	IB	IIB	IIIA	IVA	VA	VIA	VIIA	VIIIA
							1 H Hydrogen 1.0079	2 He Helium 4.00260
			5 B Boron 10.81	6 C Carbon 12.011	7 N Nitrogen 14.0067	8 O Oxygen 15.9994	9 F Fluorine 18.99840	10 Ne Neon 20.179
			13 Al Aluminum 26.98154	14 Si Silicon 28.086	15 P Phosphorus 30.97376	16 S Sulfur 32.06	17 Cl Chlorine 35.453	18 Ar Argon 39.948
28 i el 0	29 Cu Copper 63.546	30 Zn Zinc 65.38	31 Ga Gallium 69.72	32 Ge Germanium 72.59	33 As Arsenic 74.9216	34 Se Selenium 78.96	35 Br Bromine 79.904	36 Kr Krypton 83.80
46 ium 4	47 Ag Silver 107.868	48 Cd Cadmium 112.40	49 In Indium 114.82	50 Sn Tin 118.69	51 Sb Antimony 121.75	52 Te Tellurium 127.60	53 I Iodine 126.9045	54 Xe Xenon 131.30
78 um)9	79 Au Gold 196.9665	80 Hg Mercury 200.59	81 Tl Thallium 204.37	82 Pb Lead 207.2	83 Bi Bismuth 208.9804	84 Po Polonium (209)	85 At Astatine (210)	86 Rn Radon (222)

63 ım 96	64 Gd Gadolinium 157.25	65 Tb Terbium 158.9254	66 Dy Dysprosium 162.50	67 Ho Holmium 164.9304	68 Er Erbium 167.26	69 Tm Thulium 168.9342	70 Yb Ytterbium 173.04	71 Lu Lutetium 174.97
95 um)	96 Cm Curium (247)	97 Bk Berkelium (247)	98 Cf Californium (251)	99 Es Einsteinium (254)	100 Fm Fermium (257)	101 Md Mendelevium (258)	102 No Nobelium (255)	103 Lr Lawrencium (260)

Perry's Chemical Engineers' Handbook

OTHER McGRAW-HILL HANDBOOKS OF INTEREST

AMERICAN INSTITUTE OF PHYSICS *American Institute of Physics Handbook*
BAUMEISTER AND AVALLONE *Marks' Standard Handbook for Mechanical Engineers*
BRADY AND CLAUSER *Materials Handbook*
BURINGTON AND MAY *Handbook of Probability and Statistics with Tables*
CHOPEY AND HICKS *Handbook of Chemical Engineering Calculations*
CONDON AND ODISHAW *Handbook of Physics*
CONSIDINE *Process Instruments and Controls Handbook*
DAVIDSON *Handbook of Water-Soluble Gums and Resins*
DEAN *Lange's Handbook of Chemistry*
FINK AND BEATY *Standard Handbook for Electrical Engineers*
FINK AND CHRISTIANSEN *Electronics Engineers' Handbook*
GRANT *Hackh's Chemical Dictionary*
HARPER *Handbook of Plastics and Elastomers*
HARRIS AND CREDE *Shock and Vibration Handbook*
HICKS *Standard Handbook of Engineering Calculations*
HOPP AND HENNIG *Handbook of Applied Chemistry*
JURAN *Quality Control Handbook*
LYMAN *Handbook of Chemical Property Estimation Methods*
McLELLAN AND SHAND *Glass Engineering Handbook*
MAYNARD *Industrial Engineering Handbook*
PERRY *Engineering Manual*
ROHSENOW AND HARTNETT *Handbook of Heat Transfer*
ROSALER AND RICE *Standard Handbook of Plant Engineering*
SCHWARTZ *Composite Materials Handbook*
SCHWEITZER *Handbook of Separation Techniques for Chemical Engineers*
TUMA *Engineering Mathematics Handbook*
TUMA *Handbook of Physical Calculations*

PERRY'S CHEMICAL ENGINEERS' HANDBOOK

SIXTH EDITION

McGraw-Hill Book Company

New York
St. Louis
San Francisco
Auckland
Bogotá
Hamburg
London
Madrid
Mexico
Montreal
New Delhi
Panama
Paris
São Paulo
Singapore
Sydney
Tokyo
Toronto

Prepared by a staff of specialists under the editorial direction of

Late Editor
Robert H. Perry

Editor
Don W. Green
Conger-Gabel Professor of Chemical and Petroleum Engineering,
University of Kansas

Assistant Editor
James O. Maloney
Professor of Chemical Engineering,
University of Kansas

Library of Congress Cataloging in Publication Data

Main entry under title:

Perry's Chemical engineers' handook.

(McGraw-Hill chemical engineering series)
Rev. ed. of: Chemical engineers' handbook. 5th ed. 1973.
Includes bibliographical references and index.
1. Chemical engineering—Handbooks, manuals, etc.
I. Perry, Robert H., 1924–1978. II. Green, Don W.
III. Maloney, James O. IV. Chemical engineers' handbook.
V. Series.
TP151.P45 1984 660.2′8 84-837

4567890 DOW/DOW 8987

ISBN 0-07-049479-7

The editors for this book were Harold B. Crawford and Beatrice E. Eckes, the designer was Mark E. Safran, and the production supervisor was Teresa M. Leaden. It was set in Caledonia by University Graphics, Inc.

Printed and bound by R. R. Donnelley & Sons Company.

Dedicated to
Robert H. Perry

Contents

For the detailed contents of any section, consult the title page of that section. See also the alphabetical index in the back of the *Handbook*.

List of Contributors

Michael M. Abbott, Ph.D., Department of Chemical and Environmental Engineering, Rensselaer Polytechnic Institute (Section 4, Thermodynamics)

Charles M. Ambler, B.S.Ch.E. (deceased), Consultant; Director of Chemical Engineering, Sharples Division, Pennwalt Corporation (Section 19, Centrifuges)

Robert C. Amero, B.S., Staff Engineer (retired), Gulf Science and Technology Company (Section 9, Liquid Fuels, Combustion)

Eno Bagnoli, M.S., Senior Research Associate, E. I. du Pont de Nemours & Co. (Section 12, Psychrometry)

Richard Barrett, M.S., Projects Manager, Battelle Memorial Institute, Columbus Laboratories (Section 9, Steam Systems)

Kenneth J. Bell, Ph.D. , Regents Professor of Chemical Engineering, Oklahoma State University (Section 10, Thermal Design of Heat Exchangers, Condensers, Reboilers)

Richard C. Bennett, B.S.Ch.E., Division Manager, Swenson Division, Whiting Corp. (Section 19, Crystallization from Solution)

Evan Buck, M.S.Ch.E., Manager—Process Design Data and Thermodynamics, Central Engineering Department, Union Carbide Corporation (Section 3, Prediction and Correlation of Physical Properties)

Henry R. Bungay, Ph.D., Professor of Chemical and Environmental Engineering, Rensselaer Polytechnic Institute (Section 27, Biochemical Engineering)

Anthony J. Buonicore, M.Ch.E., Diplomate, AAEE, President, Buonicore-Cashman Associates, Inc. (Section 26, Air Pollution Management)

C. Edward Capes, Ph.D., Senior Research Officer and Head, Chemical Engineering Section, National Research Council, Ottawa, Canada (Section 8, Size Enlargement)

Harold F. Chambers, Jr., Ph.D., Supervisory Mechanical Engineer, Pittsburgh Energy Technology Center, U.S. Department of Energy (Section 9, Coal Liquefaction)

Ezekail L. Clark, B.S., Consultant (Section 9, Coal Gasification)

Neil H. Coates, B.S., Department Head, Energy Systems Engineering Department, The MITRE Corporation, Metrek Division (Section 9, Fluidized-Bed Combustion)

William Corder, M.S., Assistant Vice President, Consolidation Coal Company (Section 17, Sublimation)

Richard C. Corey, B.S., Department Staff (retired), Energy Systems Engineering Department, The MITRE Corporation, Metrek Division (Section 9, Solid and Gaseous Fuels, Combustion, Fuel and Energy Costs, Coal Conversion, Fired Process Equipment, Heat Transport and Regeneration)

B. B. Crocker, S.M., Distinguished Fellow—Engineering, Corporate Engineering Department, Monsanto Company (Section 18, Phase Separation)

Bruce F. Curran, S.B., Project Manager, Design Division, E. I. du Pont de Nemours & Co. (Section 6, Storage and Process Vessels)

Donald A. Dahlstrom, Ph.D., Senior Vice President, Research and Development, Eimco Process Equipment Company (Section 19, Gravity Sedimentation Operations)

J. D. Darji, M.S.Ch.E., Senior Technical Consultant, Infilco Degremont Inc. (Section 19, Ion-Exchange and Adsorption Equipment)

William M. Edwards, Ph.D., Senior Staff Research Engineer, Shell Development Company (Section 14, Mass Transfer and Gas Absorption)

Robert C. Emmett, Jr., B.S.Ch.E., Senior Process Consultant, Technology and Development, Eimco Process Equipment Company (Section 19, Gravity Sedimentation Operations)

Frank L. Evans, Jr., B.S.M.E., L.L.B. (deceased), Editor, *Hydrocarbon Processing*, Gulf Publishing Co. (Section 24, Process Machinery Drives)

J. R. Fair, Ph.D., Professor of Chemical Engineering, University of Texas (Section 18, Gas-Liquid Contacting)

R. A. Fiedler, B.S., Ch.E., Manager of Applications-Sedimentation Technology, Dorr-Oliver Inc. (Section 21, Wet Classification)

Thomas M. Flynn, Ph.D., Consultant, Cryogenic Engineering (Section 12, Cryogenic Processes)

Willard E. Fraize, Sc.D., Senior Energy Systems Engineer, Energy and Resources Division, The MITRE Corporation, Metrek Division (Section 9, Cogeneration)

Yuan C. Fu, Ph.D., Project Manager, Pittsburgh Energy Technology Center, U.S. Department of Energy (Section 9, Coal Liquefaction)

Raymond P. Genereaux, Ch.E., Chemical Engineer (retired), E. I. du Pont de Nemours & Co. (Section 6, Transport and Storage of Fluids)

W. M. Goldberger, D.Ch.E., Director of Research and Development, Superior Graphite Company (Section 21, Solid-Solid Systems)

H. A. Grabowski, B.S., Senior Engineering Consultant, C-E Environmental Systems, Combustion Engineering Inc. (Section 9, Steam Generators)

Joseph B. Gray, Ph.D., Senior Consultant (retired), Engineering Department, E. I. du Pont de Nemours and Co. (Section 19, Agitation of Low-Viscosity Particle Suspensions)

Don W. Green, Ph.D., Conger-Gabel Distinguished Professor of Chemical and Petroleum Engineering, University of Kansas (Section 1, Conversion Factors and Miscellaneous Tables)

C. Fred Gurnham, D. Eng. Sc., Consultant (Section 19, Expression)

C. Addison Hempstead, B.M.E., Senior Design Consultant (retired), E. I. du Pont de Nemours & Co. (Section 6, Process-Plant Piping)

Joseph D. Henry, Jr., Ph.D., Professor and Chairman of Chemical Engineering, West Virginia University (Section 17, Crystallization from the Melt, Separation Processes Based Primarily on Action in a Field, Novel Solid-Liquid Separation Processes)

Nevin K. Hiester, Ph.D., Associate Director, Business Intelligence Program, SRI International (Section 16, Adsorption and Ion Exchange)

W. S. Winston Ho, Ph.D., Engineering Associate, Exxon Research and Engineering Company (Section 17, Membrane Processes)

Arthur E. Hoerl, M.A., Professor, Department of Mathematical Sciences, University of Delaware (Section 2, Mathematics and Statistics)

Richard L. Hoglund, M.S., Chemical Engineer, Union Carbide Corporation—Nuclear Division (Section 17, Diffusional Separation Processes)

F. A. Holland, D.Sc., Ph.D., Chairman, Department of Chemical Engineering, University of Salford, Salford, England; Partner in Salchem Associates, Consulting Engineers (Section 25, Process Economics)

Arthur D. Holt, formerly Consultant, Processing Equipment Division, The Jeffrey Manufacturing Co. (Section 10, Thermal Design for Solids Processing; Section 11, Heat Exchangers for Solids)

Hoyt C. Hottel, S.M., Professor Emeritus of Chemical Engineering, Massachusetts Institute of Technology (Section 10, Radiation)

Arthur E. Humphrey, Ph.D., Provost, Lehigh University (Section 27, Biochemical Engineering)

Louis J. Jacobs, Jr., M.S.Ch.E., Director of Corporate Engineering Division, A. E. Staley Manufacturing Company (Section 19, Filtration)

Eric Jenett, M.S.Ch.E., Manager Process Engineering, Brown & Root, Inc. (Section 24, Power Recovery from Liquid Streams)

T. L. B. Jepsen, M.S., Min.Proc., Metallurgical Process Engineer, Basic, Inc. (Section 21, Dense-Media Separation)

Brian H. Kaye, Ph.D., Professor of Physics and Director of Institute for Fine Particles Research, Laurentian University (Section 8, Particle-Size Analysis)

Gerhard Klein, M.S., Research Engineer, Water Technology Center, University of California, Berkeley (Section 16, Absorption and Ion Exchange)

Ronald P. Klepper, B.S.Ch.E., Manager, Process Technology, Eimco Process Equipment Company (Section 19, Gravity Sedimentation Operations)

Frank S. Knoll, M.S., Min.Proc., President, Carpco, Inc. (Section 21, Electrostatic Separation)

James G. Knudsen, Ph.D., Professor of Chemical Engineering, Oregon State University (Section 10, Conduction and Convection)

Albert S. Krisher, B.S.Ch.E., Senior Fellow, Monsanto Company (Section 23, Materials of Construction)

Zdzislaw M. Kurtyka, D.Sc. (deceased), formerly Department of Chemical Engineering, The University of the West Indies, St. Augustine, Trinidad (Section 13, Azeotropy)

Robert Lemlich, Ph.D., Professor of Chemical Engineering, University of Cincinnati (Section 17, Adsorptive-Bubble Separation Methods)

M. Douglas LeVan, Ph.D., Associate Professor of Chemical Engineering, University of Virginia (Section 16, Adsorption and Ion Exchange)

Norman N. Li, Sc.D., Director, Separations Research, UOP, Inc. (Section 17, Membrane Processes)

Peter E. Liley, Ph.D., D.I.C., Professor, School of Mechanical Engineering and Center for Information and Numerical Data Analysis and Synthesis, Purdue University (Section 3, Physical and Chemical Data)

Kuang-Hui Lin, Ph.D., Research Staff, Chemical Technology Division, Oak Ridge National Laboratory (Section 4, Reaction Kinetics, Reactor Design)

John J. McKetta, Ph.D., Joe C. Walter Chair, Chemical Engineering, University of Texas (Section 2, Dimensional Analysis)

Ross E. McKinney, Sc.D., N. T. Veatch Professor of Environmental Engineering, University of Kansas (Section 26, Industrial-Wastewater Management)

Patrick M. McNeese, B.S., General Manager of Purchasing, Cities Service Oil and Gas Company (Section 22, Process Control)

James O. Maloney, Ph.D., Professor of Chemical Engineering, University of Kansas (Section 21, Liquid-Liquid Systems)

Eugene Mezey, Ph.D., Senior Chemist, Battelle Memorial Institute, Columbus Laboratories (Section 9, Electric Heating)

A. W. Michalson, B.S.Min.E., President, A. W. Michalson Company (Section 19, Ion-Exchange and Adsorption Equipment)

Shelby A. Miller, Ph.D., Senior Chemical Engineer, Argonne National Laboratory (Section 19, Leaching)

Charles J. B. Mitchell, B.S., Principal Consultant (retired), Engineering Service Division, E. I. du Pont de Nemours & Co. (Section 6, Pumping of Liquids and Gases)

D. W. Mitchell, Ph.D., Technical Director, Carpco, Inc. (Section 21, Electrostatic Separation)

Herbert A. Moak, B.S., Project Engineer, E. I. du Pont de Nemours & Co. (Section 11, Thermal Insulation)

Charles G. Moyers, Jr., Ph.D., Principal Engineer, Union Carbide Corporation—Engineering and Technical Services Division (Section 17, Crystallization from the Melt)

M. Zuhair Nashed, Ph.D., Professor of Mathematics and Professor of Electrical Engineering, University of Delaware (Section 2, general material on mathematics)

John Newman, Ph.D., Professor of Chemical Engineering, University of California, Berkeley (Section 17, Separations Based Primarily on Action in a Field, Theory of Electrical Separations)

Robert W. Norris, B.S., President, Robert W. Norris and Associates, Inc. (Section 12, Evaporative Cooling, Refrigeration)

James Y. Oldshue, Ph.D., Vice President, Mixing Technology, Mixing Equipment Company, Inc. (Section 19, Agitation of Low-Viscosity Particle Suspensions)

Carl R. Olson, M.S.E.E., Staff Planning Manager, Industrial Projects Marketing, Westinghouse Electric Corp. (Section 24, Electric Motors and Motor Controls)

Bhupendra K. Parekh, Ph.D., Research Specialist, Exxon Minerals Company (Section 21, Separation of Ultrafine Solids)

Jerry R. Peebles, M.S., Engineering Associate, Cities Service Oil and Gas Company (Section 22, Process Control)

W. R. Penney, Ph.D., Director of Corporate Process Engineering, A. E. Staley Manufacturing Company (Section 18, Gas-in-Liquid Dispersions)

Herbert A. Pohl, Ph.D., Professor of Physics, Oklahoma State University (Section 17, Separation Processes Based Primarily on Action in a Field, Dielectrophoresis)

Kent Pollock, Ph.D., Professor of Physics, Oklahoma State University (Section 17, Separation Processes Based Primarily on Action in a Field, Dielectrophoresis)

Harold F. Porter, B.S., Principal Division Consultant (retired), E. I. du Pont de Nemours and Co. (Section 20, Solids Drying and Gas-Solid Systems)

Michael E. Prudich, Ph.D., Research Engineer, Gulf Research and Development Company (Section 17, Novel Solid-Liquid Separation Processes)

Grantges J. Raymus, M.E., M.S., President, Raymus Associates, Incorporated, Packaging Consultants; Adjunct Professor and Assistant Director, Center for Packaging Engineering, Rutgers,

The State University of New Jersey (Section 7, Handling of Bulk Solids and Packaging of Solids and Liquids)

Robert C. Reid, Sc.D., Professor, Department of Chemical Engineering, Massachusetts Institute of Technology (Section 3, Prediction and Correlation of Physical Properties)

Lanny A. Robbins, Ph.D., Research Scientist, Dow Chemical Company (Section 15, Liquid-Liquid Extraction; Section 21, Co-Section Editor, Liquid-Liquid Systems)

Frank L. Rubin, B.A., B.Ch.E., Engineering Consultant, Practical Heat Transfer Consultants (Section 11, Shell-and-Tube Heat Exchangers, Other Heat Exchangers for Liquids and Gases)

Byron C. Sakiadis, Ph.D., Senior Research Fellow, Engineering Technology Laboratory, E. I. du Pont de Nemours & Co. (Section 5, Fluid and Particle Mechanics)

Adel F. Sarofim, Sc.D., Professor of Chemical Engineering and Assistant Director, Fuels Research Laboratory, Massachusetts Institute of Technology (Section 10, Radiation)

George A. Schurr, Ph.D., Consultant, E. I. du Pont de Nemours and Co. (Section 20, Solids Drying)

J. D. Seader, Ph.D., Professor of Chemical Engineering, University of Utah (Section 13, Distillation)

Konrad T. Semrau, M.S., Senior Chemical Engineer, SRI International (Section 20, Gas-Solids Separations)

Oliver W. Siebert, B.S.M.E., Senior Fellow, Monsanto Company; Adjunct Professor of Mechanical Engineering, Washington University (Section 23, Materials of Construction)

I. H. Silberberg, Ph.D., Assistant Director, Texas Petroleum Research Committee, University of Texas (Section 2, Dimensional Analysis)

Charles E. Silverblatt, M.S.Ch.E., Vice President, Technology and Development, Eimco Process Equipment Company (Section 19, Gravity Sedimentation Operations)

Julian C. Smith, B.Chem., Ch.E., Professor of Chemical Engineering, Cornell University (Section 19, Selection of a Solids-Liquid Separator)

Richard H. Snow, Ph.D., Director, National Institute for Petroleum and Energy Research (Section 8, Size Reduction and Size Enlargement)

Thomas C. Sorenson, M.B.A., Min.Eng., President, Galigher Ash (Canada) Ltd. (Section 21, Flotation)

K. S. Spiegler, Ph.D., Professor of Mechanical Engineering Emeritus, University of California, Berkeley (Section 17, Separation Processes Based Primarily on Action in a Field; Electrodialysis)

Guggilam C. Sresty, M.S., Senior Engineer, IIT Research Institute (Section 8, Crushing and Grinding Equipment)

F. C. Standiford, M.S., President, W. L. Badger Associates, Inc. (Section 10, Thermal Design of Evaporators; Section 11, Evaporators)

Paul L. Stavenger, M.S.Ch.E., Director of Technology, Dorr-Oliver Inc. (Section 21, Wet Classification)

H. Steen-Johnson, M.S.M.E., Chief Staff Engineer (retired), Elliott Co. (Section 24, Steam Turbines)

D. E. Steinmeyer, M.A., M.S., Manager, Engineering, Corporate Engineering Department, Monsanto Company (Section 18, Liquid-in-Gas Dispersions)

David Stuhlbarg, Ch.E., Heat-Transfer Consultant (Section 10, Thermal Design of Tank Coils; Section 11, Tank Coils)

David E. Stutz, B.S., Research Scientist, Battelle Memorial Institute, Columbus Laboratories (Section 9, Electric Heating)

J. S. Swearingen, Ph.D., President, Rotoflow Corp. (Section 24, Expansion Turbines)

George Tchobanoglous, Ph.D., Professor of Environmental Engineering, University of California at Davis (Section 26, Management of Industrial Solid Wastes)

Philip O. Teter, Jr., B.S., Electrical Engineer, Corps of Engineers (Section 22, Process Control)

Rich L. Thelen, B.S., M.E., Vice President, EIE Company, Inc. (Section 21, Solids Sampling)

Louis Theodore, Sc.D., Professor of Chemical Engineering, Manhattan College (Section 26, Waste Management)

Klaus D. Timmerhaus, Ph.D., Associate Dean of Engineering, Director of Engineering Research Center, University of Colorado (Section 12, Cryogenic Processes)

David B. Todd, Ph.D., Technical Director, Plastics Equipment Division, Baker Perkins Inc. (Section 19, Paste and Viscous-Material Mixing)

Robert E. Treybal, Ph.D. (deceased), Professor and Chairman, Department of Chemical Engineering, University of Rhode Island (Section 21, Liquid-Liquid Systems)

George T. Tsao, Ph.D., Director, Laboratory of Renewable Resource Engineering, Purdue University (Section 27, Biochemical Engineering)

Vincent W. Uhl, Ph.D., Professor of Chemical Engineering, University of Virginia (Section 10, Agitated Vessels)

Hendrick C. Van Ness, D.Eng., Department of Chemical and Environmental Engineering, Rensselaer Polytechnic Institute (Section 4, Thermodynamics)

Theodore Vermeulen, Ph.D. (deceased), Professor of Chemical Engineering, Faculty Scientist, Lawrence Berkeley Laboratory, and Director of the Water Technology Center, University of California, Berkeley (Section 16, Ion Exchange and Adsorption)

Edward Von Halle, Ph.D., Chemical Engineer, Union Carbide Corporation—Nuclear Division (Section 17, Diffusional Separation Processes)

F. A. Watson, M.Sc., Senior Lecturer, Department of Chemical Engineering, University of Salford, Salford, England; Partner in Salchem Associates, Consulting Engineers (Section 25, Process Economics)

Ionel Wechsler, M.S.Min. and Met., Vice President, Sala Magnetics, Inc. (Section 21, Magnetic Separation)

David F. Wells, B.S., President, David F. Wells and Associates (Section 20, Fluidized-Bed Systems)

T. C. Wherry, B.S.E.E., B.S.Ch.E., Vice President (retired), Director of Systems Research, Applied Automation, Inc.; Member, Board of Directors (Section 22, Process Control)

J. K. Wilkinson, M.Sc., Senior Lecturer, Department of Chemical Engineering, University of Salford, Salford, England; Partner in Salchem Associates, Consulting Engineers (Section 25, Process Economics)

Richard E. Worsham, B.S., Region Engineer, Cities Service Oil and Gas Company (Section 22, Process Control)

Roy M. Young, B.S.Ch.E., Process Automation Consultant, R & M Associates, Inc. (Section 22, Process Control)

Preface to the Sixth Edition

The discipline of chemical engineering has continued to contribute in numerous and important ways to worldwide industrial progress over the past decade. Significant strides have been made in process and equipment design methods. This progress has been due, in part, to factors such as increased computer utilization and availability of better materials of construction. The focus on the increased value of energy and the need for improved waste management have also affected design decisions. Basically, however, good chemical engineering practice is still founded on the sound application of the two aspects of empiricism and theory. And in both of these aspects a number of contributions have been made since publication of the fifth edition. Thus, each of the 25 sections of the fifth edition has been revised and updated, and for some areas (such as economics, distillation, extraction, and absorption) sections have been completely rewritten. Two new sections have been added to incorporate the emerging technologies of bioengineering and waste management.

The method of handling units in the *Handbook* was changed. As most engineers are aware, SI units are utilized in much of the world while U.S. customary units are still used primarily in the United States (although there has been a definite move in the United States toward greater application of the SI). To accommodate the different users, the *Handbook* has been written to the extent possible using both conventions. Tables and figures which were revised from earlier editions were not redrafted because it simply was not practical to do so. New figures and tables are generally presented in SI and, in some instances, in both sets of units. In all cases, conversion factors are provided with tables and figures to convert to the alternative convention. Numbers which appear in the text are given in both SI and U.S. customary units, and dimensional constants which appear in a number of empirical equations are also usually specified in both unit sets. In general, the editors believe the *Handbook* will be usable by persons working with either units convention.

Numerous persons assisted in the preparation of this edition. The editors in particular acknowledge Wanda S. Dekat, Georgea L. de Medina, and Guy L. Green, seniors in engineering, who performed the tedious task of preparing the index. Typing and secretarial assistance were provided by Jill A. Schoeling and Ruth R. Sleeper.

The editors especially acknowledge the contributions of Raymond Genereaux, editor of Sec.

6, "Transport and Storage of Fluids." He is the only person who has participated in the preparation of all six editions of the *Handbook*. His efforts and commitment throughout a long association with the *Handbook* are appreciated.

The untimely death of Bob Perry during the preparation of this edition was a hurtful loss. He was deeply committed to the continuation of the *Handbook* and to the quality represented in the tradition of this book. He is missed.

We also regret the loss of Frank L. Evans, Jr., and Theodore Vermeulen, section editors who made invaluable contributions to the *Handbook*.

DON W. GREEN

Perry's Chemical Engineers' Handbook

Conversion Factors and Miscellaneous Tables*

Don W. Green, Ph.D., *Conger-Gabel Distinguished Professor of Chemical and Petroleum Engineering, University of Kansas*

*Much of the material was taken from Sec. 1. of the fifth edition. The contribution of Cecil H. Chilton in developing that material is acknowledged.

CONVERSION FACTORS

TABLE 1-1 SI Base and Supplementary Quantities and Units

Quantity or "dimension"	SI unit	SI unit symbol ("abbreviation"); Use roman (upright) type
Base quantity or "dimension"		
length	meter	m
mass	kilogram	kg
time	second	s
electric current	ampere	A
thermodynamic temperature	kelvin	K
amount of substance	mole*	mol
luminous intensity	candela	cd
Supplementary quantity or "dimension"		
plane angle	radian	rad
solid angle	steradian	sr

*When the mole is used, the elementary entities must be specified; they may be atoms, molecules, ions, electrons, other particles, or specified groups of such particles.

TABLE 1-2*a* Derived Units of SI Which Have Special Names

Quantity	Unit	Symbol	Formula
frequency (of a periodic phenomenon)	hertz	Hz	1/s
force	newton	N	$(kg \cdot m)/s^2$
pressure, stress	pascal	Pa	N/m^2
energy, work, quantity of heat	joule	J	$N \cdot m$
power, radiant flux	watt	W	J/s
quantity of electricity, electric charge	coulomb	C	$A \cdot s$
electric potential, potential difference, electromotive force	volt	V	W/A
capacitance	farad	F	C/V
electric resistance	ohm	Ω	V/A
conductance	siemens	S	A/V
magnetic flux	weber	Wb	$V \cdot s$
magnetic-flux density	tesla	T	Wb/m^2
inductance	henry	H	Wb/A
luminous flux	lumen	lm	$cd \cdot sr$
illuminance	lux	lx	lm/m^2
activity (of radionuclides)	becquerel	Bq	1/s
absorbed dose	gray	Gy	J/kg

TABLE 1-2*b* Additional Common Derived Units of SI

Quantity	Unit	Symbol
acceleration	meter per second squared	m/s^2
angular acceleration	radian per second squared	rad/s^2
angular velocity	radian per second	rad/s
area	square meter	m^2
concentration (of amount of substance)	mole per cubic meter	mol/m^3
current density	ampere per square meter	A/m^2
density, mass	kilogram per cubic meter	kg/m^3
electric-charge density	coulomb per cubic meter	C/m^3
electric-field strength	volt per meter	V/m
electric-flux density	coulomb per square meter	C/m^2
energy density	joule per cubic meter	J/m^3
entropy	joule per kelvin	J/K
heat capacity	joule per kelvin	J/K
heat-flux density, irradiance	watt per square meter	W/m^2
luminance	candela per square meter	cd/m^2
magnetic-field strength	ampere per meter	A/m
molar energy	joule per mole	J/mol
molar entropy	joule per mole-kelvin	$J/(mol \cdot K)$
molar-heat capacity	joule per mole-kelvin	$J/(mol \cdot K)$
moment of force	newton-meter	$N \cdot m$
permeability	henry per meter	H/m
permittivity	farad per meter	F/m
radiance	watt per square-meter-steradian	$W/(m^2 \cdot sr)$
radiant intensity	watt per steradian	W/sr
specific-heat capacity	joule per kilogram-kelvin	$J/(kg \cdot K)$
specific energy	joule per kilogram	J/kg
specific entropy	joule per kilogram-kelvin	$J/(kg \cdot K)$
specific volume	cubic meter per kilogram	m^3/kg
surface tension	newton per meter	N/m
thermal conductivity	watt per meter-kelvin	$W/(m \cdot K)$
velocity	meter per second	m/s
viscosity, dynamic	pascal-second	$Pa \cdot s$
viscosity, kinematic	square meter per second	m^2/s
volume	cubic meter	m^3
wave number	1 per meter	1/m

TABLE 1-3 SI Prefixes

Multiplication factor	Prefix	Symbol
1 000 000 000 000 000 000 = 10^{18}	exa	E
1 000 000 000 000 000 = 10^{15}	peta	P
1 000 000 000 000 = 10^{12}	tera	T
1 000 000 000 = 10^{9}	giga	G
1 000 000 = 10^{6}	mega	M
1 000 = 10^{3}	kilo	k
100 = 10^{2}	hecto*	h
10 = 10^{1}	deka*	da
0.1 = 10^{-1}	deci*	d
0.01 = 10^{-2}	centi	c
0.001 = 10^{-3}	milli	m
0.000 001 = 10^{-6}	micro	μ
0.000 000 001 = 10^{-9}	nano	n
0.000 000 000 001 = 10^{-12}	pico	p
0.000 000 000 000 001 = 10^{-15}	femto	f
0.000 000 000 000 000 001 = 10^{-18}	atto	a

*Generally to be avoided.

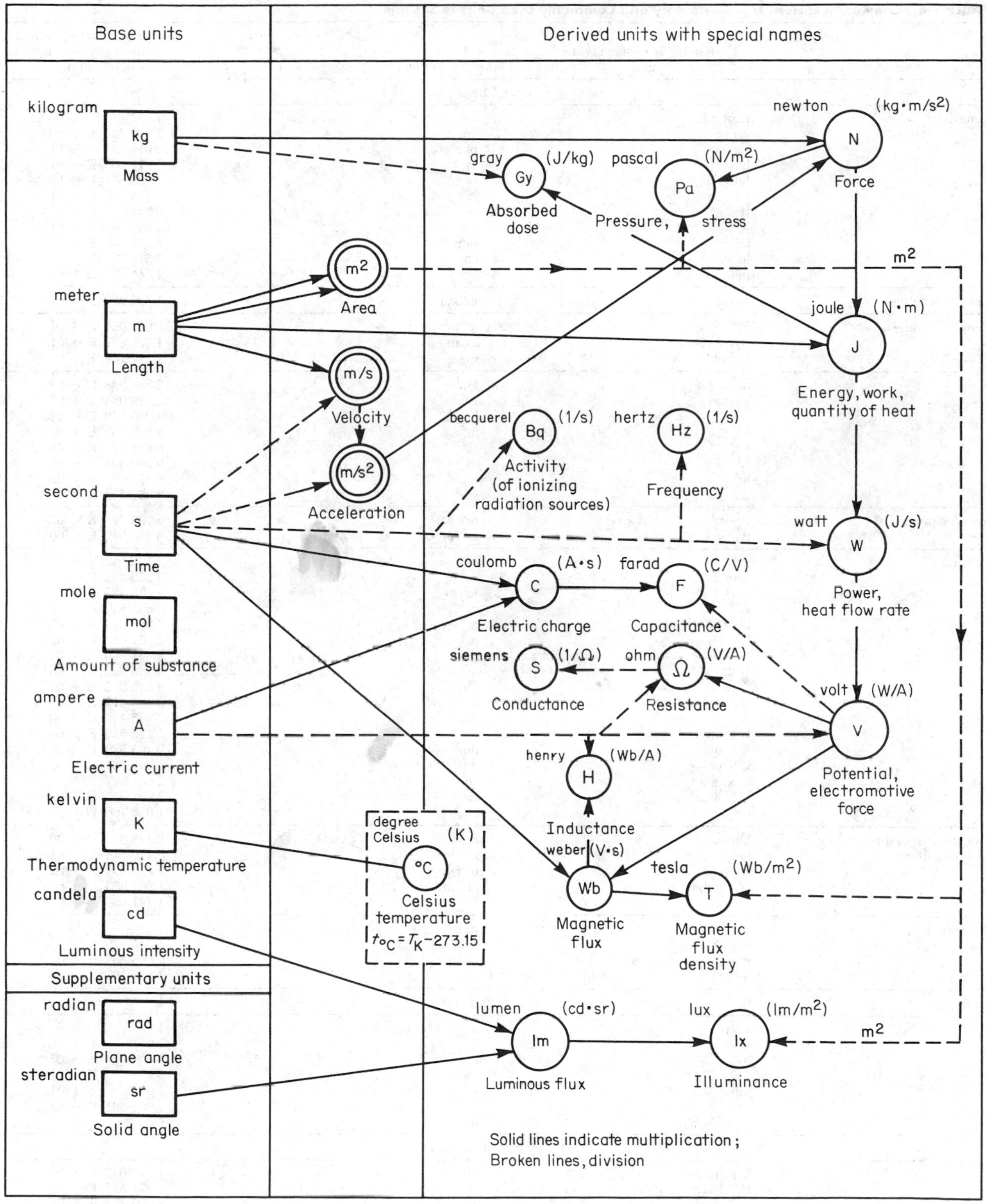

FIG. 1-1 Graphic relationships of SI units with names. *(U.S. National Bureau of Standards, LC 1078, December 1976.)*

TABLE 1-4 Conversion Factors: U.S. Customary and Commonly Used Units to SI Units

Quantity	Customary or commonly used unit	SI unit	Alternate SI unit	Conversion factor; multiply customary unit by factor to obtain SI unit
		Space,† time		
Length	naut mi	km		1.852° E + 00
	mi	km		1.609 344° E + 00
	chain	m		2.011 68° E + 01
	link	m		2.011 68° E − 01
	fathom	m		1.828 8° E + 00
	yd	m		9.144° E − 01
	ft	m		3.048° E − 01
		cm		3.048° E + 01
	in	mm		2.54° E + 01
	in	cm		2.54 E + 00
	mil	μm		2.54° E + 01
Length/length	ft/mi	m/km		1.893 939 E − 01
Length/volume	ft/U.S. gal	m/m^3		8.051 964 E + 01
	ft/ft^3	m/m^3		1.076 391 E + 01
	ft/bbl	m/m^3		1.917 134 E + 00
Area	mi^2	km^2		2.589 988 E + 00
	section	ha		2.589 988 E + 02
	acre	ha		4.046 856 E − 01
	ha	m^2		1.000 000° E + 04
	yd^2	m^2		8.361 274 E − 01
	ft^2	m^2		9.290 304° E − 02
	in^2	mm^2		6.451 6° E + 02
		cm^2		6.451 6° E + 00
Area/volume	ft^2/in^3	m^2/cm^3		5.699 291 E − 03
	ft^2/ft^3	m^2/m^3		3.280 840 E + 00
Volume	cubem	km^3		4.168 182 E + 00
	acre·ft	m^3		1.233 482 E + 03
		ha·m		1.233 482 E − 01
	yd^3	m^3		7.645 549 E − 01
	bbl (42 U.S. gal)	m^3		1.589 873 E − 02
	ft^3	m^3		2.831 685 E − 02
		dm^3	L	2.831 685 E + 01
	U.K. gal	m^3		4.546 092 E − 03
		dm^3	L	4.546 092 E + 00
	U.S. gal	m^3		3.785 412 E − 03
		dm^3	L	3.785 412 E + 00
	U.K. qt	dm^3	L	1.136 523 E + 00
	U.S. qt	dm^3	L	9.463 529 E − 01
	U.S. pt	dm^3	L	4.731 765 E − 01
	U.K. fl oz	cm^3		2.841 307 E + 01
	U.S. fl oz	cm^3		2.957 353 E + 01
	in^3	cm^3		1.638 706 E + 01
Volume/length (linear displacement)	bbl/in	m^3m		6.259 342 E + 00
	bbl/ft	m^3/m		5.216 119 E − 01
	ft^3/ft	m^3/m		9.290 304° E − 02
	U.S. gal/ft	m^3/m		1.241 933 E − 02
		L/m		1.241 933 E + 01
Plane angle	rad	rad		1
	deg (°)	rad		1.745 329 E − 02
	min (′)	rad		2.908 882 E − 04
	sec (″)	rad		4.848 137 E − 06
Solid angle	sr	sr		1
Time	year	a		1
	week	d		7.0° E + 00
	h	s		3.6° E + 03
		min		6.0° E + 01
	min	s		6.0° E + 01
		h		1.666 667 E − 02
	mμs	ns		1
		Mass, amount of substance		
Mass	U.K. ton	Mg	t	1.016 047 E + 00
	U.S. ton	Mg	t	9.071 847 E − 01
	U.K. cwt	kg		5.080 234 E + 01

TABLE 1-4 Conversion Factors: U.S. Customary and Commonly Used Units to SI Units *(Continued)*

Quantity	Customary or commonly used unit	SI unit	Alternate SI unit	Conversion factor; multiply customary unit by factor to obtain SI unit
	U.S. cwt	kg		4.535 924 E + 01
	lbm	kg		4.535 924 E − 01
	oz (troy)	g		3.110 348 E + 01
	oz (av)	g		2.834 952 E + 01
	gr	mg		6.479 891 E + 01
Amount of substance	lbm·mol	kmol		4.535 924 E − 01
	std m^3(0°C, 1 atm)	kmol		4.461 58 E − 02
	std ft^3 (60°F, 1 atm)	kmol		1.195 30 E − 03
	Enthalpy, calorific value, heat, entropy, heat capacity			
Calorific value, enthalpy (mass basis)	Btu/lbm	MJ/kg		2.326 000 E − 03
		kJ/kg	J/g	2.326 000 E + 00
		kWh/kg		6.461 112 E − 04
	cal/g	kJ/kg	J/g	4.184* E + 00
	cal/lbm	J/kg		9.224 141 E + 00
Caloric value, enthalpy (mole basis)	kcal/(g·mol)	kJ/kmol		4.184* E + 03
	Btu/(lb·mol)	kJ/kmol		2.326 000 E + 00
Calorific value (volume basis—solids and liquids)	Btu/U.S. gal	MJ/m^3	kJ/dm^3	2.787 163 E − 01
		kJ/m^3		2.787 163 E + 02
		kWh/m^3		7.742 119 E − 02
	Btu/U.K. gal	MJ/m^3	kJ/dm^3	2.320 800 E − 01
		kJ/m^3		2.320 800 E + 02
		kWh/m^3		6.446 667 E − 02
	Btu/ft^3	MJ/m^3	kJ/dm^3	3.725 895 E − 02
		kJ/m^3		3.725 895 E + 01
		kWh/m^3		1.034 971 E − 02
	cal/mL	MJ/m^3		4.184* E + 00
	(ft·lbf)/U.S. gal	kJ/m^3		3.581 692 E − 01
Calorific value (volume basis—gases)	cal/mL	kJ/m^3	J/dm^3	4.184* E + 03
	$kcal/m^3$	kJ/m^3	J/dm^3	4.184* E + 00
	Btu/ft^3	kJ/m^3	J/dm^3	3.725 895 E + 01
		kWh/m^3		1.034 971 E − 02
Specific entropy	Btu/(lbm·°R)	kJ/(kg·K)	J/(g·K)	4.186 8* E + 00
	cal/(g·K)	kJ/(kg·K)	J/(g·K)	4.184* E + 00
	kcal/(kg·°C)	kJ/(kg·K)	J/(g·K)	4.184* E + 00
Specific-heat capacity (mass basis)	kWh/(kg·°C)	kJ/(kg·K)	J/(g·K)	3.6* E + 03
	Btu/(lbm·°F)	kJ/(kg·K)	J/(g·K)	4.186 8* E + 00
	kcal/(kg·°C)	kJ/(kg·K)	J/(g·K)	4.184* E + 00
Specific-heat capacity (mole basis)	Btu/(lb·mol·°F)	kJ/(kmol·K)		4.186 8* E + 00
	cal/(g·mol·°C)	kJ/(kmol·K		4.184* E + 00
	Temperature, pressure, vacuum			
Temperature (absolute)	°R	K		5/9
	K	K		1
Temperature (traditional)	°F	°C		5/9(°F−32)
Temperature (difference)	°F	K, °C		5/9
Pressure	atm (760 mmHg at 0°C or 14,696 psi)	MPa		1.013 250*E − 01
		kPa		1.013 250*E + 02
		bar		1.013 250*E + 00
	bar	MPa		1.0* E − 01
		kPa		1.0* E + 02
	mmHg (0°C) = torr	MPa		6.894 757 E − 03
		kPa		6.894 757 E + 00
		bar		6.894 757 E − 02
	μmHg (0°C)	kPa		3.376 85 E + 00
	μ bar	kPa		2.488 4 E − 01
	mmHg = torr (0°C)	kPa		1.333 224 E − 01
	cmH_2O (4°C)	kPa		9.806 38 E − 02
	lbf/ft^2 (psf)	kPa		4.788 026 E − 02
	mHg (0°C)	Pa		1.333 224 E − 01
	bar	Pa		1.0* E − 01
	dyn/cm^2	Pa		1.0* E − 01

TABLE 1-4 Conversion Factors: U.S. Customary and Commonly Used Units to SI Units *(Continued)*

Quantity	Customary or commonly used unit	SI unit	Alternate SI unit	Conversion factor; multiply customary unit by factor to obtain SI unit
Vacuum, draft	inHg (60°F)	kPa		3.376 85 E + 00
	inH_2O (39.2°F)	kPa		2.490 82 E − 01
	inH_2O (60°F)	kPa		2.488 4 E − 01
	mmHg (0°C) = torr	kPa		1.333 224 E − 01
	cmH_2O (4°C)	kPa		9.806 38 E − 02
Liquid head	ft	m		3.048* E − 01
	in	mm		2.54* E + 01
		cm		2.54* E + 00
Pressure drop/length	psi/ft	kPa/m		2.262 059 E + 01
	Density, specific volume, concentration, dosage			
Density	lbm/ft^3	kg/m^3		1.601 846 E + 01
		g/m^3		1.601 846 E + 04
	lbm/U.S. gal	kg/m^3		1.198 264 E + 02
		g/cm^3		1.198 264 E − 01
	lbm/U.K. gal	kg/m^3		9.977 633 E + 01
	lbm/ft^3	kg/m^3		1.601 846 E + 01
		g/cm^3		1.601 846 E − 02
	g/cm^3	kg/m^3		1.0* E + 03
	lbm/ft^3	kg/m^3		1.601 846 E + 01
Specific volume	ft^3/lbm	m^3/kg		6.242 796 E − 02
		m^3/g		6.242 796 E − 05
	ft^3/lbm	dm^3/kg		6.242 796 E + 01
	U.K. gal/lbm	dm^3/kg	cm^3/g	1.002 242 E + 01
	U.S. gal/lbm	dm^3/kg	cm^3/g	8.345 404 E + 00
Specific volume (mole basis)	L/(g·mol)	$m^3/kmol$		1
	$ft^3/(lb\cdot mol)$	$m^3/kmol$		6.242 796 E − 02
Specific volume	bbl/U.S. ton	m^3/t		1.752 535 E − 01
	bbl/U.K. ton	m^3/t		1.564 763 E − 01
Yield	bbl/U.S. ton	dm^3/t	L/t	1.752 535 E + 02
	bbl/U.K. ton	dm^3/t	L/t	1.564 763 E + 02
	U.S. gal/U.S. ton	dm^3/t	L/t	4.172 702 E + 00
	U.S. gal/U.K. ton	dm^3/t	L/t	3.725 627 E + 00
Concentration (mass/mass)	wt %	kg/kg		1.0* E − 02
		g/kg		1.0* E + 01
	wt ppm	mg/kg		1
Concentration (mass/volume)	lbm/bbl	kg/m^3	g/dm^3	2.853 010 E + 00
	g/U.S. gal	kg/m^3		2.641 720 E − 01
	g/U.K. gal	kg/m^3	g/L	2.199 692 E − 01
	lbm/1000 U.S. gal	g/m^3	mg/dm^3	1.198 264 E + 02
	lbm/1000 U.K. gal	g/m^3	mg/dm^3	9.977 633 E + 01
	gr/U.S. gal	g/m^3	mg/dm^3	1.711 806 E + 01
	gr/ft^3	mg/m^3		2.288 351 E + 03
	lbm/1000 bbl	g/m^3	mg/dm^3	2.853 010 E + 00
	mg/U.S. gal	g/m^3	mg/dm^3	2.641 720 E − 01
	gr/100 ft^3	mg/m^3		2.288 351 E + 01
Concentration (volume/volume)	ft^3/ft^3	m^3/m^3		1
	bbl/(acre·ft)	m^3/m^3		1.288 931 E − 04
	vol%	m^3/m^3		1.0* E − 02
	U.K. gal/ft^3	dm^3/m^3	L/m^3	1.605 437 E + 02
	U.S. gal/ft^3	dm^3/m^3	L/m^3	1.336 806 E + 02
	mL/U.S. gal	dm^3/m^3	L/m^3	2.641 720 E − 01
	mL/U.K. gal	dm^3/m^3	L/m^3	2.199 692 E − 01
	vol ppm	cm^3/m^3		1
		dm^3/m^3	L/m^3	1.0* E − 03
	U.K. gal/1000 bbl	cm^3/m^3		2.859 403 E + 01
	U.S. gal/1000 bbl	cm^3/m^3		2.380 952 E + 01
	U.K. pt/1000 bbl	cm^3/m^3		3.574 253 E + 00
Concentration (mole/volume)	(lb·mol)/U.S. gal	$kmol/m^3$		1.198 264 E + 02
	(lb·mol)/U.K. gal	$kmol/m^3$		9.977 644 E + 01
	$(lb\cdot mol)/ft^3$	$kmol/m^3$		1.601 846 E + 01
	std ft^3 (60°F, 1 atm)/bbl	$kmol/m^3$		7.518 21 E − 03
Concentration (volume/mole)	U.S. gal/1000 std ft^3 (60°F/60°F)	$dm^3/kmol$	L/kmol	3.166 91 E + 00
	bbl/million std ft^3 (60°F/60°F)	$dm^3/kmol$	L/kmol	1.330 10 E − 01

TABLE 1-4 Conversion Factors: U.S. Customary and Commonly Used Units to SI Units *(Continued)*

Quantity	Customary or commonly used unit	SI unit	Alternate SI unit	Conversion factor; multiply customary unit by factor to obtain SI unit
		Facility throughput, capacity		
Throughput (mass basis)	U.K. ton/year	t/a		1.016 047 E + 00
	U.S. ton/year	t/a		9.071 847 E − 01
	U.K. ton/day	t/d		1.016 047 E + 00
		t/h		4.233 529 E − 02
	U.S. ton/day	t/d		9.071 847 E − 01
		t/h		3.779 936 E − 02
	U.K. ton/h	t/h		1.016 047 E + 00
	U.S. ton/h	t/h		9.071 847 E − 01
	lbm/h	kg/h		4.535 924 E − 01
Throughput (volume basis)	bbl/day	t/a		5.803 036 E + 01
		m^3/d		1.589 873 E − 01
	ft^3/day	m^3/h		1.179 869 E − 03
	bbl/h	m^3/h		1.589 873 E − 01
	ft^3/h	m^3/h		2.831 685 E − 02
	U.K. gal/h	m^3/h		4.546 092 E − 03
		L/s		1.262 803 E − 03
	U.S. gal/h	m^3/h		3.785 412 E − 03
		L/s		1.051 503 E − 03
	U.K. gal/min	m^3/h		2.727 655 E − 01
		L/s		7.576 819 E − 02
	U.S. gal/min	m^3/h		2.271 247 E − 01
		L/s		6.309 020 E − 02
Throughput (mole basis)	(lbm · mol)/h	kmol/h		4.535 924 E − 01
		kmol/s		1.259 979 E − 04
		Flow rate		
Flow rate (mass basis)	U.K. ton/min	kg/s		1.693 412 E + 01
	U.S. ton/min	kg/s		1.511 974 E + 01
	U.K. ton/h	kg/s		2.822 353 E − 01
	U.S. ton/h	kg/s		2.519 958 E − 01
	U.K. ton/day	kg/s		1.175 980 E − 02
	U.S. ton/day	kg/s		1.049 982 E − 02
	million lbm/year	kg/s		5.249 912 E + 00
	U.K. ton/year	kg/s		3.221 864 E − 05
	U.S. ton/year	kg/s		2.876 664 E − 05
	lbm/s	kg/s		4.535 924 E − 01
	lbm/min	kg/s		7.559 873 E − 03
	lbm/h	kg/s		1.259 979 E − 04
Flow rate (volume basis)	bbl/day	m^3/d		1.589 873 E − 01
		L/s		1.840 131 E − 03
	ft^3/day	m^3/d		2.831 685 E − 02
	bbl/h	L/s		3.277 413 E − 04
		m^3/s		4.416 314 E − 05
	ft^3/h	L/s		4.416 314 E − 02
		m^3/s		7.865 791 E − 06
	U.K. gal/h	L/s		7.865 791 E − 03
	U.S. gal/h	dm^3/s	L/s	1.262 803 E − 03
	U.K. gal/min	dm^3/s	L/s	1.051 503 E − 03
	U.S. gal/min	dm^3/s	L/s	7.576 820 E − 02
	ft^3/min	dm^3/s	L/s	6.309 020 E − 02
	ft^3/s	dm^3/s	L/s	4.719 474 E − 01
		dm^3/s	L/s	2.831 685 E + 01
Flow rate (mole basis)	(lb. · mol)/s	kmol/s		4.535 924 E − 01
	(lb · mol)/h	kmol/s		1.259 979 E − 04
	million scf/D	kmol/s		1.383 45 E − 02
Flow rate/length (mass basis)	lbm/(s · ft)	kg/(s · m)		1.488 164 E + 00
	lbm/(h · ft)	kg/(s · m)		4.133 789 E − 04
Flow rate/length (volume basis)	U.K. gal/(min · ft)	m^2/s	m^3/(s · m)	2.485 833 E − 04
	U.S. gal/(min · ft)	m^2/s	m^3/(s · m)	2.069 888 E − 04
	U.K. gal/(h · in)	m^2/s	m^3/(s · m)	4.971 667 E − 05
	U.S. gal/(h · in)	m^2/s	m^3/(s · m)	4.139 776 E − 05
	U.K. gal/(h · ft)	m^2/s	m^3/(s · m)	4.143 055 E − 06
	U.S. gal/(h · ft)	m^2/s	m^3/(s · m)	3.449 814 E − 06
Flow rate/area (mass basis)	lbm/(s · ft^2)	kg/(s · m^2)		4.882 428 E + 00
	lbm/(h · ft^2)	kg/(s · m^2)		1.356 230 E − 03

TABLE 1-4 Conversion Factors: U.S. Customary and Commonly Used Units to SI Units *(Continued)*

Quantity	Customary or commonly used unit	SI unit	Alternate SI unit	Conversion factor; multiply customary unit by factor to obtain SI unit
Flow rate/area (volume basis)	$ft^3/(s \cdot ft^2)$	m/s	$m^3/(s \cdot m^2)$	3.048° E − 01
	$ft^3/(min \cdot ft^2)$	m/s	$m^3/(s \cdot m^2)$	5.08° E − 03
	U.K. $gal/(h \cdot in^2)$	m/s	$m^3/(s \cdot m^2)$	1.957 349 E − 03
	U.S. $gal/(h \cdot in^2)$	m/s	$m^3/(s \cdot m^2)$	1.629 833 E − 03
	U.K. $gal/(min \cdot ft^2)$	m/s	$m^3/(s \cdot m^2)$	8.155 621 E − 04
	U.S. $gal/(min \cdot ft^2)$	m/s	$m^3/(s \cdot m^2)$	6.790 972 E − 04
	U.K. $gal/(h \cdot ft^2)$	m/s	$m^3/(s \cdot m^2)$	1.359 270 E − 05
	U.S. $gal/(h \cdot ft^2)$	m/s	$m^3/(s \cdot m^2)$	1.131 829 E − 05
		Energy, work, power		
Energy, work	therm	MJ		1.055 056 E + 02
		kJ		1.055 056 E + 05
		kWh		2.930 711 E + 01
	U.S. tonf · mi	MJ		1.431 744 E + 01
	hp · h	MJ		2.684 520 E + 00
		kJ		2.684 520 E + 03
		kWh		7.456 999 E − 01
	ch · h or CV · h	MJ		2.647 780 E + 00
		kJ		2.647 780 E + 03
		kWh		7.354 999 E − 01
	kWh	MJ		3.6° E + 00
		kJ		3.6° E + 03
	Chu	kJ		1.899 101 E + 00
		kWh		5.275 280 E − 04
	Btu	kJ		1.055 056 E + 00
		kWh		2.930 711 E − 04
	kcal	kJ		4.184° E + 00
	cal	kJ		4.184° E − 03
	ft · lbf	kJ		1.355 818 E − 03
	lbf · ft	kJ		1.355 818 E − 03
	J	kJ		1.0° E − 03
	$(lbf \cdot ft^2)/s^2$	kJ		4.214 011 E − 05
	erg	J		1.0° E − 07
Impact energy	kgf · m	J		9.806 650°E + 00
	lbf · ft	J		1.355 818 E + 00
Surface energy	erg/cm^2	mJ/m^2		1.0° E + 00
Specific-impact energy	$(kgf \cdot m)/cm^2$	J/cm^2		9.806 650°E − 02
	$(lbf \cdot ft)/in^2$	J/cm^2		2.101 522 E − 03
Power	million Btu/h	MW		2.930 711 E − 01
	ton of refrigeration	kW		3.516 853 E + 00
	Btu/s	kW		1.055 056 E + 00
	kW	kW		1
	hydraulic horsepower—hhp	kW		7.460 43 E − 01
	hp (electric)	kW		7.46° E − 01
	hp [(550 ft · lbf)/s]	kW		7.456 999 E − 01
	ch or CV	kW		7.354 999 E − 01
	Btu/min	kW		1.758 427 E − 02
	(ft · lbf)/s	kW		1.355 818 E − 03
	kcal/h	W		1.162 222 E + 00
	Btu/h	W		2.930 711 E − 01
	(ft · lbf)/min	W		2.259 697 E − 02
Power/area	$Btu/(s \cdot ft^2)$	kW/m^2		1.135 653 E + 01
	$cal/(h \cdot cm^2)$	kW/m^2		1.162 222 E − 02
	$Btu/(h \cdot ft^2)$	kW/m^2		3.154 591 E − 03
Heat-release rate, mixing power	hp/ft^3	kW/m^3		2.633 414 E + 01
	$cal/(h \cdot cm^3)$	kW/m^3		1.162 222 E + 00
	$Btu/(s \cdot ft^3)$	kW/m^3		3.725 895 E + 01
	$Btu/(h \cdot ft^3)$	kW/m^3		1.034 971 E − 02
Cooling duty (machinery)	Btu/(bhp · h)	W/kW		3.930 148 E − 01
Specific fuel consumption (mass basis)	lbm/(hp · h)	mg/J	kg/MJ	1.689 659 E − 01
		kg/kWh		6.082 774 E − 01
Specific fuel consumption (volume basis)	m^3/kWh	dm^3/MJ	mm^3/J	2.777 778 E + 02
	U.S. gal/(hp · h)	dm^3/MJ	mm^3/J	1.410 089 E + 00
	U.K. pt/(hp · h)	dm^3/MJ	mm^3/J	2.116 806 E − 01

TABLE 1-4 Conversion Factors: U.S. Customary and Commonly Used Units to SI Units *(Continued)*

Quantity	Customary or commonly used unit	SI unit	Alternate SI unit	Conversion factor; multiply customary unit by factor to obtain SI unit
Fuel consumption	U.K. gal/mi	$dm^3/100$ km	L/100 km	2.824 807 E + 02
	U.S. gal/mi	$dm^3/100$ km	L/100 km	2.352 146 E + 02
	mi/U.S. gal	km/dm^3	km/L	4.251 437 E − 01
	mi/U.K. gal	km/dm^3	km/L	3.540 064 E − 01
Velocity (linear), speed	knot	km/h		1.852* E + 00
	mi/h	km/h		1.609 344* E + 00
	ft/s	m/s		3.048* E − 01
		cm/s		3.048* E + 01
	ft/min	m/s		5.08* E − 03
	ft/h	mm/s		8.466 667 E − 02
	ft/day	mm/s		3.527 778 E − 03
		m/d		3.048* E − 01
	in/s	mm/s		2.54* E + 01
	in/min	mm/s		4.233 333 E − 01
Corrosion rate	in/year (ipy)	mm/a		2.54* E + 01
	mil/year	mm/a		2.54* E − 02
Rotational frequency	r/min	r/s		1.666 667 E − 02
		rad/s		1.047 198 E − 01
Acceleration (linear)	ft/s^2	m/s^2		3.048* E − 01
		cm/s^2		3.048* E + 01
Acceleration (rotational)	rpm/s	rad/s^2		1.047 198 E − 01
Momentum	(lbm · ft)/s	(kg · m)/s		1.382 550 E − 01
Force	U.K. tonf	kN		9.964 016 E + 00
	U.S. tonf	kN		8.896 443 E + 00
	kgf (kp)	N		9.806 650* E + 00
	lbf	N		4.448 222 E + 00
	dyn	mN		1.0 E − 02
Bending moment, torque	U.S. tonf · ft	kN · m		2.711 636 E + 00
	kgf · m	N · m		9.806 650* E + 00
	lbf · ft	N · m		1.355 818 E + 00
	lbf · in	N · m		1.129 848 E − 01
Bending moment/length	(lbf · ft)/in	(N · m)/m		5.337 866 E + 01
	(lbf · in)/in	(N · m)/m		4.448 222 E + 00
Moment of inertia	lbm · ft^2	kg · m^2		4.214 011 E − 02
Stress	U.S. $tonf/in^2$	MPa	N/mm^2	1.378 951 E + 01
	kgf/mm^2	MPa	N/mm^2	9.806 650* E + 00
	U.S. $tonf/ft^2$	MPa	N/mm^2	9.576 052 E − 02
	lbf/in^2 (psi)	MPa	N/mm^2	6.894 757 E − 03
	lbf/ft^2 (psf)	kPa		4.788 026 E − 02
	dyn/cm^2	Pa		1.0* E − 01
Mass/length	lbm/ft	kg/m		1.488 164 E + 00
Mass/area structural loading, bearing capacity (mass basis)	U.S. ton/ft^2	Mg/m^2		9.764 855 E + 00
	lbm/ft^2	kg/m^2		4.882 428 E + 00
		Miscellaneous transport properties		
Diffusivity	ft^2/s	m^2/s		9.290 304* E − 02
	m^2/s	mm^2/s		1.0* E + 06
	ft^2/h	m^2/s		2.580 64* E − 05
Thermal resistance	(°C · m^2 · h)/kcal	(K · m^2)/kW		8.604 208 E + 02
	(°F · ft^2 · h)/Btu	(K · m^2)/kW		1.761 102 E + 02
Heat flux	Btu/(h · ft^2)	kW/m^2		3.154 591 E − 03
Thermal conductivity	(cal · cm)/(s · cm^2 · °C)	W/(m · K)		4.184* E + 02
	(Btu · ft)/(h · ft^2 · °F)	W/(m · K)		1.730 735 E + 00
		(kJ · m)/(h · m^2 · K)		6.230 646 E + 00
	(kcal · m)/(h · m^2 · °C)	W/(m · K)		1.162 222 E + 00
	(Btu · in)/(h · ft^2 · °F)	W/(m · K)		1.442 279 E − 01
	(cal · cm)/(h · cm^2 · °C)	W/(m · K)		1.162 222 E − 01

TABLE 1-4 Conversion Factors: U.S. Customary and Commonly Used Units to SI Units *(Continued)*

Quantity	Customary or commonly used unit	SI unit	Alternate SI unit	Conversion factor; multiply customary unit by factor to obtain SI unit
Heat-transfer coefficient	cal/(s·cm^2·°C)	kW/(m^2·K)		4.184° E + 01
	Btu/(s·ft^2·°F)	kW/(m^2·K)		2.044 175 E + 01
	cal/(h·cm^2·°C)	kW/(m^2·K)		1.162 222 E − 02
	Btu/(h·ft^2·°F)	kW/(m^2·K)		5.678 263 E − 03
		kJ/(h·m^2·K)		2.044 175 E + 01
	Btu/(h·ft^2·°R)	kW/(m^2·K)		5.678 263 E − 03
	kcal/(h·m^2·°C)	kW/(m^2·K)		1.162 222 E − 03
Volumetric heat-transfer coefficient	Btu/(s·ft^3·°F)	kW/(m^3·K)		6.706 611 E + 01
	Btu/(h·ft^3·°F)	kW/(m^3·K)		1.862 947 E − 02
Surface tension	dyn/cm	mN/m		1
Viscosity (dynamic)	(lbf·s)/in^2	Pa·s	(N·s)/m^2	6.894 757 E + 03
	(lbf·s)/ft^2	Pa·s	(N·s)/m^2	4.788 026 E + 01
	(kgf·s)/m^2	Pa·s	(N·s)/m^2	9.806 650° E + 00
	lbm/(ft·s)	Pa·s	(N·s)/m^2	1.488 164 E + 00
	(dyn·s)/cm^2	Pa·s	(N·s)/m^2	1.0° E − 01
	cP	Pa·s	(N·s)/m^2	1.0° E − 03
	lbm/(ft·h)	Pa·s	(N·s)/m^2	4.133 789 E − 04
Viscosity (kinematic)	ft^2/s	m^2/s		9.290 304° E − 02
	in^2/s	mm^2/s		6.451 6° E + 02
	m^2/h	mm^2/s		2.777 778 E + 02
	ft^2/h	m^2/s		2.580 64° E − 05
	cSt	mm^2/s		1
Permeability	darcy	μm^2		9.869 233 E − 01
	millidarcy	μm^2		9.869 233 E − 04
Thermal flux	Btu/(h·ft^2)	W/m^2		3.152 E + 00
	Btu/(s·ft^2)	W/m^2		1.135 E + 04
	cal/(s·cm^2)	W/m^2		4.184 E + 04
Mass-transfer coefficient	(lb·mol)/[h·ft^2(lb·mol/ft^3)]	m/s		8.467 E − 05
	(g·mol)/[s·m^2(g·mol/L)]	m/s		1.0 E + 01
		Electricity, magnetism		
Admittance	S	S		1
Capacitance	μF	μF		1
Charge density	C/mm^3	C/mm^3		1
Conductance	S	S		1
	℧ (mho)	S		1
Conductivity	S/m	S/m		1
	℧/m	S/m		1
	m℧/m	mS/m		1
Current density	A/mm^2	A/mm^2		1
Displacement	C/cm^2	C/cm^2		1
Electric charge	C	C		1
Electric current	A	A		1
Electric-dipole moment	C·m	C·m		1
Electric-field strength	V/m	V/m		1
Electric flux	C	C		1
Electric polarization	C/cm^2	C/cm^2		1
Electric potential	V	V		1
	mV	mV		1
Electromagnetic moment	A·m^2	A·m^2		1
Electromotive force	V	V		1
Flux of displacement	C	C		1
Frequency	cycles/s	Hz		1
Impedance	Ω	Ω		1

TABLE 1-4 Conversion Factors: U.S. Customary and Commonly Used Units to SI Units *(Continued)*

Quantity	Customary or commonly used unit	SI unit	Alternate SI unit	Conversion factor; multiply customary unit by factor to obtain SI unit
Linear-current density	A/mm	A/mm		1
Magnetic-dipole moment	Wb·m	Wb·m		1
Magnetic-field strength	A/mm Oe gamma	A/mm A/m A/m		1 7.957 747 E + 01 7.957 747 E − 04
Magnetic flux	mWb	mWb		1
Magnetic-flux density	mT G gamma	mT T nT		1 1.0° E − 04 1
Magnetic induction	mT	mT		1
Magnetic moment	$A \cdot m^2$	$A \cdot m^2$		1
Magnetic polarization	mT	mT		1
Magnetic potential difference	A	A		1
Magnetic-vector potential	Wb/mm	Wb/mm		1
Magnetization	A/mm	A/mm		1
Modulus of admittance	S	S		1
Modulus of impedance	Ω	Ω		1
Mutual inductance	H	H		1
Permeability	μH/m	μH/m		1
Permeance	H	H		1
Permittivity	μF/m	μF/m		1
Potential difference	V	V		1
Quantity of electricity	C	C		1
Reactance	Ω	Ω		1
Reluctance	H^{-1}	H^{-1}		1
Resistance	Ω	Ω		1
Resistivity	Ω·cm Ω·m	Ω·cm Ω·m		1 1
Self-inductance	mH	mH		1
Surface density of change	mC/m^2	mC/m^2		1
Susceptance	S	S		1
Volume density of charge	C/mm^3	C/mm^3		1
		Acoustics, light, radiation		
Absorbed dose	rad	Gy		1.0° E − 02
Acoustical energy	J	J		1
Acoustical intensity	W/cm^2	W/m^2		1.0° E + 04
Acoustical power	W	W		1
Sound pressure	N/m^2	N/m^2		1.0°
Illuminance	fc	lx		1.076 391 E + 01
Illumination	fc	lx		1.076 391 E + 01
Irradiance	W/m^2	W/m^2		1
Light exposure	fc·s	lx·s		1.076 391 E + 01
Luminance	cd/m^2	cd/m^2		1
Luminous efficacy	lm/W	lm/W		1
Luminous exitance	lm/m^2	lm/m^2		1

TABLE 1-4 Conversion Factors: U.S. Customary and Commonly Used Units to SI Units *(Continued)*

Quantity	Customary or commonly used unit	SI unit	Alternate SI unit	Conversion factor; multiply customary unit by factor to obtain SI unit	
Luminous flux	lm	lm		1	
Luminous intensity	cd	cd		1	
Radiance	$W/m^2 \cdot sr$	$W/m^2 \cdot sr$		1	
Radiant energy	J	J		1	
Radiant flux	W	W		1	
Radiant intensity	W/sr	W/sr		1	
Radiant power	W	W		1	
Wavelength	Å	nm		1.0*	E − 01
Capture unit	10^{-3} cm^{-1}	m^{-1}		1.0*	E + 01
			10^{-3} cm^{-1}	1	
	m^{-1}	m^{-1}		1	
Radioactivity	Ci	Bq		3.7*	E + 10

*An asterisk indicates that the conversion factor is exact.

†Conversion factors for length, area, and volume are based on the international foot. The international foot is longer by 2 parts in 1 million than the U.S. Survey foot (land-measurement use).

NOTE: The following unit symbols are used in the table:

Unit symbol	Name	Unit symbol	Name
A	ampere	lm	lumen
a	annum (year)	lx	lux
Bq	becquerel	m	meter
C	coulomb	min	minute
cd	candela	′	minute
Ci	curie	N	newton
d	day	naut mi	U.S. nautical mile
°C	degree Celsius	Oe	oersted
°	degree	Ω	ohm
dyn	dyne	Pa	pascal
F	farad	rad	radian
fc	footcandle	r	revolution
G	gauss	S	siemens
g	gram	s	second
gr	grain	″	second
Gy	gray	sr	steradian
H	henry	St	stokes
h	hour	T	tesla
ha	hectare	t	tonne
Hz	hertz	V	volt
J	joule	W	watt
K	kelvin	Wb	weber
L, ℓ, l	liter		

NOTE: Copyright SPE-AIME, *The SI Metric System of Units and SPE's Tentative Metric Standard*, Society of Petroleum Engineers, Dallas, 1977.

TABLE 1-5 Metric Conversion Factors as Exact Numerical Multiples of SI Units

The first two digits of each numerical entry represent a power of 10. For example, the entry "—02 2.54" expresses the fact that 1 in = 2.54 × 10^{-2} m.

To convert from	To	Multiply by
abampere	ampere	+01 1.00
abcoulomb	coulomb	+01 1.00
abfarad	farad	+09 1.00
abhenry	henry	−09 1.00
abmho	mho	+09 1.00
abohm	ohm	−09 1.00
abvolt	volt	−08 1.00
acre	$meter^2$	+03 4.046 856
ampere (international of 1948)	ampere	−01 9.998 35
angstrom	meter	−10 1.00
are	$meter^2$	+02 1.00
astronomical unit	meter	+11 1.495 978
atmosphere	newton/$meter^2$	+05 1.013 25
bar	newton/$meter^2$	+05 1.00
barn	$meter^2$	−28 1.00
barrel (petroleum 42 gal)	$meter^3$	−01 1.589 873
barye	newton/$meter^2$	−01 1.00
British thermal unit (ISO/TC 12)	joule	+03 1.055 06
British thermal unit (International Steam Table)	joule	+03 1.055 04
British thermal unit (mean)	joule	+03 1.055 87
British thermal unit (thermochemical)	joule	+03 1.054 350
British thermal unit (39°F)	joule	+03 1.059 67
British thermal unit (60°F)	joule	+03 1.054 68
bushel (U.S.)	$meter^3$	−02 3.523 907
cable	meter	+02 2.194 56
caliber	meter	−04 2.54
calorie (International Steam Table)	joule	+00 4.1868
calorie (mean)	joule	+ 00 4.190 02
calorie (thermochemical)	joule	+00 4.184
calorie (15°C)	joule	+00 4.185 80
calorie (20°C)	joule	+00 4.181 90
calorie (kilogram, International Steam Table)	joule	+03 4.186 8
calorie (kilogram, mean)	joule	+03 4.190 02
calorie (kilogram, thermochemical)	joule	+03 4.184
carat (metric)	kilogram	−04 2.00
Celsius (temperature)	kelvin	$t_K = t_c + 273.15$
centimeter of mercury (0°C)	newton/$meter^2$	+03 1.333 22
centimeter of water (4°C)	newton/$meter^2$	+01 9.806 38
chain (engineer's)	meter	+01 3.048
chain (surveyor's or Gunter's)	meter	+01 2.011 68
circular mil	$meter^2$	−10 5.067 074
cord	$meter^3$	+00 3.624 556
coulomb (international of 1948)	coulomb	−01 9.998 35
cubit	meter	−01 4.572
cup	$meter^3$	−04 2.365 882
curie	disintegration/second	+10 3.70
day (mean solar)	second (mean solar)	+ 04 8.64
day (sidereal)	second (mean solar)	+04 8.616 409
degree (angle)	radian	−02 1.745 329
denier (international)	kilogram/meter	−07 1.00
dram (avoirdupois)	kilogram	− 03 1.771 845
dram (troy or apothecary)	kilogram	−03 3.887 934
dram (U.S. fluid)	$meter^3$	−06 3.696 691
dyne	newton	−05 1.00
electron volt	joule	−19 1.602 10
erg	joule	−07 1.00
Fahrenheit (temperature)	kelvin	$t_K = (5/9)(t_F + 459.67)$
Fahrenheit (temperature)	Celsius	$t_c = (5/9)(t_F - 32)$
farad (international of 1948)	farad	−01 9.995 05
faraday (based on carbon 12)	coulomb	+04 9.648 70
faraday (chemical)	coulomb	+04 9.649 57
faraday (physical)	coulomb	+04 9.652 19
fathom	meter	+00 1.828 8
fermi (femtometer)	meter	−15 1.00
fluid ounce (U.S.)	$meter^3$	−05 2.957 352
foot	meter	−01 3.048
foot (U.S. survey)	meter	−01 3.048 006
foot of water (39.2°F)	newton/$meter^2$	+03 2.988 98
footcandle	lumen/$meter^2$	+01 1.076 391
footlambert	candela/$meter^2$	+00 3.426 259
furlong	meter	+02 2.011 68
gal (galileo)	meter/$second^2$	−02 1.00
gallon (U.K. liquid)	$meter^3$	−03 4.546 087
gallon (U.S. dry)	$meter^3$	−03 4.404 883
gallon (U.S. liquid)	$meter^3$	−03 3.785 411
gamma	tesla	−09 1.00
gauss	tesla	−04 1.00
gilbert	ampere turn	−01 7.957 747
gill (U.K.)	$meter^3$	−04 1.420 652
gill (U.S.)	$meter^3$	−04 1.182 941
grad	degree (angular)	−01 9.00
grad	radian	−02 1.570 796
grain	kilogram	−05 6.479 891
gram	kilogram	−03 1.00
hand	meter	−01 1.016
hectare	$meter^2$	+04 1.00
henry (international of 1948)	henry	+00 1.000 495
hogshead (U.S.)	$meter^3$	−01 2.384 809
horsepower (550 ft lbf/s)	watt	+02 7.456 998
horsepower (boiler)	watt	+03 9.809 50
horsepower (electric)	watt	+02 7.46
horsepower (metric)	watt	+02 7.354 99
horsepower (U.K.)	watt	+02 7.457
horsepower (water)	watt	+02 7.460 43
hour (mean solar)	second (mean solar)	+03 3.60
hour (sidereal)	second (mean solar)	+03 3.590 170
hundredweight (long)	kilogram	+01 5.080 234
hundredweight (short)	kilogram	+01 4.535 923
inch	meter	−02 2.54
inch of mercury (32°F)	newton/$meter^2$	+03 3.386 389
inch of mercury (60°F)	newton/$meter^2$	+03 3.376 85
inch of water (39.2°F)	newton/$meter^2$	+02 2.490 82
inch of water (60°F)	newton/$meter^2$	+02 2.4884
joule (international of 1948)	joule	+00 1.000 165
kayser	1/meter	+02 1.00
kilocalorie (International Steam Table)	joule	+03 4.186 74
kilocalorie (mean)	joule	+03 4.190 02
kilocalorie (thermochemical)	joule	+03 4.184
kilogram mass	kilogram	+00 1.00
kilogram-force (kgf)	newton	+00 9.806 65
kilopond-force	newton	+00 9.806 65
kip	newton	+03 4.448 221
knot (international)	meter/second	−01 5.144 444
lambert	candela/$meter^2$	+04 $1/\pi$
lambert	candela/$meter^2$	+03 3.183 098
langley	joule/$meter^2$	+04 4.184
lbf (pound-force, avoidupois)	newton	+00 4.448 221
lbm (pound-mass, avoirdupois)	kilogram	−01 4.535 923
league (British nautical)	meter	+03 5.559 552
league (international nautical)	meter	+03 5.556
league (statute)	meter	+03 4.828 032
light-year	meter	+15 9.460 55
link (engineer's)	meter	−01 3.048
link (surveyor's or Gunter's)	meter	−01 2.011 68
liter	$meter^3$	−03 1.00
lux	lumen/$meter^2$	+00 1.00
maxwell	weber	−08 1.00
meter	wavelengths Kr 86	+06 1.650 763
micrometer	meter	−06 1.00
mil	meter	−05 2.54
mile (U.S. statute)	meter	+03 1.609 344
mile (U.K. nautical)	meter	+03 1.853 184
mile (international nautical)	meter	+03 1.852
mile (U.S. nautical)	meter	+03 1.852
millibar	newton/$meter^2$	+02 1.00
millimeter of mercury (0°C)	newton/$meter^2$	+02 1.333 224

TABLE 1-5 Metric Conversion Factors as Exact Numerical Multiples of SI Units *(Continued)*

The first two digits of each numerical entry represent a power of 10. For example, the entry "—02 2.54" expresses the fact that 1 in = 2.54 × 10^{-2}

To convert from	To	Multiply by
minute (angle)	radian	−04 2.908 882
minute (mean solar)	second (mean solar)	+01 6.00
minute (sidereal)	second (mean solar)	+01 5.983 617
month (mean calendar)	second (mean solar)	+06 2.628
nautical mile (international)	meter	+03 1.852
nautical mile (U.S.)	meter	+03 1.852
nautical mile (U.K.)	meter	+03 1.853 184
oersted	ampere/meter	+01 7.957 747
ohm (international of 1948)	ohm	+00 1.000 495
ounce-force (avoirdupois)	newton	−01 2.780 138
ounce-mass (avoirdupois)	kilogram	−02 2.834 952
ounce-mass (troy or apothecary)	kilogram	−02 3.110 347
ounce (U.S. fluid)	meter^3	−05 2.957 352
pace	meter	−01 7.62
parsec	meter	+16 3.083 74
pascal	newton/meter^2	+00 1.00
peck (U.S.)	meter^3	−03 8.809 767
pennyweight	kilogram	−03 1.555 173
perch	meter	+00 5.0292
phot	lumen/meter^2	+04 1.00
pica (printer's)	meter	−03 4.217 517
pint (U.S. dry)	meter^3	−04 5.506 104
pint (U.S. liquid)	meter^3	−04 4.731 764
point (printer's)	meter	−04 3.514 598
poise	(newton-second)/meter^2	−01 1.00
pole	meter	+00 5.0292
pound-force (lbf avoirdupois)	newton	+00 4.448 221
pound-mass (lbm avoirdupois)	kilogram	−01 4.535 923
pound-mass (troy or apothecary)	kilogram	−01 3.732 417
poundal	newton	−01 1.382 549
quart (U.S. dry)	meter^3	−03 1.101 220
quart (U.S. liquid)	meter^3	−04 9.463 529
rad (radiation dose absorbed)	joule/kilogram	−02 1.00
Rankine (temperature)	kelvin	$t_K = (5/9)t_R$
rayleigh (rate of photon emission)	1/second-meter^2	+10 1.00
rhe	meter^2/(newton-second)	+01 1.00
rod	meter	+00 5.0292
roentgen	coulomb/kilogram	−04 2.579 76
rutherford	disintegration/second	+06 1.00
second (angle)	radian	−06 4.848 136
second (ephemeris)	second	+00 1.000 000
second (mean solar)	second (ephemeris)	Consult American Ephemeris and Nautical Almanac
second (sidereal)	second (mean solar)	−01 9.972 695
section	meter^2	+06 2.589 988
scruple (apothecary)	kilogram	−03 1.295 978
shake	second	−08 1.00
skein	meter	+02 1.097 28
slug	kilogram	+01 1.459 390
span	meter	−01 2.286
statampere	ampere	−10 3.335 640
statcoulomb	coulomb	−10 3.335 640
statfarad	farad	−12 1.112 650
stathenry	henry	+11 8.987 554
statmho	mho	−12 1.112 650
statohm	ohm	+11 8.987 554
statute mile (U.S.)	meter	+03 1.609 344
statvolt	volt	+02 2.997 925
stere	meter^3	+00 1.00
stilb	candela/meter^2	+04 1.00
stoke	meter^2/second	−04 1.00
tablespoon	meter^3	−05 1.478 676
teaspoon	meter^3	−06 4.928 921
ton (assay)	kilogram	−02 2.916 666
ton (long)	kilogram	+03 1.016 046
ton (metric)	kilogram	+03 1.00
ton (nuclear equivalent of TNT)	joule	+09 4.20
ton (register)	meter^3	+00 2.831 684
ton (short, 2000 lb)	kilogram	+02 9.071 847
tonne	kilogram	+03 1.00
torr (0°C)	newton/meter^2	+02 1.333 22
township	meter^2	+07 9.323 957
unit pole	weber	−07 1.256 637
volt (international of 1948)	volt	+00 1.000 330
watt (international of 1948)	watt	+00 1.000 165
yard	meter	−01 9.144
year (calendar)	second (mean solar)	+07 3.1536
year (sidereal)	second (mean solar)	+07 3.155 815
year (tropical)	second (mean solar)	+07 3.155 692
year 1900, tropical, Jan., day 0, hour 12	second (ephemeris)	+07 3.155 692
year 1900, tropical, Jan., day 0, hour 12	second	+07 3.155 692

TABLE 1-6 Alphabetical Listing of Common Conversions

To convert from	To	Multiply by
Acres	Square feet	43,560
Acres	Square meters	4074
Acres	Square miles	0.001563
Acre-feet	Cubic meters	1233
Ampere-hours (absolute)	Coulombs (absolute)	3600
Angstrom units	Inches	3.937×10^{-9}
Angstrom units	Meters	1×10^{-10}
Angstrom units	Microns	1×10^{-4}
Atmospheres	Millimeters of mercury at 32°F.	760
Atmospheres	Dynes per square centimeter	1.0133×10^{6}
Atmospheres	Newtons per square meter	101,325
Atmospheres	Feet of water at 39.1°F.	33.90
Atmospheres	Grams per square centimeter	1033.3
Atmospheres	Inches of mercury at 32°F.	29.921
Atmospheres	Pounds per square foot	2116.3
Atmospheres	Pounds per square inch	14.696
Bags (cement)	Pounds (cement)	94
Barrels (cement)	Pounds (cement)	376
Barrels (oil)	Cubic meters	0.15899
Barrels (oil)	Gallons	42
Barrels (U.S. liquid)	Cubic meters	0.11924
Barrels (U.S. liquid)	Gallons	31.5
Barrels per day	Gallons per minute	0.02917
Bars	Atmospheres	0.9869
Bars	Newtons per square meter	1×10^{5}
Bars	Pounds per square inch	14.504
Board feet	Cubic feet	$1/12$
Boiler horsepower	B.t.u. per hour	33,480
Boiler horsepower	Kilowatts	9.803
B.t.u.	Calories (gram)	252
B.t.u.	Centigrade heat units (c.h.u. or p.c.u.)	0.55556
B.t.u.	Foot-pounds	777.9
B.t.u.	Horsepower-hours	3.929×10^{-4}
B.t.u.	Joules	1055.1
B.t.u.	Liter-atmospheres	10.41
B.t.u.	Pounds carbon to CO_2	6.88×10^{-5}
B.t.u.	Pounds water evaporated from and at 212°F.	0.001036
B.t.u.	Cubic foot-atmospheres	0.3676
B.t.u.	Kilowatt-hours	2.930×10^{-4}
B.t.u. per cubic foot	Joules per cubic meter	37,260
B.t.u. per hour	Watts	0.29307
B.t.u. per minute	Horsepower	0.02357
B.t.u. per pound	Joules per kilogram	2326
B.t.u. per pound per degree Fahrenheit	Calories per gram per degree centigrade	1
B.t.u. per pound per degree Fahrenheit	Joules per kilogram per degree Kelvin	4186.8
B.t.u. per second	Watts	1054.4
B.t.u. per square foot per hour	Joules per square meter per second	3.1546
B.t.u. per square foot per minute	Kilowatts per square foot	0.1758
B.t.u. per square foot per second for a temperature gradient of 1°F. per inch	Calories, gram (15°C.), per square centimeter per second for a temperature gradient of 1°C. per centimeter	1.2405
B.t.u. (60°F.) per degree Fahrenheit	Calories per degree centigrade	453.6
Bushels (U.S. dry)	Cubic feet	1.2444
Bushels (U.S. dry)	Cubic meters	0.03524
Calories, gram	B.t.u.	3.968×10^{-3}
Calories, gram	Foot-pounds	3.087
Calories, gram	Joules	4.1868
Calories, gram	Liter-atmospheres	4.130×10^{-2}
Calories, gram	Horsepower-hours	1.5591×10^{-6}
Calories. gram, per gram per degree C.	Joules per kilogram per degree Kelvin	4186.8
Calories, kilogram	Kilowatt-hours	0.0011626
Calories, kilogram per second	Kilowatts	4.185
Candle power (spherical)	Lumens	12.556
Carats (metric)	Grams	0.2
Centigrade heat units	B.t.u.	1.8
Centimeters	Angstrom units	1×10^{8}
Centimeters	Feet	0.03281
Centimeters	Inches	0.3937
Centimeters	Meters	0.01
Centimeters	Microns	10,000
Centimeters of mercury at 0°C.	Atmospheres	0.013158
Centimeters of mercury at 0°C.	Feet of water at 39.1°F.	0.4460
Centimeters of mercury at 0°C.	Newtons per square meter	1333.2
Centimeters of mercury at 0°C.	Pounds per square foot	27.845
Centimeters of mercury at 0°C.	Pounds per square inch	0.19337
Centimeters per second	Feet per minute	1.9685
Centimeters of water at 4°C.	Newtons per square meter	98.064
Centistokes	Square meters per second	1×10^{-6}
Circular mils	Square centimeters	5.067×10^{-6}
Circular mils	Square inches	7.854×10^{-7}
Circular mils	Square mils	0.7854
Cords	Cubic feet	128
Cubic centimeters	Cubic feet	3.532×10^{-5}
Cubic centimeters	Gallons	2.6417×10^{-4}
Cubic centimeters	Ounces (U.S. fluid)	0.03381
Cubic centimeters	Quarts (U.S. fluid)	0.0010567
Cubic feet	Bushels (U.S.)	0.8036
Cubic feet	Cubic centimeters	28,317
Cubic feet	Cubic meters	0.028317
Cubic feet	Cubic yards	0.03704
Cubic feet	Gallons	7.481
Cubic feet	Liters	28.316
Cubic foot-atmospheres	Foot-pounds	2116.3
Cubic foot-atmospheres	Liter-atmospheres	28.316
Cubic feet of water (60°F.)	Pounds	62.37
Cubic feet per minute	Cubic centimeters per second	472.0
Cubic feet per minute	Gallons per second	0.1247
Cubic feet per second	Gallons per minute	448.8
Cubic feet per second	Million gallons per day	0.64632
Cubic inches	Cubic meters	1.6387×10^{-5}
Cubic yards	Cubic meters	0.76456
Curies	Disintegrations per minute	2.2×10^{12}
Curies	Coulombs per minute	1.1×10^{12}
Degrees	Radians	0.017453
Drams (apothecaries' or troy)	Grams	3.888

TABLE 1-6 Alphabetical Listing of Common Conversions *(Continued)*

To convert from	To	Multiply by
Drams (avoirdupois)	Grams	1.7719
Dynes	Newtons	1×10^{-5}
Ergs	Joules	1×10^{-7}
Faradays	Coulombs (abs.)	96,500
Fathoms	Feet	6
Feet	Meters	0.3048
Feet per minute	Centimeters per second	0.5080
Feet per minute	Miles per hour	0.011364
Feet per $(\text{second})^2$	Meters per $(\text{second})^2$	0.3048
Feet of water at 39.2°F.	Newtons per square meter	2989
Foot-poundals	B.t.u.	3.995×10^{-5}
Foot-poundals	Joules	0.04214
Foot-poundals	Liter-atmospheres	4.159×10^{-4}
Foot-pounds	B.t.u.	0.0012856
Foot-pounds	Calories, gram	0.3239
Foot-pounds	Foot-poundals	32.174
Foot-pounds	Horsepower-hours	5.051×10^{-7}
Foot-pounds	Kilowatt-hours	3.766×10^{-7}
Foot-pounds	Liter-atmospheres	0.013381
Foot-pounds force	Joules	1.3558
Foot-pounds per second	Horsepower	0.0018182
Foot-pounds per second	Kilowatts	0.0013558
Furlongs	Miles	0.125
Gallons (U.S. liquid)	Barrels (U.S. liquid)	0.03175
Gallons	Cubic meters	0.003785
Gallons	Cubic feet	0.13368
Gallons	Gallons (Imperial)	0.8327
Gallons	Liters	3.785
Gallons	Ounces (U.S. fluid)	128
Gallons per minute	Cubic feet per hour	8.021
Gallons per minute	Cubic feet per second	0.002228
Grains	Grams	0.06480
Grains	Pounds	$^1/_{7000}$
Grains per cubic foot	Grams per cubic meter	2.2884
Grains per gallon	Parts per million	17.118
Grams	Drams (avoirdupois)	0.5644
Grams	Drams (troy)	0.2572
Grams	Grains	15.432
Grams	Kilograms	0.001
Grams	Pounds (avoirdupois)	0.0022046
Grams	Pounds (troy)	0.002679
Grams per cubic centimeter	Pounds per cubic foot	62.43
Grams per cubic centimeter	Pounds per gallon	8.345
Grams per liter	Grains per gallon	58.42
Grams per liter	Pounds per cubic foot	0.0624
Grams per square centimeter	Pounds per square foot	2.0482
Grams per square centimeter	Pounds per square inch	0.014223
Hectares	Acres	2.471
Hectares	Square meters	10,000
Horsepower (British)	B.t.u. per minute	42.42
Horsepower (British)	B.t.u. per hour	2545
Horsepower (British)	Foot-pounds per minute	33,000
Horsepower (British)	Foot-pounds per second	550
Horsepower (British)	Watts	745.7
Horsepower (British)	Horsepower (metric)	1.0139
Horsepower (British)	Pounds carbon to CO_2 per hour	0.175
Horsepower (British)	Pounds water evaporated per hour at 212°F	2.64
Horsepower (metric)	Foot-pounds per second	542.47
Horsepower (metric)	Kilogram-meters per second	75.0
Hours (mean solar)	Seconds	3600
Inches	Meters	0.0254
Inches of mercury at 60°F.	Newtons per square meter	3376.9
Inches of water at 60°F.	Newtons per square meter	248.84
Joules (absolute)	B.t.u. (mean)	9.480×10^{-4}
Joules (absolute)	Calories, gram (mean)	0.2389
Joules (absolute)	Cubic foot-atmospheres	0.3485
Joules (absolute)	Foot-pounds	0.7376
Joules (absolute)	Kilowatt-hours	2.7778×10^{-7}
Joules (absolute)	Liter-atmospheres	0.009869
Kilocalories	Joules	4186.8
Kilograms	Pounds (avoirdupois)	2.2046
Kilograms force	Newtons	9.807
Kilograms per square centimeter	Pounds per square inch	14.223
Kilometers	Miles	0.6214
Kilowatt-hours	B.t.u.	3414
Kilowatt-hours	Foot-pounds	2.6552×10^6
Kilowatts	Horsepower	1.3410
Knots (international)	Meters per second	0.5144
Knots (nautical miles per hour)	Miles per hour	1.1516
Lamberts	Candles per square inch	2.054
Liter-atmospheres	Cubic foot-atmospheres	0.03532
Liter-atmospheres	Foot-pounds	74.74
Liters	Cubic feet	0.03532
Liters	Cubic meters	0.001
Liters	Gallons	0.26418
Lumens	Watts	0.001496
Micromicrons	Microns	1×10^{-6}
Microns	Angstrom units	1×10^4
Microns	Meters	1×10^{-6}
Miles (nautical)	Feet	6080
Miles (nautical)	Miles (U.S. statute)	1.1516
Miles	Feet	5280
Miles	Meters	1609.3
Miles per hour	Feet per second	1.4667
Miles per hour	Meters per second	0.4470
Milliliters	Cubic centimeters	1
Millimeters	Meters	0.001
Millimeters of mercury at 0°C.	Newtons per square meter	133.32
Millimicrons	Microns	0.001
Mils	Inches	0.001
Mils	Meters	2.54×10^{-5}
Minims (U.S.)	Cubic centimeters	0.06161
Minutes (angle)	Radians	2.909×10^{-4}
Minutes (mean solar)	Seconds	60
Newtons	Kilograms	0.10197
Ounces (avoirdupois)	Kilograms	0.02835
Ounces (avoirdupois)	Ounces (troy)	0.9115
Ounces (U.S. fluid)	Cubic meters	2.957×10^{-5}
Ounces (troy)	Ounces (apothecaries')	1.000
Pints (U.S. liquid)	Cubic meters	4.732×10^{-4}
Poundals	Newtons	0.13826

TABLE 1-6 Alphabetical Listing of Common Conversions *(Continued)*

To convert from	To	Multiply by
Pounds (avoirdupois)	Grains	7000
Pounds (avoirdupois)	Kilograms	0.45359
Pounds (avoirdupois)	Pounds (troy)	1.2153
Pounds per cubic foot	Grams per cubic centimeter	0.016018
Pounds per cubic foot	Kilograms per cubic meter	16.018
Pounds per square foot	Atmospheres	4.725×10^{-4}
Pounds per square foot	Kilograms per square meter	4.882
Pounds per square inch	Atmospheres	0.06805
Pounds per square inch	Kilograms per square centimeter	0.07031
Pounds per square inch	Newtons per square meter	6894.8
Pounds force	Newtons	4.4482
Pounds force per square foot	Newtons per square meter	47.88
Pounds water evaporated from and at 212°F.	Horsepower-hours	0.379
Pound-centigrade units (p.c.u.)	B.t.u.	1.8
Quarts (U.S. liquid)	Cubic meters	9.464×10^{-4}
Radians	Degrees	57.30
Revolutions per minute	Radians per second	0.10472
Seconds (angle)	Radians	4.848×10^{-6}
Slugs	Gee pounds	1
Slugs	Kilograms	14.594
Slugs	Pounds	32.17
Square centimeters	Square feet	0.0010764

To convert from	To	Multiply by
Square feet	Square meters	0.0929
Square feet per hour	Square meters per second	2.581×10^{-5}
Square inches	Square centimeters	6.452
Square inches	Square meters	6.452×10^{-4}
Square yards	Square meters	0.8361
Stokes	Square meters per second	1×10^{-4}
Tons (long)	Kilograms	1016
Tons (long)	Pounds	2240
Tons (metric)	Kilograms	1000
Tons (metric)	Pounds	2204.6
Tons (metric)	Tons (short)	1.1023
Tons (short)	Kilograms	907.18
Tons (short)	Pounds	2000
Tons (refrigeration)	B.t.u. per hour	12,000
Tons (British shipping)	Cubic feet	42.00
Tons (U.S. shipping)	Cubic feet	40.00
Torr (mm. mercury, 0°C.)	Newtons per square meter	133.32
Watts	B.t.u. per hour	3.413
Watts	Joules per second	1
Watts	Kilogram-meters per second	0.10197
Watt-hours	Joules	3600
Yards	Meters	0.9144

TABLE 1-7 Special Tables of Conversion Factors

To convert from	To	Multiply by
h = heat-transfer coefficient		
p.c.u./(hr.)(ft.^{2})(°C.)	B.t.u./(hr.)(ft.^{2})(°F.)	1
kg.-cal./(hr.)(m.^{2})(°C.)	B.t.u./(hr.)(ft.^{2})(°F.)	0.2048
g.-cal./(sec.)(cm.^{2})(°C.)	B.t.u./(hr.)(ft.^{2})(°F.)	7380
watts/(cm.^{2})(°C.)	B.t.u./(hr.)(ft.^{2})(°F.)	1760
watts/(in.^{2})(°F.)	B.t.u./(hr.)(ft.^{2})(°F.)	490
B.t.u./(hr.)(ft.^{2})(°F.)	p.c.u./(hr.)(ft.^{2})(°C.)	1
B.t.u./(hr.)(ft.^{2})(°F.)	kg.-cal./(hr.)(m.^{2})(°C.)	4.88
B.t.u./(hr.)(ft.^{2})(°F.)	g.-cal./(sec.)(cm.^{2})(°C.)	0.0001355
B.t.u./(hr.)(ft.^{2})(°F.)	watts/(cm.^{2})(°C.)	0.000568
B.t.u./(hr.)(ft.^{2})(°F.)	watts/(in.^{2})(°F.)	0.00204
B.t.u./(hr.)(ft.^{2})(°F.)	hp./(ft.^{2})(°F.)	0.000394
B.t.u./(hr.)(ft.^{2})(°F.)	joules/(sec.)(m.^{2})(°C.)	5.678
kg.-cal./(hr.)(m.^{2})(°C.)	joules/(sec.)(m.^{2})(°C.)	1.163
watts/(m.^{2})(°C.)	joules/(sec.)(m.^{2})(°C.)	1.0
μ = viscosity		
centipoises	g./(sec.)(cm.) or poise	0.01
centipoises	lb./(sec.)(ft.)	0.000672
centipoises	lb./(hr.)(ft.)	2.42
centipoises	kg./(hr.)(m.)	3.60
centipoises	(newton)(sec.)/m.^{2}	0.001
lb./(sec.)(ft.)	(newton)(sec.)/m.^{2}	1.488
k = thermal conductivity		
g.-cal./(sec.)(cm.^{2})(°C./cm.)	B.t.u./(hr.)(ft.^{2})(°F./in.)	2903.0
watts/(cm.^{2})(°C./cm.)	B.t.u./(hr.)(ft.^{2})(°F./in.)	694.0
g.-cal./(hr.)(cm.^{2})(°C./cm.)	B.t.u./(hr.)(ft.^{2})(°F./in.)	0.8064
B.t.u./(hr.)(ft.^{2})(°F./ft.)	joules/(sec.)(m.)(°C.)	1.731
B.t.u./(hr.)(ft.^{2})(°F./in.)	joules/(sec.)(m.)(°C.)	0.1442

TABLE 1-8 Kinematic-Viscosity Conversion Formulas

Viscosity scale	Range of t, sec.	Kinematic viscosity, stokes
Saybolt Universal	$32 > t > 100$	$0.00226t - 1.95/t$
	$t > 100$	$0.00220t - 1.35/t$
Saybolt Furol	$25 > t > 40$	$0.0224t - 1.84/t$
	$t > 40$	$0.0216t - 0.60/t$
Redwood No. 1.	$34 > t > 100$	$0.00260t - 1.79/t$
	$t > 100$	$0.00247t - 0.50/t$
Redwood Admiralty	. . . .	$0.027t - 20/t$
Engler	. . . .	$0.00147t - 3.74/t$

TABLE 1-9 Values of the Gas-Law Constant

Temp. scale	Press. units	Vol. units	Wt. units	Energy units	R
Kelvin	. . . .	. . . . .	g.-moles	calories	1.9872
	. . . .	. . . . .	g.-moles	joules (abs.)	8.3144
	. . . .	. . . . .	g.-moles	joules (int.)	8.3130
	atm.	cm.^{3}	g.-moles	atm.-cm.^{3}	82.057
	atm.	liters	g.-moles	atm.-liters	0.08205
	mm. Hg	liters	g.-moles	mm. Hg-liters	62.361
	bar	liters	g.-moles	bar-liters	0.08314
	kg./cm.^{2}	liters	g.-moles	kg./(cm.^{2})(liters)	0.08478
	atm.	ft.^{3}	lb.-moles	atm.-ft.^{3}	1.314
	mm. Hg	ft.^{3}	lb.-moles	mm. Hg-ft.^{3}	998.9
	. . . .	. . . . .	lb.-moles	c.h.u. or p.c.u.	1.9872
Rankine	. . . .	. . . . .	lb.-moles	B.t.u.	1.9872
	. . . .	. . . . .	lb.-moles	hp.-hr.	0.0007805
	. . . .	. . . . .	lb.-moles	kw.-hr.	0.0005819
	atm.	ft.^{3}	lb.-moles	atm.-ft.^{3}	0.7302
	in. Hg	ft.^{3}	lb.-moles	in. Hg-ft.^{3}	21.85
	mm. Hg	ft.^{3}	lb.-moles	mm. Hg-ft.^{3}	555.0
	lb./in.^{2}abs.	ft.^{3}	lb.-moles	(lb.)(ft.^{3})/in.^{2}	10.73
	lb./ft.^{2}abs.	ft.^{3}	lb.-moles	ft.-lb.	1545.0

TABLE 1-10 United States Customary System of Weights and Measures

Linear Measure

12 inches (in.) or (″) = 1 foot (ft.) or (′)
3 feet = 1 yard (yd.)
16.5 feet, 5.5 yards = 1 rod (rd.)
5280 feet, 320 rods = 1 mile (mi.)
1 mil = 0.001 inch

Nautical:

6080.2 feet = 1 nautical mile
6 feet = 1 fathom
120 fathoms = 1 cable length
1 knot = 1 nautical mile per hour
60 nautical miles = 1° of latitude

Square Measure

144 sq. inches (sq. in.) or (in.^{2}) or (□″) = 1 sq. foot (ft.^{2}) or (□′)
9 sq. feet (ft.^{2}) (□′) = 1 sq. yard (yd.^{2})
30.25 sq. yards = 1 sq. rod, pole, or perch
160 sq. rods = {10 sq. chains, 43,560 sq. ft.} = 1 acre
640 acres = 1 sq. mile = 1 section
1 circular inch (area of circle of 1 inch diameter) = 0.7854 sq. inch
1 sq. inch = 1.2732 circular inch
1 circular mil = area of circle of 0.001 inch diameter
1,000,000 circular mils = 1 circular inch

Circular Measure

60 seconds (″) (sec.) = 1 minute (min.) or (′)
60 minutes (′) = 1 degree (°)
90 degrees (°) = 1 quadrant
360 degrees (°) = 1 circumference
57.29578 degrees = 1 radian (rad.) = 57° 17′ 44.81″

Volume Measure

Solid:

1728 cubic in. (cu. in.) (in.^{3}) = 1 cubic foot (cu. ft.)(ft.^{3})
27 cu. ft. = 1 cubic yard (cu. yd.)

Dry Measure:

2 pints = 1 quart
8 quarts = 1 peck
4 pecks = 1 bushel
1 United States Winchester bushel = 2150.42 cubic inches

Liquid:

4 gills = 1 pint (pt.)
2 pints = 1 quart (qt.)
4 quarts = 1 gallon (gal.)
7.4805 gallons = 1 cubic foot

Apothecaries' Liquid:

60 minims (min. or ♏) = 1 fluid dram or drachm
8 drams (ʒ) = 1 fluid ounce
16 ounces (oz. ℥) = 1 pint

Avoirdupois Weight

16 drams = 437.5 grains = 1 ounce (oz.)
16 ounces = 7000 grains = 1 pound (lb.)
100 pounds = 1 hundredweight (cwt.)
2000 pounds = 1 short ton; 2240 pounds = 1 long ton

Troy Weight

24 grains = 1 pennyweight (dwt.)
20 pennyweights = 1 ounce (oz.)
12 ounces = 1 pound (lb.)

Apothecaries' Weight

20 grains (gr.) = 1 scruple (℈)
3 scruples = 1 dram (ʒ)
8 drams = 1 ounce (℥)
12 ounces = 1 pound (lb.)

TABLE 1-11 Specific Gravity, Degrees Baumé, Degrees API, Degrees Twaddell, Pounds per Gallon, Pounds per Cubic Foot*

$$°Bé = 145 - \frac{145}{sp\ gr}\ (\text{heavier than } H_2O);\quad °Bé = \frac{140}{sp\ gr} - 130\ (\text{lighter than } H_2O);\quad °Tw = \frac{sp\ gr\ 60°/60°F - 1}{0.005}\quad °API = \frac{141.5}{sp\ gr} - 131.5$$

Sp. gr. 60°/60°	°Bé.	°A.P.I.	Lb. per gal. at 60°F. wt. in air	Lb. per cu. ft. at 60°F. wt. in air	Sp. gr. 60°/60°	°Bé.	°A.P.I.	Lb. per gal. at 60°F. wt. in air	Lb. per cu. ft. at 60°F. wt. in air	Sp. gr. 60°/60°	°Bé.	°A.P.I.	Lb. per gal. at 60°F. wt. in air	Lb. per cu. ft. at 60°F. wt. in air	Sp. gr. 60°/60°	°Bé.	°A.P.I.	Lb. per gal. at 60°F. wt. in air	Lb. per cu. ft. at 60°F. wt. in air
0.600	103.33	104.33	4.9929	37.350	0.700	70.00	70.64	5.8268	43.587	0.800	45.00	45.38	6.6606	49.825	0.900	25.56	25.72	7.4944	56.062
.605	101.40	102.38	5.0346	37.662	.705	68.58	69.21	5.8685	43.899	.805	43.91	44.28	6.7023	50.137	.905	24.70	24.85	7.5361	56.374
.610	99.51	100.47	5.0763	37.973	.710	67.18	67.80	5.9101	44.211	.810	42.84	43.19	6.7440	50.448	.910	23.85	23.99	7.5777	56.685
.615	97.64	98.58	5.1180	38.285	.715	65.80	66.40	5.9518	44.523	.815	41.78	42.12	6.7857	50.760	.915	23.01	23.14	7.6194	56.997
.620	95.81	96.73	5.1597	38.597	.720	64.44	65.03	5.9935	44.834	.820	40.73	41.06	6.8274	51.072	.920	22.17	22.30	7.6612	57.310
.625	94.00	94.90	5.2014	38.910	.725	63.10	63.67	6.0352	45.146	.825	39.70	40.02	6.8691	51.384	.925	21.35	21.47	7.7029	57.622
.630	92.22	93.10	5.2431	39.222	.730	61.78	62.34	6.0769	45.458	.830	38.67	38.98	6.9108	51.696	.930	20.54	20.65	7.7446	57.934
.635	90.47	91.33	5.2848	39.534	.735	60.48	61.02	6.1186	45.770	.835	37.66	37.96	6.9525	52.008	.935	19.73	19.84	7.7863	58.246
.640	88.75	89.59	5.3265	39.845	.740	59.19	59.72	6.1603	46.082	.840	36.67	36.95	6.9941	52.320	.940	18.94	19.03	7.8280	58.557
.645	87.05	87.88	5.3682	40.157	.745	57.92	58.43	6.2020	46.394	.845	35.68	35.96	7.0358	52.632	.945	18.15	18.24	7.8697	58.869
.650	85.38	86.19	5.4098	40.468	.750	56.67	57.17	6.2437	46.706	.850	34.71	34.97	7.0775	52.943	.950	17.37	17.45	7.9114	59.181
.655	83.74	84.53	5.4515	40.780	.755	55.43	55.92	6.2854	47.018	.855	33.74	34.00	7.1192	53.255	.955	16.60	16.67	7.9531	59.493
.660	82.12	82.89	5.4932	41.092	.760	54.21	54.68	6.3271	47.330	.860	32.79	33.03	7.1609	53.567	.960	15.83	15.90	7.9947	59.805
.665	80.53	81.28	5.5349	41.404	.765	53.01	53.47	6.3688	47.642	.865	31.85	32.08	7.2026	53.879	.965	15.08	15.13	8.0364	60.117
.670	78.96	79.69	5.5766	41.716	.770	51.82	52.27	6.4104	47.953	.870	30.92	31.14	7.2443	54.191	.970	14.33	14.38	8.0780	60.428
.675	77.41	78.13	5.6183	42.028	.775	50.65	51.08	6.4521	47.265	.875	30.00	30.21	7.2860	54.503	.975	13.59	13.63	8.1197	60.740
.680	75.88	76.59	5.6600	42.340	.780	49.49	49.91	6.4938	48.577	.880	29.09	29.30	7.3277	54.815	.980	12.86	12.89	8.1615	61.052
.685	74.38	75.07	5.7017	42.652	.785	48.34	48.75	6.5355	48.889	.885	28.19	28.39	7.3694	55.127	.985	12.13	12.15	8.2032	61.364
.690	72.90	73.57	5.7434	42.963	.790	47.22	47.61	6.5772	49.201	.890	27.30	27.49	7.4111	55.438	.990	11.41	11.43	8.2449	61.676
.695	71.44	72.10	5.7851	43.275	.795	46.10	46.49	6.6189	49.513	.895	26.42	26.60	7.4528	55.750	.995	10.70	10.71	8.2866	61.988
															1.000	10.00	10.00	8.3283	62.300

Sp. gr. 60°/60°	°Bé.	°Tw.	Lb. per gal. at 60°F. wt. in air	Lb. per cu. ft. at 60°F. wt. in air	Sp. gr. 60°/60°	°Bé.	°Tw.	Lb. per gal. at 60°F. wt. in air	Lb. per cu. ft. at 60°F. wt. in air	Sp. gr. 60°/60°	°Bé.	°Tw.	Lb. per gal. at 60°F. wt. in air	Lb. per cu. ft. at 60°F. wt. in air	Sp. gr. 60°/60°	°Bé.	°Tw.	Lb. per gal. at 60°F. wt. in air	Lb. per cu. ft. at 60°F wt. in air
1.005	0.72	1	8.3700	62.612	1.255	29.46	51	10.4546	78.206	1.505	48.65	101	12.5392	93.800	1.755	62.38	151	14.6238	109.394
1.010	1.44	2	8.4117	62.924	1.260	29.92	52	10.4963	78.518	1.510	48.97	102	12.5809	94.112	1.760	62.61	152	14.6655	109.705
1.015	2.14	3	8.4534	63.236	1.265	30.38	53	10.5380	78.830	1.515	49.29	103	12.6226	94.424	1.765	62.85	153	14.7072	110.017
1.020	2.84	4	8.4950	63.547	1.270	30.83	54	10.5797	79.141	1.520	49.61	104	12.6643	94.735	1.770	63.08	154	14.7489	110.329
1.025	3.54	5	8.5367	63.859	1.275	31.27	55	10.6214	79.453	1.525	49.92	105	12.7060	95.047	1.775	63.31	155	14.7906	110.641
1.030	4.22	6	8.5784	64.171	1.280	31.72	56	10.6630	79.765	1.530	50.23	106	12.7477	95.359	1.780	63.54	156	14.8323	110.953
1.035	4.90	7	8.6201	64.483	1.285	32.16	57	10.7047	80.077	1.535	50.54	107	12.7894	95.671	1.785	63.77	157	14.8740	111.265
1.040	5.58	8	8.6618	64.795	1.290	32.60	58	10.7464	80.389	1.540	50.84	108	12.8310	95.983	1.790	63.99	158	14.9157	111.577
1.045	6.24	9	8.7035	65.107	1.295	33.03	59	10.7881	80.701	1.545	51.15	109	12.8727	96.295	1.795	64.22	159	14.9574	111.889
1.050	6.91	10	8.7452	65.419	1.300	33.46	60	10.8298	81.013	1.550	51.45	110	12.9144	96.606	1.800	64.44	160	14.9990	112.200
1.055	7.56	11	8.7869	65.731	1.305	33.89	61	10.8715	81.325	1.555	51.75	111	12.9561	96.918	1.805	64.67	161	15.0407	112.512
1.060	8.21	12	8.8286	66.042	1.310	34.31	62	10.9132	81.636	1.560	52.05	112	12.9978	97.230	1.810	64.89	162	15.0824	112.824
1.065	8.85	13	8.8703	66.354	1.315	34.73	63	10.9549	81.948	1.565	52.35	113	13.0395	97.542	1.815	65.11	163	15.1241	113.136
1.070	9.49	14	8.9120	66.666	1.320	35.15	64	10.9966	82.260	1.570	52.64	114	13.0812	97.854	1.820	65.33	164	15.1658	113.448
1.075	10.12	15	8.9537	66.978	1.325	35.57	65	11.0383	82.572	1.575	52.94	115	13.1229	98.166	1.825	65.55	165	15.2075	113.760
1.080	10.74	16	8.9954	67.290	1.330	35.98	66	11.0800	82.884	1.580	53.23	116	13.1646	98.478	1.830	65.77	166	15.2492	114.072
1.085	11.36	17	9.0371	67.602	1.335	36.39	67	11.1217	83.196	1.585	53.52	117	13.2063	98.790	1.835	65.98	167	15.2909	114.384
1.090	11.97	18	9.0787	67.914	1.340	36.79	68	11.1634	83.508	1.590	53.81	118	13.2480	99.102	1.840	66.20	168	15.3326	114.696
1.095	12.58	19	9.1204	68.226	1.345	37.19	69	11.2051	83.820	1.595	54.09	119	13.2897	99.414	1.845	66.41	169	15.3743	115.007
1.100	13.18	20	9.1621	68.537	1.350	37.59	70	11.2467	84.131	1.600	54.38	120	13.3313	99.725	1.850	66.62	170	15.4160	115.318
1.105	13.78	21	9.2038	68.849	1.355	37.99	71	11.2884	84.443	1.605	54.66	121	13.3730	100.037	1.855	66.83	171	15.4577	115.630
1.110	14.37	22	9.2455	69.161	1.360	38.38	72	11.3301	84.755	1.610	54.94	122	13.4147	100.349	1.860	67.04	172	15.4993	115.943
1.115	14.96	23	9.2872	69.473	1.365	38.77	73	11.3718	85.067	1.615	55.22	123	13.4564	100.661	1.865	67.25	173	15.5410	116.255
1.120	15.54	24	9.3289	69.785	1.370	39.16	74	11.4135	85.379	1.620	55.49	124	13.4981	100.973	1.870	67.46	174	15.5827	116.567
1.125	16.11	25	9.3706	70.097	1.375	39.55	75	11.4552	85.691	1.625	55.77	125	13.5398	101.285	1.875	67.67	175	15.6244	116.879
1.130	16.68	26	9.4123	70.409	1.380	39.93	76	11.4969	86.003	1.630	56.04	126	13.5815	101.597	1.880	67.87	176	15.6661	117.191
1.135	17.25	27	9.4540	70.721	1.385	40.31	77	11.5386	86.315	1.635	56.32	127	13.6232	101.909	1.885	68.08	177	15.7078	117.503
1.140	17.81	28	9.4957	71.032	1.390	40.68	78	11.5803	86.626	1.640	56.59	128	13.6649	102.220	1.890	68.28	178	15.7495	117.814
1.145	18.36	29	9.5374	71.344	1.395	41.06	79	11.6220	86.938	1.645	56.85	129	13.7066	102.532	1.895	68.48	179	15.7912	118.126
1.150	18.91	30	9.5790	71.656	1.400	41.43	80	11.6637	87.250	1.650	57.12	130	13.7483	102.844	1.900	68.68	180	15.8329	118.438
1.155	19.46	31	9.6207	71.968	1.405	41.80	81	11.7054	87.562	1.655	57.39	131	13.7900	103.156	1.905	68.88	181	15.8746	118.740
1.160	20.00	32	9.6624	72.280	1.410	42.16	82	11.7471	87.874	1.660	57.65	132	13.8317	103.468	1.910	69.08	182	15.9163	119.062
1.165	20.54	33	9.7041	72.592	1.415	42.53	83	11.7888	88.186	1.665	57.91	133	13.8734	103.780	1.915	69.28	183	15.9580	119.374
1.170	21.07	34	9.7458	72.904	1.420	42.89	84	11.8304	88.498	1.670	58.17	134	13.9150	104.092	1.920	69.48	184	15.9996	119.686
1.175	21.60	35	9.7875	73.216	1.425	43.25	85	11.8721	88.810	1.675	58.43	135	13.9567	104.404	1.925	69.68	185	16.0413	119.998
1.180	22.12	36	9.8292	73.528	1.430	43.60	86	11.9138	89.121	1.680	58.69	136	13.9984	104.715	1.930	69.87	186	16.0830	120.309
1.185	22.64	37	9.8709	73.840	1.435	43.95	87	11.9555	89.433	1.685	58.95	137	14.0401	105.027	1.935	70.06	187	16.1247	120.621
1.190	23.15	38	9.9126	74.151	1.440	44.31	88	11.9972	89.745	1.690	59.20	138	14.0818	105.339	1.940	70.26	188	16.1664	120.933
1.195	23.66	39	9.9543	74.463	1.445	44.65	89	12.0389	90.057	1.695	59.45	139	14.1235	105.651	1.945	70.45	189	16.2081	121.245
1.200	24.17	40	9.9960	74.775	1.450	45.00	90	12.0806	90.369	1.700	59.71	140	14.1652	105.963	1.950	70.64	190	16.2498	121.557
1.205	24.67	41	10.0377	75.087	1.455	45.34	91	12.1223	90.681	1.705	59.96	141	14.2069	106.275	1.955	70.83	191	16.2915	121.869
1.210	25.17	42	10.0793	75.399	1.460	45.68	92	12.1640	90.993	1.710	60.20	142	14.2486	106.587	1.960	71.02	192	16.3332	122.181
1.215	25.66	43	10.1210	75.711	1.465	46.02	93	12.2057	91.305	1.715	60.45	143	14.2903	106.899	1.965	71.21	193	13.3749	122.493
1.220	26.15	44	10.1627	76.022	1.470	46.36	94	12.2473	91.616	1.720	60.70	144	14.3320	107.210	1.970	71.40	194	16.4166	122.804
1.225	26.63	45	10.2044	76.334	1.475	46.69	95	12.2890	91.928	1.725	60.94	145	14.3737	107.522	1.975	71.58	195	16.4583	123.116
1.230	27.11	46	10.2461	76.646	1.480	47.03	96	12.3307	92.240	1.730	61.18	146	14.4153	107.834	1.980	71.77	196	16.5000	123.428
1.235	27.59	47	10.2878	76.958	1.485	47.36	97	12.3724	92.552	1.735	61.34	147	14.4570	108.146	1.985	71.95	197	16.5417	123.740
1.240	28.06	48	10.3295	77.270	1.490	47.68	98	12.4141	92.864	1.740	61.67	148	14.4987	108.458	1.990	72.14	198	16.5833	124.052
1.245	28.53	49	10.3712	77.582	1.495	48.01	99	12.4558	93.176	1.745	61.91	149	14.5404	108.770	1.995	72.32	199	16.6250	124.364
1.250	29.00	50	10.4129	77.894	1.500	48.33	100	12.4975	93.488	1.750	62.14	150	14.5821	109.082	2.000	72.50	200	16.6667	124.676

*Prepared by Lewis V. Judson, Ph.D., Chief of Length Section of National Bureau of Standards with the advice and assistance of E. L. Peffer, B.S., A.M., late Chief of Capacity and Density Section, National Bureau of Standards.

TABLE 1-12 Temperature Conversion

General formula: °F = (°C × 9/5) + 32; °C = (°F − 32) × 5/9

C.		F.
−273.1	**−459.4**	
−268	**−450**	
−262	**−440**	
−257	**−430**	
−251	**−420**	
−246	**−410**	
−240	**−400**	
−234	**−390**	
−229	**−380**	
−223	**−370**	
−218	**−360**	
−212	**−350**	
−207	**−340**	
−201	**−330**	
−196	**−320**	
−190	**−310**	
−184	**−300**	
−179	**−290**	
−173	**−280**	
−169	**−273**	−459.4
−168	**−270**	−454
−162	**−260**	−436
−157	**−250**	−418
−151	**−240**	−400
−146	**−230**	−382
−140	**−220**	−364
−134	**−210**	−346
−129	**−200**	−328
−123	**−190**	−310
−118	**−180**	−292
−112	**−170**	−274
−107	**−160**	−256
−101	**−150**	−238
−95.6	**−140**	−220
−90.0	**−130**	−202
−84.4	**−120**	−184
−78.9	**−110**	−166
−73.3	**−100**	−148
−67.8	**−90**	−130
−62.2	**−80**	−112
−56.7	**−70**	−94
−51.1	**−60**	−76
−45.6	**−50**	−58
−40.0	**−40**	−40
−34.4	**−30**	−22
−28.9	**−20**	−4
−23.3	**−10**	14
−17.8	**0**	32

C.		F.
−17.8	**0**	32
−17.2	**1**	33.8
−16.7	**2**	35.6
−16.1	**3**	37.4
−15.6	**4**	39.2
−15.0	**5**	41.0
−14.4	**6**	42.8
−13.9	**7**	44.6
−13.3	**8**	46.4
−12.8	**9**	48.2
−12.2	**10**	50.0
−11.7	**11**	51.8
−11.1	**12**	53.6
−10.6	**13**	55.4
−10.0	**14**	57.2
−9.44	**15**	59.0
−8.89	**16**	60.8
−8.33	**17**	62.6
−7.78	**18**	64.4
−7.22	**19**	66.2
−6.67	**20**	68.0
−6.11	**21**	69.8
−5.56	**22**	71.6
−5.00	**23**	73.4
−4.44	**24**	75.2
−3.89	**25**	77.0
−3.33	**26**	78.8
−2.78	**27**	80.6
−2.22	**28**	82.4
−1.67	**29**	84.2
−1.11	**30**	86.0
−0.56	**31**	87.8
0	**32**	89.6
0.56	**33**	91.4
1.11	**34**	93.2
1.67	**35**	95.0
2.22	**36**	96.8
2.78	**37**	98.6
3.33	**38**	100.4
3.89	**39**	102.2
4.44	**40**	104.0
5.00	**41**	105.8
5.56	**42**	107.6
6.11	**43**	109.4
6.67	**44**	111.2
7.22	**45**	113.0
7.78	**46**	114.8
8.33	**47**	116.6
8.89	**48**	118.4
9.44	**49**	120.2

C.		F.
10.0	**50**	122.0
10.6	**51**	123.8
11.1	**52**	125.6
11.7	**53**	127.4
12.2	**54**	129.2
12.8	**55**	131.0
13.3	**56**	132.8
13.9	**57**	134.6
14.4	**58**	136.4
15.0	**59**	138.2
15.6	**60**	140.0
16.1	**61**	141.8
16.7	**62**	143.6
17.2	**63**	145.4
17.8	**64**	147.2
18.3	**65**	149.0
18.9	**66**	150.8
19.4	**67**	152.6
20.0	**68**	154.4
20.6	**69**	156.2
21.1	**70**	158.0
21.7	**71**	159.8
22.2	**72**	161.6
22.8	**73**	163.4
23.3	**74**	165.2
23.9	**75**	167.0
24.4	**76**	168.8
25.0	**77**	170.6
25.6	**78**	172.4
26.1	**79**	174.2
26.7	**80**	176.0
27.2	**81**	177.8
27.8	**82**	179.6
28.3	**83**	181.4
28.9	**84**	183.2
29.4	**85**	185.0
30.0	**86**	186.8
30.6	**87**	188.6
31.1	**88**	190.4
31.7	**89**	192.2
32.2	**90**	194.0
32.8	**91**	195.8
33.3	**92**	197.6
33.9	**93**	199.4
34.4	**94**	201.2
35.0	**95**	203.0
35.6	**96**	204.8
36.1	**97**	206.6
36.7	**98**	208.4
37.2	**99**	210.2
37.8	**100**	212.0

C.		F.
38	**100**	212
43	**110**	230
49	**120**	248
54	**130**	266
60	**140**	284
66	**150**	302
71	**160**	320
77	**170**	338
82	**180**	356
88	**190**	374
93	**200**	392
99	**210**	410
100	**212**	413
104	**220**	428
110	**230**	446
116	**240**	464
121	**250**	482
127	**260**	500
132	**270**	518
138	**280**	536
143	**290**	554
149	**300**	572
154	**310**	590
160	**320**	608
166	**330**	626
171	**340**	644
177	**350**	662
182	**360**	680
188	**370**	698
193	**380**	716
199	**390**	734
204	**400**	752
210	**410**	770
216	**420**	788
221	**430**	806
227	**440**	824
232	**450**	842
238	**460**	860
243	**470**	878
249	**480**	896
254	**490**	914

C.		F.
260	**500**	932
266	**510**	950
271	**520**	968
277	**530**	986
282	**540**	1004
288	**550**	1022
293	**560**	1040
299	**570**	1058
304	**580**	1076
310	**590**	1094
316	**600**	1112
321	**610**	1130
327	**620**	1148
332	**630**	1166
338	**640**	1184
343	**650**	1202
349	**660**	1220
354	**670**	1238
360	**680**	1256
366	**690**	1274
371	**700**	1292
377	**710**	1310
382	**720**	1328
388	**730**	1346
393	**740**	1364
399	**750**	1382
404	**760**	1400
410	**770**	1418
416	**780**	1436
421	**790**	1454
427	**800**	1472
432	**810**	1490
438	**820**	1508
443	**830**	1526
449	**840**	1544
454	**850**	1562
460	**860**	1580
466	**870**	1598
471	**880**	1616
477	**890**	1634
482	**900**	1652
488	**910**	1670
493	**920**	1688
499	**930**	1706
504	**940**	1724
510	**950**	1742
516	**960**	1760
521	**970**	1778
527	**980**	1796
532	**990**	1814

C.		F.
538	**1000**	1832
543	**1010**	1850
549	**1020**	1868
554	**1030**	1886
560	**1040**	1904
566	**1050**	1922
571	**1060**	1940
577	**1070**	1958
582	**1080**	1976
588	**1090**	1994
593	**1100**	2012
599	**1110**	2030
604	**1120**	2048
610	**1130**	2066
616	**1140**	2084
621	**1150**	2102
627	**1160**	2120
632	**1170**	2138
638	**1180**	2156
643	**1190**	2174
649	**1200**	2192
654	**1210**	2210
660	**1220**	2228
666	**1230**	2246
671	**1240**	2264
677	**1250**	2282
682	**1260**	2300
688	**1270**	2318
693	**1280**	2336
699	**1290**	2354
704	**1300**	2372
710	**1310**	2390
716	**1320**	2408
721	**1330**	2426
727	**1340**	2444
732	**1350**	2462
738	**1360**	2480
743	**1370**	2498
749	**1380**	2516
754	**1390**	2534
760	**1400**	2552
766	**1410**	2570
771	**1420**	2588
777	**1430**	2606
782	**1440**	2624
788	**1450**	2642
793	**1460**	2660
799	**1470**	2678
804	**1480**	2696
810	**1490**	2714

C.		F.
816	**1500**	2732
821	**1510**	2750
827	**1520**	2768
832	**1530**	2786
838	**1540**	2804
843	**1550**	2822
849	**1560**	2840
854	**1570**	2858
860	**1580**	2876
866	**1590**	2894
871	**1600**	2912
877	**1610**	2930
882	**1620**	2948
888	**1630**	2966
893	**1640**	2984
899	**1650**	3002
904	**1660**	3020
910	**1670**	3038
916	**1680**	3056
921	**1690**	3074
927	**1700**	3092
932	**1710**	3110
938	**1720**	3128
943	**1730**	3146
949	**1740**	3164
954	**1750**	3182
960	**1760**	3200
966	**1770**	3218
971	**1780**	3236
977	**1790**	3254
982	**1800**	3272
988	**1810**	3290
993	**1820**	3308
999	**1830**	3326
1004	**1840**	3344
1010	**1850**	3362
1016	**1860**	3380
1021	**1870**	3398
1027	**1880**	3416
1032	**1890**	3434
1038	**1900**	3452
1043	**1910**	3470
1049	**1920**	3488
1054	**1930**	3506
1060	**1940**	3524
1066	**1950**	3542
1071	**1960**	3560
1077	**1970**	3578
1082	**1980**	3596
1088	**1990**	3614
1093	**2000**	3632

C.		F.
1093	**2000**	3632
1099	**2010**	3650
1104	**2020**	3668
1110	**2030**	3686
1116	**2040**	3704
1121	**2050**	3722
1127	**2060**	3740
1132	**2070**	3758
1138	**2080**	3776
1143	**2090**	3794
1149	**2100**	3812
1154	**2110**	3830
1160	**2120**	3848
1166	**2130**	3866
1171	**2140**	3884
1177	**2150**	3902
1182	**2160**	3920
1188	**2170**	3938
1193	**2180**	3956
1199	**2190**	3974
1204	**2200**	3992
1210	**2210**	4010
1216	**2220**	4028
1221	**2230**	4046
1227	**2240**	4064
1232	**2250**	4082
1238	**2260**	4100
1243	**2270**	4118
1249	**2280**	4136
1254	**2290**	4154
1260	**2300**	4172
1266	**2310**	4190
1271	**2320**	4208
1277	**2330**	4226
1282	**2340**	4244
1288	**2350**	4262
1293	**2360**	4280
1299	**2370**	4298
1304	**2380**	4316
1310	**2390**	4334
1316	**2400**	4352
1321	**2410**	4370
1327	**2420**	4388
1332	**2430**	4406
1338	**2440**	4424
1343	**2450**	4442
1349	**2460**	4460
1354	**2470**	4478
1360	**2480**	4496
1366	**2490**	4514

C.		F.
1371	**2500**	4532
1377	**2510**	4550
1382	**2520**	4568
1388	**2530**	4586
1393	**2540**	4604
1399	**2550**	4622
1404	**2560**	4640
1410	**2570**	4658
1416	**2580**	4676
1421	**2590**	4694
1427	**2600**	4712
1432	**2610**	4730
1438	**2620**	4748
1443	**2630**	4766
1449	**2640**	4784
1454	**2650**	4802
1460	**2660**	4820
1466	**2670**	4838
1471	**2680**	4856
1477	**2690**	4874
1482	**2700**	4892
1488	**2710**	4910
1493	**2720**	4928
1499	**2730**	4946
1504	**2740**	4964
1510	**2750**	4982
1516	**2760**	5000
1521	**2770**	5018
1527	**2780**	5036
1532	**2790**	5054
1538	**2800**	5072
1543	**2810**	5090
1549	**2820**	5108
1554	**2830**	5126
1560	**2840**	5144
1566	**2850**	5162
1571	**2860**	5180
1577	**2870**	5198
1582	**2880**	5216
1588	**2890**	5234
1593	**2900**	5252
1599	**2910**	5270
1604	**2920**	5288
1610	**2930**	5306
1616	**2940**	5324
1621	**2950**	5342
1627	**2960**	5360
1632	**2970**	5378
1638	**2980**	5396
1643	**2990**	5414
1649	**3000**	5432

Note.—The numbers in bold-face type refer to the temperature (in either centigrade or Fahrenheit degrees) which it is desired to convert into the other scale. If converting from Fahrenheit degrees to centigrade degrees the equivalent temperature is in the left column, while if converting from degrees centigrade to degrees Fahrenheit, the equivalent temperature is in the column on the right. This table, made by Albert Sauveur, is published by permission of Mrs. Albert Sauveur.

Interpolation Factors

C.		F.	C.		F.
0.56	**1**	1.8	3.33	**6**	10.8
1.11	**2**	3.6	3.89	**7**	12.6
1.67	**3**	5.4	4.44	**8**	14.4
2.22	**4**	7.2	5.00	**9**	16.2
2.78	**5**	9.0	5.56	**10**	18.0

TABLE 1-13 Wire and Sheet-Metal Gauges*

Values in approximate decimals of an inch

As a number of gauges are in use for various shapes and metals, it is **advisable to state the thickness in thousandths when specifying gauge number.**

Gauge number	American (AWG) or Brown & Sharpe (B & S) (for nonferrous wire and sheet)†	U.S. Steel Wire (Stl WG) or Washburn & Moen or Roebling or Am. Steel & Wire [A. (steel) WG] (for steel wire)	Birmingham (BWG) (for steel wire) or Stubs Iron Wire (for iron or brass wire)‡	U.S. Standard (for sheet and plate metal, wrought iron)	Standard Birmingham (BG) (for sheet and hoop metal)	Imperial Standard Wire Gauge (SWG) (British legal standard)	Gauge number
0000000	......	0.4900	......	......	0.6666	0.500	**0000000**
000000	......	.4615	......	......	.6250	.464	**000000**
00000	......	.4305	......	......	.5883	.432	**00000**
0000	0.460	.3938	0.454	......	.5416	.400	**0000**
000	.410	.3625	.425	......	.5000	.372	**000**
00	.365	.3310	.380	......	.4452	.348	**00**
0	.325	.3065	.340	......	.3964	.324	**0**
1	.289	.2830	.300	......	.3532	.300	**1**
2	.258	.2625	.284	......	.3147	.276	**2**
3	.229	.2437	.259	0.239	.2804	.252	**3**
4	.204	.2253	.238	.224	.2500	.232	**4**
5	.182	.2070	.220	.209	.2225	.212	**5**
6	.162	.1920	.203	.194	.1981	.192	**6**
7	.144	.1770	.180	.179	.1764	.176	**7**
8	.128	.1620	.165	.164	.1570	.160	**8**
9	.114	.1483	.148	.150	.1398	.144	**9**
10	.102	.1350	.134	.135	.1250	.128	**10**
11	.091	.1205	.120	.120	.1113	.116	**11**
12	.081	.1055	.109	.105	.0991	.104	**12**
13	.072	.0915	.095	.090	.0882	.092	**13**
14	.064	.0800	.083	.075	.0785	.080	**14**
15	.057	.0720	.072	.067	.0699	.072	**15**
16	.051	.0625	.065	.060	.0625	.064	**16**
17	.045	.0540	.058	.054	.0556	.056	**17**
18	.040	.0475	.049	.0478	.0495	.048	**18**
19	.036	.0410	.042	.0418	.0440	.040	**19**
20	.032	.0348	.035	.0359	.0392	.036	**20**
21	.0285	.0317	.032	.0329	.0349	.032	**21**
22	.0253	.0286	.028	.0299	.0313	.028	**22**
23	.0226	.0258	.025	.0269	.0278	.024	**23**
24	.0201	.0230	.022	.0239	.0248	.022	**24**
25	.0179	.0204	.020	.0209	.0220	.020	**25**

Gauge number	American (AWG) or Brown & Sharpe (B & S) (for nonferrous wire and sheet)‡	U.S. Steel Wire (Stl WG) or Washburn & Moen or Roebling or Am. Steel & Wire Co. [A. (steel) WG] (for steel wire)	Birmingham (BWG) (for steel wire) or Stubs Iron Wire (for iron or brass wire)‡	U.S. Standard (for sheet and plate metal, wrought iron)	Standard Birmingham (BG) (for sheet and hoop metal)	Imperial Standard and Wire Gauge (SWG) (British legal standard)	Gauge number
26	0.0159	0.0181	0.018	0.0188	0.0196	0.018	**26**
27	.0142	.0173	.016	.0172	.0175	.0164	**27**
28	.0126	.0162	.014	.0156	.0156	.0148	**28**
29	.0113	.0150	.013	.0141	.0139	.0136	**29**
30	.0100	.0140	.012	.0125	.0123	.0124	**30**
31	.0089	.0132	.010	.0109	.0110	.0116	**31**
32	.0080	.0128	.009	.0102	.0098	.0108	**32**
33	.0071	.0118	.008	.0094	.0087	.0100	**33**
34	.0063	.0104	.007	.0086	.0077	.0092	**34**
35	.0056	.0095	.005	.0078	.0069	.0084	**35**
36	.0050	.0090	.004	.0070	.0061	.0076	**36**
37	.0045	.0085	......	.0066	.0054	.0068	**37**
38	.0040	.0080	......	.0062	.0048	.0060	**38**
39	.0035	.0075	......	......	.0043	.0052	**39**
40	.0031	.0070	......	......	.0039	.0048	**40**
41	......	.0066	......	......	.0034	.0044	**41**
42	......	.0062	......	......	.0031	.0040	**42**
43	......	.0060	......	......	.0027	.0036	**43**
44	......	.0058	......	......	.0024	.0032	**44**
45	......	.0055	......	......	.0022	.0028	**45**
46	......	.0052	......	......	.0019	.0024	**46**
47	......	.0050	......	......	.0017	.0020	**47**
48	......	.0048	......	......	.0015	.0016	**48**
49	......	.0046	......	......	.0014	.0012	**49**
50	......	.0044	......	......	.0012	.0010	**50**

Metric wire gauge is 10 times the diameter in millimeters.

*Courtesy of Dr. Lewis V. Judson with I. H. Fullmer, National Bureau of Standards.

†Sometimes used for iron wire.

‡Sometimes used for copper plate and for steel plate 12 gauge and heavier and for steel tubes.

TABLE 1-14 Fundamental Physical Constants

1 sec. = 1.00273791 sidereal seconds	sec. = mean solar second
g_0 = 9.80665 m./sec.^{2}	Definition: g_0 = standard gravity
1 liter = 0.001 cu. m.	
1 atm. = 101,325 newtons/sq. m.	Definition: atm. = standard atmosphere
1 mm. Hg (pressure) = ($^1/_{760}$) atm.	mm. Hg (pressure) = standard millimeter mercury
= 133.3224 newtons/sq. m.	
1 int. ohm = 1.000495 $\pm$ 0.000015 abs. ohm	int. = international; abs. = absolute
1 int. amp. = 0.999835 $\pm$ 0.000025 abs. amp.	amp. = ampere
1 int. coul. = 0.999835 $\pm$ 0.000025 abs. coul.	coul. = coulomb
1 int. volt = 1.000330 $\pm$ 0.000029 abs. volt	
1 int. watt = 1.000165 $\pm$ 0.000052 abs. watt	
1 int. joule = 1.000165 $\pm$ 0.000052 abs. joule	
$T_{0°C.}$ = 273.150 $\pm$ 0.010°K.	Absolute temperature of the ice point, 0°C.
$(PV)_{0°C.}^{P=0} = (RT)_{0°C.}$ = 2271.16 $\pm$ 0.04 abs. joule/mole	
= 22,414.6 $\pm$ 0.4 cu. cm. atm./mole	*PV* product for ideal gas at 0°C.
= 22.4146 $\pm$ 0.0004 liter atm./mole	
R = 8.31439 $\pm$ 0.00034 abs. joule/deg. mole	R = gas constant per mole
= 1.98719 $\pm$ 0.00013 cal./deg. mole	
= 82.0567 $\pm$ 0.0034 cu. cm. atm./deg. mole	
= 0.0820567 $\pm$ 0.0000034 liter atm./deg. mole	
ln 10 = 2.302585	ln = natural logarithm (base e)
R ln 10 = 19.14460 $\pm$ 0.00078 abs. joule/deg. mole	
= 4.57567 $\pm$ 0.00030 cal./deg. mole	
N = (6.02283 $\pm$ 0.0022) $\times 10^{23}$/mole	N = *Avogadro number*
h = (6.6242 $\pm$ 0.0044) $\times 10^{-34}$ joule sec.	h = Planck constant
c = (2.99776 $\pm$ 0.00008) $\times 10^8$ m./sec.	c = velocity of light
$(h^2/8\pi^2k)$ = (4.0258 $\pm$ 0.0037) $\times 10^{-39}$ g. sq. cm. deg.	Constant in rotational partition function of gases
$(h/8\pi^2c)$ = (2.7986 $\pm$ 0.0018) $\times 10^{-39}$ g. cm.	Constant relating wave number and moment of inertia
$Z = Nhc$ = 11.9600 $\pm$ 0.0036 abs. joule cm./mole	Z = constant relating wave number and energy per mole
= 2.85851 $\pm$ 0.0009 cal. cm./mole	
$(Z/R) = (hc/k) = c_2$ = 1.43847 $\pm$ 0.00045 cm. deg.	c_2 = second radiation constant
$\mathfrak{F}$ = 96,501.2 $\pm$ 10.0 int. coul./g.-equiv. or int. joule/int. volt g.-equiv.	$\mathfrak{F}$ = Faraday constant
= 96,485.3 $\pm$ 10.0 abs. coul./g.-equiv. or abs. joule/abs. volt g.-equiv.	
= 23,068.1 $\pm$ 2.4 cal./int. volt g.-equiv.	
= 23,060.5 $\pm$ 2.4 cal./abs. volt g.-equiv.	
e = (1.60199 $\pm$ 0.00060) $\times 10^{-19}$ abs. coul.	e = electronic charge
= (1.60199 $\pm$ 0.00060) $\times 10^{-20}$ abs. e.m.u.	
= (4.80239 $\pm$ 0.00180) $\times 10^{-10}$ abs. e.s.u.	
1 int. electron-volt/molecule = 96,501.2 $\pm$ 10 int. joule/mole	
= 23,068.1 $\pm$ 2.4 cal./mole	
1 abs. electron-volt/molecule = 96,485.3 $\pm$ 10. abs. joule/mole	
= 23,060.5 $\pm$ 2.4 cal./mole	
1 int. electron-volt = (1.60252 $\pm$ 0.00060) $\times 10^{-12}$ erg	
1 abs. electron-volt = (1.60199 $\pm$ 0.00060) $\times 10^{-12}$ erg	
hc = (1.23916 $\pm$ 0.00032) $\times 10^{-4}$ int. electron-volt cm.	Constant relating wave number and energy per molecule
= (1.23957 $\pm$ 0.00032) $\times 10^{-4}$ abs. electron-volt cm.	
k = (8.61442 $\pm$ 0.00100) $\times 10^{-5}$ int. electron-volt/deg.	k = Boltzmann constant
= (8.61727 $\pm$ 0.00100) $\times 10^{-5}$ abs. electron-volt/deg.	
= (R/N) = (1.38048 $\pm$ 0.00050) $\times 10^{-23}$ joule/deg.	
1 I.T. cal. = ($^1/_{860}$) = 0.00116279 int. watt-hr.	Definition of I.T. cal.: I.T. = International steam tables
= 4.18605 int. joule	
= 4.18674 abs. joule	
= 1.000654 cal.	cal. = thermochemical calorie
1 cal. = 4.1840 abs. joule	Definition: cal. = thermochemical calorie
= 4.1833 int. joule	
= 41.2929 $\pm$ 0.0020 cu. cm. atm.	
= 0.0412929 $\pm$ 0.0000020 liter atm.	
1 I.T. cal./g. = 1.8 B.t.u./lb.	Definition of B.t.u.: B.t.u. = I.T. British Thermal Unit
1 B.t.u. = 251.996 I.T. cal.	
= 0.293018 int. watt-hr.	
= 1054.866 int. joule	
= 1055.040 abs. joule	
= 252.161 cal.	cal. = thermochemical calorie
1 horsepower = 550 ft.-lb. (wt.)/sec.	Definition of horsepower (mechanical): lb. (wt.) = weight of 1 lb. at standard gravity
= 745.578 int. watt	
= 745.70 abs. watt	
1 in. = (1/0.3937) = 2.54 cm.	Definition of in.: in. = U.S. inch
1 ft. = 0.304800610 m.	ft. = U.S. foot (1 ft. = 12 in.)
1 lb. = 453.5924277 g.	Definition; lb. = avoirdupois pound
1 gal. = 231 cu. in.	Definition; gal. = U.S. gallon
= 0.133680555 cu. ft.	
= 3.785412 $\times 10^{-3}$ cu. m.	
= 3.785412 liter	

STATISTICAL TABLES

TABLE 1-15 Ordinates and Areas between Abscissa Values $-z$ and $+z$ of the Normal Distribution Curve

z	X	Y	A	$1-A$	z	X	Y	A	$1-A$
0	μ	0.399	0.0000	1.0000	±1.50	$\mu \pm 1.50\sigma$	0.1295	0.8664	0.1336
±0.05	$\mu \pm 0.05\sigma$	.398	.0399	0.9601	±1.55	$\mu \pm 1.55\sigma$	.1200	.8789	.1211
± .10	$\mu \pm .10\sigma$	.397	.0797	.9203	±1.60	$\mu \pm 1.60\sigma$	.1109	.8904	.1096
± .15	$\mu \pm .15\sigma$	.394	.1192	.8808	±1.65	$\mu \pm 1.65\sigma$	.1023	.9011	.0989
± .20	$\mu \pm .20\sigma$	.391	.1585	.8415	±1.70	$\mu \pm 1.70\sigma$	.0940	.9109	.0891
± .25	$\mu \pm .25\sigma$	.387	.1974	.8026	±1.75	$\mu \pm 1.75\sigma$	.0863	.9199	.0801
± .30	$\mu \pm .30\sigma$	.381	.2358	.7642	±1.80	$\mu \pm 1.80\sigma$	.0790	.9281	.0719
± .35	$\mu \pm .35\sigma$	.375	.2737	.7263	±1.85	$\mu \pm 1.85\sigma$	.0721	.9357	.0643
± .40	$\mu \pm .40\sigma$	.368	.3108	.6892	±1.90	$\mu \pm 1.90\sigma$	.0656	.9446	.0574
± .45	$\mu \pm .45\sigma$	.361	.3473	.6527	±1.95	$\mu \pm 1.95\sigma$	.0596	.9488	.0512
± .50	$\mu \pm .50\sigma$	.352	.3829	.6171	±2.00	$\mu \pm 2.00\sigma$	.0540	.9545	.0455
± .55	$\mu + .55\sigma$	.343	.4177	.5823	±2.05	$\mu \pm 2.05\sigma$	.0488	.9596	.0404
± .60	$\mu \pm .60\sigma$	.333	.4515	.5485	±2.10	$\mu \pm 2.10\sigma$	.0440	.9643	.0357
± .65	$\mu \pm .65\sigma$	.323	.4843	.5157	±2.15	$\mu \pm 2.15\sigma$	.0396	.9684	.0316
± .70	$\mu \pm .70\sigma$	.312	.5161	.4839	±2.20	$\mu \pm 2.20\sigma$	.0335	.9722	.0278
± .75	$\mu \pm .75\sigma$	.301	.5467	.4533	±2.25	$\mu \pm 2.25\sigma$	.0317	.9756	.0244
± .80	$\mu \pm .80\sigma$	.290	.5763	.4237	±2.30	$\mu \pm 2.30\sigma$	.0283	.9786	.0214
± .85	$\mu \pm .85\sigma$	.278	.6047	.3953	±2.35	$\mu \pm 2.35\sigma$	.0252	.9812	.0188
± .90	$\mu \pm .90\sigma$	.266	.6319	.3681	±2.40	$\mu \pm 2.40\sigma$	.0224	.9836	.0164
± .95	$\mu \pm .95\sigma$	.254	.6579	.3421	±2.45	$\mu \pm 2.45\sigma$	.0198	.9857	.0143
±1.00	$\mu \pm 1.00\sigma$	.242	.6827	.3173	±2.50	$\mu \pm 2.50\sigma$	.0175	.9876	.0124
±1.05	$\mu \pm 1.05\sigma$	.230	.7063	.2937	±2.55	$\mu \pm 2.55\sigma$	.0154	.9892	.0108
±1.10	$\mu \pm 1.10\sigma$	.218	.7287	.2713	±2.60	$\mu \pm 2.60\sigma$	.0136	.9907	.0093
±1.15	$\mu \pm 1.15\sigma$	.206	.7499	.2501	±2.65	$\mu \pm 2.65\sigma$	.0119	.9920	.0080
±1.20	$\mu \pm 1.20\sigma$	.194	.7699	.2301	±2.70	$\mu \pm 2.70\sigma$	.0104	.9931	.0069
±1.25	$\mu \pm 1.25\sigma$	.183	.7887	.2113	±2.75	$\mu \pm 2.75\sigma$	.0091	.9940	.0060
±1.30	$\mu \pm 1.30\sigma$	.171	.8064	.1936	±2.80	$\mu \pm 2.80\sigma$	.0079	.9949	.0051
±1.35	$\mu \pm 1.35\sigma$	.160	.8230	.1770	±2.85	$\mu \pm 2.85\sigma$	.0069	.9956	.0044
±1.40	$\mu \pm 1.40\sigma$	.150	.8385	.1615	±2.90	$\mu \pm 2.90\sigma$	.0060	.9963	.0037
±1.45	$\mu \pm 1.45\sigma$	.139	.8529	.1471	±2.95	$\mu \pm 2.95\sigma$	.0051	.9968	.0032
±1.50	$\mu \pm 1.50\sigma$	.130	.8664	.1336	±3.00	$\mu \pm 3.00\sigma$	.0044	.9973	.0027
					±4.00	$\mu \pm 4.00\sigma$	.0001	.99994	.00006
					±5.00	$\mu \pm 5.00\sigma$	.000001	.9999994	.0000006
± 0.000	μ	0.3989	.0000	1.0000	±1.036	$\mu \pm 1.036\sigma$	0.2331	0.7000	0.3000
± .126	$\mu \pm 0.126\sigma$	.3958	.1000	0.9000	±1.282	$\mu \pm 1.282\sigma$	.1755	.8000	.2000
± .253	$\mu \pm .253\sigma$	.3863	.2000	.8000	±1.645	$\mu \pm 1.645\sigma$	.1031	.9000	.1000
± .385	$\mu \pm .385\sigma$	.3704	.3000	.7000	±1.960	$\mu + 1.960\sigma$	.0584	.9500	.0500
± .524	$\mu \pm .524\sigma$	.3477	.4000	.6000	±2.576	$\mu \pm 2.576\sigma$	.0145	.9900	.0100
± .674	$\mu \pm .674\sigma$	.3178	.5000	.5000	±3.291	$\mu \pm 3.291\sigma$	.0018	.9990	.0010
± .842	$\mu \pm .842\sigma$	.2800	.6000	.4000	±3.891	$\mu \pm 3.891\sigma$	.0002	.9999	.0001

TABLE 1-16 Values of t

df	$t_{.40}$	$t_{.30}$	$t_{.20}$	$t_{.10}$	$t_{.05}$	$t_{.025}$	$t_{.01}$	$t_{.005}$
1	0.325	0.727	1.376	3.078	6.314	12.706	31.821	63.657
2	.289	.617	1.061	1.886	2.920	4.303	6.965	9.925
3	.277	.584	0.978	1.638	2.353	3.182	4.541	5.841
4	.271	.569	.941	1.533	2.132	2.776	3.747	4.604
5	.267	.559	.920	1.476	2.015	2.571	3.365	4.032
6	.265	.553	.906	1.440	1.943	2.447	3.143	3.707
7	.263	.549	.896	1.415	1.895	2.365	2.998	3.499
8	.262	.546	.889	1.397	1.860	2.306	2.896	3.355
9	.261	.543	.883	1.383	1.833	2.262	2.821	3.250
10	.260	.542	.879	1.372	1.812	2.228	2.764	3.169
11	.260	.540	.876	1.363	1.796	2.201	2.718	3.106
12	.259	.539	.873	1.356	1.782	2.179	2.681	3.055
13	.259	.538	.870	1.350	1.771	2.160	2.650	3.012
14	.258	.537	.868	1.345	1.761	2.145	2.624	2.977
15	.258	.536	.866	1.341	1.753	2.131	2.602	2.947
16	.258	.535	.865	1.337	1.746	2.120	2.583	2.921
17	.257	.534	.863	1.333	1.740	2.110	2.567	2.898
18	.257	.534	.862	1.330	1.734	2.101	2.552	2.878
19	.257	.533	.861	1.328	1.729	2.093	2.539	2.861
20	.257	.533	.860	1.325	1.725	2.086	2.528	2.845
21	.257	.532	.859	1.323	1.721	2.080	2.518	2.831
22	.256	.532	.858	1.321	1.717	2.074	2.508	2.819
23	.256	.532	.858	1.319	1.714	2.069	2.500	2.807
24	.256	.531	.857	1.318	1.711	2.064	2.492	2.797
25	.256	.531	.856	1.316	1.708	2.060	2.485	2.787
26	.256	.531	.856	1.315	1.706	2.056	2.479	2.779
27	.256	.531	.855	1.314	1.703	2.052	2.473	2.771
28	.256	.530	.855	1.313	1.701	2.048	2.467	2.763
29	.256	.530	.854	1.311	1.699	2.045	2.462	2.756
30	.256	.530	.854	1.310	1.697	2.042	2.457	2.750
40	.255	.529	.851	1.303	1.684	2.021	2.423	2.704
60	.254	.527	.848	1.296	1.671	2.000	2.390	2.660
120	.254	.526	.845	1.289	1.658	1.980	2.358	2.617
∞	.253	.524	.842	1.282	1.645	1.960	2.326	2.576

Above values refer to a single tail outside the indicated limit of t. For example, for 95 percent of the area to be between $-t$ and $+t$ in a two-tailed t distribution, use the values for $t_{0.025}$ or 2.5 percent for each tail.

TABLE 1-17 Percentiles of the χ^2 Distribution

df	Per cent									
	0.5	1	2.5	5	10	90	95	97.5	99	99.5
1	0.000039	0.00016	0.00098	0.0039	0.0158	2.71	3.84	5.02	6.63	7.88
2	.0100	.0201	.0506	.1026	.2107	4.61	5.99	7.38	9.21	10.60
3	.0717	.115	.216	.352	.584	6.25	7.81	9.35	11.34	12.84
4	.207	.297	.484	.711	1.064	7.78	9.49	11.14	13.28	14.86
5	.412	.554	.831	1.15	1.61	9.24	11.07	12.83	15.09	16.75
6	.676	.872	1.24	1.64	2.20	10.64	12.59	14.45	16.81	18.55
7	.989	1.24	1.69	2.17	2.83	12.02	14.07	16.01	18.48	20.28
8	1.34	1.65	2.18	2.73	3.49	13.36	15.51	17.53	20.09	21.96
9	1.73	2.09	2.70	3.33	4.17	14.68	16.92	19.02	21.67	23.59
10	2.16	2.56	3.25	3.94	4.87	15.99	18.31	20.48	23.21	25.19
11	2.60	3.05	3.82	4.57	5.58	17.28	19.68	21.92	24.73	26.76
12	3.07	3.57	4.40	5.23	6.30	18.55	21.03	23.34	26.22	28.30
13	3.57	4.11	5.01	5.89	7.04	19.81	22.36	24.74	27.69	29.82
14	4.07	4.66	5.63	6.57	7.79	21.06	23.68	26.12	29.14	31.32
15	4.60	5.23	6.26	7.26	8.55	22.31	25.00	27.49	30.58	32.80
16	5.14	5.81	6.91	7.96	9.31	23.54	26.30	28.85	32.00	34.27
18	6.26	7.01	8.23	9.39	10.86	25.99	28.87	31.53	34.81	37.16
20	7.43	8.26	9.59	10.85	12.44	28.41	31.41	34.17	37.57	40.00
24	9.89	10.86	12.40	13.85	15.66	33.20	36.42	39.36	42.98	45.56
30	13.79	14.95	16.79	18.49	20.60	40.26	43.77	46.98	50.89	53.67
40	20.71	22.16	24.43	26.51	29.05	51.81	55.76	59.34	63.69	66.77
60	35.53	37.48	40.48	43.19	46.46	74.40	79.08	83.30	88.38	91.95
120	83.85	86.92	91.58	95.70	100.62	140.23	146.57	152.21	158.95	163.64

For large values of degrees of freedom the approximate formula

$$\chi^2_\alpha = n\left(1 - \frac{2}{9n} + z_\alpha\sqrt{\frac{2}{9n}}\right)^3$$

where z_α is the normal deviate and n is the number of degrees of freedom, may be used. For example, $\chi^2_{.99} = 60[1 - 0.00370 + 2.326(0.06086)]^3 = 60(1.1379)^3 = 88.4$ for the 99th percentile for 60 degrees of freedom.

TABLE 1-18 *F* Distribution

Upper 5% Points ($F_{.95}$)

Degrees of freedom for denominator	Degrees of freedom for numerator: 1	2	3	4	5	6	7	8	9	10	12	15	20	24	30	40	60	120	∞
1	161	200	216	225	230	234	237	239	241	242	244	246	248	249	250	251	252	253	254
2	18.5	19.0	19.2	19.2	19.3	19.3	19.4	19.4	19.4	19.4	19.4	19.4	19.4	19.5	19.5	19.5	19.5	19.5	19.5
3	10.1	9.55	9.28	9.12	9.01	8.94	8.89	8.85	8.81	8.79	8.74	8.70	8.66	8.64	8.62	8.59	8.57	8.55	8.53
4	7.71	6.94	6.59	6.39	6.26	6.16	6.09	6.04	6.00	5.96	5.91	5.86	5.80	5.77	5.75	5.72	5.69	5.66	5.63
5	6.61	5.79	5.41	5.19	5.05	4.95	4.88	4.82	4.77	4.74	4.68	4.62	4.56	4.53	4.50	4.46	4.43	4.40	4.37
6	5.99	5.14	4.76	4.53	4.39	4.28	4.21	4.15	4.10	4.06	4.00	3.94	3.87	3.84	3.81	3.77	3.74	3.70	3.67
7	5.59	4.74	4.35	4.12	3.97	3.87	3.79	3.73	3.68	3.64	3.57	3.51	3.44	3.41	3.38	3.34	3.30	3.27	3.23
8	5.32	4.46	4.07	3.84	3.69	3.58	3.50	3.44	3.39	3.35	3.28	3.22	3.15	3.12	3.08	3.04	3.01	2.97	2.93
9	5.12	4.26	3.86	3.63	3.48	3.37	3.29	3.23	3.18	3.14	3.07	3.01	2.94	2.90	2.86	2.83	2.79	2.75	2.71
10	4.96	4.10	3.71	3.48	3.33	3.22	3.14	3.07	3.02	2.98	2.91	2.85	2.77	2.74	2.70	2.66	2.62	2.58	2.54
11	4.84	3.98	3.59	3.36	3.20	3.09	3.01	2.95	2.90	2.85	2.79	2.72	2.65	2.61	2.57	2.53	2.49	2.45	2.40
12	4.75	3.89	3.49	3.26	3.11	3.00	2.91	2.85	2.80	2.75	2.69	2.62	2.54	2.51	2.47	2.43	2.38	2.34	2.30
13	4.67	3.81	3.41	3.18	3.03	2.92	2.83	2.77	2.71	2.67	2.60	2.53	2.46	2.42	2.38	2.34	2.30	2.25	2.21
14	4.60	3.74	3.34	3.11	2.96	2.85	2.76	2.70	2.65	2.60	2.53	2.46	2.39	2.35	2.31	2.27	2.22	2.18	2.13
15	4.54	3.68	3.29	3.06	2.90	2.79	2.71	2.64	2.59	2.54	2.48	2.40	2.33	2.29	2.25	2.20	2.16	2.11	2.07
16	4.49	3.63	3.24	3.01	2.85	2.74	2.66	2.59	2.54	2.49	2.42	2.35	2.28	2.24	2.19	2.15	2.11	2.06	2.01
17	4.45	3.59	3.20	2.96	2.81	2.70	2.61	2.55	2.49	2.45	2.38	2.31	2.23	2.19	2.15	2.10	2.06	2.01	1.96
18	4.41	3.55	3.16	2.93	2.77	2.66	2.58	2.51	2.46	2.41	2.34	2.27	2.19	2.15	2.11	2.06	2.02	1.97	1.92
19	4.38	3.52	3.13	2.90	2.74	2.63	2.54	2.48	2.42	2.38	2.31	2.23	2.16	2.11	2.07	2.03	1.98	1.93	1.88
20	4.35	3.49	3.10	2.87	2.71	2.60	2.51	2.45	2.39	2.35	2.28	2.20	2.12	2.08	2.04	1.99	1.95	1.90	1.84
21	4.32	3.47	3.07	2.84	2.68	2.57	2.49	2.42	2.37	2.32	2.25	2.18	2.10	2.05	2.01	1.96	1.92	1.87	1.81
22	4.30	3.44	3.05	2.82	2.66	2.55	2.46	2.40	2.34	2.30	2.23	2.15	2.07	2.03	1.98	1.94	1.89	1.84	1.78
23	4.28	3.42	3.03	2.80	2.64	2.53	2.44	2.37	2.32	2.27	2.20	2.13	2.05	2.01	1.96	1.91	1.86	1.81	1.76
24	4.26	3.40	3.01	2.78	2.62	2.51	2.42	2.36	2.30	2.25	2.18	2.11	2.03	1.98	1.94	1.89	1.84	1.79	1.73
25	4.24	3.39	2.99	2.76	2.60	2.49	2.40	2.34	2.28	2.24	2.16	2.09	2.01	1.96	1.92	1.87	1.82	1.77	1.71
30	4.17	3.32	2.92	2.69	2.53	2.42	2.33	2.27	2.21	2.16	2.09	2.01	1.93	1.89	1.84	1.79	1.74	1.68	1.62
40	4.08	3.23	2.84	2.61	2.45	2.34	2.25	2.18	2.12	2.08	2.00	1.92	1.84	1.79	1.74	1.69	1.64	1.58	1.51
60	4.00	3.15	2.76	2.53	2.37	2.25	2.17	2.10	2.04	1.99	1.92	1.84	1.75	1.70	1.65	1.59	1.53	1.47	1.39
120	3.92	3.07	2.68	2.45	2.29	2.18	2.09	2.02	1.96	1.91	1.83	1.75	1.66	1.61	1.55	1.50	1.43	1.35	1.25
∞	3.84	3.00	2.60	2.37	2.21	2.10	2.01	1.94	1.88	1.83	1.75	1.67	1.57	1.52	1.46	1.39	1.32	1.22	1.00

Upper 1% Points ($F_{.99}$)

Degrees of freedom for denominator	Degrees of freedom for numerator: 1	2	3	4	5	6	7	8	9	10	12	15	20	24	30	40	60	120	∞
1	4052	5000	5403	5625	5764	5859	5928	5982	6023	6056	6106	6157	6209	6235	6261	6287	6313	6339	6366
2	98.5	99.0	99.2	99.2	99.3	99.3	99.4	99.4	99.4	99.4	99.4	99.4	99.4	99.5	99.5	99.5	99.5	99.5	99.5
3	34.1	30.8	29.5	28.7	28.2	27.9	27.7	27.5	27.3	27.2	27.1	26.9	26.7	26.6	26.5	26.4	26.3	26.2	26.1
4	21.2	18.0	16.7	16.0	15.5	15.2	15.0	14.8	14.7	14.5	14.4	14.2	14.0	13.9	13.8	13.7	13.7	13.6	13.5
5	16.3	13.3	12.1	11.4	11.0	10.7	10.5	10.3	10.2	10.1	9.89	9.72	9.55	9.47	9.38	9.29	9.20	9.11	9.02
6	13.7	10.9	9.78	9.15	8.75	8.47	8.26	8.10	7.98	7.87	7.72	7.56	7.40	7.31	7.23	7.14	7.06	6.97	6.88
7	12.2	9.55	8.45	7.85	7.46	7.19	6.99	6.84	6.72	6.62	6.47	6.31	6.16	6.07	5.99	5.91	5.82	5.74	5.65
8	11.3	8.65	7.59	7.01	6.63	6.37	6.18	6.03	5.91	5.81	5.67	5.52	5.36	5.28	5.20	5.12	5.03	4.95	4.86
9	10.6	8.02	6.99	6.42	6.06	5.80	5.61	5.47	5.35	5.26	5.11	4.96	4.81	4.73	4.65	4.57	4.48	4.40	4.31
10	10.0	7.56	6.55	5.99	5.64	5.39	5.20	5.06	4.94	4.85	4.71	4.56	4.41	4.33	4.25	4.17	4.08	4.00	3.91
11	9.65	7.21	6.22	5.67	5.32	5.07	4.89	4.74	4.63	4.54	4.40	4.25	4.10	4.02	3.94	3.86	3.78	3.69	3.60
12	9.33	6.93	5.95	5.41	5.06	4.82	4.64	4.50	4.39	4.30	4.16	4.01	3.86	3.78	3.70	3.62	3.54	3.45	3.36
13	9.07	6.70	5.74	5.21	4.86	4.62	4.44	4.30	4.19	4.10	3.96	3.82	3.66	3.59	3.51	3.43	3.34	3.25	3.17
14	8.86	6.51	5.56	5.04	4.70	4.46	4.28	4.14	4.03	3.94	3.80	3.66	3.51	3.43	3.35	3.27	3.18	3.09	3.00
15	8.68	6.36	5.42	4.89	4.56	4.32	4.14	4.00	3.89	3.80	3.67	3.52	3.37	3.29	3.21	3.13	3.05	2.96	2.87
16	8.53	6.23	5.29	4.77	4.44	4.20	4.03	3.89	3.78	3.69	3.55	3.41	3.26	3.18	3.10	3.02	2.93	2.84	2.75
17	8.40	6.11	5.19	4.67	4.34	4.10	3.93	3.79	3.68	3.59	3.46	3.31	3.16	3.08	3.00	2.92	2.83	2.75	2.65
18	8.29	6.01	5.09	4.58	4.25	4.01	3.84	3.71	3.60	3.51	3.37	3.23	3.08	3.00	2.92	2.84	2.75	2.66	2.57
19	8.19	5.93	5.01	4.50	4.17	3.94	3.77	3.63	3.52	3.43	3.30	3.15	3.00	2.92	2.84	2.76	2.67	2.58	2.49
20	8.10	5.85	4.94	4.43	4.10	3.87	3.70	3.56	3.46	3.37	3.23	3.09	2.94	2.86	2.78	2.69	2.61	2.52	2.42
21	8.02	5.78	4.87	4.37	4.04	3.81	3.64	3.51	3.40	3.31	3.17	3.03	2.88	2.80	2.72	2.64	2.55	2.46	2.36
22	7.95	5.72	4.82	4.31	3.99	3.76	3.59	3.45	3.35	3.26	3.12	2.98	2.83	2.75	2.67	2.58	2.50	2.40	2.31
23	7.88	5.66	4.76	4.26	3.94	3.71	3.54	3.41	3.30	3.21	3.07	2.93	2.78	2.70	2.62	2.54	2.45	2.35	2.26
24	7.82	5.61	4.72	4.22	3.90	3.67	3.50	3.36	3.26	3.17	3.03	2.89	2.74	2.66	2.58	2.49	2.40	2.31	2.21
25	7.77	5.57	4.68	4.18	3.86	3.63	3.46	3.32	3.22	3.13	2.99	2.85	2.70	2.62	2.53	2.45	2.36	2.27	2.17
30	7.56	5.39	4.51	4.02	3.70	3.47	3.30	3.17	3.07	2.98	2.84	2.70	2.55	2.47	2.39	2.30	2.21	2.11	2.01
40	7.31	5.18	4.31	3.83	3.51	3.29	3.12	2.99	2.89	2.80	2.66	2.52	2.37	2.29	2.20	2.11	2.02	1.92	1.80
60	7.08	4.98	4.13	3.65	3.34	3.12	2.95	2.82	2.72	2.63	2.50	2.35	2.20	2.12	2.03	1.94	1.84	1.73	1.60
120	6.85	4.79	3.95	3.48	3.17	2.96	2.79	2.66	2.56	2.47	2.34	2.19	2.03	1.95	1.86	1.76	1.66	1.53	1.38
∞	6.63	4.61	3.78	3.32	3.02	2.80	2.64	2.51	2.41	2.32	2.18	2.04	1.88	1.79	1.70	1.59	1.47	1.32	1.00

Interpolation should be performed using reciprocals of the degrees of freedom.

TABLE 1-19*a* Circular Segments

Angle in degrees

Central angle, degrees	Height / R	Chord / R	Height / Chord	Area / R^2	Central angle, degrees	Height / R	Chord / R	Height / Chord	Area / R^2	Central angle, degrees	Height / R	Chord / R	Height / Chord	Area / R^2
1	0.000038	0.017453	0.002177	0.0000005	61	0.138371	1.01508	0.136315	0.0950155	121	0.507576	1.74071	0.291591	0.6273404
2	.000151	.034905	.004326	.0000035	62	.142833	1.03008	.138662	.0995782	122	.515190	1.74924	.294522	.6406267
3	.000343	.052354	.006552	.0000119	63	.147360	1.04500	.141014	.1042754	123	.522841	1.75763	.297469	.6540421
4	.000609	.069799	.008725	.0000283	64	.151952	1.05984	.143373	.1091083	124	.530528	1.76590	.300429	.6675852
5	.000952	.087239	.010913	.0000554	65	.156609	1.07460	.145737	.1140780	125	.538251	1.77402	.303411	.6812546
6	.001371	.104672	.013098	.0000956	66	.161329	1.08928	.148106	.1191858	126	.546010	1.78201	.306401	.6950488
7	.001865	.122097	.015275	.0001519	67	.166114	1.10387	.150483	.1244328	127	.553802	1.78987	.309409	.7089613
8	.002436	.139513	.017461	.0002266	68	.170962	1.11839	.152864	.1298199	128	.561629	1.79759	.312434	.7230052
9	.003083	.156918	.019647	.0003226	69	.175874	1.13281	.155255	.1353483	129	.569489	1.80517	.315477	.7371642
10	.003805	.174311	.021829	.0004423	70	.180848	1.14715	.157650	.1410188	130	.577382	1.81262	.318534	.7514417
11	.004604	.191692	.024018	.0005886	71	.185885	1.16140	.160053	.1468325	131	.585307	1.81992	.321611	.7658357
12	.005479	.209057	.026207	.0007639	72	.190983	1.17557	.162460	.1527902	132	.593263	1.82709	.324704	.7803448
13	.006428	.226406	.028391	.0009708	73	.196143	1.18965	.164875	.1588927	133	.601251	1.83412	.327814	.7949670
14	.007454	.243739	.030582	.0012121	74	.201365	1.20363	.167298	.1651409	134	.609269	1.84101	.330943	.8097006
15	.008555	.261052	.032771	.0014901	75	.206647	1.21752	.169728	.1715355	135	.617317	1.84776	.334089	.8245437
16	.009732	.278346	.034963	.0018076	76	.211989	1.23132	.172164	.1780773	136	.625393	1.85436	.337255	.8394945
17	.010984	.295619	.037156	.0021671	77	.217392	1.24503	.174608	.1847666	137	.633499	1.86084	.340437	.8545511
18	.012312	.312867	.039352	.0025711	78	.222854	1.25864	.177059	.1916045	138	.641632	1.86716	.343641	.8697117
19	.013714	.330095	.041547	.0030222	79	.228375	1.27216	.179518	.1985914	139	.649793	1.87334	.346863	.8849742
20	.015192	.347296	.043744	.0035229	80	.233956	1.28558	.181985	.2057277	140	.657980	1.87939	.350103	.9003667
21	.016745	.364471	.045943	.0040756	81	.239594	1.29890	.184459	.2130141	141	.666193	1.88528	.353366	.9157968
22	.018373	.381618	.048145	.0046829	82	.245290	1.31212	.186942	.2204508	142	.674432	1.89104	.356646	.9313529
23	.020075	.398736	.050347	.0053473	83	.251044	1.32524	.189433	.2280384	143	.682695	1.89665	.359948	.9470027
24	.021852	.415823	.052551	.0060712	84	.256855	1.33826	.191932	.2357772	144	.690983	1.90211	.363272	.9627442
25	.023704	.432879	.054759	.0068570	85	.262723	1.35118	.194440	.2436676	145	.699294	1.90743	.366616	.9785754
26	.025630	.449902	.056968	.0077072	86	.268646	1.36400	.196955	.2517094	146	.707628	1.91261	.369980	.9944937
27	.027630	.466891	.059178	.0086242	87	.274626	1.37671	.199481	.2599034	147	.715985	1.91764	.373368	1.0104973
28	.029704	.483844	.061392	.0096103	88	.280660	1.38932	.202012	.2682494	148	.724363	1.92252	.376778	1.0265840
29	.031852	.500760	.063607	.0106679	89	.286750	1.40182	.204556	.2767476	149	.732762	1.92726	.380209	1.0427512
30	.034074	.517638	.065826	.0117993	90	.292893	1.41421	.207107	.2853982	150	.741181	1.93185	.383664	1.0589969
31	.036370	.534477	.068048	.0130069	91	.299091	1.42650	.209668	.2942010	151	.749620	1.93630	.387140	1.0753188
32	.038738	.551275	.070270	.0142930	92	.305342	1.43868	.212238	.3031559	152	.758078	1.94059	.390643	1.0917144
33	.041180	.568031	.072496	.0156598	93	.311645	1.45075	.214816	.3122632	153	.766555	1.94474	.394168	1.1081816
34	.043695	.584743	.074725	.0171095	94	.318002	1.46271	.217406	.3215226	154	.775049	1.94874	.397718	1.1247180
35	.046283	.601412	.076957	.0186444	95	.324410	1.47456	.220005	.3309339	155	.783560	1.95259	.401293	1.1413210
36	.048944	.618034	.079193	.0202666	96	.330869	1.48629	.222614	.3404970	156	.792088	1.95630	.404891	1.1579885
37	.051676	.634609	.081430	.0219784	97	.337380	1.49791	.225234	.3502115	157	.800632	1.95985	.408517	1.1747179
38	.054481	.651136	.083671	.0237818	98	.343941	1.50942	.227863	.3600772	158	.809191	1.96325	.412169	1.1915068
39	.057359	.667614	.085917	.0256790	99	.350552	1.52081	.230503	.3700937	159	.817765	1.96651	.415845	1.2083528
40	.060307	.684040	.088163	.0276720	100	.357212	1.53209	.233153	.3802606	160	.826352	1.96962	.419549	1.2252533
41	.063328	.700415	.090415	.0297629	101	.363922	1.54325	.235815	.3905775	161	.834952	1.97257	.423281	1.2422059
42	.066420	.716736	.092670	.0319538	102	.370680	1.55429	.238488	.4010440	162	.843566	1.97537	.427042	1.2592082
43	.069582	.733002	.094927	.0342465	103	.377485	1.56522	.241171	.4116594	163	.852191	1.97803	.430828	1.2762575
44	.072816	.749213	.097190	.0366432	104	.384339	1.57602	.243867	.4224232	164	.860827	1.98054	.434643	1.2933512
45	.076121	.765367	.099457	.0391456	105	.391239	1.58671	.246572	.4333348	165	.869474	1.98289	.438488	1.3104871
46	.079495	.781462	.101730	.0417558	106	.398185	1.59727	.249299	.4443935	166	.878131	1.98509	.442363	1.3276623
47	.082940	.797498	.104000	.0444755	107	.405177	1.60771	.252021	.4555999	167	.886797	1.98714	.446268	1.3448744
48	.086455	.813473	.106278	.0473066	108	.412215	1.61803	.254764	.4669494	168	.895472	1.98904	.450203	1.3621207
49	.090039	.829386	.108561	.0502508	109	.419297	1.62823	.257517	.4784450	169	.904154	1.99079	.454169	1.3793987
50	.093692	.845237	.110847	.0533100	110	.426424	1.63830	.260284	.4900846	170	.912844	1.99238	.458165	1.3967057
51	.097415	.861022	.113139	.0564859	111	.433594	1.64825	.263063	.5018674	171	.921541	1.99383	.462196	1.4140393
52	.101206	.876742	.115434	.0597801	112	.440807	1.65808	.265854	.5137923	172	.930244	1.99513	.466257	1.4313966
53	.105067	.892396	.117736	.0631944	113	.448063	1.66777	.268660	.5258585	173	.938952	1.99627	.470358	1.4487751
54	.108994	.907981	.120040	.0667303	114	.455361	1.67734	.271478	.5380648	174	.947664	1.99726	.474482	1.4661721
55	.112989	.923497	.122349	.0703895	115	.462700	1.68678	.274310	.5504103	175	.956381	1.99810	.478645	1.4835852
56	.117052	.938943	.124664	.0741733	116	.470081	1.69610	.277154	.5628938	176	.965101	1.99878	.482845	1.5010115
57	.121182	.954318	.126983	.0780835	117	.477501	1.70528	.280013	.5755142	177	.973823	1.99931	.487080	1.5184484
58	.125380	.969619	.129308	.0821214	118	.484962	1.71433	.282887	.5882703	178	.982548	1.99970	.491348	1.5358933
59	.129644	.984847	.131639	.0862884	119	.492462	1.72326	.285773	.6011611	179	0.991274	1.99992	.495657	1.5533435
60	.133975	1.000000	.133975	.0905860	120	.500000	1.73205	.288684	.6141847	180	1.00000	2.00000	.500000	1.5707963

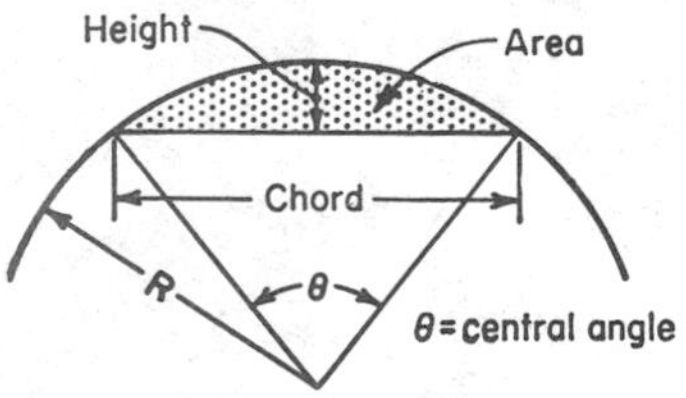

TABLE 1-19*b* Circles: Areas of Segments*

Height = h; Diameter = D; Area = A

h/D	A	h/D	A	h/D	A	h/D	A	h/D	A	h/D	A	h/D	A	h/D	A	h/D	A	h/D	A
		0.050	0.01468	0.100	0.04087	0.150	0.07387	0.200	0.11182	0.250	0.15355	0.300	0.19817	0.350	0.24498	0.400	0.29337	0.450	0.34278
0.002	0.00012	.052	.01556	.102	.04208	.152	.07531	.202	.11343	.252	.15528	.302	.20000	.352	.24689	.402	.29533	.452	.34477
.004	.00034	.054	.01646	.104	.04330	.154	.07675	.204	.11504	.254	.15702	.304	.20184	.354	.24880	.404	.29729	.454	.34676
.006	.00062	.056	.01737	.106	.04452	.156	.07819	.206	.11665	.256	.15876	.306	.20368	.356	.25071	.406	.29926	.456	.34876
.008	.00095	.058	.01830	.108	.04576	.158	.07965	.208	.11827	.258	.16051	.308	.20553	.358	.25263	.408	.30122	.458	.35075
.010	.00133	.060	.01924	.110	.04701	.160	.08111	.210	.11990	.260	.16226	.310	.20738	.360	.25455	.410	.30319	.460	.35274
.012	.00175	.062	.02020	.112	.04826	.162	.08258	.212	.12153	.262	.16402	.312	.20923	.362	.25647	.412	.30516	.462	.35474
.014	.00220	.064	.02117	.114	.04953	.164	.08406	.214	.12317	.264	.16578	.314	.21108	.364	.25839	.414	.30712	.464	.35673
.016	.00268	.066	.02215	.116	.05080	.166	.08554	.216	.12481	.266	.16755	.316	.21294	.366	.26032	.416	.30910	.466	.35873
.018	.00320	.068	.02315	.118	.05209	.168	.08704	.218	.12646	.268	.16932	.318	.21480	.368	.26225	.418	.31107	.468	.36072
.020	.00375	.070	.02417	.120	.05338	.170	.08854	.220	.12811	.270	.17109	.320	.21667	.370	.26418	.420	.31304	.470	.36272
.022	.00432	.072	.02520	.122	.05469	.172	.09004	.222	.12977	.272	.17287	.322	.21853	.372	.26611	.422	.31502	.472	.36471
.024	.00492	.074	.02624	.124	.05600	.174	.09155	.224	.13144	.274	.17465	.324	.22040	.374	.26805	.424	.31699	.474	.36671
.026	.00555	.076	.02729	.126	.05733	.176	.09307	.226	.13311	.276	.17644	.326	.22228	.376	.26998	.426	.31897	.476	.36871
.028	.00619	.078	.02836	.128	.05866	.178	.09460	.228	.13478	.278	.17823	.328	.22415	.378	.27192	.428	.32095	.478	.37071
.030	.00687	.080	.02943	.130	.06000	.180	.09613	.230	.13646	.280	.18002	.330	.22603	.380	.27386	.430	.32293	.480	.37270
.032	.00756	.082	.03053	.132	.06135	.182	.09767	.232	.13815	.282	.18182	.332	.22792	.382	.27580	.432	.32491	.482	.37470
.034	.00827	.084	.03163	.134	.06271	.184	.09922	.234	.13984	.284	.18362	.334	.22980	.384	.27775	.434	.32689	.484	.37670
.036	.00901	.086	.03275	.136	.06407	.186	.10077	.236	.14154	.286	.18542	.336	.23169	.386	.27969	.436	.32887	.486	.37870
.038	.00976	.088	.03387	.138	.06545	.188	.10233	.238	.14324	.288	.18723	.338	.23358	.388	.28164	.438	.33086	.488	.38070
.040	.01054	.090	.03501	.140	.06683	.190	.10390	.240	.14494	.290	.18905	.340	.23547	.390	.28359	.440	.33284	.490	.38270
.042	.01133	.092	.03616	.142	.06822	.192	.10547	.242	.14666	.292	.19086	.342	.23737	.392	.28554	.442	.33483	.492	.38470
.044	.01214	.094	.03732	.144	.06963	.194	.10705	.244	.14837	.294	.19268	.344	.23927	.394	.28750	.444	.33682	.494	.38670
.046	.01297	.096	.03850	.146	.07103	.196	.10864	.246	.15009	.296	.19451	.346	.24117	.396	.28945	.446	.33880	.496	.38870
.048	.01382	.098	.03968	.148	.07245	.198	.11023	.248	.15182	.298	.19634	.348	.24307	.398	.29141	.448	.34079	.498	.39070
.050	.01468	.100	.04087	.150	.07387	.200	.11182	.250	.15355	.300	.19817	.350	.24498	.400	.29337	.450	.34278	.500	.39270

**Rules for Using Table:* (1) Divide height of segment by the diameter; multiply the area in the table corresponding to the quotient, height/diameter, by the diameter squared. When segment exceeds a semicircle, its area is: Area of circle minus the area of a segment whose height is the circle diameter minus the height of the given segment. (2) To find the diameter when given the chord and the segment height: the diameter = [(½ chord)²/height] + height.

TABLE 1-20 Spheres: Segments*

h = segment height; D = sphere diameter

h/D	Segment vol. / D^3	Segment vol. / Sphere vol.	h/D	Segment vol. / D^3	Segment vol. / Sphere vol.	h/D	Segment vol. / D^3	Segment vol. / Sphere vol.	h/D	Segment vol. / D^3	Segment vol. / Sphere vol.
0.01	0.000156	0.000298	0.16	0.035923	0.068608	0.31	0.119756	0.228718	0.41	0.191877	0.366458
.02	.000619	.001184	.17	.040251	.076874	.32	.126534	.241664	.42	.199503	.381024
.03	.001385	.002646	.18	.044787	.085536	.33	.133426	.254826	.43	.207180	.395686
.04	.002446	.004672	.19	.049522	.094582	.34	.140425	.268192	.44	.214901	.410432
.05	.003796	.007250	.20	.054454	.104000	.35	.147524	.281750	.45	.222660	.425250
.06	.005429	.010368	.21	.059573	.113778	.36	.154717	.295488	.46	.230450	.440128
.07	.007338	.014314	.22	.064875	.123904	.37	.161998	.309394	.47	.238265	.455054
.08	.009517	.018176	.23	.070353	.134366	.38	.169361	.323456	.48	.246099	.470016
.09	.011960	.022842	.24	.076001	.145152	.39	.176799	.337662	.49	.253946	.485002
.10	.014661	.028000	.25	.081812	.156250	.40	.184306	.352000	.50	.261799	.500000
.11	.017613	.033638	.26	.087780	.167648						
.12	.020809	.039744	.27	.093900	.179334						
.13	.024246	.046306	.28	.100160	.191296						
.14	.027914	.053312	.29	.106560	.203522						
.15	.031809	.060750	.30	.113097	.216000						

*Given the segment height h and the sphere diameter D, first form the ratio h/D, and find from the table the value of (segment volume/D^3); then multiply this latter value by D^3, that is: (segment volume/D^3) × D^3 = segment volume.

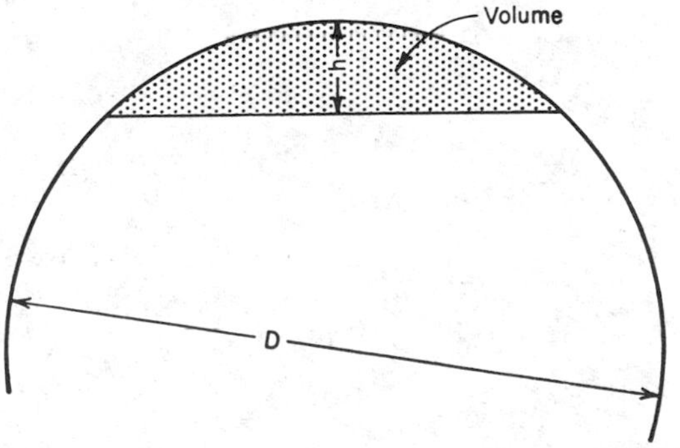

TABLE 1-21 Mathematical Signs, Symbols, and Abbreviations

$\pm$ ($\mp$)	plus or minus (minus or plus)
$:$	divided by, ratio sign
$::$	proportional sign
$<$	less than
$\nless$	not less than
$>$	greater than
$\ngtr$	not greater than
$\cong$	approximately equals, congruent
$\sim$	similar to
$\asymp$	equivalent to
$\neq$	not equal to
$\doteq$	approaches, is approximately equal to
$\propto$	varies as
∞	infinity
$\therefore$	therefore
$\sqrt{\ }$	square root
$\sqrt[3]{\ }$	cube root
$\sqrt[n]{\ }$	*n*th root
$\angle$	angle
$\perp$	perpendicular to
$\parallel$	parallel to
$\lvert x \rvert$	numerical value of x
log or $\log_{10}$	common logarithm or Briggsian logarithm
$\log_e$ or ln	natural logarithm or hyperbolic logarithm or Naperian logarithm
e	base (2.178) of natural system of logarithms
$a°$	an angle a degrees
a' a	prime, an angle a minutes
a'' a	double prime, an angle a seconds, a second
sin	sine
cos	cosine
tan	tangent
ctn or cot	cotangent
sec	secant
csc	cosecant
vers	versed sine
covers	coversed sine
exsec	exsecant
$\sin^{-1}$	anti sine or angle whose sine is
sinh	hyperbolic sine
cosh	hyperbolic cosine
tanh	hyperbolic tangent
$\sinh^{-1}$	anti hyperbolic sine or angle whose hyperbolic sine is
$f(x)$ or $\phi(x)$	function of x
Δx	increment of x
Σ	summation of
dx	differential of x
dy/dx or y'	derivative of y with respect to x
d^2y/dx^2 or y''	second derivative of y with respect to x
d^ny/dx^n	*n*th derivative of y with respect to x
$\partial y/\partial x$	partial derivative of y with respect to x
$\partial^n y/\partial x^n$	*n*th partial derivative of y with respect to x
$\frac{\partial^n y}{\partial x\, \partial y}$	*n*th partial derivative with respect to x and y
$\int$	integral of
$\int_a^b$	integral between the limits a and b
$\dot{y}$	first derivative of y with respect to time
$\ddot{y}$	second derivative of y with respect to time
Δ or ∇^2	the "Laplacian" $\left(\frac{\partial^2}{\partial x^2} + \frac{\partial^2}{\partial y^2} + \frac{\partial^2}{\partial z^2}\right)$
δ	sign of a variation
$\oint$	sign for integration around a closed path

TABLE 1-22 Greek Alphabet

Alpha	= A, α = A, a
Beta	= B, β = B, b
Gamma	= Γ, γ = G, g
Delta	= Δ, δ = D, d
Epsilon	= E, ϵ = E, e
Zeta	= Z, ζ = Z, z
Eta	= H, η = E, e
Theta	= Θ, θ = Th, th
Iota	= I, ι = I, i
Kappa	= K, κ = K, k
Lambda	= Λ, λ = L, l
Mu	= M, μ = M, m
Nu	= N, ν = N, n
Xi	= Ξ, ξ = X, x
Omicron	= O, o = O, o
Pi	= Π, π = P, p
Rho	= P, ρ = R, r
Sigma	= Σ, σ = S, s
Tau	= T, τ = T, t
Upsilon	= Υ, υ = U, u
Phi	= Φ, ϕ = Ph, ph
Chi	= X, χ = Ch, ch
Psi	= Ψ, ψ = Ps, ps
Omega	= Ω, ω = O, o

FINANCIAL TABLE

TABLE 1-23 Compound Interest Factors* (For examples demonstrating use see end of table.)

5% Compound Interest Factors

n	Single payment: Compound-amount factor. Given P, to find S: $(1+i)^n$	Single payment: Present-worth factor. Given S, to find P: $\frac{1}{(1+i)^n}$	Uniform annual series: Sinking-fund factor. Given S, to find R: $\frac{i}{(1+i)^n-1}$	Uniform annual series: Capital-recovery factor. Given P, to find R: $\frac{i(1+i)^n}{(1+i)^n-1}$	Uniform annual series: Compound-amount factor. Given R, to find S: $\frac{(1+i)^n-1}{i}$	Uniform annual series: Present-worth factor. Given R, to find P: $\frac{(1+i)^n-1}{i(1+i)^n}$
1	1.050	0.9524	1.00000	1.05000	1.000	0.952
2	1.103	.9070	0.48780	0.53780	2.050	1.859
3	1.158	.8638	.31721	.36721	3.153	2.723
4	1.216	.8227	.23201	.28201	4.310	3.546
5	1.276	.7835	.18097	.23097	5.526	4.329
6	1.340	.7462	.14702	.19702	6.802	5.076
7	1.407	.7107	.12282	.17282	8.142	5.786
8	1.477	.6768	.10472	.15472	9.549	6.463
9	1.551	.6446	.09069	.14069	11.027	7.108
10	1.629	.6139	.07940	.12950	12.578	7.722
11	1.710	.5847	.07039	.12039	14.207	8.306
12	1.796	.5568	.06283	.11283	15.917	8.863
13	1.886	.5303	.05646	.10646	17.713	9.394
14	1.980	.5051	.05102	.10102	19.599	9.899
15	2.079	.4810	.04634	.09634	21.579	10.380
16	2.183	.4581	.04227	.09227	23.657	10.838
17	2.292	.4363	.03870	.08870	25.840	11.274
18	2.407	.4155	.03555	.08555	28.132	11.690
19	2.527	.3957	.03275	.08275	30.539	12.085
20	2.653	.3769	.03024	.08024	33.066	12.462
21	2.786	.3589	.02800	.07800	35.719	12.821
22	2.925	.3418	.02597	.07597	38.505	13.163
23	3.072	.3256	.02414	.07414	41.430	13.489
24	3.225	.3101	.02247	.07247	44.502	13.799
25	3.386	.2953	.02095	.07095	47.727	14.094
26	3.556	.2812	.01956	.06956	51.113	14.375
27	3.733	.2678	.01829	.06829	54.669	14.643
28	3.920	.2551	.01712	.06712	58.403	14.898
29	4.116	.2429	.01605	.06605	62.323	15.141
30	4.322	.2314	.01505	.06505	66.489	15.372
31	4.538	.2204	.01413	.06413	70.761	15.593
32	4.765	.2099	.01328	.06328	75.299	15.803
33	5.003	.1999	.01249	.06249	80.064	16.003
34	5.253	.1904	.01176	.06176	85.067	16.193
35	5.516	.1813	.01107	.06107	90.320	16.374
40	7.040	.1420	.00828	.05828	120.800	17.159
45	8.985	.1113	.00626	.05626	159.700	17.774
50	11.467	.0872	.00478	.05478	209.348	18.256
55	14.636	.0683	.00367	.05367	272.713	18.633
60	18.679	.0535	.00283	.05283	353.584	18.929
65	23.840	.0419	.00219	.05219	456.798	19.161
70	30.426	.0329	.00170	.05170	588.529	19.343
75	38.833	.0258	.00132	.05132	756.654	19.485
80	49.561	.0202	.00103	.05103	971.229	19.596
85	63.254	.0158	.00080	.05080	1,245.087	19.684
90	80.730	.0124	.00063	.05063	1,594.607	19.752
95	103.035	.0097	.00049		2,040.694	19.806
100	131.501	.0076	.00038	.05038	2,610.025	19.848

6% Compound Interest Factors

n	Single payment: Compound-amount factor. Given P, to find S: $(1+i)^n$	Single payment: Present-worth factor. Given S, to find P: $\frac{1}{(1+i)^n}$	Uniform annual series: Sinking-fund factor. Given S, to find R: $\frac{i}{(1+i)^n-1}$	Uniform annual series: Capital-recovery factor. Given P, to find R: $\frac{i(1+i)^n}{(1+i)^n-1}$	Uniform annual series: Compound-amount factor. Given R, to find S: $\frac{(1+i)^n-1}{i}$	Uniform annual series: Present-worth factor. Given R, to find P: $\frac{(1+i)^n-1}{i(1+i)^n}$
1	1.060	0.9434	1.00000	1.06000	1.000	0.943
2	1.124	.8900	0.48544	0.54544	2.060	1.833
3	1.191	.8396	.31411	.37411	3.184	2.673
4	1.262	.7921	.22859	.28859	4.375	3.465
5	1.338	.7473	.17740	.23740	5.637	4.212
6	1.419	.7050	.14336	.20336	6.975	4.917
7	1.504	.6651	.11914	.17914	8.394	5.582
8	1.594	.6274	.10104	.16104	9.897	6.210
9	1.689	.5919	.08702	.14702	11.491	6.802
10	1.791	.5584	.07587	.13587	13.181	7.360
11	1.898	.5268	.06679	.12679	14.972	7.887
12	2.012	.4970	.05928	.11928	16.870	8.384
13	2.133	.4688	.05296	.11296	18.882	8.853
14	2.261	.4423	.04758	.10758	21.015	9.295
15	2.397	.4173	.04296	.10296	23.276	9.712
16	2.540	.3936	.03895	.09895	25.673	10.106
17	2.693	.3714	.03544	.09544	28.213	10.477
18	2.854	.3503	.03236	.09236	30.906	10.828
19	3.026	.3305	.02962	.08962	33.760	11.158
20	3.207	.3118	.02718	.08718	36.786	11.470
21	3.400	.2942	.02500	.08500	39.993	11.764
22	3.604	.2775	.02305	.08305	43.392	12.042
23	3.820	.2618	.02128	.08128	46.996	12.303
24	4.049	.2470	.01968	.07968	50.816	12.550
25	4.292	.2330	.01823	.07823	54.865	12.783
26	4.549	.2198	.01690	.07690	59.156	13.003
27	4.822	.2074	.01570	.07570	63.706	13.211
28	5.112	.1956	.01459	.07459	68.528	13.406
29	5.418	.1846	.01358	.07358	73.640	13.591
30	5.743	.1741	.01265	.07265	79.058	13.765
31	6.088	.1643	.01179	.07179	84.802	13.929
32	6.453	.1550	.01100	.07100	90.890	14.084
33	6.841	.1462	.01027	.07027	97.343	14.230
34	7.251	.1379	.00960	.06960	104.184	14.368
35	7.686	.1301	.00897	.06897	111.435	14.498
40	10.286	.0972	.00646	.06646	154.762	15.046
45	13.765	.0727	.00470	.06470	212.744	15.456
50	18.420	.0543	.00344	.06344	290.336	15.762
55	24.650	.0406	.00254	.06254	394.172	15.991
60	32.988	.0303	.00188	.06188	533.128	16.161
65	44.145	.0227	.00139	.06139	719.083	16.289
70	59.076	.0169	.00103	.06103	967.932	16.385
75	79.057	.0126	.00077	.06077	1,300.949	16.456
80	105.796	.0095	.00057	.06057	1,746.600	16.509
85	141.579	.0071	.00043	.06043	2,342.982	16.549
90	189.465	.0053	.00032	.06032	3,141.075	16.579
95	253.546	.0039	.00024	.06024	4,209.104	16.601
100	339.302	.0029	.00018	.06018	5,638.368	16.618

TABLE 1-23 Compound Interest Factors* (*Continued*)

Examples of Use of Table and Factors

Given: \$2500 is invested now at 5 per cent.
Required: Accumulated value in 10 years (*i.e.*, the amount of a given principal).

Solution:

$$S = P(1 + i)^n = \$2500 \times 1.05^{10}$$
$$\text{Compound-amount factor} = (1 + i)^n = 1.05^{10} = 1.629$$
$$S = \$2500 \times 1{,}629 = \$4062.50$$

Given: \$19,500 will be required in 5 years to replace equipment now in use.
Required: With interest available at 3 per cent, what sum must be deposited in the bank at present to provide the required capital (*i.e.*, the principal which will amount to a given sum)?

Solution:

$$P = S\frac{1}{(1 + i)^n} = \$19{,}500\frac{1}{1.03^5}$$
$$\text{Present-worth factor} = 1/(1 + i)^n = 1/1.03^5 = 0.8626$$
$$P = \$19{,}500 \times 0.8626 = \$16{,}821$$

Given: \$50,000 will be required in 10 years to purchase equipment.
Required: With interest available at 4 per cent, what sum must be deposited each year to provide the required capital (*i.e.*, the annuity which will amount to a given fund)?

Solution:

$$R = S\frac{i}{(1 + i)^n - 1} = \$50{,}000\frac{0.04}{1.04^{10} - 1}$$
$$\text{Sinking-fund factor} = \frac{i}{(1 + i)^n - 1} = \frac{0.04}{1.04^{10} - 1} = 0.08329$$
$$R = \$50.000 \times 0.08329 = \$4{,}164$$

Given: \$20,000 is invested at 10 per cent interest.
Required: Annual sum that can be withdrawn over a 20-year period (*i.e.*, the annuity provided by a given capital).

Solution:

$$R = P\frac{i(1 + i)^n}{(1 + i)^n - 1} = \$20{,}000\frac{0.10 \times 1.10^{20}}{1.10^{20} - 1}$$
$$\text{Capital-recovery factor} = \frac{i(1 + i)^n}{(1 + i)^n - 1} = \frac{0.10 \times 1.10^{20}}{1.10^{20} - 1} = 0.11746$$
$$R = \$20.000 \times 0.11746 = \$2349.20$$

Given: \$500 is invested each year at 8 per cent interest.
Required: Accumulated value in 15 years (*i.e.*, amount of an annuity).

Solution:

$$S = R\frac{(1 + i)^n - 1}{i} = \$500\frac{1.08^{15} - 1}{0.08}$$
$$\text{Compound-amount factor} = \frac{(1 - i)^n - 1}{i} = \frac{1.08^{15} - 1}{0.08} = 27.152$$
$$S = \$500 \times 27.152 = \$13{,}576$$

Given: \$8,000 is required annually for 25 years.
Required: Sum that must be deposited now at 6 per cent interest.

Solution:

$$P = R\frac{(1 + i)^n - 1}{i(1 + i)^n} = \$8000\frac{1.06^{25} - 1}{0.06 \times 1.06^{25}}$$
$$\text{Present-worth factor} = \frac{(1 + i)^n - 1}{i(1 + i)^n} = \frac{1.06^{25} - 1}{0.06 \times 1.06^{25}} = 12.783$$
$$P = \$8000 \times 12.78 = \$102{,}264$$

*Factors presented for two interest rates only. By using the appropriate formulas, values for other interest rates may be calculated.

Section 2

Mathematics*

Arthur E. Hoerl, M.A., *Professor, Department of Mathematical Sciences, University of Delaware; Fellow of the American Statistical Association; Member, American Society for Quality Control. (Mathematics and Statistics; Section Editor)*

M. Zuhair Nashed, Ph.D., *Professor of Mathematics and Professor of Electrical Engineering, University of Delaware and University of Petroleum and Minerals; Editor, Journal of Integral Equations; Executive Editor of the series Monographs and Textbooks in Pure and Applied Mathematics; Member, Editorial Board of the Journal of Nonlinear Analysis, Resultate der Mathematik, Journal of Applicable Analysis, Journal of Stochastic Analysis, and Journal of Engineering Sciences. (All general material on mathematics)*

John J. McKetta, Ph.D., *Joe C. Walter Chair, Chemical Engineering, University of Texas. (Dimensional Analysis)*

I. H. Silberberg, Ph.D., *Assistant Director, Texas Petroleum Research Committee, University of Texas. (Dimensional Analysis)*

*The contribution of William F. Ames, Georgia Institute of Technology, to material that was used from the fifth edition is acknowledged.

General References: The list of references for this section is selected to provide a broad perspective on classical and modern mathematical methods that are useful in chemical engineering. The references supplement and extend the treatment given in this section. Also included are selected references to important areas of mathematics which are not covered in the *Handbook* but which may be useful for certain areas of chemical engineering, e.g., additional topics in numerical analysis and software, optimal control and system theory, linear operators, and functional-analysis methods. Readers interested in brief summaries of theory, together with many detailed examples and solved problems on various topics of college mathematics and mathematical methods for engineers, are referred to the Schaum's Outline Series in Mathematics, published by the McGraw-Hill Book Comapny.

1. Abramowitz, M., and I. Stegun: *Handbook of Mathematical Functions with Formulas, Graphs, and Mathematical Tables,* Dover, New York, 1965.
2. Action, F. S.: *Analysis of Straight-Line Data,* Dover, New York, 1966.
3. ———: *Numerical Methods That Work,* Harper & Row, New York, 1970.
4. Adey, R. A., and C. A. Brebbia: *Basic Computational Techniques for Engineers,* Wiley, New York, 1983.
5. Alger, P.: *Mathematics for Science and Engineering,* 2d ed., McGraw-Hill, New York, 1969.
6. Allendoerfer, C. B., and C. O. Oakley: *Principles of Mathematics,* 3d ed., McGraw-Hill, New York, 1969.
7. Ames, W. F.: *Nonlinear Partial Differential Equations in Engineering,* Academic, New York, 1965.
8. ———: *Nonlinear Ordinary Differential Equations in Transport Processes,* Academic, New York, 1968.
9. ———: *Numerical Methods for Partial Differential Equations,* 2d ed., Academic, New York, 1977.
10. Angel, E., and R. Bellman: *Dynamic Programming and Partial Differential Equations,* Academic, New York, 1972.
11. Apostol, T. M.: *Mathematical Analysis,* 2d ed., Addison-Wesley, Reading, Mass., 1974.
12. Arfken, G.: *Mathematical Methods for Physicists,* 2d ed., Academic, New York, 1970.
13. Aris, R.: *The Mathematical Theory of Diffusion and Reaction in Permeable Catalysts,* vols. 1 and 2, Oxford University Press, Oxford, 1975.
14. ———: *Mathematical Modelling Techniques,* Pitman, London, 1978.
15. ——— and N. Amundson: *Mathematical Methods in Chemical Engineering,* vols. 1 and 2, Prentice-Hall, Englewood Cliffs, N.J., 1973.
16. Arya, J. C., and R. W. Lardner: *Algebra and Trigonometry with Applications,* Prentice-Hall, Englewood Cliffs, N.J., 1983.
17. Athans, M., and P. L. Falb: *Optimal Control,* McGraw-Hill, New York, 1966.
18. Atkinson, K. E.: *An Introduction to Numerical Analysis,* Wiley, New York, 1978.
19. Baker, C. T. H.: *The Numerical Treatment of Integral Equations,* Oxford University Press, New York, 1977.
20. Barrodale, I., D. K. Roberts, and B. L. Ehle: *Elementary Computer Applications,* Wiley, New York, 1971.
21. Bazaraa, M. S., and J. J. Jarvis: *Linear Programming and Network Flow,* Wiley, New York, 1977.
22. ——— and M. Shetty: *Nonlinear Programming,* Wiley, New York, 1979.
23. Beckenbach, E. F. (ed.): *Modern Mathematics for the Engineer,* 2d ed., McGraw-Hill, 1961.
24. ——— and R. E. Bellman: *Inequalities,* 3d printing, Springer-Verlag, Berlin, 1971.
25. Becker, E. B., G. F. Carey, and J. T. Oden: *Finite Elements: An Introduction,* Prentice-Hall, Englewood Cliffs, N.J., 1981 (first of a series of six volumes).
26. Bell, W. W.: *Special Functions for Scientists and Engineers,* Van Nostrand, London, 1968.
27. Bellman, R. E.: *Stability Theory of Ordinary Differential Equations,* McGraw-Hill, New York, 1953.
28. ———: *Introduction to Matrix Analysis,* McGraw-Hill, New York, 1960.
29. ——— and K. L. Cooke: *Differential-Difference Equations,* Academic, New York, 1972.
30. ——— and S. E. Dreyfus: *Applied Dynamic Programming,* Princeton University Press, Princeton, N.J., 1962.
31. Bender, E. A.: *An Introduction to Mathematical Modeling,* Wiley, New York, 1978.
32. Ben-Israel, A., and T. N. E. Greville: *Generalized Inverses: Theory and Applications,* Wiley-Interscience, New York, 1974.
33. Blum, E. K.: *Numerical Analysis and Computation: Theory and Practice,* Addison-Wesley, Reading, Mass., 1972.
34. Bodewig, E.: *Matrix Calculus,* 2d ed., Interscience, New York, 1959.
35. Boor, C. de: *A Practical Guide to Splines,* Springer-Verlag, New York, 1978.
36. Botha, J. F., and G. F. Pinder: *Fundamental Concepts in the Numerical Solution of Differential Equations,* Wiley, New York, 1983.
37. Bowman, F.: *Introduction to Bessel Functions,* Dover, New York, 1958.
38. Boyce, W. E., and R. C. Di Prima: *Elementary Differential Equations and Boundary Value Problems,* 3d ed., Wiley, New York, 1977.
39. Bradley, S. P., A. C. Hax, and T. L. Magnante: *Applied Mathematical Programming,* Addison-Wesley, Reading, Mass., 1977.
40. Brand, L.: *Advanced Calculus,* Wiley, New York, 1955.
41. ———: *Differential and Difference Equations,* Wiley, New York, 1966.
42. Brauer, F., and J. A. Nohel: *Ordinary Differential Equations,* W. A. Benjamin, New York, 1967.
43. Braun, M.: *Differential Equations and Their Applications: An Introduction to Applied Mathematics,* 3d ed., Springer-Verlag, New York, 1983.
44. Brigham, E.: *The Fast Fourier Transform,* Prentice-Hall, Englewood Cliffs, N.J., 1974.
45. Bruijn, N. G. de: *Asymptotic Methods in Analysis,* Dover, New York, 1981.
46. Bryson, A. E., and Y. C. Ho: *Applied Optimal Control,* Blaisdell Publishing Company, Waltham, Mass., 1969.
47. Brent, R.: *Algorithms for Minimization without Derivatives,* Prentice-Hall, Englewood Cliffs, N.J., 1973.
48. Buck, R. C.: *Advanced Calculus,* 3d ed., McGraw-Hill, New York, 1978.
49. Budak, B. M., A. A. Samarskii, and A. N. Tikhonov: *A Collection of Problems on Mathematical Physics,* Pergamon, Oxford, 1964.
50. Bunch, J., and D. Rose (eds.): *Sparse Matrix Computations,* Academic, New York, 1976.
51. Burden, R. L., J. D. Faires, and A. C. Reynolds: *Numerical Analysis,* 2d ed., Prindle, Weber & Schmidt, Boston, 1981.
52. Burrington, R.: *Handbook of Mathematical Tables and Formulas,* 4th ed., McGraw-Hill, New York, 1969.
53. Byrd, P., and M. Friedman: *Handbook of Elliptic Integrals for Scientists and Engineers,* 2d ed., Springer-Verlag, New York, 1971.
54. Carnahan, B., H. Luther, and J. Wilkes: *Applied Numerical Methods,* Wiley, New York, 1969.
55. Carrier, G., M. Crook, and C. Pearson: *Functions of a Complex Variable,* McGraw-Hill, New York, 1966.
56. ——— and C. Pearson: *Partial Differential Equations,* Academic, New York, 1976.
57. Carslaw, H. S.: *The Theory of Fourier Series and Integrals,* 3d ed., Dover, New York, 1930.
58. ——— and J. Jaeger: *Operational Methods in Applied Mathematics,* 2d ed., Clarendon Press, Oxford, 1948.
59. Cheney, E. W.: *Introduction to Approximation Theory,* McGraw-Hill, New York, 1966.
60. ——— and D. Kincaid: *Numerical Mathematics and Computing,* Brooks/Cole, Monterey, Calif., 1980.
61. Churchill, R. V.: *Operational Mathematics,* 3d ed., McGraw-Hill, New York, 1972.
62. ——— and J. W. Brown: *Fourier Series and Boundary Value Problems,* 3d ed., McGraw-Hill, New York, 1978.
63. ———, J. W. Brown, and R. V. Verhey: *Complex Variables and Applications,* 3d ed., McGraw-Hill, New York, 1974.
64. Clarke, F. H.: *Optimization and Nonsmooth Analysis,* Wiley, New York, 1983.
65. Coburn, L.: *Vector and Tensor Analysis,* Macmillan, New York, 1955.
66. Cochran, J. A.: *The Analysis of Linear Integral Equations,* McGraw-Hill, New York, 1972.
67. Coddington, E. A.: *An Introduction to Ordinary Differential Equations,* Prentice-Hall, Englewood Cliffs, N.J., 1961.
68. ——— and N. Levinson: *Theory of Ordinary Differential Equations,* McGraw-Hill, New York, 1955.
69. Cogan, E. J., and R. Z. Norman: *Handbook of Calculus, Difference and Differential Equations,* Prentice-Hall, Englewood Cliffs, N.J., 1958.

70. Collatz, L.: *The Numerical Treatment of Differential Equations*, 3d ed., Springer-Verlag, Berlin and New York, 1966.
71. Conte, S. D., and C. de Boor: *Elementary Numerical Analysis: An Algorithmic Approach*, 3d ed., McGraw-Hill, New York, 1980.
72. Cooper, L., and D. Steinberg: *Methods and Applications of Linear Programming*, Saunders, Philadelphia, 1974.
73. Courant, R., and D. Hilbert: *Methods of Mathematical Physics*, Interscience, New York, vol. 1, 1953; vol. 2: *Partial Differential Equations*, 1962.
74. ——— and F. John: *Introduction to Calculus and Analysis*, Wiley, New York, 1965.
75. ——— and H. Robbins: *What Is Mathematics?* Oxford University Press, New York, 1941.
76. Crandall, S.: *Engineering Analysis*, McGraw-Hill, New York, 1956.
77. Creese, T. M., and R. M. Haralick: *Differential Equations for Engineers*, McGraw-Hill, New York, 1978.
78. Cunningham, R.: *Introduction to Nonlinear Differential Equations*, McGraw-Hill, New York, 1960.
79. Dahlquist, G., and A. Bjorck: *Numerical Methods*, Prentice-Hall, Englewood Cliffs, N.J., 1974.
80. Dantzig, G. B.: *Linear Programming and Extensions*, Princeton University Press, Princeton, N.J., 1963.
81. Davis, H. T.: *Introduction to Nonlinear Differential and Integral Equations*, Dover, New York, 1962.
82. Davis, P.: *Interpolation and Approximation*, Dover, New York, 1980.
83. ——— and P. Rabinowitz: *Methods of Numerical Integration*, 2d ed., Academic, New York, 1975.
84. Denn, M. M.: *Optimization by Variational Methods*, McGraw-Hill, New York, 1969.
85. ———: *Stability of Reaction and Transport Processes*, Prentice-Hall, Englewood Cliffs, N.J., 1974.
86. Dennery, P., and A. Krzywicki: *Mathematics for Physicists*, Harper & Row, New York, 1967.
87. Dettman, J. W.: *Introduction to Linear Algebra and Differential Equations*, McGraw-Hill, New York, 1974.
88. Doetsch, G.: *A Guide to the Applications of the Laplace Transform*, Van Nostrand, Princeton, N.J., 1963.
89. Dolciani, M. P., et al.: *Modern Introductory Analysis*, Houghton Mifflin, Boston, 1970.
90. Dongarra, J. J., J. R. Bunch, C. B. Moler, and G. W. Stewart: *LINPACK Users Guide*, Society for Industrial and Applied Mathematics, Philadelphia, 1979.
91. Dorn, W. S., and D. D. McCracken: *Numerical Methods with Fortran IV Case Studies*, Wiley, New York, 1972.
92. Dreyfus, S. E.: *Dynamic Programming and the Calculus of Variations*, Academic, New York, 1965.
93. Duff, G. F. D., and D. Naylor: *Differential Equations of Applied Mathematics*, Wiley, New York, 1966.
94. Dwight, H. B.: *Tables of Integrals and Other Mathematical Data*, Macmillan, New York, 1947.
95. Dym, C. L., and E. S. Ivey: *Principles of Mathematical Modeling*, Academic, New York, 1980.
96. Elich, J., and C. J. Elich: *College Algebra with Calculator Applications*, Addison-Wesley, Boston, Mass., 1982.
97. Epstein, B.: *Partial Differential Equations*, McGraw-Hill, New York, 1962.
98. Erdelyi, A., et al.: *Higher Transcendental Functions*, vols. I, II, and III, McGraw-Hill, New York, 1953–1955.
99. Faddeev, D. K., and V. N. Faddeeva: *Computational Methods in Linear Algebra*, Freeman, San Francisco, 1963.
100. Fiacco, A. V., and G. P. McCormick: *Nonlinear Programming: Sequential Unconstrained Minimization Techniques*, Wiley, New York, 1968.
101. Finlayson, B. A.: *The Method of Weighted Residuals and Variational Principles*, Academic, New York, 1972.
102. Fisher, R. C., and A. D. Ziebur: *Integrated Algebra, Trigonometry, and Analytic Geometry*, 4th ed., Prentice-Hall, Englewood Cliffs, N.J., 1982.
103. Forsythe, G. E., M. A. Malcolm, and C. B. Moler: *Computer Methods for Mathematical Computations*, Prentice-Hall, Englewood Cliffs, N.J., 1977.
104. ——— and C. B. Moler: *Computer Solution of Linear Algebraic Systems*, Prentice-Hall, Englewood Cliffs, N.J., 1967.
105. ——— and W. R. Wasow: *Finite Difference Methods for Partial Differential Equations*, Wiley, New York, 1960.
106. Fox, L.: *Introduction to Numerical Linear Algebra*, Oxford University Press, Oxford, 1965.
107. Franklin, J. N.: *Matrix Theory*, Prentice-Hall, Englewood Cliffs, N.J., 1968.
108. Franklin, P.: *An Introduction to Fourier Methods and Laplace Transformation*, Dover, New York, 1949.
109. ———: *Functions of Complex Variable*, Prentice-Hall, Englewood Cliffs, N.J., 1958.
110. ———: *A Treatise on Advanced Calculus*, Dover, New York, 1964.
111. Freiberger, W. F. (ed.): *International Dictionary of Applied Mathematics*, Van Nostrand, New York, 1960.
112. Friedman, B.: *Principles and Techniques of Applied Mathematics*, Wiley, New York, 1957.
113. Friedman, N. A.: *Calculus and Mathematical Models*, Prindle, Weber & Schmidt, Boston, 1979.
114. Gale, D.: *The Theory of Linear Economic Models*, McGraw-Hill, New York, 1960.
115. Gantmacher, F. R.: *Applications of the Theory of Matrices*, Interscience, New York, 1959.
116. Garbow, B. S., J. M. Boyle, J. J. Dongarra, and C. B. Moler: *Matrix Eigensystem Routines—EISPACK Guide Extensions*, Springer-Verlag, Berlin and New York, 1977.
117. Gardner, K. L.: *A Programmed Vector Algebra*, Oxford University Press, Fair Lawn, N.J., 1969.
118. Gass, S. I.: *Linear Programming*, 3d ed., McGraw-Hill, New York, 1969.
119. Gear, G. W.: *Numerical Initial Value Problems in Ordinary Differential Equations*, Prentice-Hall, Englewood Cliffs, N.J., 1971.
120. Gelfand, I. M., and S. V. Fomin: *Calculus of Variations*, Prentice-Hall, Englewood Cliffs, N.J., 1963.
121. Goertzel, G., and N. Tralli: *Some Mathematical Methods of Physics*, McGraw-Hill, New York, 1960.
122. Gradshteyn, I. S., and I. M. Ryzhik: *Tables of Integrals, Series, and Products*, Academic, New York, 1980.
123. Greenberg, M. M.: *Foundations of Applied Mathematics*, Prentice-Hall, Englewood Cliffs, N.J., 1978.
124. Gregory, R., and D. Karney: *A Collection of Matrices for Testing Computational Algorithms*, Wiley, 1969.
125. Greville, T. N. E.: *Theory and Applications of Spline Functions*, Academic, New York, 1969.
126. Groetsch, C. W.: *Generalized Inverses of Linear Operators*, Marcel Dekker, New York, 1977.
127. ———: *Elements of Applicable Functional Analysis*, Marcel Dekker, New York, 1980.
128. Gustafson, R. D., and P. D. Frisk: *Plane Trigonometry*, Brooks/Cole, Monterey, Calif., 1982.
129. Haberman, R.: *Mathematical Models*, Prentice-Hall, Englewood Cliffs, N.J., 1977.
130. Hadley, G.: *Linear Programming*, Addison-Wesley, Reading, Mass., 1962.
131. ———: *Nonlinear and Dynamic Programming*, Addison-Wesley, Reading, Mass., 1964.
132. Hageman, L. A., and D. M. Young: *Applied Iterative Methods*, Academic, New York, 1981.
133. Halany, A.: *Differential Equations: Stability, Oscillations, Time Lags*, Academic, New York, 1966.
134. Hale, J. K.: *Ordinary Differential Equations*, Wiley, New York, 1969.
135. Hamming, R. W.: *Numerical Methods for Scientists and Engineers*, 2d ed., McGraw-Hill, New York, 1973.
136. Hanna, R.: *Fourier Series and Integrals of Boundary Value Problems*, Wiley, New York, 1982.
137. Hansen, A.: *Similarity Analyses of Boundary Value Probelms in Engineering*, Prentice-Hall, Englewood Cliffs, N.J., 1964.
138. Hardy, G. H., J. E. Littlewood, and G. Polya: *Inequalities*, Cambridge University Press, Cambridge, England, 1959.
139. Harvey, C. M.: *Operations Research*, Elsevier/North-Holland, New York, 1979.
140. Hastings, C., Jr.: *Approximations for Digital Computers*, Princeton University Press, Princeton, N.J., 1955.
141. Hellwig, G.: *Differential Operators of Mathematical Physics*, Addison-Wesley, Reading, Mass., 1967.
142. Henrici, P.: *Discrete Variable Methods in Ordinary Differential Equations*, Wiley, New York, 1962.
143. ———: *Applied and Computational Complex Analysis*, Wiley, New York, 1974.
144. Hetrick, D. L.: *Dynamics of Nuclear Reactors*, University of Chicago Press, Chicago, 1971.
145. Hildebrand, F. B.: *Introduction to Numerical Analysis*, 2d ed., McGraw-Hill, New York, 1974.
146. ———: *Methods of Applied Mathematics*, 2d ed., Prentice-Hall, Englewood Cliffs, N.J., 1965.
147. ———: *Advanced Calculus for Applications*, 2d ed., Prentice-Hall, Englewood Cliffs, N.J., 1976.
148. Hille, E.: *Methods in Classical and Functional Analysis*, Addison-Wesley, Reading, Mass., 1972.
149. Hodgman, R.: *Standard Mathematical Tables*, 12th ed., Chemical Rubber Company, Cleveland, 1960.

150. Holzman, A. G. (ed.): *Operations Research Support Methodology*, Marcel Dekker, New York, 1979.
151. Householder, A. S.: *The Theory of Matrices in Numerical Analysis*, Dover, New York, 1979.
152. ———: *Numerical Treatment of a Single Nonlinear Equation*, McGraw-Hill, New York, 1970; Dover, New York, 1980.
153. Hornbeck, R. W.: *Numerical Methods*, Prentice-Hall, Englewood Cliffs, N.J., 1975.
154. Hyatt, H. R., and L. Small: *Trigonometry: A Calculator Approach*, Wiley, New York, 1982.
155. Ince, E. L.: *Ordinary Differential Equations*, Dover, New York, 1956.
156. Indritz, J.: *Methods in Analysis*, Macmillan, New York, 1963.
157. Isaacson, E., and H. B. Keller: *Analysis of Numerical Methods*, Wiley, New York, 1966.
158. Jacobs, D. (ed.): *The State of the Art in Numerical Analysis*, Academic, New York, 1977.
159. ———: *Numerical Software—Needs and Availability*, Academic, New York, 1978.
160. James, G., and R. C. James: *Mathematics Dictionary*, 3d ed., Van Nostrand, New York, 1968.
161. Jeffreys, H., and B. Jeffreys: *Methods of Mathematical Physics*, 3d ed., Cambridge University Press, London, 1956.
162. Jennings, A.: *Matrix Computations for Engineers and Scientists*, Wiley, New York, 1977.
163. Johnson, R. E., and F. L. Kiokemeister: *Calculus with Analytic Geometry*, 4th ed., Allyn and Bacon, Boston, 1969.
164. Kantorovich, L. V., and G. P. Akilov: *Functional Analysis in Normed Spaces*, 2d ed., Pergamon, Oxford, 1980.
165. ——— and V. I. Krylow: *Approximate Methods of Higher Analysis*, Interscience, New York, 1958.
166. Kanwal, R. P.: *Linear Integral Equations*, Academic, New York, 1971.
167. Kaplan, W.: *Advanced Calculus*, 2d ed., Addison-Wesley, Reading, Mass., 1973.
168. Kemeny, J. G., et al.: *Introduction to Finite Mathematics*, 3d ed., Prentice Hall, Englewood Cliffs, N.J., 1975.
169. Kennedy, E.: *The History of Trigonometry: Historical Notes for the Mathematics Classroom*, National Council of Teachers of Mathematics Yearbook XXXV, Washington, 1969.
170. Kevorkian, J., and J. D. Cole: *Perturbation Methods in Applied Mathematics*, Springer-Verlag, New York, 1981.
171. Kincaid, D. R., and D. M. Young: "Survey of Iterative Methods," *Encyclopedia of Computer Science and Technology*, Marcel Dekker, New York, 1979.
172. Korn, G., and T. Korn: *Manual of Mathematics*, McGraw-Hill, New York, 1967.
173. Krasnoselskii, M. A., G. G. M. Vainikko, P. P. Zabreiko, Ya. B. Ruttiskii, and B. Ya. Stetsenko: *Approximate Solution of Operator Equations*, Wolters-Noordhoff, Groningen, Netherlands, 1972.
174. Kreyszig, E.: *Advanced Engineering Mathematics*, 5th ed., Wiley, New York, 1983.
175. ———: *Introductory Functional Analysis with Applications*, Wiley, New York, 1978.
176. Kunz, K. S.: *Numerical Analysis*, McGraw-Hill, New York, 1957.
177. Kyrala, A.: *Applied Functions of a Complex Variable*, Prentice-Hall, Englewood Cliffs, N.J., 1972.
178. Lanczos, C.: *The Variational Principles of Mechanics*, University of Toronto Press, Toronto, 1949.
179. Lapidus, L., and J. Seinfeld: *Numerical Solution of Ordinary Differential Equations*, Academic, New York, 1971.
180. Lawrence, J. D.: *A Catalog of Special Plane Curves*, Dover, New York, 1972.
181. Lawson, C. L., and R. J. Hanson: *Solving Least Squares Problems*, Prentice-Hall, Englewood Cliffs, N.J., 1974.
182. Lebeder, N. N.: *Special Functions and Their Applications*, Dover, New York, 1972.
183. LePage, W. R.: *Complex Variables and the Laplace Transform for Engineers*, McGraw-Hill, New York, 1961.
184. Lin, C. C., and L. A. Segel: *Mathematics Applied to Deterministic Problems in the Natural Sciences*, Macmillan, New York, 1974.
185. Luenberger, D. G.: *Optimization by Vector Space Methods*, Wiley, New York, 1969.
186. ———: *Introduction to Linear and Nonlinear Programming*, Addison-Wesley, Reading, Mass., 1973.
187. Luke, Y. L.: *Mathematical Functions and Their Applications*, Academic, New York, 1975.
188. Lusternik, L. A., and V. J. Sobolev: *Elements of Functional Analysis*, Wiley, New York, 1974.
189. Magnus, W., F. Oberhettinger, and R. P. Soni: *Formulas and Theorems for the Special Functions of Mathematical Physics*, 3d ed., Springer-Verlag, New York, 1966.
190. Mangasarian, O. L.: *Nonlinear Programming*, McGraw-Hill, New York, 1969.
191. Margenau, H., and G. Murphy: *The Mathematics of Physics and Chemistry*, 2d ed., Van Nostrand, Princeton, N.J., 1956.
192. Marshall, W. R., and R. L. Pigford: *The Application of Differential Equations to Chemical Engineering Problems*, Unviersity of Delaware, Newark, 1947.
193. Martin, R. H., Jr.: *Ordinary Differential Equations*, McGraw-Hill, New York, 1983.
194. McCormick, G.: *Nonlinear Programming: Theory, Algorithms, and Applications*, Wiley, New York, 1983.
195. *McGraw-Hill Encyclopedia of Science and Technology*, McGraw-Hill, New York, 1971.
196. Mickley, H. S., T. C. Sherwood, and C. E. Reed: *Applied Mathematics in Chemical Engineering*, McGraw-Hill, New York, 1957.
197. Mikhlin, S. G.: *Integral Equations*, Pergamon, New York, 1957.
198. ———: *Variational Methods in Mathematical Physics*, Pergamon, New York, 1964.
199. ———: *The Numerical Performance of Variational Methods*, Wolters-Noordhoff, Groningen, Netherlands, 1971.
200. Milne, W. E.: *Numerical Solution of Differential Equations*, Wiley, New York, 1953.
201. Mitchell, A. R.: *Computational Methods for Partial Differential Equations*, Wiley, London, 1969.
202. ——— and R. Wait: *The Finite Element Method in Partial Differential Equations*, Wiley, New York, 1977.
203. Moon, P., and D. Spencer: *Field Theory for Engineers*, Van Nostrand, Princeton, N.J., 1960.
204. ——— and ———: *Field Theory Handbook*, Springer-Verlag, Berlin, 1961.
205. Moretti, G.: *Functions of a Complex Variable*, Prentice-Hall, Englewood Cliffs, N.J., 1964.
206. Morse, P. M., and H. Feshbach: *Methods of Theoretical Physics*, McGraw-Hill, New York, 1953.
207. Murnaghan, F. D.: *Introduction to Applied Mathematics*, Wiley, New York, 1948.
208. Nashed, M. Z. (ed.): *Generalized Inverses and Applications*, Academic, New York, 1976.
209. Nayfeh, A. H.: *Perturbation Methods*, Wiley, New York, 1973.
210. ———: *Introduction to Perturbation Techniques*, Wiley, New York, 1981.
211. Naylor, A. W., and G. R. Sell: *Linear Operator Theory in Engineering and Science*, Springer-Verlag, New York, 1982.
212. Noble, B.: *Applications of Undergraduate Mathematics in Engineering*, Macmillan, New York, 1967.
213. ———: *Applied Linear Algebra*, Prentice-Hall, Englewood Cliffs, N.J., 1969.
214. Oberhettinger, F.: *Fourier Expansions: A Collection of Formulas*, Academic, New York, 1973.
215. Oden, J. T., and J. N. Reddy: *Variational Methods in Theoretical Mechanics*, 2d ed., Springer-Verlag, 1983.
216. Olmsted, J.: *Solid Analytic Geometry*, Appleton-Century, New York, 1947.
217. Orchard-Hays, W.: *Advanced Linear Programming Computing Techniques*, McGraw-Hill, New York, 1968.
218. Ortega, J. M., and W. C. Rheinboldt: *Iterative Solution of Nonlinear Equations in Several Variables*, Academic, New York, 1970.
219. Padulo, L., and M. A. Arbib: *System Theory*, Saunders, Philadelphia, 1974.
220. Papoulis, A.: *The Fourier Integral and Its Applications*, McGraw-Hill, New York, 1962.
221. Pearson, C. E.: *Handbook of Applied Mathematics*, Van Nostrand, New York, 1974.
222. Perlmutter, D. .: *Stability of Chemical Reactors*, Prentice-Hall, Englewood Cliffs, N.J., 1972.
223. Phillips, H. B.: *Vector Analysis*, Wiley, New York, 1933.
224. Pipes, L. A., and L. R. Harvill: *Applied Mathematics for Engineers and Physicists*, 3d ed., McGraw-Hill, New York, 1970.
225. Potter, M. C.: *Mathematical Methods for the Physical Sciences*, Prentice-Hall, Englewood Cliffs, N.J., 1978.
226. Powers, D. L.: *Boundary Value Problems*, Academic, New York, 1972.
227. Prenter, P. M.: *Splines and Variational Methods*, Wiley, New York, 1975.
228. Rainville, E. D.: *Special Functions*, Chelsea Publishing Company, New York, 1972.
229. Rall, L. B.: *Computational Solution of Nonlinear Operator Equations*, Wiley, New York, 1969; Dover, New York, 1981.
230. Ralston, A., and A. Rabinowitz: *A First Course in Numerical Analysis*, 2d ed., McGraw-Hill, New York, 1978.

231. Rektorys, K. (ed.): *Survey of Applicable Mathematics*, M.I.T., Cambridge, Mass., 1969.
232. Rice, J. R., (ed.): *Mathematical Software*, Academic, New York, 1971.
233. ———: *Matrix Computations and Mathematical Software*, McGraw-Hill, New York, 1981.
234. ———: *Numerical Methods, Software, and Analysis*, McGraw-Hill, New York, 1983.
235. Ritchmeyer, R., and K. Morton: *Difference Methods for Initial-Value Problems*, 2d ed., Interscience, New York, 1967.
236. Roubine, E. (ed.): *Mathematics Applied to Physics*, Springer-Verlag, New York, 1970.
237. Rudin, W.: *Principles of Mathematical Analysis*, 2d ed., McGraw-Hill, New York, 1964.
238. Saaty, T. L.: *Modern Nonlinear Equations*, Dover, New York, 1981.
239. ——— and J. Bram: *Nonlinear Mathematics*, McGraw-Hill, New York, 1964; Dover, New York, 1981.
240. Sagan, H.: *Boundary and Eigenvalue Problems in Mathematical Physics*, Wiley, New York, 1961.
241. ———: *Introduction to the Calculus of Variations*, McGraw-Hill, New York, 1969.
242. Schied, F.: *Theory and Problems of Numerical Analysis*, McGraw-Hill, New York, 1968.
243. Shampine, L., and M. Gordon: *Computer Solution of Ordinary Differential Equations*, Freeman, San Francisco, 1975.
244. Shapiro, J. F.: *Mathematical Programming: Structures and Algorithms*, Wiley, New York, 1979.
245. Shenk, A.: *Calculus and Analytic Geometry*, Goodyear Publishing Company, Santa Monica, Calif., 1977.
246. Shockley, J. E.: *Calculus and Analytic Geometry*, Saunders, Philadelphia, 1982.
247. Simmons, G. F.: *Differential Equations*, McGraw-Hill, New York, 1972.
248. Smith, G. D.: *Solution of Partial Differential Equations*, Oxford University Press, London, 1965.
249. Sneddon, I. N.: *Fourier Transforms*, McGraw-Hill, New York, 1951.
250. ———: *The Use of Integral Transforms*, McGraw-Hill, New York, 1972.
251. Sobel, M. A., and N. Lerner: *College Algebra*, Prentice-Hall, Englewood Cliffs, N.J., 1983.
252. Sokolnikoff, I. S., and R. M. Redheffer: *Mathematics of Physics and Modern Engineering*, 2d ed., McGraw-Hill, New York, 1966.
253. Spiegel, M. R.: *Applied Differential Equations*, 3d ed., Prentice-Hall, Englewood Cliffs, N.J., 1981.
254. Stakgold, I.: *Boundary Value Problems of Mathematical Physics*, vol. I and II, Macmillan, New York, 1967.
255. ———: *Green's Functions and Boundary Value Problems*, Wiley-*Interscience*, New York, 1979.
256. Stein, S. K.: *Calculus and Analytic Geometry*, 2d ed., McGraw-Hill, New York, 1977.
257. Stewart, G. W.: *Introduction to Matrix Computations*, Academic, New York, 1973.
258. Stoll, R. R., and E. T. Wong: *Linear Algebra*, Academic, New York, 1968.
259. Strang, G.: *Linear Algebra and Its Applications*, 2d ed., Academic, New York, 1980.
260. ——— and G. Fix: *An Analysis of the Finite Element Method*, Prentice-Hall, Englewood Cliffs, N.J., 1973.
261. Street, R. L.: *The Analysis and Solution of Partial Differential Equations*, Brooks/Cole, Monterey, Calif., 1973.
262. Struble, R. A.: *Nonlinear Differential Equations*, McGraw-Hill, New York, 1962.
263. Swokowski, E. W.: *Calculus with Analytic Geometry*, Prindle, Weber & Schmidt, 2d ed., Boston, 1981.
264. Taylor, A. E., and D. C. Lay: *Introduction to Functional Analysis*, 2d ed., Wiley, New York, 1980.
265. ——— and W. R. Mann: *Advanced Calculus*, 2d ed., Xerox College Publishing, Lexington, Mass., 1972.
266. Tewarson, P.: *Sparse Matrices*, Academic, New York, 1973.
267. Thomas, G. B., Jr.: *Calculus and Analytic Geometry*, 4th ed., Addison-Wesley, Reading, Mass., 1968.
268. Tiller and Tour: "Stagewise Operations–Applications of the Calculus of Finite Differences to Chemical Engineering," *Trans. Amer. Inst. Chem. Eng.*, **40**, 317–332 (1944).
269. Todd, J.: *Survey of Numerical Analysis*, McGraw-Hill, New York, 1962.
270. Tranter, C. J.: *Integral Transforms in Mathematical Physics*, Methuen, London, 1956.
271. Troutman, J. L.: *Variational Calculus with Elementary Convexity*, Springer-Verlag, New York, 1983.
272. Tychonov, A. N., and A. A. Samarskii: *Partial Differential Equations of Mathematical Physics*, Holden-Day, San Francisco, 1964.
273. Van Dyke, M.: *Perturbation Methods in Fluid Mechanics*, Academic, New York, 1964.
274. Varga, R. S.: *Matrix Iterative Analysis*, Prentice-Hall, Englewood Cliffs, N.J., 1962.
275. Vemuri, V., and W. Karplus: *Digital Computer Treatment of Partial Differential Equations*, Prentice-Hall, Englewood Cliffs, N.J., 1981.
276. Vichnevetsky, R.: *Computer Methods for Partial Differential Equations*, vol. 1: *Elliptic Equations and the Finite Element Method* (1981); vol. 2: *Initial Value Problems* (1982), Prentice-Hall, Englewood Cliffs, N.J.
277. Watson, G. N.: *Theory of Bessel Functions*, Cambridge University Press, London, 1922.
278. Weinstock, R.: *Calculus of Variations with Applications to Physics and Engineering*, Dover, New York, 1974.
279. Whipkey, K. L., and M. N. Whipkey: *The Power of Calculus*, 3d ed., Wiley, 1979.
280. Whittaker, E., and G. Watson: *A Course of Modern Analysis*, 4th ed., Cambridge University Press, London, 1958.
281. Widder, D. V.: *Advanced Calculus*, 2d ed., Prentice-Hall, Englewood Cliffs, N.J., 1961.
282. Wilde, D. J., and C. S. Beightler: *Foundations of Optimization*, Prentice-Hall, Englewood Cliffs, N.J., 1967.
283. Wilkinson, J. H.: *Rounding Errors in Algebraic Processes*, Prentice-Hall, Englewood Cliffs, N.J., 1963.
284. ———: *The Algebraic Eigenvalue Problem*, Oxford University Press, London and New York, 1965.
285. ——— and C. Reisch (eds.): *Handbook for Automatic Computation*, vol. 2, Springer-Verlag, Berlin and New York, 1971.
286. Williams, G.: *Computational Linear Algebra with Models*, Allyn and Bacon, 2d ed., Boston, 1981.
287. Willoughby, R. (ed): *Stiff Differential Systems*, Plenum, New York, 1974.
288. Wouk, A.: *A Course of Applied Functional Analysis*, Wiley-Interscience, New York, 1979.
289. Wylie, C. R.: *Plane Trigonometry*, McGraw-Hill, New York, 1955.
290. ———: *Advanced Engineering Mathematics*, 4th ed., McGraw-Hill, New York, 1975.
291. Young, D. M.: *Iterative Solution for Large Linear Systems*, Academic, New York, 1971.
292. ——— and R. T. Gregory: *A Survey of Numerical Mathematics*, vols. I and II, Addison-Wesley, Reading, Mass., 1972.
293. Zangwill, W. I.: *Nonlinear Programming: A Unified Approach*, Prentice-Hall, Englewood Cliffs, N.J., 1969.
294. Zemanian, A. H.: *Distribution Theory and Transform Analysis*, McGraw-Hill, New York, 1965.
295. Zienkiewicz, O. C.: *The Finite Element Method in Engineering Science*, McGraw-Hill, London, 1971.
296. ——— and K. Morgan: *Finite Elements and Approximations*, Wiley, New York, 1983.

REFERENCES FOR GENERAL AND SPECIFIC TOPICS

Advanced Engineering Mathematics. The references listed contain a variety of fundamental topics in mathematics for engineers. It is recommended that readers consult appropriate references in this list for background information required in the application of more advanced or specialized mathematics. *Upper undergraduate level:* 5, 12, 15, 76, 86, 121, 123, 146, 147, 161, 172, 174, 184, 191, 224, 225, 236, 252, 290. *Graduate level:* 23, 73, 112, 123, 156, 206, 207, 240, 254, 255, 280. *Mathematical tables, mathematical dictionaries, and handbooks of mathematical functions and formulas:* 1, 52, 53, 69, 94, 98, 111, 122, 149, 160, 180, 195, 204, 214, 221, 231, 285. *Mathematical modeling of physical phenomena:* 14, 15, 31, 43, 95, 129, 184, 255. *Mathematical theory of reaction, diffusion, and transport processes:* 8, 13, 15, 85, 144. *Mathematical methods in chemical engineering:* 15, 84, 192, 196, 212, 222, 268, 282. *Inequalities:* 24, 122, 138.

Vector and Tensor Analysis. 40, 48, 65, 117, 167, 174, 222.

Special Functions in Physics and Engineering. 26, 37, 98, 156, 182, 187, 189, 228, 277.

Green's Functions and Applications. 66, 68, 74, 97, 112, 123, 146, 156, 255.

Perturbation and Asymptotic Methods in Applied Mathematics. 170, 210, 211, 273.

Advanced Calculus. 11, 40, 48, 110, 167, 265, 280, 281.

Mathematical Analysis; Real Analysis. 11, 148, 237.

Approximation Theory and Interpolation. 59, 82, 83. See also general references under "Numerical Analysis and Approximate Methods."

Control and System Theory. 17, 46, 219, 239.

Functional Analysis, Linear Operators, and Their Applications to Operator Equations and Numerical Analysis. 127, 148, 164, 173, 175, 188, 208, 211, 229, 238, 264, 288.

Generalized Inverses and Least-Squares Problems. 32, 126, 181, 208.

MATHEMATICS

GENERAL

The basic problems of the sciences and engineering fall broadly into three categories:

1. *Equilibrium problems.* These are problems of steady state. In such problems the configuration of the system is to be determined. This solution does not change with time but continues indefinitely in the same pattern, hence the name "steady state." Typical chemical engineering examples include steady temperature distributions in heat conduction, equilibrium in chemical reactions, and steady diffusion problems.

2. *Eigenvalue problems.* These are extensions of equilibrium problems in which critical values of certain parameters are to be determined in addition to the corresponding steady-state configurations. The determination of eigenvalues may also arise in propagation problems. Typical chemical engineering problems include those in heat transfer and resonance in which certain boundary conditions are prescribed.

3. *Propagation problems.* These problems are concerned with predicting the subsequent behavior of a system from a knowledge of the initial state. For this reason they are often called the transient (time-varying) or unsteady-state phenomena. Chemical engineering examples include the transient state of chemical reactions (kinetics), the propagation of pressure waves in a fluid, transient behavior of an adsorption column, and the rate of approach to equilibrium of a packed distillation column.

The mathematical treatment of engineering problems involves four basic steps:

1. *Formulation.* The expression of the problem in mathematical language. That translation is based on the appropriate physical laws governing the process.

2. *Solution.* Appropriate mathematical operations are accomplished so that logical deductions may be drawn from the mathematical model.

3. *Interpretation.* Development of relations between the mathematical results and their meaning in the physical world.

4. *Refinement.* The recycling of the procedure to obtain better predictions as indicated by experimental checks.

Steps 1 and 2 are of primary interest here. The actual details are left to the various subsections, and only general approaches will be discussed.

The formulation step may result in algebraic equations, difference equations, differential equations, integral equations, or combinations of these. In any event these mathematical models usually arise from statements of physical laws such as the laws of mass and energy conservation in the form

Input of conserved quantity − output of conserved quantity + conserved quantity produced = accumulation of conserved quantity

Rate of input of conserved quantity − rate of output of conserved quantity + rate of conserved quantity produced = rate of accumulation of conserved quantity

These statements may be abbreviated by the statement

Input − output + production = accumulation

When the basic physical laws are expressed in this form, the formulation is greatly facilitated. These expressions are quite often given the names, "material balance," "energy balance," and so forth. To be a little more specific, one could write the law of conservation of energy in the steady state as

Rate of energy in − rate of energy out + rate of energy produced = 0

Many general laws of the physical universe are expressible by differential equations. Specific phenomena are then singled out from the infinity of solutions of these equations by assigning the individual initial or boundary conditions which characterize the given problem. In mathematical language one such problem, the equilibrium problem, is called a **boundary-value problem** (Fig. 2-1). Schematically the problem is characterized by a differential equation plus an open region in which the equation holds and, on the boundaries of the region, by certain conditions (boundary conditions) which are dictated by the physical problem. The solution of the equation must satisfy the differential equation inside the region and the prescribed conditions on the boundary.

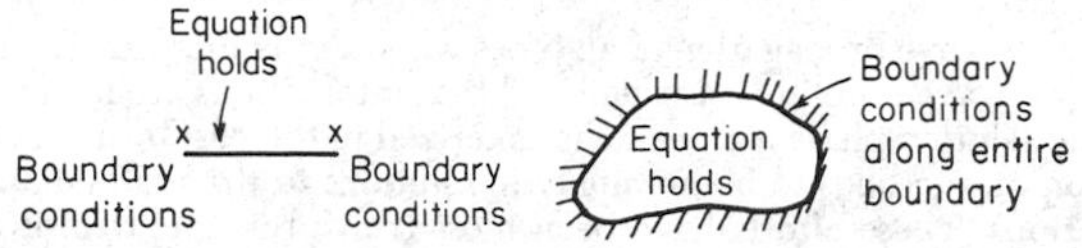

FIG. 2-1 Boundary conditions.

In mathematical language the propagation problem is known as an initial-value problem (Fig. 2-2). Schematically the problem is characterized by a differential equation plus an open region in which the equation holds. The solution of the differential equation must satisfy the initial conditions plus any "side" boundary conditions.

L. F. Richardson (1925) used descriptive phrases in illustrating the differences between these two problems. He called the equilibrium problem a **jury problem** and the propagation problem a **marching problem.** In the former the entire solution is examined and judged by a jury demanding simultaneous satisfaction of all the boundary conditions and all the internal requirements. In the latter the solution marches out from initial conditions guided in transit by "side" boundary conditions.

The description of phenomena in a "continuous" medium such as a gas or a fluid often leads to partial differential equations. In particular, phenomena of "wave" propagation are described by a class of partial differential equations called "hyperbolic," and these are essentially different in their properties from other classes such as those which describe equilibrium ("elliptic") or diffusion and heat transfer ("parabolic"). **Prototypes** are:

1. *Elliptic.* Laplace's equation

$$\partial^2 u/\partial x^2 + \partial^2 u/\partial y^2 = 0$$

Poisson's equation

$$\partial^2 u/\partial x^2 + \partial^2 u/\partial y^2 = g(x,y)$$

These do not contain the variable t (time) explicitly; accordingly their solutions represent equilibrium configurations. Laplace's equation corresponds to a "natural" equilibrium, while Poisson's equation corresponds to an equilibrium under the influence of an external force of density proportional to $g(x, y)$.

2. *Parabolic.* The heat equation

$$\partial u/\partial t = \partial^2 u/\partial x^2 + \partial^2 u/\partial y^2$$

describes nonequilibrium or propagation states of diffusion as well as heat transfer.

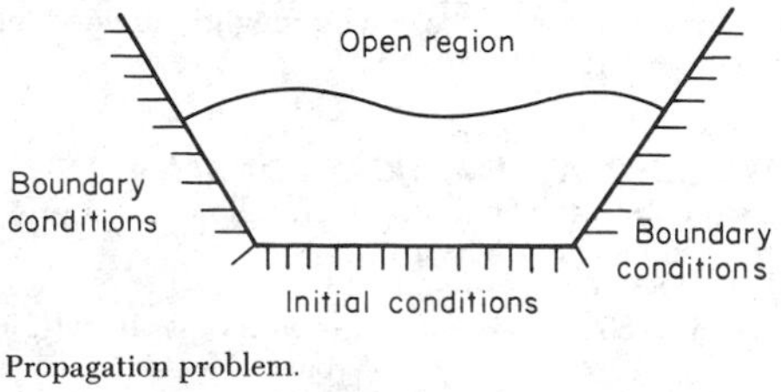

FIG. 2-2 Propagation problem.

3. *Hyperbolic.* The wave equation

$$\partial^2u/\partial t^2 = \partial^2u/\partial x^2 + \partial^2u/\partial y^2$$

describes wave propagation of all types when the assumption is made that the wave amplitude is small and that interactions are linear.

The solution phase has been characterized in the past by a concentration on methods to obtain analytic solutions to the mathematical equations. These efforts have been most fruitful in the area of the linear equations such as those just given. However, many natural phenomena are nonlinear. Various methodologies and techniques have been developed in the past two decades for the qualitative and quantitative analysis of several classes of nonlinear problems. There are, however, nonlinear phenomena for which well-known techniques are either inadequate or inapplicable. Thus formulation of the phenomena in the language generalized from the calculus does very little to help lessen our ignorance of the problem. When such situations arise, the only way open to us seems to be to retrace our steps to the time before calculus was invented. This represents the essence of an increasingly important area called numerical analysis. Thus the approach in solving a problem is to use the methods of the calculus when they represent a shortcut but to avoid them by a recourse to algebra and numerical methods when they become unmanageable.

Numerical methods almost never fail to provide an answer to any particular situation, but they can never furnish a general solution of any problem.

The mathematical details outlined here include both analytic and numerical techniques useful in obtaining solutions to problems.

Our discussion to this point has been confined to those areas in which the governing laws are well known. However, in many areas information on the governing laws is lacking, yet work must progress in these areas. Interest in the application of statistical methods to all types of problems has grown rapidly since World War II. Broadly speaking, statistical methods may be of use whenever conclusions are to be drawn or decisions made on the basis of experimental evidence. Since statistics could be defined as the technology of the scientific method, it is primarily concerned with the first two aspects of the method, namely, the performance of experiments and the drawing of conclusions from experiments. Traditionally the field is divided into two areas:

1. *Design of experiments.* When conclusions are to be drawn or decisions made on the basis of experimental evidence, statistical techniques are most useful when experimental data are subject to errors. The design of experiments may then often be carried out in such a fashion as to avoid some of the sources of experimental error and make the necessary allowances for that portion which is unavoidable. Second, the results can be presented in terms of probability statements which express the reliability of the results. Third, a statistical approach frequently forces a more thorough evaluation of the experimental aims and leads to a more definitive experiment than would otherwise have been performed.

2. *Statistical inference.* The broad problem of statistical inference is to provide measures of the uncertainty of conclusions drawn from experimental data. This area uses the theory of probability, enabling scientists to assess the reliability of their conclusions in terms of probability statements.

Both of these areas, the mathematical and the statistical, are intimately intertwined when applied to any given situation. The methods of one are often combined with the other. And both in order to be successfully used must result in the numerical answer to a problem—that is, they constitute the means to an end. Increasingly the numerical answer is being obtained from the mathematics with the aid of computing devices. These can be roughtly divided into analog and digital computers.

MISCELLANEOUS MATHEMATICAL CONSTANTS

Numerical values of the constants which follow are approximate to the number of significant digits given.

$\pi = 3.1415926536$	Pi
$e = 2.7182818285$	Napierian (natural) logarithm base
$\gamma = 0.5772156649$	Euler's constant
$\ln \pi = 1.1447298858$	Napierian (natural) logarithm of pi
$\log \pi = 0.4971498727$	Briggsian (common logarithm of pi
$\log x = 0.4342944819 \ln x$	Logarithm conversion
$\ln x = 2.302585093 \log x$	Logarithm conversion
$\sqrt{e} = 1.6487212707$	$\log e = 0.4342944819$
$\sqrt{\pi} = 1.7724538510$	$\ln 10 = 2.3025850930$
$\sqrt{\pi/2} = 1.2533141371$	$\ln 2 = 0.6931471806$
$\sqrt{\pi/3} = 1.0233267079$	$\log (\log e) = 9.637784311 - 10$
$\sqrt{2\pi} = 2.5066282746$	$\sqrt{2} = 1.4142135624$
$\sqrt{0.1} = 0.3162277660$	$\sqrt{3} = 1.7320508076$
Radian = 57.2957795131°	$\sqrt{5} = 2.2360679775$
Degree = 0.0174532925 rad	$\sqrt{6} = 2.4494897428$
Minute = 0.0002908882 rad	$\sqrt{7} = 2.6457513111$
Second = 0.0000048481 rad	$\sqrt{8} = 2.8284271248$
$\sqrt{\gamma} = 0.7597471059$	$\sqrt{10} = 3.1622776602$

THE REAL-NUMBER SYSTEM

The natural numbers, or counting numbers, are the positive integers: 1, 2, 3, 4, 5, ... The negative integers are −1, −2, −3, ...

A number in the form a/b, where a and b are integers, $b \neq 0$, is a rational number. A real number which cannot be written as the quotient of two integers is called an irrational number, e.g., $\sqrt{2}$, $\sqrt{3}$, $\sqrt{5}$, π, e, $\sqrt[3]{2}$.

There is a one-to-one correspondence between the set of real numbers and the set of points on an infinite line (coordinate line).

Order among Real Numbers; Inequalities

$a > b$ means that $a - b$ is a positive real number.
If $a < b$ and $b < c$, then $a < c$.
If $a < b$, then $a \pm c < b \pm c$ for any real number c.
If $a < b$ and $c > 0$, then $ac < bc$.
If $a < b$ and $c < 0$, then $ac > bc$.
If $a < b$ and $c < d$, then $a + c < b + d$.
If $0 < a < b$ and $0 < c < d$, then $ac < bd$.
If $a < b$ and $ab > 0$, then $1/a > 1/b$.
If $a < b$ and $ab < 0$, then $1/a < 1/b$.

Absolute Value For any real number x, $|x| = \begin{cases} x \text{ if } x \geq 0 \\ -x \text{ if } x < 0 \end{cases}$.

Properties

If $|x| = a$, where $a > 0$, then $x = a$ or $x = -a$.
$|x| = |-x|$; $-|x| \leq x \leq |x|$; $|xy| = |x|\,|y|$.
If $|x| < c$, then $-c < x < c$, where $c > 0$.
$||x| - |y|| \leq |x + y| \leq |x| + |y|$.
$\sqrt{x^2} = |x|$.

Proportions If $\dfrac{a}{b} = \dfrac{c}{d}$, then $\dfrac{a+b}{b} = \dfrac{c+d}{d}$, $\dfrac{a-b}{b} = \dfrac{c-d}{d}$, $\dfrac{a-b}{a+b} = \dfrac{c-d}{c+d}$.

Indeterminants

Form	Example*	
$(\infty)(0)$	xe^{-x}	$x \to \infty$
0^0	x^x	$x \to 0^+$
∞^0	$(\tan x)^{\cos x}$	$x \to \frac{1}{2}\pi^-$
1^∞	$(1 + x)^{1/x}$	$x \to 0^+$
$\infty - \infty$	$\sqrt{x+1} - \sqrt{x-1}$	$x \to \infty$
$\dfrac{0}{0}$	$\dfrac{\sin x}{x}$	$x \to \infty$
$\dfrac{\infty}{\infty}$	$\dfrac{e^x}{x}$	$x \to 0$

*See "Limit" in this section.

Limits of the type $0/\infty$, $\infty/0$, 0^{∞}, $\infty \cdot \infty$, $(+\infty) + (+\infty)$, and $(-\infty) + (-\infty)$ are *not* indeterminate forms.

Integral Exponents (Powers and Roots) If m and n are positive integers and a, b are numbers or functions, then the following properties hold:

$$a^{-n} = 1/a^n \qquad a \neq 0$$
$$(ab)^n = a^n b^n$$
$$(a^n)^m = a^{nm}, \qquad a^n a^m = a^{n+m}$$
$$\sqrt[n]{a} = a^{1/n} \text{ if } a > 0$$
$$\sqrt[m]{\sqrt[n]{a}} = \sqrt[mn]{a}, \; a > 0$$
$$a^{m/n} = (a^m)^{1/n} = \sqrt[n]{a^m}, \; a > 0$$
$$a^0 = 1 \; (a \neq 0)$$
$$0^a = 0 \; (a \neq 0)$$

Infinity (∞) is not a real number. It is possible to extend the real-number system by adjointing to it "∞" and "$-\infty$," and within the extended system certain operations involving $+\infty$ or $-\infty$ are possible. For example, if $0 < a < 1$, then $a^{\infty} = \lim_{x \to \infty} a^x = 0$, whereas if $a > 1$, then $a^{\infty} = \infty$, $\infty^a = \infty$ $(a > 0)$, $\infty^a = 0$ $(a < 0)$.

Care should be taken in the case of roots and fractional powers of a product; e.g., $\sqrt{xy} \neq \sqrt{x}\sqrt{y}$ if x and y are negative. This rule applies if one is careful about the domain of the functions involved; so $\sqrt{xy} = \sqrt{x}\sqrt{y}$ if $x > 0$ and $y > 0$.

Given any number $b > 0$, there is a unique function $f(x)$ defined for all real numbers x such that (1) $f(x) = b^x$ for all rational x; (2) f is increasing if $b > 1$, constant if $b = 1$, and decreasing if $0 < b < 1$. This function is called the **exponential function** b^x. For any $b > 0$, $f(x) = b^x$ is a continuous function. Also with $a,b > 0$ and x,y any real numbers, we have

$$(ab)^x = a^x b^x$$
$$b^x b^y = b^{x+y}$$
$$(b^x)^y = b^{xy}$$

The exponential function with base b can also be defined as the inverse of the logarithmic function. The most common exponential function in applications corresponds to choosing b the transcendental number e.

Logarithms $\log ab = \log a + \log b$, $a > 0$, $b > 0$.

$$\log a^n = n \log a$$
$$\log (a/b) = \log a - \log b$$
$$\log \sqrt[n]{a} = (1/n) \log a$$

Roots If a is a real number, n is a positive integer, then x is called the nth root of a if $x^n = a$. The number of nth roots is n, but not all of them are necessarily real. The principal nth root means the following: (1) if $a > 0$ the principal nth root is the unique positive root, (2) if $a < 0$, and n odd, it is the unique negative root, and (3) if $a < 0$ and n even, it is any of the complex roots. All the roots, real or complex, of a real number a may be found by De Moivre's formula or theorem (see subsection "Complex Variables"). This, however, requires that we know how to compute positive real roots of positive numbers. Tables of such roots are given in mathematical handbooks or may be computed with algorithms or calculators. Approximations can be obtained by the binomial series. Among numerical methods and iterative processes for computing roots we mention Horner's method and Newton's method.

The nth roots of a real number a are located on the circle of radius $\sqrt[n]{|a|}$ and are uniformly distributed on that circle. Thus if one can locate one of the nth roots (in particular, the principal nth root), then all the roots can be obtained geometrically.

A computing formula for roots is as follows: For $S = \sqrt[n]{X} = (X)^{1/n}$, compute $S = \exp\left(\dfrac{\ln X}{n}\right)$ **or numerically.**

Example

Square root Find $S = \sqrt{155.96}$.

Direct

```
     12  .4  9
  1 √155 .96 00        S = 12.49
    −1
   2²⌋55
     44
   24⁴⌋11 96
      −9 76
   248⁹⌋2 20 00
```

Iteration

$$S_{i+1} = \frac{1}{2}\left(\frac{X}{S_i} + S_i\right)$$

Assume $S_1 = 12$. For $X = 155.96$,

$$S_2 = \frac{1}{2}\left(\frac{155.96}{12} + 12\right) = 12.4983$$
$$S_3 = 12.488398$$
$$S_4 = 12.48839461$$

Correct to eight decimal places.

Cube root

Direct for $S = (155.96)^{1/3}$

```
               25   5 . 3 8
                 ³√155.960 000
                    125
  30*(5)(53) + 3²  |30 960
           = 7959  |23 877
 30(53)(538) + 8²  |7 083 000
         = 855484  |6 843 872
```

Iteration for $S = (X)^{1/3} = (15.596)^{1/3}$

$$S_{i+1} = \frac{2}{3} S_i + \frac{X}{3S_i^2}$$

Use $S_1 = 1$ or let $\sqrt[3]{X} = \sqrt[3]{\wedge NNN}\,(10^{P/3 + R/3})$, where $P/3$ integer and $10^{1/3} = 2.154$, $10^{2/3} = 4.642$.

Then $\sqrt[3]{\wedge NNN} = (0.15596)^{1/3}$ with $R = 2$.

The first approximation to $\sqrt[3]{\wedge NNN}$ is

$$0.349 + 1.387K - 1.236K^2 + 0.505K^3, \; K = \wedge NNN$$

For $S = (15.596)^{1/3} = (0.15596 \times 10^2)^{1/3}$,

$$S_1 = \sqrt[3]{0.15596}\,(4.642) = 0.5372(4.642) = 2.494$$
$$S_2 = \frac{2}{3}(2.494) + \frac{1}{3}\left[\frac{15.596}{(2.494)^2}\right] = 2.49846$$
$$S_3 = \frac{2}{3}(2.49846) + \frac{1}{3}\left[\frac{15.596}{(2.49846)^2}\right] = 2.4984524$$

PROGRESSIONS

Arithmetic Progression

$$\sum_{k=0}^{n-1} (a + kd) = na + \frac{1}{2} n(n-1)d = \frac{n}{2}(a + \ell)$$

where ℓ is the last term, $\ell = a + (n-1)d$.

Geometric Progression

$$\sum_{k=1}^{n} ar^{k-1} = \frac{a(r^n - 1)}{r - 1} \qquad (r \neq 1)$$

[*] Always 30 for cube root.

Arithmetic-Geometric Progression

$$\sum_{k=0}^{n-1}(a + kd)r^k = \frac{a - [a + (n - 1)d]r^n}{1 - r} + \frac{dr(1 - r^{n-1})}{(1 - r)^2} \qquad (r \neq 1)$$

$$\sum_{k=1}^{n} k^5 = \frac{1}{12} n^2(n + 1)^2(2n^2 + 2n - 1)$$

$$\sum_{k=1}^{n}(2k - 1) = n^2$$

$$\sum_{k=1}^{n}(2k - 1)^2 = \frac{1}{3} n(4n^2 - 1)$$

$$\sum_{k=1}^{n}(2k - 1)^3 = n^2(2n^2 - 1)$$

$$\gamma = \lim_{n\to\infty}\left(\sum_{m=1}^{n}\frac{1}{m} - \ln n\right) = 0.577215 \cdots$$

$$\sum_{k=1}^{n}\frac{1}{k} = \gamma + \ln n + \frac{1}{2n} - \sum_{k=2}^{\infty}\frac{A_k}{n(n + 1) \cdots (n + k - 1)}$$

where $A_k = \frac{1}{k}\int_0^1 x(1 - x)(2 - x) \cdots (k - 1 - x)\, dx$

$$A_2 = 1/12 \qquad A_3 = 1/12 \qquad A_4 = 19/80 \qquad A_5 = 9/20$$

ALGEBRAIC INEQUALITIES

Arithmetic-Geometric Inequality Let A_n and G_n denote respectively the arithmetic and the geometric means of a set of positive numbers $a_1, a_2, \ldots, a_n$. The $A_n \geq G_n$, i.e.,

$$\frac{a_1 + a_2 + \cdots + a_n}{n} \geq (a_1a_2 \cdots a_n)^{1/n}$$

The equality holds only if all of the numbers a_i are equal.

Carleman's Inequality The arithmetic and geometric means just defined satisfy the inequality

$$\sum_{r=1}^{n} G_r \leq neA_n$$

or, equivalently,

$$\sum_{r=1}^{n}(a_1a_2 \cdots a_r)^{1/r} \leq neA_n$$

where e is the best possible constant in this inequality.

Cauchy-Schwarz Inequality Let $\mathbf{a} = (a_1, a_2, \ldots, a_n)$, $\mathbf{b} = (b_1, b_2, \ldots, b_n)$, where the a_i's and b_i's are real or complex numbers. Then

$$\left|\sum_{k=1}^{n} a_k\bar{b}_k\right|^2 \leq \left(\sum_{k=1}^{n}|a_k|^2\right)\left(\sum_{k=1}^{n}|b_k|^2\right)$$

The equality holds if, and only if, the vectors **a**, **b** are linearly dependent (i.e., one vector is scalar times the other vector).

Minkowski's Inequality Let $a_1, a_2, \ldots, a_n$ and $b_1, b_2, \ldots, b_n$ be any two sets of complex numbers. Then for any real number $p > 1$,

$$\left(\sum_{k=1}^{n}|a_k + b_k|^p\right)^{1/p} \leq \left(\sum_{k=1}^{n}|a_k|^p\right)^{1/p} + \left(\sum_{k=1}^{n}|b_k|^p\right)^{1/p}$$

Hölder's Inequality Let $a_1, a_2, \ldots, a_n$ and $b_1, b_2, \ldots, b_n$ be any two sets of complex numbers, and let p and q be positive numbers with $1/p + 1/q = 1$. Then

$$\left|\sum_{k=1}^{n} a_k\bar{b}_k\right| \leq \left(\sum_{k=1}^{n}|a_k|^p\right)^{1/p}\left(\sum_{k=1}^{n}|b_k|^q\right)^{1/q}$$

The equality holds if, and only if, the sequences $|a_1|^p, |a_2|^p, \ldots, |a_n|^p$ and $|b_1|^q, |b_2|^q, \ldots, |b_n|^q$ are proportional and the argument (angle) of the complex numbers $a_k\bar{b}_k$ is independent of k. This last condition is of course automatically satisfied if $a_1, \ldots, a_n$ and $b_1, \ldots, b_n$ are positive numbers.

Lagrange's Inequality Let $a_1, a_2, \ldots, a_n$ and $b_1, b_2, \ldots, b_n$ be real numbers. Then

$$\left(\sum_{k=1}^{n} a_kb_k\right)^2 = \left(\sum_{k=1}^{n} a_k^2\right)\left(\sum_{k=1}^{n} b_k^2\right) - \sum_{1\leq k\leq j\leq n}(a_kb_j - a_jb_k)^2$$

Example Two chemical engineers John and Mary purchase stock in the same company at times $t_1, t_2, \ldots, t_n$, when the price per share is respectively $p_1, p_2, \ldots, p_n$. Their methods of investment are different, however: John purchases x shares each time, whereas Mary invests P dollars each time (fractional shares can be purchased). Who is doing better?

While one can argue intuitively that the average cost per share for Mary does not exceed that for John, we illustrate a mathematical proof using inequalities. The average cost per share for John is equal to

$$\frac{\text{Total money invested}}{\text{Number of shares purchased}} = \frac{x\sum_{i=1}^{n} p_i}{nx} = \frac{1}{n}\sum_{i=1}^{n} p_i$$

The average cost per share for Mary is

$$\frac{nP}{\sum_{i=1}^{n}\frac{P}{p_i}} = \frac{n}{\sum_{i=1}^{n}\frac{1}{p_i}}$$

Thus the average cost per share for John is the arithmetic mean of $p_1, p_2, \ldots, p_n$, whereas that for Mary is the harmonic mean of these n numbers. Since the harmonic mean is less than or equal to the arithmetic mean for any set of positive numbers and the two means are equal only if $p_1 = p_2 = \cdots = p_n$, we conclude that the average cost per share for Mary is less than than for John if two of the prices p_i are distinct. One can also give a proof based on the Cauchy-Schwarz inequality. To this end, define the vectors

$$\mathbf{a} = (p_1^{-1/2}, p_2^{-1/2}, \ldots, p_n^{-1/2}) \qquad \mathbf{b} = (p_1^{1/2}, p_2^{1/2}, \ldots, p_n^{1/2})$$

Then $\mathbf{a} \cdot \mathbf{b} = 1 + \cdots + 1 = n$, and so by the Cauchy-Schwarz inequality

$$(\mathbf{a} \cdot \mathbf{b})^2 = n^2 \leq \sum_{i=1}^{n}\frac{1}{p_i}\sum_{i=1}^{n} p_i$$

with the equality holding only if $p_1 = p_2 = \cdots = p_n$. Therefore

$$\frac{n}{\sum_{i=1}^{n}\frac{1}{p_i}} \leq \frac{\sum_{i=1}^{n} p_i}{n}$$

MENSURATION FORMULAS

Let A denote areas and V volumes in the following:

PLANE GEOMETRIC FIGURES WITH STRAIGHT BOUNDARIES

Triangles (see also "Plane Trigonometry") $A = \frac{1}{2}bh$ where b = base, h = altitude.

Rectangle $A = ab$ where a and b are the lengths of the sides.

Parallelogram (opposite sides parallel) $A = ah = ab \sin \alpha$ where a, b are the lengths of the sides, h the height, and α the angle between the sides. See Fig. 2-3.

Rhombus (equilateral parallelogram) $A = \frac{1}{2}ab$ where a, b are the lengths of the diagonals.

Trapezoid (four sides, two parallel) $A = \frac{1}{2}(a + b)h$ where the lengths of the parallel sides are a and b and h = height.

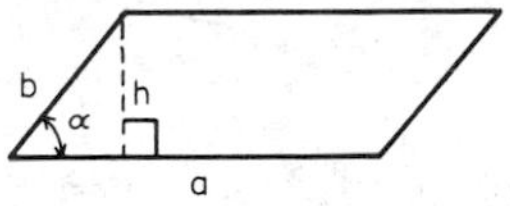

FIG. 2-3 Parallelogram.

Quadrilateral (four-sided) $A = \frac{1}{2}ab \sin \theta$ where a, b are the lengths of the diagonals and the acute angle between them is θ.

Regular Polygon of *n* Sides See Fig. 2-4.

$$A = \frac{1}{4} nl^2 \cot \frac{180°}{n} \qquad \text{where } l = \text{length of each side}$$

$$R = \frac{l}{2} \csc \frac{180°}{n} \qquad \text{where } R \text{ is the radius of the circumscribed circle}$$

$$r = \frac{l}{2} \cot \frac{180°}{n} \qquad \text{where } r \text{ is the radius of the inscribed circle}$$

$$\beta = \frac{360°}{n}$$

$$\theta = \frac{(n-2)\,180°}{n}$$

$$l = 2r \tan \frac{\beta}{2} = 2R \sin \frac{\beta}{2}$$

Inscribed and Circumscribed Circles with Regular Polygon of n Sides Let l = length of one side.

Figure	n	Area	Radius of circumscribed circle	Radius of inscribed circle
Equilateral triangle . .	3	$0.4330\, l^2$	$0.5774\, l$	$0.2887\, l$
Square	4	$1.0000\, l^2$	$0.7071\, l$	$0.5000\, l$
Pentagon	5	$1.7205\, l^2$	$0.8507\, l$	$0.6882\, l$
Hexagon	6	$2.5981\, l^2$	$1.0000\, l$	$0.8660\, l$
Heptagon	7	$3.6339\, l^2$	$1.1523\, l$	$1.0383\, l$
Octagon	8	$4.8284\, l^2$	$1.3065\, l$	$1.2071\, l$
Nonagon	9	$6.1818\, l^2$	$1.4619\, l$	$1.3737\, l$
Decagon	10	$7.6942\, l^2$	$1.6180\, l$	$1.5388\, l$

Radius r of Circle Inscribed in Triangle with Sides a, b, c

$$r = \sqrt{\frac{(s-a)(s-b)(s-c)}{s}} \qquad \text{where } s = \tfrac{1}{2}(a+b+c)$$

Radius R of Circumscribed Circle

$$R = \frac{abc}{4\sqrt{s(s-a)(s-b)(s-c)}}$$

Area of Regular Polygon of n Sides Inscribed in a Circle of Radius r

$$A = (nr^2/2) \sin (360°/n)$$

Perimeter of Inscribed Regular Polygon

$$P = 2nr \sin (180°/n)$$

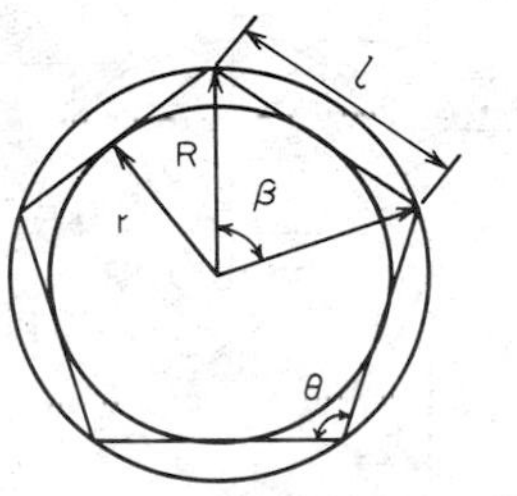

FIG. 2-4 Regular polygon.

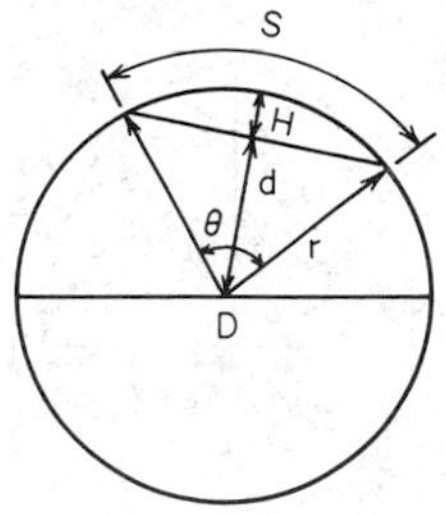

FIG. 2-5 Circle.

Area of Regular Polygon Circumscribed about a Circle of Radius r

$$A = nr^2 \tan (180°/n)$$

PLANE GEOMETRIC FIGURES WITH CURVED BOUNDARIES

Circle (Fig. 2-5) Let

C = circumference
r = radius
D = diameter
A = area
S = arc length subtended by θ
l = chord length subtended by θ
H = maximum rise of arc above chord, $r - H = d$
θ = central angle (rad) subtended by arc S

$$C = 2\pi r = \pi D \qquad (\pi = 3.14159\ldots)$$

$$S = r\theta = \tfrac{1}{2}D\theta$$

$$l = 2\sqrt{r^2 - d^2} = 2r \sin (\theta/2) = 2d \tan (\theta/2)$$

$$d = \frac{1}{2}\sqrt{4r^2 - l^2} = \frac{1}{2} l \cot \frac{\theta}{2}$$

$$\theta = \frac{S}{r} = 2 \cos^{-1} \frac{d}{r} = 2 \sin^{-1} \frac{l}{D}$$

$$A\ (\text{circle}) = \pi r^2 = \tfrac{1}{4}\pi D^2$$

$$A\ (\text{sector}) = \tfrac{1}{2} rS = \tfrac{1}{2} r^2 \theta$$

$$A\ (\text{segment}) = A\ (\text{sector}) - A\ (\text{triangle}) = \tfrac{1}{2} r^2(\theta - \sin \theta)$$

$$= r^2 \cos^{-1} \frac{r-H}{r} - (r-H)\sqrt{2rH - H^2}$$

Ring (area between two circles of radii r_1 and r_2) The circles need not be concentric, but one of the circles must enclose the other.

$$A = \pi(r_1 + r_2)(r_1 - r_2) \qquad r_1 > r_2$$

Ellipse (Fig. 2-6) Let the semiaxes of the ellipse be a and b

$$A = \pi ab$$

$$C = 4aE(k)$$

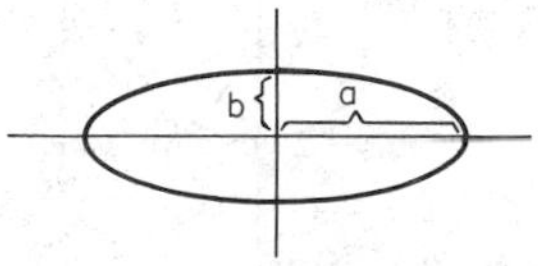

FIG. 2-6 Ellipse.

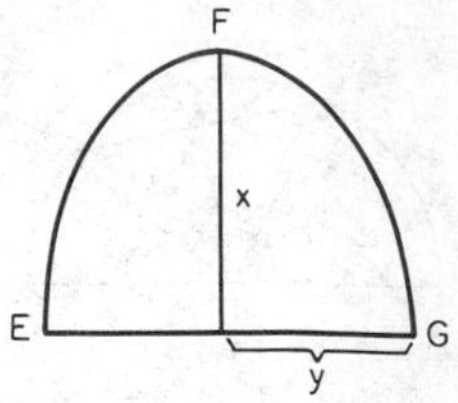

FIG. 2-7 Parabola.

where $k = 1 - (b^2/a^2)$ and $E(k)$ is the complete elliptic integral of the first kind [an approximation for the circumference $C = 2\pi \sqrt{(a^2 + b^2)/2}$].

Parabola (Fig. 2-7)

$$\text{Length of arc } EFG = \sqrt{4x^2 + y^2} + \frac{y^2}{2x} \ln \frac{2x + \sqrt{4x^2 + y^2}}{y}$$

$$\text{Area of section } EFG = \frac{4}{3}xy$$

Catenary (the curve formed by a cord of uniform weight suspended freely between two points A, B; Fig. 2-8)

$$y = \cosh(x/a)$$

Length of arc between points A and B is equal to $2a \sinh(L/a)$. Sag of the cord is $D = a \cosh(L/a) - 1$.

SOLID GEOMETRIC FIGURES WITH PLANE BOUNDARIES

Cube Volume $= a^3$; total surface area $= 6a^2$; diagonal $= a\sqrt{3}$, where a = length of one side of the cube.

Rectangular Parallelepiped Volume $= abc$; surface area $= 2(ab + ac + bc)$; diagonal $= \sqrt{a^2 + b^2 + c^2}$, where a, b, c are the lengths of the sides.

Prism Volume = (area of base) × (altitude); lateral surface area = (perimeter of right section) × (lateral edge).

Pyramid Volume = ⅓ (area of base) × (altitude); lateral area of regular pyramid = ½ (perimeter of base) × (slant height) = ½ (number of sides) (length of one side) (slant height).

Frustum of Pyramid (formed from the pyramid by cutting off the top with a plane, usually parallel to the base)

$$V = \tfrac{1}{3}(A_1 + A_2 + \sqrt{A_1 \cdot A_2})h$$

where h = altitude and A_1, A_2 are the areas of the base; lateral area of a regular figure = ½ (sum of the perimeters of base) × (slant height).

Volume and Surface Area of Regular Polyhedra with Edge l

Type of surface	Name	Volume	Surface area
4 equilateral triangles	Tetrahedron	$0.1179\ l^3$	$1.7321\ l^2$
6 squares	Hexahedron (cube)	$1.0000\ l^3$	$6.0000\ l^2$
8 equilateral triangles	Octahedron	$0.4714\ l^3$	$3.4641\ l^2$
12 pentagons	Dodecahedron	$7.6631\ l^3$	$20.6458\ l^2$
20 equilateral triangles	Icosahedron	$2.1817\ l^3$	$8.6603\ l^2$

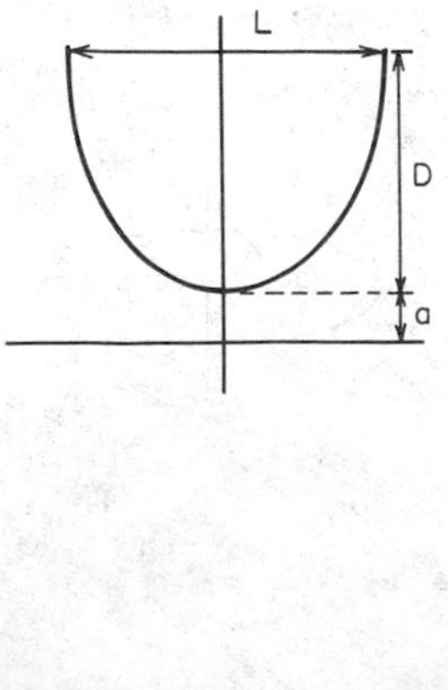

FIG. 2-8 Catenary.

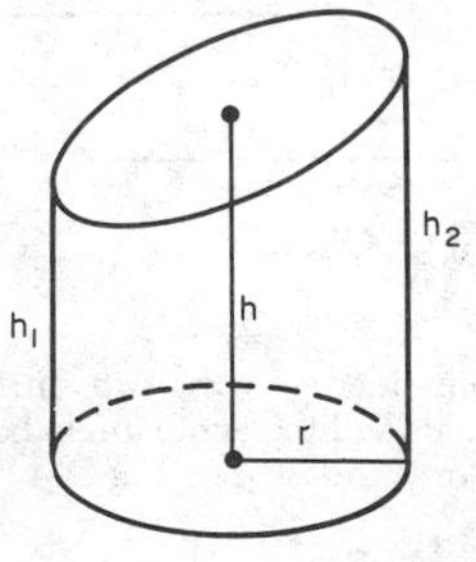

FIG. 2-9 Cylinder.

SOLIDS BOUNDED BY CURVED SURFACES

Cylinders (Fig. 2-9) V = (area of base) × (altitude); lateral surface area = (perimeter of right section) × (lateral edge).

Right Circular Cylinder $V = \pi$ (radius)^2 × (altitude); lateral surface area $= 2\pi$ (radius) × (altitude).

Truncated Right Circular Cylinder

$$V = \pi r^2 h; \text{ lateral area} = 2\pi rh$$

$$h = \tfrac{1}{2}(h_1 + h_2)$$

Hollow Cylinders Volume $= \pi h(R^2 - r^2)$, where r and R are the internal and external radii and h is the height of the cylinder.

Sphere (Fig. 2-10)

$$V \text{ (sphere)} = \tfrac{4}{3}\pi R^3 / \tfrac{1}{6}\pi D^3$$

$$V \text{ (spherical sector)} = \tfrac{2}{3}\pi R^2 h = \tfrac{1}{6}\pi h_1(3r_2^2 + h_1^2)$$

$$V \text{ (spherical segment of one base)} = \tfrac{1}{6}\pi h_1(3r_2^2 + h_1^2)$$

$$V \text{ (spherical segment of two bases)} = \tfrac{1}{6}\pi h(3r_1^2 + 3r_2^2 + h^2)$$

$$A \text{ (sphere)} = 4\pi R^2 = \pi D^2$$

$$A \text{ (zone)} = 2\pi Rh = \pi Dh$$

A (lune on the surface included between two great circles whose inclination is θ radians) $= 2R^2\theta$.

Cone V = ⅓ (area of base) × (altitude).

Right Circular Cone $V = (\pi/3) r^2 h$, where h is the altitude and r is the radius of the base; curved surface area $= \pi r \sqrt{r^2 + h^2}$, curved surface of the frustum of a right cone $= \pi(r_1 + r_2)\sqrt{h^2 + (r_1 - r_2)^2}$, where r_1, r_2 are the radii of the base and top respectively and h is the altitude; volume of the frustum of a right cone $= \pi(h/3)(r_1^2 + r_1 r_2 + r_2^2) = h/3(A_1 + A_2 + \sqrt{A_1 A_2})$, where A_1 = area of base and A_2 = area of top.

Ellipsoid $V = (\tfrac{4}{3})\pi abc$, where a, b, c are the lengths of the semiaxes.

Torus (obtained by rotating a circle of radius r about a line whose distance is $R > r$ from the center of the circle)

$$V = 2\pi^2 Rr^2 \qquad \text{Surface area} = 4\pi^2 Rr$$

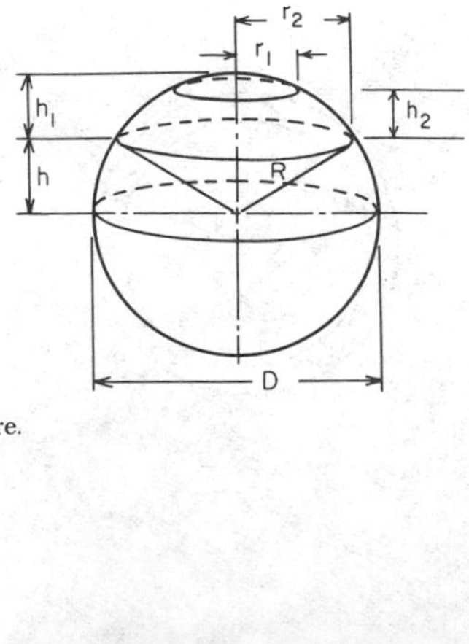

FIG. 2-10 Sphere.

Prolate Spheroid [formed by rotating an ellipse about its major axis $(2a)$]

$$\text{Surface area} = 2\pi b^2 + 2\pi(ab/e)\sin^{-1} e \qquad V = \tfrac{4}{3}\pi ab^2$$

where a, b are the major and minor axes and e = eccentricity $(e < 1)$.

Oblate Spheroid [formed by the rotation of an ellipse about its minor axis $(2b)$] Data as given previously.

$$\text{Surface area} = 2\pi a^2 + \pi\frac{b^2}{e}\ln\frac{1+e}{1-e} \qquad V = \tfrac{4}{3}\pi a^2 b$$

MISCELLANEOUS FORMULAS

See also "Differential and Integral Calculus."

Volume of a Solid Revolution (the solid generated by rotating a plane area about the x axis) $V = \pi\int_a^b [f(x)]^2\,dx$, where $y = f(x)$ is the equation of the plane curve and $a \le x \le b$.

Area of a Surface of Revolution

$$S = 2\pi\int_a^b y\,ds$$

where $ds = \sqrt{1 + (dy/dx)^2}\,dx$ and $y = f(x)$ is the equation of the plane curve rotated about the x axis to generate the surface.

Area Bounded by f(x), the x Axis, and the Lines $x = a$, $x = b$

$$A = \int_a^b f(x)\,dx \qquad (f(x) \ge 0)$$

Length of Arc of a Plane Curve

If $y = f(x)$,

$$\text{Length of arc } s = \int_a^b \sqrt{1 + \left(\frac{dy}{dx}\right)^2}\,dx$$

If $x = g(y)$,

$$\text{Length of arc } s = \int_c^d \sqrt{1 + \left(\frac{dx}{dy}\right)^2}\,dy$$

If $x = f(t)$, $y = g(t)$,

$$\text{Length of arc } s = \int_{t0}^{t1} \sqrt{\left(\frac{dx}{dt}\right)^2 + \left(\frac{dy}{dt}\right)^2}\,dt$$

In general, $(ds)^2 = (dx)^2 + (dy)^2$.

Theorems of Pappus (for volumes and areas of surfaces of revolution)

1. If a plane area is revolved about a line which lies in its plane but does not intersect the area, then the volume generated is equal to the product of the area and the distance traveled by the area's center of gravity.
2. If an arc of a plane curve is revolved about a line which lies in its plane but does not intersect the arc, then the surface area generated by the arc is equal to the product of the length of the arc and the distance traveled by its center of gravity.

These theorems are useful for determining volumes V and surface areas S of solids of revolution if the centers of gravity are known. If S and V are known, the centers of gravity may be determined.

IRREGULAR AREAS AND VOLUMES

Irregular Areas Let $y_0, y_1, \ldots, y_n$ be the lengths of a series of equally spaced parallel chords and h be their distance apart. The area of the figure is given approximately by any of the following:

$$A_T = (h/2)\,[(y_0 + y_n) + 2(y_1 + y_2 + \cdots + y_{n-1})] \qquad \text{(trapezoidal rule)}$$

$$A_S = (h/3)\,[(y_0 + y_n) + 4(y_1 + y_3 + y_5 + \cdots + y_{n-1}) + 2(y_2 + y_4 + \cdots + y_{n-2})] \qquad (n \text{ even, Simpson's rule})$$

$$A_W = (3h/8)\,[(y_0 + y_n) + 3(y_1 + y_2 + y_4 + y_5 + \cdots) + 2(y_3 + y_6 + \cdots)] \qquad \text{(Weddle's 3/8 rule)}$$

The greater the value of n, the greater the accuracy of approximation.

Irregular Volumes To find the volume, replace the y's by cross-sectional areas A_j and use the results in the preceding equations.

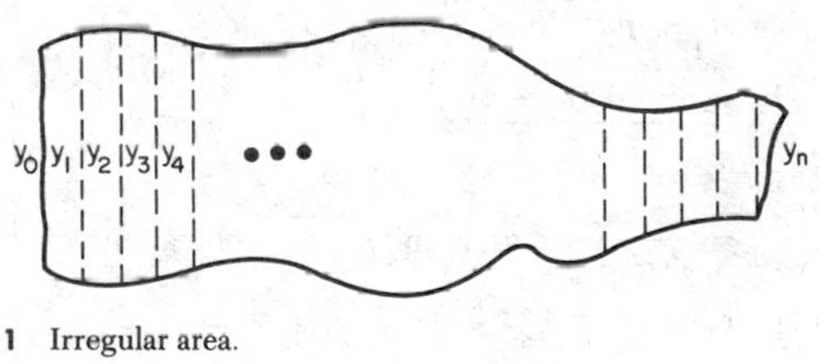

FIG. 2-11 Irregular area.

ELEMENTARY ALGEBRA

REFERENCES 6, 16, 96, 102, 117, 251. College algebra; some references with calculator applications and/or use.

OPERATIONS ON ALGEBRAIC EXPRESSIONS

An algebraic expression will here be denoted as a combination of letters and numbers such as

$$3ax - 3xy + 7x^2 + 7x^{3/2} - 2.8xy$$

Addition and Subtraction Only like terms can be added or subtracted in two algebraic expressions.

Example $(3x + 4xy - x^2) + (3x^2 + 2x - 8xy) = 5x - 4xy + 2x^2$.

Example $(2^x + 3xy - 4x^{1/2}) + (3^x + 6x - 8xy) = 2^x + 3^x + 6x - 5xy - 4x^{1/2}$.

Multiplication Multiplication of algebraic expressions is term by term, and corresponding terms are combined.

Example $(2x + 3y - 2xy)(3 + 3y) = 6x + 9y + 9y^2 - 6xy^2$.

Division This operation is analogous to that in arithmetic.

Example Divide $3e^{2x} + e^x + 1$ by $e^x + 1$.

$$\begin{array}{r|l}
 & \text{Dividend} \\
\text{Divisor } e^x + 1 & 3e^{2x} + e^x + 1 \quad \underline{3e^x - 2} \text{ quotient} \\
 & \underline{3e^{2x} + 3e^x} \\
 & -2e^x + 1 \\
 & \underline{-2e^x - 2} \\
 & +3 \text{ (remainder)}
\end{array}$$

Therefore, $3e^{2x} + e^x + 1 = (e^x + 1)(3e^x - 2) + 3$.

Operations with Zero All numerical computations (except division) can be done with zero: $a + 0 = 0 + a = a$; $a - 0 = a$; $0 - a = -a$; $(a)(0) = 0$; $a^0 = 1$ if $a \neq 0$; $0/a = 0$, $a \neq 0$. $a/0$ and $0/0$ have no meaning.

Fractional Operations

$$-\frac{x}{y} = -\left(\frac{-x}{-y}\right) = \frac{x}{-y} = \frac{-x}{y}; \frac{x}{y} = \frac{-x}{-y}; \frac{x}{y} = \frac{ax}{ay}; \text{ if } a \neq 0.$$

$$\frac{x}{y} \pm \frac{z}{y} = \frac{x \pm z}{y};$$

$$\left(\frac{x}{y}\right)\left(\frac{z}{t}\right) = \frac{xz}{yt}; (x/y)/(z/t) = \left(\frac{x}{y}\right)\left(\frac{t}{z}\right) = \frac{xt}{yz}$$

Factoring That process of analysis consisting of reducing a given expression into the product of two or more simpler expressions called *factors*. Some of the more common expressions are factored here:

(1) $(x^2 - y^2) = (x - y)(x + y)$
(2) $x^2 + 2xy + y^2 = (x + y)^2$
(3) $x^2 + ax + b = (x + c)(x + d)$ where $c + d = a$, $cd = b$
(4) $by^2 + cy + d = (ey + f)(gy + h)$ where $eg = b$, $fg + eh = c$, $fh = d$
(5) $x^2 + y^2 + z^2 + 2yz + 2xz + 2xy = (x + y + z)^2$
(6) $x^2 - y^2 - z^2 - 2yz = (x - y - z)(x + y + z)$
(7) $x^2 + y^2 + z^2 - 2xy - 2xz + 2yz = (x - y - z)^2$
(8) $x^3 - y^3 = (x - y)(x^2 + xy + y^2)$
(9) $(x^3 + y^3) = (x + y)(x^2 - xy + y^2)$
(10) $(x^4 - y^4) = (x - y)(x + y)(x^2 + y^2)$
(11) $x^5 + y^5 = (x + y)(x^4 - x^3y + x^2y^2 - xy^3 + y^4)$
(12) $x^n - y^n = (x - y)(x^{n-1} + x^{n-2}y + x^{n-3}y^2 + \cdots + y^{n-1})$

Laws of Exponents

$(a^n)^m = a^{nm}$; $a^{n+m} = a^n \cdot a^m$; $a^{n/m} = (a^n)^{1/m}$; $a^{n-m} = a^n/a^m$; $a^{1/m} = \sqrt[m]{a}$; $a^{1/2} = \sqrt{a}$; $\sqrt{x^2} = |x|$ (absolute value of x). For $x > 0$, $y > 0$, $\sqrt{xy} = \sqrt{x}\sqrt{y}$; for $x > 0$ $\sqrt[n]{x^m} = x^{m/n}$; $\sqrt[n]{1/x} = 1/\sqrt[n]{x}$

THE BINOMIAL THEOREM

If n is a positive integer,

$$(a + b)^n = a^n + na^{n-1}b + \frac{n(n-1)}{2!}a^{n-2}b^2 + \frac{n(n-1)(n-2)}{3!}a^{n-3}b^3 + \cdots + b^n = \sum_{j=0}^{n}\binom{n}{j}a^{n-j}b^j$$

where $\binom{n}{j} = \frac{n!}{j!(n-j)!}$ = number of combinations of n things taken j at a time. $n! = 1 \cdot 2 \cdot 3 \cdot 4 \cdots n$, $0! = 1$.

Example Find the sixth term of $(x + 2y)^{12}$. The sixth term is obtained by setting $j = 5$. It is

$$\binom{12}{5}x^{12-5}(2y)^5 = 1584x^7(2y)^5$$

Example $\sum_{j=0}^{14}\binom{n}{j} = (1 + 1)^{14} = 2^{14}$.

If n is not a positive integer, the sum formula no longer applies and an infinite series results for $(a + b)^n$. The coefficients are obtained from the first formulas in this case.

Example $(1 + x)^{1/2} = 1 + \frac{1}{2}x - \frac{1}{2} \cdot \frac{1}{4}x^2 + \frac{1}{2} \cdot \frac{1}{4} \cdot \frac{3}{8}x^3 \cdots$ (convergent for $x^2 < 1$).

Additional discussion is under "Infinite Series."

PROGRESSIONS

An **arithmetic progression** is a succession of terms such that each term, except the first, is derivable from the preceding by the addition of a quantity d called the common difference. All arithmetic progressions have the form $a, a + d, a + 2d, a + 3d, \ldots$. With a = first term, l = last term, d = common difference, n = number of terms, and s = sum of the terms, the following relations hold:

$$l = a + (n-1)d = -\frac{d}{2} + \sqrt{2ds + \left(a - \frac{d}{2}\right)^2}$$

$$= \frac{s}{n} + \frac{(n-1)}{2}d$$

$$s = \frac{n}{2}[2a + (n-1)d] = \frac{n}{2}(a + l) = \frac{n}{2}[2l - (n-1)d]$$

$$a = l - (n-1)d = \frac{s}{n} - \frac{(n-1)d}{2} = \frac{2s}{n} - l$$

$$d = \frac{l - a}{n - 1} = \frac{2(s - an)}{n(n-1)} = \frac{2(nl - s)}{n(n-1)}$$

$$n = \frac{l - a}{d} + 1 = \frac{2s}{l + a} = \frac{2l + d + \sqrt{(2 + d)^2 - 8ds}}{2d}$$

The **arithmetic mean or average** of two numbers a, b is $(a + b)/2$; of n numbers $a_1, \ldots, a_n$ is $(a_1 + a_2 + \cdots + a_n)/n$.

A **geometric progression** is a succession of terms such that each term, except the first, is derivable from the preceding by the multiplication of a quantity r called the **common ratio.** All such progressions have the form $a, ar, ar^2, \ldots, ar^{n-1}$. With a = first term, l = last term, r = ratio, n = number of terms, s = sum of the terms, the following relations hold:

$$l = ar^{n-1} = \frac{[a + (r-1)s]}{r} = \frac{(r-1)sr^{n-1}}{r^n - 1}$$

$$s = \frac{a(r^n - 1)}{r - 1} = \frac{a(1 - r^n)}{1 - r} = \frac{rl - a}{r - 1} = \frac{lr^n - l}{r^n - r^{n-1}}$$

$$a = \frac{l}{r^{n-1}} = \frac{(r-1)s}{r^n - 1} \quad r = \frac{s - a}{s - l} \quad \log r = \frac{\log l - \log a}{n - 1}$$

$$n = \frac{\log l - \log a}{\log r} + 1 = \frac{\log[a + (r-1)s] - \log a}{\log r}$$

The **geometric mean** of two nonnegative numbers a, b is $\sqrt{ab}$; of n numbers is $(a_1a_2 \ldots a_n)^{1/n}$.

Example Find the sum of $1 + \frac{1}{2} + \frac{1}{4} + \cdots + \frac{1}{64}$. Here $a = 1$, $r = \frac{1}{2}$, $n = 7$. Thus $s = \frac{\frac{1}{2}(\frac{1}{64}) - 1}{\frac{1}{2} - 1} = 127/64$. $s = a + ar + ar^2 + \cdots + ar^{n-1} = \frac{a}{1 - r} - \frac{ar^n}{1 - r}$. If $|r| < 1$, then $\lim_{n\to\infty} s = \frac{a}{1 - r}$, which is called the sum of the infinite geometric progression.

A progression of the form $a, (a + d)r, (a + 2d)r^2, (a + 3d)r^3$, etc., is a combined arithmetic and geometric progression. The sum of n such terms is

$$s = \frac{a - [a + (n-1)d]r^n}{1 - r} + \frac{rd(1 - r^{n-1})}{(1 - r)^2}$$

If $|r| < 1$, $\lim_{n\to\infty} s = \frac{a}{1 - r} + rd/(1 - r)^2$.

The non-zero numbers a, b, c, etc., form a **harmonic progression** if their reciprocals $1/a$, $1/b$, $1/c$, etc., form an arithmetic progression.

Example The progression $1, \frac{1}{3}, \frac{1}{5}, \frac{1}{7}, \cdots, \frac{1}{31}$ is harmonic since $1, 3, 5, 7, \ldots, 31$ form an arithmetic progression.

The **harmonic mean** of two numbers a, b is $2ab/(a + b)$.

PERMUTATIONS, COMBINATIONS, AND PROBABILITY

Each separate arrangement of all or a part of a set of things is called a **permutation.** The number of permutations of n things taken r at a time, written $P(n, r) = \frac{n!}{(n - r)!} = n(n-1)(n-2) \cdots (n - r + 1)$.

Example The permutations of a, b, c two at a time are ab, ac, ba, ca, cb, and bc. The formula is $P(3,2) = 3!/1! = 6$. The permutations of a, b, c three at a time are abc, bac, cab, acb, bca, and cba.

Each separate selection of objects that is possible irrespective of the order in which they are arranged is called a **combination.** The number of combinations of n things taken r at a time, written $C(n, r) = n!/[r!(n-r)!]$.

Example The combinations of a, b, c taken 2 at a time are ab, ac, bc; taken 3 at a time is abc.

An important relation is $r!\,C(n, r) = P(n, r)$.

If an event can occur in p ways and fail to occur in q ways, all ways being equally likely, the *probability* of its occurrence is $p/(p+q)$, and that of its failure $q/(p+q)$.

Example Two dice may be thrown in 36 separate ways. What is the probability of throwing such that their sum is 7? Seven may arise in 6 ways: 1 and 6, 2 and 5, 3 and 4, 4 and 3, 5 and 2, 6 and 1. The probability of shooting 7 is $\frac{1}{6}$.

THEORY OF EQUATIONS

Linear Equations A linear equation is one of the first degree (i.e., only the first powers of the variables are involved), and the process of obtaining definite values for the unknown is called solving the equation. Every linear equation in one variable is written $Ax + B = 0$ or $x = -B/A$. Linear equations in n variables have the form

$$\begin{aligned} a_{11}x_1 + a_{12}x_2 + \cdots + a_{1n}x_n &= b_1 \\ a_{21}x_1 + a_{22}x_2 + \cdots + a_{2n}x_n &= b_2 \\ &\vdots \\ a_{m1}x_1 + a_{m2}x_2 + \cdots + a_{mn}x_n &= b_m \end{aligned}$$

The solution of the system may then be found by elimination or matrix methods if a solution exists (see "Numerical Analysis and Approximate Methods").

Example Solution by the *elimination method:*

$$\begin{aligned} 3x_1 + 4x_2 - 5x_3 &= -2 \\ 9x_1 - 2x_2 + 3x_3 &= 1 \\ 6x_1 + 3x_2 - x_3 &= 3 \end{aligned}$$

Multiply the first row by 2 and add to the third row, leaving the first row unchanged. The result is

$$\begin{aligned} 3x_1 + 4x_2 - 5x_3 &= -2 \\ 9x_1 - 2x_2 + 3x_3 &= 1 \\ 0x_1 + 11x_2 - 11x_3 &= -1 \end{aligned}$$

Multiply the first row by -3 and add to the second row.

$$\begin{aligned} 3x_1 + 4x_2 - 5x_3 &= -2 \\ 0x_1 - 14x_2 + 18x_3 &= 7 \\ 0x_1 + 11x_2 - 11x_3 &= -1 \end{aligned}$$

Multiply the second row by $\frac{11}{14}$ and add to the third row. The result is the triangular form

$$\begin{aligned} 3x_1 + 4x_2 - 5x_3 &= -2 \\ 0x_1 - 14x_2 + 18x_3 &= 7 \\ 0x_1 + 0x_2 + \tfrac{22}{7}x_3 &= \tfrac{9}{2} \end{aligned}$$

Then from the third equation $x_3 = \frac{63}{44}$; substituting this in the second equation yields $x_2 = \frac{59}{44}$. The first equation yields $x_1 = -\frac{9}{44}$. An alternative to this method, which is equivalent to it, is to solve one of the equations for x_3 in terms of x_1 and x_2, say, the third equation, which gives $x_3 = -6x_1 + 3x_2 - 3$. This is then substituted in the second equation, yielding $-9x_1 + 7x_2 = 10$, which can then be solved for x_2, $x_2 = (10 + 9x_1)/7$. Substitution for x_3 and x_2 in the first equations gives an equation in x_1 alone.

Other methods for solving such systems will be given under "Determinants," "Matrix Algebra and Matrix Computations," and "Numerical Analysis and Approximate Methods."

Quadratic Equations Every quadratic equation in one variable is expressible in the form $ax^2 + bx + c = 0$. $a \neq 0$. This equation has two solutions, say, x_1, x_2, given by $\left.\begin{matrix} x_1 \\ x_2 \end{matrix}\right\} = \dfrac{-b \pm \sqrt{b^2 - 4ac}}{2a}$. If a, b, c are real, the **discriminant** $b^2 - 4ac$ gives the character of the roots. If $b^2 - 4ac > 0$, the roots are **real** and **unequal.** If $b^2 - 4ac < 0$, the roots are **complex conjugates.** If $b^2 - 4ac = 0$ the roots are **real** and **equal.**

Two quadratic equations in two variables can in general be solved only by numerical methods (see "Numerical Analysis and Approximate Methods"). If one equation is of the first degree, the other of the second degree, a solution may be obtained by solving the first for one unknown. This result is substituted in the second equation and the resulting quadratic equation solved.

Example Solve $x^2 + 5xy - y^2 - 15 = 0$ and $x + 2y = 10$. The second equation yields $x = 10 - 2y$. Upon substitution into the first equation the quadratic $7y^2 - 10y - 85 = 0$ is obtained. The solutions of this equation are $(5 \pm 2\sqrt{155})/7$.

Cubic Equations A cubic equation, in one variable, has the form $x^3 + bx^2 + cx + d = 0$. Every cubic equation having complex coefficients has three complex roots. If the coefficients are real numbers, then at least one of the roots must be real. The cubic equation $x^3 + bx^2 + cx + d = 0$ may be reduced by the substitution $x = y - (b/3)$ to the form $y^3 + py + q = 0$, where $p = \frac{1}{3}(3c - b^2)$, $q = \frac{1}{27}(27d - 9bc + 2b^3)$. This equation has the solutions $y_1 = A + B$, $y_2 = -\frac{1}{2}(A + B) + (i\sqrt{3}/2)(A - B)$, $y_3 = -\frac{1}{2}(A + B) - (i\sqrt{3}/2)(A - B)$, where $i^2 = -1$, $A = \sqrt[3]{-q/2 + \sqrt{R}}$, $B = \sqrt[3]{-q/2 - \sqrt{R}}$, and $R = (p/3)^3 + (q/2)^2$. If b, c, d are all real and if $R > 0$, there are one real root and two conjugate complex roots; if $R = 0$, there are three real roots, of which at least two are equal; if $R < 0$, there are three real unequal roots. If $R < 0$, these formulas are impractical. In this case, the roots are given by $y_k = \mp 2\sqrt{-p/3}\cos[(\phi/3) + 120k]$, $k = 0, 1, 2$ where $\phi = \cos^{-1}\sqrt{\dfrac{q^2/4}{-p^3/27}}$ and the upper sign applies if $q > 0$, the lower if $q < 0$.

Example $x^3 + 3x^2 + 9x + 9 = 0$ reduces to $y^3 + 6y + 2 = 0$ under $x = y - 1$. Here $p = 6$, $q = 2$, $R = 9$. Hence $A = \sqrt[3]{2}$, $B = \sqrt[3]{-4}$. The desired roots in y are $\sqrt[3]{2} - \sqrt[3]{-4}$ and $-\frac{1}{2}(\sqrt[3]{2} - \sqrt[3]{4}) \pm (i\sqrt{3}/2)(\sqrt[3]{2} + \sqrt[3]{4})$. The roots in x are $x = y - 1$.

Example $y^3 - 7y + 7 = 0$. $p = -7$, $q = 7$, $R < 0$. Hence $x_k = -\sqrt{\dfrac{28}{3}}\cos\left(\dfrac{\phi}{3} + 120k\right)$ where $\cos\phi = \sqrt{\dfrac{27}{28}}$, $\dfrac{\phi}{3} = 3°37'52''$. The roots are approximately -3.048916, 1.692020, and 1.356897.

Quartic Equations The general quartic equation $x^4 + ax^3 + bx^2 + cx + d = 0$ may be reduced to the form $y^4 + py^2 + qy + r = 0$ by the substitution $x = y - (a/4)$. Let l, m, and n denote the roots of the resolvent cubic

$$t^3 + \left(\frac{p}{2}\right)t^2 + \frac{(p^2 - 4r)t}{16} - \frac{q^2}{64} = 0$$

found by the previous method. The required roots of the reduced quartic are

$$y_1 = \pm(-\sqrt{l} - \sqrt{m} - \sqrt{n});\quad y_2 = \pm(-\sqrt{l} + \sqrt{m} + \sqrt{n})$$
$$y_3 = \pm(\sqrt{l} - \sqrt{m} + \sqrt{n});\quad y_4 = \pm(\sqrt{l} + \sqrt{m} - \sqrt{n})$$

where the upper signs are used if $q > 0$ and the lower signs if $q < 0$.

General Polynomials of the nth Degree Denote the general polynomial equation of degree n by

$$P(x) = a_0x^n + a_1x^{n-1} + \cdots + a_{n-1}x + a_n = 0$$

If $n > 4$, there is no formula which gives the roots of the general equation. However, there are some general methods which may prove useful.

Remainder Theorems When $P(x)$ is a polynomial and $P(x)$ is divided by $x - a$ until a remainder independent of x is obtained, this remainder is equal to $P(a)$.

Example $P(x) = 2x^4 - 3x^2 + 7x - 2$ when divided by $x + 1$ (here $a = -1$) results in $P(x) = (x + 1)(2x^3 - 2x^2 - x + 8) - 10$ where -10 is the remainder. It is easy to see that $P(-1) = -10$.

Factor Theorem If $P(a)$ is zero, the polynominal $P(x)$ has the factor $x - a$. In other words, if a is a root of $P(x) = 0$, then $x - a$ is a factor of $P(x)$.

If a number a is found to be a root of $P(x) = 0$, the division of $P(x)$ by $(x - a)$ leaves a polynomial of degree one less than that of the original equation, i.e., $P(x) = Q(x)(x - a)$. Roots of $Q(x) = 0$ are clearly roots of $P(x) = 0$.

Example $P(x) = x^3 - 6x^2 + 11x - 6 = 0$ has the root $+3$. Then $P(x) = (x - 3)(x^2 - 3x + 2)$. The roots of $x^2 - 3x + 2 = 0$ are 1 and 2. The roots of $P(x)$ are therefore 1, 2, 3.

Fundamental Theorem of Algebra Every polynomial of degree n has exactly n real or complex roots, counting multiplicities. If the roots of $x^n + c_1x^{n-1} + c_2x^{n-2} + \cdots + c_{n-1}x + c_n = 0$ are $r_1, r_2, r_3, \ldots, r_n$ where multiplicities may occur, then $c_1 = -\sum_{j=1}^{n} r_j$; $c_2 = \sum_{\substack{i,j=1 \\ i \neq j}}^{n} r_ir_j$ = sum of the product of the roots taken two at a time; $c_3 = -$(sum of the product of the roots taken three at a time); ... $c_n = (-1)^n r_1r_2 \cdots r_n$.

This result allows a polynomial having given numbers as roots to be formed.

Example Find a cubic equation whose roots are 1, 2, 3. $c_1 = -(1 + 2 + 3) = -6$; $c_2 = 1\cdot2 + 1\cdot3 + 2\cdot3 = 11$; $c_3 = -6$. The polynomial is $x^3 - 6x^2 + 11x - 6$.

Example $x^4 - 12x^3 + 48x^2 - 80x + 48 = 0$ has a triple root. Solve the equation. Let r be the triple root and s be the other root. Then $3r + s = 12$; $3sr + 3r^2 = 48$. Subtract $9r^2 + 3rs = 36r$ from the second equation. There results $6r^2 - 36r + 48 = 0$ or $(r - 4)(r - 2) = 0$; $r = 2$, $r = 4$. If $r = 4$, $s = 0$, which is not possible. Since $r = 4$ is extraneous, $r = 2$ is the triple root. $s = 6$ is the other root.

Every polynomial equation $a_0x^n + a_1x^{n-1} + \cdots + a_n = 0$ with *rational coefficients* may be rewritten as a polynomial, of the same degree, with *integral coefficients* by multiplying each coefficient by the least common multiple of the denominators of the coefficients.

Example The coefficients of $\frac{3}{2}x^4 + \frac{7}{3}x^3 - \frac{5}{6}x^2 + 2x - \frac{1}{6} = 0$ are rational numbers. The least common multiple of the denominators is $2 \times 3 = 6$. Therefore, the equation is equivalent to $9x^4 + 14x^3 - 5x^2 + 12x - 1 = 0$.

Rational-Root Theorem If the polynomial equation $a_0x^n + \cdots + a_n = 0$ with integral coefficients has a rational root p/q (p and q integers having no common factor > 1), then p is an exact divisor of the constant term a_n and q is an exact divisor of the leading coefficient a_0.

This result allows the computation of all rational (and integral) roots of any polynomial having integral coefficients.

Example Find all rational roots of $6y^3 + 7y^2 - 9y + 2 = 1$. The possibilities for the numerator are $\pm1, \pm2$. The possibilities for the denominator are $\pm1, \pm2, \pm3, \pm6$. Thus the possible rational roots are $\pm1, \pm2, \pm\frac{1}{2}, \pm\frac{1}{3}, \pm\frac{2}{3}, \pm\frac{1}{6}$. Trial of these values shows that -2 is a root. The reduced equation is $6y^2 - 5y + 1 = 0$, which has roots $\frac{1}{3}, \frac{1}{2}$.

Complex-Root Theorem If a polynomial equation with real coefficients has the root $a + ib$, it also has the root $a - ib$; that is, complex roots occur in conjugate pairs.

Example $x^4 - 4x^2 + 8x - 4 = 0$ has one root $1 + i$. Another root is $1 - i$. Therefore, $(x - 1 - i)(x - 1 + i) = x^2 - 2x + 2$ is a factor of the polynomial. $x^4 - 4x^2 + 8x - 4 = (x^2 - 2x + 2)(x^2 + 2x - 2)$. The other roots are $1 \pm \sqrt{3}$.

Upper Bound for the Real Roots Any number which exceeds all the roots is called an **upper bound** to the real roots. If the coefficients of a polynomial equation are all of like sign, there is no positive root. Such equations are excluded here since zero is the upper bound to the real roots. If the coefficient of the highest power of $P(x) = 0$ is negative, replace the equation by $-P(x) = 0$.

If in a polynomial $P(x) = c_0x^n + c_1x^{n-1} + \cdots + c_{n-1}x + c_n = 0$, with $c_0 > 0$, the first negative coefficient is preceded by k coefficients which are positive or zero, and if G denotes the greatest of the numerical values of the negative coefficients, then each real root is less than $1 + \sqrt[k]{G/c_0}$.

A **lower bound** to the negative roots of $P(x) = 0$ may be found by applying the rule to $P(-x) = 0$.

Example $P(x) = x^7 + 2x^5 + 4x^4 - 8x^2 - 32 = 0$. Here $k = 5$ (since 2 coefficients are zero), $G = 32$, $c_0 = 1$. The upper bound is $1 + \sqrt[5]{32} = 3$. $P(-x) = -x^7 - 2x^5 + 4x^4 - 8x^2 - 32 = 0$. $-P(-x) = x^7 + 2x^5 - 4x^4 + 8x^2 + 32 = 0$. Here $k = 3$, $G = 4$, $c_0 = 1$. The lower bound is $-(1 + \sqrt[3]{4}) \approx -2.587$. Thus all real roots r lie in the range $-2.587 < r < 3$.

Descartes Rule of Signs The number of positive real roots of a polynomial equation with real coefficients either is equal to the number v of its variations in sign or is less than v by a positive even integer. The number of negative roots of $P(x) = 0$ either is equal to the number of variations of sign of $P(-x)$ or is less than that number by a positive even integer.

Example $P(x) = x^4 + 3x^3 + x - 1 = 0$. $v = 1$; so $P(x)$ has one positive root. $P(-x) = x^4 - 3x^3 - x - 1$. Here $v = 1$; so $P(x)$has one negative root. The other two roots are complex conjugates.

Example $P(x) = x^4 - x^2 + 10x - 4 = 0$. $v = 3$; so $P(x)$ has three or one positive roots. $P(-x) = x^4 - x^2 - 10x - 4$. $v = 1$; so $P(x)$ has exactly one negative root.

Numerical methods are often used to find the roots of polynomials. A detailed discussion of these techniques is given under "Numerical Analysis and Approximate Methods."

Determinants Consider the system of two linear equations

$$a_{11}x_1 + a_{12}x_2 = b_1$$
$$a_{21}x_1 + a_{22}x_2 = b_2$$

If the first equation is multiplied by a_{22} and the second by $-a_{12}$ and the results added, we obtain

$$(a_{11}a_{22} - a_{21}a_{12})x_1 = b_1a_{22} - b_2a_{12}$$

The expression $a_{11}a_{22} - a_{21}a_{12}$ may be represented by the symbol

$$\begin{vmatrix} a_{11} & a_{12} \\ a_{21} & a_{22} \end{vmatrix} = a_{11}a_{22} - a_{21}a_{12}$$

This symbol is called a determinant of second order. The value of the square array of n^2 quantities a_{ij}, where $i = 1, \ldots, n$ is the row index, $j = 1, \ldots, n$ the column index, written in the form

$$|A| = \begin{vmatrix} a_{11} & a_{12} & a_{13} & \cdots & a_{1n} \\ a_{21} & a_{22} & \cdots & \cdots & a_{2n} \\ \vdots & & & & \\ a_{n1} & a_{n2} & a_{n3} & \cdots & a_{nn} \end{vmatrix}$$

is called a determinant. The n^2 quantities a_{ij} are called the elements of the determinant. In the determinant $|A|$ let the ith row and jth column be deleted and a new determinant be formed having $n - 1$ rows and columns. This new determinant is called the **minor** of a_{ij} denoted M_{ij}.

Example

$$\begin{vmatrix} a_{11} & a_{12} & a_{13} \\ a_{21} & a_{22} & a_{23} \\ a_{31} & a_{32} & a_{33} \end{vmatrix} \quad \text{The minor of } a_{23} \text{ is } M_{23} = \begin{vmatrix} a_{11} & a_{12} \\ a_{31} & a_{32} \end{vmatrix}.$$

The **cofactor** A_{ij}, of the element a_{ij}, is the signed minor of a_{ij} determined by the rule $A_{ij} = (-1)^{i+j}M_{ij}$.

Example $A_{23} = (-1)^5 M_{23} = -M_{23}$.

The *value* of $|A|$ is obtained by forming any of the equivalent expressions $\Sigma_{j=1}^{n} a_{ij} A_{ij}$, $\Sigma_{i=1}^{n} a_{ij}A_{ij}$, where the elements a_{ij} must be taken from a single row or a single column of A.

Example

$$\begin{vmatrix} a_{11} & a_{12} & a_{13} \\ a_{21} & a_{22} & a_{23} \\ a_{31} & a_{32} & a_{33} \end{vmatrix} = a_{31}A_{31} + a_{32}A_{32} + a_{33}A_{33}$$

$$= a_{31}\begin{vmatrix} a_{12} & a_{13} \\ a_{22} & a_{23} \end{vmatrix} - a_{32}\begin{vmatrix} a_{11} & a_{13} \\ a_{21} & a_{23} \end{vmatrix} + a_{33}\begin{vmatrix} a_{11} & a_{12} \\ a_{21} & a_{22} \end{vmatrix}$$

In general, A_{ij} will be determinants of order $n - 1$, but they may in turn be expanded by the rule.

It is easy to show that

$$\sum_{j=1}^{n} a_{ji}A_{jk} = \sum_{j=1}^{n} a_{ij}A_{jk} = \begin{cases} |A| & i = k \\ 0 & i \neq k \end{cases}$$

Fundamental Properties of Determinants

1. The value of a determinant $|A|$ is not changed if the rows and columns are interchanged.
2. If the elements of one row (or one column) of a determinant are all zero, the value of $|A|$ is zero.
3. If the elements of one row (or column) of a determinant are multiplied by the same constant factor, the value of the determinant is multiplied by this factor.
4. If one determinant is obtained from another by interchanging any two rows (or columns), the value of either is the negative of the value of the other.
5. If two rows (or columns) of a determinant are identical, the value of the determinant is zero.
6. If two determinants are identical except for one row (or column), the sum of their values is given by a single determinant obtained by adding corresponding elements of dissimilar rows (or columns) and leaving unchanged the remaining elements.

Example

$$\begin{vmatrix} 3 & 2 \\ 1 & 5 \end{vmatrix} + \begin{vmatrix} 4 & 2 \\ 7 & 5 \end{vmatrix} = 13 + 6 = 19 \qquad \text{Directly}$$

$$\begin{vmatrix} 7 & 2 \\ 8 & 5 \end{vmatrix} = 35 - 16 = 19 \qquad \text{By rule 6}$$

7. The value of a determinant is not changed if to the elements of any row (or column) are added a constant multiple of the corresponding elements of any other row (or column).
8. If all elements but one in a row (or column) are zero, the value of the determinant is the product of that element times its cofactor.

The evaluation of determinants using the definition is quite laborious. The labor can be reduced by applying the fundamental properties just outlined.

Example Evaluate

$$\begin{vmatrix} 2 & 1 & 4 & 3 \\ -1 & 4 & 2 & 1 \\ 5 & 6 & 7 & 2 \\ 1 & 3 & 4 & 5 \end{vmatrix}$$

The aim is to transform the determinant so that all elements but one in a given row (or column) are zero, *without changing the determinant value*. This may be done by utilizing property 7. Selecting the element 1 in the fourth column, add -2 times the fourth column to the third column, then -4 times the fourth column to the second column; then add the fourth column to the first column; the result is

$$|A| = \begin{vmatrix} 5 & -11 & -2 & 3 \\ 0 & 0 & 0 & 1 \\ 7 & -2 & 3 & 2 \\ 6 & -17 & -6 & 5 \end{vmatrix} = 1 \begin{vmatrix} 5 & -11 & -2 \\ 7 & -2 & 3 \\ 6 & -17 & -6 \end{vmatrix}$$

by property 8. Property 7 is now used on this 3×3 determinant. Subtract the elements of the first row from the third row. The result is

$$|A| = \begin{vmatrix} 5 & -11 & -2 \\ 7 & -2 & 3 \\ 1 & -6 & -4 \end{vmatrix}$$

Now add -7 times the third row to the second row, then -5 times the third row to the first row, resulting in

$$|A| = \begin{vmatrix} 0 & 19 & 18 \\ 0 & 40 & 31 \\ 1 & -6 & -4 \end{vmatrix} = \begin{vmatrix} 19 & 18 \\ 40 & 31 \end{vmatrix} = -131$$

The solution of n linear equations (not all b_i zero)

$$\begin{aligned} a_{11}x_1 + a_{12}x_2 + \cdots + a_{1n}x_n &= b_1 \\ a_{21}x_1 + a_{22}x_2 + \cdots + a_{2n}x_n &= b_2 \\ \vdots \qquad & \quad \vdots \\ a_{n1}x_1 + a_{n2}x_2 + \cdots + a_{nn}x_n &= b_n \end{aligned}$$

where $|A| = \begin{vmatrix} a_{11} & \cdots & a_{1n} \\ a_{21} & \cdots & a_{2n} \\ \vdots & & \\ a_{n1} & \cdots & a_{nn} \end{vmatrix} \neq 0$ has a unique solution given by $x_1 = |B_1|/|A|$, $x_2 = |B_2|/|A|, \ldots, x_n = |B_n|/|A|$, where B_k is the determinant obtained from A by replacing its kth column by $b_1, b_2, \ldots, b_n$. This technique is called **Cramer's rule.** It usually requires more labor than the method of elimination or certain numerical techniques and should not be used for computations.

Example

$$\begin{aligned} 5x_1 + 3x_2 + 3x_3 &= 48 \\ 2x_1 + 6x_2 - 3x_3 &= 18 \\ 8x_1 - 3x_2 + 2x_3 &= 21 \end{aligned}$$

The solutions are $x_1 = \dfrac{\begin{vmatrix} 48 & 3 & 3 \\ 18 & 6 & -3 \\ 21 & -3 & 2 \end{vmatrix}}{\begin{vmatrix} 5 & 3 & 3 \\ 2 & 6 & -3 \\ 8 & -3 & 2 \end{vmatrix}} = \dfrac{-693}{-231} = 3,$

$$x_2 = \frac{\begin{vmatrix} 5 & 48 & 3 \\ 2 & 18 & -3 \\ 8 & 21 & 2 \end{vmatrix}}{-231} = \frac{-1155}{-231} = 5, \quad x_3 = \frac{\begin{vmatrix} 5 & 3 & 48 \\ 2 & 6 & 18 \\ 8 & -3 & 21 \end{vmatrix}}{-231} = 6.$$

ANALYTIC GEOMETRY

REFERENCES: 102, 180, 216, 245, 246, 256, 263, 267. For solid analytic geometry, see especially Ref. 216.

Analytic geometry uses algebraic equations and methods to study geometric problems. It also permits one to visualize algebraic equations in terms of geometric curves, which frequently clarifies abstract concepts.

PLANE ANALYTIC GEOMETRY

Coordinate Systems The basic concept of analytic geometry is the establishment of a one-to-one correspondence between the points of the plane and number pairs (x, y). This correspondence may be

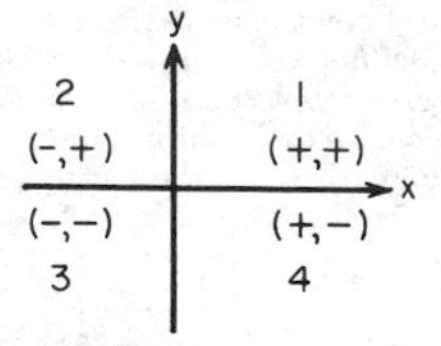

FIG. 2-12 Rectangular coordinates.

done in a number of ways. The **rectangular** or **cartesian** coordinate system consists of two straight lines intersecting at right angles (Fig. 2-12). A point is designated by (x, y), where x (the abscissa) is the distance of the point from the y axis measured parallel to the x axis, positive if to the right, negative to the left. y (ordinate) is the distance of the point from the x axis, measured parallel to the y axis, positive if above, negative if below the x axis. The **quadrants** are labeled 1, 2, 3, 4 in the drawing, the coordinates of points in the various quadrants having the depicted signs. Another common coordinate system is the **polar** coordinate system (Fig. 2-13). In this system the position of a point is designated by the pair (r, θ), $r = \sqrt{x^2 + y^2}$ being the distance to the origin $0(0,0)$ and θ being the angle the line r makes with the positive x axis (polar axis). To change from polar to rectangular coordinates, use $x = r \cos \theta$ and $y = r \sin \theta$. To change from rectangular to polar coordinates, use $r = \sqrt{x^2 + y^2}$ and $\theta = \tan^{-1} (y/x)$ if $x \neq 0$; $\theta = \pi/2$ if $x = 0$. The **distance** between two points (x_1, y_1), (x_2, y_2) is defined by $d = \sqrt{(x_1 - x_2)^2 + (y_1 - y_2)^2}$ in rectangular coordinates or by $d = \sqrt{r_1^2 + r_2^2 - 2r_1r_2 \cos (\theta_1 - \theta_2)}$ in polar coordinates. Other coordinate systems are sometimes used. For example, on the surface of a sphere latitude and longitude prove useful.

The Straight Line (Fig. 2-14) The slope m of a straight line is the tangent of the inclination angle θ made with the positive x axis. If (x_1,y_1) and (x_2, y_2) are any two points on the line, slope $= m = (y_2 - y_1)(x_2 - x_1)$. The slope of a line parallel to the x axis is zero; parallel to the y axis, it is undefined. Two lines are parallel if and only if they have the same slope. Two lines are perpendicular if and only if the product of their slopes is -1 (the exception being that case when the lines are parallel to the coordinate axes). Every equation of the type $Ax + By + C = 0$ represents a straight line, and every straight line has an equation of this form. A straight line is determined by a variety of conditions:

Given conditions	Equation of line
(1) Parallel to x axis	$y =$ constant
(2) Parallel y axis	$x =$ constant
(3) Point (x_1, y_1) and slope m	$y - y_1 = m(x - x_1$
(4) Intercept on y axis $(0, b)$, m	$y = mx + b$
(5) Intercept on x axis $(a, 0)$, m	$y = m(x - a)$
(6) Two points (x_1, y_1), (x_2, y_2)	$y - y_1 = \dfrac{y_2 - y_1}{x_2 - x_1}(x - x_1)$
(7) Two intercepts $(a, 0)$, $(0, b)$	$x/a + y/b = 1$

The angle β a line with slope m_1 makes with a line having slope m_2 is given by $\tan \beta = (m_2 - m_1)/(m_1m_2 + 1)$. A line is determined if the length and direction of the perpendicular to it (the **normal**) from the origin are given (see Fig. 2-15). Let $p =$ length of the perpendicular and α the angle that the perpendicular makes with the positive x axis. The equation of the line is $x \cos a + y \sin a = p$.

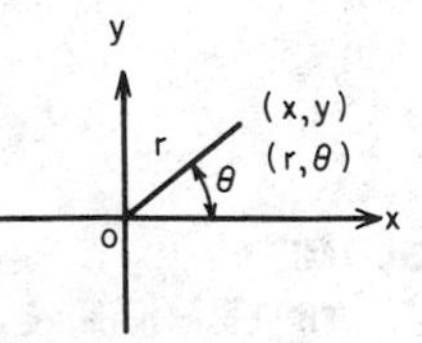

FIG. 2-13 Polar coordinates.

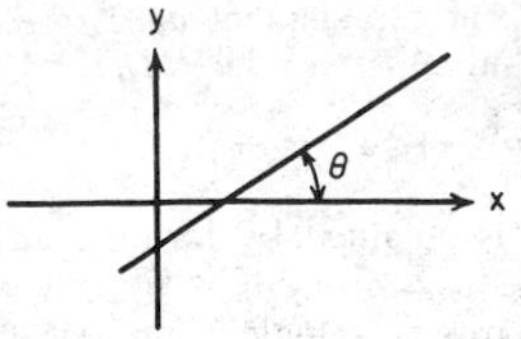

FIG. 2-14 Straight line.

The equation of a line perpendicular to a given line of slope m and passing through a point (x_1, y_1) is $y - y_1 = -(1/m)(x - x_1)$. The distance from a point (x_1, y_1) to a line with equation $Ax + by + C = 0$ is $d = \dfrac{|Ax_1 + By_1 + C|}{\sqrt{A^2 + B^2}}$.

Example If it is known that centigrade C and Fahrenheit F are linearly related and when C = 0°, F = 32°; C = 100°, F = 212°, find the equation relating C and F and that point where C = F. By using the two-point form, the equation is $F - 32 = \dfrac{212 - 32}{100 - 0}(C - 0)$ or $F = \frac{9}{5}C + 32$. Equivalently $C - 0 = \dfrac{100 - 0}{212 - 32}(F - 32)$ or $C = \frac{5}{9}(F - 32)$. Letting C = F, we have from either equation F = C = −40.

Occasionally some nonlinear algebraic equations can be reduced to linear equations under suitable substitutions or changes of variables. In other words, certain curves become the graphs of lines if the scales or coordinate axes are appropriately transformed.

Example Consider $y = bx^n$. $B = \log b$. Taking logarithms $\log y = n \log x + \log b$. Let $Y = \log y$, $X = \log x$, $B = \log b$. The equation then has the form $Y = nX + B$, which is a linear equation. Consider $y = ka^{bx}$. Taking logarithms, this becomes $\log y = \log k + (b \log a)x$, which indicates that $\log y$ and x are linearly related with slope $b \log a$ and intercept on the y axis $\log k$. Next consider $y = a + bx^n$. If the substitution $t = x^n$ is made, then the graph of y is a straight line.

Asymptotes The limiting position of the tangent to a curve as the point of contact tends to an infinite distance from the origin is called an **asymptote.** If the equation of a given curve can be expanded in a Laurent power series such that $f(x) = \Sigma_{k=0}^{n} a_kx^k + \sum_{k=0}^{n} \dfrac{b_k}{x^k}$ and $\lim_{x \to \infty} f(x) = \sum_{k=0}^{n} a_kx^k$, then the equation of the asymptote is $y = \sum_{k=0}^{n} a_kx^k$. If $n = 1$, then the asymptote is (in general oblique) a line. In this case, the equation of the asymptote may be written as

$$y = mx + b \qquad m = \lim_{x \to \infty} f'(x)$$

$$b = \lim_{x \to \infty} [f(x) - xf'(x)]$$

If $f(x) = P(x)/Q(x)$ is a rational function in which the degree of the numerator exceeds the degree of the denominator by 1, then the graph of f has an oblique asymptote which is the straight line $y = ax + b$ obtained by dividing $P(x)$ by $Q(x)$.

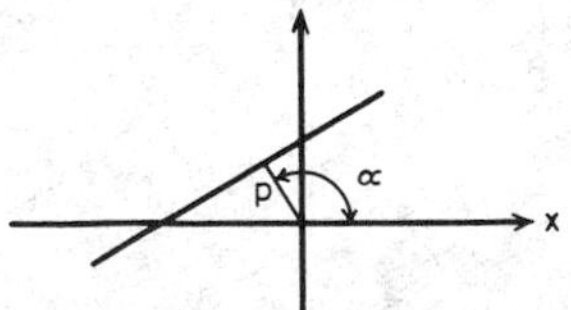

FIG. 2-15 Determination of line.

Example $f(x) = (x-1)^3/x^2 = x - 3 + (3x-1)/x^2$, so $y = x - 3$ is an asymptote to the graph of f; the graph of the equation $y = f(x)$ tends toward the line $y = x - 3$ as $x \to +\infty$ or $x \to -\infty$.

If $\lim_{x\to\infty} f(x) = b$ or $\lim_{x\to-\infty} f(x) = b$, then the line $y = b$ is called a **horizontal asymptote** of the graph of f.

The line $x = a$ is called a **vertical asymptote** of the graph of a function f if one of the limits $\lim_{x\to a+} f(x)$, $\lim_{x\to a-} f(x)$ is either ∞ or $-\infty$. Clearly, f is discontinuous at a if $x = a$ is an asymptote.

If a curve has no infinite branch (as in the case of a circle or an ellipse), there is no asymptote.

Geometric Properties of a Curve When the Equation Is Given The analysis of the properties of an equation is facilitated by the investigation of the equation by using the following techniques:

1. *Points of maximum, minimum, and inflection.* These may be investigated by means of the calculus.
2. *Symmetry.* Let $F(x, y) = 0$ be the equation of the curve.

Condition on $F(x, y)$	Symmetry
$F(x, y) = F(-x, y)$	With respect to y axis
$F(x, y) = F(x, -y)$	With respect to x axis
$F(x, y) = F(-x, -y)$	With respect to origin
$F(x, y) = F(y, x)$	With respect to the line $y = x$

Example $y^2 = x^2(x^2 - 1)$ is symmetric with respect to both axes and origin.

3. *Extent.* Only real values of x and y are considered in obtaining the points (x, y) whose coordinates satisfy the equation. The extent of them may be limited by the condition that negative numbers do not have real square roots.
4. *Intercepts.* Find those points where the curves of the function cross the coordinate axes.
5. *Asymptotes.* See preceding discussion.
6. *Direction at a point.* This may be found from the derivative of the function at a point. This concept is useful for distinguishing among a family of similar curves.

Example $y^2 = (x^2 + 1)/(x^2 - 1)$ is symmetric with respect to the x and y axis, the origin and the line $y = x$. It has the vertical asymptotes $x = \pm 1$. When $x = 0$, $y^2 = -1$; so there are no y intercepts. If $y = 0$, $(x^2 + 1)/(x^2 - 1) = 0$; so there are no x intercepts. If $|x| < 1$, y^2 is negative; so $|x| > 1$. From $x^2 = (y^2 + 1)/(y^2 - 1)$, $y = \pm 1$ are horizontal asymptotes and $|y| > 1$. As $x \to 1^+$, $y \to +\infty$; as $x \to +\infty$, $y \to +1$. The graph is given in Fig. 2-16.

Conic Sections The curves included in this group are obtained from plane sections of the cone. They include the circle, ellipse, parabola, hyperbola, and degeneratively the point and straight line. A **conic** is the locus of a point whose distance from a fixed point called the **focus** is in a constant ratio to its distance from a fixed line, called the directrix. This ratio is the **eccentricity** e. If $e = 0$, the conic is a circle; if $0 < e < 1$, the conic is an ellipse; if $e = 1$, the conic is a parabola; if $e > 1$, the conic is a hyperbola. Every conic section is representable by an equation of second degree. Conversely, every equation of second degree in two variables represents a conic. The general equation of the second degree is $Ax^2 + Bxy + Cy^2 + Dx + Ey + F = 0$. Let Δ be defined as the determinant

$$\Delta = \begin{vmatrix} 2A & B & D \\ B & 2C & E \\ D & E & 2F \end{vmatrix}$$

The table characterizes the curve represented by the equation.

	$B^2 - 4AC < 0$	$B^2 - 4AC = 0$	$B^2 - 4AC > 0$
$\Delta \neq 0$	$A\Delta < 0$ $A \neq C$, an ellipse $A\Delta < 0$ $A = C$, a circle $A\Delta > 0$, no locus	Parabola	Hyperbola
$\Delta = 0$	Point	2 parallel lines if $Q = D^2 + E^2 - 4(A + C)F > 0$ 1 straight line if $Q = 0$, no locus if $Q < 0$	2 intersecting straight lines

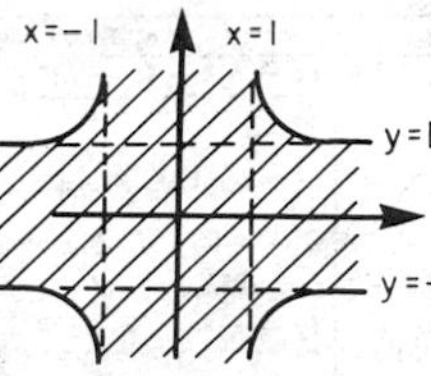

FIG. 2-16 Graph of $y^2 = (x^2 + 1)/(x^3 - 1)$.

Example $3x^2 + 4xy - 2y^2 + 3x - 2y + 7 = 0$.

$$\Delta = \begin{vmatrix} 6 & 4 & 3 \\ 4 & -4 & -2 \\ 3 & -2 & 14 \end{vmatrix} = -596 \neq 0,\ B^2 - 4AC = 40 > 0;$$ the curve is therefore a hyperbola.

To translate the axes to a new origin at (h, k) substitute for x and y in the original equation $x + h$ and $y + k$. Translation of the axes can always be accomplished to eliminate the linear terms in the second-degree equation in two variables having no xy term.

Example $x^2 + y^2 + 2x - 4y + 2 = 0$. Rewrite this as $x^2 + 2x + 1 + y^2 - 4y + 4 - 5 + 2 = 0$ or $(x + 1)^2 + (y - 2)^2 = 3$. Let $u = x + 1$, $v = y - 2$. Then $u^2 + v^2 = 3$. The axis has been translated to the new origin $(-1, 2)$.

The type of curve determined by a specific equation of the second degree can also be easily determined by reducing it to a standard form by translation and/or rotation. In the case in which the equation has no xy term, the procedure is merely to complete the squares of the terms in x and y separately.

To rotate the axes through an angle α, substitute for x the quantity $x \cos\alpha - y \sin\alpha$ and for y the quantity $x \sin\alpha + y \cos\alpha$. A rotation of the axes through $\alpha = \frac{1}{2}\cot^{-1}(A - C)/B$ will eliminate the cross-product term in the general second-degree equation.

Example Consider $3x^2 + 2xy + y^2 - 2x + 3y = 7$. A rotation of axes through $\alpha = \frac{1}{2}\cot^{-1} 1 = 22\frac{1}{2}°$ eliminates the cross-product term.

Since a proper rotation of the axes will eliminate the term Bxy, we need consider only equations of the form $Ax^2 + Cy^2 + Dx + Ey + F = 0$ in which the coefficients are those which result after the axes are rotated. Investigation is facilitated by the following table.

Condition	Conic	Equation and properties
$A = C$	Circle	$(x - h)^2 + (y - k)^2 = r^2$, center at $(h, k) = \left(-\frac{D}{2A}, -\frac{E}{2A}\right)$, radius $r = \frac{1}{2A}\sqrt{D^2 + E^2 - 4AF}$. If $D^2 + E^2 - 4AF > 0$ real circle, $= 0$ point, < 0 no locus
$A \neq 0, C = 0$	Parabola	$(x - h)^2 = 4p(y - k)$, $h = -D/2A$, $4p = -\frac{E}{A}$, $k = \frac{D^2 - 4AF}{4AE}$. $p > 0$ opens upward, $p < 0$ opens downward. Focus $(h, p + k)$, directrix $y = -p + k$
$A = 0, C \neq 0$	Parabola	$(y - k)^2 = 4p(x - h)$, $h = \frac{E^2 - 4CF}{4CD}$, $4p = -\frac{D}{C}$, $k = -\frac{E}{2C}$ if $p > 0$ opens to right, $p < 0$ opens to left. Focus $(p + h, k)$, directrix $x = -p + h$
$AC > 0, A \neq C$ $C > A$	Ellipse	$\frac{(x - h)^2}{a^2} + \frac{(y - k)^2}{b^2} = 1$, $a > b$. Here $a^2 = \frac{G}{A}$, $b^2 = \frac{G}{C}$, $G = \frac{E^2}{4C} + \frac{D^2}{4A} - F$, $h = -\frac{D}{2A}$, $k = -\frac{E}{2C}$. Major axis the x-axis, foci at $(h \pm \sqrt{a^2 - b^2}, k)$, center (h, k)

Condition	Conic	Equation and properties
$AC > 0, A \neq C$, $C < A$	Ellipse	$\frac{(x-h)^2}{b^2} + \frac{(y-k)^2}{a^2} = 1, a > b.\ a^2 = G/C$ $b^2 = G/A,\ G = \frac{E^2}{4C} + \frac{D^2}{4A} - F,\ h = -D/2A,$ $k = -E/2C$. Major axis the y-axis, foci at $(h, k \pm \sqrt{a^2 - b^2})$, center (h, k)
$AC < 0$	Hyperbola	$\frac{(x-h)^2}{a^2} - \frac{(y-k)^2}{b^2} = 1,\ a^2 = G/A,$ $b^2 = -G/C,\ G = \frac{E^2}{4C} + \frac{D^2}{4A} - F,$ $h = -D/2A,\ k = -\frac{E}{2C}$. Intersects x-axis at $\left(h \pm \frac{a}{b}\sqrt{b^2 + k^2}, 0\right)$ or $\frac{(y-k)^2}{a^2} - \frac{(x-h)^2}{b^2} = 1,\ a^2 = G/C$ $b^2 = -G/A$, all others the same. Intersects y-axis at $\left(0, k \pm \frac{a}{b}\sqrt{a^2 + h^2}\right)$.

Example $x^2 - 4y^2 + 2x + 8y - 7 = 0$ is equivalent to $(x + 1)^2/4 - (y - 1)^2 = 1$. This hyperbola opens out the x axis with intercepts on the x axis at $(-1 \pm 2\sqrt{2}, 0)$. See Fig. 2-17.

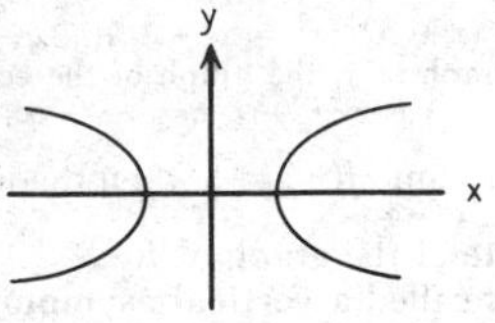

FIG. 2-17 Hyperbola.

Graphs of Polar Equations (Fig. 2-18) The polar representation of a point in the plane is (r, θ), where $0 \leq r < \infty$ and θ is the angle with the positive x axis (see under "Coordinate Systems"). $r = 0$

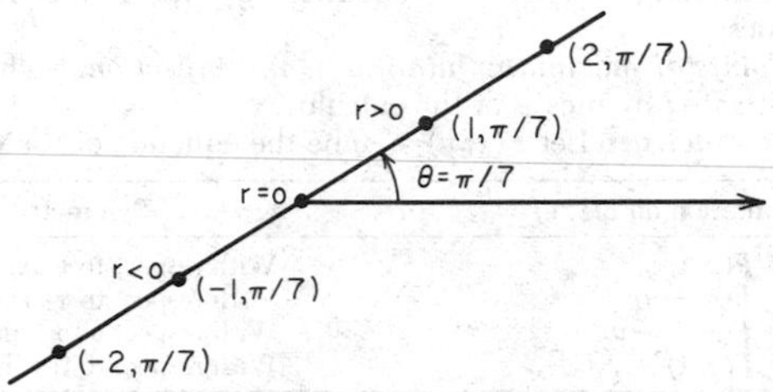

FIG. 2-18 Polar representation.

The following tabulation gives the form of the more common equations.

Polar equation	Type of curve	Representative graph
(1) $r = a$	Circle	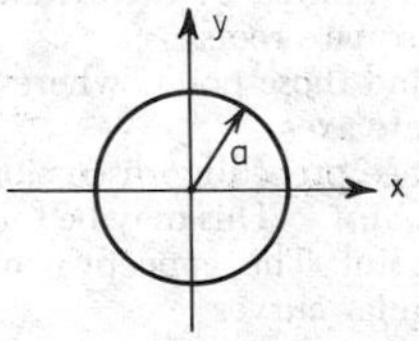 **FIG. 2-19** Circle center (0,0).
(2) $r = 2a \cos \theta$	Circle	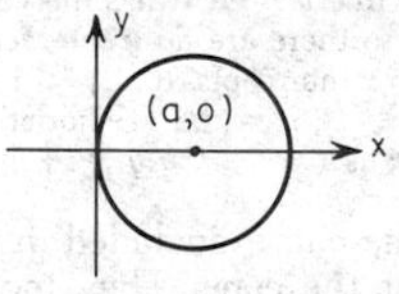 **FIG. 2-20** Circle center $(a,0)$.
(3) $r = 2a \sin \theta$	Circle	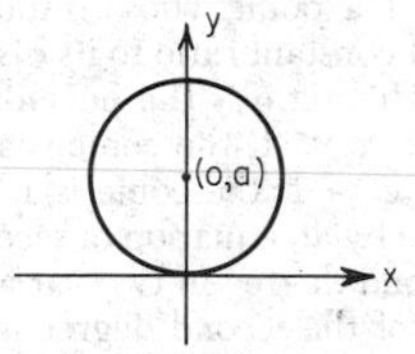 **FIG. 2-21** Circle center $(0,a)$.
(4) $r^2 - 2br \cos(\theta - \beta) + b^2 - a^2 = 0$	Circle at (b, β), radius a	
(5) $r = \frac{ke}{1 - e \cos \theta}$	$e = 1$ parabola $0 < e < 1$ ellipse $e > 1$ hyperbola	Focus at the pole, directrix $x = -k$
(6) $r\theta = a$	Hyperbolic spiral	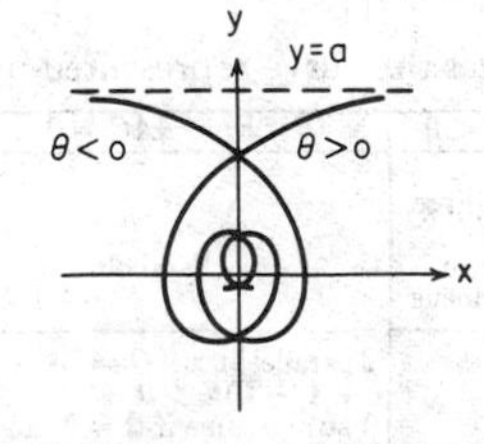 **FIG. 2-22** Hyperbolic spiral.

Polar equation	Type of curve	Representative graph
(7) $r^2 = 2a^2 \cos 2\theta$	Lemniscate	 FIG. 2-23 Lemniscate.
(8) $r = a(1 - \cos \theta)$	Cardioid	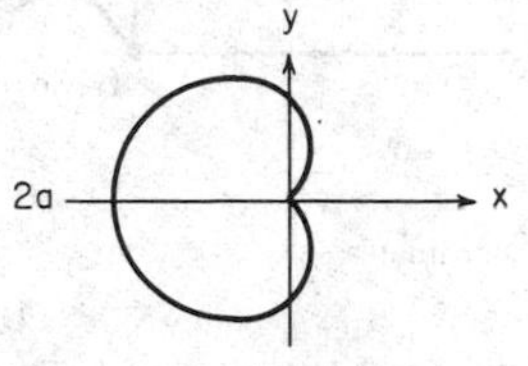 FIG. 2-24 Cardioid.
(9) $r = a \sec \theta$ $r = a \csc \theta$	Straight line parallel to y axis Straight line parallel to x axis	
(10) $r(a \cos \theta + b \sin \theta) = C$	General straight line	
(11) $r = a\theta, a > 0$	Spiral of Archimedes	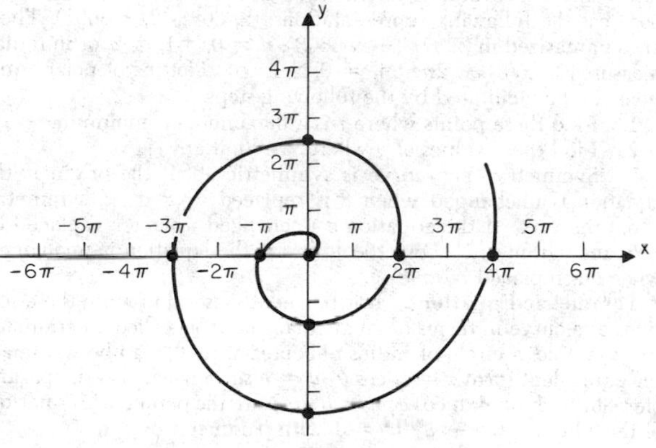 FIG. 2-25 Spiral of Archimedes.

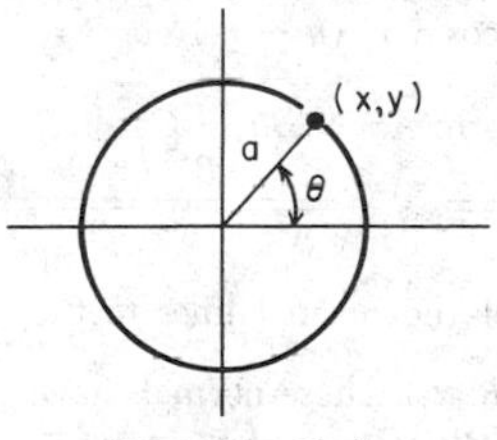

FIG. 2-26 Circle.

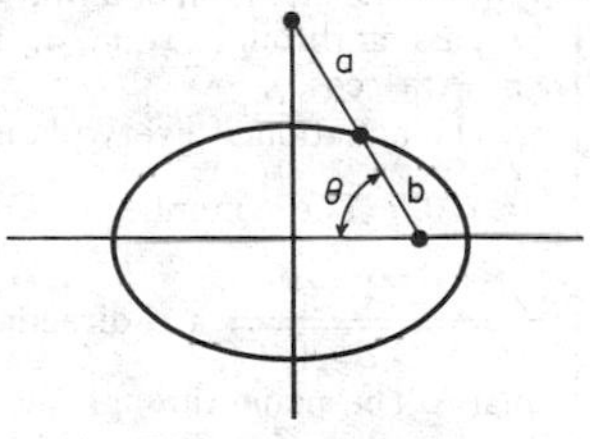

FIG. 2-27 Ellipse.

Some common equations in parametric form are given below.

(1) $(x - h)^2 + (y - k)^2 = a^2$	$x = h + a \cos \theta$ $y = k + a \sin \theta$	Circle (Fig. 2-26) Parameter is angle θ.
(2) $\dfrac{(x - h)^2}{a^2} + \dfrac{(y - k)^2}{b^2} = 1$	$x = h + a \cos \phi$ $y = k + a \sin \phi$	Ellipse (Fig. 2-27) Parameter is angle ϕ.
(3) $z^2 + y^2 = a^2$	$x = \dfrac{-at}{\sqrt{t^2 + 1}}$ $y = \dfrac{a}{\sqrt{t^2 + 1}}$	Circle Parameter is $t = \dfrac{dy}{dx}$ = slope of tangent at (x, y).
(4) $y = a \cosh \dfrac{x}{a}$	$x = a \sinh^{-1} \dfrac{s}{a}$ $y = a^2 + s^2$	Catenary (Fig. 2-28; such as hanging cable under gravity) Parameter s = arc length from $(0, a)$ to (x, y).
(5) Cycloid	$x = a(\phi - \sin \phi)$ $y = a(1 - \cos \phi)$	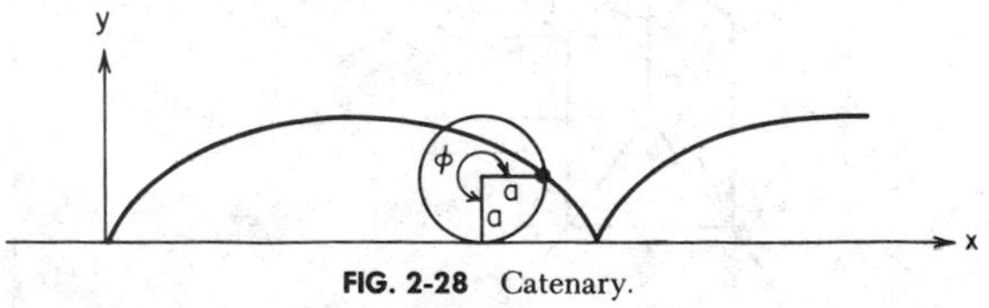 FIG. 2-28 Catenary.

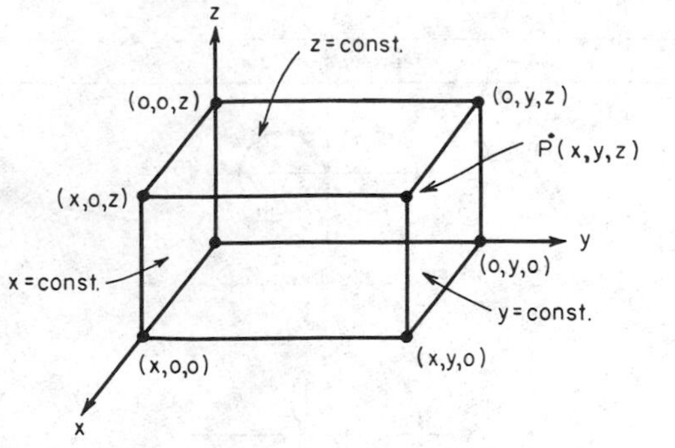

FIG. 2-29 Cartesian coordinates.

corresponds to $x = 0$, $y = 0$ regardless of θ. The same point may be represented in several different ways; thus the point $(2, \pi/3)$ or $(2, 60°)$ has the following representations: $(2, 60°)$, $(2, -300°)$. These are summarized in $(2, 60° + n\,360°)$, $n = 0, \pm 1, \pm 2$, or in radian measure $[2, (\pi/3) + 2n\pi]$, $n = 0, \pm 1, \pm 2$. Plotting of polar equations can be facilitated by the following steps:

1. Find those points where r is a maximum or minimum.
2. Find those values of θ where $r = 0$, if any.
3. Symmetry: The curve is symmetric about the origin if the equation is unchanged when θ is replaced by $\theta \pm \pi$, symmetric about the x axis if the equation is unchanged when θ is replaced by $-\theta$, and symmetric about the y axis if the equation is unchanged when θ is replaced by $\pi - \theta$.

Parametric Equations It is frequently useful to write the equations of a curve in terms of an auxiliary variable called a parameter. For example, a circle of radius a, center at (0, 0), can be written in the equivalent form $x = a \cos \theta$, $y = a \sin \phi$ where θ is the parameter. Similarly, $x = a \cos \phi$, $y = b \sin \phi$ are the parametric equations of the ellipse $x^2/a^2 + y^2/b^2 = 1$ with parameter ϕ.

SOLID ANALYTIC GEOMETRY

Coordinate Systems The commonly used coordinate systems are three in number. Others may be used in specific problems (see Morse and Feshbach, *Methods of Theoretical Physics*, McGraw-Hill, New York, 1953). The **rectangular** (cartesian) system (Fig. 2-29) consists of mutually orthogonal axes x, y, z. A triple of numbers (x, y, z) is used to represent each point. The **cylindrical** coordinate system $(r, \theta, z$; Fig. 2-30) is frequently used to locate a point in space. These are essentially the polar coordinates (r, θ) coupled with the z coordinate. As before, $x = r \cos \theta$, $y = r \sin \theta$, $z = z$ and $r^2 = x^2 + y^2$, $y/x = \tan \theta$. If r is held constant and θ and z are allowed to vary, the locus of (r, θ, z) is a right circular cylinder of radius r along the z axis. The locus of $r = C$ is a circle, and θ = constant is a plane containing the z axis and making an angle θ with the xz plane. Cylindrical coordinates are convenient to use when the problem has an axis of symmetry.

The **spherical** coordinate system is convenient if there is a point of symmetry in the system. This point is taken as the origin and the

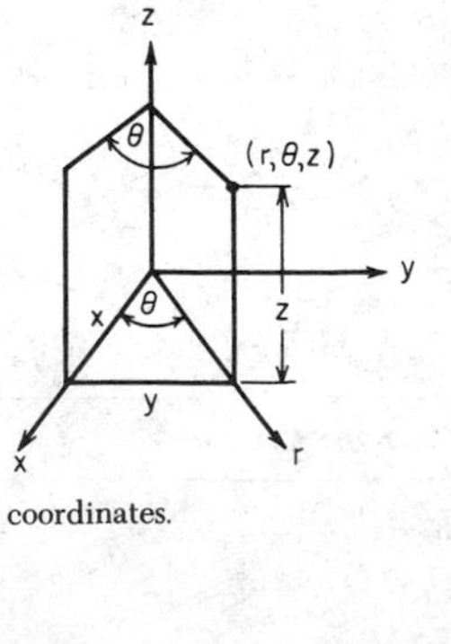

FIG. 2-30 Cylindrical coordinates.

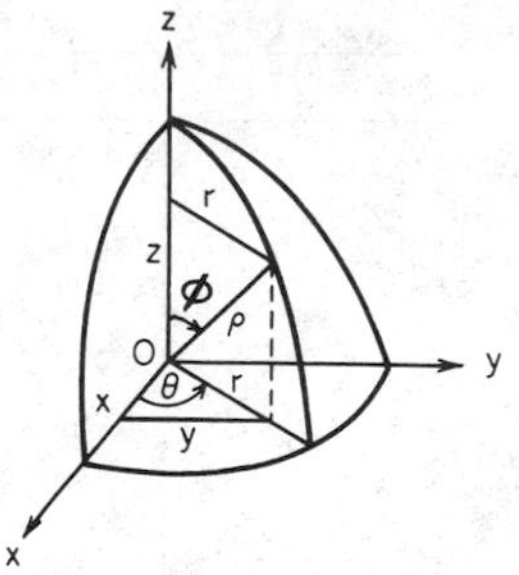

FIG. 2-31 Spherical coordinates.

coordinates (ρ, ϕ, θ) illustrated in Fig. 2-31. The relations are $x = \rho \sin \phi \cos \theta$, $y = \rho \sin \phi \sin \theta$, $z = \rho \cos \phi$ and $r = \rho \sin \phi$, $z = \rho \cos \phi$, $\theta = \theta$. θ = constant is a plane containing the z axis and making an angle θ with the xz plane. ϕ = constant is a cone with vertex at 0. ρ = constant is the surface of a sphere of radius ρ, center at the origin 0. Every point in the space may be given spherical coordinates restricted to the ranges $0 \le \phi \le \pi$, $\rho \ge 0$, $0 \le \theta < 2\pi$.

Lines and Planes The distance between two points (x_1, y_1, z_1), (x_2, y_2, z_2) is $d = \sqrt{(x_1 - x_2)^2 + (y_1 - y_2)^2 + (z_1 - z_2)^2}$. There is nothing in the geometry of three dimensions quite analogous to the slope of a line in the plane case. Instead of specifying the direction of a line by a trigonometric function evaluated for one angle, a trigonometric function evaluated for three angles is used. The angles α, β, γ that a line segment makes with the positive x, y, and z axes respectively are called the **direction angles** of the line, and $\cos \alpha$, $\cos \beta$, $\cos \gamma$ are called the **direction cosines.** Let (x_1, y_1, z_1), (x_2, y_2, z_2) be on the line. Then $\cos \alpha = (x_2 - x_1)/d$, $\cos \beta = (y_2 - y_1)/d$, $\cos \gamma = (z_2 - z_1)/d$, where d = the distance between the two points. Clearly $\cos^2 \alpha + \cos^2 \beta + \cos^2 \gamma = 1$. If two lines are specified by the direction cosines $(\cos \alpha_1, \cos \beta_1, \cos \gamma_1)$, $(\cos \alpha_2, \cos \beta_2, \cos \gamma_2)$, then the angle θ between the lines is $\cos \theta = \cos \alpha_1 \cos \alpha_2 + \cos \beta_1 \cos \beta_2 + \cos \gamma_1 \cos \gamma_2$. Thus the lines are perpendicular if and only if $\theta = 90°$ or $\cos \alpha_1 \cos \alpha_2 + \cos \beta_1 \cos \beta_2 + \cos \gamma_1 \cos \gamma_2 = 0$. The equation of a line with direction cosines $(\cos \alpha, \cos \beta, \cos \gamma)$ passing through (x_1, y_1, z_1) is $(x - x_1)/\cos \alpha = (y - y_1)/\cos \beta = (z - z_1)/\cos \gamma$.

The equation of every plane is of the form $Ax + By + Cz + D = 0$. The numbers $\dfrac{A}{\sqrt{A^2 + B^2 + C^2}}$, $\dfrac{B}{\sqrt{A^2 + B^2 + C^2}}$, $\dfrac{C}{\sqrt{A^2 + B^2 + C^2}}$ are direction cosines of the normal lines to the plane. The plane through the point (x_1, y_1, z_1) whose normals have these as direction cosines is $A(x - x_1) + B(y - y_1) + C(z - z_1) = 0$.

Example Find the equation of the plane through $(1, 5, -2)$ perpendicular to the line $(x + 9)/7 = (y - 3)/-1 = z/8$. The numbers $(7, -1, 8)$ are called **direction numbers.** They are a constant multiple of the direction cosines. $\cos \alpha = 7/114$, $\cos \beta = -1/114$, $\cos \gamma = 8/114$. The plane has the equation $7(x - 1) - 1(y - 5) + 8(z - 2) = 0$ or $7x - y + 8z + 14 = 0$.

The distance from the point (x_1, y_1, z_1) to the plane $Ax + By + Cz + D = 0$ is $d = \dfrac{|Ax_1 + By_1 + Cz_1 + D|}{\sqrt{A^2 + B^2 + C^2}}$.

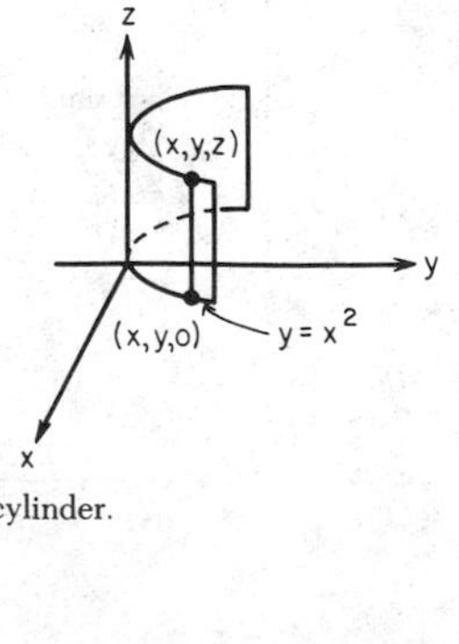

FIG. 2-32 Parabolic cylinder.

Space Curves Space curves are usually specified as the set of points whose coordinates are given parametrically by a system of equations $x = f(t)$, $y = g(t)$, $z = h(t)$ in the parameter t.

Example The equation of a straight line in space is $(x - x_1)/a = (y - y_1)/b = (z - z_1)/c$. Since all these quantities must be equal (say, to t), we may write $x = x_1 + at$, $y = y_1 + bt$, $z = z_1 + ct$, which represent the parametric equations of the line.

Example The equations $z = a \cos \beta t$, $y = a \sin \beta t$, $z = bt$, a, β, b positive constants, represent a circular helix.

Surfaces The locus of points (x, y, z) satisfying $f(x, y, z) = 0$ broadly speaking may be interpreted as a surface. The simplest surface is the **plane.** The next simplest is a **cylinder,** which is a surface generated by a straight line moving parallel to a given line and passing through a given curve.

Example The parabolic cylinder $y = x^2$ (Fig. 2-32) is generated by a straight line parallel to the z axis passing through $y = x^2$ in the plane $z = 0$.

A surface whose equation is a quadratic in the variables x, y, and z is called a **quadric surface.** Some of the more common such surfaces are tabulated and pictured in Figs. 2-33 to 2-42.

Equation	Name	Figure
(1) $\dfrac{x^2}{a^2} + \dfrac{y^2}{b^2} + \dfrac{z^2}{c^2} = 1$	Ellipsoid (sphere if $a = b = c$)	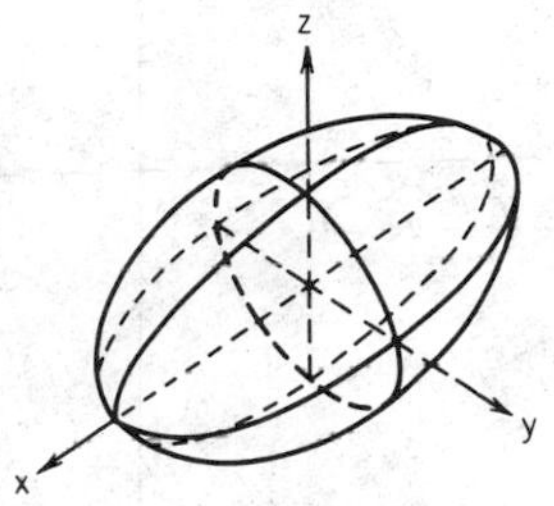 **FIG. 2-33** Ellipsoid.
(2) $\dfrac{x^2}{a^2} + \dfrac{y^2}{b^2} - \dfrac{z^2}{c^2} = 1$	Hyperboloid of one sheet	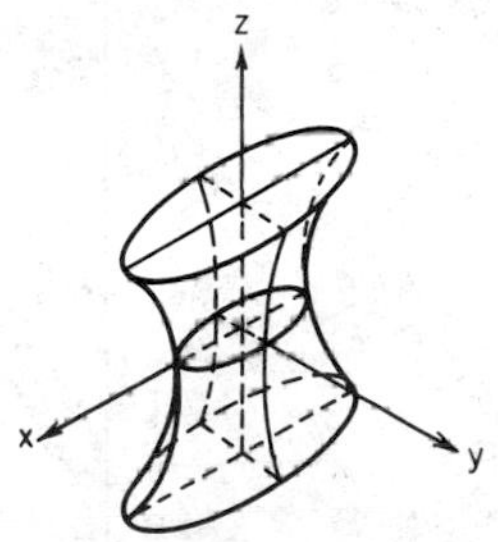 **FIG. 2-34** Hyperboloid of one sheet.
(3) $\dfrac{x^2}{a^2} + \dfrac{y^2}{b^2} - \dfrac{z^2}{c^2} = -1$	Hyperboloid of two sheets	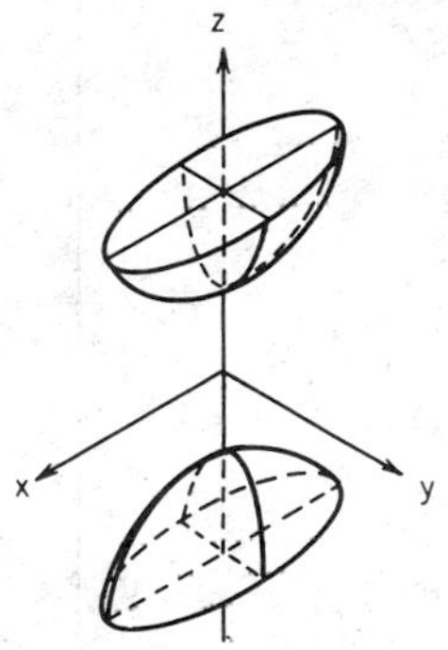 **FIG. 2-35** Hyperboloid of two sheets.
(4) $\dfrac{x^2}{a^2} + \dfrac{y^2}{b^2} - \dfrac{z^2}{c^2} = 0$	Cone	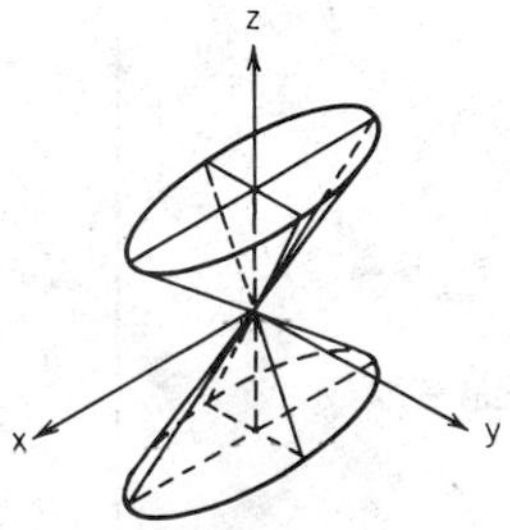 **FIG. 2-36** Cone.

Equation	Name	Figure
(5) $\frac{x^2}{a^2} + \frac{y^2}{b^2} + 2z = 0$	Elliptic paraboloid	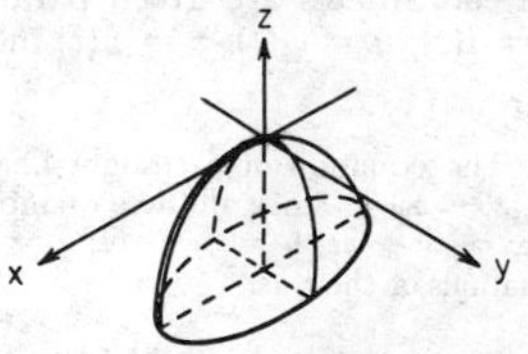 **FIG. 2-37** Elliptic paraboloid.
(6) $\frac{x^2}{a^2} - \frac{y^2}{b^2} + 2z = 0$	Hyperbolic paraboloid (saddle)	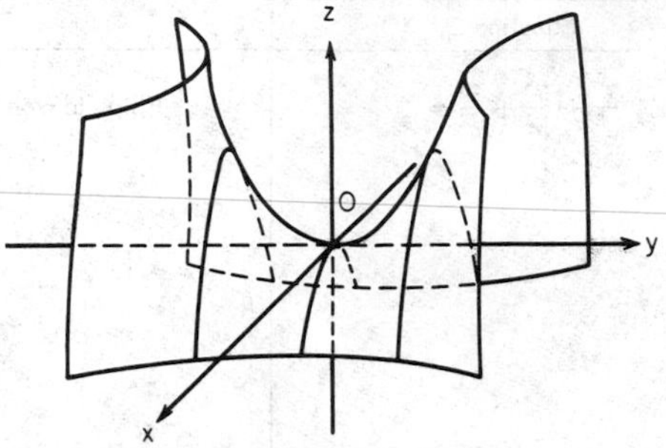 **FIG. 2-38** Hyperbolic paraboloid.
(7) $\frac{x^2}{a^2} + \frac{y^2}{b^2} = 1$	Elliptic cylinder (circular cylinder if $a = b$)	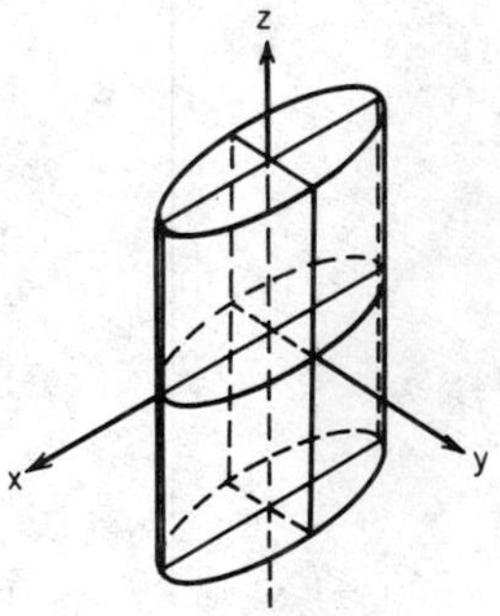 **FIG. 2-39** Elliptic cylinder.
(8) $\frac{x^2}{a^2} - \frac{y^2}{b^2} = 1$	Hyperbolic cylinder	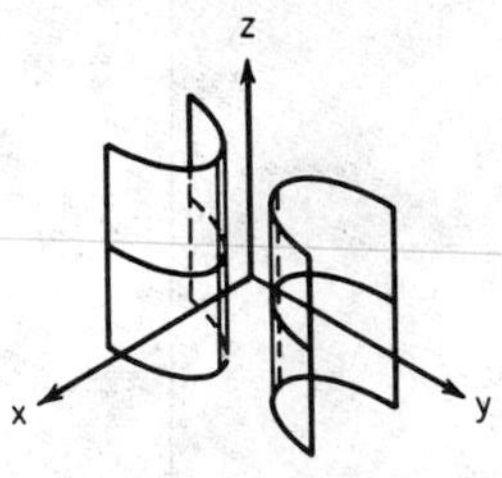 **FIG. 2-40** Hyperbolic cylinder.
(9) $y^2 + 2ax = 0$	Parabolic cylinder	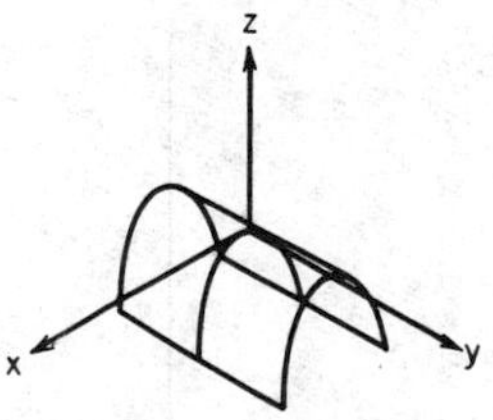 **FIG. 2-41** Parabolic cylinder.

Equation	Name	Figure
(10) $\frac{x^2}{a^2} - \frac{y^2}{b^2} = 0$	Intersecting planes	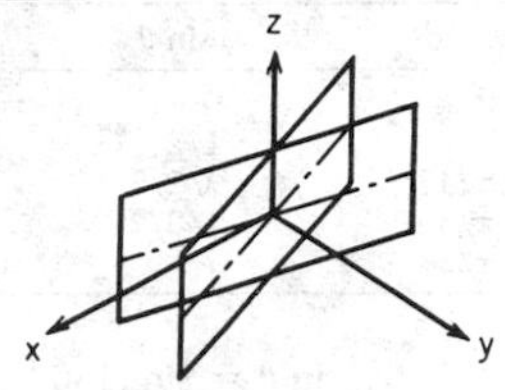 FIG. 2-42 Intersecting planes.

PLANE TRIGONOMETRY

References: 16, 102, 128, 154, 163, 169, 289. Some of these books include calculator applications, as indicated in their titles.

ANGLES

An angle is generated by the rotation of a line about a fixed center from some initial position to some terminal position. If the rotation is **clockwise,** the angle is **negative;** if it is **counterclockwise,** the angle is **positive.** Angle size is unlimited. If α, β are two angles such that $\alpha + \beta = 90°$, they are complementary; they are supplementary if $\alpha + \beta = 180°$. Angles are most commonly measured in the **sexagesimal system** or by **radian measure.** In the first system there are 360 degrees in one complete revolution; one degree = 1/90 of a right angle. The degree is subdivided into 60 minutes; the minute is subdivided into 60 seconds. In the radian system one radian is the angle at the center of a circle subtended by an arc whose length is equal to the radius of the circle. Thus 2π rad = 360°; 1 rad = 57.29578°; 1° = 0.01745 rad; 1 min = 0.00029089 rad. The advantage of radian measure is that it is *dimensionless*. The quadrants are conventionally labeled as Fig. 2-43 shows.

FUNCTIONS OF CIRCULAR TRIGONOMETRY

The trigonometric functions of angles are the ratios between the various sides of the reference triangles shown in Fig. 2-44 for the various quadrants. Clearly $r = \sqrt{x^2 + y^2} \geq 0$. The fundamental functions (see Figs. 2-45, 2-46, and 2-47) are

Sine of $\theta = \sin\theta = y/r$ Secant of $\theta = \sec\theta = r/x$

Cosine of $\theta = \cos\theta = x/r$ Cosecant of $\theta = \csc\theta = r/y$

Tangent of $\theta = \tan\theta = y/x$ Cotangent of $\theta = \cot\theta = x/y$

Magnitude and Sign of Trigonometric Functions $0 \leq \theta \leq 360°$

Function	0° to 90°	90° to 180°	180° to 270°	270° to 360°
$\sin\theta$	+0 to +1	+1 to +0	−0 to −1	−1 to −0
$\csc\theta$	+∞ to +1	+1 to +∞	−∞ to −1	−1 to −∞
$\cos\theta$	+1 to 0	−0 to −1	−1 to −0	+0 to +1
$\sec\theta$	+1 to +∞	−∞ to −1	−1 to −∞	+∞ to +1
$\tan\theta$	+0 to +∞	−∞ to −0	+0 to +∞	−∞ to −0
$\cot\theta$	+∞ to +0	−0 to −∞	+∞ to +0	−0 to −∞

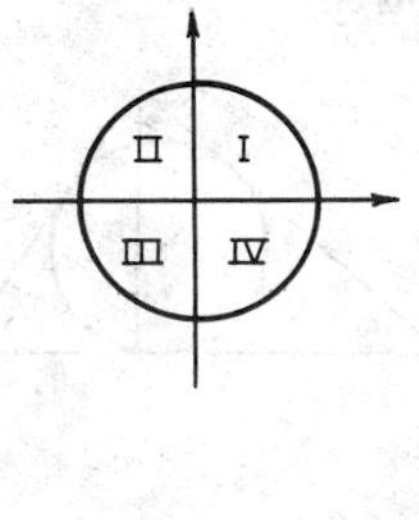

FIG. 2-43 Quadrants.

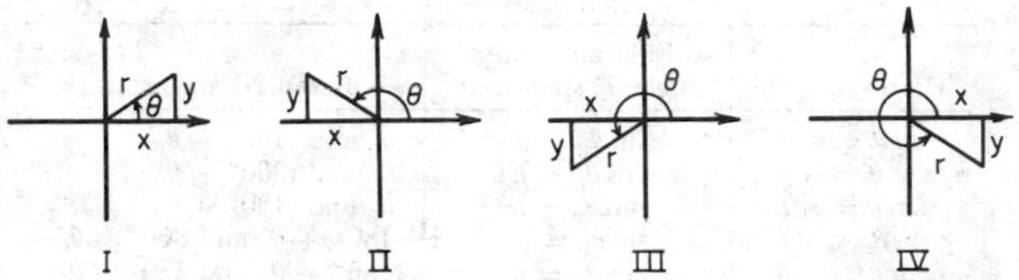

FIG. 2-44 Triangles.

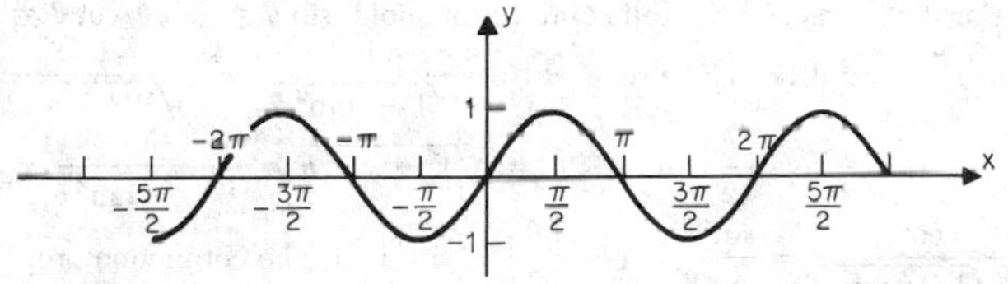

FIG. 2-45 Graph of $y = \sin x$.

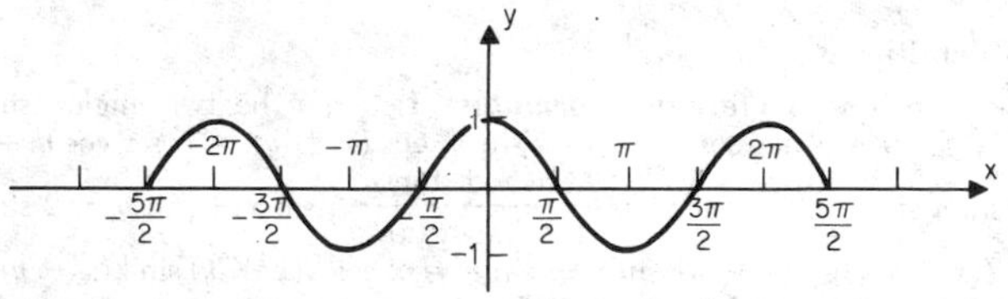

FIG. 2-46 Graph of $y = \cos x$.

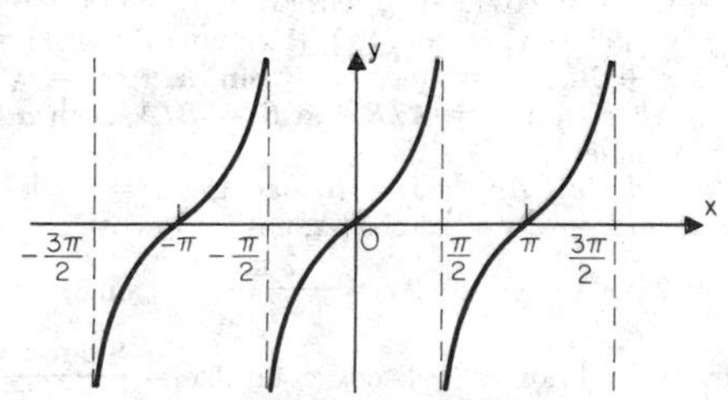

FIG. 2-47 Graph of $y = \tan x$.

Values of the Trigonometric Functions for Common Angles

$\theta°$	θ, rad	$\sin\theta$	$\cos\theta$	$\tan\theta$
0	0	0	1	0
30	$\pi/6$	$1/2$	$\sqrt{3}/2$	$\sqrt{3}/3$
45	$\pi/4$	$\sqrt{2}/2$	$\sqrt{2}/2$	1
60	$\pi/3$	$\sqrt{3}/2$	$1/2$	$\sqrt{3}$
90	$\pi/2$	1	0	$+\infty$

If $90° \le \theta \le 180°$, $\sin\theta = \sin(180° - \theta)$; $\cos\theta = -\cos(180° - \theta)$; $\tan\theta = -\tan(180° - \theta)$. If $180° \le \theta \le 270°$, $\sin\theta = -\sin(270° - \theta)$; $\cos\theta = -\cos(270° - \theta)$; $\tan\theta = \tan(270° - \theta)$. If $270° \le \theta \le 360°$, $\sin\theta = -\sin(360° - \theta)$; $\cos\theta = \cos(360° - \theta)$; $\tan\theta = -\tan(360° - \theta)$. The reciprocal properties may be used to find the values of the other functions.

If it is desired to find the angle when a function of it is given, the procedure is as follows: There will in general be two angles between 0° and 360° corresponding to the given value of the function.

Given ($a > 0$)	Find an acute angle θ_0 such that	Required angles are
$\sin\theta = +a$	$\sin\theta_0 = a$	θ_0 and $(180° - \theta_0)$
$\cos\theta = +a$	$\cos\theta_0 = a$	θ_0 and $(360° - \theta_0)$
$\tan\theta = +a$	$\tan\theta_0 = a$	θ_0 and $(180° + \theta_0)$
$\sin\theta = -a$	$\sin\theta_0 = a$	$180° + \theta_0$ and $360° - \theta_0$
$\cos\theta = -a$	$\cos\theta_0 = a$	$180° - \theta_0$ and $180° + \theta_0$
$\tan\theta = -a$	$\tan\theta_0 = a$	$180° - \theta_0$ and $360° - \theta_0$

Relations between Functions of a Single Angle $\sec\theta = 1/\cos\theta$; $\csc\theta = 1/\sin\theta$, $\tan\theta = \sin\theta/\cos\theta = \sec\theta/\csc\theta = 1/\cot\theta$; $\sin^2\theta + \cos^2\theta = 1$; $1 + \tan^2\theta = \sec^2\theta$; $1 + \cot^2\theta = \csc^2\theta$. For $0 \le \theta \le 90°$ the following results hold: $\sin\theta = \cos\theta/\cot\theta = \sqrt{1 - \cos^2\theta} = \cos\theta\,\tan\theta = \dfrac{\tan\theta}{\sqrt{1 + \tan^2\theta}} = \dfrac{1}{\sqrt{1 + \cot^2\theta}} = 2\sin\left(\dfrac{\theta}{2}\right)\cos\left(\dfrac{\theta}{2}\right)$ and $\cos\theta = \sqrt{1 - \sin^2\theta} = \dfrac{1}{\sqrt{1 + \tan^2\theta}} = \dfrac{\cot\theta}{\sqrt{1 + \cot^2\theta}} = \dfrac{\sin\theta}{\tan\theta} = \cos^2\left(\dfrac{\theta}{2}\right) - \sin^2\left(\dfrac{\theta}{2}\right)$. The cofunction property is very important. $\cos\theta = \sin(90° - \theta)$, $\sin\theta = \cos(90° - \theta)$, $\tan\theta = \cot(90° - \theta)$, $\cot\theta = \tan(90° - \theta)$, etc.

Functions of Negative Angles $\sin(-\theta) = -\sin\theta$, $\cos(-\theta) = \cos\theta$, $\tan(-\theta) = -\tan\theta$, $\sec(-\theta) = \sec\theta$, $\csc(-\theta) = -\csc\theta$, $\cot(-\theta) = -\cot\theta$.

Identities

Sum and Difference Formulas Let x, y be two angles. $\sin(x \pm y) = \sin x\cos y \pm \cos x\sin y$; $\cos(x \pm y) = \cos x\cos y \mp \sin x\sin y$; $\tan(x \pm y) = \dfrac{\tan x \pm \tan y}{1 \mp \tan x\tan y}$; $\sin x + \sin y = 2\sin ½(x + y)\cos ½(x - y)$; $\sin x - \sin y = 2\cos ½(x + y)\sin ½(x - y)$; $\cos x + \cos y = 2\cos ½(x + y)\cos ½(x - y)$; $\cos x - \cos y = -2\sin ½(x + y)\sin ½(x - y)$; $\tan x \pm \tan y = \dfrac{\sin(x \pm y)}{\cos x\cos y}$; $\sin^2 x - \sin^2 y = \cos^2 y - \cos^2 x = \sin(x + y)\sin(x - y)$; $\cos^2 x - \sin^2 y = \cos^2 y - \sin^2 x = \cos(x + y)\cos(x - y)$; $\sin(45° + x) = \cos(45° - x)$; $\sin(45° - x) = \cos(45° + x)$; $\tan(45° \pm x) = \cot(45° \mp x)$. $A\cos x + B\sin x = \sqrt{A^2 + B^2}\sin(\alpha + x) = \sqrt{A^2 + B^2}\cos(\beta - x)$ where $\tan\alpha = A/B$, $\tan\beta = B/A$; both α and β are postive acute angles.

Multiple and Half Angle Identities Let x = angle, $\sin 2x = 2\sin x\cos x$; $\sin x = 2\sin ½x\cos ½x$; $\cos 2x = \cos^2 x - \sin^2 x = 1 - 2\sin^2 x = 2\cos^2 x - 1$. $\tan 2x = \dfrac{2\tan x}{1 - \tan^2 x}$; $\sin 3x = 3\sin x - 4\sin^3 x$; $\cos 3x = 4\cos^3 x - 3\cos x$. $\tan 3x = \dfrac{3\tan x - \tan^3 x}{1 - 3\tan^2 x}$; $\sin 4x = 4\sin x\cos x - 8\sin^3 x\cos x$; $\cos 3x = 4\cos^3 x - 3\cos x$; $\cos 4x = 8\cos^4 x - 8\cos^2 x$. $\sin\left(\dfrac{x}{2}\right) = \sqrt{½(1 - \cos x)}$; $\cos\left(\dfrac{x}{2}\right) = \sqrt{½(1 + \cos x)}$; $\tan\left(\dfrac{x}{2}\right) = \sqrt{\dfrac{1 - \cos x}{1 + \cos x}} = \dfrac{\sin x}{1 + \cos x} = \dfrac{1 - \cos x}{\sin x}$.

Relations between Three Angles Whose Sum Is 180° Let x, y, z be the angles. six $x + \sin y + \sin z = 4\cos\left(\dfrac{x}{2}\right)\cos\left(\dfrac{y}{2}\right)\cos\left(\dfrac{z}{2}\right)$; $\cos x + \cos y + \cos z = 4\sin\left(\dfrac{x}{2}\right)\sin\left(\dfrac{y}{2}\right)\sin\left(\dfrac{z}{2}\right) + 1$; $\sin x + \sin y - \sin z = 4\sin\left(\dfrac{x}{2}\right)\sin\left(\dfrac{y}{2}\right)\cos\left(\dfrac{z}{2}\right)$; $\sin^2 x + \sin^2 y + \sin^2 z = 2\cos x\cos y\cos z + 2$; $\tan x + \tan y + \tan z = \tan x\tan y\tan z$; $\sin 2x + \sin 2y + \sin 2z = 4\sin x\sin y\sin z$.

INVERSE TRIGONOMETRIC FUNCTIONS

$y = \sin^{-1} x = \arcsin x$ is the angle y whose sine is x.

Example $y = \sin^{-1} ½$, y is 30°.

The complete solution of the equation $x = \sin y$ is $y = (-1)^n \sin^{-1} x + n(180°)$, $-\pi/2 \le \sin^{-1} x \le \pi/2$ where $\sin^{-1} x$ is the principal value of the angle whose sine is x. The range of principal values of the $\cos^{-1} x$ is $0 \le \cos^{-1} x \le \pi$ and $-\pi/2 \le \tan^{-1} x \le \pi/2$. If these restrictions are allowed to hold, the following formulas result: $\sin^{-1} x = \cos^{-1}\sqrt{1 - x^2} = \tan^{-1}\dfrac{x}{\sqrt{1 - x^2}} = \cot^{-1}\dfrac{\sqrt{1 - x^2}}{x} = \sec^{-1}\dfrac{1}{\sqrt{1 - x^2}} = \csc^{-1}\dfrac{1}{x} = \dfrac{\pi}{2} - \cos^{-1} x$. $\cos^{-1} x = \sin^{-1}\sqrt{1 - x^2} = \tan^{-1}\dfrac{\sqrt{1 - x^2}}{x} = \cot^{-1}\dfrac{x}{\sqrt{1 - x^2}} = \sec^{-1}\dfrac{1}{x} = \csc^{-1}\dfrac{1}{\sqrt{1 - x^2}} = \dfrac{\pi}{2} - \sin^{-1} x$. $\tan^{-1} x = \sin^{-1}\dfrac{x}{\sqrt{1 + x^2}} = \cos^{-1}\dfrac{1}{\sqrt{1 + x^2}} = \cot^{-1}\dfrac{1}{x} = \sec^{-1}\sqrt{1 + x^2} = \csc^{-1}\dfrac{\sqrt{1 + x^2}}{x}$.

RELATIONS BETWEEN ANGLES AND SIDES OF TRIANGLES

Solutions of Triangles (Fig. 2-48) Let a, b, c denote the sides and α, β, γ the angles opposite the sides in the triangle. Let $2s = a + b + c$, A = area, r = radius of the inscribed circle, R = radius of the circumscribed circle, and h = altitude. In any triangle $\alpha + \beta + \gamma = 180°$.

Law of Sines $\sin\alpha/a = \sin\beta/b = \sin\gamma/c$.

Law of Tangents $\dfrac{a + b}{a - b} = \dfrac{\tan ½(\alpha + \beta)}{\tan ½(\alpha - \beta)}$; $\dfrac{b + c}{b - c} = \dfrac{\tan ½(\beta + \gamma)}{\tan ½(\beta - \gamma)}$; $\dfrac{a + c}{a - c} = \dfrac{\tan ½(\alpha + \gamma)}{\tan ½(\alpha - \gamma)}$.

Law of Cosines $a^2 = b^2 + c^2 - 2bc\cos\alpha$; $b^2 = a^2 + c^2 - 2ac\cos\beta$; $c^2 = a^2 + b^2 - 2ab\cos\gamma$.

Other Relations In this subsection, where appropriate, two more formulas can be generated by replacing a by b, b by c, c by a, α by β, β by γ, and γ by α. $\cos\alpha = (b^2 + c^2 - a^2)/2bc$; $a = b\cos\gamma + c\cos\beta$; $\sin\alpha = \dfrac{2}{bc}\sqrt{s(s - a)(s - b)(s - c)}$; $\sin\left(\dfrac{\alpha}{2}\right) =$

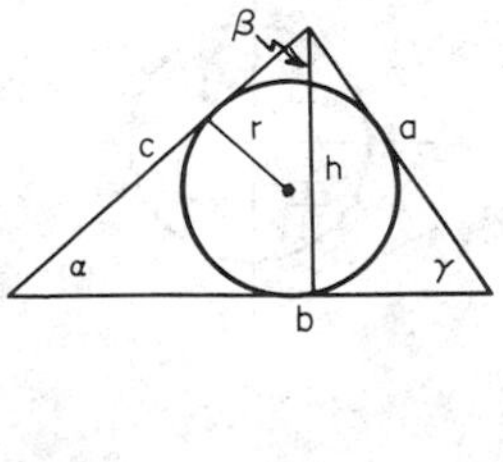

FIG. 2-48 Triangle.

$\sqrt{\dfrac{(s-b)(s-c)}{bc}}$; $\cos\left(\dfrac{\alpha}{2}\right) = \sqrt{\dfrac{s(s-a)}{bc}}$; $A = \dfrac{1}{2}bh = \dfrac{1}{2}ab\sin\gamma = \dfrac{a^2\sin\beta\sin\gamma}{2\sin\alpha} = \sqrt{s(s-a)(s-b)(s-c)} = rs$ where

$r = \sqrt{\dfrac{(s-a)(s-b)(s-c)}{s}}$. $R = a/(2\sin\alpha) = abc/4A$; $h = c\sin a = a\sin\gamma = 2rs/b$.

Example $a = 5$, $b = 4$, $\alpha = 30°$. Use the law of sines. $0.5/5 = \sin\beta/4$, $\sin\beta = \frac{2}{5}$, $\beta = 23°35'$, $\gamma = 126°25'$. So $c = \sin 126°25'/\frac{1}{10} = 10(.8047) = 8.05$.

The relations given here suffice to solve any triangle. One method for each triangle is given.

Right Triangle (Fig. 2-49) Given one side and any acute angle α or any two sides, the remaining parts can be obtained from the following formulas:

$$a = \sqrt{(c+b)(c-b)} = c\sin\alpha = b\tan\alpha$$

$$b = \sqrt{(c+a)(c-a)} = c\cos\alpha = a\cot\alpha$$

$$c = \sqrt{a^2 \pm b^2},\ \sin\alpha = \frac{a}{c},\ \cos\alpha = \frac{b}{c},\ \tan\alpha = \frac{a}{b},\ \beta = 90° - \alpha$$

$$A = \frac{1}{2}ab = \frac{a^2}{2\tan\alpha} = \frac{b^2\tan\alpha}{2} = \frac{c^2\sin 2\alpha}{4}$$

Oblique Triangles (Fig. 2-50) There are four possible cases.

1. Given b, c and the included angles α, $\frac{1}{2}(\beta+\gamma) = 90° - \frac{1}{2}\alpha$; $\tan\frac{1}{2}(\beta-\gamma) = \dfrac{b-c}{b+c}\tan\frac{1}{2}(\beta+\gamma)$; $\beta = \frac{1}{2}(\beta-\gamma)$; $\gamma = \frac{1}{2}(\beta+\gamma) - \frac{1}{2}(\beta-\gamma)$; $a = \dfrac{b\sin\alpha}{\sin\beta}$.

2. Given the three sides a b, c, $s = \frac{1}{2}(a+b+c)$; $r = \sqrt{\dfrac{(s-a)(s-b)(s-c)}{s}}$; $\tan\frac{1}{2}\alpha = \dfrac{r}{s-a}$; $\tan\frac{1}{2}\beta = \dfrac{r}{s-b}$; $\tan\frac{1}{2}\gamma = \dfrac{r}{s-c}$.

3. Given any two sides a, c and an angle opposite one of them α, $\sin\gamma = (c\sin\alpha)/a$; $\beta = 180° - \alpha - \gamma$; $b = (a\sin\beta)/(\sin\alpha)$. There may be two solutions here. γ may have two values γ_1, γ_2; $\gamma_1 < 90°$, $\gamma_2 = 180° - \gamma_1 > 90°$. If $\alpha + \gamma_2 > 180°$, use only γ_1. This case may be *impossible* if $\sin\gamma > 1$.

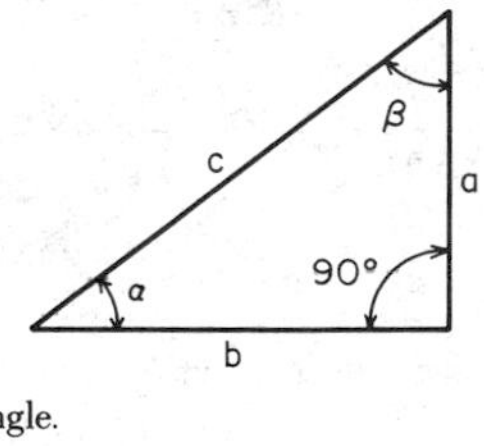

FIG. 2-49 Right triangle.

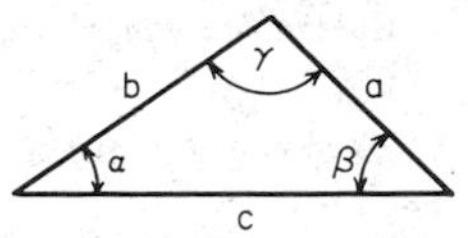

FIG. 2-50 Oblique triangle.

4. Given any side c and two angles α and β, $\gamma = 180° - \alpha - \beta$; $a = (c\sin\alpha)/(\sin\gamma)$; $b = (c\sin\beta)/(\sin\gamma)$.

HYPERBOLIC TRIGONOMETRY

The hyperbolic functions are certain combinations of exponentials e^x and e^{-x}. $\cosh x = \dfrac{e^x + e^{-x}}{2}$; $\sinh x = \dfrac{e^x - e^{-x}}{2}$; $\tanh x = \dfrac{\sinh x}{\cosh x} = \dfrac{e^x - e^{-x}}{e^x + e^{-x}}$. $\coth x = \dfrac{e^x + e^{-x}}{e^x - e^{-x}} = \dfrac{1}{\tanh x} = \dfrac{\cosh x}{\sinh x}$; $\operatorname{sech} x = \dfrac{1}{\cosh x} = \dfrac{2}{e^x + e^{-x}}$; $\operatorname{csch} x = \dfrac{1}{\sinh x} = \dfrac{2}{e^x - e^{-x}}$.

Fundamental Relationships $\sinh x + \cosh x = e^x$; $\cosh x - \sinh x = e^{-x}$; $\cosh^2 x - \sinh^2 x = 1$; $\operatorname{sech}^2 x + \tanh^2 x = 1$; $\coth^2 x - \operatorname{csch}^2 x = 1$; $\sinh 2x = 2\sinh x\cosh x$; $\cosh 2x = \cosh^2 x + \sinh^2 x = 1 + 2\sinh^2 x = 2\cosh^2 x - 1$. $\tanh 2x = (2\tanh x)/(1 + \tanh^2 x)$; $\sinh(x \pm y) = \sinh x\cosh y \pm \cosh x\sinh y$; $\cosh(x \pm y) = \cosh x\cosh y + \sinh x\sinh y$; $2\sinh^2 x/2 = \cosh x - 1$; $2\cosh^2 x/2 = \cosh x + 1$; $\sinh(-x) = -\sinh x$; $\cosh(-x) = \cosh x$; $\tanh(-x) = -\tanh x$.

When $u = a\cosh x$, $v = a\sinh x$, then $u^2 - v^2 = a^2$; which is the equation for a hyperbola. In other words, the hyperbolic functions in the parametric equations $u = a\cosh x$, $v = a\sinh x$ have the same relation to the hyperbola $u^2 - v^2 = a^2$ that the equations $u = a\cos\theta$, $v = a\sin\theta$ have to the circle $u^2 + v^2 = a^2$.

Inverse Hyperbolic Functions If $x = \sinh y$, then y is the inverse hyperbolic sine of x written $y = \sinh^{-1} x$ or arcsinh x. $\sinh^{-1} x = \log_e(x + \sqrt{x^2+1})$; $\cosh^{-1} x = \log_e(x + \sqrt{x^2-1})$ $\tanh^{-1} x = \frac{1}{2}\log_e \dfrac{1+x}{1-x}$; $\coth^{-1} x = \frac{1}{2}\log_e \dfrac{x+1}{x-1}$; $\operatorname{sech}^{-1} x = \log_e\left(\dfrac{1+\sqrt{1-x^2}}{x}\right)$; $\operatorname{csch}^{-1} = \log_e\left(\dfrac{1+\sqrt{1+x^2}}{x}\right)$.

Magnitude of the Hyperbolic Functions $\cosh x \geq 1$ with equality only for $x = 0$; $-\infty < \sinh x < \infty$; $-1 < \tanh x < 1$. $\cosh x \sim e^x$ as $x \to \infty$; $\sinh x \to e^x$ as $x \to \infty$.

APPROXIMATIONS FOR TRIGONOMETRIC FUNCTIONS

For small values of θ (θ measured in radians) $\sin\theta \approx \theta$, $\tan\theta \approx \theta$; $\cos\theta \approx 1 - (\theta^2/2)$. The following relations actually hold: $\sin\theta < \theta < \tan\theta$; $\cos\theta < \sin\theta/\theta < 1$; $\theta\sqrt{1-\theta^2} < \sin\theta < \theta$; $\cos\theta < \theta/\tan\theta < 1$; $\theta\left(1 - \dfrac{\theta^2}{2}\right) < \sin\theta < \theta$ and $\theta < \tan\theta < \dfrac{\theta}{\sqrt{1-\theta^2}}$. The behavior ratio of the functions as $\theta \to 0$ is given by the following: $\lim_{\theta\to 0} \sin\theta/\theta = 1$; $\sin\theta/\tan\theta = 1$.

DIFFERENTIAL AND INTEGRAL CALCULUS

REFERENCES: 74, 113, 163, 245, 246, 256, 263, 267, 279. See also "General References: References for General and Specific Topics—Advanced Calculus."

DIFFERENTIAL CALCULUS

An Example of Functional Notation Suppose that a storage warehouse of 16,000 ft³ is required. The construction costs per square foot are \$10, \$3, and \$2 for walls, roof, and floor respectively. What are the minimum cost dimensions? Thus, with h = height, x = width, and y = length, the respective costs are

$$\text{Walls} = 2 \times 10hy + 2 \times 10hx = 20h(y+x)$$
$$\text{Roof} = 3xy$$
$$\text{Floor} = 2xy$$

$$\text{Total cost} = 2xy + 3xy + 20h(x + y) = 5xy + 20h(x + y) \qquad (2\text{-}1)$$

and the restriction

$$\text{Total volume} = xyh \qquad (2\text{-}2)$$

Solving for h from Eq. (2-2),

$$h = \text{volume}/xy = 16{,}000/xy \qquad (2\text{-}3)$$

$$\text{Cost} = 5xy + \frac{320{,}000}{xy}(y + x) = 5xy + 320{,}000\left(\frac{1}{x} + \frac{1}{y}\right) \qquad (2\text{-}4)$$

In this form it can be shown that the minimum cost will occur for $x = y$; therefore

$$\text{Cost} = 5x^2 + 640{,}000\,(1/x)$$

By evaluation, the smallest cost will occur when $x = 40$.

$$\text{Cost} = 5(1600) + 640{,}000/40 = \$24{,}000$$

The dimensions are then $x = 40$ ft, $y = 40$ ft, $h = 16{,}000/(40 \times 40) = 10$ ft. Symbolically, the original cost relationship is written

$$\text{Cost} = f(x, y, h) = 5xy + 20h(y + x)$$

and the volume relation

$$\text{Volume} = g(x, y, h) = xyh = 16{,}000$$

In terms of the derived general relationships (2-1) and (2-2), x, y, and h are **independent variables**—cost and volume, **dependent variables.** That is, the cost and volume become fixed with the specification of dimensions. However, corresponding to the given restriction of the problem, relative to volume, the function $g(x, y, z) = xyh$ becomes a **constraint function.** In place of three independent and two dependent variables the problem reduces to two independent (volume has been constrained) and two dependent as in functions (2-3) and (2-4). Further, the requirement of minimum cost reduces the problem to three dependent variables (x, y, h) and no **degrees of freedom,** that is, freedom of independent selection.

Limits It is stated that the limit of function $f(x)$ as x approaches a (a is finite or else x is said to increase without bound) is the number N by writing

$$\lim_{x\to a} f(x) = N$$

This states that $f(x)$ can be calculated as close to N as desirable by making x sufficiently close to a. This does not put any restriction on $f(x)$ when $x = a$. More precisely, the former statement written

$$\lim_{x\to a} f(x) = N$$

means that for any given positive number ϵ a number δ can be found such that $0 < |a - x| < \delta$ implies that $|N - f(x)| < \epsilon$.

The following operations with limits (when they exist) are valid:

(1) $$\lim_{x\to a} bf(x) = b \lim_{x\to a} f(x)$$

(2) $$\lim_{x\to a} [f(x) + g(x)] = \lim_{x\to a} f(x) + \lim_{x\to a} g(x)$$

(3) $$\lim_{x\to a} [f(x)g(x)] = \lim_{x\to a} f(x) \cdot \lim_{x\to a} g(x)$$

(4) $$\lim_{x\to a}\frac{f(x)}{g(x)} = \frac{\lim_{x\to a} f(x)}{\lim_{x\to a} g(x)} \quad \text{if } \lim_{x\to a} g(x) \neq 0$$

Continuity A function $f(x)$ is **continuous** at the point $x = a$ if

$$\lim_{h\to 0} [f(a + h) - f(a)] = 0$$

Rigorously, it is stated $f(x)$ is continuous at $x = a$ if for any positive ϵ there exists a $\delta > 0$ such that $|f(a + h) - f(a)| < \epsilon$ for all x with $|x - a| < \delta$. For example, the function $(\sin x)/x$ is not continuous at $x = 0$ and therefore is said to be **discontinuous.** Discontinuities are classified into three types:

1. Removable	$y = \sin x/x$	at $x = 0$		
2. Infinite	$y = 1/x$	at $x = 0$		
3. Jump	$y = 10/(1 + e^{1/x})$	at $x = 0^+$	$y = 0^+$
		$x = 0$	$y = 0$
		$x = 0^-$	$y = 10$

Derivative The function $f(x)$ has a derivative at $x = a$, which can be denoted as $f'(a)$, if $\lim_{h\to 0} \frac{f(a + h) - f(a)}{h}$ exists. This implies continuity at $x = a$. Conversely, a function may be continuous but not have a derivative. The derivative function is

$$f'(x) = \frac{df}{dx} = \lim_{h\to 0} \frac{f(x + h) - f(x)}{h}$$

Differentiation Define $\Delta y = f(x + \Delta x) - f(x)$. Then dividing by Δx

$$\frac{\Delta y}{\Delta x} = \frac{f(x + \Delta x) - f(x)}{\Delta x}$$

Call
$$\lim_{\Delta x\to 0} \frac{\Delta y}{\Delta x} = \frac{dy}{dx}$$

then
$$\frac{dy}{dx} = \lim_{\Delta x\to 0} \frac{f(x + \Delta x) - f(x)}{\Delta x}$$

Example Find the derivative of $y = \sin x$.

$$\frac{dy}{dx} = \lim_{\Delta x\to 0} \frac{\sin(x + \Delta x) - f(x)}{\Delta x}$$

$$= \lim_{\Delta x\to 0} \frac{\sin x \cos \Delta x + \sin \Delta x \cos x - \sin x}{\Delta x}$$

$$= \lim_{\Delta x\to 0} \frac{\sin x(\cos \Delta x - 1)}{\Delta x} + \lim_{\Delta x\to 0} \frac{\sin \Delta x \cos x}{\Delta x}$$

$$= \cos x \text{ since } \lim_{\Delta x\to 0} \frac{\sin \frac{1}{2}\Delta}{\frac{1}{2}\Delta x} = 1$$

Differential Operations The following differential operations are valid: f, g, . . . are differentiable functions of x, c is a constant; e is the base of the natural logarithms.

$$\frac{dc}{dx} = 0 \qquad (2\text{-}5)$$

$$\frac{dx}{dx} = 1 \qquad (2\text{-}6)$$

$$\frac{d}{dx}(f + g) = \frac{df}{dx} + \frac{dg}{dx} \qquad (2\text{-}7)$$

$$\frac{d}{dx}(f \times g) = f\frac{dg}{dx} + g\frac{df}{dx} \qquad (2\text{-}8)$$

$$\frac{dy}{dx} = \frac{1}{dx/dy} \quad \text{if } \frac{dy}{dx} \neq 0 \qquad (2\text{-}9)$$

$$\frac{d}{dx}f^n = nf^{n-1}\frac{df}{dx} \qquad (2\text{-}10)$$

$$\frac{d}{dx}\left(\frac{f}{g}\right) = \frac{g(df/dx) - f(dg/dx)}{g^2} \qquad (2\text{-}11)$$

$$\frac{df}{dx} = \frac{df}{dv} \times \frac{dv}{dx} \text{ (chain rule)} \qquad (2\text{-}12)$$

$$\frac{df^g}{dx} = gf^{g-1}\frac{df}{dx} + f^g \ln f \frac{dg}{dx} \qquad (2\text{-}13)$$

$$\frac{da^x}{dx} = (\ln a)\, a^x \qquad (2\text{-}14)$$

Example Derive dy/dx for $x^2 + y^3 = x + xy + A$.

Here
$$\frac{d}{dx}x^2 + \frac{d}{dx}y^3 = \frac{d}{dx}x + \frac{d}{dx}xy + \frac{d}{dx}A$$

$$2x + 3y^2\frac{dy}{dx} = 1 + y + x\frac{dy}{dx} + 0$$

by rules (2-10), (2-10), (2-6), (2-8), and (2-5) respectively.

Thus $\dfrac{dy}{dx} = \dfrac{2x - 1 - y}{x - 3y^2}$

Differentials

$$de^x = e^x\,dx \qquad (2\text{-}15)$$
$$d \ln x = (1/x)\,dx \qquad (2\text{-}16)$$
$$d \log x = (\log e/x)dx \qquad (2\text{-}17)$$
$$d \sin x = \cos x\,dx \qquad (2\text{-}18)$$
$$d \cos x = -\sin x\,dx \qquad (2\text{-}19)$$
$$d \tan x = \sec^2 x\,dx \qquad (2\text{-}20)$$
$$d \cot x = -\csc^2 x\,dx \qquad (2\text{-}21)$$
$$d \sec x = \tan x \sec x\,dx \qquad (2\text{-}22)$$
$$d \csc x = -\cot x \csc x\,dx \qquad (2\text{-}23)$$
$$d \sin^{-1} x = (1 - x^2)^{-1/2}\,dx \qquad (2\text{-}24)$$
$$d \cos^{-1} = -(1 - x^2)^{-1/2}\,dx \qquad (2\text{-}25)$$
$$d \tan^{-1} x = (1 + x^2)^{-1}\,dx \qquad (2\text{-}26)$$
$$d \cot^{-1} x = -(1 + x^2)^{-1}\,dx \qquad (2\text{-}27)$$
$$d \sec^{-1} x = x^{-1}(x^2 - 1)^{-1/2}\,dx \qquad (2\text{-}28)$$
$$d \csc^{-1} x = -x^{-1}(x^2 - 1)^{-1/2}\,dx \qquad (2\text{-}29)$$
$$d \sinh x = \cosh x\,dx \qquad (2\text{-}30)$$
$$d \cosh x = \sinh x\,dx \qquad (2\text{-}31)$$
$$d \tanh x = \operatorname{sech}^2 x\,dx \qquad (2\text{-}32)$$
$$d \coth x = -\operatorname{csch}^2 x\,dx \qquad (2\text{-}33)$$
$$d \operatorname{sech} x = -\operatorname{sech} x \tanh x\,dx \qquad (2\text{-}24)$$
$$d \operatorname{csch} x = -\operatorname{csch} x \coth x\,dx \qquad (2\text{-}35)$$
$$d \sinh^{-1} x = (x^2 + 1)^{-1/2}\,dx \qquad (2\text{-}36)$$
$$d \cosh^{-1} = (x^2 - 1)^{-1/2}\,dx \qquad (2\text{-}37)$$
$$d \tanh^{-1} x = (1 - x^2)^{-1}\,dx \qquad (2\text{-}38)$$
$$d \coth^{-1} x = -(x^2 - 1)^{-1}\,dx \qquad (2\text{-}39)$$
$$d \operatorname{sech}^{-1} x = x(1 - x^2)^{-1/2}\,dx \qquad (2\text{-}40)$$
$$d \operatorname{csch}^{-1} x = -x^{-1}(x^2 + 1)^{-1/2}\,dx \qquad (2\text{-}41)$$

Example Find dy/dx for $y = \sqrt{x} \cos(1 - x^2)$.

Using

$$\frac{dy}{dx} = \sqrt{x}\frac{d}{dx}\cos(1 - x^2) + \cos(1 - x^2)\frac{d}{dx}\sqrt{x} \qquad (2\text{-}8)$$
$$\frac{d}{dx}\cos(1 - x^2) = -\sin(1 - x^2)\frac{d}{dx}(1 - x^2) \qquad (2\text{-}19)$$
$$= -\sin(1 - x^2)(0 - 2x) \qquad (2\text{-}5), (2\text{-}10)$$
$$\frac{d\sqrt{x}}{dx} = \frac{1}{2}x^{-1/2} \qquad (2\text{-}10)$$
$$\frac{dy}{dx} = 2x^{3/2}\sin(1 - x^2) + \frac{1}{2}x^{-1/2}\cos(1 - x^2)$$

Example Find the deriative of $\tan x$ with respect to $\sin x$.

$$v = \sin x$$
$$y = \tan x$$

Using

$$\frac{dx}{d \sin x} = \frac{dy}{dv} = \frac{dy}{dx}\frac{dx}{dv} \qquad (2\text{-}12)$$
$$= \frac{d \tan x}{dx}\,\frac{1}{\dfrac{d \sin x}{dx}} \qquad (2\text{-}9)$$
$$= \sec^2 x/\cos x \qquad (2\text{-}18), (2\text{-}20)$$

Very often in experimental sciences and engineering functions and their derivatives are available only through their numerical values. In particular, through measurements we may know the values of a function and its derivative only at certain points. In such cases the preceding operational rules for derivatives, including the chain rule, can be applied numerically.

Example Given the following table of values for differentiable functions f and g; evaluate the following quantities:

x	$f(x)$	$f'(x)$	$g(x)$	$g'(x)$
1	3	1	4	−4
3	0	2	4	7
4	−2	10	3	6

(1) $\dfrac{d}{dx}[f(x) + g(x)]\Big|_{x=4} = f'(4) + g'(4) = 10 + 6 = 16$

(2) $\left(\dfrac{f}{g}\right)'(1) = \dfrac{f'(1)g(1) - f(1)g'(1)}{[g(1)]^2} = \dfrac{1 \cdot 4 - 3(-4)}{(-4)^2} = \dfrac{16}{16} = 1$

(3) $\dfrac{d}{dx}[f(g(x))]\Big|_{x=4} = f'(g(4))g'(4) = f'(3)g'(4) = 2.6 = 12$

(4) $\dfrac{d}{dx}[f^3(x) - 3g(\sqrt{x})] = \left[3f^2(x)f'(x) - 3g'(\sqrt{x})\dfrac{1}{2}\dfrac{1}{\sqrt{x}}\right]_{x=1}$

$= 3f^2(1)f'(1) - 3g'(1)\dfrac{1}{2} = 3 \cdot 3^2 - 3(-4)\dfrac{1}{2} = 27 + 6 = 33$

(5) $\dfrac{d}{dx}[g(\cot x)]\Big|_{x=\pi/4} = g'(\cot x)\dfrac{d}{dx}\cot x\Big|_{x=\pi/4}$

$= g'(\cot x)(-\csc^2 x)\Big|_{x=\pi/4} = -g'(1)\csc^2\dfrac{\pi}{4} = -(-4) \cdot 2 = 8$

Higher Differentials The first derivative of $f(x)$ with respect to x is denoted by f' or df/dx. The derivative of the first derivative is called the second derivative of $f(x)$ with respect to x and is denoted by f'', $f^{(2)}$, or d^2f/dx^2; and similarly for the higher-order derivatives.

Example Given $f(x) = 3x^3 + 2x + 1$, calculate all derivative values at $x = 3$.

$$\frac{df(x)}{dx} = 9x^2 + 2 \qquad x = 3,\ f'(3) = 9(9) + 2 = 83$$
$$\frac{d^2f(x)}{dx^2} = 18x \qquad x = 3,\ f''(3) = 18(3) = 54$$
$$\frac{d^3f(x)}{dx^3} = 18 \qquad x = 3,\ f'''(3) = 18$$
$$\frac{d^nf(x)}{dx^n} = 0 \qquad \text{for } n \geq 4$$

If $f'(x) > 0$ on (a, b), then f is increasing on (a, b). If $f'(x) < 0$ on (a, b), then f is decreasing on (a, b).

The graph of a function $y = f(x)$ is concave up if f' is increasing on (a, b); it is concave down if f' is decreasing on (a, b).

If $f''(x)$ exists on (a, b) and if $f''(x) > 0$, then f is concave up on (a, b). If $f''(x) < 0$, then f is concave down on (a, b).

An inflection point is a point at which a function changes the direction of its concavity.

Indeterminate Forms: L'Hospital's Theorem Forms of the type $0/0$, ∞/∞, $0 \times \infty$, etc., are called indeterminates. To find the limiting values that the corresponding functions approach, L'Hospital's theorem is useful: If two functions $f(x)$ and $g(x)$ both become zero at $x = a$, then the limit of their quotient is equal to the limit of the quotient of their separate derivatives, if the limit exists or is $+\infty$ or $-\infty$.

Example Find $\lim\limits_{n \to 0} \dfrac{\sin x}{x}$.

Here $\lim\limits_{x \to 0} \dfrac{\sin x}{x} = \lim\limits_{x \to 0} \dfrac{d \sin x}{dx} = \lim\limits_{x \to 0} \dfrac{\cos x}{1} = 1$

Example Find $\lim\limits_{x \to \infty} \dfrac{(1.1)^x}{x^{1000}}$.

$$\lim_{x \to \infty} \frac{(1.1)^x}{x^{1000}} = \lim_{x \to \infty} \frac{d(1.1)^x}{dx^{1000}} = \lim_{x \to \infty} \frac{(\ln 1.1)(1.1)^x}{1000xx^{999}}$$

Obviously $\lim_{x\to\infty} \frac{1.1^x}{x^{1000}} = \infty$, since repeated application of the rule will reduce the denominator to a finite number 1000! while the numerator remains infinitely large.

Example Find $\lim_{x\to\infty} x^3 e^{-x}$.

$$\lim_{x\to\infty} x^3 e^{-x} = \lim_{x\to\infty} \frac{x^3}{e^x} = \lim_{x\to\infty} \frac{6}{e^x} = 0$$

Example Find $\lim (1 - x)^{1/x}$.

Let
$$y = (1 - x)^{1/x}$$
$$\ln y = (1/x) \ln (1 - x)$$
$$\lim_{x\to 0} (\ln y) = \lim_{x\to 0} \frac{\ln(1 - x)}{x} = -1$$

Therefore,
$$\lim_{x\to 0} y = e^{-1}$$

Example Find $\lim_{x\to 1} \left(\frac{1}{x^2 - 1} - \frac{1}{x - 1}\right)$

$$= \lim_{x\to 1} \frac{-x^2 + x}{x^3 - x^2 - x + 1} = \lim_{x\to 1} \frac{-2}{6x - 2} = -\frac{1}{2}$$

Partial Derivative The abbreviation $z = f(x, y)$ means that z is a function of the two variables x and y. The derivative of z with respect to x, treating y as a constant, is called the partial derivative with respect to x and is usually denoted as $\partial z/\partial x$ or $\partial f(x, y)/\partial x$ or simply f_x. Partial differentiation, like full differentiation, is quite simple to apply. Conversely, the solution of partial differential equations is appreciably more difficult than that of differential equations.

Example Find $\partial z/\partial x$ and $\partial z/\partial y$ for $z = ye^{x^2} + xe^y$.

$$\frac{\partial z}{\partial x} = y\frac{\partial e^{x^2}}{\partial x} + e^y \frac{\partial x}{\partial x} = 2xye^{x^2} + e^y \qquad \frac{\partial z}{\partial y} = e^{x^2}\frac{\partial y}{\partial y} + x\frac{\partial e^y}{\partial y} = e^{x^2} + xe^y$$

Order of Differentiation It is generally true that the order of differentiation is immaterial for any number of differentiations or variables provided the function and the appropriate derivatives are continuous. For $z = f(x, y)$ it follows:

$$\frac{\partial^3 f}{\partial y^2\, \partial x} = \frac{\partial^3 f}{\partial y\, \partial x\, \partial y} = \frac{\partial^3 f}{\partial x\, \partial y^2}$$

General Form for Partial Differentiation

1. Given $f(x, y) = 0$ and $x = g(t)$, $y = h(t)$.

Then
$$\frac{df}{dt} = \frac{\partial f}{\partial x}\frac{dx}{dt} + \frac{\partial f}{\partial y}\frac{dy}{dt}$$
$$\frac{d^2 f}{dt^2} = \frac{\partial^2 f}{\partial x^2}\left(\frac{dx}{dt}\right)^2 + 2\frac{\partial^2 f}{\partial x\, \partial y}\frac{dx}{dt}\frac{dy}{dt} + \frac{\partial^2 f}{\partial y^2}\left(\frac{dy}{dt}\right)^2 + \frac{\partial f}{\partial x}\frac{d^2 x}{dt^2} + \frac{\partial f}{\partial y}\frac{d^2 y}{dt^2}$$

Example Find df/dt for $f = xy$, $x = \rho \sin t$, $y = \rho \cos t$.

$$\frac{df}{dt} = \frac{\partial(xy)}{\partial x}\left(\frac{d\,\rho \sin t}{dt}\right) + \frac{\partial(xy)}{\partial y}\left(\frac{d\,\rho \cos t}{dt}\right)$$
$$= y(\rho \cos t) + x(-\rho \sin t)$$
$$= \rho^2 \cos^2 t - \rho^2 \sin^2 t$$

2. Given $f(x, y) = 0$ and $x = g(t, s)$, $y = h(t, s)$.

Then
$$\frac{\partial f}{\partial t} = \frac{\partial f}{\partial x}\frac{\partial x}{\partial t} + \frac{\partial f}{\partial y}\frac{\partial y}{\partial t}$$
$$\frac{\partial f}{\partial s} = \frac{\partial f}{\partial x}\frac{\partial x}{\partial s} + \frac{\partial f}{\partial y}\frac{\partial y}{\partial x}$$

Differentiation of Composite Function

Rule 1. Given $f(x, y) = 0$, then $\frac{dy}{dx} = -\frac{\partial f/\partial x}{\partial f/\partial y}\left(\frac{\partial f}{\partial y} \neq 0\right)$.

Example Find dy/dx for $x^2 + y^3 = x + xy + A$. Let $f(x, y) = x^2 + y^3 - x - xy - A = 0$.

$$\partial f/\partial x = 2x - 1 - y \qquad \partial f/\partial y = 3y^2 - x$$
$$dy/dx = (2x - 1 - y)/(3y^2 - x)$$

Rule 2. Given $f(u) = 0$ where $u = g(x)$, then

$$\frac{df}{dx} = f'(u)\frac{du}{dx}$$
$$\frac{d^2 f}{dx^2} = f''(u)\left(\frac{du}{dx}\right)^2 + f'(u)\frac{d^2 u}{dx^2}$$

Example Find df/dx for $f = \sin^2 u$ and $u = \sqrt{1 - x^2}$

$$\frac{df}{dx} = \frac{d \sin^2 u}{du}\frac{d\sqrt{1 - x^2}}{dx}$$
$$= 2 \sin u \cos u \left(\frac{1}{2}\right)(-2x)(1 - x^2)^{-1/2}$$
$$= 2\frac{\sqrt{1 - u^2}}{u} \sin u \cos u$$

Rule 3. Given $f(u) = 0$ where $u = g(x,y)$, then

$$\frac{\partial f}{\partial x} = f'(u)\frac{\partial u}{\partial x} \qquad \frac{\partial f}{\partial y} = f'(u)\frac{\partial u}{\partial y}$$
$$\frac{\partial^2 f}{\partial x^2} = f''\left(\frac{\partial u}{\partial x}\right)^2 + f'\frac{\partial^2 u}{\partial x^2}$$
$$\frac{\partial^2 f}{\partial x\, \partial y} = f''\frac{\partial u}{\partial x}\frac{\partial u}{\partial y} + f'\frac{\partial^2 u}{\partial x\, \partial y}$$
$$\frac{\partial^2 f}{\partial y^2} = f''\left(\frac{\partial u}{\partial y}\right)^2 + f'\frac{\partial^2 u}{\partial y^2}$$

INTEGRAL CALCULUS

Indefinite Integral If $f'(x)$ is the derivative of $f(x)$, an antiderivative of $f'(x)$ is $f(x)$. Symbolically, the indefinite integral of $f'(x)$ is

$$\int f'(x)\, dx = f(x) + c$$

where c is an arbitrary constant to be determined by the problem. By virtue of the known formulas for differentiation the following relationships hold (a is a constant):

$$\int (du + dv + dw) = \int du + \int dv + \int dw \tag{2-42}$$
$$\int a\, dv = a\int dv \tag{2-43}$$
$$\int v^n\, dv = \frac{v^{n+1}}{n+1} + c\ (n \neq -1) \tag{2-44}$$
$$\int \frac{dv}{v} = \ln|v| + c \tag{2-45}$$
$$\int a^v\, dv = \frac{a^v}{\ln a} + c \tag{2-46}$$
$$\int e^v\, dv = e^v + c \tag{2-47}$$
$$\int \sin v\, dv = -\cos v + c \tag{2-48}$$
$$\int \cos v\, dv = \sin v + c \tag{2-49}$$
$$\int \sec^2 v\, dv = \tan v + c \tag{2-50}$$
$$\int \csc^2 v\, dv = -\cot v + c \tag{2-51}$$
$$\int \sec v \tan v\, dv = \sec v + c \tag{2-52}$$

$$\int \csc v \cot v \, dv = -\csc v + c \qquad (2\text{-}53)$$

$$\int \frac{dv}{v^2 + a^2} = \frac{1}{a}\tan^{-1}\frac{v}{a} + c \qquad (2\text{-}54)$$

$$\int \frac{dv}{\sqrt{a^2 - v^2}} = \sin^{-1}\frac{v}{a} + c \qquad (2\text{-}55)$$

$$\int \frac{dv}{v^2 - a^2} = \frac{1}{2a}\ln\left|\frac{v-a}{v+a}\right| + c \qquad (2\text{-}56)$$

$$\int \frac{dv}{\sqrt{v^2 \pm a^2}} = \ln|v + \sqrt{v^2 \pm a^2}| + c \qquad (2\text{-}57)$$

$$\int \sec v \, dv = \ln(\sec v + \tan v) + c \qquad (2\text{-}58)$$

$$\int \csc v \, dv = \ln(\csc v - \cot v) + c \qquad (2\text{-}59)$$

Example Derive $\int a^v\,dv = (a^v/\ln a) + c$. By reference to the differentiation formula $da^v/dv = a^v \ln a$, or in the more usable form $d(a^v/\ln a) = a^v\,dv$, let $f' = a^v\,dv$; then $f = a^v/\ln a$ and hence $\int a^v dv = (a^v/\ln a) + c$.

Example Find $\int(3x^2 + e^x - 10)\,dx$. Using

$$\int(3x^2 + e^x - 10)\,dx = 3\int x^2\,dx + \int e^x\,dx - 10\int dx \qquad (2\text{-}42)$$

$$= x^3 + e^x - 10x + c \qquad (2\text{-}44), (2\text{-}47)$$

Example Find $\int \frac{7x\,dx}{2 - 3x^2}$. Let

$$v = 2 - 3x^2$$
$$dv = -6x\,dx$$

Thus

$$\int \frac{7x\,dx}{2-3x^2} = 7\int \frac{x\,dx}{2-3x^2} = -\frac{7}{6}\int \frac{-6x\,dx}{2-3x^2}$$
$$= -\frac{7}{6}\int \frac{dv}{v}$$
$$= -\frac{7}{6}\ln|v| + c$$
$$= -\frac{7}{6}\ln|2 - 3x^2| + c$$

Example: Constant of Integration By definition the derivative of x^3 is $3x^2$, and x^3 is therefore the integral of $3x^2$. However, if $f = x^3 + 10$, it follows that $f' = 3x^2\,dx$, and $x^3 + 10$ is therefore also the integral of $3x^2$. For this reason the constant c in $\int 3x^2\,dx = x^3 + c$ must be determined by the problem conditions, i.e., the value of f for a specified x.

Methods of Integration In practice it is rare when generally encountered functions can be directly integrated. For example, the integrand in $\int \sqrt{\sin x}\,dx$ which appears quite simple has no elementary function whose derivative is $\sqrt{\sin x}$. In general, there is no explicit way of determining whether a particular function can be integrated into an elementary form. As a whole, integration is a trial-and-error proposition which depends on the effort and ingenuity of the practitioner. The following are general procedures which can be used to find the elementary forms of the integral when they exist. When they do not exist or cannot be found either from tabled integration formulas or directly, the only recourse is series expansion as illustrated later. Indefinite integrals cannot be solved numerically unless they are redefined as definite integrals (see "Definite Integral"), i.e., $F(x) = \int f(x)\,dx$, indefinite, whereas $F(x) = \int_a^x f(t)\,dt$, definite.

Direct Formula Many integrals can be solved by transformation in the integrand to one of the forms given previously.

Example Find $\int x^2\sqrt{3x^3 + 10}\,dx$. Let $v = 3x^3 + 10$ for which $dv = 9x^2\,dx$. Thus

$$\int x^2\sqrt{3x^3+10}\,dx = \int (3x^3+10)^{1/2}(x^2\,dx)$$
$$= \frac{1}{9}\int (3x^3+10)^{1/2}(9x^2\,dx)$$
$$= \frac{1}{9}\int v^{1/2}\,dv$$
$$= \frac{1}{9}\frac{v^{3/2}}{3/2} + c \qquad \text{[by Eq. (2-44)]}$$
$$= \frac{2}{27}(3x^3+10)^{3/2} + c$$

Trigonometric Substitution This technique is particularly well adapted to integrands in the form of radicals. For these the function is transformed into a trigonometric form. In the latter form they may be more easily recognizable relative to the identity formulas. These functions and their transformations are

$$\sqrt{x^2 - a^2} \quad \text{Let } x = a\sec\theta$$
$$\sqrt{x^2 + a^2} \quad \text{Let } x = a\tan\theta$$
$$\sqrt{a^2 - x^2} \quad \text{Let } x = a\sin\theta$$

Example Find $\int \frac{\sqrt{4 - 9x^2}}{x^2}\,dx$. Let $x = \frac{2}{3}\sin\theta$; then $dx = \frac{2}{3}\cos\theta\,d\theta$.

$$3\int \frac{\sqrt{\left(\frac{2}{3}\right)^2 - x^2}}{x^2}\,dx = 3\int \frac{\frac{2}{3}\sqrt{1-\sin^2\theta}}{\left(\frac{2}{3}\right)^2\sin^2\theta}\left(\frac{2}{3}\cos\theta\,d\theta\right)$$
$$= 3\int \frac{\cos^2\theta}{\sin^2\theta}\,d\theta$$
$$= 3\int \cot^2\theta\,d\theta$$
$$= -3\cot\theta - \theta + c \text{ by trigonometric transform}$$
$$= -\frac{\sqrt{4-9x^2}}{x} - 3\sin^{-1}\frac{3}{2}x + c \text{ in terms of } x$$

Algebraic Substitution Functions containing elements of the type $(a + bx)^{1/n}$ are best handled by the algebraic transformation $y^n = a + bx$.

Example Find $\int \frac{x\,dx}{(3+4x)^{1/4}}$. Let $3 + 4x = y^4$; then $4dx = 4y^3\,dy$ and

$$\int \frac{x\,dx}{(3+4x)^{1/4}} = \int \frac{\frac{y^4 - 3}{4}y^3\,dy}{y}$$
$$= \frac{1}{4}\int y^2(y^4 - 3)\,dy$$
$$= \frac{1}{4}\frac{y^7}{7} - \frac{3}{4}\frac{y^3}{3} + c$$
$$= \frac{7}{4}(3+4x)^{7/4} - \frac{1}{4}(3+4x)^{3/4} + c$$

General The number of possible transformations one might use are unlimited. No specific overall rules can be given. Success in handling integration problems depends primarily upon experience and ingenuity. The following example illustrates the extent to which alternative approaches are possible.

Example Find $\int \frac{dx}{e^x - 1}$. Let $e^x = y$; then $e^x\,dx = dy$ or $dx = 1/y\ dy$.

$$\int \frac{dx}{e^x - 1} = \int \frac{(1/y)\,dy}{y - 1} = \int \frac{dy}{y^2 - y} = \ln\frac{y-1}{y} = \ln\frac{e^x - 2}{e^x}$$

Partial Fractions **Rational functions** are of the type $f(x)/g(x)$ where $f(x)$ and $g(x)$ are polynomial expressions of degrees m and n respectively. If the degree of f is higher than g, perform the algebraic division—the remainder will then be at least one degree less than the denominator. Consider the following types:

Type 1. Reducible denominator to linear unequal factors. For example,

$$\frac{1}{x^3 - x^2 - 4x + 4}$$
$$= \frac{1}{(x+2)(x-2)(x-1)}$$
$$= \frac{A}{x+2} + \frac{B}{x-2} + \frac{C}{x-1}$$
$$= \frac{A(x-2)(x-1) + B(x+2)(x-1) + C(x+2)(x-2)}{(x+2)(x-2)(x-1)}$$
$$= \frac{x^2(A+B+C) + x(-3A+B) + (2A-2B-4C)}{(x+2)(x-2)(x-1)}$$

or by equating coefficients and solving for A, B, and C

$$= \frac{1}{12(x+2)} + \frac{1}{4(x-2)} - \frac{1}{3(x-1)}$$

Hence

$$\int \frac{dx}{x^3 - x^2 - 4x + 4} = \int \frac{dx}{12(x+2)} + \int \frac{dx}{4(x-2)} - \int \frac{dx}{3(x-1)}$$

Type 2. Reducible denominator to linear but some equal factors. For example,

$$\frac{x^2 + 6x - 1}{(x-3)^2(x-1)} = \frac{A(x-3)(x-1) + B(x-1) + C(x-3)^2}{(x-3)^2(x-1)}$$

equating as before

$$= -\frac{1}{2(x-3)} + \frac{13}{(x-3)^2} + \frac{3}{2(x-1)}$$

Type 3. Reducible denominator to quadratic factors.

$$\frac{8x^2 + 3}{(x^2 + x + 1)(x-2)} = \frac{Ax + B}{x^2 + x + 1} + \frac{C}{x-2)}$$
$$= \frac{Ax(x-2) + B(x-2) + C(x^2 + x + 1)}{(x^2 + x + 1)(x-2)}$$

by equating coefficients and solving for A, B, and C

$$= \frac{3x+1}{x^2 + x + 1} + \frac{5}{x-2}$$

Parts An extremely useful formula for integration is the relation

$$d(uv) = u\,dv + v\,du$$

and
$$uv = \int u\,dv + \int v\,du$$

or
$$\int u\,dv = uv - \int v\,du$$

No general rule for breaking an integrand can be given. Experience alone limits the use of this technique. It is particularly useful for trigonometric and exponential functions.

Example Find $\int xe^x\,dx$. Let

$$u = x \quad \text{and} \quad dv = e^x\,dx$$
$$du = dx \qquad v = e^x$$

Therefore

$$\int xe^x\,dx = xe^x - \int e^x\,dx$$
$$= xe^x - e^x + c$$

Example Find $\int e^x \sin x\,dx$. Let

$$u = e^x \qquad dv = \sin x\,dx$$
$$du = e^{x\,dx} \qquad v = -\cos x$$

$$\int e^x \sin x\,dx = -e^x \cos x + \int e^x \cos x\,dx$$

Again

$$u = e^x \qquad dv = \cos x\,dx$$
$$du = e^x\,dx \qquad v = \sin x$$

$$\int e^x \sin x\,dx = -e^x \cos x + e^x \sin x - \int e^x \sin x\,dx + c$$
$$= (e^x/2)(\sin x - \cos x) + c$$

Series Expansion When an explicit function cannot be found, the integration can sometimes be carried out by a series expansion.

Example Find $\int e^{-x^2}\,dx$. Since

$$e^{-x^2} = 1 - x^2 + \frac{x^4}{2!} - \frac{x^6}{3!} + \cdots$$

$$\int e^{-x^2}\,dx = \int dx - \int x^2\,dx + \int \frac{x^4}{2!}\,dx - \int \frac{x^6}{3!}\,dx + \cdots$$
$$= x - \frac{x^3}{3} + \frac{x^5}{5.2!} - \frac{x^7}{7.3!} + \cdots \qquad \text{for all } x$$

Definite Integral The concept and derivation of the definite integral are completely different from those for the indefinite integral. These are by definition different types of operations. However, the formal operation $\int$ as it turns out treats the integrand in the same way for both.

Consider the function $f(x) = 10 - 10e^{-2x}$. Definite $x_1 = a$ and $x_n = b$, and suppose it is desirable to compute the area between the curve and the coordinate axis $y = 0$ and bounded by $x_1 = a$, $x_n = b$. Obviously, by a sufficiently large number of rectangles this area could be approximated as closely as desired by the formula

$$\sum_{i=1}^{n-1} f(\xi_i)(x_{i+1} - x_i) = f(\xi_1)(x_2 - a) + f(\xi_2)(x_3 - x_2)$$
$$+ \cdots + f(\xi_{n-1})(b - x_{n-1}) \qquad x_{i-1} \le \xi_{i-1} \le x_i$$

The definite integral of $f(x)$ is defined as

$$\int_a^b f(x)\,dx = \lim_{n\to\infty} \sum_{i=1}^{n} f(\xi_i)(x_{i+1} - x_i)$$

where the points $x_1, x_2, \ldots, x_n$ are equally spaced. For a rigorous definition of the definite integral the references should be consulted.

Thus, the value of a definite integral depends on the limits a, b and any selected variable coefficients in the function but not on the **dummy variable** of integration x. Symbolically

$$F(x) = \int f(x)\,dx \qquad \text{indefinite integral where } dF/dx = f(x)$$

$$\text{or } F(a, b) = \int_a^b f(x)\,dx \qquad \text{definite integral}$$

$$F(\alpha) = \int_a^b f(x, \alpha)\,dx$$

There are certain restrictions of the integration definition, "The function $f(x)$ must be continuous in the finite interval (a, b) with at most a finite number of finite discontinuities," which must be observed before integration formulas can be generally applied. Two of these restrictions give rise to so-called **improper integrals** and require special handling. These occur when

1. The limits of integration are not both finite, i.e., $\int_0^\infty e^{-x}\,dx$.
2. The function becomes infinite within the interval of integration, i.e., $\int_0^1 \frac{1}{\sqrt{x}}\,dx$.

Techniques for determining when integration is valid under these conditions are available in the references. However, the following simplified rules will, in general, serve as a guide for most practical applications.

Rule 1. For the integral $\int_0^\infty \frac{\phi(x)}{x^n}\,dx$, if $\phi(x)$ is bounded, the integral will converge for $n > 1$ and not converge for $n \le 1$.

It is easily seen that $\int_0^\infty e^{-x}\,dx$ converges by reference to the summation formula and noting $1/x^2 > 1/e^x > 0$ for large x.

Rule 2. For the integral $\int_a^b \frac{\phi(x)}{(a-x)^n}\,dx$, if $\phi(x)$ is bounded, the integral will converge for $n < 1$ and diverge for $n \ge 1$. Thus $\int_0^1 \frac{1}{\sqrt{x}}\,dx$ will converge (exist) since $1/2 < n < 1$.

Properties The fundamental theorem of calculus states

$$\int_a^b f(x)\,dx = F(b) - F(a) \tag{2-60}$$

where $\quad dF(x)/dx = f(x)$

Other properties of the definite integral are

$$\int_a^b c[f(x)\,dx] = c\int_a^b f(x)\,dx \tag{2-61}$$

$$\int_a^b [f_1(x) + f_2(x)]\,dx) = \int_a^b f_1(x)\,dx + \int_a^b f_2(x)\,dx \tag{2-62}$$

$$\int_a^b f(x)\,dx = -\int_b^a f(x)\,dx \tag{2-63}$$

$$\int_a^b f(x)\,dx = \int_a^c f(x)\,dx + \int_c^b f(x)\,dx \tag{2-64}$$

$$\int_a^b f(x)\,dx = (b-a)f(\xi) \qquad \text{for some } \xi \text{ in } (a, b) \tag{2-65}$$

$$\frac{\partial}{\partial b}\int_a^b f(x)\,dx = f(b) \tag{2-66}$$

$$\frac{\partial}{\partial a}\int_a^b f(x)\,dx = -f(a) \tag{2-67}$$

$$\frac{dF(\alpha)}{d\alpha} = \int_a^b \frac{\partial f(x,\alpha)}{\partial \alpha}\,dx \text{ if } a \text{ and } b \text{ are constant} \tag{2-68}$$

$$\int_a^b dx \int_c^d f(x,\alpha)d\alpha = \int_c^d d\alpha \int_a^b f(x,\alpha)\,dx \tag{2-69}$$

Example Find $\int_0^{\pi/2} \sin x\,dx$.

$$\int_0^{\pi/2} \sin x\,dx = \Big[-\cos x\Big]_0^{\pi/2} = -\left(\cos\frac{\pi}{2} - \cos 0\right) = 1$$

since $\qquad -d\cos x/dx = \sin x$

Example Find $\int_0^2 \frac{dx}{(x-1)^2}$. Direct application of the formula would yield the incorrect value

$$\int_0^2 \frac{dx}{(x-1)^2} = \left[-\frac{1}{x-1}\right]_0^2 = -2$$

It should be noted that $f(x) = 1/(x-1)^2$ becomes unbounded as $x \to 1$ and by Rule 2 the integral diverges and hence is said not to exist.

Methods of Integration All the methods of integration available for the indefinite integral can be used for definite integrals. In addition, several others are available for the latter integrals and are indicated below.

Change of Variable This substitution is basically the same as previously indicated for indefinite integrals. However, for definite integrals the limits of integration must also be changed: i.e., for $x = \phi(t)$,

$$\int_a^b f(x)\,dx = \int_{t_0}^{t_1} f[\phi(t)\phi'(t)\,dt \tag{2-70}$$

where $t = t_0$ when $x = a$
$\qquad t = t_1$ when $x = b$

Example Find $\int_0^4 \sqrt{16 - x^2}\,dx$. Let

$$x = 4\sin\theta \qquad x = 0 \quad \theta = 0$$
$$dx = 4\cos\theta\,d\theta \qquad x = 4 \quad \theta = \pi/2$$

Then

$$\int_0^4 \sqrt{16 - x^2}\,dx = 16\int_0^{\pi/2} \cos^2\theta\,d\theta = 16[\tfrac{1}{2}\theta + \tfrac{1}{4}\sin 2\theta]_0^{\pi/2} = 4\pi$$

Differentiation Here the application of the general rules for differentiating under the integral sign may be useful.

Example Find $\phi(\alpha) = \int_0^\infty \frac{e^{-\alpha x}\sin x}{x}\,dx$ $(\alpha > 0)$. Since this is a continuous function of α, it may be differentiated under the integral sign

$$\frac{d\phi}{d\alpha} = -\int_0^\infty e^{-\alpha x}\sin x\,dx$$
$$= 1/(1 + \alpha^2)$$
$$\phi(\alpha) = -\tan^{-1}\alpha + c$$

and since $\phi(\alpha) \to 0$ as $\alpha \to \infty$,

$$c = \pi/2$$
$$\phi(\alpha) = -\tan^{-1}\alpha + \pi/2$$

Integration It is sometimes useful to generate a double integral to solve a problem. By this approach the fundamental theorem indicated by Eq. (2-69) can be used.

Example Find $\int_0^1 \frac{x^b - x^a}{\ln x}\,dx$. Consider $\int_0^1 x^\alpha\,dx = \frac{1}{\alpha+1}$ $(\alpha > -1)$. Then multiplying both sides by $d\alpha$ and integrating between a and b,

$$\int_a^b d\alpha \int_0^1 x^\alpha\,dx = \int_a^b \frac{d\alpha}{\alpha+1} = \ln\left|\frac{b+1}{a+1}\right|$$

But also

$$\int_a^b d\alpha \int_0^1 x^\alpha\,dx = \int_0^1 dx \int_a^b x^\alpha\,d\alpha = \int_0^1 \frac{x^b - x^a}{\ln x}\,dx$$

Therefore

$$\int_0^1 \frac{x^b - x^a}{\ln x}\,dx = \ln\left|\frac{b+1}{a+1}\right|$$

Complex Variable Certain definite integrals can be evaluated by the technique of complex variable integration. This is described in the references for "Complex Variables."

Numerical Because of the property of definite integrals another method for obtaining their solution is available which cannot be applied to indefinite integrals. This involves a numerical approximation based on the previously outlined summation definition:

$$\lim_{n\to\infty} \sum_1^{n-1} f(\xi_i)(x_{i+1} - x_i) = \int_a^b f(x)\,dx \tag{2-71}$$

where $x_1 = a \quad$ and $\quad x_n = b$

Examples of this procedure are given in the subsection "Numerical Analysis and Approximate Methods."

INFINITE SERIES

REFERENCES: 11, 40, 48, 122, 123, 167. For asymptotic series and asymptotic methods, see Refs. 45, 123.

DEFINITIONS

A succession of numbers or terms which are formed according to some definite rule is called a **sequence.** The indicated sum of the terms of a sequence is called a **series.** A series of the form $a_0 + a_1 (x - c) + a_2 (x - c)^2 + \cdots + a_n (x - c)^n + \cdots$ is called a **power series.**

Consider the sum of a finite number of terms in the geometric series (a special case of a power series).

$$S_n = a + ar + ar^2 + ar^3 + \cdots + ar^{n-1} \qquad (2\text{-}72)$$

It has been shown that for any number of terms n their sum is equal to the following:

$$S_n = a\frac{1 - r^n}{1 - r}$$

In this form the geometric series is assumed finite.

In the form of Eq. (2-72) it can further be defined that the terms in the series be nonending and therefore an infinite series.

$$S = a + ar + ar^2 + \cdots + ar^n + \cdots \qquad (2\text{-}73)$$

However, the defined sum of the terms [Eq. (2-72)]

$$S_n = a\frac{1 - r^n}{1 - r} \qquad r \neq 1$$

while valid for any finite value of r and n now takes on a different interpretation. In this sense it is necessary to consider the limit of S_n as n increases indefinitely:

$$S = \lim_{n\to\infty} S_n$$
$$= a \lim_{n\to\infty} \frac{1 - r^n}{1 - r}$$

For this, it is stated the infinite series **converges** if the limit of S_n approaches a fixed finite value as n approaches infinity. Otherwise, the series is **divergent.**

On this basis an analysis of $S = a\lim_{n\to\infty} \frac{1 - r^n}{1 - r}$ shows that if r is less than 1 but greater than -1, the infinite series is convergent. For values outside of the range $-1 < r < 1$ the series is divergent because the sum is not defined. The range $-1 < r < 1$ is called the **region of convergence.** (We assume $a \neq 0$.)

Consider the divergence of Eq. (2-73) when $r = -1$ and $+1$. For the former case $r = -1$,

$$S = a + a(-1) + a(-1)^2 + a(-1)^3 + \cdots + a(-1)^n + \cdots$$
$$= a - a + a - a + a - \cdots$$

and for which

$$S = a \lim_{n\to\infty} \frac{1 - r^n}{1 - r}$$
$$= a \lim_{n\to\infty} \frac{1 - (-1)^n}{1 + 1} \qquad \text{undefined limit (if } a \neq 0)$$

Since the limit sum does not exist, the series is divergent. This is defined as a bounded or **oscillating divergent series.** Similarly for the value $r = +1$,

$$S = a + a(1) + a(1)^2 + a(1)^3 + \cdots + a(1)^n + \cdots$$
$$S = a + a + a + a + \cdots + a + \cdots \qquad (a \neq 0)$$

The series is also divergent but defined as an **unbounded divergent series.**

There are also two types of convergent series. Consider the new series

$$S = 1 - \frac{1}{2} + \frac{1}{3} - \frac{1}{4} + \cdots + (-1)^{n+1}\frac{1}{n} + \cdots \qquad (2\text{-}74)$$

It can be shown that the series (2-74) does converge to the value $S = \log 2$. However, if each term is replaced by its absolute value, the series becomes unbounded and therefore divergent (unbounded divergent):

$$S = 1 + \frac{1}{2} + \frac{1}{3} + \frac{1}{4} + \frac{1}{5} + \cdots \qquad (2\text{-}75)$$

In this case the series (2-74) is defined as a **conditionally convergent** series. If the replacement series of absolute values also converges, the series is defined to **converge absolutely.**

Series (2-74) is further defined as an **alternating series,** while series (2-75) is referred to as a **positive series.**

OPERATIONS WITH INFINITE SERIES

1. The convergence or divergence of an infinite series is unaffected by the removal of a finite number of finite terms. This is a trivial theorem but useful to remember, especially when using the comparison test to be described in the subsection "Tests for Convergence and Divergence."

2. If a series is conditionally convergent, its sums can be made to have any arbitrary value by a suitable rearrangement of the series; it can in fact be made divergent or oscillatory (Riemann's theorem). This seemingly paradoxical theorem can be illustrated by the following example.

Example

$$S = 1 - \frac{1}{2} + \frac{1}{3} - \frac{1}{4} + \frac{1}{5} - \frac{1}{6} + \cdots$$

The series is rearranged so that each positive term is followed by two negative terms:

$$t = 1 - \frac{1}{2} - \frac{1}{4} + \frac{1}{3} - \frac{1}{6} - \frac{1}{8} + \frac{1}{5} - \frac{1}{10} - \frac{1}{12} + \cdots$$

Define t_{3n} for the first $3n$ terms in the series

$$t_{3n} = \left(1 - \frac{1}{2}\right) - \frac{1}{4} + \left(\frac{1}{3} - \frac{1}{6}\right) - \frac{1}{8} + \cdots + \left(\frac{1}{2n - 1} - \frac{1}{4n - 2}\right) - \frac{1}{4n}$$
$$= \frac{1}{2} - \frac{1}{4} + \frac{1}{6} - \frac{1}{8} + \cdots + \frac{1}{4n - 2} - \frac{1}{4n}$$
$$= \frac{1}{2}\left(1 - \frac{1}{2} + \frac{1}{3} - \frac{1}{4} + \cdots + \frac{1}{2n - 1} - \frac{1}{2n}\right)$$
$$= \frac{1}{2} S_{2n}$$

where S_{2n} is the sum of the first $2n$ terms of the original series. Thus

$$\lim_{n\to\infty} t_{3n} = \lim_{n\to\infty} \frac{1}{2} S_{2n}$$
$$= \frac{1}{2} S$$

and since $\lim t_{3n+2} = \lim t_{3n+1} = \lim t_{3n}$, it follows the sum of the series t is (½) S. Hence a rearrangement of the terms of an alternating series alters the sum of the series.

3. A series of positive terms, if convergent, has a sum independent of the order of its terms; but if divergent, it remains divergent however its terms are rearranged.

4. An oscillatory series can always be made to converge by grouping the terms in brackets.

Example Consider the series

$$1-\frac{1}{2}+\frac{2}{3}-\frac{3}{4}+\frac{4}{5}-\frac{5}{6}+\cdots$$

which oscillates between the values 0.306 and 1.306. However, the series

$$\left(1-\frac{1}{2}\right)+\left(\frac{2}{3}-\frac{3}{4}\right)+\left(\frac{4}{5}-\frac{5}{6}\right)+\cdots=\frac{1}{2}-\frac{1}{12}-\frac{1}{30}-\frac{1}{56}-\cdots$$

$$\cong 0.306\cdots \quad \text{and} \quad 1-\left(\frac{1}{2}-\frac{2}{3}\right)-\left(\frac{3}{4}-\frac{4}{5}\right)-\left(\frac{5}{6}-\frac{6}{7}\right)+\cdots$$

$$=1+\frac{1}{6}+\frac{1}{20}+\frac{1}{42}+\cdots=1.306\cdots$$

5. A power series can be inverted, provided the first-degree term is not zero. Given

$$y=b_1x+b_2x^2+b_3x^3+b_4x^4+b_5x^5+b_6x^6+b_7x^7+\cdots$$

then

$$x=B_1y+B_2y^2+B_3y^3+B_4y^4+B_5y^5+B_6y^6+B_7y^7+\cdots$$

where

$$B_1=1/b_1$$
$$B_2=b_2/b_1^3$$
$$B_3=(1/b_1^5)(2b_2^2-b_1b_3)$$
$$B_4=(1/b_1^7)(5b_1b_2b_3-b_1^2b_4-5b_2^3)$$

Additional coefficients are available in the references.

6. Two series may be added or subtracted term by term provided each is a convergent series. The joint sum is equal to the sum (or difference) of the individuals.

7. The sum of two divegent series can be convergent. Similarly, the sum of a convergent series and a divergent series must be divergent.

Example Given

$$\sum_{n=1}^{\infty}\left(\frac{1+n}{n^2}\right)=\frac{2}{1}+\frac{3}{4}+\frac{4}{9}+\frac{5}{16}+\cdots \qquad \text{a divergent series}$$

$$\sum_{n=1}^{\infty}\left(\frac{1-n}{n^2}\right)=-\frac{1}{4}-\frac{2}{9}-\frac{3}{16}+\cdots \qquad \text{a divergent series}$$

However,

$$\sum\left(\frac{1+n}{n^2}\right)+\sum\left(\frac{1-n}{n^2}\right)=\sum\left(\frac{1+n+1-n}{n^2}\right)$$

$$=2\sum\frac{1}{n^2} \qquad \text{convergent}$$

8. A power series may be integrated term by term to represent the integral of the function within an interval of the region of convergence. If $f(x)=a_0+a_1x+a_2x^2+\cdots$, then

$$\int_{x1}^{x2}f(x)\,dx=\int_{x1}^{x2}a_0\,dx+\int_{x1}^{x2}a_1x\,dx+\int_{x1}^{x2}a_2x^2\,dx+\cdots$$

9. A power series may be differentiated term by term and represents the function $df(x)/dx$ within the same region of convergence as $f(x)$.

TESTS FOR CONVERGENCE AND DIVERGENCE

In general, the problem of determining whether a given series will converge or not can require a great deal of ingenuity and resourcefulness. There is no all-inclusive test which can be applied to all series. As the only alternative, it is necessary to apply one or more of the developed theorems in an attempt to ascertain the convergence or divergence of the series under study. The following defined tests are given in relative order of effectiveness:

1. Comparison Test. A series will converge if the absolute value of each term (with or without a finite number of terms) is less than the corresponding term of a known convergent series. Similarly, a positive series is divergent if it is termwise larger than a known divergent series of positive terms.

Example: Need for Comparison with a Positive Series

Given: $$1-\frac{1}{1\times 2}+\frac{1}{2}-\frac{1}{2\times 3}+\frac{1}{3}-\frac{1}{3\times 4}+\cdots$$

Reference: $$1-1+\frac{1}{2}-\frac{1}{2}+\frac{1}{3}-\frac{1}{3}+\cdots$$

The reference series is known to converge to zero. The absolute value of each term of the reference series is numerically greater than or equal to that of the given series, and the former series converges to zero. However, the given series does diverge since

$$1-\frac{1}{1\times 2}=\frac{1}{2}$$
$$\frac{1}{2}-\frac{1}{2\times 3}=\frac{1}{3}$$
$$\frac{1}{3}-\frac{1}{3\times 4}=\frac{1}{4} \qquad \text{etc.}$$

or

$$\frac{1}{2}+\frac{1}{3}+\frac{1}{4}+\frac{1}{5}+\cdots$$

Example: Convergent Series

Given: $$1+\frac{2}{3}+\frac{3}{3^2}+\frac{4}{3^3}+\frac{5}{3^4}+\cdots+\frac{n}{3^{n-1}}+\cdots$$

Reference: $$1+\frac{1}{2}+\frac{1}{4}+\frac{1}{8}+\frac{1}{16}+\cdots+\frac{1}{2^{n-1}}\cdots$$

$$\left(\text{Geometrical series with } a=1,\ r=\frac{1}{2}\right)$$

Since the convergence or divergence of a series is unaffected by neglecting a finite number of terms, the comparison test for the given series can be made without the first term:

$$\begin{array}{cccccccc} 1 & \frac{2}{3} & \frac{3}{3^2} & \frac{4}{3^3} & \frac{5}{3^4} & \cdots & \left(\frac{1}{3}\right)^{n-1} \\ & \updownarrow & \updownarrow & \updownarrow & \updownarrow & & \updownarrow \\ & 1 & \frac{1}{2} & \frac{1}{4} & \frac{1}{8} & \cdots & \frac{1}{2^{n-2}} \end{array}$$

In this case $\frac{2}{3}<1,\ \frac{3}{9}<\frac{1}{2},\ \frac{4}{27}<\frac{1}{4},\ \frac{5}{81}<\frac{1}{8}$, and similarly for all higher-order terms since

$$\frac{n}{3^{n-1}}<\frac{1}{2^{n-2}} \qquad \text{for } n>2$$

Therefore, since the comparison series converges, the given series must also converge. Further, the limit of the given series must be less than 3 as follows:

$$S=1+\frac{1}{2}+\frac{1}{4}+\frac{1}{8}+\cdots=\frac{1}{1-r}=2$$

$$1+\frac{2}{3}+\frac{3}{3^2}+\frac{4}{3^3}+\cdots<1+S=3 \qquad \text{where } S=2$$

2. *n*th-Term Test. A series is divergent if the nth term of the series does not approach zero as n becomes increasingly large.

Example: Divergent Series Given

$$2-\frac{2^2}{2^4}+\frac{2^3}{3^4}-\frac{2^4}{4^4}+\frac{2^5}{5^4}+\cdots+(-1)^{n+1}\frac{2^n}{n^4}$$

A check calculation of the first few terms would indicate a rapid decrease in numerical value for the successive terms. This could lead to the false impression that the series was convergent. This theorem is a negative test (a test of whether the series is divergent only) but does show, as in this example, the divergence of a series, i.e.,

$$\lim_{n\to\infty}\frac{2^n}{n^4} = \lim_{n\to\infty}\frac{(d^4/dn^4)(2^n)}{(d^4/dn^4)(n^4)}$$

$$= \lim_{n\to\infty}\frac{(\ln 2)^4 2^n}{24} \quad \text{by L'Hospital's theorem}$$

(see "Indeterminate Forms")

$$= \infty$$

Example: Inconclusive Test Given

$$\ln\frac{1}{2} + \ln\frac{2}{3} + \ln\frac{3}{4} + \ln\frac{4}{5} + \cdots + \ln\frac{n}{1+n} + \cdots$$

Under the conditions of the *n*th-term test, application to this series is inconclusive. Although $\lim_{n\to\infty} \ln[n/(1+n)]$ does equal zero, the converse conclusion based on the theorem cannot be inferred. In this example the series is actually divergent but cannot be demonstrated as divergent by this test.

3. Ratio Test. If the absolute ratio of the $(n+1)$ term divided by the *n*th term as *n* becomes unbounded approaches

a. a number less than 1, the series is convergent.
b. a number greater than 1, the series is divergent.
c. a number equal to 1, the test is inconclusive.

Example: Convergent Series Given

$$1 + \frac{2^5}{2!} + \frac{3^5}{3!} + \frac{4^5}{4!} + \cdots + \frac{n^5}{n!} + \frac{(n+1)^5}{(n+1)!} + \cdots$$

Applying the ratio test,

$$\lim_{n\to\infty}\frac{(n+1)^5/(n+1)!}{n^5/n!} = \lim_{n\to\infty}\left(\frac{n+1}{n}\right)^5 \frac{n!}{(n+1)!}$$

$$= \lim_{n\to\infty}\left(\frac{n+1}{n}\right)^5\left(\frac{1}{n+1}\right)$$

$$= \lim_{n\to\infty}\left(\frac{n+1}{n}\right)^5 \lim_{n\to\infty}\left(\frac{1}{n+1}\right)$$

$$= 1 \times 0 = 0$$

Therefore the series converges under statement *a*.

Example: Divergent Series Given

$$\frac{2}{1\times 2} + \frac{2^2}{2\times 3} + \frac{2^3}{3\times 4} + \cdots + \frac{2^n}{n(n+1)} + \frac{2^{n+1}}{(n+1)(n+2)} + \cdots$$

By the ratio test

$$\lim_{n\to\infty}\frac{2^{n+1}/(n+1)(n+2)}{2^n/(n)(n+1)} = \lim_{n\to\infty}\frac{2^{n+1}}{2^n}\frac{n(n+1)}{(n+2)(n+1)}$$

$$= \lim_{n\to\infty} 2 \lim_{n\to\infty}\frac{n}{n+2}$$

$$= 2 \times 1 = 2$$

Thus by statement *b* the series is divergent.

Example: Inconclusive Test Given

$$1 + \frac{1}{2^r} + \frac{1}{3^r} + \frac{1}{4^r} + \cdots + \frac{1}{n^r} + \frac{1}{(n+1)^r} + \cdots$$

Again

$$\lim_{n\to\infty}\frac{1/(n+1)^r}{1/n^r} = \lim_{n\to\infty}\left(\frac{n}{n+1}\right)^r$$

$$= 1 \text{ for any value of } r$$

Under *c* the ratio test fails and no conclusion can be stated. However, as will be indicated, this series converges if $r > 1$ and diverges if $r \le 1$.

4. Summation Test. If the partial summation S_n of a series converges as *n* becomes unbounded, the series converges, and diverges if S_n diverges.

Example: Convergent Series Given

$$1 + \frac{2}{3} + \frac{2^2}{3^2} + \frac{2^3}{3^3} + \cdots + \frac{2^n}{3^n} + \cdots$$

From the geometric-progression formula

$$S_n = 1 + \frac{2}{3} + \frac{2^2}{3^2} + \cdots + \left(\frac{2}{3}\right)^{n-1}$$

$$= \frac{1-(2/3)^n}{1-2/3}$$

and

$$= 3[1-(2/3)^n]$$

$$\lim_{n\to\infty} S_n = 3$$

Hence the series converges.

Example: *Divergent Series* Given

$$\ln\frac{1}{1+1} + \ln\frac{2}{2+1} + \ln\frac{3}{3+1} + \cdots + \ln\frac{n}{n+1} + \cdots$$

The partial sum S_n is given by

$$S_n = \ln\frac{1}{1+1} + \ln\frac{2}{2+1} + \cdots + \ln\frac{n}{n+1}$$

$$= \ln\left(\frac{1}{1+1}\cdot\frac{2}{2+1}\cdots\frac{n}{n+1}\right)$$

$$= \ln\frac{1}{n+1}$$

and since $\lim_{n\to\infty} S_n = -\infty$, the series diverges.

Example: Divergent Series Given

$$1 + \frac{1}{2} + \frac{1}{3} + \frac{1}{4} + \frac{1}{5} + \cdots + \frac{1}{n} + \cdots$$

It can be shown for

$$S_n = 1 + \frac{1}{2} + \frac{1}{3} + \cdots + \frac{1}{n}$$

that

$$S_n > \ln n + \frac{1}{2n} \quad \text{for all } n$$

Therefore, since $\lim_{n\to\infty} [\ln n + (1/2n)]$ diverges, the given series also diverges.

5. Alternating-Series Test. If the terms of a series are alternately positive and negative and never increase in numerical value, the series will converge, provided that the terms tend to zero as a limit.

6. Cauchy's Root Test. If the *n*th root of the absolute value and the *n*th term, as *n* becomes unbounded, approaches

a. a number less than 1, the series is convergnet.
b. a number greater than 1, the series is divergent.
c. a number equal to 1, the test is inconclusive.

Example: Convergent Series Given

$$\frac{1}{2} + \left(\frac{2}{2+1}\right)^{2^2} + \left(\frac{3}{3+1}\right)^{3^2} + \cdots + \left(\frac{n}{n+1}\right)^{n^2} + \cdots$$

The application of the *n*th root test results in

$$\lim_{n\to\infty}\sqrt[n]{\left(\frac{n}{n+1}\right)^{n^2}} = \lim_{n\to\infty}\left[\left(\frac{n}{n+1}\right)^{n^2}\right]^{1/n}$$

$$= \lim_{n\to\infty}\left(\frac{n}{n+1}\right)^n$$

$$= 1/e$$

Since $1/e$ is less than 1, the series converges.

Example: Divergent Series Given

$$1 + \frac{3^2}{2^{10}} + \frac{3^3}{3^{10}} + \frac{3^4}{4^{10}} + \cdots \frac{3^n}{n^{10}} + \cdots$$

By the root test the *n*th-root limit of the *n*th term is taken as follows:

$$\lim_{n\to\infty}\sqrt[n]{\frac{3^n}{n^{10}}} = \lim_{n\to\infty}\left(\frac{3^n}{n^{10}}\right)^{1/n}$$

$$= \lim_{n\to\infty}(3)\lim_{n\to\infty}(n^{-10/n})$$

$$= 3$$

Therefore the series diverges. The fact that $\lim_{n\to\infty} (n^{-10/n}) = 1$ is easily verified by considering

$$X = n^{10/n}$$

$$\log X = \frac{10}{n} \log n$$

$$\lim_{n\to\infty} \log X = \lim_{n\to\infty} \frac{10}{n} \log n = 0$$

Hence $$\lim_{n\to\infty} (n^{10/n}) = 1$$

Example: Inconclusive Test Given

$$1 + \frac{1}{2^r} + \frac{1}{3^r} + \frac{1}{4^r} + \cdots + \frac{1}{n^r} + \cdots$$

By the root test

$$\lim_{n\to\infty} \sqrt[n]{1/n^r} = \lim_{n\to\infty} 1/n^{r/n}$$

$$= 1 \text{ for all } r$$

and therefore the test result is inconclusive. It will be shown that the series does converge if $r > 1$.

7. Maclaurin's Integral Test. Suppose Σa_n is a series of positive terms and f is a continuous decreasing function such that $f(x) \geq 0$ for $1 \leq x < \infty$ and $f(n) = a_n$. Then the series and the improper integral $\int_1^\infty f(x)\,dx$ either both converge or both diverge.

Example Given

$$1 + \frac{1}{2^r} + \frac{1}{3^r} + \frac{1}{4^r} + \cdots + \frac{1}{n^r} + \cdots$$

For this example other tests are useful for certain regions of r, but all break down when $r > 1$. The determination of the region of convergence in this case will be derived on this basis.

It follows that if $r \leq 0$, then the criterion $\lim_{n\to\infty} a_n = 0$ is not met by test 2 and therefore the series diverges for $r \leq 0$.

Define $f(x) = 1/x^r$ which is continuous, decreasing, and the $\lim_{x\to\infty} 1/x^r = 0$ for $r > 0$. Thus, for $r \neq 1$*

$$\int_1^\infty \frac{1}{x^r} = \lim_{n\to\infty} \int_1^n \frac{1}{x^r}\,dx$$

$$= \lim_{n\to\infty} \left[\left(\frac{1/1-r}{x^{r-1}}\right)\Big|_1^n\right]$$

$$= \lim_{n\to\infty} \left[\frac{1}{1-r}\frac{1}{n^{r-1}} - \frac{1}{1-r}\right]$$

$$= \lim_{n\to\infty} \left[\frac{1}{r-1}\left(1 - \frac{1}{n^{r-1}}\right)\right]$$

The limit exists (finite) when $r > 1$, and therefore the series converges. For $0 < r < 1$ the limit does not exist and the series diverges. Further, when $r = 1$,

$$\int_1^\infty \frac{1}{x}\,dx = \lim_{n\to\infty} \int_1^n \frac{1}{x}\,dx$$

$$= \lim_{n\to\infty} [\ln n]$$

$$= \infty \quad \text{which implies divergence in the case } r = 1$$

Other tests of a more specialized and specific nature are available. However, these require more detailed mathematical definitions and recourse to advanced concepts such as upper and lower bounds. The indicated references should be consulted for details on their application and use.

*General integration identity not valid for $r = 1$.

SERIES SUMMATION AND IDENTITIES

Sums for the First n Numbers to Integer Powers

$$\sum_{j=1}^{n} j = \frac{n(n+1)}{2} = 1 + 2 + 3 + 4 + \cdots + n$$

$$\sum_{j=1}^{n} j^2 = \frac{n(n+1)(2n+1)}{6} = 1^2 + 2^2 + 3^2 + 4^2 + \cdots + n^2$$

$$\sum_{j=1}^{n} j^3 = \frac{n^2(n+1)^2}{4} = 1^3 + 2^3 + 3^3 + \cdots + n^3$$

$$\sum_{j=1}^{n} j^4 = \frac{n(n+1)(2n+1)(3n^2+3n-1)}{30} = 1^4 + 2^4 + 3^4 + \cdots + n^4$$

Example Find $1^4 + 2^4 + 3^4 + \cdots + 1000^4$.

$$\sum_{1}^{1000} n^4 = \frac{1000(1001)(2001)(3{,}000{,}000 + 3000 - 1)}{30}$$

$$= 200{,}500{,}333{,}333{,}300$$

Example Find $1^4 + 3^4 + 5^4 + 7^4 + \cdots + 999^4 = \sum_{1}^{500} (2n-1)^4$.

$$\sum_{1}^{500} (2n-1)^4 = 1^4 + 3^4 + 5^4 + \cdots + 999^4$$

$$= 1^4 + 2^4 + 3^4 + 4^4 + 5^4 + \cdots + 1000^4 - (2^4 + 4^4 + 6^4 + \cdots + 1000^4)$$

$$= \sum_{1}^{1000} n^4 - \sum_{1}^{500} (2n)^4$$

$$= \sum_{1}^{1000} n^4 - 2^4 \sum_{1}^{500} n^4$$

$$= \frac{1000(1001)(2001)(3{,}000{,}000 + 3000 - 1)}{30} - 16\,\frac{500(501)(1001)(751{,}499)}{30}$$

$$= 99{,}999{,}666{,}666{,}900$$

Example Find $1^4 + 1.25^4 + 1.50^4 + 1.75^4 + \cdots + 100^4$.

$$\sum_{1}^{397} \left(\frac{3}{4} + \frac{1}{4}n\right)^4 = \sum_{4}^{400} \left(\frac{n}{4}\right)^4$$

$$= \frac{1}{4^4} \sum_{4}^{400} n^4 = \frac{1}{4^4}\left[\sum_{1}^{400} n^4 - 1^4 - 2^4 - 3^4\right]$$

$$= \frac{1}{256}\left[\frac{400(401)(801)(481{,}199)}{30} - 98\right]$$

$$= 8{,}050{,}083{,}332\,\frac{115}{128}$$

Arithmetic Progression

$$\sum_{k=1}^{n} [a + (k-1)d] = a + (a+d) + (a+2d) + (a+3d) + \cdots + [a + (n-1)d]$$

$$= na + \frac{1}{2}n(n-1)d$$

Example Find $1 + 6 + 11 + 16 + 21 + \cdots + 201$.

$$\sum_{1}^{41} [1 + (n-1)5] = 41(1) + \frac{1}{2}(41)(40)5$$

$$= 4141$$

Geometric Progression

$$\sum_{j=1}^{n} ar^{j-1} = a + ar + ar^2 + ar^3 + \cdots + ar^{n-1}$$

$$= a\frac{1-r^n}{1-r} \qquad r \neq 1$$

Example Find $\Sigma(ar^n)$ for $a = 2$, $r = ⅛$, $n = 51$.

$$\sum_{1}^{51} 2\left(\frac{1}{8}\right)^{n-1} = 2\frac{1-(⅛)^{51}}{1-⅛}$$

Harmonic Progression

$$\sum_{k=0}^{n} \frac{1}{a+kd} = \frac{1}{a} + \frac{1}{a+d} + \frac{1}{a+2d} + \frac{1}{a+3d} + \frac{1}{a+3d} + \cdots + \frac{1}{a+nd}$$

The reciprocals of the terms of the arithmetic-progression series are called harmonic progression. No general summation formulas are available for this series.

Binomial Series

$$(x+y)^n = x^n + nx^{n-1}y + \frac{n(n-1)}{2!}x^{n-2}y^2 + \frac{n(n-1)(n-2)}{3!}x^{n-3}y^3 + \cdots + \frac{n!}{(n-r)!r!}x^{n-r}y^r + \cdots + y^n$$

$$(1 \pm x)^n = 1 \pm nx + \frac{n(n-1)}{2!}x^2 \pm \frac{n(n-1)(n-2)}{3!}x^3 + \cdots \quad (x^2 < 1)$$

Example Find the coefficient of the term x^7y^{12} in $(x+y)^{19}$.

$$\frac{n!}{(n-r)!r!} = \frac{19!}{7!12!} = \frac{19 \times 18 \times \cdots \times 13}{7 \times 6 \times 5 \times \cdots \times 1} = 50{,}388$$

Example Find the sum of $1 - \frac{1}{3}x - \frac{1 \times 2}{3 \times 6}x^2 - \frac{1 \times 2 \times 5}{3 \times 6 \times 9}x^3 + \cdots$. By identification to the second form of the binomial series with $n = ⅓$; then

$$1 - \frac{1}{3}x - \frac{1 \times 2}{3 \times 6}x^2 - \frac{1 \times 2 \times 5}{3 \times 6 \times 9}x^3 + \cdots = (1-x)^{1/3}$$

which converges for $x^2 < 1$.

Example Find $\sqrt{5}$ to 6 decimal places.

$$\sqrt{5} = 2\sqrt{1 + ¼} = 2\left(1 + \frac{1}{4}\right)^{1/2}$$

$$= 2\left[1 + \frac{1}{2(4)} - \frac{1}{8(16)} + \frac{1}{16(64)} - \frac{5}{128(256)} + \cdots\right]$$

1	1.00000000
1/2(4)	0.125
1/8(16)	−0.0078125
1/16(64)	0.00097656
5/128(256)	−0.00015259
7/256(1024)	0.00002670
21/1024(4096)	−0.00000501
33/2086(16384)	0.00000097
429/33376(65536)	−0.00000020
715/66752(262144)	0.00000001
	1.11803396 × 2 = 2.23606792 ≈ $\sqrt{5}$

Taylor's Series

$$f(x+h) = f(h) + xf'(h) + \frac{x^2}{2!}f''(h) + \frac{x^3}{3!}f'''(h) + \cdots$$

or

$$f(x) = f(x_0) + f'(x_0)(x - x_0) + \frac{f''(x_0)}{2!}(x-x_0)^2 + \frac{f'''(x_0)}{3!}(x-x_0)^3 + \cdots$$

Example Find the Taylor expansion of the polynomial $f(x) = 3 + 5x - 2x^2 - 2x^3$ about $x_0 = 1$. $f'(x = 5 - 4x - 6x^2$, $f''(x) = -4 - 12x$, $f'''(x) = -12$, $f^{(n)}(x) = 0$ for $n \geq 4$. Thus, $f(1) = 4$, $f'(1) = -5$, $f''(1) = -16$, $f'''(1) = -12$, and $f(x) = 4 - 5(x-1) - 8(x-1)^2 - 2(x-1)^3$.

Example Find a series expansion for $f(x) = \ln(1+x)$ about $x_0 = 0$.

$f'(x) = (1+x)^{-1}$, $f''(x) = -(1+x)^{-2}$, $f'''(x) = 2(1+x)^{-3}$, etc.

thus $f(0) = 0$, $f'(0) = 1$, $f''(0) = -1$, $f'''(1) = 2$, etc.

$$\ln(x+1) = x - \frac{x^2}{2} + \frac{x^3}{3} - \frac{x^4}{4} + \cdots + (-1)^{n+1}\frac{x^n}{n} + \cdots$$

which converges for $-1 < x \leq 1$.

Maclaurin's Series

$$f(x) = f(0) + xf'(0) + \frac{x^2}{2!}f''(0) + \frac{x^3}{3!}f'''(0) + \cdots$$

This is simply a special case of Taylor's series when h is set to zero.

Example Compute $\sqrt[3]{e}$ to five places.

$$e^x = 1 + x + \frac{x^2}{2!} + \frac{x^3}{3!} + \cdots$$

$$e^{1/3} = 1 + \frac{1}{3} + \frac{1}{(3^2)2!} + \frac{1}{(3^3)3!} + \frac{1}{(3^4)4!} + \text{remainder}$$

$$= 1.39557 + R_5$$

The remainder R_5 can be estimated by a derived formula

$$R_n = \frac{x^n}{n!}f^{(n)}(x_0) \qquad x_0 \text{ such that } 0 < x_0 < x \text{ or } x < x_0 < 0 \text{ if } x < 0$$

which in this case is

$$R_5 = \frac{(⅓)^5}{5!}e^{1/3}$$

since

$$e^{1/3} < 3^{1/3}$$

$$R_5 < \frac{\sqrt[3]{3}}{3^5 5!} \cong 0.00005$$

Therefore $e^{1/3} \approx 1.3956$ to 5 places.

Exponential Series

$$e^x = 1 + x + \frac{x^2}{2!} + \frac{x^3}{3!} + \cdots + \frac{x^n}{n!} + \cdots \qquad -\infty < x < \infty$$

Logarithmic Series

$$\ln x = \frac{x-1}{x} + \frac{1}{2}\left(\frac{x-1}{x}\right)^2 + \frac{1}{3}\left(\frac{x-1}{x}\right)^3 + \cdots \qquad (x > ½)$$

$$\ln x = 2\left(\frac{x-1}{x+1}\right) + \frac{1}{3}\left(\frac{x-1}{x+1}\right)^3 + \cdots \qquad (x > 0)$$

Trigonometric Series*

$$\sin x = x - \frac{x^3}{3!} + \frac{x^5}{5!} - \frac{x^7}{7!} + \cdots \qquad -\infty < x < \infty$$

$$\cos x = 1 - \frac{x^2}{2!} + \frac{x^4}{4!} - \frac{x^6}{6!} + \cdots \qquad -\infty < x < \infty$$

$$\sin^{-1} x = x + \frac{x^3}{6} + \frac{1}{2}\cdot\frac{3}{4}\cdot\frac{x^5}{5} + \frac{1}{2}\cdot\frac{3}{4}\cdot\frac{5}{6}\cdot\frac{x^7}{7} + \cdots \qquad (x^2 < 1)$$

$$\tan^{-1} x = x - \frac{1}{3}x^3 + \frac{1}{5}x^5 - \frac{1}{7}x^7 + \cdots \qquad (x^2 < 1)$$

Miscellaneous Infinite Series†

$$e = 1 + \frac{1}{1} + \frac{1}{2!} + \frac{1}{3!} + \frac{1}{4!} + \cdots + \frac{1}{n!} + \cdots$$

$$\pi = 4\left(1 - \frac{1}{3} + \frac{1}{5} - \frac{1}{7} + \frac{1}{9} - \frac{1}{11} + \cdots + \frac{(-1)^{n+1}}{2n-1} + \cdots\right)$$

$$\pi^2 = 6 + \frac{6}{2^2} + \frac{6}{3^2} + \frac{6}{4^2} + \frac{6}{5^2} + \cdots + \frac{6}{n^2} + \cdots$$

$$\ln 2 = 1 - \frac{1}{2} + \frac{1}{3} - \frac{1}{4} + \frac{1}{5} + \cdots + (-1)^n \frac{1}{n+1} + \cdots$$

$$\sqrt{10} = 3\left[1 + \frac{1}{2(9)} + \frac{1}{2!2^2 9^2} + \frac{3}{3!2^3 9^3} + \frac{3.5}{4!2^4 9^4} + \cdots\right.$$

$$2 = 1 + \frac{1}{2} + \frac{1}{4} + \frac{1}{8} + \frac{1}{16} + \cdots + \frac{1}{2^n} + \cdots$$

Partial Sums of Infinite Series, and How They Grow Calculus textbooks devote much space to tests for convergence and divergence of series that are of little practical value, since a convergent series either converges rapidly, in which case almost any test (among those presented in the preceding subsections) will do; or it converges slowly, in which case it is not going to be of much use unless there is some way to get at its sum without adding up an unreasonable number of terms. In what follows we discuss topics connected with the problem of finding out, as accurately as possible, how fast a convergent series converges and how fast a divergent series diverges. The presentation follows an article by R. P. Boas [Am. Math Mon., 84 (1977); see also references cited therein].

Computing partial sums to achieve a desired approximation to the sum of a convergent series may require adding up millions of terms. At present it is possible to add a million terms on a computer in a few minutes. However 10^{12} terms seem beyond the range of current computers. If we count a term per microsecond, 10^{12} terms would take about 12 days; 10^{15} terms, about 32 years.

Example Power series typically converge rather fast. Consider, for example, a geometric series with ratio $x > 0$ close to 1. If we want to take enough terms so that the error in the sum is less than ϵ, we need $N+1$ terms, where N is large enough to make $x^{N+1} < \epsilon(1 - x)$. For example, when $\epsilon = 0.005$ (i.e., to two-decimal accuracy) and $x = 0.95$, we still need only 162 terms. If we need much greater accuracy, say, $\epsilon = 5 \times 10^{-100}$, we will need 4517 terms, which is trivial for a high-speed computer. However, as soon as we get away from power series, the situation deteriorates. For example, if we want to add terms of $\Sigma_{n=1}^{\infty} 1/n^2$ until the error is less than 0.005, we need 200 terms, more than for $\Sigma(0.95)^n$; and each additional decimal place demands 10 times as many terms as the previous one. To get 10 decimal places we should have to add 2×10^{10} terms, which is rather impractical. In this example, we know that $\Sigma_{n=1}^{\infty} 1/n^2 = \pi^2/6$, and we can compute the sum very accurately. However, similar statements can be made about rates of convergence even if we do not know the sum. We show how such computations can be made for a large class of series. To dramatize how slow convergence can be, the series $\Sigma_{n=2}^{\infty} n^{-1}(\log n)^{-2}$ would require about 10^{87} terms to get its sum to two decimal places, but its sum is known, by indirect methods, to be approximately 2.10974.

Euler-Maclaurin Formula Consider a series of the form $\Sigma f(n)$, where f is positive and decreasing, $|f'|$ also decreases, and f is the derivative of a function whose values can be easily calculated. For a convergent series of this form, the Euler-Maclaurin formula gives the following expression for the remainders of the series:

$$R_n \equiv \sum_{k=n+1}^{\infty} f(k) = \int_{n+1/2}^{\infty} f(t)\,dt + \int_{n+1/2}^{\infty} P_1(t)f'(t)\,dt$$

where $P_1(t) = t - [t] - ½$ ($[t]$ being the integer part of t, e.g., $[3.6] = 3$). It is easy to show that the second integral is negative, and that the remainder R_n satisfies

$$\int_{n+1/2}^{\infty} f(t)\,dt > R_n > \int_{n+1/2}^{\infty} f(t)\,dt + \frac{1}{8}f'\left(n + \frac{1}{2}\right)$$

Consequently, we have $R_n < \epsilon$ together with $R_{n-1} > \epsilon$ (so that precisely n terms of the series are required to get a remainder less than ϵ), provided we have both

$$\int_{n+1/2}^{\infty} f(t)\,dt < \epsilon$$

and

$$\int_{n-1/2}^{\infty} f(t)dt + \frac{1}{8}f'\left(n - \frac{1}{2}\right) > \epsilon$$

If $y = \psi(x)$ is the inverse of $x = \int_h^{\infty} f(t)\,dt$, then R_n is nearly equal to some specified $\epsilon > 0$ when n is nearly equal to $\psi(\epsilon)$.

Example $f(k) = 1/k^2$. Then $\int_{n-1/2}^{\infty} f(t)\,dt = \frac{2}{2n-1}$, and the remainder will be less than ϵ for the first time at n, provided $\frac{2}{2n-1} < \epsilon$ and $\frac{2}{2n-1} - \frac{2}{(2n-1)^3} > \epsilon$. If $\epsilon = 0.005$, we get $n > 199.5$, and with this value of n the second inequality is satisfied. This calculation illustrates the fallacy of the advice that is sometimes given, namely, that in calculating the sum of a series one can stop when the terms become smaller than the admissible error. In this example, the 200th term is less than 4×10^{-5}, but the error in the sum up to that point is much larger (close to 5×10^{-3}).

There are also more elaborate versions of the Euler-Maclaurin formula (than the simplest version just discussed) that can also be used to determine how fast a convergent series converges or how fast a divergent series diverges. The Euler-Maclaurin formula enables us to approximate the partial sums of a series (if the series has sufficiently simple structure) by integrals, with errors that can be estimated. The more refined versions of the Euler-Maclaurin involve the Bernoulli numbers.

Asymptotic Series Certain divergent series are useful in the asymptotic analysis of functions. A divergent series of the form

$$a_0 + a_1/x + a_2/x^2 + \cdots + a_n/x^n + \cdots$$

is called an asymptotic representation of the function f if

$$\lim_{n\to\infty} x^n\,[f(x) - S_{n+1}(x)] = 0$$

for any value of n, where S_{n+1} is the sum of the first $(n+1)$ terms of the series.

If a function has an asymptotic-series expansion, then such a series is unique. An asymptotic series may be integrated term by term, but in general it cannot be differentiated term by term.

*tan x series has awkward coefficients and should be computed as $\left[(\text{sign})\frac{\sin x}{\sqrt{1 - \sin^2 x}}\right]$.

†These series are useful for testing general-purpose computer programs.

COMPLEX VARIABLES

REFERENCES: *General.* 55, 63, 109, 167, 205. *Applied and computational complex analysis.* 143, 177, 183.

Numbers of the form $z = x + iy$, where x and y are real, $i^2 = -1$, are called **complex numbers.** The numbers $z = x + iy$ are representable in the plane as shown in Fig. 2-51. The following definitions and terminology are used:

1. Distance $OP = r =$ **modulus** of z written $|z|$. $|z| = \sqrt{x^2 + y^2}$.
2. x is the **real part** of z.
3. y is the **imaginary part** of z.
4. The angle θ, $0 \le \theta < 2\pi$, measured counterclockwise from the positive x axis to OP is the **argument** of z. $\theta = \arctan y/x = \arcsin y/r = \arccos x/r$ if $x \neq 0$, $\theta = \pi/2$ if $x = 0$ and $y > 0$.
5. The numbers r, θ are the **polar coordinates** of z.
6. $\bar{z} = x - iy$ is the complex **conjugate** of z.

ALGEBRA

Let $z_1 = x_1 + iy_1$, $z_2 = x_2 + iy_2$.

Equality $z_1 = z_2$ if and only if $x_1 = x_2$ and $y_1 = y_2$.

Addition $z_1 + z_2 = (x_1 + x_2) + i(y_1 + y_2)$.

Subtraction $z_1 - z_2 = (x_1 - x_2) + i(y_1 - y_2)$.

Multiplication $z_1 \cdot z_2 = (x_1x_2 - y_1y_2) + i(x_1y_2 + x_2y_1)$.

Division $z_1/z_2 = \dfrac{x_1x_2 + y_1y_2}{x_2^2 + i\,y_2^2} + i\dfrac{x_2y_1 - x_1y_2}{x_2^2 + y_2^2}$ $z_2 \neq 0$.

SPECIAL OPERATIONS

$z\bar{z} = x^2 + y^2 = |z|^2$; $\overline{z_1 \pm z_2} = \bar{z}_1 \pm \bar{z}_2$; $\bar{\bar{z}}_1 = z_1$; $\overline{z_1z_2} = \bar{z}_1\bar{z}_2$; $|z_1 \cdot z_2| = |z_1| \cdot |z_2|$; $\arg(z_1 \cdot z_2) = \arg z_1 + \arg z_2$; $\arg(z_1/z_2) = \arg z_1 - \arg z_2$; $i^{4n} = 1$ for n any integer; $i^{2n} = -1$ where n is any odd integer; $z + \bar{z} = 2x$; $z - \bar{z} = 2iy$.

Every complex quantity can be expressed in the form $x + iy$.

Example $(1 + i)^2 + (2 - i)^2 = 1 + 2i + i^2 + 4 - 4i + i^2 = 5 - 2i + 2i^2 = 3 - 2i$ since $i^2 = -1$.

Example $\dfrac{1 + 2i}{3 - i} = \dfrac{(1 + 2i)(3 + i)}{(3 - i)(3 + i)} = \dfrac{3 + 7i + 2i^2}{9 - i^2} = \dfrac{1 + 7i}{10} = \dfrac{1}{10} + \dfrac{7}{10}i$.

Example $\dfrac{3 + i}{i^3} = \dfrac{i(3 + i)}{i^4} = i(3 + i) = -1 + 3i$.

TRIGONOMETRIC REPRESENTATION

By referring to Fig. 2-51 there results $x = r\cos\theta$, $y = r\sin\theta$ so that $z = x + iy = r(\cos\theta + i\sin\theta)$, which is called the **polar form** of the complex number. $\cos\theta + i\sin\theta = e^{i\theta}$. Hence $z = x + iy = re^{i\theta}$. $\bar{z} = x - iy = re^{-i\theta}$. Two important results from this are $\cos\theta = (e^{i\theta} + e^{-i\theta})/2$ and $\sin\theta = (e^{i\theta} - e^{-i\theta})/2i$. Let $z_1 = r_1e^{i\theta_1}$, $z_2 = r_2e^{i\theta_2}$. This form is convenient for multiplication for $z_1z_2 = r_1r_2e^{i(\theta_1+\theta_2)}$ and for division for $z_1/z_2 = (r_1/r_2)e^{i(\theta_1-\theta_2)}$, $z_2 \neq 0$.

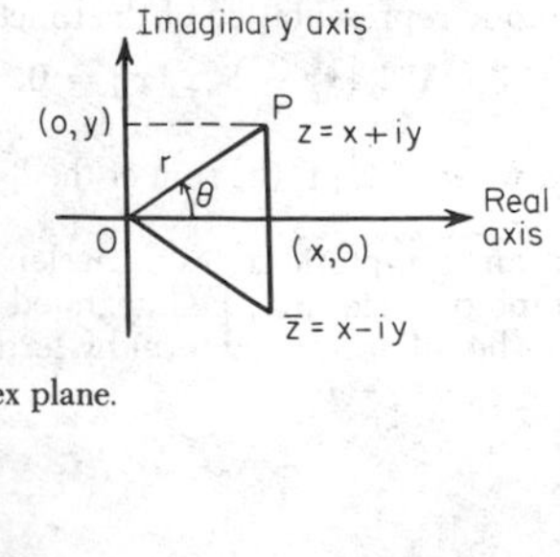

FIG. 2-51 Complex plane.

Example $1 + 2i = \sqrt{5}(\cos 63°26' + i\sin 63°26') = \sqrt{5}e^{i(1.10713)}$, where 1.10713 rad = 63°26′.

Example $\dfrac{1 + i}{3i} = \dfrac{\sqrt{2}}{3}e^{i(\pi/4-\pi/2)} = \dfrac{\sqrt{2}}{3}e^{-i\pi/4}$.

POWERS AND ROOTS

If n is a positive integer, $z^n = (re^{i\theta})^n = r^ne^{in\theta} = r^n(\cos n\theta + i\sin n\theta)$.

Example $(1 + i)^8 = (\sqrt{2})^8e^{i8(\pi/4)} = 16\,e^{2\pi i} = 16$.

Example

$$(\sqrt{3} - i)^5 = 2^5e^{i5(11\pi/6)} = 32\left(\cos\frac{55\pi}{6} + i\sin\frac{55\pi}{6}\right) = 32\left(-\frac{\sqrt{3}}{2} - \frac{i}{2}\right) = -16\sqrt{3} - 16i$$

If n is a positive integer,

$$z^{1/n} = r^{1/n}e^{i[(\theta+2k\pi)/n]} = r^{1/n}\left[\cos\left(\frac{\theta + 2k\pi}{n}\right) + i\sin\left(\frac{\theta + 2k\pi}{n}\right)\right]$$

and selecting values of $k = 0, 1, 2, 3, \ldots, n - 1$ give the n distinct values of $z^{1/n}$. The n roots of a complex quantity are uniformly spaced around a circle, with radius $r^{1/n}$, in the complex plane in a symmetric fashion.

Example Find the three cube roots of -8. Here $r = 8$, $\theta = \pi$. The roots are $z_0 = 2(\cos\pi/3 + i\sin\pi/3) = 1 + i\sqrt{3}$, $z_1 = 2(\cos\pi + i\sin\pi) = -2$, $z_2 = 2(\cos 5\pi/3 + i\sin 5\pi/3) = 1 - i\sqrt{3}$.

ELEMENTARY COMPLEX FUNCTIONS

Polynomials A polynomial in z, $a_nz^n + a_{n-1}z^{n-1} + \cdots + a_0$, where n is a positive integer, is simply a sum of complex numbers times integral powers of z which have already been defined. Every polynomial of degree n has precisely n complex roots provided each multiple root of multiplicity m is counted m times.

Exponential Functions The exponential function e^z is defined by the equation $e^z = e^{x+iy} = e^x \cdot e^{iy} = e^x(\cos y + i\sin y)$. Properties: $e^0 = 1$; $e^{z_1} \cdot e^{z_2} = e^{z_1+z_2}$; $e^{z_1}/e^{z_2} = e^{z_1-z_2}$; $e^{z+2k\pi i} = e^z$.

Trigonometric Functions $\sin z = (e^{iz} - e^{-iz})/2i$; $\cos z = (e^{iz} + e^{-iz})/2$; $\tan z = \sin z/\cos z$; $\cot z = \cos z/\sin z$; $\sec z = 1/\cos z$; $\csc z = 1/\sin z$. Fundamental identities for these functions are the same as their real counterparts. Thus $\cos^2 z + \sin^2 z = 1$, $\cos(z_1 \pm z_2) = \cos z_1 \cos z_2 \mp \sin z_1 \sin z_2$, $\sin(z_1 \pm z_2) = \sin z_1 \cos z_2 \pm \cos z_1 \sin z_2$. The sine and cosine of z are **periodic functions of period** 2π; thus $\sin(z + 2\pi) = \sin z$. For computation purposes $\sin z = \sin(x + iy) = \sin x \cosh y + i\cos x \sinh y$, where $\sin x$, $\cosh y$, etc., are the real trigonometric and hyperbolic functions. Similarly, $\cos z = \cos x \cosh y - i\sin x \sinh y$. If $x = 0$ in the results given, $\cos iy = \cosh y$, $\sin iy = i\sinh y$.

Example $\cos(1 + 2i) = \cos 1 \cosh 2 - i\sin 1 \sinh 2 = (0.54030)(3.7622) - i(0.84147)(3.6269) = 2.033 - i3.052$.

Example Find all solutions of $\sin z = 3$. From previous data $\sin z = \sin x \cosh y + i\cos x \sinh y = 3$. Equating real and imaginary parts $\sin x \cosh y = 3$, $\cos x \sinh y = 0$. The second equation can hold for $y = 0$ or for $x = \pi/2, 3\pi/2, \ldots$. If $y = 0$, $\cosh 0 = 1$ and $\sin x = 3$ is impossible for real x. Therefore, $x = \pm\pi/2, \pm 3\pi/2, \ldots \pm(2n + 1)\pi/2$, $n = 0, \pm 1, \pm 2, \ldots$. However, $\sin 3\pi/2 = -1$ and $\cosh y \ge 1$. Hence $x = \pi/2, 5\pi/2, \ldots$. The solution is $z = [(4n + 1)\pi]/2 + i\cosh^{-1}3$, $n = 0, 1, 2, 3, \ldots$.

Example Find all solutions of $e^z = -i$. $e^z = e^x(\cos y + i \sin y) = -i$. Equating real and imaginary parts gives $e^x \cos y = 0$, $e^x \sin y = -1$. From the first $y = \pm \pi/2, \pm 3\pi/2, \ldots$ But $e^x > 0$. Therefore, $y = 3\pi/2, 7\pi/2, -\pi/2, \ldots$ Then $x = 0$. The solution is $z = i[(4n + 3)\pi]/2$.

Two important facets of these functions should be recognized. First, the sin z is *unbounded;* and, second, e_z takes *all* complex values *except* 0.

Hyperbolic Functions $\sinh z = (e^z - e^{-z})/2$; $\cosh z = (e^z + e^{-z})/2$; $\tanh z = \sinh z/\cosh z$; $\coth z = \cosh z/\sinh z$; $\operatorname{csch} z = 1/\sinh z$; $\operatorname{sech} z = 1/\cosh z$. Fundamental identities for these functions are the same as for their real counterparts. $\cosh^2 z - \sinh^2 z = 1$; $\sinh(z_1 + z_2) = \sinh z_1 \cosh z_2 + \cosh z_1 \sinh z_2$; $\cosh(z_1 + z_2) = \cosh z_1 \cosh z_2 + \sinh z_1 \sinh z_2$; $\cosh z + \sinh z = e^z$; $\cosh z + \sinh z = e^{-z}$. The hyperbolic sine and hyperbolic cosine are **periodic functions** with the imaginary period $2\pi i$. That is, $\sinh(z + 2\pi i) = \sinh z$.

Logarithms The logarithm of z, $\log z = \log |z| + i(\theta + 2n\pi)$, where $\log |z|$ is taken to the base e and θ is the **principal argument** of z, that is, the particular argument lying in the interval $0 \le \theta < 2\pi$. The logarithm of z is infinitely many valued. If $n = 0$, the resulting logarithm is called the **principal value.** The familiar laws $\log z_1z_2 = \log z_1 + \log z_2$, $\log z_1/z_2 = \log z_1 - \log z_2$, $\log z^n = n \log z$ hold for the principal value.

Example $\log(1 + i) = \log \sqrt{2} + i\left(\frac{\pi}{4} + 2n\pi\right)$.

General powers of z are defined by $z^\alpha = e^{\alpha \log z}$. Since $\log z$ is infinitely many valued, so too is z^α unless α is a rational number.

Example $i^i = e^{i\log i} = e^{i[\log|i| + i(\pi/2 + 2n\pi)]} = e^{-(\pi/2 + 2n\pi)}$. Thus i^i is real with principal value $(n = 0) = e^{-\pi/2}$.

Example $(\sqrt{2})^{1+i} = e^{(1+i)\log\sqrt{2}} = e^{\log\sqrt{2}}.\ e^{i\log\sqrt{2}} = \sqrt{2}.\ (\cos \log \sqrt{2} + i \sin \log \sqrt{2}) = \sqrt{2}[\cos(0.3466) + i \sin(0.3466)]$.

Inverse Trigonometric Functions $\cos^{-1} z = -i \log(z \pm \sqrt{z^2 - 1})$; $\sin^{-1} z = -i \log(iz \pm \sqrt{1 - z^2})$; $\tan^{-1} z = \frac{i}{2} \log \left(\frac{i + z}{i - z}\right)$. These functions are infinitely many valued.

Inverse Hyperbolic Functions $\cosh^{-1} z = \log(z \pm \sqrt{z^2 - 1})$; $\sinh^{-1} z = \log(z \pm \sqrt{z^2 + 1})$; $\tanh^{-1} z = \frac{1}{2} \log \left(\frac{1 + z}{1 - z}\right)$.

COMPLEX FUNCTIONS (ANALYTIC)

In the real-number system a greater than $b(a > b)$ and b less than $c(b < c)$ define an order relation. These relations have no meaning for complex numbers. The absolute value is used for ordering. Some important relations follow: $|z| \ge x$; $|z| \ge y$; $|z_1 + z_2| \le |z_1| + |z_2|$; $|z_1 - z_2| \ge ||z_1| - |z_2||$; $|z| \ge (|x| + |y|)/\sqrt{2}$. Parts of the complex plane, commonly called **regions** or **domains,** are described by using inequalities.

Example $|z - 3| \le 5$. This is equivalent to $\sqrt{(x - 3)^2 + y^2} \le 5$, which is the set of all points within and on the circle, centered at $x = 3$, $y = 0$ of radius 5.

Example $|z - 1| \le x$ represents the set of all points inside and on the parabola $2x = y^2 + 1$ or, equivalently, $2x \ge y^2 + 1$.

Functions of a Complex Variable If $z = x + iy$, $w = u + iv$ and if for each value of z in some region of the complex plane one or more values of w are defined, then w is said to be a function of z, $w = f(z)$. Some of these functions have already been discussed, e.g., $\sin z$, $\log z$. All functions are reducible to the form $w = u(x, y) + iv(x, y)$, where u, v are real functions of the real variables x and y.

Example $z^3 = (x + iy)^3 = x^3 + 3x^2(iy) + 3x(iy)^2 + (iy)^3 = (x^3 - 3xy^2) + i(3x^2y - y^3)$.

Example $\cos z = \cos x \cosh y - i \sin x \sinh y$.

Differentiation The *derivative* of $w = f(z)$ is $\frac{dw}{dz} = \lim_{\Delta z \to 0} \frac{f(z + \Delta z) - f(z)}{\Delta z}$, and for the derivative to exist the limit must be the same no matter how Δz approaches zero. If w_1, w_2 are differentiable functions of z, the following rules apply:

$$\frac{d(w_1 \pm w_2)}{dz} = \frac{dw_1}{dz} \pm \frac{dw_2}{dz} \qquad \frac{d(w_1w_2)}{dz} = w_2\frac{dw_1}{dz} + w_1\frac{dw_2}{dz}$$

$$\frac{d(w_1/w_2)}{dz} = \frac{w_2(dw_1/dz) - w_1(dw_2/dz)}{w_2^2}$$

and

$$\frac{dw_1^n}{dz} = nw_1^{n-1}\frac{dw_1}{dz}$$

For $w = f(z)$ to be differentiable, it is necessary that $\partial u/\partial x = \partial v/\partial y$ and $\partial v/\partial x = {}^-\partial u/\partial y$. The last two equations are called the Cauchy-Riemann equations. The derivative $\frac{dw}{dz} = \frac{\partial u}{\partial x} + i\frac{\partial v}{\partial x} = \frac{\partial v}{\partial y} - i\frac{\partial u}{\partial y}$. If $f(z)$ possesses a derivative at z_o and at every point in some neighborhood of z_0, then $f(z)$ is said to be analytic at z_0. If the Cauchy-Riemann equations are satisfied and u, v, $\frac{\partial u}{\partial x}, \frac{\partial u}{\partial y}, \frac{\partial v}{\partial x}, \frac{\partial v}{\partial y}$ are continuous in a region of the complex plane, then $f(z)$ is analytic in that region.

Example $w = z\bar{z} = x^2 + y^2$. Here $u = x^2 + y^2$, $v = 0$. $\partial u/\partial x = 2x$, $\partial u/\partial y = 2y$, $\partial v/\partial x = \partial v/\partial y = 0$. These are continuous everywhere, but the Cauchy-Riemann equations hold only at the origin. Therefore, w is nowhere analytic, but it is differentiable at $z = 0$ only.

Example $w = e^z = e^x \cos y + ie^x \sin y$. $u = e^x \cos y$, $v = e^x \sin y$. $\partial u/\partial x = e^x \cos y$, $\partial u/\partial y = -e^x \sin y$, $\partial v/\partial x = e^x \sin y$, $\partial v/\partial y = e^x \cos y$. The continuity and Cauchy-Riemann requirements are satisfied for all finite z. Hence e^z is analytic (except at ∞) and $dw/\partial z = \partial u/\partial x + i(\partial v/\partial x) = e^z$.

Example $w = \frac{1}{z} = \frac{x - iy}{x^2 + y^2} = \frac{x}{x^2 + y^2} - i\frac{y}{x^2 + y^2}$. It is easy to see that dw/dz exists except at $z = 0$. Thus $1/z$ is analytic except at $z = 0$.

Singular Points If $f(z)$ is analytic in a region except at certain points, those points are called singular points.

Example $1/z$ has a singular point at zero.

Example $\tan z$ has singular points at $z = \pm(2n + 1)(\pi/2)$, $n = 0, 1, 2, \ldots$

The derivatives of the common functions, given earlier, are the same as their real counterparts.

Example $(d/dz)(\log z) = 1/z$, $(d/dz)(\sin z) = \cos z$.

Harmonic Functions Both the *real* and the *imaginary* parts of any analytic function $f = u + iv$ satisfy Laplace's equation $\partial^2\phi/\partial x^2 + \partial^2\phi/\partial y^2 = 0$. A function which possesses continuous second partial derivatives and satisfies Laplace's equation is called a harmonic function.

Example $e^z = e^x \cos y + ie^x \sin y$. $u = e^x \cos y$, $\partial u/\partial x = e^x \cos y$, $\partial^2u/\partial x^2 = e^x \cos y$, $\partial u/\partial y = -e^x \sin y$, $\partial^2u/\partial y^2 = -e^x \cos y$. Clearly $\partial^2u/\partial x^2 + \partial^2u/\partial y^2 = 0$. Similarly, $v = e^x \sin y$ is also harmonic.

If $w = u + iv$ is analytic, the curves $u(x, y) = c$ and $v(x, y) = k$ intersect at right angles, if $w'(z) \ne 0$.

Example $z^3 = (x^3 - 3xy^2) + i(3x^2y - y^3)$. Set $u = x^3 - 3xy^2 = c$, $v = 3x^2y - y^3 = k$. By implicit differentiation there results, respectively, $dy/dx = (x^2 - y^2)/2xy$, $dy/dx = 2xy/(y^2 - x^2)$, which are clearly negative reciprocals, the condition for perpendicularity.

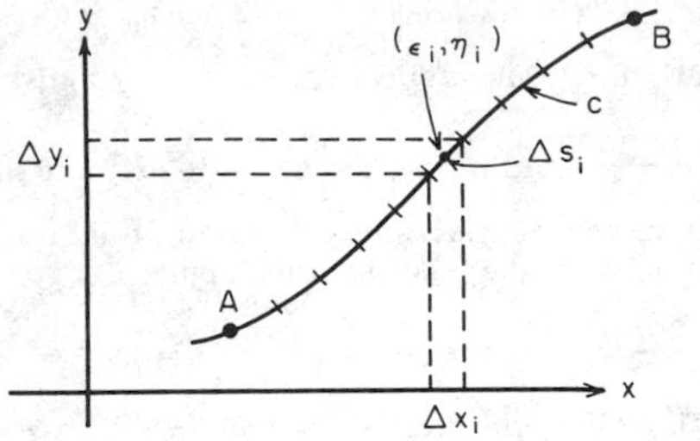

FIG. 2-52 Line integral.

Integration In much of the work with complex variables a simple extension of integration called **line** or **curvilinear** integration is of fundamental importance. Since any complex line integral can be expressed in terms of real line integrals, we define only real line integrals. Let $F(x,y)$ be a real, continuous function of x and y and c be any continuous curve of finite length joining the points A and B (Fig. 2-52). $F(x,y)$ is not related to the curve c. Divide c up into n segments, Δs_i, whose projection on the x axis is Δx_i and on the y axis is Δy_i. Let (ϵ_i, η_i) be the coordinates of an arbitrary point on Δs_i. The limits of the sums

$$\lim_{\Delta s_i \to 0} \sum_{i=1}^{n} F(\epsilon_i, \eta_i)\, \Delta s_i = \int_c F(x, y)\, ds$$

$$\lim_{\Delta s_i \to 0} \sum_{i=1}^{n} F(\epsilon_i, \eta_i)\, \Delta x_i = \int_c F(x, y)\, dx$$

$$\lim_{\Delta s_i \to 0} \sum_{i=1}^{n} F(\epsilon_i, \eta_i)\, \Delta_{yi} = \int_c F(x, y)\, dy$$

are known as line integrals. Much of the initial strangeness of these integrals will vanish if it be observed that the ordinary definite integral $\int_a^b f(x)\, dx$ is just a line integral in which the curve c is a line segment on the x axis and $F(x, y)$ is a function of x alone. The evaluation of line integrals can be reduced to evaluation of ordinary integrals.

Example $\int_c y(1 + x)\, dy$, where c: $y = 1 - x^2$ from $(-1, 0)$ to $(1, 0)$. Clearly $y = 1 - x^2$, $dy = -2x\, dx$. Thus $\int_c y(1 + x)\, dy = -2\int_{-1}^{1} (1 - x^2)(1 + x)x\, dx = -8/15$.

Example $\int_c x^2y\, ds$, c is the square whose vertices are (0, 0), (1, 0), (1, 1), (0, 1). $ds = \sqrt{dx^2 + dy^2}$. When $dx = 0$, $ds = dy$. From (0, 0) to (1, 0), $y = 0$, $dy = 0$. Similar arguments for the other sides give $\int_c x^2y\, ds = \int_0^1 0.x^2\, dx + \int_0^1 y\, dy + \int_1^0 x^2\, dx + \int_1^0 0.y\, dy = 1/2 - 1/3 = 1/6$.

Let $f(z)$ be any function of z, analytic or not, and c any curve as above. The complex integral is calculated as $\int_c f(z)\, dz = \int_c (u\, dx - v\, dy) + i\int_c (v\, dx + u\, dy)$, where $f(z) = u(x, y) + iv(x, y)$. Properties of line integrals are the same as those for ordinary integrals. That is, $\int_c [f(z) \pm g(z)]\, dz = \int_c f(z)\, dz \pm \int_c g(z)\, dz$; $\int_c kf(z)\, dz = k\int_c f(z)\, dz$ for any constant k, etc.

Example $\int_c (x^2 + iy)\, dz$ along c: $y = x$, 0 to $1 + i$. This becomes $\int_c (x^2 + iy)\, dz = \int_c (x^2\, dx - y\, dy) + i\int_c (y\, dx + x^2\, dy) = \int_0^1 x^2\, dx - \int_0^1 x\, dy + i\int_0^1 x\, dx + i\int_0^1 x^2\, dx = 1/6 + 5i/6$.

Conformal Mapping Every function of a complex variable $w = f(z) = u(x, y) + iv(x, y)$ transforms the x, y plane into the u, v plane in some manner. A conformal transformation is one in which angles between curves are preserved in *magnitude* and *sense.* Every analytic funcion, except at those points where $f'(z) = 0$, is a conformal transformation. See Fig. 2-53.

Example $w = z^2$. $u + iv = (x^2 - y^2) + 2ixy$ or $u = x^2 - y^2$, $v = 2xy$. These are the transformation equations between the (x, y) and (u, v) planes. Lines parallel to the x axis, $y = c_1$ map into curves in the u, v plane with parametric equations $u = x^2 - c_1^2$, $v = 2c_1x$. Eliminating x, $u = (v^2/4c_1^2) - c_1^2$, which represents a family of parabolas with the origin of the w plane as focus, the line $v = 0$ as axis and opening to the right. Similar arguments apply to $x = c_2$.

The principles of complex variables are useful in the solution of a variety of applied problems. See the references for additional information.

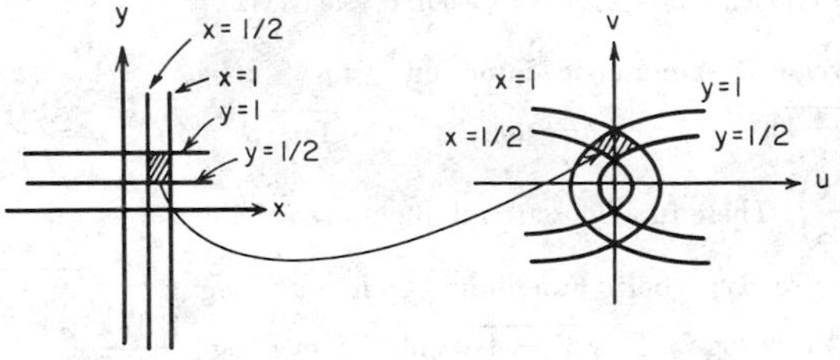

FIG. 2-53 Conformal transformation.

DIFFERENTIAL EQUATIONS

REFERENCES: *Ordinary Differential Equations. Elementary level:* 38, 42, 43, 67, 69, 77, 87, 193, 247, 252. *Intermediate level:* 29, 38, 41, 42, 78, 81, 87, 93, 147, 262. *Theory and advanced topics:* 27, 68, 93, 133, 134, 155, 238, 239. *Applications:* 7, 81, 87, 247.

Partial Differential Equations. Elementary level and solution methods: 7, 38, 56, 62, 137, 147, 161, 203, 226, 240, 248, 261. *Theory and advanced level:* 73, 97, 112, 141, 272.

See also "Numerical Analysis and Approximate Methods" and "General References: References for General and Specific Topics—Advanced Engineering Mathematics" for additional references on topics in ordinary and partial differential equations.

The natural laws in any scientific or technological field are not regarded as precise and definitive until they have been expressed in mathematical form. Such a form, often an equation, is a relation between the quantity of interest, say, product yield, and independent variables such as time and temperature upon which yield depends. When it happens that this equation involves, besides the function itself, one or more of its derivatives it is called a **differential equation.**

Example The homogeneous bimolecular reaction $A + B \xrightarrow{k} C$ is characterized by the differential equation $dx/dt = k(a - x)(b - x)$, where a = initial concentration of A, b = initial concentration of B, and $x = x(t)$ = concentration of C as a function of time t.

Example The differential equation of heat conduction in a moving fluid with velocity components v_x, v_y is

$$\frac{\partial u}{\partial t} + v_x\frac{\partial u}{\partial x} + v_y\frac{\partial u}{\partial y} = \frac{K}{\rho c_p}\left(\frac{\partial^2 u}{\partial x^2} + \frac{\partial^2 u}{\partial y^2}\right)$$

where $u = u(x, y, t)$ = temperature, K = thermal conductivity, ρ = density, and c_p = specific heat at constant pressure.

ORDINARY DIFFERENTIAL EQUATIONS

When the function involved in the equation depends upon only one variable, its derivatives are ordinary derivatives and the differential equation is called an **ordinary differential equation.** When the function depends upon several independent variables, then the equation is called a **partial differential equation.** The theories of ordinary and partial differential equations are quite different. In almost every respect the latter is more difficult.

Whichever the type, a differential equation is said to be of nth order if it involves derivatives of order n but no higher. The equation in the first example is of first order and that in the second example of second order. The **degree** of a differential equation is the power to which the derivative of the highest order is raised after the equation has been cleared of fractions and radicals in the dependent variable and its derivatives.

A relation between the variables, involving no derivatives, is called a solution of the differential equation if this relation, when substituted in the equation, satisfies the equation. A solution of an ordinary differential equation which includes the maximum possible number of "arbitrary" constants is called the **general solution.** The maximum number of "arbitrary" constants is exactly equal to the order of the differential equation. If any set of specific values of the constants is chosen, the result is still a solution called a **particular solution.**

Example The general solution of $(d^2x/dt^2) + k^2x = 0$ is $x = A \cos kt + B \sin kt$, where A, B are arbitrary constants. A particular solution is $x = \frac{1}{2} \cos kt + 3 \sin kt$.

In the case of some equations still other solutions exist called singular solutions. A **singular solution** *is any solution of the differential equation which is not included in the general solution.*

Example $y = x(dy/dx) - \frac{1}{4}(dy/dx)^2$ has the general solution $y = cx - \frac{1}{4}c^2$, where c is an arbitrary constant; $y = x^2$ is a singular solution, as is easily verified.

Ordinary Differential Equations of the First Order

Equations with Separable Variables Every differential equation of the first order and of the first degree can be written in the form $M(x, y)\,dx + N(x, y)\,dy = 0$. If the equation can be transformed so that M does not involve y and N does not involve x, then the variables are said to be separated. The solution can then be obtained by **quadrature,** which means that $y = \int f(x)\,dx + c$, which may or may not be expressible in simpler form.

Example Two liquids A and B are boiling together in a vessel. Experimentally it is found that the ratio of the rates at which A and B are evaporating at any time is proportional to the ratio of the amount of A (say, x) to the amount of B (say, y) still in the liquid state. This physical law is expressible as $(dy/dt)/(dx/dt) = ky/x$ or $dy/dx = ky/x$, where k is a proportionality constant. This equation may be written $dy/y = k(dx/x)$, in which the variables are separated. The solution is $\ln y = k \ln x + \ln c$ or $y = cx^k$

Exact Equations The equation $M(x, y)\,dx + N(x, y)\,dy = 0$ is exact if and only if $\partial M/\partial y = \partial N/\partial x$. In this case there exists a function $w = f(x, y)$ such that $\partial f/\partial x = M$, $\partial f/\partial y = N$, and $f(x, y) = C$ is the required solution. $f(x, y)$ is found as follows: treat y as though it were constant and evaluate $\int M(x, y)\,dx$. Then treat x as though it were constant and evaluate $\int N(x, y)\,dy$. The sum of all unlike terms in these two integrals (including no repetitions) is $f(x, y)$.

Example $(2xy - \cos x)\,dx + (x^2 - 1)\,dy = 0$ is exact for $\partial M/\partial y = 2x$, $\partial N/\partial x = 2x$. $\int M dx = \int(2xy - \cos x)\,dx = x^2y - \sin x$, $\int N\,dy = \int(x^2 - 1)\,dy = x^2y - y$. The solution is $x^2y - \sin x - y = C$, as may easily be verified.

Linear Equations A differential equation is said to be linear when it is of first degree in the dependent variable and its derivatives. The general linear first-order differential equation has the form $dy/dx + P(x)y = Q(x)$. Its general solution is

$$y = e^{-\int P dx}\left[\int Q e^{\int P dx}\,dx + C\right]$$

Example A tank initially holds 200 gal of a salt solution in which 100 lb is dissolved. Six gallons of brine containing 4 lb of salt run into the tank per minute. If mixing is perfect and the output rate is 4 gal/min, what is the amount A of salt in the tank at time t? The differential equation of A is $dA/dt + [2/(100 + t)]\,A = 24$. Its general solution is $(100 + t)^2A = 8(100 + t)^3 + C$. At $t = 0$, $A = 100$; so the particular solution is $(100 + t)^2A = 8(100 + t)^3 - 7(100)^3$.

Equations Reducible to Linear Equations The Bernoulli equation $(dy/dx) + P(x)y = y^nQ(x)$, $n \neq 0, 1$ is reducible to a linear equation by the substitution $z = y^{-n+1}$. The result of this is $[1/(1 - n)](dz/dx) + zP(x) = Q(x)$, which is readily integrated by techniques of the previous section. If $n = 0$ or 1, the equation is already linear.

Example $\dfrac{dy}{dx} = \dfrac{xy^2 + 2y}{x}$. This is equivalent to $\dfrac{dy}{dx} - \dfrac{2}{x}y = y^2$. Set $z = y^{-1}$. $\dfrac{dy}{dx} = -y^2\dfrac{dz}{dx}$. The equation becomes $\dfrac{dz}{dx} + \dfrac{2}{x}z = -1$.

The "Riccati" equation in the form $(dy/dx) + ay^2 + Q(x)y + R(x) = 0$ can be reduced to the second-order linear differential equation $(d^2u/dx^2) + Q(x)(du/dx) + aR(x)u = 0$ by the substitution $u = e^{a\int y dx}$. The equation for u is then solved by techniques for higher-order equations.

Homogeneous Equations A function $g(x, y)$ is said to be homogeneous of degree m if for any number r, $g(rx, ry) = r^mg(x, y)$, e.g., $x^2y^2\,dx + xy^3\,dx = 0$ is homogeneous of degree 4 since x^2y^2 and xy^3 both have the required property. $\cos(xy)\,dx + x^2y^3\,dy = 0$ is not homogeneous. If the differential equation is homogeneous, then either of the substitutions $y = vx$ or $x = vy$ will reduce the equation to the variables-separable case. Let $y = vx$. Then $dy/dx = v + x(dv/dx)$, so that $dy/dx = f(x, y)$ becomes $v + x(dv/dx) = x^mF(v)$ or $dv/[F(v) - v] = dx/x^{1-m}$.

Example $(xe^{y/x} + y)\,dx - x\,dy = 0$ is equivalent to $dy/dx = e^{y/x} + (y/x)$. Let $y = vx$. Then $F(v) = e^v + v$ and $dv = dx/x$. The solution is $\ln x + e^{-y/x} = C$.

Equations Reducible to Homogeneous Equations The differential equation of the form $\dfrac{dy}{dx} = f\left(\dfrac{ax + by + c}{dx + ey + g}\right)$ is reducible to a homogeneous equation by the substitution $u = ax + by + c$, $v = dx + ey + g$ for $ae - bd \neq 0$ and by $z = ax + by$, $w = dx + ey$ if $ae - bd = 0$.

Example $\dfrac{dy}{dx} = \left(\dfrac{6x + 4y - 3}{3x + y - 1}\right)^2$. Since $ae - bd = -6 \neq 0$, set $u = 6x + 4y - 3$, $v = 3x + y - 1$. $du = 6dx + 4dy$, $dv = 3dx + dy$, $dx = \frac{2}{3}dv - \frac{1}{6}du$, $dy = +\frac{1}{2}du - dv$. Therefore, $(u^2 + 3v^2)\,du = (4u^2 + 6v^2)\,dv$, which is homogeneous.

Equations Linear in f(y) $f'(y)(dy/dx) + Pf(y) = Q$, where P, Q are functions of x alone, is linear in $f(y)$.

Example $e^y(dy/dx) - e^y = x$. Let $v = e^y$, $dv/dx = e^y(dy/dx)$. The equation becomes $(dv/dx) - v = x$, which is a linear equation with solution $v = ce^x - x - 1$ or $y = \ln(ce^x - x - 1)$. The last few subsections have emphasized the importance of substitutions in solving differential equations.

Equations of the First Order but Not First Degree A general differential equation of the first order is a relation $F(x, y, dy/dx) = 0$. If F is not of first degree in dy/dx, this equation may be satisfied by several solution curves through a given point.

Equations Solvable for dy/dx After solving for dy/dx, treat each of the solutions as it occurs under some case studied previously.

Example $\left(\dfrac{dy}{dx}\right)^2 - x\dfrac{dy}{dx} - 2y\dfrac{dy}{dx} + 2xy = 0$. This is equivalent to $\left(\dfrac{dy}{dx} - 2y\right)\left(\dfrac{dy}{dx} - x\right) = 0$ or $\dfrac{dy}{dx} = x$, $\dfrac{dy}{dx} = 2y$. These equations have the solution $y = ce^{2x}$ and $y = x^2/2 + c$.

Equations Solvable for y Solution of the general equation for y gives one or more equations of the form $y = f(x, dy/dx)$. To solve this differentiate with respect to x and setting $p = dy/dx$ there results $p = \partial f/\partial x + (\partial f/\partial p)(dp/dx)$. The result is a differential equation of the first order and first degree in x and p. A solution may be found in the form $g(x, p, c) = 0$ and p eliminated between g and f to give the solution.

Example Clairaut's equation, written $y = xp + h(p)$, is an important case. Differentiating with respect to x, $p = p + [x + h'(p)]\,(dp/dx)$, and this is satisfied only if $x = -h'(p)$ or $dp/dx = 0$. The latter gives $p = c$, which substituted in the equation yields the solution $y = cx + f(c)$. Turning to $h'(p) = -x$ substitution in the original equation gives a parametric solution $y = -ph'(p) + h(p)$, $x = -h'(p)$. For example, $y = xp + p^2$ yields $(x + 2p)p = 0$ with solutions $y = cx + c^2$ and $y = -(x^2/4) + c$. Notice that the latter solution is singular.

Equations Solvable for x The process for such equations is analogous to the preceding discussion except that there results

$$x = f(y, p) \quad \text{and} \quad \frac{dx}{dy} = \frac{1}{p} = \frac{\partial f}{\partial y} + \frac{\partial f}{\partial p}\frac{dp}{dy}$$

Ordinary Differential Equations of Higher Order The higher-order differential equations, especially those of order 2, are of great importance because of physical situations describable by them.

Equation $y^{(n)} = f(x)$ Such a differential equation can be solved by n integrations. The solution will contain n arbitrary constants.

Second-Order Equations: Dependent Variable Missing Such an equation is of the form $F\left(x, \frac{dy}{dx}, \frac{d^2y}{dx^2}\right) = 0$. It can be reduced to a first-order equation by substituting $p = dy/dx$ and $dp/dx = d^2y/dx^2$.

Example A body of weight w lb falls from rest in a medium offering resistance proportional to the square of the velocity. If the limiting velocity (when acceleration = 0) is V and y acts positive downward, the differential equation of motion is $\frac{d^2y}{dt^2} = g - \frac{g}{w}\lambda\left(\frac{dy}{dt}\right)^2$ by Newton's laws. If $p = dy/dt$, the equation becomes $\frac{dp}{dt} = g\left(1 - \frac{\lambda}{w}p^2\right)$. This equation is of the separable-variables type. Two integrations and the use of the limiting velocity give the solution $y = (V^2/g) \ln \cosh (gt/V)$.

Second-Order Equations: Independent Variable Missing Such an equation is of the form $F\left(y, \frac{dy}{dx}, \frac{d^2y}{dx^2}\right) = 0$. Set $\frac{dy}{dx} = p$, $\frac{d^2y}{dx^2} = p\frac{dp}{dy}$. The result is a first-order equation in p, $F\left(y, p, p\frac{dp}{dy}\right) = 0$.

Example The capillary curve for one vertical plate is given by $\frac{d^2y}{dx^2} = \frac{4y}{c^2}\left[1 + \left(\frac{dy}{dx}\right)^2\right]^{3/2}$. Its solution by this technique is $x + \sqrt{c^2 - y^2} - \sqrt{c^2 - h_0^2} = \frac{c}{2}\left(\cosh^{-1}\frac{c}{y} - \cosh^{-1}\frac{c}{h_0}\right)$, where c, h_0 are physical constants.

Example $yy'' + 2(y')^2 = 0$. $yp(dp/dy) + 2p^2 = 0$ or $y(dp/dy) + 2p = 0$ and $p = 0$. The first when integrated twice gives $y^3 = c_1x + c_2$. This solution clearly includes the second $y = x + c$ as a special case.

Linear Differential Equations with Constant Coefficients and Right-Hand Member Zero (Homogeneous) The solution of $y'' + ay' + by = 0$ depends upon the nature of the roots of the characteristic equation $m^2 + am + b = 0$ obtained by substituting the trial solution $y = e^{mx}$ in the equation.

Distinct Real Roots If the roots of the characteristic equation are distinct real roots, r_1 and r_2, say, the solution is $y = Ae^{r_1x} + Be^{r_2x}$, where A and B are arbitrary constants.

Example $y'' + 4y' + 3 = 0$. The characteristic equation is $m^2 + 4m + 3 = 0$. The roots are -3 and -1, and the general solution is $y = Ae^{-3x} + Be^{-x}$.

Multiple Real Roots If $r_1 = r_2$, the solution of the differential equation is $y = e^{r_1x}(A + Bx)$.

Example $y'' + 4y' + 4 = 0$. The characteristic equation is $m^2 + 4m + 4 = 0$ with roots -2 and -2. The solution is $y = e^{-2x}(A + Bx)$.

Complex Roots If the characteristic roots are $p \pm iq$, then the solution is $y = e^{px}(A \cos qx + B \sin qx)$.

Example The differential equation $My'' + Ay' + ky = 0$ represents the vibration of a linear system of mass M, spring constant k, and damping constant A. If $A < 2\sqrt{kM}$, the roots of the characteristic equation $Mm^2 + Am + k = 0$ are complex $-\frac{A}{2M} \pm i\sqrt{\frac{k}{M} - \left(\frac{A}{2M}\right)^2}$ and the solution is $y = e^{-(At/2M)}$

$$\left\{c_1 \cos\left(\sqrt{\frac{k}{M} - \left(\frac{A}{2M}\right)^2}\right)t + c_2 \sin\left(\sqrt{\frac{k}{M} - \left(\frac{A}{2M}\right)^2}\right)t\right\}.$$

This solution is oscillatory, representing undercritical damping.

All these results generalize to homogeneous linear differential equations with constant coefficients of order higher than 2. These equations (especially of order 2) have been much used because of the ease of solution. Oscillations, electric circuits, diffusion processes, and heat-flow problems are a few examples for which such equations are useful.

Linear Nonhomogeneous Differential Equations

Linear Differential Equations Right-Hand Member $f(x) \neq 0$ Again the specific remarks for $y'' + ay' + by = f(x)$ apply to differential equations of similar type but higher order. We shall discuss two general methods.

Method of Undetermined Coefficients Use of this method is limited to equations exhibiting both constant coefficients and particular forms of the function $f(x)$. In most cases $f(x)$ will be a sum or product of functions of the type constant, x^n (n a positive integer), e^{mx}, $\cos kx$, $\sin kx$. When this is the case, the solution of the equation is $y = H(x) + P(x)$, where $H(x)$ is a solution of the homogeneous equations found by the method of the preceding subsection and $P(x)$ is a particular integral found by using the following table subject to these conditions: (1) When $f(x)$ consists of the sum of several terms, the appropriate form of $P(x)$ is the sum of the particular integrals corresponding to these terms individually. (2) When a term in any of the trial integrals listed is already a part of the homogeneous solution, the indicated form of the particular integral is multiplied by x.

Form of Particular Integral

If $f(x)$ is	Then $P(x)$ is
a (constant)	A (constant)
ax^n	$A_nx^n + A_{n-1}x^{n-1} + \cdots A_1x + A_0$
ae^{rx}	Be^{rx}
$c \cos kx$, $d \sin kx$	$A \cos kx + B \sin kx$
$gx^ne^{rx} \cos kx$, $hx^ne^{rx} \sin kx$	$(A_nx^n + \cdots + A_0)e^{rx} \cos kx + (B_nx^n + \cdots + B_0)e^{rx} \sin kx$

Since the form of the particular integral is known, the constants may be evaluated by substitution in the differential equation.

Example $y'' + 2y' + y = 3e^{2x} - \cos x + x^3$. The characteristic equation is $(m + 1)^2 = 0$ so that the homogeneous solution is $y = (c_1 + c_2x)e^{-x}$. To find a particular solution we use the trial solution from the table, $y = a_1e^{2x} + a_2 \cos x + a_3 \sin x + a_4x^3 + a_5x^2 + a_6x + a_7$. Substituting this in the differential equation collecting and equating like terms, there results $a_1 = 1/3$, $a_2 = 0$, $a_3 = -1/2$, $a_4 = 1$, $a_5 = -6$, $a_6 = 18$, and $a_7 = -24$. The solution is $y = (c_1 + c_2x)e^{-x} + \frac{1}{3}e^{2x} - \frac{1}{2}\sin x + x^3 - 6x^2 + 18x - 24$.

Method of Variation of Parameters This method is applicable to any linear equation. The technique is developed for a second-order equation but immediately extends to higher order. Let the equation be $y'' + a(x)y' + b(x)y = R(x)$ and let the solution of the homogeneous equation, found by some method, be $y = c_1f_1(x) + c_2f_2(x)$. It is now assumed that a particular integral of the differential equation is of the form $P(x) = uf_1 + vf_2$ where u, v are functions of x to be determined by two equations. One equation results from the requirement that $uf_1 + vf_2$ satisfy the differential equation, and the other is a degree of freedom open to the analyst. The best choice proves to be

$$u'f_1 + v'f_2 = 0 \quad \text{and} \quad u'f'_1 + v'f'_2 = R(x)$$

Then
$$u' = \frac{du}{dx} = -\frac{f_2}{f_1 f'_2 - f_2 f'_1} R(x)$$
$$v' = \frac{dv}{dx} = \frac{f_1}{f_1 f'_2 - f_2 f'_1} R(x)$$

and since f_1, f_2, and R are known u, v may be found by direct integration.

Example $(1 - x^2)\dfrac{d^2y}{dx^2} - \dfrac{1}{x}\dfrac{dy}{dx} = x$. The homogeneous equation $(1 - x^2)\dfrac{d^2y}{dx^2} - \dfrac{1}{x}\dfrac{dy}{dx} = 0$ reduces to $\dfrac{dp}{p} = \dfrac{dx}{x(1 - x^2)}$ when we set $dy/dx = p$. Upon integrating twice, $y = c_1\sqrt{x^2 - 1} + c_2$ is the homogeneous solution. Now assume that the particular solution has the form $y = u\sqrt{x^2 - 1} + v$. The equations for u and v become

$$u' = du/dx = 1$$
$$v' = \frac{dv}{dx} = -\frac{1}{\sqrt{x^2 - 1}}$$

so that $u = x$ and $v = \ln(x + \sqrt{x^2 - 1})$. The complete solution is $y = c_1\sqrt{x^2 - 1} + c_2 + x\sqrt{x^2 - 1} + \ln(x + \sqrt{x^2 - 1})$.

SPECIAL DIFFERENTIAL EQUATIONS

Euler's Equation The linear equation $x^n y^{(n)} + a_1 x^{n-1} y^{(n-1)} + \cdots + a_{n-1}xy' + a_n y = R(x)$ can be reduced to a linear equation with constant coefficients by the change of variable $x = e^t$. To solve the homogeneous equation substitute y-x^r into it, cancel the powers of x, which are the same for all terms, and solve the resulting polynomial for r. In case of multiple or complex roots there results the form $y = x^r(\log x)^r$ and $y = x^\alpha[\cos(\beta \log x) + i \sin(\beta \log x)]$.

Example Solve $x^2y'' - 2y = 0$. By setting $y = x^r$, $x^r[r(r - 1) - 2] = 0$. The roots of $r^2 - r - 2 = 0$ are $r = 2, -1$. The general solution is $y = Ax^2 + B/x$.

The equation $(ax + b)^n y^{(n)} + a_1(ax + b)^{n-1}y^{(n-1)} + \cdots + a_n y = R(x)$ can be reduced to the Euler form by the substitution $ax + b = z$. It may be treated without change of variable, the homogeneous equation having solutions of the form $y = (ax + b)^r$.

Bessel's Equation The linear equation $x^2(d^2y/dx^2) + (1 - 2\alpha)x(dy/dx) + [\beta^2\gamma^2x^{2\gamma} + (\alpha^2 - p^2\gamma^2)]y = 0$ is the general Bessel equation. By series methods, not to be discussed here, this equation can be shown to have the solution

$$y = Ax^\alpha J_p(\beta x^\gamma) + Bx^\alpha J_{-p}(\beta x^\gamma) \qquad p \text{ not an integer or zero}$$
$$y = Ax^\alpha J_p(\beta x^\gamma) + Bx^\alpha Y_p(\beta x^\gamma) \qquad p \text{ an integer}$$

where
$$J_p(x) = \left(\frac{x}{2}\right)^p \sum_{k=0}^{\infty} \frac{(-1)^k(x/2)^{2k}}{k!\,\Gamma(p + k + 1)}$$
$$J_{-p}(x) = \left(\frac{x}{2}\right)^{-p} \sum_{k=0}^{\infty} \frac{(-1)^k(x)/2)^{2k}}{k!\,\Gamma(k + 1 - p)} \qquad p \text{ not an integer}$$
and
$$\Gamma(n) = \int_0^\infty x^{-1}e^{n-x}\,dx \qquad n > 0$$

is the gamma function. For p an integer

$$J_p(x) = \left(\frac{x}{2}\right)^p \sum_{k=0}^{\infty} \frac{(-1)^k(x/2)^{2k}}{k!(p + k)!}$$

(Bessel function of the first kind of order p)

$$Y_p(x) = \frac{2}{\pi}\left\{\left(\ln\frac{x}{2} + \gamma\right)J_p(x) - \frac{1}{2}\sum_{k=0}^{n-1}\frac{(p - k - 1)!(x/2^{2k-p})}{k!} + \frac{1}{2}\sum_{k=0}^{\infty}(-1)^{k+1}[\phi(k) + \phi(k + p)]\frac{(x/2)^{2k+p}}{k!(p + k)!}\right\}$$

where $\gamma = 0.5772157\ldots$ = Euler's constant and $\phi(k) = \sum_{m=1}^{k} 1/m$, $k \geq 1$, $\phi(0) = 0$. The series converge for all x. Much of the importance of Bessel's equation and Bessel functions lies in the fact that the solutions of numerous linear differential equations can be expressed in terms of them.

Example $d^2y/dx^2 + [9x - (63/4x^2)]y = 0$. In general form this is $x^2(d^2y/dx^2) + (9x^3 - 63/4)y = 0$. Thus $\alpha = 1/2$, $\gamma = 3/2$, $\beta = 2$, $p = 8/3$. The solution is (since $p \neq$ integer) $y = Ax^{1/2}J_{8/3}(2x^{3/2}) + Bx^{1/2}J_{-8/3}(2x^{3/2})$. Tables are available for the evaluation of many of these functions.

Example The heat flow through a wedge-shaped fin is characterized by the equation $x^2(d^2y/dx^2) + x(dy/dx) - axy = 0$, where $y = T - T_{\text{air}}$, α is a combination of physical constants, and x = distance from fin end. By comparing this with the standard equation, there results $\alpha = 0$, $p = 0$, $\gamma = 1/2$, $\beta^2 = -4a^2$ or $\beta = 2ai$. The solution is $y = AJ_0(2ai\sqrt{x}) + BY_0(2ai\sqrt{x})$.

Legendre's Equation The Legendre equation $(1 - x^2)y'' - 2xy' + n(n + 1)y = 0$, $n \geq 0$, has the solution $y = Au_n(x) + Bv_n(x)$ for n not an integer where

$$u_n(x) = 1 - \frac{n(n+1)}{2!}x^2 + \frac{n(n-2)(n+1)(n+3)}{4!}x^4 - \frac{n(n-2)(n-4)(n+1)(n+3)(n+5)}{6!}x^6 + \cdots$$
$$v_n(x) = x - \frac{(n-1)(n+2)}{3!}x^3 + \frac{(n-1)(n+2)(n+4)}{5!}x^5 - \cdots$$

If n is an even integer or zero, u_n is a polynomial in x. If n is an odd integer, then v_n is a polynomial. The interval of convergence for the series is $-1 < x < 1$. If n is an integer, set $P_n(x) = \dfrac{u_n(x)}{u_n(1)}$ (n even or zero), $P_n = \dfrac{v_n(x)}{v_n(1)}$ (n odd). The polynomials P_n are the so-called Legendre polynomials, $P_0(x) = 1$, $P_1(x) = x$, $P_2(x) = \frac{1}{2}(3x^2 - 1)$, $P_3(x) = \frac{1}{2}(5x^3 - 3x)$, ... Designate by $Q_n(x)$ the Legendre functions of the second kind where $Q_n(x) = -v_n(1)\,u_n(x)$ for n odd and $Q_n(x) = u_n(1)\,v_n(x)$ for n even. The solution of Legendre's equation for integer n is $y = AP_n(x) + BQ_n(x)$.

Laguerre's Equation The Laguerre equation $x(d^2y/dx^2) + (c - x)(dy/dx) - ay = 0$ is satisfied by the confluent hypergeometric function of Kummer, $M(a, c; x)$. If c is not an integer, $y = AM(a, c; x) + Bx^{1-c}M(1 + a - c, 2 - c; x)$. If $c = 1$, $a = -n$, n a positive integer or zero, one solution is the Laguerre polynomial $AL_n(x)$; $L_0 = 1$, $L_1 = 1 - x$, $L_2 = 2 - 4x + x^2$, $L_3 = 6 - 18x + 9x^2 - x^3$, $L_4 = 24 - 96x + 72x^2 - 16x^3 + x^4$, $L_{r+1} = (1 + 2r - x)L_r - r^2L_{r-1}$. If $c = k + 1$, $a = k - n$ where k and n are integers; one solution is the associated Laguerre polynomial $y = A(d^k/dx^k)L_n(x)$ if $k \leq n$.

Example $xy'' + (1 - x)y' + 3y = 0$. Here $c = 1$, $a = -3$. One solution is $y = AL_3 = A(6 - 18x + 9x^2 - x^3)$.

Hermite's Equation The Hermite equation $y'' - 2xy' + 2ny = 0$ is satisfied by the Hermite polynomial of degree n, $y = AH_n(x)$ if n is a positive integer or zero. $H_0(x) = 1$, $H_1(x) = 2x$, $H_2(x) = 4x^2 - 2$, $H_3(x) = 8x^3 - 12x$, $H_4(x) = 16x^4 - 48x^2 + 12$, $H_{r+1}(x) = 2xH_r(x) - 2rH_{r-1}(x)$.

Example $y'' - 2xy' + 6y = 0$. Here $n = 3$; so $y = AH_3 = A(8x^3 - 12x)$ is a solution.

Chebyshev's Equation The equation $(1 - x^2)y'' - xy' + n^2y = 0$ for n a positive integer or zero is satisfied by the nth Chebyshev polynomial $y = AT_n(x)$. $T_0(x) = 1$, $T_1(x) = x$, $T_2(x) = 2x^2 - 1$, $T_3(x) = 4x^3 - 3x$, $T_4(x) = 8x^4 - 8x^2 + 1$; $T_{r+1}(x) = 2xT_r(x) - T_{r-1}(x)$.

Example $(1 - x^2)y'' - xy' + 36y = 0$. Here $n = 6$. A solution is $y = AT_6(x) = 2xT_5(x) - T_4(x) = 2x(2xT_4 - T_3) - T_4 = 32x^6 - 48x^4 + 18x^2 - 1$. Further details on these special equations and others can be found in the literature.

PARTIAL DIFFERENTIAL EQUATIONS

The analysis of situations involving two or more independent variables frequently results in a partial differential equation.

Example The equation $\partial T/\partial t = K(\partial^2 T/\partial x^2)$ represents the unsteady one-dimensional conduction of heat.

Example The equation for the unsteady transverse motion of a uniform beam clamped at the ends is $\dfrac{\partial^4 y}{\partial x^4} + \dfrac{\rho}{EI}\dfrac{\partial^2 y}{\partial t^2} = 0$.

Example The expansion of a gas behind a piston is characterized by the simultaneous equations $\dfrac{\partial u}{\partial t} + u\dfrac{\partial u}{\partial x} + \dfrac{c^2}{\rho}\dfrac{\partial \rho}{\partial x} = 0$ and $\dfrac{\partial \rho}{\partial t} + u\dfrac{\partial \rho}{\partial x} + \rho\dfrac{\partial u}{\partial x} = 0$.

Example The heating of a diathermanous solid is characterized by the equation $\alpha\,(\partial^2\theta/\partial x^2) + \beta e^{-\gamma x} = \partial\theta/\partial t$.

The partial differential equation $\partial^2 f/\partial x\,\partial y = 0$ can be solved by two integrations yielding the solution $f = g(x) + h(y)$, where $g(x)$ and $h(y)$ are arbitrary differentiable functions. This result is an example of the fact that the general solution of partial differential equations involves arbitrary functions in contrast to the solution of ordinary differential equations, which involve only arbitrary constants. A number of methods are available for finding the general solution of a partial differential equation. In most applications of partial differential equations the general solution is of limited use. In such applications the solution of a partial differential equation must satisfy both the equation and certain auxiliary conditions called **initial** and/or **boundary** conditions, which are dictated by the problem. Examples of these include those in which the wall temperature is a fixed constant $T(x_0) = T_0$, there is no diffusion across a nonpermeable wall, and the like. In ordinary differential equations these auxiliary conditions allow definite numbers to be assigned to the constants of integration. In partial differential equations the boundary conditions demand that the arbitrary functions resulting from integration assume specific forms. Except for a few cases (some first-order equations, D'Alembert's solution of the wave equation, and others) a procedure which first determines the arbitrary functions and then specializes them to fit the boundary conditions is usually not feasible. A more fruitful attack is to determine directly a set of particular solutions and then combine them so that the boundary conditions are satisfied. The only area in which much analysis has been accomplished is for linear homogeneous partial differential equations. Such equations have the property that if $f_1, f_2, \ldots, f_n, \ldots$ are individually solutions, then the function $f = \sum_{i=1}^{\infty} f_i$ is also a solution, provided the series converges and is differentiable up to the order (termwise) of the equation.

Quasilinear Partial Differential Equations of First Order By this is meant that the partial derivatives appear to the first degree only, while the coefficients may be functions of all the independent variables and the dependent variable. The general solution of the partial differential equation $P(x, y, z)(\partial z/\partial x) + Q(x, y, z)(\partial z/\partial y) = R(x, y, z)$ is $F(u, v) = 0$, where F is an arbitrary function arising from a relation $F(c_1, c_2) = 0$ and $u(x, y, z) = c_1$, $v(x, y, z) = c_2$ form a solution of the system $dx/P = dy/Q = dz/R$. If in addition it is desired to have the solution surface pass through a given curve (auxiliary condition), say, C, whose parametric equations are $x = x(t)$, $y = y(t)$, $z = z(t)$, where t is a parameter, then the particular solution of $u = c_1$, $v = c_2$ must be such that $u[x(t), y(t), z(t)] = c_1$, $v[x(t), y(t), z(t)] = c_2$. From these two equations eliminate the variable t to obtain a relationship $F(c_1, c_2) = 0$. The solution is then $F(u, v) = 0$.

Example Find the solution of $y(\partial z/\partial x) + zx(\partial z/\partial y) = 2xy$ passing through the circle $z = 0$, $x^2 + y^2 = 1$. The curve c in parametric form is $x = \cos t$, $y = \sin t$, $z = 0$. The equations $dx/y = dy/zx = dz/2xy$ have the solutions $u(x, y, z) = x^2 - z = c_1$, and $v(x, y, z) = y^2 - (z^2/2) = c^2$. Substituting the parametric equations for c into these results yields $c_1 + c_2 = 1$, upon elimination of t. The desired solution is then $x^2 - z + y^2 - (z^2/2) = 1$.

The general techniques applicable to the pfaffian equation $P\,dx + Q\,dy + R\,dx = 0$ and $dx/P = dy/Q = dz/R$ may be found in the literature. Generalization of this method for nonlinear partial differential equations of first order have been worked out. Cauchy's method of characteristics, Charpit's method, and Jacobi's method can be found in the literature.

Partial Differential Equations of Second and Higher Order Many of the applications to scientific problems fall naturally into partial differential equations of second order, although there are important exceptions in elasticity, vibration theory, and elsewhere.

Phenomena of **propagation** such as vibrations are characterized by equations of "hyperbolic" type which are essentially different in their properties from other classes such as those which describe equilibrium (elliptic) or unsteady diffusion and heat transfer (parabolic). Prototypes are as follows:

Elliptic Laplace's equation $\partial^2 u/\partial x^2 + \partial^2 u/\partial y^2 = 0$ and Poisson's equation $\partial^2 u/\partial x^2 + \partial^2 u/\partial y^2 = g(x, y)$ do not contain the variable time explicitly and consequently represent equilibrium configurations. Laplace's equation is satisfied by the gravitational potential function at points in space not occupied by mass as well as by static electric or magnetic potential at points free from electric charges or magnetic poles. Other important functions satisfying Laplace's equation are the velocity potential of the irrotational motion of an incompressible fluid, used in hydrodynamics; the steady temperature at points in a homogeneous solid, and the steady state of diffusion through a homogeneous body. The gravitational potential V at points occupied by mass of density d satisfies Poisson's equation $\partial^2 V/\partial x^2 + \partial^2 V/\partial y^2 + \partial^2 V/\partial z^2 = -4\pi d$.

Parabolic The heat equation $\partial T/\partial t = \partial^2 T/\partial x^2 + \partial^2 T/\partial y^2$ represents nonequilibrium or unsteady states of heat conduction and diffusion.

Hyperbolic The wave equation $\partial^2 u/\partial t^2 = c^2(\partial^2 u/\partial x^2 + \partial^2 u/\partial y^2)$ represents wave propagation of many varied types.

The solution of problems involving partial differential equations often revolves about an attempt to reduce the partial differential equation to one or more ordinary differential equations. The solutions of the ordinary differential equations are then combined (if possible) so that the boundary conditions as well as the original partial differential equation are simultaneously satisfied. Three of these techniques are illustrated.

Similarity Variables The physical meaning of the term "similarity" relates to internal similitude, or self-similitude. Thus, similar solutions in boundary-layer flow over a horizontal flat plate are those for which the horizontal component of velocity u has the property that two velocity profiles located at different coordinates x differ only by a scale factor. The mathematical interpretation of the term similarity is a transformation of variables carried out so that a reduction in the number of independent variables is achieved. There are essentially two methods for finding similarity variables, "separation of variables" (not the classical concept) and the use of "continuous transformation groups." The basic theory is available in Hansen or Ames (see the references).

Example The equation $\partial\theta/\partial x = (A/y)(\partial^2\theta/\partial y^2)$ with the boundary conditions $\theta = 0$ at $x = 0$, $y > 0$; $\theta = 0$ at $y = \infty$, $x > 0$; $\theta = 1$ at $y = 0$, $x > 0$ represents the nondimensional temperature θ of a fluid moving past an infinitely wide flat plate immersed in the fluid. Turbulent transfer is neglected, as is molecular transport except in the y direction. It is now assumed that the equation and the boundary conditions can be satisfied by a solution of the form $\theta = f(y/x^n) = f(u)$, where $\theta = 0$ at $u = \infty$ and $\theta = 0$ at $u = 0$. The purpose here is to replace the independent variables x and y by the single variable u when it is hoped that a value of n exists which will allow x and y to be completely eliminated in the equation. In this case since $u = y/x^n$, there results after some calculation $\partial\theta/\partial x = -(nu/x)(d\theta/du)$, $\partial^2\theta/\partial y^2 = (1/x^{2n})(d^2\theta/du^2)$, and when these are substituted in the equation, $-(1/x)nu(d\theta/du) = (1/x^{3n})(A/u)(d^2\theta/du^2)$. For this to be a function of u only, choose $n = ⅓$. There results $(d^2\theta/du^2) + (u^2/3A)(d\theta/du) = 0$. Two integrations and use of the boundary conditions for this ordinary differential equation give the solution

$$\theta = \int_u^{\infty} \exp(-u^3/9A)\,du \Big/ \int_0^{\infty} \exp(-u^3/9A)\,du$$

Group Method Many of the elementary similarity variables are easily constructed from the following rules.

General Theory Let Σ be a system of partial differential equations given by $\phi_j = 0$ $(j = 1, 2, \ldots, n)$ in which x_i $(i = 1, \ldots, m)$ and y_j $(j = 1, 2, \ldots, n)$ are the independent and dependent variables, respectively. Let a group Γ_1 consisting of a set of the transformations be defined as $\bar{x}_1 = a^{\alpha_1}x_1$, $\bar{x}_r = a^{\alpha_r}x_r$, $r = 2, \ldots, m$; $\bar{y}_j = a^{\gamma_j}y_j$, $j = 1, \ldots, n$, where the parameter $a \neq 0$ is real and the α_r, γ_j are to be determined from the condition that the system Σ be constant conformally invariant under Γ_1. A set of functions $\phi_j(x_i)$ is said to be "conformally invariant" under a one-parameter group $\bar{x}_i \to x_i$ if $\phi_j(\bar{x}_i = f_j(\bar{x}_i; a)\phi_j(\bar{x}_i)$, $i = 1, 2, \ldots, n$, $j = 1, 2, \ldots, m$, where the $\phi_j(\bar{x}_i)$ are exactly the same functions of the $\bar{x}_i$ as the ϕ_j are of the x_i. These functions will be said to be "constant conformally invariant" if $f_j(\bar{x}_i; a)$ are independent of the $\bar{x}_i$ and "absolutely invariant" if $f_j \equiv 1$ for all j.

The requirement of constant conformal invariance gives rise to a set of simultaneous equations, nontrivial solutions of which generate similarity variables which are the invariants of Γ_1.

Part 1: Simplest Case Let x_1 be the independent variable to appear in the denominator. There are two cases: If $\alpha_1 \neq 0$, the invariants (similarity variables) of the linear group Γ_1 are $n_r = x_r/x_1^{\beta_r}$, $\beta_r = \alpha_r/\alpha_1$, $r = 2, \ldots, m$, and $f_j(n_2, n_3, \ldots, n_m) = y_j(x_1, x_2, \ldots, x_m)/(x_1)^{\gamma_j/\alpha_1}$, $j = 1, 2, \ldots, n$. If $\alpha_1 = 0$ and the simultaneous equations have a nontrivial solution, then we may choose the special group Γ_2 consisting of $\bar{x}_1 = x_1 + \ln a$, $\bar{x}_r = a^{\alpha_r}x_r$, $\bar{y}_j = a^{\gamma_j}y_j$ and the invariants (similarity variables) of the group are $\eta_r = x_r/\exp(\alpha_r x_1)$, $r = 2, \ldots, m$; $f_j(\eta_2, \ldots, \eta_m) = y_j(x_1, \ldots, x_m)/\exp(\gamma_j x_1)$, $j = 1, 2, \ldots, n$.

Part 2: Extension of One-Parameter Group An extension of the simplest case for the elimination of two or more variables is given here. We suppose two variables are to be eliminated, say, x_1 and x_2. Let Γ_1 be the one-parameter linear group of Part 1. If $\alpha_1 = \alpha_2 \neq 0$, the invariants of Γ_1 are $\eta_r = x_r/(bx_1 + cx_2)^{\alpha_r/\alpha_1}$, $r = 3, \ldots, m$; and $f_j(\eta_3, \ldots, \eta_m) = y_j(x_1, \ldots, x_m)/(bx_1 + cx_2)^{\gamma_j/\alpha_1}$, where b and c are arbitrary constants.

If $\alpha_1 = \alpha_2 = 0$, the invariants (similarity variables) are $\eta_r = x_r/\exp[(bx_1 + cx_2 + d)\alpha_r]$ and $f_j(\eta_3, \ldots, \eta_m) = y_j(x_1, \ldots, x_m/\exp[(bx_1 + cx_2 + d)\gamma_j]$.

Part 3: A Two-Parameter Group Instead of the group Γ_1 we may use the two-parameter group Γ_2 consisting of Γ_2: $\bar{x}_1 = a^{\alpha_1}x_1$, $\bar{x}_2 = b^{\beta_2}x_2$, $\bar{x}_r = a^{\alpha_r}b^{\beta_r}x_r$, $r = 3, \ldots, m$ and $\bar{y}_j = a^{\gamma_j}b^{\delta_j}y_j$, $j = 1, 2, \ldots, m$. The requirement that the system of partial differential equations by constant conformally invariant under Γ_2 generates the similarity variables. Four cases arise:

(1) $\alpha_1 \neq 0$, $\beta_2 \neq 0$

$$\eta_r = \frac{x_r}{x_1^{\alpha_r/\alpha_1}\,x_2^{\beta_r/\beta_2}} \qquad r = 3, \ldots, m$$

$$f_j(\eta_3, \ldots, \eta_m) = \frac{y_j(x_1, \ldots, x_m)}{x_1^{\gamma_j/\alpha_1}\,x_2^{\delta_j/\beta_2}} \qquad j = 1, \ldots, n$$

(2) $\alpha_1 \neq 0$, $\beta_2 = 0$

$$\eta_r = \frac{x_r}{x_1^{\alpha_r/\alpha_1}\exp(\beta_r x_2)}$$

$$f_j = \frac{y_j}{x_1^{\gamma_j/\alpha_1}\exp(\delta_j x_2)}$$

(3) $\alpha_1 = 0$, $\beta_2 \neq 0$

$$\eta_r = \frac{x_r}{\exp(\alpha_r x_1)x_2^{\beta_r/\beta_2}}$$

$$f_j = \frac{y_j}{\exp(\gamma_j x_1)x_2^{\delta_j/\beta_2}}$$

(4) $\alpha_1 = \beta_2 = 0$

$$\eta_r = \frac{x_r}{\exp(\alpha_r x_1 + \beta_r x_2)}$$

$$f_j = \frac{y_j}{\exp(\gamma_j x_1 + \delta_j x_2)}$$

Example The equations for transient, incompressible two-dimensional boundary-layer flow are $u_t + uu_x + vu_y = -\rho^{-1}p_x + \nu u_{yy}$, $u_x + v_y = 0$ with the boundary conditions $u = v = 0$ for $y = 0$; $u \to u_e(x, t)$ for $y \to \infty$. Introducing the stream function $\psi(u = \psi_y, v = -\psi_x)$ and applying Bernoulli's equation for flow outside the boundary layer $[-(1/\rho)p_x = u_{et} + u_e u_{ex}]$, we get the equation for ψ as $\psi_{yt} + \psi_y\psi_{xy} - \psi_x\psi_{yy} = u_{et} + u_e u_{ex} + \nu\psi_{yyy}$ with boundary conditions $\psi_x = \psi_y = 0$ for $y = 0$; $\psi_y \to u_e(x, t)$ for $y \to \infty$. Choose the linear group $\bar{x} = a^m x$, $\bar{y} = a^n y$, $\bar{t} = a^r t$, $\bar{\psi} = a^s\psi$, $\bar{u}_e = a^p u_e$ with $a \neq 0$. The quantities m, n, r, s, p are real numbers to be determined. Upon setting this group into the equation for the stream function, we get $a^{n+r-s}\bar{\psi}_{\bar{y}\bar{t}} + a^{m+2n-2s}(\bar{\psi}_{\bar{y}}\bar{\psi}_{\bar{x}\bar{y}} - \bar{\psi}_{\bar{x}}\bar{\psi}_{\bar{y}\bar{y}}) = a^{r-p}\bar{u}_{e\bar{t}} + a^{m-2p}\bar{u}_e\bar{u}_{e\bar{x}} + \nu a^{3n-s}\bar{\psi}_{\bar{y}\bar{y}\bar{y}}$. The requirement of constant conformal invariance requires that these coefficients all be the same—that is to say, the system of equations for m, n, r, s, and p is $n + r - s = m + 2n - 2s = r - p = m - 2p = 3n - s$. We now choose to eliminate t (we could just as easily eliminate x or y first) so that the solutions for $r \neq 0$ are equivalent to the solutions (division by r is convenient since the invariants have exponents of the form s/r, n/r) of $n/r + 1 - s/r = m/r + 2(n/r) - 2(s/r) = 1 - p/r = m/r - 2(p/r) = 3(n/r) - s/r$. The solution of this system can proceed along a variety of lines. We attempt to solve in (nontrivially) by setting $m/r = A$ where A is an arbitrary constant. Then $p/r = m/r - 1 = A - 1$; and successively $s/r = A - \tfrac{1}{2}$, and $n/r = \tfrac{1}{2}$. Other alternatives are possible but will result in the same system multiplied by an arbitrary constant. From Part 1 the corresponding invariants of the group, which are the similarity variables, are given by $\xi = x/t^A$, $\eta = y/t^{1/2}$, $f(\xi, \eta) = \psi(x, y, t)/t^{A-1/2}$, $V(\xi) = u_e(x, t)/t^{A-1}$. For similarity in t, the velocity of the outer flow must be of the form $u_e(x, t) = V(\xi)t^{A-1} = V(x/t^A)t^{A-1}$.

The partial differential equation for f is obtained by substituting the similarity variables into the equation and boundary conditions. Carrying out the differentiations gives

$$\nu f_{\eta\eta\eta} + \frac{1}{2}\eta f_{\eta\eta} - (A - 1)f_\eta + A\xi f_{\xi\eta} + f_\xi f_{\eta\eta} - f_\eta f_{\xi\eta}$$
$$+ (A - 1)V(\xi) - A\xi(dV/d\xi) + V(dV/d\xi) = 0$$

with $\partial f/\partial\xi = \partial f/\partial\eta = 0$ for $\eta = 0$; $\partial f/\partial\eta \to V(\xi)$ for $\eta \to \infty$. A further reduction can be accomplished by a reapplication of the procedure.

Additional examples and a bibliography leading to the deductive theory of similarity variables are available in Ames, "Recent Developments in the Nonlinear Equations of Transport," *Ind. Eng. Chem. Fundam.* (February 1969).

Separation of Variables This is a powerful, well-utilized method which is applicable in certain circumstances. It consists of assuming that the solution for a partial differential equation (illustrated for order 2) has the form $U = f(x)g(y)$. If it is then possible to obtain an ordinary differential equation on one side of the equation depending only on x and on the other side only on y, the partial differential equation is said to be **separable** in the variables x, y. If this is the case, one side of the equation is a function of x alone and the other of y alone. The two can be equal only if each is a constant, say λ. Thus the problem has again been reduced to the solution of ordinary differential equations.

Example Laplace's equation $\partial^2V/\partial x^2 + \partial^2V/\partial y^2 = 0$ plus the boundary conditions $V(0, y) = 0$, $V(l, y) = 0$, $V(x, \infty) = 0$, $V(x, 0) = f(x)$ represents the steady-state potential in a thin plate (in z direction) of infinite extent in the y direction and of width l in the x direction. A potential $f(x)$ is impressed (at $y = 0$) from $x = 0$ to $x = 1$, and the sides are grounded. To obtain a solution of this boundary-value problem assume $V(x, y) = f(x)g(y)$. Substitution in the differential equation yields $f''(x)g(y) + f(x)g''(y) = 0$, or $g''(y)/g(y) = -f''(x)/f(x) = \lambda^2$ (say). This system becomes $g''(y) - \lambda^2 g(y) = 0$ and $f''(x) + \lambda^2 f(x) = 0$. The solutions of these ordinary differential equations are respectively $g(y) = Ae^{\lambda y} + Be^{-\lambda y}$, $f(x) = C\sin\lambda x + D\cos\lambda x$. Then $f(x)g(y) = (Ae^{\lambda y} + Be^{-\lambda y})(C\sin\lambda x + D\cos\lambda x)$. Now $V(0, y) = 0$ so that $f(0)g(y) = (Ae^{\lambda y} + Be^{-\lambda y})D \equiv 0$ for all y. Hence $D = 0$. The solution then has the form $\sin\lambda x\,(Ae^{\lambda y} + Be^{-\lambda y})$ where the multiplicative constant C has been eliminated. Since $V(l, y) = 0$, $\sin\lambda l(Ae^{\lambda y} + Be^{-\lambda y}) \equiv 0$. Clearly the bracketed function of y is not zero, for the solution would then be the identically zero solution. Hence $\sin\lambda l = 0$ or $\lambda_n = n\pi/l$, $n = 1, 2, \ldots$ where $\lambda_n = n$th eigenvalue.

The solution now has the form $\sin(n\pi x/l)(Ae^{n\pi y/l} + Be^{-n\pi y/l})$. Since $V(x, \infty) = 0$, A must be taken to be zero because e^y becomes arbitrarily large as $y \to \infty$. The solution then reads $B_n \sin(n\pi x/l)e^{-n\pi y/l}$, where B_n is the multiplicative constant. The differential equation is linear and homogeneous so that $\sum_{n=1}^{\infty} B_n e^{-n\pi y/l} \sin(n\pi x/l)$ is also a solution. Satisfaction of the last boundary condition is ensured by taking $B_n = \frac{2}{l}\int_0^l f(x) \sin(n\pi x/l)\, dx$ = Fourier sine coefficients of $f(x)$. Further, convergence and differentiability of this series are established quite easily. Thus the solution is

$$V(x, y) = \sum_{n=1}^{\infty} B_n e^{-n\pi y/l} \sin\frac{n\pi x}{l}$$

Integral-Transform Method A number of integral transforms are used in the solution of differential equations. Only one, the Laplace transform, will be discussed here [for others, see "Integral Transforms (Operational Methods)"]. The one-sided Laplace transform indicated by $L[f(t)]$ is defined by the equation $L[f(t)] = \int_0^\infty f(t)e^{-st}\, dt$. It has numerous important properties. The ones of interest here are $L[f'(t)] = sL[f(t)] - f(0)$; $L[f''(t)] = s^2L[f(t)] - sf(0) - f'(0)$; $L[f^{(n)}(t)] = s^nL[f(t)] - s^{n-1}f(0) - s^{n-2}f'(0) - \cdots - f^{(n-1)}(0)$ for ordinary derivatives. For partial derivatives an indication of which variable is being transformed avoids confusion. Thus, if $y = y(x, t)$, $L_t\left[\frac{\partial y}{\partial t}\right] = sL[y(x, t)] - y(x, 0)$, whereas $L_t\left[\frac{\partial y}{\partial x}\right] = \frac{dL_t[y(x, t)]}{dx}$, since $L[y(x, t)]$ is "really" only a function of x. Otherwise the results are similar. These facts coupled with the linearity of the transform, i.e., $L[af(t) + bg(t)] = aL[f(t)] + bL[g(t)]$, make it a useful device in solving some linear differential equations.

Its use reduces the solution of ordinary differential equations to the solution of algebraic equations for $L[y]$. The solution of partial differential equations is reduced to the solution of ordinary differential equations. In both situations the inverse transform must be obtained either from tables, of which there are several, or by use of complex inversion methods.

Example $y'' - 3y' + 2y = 0$. Initial conditions at $t = 0$, $y = 0$, $y' = 1$. $L[y''] = s^2L[y] - 1$. $L[y'] = sL[y]$. Hence $L[y'' - 3y' + 2y] = s^2L[y] - 1 - 3sL[y] + 2L[y] = 0$. Solving for $L[y]$ there results in $L[y] = \frac{1}{(s-2)(s-1)} = \frac{1}{s-2} - \frac{1}{s-1}$. From the tables the function y having this transform is $y = e^{2t} - e^t$.

Example The equation $\partial c/\partial t = D(\partial^2 c/\partial x^2)$ represents the diffusion in a semi-infinite medium, $x \geq 0$. Under the boundary conditions $c(0, t) = c_0$, $c(x, 0) = 0$ find a solution of the diffusion equation. By taking the Laplace transform of both sides with respect to t,

$$\int_0^\infty e^{-st}\frac{\partial^2 c}{\partial x^2}\, dt = \frac{1}{D}\int_0^\infty e^{-st}\frac{\partial c}{\partial t}\, dt$$

or
$$d^2F/dx^2 = (1/D)sF - c(x, 0) = sF/D$$

where $F(x, s) = L_t[c(x, t)]$. Hence

$$d^2F/dx^2 - (s/D)F = 0$$

The last boundary condition transforms into $F(0, s) = c_0/s$. Finally the solution of the ordinary differential equation for F subject to $F(0, s) = c_0/s$ and F remains finite as $x \to \infty$ is $F(x, s) = (c_0/s)e^{-\sqrt{(s/D)}x}$. Reference to a table shows that the function having this as its Laplace transform is

$$c(x, t) = c_0\left[1 - \frac{2}{\sqrt{\pi}}\int_0^{x/2\sqrt{Dt}} e^{-u^2}\, du\right]$$

DIFFERENCE EQUATIONS

References: 29, 41, 69, 238, 268.

Certain situations are such that the independent variable does not vary continuously but has meaning only for discrete values. Typical illustrations occur in the stagewise processes found in chemical engineering such as distillation, staged extraction systems, and absorption columns. In each of these the operation is characterized by a finite between-stage change of the dependent variable in which the independent variable is the integral number of the stage. The importance of difference equations is twofold: (1) to analyze problems of the type described and (2) to obtain approximate solutions of problems which lead, in their formulation, to differential equations. In this subsection only problems of analysis are considered; the application to approximate solutions is considered under "Numerical Analysis and Approximate Methods."

ELEMENTS OF THE CALCULUS OF FINITE DIFFERENCES

Let $y = f(x)$ be defined for discrete equidistant values of x, which will be denoted by x_n. The corresponding value of y will be written $y_n = f(x_n)$. The first forward difference of $f(x)$ denoted by $\Delta f(x) = f(x + h) - f(x)$ where $h = x_n - x_{n-1}$ = interval length.

Example Let $f(x) = x^2$. Then $\Delta f(x) = (x + h)^2 - x^2 = 2hx + h^2$.

The second forward difference is obtained by taking the difference of the first; thus $\Delta\Delta f(x) = \Delta^2 f(x) = \Delta f(x + h) - \Delta f(x) = f(x + 2h) - 2f(x + h) + f(x)$.

Example $f(x) = x^2$, $\Delta^2 f(x) = \Delta[\Delta f(x)] = \Delta 2hx + \Delta h^2 = 2h(x + h) - 2hx + h^2 - h^2 = 2h^2$.

Similarly the nth forward difference is defined by the relation $\Delta^n f(x) = \Delta[\Delta^{n-1}f(x)]$. Other difference relations are also quite useful. Some of these are $\nabla f(x) = f(x) - f(x - h)$, which is called the backward difference, and $\delta f(x) = f[x + (h/2)] - f[x - (h/2)]$, called the central difference. Some properties of the operator Δ are quite important. If C is any constant, $\Delta C = 0$; if $f(x)$ is any function of period h, $\Delta f(x) = 0$ (in fact, periodic functions of period h play the same role here as constants do in the differential calculus); $\Delta[f(x) + g(x)] = \Delta f(x) + \Delta g(x)$; $\Delta^m[\Delta^n f(x)] = \Delta^{m+n}f(x)$; $\Delta[f(x)g(x)] = f(x)\Delta g(x) + g(x + h)\,\Delta f(x)$; $\Delta\left[\frac{f(x)}{g(x)}\right] = \frac{g(x)\,\Delta f(x) - f(x)\,\Delta g(x)}{g(x)g(x + h)}$.

Example $\Delta(x \sin x) = x\Delta \sin x + \sin(x + h)\,\Delta x = 2\sin(h/2)\cos[x + (h/2)] + h\sin(x + h)$.

DIFFERENCE EQUATIONS

A difference equation is a relation between the differences and the independent variable, $\phi(\Delta^n y, \Delta^{n-1}y, \ldots, \Delta y, y, x) = 0$, where ϕ is some given function. The general case in which the interval between the successive points is any real number h, instead of 1, can be reduced to that with interval size 1 by the substitution $x = hx'$. Hence all further difference-equation work will assume the interval size between successive points is 1.

Example $f(x + 1) - (\alpha + 1)f(x) + \alpha f(x - 1) = 0$. Common notation usually is $y_x = f(x)$. This equation is then written $y_{x+1} - (\alpha + 1)y_x + \alpha y_{x-1} = 0$.

Example $y_{x+2} + 2y_x y_{x+1} + y_x = x^2$.

Example $y_{x+1} - y_x = 2^x$.

The order of the difference equation is the difference between the largest and smallest arguments when written in the form of the second example. The first and second examples are both of order 2, while the third example is of order 1. A linear difference equation involves no products or other nonlinear functions of the dependent variable and its differences. The first and third examples are linear, while the second example is nonlinear.

A solution of a difference equation is a relation between the variables which satisfies the equation. If the difference equation is of order n, the general solution involves n arbitrary constants. The techniques for solving difference equations resemble techniques used for differential equations.

Equation $\Delta^n y = a$ The solution of $\Delta^n y = a$, where a is a constant, is a polynomial of degree n plus an arbitrary periodic function of period 1. That is, $y = (ax^n/n!) + c_1x^{n-1} + c_2x^{n-2} + \cdots + c_n + f(x)$, where $f(x + 1) = f(x)$.

Example $\Delta^3 y = 6$. The solution is $y = x^3 + c_1x^2 + c_2x + c_3 + f(x)$; c_1, c_2, c_3 are arbitrary constants, and $f(x)$ is an arbitrary periodic function of period 1.

Equation $y_{x+1} - y_x = \phi(x)$ This equation states that the first difference of the unknown function is equal to the given function $\phi(x)$. The solution by analogy with solving the differential equation $dy/dx = \phi(x)$ by integration is obtained by "finite integration" or summation. When there are only a finite number of data points, this is easily accomplished by writing $y_x = y_0 t \sum_{t=1}^{x} \phi(t-1)$, where the data points are numbered from 1 to x. This is the only situation considered here.

Examples If $\phi(x) = 1$, $y_x = x$. If $\phi(x) = x$, $y_x = [x(x-1)]/2$. If $\phi(x) = a^x$, $a \neq 0$, $y_x = a^x/(a-1)$. In all cases $y_0 = 0$.

Other examples may be evaluated by using summation, that is, $y_2 = y_1 + \phi(1)$, $y_3 = y_2 + \phi(2) = y_1 + \phi(1) + \phi(2)$, $y_4 = y_3 + \phi(3) = y_1 + \phi(1) + \phi(2) + \phi(3), \ldots, y_x = y_1 + \sum_{t=1}^{x-1} \phi(t)$.

Example $y_{x+1} - ry_x = 1$, r constant, $x > 0$ and $y_0 = 1$. $y_1 = 1 + r$, $y_2 = 1 + r + r^2, \ldots, y_x = 1 + r + \cdots + r^x = (1 - r^{x+1})/(1 - r)$ for $r \neq 1$ and $y_x = 1 + x$ for $r = 1$.

Linear Difference Equations The linear difference equation of order n has the form $P_n y_{x+n} + P_{n-1}y_{x+n-1} + \cdots + P_1 y_{x+1} + P_0 y_x = Q(x)$ with $P_n \neq 0$ and $P_0 \neq 0$ and P_j; $j = 0, \ldots, n$ are functions of x.

Constant Coefficient and $Q(x) = 0$ (Homogeneous) The solution is obtained by trying a solution of the form $y_x = c\beta^x$. When this trial solution is substituted in the difference equation, a polynomial of degree n results for β. If the solutions of this polynomial are denoted by $\beta_1, \beta_2, \ldots, \beta_n$ then the following cases result: (1) if all the β_j's are real and unequal, the solution is $y_x = \sum_{j=1}^{n} c_j\beta_j^x$, where the $c_1, \ldots, c_n$ are arbitrary constants; (2) if the roots are real and repeated, say, β_j has multiplicity m, then the partial solution corresponding to β_j is $\beta_j^x(c_1 + c_2x + \cdots + c_mx^{m-1})$; (3) if the roots are complex conjugates, say, $a + ib = pe^{i\theta}$ and $a - ib = pe^{-i\theta}$, the partial solution corresponding to this pair is $p^x(c_1 \cos \theta x + c_2 \sin \theta x)$; and (4) if the roots are multiple complex conjugates, say, $a + ib = pe^{i\theta}$ and $a - ib = pe^{-i\theta}$ are m-fold, then the partial solution corresponding to these is $p^x[(c_1 + c_2x + \cdots + c_mx^{m-1}) \cos \theta x + (d_1 + d_2x + \cdots + d_mx^{m-1}) \sin \theta x]$.

Example The equation $y_{x+1} - (\alpha + 1)y_x + \alpha y_{x-1} = 0$, $y_0 = c_0$ and $y_{m+1} = x_{m+1}/k$ represents the steady-state composition of transferable material in the raffinate stream of a staged countercurrent liquid–liquid extraction system. Clearly y is a function of the stage number x. α is a combination of system constants. By using the trial solution $y_x = c\beta^x$, there results $\beta^2 - (\alpha + 1)\beta + \alpha = 0$, so that $\beta_1 = 1$, $\beta_2 = \alpha$. The general solution is $y_x = c_1 + c_2\alpha^x$. By using the side conditions, $c_1 = c_0 - c_2$, $c_2 = (y_{m+1} - c_0)(\alpha^{m+1} - 1)$. The desired solution is $(y_x - c_0)(y_{m+1} - c_0) = (\alpha^x - 1)(\alpha^{m+1} - 1)$.

Example $y_{x+3} - 3y_{x+2} + 4y_x = 0$. By setting $y_x = c\beta^x$, there results $\beta^3 - 3\beta^2 + 4 = 0$ or $\beta_1 = -1$, $\beta_2 = 2$, $\beta_3 = 2$. The general solution is $y_x = c_1(-1)^x + 2^x(c_2 + c_3x)$.

Example $y_{x+1} - 2y_x + 2y_{x-1} = 0$. $\beta_1 = 1 + i$, $\beta_2 = 1 - i$. $p = \sqrt{1+1} = \sqrt{2}$, $\theta = \pi/4$. The solution is $y_x = 2^{x/2}[c_1 \cos(x\pi/4) + c_2 \sin(x\pi/4)]$.

Constant Coefficients and $Q(x) \neq 0$ (Nonhomogeneous) In this case the general solution is found by first obtaining the homogeneous solution, say, y_x^H and adding to it any particular solution with $Q(x) \neq 0$, say, y_x^P. There are several means of obtaining the particular solution.

Method of Undetermined Coefficients If $Q(x)$ is a product or linear combination of products of the functions e^{bx}, a^x, x^p (p a positive integer or zero) cos cx and sin cx, this method may be used. The "families" $[a^x]$, $[e^{bx}]$, [sin cx, cos cx] and $[x^p, x^{p-1}, \ldots, x, 1]$ are defined for each of the above functions in the following way: The family of a term f_x is the set of all functions of which f_x and all operations of the form a^{x+y}, $\cos c(x + y)$, $\sin c(x + y)$, $(x + y)^p$ on f_x and their linear combinations result in. The technique involves the following steps: (1) Solve the homogeneous system. (2) Construct the family of each term. (3) If the family has no representative in the homogeneous solution, assume y_x^P is a linear combination of the families of each term and determine the constants so that the equation is satisfied. (4) If a family has a representative in the homogeneous solution, multiply each member of the family by the smallest integral power of x for which all such representatives are removed and revert to step 3.

Example $y_{x+1} - 3y_x + 2y_{x-1} = 1 + a^x$. $a \neq 0$. The homogeneous solution is $y_x^H = c_1 + c_2 2^x$. The family of 1 is 1 and of a^x is a^x. However, 1 is a solution of the homogeneous system. Therefore, try $y_x^P = Ax + Ba^x$. Substituting in the equation there results $y_x = c_1 + c_2 2^x - x + \dfrac{a}{(a-1)(a-2)}a^x$, $a \neq 1$, $a \neq 2$. If $a = 1$, $y_x = c_1 + c_2 2^x - x$. If $a = 2$, $y_x = c_1 + c_2 2^x - x + x2^x$.

Example The family of $x^2 3^x$ is $[x^2 3^x, x3^x, 3^x]$.

Method of Variation of Parameters This technique is applicable to general linear difference equations. It is illustrated for the second-order system $y_{x+2} + Ay_{x+1} + By_x = \phi(x)$. Assume that the homogeneous solution has been found by some technique and write $y_x^H = c_1u_x + c_2v_x$. Assume that a particular solution $y_x^P = D_xu_x + E_xv_x$. E_x and D_x can be found by solving the equations:

$$E_{x+1} - E_x = \frac{u_{x+1}\phi(x)}{u_{x+1}v_{x+2} - u_{x+2}v_{x+1}}$$

$$D_{x+1} - D_x = \frac{v_{x+1}\phi(x)}{v_{x+1}u_{x+2} - v_{x+2}u_{x+1}}$$

by summation. The general solution is then $y_x = y_x^P + y_x^H$.

Example $y_{x+2} - (2\cos\alpha)y_{x+1} + y_x = \sqrt{x}$ $x \geq 0$. The homogeneous solution is $y_x^H = c_1 \cos \alpha x + c_2 \sin \alpha x$. With $u_x = \cos \alpha x$, $v_x = \sin \alpha x$ it is found that $E_{x+1} - E_x = \dfrac{[\cos \alpha(x+1)]\sqrt{x}}{\sin \alpha}$, $D_{x+1} - D_x = \dfrac{-\sqrt{x}\sin \alpha(x+1)}{\sin \alpha}$. Thus for $\sin \alpha \neq 0$, $E_0 = 0$, $D_0 = 0$.

$$D_x = -\sum_{n=0}^{x} \frac{\sqrt{n}\sin n\alpha}{\sin \alpha} \qquad E_x = \sum_{n=0}^{x} \frac{\sqrt{n}\cos n\alpha}{\sin \alpha}$$

The general solution is $y_x = c_1 \cos \alpha x + c_2 \sin \alpha x - \cos \alpha x \sum_{n=0}^{x} \dfrac{\sqrt{n}\sin \alpha n}{\sin \alpha} + \sin \alpha x \sum_{n=0}^{x} \dfrac{\sqrt{n}\cos \alpha n}{\sin \alpha}$.

Variable Coefficients The method of variation of parameters applies equally well to the linear difference equation with variable coefficients. Techniques are therefore needed to solve the homogeneous system with variable coefficients.

Equation $y_{x+1} - a_xy_x = 0$ By assuming that this equation is valid for $x \geq 0$ and $y_0 = c$, the solution is $y_x = c\prod_{n=1}^{x} a_{x-1}$.

Example $y_{x+1} + \frac{x+2}{x+1} y_x = 0$. The solution is

$$y_x = c \prod_{n=1}^{x} \left(-\frac{n+1}{n} \right) = c(-1)^x \frac{2}{1} \cdot \frac{3}{2} \cdots \frac{x+1}{x} = (-1)^x c(x+1)$$

Example $y_{x+1} - xy_x = 0$. The solution is $y_x = c(x-1)!$

Reduction of Order If one homogeneous solution, say, u_x, can be found by inspection or otherwise, an equation of lower order can be obtained by the substitution $v_x = y_x/u_x$. The resultant equation must be satisfied by v_x = constant or $\Delta v_x = 0$. Thus the equation will be of reduced order if the new variable $U_x = \Delta(y_x/u_x)$ is introduced.

Example $(x+2)y_{x+2} - (x+3)y_{x+1} + y_x = 0$. By observation $u_x = 1$ is a solution. Set $U_x = \Delta y_x = y_{x+1} - y_x$. There results $(x+2)U_{x+1} - U_x = 0$, which is of degree one lower than the original equation. The complete solution for y_x is finally $y_x = c_0 \sum_{n=0}^{x} \frac{1}{n!} + c_1$.

Factorization If the difference equation can be factored, then the general solution can be obtained by solving two or more successive equations of lower order. Consider $y_{x+2} + A_x y_{x+1} + B_x y_x = \phi(x)$. If there exists a_x, b_x such that $a_x + b_x = -A_x$ and $a_x b_x = B_x$, then the difference equation may be written $y_{x+2} - (a_x + b_x)\, y_{x+1} + a_x b_x y_x = \phi(x)$. First solve $U_{x+1} - b_x U_x = \phi(x)$ and then $y_{x+1} - a_x y_x = U_x$.

Example $y_{x+2} - (2x+1)y_{x+1} + (x^2+x)y_x = 0$. Set $a_x = x$, $b_x = x + 1$. Solve $u_{x+1} - (x+1)u_x = 0$ and then $y_{x+1} - xy_x = u_x$.

Substitution If it is possible to rearrange a difference equation so that it takes the form $af_{x+2}y_{x+2} + bf_{x+1}y_{x+1} + cf_x y_x = \phi(x)$ with a, b, c constants, then the substitution $u_x = f_x y_x$ reduces the equation to one with constant coefficients.

Example $(x+2)^2 y_{x+2} - 3(x+1)^2 y_{x+1} + 2x^2 y_x = 0$. Set $u_x = x^2 y_x$. The equation becomes $u_{x+2} - 3u_{x+1} + 2u_x = 0$, which is linear and easily solved by previous methods

The substitution $u_x = y_x/f_x$ reduces $af_x f_{x+2} + bf_x f_{x+2} y_{x+1} + cf_{x+1} f_{x+2} y_x = \phi(x)$ to an equation with constant coefficients.

Example $x(x+1)y_{x+2} + 3x(x+2)y_{x+1} - 4(x+1)(x+2)y_x = x$. Set $u_x = y_x/f_x = y_x/x$. Then $y_x = xu_x$, $y_{x+1} = (x+1)u_{x+1}$ and $y_{x+2} = (x+2)u_{x+2}$. Substitution in the equation yields $x(x+1)(x+2)u_{x+2} + 3x(x+2)(x+1)u_{u+1} - 4x(x+1)(x+2)u_x = x$ or $u_{x+2} + 3u_{x+1} - 4u_x = [1/(x+1)]\,(x+2)$, which is a linear equation with constant coefficients.

Nonlinear Difference Equations: Riccati Difference Equation The Riccati equation $y_{x+1}y_x + ay_{x+1} + by_x + c = 0$ is a nonlinear difference equation which can be solved by reduction to linear form. Set $y = z + h$. The equation becomes $z_{x+1}z_x + (h+a)z_{x+1} + (h+b)z_x + h^2 + (a+b)h + c = 0$. If h is selected as a root of $h^2 + (a+b)h + c = 0$ and the equation is divided by $z_{x+1}z_x$ there results $[(h+b)/z_{x+1}] + [(h+a)/z_x] + 1 = 0$. This is a linear equation with constant coefficients. The solution is

$$y_x = h + \frac{1}{c\left[-\dfrac{a+h}{b+h}\right]^x - \dfrac{1}{(a+h)+(b+h)}}$$

Example This equation is obtained in distillation problems, among others, in which the number of theoretical plates is required. If the relative volatility is assumed to be constant, the plates are theoretically perfect, and the molal liquid and vapor rates are constant, then a material balance around the nth plate of the enriching section yields a Riccati difference equation.

INTEGRAL EQUATIONS

REFERENCES: 66, 73, 81, 146, 156, 166, 197, 238, 255. See also "Numerical Analysis and Approximate Methods."

An integral equation is any equation in which the unknown function appears under the sign of integration and possibly outside the sign of integration. Such equations are important in mathematical applications to physical problems because it is often possible to restate a differential equation, together with its boundary conditions, as a single integral equation.

The simplest type of integral equation arises from the integration of the differential equation $dy/dx = f(x, y)$ with the initial condition $y = y_0$ when $x = x_0$. This is equivalent to $dy = f(x, y)\, dx$ or $\int_{y_0}^{y} dy = \int_{x_0}^{x} f(x, y)\, dx$. Upon integrating there results $y = y_0 + \int_{x_0}^{x} f(x, y)\, dx$. This type of equation can be solved by a process of **successive approximations** often called the **Picard method,** which is useful for obtaining approximate solutions of differential equations. The iteration procedure is to select a first approximation for y, say, $y = y_0$, then $y^{(1)} = y_0 + \int_{x_0}^{x} f(x, y_0)\, dx$. After integrating, $y^{(2)} = y_0 + \int_{x_0}^{x} f(x, y^{(1)})\, dx$, and in general $y^{(n)} = y_0 + \int_{x_0}^{x} f(x, y^{(n-1)})\, dx$.

Example $dy/dx = x^2 + y$. $x_0 = 0$, $y_0 = 1$. This problem is equivalent to the integral equation $y = 1 + \int_0^x (x^2 + y)\, dx$. Let the initial approximation for y be 1. Then $y^{(1)} = 1 + \int_0^x (x^2 + 1)\, dx = 1 + x + \frac{x^3}{3}$. $y^{(2)} = 1 + \int_0^x [x^2 + y^{(1)}]\, dx = 1 + \int_0^x \left[x^2 + 1 + x + \frac{x^3}{3} \right] dx = 1 + x + \frac{x^2}{2} + \frac{x^3}{3} + \frac{x^4}{12}$, etc.

CLASSIFICATION OF INTEGRAL EQUATIONS

$\int_a^b K(x, t)u(t)\, dt = f(x)$, where $f(x)$, $K(x, t)$, a and b are known and u is to be determined, is a linear equation of the first kind of **Fredholm type.** $K(x, t)$ is called the **kernel** function of the equation. If u, the unknown function, occurs only to the first power, the equation is said to be linear.

$\int_a^x K(x, t)u(t)\, dt = f(x)$ is a linear integral equation of the first kind of **Volterra type.**

An equation of the form $u(x) = f(x) + \int_a^b K(x, t)u(t)\, dt$ is said to be a linear integral equation of the second kind. If b is constant, it is of Fredholm type. If $b = x$, it is of Volterra type. If $f(x)$ is identically zero, then $u(x) = \int_a^b K(x, t)u(t)\, dt$ is the homogeneous linear integral equation of the second kind. Sometimes a parameter λ is introduced; thus $u(x) = f(x) + \lambda \int_a^b K(x, t)u(t)\, dt$, which facilitates the solution and may take on various values in a particular problem.

The equation $\phi(x)u(x) = f(x) + \int_a^b K(x, t)u(t)\, dt$ is the linear integral equation of the third kind. If b = constant, it is of Fredholm type, and if $b = x$, it is of Volterra type.

If the unknown function u appears in the equation in any way except to the first power, the integral equation is said to be nonlinear. The equation $u(x) = f(x) + \int_a^b K(x, t)[u(t)]^{3/2}\, dt$ is nonlinear. The differential equation $du/dx = g(x, u)$ is equivalent to the nonlinear integral equation $u(x) = c + \int_a^x g[t, u(t)]\, dt$.

An integral equation is said to be singular when either one or both of the limits of integration become infinite or if $K(x, t)$ becomes infinite for one or more points of the interval under discussion.

Example $u(x) = x + \int_0^\infty \cos(xt)\, u(t)\, dt$ and $f(x) = \int_0^x \frac{u(t)}{x-t}\, dt$ are both singular. The kernel of the first equation is $\cos(xt)$, and that of the second is $(x-t)^{-1}$.

RELATION TO DIFFERENTIAL EQUATIONS

The **Leibnitz rule** $\frac{d}{dx}\int_{a(x)}^{b(x)} F(x,t)\, dt = \int_{a(x)}^{b(x)} \frac{\partial F(x,t)}{\partial x}\, dt + F[x, b(x)]\frac{db}{dx} - F[x, a(x)]\frac{da}{dx}$ is frequently useful for differentiation of an integral involving a parameter. If $I_n(x) = \int_a^x (x-t)^{n-1} f(t)\, dt$, then $d^n I_n/dx^n = (n-1)!\, f(x)$, and since $I_n(a) = 0$, $n \geq I$, it follows that $I_n(a) = [dI_n(a)/dx] = [d^2 I_n(a)/dx^2] = \cdots = [d^{n-1} I_n(a)/dx^{n-1}] = 0$. Hence $\frac{I_n(x)}{(n-1)!} = \frac{1}{(n-1)!}\int_a^x (x-t)n^{-1} f(t)\, dt = \underbrace{\int_a^x \cdots \int_a^x}\, f(x)\, \underbrace{dx \cdots dx}_{n \text{ times}}$. This result can be used to show the equivalence of the initial-value problem consisting of the second-order differential equation $d^2y/dx^2 + A(x)(dy/dx) + B(x)y = f(x)$ together with the prescribed initial conditions $y(a) = y_0$, $y'(a) = y_0'$ to the integral equation

$$y(x) = -\int_a^x \{A(t) + (x-t)[B(t) - A'(t)]\} y(t)\, dt$$

$$+ \int_a^x (x-t) f(t)\, dt + [A(a)y_0 + y_0'](x-a) + y_0$$

or equivalently $y(x) = \int_a^x K(x,t) y(t)\, dt + F(x)$

where $\quad K(x,t) = (t-x)[B(t) - A'(t)] - A(t)$

and $F(x) = \int_a^x (x-t) f(t)\, dt + [A(a)y_0 + y_0'](x-a) + y_0$

This integral equation is a **Volterra equation** of the second kind. Thus the initial-value problem is equivalent to a Volterra integral equation of the second kind.

Example $d^2y/dx^2 + x^2(dy/dx) + xy = x$, $y(0) = 1$, $y'(0) = 0$. Here $A(x) = x^2$, $B(x) = x$, $f(x) = x$. The equivalent integral equation is $y(x) = \int_0^x K(x,t)\, y(t)\, dt + F(x)$ where $K(x,t) = t(x-t) - t^2$ and $F(x) = \int_0^x (x-t)t\, dt + 1 = x^3/6 + 1$. Combining these $y(x) = \int_0^x t[x - 2t]y(t)\, dt + x^3/6 + 1$.

The expression for $I_n(x)/(n-1)!$ can also be used to show the equivalence of boundary-value problems to Fredholm integral equations of the second kind. For example, the problem $(d^2y/dx^2) + \lambda y = 0$ with $y(0) = 0$, $y(a) = 0$ is equivalent to the integral equation $y(x) = \lambda \int_0^a K(x,t)y(t)\, dt$, where $K(x,t) = (t/a)(a-x)$ when $t < x$ and $K(x,t) = (x/a)(a-t)$ when $t > x$.

The differential equation may be recovered from the integral equation by differentiating the integral equation by using the Leibnitz rule.

METHODS OF SOLUTION

To solve an integral equation of any type is to find the unknown function. In general the solution of integral equations by exact analytic methods is not easy. Often approximate or numerical methods must be resorted to. A few exact and approximate methods are considered here. Numerical methods are left for the discussion of numerical analysis.

Equations of Convolution Type The equation $u(x) = f(x) + \lambda \int_0^x K(x-t)u(t)\, dt$ is a special case of the linear integral equation of the second kind of Volterra type. The integral part is the convolution integral discussed under "Integral Transforms (Operational Methods)"; so the solution can be accomplished by Laplace transforms; $L[u(x)] = L[f(x)] + \lambda L[u(x)]L[K(x)]$ or $L[u(x)] = \frac{L[f(x)]}{1 - \lambda L[K(x)]}$, $u(x) = L^{-1}\left[\frac{L[f(x)]}{1 - \lambda L[K(x)]}\right]$. Equations of the type considered here occur quite frequently in practice in what can be called "cause-and-effect" systems.

Example In a certain linear system, the effect $E(t)$ due to a cause $C = \lambda E$ at time τ is a function only of the elapsed time $t - \tau$. If the system has the activity level 1 at time $t < 0$, the cause λE and effect (E) relation is given by the integral equation $E(t) = 1 + \lambda \int_0^t K(t-\tau)E(\tau)\, d\tau$. Let $K(t-\tau) = t - \tau$. Then $E(t) = 1 + \lambda \int_0^t (t-\tau)E(\tau)\, d\tau$. By using the transform method, $E(t) = L^{-1}\left[\frac{L[1]}{1 - \lambda L[K(t)]}\right] = L^{-1}\left[\frac{1/p}{1 - \lambda/p^2}\right] = L^{-1}\left[\frac{p}{p^2 - \lambda}\right] = \cosh\sqrt{\lambda}t$.

General Abel Equation The Volterra equation $F(x) = \int_0^x \frac{y(t)}{(x-t)^\alpha}\, dt$, $0 < \alpha < 1$ is known as Abel's equation. Its solution is $y(x) = \frac{\sin \alpha\pi}{\pi}\frac{d}{dx}\int_0^x \frac{F(t)\, dt}{(x-t)^{1-\alpha}}$.

Equations with Separable Kernels The integral equation of second kind of Fredholm type $y(x) = f(x) + \int_a^b K(x,t)y(t)\, dt$ is said to have a separable kernel if $K(x,t) = \sum_{i=1}^n U_i(x)V_i(t)$. For example, if $K(x,t) = u(x)v(t)$, the solution in this case is $y(x) = f(x) + \lambda \frac{\int_a^b K(x,t)f(t)\, dt}{1 - \lambda \int_a^b K(x,x)\, dx}$.

Example $y(x) = x + \lambda \int_0^1 xty(t)\, dt$. The solution is $y(x) = f(x) + \lambda \frac{\int_0^1 (xt)t\, dt}{1 - \lambda\int_0^1 x^2\, dx} = x + \lambda \frac{x/3}{1 - (\lambda/3)} = \frac{3x}{3 - \lambda}$, $\lambda \neq 3$.

Method of Successive Approximations Consider the equation $y(x) = f(x) + \lambda \int_a^b K(x,t)y(t)\, dt$. In this method a unique solution is obtained in sequence form as follows: Substitute in the right-hand member of the equation $y_0(t)$ for $y(t)$. Upon integration there results $y_1(t) = f(x) + \lambda \int_a^b K(x,t)y_0(t)\, dt$. Continue in like manner by replacing y_0 by y_1, y_1 by y_2, etc. A series of functions $y_0(x)$, $y_1(x)$, $y_2(x)$, . . . are obtained which satisfy the equations

$$y_n(x) = f(x) + \lambda \int_a^b K(x,t)y_{n-1}(t)\, dt$$

Then $y_n(x) = f(x) + \lambda \int_a^b K(x,t)f(t)\, dt + \lambda^2 \int_a^b K(x,t) \int_a^b K(t,t_1)f(t_1)\, dt_1\, dt + \lambda^3 \int_a^b K(x,t)\int_a^b K(t,t_1)\int_a^b K(t_1,t_2)f(t_2)\, dt_2\, dt_1\, dt + \cdots + R_n$, where R_n is the remainder, and $|R_n| \leq |\lambda^n| \begin{pmatrix}\text{max. } y_0 \\ a \leq x \leq b\end{pmatrix} M^n(b-a)^n$, where M = maximum value of $|K|$ in the rectangle $a \leq t \leq b$, $a \leq x \leq b$. If $|\lambda| M(b-a) < 1$, $\lim_{n\to\infty} R_n = 0$. Then $y_n(x) \to y(x)$, which is the unique solution.

Example Consider the equation $y(x) = 1 + \lambda \int_0^1 (1 - 3xt)y(t)\, dt$.

$$y(x) = 1 + \lambda \int_0^1 (1-3xt)\, dt + \lambda^2 \int_0^1 (1-3xt) \int_0^1 (1 - 3tt_1)\, dt_1\, dt + \cdots$$

$$= 1 + \lambda\left(1 - \frac{3}{2}x\right) + \lambda^2\frac{1}{4} + \frac{1}{4}\lambda^3\left(1 - \frac{3}{2}x\right) + \frac{\lambda^4}{16} + \frac{1}{16}\lambda^5\left(1 - \frac{3}{2}x\right) + \cdots$$

$$= \left(1 + \frac{\lambda^2}{4} + \frac{\lambda^4}{16} + \cdots\right)\left(1 + \lambda\left(1 - \frac{3}{2}x\right)\right)$$

$$= \frac{1 + \lambda\left(1 - \frac{3}{2}x\right)}{1 - \frac{1}{4}\lambda^2} \quad |\lambda| < 2$$

INTEGRAL TRANSFORMS (OPERATIONAL METHODS)

References: 57, 58, 61, 62, 88, 108, 136, 214, 224, 249, 250, 270, 294.

The term "operational method" implies a procedure of solving differential and difference equations by which the boundary or initial conditions are automatically satisfied in the course of the solution. The technique offers a very powerful tool in the applications of mathematics.

Most integral transforms are special cases of the equation $g(s) = \int_a^b f(t)K(s, t)\, dt$ in which $g(s)$ is said to be the **transform** of $f(t)$ and $K(s, t)$ is called the **kernel** of the transform. A tabulation of the more important kernels and the interval (a, b) of applicability follows. The first three transforms are considered here.

Name of transform	(a, b)	$K(s, t)$
Laplace	$(0, \infty)$	e^{-st}
Fourier	$(-\infty, \infty)$	$\frac{1}{\sqrt{2\pi}} e^{-ist}$
Fourier cosine	$(0, \infty)$	$\sqrt{\frac{2}{\pi}} \cos st$
Fourier sine	$(0, \infty)$	$\sqrt{\frac{2}{\pi}} \sin st$
Mellin	$(0, \infty)$	t^{s-1}
Hankel	$(0, \infty)$	$tJ_\nu(st),\ \nu \geq -\frac{1}{2}$

LAPLACE TRANSFORM

The Laplace transform of a function $f(t)$ is defined by $F(s) = L\{f(t)\} = \int_0^\infty e^{-st}f(t)\, dt$, where s is a complex variable. Note that the transform is an improper integral and therefore may not exist for all continuous functions and all values of s. It is to be understood that we restrict consideration to those values of s and those functions f for which this improper integral converges.

The function $L[f(t)] = g(s)$ is called the **direct transform,** and $L^{-1}[g(s)] = f(t)$ is called the **inverse transform.** Both the direct and the inverse transforms are tabulated for many often-occurring functions. In general, $L^{-1}[g(s)] = \frac{1}{2\pi i}\int_{\alpha-i\infty}^{\alpha+i\infty} e^{st}g(s)\, ds$, and to evaluate this integral requires a knowledge of complex variables, the theory of residues, and contour integration.

A function is said to be piecewise continuous on an interval if it has only a finite number of finite (or jump) discontinuities. A function f on $0 < t < \infty$ is said to be of exponential growth at infinity if there exist constants M and α such that $|f(t)| \leq Me^{\alpha t}$ for sufficiently large t.

Sufficient Conditions for the Existence of Laplace Transform Suppose f is a function which is (1) piecewise continuous on every finite interval $0 < t < T$, (2) of exponential growth at infinity, and (3) $\int_0^\delta |f(t)|\, dt$ exist (finite) for every finite $\delta > 0$. Then the Laplace transform of f exists for all complex numbers s with sufficiently large real part.

Note that condition 3 is automatically satisfied if f is assumed to be piecewise continuous on every finite interval $0 \leq t < T$. The function $f(t) = t^{-1/2}$ is not piecewise continuous on $0 \leq t \leq T$ but satisfies conditions 1 to 3.

Let Λ denote the class of all fuctions on $0 < t < \infty$ which satisfy conditions 1 to 3.

Example Let $f(t)$ be the Heaviside step function at $t = t_0$; i.e., $f(t) = 0$ for $t \leq t_0$, and $f(t) = 1$ for $t > t_0$. Then $L\{f(t)\} = \int_{t_0}^\infty e^{-st}\, dt = \lim_{T\to\infty}\int_{t_0}^T e^{-st}\, dt = \lim_{T\to\infty} \frac{1}{s}(e^{-st_0} - e^{-sT}) = \frac{e^{-st_0}}{s}$ provided $s > 0$.

Example Let $f(t) = e^{5t}$, $t \geq 0$. Then $L\{e^{st}\} = \int_0^\infty e^{-(s-5)}\, dt = 1/(s-5)$, provided Re $s > 5$.

Example Let $f(t) = e^{at}$, $t \geq 0$, where a is a real number. Then $L\{e^{at}\} = \int_0^\infty e^{-(s-a)}\, dt = 1/(s-a)$, provided Re $s > a$.

Properties of the Laplace Transform

1. The Laplace transform is a linear operator: $L\{af(t) + bg(t)\} = aL\{f(t)\} + bL\{g(t)\}$ for any constants a, b and any two functions f and g whose Laplace transforms exist.
2. The Laplace transform of a real-valued function is real for real s. If $f(t)$ is a complex-valued function, $f(t) = u(t) + iv(t)$, where u and v are real, then $L\{f(t)\} = L\{u(t)\} + iL\{v(t)\}$. Thus $L\{u(t)\}$ is the real part of $L\{f(t)\}$, and $L\{v(t)\}$ is the imaginary part of $L\{f(t)\}$.
3. The Laplace transform of a function in the class Λ has derivatives of all orders, and $L\{t^kf(t)\} = (-1)^k d^kF(s)/ds^k$, $k = 1, 2, 3, \ldots$.

Example $\int_0^\infty e^{-st}\sin at\, dt = \frac{a}{s^2+a^2}$, $s > 0$. By property 3, $\frac{2as}{(s^2+a^2)^2} = \int_0^\infty e^{-st}\, t \sin t\, dt = L\{t \sin t\}$.

Example By applying property 3 with $f(t) = 1$ and using the preceding results, we obtain $L\{t^k\} = (-1)^k \frac{d^k}{ds^k}\left(\frac{1}{s}\right) = \frac{k!}{s^{k+1}}$, provided Re $s > 0$; $k = 1, 2, \ldots$. Similarly, we obtain $L\{t^ke^{at}\} = (-1)^k \frac{d^k}{ds^k}\left(\frac{1}{s-a}\right) = \frac{k!}{(s-a)^{k+1}}$.

4. Frequency-shift property (or, equivalently, the transform of an exponentially modulated function). If $F(s)$ is the Laplace transform of a function $f(t)$ in the class Λ, then for any constant a, $L\{e^{at}f(t)\} = F(s-a)$.

Example $L\{te^{-at}\} = \frac{1}{(s+a)^2}$, $s > 0$.

5. Time-shift property. Let $u(t-a)$ be the unit step function at $t = a$. Then $L\{f(t-a)u(t-a)\} = e^{-as}F(s)$.
6. Transform of a derivative. Let f be a differentiable function such that both f and f' belong to the class Λ. Then $L\{f'(t)\} = sF(s) - f(0)$.
7. Transform of a higher-order derivative. Let f be a function which has continuous derivatives up to order n on $(0, \infty)$, and suppose that f and its derivatives up to order n belong to the class Λ. Then $L\{f^{(j)}(t)\} = s^jF(s) - s^{j-1}f(0) - s^{j-2}f'(0) - \cdots - sf^{(j-2)}(0) - f^{(j-1)}(0)$ for $j = 1, 2, \ldots, k$.

Example

$$L\{f''(t)\} = s^2L\{f(t)\} - sf(0) - f'(0)$$
$$L\{f'''(t)\} = s^3L\{f(t)\} - s^2f(0) - sf'(0) - f''(0)$$

Example Solve $y'' + y = 2e^t$, $y(0) = y'(0) = 2$. $L[y''] = -y'(0) - sy(0) + s^2L[y] = -2 - 2s + s^2L[y]$. Thus $-2 - 2s + s^2L[y] + L[y] = 2L[e^t] = \frac{2}{s-1}$. $L[y] = \frac{2s^2}{(s-1)(s^2+1)} = \frac{1}{s-1} + \frac{s}{s^2+1} + \frac{1}{s^2+1}$. Hence $y = e^t + \cos t + \sin t$.

A short table of very common Laplace transforms and inverse transforms follows. The references include more detailed tables. NOTE: $\Gamma(n+1) = \int_0^\infty x^ne^{-x}\, dx$ (gamma function); $J_n(t)$ = Bessel function of the first kind of order n.

$f(t)$	$g(s)$	$f(t)$	$g(s)$
1	$1/s$	$e^{-at}(1-at)$	$\frac{s}{(s+a)^2}$
t^n, $(na + \text{integer})$	$\frac{n!}{s^{n+1}}$	$\frac{t \sin at}{2a}$	$\frac{s}{(s^2+a^2)^2}$
t^n, $n \neq + \text{integer}$	$\frac{\Gamma(n+1)}{s^{n+1}}$	$\frac{1}{2a^2} \sin at \sinh at$	$\frac{s}{s^4+4a^4}$
$\cos at$	$\frac{s}{s^2+a^2}$	$\cos at \cosh at$	$\frac{s^3}{s^4+4a^4}$
$\sin at$	$\frac{a}{s^2+a^2}$	$\frac{1}{2a}(\sinh at + \sin at)$	$\frac{s^2}{s^4-a^4}$
$\cosh at$	$\frac{s}{s^2-a^2}$	$\frac{1}{2}(\cosh at + \cos at)$	$\frac{s^3}{s^4-a^4}$
$\sinh at$	$\frac{a}{s^2-a^2}$	$\frac{\sin at}{t}$	$\tan^{-1}\frac{a}{s}$
e^{-at}	$\frac{1}{s+a}$	$J_0(at)$	$\frac{1}{\sqrt{s^2+a^2}}$
$e^{-bt}\cos at$	$\frac{s+b}{(s+b)^2+a^2}$	$\frac{n}{a^n}\frac{J_n(at)}{t}$	$\frac{1}{(\sqrt{s^2+a^2}+s)^n}$
$e^{-bt}\sin a$	$\frac{a}{(s+b)^2+a^2}$	$J_0(2\sqrt{at})$	$\frac{1}{s}e^{-a/s}$

8. $L\left[\int_a^t f(t)\,dt\right] = \frac{1}{s}L[f(t)] + \frac{1}{s}\int_a^0 f(t)\,dt.$

Example Find $f(t)$ if $L[f(t)] = \frac{1}{s^2}\left[\frac{1}{s^2-a^2}\right]$. $L\left[\frac{1}{a}\sinh at\right] = \frac{1}{s^2-a^2}$. Therefore, $f(t) = \int_0^t \left[\int_0^t \frac{1}{a}\sinh at\,dt\right] dt = \frac{1}{a^2}\left[\frac{\sinh at}{a} - t\right]$.

9. $L\left[\frac{f(t)}{t}\right] = \int_s^\infty g(s)\,ds$; $L\left[\frac{f(t)}{t^k}\right] = \underbrace{\int_s^\infty \cdots \int_s^\infty}_{k \text{ integrals}} g(s)\,(ds)^k$.

Example $L\left[\frac{\sin at}{t}\right] = \int_s^\infty L[\sin at]\,ds = \int_s^\infty \frac{a\,ds}{s^2+a^2} = \cot^{-1}\frac{s}{a}$.

10. The unit step function $u(t-a) = \begin{cases} 0 & t < a \\ 1 & t > a \end{cases}$. $L[u(t-a)] = e^{-as}/s$.

11. The unit impulse function is $\delta(a) = u'(t-a) = \begin{cases} \infty & \text{at } t = a \\ 0 & \text{elsewhere} \end{cases}$. $L[u'(t-a)] = e^{-as}$.

12. $L^{-1}[e^{-as}g(s)] = f(t-a)u(t-a)$(second shift theorem).

13. If $f(t)$ is periodic of period b, i.e., $f(t+b) = f(t)$, then $L[f(t)] = \left[\frac{1}{1-e^{-bs}}\right]\int_0^b e^{-st}f(t)\,dt.$

Example The partial differential equations relating gas composition to position and time in a gas chromatograph are $\partial y/\partial n + \partial x/\partial\theta = 0$, $\partial y/\partial n = x - y$, where $x = mx'$, $n = (k_G aP/G_m)h$, $\theta = (mk_G aP/\rho_B)t$ and G_M = molar velocity, y = mole fraction of the component in the gas phase, ρ_B = bulk density, h = distance from the entrance, P = pressure, k_G = mass-transfer coefficient, and m = slope of the equilibrium line. These equations are equivalent to $\partial^2 y/\partial n\,\partial\theta + \partial y/\partial n + \partial y/\partial\theta = 0$, where the boundary conditions considered here are $y(n, \theta) = y(0, \theta) = 0$ and $x(n, 0) = y(n, 0) + (\partial y/\partial n)(n, 0) = \delta(0)$ (see property 11). The problem is conveniently solved by using the Laplace transform of y with respect to n; write $g(s, \theta) = \int_0^\infty e^{-ns}y(n, \theta)\,dn$. Operating on the partial differential equation gives $s(dg/d\theta) - (\partial y/\partial\theta)(0, \theta) + sg - y(0, \theta) + dg/d\theta = 0$ or $(s+1)(dg/d\theta) + sg = (\partial y/\partial\theta)(0, \theta) + y(0, \theta) = 0$. The second boundary condition gives $g(s, 0) + sg(s, 0) - y(0, 0) = 1$ or $g(s, 0) + sg(s, 0) = 1$ ($L[\delta(0)] = 1$). A solution of the ordinary differential equation for g consistent with this second condition is $g(s, \theta) = \frac{1}{s+1}e^{-(s\theta/s+1)}$. Inversion of this transform gives the solution $y(n, \theta) = e^{-(n+\theta)}I_0(2\sqrt{n\theta})$ where I_0 = zero-order Bessel function of an imaginary argument. For large u, $I_n(u) \backsim e^u/2\sqrt{\pi u}$. Hence for large n,

$$y(n, \theta) \backsim \frac{\exp[-(\sqrt{\theta} - \sqrt{n})^2]}{2\pi^{1/2}(n\theta)^{1/4}}$$

or for sufficiently large n, the peak concentration occurs near $\theta = n$.

Other applications of Laplace transforms are given under "Differential Equations."

CONVOLUTION INTEGRAL

The convolution integral (faltung) of two functions $f(t)$, $r(t)$ is $x(t) = f(t)^\circ r(t) = \int_0^t f(\tau)r(t-\tau)\,d\tau$.

Example $t^\circ \sin t = \int_0^t \tau \sin(t-\tau)\,d\tau = t - \sin t$.

14. $L[f(t)]L[h(t)] = L[f(t)^\circ h(t)]$

FOURIER TRANSFORM

The Fourier transform is given by $F[f(t)] = \frac{1}{\sqrt{2\pi}}\int_{-\infty}^\infty f(t)e^{-ist}\,dt = g(s)$ and its inverse by $F^{-1}[g(s)] = \frac{1}{\sqrt{2\pi}}\int_{-\infty}^\infty g(s)e^{ist}\,dt = f(t)$. In brief, the condition for the Fourier transform to exist is that $\int_{-\infty}^\infty |f(t)|\,dt < \infty$, although certain functions may have a Fourier transform even if this is violated.

Example The function in Fig. 2-54, $f(t) = \begin{cases} 1 & -a \le t \le a \\ 0 & \text{elsewhere} \end{cases}$ has $F[f(t)] = \int_{-a}^a e^{-ist}\,dt = \int_0^a e^{ist}\,dt + \int_0^a e^{-ist}\,dt = 2\int_0^a \cos st\,dt = \frac{2\sin sa}{s}$.

Properties of the Fourier Transform Let $F[f(t)] = g(s)$; $F^{-1}[g(s)] = f(t)$.

1. $F[f^{(n)}(t)] = (is)^n F[f(t)$.
2. $F[af(t) + bh(t)] = aF[f(t)] + bF[h(t)]$.
3. $F[f(-t)] = g(-s)$.
4. $F[f(at)] = \frac{1}{a}g\left(\frac{s}{a}\right)$, $a > 0$.
5. $F[e^{-iwt}f(t)] = g(s+w)$.
6. $F[f(t+t_1)] = e^{ist}t_1 g(s)$.
7. $F[f(t)] = G(is) + G(-is)$ if $f(t) = f(-t)$ (f even)
   $F[f(t)] = G(is) - G(-is)$ if $f(t) = -f(-t)$ (f odd)

where $G(s) = L[f(t)]$. This result allows the use of the Laplace-transform tables to obtain the Fourier transforms.

Example Find $F[e^{-a|t|}]$ by property 7. $e^{-a|t|}$ is even. So $L[e^{-at}] = 1/(s+a)$. Therefore, $F[e^{-a|t|}] = 1/(is+a) + 1/(-is+a) = 2a/(s^2+a^2)$.

Tables of this transform may be found in *Tables of Integral Transforms*, Bateman Manuscript Project, vol. 1, McGraw-Hill, New York, 1954. See Ref. 98.

FOURIER COSINE TRANSFORM

The Fourier cosine transform is given by $F_c[f(t)] = g(s) = \sqrt{\frac{2}{\pi}}\int_0^\infty f(t)\cos st\,dt$ and its inverse by $F_c^{-1}[g(s)] = f(t) =$

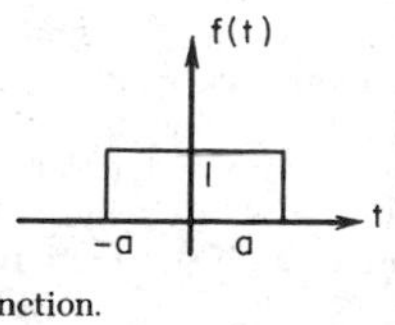

FIG. 2-54 Rectangular function.

$\sqrt{\frac{2}{\pi}} \int_0^\infty g(s) \cos st \, ds$. The Fourier sine transform F_s is obtainable by replacing the cosine by the sine in these integrals.

Example $F_c[f(t)]$, $f(t) = \begin{cases} 1 & 0 < t < a \\ 0 & a < t < \infty \end{cases}$ $F_c[f(t)] = \sqrt{\frac{2}{\pi}} \int_0^a \cos st \, dt = \sqrt{\frac{2}{\pi}} \frac{\sin as}{s}$.

Properties of the Fourier Cosine Transform $F_c[f(t)] = g(s)$.

1. $F_c[af(t) + bh(t)] = aF_c[f(t)] + bF_c[h(t)]$.
2. $F_c[f(at)] = (1/a)g(s/a)$.
3. $F_c[f(at)\cos bt] = \frac{1}{2a}\left[g\left(\frac{s+b}{a}\right) + g\left(\frac{s-b}{a}\right)\right]$, $a, b > 0$.
4. $F_c[t^{2n}f(t)] = (-1)^n (d^{2n}g/ds^{2n})$.
5. $F_c[t^{2n+1}f(t)] = (-1)^n (d^{2n+1}/ds^{2n+1}) F_s[f(t)]$.

A short table of Fourier cosine transforms follows. More extensive tables can be found in Ref. 98.

$f(t)$		$\frac{g(s)}{\sqrt{2/\pi}}$
t $2 - t$ 0	$0 < t < 1$ $1 < t < 2$ $2 < t < \infty$	$\frac{1}{s^2}[2 \cos s - 1 - \cos 2s]$
$t^{-1/2}$		$\pi^{1/2}(2s)^{-1/2}$
0 $(t - a)^{-1/2}$	$0 < t < a$ $a < t < \infty$	$\pi^{1/2}(2s)^{-1/2}[\cos as - \sin as]$
$(t^2 + a^2)^{-1}$		$\frac{1}{2}\pi a^{-1}e^{-as}$
e^{-at}	$a > 0$	$\frac{a}{s^2 + a^2}$
e^{-at^2}	$a > 0$	$\frac{1}{2}\pi^{1/2}a^{-1/2}e^{-s^2/4a}$
$\frac{\sin at}{t}$	$a > 0$	$\begin{cases} \pi/2 & s < a \\ \pi/4 & s = a \\ 0 & s > a \end{cases}$

Example The temperature θ in the semi-infinite rod $0 \le x < \infty$ is determined by the differential equation $\partial\theta/\partial t = k(\partial^2\theta/\partial x^2)$ and the condition $\theta = 0$ when $t = 0$, $x \ge 0$; $\partial\theta/\partial x = -\mu =$ constant when $x = 0$, $t > 0$. By using the Fourier cosine transform a solution may be found as $\theta(x, t) = \frac{2\mu}{\pi} \int_0^\infty \frac{\cos px}{p}(1 - e^{-kp^2t})\, dp$ (see Sneddon in the references).

MATRIX ALGEBRA AND MATRIX COMPUTATIONS

REFERENCES: *Matrix algebra and linear algebra.* 28, 34, 87, 213, 258, 286; *advanced topics:* 107, 115, 151. *Matrix computations and numerical linear algebra.* 50, 51, 54, 60, 79, 99, 103, 104, 106, 151, 162, 171, 181, 214, 257, 258, 259, 266, 283, 284, 286. *Applications of matrices.* 115, 209, 213, 253, 259, 266. *Packages and users' guides for matrix computations.* 90, 103, 116, 124. *Numerical and mathematical software for matrix computations.* 159, 232, 233, 234.

MATRIX ALGEBRA

Matrices A rectangular array of mn quantities, arranged in m rows and n columns

$$A = (a_{ij}) = \begin{bmatrix} a_{11} & \cdots & a_{1n} \\ a_{21} & \cdots & a_{2n} \\ \vdots & & \\ a_{m1} & \cdots & a_{mn} \end{bmatrix}$$

is called a matrix. The elements a_{ij} may be real or complex. The notation a_{ij} means the element in the ith row and jth column, i is called the row index, j the column index. If $m = n$ the matrix is said to be square and of order n. A matrix, even if it is square, does not have a numerical value, as a determinant does. However, if the matrix A is square, a determinant can be formed which has the same elements as the matrix A. This is called the determinant of the matrix and is written det (A) or $|A|$. If A is square and det $(A) \ne 0$, A is said to be nonsingular; if det $(A) = 0$, A is said to be singular. A matrix A has rank r if and only if it has a nonvanishing determinant of order r and no nonvanishing determinant of order $> r$.

Equality of Matrices Let $A = (a_{ij})$, $B = (b_{ij})$. Two matrices A and B are *equal* $(=)$ if and only if they are identical; that is, they have the same number of rows and the same number of columns and equal corresponding elements ($a_{ij} = b_{ij}$ for all i and j).

Addition and Subtraction The operations of addition $(+)$ and subtraction $(-)$ of two or more matrices are possible if and only if they have the same number of rows and columns. Thus $A \pm B = (a_{ij} \pm b_{ij})$; i.e., addition and subtraction are of corresponding elements.

Example

$$\begin{bmatrix} 3 & 1 & 4 & -1 \\ 2 & 5 & 3 & 2 \\ 0 & 6 & 2 & -5 \end{bmatrix} + \begin{bmatrix} 2 & 6 & 1 & 4 \\ 1 & -1 & 0 & 1 \\ 3 & 2 & 1 & 0 \end{bmatrix} = \begin{bmatrix} 5 & 7 & 5 & 3 \\ 3 & 4 & 3 & 3 \\ 3 & 8 & 3 & -5 \end{bmatrix}$$

Transposition The matrix obtained from A by interchanging the rows and columns of A is called the **transpose** of A, written A' or A^T.

Example

$$A = \begin{bmatrix} 1 & 3 & 4 \\ 2 & 1 & 6 \end{bmatrix}, \quad A^T = \begin{bmatrix} 1 & 2 \\ 3 & 1 \\ 4 & 6 \end{bmatrix}$$

Note that $(A^T)^T = A$.

Multiplication Let $A = (a_{ij})$, $i = 1, \ldots, m_1$; $j = 1, \ldots, m_2$. $B = (b_{ij})$. $i = 1, \ldots, n_1$; $j = 1, \ldots, n_2$. The product AB is defined if and only if the number of columns of A (m_2) equals the number of rows of $B(n_1)$, i.e., $n_1 = m_2$. For two such matrices the product $P = AB$ is defined by summing the element by element products of a row of A by a column of B.

This is the row by column rule. Thus $p_{ij} = \sum_{k=1}^{n_1} a_{ik}b_{kj}$. The resulting matrix has m_1 rows and n_2 columns.

Example

$$\begin{bmatrix} 3 & 2 \\ 1 & 1 \\ 5 & 4 \end{bmatrix} \begin{bmatrix} 0 & 1 & 5 & 6 \\ -2 & 0 & 1 & 3 \end{bmatrix} = \begin{bmatrix} -4 & 3 & 17 & 24 \\ -2 & 1 & 6 & 9 \\ -8 & 5 & 29 & 42 \end{bmatrix}$$

It is helpful to remember that the element p_{ij} is formed from the ith row of the first matrix and the jth column of the second matrix. The matrix product is not commutative. That is, $AB \ne BA$ in general.

Example

$$\begin{bmatrix}2 & 1\\ 3 & -2\end{bmatrix}\begin{bmatrix}1 & 4\\ 0 & 2\end{bmatrix} = \begin{bmatrix}2 & 10\\ 3 & 8\end{bmatrix} = AB$$

$$\begin{bmatrix}1 & 4\\ 0 & 2\end{bmatrix}\begin{bmatrix}2 & 1\\ 3 & -2\end{bmatrix} = \begin{bmatrix}14 & -7\\ 6 & -4\end{bmatrix} = BA$$

Inverse of a Matrix A square matrix A is said to have an inverse if there exists a matrix B such that $AB = BA = I$, where I is the identity matrix of order n.

$$\begin{bmatrix}1 & 0 \cdots 0\\ 0 & 1 \cdots\\ \vdots & 1 \quad 0\\ 0 & \cdots 0 \quad 1\end{bmatrix}$$

The inverse B is a square matrix of the order of A, designated by A^{-1}. Thus $AA^{-1} = A^{-1}A = I$. A square matrix A has an inverse if and only if A is nonsingular.

Certain relations are important:

(1) $(AB)^{-1} = B^{-1}A^{-1}$

(2) $(AB)^T = B^TA^T$

(3) $(A^{-1})^T = (A^T)^{-1}$

(4) $(ABC)^{-1} = C^{-1}B^{-1}A^{-1}$

Scalar Multiplication Let c be any real or complex number. Then $cA = (ca_{ij})$.

Example

$$3\begin{bmatrix}1 & 4 & 2\\ 3 & 0 & -1\end{bmatrix} = \begin{bmatrix}3 & 12 & 6\\ 9 & 0 & -3\end{bmatrix}$$

Adjugate Matrix of a Matrix Let A_{ij} denote the cofactor of the element a_{ij} in the determinant of the matrix A. The matrix B^T where $B = (A_{ij})$ is called the adjugate matrix of A written adj $A = B^T$. Then $A^{-1} = \text{adj } A/|A|$. This definition may be used to calculate A^{-1}. However, it is very laborious and the inversion is usually accomplished by numerical techniques shown under "Numerical Analysis and Approximate Methods."

Example Let $A = \begin{bmatrix}3 & 0 & -1\\ -1 & 2 & 1\\ 3 & 6 & 3\end{bmatrix}$ Form $B = (A_{ij})$, $B =$

$$\begin{bmatrix}0 & 6 & -12\\ -6 & 12 & -18\\ 2 & -2 & 6\end{bmatrix}, \text{ adj } A = B^T = \begin{bmatrix}0 & -6 & 2\\ 6 & 12 & -2\\ -12 & -18 & 6\end{bmatrix}; \ |A| = 12.$$

$$A^{-1} = \frac{\text{adj } A}{|A|} = \begin{bmatrix}0 & -\frac{1}{2} & \frac{1}{6}\\ \frac{1}{2} & 1 & -\frac{1}{6}\\ -1 & -\frac{3}{2} & \frac{1}{2}\end{bmatrix}$$

Linear Equations in Matrix Form Every set of n nonhomogeneous linear equations in n unknowns

$$\begin{aligned}a_{11}x_1 + a_{12}x_2 + \cdots + a_{1n}x_n &= b_1\\ a_{21}x_1 + a_{22}x_2 + \cdots + a_{2n}x_n &= b_2\\ \vdots \qquad & \quad \vdots\\ a_{n1}x_1 + a_{n2}x_2 + \cdots + a_{nn}x_n &= b_n\end{aligned}$$

can be written in matrix form as $AX = B$, where $A = (a_{ij})$, $X' = [x_1 \cdots x_n]$, and $B^1 = [b_1 \cdots b_n]$. The solution for the unknowns is $X = A^{-1}B$.

Example

$$\begin{aligned}3x_1 + 0x_2 - x_3 &= 4\\ -x_1 + 2x_2 + x_3 &= -2\\ 3x_1 + 6x_2 + 3x_3 &= 6\end{aligned}$$

in matrix form is written

$$\begin{bmatrix}3 & 0 & -1\\ -1 & 2 & 1\\ 3 & 6 & 3\end{bmatrix}\begin{bmatrix}x_1\\ x_2\\ x_3\end{bmatrix} = \begin{bmatrix}4\\ -2\\ 6\end{bmatrix}$$

The solution is $\begin{bmatrix}x_1\\ x_2\\ x_3\end{bmatrix} = \begin{bmatrix}0 & -\frac{1}{2} & \frac{1}{6}\\ \frac{1}{2} & 1 & -\frac{1}{6}\\ -1 & -\frac{3}{2} & \frac{1}{2}\end{bmatrix}\begin{bmatrix}4\\ -2\\ 6\end{bmatrix} = \begin{bmatrix}2\\ -1\\ 2\end{bmatrix}$; that is, $x_1 = 2$, $x_2 = -1$, $x_3 = 2$.

Special Square Matrices

1. A triangular matrix is a matrix all of whose elements above or below the main diagonal (set of elements $a_{11}, \ldots, a_{nn}$) are zero. If A is triangular, $\det(A) = a_{11} \cdot a_{22} \ldots a_{nn}$.
2. A diagonal matrix is one such that all elements both above and below the main diagonal are zero (i.e., $a_{ij} = 0$ for all $i \neq j$). If all diagonal elements are equal, the matrix is called scalar. If A is diagonal, $A = (a_{ij})$, $A^{-1} = (1/a_{ij})$.
3. If $a_{ij} = a_{ji}$ for all i and j (i.e., $A = A^T$), the matrix is symmetric.
4. If $a_{ij} = -a_{ji}$ for $i \neq j$ but the a_{ij} are not all zero, the matrix is skew.
5. If $a_{ij} = -a_{ji}$ for all i and j (i.e., $a_{ii} = 0$), the matrix is skew symmetric.
6. If $A^T = A^{-1}$, the matrix A is orthogonal.
7. If the matrix $A^* = (\bar{a}_{ij})^T$, $\bar{a}_{ij}$ = complex conjugate of a_{ij}, A^* is the hermitian conjugate of A.
8. If $A = A^{-1}$, A is involutory.
9. If $A = A^*$, A is hermitian.
10. If $A = -A^*$, A is skew hermitian.
11. If $A^{-1} = A^*$, A is unitary.

If A is any matrix, then AA^T and A^TA are square symmetric matrices, usually of different order.

Example Let $A = \begin{bmatrix}5 & 1 & 3 & 0\\ 3 & 4 & 1 & 5\\ 2 & -2 & 0 & 1\end{bmatrix}$, $A^T = \begin{bmatrix}5 & 3 & 2\\ 1 & 4 & -2\\ 3 & 1 & 0\\ 0 & 5 & 1\end{bmatrix}$,

$$AA^T = \begin{bmatrix}35 & 22 & 8\\ 22 & 51 & 3\\ 8 & 3 & 9\end{bmatrix}, \ A^TA = \begin{bmatrix}38 & 13 & 18 & 17\\ 13 & 21 & 7 & 18\\ 18 & 7 & 10 & 5\\ 17 & 18 & 5 & 26\end{bmatrix}$$

Matrix Calculus

Differentiation Let the elements of $A = [a_{ij}(t)]$ be differentiable functions of t. Then $\dfrac{dA}{dt} = \left[\dfrac{da_{ij}(t)}{dt}\right]$.

Example $A = \begin{bmatrix}\sin t & \cos t\\ -\cos t & \sin t\end{bmatrix}$, $\dfrac{dA}{dt} = \begin{bmatrix}\cos t & -\sin t\\ \sin t & \cos t\end{bmatrix}$.

Integration The integral $\int A\, dt = [\int a_{ij}(t)\, dt]$.

Example $A = \begin{bmatrix}t & 2\\ t^2 & e^t\end{bmatrix}$, $\int A\, dt = \begin{bmatrix}t^2/2 & 2t\\ t^3/3 & e^t\end{bmatrix}$.

The matrix $B = \lambda I - A$ is called the characteristic (eigen) matrix of A. Here A is square of order n, λ is a scalar parameter, and I is the $n \times n$ identity. det B = det $(\lambda I - A) = 0$ is the characteristic (eigen) equation for A. The characteristic equation is always of the same degree as the order of A. The roots of the characteristic equation are called the eigenvalues of A.

Example

$$A = \begin{bmatrix}1 & 2\\ 3 & 8\end{bmatrix} \quad B = \begin{bmatrix}\lambda & 0\\ 0 & \lambda\end{bmatrix} - \begin{bmatrix}1 & 2\\ 3 & 8\end{bmatrix} = \begin{bmatrix}\lambda - 1 & -2\\ -3 & \lambda - 8\end{bmatrix}$$

is the characteristic matrix and $f(\lambda) = \det(B) = \det(\lambda I - A) = (\lambda - 1)(\lambda - 8) - 6 = \lambda^2 - 9\lambda + 2 = 0$ is the characteristic equation. The eigenvalues of A are the roots of $\lambda^2 - 9\lambda + 2 = 0$, which are $(9 \pm \sqrt{73})/2$.

A nonzero matrix X_i, which has one column and n rows, called a column vector satisfying the equation

$$(\lambda I - A)X_i = 0$$

and associated with the ith characteristic root λ_i is called an eigenvector.

Vector and Matrix Norms To carry out error analysis for approximate and iterative methods for the solutions of linear systems, one needs notions for vectors in R^n and for matrices that are analogous to the notion of length of a geometric vector. Let R^n denote the set of all vectors with n components, $x = (x_1, \ldots, x_n)$. In dealing with matrices it is convenient to treat vectors in R^n as columns, and so $x = (x_1, \ldots, x_n)^T$; however, we shall here write them simply as row vectors. A norm on R^n is a real-valued function f defined on R^n with the following properties:

1. $f(x) \geq 0$ for all $x \in R^n$.
2. $f(x) = 0$ if and only if $x = (0, 0, \ldots, 0)$.
3. $f(\alpha x) = |\alpha| f(x)$ for all real numbers α and $x \in R^n$.
4. $f(x + y) \leq f(x) + f(y)$ for all $x, y \in R^n$.

The usual notation for a norm is $f(x) = \|x\|$.

MATRIX COMPUTATIONS

The principal topics in linear algebra involve systems of linear equations, matrices, vector spaces, linear transformations, eigenvalues and eigenvectors, and least-squares problems. Space does not allow an overview of these topics with particular reference to computational methods; however, we shall discuss some of the elementary ideas and computational techniques.

Elementary Row Operations Gaussian elimination involves the application of elementary row operations to the augmented matrix of the linear system. For any $m \times n$ matrix, there are three types of elementary row operations: (1) multiplication of any row by a *nonzero* number, (2) interchange of two rows, and (3) addition to a certain row a constant times another row. Note that these operations are reversible; thus when they are performed on the augmented matrix of a system of linear equations, they produce a system which is equivalent to the original one.

Echelon Form Every matrix can be reduced to an echelon form by elementary row operations. A matrix is said to be in echelon form if it satisfies three properties: (1) the nonzero rows come first, so if there are rows consisting entirely of zeros, they would be grouped together at the bottom of the matrix; (2) the first nonzero entry (in a nonzero row) is called the pivot, and all entries below a pivot are zeros; and (3) in any two consecutive nonzero rows, the pivot in the lower row is farther to the right than the pivot in the higher order. Thus an echelon form is an upper triangular matrix in which the pivots sit on a staircase but not necessarily on the main diagonal. Shown here are two typical echelon matrices in which the pivots are circled, whereas the other starred entries may or may not be zero.

$$\begin{bmatrix} \circledast & * & * & * \\ 0 & 0 & \circledast & * \\ 0 & 0 & 0 & 0 \\ 0 & 0 & 0 & 0 \\ 0 & 0 & 0 & 0 \end{bmatrix} \qquad \begin{bmatrix} \circledast & * & * & * & * & * \\ 0 & \circledast & * & * & * & * \\ 0 & 0 & \circledast & * & * & * \\ 0 & 0 & 0 & 0 & 0 & \circledast \\ 0 & 0 & 0 & 0 & 0 & 0 \end{bmatrix}$$

An echelon form is not unique. If an echelon form satisfies the additional property that (4) each column that contains a pivot has zeros everywhere else, then it is called a *reduced echelon form*, e.g.,

$$\begin{bmatrix} \circledast & * & 0 & * \\ 0 & 0 & \circledast & * \\ 0 & 0 & 0 & 0 \end{bmatrix}$$

Some authors require in these definitions that the "pivots" be 1, which of course can be achieved by elementary row operations.

Elementary Matrices Any matrix obtained from the identity matrix (of a given order) by performing a single elementary row operation on it is called an elementary matrix.

Examples $E_1 = \begin{bmatrix} 5 & 0 \\ 0 & 1 \end{bmatrix}$, $E_2 = \begin{bmatrix} 1 & 0 \\ -3 & 1 \end{bmatrix}$, $E_3 = \begin{bmatrix} 0 & 1 \\ 1 & 0 \end{bmatrix}$. The row operations performed on $I = \begin{bmatrix} 1 & 0 \\ 0 & 1 \end{bmatrix}$ to obtain these matrices are respectively $5R_1$, $R_2 - 3R_1$ (i.e., subtracting from the second row 3 times the first row), and $R_1 \leftrightarrow R_2$ (interchange of rows). In contrast, $\begin{bmatrix} 2 & 0 \\ -3 & 1 \end{bmatrix}$ is not an elementary matrix since it is obtained from the identity matrix by two elementary operations (one operation does not suffice).

A **permutation matrix** is a matrix obtained from I by interchanging two or more rows. Clearly every elementary matrix is invertible, and its inverse is obtained by performing the inverse of the same elementary row operation on the identity matrix.

Example To obtain the inverse of E_2 in the preceding example we add to the second row of I 3 times the first row, so $E_2^{-1} = \begin{bmatrix} 1 & 0 \\ 3 & 1 \end{bmatrix}$.

A Bookkeeping Technique That Is Useful in Performing Row Operations on a Computer If A is any matrix and E is the elementary row matrix obtained from I by performing an elementary row operation on I, then EA is the same as the matrix obtained from A by performing the same elementary row operation on A.

Computation of A^{-1} by Using Elementary Row Operations Suppose that A is a square matrix that is reducible by elementary row operations to the identity matrix I. Then $E_p \cdots E_2E_1A = I$ for some elementary matrices $E_1, E_2, \ldots, E_p$. Let $B = E_p \cdots E_2E_1$. Then $BA = I$; thus A^{-1} is precisely B. However, rather than performing explicitly the multiplication of the matrices in B, the bookkeeping technique is invoked, and one does the bookkeeping on the identity as the elementary row operations on A are performed. Thus the procedure consists of augmenting the matrix A by I and performing elementary row operations on $[A : I]$ to reduce it to $[I : B]$. Then $B = A^{-1}$.

Example Show that the matrix

$$A = \begin{bmatrix} 1 & 2 & 3 \\ 4 & 5 & 6 \\ 7 & 8 & 10 \end{bmatrix}$$

is invertible, and compute its inverse by using elementary row operations.

$$[A : I_3] = \left[\begin{array}{ccc|ccc} 1 & 2 & 3 & 1 & 0 & 0 \\ 4 & 5 & 6 & 0 & 1 & 0 \\ 7 & 8 & 10 & 0 & 0 & 1 \end{array}\right] \xrightarrow[R_3 - 7R_1]{R_2 - 4R_1} \left[\begin{array}{ccc|ccc} 1 & 2 & 3 & 1 & 0 & 0 \\ 0 & -3 & -6 & -4 & 1 & 0 \\ 0 & -6 & -11 & -7 & 0 & 0 \end{array}\right]$$

$$\xrightarrow{R_3 - 2R_2} \left[\begin{array}{ccc|ccc} 1 & 2 & 3 & 1 & 0 & 0 \\ 0 & 1 & 2 & 4/3 & -1/3 & 0 \\ 0 & 0 & 1 & 1 & -2 & 1 \end{array}\right] \xrightarrow[R_2 - 2R_3]{R_1 - 3R_3} \left[\begin{array}{ccc|ccc} 1 & 2 & 0 & -2 & 6 & -3 \\ 0 & 1 & 0 & -2/3 & 11/3 & -2 \\ 0 & 0 & 1 & 1 & -1 & 1 \end{array}\right]$$

$$\xrightarrow{R_1 - 2R_2} \left[\begin{array}{ccc|ccc} 1 & 0 & 0 & -2/3 & -4/3 & 1 \\ 0 & 1 & 0 & -2/3 & 11/3 & -2 \\ 0 & 0 & 1 & 1 & -2 & 1 \end{array}\right]$$

$$\text{Thus } A^{-1} = \frac{1}{3}\begin{bmatrix} -2 & -4 & 3 \\ -2 & 11 & -6 \\ 3 & -6 & 3 \end{bmatrix}$$

Here we have indicated the elementary row operations performed in each particular transition; e.g., in the last step $R_1 - 2R_2$ means subtraction from the first row (of the matrix in the previous step) twice the second row.

The procedure illustrated in the example fails only if A is singular (noninvertible). For any $n \times n$ matrix the following statements are equivalent: (1) A is invertible, (2) A is row-equivalent (i.e., it can be reduced by elementary row operations) to I, and (3) the only solution of the homogeneous system $Ax = 0$ is the trivial solution (the zero vector). Thus if we start with a noninvertible matrix, then it will not be possible to complete the reduction to $[I\vdots B]$, since such a reduction does not exist; this will be discovered by the appearance of a zero row in the process.

Example In the matrix A shown previously, if we replace the entry 10 by 9, a singular matrix is obtained. Applying the previous procedure leads to

$$\left[\begin{array}{ccc|ccc} 1 & 2 & 3 & 1 & 0 & 0 \\ 0 & 1 & 2 & 4/3 & -1/3 & 0 \\ 0 & 0 & 0 & \circ & \circ & \circ \end{array}\right]$$

so we stop (declaring that A is noninvertible).

Example

1. Find an echelon form for the matrix A for all possible choices of the parameter c.

$$A = \begin{bmatrix} 1 & 3 & -2 & 0 \\ 0 & 1 & 2 & 1 \\ 2 & 6 & c & 3 \end{bmatrix}$$

2. Find the reduced-echelon form of the matrix A when $c = -4$.
3. Find all solutions of the homogeneous system

$$x_1 + 3x_2 - 2x_3 = 0$$
$$x_2 + 2x_3 + x_4 = 0$$
$$2x_1 + 6x_2 - x_3 + 3x_4 = 0$$

Note that the coefficient matrix is the same as the matrix A in part 1 with $c = -1$.

Solution. Subtracting twice the first row from the third row, we obtain

$$\begin{bmatrix} 1 & 3 & -2 & 0 \\ 0 & 1 & 2 & 1 \\ 0 & 0 & c+4 & 3 \end{bmatrix}$$

Thus for $c \neq -4$, an echelon form for A is

$$\begin{bmatrix} 1 & 3 & -2 & 0 \\ 0 & 1 & 2 & 1 \\ 0 & 0 & 1 & \dfrac{3}{c+4} \end{bmatrix}$$

For $c = -4$ an echelon form for A is

$$\begin{bmatrix} 1 & \textcircled{3} & -2 & 0 \\ 0 & 1 & 2 & \textcircled{1} \\ 0 & 0 & 0 & 1 \end{bmatrix}$$

To obtain the reduced-echelon form we must transform the matrix so that the circled entries become zero. To this end, we subtract the third row from the second row and then, using the new second row, we subtract from the first row 3 times the new second row. These elementary row operations lead to the matrix

$$\begin{bmatrix} 1 & 0 & -8 & 0 \\ 0 & 1 & 2 & 0 \\ 0 & 0 & 0 & 1 \end{bmatrix}$$

4. An echelon form for A when $c = -1$ is

$$\begin{bmatrix} 1 & 3 & -2 & 0 \\ 0 & 1 & 2 & 1 \\ 0 & 0 & 1 & 1 \end{bmatrix} \rightarrow \begin{bmatrix} 1 & 3 & -2 & 0 \\ 0 & 1 & 0 & -1 \\ 0 & 0 & 1 & 1 \end{bmatrix}$$

Thus $x_3 + x_4 = 0$. Let $x_4 = t$. Then $x_3 = -t$. Also, $x_2 - x_4 = 0$, so $x_2 = t$. Finally, $x_1 + 3x_2 - 2x_3 = 0$, so $x_1 = -5t$. Thus the general solution is of the form $x = t(-5, 1, -1, 1)$ for any real number t.

LU Factorization of a Matrix To every $m \times n$ matrix A there exists a permutation matrix P, a lower triangular matrix L with unit diagonal elements, and an $m \times n$ (upper triangular) echelon matrix U such that $PA = LU$. The Gauss elimination is in essence an algorithm to determine U, P, and L. The permutation matrix P may be needed since it may be necessary in carrying out the Gauss elimination to interchange two rows of A to produce a (nonzero) pivot, such as if we start with $A = \begin{bmatrix} 0 & 2 \\ 1 & 6 \end{bmatrix}$. If A is a square matrix and if principal submatrices of A are all nonsingular, then we may choose P as the identity in the preceding factorization and obtain $A = LU$. This factorization is unique if L is normalized (as assumed previously), so that it has unit elements on the main diagonal. (Recall that the principal submatrices of $A = [a_{ij}]_{n\times n}$ are the matrices $[a_{ij}]_{k\times k}$, i.e., $i, j = 1, \ldots, k$, and $k = 1, 2 \ldots, n$). The LU factorization is illustrated in the following example by using Gauss elimination. Some applications of such factorizations in computational methods are also indicated.

Example. $A = \begin{bmatrix} 1 & 2 & 4 \\ -1 & 1 & 2 \\ 1 & -1 & 1 \end{bmatrix}$ can be factored in the form $A = LU$, whereas $B = \begin{bmatrix} 1 & -1 & 1 \\ -1 & 1 & 2 \\ 1 & 2 & 4 \end{bmatrix}$ requires a permutation P so that $PB = LU$. This can be easily detected in the process of Gauss elimination. Note that $A = PB$, where $P = \begin{bmatrix} 0 & 0 & 1 \\ 0 & 1 & 0 \\ 1 & 0 & 0 \end{bmatrix}$

Example. Let $C = \begin{bmatrix} 1 & 1 & -1 \\ 2 & -1 & 1 \\ 4 & 1 & 2 \end{bmatrix}$; then

$$C \xrightarrow[R3-4R1]{R2-2R1} \begin{bmatrix} 1 & 1 & -1 \\ 0 & -3 & 3 \\ 0 & -3 & 6 \end{bmatrix} \xrightarrow{R3-R2} \begin{bmatrix} 1 & 1 & -1 \\ 0 & -3 & 3 \\ 0 & 0 & 3 \end{bmatrix} = U$$

Note that this case required three elementary operations all of which were of type 3, i.e., the addition to a given row of a "multiple" of another row. No interchange of rows was required, so $P = I$. Denoting the performed elementary operations by E_1, E_2, and E_3, we have $E_3E_2E_1C = U$, so $C = (E_3E_2E_1)^{-1} U = E_1^{-1} E_2^{-1} E_3^{-1} U = LU$, where $L = E_1^{-1}E_2^{-1}E_3^{-1}$. Thus L is obtained from the identity matrix by applying in reverse order the inverse of the operations performed in the reduction of A to an upper triangular matrix. That is,

$$\begin{bmatrix} 1 & 0 & 0 \\ 0 & 1 & 0 \\ 0 & 0 & 1 \end{bmatrix} \xrightarrow{R3+R2} \begin{bmatrix} 1 & 0 & 0 \\ 0 & 1 & 0 \\ 0 & 1 & 1 \end{bmatrix} \xrightarrow[R2+2R1]{R3+4R1} \begin{bmatrix} 1 & 0 & 0 \\ 2 & 1 & 0 \\ 4 & 1 & 1 \end{bmatrix} = L$$

Solution of $Ax = b$ by Using LU Factorization Suppose that the indicated system is compatible and that $A = LU$ (the case $PA = LU$ is similarly handled and amounts to rearranging the equations). Let $z = Ux$. Then $Ax = LUx = b$ implies that $Lz = b$. Thus to solve $Ax = b$ we first solve $Lz = b$ for z and then solve $Ux = z$ for x. This procedure does not require that A be invertible and can be used to determine all solutions of a compatible system $Ax = b$. Note that the systems $Lz = b$ and $Ux = z$ are both in triangular forms and thus can be easily solved.

Example This is illustrated by solving $Ax = b$, where A is the nonsingular matrix in the previous example and $b = (b_1, b_2, b_3)^T$. Then $Lz = b$ is of the form

$$\begin{bmatrix} 1 & 0 & 0 \\ 2 & 1 & 0 \\ 4 & 1 & 1 \end{bmatrix} \begin{bmatrix} z_1 \\ z_2 \\ z_3 \end{bmatrix} = \begin{bmatrix} b_1 \\ b_2 \\ b_3 \end{bmatrix}$$

Forward substitutions lead to $z_1 = b_1$, $z_2 = b_2 - 2b_1$ and $z_3 = b_3 - z_2 - 4z_1 = b_3 - b_2 - 2b_1$. Now we solve the upper triangular system $Ux = z$, or

$$\begin{bmatrix} 1 & 1 & -1 \\ 0 & -3 & 3 \\ 0 & 0 & 3 \end{bmatrix} \begin{bmatrix} x_1 \\ x_2 \\ x_3 \end{bmatrix} = \begin{bmatrix} b_1 \\ b_2 - 2b_1 \\ b_3 - b_2 - 2b_1 \end{bmatrix}$$

Backward substitutions lead to $x_3 = \frac{1}{3}(b_3 - b_2 - 2b_1)$, $x_2 = \frac{1}{3}(-2b_2 + b_3)$, $x_1 = \frac{1}{3}(b_1 + b_2)$, so

$$\begin{bmatrix} x_1 \\ x_2 \\ x_3 \end{bmatrix} = \frac{1}{3}\begin{bmatrix} 1 & -1 & 0 \\ 0 & -2 & 1 \\ -2 & -1 & 1 \end{bmatrix}\begin{bmatrix} b_1 \\ b_2 \\ b_3 \end{bmatrix}$$

This implies that

$$A^{-1} = \frac{1}{3}\begin{bmatrix} 1 & -1 & 0 \\ 0 & -2 & 1 \\ -2 & -1 & 1 \end{bmatrix}$$

Computationally, to obtain A^{-1} by using LU factorization of A, we determine the column vectors of A one at a time. For instance, to determine the second column of A^{-1}, we solve the system $Ax = e_2$ where $e_2 = (0, 1, 0)^T$. Solving the triangular systems $Lz = e_2$ and $Ux = z$, we get $x = \frac{1}{3}(-1, -2, -1)^T$, which is the second column of A^{-1}.

Pivoting in Gauss Elimination It might seem that the Gauss elimination completely disposes of the problem of finding solutions of linear systems, and theoretically it does. In practice, however, things are not so simple.

Example. Assume three-decimal floating arithmetic (i.e., only the three most significant digits of any number are retained), and solve the following system by Gauss elimination:

$$0.000100x_1 + 1.00x_2 = 1.00$$
$$1.00x_1 + 1.00x_2 = 2.00$$

We obtain

$$0.100 \times 10^{-3}x_1 + 0.100 \times 10^1 x_2 = 0.100 \times 10^1$$
$$-0.100 \times 10^5 x_2 = -0.100 \times 10^5$$

so that $x_2 = 1.00$ and $x_1 = 0.00$.

We check our solution by computing the residual vector $\mathbf{r} = b - Ax$:

$$r_1 = 0.100 \times 10^1 - 0.100 \times 10^{-3}x_1 - 0.100 \times 10^1 x_2 = 0.00$$
$$r_2 = 0.200 \times 10^1 - 0.100 \times 10^1 x_1 - 0.100 \times 10^1 x_2 = 0.100 \times 10^1$$

The fact that $r_2 = 1$ indicates that our "solution" is not very good. Indeed the exact solution of the system is $x_1 = 1.00010$ and $x_2 = 0.99990$, so the result computed by Gauss elimination is pretty bad.

Now reverse the order of the equations (that is, pivot) and solve

$$0.100 \times 10^1 x_1 + 0.100 \times 10^1 x_2 = 0.200 \times 10^1$$
$$0.100 \times 10^1 x_2 = 0.100 \times 10^1$$

so that $x_2 = 1.00$ and $x_1 = 1.00$. In this case the residual vector is $r_1 = 0.00$ and $r_2 = 0.100 \times 10^{-3}$, a considerable improvement over the previous result. In fact, the solution is as good as one could hope for by using three-digit arithmetic.

The moral of the preceding example is that the order of equations can make a large difference in how good an answer is obtained. It should be clear that the poor results in the first case are caused by having the large multiplier $(0.100 \times 10^1)/(0.100 \times 10^{-3})$, which resulted from dividing by a relatively small a_{11}. It is not enough just to avoid zero "pivots"; one must also avoid using pivots that are relatively small.

This magnification of errors can be reduced if we arrange that the pivot at any stage is larger in magnitude than any remaining element in the column. If this is done, the multipliers will then be less than or equal to 1 in magnitude. Gauss elimination modified in this manner is called **pivotal condensation or partial pivoting.**

Example The following system will be solved by pivotal condensation by using three-decimal floating arithmetic.

$$1.01x_1 + 0.98x_2 = 1.99$$
$$1.00x_1 + 0.99x_2 = 1.99$$

Then,

$$0.101 \times 10^1 x_1 + 0.980 \times 10^0 x_2 = 0.199 \times 10^1$$
$$0.200 \times 10^{-1} x_2 = 0.300 \times 10^{-1}$$

Or

$$x_2 = 1.50 \qquad x_1 = 0.515$$

The residuals are

$$r_1 = 0.00 \qquad r_2 = 0.01$$

which seems to be acceptable. Notice, however, that the actual solution is $x_1 = x_2 = 1$, so this answer is very bad. This system is inherently nasty, as can be seen by considering the system with the right side changed to $b^T = (2.02, 2.00)$, surely a "small" change. Now the actual solution is $x_1 = x_2 = 2$; a 1 percent change in b produces a 100 percent change in x. The trouble in this case resides in the fact that the coefficient matrix is nearly singular (its determinant is almost zero), and so solutions are highly sensitive to errors. Such problems are said to be poorly conditioned and require special handling.

NUMERICAL APPROXIMATIONS TO SOME EXPRESSIONS

APPROXIMATION IDENTITIES

For the following relationships the sign $\cong$ means approximately equal to, when X is small:

Approximation	Approximation
$\dfrac{1}{1 \pm X} \cong 1 \mp X$	$\sqrt{1 \pm X} \cong 1 \pm \dfrac{X}{2}$
$\dfrac{1 + Y}{1 \mp X} \cong 1 + Y \pm X$	$(1 \pm X)^{-n} \cong 1 \mp nX$
$(1 \pm X)^n \cong 1 \pm nX$	$(1 \pm X)^{-1/2} \cong 1 \mp \dfrac{X}{2}$
$(a \pm X)^2 \cong a^2 \pm 2aX$	$e^x \cong 1 + X$
$\sin X \cong X(X \text{ rad})$	$\tan X \cong X$
$\sqrt{Y(Y + X)} \cong \dfrac{2Y + X}{2}$	$\sqrt{Y^2 + X^2} \cong Y + \dfrac{X^2}{2Y}\left(\dfrac{X}{Y}\text{ small}\right)$

SUMMATION AND APPROXIMATION

$$\sum_1^m \sqrt{n} \cong \frac{2}{3}m^{3/2} + \frac{\sqrt{m}}{2} - 0.245$$

$$n! \cong e^{-n} n \sqrt{2\pi n}$$

$$n! \cong \sqrt{2\pi}\left\{\frac{\sqrt{n^2 + n + \frac{1}{6}}}{e}\right\}^{n+1/2}$$

$$\sqrt{X} \cong \frac{1}{2}\left[\frac{X}{S} + S\right] \qquad S = \sqrt{[X]}$$

$[X]$ = smallest integer containing X, or S is a tabled value. Precision of $\sqrt{X}$ is roughly twice as many significant digits as $S \cong \sqrt{X}$. With a three-place square-root table the root of any number can be calculated to six places with one application of the preceding correction: Given $X = 105.53196$ from $\sqrt{106} = 10.29563$ from a table

$$\sqrt{105.53196} \cong 0.5\left[\frac{105.53196}{10.29563} + 10.29563\right] \cong 10.2729$$

A second application will give the root to eight decimal places.

NUMERICAL ANALYSIS AND APPROXIMATE METHODS

REFERENCES: *General.* 3, 4, 18, 20, 33, 51, 54, 60, 71, 79, 91, 132, 135, 140, 145, 153, 157, 158, 165, 176, 218, 230, 242, 269, 292. Textbooks which cover at an introductory level a variety of topics that constitute a core of numerical methods for practicing engineers. *Numerical solution of nonlinear equations or systems; matrix iterative methods.* 152, 171, 218, 229, 274, 291. *Numerical solution of ordinary differential equations.* 70, 119, 123, 142, 179, 200, 243, 287. *Numerical solutions of integral equations.* 19, 127, 146, 164, 165, 173, 199, 208. *Numerical solutions of partial differential equations.* 9, 10, 70, 105, 123, 132, 157, 201, 202, 235, 275, 276. *Spline functions and applications.* 33, 35, 51, 60, 125, 227. *Finite elements and applications.* 25, 199, 202, 260, 276, 295, 296. *Fast Fourier transforms.* 44, 51, 135, 230.

INTRODUCTION

The goal of approximate and numerical methods is to provide convenient techniques for obtaining useful information from mathematical formulations of physical problems. Often this mathematical statement is not solvable by analytical means. Or perhaps analytic solutions are available but in a form that is inconvenient for direct interpretation numerically. In the first case it is necessary either to attempt to approximate the problem satisfactorily by one which will be amenable to analysis, to obtain an approximate solution to the original problem by numerical means, or to use the two techniques in combination.

Numerical techniques therefore do not yield exact results in the sense of the mathematician. Since most numerical calculations are inexact, the concept of error is an important feature. The error associated with an approximate value is defined as

$$\text{True value} = \text{approximate value} + \text{error}$$

The four sources of error are as follows:

1. *Gross errors.* These result from unpredictable human, mechanical, or electrical mistakes.

2. *Round-off errors.* These are the consequence of using a number specified by m correct digits to approximate a number which requires more than m digits for its exact specification. For example, approximate the irrational number $\sqrt{2}$ by 1.414. Such errors are often present in experimental data, in which case they may be called inherent errors, due either to empiricism or to the fact that the computer dictates the number of digits. Such errors may be especially damaging in areas such as matrix inversion or the numerical solution of partial differential equations when the number of algebraic operations is extremely large.

3. *Truncation errors.* These errors arise from the substitution of a finite number of steps for an infinite sequence of steps which would yield the exact result. To illustrate this error consider the infinite series for $e^{-x} \cdot e^{-x} = 1 - x + x^2/2 - x^3/6 + E_T(x)$, where E_T is the truncation error, $E_T = (1/24)e^{-\epsilon}x^4$, $0 < \epsilon < x$. If x is positive, ϵ is also positive. Hence $e^{-\epsilon} < 1$. The approximation $e^{-x} \approx 1 - x + x^2/2 - x^3/6$ is in error by a positive amount smaller than $(1/24)x^4$.

4. *Inherited errors.* These arise as a result of errors occurring in the previous steps of the computational algorithm.

The study of errors in a computation is related to the theory of probability. In what follows a relation for the error will be given in certain instances.

NUMERICAL SOLUTION OF LINEAR EQUATIONS AND ASSOCIATED PROBLEMS

The methods described here are concerned with a set of n linear equations in n unknowns $x_1, x_2, \ldots, x_n$ expressed in the form

$$\begin{aligned} a_{11}x_1 + a_{12}x_2 + a_{13}x_3 + \cdots + a_{1n}x_n &= b_1 \\ a_{21}x_1 + a_{22}x_2 + a_{23}x_3 + \cdots + a_{2n}x_n &= b_2 \\ \cdots\cdots\cdots\cdots\cdots\cdots\cdots\cdots \\ a_{n1}x_1 + a_{n2}x_2 + a_{n3}x_3 + \cdots + a_{nn}x_n &= b_n \end{aligned} \tag{2-76}$$

where the n^2 coefficients a_{ij} and the n right-hand members are given. Equations (2-76) may be written in matrix form as

$$AX = B \tag{2-77}$$

where

$$A = \begin{bmatrix} a_{11}a_{12} \cdots a_{1n} \\ a_{21}a_{22} \cdots a_{2n} \\ \cdots\cdots\cdots \\ a_{n1}a_{n2} \cdots a_{nn} \end{bmatrix} \quad X = \begin{bmatrix} x_1 \\ x_2 \\ \vdots \\ x_n \end{bmatrix} \quad B = \begin{bmatrix} b_1 \\ b_2 \\ \vdots \\ b_n \end{bmatrix}$$

and in the terminology a_{ij}, i = row index, j = column index. The problem of determining the values of $x_1, x_2, \ldots, x_n$ satisfying Eqs. (2-76) may be accomplished numerically from the form (2-76) or from (2-77) by matrix-inversion techniques. In either case the methods are **direct** (meaning "once through") or **iterative** (repeated) procedures.

Direct Methods for Solving Eqs. (2-76) Suppose that not all the b_j's are zero and that the determinant of $A \neq 0$. Then Eqs. (2-76) have a unique nontrivial solution.

Gauss Reduction This method is the simplest practical method for solving Eqs. (2-76). It consists of dividing the first equation by a_{11} (if $a_{11} = 0$, reorder the equations) and using the result to eliminate x_1 from all succeeding equations. Next, the modified second equation is divided by a'_{22} (if $a'_{22} = 0$, a renumbering of equations and/or variables may again be necessary), and the resulting equation is used to eliminate x_2 from the succeeding equations. This elimination is done n times. The result is of the "triangular" form.

$$\begin{aligned} x_1 + a'_{12}x_2 + a'_{13}x_3 + \cdots + a'_{1n}x_n &= b'_1 \\ x_2 + a'_{23}x_3 + \cdots + a'_{2n}x_n &= b'_2 \\ \cdots\cdots\cdots\cdots\cdots\cdots \\ x_{n-1} + a'_{n-1,n}x_n &= b'_{n-1} \\ x_n &= b'_n \end{aligned}$$

where the a'_{ij} and b'_j represent the specific numerical values obtained by the preceding process. The solution is then obtained by working backward from the last equation. The procedure is illustrated as it would be done in practice in which only the coefficient array is recorded.

Example

$$\begin{aligned} 18.7492x_1 + 6.0832x_2 - 4.8742x_3 &= 18.4666 \\ 6.0832x_1 + 12.3664x_2 + 2.4326x_3 &= 16.4098 \\ -4.8742x_1 + 2.4326x_2 + 16.8858x_3 &= 7.8678 \end{aligned}$$

18.7492	6.0832	−4.8742	18.4666
6.0832	12.3664	2.4326	16.4098
−4.8742	2.4326	16.8858	7.8678

Division by 18.7492 in the first row and then elimination of x_1 yields

1	0.32445	−0.25997	0.98493
	5.19635	2.00702	5.20914
	2.00702	7.80933	6.33427

Then division of the new second row by 5.19635 and elimination of x_2 in the third equation yields (writing down only the new second and third equations)

1	0.38624	1.00246
	7.03414	4.32234

Finally, dividing the last line by 7.03414 there results $x_3 = 0.61448$. Substitution of this into the previous line yields $x_2 = 0.76512$ and finally $x_1 = 0.89643$. The final result must be checked in each of the original equations.

The Gauss reduction is the basic procedure from which all other direct procedures have evolved. Its disadvantage is the recording of the new arrays and the possibility of gross errors. A modification using the kth equation, at the kth stage, to eliminate x_k from the preceding as well as the following equation gives a final diagonal form of the array. The solution is then obtained immediately. This procedure is called the Gauss-Jordan reduction. Another modification called the Gauss-Doolittle reduction is specifically for symmetric equations. (These and the checks on the methods are outlined in the references; these are especially well done in Bodewig.)

Crout Reduction A modification of the Gauss procedure which is well adapted for use on desk calculators and digital computers is a method devised by Crout. Recording of intermediate steps is minimized in this procedure. The Crout algorithm is summarized by the equations

$$\begin{aligned} a'_{ij} &= a_{ij} - \sum_{k=1}^{j-1} a'_{ik}a'_{kj} \qquad i \geq j \\ a'_{ij} &= \frac{1}{a'_{ii}}\left[a_{ij} - \sum_{k=1}^{i-1} a'_{ik}a'_{kj}\right] \quad i < j \\ b'_i &= \frac{1}{a'_{ii}}\left[b_i - \sum_{k=1}^{i-1} a'_{ik}b'_k\right] \end{aligned} \tag{2-78}$$

and finally the solution

$$x_i = b'_i - \sum_{k=i+1}^{n} a'_{ik}x_k \tag{2-79}$$

and i and j run from 1 to n unless other restrictions are present. An excellent discussion of this technique is found in Hildebrand (145).

Example

$$\begin{aligned} 2x_1 - 3x_2 + x_3 &= 1 \\ x_1 + 2x_2 + 3x_3 &= 3 \\ 4x_1 - x_2 - x_3 &= 1 \end{aligned}$$

From Eq. (2-78)

$$a'_{11} = a_{11} = 2 \quad a'_{12} = \frac{1}{a'_{11}}[a_{12}] = -\frac{3}{2}$$

$$a'_{21} = a_{21} = 1 \quad a'_{22} = a_{22} - a'_{21}a'_{12} = \frac{7}{2}$$

$$a'_{31} = a_{31} = 4 \quad a'_{13} = \frac{a_{13}}{a'_{11}} = \frac{1}{2}$$

$$a'_{23} = \frac{1}{a'_{22}}[a_{23} - a'_{21}a'_{13}] = \frac{5}{7}$$

$$a'_{33} = a_{33} - \sum_{k=1}^{2} a'_{3k}a'_{k3} = a_{33} - a'_{31}a'_{13} - a'_{32}a'_{23} = -\frac{46}{7}$$

$$b'_1 = \frac{b_1}{a'_{11}} = \frac{1}{2} \qquad b'_2 = \frac{1}{a'_{22}}[b_2 - a'_{21}b'_1] = \frac{5}{7}$$

$$b'_3 = \frac{1}{a'_{33}}[b_3 - a'_{31}b'_1 - a'_{32}b'_2] = \frac{16}{23}$$

$$a'_{32} = a_{32} - a'_{31}a'_{12} = +5$$

Hence by Eq. (2-79)

$$x_3 = b'_3 = 16/23 \qquad x_2 = b'_2 - a'_{23}x_3 = 5/23$$

$$x_1 = b'_1 - a'_{12}x_2 - a'_{13}x_3 = 11/23$$

Iterative Methods for Solving Eqs. (2-76) In certain systems, for example, in the least-squares problems of statistics, it often happens that the diagonal elements (the elements a_{ii}) of Eqs. (2-76) dominate strongly over the other elements. In these cases iterative methods may be used to solve the linear system [Eqs. (2-76)]. The more the diagonal terms dominate, the more rapidly the process converges and is in many cases superior to the direct processes.

Iteration in Total Steps Referring to the linear system [Eqs. (2-76)], the first set of approximate values is obtained by taking into account only the dominant diagonal terms in each equation. The approximate values are then inserted into the full system to obtain the second approximation, and so on. If the system has been rewritten so that the diagonal terms dominate, then the procedure is to rewrite it as

$$\begin{aligned} x_1 &= (1/a_{11})(b_1 - a_{12}x_2 - a_{13}x_3 - \cdots - a_{1n}x_n) \\ x_2 &= (1/a_{22})(b_2 - a_{21}x_1 - a_{23}x_3 - \cdots - a_{2n}x_n) \\ &\cdots\cdots\cdots\cdots\cdots\cdots \\ x_n &= (1/a_{nn})(b_n - a_{n1}x_1 - a_{n2}x_2 - \cdots - a_{n\,n-1}x_{n-1}) \end{aligned} \tag{2-80}$$

The initial approximation is

$$x_1^{(0)} = b_1/a_{11} \qquad x_2^{(0)} = b_2/a_{22} \quad \cdots \quad x_n^{(0)} = b_n/a_{nn} \tag{2-81}$$

The next approximation is obtained by inserting the initial approximations in Eq. (2-80) and repeating until the successive approximations agree to within a specified tolerance.

Iteration in Single Steps In this method a diagonal unknown, say, x_1, is computed approximately, all others being neglected. This value is inserted into all other equations, and from one of them an approximation for a second diagonal element is obtained, and so forth. Thus at every step all unknowns are computed by means of all components already known.

Example

$$\begin{aligned} 10x_1 + x_2 - x_3 &= 2 \\ 2x_1 + 15x_2 - 3x_3 &= 6 \\ 3x_1 - x_2 + 20x_3 &= -4 \end{aligned}$$

$x_1^{(0)} = 2/10 = 1/5$, then $15x_2^{(0)} = 6 - 2/5$ $\quad x_2^{(0)} = 28/75$ $\quad 20x_3^{(0)} = -4 - 3/5 + 28/75 = -347/75$ $\quad x_3^{(0)} = -347/75 \times 20$, and so forth.

Relaxation These procedures do not readily lend themselves to mechanization but are useful for hand computation. Their great facility is again concerned with linear systems having dominant diagonal elements. The ingenuity of the computer is important in this procedure, as will be seen. The first step in the computation is to define **residuals** $R_1, R_2, \ldots, R_n$ from the system (2-76) by the equations

$$\begin{aligned} R_1 &= b_1 - a_{11}x_1 - a_{12}x_2 - \cdots - a_{1n}x_n \\ R_2 &= b_2 - a_{21}x_1 - a_{22}x_2 - \cdots - a_{2n}x_n \\ &\cdots\cdots\cdots\cdots\cdots\cdots \\ R_n &= b_n - a_{n1}x_1 - a_{n2}x_2 - \cdots - a_{nn}x_n \end{aligned} \tag{2-82}$$

The unknowns are then estimated, say, by the system (2-81), and the corresponding residuals are calculated, after which the estimated values of the unknowns are successively modified, one or more at a time, such that the magnitudes of all residuals are reduced approximately to zero. Since the residuals are known at each step, one usually focuses attention on the residual of largest magnitude to reduce it to zero. A helpful hint here is to note in Eqs. (2-82) that if x_i is increased by 1 and all other unknowns are held fixed, then R_j decreases by a_{ji}.

Example A typical sequence of relaxations is shown below for the system

$$\begin{aligned} 9.37x_1 + 3.04x_2 - 2x_3 &= 3 \\ 2.05x_1 + 7.22x_2 + 1.22x_3 &= 6 \\ -1.41x_1 + 1.83x_2 + 6.30x_3 &= -4 \end{aligned}$$

$$\begin{aligned} x_1^{(0)} &= 3/9.37 = 0.32 & R_1 &= 3 - 9.37x_1 - 3.04x_2 + 2x_3 \\ x_2^{(0)} &= 6/7.22 = 0.83 & R_2 &= 6 - 2.05x_1 - 7.22x_2 - 1.22x_3 \\ x_3^{(0)} &= -4/6.3 = -0.63 & R_3 &= -4 + 1.41x_1 - 1.83x_2 - 6.30x_3 \end{aligned}$$

Step no.	x_1	x_2	x_3	R_1	R_2	R_3
0	0.32	0.83	−0.63	−3.78	0.11	−1.07

The largest residual in magnitude, at this stage, is $R_1 = -3.78$. Thus we decrease x_1 by a convenient amount, usually of the size $R_1/a_{11} = -3.78/9.37 \approx -0.4$, and recompute the residuals.

Step no.	x_1	x_2	x_3	R_1	R_2	R_3
1	−0.08	0.83	−0.63	−0.03	0.93	−1.63

At this stage $R_3 = -1.63$ is largest. Decrease x_3 by $R_3/a_{33} = -1.63/6.30 \approx -0.27$ and proceed.

2	−0.08	0.83	−0.9	−0.57	1.26	0.07

Iterative procedures are described in detail in Refs. [34], [51], and [274] and in the general references given.

Matrix Inversion In some problems, such as those encountered in statistical regression analysis, it is essential that the system (2-76) be solved by matrix inversion of Eq. (2-77). Thus $X = A^{-1}B$, where A^{-1} is the inverse of A, defined under "Elementary Algebra."

The number of methods for inverting matrices are many and varied. The methods previously described may be continued to obtain the inverse of the matrix. In addition, two procedures which are admirably suited to use on large-scale digital computers are given.

Modified-Square-Root Method (Choleski) This is a direct method for inverting symmetric positive definite matrices. As will be seen, the division by the square root of the leading element reduces the round-off errors, and the square rooting of all elements brings all matrix elements nearer in value, which is helpful. The matrix A is inverted by the following sequence of calculations where

$$A^{-1} = \begin{bmatrix} c_{11} & c_{12} & c_{13} & \cdots & c_{1n} \\ c_{21} & c_{22} & \cdots & \cdots & c_{2n} \\ \cdots & \cdots & \cdots & \cdots & \cdots \\ c_{n1} & c_{n2} & \cdots & \cdots & c_{nn} \end{bmatrix}$$

Since A is symmetric, A^{-1} is also symmetric, so that $c_{ij} = c_{ji}$. Hence only c_{ij}, $i \geq j$ need be calculated. In general, calculate

$$\begin{aligned} t_{ii} &= \sqrt{a_{ii} - \sum_{k=1}^{i-1} t_{ki}^2} \\ t_{ij} &= \frac{a_{ij} - \sum_{k=1}^{i-1} t_{ki}t_{kj}}{t_{ii}} \qquad i \neq j,\ i > j \\ c_{ii} &= \frac{1}{t_{ii}^2} - \frac{\sum_{k=i+1}^{n} t_{ik}c_{ik}}{t_{ii}} \\ c_{ij} &= -\frac{\sum_{k=j}^{n} t_{ik}c_{kj}}{t_{ii}} \quad \text{and} \quad c_{ij} = c_{ji} \end{aligned} \tag{2-83}$$

These equations are illustrated here for a 4 × 4 matrix.

$$t_{11} = \sqrt{a_{11}} \quad t_{12} = a_{12}/t_{11} \quad t_{13} = a_{13}/t_{11} \quad t_{14} = a_{14}/t_{11}$$

$$t_{22} = \sqrt{a_{22} - t_{12}^2} \quad t_{23} = \frac{a_{23} - t_{12}t_{13}}{t_{22}}$$

$$t_{24} = \frac{a_{24} - t_{12}t_{14}}{t_{22}} \quad t_{33} = \sqrt{a_{33} - t_{13}^2 - t_{23}^2}$$

$$t_{34} = \frac{a_{34} - t_{13}t_{14} - t_{23}t_{24}}{t_{33}} \quad t_{44} = \sqrt{a_{44} - t_{14}^2 - t_{24}^2 - t_{34}^2}$$

and finally it is necessary to calculate the c_{ij} in reverse order:

$$c_{44} = \frac{1}{t_{44}^2} \quad c_{34} = \frac{-t_{34}c_{44}}{t_{33}} \quad c_{24} = \frac{-t_{23}c_{34} - t_{24}c_{44}}{t_{22}}$$

$$c_{14} = \frac{-t_{12}c_{24} - t_{13}c_{34} - t_{14}c_{44}}{t_{11}} \quad c_{33} = \frac{1}{t_{33}^2} - \frac{t_{34}c_{34}}{t_{33}}$$

$$c_{23} = \frac{-t_{23}c_{33} - t_{24}c_{34}}{t_{22}}, \ldots$$

$$c_{11} = \frac{1}{t_{11}^2} - \frac{t_{12}c_{12} + t_{13}c_{13} + t_{14}c_{14}}{t_{11}}$$

Smith Algorithm This procedure is again a direct procedure which basically consists of a factorization of the matrix A until it is reduced to the identity by elementary row and column operations. The algorithm is outlined in Bodewig (34).

NUMERICAL SOLUTION OF NONLINEAR EQUATIONS IN ONE VARIABLE

Special Methods for Polynomials Consider a polynomial equation of degree n:

$$P(x) = a_0x^n + a_1x^{n-1} + a_2x^{n-2} + \cdots + a_{n-1}x + a_n = 0 \tag{2-84}$$

with real coefficients. $P(x)$ has exactly n roots, which may be real or complex. If all the coefficients of $P(x)$ are integers, then any rational root, say, r/s (r, s integers, having no common divisors) of $P(x)$, must be such that r is an integral divisor of a_n and s is an integral divisor of a_0. Further, any polynomial with rational coefficients may be converted into one with integral coefficients by multiplying by the lowest common multiple of the denominators of the coefficients.

Example $3x^4 - \frac{5}{3}x^2 + \frac{1}{5}x - 2 = 0$. The lowest common multiple of the denominator is 15. Thus multiplying by 15 (which does not change the roots) gives $45x^4 - 25x^2 + 3x - 30 = 0$. The only possible rational roots r/s are such that r may have the values ±30, ±15, ±10, ±6, ±3, ±2, ±1. s may have the values ±45, ±15, ±9, ±5, ±3, ±1. The possible rational roots may then be formed from all possible quotients, having no common factor.

In addition to these results, one can obtain an upper and lower bound for the real roots by the following device: If $a_0 > 0$ in Eq. (2-84) and if in Eq. (2-84) the first negative coefficient is preceded by k coefficients which are positive or zero, and if G is the greatest of the absolute values of the negative coefficients, then each real root is less than $1 + \sqrt[k]{G/c_0}$.

Example $P(x) = x^5 + 3x^4 - 7x^2 - 40x + 2 = 0$. Here $a_0 = 1$, $G = 40$, and $k = 3$ since we must supply 0 as the coefficient for x^3. Thus $1 + \sqrt[3]{40} \approx 4.42$ is an upper bound for the real roots.

A lower bound to the real roots may be found by applying the criterion to the equation $P(-x)$.

Example $P(-x) = -x^5 + 3x^4 - 7x^2 + 40x + 2 = 0$, which is equivalent to $x^5 - 3x^4 + 7x^2 - 40x - 2 = 0$ since a_0 must be +. Then $a_0 = 1$, $G = 40$, and $k = 1$. Hence $-(1 + 40) = -41$ is a lower bound. Thus all real roots $-41 < r < 4.42$.

One last result is helpful in getting an estimate of how many positive and negative real roots there are.

Descartes Rule The number of positive real roots of a polynomial with real coefficients is either equal to the number of changes in sign v or is less than v by a positive even integer. The number of negative roots of $f(x)$ is either equal to the number of variations of sign of $f(-x)$ or is less than this by a positive even integer.

Example $f(x) = x^4 - 13x^2 + 4x - 2 = 0$ has three changes in sign; therefore, there are either three or one positive roots. $f(-x) = x^4 - 13x^2 - 4x - 2$ has one change in sign. Therefore, there is one negative root.

More information on properties of polynomials and special techniques may be found in Dickson, *New First Course in the Theory of Equations*, Wiley, New York, 1939; and MacDuffee, *Theory of Equations*, Wiley, New York, 1954. See also Ref. 71.

Graeffe Root-Squaring Technique This is an iterative method for finding the roots of the algebraic equation

$$f(x) = a_0x^p + a_1x^{p-1} + \cdots + a_{p-1}x + a_p = 0 \tag{2-85}$$

If the roots are $r_1, r_2, r_3, \ldots$, then one can write

$$S_p = r_1^p(1 + r_2^p/r_1^p + r_3^p/r_1^p + \cdots) \tag{2-86}$$

and if one root is larger than all the others, say, r_1, then for p large enough all terms (other than 1) would become negligible and thus

$$S_p \approx r_1^p$$

or

$$\lim_{p\to\infty} S_p^{1/p} = r_1$$

The Graeffe procedure provides an efficient way for computing S_p of Eq. (2-86) via a sequence of equations such that the roots of each equation are the squares of the roots of the preceding equations in the sequence. This serves the purpose of ultimately obtaining an equation whose roots are so widely separated in magnitude that they may be read approximately from the equation by inspection. The basic procedure is illustrated for a polynomial of degree 4:

$$f(x) = a_0x^4 + a_1x^3 + a_2x^2 + a_3x + a_4 = 0 \qquad (2\text{-}87)$$

Rewrite Eq. (2-87) as

$$a_0x^4 + a_2x^2 + a_4 = -a_1x^3 - a_3x$$

and square both sides so that upon grouping

$$a_0^2x^8 + (2a_0a_2 - a_1^2)x^6 + (2a_0a_4 - 2a_1a_3 + a_2^2)x^4 + (2a_2a_4 - a_3^2)x^2 + a_4^2 = 0 \qquad (2\text{-}88)$$

Since this involves only even powers of x, we may set $y = x^2$ and rewrite Eq. (2-88) as

$$a_0^2y^4 + (2a_0a_2 - a_1^2)y^3 + (2a_0a_4 - 2a_1a_3 + a_2^2)y^2 + (2a_2a_4 - a_3^2)y + a_4^2 = 0$$

whose roots are the squares of the original equation. If we repeat this process again, the new equation has roots which are the fourth power, etc. After p such operations the roots are 2^p (original roots). Then $a_1/a_0 = -\sum_{i=1}^{4} r_i$, $a_1^{(1)}/a_0^{(1)} = -\Sigma r_i^2, \ldots, a_1^{(p)}/a_0^{(p)} = -\Sigma r_i^{2p}$. If the roots are all distinct and r_1 is the largest in magnitude, then eventually $r_1^{2p} \approx -a_1^{(p)}/a_0^{(p)}$. And if r_2 is the next largest in magnitude, then $r_2^{2p} \approx a_2^{(p)}/a_1^{(p)}$. And in general $a_n^{(p)}/a_{n-1}^{(p)} \approx (-1)^n r_n^{2p}$. This procedure is easily generalized to polynomials of arbitrary degree and specialized to the case of multiple and complex roots. An excellent discussion of these ideas is found in Householder and Hildebrand (see references). The signs of the roots are undetermined, but these may be obtained by trial of both possibilities.

Example $f(x) = x^4 - 7x^3 + 9x^2 + 7x - 10 = 0.$

Step p	$a_0^{(p)}$	$a_1^{(p)}$	$a_2^{(p)}$	$a_3^{(p)}$	$a_4^{(p)}$
0	1	−7	9	7	−10
1	1	−31	159	−229	100
2	1	−643	11,283	−20,641	10,000

Thus $\quad r_1^4 \approx -a_1^{(2)}/a_0^{(2)} = 643 \quad r_1 \approx +5.04$

$\quad r_2^4 \approx -a_2^{(2)}/a_1^{(2)} = 17.6 \quad r_2 \approx +2.05$

and so forth. The actual roots are +5, +2, +1, and −1.

Other methods include Bernoulli iteration, Bairstow iteration, Lin iteration, and so forth. These may be found in the cited literature. In addition the methods given in the following subsections may be used for the numerial solution of polynomials.

General Methods for Nonlinear Equations in One Variable

Successive Substitutions Let $f(x) = 0$ be the nonlinear equation to be solved. If this is rewritten as $x = F(x)$, then an iterative scheme can be set up in the form $x_{k+1} = F(x_k)$. To start the iteration an initial guess must be obtained graphically or otherwise. The convergence or divergence of the procedure depends upon the method of writing $x = F(x)$, of which there will usually be several forms. A general rule to ensure convergence cannot be given. However, if a is a root of $f(x) = 0$, and if $|F'(a)| < 1$, then for any initial approximation sufficiently close to a, the method converges to a. This process is called **first order** because the error in x_{k+1} is proportional to the first power of the error in x_k for large k.

Example $f(x) = x^3 - x - 1 = 0$. A rough plot shows a real root of approximately 1.3. The equation can be written in the form $x = F(x)$ in several ways such as $x = x^3 - 1$, $x = 1/(x^2 - 1)$, and $x = (1 + x)^{1/3}$. In the first case $F'(x) = 3x^2 = 5.07$ at $x = 1.3$, in the second $F(1.3) = 5.46$, and only in the third case is $F'(1.3) < 1$. Hence only the third iterative process has a chance to converge. This is illustrated in the following table.

Iteration Table

Step k	$x = \frac{1}{x^2 - 1}$	$x = x^3 - 1$	$x = (1 + x)^{1/3}$
0	1.3	1.3	1.3
1	1.4493	1.197	1.32
2	0.9087	0.7150	1.3238
3	−5.737	−0.6345	1.3247
4	....	....	1.3247

Methods of Perturbation Let $f(x) = 0$ be the equation. In general, the iterative relation is

$$x_{k+1} = x_k - [f(x_k)/\alpha_k] \qquad (2\text{-}89)$$

where the iteration begins with x_0 as an initial approximation and α_k as some functional.

Newton-Raphson procedure. This variant chooses $\alpha_k = f'(x_k)$ where $f' = df/dx$ and geometrically consists of replacing the graph of $f(x)$ by the tangent line at $x = x_k$ in each successive step. If $f'(x)$ and $f''(x)$ have the same sign throughout an interval $a \le x \le b$ containing the solution, with $f(a)$, $f(b)$ of opposite signs, then the process converges starting from any x_0 in the interval $a \le x \le b$. The process is second order.

Example

$$f(x) = x - 1 + \frac{(0.5)^x - 0.5}{0.3}$$

$$f'(x) = 1 - 2.3105[0.5]^x$$

An approximate root (obtained graphically) is 2.

Step k	x_k	$f(x_k)$	$f'(x_k)$
0	2	0.1667	0.4224
1	1.6054	0.0342	0.2407
2	1.4632	0.0055	0.1620

Method of false position. This variant is commenced by finding x_0 and x_1 such that $f(x_0)$, $f(x_1)$ are of opposite signs. Then α_1 = slope of secant line joining $[x_0, f(x_0)]$ and $[x_1, f(x_1)]$ so that

$$x_2 = x_1 - \frac{x_1 - x_0}{f(x_1) - f(x_0)} f(x_1)$$

In each following step α_k is the slope of the line joining $[x_k, f(x_k)]$ to the most recently determined point where $f(x_j)$ has the opposite sign from that of $f(x_k)$. This method is of first order.

Method of Wegstein This is a variant of the method of successive substitutions which forces and/or accelerates convergence. The iterative procedure $x_{k+1} = F(x_k)$ is revised by setting $\hat{x}_{k+1} = F(x_k)$ and then taking $x_{k+1} = qx_k + (1 - q)\hat{x}_{k+1}$, where q is a suitably chosen number which may be taken as constant throughout or may be adjusted at each step. Wegstein found that suitably chosen q's are related to the basic process as follows:

Behavior of successive substitution process	Range of optimum q
Oscillatory convergence	$0 < q < ½$
Oscillatory divergence	$½ < q < 1$
Monotonic convergence	$q < 0$
Monotonic divergence	$1 < q$

At each step q may be calculated to give a locally optimum value by setting

$$q = \frac{\hat{x}_{k+2} - \hat{x}_{k+1}}{\hat{x}_{k+2} - 2\hat{x}_{k+1} + \hat{x}_k}$$

Numerical Solution of Simultaneous Nonlinear Equations The techniques illustrated here will be demonstrated for two simultaneous equations $f(x, y) = 0$, $g(x, y) = 0$. They immediately generalize to more than two simultaneous equations.

Method of Successive Substitutions The two simultaneous equations can be written in various ways in equivalent forms

$$\begin{aligned} x &= F(x, y) \\ y &= G(x, y) \end{aligned} \tag{2-90}$$

And the method of successive substitutions can be based on

$$\begin{aligned} x_{k+1} &= F(x_k, y_k) \\ y_{k+1} &= G(x_k, y_k) \end{aligned}$$

Again the procedure is of the first order and a sufficient condition for convergence is

$$\left|\frac{\partial F}{\partial x}\right| + \left|\frac{\partial F}{\partial y}\right| < 1 \qquad \left|\frac{\partial G}{\partial x}\right| + \left|\frac{\partial G}{\partial y}\right| < 1$$

in the iteration neighborhood of the true solution.

Newton-Raphson Procedure Using $f(x, y) = 0$ and $g(x, y) = 0$ for the two simultaneous equations, start from an approximation, say, (x_0, y_0), obtained graphically or from a two-way table. Then solve successively the linear equations

$$\begin{aligned} \Delta x_k \frac{\partial f}{\partial x}(x_k, y_k) + \Delta y_k \frac{\partial f}{\partial y}(x_k, y_k) &= -f(x_k, y) \\ \Delta x_k \frac{\partial g}{\partial x}(x_k, y_k) + \Delta y_k \frac{\partial g}{\partial y}(x_k, y_k) &= -g(x_k, y_k) \end{aligned} \tag{2-91}$$

for Δx_k and Δy_k. Then the $k + 1$ approximation is given from $x_{k+1} = x_k + \Delta x_k$, $y_{k+1} = y_k + \Delta y_k$. A modification consists in solving Eqs. (2-91) with (x_k, y_k) replaced by (x_0, y_0) (or other suitable pair later on in the iteration) in the derivatives. This means that the derivatives (and therefore the coefficients of Δx_k, Δy_k) are independent of k. Hence the results become

$$\begin{aligned} \Delta x_k &= \frac{-f(x_k, y_k)(\partial g/\partial y)(x_0, y_0) + g(x_k, y_k)(\partial f/\partial y)(x_0, y_0)}{(\partial f/\partial x)(x_0, y_0)(\partial g/\partial y)(x_0, y_0) - (\partial f/\partial y)(x_0, y_0)(\partial g/\partial x)(x_0, y_0)} \\ \Delta y_k &= \frac{-g(x_k, y_k)(\partial f/\partial x)(x_0, y_0) + f(x_k, y_k)(\partial g/\partial x)(x_0, y_0)}{(\partial f/\partial x)(x_0, y_0)(\partial g/\partial y)(x_0, y_0) - (\partial f/\partial y)(x_0, y_0)(\partial g/\partial x)(x_0, y_0)} \end{aligned} \tag{2-92}$$

and $x_{k+1} = \Delta x_k + x_k$, $y_{k+1} = \Delta y_k + y_k$. Such an alteration of the basic technique reduces the rapidity of convergence in general.

Example

$$\begin{aligned} f(x, y) &= 4x^2 + 6x - 4xy + 2y^2 - 3 \\ g(x, y) &= 2x^2 - 4xy + y^2 \end{aligned}$$

By plotting one of the approximate roots is found to be $x_0 = 0.4$, $y_0 = 0.3$. At this point there results $\partial f/\partial x = 8$, $\partial f/\partial y = -0.4$, $\partial g/\partial x = 0.4$, and $\partial g/\partial y = -1$. Hence from Eqs. (2-92)

$$\begin{aligned} x_{k+1} &= x_k + \Delta x_k = x_k + \frac{-f(x_k, y_k) - 0.4g(x_k, y_k)}{8(-1) - (-0.4)(0.4)} \\ &= x_k - 0.12755f(x_k, y_k) - 0.05102g(x_k, y_k) \end{aligned}$$

and $\quad y_{k+1} = y_k - 0.05102f(x_k, y_k) + 1.02041g(x_k, y_k)$

The first few iteration steps are as follows:

Step k	x_k	y_k	$f(x_k, y_k)$	$g(x_k, y_k)$
0	0.4	0.3	−0.26	0.07
1	0.43673	0.24184	0.078	0.0175
2	0.42672	0.25573	−0.0170	−0.007
3	0.42925	0.24943	0.0077	0.0010

Method of Continuity In the case of n equations in n unknowns, when n is large, determining the approximate solution may involve considerable effort. In such a case the method of continuity is admirably suited for use on either digital or analog computers. It consists basically of the introduction of an extra variable into the n equations

$$f_i(x_1, x_2, \ldots, x_n) = 0 \qquad i = 1, \ldots, n \tag{2-93}$$

and replacing them by

$$f_i(x_1, x_2, \ldots, x_n, \lambda) = 0 \qquad i = 1, \ldots, n \tag{2-94}$$

where λ is introduced in such a way that the functions (2-94) depend in a simple way upon λ and reduce to an easily solvable system for $\lambda = 0$ and to the original equations (2-93) for $\lambda = 1$. A system of ordinary differential equations, with independent variable λ, is then constructed by differentiating Eqs. (2-94) with respect to λ. There results

$$\sum_{j=1}^{n} \frac{\partial f_i}{\partial x_j}\frac{dx_j}{d\lambda} + \frac{\partial f_i}{\partial \lambda} = 0 \tag{2-95}$$

where $x_1, \ldots, x_n$ are considered as functions of λ. Equations (2-95) are integrated, with initial conditions obtained from Eqs. (2-94) with $\lambda = 0$, from $\lambda = 0$ to $\lambda = 1$. If the solution can be continued to $\lambda = 1$, the values of $x_1, \ldots, x_n$ for $\lambda = 1$ will be a solution of the original equations. If the integration becomes infinite, the parameter λ must be introduced in a different fashion. Integration of the differential equations (which are usually nonlinear in λ) may be accomplished on an analog computer or by digital means by using techniques described under "Numerical Solution of Ordinary Differential Equations."

Example

$$\begin{aligned} f(x, y) &= 1 + x + y - x^2 + 8xy + y^3 = 0 \\ g(x, y) &= 1 + 2x + 3y + x^2 + xy - ye^x = 0 \end{aligned}$$

Introduce λ as

$$\begin{aligned} f(x, y, \lambda) &= (2 + x + y) + \lambda(-x^2 + 8xy + y^3) = 0 \\ g(x, y, \lambda) &= (1 + 2x - 3y) + \lambda(x^2 + xy - ye^x) = 0 \end{aligned}$$

For $\lambda = 1$ these reduce to the original equations but for $\lambda = 0$ they are the linear system

$$\begin{aligned} x + y &= -2 \\ 2x - 3y &= -1 \end{aligned}$$

which has the unique solution $x = -1.4$, $y = -0.6$. The differential equations (2-95) become in this case

$$\begin{aligned} \frac{\partial f}{\partial x}\frac{dx}{d\lambda} + \frac{\partial f}{\partial y}\frac{dy}{d\lambda} &= -\frac{\partial f}{\partial \lambda} \\ \frac{\partial g}{\partial x}\frac{dx}{d\lambda} + \frac{\partial g}{\partial y}\frac{dy}{d\lambda} &= -\frac{\partial g}{\partial \lambda} \end{aligned} \tag{2-96}$$

or

$$\frac{dx}{d\lambda} = \frac{\dfrac{\partial f}{\partial y}\dfrac{\partial g}{\partial \lambda} - \dfrac{\partial f}{\partial \lambda}\dfrac{\partial g}{\partial y}}{\dfrac{\partial f}{\partial x}\dfrac{\partial g}{\partial y} - \dfrac{\partial f}{\partial y}\dfrac{\partial g}{\partial x}}$$

$$\frac{dy}{d\lambda} = \frac{\dfrac{\partial f}{\partial \lambda}\dfrac{\partial g}{\partial x} - \dfrac{\partial f}{\partial x}\dfrac{\partial g}{\partial \lambda}}{\dfrac{\partial f}{\partial x}\dfrac{\partial g}{\partial y} - \dfrac{\partial f}{\partial y}\dfrac{\partial g}{\partial x}}$$

where

$$\begin{aligned} \partial f/\partial x &= 1 - 2\lambda x + 8\lambda y \\ \partial f/\partial y &= 1 + 8\lambda x + 3\lambda y^2 \\ \partial g/\partial x &= 2 + 2\lambda x + \lambda y - \lambda y e^x \\ \partial g/\partial y &= -3 + \lambda x - \lambda e^x \end{aligned}$$

Equations (2-96) are integrated in λ (starting with $x = -1.4$, $y = -0.6$ at $\lambda = 0$) to $\lambda = 1$. The values of x, y at $\lambda = 1$ constitute the solution.

Other Methods Other methods can be found in the literature. To be especially mentioned are methods of steepest descent (see Housholder in the references) and relaxation methods (see Southwell,

Relaxation Methods in Theoretical Physics, Oxford, New York, 1946) and Refs. 71 and 218.

INTERPOLATION AND FINITE DIFFERENCES

The practicing engineer finds it constantly necessary to refer to tables as sources of information. Consequently interpolation, or that procedure of "reading between the lines of the table," is a necessary topic in numerical analysis.

Linear Interpolation If a function $f(x)$ is approximately linear in a certain range, then the ratio $\frac{f(x_1) - f(x_0)}{x_1 - x_0} = f[x_0, x_1]$ is approximately independent of x_0, x_1 in the range. The linear approximation to the function $f(x)$, $x_0 < x < x_1$ then leads to the interpolation formula

$$\begin{aligned} f(x) &\approx f(x_0) + (x - x_0)f[x_0, x_1] \\ &\approx f(x_0) + \frac{x - x_0}{x_1 - x_0}[f(x_1) - f(x_0)] \qquad (2\text{-}97) \\ &\approx \frac{1}{x_1 - x_0}[(x_1 - x)f(x_0) - (x_0 - x)f(x_1)] \end{aligned}$$

Example Find cosh 0.83 by linear interpolation given cosh 0.8 and cosh 0.9.

x_i	$f(x_i)$	$x_i - 0.83$
0.8	1.33743	−0.03
0.9	1.43309	+0.07

$$f(0.83) \approx 1/0.10[(0.07)(1.33743) - (-0.03)(1.43309)]$$
$$f(0.83) \approx 1.36613$$

Since the true five-place value is 1.36468, it is seen that here linear interpolation gives three significant figures.

Divided Differences of Higher Order and Higher-Order Interpolation The first-order divided difference $f[x_0, x_1]$ was defined previously. Divided differences of second and higher order are defined iteratively by

$$\begin{aligned} f[x_0, x_1, x_2] &= \frac{f[x_1, x_2] - f[x_0, x_1]}{x_2 - x_0} \\ &\vdots \qquad\qquad (2\text{-}98) \\ f[x_0, x_1, \ldots, x_k] &= \frac{f[x_1, \ldots, x_k] - f[x_0, x_1, \ldots, x_{k-1}]}{x_k - x_0} \end{aligned}$$

and a convenient form for computational purposes is

$$f[x_0, x_1, \ldots, x_k] = \sum_{j=0}^{k}{}' \frac{f(x_j)}{(x_j - x_0)(x_j - x_1) \cdots (x_j - x_k)}$$

for any $k \geq 0$, where the $'$ means that the term $(x_j - x_j)$ is omitted in the denominator. For example,

$$f[x_0, x_1, x_2] = \frac{f(x_0)}{(x_0 - x_1)(x_0 - x_2)} + \frac{f(x_1)}{(x_1 - x_0)(x_1 - x_2)} + \frac{f(x_2)}{(x_2 - x_0)(x_2 - x_1)}$$

If the accuracy afforded by a linear approximation is inadequate, a generally more accurate result may be based upon the assumption that $f(x)$ may be approximated by a polynomial of degree 2 or higher over certain ranges. This assumption leads to Newton's fundamental interpolation formula with divided differences

$$\begin{aligned} f(x) &\approx f(x_0) + (x - x_0)f[x_0, x_1] + (x - x_0)(x - x_1)f[x_0, x_1, x_2] \\ &\quad + \cdots + (x - x_0)(x - x_1) \cdots (x - x_{n-1})f[x_0, x_1, \ldots, x_n] \\ &\quad + E_n(x) \qquad (2\text{-}99) \end{aligned}$$

where $E_n(x) = \text{error} = \frac{1}{(n+1)!} f^{n+1)}(\epsilon)\pi(x)$

where minimum $(x_0, \ldots, x) < \epsilon <$ maximum $(x_0, x_1, \ldots, x_n, x)$ and $\pi(x) = (x - x_0)(x - x_1) \cdots (x - x_n)$. In order to use Eq. (2-99) most effectively one may first form a divided-difference table. For example, for third-order interpolation the difference table is

x_0	$f(x_0)$			
		$f[x_0, x_1]$		
x_1	$f(x_1)$		$f[x_0, x_1, x_2]$	
		$f[x_1, x_2]$		$f[x_0, x_1, x_2, x_3]$
x_2	$f(x_2)$		$f[x_1, x_2, x_3]$	
		$f[x_2, x_3]$		
x_3	$f(x_3)$			

where each entry is given by taking the difference between diagonally adjacent entries to the left, divided by the abscissas corresponding to the ordinates intercepted by the diagonals passing through the calculated entry.

Example Calculate by third-order interpolation the value of cosh 0.83, given cosh 0.60, cosh 0.80, cosh 0.90, and cosh 1.10.

$x_0 = 0.60$	1.18547			
		0.7598		
$x_1 = 0.80$	1.33743		0.6560	
		0.9566		0.1586
$x_2 = 0.90$	1.43309		0.7353	
		1.1772		
$x_3 = 1.10$	1.66852			

From Eq. (2-99) with $n = 3$ we have cosh 0.83 ≈ 1.18547 + (0.23)(0.7598) + (0.23)(0.03)(0.6560) + (0.23)(0.03)(−0.07)(0.1586) = 1.36464, which varies from the true value by 0.00004.

Equally Spaced Forward Differences Formulas (2-98) and (2-99) are given in general form for unequally spaced ordinates. If the ordinates are *equally spaced,* i.e., $x_j - x_{j-1} = \Delta x$ for all j, then the first differences are denoted by $\Delta f(x_0) = f(x_1) - f(x_0)$ or $\Delta y_0 = y_1 - y_0$, where $y = f(x)$. The differences of these first differences, called second differences, are denoted by $\Delta^2 y_0, \Delta^2 y_1, \ldots, \Delta^2 y_n$. Thus

$$\Delta^2 y_0 = \Delta y_1 - \Delta y_0 = y_2 - y_1 - y_1 + y_0 = y_2 - 2y_1 + y_0$$

And in general

$$\Delta^j y_0 = \sum_{n=0}^{j} (-1)^n \binom{j}{n} y_{j-n}$$

where $\binom{j}{n} = \frac{j!}{n!(j-n)!}$ = binomial coefficients.

Thus

$$\begin{aligned} \Delta^4 y_0 &= \sum_{n=0}^{4} (-1)^n \binom{4}{n} y_{4-n} = \binom{4}{0} y_4 - \binom{4}{1} y_3 + \binom{4}{2} y_2 \\ &\quad - \binom{4}{3} y_1 + \binom{4}{4} y_0 = y_4 - 4y_3 + 6y_2 - 4y_1 + y_0 \end{aligned}$$

A horizontal-difference table is convenient here. This is illustrated for a particular example. Note that decimal points are often omitted.

x	y	Δy	$\Delta^2 y$	$\Delta^3 y$	$\Delta^4 y$	$\Delta^5 y$
1.5	2.129					
1.6	2.376	247				
1.7	2.645	269	22			
1.8	2.942	297	28	6		
1.9	3.268	326	29	1	−5	
2.0	3.627	359	33	4	3	8

Example If y = polynomial of degree n

$$= a_n x^n + a_{n-1}x^{n-1} + \cdots + a_0 \qquad \Delta x = \text{constant}$$

then Δy = polynomial of degree $n - 1$

$$\vdots$$

$$\Delta^j y = \text{polynomial of degree } n - j \qquad j \le n$$

$$\Delta^n y = a_n (\Delta x)^n n!$$

That is, if the values of the independent variable are all separated by equal intervals, then the nth differences of a polynomial of the nth degree are constant. Conversely, if the nth differences of a tabulated function are constant when the independent variable is separated by equal intervals, then the function is a polynomial of degree n. This result is quite useful, for it allows one to select a polynomial of appropriate degree to use to fit data.

Example The data* in the following table when differentiated show that the approximate relation between x and y is a cubic, since the third differences are sensibly constant.

x	y	Δy	$\Delta^2 y$	$\Delta^3 y$
0	2.105			
0.2	2.808	703		
0.4	3.614	806	103	
0.6	4.604	990	184	81
0.8	5.857	1253	263	79
1.0	7.451	1594	341	78
1.2	9.467	2016	422	81
1.4	11.985	2518	502	80

Lagrange Interpolation Formulas The Newton formulas (2-99) are expressed in terms of divided differences. It is often useful to have interpolation formulas expressed explicitly in terms of the ordinates involved. This is accomplished by the Lagrange interpolation polynomial of degree n

$$y(x) = \sum_{j=0}^{n} \frac{\pi(x)}{(x - x_j)\pi'(x_j)} f(x_j) \qquad (2\text{-}100)$$

where $\pi(x) = (x - x_0)(x - x_1) \cdots (x - x_n)$

$\pi'(x_j) = (x_j - x_0)(x_j - x_1) \cdots (x_j - x_n)$

where $(x_j - x_j)$ is the omitted factor. Thus

$$f(x) = y(x) + E_n(x)$$

$$E_n(x) = \frac{1}{(n+1)!}\pi(x) f^{(n+1)}(\epsilon)$$

Example The interpolation polynomial of degree 3 is

$$y(x) = \frac{(x - x_1)(x - x_2)(x - x_3)}{(x_0 - x_1)(x_0 - x_2)(x_0 - x_3)} f(x_0) + \frac{(x - x_0)(x - x_2)(x - x_3)}{(x_1 - x_0)(x_1 - x_2)(x_1 - x_3)} f(x_1) + \frac{(x - x_0)(x - x_1)(x - x_3)}{(x_2 - x_0)(x_2 - x_1)(x_2 - x_3)} f(x_2) + \frac{(x - x_0)(x - x_1)(x - x_2)}{(x_3 - x_0)(x_3 - x_1)(x_3 - x_2)} f(x_3)$$

Thus directly from the data

x	0	1	3	4
$f(x)$	1	1	−1	2

we have as an interpolation polynomial $y(x)$ for $f(x)$

$$y(x) = 1 \cdot \frac{(x-1)(x-3)(x-4)}{(0-1)(0-3)(0-4)} + 1\frac{(x-0)(x-3)(x-4)}{(1-0)(1-3)(1-4)} - 1 \cdot \frac{(x-0)(x-1)(x-4)}{(3-0)(3-1)(3-4)} + 2\frac{(x-0)(x-1)(x-3)}{(4-0)(4-1)(4-3)}$$

*Reprinted by permission from I. S. and E. S. Sokolnikoff, *Higher Mathematics for Engineers and Physicists*, McGraw-Hill, New York, 1941.

Other Difference Methods (Equally Spaced Ordinates); Backward Differences The backward differences denoted by

$$\nabla f(x) = f(x) - f(x - h)$$

$$\nabla^2 f(x) = \nabla f(x) - \nabla f(x - h) \cdots$$

$$\nabla f^n(x) = \nabla^{n-1} f(x) - \nabla^{n-1} f(x - h)$$

are useful for calculation near the end of tabulated data.

Central Differences The central difference denoted by

$$\delta f(x) = f\left(x + \frac{h}{2}\right) - f\left(x - \frac{h}{2}\right)$$

$$\delta^n f(x) = \delta^{n-1} f\left(x + \frac{h}{2}\right) - \delta^{n-1} f\left(x - \frac{h}{2}\right)$$

is useful for calculating at the interior points of tabulated data.

Also to be found in the literature already cited are gaussian, Stirling, Bessel, Everett, and Comrie differences, and so forth.

Inverse Interpolation This is the process of finding the value of the independent variable or abscissa corresponding to a given value of the function when the latter is between two tabulated values of the abscissa. One method of accomplishing this is to use Lagrange's interpolation formula in the form $x = \psi(y) = \sum_{j=0}^{n} \frac{\pi(y)}{(y - y_j)\pi'(y_j)} x_j$, where x is expressed as a function of y. See Refs. 18, 51, and 79.

NUMERICAL DIFFERENTIATION

Numerical differentiation should be avoided whenever possible, particularly when data are empirical and subject to appreciable observation errors. Errors in data can affect numerical derivatives quite strongly; i.e., differentiation is a roughening process. When such a calculation must be made, it is usually desirable first to smooth the data to a certain extent.

Use of Interpolation Formula If the data are given over equidistant values of the independent variable x, an interpolation formula such as the Newton formula (see Hildebrand in references) may be used and the resulting formula differentiated analytically. If the independent variable is not at equidistant values, then Lagrange's formulas must be used. By differentiating three- and five-point Lagrange interpolation formulas the following differentiation formulas result for equally spaced tabular points:

Three-Point Formulas Let x_0, x_1, x_2 be the three points.

$$f'(x_0) = \frac{1}{2h}[-3f(x_0) + 4f(x_1) - f(x_2)] + \frac{h^2}{3} f'''(\epsilon)$$

$$f'(x_1) = \frac{1}{2h}[-f(x_0) + f(x_2)] - \frac{h^2}{6} f'''(\epsilon) \qquad (2\text{-}101)$$

$$f'(x_2) = \frac{1}{2h}[f(x_0) - 4f(x_1) + 3f(x_2)] + \frac{h^2}{3} f'''(\epsilon)$$

where the last term is an error term $\min_j x_j < \epsilon < \max_j x_j$.

Five-Point Formulas Let x_0, x_1, x_2, x_3, x_4 be the five values of the equally-spaced independent variable and $f_i = f(x_i)$.

$$f'(x_0) = \frac{1}{12h}[-25f_0 + 48f_1 - 36f_2 + 16f_3 - 3f_4] + \frac{h^4}{5} f^{(V)}(\epsilon)$$

$$f'(x_1) = \frac{1}{12h}[-3f_0 - 10f_1 + 18f_2 - 6f_3 + f_4] - \frac{h^4}{20} f^{(V)}(\epsilon)$$

$$f'(x_2) = \frac{1}{12h}[f_0 - 8f_1 + 8f_3 - f_4] + \frac{h^4}{30} f^{(V)}(\epsilon) \qquad (2\text{-}102)$$

$$f'(x_3) = \frac{1}{12h}[-f_0 + 6f_1 - 18f_2 + 10f_3 + 3f_4] - \frac{h^4}{20} f^{(V)}(\epsilon)$$

$$f'(x_4) = \frac{1}{12h}[3f_0 - 16f_1 + 36f_2 - 48f_3 + 25f_4] + \frac{h^4}{5} f^{(V)}(\epsilon)$$

and the last term is again an error term.

Smoothing Techniques These techniques involve the approximation of the tabular data by a least-squares fit of the data by using some known functional form, usually a polynomial (for the concept of least squares see "Statistics"). In place of approximating $f(x)$ by a single least-squares polynomial of degree n over the entire range of the tabulation, it is often desirable to replace each tabulated value by the value taken on by a least-squares polynomial of degree n relevant to a subrange of $2M + 1$ points centered, when possible, at the point for which the entry is to be modified. Thus each smoothed value replaces a tabulated value. Let $f_j = f(x_j)$ be the tabular points and y_j = smoothed values.

First-Degree Least Squares with Three Points

$$y_0 = \tfrac{1}{6}[5f_0 + 2f_1 - f_2]$$
$$y_1 = \tfrac{1}{3}[f_0 + f_1 + f_2]$$
$$y_2 = \tfrac{1}{6}[-f_0 + 2f_1 + 5f_2]$$

First-Degree Least Squares with Five Points

$$y_0 = \tfrac{1}{5}[3f_0 + 2f_1 + f_2 - f_4]$$
$$y_1 = \tfrac{1}{10}[4f_0 + 3f_1 + 2f_2 + f_3]$$
$$y_2 = \tfrac{1}{5}[f_0 + f_1 + f_3 + f_4]$$
$$y_3 = \tfrac{1}{10}[f_0 + 2f_1 + 3f_1 + 4f_3]$$
$$y_4 = \tfrac{1}{5}[-f_0 + f_2 + 2f_3 + 3f_4]$$

Thus, for example, if first-degree five-point least squares are used, the central formula is employed for all values except the first two and the last two, for which the off-center formulas are used.

Third-Degree Least Squares with Seven Points

$$y_0 = \tfrac{1}{42}[39f_0 + 8f_1 - 4f_2 - 4f_3 + f_4 + 4f_5 - 2f_6]$$
$$y_1 = \tfrac{1}{42}[8f_0 + 19f_1 + 16f_2 + 6f_3 - 4f_4 - 7f_5 + 4f_6]$$
$$y_2 = \tfrac{1}{42}[-4f_0 + 16f_1 + 19f_2 + 12f_3 + 2f_4 - 4f_5 + f_6]$$
$$y_3 = \tfrac{1}{21}[-2f_0 + 3f_1 + 6f_2 + 7f_3 + 6f_4 + 3f_5 - 2f_6] \qquad (2\text{-}103)$$
$$y_4 = \tfrac{1}{42}[f_0 - 4f_1 + 2f_2 + 12f_3 + 19f_4 + 16f_5 - 4f_6]$$
$$y_5 = \tfrac{1}{42}[4f_0 - 7f_1 - 4f_2 + 6f_3 + 16f_4 + 19f_5 + 8f_6]$$
$$y_6 = \tfrac{1}{42}[-2f_0 + 4f_1 + f_2 - 4f_3 - 4f_4 + 8f_5 + 39f_6]$$

Additional smoothing formulas may be found in the references. After the data have been smoothed, any of the interpolation polynomials or an appropriate least-squares polynomial may be fitted and the results used to obtain the derivative.

Example

x	0	1	2	3	4	5	6	7	8
$f(x)$	54	145	227	359	401	342	259	112	65

These data are smoothed with Formulas (2-103) as follows (two points are illustrated): To smooth the first point use the first formula of (2-103).

$$y_0 = \tfrac{1}{42}[39(54) + 8(145) - 4(227) - 4(359) + 401 + 4(342) - 2(259)]$$
$$= 51.7$$
$$y_2 = \tfrac{1}{42}[-4(54) + 16(145) + 19(227) + 12(359) + 2(401) - 4(342) + 259]$$
$$= 248.05$$

Least-Squares Methods

Parabolic For five evenly spaced neighboring abscissas labeled $x_{-2}, x_{-1}, x_0, x_1, x_2$ and their ordinates $f_{-2}, f_{-1}, f_0, f_1, f_2$ assume that a parabola is fit by least squares. There results for all interior points, except the first and last two points of the data, the formula for the numerical derivative

$$f_0' = 1/10h[-2f_{-2} - f_{-1} + f_1 + 2f_2]$$

For the first two data points designated by 0 and h

$$f'(0) = 1/20h[-21f(0) + 13f(h) + 17f(2h) - 9f(3h)]$$
$$f'(h) = 1/20h[-11f(0) + 3f(h) + 7f(2h) + f(3h)]$$

and for the last two given by $\alpha - h, \alpha$

$$f'(\alpha - h) = 1/20h[11f(\alpha) - 3f(\alpha - h) - 7f(\alpha - 2h) - f(\alpha - 3h)]$$
$$f'(\alpha) = 1/20h[+21f(\alpha) - 13f(\alpha - h) - 17f(\alpha - 2h) + 9f(\alpha - 3h)]$$

Quartic (Douglas-Avakian) A fourth-degree polynomial $y = a + bx + cx^2 + dx^3 + ex^4$ is fitted to seven adjacent equidistant points (spacing h) after a translation of coordinates has been made so that $x = 0$ corresponds to the central point of the seven. Thus these may be called $-3h, -2h, -h, 0, h, 2h, 3h$. Let k = coefficient of h for the seven points. That is, in $-3h$, $k = -3$. Then the coefficients for the polynomial are

$$a = \frac{524\Sigma f(kh) - 245\Sigma k^2 f(kh) + 21\Sigma k^4 f(kh)}{924}$$
$$b = \frac{397\Sigma kf(kh)}{1512h} - \frac{7\Sigma k^3 f(kh)}{216h}$$
$$c = \frac{-840\Sigma f(kh) + 679\Sigma k^2 f(kh) - 67\Sigma k^4 f(kh)}{3168h^2}$$
$$d = \frac{-7\Sigma kf(kh) + \Sigma k^3 f(kh)}{216h^3}$$
$$e = \frac{72\Sigma f(kh) - 67\Sigma k^2 f(kh) + 7\Sigma k^4 f(kh)}{3168h^4}$$

where all summations run from $k = -3$ to $k = +3$ and $f(kh)$ = tabular value at kh. The slope of the polyomial at $x = 0$ is $dy/dx = b$.

Example Find the constants and dy/dz at $z = 3$ for the data*

$f(z)$	2	3	2	−1	−2	−2	−1
z	0	1	2	3	4	5	6

Here $h = 1$. Set $x = z - 3$, which moves the origin to the midpoint at $z = 3$. An aid to the hand calculation follows:

z	y	x	k	$kf(kh)$	$k^2f(kh)$	$k^3f(kh)$	$k^4f(kh)$
0	2	−3	−3	−6	18	−54	162
1	3	−2	−2	−6	12	−24	48
2	2	−1	−1	−2	2	−2	2
3	−1	0	0	0	0	0	0
4	−2	1	1	−2	−2	−2	−2
5	−2	2	2	−4	−8	−16	−32
6	−1	3	3	−3	−9	−27	−81
Σ				−23	13	−125	97

Thus:

$$a = \frac{524(1) - 245(13) + 21(97)}{924} = -0.68$$
$$b = \frac{397(-23)}{1512} - \frac{7(-125)}{216} = -1.99$$

and so forth. The slope at $x = 0$ ($z = 3$) is $dy/dx = -1.99 = b$.

NUMERICAL INTEGRATION

A multitude of formulas have been developed to accomplish numerical integration, which consists of computing the value of a definite integral from a set of numerical values of the integrand.

Newton-Cotes Integration Formulas (Equally Spaced Ordinates) for Functions of One Variable The definite integral $\int_a^b f(x)\,dx$ is to be evaluated.

Trapezoidal Rule This formula consists of subdividing the interval $a \le x \le b$ into n subintervals a to $a + h$, $a + h$ to $a + 2h$, . . . and replacing the graph of $f(x)$ by the result of joining the ends of

*Reprinted by permission from Mickley, Sherwood, and Reed, *Applied Mathematics in Chemical Engineering*, McGraw-Hill, New York, 1957.

adjacent ordinates by line segments. If $f_j = f(x_j) = f(a + jh)$, $f_0 = f(a)$, $f_n = f(b)$, the integration formula is

$$\int_a^b f(x)\,dx = \frac{h}{2}[f_0 + 2f_1 + 2f_2 + \cdots + 2f_{n-1} + f_n] + E_n$$

$$\text{where}\qquad |E_n| = \frac{nh^3}{12}|f''(\epsilon)| = \frac{(b-a)^3}{12n^2}|f''(\epsilon)| \qquad a < \epsilon < b$$

This procedure is not of high accuracy. However, if $f''(x)$ is continuous in $a < x < b$, the error goes to zero as $1/n^2$, $n \to \infty$.

Parabolic Rule (Simpson's Rule) This procedure consists of subdividing the interval $a < x < b$ into $n/2$ subintervals, each of length $2h$, where n is an *even* integer. By using the notation as above the integration formula is

$$\int_a^b f(x)\,dx = \frac{h}{3}[f_0 + 4f_1 + 2f_2 + 4f_3 + \cdots + 4f_{n-3} + 2f_{n-2} + 4f_{n-1} + f_n] + E_n$$

$$\text{where}\qquad |E_n| = \frac{nh^5}{180}|f^{(\text{IV})}(\epsilon)| = \frac{(b-a)^5}{180n^4}|f^{(\text{IV})}(\epsilon)| \qquad a < \epsilon < b$$

This method approximates $f(x)$ by a parabola on each subinterval. This rule is generally more accurate than the trapezoidal rule. It is the most widely used integration formula.

Weddle's Rule This procedure consists of subdividing the integral $a < x < b$ into $n/6$ subintervals, each of length $6h$, where n is a multiple of 6. By using the notation from the trapezoidal rule there results

$$\int_a^b f(x)\,dx = \frac{3h}{10}[f_0 + 5f_1 + f_2 + 6f_3 + f_4 + 5f_5 + 2f_6 + 5f_7 + f_8 + \cdots + 6f_{n-3} + f_{n-2} + 5f_{n-1} + f_n] + E_n$$

Note that the coefficients of f_j follow the rule 1, 5, 1, 6, 1, 5, 2, 5, 1, 6, 1, 5, 2, 5, etc. This procedure consists of approximating $f(x)$ by a polynomial of degree 6 on each subinterval. Here

$$E_n = nh^7/1400[10f^{(6)}(\epsilon_1) + 9h^2f^{(8)}(\epsilon_2)]$$

Two-Dimensional Formula Formulas for two-way integration over a rectangle, circle, ellipse, and so forth, may be developed by a double application of one-dimensional integration formulas. The two-dimensional generalization of the parabolic rule is given here. Consider the iterated integral $\int_a^b \int_c^d f(x, y)\,dx\,dy$. Subdivide $c < x < d$ into m (even) subintervals of length $h = (d - c)/m$, and $a < y < b$ into n (even) subintervals of length $k = (b - a)/n$. This gives a subdivision of the rectangle $a \le y \le b$, $c \le x \le d$ into subrectangles. Let $x_j = c + jh$, $y_j = a + jk$, $f_{i,j} = f(x_i, y_j)$. Then $\int_a^b \int_c^d f(x, y)\,dx\,dy = \frac{hk}{9}[(f_{0,0} + 4f_{1,0} + 2f_{2,0} + \cdots + f_{m,0}) + 4(f_{0,1} + 4f_{1,1} + 2f_{2,1} + \cdots + f_{m,1}) + 2(f_{0,2} + 4f_{1,2} + 2f_{2,2} + \cdots + f_{m,2}) + \cdots + (f_{0,n} + 4f_{1,n} + 2f_{2,n} + \cdots + f_{m,n})] + E_{m,n}$ where $E_{m,n} = -\frac{hk}{90}\left[mh^4 \frac{\partial^4 f(\epsilon_1, \eta_1)}{\partial x^4} + nk^4 \frac{\partial^4 f(\epsilon_2, \eta_2)}{\partial y^4}\right]$ and ϵ_1, ϵ_2 lie in $c < x < d$, η_1, η_2 lie in $a < y < b$.

In addition, the following literature is cited: Tyler, "Numerical Integration of Functions of Several Variables," *Can. J. Math.*, **5**, 393 (1953); and Davis and Rabinowitz, "Monte Carlo Experiments in Computing Multiple Integrals," *Mathematical Tables and Other Aids to Computation*, vol. 10, 1956).

Gaussian Integration Formulas (Unequally Spaced Abscissas) These formulas are capable of yielding comparable accuracy with fewer ordinates than the equally spaced formulas. The ordinates are obtained by optimizing the distribution of the abscissas rather than by arbitrary choice. For the details of these formulas Hildebrand (145) is an excellent reference.

NUMERICAL SOLUTION OF ORDINARY DIFFERENTIAL EQUATIONS

A variety of methods have been devised to solve ordinary differential equations numerically. The general references contain some information. More specific information may be found in Collatz, *The Numerical Treatment of Differential Equations*, Springer, Berlin, 1960; Hamming, *Numerical Methods for Scientists and Engineers*, McGraw-Hill, New York, 1962; Fox (ed.), *Numerical Solution of Ordinary and Partial Differential Equations*, Pergamon, Oxford, 1962; Fox, *Numerical Solution of Two-Point Boundary Problems*, Oxford, London, 1957; and Henrici, *Discrete Variable Methods in Ordinary Differential Equations*, Wiley, New York, 1962. By a numerical solution of a differential equation is meant a table of values of the dependent variable and its derivatives over only a limited portion of the range of the independent variable. Every differential equation of order n can be written as n first-order equations. Therefore, the methods given in this subsection will be for first-order equations, and the generalization to simultaneous systems will be subsequently discussed. No single numerical method is applied to every ordinary differential equation or for that matter to every member of the much smaller class of linear ordinary differential equations. There are a large variety of problem types; so a variety of methods is required. Since computer "libraries" have only one or two canned routines, there is a tendency among the less experienced to apply those programs on all problems. One should be forewarned that such a practice can generate erroneous results.

Modified Euler Method This method is simple and yields modest accuracy. If extreme accuracy is desired, a more sophisticated method should be selected. Let the first-order differential equation be $dy/dx = f(x, y)$ with the initial condition (x_0, y_0), that is, $y = y_0$ when $x = x_0$. The procedure is as follows:

Step 1. From the given initial conditions (x_0, y_0) compute $y_0' = f(x_0, y_0)$ and $y_0'' = \frac{\partial f(x_0, y_0)}{\partial x} + \frac{\partial f(x_0, y_0)}{\partial y} y_0'$. Then determine $y_0 = y_0 + hy_0' + (h^2/2)y_0''$, where h = subdivision of the independent variable.

Step 2. Determine $y_1' = f(x_1, y_1)$. $(x_1 = x_0 + h)$. These prepare us for:

Predictor Steps

Step 3. For $n \ge 1$ calculate $(y_{n+1})_1 = y_{n-1} + 2hy_n'$.

Step 4. Calculate $(y_{n+1}')_1 = f[x_{n+1}, (y_{n+1})_1]$.

Corrector Steps

Step 5. Calculate $(y_{n+1})_2 = y_n + (h/2)[(y_{n+1}')_1 + y_n']$, where y_n, y_n' without the subscripts are the previous values obtained by this process (or by steps 1 and 2).

Step 6. $(y_{n+1}')_2 = f[x_{n+1}, (y_{n+1})_2]$.

Step 7. Repeat the corrector steps 5 and 6 if necessary until the desired accuracy is produced in y_{n+1}, y_{n+1}'.

Example Consider the equation $y' = 2y^2 + x$ with the initial conditions $y_0 = 1$ when $x_0 = 0$. Let $h = 0.1$. A few steps of the computation are illustrated.

Step no.	
1	$y_0' = 2y_0^2 + x_0 = 2$
	$y_0'' = 1 + 4y_0y_0' = 1 + 8 = 9$
	$y_1 = 1 + (0.1)(2) + [(0.1)^2/2]9 = 1.245$
2	$y_1' = 2y_1^2 + x_1 = 3.100 + 0.1 = 3.210$
3	$(y_2)_1 = y_0 + 2hy'_1 = 1 + 2(0.1)3.210 = 1.642$
4	$(y_2')_1 = 2(y_2)_1^2 + x_2 = 5.592$
5	$(y_2)_2 = y_1 + (0.1/2)[(y_2')_1 + y_1'] = 1.685$
6	$(y_2')_2 = 2(y_2)_2^2 + x_2 = 5.878$
5 (repeat)	$(y_2)_3 = y_1 + (0.05)[(y_2')_2 + y_1'] = 1.699$
6 (repeat)	$(y_2')_3 = 2(y_2)_3^2 + x_2 = 5.974$

and so forth. This procedure may be programmed for a computer. A discussion of the truncation error of this process may be found in Milne (see references).

Modified Adam's Method The procedure given here was developed by retaining third differences. It can then be considered as a more exact predictor-corrector method than the Euler method. The procedure is as follows for $dy/dx = f(x, y)$ and h = interval size:

Steps 1 and 2. The same as in the Euler method.

Predictor Steps

Step 3. $(y_{n+1})_1 = y_n + (h/24)[55y'_n - 59y'_{n-1} + 37y'_{n-2} - 9y'_{n-3}]$ where y'_n, y'_{n-1}, etc., are calculated in step 1.

Step 4. $(y'_{n+1})_1 = f[x_{n+1}, (y_{n+1})_1]$.

Corrector Steps

Step 5. $(y_{n+1})_2 = y_n + \dfrac{h}{24}[9(y'_{n+1})_1 + 19y'_n - 5y'_{n-1} + y'_{n-2}]$.

Step 6. $(y'_{n+1})_2 = f[x_{n+1}, (y_{n+1})_2]$.

Step 7. Iterate steps 5 and 6 if necessary.

Runge-Kutta Methods These methods are self-starting and are inherently stable. Kopal (*Numerical Analysis*, Wiley, New York, 1955) is a good reference for their derivation and discussion. Third- and fourth-order procedures are given here for $dy/dx = f(x, y)$, h = interval size.

Third Order (Error ≈ h^4)

$$k_0 = hf(x_n, y_n)$$
$$k_1 = hf(x_n + \tfrac{1}{2}h, y_n + \tfrac{1}{2}k_0)$$
$$k_2 = hf(x_n + h, y_n + 2k_1 - k_0)$$

and

$$y_{n+1} = y_n + \tfrac{1}{6}(k_0 + 4k_1 + k_2)$$

for all $n \geq 0$, with initial condition (x_0, y_0).

Fourth Order (Error ≈ h^5)

$$k_0 = hf(x_n, y_n)$$
$$k_1 = hf(x_n + \tfrac{1}{2}h, y_n + \tfrac{1}{2}k_0)$$
$$k_2 = hf(x_n + \tfrac{1}{2}h, y_n + \tfrac{1}{2}k_1)$$
$$k_3 = hf(x_n + h, y_n + k_2)$$
$$y_{n+1} = y_n + \tfrac{1}{6}(k_0 + 2k_1 + 2k_2 + k_3)$$

Example: Third Order Let $dy/dx = x - 2y$ with initial condition $y_0 = 1$ when $x_0 = 0$ and let $h = 0.1$. Clearly $x_n = nh$. To calculate y_1 proceed as follows:

$$k_0 = 0.1[x_0 - 2y_0] = -0.2$$
$$k_1 = 0.1[0.05 - 2(1 - 0.1)] = -0.175$$
$$k_2 = 0.1[0.1 - 2(1 - 0.35 + 0.2)] = -0.16$$
$$y_1 = 1 + \tfrac{1}{6}(-0.2 - 0.7 - 0.16) = 0.8234$$

Equations of Higher Order and Simultaneous Differential Equations Any differential equation of second or higher order can be reduced to a simultaneous system of first-order equations by the introduction of auxiliary variables. Consider the equations

$$d^2x/dt^2 + xy(dx/dt) + z = e^z$$
$$d^2y/dt^2 + zy(dy/dt) = 7 + t^2$$
$$d^2z/dt^2 + xz(dz/dt) + x = e^x$$

If the new variables $x_1 = x$, $x_2 = y$, $x_3 = z$, $x_4 = dx_1/dt$, $x_5 = dx_2/dt$, $x_6 = dx_3/dt$, the equations become

$$dx_1/dt = x_4$$
$$dx_2/dt = x_5$$
$$dx_3/dt = x_6$$
$$dx_4/dt = -x_1x_2x_4 - x_3 + e^{x_3}$$
$$dx_5/dt = -x_3x_2x_5 + 7 + t^2$$
$$dx_6/dt = -x_1x_3x_6 - x_1 + e^{x_1}$$

which is a system of the general form

$$dx_i/dt = f_i(t, x_1, x_2, x_3, \ldots, x_n)$$

$i = 1, 2, \ldots, n$. Such systems may be solved by application simultaneously of any of the preceding numerical techniques. A Runge-Kutta method for

$$dx/dt = f(t, x, y)$$
$$dy/dt = g(t, x, y)$$

is given in the following paragraph. The fourth-order procedure is shown.

Starting at the initial conditions x_0, y_0, t_0, the next values x_1, y_1 are computed via the following equations (where $\Delta t = h$, $t_j = h + t_{j-1}$):

$$k_0 = hf(t_0, x_0, y_0) \qquad l_0 = hg(t_0, x_0, y_0)$$
$$k_1 = hf\left(t_0 + \frac{h}{2}, x_0 + \frac{k_0}{2}, y_0 + \frac{l_0}{2}\right)$$
$$l_1 = hg\left(t_0 + \frac{h}{2}, x_0 + \frac{k_0}{2}, y_0 + \frac{l_0}{2}\right)$$
$$k_2 = hf\left(t_0 + \frac{h}{2}, x_0 + \frac{k_1}{2}, y_0 + \frac{l_1}{2}\right)$$
$$l_2 = hg\left(t_0 + \frac{h}{2}, x_0 + \frac{k_1}{2}, y_0 + \frac{l_1}{2}\right)$$
$$k_3 = hf(t_0 + h, x_0 + k_2, y_0 + l_2)$$
$$l_3 = hg(t_0 + h, x_0 + k_2, y_0 + l_2)$$

and

$$x_1 = x_0 + \tfrac{1}{6}(k_0 + 2k_1 + 2k_2 + k_3)$$
$$y_1 = y_0 + \tfrac{1}{6}(l_0 + 2l_1 + 2l_2 + l_3)$$

To continue the computation, replace t_0, x_0, y_0, in the preceding formulas by $t_1 = t_0 + h$, x_1, y_1 just calculated. Extension of this method to more than two equations follows precisely the same pattern.

NUMERICAL SOLUTION OF INTEGRAL EQUATIONS

In this subsection is considered a method of solving numerically the Fredholm integral equation of the second kind:

$$u(x) = f(x) + \lambda \int_a^b k(x, t)u(t)\, dt \qquad \text{for } u(x) \qquad (2\text{-}104)$$

The method discussed arises because a definite integral can be closely approximated by any of several numerical integration formulas (each of which arises by approximating the function by some polynomial over an interval). Thus the definite integral in Eq. (2-104) can be replaced by an integration formula, and Eq. (2-104) may be written

$$u(x) = f(x) + \lambda(b - a)\left[\sum_{i=1}^{n} c_i k(x, t_i)u(t_i)\right] \qquad (2\text{-}105)$$

where $t_1, \ldots, t_n$ are points of subdivision of the t axis, $a \leq t \leq b$, and the c's are coefficients whose values depend upon the type of numerical integration formula used. Now Eq. (2-105) must hold for all values of x, $a \leq x \leq b$; so it must hold for $x = t_1$, $x = t_2, \ldots, x = t_n$. Substituting for x successively $t_1, t_2, \ldots, t_n$ and setting $u(t_i) = u_i$, $f(t_i) = f_i$, we get n linear algebraic equations for the n unknowns $u_1, \ldots, u_n$. That is,

$$u_i = f_i + (b - a)[c_1k(t_i, t_1)u_1 + c_2k(t_i, t_2)u_2 + \cdots + c_nk(t_i, t_n)u_n] \qquad i = 1, 2, \ldots, n$$

These u_j may be solved for by the methods under "Numerical Solution of Linear Equations and Associated Problems" and substituted into Eq. (2-105) to yield an approximate solution for Eq. (2-104).

Example Solve numerically $u(x) = x + \tfrac{1}{2}\int_0^1 (t + x)u(t)\, dt$. In this example $a = 0$, $b = 1$. Take $n = 3$, $t_1 = 0$, $t_2 = \tfrac{1}{2}$, $t_3 = 1$. Then Eq. (2-105) takes the form (for which we have used the parabolic rule)

$$u(x) = x + (1/3)\left[\frac{1/2}{3}(t_1 + x)u(t_1) + 4(t_2 + x)u(t_2) + (t_3 + x)u(t_3)\right]$$
$$= x + (1/18)[(t_1 + x)u(t_1) + 4(t_2 + x)u(t_2) + (t_3 + x)u(t_3)]$$

This must hold for all x, $0 \le x \le 1$. Hence take $x = t_1 = 0$, $x = t_2 = \frac{1}{2}$, $x = t_3 = 1$. Thus

$$\begin{aligned} u(t_1) &= t_1 + \tfrac{1}{18}[2t_1u(t_1) + 4(t_2 + t_1)u(t_2) + (t_3 + t_1)u(t_3)] \\ u(t_2) &= t_2 + \tfrac{1}{18}[(t_1 + t_2)u(t_1) + 4(2t_2)u(t_2) + (t_3 + t_2)u(t_3)] \qquad (2\text{-}106) \\ u(t_3) &= t_3 + \tfrac{1}{18}[(t_1 + t_3)u(t_1) + 4(t_2 + t_3)u(t_2) + 2t_3u(t_3)] \end{aligned}$$

By setting in the values of t_1, t_2, t_3 and $u(t_i) = u_i$, Eqs. (2-106) become

$$\begin{aligned} 18u_1 - 2u_2 - u_3 &= 0 \\ -u_1 + 28u_2 - 3u_3 &= 18 \\ -u_1 - 6u_2 + 16u_3 &= 18 \end{aligned}$$

with the solution $u_1 = \frac{12}{11}$, $u_2 = \frac{57}{11}$, $u_3 = \frac{102}{11}$. Thus

$$\begin{aligned} u(x) &= x + \tfrac{1}{18}[x\tfrac{12}{11} + 4(\tfrac{1}{2} + x)\tfrac{57}{11} + (1 + x)\tfrac{102}{11}] \\ &= \tfrac{90}{11}x + \tfrac{12}{11} \end{aligned}$$

Because of the work involved in solving large systems of simultaneous linear equations it is desirable that only a small number of u's be computed. Thus the gaussian integration formulas are useful because of the economy they offer. See references on numerical solutions of integral equations.

NUMERICAL SOLUTION OF PARTIAL DIFFERENTIAL EQUATIONS

Computational methods for partial differential equations have evolved from finite-difference approximations for the partial derivatives and, in the case of hyperbolic systems, from characteristics. Partial derivatives can be approximated by finite differences in many ways, depending upon accuracy requirements. Some of the more common operators and their approximations are given in Fig. 2-55. The details of the theory are available in Forsythe and Wasow, *Finite Difference Methods for Partial Differential Equations*, Wiley, New York, 1960; Fox (ed.), *Numerical Solution of Ordinary and Partial Differential Equations*, Pergamon, Oxford, 1962; Richtmyer and Morton, *Difference Methods for Initial Value Problems*, Interscience, New York, 1967; Ames, *Nonlinear Partial Differential Equations in Engineering*, Academic, New York, 1965; Wachspress, *Iterative Solution of Elliptic Systems*, Prentice-Hall, Englewood Cliffs, N.J., 1966; Ames, *Numerical Methods for Partial Differential Equations*, Nelson, London, 1969; and others cited at the beginning of this section.

The techniques will be introduced by two examples, followed by an error discussion and then by a last example.

Example: Typical Linear-Diffusion Problem Consider the problem

$$\begin{aligned} u_t &= u_{xx},\ 0 < x < 1,\ 0 < t \le T \\ u(x, 0) &= f(x),\ 0 < x < 1 \\ u(0, t) &= g(t),\ 0 < t \le T \\ u(1, t) &= h(t),\ 0 < t \le T \end{aligned}$$

A finite-difference analog for this problem is developed by introducing a net whose mesh points are denoted by $x_i = ih$, $t_j = jk$, where $i = 0, 1, 2, \ldots, M$; $j = 0, 1, \ldots, N$ with $h = \Delta x = 1/M$, $k = \Delta t = T/N$. The boundaries are specified by $i = 0$ and $i = M$ and any "false" boundaries by $i = -1, -2, \ldots, i = M + 1, M + 2, \ldots,$ and so forth. The initial line is denoted by $j = 0$, and the discrete approximation at $x_i = ih$, $t_j = jk$ is designated $U_{i,j}$.

If an approximate solution $U_{i,j}$ is assumed to be known at all the mesh points up to the t_j, a method must be specified to advance the solution to time t_{j+1}. The value of $U_{i,j+1}$ at $x = 0$ and $x = 1$ should be selected as those boundary conditions specified above—that is,

$$U_{0,j+1} = g(t_{j+1}),\ U_{M,j+1} = h(t_{j+1})$$

At other points $0 < i < M$, the partial differential equation will be replaced by some difference equation. The simplest replacement consists in approximating the space derivative by a centered second difference and the time derivative by a forward difference at (x_i, t_j). The resulting difference equation is

$$\frac{1}{k}(U_{i,j+1} - U_{i,j}) = \frac{1}{h^2}(u_{i+1,j} - 2U_{i,j} + U_{i-1,j}),\ i = 1, \ldots, M - 1$$

Upon solving for $U_{i,j+1}$ we obtain the *explicit* equation for "marching" ahead in time

$$U_{i,j+1} = rU_{i-1,j} + (1 - 2r)U_{i,j} + rU_{i+1,j} \qquad i = 1, \ldots, M - 1$$

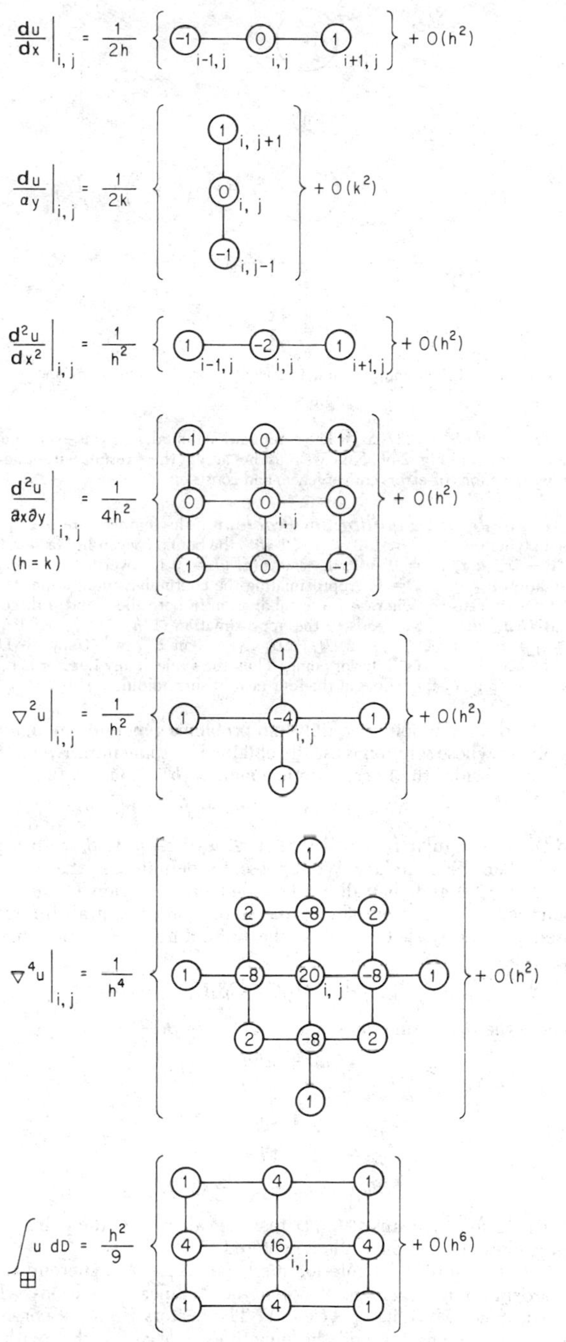

FIG. 2-55 Computational molecules.

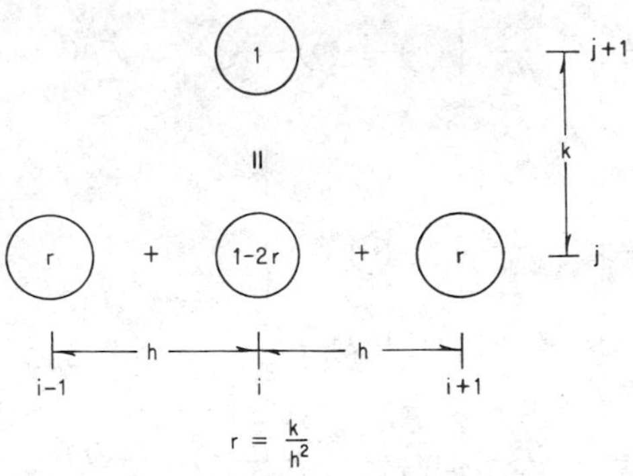

FIG. 2-56 Explicit computational molecules for diffusion equation.

where $r = k/h^2 = \Delta t/(\Delta x)^2$. The computational model for this explicit equation is shown in Fig. 2-56. Now we must not accept these results without examining questions of error, convergence, and consistency.

Example: An Equilibrium Problem The temperature $u(x, y)$ in a plate lying on $0 < x < 1, 0 < y < 1$ with the boundary conditions $u(x, 0) = x(1 - x)$, $u(x, 1) = 0$, $u(0, y) = 0$, $u(a, y) = 0$ is governed by Laplace's equation $u_{xx} + u_{yy} = 0$. Approximating the continuous integration domain with a discrete net where h is the subdivision in both the x and y directions and $U(ih, jh) = U_{i,j}$ leads to the approximation $(1/h^2)(U_{i+1,j} - 2U_{i,j} + U_{i-1,j}) + (1/h^2)(U_{i,j+1} - 2\,U_{i,j} + U_{i,j-1}) = 0$ or $U_{i,j} = \frac{1}{4}(U_{i+1,j} + U_{i-1,j} + U_{i,j+1} + U_{i,j-1})$ at interior points. Thus, the value at any interior net point is the average of the values at the four nearest surrounding points.

The discretization of equilibrium problems generates an algebraic problem whose solution is usually obtained by some iterative method. Let us consider the linear elliptic equation $(b^2 - 4ac < 0)$

$$au_{xx} + cu_{yy} + du_x + eu_y + fu = g(x, y)$$

in the rectangular region $R: 0 \le x \le \alpha$, $0 \le y \le \beta$, with u prescribed on the boundary. We suppose, for definiteness, that $a > 0$, $c > 0$, $f \le 0$ and that all are bounded in the region R and on its boundary B. Upon employing the second-order central differences used previously, with $h = k$, the finite-difference approximation becomes

$$\beta_1 U_{i+1,j} + \beta_2 U_{i-1,j} + \beta_3 U_{i,j+1} + \beta_4 U_{i,j-1} - \beta_0 U_{i,j} = h^2 g_{ij}$$

where the β_i are functions of $x_i = ih$, $y_j = jh$, given by

$$\beta_1 = a_{ij} + \tfrac{1}{2}hd_{ij}$$

$$\beta_2 = a_{ij} - \tfrac{1}{2}hd_{ij}$$

$$\beta_3 = c_{ij} + \tfrac{1}{2}he_{ij}$$

$$\beta_4 = c_{ij} - \tfrac{1}{2}he_{ij}$$

$$\beta_0 = 2(a_{ij} + c_{ij} - \tfrac{1}{2}h^2 f_{ij})$$

The notation a_{ij} means $a(ih, jh)$, that is, evaluation at the point where the computational molecule is centered.

Let the number of interior mesh points be N, whereupon the approximation generates a system of N linear equations whose matrix form we write as $AU = V$. The vectors **U** and **V** consist of the N unknowns $U_{i,j}$ and the quantities $-h^2 g_{ij}$ together with the boundary values respectively. The matrix A has real coefficients whose main-diagonal elements are the β_0 and whose off-diagonal elements are the *negatives* of the β_i which do not originate from the boundary points.

In practice, the solution of $AU = \mathbf{V}$ is usually accomplished by some iterative method (see Bodewig; Varga; or Ames in the references), of which the Gauss-Seidel (iteration by single steps) accelerated by successive overrelaxation (SOR) and an alternating-direction implicit (ADI) method are typical. The first of these requires an ordering U_i, $i = 1, \ldots, N$, of the unknowns, which we shall assume to be from the bottom of the net to the top, moving from left to right—that is, $U_{1,1}, U_{1,2}, \ldots, U_{2,1}, \ldots$. The SOR technique is defined by means of the relation

$$U_i^{(k)} = U_i^{(k-1)} + \omega[\overline{U}_i^{(k)} - U_i^{(k-1)}] = (1 - \omega)U_i^{(k-1)} + \omega\overline{U}_i^{(k)}$$

where ω is the overrelaxation parameter $(1 \le \omega < 2)$ and $\overline{U}_i^{(k)}$ are the components of the kth Gauss-Seidel iteration. Thus, the accepted value at step k is extrapolated from the Gauss-Seidel value and the previous accepted value. Upon setting the SOR relation in our sample problem, we have the algebraic form of SOR

$$U_i^{(k)} = (1 - \omega)U_i^{(k-1)} + \omega\left[\sum_{j=1}^{i-1} g_{ij}U_j^{(k)} + \sum_{j=i+1}^{N} g_{ij}U_j^{(k-1)} + c_i\right]$$

or the matrix form

$$U^{(k)} = (D + \omega R)^{-1}\{[(1 - \omega)D - \omega S]U^{(k-1)} + \omega V\}$$

and $A = R + D + S$, where R, D, and S are matrices having the same elements as A, respectively below, on, and above the main diagonal with zeros elsewhere.

The ADI methods have proved highly effective and versatile. Birkhoff, Varga, and Young, Alt and Rubinoff (eds.), *Advances in Computers*, Academic, New York, 1962, p. 189, summarize ADI methods. For recent results see Refs. 9, 132, 171, and 291. Basically, a complete ADI iteration consists of a first half in the row direction followed by a second half in the column direction. We illustrate the Peaceman-Rachford ADI technique for Laplace's equation beginning with the equal mesh discretization of the preceding example.

The iteration proceeds from $U_{i,j}^{(k)}$ to the determination of

$$U_{i,j}^{[k+(1/2)]} = U_{i,j}^{(k)} + \rho_k[U_{i+1,j}^{[k+(1/2)]} + U_{i-1,j}^{[k+(1/2)]} - 2U_{i,j}^{[k+(1/2)]}$$
$$+ \rho_k[U_{i,j+1}^{(k)} + U_{i,j-1}^{(k)} - 2U_{i,j}^{(k)}]$$

by a single-row (line) iteration followed by a single-column iteration determined from

$$U_{i,j}^{(k+1)} = U_{i,j}^{[k+(1/2)]} + \rho_k[U_{i+1,j}^{[k+(1/2)]} + U_{i-1,j}^{[k+(1/2)]} - 2U_{i,j}^{[k+(1/2)]}]$$
$$+ \rho_k[U_{i,j+1}^{(k+1)} + U_{i,j-1}^{(k+1)} - 2U_{i,j}^{(k+1)}]$$

The quantities ρ_k, called **iteration parameters,** usually depend upon k. The same values should be employed for both parts of the iteration. Suggested values of the ρ_k (and SOR ω) are given in the references.

The presence of round-off error, truncation error, or other computational error may lead to numerical instability. A finite-difference process within the semi-infinite strip $0 < x < 1$, $t > 0$ is called stepwise unstable if for a fixed network and fixed homogeneous boundary conditions there exist initial disturbances for which the finite-difference solutions $U_{i,j}$ become unbounded as $j \to \infty[U_{i,j} = U(i\,\Delta x, j\,\Delta t)]$. Other, more precise definitions are available. A finite-difference scheme $U_{i,j}$ is said to converge if $U(P)$ converges to the solution $u(P)$, of the continuous problem, with the same boundary values, as the mesh is successively refined. A finite-difference approximation is said to be consistent with the approximated differential equation if the local truncation error tends to zero as the interval sizes go to zero in any way. Thus, a successful numerical method must be *consistent, stable,* and *convergent,* for otherwise the computed results may be meaningless.

Example The quasilinear (highest-order derivatives are linear) system $u_{xx} + f(x, t, u)u_x + g(x, t, u) = p(x, t, u)u_t$ is representative of nonlinear diffusion-reaction systems. With $\delta_x^2 U_{i,j+1} = U_{i+1,j+1} - 2U_{i,j+1} + U_{i-1,j+1}$ and $\mu\delta_x U_{i,j+1} = \frac{1}{2}(U_{i+1,j+1} - U_{i-1,j+1})$, the finite-difference approximation

$$(1/h^2)\delta_x^2 U_{i,j+1} + (1/h)f[ih, (j + 1)k, U_{i,j}]\mu\delta_x U_{i,j+1} + g[ih, (j + 1)k, U_{i,j}]$$
$$= p[ih, (j + 1)k, U_{i,j}](U_{i,j+1} - U_{i,j})/k$$

contains $U_{i,j+1}$ only linearly. Thus, the algebraic problem is linear and tridiagonal at each time step. Note that Burger's equation $u_t + uu_x = vu_{xx}$ is an example of this form.

Linear equations with tridiagonal matrices (nonzero elements only on the main diagonal and those above and below it) occur so frequently that the explicit (Thomas) tridiagonal algorithm is given here. The general tridiagonal system of n equations may be expressed as

$$b_1u_1 + c_1u_2 = d_1$$
$$a_iu_{i-1} + b_iu_i + c_iu_{i+1} = d_i,\ i = 2, 3, \ldots, n-1$$
$$a_nu_{n-1} + b_nu_n = d_n$$

The gaussian elimination process—that is, successive subtraction of a suitable multiple of each equation from the following equations—transforms the system into a simpler one of upper bidiagonal form. The coefficients of this new system we designate by a_i', b_i', c_i', d_i', and in particular we note that $a_i' = 0$, $i = 2, 3, \ldots, n$, $b_i' = 1$, $i = 1, 2, \ldots, n$. The coefficients c_i', d_i' are calculated successively from the relations $c_1' = c_1/b_1$, $d_1' = d_1/b_1$, $c_{i+1}' = c_{i+1}/(b_{i+1} - a_{i+1}c_i')$, $d_{i+1}' = (d_{i+1} - a_{i+1}d_i')/(b_{i+1} - a_{i+1}c_i')$, $i = 1, 2, \ldots, n-1$, and of course, $c_n = 0$. Having completed the elimination, we examine the new system and see that the nth equation is now $u_n = d_n'$. Substituting this value into the $(n-1)$st equation $u_{n-1} + c_{n-1}'u_n = d_n'$, we have $u_{n-1} = d_n' - c_{n-1}'u_n$. Thus, starting with u_n, we have successively the solution for u_i as $u_i = d_i' - c_i'u_{i+1}$, $i = n-1, n-2, \ldots, 1$. When performing this calculation, we must be sure that $b_1 \neq 0$ and that $b_{i+1} - a_{i+1}c_i' \neq 0$. If $b_1 = 0$, we can solve for u_2 and thus reduce the system size by solving for u_{i+2}. A serious source of round-off error can arise if $|b_{i+1} - a_{i+1}c_i'|$ is small.

SPLINE FUNCTIONS

Let $x_1, x_2, \ldots, x_n$ be given real numbers with $x_1 < x_2 < \cdots < x_n$. A spline function $S(x)$ with knots $x_1, x_2, \ldots, x_n$ is a function (defined for all real numbers) consisting of polynomial pieces on each of the subintervals (x_i, x_{i+1}) for $i = 0, 1, \ldots, n$ (where we let $x_0 = -\infty$ and $x_{n+1} = +\infty$), such that the pieces are joined together with certain smoothness conditions. A spline function of degree m is a function defined on the entire real line and satisfying the following two properties: (1) in each subinterval (x_i, x_{i+1}) for $i = 0, 1, \ldots, n$, $S(x)$ is given by some polynomial of degree m or less; and (2) $S(x)$ and its derivatives of orders $1, 2, \ldots, m-1$ are continuous everywhere. Often we use splines on $[a, b]$ rather than on $(-\infty, \infty)$. A simple example is a spline function of degree 1 which is precisely a continuous polygonal function; i.e., the polynomial pieces in this case are straight-line segments which are joined together to achieve continuity on the entire interval. A spline function of degree 1 with knots $x_1, x_2, \ldots, x_n$ can be written in the form $S(x) = a_ix + b_i$ for $x_i \leq x \leq x_{i+1}$, $i = 1, 2, \ldots, n-1$ with $S(x_i) = S(x_{i+1})$ for all $i = 2, \ldots, n-1$. The spline functions of degree 1 can be used for piecewise linear interpolation of a set of n points in the plane with distinct abscissas: $(x_1, y_1), (x_2, y_2), \ldots, (x_n, y_n)$, where $x_1 < x_2 < \cdots < x_n$. The resulting interpolating spline S of degree 1 can be written in the form

$$S(x) = y_i + m_i(x - x_i),\ x \in [x_i, x_{i+1}],\ i = 1, 2, \ldots, n-1.$$

While, in general, a spline function is given by different polynomials in adjoining subintervals (x_{i-1}, x_i) and (x_i, x_{i+1}), the definition does not require this; thus a spline might be a single polynomial on the interval $(-\infty, \infty)$. However, it will be convenient to exclude this case and talk about splines only when the function is given by different polynomials on at least two subintervals.

Higher-degree splines are used whenever smoothness is needed in the approximating function. For example, if we wish the interpolating function to have a continuous first derivative, then obviously splines of degree 1 will not do the job since at each knot the slope of such a spline can be changed abruptly. If the approximating or interpolating spline is to have a continuous mth derivative, a spline of degree at least $m + 1$ is needed. The choice of degree 3 is frequently made in applications, and the resulting splines are called **cubic splines.** In this case the pieces of the graphs are cubic polynomials joined together in such a way that the resulting spline function has continuous second derivatives everywhere. The spline in this case will have discontinuities in the third derivative; however, such discontinuities cannot be detected visually, and the graph of the function will appear smooth. For a drafter, a spline is a mechanical device (used to draw a smooth curve) consisting of a strip or rod of some flexible material to which weights are attached, so that it can be constrained through or near certain plotted points on a graph. The term "spline function" is intended to suggest that the graph of such a function is similar to a curve drawn by a mechanical spline; indeed, to a first order of approximation the curve plotted by a mechanical spline is a cubic spline function. Cubic spline functions are simple to calculate and use in interpolation and approximation problems. Cubic splines suffice in practice for most problems.

Application of Splines Polynomials have long been widely used to approximate other functions in interpolation, in solutions of boundary value problems, etc., because of their simple mathematical properties. However, numerical difficulties arise when accuracy requires the use of polynomials of moderately high degree (say, 200). For example, a polynomial of high degree fitted to a large number of data points tends to exhibit severe oscillatory behavior that is not present in a curve drawn with a spline or a french curve. Theory and numerical experience suggest that as a general rule a spline function is a more suitable and adaptable approximating function than a polynomial involving a comparable number of parameters. Note that a spline of order m involves more parameters than a polynomial of degree m. The derivative of a spline of order m is another spline of order $m - 1$, and similarly for indefinite integrals.

Let $S_m(x_1, x_2, \ldots, x_n)$ denote the set of all spline functions of degree m having knots $x_1, x_2, \ldots, x_n$, and let P_m denote the class of all polynomials of degree at most m. The truncated power function

$$x_+^m = \begin{cases} x^m & \text{if } x > 0 \\ 0 & \text{if } x \leq 0 \end{cases}$$

is a spline function; for $m = 0$ it is simply the Heaviside step function. Any function S in $S_m(x_1, \ldots, x_n)$ has a unique representation of the form

$$S(x) = p(x) + \sum_{j=1}^{n} c_j(x - x_j)_+^m$$

for some p in P_m and constants $c_1, \ldots, c_n$.

A spline of odd order $2k - 1$ with the knots $x_1, x_2, \ldots, x_n$ is called a **natural spline** if it is given in each of the two intervals $(-\infty, x_1)$, (x_n, ∞) by some polynomial of degree $k - 1$ (rather than $2k - 1$) or less.

Interpolation by Natural Cubic Splines Let $(x_1, y_1), (x_2, y_2), \ldots, (x_n, y_n)$ be data points, where the knots $x_1, x_2, \ldots, x_n$ are assumed to be distinct, and arranged in ascending order. Then there exists a unique natural cubic spline interpolant to the data, i.e., a unique cubic spline S satisfying the free boundary conditions $S''(x_1) = S''(x_n) = 0$ and interpolating the data: $S(x_i) = y_i$, $i = 1, 2, \ldots, n$. If one does not impose the free boundary conditions, there would be an infinite number of cubic splines that interpolate the given data.

Natural cubic splines are the best functions to employ for curve fitting or interpolation in the sense that they provide the smoothest interpolating function for the data points. More precisely, the following result holds. Let S be the natural spline function that interpolates the data points. Let f be any function that interpolates the data, and assume that f'' is continuous on an open interval (a, b) that contains the knots. Then

$$\int_a^b [S''(x)]^2\,dx \leq \int_a^b [f''(x)]^2\,dx$$

For interpolation at the knots, odd-degree splines behave better than even-degree splines. If one is interested in interpolation at points other than the knots (of the splines), then even-degree splines can be used effectively.

Algorithm for Interpolating by Natural Cubic Splines Suppose that we are given a table of function values

x	x_1	x_2	$\cdots$	x_n
y	y_1	y_2		y_n

and we wish to interpolate by a natural cubic spline whose knots coincide with the values of the x_i's in the table. It is assumed that these values are arranged in increasing order. Thus the interpolating spline S must satisfy the following conditions:

$$S(x_i) = y_i \qquad 1 \leqq i \leqq n$$
$$\lim_{x\to x_t^-} S^{(k)}(x) = \lim_{x\to x_t^+} S^{(k)}(x) \qquad 0 \leqq k \leqq 2,\ 2 \leqq i \leqq n-1$$
$$S''(x_1) = S''(x_n) = 0$$

It can be shown that on the interval $[x_i, x_{i+1}]$ the required spline is of the form

$$S_i(x) = \frac{z_{i+1}}{6h_i}(x - x_i)^3 + \frac{z_i}{6h_i}(x_{i+1} - x)^3 + \left(\frac{y_{i+1}}{h_i} - \frac{z_{i+1}h_i}{6}\right)(x - x_i) + \left(\frac{y_i}{h_i} - \frac{z_i h_i}{6}\right)(x_{i+1} - x)$$

where $h_i = x_{i+1} - x_i$ and $z_i = S''(t_i)$. The values of z_i can be determined by imposing the condition on the continuity of S'. This leads to the system of $n - 2$ equations for the values $z_2, \ldots, z_{n-1}$:

$$h_{i-1}z_{i-1} + 2(h_{i-1} + h_i)z_i + h_i z_{i+1} = (6/h_i)(y_{i+1} - y_i) - (6/h_{i-1})(y_i - y_{i-1})$$

for $2 \leqq i \leqq n - 1$. Note that $z_1 = z_n = 0$. This system of linear equations is symmetric and tridiagonal, and so the values z_i can be calculated by using available subroutines for solving tridiagonal symmetric matrix equations.

Interpolation with Quadratic Splines A useful approximation scheme consists of interpolation with quadratic splines, in which the nodes for interpolation are chosen to be the first and the last knot and the midpoints between the knots. Recall that the knots are, by definition, the points where the spline function is allowed to change from one polynomial to another, whereas the nodes are the points where the values of the spline are specified. In the case of natural cubic splines, the nodes were taken to be the same as the knots. However, in some schemes it may be advantageous to choose the nodes as noted previously.

Suppose that the knots $a = x_1 < x_2 < \cdots < x_n = b$ have been specified, and let the nodes be the points

$$t_0 = x_1 \qquad t_n = x_n$$
$$t_j = (1/2)(x_j + x_{j+1}) \qquad j = 1, 2, \ldots, n-1.$$

The problem is to find a quadratic spline S which interpolates the data points at the nodes, i.e., $S(t_j) = y_j$, $j = 0, 1, \ldots, n$. The knots determine $n - 1$ subintervals, and in each of them S can be a different quadratic polynomial, but since S is a quadratic spline, its first derivative S' should be continuous. Let S be equal to the quadratic polynomial S_i on the interval $[x_i, x_{i+1}]$. Then to satisfy the required properties it is easy to show that S_i must have the form

$$S_i(x) = y_i + \frac{1}{2}(z_{i+1} + z_i)(x - t_i) + \frac{1}{2h_i}(z_{i+1} - z_i)(x - t_i)^2$$

where $h_i = x_{i+1} - x_i$ and $z_i = S'(x_i)$, $i = 1, 2, \ldots, n$, are solutions of the system of linear equations in the matrix form.

$$\begin{bmatrix} 3h_1 & h_1 & & & \mathbf{0} \\ h_1 & 3(h_1 + h_2) & h_2 & & \\ & h_2 & 3(h_2 + h_3) & h_3 & \\ & & \ddots & \ddots & \ddots \\ \mathbf{0} & & h_{n-2} & 3(h_{n-2} + h_{n-1}) & h_{n-1} \\ & & & h_{n-1} & 3h_{n-1} \end{bmatrix} \times \begin{bmatrix} z_1 \\ z_2 \\ z_3 \\ \vdots \\ z_{n-1} \\ z_n \end{bmatrix} = 8 \begin{bmatrix} y_1 - y_0 \\ y_2 - y_1 \\ y_3 - y_2 \\ \vdots \\ y_{n-1} - y_{n-2} \\ y_n - y_{n-1} \end{bmatrix}$$

B Splines The B splines form a basis for the set of all splines. They have also been referred to as "bell" splines because of their shape (for large-order splines).

The B splines of degree 0 are defined by

$$B_i^0(x) = \begin{cases} 1 & \text{if } x_i \leqq x \leqq x_{i+1} \\ 0 & \text{otherwise} \end{cases}$$

All the higher-degree B splines are generated from B_i^0 by a simple recursive definition

$$B_i^k(x) = \left(\frac{x - x_i}{x_{i+k} - x_i}\right) B_i^{k-1}(x) + \left(\frac{x_{i+k+1} - x}{x_{i+k+1} - x_{i+1}}\right) B_{i+1}^{k-1}(x)$$

where $k = 1, 2, 3, \ldots$ and $i = 0, \pm 1, \pm 2, \ldots$. For example, the B splines of degree 1 are given by

$$B_i^1(x) = \left(\frac{x - x_i}{x_{i+1} - x_i}\right) B_i^0(x) + \left(\frac{x_{i+2} - x}{x_{i+2} - x_{i+1}}\right) B_{i+1}^0(x)$$
$$= \begin{cases} 1 & \text{if } x \leqq x_i \text{ or } x \geqq x_{i+2} \\ \dfrac{x - x_i}{x_{i+1} - x_i} & \text{if } x_i < x < x_{i+1} \\ \dfrac{x_{i+2} - x}{x_{i+2} - x_{i+1}} & \text{if } x_{i+1} \leqq x < x_{i+2} \end{cases}$$

The graphs of B_i^0 and B_i^1 are shown in Fig. 2-57.

Finite Element Method in One Dimension:

Example Consider the following two-point boundary value problem (BVP)

$$-y'' + y = f(x) \qquad 0 < x < 1$$
$$y(0) = y(1) = 0$$

We seek approximations to the solution by Galerkin's method by using, for simplicity, B splines of order 1. Thus let $\{x_j\}_{j=0}^{n+1}$ be a sequence of $n + 2$ points (which we take for simplicity to be equally spaced) with $x_0 = 0$ and $x_{n+1} = 1$. Let $h = \dfrac{1}{n+1}$ and $x_j = jh$. The B splines of order 1 are defined by

$$\phi_j(x) = \begin{cases} 0 & x \leq x_{j-1} \\ (x - x_{j-1})/h & x_{j-1} \leq x \leq x_j \\ (x_j - x)/h & x_j \leq x \leq x_{j+1} \\ 0 & x \geq x_{j+1} \end{cases}$$

The finite-element method with these splines consists of approximating the exact solution of the preceding BVP by a linear combination of the ϕ_j's, i.e., by a function of the form

$$u(x) = \sum_{j=1}^{n} c_j\phi_j(x)$$

To determine the coefficients we first recast the BVP in a variational form. Multiplying both sides of the differential equation $-y'' + y = f$ by a "testing" function ϕ in $C'[0, 1]$, the space of continuously differentiable functions, with $\phi(0) = \phi(1) = 0$, we obtain

$$\begin{cases} \displaystyle\int_0^1 (y'\phi' + y\phi)\,dx = \int_0^1 f\phi\,dx \\ \text{for all } \phi \epsilon C'[0, 1] \text{ with } \phi(0) = \phi(1) = 0 \\ y(0) = y(1) = 0 \end{cases}$$

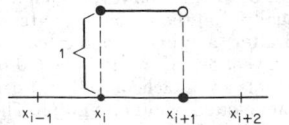

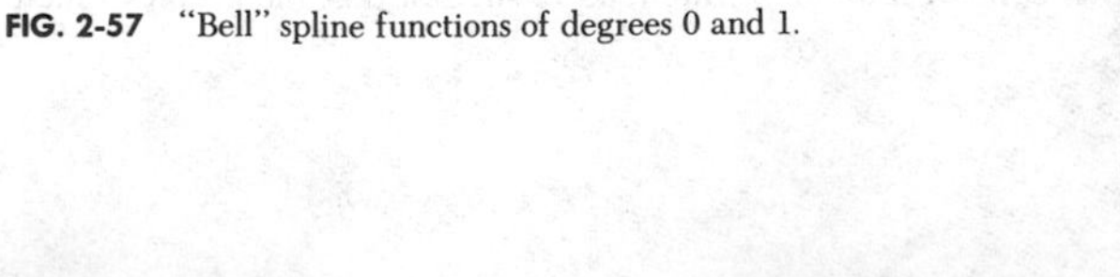

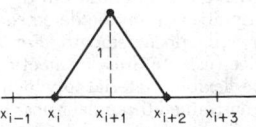

FIG. 2-57 "Bell" spline functions of degrees 0 and 1.

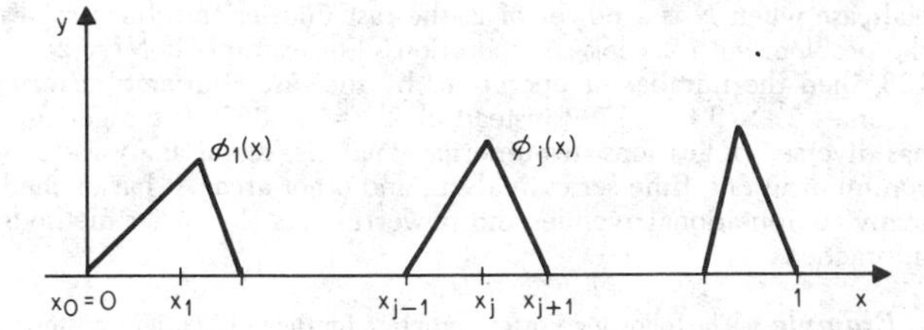

FIG. 2-58 *B splines of order 1.*

Now we apply the Galerkin method and replace the preceding equation by the following approximating equations

$$\int_0^1 (u'\phi_i' + u\phi_i)\,dx = \int_0^1 f\phi_i\,dx \qquad i = 1, 2, \ldots, n$$

or

$$\sum_{j=1}^{n} \left[\int_0^1 (\phi_j'\phi_i' + \phi_j\phi_i)\,dx \right] c_j = \int_0^1 f\phi_i\,dx,\ i = 1, \ldots, n$$

We can write these equations in matrix form $Ac = b$, where $A = (a_{ij})$ with $a_{ij} = \int_0^1 (\phi_j'\phi_i' + \phi_j\phi_i)\,dx$, $c = (c_j)$, $b = (b_j)$ with $b_j = \int_0^1 f\phi_j\,dx$. Thus the problem of approximating the solution to the BVP is reduced to solving a linear system which involves a symmetric and tridiagonal system. Note again the advantage of using B splines, which are easy to compute with and have led to this tridiagonal system, since ϕ_j vanishes outside the subinterval $[x_{j-1}, x_{j+1}]$. (See Fig. 2-58.)

Algorithm: finite element method with B splines. Store the tridiagonal matrix A in a $n \times 3$ matrix:

For $i = 1, \ldots, n$

$a_{i1} = a_{i3} = -1/h + h/6$

$a_{i2} = 2/h + 2h/3$

Compute vector **b** by Simpson's rule, with step size $h/2$:

For $i = 1, \ldots, n$

$b_i \cong [f(x_i - h/2) + f(x_i) + f(x_i + h/2)]h/3$

Solve the system $Ac = b$ by the gaussian elimination method (tridiagonal matrix):

Triangulation:

For $k = 1, \ldots, n - 1$

$p = a_{k+1,k}/a_{kk}$

$b_{k+1} = b_{k+1} - pb_k$

Back substitution:

For $k = n - 1, \ldots, 1$

$c_k = (b_k - a_{k3}c_{k+1})/a_{k2}$

FAST FOURIER TRANSFORM (FFT) ALGORITHM OR THE COOLEY-TUKEY ALGORITHM FOR COMPUTATION OF INTERPOLATORY TRIGONOMETRIC POLYNOMIALS

Suppose that we are given a set of $2m$ points in the plane, say, (x_0, y_0), (x_1, y_1), . . . , (x_{2m-1}, y_{2m-1}) with the first elements in the pairs equally spaced within an interval. To be specific we take the interval to be $[-\pi, \pi]$, so then

$$x_j = -\pi + (j/m)\pi \quad \text{for } j = 0, 1, \ldots, 2m - 1$$

If this were not the case, a simple transformation $z = \alpha x + \beta$ could be used to translate the given interval into $[-\pi, \pi]$. For example, suppose that the given interval is [0, 4]. When $x = 0$, $z = -\pi$, so $\beta = -\pi$; when $x = 3$, $z = \pi$, so $\alpha = \pi/2$, and the required transformation is $z_j = (\pi/2)(x_j - 2)$.

For a fixed $n < m$, consider the set of function $F_n = \{\phi_0, \phi_1, \ldots, \phi_{2n-1}\}$ where

$$\phi_0(x) = 1/\sqrt{2m}$$

$$\phi_k(x) = (1/\sqrt{m}) \cos kx \qquad k = 1, 2, \ldots, n$$

$$\phi_{n+k}(x) = (1/\sqrt{m}) \sin kx \qquad k = 1, 2, \ldots, n - 1$$

We consider the problem of determining the trigonometric polynomial S_n, which is in the span of the set F_n (i.e., it is a linear combination of the functions in F_n) and which provides the best least-squares fit to the $2m$ paired data points $\{(x_j, y_j)\}_{j=0}^{2m-1}$; in other words, the problem to minimize the function

$$E(S_n) = \sum_{j=0}^{2m-1} \{y_j - S_n(x_j)\}^2$$

or to compute the constants (coefficients of S_n).

A celebrated paper by J. W. Cooley and J. W. Tukey, published in 1965, describes a method for calculating these constants which requires only $2m\ell n(2m)$ multiplications and $2m\ell n(2m)$ additions, provided m is chosen in an appropriate manner (e.g., m is factored into powers of 2). This method entails a drastic reduction in the number of calculations (e.g., 7000 calculations compared with 1 million in the direct method for data of 1000 points).

The Algorithm For the data points (x_0, y_0), (x_1, y_1), . . . , (x_{N-1}, y_{N-1}), where $N = 2^p$ and $x_j = -\pi + 2j\pi/N$ for $j = 0, 1, \ldots, N - 1$, the problem is to compute the approximation which minimizes

$$\sum_{j=0}^{2m-1} \left\{ y_j - \left[\frac{a_0}{\sqrt{2m}} + \frac{a_n}{\sqrt{m}} \cos nx_j + \frac{1}{\sqrt{m}} \sum_{k=1}^{n-1} (a_k \cos kx_j + a_{n+k} \sin kx_j) \right] \right\}^2$$

over all possible choices of the constants $a_0, a_1, \ldots, a_{2n-1}$. The best least-squares approximation in this case is given by the interpolatory trigonometric polynomial

$$S_m(x) = \frac{a_0 + a_m \cos mx}{2} + \sum_{j=1}^{m-1} [a_j \cos jx + a_{m+j} \sin jx]$$

While trigometric polynomials of high order can produce satisfactory results for a large number of equally spaced data points, their use until two decades ago had been inhibited by the large number of arithmetic operations required. The interpolation of $2m$ data points requires approximately $(2m)^2$ multiplications and $(2m)^2$ additions by the direct calculation presented here.

Fast Fourier Transform Algorithm

$$F(x) = \sum_{k=0}^{N-1} c_k e^{ik\pi x} = \sum_{k=1}^{N-1} c_k(\cos k\pi x + i \sin k\pi x)$$

where $i = \sqrt{-1}$.

Input: N; $y_0, y_1, \ldots, y_{N-1}$.

1. Set $M = N/2$, $p = \log_2 N$, $q = p - 1$, $w = e^{2\pi i/N}$.
2. For $j = 0, 1, \ldots, N - 1$, set $c_j = y_j$.
3. For $j = 1, 2, \ldots, M$ set $u_j = w^j$, $u_{j+M} = -u_j$.
4. Set $K = 0$, $u_0 = 1$.
5. For $K = 1, 2, \ldots, p$ do steps 6 to 12.
6. While $K < N - 1$, do steps 7 to 11.
7. For $j = 1, 2, \ldots, M$ do steps 8 to 10.
8. Suppose K is of the form $K = k_{p-1}2^{p-1} + k_{p-2}2^{p-2} + \cdots + k_1 2 + k_0$; set $K_1 = k_{p-1}2^{p-q-1} + \cdots + k_{q+1}2 + k_q$; $K_2 = k_q 2^{p-1} + k_{q+1}2^{p-1} + \cdots + k_{p-1}2^q$.
9. Set $v = c_{K+M}u_{K2}$. $c_{K+M} = c_K - v$ and $c_K = c_K + v$.
10. Set $K = K + 1$.
11. Set $K = K + M$.
12. Set $K = 0$; $M = M/2$; $q = q - 1$.
13. While $K < N - 1$, do steps 14 to 16.
14. Let K be factored as in step 8. Set $j = k_0 2^{p-1} + k_1 2^{p-2} + \cdots + k_{p-2}2 + k_{p-1}$.
15. If $j > k$, then interchange c_j and c_k.
16. Set $K = K + 1$.
17. For $j = 0, 1, \ldots, N - 1$ set $c_j = 2c_j/N$.
18. Output $(c_0, c_1, \ldots, c_N)$; stop.

Discrete Fourier Transform (DFT) The summation of the type considered in the preceding subsection arises in many applications besides least-squares fit of data points by trigonometric polynomials. It is a special case of the discrete Fourier transform. Let $\{g_k\}_{k=0}^{N-1}$ be a set of complex numbers and let

$$G_j = \sum_{k=0}^{N-1} g_k e^{2\pi ijk/N} \qquad j = 0, 1, \ldots, N-1$$
$$= \sum_{k=0}^{N-1} g_k z^k$$

where $z = e^{2\pi i/N}$. The sequence $\{G_j\}$ is called the discrete Fourier transform of the sequence $\{g_j\}$ by analogy with the (continuous) Fourier transform

$$G(x) = \int_{-\infty}^{\infty} g(t)e^{2\pi ixt}\,dt$$

Just as the continuous Fourier transform can be inverted, so can the discrete Fourier transform to yield

$$g_k = \frac{1}{N}\sum_{j=0}^{N-1} G_j z^{-jk} \qquad k = 0, 1, \ldots, N-1$$

Direct calculation of the inverse discrete Fourier transform via this formula is very cumbersome for large N; in addition, it could lead to meaningless numerical results when round-off errors are present. The fast Fourier transform algorithm is very helpful in such computations.

Consider the problem of computing the Fourier coefficients $c_0, c_1, \ldots, c_{N-1}$ for a function $f(x) = \Sigma_{j=0}^{N-1} c_j e^{ijx}$ whose values are known at the points $2\pi k/N$, $k = 0, 1, 2, \ldots, N-1$. Then

$$c_j = N^{-1}\sum_{k=1}^{N-1} f\left(\frac{2\pi k}{N}\right) e^{-ij2\pi k/N} \qquad j = 0, 1, \ldots, N-1$$

Set $w = e^{-2\pi i/N}$, $a_k = f\left(\frac{2\pi k}{N}\right)N^{-1}$. Then we can state the problem as follows: For $j = 0, 1, \ldots, N-1$, compute $c_j = \Sigma_{k=0}^{N-1} a_k w^{jk}$, where $w^N = 1$. With the fast Fourier method, one needs only $N(r_1 + r_2 + \cdots + r_p)$ operations if $r_1 r_2 \cdots r_p = N$. (By an "operation" we mean one complex multiplication and one complex addition.) In contrast, if one solves this problem by conventional methods (say, by using Horner's scheme), one needs N operations per coefficient, i.e., N^2 operations. In the important (and often realized) special case when N is a power of 2, the fast Fourier transform solves the problem with $2N \log_2 N$ operations. For example if $N = 2^7 = 128$, then the number of operations by the fast Fourier transform becomes $2^7 \times 14 = 1792$ instead of $2^{14} = 16{,}383$. The algorithm has diverse applications in numerical analysis, signal analysis, telecommunications, time-series analysis, and other areas. It has opened many computational avenues and powerful uses of Fourier methods in practice.

Example The following Fortran program for the fast Fourier transform, for the case in which N is a power of 2, is quoted from Cooley, Lewis, and Welch, *IEEE Trans.*, **E-12**(1) (March 1965).

```
    SUBROUTINE FFT(A,M)
    COMPLEX A(1),U,W,T
    N = 2**M
    NV2 = N/2
    NM1 = N - 1
    J = 1
    DO 7 I = 1,NM1
    IF(I.GE.J) GO TO 5
    T = A(J)
    A(J) = A(I)
    A(I) = T
  5 K = NV2
  6 IF(K.GE.J) GO TO 7
    J = J - K
    K = K/2
    GO TO 6
  7 J = J + K
    DO 20 L = 1,M
    LE = 2**L
    LE1 = LE/2
    U = (1.,0.)
    ANG = 3.14159265358979/LE1
    W = CMPLX(COS(ANG),SIN(ANG))
    DO 20 J = 1,LE1
    DO 10 I = J,N,LE
    IP = I + LE1
    T = A(IP)*U
    A(IP) = A(I) - T
 10 A(I) = A(I) + T
 20 U = U*W
    RETURN
    END
```

CALCULUS OF VARIATIONS

REFERENCES: *General.* 73, 75, 84, 92, 120, 146, 241, 271, 278. *Variational methods in mechanics and mathematical physics; numerical methods.* 84, 101, 178, 198, 199, 215, 227, 278.

SOME CLASSICAL EXAMPLES AND FORMULATION OF THE "SIMPLEST PROBLEM" IN THE CALCULUS OF VARIATIONS

Jean Bernoulli in 1696 proposed the following problem: Find the curve connecting two points in a vertical plane along which a particle falling from the higher point under the influence of gravity will reach the lower point in minimum time. The problem is called the brachistochrone problem (in Greek, *brachys* = minimum, *chronos* = time). To formulate this problem mathematically, it is convenient to choose the axes and locate the points as shown in Fig. 2-59. Let Γ denote a curve joining P and Q, and let ds denote the element of arc length along this curve. The time that the particle takes to fall along this curve is

$$T = \int dt = \int_\Gamma \frac{ds}{v(x, y)} = \int_0^b \frac{\sqrt{1 + (y')^2}}{v(x, y)}\,dx$$

where $v(x, y)$ is the velocity of the particle at the point (x, y). Assume no loss of energy due to friction, so that by principle of conservation of energy the sum of the kinetic and potential energies must be constant at all points on the curve. Thus

$$\tfrac{1}{2}mv_0^2 + mg\alpha = \tfrac{1}{2}mv^2(x, y) + mgy$$

where v_0 is the initial velocity and m is the mass of the particle. Solving for $v(x, y)$ and substituting it in the foregoing expression for T, we obtain

$$T = \int_0^b \frac{\sqrt{1 + (y')^2}}{\sqrt{v_0^2 + 2g(\alpha - y)}}\,dx$$

Note that T is a functional of y: to each curve y we get a unique value of T. The problem is to find among all curves joining the points

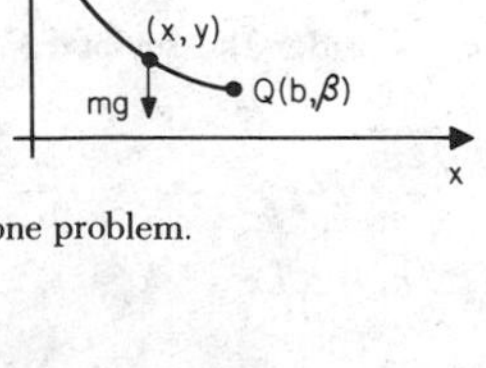

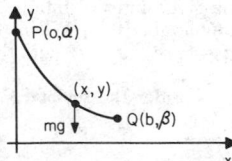

FIG. 2-59 Brachischrone problem.

P and Q [i.e., satisfying $y(0) = \alpha$, $y(b) = \beta$, where α and β are prescribed] that curve which minimizes $T[y]$. The solution turns out to be a cycloid.

Historically, the brachistochrone problem was the first minimization problem for which the calculus of Newton and Leibniz was not adequate. For in the case of a minimization problem in ordinary calculus we are given a function and asked to find the *points* (if any) in its domain at which the function attains a minimum. In contrast, a basic problem in the calculus of variations is to find *among all curves* satisfying certain conditions that curve which minimizes (or maximizes) a given functional. It is expected, therefore, that a solution to this type of problem would require the development of a calculus of functionals paralleling the classical theory of functions of real variables.

As is well known, the shortest distance between two points in a plane is along a straight line. This provides the solution to the following variational problem. Among all plane curves joining two points find the curve $\hat{y} = \phi(x)$ which minimizes the functional $J[y] = \int_a^b \sqrt{1 + (y')^2}\, dx$. More interesting generalizations of this problem involve finding the shortest distance between two points lying on a prescribed surface and, in particular, the curve on the prescribed surface along which the shortest distance is attained. For example, the shortest distance between two points on a sphere is along an arc on a great circle.

The least-surface-area problem, or the problem of Plateau, is to find among all curves joining two points (x_1, y_1) and (x_2, y_2) the one which generates the surface of area of revolution when the curves are revolved about the x axis. Mathematically the problem is to minimize the functional

$$S[y] = \int_\Gamma 2\pi y ds = 2\pi \int_{x1}^{x2} y\sqrt{1 + (y')^2}\, dx$$

This is an interesting problem since it turns out that there are three possibilities, depending on the relative positions of the points.

Fermat's principle states that a light ray travels along that path which yields the minimum transit time. This provides a **variational principle** in geometric optics which enables us to find the path along which a light ray propagates in an inhomogeneous medium. For instance, if we have a plane transparent medium in which the index of refraction at each point is $n(x, y)$, then the velocity of light in that medium is $v(x, y) = c/n(x, y)$, where c is a constant. The transit time (to be minimized) is then

$$T[y] = \frac{1}{c}\int_{x1}^{x2} n(x, y)\sqrt{1 + (y')^2}\, dx$$

The foregoing examples are special cases of the variational problem of finding among all functions which have a continuous first derivative on $[a, b]$ and satisfy the prescribed boundary conditions

$$(BC) \qquad y(a) = \alpha \qquad y(b) = \beta$$

the function which minimizes the functional

$$(°) \qquad J[y] = \int_a^b f(x, y(x), y'(x))\, dx$$

where $f(x, y, z)$ is a given function which has continuous first partial derivatives. This problem is often called the "simplest problem in the calculus of variations."

METHOD OF FINITE DIFFERENCES IN THE CALCULUS OF VARIATIONS

Suppose that among all functions in the class $C^2[a, b]$ of twice continuously differentiable functions which satisfy the prescribed boundary conditions (BC) the function $\hat{y}$ yields a relative minimum of the functional $J[y]$ given in (°), i.e., $J[y] \geq J[\hat{y}]$ for all y in some "neighborhood" of $\hat{y}$. What equation does $\hat{y}$ satisfy? (Here we assume additional smoothness of the functions y and f for convenience.) Euler in 1744 determined the necessary condition for a function to extremize a given functional. He accomplished this by using the method of finite difference and the calculus of functions of n variables. His approach provides a method for approximating the extremizing function and is particularly suitable for modern computational procedures. We partition the interval $[a, b]$ into $n + 1$ subintervals using the points $a = x_0 < x_1 < \cdots < x_{n+1} = b$, approximating the first derivative of y by a difference quotient:

$$y'(x_i) \approx \frac{y(x_{i+1}) - y(x_i)}{x_{i+1} - x_i}$$

and an integral by a finite sum:

$$\int_a^b g(x)\, dx \approx \sum_{i=0}^{n} g(x_i)(x_{i+1} - x_i)$$

For simplicity we take equal subdivision and let $h = (b - a)/(n + 1) = x_{i+1} - x_i$ for all $i = 0, 1, \ldots, n$. We denote $y(x_i)$ by y_i. Then the functional $J[y]$ may be approximated as follows:

$$J[y] \approx \sum_{i=0}^{n} f\left(x_i, y_i, \frac{y_{i+1} - y_i}{h}\right) h =: I(y_1, y_2, \ldots, y_n)$$

There are only n unknowns $y_1, \ldots, y_n$ since $y_0 = \alpha$ and $y_{n+1} = \beta$. The problem thus reduces to ordinary calculus. If $\bar{y}_1, \ldots, \bar{y}_n$ minimizes $I(y_1, \ldots, y_n)$, then

$$\left.\frac{\partial I}{\partial y_k}\right|_{(\bar{y}_1, \ldots, \bar{y}_n)} = 0 \quad \text{for } k = 1, 2, \ldots, n$$

To compute $\partial I/\partial y_k$ for a fixed k, note that only two terms in the above sum contains terms in y_k:

$$I(y_1, \cdots, y_n) = \cdots + f\left(x_k, y_{k-1}, \frac{y_k - y_{k-1}}{h}\right) h + f\left(x_k, y_k, \frac{y_{k+1} - y_k}{h}\right) h + \cdots$$

Thus by using the chain rule we obtain

$$f_2\left(x_k, y_k, \frac{y_{k+1} - y_k}{h}\right) - \frac{1}{h}\left[f_3\left(x_k, y_k, \frac{y_{k+1} - y_k}{h}\right) - f_3\left(x_{k-1}, y_{k-1}, \frac{y_k - y_{k-1}}{h}\right)\right] = 0 \quad \text{for } k = 1, 2, \ldots, n$$

Euler recognized that this system of (nonlinear) equations is a finite difference approximation of the differential equation

$$f_y(x, y, y') - (d/dx)\, f_{y'}(x, y, y') = 0$$

This latter equation is known as the Euler equation or the Euler-Lagrange equation. It may be obtained as the limiting case of the discrete system as the mesh size h tends to zero. Then y_i tends to $y(x)$, where x is a point in $[a, b]$, and as $h \to 0$, $f_2\left(x_k, y_k, \frac{y_{k+1} - y_k}{h}\right)$ tends to $f_2(x, y(x), y'(x))$, $\frac{1}{h}\left[f_3\left(x_k, y_k, \frac{y_{k+1} - y_k}{h}\right) - f_3\left(x_{k-1}, y_{k-1}, \frac{y_k - y_{k-1}}{h}\right)\right]$ tends to $(d/dx)f_{y'}(x, y(x), y'(x))$, assuming f is sufficiently smooth.

This finite difference method shows that the problem of extremizing the functional can be approximated by a problem of finding an extremum of a function of n real variables. We may interpret this procedure also as a method for finding a polygonal line approximation to the extremizing function: the polygonal with vertices (a, α), $(x_1, \bar{y}_1), \ldots, (x_n, \bar{y}_n), (x_{n+1}, \beta)$ is an approximation to the curve $\hat{y}$ which extremizes the functional $J[y]$. In this procedure, the solution of the discrete system provides an approximation of $\hat{y}$ at the knots $x_1, x_2, \ldots, x_n$. An approximate solution for all x in $[a, b]$ may be obtained by interpolation (e.g., by polynomial or spline function interpolation). It should be pointed out, however, that the question of convergence of the approximate extremizing functions obtained by this finite difference method together with interpolation to the exact extremizing function is more subtle.

Example $J[y] = \int_0^3 \{(y')^2 + xy\}\,dx,\ y(0) = 0,\ y(3) = 4$. Let $n = 2$, so $h = (3 - 0/2 + 1) = 1,\ x_0 = x,\ x_1 = 1,\ x_2 = 2,\ x_3 = 3$.

$$y'(x_i) \approx \frac{y_{i+1} - y_i}{h} = y_{i+1} - y_i$$

$$J[y] \approx \sum_{i=0}^{2} f(x_i, y_i, y_{i+1} - y_i) =: \phi(y_1, y_2)$$

$$\phi(y_1, y_2) = (y_1 - y_0)^2 + 0y_0 + (y_2 - y_1)^2 + y_1 + (y_3 - y_2)^2 + 2y_2$$

Substituting the values $y_0 = 0$ and $y_3 = 4$, we obtain

$$\phi(y_1, y_2) = y_1^2 + (y_2 - y_1)^2 + y_1 + (4 - y_2)^2 + 2y_2$$

Setting $\partial\phi/\partial y = 2\phi/\partial y_2 = 0$, we obtain

$$4y_1 - 2y_2 + 1 = 0 \qquad \text{and} \qquad -y_1 + 2y_2 - 3 = 0$$

This system has the solution $y_1 = 2/3$ and $y_2 = 11/6$. The exact solution may be obtained by solving the Euler equation $fy - (d/dx)fy' = 0$, where in this case $f(x, y, y') = (y')^2 + xy$, so $f_y = x$, $f_{y'} = 2y'$, and $x - (d/dx)(2y') = 0$. So $y' = x^2/4 + c$, $y(x) = x^3/12 + cx + c_1$. Since $y(0) = 0$, $c_1 = 0$. Now $y(3) = 4$, so $4 = 27/12 + 3c$, or $c = 7/12$. The exact solution which minimizes $J[y]$ and satisfies the prescribed boundary conditions is $y(x) = x^3/12 + 7x/12$.

For problems of the types just discussed consider the integral

$$I = \int_a^b f\left(x, y, \frac{dy}{dx}\right) dx$$

taken along the curve $y = g(x)$ between the points A and B in a plane. Then the integral I has a maximum or a minimum value along a curve c if $y = g(x)$ is a solution of the equation

$$\partial f/\partial y - (d/dx)\,(\partial f/\partial y') = 0$$

Example By elementary calculus the length of the curve $y = g(x)$ is given by

$$S = \int_A^B \sqrt{dx^2 + dy^2} = \int_A^B \sqrt{1 + \left(\frac{dy}{dx}\right)^2}\,dx$$

Here

$$f = \sqrt{1 + \left(\frac{dy}{dx}\right)^2},\quad \frac{\partial f}{\partial y} = 0,\quad \frac{\partial f}{\partial y'} = \frac{y'}{\sqrt{1 + y'^2}}$$

By the equation preceding the example,

$$0 - \frac{d}{dx}\frac{y'}{\sqrt{1 + y'^2}} = 0$$

$$\frac{\sqrt{1 + y'^2}(y'') - y'(y'y''/\sqrt{1 + y'^2})}{1 + y'^2} = 0$$

or $y'' = 0$. For this $y = ax + b$ is the curve with minimum length between A and B. The constants a and b are used to make $y = f(x)$ pass through the points A and B.

Example The area of revolution is given by

$$R = 2\pi \int_A^B y\,ds = 2\pi \int_A^B y\sqrt{1 + y'^2}\,dx$$

where $\quad f = y\sqrt{1 + y'^2} \quad \partial f/\partial y = \sqrt{1 + y'^2} \quad \partial f/\partial y' = \dfrac{yy'}{\sqrt{1 + y'^2}}$

By the equation preceding the first example,

$$\sqrt{1 + y'^2} - \frac{d}{dx}\frac{yy'}{\sqrt{1 + y'^2}} = 0$$

$$\sqrt{1 + y'^2} - \frac{\sqrt{1 + y'^2}(yy'' + y'^2) - yy'(y'y''/\sqrt{1 + y'^2})}{(1 + y'^2)} = 0$$

$$(1 + y'^2) - \frac{1 + y'^2)(yy'' + y'^2) - yy'^2y''}{(1 + y'^2)} = 0$$

$$(1 + y'^2)^2 - y'^2 - yy'' - y'^4 = 0$$

$$1 + y'^2 - yy'' = 0$$

By setting $y' = \theta$ and $y'' = \theta(d\theta/dy)$ the equation reduces to

$$\theta\,d\theta/(1 + \theta^2) = dy/y$$

which has the solution $y = c \cosh (x - d)/c$. The constants c and d are used to pass the curve through the points A and B.

References should be consulted for solving constrained variational problems. Some examples of Euler-Lagrange equations are shown in Table 2-1.

TABLE 2-1 Euler-Lagrange Equations and Natural Boundary Conditions for Some Standard Functionals in the Calculus of Variations

Functional	Euler-Lagrange equations	Natural boundary conditions
1. $\int_a^b f(x, y(x), y'(x))\,dx$	$f_y - \frac{d}{dx} f_{y'} = 0$	1. $f_{y'}(a, y(a), y'(a)) = 0$, $f_{y'}(b, y(b), y(b), y'(b)) = 0$
2. $\int_a^b f(x, y(x), y'(x), y''(x))\,dx$	$f_y - \frac{d}{dx} f_{y'} + \frac{d^2}{dx^2} f_{y''} = 0$	2. $f_{y''}(a, y(a), y'(a)) = 0$ and $\left(f_{y'} - \frac{d}{dx} f_{y''}\right)\Big\vert_{x=a} = 0$ (similar two conditions at $x = b$)
3. $\int_a^b f(x, y(x), z(x), y'(x), z'(x))\,dx$ (one independent variable x, two unknown functions)	$f_y - \frac{d}{dx} f_{y'} = 0$; $f_z - \frac{d}{dz} f_{z'} = 0$	3. $f_{y'}\Big\vert_{x=a} = f_{z'}\Big\vert_{x=a} = 0$ (similar conditions at $x = b$)
4. $\iint_\Omega f(x, y, u(x, y), u_x(x, y), u_y(x, y))\,dx\,dy$	$f_u - \frac{\partial}{\partial x} f_{u_x} - \frac{\partial}{\partial y} f_{u_y} = 0$	
5. $\iint_\Omega f(x, y, u(x, y), u_x(x, y), u_y, u_{xx}, u_{xy}, u_{yy})\,dx\,dy$	$f_u - \frac{\partial}{\partial x} f_{u_x} - \frac{\partial}{\partial y} f_{u_y} + \frac{\partial^2 f_{u_{xx}}}{\partial x^2} + \frac{\partial^2 f_{u_{xy}}}{\partial x \partial y} + \frac{\partial^2 f_{u_{yy}}}{\partial y^2} = 0$	

OPTIMIZATION

References: *General.* 39, 47, 84, 186, 282. Textbooks that cover a variety of topics in optimization theory and algorithms and mathematical programming. *Linear programming and applications.* 21, 72, 80, 114, 118, 130, 186, 217. *Nonlinear programming; theory and algorithms.* 22, 100, 131, 186, 190, 194, 293. *Network flows.* 21, 150. *Dynamic programming.* 30, 92, 131. *Advanced topics in optimization theory and algorithms.* 22, 64, 185, 194, 271. *Operations research.* 139, 150, 244.

By **optimization** we shall mean finding the best way to do things—that is, to achieve an optimum. "Optimum" is a technical term connoting quantitative measurement and mathematical analysis. Many phases of optimization theory have been known for centuries. But huge and tedious computations prevented their practical application until the age of computers opened up new vistas.

SEARCH METHODS

Suppose that it is desired to determine what settings of the independent variables will yield the optimal (maximum or minimum) value of a dependent variable when the **functional dependence** is not known. Assume further that the total number n of performable experiments is limited. Any set of instructions for placing the n experiments is called a **search plan.** Any investigation seeking the optimal value of an unknown function is a **search problem.** Search problems are classifiable according to the number of independent variables and whether they are error-free (deterministic) or subject to random error (stochastic). Search plans fall naturally into two mutually exclusive categories, simultaneous and sequential. Any plan specifying the location of every experiment before any results are known will bear the former name. A plan permitting the analyst to base future experiments on past outcomes is called sequential. The advantage of sequential plans over simultaneous plans increases exponentially with the number of experiments. Statistical experimental design is usually concerned more with simultaneous than with sequential procedures.

Single-Variable Search Consider two experimental settings x_1 and x_2 with $x_1 < x_2$ (on some scale) with outputs y_1 and y_2. The output y is unimodal if $x_2 < x^\circ$ implies $y_1 < y_2$, and if $x_1 > x^\circ$ implies $y_1 > y_2$, where y° is the optimum occurring at x°. In other words, if the points are both on the same side of the optimum, the one nearer gives the higher value of y. A unimodal function does not have to have a derivative or be continuous. Choice of an optimal search plan requires an a priori criterion of search effectiveness—that is, a measure that can be evaluated from examining the search plan alone, before any experiments are carried out. With n experiments, let the end points of the search interval be denoted x_0 and x_{n+1}. In general, the optimum x° will lie, say, on $x_{K-1} \le x^\circ \le x_{K+1}$. Define $l_n = x_{K+1} - x_{K-1}$ as the interval of uncertainty after n experiments. Then $L_n(x_k) = \max\limits_{1\le K\le n} [l_n(x_k, K)]$ depends only on the search plan. Then seek among all possible plans that one making L_n as small as possible. One then finds the optimal plan x_n° by a minimax scheme $L_n^\circ = \min\limits_{x_k} \max\limits_{1\le K\le n} [l_n(x_k, K)]$,$[L_n^\circ = L_n(x_k^\circ)]$. This is a meaningful, nontrivial search plan.

While simultaneous searches may be necessary, sequential ones are more efficient; so we indicate two briefly—**dichotomous** and **Fibonacci.** By elementary transformation, the search interval can be transformed to $x_0 = 0$, $x_{n+1} = 1$. With two experiments x_1, x_2, $0 \le x_1 < x_2 \le 1$, the maximum level of uncertainty is $L_2 = \max(x_2, 1 - x_1)$ so that $L_2^\circ = ½ + \epsilon/2$. That is, we should place both experiments at the center of the interval as close together as possible. That is, $x_1 = ½ - \epsilon/2$, $x_2 = ½ + \epsilon/2$. Then place the third and fourth experiments in the middle of the remaining interval of length $(1 + \epsilon)/2$. The placement of the third and fourth, and fifth and sixth, experiments is shown in Fig. 2-60. Here ϵ is the least separation between two experiments for which a difference in the y's can be detected. After the third and fourth experiments, the interval of uncertainty is ¼$(1 + 3\epsilon)$. In general, after n experiments (n even) one can locate the optimum within an interval of length $2^{-n/2} + (1 - 2^{-n/2})\epsilon$. Consequently, the effectiveness of the dichotomous search grows exponentially with n.

An even better method is the Fibonacci search. With $F_0 = F_1 = 1$, the difference equation $F_k = F_{k-1} + F_{k-2}$, $k = 2, 3, \ldots$ defines the Fibonacci numbers. With the number of experiments n fixed, the first experiment must be placed $L_2^\circ = F_n^{-1}[F_{n-1} + (-1)^n\epsilon]$ units from one end of the original unit interval of uncertainty. Because of the method's symmetry, it does not matter which end is chosen. The succeeding points are chosen according to the optimal search plan $L_{n-k}^\circ = F_n^{-1}[(F_{k+1}F_{n-2} - F_{k-1}F_n)\epsilon]$. After n experiments, one can locate the optimum within an interval of length $F_n^{-1}(1 + F_{n-2}\epsilon)$. With a dichotomous sequential search of 20 experiments, the reduction rate is 1 to 1024. For a Fibonacci search it is 1 to 10,946.

Other effective search methods are those of lattice and random search.

Multivariable Searches Multivariable problems have a structure entirely different from that of a single variable (Bellman). Thus, single-variable methods are not extended to the multivariable case without considerable trouble. This curse of dimensionality makes the unimodal assumption less plausible: measures of search effectiveness independent of the experimenters' luck are difficult to develop, and the size of multidimensional space forces us to seek regions of uncertainty which are very small parts of the original experimental region. A good introduction to various of these methods, together with pitfalls such as ridges, is contained in Wilde (see references).

CONSTRAINTS

In addition, the physical problem at hand may impose constraints on the variables. Constraints may take many forms, such as restrictions on variable range (e.g., $a < x < b$), equality constraints (e.g., $x^2 + y^2 = 2$), inequality constraints (e.g., $x^2 + y^2 < 2$), and the like.

For unconstrained problems we have the following rules:

1. If $y = f(x)$ and $dy/dx = 0$ at a point P, then at that point, f has (a) a local maximum if $d^2y/dx^2 < 0$ at P; (b) a local minimum if $d^2y/dx^2 > 0$ at P; and (c) a saddle (inflection) point if $d^2y/dx^2 = 0$ but $d^3y/dx^3 \ne 0$ at P. If $d^3y/dx^3 = 0$ at P and $d^4y/dx^4 < 0$, there is a local maximum at P, etc.

2. If $z = f(x, y)$ and if $f_x = f_y = 0$ at P, then at that point f has (a) a local maximum if $f_{xx} < 0$, $f_{xx}f_{yy} > f_{xy}^2$ at P; (b) a local minimum if $f_{xx} > 0$, $f_{xx}f_{yy} > f_{xy}^2$ at P; (c) a saddle point if $f_{xx}f_{yy} < f_{xy}^2$ at P; and (d) if $f_{xx}f_{yy} = f_{xy}^2$ at P, further investigation is necessary.

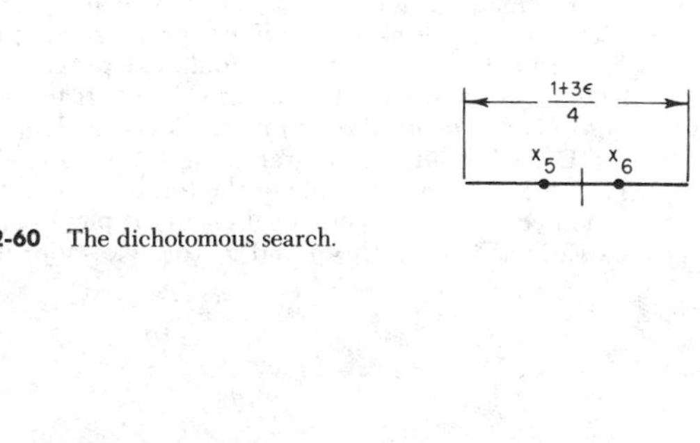

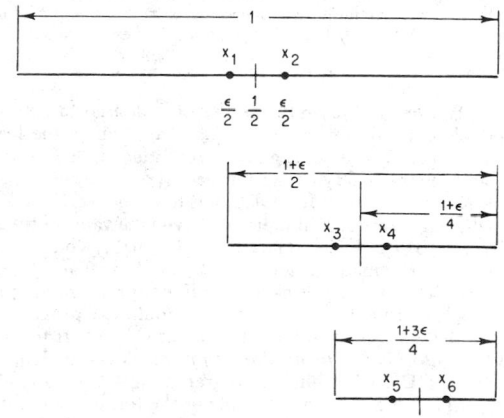

FIG. 2-60 The dichotomous search.

3. Let $z = f(x_1, x_2, \ldots, x_n)$, $\partial z/\partial x_i = z'_i$, $\partial^2 z/\partial x_i \partial x_j = z''_{i,j}$ and denote the hessian matrix $H = \begin{bmatrix} z''_{11}\, z''_{12} \cdots z''_{1n} \\ \cdots \\ z''_{n1}\, z''_{n2} \cdots z''_{nn} \end{bmatrix}$. If $z'_i = 0$, for all i, at a point P, then at that point f has (a) a local minimum if H is positive definite at P; (b) a local maximum if H is negative definite at P; (c) a valley (ridge) if H is semidefinite at P [i.e., nonnegative (nonpositive), vanishing for dx along the valley (ridge)]; and (d) a saddle if H is indefinite at P (positive, negative, or zero).

Local maxima or minima may also be attained at a point where the first derivatives fail to exist. When constraints are placed on a problem, e.g., a region with a finite boundary, it may happen that the optimum occurs at a boundary point at which derivatives may or may not exist and may or may not differ from zero.

Since constrained problems are of a very wide variety, only a few of the most common methods can be included here.

Equality Constraints When the objective function must satisfy side conditions given as equations relating the independent variables, the optimum must necessarily lie on a boundary of the feasible region. Various methods (Wilde and Beightler; see references) have been devised to transform the equality-constrained problem into one having the optimum inside the feasible region. Typical of these is that of Lagrange multipliers. The technique is well illustrated by the problem of maximizing or minimizing a function $f(x, y, z)$ subject to two (equality) constraints of the form $g(x, y, z) = 0$ and $h(x, y, z) = 0$, where g and h are not functionally dependent, so that the constraints are neither equivalent nor incompatible. Upon introduction of the arbitrary Lagrange multipliers λ_1 and λ_2, one obtains the three equations $f_x + \lambda_1 g_x + \lambda_2 h_x = 0$, $f_y + \lambda_1 g_y + \lambda_2 h_y = 0$, and $f_z + \lambda_1 g_z + \lambda_2 h_z = 0$ at (x_0, y_0, z_0), which together with the conditions $g = 0$, $h = 0$ at (x_0, y_0, z_0) constitute five equations in the five unknowns x_0, y_0, z_0, λ_1, and λ_2. The first three equations are easily remembered, for they are the necessary conditions, $\psi_x = 0$, $\psi_y = 0$, and $\psi = 0$, that $\psi = f + \lambda_1 g + \lambda_2 h$ attains a local maximum or minimum when no constraints are imposed.

More generally, if $f(x_1 \ldots, x_n)$ is to be optimized subject to m ($m < n$) independent constraints $g_i = 0$, $i = 1, \ldots, m$, form the auxiliary function $\psi = f + \sum_{i=1}^m \lambda_i g_i$ and write down the necessary conditions for ψ to attain a local maximum or minimum with no constraints.

Example The maximum value of the function $f(x, y, z) = x^2y^2z^2$ subject to the equality constraint $x^2 + y^2 + z^2 = c^2$ must occur on that spherical surface. According to the method, we introduce a Lagrange multiplier λ and find the maximum of the unconstrained problem defined by the auxiliary function $\psi = x^2y^2z^2 + \lambda(x^2 + y^2 + z^2 - c^2)$. By calculating ψ_x, ψ_y, and ψ_z and equating to zero, we obtain $xy^2z^2 + \lambda x = 0$, $x^2yz^2 + \lambda y = 0$, and $x^2y^2z + \lambda z = 0$. Together with the constraint, these provide four equations for the unknowns x_0, y_0, z_0, and λ. The solution $x = y = z = 0$ is excluded, since f takes its minimum value, zero, there. The other solutions are $x^2 = y^2 = z^2$, $\lambda = -x^4$. From the constraint it then follows that $x = \pm c/\sqrt{3}$, $y = \pm c/\sqrt{3}$, $z = \pm c/\sqrt{3}$, and $\lambda = -c^4/9$. At all these points, f takes the maximum value $c^6/27$.

Inequality Constraints Suppose that it is desired to optimize a differentiable objective function $y(x_1, \ldots, x_N)$, where the feasible domain is defined by the nonnegativity conditions $x_i \geq 0$, $i = 1, 2, \ldots, N$, and the inequality constraints are $f_k(x_1, \ldots, x_N) \geq 0$, $k = 1, 2, \ldots, K$. An equivalent formulation is to introduce K slack variables f_k which measure the difference between the value of the function $f_k(x_1, \ldots, x_N)$ and zero. Thus, the K inequality constraints are replaced by the K equations $f_k(x_1, \ldots, x_N) - f_k = 0$ and $f_k \geq 0$, $k = 1, \ldots, K$. This problem is called the **linear-programming** problem when all the functions are linear and **nonlinear programming** otherwise. Methods for these important areas will be left for the literature [see also Gass, *Linear Programming*, McGraw-Hill, New York, 1964; and Dorn, "Nonlinear Programming—A Survey," *Manage. Sci.*, **9**, 171 (1963)]. As an example of the formulation, suppose that the operation cost y of a hypothetical chemical plant depends on the process variables x_1 (pressure) and x_2 (recycle ratio) in the form $y = 100x_1 + 2 \times 10^8 x_1^{-1} x_2^{-1} + 3 \times 10^4 x_2$, with $x_1 > 0$, $x_2 > 0$. If the inequality constraint is $x_1 x_2 \geq 900$, it must be placed in the proper form $f_1(x_1, x_2) = x_1 x_2 - 900 \geq 0$. The alternate form involves introducing a nonnegative slack variable f_1 such that $f_1(x_1, x_2) - f_1 = x_1 x_2 - 900 - f_1 = 0$, $f_1 \geq 0$.

VARIATIONAL CALCULUS METHODS

The elementary concepts from this subject were discussed under "Differential and Integral Calculus." Variational problems constitute an important class of optimization problems in which one seeks to determine one or more functions, subject to certain constraints, so as to optimize certain definite integrals. The integrand of these integrals depends upon the unknown function or functions and/or certain of their derivatives. One of the simple examples concerns the desire to optimize an integral of the form $I = \int_a^b f(x, u, u')dx$ subject to the constraints $u(a) = A$ and $u(b) = B$, where a, b, A, and B are given constants. The optimum function u must satisfy the Euler equation $(d/dx)(\partial F/\partial u') - \partial F/\partial u = 0$, which can be written in the expanded form $F_{u'u'}(d^2u/dx^2) + F_{uu'}(du/dx) + (F_{xu'} - F_u) = 0$. These equations are generally quasilinear (linear in the highest-order derivative). The Euler equation is only a necessary condition, which must be satisfied by u if it is to qualify. The development of sufficient conditions is much more difficult.

If $I = \int_a^b F(x; u_1, \ldots, u_n; u'_1, \ldots, u'_n)\, dx$ where $u_i(a)$ and $u_i(b)$, $i = 1, \ldots, n$ are prescribed, the functions $u_1, \ldots, u_n$ are sought which optimize I subject to the prescribed end conditions. The necessary condition is the Euler equation $(d/dx)(\partial F/\partial u'_r) - (\partial F/\partial u_r) = 0$, $r = 1, 2, \ldots, n$.

Example The Euler equations associated with the integral $\int_a^b [(u'_1)^2 + (u'_2)^2 - 2u_1u_2 + 2xu_1]\, dx$ are $(d/dx)(2u'_1) - (-2u_2 + 2x) = 0$ and $(d/dx)(2u'_2) - (-2u_1) = 0$ or $u''_1 + u_2 = x$ and $u''_2 + u_1 = 0$.

If $I = \int_a^b F(x, u, u', u'')\, dx$, $u(a)$, $u(b)$ prescribed, then the function u generating the optimum value of I must satisfy the Euler equation $(d^2/dx^2)(\partial F/\partial u'') - (d/dx)(\partial F/\partial u') + (\partial F/\partial u) = 0$.

If $I = \iint_R F(x, y, u, u_x, u_y)\, dx\, dy$ with u prescribed on the closed boundary C of R, then the function u generating the optimum value of I must satisfy the Euler equation $(\partial/\partial x)(\partial F/\partial u_x) + (\partial/\partial y)(\partial F/\partial u_y) - (\partial F/\partial u) = 0$.

Suppose that $I = \int_a^b F(x, u, u')\, dx$ is to be optimized, where $u(a)$ and $u(b)$ are prescribed as before, but an additional integral constraint $\int_a^b G(x, u, u')\, dx = K$ must be satisfied. In this case the appropriate Euler equation is found by replacing F in the Euler equation above by $H = F + \lambda G$ and solving the unconstrained problem. λ is of the same nature as a Lagrange multiplier.

Example $I = \int_0^1 (u')^2\, dx$ subject to $u(0) = 0$, $u(1) = 0$ and the integral constraint $\int_0^1 u\, dx = 1$. Let $H = (u')^2 + \lambda u$ whose Euler equation is $2u'' - \lambda = 0$. Solving this and setting $u(0) = 0$ and $u(1) = 0$, there follows $u = 6x(1 - x)$, $\lambda = -24$.

PROGRAMMING

The general optimization problem is that of finding an optimum in a feasible region defined by equations and inequalities. When there are no equations but only strict inequalities satisfied at the optimum, it can be found by setting first derivatives to zero. When a known set of equations but no inequalities constrain the optimum, it is locatable by an indirect approach using decision derivatives. In the general problem one cannot predict which inequality constraints will be "tight" (satisfied as strict equalities) and which "loose" (satisfied as strict inequalities) at the optimum. Optimization problems of this type are called **mathematical-programming** problems. In this section we shall discuss dynamic programming (Bellman) and linear programming.

Dynamic Programming A large number of interesting and significant scientific, production, and management problems can be classified as multistage decision processes. Among the more useful ways of furnishing numerical answers for such problems is the theory of dynamic programming due to R. E. Bellman and his collaborators. This approach, based on the use of **functional equations** and the **principle of optimality,** was developed to employ the capabilities of the digital computer. A typical problem from operations research will be used to discuss the concepts.

Consider the problem of maximizing the function $R(x_1, x_2, \ldots, x_n) = \Sigma_{i=1}^{n} g_i(x_i)$ over the region $x_i \geq 0$, $\Sigma_{i=1}^{n} x_i = x$. To treat this problem, it is first embedded in a family of allocation processes. Second, in place of considering a particular quantity of resources and a fixed number of activities, suppose that we examine the entire family of such problems in which x may assume any positive value and n is any positive integer. What seems to be a static process is artificially imbued with a timelike property by requiring the allocations to be made one at a time. First a quantity of resources is assigned to the nth activity (stage), then to the $(n - 1)$st activity, etc. In this fashion the process is made dynamic.

The mathematical model follows from these assumptions. Since the maximum of R depends upon x and n, this dependence is made explicit by defining the sequence $[f_n(x)]$, $n = 1, 2, \ldots$, $x \geq 0$ as $f_n(x) = \max\limits_{[x_i]} R(x_1, x_2, \ldots, x_n)$, $x_i \geq 0$, $\Sigma_{i=1}^{n} x_i = x$. The function $f_n(x)$ is the optimal return from an allocation of the quantity of resources x to n activities. Clearly $f_n(0) = 0$ provided $g_i(0) = 0$ for each i and $f_1(x) = g_1(x)$, $x \geq 0$. Next a recurrence relation between $f_n(x)$ and $f_{n-1}(x)$, for arbitrary n and x, is obtained. Let x_n, $0 \leq x_n \leq x$ be the allocation to the nth activity. The remaining resources $x - x_n$ will be used to obtain a maximum return from the remaining $n - 1$ activities. The optimal return for $n - 1$ activities, starting with resources $x - x_n$, is $f_{n-1}(x - x_n)$. Thus the initial allocation of x_n to the nth activity generates a total return of $g_n(x_n) + f_{n-1}(x - x_n)$ from the nth activity. An optimal choice of x_n maximizes this function. Thus with $f_1(x) = g_1(x)$ the basic functional equation becomes

$$f_n(x) = \max_{0 \leq x_n \leq x} [g_n(x) + f_{n-1}(x - x_n)] \qquad (2\text{-}107)$$

$n = 2, 3, \ldots$. The foregoing result utilizes the very general concept or principle of optimality: An optimal policy has the property that whatever the initial state and initial decision are, the remaining decisions must constitute an optimal policy with regard to the state resulting from the first decision process. This simple property of multistage decision processes is the cornerstone of dynamic programming.

Extension of this concept to more complex problems follows easily. For example, if there are two types of resources in quantities x and y and the return (utility) function is $g_i(x_i, y_i)$ from the ith activity due to allocations x_i and y_i, then the mathematical problem is that of maximizing the function of $2n$ variables

$$R(x_1, x_2, \ldots, x_n; y_1, \ldots, y_n) = \sum_{i=1}^{n} g_i(x_i, y_i)$$

subject to the two constraints $\Sigma_{i=1}^{n} x_i = x$, $x_i \geq 0$, $\Sigma_{i=1}^{n} y_i = y$, $y_i \geq 0$. With $f_n(x, y) = \max\limits_{(x,y)} R_n$ the recurrence relation is

$$f_n(x, y) = \max_{0 \leq x_n \leq x} \max_{0 \leq y_n \leq y} [g_n(x_n, y_n) + f_{n-1}(x - x_n, y - y_n)] \qquad (2\text{-}108)$$

and $f_1(x, y) = g_1(x, y)$. Further extensions, applications, and computational algorithms are found in the cited literature. Additionally, we recommend Bellman and Dreyfus, *Applied Dynamic Programming*, Princeton University Press, Princeton, N.J., 1962, for its lucid account of the subject. See also Howard, *Dynamic Programming and Markov Processes*, Wiley, New York, 1960.

Linear Programming The combined term "linear programming" is given to any method for finding where a given linear function of several variables takes on an extreme value, and what that value is, when the variables are nonnegative and are constrained by linear equalities or inequalities. A very general problem consists of maximizing $f = \Sigma_{j=1}^{n} c_j x_j$ subject to the constraints $x_j \geq 0$ $(j = 1, 2, \ldots, n)$ and $\Sigma_{j=1}^{n} a_{ij} x_j \leq b_i$ $(i = 1, 2, \ldots, m)$. With S the set of all points whose coordinates x_j satisfy all the constraints, we must ask three questions: (1) Are the constraints *consistent?* If not, S is empty and there is no solution. (2) If S is not empty, does the function f bcome *unbounded* on S? If so, the problem has no solution. If not, then there is a point P of S that is optimal in the sense that if Q is any point of S then $f(Q) \leq f(P)$. (3) How can we find P?

The simplex algorithm (see Ficken, *The Simplex Algorithm of Linear Programming*, Holt, New York, 1961, for an elementary account), in a sense, prepares the problem before calculation in such a way that favorable answers to these questions are tentatively assumed for the given problem and can be guaranteed for the prepared problem. The calculations then reveal whether or not those assumptions are justified for the given problem. The simplex algorithm terminates automatically, yielding full information on the given problem and so-called dual problem. The dual of the general problem of linear programming is to minimize $g(u_1, \ldots, u_m) = \Sigma_{i=1}^{m} u_i b_i$ subject to $u_i \geq 0$ $(i = 1, 2, \ldots, m)$ and $\Sigma_{i=1}^{m} u_i a_{ij} \geq c_j (j = 1, 2, \ldots, n)$. Let A be the matrix $[a_{ij}]$, $c = [c_j]$, $U = [u_i]$ be row vectors, and $B = [b_i]^T$, $X = [x_j]^T$ be column vectors. In matrix form the original (primal) problem is to maximize $f(X) = CX$ subject to $X \geq 0$, $AX \leq B$. The dual is to minimize $g(U) = UB$ subject to $U \geq 0$, $UA \geq C$.

Example Maximize $3x + 4y$ subject to the constraints $x \geq 0$, $y \geq 0$, $2x + 5y \leq 8$ and $4x + 3y \leq 10$. The dual problem is to minimize $8u + 10v$ subject to the constraints $u \geq 0$, $v \geq 0$, $2u + 4v \geq 3$, and $5u + 3v \geq 4$.

Example (From Symonds, *Linear Programming: The Solution of Refinery Problems*, Esso Standard Oil Co., New York, 1955). Maximize $5P + 4V - 3F$ subject to $P \geq 0$, $V \geq 0$, $F \geq 0$ and $2P + V - 2F \leq 0$, $P + V - 3F \leq 0$, $P + V \leq 1000$.

Simplex Method

1. *Original problem.* Let the column vector $[x_j]^T = x$ $(j = 1, 2, \ldots, n)$ and the row vector $[c_j] = c$. To maximize $f(x) = \Sigma_{j=1}^{n} c_j x_j = cx$ subject to the n constraints $x_j \geq 0$ $(j = 1, \ldots, n)$ and m further constraints h_i: $\Sigma_{j=1}^{n} a_{ij} x_j \,{}^\circ{}_i\, b_i$ $(i = 1, 2, \ldots, m)$where ${}^\circ{}_i$ can be $\geq$ = or $\leq$. If any $b_i < 0$, multiply h_i by -1; thus we may assume $b_i \geq 0$. We suppose the m constraints have been arranged so that ${}^\circ{}_i$ is $\geq$ for $i = 1, \ldots, g$; ${}^\circ{}_i$ is = for $i = g + 1, \ldots, g + e$; ${}^\circ{}_i$ is $\leq$ for $i = g + e + 1, \ldots, g + e + l = m$.

2. *Adjusted original problem.* Introduce $m + g$ further variables with associated constraints and coefficients for use in f. Thus, replacing j by $j + m$, f becomes $f(x) = \Sigma_{j=m+1}^{m+n} c_j x_j$ and constraints $x_j \geq 0$ and h_i: $\Sigma_{j=m+1}^{m+n} a_{ij} x_j \,{}^\circ{}_i\, b_i$, $i = 1, \ldots, m$.

3. *Prepared problem.* For $i = 1, \ldots, g$ replace h_i by H_i: $x_i + \Sigma_{j=m+1}^{m+n} a_{ij} x_j\, x_{m+n+i} = b_i$, define $c_i = -M$($M > 0$ and "large") and $C_{m+n+i} = 0$, and add the constraints $x_i \geq 0$, $x_{m+n+i} \geq 0$. For $i = g + e + 1, \ldots, m$ replace h_i by H_i. $x_i + \Sigma_{j=m+1}^{m+n} a_{ij} x_j = b_i$, define $c_i = 0$, and adjoin $x_i \geq 0$. Let J run from 1 to $N = n + m + g$; put $X = [x_J]^T$ and $j = m + 1, \ldots, m + n$ The new function to be maximized is $f(X) = \Sigma_{J=1}^{N} c_J \mathbf{X}_J$. Actually this is $f(X) = -M \Sigma_{i=1}^{g+e} x_i + \Sigma_{j=m+1}^{vm+n} c_j x_j$, for all other coefficients are zero. Thus for $J = g + e + 1, \ldots, m$ and $m + n + 1, \ldots, N$ the variables x_j make no contribution to f. They are called **slack variables,** since they take up the slack permitted by the inequalities ($\leq$ and $\geq$) in h_i. Any variable x_i, $i = 1, \ldots, g + e$ whose value is not zero gives rise to a large negative term $-Mx_i$. Such a term will keep $f(X)$ less than it would be with that $x_i = 0$. The effect of $c_i = -M(i = 1, \ldots, g + e)$ is to make it likely that the optimal solution will have the **artificial variables** $\mathbf{x_i} = 0(\mathbf{i} = 1, \ldots, \mathbf{g} + \mathbf{e})$.

The prepared problem now has the form—maximize $f(X) = \Sigma_{J=1}^{N} c_J x_J$ subject to $x_J \geq 0$ and H_i: $\Sigma_{J=1}^{N} a_{iJ} x_J = (i = 1, \ldots, m)$, where $b_i \geq 0$, $a_{i\beta} = \delta_{i\beta} = \begin{cases} 0i \neq \beta \\ 1i \neq \beta \end{cases}$ $(\beta = 1, \ldots, m)$, $a_{i,m+n+\beta} = -\delta_{i\beta}$ $(\beta = 1, \ldots, g)$, and a_{ij} came from h_i.

The set of feasible points S_P (points satisfying all constraints) for the prepared problem is not empty, and $f(X)$ is bounded above on S_P.

Example We shall use the preceding example for this and in the following discussion. Here $m = 3$. The numbers $b_1 = 0$, $b_2 = 0$, and $b_3 = 1000$ are all ≥ 0. Neither ≥ 0 nor $= 0$ occurs in the constraints; so g and e are both zero, and there is no need to reorder the constraints. The adjusted original problem, arising from a renaming of the variables as $P = x_4$, $V = x_5$, and $F = x_6$, becomes—maximize $f(x) = 5x_4 + 4x_5 - 3x_6$ subject to $x_j \geq 0$ (j = 4, 5, 6), $2x_4 + x_5 - 2x_6 \leq 0$, $x_4 + x_5 - 3x_6 \leq 0$, $x_4 + x_5 \leq 1000$.

To obtain the prepared problem, we note that each constraint contains $\leq$ and three slack variables x_1, x_2, and x_3 are needed together with coefficients $c_1 = c_2 = c_3 = 0$ and constraints $x_1 \geq 0$, $x_2 \geq 0$, and $x_3 \geq 0$. Thus the prepared problem is—maximize $5x_4 + 4x_5 - 3x_6$ subject to $x_J \geq 0$ (J = 1, . . . , 6), $x_1 + 2x_4 + x_5 - 2x_6 = 0$, $x_2 + x_4 + x_5 - 3x_6 = 0$, and $x_3 + x_4 + x_5 = 1000$.

Computational Tableaux To start the simplex procedure we arrange the problem matrix in a tableau (array) of the form

			$c_1 \cdots\cdots$	$c_J \cdots\cdots$	c_N
C_Q	A_Q	B	$A_1 \cdots\cdots$	$A_J \cdots\cdots$	A_N
c_{q1}	A_{q1}	x_{q1}	a_{11}	a_{1J}	a_{1N}
$\vdots$	$\vdots$	$\vdots$	$\vdots$	$\vdots$	$\vdots$
c_{qm}	A_{qm}	x_{qm}	a_{m1}	a_{mJ}	a_{mN}
	z_J	C_QB	$C_QA_1 \cdots$	C_QA_J	C_QA_N
$z_J - c_J$			$z_1 - c_1 \cdots$	$z_J - c_J \cdots$	$z_N - c_N$

where $Q = (q_i)$ (i = 1, . . . , m) denotes a set of m distinct indices $q_1, \ldots, q_m$ chosen from 1 to N. The m vectors $\mathbf{A}_Q$ will always be independent and therefore a basis for the space of m-component column vectors; they determine the entire array. The numbers a_{iJ} are the components of A_J with respect to $\mathbf{A}_Q$; that is, $A_J = \Sigma_{i=1}^{m} A_{qi}a_{iJ}$. For i, β = 1, . . . , m, $a_{iq\beta} = \delta_{qi\alpha}$ and also for the q_βth column has 1 in the β row and zero elsewhere. With $x_J = 0$ for $J \epsilon R$ (the residual set, J not in Q), the X with x_{qi} in the B column has $B = \Sigma_i A_{qi}x_{qi}$ and $x_{qi} \geq 0$. Since the vectors $\mathbf{A}_Q$ are independent, X is not only feasible but extreme. Further $f(X) = \Sigma_i c_{qi}x_{qi} = C_QX$.

The first table follows directly from the prepared problem. The set Q is 1, . . . , m, $q_i = i$, $x_{qi} = b_i$, and $c_{qi} = -M$ for i = 1, . . . , $g + e$, while $c_{qi} = 0$ for $i = g + e + 1, \ldots, m$ and $f(X) = -M \Sigma_{i=1}^{g+e} b_i$. Here and later the M's in z_j and $z_j - c_j$ may be carried along separately and $z_{qi} = c_{qi}$ (i = 1, . . . , m).

A necessary and sufficient condition for the optimality of **X**: If $z_J - c_J \geq 0$ (J = 1, . . . , N) in a tableau, then the vector **X** determined by x_{qi}, with $x_J = 0$ ($J \epsilon R$), is optimal.

Once an initial computational tableau has been constructed, the simplex algorithm calls for the successive application (iteration) of the following steps:

1. Test the $z_J - c_J$ elements to determine whether a maximum solution has been found, i.e., whether $z_J - c_J \geq 0$ for all J.
2. Select the vector to be introduced (into the basis) if some $z_J - c_J < 0$; i.e., select the vector with minimum $z_J - c_J$.
3. Select the vector to be eliminated (from the basis) to ensure feasibility of the new solution, i.e., the vector with min (x_{qi}/a_{ik}) for those $a_{ik} > 0$ (k corresponds to the vector of step 2). If all $a_{ik} \leq 0$, the solution is unbounded.
4. Transform the tableau by an elimination procedure (see the references cited).

Example The tableau (two steps) for the preceding example, is given here.

Tableau

		c_j		0	0	0	5	4	−3	
C_Q		A_Q	B	A_1	A_2	A_3	A_4	A_5	A_6	
0		A_1	0	1	0	0	2	1	−2	$\min(z_J - c_J) = -5$; so
0	out	A_2	0	0	1	0	1	1	−3	A_4 in; min $B/A_4^+ = 0$
0		A_3	1000	0	0	1	1	1	0	twice; A_2 out
		z_J	0	0	0	0	0	0	0	
$z_J - c_J$				0	0	0	−5	−4	3	
0	out	A_1	0	1	−2	0	0	−1	4	
5		A_4	0	0	1	0	1	1	−3	
0		A_3	1000	0	−1	1	0	0	3	
		z_J	0	0	5	0	5	5	−15	
$z_J - c_J$ etc.				0	5	0	0	1	−12	

STATISTICS

GENERAL REFERENCES: *Probability*. Parsen, *Modern Probability Theory and Its Applications*, Wiley, New York, 1960. Uspensky, *Introduction to Mathematical Probability*, McGraw-Hill, New York, 1937. *General statistics*. Hald, *Statistical Theory with Engineering Applications*, Wiley, New York, 1952. Hamburg, *Statistical Analysis for Decision Making*, 2d ed., Harcourt, New York, 1977. Lapin, *Statistics for Modern Business Decisions*, 2d ed., Harcourt, New York, 1978. *Mathematical statistics*. Kendall and Stuart, *The Advanced Theory of Statistics*, vol. 1, 4th ed., 1977, vol. 2, 4th ed., 1979, vol. 3, 3d ed., 1976, Hafner, New York. Mood, Graybill, and Boes, *Introduction to the Theory of Statistics*, 3d ed., McGraw-Hill, New York, 1974. *Regression*. Chatterjee and Price, *Regression Analysis by Example*, Wiley, New York, 1977. Draper and Smith, *Applied Regression Analysis*, 2d ed., Wiley, New York, 1981. Weisberg, *Applied Linear Regression*, Wiley, New York, 1980. *Design*. Box, Hunter, and Hunter, *Statistics for Experimenters*, Wiley, New York, 1978. Hicks, *Fundamental Concepts in the Design of Experiments*, Holt, New York, 1965.

INTRODUCTION

Statistics represents a body of knowledge which enables one to deal with quantitative data reflecting any degree of uncertainty. There are six basic aspects of applied statistics. These are:

1. Type of data
2. Random variables
3. Models
4. Parameters
5. Sample statistics
6. Characterization of chance occurrences

From these can be developed strategies and procedures for dealing with (1) estimation and (2) inferential statistics. The following has been directed more toward inferential statistics because of its broader utility.

Detailed illustrations and examples are used throughout to develop basic statistical methodology for dealing with a broad area of applications. However, in addition to this material, there are many specialized topics as well as some very subtle areas which have not been discussed. The references should be used for more detailed information. It is hoped that through a study of this material potential users will be able to recognize the broader implications of specialized applications and request assistance when it is required.

The following represents a simplified introduction to the aspects and utility of statistical methodology that will serve as a broad overview of the concepts and strategy of statistics.

Type of Data In general, statistics deals with two types of data: counts and measurements. Counts represent the number of discrete outcomes, such as the number of defective parts in a shipment, the number of lost-time accidents, and so forth. Measurement data are treated as a continuum. For example, the tensile strength of a synthetic yarn theoretically could be measured to any degree of precision. A subtle aspect associated with count and measurement data is that some types of count data can be dealt with through the application of techniques which have been developed for measurement data alone. This ability is due to the fact that some simplified measurement statistics serve as an excellent approximation for the more tedious count statistics. This will be illustrated.

Random Variables Applied statistics deals with quantitative data. In tossing a fair coin the successive outcomes would tend to be different, with heads and tails occurring randomly over a period of time. Given a long strand of synthetic fiber, the tensile strength of successive samples would tend to vary significantly from sample to sample.

Counts and measurements are characterized as random variables, that is, observations which are susceptible to chance. Virtually all quantitative data are susceptible to chance in one way or another.

Models Part of the foundation of statistics consists of the mathematical models which characterize an experiment. The models themselves are mathematical ways of describing the probability, or relative likelihood, of observing specified values of random variables. For example, in tossing a coin once, a random variable x could be defined by assigning to x the value 1 for a head and 0 for a tail. Given a fair coin, the probability of observing a head on a toss would be a .5, and similarly for a tail. Therefore, the mathematical model governing this experiment can be written as

x	$P(x)$
0	.5
1	.5

where $P(x)$ stands for what is called a **probability function.** This term is reserved for count data, in that probabilities can be defined for particular outcomes.

The probability function that has been displayed is a very special case of the more general case, which is called the **binomial probability distribution.**

For measurement data which are considered continuous, the term "relative probability" or "density" is used. For example, consider a spinner wheel which conceptually can be thought of as being marked off on the circumference infinitely precisely from 0 up to, but not including, 1. In spinning the wheel, the probability of the wheel's stopping at a specified marking point at any particular x value, where $0 \le x < 1$, is zero, for example, stopping at the value $x = \sqrt{.5}$. For the spinning wheel, the model or density function would be defined by $f(x) = 1$ for $0 \le x < 1$. Graphically, this is shown in Fig. 2-61. The relative-probability concept refers to the fact that density reflects the relative likelihood of occurrence; in this case, each number between 0 and 1 is equally likely.

For measurement data, probability is defined by the area under the curve between specified limits. A density function always must have a total area of 1.

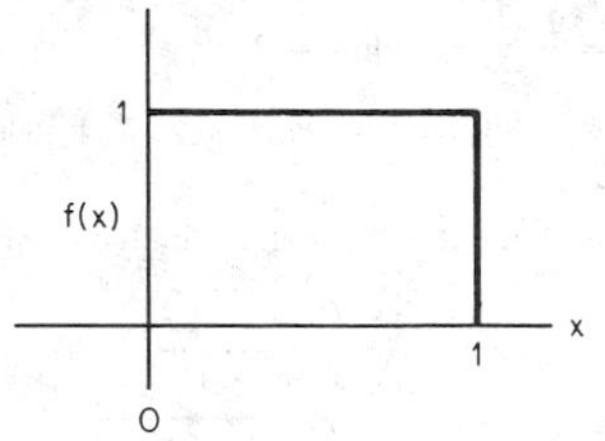

FIG. 2-61 Density function.

Example For the density of Fig. 2-61 the

$$P[0 \le x \le .4] = .4$$
$$P[.2 \le x \le .9] = .7$$
$$P[.6 \le x < 1] = .4$$

and so forth. Since the probability associated with any particular point value is zero, it makes no difference whether the limit point is defined by a closed interval ($\le$ or $\ge$) or an open interval ($<$ or $>$).

Many different types of models are used as the foundation for statistical analysis. These models are also referred to as **populations.**

Parameters As a way of characterizing probability functions and densities, certain types of quantities called parameters can be defined. For example, the center of gravity of the distribution is defined to be the **population mean,** which is designated as μ. For the coin toss $\mu = .5$, which corresponds to the average value of x; i.e., for half of the time x will take on a value 0 and for the other half a value 1. The average would be .5. For the spinning wheel, the average value would also be .5.

Another parameter is called the **standard deviation,** which is designated as σ. The square of the standard deviation is used frequently and is called the **popular variance,** σ^2. Basically, the standard deviation is a quantity which measures the spread or dispersion of the distribution from its mean μ. If the spread is broad, then the standard deviation will be larger than if it were more constrained. In mechanics the variance is the moment of inertia about a line normal to the x axis through the center of gravity μ.

For specified probability and density functions, the respective means and variances are defined by the following:

Probability functions	Density functions
$E(x) = \mu = \sum_x x\,P(x)$	$E(x) = \mu = \int_x x\,f(x)\,dx$
$\text{Var}(x) = \sigma^2 = \sum_x (x-\mu)^2\,P(x)$	$\text{Var}(x) = \sigma^2 = \int_x (x-\mu)^2 f(x)\,dx$

where $E(x)$ is defined to be the expected or average value of x.

Sample Statistics Many types of sample statistics will be defined. Two very special types are the **sample mean,** designated as $\bar{x}$, and the **sample standard deviation,** designated as s. These are, by definition, random variables. Parameters like μ and σ are not random variables; they are fixed constants.

Example In an experiment, six random numbers (rounded to four decimal places) were observed from the uniform distribution $f(x) = 1$ for $0 \le x < 1$:

.1009
.3754
.0842
.9901
.1280
.6606

The sample mean corresponds to the arithmetic average of the observations, which will be designated as x_1 through x_6, where

$$\bar{x} = \frac{1}{n}\sum_{i=1}^{n} x_i \text{ with } n = 6$$
$$= (1/6)(.1009 + .3754 + \cdots + .6606)$$
$$= .3899$$

The sample standard deviation s is defined by the computation

$$s = \sqrt{\frac{\Sigma(x_i - \bar{x})^2}{n-1}}$$
$$= \sqrt{\frac{n\Sigma x_i^2 - (\Sigma x_i)^2}{n(n-1)}}$$

In effect, this represents the root of a statistical average of the squares. The devisor quantity $(n - 1)$ will be referred to as the **degrees of freedom.**

The sample value of the standard deviation for the data given is .3686.

The following is a tabulation of the deviations $(x_i - \bar{x}_j)$ for the data:

x	$x - \bar{x}$
.1009	−.2890
.3754	−.0145
.0842	−.3057
.9901	.6002
.1280	−.2619
.6606	.2707
$\bar{x}$ = .3899	s = .3686

In effect, the standard deviation quantifies the relative magnitude of the deviation numbers, i.e., a special type of "average" of the distance of points from their center. In statistical theory, it turns out that the corresponding variance quantities s^2 have remarkable properties which make possible broad generalities for sample statistics and therefore also their counterparts, the standard deviations.

For the corresponding population, the parameter values are μ = .50 and σ = .2887. If instead of using individual observations only averages of six were reported, then the corresponding population parameter values would be μ = .50 and $\sigma_{\bar{x}} = \sigma/\sqrt{6} = .1179$. The corresponding variance for an average will be written occasionally as Var $(\bar{x})$ = var $(x)/n$. In effect, the variance of an average is inversely proportional to the sample size n, which reflects the fact that sample averages will tend to cluster about μ much more closely than individual observations. This is illustrated in greater detail under "Measurement Data and Sampling Densities."

Characterization of Chance Occurrences To deal with a broad area of statistical applications, it is necessary to characterize the way in which random variables will vary by chance alone. The basic foundation for this characteristic is laid through a density called the gaussian, or normal, distribution.

Determining the area under the normal curve is a very tedious procedure. However, by standardizing a random variable that is normally distributed, it is possible to relate all normally distributed random variables to one table. The standardization is defined by the identity $z = (x - \mu)/\sigma$, where z is called the **unit normal.** Further, it is possible to standardize the sampling distribution of averages $\bar{x}$ by the identity $z = (\bar{x} - \mu)/(\sigma/\sqrt{n})$.

A remarkable property of the normal distribution is that, almost regardless of the distribution of x, sample averages $\bar{x}$ will approach the gaussian distribution as n gets large. Even for relatively small values of n, of about 10, the approximation in most cases is quite close. For example, sample averages of size 10 from the uniform distribution will have essentially a gaussian distribution.

Also, in many applications involving count data, the normal distribution can be used as a close approximation. In particular, the approximation is quite close for the binomial distribution within certain guidelines.

ENUMERATION DATA AND PROBABILITY DISTRIBUTIONS

Introduction Many types of statistical applications are characterized by enumeration data in the form of counts. Examples are the number of lost-time accidents in a plant, the number of defective items in a sample, and the number of items in a sample that fall within several specified categories.

The sampling distribution of count data can be characterized through probability distributions. In many cases, count data are appropriately interpreted through their corresponding distributions. However, in other situations analysis is greatly facilitated through distributions which have been developed for measurement data. Examples of each will be illustrated in the following subsections.

Binomial Probability Distribution

Nature Consider an experiment in which each outcome is classified into one of two categories, one of which will be defined as a success and the other as a failure. Given that the probability of success p is constant from trial to trial, then the probability of observing a specified number of successes x in n trials is defined by the binomial distribution. The sequence of outcomes is called a **Bernoulli process.**

Nomenclature

n = total number of trials
x = number of successes in n trials
p = probability of observing a success on any one trial
$\hat{p} = x/n$, the proportion of successes in n trials

Probability Law

$$P(x) = P\left(\frac{x}{n}\right) = \binom{n}{x} p^x(1-p)^{n-x} \qquad x = 0, 1, 2, \ldots, n$$

where $\binom{n}{x} = \dfrac{n!}{x!(n-x)!}$.

Properties

$$E(x) = np \qquad \text{Var}(x) = np(1-p)$$
$$E(\hat{p}) = p \qquad \text{Var}(\hat{p}) = p(1-p)/n$$

Example For a very special coin, the probability of observing a head (the defined success) on a single toss is 0.4. What is the probability distribution for the number of heads in five tosses of the coin?

x	$x/5$	$P(x)$	
0	0	.07776	$E(x)$ = 2.0
1	.2	.25920	Var(x) = 1.2
2	.4	.34560	
3	.6	.23040	$E(x/n)$ = 0.4
4	.8	.07680	Var(x/n) = 0.048
5	1.0	.01024	
		1.	

Example A process has been producing defective items at a rate of 6 percent. If 20 items are selected randomly during the day, what is the probability that at most 2 will be defective?

$$P(x \le 2) = P(x = 0) + P(x = 1) + P(x = 2)$$

x	$P(x)$
0	.29011
1	.37035
2	.22457
	.88503, or roughly an 88.5 percent chance

Geometric Probability Distribution

Nature Consider an experiment in which each outcome is classified into one of two categories, one of which will be defined as a success and the other as a failure. Given that the probability of success p is constant from trial to trial, then the probability of observing the first success on the xth trial is defined by the geometric distribution.

Nomenclature

p = probability of observing a success on any one trial
x = the number of trials under the first success

Probability Law

$$P(x) = p(1-p)^{x-1} \qquad x = 1, 2, 3, \ldots$$

Properties

$$E(x) = 1/p \qquad \text{Var}(x) = (1 - p)/p^2$$

Example What is the probability distribution for the number of tosses until the first head is observed?

x	$P(x)$
1	1/2
2	1/4
3	1/8
.	.
.	.
.	.
x	$(1/2)^x$
.	.
.	.
.	.

Example What is the probability in roulette that red will not occur on the next three spins of the wheel? Since

$$p = 18/38 \text{ (36 numbers plus 0, 00, of which 18 are red)}$$
$$P(x \le 4) = 1 - P(x = 1) - P(x = 2) - P(x = 3)$$
$$= 1 - (18/38) - (18/38)(20/38) - (18/38)(20/38)(20/38)$$
$$= .1458$$

This can also be determined by the binomial where $p = 18/38$ and the probability of no successes is computed from

$$P(x = 0) = \binom{3}{0}\left(\frac{18}{38}\right)^0\left(\frac{20}{38}\right)^3 = .1458$$

Poisson Probability Distribution

Nature In monitoring a moving threadline, one criterion of quality would be the frequency of broken filaments. These can be identified as they occur through the threadline by a broken-filament detector mounted adjacent to the threadline. In this context, the random occurrences of broken filaments can be modeled by the Poisson distribution. This is called a Poisson process and corresponds to a probabilistic description of the frequency of defects or, in general, what are called arrivals at points on a continuous line or in time. Other examples include:

1. The number of cars (arrivals) that pass a point on a high-speed highway between 10:00 and 11:00 A.M. on Wednesdays
2. The number of customers arriving at a bank between 10:00 and 10:10 A.M.
3. The number of telephone calls received through a switchboard between 9:00 and 10:00 A.M.
4. The number of insurance claims that are filed each week
5. The number of spinning machines that break down during 1 day at a large plant.

Nomenclature

x = total number of arrivals in a total length L or total period T

a = average rate of arrivals for a unit length or unit time

$\lambda = aL$ = expected or average number of arrivals for the total length L

$\lambda = aT$ = expected or average number of arrivals for the total time T

Probability Law Given that a is constant for the total length L or period T, the probability of observing x arrivals in some period L or T is given by

$$P(x) = \frac{\lambda^x}{x!}e^{-\lambda} \qquad x = 0, 1, 2, \ldots$$

Properties

$$E(x) = \lambda \qquad \text{Var}(x) = \lambda$$

Example The number of broken filaments in a threadline has been averaging .015 per yard. What is the probability of observing exactly two broken filaments in the next 100 yd? In this example, $a = .015$/yd and $L = 100$ yd; therefore $\lambda = (.015)(100) = 1.5$:

$$P(x = 2) = \frac{(1.5)^2}{2!}e^{-1.5} = .2510$$

Example A commercial item is sold in a retail outlet as a unit product. In the past, sales have averaged 10 units per month with no seasonal variation. The retail outlet must order replacement items 2 months in advance. If the outlet starts the next 2-month period with 25 items on hand, what is the probability that it will stock out before the end of the second month?

Given $a = 10$/month, then $\lambda = 10 \times 2 = 20$ for the total period of 2 months:

$$P(x \ge 26) = \sum_{26}^{\infty} P(x) = 1 - \sum_{0}^{25} P(x)$$

$$\sum_{0}^{25} \frac{20^x}{x!}e^{-20} = e^{-20}\left[1 + \frac{20}{1} + \frac{20^2}{2!} + \cdots + \frac{20^{25}}{25!}\right]$$

$$= .887815$$

Therefore $P(x \ge 26) = .112185$, or roughly an 11 percent chance of a stockout.

Example For a Bernoulli process, it can be shown that as n gets large and p gets small in such a way that $\lambda = np$ remains constant, the binomial distribution converges to the Poisson distribution. In other words, the Poisson distribution can be a computationally convenient approximation to the binomial distribution for large $n(\ge 100)$ and small $p(\le .10)$. For example, given $\lambda = 1.28$,

	Binomial				
x	$n = 10$ $p = .128$	40 .032	160 .008	640 .002	Poisson, $\lambda = 1.28$
0	.2542	.2723	.2766	.2777	.2780
1	.3731	.3600	.3569	.3561	.3559
2	.2465	.2321	.2288	.2280	.2278
3	.0965	.0972	.0972	.0972	.0972
4	.0248	.0297	.0308	.0310	.0311
5	.0044	.0071	.0077	.0079	.0080
6	.0005	.0014	.0016	.0017	.0017
7	0	.0002	.0003	.0003	.0003

Hypergeometric Probability Distribution

Nature In an experiment in which one samples from a relatively small group of items, each of which is classified in one of two categories, A or B, the hypergeometric distribution can be defined. One example is the probability of drawing two red and two black cards from a deck of cards. The hypergeometric distribution is the analog of the binomial distribution when successive trials are not independent, i.e., when the total group of items is not infinite.

Nomenclature

N = total group size

n = sample group size

X = number of items in the total group with a specified attribute A

$N - X$ = number of items in the total group with the other attribute B

x = number of items in the sample with a specified attribute A

$n - x$ = number of items in the sample with the other attribute B

	Population	Sample
Category A	X	x
Category B	$N - X$	$n - x$
Total	N	n

Probability Law

$$P(x) = \binom{N - X}{n - x}\binom{X}{x} \Big/ \binom{N}{n}$$

Example What is the probability that an appointed special committee of 4 has no female members when the members are randomly selected from a candidate group of 10 males and 7 females?

$$P(x = 0) = \frac{\binom{10}{4}\binom{7}{0}}{\binom{17}{4}} = .0882$$

Example A bin contains 300 items, of which 240 are good and 60 are defective. In a sample of 6 what is the probability of selecting 4 good and 2 defective items by chance?

$$P(x) = \frac{\binom{240}{4}\binom{60}{2}}{\binom{300}{6}} = .2478$$

Multinomial Distribution

Nature For an experiment in which successive outcomes can be classified into two or more categories and the probabilities associated with the respective outcomes remain constant, then the experiment can be characterized through the multinomial distribution.

Nomenclature

n = total number of trials

k = total number of distinct categories

p_j = probability of observing category j on any one trial, $j = 1, 2, \ldots, k$

x_j = total number of occurrences in category j in n trials

Probability Law

$$P(x_1, x_2, \ldots, x_k) = \frac{n!}{x_1!x_2! \ldots x_k!} p_1^{x_1} p_2^{x_2} \ldots p_k^{x_k}$$

Example In tossing a die 12 times, what is the probability that each face value will occur exactly twice?

$$p(2, 2, 2, 2, 2, 2) = \frac{12!}{2!2!2!2!2!2!}\left(\frac{1}{6}\right)^2\left(\frac{1}{6}\right)^2\left(\frac{1}{6}\right)^2\left(\frac{1}{6}\right)^2\left(\frac{1}{6}\right)^2\left(\frac{1}{6}\right)^2 = .003438$$

MEASUREMENT DATA AND SAMPLING DENSITIES

Introduction The following example data are used throughout this subsection to illustrate concepts. Consider, for the purpose of illustration, that five synthetic-yarn samples have been selected randomly from a production line and tested for tensile strength on each of 20 production days. For this, assume that each group of five corresponds to a day, Monday through Friday, for a period of 4 weeks:

Monday 1	Tuesday 2	Wednesday 3	Thursday 4	Friday 5	Groups of 25 pooled
36.48	38.06	35.28	36.34	36.73	
35.33	31.86	36.58	36.25	37.17	
35.92	33.81	38.81	30.46	33.07	
32.28	30.30	33.31	37.37	34.27	
31.61	35.27	33.88	37.52	36.94	
$\bar{x}$ = 34.32	33.86	35.57	35.59	35.64	35.00
s = 2.22	3.01	2.22	2.92	1.85	2.40
6	7	8	9	10	
38.67	36.62	35.03	35.80	36.82	
32.08	33.05	36.22	33.16	36.49	
33.79	35.43	32.71	35.19	32.83	
32.85	36.63	32.52	32.91	32.43	
35.22	31.46	27.23	35.44	34.16	
$\bar{x}$ = 34.52	34.64	32.74	34.50	34.54	34.19
s = 2.60	2.30	3.46	1.36	2.03	2.35

Monday 11	Tuesday 12	Wednesday 13	Thursday 14	Friday 15	Groups of 25 pooled
39.63	34.52	36.05	36.64	31.57	
34.38	37.39	35.36	31.18	36.21	
36.51	34.16	35.00	36.13	33.84	
30.00	35.76	33.61	37.51	35.01	
39.64	37.63	36.98	39.05	34.95	
$\bar{x}$ = 36.03	35.89	35.40	36.10	34.32	35.55
s = 4.04	1.59	1.25	2.96	1.75	2.42
16	17	18	19	20	
37.68	35.97	33.71	35.61	36.65	
36.38	35.92	32.34	37.13	37.91	
38.43	36.51	33.29	31.37	42.18	
39.07	33.89	32.81	35.89	39.25	
33.06	36.01	37.13	36.33	33.32	
$\bar{x}$ = 36.92	35.66	33.86	35.27	37.86	35.91
s = 2.38	1.02	1.90	2.25	3.27	2.52

Pooled sample of 100: $\bar{x}$ = 35.16 s = 2.47

Even if the process were at steady state, tensile strength, a key property, would still reflect some variation. Steady state, or stable operation of any process, has associated with it a characteristic variation. Superimposed on this is the testing method, which is itself a process with its own characteristic variation. The observed variation is a composite of these two variations.

Assume that the table represents "typical" production-line performance. The numbers themselves have been generated on a computer and represent random observations from a population with μ = 35 and a population standard deviation σ = 2.45. The sample values reflect the way in which tensile strength can vary by chance alone. In practice, a production supervisor unschooled in statistics but interested in high tensile performance would be despondent on the eighth day and exuberant on the twentieth day. If the supervisor were more concerned with uniformity, the lowest and highest points would have been on the eleventh and seventeenth days.

An objective of statistical analysis is to serve as a guide in decision making in the context of normal variation. In the case of the production supervisor, it is to make a decision, with a high probability of being correct, that something has in fact changed the operation.

Suppose that an engineering change has been made in the process and five new tensile samples have been tested with the results:

36.81	
38.34	$\bar{x}$ = 37.14
34.87	s = 1.85
39.58	
36.12	

In this situation, management would inquire whether the product has been improved by increased tensile strength. To answer this question, in addition to a variety of analogous questions, it is necessary to have some type of scientific basis upon which to draw a conclusion.

A scientific basis for the evaluation and interpretation of data is contained in the accompanying table descriptions. These tables characterize the way in which sample values will vary by chance alone in the context of individual observations, averages, and variances.

Table number, Sec. 1	Designated symbol	Variable	Sampling distribution of
1-15	z	$\dfrac{x-\mu}{\sigma}$	Observations*
1-15	z	$\dfrac{\bar{x}-\mu}{\sigma/\sqrt{n}}$	Averages
1-16	t	$\dfrac{\bar{x}-\mu}{s/\sqrt{n}}$	Averages when σ is unknown*
1-17	χ^2	$(s^2/\sigma^2)(\text{df})$	Variances*
1-18	F	s_1^2/s_2^2	Ratio of two independent sample variances*

*When sampling from a gaussian distribution.

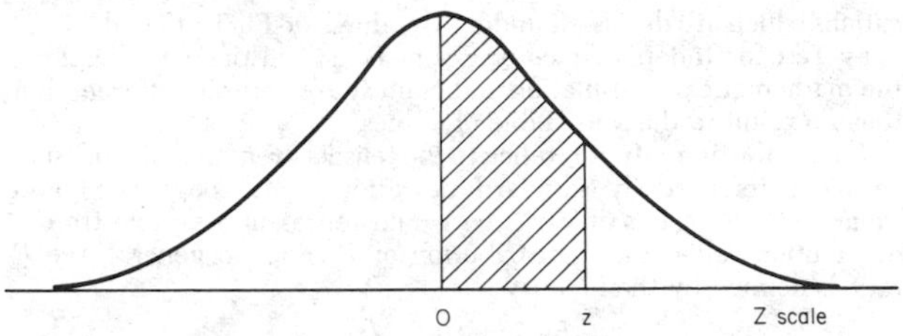

FIG. 2-62 Normal distribution.

Normal Distribution of Observations Many types of data follow what is called the gaussian, or bell-shaped, curve; this is especially true of averages. Basically, the gaussian curve is a purely mathematical function which has very special properties. However, owing to some mathematically intractable aspects primary use of the function is restricted to tabulated values.

Basically, the tabled values represent area (proportions or probability) associated with a scaling variable designated by Z in Fig. 2-62. The normal curve is centered at 0, and for particular values of Z, designated as z, the tabulated numbers represent the corresponding area under the curve between 0 and z. For example, between 0 and 1 the area is .3413. The area between 0 and 2 is .4772; therefore, the area between 1 and 2 is .4772 − .3413 = .1359.

Also, since the normal curve is symmetric, areas to the left can be determined in exactly the same way. For example, the area between − 2 and +1 would include the area between −2 and 0, .4772 (the same as 0 to 2), plus the area between 0 and 1, .3413, or a total area of .8185.

Any types of observation which are applicable to the normal curve can be transformed to Z values by the relationship $z = (x - \mu)/\sigma$ and, conversely, Z values to x values by $x = \mu + \sigma z$, as shown in Fig. 2-63. For example, for tensile strength, with $\mu = 35$ and $\sigma = 2.45$, this would dictate $z = (x - 35)/2.45$ and $x = 35 + 2.45z$.

Example What proportion of tensile values will fall between 34 and 36?

$$z_1 = (34 - 35)/2.45 = -.41 \qquad z_2 = (36 - 35)/2.45 = .41$$

$$P[-.41 \le z \le .41] = .3182, \text{ or roughly 32 percent}$$

Example What midrange of tensile values will include 95 percent of the sample values? Since $P[-1.96 \le z \le 1.96] = .95$, the corresponding values of x are

$$x_1 = 35 - 1.96(2.45) = 30.2$$

$$x_2 = 35 + 1.96(2.45) = 39.8$$

or

$$P[30.2 \le x \le 39.8] = .95$$

Normal Distribution of Averages An examination of the tensile-strength data previously tabulated would show that the range (largest minus the smallest) of tensile strength within days averages 5.72. The average range in $\bar{x}$ values within each week is 2.37, while the range in the four weekly averages is 1.72. This reflects the fact that averages tend to be less variable in a predictable way. Given that the variance of x is $\text{var}(x) = \sigma^2$, then the variance of $\bar{x}$ based on n observations is $\text{var}(\bar{x}) = \sigma^2/n$.

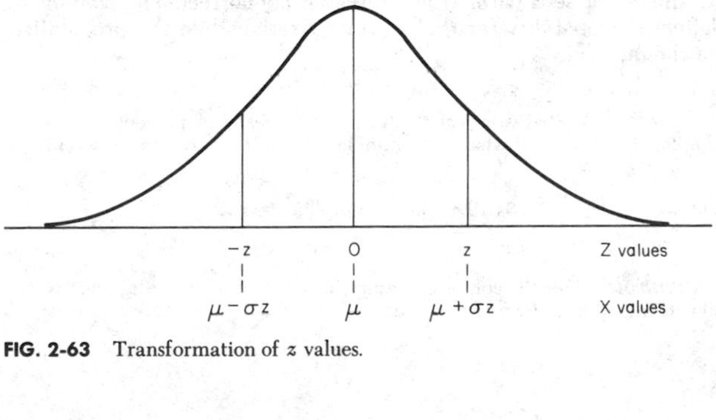

FIG. 2-63 Transformation of z values.

For averages of n observations, the corresponding relationship for the Z-scale relationship is

$$z = (\bar{x} - \mu/\sigma/\sqrt{n}) \qquad \text{or} \qquad \bar{x} = \mu + \frac{\sigma}{\sqrt{n}} z$$

Example What proportion of daily tensile averages will fall between 34 and 36?

$$z_1 = (34 - 35)/(2.45/\sqrt{5}) = -.91 \qquad z_2 = (36 - 35)/(2.45/\sqrt{5}) = .91$$

$$P[-.91 \le z \le .91] = .6372, \text{ or roughly 64 percent}$$

Example What midrange of daily tensile averages will include 95 percent of the sample values?

$$x_1 = 35 - 1.96(2.45/\sqrt{5}) = 32.85$$

$$x_2 = 35 + 1.96(2.45/\sqrt{5}) = 37.15$$

or

$$P[32.85 \le \bar{x} \le 37.15] = .95$$

Example What proportion of weekly tensile averages will fall between 34 and 36?

$$z_1 = (34 - 35)/(2.45/\sqrt{25}) = -2.04$$

$$z_2 = (36 - 35)/(2.45/\sqrt{25}) = 2.04$$

$$P[-2.04 \le z \le 2.04] = .9586, \text{ or roughly 96 percent}$$

Distribution of Averages The normal curve relies on a knowledge of σ, or in special cases, when it is unknown, s can be used with the normal curve as an approximation when $n > 30$. For example, with $n > 30$ the intervals $\bar{x} \pm s$ and $\bar{x} \pm 2s$ will include roughly 68 and 95 percent of the sample values respectively when the distribution is normal.

In applications sample sizes are usually small and σ unknown. In these cases, the t distribution can be used where

$$t = (\bar{x} - \mu)/(s/\sqrt{n}) \qquad \text{or} \qquad \bar{x} = \mu + ts/\sqrt{n}$$

The t distribution is also symmetric and centered at zero. It is said to be robust in the sense that even when the individual observations x are not normally distributed, sample averages of x have distributions which tend toward normality as n gets large. Even for small n of 5 through 10, the approximation is usually relatively accurate.

In reference to the tensile-strength table, consider the summary statistics $\bar{x}$ and s by days. For each day, the t statistic could be computed. If this were repeated over an extensive simulation and the resultant t quantities plotted in a frequency distribution, they would match the corresponding distribution of t values summarized in Table 1-16 in Sec. 1.

Since the t distribution relies on the sample standard deivation s, the resultant distribution will differ acording to the sample size n. To designate this difference, the respective distributions are classified according to what are called the degrees of freedom and abbreviated as df. In simple problems, the df are just the sample size minus 1. In more complicated applications the df can be different. In general, degrees of freedom are the number of quantities minus the number of constraints. For example, four numbers in a square which must have row and column sums equal to zero have only one df, i.e., four numbers minus three constraints (the fourth constraint is redundant). For example, pick a, then

$$\begin{array}{cc} a & -a = 0 \\ \underline{-a} & \underline{a} = 0 \\ 0 & 0 \end{array}$$

Example For a sample size $n = 5$, what values of t define a midarea of 90 percent. For 4 df the tabled value of t corresponding to a midarea of 90 percent is 2.132; i.e., $P[-2.132 \le t \le 2.132] = .90$.

Example For a sample size $n = 25$, what values of t define a midarea of 95 percent? For 24 df the tabled value of t corresponding to a midarea of 95 percent is 2.064; i.e., $P[-2.064 \le t \le 2.064] = .95$.

Example What is the sample value of t for the first day of tensile data?

$$\text{Sample } t = (34.32 - 35)/(2.22/\sqrt{5}) = -.68$$

Note that on the average 90 percent of all such sample values would be expected to fall within the interval ± 2.132.

t Distribution for the Difference in Two Sample Means

Population Variances Are Equal The t distribution can be readily extended to the difference in two sample means when the respective populations have the same variance σ^2:

$$t = \frac{(\bar{x}_1 - \bar{x}_2) - (\mu_1 - \mu_2)}{s_p \sqrt{1/n_1 + 1/n_2}}$$

where s_p^2 is a pooled variance defined by

$$s_p^2 = \frac{(n_1 - 1)s_1^2 + (n_2 - 1)s_2^2}{(n_1 - 1) + (n_2 - 1)}$$

In this application, the t distribution has $(n_1 + n_2 - 2)$ df.

Population Variances Are Unequal When population variances are unequal, an approximate t quantity can be used:

$$t = \frac{(\bar{x}_1 - \bar{x}_2) - (\mu_1 - \mu_2)}{\sqrt{a + b}}$$

with $a = s_1^2/n_1 \qquad b = s_2^2/n_2$

and
$$\text{df} = \frac{(a + b)^2}{a^2/(n_1 - 1) + b^2/(n_2 - 1)}$$

Chi-Square Distribution For some industrial applications, product uniformity is of primary importance. The sample standard deviation s is most often used to characterize uniformity. In dealing with this problem, the chi-square distribution can be used where $\chi^2 = (s^2/\sigma^2)$ (df). The chi-square distribution is a family of distributions which are defined by the degrees of freedom associated with the sample variance. For most applications, df is equal to the sample size minus 1.

In terms of the tensile-strength table previously given, the respective chi-square sample values for the daily, weekly, and monthly figures could be computed. The corresponding df would be 4, 24, and 99 respectively. These numbers would represent sample values from the respective distributions which are summarized in Table 1-17 in Sec. 1.

In a manner similar to the use of the t distribution, chi square can be interpreted in a direct probabilistic sense corresponding to a midarea of $(1 - \alpha)$:

$$P[\chi_1^2 \le (s^2/\sigma^2)(\text{df}) \le \chi_2^2] = 1 - \alpha$$

where χ_1^2 corresponds to a lower-tail area of $\alpha/2$ and χ_2^2 an upper-tail area of $\alpha/2$.

The basic underlying assumption for the mathematical derivation of chi square is that a random sample was selected from a normal distribution with variance σ^2. When the population is not normal but skewed, square probabilities could be substantially in error.

Example On the basis of a sample size $n = 5$, what midrange of values will include the sample ratio s/σ with a probability of 90 percent?

$$P[.484 \le (s^2/\sigma^2)(4) \le 11.1] = .90$$

or
$$P[.35 \le s/\sigma \le 6.66] = .90$$

Example On the basis of a sample size $n = 25$, what midrange of values will include the sample ratio s/σ with a probability of 90 percent?

$$P[12.4 \le (s^2/\sigma^2)(24) \le 39.4] = .90$$

or
$$P[.72 \le s/\sigma \le 1.28] = .90$$

This states that the sample standard deviation will be at least 72 percent and not more than 128 percent of the population variance 90 percent of the time. Conversely, 10 percent of the time the standard deviation will underestimate or overestimate the population standard deviation by the corresponding amount. Even for samples as large as 25, the relative reliability of a sample standard deviation is poor.

The chi-square distribution can be applied to other types of application which are of an entirely different nature. These include applications which are discussed under "Goodness-of-Fit Test" and "Two-Way Test for Independence of Count Data." In these applications, the mathematical formulation and context are entirely different, but they do result in the same table of values.

***F* Distribution** In reference to the tensile-strength table, the successive pairs of daily standard deviations could be ratioed and squared. These ratios of variance would represent a sample from a distribution called the F distribution or F ratio. In general, the F ratio is defined by the identity

$$F(\gamma_1, \gamma_2) = s_1^2/s_2^2$$

where γ_1 and γ_2 correspond to the respective df's for the sample variances. In statistical applications, it turns out that the primary area of interest is found when the ratios are greater than 1. For this reason, most tabled values are defined for an upper-tail area. However, defining F_2 to be that value corresponding to an upper-tail area of $\alpha/2$, then F_1 for a lower-tail area of $\alpha/2$ can be determined through the identity

$$F_1(\gamma_1, \gamma_2) = 1/F_2(\gamma_2, \gamma_1)$$

The F distribution, similar to the chi square, is sensitive to the basic assumption that sample values were selected randomly from a normal distribution

Example For two sample variances with 4 df each, what limits will bracket their ratio with a midarea probability of 90 percent?

$$P[1/6.39 \le s_1^2/s_2^2 \le 6.39] = .90$$

or
$$P[.40 \le s_1/s_2 \le 2.53] = .90$$

Confidence Interval for a Mean For the daily sample tensile-strength data with 4 df it is known that $P[-2.132 \le t \le 2.132] = .90$. This states that 90 percent of all samples will have sample t values which fall within the specified limits. In fact, for the 20 daily samples exactly 16 do fall within the specified limits (note that the binomial with $n = 20$ and $p = .90$ would describe the likelihood of exactly none through 20 falling within the prescribed limits—the sample of 20 is only a sample).

Consider the new daily sample (with $n = 5$, $\bar{x} = 37.14$, and $s = 1.85$) which was observed after a process change. In this case, the same probability holds. However, in this instance the sample value of t cannot be computed, since the new μ, under the process change, is not known. Therefore $P[-2.132 \le (37.14 - \mu)/(1.85/\sqrt{5}) \le 2.132] = .90$. In effect, this identity limits the magnitude of possible values for μ. The magnitude of μ can be only large enough to retain the t quantity above -2.132 and small enough to retain the t quantity below $+2.132$. This can be found by rearranging the quantities within the bracket; i.e., $P[35.78 \le \mu \le 38.90] = .90$. This states that we are 90 percent sure that the interval from 35.78 to 38.90 includes the unknown parameter μ.

In general,

$$P\left[\bar{x} - t\frac{s}{\sqrt{n}} \le \mu \le \bar{x} + t\frac{s}{\sqrt{n}}\right] = 1 - \alpha$$

where t is defined for an upper-tail area of $\alpha/2$ with $(n - 1)$ df. In this application, the interval limits $(\bar{x} + t\, s/\sqrt{n})$ are random variables which will cover the unknown parameter μ with probability $(1 - \alpha)$. The converse, that we are $100\,(1 - \alpha)$ percent sure that the parameter value is within the interval, is not correct. This statement defines a probability for the parameter rather than the probability for the interval.

Example What values of t define the midarea of 95 percent for weekly samples of size 25, and what is the sample value of t for the second week?

$$P[-2.064 \le t \le 2.064] = .95$$

and
$$(34.19 - 35)/(2.35/\sqrt{25}) = 1.72.$$

Example For the composite sample of 100 tensile strengths, what is the 90 percent confidence interval for μ?

$$P\left[35.16 - 1.645\frac{2.47}{\sqrt{100}} < \mu < 35.16 + 1.645\frac{2.47}{\sqrt{100}}\right] = .90$$

or $$P[34.75 \le \mu \le 35.57] = .90$$

Confidence Interval for the Difference in Two Population Means The confidence interval for a mean can be extended to include the difference between two population means. This interval is based on the assumption that the respective populations have the same variance σ^2:

$$(\bar{x}_1 - \bar{x}_2) - ts_p\sqrt{1/n_1 + 1/n_2} \le \mu_1 - \mu_2 \le (\bar{x}_1 - \bar{x}_2) + ts_p\sqrt{1/n_1 + 1/n_2}$$

Example Compute the 95 percent confidence interval based on the original 100-point sample and the subsequent 5-point sample:

$$s_p^2 = \frac{99(2.47)^2 + 4(1.85)^2}{103} = 5.997$$

or $$s_p = 2.45$$

With 103 df and $\alpha = .05$, $t = \pm 1.96$. Therefore

$$(35.16 - 37.14) \pm 1.96(2.45)\sqrt{1/100 + 1/5} = -1.98 \pm 2.20$$

or $$-4.18 \le (\mu_1 - \mu_2) \le .22$$

Note that if the respective samples had been based on 52 observations each rather than 100 and 5, the uncertainty factor would have been $\pm$ .94 rather than the observed $\pm$2.20. The interval width tends to be minimum when $n_1 = n_2$.

Confidence Interval for a Proportion Suppose that in the previous tensile-strength data sample packages testing below 32 are defined as unacceptable. For the total sample of 100, there would be 10 classified as unacceptable. On the basis of a sample proportion $\hat{p}$, a 100 $(1 - \alpha)$ percent confidence interval for p, the population proportion can be derived for large samples when both $np > 5$ and $n(1 - p) > 5$.

$$\hat{p} \pm t\sqrt{\hat{p}(1 - \hat{p})/n}$$

Example For the sample data, compute a 90 percent confidence interval.

$$10/100 \pm 1.645\sqrt{(.1)(.9)/100} = .1 \pm .049$$

or $$P[.051 \le p \le .149] = .90$$

Confidence Interval for a Variance The chi-square distribution can be used to derive a confidence interval for a population variance σ^2 when the parent population is normally distributed. For a $100(1 - \alpha)$ percent confidence interval

$$\frac{(\text{df})s^2}{\chi_2^2} \le \sigma^2 \le \frac{(\text{df})s^2}{\chi_1^2}$$

where χ_1^2 corresponds to a lower-tail area of $\alpha/2$ and χ_2^2 to an upper-tail area of $\alpha/2$.

Example For the first week of tensile-strength samples compute the 90 percent confidence interval for σ^2 (df = 24, corresponding to $n = 25$):

$$\frac{24(2.40)^2}{36.4} \le \sigma^2 \le \frac{24(2.40)^2}{13.8}$$

$$3.80 \le \sigma^2 \le 10.02$$

or $$1.95 \le \sigma \le 3.17$$

TESTS OF HYPOTHESIS

General Nature of Tests The general nature of tests can be illustrated with a simple example. In a court of law, when a defendant is charged with a crime, the judge instructs the jury initially to presume that the defendant is innocent of the crime. The jurors are then presented with evidence and counterargument as to the defendant's guilt or innocence. If the evidence suggests beyond a reasonable doubt that the defendant did, in fact, commit the crime, they have been instructed to find the defendant guilty; otherwise, not guilty. The burden of proof is on the prosecution.

Jury trials represent a form of decision making. In statistics, an analogous procedure for making decisions falls into an area of statistical inference called **hypothesis testing.**

Suppose that a company has been using a certain supplier of raw materials in one of its chemical processes. A new supplier approaches the company and states that its material, at the same cost, will increase the process yield. If the new supplier has a good reputation, the company might be willing to run a limited test. On the basis of the test results it would then make a decision to change suppliers or not. Good management would dictate that an improvement must be demonstrated (beyond a reasonable doubt) for the new material. That is, the burden of proof is tied to the new material. In setting up a test of hypothesis for this application, the initial assumption would be defined as a null hypothesis and symbolized as H_0. The null hypothesis would state that yield for the new material is no greater than for the conventional material. The symbol μ_0 would be used to designate the known current level of yield for the standard material and μ for the unknown population yield for the new material. Thus, the null hypothesis can be symbolized as H_0: $\mu \le \mu_0$.

The alternative to H_0 is called the alternative hypothesis and is symbolized as H_1: $\mu > \mu_0$.

Given a series of tests with the new material, the average yield $\bar{x}$ would be compared with μ_0. If $\bar{x} < \mu_0$, the new supplier would be dismissed. If $\bar{x} > \mu_0$, the question would be: Is it sufficiently greater in the light of its corresponding reliability, i.e., beyond a reasonable doubt? If the confidence interval for μ included μ_0, the answer would be no, but if it did not include μ_0, the answer would be yes. In this simple application, the formal test of hypothesis would result in the same conclusion as that derived from the confidence interval. However, the utility of tests of hypothesis lies in their generality, whereas confidence intervals are restricted to a few special cases.

Test of Hypothesis for a Mean Procedure

Nomenclature

- μ = mean of the population from which the sample has been drawn
- σ = standard deviation of the population from which the sample has been drawn
- μ_0 = base or reference level
- H_0 = null hypothesis
- H_1 = alternative hypothesis
- α = significance level, usually set at .10, .05, or .01
- t = tabled t value corresponding to the significance level α. For a two-tailed test, each corresponding tail would have an area of $\alpha/2$, and for a one-tailed test, one tail area would be equal to α. If σ^2 is known, then z would be used rather than the t.
- $t = (\bar{x} - \mu_0)/(s/\sqrt{n})$ = sample value of the test statistic.

Assumptions

1. The n observations $x_1, x_2, \ldots, x_n$ have been selected randomly.
2. The population from which the observations were obtained is normally distributed with an unknown mean μ and standard deviation σ. In actual practice, this is a robust test, in the sense that in most types of problems it is not sensitive to the normality assumption when the sample size is 10 or greater.

Test of Hypothesis

1. Under the null hypothesis, it is assumed that the sample came from a population whose mean μ is equivalent to some base or reference designated by μ_0. This can take one of three forms:

Form 1	Form 2	Form 3
H_0: $\mu = \mu_0$	H_0: $\mu \le \mu_0$	H_0: $\mu \ge \mu_0$
H_1: $\mu \ne \mu_0$	H_1: $\mu > \mu_0$	H_1: $\mu < \mu_0$
Two-tailed test	Upper-tailed test	Lower-tailed test

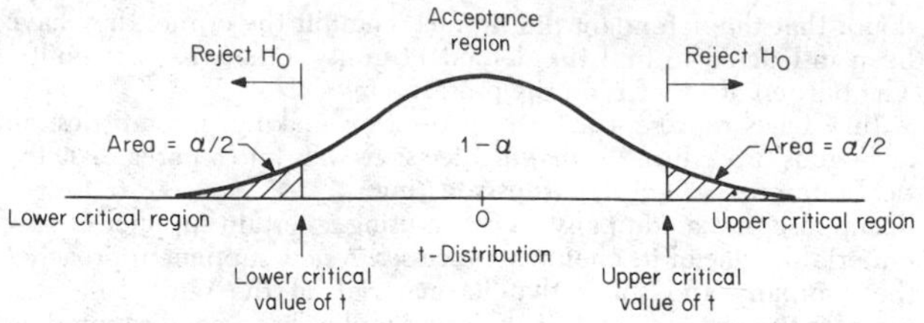

FIG. 2-64 Acceptance region.

2. If the null hypothesis is assumed to be true, say, in the case of a two-sided test, form 1, then the distribution of the test statistic t is known. Given a random sample, one can predict how far its sample value of t might be expected to deviate from zero (the midvalue of t) by chance alone. If the sample value of t does, in fact, deviate too far from zero, then this is defined to be sufficient evidence to refute the assumption of the null hypothesis. It is consequently rejected, and the converse or alternative hypothesis is accepted.
3. The rule for accepting H_0 is specified by selection of the α level as indicated in Fig. 2-64. For forms 2 and 3 the α area is defined to be in the upper or the lower tail respectively.
4. The decision rules for each of the three forms are defined as follows: If the sample t falls within the acceptance region, accept H_0 for lack of contrary evidence. If the sample t falls in the critical region, reject H_0 at a significance level of 100α percent.

Example
Application. In the past, the yield for a chemical process has been established at 89.6 percent with a standard deviation of 3.4 percent. A new supplier of raw materials will be used and tested for 7 days.
Procedure
1. The standard of reference is $\mu_0 = 89.6$ with a known $\sigma = 3.4$.
2. It is of interest to demonstrate whether an increase in yield is achieved with the new material; therefore,

$$H_0: \mu \leq 89.6 \qquad H_1: \mu > 89.6$$

3. Select $\alpha = .05$, and since σ is known (the new material would not affect the day-to-day variability in yield), the test statistic would be z with a corresponding critical value $cv(z) = 1.645$.
4. The decision rule:

$$\text{Accept } H_0 \text{ if sample } z < 1.645$$
$$\text{Reject } H_0 \text{ if sample } z > 1.645$$

5. A 7-day test was carried out, and daily yields averaged 91.6 percent with a sample standard deviation $s = 3.6$ (this is not needed for the test of hypothesis).
6. For the data sample $z = (91.6 - 89.6)/(3.4/\sqrt{7}) = 1.56$.
7. Since the sample $z < cv(z)$, accept the null hypothesis for lack of contrary evidence; i.e., an improvement has not been demonstrated beyond a reasonable doubt (see "Power").

Example
Application. In the past, the break strength of a synthetic yarn has averaged 34.6 lb. The first-stage draw ratio of the spinning machines has been increased. Production management wants to determine whether the break strength has changed under the new condition.
Procedure
1. The standard of reference is $\mu_0 = 34.6$.
2. It is of interest to demonstrate whether a change has occurred; therefore,

$$H_0: \mu = 34.6 \qquad H_1: \mu \neq 34.6$$

3. Select $\alpha = .05$, and since with the change in draw ratio the uniformity might change, the sample standard deviation would be used, and therefore t would be the appropriate test statistic.
4. A sample of 21 ends was selected randomly and tested on an Instron with the results $\bar{x} = 35.55$ and $s = 2.041$.
5. For 20 df and a two-tailed α level of 5 percent, the critical values of t are given by $\pm$ 2.086 with a decision rule:

$$\text{Accept } H_0 \text{ if } -2.086 < \text{sample } t < 2.086$$
$$\text{Reject } H_0 \text{ if sample } t < -2.086 \text{ or } > 2.086$$

6. For the data sample $t = (35.55 - 34.6)/(2.041/\sqrt{21}) = 2.133$.
7. Since $2.133 > 2.086$, reject H_0 and accept H_1. It has been demonstrated that an improvement in break strength has been achieved.

Power In testing a null hypothesis in a particular application, the null hypothesis is, in fact, either true or false. The analyst never knows in an absolute sense whether it is true or false but can only make a judgmental statement by accepting or rejecting the null hypothesis on the basis of the evidence contained in the sample data. However, whatever the application, the null hypothesis is either true or false. From a probabilistic standpoint, the analyst can never be absolutely sure of a decision which is based on evidence alone. However, it is possible to make theoretical probabilistic statements about the analysis if certain assumptions are made.

In a given application the analyst will draw a conclusion of either accepting or rejecting H_0 on the basis of the evidence. This potential action can be compared with the absolute nature of H_0 as follows:

Conclusion based on a sample	Absolute nature: H_0 is true	Absolute nature: H_0 is false
Accept H_0	$1 - \alpha$	β = type II error
Reject H_0	α = type I error	Power = $1 - \beta$

As indicated in the table, the analyst might make one of two mistakes based on the evidence of the sample. The analyst might reject H_0 when in actual fact it is true, a type I error, or conversely might accept H_0 when it is false, a type II error. Symbolically,

$$\alpha = P(\text{rejecting } H_0 \mid H_0 \text{ is true})$$
$$\beta = P(\text{accepting } H_0 \mid H_0 \text{ is false})$$

In tossing a fair coin, there is some remote chance that 10 successive heads will be observed. If a bystander observes this, it might be concluded, falsely, that the coin is two-headed. This corresponds to a type I error, i.e., rejection of the notion that a fair coin has been used when in fact a fair coin has been used. Conversely, in a class experiment in which 9 fair coins and 2 two-headed coins ($\mu = 6.5$) are tossed simultaneously 10 times, there is a relatively high chance that the resultant confidence interval based on the experimental results will include the assumed and false value 5.5. If this occurs (and it will happen on occasion), the students will accept the experimental evidence and suspect nothing. This will be a false conclusion because the true value of $\mu = 6.5$.

What if the group of coins is tossed 20 times rather than 10 times? If they are tossed 20 times, there is a 91 percent chance that the resultant confidence interval will not include 5.5, while on the basis of 10 tosses this probability is only a 68 percent chance. This type of probability is called **power.** For this contrived example, only the experimenter knows that $\mu_0 = 5.5$ is false and that the actual $\mu = 6.5$. Therefore, the chance of rejecting $\mu_0 = 5.5$, the power, can be computed where

$$\text{Power} = P(\text{rejecting } H_0{:}\mu = 5.5 \mid \mu = 6.5)$$

In this example, power can be computed because the true μ is known. For an industrial application power can be computed only for suggested alternative values. In other words, the experimenter states the threshold value that is to be detected. Given this, it is then possible to determine how many samples will be required in order to achieve a specified power.

Example Consider the following real application. A paper company receives 200,000 cartons from a supplier. It has been agreed originally that the shipment of cartons should contain no more than 10 percent defective items in regard to a certain type of defect. In practice, the quality-assurance group could sample some n cartons, selected randomly, and either accept or reject the shipment according to the number of defective items found in the sample. The corresponding hypotheses would be

$$H_0: p \leq .10 \qquad H_1: p > .10$$

Suppose that the members of the group examine one carton. A rule could be that if the sampled carton is defective, they are to send the shipment back; otherwise they are to accept the shipment. The probability of returning or accepting the shipment will depend on the true proportion of defective items in the shipment (it is assumed that defective items are randomly distributed in the shipment). For example,

True proportion of defects	One-sample probability of	
	Accepting defects = 0	Rejecting defects = 1
.01	.99	.01
.05	.95	.05
.10	.90	.10
.20	.80	.20
.50	.50	.50

This would be totally unsatisfactory to the consumer. The rule would dictate that it would be compulsory for the consumer to sample more than 1 carton. If the group sampled 10 cartons under a new rule of accepting the shipment if at most 1 defective carton was found in the shipment and of returning the shipment if 2 or more were found, then the following probabilities would be defined:

	True proportion of defects	10-sample probability of	
		Accepting defects ≤ 1	Rejecting defects > 1
H_0 true	.05	.914	.086
	.10	.736	.264
H_0 false	.15	.544	.456
	.20	.374	.626
	.25	.244	.756
	.30	.149	.851

This table can be interpreted in the sense of type I, type II, and power probabilities:

H_0 true: For $p \leq .10$ $\quad \alpha \leq .264$

H_0 false: For $p = .15$ $\quad \beta = 544$ $\quad$ Power $= .456$

For $p = .20$ $\quad \beta = .374$ $\quad$ Power $= .626$

etc.

Suppose that the sample size is fixed at $n = 100$ with the decision rule set at 10 defects:

	True proportion of defects	100-sample probability of	
		Accepting defects ≤ 10	Rejecting defects > 10
H_0 true	.05	.989	.011
	.10	.583	.417
H_0 false	.15	.099	.901
	.20	.006	.994

Obviously, the rule of accepting or rejecting a shipment based on a sample of $n = 100$ is poor for the producer. Given that the true proportion is $p = .09999$, an acceptable lot, the probability is almost a 42 percent chance that the shipment will be rejected. Therefore, it is standard practice to allow more than 10 percent defects in the sample. If the type I error is fixed at, say, $\alpha = .04$, then the consumer agrees to accept the shipment if 15 or fewer defective items are found in the sample. The cutoff of 15 arises because the probability of observing by chance 16 or more defects, given $n = 100$ and $p = .10$, is .04—the producer's risk.

Now, working under conventional rules, what happens to the type II error and the corresponding power when the true proportion is greater than 10 percent?

	True proportion of defects	100-sample probability of	
		Accepting defects ≤ 15	Rejecting defects > 15
H_0 true	.05	1.000	.000
	.10	.960	.040
H_0 false	.15	.568	.432
	.20	.129	.871

This seems to be an intolerable situation for the consumer. The risk is a 56.8 percent chance that it will accept a shipment which contains 15 percent defective items. What can be done? The simple solution is to increase the sample size. Suppose that n is set at 1000 with the decision rule set at 115 defects:

	True proportion of defects	1000-sample probability of	
		Accepting defects ≤ 115	Rejecting defects > 115
H_0 true	.05	1.000	.000
	.10	.950	.050
H_0 false	.11	.712	.288
	.12	.330	.670
	.13	.087	.913
	.14	.013	.987
	.15	.001	.999

Even with $n = 1000$ the probability dictates that there is a 71.2 percent chance that the consumer will accept a shipment which contains 11 percent defective items. Even if the sample is increased to $n = 5000$,

$$\beta = P(\text{accepting shipment} \mid n = 5000;\ p = .11;\ \alpha = .05)$$
$$= .242$$

or $\quad$ Power $= .758$

Management must decide what it is willing to live with. To control its destiny it can specify two of three quantities, given an α level, and the third quantity is then defined:

1. Power
2. The true maximum p level which management is willing to live with
3. The sample size n

If management fixes p close to .10 and power close to 1, then the required n will be gigantic. If p is relatively large conpared with .10 and power is, say, 80 percent, then n will be realistic, as shown in Fig. 2-65.

The same type of principle holds for any kind of experimental work. How large a difference do you wish to detect and with what power? The answer will fix the number of experiments that you must carry out and pay for.

Two-Population Test of Hypothesis for Means

Nature Two samples were selected from different locations in a plastic-film sheet and measured for thickness. The thickness of the respective samples was measured at 10 close but equally spaced points in each of the samples. It was of interest to compare the average thickness of the respective samples to detect whether they were significantly different. That is, was there a significant variation in thickness between locations?

From a modeling standpoint statisticians would define this problem as a two-population test of hypothesis. They would define the respective sample sheets as two populations from which 10 sample thickness determinations were measured for each.

In order to formalize a comparison between populations based on their respective samples, it is necessary to have some basis of comparison. This basis is predicated on the distribution of the t statistic. In effect, the t statistic characterizes the way in which two sample means from two separate populations will tend to vary by chance

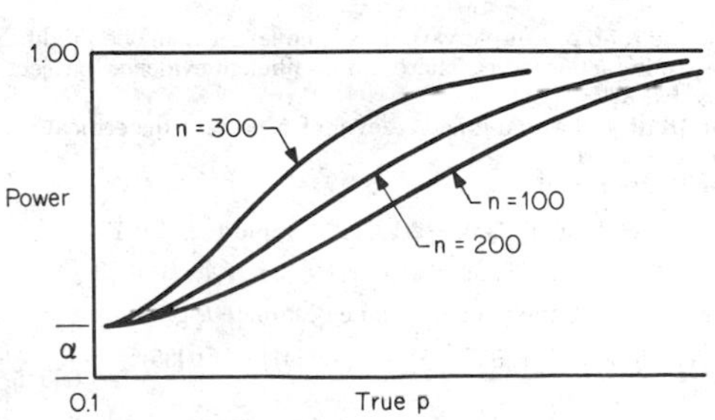

FIG. 2-65 Power versus true P.

alone when the population means and variances are equal. Consider the following:

Population 1		Population 2	
Normal	Sample 1	Normal	Sample 2
μ_1	n_1	μ_2	n_2
	$\bar{x}_1$		$\bar{x}_2$
σ_1^2	s_1^2	σ_2^2	s_2^2

Consider the hypothesis $\mu_1 = \mu_2$. If, in fact, the hypothesis is correct, i.e., $\mu_1 = \mu_2$ (under the condition $\sigma_1^2 = \sigma_2^2$), then the sampling distribution of $(\bar{x}_1 - \bar{x}_2)$ is predictable through the t distribution. The observed sample values then can be compared with the corresponding t distribution. If the sample values are reasonably close (as reflected through the α level), that is, $\bar{x}_1$ and $\bar{x}_2$ are not "too different" from each other on the basis of the t distribution, the null hypothesis would be accepted. Conversely, if they deviate from each other "too much" and the deviation is therefore not ascribable to chance, the conjecture would be questioned and the null hypothesis rejected.

Example

Application. Two samples were selected from different locations in a plastic-film sheet. The thickness of the respective samples was measured at 10 close but equally spaced points.

Procedure

1. Demonstrate whether the thicknesses of the respective sample locations are significantly different from each other; therefore,

$$H_0: \mu_1 = \mu_2 \qquad H_1: \mu_1 \neq \mu_2$$

2. Select $\alpha = .05$.
3. Summarize the statistics for the respective samples:

Sample 1		Sample 2	
1.473	1.367	1.474	1.417
1.484	1.276	1.501	1.448
1.484	1.485	1.485	1.469
1.425	1.462	1.435	1.474
1.448	1.439	1.348	1.452
$\bar{x}_1 = 1.434$	$s_1 = .0664$	$\bar{x}_2 = 1.450$	$s_2 = .0435$

4. As a first step, the assumption for the standard t test, that $\sigma_1^2 = \sigma_2^2$, can be tested through the F distribution. For this hypothesis, H_0: $\sigma_1^2 = \sigma_2^2$ would be tested against H_1: $\sigma_1^2 \neq \sigma_2^2$. Since this is a two-tailed test and conventionally only the upper tail for F is published, the procedure is to use the largest ratio and the corresponding ordered degrees of freedom. This achieves the same end result through one table. However, since the largest ratio is arbitrary, it is necessary to define the true α level as twice the value of the tabled value. Therefore, by using the tabled $\alpha = .05$ the corresponding critical value for $F(9,9) = 3.18$ would be for a true $\alpha = .10$. For the sample,

$$\text{Sample } F = (.0664/.0435)^2 = 2.33$$

Therefore, the ratio of sample variances is no larger than one might expect to observe when in fact $\sigma_1^2 = \sigma_2^2$. There is not sufficient evidence to reject the null hypothesis that $\sigma_1^2 = \sigma_2^2$.

5. For 18 df and a two-tailed α level of 5 percent the critical values of t are given by ± 2.101.
6. The decision rule:

Accept H_0 if $-2.101 \leq \text{sample } t \leq 2.101$

Reject H_0 otherwise

7. For the sample the pooled variance estimate is given by

$$s_p^2 = \frac{9(.0664)^2 + 9(.0435)^2}{9 + 9} = \frac{(.0664)^2 + (.0435)^2}{2} = .00315$$

or $\qquad s_p = .056$

8. The sample statistic value of t is

$$\text{Sample } t = \frac{1.434 - 1.450}{.056\sqrt{1/10 + 1/10}} = -.64$$

9. Since the sample value of t falls within the acceptance region, accept H_0 for lack of contrary evidence; i.e., there is insufficient evidence to demonstrate that thickness differs between the two selected locations.

Example

Application. An efficiency expert carried out a driving test to determine whether there was a significant cost differential between standard and high-test gasoline for city driving. For his study the most viable procedure would have been to drive through an equal number of tankloads of gasoline in an alternating sequence. However, from a practical standpoint he would be unable to drive through a total full tankload, because this would have resulted in his being stranded in traffic. Since he would not drive through a full tank, a residue from the previous gasoline type would remain in subsequent filling. Therefore the expert opted to run the following sequence: (1) standard, 5 tanks; (2) high-test, 10 tanks; and (3) standard, 5 tanks. By using this procedure he would minimize the residue problem. In addition, any time tendency for the car to get out of tune would be reasonably spread across the two types of gasoline.

Procedure

1. It was of interest to demonstrate whether one type of gasoline was economically more efficient than the other. For city driving the engineer estimated his mileage at about 25 mi/gal with a standard brand. Therefore, using his current pricings of \$1.298/gal for standard and \$1.388/gal for high-test, he would need to achieve 26.73 mi/gal with high-test to break even; i.e.,

$$129.8/25 = 138.8/26.73 = 5.19 \text{ cents/mi}$$

Therefore, the null hypothesis would be defined by

$$H_0: \mu_1 = \mu_2 - 1.73 \text{ with } H_1: \mu_1 \neq \mu_2 - 1.73$$

where μ_1 would be the population mean for the standard brand and μ_2 for the high-test.

2. Select $\alpha = .05$.
3. For this application it was assumed that $\sigma_1^2 = \sigma_2^2$. Basically, variation in gasoline mileage is attributable to the number of stops and starts, traffic congestion, total idling time, and so forth, rather than to the relatively minuscule nonuniformity of gasoline. The test itself was carried out over a period during which driving conditions were essentially constant but random day by day.
4. Each trial was started with a full tank, at which time the mileage was noted. When the tank became almost empty, it was filled with the gasoline type dictated by the schedule and the mileage noted again. The total differential mileage was divided by the gallons to fill and the computed ratio determining miles per gallon. For the 20 trials:

Standard		High-test	
23.2	24.5	27.1	31.4
25.6	27.7	26.0	26.4
26.6	23.1	27.2	29.0
26.7	24.0	30.3	28.8
28.8	24.8	26.0	26.6
$\bar{x}_1 = 25.5$	$s_1 = 1.92$	$\bar{x}_2 = 27.88$	$s_2 = 1.89$

5. With $\alpha = .05$ and 18 df the critical values of t are ± 2.101; therefore, the decision rule is given by

Accept H_0 if $-2.101 \leq \text{sample } t \leq 2.101$

Otherwise reject H_0 and accept H_1

6. For the sample the pooled variance estimate is given by ($9 \text{ df}_1 = 9 \text{ df}_2$):

$$s_p^2 = \frac{(1.92)^2 + (1.89)^2}{2} = 3.63 \qquad \text{or} \qquad s_p = 1.91$$

7. The sample t statistic:

$$\text{Sample } t = \frac{(25.5 - 27.88) - (-1.73)}{1.91\sqrt{1/10 + 1/10}} = -.76$$

8. Since the sample value of t falls within the acceptance region, accept H_0 for lack of contrary evidence. Since the efficiency expert would continue

to drive his car, he decided to switch to high-test even though no significant difference was detected. This decision was based on the fact that the sample mileage differential, 2.38, was greater than the breakeven point of 1.73 and no other considerations favored either standard or high-test. In addition, this choice would require fewer stops at the gas station.

Test of Hypothesis for Paired Observations

Nature In some types of applications, associated pairs of observations are defined. For example, (1) pairs of samples from two populations are treated in the same way, or (2) two types of measurements are made on the same unit. For applications of this type, it is not only more effective but necessary to define the random variable as the difference between the pairs of observations. The difference numbers can then be tested by the standard t distribution.

Examples of the two types of applications are as follows:

1. *Sample treatment*

a. Two types of metal specimens buried in the ground together in a variety of soil types to determine corrosion resistance

b. Wear-rate test with two different types of tractor tires mounted in pairs on n tractors for a defined period of time

2. *Same unit*

a. Blood-pressure measurements made on the same individual before and after the administration of a stimulus

b. Smoothness determinations on the same film samples at two different testing laboratories

Test of Hypothesis for Matched Pairs: Procedure

Nomenclature

d_i = sample difference between the ith pair of observations
s = sample standard deviation of differences
μ = population mean of differences
σ = population standard deviation of differences
μ_0 = base or reference level of comparison
H_0 = null hypothesis
H_1 = alternative hypothesis
α = significance level
t = tabled value with $(n-1)$ df
t = $(\bar{d} - \mu_0/s/\sqrt{n})$, the sample value of t

Assumptions

1. The n pairs of samples have been selected and assigned for testing in a random way.

2. The population of differences is normally distributed with a mean μ and variance σ^2. As in the previous application of the t distribution, this is a robust procedure, i.e., not sensitive to the normality assumption if the sample size is 10 or greater in most situations.

Test of Hypothesis

1. Under the null hypothesis, it is assumed that the sample came from a population whose mean μ is equivalent to some base or reference level designated by μ_0. For most applications of this type, the value of μ_0 is defined to be zero; that is, it is of interest generally to demonstrate a difference not equal to zero. The hypothesis can take one of three forms:

Form 1	Form 2	Form 3
H_0: $\mu = \mu_0$	H_0: $\mu \le \mu_0$	H_0: $\mu \ge \mu_0$
H_1: $\mu \ne \mu_0$	H_1: $\mu > \mu_0$	H_1: $\mu < \mu_0$
Two-tailed test	Upper-tailed test	Lower-tailed test

2. If the null hypothesis is assumed to be true, say, in the case of a lower-tailed test, form 3, then the distribution of the test statistic t is known under the null hypothesis that limits $\mu = \mu_0$. Given a random sample, one can predict how far its sample value of t might be expected to deviate from zero by chance alone when $\mu = \mu_0$. If the sample value of t is too small, as in the case of a negative value, then this would be defined as sufficient evidence to reject the null hypothesis.

3. Select α.

4. The critical values or value of t would be defined by the tabled value of t with $(n-1)$ df corresponding to a tail area of α. For a two-tailed test, each tail area would be $\alpha/2$, and for a one-tailed test there would be an upper-tail or a lower-tail area of α corresponding to forms 2 and 3 respectively.

5. The decision rule for each of the three forms would be to reject the null hypothesis if the sample value of t fell in that area of the t distribution defined by α, which is called the critical region. Otherwise, the alternative hypothesis would be accepted for lack of contrary evidence.

Example

Application. Pairs of pipes have been buried in 11 different locations to determine corrosion on nonbituminous pipe coatings for underground use. One type includes a lead-coated steel pipe and the other a bare steel pipe.

Procedure

1. The standard of reference is taken as $\mu_0 = 0$, corresponding to no difference in the two types.

2. It is of interest to demonstrate whether either type of pipe has a greater corrosion resistance than the other. Therefore,

$$H_0: \mu = 0 \qquad H_1: \mu \ne 0$$

3. Select $\alpha = .05$. Therefore, with $n = 11$ the critical values of t with 10 df are defined by $t = \pm 2.228$.

4. The decision rule:

Accept H_0 if $-2.228 \le$ sample $t \le 2.228$

Reject H_0 otherwise

5. The sample of 11 pairs of corrosion determinations and their differences are as follows:

Soil type	Lead-coated steel pipe	Bare steel pipe	d = difference
A	27.3	41.4	−14.1
B	18.4	18.9	−0.5
C	11.9	21.7	−9.8
D	11.3	16.8	−5.5
E	14.8	9.0	5.8
F	20.8	19.3	1.5
G	17.9	32.1	−14.2
H	7.8	7.4	0.4
I	14.7	20.7	−6.0
J	19.0	34.4	−15.4
K	65.3	76.2	−10.9

6. The sample statistics:

$$\bar{d} = -6.245 \quad s^2 = \frac{11\Sigma d^2 - (\Sigma d)^2}{11 \times 10} = 52.56$$

or

$$s = 7.25$$

$$\text{Sample } t = (-6.245 - 0)/(7.25/\sqrt{11})$$

$$= -2.86$$

7. Since the sample t of $-2.86 <$ tabled t of -2.228, reject H_0 and accept H_1; that is, it has been demonstrated that, on the basis of the evidence, lead-coated steel pipe has a greater corrosion resistance than bare steel pipe.

Example

Application. A stimulus was tested for its effect on blood pressure. Ten men were selected randomly, and their blood pressure was measured before and after the stimulus was administered. It was of interest to determine whether the stimulus had caused a significant increase in the blood pressure.

Procedure

1. The standard of reference was taken as $\mu_0 \le 0$, corresponding to no increase.

2. It was of interest to demonstrate an increase in blood pressure if in fact an increase did occur. Therefore,

$$H_0: \mu_0 \le 0 \qquad H_1: \mu_0 > 0$$

3. Select $\alpha = .05$. Therefore, with $n = 10$ the critical value of t with 9 df is defined by $t = 1.833$.

4. The decision rule:

Accept H_0 if sample $t < 1.833$

Reject H_0 if sample $t > 1.833$

5. The sample of 10 pairs of blood pressure and their differences were as follows:

Individual	Before	After	d = difference
1	138	146	8
2	116	118	2
3	124	120	−4
4	128	136	8
5	155	174	19
6	129	133	4
7	130	129	−1
8	148	155	7
9	143	148	5
10	159	155	−4

6. The sample statistics:

$$\bar{d} = 4.4 \qquad s = 6.85$$

$$\text{Sample } t = (4.4 - 0)/(6.85/\sqrt{10}) = 2.03$$

7. Since the sample $t = 2.03 >$ critical $t = 1.833$, reject the null hypothesis. It has been demonstrated that the population of men from which the sample was drawn tend, as a whole, to have an increase in blood pressure after the stimulus has been given. The distribution of differences d seems to indicate that the degree of response varies by individuals.

Test of Hypothesis for a Proportion

Nature Some types of statistical applications deal with counts and proportions rather than measurements. Examples are (1) the proportion of workers in a plant who are out sick, (2) lost-time worker accidents per month, (3) defective items in a shipment lot, and (4) preference in consumer surveys.

The procedure for testing the significance of a sample proportion follows that for a sample mean. In this case, however, owing to the nature of the problem the appropriate test statistic is Z. This follows from the fact that the null hypothesis requires the specification of the goal or reference quantity p_0, and since the distribution is a binomial proportion, the associated variance is $[p_0(1 - p_0)]n$ under the null hypothesis. The primary requirement is that the sample size n satisfy normal approximation criteria for a binomial proportion, roughly $np > 5$ and $n(1 - p) > 5$.

Test of Hypothesis for a Proportion: Procedure

Nomenclature

p = mean proportion of the population from which the sample has been drawn

p_0 = base or reference proportion

$[p_0(1 - p_0)]/n$ = base or reference variance

$\hat{p} = x/n$ = sample proportion, where x refers to the number of observations out of n which have the specified attribute

H_0 = assumption or null hypothesis regarding the population proportion

H_1 = alternative hypothesis

α = significance level, usually set at .10, .05, or .01

z = Tabled Z value corresponding to the significance level α. The sample sizes required for the z approximation according to the magnitude of p_0 are given in Sec. 1.

$z = (\hat{p} - p_0)/\sqrt{p_0(1 - p_0)/n}$, the sample value of the test statistic

Assumptions

1. The n observations have been selected randomly.
2. The sample size n is sufficiently large to meet the requirement for the Z approximation.

Test of Hypothesis 1. Under the null hypothesis, it is assumed that the sample came from a population with a proportion p_0 of items having the specified attribute. For example, in tossing a coin the population could be thought of as having an unbounded number of potential tosses. If it is assumed that the coin is fair, this would dictate $p_0 = 1/2$ for the proportional number of heads in the population. The null hypothesis can take one of three forms:

Form 1	Form 2	Form 3
H_0: $p = p_0$	H_0: $p \leq p_0$	H_0: $p \geq p_0$
H_1: $p \neq p_0$	H_1: $p > p_0$	H_1: $p < p_0$
Two-tailed test	Upper-tailed test	Lower-tailed test

2. If the null hypothesis is assumed to be true, then the sampling distribution of the test statistic Z is known. Given a random sample, it is possible to predict how far the sample proportion x/n might deviate from its assumed population proportion p_0 through the Z distribution. When the sample proportion deviates too far, as defined by the significance level α, this serves as the justification for rejecting the assumption, that is, rejecting the null hypothesis.

3. The decision rule is given by

Form 1: Accept H_0 if lower critical $z <$ sample $z <$ upper critical z
Reject H_0 otherwise

Form 2: Accept H_0 if sample $z <$ upper critical z
Reject H_0 otherwise

Form 3: Accept H_0 if lower critical $z <$ sample z
Reject H_0 otherwise

Example

Application. A company has received a very large shipment of rivets. One product specification required that no more than 2 percent of the rivets have diameters greater than 14.28 mm. Any rivet with a diameter greater than this would be classified as defective. A random sample of 600 was selected and tested with a go–no-go gauge. Of these, 16 rivets were found to be defective. Is this sufficient evidence to conclude that the shipment contains more than 2 percent defective rivets?

Procedure

1. The quality goal is $p \leq .02$. It would be assumed initially that the shipment meets this standard; i.e., H_0: $p \leq .02$.
2. The assumption in step 1 would first be tested by obtaining a random sample. Under the assumption that $p \leq .02$, the distribution for a sample proportion would be defined by the z distribution. This distribution would define an upper bound corresponding to the upper critical value for the sample proportion. It would be unlikely that the sample proportion would rise above that value if, in fact, $p \leq .02$. If the observed sample proportion exceeds that limit, corresponding to what would be a very unlikely chance outcome, this would lead one to question the assumption that $p \leq .02$. That is, one would conclude that the null hypothesis is false. To test, set

$$H_0: p \leq .02 \qquad H_1: p > .02$$

3. Select $\alpha = .05$.
4. With $\alpha = .05$, the upper critical value of Z = 1.645.
5. The decision rule:

Accept H_0 if sample $z < 1.645$
Reject H_0 if sample $z > 1.645$

6. The sample z is given by

$$\text{Sample } z = \frac{(16/600) - .02}{\sqrt{(.02)(.98)/600}} = 1.17$$

7. Since the sample $z < 1.645$, accept H_0 for lack of contrary evidence; there is not sufficient evidence to demonstrate that the defect proportion in the shipment is greater than 2 percent.

Test of Hypothesis for Two Proportions

Nature In some types of engineering and management-science problems, we may be concerned with a random variable which represents a proportion, for example, the proportional number of defective items per day. The method described previously relates to a single proportion. In this subsection two proportions will be considered.

A certain change in a manufacturing procedure for producing component parts is being considered. Samples are taken by using both the existing and the new procedures in order to determine whether the new procedure results in an improvement. In this application, it is of interest to demonstrate statistically whether the population proportion p_2 for the new procedure is less than the population proportion p_1 for the old procedure on the basis of a sample of data.

Test of Hypothesis for Two Proportions: Procedure

Nomenclature

p_1 = population 1 proportion
p_2 = population 2 proportion
n_1 = sample size from population 1
n_2 = sample size from population 2
x_1 = number of observations out of n_1 that have the designated attribute
x_2 = number of observations out of n_2 that have the designated attribute
$\hat{p}_1$ = x_1/n_1, the sample proportion from population 1
$\hat{p}_2$ = x_2/n_2, the sample proportion from population 2
α = significance level
H_0 = null hypothesis
H_1 = alternative hypothesis
z = tabled Z value corresponding to the stated significance level α

$$z = \frac{\hat{p}_1 - \hat{p}_2}{\sqrt{\hat{p}_1(1-\hat{p}_1)/n_1 + \hat{p}_2(1-\hat{p}_2)/n_2}}, \text{ the sample value of Z}$$

Assumptions

1. The respective two samples of n_1 and n_2 observations have been selected randomly.
2. The sample sizes n_1 and n_2 are sufficiently large to meet the requirement for the Z approximation; i.e., $x_1 > 5$, $x_2 > 5$.

Test of Hypothesis

1. Under the null hypothesis, it is assumed that the respective two samples have come from populations with equal proportions $p_1 = p_2$. Under this hypothesis, the sampling distribution of the corresponding Z statistic is known. On the basis of the observed data, if the resultant sample value of Z represents an unusual outcome, that is, if it falls within the critical region, this would cast doubt on the assumption of equal proportions. Therefore, it will have been demonstrated statistically that the population proportions are in fact not equal. The various hypotheses can be stated:

Form 1	Form 2	Form 3
H_0: $p_1 = p_2$	H_0: $p_1 \le p_2$	H_0: $p_1 \ge p_2$
H_1: $p_1 \ne p_2$	H_1: $p_1 > p_2$	H_1: $p_1 < p_2$
Two-tailed test	Upper-tailed test	Lower-tailed test

2. The decision rule for form 1 is given by

Accept H_0 if lower critical $z <$ sample $z <$ upper critical z
Reject H_0 otherwise

Example

Application. A certain engineering change was made in a manufacturing procedure for component parts. Samples were taken during the last week of operations with the old procedure and during the first week of operations with the new procedure. It is of interest to determine whether the proportional numbers of defects for the respective populations differ on the basis of the sample information.

Procedure

1. The hypotheses are

$$H_0: p_1 = p_2 \qquad H_1: p_1 \ne p_2$$

2. Select $\alpha = .05$. Therefore, the critical values of z are ± 1.96.
3. For the samples, 75 out of 1720 parts from the previous procedure and 80 out of 2780 parts under the new procedure were found to be defective; therefore,

$$\hat{p}_1 = 75/1720 = .0436 \qquad \hat{p}_2 = 80/2780 = .0288$$

4. The decision rule:

Accept H_0 if $-1.96 \le$ sample $Z \le 1.96$
Reject H_0 otherwise

5. The sample statistic:

$$\text{Sample } z = \frac{.0436 - .0288}{\sqrt{(.0436)(.9564)/1720 + (.0288)(.9712)/2780}} = 2.53$$

6. Since the sample z of $2.53 >$ tabled z of 1.96, reject H_0 and conclude that the new procedure has resulted in a reduced defect rate.

Goodness-of-Fit Test

Nature A standard die has six sides numbered from 1 to 6. If one were really interested in determining whether a particular die was well balanced, one would have to carry out an experiment. To do this, it might be decided to count the frequencies of outcomes, 1 through 6, in tossing the die N times. On the assumption that the die is perfectly balanced, one would expect to observe $N/6$ occurrences each for 1, 2, 3, 4, 5, and 6. However, chance dictates that exactly $N/6$ occurrences each will not be observed. For example, given a perfectly balanced die, the probability is only 1 chance in 65 that one will observe 1 outcome each, for 1 through 6, in tossing the die 6 times. Therefore, an outcome different from 1 occurrence each can be expected. Conversely, an outcome of six 3s would seem to be too unusual to have occurred by chance alone.

Some industrial applications involve the concept outlined here. The basic idea is to test whether or not a group of observations follows a preconceived distribution. In the case cited, the distribution is uniform; i.e., each face value should *tend* to occur with the same frequency.

Goodness-of-Fit Test: Procedure

Nomenclature Each experimental observation can be classified into one of r possible categories or cells.

r = total number of cells
O_j = number of observations occurring in cell j
E_j = expected number of observations for cell j based on the preconceived distribution
N = total number of observations
f = degrees of freedom for the test. In general, this will be equal to $(r - 1)$ minus the number of statistical quantities on which the E_j's are based (see the examples which follow for details).

Assumptions

1. The observations represent a sample selected randomly from a population which has been specified.
2. The number of expectation counts E_j within each category should be roughly 5 or more. If an E_j count is significantly less than 5, that cell should be pooled with an adjacent cell.

Computation for E_j On the basis of the specified population, the probability of observing a count in cell j is defined by p_j. For a sample of size N, corresponding to N total counts, the expected frequency is given by $E_j = Np_j$.

Test Statistics: Chi Square

$$\chi^2 = \sum_{j=1}^{r} \frac{(O_j - E_j)^2}{E_j} \qquad \text{with } f \text{ df}$$

Test of Hypothesis

1. H_0: The sample came from the specified theoretical distribution

   H_1: The sample did not come from the specified theoretical distribution
2. For a stated level of α,

   Reject H_0 if sample $\chi^2 >$ tabled χ^2

   Accept H_0 if sample $\chi^2 <$ tabled χ^2

Example

Application. An individual counted the number of occurrences of 1s through 6s in tossing a die 600 times.

Nomenclature
$r = 6$
O_j = observed frequencies (see table following)
E_j = 600/6 = 100 since it is assumed that each outcome is equally likely with probability 1/6
$f = (6 - 1) = 5$, since there are no statistical quantities computed from the sample, with only the total count N, based on the sample size, being used

Computation
1. The following data are given:

Cell	O_j	E_j
1	89	100
2	106	100
3	112	100
4	96	100
5	109	100
6	88	100
Total	600	600

2. The hypotheses:

H_0: the die is perfectly balanced

H_1: the die is not perfectly balanced

3. Assume $\alpha = .05$; therefore, the critical value of $\chi^2(5) = 11.07$.
4. The decision rule:

Reject H_0 if sample $\chi^2 > 11.07$

Accept H_0 otherwise

5. Sample statistic:

$$\text{Sample } \chi^2 = \frac{(89-100)^2}{100} + \frac{(106-100)^2}{100} + \cdots + \frac{(88-100)^2}{100}$$
$$= 5.42$$

6. Conclusion: Accept H_0 since sample $\chi^2 < 11.07$; i.e., there is not sufficient evidence to indicate that the die is out of balance.

Example

Application. A production-line product is rejected if one of its characteristics does not fall within specified limits. The standard goal is that no more than 2 percent of the production should be rejected.

Computation
1. Of 950 units produced during the day, 28 units were rejected.
2. The hypotheses:

H_0: the process is in control

H_1: the process is not in control

3. Assume that $\alpha = .05$; therefore, the critical value of $\chi^2(1) = 3.84$. One degree of freedom is defined since $(r - 1) = 1$, and no statistical quantities have been computed for the data.
4. The decision rule:

Reject H_0 if sample $\chi^2 > 3.84$

Accept H_0 otherwise

5. Since it is assumed that $p = .02$, this would dictate that in a sample of 950 there would be on the average (.02)(950) = 19 defective items and 931 acceptable items:

Category	Observed O_j	Expectation $E_j = 950p_j$
Acceptable	922	931
Not acceptable	28	19
Total	950	950

$$\text{Sample } \chi^2 = \frac{(922-931)^2}{931} + \frac{(28-19)^2}{19}$$
$$= 4.35 \text{ with critical } \chi^2 = 3.84$$

6. Conclusion. Since the sample value exceeds the critical value, it would be concluded that the process is not in control.

Example

Application. A frequency count of workers was tabulated according to the number of defective items that they produced. An unresolved question is whether the observed distribution is a Poisson distribution. That is, do observed and expected frequencies agree within chance variation?

Computation
1. The hypotheses:

H_0: there are no significant differences, in number of defective units, between workers

H_1: there are significant differences

2. Assume that $\alpha = .05$.
3. Test statistic:

No. of defective units	O_j		E_j	
0	3	10	2.06	8.70 pool
1	7		6.64	
2	9		10.73	
3	12		11.55	
4	9		9.33	
5	6		6.03	
6	3	6	3.24	5.66 pool
7	2		1.50	
8	0		.60	
9	1		.22	
≥10	0		.10	
Sum	52		52	

The expectation numbers E_j were computed as follows: For the Poisson distribution, $\lambda = E(x)$; therefore, an estimate of λ is the average number of defective units per worker, i.e., $\hat{\lambda} = (1/52)(0 \times 3 + 1 \times 7 + \cdots + 9 \times 1) = 3.23$. Given this approximation, the probability of no defective units for a worker would be $(3.23)^0/0!)e^{-3.23} = .0396$. For the 52 workers, the number of workers producing no defective units would have an expectation $E = 52(0.396) = 2.06$, and so forth.

The sample chi-square value is computed from

$$\chi^2 = \frac{(10-8.70)^2}{8.70} + \frac{(9-10.73)^2}{10.73} + \cdots + \frac{(6-5.66)^2}{5.66}$$
$$= .58$$

4. The critical value of χ^2 would be based on four degrees of freedom. This corresponds to $(r - 1) - 1$, since one statistical quantity $\hat{\lambda}$ was computed from the sample and used to derive the expectation numbers.
5. The critical value of $\chi^2(4) = 9.49$ with $\alpha = .05$; therefore, accept H_0.

Example

Application. In a U.S. Department of Agriculture study, *Women's Measurements for Garment and Pattern Construction* (Misc. Publ. 454, 1941), 59 measurements were made on a reported sample of 10,042 white women living in the United States. One measurement, stature, was measured to the nearest inch and tabulated into 16 categories. In this study, it was of interest to test whether stature is normally distributed.

Computation
1. The hypotheses:

H_0: stature is normally distributed

H_1: stature is not normally distributed

2. Assume that $\alpha = .05$.
3. For the sample $\bar{x} = 63.66$, $s = 2.485$.
4. Under the normality assumption, tabled values from the normal table would be used to determine the p_j numbers. These numbers are then multiplied by N to obtain the corresponding expectations $E_j = Np_j$.
5. To obtain the conversion from height values to Z values, the parameters μ and σ must be specified. Since these values are unknown for the population, women's stature, the sample estimates $\bar{x}$ and s are used. In this application, these would correspond to two statistical quantities, and therefore the resultant degrees of freedom would be $f = (16 - 1) - 2 = 13$.
6. The sample data are:

Stature Frequencies for Women

Stature, in	Z value	P_j	E_j	O_j
56 and below		.0020	20	29
	−2.882			
57		.0046	46	59
	−2.479			
58		.0123	124	127
	−2.077			
59		.0281	283	248
	−1.674			
60		.0547	549	550
	−1.272			
61		.0906	910	857
	−.869			
62		.1280	1,285	1,255
	−.467			
63		.1540	1,547	1,606
	−.064			
64		.1580	1,587	1,661
	.338			
65		.1382	1,387	1,442
	.741			
66		.1030	1,034	972
	1.143			
67		.0654	657	640
	1.545			
68		.0354	356	337
	1.948			
69		.0163	164	153
	2.350			
70		.0064	65	75
	2.753			
71 and above		.0030	30	31
			10,044*	10,042

*Rounding error.

As an example of the computation for the E_j numbers, consider the stature row corresponding to 61 in. By definition, this row includes the proportion of women between 60.5 and 61.5 in; therefore, as Z numbers

$$\text{For } 60.5\ z = (60.5 - 63.66)/2.485 = -1.674$$
$$\text{For } 61.5\ z = (61.5 - 63.66)/2.485 = -1.272$$

The area under the normal curve below $z = -1.674$ is .0470, and the area below $z = -1.272$ is .1017; therefore, the area between is the difference .0547. Since there are 10,042 women in the sample, it would be expected that (.0547)(10,042), or 549, women would fall within the specified limits of 60.5 to 61.5 in.

7. Sample χ^2:

$$\chi^2 = \frac{(20-29)^2}{20} + \frac{(46-59)^2}{46} + \cdots + \frac{(30-31)^2}{30}$$
$$= 31.28$$

8. With $\alpha = .05$, the critical value of $\chi^2(13) = 26.30$. Therefore, reject the null hypothesis; i.e., there is significant deviation from normality.

Two-Way Test for Independence for Count Data

Nature When individuals or items are observed and classified according to two different criteria, the resultant counts can be statistically analyzed. For example, a group of men might be randomly selected and classified according to eye and hair color. Patients in a hospital could be surveyed and classified according to whether or not they have lung cancer and are or are not smokers.

Count data, based on a random selection of individuals or items which are classified according to two different criteria, can be statistically analyzed through the χ^2 distribution. The purpose of this analysis is to determine whether the respective criteria are dependent. That is, do blue-eyed people tend to be light-haired, for example? Do smokers tend to have a higher rate of lung cancer than nonsmokers?

Two-Way Test for Independence for Count Data: Procedure

Nomenclature

1. Each observation is classified into each of two categories:

a. The first one into 2, 3, . . . , or r categories

b. The second one into 2, 3, . . . , or c categories

2. O_{ij} = number of observations (observed counts) in cell (i,j) with

$$i = 1, 2, \ldots, r$$
$$j = 1, 2, \ldots, c$$

3. N = total number of observations

4. E_{ij} = computed number for cell (i,j) which is an expectation based on the assumption that the two characteristics are independent

5. R_i = subtotal of counts in row i

6. C_j = subtotal of counts in column j

7. α = significance level

8. H_0 = null hypothesis

9. H_1 = alternative hypothesis

10. χ^2 = critical value of χ^2 corresponding to the significance level α and $(r-1)(c-1)$ df

11. Sample $\chi^2 = \sum_{i,j}^{c,r} \frac{(O_{ij} - E_{ij})^2}{E_{ij}}$

Assumptions

1. The observations represent a sample selected randomly from a large total population.

2. The number of expectation counts E_{ij} within each cell should be approximately 2 or more for arrays 3 × 3 or larger. If any cell contains a number smaller than 2, appropriate rows or columns should be combined to increase the magnitude of the expectation count. For arrays 2 × 2, approximately 4 or more are required. If the number is less than 4, the exact Fisher test should be used.

Test of Hypothesis Under the null hypothesis, the classification criteria are assumed to be independent, i.e.,

H_0: the criteria are independent

H_1: the criteria are not independent

For the stated level of α,

Reject H_0 if sample $\chi^2 >$ tabled χ^2

Accept H_0 otherwise

Computation for E_{ij} Compute E_{ij} across rows or down columns by using either of the following identities:

$$E_{ij} = C_j\left(\frac{R_i}{N}\right) \quad \text{across rows}$$

$$E_{ij} = R_i\left(\frac{C_j}{N}\right) \quad \text{down columns}$$

Sample χ^2 Value

$$\chi^2 = \sum_{i,j} \frac{(O_{ij} - E_{ij})^2}{E_{ij}}$$

In the special case of $r = 2$ and $c = 2$, a more accurate and simplinied formula which does not require the direct computation of E_{ij} can be used:

$$\chi^2 = \frac{[|O_{11}O_{22} - O_{12}O_{21}| - \tfrac{1}{2}N]^2 N}{R_1R_2C_1C_2}$$

Example

Application. A survey of 100 patients at a local Veterans Administration hospital was conducted to determine whether there is an association between smoking and lung cancer.

Procedure

1. It was of interest to demonstrate whether an association exists between smoking and lung cancer; therefore the following assumption was made:

H_0: smoking and lung cancer are independent

H_1: they are not independent

2. The sample of 100 patients was classified according to the two criteria:

		Lung cancer		
		Yes	No	R_i
Smoker	Yes	16	27	= 43
	No	4	53	= 57
	C_j	20	80	100

3. Select $\alpha = .05$; therefore with $(r - 1)(c - 1) = 1$ df the critical value of χ^2 is 3.84.
4. The decision rule:

Accept H_0 if sample $\chi^2 < 3.84$

Reject H_0 otherwise

5. The sample value of χ^2 by using the special formula is

$$\text{Sample } \chi^2 = \frac{[|16 \times 53 - 4 \times 27| - 50]^2 100}{(43)(57)(20)(80)}$$

$$= 12.14$$

6. Since the sample χ^2 of 12.14 > tabled χ^2 of 3.84, reject H_0 and accept H_1. We conclude that in the population from which these data are drawn, i.e., patients in a veterans hospital, there appears to be a significant statistical dependence between smoking and lung canrer. However, this dependence does not imply cause and effect. These data suggest that in this population there appears to be a greater proportion of lung-cancer victims among smokers than among nonsmokers. This association may be caused by some other factor, say, a physiological one, that increases a person's chance of both becoming a smoker and contracting lung cancer.

Example

Application. A market research study was carried out to relate the subjective "feel" of a consumer product to consumer preference. In other words, is the consumer's preference for the product associated with the feel of the product, or is the preference independent of the product feel?

Procedure

1. It was of interest to demonstrate whether an association exists between feel and preference; therefore, assume

H_0: feel and prefeence are independent

H_1: they are not independent

2. A sample of 200 people was asked to classify the product according to two criteria:

a. Liking for this product
b. Liking for the feel of the product

		Like feel		
		Yes	No	R_i
Like product	Yes	114	13	= 127
	No	55	18	= 73
	C_j	169	31	200

3. Select $\alpha = .05$; therefore, with $(r - 1)(c - 1) = 1$ df, the critical value of χ^2 is 3.84.
4. The decision rule:

Accept H_0 if sample $\chi^2 < 3.84$

Reject H_0 otherwise

5. The sample value of χ^2 by using the special formula is

$$\text{Sample } \chi^2 = \frac{[|114 \times 18 - 13 \times 55| - 100]^2 200}{(169)(31)(127)(73)}$$

$$= 6.30$$

6. Since the sample χ^2 of 6.30 > tabled χ^2 of 3.84, reject H_0 and accept H_1. The relative proportionality of $E_{11} = 169(127/200) = 107.3$ to the observed 114 compared with $E_{22} = 31(73/200) = 11.3$ to the observed 18 suggests that when the consumer likes the feel, the consumer tends to like the product, and conversely for not liking the feel. The proportions 169/200 = 84.5 percent and 127/200 = 63.5 percent suggest further that there are other attributes of the product which tend to nullify the beneficial feel of the product.

Example

Application. J. F. Tocher [*Biometrika*, **5**, 289–300 (1970)] classified 4235 male inmates of asylums in Scotland by eye and hair color.

Procedure

1. It was of interest to demonstrate whether an association exists between eye and hair color; therefore, assume

H_0: eye and hair color are independent

H_1: they are not independent

2. The sample data:

	Hair color				
Eye color	Red	Fair	Medium	Dark	R_i
Light	37	175	1345	346	= 1903
Medium	20	77	788	497	= 1382
Dark	9	23	389	529	= 950
C_j	66	275	2522	1372	4235

3. Select $\alpha = .05$; therefore, with $(r - 1)(c - 1) = 6$ df the critical value of χ^2 is 12.6.
4. The decision rule:

Accept H_0 if sample $\chi^2 < 12.6$

Reject H_0 otherwise

5. The sample value of χ^2 is determined through the computations for the expectations E_{ij}. In effect, these numbers represent the expected counts within cells if an equal proportionality according to the corresponding row and column proportions is assumed for that cell. For example, with $i = 1$ and $j = 3$ the sample proportion of light-eyed inmates is 1903/4235 and for medium hair color 2522/4235. Under the hypothesis that they are independent, the proportionality of light-eyed and medium-haired inmates would be for cell (1,3)

$$\left(\frac{1903}{4235}\right)\left(\frac{2522}{4235}\right) = .2676$$

Since there are 4235 total inmates, the number of inmates with this combination characteristic should be roughly 4235 × .2676 = 1133.3. This computation reduces to

$$E_{ij} = C_j\left(\frac{R_i}{N}\right) \quad \text{across rows}$$

$$E_{ij} = R_i\left(\frac{C_j}{N}\right) \quad \text{down columns}$$

The corresponding values for E_{ij} are

29.7	123.6	1133.3	616.5
21.5	89.7	823.0	447.7
14.8	61.7	565.7	307.8

with

$$\text{Sample } \chi^2 = 430.93$$

6. Since the sample value of χ^2 exceeds the critical value, we would conclude that there is some association between hair and eye color. A comparison of the pairs of values O_{ij} and E_{ij} suggests a much larger proportionality of dark eyes with dark hair; i.e., dark eyes tend to be associated with dark hair. This has been demonstrated only for the population of male asylum inmates in Scotland around the turn of the twentieth century.

NOTE. Since the row and column subtotals must match the sample subtotals, the χ^2 test has $(r - 1)(c - 1)$ df. In effect, once all the numbers except one have been specified in the first column, the last number becomes fixed because the subtotal is fixed. This would correspond to $(r - 1)$ df, and similarly for each additional column up to, but not including, the last column. The last column is fixed by the row subtotals and hence $(r - 1)(c - 1)$ total df.

One-Way Analysis of Variance

Nature In a previous subsection, the statistical procedure for testing for a significant difference between two sample means was discussed. For three or more sample means this procedure is generalized through an analysis of variance. For the special case of two sample means, the statistical conclusion would, in fact, be the same even though two different test statistics would be used. For two sample means, the t statistic was used, and for the generalization the F statistic will be used. For the special case of two samples, $F(1,\gamma_2) = t^2(\gamma_2)$.

Consider a reported experiment in which 14 samples were taken and, at random, referred to five groups (of predetermined size for other considerations). These respective groups were then individually processed through five different treatments. It was of interest to determine whether the respective treatments resulted in differences in strength levels across the sample groups. For the sample data the averages were as follows:

	Treatment type				
	1	2	3	4	5
	17.3	15.4	18.1	17.9	17.1
	18.3	17.4	16.4	19.5	
	17.6	17.1		20.0	
	18.4				
	18.3				
Column average	17.98	16.63	17.25	19.13	17.10

Overall average = 17.77

As reflected through the column averages, there seem to be relatively large differences associated with the respective treatments. However, on the other side of the balance, there seem also to be relatively large differences within the treatments. The analysis of variance weighs the relative variation between column means against the variation within columns to determine whether the variation between column means is greater than one might expect on the basis of the variation within. For example, if the data had been as shown in the accompanying table, it would be clear that virtually all pairs

	Treatment type				
	1	2	3	4	5
	18.0	16.7	17.3	19.3	17.1
	17.9	16.6	17.2	19.1	
	17.8	16.6		19.0	
	18.1				
	18.1				
Column average	17.98	16.63	17.25	19.13	17.1

Overall average = 17.77

are significantly different. This could be easily demonstrated even in terms of the previous technique for the two-population t statistic. The reason for this is based on the magnitude of the differences in sample means relative to the magnitude of the pooled standard deviation $s_p \approx .1$. For example, the t value for the first pair is $t = (17.98 - 16.63)/(.1\sqrt{1/5 + 1/3}) = 18.49$.

For the actual data, the significance of the relative differences across treatents compared with the variation within treatments is not so clear. The F statistic can be used to quantify this basic idea. That is, on the basis of the variation within treatments, is the variation between treatments sufficiently large to conclude that it is not ascribable to chance? The degree of variation within columns will be quantified by a variance which, in the statistical literature, is labeled the error mean square (EMS). Another type of mean square will be computed on the basis of the column averages. If the magnitude of true differences between the treatments is relatively large, then the variation across column averages will tend to be large. This will result in a large significant F ratio when the variation across columns is divided by the variation within columns.

One-Way Analysis of Variance: Procedure

Nomenclature Given an array,

	Treatment level					
	1	2	...	j	...	k
	y_{11}	y_{12}		y_{1j}		y_{1k}
	y_{21}	y_{22}		y_{2j}		y_{2k}
	$\vdots$	$\vdots$		$\vdots$		$\vdots$
Sample size	n_1	n_2		n_j		n_k
Column sum	S_1	S_2		S_j		S_k
Column average	$\bar{y}_1$	$\bar{y}_2$		$\bar{y}_j$		$\bar{y}_k$

k = number of treatments (columns)
$j = 1, 2, 3, \ldots, k$, the column index
n_j = number of observations in jth column.
$n = n_1 + n_2 + \cdots + n_k$, the total number of observations
S_j = sum of observations in column j
$S = S_1 + S_2 + \cdots + S_k$, the total sum of observations
$\bar{y}_j = S_j/n_j$, the mean of column j
$\bar{y} = S/n$, the overall mean

Linear Model The analysis of variance is predicated on modeling the sample data through a linear model which for a one-way analysis is defined by

$$y_{ij} = \mu + \alpha_j + \epsilon_{ij}$$

where μ = overall population means
α_j = jth treatment effect. These can be interpretated as relatives with their corresponding sum $\Sigma_1^k \alpha_j = 0$ (a fixed-effects model). The purpose of this is to quantify the relative magnitude of differences across treatment levels.
ϵ_{ij} = random error associated with ith observation under the jth treatment level

Assumptions

1. The respective observations y_{ij} are random and independent.
2. The model is a true representation of the observations.
3. The random errors ϵ_{ij} are normally distributed with a population mean 0 and variance σ^2. An assumption for this procedure is that the variances within the respective columns are equal to σ^2.

Analysis-of-Variance Table

1. The total sum of squared deviations TSS $= \Sigma(y_{ij} - \bar{y})^2$ can be partitioned into two components:

a. The sum of squared deviations within columns, called the error sum of squares (ESS)

b. The sum of squared deviations between columns (CSS)

Therefore, TSS = ESS + CSS.

2. From the standpoint of a summary display, the respective sums of squared deviations (SS) can be arranged in an analysis-of-variance table with their respective degrees of freedom (df):

Source	SS	df	MS	*F* ratio
Columns		$k - 1$		
Error		$n - k$		
Total		$n - 1$		

The fourth column in the table is labeled MS, an abbreviation for mean square (it is actually a special type of variance), and is equal to the corresponding SS divided by the df. The F ratio is defined to be the column mean square (CMS) divided by the error mean square (EMS).

Computational Procedure

$$\text{TSS} = \sum_{i,j} (y_{ij} - \bar{y})^2 = \Sigma y_{ij}^2 - S^2/n$$

$$\text{ESS} = \sum_{j=1}^{k}\left[\sum_{i=1}^{n_j} (y_{ij} - \bar{y}_j)^2\right] = \sum_{j=1}^{k}\left[\sum_{i=1}^{n_j} y_{ij}^2 - S_j^2/n_j\right]$$

and $$\text{ESS}_j = \sum_{i=1}^{n_j} (y_{ij} - \bar{y}_j)^2$$

$$\text{CSS} = \sum_{j=1}^{k} n_j(\bar{y}_j - \bar{y})^2 = \sum_{j=1}^{k} S_j^2/n_j - S^2/n$$

Test of Hypothesis

H_0: $\alpha_j = 0$ for all j
H_1: not all α_j are zero

Under the null hypothesis, it is assumed that the treatment effects are zero; i.e., there are no differences between treatments corresponding to all α's equal to zero. Given that the null hypothesis is true, then the sampling distribution of the ratio of CMS over EMS is an F distribution. That is, the ratio represents the chance distribution of two independent sample variances both of which have a population variance σ^2. For a stated α level, the critical value of $F(k-1, n-k)$ can be found in the table. If the sample value of F exceeds the tabled value, this would be interpreted as an unusual outcome, and therefore the null hypothesis would be questioned and subsequently rejected.

Decision Rule Accept H_0 if sample $F <$ tabled F; reject H_0 otherwise.

Example

Application. Fourteen samples were randomly assigned to five different treatment conditions according to a prescribed schedule. After treatment, the strength levels of the respective samples were tested. It was of interest to determine whether the types of treatment affected the resultant strength level of the samples.

Procedure

1. The hypotheses:

$$H_0\text{: } \alpha_j = 0 \text{ for } j = 1, 2, 3, 4, \text{ and } 5$$
$$H_1\text{: not all } \alpha_j \text{ are zero}$$

2. Select $\alpha = .05$; therefore, with $(k-1) = 4$ df for the CMS and $(n-k) = 9$ df for the EMS the critical value of $F(4,9) = 3.63$.

3. The decision rule:

Accept H_0 if sample $F < 3.63$

Reject H_0 otherwise

4. The sample data and associated summaries:

	Treatment				
	1	2	3	4	5
	17.3	15.4	18.1	17.9	17.1
	18.3	17.4	16.4	19.5	
	17.6	17.1		20.0	
	18.4				
	18.3				
$n_j =$	5	3	2	3	1
$S_j =$	89.9	49.9	34.5	57.4	17.1
$\bar{y}_j =$	17.98	16.63	17.25	19.13	17.10
$\text{ESS}_j =$	.988	2.327	1.445	2.407	

Overall: $n = 14$ $\quad S = 248.8 \quad \bar{y} = 17.77$

5. The total sum of squared deviations was computed by the formula

$$\text{TSS} = \sum_{i,j} y_{ij}^2 - S^2/n$$
$$= 4439.36 - (248.8)^2/14$$
$$= 17.829$$

6. The within or error sum of squared deviations was computed by summing the ESS_j across columns, where $\text{ESS}_j = \Sigma_{i=1}^{n_j} y_{ij}^2 - S_j^2/n_j$ within columns with ESS = 7.167.

7. The column sum of squared deviations can be obtained by differences through the partition identity, i.e.,

$$\text{CSS} = \text{TSS} - \text{ESS}$$
$$= 17.829 - 7.167$$
$$= 10.662$$

However, for purposes of generality, the CSS will be determined in two different ways:

a. The awkward but conceptual formula:

$$\text{CSS} = \sum_{j=1}^{k} n_j(\bar{y}_j - \bar{y})^2$$
$$= 5(17.98 - 17.77)^2 + 3(16.63 - 17.77)^2$$
$$+ 2(17.25 - 17.77)^2 + 3(19.13 - 17.77)^2$$
$$+ 1(17.10 - 17.77)^2$$
$$= 10.662$$

This corresponds to the computation of the sum of squared deviations about a mean, correcting for the fact that the individual numbers represent averages themselves. Therefore, the individual terms are corrected for the corresponding sample sizes through the identity (under the null hypothesis)

$$\text{Var}(\bar{x}_j) = \sigma^2/n_j$$

or

$$\sigma^2 = n_j \text{ var}(\bar{x}_j)$$

b. The convenient computational formula:

$$\text{CSS} = \Sigma S_j^2/n_j - S^2/n$$
$$= \frac{(89.9)^2}{5} + \frac{(49.9)^2}{3} + \frac{(34.5)^2}{2} + \frac{(57.4)^2}{3}$$
$$+ \frac{(17.1)^2}{1} - \frac{(248.8)^2}{14}$$
$$= 10.662$$

8. The analysis of variance:

Source	SS	df	MS	F ratio
Between columns	10.662	4	2.666	3.35
Error	7.167	9	.796	
Total	17.829	13		

9. The decision: Since the sample value of $F = 3.35 <$ tabled value of $F = 3.63$, we must conclude that there is not *sufficient* evidence to reject the null hypothesis. This example points out the fact that when an experiment is less than ideal, an unsatisfying solution may result. The statistical results seem to indicate that there are true differences between the treatments, but unfortunately they can not be demonstrated. As in the case of two populations, the optimum design would require an equal number of samples within each treatment column. If the schedule had permitted three samples within each treatment ($n = 15$) and the same column averages had resulted

$$\bar{y}_1 = 17.98$$
$$\bar{y}_2 = 16.63$$
$$\bar{y}_3 = 17.25$$
$$\bar{y}_4 = 19.13$$
$$\bar{y}_5 = 17.10$$

with the same EMS of .796 (without any of the basic quantities being changed), the corresponding sample F ratio would have been 3.57 with a critical $F(4,10) = 3.48$ rather than the observed value 3.35 with a critical $F(4,9) = 3.63$. In other words, with a more efficient design the resultant conclusion would have demonstrated a significant difference without the recourse of an additional completely new test.

Example

Application. A manufacturer of paper used for paper towels was interested in improving the tensile strength of the product. Since it was suspected that the tensile strength was a function of the concentration of an additive in the pulp, it was decided to investigate five different levels of concentrations with five samples selected randomly under each level.

Procedure

1. The hypotheses:

$$H_0\text{: } \alpha_j = 0 \text{ for } j = 1, 2, 3, 4, \text{ and } 5$$
$$H_1\text{: not all } \alpha_j \text{ are zero}$$

2. Select $\alpha = .05$; therefore, with $(k-1) = 4$ df for the CMS and $n - k = 20$ df for the EMS the critical value of $F(4, 20) = 2.87$.

3. The decision rule:

Accept H_0 if sample $F < 2.87$

Reject H_0 otherwise

4. The sample data and associated summaries for the tensile-strength determinations:

	Additive concentration				
	0%	2%	4%	6%	10%
	17	16	22	20	22
	14	23	24	21	20
	15	17	18	19	17
	18	20	18	24	20
	19	19	23	25	19
$n_j =$	5	5	5	5	5
$S_j =$	83	95	105	109	98
$\Sigma y_{ij}^2 =$	1395	1835	2237	2403	1934
$\bar{y}_j =$	16.6	19.0	21.0	21.8	19.6
$\text{ESS}_j =$	17.2	30.0	32.0	26.8	13.2

Overall: $n = 25 \quad S = 490 \quad \bar{y} = 19.60$

where S_j corresponds to the column sum and s_j^2 to the within-column variance.

5. For the two-population test of hypothesis, the assumption $\sigma_1^2 = \sigma_2^2$ was made. This was tested as a preliminary to the main test through the F ratio of sample variances s_1^2 and s_2^2. Also, in the analysis of variance it is assumed that the population variances σ_j^2 within columns are all equal to σ^2. The generalization of a two-population F test for the equality of variances is called Bartlett's test, which is summarized in the following material. Application of Bartlett's test to the sample data in this example leads to accepting the null hypothesis H_0: $\sigma_j^2 = \sigma^2$ across j.

6. The total sum of squared deviations was computed by the formula

$$\text{TSS} = \Sigma y_{ij}^2 - S^2/n = 9804 - (490)^2/25 = 200.00$$

7. The within or error sum of squared deviations was computed by summing the ESS_j across columns:

$$\text{ESS} = 17.2 + 30.0 + 32.0 + 26.8 + 13.2 = 119.2$$

8. The between-column sum of squared deviations was computed by

$$\text{CSS} = \frac{83^2}{5} + \frac{95^2}{5} + \frac{105^2}{5} + \frac{109^2}{5} + \frac{98^2}{5} - \frac{490^2}{25} = 80.8$$

9. The analysis of variance:

Source	SS	df	MS	Sample F ratio
Between columns	80.8	4	20.20	3.39
Error	119.2	20	5.96	
Total	200.0	24		

10. The decision: Since the sample value of $F = 3.39 >$ tabled $F = 2.87$, we would reject the hypothesis and conclude that there are at least two values of α_j different from zero (since they are only relatives, they are defined to sum to zero, and therefore if one is different from zero, one or more others also must be different from zero).

For this application, it is of interest to go to the second stage and estimate the respective quantities α_j. On the basis of statistical theory and the fact that the estimates are based on a sample set of data, the nomenclature $\hat{\alpha}_j$ is used for these estimates. They are obtained through the identity $\hat{\alpha}_j = \bar{y}_j - \bar{y}$. Thus

$$\hat{\alpha}_1 = -3.0$$
$$\hat{\alpha}_2 = -\ .6$$
$$\hat{\alpha}_3 = 1.4$$
$$\hat{\alpha}_4 = 2.2$$
$$\hat{\alpha}_5 = 0$$

A third-stage analysis can be applied to determine which pairs, of all possible pairs, are significantly different. One method for carrying out this analysis is called the Duncan multiple-range test (Duncan, "Multiple Range and Multiple F Tests," *Biometrics*, no. 11, 1–42, 1956). For this, arrange the respective estimates in increasing order:

$$\hat{\alpha}_1 = -3.0$$
$$\hat{\alpha}_2 = -\ .6$$
$$\hat{\alpha}_5 = 0$$
$$\hat{\alpha}_3 = 1.4$$
$$\hat{\alpha}_4 = 2.2$$

The results of this test indicate significant differences between only two pairs. The test itself weighs the relative magnitude of the difference between successive pairs relative to the reliability of the difference, in a way similar to the conceptual idea of a confidence interval between two population means. See an ensuing subsection for a discussion of this procedure. In addition, see the second example under "Multiple Regression" for a more viable approach to the analysis.

Example

Application. Five samples, cut ⅝ by 2¾ in, were selected in the cross direction (CD) of a 1-mil polyolefin-based film and measured for thickness at 10 equally spaced locations in the 2¾-in direction. It was of interest to characterize the CD variation relative to the variation within the samples.

Procedure

1. The hypotheses:

$$H_0\text{: } \alpha_j = 0 \text{ for } j = 1, 2, 3, 4, \text{ and } 5$$
$$H_1\text{: not all } \alpha_j \text{ are zero}$$

2. Select $\alpha = .05$. Therefore, with $(k - 1) = 4$ df for the mean square between (MSB) and $(n - k) = 45$ df for the mean square error (MSE) the critical value of F (4, 45) = 2.58.

3. The decision rule:

Accept H_0 if sample $F < 2.58$

Reject H_0 if sample $F > 2.58$

4. The sample data and associated summaries for the thickness measurements:

	Sample number				
	1	2	3	4	5
	1.014	1.031	.991	1.004	.964
	1.001	1.033	.987	.983	.956
	.998	1.024	.975	.996	.957
	1.013	1.021	.970	.992	.967
	1.014	1.038	.974	.997	.968
	1.008	1.049	.980	.999	.968
	1.012	1.059	.960	1.004	.977
	.970	1.054	.960	.953	.982
	.964	1.056	.982	1.004	.981
	.956	1.057	.971	.997	.988
n_j	10	10	10	10	10
S_j	9.950	10.422	9.750	9.929	9.708
$\bar{y}_j$	.995	1.042	.975	.993	.971
SSE_j	.00466	.00189	.00097	.00214	.00105
s_j	.023	.014	.010	.015	.011

Overall: $n = 50$ $S = 49.759$ $\bar{y} = .9952$

5. The Bartlett test of H_0: $\sigma_1^2 = \cdots = \sigma_5^2 = \sigma^2$ indicates that there is insufficient evidence to reject this hypothesis; i.e., the differences across the five sample variances are no larger than one might expect by chance alone at an α level of 5 percent.

6. The total sum of squared deviations was computed by the formula

$$\text{SST} = \Sigma y_{ij}^2 - S^2/n = .04289$$

7. The within or error sum of squared deviations was computed by summing the SSE_j across the columns:

$$\text{SSE} = .00466 + .00189 + .00097 + .00214 + .00105 = .01071$$

8. The between-column sum of squared deviations was computed by

$$\text{SSB} = \frac{(9.950)^2}{10} + \frac{(10.422)^2}{10} + \frac{(9.750)^2}{10} + \frac{(9.929)^2}{10} + \frac{(9.708)^2}{10} - \frac{49.759)^2}{50} = .03218$$

9. The analysis of variance:

Source	SS	df	MS	Sample F ratio
Between columns	.03218	4	.008045	33.8
Error	.01071	45	.000238	
Total	.04289	49		

10. The decision: Since the sample value of $F = 33.8 >$ tabled $F = 2.58$, the null hypothesis would be rejected.

11. For this application it was of interest to characterize the true variation in the CD direction, i.e., to estimate the true CD variance independently of the inherent short-term-error variance σ^2. Since the expectation for the MSB is $\sigma^2 + n_j\sigma_T^2$,

$$\hat{\sigma}^2 + 10\hat{\sigma}_T^2 = .008045$$

where

$$\hat{\sigma}^2 = .000238$$

Therefore,

$$\hat{\sigma}_T^2 = (1/10)(0.008045 - .000238) = .000781$$

or

$$\hat{\sigma} = .0154 \qquad \hat{\sigma}_T = .0279$$

In other words, there is roughly twice as much true variation in the CD direction as there is within sample variation.

Bartlett's Test for Equal Variances In the preceding example it was concluded that there was insufficient evidence to reject the null hypothesis H_0: $\sigma_1^2 = \sigma_2^2 = \cdots = \sigma_5^2$. This test was based on a procedure developed by Bartlett [*Proc. R. Soc.*, 273–275 (1937)].

Given k sample variances, $s_1^2, s_2^2, \ldots, s_k^2$, from samples of size $n_1, n_2, \ldots, n_k$ with $N = \Sigma n_j$, compute the pooled estimate

$$s_p^2 = \frac{1}{N-k} \sum_1^k (n_j - 1)s_j^2$$

Define

$$q = (N-k) \ln s_p^2 - \Sigma[(n_j - 1) \ln s_j^2]$$

and
$$h = 1 + \frac{1}{3(k-1)}\left(\left(\sum_1^k \frac{1}{n_j - 1}\right) - \frac{1}{N-k}\right)$$

Then $q/h \simeq \chi^2(k-1)$, i.e., $(k-1)$ df, corresponding to an upper-tail test.

Example For the problem statement and data refer to the preceding example.

1. Data summary:

Sample	n_j	s_j
1	10	.023
2	10	.014
3	10	.010
4	10	.015
5	10	.011
	$N = 50$	

2. Hypothesis:

$$H_0: \sigma_1^2 = \sigma_2^2 = \sigma_3^2 = \sigma_4^2 = \sigma_5^2$$
$$H_1: \text{not all } \sigma_j^2 \text{ are equal}$$

3. Significance level: Pick $\alpha = .05$; therefore, critical value of $\chi^2(4) = 9.49$.

4. Pooled variance:

$$s_p^2 = \frac{9(.023)^2 + 9(.014)^2 + \cdots + 9(.011)^2}{45}$$
$$= \frac{(.023)^2 + (.014)^2 + \cdots + (.011)^2}{5}$$
$$= .0002342$$

5. Compute q:

$$q = (50-5) \ln(.0002342) - [9 \ln(.023)^2 + 9 \ln(.014)^2 + \cdots + 9 \ln(.011)^2]$$
$$= -376.17 - (-384.40)$$
$$= 8.23$$

6. Compute h:

$$h = 1 + \frac{1}{3(4)}\left(\frac{1}{9} + \frac{1}{9} + \cdots + \frac{1}{9} - \frac{1}{45}\right)$$
$$= 1.0444$$

7. Test of hypothesis: The sample value of the chi-square approximation is

$$\text{Sample } \chi^2 = 8.23/1.0444$$
$$= 7.88$$

Since sample $\chi^2 = 7.88 <$ tabled $\chi^2 = 9.49$, accept the null hypothesis; i.e., there is insufficient evidence to reject H_0.

Example For the problem statement and data refer to the first example under "One-Way Analysis of Variance: Procedure."

1. Data summary:

Sample	n_j	s_j^2
1	5	.2470
2	3	1.1635
3	2	1.4450
4	3	1.2035
5	1	Indeterminate (0/0)
		$N = 13$

2. Hypothesis:

$$H_0: \sigma_1^2 = \sigma_2^2 = \sigma_3^2 = \sigma_4^2$$
$$H_1: \text{not all } \sigma_j^2 \text{ are equal}$$

3. Significance level. Pick $\alpha = .05$; therefore, the upper critical value of $\chi^2(3) = 7.81$.

4. Pooled variance:

$$s_p^2 = \frac{4(.2470) + 2(1.1635) + 1(1.4450) + 2(1.2035)}{9}$$
$$= .7963$$

5. Compute q:

$$q = (13-4) \ln(.7963) - [4 \ln(.2470) + 2 \ln(1.1635) + \ln(1.445) + 2(1.2035)]$$
$$= -2.050 - (-4.552)$$
$$= 2.502$$

6. Compute h:

$$h = 1 + \frac{1}{3(3)}\left(\frac{1}{4} + \frac{1}{2} + \frac{1}{1} + \frac{1}{2} - \frac{1}{9}\right)$$
$$= 1.238$$

7. Test of hypothesis: The sample value of the chi-square approximation is

$$\text{Sample } \chi^2 = 2.502/1.238$$
$$= 2.02$$

Since sample $\chi^2 = 2.02 <$ tabled $\chi^2 = 7.81$, accept the null hypothesis.

NOTE: Bartlett's test is an approximate χ^2 test and suffers from two limitations. First, the test is sensitive to an assumption that observations within sample groups are normally distributed. When observations deviate substantially from normality, the true type I error α can be substantially different from the assumed value used in the chi-square table. Second, the power for Bartlett's test tends to be small. In statistical theory, it is interpreted as a relatively weak test of hypothesis. There are other specialized tests but no general test other than Bartlett's test.

Duncan Multiple-Range Test In the second example presented under "One-Way Analysis of Variance," the sample estimates

$$\hat{\alpha}_1 = -3.0$$
$$\hat{\alpha}_2 = -.6$$
$$\hat{\alpha}_3 = 1.4$$
$$\hat{\alpha}_4 = 2.2$$
$$\hat{\alpha}_5 = 0$$

were analyzed, through the Duncan multiple-range test, for significant differences. It was concluded that only two pairs of estimates were significantly different: $(\hat{\alpha}_4, \hat{\alpha}_1)$ and $(\hat{\alpha}_3, \hat{\alpha}_1)$.

Example The following is a summary of the steps to be taken to demonstrate significant differences:

1. Arrange the $\hat{\alpha}$'s in order from low to high:

$$\hat{\alpha}_1 = -3.0$$
$$\hat{\alpha}_2 = -.6$$
$$\hat{\alpha}_5 = 0$$
$$\hat{\alpha}_3 = 1.4$$
$$\hat{\alpha}_4 = 2.2$$

2. Enter the analysis-of-variance table and tabulate the error mean square with its degrees of freedom: $\hat{\sigma}^2 = 5.96$ with 20 df.

3. Compute the standard error for the respective $\hat{\alpha}$'s, i.e.,

$$\text{Standard error } \hat{\alpha}_j = \sqrt{\hat{\sigma}^2/n_j}$$
$$= \sqrt{5.96/5}$$
$$= 1.09$$

4. Enter Duncan's table (Table 2-1) of significant ranges at the α level desired ($\alpha = .05$ in this example) by using the df for the error (20) and $p =$

TABLE 2-1 Significant Ranges: Duncan Multiple-Range Test*

n_2/p	2	3	4	5	6	7	8	9	10	12	14	16	18	20	50	100
	Significant ranges for a 1 percent level new† multiple-range test															
1	90.0	90.0	90.0	90.0	90.0	90.0	90.0	90.0	90.0	90.0	90.0	90.0	90.0	90.0	90.0	90.0
2	14.0	14.0	14.0	14.0	14.0	14.0	14.0	14.0	14.0	14.0	14.0	14.0	14.0	14.0	14.0	14.0
3	8.26	8.5	8.6	8.7	8.8	8.9	8.9	9.0	9.0	9.0	9.1	9.2	9.3	9.3	9.3	9.3
4	6.51	6.8	6.9	7.0	7.1	7.1	7.2	7.2	7.3	7.3	7.4	7.4	7.5	7.5	7.5	7.5
5	5.70	5.96	6.11	6.18	6.26	6.33	6.40	6.44	6.5	6.6	6.6	6.7	6.7	68	6.8	6.8
6	5.24	5.51	5.65	5.73	5.81	5.88	5.95	6.00	6.0	6.1	6.2	6.2	6.3	6.3	6.3	6.3
7	4.95	5.22	5.37	5.45	5.53	5.61	5.69	5.73	5.8	5.8	5.9	5.9	6.0	6,0	6.0	6.0
8	4.74	5.00	5.14	5.23	5.32	5.40	5.47	5.51	5.5	5.6	5.7	5.7	5.8	5.8	5.8	5.8
9	4.60	4.86	4.99	5.08	5.17	5.25	5.32	5.36	5.4	5.5	5.5	5.6	5.7	5.7	5.7	5.7
10	4.48	4.73	4.88	4.96	5.06	5.13	5.20	5.24	5.28	5.36	5.42	5.48	5.54	5.55	5.55	5.55
11	4.39	4.63	4.77	4.86	4.94	5.01	5.06	5.12	5.15	5.24	5.28	5.34	5.38	5.39	5.39	5.39
12	4.32	4.55	4.68	4.76	4.84	4.92	4.96	5.02	5.07	5.13	5.17	5.22	5.24	5.26	5.26	5.26
13	4.26	4.48	4.62	4.69	4.74	4.84	4.88	4.94	4.98	5.04	5.08	5.13	5.14	5.15	5.15	5.15
14	4.21	4.42	4.55	4.63	4.70	4.78	4.83	4.87	4.91	4.96	5.00	5.04	5.06	5.07	5.07	5.07
15	4.17	4.37	4.50	4.58	4.64	4.72	4.77	4.81	4.84	4.90	4.94	4.97	4.99	5.00	5.00	5.00
16	4.13	4.34	4.45	4.54	4.60	4.67	4.72	4.76	4.79	4.84	4.88	4.91	4.93	4.94	4.94	4.94
17	4.10	4.30	4.41	4.50	4.56	4.63	4.68	4.73	4.75	4.80	4.83	4.86	4.88	4.89	4.89	4.89
18	4.67	4.27	4.38	4.46	4.43	4.59	4.64	4.68	4.71	4.76	4.79	4.82	4.84	4.85	4.85	4.85
19	4.05	4.24	4.35	4.43	4.50	4.56	4.61	4.64	4.67	4.72	4.76	4.79	4.81	4.82	4.82	4.82
20	4.02	4.22	4.33	4.40	4.47	4.53	4.58	4.61	4.65	4.69	4.73	4.76	4.78	4.79	4.79	4.79
22	3.99	4.17	4.28	4.36	4.42	4.48	4.53	4.57	4.60	4.65	4.68	4.71	4.74	4.75	4.75	4.75
24	3.96	4.14	4.24	4.33	4.39	4.44	4.49	4.53	4.57	4.62	4.64	4.67	4.70	4.72	4.74	4.74
26	3.93	4.11	4.21	4.30	4.36	4.41	4.46	4.50	4.53	4.58	4.62	4.65	4.67	4.69	4.73	4.73
28	3.91	4.08	4.18	4.28	4.34	4.39	4.43	4.47	4.51	4.56	4.60	4.62	4.65	4.67	4.72	4.72
30	3.89	4.06	4.16	4.22	4.32	4.36	4.41	4.45	4.48	4.54	4.58	4.61	4.63	4.65	4.71	4.71
40	3.82	3.99	4.10	4.17	4.24	4.30	4.34	4.37	4.41	4.46	4.51	4.54	4.57	4.59	4.69	4.69
60	3.76	3.92	4.03	4.12	4.17	4.23	4.27	4.31	4.34	4.39	4.44	4.47	4.50	4.53	4.66	4.66
100	3.71	3.86	3.98	4.06	4.11	4.17	4.21	4.25	4.29	4.35	4.38	4.42	4.45	4.48	4.64	4.65
∞	3.64	3.80	3.90	3.98	4.04	4.09	4.14	4.17	4.20	4.26	4.31	4.34	4.38	4.41	4.60	4.68
	Significant studentized ranges for a 5 percent level new† multiple-range test															
1	18.0	18.0	18.0	18.0	18.0	18.0	18.0	18.0	18.0	18.0	18.0	18.0	18.0	18.0	18.0	18.0
2	6.09	6.09	6.09	6.09	6.09	6.09	6.09	6.09	6.09	6.09	6.09	6.09	6.09	6.09	6.09	6.09
3	4.50	4.50	4.50	4.50	4.50	4.50	4.50	4.50	4.50	4.50	4.50	4.50	4.50	4.50	4.50	4.50
4	3.93	4.01	4.02	4.02	4.02	4.02	4.02	4.02	4.02	4.02	4.02	4.02	4.02	4.02	4.02	4.02
5	3.64	3.74	3.79	3.83	3.83	3.83	3.83	3.83	3.83	3.83	3.83	3.83	3.83	3.83	3.83	3.83
6	3.46	3.58	3.64	3.68	3.68	3.68	3.68	3.68	3.68	3.68	3.68	3.68	3.68	3.68	3.68	3.68
7	3.35	3.47	3.54	3.58	3.60	3.61	3.61	3.61	3.61	3.61	3.61	3.61	3.61	3.61	3.61	3.61
8	3.26	3.39	3.47	3.52	3.55	3.56	3.56	3.56	3.56	3.56	3.56	3.56	3.56	3.56	3.56	3.56
9	3.20	3.34	3.41	3.47	3.50	3.52	3.52	3.52	3.52	3.52	3.52	3.52	3.52	3.52	3.52	3.52
10	3.15	3.30	3.37	3.43	3.46	3.47	3.47	3.47	3.47	3.47	3.47	3.47	3.47	3.48	3.48	3.38
11	3.11	3.27	3.35	3.39	3.43	3.44	3.45	3.46	3.46	3.46	3.46	3.46	3.47	3.48	3.48	3.48
12	3.08	3.23	3.33	3.36	3.40	3.42	3.44	3.44	3.46	3.46	3.46	3.46	3.47	3.48	3.48	3.48
13	3.06	3.21	3.30	3.35	3.38	3.41	3.42	3.44	3.45	3.45	3.46	3.46	3.47	3.47	3.47	3.47
14	3.03	3.18	3.27	3.33	3.37	3.39	3.41	3.42	3.44	3.45	3.46	3.46	3.47	3.47	3.47	3.47
15	3.01	3.16	3.25	3.31	3.36	3.38	3.40	3.42	3.43	3.44	3.45	3.46	3.47	3.47	3.47	3.47
16	3.00	3.15	3.23	3.30	3.34	3.37	3.39	3.41	3.43	3.44	3.45	3.46	3.47	3.47	3.47	3.47
17	2.98	3.13	3.22	3.28	3.33	3.36	3.38	3.40	3.42	3.44	3.45	3.46	3.47	3.47	3.47	3.47
18	2.97	3.12	3.21	3.27	3.32	3.35	3.37	3.39	3.41	3.43	3.45	3.46	3.47	3.47	3.47	3.47
19	2.96	3.11	3.19	3.26	3.31	3.35	3.37	3.39	3.41	3.43	3.44	3.46	3.47	3.47	3.47	3.47
20	2.95	3.10	3.18	3.25	3.30	3.34	3.36	3.38	3.40	3.43	3.44	3.46	3.47	3.47	3.47	3.47
22	2.93	3.08	3.17	3.24	3.29	3.32	3.35	3.37	3.39	3.42	3.44	3.45	3.46	3.47	3.47	3.47
24	2.92	3.07	3.15	3.22	3.28	3.31	3.34	3.37	3.38	3.41	3.44	3.45	3.46	3.47	3.47	3.47
26	2.91	3.06	3.14	3.21	3.27	3.30	3.34	3.36	3.38	3.41	3.43	3.45	3.46	3.47	3.47	3.47
28	2.90	3.04	3.13	3.20	3.26	3.30	3.33	3.35	3.37	3.40	3.43	3.45	3.46	3.47	3.47	3.47
30	2.89	3.04	3.12	3.20	3.25	3.29	3.32	3.35	3.37	3.40	3.43	3.44	3.46	3.47	3.47	3.47
40	2.86	3.01	3.10	3.17	3.22	3.27	3.30	3.33	3.35	3.39	3.42	3.44	3.46	3.47	3.47	3.47
60	2.83	2.98	3.08	3.14	3.20	3.24	3.28	3.31	3.33	3.37	3.40	3.43	3.45	3.47	3.48	3.48
100	2.80	2.95	3.05	3.12	3.18	3.22	3.26	3.29	3.32	3.36	3.40	3.42	3.45	3.47	3.53	3.53
∞	2.77	2.92	3.02	3.09	3.15	3.19	3.23	3.26	3.29	3.34	3.38	3.41	3.44	3.47	3.61	3.67

*From D. B. Duncan, "Multiple Range and Multiple F Tests," *Biometrics*, **11** (1956). With permission from the Biometric Society.
†Using special protection levels based on degrees of freedom.

2, 3, . . . , k, where k corresponds to the total number of treatment levels, five in this example:

p	Duncan's range numbers
2	2.95
3	3.10
4	3.18
5	3.25

5. Multiply the Duncan range numbers by the standard error from step 3:

p	Duncan × standard error
2	3.22
3	3.38
4	3.47
5	3.54

6. Test the observed ranges against their corresponding critical values with the decision rule that if the observed difference is larger than the critical value, it is significantly different; otherwise, not.

p	Quantities	Difference	Critical value	Conclusion
5	$\hat{\alpha}_4 - \hat{\alpha}_4$	5.2	3.54	Significant difference
4	$\hat{\alpha}_3 - \hat{\alpha}_1$	4.4	3.47	Significant difference
3	$\hat{\alpha}_5 - \hat{\alpha}_1$	3.0	3.38	Not
2	$\hat{\alpha}_2 - \hat{\alpha}_1$	2.4	3.22	Not
4	$\hat{\alpha}_4 - \hat{\alpha}_2$	2.8	3.47	Not
3	$\hat{\alpha}_3 - \hat{\alpha}_2$	2.0	3.38	Not
2	$\hat{\alpha}_5 - \hat{\alpha}_2$	.6	3.22	Not
3	$\hat{\alpha}_4 - \hat{\alpha}_5$	2.2	3.38	Not
2	$\hat{\alpha}_3 - \hat{\alpha}_5$	1.4	3.22	Not
2	$\hat{\alpha}_4 - \hat{\alpha}_3$	.8	3.22	Not

7. The sample $\hat{\alpha}$ can be plotted and lines drawn under significant differences to illustrate the conclusions (see Fig. 2-66). In terms of the original concentration levels, only the 6 and 4 percent levels are significantly different from the 0 percent level.

Two-Way Analysis of Variance

Nature An analysis of corrosion rates for two types of pipes buried in a variety of soils was discussed under "Test of Hypothesis for Matched Pairs." The method of analysis was called **paired observations.** In that example, it was possible to evaluate the relative corrosion rate between the two types by analyzing the differences within soil types. An examination of the data indicates that by working with differences it was possible to eliminate the direct influence of soil types. However, on the basis of an examination of soil types, for example, A, H, and K, it is apparent that the type of soil has a direct influence on the corrosion rate by itself.

If several other types of pipe had been included in this application, the use of paired comparisons would not have been appropriate, since the technique is restricted to two treatments (types of pipe). If, for example, six types of pipe had been included in the study, one might be tempted to make all comparisons two at a time. However, as a statistical procedure this comparison across 15 pairs is not valid. In addition, in many types of applications it is equally important to demonstrate the magnitude of these row effects (soil effects). In some of these the row effects might not be statistically significant, and it would be important to quantify this.

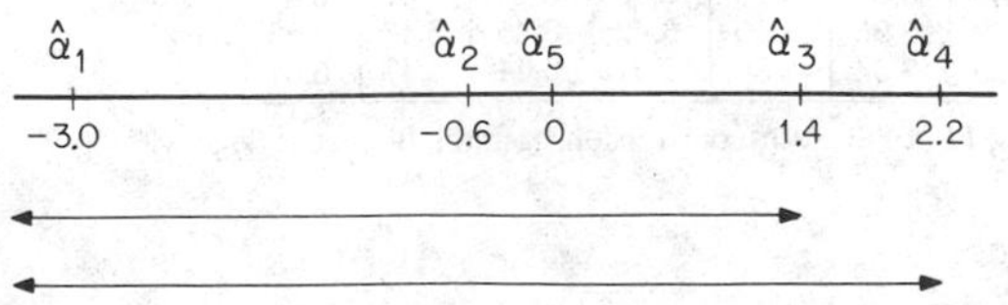

FIG. 2-66 Graphical illustrations of conclusions.

The technique of evaluating applications involving two different treatment effects, such as pipe and soil types, is called **two-way analysis.** For these applications, there are two distinct ways of modeling the experiment, with and without what is called interaction. In the corrosion example, if the relative rate of corrosion between pipes tended to vary according to soil type, this would reflect interaction. As an illustration, see the following table.

	Noninteractive			Interactive		
Soil type	Steel	Bare	Difference	Steel	Bare	Difference
1	7.6	10.4	−2.8	7.6	20.7	−13.1
2	24.5	26.3	−1.8	24.5	25.3	− .8
3	52.6	56.3	−3.7	52.6	42.6	10.0
4	79.8	81.4	−1.6	79.8	60.3	19.5

Interaction reflects the fact that the relative differences, in this case the individual differences, will change in characteristic according to the level of the corresponding row. For the interactive data shown in the table, the relative differences between pipes change according to the soil type. If the soil is relatively noncorrosive, the steel pipe is preferable, and, correspondingly, if the soil is relatively more corrosive, the bare pipe is preferable.

In dealing with an interactive system, it is necessary to have replicated observations, namely, repeated observations or samples within the row and column combinations.

Alternative modeling structures will be illustrated later in this subsection through two examples, noninteractive in the first example and interactive in the second.

Two-Way Analysis of Variance: Procedure

Nomenclature

Treatment type II	Treatment type I: level 1	2	3	. . .	c
1	y_{111} y_{112} ↓	y_{121} y_{122} ↓	y_{131} y_{132} ↓		y_{1c1} y_{1c1} ↓
2	y_{211} y_{212} ↓	y_{221} y_{222} ↓	y_{231} y_{232} ↓		y_{2c1} y_{2c2} ↓
r	y_{r11} y_{r12} ↓	y_{r21} y_{r22} ↓	y_{r31} y_{r32} ↓		y_{rc1} y_{rc2} ↓

c = number of column treatments indexed as j
r = number of row treatments indexed as i
m = number of replications within each row and column combination indexed as k. Unless this is the same number within each (i,j), a more sophisticated analysis must be used. In many applications a noninteractive system can be assumed on the basis of the nature of the study. In this case $m = 1$, corresponding to one observation for each (i,j) combination
S_j = sum of observations in column j
S_i = sum of observations in row i
S_k = sum of observations in row and column (i,j)
S = sum of all observations
n_j = number of observations in column j
n_i = number of observations in row i
n = total number of observations

Linear Model There are two types of model:

Noninteractive

$$y_{ijk} = \mu + \alpha_i + \beta_j + \epsilon_{ijk}$$
(with $k \equiv 1$ usually)

Interactive

$$y_{ijk} = \mu + \alpha_i + \beta_j + (\alpha\beta)_{ij} + \epsilon_{ijk}$$

In general, if the experiment has been replicated ($m > 1$), then the analysis should be carried out under the interactive format as an independent control. The respective terms are defined as:

μ = overall population mean
α_i = the ith-row treatment effect (with a fixed-effects model stipulation $\Sigma\alpha_i = 0$)
β_j = jth-column treatment effect (with a fixed-effects model stipulation $\Sigma\beta_j = 0$)
$(\alpha\beta)_{ij}$ = (i,j) interactive effect [with a fixed-effects model stipulation $\Sigma(\alpha\beta)_{ij} = 0$]
ϵ_{ijk} = random error

Assumptions

1. The respective observations y_{ijk} are random and independent.
2. The model is a true representation of the observations.
3. The random errors ϵ_{ijk} are normally distributed with a population mean zero and variance σ^2.

Analysis-of-Variance Table

1. The total sum of squared deviations TSS $= \Sigma(y_{ijk} - \bar{y})^2$ can be partitioned into four components:

$$\text{TSS} = \text{CSS} + \text{RSS} + \text{ISS} + \text{ESS}$$

where CSS = column sum of squared deviations
RSS = row sum of squared deviations
ISS = interaction sum of squared deviations
ESS = error sum of squared deviations

2. Analysis-of-variance table:

Source	SS	df	MS	F ratio
Rows		$r - 1$		
Columns		$c - 1$		
Interaction ($r \times c$)		$(r - 1)(c - 1)$		
Error		$rc(m - 1)$		
Total		$n - 1$		

When $m = 1$, the interaction source is labeled the error since the error source cannot be estimated directly.

Computational Procedure

$$\text{TSS} = \sum_{i,j,k} y_{ijk}^2 - S^2/n$$

$$\text{RSS} = \sum_{i=1}^{r} S_i^2/n_i - S^2/n$$

$$\text{CSS} = \sum_{j=1}^{c} S_j^2/n_j - S^2/n$$

$$\text{ESS} = \sum_{i,j}\left[\sum_{l}^{m} y_{ijk}^2 - S_k^2/m\right]$$

$$\text{ISS} = \text{TSS} - \text{RSS} - \text{CSS} - \text{ESS}$$

There is a direct computational procedure for ISS which is somewhat complicated. It is not required for a two-way analysis but is necessary for higher-order analyses of variance. For these, the recommended texts included in "General References: Design" should be consulted.

Test of Hypothesis The test of hypothesis for the fixed-effects model can include three tests:

Rows	Columns	$r \times c$
H_0: $\alpha_i = 0$ H_1: not all α_i are zero	H_0: $\beta_j = 0$ H_1: not all β_j are zero	H_0: $(\alpha\beta)_{ij} = 0$ H_1: not all $(\alpha\beta)_{ij}$ are zero

Decision Rule Accept H_0 if sample $F <$ tabled F; otherwise reject it under the stated guidelines.

Example

Application. Pairs of two different types of pipes were buried in 11 different locations to determine their relative corrosion resistance. This will be analyzed through a noninteractive model formulation.

Procedure

1. It is of interest to estimate the degree of difference in corrosion rate between pipe types and between soil types. For this purpose, the statistical significance of the respective two sources of variation is tested through the null hypothesis:

Test 1 H_0: $\beta_j = 0$
H_1: not all $\beta_j = 0$
Test 2 H_0: $\alpha_i = 0$
H_1: not all $\alpha_i = 0$

2. Select $\alpha = .05$. Therefore, with $c = 2$ the column test for β_j has 1 df for the numerator, and with $r = 11$ the row test for α_i has 10 df for the numerator. The denominator mean-square estimate is based on $(r - 1)(c - 1)$, or 11 df, since no replications are available. The corresponding critical values for the F ratio are $F(1,11) = 4.84$ and $F(10,11) = 2.85$. The critical value of F is larger in the former test, given only 1 df for the numerator in that test while 10 df for the numerator are available in the latter test. This is due to the fact that the numerator mean square has only 1 df and therefore is a less reliable estimate for the column effect than the row-effect mean square, which has 10 df.

3. The decision rule:

Accept H_0 if sample $F <$ critical value F
Reject H_0 otherwise

4. The sample of 11 pairs of corrosion determinations and respective partial sums S_i and S_j is as follows:

Soil type	Coated steel pipe	Bare pipe	S_i
1	27.3	41.4	68.7
2	18.4	18.9	37.3
3	11.9	21.7	33.6
4	11.3	16.8	28.1
5	14.8	9.0	23.8
6	20.8	19.3	40.1
7	17.9	32.1	50.0
8	7.8	7.4	15.2
9	14.7	20.7	35.4
10	19.0	34.4	53.4
11	65.3	76.2	141.5
S_j	229.2	297.9	527.1

5. The respective quantities for the analysis-of-variance table are computed by

$$\begin{aligned}\text{TSS} &= \Sigma y^2 - S^2/n\\ &= (27.3^2 + 41.4^2 + \cdots + 76.2^2) - 527.1^2/22\\ &= 37062.01 - 527.1^2/22\\ &= 6379.67\\ \text{CSS} &= \Sigma S_j^2/n_j - S^2/n\\ &= 229.2^2/11 + 297.9^2/11 - 527.1^2/22\\ &= 214.53\\ \text{RSS} &= \Sigma S_i^2/n_i - S^2/n\\ &= 68.7^2/2 + 37.3^2/2 + \cdots + 141.5^2/2 - 527.1^2/22\\ &= 5902.17\\ \text{ESS} &= 6379.67 - 214.53 - 5902.17\\ &= 262.97\end{aligned}$$

6. The analysis of variance:

Source	SS	df	MS	F ratio
Columns: pipes	214.53	1	214.5	8.16
Rows: soils	5902.17	10	590.2	22.44
Error	262.97	10	26.3	
Total	6379.67	21		

7. Since both sample F ratios exceed their respective critical values, the corresponding null hypotheses would be rejected. Also, it should be noted that for the pipe test the sample F ratio of 8.16 is the square of the sample $t = -2.857$, corresponding to the paired test of hypothesis. In this special case, with $r = 2$, the t and F tests are the same, since in the former test the effect of soil type has been removed by using paired differences.

8. Given a significant difference, the analysis can be extended to estimation of the respective effects. For example, under the assumptions the error-mean-

square observation is an estimate of σ^2. This is written $\hat{\sigma}^2 = 26.3$. Similarly, an estimate of the magnitude of the respective group values of α_i and β_j is sometimes computed (even though the corresponding σ^2's are not true variance quantities). By defining

$$\sigma_\alpha^2 = \frac{1}{c-1}\,\Sigma\alpha_j^2 \quad (\Sigma\alpha_j = 0)$$

and

$$\sigma_\beta^2 = \frac{1}{r-1}\,\Sigma\beta_i^2 \quad (\Sigma\beta_i = 0)$$

the following identities can be developed:

$$E(\text{CMS}) = \sigma^2 + r\sigma_\alpha^2$$
$$E(\text{RMS}) = \sigma^2 + c\sigma_\beta^2$$

where E(CMS) or E(RMS) denotes that on the average (over a great many repeated experiments) the sample mean squares will be equal to the right side. On the basis of one experiment, an estimate can be determined through the sample values:

$$\hat{\sigma}_\alpha^2 = 1/r(\text{sample CMS} - \text{sample EMS})$$
$$= \tfrac{1}{11}(214.5 - 26.3)$$
$$= 17.1$$

and

$$\hat{\sigma}_\beta^2 = 1/c(\text{sample RMS} - \text{sample EMS})$$
$$= \tfrac{1}{2}(590.2 - 26.3)$$
$$= 282.0$$

These numbers reflect the fact that variation is appreciably greater between soil types than between pipe types.

Example

Application. A test was carried out to evaluate three additive types at two addition points in the process of making paper. The additives serve the purpose of improving wet strength. The object of the study was to determine what effect starch has on wet-strength additives.

Procedure

1. In this study, it was anticipated that the additive types and addition points might be interactive, and therefore three replications were carried out with each of the 3 × 2 = 6 combinations of tests.
2. With an interactive model, the corresponding null hypotheses are

Test 1	H_0: $\alpha_i = 0$	
	H_1: not all α_i are zero	
Test 2	H_0: $\beta_j = 0$	
	H_1: not all β_j are zero	
Test 3	H_0: $(\alpha\beta)_{ij} = 0$	
	H_1: not all $(\alpha\beta)_{ij}$ are zero	

3. Select $\alpha = .05$. With

$$(c-1) = 2 \text{ df for test 1}$$
$$(r-1) = 1 \text{ df for test 2}$$
$$(r-1)(c-1) = 2 \text{ df for test 3}$$

and

$$rc(m-1) = 12 \text{ df for the denominator}$$

the corresponding critical values of F are

Test 1 $F(2,12) = 3.89$
Test 2 $F(1,12) = 4.75$
Test 3 $F(2,12) = 3.89$

4. The decision rule:

Accept H_0 if sample $F <$ critical value F
Reject it otherwise

5. The sample of 18 trials and the corresponding partial sums:

	Additive			
Addition point	1	2	3	S_i
1	.51	.56	.53	
	.57	.60	.53	4.94
	.53	.55	.56	
2	.53	.53	.83	
	.49	.59	.74	5.53
	.55	.58	.69	
S_j	3.18	3.41	3.88	10.47

For the S_k the respective quantities are 1.61, 1.71, 1.62, 1.57, 1.70, and 2.26.

6. For the analysis of variance:

$$\text{TSS} = \Sigma y^2 - S^2/n$$
$$= .51^2 + .57^2 + \cdots + .69^2 - 10.47^2/18$$
$$= .12885$$

$$\text{CSS} = \Sigma S_j^2/n_j - S^2/n$$
$$= 3.18^2/6 + 3.41^2/6 + 3.88^2/6 - 10.47^2/18$$
$$= .04243$$

$$\text{RSS} = \Sigma S_i^2/n_i - S^2/n$$
$$= 4.94^2/9 + 5.53^2/9 - 10.47^2/18$$
$$= .01934$$

$$\text{ESS} = \sum_{i,j}\left[\sum_{1}^{m} y_{ijk}^2 - S_k^2/m\right]$$
$$= [.51^2 + .57^2 + .53^2 - 1.61^2/3] + [.56^2 + .60^2 + .55^2 - 1.71^2/3]$$
$$+ \cdots + [.83^2 + .74^2 + .69^2 - 2.26^2/3]$$
$$= .01787$$

$$\text{ISS} = .12885 - .04243 - .01934 - .01787$$
$$= .04921$$

7. The analysis of variance:

Source	SS	df	MS	*F* ratio
Column—additive	.04243	2	.021215	14.2
Row—addition point	.01934	1	.01934	13.0
$c \times r$	.04921	2	.024605	16.5
Error	.01787	12	.001489	
Total	.12885	17		

8. Since each sample F ratio exceeds its corresponding critical value, the corresponding null hypotheses would be rejected. On the basis of the data, it is obvious that the best combination for increased wet strength is to use additive 3 at the second addition point.

SIMPLE LINEAR LEAST SQUARES

Nature In the pipe-corrosion study (see "Test of Hypothesis for Matched Pairs") it was demonstrated that soil type influenced the corrosion rate across both types of pipe. If a measure of acidity such as pH had been determined, it would have been of interest to relate the pH level to the rate of corrosion.

As a first step, the rate of corrosion averaged over the two types of pipe could have been plotted against the corresponding pH level for each soil type. This type of plot is called a **scatter plot.** In such an application, it would be of interest to determine the way in which the corrosion rate changes according to the soil pH. For plotting, the corrosion rate would be defined on the ordinate, or y axis, and pH on the abscissa, or x axis.

If the trend between corrosion rate and pH were linear, a straight line could be fitted to the data. The "best" method for fitting the line would be through the principle of least squares. Under this principle, the line would have the property of lying as close as possible to the sample points. For statistical purposes, "close" is defined in the sense that the sum of squared distances between the points and their corresponding line values would be as small as possible. Graphically, for a sample of four points, the distance d_i between the y_i values and their corresponding line values would appear as shown in Fig. 2-67.

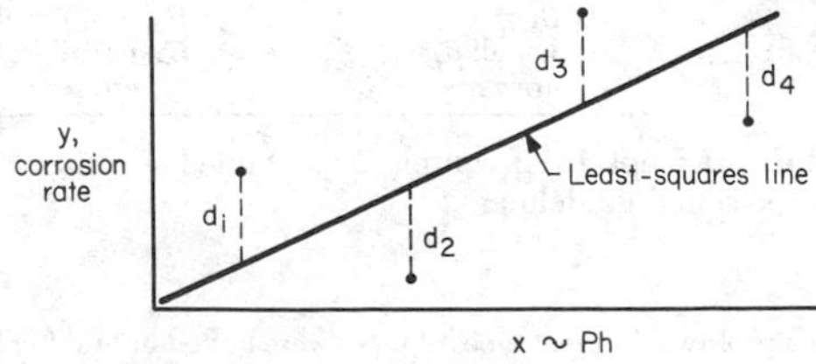

FIG. 2-67 Least-squares fit.

It can be shown that this is, in fact, the best line that can be computed, given a few basic assumptions.

Simple Linear Least Squares: Procedure

Nomenclature (x_i, y_i) is the ith pair of observations, where x_i is defined as the abscissa, or horizontal, variable and y_i the ordinate, or vertical, variable. $Y = \beta_0 + \beta_1 x$ is the true linear relationship between x and the true value Y. $y_i = Y_i + \epsilon_i$, the sample value y_i, is equal to the corresponding true value Y_i, for a stated x_i, plus some type of measurement error or distribution deviation ϵ_i. $\hat{Y} = \hat{\beta}_0 + \hat{\beta}_1 x$ is the least-squares model based on the sample data.

Assumptions

1. The measurement or distribution deviations ϵ_i, also called the **random errors,** are independent and normally distributed with a population mean zero and variance σ^2. For the principle of least squares, the normality assumption is not required; however, for the purpose of inferential statistics, it is required.
2. The abscissa variable x is measured without error. The variable x is also called the independent variable in the sense that it is specified, and the corresponding value y_i, the dependent variable, is measured with some degree of distribution ϵ. For example, for a random selection of male adults, their heights and weights could be measured. Here the independent variable x would be defined as the height and the dependent variable y as the weight. For the sample, the error term would account for the two types of deviation: (1) whether the ith individual tended to be overweight or underweight relative to the population as a whole and (2), at that point in time, whether he might be going into or out of a diet-control program.
3. The association between x and y is linear.

Scatter Plot It is usually advisable to plot the observed pairs of observations to (1) support the linearity assumption and (2) detect potential outliers. Suspect or supported outliers can be omitted from the least-squares "fit" and then subsequently tested on the basis of the least-squares fit.

Computations Through the principle of least squares, the minimization of the quantity ESS $= \Sigma(y_i - \hat{\beta}_0 - \hat{\beta}_1 x_i)^2$ defines two normal equations in $\hat{\beta}_0$ and $\hat{\beta}_1$:

$$(n)\hat{\beta}_0 + (\Sigma x_i)\hat{\beta}_1 = \Sigma y_i$$
$$(\Sigma x_i)\hat{\beta}_0 + (\Sigma x_i^2)\hat{\beta}_1 = \Sigma x_i y_i$$

The simultaneous solution to these equations is given by

$$\hat{\beta}_1 = \frac{n\Sigma x_i y_i - (\Sigma x_i)(\Sigma y_i)}{n\Sigma x_i^2 - (\Sigma x_i)(\Sigma x_i)}$$
$$\hat{\beta}_0 = \bar{y} - \hat{\beta}_1 \bar{x}$$

The error sum of squares can be computed from the identity ESS $= \Sigma y_i^2 - \hat{\beta}_0 \Sigma y_i - \hat{\beta}_1 \Sigma x_i y_i$.

Analysis of Variance

1. The reliability of the fit is measured by the significance of the signal-to-noise ratio:

$$F \text{ ratio} = \text{signal/noise} = (n - 2)\frac{\text{accounted variation in } y}{\text{unaccounted variation in } y}$$

where

$$\text{Unaccounted variation (error SS)} = \text{ESS}$$
$$\text{Accounted variation (regression SS)} = \Sigma(\hat{Y}_i - \bar{y})^2 = \hat{\beta}_0 \Sigma y_i + \hat{\beta}_1 \Sigma x_i y_i - (1/n)(\Sigma y_i)^2$$

These computations are based on what is called the partition theorem $\Sigma(y_i - \bar{y})^2 = \Sigma(y_i - \hat{Y}_i)^2 + \Sigma(\hat{Y}_i - \bar{y})^2$. The term on the left corresponds to the total variation in y_i when x_i is not considered. The first term on the right defines the unexplained variation in y_i when x_i is used to predict y_i. Therefore, the second term on the right is the variation in y_i that has been accounted for through x_i. It is called the **regression sum of squares.**

2. As a summary, the quantities given previously are usually displayed in an array called an analysis-of-variance table:

Source	SS	df	MS	*F* ratio
Regression		1		
Error		$n - 2$		
Total		$n - 1$		

3. The extent to which there is an association between x and y also can be expressed by the sample coefficient of determination

$$r^2 = \frac{\text{regression SS}}{\text{total SS}}$$

and the corresponding correlation coefficient

$$r = (\text{sign of } \hat{\beta}_1)\sqrt{\frac{\text{RSS}}{\text{TSS}}}$$

Also the correlation coefficient r can be computed directly by

$$r = \frac{n\Sigma x_i y_i - (\Sigma x_i)(\Sigma y_i)}{\sqrt{(n\Sigma y_i^2 - \Sigma y_i \Sigma y_i)(n\Sigma x_i^2 - \Sigma x_i \Sigma x_i)}}$$

4. The significance of the fit is determined by the magnitude of the sample F ratio compared with the tabled critical value $F(1, n - 2)$ for the selected α level. The null and alternative hypotheses are

$$H_0: \beta_1 = 0 \qquad H_1: \beta_1 \neq 0$$

5. The decision rule:

Accept H_0 if sample $F <$ tabled F
Reject H_0 otherwise

6. The residual mean square also is called the residual variance and is designated as $\hat{\sigma}^2$. It serves as an estimator for σ^2.

Reliability of the Slope Estimator $\hat{\beta}_1$ The $100(1 - \alpha)$ percent confidence interval for $\hat{\beta}_1$ as an estimator for β_1 is given by

$$\hat{\beta}_1 - t\hat{\sigma}\sqrt{\frac{1}{\Sigma(x_i - \bar{x})^2}} < \beta_1 < \hat{\beta}_1 + t\hat{\sigma}\sqrt{\frac{1}{\Sigma(x_i - \bar{x})^2}}$$

where t is the tabled value corresponding to $(n - 2)$ df and an upper-tail area of $\alpha/2$.

Reliability for a New Prediction Let $x°$ be a new value of x; then $\hat{Y}°$ would be an estimator of both $Y°$ (where the true line is defined for that value of $x°$) and $y°$ (the corresponding sample value of $y°$ associated with $x°$), where $\hat{Y}° = \hat{\beta}_0 + \hat{\beta}_1 x°$. Then the $100(1 - \alpha)$ percent confidence intervals are

$$\hat{Y}° - t\hat{\sigma}\left[\frac{1}{n} + \frac{(x° - \bar{x})^2}{\Sigma(x_i - \bar{x})^2}\right]^{1/2} < Y° < \hat{Y}° + t\hat{\sigma}\left[\frac{1}{n} + \frac{(x° - \bar{x})^2}{\Sigma(x_i - \bar{x})^2}\right]^{1/2}$$

and

$$\hat{Y}° - t\hat{\sigma}\left[1 + \frac{1}{n} + \frac{(x° - \bar{x})^2}{\Sigma(x_i - \bar{x})^2}\right]^{1/2} < y° < \hat{Y}° - t\hat{\sigma}\left[1 + \frac{1}{n} + \frac{(x° - \bar{x})^2}{\Sigma(x_i - \bar{x})^2}\right]^{1/2}$$

Example

Application. Brenner (*Magnetic Method for Measuring the Thickness of Non-magnetic Coatings on Iron and Steel,* National Bureau of Standards, RP1081, March 1938) suggests an alternative way of measuring the thickness of nonmagnetic coatings of galvanized zinc on iron and steel. This procedure is based on a nondestructive magnetic method as a substitute for the standard destructive stripping method. A random sample of 11 pieces was selected and measured by both methods.

Nomenclature. The calibration between the magnetic and the stripping methods can be determined through the model

$$y = \beta_0 + \beta_1 x + \epsilon$$

where x = strip-method determination
y = magnetic-method determination

The least-squares model would then be inverted to $(y - \hat{\beta}_0)/\hat{\beta}_1$ as an estimator for the "true" thickness. In this example, it would be assumed that the thickness determination by the strip method is essentially without error while the determination by the magnetic method is an approximation.

Null hypothesis

1. The hypotheses:

$$H_0: \beta_1 = 0 \qquad H_1: \beta_1 \neq 0$$

2. Select $\alpha = .05$. Therefore, with 1 df for the numerator MS and 9 df for the denominator MS, the critical value of $F(1,9) = 5.12$.

3. The decision rule:

Accept H_0 if sample $F < 5.12$
Reject H_0 otherwise

Sample data

Thickness, 10^{-5} In

Stripping method, x	Magnetic method, y
104	85
114	115
116	105
129	127
132	120
139	121
174	155
312	250
338	310
465	443
720	630

Computations. The normal equations are defined by

$$(n)\hat{\beta}_0 + (\Sigma x)\hat{\beta}_1 = \Sigma y$$
$$(\Sigma x)\hat{\beta}_0 + (\Sigma x^2)\hat{\beta}_1 = \Sigma xy$$

For the sample

$$11\hat{\beta}_0 + 2743\hat{\beta}_1 = 2461$$
$$2743\hat{\beta}_0 + 1{,}067{,}143\hat{\beta}_1 = 952{,}517$$

with $\Sigma y^2 = 852{,}419$.

The solution to the normal equations is given by

$$\hat{\beta}_0 = 3.19955 \qquad \hat{\beta}_1 = .884362$$

The error sum of squares can be computed from the formula

$$\text{ESS} = \Sigma y^2 - \hat{\beta}_0 \Sigma y - \hat{\beta}_1 \Sigma xy$$

if a sufficient number of significant digits is retained (usually six or seven digits are sufficient). Otherwise, the computational formula ESS $= \Sigma(y_i - \hat{Y}_i)^2$ should be used. Here

$$\text{ESS} = 852{,}419 - (3.19955)(2461) - .884362(952{,}517)$$
$$= 2175.07$$

Analysis of variance

Source	SS	df	MS	F ratio
Regression	299,651	1	299,651	1240
Error	2,175	9	241.7	
Total	301,826	10		

On the basis of the sample F ratio, there is no question as to the association between the two methods. The correlation coefficient, another measure of association, is $r = .9964$.

The calibration. To predict the "true" thickness on the basis of the magnetic method alone, the inverted model

(Magnetic thickness − 3.20)/.884362 = 1.1308 (magnetic thickness) − 3.62

would be used. To assess the reliability of this prediction, the corresponding residuals, where

Residual = true thickness − predicted thickness

should be computed:

True	Predicted	Residual
104	92.5	11.5
114	126.4	−12.4
116	115.1	.9
129	140.0	−11.0
132	132.1	− .1
139	133.2	5.8
174	171.7	2.3
312	279.1	32.9
338	346.9	− 8.9
465	497.3	−32.3
720	708.8	11.2

The standard deviation for the residuals equals 17.6. The personnel responsible for reporting thickness need to assess the relative reliability of the alternative method, balanced against the inherent cost for the more exact destructive test, in deciding whether it should be used.

Example

Application. The last step in a process for the manufacture of a chemical intermediate is grinding. As for most materials subjected to a grinding mechanism, the particle-size distribution tends to follow a log-normal distribution; that is, the log of particle size is normally distributed. This results in a straight line on log-normal-probability paper, the cumulative distribution being plotted against the log of particle size. A sample of particles was screened with six different mesh sizes.

Nomenclature. For the purposes of nomenclature, define

x = log (sieve opening) for each corresponding screen size
y = z-scale value for the corresponding cumulative proportion of particles which are larger than the corresponding screen size (the cumulative scale on normal probability paper corresponds to a linear scale of z values)

Then $y = \beta_0 + \beta_1 x + \epsilon$.

Null hypothesis

1. The hypotheses:

$$H_0: \beta_1 = 0 \qquad H_1: \beta_1 \neq 0$$

2. Select $\alpha = .05$. Therefore, with 1 and 4 df the critical value is $F(1,4) = 7.71$.

3. The decision rule:

Accept H_0 if sample $F < 7.71$
Reject H_0 otherwise

Sample data

Mesh size	Sieve opening	x log	% cumulative proportion	y z score
30	595	2.775	9.43	−1.32
50	297	2.473	48.86	− .03
100	149	2.173	70.77	.84
140	105	2.021	79.97	.84
200	74	1.869	88.66	1.21
325	44	1.643	93.62	1.52

The z scores correspond to the cumulative areas in Table 1-15 in Sec. 1; i.e., a $z = -1.32$ for a cumulative area of .0943.

Computations. The normal equations for the sample data are

$$6\hat{\beta}_0 + 12.954\hat{\beta}_1 = 2.77$$
$$12.954\hat{\beta}_0 + 28.815\hat{\beta}_1 = 3.9145$$
$$\Sigma y^2 = 6.5259$$

The solution to the normal equations is given by

$$\hat{\beta}_0 = 5.7258 \qquad \hat{\beta}_1 = -2.4382$$

The error sum of squares ESS = .2098.

Analysis of variance

Source	SS	df	MS	*F* ratio
Regression	5.0373	1	5.0375	96.
Error	.2098	4	.05245	
Total	5.2471	5		

This is a statistically significant fit.

Comment. On the basis of a scatter plot, it might be concluded that the cumulative proportion with a 30-mesh screen might be an outlier. Since the data are as reported, it is difficult to assess the validity of this point. This could represent a blunder in the screening process (an actual 19.43 percent rather than the reported 9.43 percent, for example), a poor distribution of the larger-size particles, or possibly a screen that is not in calibration.

A reworking of the remaining five data points leads to $\hat{Y} = 4.7048 - 1.9092x$ and the corresponding analysis of variance:

Source	SS	df	MS	*F* ratio
Regression	1.4300	1	1.4300	544
Error	.0079	3	.00263	
Total	1.4379	4		

The corresponding reduction in the error mean square from .052 to .0026 reflects the magnitude of influence of the suspect point.

A way of assessing the reliability of the suspect point is to compute the confidence interval for that point on the basis of the model presented. With $\alpha = .10$, the corresponding $t = \pm 2.353$ with 3 df; therefore, with $\bar{x} = 2.038$, $\Sigma(x - \bar{x})^2 = .3923$, and

$$\hat{Y}^\circ = 4.7048 - 1.9092(2.775)$$
$$= -.59$$

The 90 percent confidence interval for y° is given by

$$-.59 - 2.353(.00263)^{1/2}\left[1 + \frac{1}{5} + \frac{(2.775 - 2.038)^2}{.3923}\right]^{1/2} < y^\circ$$
$$< -.59 + 2.353(.00263)^{1/2}\left[1 + \frac{1}{5} + \frac{(2.775 - 2.038)^2}{.3923}\right]^{1/2}$$

This reduces to $-.78 < y^\circ < -.40$. The sample value of -1.32 is well out of bounds from this interval.

The final decision should rest with the engineer responsible for the particle-size determinations.

This example illustrates the point that a significant fit does not demonstrate the assumptions for the fitting. That is, with the six data points the corresponding $F = 96$, whereas the particle-size distribution is not really log-normal over the full range of sizes.

SIMPLE NONLINEAR REGRESSION

Nature In modeling a relationship based on n pairs of observations (x_i, y_i), it sometimes turns out that the relationship is curvilinear. When the trend is curvilinear, the most appropriate model for a particular set of data can be one of two types, defined according to the way in which the unknown coefficients occur. If the coefficients can be added linearly to the model, it is said to be a linear model. Conversely, if the coefficients cannot be added linearly, the model is called a nonlinear model. Examples of these two types are as follows:

Linear type

$$y = a + bx + cx^2$$
$$y = a + bx + c \ln x$$
$$y = a + b/x + c/x^2$$

Nonlinear type

$$y = a + be^{cx}$$
$$y = a + bx^c$$
$$y = a + b/(c + x)$$

For the former, the principle of least squares can be applied directly. This procedure is an extension of simple linear regression and is discussed under "Multiple Regression." For nonlinear regression, the principle of least squares cannot be applied directly because the resultant normal equations would contain unknown coefficients in a nonlinear form and could not be solved directly. Therefore, it is necessary to use some iterative computational procedure. One procedure is called the method of false position, which will be illustrated in the subsequent examples.

There are no analytic methods for determining the most appropriate model for a particular set of data. In many cases, however, reasonable candidate models can be selected on the basis of the nature of the application. For example, in relating the temperature to the elapsed time of a fluid cooling in the atmosphere, a model which has an asymptotic property would be the appropriate model ($\text{TEMP} = a + be^{-c\text{TIME}}$ where a represents the asymptotic temperature corresponding to $t \to \infty$).

When the functional form is unknown, general trend curves can be used as an initial screening guide. A variety of candidate functions should be fitted to the data to develop a viable and appropriate relationship. The most effective screening can be achieved when replicated observations are available. In these cases, it is possible to assess directly the viability of a particular model form through what is called a **lack-of-fit analysis.** This is illustrated in the second example in this subsection.

The usual practice in these applications is to concentrate on model development and computation rather than on statistical aspects. In general, nonlinear regression should be applied only to problems in which there is a well-defined, clear association between the two variables; therefore, a test of hypothesis on the significance of the fit would be somewhat ludicrous. In addition, the generalization of the theory for the associated confidence intervals for nonlinear coefficients is not well developed.

Simple Nonlinear Regression: Procedure

Nomenclature n pairs of observations (x_i, y_i) are defined where

x_i = ith observation of the independent variable. This is considered independent in the sense of being specified and measured without error.

y_i = ith observation of the dependent variable. This is also defined as the response and is considered a random variable which reflects some degree of error in the sense that

$$y_i = Y_i + \epsilon_i$$

where Y_i is the true value at the defined level x_i and ϵ_i corresponds to some measurement or distribution error with constant variance σ^2 over the range of x.

Y = $f(x)$, the particular model form selected for the least-squares fit.

Screening Procedure There are no analytical methods for determining the correct functional form for a particular set of data. In many cases, however, reasonable candidate models can be selected on the basis of the nature of the application. If no guideline is available, the following general procedure can be used:

1. Select a candidate model.
2. Compute the iterative least-squares solution and the residual for each point.
3. Plot the residuals against their corresponding x values. If there is no discernible trend, the functional form is probably adequate. If a trend is indicated, an alternative model should be tried.

Computational Procedure There are general computer programs available to handle a wide variety of functional forms. However, if desired, a fairly simple computational procedure can be used with most small programmable calculators. The algorithm, called the **method of false position,** is based on the following technique:

1. Given a particular functional form, define a new independent variable z which includes the nonlinear coefficient. For example, with

$$y = a + be^{cx} \quad \text{let} \quad z = e^{cx}$$
$$y = a + bx^c \quad \text{let} \quad z = x^c$$
$$y = a + b/(c + x) \quad \text{let} \quad z = 1/(c + x)$$

2. Pick an appropriate (or any) value for the nonlinear coefficient and compute the z_i values over the data set.

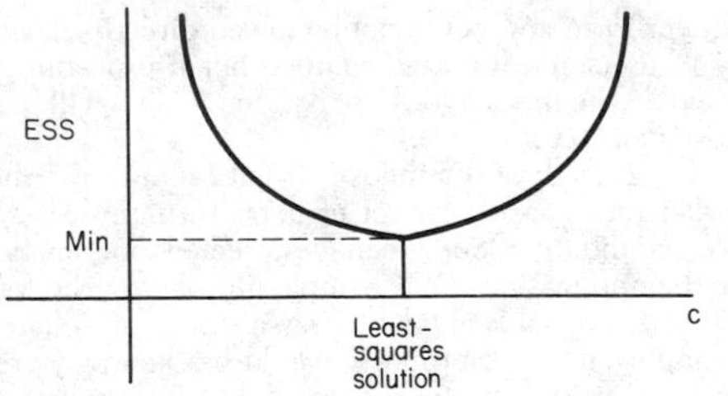

FIG. 2-68 Minimization of error sum of squares.

3. Compute the simple linear-regression solution and the error sum of squares (ESS) for the particular value of the nonlinear coefficient c.
4. Try another value of c and repeat steps 2 and 3. Vary c until the ESS is minimal. Almost always a plot of ESS against c will be a simple convex function as shown in Fig. 2-68.

Then the following successive approximation formula can be used, given three equally spaced values of c and corresponding values of ESS which are not too far removed from the minimum:

$$c_5 = c_2 + \frac{\Delta c}{2} \frac{\text{ESS}_1 - \text{ESS}_3}{\text{ESS}_1 - 2\text{ESS}_2 + \text{ESS}_3}$$

The new guess c_5 can be used as the new center for three additional values. In general, the c-interval width can be reduced by 2.

Reliability of the Fit A useful quantity to characterize the fit is the standard error of the estimate $\hat{\sigma}$, which quantifies the relative scatter of observations about the derived least-squares curve. For this $\hat{\sigma}^2 = \text{ESS}/(n - p - 1)$, where p represents the number of variable terms. As a guide in comparing alternative models fitted to the same data, their respective standard errors can be used.

Example

Application. Data were collected on the cooling of water in the atmosphere as a function of time.

Sample data

Time x	Temperature y
0	92.0
1	85.5
2	79.5
3	74.5
5	67.0
7	60.5
10	53.5
15	45.0
20	39.5

Model form. On the basis of the nature of the data, an exponential model was selected initially to represent the trend $y = a + be^{cx}$. In this example, the resultant temperature would approach as an asymptotic (a with c negative) the wet-bulb temperature of the surrounding atmosphere. Unfortunately, this temperature was not reported.

Computations. Since an asymptotic trend is indicated, a reasonable initial guess for c was set at $-.20$. Corresponding to this value, z_i was defined as follows:

x	$z = e^{-.20x}$	y
0	1.0000	92.0
1	.8187	85.5
2	.6703	79.5
3	.5488	74.5
5	.3679	67.0
7	.2466	60.5
10	.1353	53.5
15	.0498	45.0
20	.0183	39.5

By using the computational procedure outlined in the previous subsection, the following normal equations were defined:

$$9a + 3.8558b = 597.0$$
$$3.8558a + 2.6381b = 305.9498$$

for which $a = 44.5336$, $b = 50.8840$, and ESS $= 86.9767$.

A second try with $c = -.10$ resulted in a new ESS of 3.1552. This would dictate a least-squares solution in the neighborhood of about $-.10$. Two other values were tried which are summarized here:

c	ESS
−.10	3.1552
−.11	1.8321
−.12	3.2481

No practical improvement could be achieved by refining the value of c from $-.11$. The following is a summary of fit:

x	$z = e^{-.11x}$	y	$\hat{Y}$	Residual
0	1.0000	92.0	91.44	.56
1	.8958	85.5	85.42	.08
2	.8025	79.5	80.02	−.52
3	.7189	74.5	75.19	−.69
5	.5769	67.0	66.97	.03
7	.4630	60.5	60.38	.12
10	.3329	53.5	52.85	.65
15	.1920	45.0	44.71	.29
20	.1108	39.5	40.01	−.51
				ESS = 1.8321

$$\hat{\sigma} = \left(\frac{1.8321}{9 - 3}\right)^{1/2} = .55$$

and

$$\hat{Y} = 33.60 + 57.845e^{-.11x}$$

An examination of the residuals indicates a wavering trend which suggests that the exponential model does not truly represent the temperature trend over time. If the model were a true model, the residuals would tend to be random over time.

Alternative model form. A model of the form $y = a + b/(c + x)$ was also considered. For this model, the computational form $y = a + bz$ was defined.

As an initial scan, the error sum of squares was computed with the following values of c:

c	ESS
10	1.4490
11	.2392
12	.5134

Using the successive approximation formula

$$c_5 = c_2 + \frac{\Delta c}{2} \frac{\text{ESS}_1 - \text{ESS}_3}{\text{ESS}_1 - 2\text{ESS}_2 + \text{ESS}_3}$$

resulted in $c_5 = 11.3152$. This was rounded to 11.3, and a new $\Delta c = .1$ was used:

c	ESS
11.2	.1924
11.3	.1897
11.4	.2002

The value of $c = 11.3$ was considered sufficient. The corresponding model is given by

$$\hat{Y} = 9.7814 + \frac{929.06}{11.3 + x}$$

with the following:

x	y	$\hat{Y}$	Residual
0	92.0	92.00	0
1	85.5	85.31	.19
2	79.5	79.64	−.14
3	74.5	74.75	−.25
5	67.0	66.78	.22
7	60.5	60.55	−.05
10	53.5	53.40	.10
15	45.0	45.11	−.11
20	39.5	39.46	.04
			$\hat{\sigma}$ = .179

With this model, no discernible trend in the residuals is indicated over the range of data.

It should be noted that in this example the exponential model has an asymptote of 33.60 while the rational function has a value of 9.78. Water will cool to the wet-bulb temperature of the surrounding atmosphere, but unfortunately that temperature was not reported in the original study.

Example

Application. An experiment was carried out to investigate the way in which the tensile strength of cement varies with curing time.

Sample data. y = tensile strength, kg/cm^2.

Curing Time, Days (x)

1	2	3	7	28
13.0	21.9	29.8	32.4	41.8
13.3	24.5	28.0	30.4	42.6
11.8	24.7	24.1	34.5	40.3
		24.2	33.1	35.7
		26.2	35.7	37.3

Model form. On the basis of previous knowledge, the functional relationship between curing time and tensile strength is defined by $y = ae^{-b/x}$.

Computations. In this particular application, replicated observations are available, and therefore it is possible to make two separate types of analysis: (1) the assumption of constant variance can be checked, and (2) a lack-of-fit determination can be made. The lack-of-fit determination serves as a way of assessing the viability of the model.

The model could be fitted in its standard form or as a linear log function $\ln y = \ln a - b/x$, where it is assumed that $\ln y = (\ln Y) + \epsilon$. For the first analysis, the variance was computed within days of curing for both the arithmetic tensile strength and its natural logarithm:

Curing time	df	Within variance: Arithmetic	ln
1	2	.63	.0040
2	2	2.44	.0045
3	4	6.07	.0085
7	4	4.11	.0038
8	4	8.70	.0057

Since the arithmetic within variances indicates an increase with curing time which is not reflected in the log form, the second model form is the appropriate one. If both had indicated a trend, then a weighted least-squares solution would have been required.

In log form, the least-squares procedure reduces to the simple least-squares solution:

$$\ln \hat{Y} = 3.6878 - 1.14553/t$$

The second analysis, lack of fit, can be carried out through the analysis of variance. For this, the error sum of squares is partitioned into two components, a lack of fit and a random sum of squares. The random part is defined by the sum of squares within curing time. This quantity reflects the pure error in the tensile-strength determinations independently of the model being fitted. The corresponding lack of fit represents the difference between the total error sum of squares minus the random part.

Analysis of variance

Source	SS	df	MS	F ratio
Regression	2.6780	1	2.678	481.
Error	.1085	19		
Lack of fit	.0194	3	.00647	1.16
Random	.0891	16	.00557	
Total	2.7865	20		

The random sum of squares was computed by cumulating the product of the within variances times their corresponding df:

$$\text{Random} = 2x.0040 + 2x.0045 + \cdots + 4x.0057 = .0891$$

Since the lack-of-fit MS divided by the random MS is small (1.16), the functional form is adequate in representing the trend between tensile strength and curing time. If this had not been the case, an alternative model would have been required.

MULTIPLE REGRESSION

Nature Multiple regression is an extension of simple linear regression when more terms are added. The basic principle of least squares can be readily applied regardless of the number of linear-coefficient terms.

Multiple Regression: Procedure

Basic Nomenclature n sets of observations ($x_{i1}, x_{i2}, x_{i3}, \ldots, x_{ip}, y_i$) are defined where

x_{ij} = ith observation for independent variable j. Any of the independent variables may represent functions of the other independent variables, i.e., $x_2 = x_1^2$, $x_3 = x_1x_2$, etc.

y_i = ith observation of the dependent variable. This is also defined as the response and is considered a random variable which reflects some degree of error in the sense that $y_i = Y_i + \epsilon_i$, where Y_i is the true value and ϵ_i corresponds to some measurement or distribution error with constant variance σ^2 over the range of data.

$Y = \beta_0 + \beta_1x_1 + \cdots + \beta_px_p$, the true model form which has been specified over the range of data in the sample

$\hat{Y} = \hat{\beta}_0 + \hat{\beta}_1x_1 + \hat{\beta}_2x_2 + \cdots + \hat{\beta}_px_p$, the least-squares model

Matrix Notation For the generalization of multiple regression, it is necessary to define matrix notation. X is an $n \times (p + 1)$ array of observations:

x_0	x_1	x_2	...	x_p
1	x_{11}	x_{12}		x_{1p}
1	x_{21}	x_{22}		x_{2p}
1	x_{31}	x_{32}		x_{3p}
.	.	.		.
.	.	.		.
.	.	.		.
1	x_{n1}	x_{n2}		x_{np}

The x columns can be mean-corrected without any loss of generality, in which case $\hat{\beta}_0 = \bar{y}$.

X' = transpose of X, where each column of X defines a corresponding row in X'

y = column vector (a one-column matrix):

$$\mathbf{y} = \begin{matrix} y_1 \\ y_2 \\ \vdots \\ y_n \end{matrix}$$

$X'X$ = $(p + 1) \times (p + 1)$ product matrix

$(X'X)^{-1}$ = inverse of the product matrix with diagonal elements c_{jj}

$\hat{\beta} = (X'X)^{-1}X'\mathbf{y}$ = least-squares-solution vector

$\hat{\mathbf{Y}} = X\hat{\beta}$, the least-squares estimator for $\mathbf{Y}$ (the column of true values)

ESS $= (\mathbf{y} - \hat{\mathbf{Y}})'(\mathbf{y} - \hat{\mathbf{Y}}) = \mathbf{y}'\mathbf{y} - \hat{\beta}'X'\mathbf{y}$, the residual sum of squares

TSS $= (\mathbf{y} - \bar{\mathbf{y}})^1/(\mathbf{y} - \bar{y})$, the total sum of squares

RSS $=$ TSS $-$ ESS, the regression sum of squares

$R^2 = \dfrac{\text{RSS}}{\text{TSS}}$, the multiple coefficient of determination

$\text{Var}(\hat{\beta}_j) = \sigma^2 c_{jj}$

Computational Procedure Multiple regressions are run on computers. For these runs, the model form and sample data are specified by the analyst. The computer output will include four important sets of computational summaries:

1. Regression coefficients
2. Residuals
3. Analysis of variance
4. t values for regression coefficients

Example

Application. Merriman ("The Method of Least Squares Aplied to a Hydraulic Problem," *J. Franklin Inst.*, 233–241, October 1877) reported on a study of stream velocity as a function of relative depth of the stream.

Sample data

Depth*	Velocity, y, ft/s
0	3.1950
.1	3.2299
.2	3.2532
.3	3.2611
.4	3.2516
.5	3.2282
.6	3.1807
.7	3.1266
.8	3.0594
.9	2.9759

*As a fraction of total depth.

Model. Owing to the curvature of velocity with depth, a quadratic model was specified:

$$\text{Velocity} = \beta_0 + \beta_1 x_1 + \beta_2 x_2$$

where $x_2 = x_1^2$.

Normal equations. In matrix notation, normal equations are defined by $X'X\hat{\beta} = X'y$. In summation notation, the three normal equations are defined by

$$(n)\hat{\beta}_0 + (\Sigma x_1)\hat{\beta}_1 + (\Sigma x_2)\hat{\beta}_2 = \Sigma y$$
$$(\Sigma x_1)\hat{\beta}_0 + (\Sigma x_1^2)\hat{\beta}_1 + (\Sigma x_1 x_2)\hat{\beta}_2 = \Sigma x_1 y$$
$$(\Sigma x_2)\hat{\beta}_0 + (\Sigma x_1 x_2)\hat{\beta}_1 + (\Sigma x_2^2)\hat{\beta}_2 = \Sigma x_2 y$$

For the sample data, the normal equations are

$$10\hat{\beta}_0 + 4.5\hat{\beta}_1 + 2.85\hat{\beta}_2 = 31.7616$$
$$4.5\hat{\beta}_0 + 2.85\hat{\beta}_1 + 2.025\hat{\beta}_2 = 14.08957$$
$$2.85\hat{\beta}_0 + 2.025\hat{\beta}_1 + 1.5333\hat{\beta}_2 = 8.828813$$

The algebraic solution to the simultaneous equations is

$$\hat{\beta}_0 = 3.19513 \qquad \hat{\beta}_1 = .4425 \qquad \hat{\beta}_2 = -.7653$$

The inverse of the product matrix

$$X'X = \begin{pmatrix} 10 & 4.5 & 2.85 \\ 4.5 & 2.85 & 2.025 \\ 2.85 & 2.025 & 1.5333 \end{pmatrix}$$

is

$$(X'X)^{-1} = \begin{pmatrix} .6182 & -2.5909 & 2.2727 \\ -2.5909 & 16.5530 & -17.0455 \\ 2.2727 & -17.0455 & 18.9394 \end{pmatrix}$$

Residuals. On the basis of the least-squares fit, residuals were computed for the sample data:

Depth	y	$\hat{Y}$	Residual
0	3.1950	3.1951	−.0001
.1	3.2299	3.2317	−.0018
.2	3.2532	3.2530	.0002
.3	3.2611	3.2590	.0021
.4	3.2516	3.2497	.0019
.5	3.2282	3.2251	.0031
.6	3.1807	3.1851	−.0044
.7	3.1266	3.1299	−.0033
.8	3.0594	3.0594	0
.9	2.9759	2.9735	.0024

The wavering trend in the residuals suggests that the quadratic model does not fully represent the functional relationship between velocity and depth. Several alternative functional forms were tried without success.

Analysis of variance

Source	SS	df	MS	F ratio
Regression	.080949	2	.04047	4976.
Error	.000057	7	.0000081	
Total	.081006	9		

$$\hat{\sigma} = \sqrt{.0000081} = .0029$$
$$R^2 = .080949/.081006 = .9993$$

where

$$\Sigma(y - \bar{y})^2 = .081006$$
$$\Sigma(y - \hat{Y})^2 = .000057$$
$$\Sigma(\hat{Y} - \bar{y})^2 = .080949$$

Test of hypothesis. A test of hypothesis for the fit as a whole can be carried out with

$$H_0\text{: } Y = \beta_0 \text{ (Y is independent of } x)$$
$$H_1\text{: } Y \neq \beta_0$$

Given $\alpha = .05$, the critical value of $F(2,7) = 4.75$. On the basis of the sample value $F = 4976$. The null hypothesis is rejected. However, it should be noted that this test does not validate the model; rather it reflects the existence of an association between velocity and depth.

t values. A sample t value can be computed for each regression coefficient j through the identity $t_j = \hat{\beta}_j/(\hat{\sigma}\sqrt{c_{jj}})$, where c_{jj} is the (j,j) element in the inverse. For the two variables x_1 and x_2,

Coefficient	c_{jj}	Sample t value
.4425	16.55	37.5
−.7653	18.94	−60.6

Computational note. From a computational standpoint, it is usually advisable to define the variables in deviation units. For example, in the problem presented, let

$$x_1 = \text{depth} - \overline{\text{depth}}$$
$$= \text{depth} - .45$$

For expansion terms such as a square, define

$$x_2 = x_1^2 - \overline{x_1^2} \qquad (\overline{x_1^2} = .0825)$$

For the previous sample data,

Deviation units	
x_1	x_2
−.45	.12
−.35	.04
−.25	−.02
−.25	−.02
−.15	−.06
−.05	−.08
.05	−.08
.15	−.06
.25	−.02
.35	.04
.45	.12

The resultant analysis-of-variance tables will remain exactly the same. However, the corresponding coefficient t value for the linear coefficient will usually be improved. This is an idiosyncrasy of regression modeling. With the coded data presented, the least-squares solution is given by

$$\hat{Y} = 3.17616 - .2462x_1 - .7653x_2$$

with a corresponding t value for $\hat{\beta}_1 = -.2462$ of $t = -63.63$.

When expansion terms are used but not expanded about the mean, the corresponding t values for the generating terms should not be used. For example, if $x_3 = x_1x_2$ is used rather than the correct expansion $(x_1 - \bar{x}_1)(x_2 - \bar{x}_2)$, then the corresponding t values for x_1 and x_2 should not be used.

Example

Application. In the section "One-Way Analysis of Variance" three examples were illustrated. For two of these, the first and third, the only method of analysis was the method that was used. For the second example a more viable method would be multiple regression. The reason for this is associated with the definition of the treatment levels. For this example, the treatment levels are quantitatively ordered; that is, they define successive levels of treatment. In the other two examples cited the respective levels (types) represent nominal data which do not have any type of ordering.

Model. Define

$$y = \beta_0 + \beta_1(x - 4.) + \beta_2(x - 4.)^2 + \epsilon$$

where y = strength strength
x = additive level in percent

Normal equations. For the sample data the normal equations are

$$25\hat{\beta}_0 + 0\hat{\beta}_1 + 200\,\hat{\beta}_2 = 490$$
$$0\hat{\beta}_0 + 200\hat{\beta}_1 + 0\hat{\beta}_2 = 88$$
$$200\hat{\beta}_0 + 0\hat{\beta}_1 + 2720\hat{\beta}_2 = 3712$$

The algebraic solution is given by

$$\hat{\beta}_0 = 21.09 \qquad \hat{\beta}_1 = 0.44 \qquad \hat{\beta}_2 = -.1857$$

with an ESS = 122.65.

Analysis of variance. For the analysis of variance the lack-of-fit component (see the second example in the preceding subsection) can be computed since five treatment levels were defined (5 − 3 = 2 df for lack of fit) and, in addition, experimental points were replicated wihin the treatment levels:

Source	SS	df	MS
Regression	77.35	2	38.675
Error	122.65	22	5.575
Lack of fit	3.45	2	1.725
Random	119.20	20	5.960
Total	200.00	24	

Test of hypothesis. The *adequacy* of the quadratic model can be tested through the F ratio for the lack of fit and random mean square. For this, select an α level of .10 which defines a critical $F(2,20) = 2.59$ for the test of

$$H_0\colon Y = \beta_0 + \beta_1 x + \beta_2 x^2 \qquad \text{with} \qquad H_1\colon Y \neq \beta_0 + \beta_1 x + \beta_2 x^2$$

Since the sample $F = .29 <$ the critical F of 2.59, accept the null hypothesis for lack of contrary evidence. Therefore, the quadratic model is adequate to represent the change in tensile strength with additive concentration. It should be noted that the term "adequate" is defined in the sense of a sufficient condition, not a necessary condition. That is, it is possible to pass this test (accept H_0) when, in fact, $Y = \beta_0 + \beta_1 x$ or even $Y = \beta_0$. Also, this test does not validate the relationship when the null hypothesis is accepted. Rather, within the relative reliability of the tensile-strength determinations the quadratic model is adequate to represent the trend.

The corresponding test for the regression can be defined by pooling the respective error components (there are valid arguments for and against pooling—it is not a universally accepted procedure). This is given by the error mean square. For this test, at an α level of 5 percent the critical value of $F(2,22) = 3.44$ with

$$H_0\colon Y = \beta_0 \qquad H_1\colon Y \neq \beta_0$$

Therefore, with a sample $F = 9.64 > 3.44$, reject the null hypothesis and accept H_1.

Example

Application. Mills (*Statistical Methods*, Holt, New York, 1938, pp. 531–542) illustrated the theory of multiple regression through an agricultural example. In this study, the author related three monthly summer temperatures to the corn yield in Kansas for the years 1890–1933. For the purposes of the study, the author states that he used monthly temperatures averaged over three government stations, located in Dodge City, Concordia, and Iola, for June, July, and August.

There evidently was some misunderstanding by Mills regarding the temperature data. According to the U.S. *Monthly Weather Statistics*, the monthly figures correspond to temperatures averaged over the whole state for the years 1890–1922 and averages for the aforementioned three cities for the period 1923–1933. In actuality, differences between the two sets of numbers for this latter period are relatively minor, although these differences do influence the subsequent statistical analysis. For the early period, no comparison is possible since temperature data were not recorded in Iola prior to 1914.

To expand on Mills's original analysis, precipitation data by months were included in the analysis.

Corn Yield in Kansas

Year	Average June temperature	Average July temperature	Average August temperature	June rain	July rain	August rain	Yield
1890	77.6	83.1	76.1	2.7	1.3	4.2	15.6
1	70.7	74.0	75.1	6.3	4.6	1.3	26.7
2	73.4	77.5	76.5	1.9	3.3	3.4	24.5
3	74.7	79.5	73.8	3.9	3.5	2.7	21.3
4	74.2	77.8	78.0	4.9	1.6	.7	11.2
5	71.7	74.9	76.0	5.0	5.5	5.2	24.3
6	74.1	78.1	78.7	4.2	4.6	2.7	28.0
7	76.6	80.2	76.0	3.6	3.1	3.1	18.0
8	75.0	77.7	78.2	4.4	3.0	2.5	16.0
9	73.9	76.2	80.6	5.4	5.4	1.9	27.0
1900	74.9	77.9	81.0	3.6	3.3	2.1	19.0
1	77.3	85.0	79.1	2.3	1.8	2.6	7.8
2	70.9	76.8	78.2	5.8	4.2	5.5	29.9
3	67.2	78.3	75.3	2.4	3.0	4.8	25.6
4	70.4	75.6	74.6	6.6	6.0	3.3	20.9
5	75.5	74.5	78.7	3.8	5.4	1.8	27.7
6	71.8	73.8	76.3	3.7	4.6	3.8	28.9
7	72.0	78.4	78.1	4.8	3.8	3.2	22.1
8	72.1	75.8	76.2	7.9	3.4	3.8	22.0
9	73.1	78.1	80.1	5.5	5.4	1.2	19.9
1910	72.2	79.5	75.7	1.9	2.0	4.4	19.0
1	80.5	78.6	76.4	.6	4.0	3.6	14.5
2	69.3	79.9	77.4	4.0	2.6	4.3	23.0
3	74.2	82.1	84.2	2.4	1.3	.5	3.2
4	78.2	79.9	78.2	4.0	2.7	3.3	18.5
5	69.2	74.0	70.1	6.5	5.9	4.6	31.0
6	70.3	81.2	79.6	6.0	.8	2.1	10.0
7	72.8	80.8	73.4	2.1	1.5	4.9	13.0
8	78.4	78.3	82.3	1.3	2.6	2.3	7.1
9	72.3	80.2	78.3	4.4	1.9	1.8	15.2
1920	72.8	77.6	72.9	2.9	3.2	4.3	26.5
1	74.4	79.2	78.6	5.0	3.1	3.3	22.2
2	75.2	77.0	80.1	2.3	4.9	1.5	19.3
3	73.0	78.7	78.2	6.1	3.2	2.4	21.7
4	74.1	75.6	78.8	2.2	3.7	4.1	21.7
5	77.6	79.8	77.2	4.0	3.2	3.2	16.6
6	72.7	78.8	79.4	2.7	2.5	2.8	11.0
7	71.2	77.2	71.8	5.9	3.7	6.5	30.0
8	67.3	77.5	76.9	7.4	4.4	3.1	27.0
9	72.1	78.9	78.9	4.0	3.5	2.1	17.5
1930	72.7	81.4	79.7	3.9	2.0	2.8	12.0
1	77.8	80.4	76.0	2.2	2.3	3.3	17.5
2	73.4	81.6	79.3	5.7	3.1	2.9	18.5
3	80.7	81.4	76.7	.9	2.8	5.2	11.5
Averages	73.6	78.5	77.4	4.0	3.4	3.2	19.6

Sample data. The accompanying table includes a summary of temperature, precipitation, and corn yield for the period 1890–1933.

Preliminary analysis. As a preliminary analysis for modeling, 12 key years were selected: the 4 lowest-yield years, a middle 4, and the 4 highest-yield years. The following is a summary of these data:

Year	June temperature	July temperature	August temperature	June rain	July rain	August rain	Yield
			Lowest yield				
1913	74	82	84	2	1	½	3.2
1918	78	78	82	1	3	2	7.1
1901	77	85	79	2	2	3	7.8
1926	73	79	79	3	2	3	11.0
Average		79°			2 in		7.2

Year	June tempera-ture	July tempera-ture	August tempera-ture	June rain	July rain	August rain	Yield
			Middle yield				
1897	77	80	76	4	3	3	18.0
1922	75	77	80	2	5	2	19.3
1909	73	78	80	6	5	1	19.9
1893	75	80	74	4	4	3	21.3
Average		77°			3½ in		19.6
			Highest yield				
1906	72	74	76	4	5	4	28.9
1902	71	77	78	6	4	6	29.9
1927	71	77	72	6	4	6	30.0
1915	69	74	70	7	6	5	31.0
Average		73°			5 in		30.0
44-year averages		76½°			3½ in		19.6

As reflected by the data, corn yield seems to increase with increased precipitation and decreased temperature. The highest yields occur when both are simultaneously at their favorable extremes, and conversely for the lowest yields. The intermediate yields seem to occur in various combinations of favorable and unfavorable deviations from the temperature and rain means.

Regression model. To evaluate the relative associations of temperature and precipitation with corn yield, the following regression model was assumed:

$$Y = \beta_0 + \beta_1 x_1 + \beta_2 x_2 + \beta_3 x_3 + \beta_4 x_4 + \beta_5 x_5 + \beta_6 x_6$$

where $x_1 - x_3$ corresponds to the successive monthly temperatures and $x_4 - x_6$ to the successive monthly precipitations.

Least-squares solution. The least-squares solution resulted in the model

$$\hat{Y} = 105.79 - .3905(\text{June } T) - .5660(\text{July } T) - .3341(\text{August } T) + .4132(\text{June } R) + 2.1664(\text{July } R) + 1.2463(\text{August } R)$$

Residuals. A summary of residuals for the 44 years of data is presented in the accompanying table. The sum of squared residuals, ESS = 493.36 wih 44 − 7 = 37 df, results in a standard error of $\hat{\sigma}$ = 3.65.

Residuals for Six-Term Regression

Year	y	$\hat{Y}$	Residual
1890	15.6	12.2	3.4
1	26.7	25.4	1.3
2	24.5	19.9	4.6
3	21.3	19.5	1.8
4	11.2	13.1	−1.9
5	24.3	30.5	−6.2
6	28.0	21.4	6.6
7	18.0	17.2	.8
8	16.0	17.8	−1.8
9	27.0	23.2	3.8
1900	19.0	16.6	2.4
1	7.8	9.2	−1.4
2	29.9	26.9	3.0
3	25.6	23.5	2.1
4	20.9	30.4	−9.5
5	27.7	23.4	4.3
6	28.9	26.7	2.2
7	22.1	21.4	.7
8	22.0	24.6	−2.6
9	19.9	21.7	−1.8
1910	19.0	17.9	1.1
1	14.5	17.7	−3.2
2	23.0	20.3	2.7
3	3.2	6.6	−3.4
4	18.5	15.5	3.0
5	31.0	34.7	−3.7
6	10.0	12.6	−2.6
7	13.0	17.3	−4.3
8	7.1	12.4	−5.3
9	15.2	14.2	1.0
1920	26.5	22.6	3.9
1	22.2	18.5	3.7
2	19.3	19.5	−.2
3	21.7	19.1	2.6
4	21.7	21.8	−.1
5	16.6	17.1	−.5
6	11.0	16.3	−5.3
7	30.0	28.9	1.1
8	27.0	26.4	.6
9	17.5	18.5	−1.0
1930	12.0	14.1	−2.1
1	17.5	14.5	3.0
2	18.5	17.1	1.4
3	11.5	15.5	−4.0

An examination of the residuals indicates that for the year 1904 the residual −9.5 might be excessive. A rough criterion is given by the magnitude of the ratio of the residual divided by the standard error, −9.5/3.65 = −2.60. As a rough rule, a standardized residual (residual over standard error) should be within about ±2.50. (See Anscombe, "Examination of Residuals," *Proc. 4th Berkeley Symp. Math. Stat. Prob. I*, 1961, pp. 1–36, for a more rigorous statistical procedure.) An examination of government documents did not disclose any reason, such as insect infestation, for the excessively low yield in 1904. Conversely, an examination of the temperature and precipitation data for that year reflects the fact that the corn yield should have been higher. Also, an examination of the corn yield in 1904 in the surrounding states indicates a relatively high yield year. Therefore, this year was considered an outlier and the regression rerun with 43 years of data. For applications of this type, outlier points can exert a relatively high degree of leverage on the resultant least-squares solution. For example, for the total ESS, 493 over 44 years, this one year accounts for more than 100 units.

The least-squares solution for the 43 years of data is

$$\hat{Y} = 102.43 - .4449(\text{June } T) - .3737(\text{July } T) - .4527(\text{August } T) + .4741(\text{June } R) + 2.7532(\text{July } R) + 1.0874(\text{August } R)$$

with a corresponding ESS = 384 with $\hat{\sigma}$ = 3.27. The prediction for 1904 by using the revised model is $\hat{Y}$ = 32.33 with a standardized residual value = (20.9 − 32.33)/3.27 = −3.50. This is a more viable criterion, since with the outlier point in the data set, least squares will tend to overfit that point at the expense of the other observations.

Analysis of variance. The following is the summary analysis-of-variance table for the 43-point regression model:

Source	SS	df	MS	F ratio
Regression	1580.2	6	263.36	24.7
Error	384.5	36	10.68	
Total	1964.6	42		
	R^2 = .80			

Test of hypothesis. For the model fit as a whole

$$H_0\colon Y = \beta_0 \qquad H_1\colon Y \neq \beta_0$$

With α = .05, the critical value of $F(6,36)$ = 2.39; therefore, on the basis of a sample value of F = 24.7, the null hypothesis is rejected. It has been demonstrated that there is an association between temperature and precipitation with corn yield.

Coefficient t values. As a final summary, sample t values should be displayed:

Coefficient	Estimate	Standard deviation	t value
June temperature	−.4449	.2188	−2.03
July temperature	−.3737	.3541	−1.06
August temperature	−.4527	.2636	−1.72
June rainfall	.4741	.3838	1.24
July rainfall	2.7532	.6636	4.15
August rainfall	1.0874	.5223	2.08

The t values reflect the relative contribution of the respective variables in predicting corn yield. In general, they can be tested statistically through the t distribution with $(n - p - 1)$ df corresponding to the df of the error standard deviation. However, the viability of variable screening, on the basis of statistical theory, is not well developed. In general, the power (the probability of rejecting the null hypothesis $H_0\colon \beta_j = 0$ when it is false) tends to be relatively small in regression applications. Therefore, the extent to which variable screening is implemented depends on the requirements, objectives, and nature of the regression application. In this particular application, it would be recommended that all terms be retained owing to the nature of the problem; i.e., if July precipitation and June temperature are significant, then the other comparable variables also should be important.

DIMENSIONAL ANALYSIS

General References: Boucher and Alves, "Dimensionless Numbers," *Chem. Eng. Prog.*, **55**, 55–64 (1960). Bridgman, *Dimensional Analysis*, Yale University Press, New Haven, Conn., 1937. Buckingham, "On Physically Similar Systems: Illustrations of the Use of Dimensional Analysis," *Phys. Rev.*, **4**, 345 (1914). Klinkenberg, *Chem. Eng. Sci.*, **4**, 130–140, 167–177 (1955). Langhaar, *Dimensional Analysis and Theory of Models*, Wiley, New York, 1951. Murphy, "Dimensional Analysis," *Bull. VPI*, **42**(6) (1949). Porter, *The Method of Dimensions*, Methuen, London, 1933. Rayleigh, "The Principle of Similitude," *Nature*, **95**, 66 (1915). Silberberg and McKetta, "Learning How to Use Dimensional Analysis," *Pet. Refiner*, **32**(4), 179; (5), 147; (6), 101; (7), 129 (1953). Van Driest, "On Dimensional Analysis and the Presentation of Data in Fluid Flow Problems," *J. Appl. Mech.*, **68**, A-34 (March 1946).

Dimensional analysis is simply a mathematical tool. It is not necessary to be a mathematician, however, to be able to use and to understand dimensional analysis. A knowledge of its use will enable engineers to save considerable time in planning experiments and in correlating results of an experiment or using correlations prepared by others.

In order to apply dimensional analysis to a situation, engineers need know only the variables believed to be involved and their dimensions. They need use only these along with fundamental algebra to save time, trouble, and expense in experimental investigation or correlation of a problem. In some cases, without any experimental work at all they can tell from dimensional analysis whether a suspected variable is really involved in a particular problem.

In all cases, dimensional analysis will reduce the number of experimental variables to be correlated, and often it will point out the best experimental approach to the problem. It will not give quantitative information, however; experiment must still be relied upon for that purpose.

THE SIMPLEST PROBLEMS

When only two variables are to be correlated, a simple plot of one against the other results in a single curve. When the number of variables is increased to three, then values of the first are plotted against corresponding values of the second for several constant values of the third, producing a set of curves. The addition of a fourth or more variables greatly complicates both the experimental and the correlation work.

For an example of the value of dimensional analysis in a simple problem involving four variables, this hypothetical case is considered: Assume that a problem exists in which the only variables are two forces, F and P, and two lengths, X and Y.

Variable Correlation The correlation of this simple problem by plotting one variable against another at constant values of the third and fourth variables would be anything but simple. In the experimental work F would have to be held constant at $F = F_1$ while the variation of Y with X was determined at constant values of $P(P = P_1, P_2, P_3$, etc.). Then F would have to be held at $F = F_2$ while the variation of Y with X was again determined for the various constant values of P. This procedure would have to be repeated again and again until the range of values of F had been covered.

Dimensional-Analysis Correlations Since in this simple problem the two variables F and P have the dimension of force and the two variables X and Y have the dimension of length, we would have two dimensionless groups F/P and X/Y. Application of the basic princple of dimensional analysis would show that the plot of F/P versus X/Y would yield a single curve. It is then easily seen that the saving in experimental work would be tremendous since no single variable would have to be held constant. Only the dimensionless group X/Y would need to be varied, and the corresponding value of F/P would be measured.

Admittedly such simple problems are rare in chemical engineering studies. A more typical one would involve more variables with varying degrees of dimensional complexity in the individual variables. Such a problem is the one involving pressure drop accompanying the flow of a fluid through a closed circuit. In this problem suppose that seven variables are considered: pressure drop, fluid velocity, fluid density, fluid viscosity, conduit length, conduit diameter, and conduit wall roughness. Application of dimensional analysis to this problem reduces the number of quantities to be correlated from seven variables to four dimensionless groups of variables, and a consideration of the mechanics of fluids further reduces the number to three dimensionless groups. A single chart then becomes sufficient for the general correlation of the seven original variables.

RAYLEIGH METHOD

Rayleigh's method is based upon the premise that if n quantities, Q_1, Q_2, Q_3, . . . , Q_n, are involved in a certain physical phenomenon, for the purpose of the dimensional analysis their mutual dependence may be expressed as a power product of the following type:

$$Q_1 = KQ_2^{a2}Q_3^{a3} \cdots Q_n^{an} \tag{2-109}$$

where K is a dimensionless constant.

In this equation, Q_1 might be construed to be the quantity which is of principal interest, although such an interpretation is entirely unessential to the method. The Q's would include all the variables known to enter into the particular phenomenon and, in addition, all dimensional constants either demanded by the dimensional system employed or otherwise known to be involved. The requirement of dimensional homogeneity places some restrictions upon the values that the $n - 1$ constants a_2, a_3, . . . , a_n may have. If the n variables and dimensional constants consist of r primary dimensions, then there exists a maximum of r conditions which the constant exponents of Eq.(2-109) must satisfy. The word "maximum" is advisedly used, for in some cases, depending upon the dimensional system used or the dimensional nature of the quantities involved, two or more of the r conditions may in effect be identical, thereby reducing the number of actual conditions to a value less than r. Consequently, of the $n - 1$ constants, a minimum of $n - 1 - r$ is not restricted by the requirement of dimensional homogeneity. The final result of dimensional analysis by the Rayleigh method is an arrangement of the n quantities into such a form that a dimensionless product or group containing Q_1 is equated to the product of a minimum of $n - 1 - r$ other dimensionless groups, each raised to the power represented by one of the $n - 1 - r$ unrestricted exponents. As a consequence, there results an arrangement of the n quantities into a minimum of $n - 1 - r + 1 = n - r$ dimensionless groups.

Example: Rayleigh Method—Pressure Drop by Friction in Fluid Flow When a fluid is flowing in a small straight length of dL of a pipe with internal diameter D, a pressure drop $-dp_f$ occurs as a result of friction. The variables believed to be involved and their dimensions in the engineering system* are as follows:

Pressure drop $= -dp_f = (F/L^2)$

Pipe internal diameter $= D = (L)$

Pipe length $= dL = (L)$

*Any of the systems of dimensions could have been chosen. The engineering system is used here only because most readers are more familiar with it.

Pipe roughness $= \epsilon = (L)$
Pipe linear velocity $= V = (L/\theta)$
Fluid absolute viscosity $= \mu = (M/L\theta)$
Fluid density $= \rho = (M/L^3)$
Dimensional constant $= g_c = (ML/F\theta^2)$

An inspection of the dimensions of these quantities indicates that all are dimensionally combinable and should appear in the result of the dimensional analysis. Therefore, since there are eight quantities and four primary dimensions, a minimum of three ($n - r - 1 = 3$) unrestricted exponents and a minimum of four ($n - r = 4$) dimensionless groups may be expected.

The basic equation relating the variables is

$$-dp_f = KD^a(dL)^b\epsilon^c V^d\rho^e\mu^h g_c^j \tag{2-110}$$

The corresponding dimensional equation is

$$\frac{F}{L^2} = (L)^a(L)^b(L)^c\left(\frac{L}{\theta}\right)^d\left(\frac{M}{L^3}\right)^e\left(\frac{M}{L\theta}\right)^h\left(\frac{ML}{F\theta^2}\right)^j \tag{2-111}$$

If the condition of dimensional homogeneity is applied, there results the following set of equations:

$$\begin{aligned} \Sigma F &= 0{:}\ 1 = -j \\ \Sigma M &= 0{:}\ 0 = e + h + j \\ \Sigma L &= 0{:}\ -2 = a + b + c + d - 3e - h + j \\ \Sigma\theta &= 0{:}\ 0 = -d - h - 2j \end{aligned} \tag{2-112}$$

No two of these four equations are identical, nor may any be obtained from a linear combination of two others. Therefore, the number of conditions is the maximum ($r = 4$), and the minimum number of dimensionless groups may be expected ($n - r = 4$). Since there are four independent equations relating the seven constants, the values of four of the constants may be determined in terms of the remaining three. If we let these three unrestricted constants be b, c, and h, the solutions to Eqs. (2-112) then become

$$\begin{aligned} a &= -b - c - h \\ d &= 2 - h \\ e &= 1 - h \\ j &= -1 \end{aligned} \tag{2-113}$$

The insertion of the solutions represented by Eqs. (2-113) into Eq. (2-111) results in the expression

$$-dp_f = KD^{-b-c-h}(dL)^b\epsilon^c V^{2-h}\rho^{1-h}\mu^h g_c^{-1} \tag{2-114}$$

or

$$\frac{g_c(-dp_f)}{\rho V^2} = K(dL/D)^b(DV\rho/\mu)^{-h}(\epsilon/D)^c \tag{2-115}$$

The seven variables plus g_c are now correlated in the form of four dimensionless groups. From theoretical considerations, it can be shown that the pressure drop $-dp_f$ must be directly proportional to the pipe length dL; consequently, the exponent b must have the value of unity. The dimensionless group $g_c(-dp_f)/\rho V^2$ is called the Euler number, N_{Eu}. The dimensionless group $DV\rho/\mu$ is called the Reynolds number, N_{Re}. Both of these numbers have general application in fluid flow. The dimensionless group ϵ/D is called the roughness factor of the pipe.

For reasons which will be discussed briefly later, in place of Eq. (2-115) with $b = 1$ there may be written the more general functional equation

$$-dp_f/\rho = (V^2/g_c)(dL/D)\Phi(N_{\text{Re}}, \epsilon/D) \tag{2-116}$$

If the function $\Phi(N_{\text{Re}}, \epsilon/D)$ is given the symbol $f/2$, Eq. (2-116) may be written as

$$-dp_f/\rho = f(V^2/2g_c)(dL/D) \tag{2-116a}$$

The function of the Reynolds number and the roughness factor denoted by the symbol f is called the friction factor.

In this example the exponents b, c, and h were purposely selected as the unrestricted exponents in order to obtain the results in the conventional form. Three other exponents might have been so selected, with the obvious exception of j, which must have the value of -1 for this example. Inspection of the equations for ΣM and $\Sigma\theta$ also shows that, of the exponents d, e, and h, no more than one may be selected as an unrestricted exponent. For example, the selection of a, b, and d as the unrestricted exponents would have resulted in the following four dimensionless groups:

$$\frac{g_c(-dp_f)\epsilon^2\rho}{\mu^2} \qquad \frac{\epsilon}{D} \qquad \frac{dL}{D} \qquad \frac{\epsilon V\rho}{\mu}$$

The appearance of ϵ, a quantity which is difficult to determine with exactness, in three of these four dimensionless groups is undesirable. If $\epsilon V\rho/\mu$ is divided by ϵ/D, the Reynolds number $DV\rho/\mu$ results. If the group $g_c(-dp_f)\epsilon^2\rho/\mu^2$ is divided by $(\epsilon V\rho/\mu)^2$, the Euler number $g_c(-dp_f)/\rho V^2$ results. In this manner, the results of dimensionless analysis may be converted from dimensionless groups that are not desired to groups that are more convenient and desirable.

The transformation of $(DV\rho/\mu)^{-h}(\epsilon/D)^c$ in Eq. (2-115) to an unspecified function of these two dimensionless groups $\Phi(N_{\text{Re}}, \epsilon/D)$ in Eq. (2-116) is a definite feature of the Rayleigh method. Solutions of the type of Eq. (2-115) may be regarded as special and may be summed up to give a more general solution. The requirement of dimensional homogeneity for the general solution is equivalent to a requirement of dimensional homogeneity for each special solution. Furthermore, since no restrictions are placed upon the constant K or the exponents in Eq. (2-115), the general solution may be regarded as a general function, as in Eq.(2-116).

The justification of the transition from a power-product equation to a general functional expression in the Rayleigh method implies a restriction upon the nature of the general function. If Rayleigh's method is to be taken at its face value, the function $\Phi(N_{\text{Re}}, \epsilon/D)$ in the first example must be representable as a product of two functions $\Phi_1(N_{\text{Re}})$ and $\Phi_2(\epsilon/D)$. In a more general case, suppose that an analysis by the Rayleigh method indicated that the dimensionless group P is a function of the dimensionless groups R and S, or

$$P = \Phi(R, S) \tag{2-117}$$

The implicit restriction, however, is that the function be representable as

$$P = \Phi_1(R)\Phi_2(S) \tag{2-117a}$$

and not as

$$P = \Phi_3(R) + \Phi_4(S) \tag{2-117b}$$

nor as

$$P = [\Phi_5(R)]^{\Phi_6(S)} \tag{2-117c}$$

If this apparent restriction in Rayleigh's method on the algebraic form in which the solution may be expressed were a general restriction in dimensional analysis, it would indeed be serious. Fortunately, Rayleigh's method may be rigorously regarded as nothing more than an algebraic procedure for determining the dimensionless groups that constitute a *complete* set for the group of quantities being considered. Any implication of restrictions on the nature of the functions of these groups is merely a trivial consequence of the algebraic procedure.

BUCKINGHAM PI METHOD

Although this method was introduced in 1914, no rigorous proof was presented until Langhaar's contribution in 1951. He stated the pi theorem as follows:

> If an equation is dimensionally homogeneous, it can be reduced to a relationship among a complete set of dimensionless products. . . . A set of dimensionless products of given variables is complete if each product in the set is independent of the others, and every other dimensionless product of the variables is a product of powers of dimensionless products in the set.

The dependence of n quantities may be expressed in the form of

$$Q_1^{a1}Q_2^{a2}Q_3^{a3}\cdots Q_m^{am}\cdots Q_n^{an} = \pi_i \quad i = 1, 2, 3, \ldots, p \tag{2-118}$$

The significance of m and p will be explained shortly. No dimensionless constant is needed in Eq. (2-118). In the Rayleigh method, the minimum number of dimensionless groups possible from n quantities described by r dimensions was $n - r$. In the Buckingham method, it is necessary to know beforehand how many dimensionless groups will constitute a complete set. Consequently, the symbol m is introduced and defined as the number of **restrictions** placed upon Eq. (2-118) by virtue of the requirement of dimensional homogeneity. Furthermore, m will have as its maximum value the number of primary dimensions r. If there are m restrictions on the values of a_1, a_2, a_3, . . . , a_n in Eq. (2-118), there will be $n - m$ unrestricted exponents, and consequently $n - m$ dimensionless groups will result.

Langhaar has proved that the number of dimensionless groups p constituting a complete set for n quantities is given by the equation

$$p = n - m \qquad (2\text{-}119)$$

According to Buckingham's theorem, the p dimensionless groups $\pi_1, \pi_2, \pi_3, \ldots, \pi_p$ are then related by the general functional equation

$$\phi(\pi_1, \pi_2, \pi_3, \ldots, \pi_p) = 0 \qquad (2\text{-}120)$$

This equation does no more than state that the particular phenomenon may now be described as rigorously and accurately in terms of the complete set of p dimensionless groups as it might be in terms of the n quantities involved in the phenomenon.

The evaluation of m has been the chief difficulty in the use of the Buckingham method. Van Driest stated that m could be considered as the maximum number of quantities involved in the problem that could be combined without forming a dimensionless group. Langhaar has rigorously proved this to be true and has further shown that m is in reality the rank of the matrix formed from the exponents of the dimensions of the quantities. This use of matrices in connection with Buckingham's method is convenient, but it is by no means essential to the method. The determination of m as the maximum number of variables that will not combine to form a dimensionless group is entirely satisfactory even though sometimes rather tedious. A simple demonstration of the fact that the maximum value of m is the number of primary dimensions results from a consideration of the five variables force, mass, length, time, and temperature with engineering-system dimensions. Obviously these five variables cannot possibly form a dimensionless group, but the inclusion of any other variable describable by the dimensions F, M, L, θ, and T will permit the formation of a dimensionless group.

Since there are by definition m restrictions upon the values of the constant exponents $a_1, a_2, a_3, \ldots, a_m, \ldots, a_n$ in Eq. (2-118), let the constants $a_1, a_2, a_3, \ldots, a_m$ be those so restricted. The constants $a_{m+1}, a_{m+2}, \ldots, a_n$ may be assigned any value, as they represent the $n - m$ unrestricted constants. To determine $\pi_1(i = 1)$ from Eq. (2-118), let $a_{m+1} = 1$ and $a_{m+2}, a_{m+3}, \ldots, a_n$ all equal zero. Similarly, to determine $\pi_2(i = 2)$, let $a_{m+2} = 1$ and $a_{m+1}, a_{m+3}, \ldots, a_n$ all equal zero. If this procedure is followed for all the $n - m$ unrestricted constants, the following set of equations will result:

$$\begin{aligned} \pi_1 &= Q_1^{a1}Q_2^{a2}Q_3^{a3} \cdots Q_m^{am}Q_{m+1} \\ \pi_2 &= Q_1^{a1}Q_2^{a2}Q_3^{a3} \cdots Q_m^{am}Q_{m+2} \\ &\vdots \\ \pi_p &= Q_1^{a1}Q_2^{a2}Q_3^{a3} \cdots Q_m^{am}mQ_n \end{aligned} \qquad (2\text{-}121)$$

This set of equations represents the equations that must be constructed from the quantities in any problem when Buckingham's method is used. The same m quantities appear in all p equations. In order to avoid difficulty or a complete impasse in the solutions of these equations, it is essential that these m repeating quantities be such that they alone are not capable of forming a dimensionless group.

Example: Buckingham Pi Method—Heat-Transfer Film Coefficient It is desired to determine a complete set of dimensionless groups with which to correlate experimental data on the film coefficient of heat transfer between the walls of a straight conduit with circular cross section and a fluid flowing in that conduit. The variables and the dimensional constant believed to be involved and their dimensions in the engineering system are given below:

Film coefficient = h = $(F/L\theta T)$
Conduit internal diameter = D = (L)
Fluid linear velocity = V = (L/θ)
Fluid density = ρ = (M/L^3)
Fluid absolute viscosity = μ = $(M/L\theta)$
Fluid thermal conductivity = k = $(F/\theta T)$
Fluid specific heat = c_p = (FL/MT)
Dimensional constant = g_c = $(ML/F\theta^2)$

The first step in the solution requires the determination of m. Since five primary dimensions ($FML\theta T$) are involved, the maximum value that m might have is 5. A few trials will show that there are several sets of five quantities which do not form dimensionless groups. Since m therefore has the value 5, then $n = 8$ quantities listed should produce a complete set of $p = n - m = 3$ dimensionless groups according to Eq. (2-119).

Let the five repeating quantities that do not form a dimensionless group be D, V, μ, k, and g_c. According to Eq. (2-121), the following equations represent the three dimensionless groups:

$$\pi_1 = D^aV^b\mu^dk^eg_c^fh \qquad (2\text{-}122)$$
$$\pi_2 = D^aV^b\mu^dk^eg_c^fc_p \qquad (2\text{-}123)$$
$$\pi_3 = D^aV^b\mu^dk^eg_c^f\rho \qquad (2\text{-}124)$$

In these equations, h was included with the five repeating variables to form π_1. Similarly, c_p was included to form π_2, and ρ was included to form π_3.

The procedure from this point is identical to that in the Rayleigh method with the exception that no undetermined exponents will be found, as the number of restrictions on each equation and the number of exponents are the same. This is a general characteristic of the Buckingham method. The dimensional equation corresponding to Eq. (2-122) is

$$F^\circ M^\circ L^\circ \theta^\circ T^\circ = (L)^a(L/\theta)^b(M/L\theta)^d(F/\theta T)^e(ML/F\theta^2)^f(F/L\theta T) \qquad (2\text{-}125)$$

Applying the condition of dimensional homogeneity, we have

$$\begin{aligned} \Sigma F &= 0{:}\ 0 = e - f + 1 \\ \Sigma M &= 0{:}\ 0 = d + f \\ \Sigma L &= 0{:}\ 0 = a + b - d + f - 1 \\ \Sigma \theta &= 0{:}\ 0 = -b - d - e - 2f - 1 \\ \Sigma T &= 0{:}\ 0 = -e - 1 \end{aligned} \qquad (2\text{-}126)$$

The solutions to Eqs. (2-126) are

$$a = 1,\quad b = 0,\quad d = 0,\quad e = -1,\quad f = 0 \qquad (2\text{-}127)$$

Equation (2-122) then becomes

$$\pi_1 = D^1V^0\mu^0k^{-1}g_c^0h \qquad (2\text{-}128)$$
$$\pi_1 = hD/k$$

In a similar manner, the solutions to Eqs. (2-123) and (2-124) are found to be

$$\pi_2 = c_p\mu/k \qquad (2\text{-}129)$$
$$\pi_3 = DV\rho/\mu \qquad (2\text{-}130)$$

The dimensionless group hD/k is called the Nusselt number, N_{Nu}, and the group $c_p\mu/k$ is the Prandtl number, N_{Pr}. The group $DV\rho/\mu$ is the familiar Reynolds number, N_{Re}, encountered in fluid-friction problems. These three dimensionless groups are frequently used in heat-transfer-film-coefficient correlations. Functionally, their relation may be expressed as

$$\phi(N_{Nu}, N_{Pr}, N_{Re}) = 0 \qquad (2\text{-}131)$$

or as

$$N_{Nu} = \phi_1(N_{Pr}, N_{Re}) \qquad (2\text{-}132)$$

It has been found that these dimensionless groups may be correlated well by an equation of the type

$$hD/k = K(c_p\mu/k)^a(DV\rho/\mu)^b \qquad (2\text{-}133)$$

in which K, a, and b are experimentally determined dimensionless constants. However, any other type of algebraic expression or perhaps simply a graphical relation among these three groups that accurately fits the experimental data would be an equally valid manner of expressing Eq. (2-131).

Naturally, other dimensionless groups might have been obtained in the example by employing a different set of five repeating quantities that would not form a dimensionless group among themselves. Some of these groups may be found among those presented in Table 2-2. Such a complete set of three dimensionless groups might consist of Stanton, Reynolds, and Prandtl numbers or of Stanton, Peclet, and Prandtl numbers. Also, such a complete set different from that obtained in the preceding example will result from a multiplication of appropriate powers of the Nusselt, Prandtl, and Reynolds numbers. For such a set to be complete, however, it must satisfy the con-

TABLE 2-2 Dimensionless Groups in the Engineering System of Dimensions*

Biot number	N_{Bi}	hL/k
Condensation number	N_{Co}	$(h/k)(\mu^2/\rho^2 g)^{1/3}$
Number used in condensation of vapors	N_{Cv}	$L^3\rho^2 g\lambda/k\mu\Delta t$
Euler number	N_{Eu}	$g_c(-dp)/\rho V^2$
Fourier number	N_{Fo}	$k\theta/\rho c L^2$
Froude number	N_{Fr}	V^2/Lg
Graetz number	N_{Gz}	wc/kL
Grashof number	N_{Gr}	$L^3\rho^2\beta g\Delta t/\mu^2$
Mach number	N_{Ma}	V/V_a
Nusselt number	N_{Nu}	hD/k
Peclet number	N_{Pe}	$DV\rho c/k$
Prandtl number	N_{Pr}	$c\mu/k$
Reynolds number	N_{Re}	$DV\rho/\mu$
Schmidt number	N_{Sc}	$\mu/\rho D_v$
Stanton number	N_{St}	$h/cV\rho$
Weber number	N_{We}	$LV^2\rho/\sigma g_c$

* For additional groups see work of Boucher and Alves.

dition that each of the three dimensionless groups be independent of the other two.

Although it is not directly necessary in the Buckingham method to know how many quantities are really involved (dimensionally combinable) in the problem, it is essential to the method that the correct estimation of the value of m be made. Since the inclusion of a quantity that is not dimensionally combinable will increase the value of m, thereby increasing the number of repeating quantities in the equations of the type of Eq. (2-121), some difficulty might be anticipated. However, all dimensionally noncombinable quantities must certainly be included in the largest sets of quantities characterized by the inability to form a dimensionless group among themselves. In such a case, when the equations of the type of Eqs. (2-121) are constructed, these quantities are included among the m repeating quantities but, being noncombinable, drop out from each dimensionless group being formed. This was the case with g_c in the second example. Since both n and m are increased by 1 as a result of the inclusion of each noncombinable quantity, neither the prediction of the number of dimensionless groups nor the results of the analysis are affected in any way. Thus, the problem is handled automatically by the determination of m as the largest number of quantities that are not capable of forming a dimensionless group and by the use of this set of quantities as the m repeating quantities in Eqs.(2-121). Similarly, no difficulties arise when the method employing matrices is used.

By virtue of the inclusion of dimensionally noncombinable g_c in the second example, the value of m resulting was the maximum value, the same as the number of primary dimensions ($FML\theta T$). Had g_c been omitted (an omission which the results of the analysis justified), the value of m would have been 4, or 1 less than the number of primary dimensions. Inspection of Eqs. (2-126) for ΣF and ΣT shows that if f is assigned the value of zero (equivalent to omission of g_c), these two equations become identical. Furthermore, inspection of the dimensions of the quantities excluding g_c shows that F/T is in reality a primary dimension for that set of variables. In other words, a four-dimensional system (F/T, M, L, θ) is adequate for the problem. Dimensions need not have physical significance, so that F/T is a perfectly permissible primary dimension for this problem, and actually the value of m was the number of primary dimensions. It would be misleading, however, not to stress the fact that all cases in which m is apparently less than the number of primary dimensions may not be so explained. The four quantities g_c, $-dp_f$, ρ, and V, as defined for the first example, are described in the engineering system by the four dimensions F, M, L, and θ. However, $m = 3$ for this case, as these four quantities do combine to form the Euler number.

A lack of space necessarily limits the discussion on this important subject. It is suggested that beginners follow the comprehensive discussion of Silberberg and McKetta in the four-part article in *Petroleum Refiner* in 1953. Those more familiar with the subject should refer to Langhaar (op. cit.) for a more thorough and sophisticated treatment.

Section 3

Physical and Chemical Data

Peter E. Liley, Ph.D., D.I.C., *School of Mechanical Engineering and Center for Information and Numerical Data Analysis and Synthesis, Purdue University; Member, Institute of Physics, London. (Physical and Chemical Data)*

Robert C. Reid, Sc.D., *Department of Chemical Engineering, Massachusetts Institute of Technology. (Prediction and Correlation of Physical Properties)*

Evan Buck, M.S.Ch.E., *Manager—Process Design Data and Thermodynamics, Central Engineering Department, Union Carbide Corporation. (Prediction and Correlation of Physical Properties)*

DENSITIES OF AQUEOUS ORGANIC SOLUTIONS

DENSITIES OF MISCELLANEOUS MATERIALS

SOLUBILITIES

THERMAL EXPANSION

JOULE-THOMSON EFFECT

CRITICAL CONSTANTS

COMPRESSIBILITIES

LATENT HEATS

SPECIFIC HEATS OF PURE COMPOUNDS

SPECIFIC HEATS OF AQUEOUS SOLUTIONS

SPECIFIC HEATS OF MISCELLANEOUS MATERIALS

HEATS AND FREE ENERGIES OF FORMATION

HEATS OF COMBUSTION

HEATS OF SOLUTION

THERMODYNAMIC PROPERTIES

TRANSPORT PROPERTIES

PREDICTION AND CORRELATION OF PHYSICAL PROPERTIES

GENERAL REFERENCES

Considerations of reader interest, space availability, the system or systems of units employed, copyright considerations, etc., have all influenced the revision of material in previous editions for the present edition. Reference is made at numerous places to various specialized works and also, when appropriate, to more general works. A listing of general works may be useful to readers in need of further information.

Canjar and Manning, *Thermodynamic Properties and Reduced Correlations for Gases,* Gulf, Houston, 1967. D'Ans and Lax, *Handbook for Chemists and Physicists* (in German), Springer-Verlag, Berlin (often referred to as D'Ans-Lax). El-Sabban and Scott, U.S. Bur. Mines Bull. 654, 1970; review of chemical thermodynamic properties of 25 organic compounds and list of 27 other general references, plus specific references. Gallant, *Ser. Hydrocarbon Process.*, **44**(7), (1965–1969), listing physical properties of hydrocarbons. Hala, Pick, et al., *Vapor Liquid Equilibria*, Pergamon, New York, 1958. *Handbook of Fundamentals,* American Society of Heating, Refrigerating and Air-Conditioning Engineers, New York, 1972. Hilsenrath et al., *Tables of Thermal Properties of Gases,* NBS Circ. 564, 1955. Hultgren, Orr, et al., *Selected Values of Thermodynamic Properties of Metals and Alloys*, Wiley, New York, 1963. *International Critical Tables,* McGraw-Hill, New York. Janz, *Estimation of Thermodynamic Properties of Organic Compounds,* Academic, New York, 1958. Jordan, *Vapor Pressure of Organic Compounds,* Interscience, New York, 1954. Kaye and Laby, *Tables of Physical and Chemical Constants,* 12th ed., New York, 1960, and later editions. King, *Phase Equilibrium in Mixtures,* Pergamon, New York, 1969. Kubaschewski and Evans, *Metallurgical Thermochemistry,* Wiley, New York, 1956. Landolt-Börnstein, *Eigenschaften der Materie in Ihren Aggregatzuständen,* two volumes on transport phenomena, 1968 and 1969; other volumes should also be consulted for thermodynamic properties, magnetic properties, etc. Linke and Seidell, *Solubilities of Inorganic and Metal-Organic Compounds,* various volumes, Van Nostrand, Princeton, N.J. Nesmeyanov, *Vapor Pressure of the Elements,* Academic, New York, 1963; and Moscow original, 1961. Reid and Sherwood, *Properties of Gases and Liquids,* McGraw-Hill, New York, 1966. Reisman, *Phase Equilibria,* Academic, New York, 1970. *Selected Values of Chemical Thermodynamic Properties,* NBS Circ. 500, plus additional Tech. Notes TN 270-1, 1965 ff. *Selected Values of Properties of Hydrocarbons and Related Compounds,* American Petroleum Institute Research Proj. 44, Carnegie Institute of Technology, Pittsburgh; continued at Thermodynamics Research Center, Texas A&M University, College Station, 1972. Stephen and Stephen, *Solubilities of Inorganic and Organic Compounds,* Macmillan, New York, 1963. Stull, Prophet, et al., *JANAF Thermochemical Tables,* 2d ed., NSRDS-NBS-37, 1971. Techo, "Bibliography of Thermodynamic Networks of Pure Substances," M.S. thesis, Georgia Institute of Technology, Atlanta, 1958. *Thermodynamic Charts for 13 Materials,* Institute of International Refrigeration, 177 Boulevard Malesherbes, Paris. *Thermodynamic Properties of Refrigerants,* New York, 1968. *Thermophysical Properties of Refrigerants,* American Society of Heating, Refrigerating and Air-Conditioning Engineers, New York, 1973. Timmermans, *Physico-Chemical Constants of Pure Organic Compounds,* Elsevier, Amsterdam, 1950. Touloukian et al., *Thermophysical Properties of Matter,* TPRC Data Ser., Plenum, New York; thermal conductivity, specific heat, radiative properties of solids, liquids, and gases, viscosity.

PHYSICAL PROPERTIES OF PURE SUBSTANCES

TABLE 3-1 Physical Properties of the Elements and Inorganic Compounds*

Abbreviations Used in the Table

a., acid
A., specific gravity with reference to air = 1
abs., absolute
ac., acetic acid
act., acetone
al., 95 per cent ethyl alcohol
alk, alkali (*i.e.*, aq. NaOH or KOH)
am., amyl (C_5H_{11})
amor., amorphous
anh., anhydrous
aq., aqueous or water
aq. reg., aqua regia
atm., atmosphere or 760 mm. of mercury pressure
bk., black
brn., brown
bz., benzene
c., cold
cb., cubic
cc, cubic centimeter
chl., chloroform
col., colorless or white
conc., concentrated
cr., crystals or crystalline
d., decomposes
D., specific gravity with reference to hydrogen = 1
d. 50, decomposes at 50°C.; 50 d., melts at 50°C. with decomposition
delq., deliquescent
dil., dilute
dk., dark
eff., effloresces or efflorescent
et., ethyl ether
expl., explodes
gel., gelatinous
gly., glycerol, (glycerin)
gn., green
h., hot
hex., hexagonal
hyg., hygroscopic
i., insoluble
ign., ignites
lq., liquid
lt., light
m. al., methyl alcohol
mn., monoclinic
nd., needles
NH_3, liquid ammonia
NH_4OH, ammonium hydroxide solution
oct., octahedral
or., orange
pd., powder
pl., plates
pr., prisms or prismatic
pyr., pyridine
rhb., rhombic (orthorhombic)
s., soluble
satd., saturated
sl., slightly
soln., solution
subl., sublimes
sulf., sulfides
tart. a., tartaric acid
tet., tetragonal
tr., transition
tri., triclinic
trig., trigonal
v., very
vac., in vacuo
vl., violet
volt., volatile or volatilizes
wh., white
yel., yellow
∞, soluble in all proportions
<, less than
>, greater than
42±, about or near 42
$-3H_2O$, 100, loses 3 moles of water per formula weight at 100°C.

Formula weights are based upon the International Atomic Weights of 1941 and are computed to the nearest hundredth.

Refractive index, where given for a uniaxial crystal, is for the ordinary (ω) ray; where given for a biaxial crystal, the index given is for the median (β) value. Unless otherwise specified, the index is given for the sodium D-line (λ = 589.3 mμ).

Specific gravity values are given at room temperatures (15° to 20°C.) unless otherwise indicated by the small figures which follow the value: thus, "$5.6\frac{18°}{4}$" indicates a specific gravity of 5.6 for the substance at 18°C. referred to water at 4°C. In this table the values for the specific gravity of gases are given with reference to air (A) = 1, or hydrogen (D) = 1.

Melting point is recorded in a certain case as "82 d." and in some other case as "d. 82," the distinction being made in this manner to indicate that the former is a melting point with decomposition at 82°C., while in the latter decomposition only occurs at 82°C. Where a value such as "$-2H_2O$, 82" is given it indicates loss of 2 moles of water per formula weight of the compound at a temperature of 82°C.

Boiling point is given at atmospheric pressure (760 mm, of mercury) unless otherwise indicated; thus, "$82^{15 mm.}$" indicates the boiling point is 82°C. when the pressure is 15 mm.

Solubility is given in parts by weight (of the formula shown at the extreme left) per 100 parts by weight of the solvent; the small superscript indicates the temperature. In the case of gases the solubility is often expressed in some manner as "$5^{10°}$ cc" which indicates that at 10°C., 5 cc. of the gas are soluble in 100 g. of the solvent. The symbols of the common mineral acids: H_2SO_4, HNO_3, HCl, etc., represent dilute aqueous solutions of these acids. See also special tables on Solubility.

REFERENCES: The information given in this table has been collected mainly from the following sources: Mellor, "A Comprehensive Treatise on Inorganic and Theoretical Chemistry," Longmans, New York, 1922. Abegg, "Handbuch der anorganischen Chemie," S. Hirzel, Leipzig, 1905. Gmelin-Kraut, "Handbuch der anorganischen Chemie," 7th ed., Carl Winter, Heidelberg; 8th ed., Verlag Chemie, Berlin, 1924. Friend, "Textbook of Inorganic Chemistry," Griffin, London, 1914. Winchell, "Microscopic Character of Artificial Inorganic Solid Substances or Artificial Minerals," Wiley, New York, 1931. "International Critical Tables," McGraw-Hill, New York, 1926. "Tables annuelles internationales de constants et donnes numeriques," McGraw-Hill, New York. "Annual Tables of Physical Constants and Numerical Data," National Research Council, Princeton, N.J., 1943. Comey and Hahn, "A Dictionary of Chemical Solubilities," Macmillan, New York, 1921. Seidell, "Solubilities of Inorganic and Metal Organic Compounds," Van Nostrand, New York, 1940.

Name	Formula	Formula weight	Color, crystalline form and refractive index	Specific gravity	Melting point, °C.	Boiling point, °C.	Solubility in 100 parts: Cold water	Hot water	Other reagents
Aluminum	Al	26.97	silv., cb.	$2.70^{20°}$	660	2056	i.	i.	s. HCl, H_2SO_4, alk.
acetate, normal	$Al(C_2H_3O_2)_3$	204.10	wh. pd.	...	d. 200	...	s.	d.	
acetate, basic	$Al(OH)(C_2H_3O_2)_2$	162.07	wh., amor.	...	d.	...	i.	...	s.a.; i. NH_4 salts
bromide	$AlBr_3$	266.72	trig.	$3.01\frac{25°}{4}$	97.5	268	s.	...	s. al., act., CS_2
bromide	$AlBr_3.6H_2O$	374.82	col., delq. cr.	...	d. 100	...	s.	s.	s. al., CS_2
carbide	Al_4C_3	143.91	yel., hex., 2.70	2.95	d. >2200	...	d. to CH_4	...	s. a.; i. act.
chloride	$AlCl_3$	133.34	wh., delq., hex.	$2.44\frac{25°}{4}$	$194^{5.2 atm.}$	$182.7^{752 mm}$; subl. 178	$69.87^{15°}$	s. d.	s. et., chl., CCl_4; i. bz.
chloride	$AlCl_3.6H_2O$	241.44	col., delq., trig., 1.560	...	...	...	400	v. s.	50 al.; s. et.
fluoride (fluellite)	$AlF_3.H_2O$	101.99	col., rhb., 1.490	2.17	d.	...	sl. s.		
fluoride	$Al_2F_6.7H_2O$	294.05	wh., cr. pd.	...	$-4H_2O$, 120	$-6H_2O$, 250	i.	sl. s.	
hydroxide	$Al(OH)_3$	77.99	wh., mn.	2.42	$-2H_2O$, 300	...	$0.000104^{18°}$	i.	s. a., alk.; i. a.
nitrate	$Al(NO_3)_3.9H_2O$	375.14	rhb., delq.	...	73	d. 134	v. s.	v. s. d.	s. al., CS_2
nitride	Al_2N_2	81.96	yel., hex.	$3.05\frac{25°}{4}$	$2150^{4 atm.}$	d. >1400	d. slowly	...	s. alk. d.
oxide	Al_2O_3	101.94	col., hex., 1.67–8	3.99	1999 to 2032	...	i.	i.	v. sl. s. a., alk.
oxide (corundum)	Al_2O_3	101.94	wh., trig., 1.768	4.00	1999 to 2032	2210	i.	i.	v. sl. s. a., alk.
phosphate	$AlPO_4$	121.95	col., hex.	2.59	...	...	i.	i.	s. a., alk.; i. ac.
potassium silicate (muscovite)	$3Al_2O_3.K_2O.6SiO_2.2H_2O$	796.40	mn., 1.590	2.9	d.	...	i.		
potassium silicate (orthoclase)	$Al_2O_3.K_2O.6SiO_2$	556.49	col., mn., 1.524	2.56	1450 (1150)	...	i.		

Aluminum potassium tartrate	$AlK(C_4H_4O_6)_2$	362.21	col.	...	...	...	s.	s.	
sodium fluoride (cryolite)	$AlF_3.3NaF$	209.96	wh., mn., 1.3389	2.90	1000	...	sl. s.	...	i. HCl
sodium silicate	$Al_2O_3.Na_2O.6SiO_2$	524 29	col., tri., 1.529	2.61	1100	...	i.	i.	d. a.
sulfate	$Al_2(SO_4)_3$	342.12	wh. cr.	2.71	d. 770	...	$31.3^{0°}$	$89^{100°}$	
Alum, ammonium (tschermigite)	$Al_2(SO_4)_3.(NH_4)_2SO_4.24H_2O$	906.64	col., oct., 1.4594	$1.64\frac{20°}{4}$	93.5	$-20H_2O$, 120; $-24H_2O$, 200	$3.9^{0°}$	∞ 100°	i. al.
ammonium chrome	$Cr_2(SO_4)_3.(NH_4)_2SO_4.24H_2O$	956.72	gn. or vl., oct., 1.4842	1.72	100 d.	...	$21.2^{25°}$	...	s. al.
ammonium iron	$Fe_2(SO_4)_3.(NH_4)_2SO_4.24H_2O$	964.40	vl., oct., 1.485	1.71	40	...	$124^{25°}$	...	i. al.
potassium (kalinite)	$Al_2(SO_4)_3.K_2SO_4.24H_2O$	948.76	col., mn., 1.4564	$1.76\frac{26°}{4}$	92	$-18H_2O$, 64.5	$5.7^{0°}$	$\infty^{93°}$	
potassium chrome	$Cr_2(SO_4)_3.K_2SO_4.24H_2O$	998.84	red or gn., cb., 1.4814	1.83	89	...	20	50	i. al.
sodium	$Al_2(SO_4)_3.Na_2SO_4.24H_2O$	916.56	col., oct., 1.4388	$1.675\frac{20°}{4}$	61	...	$106.4^{0°}$	$121.7^{45°}$	i. al.
Ammonia†	NH_3	17.03	col. gas, 1.325 (lq.)	$0.817^{-79°}$ 0.5971 (A)	−77.7	−33.4	$89.9^{0°}$	$7.4^{96°}$	$14.8^{20°}$ al.; s. et.
Ammonium acetate	$NH_4C_2H_3O_2$	77.08	wh., hyg. cr.	1.073	114	d.	$148^{4°}$	...	s. al.; sl. s. act.
auricyanide	$NH_4CN.Au(CN)_3.H_2O$	337.33	pl.	...	d. 200	...	s.	v. s.	i. al.
bicarbonate	NH_4HCO_3	79.06	mn. or rhb., 1.5358	1.573	d. 35–60	...	$11.9^{0°}$	$27^{30°}$	i. al.
bromide	NH_4Br	97.96	col., cb., 1.7108	$2.327\frac{15°}{4}$	subl. 542	...	$68^{10°}$	$145.6^{100°}$	s. al., et., act.
carbonate	$(NH_4)_2CO_3.H_2O$	114.11	col. pl.	...	d. 58	...	$100^{15°}$	...	i. al., CS_2, NH_3
carbonate, carbamate	$NH_4HCO_3.NH_2CO_2NH_4$‡	157.11	wh. cr.	...	subl.	...	$25^{15°}$	$67^{65°}$	
carbonate, sesqui-	$(NH_4)_2CO_3.2NH_4HCO_3.H_2O$	272.22	wh.	...	d.	...	$20^{15°}$	$50^{49°}$	
chloride (salammoniac)	NH_4Cl	53.50	wh., cb., 1.639, 1.6426	$1.53^{17°}$	d. 350	subl. 520	$29.4^{0°}$	$77.3^{100°}$	s. NH_3; sl. s. al., m. al.
chloroplatinate	$(NH_4)_2PtCl_6$	444.05	yel., cb.	3.065	d.	...	$0.7^{15°}$	$1.25^{100°}$	0.005 al.
chloroplatinite	$(NH_4)_2PtCl_4$	373.14	tet.	...	d.	...	s.	v. s.	
chlorostannate	$(NH_4)_2SnCl_6$	367.52	pink., cb.	2.4	...	...	$33.3^{15°}$		
chromate	$(NH_4)_2CrO_4$	152.09	yel., mn.	$1.91^{12°}$	d. 180	...	$40.5^{30°}$	d.	sl. s. act., NH_3; i. al.
cyanide	NH_4CN	44.06	col., cb.	$0.79^{100°}$ (A)	36	...	s.	v. s.	s. al.
dichromate	$(NH_4)_2Cr_2O_7$	252.10	or., mn.	2.15	d. 185	...	$47.2^{30°}$	v. s.	s. al.; i. act.
ferrocyanide	$(NH_4)_4Fe(CN)_6.6H_2O$	392.21	mn.	...	d.	...	s.	...	i. al.
fluoride	NH_4F	37.04	wh., hex.	...	...	...	v. s.	d.	s. al.; i. NH_3
fluoride, acid	$NH_4F.HF$	57.05	wh., rhb., 1.390	$2.21\frac{12}{12}$	...	...	v. s.		
formate	HCO_2NH_4	63.06	col., mn., delq.	1.266	114–116	d. 180; subl. in vac.	$102^{0°}$	$531^{80°}$	s. al.
hydrosulfide	NH_4HS	51.11	col., rhb.	...	d.	subl. 120	v. s.	...	s. al.
hydroxide	NH_4OH	35.05	in soln. only	...	...	...	s.		
molybdate	$(NH_4)_2MoO_4$	196.03	mn.	2.27	d.	...	d.	d.	i. al., NH_3
molybdate, hepta-	$(NH_4)_6Mo_7O_{24}.4H_2O$‡	1235.95	col., mn.	...	...	...	$44^{25°}$	...	i. al.
nitrate (α), stable −16° to 32°	NH_4NO_3	80.05	col., tet., 1.611	$1.66\frac{25°}{4}$	169.6	d. 210	$118.3^{0°}$	$241.8^{30°}$	
nitrate (β), stable 32° to 84°	NH_4NO_3	80.05	col., rhb. or mn.	$1.725\frac{25°}{4}$	...	d. 210	$365.8^{35°}$	$580^{80°}$	$3.8^{20°}$ al., $17.1^{20°}$ m. al.; v. s. NH_3
nitrite	NH_4NO_2	64.05	wh. nd.	1.69	expl.	...	s.	d.	s. al.
osmochloride	$(NH_4)_2OsCl_6$	439.02	cb.	$2.93\frac{20°}{4}$					
oxalate	$(NH_4)_2C_2O_4.H_2O$	142.12	col., rhb.	1.501	...	...	$2.5^{0°}$	$11.8^{50°}$	sl. s. al.; i. NH_3
oxalate, acid	$NH_4HC_2O_4.H_2O$	125.08	col., trimetric	1.556	d.	...	s.		
perchlorate	NH_4ClO_4	117.50	col., rhb., 1.4833	1.95	d.	...	$10.9^{0°}$	$46.9^{100°}$	$2^{20°}$ al.; s. act.; i. et.
persulfate	$(NH_4)_2S_2O_8$	228.20	wh., mn., 1.5016	1.98	d. 120	...	$58.2^{0°}$	d.	
phosphate, monobasic	$NH_4H_2PO_4$	115.04	col., tet., 1.5246	$1.803\frac{19°}{4}$	...	...	$22.7^{0°}$	$173.2^{100°}$	i. ac.
phosphate, dibasic	$(NH_4)_2HPO_4$	132.07	col., mn., 1.53	1.619	...	...	$131^{15°}$	...	i. act.
phosphate, meta-	$(NH_4)_4P_4O_{12}$	388.08	col., mn.	2.21	...	...	s.		

*By N. A. Lange, Ph.D., Handbook Publishers, Inc., Sandusky, Ohio. Abridged from table of Physical Constants of Inorganic Compounds in Lange, "Handbook of Chemistry."
† See special tables.
‡ Usual commercial form.

TABLE 3-1 Physical Properties of the Elements and Inorganic Compounds* *(Continued)*

Name	Formula	Formula weight	Color, crystalline form and refractive index	Specific gravity	Melting point, °C.	Boiling point, °C.	Solubility in 100 parts: Cold water	Hot water	Other reagents
Ammonium phosphomolybdate	$(NH_4)_3PO_4.12MoO_3.3H_2O$ (?)	1930.55	yel.	...	d.	...	$0.03^{15°}$	i.	s. alk.; i. al., HNO_3
silicofluoride	$(NH_4)_2SiF_6$	178.14	cb., 1.3696	2.01	...	subl.	$18.5^{17.5°}$	55.5	s. al.; i. act.
sulfamate	$NH_4.SO_3NH_2$	114.12	col. pl.	...	132	d. 160	$134^{0°}$	$357^{60°}$	
sulfate (mascagnite)	$(NH_4)_2SO_4$	132.14	col., rhb., 1.5230	$1.769\frac{20°}{4}$	235 d.	...	$70.6^{0°}$	$103.3^{100°}$	i. al., act., CS_2
sulfate, acid	NH_4HSO_4	115.11	col., rhb., 1.480	1.78	146.9	490	100	...	v. sl. s. al.; i. act.
sulfide	$(NH_4)_2S$	68.14	yel.-wh.	...	d.	...	v. s.	...	$120^{25°}$ NH_3
sulfide, penta-	$(NH_4)_2S_5$	196.38	or.-red pr.	...	...	...	s.		
sulfite	$(NH_4)_2SO_3.H_2O$	134.16	col., mn.	1.41	d.	...	$100^{12°}$	...	i. al., act.
sulfite, acid	NH_4HSO_3	99.11	rhb.	$2.03\frac{12°}{4}$	d.	...	s.		
tartrate	$(NH_4)_2C_4H_4O_6$	184.15	col., mn.	1.60	d.	...	$45^{0°}$	$87^{60°}$	sl. s. al.
thiocyanate	NH_4CNS	76.12	col., mn., 1.685±	1.305	149.6	d. 170	$120^{0°}$	$170^{20°}$	s. al., act., NH_3, SO_2
vanadate, meta-	NH_4VO_3	116.99	col. cr.	2.326	d.	...	$0.44^{18°}$	$3.05^{70°}$	i. al., NH_4Cl
Antimony	Sb	121.76	tin wh., trig.	$6.684^{25°}$	630.5	1380	i.	i.	s. aq. reg., h. conc. H_2SO_4
chloride, tri- (butter of antimony)*	$SbCl_3$	228.13	col., rhb., delq.	$3.14\frac{20°}{4}$	73.4	220.2	$601.6^{0°}$	$\infty^{72°}$	s. al., HCl, HBr, $H_2C_4H_4O_6$
oxide, tri- (valentinite)	Sb_2O_3	291.52	rhb., 2.35	5.67	656	1570	v. sl. s.	sl. s.	s. HCl, KOH, $H_2C_4H_4O_6$
oxide, tri- (senarmontite)	Sb_2O_3	291.52	cb., 2.087	5.2	652				
sulfide, tri- (stibnite)	Sb_2S_3	339.70	bk., rhb., 4.046	4.64	550	...	$0.00017^{18°}$	d.	s. HCl; alk., NH_4HS, K_2S; i. ac.
sulfide, penta-	Sb_2S_5	403.82	golden	$4.120^{0°}$	−2S, 135	...	i.	i.	s. HCl, alk., NH_4HS
telluride, tri-	Sb_2Te_3	626.35	gray	...	629				
Antimonyl potassium tartrate (tartar emetic)	$(SbO)KC_4H_4O_6.\frac{1}{2}H_2O$	333.94	wh., rhb.	2.60	$-\frac{1}{2}H_2O$, 100	...	$5.26^{8.7°}$	$35.7^{100°}$	s. gly.; i. al.
sulfate, normal	$(SbO)_2SO_4$	371.58	wh. pd.	4.89	...	...	d.	d.	
sulfate, basic	$(SbO)_2SO_4.Sb_2(OH)_4$	683.13	wh. pd.	...	...	...	i.	d.	$5.15^{15°}$ gly.
Argon	A	39.94	col. gas	$1.65^{-233°}$; $1.402^{-185.7°}$; 1.38 (A)	−189.2	−185.7	$5.6^{0°}$ cc	$2.23^{50°}$ cc	$24^{25°}$ cc al.
Arsenic (crystalline)(α)	As_4	299.64	met., hex.	$5.727^{14°}$	$814^{36 atm.}$	subl. 615	i.	i.	s. HNO_3
Arsenic (black)(β)	As_4	299.64	bk., amor.	$4.7^{20°}$	...	...	i.	i.	s. HNO_3, aq. reg., aq. Cl_2, h. alk.
Arsenic (yellow)(γ)	As_4	299.64	yel., cb.	$2.0^{20°}$	d. 358				
acid, ortho-	$H_3AsO_4.\frac{1}{2}H_2O$	150.94	col., hyg.	2.0–2.5	35.5	$-H_2O$, 160	16.7	50	s. alk.
acid, meta-	$HAsO_3$	123.92	wh., hyg.	...	d.		d. to form H_3AsO_4	H_3AsO_4	
acid, pyro-	$H_4As_2O_7$	265.85	col.	...	d. 206		d. to form H_3AsO_4	H_3AsO_4	
pentoxide	As_2O_5	229.82	wh., amor.	4.086	...	d.	$59.5^{0°}$	$76.7^{100°}$	s. alk., al.
sulfide, di- (realgar)	As_2S_2	213.94	red, mn., 2.68	(α)$3.506^{19°}$; (β)$3.254^{19°}$	(α)tr. 267; (β)307	565	i.	d.	s. K_2S, $NaHCO_3$
sulfide, penta-	As_2S_5	310.12	yel.	...	...	d. 500	$0.00013^{60°}$	i.	s. HNO_3, alk.
Arsenious chloride (butter of arsenic)	$AsCl_3$	181.28	oily lq.	lq. 2.163	−18	130	d.	d.	s. HCl, HBr, PCl_3
hydride (arsine)	AsH_3	77.93	col. gas	2.695 (A)	−113.5	−55; d. 230	20 cc	sl. s.	sl. s. alk.
oxide (arsenolite)	As_2O_3	197.82	col., cb., fibrous, 1.755	$3.865\frac{25°}{4}$	subl.	...	sl. s.	sl. s.	i. al., et.
oxide (claudetite)	As_2O_3	197.82	col., mn., 1.92	3.85	subl.	...	sl. s.	sl. s.	i. al., et.
oxide	As_2O_3	197.82	amor. or vitreous	3.738	315	...	$1.2^{10°}$	$2.93^{40°}$	s. HCl, alk., Na_2CO_3; i. al., et.
Auric chloride	$AuCl_3.2H_2O$	339.60	or. cr.	...	d.	...	v. s.	v. s.	s. HCl, al., et.; sl. s. NH_3
cyanide	$Au(CN)_3.6H_2O$	383.35	...	...	d. 50	...	v. s.	v. s.	s. al.
Aurous chloride	AuCl	232.66	yel. cr.	7.4	$AuCl_3$, 170	d. 290	d.	d.	s. HCl, HBr; d. al.
cyanide	AuCN	223.22	yel. cr.	...	d.	...	i.	i.	s. KCN; i. al., et.
Cf. also under *Gold*									
Barium	Ba	137.36	silv. met.	3.5	850	1140	d.	d.	s. a.; d. al.
acetate	$Ba(C_2H_3O_2)_2$	255.45	col.	2.468	...	...	$58.8^{0°}$	$75.0^{100°}$	
acetate	$Ba(C_2H_3O_2)_2.H_2O$	273.46	wh., tri. pr., 1.517	2.19	$-H_2O$, 41	...	$75^{30°}$(anh.)	$79^{40°}$(anh.)	i. al.
bromide	$BaBr_2$	297.19	col.	$4.781\frac{24°}{4}$	847	d.	$98^{0°}$	$149^{100°}$	v. s. m. al.; v. sl. s. act.
bromide	$BaBr_2.2H_2O$	333.22	col., mn., 1.7266	3.69	$-2H_2O$, 100	d.	v. s.	v. s.	s. al.
carbonate (witherite)	$BaCO_3$	197.37	wh., rhb., 1.676	4.29	tr. 811 to α	d. 1450	$0.0022^{18°}$	$0.0065^{100°}$	s. a.; i. al.
carbonate (α)	$BaCO_3$	197.37	wh., hex.	...	tr. 982 to β				
carbonate (β)	$BaCO_3$	197.37	wh.	...	$1740^{90 atm}$	...	$0.0022^{18°}$	$0.0065^{100°}$	s. a.; i. al.

Barium chlorate	$Ba(ClO_3)_2$	304.27	col.		414		$20.35^{0°}$	$84.8^{80°}$	
chlorate	$Ba(ClO_3)_2.H_2O$†	322.29	col.; mn., 1.577	3.179	d. 120		s.	s.	sl. s. al., act.
chloride	$BaCl_2$	208.27	col.; mn., 1.7361	$3.856\frac{24}{4}^{°}$	tr. 925	1560	$31^{0°}$	$59^{100°}$	sl. s. HCl, HNO_3; i. al.
chloride	$BaCl_2$	208.27	col., cb.		962	1560			
chloride	$BaCl_2.2H_2O$†	244.31	col., mn., 1.646	$3.097\frac{24}{4}^{°}$	$-2H_2O$, 100		$39.3^{0°}$	$76.8^{100°}$	sl. s. HCl, HNO_3; i. al.
hydroxide	$Ba(OH)_2$	171.38	col., mn.	4.495			$1.67^{0°}$	$101.4^{80°}$	
hydroxide	$Ba(OH)_2.8H_2O$	315.50	col., mn., 1.5017	$2.188^{16°}$	77.9	$-8H_2O$, 550	$5.6^{15°}$		v. sl. s. al.; i. et.
nitrate (nitrobarite)	$Ba(NO_3)_2$	261.38	col., cb., 1.572	$3.244^{23°}$	592	d.	$5.0^{0°}$	$34.2^{100°}$	sl. s. a.; i. al.
oxalate	BaC_2O_4	225.38	wh. cr.	2.658			$0.0016^{8°}$	$0.0024^{24°}$	s. a., NH_4Cl; i. al.
oxide	BaO	153.36	col., cb., 1.98	5.72	1923	2000±	$1.5^{0°}$	$90.8^{80°}$	s. HCl, HNO_3, abs. al.; i. NH_3, act.
peroxide	BaO_2†	169.36	gray or wh. pd.	4.958	−O, 800		v. sl. s.	d.	s. dil. a.; i. act.
peroxide	$BaO_2.8H_2O$	313.49	pearly sc.		$-8H_2O$, 100		0.168	d.	s. dil. a.; i. al., et., act.
phosphate, monobasic	$BaH_4(PO_4)_2$	331.35	tri.	$2.9^{4°}$			d.	d.	s. a.
phosphate, dibasic	$BaHPO_4$	233.35	wh., rhb. nd., 1.635	$4.165^{18°}$			0.015		s. a., NH_4 salts
phosphate, tribasic	$Ba_3(PO_4)_2$	602.04	wh., cb.	$4.1^{16°}$			i.		s. a.
phosphate, pyro-	$Ba_2P_2O_7$	448.68	wh., rhb.	$3.9^{20°}$			0.01		s. a., NH_4 salts
silicofluoride	$BaSiF_6$	279.42	pr.	$4.279^{15°}$			$0.026^{17°}$	$0.09^{100°}$	sl. s. HCl, NH_4Cl; i. al.
sulfate (barite, barytes)	$BaSO_4$	233.42	col., rhb., 1.636	$4.499^{15°}$	1580 d.	tr. to mn. 1149	$0.000115^{0°}$	$0.000285^{30°}$	s. conc. H_2SO_4; 0.006, 3% HCl
sulfide, mono-	BaS	169.42	col., cb., 2.155	$4.25^{15°}$			d.	d.	d. HCl; i. al.
sulfide, tri-	BaS_3	233.54	yel.-gn.		d. 400		s.	s.	
sulfide, tetra-	$BaS_4.2H_2O$	301.63	red, rhb.	$2.988^{20°}$	d. 200		$41^{15°}$	v. s.	i. al., CS_2
Beryllium (glucinum)	Be(Gl)	9.02	gray, met., hex.	1.816	1284	2767	i.	sl. s. d.	s. dil. a., alk.
Bismuth	Bi	209.00	silv. wh. or reddish, hex.	$9.80^{20°}$	271	1450	i.	i.	s. aq. reg., conc. H_2SO_4; HNO_3
carbonate, sub-	$Bi_2O_3.CO_2.H_2O$	528.03	wh. pd.	6.86	d.		i.	i.	s. a.
chloride, di-	$BiCl_2$(?)	279.91	bk. nd.	4.86	163	d. 300	d.		
chloride, tri-	$BiCl_3$†	315.37	wh. cr.	4.75	230	447	d.		s. al.
nitrate	$Bi(NO_3)_3.5H_2O$	485.10	col., tri.	2.82	d. 30	$-5H_2O$, 80	d.		$42^{19°}$ act.; s. a.; i. al.
nitrate, sub-	$BiONO_3.H_2O$	305.02	hex. pl.	$4.928^{15°}$	d. 260		i.	i.	s. a.
oxide, tri-	Bi_2O_3	466.00	yel., rhb.	8.9	820	1900±	i.	i.	s. a.
oxide, tri-	Bi_2O_3	466.00	yel., tet.	8.55	860		i.	i.	s. a.
oxide, tri-	Bi_2O_3	466.00	yel., cb.	8.20	tr. 704		i.	i.	s. a.
oxychloride	BiOCl	260.46	wh., amor.	$7.72^{15°}$			sl. s.	sl. s.	s. a.; i. act., NH_3, $H_2C_4H_4O_6$
Boric acid	H_3BO_3	61.84	wh., tri.	$1.435^{15°}$	185 d.		$2.66^{0°}$	$40.2^{100°}$	$22.2^{20°}$ gly., $0.24^{25°}$ et.; s. al.
Boron	B	10.82	gray or bk., amor. or mn.	2.32	2300	2550	i.	i.	s. HNO_3; i. al.
carbide	B_4C	55.29	bk. cr.	2.54	2450	>3500	i.	i.	i. a.
oxide	B_2O_3	69.64	col. glass, 1.459	1.85	577	>1500	$1.1^{0°}$	$15.7^{100°}$	s. a., al., gly.
oxide (sassolite)	$B_2O_3.3H_2O$	123.69	tri., 1.456	1.49	d.		sl. s.	s.	
Bromic acid	$HBrO_3$	128.92	col.; in soln. only		d. 100		v. s.	d.	
Bromine	Br_2	159.83	rhb., or red lq.	$3.119^{20°}$; 5.87 (A)	−7.2	58.78	$4.22^{0°}$	$3.13^{30°}$	s. al., et., alk., CS_2
hydrate	$Br_2.10H_2O$	339.99	red, oct.		d. 6.8		s.		
Cadmium	Cd	112.41	silv. met., hex.	$8.65^{20°}$	320.9	767	i.	i.	s. a., NH_4NO_3
acetate	$Cd(C_2H_3O_2)_2$	230.50	col.	2.341	256	d.	v. s.		s. m. al.
acetate	$Cd(C_2H_3O_2)_2.2H_2O$†	266.53	col., mn.	2.01	$-H_2O$, 130		v. s.		s. al.
carbonate	$CdCO_3$	172.42	wh., trig.	$4.258^{4°}$	d. <500		i.	i.	s. a., KCN, NH_4 salts; i. NH_3
chloride	$CdCl_2$	183.32	wh., cb.	$4.047\frac{25}{4}$	568	960	$90^{0°}$	$147^{100°}$	$1.52^{15°}$ al.; i. et., act.
chloride	$CdCl_2.2½H_2O$	228.36	col., mn., 1.6513	3.327	tr. 34		$168^{20°}$	$180^{100°}$	$2.05^{15°}$ m. al.
cyanide	$Cd(CN)_2$	164.45			d. >200		$0.0247^{15°}$		s. a.; NH_4OH, KCN
hydroxide	$Cd(OH)_2$	146.43	wh., trig.	$4.79\frac{15}{4}^{°}$	d. 300		$0.00026^{25°}$		s. a., NH_4 salts; i. alk.
nitrate	$Cd(NO_3)_2$	236.43	col.		350		$109.7^{0°}$	$326^{59.5°}$	v. s. a.
nitrate	$Cd(NO_3)_2.4H_2O$†	308.49	col. nd.	$2.455\frac{17}{4}^{°}$	59.4	132	$215^{0°}$		s. al., NH_3; i. HNO_3
oxide	CdO	128.41	brn., cb.	8.15			i.	i.	s. a., NH_4 salts; i. alk.
oxide	CdO	128.41	brn., amor, 2.49	6.95	d. 900–1000		i.	i.	s. a., NH_4 salts; i. alk.
oxide, sub-	Cd_2O	240.82	gn., amor.	$8.192\frac{18}{4}^{°}$	d.				d. a., alk.

* Usually the solution.
† Usual commercial form.

TABLE 3-1 Physical Properties of the Elements and Inorganic Compounds *(Continued)*

Name	Formula	Formula weight	Color, crystalline form and refractive index	Specific gravity	Melting point, °C.
Cadmium sulfate	$CdSO_4$	208.47	rhb.	$4.691\frac{24°}{4}$	1000
sulfate	$CdSO_4.H_2O$	226.49	mn.	$3.786^{20°}$	tr. 108
sulfate	$3CdSO_4.8H_2O$*	769.54	col., mn., 1.565	3.09	tr. 41.5
sulfate	$CdSO_4.4H_2O$	280.53	col.	3.05	...
sulfate	$CdSO_4.7H_2O$	334.58	mn.	$2.48\frac{20°}{4}$	tr. 4
sulfide (greenockite)	CdS	144.47	yel.-or., hex., 2.506	4.58	1750^{100atm}
Calcium	Ca	40.08	silv. met., cb.	$1.55^{20°}$	810
acetate	$Ca(C_2H_3O_2)_2.H_2O$	176.18	wh. nd.	...	d.
aluminate	$Ca(AlO_2)_2$	158.02	col., rhb. or mn.	$3.67^{20°}$	1600
aluminum silicate (anorthite)	$CaO.Al_2O_3.2SiO_2$	278.14	tri., 1.5832	2.765	1551
arsenate	$Ca_3(AsO_4)_2$	398.06	wh. pd.	...	...
bromide	$CaBr_2$	199.91	delq. nd.	$3.353\frac{25°}{4}$	760
carbonate (aragonite)	$CaCO_3$	100.09	col., rhb., 1.6809	2.93	d. 825
carbonate (calcite)	$CaCO_3$	100.09	col., hex., 1.550	$2.711\frac{25°}{4}$	$1339^{103atm.}$
chloride (hydrophilite)	$CaCl_2$*	110.99	wh., delq., cb, 1.52	$2.152\frac{15°}{4}$	772
chloride	$CaCl_2.H_2O$	129.01	col., delq.	...	...
chloride	$CaCl_2.6H_2O$	219.09	col., trig., 1.417	$1.68^{17°}$	29.92
citrate	$Ca_3(C_6H_5O_7)_2.4H_2O$	570.50	col. nd.	...	$-2H_2O$, 130
cyanamide	$CaCN_2$	80.11	col., rhombohedral	...	...
ferrocyanide	$Ca_2Fe(CN)_6.12H_2O$	508.31	yel., tri., 1.5818	1.7	...
fluoride (fluorite)	CaF_2	78.08	wh., cb., 1.4339	$3.180^{20°}$	1330
formate	$Ca(HCO_2)_2$	130.12	col., rhb.	2.015	d.
hydride	CaH_2	42.10	wh. cr. or pd.	1.7	d. 675
hydroxide	$Ca(OH)_2$	74.10	col., hex., 1.574	2.2	$-H_2O$, 580
hypochlorite	$Ca(ClO)_2.4H_2O$	215.06	wh., feathery cr.	...	d.
hypophosphate	$Ca_2P_2O_6.2H_2O$	274.15	granular	...	$-2H_2O$, 200
lactate	$Ca(C_3H_5O_3)_2.5H_2O$	308.30	col., eff.	...	$-3H_2O$, 100
magnesium carbonate (dolomite)	$CaO.MgO.2CO_2$	184.42	trig., 1.68174	2.872	d. 730–760
magnesium silicate (diopside)	$CaO.MgO.2SiO_2$	216.52	wh., mn.	3.3	1391
nitrate (nitrocalcite)	$Ca(NO_3)_2$	164.10	col., cb.	2.36	561
nitrate	$Ca(NO_3)_2.4H_2O$*	236.16	col., mn., 1.498	1.82	42.7
nitride	Ca_3N_2	148.26	brn. cr.	$2.63^{17°}$	900
nitrite	$Ca(NO_2)_2.H_2O$	150.11	delq., hex.	$2.23^{34°}$	...
oxalate	CaC_2O_4	128.10	col., cb.	$2.2^{4°}$	d.
oxalate	$CaC_2O_4.H_2O$	146.12	col.	2.2	$-H_2O$, 200
oxide	CaO	56.08	col., cb., 1.837	3.32	2570
peroxide	$CaO_2.8H_2O$	216.21	pearly, tet.	...	$-8H_2O$, 100
phosphate, monobasic	$CaH_4(PO_4)_2.H_2O$	252.09	wh., tri.	$2.220\frac{16°}{4}$	$-H_2O$, 100
phosphate, dibasic	$CaHPO_4.2H_2O$	172.10	wh., mn. pl.	$2.306\frac{16°}{4}$	d.
phosphate, tribasic	$Ca_3(PO_4)_2$	310.20	wh., amor.	3.14	1670
phosphate, meta-	$Ca(PO_3)_2$	198.04	wh., tet., 1.588	2.82	975
phosphate, pyro-	$Ca_2P_2O_7$	254.12	col., biaxial, 1.60	3.09	1230
phosphate, pyro- (brushite)	$Ca_2P_2O_7.5H_2O$	344.20	wh., mn.	2.25	...
phosphide	Ca_3P_2	182.20	red cr.	$2.51^{15°}$	>1600]
silicate (α) (pseudowollastonite)	$CaSiO_3$	116.14	col., pseudo hex., 1.6150 or mn.(?)	2.905	1540
silicate (β) (wollastonite)	$CaSiO_3$	116.14	col., mn., 1.610	2.915	tr. 1190 to α
sulfate (anhydrite)	$CaSO_4$	136.14	col., rhb., 1.576, or mn, 1.50	2.96	1450(mn.)
sulfate (gypsum)	$CaSO_4.2H_2O$	172.17	col., mn., 1.5226	2.32	$-1\frac{1}{2}H_2O$, 128
sulfhydrate	$Ca(SH)_2.6H_2O$	214.31	col. pr.	...	d. 15
sulfide (oldhamite)	CaS	72.14	col., cb.	$2.8^{15°}$	...
sulfite	$CaSO_3.2H_2O$	156.17	wh., cr., 1.595	...	$-2H_2O$, 100
tartrate	$CaC_4H_4O_6.4H_2O$	260.22	col., rhb.	...	d.
thiocyanate	$Ca(CNS)_2.3H_2O$	210.28	wh., delq. cr.	...	...
thiosulfate	$CaS_2O_3.6H_2O$	260.30	col., tri., 1.56	$1.873^{16°}$	d.
tungstate (scheelite)	$CaWO_4$	288 00	wh., tet., 1.9200	6.06	...

Name	Boiling point, °C.	Solubility in 100 parts: Cold water	Hot water	Other reagents
Cadmium sulfate	...	$76.5^{0°}$	$60.8^{100°}$	i. act., NH_3
sulfate	...	s.	s.	
sulfate	...	$114.2^{0°}$	$127.6^{60°}$	i. al.
sulfate	...	s.	s.	i. al.
sulfate	...	$350^{-5°}$	...	i. al.
sulfide (greenockite)	subl. in N_2, 980	0.000001	Colloidal	s. a.; v. s. NH_4OH
Calcium	1200 ± 30	d.	d.	s. a.; sl. s. al.
acetate	...	$52^{0°}$	$45.5^{80°}$	sl. s. al.
aluminate	...	d.	...	s. HCl
aluminum silicate (anorthite)				
arsenate	...	$0.013^{25°}$	i.	s. dil. a.
bromide	810	$125^{0°}$	$312^{105°}$	s. al., act.; sl. s. NH_3
carbonate (aragonite)	...	$0.0012^{20°}$†	$0.002^{100°}$	s. a., NH_4Cl
carbonate (calcite)	...	$0.0014^{25°}$	$0.002^{100°}$	s. a., NH_4Cl
chloride (hydrophilite)	>1600	$59.5^{0°}$	$347^{260°}$	s. al.
chloride	...	s.	s.	s. al.
chloride	$-6H_2O$, 200	v. s.	v. s.	s. al.
citrate	$-4H_2O$, 185	$0.085^{18°}$	$0.096^{25°}$	$0.0065^{18°}$ al.
cyanamide	...	s. d.	d.	
ferrocyanide	...	s.	$150^{90°}$	i. al.
fluoride (fluorite)	...	$0.0016^{18°}$	$0.0017^{26°}$	sl. s. a.
formate	...	$16.1^{0°}$	18.4^{100}	i. al., et.
hydride	...	d.	...	d. a.; i. bz.
hydroxide	...	$0.185^{0°}$	$0.077^{100°}$	s. NH_4Cl
hypochlorite	...	delq.; d.	d.	d. a.
hypophosphate	...	i.	...	s. HCl, $H_4P_2O_6$
lactate	...	10.5	∞	∞ h. al.; i. et.
magnesium carbonate (dolomite)	...	$0.032^{18°}$		
magnesium silicate (diopside)	...	i.	i.	
nitrate (nitrocalcite)	...	$102^{0°}$	$376^{151°}$	$14^{15°}$ al.; s. amyl al., NH_3
nitrate	...	$266^{0°}$	v. s.	
nitride	...	d.	d.	s. dil. a.; i. abs. al.
nitrite	...	$77^{0°}$	$417^{90°}$	s. 90% al.
oxalate	...	$0.00067^{13°}$	$0.0014^{95°}$	s. a.; i. ac.
oxalate	...	i.	i.	s. a.; i. ac
oxide	2850	Forms $Ca(OH)_2$	...	s. a.; i. al.
peroxide	expl. 275	sl. s.	d.	s. a. d.; i. al., et.
phosphate, monobasic	d. 200	...	d.	
phosphate, dibasic	...	$0.02^{24.5°}$	$0.075^{100°}$	
phosphate, tribasic	...	0.0025	d.	s. a.; i. al.; ac.
phosphate, meta-	...	i.	i.	i. a.
phosphate, pyro-	...	i.	...	s. a.
phosphate, pyro- (brushite)	...	sl. s.	...	s. a.; i. NH_4Cl
phosphide	...	d.	...	s. dil. a.; i. al., et.
silicate (α) (pseudowollastonite)	...	$0.0095^{17°}$	...	s. HCl
silicate (β) (wollastonite)				
sulfate (anhydrite)	tr. 1193 to rhb.	$0.298^{20°}$	$0.1619^{100°}$	s. a., $Na_2S_2O_3$, NH_4 salts
sulfate (gypsum)	$-2H_2O$, 163	$0.223^{0°}$	$0.257^{50°}$	s. a., gly., $Na_2S_2O_3$, NH_4 salts
sulfhydrate	...	v. s.	v. s.	s. al.
sulfide (oldhamite)	...	d.	d.	s. a.
sulfite	d. 650	$0.0043^{18°}$	$0.0027^{90°}$	s. H_2SO_3
tartrate	...	$0.037^{0°}$	$0.22^{85°}$	sl. s. al.
thiocyanate	...	s.	v. s.	v. s. al.
thiosulfate	...	$71.2^{9°}$	d.	i. al.
tungstate (scheelite)	...	0.2	...	s. NH_4Cl; i. a.

Carbon, *Cf.* table of organic compounds									
Carbon, amorphous	C	12.01	bk., amor.	1.8–2.1	>3500	4200	i.	i.	i. a., alk.
Carbon, diamond	C	12.01	col., cb., 2.4195	$3.51^{20°}$	>3500	4200	i.	i.	i. a., alk.
Carbon, graphite	C	12.01	bk., hex.	$2.26^{20°}$	>3500	4200	i.	i.	i. a., alk.
dioxide	CO_2	44.01	col. gas	lq. $1.101^{-37°}$; 1.53 (A); solid $1.56^{-79°}$	$-56.6^{5.2 atm.}$	subl. −78.5	$179.7^{0°}$ cc	$90.1^{20°}$ cc	s. a., alk.
disulfide	CS_2	76.13	col. lq.	lq. $1.261\frac{22°}{20}$; 2.63 (A)	−108.6	46.3	$0.2^{0°}$	$0.014^{50°}$	s. al., et.
monoxide	CO	28.01	col., poisonous, odorless gas	lq. $0.814^{-\frac{195°}{4}}$; 0.968 (A)	−207	−192	$0.0044^{0°}$; $3.5^{0°}$ cc	$0.0018^{50°}$ $2.32^{20°}$ cc	s. al., Cu_2Cl_2
oxychloride (phosgene)	$COCl_2$	98.92	poisonous gas	$1.392\frac{19°}{4}$	−104	8.2^{756mm}	v. s. sl. d.	d.	s. ac., CCl_4, bz.; d.a.
oxysulfide	COS	60.07	gas	lq. $1.24^{-87°}$; 2.10 (A)	−138.2	-50.2^{760mm}	$133^{0°}$ cc	$40.3^{30°}$ cc	v. s. alk., al.
suboxide	C_3O_2	68.03	gas	lq. $1.114^{0°}$	−107	7^{761mm}	d.	...	s. et.
thionyl chloride	$CSCl_2$	114.98	yel.-red lq.	$1.509^{15°}$	...	73.5			
Ceric hydroxide	$2CeO_2.3H_2O$	398.31	yel., gelatinous	...	...	...	...	...	s. a.; sl. s. alk. carb.; i. alk
hydroxynitrate	$Ce(OH)(NO_3)_3.3H_2O$	397.21	red, mn.	...	...	...	d.		
oxide	CeO_2	172.13	wh. or pa. yel., cb.	7.3	1950	...	i.	i.	s. H_2SO_4, HCl
sulfate	$Ce(SO_4)_2.4H_2O$	404.31	yel., rhb.	3.91	...	...	s. d.	...	s. dil. H_2SO_4
Cerium	Ce	140.13	steel gray, cb. or hex.	$6.9^{20°}$ cb.; 6.7 hex.	645	1400	i.	Slowly oxidized	s. dil. a.; i. al.
Cerous sulfate	$Ce_2(SO_4)_3$	568.44	wh., mn. or rhb.	3.91	...	...	$18.9^{80°}$	$0.4^{100°}$	
sulfate	$Ce_2(SO_4)_3.8H_2O$	712.57	tri.	$2.886^{17°}$	$-8H_2O$, 630	...	$25^{0°}$	$7.6^{40°}$	
Cesium	Ce	132.91	silv. met., hex.	$1.90^{20°}$	28.5	670	d.	...	s. a., al., NH_3
Chloric acid	$HClO_3.7H_2O$	210.58	lq.	$1.282^{14.2°}$	<−20	d. 40	v. s.		
Chlorine	Cl_2	70.91	rhb., or gn.-yel. gas	lq. $1.56^{-33.6°}$; $2.49^{0°}$ (A)	−101.6	−34.6	$1.46^{0°}$; $310^{10°}$ cc	$0.57^{30°}$; $177^{30°}$ cc	s. alk.
hydrate	$Cl_2.8H_2O$	215.04	rhb.	1.23	d. 9.6	...	s.	...	s. alk.
Chloroplatinic acid	$H_2PtCl_6.6H_2O$	518.08	red-brn., delq.	2.431	60	...	v. s.	v. s.	s. al., et.
Chlorostannic acid	$H_2SnCl_6.6H_2O$	441.55	delq.	$1.971^{28°}$	19.2	...	s.		
Chlorosulfonic acid	$HO.SO_2.Cl$	116.52	col. lq.	$1.787^{25°}$	−80	151.5^{765mm}	d.	...	d. al.; i. CS_2
Chromic acetate	$Cr_2(C_2H_3O_2)_6.2H_2O$	494.32	gn.	...	...	...	s.	...	$4.76^{15°}$ m. al.
chloride	$CrCl_3$	158.38	pink, trig.	$2.757^{15°}$	...	1200–1500 d.	i.§	sl. s.	i. a., act., CS_2
chloride	$CrCl_3.6H_2O$*	266.48	vl. or gn., hex. pl.	$1.835\frac{25°}{4}$	subl. 83	...	v. s. d.	...	s. al.; i. et.
fluoride	CrF_3	109.01	gn., rhb.	3.8	>1000	d.	i.	...	sl. s. a.; i. al., NH_3
hydroxide	$Cr(OH)_3$	103.03	gn. or blue, gelatinous	...	...	...	i.	...	s. a., alk.; sl. s. NH_3
hydroxide	$Cr(OH)_3.2H_2O$	139.07	gn.	...	$-2H_2O$, 100	...	i.	i.	s. a., alk.
nitrate	$Cr(NO_3)_3.9H_2O$*	400.18	purple pr.	...	36.5	d. 100	s.	s.	s. a., alk., al., act.
nitrate	$Cr(NO_3)_3.7\frac{1}{2}H_2O$	373.15	purple, mn.	...	100	d.	s.	s.	
oxide	Cr_2O_3	152.02	dark gn., hex.	5.21	1900	...	i.	i.	sl. s. a.
sulfate	$Cr_2(SO_4)_3$	392.20	rose pd.	3.012	...	...	i.§	...	i. a.
sulfate	$Cr_2(SO_4)_3.5H_2O$	482.28	gn.	...	...	...	s.	...	s. al., H_2SO_4
sulfate	$Cr_2(SO_4)_3.15H_2O$	662.44	vl.	$1.867^{17°}$	100	$-10H_2O$, 100	s.	d. 67°	sl. s. al.
sulfate	$Cr_2(SO_4)_3.18H_2O$	716.49	vl., cb., 1.564	$1.7^{22°}$	...	$-12H_2O$, 100	$120^{20°}$	d.	s. al.
sulfide	Cr_2S_3	200.02	brn.-bk. pd.	$3.77^{19°}$	−S, 1350	...	i.	d.	s. h. HNO_3
Chromium	Cr	52.01	gray, met., cb.	7.1	1615	2200	i.	i.	s. HCl, dil. H_2SO_4; i. HNO_3
trioxide (chromic acid)	CrO_3	100.01	red, rhb.	2.70	197 d.	...	$164.9^{0°}$	$206.7^{100°}$	s. H_2SO_4, al., et.
Chromous chloride	$CrCl_2$	122.92	wh., delq.	2.75	...	...	v. s.	v. s.	sl. s. al.; i. et.
hydroxide	$Cr(OH)_2$	86.03	yel.-brn.	...	d.	...	d.	...	s. conc. a.
oxide	CrO	68.01	bk. pd.	...	...	...	i.	i.	i. dil. HNO_3
sulfate	$CrSO_4.7H_2O$	274.18	blue	...	...	...	12.35^{0}	...	sl. s. al.
sulfide (daubrelite)	CrS	84.07	bk. pd.	3.97	1550	...	i.	...	v. s. a.
Chromyl chloride	CrO_2Cl_2	154.92	dark red lq.	1.92	−96.5	117.6	d.	...	s. et.
Cobalt	Co	58.94	silv. met., cb.	$8.9^{20°}$	1480	2900	i.	i.	s. a.
carbonyl	$Co(CO)_4$	170.98	or. cr.	$1.73^{18°}$	51	d. 52	i.	d.	s. al., et., CS_2
sulfide, di-	CoS_2	123.06	bk., cb.	4.269	...	...	i.	...	s. HNO_3, aq. reg.
Cobaltic chloride	$CoCl_3$	165.31	red cr.	2.94	subl.	...	s.	s.	
chloride, dichro	$Co(NH_3)_3Cl_3.H_2O$	234.42	...	...	...	...	s.	...	s. a., al.
chloride, luteo	$Co(NH_3)_6Cl_3$	267.50	or., mn.	$1.7016^{20°}$	...	...	$4.26^{0°}$	$12.74^{46.6°}$	i. al., NH_4OH
chloride, praseo	$Co(NH_3)_4Cl_3.H_2O$	251.46	gn., rhb.	1.847	...	...	v. s.	...	s. a.; i. al.

* Usual commercial form.

† The solubility of $CaCO_3$ in H_2O is greatly increased by increasing the amount of CO_2 in the H_2O.

§ Also a soluble modification.

TABLE 3-1 Physical Properties of the Elements and Inorganic Compounds *(Continued)*

Name	Formula	Formula weight	Color, crystalline form and refractive index	Specific gravity	Melting point, °C.	Boiling point, °C.	Solubility in 100 parts: Cold water	Hot water	Other reagents
Cobaltic chloride, purpureo	$Co(NH_3)_5Cl_3$	250.47	rhb.	$1.819\frac{25°}{25}$	......	......	$0.232^{0°}$	$1.031^{46.6°}$	i. al.
chloride, roseo	$Co(NH_3)_5Cl_3.H_2O$	268.49	brick red	......	d. 100	......	$16.12^{0°}$	$24.87^{16°}$	sl. s. HCl
hydroxide	$Co(OH)_3$	109.96	bk.	......	$-1\frac{1}{2}H_2O$, 100	......	i.	i.	s. a.; i. al.
oxide	Co_2O_3	165.88	bk.	5.18	d. 900	......	i.	i.	s. a.
sulfate	$Co_2(SO_4)_3$	406.06	blue cr.	......	......	......	d.	......	s. H_2SO_4
sulfide	Co_2S_3	214.06	bk. cr.	4.8	......	......	i.	......	d. a.
Cobalto-cobaltic oxide	Co_3O_4	240.82	bk., cb.	6.07	......	......	i.	i.	s. H_2SO_4; i. HCl, HNO_3
Cobaltous acetate	$Co(C_2H_3O_2)_2.4H_2O$	249.09	red-vl., mn., 1.542	$1.7053^{18.7°}$	$-4H_2O$, 140	......	s.	s.	s. a., al.
chloride	$CoCl_2$	129.85	blue cr.	3.356	subl.	1049	$45^{7°}$	$105^{96°}$	31 al.; 8.6 act.
chloride	$CoCl_2.6H_2O$*	237.95	red, mn.	$1.924\frac{25°}{25}$	86	$-6H_2O$, 110	$116.5^{0°}$	$177^{80°}$	v. s. et., act.
nitrate	$Co(NO_3)_2.6H_2O$	291.05	red, mn., 1.4	$1.883\frac{25°}{25}$	<100	d.	$84.03^{0°}$(anh.)	$334.9^{90°}$(anh.)	$100^{12.5°}$ al.; s. act.; sl. s. NH_3
oxide	CoO	74.94	brn., cb.	5.68	d. 1800	......	i.	i.	s. a., NH_4OH; i. al.
sulfate	$CoSO_4$	155.00	red pd.	$3.710^{25°}$	d. 880	......	$25.6^{0°}$	$83^{100°}$	$1.04^{18°}$ m. al.; i. NH_3
sulfate	$CoSO_4.H_2O$	173.02	red pd., mn.(?), 1.639	3.13	d.	......	s.	s.	
sulfate (biebeorite)	$CoSO_4.7H_2O$*	281.11	red, mn., 1.483	$1.948\frac{25°}{25}$	96.8	$-7H_2O$, 420	$33^{20°}$	s.	$2.5^{3°}$ al.
sulfide (syeporite)	CoS	91.00	brn. nd.	$5.45^{18°}$	>1100	......	$0.00038^{18°}$	......	s. a., aq. reg.
Copper	Cu	63.57	yel.-red met., cb.	$8.92^{20°}$	1083	2300	i.	i.	s. HNO_3, h. H_2SO_4
Cupric acetate	$Cu(C_2H_3O_2)_2$	181.66	......	$1.930\frac{20°}{4}$	......	......	s.		
acetate	$Cu(C_2H_3O_2)_2.H_2O$	199.67	dark gn., mn.	1.882	115	240 d.	7.2	20	7 al.; s. et., gly.
aceto-arsenite (Paris green)	$(CuOAs_2O_3)_3.Cu(C_2H_3O_2)_2$*	1013.83	gn.	......	......	......	i.	......	s. a., NH_4OH
ammonium chloride	$CuCl_2.2NH_4Cl.2H_2O$	277.51	blue, tet., 1.670, 1.744	1.98	d. 110	......	$33.8^{0°}$	$99.3^{80°}$	s. a.
ammonium sulfate	$CuSO_4.4NH_3.H_2O$	245.77	blue, rhb.	1.81	d. 150	......	$18.05^{21.5°}$	d.	i. al.
carbonate, basic (azurite)	$2CuCO_3.Cu(OH)_2$	344.75	blue, mn., 1.758	3.88	d. 220	......	i.	d.	s. NH_4OH, h. aq. $NaHCO_3$
carbonate, basic (malachite)	$CuCO_3.Cu(OH)_2$	221.17	dark gn.; mn., 1.875	3.9	d.	......	i.	d.	s. KCN; 0.03 aq. CO
chloride (eriochalcite)	$CuCl_2$	134.48	brn.-yel. pd.	3.054	498	Forms Cu_2Cl_2 993	$70.7^{0°}$	$107.9^{100°}$	$53^{15°}$ al.; $68^{15°}$ m. al.
chloride	$CuCl_2.2H_2O$	170.52	gn., rhb., 1.684	$2.39^{22.4°}$	$-2H_2O$, 110	d.	$110.4^{0°}$	$192.4^{100°}$	s. al.; et., NH_4Cl
chromate, basic	$CuCrO_4.2CuO.2H_2O$	374.75	yel.-brn.	......	$-2H_2O$, 260	......	i.	......	s. HNO_3, NH_4OH
cyanide	$Cu(CN)_2$	115.61	yel.-gn.	......	d.	......	i.	......	s. KCN, C_5H_5N
dichromate	$CuCr_2O_7.2H_2O$	315.62	bk., tri.	$2.286^{18°}$	$-2H_2O$, 100	......	sl. s.	d.	s. a.; NH_4OH
ferricyanide	$Cu_3[Fe(CN)_6]_2$	614.63	yel.-gn.	......	......	......	i.	......	s. NH_4OH; i. HCl
ferrocyanide	$Cu_2Fe(CN)_6.7H_2O$	465.21	red-brn.	......	......	......	i.	i.	s. NH_4OH; i. a., NH_3
formate	$Cu(HCO_2)_2$	153.61	blue, mn.	1.831	......	......	12.5	d.	0.25 al.
hydroxide	$Cu(OH)_2$	97.59	blue, gelatinous	3.368	$-H_2O$	......	i.	d.	s. a., NH_4OH, KCN, al.
lactate	$Cu(C_3H_5O_3)_2.2H_2O$	277.74	dark blue, mn.	......	......	......	16.7	$45^{100°}$	sl. s. al.
nitrate	$Cu(NO_3)_2.3H_2O$*	241.63	blue, delq.	$2.047^{3.9°}$	114.5	$-HNO_3$, 170	$381^{40°}$	$666^{80°}$	$100^{12.5°}$ al.
nitrate	$Cu(NO_3)_2.6H_2O$	295.68	blue, rhb.	2.074	$-3H_2O$, 26.4	......	$243.7^{0°}$	∞	s. al.
oxide (paramelaconite)	CuO	79.57	bk., cb..	6.40	d. 1026	......	i.	i.	s. a.; KCN, NH_4Cl
oxide (tenorite)	CuO	79.57	bk., tri., 2.63	6.45	d. 1026	......	i.	i.	s. a.; KCN, NH_4Cl
oxychloride	$CuCl_2.2CuO.4H_2O$	365.69	blue-gn.	......	$-3H_2O$, 140	......	i.	......	s. a.
phosphide	Cu_3P_2	252.67	bk.	6.35	d.	......	i.	......	s. HNO_3; i. HCl
sulfate (hydrocyanite)	$CuSO_4$	159.63	gn.-wh., rhb., 1.733	$3.606^{15°}$	d. >600	Forms CuO, 650	$14.3^{0°}$	$75.4^{100°}$	i. al.
sulfate (blue vitriol or chalcanthite)	$CuSO_4.5H_2O$*	249.71	blue, tri., 1.5368	$2.286\frac{15.6°}{4}$	$-4H_2O$, 110	$-5H_2O$, 250	$24.3^{0°}$	$205^{100°}$	$1.1^{3°}$ al.
sulfide (covellite)	CuS	95.63	blue, hex. or mn., 1.45	4.6	tr. 103	d. 220	$0.000033^{18°}$	......	s. HNO_3, KCN
tartate	$CuC_4H_4O_6.3H_2O$	265.69	l+ gn. pd.	......	d.	......	$0.02^{15°}$	$0.14^{85°}$	s. a.; KOH
Cuprous ammonium iodide	$CuI.NH_4I.H_2O$	353.47	rhb. pl.	......	......	......	d.	......	s. NH_4I
carbonate	Cu_2CO_3	187.15	yel.	4.4	d.	......	i.	i.	s. a.; NH_4OH
chloride (nantokite)	Cu_2Cl_2	198.05	wh., cb., 1.973	3.53	422	1366	$1.52^{25°}$	......	s. HCl, NH_4OH, al.
cyanide	$Cu_2(CN)_2$	179.16	wh., mn.	2.9	474.5	d.	i.	i.	s. KCN, HCl, NH_4OH; sl. s. NH_3
ferricyanide	$Cu_3Fe(CN)_6$	402.67	brn.-red	......	......	......	i.	......	s. NH_4OH; i. HCl
ferrocyanide	$Cu_4Fe(CN)_6$	466.24	brn.-red	......	......	......	i.	......	s. NH_4OH; i. NH_4Cl
fluoride	Cu_2F_2	165.14	red cr.	......	908	subl. 1100	i.	......	s. HF, HCl, HNO_3; i. al.
hydroxide	CuOH	80.58	yel.	3.4	$-\frac{1}{2}H_2O$, 360	......	i.	i.	s. a., NH_4OH
oxide (cuprite)	Cu_2O	143.14	red, cb., 2.705	6.0	1235	$-O$, 1800	i.	i.	s. HCl, NH_4Cl, NH_4OH

Cuprous phosphide	Cu_6P_2	443.38	gray-bk.	6.4 to 6.8	....	....	i.	....	s. HNO_3; i. HCl
sulfide (chalcocite)	Cu_2S	159.20	bk., rhb.	5.6	1100	....	$0.0005^{18°}$	....	s. HNO_3, NH_4OH; i. act.
sulfide	Cu_2S	159.20	bk., cb.	5.80	1130	....	$0.0005^{18°}$	....	s. HNO_3, NH_4OH; i. act.
Cyanogen	C_2N_2	52.02	poisonous gas	lq. $0.866^{-17.2°}$; 1.806 (A)	−34.4	−20.5	$450^{20°}$ cc	....	$2300^{20°}$ cc al.; $500^{18°}$ cc et.
Cyanogen compounds, *Cf.* table of organic compounds									
Ferric acetate, basic	$Fe(OH)(C_2H_3O_2)_2$	190.95	brn., amor.	....	....	....	i.	....	s. a.; al.
ammonium sulfate, *Cf.* Alum									
chloride (molysite)	$FeCl_3$	162.22	bk.-brn., hex. delq.	$2.804^{11°}$	282	315	$74.4^{0°}$	$535.8^{100°}$	v. s. al.; et. +HCl
chloride	$FeCl_3.6H_2O$*	270.32	red-yel., delq.	....	37	280	$246^{0°}$	∞	s. al., act., gly.
ferrocyanide (Prussian blue)	$Fe_4[Fe(CN)_6]_3$	859.27	dark blue	....	d.	....	i.	d.	s. HCl, conc. H_2SO_4; i. al., et.
hydroxide	$Fe(OH)_3$	106.87	red-brn.	3.4 to 3.9	−1½H_2O, 500	....	i.	i.	s. a.; i. al., et.
lactate	$Fe(C_3H_5O_3)_3$	323.06	brn., amor., delq.	....	....	....	v. s.	v. s.	i. et.
nitrate	$Fe(NO_3)_3.6H_2O$	349.97	rhb., delq.	$1.684^{20°}$	35	d.	$150^{0°}$	∞	s. al., act.
oxide (hematite)	Fe_2O_3	159.70	red or bk., trig., 3.042	5.12	1560 d.	....	i.	....	s. HCl
sulfate	$Fe_2(SO_4)_3$	399.88	rhb., 1.814	$3.097^{18°}$	d. 480	....	sl. s.	d.	i. H_2SO_4, NH_3
sulfate (coquimbite)	$Fe_2(SO_4)_3.9H_2O$	562.02	yel., trig.	2.1	....	....	440	d.	s. abs. al.
Ferroso-ferric chloride	$FeCl_2.2FeCl_3.18H_2O$	775.49	yel., delq.	....	d. 50	....	s.	s.	
ferricyanide (Prussian green)	$Fe'''_4Fe''_3[Fe(CN)_6]_6$	1662.70	gn.	....	d. 180	....	i.	....	s. d. h. HCl
oxide (magnetite; magnetic iron oxide)	Fe_3O_4	231.55	bk., cb., 2.42	5.2	1538 d.	....	i.	i.	i. al.
oxide, hydrated	$Fe_3O_4.4H_2O$	303.61	bk.	....	d.	....	i.	i.	s. a.
Ferrous ammonium sulfate	$FeSO_4.(NH_4)_2SO_4.6H_2O$	392.15	blue-gn., mn., 1.4915	1.864	d.	....	$18^{0°}$	$100^{75°}$	i. al.
chloride (lawrencite)	$FeCl_2$	126.76	gn.-yel., hex., 1.567	2.7	....	delq.	$64.4^{10°}$	$105.7^{100°}$	100 al.; s. act.; i. et.
chloroplatinate	$FePtCl_6.6H_2O$	571.92	yel., hex.	2.714	....	....	v. s.	v. s.	
ferricyanide (Turnbull's blue)	$Fe_3[Fe(CN)_6]_2$	591.47	dark blue	....	d.	....	i.	....	i. dil. a., al.
ferrocyanide	$Fe_2Fe(CN)_6$	323.66	blue-wh., amor.	....	....	....	i.		
formate	$Fe(HCO_2)_2.2H_2O$	181.92	....	....	d.	....	sl. s.		
hydroxide	$Fe(OH)_2$	89.87	lt. gn.	3.4	....	....	0.00067	....	s. a., NH_4Cl
nitrate	$Fe(NO_3)_2.6H_2O$	287.96	cr.	....	60.5	....	$200^{0°}$	$300^{25°}$	
oxide	FeO	71.85	bk.	5.7	1420	....	i.	i.	s. a.; i. alk.
phosphate (vivianite)	$Fe_3(PO_4)_2.8H_2O$	501.64	blue, mn., 1.592, 1.603	2.58	....	....	i.	i.	s. a.; i. ac.
silicate	$FeSiO_3$	131.91	mn.	3.5	1550				
sulfate (siderotilate)	$FeSO_4.5H_2O$	241.99	gn., tri., 1.536	2.2	....	−5H_2O, 300	s.	s.	i. al.
sulfate (copperas)	$FeSO_4.7H_2O$*	278.02	blue-gn., mn.	$1.899^{14.8°}$	64	−7H_2O, 300	$32.8^{0°}$	$149^{50°}$	i. al.
sulfide	FeS	87.91	bk., hex.	4.84	1193	d.	$0.000616^{18°}$	....	s. a.; i. NH_3
Cf. also under iron									
Fluoboric acid	HBF_4	87.83	col. lq.	....	....	130 d.	∞	∞	s. al.
Fluorine	F_2	38.00	gn.-yel. gas	lq. $1.51^{-187°}$; $1.31^{15°}$ (A)	−223	−187	d.		
Fluosilicic acid	H_2SiF_6	144.08	....	....	....	....	s.	s.	
Gadolinium	Gd	156.9							
Gallium bromide	$GaBr_3$	309.47	delq. cr.	....	....	....	s.	s.	
Glucinum *Cf.* Beryllium									
Gold	Au	197.20	yel. met., cb.	$19.3^{20°}$	1063	2600	i.	i.	s. aq. reg., KCN; i. a.
Gold, colloidal	Au	197.20	blue to vl.	....	....	....	s.	....	s. aq. reg., KCN; i. a.
Gold salts *Cf.* under Auric and Aurous									
Hafnium	Hf	178.6	hex.	12.1	>1700	>3200(?)			
Helium	He	4.00	col. gas	0.1368 (A)	<−272.2	−268.9	$0.97^{0°}$ cc	$1.08^{50°}$ cc	Absorbed by Pt
Hydrazine	N_2H_4	32.05	col. lq.	$1.011^{\frac{15}{4}°}$	1.4	113.5	∞	∞	s. al.
formate	$N_2H_4.2HCO_2H$	124.10	eb.	....	128	....	s.		
hydrate	$N_2H_4.H_2O$	50.06	col.	$1.03^{21°}$	−40	$118.5^{739.5mm}$	∞	∞	∞ al.; i. et.
hydrochloride	$N_2H_4.HCl$	68.51	yel. lq.	....	....	....	v. s.	v. s.	sl. s. al.
hydrochloride, di-	$N_2H_4.2HCl$	104.98	wh., cb.	1.42	198	....	s.	v. s.	s. al.
nitrate	$N_2H_4.HNO_3$	95.06	cr.	....	70.7	subl. 140	$174.9^{10°}$	v. s.	
nitrate, di-	$N_2H_4.2HNO_3$	158.08	nd.	....	104	d.	v. s.		
sulfate	$N_2H_4.½H_2SO_4$	81.09	delq. pl.	....	85	....	v. s.	....	i. al.
surate	$N_2H_4.H_2SO_4$	130.12	rhb.	1.378	254	....	$3.055^{22°}$	$27.65^{60°}$	v. sl. s. abs. al.
Hydrazoic acid (azoimide)	HN_3	43.03	col. lq.	....	−80	37	∞	∞	∞ al.
Hydriodic acid	HI	127.93	col. gas	$4.4^{0°}$ (A)	−50.8	−35.5	$42500^{10°}$ cc	v. s.	s. al.
Hydriodic acid	$HI.H_2O$	145.94	col. lq.	$1.7^{15°}$	....	127^{774mm}	∞	....	∞ al.
Hydriodic acid	$HI.2H_2O$	163.96	col. lq.	....	−43	....	∞	....	s. al.
Hydriodic acid	$HI.3H_2O$	181.98	col. lq.	....	−48	....	∞	....	s. al.
Hydriodic acid	$HI.4H_2O$	199.99	col. lq.	....	−36.5	....	∞	....	s. al.
Hydrobromic acid	HBr	80.92	col. gas; 1.325 (lq.)	$2.71^{0°}$ (A)	−86	−67	$221^{0°}$	$130^{100°}$	s. al.

* Usual commercial form.

TABLE 3-1 Physical Properties of the Elements and Inorganic Compounds *(Continued)*

Name	Formula	Formula weight	Color, crystalline form and refractive index	Specific gravity	Melting point, °C.	Boiling point, °C.	Solubility in 100 parts: Cold water	Hot water	Other reagents
Hydrobromic acid	$HBr.H_2O$	98.94	col. lq.	1.78	...	...	...	...	Stable at −15.5° and 1 atm., and at −11.3° and 2.5 atm.
Hydrobromic acid	HBr (47.8% in H_2O)	80.92	col. lq.	1.486	...	126	∞	...	s. al.
Hydrobromic acid	$HBr.2H_2O$	116.96	wh. cr.	$2.11^{-15°}$	−11	...	s.	s.	
Hydrochloric acid	HCl*	36.47	col. gas; 1.256 (lq.)	$1.268^{0°}$ (A)	−111	−85	$82.3^{0°}$	$56.1^{60°}$	s. al., et.
Hydrochloric acid	HCl (45.2% in H_2O)	36.47	col. lq.	1.48	−15.35	...	∞	...	s. al.
Hydrochloric acid	$HCl.2H_2O$	72.50	col. lq.	$1.46^{-18.3°}_{4}$	0	d.	∞	...	s. al.
Hydrochloric acid	$HCl.3H_2O$	90.51	col. lq.	...	−24.4	d.	∞	...	s. al.
Hydrocyanic acid (prussic acid)	HCN	27.03	poisonous gas or col. lq., 1.254	$0.697^{18°}$	−14	26	∞	...	∞ al., et.
Hydrofluoric acid	HF	20.01	gas or col. lq.	$0.988^{13.6°}$	−83	19.4	∞ 0° to 19.4°	v. s.	
Hydrofluoric acid	HF (35.35% in H_2O)	20.01	col. lq.	1.15	−35	120	v. s.		
Hydrogen	H_2	2.016	col. gas or cb.	lq. $0.0709^{-252.7°}$ 0.06948 (A)	−259.1	−252.7	$2.1^{0°}$ cc	$0.85^{80°}$ cc	sl. s. Fe, Pd, Pt
peroxide	H_2O_2†	34.02	col. lq., 1.333	$1.438^{20°}_{4}$	−0.89	151.4^{760mm}	∞	...	s. a., et.; i. petr. et
selenide	H_2Se	81.22	col. gas	$2.12^{-42°}$	−64	−42	$377^{4°}$ cc	$270^{22.5°}$ cc	s. CS_2, $COCl_2$
sulfide	H_2S	34.08	col. gas	1.1895 (A)	−82.9	−59.6	$437^{0°}$ cc	$186^{40°}$ cc	$9.54^{15°}$ cc al.; s. CS_2
Hydroxylamine	NH_2OH	33.03	rhb., delq.	$1.35^{18°}$	34	56.5^{22mm}	s.	d.	s. a., al.
hydrochloride	$NH_2OH.HCl$	69.50	col., mn.	$1.67^{17°}$	151	d.	$83.3^{17°}$	v. s.	s. al.; i. et.
nitrate	$NH_2OH.HNO_3$	96.05	col. cr.	...	48	d. <100	v. s.	d.	v. s. abs. al.
sulfate	$NH_2OH.\frac{1}{2}H_2SO_4$	82.07	col., mn.	...	170 d.	...	$32.9^{0°}$	$68.5^{90°}$	v. sl. s. al.; i. et., abs. al.
Hypobromous acid	HBrO	96.92	yel.	...	...	40^{50mm}	s.	d.	
Illinium	Il	146(?)							
Indium	In	114.76	soft, tet. met.	$7.3^{20°}$	155	1450	i.	i.	s. a.
Iodic acid	HIO_3	175.93	col., rhb.	$4.629^{0°}$	110 d.	...	$286^{0°}$	$576^{101°}$	v. s. 87% al.; i. abs. al., et., chl.
Iodine	I_2	253.84	blue-bk., rhb.	$4.93^{20°}$	113.5	184.35	$0.0162^{0°}$	$0.09566^{60°}$	s. al., KI, et.
oxide, penta-	I_2O_5	333.84	wh., trimetric	$4.799^{25°}_{4}$	d. 300	...	$187.4^{13°}$	...	i. abs. al., et., chl.
Iodoplatinic acid	$H_2PtI_6.9H_2O$	1120.91	brn., delq. mn.	...	...	...	s. d.		
Iridium	Ir	193.10	wh. met., cb.	$22.4^{20°}$	2350	>4800	i.	i.	sl. s. aq. reg., aq. Cl_2
Iron, cast‡	Fe	55.85	gray	7.03	1275	...	i.	i.	s. a.; i. alk.
pure	Fe	55.85	silv. met., cb.	$7.86^{20°}$	1535	3000	i.	i.	s. a.; i. alk.
steel	Fe	55.85	silv. gray	7.6 to 7.8	1375	...	i.	i.	s. a.; i. alk.
white pig	Fe	55.85	gray	7.6 to 7.8	1075	...	i.	i.	s. a.; i. alk.
wrought	Fe	55.85	gray	7.86	1505	...	i.	i.	s. a.; i. alk.
carbide (cementite)	Fe_3C	179.56	pseudo hex.	7.4	1837	...	i.	i.	s. a.
carbonyl	$Fe(CO)_5$	195.90	pa. yel. lq.	$1.457^{21°}$	−21	102.5^{760mm}	i.	...	s. al., H_2SO_4, alk.
nitride	Fe_2N	125.71	gray	6.35	d. >560	...	d.	...	s. HCl, H_2SO_4
silicide	FeSi	83.91	yel.-gray, oct.	$6.1^{20°}_{4}$	...	...	i.	i.	i. aq. reg.
sulfide, di- (marcasite)	FeS_2	119.97	yel., rhb.	4.87	tr. 450	d.	0.00049	...	i. dil. a.
sulfide, di- (pyrite)	FeS_2	119.97	yel., cb.	5.0	1171	d.	0.0005	...	i. dil. a.
sulfide (pyrrhotite)	Fe_7S_8	647.43	hex.	$4.6^{20°}_{4}$	d. >700	...	i.		
Cf. also under ferric and ferrous									
Krypton	Kr	83.70	col. gas	2.818 (A)	−169	−151.8	$11.05^{0°}$ cc	$3.57^{60°}$ cc	sl. s. al., bz.
Lanthanum	La	138.92	lead gray	$6.15^{20°}$	826	1800	d.	...	s. a.
Lead	Pb	207.21	silv. met., cb.	$11.337^{20°}_{20}$	327.5	1620	i.	i.	s. HNO_3; i. c. HCl, H_2SO_4
acetate	$Pb(C_2H_3O_2)_2$	325.30	wh. cr.	$3.251^{20°}_{4}$	280	...	$19.7^{0°}$	$221^{50°}$	s. gly.; v. sl. s. al.
acetate (sugar of lead)	$Pb(C_2H_3O_2)_2.3H_2O$§	379.35	wh., mn.	2.55	$-3H_2O$, 75	...	$45.64^{15°}$	$200^{100°}$	s. gly.; sl. s. al.
acetate	$Pb(C_2H_3O_2)_2.10H_2O$	505.46	wh., rhb.	1.689	22	...	s.	s.	
acetate, basic	$Pb_2(C_2H_3O_2)_3OH$	608.56	wh.	...	...	...	v. s.	...	sl. s. al.
acetate, basic	$Pb(C_2H_3O_2)_2.Pb(OH)_2.H_2O$	584.54	wh. nd.	...	...	...	v. s.	...	s. al.
acetate, basic	$Pb(C_2H_3O_2)_2.2Pb(OH)_2$	807.75	wh. nd.	...	...	...	5.55	18.2	s. al.
arsenate, monobasic	$PbH_4(AsO_4)_2$	489.06	tri., 1.82	$4.46^{15°}$	d. 140	...	d.	...	s. HNO_3
arsenate, dibasic (schultenite)	$PbHAsO_4$	347.13	wh., mn., 1.9097	5.94	d. >200	$-H_2O$, 280	i.	sl. s.	s. HNO_3, NaOH
arsenate, meta-	$Pb(AsO_3)_2$	453.03	hex.	$6.42^{15°}$	...	...	d.	...	s. HNO_3
arsenate, pyro-	$Pb_2As_2O_7$	676.24	rhb., 2.03	$6.85^{15°}_{15}$	802	...	i.	d.	s. HCl, HNO_3; i. ac.

Lead azide	PbN_6	291.26	col. nd.	...	expl. 350	...	i.	$0.05^{100°}$	v. s. ac.; i. NH_4OH
bromide	$PbBr_2$	367.05	col., rhb.	6.66	373	918	$0.4554^{0°}$	$4.75^{100°}$	s. a., KBr.; sl. s. NH_3; i. al.
carbonate (cerussite)	$PbCO_3$	267.22	wh., rhb., 2.0763	6.6	d. 315	...	$0.00011^{20°}$	d.	s. a., alk.; i. NH_3, al.
carbonate, basic (hydrocerussite; white lead)	$2PbCO_3.Pb(OH)_2$§	775.67	wh., hex.	6.14	d. 400	...	i.	i.	s. ac.; sl. s. aq. CO_2
chloride (cotunnite)	$PbCl_2$	278.12	wh., rhb., 2.2172	5.80	501	954^{760mm}	$0.673^{0°}$	$3.34^{100°}$	sl. s. dil. HCl, NH_3; i. al.
chromate (crocoite)	$PbCrO_4$	323.22	yel., mn., 2.42	6.12	844	d.	$0.000007^{20°}$	i.	s. a., alk.; i. NH_3, ac.
chromate, basic	$PbCrO_4.PbO$	546.43	or.-yel. nd.	...	...	...	i.	i.	s. a., alk.
formate	$Pb(HCO_2)_2$	297.25	wh., rhb.	4.56	d. 190	...	$1.6^{16°}$	$18^{100°}$ d.	i. al.
hydroxide	$3PbO.H_2O$	687.65	cb.	7.592	$-H_2O$, 130	...	0.014	...	s. a., alk.
nitrate	$Pb(NO_3)_2$	331.23	col., cb. or mn., 1.7815	4.53	d. 470	...	$38.8^{0°}$	$138.8^{100°}$	$8.8^{22°}$ al.
oxide, sub-	Pb_2O	430.42	bk., amor.	8.34	d. red heat	...	i.	i.	s. a., alk.
oxide, mono- (litharge)	PbO	223.21	yel., tet.	9.53	888	...	$0.0068^{18°}$	...	s. alk., PbAc, NH_4Cl, $CaCl_2$
oxide, mono (massicotite)	PbO	223.21	yel., rhb., 2.61	8.0					
oxide, mono-	PbO	223.21	amor.	9.2 to 9.5	...	...	i.	i.	s. alk., PbAc, NH_4Cl, $CaCl_2$
oxide, red (minium)	Pb_3O_4	685.63	red, amor.	9.1	d. 500	...	i.	i.	s. ac., h. HCl
oxide, sesqui-	Pb_2O_3	462.42	red-yel., amor.	...	d. 360	...	i.	i.	s. a., alk.
oxide, di- (plattnerite)	PbO_2	239.21	brn., tet., 2.229	9.375	d. 290	...	i.	i.	s. ac., h. alk.; i. al.
silicate	$PbSiO_3$	283.27	col., mn., 1.961	6.49	766	...	i.	...	s. a.
sulfate (anglesite)	$PbSO_4$	303.27	wh., mn. or rhb 1.8823	6.2	1170	...	$0.0028^{0°}$	$0.0056^{40°}$	s. conc. a., NH_4 salts; i. al.
sulfate, acid	$Pb(HSO_4)_2.H_2O$	419.36	cr.	...	d.	...	$0.0001^{18°}$	...	sl. s. H_2SO_4
sulfate, basic (lanarkite)	$PbSO_4.PbO$	526.48	col., mn.	6.92	977	...	$0.0044^{18°}$	...	sl. s. H_2SO_4
sulfide (galena)	PbS	239.27	lead gray, cb., 3.912	7.5	1120	...	$0.00009^{18°}$	i.	s. a.; i. alk.
thiocyanate	$Pb(CNS)_2$	323.37	col., mn.	3.82	d. 190	...	$0.05^{20°}$	s.	s. KCNS, HNO_3
Lithium	Li	6.94	silv. met. cb.	$0.53^{20°}$	186	1336 ± 5	d.	d.	s. a., NH_3
benzoate	$LiC_7H_5O_2$	128.05	wh. leaflets	...	...	...	$33^{25°}$	$40^{100°}$	$7.7^{25°}$, $10^{78°}$ al.
bromide	LiBr	86.86	wh., delq., cb., 1.784	$3.464\frac{25°}{4}$	547	1265	$143^{0°}$ ($2H_2O$)	$266^{100°}$ ($1H_2O$)	s. al., act.
bromide	$LiBr.2H_2O$	122.89	wh. pr.	...	44	...	$246^{20°}$	...	s. al.
carbonate	Li_2CO_3	73.89	col., mn., 1.567	$2.11^{0°}$	618	d.	$1.54^{0°}$	$0.72^{100°}$	s. dil. a.; i. al., act., NH_3
chloride	LiCl	42.40	wh., delq., cb., 1.662	$2.068\frac{25°}{4}$	614	1360	$67^{0°}$	$127.5^{100°}$	$2.48^{15°}$ al.; s. et.
citrate	$Li_3C_6H_5O_7.4H_2O$	281.98	wh. cr.	...	d.	...	$61.2^{15°}$	$66.7^{100°}$	sl. s. al., et.
fluoride	LiF	25.94	wh., cb., 1.3915	$2.295^{21.5°}$	870	1670	$0.27^{18°}$	$0.135^{35°}$	s. HF; i. act.
formate	$LiHCO_2.H_2O$	69.97	col., rhb.	1.46	$-H_2O$, 94	...	$49.2^{0°}$	$346.6^{104°}$	sl. s. al., et.
hydride	LiH	7.95	wh., cb.	0.820	680	...	d.	...	i. et.
hydroxide	LiOH	23.95	wh. cr.	2.54	445	925±	$12.7^{0°}$	$17.5^{100°}$	sl. s. al.
hydroxide	$LiOH.H_2O$	41.96	col., mn.	1.83	...	d.	$22.3^{10°}$	$26.8^{80°}$	sl. s. al.
nitrate	$LiNO_3$	68.95	col., trig., 1.735	2.38	261	...	$53.4^{0°}$	$194^{70°}$	s. al., NH_3
nitrate	$LiNO_3.3H_2O$	123.00	col.	...	29.88	...	v. s.	∞	
oxide	Li_2O	29.88	col., 1.644	$2.013\frac{25°}{4}$	...	subl. <1000	forms LiOH		
phosphate, monobasic	LiH_2PO_4	103.94	col.	2.461	>100				
phosphate, tribasic	Li_3PO_4	115.80	wh., rhb.	$2.537^{17.5°}$	837	...	$0.034^{18°}$	v. sl. s.	s. a., NH_4Cl; i. act.
phosphate, tribasic	$Li_3PO_4.12H_2O$	331.99	wh., trig.	1.645	100	...	v. sl. s.	v. sl. s.	
salicylate	$LiC_7H_5O_3$	144.05	col.	...	d.	...	$128^{25°}$	...	v. s. al.
sulfate	Li_2SO_4	109.94	col., mn., 1.465	2.22	860	...	$35.34^{0°}$	$29.9^{100°}$	i. act., 80% al.
sulfate	$Li_2SO_4.H_2O$§	127.96	col., mn., 1.477	2.06	$-H_2O$, 130	...	$43.6^{0°}$	$35^{100°}$	i. 80% al.
sulfate, acid	$LiHSO_4$	104.01	pr.	$2.123^{13°}$	170.5	...	d.		
Lutecium	Lu	174.99							
Magnesium	Mg	24.32	silv. met., hex.	$1.74^{20°}$	651	1110	i.	sl. s. d.	s. a., NH_4 salts
acetate	$Mg(C_2H_3O_2)_2$	142.41	wh.	1.42	323	...	v. s.	v. s.	$5.25^{15°}$ m. al.
acetate	$Mg(C_2H_3O_2)_2.4H_2O$§	214.47	wh., mn. pr., 1.491	1.454	80	...	v. s.	v. s.	v. s. al.
aluminate (spinel)	$MgO.Al_2O_3$	142.26	col. cb., 1.718–23	3.6	2135	...	i.	...	v. sl. s. dil. HCl; i. dil. HNO_3
ammonium chloride	$MgCl_2.NH_4Cl.6H_2O$	256.83	wh., rhb., delq.	1.456	$-4H_2O$, 195	...	16.7	s.	
ammonium phosphate (struvite)	$MgNH_4PO_4.6H_2O$	245.44	col., rhb., 1.496	1.715	d. 100	...	$0.0231^{0°}$	$0.0195^{80°}$	s. a.; i. al.
ammonium sulfate (boussingaultite)	$MgSO_4.(NH_4)_2SO_4.6H_2O$	360.62	col., mn.	1.72	>120	...	$16.86^{0°}$	$130^{100°}$	
benzoate	$Mg(C_7H_5O_2)_2.3H_2O$	320.59	wh. pd.	...	$-3H_2O$, 110]	...	$4.5^{25°}$ (anh.)	s.	s. act.
carbonate (magnesite)	$MgCO_3$	83.43	wh., trig. 1.700	3.037	d. 350	...	0.0106	...	s. a., aq. CO_2; i. act., NH_3
carbonate (nesquehonite)	$MgCO_3.3H_2O$	138.38	col., rhb., 1.501	1.852	$-H_2O$, 100	...	$0.1518^{19°}$	d.	s. a., aq. CO_2
carbonate, basic (hydromagnesite)	$3MgCO_3.Mg(OH)_2.3H_2O$	365.37	wh., rhb., 1.530	2.16	d.	...	0.04	0.011	s. a., NH_4 salts; i. al.

* Usual commercial form about 31 per cent.
† Usual commercial forms 3 or 30 per cent.
‡ See also a table of alloys.
§ Usual commercial form.

TABLE 3-1 Physical Properties of the Elements and Inorganic Compounds *(Continued)*

Name	Formula	Formula weight	Color, crystalline form and refractive index	Specific gravity	Melting point, °C.	Boiling point, °C.	Solubility in 100 parts: Cold water	Hot water	Other reagents
Magnesium chloride (chloromagnesite)	$MgCl_2$	95.23	col., hex., 1.675	$2.325^{25°}$	712	1412	$52.8^{0°}$	$73^{100°}$	50 al.
chloride (bischofite)	$MgCl_2.6H_2O$*	203.33	wh., delq., mn., 1.507	1.56	118 d.	d.	$281^{0°}$	$918^{100°}$	50 al.
hydroxide (brucite)	$Mg(OH)_2$	58.34	wh., trig., 1.5617	2.4	d.	...	$0.0009^{18°}$	...	s. NH_4 salts, dil. a.
nitride	Mg_3N_2	100.98	gn.-yel., amor.	...	d.	...	i.	d.	s. a.; i. al.
oxide (magnesia; periclase)	MgO	40.32	col., cb., 1.7364	3.65	2800	3600	0.00062	...	s. a., NH_4 salts; i. al.
perchlorate	$Mg(ClO_4)_2$*	223.23	wh., delq.	$2.60^{25°}$	d.	...	$99.6^{25°}$	v. s.	24^{25} al., $51.8^{25°}$ m. al.; 0.29 et.
peroxide	MgO_2	56.32	wh. pd.	...	expl. 275	...	i.	i.	s. a.
phosphate, pyro-	$Mg_2P_2O_7$	222.60	col., mn., 1.604	$2.598^{22°}$	1383	...	i.	i.	s. a.; i. alk.
phosphate, pyro-	$Mg_2P_2O_7.3H_2O$	276.65	wh., amor.	2.56	$-3H_2O$, 100	...	i.	sl. s.	s. a.; i. al.
potassium chloride (carnallite)	$MgCl_2.KCl.6H_2O$	277.88	delq., rhb., 1.475	$1.60^{19.4°}_{4}$	265	...	$64.5^{19°}$ d.	d.	d. al.
potassium sulfate (picromerite)	$MgSO_4.K_2SO_4.6H_2O$	402.73	mn., 1.4629	2.15	d. 72	...	$19.26^{0°}$	$81.7^{75°}$	
silicofluoride	$MgSiF_6.6H_2O$	274.48	col., trig., 1.3439	$1.788^{17.5°}_{4}$	d.	...	$64.8^{17.5°}$	s.	d. HF
sodium chloride	$MgCl_2.NaCl.H_2O$	171.70	col.	...	...	...	s.	s.	
sulfate	$MgSO_4$	120.38	col.	2.66	1185	...	$26.9^{0°}$	$68.3^{100°}$	s. al.
sulfate (epsom salt; epsomite)	$MgSO_4.7H_2O$*	246.49	col., rhb., 1.4554	1.68	70 d.	...	$72.4^{0°}$	$178^{40°}$	s. al.
Manganese	Mn	54.93	gray-pink met.	$7.2^{20°}$	1260	1900	d.	...	s. dil. a.
acetate	$Mn(C_2H_3O_2)_2$	173.02	...	$1.74^{20°}_{4}$	...	...	s.	s.	
acetate	$Mn(C_2H_3O_2)_2.4H_2O$*	245.08	pa. pink, mn.	1.589	...	...	s.	$64.5^{50°}$	s. al., m. al.
carbonate (rhodocrosite)	$MnCO_3$	114.94	rose, trig., 1.817	3.125	d.	...	$0.0065^{25°}$	...	s. aq. CO_2, dil. a.; i. NH_3, al.
chloride (scacchite)	$MnCl_2$	125.84	rose, delq., cb.	$2.977^{25°}_{4}$	650	1190	$63.4^{0°}$	$123.8^{106°}$	s. al.; i. et., NH_3
chloride	$MnCl_2.4H_2O$*	197.91	rose red, delq., mn. 1.575	2.01	58.0	$-H_2O$, 106; $-4H_2O$, 200	$151^{8°}$	∞	s. al.; i. et.
chloride, per-	$MnCl_4$	196.76	gn.	...	...	...	s.	s.	s. al., et.
hydroxide (ous) (pyrochroite)	$Mn(OH)_2$	88.95	wh., trig.	$3.258^{13°}$	d.	...	$0.002^{20°}$	i.	s. a., NH_4 salts; i. alk.
hydroxide (ic) (manganite)	$Mn_2O_3.H_2O$	175.88	brn., rhb., 2.24	3.258	d.	...	i.	i.	s. h. H_2SO_4
nitrate	$Mn(NO_3)_2.6H_2O$	287.04	rose red, mn.	$1.82^{21°}$	25.8	129.5	$426^{0°}$	∞	v. s. al.
oxide (ous) (manganosite)	MnO	70.93	gray-gn., cb., 2.16	5.18	1650	...	i.	i.	s. a., NH_4Cl
oxide (ic)	Mn_2O_3	157.86	brn.-bk., cb.	4.81	−O, 1080	...	i.	i.	s. a.; i. act.
oxide, di- (pyrolusite; polianite)	MnO_2*	86.93	bk., rhb.	5.026	−O, >230	...	i.	i.	s. HCl; i. HNO_3, act.
sulfate (ous)	$MnSO_4$	150.99	red-wh.	3.235	700	d. 850	$53^{0°}$	$73^{50°}$	s. al.; i. et.
sulfate (ous) (szmikite)	$MnSO_4.H_2O$	169.01	pa. pink, mn., 1.595	2.87	Stable 57 to 117	...	$98.47^{48°}$	$79.77^{100°}$	
sulfate (ous)	$MnSO_4.2H_2O$	187.02	...	$2.526^{15°}$	Stable 40 to 57	...	$85.27^{35°}$	$106.8^{55°}$	
sulfate (ous)	$MnSO_4.3H_2O$	205.04	...	$2.356^{15°}$	Stable 30 to 40	...	$74.22^{5°}$	$99.31^{57°}$	
sulfate (ous)	$MnSO_4.4H_2O$*	223.05	pink, rhb. or mn., 1.518	2.107	Stable 18 to 30	$-4H_2O$, 450	$136^{16°}$	$169^{50°}$	i. al.
sulfate (ous)	$MnSO_4.5H_2O$	241.07	pink, tri., 1.508	$2.103^{15°}$	Stable 8 to 18	...	$142^{5°}$	$200^{35°}$	
sulfate (ous)	$MnSO_4.6H_2O$	259.09	...	...	Stable −5 to +8	...	$204^{0°}$	$247^{9°}$	
sulfate (ous)	$MnSO_4.7H_2O$	277.10	pink, mn. or rhb.	2.092	Stable −10 to −5; 19 d.	$-7H_2O$, 280	$176^{0°}$	$251^{14°}$	
sulfate (ic)	$Mn_2(SO_4)_3$	398.04	gn., delq. cr.	3.24	d. 160	...	v. s.	d.	s. HCl, dil. H_2SO_4; i. conc. H_2SO_4, HNO_3
Masurium	Ma	98–99.5	...	11.5	2300 (?)				
Mercuric acetate	$Hg(C_2H_3O_2)_2$	318.70	wh. pl.	3.270	d.	...	$25^{10°}$	$100^{100°}$	s. al. sl. d.
bromide	$HgBr_2$	360.44	wh., rhb.	6.053	237	322	$0.5^{20°}$	$25^{100°}$	$25.2^{0°}$ al.; v. sl. s. et.
carbonate, basic	$HgCO_3.2HgO$	693.84	brn.-red	...	...	...	i.	...	s. aq. CO_2, NH_4Cl
chloride (corrosive sublimate)	$HgCl_2$	271.52	wh., rhb., 1.859	5.44	277	304	$3.6^{0°}$	$61.3^{100°}$	$33^{25°}$ 99% al.; 33 et.
fulminate	$Hg(CNO)_2$	284.65	cb.	4.42	expl.	...	sl. s.	...	s. NH_4OH, al.
hydroxide	$Hg(OH)_2$	234.63		...	$-H_2O$, 175	...	i.	i.	s. a.
oxide (montroydite)	HgO	216.61	yel. or red, rhb., 2.5	11.14	d. 100	...	$0.0052^{25°}$	$0.041^{100°}$	s. a.; i. al.
oxychloride (kleinite)	$HgCl_2.3HgO$	921.35	yel., hex.	7.93	d. 260	...	i.	d.	s. HCl
silicofluoride, basic	$HgSiF_6.HgO.3H_2O$	613.33	yel. nd.	...	...	...	d.	...	s. a.
sulfate	$HgSO_4$	296.67	wh., rhb.	6.47	d.	...	d.	...	s. a.; i. al., act., NH_3
sulfate, basic (turpeth)	$HgSO_4.2HgO$	729.89	yel., tet.	6.44	...	...	0.005	$0.167^{100°}$	s. a.; i. al.
Mercurous acetate	$HgC_2H_3O_2$	259.65	wh. sc.	...	d.	...	$0.75^{13°}$	d.	s. H_2SO_4, HNO_3; i. al.
bromide	HgBr	280.53	wh., tet.	7.307	subl. 345	...	7×10^{-9}	i.	s. a.; i. al., act.
carbonate	Hg_2CO_3	461.23	yel. pd.	...	d. 130	...	i.	d.	s. NH_4Cl
chloride (calomel)	HgCl	236.07	wh., tet., 1.9733	7.150	302	383.7	$0.0014^{0°}$	$0.0007^{43°}$	s. aq. reg., $Hg(NO_3)_2$; sl. s. HNO_3, HCl; i. al., et.
iodide	HgI	327.53	yel., tet.	7.70	290 d.	subl. 140; 310d.	2×10^{-8}	v. sl. s.	s. KI; i. al.
nitrate	$HgNO_3.H_2O$	280.63	wh. mn.	$4.785^{3.9°}$	70	expl.	v. s.	d.	s. HNO_3; i. al., et.

Mercurous oxide	Hg_2O	417.22	bk.	9.8	d. 100		i.	0.0007	s. h. ac.; i. alk., dil. HCl, NH_3
sulfate	Hg_2SO_4	497.28	wh., mn.	7.56	d.		$0.055^{16.5°}$	$0.092^{100°}$	s. H_2SO_4, HNO_3
Mercury†	Hg	200.61	silv. lq. or hex.(?)	$13.546^{20°}$	−38.87	356.9	i.	i.	s. HNO_3; i. HCl
Molybdenum	Mo	95.95	gray, cb.	10.2	2620 ± 10	3700	i.	i.	s. h. conc. H_2SO_4; i. HCl, HF, NH_3, dil. H_2SO_4, Hg
chloride, di-	$MoCl_2$	166.85	yel., amor.	$3.714\frac{25°}{4}$	d.		i.	i.	s. HCl, H_2SO_4, NH_4OH, al., et.
chloride, tri-	$MoCl_3$	202.32	dark red pd.	$3.578\frac{25°}{4}$	d.		i.	d.	s. HNO_3, H_2SO_4; v. sl. s. al., et.
chloride, tetra-	$MoCl_4$	237.78	brn., delq.		volt.	d.	s.	d.	s. HNO_3, H_2SO_4; sl. s. al., et.
chloride, penta-	$MoCl_5$	273.24	bk. cr.	$2.928\frac{25°}{4}$	194	268	s.	d.	s. HNO_3, H_2SO_4; i. abs. al., et.
oxide, tri- (molybdite)	MoO_3	143.95	col., rhb.	$4.50^{19.5°}$	795	subl.	$0.107^{18°}$	$2.106^{70°}$	s. a., NH_4OH
sulfide, di- (molybdenite)	MoS_2	160.07	bk., hex., 4.7	$4.80^{14°}$	1185		i.	i.	s. H_2SO_4, aq. reg.
sulfide, tri-	MoS_3	192.13	red-brn.		d.		sl. s.	s.	s. alk. sulfides
sulfide, tetra-	MoS_4	224.19	brn. pd.		d.		i.	i.	s. alk. sulfides; i. NH_3
Molybdic acid	H_2MoO_4	161.97	yel-wh., hex.		d. 115		v. sl. s.	sl. s.	s. NH_4OH, H_2SO_4; i. NH_3
Molybdic acid	$H_2MoO_4.H_2O$	179.98	yel., mn.	$3.124^{15°}$	$-H_2O$, 70	$-2H_2O$, 200	$0.133^{18°}$	$2.13^{70°}$	s. a., NH_4OH, NH_4 salts
Neodymium	Nd	144.27	yellowish	$6.9^{20°}$	840		d.		
Neon	Ne	20.18	col. gas	lq. $1.204^{-245.9°}$ 0.674 (A)	−248.67	−245.9	$2.6^{0°}$ cc	$1.1^{45°}$ cc	s. lq. O_2, al., act., bz.
Neptunium	Np^{239}	239			Produced by Neutron bombardment of U^{238}				
Nickel	Ni	58.69	silv. met., cb.	8.90^{20}	1452	2900	i.	i.	s. dil. HNO_3; sl. s. H_2SO_4, HCl; i. NH_3
acetate	$Ni(C_2H_3O_2)_2$	176.78	gn. pr.	1.798	d.		16.6		i. al.
ammonium chloride	$NiCl_2.NH_4Cl.6H_2O$	291.20	gn., delq., mn.	1.645			$150^{25°}$	v. s.	
ammonium sulfate	$NiSO_4.(NH_4)_2SO_4.6H_2O$	394.99	blue-gn., mn., 1.5007	1.923			$2.5^{8.5°}$	$39.2^{85°}$	v. sl. s. $(NH_4)_2SO_4$
bromate	$Ni(BrO_3)_2.6H_2O$	422.62	gn., cb.	2.575	d.		28		s. NH_4OH
bromide	$NiBr_2$	218.52	yel., delq.	$4.64\frac{28°}{4}$	d.		$112.8^{0°}$	$156^{100°}$	s. al., et., NH_4OH
bromide	$NiBr_2.3H_2O$	272.57	gn., delq.		$-3H_2O$, 200		$199^{0°}$	$316^{100°}$	s. al., et., NH_4OH
bromide, ammonia	$NiBr_2.6NH_3$	320.71	vl. pd.	1.837			v. s.	d.	i. c. NH_4OH
bromoplatinate	$NiPtBr_6.6H_2O$	841.51	trig.	3.715					
carbonate	$NiCO_3$	118.70	lt. gn., rhb.		d.		$0.0093^{25°}$	i.	s. a.
carbonate, basic	$2NiCO_3.3Ni(OH)_2.4H_2O$	587.58	lt. gn.		d.		i.	d.	s. a., NH_4 salts
carbonyl	$Ni(CO)_4$	170.73	lq.	$1.31^{17°}$	−25	43^{751mm}	$0.018^{9.8°}$	i.	s. aq. reg., HNO_3, al., et
chloride	$NiCl_2$	129.60	yel., delq.	3.544	subl.	973	$53.8^{0°}$	$87.6^{100°}$	s. NH_4OH, al.; i. NH_3
chloride	$NiCl_2.6H_2O$*	237.70	gn., delq., mn., 1.57±				180	v. s.	v. s. al.
chloride, ammonia	$NiCl_2.6NH_3$	231.80					s.	d.	s. NH_4OH; i. al.
cyanide	$Ni(CN)_2.4H_2O$	182.79	gn. pl.		$-4H_2O$, 200	d.	i.	i.	s. KCN; i. dil. KCl
dimethylglyoxime	$NiC_8H_{14}O_4N_4$	288.91	scarlet red cr.		subl. 250		i.	i.	s. abs. al., a.; i. ac., NH_4OH
formate	$Ni(HCO_2)_2.2H_2O$	184.76	gn. cr.	2.154	d.		s.		
hydroxide (ic)	$Ni(OH)_3$	109.71	bk.		d.		i.	i.	s. a., NH_4OH, NH_4Cl
hydroxide (ous)	$Ni(OH)_2.\frac{1}{4}H_2O$	97.21	lt. gn.	4.36	d.		v. sl. s.	v. sl. s.	s. a., NH_4OH; i. alk.
nitrate	$Ni(NO_3)_2.6H_2O$	290.80	gn., mn.	2.05	56.7	136.7	$243.0^{0°}$	$\infty^{56.7°}$	s. NH_4OH; i. abs. al.
nitrate, ammonia	$Ni(NO_3)_2.4NH_3.2H_2O$	286.87					v. s.		i. al.
oxide, mono- (bunsenite)	NiO	74.69	gn.-bk., cb., 2.37	7.45	Forms Ni_2O_3 at 400		i.	i.	s. a., NH_4OH
potassium cyanide	$Ni(CN)_2.2KCN.H_2O$	258.97	red yel., mn.	$1.875^{11°}$	$-H_2O$, 100		s.		d. a.
sulfate	$NiSO_4$	154.75	yel., cb.	3.68	$-SO_3$, 840		$27.2^{0°}$	$76.7^{100°}$	i. al., et., act.
sulfate	$NiSO_4.6H_2O$*	262.85	gn. mn. or blue, tet., 1.5109	2.07	tr. 53.3	$-6H_2O$, 280	$131^{50°}$	$280^{100°}$	v. s. NH_4OH, al.
sulfate (morenosite)	$NiSO_4.7H_2O$	280.86	gn., rhb., 1.4893	1.948	98–100	$-6H_2O$, 103	$63.5^{0°}$	$117.8^{80°}$	s. al.
Nitric acid	HNO_3	63.02	col. lq.	1.502	−42	86	∞	∞	expl. with al.
Nitric acid	$HNO_3.H_2O$	81.03	col. lq.		−38		∞	∞	d. al.
Nitric acid	$HNO_3.3H_2O$	117.06	col. lq.		−18.5		$263^{-20°}$	∞	d. al.
Nitro acid sulfite	NO_2HSO_3	127.08	col., rhb.		73 d.		d.		s. H_2SO_4
Nitrogen	N_2	28.02	col. gas or cb. cr.	$1.026^{-252.5°}$ $0.808^{-195.8°}$ 12.50° (D)	−209.86	−195.8	$2.35^{0°}$ cc	$1.55^{20°}$ cc	sl. s. al.

* Usual commercial form.
† See also special table 48.

TABLE 3-1 Physical Properties of the Elements and Inorganic Compounds *(Continued)*

Name	Formula	Formula weight	Color, crystalline form and refractive index	Specific gravity	Melting point, °C.	Boiling point, °C.	Solubility in 100 parts: Cold water	Hot water	Other reagents
Nitrogen oxide, mono- (ous)	N_2O	44.02	col. gas	lq. $1.226^{-89°}$; 1.530 (A)	−102.3	−90.7	$130.52^{0°}$ cc	$60.82^{24°}$ cc	s. H_2SO_4, al.
oxide, di- (ic)	NO or $(NO)_2$	30.01 (60.02)	col. gas	lq. $1.269^{-150.2°}$; 1.0367 (A)	−161	−151	$7.34^{0°}$ cc	$0.0^{100°}$	26.6 cc al.; 3.5 cc H_2SO_4; s. aq. $FeSO_4$
oxide, tri-	N_2O_3	76.02	red-brn. gas or blue lq. or solid	$1.447^{2°}$	−102	3.5	s.	....	s. a., et.
oxide, tetra- (per- or di-)	NO_2 or $(NO_2)_2$	46.01 (92.02)	yel. lq., col. solid, red-brn. gas	$1.448^{20°}$	−9.3	21.3	d.	....	s. HNO_3, H_2SO_4, chl., CS_2
oxide, penta-	N_2O_5	108.02	wh., rhb.	$1.63^{18°}$	30	47	s.	Forms HNO_3	
oxybromide	NOBr	109.92	brn. lq.	>1.0	−55.5	−2	d.		
oxychloride	NOCl	65.47	red-yel. lq. or gas	$1.417^{-12°}$; 2.31 (A)	−64.5	−5.5	d.	....	s. fuming H_2SO_4
Nitroxyl chloride	NO_2Cl	81.47	yel.-brn. gas	lq. $1.32^{14°}$	<−30	5	d		
Osmium	Os	190.2	blue, hex.	$22.48^{20°}$	2700	>5300	i.	i.	sl. s. aq. reg., HNO_3; i. NH_3
chloride, di-	$OsCl_2$	261.11	gn., delq.	....	....	....	s. d.	....	s. NaCl, al., et.
chloride, tri-	$OsCl_3$	296.57	brn., cb.	....	d. 560–600	....	sl. s.	....	s. a., alk., al.; sl. s. et.
chloride, tetra-	$OsCl_4$	332.03	red-yel. nd.	....	....	....	s. d.	....	s. HCl, al.
Oxygen	O_2	32.00	col. gas or hex. solid	$1.14^{-183°}$; $1.426^{-252.5°}$; 1.1053 (A)	−218.4	−183	$4.89^{0°}$ cc	$2.6^{30°}$ cc; $1.7^{100°}$ cc	sl. s. al., s. fused Ag
Ozone	O_3	48.00	col. gas	$1.71^{-183°}$; $3.03^{-80°}$; 1.658 (A)	−251	−112	$0.494^{0°}$ cc	$0^{60°}$ cc	s. oil turp., oil cinn.
Palladium	Pd	106.70	silv. met., cb.	$12.0^{20°}$; $11^{1550°}$	1555	2200	i.	i.	s. aq. reg., h. H_2SO_4; i. NH_3
bromide (ous)	$PdBr_2$	266.53	brn.	....	....	....	i.	i.	s. HBr
chloride	$PdCl_2$	177.61	brn., cb.	....	500 d.	....	s.	s.	s. HCl, act., al.
chloride	$PdCl_2.2H_2O$	213.65	brn. pr.	....	....	....	s.	s.	s. HCl, act., al.
cyanide	$Pd(CN)_2$	158.74	yel.	....	d.	....	i.	i.	s. HCN, KCN, NH_4OH; i. dil. a.
hydride	Pd_2H	214.41	met.	11.06	d.				
Palladous dichlorodiammine	$Pd(NH_3)_2Cl_2$	211.68	red or yel., tet.	2.5	....	....	s.	....	s. a., NH_4OH
Perchloric acid	$HClO_4$	100.46	unstable, col. lq	$1.768\frac{22}{4}°$	−112	16^{18mm}	s.		
Perchloric acid	$HClO_4.H_2O$	118.48	fairly stable nd.	1.88	50	d.	s.		
Perchloric acid	$HClO_4.2H_2O$* 73.6% anh.	136.50	stable lq., col.	$1.71\frac{25}{4}°$	−17.8	200	v. s.	....	s. al.
Periodic acid	HIO_4	191.93	wh. cr.	....	d. 138	subl. 110	s.		
Periodic acid	$HIO_4.2H_2O$	227.96	delq., mn.	....	d. 110	....	v. s.	v. s.	sl. s. al.; et.
Permanganic acid	$HMnO_4$	119.94	exists only in solution	....		....	v. s.	d.	d. al.
Permolybdic acid	$HMoO_4.2H_2O$	196.99	wh. cr.	....	....	....	v. s.	v. s.	
Persulfuric acid	$H_2S_2O_8$	194.14	hyg. cr.	....	<60	....	v. s.	d.	
Phosphamic acid	$PONH_2.(OH)_2$	97.02	cb.	....	d.	....	v. s.	v. s. d.	i. al.
Phosphatomolybdic acid	$H_7P(Mo_2O_7)_6.28H_2O$	2365.88	yel. cb.	....	78	$-25H_2O$, 140	s.	....	s. HNO_3
Phosphine	PH_3	34.00	col. gas	lq. $0.746^{-90°}$; 1.146 (A)	−132.5	−85	$26^{17°}$ cc	$i.^{100°}$	s. $CuCl_2$, al.; et.
Phosphonium chloride	PH_4Cl	70.47	wh., cb.	....	28^{46atm}	subl.	d.		
Phosphoric acid, hypo-	$H_4P_2O_6$	161.99	cr.	....	55	d. 70	s.	$450^{62°}$	
Phosphoric acid, meta-	HPO_3	79.99	vitreous, delq.	2.2–2.5	subl.	....	s.	Forms H_3PO_4	i. lq. CO_2
Phosphoric acid, ortho-	H_3PO_4†	98.00	col., rhb.	$1.834^{18.2°}$	42.35	$-\frac{1}{2}H_2O$, 213	$2340^{26°}$	v. s.	s. al.
Phosphoric acid, pyro-	$H_4P_2O_7$	177.99	wh. nd.	....	61	....	$800^{23°}$	Forms H_3PO_4	v. s. al., et.
Phosphorous acid, hypo-	H_3PO_2	66.00	syrupy	$1.493^{18.8°}$	26.5	d.	∞	∞	
Phosphorous acid, ortho-	H_3PO_3	82.00	col.	$1.651^{21.2°}$	74	d. 200	$307.3^{0°}$	$730^{40°}$	
Phosphorous acid, pyro-	$H_4P_2O_5$	145.99	nd.	....	38	d. 130	d.		
Phosphorus, black	P_4	123.92	rhombohedral	2.69	....	ign. in air, 400	i.	i.	i. CS_2
Phosphorus, red	P_4	123.92	red, cb.	$2.20^{20°}$	590^{43atm}	ign. in air, 725	i.	i.	s. alk.; i. CS_2, NH_3; et.
Phosphorus, yellow	P_4	123.92	yel., hex., 2.1168	$1.82^{20°}$; lq. $1.745^{44.5°}$	44.1; ign. 34	280	0.0003	sl. s.	0.4 al.; $1000^{10°}$ CS_2; $1.5^{0°}$, $10^{81°}$ bz.; s. NH_3
chloride, tri-	PCl_3	137.35	col., fuming lq.	$1.574\frac{20.8}{4}$	−111.8	75.95^{760mm}	d.	....	s. et., chl., CS_2
chloride, penta-	PCl_5	208.27	delq., tet.	solid 1.6; $3.60^{295°}$ (A)	148 under pressure	subl. 160	d.	....	s. CS_2, C_6H_5COCl
oxide, penta-	P_2O_5	141.96	wh., delq., amor.	2.387	subl. 250	....	Forms H_3PO_4	v. s.	s. H_2SO_4; i. NH_3, act.
oxychloride	$POCl_3$	153.35	col., fuming lq.	1.675	2	107.2^{760mm}	d.	....	d. al.

Phosphotungstic acid	$P_2O_5.2WO_3.42H_2O$	3681.67	yel.-gn. cr.	...	...	...	s.	...	s. al., et.
Platinum	Pt	195.23	silv. met., cb.	$21.45^{20°}$; lq. $19^{1755°}$	1755	4300	i.	i.	s. aq. reg., fused alk.
chloride (ic)	$PtCl_4$	337.06	brn.	...	d. 370	...	$140^{25°}$	v. s.	s. al., act.; sl. s. NH_3; i. et.
chloride (ous)	$PtCl_2$	266.14	brn.	$5.87^{11°}$	d. 581	...	i.	i.	s. HCl, NH_4OH; sl. s. NH_3; i. al., et.
chloride (ic)	$PtCl_4.8H_2O$	481.19	red, mn.	2.43	$-4H_2O$, 100	...	v. s.	v. s.	s. al., et.
cyanide (ous)	$Pt(CN)_2$	247.27	yel.-brn.	...	...	...	i.	i.	i. alk.
Plutonium	Pu	238	...	Produced by deuteron bombardment on U^{238}					
Plutonium	Pu	239	...	Produced by neutron bombardment on U^{238}					
Potassium	K	39.10	silv. met., cb.	$0.86^{20°}$; lq. $0.83^{62°}$	62.3	760	d.	Forms KOH	s. a., al., Hg
acetate	$KC_2H_3O_2$	98.14	wh. pd.	1.8	292	...	$217^{0°}$	$396^{90°}$	33 al.; i. et.
acetate, acid	$KH(C_2H_3O_2)_2$	158.19	delq. nd. or pl.	...	148	d. 200	d.	...	s. ac.
aluminate	$K_2(AlO_2)_2.3H_2O$	250.18	cr.	...	...	...	s.	d.	s. alk.; i al.
amide	KNH_2	55.12	yel.-grn.	...	338	subl. 400	d.	...	d. al.; $3.6^{25°}$ NH_3
arsenate (monobasic)	KH_2AsO_4	180.02	col., tet., 1.5674	2.867	288	...	$18.87^{6°}$	v. s.	i. al.
auricyanide	$KAu(CN)_4.1.5H_2O$	367.39	pl.	...	d. 200	...	s.	v. s.	s. al.
aurocyanide	$KAu(CN)_2$	288.33	rhb.	...	...	...	14.3	$200^{100°}$	sl. s. al.; i. et.
bicarbonate	$KHCO_3$	100.11	mn., 1.482	2.17	d. 100–200	...	$22.4^{0°}$	$60^{60°}$	i. satd. K_2CO_3, al.
bisulfate	$KHSO_4$	136.16	rhb., or mn., 1.480	2.35	210	d.	$36.3^{0°}$	$121.6^{100°}$	d. al.
bromate	$KBrO_3$	167.01	trig.	$3.27^{17.5°}$	370 d.	...	$3.1^{10°}$	$49.75^{100°}$	sl. s. al.; i. act.
bromide	KBr	119.01	col., cb., 1.5594	$2.75^{25°}$	730	1380	$53.5^{0°}$	$104^{100°}$	sl. s. al., et.
carbonate	K_2CO_3	138.20	wh., delq. pd., 1.531	2.29	891	d.	$105.5^{0°}$	$156^{100°}$	i. al.
carbonate	$K_2CO_3.2H_2O$	174.23	rhb.	2.043	...	...	$183^{0°}$	$331^{100°}$	
carbonate	$2K_2CO_3.3H_2O$	330.45	mn.	2.13	...	...	$129.4^{0°}$	$268^{100°}$	
chlorate	$KClO_3$	122.56	col., mn., 1.5167	2.32	368	d. 400	$3.3^{0°}$	$57^{100°}$	0.83 al.; s. alk.
chloride (sylvite)	KCl	74.56	col., cb., 1.4904	1.988	790	1500	$27.6^{0°}$	$56.7^{100°}$	s. al., alk.
chloroplatinate	K_2PtCl_6	486.16	yel., cb., 1.825±	3.499	d. 250	...	$0.74^{0°}$	$5.2^{100°}$	i. al., et.
chromate (tarapacaite)	K_2CrO_4	194.20	yel., rhb., 1.7261	$2.732^{18°}$	975	...	$58.0^{0°}$	$75.6^{100°}$	i. al.
cyanate	KCNO	81.11	wh., tet.	2.048	...	...	s.	d.	v. sl. s. al.
cyanide	KCN	65.11	wh., cb., delq., 1.410	$1.52^{16°}$	634.5	...	s.	$122.2^{103.3°}$	s. gly.; $0.9^{19.5°}$ al.; 1.3 h. al.
dichromate	$K_2Cr_2O_7$	294.21	red, tri.	2.69	398	d.	$4.9^{0°}$	$80^{100°}$	i. al.
ferricyanide	$K_3Fe(CN)_6$	329.25	red, mn. pr., 1.5689	1.84	d.	...	$33^{4.4°}$	$77.5^{100°}$	s. act.; sl. s. al.; i. NH_3
ferrocyanide	$K_4Fe(CN)_6.3H_2O$	422.39	yel., mn., 1.5772	$1.853^{17°}$	$-3H_2O$, 70	...	$27.8^{12.2°}$	$90.6^{96.3°}$	s. act.; i. NH_3, al., et.
formate	$KHCO_2$	84.11	col., rhb.	1.91	167.5	d.	$331^{18°}$	$657^{90°}$	sl. s. al.; i. et.
hydride	KH	40.10	cb., 1.453	0.80	d.	...	d.	...	i. et., bz., CS_2
hydrosulfide	KHS	72.16	wh., delq., rhb.	2.0	455	...	s.	s. d.	s. al.
hydroxide	KOH	56.10	wh., delq., rhb.	2.044	380	1320	$97^{0°}$	$178^{100°}$	v. s. al., et.; i. NH_3
iodate	KIO_3	214.02	col., mn.	3.89	560	...	$4.73^{0°}$	$32.2^{100°}$	s. KI; i. al., NH_3
iodide	KI	166.02	wh., cb., 1.6670	3.13	723	1330	$127.5^{0°}$	$208^{100°}$	$4^{20°}$ al.; s. NH_3; sl. s. et.
iodide, tri-	KI_3	419.86	dark blue, delq., mn.	3.498	45	d. 225	v. s.	...	s. KI, al.
iodoplatinate	K_2PtI_6	1034.94	cb.	5.18	...	...	s.		
manganate	K_2MnO_4	197.12	gn., rhb.	...	d. 190	...	d.	...	s. KOH
metabisulfite	$K_2S_2O_5$	222.31	mn., pl.	...	d. 150	...	$25^{0°}$	$120^{94°}$	sl. s. al.; i. et.
nitrate (saltpeter)	KNO_3	101.10	col., rhb., 1.5038	$2.11^{10.6°}$	tr. 129; 333	d. 400	$13.3^{0°}$	$246^{100°}$	$0.1^{80°}$ al.; i. et.
nitrite	KNO_2	85.10	pr.	1.915	297	d. 350	$281^{0°}$	$413^{100°}$	v. s. NH_3; sl. s. al.
oxalate	$K_2C_2O_4.H_2O$	184.23	wh., mn.	2.13	d.	...	$28.7^{0°}$	$83.2^{100°}$	
oxalate, acid	KHC_2O_4‡	128.12	mn., 1.545	2.0	d.	...	$14.3^{50°}$	$48.1^{100°}$	
oxalate, acid	$KHC_2O_4.\frac{1}{2}H_2O$	137.13	trimetric	...	d.	...	$2.2^{0°}$	$51.5^{100°}$	
oxide	K_2O	94.19	wh., cb.	$2.32\frac{20°}{4}$	...	...	Forms KOH	v. s.	s. al., et.
perchlorate	$KClO_4$	138.55	col., rhb., 1.4737	$2.524\frac{11°}{4}$*	d. 400	...	$0.75^{0°}$	$21.8^{100°}$	$0.105^{20°}$ m. al.; i. et.
permanganate	$KMnO_4$	158.03	purple, rhb.	2.703	d. <240	...	$2.83^{0°}$	$32.35^{75°}$	s. H_2SO_4; d. al.
persulfate	$K_2S_2O_8$	270.31	wh., tri., 1.4669	...	d. <100	...	$1.77^{0°}$	$10^{40°}$	i. al.
phosphate, monobasic	KH_2PO_4	136.09	col., delq., tet., 1.5095	2.338	256	...	$14.8^{0°}$	$83.5^{90°}$	i. al.
phosphate, dibasic	K_2HPO_4	174.18	wh., delq.	...	d.	...	$33^{25°}$	v. s.	sl. s. al.
phosphate, tribasic	K_3PO_4	212.27	wh., rhb.	$2.564^{17°}$	1340	...	$193.1^{25°}$	v. s.	i. al.
phosphate, meta-	KPO_3	118.08	wh. pd.	$2.258^{14.5°}$	tr. 450; 798	1320	s.	s.	
phosphate, meta-	$K_4P_4O_{12}.2H_2O$	508.34	amor.	$2.264^{14.5°}$	$-2H_2O$, 100	d.	s.	83	s. a.
phosphate, pyro-	$K_4P_2O_7.3H_2O$	384.39	delq.	2.33	$-2H_2O$, 180	$-3H_2O$, 300	s.	v. s.	i. al.
phthalate, acid	$KHC_8H_4O_4$	204.22	wh. cr.	1.63	d.	...	$10.2^{25°}$	36	
platinocyanide	$K_2Pt(CN)_4.3H_2O$	431.54	yel., rhb., 1.62±	$2.45^{16°}$	...	...	sl. s.	v. s.	s. al., et.
silicate	K_2SiO_3	154.25	hyg. 1.521±	...	976	...	s.	s.	i. al.
silicate, tetra-	$K_2Si_4O_9.H_2O$	352.45	rhb., 1.530	2.417	d. 400	...	s.	s.	i. al.
sulfate (arcanite)	K_2SO_4	174.25	col., rhb., 1.4947	2.662	tr. 588	...	$7.35^{0°}$	$24.1^{100°}$	i. al., act., CS_2

* One commercial form 70 to 72 per cent.

† Common commercial form 85 per cent H_3PO_4 in aqueous solution.

‡ Usual commercial form.

TABLE 3-1 Physical Properties of the Elements and Inorganic Compounds *(Continued)*

Name	Formula	Formula weight	Color, crystalline form and refractive index	Specific gravity	Melting point, °C.	Boiling point, °C.	Solubility in 100 parts: Cold water	Hot water	Other reagents
Potassium sulfate, pyro-	$K_2S_2O_7$	254.31	col.	2.277	300	....	s.	d.	
sulfide, mono-	$K_2S.5H_2O$	200.33	rhb., delq.	....	60	$-3H_2O$, 150	s.	....	s. al., gly.; i. et.
sulfite	$K_2SO_3.2H_2O$	194.28	wh., rhb.	....	d.	....	100	>100	sl. s. al.; i. NH_3
sulfite, acid	$KHSO_3$	120.16	wh., mn.	....	d. 190	....	$45.5^{15°}$	$91.5^{75°}$	i. abs. al.
tartrate	$K_2C_4H_4O_6.\frac{1}{2}H_2O$	235.27	col., mn., 1.526	1.98	....	d.	$12.5^{17.5°}$	$278^{100°}$	sl. s. al.
tartrate, acid	$KHC_4H_4O_6$*	188.18	col., rhb.	1.956	....	....	$0.37^{0°}$	$6.1^{100°}$	s. a., alk.; i. al., ac.
thiocyanate	KCNS	97.17	col.,delq., mn., 1.660±	1.886	172.3	d. 500	$177^{0°}$	$217^{20°}$	$20.8^{22°}$ act.; s. al.
thiosulfate	$K_2S_2O_3$	190.31	col., cb.	....	d. 400	....	$96.1^{0°}$	$311.2^{90°}$	
thiosulfate	$3K_2S_2O_3.H_2O$	588.95	delq., mn.	2.23	$-H_2O$, 180	d.	....	....	i. al.
Praseodymium	Pr	140.92	yel.	$6.5^{20°}$	940	....	d.		
Radium	Ra	226.05	wh., met.	5?	960	1140	d. $+H_2$	....	d. a.
bromide	$RaBr_2$	385.88	wh., mn.	5.79	728	subl. 900	$70^{20°}$	s.	s. al.
Radon (Niton)	Rn	222.0	gas	lq. 5.5; 111 (D)	−71	−62	$51^{0°}$ cc	$8.5^{50°}$ cc	
Rhenium	Re	186.31	hex.	....	3440	....	....	....	i. HF, HCl; s. H_2SO_4; HNO_3
Rhodium	Rh	102.91	gray-wh.; cb.	12.5	1955	>2500	i.	i.	sl. s. aq. reg., a.
chloride	$RhCl_3$	209.28	red	....	d. 450	subl. 800±	i.	i.	v. sl. s. alk.; i. aq. reg., a.
chloride	$RhCl_3.4H_2O$	281.35	dark red	....	....	....	v. s.	....	s. HCl, al.; i. et.
Rubidium	Rb	85.48	silv. wh.	lq. $1.475^{38.5}$; $1.53^{20°}$	38.5	700	d.	....	s. a., al.
Ruthenium	Ru	101.70	bk., porous	8.6	>1950	....	i.	i.	sl. s. aq. reg.; a.
Ruthenium	Ru	101.70	gray, hex.	$12.2^{20°}$	2450	>2700	i.	i.	
Samarium	Sm (also Sa)	150.43	....	7.7	>1300				
Scandium	Sc	45.10	....	2.5?	1200	2400			
Selenic acid	H_2SeO_4	144.98	hex. pr.	$2.950\frac{15°}{4}$	58	260	$1300^{30°}$	$\infty^{60°}$	s. H_2SO_4; d. al.; i. NH_3
Selenic acid	$H_2SeO_4.H_2O$	162.99	nd.	$2.627\frac{15°}{4}$	26	205	v. s.		
Selenium	Se_8	631.68	red pd., amor., 2.92	$4.26^{25°}$	50	688	i.	i.	s. CS_2, H_2SO_4, CH_2I_2
Selenium	Se_8	631.68	gray, trig., 3.00; red, hex.	4.80; 4.50	220	688	i.	i.	s. CS_2, H_2SO_4
Selenium	Se_8	631.68	steel gray	$4.8^{25°}$	217	688	i.	i.	i. CS_2; s. H_2SO_4
Selenous acid	H_2SeO_3	128.98	hex.	$3.004\frac{15°}{4}$	d.	....	$90^{0°}$	$400^{90°}$	v. s. al.; i. NH_3
Silicic acid, meta-	H_2SiO_3	78.08	amor., 1.41	2.1–2.3	....	....	i.	i.	s. alk.; i. NH_4Cl
Silicic acid, ortho-	H_4SiO_4	96.09	amor.	$1.576^{17°}$	....	....	sl. s.	sl. s.	s. alk.; i. NH_4Cl
Silicon, crystalline	Si	28.06	gray, cb., 3.736	$2.4^{20°}$	1420	2600	i.	i.	s. HNO_3 + HF, Ag; sl. s. Pb, Zn; i. HF
Silicon, graphitic	Si	28.06	cr.	2.0–2.5	....	2600	i.	i.	s. HNO_3 + HF, fused alk.; i. HF.
Silicon, amorphous	Si	28.06	brn., amor.	2	....	2600	i.	i.	s. HF, KOH
carbide	SiC	40.07	blue-bk., trig., 2.654	3.17	>2700	subl. 2200	i.	i.	s. fused alk.; i. a.
chloride, tri-	Si_2Cl_6	268.86	lf. or lq.	$1.58^{0°}$	−1	144^{760mm}	d.	....	d. alk.
chloride, tetra-	$SiCl_4$	169.89	col., fuming lq., 1.412	1.50	−70	57.6	d.	....	d. conc. H_2SO_4, al.
fluoride	SiF_4	104.06	gas	3.57 (A)	−95.7	-65^{1810mm}	v. s. d.	....	s. HNO_3, al., et.
hydride (silane)	SiH_4	32.09	col. gas	lq. $0.68^{-185°}$	−185	-112^{760mm}	i.	....	i. al., et.; d. KOH
oxide, di- (opal)	$SiO_2.xH_2O$	......	iridescent, amor.	2.2	1600–1750	subl. 1750	i.	i.	s. HF, h. alk., fused $CaCl_2$
oxide, di- (cristobalite)	SiO_2	60.06	col., cb. or tet., 1.487	2.32	1710	2230	i.	i.	s. HF; i. alk.
oxide, di- (lechatelierite)	SiO_2	60.06	....	2.20	....	2230	i.	i.	s. HF; i. alk.
oxide, di- (quartz)	SiO_2	60.06	hex., 1.5442	$2.650^{20°}$	tr. <1425	2230	i.	i.	s. HF; i. alk.
oxide, di- (tridymite)	SiO_2	60.06	trig., rhb., 1.469	2.26	tr. 1670	2230	i.	....	s. HF; i. alk.
Silver	Ag	107.88	silv. met., cb.	$10.5^{20°}$	960.5	1950	i.	i.	s. HNO_3, h. H_2SO_4; i. alk.
bromide (bromyrite)	AgBr	187.80	pa. yel., cb., 2.252	$6.473\frac{25°}{4}$	434	d. 700]	$0.00002^{20°}$	$0.00037^{100°}$	$0.51^{18°}$ NH_4OH; s. KCN, $Na_2S_2O_3$
carbonate	Ag_2CO_3	275.77	yel. pd.	6.077	218 d.	....	$0.003^{20°}$	$0.05^{100°}$	s. NH_4OH, $Na_2S_2O_3$; i. al.
chloride (cerargyrite)	AgCl	143.34	wh., cb., 2.071	5.56	455	1550	$0.000089^{10°}$	$0.00217^{100°}$	s. NH_4OH, KCN; sl. s. HCl
cyanide	AgCN	133.90	wh., 1.685±	3.95	$-(CN)_2$, 320	....	$0.000022^{20°}$	....	s. NH_4OH, KCN, HNO_3
nitrate (lunar caustic)	$AgNO_3$	169.89	col., rhb., 1.744	$4.352\frac{19°}{4}$	212	444 d.	$122^{0°}$	$952^{100°}$	s. gly.; v. sl. s. al.
Sodium	Na	22.997	silv. met, cb.	$0.97^{20°}$	97.5	880	d., forms NaOH	....	i. bz.; d. al.
acetate	$NaC_2H_3O_2$	82.04	wh., mn., 1.464	1.528	324	....	$46.5^{20°}$	$170^{100°}$	$2.1^{18°}$ al.
acetate	$NaC_2H_3O_2.3H_2O$	136.09	wh., mn.	1.45	58	$-3H_2O$, 120	v. s.	v. s.	$7.8^{25°}$ abs. al.
aluminate	$NaAlO_2$	81.97	amor.	....	1650	....	s.	v. s.	i. al.
amide	$NaNH_2$	39.02	olive gn.	....	210	400	d.	....	d. al.

Sodium ammonium phosphate	$NaNH_4HPO_4.4H_2O$	209.09	col., mn.	1.574	79 d.		16.7	100	i. al.
antimonate, meta-	$2NaSbO_3.7H_2O$	511.63	cb.				$0.031^{12.8\circ}$		sl. s. al., NH_4 salts; i. ac.
arsenate	$Na_3AsO_4.12H_2O$	424.09	hex., 1.4589	1.759	86.3		$26.7^{17\circ}$		1.67 al., $50^{15\circ}$ gly.
arsenate, acid (monobasic)	$NaH_2AsO_4.H_2O$	181.94	rhb., 1.5535	2.535	d. 100		s.		
arsenate, acid (dibasic)	$Na_2HAsO_4.7H_2O$*	312.02	col., mn., 1.4658	1.871	125	$-7H_2O$, 100	$61^{15\circ}$	v. s.	sl. s. al.
arsenate, acid (dibasic)	$Na_2HAsO_4.12H_2O$	402.10	mn., 1.4496	1.72	28	$-12H_2O$, 100	$5.59^{0.1\circ}$	$140.7^{30\circ}$	sl. s. al.
arsenite, acid	Na_2HAsO_3	169.91	col.	1.87			v. s.		
benzoate	$NaC_7H_5O_2$	144.11	col. cr.				$62.5^{25\circ}$	$76.9^{100\circ}$	$2.3^{25\circ}$, $8.3^{78\circ}$ al.
bicarbonate	$NaHCO_3$	84.01	wh., mn., 1.500	2.20	$-CO_2$, 270		$6.9^{0\circ}$	$16.4^{60\circ}$	i. al.
bifluoride	$NaHF_2$	62.00	col. cr.		d.		$3.7^{20\circ}$	s.	
bisulfate	$NaHSO_4$	120.06	col., tri.	2.742	>315	d., $-H_2O$	$50^{0\circ}$	$100^{100\circ}$	d. al.; i. NH_3
bisulfite	$NaHSO_3$	104.06	col., mn., 1.526	1.48	d.		sl. s.	s.	i. al., act.
borate, tetra-	$Na_2B_4O_7$	201.27		2.367	741		$1.3^{0\circ}$	$8.79^{40\circ}$	i. al.
borate, tetra	$Na_2B_4O_7.5H_2O$	291.35	col., rhb., 1.461	1.815			$22^{62\circ}$ (anh.)	$52.3^{100\circ}$ (anh.)	
borate, tetra- (borax)	$Na_2B_4O_7.10H_2O$*	381.43	wh., mn., 1.4694	1.73	75	$-10H_2O$, 200	$1.3^{0.5}$ (anh).	$20.3^{60\circ}$ (anh.)	s. gly.; i. abs. al.
bromate	$NaBrO_3$	150.91	col., cb.	$3.339^{17.5\circ}$	381		$27.5^{0\circ}$	$90.9^{100\circ}$	i. al.
bromide	NaBr	102.91	col., cb., 1.6412	$3.205^{17.5\circ}$	755	1390	$90^{20\circ}$	$121^{100\circ}$	sl. s. al.
bromide	$NaBr.2H_2O$	138.95	col., mn.	2.176	50.7		$79.5^{0\circ}$ (anh.)	$118.3^{80\circ}$ (anh.)	sl. s. al.
carbonate (soda ash)	Na_2CO_3	106.00	wh. pd., 1.535	2.533	851	d.	$7.1^{0\circ}$	$48.5^{104\circ}$	i. al., et.
carbonate	$Na_2CO_3.H_2O$	124.02	wh., rhb., 1.506-1.509	1.55	$-H_2O$, 100		s.	s.	s. gly.; i. al., et.
carbonate	$Na_2CO_3.7H_2O$	232.12	rhb. or trig.	1.51	d. 35.1		s.	s.	
carbonate (sal soda)	$Na_2CO_3.10H_2O$	286.16	wh., mn., 1.425	1.46			$21.5^{0\circ}$	$238^{30\circ}$	i. al.
carbonate, sesqui- (trona)	$Na_3H(CO_3)_2.2H_2O$	226.05	wh., mn., 1.5073	2.112	d.		$13^{0\circ}$	$42^{100\circ}$	
chlorate	$NaClO_3$	106.45	wh., cb., or trig., 1.5151	$2.490^{15\circ}$	248	d.	$79^{0\circ}$	$230^{100\circ}$	s. al.
chloride	NaCl	58.45	col., cb., 1.5443	2.163	800.4	1413	$35.7^{0\circ}$	$39.8^{100\circ}$	sl. s. al.; i. conc. HCl
chromate	Na_2CrO_4	162.00	yel., rhb.	2.723	392		$32^{0\circ}$	$126^{100\circ}$	
chromate	$Na_2CrO_4.10H_2O$	342.16	yel., delq., mn.	1.483	19.9		v. s.	∞	sl. s. al.
citrate	$2Na_3C_6H_5O_7.11H_2O$	714.36	wh., rhb.	$1.857^{\frac{23}{4}.5\circ}$	$-11H_2O$, 150	d.	$91^{25\circ}$	$250^{100\circ}$	i. al.
cyanide	NaCN	49.02	wh., cb., 1.452		563.7	1496	$48^{10\circ}$	$82^{35\circ}$	s. NH_3; sl. s. al.
dichromate	$Na_2Cr_2O_7.2H_2O$	298.05	red, mn., 1.6994	$2.52^{13\circ}$	$-2H_2O$, 84.6; 356 (anh.)	d. 400	$238^{0\circ}$	$508^{80\circ}$	
ferricyanide	$Na_3Fe(CN)_6.H_2O$	298.97	red, delq.				$18.9^{0\circ}$	$67^{100\circ}$	i. al.
ferrocyanide	$Na_4Fe(CN)_6.10H_2O$	484.11	yel., mn.	1.458			$17.9^{20\circ}$ (anh.)	$63^{98.5\circ}$ (anh.)	i. al.
fluoride (villiaumite)	NaF	42.00	tet., 1.3258	2.79	992		$4^{0\circ}$	$5^{100\circ}$	v. sl. s. al.
formate	$NaHCO_2$	68.01	wh., mn.	1.919	253		$44^{0\circ}$	$160^{100\circ}$	sl. s. al.; i. et.
hydride	NaH	24.005	silv. nd., 1.470	0.92	d. 800		d.		i. bz., CS_2, CCl_4, NH_3; s. molten metal
hydrosulfide	$NaSH.2H_2O$	92.10	col., delq., nd.		d.		s.	s.	s. al.; d. a.
hydrosulfide	$NaSH.3H_2O$	110.11	rhb.		22	d.	s.	s.	s. al.; d. a.
hydrosulfite	$Na_2S_2O_4.2H_2O$	210.15	col. cr.		d.		$22^{20\circ}$	d.	i. al.
hydroxide	NaOH	40.00	wh., delq.	2.130	318.4	1390	$42^{0\circ}$	$347^{100\circ}$	v. s. al., et., gly.; i. act.
hydroxide	$NaOH.3\frac{1}{2}H_2O$	103.06	mn.		15.5		s.	v. s.	
hypochlorite	NaOCl	74.45	pa. yel., in soln. only		d.		$26^{0\circ}$	$158^{56\circ}$	
iodide	NaI*	149.92	col., cb., 1.7745	$3.667^{0\circ}$	651	1300	$158.7^{0\circ}$	$302^{100\circ}$	v. s. al., act.
iodide	$NaI.2H_2O$	185.95	col., mn.	2.448			v. s.	v. s.	v. s. NH_3
lactate	$NaC_3H_5O_3$	112.07	col., amor.		d.		v. s.	v. s.	s. al.; i. et.
nitrate (soda niter)	$NaNO_3$	85.01	col., trig., 1.5874	2.257	308	d. 380	$73^{0\circ}$	$180^{100\circ}$	s. NH_3; sl. s. gly., al.
nitrite	$NaNO_2$	69.01	pa. yel., rhb.	$2.168^{0\circ}$	271	d. 320	$72.1^{0\circ}$	$163.2^{100\circ}$	$0.3^{20\circ}$ et.; 0.3 abs. al.; $4.4^{20\circ}$ m. al.; v. s. NH_3
oxide	Na_2O	61.99	wh., delq.	2.27	subl.		Forms NaOH		d. al.
perborate	$NaBO_3.H_2O$	99.83	wh. pd.		d. 40		sl. s.	d.	s. gly., alk.
perchlorate	$NaClO_4$	122.45	rhb., 1.4617		482 d.		$170^{0\circ}$	$320^{100\circ}$	s. al.; 51 m. al.; 52 act.; i. et.
perchlorate	$NaClO_4.H_2O$	140.47	hex.	2.02	d. 130		$209^{15\circ}$	$284^{50\circ}$	s. al.
peroxide	Na_2O_2*	77.99	yel.-wh. pd.	2.805	d.		s. d.	d.	s. dil. a.
peroxide	$Na_2O_2.8H_2O$	222.12	wh., hex.		d. 30		s. d.	d.	
phosphate, monobasic	$NaH_2PO_4.H_2O$*	138.01	col., rhb., 1.4852	2.040	$-H_2O$, 100	d. 200	$71^{0\circ}$	$390^{83\circ}$	i. al.
phosphate, monobasic	$NaH_2PO_4.2H_2O$	156.03	col., rhb., 1.4629	1.91	60		$91.1^{0\circ}$	$308^{40\circ}$	
phosphate, dibasic	$Na_2HPO_4.7H_2O$	268.09	col., mn., 1.4424	1.679	d.		$185^{40\circ}$	$2000^{100\circ}$	
phosphate, dibasic	$Na_2HPO_4.12H_2O$	358.17	col., mn., 1.4361	1.52	34.6	$-12H_2O$, 180	$4.3^{0\circ}$	$76.7^{30\circ}$	i. al.
phosphate, tribasic	Na_3PO_4	163.97	wh.	$2.537^{17.5\circ}$	1340		$4.5^{0\circ}$	$77^{100\circ}$	
phosphate, tribasic	$Na_3PO_4.12H_2O$*	380.16	wh., trig., 1.4458	1.62	73.4	$-11H_2O$, 100	$28.3^{15\circ}$	∞	i. CS_2
phosphate, meta-	$Na_4P_4O_{12}$	407.91	col.	2.476	616 d.		s.	s.	s. a., alk.
phosphate, pyro-	$Na_4P_2O_7$*	265.95	wh.	2.45	988		$2.26^{0\circ}$	$45^{96\circ}$	d. a.
phosphate, pyro-	$Na_4P_2O_7.10H_2O$	446.11	mn., 1.4525	1.82	d.		$5.4^{0\circ}$	$93^{100\circ}$	i. al., NH_3
phosphate (pyrodisodium)	$Na_2H_2P_2O_7$	221.97	col., mn., 1.510	1.862	d. 220		$4.5^{0\circ}$	$21^{40\circ}$	
phosphate (pyrodisodium)	$Na_2H_2P_2O_7.6H_2O$	330.07	col., mn., 1.4645	1.848			$6.9^{0\circ}$	$36^{40\circ}$	
potassium tartrate	$NaKC_4H_4O_6.4H_2O$	282.23	rhb., 1.493	1.790	70 to 80	$-4H_2O$, 215	$26^{0\circ}$	$66^{26\circ}$	sl. s. al.
silicate, meta-	Na_2SiO_3	122.05	col., rhb., 1.520		1088		s.	s. d.	i. Na or K salts, al.

*Usual commercial form.

TABLE 3-1 Physical Properties of the Elements and Inorganic Compounds *(Continued)*

Name	Formula	Formula weight	Color, crystalline form and refractive index	Specific gravity	Melting point, °C.	Boiling point, °C.	Solubility in 100 parts: Cold water	Hot water	Other reagents
Sodium silicate, meta-	$Na_2SiO_3.9H_2O$	284.20	rhb.		47	$-6H_2O$, 100	v. s.	v. s.	$29^{18°}$, ½N NaOH
silicate, ortho-	Na_4SiO_4	184.05	col., hex., 1.530		1018		s.	s.	
silicofluoride	Na_2SiF_6	188.05	wh., hex., 1.312	2.679	d.		$0.44^{0°}$	$2.45^{100°}$	i. al.
stannate	$Na_2SnO_3.3H_2O$	266.74	hex. tablets		d. 140		$50^{0°}$	$67^{50°}$	i. al., act.
sulfate (thenardite)	Na_2SO_4	142.05	col., rhb., 1.477	2.698	tr. 100 to mn.		$5^{0°}$	$42^{100°}$	i. al.
sulfate	Na_2SO_4	142.05	col., mn.		tr. 500 to hex.		$48.8^{40°}$	$42.5^{100°}$	d. HI; s. H_2SO_4
sulfate	Na_2SO_4	142.05	col., hex.		884		$19.4^{20°}$	$45.3^{60°}$	
sulfate	$Na_2SO_4.7H_2O$	268.17	tet.				$44.9^{0°}$	$202.6^{26°}$	
sulfate (Glauber's salt)	$Na_2SO_4.10H_2O$	322.21	col., mn., 1.396	1.464	32.4	$-10H_2O$, 100	$36^{15°}$	$412^{34°}$	i. al.
sulfide, mono-	Na_2S	78.05	pink or wh., amor.	1.856			$15.4^{10°}$	$57.3^{90°}$	sl. s. al.; i. et.
sulfide, tetra-	Na_2S_4	174.23	yel., cb.		275		s.	s.	s. al.
sulfide, penta-	Na_2S_5	206.29	yel.		251.8		s.	s.	s. al.
sulfite	Na_2SO_3	126.05	hex. pr., 1.565	$2.633\frac{15°}{4}$	d.		$13.9^{0°}$	$28.3^{84°}$	i. al., NH
sulfite	$Na_2SO_3.7H_2O$	252.17	mn.	1.561	$-7H_2O$, 150	d.	$34.7^{2°}$	$67.8^{18°}$	i. al.
tartrate	$Na_2C_4H_4O_6.2H_2O$	230.10	rhb.	1.818			$29^{6°}$	$66^{43°}$	i. al.
thiocyanate	NaCNS	81.08	delq., rhb., 1.625±		287		$110^{10°}$	$225^{100°}$	v. s. al.
thiosulfate	$Na_2S_2O_3$	158.11	mn.	1.667			$50^{0°}$	$231^{80°}$	
thiosulfate (hypo)	$Na_2S_2O_3.5H_2O$*	248.19	mn. pr., 1.5079	1.685	d. 48.0		$74.7^{0°}$	$301.8^{60°}$	s. NH_3; v. sl. s. al.
tungstate	Na_2WO_4	293.91	wh., rhb.	4.179	692		$57.5^{80°}$	$97^{100°}$	
tungstate	$Na_2WO_4.2H_2O$*	329.95	wh., rhb.	3.245	$-2H_2O$, 100		$88^{0°}$	$123.5^{100°}$	sl. s. NH_3; i. a., al.
tungstate, para-	$Na_6W_7O_{24}.16H_2O$	2097.68	wh., tri.	$3.987^{14°}$	$-16H_2O$, 300		8	d.	
uranate	Na_2UO_4	348.06	yel.				i.	i.	s. alk. carb., dil. a.
vanadate	$Na_3VO_4.16H_2O$	472.20	col. nd.		866 (anh.)		v. s.	d.	i. al.
vanadate, pyro-	$Na_4V_2O_7$	305.89	hex.		654		s.		i. al.
Stannic chloride	$SnCl_4$	260.53	col., fuming lq.	2.226	−30.2	114.1	s.	d.	s. abs. al., act., NH_3; s. ∞ CS_2
oxide (cassiterite)	SnO_2	150.70	wh., tet., 1.9968	7.0	1127		i.	i.	s. conc. H_2SO_4; i. alk., NH_4OH, NH_3
sulfate	$Sn(SO_4)_2.2H_2O$	346.85	col., delq., hex.				v. s.	d.	s. dil. H_2SO_4, HCl; d. abs. al.
Stannous bromide	$SnBr_2$	278.53	yel., rhb.	$5.12^{17°}$	215.5	620	s.	d.	s. C_6H_5N
chloride	$SnCl_2$	189.61	wh., rhb.		246.8	623	$83.9^{0°}$	$269.8^{15°}$	s. alk., abs. al., et.
chloride (tin salt)	$SnCl_2.2H_2O$*	225.65	wh., tri.	$2.71^{15.5°}$	37.7	d.	$118.7^{0°}$	∞	s. tart. a., alk., al.
sulfate	$SnSO_4$	214.76	wh. cr.		$-SO_2$, 360		$19^{19°}$	$18^{100°}$	s. H_2SO_4
Strontium	Sr	87.63	silv. met.	2.6	800	1150	d.	Forms $Sr(OH)_2$	s. al., a.
acetate	$Sr(C_2H_3O_2)_2$	205.72	wh. cr.	2.099		d.	$36.9^{0°}$	$36.4^{97°}$	$0.26^{15°}$ m. al.
carbonate (strontianite)	$SrCO_3$	147.64	wh., rhb., 1.664	3.70	1497^{60atm}	$-CO_2$, 1350	$0.0011^{18°}$	$0.065^{100°}$	s. a., NH_4 salts, aq. CO_2
chloride	$SrCl_2$	158.54	wh., cb., 1.6499	3.052	873		$43.5^{0°}$	$100.8^{100°}$	v. sl. s. act., abs. al.; i. NH_3
chloride	$SrCl_2.6H_2O$*	266.64	wh., rhb., 1.5364	$1.933^{17°}$	$-4H_2O$, 61	$-6H_2O$, 100	$104^{0°}$	$198^{40°}$	
hydroxide	$Sr(OH)_2$	121.65	wh., delq.	3.625	375		$0.41^{0°}$	$21.83^{100°}$	s. NH_4Cl
hydroxide	$Sr(OH)_2.8H_2O$*	265.77	col., tet., 1.499	1.90	$-7H_2O$ in dry air		$0.90^{0°}$	$47.7^{100°}$	s. NH_4Cl; i. act.
nitrate	$Sr(NO_3)_2$*	211.65	col., cb., 1.5878	2.986	570		$40^{0°}$	$100^{89°}$	s. NH_3; 0.012 abs. al.
nitrate	$Sr(NO_3)_2.4H_2O$	283.71	wh., mn.	2.2			$62.2^{0°}$	$124^{20°}$	i. HNO_3
oxide (strontia)	SrO	103.63	col., cb., 1.870	4.7	2430		Forms $Sr(OH)_2$		sl. s. al.; i. et.
peroxide	SrO_2	119.63	wh. pd.		d.		$0.008^{20°}$	d.	s. al., NH_4Cl; i. act.
peroxide	$SrO_2.8H_2O$	263.76	wh. cr.		$-8H_2O$, 100	d.	$0.018^{20°}$	d.	s. al.; i. NH_4OH
sulfate (celestite)	$SrSO_4$	183.69	col., rhb., 1.6237	3.96	1580 d.		$0.0113^{0°}$	$0.0114^{32°}$	sl. s. a.; i. dil. H_2SO_4, al.
sulfate, acid	$Sr(HSO_4)_2$	281.77	col., granular		d.		d.		$14^{70°}$ H_2SO_4
Sulfamic acid	NH_2SO_3H	97.09	wh., rhb.	$2.03\frac{12°}{4}$	205 d.		$20^{0°}$	$40^{70°}$	sl. s. al., act.; i. et.
Sulfur, amorphous	S	32.06	pa. yel. pd., 2.0–2.9	2.046	120	444.6	i.	i.	sl. s. CS_2
Sulfur, monoclinic	S_8	256.48	pa. yel., mn.	1.96	119.0	444.6	i.	i.	s. CS_2, al.
Sulfur, rhombic	S_8	256.48	pa. yel., rhb.	2.07	112.8	444.6	i.	i.	$24^{0°}$, $181^{55°}$ CS_2
Sulfur bromide, mono-	S_2Br_2	223.95	red, fuming lq.	2.635	−46	$54^{0.18mm}$	d.		
chloride, mono-	S_2Cl_2	135.03	red-yel. lq.	1.687	−80	138	d.		s. CS_2, et., bz.
chloride, di-	SCl_2	102.97	dark red fuming lq.	$1.621\frac{15°}{15}$	−78	59	d.		d. al.
chloride, tetra-	SCl_4	173.89	yel.-brn. lq.		−30	d. > −20	d.		
oxide, di-	SO_2	64.06	col. gas	lq., $1.434^{0°}$; 2.264 (A)	−75.5	−10.0	$22.8^{0°}$	$4.5^{50°}$	s. H_2SO_4; al., ac.
oxide, tri-(α)	SO_3	80.06	col. pr.	lq., 1.923; 2.75 (A)	16.83	44.6	d.		s. H_2SO_4
oxide, tri-(β)	$(SO_3)_2$	160.12	col., silky, nd.	$1.97^{20°}$	50		Forms H_2SO_4		s. H_2SO_4

Sulfuric acid	H_2SO_4*	98.08	col., viscous lq.	$1.834\frac{18°}{4}$	10.49	d. 340	∞	∞	d. al.
Sulfuric acid	$H_2SO_4.H_2O$	116.09	pr. or lq.	$1.842\frac{15°}{4}$	8.62	290	∞	∞	d. al.
Sulfuric acid	$H_2SO_4.2H_2O$	134.11	col. lq.	$1.650\frac{0°}{4}$	−38.9	167	∞	∞	d. al.
Sulfuric acid, pyro-	$H_2S_2O_7$	178.14	cr.	$1.9^{20°}$	35	d.	d.	...	d. al.
Sulfuric oxychloride	SO_2Cl_2	134.97	col. lq.	$1.667\frac{20°}{4}$	−54.1	69.1^{760mm}	d.	...	s. ac.; d. al.
Sulfurous oxybromide	$SOBr_2$	207.89	or.-yel. lq.	$2.68^{18°}$	−50	68^{40mm}	d.	...	s. bz., CS_2, CCl_4; d. act.
oxychloride	$SOCl_2$	118.97	col. lq.	1.638	−104.5	78.8	d.	...	s. bz., chl.
Tantalum	Ta	180.88	bk.-gray, cb.	16.6	2850	>4100	i.	i.	s. fused alk., HF; i. HCl, HNO_3, H_2SO_4
Tellurium	Te	127.61	met., hex.	(α) 6.24; (β) 6.00	452	1390	i.	i.	s. H_2SO_4, HNO_3, KCN, KOH, aq. reg.; i. CS_2
Terbium	Tb	159.20							
Thallium	Tl	204.39	blue-wh., tet.	11.85	303.5	1650	i.	i.	s. HNO_3, H_2SO_4; i. NH_3
acetate	$TlC_2H_3O_2$	263.43	silky nd.	3.68	110	...	v. s.	...	v. s. al.
chloride, mono-	TlCl	239.85	wh., cb.	7.00	430	806	$0.21^{0°}$	$1.8^{100°}$	sl. s. HCl; i. al., NH_4OH
chloride, sesqui-	Tl_2Cl_3	515.15	yel., hex.	5.9	400–500	d.	$0.26^{15°}$	$1.9^{100°}$	
chloride, tri-	$TlCl_3$	310.76	hex. pl.	...	25	d.	v. s.	...	s. al., et.
chloride, tri-	$TlCl_3.4H_2O$	382.83	nd.	...	37	$-4H_2O$, 100	$86.2^{17°}$	d.	s. al., et.
sulfate (ic)	$Tl_2(SO_4)_3.7H_2O$	823.07	lf.	...	$-6H_2O$, 200	d.	d.	d.	s. dil. H_2SO_4
sulfate (ous)	Tl_2SO_4	504.84	col., rhb., 1.8671	6.77	632	d.	$2.70^{0°}$	$18.45^{100°}$	
sulfate, acid	$TlHSO_4$	301.46	trimorphous	...	115 d.	...	...	...	v. sl. s. dil. H_2SO_4
Thio, *Cf.* sulfo or sulfur									
Thorium	Th	232.12	cb.	11.2	1845	>3000	i.	i.	s. HCl, H_2SO; sl. s. HNO_3; i. HF, alk.
oxide, di- (thorianite)	ThO_2	264.12	wh., cb.	9.69	>2800	4400	i.	...	s. h. H_2SO_4; i. alk.
sulfate	$Th(SO_4)_2$	424.24	...	$4.225^{17°}$	...	...	$0.74^{0°}$	$5.22^{60°}$	
sulfate	$Th(SO_4)_2.9H_2O$	586.38	mn. pr.	2.77	$-9H_2O$, 400	...	sl. s.	sl. s.	
Thulium	Tm	169.40	...		...	...	i.	i.	
Tin	Sn	118.70	silv. met., tet.	7.31	231.85	2260	i.	i.	s. HCl, H_2SO_4, dil. HNO_3, h. aq KOH
Tin	Sn	118.70	gray, cb.	5.750	Stable −163 to +18	2260	i.	i.	s. a., h. alk. solns.
Tin salts, *Cf.* stannic and stannous									
Titanic acid	H_2TiO_3	97.92	wh. pd.	...	...	...	i.	i.	s. alk.; v. sl. s. dil. a.; i. al.
Titanium	Ti	47.90	dark gray, cb.	$4.50^{17.5°}$	1800	>3000	i.	d.	s. a.
chloride, di-	$TiCl_2$	118.81	bk., delq.	...	Unstable in air	...	d.	...	i. CS_2, et., chl.
chloride, tri-	$TiCl_3$	154.27	vl., delq.	...	d. 440	...	s.	s.	
chloride, tetra-	$TiCl_4$*	189.73	col. lq.	lq., 1.726	−30	136.4	s.	d.	s. dil. HCl
oxide, di- (anatase)	TiO_2	79.90	brn. or bk., tet., 2.534–2.564	3.84	...	...	i.	i.	sl. s. alk.
oxide, di- (brookite)	TiO_2	79.90	brn. or bk., rhb., 2.586	4.17	...	...	i.	i.	
oxide, di- (rutile)	TiO_2	79.90	col. if pure, tet., 2.615	4.26	1640 d.	<3000	i.	i.	s. H_2SO_4, alk.
Tungsten	W	183.92	gray-bk., cb.	19.3	3370	5900	i.	i.	s. h. conc. KOH; sl. s. NH_3, HNO_3, aq. reg.
carbide	WC	195.93	gray pd., cb.	$15.7^{18°}$	2777	6000	i.	i.	s. F_2; i. a.
carbide	W_2C	379.85	iron gray	$16.06^{18°}$	2877	6000	i.	i.	s. h. HNO_3; sl. s. HCl, H_2SO_4
oxide, tri-	WO_3	231.92	yel., rhb.	7.16	>2130	...	i.	i.	s. alk.; i. a.
Tungstic acid (tungstite)	H_2WO_4	249.94	yel., rhb. 2.24	5.5	$-\frac{1}{2}H_2O$, 100; 1473	...	i.	sl. s.	s. HF, alk., NH_3
Uranic acid	H_2UO_4	304.09	yel. pd.	$5.926^{15°}$	$-H_2O$, 250 to 300	...	i.	i.	s. a., alk. carb.; i. alk.
Uranium	U	238.07	wh. cr.	$18.485\frac{13°}{4}$	1133	3500	i.	i.	s. a.; i. alk.
carbide	U_2C_3	512.14	cr.	11.28	2400	...	d.	...	d. a.
oxide, di- (uraninite)	UO_2	270.07	bk., rhb.	10.9	2176	...	i.	i.	s. HNO_3, conc. H_2SO_4
oxide (pitchblende)	U_3O_8	842.21	olive gn.	7.31	d.	...	i.	i.	s. HNO_3, H_2SO_4
sulfate (ous)	$U(SO_4)_2.4H_2O$	502.25	gn., rhb.	...	$-4H_2O$, 300	...	$23^{11°}$	$9^{63°}$	s. dil. a.
Uranyl acetate	$UO_2(C_2H_3O_2)_2.2H_2O$	424.19	yel., rhb.	$2.89^{15°}$	$-2H_2O$, 110	...	$9.2^{17°}$	d.	s. al., act.
carbonate (rutherfordine)	UO_2CO_3	330.08	tet.	5.6					
nitrate	$UO_2(NO_3)_2.6H_2O$	502.18	yel., rhb., 1.4967	2.807	60.2	118	$170.3^{0°}$	$\infty^{60°}$	v. s. ac., al., et.; i. dil. alk.
sulfate	$UO_2SO_4.3H_2O$	420.18	yel. cr.	$3.28^{16.5°}$	d. 100	...	$18.9^{13.2°}$	$230^{25°}$	4 al.; s. a.
Vanadic acid, meta-	HVO_3	99.96	yel. scales	...	...	...	i.	...	s. a., alk.; i. NH_3

* Usual commercial form.

TABLE 3-1 Physical Properties of Elements and Inorganic Compounds *(Concluded)*

Name	Formula	Formula weight	Color, crystalline form and refractive index	Specific gravity	Melting point, °C.	Boiling point, °C.	Solubility in 100 parts: Cold water	Hot water	Other reagents
Vanadic acid, pyro-	$H_4V_2O_7$	217.93	pa. yel., amor.	......	......	......	i.	......	s. a., alk., NH_4OH
Vanadium	V	50.95	lt. gray, cb.	5.96	1710	3000	i.	i.	s. HNO_3, H_2SO_4; i. aq., alk.
chloride, di-	VCl_2	121.86	gn., hex., delq.	$3.23^{18°}$	......	......	s.	d.	s. al., et.
chloride, tri-	VCl_3	157.23	pink, tabular, delq.	$3.00^{18°}$	d.	......	s.	d.	s. abs. al., et.
chloride, tetra-	VCl_4	192.78	red lq.	$1.816^{30°}$	−109	148.5^{755mm}	s. d.	......	s. abs. al., et., chl., ac.
oxide, di-	V_2O_2	133.90	lt. gray cr.	3.64	ign.	......	i.	i.	s. a.
oxide, tri-	V_2O_3	149.90	bk. cr.	$4.87\frac{18°}{4}$	1970	......	sl. s.	s.	s. HNO_3, HF, alk.
oxide, tetra-	V_2O_4	165.90	blue cr.	4.399	1967	......	i.	i.	s. a., alk.
oxide, penta-	V_2O_5	181.90	red-yel., rhb.	$3.357\frac{18°}{4}$	800	d. 1750	$0.8^{20°}$	......	s. a., alk.; i. abs. al.
oxychloride, mono-	VOCl	102.41	brn. pd.	2.824	......	......	i.	......	v. s. HNO_3
Vanadyl chloride	$(VO)_2Cl$	169.36	yel. cr.	3.64	d. in air	......	i.	......	s. HNO_3
chloride, di-	$VOCl_2$	137.86	gn., delq.	$2.88^{13°}$	......	......	d.	......	s. abs. al., dil. HNO_3
chloride, tri-	$VOCl_3$	173.32	yel. lq.	1.829	<−15	127.19	s. d.	......	s. al., et., ∞Br_2
Water†	H_2O	18.016	col. lq., $1.33300^{20°}$; hex. solid, 1.309	$1.00^{4°}$ (lq.); $0.915^{0°}$ (ice)	0	100	......	......	∞al.; sl. s. et.
Water, heavy	D_2O	20.029	col. lq., $1.32844^{20°}$	$1.107^{20°}$	3.82	101.42	∞	∞	∞ al.; sl. s. et.
Xenon	Xe	131.30	col. gas	lq., $3.06^{-109.1}$ $2.7^{-140°}$ 4.53 (A)	−140	−109.1	24.2$^{0°}$ cc	$7.3^{50°}$ cc	
Ytterbium	Yb	173.04							
Yttrium	Y	88.92	dark gray, hex.	5.51	1490	2500	sl. d.	d.	v. s. dil. a., h. KOH
Zinc	Zn	65.38	silv. met., hex.	7.140	419.4	907	i.	i.	s. a., ac., alk.
acetate	$Zn(C_2H_3O_2)_2$	183.47	mn.	1.840	242	subl. in vac.	$30^{25°}$	$44.6^{100°}$	$2.8^{25°}$, $166^{79°}$ al.
acetate	$Zn(C_2H_3O_2)_2.2H_2O$*	219.50	wh., mn., 1.494	1.735	237	$-2H_2O$, 100	$40^{25°}$	$66.6^{100°}$	v. s. al.
bromide	$ZnBr_2$	225.21	rhb.	$4.219^{4°}$	394	650	$390^{0°}$	$670^{100°}$	v. s. NH_4OH, al., et.
carbonate	$ZnCO_3$	125.39	wh., trig., 1.818	4.42	$-CO_2$, 300	......	$0.001^{15°}$	......	s. a., alk., NH_4 salts; i. act., NH_3
chloride	$ZnCl_2$	136.29	wh., delq., 1.687, uniaxial	$2.91\frac{25°}{4}$	283	732	$432^{25°}$	$615^{100°}$	$100^{12.5°}$ al.; v. s. et.; i. NH_3
cyanide	$Zn(CN)_2$	117.42	col., rhb.	......	d. 80	......	$0.0005^{18°}$	sl. s.	s. KCN, NH_3, alk.; i. al.
hydroxide	$Zn(OH)_2$	99.40	col., rhb.	3.053	d. 125	......	$0.00052^{18°}$	......	s. a., alk., NH_4OH
iodide	ZnI_2	319.22	cb.	$4.666\frac{14.2°}{4}$	446	624	$430^{0°}$	$510^{100°}$	s. a., al., NH_3, aq. $(NH_4)_2CO_3$
nitrate	$Zn(NO_3)_2.6H_2O$	297.49	col., tet.	$2.065\frac{14°}{4}$	36.4	$-6H_2O$, 105	324.5	$∞^{36.4°}$	v. s. al.
oxide (zincite)	ZnO	81.38	wh., hex., 2.004	5.606	>1800	......	$0.00042^{18°}$	......	s. a., alk., NH_4Cl; i. NH_3
oxide	ZnO	81.38	wh., amor.	5.47	>1800	......	$0.00042^{18°}$		
peroxide	ZnO_2	97.38	yel.	1.571	expl. 212	......	0.0022	......	i. NH_4OH; d. a.
phosphide	Zn_3P_2	258.10	steel gray, cb.	$4.55\frac{13°}{4}$	>420	1100	i.	......	s. dil. a.
silicate	$ZnSiO_3$	141.44	hex. or rhb.; glass, 1.650	3.52	1437	......	i.		
sulfate (zincosite)	$ZnSO_4$	161.44	wh., rhb., 1.669	$3.74\frac{15°}{4}$	d. 740	......	$42^{0°}$	$61^{100°}$	sl. s. al.; s. gly.
sulfate	$ZnSO_4.H_2O$	179.46	col.	$3.28\frac{15°}{4}$	d. 238	......	s.	$89.5^{100°}$	
sulfate	$ZnSO_4.6H_2O$	269.54	mn.	$2.072\frac{15°}{4}$	$-5H_2O$, 70	......	s.	s.	sl. s. al.; i. act.; NH_3
sulfate (goslarite)	$ZnSO_4.7H_2O$*	287.55	rhb., 1.4801	$1.966^{16.5°}$	tr. 39	$-7H_2O$, 280	115.2$^{0°}$	$653.6^{100°}$	sl. s. al.; i. act.; NH_3
sulfide (α) (wurzite)	ZnS	97.44	wh., hex., 2.356	4.087	1850^{150atm}	subl. 1185	$0.00069^{18°}$	i.	v. s. a.; i. ac.
sulfide (β) (sphalerite)	ZnS	97.44	wh., cb.; glass (?) 2.18–2.25	$4.102\frac{25°}{4}$	tr. 1020	......	i.	i.	s. a.
sulfide (blende)	ZnS	97.44	wh., granular	4.04	......	......	i.	i.	v. s. a.; i. ac.
sulfite	$ZnSO_3.2½H_2O$	190.48	mn.	......	$-2½H_2O$, 100	d. 200	0.16	d.	s. H_2SO_3, NH_4OH; i. al.
Zirconium	Zr	91.22	cb., pd. ign. easily	6.4	1700	>2900	i.	i.	s. HF, aq. reg.; sl. s. a.
oxide, di- (baddeleyite)	ZrO_2	123.22	yel. or brn., mn., 2.19	5.49	2700	......	i.	i.	s. H_2SO_4, HF
oxide, di- (free from Hf)	ZrO_2	123.22	wh., mn.	5.73	......	4300	i.	i.	s. H_2SO_4, HF

* Usual commercial form.

† *Cf.* special tables on water and steam, Nos. 3-3, 3-4, 3-5, 3-45, 3-46, 3-49, 3-259, 3-260, 3-261, 3-267, and 3-275.

NOTE: °F = ⁹⁄₅ °C + 32.

TABLE 3-2 Physical Properties of Organic Compounds*

Abbreviations Used in the Table

(A), density referred to air
al., ethyl alcohol
amor., amorphous
aq., aqua, water
brn., brown
bz., benzene
c., cubic
cc., cubic centimeter
chl., chloroform
col., colorless
cr., crystalline
d., decomposes
d-, dextrorotatory
dl-, dextro-laevorotatory
et., ethyl ether
expl., explodes
gn., green
h., hot
hex., hexagonal
i-, iso-, containing the group $(CH_3)_2CH$-
i., insoluble
ign., ignites
l-, laevorotatory
lf., leaflets
lq., liquid
m-, meta
mn., monoclinic
n-, normal
nd., needles
o-, ortho
or., orange
p-, para
pd., powder
pet., petroleum ether
pl., plates
pr., prisms
rhb., rhombic
s., soluble
s-, sec-, secondary
silv., silvery
sl., slightly
subl., sublimes
sym., symmetrical
t-, tertiary
tet., tetragonal
tri., triclinic
uns., unsymmetrical
v., very
v. s., very soluble
v. sl. s., very slightly soluble
wh., white
yel., yellow
(+), right rotation
>, greater than
<, less than
∞, infinitely

This table of the physical properties includes the organic compounds of most general interest. For the properties of other organic compounds, reference must be made to larger tables in Lange's "Handbook of Chemistry" (Handbook Publishers), "Handbook of Chemistry and Physics" (Chemical Rubber Publishing Co.), Van Nostrand's "Chemical Annual," "International Critical Tables" (McGraw-Hill), and similar works.

The **molecular weights** are based on the 1941 atomic weight values. The **densities** are given for the temperature indicated and are usually referred to water at 4°C., *e.g.*, $1.028^{95/4}$ a density of 1.028 at 95°C. referred to water at 4°C., the 4 being omitted when it is not clear whether the reference is to water at 4°C. or at the temperature indicated by the upper figure. The melting and boiling points given have been selected from available data as probably the most accurate. The **solubility** is given in grams of the substance in 100 g. of the solvent. In the case of gases, the solubility is often expressed in some manner as "5^{10} cc." which indicates that, at 10°C., 5 cc. of the gas are soluble in 100 g. of the solvent.

Name	Synonym	Formula	Formula weight	Form and color	Specific gravity	Melting point, °C.	Boiling point, °C.	Solubility in 100 parts: Water	Alcohol	Ether
Abietic acid	sylvic acid, abietinic acid	$C_{20}H_{30}O_2$	302.44	lf.	....	182	....	i.	v. s.	v. s.
Acenaphthene	naphthylene ethylene	$C_{10}H_6(CH_2)_2$	154.20	rhb./al.	$1.069^{95/95}$	95	278–9	i.	s. h.	s. chl.
Acetal	acetaldehyde diethylacetal	$CH_3CH(OC_2H_5)_2$	118.17	lq.	$0.821^{22/4}$	....	102.2	6^{25}	∞	∞
Acet-aldehyde	ethanal	CH_3CHO	44.05	col. lq.	$0.783^{18/4}$	−123.5	20.2	∞	∞	∞
-aldehyde, par-	paraldehyde	$(C_2H_4O)_3$	132.16	col. cr.	$0.994^{20/4}$	10.5–12	124.4^{752}	12^{18}	∞	∞
-aldehyde ammonia	....	$CH_3CHOHNH_2$	61.08	col. cr.	....	97	100–10 d.	v. s.	v. s.	sl. s.
-amide	ethanamide	CH_3CONH_2	59.07	col. cr.	1.159	81(69.4)	222	s.	s.	v. sl. s.
-anilide	antifebrin	$C_6H_5NHCOCH_3$	135.16	rhb./al.	1.21^{4}	113–4	305	0.5^{6}	21^{20}	7^{25}
-phenetidide (o-)	o-ethoxyacetanilide	$CH_3CONHC_6H_4OC_2H_5$	179.21	lf./al.	....	79	>250	i.	s.	
(m-)	acetyl-m-phenetidine	$CH_3CONHC_6H_4OC_2H_5$	179.21	lf./al.	....	96–7	....	sl. s.	s.	
-toluidide (o-)	N-tolylacetamide	$CH_3C_6H_4NHCOCH_3$	149.19	rhb.	1.168^{15}	110	296	0.86^{19}	s.	s.
(p-)	N-tolylacetamide	$CH_3C_6H_4NHCOCH_3$	149.19	rhb. or mn.	1.212^{15}	153	306–7	0.09^{22}	10^{25}	s.
Acetic acid	ethanoic acid, vinegar acid	CH_3CO_2H	60.05	col. lq.	$1.049^{20/4}$	16.7	118.1	∞	∞	∞
anhydride	acetyl oxide, acetic oxide	$(CH_3CO)_2O$	102.09	col. lq.	$1.082^{20/4}$	−73	139.6	12 c.	∞	∞
nitrile	methyl cyanide	CH_3CN	41.05	col. lq.	$0.783^{20/4}$	−41	81.6–2.0	∞	∞	∞
Acetone	propanone, dimethyl ketone	CH_3COCH_3	58.08	col. lq.	$0.792^{20/4}$	−94.6	56.5	∞	∞	∞
Acetonyl urea	dimethyl hydantoin	$<NHCONHCOC>(CH_3)_2$	128.13	tri./al.	....	175	subl.	s.	s.	s.
Acetophenone benzoyl hydride	methyl-phenyl ketone	$CH_3COC_6H_5$	120.14	lf.	$1.033^{15/15}$	20.5	202.3^{749}	i.	s.	s.
Acetyl-chloride	ethanoyl chloride	CH_3COCl	78.50	col. lq.	$1.105^{20/4}$	−112.0	51–2	d.	d.	∞
-phenylenediamine (-p)	amino-acetanilide (p)	$C_2H_3ONHC_6H_4NH_2$	150.18	nd./aq.	....	162	....	s. h.	v. s.	v. s.
Acetylene	ethyne; ethine	$HC:CH$	26.04	col. gas	(A) 0.906	-81.5^{891}	-84^{760}	100 cc.^{18}	600 cc.^{18}	
dichloride (cis)	1,2-dichloroethene	$CHCl:CHCl$	96.95	col. lq.	$1.291^{15/4}$	−80.5	60.3	0.35^{20}	∞	∞
(trans)	dioform	$CHCl:CHCl$	96.95	col. lq.	$1.265^{15/4}$	−50	48.4	0.63^{20}	∞	∞
Aconitic acid	equisetic acid; citridic acid	$C_3H_3(CO_2H)_3$	174.11	cr./aq.	....	192 d.	....	33^{15}	sl. s.	v. sl. s.
Acridine	....	$C_6H_4 < (CH)(N) > C_6H_4$	179.21	rhb./aq. al.	....	110–1	346	sl. s. h.	s.	s.
Acrolein ethylene aldehyde	acrylic aldehyde, propenal	$CH_2:CH.CHO$	56.06	col. lq.	$0.841^{20/4}$	−87.7	52.5	40	s.	s.
Acrylic acid	propenoic acid	$CH_2:CH.CO_2H$	72.06	col. lq.	$1.062^{16/4}$	12–13	141–2	∞	∞	
nitrile	vinyl cyanide	$CH_2:CH.CN$	53.06	col. lq.	0.811^{20}	−82	78–9	s.		
Adipic acid	hexandioc acid, adipinic acid	$(CH_2CH_2CO_2H)_2$	146.14	mn. pr.	$1.360^{25/4}$	151–3	265^{10}	1.4^{15}	v. s.	0.6^{15}
amide	....	$(CH_2CH_2CONH_2)_2$	144.17	cr. pd.	....	226–7	....	0.4^{12}		
nitrile	tetramethylene	$(CH_2CH_2CN)_2$	108.14	col. oil	$0.951^{19/19}$	1	295	v. sl. s.	s.	v. sl. s.
Adrenaline (1-) (3,4,1)	1-suprarenine	$C_6H_3(OH)_2(CHOHCH_2NHCH_3)$	183.20	col. pd.	....	d. 207–11	....	0.03^{20}	v. sl. s.	i.
Alanine (α) (dl-)	....	$CH_3CH(NH_2)CO_2H$	89.09	nd./aq.	....	295 d.	subl. >200	22^{17}	v. sl. s.	i.
Aldol acetaldol,	2-hydroxybutyraldehyde	$CH_3CH(OH)CH_2CO_2H$	88.10	col. lq.	$1.103^{20/4}$	....	83^{20}	∞	∞	s.
Alizarin	Anthraquinoic acid,	$C_6H_4(CO)_2C_6H_2(OH)_2$	240.20	red rhb.	....	289–90	430	$0\ 03^{100}$	v. s.	v. s.
Allyl alcohol	propen-1-ol-3,propenyl alcohol	$CH_2:CH.CH_2OH$	58.08	col. lq.	$0.854^{20/4}$	−129	96.6	∞	∞	∞
bromide	3-bromo-propene-1	$CH_2:CH.CH_2Br$	120.99	lq.	$1.398^{20/4}$	−119.4	$70\text{–}1^{753}$	i.	∞	∞
chloride	3-chloro-propene-1	$CH_2:CH.CH_2Cl$	76.53	col. lq.	$0.938^{20/4}$	−136.4	44.6	<0.1	∞	∞
thiocyanate (i)	mustard oil	$CH_2:CH.CH_2NCS$	99.15	col. oil	$1.013^{20/4}$	−80	152	0.2	∞	∞
thiourea	thiosinamide	$CH_2:CH.CH_2NHCSNH_2$	116.18	col. pr.	$1.219^{20/20}$	77–8	....	3^{0}	s.	v. sl. s.
Aluminum ethoxide	....	$Al(OCH_2CH_3)_3$	164.15	pd.	$1.142^{20/0}$	150–60	$200\text{–}5^{10}$	d.	i.	v. sl. s.
Amino-anthraquinone (α)	....	$C_6H_4(CO)_2C_6H_3NH_2$	223.22	red nd.	....	256	subl.	i.	s.	s.
(β)	....	$C_6H_4(CO)_2C_6H_3NH_2$	223.22	red nd.	....	302	subl.	i.	s.	i.
-azobenzene	....	$C_6H_5N:N.C_6H_4NH_2$	197.23	yel. mn.	....	126–7	225^{120}	sl. s. h.	s. h.	s.
-benzoic acid (m-)	....	$H_2N.C_6H_4CO_2H$	137.13	nd./aq.	$1.511^{4°}$	173–4	....	v. sl. s.	2^{10}	1.8^{6}
(p-)	aminodracylic acid	$H_2N.C_6H_4CO_2H$	137.13	mn. pr.	....	187–8	....	0.3^{18}	11^{10}	8.2^{6}

* By N. A. Lange, Ph.D, Handbook Publishers, Inc., Sandusky, Ohio. Abridged from table of Physical Constants of Organic Compounds in Lange's "Handbook of Chemistry."

TABLE 3-2 Physical Properties of Organic Compounds *(Continued)*

Name	Synonym	Formula	Formula weight	Form and color	Specific gravity	Melting point, °C.	Boiling point, °C.	Solubility in 100 parts: Water	Alcohol	Ether
Amino-diphenylamine (p-)	....	$H_2N.C_6H_4NH.C_6H_5$	184.23	nd./aq. al.	....	67	354	sl. s.	s.	s.
-G-acid (2-)(6-,8-), Na_2 salt	....	$C_{10}H_5(NH_2)(SO_3Na)_2$	347.28	....	....	....	....	v. sl. s.		
-mono-potassium salt	....	$C_{10}H_5(NH_2)S_2O_6HK$	341.39	....	....	....	....	12.8^{20}		
-sodium salt	....	$C_{10}H_5(NH_2)S_2O_6HNa$	325.29	....	....	....	....	2.7^{18}		
-J-acid (2-)(5-,7-)	....	$C_{10}H_5(NH_2)(SO_3H)_2$	303.30	....	....	....	....	10.0^{20}		
-mono-potassium salt	....	$C_{10}H_5(NH_2)S_2O_6HK$	341.39	....	....	....	....	3.4^{18}		
-naphthol sulfonic (1-,2-,4-)(α-)	....	$C_{10}H_5OHNH_2SO_3H\frac{1}{2}H_2O$	248.25	....	....	....	....	v. s.		
(1-,8-,4-)	....	$NH_2(OH)C_{10}H_5SO_3H$	239.24	....	....	....	....	v. sl. s.		
-phenol (o-)	2-aminophenol	$H_2N.C_6H_4.OH$	109.12	col. nd.	....	173	subl.	1.7^{0}	4.3^{0}	v. s.
(m-)	3-aminophenol	$H_2N.C_6H_4.OH$	109.12	pr.	....	122–3	....	2.6^{0}	s.	sl. s.
(p-)	p-hydroxyaniline	$H_2N.C_6H_4.OH$	109.12	lf.	....	184–6 d.	subl.	1.1^{0}	4^{0}	i. bz.
-toluene sulfonic acid (1-,2-,3-)	....	$C_6H_3(CH_3)(NH_2)SO_3H$	187.21	nd.	....	....	....	0.97^{11}		
(1-,4-,2-)	....	$C_6H_3(CH_3)(NH_2)SO_3H.H_2O$	205.23	mn.	....	d.	....	0.5^{20}	i.	
(1-,4-,3-)	....	$C_6H_3(CH_3)(NH_2)SO_3H.\frac{1}{2}H_2O$	196.22	nd.	....	....	....	0.47		
(1-,2-,5-)	....	$C_6H_3(CH_3)(NH_2)SO_3H.H_2O$	205.23	tri./aq.	....	$-H_2O$, 120	....	3^{11}	i.	
Amyl acetate (n-)	....	$CH_3CO_2CH_2(CH_2)_3CH_3$	130.18	col. lq.	$0.879^{20/20}$	−70.8	148.4^{737}	v. sl. s.	∞	∞
(i-)	common amyl acetate	$CH_3CO_2CH_2CH_2CH(CH_3)_2$	130.18	col. lq.	$0.876^{15/4}$	....	142^{757}	0.3^{15}	∞	∞
		$CH_3CO_2CH_2CH(CH_3)C_2H_5$	130.18	col. lq.	0.880^{13}	....	141–2	v. sl. s.	∞	∞
(s-)	α-Me-Bu-acetate	$CH_3CO_2CH(CH_3)CH_2C_2H_5$	130.18	col. lq.	$0\ 922^{0}$	....	133.5	sl. s.	∞	∞
(s-)	di Et-carbinol acetate	$CH_3CO_2CH(C_2H_5)_2$	130.18	col. lq.	$0.871^{20/4}$	....	133	sl. s.	∞	∞
(t-)	....	$CH_3CO_2C(CH_3)_2C_2H_5$	130.18	col. lq.	$0\ 874^{19}$	....	124.5^{749}	v. sl. s.	∞	∞
alcohol (n-) fusel oil,	pentanol-1	$CH_3(CH_2)_3CH_2OH$	88.15	col. lq.	$0.817^{20/20}$	−78 5	137.9	2.7^{22}	∞	∞
(s-,n-) methyl-propyl carbinol,	pentanol-2	$C_2H_5CH_2CH(OH)CH_3$	88.15	col. lq.	$0.810^{20/20}$	....	119.5	4^{20}	∞	∞
(prim.-,i-) isobutyl carbinol,..	2-methyl-butanol-4	$(CH_3)_2CHCH_2CH_2OH$	88.15	col. lq.	$0.813^{15/4}$	−117.2	132.0	2^{14}	∞	∞
		$(C_2H_5)_2CHOH$	88.15	col. lq.	$0.815^{25/4}$	....	115.6	5.5^{30}	∞	∞
(s-,i-)	2-methyl-butanol-3	$(CH_3)_2CHCH(OH)CH_3$	88.15	col. lq.	0.819^{19}	....	113–4	2.8^{30}	∞	∞
(t-)	2-methyl-butanol-2	$(CH_3)_2C(OH)C_2H_5$	88.15	col. lq.	$0\ 809^{20/4}$	−11.9	102	sl. s.	s.	s.
		$(CH_3)_3CCH_2OH$	88.15	cr.	....	52–3	113–4	sl. s.	∞	∞
(d-)	active amyl alcohol	$C_2H_5CH(CH_3)CH_2OH$	88.15	col. lq.	$0.816^{20/4}$	....	128	3.6^{30}	∞	∞
-amine (n-)	....	$CH_3(CH_2)_4NH_2$	87.16	col. lq.	0.766^{19}	−55	103–4	s.	s.	s.
(s-,n-)	....	$(C_3H_7)(CH_3)CHNH_2$	87.16	col. lq.	$0.749^{20/4}$	....	91–2	∞	∞	∞
(i-)	....	$(CH_3)_2CH(CH_2)_2NH_2$	87.16	col. lq.	$0.751^{18/4}$	....	95	∞	∞	∞
(t-)	....	$(C_2H_5)(CH_3)_2CNH_2$	87.16	col. lq.	$0.731^{25/4}$	−105	77–8	∞	∞	∞
....	1-NH₂-2-Me-butane	$C_2H_5CH(CH_3)CH_2NH_2$	87.16	col. lq.	0.755^{18}	....	95–6	∞	∞	∞
....	3-amino pentane	$(C_2H_5)_2CHNH_2$	87.16	col. lq.	$0.749^{20/4}$	....	90–1	∞	∞	∞
....	3-NH_2-2-Me-butane	$(CH_3)_2CHCH(CH_3)NH_2$	87.16	col. lq.	0.757^{18}	....	83–4	∞	∞	∞
aniline (i-)	....	$C_6H_5NHC_5H_{11}$	163.25	lq.	$0.928^{15/4}$	....	254.5	i.		
benzoate (i-)	....	$C_6H_5CO_2C_5H_{11}$	192.25	col. lq.	$0.992^{14/14}$	....	261^{746}	i.	∞	∞
bromide (n-)	1-bromopentane	$CH_3(CH_2)_3CH_2Br$	151.05	col. lq.	$1.218^{20/4}$	−95	129.7	i.	s.	
(i-)	4-Br-2-Me-butane	$(CH_3)_2CH(CH_2)_2Br$	151.05	col. lq.	$1.220^{17/15}$	....	120^{745}	0.02^{16}	s.	s.
(t-)	2-Br-2-Me-butane	$(CH_3)_2C(Br)C_2H_5$	151.05	lq.	$1.216^{19/0}$	....	108^{765}	i.	s.	s.
n-butyrate (n-)	....	$C_2H_5CH_2CO_2(CH_2)_4CH_3$	158.23	col. lq.	$0.871^{15/4}$	−73.2	186.4	0.05^{50}	∞	∞
(i-)	....	$C_2H_5CH_2CO_2.C_5H_{11}$	158.23	col. lq.	$0.866^{19/15}$	....	178.6	i.	∞	∞
(t-)	....	$C_3H_7CO_2C(CH_3)_2C_2H_5$	158.23	col. lq.	$0.865^{15/0}$	....	164	sl. s.	∞	∞
i-butyrate (i-)	....	$(CH_3)_2CHCO_2C_5H_{11}$	158.23	lq.	$0.876^{0/4}$	....	168.8	i.	s.	s.
chloride (n-)	1-chloropentane	$CH_3(CH_2)_3CH_2Cl$	106.60	col. lq.	$0.878^{20/4}$	−99	108.4	i.	s.	s.
(s-)	2-chloropentane	$C_2H_5CH_2CHClCH_3$	106.60	lq.	$0.870^{20/4}$	....	96.7	i.	s.	s.
(s-)	3-chloropentane	$(C_2H_5)_2CHCl$	106.60	col. lq.	0.895^{21}	....	97.3	i.	∞	∞
(i-)	4-Cl-2-Me-butane	$(CH_3)_2CH(CH_2)_2Cl$	106.60	col. lq.	$0.893^{20/4}$	....	99.7^{758}	i.	s.	∞
(s-,i-)	3-Cl-2-Me-butane	$(CH_3)CHCHClCH_3$	106.60	lq.	0.883^{0}	....	91^{753}	i.	s.	s.
(t-)	2-Cl-2-Me-butane	$(CH_3)_2CClC_2H_5$	106.60	lq.	$0.871^{20/4}$	−72.9	85.7	i.	s.	s.
....	1-Cl-2-Me-butane	$(CH_3)(C_2H_5)CHCH_2Cl$	106.60	lq.	$0.881^{17.5}$	....	98–9	i.	s.	s.
i-cyanide (i-)	iso-caproic iso-nitrile	$(CH_3)_2CH(CH_2)_2NC$	97.16	lq.	....	....	137–9	i.	s.	s.
formate (n-)	....	$HCO_2CH_2(CH_2)_3CH_3$	116.16	lq.	0.902^{0}	−73.5	132	v. sl. s.	∞	∞
(i-)	....	$HCO_2CH_2CH_2CH(CH_3)_2$	116.16	lq.	$0.882^{20/4}$	−93.5	123.5	0.3^{22}	∞	∞
iodide (n-)	1-iodopentane	$CH_3(CH_2)_3CH_2I$	198.06	lq.	$1.510^{20/4}$	−86	157.0	i.	s.	∞
(i-)	4-I-2-Me-butane	$(CH_3)_2CHCH_2CH_2I$	198.06	lq	$1.515^{18/4}$	....	147^{765}	i.	∞	∞
(s-,n-)	2-iodopentane	$C_2H_5CH_2CHICH_3$	198.06	lq.	$1.507^{17/4}$	....	144–5	i.	∞	∞
(t-)	2-I-2-Me-butane	$(CH_3)_2CIC_2H_5$	198.06	lq.	$1.471^{19/15}$	....	127^{765}	i.	∞	∞
		$C_2H_5CH(CH_3)CH_2I$	198.06	lq.	$1.524^{20/4}$	....	148	i.	∞	∞
mercaptan (n-)	pentanthiol-1	$CH_3(CH_2)_3CH_2SH$	104.21	lq.	0.857^{20}	....	126^{767}	i.	∞	∞
(n-)	pentanthiol-3	$(C_2H_5)_2CHSH$	104.21	col. lq.	....	....	105	i.	∞	∞
(i-)	2-Me-butanthiol-4	$(CH_3)_2CH(CH_2)_2SH$	104.21	lq.	$0.835^{20/4}$	....	120	i.	∞	∞
phenol (t-)(p-)	pentaphen	$C_5H_{11}.C_6H_4OH$	164.24	cr.	....	93	265–7	sl. s.	s.	s.

propionate (n-)	....	$C_2H_5CO_2(CH_2)_4CH_3$	144.21	lq.	$0.876^{15/4}$	−73.1	168.7	i.	∞	∞
(i-)	....	$C_2H_5CO_2(CH_2)_2CH(CH_3)_2$	144.21	col. lq.	$0.870^{20/4}$	....	160.2	0.1^{25}	∞	∞
(act.)	....	$C_2H_5CO_2C_5H_{11}$	144.21	col. lq.	$0.866^{20/4}$	....	58^{16}	v. sl. s.	∞	∞
salicylate (n-)	....	$HOC_6H_4CO_2C_5H_{11}$	208.25	lq.	1.065^{15}	....	265	i.	∞	∞
Amyl i-valerate (i)	....	$C_4H_9CO_2C_5H_{11}$	172.26	col. lq.	$0.858^{20/15}$	....	194	v. sl. s.	∞	∞
(t-)	....	$C_4H_9CO_2C_5H_{11}$	172.26	col. lq.	$0.861^{14/0}$	....	173–4	sl. s.	s.	s.
Amylene (n-)(α-)	pentene-1	$C_2H_5CH_2CH{:}CH_2$	70.13	lq.	0.644^{20}	....	30–1	i.	∞	∞
(i-)	2-methyl-butene-3	$(CH_3)_2CHCH{:}CH_2$	70.13	col. lq.	0.632^{15}	−135	20.5^{771}	i.	∞	∞
(α-)	2-methyl-butene-1	$(C_2H_5)(CH_3)C{:}CH_2$	70.13	col. lq.	$0.667^{0/0}$	....	$31–2^{758}$	i.	∞	∞
(-n)(β-)	pentene-2	$C_2H_5CH{:}CHCH_3$	70.13	col. lq.	$0.650^{20/4}$	−139	36.4	v. sl. s.	∞	∞
(i-)(β-)	2-methyl-butene-2	$(CH_3)_2C{:}CHCH_3$	70.13	col. lq.	$0.663^{19/4}$	−124	37–8	i.	s.	∞
Anethole (p-)	p-propenyl anisole	$CH_3CH{:}CH.C_6H_4OCH_3$	148.20	lf./al.	$0.991^{20/20}$	22.5	235.3	v. sl. s.	s.	∞
Anhydroformald-aniline	methylene aniline	$(CH_2NC_6H_5)_3$	315.40	pr./al.	....	143	185	i.	sl. s.	s.
Aniline	amino benzene, phenyl amine, cyanol	$C_6H_5NH_2$	93.12	col. oil	$1.022^{20/4}$	−6.2	184.4	3.6^{18}	∞	∞
hydrochloride	aniline salt, aniline chloride	$C_6H_5NH_2.HCl$	129.59	cr.	1.222^{4}	198	245	18^{15}	s.	i.
nitrate	....	$C_6H_5NH_2.HNO_3$	156.14	rhb.	1.356^{4}	d. 190	....	s.	s.	sl. s.
sulfate	....	$(C_6H_5NH_2)_2.H_2SO_4$	284.32	lf./al.	1.377^{4}	d.	....	5^{14}	sl. s.	i.
Anisal-acetone (p-)	MeO-benzalacetone	$CH_3OC_6H_4CH{:}CHCOCH_3$	176.22	lf./et.	....	73–4	....	i.	v. s.	v. s.
Anisic acid (p-)	....	$CH_3OC_6H_4CO_2H$	152.14	mn./aq.	1.385^{4}	184.2	275–80	0.03^{19}	v. s.	v. s.
aldehyde (p-)	....	$CH_3OC_6H_4CHO$	136.14	col. oil	$1.123^{20/4}$	2.5	247–8	v. sl. s.	∞	∞
Anisidine (o-)	2-amino-anisole	$CH_3OC_6H_4NH_2$	123.15	col. lq.	$1.098^{15/15}$	5.2	225	v. sl. s.	∞	∞
(m-)	MeO-aniline(m)	$CH_3OC_6H_4NH_2$	123.15	oil	$1.096^{20/4}$	<−12	251	v. sl. s.	s.	s.
(p-)	4-amino anisole	$CH_3OC_6H_4NH_2$	123.15	pl./aq.	$1.089^{55/55}$	57.2	243	s. h.	s.	s.
Anisole	methyl phenyl ether	$CH_3OC_6H_5$	108.13	col. lq.	$0.990^{22/4}$	−37.3	154–5	i.	s.	s.
Anthracene	paranaphthalene, anthracin green oil	$C_6H_4{:}(CH)_2{:}C_6H_4$	178.22	col. mn.	$1.25^{27/4}$	217–8	340–2	i.	1.5^{20}	
Anthramine (α)	α-amino-anthracene	$C_6H_4{:}(CH)_2{:}C_6H_3NH_2$	193.24	yel./al.	....	130±	....	i.	s.	
(β)	β-amino-anthracene	$C_6H_4{:}(CH)_2{:}C_6H_3NH_2$	193.24	yel./al.	....	238	subl.	i.	sl. s.	sl. s.
Anthranil	....	$C_6H_4{:}(NH)CO$	119.12	col. oil	$1.187^{15/4}$	<−18	d. >215	sl. s. h.	s.	s.
Anthranilic acid (o-)	....	$H_2NC_6H_4CO_2H$	137.13	col. rhb.	....	144–5	subl.	0.35^{14}	11^{10}	16^{7}
Anthrapurpurin (1-,2-,7-)	....	$C_{14}H_5O_2(OH)_3$	256.20	or. nd./al.	....	369	462	sl. s. h.	v. s. h.	sl. s.
Anthraquinone	diphenyleneketone, dihydrodiketoanthracene	$C_6H_4{:}(CO)_2{:}C_6H_4$	208.20	yel. rhb.	$1.438^{20/4}$	286	379–81	i.	0.05^{18}	v. sl. s.
disulfonate Na_2 (1-,5-)	ρ-anthraquinone disulfonate	$C_{14}H_6O_2(SO_3Na)_2.5H_2O$	502.38	yel. lf.	....	....	....	v. s.	i.	i.
(1-,8-)	x-anthraquinone disulfonate	$C_{14}H_6O_2(SO_3Na)_2.4H_2O$	484.37	yel. pr.	....	....	....	sl. s.		
(2-,6-)	....	$C_{14}H_6O_2(SO_3Na)_2.7H_2O$	538.41	col. cr.	....	....	....	3.9^{20}		
(2-,7-)	....	$C_{14}H_6O_2(SO_3Na)_2.4H_2O$	484.37	cr.	....	....	....	30.5^{20}	v. sl. s.	i.
sulfonate Na (1-)	....	$C_{14}H_7O_2SO_3Na$	310.25	yel. lf.	....	....	....	0.53^{20}	i.	i.
(2-)	....	$C_{14}H_7O_2SO_3Na$	310.25	silv. lf.	....	....	....	0.84^{25}	i.	i.
Anthrarufin (1-,5-)	....	$C_{14}H_6O_3(OH)_2$	240.20	yel. lf.	....	280	subl.	i.	sl. s.	s.
Antipyrene	1-ph-2,3-diMepyrazolone-5	$C_{11}H_{12}ON_2$	188.22	mn./aq.	$1.088^{113/4}$	113(109)	319^{174}	100^{25}	100	sl. s.
Apiole	1-allyl-2, 5-diMeO-3,4 methylenedioxybenzene	$C_{12}H_{14}O_2$	222.23	col. nd.	$1.02^{20/4}$	30	294	i	s.	s.
Arabinose (α)(d- or l-)	....	$CH_2OH(CHOH)_3CHO$	150.13	rhb. pr.	$1.585^{20/4}$	159.5	....	46^{0}	$0.5^{9°}$	i.
(dl-)	....	$CH_2OH(CHOH)_3CHO$	150.13	....	....	164.5	....	16.9^{10}		
Arachidic acid	eicosanoic acid	$CH_3(CH_2)_{18}CO_2H$	312.52	col. lf.	....	77	328	i.	s. h.	v. s.
Arsanilic acid (p-)	....	$H_2N.C_6H_4.AsO_3H_2$	217.04	nd./aq.	....	232	....	v. s. h.	v. s. h.	i.
Asparagine (l-)	....	$HO_2C.C_2H_3(NH_2).CONH_2$	132.12	rhb.	$1.543^{15/4}$	227–35	d. 235	3.1^{28}	i. c.	
Aspirin (o-)	....	$CH_3CO_2.C_6H_4.OH$	180.15	nd./aq.	....	135–6	....	1^{37}	s.	5^{20}
Atropic acid	α-phenyl acrylic acid	$C_6H_5C({:}CH_2).CO_2H$	148.15	nd./aq.	....	106–7	267 d.	0.1 c.	s.	s.
Auramine	4,4′-dimethylaminobenzophenomide	$[(CH_3)_2NC_6H_4]_2C{:}NH$	267.36	col./al.	....	136	....	i.	7^{20}	2.3^{20}
Aurine, coralline (4-,4′-)	....	$(HOC_6H_4)_2C{:}C_6H_4{:}O$	290.30	red	....	310 d.	....	i.	s.	s.
Azo-anisole (2-,2′-)	diMeO-azobenzene	$(CH_3O.C_6H_4N{:})2$	242.27	or. pr.	....	153	....	i.	s.	s.
benzene	diphenyldiimide	$C_6H_5N{:}N.C_6H_5$	182.22	or. mn.	$1.203^{20/4}$	68	297	i.	4.2^{20}	
Azoxybenzene	....	$(C_6H_5)_2N_2O$	198.22	yel. rhb.	$1.248^{20/20}$	36	d.	i.	11.4^{15}	
Barbituric acid	malonyl urea	$CO{:}(NHCO)_2{:}CH_2.2H_2O$	164.12	col./aq.	....	d. 245	....	s. h.	sl. s.	s.
Benzal acetone	Me-cinnamyl ketone	$C_6H_5CH{:}CHCOCH_3$	146.18	pl.	$1.035^{20/20}$	41–2	260–2	i.	s.	s.
Benzaldehyde	artificial almond oil	C_6H_5CHO	106.12	col. lq.	$1.046^{20/4}$	−26	179	0.3	∞	∞
Benzamide	....	$C_6H_5CONH_2$	121.13	col. pr.	1.341	130	290	1.35^{25}	17^{25}	sl. s.
Benzanilide	....	$C_6H_5CONHC_6H_5$	197.23	lf./al.	1.31^{4}	163	$117–9^{10}$	i.	4^{30}	sl. s.
Benzene	benzol, phenyl hydride, cyclohexatriene	C_6H_6	78.11	col. lq.	$0.879^{20/4}$	5.5	80.1	0.07^{22}	s.	∞
sulfinic acid	....	$C_6H_5SO_2H$	142.17	pr./aq.	....	83–4	d. >100	v. s. h.	v. s.	v. s.
sulfonic acid	....	$C_6H_5SO_3H$	158.17	col. nd.	....	65–6	d.	v. s.	v. s.	i.
sulfonic amide	benzene sulfonamide	$C_6H_5SO_2NH_2$	157.18	mn./aq.	....	156	....	0.43^{16}	v. s.	v. s.
sulfonic chloride	benzene sulfonyl chloride	$C_6H_5SO_2Cl$	176.62	cr.	$1.384^{15/15}$	14.5	251.5	i.	v. s.	s.
Benzidine (4-,4′-)	....	$NH_2.C_6H_4.C_6H_4.NH_2$	184.23	cr./aq.	....	128–9	400^{740}	1 h.	1 h.	2
disulfonic acid (2-,2′-)	....	$(.C_6H_3(NH_2)SO_3H)_2.3H_2O$	398.40	pr./aq.	....	d. >175	....	0.09^{25}	i.	i.
(3-,3′-)	....	$(.C_6H_3(NH_2)SO_3H)_2$	344.35	....	....	....	....	v. sl. s.		
Benzil	dibenzoyl	$C_6H_5CO.COC_6H_5$	210.22	pr.	1.23^{15}	95	348 d.	i.	v. s.	v. s.
Benzoic acid	....	$C_6H_5CO_2H$	122.12	mn. pr.	$1.266^{15/4}$	121.7	249.2	0.2^{17}	46^{15}	66^{15}
anhydride	....	$(C_6H_5CO)_2O$	226.22	rhb./et.	$1.199^{15/4}$	42	360	i.	s.	s.
nitrile	phenyl cyanide	C_6H_5CN	103.12	col. lq.	$1.001^{25/6}$	−12.9	190.7	1^{100}	∞	∞

TABLE 3-2 Physical Properties of Organic Compounds *(Continued)*

Name	Synonym	Formula	Formula weight	Form and color	Specific gravity	Melting point, °C.	Boiling point, °C.	Solubility in 100 parts: Water	Alcohol	Ether
Benzoin (dl-)		$C_6H_5CO.CHOHC_6H_5$	212.24	mn.	....	133–7	344^{768}	v. sl. s.	s. h.	sl. s.
Benzophenone	diphenyl ketone	$C_6H_5COC_6H_5$	182.21	col. rhb.	1.083^{54}	48.5	305.4	i.	6.5^{15}	15^{18}
Benzotrichloride	phenyl chloroform	$C_6H_5CCl_3$	195.48	col. lq.	1.380^{14}	−4.75	220.7	i.	s.	s.
Benzoyl-benzoic acid (o-)		$C_6H_5COC_6H_4CO_2H.H_2O$	244.24	tri./aq.	....	93(128)	....	sl. s.		
-chloride		C_6H_5COCl	140.57	col. lq.	$1.212^{20/4}$	−0.5	197.2	d.	d. h.	∞
-peroxide		$(C_6H_5CO)_2O_2$	242.22	rhb./et.	....	108 d.	expl.	i.	s. h.	s.
Benzyl acetate		$CH_3CO_2CH_2C_6H_5$	150.17	col. lq.	1.057^{17}	−51.5	213.5	i.	∞	∞
alcohol	phenyl carbinol	$C_6H_5CH_2OH$	108.13	col. lq.	$1.043^{20/4}$	−15.3	204.7	4^{17}	∞	∞
amine	ω-amino toluene	$C_6H_5CH_2NH_2$	107.15	lq.	$0.982^{20/4}$	....	184.5	∞	∞	∞
aniline	phenyl-benzylamine	$C_6H_5CH_2NHC_6H_5$	183.24	mn. pr.	$1.065^{25/25}$	37–8	306^{759}	i.	......	s.
benzoate		$C_6H_5CO_2CH_2C_6H_5$	212.24	nd.	$1.12^{20/4}$	21	323–4	i.	∞	∞
butyrate		$C_2H_5CH_2CO_2CH_2C_6H_5$	178.22	col. lq.	$1.016^{16/18}$	....	238–40	i.	v. s.	v. s.
chloride	ω-chlorotoluene	$C_6H_5CH_2Cl$	126.58	col. lq.	$1.100^{20/20}$	−39	179.4	i.	∞	∞
ether	dibenzyl ether	$(C_6H_5CH_2)_2O$	198.25	lq.	1.036^{16}	....	295–8	i.	s. h.	s.
formate		$HCO_2CH_2C_6H_5$	136.14	col. lq.	1.081^{23}	3.6	$202–3^{747}$	i.	s.	∞
propionate		$C_2H_5CO_2CH_2C_6H_5$	164.20	lq.	$1.036^{16/17}$	....	220–2	i.		
Berberonic acid (2-,4-,5-)		$C_5H_2N(CO_2H)_3.2H_2O$	247.16	tri.	....	243	....	v. sl. s.	sl. s. h.	i.
Biuret	allophanamide	$NH(CONH_2)_2$	103.08	nd./al.	....	192–3 d.	....	1.3^{0}	s.	
Borneol (dl-)		$C_{10}H_{17}OH$	154.24	col. cr.	$1.011^{20/4}$	210.5	subl.	v. sl. s.		
(d- or l-)		$C_{10}H_{17}OH$	154.24	col. cr.	$1.011^{20/4}$	208–9	212–3	v. sl. s.	v. s.	v. s.
(iso-)		$C_{10}H_{17}OH$	154.24	col. cr.	....	212	....	i.		
Bornyl acetate (d-)		$CH_3CO_2C_{10}H_{17}$	196.28	rhb./pet.	0.991^{15}	29	226–7	i.	s.	s.
Bromo-aniline (p-)		$BrC_6H_4NH_2$	172.03	rhb.	1.8^{20}	63–4	....	i. c.	v. s.	v. s.
-benzene	phenyl bromide	C_6H_5Br	157.02	col. lq.	$1.495^{20/4}$	−30.6	156.2	i.	s.	∞
-camphor (3-)(d-)	α-bromocamphor	$BrC_{10}H_{15}O$	231.14	cr.	$1.449^{20/4}$	77–8	274	i.	20^{28}	v. s.
-diphenyl (p-)		$BrC_6H_4.C_6H_5$	233.11	cr./al.	....	90–1	310	i.	s.	34^{25}
-naphthalene (α-)	α-naphthyl bromide	$C_{10}H_7Br$	207.07	col. oil	$1.482^{20/4}$	5–6	281.1	i.	s.	∞
(β-)	β-naphthyl bromide	$C_{10}H_7Br$	207.07	lf./al.	1.605^{0}	59	281–2	i.	6^{20}	v. s.
-phenol (o-)		BrC_6H_4OH	173.02	col. lq.	1.553^{80}	5.6	194–5	s.	s.	∞
(m-)		BrC_6H_4OH	173.02	cr.	....	32–3	236–7	....	s.	s.
(p-)		BrC_6H_4OH	173.02	tet. cr.	1.588^{80}	63.5	238	1.4^{15}	v. s.	v. s.
-styrene (ω)(1)		$C_6H_5CH:CHBr$	183.05	lq.	$1.422^{20/4}$	7	221	i.	∞	∞
(2)		$C_6H_5CH:CHBr$	183.05	lq.	$1.427^{20/4}$	−7.5	108^{26}	i.	∞	∞
-toluene (o-)	o-tolyl bromide	$CH_3.C_6H_4Br$	171.04	col. lq.	$1.422^{20/4}$	−28	181.8	i.	s.	$∞^{25}$
(m-)		$CH_3.C_6H_4Br$	171.04	col. lq.	$1.410^{20/4}$	−39.8	183.7	i.	s.	s.
(p-)		$CH_3.C_6H_4Br$	171.04	cr./al.	$1.390^{20/4}$	28.5	184–5	i.	s.	$∞^{25}$
Bromoform	tribromo-methane	$CHBr_3$	252.77	col. lq.	$2.890^{20/4}$	8–9	150.5	0.1 c.	∞	∞
Butadiene (1-,2-)	methyl-allene	$CH_3CH:C:CH_2$	54.09	lq.	....	....	18–9	i.	∞	∞
(1-,3-)	erythrene	$CH_2:CHCH:CH_2$	54.09	col. gas	$0.621^{20/4}$	−108.9	−4.41	i.	∞	∞
Butadienyl acetylene		$CH_2:(CH)_2:CH.C:CH$	78.11	col. lq.	$0.773^{20/4}$	....	83–6	i.		
Butane	diethyl	$CH_3CH_2CH_2CH_3$	58.12	col. gas	0.60^{0}	−135	−0.6	i.	s.	s.
(i-)	trimethyl-methane	$(CH_3)_2CHCH_3$	58.12	col. gas	0.60^{0}	−145	−10	i.	s.	s.
Butyl acetate (n-)		$CH_3CO_2(CH_2)_2C_2H_5$	116.16	col. lq.	0.882^{20}	−76.3	125^{740}	0.7	∞	∞
(s-)		$CH_3CO_2CH(CH_3)C_2H_5$	116.16	col. lq.	$0.865^{25/4}$	....	112^{744}	i.	∞	∞
(i-)		$CH_3CO_2CH_2CH(CH_3)_2$	116.16	col. lq.	$0.871^{20/4}$	−98.9	118	0.6^{25}	∞	∞
(tert-)		$CH_3CO_2C(CH_3)_3$	116.16	col. lq.	$0.866^{20/4}$	....	$95–6^{750}$	i.	∞	∞
alcohol (n-)	butanol-1	$C_2H_5CH_2CH_2OH$	74.12	col. lq.	$0.810^{20/4}$	−79.9	117	9^{15}	∞	∞
(s-)	butanol-2	$C_2H_5CH(OH)CH_3$	74.12	col. lq.	$0.808^{20/4}$	−114.7	99.5	12.5^{20}	∞	∞
(i-)	2-methyl-propanol-1	$(CH_3)_2CHCH_2OH$	74.12	col. lq.	$0.805^{17.5}$	−108	107–8	10^{15}	∞	∞
(tert-)	2-methyl-propanol-2	$(CH_3)_3COH$	74.12	lq.	0.779^{26}	25.5	82.9	∞	∞	∞
amine (n-)		$C_2H_5CH_2CH_2NH_2$	73.14	col. lq.	$0.739^{25/4}$	−50	77.8	∞	∞	∞
(s-)		$C_2H_5CH(NH_2)CH_3$	73.14	col. lq.	$0.724^{20/4}$	−104	66^{772}	∞	∞	∞
(i-)		$(CH_3)_2CHCH_2NH_2$	73.14	col. lq.	$0.732^{20/20}$	−85	68–9	∞	∞	∞
(t-)		$(CH_3)_3CNH_2$	73.14	col. lq.	$0.698^{18/4}$	−67.5	45.2	....	∞	
p-aminophenol (N)(n)		$C_4H_9NH.C_6H_4.OH$	165.23	....	....	71	....	i.		
(N)(i-)		$C_4H_9NH.C_6H_4.OH$	165.23	....	....	79	....	i.		
aniline (n-)		$C_4H_9NHC_6H_5$	149.23	lq.	....	....	235^{720}	i.	v. s.	v. s.
(i-)		$C_4H_9NHC_6H_5$	149.23	oil	$0.940^{20/4}$	....	231–2	0.01^{15}	v. s.	v. s.
arsonic acid (n-)		$C_4H_9AsO(OH)_2$	182.04	col. lf.	....	158–9	....	s.	s.	i.
benzoate (n-)		$C_6H_5CO_2C_4H_9$	178.22	col. oil	$1.005^{25/25}$	−22	249–50	i.	s.	s.
(i-)		$C_6H_5CO_2C_4H_9$	178.22	col. oil	$0.997^{25/25}$	....	241.5	i.	∞	∞
bromide (n-)	1-bromo-butane	$C_2H_5CH_2CH_2Br$	137.03	lq.	$1.277^{20/4}$	−112.4	101.6	0.06^{16}	∞	∞
(s-)	2-bromo-butane	$C_2H_5CH(Br)CH_3$	137.03	lq.	$1.251^{25/4}$	−112	91.3	i.		
(i-)	1-Br-2-Me-propane	$(CH_3)_2CHCH_2Br$	137.03	lq.	$1.258^{25/4}$	−118.5	91.5	0.06^{18}	∞	∞
(t-)	2-Br-2-Me-propane	$(CH_3)_3CBr$	137.03	lq.	$1.211^{20/4}$	−16.2	73.3	i.	∞	∞

butyrate (n-)(n-)	....	$C_2H_5CH_2CO_2CH_2CH_2C_2H_5$	144.21	col. lq.	$0.872^{20/20}$	....	165.7^{736}	i.	∞	∞
(n-)(i-)	....	$C_2H_5CH_2CO_2CH_2CH(CH_3)_2$	144.21	col. lq.	$0.863^{18/4}$	....	156.9	i.	∞	∞
(i-)(i-)	....	$(CH_3)_2CHCO_2CH_2CH(CH_3)_2$	144.21	col. lq.	$0.875^{0/4}$	−80.7	148–9	i.	∞	∞
caproate	....	$CH_3(CH_2)_4CO_2C_4H_9$	172.26	col. lq.	$0.882^{0/0}$	....	204.3	i.		
carbamate (i-)	....	$NH_2CO_2CH_2CH(CH_3)_2$	117.15	col. lf.	$0.956^{76/4}$	65	206–7	i.	s.	s.
cellosolve (n-)	2-BuO-ethanol-1	$C_4H_9OCH_2CH_2OH$	118.17	col. lq.	$0.903^{20/4}$	....	171.2	∞	∞	∞
chloride (n-)	1-chloro-butane	$C_2H_5CH_2CH_2Cl$	92.57	col. lq.	0.887^{20}	−123.1	77.9^{763}	0.07^{18}	∞	∞
(s-)	2-chloro-butane	$C_2H_5.CHCl.CH_3$	92.57	col. lq.	$0.871^{20/4}$	−131	67.8^{767}	i.	∞	∞
(i-)	1-Cl-2-Me-propane	$(CH_3)_2CHCH_2Cl$	92.57	col. lq.	0.884^{15}	−131.2	68.9	i.	∞	∞
(t-)	2-Cl-2-Me-propane	$(CH_3)_3CCl$	92.57	col. lq.	0.847^{15}	−26.5	51–2	i.	∞	∞
dimethylbenzene (t-)(1-,3-,5-)	....	$(CH_3)_3C.C_6H_3:(CH_3)_2$	162.26	col. lq.	....	....	$200–2^{147}$	i.		
formate (n-)	....	$HCO_2CH_2CH_2C_2H_5$	102.13	lq.	0.911^{0}	....	106.9	v. sl. s.	∞	∞
(s-)	....	$HCO_2CH(CH_3)C_2H_5$	102.13	lq.	$0.882^{20/4}$	....	97	sl. s.	∞	∞
(i-)	....	$HCO_2CH_2CH(CH_3)_2$	102.13	lq.	$0.885^{20/4}$	−95.3	98.2	1.1^{22}	∞	∞
furoate (n-)	....	$OC_4H_3CO_2C_4H_9$	168.19	col. lq.	$1.056^{20/4}$	....	$118–20^{25}$	i.	∞	∞
iodide (n-)	1-iodo-butane	$C_2H_5CH_2CH_2I$	184.03	lq.	$1.617^{20/4}$	−103.5	129.9	i.	∞	∞
(s-)	2-iodo-butane	$C_2H_5CHICH_3$	184.03	lq.	1.595^{20}	−104	118–9	i.	∞	∞
(i-)	1-iodo-2-Me-propane	$(CH_3)_2CHCH_2I$	184.03	lq.	$1.606^{20/4}$	−90.7	120	i.	∞	∞
(t-)	2-iodo-2-Me-propane	$(CH_3)_3CI$	184.03	lq.	$1.370^{19/15}$	−34	99	i.	∞	∞
lactate (n-)	....	$CH_3CH(OH)CO_2C_4H_9$	146.18	col. lq.	0.968	....	$75–6^{6}$	sl. s.	∞	∞
mercaptan (n-)	butanthiol-1	$C_2H_5CH_2CH_2SH$	90.18	col. lq.	$0.837^{25/4}$	−116	97–8	sl. s.	v. s.	v. s
(i-)	2-Me-propanthiol-1	$(CH_3)_2CHCH_2SH$	90.18	lq.	$0.836^{20/4}$	<−79	88	v. sl. s.	s.	s.
(t-)	....	$(CH_3)_3CSH$	90.18	lq	....	....	65–7			
methacrylate (n-)	....	$CH_2:C(CH_3)CO_2C_4H_9$	142.19	lq.	$0.889^{15.6}$	....	155	i		
(i-)	....	$CH_2:C(CH_3)CO_2C_4H_9$	142.19	lq.	$0.889^{15.6}$	....	155	i.		
phenol (p-)(t-)	....	$(CH_3)_3C.C_6H_4.OH$	150.21	nd./aq.	$0.908^{112/4}$	99	236–8	sl. s.	s.	s.
propionate (n-)	....	$C_2H_5CO_2C_4H_9$	130.18	col. lq.	0.883^{15}	−89.55	146	i.	∞	∞
(s-)	....	$C_2H_5CO_2C_4H_9$	130.18	col. lq.	$0.866^{20/4}$	....	132.5	i.	∞	∞
(i-)	....	$C_2H_5CO_2C_4H_9$	130.18	col. lq.	$0.888^{0/4}$	−71.4	136.8	i.	∞	∞
stearate (n-)	....	$CH_3(CH_2)_{16}CO_2C_4H_9$	340.57	col. lq.	$0.855^{25/25}$	27.5	$220–5^{25}$	0.3^{25}	s.	s.
(i-)	....	$CH_3(CH_2)_{16}CO_2C_4H_9$	340.57	wax	....	25	....	i.		
iso-thiocyanate (n-)	butyl mustard oil	$C_2H_5CH_2CH_2.N:CS$	115.19	lq.	0.956^{11}	....	165^{724}	i.	s.	s.
(i-)	iso-Bu mustard oil	$(CH_3)_2CHCH_2.N:CS$	115.19	lq.	$0.964^{14/4}$	....	162	i.	s.	s.
(s-)(d-)	....	$C_4H_9.N:CS$	115.19	lq.	$0.943^{20/4}$	....	159–63	i.	s.	s.
(t-)	....	$(CH_3)_3C.N:CS$	115.19	lq.	0.919^{10}	10.5	140^{770}	i.	s.	s.
valerate (n-)(n-)	....	$CH_3(CH_2)_3CO_2(CH_2)_3CH_3$	158.23	lq.	$0.870^{15/4}$	−93	186	v. sl. s.	∞	∞
(i-)(n-)	....	$(CH_3)_2CHCH_2CO_2(CH_2)_3CH_3$	158.23	lq.	$0.862^{25/4}$	....	168.8	i.	∞	∞
(i-)(s-)	....	$(CH_3)_2CHCH_2CO_2C_4H_9$	158.23	col. lq.	$0.848^{20/4}$	....	$163–4^{752}$	i.	∞	∞
(i-)(i-)	....	$C_4H_9CO_2C_4H_9$	158.23	col. lq.	$0.874^{0/4}$	....	168.7	i.	∞	∞
Butylene (α-)	butene-1	$C_2H_5CH:CH_2$	56.10	col. gas	0.6^{9}	−130	-5^{758}	i.	v. s.	v. s.
(β-)	butene-2	$CH_3CH:CHCH_3$	56.10	col. gas	....	−127	3^{746}			
Butyraldehyde (n-)	....	$CH_3CH_2CH_2CHO$	72.10	col. lq.	$0.817^{20/4}$	−99	75.7	4	∞	∞
(i-)	2-Me-propanol	$(CH_3)_2CHCHO$	72.10	col. lq.	$0.794^{20/4}$	−65.9	64^{757}	11^{20}	∞	∞
Butyric acid (n-)	butanoic acid	$C_2H_5CH_2CO_2H$	88.10	col. lq.	$0.964^{20/4}$	−4.7	163.5^{757}	∞	∞	∞
(i-)	2-Me-propanoic acid	$(CH_3)_2CHCO_2H$	88.10	col. lq.	$0.949^{20/4}$	−47	154.5	20^{20}	∞	∞
amide (n-)	n-butyramide	$C_2H_5CH_2CONH_2$	87.12	rhb.	1.032	115–6	216	16.3^{15}	s.	sl. s.
(i-)	iso-butyramide	$(CH_3)_2CHCONH_2$	87.12	mn. pl.	1.013	129–30	216–20	v. s.	s.	sl. s.
anhydride (n-)	....	$(C_2H_5CH_2CO)_2O$	158.19	col. lq.	$0.968^{20/20}$	−75	199.5	d.	d	∞
(i-)	....	$[(CH_3)_2CHCO]_2O$	158.19	col. lq.	$0.950^{25/4}$	−53.5	181.5^{734}	d.	d	∞
anilide (n-)	n-butyranilide	$C_3H_7CONHC_6H_5$	163.21	mn. pr.	1.134	92	189^{15}	i.	s.	s.
Caffeic acid (3-,4-)	....	$(HO)_2C_6H_3C_2H_2CO_2H$	180.15	yel./aq.	....	195–213	d.	s. h.	s.	sl. s.
Caffeine	....	$C_8H_{10}O_2N_4.H_2O$	212.21	nd./al.	1.23^{19}	237	subl.	2	2	0.3
Camphene (dl-)	....	$C_{10}H_{16}$	136.23	cr.	0.822^{78}	50	160	i.	s.	s.
(d- or l-)	....	$C_{10}H_{16}$	136.23	cr.	$0.845^{50/4}$	42.7	159.6	i.	s.	s.
Camphor (d-)	....	$C_{10}H_{16}O$	152.23	trig.	$0.999^{9/9}$	178–9	209.1^{759}	0.1	120^{12}	v. s.
Camphoric acid (d-)	....	$C_8H_{14}(CO_2H)_2$	200.23	mn.	1.186	187	....	0.6^{12}	s.	
Cantharidine	....	$C_{10}H_{12}O_4$	196.20	cr.	....	212	....	0.003		
Capric acid	decanoic acid	$CH_3(CH_2)_8CO_2H$	172.26	col. nd.	0.889^{37}	31.5	268–70	0.003	s.	s.
Caproic acid (n-)	hexanoic acid	$CH_3(CH_2)_4CO_2H$	116.16	oily lq.	$0.922^{20/4}$	−1.5	202^{761}	1.1^{20}	s.	s.
(i-)	2-Me-pentanoic-5 acid	$(CH_3)_2CH(CH_2)_2.CO_2H$	116.16	col. oil	$0.925^{20/4}$	−35	207.7	v. sl. s.	s.	s.
Caprylic acid (n-)	octanoic acid	$CH_3(CH_2)_6CO_2H$	144.21	col. lf.	$0.910^{20/4}$	16	237.5	0.07^{15}	s.	s.
Carbazole	diphenylenelimine, dibenzopyrrole	$(C_6H_4)_2NH$	167.20	lf.	....	244.8	354.8	i.	0.92^{14}	sl. s.
Carbitol	diethylene glycol mono-Et ether	$C_2H_5O(CH_2)_2O(CH_2)_2OH$	134.17	col. lq.	$0.990^{20/20}$	....	201.9	∞	v. s.	s.
Carbon disulfide	....	CS_2	76.13	col. lq.	$1.263^{20/4}$	−108.6	46.3	0.2^{0}	∞	∞
monoxide	....	CO	28.01	col. gas	$0.81^{-195/4}$	−207	−192	3.5^{0} cc.	s.	
suboxide	....	OC:C:CO	68.03	gas	1.114^{0}	−107	7^{761}	d.	....	s.
tetrabromide	tetrabromomethane	CBr_4	331.67	col. mn.	3.42	90.1(48)	189.5	0.02^{30}	s.	s.
tetrachloride	tetrachloromethane	CCl_4	153.84	col. lq.	$1.595^{20/4}$	−22.6	76.8	0.08^{20}	∞	∞
tetrafluoride	tetrafluoromethane	CF_4	88.01	gas	....	....	−128	sl. s.		

TABLE 3-2 Physical Properties of Organic Compounds *(Continued)*

Name	Formula	Formula weight	Form and color	Specific gravity	Melting point, °C.	Boiling point, °C.	Solubility in 100 parts		
							Water	Alcohol	Ether
Carbonyl sulfide	COS	60.07	col. gas	1.24^{-87}	−138.2	-50.2^{760}	80^{14} cc.	s.	s.
Carminic acid	$C_{22}H_{20}O_{13}$	492.40	red pd.	....	d. 136	....	s.	s.	v. sl. s.
Carvacrol (1-,2-,4-)	$CH_3C_6H_3(OH)CH(CH_3)_2$	150.21	col. lq.	$0.977^{20/4}$	0.5	238	v. sl. s.	∞	∞
Carvacrylamine (2-,1-,4-)	$H_2NC_6H_3(CH_3)C_3H_7$	149.23	oil	0.994^{20}	−16	241	v. sl. s.	s.	s.
Carvone (d-)	$C_{10}H_{14}O$	150.21	col. lq.	$0.961^{20/4}$	....	230^{755}	i.	∞	∞
Cellosolve	$C_2H_5O(CH_2)_2OH$	90.12	col. lq.	$0.931^{20/4}$	−70	135.1	∞	∞	∞
acetate	$CH_3CO_2CH_2CH_2OC_2H_5$	132.16	col. lq.	$0.975^{20/4}$	....	156.3	22	∞	∞
Cellulose	$(C_6H_{10}O_5)x$	162.14	amor.	1.3–1.4	....	....	i.	i.	i.
Cetyl acetate	$CH_3CO_2(CH_2)_{15}CH_3$	284.47	nd.	0.858^{20}	22–3	200^{15}	i.	v. sl. s. c.	
alcohol	$CH_3(CH_2)_{14}CH_2OH$	242.43	lf.	$0.818^{50/4}$	49–50	189.5^{15}	i.	s.	s.
Chloral	$CCl_3.CHO$	147.40	col. lq.	$1.505^{25/4}$	−57±	97.6^{768}	v. s.	∞	∞
hydrate	$CCl_3.CH(OH)_2$	165.42	mn. pr.	$1.619^{50/4}$	51.7±	d. 98	474^{17}	v. s.	s.
Chloranil	$OC:(CCl.CCl)_2:CO$	245.89	yel./bz.	....	290	subl.	i.	i. c.	i. c.
Chloretone	$Cl_3C.C(OH)(CH_3)_2$	177.47	col. cr.	....	97	167	0.8 c.	111	s.
Chloro-acetanilide (p-)	$CH_3CO_2NHC_6H_4Cl$	169.61	rhb.	1.385^{22}	175–6	....	sl. s.	s.	v. s.
-acetic acid	$ClCH_2CO_2H$	94.50	col. cr.	$1.58^{20/20}$	61.2	189.5	v. s.	s.	s.
-acetone	CH_3COCH_2Cl	92.53	col. lq.	1.162^{16}	−44.5	121	∞	∞	∞
-acetophenone (ω-)	$C_6H_5COCH_2Cl$	154.59	rhb.	1.324^{15}	58–9	245–7	0.11	v. s.	v. s.
-acetyl chloride	$ClCH_2COCl$	112.95	col. lq.	$1.498^{20/20}$	....	105	d.	d.	
-aniline (o-)	$ClC_6H_4NH_2$	127.57	lq.	$1.213^{20/4}$	0	210.5	i.	....	s.
(m-)	$ClC_6H_4NH_2$	127.57	lq.	$1.216^{20/4}$	−10.4	230^{767}	i.	....	s.
(p-)	$ClC_6H_4NH_2$	127.57	rhb.	1.427^{19}	70–1	230–1	s. h.	s.	s.
-anthraquinone (1-)	$C_6H_4(CO)_2C_6H_3Cl$	242.65	yel. nd.	....	162	subl.	i.	sl. s. h.	
(2-)	$C_6H_4(CO)_2C_6H_3Cl$	242.65	nd./al.	....	208–9	....	i.		
-benzaldehyde (o-)	ClC_6H_4CHO	140.57	nd.	1.29^{8}	11	208^{748}	v. sl. s.	v. s.	v. s.
(m-)	ClC_6H_4CHO	140.57	pr.	1.250^{15}	17–8	213–4	v. sl. s.	v. s.	v. s.
(p-)	ClC_6H_4CHO	140.57	pr.	1.196^{61}	47.8	213^{748}	s. h.	v. s.	v. s.
-benzene	C_6H_5Cl	112.56	col. lq.	$1.107^{20/4}$	−45.2	132.1	0.049^{20}	∞	∞
-benzoic acid (o-)	$ClC_6H_4CO_2H$	156.57	mn./aq.	$1.544^{25/4}$	141–2	....	0.208^{25}	s.	s.
(m-)	$ClC_6H_4CO_2H$	156.57	pr.	$1.496^{25/4}$	158	....	0.041^{25}	s.	s.
(p-)	$ClC_6H_4CO_2H$	156.57	tri.	1.541^{24}	242–3	subl.	0.008^{25}	s.	s.
-buta-1,3-diene (2-)	$CH_2:CCl.CH:CH_2$	88.54	col. lq.	$0.958^{20/20}$	....	59.4	v. sl. s.	∞	∞
(1-)	$CH_2:CH.CH:CHCl$	88.54	col. lq.	$0.965^{20/20}$	....	69	v. sl. s.	∞	∞
-buta-1,2-diene (4-)	$CH_2:C:CH.CH_2Cl$	88.54	col. lq.	$0.991^{20/20}$	....	88	d.		
-dimethylhydantoin	$—C(CH_3)_2N(Cl)CON(Cl)CO—$	197.03	....	$1.5^{20/20}$	130	....	0.21^{25}		
-dinitrobenzene (α)(1-,2-)(4-)	$ClC_6H_3(NO_2)_2$	202.56	cr./et.	....	39(36)	315 d.	i.	v. s. h.	v. s.
(α)(1-,3-)(4-)	$ClC_6H_3(NO_2)_2$	202.56	rhb./et.	1.697^{22}	53(43)	315 d.	i.	s. h.	s.
-diphenyl (o-)	$C_6H_5.C_6H_4Cl$	188.65	cr.	....	34	267–8	i.		
(m-)	$C_6H_5.C_6H_4Cl$	188.65	cr.	....	89	284–5	i.		
(p-)	$C_6H_5.C_6H_4Cl$	188.65	lf.	....	77.5	282	i.		
-hydroquinone	$ClC_6H_3(OH)_2$	144.56	mn.	....	106	263 sl. d.	v. s.	v. s.	v. s.
-naphthalene (α-)	$C_{10}H_7Cl$	162.61	col. lq.	$1.194^{20/4}$	−20	259.3	i.	s.	∞
(β-)	$C_{10}H_7Cl$	162.61	lf./al.	1.266^{16}	56–7	264^{751}	i.	v. s.	v. s.
-nitrobenzene (o-)	$ClC_6H_4NO_2$	157.56	mn. nd.	$1.305^{80/4}$	32.5	245.5^{753}	i.	s. h.	s.
(m-)	$ClC_6H_4NO_2$	157.56	yel./al.	$1.343^{50/4}$	44.4(24)	235.6	i.	v. s. h.	v. s.
(p-)	$ClC_6H_4NO_2$	157.56	mn. pr.	1.298^{91}	83–4	242^{761}	i.	v. s. h.	v. s.
-nitrotoluene (2-,4-)	$CH_3C_6H_3(NO_2)(Cl)$	171.56	cr.	1.256^{80}	38.2	240^{718}	i.		
(2-,6-)	$CH_3C_6H_3(NO_2)(Cl)$	171.56	cr.	....	37.5	238	i.		
-phenol (o-)	ClC_6H_4OH	128.56	col. lq.	$1.241^{18/15}$	7(0)	175–6	2.85^{20}	s.	s.
(m-)	ClC_6H_4OH	128.56	nd.	1.268^{25}	32–3	214	2.60^{20}	s.	s.
(p-)	ClC_6H_4OH	128.56	nd.	$1.306^{20/4}$	41–3	217	2.71^{20}	v. s.	v. s.
-propionic acid (α)(dl-)	$CH_3.CHCl.CO_2H$	108.53	col. lq.	1.306^{9}	<−20	186	∞	∞	∞
-toluene (o-)	$CH_3.C_6H_4Cl$	126.58	col. lq.	$1.082^{20/4}$	−34	159.5	i.	s.	∞
(m-)	$CH_3.C_6H_4Cl$	126.58	col. lq.	$1.072^{20/4}$	−47.8	161.6	i.	s.	∞
(p-)	$CH_3.C_6H_4Cl$	126.58	col. lq.	$1.070^{20/4}$	7.5	162.2	i.	s.	∞
Chloroform	$CHCl_3$	119.39	col. lq.	1.489^{20}	−63.5	61.2	0.82^{20}	∞	∞
Chlorophyll (α-)	$C_{55}H_{72}O_5N_4Mg$	893.48	....	....	....	d.	i.	s.	
Chloropicrin	Cl_3CNO_2	164.39	lq.	$1.651^{23/4}$	−64	112.3^{766}	0.17^{18}	s.	s.
Cholesterol	$C_{27}H_{45}OH.H_2O$	404.65	rhb./al.	1.067	149–51	subl.	0.26^{20}	1.1^{17}	18
Chrysene	$C_{18}H_{12}$	228.28	col. rhb.	....	253–4	448	i.	0.1^{16}	v. sl. s.
Chrysoidine (2-,4-)	$C_6H_5.N:N.C_6H_3(NH_2)_2$	212.25	yel. cr.	....	117.5	....	sl. s. h.	s.	s.
Chrysophanic acid	$C_{14}H_5(OH)_2(CH_3)O_2$	254.23	yel./al.	....	195	subl.	i. c.	s. h.	sl. s.
Cinchomeronic acid (3-,4-)	$C_5H_3N(CO_2H)_2$	167.12	cr./HCl	....	258–9 d.	subl. d.	v. sl. s.	sl. s.	i.
Cineole, eucalyptole	$C_{10}H_{18}O$	154.24	col. oil	0.927^{20}	1.5	176–7	1.9^{15}	∞	∞
Cinnamic acid (cis-)	$C_6H_5CH:CHCO_2H$	148.15	mn. pr.	1.284^{4}	68	125^{19}			
(trans-)	$C_6H_5CH:CHCO_2H$	148.15	mn. pr.	1.245	133	300	0.04^{18}	24^{20}	v. s.

aldehyde	$C_6H_5CH{:}CHCHO$	132.15	lq.	1.110$^{20/20}$	−7.5	252 sl. d.	v. sl. s.	s.	∞
Cinnamyl alcohol	$C_6H_5CH{:}CHCH_2OH$	134.17	nd.	1.040$^{35/35}$	33	257.5	sl. s.	v. s.	v. s.
cinnamate	$C_8H_7CO_2C_9H_9$	264.31	nd. or pr.	1.085$^{16.5}$	44	....	i.	4 c.	33
Citraconic acid (cis-)	$CH_3C(CO_2H){:}CHCO_2H$	130.10	nd.	1.617	92–3	....	360^{25}	s.	s.
Citral (α)	$C_9H_{15}CHO$	152.23	col. oil	0.890$^{17/4}$	....	229	i.	∞	∞
Citric acid	$C_3H_4(OH)(CO_2H)_3$	192.12	cr.	1.542$^{20/4}$	153	d.	207.7^{25}	76^{15}	2^{15}
Citronellal (d-)	$C_9H_{17}.CHO$	154.24	col. oil	0.855$^{17.5}$	....	204–8	v. sl. s.	∞	∞
Citronellol (d-)	$C_{10}H_{20}O$	156.26	col. oil	0.848$^{20/4}$	....	224–5	v. sl. s.	∞	∞
Coniine (d-)(2-)	$C_3H_7.C_5H_{10}N$	127.22	col. lq.	0.847^{17}	−2	166–7	1.1	v. s.	v. s.
Coumaric acid (o-)	$HOC_6H_4CH{:}CHCO_2H$	164.15	nd./aq.	....	207–8	subl.	sl. s. c.	s.	v. sl. s.
(p-)	$HOC_6H_4CH{:}CHCO_2H$	164.15	cr./aq.	....	206–7 d.	....	s. h.	v. s. h.	v. s.
Coumarin	$C_9H_6O_2$	146.14	rhb./et.	0.935$^{20/4}$	70	290–1	0.3 c.	v. s.	s.
Coumarone	C_8H_6O	118.13	oil	1.078$^{15/15}$	<−18	173–4	i.	....	s.
Creatine	$C_4H_9N_3O_2.H_2O$	149.15	mn./aq.	....	295	....	1.4^{18}	0.01^{17}	i.
Creatinine	$C_4H_7N_3O$	113.12	mn.	....	260 d.	....	8.7^{16}	1^{16}	
Creosol (3-,1-,4-)	$CH_3O.C_6H_3(CH_3)OH$	138.16	pr.	1.092$^{20/20}$	5.5	221–2^{765}	v. sl. s.	∞	∞
Cresidine (1-,2-,4-)	$CH_3(NH_2)C_6H_3.OCH_3$	137.18	nd./pet.	....	93–4	235	v. sl. s.	s.	s.
Cresol (o-)	$CH_3C_6H_4OH$	108.13	cr.	1.048$^{20/4}$	30.8	190.8	2.5	∞^{30}	∞^{30}
(m-)	$CH_3C_6H_4OH$	108.13	lq.	1.034$^{20/4}$	10.9	202.8	0.5	∞	∞
(p-)	$CH_3C_6H_4OH$	108.13	pr.	1.035$^{20/4}$	35–6	202	1.8	∞^{36}	∞^{36}
Cresyl benzoate (o-)	$C_6H_5CO_2C_6H_4CH_3$	212.24	lq.	....	....	308	i.		
(m-)	$C_6H_5CO_2C_6H_4CH_3$	212.24	cr.	....	55	314	i.		
(p-)	$C_6H_5CO_2C_6H_4CH_3$	212.24	cr.	....	71.5	316	i.		
Crotonic acid (α-)	$CH_3CH{:}CHCO_2H$	86.09	col. mn.	0.964$^{79.7}$	72	189	8.3^{15}		
acid (β-)(cis-)	$CH_3CH{:}CHCO_2H$	86.09	nd.	1.031$^{15/4}$	15.5	170–1 d.	∞^{25}	s.	
aldehyde (α)	$CH_3CH{:}CHCHO$	70.09	col. lq.	0.853$^{20/20}$	−69	102.2	18	∞	∞
Cumene	$C_6H_5CH(CH_3)_2$	120.19	col. lq.	0.862$^{20/4}$	−96.9	152.5	i.	∞	∞
Cumic acid (p-)	$(CH_3)_2CH.C_6H_4CO_2H$	164.20	tri.	1.162^{4}	116–7	subl.	0.02^{25}	s.	s.
Cumidine (p-)	$(CH_3)_2CH.C_6H_4NH_2$	135.20	lq.	0.953	<−20	225^{761}	i.		
Cyanamide	$H_2N.CN$	42.04	col. nd.	1.073$^{48/4}$	44–5	140^{19}	v. s.	v. s.	v. s.
Cyanic acid	HOCN or HNCO	43.03	gas	1.140^{0}	−80	−64^{0}	sl. s.	....	s.
Cyanoacetic acid	$CH_2(CN)CO_2H$	85.06	col. lq.	....	65–6	108$^{0.2}$	s.	s.	s.
Cyanogen	$(CN)_2$	52.04	col. gas	0.866^{17}	−34.4	−21	450^{20} cc.	2300^{20} cc.	500^{20} cc.
bromide	BrCN	105.93	nd.	2.015$^{20/4}$	52	61.3^{750}	s.	s.	s.
chloride	ClCN	61.48	gas	1.222^{0}	−6.5	12.5–13	2500^{20} cc.	v. s.	5000^{20} cc.
Cyanuric acid	$C_3H_3O_3N_3.2H_2O$	165.11	mn./aq.	1.768$^{0/4}$	>360	d.	0.27^{17}	0.1^{22}	
Cyclo-butane	$CH_2<(CH_2)_2>CH_2$	56.10	col. gas	0.703$^{0/4}$	−50	11–12^{726}	i.	v. s.	
-heptane	$CH_2<(CH_2CH_2CH_2)_2>$	98.18	oil	0.810$^{20/4}$	−12	118–20	i.		
-hexane	$CH_2<(CH_2CH_2)_2>CH_2$	84.16	col. lq.	0.779$^{20/4}$	6.5	80–1	i.	∞	∞
-hexanol	$CH_2<(CH_2CH_2)_2>CHOH$	100.16	col. nd.	0.962$^{20/4}$	23.9	160–1	3.6^{20}	s.	s.
-hexanone	$CH_2<(CH_2CH_2)_2>CO$	98.14	col. oil	0.947$^{19/4}$	−45	155–6	s.	s.	s.
-hexene	$(.CH_2.CH_2CH{:})_2$	82.14	lq.	0.810$^{20/4}$	−103.7	83.3	v. sl. s.	v. s.	v. s.
-hexyl acetate	$CH_3CO_2C_6H_{11}$	142.19	oil	0.985$^{0/4}$	....	174^{750}	i.	∞	∞
amine	$CH_2<(CH_2CH_2)_2>CHNH_2$	99.17	col. lq.	0.865$^{20/0}$	....	134	i.	s.	
bromide	$CH_2<(CH_2CH_2)_2>CHBr$	163.06	col. lq.	1.324$^{20/20}$	....	165^{714}	i.	s.	s.
chloride	$CH_2<(CH_2CH_2)_2>CHCl$	118.61	col. lq.	0.977$^{18/4}$	−43.9	142	i.	....	∞
-pentadiene (1-,3-)	$CH_2<(CH{:}CH)_2>$	66.10	col. lq.	0.805$^{19/4}$	−85	41–2	i.	∞	∞
-pentane	$CH_2<(CH_2CH_2)_2>$	70.13	col. oil	0.745$^{20/4}$	−93.3	49–50	i.		
-pentanone	$<(CH_2CH_2)_2>CO$	84.11	col. oil	0.948^{20}	−58.2	129–30	v. sl. s.		
-propane	$<CH_2CH_2CH_2>$	42.08	col. gas	0.720^{-79}	−126.6	−34^{749}	i.	s.	s.
Cymene (o-)	$CH_3.C_6H_4CH(CH_3)_2$	134.21	col. lq.	0.875$^{20/4}$	....	177	i.	s.	s.
(m-)	$CH_3.C_6H_4CH(CH_3)_2$	134.21	col. lq.	0.862^{20}	<−25	175–6	i.	s.	s.
(p-)	$CH_3.C_6H_4CH(CH_3)_2$	134.21	col. lq.	0.857$^{20/4}$	−73.5	176–7	i.	s.	s.
Cystine (l-)	$[.SCH_2CH(NH_2)CO_2H]_2$	240.29	pl.	....	d. 258–61	....	0.01^{19}	i.	
Dambose	$C_6H_6(OH)_6$	180.16	mn./aq.	1.752	253	319^{15}	2^{12}	i.	i.
Decahydronaphthalene (cis-)	$C_{10}H_{18}$	138.24	lq.	0.895$^{18/4}$	−51	193.3	i.	s.	s.
(trans-)	$C_{10}H_{18}$	138.24	lq.	0.872$^{20/4}$	−32	185.3	i.	s.	s.
Decane (n-)	$CH_3(CH_2)_8CH_3$	142.28	col. lq.	0.730^{2}	−29.7	174.0	i.	∞	∞
Decyl alcohol	$CH_3(CH_2)_8CH_2OH$	158.28	col. oil	0.830$^{20/4}$	7	232.9	i.	s.	
Dextrin	$(C_6H_{10}O_5)x$	162.14	amor.	1.038	....	....	s.	i.	i.
Diacetone alcohol	$(CH_3)_2C(OH).CH_2COCH_3$	116.16	lq.	0.931^{25}	−47	167.9	∞	∞	∞
Diamino-benzophenone (4-,4′-)	$H_2NC_6H_4COC_6H_4NH_2$	212.24	yel. nd.	....	237–9	....	sl. s. h.	s.	s.
-diphenylamine (4-,4′-)	$H_2NC_6H_4NHC_6H_4NH_2$	199.25	lf./aq.	....	158	d.	sl. s.	s.	s.
-diphenylmethane (4-,4′-)	$H_2NC_6H_4CH_2C_6H_4NH_2$	198.26	nd./aq.	....	93–4	249–53^{15}	sl. s. c.	s.	s.
-diphenylurea (4-,4′-)	$(H_2NC_6H_4NH)_2CO$	242.28	cr.	....	subl. 310	....	v. sl. s.		
Diamyl-amine (i-)	$[(CH_3)_2CHCH_2CH_2]_2NH$	157.29	col. lq.	0.767$^{21/4}$	−44	188–90	sl. s.	s.	∞
ether (n-)	$(C_2H_5CH_2CH_2CH_2)_2O$	158.28	col. lq.	0.774$^{20/4}$	−69	190	i.	∞	∞
(i-)	$[(CH_3)_2CH(CH_2)_2]_2O$	158.28	col. lq.	0.777$^{20/4}$	....	173.4	i.	∞	∞

TABLE 3-2 Physical Properties of Organic Compounds *(Continued)*

Name	Formula	Formula weight	Form and color	Specific gravity	Melting point, °C.	Boiling point, °C.	Solubility in 100 parts: Water	Alcohol	Ether
Diamyl ketone (i-)	$[(CH_3)_2CHCH_2CH_2]_2CO$	170.29	yel. oil	$0.821^{25/4}$	14.6	228	i.	s.	s.
phthalate (n-)	$C_6H_4(CO_2C_5H_{11})_2$	306.39	col. lq.	....	....	$204-6^{11}$			
(i-)	$C_6H_4(CO_2C_5H_{11})_2$	306.39	col. lq.	1.03	....	225^{40}	i.	s.	s.
tartrate (i-)	$(HOCH.CO_2C_5H_{11})_2$	290.35	lq.	$1.063^{15/4}$	....	195^{16}	i.		
Dianisidine (o-)(4-,3-)$_2$	$[NH_2(OCH_3)C_6H_3.]_2$	244.28	col. lf.	....	131.5	....	i.	s.	s.
Diazo-aminobenzene	$C_6H_5N{:}N.NHC_6H_5$	197.23	yel. lf.	....	96–8	expl.	i.	s. h.	v. s.
-aminotoluene (2-,2′-)	$C_7H_7N{:}N.NHC_7H_7$	225.28	or. cr.	....	51	....	0.05		
-methane	$CH_2{:}N_2$	42.04	gas	....	−145	−23	d.	....	s.
Dibenzothiazyl-disulfide (2-,2′-)	$(C_6H_4NSC)_2S_2$	232.46	cr.	1.50	180	d.	i.		
Dibenzoyl methane	$(C_6H_5CO)_2CH_2$	224.25	rhb./al.	....	78	$219-21^{18}$	i.	4.4^{20}	s.
Dibenzyl-amine	$(C_6H_5CH_2)_2NH$	197.27	col. oil	$1.028^{25/25}$	−26	$268-71^{250}$	i.	s.	s.
-aniline	$C_6H_5N(CH_2C_6H_5)_2$	273.36	pr./al.	....	70–1	>300	i.	v. s. h.	s.
ketone	$(C_6H_5CH_2)_2CO$	210.26	cr.	....	34–5	330.6	i.		
phthalate (o-)	$C_6H_4(CO_2CH_2C_6H_5)_2$	346.36	pr./al.	....	42–3	274^{12}	v. sl. s.	s.	s.
succinate	$(.CH_2CO_2CH_2C_6H_5)_2$	298.32	lf./al.	....	45–6	238^{14}	i.	s.	s.
Dibromo-benzene (o-)	$C_6H_4Br_2$	235.92	col. lq.	$1.956^{20/4}$	1.8	221–2	i.	s.	s.
(m-)	$C_6H_4Br_2$	235.92	col. lq.	$1.952^{20/4}$	−6.9	219^{755}	i.	s.	s.
(p-)	$C_6H_4Br_2$	235.92	pl./al.	2.261^{18}	87–8	218.6^{758}	i.	1.6	71^{25}
-diphenyl (4-,4′-)	$BrC_6H_4.C_6H_4Br$	312.02	mn. pr.	1.897	164–5	355–60	i.	v. sl. s. h.	
Dibutyl-adipate (n-)	$(.CH_2CH_2CO_2C_4H_9)_2$	258.35	col. lq.	$0.965^{20/4}$	−38	183^{14}	i.	∞	∞
(i-)	$(.CH_2CH_2CO_2C_4H_9)_2$	258.35	col. lq.	0.950^{25}	−20	278–80	i.		
-amine (n-)	$(C_2H_5CH_2CH_2)_2NH$	129.24	col. lq.	$0.768^{20/20}$	....	159^{761}	∞	∞	∞
(i-)	$[(CH_3)_2CHCH_2]_2NH$	129.24	col. lq.	$0.741^{25/4}$	−70	139–40	v. sl. s.	s.	s.
-p-aminophenol (s-)	$(C_4H_9)_2N.C_6H_4OH$	221.33	lq.	....	....	170^{10}	i.		
-aniline (n-)	$C_6H_5N(C_4H_9)_2$	205.33	lq.	....	....	262.8	i.	∞	∞
carbonate (n-)	$CO(OC_4H_9)_2$	174.23	col. lq.	$0.924^{20/4}$	....	207^{740}	i.	s.	
(i-)	$CO(OC_4H_9)_2$	174.23	col. lq.	0.919^{15}	....	190	i.		
(s-)	$CO(OC_4H_9)_2$	174.23	col. lq.	....	....	178–80			
ether (n-)	$(C_2H_5CH_2CH_2)_2O$	130.22	lq.	$0.769^{20/20}$	−98	142.4	<0.05	∞	∞
(i-)	$[(CH_3)_2CHCH_2]_2O$	130.22	lq.	0.762^{15}	....	122.5	i.	∞	∞
(s-)	$[C_2H_5(CH_3)CH]_2O$	130.22	lq.	0.756^{21}	....	121	i.	∞	∞
ketone (n-)	$(C_2H_5CH_2CH_2)_2CO$	142.23	lq.	$0.827^{18/4}$	−5.9	187.7	i.	s.	v. s.
(i-)	$[(CH_3)_2CHCH_2]_2CO$	142.23	oil	$0.805^{21/4}$	....	168.1	<0.06	∞	∞
malate (l-)(n-)	$C_2H_4O(CO_2C_4H_9)_2$	246.30	lq.	$1.038^{20/4}$	....	$170-1^{13}$	v. sl. s.		
oxalate (n-)	$(.CO_2C_4H_9)_2$	202.24	col. lq.	$0.986^{20/4}$	−29.6	245.5	i.	s.	s.
phthalate (n-)	$C_6H_4(CO_2C_4H_9)_2$	278.34	col. lq.	1.045^{21}	....	340	0.04^{25}	∞	∞
tartrate (d-)(n-)	$(CHOHCO_2C_4H_9)_2$	262.30	pr.	1.098^{15}	22–2.5	$200-3^{18}$	i.		
(d-)(i-)	$(CHOHCO_2C_4H_9)_2$	262.30	cr.	$1.031^{75/4}$	73–4	323–5	v. sl. s.		
Dichloro-acetic acid	$Cl_2CH.CO_2H$	128.95	lq.	$1.560^{25/25}$	9.7(−4)	194.4	∞	∞	∞
-acetone (αα-)	$Cl_2CHCOCH_3$	126.98	lq.	1.234^{15}	....	120	v. sl. s.	s.	s.
-aniline (2-,5-)	$Cl_2C_6H_3NH_2$	162.02	nd.	....	50	251	v. sl. s.	s.	s.
-anthraquinone (1-,3-)	$C_6H_4{:}(CO)_2{:}C_6H_2Cl_2$	277.10	yel. nd.	....	208–9	....	i.	i.	
(1-,4-)	$C_6H_4{:}(CO)_2{:}C_6H_2Cl_2$	277.10	yel. nd.	....	187.5	....	i.	v. sl. s.	v. sl. s.
(1-,5-)	$C_6H_3Cl{:}(CO)_2{:}C_6H_3Cl$	277.10	yel. nd.	....	251	....	i.	sl. s.	
(1-,6-)	$C_6H_3Cl{:}(CO)_2{:}C_6H_3Cl$	277.10	yel. nd.	....	203–4	....	i.		
(1-,8-)	$C_6H_3Cl{:}(CO)_2{:}C_6H_3Cl$	277.10	yel. nd.	....	202–3	....	i.	sl. s.	
(2-,3-)	$C_6H_4{:}(CO)_2{:}C_6H_2Cl_2$	277.10	yel. nd.	....	268–70	....	i.	sl. s.	
(2-,6-)	$C_6H_3Cl{:}(CO)_2{:}C_6H_3Cl$	277.10	yel. nd.	....	282	....	i.		
(2-,7-)	$C_6H_3Cl{:}(CO)_2{:}C_6H_3Cl$	277.10	yel. nd.	....	210–11	....	i.		
-benzene (o-)	$C_6H_4Cl_2$	147.01	col. lq.	$1.305^{20/4}$	−17.6	179	i.	∞	∞
(m-)	$C_6H_4Cl_2$	147.01	col. lq.	$1.288^{20/4}$	−24.8	172^{766}	i.	s.	s.
(p-)	$C_6H_4Cl_2$	147.01	col. mn.	1.458^{21}	53	174^{764}	i.	v. s.	v. s.
-butane (n-)(1-,4-)	$ClCH_2(CH_2)_2CH_2Cl$	127.02	lq.	....	−38.7	161–3			
-diphenyl (4-,4′-)	$ClC_6H_4.C_6H_4Cl$	223.10	pr.	$1.442^{0/4}$	148	315–9	i.	v. sl. s.	4^{25}
-ethane (1-,2-)	$ClCH_2.CH_2Cl$	98.97	col. lq.	$1.256^{20/20}$	−35.3	83.7	0.9^{0}	∞	∞
-naphthalene (β-)(1-,4-)	$C_{10}H_6Cl_2$	197.06	nd./al.	$1.300^{76/4}$	67–8	$286-7^{740}$	i.	v. sl. s.	
(γ-)(1-,5-)	$C_{10}H_6Cl_2$	197.06	lf./al.	....	107	subl.	i.	s.	s.
-nitrobenzene (2-,5-)	$Cl_2C_6H_3NO_2$	192.01	tri./al.	1.669^{22}	54.6	266	i.	v. s. h.	
-pentane (1-,5-)	$ClCH_2(CH_2)_3CH_2Cl$	141.04	col. lq.	$1.094^{25/4}$	....	180–1	i.	s.	s.
-phenol (2-,4-)	$Cl_2C_6H_3OH$	163.01	nd.	$1.383^{60/25}$	45	209–10	0.45^{20}	v. s.	v. s.
Dichloramine T (p-)	$CH_3C_6H_4SO_2NCl_2$	240.11	cr.	....	83	....	sl. s.		
Dicyandiamide	$H_2N.C({:}NH).NH.CN$	84.08	mn. pl.	1.40^{14}	207–8	d.	2.3^{13}	1.3^{13}	0.01^{13}
Diethanolamine	$HN(CH_2CH_2OH)_2$	105.14	pr.	$1.097^{20/4}$	28	270^{748}	∞	∞	v. sl. s.
Diethyl adipate	$(.CH_2CH_2CO_2C_2H_5)_2$	202.24	col. lq.	$1.009^{20/4}$	−21	$239-41^{761}$	0.43^{30}	s.	s.
-amine	$(C_2H_5)_2NH$	73.14	col. lq.	$0.712^{15/15}$	−38.9	55.5^{759}	v. s.	∞	∞
-aminophenol (m-)	$(C_2H_5)_2N.C_6H_4.OH$	165.23	rhb.	....	78	276–80	s.		

-aniline	$(C_2H_5)_2NC_6H_5$	149.23	oil	$0.934^{20/4}$	−34.4	216	1.4^{12}	s.	s.
sulfonic acid (m-)	$(C_2H_5)_2NC_6H_4SO_3H$	229.29	cr.	....	270 d.	....	s.		
carbonate	$OC(OC_2H_5)_2$	118.13	col. lq.	$0.975^{20/4}$	−43	126^{759}	i.	∞	∞
diethyl malonate	$(C_2H_5)_2C(CO_2C_2H_5)_2$	216.27	col. lq.	$0.985^{20/4}$	....	230	i.	∞	∞
Diethyl dimethyl malonate	$(CH_3)_2C(CO_2C_2H_5)_2$	188.22	col. lq.	$0.994^{25/25}$	....	196.7	i.	∞	∞
glutarate	$CH_2(CH_2CO_2C_2H_5)_2$	188.22	syrup	1.025^{21}	−24	237	0.88^{20}	v. s.	s.
ketone	$(C_2H_5)_2CO$	86.13	col. lq.	$0.816^{19/4}$	−42	101.7	4.7^{20}	∞	∞
malonate	$CH_2(CO_2C_2H_5)_2$	160.17	col. lq.	$1.055^{20/4}$	−49.8	198.9	2.08^{20}	∞	∞
-malonic acid	$(C_2H_5)_2C(CO_2H)_2$	160.17	pr./aq.	....	125	d. 170–80	65^{16}	v. s.	v. s.
-naphthylamine (α-)	$C_{10}H_7N(C_2H_5)_2$	199.28	col. oil	1.005	....	285–90	i.	∞	∞
(β-)	$C_{10}H_7N(C_2H_5)_2$	199.28	col. oil	1.026	....	318	i.	∞	∞
oxalate	$(.CO_2C_2H_5)_2$	146.14	col. lq.	$1.079^{20/4}$	−40.6	186	v. sl. s.	∞	∞
phthalate (o-)	$C_6H_4(CO_2C_2H_5)_2$	222.23	col. lq.	$1.121^{25/25}$	....	298–9	i.	∞	∞
sulfate	$O_2S(OC_2H_5)_2$	154.18	col. lq.	$1.172^{25/4}$	−25	210	i.	s.	∞
sulfide	$(C_2H_5)_2S$	90.18	col. lq.	$0.837^{20/4}$	−99.5	$92–3^{754}$	0.31^{20}	∞	∞
tartrate (d-)	$(CHOH.CO_2C_2H_5)_2$	206.19	lq.	$1.204^{20/4}$	17	280	sl. s.	∞	∞
-toluidine (o-)	$CH_3.C_6H_4.N(C_2H_5)_2$	163.25	lq.	....	....	$208–9^{755}$	i.	s.	s.
(m-)	$CH_3.C_6H_4.N(C_2H_5)_2$	163.25	lq.	....	....	231–2	i.	s.	s.
(p-)	$CH_3.C_6H_4.N(C_2H_5)_2$	163.25	lq.	$0.924^{15.5}$	....	228–9	i.		
Diethyleneglycol dinitrate	$O(CH_2CH_2ONO_2)_2$	196.12	lq.	$1.377^{25/4}$	−11.3	....	i.		
Difluorodichloromethane	F_2CCl_2	120.92	gas	1.486^{-30}	−155	−29.2	5.7 cc.[26]	s.	s.
Diglycerol	$[(HO)_2C_3H_5]_2O$	166.17	lq.	....	....	$220–30^{10}$	s. h.	....	i.
Dihydroxy-dinaphthyl (α-)	$(HO.C_{10}H_6.)_2$	286.31	pl./al.	....	300	....	i.	s.	v. s.
(-2,-2′,-1,-1′)	$(HO.C_{10}H_6.)_2$	286.31	nd./al.	....	218	subl.	i.	s.	v. s.
-diphenyl (4-,4′-)	$(HO.C_6H_4.)_2$	186.20	rhb./al.	1.25	270–2	subl.	sl. s.	v. s.	v. s.
-ethyl formal (β-)	$CH_2(OCH_2CH_2OH)_2$	136.15	lq.	1.154^{25}	−5.3	264	∞		
-naphthalene (1-,5-)	$C_{10}H_6(OH)_2$	160.16	pr./aq.	....	258–60	d.	sl. s.	s.	v. s.
(1-,8-)	$C_{10}H_6(OH)_2$	160.16	nd.	....	140	....	sl. s. h.	....	v. s.
Dimethoxy-benzene (p-)	$(CH_3O)_2C_6H_4$	138.16	lf.	$1.053^{56/55}$	56	212.6	v. sl. s.	v. s.	v. s.
-diphenylamine (4-,4′-)	$HN(C_6H_4OCH_3)_2$	229.26	cr.	....	103	....	i.		
-ethyl adipate	$(CH_2)_4(CO_2C_2H_4OCH_3)_2$	262.30	lq.	$1.075^{15.6}$	....	$145–50^{2}$	5		
Dimethyl adipate	$[(CH_2)_2CO_2CH_3]_2$	174.19	col. lq.	$1.063^{20/4}$	10–1	115^{18}	i.		
-amine	$(CH_3)_2NH$	45.08	col. lq.	$0.680^{0/4}$	−96	7.4	v. s.	s.	s.
-aminoazobenzene (p-)	$C_6H_5N:N.C_6H_4N(CH_3)_2$	225.28	yel./al.	....	116–7	d.	i.	s.	s.
-aminoethanol	$(CH_3)_2NCH_2CH_2OH$	89.14	col. lq.	$0.887^{20/4}$	....	135^{758}	∞		
-aminophenol (m-)	$(CH_3)_2NC_6H_4OH$	137.18	nd.	....	85	265–8	sl. s. h.	s.	s.
-aniline	$(CH_3)_2NC_6H_5$	121.18	yel. lq.	$0.956^{20/4}$	2.5	193	i.	s.	s.
sulfonic acid (m-)	$(CH_3)_2NC_6H_4SO_3H$	201.24	cr.	....	d. 266	....	s.		
(p-)	$(CH_3)_2NC_6H_4SO_3H.H_2O$	219.25	pr.	....	257	....	s. h.	v. sl. s.	v. sl. s.
carbonate	$OC(OCH_3)_2$	90.08	col. lq.	$1.070^{20/4}$	0.5	89–90	i.	∞	∞
ether	CH_3OCH_3	46.07	gas	....	−138.5	−23.7	3700 cc.[18]	s.	s.
-formamide	$HCON(CH_3)_2$	73.09	lq.	0.945^{25}	−58.3	152.8	∞		
fumarate	$(:CHCO_2CH_3)_2$	144.12	col. tri.	....	102	192	i.	sl. s.	sl. s.
glutarate	$(CH_2)_3(CO_2CH_3)_2$	160.17	lq.	$1.089^{15.6}$	−37	130^{50}			
glyoxime	$(CH_3.C:NOH)_2$	116.12	col. cr.	....	240–6	....	0.06^{20}	v. s.	v. s.
-naphthalene (1-,4-)	$C_{10}H_6(CH_3)_2$	156.22	lq.	$1.016^{20/4}$	<−18	264–6	i.		
(2-,3-)	$C_{10}H_6(CH_3)_2$	156.22	lf./al.	....	104	265^{767}	i.	sl. s.	
-naphthylamine (α-)	$C_{10}H_7N(CH_3)_2$	171.23	col. oil	1.042^{20}	....	274.5^{711}	i.	s.	s.
(β-)	$C_{10}H_7N(CH_3)_2$	171.23	col. cr.	$1.039^{70/70}$	46	304–5	i.	s.	s.
oxalate	$(.CO_2CH_3)_2$	118.09	col. mn.	1.148^{54}	54	163.3	6	s.	s.
phthalate (o-)	$C_6H_4(CO_2CH_3)_2$	194.18	col. lq.	$1.189^{25/25}$	....	280^{734}	0.43		
sulfate	$(CH_3O)_2SO_2$	126.13	col. oil	$1.352^{0/4}$	−26.8	188.3	v. sl. s.	∞	∞
sulfide	$(CH_3)_2S$	62.13	oil	$0.846^{21/4}$	−83.2	37.3	i.	s.	s.
tartrate (d-)	$(CHOH.CO_2CH_3)_2$	178.14	cr.	$1.328^{20/4}$	61.5	280	s.	200^{15}	
-vinyl-ethenyl carbinol	$(CH_3)_2COH.C{:}C.CH{:}CH_2$	110.15	lq.	$0.887^{20/4}$	....	150	6^{29}		
Dinaphthyl (αα-)	$C_{10}H_7.C_{10}H_7$	254.31	lf./al.	....	160	$240–4^{12}$	i.	s. h.	s.
-methane (αα′-)	$(C_{10}H_7)_2CH_2$	268.34	pr./al.	....	109	>360	i.	0.8 c.	v. s.
(β,β′-)	$(C_{10}H_7)_2CH_2$	268.34	nd./al.	....	92	....	i.	s.	
Dinitro-anisole (1-)(2-,4-)	$CH_3OC_6H_3(NO_2)_2$	198.13	col. mn.	1.341^{20}	94–5	....	sl. s. h.	1.5^{20}	
-benzene (o-)	$C_6H_4(NO_2)_2$	168.11	col. mn.	1.59^{18}	117–8	319^{774}	0.01 c.	1.9^{21}	
(m-)	$C_6H_4(NO_2)_2$	168.11	col. rhb.	$1.575^{20/4}$	89.8	300–2	0.3^{99}	3^{20}	
(p-)	$C_6H_4(NO_2)_2$	168.11	col. mn.	1.625^{18}	173–4	299^{777}	0.18^{100}	0.18^{21}	
sulfonic acid (2-,4-)(1-)	$(NO_2)_2C_6H_3SO_3H.3H_2O$	302.22	pr.	....	106–8	....	s.	s.	v. sl. s.
-benzoic acid (2-,4-)	$(NO_2)_2C_6H_3CO_2H$	212.12	cr./aq.	....	179–80	....	1.85^{25}	s.	
(3-,5-)	$(NO_2)_2C_6H_3CO_2H$	212.12	mn. pr.	....	204–5	subl.	s h.	v. s.	sl. s.
-benzophenone (4-,4′-)	$(NO_2C_6H_4)_2CO$	272.21	col. nd.	....	189	....	i.		
-diphenyl (4-,4′-)	$(NO_2C_6H_4)_2$	244.20	nd./al.	1.445	233	....	i.	1.5^{20}	
(2-,4′-)	$(NO_2C_6H_4)_2$	244.20	mn.	1.474	93.5	....	i.	v. s. h.	
-naphthalene (1-,5-)	$C_{10}H_6(NO_2)_2$	218.16	nd.	....	216	subl.	i.		
(1-,8-)	$C_{10}H_6(NO_2)_2$	218.16	rhb.	....	170–2	d.	i.	0.2 c.	

TABLE 3-2 Physical Properties of Organic Compounds *(Continued)*

Name	Formula	Formula weight	Form and color	Specific gravity	Melting point, °C.	Boiling point, °C.	Solubility in 100 parts: Water	Alcohol	Ether
Dinitro-phenol (2-,3-)	$(NO_2)_2C_6H_3OH$	184.11	yel. mn.	1.681^{20}	144–5	....	sl. s.	v. s. h.	v. s.
(2-,4-)	$(NO_2)C_6H_3OH$	184.11	yel. rhb.	1.683^{24}	114–5	subl.	0.5 c.	4^{20}	v. s. h.
(2-,6-)	$(NO_2)C_6H_3OH$	184.11	yel. rhb.	....	63–4	....	s. h.	s. h.	s.
-salicylic acid (3-,5-)	$(NO_2)_2C_6H_2(OH)CO_2H.H_2O$	246.13	pl./aq.	....	173 d.	....	s. c.	v. s.	v. s.
-stilbene (4-,4′-)	$(NO_2C_6H_4CH:)_2$	270.24	yel. lf.	....	210–6	....	i.	v. sl. s.	v. sl. s.
-toluene (2-,4-)	$(NO_2)_2C_6H_3CH_3$	182.13	nd.	1.321^{71}	70	300	0.03^{22}	1.2^{15}	9^{15}
(3-,4-)	$(NO_2)_2C_6H_3CH_3$	182.13	nd.	1.259^{111}	60–1	....	i.		
(3-,5-)	$(NO_2)_2C_6H_3CH_3$	182.13	mn. pr.	1.277^{111}	92–3	subl.	sl. s.	s. h.	s.
Dioxane	$O < (CH_2.CH_2)_2 > O$	88.10	col. lq.	$1.033^{20/4}$	9.5–10.5	101.1	∞	s.	s.
Dipentene	$C_{13}H_{10}$	136.23	col. lq.	0.865^{18}	....	178	i.		
Diphenyl	$C_6H_5.C_6H_5$	154.20	col. mn.	$0.992^{73/4}$	69–70	254.9	i.	10^{20}	6.6^{20}
-amine	$C_6H_5NHC_6H_5$	169.22	col. mn.	$1.160^{20/20}$	52.9	302	0.03^{25}	$56^{19.5}$	s.
carbonate	$O(COC_6H_5)_2$	214.21	nd./al.	1.272^{14}	80	302–6	i.	v. s.	s.
-chloroarsine	$(C_6H_5)_2AsCl$	264.57	rhb.	1.583^{40}	43–4	d. 327	0.2 d.	20	s.
-ethane	$(C_6H_5CH_2.)_2$	182.25	col. pr.	$0.978^{50/50}$	52–3	284	i.	s.	v. s.
ether	$C_6H_5OC_6H_5$	170.20	col. rhb.	1.073^{20}	27	259	v. sl. s.	s.	∞
guanidine	$(C_6H_5NH)_2C:NH$	211.26	mn./al.	....	147–8	d. >170	v. sl. s.	9^{20}	sl. s.
-methane	$(C_6H_5)_2CH_2$	168.23	col. pr.	$1.001^{26/4}$	26–7	265	i.	v. s.	v. s.
phenylenediamine (p-)	$(C_6H_5NH)_2C_6H_4$	260.32	cr.	....	152	....	i.		
succinate	$(.CH_2CO_2C_6H_5)_2$	270.27	lf./al.	....	122–3	330	i.	....	s.
sulfide	$(C_6H_5)_2S$	186.26	col. lq.	$1.119^{15/15}$	<−40	296–7	i.	s. h.	∞
sulfone	$(C_6H_5)_2SO_2$	218.26	nd./aq.	$1.248^{25/4}$	128–9	379	sl. s. h.	s. h.	
urea (uns.)	$(C_6H_5)_2NCONH_2$	212.24	rhb.	1.276	189	....	v. sl. s.	s.	s.
Diphenylene oxide	$< (C_6H_4)_2O$	168.18	lf./al.	....	86–7	287–8	i.	s. h.	v. s.
Dipropyl adipate (n-)	$(.CH_2CH_2CO_2C_3H_7)_2$	230.30	col. lq.	$0.979^{20/4}$	−20.3	$143–5^{10}$	i.	s.	s.
-amine (n-)	$(C_2H_5CH_2)_2NH$	101.19	col. lq.	$0.739^{20/4}$	−39.6	110–1	s.	∞	∞
(i-)	$[(CH_3)_2CH]_2NH$	101.19	col. lq.	0.722^{22}	−61	83.5^{743}	s.	s.	
aniline (n-)	$C_6H_5N(C_3H_7)_2$	177.28	yel. oil	0.910^{20}	....	245.4	i.	s.	s.
carbonate (n-)	$O(COCH_2C_2H_5)_2$	146.18	col. lq.	0.968^{22}	....	168.2	v. sl. s.		
ether (n-)	$(C_2H_5CH_2)_2O$	102.17	col. lq.	$0.744^{21/0}$	−122	91	sl. s.	∞	∞
(i-)	$[(CH_3)_2CH]_2O$	102.17	col. lq.	$0.725^{21/0}$	−60	69	0.2	∞	∞
ketone (n-)	$(C_2H_5CH_2)_2CO$	114.18	col. lq.	$0.822^{20/4}$	−32.6	144.2	0.43	∞	∞
(i-)	$[(CH_3)_2CH]_2CO$	114.18	col. lq.	$0.806^{20/4}$	....	123.7	v. sl. s.	∞	∞
oxalate (n-)	$(CO_2CH_2C_2H_5)_2$	174.19	col. lq.	$1.038^{0/0}$	−51.7	213.5	d. h.		
(i-)	$[CO_2CH(CH_3)_2]_2$	174.19	col. lq.	....	....	190			
Disalicylal ethylenediamine	$[HOC_6H_4CH:NCH_2.]_2$	268.30	cr.	1.34	125–6	....	0.03^{28}		
Ditolyl guanidine (o-)	$(C_7H_7NH)_2C:NH$	239.31	cr.	$1.10^{20/4}$	178–9	....	v. sl. s.	s. h.	s.
Divinyl acetylene	$(H_2C:CH.C:)_2$	78.11	lq.	$0.776^{20/4}$	....	85	i.		
Docosane (n-)	$CH_3(CH_2)_{20}CH_3$	310.59	cr.	$0.778^{44/4}$	44.5	224.5^{15}	i.	4 h.	v. s.
Dodecane (n-)	$CH_3(CH_2)_{10}CH_3$	170.33	lq.	$0.751^{20/4}$	−9.6	214.5	i.	v. s.	v. s.
Dulcitol	$CH_2OH(CHOH)_4CH_2OH$	182.17	mn.	1.466^{15}	189	$290–5^{3}$	3.2^{15}	v. sl. s.	i.
Durene (1-,2-,4-,5-)	$(CH_3)_4C_6H_2$	134.21	mn.	$0.838^{81/4}$	79–80	193–5	i.	s.	s.
Elaidic acid	$C_8H_{17}CH:CH(CH_2)_7CO_2H$	282.45	lf./al.	$0.851^{79/4}$	51–2	288^{100}	i.	v. s.	v. s.
Eosine	$C_{20}H_8O_5Br_4$	647.93	col. cr.	....	....	....	i.	s.	
Ephedrine (l-)	$C_6H_5CHOHCH(CH_3)NHCH_3$	165.23	cr./et.	....	40	255	5	500	s.
Epichlorhydrin (α-)	$C_2H_3O.CH_2Cl$	92.53	lq.	$1.183^{25/25}$	−25.6	117^{756}	<5	∞	∞
Epidichlorohydrin (α-)	$CH_2:CCl.CH_2Cl$	110.98	col. lq.	1.204^{25}	....	94	i.	∞	∞
Erythritol (dl-)	$CH_2OH(CHOH)_2CH_2OH$	122.12	tet. pr.	$1.451^{20/4}$	126	329–31	60	sl. s. c.	i.
tetranitrate	$C_4H_6(ONO_2)_4$	302.12	lf./al.	....	61	expl.	i. c.	s.	s.
Ethane	CH_3CH_3	30.07	col. gas	0.546^{-88}	−172	−88.6	4.7 $cc.^{20}$	150 cc.	
Ethanol-amine	$HOCH_2CH_2NH_2$	61.08	col. oil	1.022^{20}	10.5	171^{757}	∞	∞	1
formamide	$HCONHCH_2CH_2OH$	89.09	lq.	1.169^{25}	<−40	d.	∞		
Ether	$(CH_3CH_2)_2O$	74.12	col. lq.	$0.708^{25/4}$	−116.3	34.6	7.5^{20}	∞	
Ethyl abietate	$C_{19}H_{29}CO_2C_2H_5$	330.49	lq.	$1.020^{20/20}$	....	200^{4}	i.		
acetate	$CH_3CO_2C_2H_5$	88.10	col. lq.	$0.901^{20/4}$	−82.4	77.1	8.5^{15}	∞	∞
acetoacetate	$CH_3COCH_2CO_2C_2H_5$	130.14	col. lq.	$1.025^{20/4}$	−45	180^{755}	13^{17}	∞	∞
alcohol	CH_3CH_2OH	46.07	col. lq.	$0.789^{20/4}$	−112	78.4	∞	....	∞
-amine	$C_2H_5NH_2$	45.08	col. lq.	$0.689^{15/15}$	−80.6	16.6	∞	∞	∞
hydrochloride	$C_2H_5NH_2.HCl$	81.55	mn.	1.216	108–9	....	240^{17}	v. s.	i.
aniline	$C_6H_5NHC_2H_5$	121.18	lq.	$0.963^{20/4}$	−63.5	204	i.	∞	∞
sulfonic acid (m-)	$C_2H_5NHC_6H_4SO_3H$	201.24	nd./aq.	....	d. 294	....	2.15^{15}		
anisate (p-)	$CH_3OC_6H_4CO_2C_2H_5$	180.20	lq.	$1.103^{25/25}$	7–8	269–70	i.	s.	s.
anthranilate (o-)	$NH_2C_6H_4CO_2C_2H_5$	165.19	cr.	$1.117^{20/4}$	13	266–8	v. sl. s.	s.	s.
benzene	$C_6H_5.C_2H_5$	106.16	col. lq.	$0.867^{20/4}$	−94.4	136.2	0.01^{15}	∞	∞
benzoate	$C_6H_5CO_2C_2H_5$	150.17	col. lq.	$1.052^{15/15}$	−34.6	211–2	0.08^{20}	∞	∞
-benzyl-aniline	$C_6H_5N(C_2H_5)CH_2C_6H_5$	211.29	yel. oil	$1.034^{18.5}$	....	285^{10}	i.	18	∞

bromide	C_2H_5Br	108.98	col. lq.	$1.431^{20/4}$	−117.8	38.4	1.06^{0}	∞	∞
butyrate (n-)	$C_2H_5CH_2CO_2C_2H_5$	116.16	col. lq.	$0.879^{20/4}$	−93.3	120–1	0.68^{25}	∞	∞
(i-)	$(CH_3)_2CHCO_2C_2H_5$	116.16	col. lq.	$0.871^{20/4}$	−88.2	110–1	sl. s.	∞	∞
caprate (n-)	$CH_3(CH_2)_8CO_2C_2H_5$	200.31	lq.	0.859^{28}	−20	244.6^{758}	0.002^{20}	∞	∞
Ethyl caproate (n-)	$CH_3(CH_2)_4CO_2C_2H_5$	144.21	col. lq.	$0.873^{20/20}$	−67.5	$165–6^{736}$	i.	∞	∞
caprylate (n-)	$CH_3(CH_2)_6CO_2C_2H_5$	172.26	col. lq.	0.878^{17}	−45	$207–8^{753}$	i.	∞	∞
chloride	CH_3CH_2Cl	64.52	col. lq.	$0.917^{6/6}$	−139	13	0.45^{0}	∞	∞
chloroacetate	$ClCH_2CO_2C_2H_5$	122.55	col. lq.	$1.159^{20/4}$	−26	144	i.	∞	∞
chlorocarbonate	$ClCO_2CH_2CH_3$	108.53	col. lq.	$1.138^{20/4}$	−80.6	94–5	d.	∞	∞
cinnamate (trans-)	$C_6H_5CH:CHCO_2H$	176.21	col. lq.	$1.049^{20/4}$	12	271	i.	∞	∞
cyanoacetate	$CH_2(CN)CO_2C_2H_5$	113.11	col. lq.	$1.062^{20/4}$	−22.5	208^{753}	2^{25}	∞	∞
formate	$HCO_2CH_2CH_3$	74.08	col. lq.	$0.923^{20/4}$	−79	54^{760}	11^{18}	∞	∞
furoate (α)	$OC_4H_3CO_2C_2H_5$	140.13	lf.	$1.117^{21/4}$	34	195^{766}	i.	∞	∞
heptoate	$CH_3(CH_2)_5CO_2C_2H_5$	158.23	col. lq.	$0.872^{20/20}$	−66.1	187–8	0.029^{20}	∞	∞
hypochlorite	$ClOCH_2CH_3$	80.52	yel. lq.	$1.013^{-6/4}$	expl.	36^{752}	....	....	∞
iodide	CH_3CH_2I	155.98	col. lq.	$1.933^{20/4}$	−105	72.4	0.4^{20}	∞	∞
lactate	$CH_3CH(OH)CO_2C_2H_5$	118.13	oil	$1.030^{25/4}$	....	155	∞	∞	∞
laurate	$CH_3(CH_2)_{10}CO_2C_2H_5$	228.36	oil	$0.868^{13/4}$	−10.7	269	i.	s.	∞
mercaptan	CH_3CH_2SH	62.13	lq.	$0.839^{20/4}$	−121	36–7	1.5	s.	s.
methacrylate	$CH_2:C(CH_3)CO_2C_2H_5$	114.14	col. lq.	$0.913^{15.6}$	....	118	i.	s.	s.
naphthylamine (α-)	$C_{10}H_7NHC_2H_5$	171.23	oil	$1.060^{20/4}$	....	303^{723}	i.	s.	s.
naphthyl ether (α-)	$C_{10}H_7OC_2H_5$	172.22	cr.	$1.061^{20/20}$	5.5	276.4	i.	s.	s.
nitrate	$C_2H_5ONO_2$	91.07	col. lq.	$1.100^{25/4}$	−102	87–8	1.3^{55}	∞	∞
nitrite	C_2H_5ONO	75.07	lq.	$0.900^{15.5}$	....	17	v. sl. s.	∞	∞
oleate	$C_{17}H_{33}CO_2C_2H_5$	310.50	oil	0.867^{25}	<−15	$216–8^{15}$	i.	∞	∞
palmitate	$CH_3(CH_2)_{14}CO_2C_2H_5$	284.47	col. nd.	$0.858^{25/4}$	24–5	191^{10}	i.	s.	s.
pelargonate	$CH_3(CH_2)_7CO_2C_2H_5$	186.29	col. lq.	$0.866^{17.5}$	−44.5	$227–8^{757}$	i.	∞	∞
propionate	$CH_3CH_2CO_2C_2H_5$	102.13	col. lq.	$0.891^{20/4}$	−72.6	99.1	2.4^{20}	∞	∞
salicylate (o-)	$HOC_6H_4CO_2C_2H_5$	166.17	col. lq.	$1.136^{15/4}$	1.3	233–4	i.	∞	∞
stearate	$CH_3(CH_2)_{16}CO_2C_2H_5$	312.52	col. cr.	$0.848^{36.3}$	33.4(31)	201^{10}	i.	s.	s.
toluate (o-)	$CH_3.C_6H_4CO_2C_2H_5$	164.20	lq.	$1.032^{25/25}$	<−10	227	i.	∞	∞
(m-)	$CH_3.C_6H_4CO_2C_2H_5$	164.20	lq.	$1.030^{20/20}$	....	231^{750}	i.	∞	∞
toluene sulfonate (p-)	$CH_3.C_6H_4SO_3C_2H_5$	200.25	pr./al.	$1.166^{48/4}$	33–4	221.3	i.	s.	s.
toluidine (o-)	$CH_3.C_6H_4NHC_2H_5$	135.20	lq.	$0.948^{25/4}$	<−15	215–6	i.		
(p-)	$CH_3.C_6H_4NHC_2H_5$	135.20	lq.	$0.942^{25/4}$	....	217	i.		
urea	$C_2H_5NH.CO.NH_2$	88.11	nd.	1.213^{18}	92	....	v. s.	80	i.
valerate (n-)	$CH_3(CH_2)_3CO_2C_2H_5$	130.18	col. lq.	0.877^{20}	−91.2	145.5	0.24^{25}	∞	∞
(i-)	$(CH_3)_2CH(CH_2)_2CO_2C_2H_5$	130.18	col. lq.	$0.867^{20/4}$	−99.3	135	0.17^{20}	∞	∞
Ethylal	$CH_2(OC_2H_5)_2$	104.15	lq.	$0.824^{25/4}$	−66.5	89	9^{18}	∞	∞
Ethylene	$H_2C:CH_2$	28.05	col. gas	$0.57^{-102/4}$	−169	−103.9	26 cc.^{0}	360 cc.	s.
bromide	$BrCH_2.CH_2Br$	187.88	col. lq.	$2.180^{20/4}$	10	131.5	0.43^{30}	∞	∞
bromohydrin	$BrCH_2.CH_2OH$	124.98	col. lq.	$1.772^{20/4}$	....	150.3	sl. s.	s.	
chlorobromide	$ClCH_2.CH_2Br$	143.43	lq.	1.689^{19}	−16.6	106.7	0.69^{30}		
chlorohydrin	$ClCH_2.CH_2OH$	80.52	col. lq.	$1.213^{20/4}$	−69	128.8	∞	∞	∞
diamine	$H_2NCH_2.CH_2NH_2$	60.10	col. lq.	$0.900^{20/20}$	8.5	117.2	∞	∞	0.3
oxide	$<(CH_2)_2>O$	44.05	lq.	$0.887^{7/4}$	−111.3	13.5^{747}	∞	∞	v. s.
Ethylidene diacetate	$CH_3CH(O_2CCH_3)_2$	146.14	col. lq.	1.061^{12}	18.85	168^{740}	sl. s.	∞	
Eugenol (1-,4-,3-)	$C_3H_5.C_6H_3(OH)OCH_3$	164.20	oil	$1.070^{15/15}$	10.3	253.5	v. sl. s.	∞	∞
(i-)(1-,3-,4-)	$C_3H_5.C_6H_3(OCH_3)OH$	164.20	oil	$1.091^{15/15}$	−10	267.5	v. sl. s.	∞	∞
Fenchyl alcohol (dl-)	$C_{10}H_{17}OH$	154.24	col. cr.	0.935^{40}	35	201	sl. s.		
(d-)(α-)	$C_{10}H_{17}OH$	154.24	col. pr.	$0.964^{20/4}$	45–7	201–2	sl. s.	s.	s.
(i-)(l-)	$C_{10}H_{17}OH$	154.24	col. cr.	0.961	61–2	201–2	i.		
Ferric dimethyl-dithiocarbamate	$Fe[SSCN(CH_3)_2]_3$	416.47	cr.	....	d. 100–30	ign. >150	v. sl. s.		
Fluorene	$(C_6H_4)_2>CH_2$	166.21	cr./al.	$1.203^{0/4}$	115–6	293–5	i.	s. h.	s.
Fluorescein	$C_{20}H_{12}O_5$	332.30	yel. red	....	d. >290	....	v. sl. s. h.	s. h.	
Fluoro-dichloromethane	$FCHCl_2$	102.93	gas	1.426^{0}	−127	14.5	i.	s.	s.
-trichloromethane	Cl_3CF	137.38	col. lq.	$1.494^{17.2}$	....	24.9	i.	∞	∞
Formaldehyde	$HCHO$	30.03	gas	0.815^{-20}	−92	−21	v. s.	v. s.	v. s.
(m-)	$(CH_2O)_3$	90.08	wh.	1.17^{65}	64	114.5^{759}	21^{25}	s.	s.
(p-)	$(CH_2O)_x.xH_2O$	(30.03)	amor.	....	150–60	subl.	$20–30^{18}$	i.	i.
Formamide	$HCONH_2$	45.04	lq.	$1.139^{20/4}$	2	193	∞	∞	v. sl. s.
Formanilide	$HCONHC_6H_5$	121.13	mn.	$1.147^{15/15}$	47	216^{120}	sl. s.	v. s.	s.
Formic acid	HCO_2H	46.03	col. lq.	$1.220^{20/4}$	8.6	100.8	∞	∞	
Fructose	$CH_2OH(CHOH)_3COCH_2OH$	180.16	nd./aq.	$1.669^{17.5}$	95–105	....	v. s.	8^{18}	
Fuchsin	$C_{20}H_{19}N_3HCl$	337.84	red	1.22	d. >200	....	0.3	s.	i.
Fulminic acid	$C:NOH$	43.03							
Fumaric acid (trans-)	$HO_2CCH:CHCO_2H$	116.07	col. pr.	$1.635^{20/4}$	286–7	290	0.7^{17}	5.8^{30}	0.7^{25}
Furfural	$C_4H_3O.CHO$	96.08	lq.	$1.159^{20/4}$	−38.7	161.7^{760}	9.1^{13}	∞	∞
Furfuran	C_4H_4O	68.07	col. lq.	$0.937^{20/4}$	....	$31–2^{756}$	i.	s.	s.

TABLE 3-2 Physical Properties of Organic Compounds *(Continued)*

Name	Formula	Formula weight	Form and color	Specific gravity	Melting point, °C.	Boiling point, °C.	Solubility in 100 parts: Water	Alcohol	Ether
Furfuryl acetate	$CH_3CO_2CH_2C_4H_3O$	140.13	col. oil	$1.118^{20/4}$	....	175–7	i.	s.	s.
alcohol	$C_4H_3O.CH_2OH$	98.10	oil	$1.129^{25/4}$	....	169.5^{752}	∞	s.	s.
butyrate	$C_3H_7CO_2CH_2.C_4H_3O$	168.19	col. lq.	$1.053^{20/4}$	....	212–3	v. sl. s.	s.	∞
propionate	$C_2H_5CO_2CH_2.C_4H_3O$	154.16	col. lq.	$1.109^{20/4}$	....	195–6	v. sl. s.	s.	∞
Furoic acid	$C_4H_3O.CO_2H$	112.08	mn. pr.	....	133–4	230–2	3.6^{15}	s.	s.
G-acid, K salt (2-)(6-,8-)	$HOC_{10}H_5(SO_3K)_2$	380.46	cr.	....	....	....	8^{25}		
Na salt (2-)(6-,8-)	$HOC_{10}H_5(SO_3Na)_2$	348.26	cr.	....	....	....	34^{20}		
Galactose (d-)(α-)	$C_5H_{11}O_5.CHO$	180.16	pr.	....	165.5	....	10.3^{0}	0.6^{40}	
Gallic acid (3-,4-,5-)	$(HO)_3C_6H_2CO_2H.H_2O$	188.13	mn./aq.	$1.694^{4/4}$	d. 220	....	1^{13}	28^{15}	2.5^{15}
Gamma acid (2-,8-,6-)	$C_{10}H_5(NH_2)(OH)SO_3H$	239.24	cr.						
Geraniol	$C_9H_{15}CH_2OH$	154.24	col. lq.	0.883^{15}	<−15	230	i.	∞	∞
Glucose (d-)(α-)	$C_5H_{11}O_5.CHO$	180.16	rhb.	1.544^{25}	146	....	$82^{17.5}$	sl. s.	i.
(d-)(β-)	$C_6H_{12}O_6.H_2O$	198.17	cr.	$1.562^{18/4}$	150	....	154^{15}		
Glucuronic acid	$CHO(CHOH)_4CO_2H$	194.14	cr.	....	154	d.	v. s.		
Glutam(in)ic acid (dl-)	$[.CHNH_2(CH_2)_2.](CO_2H)_2$	147.13	cr./aq.	1.460	199 d.	....	1.5^{20}	v. sl. s.	v. sl. s.
Glutaric acid	$CH_2(CH_2CO_2H)_2$	132.11	col. cr.	1.429^{15}	97.5	200^{20}	63.9^{20}	v. s.	v. s.
Glycerol	$CH_2OH.CHOH.CH_2OH$	92.09	col. lq.	$1.260^{50/4}$	17.9	290	∞	∞	i.
acetate (mono-)	$C_5H_{10}O_4$	134.13	col. oil	$1.20^{20/4}$	....	158^{165}	v. s.	v. s.	sl. s.
(di-)	$(CH_3CO_2)_2C_3H_5OH$	176.17	col. lq.	$1.178^{15/15}$	40	$175–6^{40}$	s.	s.	sl. s.
nitrate (mono-) (α-)	$CH_2OH.CHOH.CH_2NO_3$	137.09	col. pr.	1.40^{15}	58–9	155–60	70^{15}	v. s.	v. sl. s.
(β-)	$CH_2OH.CHNO_3.CH_2OH$	137.09	lf.	1.40^{15}	54	155–60	....	v. s.	sl. s.
dinitrate (1-,3-)	$CHOH(CH_2ONO_2)_2$	182.09	oil	1.47^{15}	<−30	$146–8^{15}$	....	v. s.	v. s.
Glyceryl triacetate	$(CH_3CO_2)_3C_3H_5$	218.20	col. lq.	$1.161^{17/4}$	−78	258–9	7.17^{15}	∞	∞
tribenzoate	$(C_6H_5CO_2)_3C_3H_5$	404.40	nd.	1.228^{12}	75–6	d.	i.	s. h.	s.
tributyrate	$(C_2H_5CH_2CO_2)_3C_3H_5$	302.36	col. lq.	$1.032^{20/4}$	<−75	305–9	i.	s.	s.
tricaprate	$[CH_3(CH_2)_8CO_2]_3C_3H_5$	554.83	col. cr.	$0.921^{40/4}$	31(25)	....	i.	s. h.	v. s.
tricaproate	$[CH_3(CH_2)_4CO_2]_3C_3H_5$	386.51	col. lq.	$0.987^{20/4}$	−25	....	i.	s.	s.
tricaprylate	$[CH_3(CH_2)_6CO_2]_3C_3H_5$	470.67	col. lq.	$0.954^{20/4}$	8.3(−21)	....	i.	s.	s.
trilaurate	$[CH_3(CH_2)_{10}CO_2]_3C_3H_5$	638.98	col. nd.	$0.894^{60/4}$	45–6	....	i.	sl. s. c.	v. s.
trimyristate	$[CH_3(CH_2)_{12}CO_2]_3C_3H_5$	723.14	lf.	$0.885^{60/6}$	56.5	....	i.	s.	
trinitrate	$CH_2NO_3.CHNO_3.CH_2NO_3$	227.09	yel. oil	1.601^{15}	13.3(2)	160^{15}	0.18^{20}	50^{20}	∞
trinitrite	$CH_2NO_2.CHNO_2.CH_2NO_2$	179.09	yel. lq.	$1.291^{10/16}$	....	150 sl. d.	d.	d.	s.
trioleate	$(C_{17}H_{33}CO_2)_3C_3H_5$	885.40	col. oil	0.915^{15}	−4	240^{18}	i.	sl. s.	v. s.
tripalmitate	$[CH_3(CH_2)_{14}CO_2]_3C_3H_5$	807.29	col. nd.	$0.866^{80/4}$	65.1	$310–20^{0.1}$	i.	0.004^{21}	v. s.
tristearate	$[CH_3(CH_2)_{16}CO_2]_3C_3H_5$	891.45	col. pr.	$0.862^{80/4}$	70.8(55)	....	i.	s. h.	s. h.
Glycide	$C_2H_3O.CH_2OH$	74.08	col. lq.	$1.114^{16/16}$	....	166 sl. d.	∞	∞	∞
Glycine, Glycocoll	$NH_2CH_2.CO_2H$	75.07	mn.	1.161	232–6 d.	....	23 c.	0.1 c.	i.
Glycol	$CH_2OH.CH_2OH$	62.07	col. lq.	$1.113^{19/4}$	−15.6	197.4	∞	∞	1.0
diacetate	$(CH_3CO_2CH_2.)_2$	146.14	col. lq.	$1.109^{14/4}$	−31	190.5	14.3^{22}	∞	∞
dibenzoate	$(C_6H_5CO_2CH_2.)_2$	270.27	rhb./et.	....	73–4	>360	i.	....	s.
dibutyrate	$(C_3H_7CO_2CH_2.)_2$	202.24	col. lq.	1.024^{0}	....	240	i.	v. s.	v. s.
dicaprylate	$(C_7H_{15}CO_2CH_2.)_2$	314.45	lq.	....	22	....	i.		
diformate	$(HCO_2CH_2.)_2$	118.09	lq.	....	....	174	v. sl. s.		
dilaurate	$(C_{11}H_{23}CO_2CH_2.)_2$	426.66	amor.	....	52–4	188^{20}	i.	v. s.	v. s.
dinitrate	$(O_2NO.CH_2.)_2$	152.07	yel. lq.	$1.482^{21/2}$	−20	expl. 114	0.92^{25}	s.	∞
dinitrite	$(ONO.CH_2.)_2$	120.07	lq.	1.216^{0}	<−15	96–8	i.	s. d.	s.
dipalmitate	$(C_{15}H_{31}CO_2CH_2.)_2$	538.87	nd.	....	71–2	$260^{0.1}$	i.	s.	s.
dipropionate	$(C_2H_5CO_2CH_2.)_2$	174.19	lq.	1.045^{25}	....	211–2	sl. s.	∞	∞
ether	$(HO.CH_2CH_2)_2O$	106.12	lq.	$1.118^{20/20}$	−10.5	244.8	∞	∞	i.
formal	$< O.CH_2CH_2OCH_2 >$	74.08	lq.	$1.060^{20/4}$	....	75–6	∞		
formate (mono-)	$HCO_2CH_2CH_2OH$	90.08	lq.	$1.199^{15/4}$	....	180	∞		
Glycolic acid	$HOCH_2CO_2H$	76.05	nd./aq.	....	79(63)	d.	∞	90^{25}	v. s.
Guaiacol (o-)	$CH_3O.C_6H_4OH$	124.13	pr.	$1.140^{15/15}$	28.3	205	1.7^{15}	v. s.	v. s.
Guanidine	$NH:C(NH_2)_2$	59.07	col. cr.	....	50	....	v. s.	s.	
H-acid, Na salt (1-,8-,3-,6-)	$C_{10}H_8O_7NS_2Na.1\frac{1}{2}H_2O$	368.31	cr.	....	....	....	0.17^{20}		
Heptacosane (n-)	$CH_3(CH_2)_{25}CH_3$	380.72	col. cr.	$0.780^{60/4}$	59.5	270^{15}	i.		
Heptane (n-)	$CH_3(CH_2)_5CH_3$	100.20	col. lq.	$0.684^{20/4}$	−90.6	98.4^{760}	0.005^{15}	sl. s.	∞
(i-)	$(CH_3)_2CH(CH_2)_3CH_3$	100.20	col. lq.	$0.679^{20/4}$	−118.2	90.0	i.	s.	∞
	$C_3H_7.CH(CH_3).C_2H_5$	100.20	col. lq.	$0.687^{20/4}$	−119.4	91.8	i.	s.	∞
	$(CH_3)_3C.CH_2.C_2H_5$	100.20	col. lq.	$0.674^{20/4}$	−125	79.1	i.	s.	∞
	$[(CH_3)_2CH]_2CH_2$	100.20	col. lq.	$0.675^{20/4}$	−119.4	80.8	i.	s.	∞
	$(CH_3)_2C(C_2H_5)_2$	100.20	col. lq.	$0.693^{20/4}$	−135.0	86.0	i.	s.	∞
	$(C_2H_5)_3CH$	100.20	col. lq.	$0.698^{20/4}$	−118.7	93.5	i.	s.	∞
	$(CH_3)_3C.CH(CH_3)_2$	100.20	col. lq.	$0.690^{20/4}$	−25	80.8	i.	s.	∞
Heptoic acid	$CH_3(CH_2)_5CO_2H$	130.18	col. lq.	0.918^{20}	−10	221–2	0.25^{15}	s.	s.
aldehyde	$CH_3(CH_2)_5CHO$	114.18	col. lq.	$0.850^{20/4}$	−42	155	0.02^{20}	∞	∞

Heptyl acetate (n-)	$CH_3CO_2CH_2(CH_2)_5CH_3$	158.24	col. lq.	$0.874^{16/16}$	....	191.5^{759}	i.	s.	s.
alcohol (n-)	$CH_3(CH_2)_5CH_2OH$	116.20	col. lq.	$0.824^{20/4}$	34.6	175^{756}	0.18^{25}	∞	∞
	$[(CH_3)_2CH]_2CHOH$	116.20	col. lq.	$0.829^{20/4}$	....	140	v. sl. s.	∞	∞
	$(C_2H_5.CH_2.CH_2)_2CHOH$	116.20	lq.	$0.820^{20/4}$	−37	156	i.	s.	s.
mercaptan	$CH_3CH(SH).C_5H_{11}$	132.26	lq.	0.835^{20}	....	$174–5^{765}$	i.		
Hexachloro-benzene	C_6Cl_6	284.80	mn.	2.044^{24}	228–31	309^{742}	i.	v. sl. s. h.	s. h.
-ethane	$CCl_3.CCl_3$	236.76	rhb.	$2.091^{20/4}$	186–7	186^{777}	0.005^{22}	v. s.	v. s.
Hexacosane (n-)	$CH_3(CH_2)_{24}CH_3$	366.69	cr.	$0.779^{57/4}$	56.6	262^{15}	i.	v. sl. s.	
Hexadecane (n-)	$CH_3(CH_2)_{14}CH_3$	226.43	lf.	$0.774^{20/4}$	18.5	287.5	i.	∞	∞
Hexaethylbenzene	$C_6(C_2H_5)_6$	246.42	pr./al.	$0.831^{130/4}$	130	298.3	i.	0.75^{25}	8^{25}
Hexamethylbenzene	$C_6(CH_3)_6$	162.26	pl./al.	....	166	265	i.	0.2^{0}	v. s.
Hexamethylene-diamine	$NH_2(CH_2)_6NH_2$	116.20	lf.	....	42	204–5	v. s.	s.	
-diisocyanate	$OCN(CH_2)_6NCO$	168.19	lq.	1.04^{28}	....	$143–4^{20}$	d.	d.	
-glycol	$HO(CH_2)_6OH$	118.17	nd./aq.	....	42	250	s.	s.	sl. s. h.
tetramine	$(CH_2)_6N_4$	140.19	col. rhb.	....	subl.	....	81^{12}	3	v. sl. s.
Hexane (n-)	$CH_3(CH_2)_4CH_3$	86.17	col. lq.	$0.659^{20/4}$	−94	69	0.014^{15}	50^{33}	∞
(i-)	$(CH_3)_2CH(CH_2)_2CH_3$	86.17	lq.	$0.654^{20/4}$	−153.7	60.2	i.	....	s.
(neo-)	$(CH_3)_3C.C_2H_5$	86.17	lq.	$0.649^{20/20}$	−98.2	49.7	i.	....	s.
	$(CH_3)_2CH.CH(CH_3)_2$	86.17	lq.	$0.662^{20/4}$	−129.8	58.0^{760}	i.	....	s.
	$(C_2H_5)_2CHCH_3$	86.17	lq.	$0.664^{20/4}$	−118	63.2	i.	....	s.
Hexyl acetate (n-)	$CH_3CO_2(CH_2)_5CH_3$	144.21	col. lq.	$0.890^{0/0}$	....	169.2	i.	v. s.	v. s.
alcohol (n-)	$CH_3(CH_2)_4CH_2OH$	102.17	col. lq.	$0.820^{20/20}$	−51.6	157.2	0.6^{20}	∞	∞
	$(CH_3)_2CH.C(CH_3)_2OH$	102.17	lq.	$0.821^{20/0}$	−14	120–1	v. sl. s.	∞	∞
	$(CH_3)_2COH.CH_2C_2H_5$	102.17	lq.	$0.809^{20/4}$	−107	123^{762}	v. sl. s.	∞	∞
formate (n-)	$HCO_2CH_2(CH_2)_4CH_3$	130.18	lq.	0.898^{0}	....	153.6	....	∞	∞
resorcinol (2-,4-)	$CH_3(CH_2)_5C_6H_3(OH)_2$	194.26	col. nd.	....	68–70	179^{7}	0.05	v. s.	s.
Hippuric acid	$C_6H_5CONHCH_2CO_2H$	179.17	rhb.	$1.371^{20/4}$	187–8	d.	0.4^{20}	s. h.	0.25^{18}
Histidine (l-)	$C_6H_9O_2N_3$	155.16	lf./aq.	....	d. 287	....	s.	v. sl. s.	i.
Homophthalic acid (o-)	$HO_2C.C_6H_4.CH_2CO_2H$	180.15	cr./aq.	....	175–80	....	s. h.	v. s.	sl. s.
Hydracrylic acid	$HOCH_2CH_2CO_2H$	90.08	syrup	....	....	d.			
Hydro-cyanic acid	HCN	27.03	lq.	0.697^{18}	−12	25–6	∞	∞	∞
-quinone (p-)	$C_6H_4(OH)_2$	110.11	cr.	1.332^{15}	170.3	285^{730}	6^{15}	v. s.	v. s.
Hydroxy-benzaldehyde (p-)	$HO.C_6H_4.CHO$	122.12	nd./aq.	1.129^{130}	116–7	subl.	1.38^{31}		
-benzanilide (o-)	$HO.C_6H_4.CONHC_6H_5$	213.23	pr./al.	....	135	d.	v. sl. s. h.	s.	s.
-quinoline (2-)(α-)	$C_9H_6N.OH$	145.15	pr./al.	....	199–200	subl.	s. h.	v. s.	v. s.
(8-)(o-)	$C_9H_6N.OH$	145.15	pr.	....	75–6	266.6^{752}	v. sl. s. c.	s.	sl. s.
Indigo	$[C_6H_4(CO)(NH)C:]_2$	262.26	cr.	1.35	390–2	subl.	i.	i.	i.
White	$C_{16}H_{12}O_2N_2$	264.27	gray	....	....	....	i.	s.	s.
Indole	C_8H_7N	117.14	lf./aq.	....	52	253–4	s. h.	s. h.	s.
Indoxyl	C_8H_6NOH	133.14	yel. pr.	....	85	110	s.	s.	s.
Iodo-benzene	C_6H_5I	204.02	col. lq.	$1.824^{25/4}$	−28.5	188.6	0.034^{20}	s.	∞
-phenol (p-)	IC_6H_4OH	220.02	nd./aq.	1.857^{112}	93–4	d.	sl. s.	v. s.	v. s.
Iodoform	HCI_3	393.78	yel. hex.	4.008^{17}	119	subl.	0.01^{25}	1.5^{17}	13.6^{25}
Ionone (α-)	$C_{10}H_{16}:CHCOCH_3$	192.29	col. oil	0.930^{20}	....	136.1^{17}	sl. s.	∞	∞
(β-)	$C_{10}H_{16}:CHCOCH_3$	192.29	col. oil	0.944^{20}	....	140^{18}	sl. s.	∞	∞
Irone (β-)	$C_{14}H_{22}O$	206.32	col. oil	0.939^{20}	....	144^{16}	v. sl. s.	∞	v. s.
Isatin	$C_6H_4 < (CO)(N) > COH$	147.13	yel. red	....	200–1	subl.	s. h.	v. s. h.	sl. s.
Isoprene	$CH_2:CH.C(CH_3):CH_2$	68.11	col. lq.	$0.681^{20/4}$	−120	34	i.	∞	∞
Ketene	$H_2C:CO$	42.04	col. gas	....	−151	−56	d.	d.	s.
Koch acid (1-)(3-,6-,8-)	$C_{10}H_4(NH_2)S_3O_9HNa_2$	427.34	cr.	....	....	....	7.2^{20}		
Lactic acid (dl-)	$CH_3CH(OH)CO_2H$	90.08	hyg.	$1.249^{15/4}$	16.8	122^{14}	∞	∞	∞
anhydride	$C_6H_{10}O_5$	162.14	yel. oil	....	....	d. 250	v. sl. s.	s.	s.
Lactide (dl-)	$C_6H_8C_4$	144.12	tri./al.	$0.862^{10/4}$	124.5	255^{757}	v. sl. s.	v. sl. s. c.	
Lactose	$C_{12}H_{22}O_{11}.H_2O$	360.31	col. rhb.	1.525^{20}	202	d.	17^{10}	i.	i.
Lauric acid	$CH_3(CH_2)_{10}CO_2H$	200.31	col. nd.	$0.869^{50/4}$	48(44)	225^{100}	i.	s.	s.
Laurone	$[CH_3(CH_2)_{10}]_2CO$	338.60	pl.	$0.809^{69/4}$	69–70	....	i.	i. c.	
Lauryl alcohol	$CH_3(CH_2)_{10}CH_2OH$	186.33	lf.	$0.831^{24/4}$	24	255–9	i.	s.	s.
Lead tetraethyl	$Pb(CH_2CH_3)_4$	323.45	col. lq.	$1.659^{18/4}$	−136	152^{291}	i.	sl. s.	∞
tetramethyl	$Pb(CH_3)_4$	267.35	col. lq.	$1.995^{20/4}$	−27.5	110^{760}	i.	∞	∞
Lecithin (protagon)	$C_{42}H_{48}O_9PN$	778.08	wax	....	150–200 d.	....	i.	s. h.	s. h.
Lepidine (py-4)	$C_9H_6N.CH_3$	143.18	lq.	1.086^{20}	9–10	261–3	sl. s.	∞	∞
Leucine (l-)	$(CH_3)_2CHCH_2CH(NH_2)CO_2H$	131.17	cr.	1.293^{18}	295	subl.	2.2^{18}		
Levulinic acid	$CH_3CO(CH_2)_2CO_2H$	116.11	lf.	$1.140^{20/20}$	33.5	245–6	v. s.	v. s.	v. s.
Limonene (d- or l-)	$C_{10}H_{16}$	136.23	lq.	$0.842^{20/4}$	−96.9	177	i.	∞	∞
Linalool (d- or l-)	$C_{10}H_{17}OH$	154.24	col. oil	0.868^{20}	....	198–200	v. sl. s.	s.	∞
Linalyl acetate	$CH_3CO_2C_{10}H_{17}$	196.28	col. lq.	0.895^{20}	....	220^{762} d.	v. sl. s.	∞	∞
Linoleic acid	$C_{17}H_{31}CO_2H$	280.44	yel. oil	$0.903^{18/4}$	−9.5	$229–30^{16}$	i.	∞	∞
Maleic acid	$HO_2C.CH:CH.CO_2H$	116.07	mn.	1.609	130.5	135 d.	79^{25}	70^{30}	8^{25}
anhydride	$< (.CHCO)_2 > O$	98.06	cr.	1.5	57–60	202	16.3^{30}		
Malic acid (dl-)	$HO_2CCH_2CH(OH)CO_2H$	134.09	col. cr.	$1.601^{20/4}$	128–9	150 d.	144^{26}	v. s.	v. s.
(d- or l-)	$HO_2CCH_2CH(OH)CO_2H$	134.09	col. cr.	$1.595^{20/4}$	99–100	140 d.	v. s.	v. s.	8.4^{15}
Malonic acid	$H_2C(CO_2H)_2$	104.06	col. tri.	1.631^{15}	130–5 d.	....	138^{16}	42^{25}	8^{15}

TABLE 3-2 Physical Properties of Organic Compounds *(Continued)*

Name	Formula	Formula weight	Form and color	Specific gravity	Melting point, °C.	Boiling point, °C.	Solubility in 100 parts: Water	Alcohol	Ether
Maltose	$C_{12}H_{22}O_{11}.H_2O$	360.31	col. nd.	1.540^{17}	d.	....	108^{25}	v. sl. s. c.	i.
Mandelic acid (dl-)	$C_6H_5CH(OH)CO_2H$	152.14	rhb./aq.	$1.300^{20/4}$	118.1	d.	16^{20}	s.	s.
Mannitol (d-)	$CH_2OH(CHOH)_4CH_2OH$	182.17	col. rhb.	$1.489^{20/4}$	166	$290–5^3$	13^{14}	0.01^{14}	i.
Mannose (d-)	$CH_2OH(CHOH)_4CHO$	180.16	rhb.	$1.539^{20/4}$	132	....	248^{17}	v. sl. s.	i.
Margaric acid	$CH_3(CH_2)_{15}CO_2H$	270.44	col. pl.	0.853^{60}	60–1	227^{100}	i.	32^{28}	v. s.
Mellitic acid	$C_6(CO_2H)_6$	342.17	nd./al.	....	286–8	d.	v. s.	v. s.	
Menthol (l-)(α-)	$C_{10}H_{19}OH$	156.26	col. cr.	$0.890^{15/15}$	42–3	212	0.04 c.	v. s.	v. s.
Mercapto-benzothiazole (2-)	$< C_6H_4N{:}C(SH)S >$	167.24	nd.	$1.42^{20/4}$	179	d.	i.	s.	sl. s.
-thiazoline (2-)	$< CH_2N{:}C(SH)SCH_2 >$	119.20	cr.	1.50	106	....	1.6^{60}		
Mercuric cyanide	$Hg(CN)_2$	252.65	cr.	4.003^{22}	d. 320	....	12.5^{15}		
fulminate	$Hg(ONC)_2.\frac{1}{2}H_2O$	293.65	cr./aq.	4.4	expl.	....	0.07^{12}	s.	
Mesityl oxide	$(CH_3)_2C{:}CHCOCH_3$	98.14	lq.	$0.858^{20/4}$	−59	130^{750}	3^{20}	∞	∞
Mesitylene (1-,3-,5-)	$C_6H_3(CH_3)_3$	120.19	col. lq.	$0.865^{20/4}$	−45(−52)	164.8	i.	s.	∞
Metanilic acid (m-)	$H_2NC_6H_4SO_3H$	173.18	col. nd.	....	d.	....	2^{15}	v. sl. s.	v. sl. s.
Methane	CH_4	16.04	gas	0.415^{-164}	−182.6	−161.4	0.4^{20} cc.	47^{20} cc.	104^{10} cc.
Methoxy-methoxyethanol	$CH_3(OCH_2)_2CH_2OH$	106.12	lq.	1.038^{25}	<−70	167.5	∞		
Methyl acetate	$CH_3CO_2CH_3$	74.08	col. lq.	$0.924^{20/4}$	−98.7	57.1	33^{22}	∞	∞
acrylic acid (α-)	$CH_2{:}C(CH_3)CO_2H$	86.09	pr.	$1.015^{20/4}$	15–16	161–3	s. h.	∞	∞
alcohol	CH_3OH	32.04	col. lq.	$0.792^{20/4}$	−97–8	64.7	∞	∞	∞
-amine	CH_3NH_2	31.06	col. gas	0.699^{-11}	−92.5	-6.7^{758}	v. s.	v. s.	
-amine hydrochloride	$CH_3NH_2.HCl$	67.52	pl./al.	1.23	226–8	230^{15}	v. s.	23 h.	i.
aniline	$C_6H_5NHCH_3$	107.15	lq.	$0.989^{20/4}$	−57	195.5	0.01^{25}	s.	∞
anthracene (α-)	$C_6H_4{:}(CH)_2{:}C_6H_3CH_3$	192.25	lf./al.	$1.047^{99.4}$	86	....	i.		
(β-)	$C_6H_4{:}(CH)_2{:}C_6H_3CH_3$	192.25	col. lf.	$1.181^{0/4}$	207	....	i.	v. sl. s.	v. sl. s.
anthranilate (o-)	$NH_2C_6H_4CO_2CH_3$	151.16	col. lq.	$1.168^{19/4}$	24	135.5^{15}	sl. s.	s.	s.
anthraquinone (2-)	$C_6H_4{:}(CO)_2{:}C_6H_3CH_3$	222.23	col. nd.	....	176–7	subl.	i.	s.	s.
benzoate	$C_6H_5CO_2CH_3$	136.14	col. lq.	$1.087^{25/25}$	−12.5	198–9	0.02^{30}	∞	∞
benzylaniline	$C_6H_5N(CH_3)CH_2C_6H_5$	197.27	lq.	....	9.2	305–6	i.	s.	s.
bromide	CH_3Br	94.95	gas	$1.732^{0/0}$	−93	4.5^{758}	v. sl. s.	s.	s.
butyrate (n-)	$CH_3(CH_2)_2CO_2CH_3$	102.13	col. lq.	$0.898^{20/4}$	<−95	102.3	1.7	∞	∞
(i-)	$(CH_3)_2CHCO_2CH_3$	102.13	col. lq.	$0.891^{20/4}$	−84.7	92.6	v. sl. s.	∞	∞
caprate	$CH_3(CH_2)_8CO_2CH_3$	186.29	lq.	....	−18	223–4	i.	∞	∞
caproate (n-)	$CH_3(CH_2)_4CO_2CH_3$	130.18	col. lq.	$0.904^{0/0}$	....	149.5	i.	∞	∞
caprylate	$CH_3(CH_2)_6CO_2CH_3$	158.23	col. lq.	0.887^{18}	−40	192–4	i.	∞	∞
cellosolve	$CH_3OCH_2CH_2OH$	76.09	col. lq.	$0.965^{20/4}$	....	124–5	∞	∞	∞
chloride	CH_3Cl	50.49	gas	0.952^0	−97.7	−24	280^{16} cc.	v. s.	v. s.
chloroacetate	$ClCH_2CO_2CH_3$	108.53	col. lq.	$1.236^{20/4}$	−32.7	130^{740}	v. sl. s.	∞	∞
chloroformate	$ClCO_2CH_3$	94.50	col. lq.	1.236^{15}	....	71–2	d.	∞	∞
cinnamate	$C_6H_5CH{:}CHCO_2CH_3$	162.18	cr.	$1.042^{36/0}$	33.4	263	i.	v. s.	v. s.
cyclohexane	$CH_2 < (CH_2CH_2)_2 > CHCH_3$	98.18	col. lq.	$0.769^{20/4}$	−126.3	101	i.	s.	s.
ethyl carbonate	$CH_3O.CO.OC_2H_5$	104.10	lq.	1.002^{27}	−14.5	109.2	i.	∞	∞
ethyl ketone	$CH_3.CO.C_2H_5$	72.10	col. lq.	$0.805^{20/4}$	−85.9	79.6	35^{10}	∞	∞
ethyl oxalate	$CH_3OCO.CO_2C_2H_5$	132.11	lq.	$1.156^{0/0}$	....	173.7	i.	v. s.	v. s.
formate	HCO_2CH_3	60.05	lq.	$0.974^{20/4}$	−99.8	32	30^{20}	∞	
furoate	$C_4H_3O.CO_2CH_3$	126.11	col. lq.	$1.179^{21/4}$	....	181.3	i.	∞	∞
glucamine	$CH_2OH(CHOH)_4CH_2NHCH_3$	195.21							
glycolate	$HOCH_2CO_2CH_3$	90.08	lq.	1.168^{18}	....	151.2			
heptoate	$CH_3(CH_2)_5CO_2CH_3$	144.21	lq.	$0.881^{15/4}$	....	172–3	i.		
hypochlorite	$ClOCH_3$	66.49	gas	....	....	12^{726}			
iodide	CH_3I	141.95	col. lq.	$2.279^{20/4}$	−64.4	42.4	1.8^{15}	∞	∞
lactate	$CH_3CH(OH)CO_2CH_3$	104.10	lq.	1.090^{19}	....	144.8	∞	s.	s.
laurate	$CH_3(CH_2)_{10}CO_2CH_3$	214.34	lq.	....	5	148^{18}	i.		
mercaptan	CH_3SH	48.10	gas	0.896^0	−121	5.8^{752}	s.	v. s.	v. s.
methacrylate	$CH_2{:}C(CH_3)CO_2CH_3$	100.11	lq.	$0.950^{15.6}$	−48	100.3	i.		
myristate	$CH_3(CH_2)_{12}CO_2CH_3$	242.39	cr./al.	....	18–9	295^{715}	i.		
naphthalene (α-)	$C_{10}H_7CH_3$	142.19	oil	$1.025^{14/4}$	−19	244.6	i.	v. s.	v. s.
(β-)	$C_{10}H_7CH_3$	142.19	mn.	$0.994^{40/4}$	35–6	241–2	i.	v. s.	v. s.
nitrate	CH_3ONO_2	77.04	lq.	1.203^{25}	expl.	65	sl. s.	s.	s.
nitrite	CH_3ONO	61.04	gas	0.991^{15}	....	−12	....	s.	s.
nonyl ketone (n-)	$CH_3(CH_2)_8COCH_3$	170.29	col. oil	$0.828^{20/20}$	13.5	228	i.	s.	s.
oleate	$C_{17}H_{33}CO_2CH_3$	296.48	oil	0.879^{18}	....	$190–1^{10}$	i.	∞	∞
orange	$(CH_3)_2NC_6H_4N_2C_6H_4SO_3Na$	327.33	red pd.	....	....	....	0.2 c.		
palmitate	$CH_3(CH_2)_{14}CO_2CH_3$	270.44	col. cr.	....	30–1	196^{15}	i.	s.	s.
phosphine	CH_3PH_2	48.03	gas	....	....	-14^{759}	i.	sl. s.	
propionate	$CH_3CH_2CO_2CH_3$	88.10	col. lq.	$0.915^{20/4}$	−87.5	79.7	0.5^{20}	∞	∞
propyl ketone (n-)	$CH_3COCH_2CH_2CH_3$	86.13	col. lq.	$0.812^{15/15}$	−77.8	102	v. sl. s.	∞	∞
salicylate (o-)	$HO.C_6H_4CO_2CH_3$	152.14	col. lq.	$1.182^{25/25}$	−8.3	222.2	0.07^{30}	∞	∞

stearate	$CH_3(CH_2)_{16}CO_2CH_3$	298.49	col. cr.	....	38–9	215^{15}	i.	s.	s.
toluate (o-)	$CH_3.C_6H_4CO_2CH_3$	150.17	col. lq.	1.073^{15}	<−50	213	i.	∞	∞
(m-)	$CH_3.C_6H_4CO_2CH_3$	150.17	col. lq.	1.066^{15}	....	215	i.		
(p-)	$CH_3.C_6H_4CO_2CH_3$	150.17	cr.	....	33–4	217	i.	v. s.	v. s.
Methyl toluidine (o-)	$CH_3.C_6H_4NHCH_3$	121.18	lq.	0.973^{15}	....	206–7	i.	∞	∞
(m-)	$CH_3.C_6H_4NHCH_3$	121.18	lq.	....	....	206–7	i.	∞	∞
(p-)	$CH_3.C_6H_4NHCH_3$	121.18	lq.	$0.935^{55/4}$	....	211^{761}	i.	∞	∞
valerate (n-)	$CH_3(CH_2)_3CO_2CH_3$	116.16	lq.	$0.895^{15/4}$	−91	127.3	v. sl. s.	∞	∞
(i-)	$(CH_3)_2CHCH_2CO_2CH_3$	116.16	col. lq.	$0.881^{20/4}$	....	$116–7^{764}$	v. sl. s.	∞	∞
vinyl ketone	$CH_3COCH:CH_2$	70.09	lq.	$0.836^{20/4}$	....	81	>85		
Methylal	$HCH(OCH_3)_2$	76.09	col. lq.	$0.866^{15/4}$	−104.8	42–3	33	∞	∞
Methylene-bis-(phenyl-4-isocyanate)	$(OCN.C_6H_4)_2CH_2$	250.25	lq.	1.222^{30}	....	$210–2^{13}$	d.	d.	
bromide	CH_2Br_2	173.86	col. lq.	$2.495^{20/4}$	−52.8	98.5^{756}	1.17^{0}	∞	∞
chloride	CH_2Cl_2	84.94	col. lq.	$1.336^{20/4}$	−96.7	40–1	2^{20}	∞	∞
dianiline	$(C_6H_5NH)_2CH_2$	198.26	cr.	....	65	208–9 d.	i.	s.	s.
iodide	CH_2I_2	267.87	col. lq.	$3.325^{20/4}$	5.7	180 d.	1.4^{20}	∞	∞
Michler's hydrol (p-,p'-)	$[(CH_3)_2NC_6H_4]_2CHOH$	270.36	gn.	....	96–7	....	i.	s. h.	s.
ketone	$[(CH_3)_2NC_6H_4]_2CO$	268.35	lf./al.	....	174	>360 d.	i.	sl. s.	v. sl. s.
Morphine	$C_{17}H_{19}O_3N.H_2O$	303.35	pr./al.	1.317	254 d.	....	0.02^{20}	sl. s.	s.
Mucic acid	$(.CHOHCHOHCO_2H)_2$	210.14	pd.	....	206–14	....	0.33^{14}	i.	i.
Mustard gas	$(ClCH_2.CH_2)_2S$	159.08	oil	$1.275^{20/4}$	13–4	217	0.07^{25}	s.	s.
Myricyl alcohol	$C_{31}H_{63}OH(?)$	452.82	cr.	0.777^{95}	88	....	i.	v. sl. s.	v. s.
Myristic acid	$CH_3(CH_2)_{12}CO_2H$	228.36	col. lf.	$0.853^{70/4}$	57–8	250.5^{100}	i.	v. s.	v. s.
Myristyl alcohol	$CH_3(CH_2)_{12}CH_2OH$	214.38	cr.	$0.824^{38/4}$	38	167^{15}	<0.02	sl. s.	s.
Naphthalene	$C_{10}H_8$	128.16	pl./al.	$1.145^{20/4}$	80.2	217.9	0.003^{25}	9.5^{20}	v. s.
disulfonic acid (1-,5-)	$C_{10}H_6(SO_3H)_2$	288.28	lf.	....	d.	....	102^{20}	s.	i.
(1-,6-)	$C_{10}H_6(SO_3H)_2$	288.28	cr.	....	d. 125	....	164^{20}	s.	i.
sulfonic acid (α-)	$C_{10}H_7SO_3H.2H_2O$	244.26	cr.	....	90	....	v. s.	v. s.	sl. s.
(β-)	$C_{10}H_7SO_3H.H_2O$	226.24	cr.	....	125	....	77^{30}		
Naphthasultam (1-,8-)	$C_{10}H_7O_2NS$	205.22	nd.	....	177–8	....	s. h.	sl. s.	s.
disulfonate Na (1-,8-)	$C_{10}H_5O_8NS_3Na_2.2H_2O$	445.35	cr.	....	....	....	v. s.		
(2-,4-)	$C_{10}H_4O_8NS_3Na_3.8\frac{1}{2}H_2O$	584.45	lf.	....	....	....	v. s.	sl. s.	
Naphthoic acid (α-)	$C_{10}H_7CO_2H$	172.17	nd.	....	160–1	300	v. sl. s. h.	s. h.	s.
(β-)	$C_{10}H_7CO_2H$	172.17	mn.	$1.077^{100/4}$	184	>300	0.007^{25}	s.	s.
Naphthol (α-)	$C_{10}H_7OH$	144.16	mn.	1.224^{4}	96	278–80	sl. s. h.	v. s.	v. s.
(β-)	$C_{10}H_7OH$	144.16	mn.	1.217^{4}	122–3	285–6	0.074^{25}	v. s.	v. s.
sulfonic acid (α-)(1-,2-)	$HO.C_{10}H_6SO_3H$	224.22	pl./aq.	....	>250	....	v. s. h.	....	i.
(β-)(2-,6-)	$HO.C_{10}H_6SO_3H$	224.22	lf.	....	125	....	v. s.	v. s.	
Naphthyl acetate (α-)	$CH_3CO_2C_{10}H_7$	186.20	nd./al.	....	46–9	....	sl. s. h.	s.	s.
(β-)	$CH_3CO_2C_{10}H_7$	186.20	nd./al.	....	69–70	....	i.	s.	s.
amine (α-)	$C_{10}H_7NH_2$	143.18	rhb.	$1.123^{25/25}$	50	300.8	0.17 c.	v. s.	v. s.
(β-)	$C_{10}H_7NH_2$	143.18	lf./aq.	$1.061^{98/4}$	111–2	306.1	v. s. h.	s.	s.
amine hydrochloride (α-)	$C_{10}H_7NH_2.HCl$	179.65	nd.	....	....	subl.	3.8^{20}	s.	s.
(β-)	$C_{10}H_7NH_2.HCl$	179.65	lf.	....	....	....	v. s.	v. s.	
amine sulfonic acid (1-,4-)	$NH_2.C_{10}H_6.SO_3H$	223.24	nd.	....	d.	....	0.2^{100}	i.	i.
(1-,5-)	$NH_2.C_{10}H_6.SO_3H.H_2O$	241.26	cr.	....	....	....	sl. s.		
(1-,7-)	$NH_2.C_{10}H_6.SO_3H.H_2O$	241.26	cr.	....	....	....	0.46^{25}		
(1-,8-)	$NH_2.C_{10}H_6.SO_3H.H_2O$	241.26	cr.	....	....	....	0.42^{100}		
(2-,5-)	$NH_2.C_{10}H_6.SO_3H$	223.24	cr.	....	....	....	0.08		
(2-,6-)	$NH_2.C_{10}H_6.SO_3H.H_2O$	241.26	cr.	....	....	....	0.38^{100}		
(2-,7-)	$NH_2.C_{10}H_6.SO_3H.H_2O$	241.26	cr.	....	....	....	0.28^{100}		
isocyanate (α-)	$C_{10}H_7N:CO$	169.17	col. lq.	1.18	....	269–70	d.	s.	s.
Nicotine	$C_{10}H_{14}N_2$	162.23	oil	$1.009^{20/4}$	<−80	246^{730}	s.	∞	∞
Nicotinic acid (3-)	$C_5H_4NCO_2H$	123.11	nd./al.	....	235.2	subl.	s. h.	s. h.	v. sl. s.
(i-)(4-)	$C_5H_4NCO_2H$	123.11	nd./aq.	....	317	d.	s. h.	sl. s. h.	v. sl. s.
Nitro-acetanilide (p-)	$CH_3CONHC_6H_4NO_2$	180.16	rhb.	....	215–6	....	s. h.	s.	s.
-acetophenone (m-)	$CH_3COC_6H_4NO_2$	165.14	nd.	....	80–1	202	i.	s.	
-aminoanisole (4-,1-,2-)	$NO_2.C_6H_3(OCH_3)NH_2$	168.15	red nd.	1.207^{156}	118	....	i.	s.	
(5-,1-,2-)	$NO_2.C_6H_3(OCH_3)NH_2$	168.15	yel. nd.	1.211^{156}	139–40	....			
(3-,1-,4-)	$NO_2.C_6H_3(OCH_3)NH_2$	168.15	red	....	123	....	sl. s.	s.	s.
-aminophenol (4-,2-,1-)	$NO_2.C_6H_3(NH_2)OH$	154.12	or. pr.	....	142–3	....	sl. s. c.	v. s.	v. s.
-aniline (o-)	$NO_2.C_6H_4NH_2$	138.12	yel. rhb.	1.442^{15}	71.5	284.1	s. h.	v. s.	v. s.
(m-)	$NO_2.C_6H_4NH_2$	138.12	yel. rhb.	1.43	114	306.4	0.11^{20}	7.1^{20}	7.9^{20}
(p-)	$NO_2.C_6H_4NH_2$	138.12	yel. mn.	1.437^{14}	146–7	331.7	0.08^{19}	5.8^{20}	6.1^{20}
-anisole (o-)	$CH_3OC_6H_4NO_2$	153.13	col. cr.	$1.254^{20/4}$	9.4	272–3	0.17^{30}	∞	∞
(p-)	$CH_3OC_6H_4NO_2$	153.13	pr./al.	1.233^{20}	54	274	0.06^{30}	v. s.	v. s.
-anthraquinone (α-)	$C_6H_4:(CO)_2:C_6H_3NO_2$	253.20	nd.	....	230	270^{7}	i.	sl. s.	v. sl. s.
-anthraquinone sulfonic acid (1-,5-)	$NO_2.C_{14}H_6O_2.SO_3H$	333.26	yel. cr.	....	....	....	s.	i.	i.
-benzal chloride (m-)	$NO_2.C_6H_4.CHCl_2$	206.03	mn.	....	65	....	i.	v. s. h.	v. s.
-benzaldehyde (m-)	$NO_2.C_6H_4CHO$	151.12	nd./aq.	....	58	164^{23}	1.95^{112}	v. s. h.	v. s.

TABLE 3-2 Physical Properties of Organic Compounds *(Continued)*

Name	Formula	Formula weight	Form and color	Specific gravity	Melting point, °C.	Boiling point, °C.	Solubility in 100 parts: Water	Alcohol	Ether
Nitro-benzene	$C_6H_5NO_2$	123.11	yel. lq.	$1.205^{18/4}$	5.7	210.9	0.19^{20}	v. s.	∞
-benzidine (2-)	$NH_2C_6H_4C_6H_3(NH_2)NO_2$	229.23	red nd.	....	143	....	sl. s. h.		
-benzoic acid (o-)	$NO_2.C_6H_4.CO_2H$	167.12	tri./aq.	$1.575^{4/4}$	147.5	....	0.65^{20}	28^{11}	22^{11}
(m-)	$NO_2.C_6H_4.CO_2H$	167.12	mn.	$1.494^{4/4}$	140–1	....	0.24^{165}	3^{12}	25^{10}
(p-)	$NO_2.C_6H_4.CO_2H$	167.12	yel. mn.	$1.550^{32/4}$	240–2	subl.	0.02^{15}	0.9^{10}	2.2^{18}
-benzyl alcohol (m-)	$NO_2.C_6H_4.CH_2OH$	153.13	cr.	....	27	$175–80^{3}$			
-benzyl bromide (p-)	$NO_2.C_6H_4CH_2Br$	216.04	nd./al.	....	99–100	....	i.	2^{19}	v. s.
-chlorotoluene (1-,2-,6-)	$CH_3.C_6H_3(NO_2)Cl$	171.58	cr.	....	37.5	238	i.		
-cresol (1-,3-,4-)	$CH_3.C_6H_3(NO_2)OH$	153.13	yel.	$1.240^{39/4}$	32	125^{22}	v. sl. s.	v. s.	v. s.
-cymene (1-,2-,4-)	$CH_3.C_6H_3(NO_2)CH(CH_3)_2$	179.21	oil	$1.067^{20/4}$	....	152^{15}	i.		
-dimethylaniline (o-)	$NO_2.C_6N_4NHCH_3$	166.18	yel. oil	$1.179^{20/4}$	....	$151–3^{30}$	v. sl. s.	v. s.	v. s.
(m-)	$NO_2.C_6H_4NHCH_3$	166.18	red mn.	1.313^{17}	60–1	280–5	i.	s.	s.
(p-)	$NO_2.C_6H_4NHCH_3$	166.18	yel. nd.	....	163–4	....	i.	s. h.	
-diphenyl (o-)	$C_6H_5.C_6H_4NO_2$	199.20	rhb.	1.44	37	320	i.	s.	v. s.
(p-)	$C_6H_5.C_6H_4NO_2$	199.20	nd./al.	....	113–4	340	i.	sl. s. c.	v. s.
-diphenylamine (o-)	$C_6H_5.NH.C_6H_4NO_2$	214.22	or. cr.	....	75–6				
-guanidine	$H_2NC(NH)NHNO_2$	104.07	nd./aq.	....	246–7	....	9^{100}	sl. s.	v. sl. s.
-naphthalene (α-)	$C_{10}H_7NO_2$	173.16	yel./al.	1.223^{62}	59–60	304	i.	s.	s.
(β-)	$C_{10}H_7NO_2$	173.16	col./al.	....	79	165^{15}	i.	v. s.	v. s.
-phenol (o-)	$NO_2.C_6H_4.OH$	139.11	yel. mn.	1.295^{45}	44–5	214.5	1.08^{100}	v. s.	v. s.
(m-)	$NO_2.C_6H_4.OH$	139.11	col. mn.	1.485^{20}	96–7	194^{70}	1.35^{20}	v. s.	s.
(p-)	$NO_2.C_6H_4.OH$	139.11	yel. pr.	1.479^{20}	113–4	subl.	1.6^{25}	v. s.	v. s.
-phenol sulfonic acid (1-,4-,2-)	$HO.C_6H_3(NO_2)SO_3H.3H_2O$	273.22	nd.	....	d. 110	....	v. s.	v. s.	sl. s
(1-,2-,4-)	$HO.C_6H_3(NO_2)SO_3H.3H_2O$	273.22	nd./aq.	....	51.5	....	v. s.	v. s.	
-phthalic acid (3-)	$NO_2.C_6H_3(CO_2H)_2$	211.13	yel./aq.	....	222	....	2.05^{25}	v. s. h.	sl. s.
(4-)	$NO_2.C_6H_3(CO_2H)_2$	211.13	yel. cr.	....	164–5	....	v. s.	v. s.	s.
-toluene (o-)	$CH_3.C_6H_4NO_2$	137.13	yel. lq.	$1.163^{20/4}$	−4.1	222.3	0.07^{30}	∞	∞
(m-)	$CH_3.C_6H_4NO_2$	137.13	lq.	$1.160^{18/4}$	15–16	230–1	0.05^{30}	∞	∞
(p-)	$CH_3.C_6H_4NO_2$	137.13	rhb.	$1.139^{55/65}$	51.9	237.7	0.04^{30}	8.6^{15}	80.8^{15}
-toluene sulfonic acid (1-,4-,2-)	$CH_3.C_6H_3(NO_2)SO_3H.2H_2O$	253.23	pl./aq.	....	130	....	47.7^{23}	v. s.	v. s.
-toluidine (4-,1-,2-)	$NO_2.C_6H_3(CH_3)NH_2$	152.15	yel. mn.	1.365^{15}	105–7	....	v. sl. s.	s.	s.
(3-,1-,4-)	$NO_2.C_6H_3(CH_3)NH_2$	152.15	red mn.	1.312^{17}	116–7	....	sl. s. h.	s.	
Nitron	$C_{20}H_{16}N_4$	312.36	yel. lf.	....	189–90 d.	....	i.	s. h.	v. sl. s.
Nitroso-dimethylaniline (p-)	$ON.C_6H_4N(CH_3)_2$	150.18	gn. tri.	....	86–7	....	i.	s.	s.
-naphthol (β-)(1-)	$ON.C_{10}H_6OH$	173.16	brn. pr.	....	109.5	....	0.1^{20}	2.4^{18}	
Nonadecane (n-)	$CH_3(CH_2)_{17}CH_3$	268.51	cr.	$0.777^{32/4}$	32	330	i.	sl. s.	s.
Nonane (n-)	$CH_3(CH_2)_7CH_3$	128.25	col. lq.	$0.718^{20/4}$	−53.7	150.5^{759}	i.	sl. s.	s.
Octadecane (n-)	$CH_3(CH_2)_{16}CH_3$	254.48	cr.	$0.775^{28/4}$	28	317	i.	sl. s.	s.
Octane (n-)	$CH_3(CH_2)_6CH_3$	114.22	col. lq.	$0.703^{20/4}$	−56.5	125.7	0.002^{16}	sl. s.	s.
(iso-)	$(CH_3)_3CCH_2CH(CH_3)_2$	114.22	col. lq.	$0.692^{20/4}$	−107.4	99.3^{760}	i.	sl. s.	s.
Octyl acetate (n-)	$CH_3CO_2CH_2(CH_2)_6CH_3$	172.26	col. lq.	$0.885^{0/4}$	−38.5	210	i.	s.	s.
(sec-)	$CH_3CO_2CH(CH_3)C_6H_{13}$	172.26	col. lq.	$0.863^{14/4}$	....	195	i.	s.	s.
alcohol (n-)	$CH_3(CH_2)_6CH_2OH$	130.22	col. lq.	$0.827^{20/4}$	−16	194–5	0.054^{25}	∞	∞
(sec-)	$CH_3(CH_2)_5CH(OH)CH_3$	130.22	col. lq.	$0.822^{20/4}$	−38.6	179–80	0.096^{25}	∞	∞
Octylene (n-)	$CH_3(CH_2)_5CH:CH_2$	112.21	lq.	$0.721^{18/4}$	....	126	i.	∞	∞
Oleic acid	$C_8H_{17}CH:CH(CH_2)_7CO_2H$	282.45	col. nd.	$0.854^{78/4}$	14	$285–6^{100}$	i.	∞	∞
Orcinol (1-,3-,5-)	$(HO)_2C_6H_3.CH_3$	124.13	pr./bz.	1.290^{4}	107–8	287–90	v. s.	v. s.	v. s.
Oxalic acid	$HO_2C.CO_2H.2H_2O$	126.07	col. mn.	$1.653^{19/4}$	101.5	subl.	s.	s.	1.3
Palmitic acid	$CH_3(CH_2)_{14}CO_2H$	256.42	col. pl.	$0.849^{70/4}$	63–4	271.5^{100}	i.	9^{20}	s.
Pelargonic acid	$CH_3(CH_2)_7CO_2H$	158.23	col. oil	$0.906^{20/4}$	12.5	253–4	v. sl. s.	s.	s.
Penta-chloroethane	$CHCl_2.CCl_3$	202.31	col. lq.	$1.671^{25/4}$	−22	162	0.05^{20}	∞	∞
-decane (n-)	$CH_3(CH_2)_{13}CH_3$	212.41	col. lq.	$0.770^{20/4}$	10	270.5	i.	v. s.	v. s.
-erythritol	$C(CH_2OH)_4$	136.15	cr.	....	262	276^{30}	5.6^{15}	v. sl. s.	i.
Pentandiol	$HOCH_2(CH_2)_3CH_2OH$	104.15	lq.	$0.994^{20/4}$	....	239.4	∞		
Pentane (n-)	$CH_3(CH_2)_3CH_3$	72.15	col. lq.	$0.630^{18/4}$	−129.7	36.3	0.036^{16}	∞	∞
(i-)	$(CH_3)_2CHCH_2CH_3$	72.15	col. lq.	0.621^{19}	−160.0	27.95	i.	∞	∞
(neo-)	$(CH_3)_2C(CH_3)_2$	72.15	col. lq.	$0.613^{20/4}$	−20	9.5	i.	s.	s.
Phenacetin	$C_2H_5OC_6H_4NHCOCH_3$	179.21	col. mn.	....	134–5	d.	0.7^{20}	40 h.	1.6^{25}
Phenanthrene	$<(C_6H_4CH)_2>$	178.22	pl./al.	1.179^{25}	99–100	340	i.	10 h.	v. s.
Phenetidine (o-)	$C_2H_5O.C_6H_4.NH_2$	137.18	oil	....	<−21	228–9	i.	s.	s.
(p-)	$C_2H_5O.C_6H_4.NH_2$	137.18	lq.	1.061^{15}	3–4	254–5	i.	s.	s.
Phenetole	$C_2H_5O.C_6H_5$	122.16	col. lq.	$0.967^{20/4}$	−30.2	172	i.	∞	∞
Phenol	C_6H_5OH	94.11	col. nd.	$1.071^{25/4}$	42–3	181.4	8.2^{15}	∞	∞
-phthalein	$C_{20}H_{14}O_4$	318.31	col. rhb.	$1.299^{25/4}$	261–2	....	0.2^{20}	10^{25}	5.9 c.
-sulfonic acid (o-)	$HO.C_6H_4SO_3H.\frac{3}{4}H_2O$	187.68	cr.	....	50 d.	....	v. s.	v. s.	
Phenyl acetaldehyde	$C_6H_5CH_2CHO$	120.14	lq.	1.025^{20}	....	193–4	v. sl. s.	∞	∞
acetic acid	$C_6H_5CH_2CO_2H$	136.14	lf.	$1.081^{80/4}$	76–7	265.5	1.66^{20}	v. s.	v. s.

-acetylene	$C_6H_5C:CH$	102.13	col. lq.	$0.930^{20/4}$	−43	142–3	i.	∞	∞
aniline (o-)	$C_6H_5.C_6H_4.NH_2$	169.22	cr.	....	45–6	299^{760}	v. sl. s.	s.	s.
(p-)	$C_6H_5.C_6H_4.NH_2$	169.22	lf.	....	50–2	302	s. h.	s.	s.
Phenyl-ethyl alcohol	$C_6H_5CH_2CH_2OH$	122.16	col. oil	$1.023^{18/4}$	....	$219\text{–}21^{750}$	1.6^{20}	s.	∞
-glycine	$C_6H_5NHCH_2CO_2H$	151.16	cr.	....	127	....	s.	s.	sl. s.
-hydrazine	$C_6H_5NH.NH_2$	108.14	yel. oil	$1.097^{28/4}$	19.6	243.5	sl. s. h.	∞	∞
-hydrazine sulfonic acid (p-)	$H_2NNHC_6H_4SO_3H$	188.20	cr./al.	....	286	....	0.6^{12}	sl. s.	
isocyanate	$C_6H_5N:CO$	119.12	lq.	$1.096^{20/4}$	....	166^{769}	d.	d.	v. s.
-methylpyrazolone (3-)(N-)	$C_4H_5ON_2.C_6H_5$	174.20	pr./aq.	....	128	191^{17}	1^{20}	v. s. h.	v. sl. s.
-mustard oil	$C_6H_5N:CS$	135.18	col. lq.	$1.138^{15/15}$	−21	219–20	i.	s.	s.
naphthalene (α-)	$C_{10}H_7.C_6H_5$	204.26	waxy	....	45	336–7	i.	v. s.	v. s.
(β-)	$C_{10}H_7.C_6H_5$	204.26	lf./al.	....	102.5	345–6	i.	sl. s.	sl. s.
naphthylamine (α-)	$C_{10}H_7NHC_6H_5$	219.27	pr./al.	1.17	62	335^{258}	0.08^{60}	s.	s.
(β-)	$C_{10}H_7NHC_6H_5$	219.27	rhb.	1.18	107–8	399.5	0.4^{60}	v. s. h.	v. s. h.
phenol (o-)	$C_6H_5.C_6H_4OH$	170.20	nd.	....	56–7	275	i.	s.	s.
(p-)	$C_6H_5.C_6H_4OH$	170.20	nd.	....	164–5	305–8	i.	s.	s.
propyl alcohol (γ-)	$C_6H_5(CH_2)_3OH$	136.19	oil	$1.008^{20/4}$	<−18	235–7	sl. s.	∞	∞
quinoline (2-)(α-)	$C_6H_5.C_9H_6N$	205.25	nd.	....	86	363	sl. s.	s. h.	s.
(8-)(0-)	$C_6H_5.C_9H_6N$	205.25	lq.	....	....	283^{187}	sl. s.	s.	s.
salicylate, salol	$HO.C_6H_4CO_2C_6H_5$	214.21	rhb./al.	$1.250^{20/4}$	42–3	$172\text{–}3^{12}$	0.015^{25}	v. s.	s.
stearate	$CH_3(CH_2)_{16}CO_2C_6H_5$	360.56	cr.	....	52	267^{15}	i.		
urethane	$C_6H_5NHCO_2C_2H_5$	165.19	pl./al.	$1.106^{30/4}$	52–3	237–8	i. c.	s.	s.
Phenylene-diamine (o-)	$C_6H_4(NH_2)_2$	108.14	lf./aq.	....	103–4	256–8	733^{81}	v. s.	v. s.
(m-)	$C_6H_4(NH_2)_2$	108.14	rhb.	$1.139^{15/15}$	62.8	284–7	35.1^{25}	v. s.	s.
(p-)	$C_6H_4(NH_2)_2$	108.14	mn.	....	140	267	669^{107}	s.	s.
Phloroglucinol (1-,3-,5-)	$C_6H_3(OH)_3.2H_2O$	162.14	rhb.	....	117	subl.	1.13^{25}	v. s.	v. s.
Phorone	$[(CH_3)_2C:CH]_2CO$	138.20	yel. pr.	$0.885^{20/4}$	28	197.2^{743}	0.1^{50}	s.	s.
Phosgene	$OCCl_2$	98.92	gas	$1.392^{19/4}$	−104	8.2^{756}	v. sl. s.		
Phthalic acid (o-)	$C_6H_4(CO_2H)_2$	166.13	mn./aq.	$1.593^{20/4}$	208	d.	0.70^{25}	12^{18}	0.68^{15}
(m-)(iso-)	$C_6H_4(CO_2H)_2$	166.13	nd./aq.	....	330	subl.	0.2^{100}	s.	
anhydride (o-)	$C_6H_4 < (CO)_2 > O$	148.11	rhb.	1.527^{4}	130.8	284.5	v. sl. s.	s.	sl. s.
nitrile (o-)	$C_6H_4(CN)_2$	128.13	cr.	....	141	....	sl. s. c.		
Phthalide	$C_6H_4(CH_2)(CO) > O$	134.13	nd./aq.	$1.164^{99/4}$	73(65)	290	v. sl. s.	s.	
Phthalimide (o-)	$C_6H_4 < (CO)_2 > NH$	147.13	cr./et.	....	238	subl.	0.04^{25}	5	s. h.
Picoline (α-)	$C_5H_4N.CH_3$	93.12	col. lq.	$0.950^{15/4}$	−70	128.8	v. s.	∞	∞
(β-)	$C_5H_4N.CH_3$	93.12	col. lq.	$0.961^{15/4}$	....	143.5	∞	∞	∞
(γ-)	$C_5H_4N.CH_3$	93.12	lq.	$0.957^{15/4}$	....	143.1	∞	∞	∞
Picramic acid (1-,2-,4-,6-)	$HO.C_6H_2(NH_2)(NO_2)_2$	199.12	red nd.	....	169	....	0.14^{22}	s.	sl. s.
Picric acid (2-,4-,6-)	$HO.C_6H_2(NO_2)_3$	229.11	yel. rhb.	$1.763^{20/4}$	121.8	expl.	1.23^{20}	6^{20}	1^{13}
Picryl chloride (2-,4-,6-)	$ClC_6H_2(NO_2)_3$	247.56	yel. mn.	1.797^{20}	83	d.	0.018^{15}	4.8^{17}	7^{17}
Pinacol	$[(CH_3)_2C.OH]_2$	118.17	col. nd.	0.967^{15}	43(38)	$171\text{–}2^{739}$	sl. s. c.	v. s.	v. s.
Pinacoline	$CH_3COC(CH_3)_3$	100.16	col. lq.	0.800^{16}	−52.5	106.2	2.5^{15}	s.	s.
Pinene (α-)(dl-)	$C_{10}H_{16}$	136.23	col. lq.	$0.878^{20/4}$	−55	154–6	v. sl. s.	s.	∞
hydrochloride	$C_{10}H_{17}Cl$	172.69	lf.	....	131–2	207–8	i.	33	s.
Pinol (dl-)	$C_{10}H_{16}O$	152.23	lq.	$0.953^{20/20}$	....	183–4	....	s.	s.
Piperidine	$CH_2 < (CH_2CH_2)_2 > NH$	85.15	lq.	$0.860^{20/4}$	−9	106	∞	∞	
carboxylic acid (α-)(dl-)	$HO_2C.CH < (CH_2CH_2)_2 > NH$	129.16	cr.	....	264	....	s.		
Piperidinium pentamethylene dithiocarbamate	$(CH_2)_5CS_2H.HN(CH_2)_5$	232.41	cr.	1.13	175	....	6^{28}		
Propane	$CH_3CH_2CH_3$	44.09	gas	$0.585^{-45/4}$	−187.1	−42.2	6.5^{18} cc.	s.	v. s.
Propionic acid	$CH_3CH_2CO_2H$	74.08	col. lq.	$0.992^{20/4}$	−22	141.1	∞	∞	∞
aldehyde	CH_3CH_2CHO	58.08	col. lq.	$0.807^{20/4}$	−81	49.5^{740}	20^{20}	∞	∞
anhydride	$(CH_3CH_2CO)_2O$	130.14	col. lq.	$1.012^{20/4}$	−45	168.8^{780}	d.	d.	
Propyl acetate (n-)	$CH_3COCH_2CH_2CH_3$	102.13	col. lq.	$0.886^{20/4}$	−92.5	101.6	1.6^{16}	∞	∞
(i-)	$CH_3CO_2CH(CH_3)_2$	103.13	col. lq.	$0.874^{20/20}$	−73.4	88.4	3^{20}	∞	∞
alcohol (n-)	$CH_3CH_2CH_2OH$	60.09	col. lq.	$0.804^{20/4}$	−127	97.8	∞	∞	∞
(i-)	$(CH_3)_2CHOH$	60.09	col. lq.	$0.789^{20/4}$	−85.8	82.5	∞	∞	∞
amine (n-)	$CH_3CH_2CH_2NH_2$	59.11	col. lq.	$0.718^{20/20}$	−83	$49\text{–}50^{781}$	∞	∞	∞
(i-)	$(CH_3)_2CHNH_2$	59.11	col. lq.	$0.694^{15/4}$	−101	33–4	∞	∞	∞
aniline (n-)	$C_6H_5NHCH_2CH_2CH_3$	135.20	lq.	0.949^{18}	....	222	i.	v. s.	v. s.
benzoate (n-)	$C_6H_5CO_2CH_2CH_2CH_3$	164.20	col. lq.	$1.021^{25/25}$	−51.6	231	i.	s.	s.
(i-)	$C_6H_5CO_2CH(CH_3)_2$	164.20	col. lq.	$1.010^{25/25}$	....	218.5	i.	s.	s.
bromide (n-)	$CH_3CH_2CH_2Br$	123.00	col. lq.	$1.353^{20/4}$	−109.9	70.8	0.25^{20}	∞	∞
(i-)	$(CH_3)_2CHBr$	123.00	col. lq.	$1.310^{20/4}$	−89	60	0.32^{20}	∞	∞
n-butyrate (n-)	$C_2H_5CH_2CO_2CH_2C_2H_5$	130.18	col. lq.	0.879^{15}	−95.2	142.7	0.17^{17}	∞	∞
i-butyrate (n-)	$(CH_3)_2CHCO_2CH_2C_2H_5$	130.18	col. lq.	$0.884^{0/4}$	....	134–5	v. sl. s.		
n-butyrate (i-)	$C_2H_5CH_2CO_2CH(CH_3)_2$	130.18	col. lq.	0.865^{18}	....	128	v. sl. s.		
i-butyrate (i-)	$(CH_3)_2CHCO_2CH(CH_3)_2$	130.18	col. lq.	$0.869^{0/4}$	....	120.8	v. sl. s.		
chloride (n-)	$CH_3CH_2CH_2Cl$	78.54	col. lq.	$0.890^{20/4}$	−122.8	46.4	0.27^{20}	∞	∞
(i-)	$(CH_3)_2CHCl$	78.54	col. lq.	0.859^{20}	−117	36.5	0.31^{20}	∞	∞

TABLE 3-2 Physical Properties of Organic Compounds *(Continued)*

Name	Formula	Formula weight	Form and color	Specific gravity	Melting point, °C.	Boiling point, °C.	Solubility in 100 parts: Water	Alcohol	Ether
Propyl formate (n-)	$HCO_2CH_2CH_2CH_3$	88.10	col. lq.	$0.901^{20/4}$	−92.9	81.3	2.2^{22}	∞	∞
(i-)	$HCO_2CH(CH_3)_2$	88.10	col. lq.	$0.873^{20/4}$	....	$68–71^{751}$	2.1^{22}	∞	∞
furoate (n-)	$C_4H_3O.CO_2C_3H_7$	154.16	col. lq.	$1.075^{26/4}$	....	211	v. sl. s.	s.	∞
lactate (n-)	$CH_3CH(OH)CO_2CH_2C_2H_5$	132.16	col. lq.	....	....	$122–3^{150}$	s.	s.	s.
(i-)	$CH_3CH(OH)CO_2CH(CH_3)_2$	132.16	col. lq.	....	....	167.5	s.	s.	s.
mercaptan (n-)	$CH_3CH_2CH_2SH$	76.15	lq.	$0.836^{25/4}$	−112	67–8	v. sl. s.	s.	s.
(i-)	$(CH_3)_2CHSH$	76.15	lq.	$0.809^{25/4}$	−130.7	58–60	v. sl. s.	∞	∞
propionate (n-)	$C_2H_5CO_2CH_2C_2H_5$	116.16	col. lq.	$0.883^{20/4}$	−76	122–3	0.56^{25}	∞	∞
(i-)	$C_2H_5CO_2CH(CH_3)_2$	116.16	col. lq.	0.893^{0}	....	$109–11^{750}$	0.6^{25}	∞	∞
thiocyanate (i-)	$(CH_3)_2CH.CNS$	101.16	lq.	0.963^{20}	....	$152–3^{754}$	i.	∞	∞
n-valerate (n-)	$CH_3(CH_2)_3CO_2CH_2C_2H_5$	144.21	lq.	0.874^{15}	−70.7	67.5	i.	∞	∞
i-valerate (n-)	$(CH_3)_2CHCH_2CO_2C_3H_7$	144.21	col. lq.	$0.863^{20/4}$	....	155.9	i.	∞	∞
i-valerate (i-)	$(CH_3)_2CHCH_2CO_2C_3H_7$	144.21	col. lq.	0.854^{17}	....	142^{756}			
Propylene	$CH_3CH:CH_2$	42.08	gas	$0.609^{-47/4}$	−185	-48^{749}	44.6 cc.	1200 cc.	-
bromide	$CH_3CHBrCH_2Br$	201.91	col. lq.	$1.933^{20/4}$	−55.5	141.6	0.25^{20}	s.	v. s.
chlorohydrin	$CH_3CHClCH_2OH$	94.54	col. lq.	1.103^{20}	....	133–4	s.	s.	s.
chloride	$CH_3CHClCH_2Cl$	112.99	col. lq.	$1.159^{20/20}$	<−70	96.8	0.27^{20}	v. s.	v. s.
glycol	$CH_3CH(OH)CH_2OH$	76.09	col. oil	$1.040^{19.4}$	....	188–9	∞	∞	8
oxide	$CH_3(CHCH_2)O$	58.08	col. lq.	$0.831^{20/20}$	....	35	33^{20}	∞	∞
Protocatechuic acid (3-,4-)	$(HO)_2C_6H_3CO_2H.H_2O$	172.13	nd./aq.	$1.542^{4/4}$	199 d.	....	1.82^{14}	v. s.	s.
Pulegol (iso-)(d-)	$C_{10}H_{17}OH$	154.24	col. lq.	$0.911^{20/4}$	....	$86–9^{10}$	v. sl. s.		
Pulegone	$C_{10}H_{16}O$	152.23	col. lq.	$0.932^{20/20}$	....	224^{754}	i.	∞	∞
Pyrazole	—NH.N:CH.CH:CH—	68.08	nd./et.	....	70	186–8	s.	s.	s.
Pyrazoline	—NH.N:CH.CH_2CH_2—	70.09	lq.	....	....	144	∞	∞	sl. s.
Pyrazolone	—NH.CO.CH_2CH:N—	84.08	nd.	....	165	subl. d.	s.	v. s.	v. sl. s.
Pyrene	$C_{16}H_{10}$	202.24	yel. pr.	$1.277^{0/4}$	149–50	>360	i.	3 h.	v. s.
Pyridazine	$N_2 < (CHCH)_2 >$	80.09	lq.	$1.107^{20/4}$	−8	208	∞	s.	s.
Pyridine	$CH < (CHCH)_2 > N$	79.10	col. lq.	$0.982^{20/4}$	−42	115–6	∞	∞	s.
Pyrocatechol (o-)	$C_6H_4(OH)_2$	110.11	nd./aq.	1.344^{4}	104–5	240–5	45.1^{20}	v. s.	v. s.
Pyrogallol (1-,2-,3-)	$C_6H_3(OH)_3$	126.11	nd.	1.453^{4}	133–4	309	40^{13}	s.	s.
Pyrone	$CO < (CHCH)_2 > O$	96.08	cr.	$1.190^{40.3}$	32.5	215–7	v. sl. s.	s.	v. s.
Pyrrole	$< (CH:CH)_2 > NH$	67.09	lq.	$0.948^{20/4}$	....	131	i.	s.	s.
Pyrrolidine	$< (CH_2.CH_2)_2 > NH$	71.12	lq.	$0.852^{22.5}$	....	87–8	∞	∞	∞
Pyrroline	$< (CH.CH_2)_2 > NH$	69.10	lq.	$0.910^{20/4}$	....	90–1	v. s.	∞	∞
Pyruvic acid	CH_3COCO_2H	88.06	col. lq.	$1.267^{20/4}$	13.6	165	∞	∞	∞
Quercitrin	$C_{21}H_{20}O_{11}.2H_2O$	484.40	yel. nd.	....	182–5	....	0.04^{20}	s.	sl. s.
Quinaldine (py-2)	$CH_3.C_9H_6N$	143.18	lq.	$1.059^{20/4}$	−1	$244–5^{750}$	v. sl. s.	....	s.
Quinoline	C_9H_7N	129.15	lq.	1.095^{20}	−15	237.1^{747}	6	∞	∞
(iso-)	C_9H_7N	129.15	pl.	$1.099^{21/4}$	24.6	240.5^{763}	sl. s.	....	s.
-diol (1-,3-)	—C_6H_4CH:C(OH)N:C(OH)—	161.15	cr.	....	237	....	v. sl. s.		
Quinone (p-)	$CO < (CHCH)_2 > CO$	108.09	yel. mn.	$1.318^{20/4}$	115.7	subl.	sl. s. h.	s.	s.
R-acid Ca salt (2-)(3-,6-)	$HOC_{10}H_5(SO_3)_2Ca$	342.35	cr.	....	....	....	30.6^{25}		
K salt	$HOC_{10}H_5(SO_3K)_2$	380.46	cr.	....	....	....	29.5^{25}		
Na salt	$HOC_{10}H_5(SO_3Na)_2$	348.26	cr.	....	....	....	25.2^{25}		
Raffinose	$C_{18}H_{32}O_{16}.5H_2O$	594.52	cr./aq.	1.465^{0}	119	d. 130	14.3^{20}	0.1^{20}	
Resorcinol (m-)	$C_6H_4(OH)_2$	110.11	col. rhb.	1.272^{15}	110.7	276.5	147^{12}	v. s.	v. s.
Retene	$C_{18}H_{18}$	234.32	lf./al.	1.13^{16}	98–9	390–4	i.	69 h.	v. s. h.
Rhamnose (β-)	$CH_3(CHOH)_4CHO.H_2O$	182.17	col. mn.	$1.471^{20/4}$	126	....	60.8^{21}	....	i.
Ricinoleic acid	$C_{17}H_{32}(OH)CO_2H$	298.45	lq.	0.954^{16}	4–5	$226–8^{10}$	i.	∞	∞
Rosaniline	$C_{20}H_{21}ON_3$	319.39	col. nd.	....	186 d.	....	v. sl. s.	sl. s.	i.
Rosolic acid	$C_{20}H_{16}O_3$	304.33	red lf.	....	308–10 d.	....	0.12^{25}	v. s. h.	sl. s.
Saccharin	$C_6H_4(CO)(SO_2) > NH$	183.18	mn.	....	225–8	subl.	0.4^{25}	3.1 c.	1.05 c.
Safrole (1-,3-,4-)	$CH_2:CHCH_2.C_6H_3:O_2CH_2$	162.18	col. mn.	$1.100^{20/4}$	11.2	233–4	i.	s.	∞
(iso-)(1-,3-,4-)	$CH_3.CH:CH.C_6H_3:O_2CH_2$	162.18	col. lq.	$1.122^{20/4}$	6–7	252–3	i.	∞	∞
Salicylic acid (o-)	$HO.C_6H_4.CO_2H$	138.12	mn.	$1.443^{20/4}$	159	211^{20}	0.2^{23}	49^{15}	51^{15}
aldehyde (o-)	$HO.C_6H_4.CHO$	122.12	col. oil	$1.153^{25/4}$	−7	196.5	1.7^{86}	∞	∞
Saligenin	$HO.C_6H_4.CH_2OH$	124.13	rhb./aq.	1.161^{25}	86–7	subl.	6.6^{15}	v. s.	v.s.
Schaeffer's salt, Ca	$(HOC_{10}H_6SO_3)_2Ca.5H_2O$	576.59	cr.	....	....	....	4.76^{20}		
K	$HOC_{10}H_6SO_3K$	262.31	cr.	....	....	....	3.46^{25}		
Na	$HOC_{10}H_6SO_3Na$	246.21	cr.	....	....	....	6.29^{25}		
Semicarbazide	$NH_2.CO.NH.NH_2$	75.07	pr./al.	....	96	....	v. s.	v. s.	i.
hydrochloride	$NH_2.CO.NH.NH_3Cl$	111.54	pr.	....	173 d.	....	v. s.	sl. s.	i.
Skatole (3-)	$CH_3.C_8H_6N$	131.17	lf.	....	95	$265–6^{755}$	0.05 c.	s.	s.
Sodium methylate	CH_3ONa	54.03	pd.	....	d. 300	....	d.		
Sorbitol	$[CH_2OH(CHOH)_2]_2$	182.17	cr.	....	110–2	....	v. s.	v. s. h.	
Sorbose (d- or l-)	$C_6H_{12}O_6$	180.16	rhb.	1.654^{15}	165	....	55^{17}	sl. s.	
Starch	$(C_6H_{10}O_5)x$	162.14	amor.	1.50^{21}	d.	....	i.	i.	i

Stearic acid	$CH_3(CH_2)_{16}CO_2H$	284.47	mn.	$0.847^{69.3}$	70–1	291^{110}	0.03^{25}	2^{20}	6
amide	$CH_3(CH_2)_{16}CONH_2$	283.48	col. cr.	....	108–9	251^{12}	i.	s. h.	s. h.
Styrene	$C_6H_5CH:CH_2$	104.14	col. lq.	$0.903^{20/4}$	−31	145–6	v. sl. s.	∞	∞
Suberic acid	$HO_2C(CH_2)_6CO_2H$	174.19	nd./aq.	$1.266^{25/4}$	140–4	279^{100}	0.14^{16}	s.	0.8^{15}
Succinic acid	$HO_2C(CH_2)_2CO_2H$	118.09	col. mn.	$1.572^{25/4}$	189–90	235 d.	6.8^{20}	9.9^{15}	1.2^{15}
Sucrose	$C_{12}H_{22}O_{11}$	342.30	col. mn.	1.588^{15}	170–86 d.	....	179^{0}	0.9	i.
Sulfanilic acid (p-)	$H_2N.C_6H_4.SO_3H$	173.18	col. cr.	....	d. >280	....	0.8^{10}	v. sl. s.	v. sl. s.
Sylvestrene (d-)	$C_{10}H_{16}$	136.23	lq.	$0.863^{20/4}$	....	176–7			
Tartaric acid (meso-)	$(CHOHCO_2H)_2$	150.09	cr.	1.737	159–60	....	120^{15}		
(racemic)	$(CHOHCO_2H)_2.H_2O$	168.10	tri.	$1.697^{20/4}$	205–6	....	20.6^{20}	2^{0}	0.09
(d- or l-)	$(CHOHCO_2H)_2$	150.09	mn.	$1.760^{20/4}$	168–70	d.	139^{20}	25^{15}	0.4^{15}
Tartronic acid	$CH(OH)(CO_2H)_2.\frac{1}{2}H_2O$	129.07	pr./aq.	....	d. 155–8	subl.	v. s.	v. s.	i.
Terephthalic acid (p-)	$C_6H_4(CO_2H)_2$	166.13	cr.	1.510	subl.	....	0.001 c.	sl. s. h.	i.
Terpin hydrate (cis-)	$C_{10}H_{20}O_2.H_2O$	190.28	rhb.	....	117	d.	0.4^{15}	10^{15}	1^{15}
Terpineol (α-)(d- or l-)	$C_{10}H_{18}O$	154.24	col. cr.	0.935^{15}	38–40	219–21	i.	v. s.	v. s.
(dl-)	$C_{10}H_{18}O$	154.24	col. cr.	$0.935^{20/20}$	35	$218–9^{752}$	i.	v. s.	v. s.
Terpinyl acetate (α-)(dl-)	$CH_3CO_2.C_{10}H_{17}$	196.28	lq.	$0.966^{20/4}$	<−50	220 d.	i.	20	
Tetrabromo-ethane (sym)	$Br_2CH.CHBr_2$	345.70	col. lq.	$2.964^{20/4}$	−1.0	151^{54}	i.	∞	∞
(uns)	$Br_3C.CH_2Br$	345.70	col. lq.	$2.875^{20/4}$	0	104^{13}	....	s.	
Tetrachloro-ethane (sym)	$Cl_2CH.CHCl_2$	167.86	col. lq.	$1.600^{20/4}$	−36	146.3	0.29^{20}	∞	∞
(uns)	$Cl_3C.CH_2Cl$	167.86	lq.	$1.588^{20/4}$	....	129–30	i.	∞	∞
-ethylene	$Cl_2C:CCl_2$	165.85	col. lq.	$1.624^{15/4}$	−19	120.8	0.02^{20}	∞	∞
Tetracosane (n-)	$CH_3(CH_2)_{22}CH_3$	338.64	cr.	$0.779^{51/4}$	51.1	324	....	....	s.
Tetradecane (n-)	$CH_3(CH_2)_{12}CH_3$	198.38	col. lq.	$0.765^{20/4}$	5.5	252.5	i.	v. s.	v. s.
Tetraethyl-thiuram disulfide	$[(C_2H_5)_2NCS]_2S_2$	296.52	cr.	1.17	70	....	i.		
Tetrafluoro-ethylene	$F_2C:CF_2$	100.02	gas	1.58^{-78}	−142.5	−76.3	0.01^{30}		
Tetrahydro-furan	$—CH_2(CH_2)_2CH_2.O—$	72.10	col. lq.	$0.888^{21/4}$	−65	65–6	s.	s.	s.
-furfuryl alcohol	$C_4H_7O.CH_2OH$	102.13	col. lq.	$1.050^{20/4}$	....	$177–8^{743}$	∞	∞	∞
-pyran	$—CH_2(CH_2)_3CH_2.O—$	86.13	lq.	$0.881^{20/4}$	....	88	s.		
Tetralin	$—C_6H_4CH_2(CH_2)_2CH_2—$	132.20	col. lq.	$0.973^{18/4}$	−31	206^{764}	i.	s.	s.
Tetramethyl-thiuram disulfide	$[(CH_3)_2NCS]_2S_2$	240.41	cr.	1.29	155–6	....	i.		
Tetryl (2-,4-,6-)	$(NO_2)_3C_6H_2.N(CH_3)NO_2$	287.15	yel. mn.	1.57^{19}	130.5	expl.	i.	s. h.	s.
Theobromine	$C_7H_8O_2N_4$	180.17	rhb.	....	330	....	0.06^{15}	0.06 c.	0.03 h.
Thio-acetic acid	$CH_3.CO.SH$	76.11	yel. lq.	1.074^{10}	<−17	93	s.	∞	∞
-aniline (4-, 4′-)	$(NH_2.C_6H_4)_2S$	216.29	nd./aq.	....	108	....	sl. s. h.	s.	s.
-carbanilide	$(C_6H_5.NH)_2CS$	228.30	rhb./al.	1.3^{24}	154	d.	i.	v. s.	v. s.
-naphthol (β-)	$C_{10}H_7.SH$	160.22	cr./al.	....	81	286–8	v. sl. s.	v. s.	v. s.
-phenol	$C_6H_5.SH$	110.17	col. lq.	$1.074^{23/4}$	....	168–9	v. sl. s.	v. s.	∞
-salicylic acid (o-)	$HS.C_6H_4.CO_2H$	154.18	yel. nd.	....	164	subl.	sl. s. h.	s.	
-urea	$NH_2.CS.NH_2$	76.12	rhb./al.	$1.405^{20/4}$	180–2	d.	9.2^{13}	s.	sl. s.
Thiophene	$<(CH:CH)_2>S$	84.13	col. lq.	$1.070^{15/4}$	−30	84	i.	s.	
Thymol (5-,2-,1-)	$(CH_3)(C_3H_7)C_6H_3OH$	150.21	cr.	$0.972^{25/25}$	51.5	232^{752}	0.09^{19}	v. s.	v. s.
Tolidine (0-)(3-,3′-,4-,4′-)	$[CH_3(NH_2)C_6H_3]_2$	212.28	lf.	....	128–9	....	v. sl. s.	s.	s.
Toluene	$C_6H_5.CH_3$	92.13	col. lq.	$0.866^{20/4}$	−95	110.8	0.05^{16}	s.	∞
sulfonic acid (o-)	$CH_3.C_6H_4SO_3H.2H_2O$	208.23	cr.	....	d.	128.8^{0}	v. s.	s.	
(p-)	$CH_3.C_6H_4SO_3H.H_2O$	190.21	mn.	....	104–5	$146–7^{0}$	v. s.	s.	
sulfonic amide (p-)	$CH_3.C_6H_4SO_2NH_2$	171.21	mn.	....	137	....	0.2^{9}	7.4^{5}	
sulfonic chloride (p-)	$CH_3.C_6H_4.SO_2Cl$	190.64	tri.	....	69	134.5^{10}	i.	s.	s.
Toluic acid (o-)	$CH_3.C_6H_4.CO_2H$	136.14	cr./aq.	$1.062^{115/4}$	104–5	259^{751}	2.17^{100}	v. s.	
(m-)	$CH_3.C_6H_4.CO_2H$	136.14	pr./aq.	$1.054^{112/4}$	110–1	263	1.6^{100}	v. s.	v. s.
(p-)	$CH_3.C_6H_4.CO_2H$	136.14	cr./aq.	....	179–80	274–5	1.3^{100}	v. s.	v. s.
Toluidine (o-)	$CH_3.C_6H_4.NH_2$	107.15	col. lq.	$0.999^{20/4}$	−16.3	199.7	1.5^{25}	∞	∞
(m-)	$CH_3.C_6H_4.NH_2$	107.15	col. lq.	$0.989^{20/4}$	−31.5	203.3	sl. s.	∞	∞
(p-)	$CH_3.C_6H_4.NH_2$	107.15	cr.	$1.046^{20/4}$	44–5	200.3	0.74^{21}	v s.	v. s.
hydrochloride (o-)	$CH_3.C_6H_4.NH_3Cl$	143.62	mn. pr.	....	218–20	242	s.	sl. s.	
sulfonic acid (1-,2-,3-)	$CH_3(NH_2)C_6H_3SO_3H$	187.21	cr.	....	....	....	0.97^{11}		
Toluylenediamine (1-,2-,4-)	$CH_3.C_6H_3(NH_2)_2$	122.17	rhb.	....	99	283–5	s. h.	s.	s.
Tolylene diisocyanate (1-,2-,4-)	$CH_3.C_6H_3(NCO)_2$	174.15	lq.	1.23^{28}	....	134.5^{20}	d.	d.	
Trehalose	$C_{12}H_{22}O_{11}.2H_2O$	378.33	rhb./al.	....	97	....	s. h.	sl. s. h.	i.
Triamylamine (n-)	$[CH_3(CH_2)_3CH_2]_3N$	227.42	lq.	....	....	240–5	i.		
(i-)	$[(CH_3)_2CH(CH_2)_2]_3N$	227.42	col. lq.	$0.786^{20/4}$	....	235	i.		
Tributyl-amine (n-)	$[CH_3(CH_2)_2CH_2]_3N$	185.34	col. lq.	$0.778^{20/20}$	....	216.5^{761}	i.	s.	∞
phosphite	$[CH_3(CH_2)_3O]_3P$	250.32	lq.	$0.925^{20/4}$	....	$122–3^{12}$	i.		
Trichloro-acetic acid	$Cl_3C.CO_2H$	163.40	cr.	$1.617^{46/15}$	58	195.5^{754}	120^{25}	s.	s.
-benzene (s-)(1-,3-,5-)	$C_6H_3Cl_3$	181.46	nd.	....	63.5	208.5^{764}	i.	sl. s.	
-ethane (1-,1-,1-)	$Cl_3C.CH_3$	133.42	lq.	$1.325^{26/4}$	....	74.1	i.	∞	∞
-ethylene	$Cl_2C:CHCl$	131.40	col. lq.	$1.466^{20/20}$	−73	87.2	0.1^{25}	∞	∞
-phenol	$Cl_3C_6H_2OH$	197.46	nd.	$1.490^{75/4}$	68–9	246	0.09^{25}	v. s.	v. s.
Tricosane (n-)	$CH_3(CH_2)_{21}CH_3$	324.61	lf.	$0.779^{48/4}$	47.7	234^{15}	i.		
Tricresyl phosphate (o-)	$OP(OC_6H_4CH_3)_3$	368.36	lq.	....	....	....	i.		
Tridecane (n-)	$CH_3(CH_2)_{11}CH_3$	184.35	col. lq.	$0.757^{20/4}$	−6.2	234	i.	v. s.	v. s.
Triethanol amine	$(HOCH_2CH_2)_3N$	149.19	col. lq.	$1.126^{20/20}$	20–1	$277–9^{150}$	∞	∞	sl. s.

TABLE 3-2 Physical Properties of Organic Compounds *(Concluded)*

Name	Formula	Formula weight	Form and color	Specific gravity	Melting point, °C.	Boiling point, °C.	Solubility in 100 parts: Water	Alcohol	Ether
Triethyl-amine	$(CH_3CH_2)_3N$	101.19	col. oil	$0.729^{20/20}$	−114.8	89.4	∞ > 19^{0}	∞	∞
-benzene (1-,3-,5-)	$(C_2H_5)_3C_6H_3$	162.26	lq.	$0.861^{20/4}$	....	215	i.	s.	s.
(1-,2-,4-)	$(C_2H_5)_3C_6H_3$	162.26	lq.	$0.882^{17/4}$	....	$217\text{–}8^{755}$	i.	s.	s.
borate	$B(OCH_2CH_3)_3$	146.00	lq.	$0.864^{20/20}$	....	120	d.		
citrate	$HOC_3H_4(CO_2C_2H_5)_3$	276.28	oil	$1.137^{20/4}$	....	294	i.	∞	∞
Triethylene glycol	$(.CH_2OCH_2CH_2OH)_2$	150.17	col. lq.	$1.125^{20/20}$	−5	290	∞	∞	v. sl. s.
Trifluoro-chloromethane	CF_3Cl	104.47	gas	1.726^{-130}	−182	−80			
chloroethylene	$F_2C:CFCl$	116.48	gas	....	−157.5	−27.9	d.		
-trichloroethane	$Cl_2CF.CClF_2$	187.39	lq.	$1.576^{20/4}$	−35	47.6	i.	∞	∞
Trimethoxybutane (1-,3-,3-)	$CH_2(OCH_3)CH_2C(OCH_3)_2CH_3$	148.20	lq.	0.932	....	$63\text{–}5^{25}$	d.		
Trimethylamine	$(CH_3)_3N$	59.11	gas	0.662^{-5}	−124	3.5	41^{19}	s.	s.
Trimethylene bromide	$BrCH_2CH_2CH_2Br$	201.91	lq.	$1.987^{15/4}$	−34.4	167.5	0.17^{30}	s.	s.
chloride	$ClCH_2CH_2CH_2Cl$	112.99	lq.	1.201^{15}	....	123–5	0.27^{25}	s.	s.
glycol	$HOCH_2CH_2CH_2OH$	76.09	oil	$1.060^{20/4}$	....	214	∞	∞	
Trinitro-benzene (1-,3-,5-)	$C_6H_3(NO_2)_3$	213.11	col. rhb.	$1.688^{20/4}$	121	d.	0.03^{15}	1.9^{18}	1.5^{18}
-benzoic acid (2-,4-,6-)	$(NO_2)_3C_6H_2CO_2H$	257.12	rhb./aq.	....	210–20 d.	....	2.05^{24}		
-tert-butylxylene	$(NO_2)_3C_6(CH_3)_2C_4H_9$	297.26	nd./al.	....	110	....	i.	sl. s.	s.
-naphthalene (α-)(1-,3-,5-)	$C_{10}H_5(NO_2)_3$	263.16	rhb.	....	122–3	....	i.	s.	
(β-)(1-,3-,8-)	$C_{10}H_5(NO_2)_3$	263.16	cr./al.	....	218–9	....	0.02^{100}	0.05^{23}	0.13^{15}
(γ-)(1-,4-,5-)	$C_{10}H_5(NO_2)_3$	263.16	yel. cr.	....	148–9	....	i.	0.11^{19}	0.4^{19}
-phenol (2-,3-,6-)	$(NO_2)_3C_6H_2OH$	229.11	nd.	....	117–8	....	s. h.	v. s.	v. s.
-toluene (β-)(2-,3-,4-)	$CH_3C_6H_2(NO_2)_3$	227.13	cr.	$1.620^{20/4}$	112	expl.	i.	sl. s. c.	s.
(γ-)(2-,4-,5-)	$CH_3C_6H_2(NO_2)_3$	227.13	yel. pl.	$1.620^{20/4}$	104	expl.	i.	s. h.	v. s.
(α-)(2-,4-,6-)	$CH_3C_6H_2(NO_2)_3$	227.13	cr./al.	1.654	80.8	expl.	0.01^{20}	1.5^{22}	5^{33}
Trional	$C_2H_5(CH_3)C(SO_2C_2H_5)_2$	242.34	pl./al.	$1.199^{85/4}$	76	d.	0.3^{15}	5^{0}	6.6^{15}
Triphenyl-arsine	$(C_6H_5)_3As$	306.21	pl.	1.306	59–60	>360	i.	s.	v. s.
carbinol	$(C_6H_5)_3COH$	260.32	cr.	$1.188^{20/4}$	162.5	>360	i.	v. s.	v. s.
guanidine (α-)	$C_6H_5N:C(NHC_6H_5)_2$	287.35	rhb./al.	1.13	144–5	d.	i.	4^{0}	
methane	$(C_6H_5)_3CH$	244.32	cr.	$1.014^{99/4}$	93.4	359^{754}	i.	v. s. h.	v. s.
methyl	$(C_6H_5)_3C$ . . .	243.31	col. cr.	....	145–7	d.	i.	sl. s. h.	
phosphate	$OP(OC_6H_5)_3$	326.28	pr./al.	$1.206^{58/4}$	49–50	245^{11}	i.	155^{25}	v. s.
Tripropylamine (n-)	$(CH_3CH_2CH_2)_3N$	143.27	col. lq.	$0.757^{20/4}$	−93.5	156.5	v. sl. s.	∞	∞
Undecane (n-)	$CH_3(CH_2)_9CH_3$	156.30	col. lq.	$0.741^{20/4}$	−25.6	194.5	i.	∞	∞
Urea	$H_2N.CO.NH_2$	60.06	col. pr.	$1.335^{20/4}$	132.7	d.	100^{17}	20^{20}	sl. s.
nitrate	$CO(NH_2)_2.HNO_3$	123.07	col. mn.	....	152 d.	....	v. s. h.	s.	
Uric acid	$C_5H_4O_3N_4$	168.11	cr.	1.893^{20}	d.	....	0.06 h.	i.	i.
Valeric acid (n-)	$C_2H_5CH_2CH_2CO_2H$	102.13	col. lq.	$0.939^{20/4}$	−34.5	187	3.3^{16}	∞	∞
(i-)	$(CH_3)_2CHCH_2CO_2H$	102.13	col. lq.	$0.931^{20/20}$	−37.6	176	4.2^{20}	∞	∞
aldehyde (n-)	$C_2H_5CH_2CH_2CHO$	86.13	lq.	0.819^{11}	−92	103.4	v. sl. s.	s.	s.
(i-)	$(CH_3)_2CHCH_2CHO$	86.13	col. lq.	0.803^{17}	−51	92.5	sl. s.	s.	s.
amide (n-)	$C_2H_5CH_2CH_2CONH_2$	101.15	mn. pl.	1.023	106	....	v. s.	v. s.	v. s.
(i-)	$(CH_3)_2CHCH_2CONH_2$	101.15	mn.	$0.965^{20/4}$	135–7	232	s.	s.	s.
Vanillic acid (3-,4-,1-)	$CH_3O(OH)C_6H_3CO_2H$	168.14	nd./aq.	....	207	subl.	0.12^{14}	v. s.	v. s.
alcohol (3-,4-,1-)	$CH_3O(OH)C_6H_3CH_2OH$	154.16	mn./aq.	....	115	d.	v. s. h.	v. s.	v. s.
Vanillin (3-,4-,1-)	$CH_3O(OH)C_6H_3CHO$	152.14	mn.	1.056	81–2	285	1^{14}	v. s.	v. s.
Veratrole (o-)	$C_6H_4(OCH_3)_2$	138.16	cr.	$1.091^{15/15}$	22.5	207.1	v. sl. s.	s.	s.
Vinyl acetate	$CH_3CO_2CH:CH_2$	86.09	col. lq.	$0.932^{20/4}$	<−60	72–3	2^{20}	∞	∞
(poly-)	$(CH_3CO_2CH:CH_2)x$	(86.09)	....	1.19^{20}	100–25	....	i.		
acetic acid	$CH_2:CH.CH_2CO_2H$	86.09	col. lq.	$1.013^{15/15}$	−39	163	s.	∞	∞
acetylene	$CH_2:CH.C:CH$	52.07	gas	$0.705^{1.5}$	....	5.5	$0.67^{0.8}$		
alcohol	$CH_2:CHOH$	44.06							
(poly-)	$(CH_2:CHOH)x$	(44.06)	....	1.3^{20}	d. >200	....	s.		
chloride	$CH_2:CHCl$	62.50	gas	$0.908^{25/25}$	−160	−12	sl. s.	s.	v. s.
propionate	$C_2H_5CO_2CH:CH_2$	100.11	lq.	....	....	93–5	v. sl. s.		
Xylene (o-)	$C_6H_4(CH_3)_2$	106.16	col. lq.	$0.881^{20/4}$	−25	144	i.	s.	∞
(m-)	$C_6H_4(CH_3)_2$	106.16	col. lq.	$0.867^{17/4}$	−47.4	139.3	i.	s.	∞
(p-)	$C_6H_4(CH_3)_2$	106.16	col. lq.	$0.861^{20/4}$	13.2	138.5	i.	s.	v. s.
sulfonic acid (1-,4-,2-)	$(CH_3)_2C_6H_3SO_3H.2H_2O$	222.25	col. lf.	....	86	$149^{0.1}$	s.		
Xylidine (1:2)(3-)	$(CH_3)_2C_6H_3NH_2$	121.18	lq.	0.991^{15}	<−15	223	v. sl. s.	s.	s.
(1:2)(4-)	$(CH_3)_2C_6H_3NH_2$	121.18	pr.	$1.076^{17.5}$	49–50	224–6	v. sl. s.		
(1:3)(2-)	$(CH_3)_2C_6H_3NH_2$	121.18	lq.	0.980^{15}	....	216–7	v. sl. s.		
(1:3)(4-)	$(CH_3)_2C_6H_3NH_2$	121.18	lq.	$0.978^{20/4}$	....	213–4	v. sl. s.		
(1:3)(5-)	$(CH_3)_2C_6H_3NH_2$	121.18	oil	$0.972^{20/4}$	....	221–2	v. sl. s.		
(1:4)(2-)	$(CH_3)_2C_6H_3NH_2$	121.18	oil	$0.979^{21/4}$	15.5	215^{739}	v. sl. s.		
Xylose (l-)(+)	$CH_2OH(CHOH)_3CHO$	150.13	nd.	1.535^{0}	153–4	....	117^{20}	v. sl. s.	i.
Xylylene dichloride (p-)	$C_6H_4(CH_2Cl)_2$	175.06	mn.	1.417^{0}	100.5	240–5 d.	i.	s.	v. sl. s.
Zinc diethyl	$Zn(CH_2CH_3)_2$	123.50	col. lq.	1.182^{18}	−28	118	d.	d.	
dimethyl	$Zn(CH_3)_2$	95.45	col. lq.	1.386^{11}	−40	46	d.	d.	
dimethyl-dithiocarbamate	$Zn[S_2CN(CH_3)_2]_2$	305.79	....	$2.00^{40/4}$	248–50	....	i.		

NOTE: °F = 9/5 °C + 32.

VAPOR PRESSURES OF PURE SUBSTANCES

UNITS CONVERSIONS

For this subsection, the following units conversions are applicable:

°F = 9/5 °C + 32.

To convert millimeters of mercury to pounds-force per square inch, multiply by 0.01934.

ADDITIONAL REFERENCES

Additional compilations of vapor-pressure data include Boublik, Fried, and Hala, *The Vapor Pressures of Pure Substances*, Elsevier, Amsterdam, 1984. See also Hirata, Ohe, and Nagahama, *Computer Aided Data Book of Vapor-Liquid Equilibria*, Kodansha/Elsevier, Tokyo, 1975; Weishaupt, *Landolt-Börnstein New Series Group IV*, vol. 3: *Thermodynamic Equilibria of Boiling Mixtures*, Springer-Verlag, Berlin, 1975; Wichterle, Linek, and Hala, *Vapor-Liquid Equilibrium Data Bibliography*, Elsevier, Amsterdam, 1973; suppl. 1, 1976; suppl. 2, 1982.

TABLE 3-3 Vapor Pressure of Water Ice from −15 to 0°C*

mmHg

t, °C.	0.0	0.1	0.2	0.3	0.4	0.5	0.6	0.7	0.8	0.9
−14	1.361	1.348	1.336	1.324	1.312	1.300	1.288	1.276	1.264	1.253
−13	1.490	1.477	1.464	1.450	1.437	1.424	1.411	1.399	1.386	1.373
−12	1.632	1.617	1.602	1.588	1.574	1.559	1.546	1.532	1.518	1.504
−11	1.785	1.769	1.753	1.737	1.722	1.707	1.691	1.676	1.661	1.646
−10	1.950	1.934	1.916	1.899	1.883	1.866	1.849	1.833	1.817	1.800
− 9	2.131	2.112	2.093	2.075	2.057	2.039	2.021	2.003	1.985	1.968
− 8	2.326	2.306	2.285	2.266	2.246	2.226	2.207	2.187	2.168	2.149
− 7	2.537	2.515	2.493	2.472	2.450	2.429	2.408	2.387	2.367	2.346
− 6	2.765	2.742	2.718	2.695	2.672	2.649	2.626	2.603	2.581	2.559
− 5	3.013	2.987	2.962	2.937	2.912	2.887	2.862	2.838	2.813	2.790
− 4	3.280	3.252	3.225	3.198	3.171	3.144	3.117	3.091	3.065	3.039
− 3	3.568	3.539	3.509	3.480	3.451	3.422	3.393	3.364	3.336	3.308
− 2	3.880	3.848	3.816	3.785	3.753	3.722	3.691	3.660	3.630	3.599
− 1	4.217	4.182	4.147	4.113	4.079	4.045	4.012	3.979	3.946	3.913
− 0	4.579	4.542	4.504	4.467	4.431	4.395	4.359	4.323	4.287	4.252

* For data at 0(0.2)–30(2)–98°C. see p. 2324, "Handbook of Chemistry and Physics," 40th ed., Chemical Rubber Publishing Co.

TABLE 3-4 Vapor Pressure of Liquid Water from −16 to 0°C*

mmHg

t, °C.	0.0	0.1	0.2	0.3	0.4	0.5	0.6	0.7	0.8	0.9
−15	1.436	1.425	1.414	1.402	1.390	1.379	1.368	1.356	1.345	1.334
−14	1.560	1.547	1.534	1.522	1.511	1.497	1.485	1.472	1.460	1.449
−13	1.691	1.678	1.665	1.651	1.637	1.624	1.611	1.599	1.585	1.572
−12	1.834	1.819	1.804	1.790	1.776	1.761	1.748	1.734	1.720	1.705
−11	1.987	1.971	1.955	1.939	1.924	1.909	1.893	1.878	1.863	1.848
−10	2.149	2.134	2.116	2.099	2.084	2.067	2.050	2.034	2.018	2.001
− 9	2.326	2.307	2.289	2.271	2.254	2.236	2.219	2.201	2.184	2.167
− 8	2.514	2.495	2.475	2.456	2.437	2.418	2.399	2.380	2.362	2.343
− 7	2.715	2.695	2.674	2.654	2.633	2.613	2.593	2.572	2.553	2.533
− 6	2.931	2.909	2.887	2.866	2.843	2.822	2.800	2.778	2.757	2.736
− 5	3.163	3.139	3.115	3.092	3.069	3.046	3.022	3.000	2.976	2.955
− 4	3.410	3.384	3.359	3.334	3.309	3.284	3.259	3.235	3.211	3.187
− 3	3.673	3.647	3.620	3.593	3.567	3.540	3.514	3.487	3.461	3.436
− 2	3.956	3.927	3.898	3.871	3.841	3.813	3.785	3.757	3.730	3.702
− 1	4.258	4.227	4.196	4.165	4.135	4.105	4.075	4.045	4.016	3.986
− 0	4.579	4.546	4.513	4.480	4.448	4.416	4.385	4.353	4.320	4.289

* Computed from the above table with the aid of the thermodynamic equation

$$\log_{10} \frac{p_w}{p_i} = \frac{-1.1489t}{273.1 + t} - 1.330 \times 10^{-5}t^2 + 9.084 \times 10^{-8}t^3$$

TABLE 3-5 Vapor Pressure of Liquid Water from 0 to 100°C*

mmHg

t, °C.	0.0	0.1	0.2	0.3	0.4	0.5	0.6	0.7	0.8	0.9
0	4.579	4.613	4.647	4.681	4.715	4.750	4.785	4.820	4.855	4.890
1	4.926	4.962	4.998	5.034	5.070	5.107	5.144	5.181	5.219	5.256
2	5.294	5.332	5.370	5.408	5.447	5.486	5.525	5.565	5.605	5.645
3	5.685	5.725	5.766	5.807	5.848	5.889	5.931	5.973	6.015	6.058
4	6.101	6.144	6.187	6.230	6.274	6.318	6.363	6.408	6.453	6.498
5	6.543	6.589	6.635	6.681	6.728	6.775	6.822	6.869	6.917	6.965
6	7.013	7.062	7.111	7.160	7.209	7.259	7.309	7.360	7.411	7.462
7	7.513	7 565	7.617	7.669	7.722	7.775	7.828	7.882	7.936	7.990
8	8.045	8.100	8.155	8.211	8.267	8.323	8.380	8.437	8.494	8.551
9	8.609	8.668	8.727	8.786	8.845	8.905	8.965	9.025	9.086	9.147

TABLE 3-5 Vapor Pressure of Liquid Water from 0 to 100°C* (*Concluded*)

t, °C.	0.0	0.1	0.2	0.3	0.4	0.5	0.6	0.7	0.8	0.9
10	9.209	9.2 1	9.333	9.395	9.458	9.521	9.585	9.649	9.714	9.779
11	9.844	9.910	9.976	10.042	10.109	10.176	10.244	10.312	10.380	10.449
12	10.518	10.588	10.658	10.728	10.799	10.870	10.941	11.013	11.085	11.158
13	11.231	11.305	11.379	11.453	11.528	11.604	11.680	11.756	11.833	11.910
14	11.987	12.065	12.144	12.223	12.302	12.382	12.462	12.543	12.624	12.706
15	12.788	12.870	12.953	13.037	13.121	13.205	13.290	13.375	13.461	13.547
16	13.634	13.721	13.809	13.898	13.987	14.076	14.166	14.256	14.347	14.438
17	14.530	14.622	14.715	14.809	14.903	14.997	15.092	15.188	15.284	15.380
18	15.477	15.575	15.673	15.772	15.871	15.971	16.071	16.171	16.272	16.374
19	16.477	16.581	16.685	16.789	16.894	16.999	17.105	17.212	17.319	17.427
20	17.535	17.644	17.753	17.863	17.974	18.085	18.197	18.309	18.422	18.536
21	18.650	18.765	18.880	18.996	19.113	19.231	19.349	19.468	19.587	19.707
22	19.827	19.948	20.070	20.193	20.316	20.440	20.565	20.690	20.815	20.941
23	21.068	21.196	21.324	21.453	21.583	21.714	21.845	21.977	22.110	22.243
24	22.377	22.512	22.648	22.785	22.922	23.060	23.198	23.337	23.476	23.616
25	23.756	23.897	24.039	24.182	24.326	24.471	24.617	24.764	24.912	25.060
26	25.209	25.359	25.509	25.660	25.812	25.964	26.117	26.271	26.426	26.582
27	26.739	26.897	27.055	27.214	27.374	27.535	27.696	27.858	28.021	28.185
28	28.349	28.514	28.680	28.847	29.015	29.184	29.354	29.525	29.697	29.870
29	30.043	30.217	30.392	30.568	30.745	30.923	31.102	31.281	31.461	31.642
30	31.824	32.007	32.191	32.376	32.561	32.747	32.934	33.122	33.312	33.503
31	33.695	33.888	34.082	34.276	34.471	34.667	34.864	35.062	35.261	35.462
32	35.663	35.865	36.068	36.272	36.477	36.683	36.891	37.099	37.308	37.518
33	37 729	37.942	38.155	38.369	33.584	38.801	39.018	39.237	39.457	39.677
34	39.898	40.121	40.344	40.569	40.796	41.023	41.251	41.480	41.710	41.942
35	42.175	42.409	42.644	42.880	43.117	43.355	43.595	43.836	44.078	44.320
36	44.563	44.808	45.054	45.301	45.549	45.799	46.050	46.302	46.556	46.811
37	47.067	47.324	47.582	47.841	48.102	48.364	48.627	48.891	49.157	49.424
38	49.692	49.961	50.231	50.502	50.774	51.048	51.323	51.600	51.879	52.160
39	52.442	52.725	53.009	53.294	53.580	53.867	54.156	54.446	54.737	55.030
40	55.324	55.61	55.91	56.21	56.51	56.81	57.11	57.41	57.72	58.03
41	58.34	58.65	58.96	59.27	59.58	59.90	60.22	60.54	60.86	61.18
42	61.50	61.82	62.14	62.47	62.80	63.13	63.46	63.79	64.12	64.46
43	64.80	65.14	65.48	65.82	66.16	66.51	66.86	67.21	67.56	67.91
44	68.26	68.61	68.97	69.33	69.69	70.05	70.41	70.77	71.14	71.51
45	71.88	72.25	72.62	72.99	73.36	73.74	74.12	74.50	74.88	75.26
46	75.65	76.04	76.43	76.82	77.21	77.60	78.00	78.40	78.80	79.20
47	79.60	80.00	80.41	80.82	81.23	81.64	82.05	82.46	82.87	83.29
48	83.71	84.13	84.56	84.99	85.42	85.85	86.28	86.71	87.14	87.58
49	88.02	88.46	88.90	89.34	89.79	90.24	90.69	91.14	91.59	92.05

t, °C.	0	1	2	3	4	5	6	7	8	9
50	92.51	97.20	102.09	107.20	112.51	118.04	123.80	129.82	136.08	142.60
60	149.38	156.43	163.77	171.38	179.31	187.54	196.09	204.96	214.17	223.73
70	233.7	243.9	254.6	265.7	277.2	289.1	301.4	314.1	327.3	341.0
80	355.1	369.7	384.9	400.6	416.8	433.6	450.9	468.7	487.1	506.1
90	525.76	527.76	529.77	531.78	533.80	535.82	537.86	539.90	541.95	544.00
91	546.05	548.11	550.18	552.26	554.35	556.44	558.53	560.64	562.75	564.87
92	566.99	569.12	571.26	573.40	575.55	577.71	579.87	582.04	584.22	586.41
93	588.60	590.80	593.00	595.21	597.43	599.66	601.89	604.13	606.38	608.64
94	610.90	613.17	615.44	617.72	620.01	622.31	624.61	626.92	629.24	631.57
95	633.90	636.24	638.59	640.94	643.30	645.67	648.05	650.43	652.82	655.22
96	657.62	660.03	662.45	664.88	667.31	669.75	672.20	674.66	677.12	679.69
97	682.07	684.55	687.04	689.54	692.05	694.57	697.10	699.63	702.17	704.71
98	707.27	709.83	712.40	714.98	717.56	720 15	722.75	725.36	727.98	730.61
99	733.24	735.88	738.53	741.18	743.85	746.52	749.20	751.89	754.58	757.29
100	760.00	762.72	765.45	768.19	770.93	773.68	776.44	779.22	782.00	784.78
101	787.57	790.37	793.18	796.00	798.82	801.66	804.50	807.35	810.21	813.08

*From the Physikalisch-technische Reichsanstalt, Holborn, Scheel, and Henning, "Wärmetabellen," Friedrich Vieweg & Sohn, Brunswick, 1909. By permission. For data at 50(0.2) 101.8°C., see "Handbook of Chemistry and Physics," 40th ed., p. 2326, Chemical Rubber Publishing Co. For a tabulation of temperature for pressures 700(1)779 mm. Hg, see Atack, "Handbook of Chemical Data," p. 117, Reinhold, New York, 1957. For a tabulation of pressure for 105(5)200(10)370°C., see Atack, p. 134, and for 100(1)374°C., see "Handbook of Chemistry and Physics," 40th ed., pp. 2328–2330, Chemical Rubber Publishing Co.

TABLE 3-6 Vapor Pressures of Inorganic Compounds, above 1 atm*

Compound		Pressure, atm.									Critical point	
Name	Formula	1	2	5	10	20	30	40	50	60	t_c, °C.	P_c, atm.
		Temperature, °C.										
Ammonia	NH_3	−33.6	−18.7	+4.7	25.7	50.1	66.1	78.9	89.3	98.3	132.4	111.5
Carbon monoxide	CO	−191.3	−183.5	−170.7	−161.0	−149.7	−141.9	....	....	....	−138.7	34.6
dioxide	CO_2	−78.2	−69.1	−56.7	−39.5	−18.9	−5.3	+5.9	14.9	22.4	31.1	73.0
disulfide	CS_2	46.5	69.1	104.8	136.3	175.5	201.5	222.8	240.0	256.0	273.0	72.9
Chlorine	Cl_2	−33.8	−16.9	+10.3	35.6	65.0	84.8	101.6	115.2	127.1	144.0	76.1
para-Hydrogen	H_2	−252.5	−250.2	−246.0	−241.8	....	....	....	....	....	−240.0	12.80
Hydrogen bromide	HBr	−66.5	−51.5	−29.1	−8.4	+16.8	33.9	48.1	60.0	70.6	90.0	84.4
chloride	HCl	−84.8	−71.4	−50.5	−31.7	−8.8	+5.9	17.8	27.9	36.2	51.4	81.6
cyanide	HCN	25.9	45.8	75.8	102.7	135.0	153.8	169.9	183.5	....	183.5	50.0
Water	H_2O	100.0	120.1	152.4	180.5	213.1	234.6	251.1	264.7	276.5	374.2	218.0
Hydrogen sulfide	H_2S	−60.4	−45.9	−22.3	−0.4	+25.5	41.9	55.8	66.7	76.3	100.3	88.9
Krypton	Kr	−152.0	−143.5	−130.0	−118.0	−101.7	−88.8	−78.4	−66.5	....	−63	54
Nitrogen	N_2	−195.8	−189.2	−179.1	−169.8	−157.6	−148.3	....	....	....	−147.2	33.5
Oxygen	O_2	−183.1	−176.0	−164.5	−153.2	−140.0	−130.7	−124.1	....	....	−118.9	49.7
Sulfur dioxide	SO_2	−10.0	+6.3	32.1	55.5	83.8	102.6	118.0	130.2	141.7	157.2	77.7
trioxide	SO_3	44.8	60.0	82.5	104.0	138.0	157.8	175.0	187.8	198.0	218.3	83.6

* Compiled from the extended tables published by D. R. Stull in *Ind. Eng. Chem.*, **39**, 517 (1947).

TABLE 3-7 Vapor Pressures of Inorganic Compounds, up to 1 atm*

Compound: Name	Formula	Pressure, mm. Hg: 1	5	10	20	40	60	100	200	400	760	Melting point, °C.
		Temperature, °C.										
Aluminum	Al	1284	1421	1487	1555	1635	1684	1749	1844	1947	2056	660
borohydride	$Al(BH_4)_3$	......	−52.2	−42.9	−32.5	−20.9	−13.4	−3.9	+11.2	28.1	45.9	−64.
bromide	$AlBr_3$	81.3	103.8	118.0	134.0	150.6	161.7	176.1	199.8	227.0	256.3	97.
chloride	Al_2Cl_6	100.0	116.4	123.8	131.8	139.9	145.4	152.0	161.8	171.6	180.2	192.4
fluoride	AlF_3	1238	1298	1324	1350	1378	1398	1422	1457	1496	1537	1040
iodide	AlI_3	178.0	207.7	225.8	244.2	265.0	277.8	294.5	322.0	354.0	385.5	
oxide	Al_2O_3	2148	2306	2385	2465	2549	2599	2665	2766	2874	2977	2050
Ammonia	NH_3	−109.1	−97.5	−91.9	−85.8	−79.2	−74.3	−68.4	−57.0	−45.4	−33.6	−77.7
heavy	ND_3	......	......	......	......	......	−74.0	−67.4	−57.0	−45.4	−33.4	−74.0
Ammonium bromide	NH_4Br	198.3	234.5	252.0	270.6	290.0	303.8	320.0	345.3	370.8	396.0	
carbamate	$N_2H_6CO_2$	−26.1	−10.4	−2.9	+5.3	14.0	19.6	26.7	37.2	48.0	58.3	
chloride	NH_4Cl	160.4	193.8	209.8	226.1	245.0	256.2	271.5	293.2	316.5	337.8	520
cyanide	NH_4CN	−50.6	−35.7	−28.6	−20.9	−12.6	−7.4	−0.5	+9.6	20.5	31.7	36
hydrogen sulfide	NH_4HS	−51.1	−36.0	−28.7	−20.8	−12.3	−7.0	0.0	+10.5	21.8	33.3	
iodide	NH_4I	210.9	247.0	263.5	282.8	302.8	316.0	331.8	355.8	381.0	404.9	
Antimony	Sb	886	984	1033	1084	1141	1176	1223	1288	1364	1440	630.5
tribromide	$SbBr_3$	93.9	126.0	142.7	158.3	177.4	188.1	203.5	225.7	250.2	275.0	96.6
trichloride	$SbCl_3$	49.2	71.4	85.2	100.6	117.8	128.3	143.3	165.9	192.2	219.0	73.4
pentachloride	$SbCl_5$	22.7	48.6	61.8	75.8	91.0	101.0	114.1	......	......	......	2.8
triiodide	SbI_3	163.6	203.8	223.5	244.8	267.8	282.5	303.5	333.8	368.5	401.0	167
trioxide	Sb_4O_6	574	626	666	729	812	873	957	1085	1242	1425	656
Argon	A	−218.2	−213.9	−210.9	−207.9	−204.9	−202.9	−200.5	−195.6	−190.6	−185.6	−189.2
Arsenic	As	372	416	437	459	483	498	518	548	579	610	814
Arsenic tribromide	$AsBr_3$	41.8	70.6	85.2	101.3	118.7	130.0	145.2	167.7	193.6	220.0	
trichloride	$AsCl_3$	−11.4	+11.7	+23.5	36.0	50.0	58.7	70.9	89.2	109.7	130.4	−18
trifluoride	AsF_3	......	......	......	......	−2.5	+4.2	13.2	26.7	41.4	56.3	−5.9
pentafluoride	AsF_5	−117.9	−108.0	−103.1	−98.0	−92.4	−88.5	−84.3	−75.5	−64.0	−52.8	−79.8
trioxide	As_2O_3	212.5	242.6	259.7	279.2	299.2	310.3	332.5	370.0	412.2	457.2	312.8
Arsine	AsH_3	−142.6	−130.8	−124.7	−117.7	−110.2	−104.8	−98.0	−87.2	−75.2	−62.1	−116.3
Barium	Ba	......	984	1049	1120	1195	1240	1301	1403	1518	1638	850
Beryllium borohydride	$Be(BH_4)_2$	+1.0	19.8	28.1	36.8	46.2	51.7	58.6	69.0	79.7	90.0	123
bromide	$BeBr_2$	289	325	342	361	379	390	405	427	451	474	490
chloride	$BeCl_2$	291	328	346	365	384	395	411	435	461	487	405
iodide	BeI_2	283	322	341	361	382	394	411	435	461	487	488
Bismuth	Bi	1021	1099	1136	1177	1217	1240	1271	1319	1370	1420	271
tribromide	$BiBr_3$	......	261	282	305	327	340	360	392	425	461	218
trichloride	$BiCl_3$	......	242	264	287	311	324	343	372	405	441	230
Diborane hydrobromide	B_2H_5Br	−93.3	−75.3	−66.3	−56.4	−45.4	−38.2	−29.0	−15.4	0.0	+16.3	−104.2
Borine carbonyl	BH_3CO	−139.2	−127.3	−121.1	−114.1	−106.6	−101.9	−95.3	−85.5	−74.8	−64.0	−137.0
triamine	$B_3N_3H_6$	−63.0	−45.0	−35.3	−25.0	−13.2	−5.8	+4.0	18.5	34.3	50.6	−58.2
Boron hydrides												
dihydrodecaborane	$B_{10}H_{14}$	60.0	80.8	90.2	100.0	117.4	127.8	142.3	163.8	......	......	99.6
dihydrodiborane	B_2H_6	−159.7	−149.5	−144.3	−138.5	−131.6	−127.2	−120.9	−111.2	−99.6	−86.5	−169
dihydropentaborane	B_5H_9	......	−40.4	−30.7	−20.0	−8.0	−0.4	+9.6	24.6	40.8	58.1	−47.0
tetrahydropentaborane	B_5H_{11}	−50.2	−29.9	−19.9	−9.2	+2.7	10.2	20.1	34.8	51.2	67.0	
tetrahydrotetraborane	B_4H_{10}	−90.9	−73.1	−64.3	−54.8	−44.3	−37.4	−28.1	−14.0	+0.8	16.1	−119.9
Boron tribromide	BBr_3	−41.4	−20.4	−10.1	+1.5	14.0	22.1	33.5	50.3	70.0	91.7	−45
trichloride	BCl_3	−91.5	−75.2	−66.9	−57.9	−47.8	−41.2	−32.4	−18.9	−3.6	+12.7	−107
trifluoride	BF_3	−154.6	−145.4	−141.3	−136.4	−131.0	−127.6	−123.0	−115.9	−108.3	−100.7	−126.8
Bromine	Br_2	−48.7	−32.8	−25.0	−16.8	−8.0	−0.6	+9.3	24.3	41.0	58.2	−7.3
pentafluoride	BrF_5	−69.3	−51.0	−41.9	−32.0	−21.0	−14.0	−4.5	+9.9	25.7	40.4	−61.4
Cadmium	Cd	394	455	484	516	553	578	611	658	711	765	320.9
chloride	$CdCl_2$	......	618	656	695	736	762	797	847	908	967	568
fluoride	CdF_2	1112	1231	1286	1344	1400	1436	1486	1561	1651	1751	520
iodide	CdI_2	416	481	512	546	584	608	640	688	742	796	385
oxide	CdO	1000	1100	1149	1200	1257	1295	1341	1409	1484	1559	
Calcium	Ca	......	926	983	1046	1111	1152	1207	1288	1388	1487	851
Carbon (graphite)	C	3586	3828	3946	4069	4196	4273	4373	4516	4660	4827	
dioxide	CO_2	−134.3	−124.4	−119.5	−114.4	−108.6	−104.8	−100.2	−93.0	−85.7	−78.2	−57.5
disulfide	CS_2	−73.8	−54.3	−44.7	−34.3	−22.5	−15.3	−5.1	+10.4	28.0	46.5	−110.8
monoxide	CO	−222.0	−217.2	−215.0	−212.8	−210.0	−208.1	−205.7	−201.3	−196.3	−191.3	−205.0
oxyselenide	COSe	−117.1	−102.3	−95.0	−86.3	−76.4	−70.2	−61.7	−49.8	−35.6	−21.9	
oxysulfide	COS	−132.4	−119.8	−113.3	−106.0	−98.3	−93.0	−85.9	−75.0	−62.7	−49.9	−138.8
selenosulfide	CSeS	−47.3	−26.5	−16.0	−4.4	+8.6	17.0	28.3	45.7	65.2	85.6	−75.2
subsulfide	C_3S_2	14.0	41.2	54.9	69.3	85.6	96.0	109.9	130.8	......	......	+0.4
tetrabromide	CBr_4	......	......	......	......	96.3	106.3	119.7	139.7	163.5	189.5	90.1
tetrachloride	CCl_4	−50.0	−30.0	−19.6	−8.2	+4.3	12.3	23.0	38.3	57.8	76.7	−22.6
tetrafluoride	CF_4	−184.6	−174.1	−169.3	−164.3	−158.8	−155.4	−150.7	−143.6	−135.5	−127.7	−183.7
Cesium	Cs	279	341	375	409	449	474	509	561	624	690	28.5
bromide	CsBr	748	838	887	938	993	1026	1072	1140	1221	1300	636
chloride	CsCl	744	837	884	934	989	1023	1069	1139	1217	1300	646
fluoride	CsF	712	798	844	893	947	980	1025	1092	1170	1251	683
iodide	CsI	738	828	873	923	976	1009	1055	1124	1200	1280	621
Chlorine	Cl_2	−118.0	−106.7	−101.6	−93.3	−84.5	−79.0	−71.7	−60.2	−47.3	−33.8	−100.7
fluoride	ClF	......	−143.4	−139.0	−134.3	−128.8	−125.3	−120.8	−114.4	−107.0	−100.5	−145
trifluoride	ClF_3	......	−80.4	−71.8	−62.3	−51.3	−44.1	−34.7	−20.7	−4.9	+11.5	−83
monoxide	Cl_2O	−98.5	−81.6	−73.1	−64.3	−54.3	−48.0	−39.4	−26.5	−12.5	+2.2	−116
dioxide	ClO_2	......	......	−59.0	−51.2	−42.8	−37.2	−29.4	−17.8	−4.0	+11.1	−59
heptoxide	Cl_2O_7	−45.3	−23.8	−13.2	−2.1	+10.3	+18.2	29.1	44.6	62.2	78.8	−91
Chlorosulfonic acid	HSO_3Cl	32.0	53.5	64.0	75.3	87.6	95.2	105.3	120.0	136.1	151.0	−80
Chromium	Cr	1616	1768	1845	1928	2013	2067	2139	2243	2361	2482	1615
carbonyl	$Cr(CO)_6$	36.0	58.0	68.3	79.5	91.2	98.3	108.0	121.8	137.2	151.0	
oxychloride	CrO_2Cl_2	−18.4	+3.2	13.8	25.7	38.5	46.7	58.0	75.2	95.2	117.1	
Cobalt chloride	$CoCl_2$	......	......	......	......	770	801	843	904	974	1050	735
nitrosyl tricarbonyl	$Co(CO)_3NO$	......	......	......	−1.3	+11.0	18.5	29.0	44.4	62.0	80.0	−11
Columbium fluoride	CbF_5	......	......	86.3	103.0	121.5	133.2	148.5	172.2	198.0	225.0	75.5
Copper	Cu	1628	1795	1879	1970	2067	2127	2207	2325	2465	2595	1083
Cuprous bromide	Cu_2Br_2	572	666	718	777	844	887	951	1052	1189	1355	504
chloride	Cu_2Cl_2	546	645	702	766	838	886	960	1077	1249	1490	422
iodide	Cu_2I_2	......	610	656	716	786	836	907	1018	1158	1336	605
Cyanogen	C_2N_2	−95.8	−83.2	−76.8	−70.1	−62.7	−57.9	−51.8	−42.6	−33.0	−21.0	−34.4
bromide	CNBr	−35.7	−18.3	−10.0	−1.0	+8.6	14.7	22.6	33.8	46.0	61.5	58
chloride	CNCl	−76.7	−61.4	−53.8	−46.1	−37.5	−32.1	−24.9	−14.1	−2.3	+13.1	−6.5
fluoride	CNF	−134.4	−123.8	−118.5	−112.8	−106.4	−102.3	−97.0	−89.2	−80.5	−72.6	

* Compiled from the extended tables published by D. R. Stull in *Ind. Eng. Chem.*, **39**, 517 (1947).

TABLE 3-7 Vapor Pressures of Inorganic Compounds, up to 1 atm *(Continued)*

Compound		Pressure, mm. Hg										Melting point, °C.
Name	Formula	1	5	10	20	40	60	100	200	400	760	
		Temperature, °C.										
Deuterium cyanide	DCN	−68.9	−54.0	−46.7	−38.8	−30.1	−24.7	−17.5	−5.4	+10.0	26.2	−12
Fluorine	F_2	−223.0	−216.9	−214.1	−211.0	−207.7	−205.6	−202.7	−198.3	−193.2	−187.9	−223
oxide	F_2O	−196.1	−186.6	−182.3	−177.8	−173.0	−170.0	−165.8	−159.0	−151.9	−144.6	−223.9
Germanium bromide	$GeBr_4$	......	43.3	56.8	71.8	88.1	98.8	113.2	135.4	161.6	189.0	26.1
chloride	$GeCl_4$	−45.0	−24.9	−15.0	−4.1	+8.0	16.2	27.5	44.4	63.8	84.0	−49.5
hydride	GeH_4	−163.0	−151.0	−145.3	−139.2	−131.6	−126.7	−120.3	−111.2	−100.2	−88.9	−165
Trichlorogermane	$GeHCl_3$	−41.3	−22.3	−13.0	−3.0	+8.8	16.2	26.5	41.6	58.3	75.0	−71.1
Tetramethylgermane	$Ge(CH_3)_4$	−73.2	−54.6	−45.2	−35.0	−23.4	−16.2	−6.3	+8.8	26.0	44.0	−88
Digermane	Ge_2H_6	−88.7	−69.8	−60.1	−49.9	−38.2	−30.7	−20.3	−4.7	+13.3	31.5	−109
Trigermane	Ge_3H_8	−36.9	−12.8	−0.9	+11.8	26.3	35.5	47.9	67.0	88.6	110.8	−105.6
Gold	Au	1869	2059	2154	2256	2363	2431	2521	2657	2807	2966	1063
Helium	He	−271.7	−271.5	−271.3	−271.1	−270.7	−270.6	−270.3	−269.8	−269.3	−268.6	
para-Hydrogen	H_2	−263.3	−261.9	−261.3	−260.4	−259.6	−258.9	−257.9	−256.3	−254.5	−252.5	−259.1
Hydrogen bromide	HBr	−138.8	−127.4	−121.8	−115.4	−108.3	−103.8	−97.7	−88.1	−78.0	−66.5	−87.0
chloride	HCl	−150.8	−140.7	−135.6	−130.0	−123.8	−119.6	−114.0	−105.2	−95.3	−84.8	−114.3
cyanide	HCN	−71.0	−55.3	−47.7	−39.7	−30.9	−25.1	−17.8	−5.3	+10.2	25.9	−13.2
fluoride	H_2F_2	......	−74.7	−65.8	−56.0	−45.0	−37.9	−28.2	−13.2	+2.5	19.7	−83.7
iodide	HI	−123.3	−109.6	−102.3	−94.5	−85.6	−79.8	−72.1	−60.3	−48.3	−35.1	−50.9
oxide (water)	H_2O	−17.3	+1.2	11.2	22.1	34.0	41.5	51.6	66.5	83.0	100.0	0.0
sulfide	H_2S	−134.3	−122.4	−116.3	−109.7	−102.3	−97.9	−91.6	−82.3	−71.8	−60.4	−85.5
disulfide	HSSH	−43.2	−24.4	−15.2	−5.1	+6.0	12.8	22.0	35.3	49.6	64.0	−89.7
selenide	H_2Se	−115.3	−103.4	−97.9	−91.8	−84.7	−80.2	−74.2	−65.2	−53.6	−41.1	−64
telluride	H_2Te	−96.4	−82.4	−75.4	−67.8	−59.1	−53.7	−45.7	−32.4	−17.2	−2.0	−49.0
Iodine	I_2	38.7	62.2	73.2	84.7	97.5	105.4	116.5	137.3	159.8	183.0	112.9
heptafluoride	IF_7	−87.0	−70.7	−63.0	−54.5	−45.3	−39.4	−31.9	−20.7	−8.3	+4.0	5.5
Iron	Fe	1787	1957	2039	2128	2224	2283	2360	2475	2605	2735	1535
pentacarbonyl	$Fe(CO)_5$	......	−6.5	+4.6	16.7	30.3	39.1	50.3	68.0	86.1	105.0	−21
Ferric chloride	Fe_2Cl_6	194.0	221.8	235.5	246.0	256.8	263.7	272.5	285.0	298.0	319.0	304
Ferrous chloride	$FeCl_2$	......	......	700	737	779	805	842	897	961	1026	
Krypton	Kr	−199.3	−191.3	−187.2	−182.9	−178.4	−175.7	−171.8	−165.9	−159.0	−152.0	−156.7
Lead	Pb	973	1099	1162	1234	1309	1358	1421	1519	1630	1744	327.5
bromide	$PbBr_2$	513	578	610	646	686	711	745	796	856	914	373
chloride	$PbCl_2$	547	615	648	684	725	750	784	833	893	954	501
fluoride	PbF_2	......	861	904	950	1003	1036	1080	1144	1219	1293	855
iodide	PbI_2	479	540	571	605	644	668	701	750	807	872	402
oxide	PbO	943	1039	1085	1134	1189	1222	1265	1330	1402	1472	890
sulfide	PbS	852	928	975	1005	1048	1074	1108	1160	1221	1281	1114
Lithium	Li	723	828	881	940	1003	1042	1097	1178	1273	1372	186
bromide	LiBr	748	840	888	939	994	1028	1076	1147	1226	1310	547
chloride	LiCl	783	880	932	987	1045	1081	1129	1203	1290	1382	614
fluoride	LiF	1047	1156	1211	1270	1333	1372	1425	1503	1591	1681	870
iodide	LiI	723	802	841	883	927	955	993	1049	1110	1171	446
Magnesium	Mg	621	702	743	789	838	868	909	967	1034	1107	651
chloride	$MgCl_2$	778	877	930	988	1050	1088	1142	1223	1316	1418	712
Manganese	Mn	1292	1434	1505	1583	1666	1720	1792	1900	2029	2151	1260
chloride	$MnCl_2$	......	736	778	825	879	913	960	1028	1108	1190	650
Mercury	Hg	126.2	164.8	184.0	204.6	228.8	242.0	261.7	290.7	323.0	357.0	−38.9
Mercuric bromide	$HgBr_2$	136.5	165.3	179.8	194.3	211.5	221.0	237.8	262.7	290.0	319.0	237
chloride	$HgCl_2$	136.2	166.0	180.2	195.8	212.5	222.2	237.0	256.5	275.5	304.0	277
iodide	HgI_2	157.5	189.2	204.5	220.0	238.2	249.0	261.8	291.0	324.2	354.0	259
Molybdenum	Mo	3102	3393	3535	3690	3859	3964	4109	4322	4553	4804	2622
hexafluoride	MoF_6	−65.5	−49.0	−40.8	−32.0	−22.1	−16.2	−8.0	+4.1	17.2	36.0	17
oxide	MoO_3	734	785	814	851	892	917	955	1014	1082	1151	795
Neon	Ne	−257.3	−255.5	−254.6	−253.7	−252.6	−251.9	−251.0	−249.7	−248.1	−246.0	−248.7
Nickel	Ni	1810	1979	2057	2143	2234	2289	2364	2473	2603	2732	1452
carbonyl	$Ni(CO)_4$	......	......	......	......	−23.0	−15.9	−6.0	+8.8	25.8	42.5	−25
chloride	$NiCl_2$	671	731	759	789	821	840	866	904	945	987	1001
Nitrogen	N_2	−226.1	−221.3	−219.1	−216.8	−214.0	−212.3	−209.7	−205.6	−200.9	−195.8	−210.0
Nitric oxide	NO	−184.5	−180.6	−178.2	−175.3	−171.7	−168.9	−166.0	−162.3	−156.8	−151.7	−161
Nitrogen dioxide	NO_2	−55.6	−42.7	−36.7	−30.4	−23.9	−19.9	−14.7	−5.0	+8.0	21.0	−9.3
Nitrogen pentoxide	N_2O_5	−36.8	−23.0	−16.7	−10.0	−2.9	+1.8	7.4	15.6	24.4	32.4	30
Nitrous oxide	N_2O	−143.4	−133.4	−128.7	−124.0	−118.3	−114.9	−110.3	−103.6	−96.2	−85.5	−90.9
Nitrosyl chloride	NOCl	......	......	......	......	−60.2	−54.2	−46.3	−34.0	−20.3	−6.4	−64.5
fluoride	NOF	−132.0	−120.3	−114.3	−107.8	−100.3	−95.7	−88.8	−79.2	−68.2	−56.0	−134
Osmium tetroxide (yellow)	OsO_4	3.2	22.0	31.3	41.0	51.7	59.4	71.5	89.5	109.3	130.0	56
(white)	OsO_4	−5.6	+15.6	26.0	37.4	50.5	59.4	71.5	89.5	109.3	130.0	42
Oxygen	O_2	−219.1	−213.4	−210.6	−207.5	−204.1	−201.9	−198.8	−194.0	−188.8	−183.1	−218.7
Ozone	O_3	−180.4	−168.6	−163.2	−157.2	−150.7	−146.7	−141.0	−132.6	−122.5	−111.1	−251
Phosgene	$COCl_2$	−92.9	−77.0	−69.3	−60.3	−50.3	−44.0	−35.6	−22.3	−7.6	+8.3	−104
Phosphorus (yellow)	P	76.6	111.2	128.0	146.2	166.7	179.8	197.3	222.7	251.0	280.0	44.1
(violet)	P	237	271	287	306	323	334	349	370	391	417	590
tribromide	PBr_3	7.8	34.4	47.8	62.4	79.0	89.8	103.6	125.2	149.7	175.3	−40
trichloride	PCl_3	−51.6	−31.5	−21.3	−10.2	+2.3	10.2	21.0	37.6	56.9	74.2	−111.8
pentachloride	PCl_5	55.5	74.0	83.2	92.5	102.5	108.3	117.0	131.3	147.2	162.0	
Phosphine	PH_3	......	......	......	......	−129.4	−125.0	−118.8	−109.4	−98.3	−87.5	−132.5
Phosphonium bromide	PH_4Br	−43.7	−28.5	−21.2	−13.3	−5.0	+0.3	7.4	17.6	28.0	38.3	
chloride	PH_4Cl	−91.0	−79.6	−74.0	−68.0	−61.5	−57.3	−52.0	−44.0	−35.4	−27.0	−28.5
iodide	PH_4I	−25.2	−9.0	−1.1	+7.3	16.1	21.9	29.3	39.9	51.6	62.3	
Phosphorus trioxide	P_4O_6	......	39.7	53.0	67.8	84.0	94.2	108.3	129.0	150.3	173.1	22.5
pentoxide	P_4O_{10}	384	424	442	462	481	493	510	532	556	591	569
oxychloride	$POCl_3$	......	......	2.0	13.6	27.3	35.8	47.4	65.0	84.3	105.1	2
thiobromide	$PSBr_3$	50.0	72.4	83.6	95.5	108.0	116.0	126.3	141.8	157.8	175.0	38
thiochloride	$PSCl_3$	−18.3	+4.6	16.1	29.0	42.7	51.8	63.8	82.0	102.3	124.0	−36.2
Platinum	Pt	2730	3007	3146	3302	3469	3574	3714	3923	4169	4407	1755
Potassium	K	341	408	443	483	524	550	586	643	708	774	62.3
bromide	KBr	795	892	940	994	1050	1087	1137	1212	1297	1383	730
chloride	KCl	821	919	968	1020	1078	1115	1164	1239	1322	1407	790
fluoride	KF	885	988	1039	1096	1156	1193	1245	1323	1411	1502	880
hydroxide	KOH	719	814	863	918	976	1013	1064	1142	1233	1327	380
iodide	KI	745	840	887	938	995	1030	1080	1152	1238	1324	723
Radon	Rn	−144.2	−132.4	−126.3	−119.2	−111.3	−106.2	−99.0	−87.7	−75.0	−61.8	−71
Rhenium heptoxide	Re_2O_7	212.5	237.5	248.0	261.0	272.0	280.0	289.0	307.0	336.0	362.4	296

TABLE 3-7 Vapor Pressures of Inorganic Compounds, up to 1 atm *(Concluded)*

Compound		Pressure, mm. Hg										Melting point, °C.
Name	Formula	1	5	10	20	40	60	100	200	400	760	
		Temperature, °C.										
Rubidium	Rb	297	358	389	422	459	482	514	563	620	679	38.5
bromide	RbBr	781	876	923	975	1031	1066	1114	1186	1267	1352	682
chloride	RbCl	792	887	937	990	1047	1084	1133	1207	1294	1381	715
fluoride	RbF	921	982	1016	1052	1096	1123	1168	1239	1322	1408	760
iodide	RbI	748	839	884	935	991	1026	1072	1141	1223	1304	642
Selenium	Se	356	413	442	473	506	527	554	594	637	680	217
dioxide	SeO_2	157.0	187.7	202.5	217.5	234.1	244.6	258.0	277.0	297.7	317.0	340
hexafluoride	SeF_6	−118.6	−105.2	−98.9	−92.3	−84.7	−80.0	−73.9	−64.8	−55.2	−45.8	−34.7
oxychloride	$SeOCl_2$	34.8	59.8	71.9	84.2	98.0	106.5	118.0	134.6	151.7	168.0	8.5
tetrachloride	$SeCl_4$	74.0	96.3	107.4	118.1	130.1	137.8	147.5	161.0	176.4	191.5	
Silicon	Si	1724	1835	1888	1942	2000	2036	2083	2151	2220	2287	1420
dioxide	SiO_2	......	......	1732	1798	1867	1911	1969	2053	2141	2227	1710
tetrachloride	$SiCl_4$	−63.4	−44.1	−34.4	−24.0	−12.1	−4.8	+5.4	21.0	38.4	56.8	−68.8
tetrafluoride	SiF_4	−144.0	−134.8	−130.4	−125.9	−120.8	−117.5	−113.3	−170.2	−100.7	−94.8	−90
Trichlorofluorosilane	$SiFCl_3$	−92.6	−76.4	−68.3	−59.0	−48.8	−42.2	−33.2	−19.3	−4.0	+12.2	−120.8
Iodosilane	SiH_3I	......	−53.0	−47.7	−33.4	−21.8	−14.3	−4.4	+10.7	27.9	45.4	−57.0
Diiodosilane	SiH_2I_2	......	3.8	18.0	34.1	52.6	64.0	79.4	101.8	125.5	149.5	−1.0
Disiloxan	$(SiH_3)_2O$	−112.5	−95.8	−88.2	−79.8	−70.4	−64.2	−55.9	−43.5	−29.3	−15.4	−144.2
Trisilane	Si_3H_8	−68.9	−49.7	−40.0	−29.0	−16.9	−9.0	+1.6	17.8	35.5	53.1	−117.2
Trisilazane	$(SiH_3)_3N$	−68.7	−49.9	−40.4	−30.0	−18.5	−11.0	−1.1	+14.0	31.0	48.7	−105.7
Tetrasilane	Si_4H_{10}	−27.7	−6.2	+4.3	15.8	28.4	36.6	47.4	63.6	81.7	100.0	−93.6
Octachlorotrisilane	Si_3Cl_8	46.3	74.7	89.3	104.2	121.5	132.0	146.0	166.2	189.5	211.4	
Hexachlorodisiloxane	$(SiCl_3)_2O$	−5.0	17.8	29.4	41.5	55.2	63.8	75.4	92.5	113.6	135.6	−33.2
Hexachlorodisilane	Si_2Cl_6	+4.0	27.4	38.8	51.5	65.3	73.9	85.4	102.2	120.6	139.0	−1.2
Tribromosilane	$SiHBr_3$	−30.5	−8.0	+3.4	16.0	30.0	39.2	51.6	70.2	90.2	111.8	−73.5
Trichlorosilane	$SiHCl_3$	−80.7	−62.6	−53.4	−43.8	−32.9	−25.8	−16.4	−1.8	+14.5	31.8	−126.6
Trifluorosilane	$SiHF_3$	−152.0	−142.7	−138.2	−132.9	−127.3	−123.7	−118.7	−111.3	−102.8	−95.0	−131.4
Dibromosilane	SiH_2Br_2	−60.9	−40.0	−29.4	−18.0	−5.2	+3.2	14.1	31.6	50.7	70.5	−70.2
Difluorosilane	SiH_2F_2	−146.7	−136.0	−130.4	−124.3	−117.6	−113.3	−107.3	−98.3	−87.6	−77.8	
Monobromosilane	SiH_3Br	......	−85.7	−77.3	−68.3	−57.8	−51.1	−42.3	−28.6	−13.3	+2.4	−93.9
Monochlorosilane	SiH_3Cl	−117.8	−104.3	−97.7	−90.1	−81.8	−76.0	−68.5	−57.0	−44.5	−30.4	
Monofluorosilane	SiH_3F	−153.0	−145.5	−141.2	−136.3	−130.8	−127.2	−122.4	−115.2	−106.8	−98.0	
Tribromofluorosilane	$SiFBr_3$	−46.1	−25.4	−15.1	−3.7	+9.2	17.4	28.6	45.7	64.6	83.8	−82.5
Dichlorodifluorosilane	SiF_2Cl_2	−124.7	−110.5	−102.9	−94.5	−85.0	−78.6	−70.3	−58.0	−45.0	−31.8	−139.7
Trifluorobromosilane	SiF_3Br	......	......	......	......	......	......	......	−69.8	−55.9	−41.7	−70.5
Trifluorochlorosilane	SiF_3Cl	−144.0	−133.0	−127.0	−120.5	−112.8	−108.2	−101.7	−91.7	−81.0	−70.0	−142
Hexafluorodisilane	Si_2F_6	−81.0	−68.8	−63.1	−57.0	−50.6	−46.7	−41.7	−34.2	−26.4	−18.9	−18.6
Dichlorofluorobromosilane	$SiFCl_2Br$	−86.5	−68.4	−59.0	−48.8	−37.0	−29.0	−19.5	−3.2	+15.4	35.4	−112.3
Dibromochlorofluorosilane	$SiFClBr_2$	−65.2	−45.5	−35.6	−24.5	−12.0	−4.7	+6.3	23.0	43.0	59.5	−99.3
Silane	SiH_4	−179.3	−168.6	−163.0	−156.9	−150.3	−146.3	−140.5	−131.6	−122.0	−111.5	−185
Disilane	Si_2H_6	−114.8	−99.3	−91.4	−82.7	−72.8	−66.4	−57.5	−44.6	−29.0	−14.3	−132.6
Silver	Ag	1357	1500	1575	1658	1743	1795	1865	1971	2090	2212	960.5
chloride	AgCl	912	1019	1074	1134	1200	1242	1297	1379	1467	1564	455
iodide	AgI	820	927	983	1045	1111	1152	1210	1297	1400	1506	552
Sodium	Na	439	511	549	589	633	662	701	758	823	892	97.5
bromide	NaBr	806	903	952	1005	1063	1099	1148	1220	1304	1392	755
chloride	NaCl	865	967	1017	1072	1131	1169	1220	1296	1379	1465	800
cyanide	NaCN	817	928	983	1046	1115	1156	1214	1302	1401	1497	564
fluoride	NaF	1077	1186	1240	1300	1363	1403	1455	1531	1617	1704	992
hydroxide	NaOH	739	843	897	953	1017	1057	1111	1192	1286	1378	318
iodide	NaI	767	857	903	952	1005	1039	1083	1150	1225	1304	651
Strontium	Sr	......	847	898	953	1018	1057	1111	1192	1285	1384	800
Strontium oxide	SrO	2068	2198	2262	2333	2410	......	......	......	......	......	2430
Sulfur	S	183.8	223.0	243.8	264.7	288.3	305.5	327.2	359.7	399.6	444.6	112.8
monochloride	S_2Cl_2	−7.4	+15.7	27.5	40.0	54.1	63.2	75.3	93.5	115.4	138.0	−80
hexafluoride	SF_6	−132.7	−120.6	−114.7	−108.4	−101.5	−96.8	−90.9	−82.3	−72.6	−63.5	−50.2
Sulfuryl chloride	SO_2Cl_2	......	−35.1	−24.8	−13.4	−1.0	+7.2	17.8	33.7	51.3	69.2	−54.1
Sulfur dioxide	SO_2	−95.5	−83.0	−76.8	−69.7	−60.5	−54.6	−46.9	−35.4	−23.0	−10.0	−73.2
trioxide (α)	SO_3	−39.0	−23.7	−16.5	−9.1	−1.0	+4.0	10.5	20.5	32.6	44.8	16.8
trioxide (β)	SO_3	−34.0	−19.2	−12.3	−4.9	+3.2	8.0	14.3	23.7	32.6	44.8	32.3
trioxide (γ)	SO_3	−15.3	−2.0	+4.3	11.1	17.9	21.4	28.0	35.8	44.0	51.6	62.1
Tellurium	Te	520	605	650	697	753	789	838	910	997	1087	452
chloride	$TeCl_4$	......	......	233	253	273	287	304	330	360	392	224
fluoride	TeF_6	−111.3	−98.8	−92.4	−86.0	−78.4	−73.8	−67.9	−57.3	−48.2	−38.6	−37.8
Thallium	Tl	825	931	983	1040	1103	1143	1196	1274	1364	1457	3035
Thallous bromide	TlBr	......	490	522	559	598	621	653	703	759	819	460
chloride	TlCl	......	487	517	550	589	612	645	694	748	807	430
iodide	TlI	440	502	531	567	607	631	663	712	763	823	440
Thionyl bromide	$SOBr_2$	−6.7	+18.4	31.0	44.1	58.8	68.3	80.6	99.0	119.2	139.5	−52.2
Thionyl chloride	$SOCl_2$	−52.9	−32.4	−21.9	−10.5	+2.2	10.4	21.4	37.9	56.5	75.4	−104.5
Tin	Sn	1492	1634	1703	1777	1855	1903	1968	2063	2169	2270	231.9
Stannic bromide	$SnBr_4$	......	58.3	72.7	88.1	105.5	116.2	131.0	152.8	177.7	204.7	31.0
Stannous chloride	$SnCl_2$	316	366	391	420	450	467	493	533	577	623	246.8
Stannic chloride	$SnCl_4$	−22.7	−1.0	+10.0	22.0	35.2	43.5	54.7	72.0	92.1	113.0	−30.2
iodide	SnI_4	......	156.0	175.8	196.2	218.8	234.2	254.2	283.5	315.5	348.0	144.5
hydride	SnH_4	−140.0	−125.8	−118.5	−111.2	−102.3	−96.6	−89.2	−78.0	−65.2	−52.3	−149.9
Tin tetramethyl	$Sn(CH_3)_4$	−51.3	−31.0	−20.6	−9.3	+3.5	11.7	22.8	39.8	58.5	78.0	
trimethyl-ethyl	$Sn(CH_3)_3.C_2H_5$	−30.0	−7.6	+3.8	16.1	30.0	38.4	50.0	67.3	87.6	108.8	
trimethyl-propyl	$Sn(CH_3)_3.C_3H_7$	−12.0	+10.7	21.8	34.0	48.5	57.5	69.8	88.0	109.6	131.7	
Titanium chloride	$TiCl_4$	−13.9	+9.4	21.3	34.2	48.4	58.0	71.0	90.5	112.7	136.0	−30
Tungsten	W	3990	4337	4507	4690	,4886	5007	5168	5403	5666	5927	3370
Tungsten hexafluoride	WF_6	−71.4	−56.5	−49.2	−41.5	−33.0	−27.5	−20.3	−10.0	+1.2	17.3	−0.5
Uranium hexafluoride	UF_6	−38.8	−22.0	−13.8	−5.2	+4.4	10.4	18.2	30.0	42.7	55.7	69.2
Vanadyl trichloride	$VOCl_3$	−23.2	+0.2	12.2	26.6	40.0	49.8	62.5	82.0	103.5	127.2	
Xenon	Xe	−168.5	−158.2	−152.8	−147.1	−141.2	−137.7	−132.8	−125.4	−117.1	−108.0	−111.6
Zinc	Zn	487	558	593	632	673	700	736	788	844	907	419.4
chloride	$ZnCl_2$	428	481	508	536	566	584	610	648	689	732	365
fluoride	ZnF_2	970	1055	1086	1129	1175	1207	1254	1329	1417	1497	872
diethyl	$Zn(C_2H_5)_2$	−22.4	0.0	+11.7	24.2	38.0	47.2	59.1	77.0	97.3	118.0	−28
Zirconium bromide	$ZrBr_4$	207	237	250	266	281	289	301	318	337	357	450
chloride	$ZrCl_4$	190	217	230	243	259	268	279	295	312	331	437
iodide	ZrI_4	264	297	311	329	344	355	369	389	409	431	499

TABLE 3-8 Vapor Pressures of Organic Compounds, up to 1 atm*

Compound		Pressure, mm. Hg										Melting point, °C.
Name	Formula	1	5	10	20	40	60	100	200	400	760	
		Temperature, °C.										
Acenaphthalene	$C_{12}H_{10}$	.....	114.8	131.2	148.7	168.2	181.2	197.5	222.1	250.0	277.5	95
Acetal	$C_6H_{14}O_2$	−23.0	−2.3	+8.0	19.6	31.9	39.8	50.1	66.3	84.0	102.2	
Acetaldehyde	C_2H_4O	−81.5	−65.1	−56.8	−47.8	−37.8	−31.4	−22.6	−10.0	+4.9	20.2	−123.5
Acetamide	C_2H_5NO	65.0	92.0	105.0	120.0	135.8	145.8	158.0	178.3	200.0	222.0	81
Acetanilide	C_8H_9NO	114.0	146.6	162.0	180.0	199.6	211.8	227.2	250.5	277.0	303.8	113.5
Acetic acid	$C_2H_4O_2$	−17.2	+6.3	17.5	29.9	43.0	51.7	63.0	80.0	99.0	118.1	16.7
anhydride	$C_4H_6O_3$	1.7	24.8	36.0	48.3	62.1	70.8	82.2	100.0	119.8	139.6	−73
Acetone	C_3H_6O	−59.4	−40.5	−31.1	−20.8	−9.4	−2.0	+7.7	22.7	39.5	56.5	−94.6
Acetonitrile	C_2H_3N	−47.0	−26.6	−16.3	−5.0	+7.7	15.9	27.0	43.7	62.5	81.8	−41
Acetophenone	C_8H_8O	37.1	64.0	78.0	92.4	109.4	119.8	133.6	154.2	178.0	202.4	20.5
Acetyl chloride	C_2H_3OCl	−50.0	−35.0	−27.6	−19.6	−10.4	−4.5	+3.2	16.1	32.0	50.8	−112.0
Acetylene	C_2H_2	−142.9	−133.0	−128.2	−122.8	−116.7	−112.8	−107.9	−100.3	−92.0	−84.0	−81.5
Acridine	$C_{13}H_9N$	129.4	165.8	184.0	203.5	224.2	238.7	256.0	284.0	314.3	346.0	110.5
Acrolein (2-propenal)	C_3H_4O	−64.5	−46.0	−36.7	−26.3	−15.0	−7.5	+2.5	17.5	34.5	52.5	−87.7
Acrylic acid	$C_3H_4O_2$	+3.5	27.3	39.0	52.0	66.2	75.0	86.1	103.3	122.0	141.0	14
Adipic acid	$C_6H_{10}O_4$	159.5	191.0	205.5	222.0	240.5	251.0	265.0	287.8	312.5	337.5	152
Allene (propadiene)	C_3H_4	−120.6	−108.0	−101.0	−93.4	−85.2	−78.8	−72.5	−61.3	−48.5	−35.0	−136
Allyl alcohol (propen-1-ol-3)	C_3H_6O	−20.0	+0.2	10.5	21.7	33.4	40.3	50.0	64.5	80.2	96.6	−129
chloride (3-chloropropene)	C_3H_5Cl	−70.0	−52.0	−42.9	−32.8	−21.2	−14.1	−4.5	10.4	27.5	44.6	−136.4
isopropyl ether	$C_6H_{12}O$	−43.7	−23.1	−12.9	−1.8	+10.9	18.7	29.0	44.3	61.7	79.5	
isothiocyanate	C_4H_5NS	−2.0	+25.3	38.3	52.1	67.4	76.2	89.5	108.0	129.8	150.7	−80
n-propyl ether	$C_6H_{12}O$	−39.0	−18.2	−7.9	+3.7	16.4	25.0	35.8	52.6	71.4	90.5	
4-Allylveratrole	$C_{11}H_{14}O_2$	85.0	113.9	127.0	142.8	158.3	169.6	183.7	204.0	226.2	248.0	
iso-Amyl acetate	$C_7H_{14}O_2$	0.0	+23.7	35.2	47.8	62.1	71.0	83.2	101.3	121.5	142.0	
n-Amyl alcohol	$C_5H_{12}O$	+13.6	34.7	44.9	55.8	68.0	75.5	85.8	102.0	119.8	137.8	
iso-Amyl alcohol	$C_5H_{12}O$	+10.0	30.9	40.8	51.7	63.4	71.0	80.7	95.8	113.7	130.6	−117.2
sec-Amyl alcohol (2-pentanol)	$C_5H_{12}O$	+1.5	22.1	32.2	42.6	54.1	61.5	70.7	85.7	102.3	119.7	
tert-Amyl alcohol	$C_5H_{12}O$	−12.9	+7.2	17.2	27.9	38.8	46.0	55.3	69.7	85.7	101.7	−11.9
sec-Amylbenzene	$C_{11}H_{16}$	29.0	55.8	69.2	83.8	100.0	110.4	124.1	145.2	168.0	193.0	
iso-Amyl benzoate	$C_{12}H_{16}O_2$	72.0	104.5	121.6	139.7	158.3	171.4	186.8	210.2	235.8	262.0	
bromide (1-bromo-3-methylbutane)	$C_5H_{11}Br$	−20.4	+2.1	13.6	26.1	39.8	48.7	60.4	78.7	99.4	120.4	
n-butyrate	$C_9H_{18}O_2$	21.2	47.1	59.9	74.0	90.0	99.8	113.1	133.2	155.3	178.6	
formate	$C_6H_{12}O_2$	−17.5	+5.4	17.1	30.0	44.0	53.3	65.4	83.2	102.7	123.3	
iodide (1-iodo-3-methylbutane)	$C_5H_{11}I$	−2.5	+21.9	34.1	47.6	62.3	71.9	84.4	103.8	125.8	148.2	
isobutyrate	$C_9H_{18}O_2$	14.8	40.1	52.8	66.6	81.8	91.7	104.4	124.2	146.0	168.8	
Amyl isopropionate	$C_8H_{16}O_2$	+8.5	33.7	46.3	60.0	75.5	85.2	97.6	117.3	138.4	160.2	
iso-Amyl isovalerate	$C_{10}H_{20}O_2$	27.0	54.4	68.6	83.8	100.6	110.3	125.1	146.1	169.5	194.0	
n-Amyl levulinate	$C_{10}H_{18}O_3$	81.3	110.0	124.0	139.7	155.8	165.2	180.5	203.1	227.4	253.2	
iso-Amyl levulinate	$C_{10}H_{18}O_3$	75.6	104.0	118.8	134.4	151.7	162.6	177.0	198.1	222.7	247.9	
nitrate	$C_5H_{11}NO_3$	+5.2	28.8	40.3	53.5	67.6	76.3	88.6	106.7	126.5	147.5	
4-tert-Amylphenol	$C_{11}H_{16}O$	.....	109.8	125.5	142.3	160.3	172.6	189.0	213.0	239.5	266.0	93
Anethole	$C_{10}H_{12}O$	62.6	91.6	106.0	121.8	139.3	149.8	164.2	186.1	210.5	235.3	22.5
Angelonitrile	C_5H_7N	−8.0	+15.0	28.0	41.0	55.8	65.2	77.5	96.3	117.7	140.0	
Aniline	C_6H_7N	34.8	57.9	69.4	82.0	96.7	106.0	119.9	140.1	161.9	184.4	−6.2
2-Anilinoethanol	$C_8H_{11}NO$	104.0	134.3	149.6	165.7	183.7	194.0	209.5	230.6	254.5	279.6	
Anisaldehyde	$C_8H_8O_2$	73.2	102.6	117.8	133.5	150.5	161.7	176.7	199.0	223.0	248.0	2.5
o-Anisidine (2-methoxyaniline)	C_7H_9NO	61.0	88.0	101.7	116.1	132.0	142.1	155.2	175.3	197.3	218.5	5.2
Anthracene	$C_{14}H_{10}$	145.0	173.5	187.2	201.9	217.5	231.8	250.0	279.0	310.2	342.0	217.5
Anthraquinone	$C_{14}H_8O_2$	190.0	219.4	234.2	248.3	264.3	273.3	285.0	314.6	346.2	379.9	286
Azelaic acid	$C_9H_{16}O_4$	178.3	210.4	225.5	242.4	260.0	271.8	286.5	309.6	332.8	356.5	106.5
Azelaldehyde	$C_9H_{18}O$	33.3	58.4	71.6	85.0	100.2	110.0	123.0	142.1	163.4	185.0	
Azobenzene	$C_{12}H_{10}N_2$	103.5	135.7	151.5	168.3	187.9	199.8	216.0	240.0	266.1	293.0	68
Benzal chloride (α,α-Dichlorotoluene)	$C_7H_6Cl_2$	35.4	64.0	78.7	94.3	112.1	123.4	138.3	160.7	187.0	214.0	−16.1
Benzaldehyde	C_7H_6O	26.2	50.1	62.0	75.0	90.1	99.6	112.5	131.7	154.1	179.0	−26
Benzanthrone	$C_{17}H_{10}O$	225.0	274.5	297.2	322.5	350.0	368.8	390.0	426.5	.....	.....	174
Benzene	C_6H_6	−36.7	−19.6	−11.5	−2.6	+7.6	15.4	26.1	42.2	60.6	80.1	+5.5
Benzenesulfonylchloride	$C_6H_5ClO_2S$	65.9	96.5	112.0	129.0	147.7	158.2	174.5	198.0	224.0	251.5	14.5
Benzil	$C_{14}H_{10}O_2$	128.4	165.2	183.0	202.8	224.5	238.2	255.8	283.5	314.3	347.0	95
Benzoic acid	$C_7H_6O_2$	96.0	119.5	132.1	146.7	162.6	172.8	186.2	205.8	227.0	249.2	121.7
anhydride	$C_{14}H_{10}O_3$	143.8	180.0	198.0	218.0	239.8	252.7	270.4	299.1	328.8	360.0	42
Benzoin	$C_{14}H_{12}O_2$	135.6	170.2	188.1	207.0	227.6	241.7	258.0	284.4	313.5	343.0	132
Benzonitrile	C_7H_5N	28.2	55.3	69.2	83.4	99.6	109.8	123.5	144.1	166.7	190.6	−12.9
Benzophenone	$C_{13}H_{10}O$	108.2	141.7	157.6	175.8	195.7	208.2	224.4	249.8	276.8	305.4	48.5
Benzotrichloride (α,α,α-Trichlorotoluene)	$C_7H_5Cl_3$	45.8	73.7	87.6	102.7	119.8	130.0	144.3	165.6	189.2	213.5	−21.2
Benzotrifluoride (α,α,α-Trifluorotoluene)	$C_7H_5F_3$	−32.0	−10.3	+0.4	12.2	25.7	34.0	45.3	62.5	82.0	102.2	−29.3
Benzoyl bromide	C_7H_5BrO	47.0	75.4	89.8	105.4	122.6	133.4	147.7	169.2	193.7	218.5	0
chloride	C_7H_5ClO	32.1	59.1	73.0	87.6	103.8	114.7	128.0	149.5	172.8	197.2	−0.5
nitrile	C_8H_5NO	44.5	71.7	85.5	100.2	116.6	127.0	141.0	161.3	185.0	208.0	33.5
Benzyl acetate	$C_9H_{10}O_2$	45.0	73.4	87.6	102.3	119.6	129.8	144.0	165.5	189.0	213.5	−51.5
alcohol	C_7H_8O	58.0	80.8	92.6	105.8	119.8	129.3	141.7	160.0	183.0	204.7	−15.3
Benzylamine	C_7H_9N	29.0	54.8	67.7	81.8	97.3	107.3	120.0	140.0	161.3	184.5	
Benzyl bromide (α-bromotoluene)	C_7H_7Br	32.2	59.6	73.4	88.3	104.8	115.6	129.8	150.8	175.2	198.5	−4
chloride (α-chlorotoluene)	C_7H_7Cl	22.0	47.8	60.8	75.0	90.7	100.5	114.2	134.0	155.8	179.4	−39
cinnamate	$C_{16}H_{14}O_2$	173.8	206.3	221.5	239.3	255.8	267.0	281.5	303.8	326.7	350.0	39
Benzyldichlorosilane	$C_7H_8Cl_2Si$	45.3	70.2	83.2	96.7	111.8	121.3	133.5	152.0	173.0	194.3	
Benzyl ethyl ether	$C_9H_{12}O$	26.0	52.0	65.0	79.6	95.4	105.5	118.9	139.6	161.5	185.0	
phenyl ether	$C_{13}H_{12}O$	95.4	127.7	144.0	160.7	180.1	192.6	209.2	233.2	259.8	287.0	
isothiocyanate	C_8H_7NS	79.5	107.8	121.8	137.0	153.0	163.8	177.7	198.0	220.4	243.0	
Biphenyl	$C_{12}H_{10}$	70.6	101.8	117.0	134.2	152.5	165.2	180.7	204.2	229.4	254.9	69.5
1-Biphenyloxy-2,3-epoxypropane	$C_{15}H_{14}O_2$	135.3	169.9	187.2	205.8	226.3	239.7	255.0	280.4	309.8	340.0	
d-Bornyl acetate	$C_{12}H_{20}O_2$	46.9	75.7	90.2	106.0	123.7	135.7	149.8	172.0	197.5	223.0	29
Bornyl n-butyrate	$C_{14}H_{24}O_2$	74.0	103.4	118.0	133.8	150.7	161.8	176.4	198.0	222.2	247.0	
formate	$C_{11}H_{18}O_2$	47.0	74.8	89.3	104.0	121.2	131.7	145.8	166.4	190.2	214.0	
isobutyrate	$C_{14}H_{24}O_2$	70.0	99.8	114.0	130.0	147.2	157.6	172.2	194.2	218.2	243.0	
propionate	$C_{13}H_{22}O_2$	64.6	93.7	108.0	123.7	140.4	151.2	165.7	187.5	211.2	235.0	
Brassidic acid	$C_{22}H_{42}O_2$	209.6	241.7	256.0	272.9	290.0	301.5	316.2	336.8	359.6	382.5	61.5
Bromoacetic acid	$C_2H_3BrO_2$	54.7	81.6	94.1	108.2	124.0	133.8	146.3	165.8	186.7	208.0	49.5
4-Bromoanisole	C_7H_7BrO	48.8	77.8	91.9	107.8	125.0	136.0	150.1	172.7	197.5	223.0	12.5

* Compiled from the extended tables published by D. R. Stull in *Ind. Eng. Chem.*, **39**, 517 (1947). For information on fuels see Hibbard, *N.A.C.A. Research Mem.* E56I21, 1956. For methane see Johnson (ed.), WADD-TR-60-56, 1960.

TABLE 3-8 Vapor Pressures of Organic Compounds, up to 1 atm *(Continued)*

Compound		Pressure, mm. Hg										Melting point, °C.
Name	Formula	1	5	10	20	40	60	100	200	400	760	
		Temperature, °C.										
Bromobenzene	C_6H_5Br	+2.9	27.8	40.0	53.8	68.6	78.1	90.8	110.1	132.3	156.2	−30.7
4-Bromobiphenyl	$C_{12}H_9Br$	98.0	133.7	150.6	169.8	190.8	204.5	221.8	248.2	277.7	310.0	90.5
1-Bromo-2-butanol	C_4H_9BrO	23.7	45.4	55.8	67.2	79.5	87.0	97.6	112.1	128.3	145.0	
1-Bromo-2-butanone	C_4H_7BrO	+6.2	30.0	41.8	54.2	68.2	77.3	89.2	107.0	126.3	147.0	
cis-1-Bromo-1-butene	C_4H_7Br	−44.0	−23.2	−12.8	−1.4	+11.5	19.8	30.8	47.8	66.8	86.2	
trans-1-Bromo-1-butene	C_4H_7Br	−38.4	−17.0	−6.4	+5.4	18.4	27.2	38.1	55.7	75.0	94.7	−100.3
2-Bromo-1-butene	C_4H_7Br	−47.3	−27.0	−16.8	−5.3	+7.2	15.4	26.3	42.8	61.9	81.0	−133.4
eis-2-Bromo-2-butene	C_4H_7Br	−39.0	−17.9	−7.2	+4.6	17.7	26.2	37.5	54.5	74.0	93.9	−111.2
trans-2-Bromo-2-butene	C_4H_7Br	−45.0	−24.1	−13.8	−2.4	+10.5	18.7	29.9	46.5	66.0	85.5	−114.6
1,4-Bromochlorobenzene	C_6H_4BrCl	32.0	59.5	72.7	87.8	103.8	114.8	128.0	149.5	172.6	196.9	
1-Bromo-1-chloroethane	C_2H_4BrCl	−36.0	−18.0	−9.4	0.0	+10.4	17.0	28.0	44.7	63.4	82.7	16.6
1-Bromo-2-chloroethane	C_2H_4BrCl	−28.8	−7.0	+4.1	16.0	29.7	38.0	49.5	66.8	86.0	106.7	−16.6
2-Bromo-4,6-dichlorophenol	$C_6H_3BrCl_2O$	84.0	115.6	130.8	147.7	165.8	177.6	193.2	216.5	242.0	268.0	68
1-Bromo-4-ethyl benzene	C_8H_9Br	30.4	42.5	74.0	90.2	108.5	121.0	135.5	156.5	182.0	206.0	−45.0
(2-Bromoethyl)-benzene	C_8H_9Br	48.0	76.2	90.5	105.8	123.2	133.8	148.2	169.8	194.0	219.0	
2-Bromoethyl 2-chloroethyl ether	C_4H_8BrClO	36.5	63.2	76.3	90.8	106.6	116.4	129.8	150.0	172.3	195.8	
(2-Bromoethyl)-cyclohexane	$C_8H_{15}Br$	38.7	66.6	80.5	95.8	113.0	123.7	138.0	160.0	186.2	213.0	
1-Bromoethylene	C_2H_3Br	−95.4	−77.8	−68.8	−58.8	−48.1	−41.2	−31.9	−17.2	−1.1	+15.8	−138
Bromoform (tribromomethane)	$CHBr_3$	......	22.0	34.0	48.0	63.6	73.4	85.9	106.1	127.9	150.5	8.5
1-Bromonaphthalene	$C_{10}H_7Br$	84.2	117.5	133.6	150.2	170.2	183.5	198.8	224.2	252.0	281.1	5.5
2-Bromo-4-phenylphenol	$C_{12}H_9BrO$	100.0	135.4	152.3	171.8	193.8	207.0	224.5	251.0	280.2	311.0	95
3-Bromopyridine	C_5H_4BrN	16.8	42.0	55.2	69.1	84.1	94.1	107.8	127.7	150.0	173.4	
2-Bromotoluene	C_7H_7Br	24.4	49.7	62.3	76.0	91.0	100.0	112.0	133.6	157.3	181.8	−28
3-Bromotoluene	C_7H_7Br	14.8	50.8	64.0	78.1	93.9	104.1	117.8	138.0	160.0	183.7	39.8
4-Bromotoluene	C_7H_7Br	10.3	47.5	61.1	75.2	91.8	102.3	116.4	137.4	160.2	184.5	28.5
3-Bromo-2,4,6-trichlorophenol	$C_6H_2BrCl_3O$	112.4	146.2	163.2	181.8	200.5	213.0	229.3	253.0	278.0	305.8	
2-Bromo-1,4-xylene	C_8H_9Br	37.5	65.0	78.8	94.0	110.6	121.6	135.7	156.4	181.0	206.7	+9.5
1,2-Butadiene (methyl allene)	C_4H_6	−89.0	−72.7	−64.2	−54.9	−44.3	−37.5	−28.3	−14.2	+1.8	18.5	
1,3-Butadiene	C_4H_6	−102.8	−87.6	−79.7	−71.0	−61.3	−55.1	−46.8	−33.9	−19.3	−4.5	−108.9
n-Butane	C_4H_{10}	−101.5	−85.7	−77.8	−68.9	−59.1	−52.8	−44.2	−31.2	−16.3	−0.5	−135
iso-Butane (2-methylpropane)	C_4H_{10}	−109.2	−94.1	−86.4	−77.9	−68.4	−62.4	−54.1	−41.5	−27.1	−11.7	−145
1,3-Butanediol	$C_4H_{10}O_2$	22.2	67.5	85.3	100.0	117.4	127.5	141.2	161.0	183.8	206.5	77
1,2,3-Butanetriol	$C_4H_{10}O_3$	102.0	132.0	146.0	161.0	178.0	188.0	202.5	222.0	243.5	264.0	
1-Butene	C_4H_8	−104.8	−89.4	−81.6	−73.0	−63.4	−57.2	−48.9	−36.2	−21.7	−6.3	−130
cis-2-Butene	C_4H_8	−96.4	−81.1	−73.4	−64.6	−54.7	−48.4	−39.8	−26.8	−12.0	+3.7	−138.9
trans-2-Butene	C_4H_8	−99.4	−84.0	−76.3	−67.5	−57.6	−51.3	−42.7	−29.7	−14.8	+0.9	−105.4
3-Butenenitrile	C_4H_5N	−19.6	+2.9	14.1	26.6	40.0	48.8	60.2	78.0	98.0	119.0	
iso-Butyl acetate	$C_6H_{12}O_2$	−21.2	+1.4	12.8	25.5	39.2	48.0	59.7	77.6	97.5	118.0	−98.9
n-Butyl acrylate	$C_7H_{12}O_2$	−0.5	+23.5	35.5	48.6	63.4	72.6	85.1	104.0	125.2	147.4	−64.6
alcohol	$C_4H_{10}O$	−1.2	+20.0	30.2	41.5	53.4	60.3	70.1	84.3	100.8	117.5	−79.9
iso-Butyl alcohol	$C_4H_{10}O$	−9.0	+11.6	21.7	32.4	44.1	51.7	61.5	75.9	91.4	108.0	−108
sec-Butyl alcohol	$C_4H_{10}O$	−12.2	+7.2	16.9	27.3	38.1	45.2	54.1	67.9	83.9	99.5	−114.7
tert-Butyl alcohol	$C_4H_{10}O$	−20.4	−3.0	+5.5	14.3	24.5	31.0	39.8	52.7	68.0	82.9	25.3
iso-Butyl amine	$C_4H_{11}N$	−50.0	−31.0	−21.0	−10.3	+1.3	8.8	18.8	32.0	50.7	68.6	−85.0
n-Butylbenzene	$C_{10}H_{14}$	22.7	48.8	62.0	76.3	92.4	102.6	116.2	136.9	159.2	183.1	−88.0
iso-Butylbenzene	$C_{10}H_{14}$	14.1	40.5	53.7	67.8	83.3	93.3	107.0	127.2	149.6	172.8	−51.5
sec-Butylbenzene	$C_{10}H_{14}$	18.6	44.2	57.0	70.6	86.2	96.0	109.5	128.8	150.3	173.5	−75.5
tert-Butylbenzene	$C_{10}H_{14}$	13.0	39.0	51.7	65.6	80.8	90.6	103.8	123.7	145.8	168.5	−58
iso-Butyl benzoate	$C_{11}H_{14}O_2$	64.0	93.6	108.6	124.2	141.8	152.0	166.4	188.2	212.8	237.0	
n-Butyl bromide (1-bromobutane)	C_4H_9Br	−33.0	−11.2	−0.3	+11.6	24.8	33.4	44.7	62.0	81.7	101.6	−112.4
iso-Butyl n-butyrate	$C_8H_{16}O_2$	+4.6	30.0	42.2	56.1	71.7	81.3	94.0	113.9	135.7	156.9	
carbamate	$C_5H_{11}NO_2$	......	83.7	96.4	110.1	125.3	134.6	147.2	165.7	186.0	206.5	65
Butyl carbitol (diethylene glycol butyl ether)	$C_8H_{18}O_3$	70.0	95.7	107.8	120.5	135.5	146.0	159.8	181.2	205.0	231.2	
n-Butyl chloride (1-chlorobutane)	C_4H_9Cl	−49.0	−28.9	−18.6	−7.4	+5.0	13.0	24.0	40.0	58.8	77.8	−123.1
iso-Butyl chloride	C_4H_9Cl	−53.8	−34.3	−24.5	−13.8	−1.9	+5.9	16.0	32.0	50.0	68.9	−131.2
sec-Butyl chloride (2-Chlorobutane)	C_4H_9Cl	−60.2	−39.8	−29.2	−17.7	−5.0	+3.4	14.2	31.5	50.0	68.0	−131.3
tert-Butyl chloride	C_4H_9Cl	......	......	......	......	−19.0	−11.4	−1.0	+14.6	32.6	51.0	−26.5
sec-Butyl chloroacetate	$C_6H_{11}ClO_2$	17.0	41.8	54.6	68.2	83.6	93.0	105.5	124.1	146.0	167.8	
2-tert-Butyl-4-cresol	$C_{11}H_{16}O$	70.0	98.0	112.0	127.2	143.9	153.7	167.0	187.8	210.0	232.6	
4-tert-Butyl-2-cresol	$C_{11}H_{16}O$	74.3	103.7	118.0	134.0	150.8	161.7	176.2	197.8	221.8	247.0	
iso-Butyl dichloroacetate	$C_6H_{10}Cl_2O_2$	28.6	54.3	67.5	81.4	96.7	106.6	119.8	139.2	160.0	183.0	
2,3-Butylene glycol (2,3-butanediol)	$C_4H_{10}O_2$	44.0	68.4	80.3	93.4	107.8	116.3	127.8	145.6	164.0	182.0	22.5
2-Butyl-2-ethylbutane-1,3-diol	$C_{10}H_{22}O_2$	94.1	122.6	136.8	151.2	167.8	178.0	191.9	212.0	233.5	255.0	
2-tert-Butyl-4-ethylphenol	$C_{12}H_{18}O$	76.3	106.2	121.0	137.0	154.0	165.4	179.0	200.3	223.8	247.8	
n-Butyl formate	$C_5H_{10}O_2$	−26.4	−4.7	+6.1	18.0	31.6	39.8	51.0	67.9	86.2	106.0	
iso-Butyl formate	$C_5H_{10}O_2$	−32.7	−11.4	−0.8	+11.0	24.1	32.4	43.4	60.0	79.0	98.2	−95.3
sec-Butyl formate	$C_5H_{10}O_2$	−34.4	−13.3	−3.1	+8.4	21.3	29.6	40.2	56.8	75.2	93.6	
sec-Butyl glycolate	$C_6H_{12}O_3$	28.3	53.6	66.0	79.8	94.2	104.0	116.4	135.5	155.6	177.5	
iso-Butyl iodide (1-iodo-2-methylpropane)	C_4H_9I	−17.0	+5.8	17.0	29.8	42.8	51.8	63.5	81.0	100.3	120.4	−90.7
isobutyrate	$C_8H_{16}O_2$	+4.1	28.0	39.9	52.4	67.2	75.9	88.0	106.3	126.3	147.5	−80.7
isovalerate	$C_9H_{18}O_2$	16.0	41.2	53.8	67.7	82.7	92.4	105.2	124.8	146.4	168.7	
levulinate	$C_9H_{16}O_3$	65.0	92.1	105.9	120.2	136.2	147.0	160.2	181.8	205.5	229.9	
naphthylketone (1-isovaleronaphthone)	$C_{15}H_{16}O$	136.0	167.9	184.0	201.6	219.7	231.5	246.7	269.7	294.0	320.0	
2-sec-Butylphenol	$C_{10}H_{14}O$	57.4	86.0	100.8	116.1	133.4	143.9	157.3	179.7	203.8	228.0	
2-tert-Butylphenol	$C_{10}H_{14}O$	56.6	84.2	98.1	113.0	129.2	140.0	153.5	173.8	196.3	219.5	
4-iso-Butylphenol	$C_{10}H_{14}O$	72.1	100.9	115.5	130.3	147.2	157.0	171.2	192.1	214.7	237.0	
4-sec-Butylphenol	$C_{10}H_{14}O$	71.4	100.5	114.8	130.3	147.8	157.9	172.4	194.3	217.6	242.1	
4-tert-Butylphenol	$C_{10}H_{14}O$	70.0	99.2	114.0	129.5	146.0	156.0	170.2	191.5	214.0	238.0	99
2-(4-tert-Butylphenoxy)ethyl acetate	$C_{14}H_{20}O_3$	118.0	150.0	165.8	183.3	201.5	212.8	228.0	250.3	277.6	304.4	
4-tert-Butylphenyl dichlorophosphate	$C_{10}H_{13}Cl_2O_2P$	96.0	129.6	146.0	164.0	184.3	197.2	214.3	240.0	268.2	299.0	
tert-Butyl phenyl ketone (pivalophenone)	$C_{11}H_{14}O$	57.8	85.7	99.0	114.3	130.4	140.8	154.0	175.0	197.7	220.0	
iso-Butyl propionate	$C_7H_{14}O_2$	−2.3	+20.9	32.3	44.8	58.5	67.6	79.5	97.0	116.4	136.8	−71
4-tert-Butyl-2,5-xylenol	$C_{12}H_{18}O$	88.2	119.8	135.0	151.0	169.8	180.3	195.0	217.5	241.3	265.3	
4-tert-Butyl-2,6-xylenol	$C_{12}H_{18}O$	74.0	103.9	119.0	135.0	152.2	163.6	176.0	196.0	217.8	239.8	
6-tert-Butyl-2,4-xylenol	$C_{12}H_{18}O$	70.3	100.2	115.0	131.0	148.5	158.2	172.0	192.3	214.2	236.5	
6-tert-Butyl-3,4-xylenol	$C_{12}H_{18}O$	83.9	113.6	127.0	143.0	159.7	170.0	184.0	204.5	226.7	249.5	
Butyric acid	$C_4H_8O_2$	25.5	49.8	61.5	74.0	88.0	96.5	108.0	125.5	144.5	163.5	−74

TABLE 3-8 Vapor Pressures of Organic Compounds, up to 1 atm *(Continued)*

Compound		Pressure, mm. Hg										Melting point, °C.
		1	5	10	20	40	60	100	200	400	760	
Name	Formula	Temperature, °C.										
iso-Butyric acid	$C_4H_8O_2$	14.7	39.3	51.2	64.0	77.8	86.3	98.0	115.8	134.5	154.5	−47
Butyronitrile	C_4H_7N	−20.0	+2.1	13.4	25.7	38.4	47.3	59.0	76.7	96.8	117.5	
iso-Valerophenone	$C_{11}H_{14}O$	58.3	87.0	101.4	116.8	133.8	144.6	158.0	180.1	204.2	228.0	
Camphene	$C_{10}H_{16}$	...	...	47.2	60.4	75.7	85.0	97.9	117.5	138.7	160.5	50
Campholenic acid	$C_{10}H_{16}O_2$	97.6	125.7	139.8	153.9	170.0	180.0	193.7	212.7	234.0	256.0	
d-Camphor	$C_{10}H_{16}O$	41.5	68.6	82.3	97.5	114.0	124.0	138.0	157.9	182.0	209.2	178.5
Camphylamine	$C_{10}H_{19}N$	45.3	74.0	83.7	97.6	112.5	122.0	134.6	153.0	173.8	195.0	
Capraldehyde	$C_{10}H_{20}O$	51.9	78.8	92.0	106.3	122.2	132.0	145.3	164.8	186.3	208.5	
Capric acid	$C_{10}H_{20}O_2$	125.0	142.0	152.2	165.0	179.9	189.8	200.0	217.1	240.3	268.4	31.5
n-Caproic acid	$C_6H_{12}O_2$	71.4	89.5	99.5	111.8	125.0	133.3	144.0	160.8	181.0	202.0	−1.5
iso-Caproic acid	$C_6H_{12}O_2$	66.2	83.0	94.0	107.0	120.4	129.6	141.4	158.3	181.0	207.7	−35
iso-Caprolactone	$C_6H_{10}O_2$	38.3	66.4	80.3	95.7	112.3	123.2	137.2	157.8	182.1	207.0	
Capronitrile	$C_6H_{11}N$	9.2	34.6	47.5	61.7	76.9	86.8	99.8	119.7	141.0	163.7	
Capryl alcohol (2-octanol)	$C_8H_{18}O$	32.8	57.6	70.0	83.3	98.0	107.4	119.8	138.0	157.5	178.5	−38.6
Caprylaldehyde	$C_8H_{16}O$	73.4	92.0	101.2	110.2	120.0	126.0	133.9	145.4	156.5	168.5	
Caprylic acid (octanoic acid)	$C_8H_{16}O_2$	92.3	114.1	124.0	136.4	150.6	160.0	172.2	190.3	213.9	237.5	16
Caprylonitrile	$C_8H_{15}N$	43.0	67.6	80.4	94.6	110.6	121.2	134.8	155.2	179.5	204.5	
Carbazole	$C_{12}H_9N$	...	...	...	...	...	248.2	265.0	292.5	323.0	354.8	244.8
Carbon dioxide	CO_2	−134.3	−124.4	−119.5	−114.4	−108.6	−104.8	−100.2	−93.0	−85.7	−78.2	−57.5
disulfide	CS_2	−73.8	−54.3	−44.7	−34.3	−22.5	−15.3	−5.1	+10.4	28.0	46.5	−110.8
monoxide	CO	−222.0	−217.2	−215.0	−212.8	−210.0	−208.1	−205.7	−201.3	−196.3	−191.3	−205.0
oxyselenide (carbonyl selenide)	$COSe$	−117.1	−102.3	−95.0	−86.3	−76.4	−70.2	−61.7	−49.8	−35.6	−21.9	
oxysulfide (carbonyl sulfide)	COS	−132.4	−119.8	−113.3	−106.0	−98.3	−93.0	−85.9	−75.0	−62.7	−49.9	−138.8
tetrabromide	CBr_4	...	...	...	...	96.3	106.3	119.7	139.7	163.5	189.5	90.1
tetrachloride	CCl_4	−50.0	−30.0	−19.6	−8.2	+4.3	12.3	23.0	38.3	57.8	76.7	−22.6
tetrafluoride	CF_4	−184.6	−174.1	−169.3	−164.3	−158.8	−155.4	−150.7	−143.6	−135.5	−127.7	−183.7
Carvacrol	$C_{10}H_{14}O$	70.0	98.4	113.2	127.9	145.2	155.3	169.7	191.2	213.8	237.0	+0.5
Carvone	$C_{10}H_{14}O$	57.4	86.1	100.4	116.1	133.0	143.8	157.3	179.6	203.5	227.5	
Chavibetol	$C_{10}H_{12}O_2$	83.6	113.3	127.0	143.2	159.8	170.7	185.5	206.8	229.8	254.0	
Chloral (trichloroacetaldehyde)	C_2HCl_3O	−37.8	−16.0	−5.0	+7.2	20.2	29.1	40.2	57.8	77.5	97.7	−57
hydrate (trichloroacetaldehyde hydrate)	$C_2H_3Cl_3O_2$	−9.8	+10.0	19.5	29.2	39.7	46.2	55.0	68.0	82.1	96.2	51.7
Chloranil	$C_6Cl_4O_2$	70.7	89.3	97.8	106.4	116.1	122.0	129.5	140.3	151.3	162.6	290
Chloroacetic acid	$C_2H_3ClO_2$	43.0	68.3	81.0	94.2	109.2	118.3	130.7	149.0	169.0	189.5	61.2
anhydride	$C_4H_4Cl_2O_3$	67.2	94.1	108.0	122.4	138.2	148.0	159.8	177.8	197.0	217.0	46
2-Chloroaniline	C_6H_6ClN	46.3	72.3	84.8	99.2	115.6	125.7	139.5	160.0	183.7	208.8	0
3-Chloroaniline	C_6H_6ClN	63.5	89.8	102.0	116.7	133.6	144.1	158.0	179.5	203.5	228.5	−10.4
4-Chloroaniline	C_6H_6ClN	59.3	87.9	102.1	117.8	135.0	145.8	159.9	182.3	206.6	230.5	70.5
Chlorobenzene	C_6H_5Cl	−13.0	+10.6	22.2	35.3	49.7	58.3	70.7	89.4	110.0	132.2	−45.2
2-Chlorobenzotrichloride (2-α,α,α-tetrachlorotoluene)	$C_7H_4Cl_4$	69.0	101.8	117.9	135.8	155.0	167.8	185.0	208.0	233.0	262.1	28.7
2-Chlorobenzotrifluoride (2-chloro-α,α,α-trifluorotoluene)	$C_7H_4ClF_3$	0.0	24.7	37.1	50.6	65.9	75.4	88.3	108.3	130.0	152.2	−6.0
2-Chlorobiphenyl	$C_{12}H_9Cl$	89.3	109.8	134.7	151.2	169.9	182.1	197.0	219.6	243.8	267.5	34
4-Chlorobiphenyl	$C_{12}H_9Cl$	96.4	129.8	146.0	164.0	183.8	196.0	212.5	237.8	264.5	292.9	75.5
α-Chlorocrotonic acid	$C_4H_5ClO_2$	70.0	95.6	108.0	121.2	135.6	144.4	155.9	173.8	193.2	212.0	
Chlorodifluoromethane	$CHClF_2$	−122.8	−110.2	−103.7	−96.5	−88.6	−83.4	−76.4	−65.8	−53.6	−40.8	−160
Chlorodimethylphenylsilane	$C_8H_{11}ClSi$	29.8	56.7	70.0	84.7	101.2	111.5	124.7	145.5	168.6	193.5	
1-Chloro-2-ethoxybenzene	C_8H_9ClO	45.8	72.8	86.5	101.5	117.8	127.8	141.8	162.0	185.5	208.0	
2-(2-Chloroethoxy) ethanol	$C_4H_9ClO_2$	53.0	78.3	90.7	104.1	118.4	127.5	139.5	157.2	176.5	196.0	
bis-2-Chloroethyl acetacetal	$C_6H_{12}Cl_2O_2$	56.2	83.7	97.6	112.2	127.8	138.0	150.7	169.8	190.5	212.6	
1-Chloro-2-ethylbenzene	C_8H_9Cl	17.2	43.0	56.1	70.3	86.2	96.4	110.0	130.2	152.2	177.6	−80.2
1-Chloro-3-ethylbenzene	C_8H_9Cl	18.6	45.2	58.1	73.0	89.2	99.6	113.6	133.8	156.7	181.1	−53.3
1-Chloro-4-ethylbenzene	C_8H_9Cl	19.2	46.4	60.0	75.5	91.8	102.0	116.0	137.0	159.8	184.3	−62.6
2-Chloroethyl chloroacetate	$C_4H_6Cl_2O_2$	46.0	72.1	86.0	100.0	116.0	126.2	140.0	159.8	182.2	205.0	
2-Chloroethyl 2-chloroisopropyl ether	$C_5H_{10}Cl_2O$	24.7	50.1	63.0	77.2	92.4	102.2	115.8	135.7	156.5	180.0	
2-Chloroethyl 2-chloropropyl ether	$C_5H_{10}Cl_2O$	29.8	56.5	70.0	84.8	101.5	111.8	125.6	146.3	169.8	194.1	
2-Chloroethyl α-methylbenzyl ether	$C_{10}H_{13}ClO$	62.3	91.4	106.0	121.8	139.6	150.0	164.8	186.3	210.8	235.0	
Chloroform (trichloromethane)	$CHCl_3$	−58.0	−39.1	−29.7	−19.0	−7.1	+0.5	10.4	25.9	42.7	61.3	−63.5
1-Chloronaphthalene	$C_{10}H_7Cl$	80.6	104.8	118.6	134.4	153.2	165.6	180.4	204.2	230.8	259.3	−20
4-Chlorophenethyl alcohol	C_8H_9ClO	84.0	114.3	129.0	145.0	162.0	173.5	188.1	210.0	234.5	259.3	
2-Chlorophenol	C_6H_5ClO	12.1	38.2	51.2	65.9	82.0	92.0	106.0	126.4	149.8	174.5	7
3-Chlorophenol	C_6H_5ClO	44.2	72.0	86.1	101.7	118.0	129.4	143.0	164.8	188.7	214.0	32.5
4-Chlorophenol	C_6H_5ClO	49.8	78.2	92.2	108.1	125.0	136.1	150.0	172.0	196.0	220.0	42
2-Chloro-3-phenylphenol	$C_{12}H_9ClO$	118.0	152.2	169.7	186.7	207.4	219.6	237.0	261.3	289.4	317.5	+6
2-Chloro-6-phenylphenol	$C_{12}H_9ClO$	119.8	153.7	170.7	189.8	208.2	220.0	237.1	261.6	289.5	317.0	
Chloropicrin (trichloronitromethane)	CCl_3NO_2	−25.5	−3.3	+7.8	20.0	33.8	42.3	53.8	71.8	91.8	111.9	−64
1-Chloropropene	C_3H_5Cl	−81.3	−63.4	−54.1	−44.0	−32.7	−25.1	−15.1	+1.3	18.0	37.0	−99.0
2-Chloropyridine	C_5H_4ClN	13.3	38.8	51.7	65.8	81.7	91.6	104.6	125.0	147.7	170.2	
3-Chlorostyrene	C_8H_7Cl	25.3	51.3	65.2	80.0	96.5	107.2	121.2	142.2	165.7	190.0	
4-Chlorostyrene	C_8H_7Cl	28.0	54.5	67.5	82.0	98.0	108.5	122.0	143.5	166.0	191.0	−15.0
1-Chlorotetradecane	$C_{14}H_{29}Cl$	98.5	131.8	148.2	166.2	187.0	199.8	215.5	240.3	267.5	296.0	+0.9
2-Chlorotoluene	C_7H_7Cl	+5.4	30.6	43.2	56.9	72.0	81.8	94.7	115.0	137.1	159.3	
3-Chlorotoluene	C_7H_7Cl	+4.8	30.3	43.2	57.4	73.0	83.2	96.3	116.6	139.7	162.3	
4-Chlorotoluene	C_7H_7Cl	+5.5	31.0	43.8	57.8	73.5	83.3	96.6	117.1	139.8	162.3	+7.3
Chlorotriethylsilane	$C_6H_{15}ClSi$	−4.9	+19.8	32.0	45.5	60.2	69.5	82.3	101.6	123.6	146.3	
1-Chloro-1,2,2-trifluoroethylene	C_2ClF_3	−116.0	−102.5	−95.9	−88.2	−79.7	−74.1	−66.7	−55.0	−41.7	−27.9	−157.5
Chlorotrifluoromethane	$CClF_3$	−149.5	−139.2	−134.1	−128.5	−121.9	−117.3	−111.7	−102.5	−92.7	−81.2	
Chlorotrimethylsilane	C_3H_9ClSi	−62.8	−43.6	−34.0	−23.2	−11.4	−4.0	+6.0	21.9	39.4	57.9	
trans-Cinnamic acid	$C_9H_8O_2$	127.5	157.8	173.0	189.5	207.1	217.8	232.4	253.3	276.7	300.0	133
Cinnamyl alcohol	$C_9H_{10}O$	72.6	102.5	117.8	133.7	151.0	162.0	177.8	199.8	224.6	250.0	33
Cinnamylaldehyde	C_9H_8O	76.1	105.8	120.0	135.7	152.2	163.7	177.7	199.3	222.4	246.0	−7.5
Citraconic anhydride	$C_5H_4O_3$	47.1	74.8	88.9	103.8	120.3	131.3	145.4	165.8	189.8	213.5	
cis-α-Citral	$C_{10}H_{16}O$	61.7	90.0	103.9	119.4	135.9	146.3	160.0	181.8	205.0	228.0	
d-Citronellal	$C_{10}H_{18}O$	44.0	71.4	84.8	99.8	116.1	126.2	140.1	160.0	183.8	206.5	
Citronellic acid	$C_{10}H_{18}O_2$	99.5	127.3	141.4	155.6	171.9	182.1	195.4	214.5	236.6	257.0	
Citronellol	$C_{10}H_{20}O$	66.4	93.6	107.0	121.5	137.2	147.2	159.8	179.8	201.0	221.5	
Citronellyl acetate	$C_{12}H_{22}O_2$	74.7	100.2	113.0	126.0	140.5	149.7	161.0	178.8	197.8	217.0	
Coumarin	$C_9H_6O_2$	106.0	137.8	153.4	170.0	189.0	200.5	216.5	240.0	264.7	291.0	70

TABLE 3-8 Vapor Pressures of Organic Compounds, up to 1 atm *(Continued)*

Compound		Pressure, mm. Hg										Melting point, °C.
Name	Formula	1	5	10	20	40	60	100	200	400	760	
		Temperature, °C.										
o-Cresol (2-cresol; 2-methylphenol)	C_7H_8O	38.2	64.0	76.7	90.5	105.8	115.5	127.4	146.7	168.4	190.8	30.8
m-Cresol (3-cresol; 3-methylphenol)	C_7H_8O	52.0	76.0	87.8	101.4	116.0	125.8	138.0	157.3	179.0	202.8	10.9
p-Cresol (4-cresol; 4-methylphenol)	C_7H_8O	53.0	76.5	88.6	102.3	117.7	127.0	140.0	157.3	179.4	201.8	35.5
cis-Crotonic acid	$C_4H_6O_2$	33.5	57.4	69.0	82.0	96.0	104.5	116.3	133.9	152.2	171.9	15.5
trans-Crotonic acid	$C_4H_6O_2$	......	......	80.0	93.0	107.8	116.7	128.0	146.0	165.5	185.0	72
cis-Crotononitrile	C_4H_5N	−29.0	−7.1	+4.0	16.4	30.0	38.5	50.1	68.0	88.0	108.0	
trans-Crotononitrile	C_4H_5N	−19.5	+3.5	15.0	27.8	41.8	50.9	62.8	81.1	101.5	122.8	
Cumene	C_9H_{12}	+2.9	26.8	38.3	51.5	66.1	75.4	88.1	107.3	129.2	152.4	−96.0
4-Cumidene	$C_9H_{13}N$	60.0	88.2	102.2	117.8	134.2	145.0	158.0	180.0	203.2	227.0	
Cuminal	$C_{10}H_{12}O$	58.0	87.3	102.0	117.9	135.2	146.0	160.0	182.8	206.7	232.0	
Cuminyl alcohol	$C_{10}H_{14}O$	74.2	103.7	118.0	133.8	150.3	161.7	176.2	197.9	221.7	246.6	
2-Cyano-2-n-butyl acetate	$C_7H_{11}NO_2$	42.0	68.7	82.0	96.2	111.8	121.5	133.8	152.2	173.4	195.2	
Cyanogen	C_2N_2	−95.8	−83.2	−76.8	−70.1	−62.7	−57.9	−51.8	−42.6	−33.0	−21.0	−34.4
bromide	$CBrN$	−35.7	−18.3	−10.0	−1.0	+8.6	14.7	22.6	33.8	46.0	61.5	58
chloride	$CClN$	−76.7	−61.4	−53.8	−46.1	−37.5	−32.1	−24.9	−14.1	−2.3	+13.1	−6.5
iodide	CIN	25.2	47.2	57.7	68.6	80.3	88.0	97.6	111.5	126.1	141.1	
Cyclobutane	C_4H_8	−92.0	−76.0	−67.9	−58.7	−48.4	−41.8	−32.8	−18.9	−3.4	+12.9	−50
Cyclobutene	C_4H_6	−99.1	−83.4	−75.4	−66.6	−56.4	−50.0	−41.2	−27.8	−12.2	+2.4	
Cyclohexane	C_6H_{12}	−45.3	−25.4	−15.9	−5.0	+6.7	14.7	25.5	42.0	60.8	80.7	+6.6
Cyclohexaneethanol	$C_8H_{16}O$	50.4	77.2	90.0	104.0	119.8	129.8	142.7	161.7	183.5	205.4	
Cyclohexanol	$C_6H_{12}O$	21.0	44.0	56.0	68.8	83.0	91.8	103.7	121.7	141.4	161.0	23.9
Cyclohexanone	$C_6H_{10}O$	+1.4	26.4	38.7	52.5	67.8	77.5	90.4	110.3	132.5	155.6	−45.0
2-Cyclohexyl-4,6-dinitrophenol	$C_{12}H_{14}N_2O_5$	132.8	161.8	175.9	191.2	206.7	216.0	229.0	248.7	269.8	291.5	
Cyclopentane	C_5H_{10}	−68.0	−49.6	−40.4	−30.1	−18.6	−11.3	−1.3	+13.8	31.0	49.3	−93.7
Cyclopropane	C_3H_6	−116.8	−104.2	−97.5	−90.3	−82.3	−77.0	−70.0	−59.1	−46.9	−33.5	−126.6
Cymene	$C_{10}H_{14}$	17.3	43.9	57.0	71.1	87.0	97.2	110.8	131.4	153.5	177.2	−68.2
cis-Decalin	$C_{10}H_{18}$	22.5	50.1	64.2	79.8	97.2	108.0	123.2	145.4	169.9	194.6	−43.3
trans-Decalin	$C_{10}H_{18}$	−0.8	+30.6	47.2	65.3	85.7	98.4	114.6	136.2	160.1	186.7	−30.7
Decane	$C_{10}H_{22}$	16.5	42.3	55.7	69.8	85.5	95.5	108.6	128.4	150.6	174.1	−29.7
Decan-2-one	$C_{10}H_{20}O$	44.2	71.9	85.8	100.7	117.1	127.8	142.0	163.2	186.7	211.0	+3.5
1-Decene	$C_{10}H_{20}$	14.7	40.3	53.7	67.8	83.3	93.5	106.5	126.7	149.2	172.0	
Decyl alcohol	$C_{10}H_{22}O$	69.5	97.3	111.3	125.8	142.1	152.0	165.8	186.2	208.8	231.0	+7
Decyltrimethylsilane	$C_{13}H_{30}Si$	67.4	96.4	111.0	126.5	144.0	154.3	169.5	191.0	215.5	240.0	
Dehydroacetic acid	$C_8H_8O_4$	91.7	122.0	137.3	153.0	171.0	181.5	197.5	219.5	244.5	269.0	
Desoxybenzoin	$C_{14}H_{12}O$	123.3	156.2	173.5	192.0	212.0	224.5	241.3	265.2	293.0	321.0	60
Diacetamide	$C_4H_7NO_2$	70.0	95.0	108.0	122.6	138.2	148.0	160.6	180.8	202.0	223.0	78.5
Diacetylene (1,3-butadiyne)	C_4H_2	−82.5	−68.0	−61.2	−53.8	−45.9	−41.0	−34.0	−20.9	−6.1	+9.7	−34.9
Diallyldichlorosilane	$C_6H_{10}Cl_2Si$	+9.5	34.8	47.4	61.3	76.4	86.3	99.7	119.4	142.0	165.3	
Diallyl sulfide	$C_6H_{10}S$	−9.5	+14.4	26.6	39.7	54.2	63.7	75.8	94.8	116.1	138.6	−83
Diisoamyl ether	$C_{10}H_{22}O$	18.6	44.3	57.0	70.7	86.3	96.0	109.6	129.0	150.3	173.4	
oxalate	$C_{12}H_{22}O_4$	85.4	116.0	131.4	147.7	165.7	177.0	192.2	215.0	240.0	265.0	
sulfide	$C_{10}H_{22}S$	43.0	73.0	87.6	102.7	120.0	130.6	145.3	166.4	191.0	216.0	
Dibenzylamine	$C_{14}H_{15}N$	118.3	149.8	165.6	182.2	200.2	212.2	227.3	249.8	274.3	300.0	−26
Dibenzyl ketone (1,3-diphenyl-2-propanone)	$C_{15}H_{14}O$	125.5	159.8	177.6	195.7	216.6	229.4	246.6	272.3	301.7	330.5	34.5
1,4-Dibromobenzene	$C_6H_4Br_2$	61.0	79.3	87.7	103.6	120.8	131.6	146.5	168.5	192.5	218.6	87.5
1,2-Dibromobutane	$C_4H_8Br_2$	7.5	33.2	46.1	60.0	76.0	86.0	99.8	120.2	143.5	166.3	−64.5
dl-2,3-Dibromobutane	$C_4H_8Br_2$	+5.0	30.0	41.6	56.4	72.0	82.0	95.3	115.7	138.0	160.5	
meso-2,3-Dibromobutane	$C_4H_8Br_2$	+1.5	26.6	39.3	53.2	68.0	78.0	91.7	111.8	134.2	157.3	−34.5
1,2-Dibromodecane	$C_{10}H_{20}Br_2$	95.7	123.6	137.3	151.0	167.4	177.5	190.2	209.6	229.8	250.4	
Di(2-bromoethyl) ether	$C_4H_8Br_2O$	47.7	75.3	88.5	103.6	119.8	130.0	144.0	165.0	188.0	212.5	
α,β-Dibromomaleic anhydride	$C_4H_2Br_2O_3$	50.0	78.0	92.0	106.7	123.5	133.8	147.7	168.0	192.0	215.0	
1,2-Dibromo-2-methylpropane	$C_4H_8Br_2$	−28.8	−3.0	+10.5	25.7	42.3	53.7	68.8	92.1	119.8	149.0	−70.3
1,3-Dibromo-2-methylpropane	$C_4H_8Br_2$	14.0	40.0	53.0	67.5	83.5	93.7	107.4	117.8	150.6	174.6	
1,2-Dibromopentane	$C_5H_{10}Br_2$	19.8	45.4	58.0	72.0	87.4	97.4	110.1	130.2	151.8	175.0	
1,2-Dibromopropane	$C_3H_6Br_2$	−7.0	+17.3	29.4	42.3	57.2	66.4	78.7	97.8	118.5	141.6	−55.5
1,3-Dibromopropane	$C_3H_6Br_2$	+9.7	35.4	48.0	62.1	77.8	87.8	101.3	121.7	144.1	167.5	−34.4
2,3-Dibromopropene	$C_3H_4Br_2$	−6.0	+17.9	30.0	43.2	57.8	67.0	79.5	98.0	119.5	141.2	
2,3-Dibromo-1-propanol	$C_3H_6Br_2O$	57.0	84.5	98.2	113.5	129.8	140.0	153.0	173.8	196.0	219.0	
Diisobutylamine	$C_8H_{19}N$	−5.1	+18.4	30.6	43.7	57.8	67.0	79.2	97.6	118.0	139.5	−70
2,6-Ditert-butyl-4-cresol	$C_{15}H_{24}O$	85.8	116.2	131.0	147.0	164.1	175.2	190.0	212.8	237.6	262.5	
4,6-Ditert-butyl-2-cresol	$C_{15}H_{24}O$	86.2	117.3	132.4	149.0	167.4	179.0	194.0	217.5	243.4	269.3	
4,6-Ditert-butyl-3-cresol	$C_{15}H_{24}O$	103.7	135.2	150.0	167.0	185.3	196.1	211.0	233.0	257.1	282.0	
2,6-Ditert-butyl-4-ethylphenol	$C_{16}H_{26}O$	89.1	121.4	137.0	154.0	172.1	183.9	198.0	220.0	244.0	268.6	
4,6-Ditert-butyl-3-ethylphenol	$C_{16}H_{26}O$	111.5	142.6	157.4	174.0	192.3	204.4	218.0	241.7	264.6	290.0	
Diisobutyl oxalate	$C_{10}H_{18}O_4$	63.2	91.2	105.3	120.3	137.5	147.8	161.8	183.5	205.8	229.5	
2,4-Ditert-butylphenol	$C_{14}H_{22}O$	84.5	115.4	130.0	146.0	164.3	175.8	190.0	212.5	237.0	260.8	
Dibutyl phthalate	$C_{16}H_{22}O_4$	148.2	182.1	198.2	216.2	235.8	247.8	263.7	287.0	313.5	340.0	
sulfide	$C_8H_{18}S$	+21.7	51.8	66.4	80.5	96.0	105.8	118.6	138.0	159.0	182.0	−79.7
Diisobutyl d-tartrate	$C_{12}H_{22}O_6$	117.8	151.8	169.0	188.0	208.5	221.6	239.5	264.7	294.0	324.0	73.5
Dicarvacryl-mono-(6-chloro-2-xenyl) phosphate	$C_{32}H_{34}ClO_4P$	204.2	234.5	249.3	264.5	280.5	290.7	304.9	323.8	342.0	361.0	
Dicarvacryl-2-tolyl phosphate	$C_{27}H_{33}O_4P$	180.2	209.3	221.8	237.0	251.5	260.3	272.5	290.0	309.8	330.0	
Dichloroacetic acid	$C_2H_2Cl_2O_2$	44.0	69.8	82.6	96.3	111.8	121.5	134.0	152.3	173.7	194.4	9.7
1,2-Dichlorobenzene	$C_6H_4Cl_2$	20.0	46.0	59.1	73.4	89.4	99.5	112.9	133.4	155.8	179.0	−17.6
1,3-Dichlorobenzene	$C_6H_4Cl_2$	12.1	39.0	52.0	66.2	82.0	92.2	105.0	125.9	149.0	173.0	−24.2
1,4-Dichlorobenzene	$C_6H_4Cl_2$	......	......	54.8	69.2	84.8	95.2	108.4	128.3	150.2	173.9	53.0
1,2-Dichlorobutane	$C_4H_8Cl_2$	−23.6	−0.3	+11.5	24.5	37.7	47.8	60.2	79.7	100.8	123.5	
2,3-Dichlorobutane	$C_4H_8Cl_2$	−25.2	−3.0	+8.5	21.2	35.0	43.9	56.0	74.0	94.2	116.0	−80.4
1,2-Dichloro-1,2-difluoroethylene	$C_2Cl_2F_2$	−82.0	−65.6	−57.3	−48.3	−38.2	−31.8	−23.0	−10.0	+5.0	20.9	−112
Dichlorodifluoromethane	CCl_2F_2	−118.5	−104.6	−97.8	−90.1	−81.6	−76.1	−68.6	−57.0	−43.9	−29.8	
Dichlorodiphenyl silane	$C_{12}H_{10}Cl_2Si$	109.6	142.4	158.0	176.0	195.5	207.5	223.8	248.0	275.5	304.0	
Dichlorodiisopropyl ether	$C_6H_{12}Cl_2O$	29.6	55.2	68.2	82.2	97.3	106.9	119.7	139.0	159.8	182.7	
Di(2-chloroethoxy) methane	$C_5H_{10}Cl_2O_2$	53.0	80.4	94.0	109.5	125.5	135.8	149.6	170.0	192.0	215.0	
Dichloroethoxymethylsilane	$C_3H_8Cl_2OSi$	−33.8	−12.1	−1.3	+11.3	24.4	32.6	44.1	61.0	80.3	100.6	
1,2-Dichloro-3-ethylbenzene	$C_8H_8Cl_2$	46.0	75.0	90.0	105.9	123.8	135.0	149.8	172.0	197.0	222.1	−40.8
1,2-Dichloro-4-ethylbenzene	$C_8H_8Cl_2$	47.0	77.2	92.3	109.6	127.5	139.0	153.3	176.0	201.7	226.6	−76.4
1,4-Dichloro-2-ethylbenzene	$C_8H_8Cl_2$	38.5	68.0	83.2	99.8	118.0	129.0	144.0	166.2	191.5	216.3	−61.2
cis-1,2-Dichloroethylene	$C_2H_2Cl_2$	−58.4	−39.2	−29.9	−19.4	−7.9	−0.5	+9.5	24.6	41.0	59.0	−80.5
trans-1,2-Dichloro ethylene	$C_2H_2Cl_2$	−65.4	−47.2	−38.0	−28.0	−17.0	−10.0	−0.2	+14.3	30.8	47.8	−50.0

TABLE 3-8 Vapor Pressures of Organic Compounds, up to 1 atm *(Continued)*

Compound		Pressure, mm. Hg										Melting point, °C.
Name	Formula	1	5	10	20	40	60	100	200	400	760	
		Temperature, °C.										
Di(2-chloroethyl) ether	$C_4H_8Cl_2O$	23.5	49.3	62.0	76.0	91.5	101.5	114.5	134.0	155.4	178.5	
Dichlorofluoromethane	$CHCl_2F$	−91.3	−75.5	−67.5	−58.6	−48.8	−42.6	−33.9	−20.9	−6.2	+8.9	−135
1,5-Dichlorohexamethyltrisiloxane	$C_6H_{18}Cl_2O_2Si_3$	26.0	52.0	65.1	79.0	94.8	105.0	118.2	138.3	160.2	184.0	−53.0
Dichloromethylphenylsilane	$C_7H_8Cl_2Si$	35.7	63.5	77.4	92.4	109.5	120.0	134.2	155.5	180.2	205.5	
1,1-Dichloro-2-methylpropane	$C_4H_8Cl_2$	−31.0	−8.4	+2.6	14.6	28.2	37.0	48.2	65.8	85.4	106.0	
1,2-Dichloro-2-methylpropane	$C_4H_8Cl_2$	−25.8	−4.2	+6.7	18.7	32.0	40.2	51.7	68.9	87.8	108.0	
1,3-Dichloro-2-methylpropane	$C_4H_8Cl_2$	−3.0	+20.6	32.0	44.8	58.6	67.5	78.8	96.1	115.4	135.0	
2,4-Dichlorophenol	$C_6H_4Cl_2O$	53.0	80.0	92.8	107.7	123.4	133.5	146.0	165.2	187.5	210.0	45.0
2,6-Dichlorophenol	$C_6H_4Cl_2O$	59.5	87.6	101.0	115.5	131.6	141.8	154.6	175.5	197.7	220.0	
α,α-Dichlorophenylacetonitrile	$C_8H_5Cl_2N$	56.0	84.0	98.1	113.8	130.0	141.0	154.5	176.2	199.5	223.5	
Dichlorophenylarsine	$C_6H_5AsCl_2$	61.8	100.0	116.0	133.1	151.0	163.2	178.9	202.8	228.8	256.5	
1,2-Dichloropropane	$C_3H_6Cl_2$	−38.5	−17.0	−6.1	+6.0	19.4	28.0	39.4	57.0	76.0	96.8	
2,3-Dichlorostyrene	$C_8H_6Cl_2$	61.0	90.1	104.6	120.5	137.8	149.0	163.5	185.7	210.0	235.0	
2,4-Dichlorostyrene	$C_8H_6Cl_2$	53.5	82.2	97.4	111.8	129.2	140.0	153.8	176.0	200.0	225.0	
2,5-Dichlorostyrene	$C_8H_6Cl_2$	55.5	83.9	98.2	114.0	131.0	142.0	155.8	178.0	202.5	227.0	
2,6-Dichlorostyrene	$C_8H_6Cl_2$	47.8	75.7	90.0	105.5	122.4	133.3	147.6	169.0	193.5	217.0	
3,4-Dichlorostyrene	$C_8H_6Cl_2$	57.2	86.0	100.4	116.2	133.7	144.6	158.2	181.5	205.7	230.0	
3,5-Dichlorostyrene	$C_8H_6Cl_2$	53.5	82.2	97.4	111.8	129.2	140.0	153.8	176.0	200.0	225.0	
1,2-Dichlorotetraethylbenzene	$C_{14}H_{20}Cl_2$	105.6	138.7	155.0	172.5	192.2	204.8	220.7	245.6	272.8	302.0	
1,4-Dichlorotetraethylbenzene	$C_{14}H_{20}Cl_2$	91.7	126.1	143.8	162.0	183.2	195.8	212.0	238.5	265.8	296.5	
1,2-Dichloro-1,1,2,2-tetrafluoroethane	$C_2Cl_2F_4$	−95.4	−80.0	−72.3	−63.5	−53.7	−47.5	−39.1	−26.3	−12.0	+3.5	−94
Dichloro-4-tolylsilane	$C_7H_8Cl_2Si$	46.2	71.7	84.2	97.8	113.2	122.6	135.5	153.5	175.2	196.3	
3,4-Dichloro-α,α,α-trifluorotoluene	$C_7H_3Cl_2F_3$	11.0	38.3	52.2	67.3	84.0	95.0	109.2	129.0	150.5	172.8	−12.1
Dicyclopentadiene	$C_{10}H_8$	......	34.1	47.6	62.0	77.9	88.0	101.7	121.8	144.2	166.6	32.9
Diethoxydimethylsilane	$C_6H_{16}O_2Si$	−19.1	+2.4	13.3	25.3	38.0	46.3	57.6	74.2	93.2	113.5	
Diethoxydiphenylsilane	$C_{16}H_{20}O_2Si$	111.5	142.8	157.6	174.3	193.2	205.0	220.0	243.8	259.7	296.0	
Diethyl adipate	$C_{10}H_{18}O_4$	74.0	106.6	123.0	138.3	154.6	165.8	179.0	198.2	219.1	240.0	−21
Diethylamine	$C_4H_{11}N$	......	......	−33.0	−22.6	−11.3	−4.0	+6.0	21.0	38.0	55.5	−38.9
N-Diethylaniline	$C_{10}H_{15}N$	49.7	78.0	91.9	107.2	123.6	133.8	147.3	168.2	192.4	215.5	−34.4
Diethyl arsanilate	$C_{10}H_{16}AsNO_3$	38.0	62.6	74.8	88.0	102.6	111.8	123.8	141.9	161.0	181.0	
1,2-Diethylbenzene	$C_{10}H_{14}$	22.3	48.7	62.0	76.4	92.5	102.6	116.2	136.7	159.0	183.5	−31.4
1,3-Diethylbenzene	$C_{10}H_{14}$	20.7	46.8	59.9	74.5	90.4	100.7	114.4	134.8	156.9	181.1	−83.9
1,4-Diethylbenzene	$C_{10}H_{14}$	20.7	47.1	60.3	74.7	91.1	101.3	115.3	136.1	159.0	183.8	−43.2
Diethyl carbonate	$C_5H_{10}O_3$	−10.1	+12.3	23.8	36.0	49.5	57.9	69.7	86.5	105.8	125.8	−43
cis-Diethyl citraconate	$C_9H_{14}O_4$	59.8	88.3	103.0	118.2	135.7	146.2	160.0	182.3	206.5	230.3	
Diethyl dioxosuccinate	$C_8H_{10}O_6$	70.0	98.0	112.0	126.8	143.8	153.7	167.7	188.0	210.8	233.5	
Diethylene glycol	$C_4H_{10}O_3$	91.8	120.0	133.8	148.0	164.3	174.0	187.5	207.0	226.5	244.8	
Diethyleneglycol-bis-chloroacetate	$C_8H_{12}Cl_2O_5$	148.3	180.0	195.8	212.0	229.0	239.5	252.0	271.5	291.8	313.0	
Diethylene glycol dimethyl ether Di(2-methoxyethyl) ether	$C_6H_{14}O_3$	13.0	37.6	50.0	63.0	77.5	86.8	99.5	118.0	138.5	159.8	
glycol ethyl ether	$C_6H_{14}O_3$	45.3	72.0	85.8	100.3	116.7	126.8	140.3	159.0	180.3	201.9	
Diethyl ether	$C_4H_{10}O$	−74.3	−56.9	−48.1	−38.5	27.7	−21.8	−11.5	+2.2	17.9	34.6	−116.3
ethylmalonate	$C_9H_{16}O_4$	50.8	77.8	91.6	106.0	122.4	132.4	146.0	166.0	188.7	211.5	
fumarate	$C_8H_{12}O_4$	53.2	81.2	95.3	110.2	126.7	137.7	151.1	172.2	195.8	218.5	+0.6
glutarate	$C_9H_{16}O_4$	65.6	94.7	109.7	125.4	142.8	153.2	167.8	189.5	212.8	237.0	
Diethylhexadecylamine	$C_{20}H_{43}N$	139.8	175.8	194.0	213.5	235.0	248.5	265.5	292.8	324.6	355.0	
Diethyl itaconate	$C_9H_{14}O_4$	51.3	80.2	95.2	111.0	128.2	139.9	154.3	177.5	203.1	227.9	
ketone (3-pentanone)	$C_5H_{10}O$	−12.7	+7.5	17.2	27.9	39.4	46.7	56.2	70.6	86.3	102.7	−42
malate	$C_8H_{14}O_5$	80.7	110.4	125.3	141.2	157.8	169.0	183.9	205.3	229.5	253.4	
maleate	$C_8H_{12}O_4$	57.3	85.6	100.0	115.3	131.8	142.4	156.0	177.8	201.7	225.0	
malonate	$C_7H_{12}O_4$	40.0	67.5	81.3	95.9	113.3	123.0	136.2	155.5	176.8	198.9	−49.8
mesaconate	$C_9H_{14}O_4$	62.8	91.0	105.3	120.3	137.3	147.9	161.6	183.2	205.8	229.0	
oxalate	$C_6H_{10}O_4$	47.4	71.8	83.8	96.8	110.6	119.7	130.8	147.9	166.2	185.7	−40.6
phthalate	$C_{12}H_{14}O_4$	108.8	140.7	156.0	173.6	192.1	204.1	219.5	243.0	267.5	294.0	
sebacate	$C_{14}H_{26}O_4$	125.3	156.2	172.1	189.8	207.5	218.4	234.4	255.8	280.3	305.5	1.3
2,5-Diethylstyrene	$C_{12}H_{16}$	49.7	78.4	92.6	108.5	125.8	136.8	151.0	173.2	198.0	223.0	
Diethyl succinate	$C_8H_{14}O_4$	54.6	83.0	96.6	111.7	127.8	138.2	151.1	171.7	193.8	216.5	−20.8
isosuccinate	$C_8H_{14}O_4$	39.8	66.7	80.0	94.7	111.0	121.4	134.8	155.1	177.7	201.3	
sulfate	$C_4H_{10}O_4S$	47.0	74.0	87.7	102.1	118.0	128.6	142.5	162.5	185.5	209.5	−25.0
sulfide	$C_4H_{10}S$	−39.6	−18.6	−8.0	+3.5	16.1	24.2	35.0	51.3	69.7	88.0	−99.5
sulfite	$C_4H_{10}O_3S$	10.0	34.2	46.4	59.7	74.2	83.8	96.3	115.8	137.0	159.0	
d-Diethyl tartrate	$C_8H_{14}O_6$	102.0	133.0	148.0	164.2	182.3	194.0	208.5	230.4	254.8	280.0	17
dl-Diethyl tartrate	$C_8H_{14}O_6$	100.0	131.7	147.2	163.8	181.7	193.2	208.0	230.0	254.3	280.0	
3,5-Diethyltoluene	$C_{11}H_{16}$	34.0	61.5	75.3	90.2	107.0	117.7	131.7	152.4	176.5	200.7	
Diethylzinc	$C_4H_{10}Zn$	−22.4	0.0	+11.7	24.2	38.0	47.2	59.1	77.0	97.3	118.0	−28
l-Dihydrocarvone	$C_{10}H_{16}O$	46.6	75.5	90.0	106.0	123.7	134.7	149.7	171.8	197.0	223.0	
Dihydrocitronellol	$C_{10}H_{22}O$	68.0	91.7	103.0	115.0	127.6	136.7	145.9	160.2	176.8	193.5	
1,4-Dihydroxyanthraquinone	$C_{14}H_8O_4$	196.7	239.8	259.8	282.0	307.4	323.3	344.5	377.8	413.0	450.0	194
Dimethylacetylene (2-butyne)	C_4H_6	−73.0	−57.9	−50.5	−42.5	−33.9	−27.8	−18.8	−5.0	+10.6	27.2	−32.5
Dimethylamine	C_2H_7N	−87.7	−72.2	−64.6	−56.0	−46.7	−40.7	−32.6	−20.4	−7.1	+7.4	−96
N,N-Dimethylaniline	$C_8H_{11}N$	29.5	56.3	70.0	84.8	101.6	111.9	125.8	146.5	169.2	193.1	+2.5
imethyl arsanilate	$C_8H_{12}AsNO_3$	15.0	39.6	51.8	65.0	79.7	88.6	101.0	119.8	140.3	160.5	
Di(α-methylbenzyl) ether	$C_{16}H_{18}O$	96.7	128.3	144.0	160.3	179.6	191.5	206.8	229.7	254.8	281.0	
2,2-Dimethylbutane	C_6H_{14}	−69.3	−50.7	−41.5	−31.1	−19.5	−12.1	−2.0	+13.4	31.0	49.7	−99.8
2,3-Dimethylbutane	C_6H_{14}	−63.6	−44.5	−34.9	−24.1	−12.4	−4.9	+5.4	21.1	39.0	58.0	−128.2
Dimethyl citraconate	$C_7H_{10}O_4$	50.8	78.2	91.8	106.5	122.6	132.7	145.8	165.8	188.0	210.5	
1,1-Dimethylcyclohexane	C_8H_{16}	−24.4	−1.4	+10.3	23.0	37.3	45.7	57.9	76.2	97.2	119.5	−34
cis-1,2-Dimethylcyclohexane	C_8H_{16}	−15.9	+7.3	18.4	31.1	45.3	54.4	66.8	85.6	107.0	129.7	−50.0
crans-1,2-Dimethylcyclohexane	C_8H_{16}	−21.1	+1.7	13.0	25.6	39.7	48.7	61.0	79.6	100.9	123.4	−88.0
crans-1,3-Dimethylcyclohexane	C_8H_{16}	−19.4	+3.4	14.9	27.4	41.4	50.4	62.5	81.0	102.1	124.4	−92.0
cis-1,3-Dimethylcyclohexane	C_8H_{16}	−22.7	0.0	+11.2	23.6	37.5	46.4	58.5	76.9	97.8	120.1	−76.2
cis-1,4-Dimethylcyclohexane	C_8H_{16}	−20.0	+3.2	14.5	27.1	41.1	50.1	62.3	80.8	101.9	124.3	−87.4
trans-1,4-Dimethylcyclohexane	C_8H_{16}	−24.3	−1.7	+10.1	22.6	36.5	45.4	57.6	76.0	97.0	119.3	−36.9
Dimethyl ether	C_2H_6O	−115.7	−101.1	−93.3	−85.2	−76.2	−70.4	−62.7	−50.9	−37.8	−23.7	−138.5
2,2-Dimethylhexane	C_8H_{18}	−29.7	−7.9	+3.1	15.0	28.2	36.7	48.2	65.7	85.6	106.8	
2,3-Dimethylhexane	C_8H_{18}	−23.0	−1.1	+9.9	22.1	35.6	44.2	56.0	73.8	94.1	115.6	
2,4-Dimethylhexane	C_8H_{18}	−26.9	−5.3	+5.2	17.2	30.5	39.0	50.6	68.1	88.2	109.4	
2,5-Dimethylhexane	C_8H_{18}	−26.7	−5.5	+5.3	17.2	30.4	38.9	50.5	68.0	87.9	109.1	−90.7

TABLE 3-8 Vapor Pressures of Organic Compounds, up to 1 atm *(Continued)*

Compound		Pressure, mm. Hg										Melting point, °C.
Name	Formula	1	5	10	20	40	60	100	200	400	760	
		Temperature °C.										
3,3-Dimethylhexane	C_8H_{18}	−25.8	−4.4	+6.1	18.2	31.7	40.4	52.5	70.0	90.4	112.0	
3,4-Dimethylhexane	C_8H_{18}	−22.1	+0.2	11.3	23.5	37.1	45.8	57.7	75.6	96.0	117.7	
Dimethyl itaconate	$C_7H_{10}O_4$	69.3	94.0	106.6	119.7	133.7	142.6	153.7	171.0	189.8	208.0	38
l-Dimethyl malate	$C_6H_{10}O_5$	75.4	104.0	118.3	133.8	150.1	160.4	175.1	196.3	219.5	242.6	
Dimethyl maleate	$C_6H_8O_4$	45.7	73.0	86.4	101.3	117.2	127.1	140.4	160.0	182.2	205.0	
malonate	$C_5H_8O_4$	35.0	59.8	72.0	85.0	100.0	109.7	121.9	140.0	159.8	180.7	−62
trans-Dimethyl mesaconate	$C_7H_{10}O_4$	46.8	74.0	87.8	102.1	118.0	127.8	141.5	161.0	183.5	206.0	
2,7-Dimethyloctane	$C_{10}H_{22}$	+6.3	30.5	42.3	55.8	71.2	80.8	93.9	114.0	136.0	159.7	−52.8
Dimethyl oxalate	$C_4H_6O_4$	20.0	44.0	56.0	69.4	83.6	92.8	104.8	123.3	143.3	163.3	
2,2-Dimethylpentane	C_7H_{16}	−49.0	−28.7	−18.7	−7.5	+5.0	13.0	23.9	40.3	59.2	79.2	−123.7
2,3-Dimethylpentane	C_7H_{16}	−42.0	−20.8	−10.3	+1.1	13.9	22.1	33.3	50.1	69.4	89.8	−135
2,4-Dimethylpentane	C_7H_{16}	−48.0	−27.4	−17.1	−5.9	+6.5	14.5	25.4	41.8	60.6	80.5	−119.5
3,3-Dimethylpentane	C_7H_{16}	−45.9	−25.0	−14.4	−2.9	+9.9	18.1	29.3	46.2	65.5	86.1	−135.0
2,3-Dimethylphenol (2,3-xylenol)	$C_8H_{10}O$	56.0	83.8	97.6	112.0	129.2	139.5	152.2	173.0	196.0	218.0	75
2,4-Dimethylphenol (2,4-xylenol)	$C_8H_{10}O$	51.8	78.0	91.3	105.0	121.5	131.0	143.0	161.5	184.2	211.5	25.5
2,5-Dimethylphenol (2,5-xylenol)	$C_8H_{10}O$	51.8	78.0	91.3	105.0	121.5	131.0	143.0	161.5	184.2	211.5	74.5
3,4-Dimethylphenol (3,4-xylenol)	$C_8H_{10}O$	66.2	93.8	107.7	122.0	138.0	148.0	161.0	181.5	203.6	225.2	62.5
3.5-Dimethylphenol (3,5-xylenol)	$C_8H_{10}O$	62.0	89.2	102.4	117.0	133.3	143.5	156.0	176.2	197.8	219.5	68
Dimethylphenylsilane	$C_8H_{12}Si$	+5.3	30.3	42.6	56.2	71.4	81.3	94.2	114.2	136.4	159.3	
Dimethyl phthalate	$C_{10}H_{10}O_4$	100.3	131.8	147.6	164.0	182.8	194.0	210.0	232.7	257.8	283.7	
3,5-Dimethyl-1,2-pyrone	$C_7H_8O_2$	78.6	107.6	122.0	136.4	152.7	163.8	177.5	198.0	221.0	245.0	51.5
4,6-Dimethylresorcinol	$C_8H_{10}O_2$	49.0	76.8	90.7	105.8	122.5	133.2	147.3	167.8	192.0	215.0	
Dimethyl sebacate	$C_{12}H_{22}O_4$	104.0	139.8	156.2	175.8	196.0	208.0	222.6	245.0	269.6	293.5	38
2,4-Dimethylstyrene	$C_{10}H_{12}$	34.2	61.9	75.8	90.8	107.7	118.0	132.3	153.2	177.5	202.0	
2,5-Dimethylstyrene	$C_{10}H_{12}$	29.0	55.9	69.0	84.0	100.2	110.7	124.7	145.6	168.7	193.0	
α,α-Dimethylsuccinic anhydride	$C_6H_8O_3$	61.4	88.1	102.0	116.3	132.3	142.4	155.3	175.8	197.5	219.5	
Dimethyl sulfide	C_2H_6S	−75.6	−58.0	−49.2	−39.4	−28.4	−21.4	−12.0	+2.6	18.7	36.0	−83.2
d-Dimethyl tartrate	$C_6H_{10}O_6$	102.1	133.2	148.2	164.3	182.4	193.8	208.8	230.5	255.0	280.0	61.5
dl-Dimethyl tartrate	$C_6H_{10}O_6$	100.4	131.8	147.5	164.0	182.4	193.8	209.5	232.3	257.4	282.0	89
N,N-Dimethyl-2-toluidine	$C_9H_{13}N$	28.8	54.1	66.2	80.2	95.0	105.2	118.1	138.3	161.5	184.8	−61
N,N-Dimethyl-4-toluidine	$C_9H_{13}N$	50.1	74.3	86.7	100.0	116.3	126.4	140.3	161.6	185.4	209.5	
Di(nitrosomethyl) amine	$C_2H_5N_3O_2$	+3.2	27.8	40.0	53.7	68.2	77.7	90.3	110.0	131.3	153.0	
Diosphenol	$C_{10}H_{16}O_2$	66.7	95.4	109.0	124.0	141.2	151.3	165.6	186.2	209.5	232.0	
1,4-Dioxane	$C_4H_8O_2$	−35.8	−12.8	−1.2	+12.0	25.2	33.8	45.1	62.3	81.8	101.1	10
Dipentene	$C_{10}H_{16}$	14.0	40.4	53.8	68.2	84.3	94.6	108.3	128.2	150.5	174.6	
Diphenylamine	$C_{12}H_{11}N$	108.3	141.7	157.0	175.2	194.3	206.9	222.8	247.5	274.1	302.0	52.9
Diphenyl carbinol (benzhydrol)	$C_{13}H_{12}O$	110.0	145.0	162.0	180.9	200.0	212.0	227.5	250.0	275.6	301.0	68.5
chlorophosphate	$C_{12}H_{10}ClPO_3$	121.5	160.5	182.0	203.8	227.9	244.2	265.0	299.5	337.2	378.0	
disulfide	$C_{12}H_{10}S_2$	131.6	164.0	180.0	197.0	214.8	226.2	241.3	262.6	285.8	310.0	61
1,2-Diphenylethane (dibenzyl)	$C_{14}H_{14}$	86.8	119.8	136.0	153.7	173.7	186.0	202.8	227.8	255.0	284.0	51.5
Diphenyl ether	$C_{12}H_{10}O$	66.1	97.8	114.0	130.8	150.0	162.0	178.8	203.3	230.7	258.5	27
1,1-Diphenylethylene	$C_{14}H_{12}$	87.4	119.6	135.0	151.8	170.8	183.4	198.6	222.8	249.8	277.0	
trans-Diphenylethylene	$C_{14}H_{12}$	113.2	145.8	161.0	179.8	199.0	211.5	227.4	251.7	278.3	306.5	124
1,1-Diphenylhydrazine	$C_{12}H_{12}N_2$	126.0	159.3	176.1	194.0	213.5	225.9	242.5	267.2	294.0	322.2	44
Diphenylmethane	$C_{13}H_{12}$	76.0	107.4	122.8	139.8	157.8	170.2	186.3	210.7	237.5	264.5	26.5
Diphenyl sulfide	$C_{12}H_{10}S$	96.1	129.0	145.0	162.0	182.8	194.8	211.8	236.8	263.9	292.5	
Diphenyl-2-tolyl thiophosphate	$C_{18}H_{17}O_3PS$	159.7	179.8	201.6	215.5	230.6	240.4	252.5	270.3	290.0	310.0	
1,2-Dipropoxyethane	$C_8H_{18}O_2$	−38.8	−10.3	+5.0	22.3	42.3	55.8	74.2	103.8	140.0	180.0	
1,2-Diisopropylbenzene	$C_{12}H_{18}$	40.0	67.8	81.8	96.8	114.0	124.3	138.7	159.8	184.3	209.0	
1,3-Diisopropylbenzene	$C_{12}H_{18}$	34.7	62.3	76.0	91.2	107.9	118.2	132.3	153.7	177.6	202.0	−105
Dipropylene glycol	$C_6H_{14}O_3$	73.8	102.1	116.2	131.3	147.4	156.5	169.9	189.9	210.5	231.8	
Dipropyleneglycol monobutyl ether	$C_{10}H_{22}O_3$	64.7	92.0	106.0	120.4	136.3	146.3	159.8	180.0	203.8	227.0	
isopropyl ether	$C_9H_{20}O_3$	46.0	72.8	86.2	100.8	117.0	126.8	140.3	160.0	183.1	205.6	
Di-n-propyl ether	$C_6H_{14}O$	−43.3	−22.3	−11.8	0.0	+13.2	21.6	33.0	50.3	69.5	89.5	−122
Diisopropyl ether	$C_6H_{14}O$	−57.0	−37.4	−27.4	−16.7	−4.5	+3.4	13.7	30.0	48.2	67.5	−60
Di-n-propyl ketone (4-heptanone)	$C_7H_{14}O$	23.0	44.4	55.0	66.2	78.1	85.8	96.0	111.2	127.3	143.7	−32.6
Di-n-propyl oxalate	$C_8H_{14}O_4$	53.4	80.2	93.9	108.6	124.6	134.8	148.1	168.0	190.3	213.5	
Diisopropyl oxalate	$C_8H_{14}O_4$	43.2	69.0	81.9	95.6	110.5	120.0	132.6	151.2	171.8	193.5	
Di-n-propyl succinate	$C_{10}H_{18}O_4$	77.5	107.6	122.2	138.0	154.8	166.0	180.3	202.5	226.5	250.8	
Di-n-propyl d-tartrate	$C_{10}H_{18}O_6$	115.6	147.7	163.5	180.4	199.7	211.7	227.0	250.1	275.6	303.0	
Diisopropyl d-tartrate	$C_{10}H_{18}O_6$	103.7	133.7	148.2	164.0	181.8	192.6	207.3	228.2	251.8	275.0	
Divinyl acetylene (1,5-hexadiene-3-yne)	C_6H_6	−45.1	−24.4	−14.0	−2.8	+10.0	18.1	29.5	46.0	64.4	84.0	
1,3-Divinylbenzene	$C_{10}H_{10}$	32.7	60.0	73.8	88.7	105.5	116.0	130.0	151.4	175.2	199.5	−66.9
Docosane	$C_{22}H_{46}$	157.8	195.4	213.0	233.5	254.5	268.3	286.0	314.2	343.5	376.0	44.5
n-Dodecane	$C_{12}H_{26}$	47.8	75.8	90.0	104.6	121.7	132.1	146.2	167.2	191.0	216.2	−9.6
1-Dodecene	$C_{12}H_{24}$	47.2	74.0	87.8	102.4	118.6	128.5	142.3	162.2	185.5	208.0	−31.5
n-Dodecyl alcohol	$C_{12}H_{26}O$	91.0	120.2	134.7	150.0	167.2	177.8	192.0	213.0	235.7	259.0	24
Dodecylamine	$C_{12}H_{27}N$	82.8	111.8	127.8	141.6	157.4	168.0	182.1	203.0	225.0	248.0	
Dodecyltrimethylsilane	$C_{15}H_{34}Si$	91.2	122.1	137.7	153.8	172.1	184.2	199.5	222.0	248.0	273.0	
Elaidic acid	$C_{18}H_{34}O_2$	171.3	206.7	223.5	242.3	260.8	273.0	288.0	312.4	337.0	362.0	51.5
Epichlorohydrin	C_3H_5ClO	−16.5	+5.6	16.6	29.0	42.0	50.6	62.0	79.3	98.0	117.9	−25.6
1,2-Epoxy-2-methylpropane	C_4H_8O	−69.0	−50.0	−40.3	−29.5	−17.3	−9.7	+1.2	17.5	36.0	55.5	
Erucic acid	$C_{22}H_{42}O_2$	206.7	239.7	254.5	270.6	289.1	300.2	314.4	336.5	358.8	381.5	33.5
Estragole (p-methoxy allyl benzene)	$C_{10}H_{12}O$	52.6	80.0	93.7	108.4	124.6	135.2	148.5	168.7	192.0	215.0	
Ethane	C_2H_6	−159.5	−148.5	−142.9	−136.7	−129.8	−125.4	−119.3	−110.2	−99.7	−88.6	−183.2
Ethoxydimethylphenylsilane	$C_{10}H_{16}OSi$	36.3	63.1	76.2	91.0	107.2	127.5	131.4	151.5	175.0	199.5	
Ethoxytrimethylsilane	$C_5H_{14}OSi$	−50.9	−31.0	−20.7	−9.8	+3.7	11.5	22.1	38.1	56.3	75.7	
Ethoxytriphenylsilane	$C_{20}H_{20}OSi$	167.0	198.2	213.5	230.0	247.0	258.3	273.5	295.0	319.5	344.0	
Ethyl acetate	$C_4H_8O_2$	−43.4	−23.5	−13.5	−3.0	+9.1	16.6	27.0	42.0	59.3	77.1	−82.4
acetoacetate	$C_6H_{10}O_3$	28.5	54.0	67.3	81.1	96.2	106.0	118.5	138.0	158.2	180.8	−45
Ethylacetylene (1-butyne)	C_4H_6	−92.5	−76.7	−68.7	−59.9	−50.0	−43.4	−34.9	−21.6	−6.9	+8.7	−130
Ethyl acrylate	$C_5H_8O_2$	−29.5	−8.7	+2.0	13.0	26.0	33.5	44.5	61.5	80.0	99.5	−71.2
α-Ethylacrylic acid	$C_5H_8O_2$	47.0	70.7	82.0	94.4	108.1	116.7	127.5	144.0	160.7	179.2	
α-Ethylacrylonitrile	C_5H_7N	−29.0	−6.4	+5.0	17.7	31.8	40.6	53.0	71.6	92.2	114.0	
Ethyl alcohol (ethanol)	C_2H_6O	−31.3	−12.0	−2.3	+8.0	19.0	26.0	34.9	48.4	63.5	78.4	−112
Ethylamine	C_2H_7N	−82.3	−66.4	−58.3	−48.6	−39.8	−33.4	−25.1	−12.3	+2.0	16.6	−80.6
4-Ethylaniline	$C_8H_{11}N$	52.0	80.0	93.8	109.0	125.7	136.0	149.8	170.6	194.2	217.4	−4
N-Ethylaniline	$C_8H_{11}N$	38.5	66.4	80.6	96.0	113.2	123.6	137.3	156.9	180.8	204.0	−63.5

TABLE 3-8 Vapor Pressures of Organic Compounds, up to 1 atm *(Continued)*

Compound: Name	Compound: Formula	Pressure, mm. Hg: 1	5	10	20	40	60	100	200	400	760	Melting point, °C.
		Temperature, °C.										
2-Ethylanisole	$C_9H_{12}O$	29.7	55.9	69.0	83.1	98.8	109.0	122.3	142.1	164.2	187.1	
3-Ethylanisole	$C_9H_{12}O$	33.7	60.3	73.9	88.5	104.8	115.5	129.2	149.7	172.8	196.5	
4-Ethylanisole	$C_9H_{12}O$	33.5	60.2	73.9	88.5	104.7	115.4	128.4	149.2	172.3	196.5	
Ethylbenzene	C_8H_{10}	−9.8	+13.9	25.9	38.6	52.8	61.8	74.1	92.7	113.8	136.2	−94.9
Ethyl benzoate	$C_9H_{10}O_2$	44.0	72.0	86.0	101.4	118.2	129.0	143.2	164.8	188.4	213.4	−34.6
benzoylacetate	$C_{11}H_{12}O_3$	107.6	136.4	150.3	166.8	181.8	191.9	205.0	223.8	244.7	265.0	
bromide	C_2H_5Br	−74.3	−56.4	−47.5	−37.8	−26.7	−19.5	−10.0	+4.5	21.0	38.4	−117.8
α-bromoisobutyrate	$C_6H_{11}BrO_2$	10.6	35.8	48.0	61.8	77.0	86.7	99.8	119.7	141.2	163.6	
n-butyrate	$C_6H_{12}O_2$	−18.4	+4.0	15.3	27.8	41.5	50.1	62.0	79.8	100.0	121.0	−93.3
isobutyrate	$C_6H_{12}O_2$	−24.3	−2.4	+8.4	20.6	33.8	42.3	53.5	71.0	90.0	110.0	−88.2
Ethylcamphoronic anhydride	$C_{11}H_{16}O_5$	118.2	149.8	165.0	181.8	199.8	211.5	226.6	248.5	272.8	298.0	
Ethyl isocaproate	$C_8H_{16}O_2$	11.0	35.8	48.0	61.7	76.3	85.8	98.4	117.8	139.2	160.4	
carbamate	$C_3H_7NO_2$	......	65.8	77.8	91.0	105.6	114.8	126.2	144.2	164.0	184.0	49
carbanilate	$C_9H_{11}NO_2$	107.8	131.8	143.7	155.5	168.8	177.3	187.9	203.8	220.0	237.0	52.5
Ethylcetylamine	$C_{18}H_{39}N$	133.2	168.2	186.0	205.5	226.5	239.8	256.8	283.3	313.0	342.0	
Ethyl chloride	C_2H_5Cl	−89.8	−73.9	−65.8	−56.8	−47.0	−40.6	−32.0	−18.6	−3.9	+12.3	−139
chloroacetate	$C_4H_7ClO_2$	+1.0	25.4	37.5	50.4	65.2	74.0	86.0	103.8	123.8	144.2	−26
chloroglyoxylate	$C_4H_5ClO_3$	−5.1	+18.0	29.9	42.0	56.0	65.2	76.6	94.5	114.7	135.0	
α-chloropropionate	$C_5H_9ClO_2$	+6.6	30.2	41.9	54.3	68.2	77.3	89.3	107.2	126.2	146.5	
trans-cinnamate	$C_{11}H_{12}O_2$	87.6	108.5	134.0	150.3	169.2	181.2	196.0	219.3	245.0	271.0	12
3-Ethylcumene	$C_{11}H_{16}$	28.3	55.5	68.8	83.6	99.9	110.2	124.3	145.4	168.2	193.0	
4-Ethylcumene	$C_{11}H_{16}$	31.5	58.4	72.0	86.7	103.3	113.8	127.2	148.3	171.8	195.8	
Ethyl cyanoacetate	$C_5H_7NO_2$	67.8	93.5	106.0	119.8	133.8	142.1	152.8	169.8	187.8	206.0	
Ethylcyclohexane	C_8H_{16}	−14.5	+9.2	20.6	33.4	47.6	56.7	69.0	87.8	109.1	131.8	−111.3
Ethylcyclopentane	C_7H_{14}	−32.2	−10.8	−0.1	+11.7	25.0	33.4	45.0	62.4	82.3	103.4	−138.6
Ethyl dichloroacetate	$C_4H_6Cl_2O_2$	9.6	34.0	46.3	59.5	74.0	83.6	96.1	115.2	135.9	156.5	
N,N-diethyloxamate	$C_8H_{15}NO_3$	76.0	106.3	121.7	137.7	154.4	166.0	180.3	202.8	226.5	252.0	
N-Ethyldiphenylamine	$C_{14}H_{15}N$	98.3	130.2	146.0	162.8	182.0	193.7	209.8	233.0	258.8	286.0	
Ethylene	C_2H_4	−168.3	−158.3	−153.2	−147.6	−141.3	−137.3	−131.8	−123.4	−113.9	−103.7	−169
Ethylene-bis-(chloroacetate)	$C_6H_8Cl_2O_4$	112.0	142.4	158.0	173.5	191.0	201.8	215.0	237.3	259.5	283.5	
Ethylene chlorohydrin (2-chloroethanol)	C_2H_5ClO	−4.0	+19.0	30.3	42.5	56.0	64.1	75.0	91.8	110.0	128.8	−69
diamine (1,2-ethanediamine)	$C_2H_8N_2$	−11.0	+10.5	21.5	33.0	45.8	53.8	62.5	81.0	99.0	117.2	8.5
dibromide (1,2-dibromethane)	$C_2H_4Br_2$	−27.0	+4.7	18.6	32.7	48.0	57.9	70.4	89.8	110.1	131.5	10
dichloride (1,2-dichloroethane)	$C_2H_4Cl_2$	−44.5	−24.0	−13.6	−2.4	+10.0	18.1	29.4	45.7	64.0	82.4	−35.3
glycol (1,2-ethanediol)	$C_2H_6O_2$	53.0	79.7	92.1	105.8	120.0	129.5	141.8	158.5	178.5	197.3	−15.6
glycol diethyl ether (1,2-diethoxyethane)	$C_6H_{14}O_2$	−33.5	−10.2	+1.6	14.7	29.7	39.0	51.8	71.8	94.1	119.5	
glycol dimethyl ether (1,2-dimethoxyethane)	$C_4H_{10}O_2$	−48.0	−26.2	−15.3	−3.0	+10.7	19.7	31.8	50.0	70.8	93.0	
glycol monomethyl ether (2-methoxyethanol)	$C_3H_8O_2$	−13.5	+10.2	22.0	34.3	47.8	56.4	68.0	85.3	104.3	124.4	
oxide	C_2H_4O	−89.7	−73.8	−65.7	−56.6	−46.9	−40.7	−32.1	−19.5	−4.9	+10.7	−111.3
Ethyl α-ethylacetoacetate	$C_8H_{14}O_3$	40.5	67.3	80.2	94.6	110.3	120.6	133.8	153.2	175.6	198.0	
fluoride	C_2H_5F	−117.0	−103.8	−97.7	−90.0	−81.8	−76.4	−69.3	−58.0	−45.5	−32.0	
formate	$C_3H_6O_2$	−60.5	−42.2	−33.0	−22.7	−11.5	−4.3	−5.4	20.0	37.1	54.3	−79
2-furoate	$C_7H_8O_3$	37.6	63.8	77.1	91.5	107.5	117.5	130.4	150.1	172.5	195.0	34
glycolate	$C_4H_8O_3$	14.3	38.8	50.5	63.9	78.1	87.6	99.8	117.8	138.0	158.2	
3-Ethylhexane	C_8H_{18}	−20.0	+2.1	12.8	25.0	38.5	47.1	58.9	76.7	97.0	118.5	
2-Ethylhexyl acrylate	$C_{11}H_{20}O_2$	50.0	77.7	91.8	106.3	123.7	134.0	147.9	168.2	192.2	216.0	
Ethylidene chloride (1,1-dichloroethane)	$C_2H_4Cl_2$	−60.7	−41.9	−32.3	−21.9	−10.2	−2.9	+7.2	22.4	39.8	57.4	−96.7
fluoride (1,1-difluoroethane)	$C_2H_4F_2$	−112.5	−98.4	−91.7	−84.1	−75.8	−70.4	−63.2	−52.0	−39.5	−26.5	−117
Ethyl iodide	C_2H_5I	−54.4	−34.3	−24.3	−13.1	−0.9	+7.2	18.0	34.1	52.3	72.4	−105
Ethyl l-leucinate	$C_8H_{17}NO_2$	27.8	57.3	72.1	88.0	106.0	117.8	131.8	149.8	167.3	184.0	
Ethyl levulinate	$C_7H_{12}O_3$	47.3	74.0	87.3	101.8	117.7	127.6	141.3	160.2	183.0	206.2	
Ethyl mercaptan (ethanethiol)	C_2H_6S	−76.7	−59.1	−50.2	−40.7	−29.8	−22.4	−13.0	+1.5	17.7	35.0	−121
Ethyl methylcarbamate	$C_4H_9NO_2$	26.5	51.0	63.2	76.1	91.0	100.0	112.0	130.0	149.8	170.0	
Ethyl methyl ether	C_3H_8O	−91.0	−75.6	−67.8	−59.1	−49.4	−43.3	−34.8	−22.0	−7.8	+7.5	
1-Ethylnaphthalene	$C_{12}H_{12}$	70.0	101.4	116.8	133.8	152.0	164.1	180.0	204.6	230.8	258.1	−27
Ethyl α-naphthyl ketone (1-propionaphthone)	$C_{13}H_{12}O$	124.0	155.5	171.0	188.1	206.9	218.2	233.5	255.5	280.2	306.0	
Ethyl 3-nitrobenzoate	$C_9H_9NO_4$	108.1	140.2	155.0	173.6	192.6	205.0	220.3	244.6	270.6	298.0	47
3-Ethylpentane	C_7H_{16}	−37.8	−17.0	−6.8	+4.7	17.5	25.7	36.9	53.8	73.0	93.5	−118.6
4-Ethylphenetole	$C_{10}H_{14}O$	48.5	75.7	89.5	103.8	119.8	129.8	143.5	163.2	185.7	208.0	
2-Ethylphenol	$C_8H_{10}O$	46.2	73.4	87.0	101.5	117.9	127.9	141.8	161.6	184.5	207.5	−45
3-Ethylphenol	$C_8H_{10}O$	60.0	86.8	100.2	114.5	130.0	139.8	152.0	171.8	193.3	214.0	−4
4-Ethylphenol	$C_8H_{10}O$	59.3	86.5	100.2	115.0	131.3	141.7	154.2	175.0	197.4	219.0	46.5
Ethyl phenyl ether (phenetole)	$C_8H_{10}O$	18.1	43.7	56.4	70.3	86.6	95.4	108.4	127.9	149.8	172.0	−30.2
Ethyl propionate	$C_5H_{10}O_2$	−28.0	−7.2	+3.4	14.3	27.2	35.1	45.2	61.7	79.8	99.1	−72.6
Ethyl propyl ether	$C_5H_{12}O$	−64.3	−45.0	−35.0	−24.0	−12.0	−4.0	+6.8	23.3	41.6	61.7	
Ethyl salicylate	$C_9H_{10}O_3$	61.2	90.0	104.2	119.3	136.7	147.6	161.5	183.7	207.0	231.5	1.3
3-Ethylstyrene	$C_{10}H_{12}$	28.3	55.0	68.3	82.8	99.2	109.6	123.2	144.0	167.2	191.5	
4-Ethylstyrene	$C_{10}H_{12}$	26.0	52.7	66.3	80.8	97.3	107.6	121.5	142.0	165.0	189.0	
Ethylisothiocyanate	C_3H_5NS	13.2	+10.6	22.8	36.1	50.8	59.8	71.9	90.0	110.1	131.0	−5.9
2-Ethyltoluene	C_9H_{12}	9.4	34.8	47.6	61.2	76.4	86.0	99.0	119.0	141.4	165.1	
3-Ethyltoluene	C_9H_{12}	7.2	32.3	44.7	58.2	73.3	82.9	95.9	115.5	137.8	161.3	−95.5
4-Ethyltoluene	C_9H_{12}	7.6	32.7	44.9	58.5	73.6	83.2	96.3	116.1	136.4	162.0	
Ethyl trichloroacetate	$C_4H_5Cl_3O_2$	20.7	45.5	57.7	70.6	85.5	94.4	107.4	125.8	146.0	167.0	
Ethyltrimethylsilane	$C_5H_{14}Si$	−60.6	−41.4	−31.8	−21.0	−9.0	−1.2	+9.2	25.0	42.8	62.0	
Ethyltrimethyltin	$C_5H_{14}Sn$	−30.0	−7.6	+3.8	16.1	30.0	38.4	50.0	67.3	87.6	108.8	
Ethyl isovalerate	$C_7H_{14}O_2$	−6.1	+17.0	28.7	41.3	55.2	64.0	75.9	93.8	114.0	134.3	−99.3
2-Ethyl-1,4-xylene	$C_{10}H_{14}$	25.7	52.0	65.6	79.8	96.0	106.2	120.0	140.2	163.1	186.9	
4-Ethyl-1,3-xylene	$C_{10}H_{14}$	26.3	53.0	66.4	80.6	97.2	107.4	121.2	141.8	164.4	188.4	
5-Ethyl-1,3-xylene	$C_{10}H_{14}$	22.1	48.8	62.1	76.5	92.6	103.0	116.5	137.4	159.6	183.7	
Eugenol	$C_{10}H_{12}O_2$	78.4	108.1	123.0	138.7	155.8	167.3	182.2	204.7	228.3	253.5	
iso-Eugenol	$C_{10}H_{12}O_2$	86.3	117.0	132.4	149.0	167.0	178.2	194.0	217.2	242.3	267.5	−10
Eugenyl acetate	$C_{12}H_{14}O_3$	101.6	132.3	148.0	164.2	183.0	194.0	209.7	232.5	257.4	282.0	295
Fencholic acid	$C_{10}H_{16}O_2$	101.7	128.7	142.3	155.8	171.8	181.5	194.0	215.0	237.8	264.1	19
d-Fenchone	$C_{10}H_{16}O$	28.0	54.7	68.3	83.0	99.5	109.8	123.6	144.0	166.8	191.0	5
dl-Fenchyl alcohol	$C_{10}H_{18}O$	45.8	70.3	82.1	95.6	110.8	120.2	132.3	150.0	173.2	201.0	35
Fluorene	$C_{13}H_{10}$	......	129.3	146.0	164.2	185.2	197.8	214.7	240.3	268.6	295.0	113
Fluorobenzene	C_6H_5F	−43.4	−22.8	−12.4	−1.2	+11.5	19.6	30.4	47.2	65.7	84.7	−42.1

TABLE 3-8 Vapor Pressures of Organic Compounds, up to 1 atm *(Continued)*

Compound		Pressure, mm. Hg										Melting point, °C.
Name	Formula	1	5	10	20	40	60	100	200	400	760	
		Temperature, °C.										
2-Fluorotoluene	C_7H_7F	−24.2	−2.2	+8.9	21.4	34.7	43.7	55.3	73.0	92.8	114.0	−80
3-Fluorotoluene	C_7H_7F	−22.4	−0.3	+11.0	23.4	37.0	45.8	57.5	75.4	95.4	116.0	−110.8
4-Fluorotoluene	C_7H_7F	−21.8	+0.3	11.8	24.0	37.8	46.5	58.1	76.0	96.1	117.0	
Formaldehyde	CH_2O	......	......	−88.0	−79.6	−70.6	−65.0	−57.3	−46.0	−33.0	−19.5	−92
Formamide	CH_3NO	70.5	96.3	109.5	122.5	137.5	147.0	157.5	175.5	193.5	210.5	
Formic acid	CH_2O_2	−20.0	−5.0	+2.1	10.3	24.0	32.4	43.8	61.4	80.3	100.6	8.2
trans-Fumaryl chloride	$C_4H_2Cl_2O_2$	+15.0	38.5	51.8	65.0	79.5	89.0	101.0	120.0	140.0	160.0	
Furfural (2-furaldehyde)	$C_5H_4O_2$	18.5	42.6	54.8	67.8	82.1	91.5	103.4	121.8	141.8	161.8	
Furfuryl alcohol	$C_5H_6O_2$	31.8	56.0	68.0	81.0	95.7	104.0	115.9	133.1	151.8	170.0	
Geraniol	$C_{10}H_{18}O$	69.2	96.8	110.0	125.6	141.8	151.5	165.3	185.6	207.8	230.0	
Geranyl acetate	$C_{12}H_{20}O_2$	73.5	102.7	117.9	133.0	150.0	160.3	175.2	196.3	219.8	243.3	
Geranyl n-butyrate	$C_{14}H_{24}O_2$	96.8	125.2	139.0	153.8	170.1	180.2	193.8	214.0	235.0	257.4	
Geranyl isobutyrate	$C_{14}H_{24}O_2$	90.9	119.6	133.0	147.9	164.0	174.0	187.7	207.6	228.5	251.0	
Geranyl formate	$C_{11}H_{18}O_2$	61.8	90.3	104.3	119.8	136.2	147.2	160.7	182.6	205.8	230.0	
Glutaric acid	$C_5H_8O_4$	155.5	183.8	196.0	210.5	226.3	235.5	247.0	265.0	283.5	303.0	97.5
Glutaric anhydride	$C_5H_6O_3$	100.8	133.3	149.5	166.0	185.5	196.2	212.5	236.5	261.0	287.0	
Glutaronitrile	$C_5H_6N_2$	91.3	123.7	140.0	156.5	176.4	189.5	205.5	230.0	257.3	286.2	
Glutaryl chloride	$C_5H_6Cl_2O_2$	56.1	84.0	97.8	112.3	128.3	139.1	151.8	172.4	195.3	217.0	
Glycerol	$C_3H_8O_3$	125.5	153.8	167.2	182.2	198.0	208.0	220.1	240.0	263.0	290.0	17.9
Glycerol dichlorohydrin (1,3-dichloro-2-propanol)	$C_3H_6Cl_2O$	28.0	52.2	64.7	78.0	93.0	102.0	114.8	133.3	153.5	174.3	
Glycol diacetate	$C_6H_{10}O_4$	38.3	64.1	77.1	90.8	106.1	115.8	128.0	147.8	168.3	190.5	−31
Glycolide (1,4-dioxane-2,6-dione)	$C_4H_4O_4$	......	103.0	116.6	132.0	148.6	158.2	173.2	194.0	217.0	240.0	97
Guaicol (2-methoxyphenol)	$C_7H_8O_2$	52.4	79.1	92.0	106.0	121.6	131.0	144.0	162.7	184.1	205.0	28.3
Heneicosane	$C_{21}H_{44}$	152.6	188.0	205.4	223.2	243.4	255.3	272.0	296.5	323.8	350.5	40.4
Heptacosane	$C_{27}H_{56}$	211.7	248.6	266.8	284.6	305.7	318.3	333.5	359.4	385.0	410.6	59.5
Heptadecane	$C_{17}H_{36}$	115.0	145.2	160.0	177.7	195.8	207.3	223.0	247.8	274.5	303.0	22.5
Heptaldehyde (enanthaldehyde)	$C_7H_{14}O$	12.0	32.7	43.0	54.0	66.3	74.0	84.0	102.0	125.5	155.0	−42
n-Heptane	C_7H_{16}	−34.0	−12.7	−2.1	+9.5	22.3	30.6	41.8	58.7	78.0	98.4	−90.6
Heptanoic acid (enanthic acid)	$C_7H_{14}O_2$	78.0	101.3	113.2	125.6	139.5	148.5	160.0	179.5	199.6	221.5	−10
1-Heptanol	$C_7H_{16}O$	42.4	64.3	74.7	85.8	99.8	108.0	119.5	136.6	155.6	175.8	34.6
Heptanoyl chloride (enanthyl chloride)	$C_7H_{13}ClO$	34.2	54.6	64.6	75.0	86.4	93.5	102.7	116.3	130.7	145.0	
2-Heptene	C_7H_{14}	−35.8	−14.1	−3.5	+8.3	21.5	30.0	41.3	58.6	78.1	98.5	
Heptylbenzene	$C_{13}H_{20}$	64.0	94.6	110.0	126.0	144.0	154.8	170.2	193.3	217.8	244.0	
Heptyl cyanide (enanthonitrile)	$C_7H_{13}N$	21.0	47.8	61.6	76.3	92.6	103.0	116.8	137.7	160.0	184.6	
Hexachlorobenzene	C_6Cl_6	114.4	149.3	166.4	185.7	206.0	219.0	235.5	258.5	283.5	309.4	230
Hexachloroethane	C_2Cl_6	32.7	49.8	73.5	87.6	102.3	112.0	124.2	143.1	163.8	185.6	186.6
Hexacosane	$C_{26}H_{54}$	204.0	240.0	257.4	275.8	295.2	307.8	323.2	348.4	374.6	399.8	56.6
Hexadecane	$C_{16}H_{34}$	105.3	135.2	149.8	164.7	181.3	193.2	208.5	231.7	258.3	287.5	18.5
1-Hexadecene	$C_{16}H_{32}$	101.6	131.7	146.2	162.0	178.8	190.8	205.3	226.8	250.0	274.0	4
n-Hexadecyl alcohol (cetyl alcohol)	$C_{16}H_{34}O$	122.7	158.3	177.8	197.8	219.8	234.3	251.7	280.2	312.7	344.0	49.3
n-Hexadecylamine (cetylamine)	$C_{16}H_{35}N$	123.6	157.8	176.0	195.7	215.7	228.8	245.8	272.2	300.4	330.0	
Hexaethylbenzene	$C_{18}H_{30}$	......	134.3	150.3	168.0	187.7	199.7	216.0	241.7	268.5	298.3	130
n-Hexane	C_6H_{14}	−53.9	−34.5	−25.0	−14.1	−2.3	+5.4	15.8	31.6	49.6	68.7	−95.3
1-Hexanol	$C_6H_{14}O$	24.4	47.2	58.2	70.3	83.7	92.0	102.8	119.6	138.0	157.0	−51.6
2-Hexanol	$C_6H_{14}O$	14.6	34.8	45.0	55.9	67.9	76.0	87.3	103.7	121.8	139.9	
3-Hexanol	$C_6H_{14}O$	+2.5	25.7	36.7	49.0	62.2	70.7	81.8	98.3	117.0	135.5	
1-Hexene	C_6H_{12}	−57.5	−38.0	−28.1	−17.2	−5.0	+2.8	13.0	29.0	46.8	66.0	−98.5
n-Hexyl levulinate	$C_{11}H_{20}O_3$	90.0	120.0	134.7	150.2	167.8	179.0	193.6	215.7	241.0	266.8	
n-Hexyl phenyl ketone (enanthophenone)	$C_{13}H_{18}O$	100.0	130.3	145.5	161.0	178.9	189.8	204.2	225.0	248.3	271.3	
Hydrocinnamic acid	$C_9H_{10}O_2$	102.2	133.5	148.7	165.0	183.3	194.0	209.0	230.8	255.0	279.8	48.5
Hydrogen cyanide (hydrocyanic acid)	CHN	−71.0	−55.3	−47.7	−39.7	−30.9	−25.1	−17.8	−5.3	+10.2	25.9	−13.2
Hydroquinone	$C_6H_6O_2$	132.4	153.3	163.5	174.6	192.0	203.0	216.5	238.0	262.5	286.2	170.3
4-Hydroxybenzaldehyde	$C_7H_6O_2$	121.2	153.2	169.7	186.8	206.0	217.5	233.5	256.8	282.6	310.0	115.5
α-Hydroxyisobutyric acid	$C_4H_8O_3$	73.5	98.5	110.5	123.8	138.0	146.4	157.7	175.2	193.8	212.0	79
α-Hydroxybutyronitrile	C_5H_9NO	41.0	65.8	77.8	90.7	104.8	113.9	125.0	142.0	159.8	178.8	
4-Hydroxy-3-methyl-2-butanone	$C_5H_{10}O_2$	44.6	69.3	81.0	94.0	108.2	117.4	129.0	146.5	165.5	185.0	
4-Hydroxy-4-methyl-2-pentanone	$C_6H_{12}O_2$	22.0	46.7	58.8	72.0	86.7	96.0	108.2	126.8	147.5	167.9	−47
3-Hydroxypropionitrile	C_3H_5NO	58.7	87.8	102.0	117.9	134.1	144.7	157.7	178.0	200.0	221.0	
Indene	C_9H_8	16.4	44.3	58.5	73.9	90.7	100.8	114.7	135.6	157.8	181.6	−2
Iodobenzene	C_6H_5I	24.1	50.6	64.0	78.3	94.4	105.0	118.3	139.8	163.9	188.6	−28.5
Iodononane	$C_9H_{19}I$	70.0	96.2	109.0	123.0	138.1	147.7	159.8	179.0	199.3	219.5	
2-Iodotoluene	C_7H_7I	37.2	65.9	79.8	95.6	112.4	123.8	138.1	160.0	185.7	211.0	
α-Ionone	$C_{13}H_{20}O$	79.5	108.8	123.0	139.0	155.6	166.3	181.2	202.5	225.2	250.0	
Isoprene	C_5H_8	−79.8	−62.3	−53.3	−43.5	−32.6	−25.4	−16.0	−1.2	+15.4	32.6	−146.7
Lauraldehyde	$C_{12}H_{24}O$	77.7	108.4	123.7	140.2	157.8	168.7	184.5	207.8	231.8	257.0	44.5
Lauric acid	$C_{12}H_{24}O_2$	121.0	150.6	166.0	183.6	201.4	212.7	227.5	249.8	273.8	299.2	48
Levulinaldehyde	$C_5H_8O_2$	28.1	54.9	68.0	82.7	98.3	108.4	121.8	142.0	164.0	187.0	
Levulinic acid	$C_5H_8O_3$	102.0	128.1	141.8	154.1	169.5	178.0	190.2	208.3	227.4	245.8	33.5
d-Limonene	$C_{10}H_{16}$	14.0	40.4	53.8	68.2	84.3	94.6	108.3	128.5	151.4	175.0	−96.9
Linalyl acetate	$C_{12}H_{20}O_2$	55.4	82.5	96.0	111.4	127.7	138.1	151.8	173.3	196.2	220.0	
Maleic anhydride	$C_4H_2O_3$	44.0	63.4	78.7	95.0	111.8	122.0	135.8	155.9	179.5	202.0	58
Menthane	$C_{10}H_{20}$	+9.7	35.7	48.3	62.7	78.3	88.6	102.1	122.7	146.0	169.5	
l-Menthol	$C_{10}H_{20}O$	56.0	83.2	96.0	110.3	126.1	136.1	149.4	168.3	190.2	212.0	42.5
Menthyl acetate	$C_{12}H_{22}O_2$	57.4	85.8	100.0	115.4	132.1	143.2	156.7	178.8	202.8	227.0	
benzoate	$C_{17}H_{24}O_2$	123.2	154.2	170.0	186.3	204.3	215.8	230.4	253.2	277.1	301.0	54.5
formate	$C_{11}H_{20}O_2$	47.3	75.8	90.0	105.8	123.0	133.8	148.0	169.8	194.2	219.0	
Mesityl oxide	$C_6H_{10}O$	−8.7	+14.1	26.0	37.9	51.7	60.4	72.1	90.0	109.8	130.0	−59
Methacrylic acid	$C_4H_6O_2$	25.5	48.5	60.0	72.7	86.4	95.3	106.6	123.9	142.5	161.0	15
Methacrylonitrile	C_4H_5N	−44.5	−23.3	−12.5	−0.6	+12.8	21.5	32.8	50.0	70.3	90.3	
Methane	CH_4	−205.9	−199.0	−195.5	−191.8	−187.7	−185.1	−181.4	−175.5	−168.8	−161.5	−182.5
Methanethiol	CH_4S	−90.7	−75.3	−67.5	−58.8	−49.2	−43.1	−34.8	−22.1	−7.9	+6.8	−121
Methoxyacetic acid	$C_3H_6O_3$	52.5	79.3	92.0	106.5	122.0	131.8	144.5	163.5	184.2	204.0	
N-Methylacetanilide	$C_9H_{11}NO$	......	103.8	118.6	135.1	152.2	164.2	179.8	202.3	227.4	253.0	102
Methyl acetate	$C_3H_6O_2$	−57.2	−38.6	−29.3	−19.1	−7.9	−0.5	+9.4	24.0	40.0	57.8	−98.7
acetylene (propyne)	C_3H_4	−111.0	−97.5	−90.5	−82.9	−74.3	−68.8	−61.3	−49.8	−37.2	−23.3	−102.7
acrylate	$C_4H_6O_2$	−43.7	−23.6	−13.5	−2.7	+9.2	17.3	28.0	43.9	61.8	80.2	
alcohol (methanol)	CH_4O	−44.0	−25.3	−16.2	−6.0	+5.0	12.1	21.2	34.8	49.9	64.7	−97.8
Methylamine	CH_5N	−95.8	−81.3	−73.8	−65.9	−56.9	−51.3	−43.7	−32.4	−19.7	−6.3	−93.5

TABLE 3-8 Vapor Pressures of Organic Compounds, up to 1 atm *(Continued)*

Compound: Name	Formula	Pressure, mm. Hg: 1	5	10	20	40	60	100	200	400	760	Melting point, °C.
		Temperature, °C.										
N-Methylaniline	C_7H_9N	36.0	62.8	76.2	90.5	106.0	115.8	129.8	149.3	172.0	195.5	−57
Methyl anthranilate	$C_8H_9NO_2$	77.6	109.0	124.2	141.5	159.7	172.0	187.8	212.4	238.5	266.5	24
benzoate	$C_8H_8O_2$	39.0	64.4	77.3	91.8	107.8	117.4	130.8	151.4	174.7	199.5	−12.5
2-Methylbenzothiazole	C_8H_7NS	70.0	97.5	111.2	125.5	141.2	150.4	163.9	183.2	204.5	225.5	15.4
α-Methylbenzyl alcohol	$C_8H_{10}O$	49.0	75.2	88.0	102.1	117.8	127.4	140.3	159.0	180.7	204.0	
Methyl bromide	CH_3Br	−96.3	−80.6	−72.8	−64.0	−54.2	−48.0	−39.4	−26.5	−11.9	+3.6	−93
2-Methyl-1-butene	C_5H_{10}	−89.1	−72.8	−64.3	−54.8	−44.1	−37.3	−28.0	−13.8	+2.5	20.2	−135
2-Methyl-2-butene	C_5H_{10}	−75.4	−57.0	−47.9	−37.9	−26.7	−19.4	−9.9	+4.9	21.6	38.5	−133
Methyl isobutyl carbinol (2-methyl-4-pentanol)	$C_6H_{14}O$	−0.3	+22.1	33.3	45.4	58.2	67.0	78.0	94.9	113.5	131.7	
n-butyl ketone (2-hexanone)	$C_6H_{12}O$	+7.7	28.8	38.8	50.0	62.0	69.8	79.8	94.3	111.0	127.5	−56.9
isobutyl ketone (4-methyl-2-pentanone)	$C_6H_{12}O$	−1.4	+19.7	30.0	40.8	52.8	60.4	70.4	85.6	102.0	119.0	−84.7
n-butyrate	$C_5H_{10}O_2$	−26.8	−5.5	+5.0	16.7	29.6	37.4	48.0	64.3	83.1	102.3	
isobutyrate	$C_5H_{10}O_2$	−34.1	−13.0	−2.9	+8.4	21.0	28.9	39.6	55.7	73.6	92.6	−84.7
caprate	$C_{11}H_{22}O_2$	63.7	93.5	108.0	123.0	139.0	148.6	161.5	181.6	202.9	224.0	−18
caproate	$C_7H_{14}O_2$	+5.0	30.0	42.0	55.4	70.0	79.7	91.4	109.8	129.8	150	
caprylate	$C_9H_{18}O_2$	34.2	61.7	74.9	89.0	105.3	115.3	128.0	148.1	170.0	193.0	−40
chloride	CH_3Cl	......	−99.5	−92.4	−84.8	−76.0	−70.4	−63.0	−51.2	−38.0	−24.0	−97.7
chloroacetate	$C_3H_5ClO_2$	−2.9	19.0	30.0	41.5	54.5	63.0	73.5	90.5	109.5	130.3	−31.9
cinnamate	$C_{10}H_{10}O_2$	77.4	108.1	123.0	140.0	157.9	170.0	185.8	209.6	235.0	263.0	33.4
α-Methylcinnamic acid	$C_{10}H_{10}O_2$	125.7	155.0	169.8	185.2	201.8	212.0	224.8	245.0	266.8	288.0	
Methylcyclohexane	C_7H_{14}	−35.9	−14.0	−3.2	+8.7	22.0	30.5	42.1	59.6	79.6	100.9	−126.4
Methylcyclopentane	C_6H_{12}	−53.7	−33.8	−23.7	−12.8	−0.6	+7.2	17.9	34.0	52.3	71.8	−142.4
Methylcyclopropane	C_4H_8	−96.0	−80.6	−72.8	−64.0	−54.2	−48.0	−39.3	−26.0	−11.3	+4.5	
Methyl n-decyl ketone (n-dodecan-2-one)	$C_{12}H_{24}O$	77.1	106.0	120.4	136.0	152.4	163.8	177.5	199.0	222.5	246.5	
dichloroacetate	$C_3H_4Cl_2O_2$	3.2	26.7	38.1	50.7	64.7	73.6	85.4	103.2	122.6	143.0	
N-Methyldiphenylamine	$C_{13}H_{13}N$	103.5	134.0	149.7	165.8	184.0	195.4	210.1	232.8	257.0	282.0	−7.6
Methyl n-dodecyl ketone (2-tetradecanone)	$C_{14}H_{28}O$	99.3	130.0	145.5	161.3	179.8	191.4	206.0	228.2	253.3	278.0	
Methylene bromide (dibromomethane)	CH_2Br_2	−35.1	−13.2	−2.4	+9.7	23.3	31.6	42.3	58.5	79.0	98.6	−52.8
chloride (dichloromethane)	CH_2Cl_2	−70.0	−52.1	−43.3	−33.4	−22.3	−15.7	−6.3	+8.0	24.1	40.7	−96.7
Methyl ethyl ketone (2-butanone)	C_4H_8O	−48.3	−28.0	−17.7	−6.5	+6.0	14.0	25.0	41.6	60.0	79.6	−85.9
2-Methyl-3-ethylpentane	C_8H_{18}	−24.0	−1.8	+9.5	21.7	35.2	43.9	55.7	73.6	94.0	115.6	−114.5
3-Methyl-3-ethylpentane	C_8H_{18}	−23.9	−1.4	+9.9	22.3	36.2	45.0	57.1	75.3	96.2	118.3	−90
Methyl fluoride	CH_3F	−147.3	−137.0	−131.6	−125.9	−119.1	−115.0	−109.0	−99.9	−89.5	−78.2	
formate	$C_2H_4O_2$	−74.2	−57.0	−48.6	−39.2	−28.7	−21.9	−12.9	+0.8	16.0	32.0	−99.8
α-Methylglutaric anhydride	$C_6H_8O_3$	93.8	125.4	141.8	157.7	177.5	189.9	205.0	229.1	255.5	282.5	
Methyl glycolate	$C_3H_6O_3$	+9.6	33.7	45.3	58.1	72.3	81.8	93.7	111.8	131.7	151.5	
2-Methylheptadecane	$C_{18}H_{38}$	119.8	152.0	168.7	186.0	204.8	216.3	231.5	254.5	279.8	306.5	
2-Methylheptane	C_8H_{18}	−21.0	+1.3	12.3	24.4	37.9	46.6	58.3	76.0	96.2	117.6	−109.5
3-Methylheptane	C_8H_{18}	−19.8	+2.6	13.3	25.4	38.9	47.6	59.4	77.1	97.4	118.9	−120.8
4-Methylheptane	C_8H_{18}	−20.4	+1.5	12.4	24.5	38.0	46.6	58.3	76.1	96.3	117.7	−121.1
2-Methyl-2-heptene	C_8H_{16}	−16.1	+6.7	17.8	30.4	44.0	52.8	64.6	82.3	102.2	122.5	
6-Methyl-3-hepten-2-ol	$C_8H_{16}O$	41.6	65.0	76.7	89.3	102.7	111.5	122.6	139.5	156.6	175.5	
6-Methyl-5-hepten-2-ol	$C_8H_{16}O$	41.9	66.0	77.8	90.4	104.0	112.8	123.8	140.0	156.6	174.3	
2-Methylhexane	C_7H_{16}	−40.4	−19.5	−9.1	+2.3	14.9	23.0	34.1	50.8	69.8	90.0	−118.2
3-Methylhexane	C_7H_{16}	−39.0	−18.1	−7.8	+3.6	16.4	24.5	35.6	52.4	71.6	91.9	
Methyl iodide	CH_3I	......	−55.0	−45.8	−35.6	−24.2	−16.9	−7.0	+8.0	25.3	42.4	−64.4
laurate	$C_{13}H_{26}O_2$	87.8	117.9	133.2	149.0	166.0	176.8	190.8	......	......	......	5
levulinate	$C_6H_{10}O_3$	39.8	66.4	79.7	93.7	109.5	119.3	133.0	153.4	175.8	197.7	
methacrylate	$C_5H_8O_2$	−30.5	−10.0	+1.0	11.0	25.5	34.5	47.0	63.0	82.0	101.0	
myristate	$C_{15}H_{30}O_2$	115.0	145.7	160.8	177.8	195.8	207.5	222.6	245.3	269.8	295.8	18.5
α-naphthyl ketone (1-acetonaphthone)	$C_{12}H_{10}O$	115.6	146.3	161.5	178.4	196.8	208.6	223.8	246.7	270.5	295.5	
β-naphthyl ketone (2-acetonaphthone)	$C_{12}H_{10}O$	120.2	152.3	168.5	185.7	203.8	214.7	229.8	251.6	275.8	301.0	55.5
n-nonyl ketone (undecan-2-one)	$C_{11}H_{22}O$	68.2	95.5	108.9	123.1	139.0	148.6	161.0	181.2	202.3	224.0	15
palmitate	$C_{17}H_{34}O_2$	134.3	166.8	184.3	202.0	......	......	......	......	......	......	30
n-pentadecyl ketone (2-heptadecanone)	$C_{17}H_{34}O$	129.6	161.6	178.0	196.4	214.3	226.7	242.0	265.8	291.7	319.5	
2-Methylpentane	C_6H_{14}	−60.9	−41.7	−32.1	−21.4	−9.7	−1.9	+8.1	24.1	41.6	60.3	−154
3-Methylpentane	C_6H_{14}	−59.0	−39.8	−30.1	−19.4	−7.3	+0.1	10.5	26.5	44.2	63.3	−118
2-Methyl-1-pentanol	$C_6H_{14}O$	15.4	38.0	49.6	61.6	74.7	83.4	94.2	111.3	129.8	147.9	
2-Methyl-2-pentanol	$C_6H_{14}O$	−4.5	+16.8	27.6	38.8	51.3	58.8	69.2	85.0	102.6	121.2	−103
Methyl n-pentyl ketone (2-heptanone)	$C_7H_{14}O$	19.3	43.6	55.5	67.7	81.2	89.8	100.0	116.1	133.2	150.2	
phenyl ether (anisole)	C_7H_8O	+5.4	30.0	42.2	55.8	70.7	80.1	93.0	112.3	133.8	155.5	−37.3
2-Methylpropene	C_4H_8	−105.1	−96.5	−81.9	−73.4	−63.8	−57.7	−49.3	−36.7	−22.2	−6.9	−140.3
Methyl propionate	$C_4H_8O_2$	−42.0	−21.5	−11.8	−1.0	+11.0	18.7	29.0	44.2	61.8	79.8	−87.5
4-Methylpropiophenone	$C_{10}H_{12}O$	59.6	89.3	103.8	120.2	138.0	149.3	164.2	187.4	212.7	238.5	
2-Methylpropionyl bromide	C_4H_7BrO	13.5	38.4	50.6	64.1	79.4	88.8	101.6	120.5	141.7	163.0	
Methyl propyl ether	$C_4H_{10}O$	−72.2	−54.3	−45.4	−35.4	−24.3	−17.4	−8.1	+6.0	22.5	39.1	
n-propyl ketone (2-pentanone)	$C_5H_{10}O$	−12.0	+8.0	17.9	28.5	39.8	47.3	56.8	71.0	86.8	103.3	−77.8
isopropyl ketone (3-Methyl-2-butanone)	$C_5H_{10}O$	−19.9	−1.0	+8.3	18.3	29.6	36.2	45.5	59.0	73.8	88.9	−92
2-Methylquinoline	$C_{10}H_9N$	75.3	104.0	119.0	134.0	150.8	161.7	176.2	197.8	211.7	246.5	−1
Methyl salicylate	$C_8H_8O_3$	54.0	81.6	95.3	110.0	126.2	136.7	150.0	172.6	197.5	223.2	−8.3
α-Methyl styrene	C_9H_{10}	7.4	34.0	47.1	61.8	77.8	88.3	102.2	121.8	143.0	165.4	−23.2
4-Methyl styrene	C_9H_{10}	16.0	42.0	55.1	69.2	85.0	95.0	108.6	128.7	151.2	175.0	
Methyl n-tetradecyl ketone (2-hexadecanone)	$C_{16}H_{32}O$	109.8	151.5	167.3	184.6	203.7	215.0	230.5	254.4	279.8	307.0	
thiocyanate	C_2H_3NS	−14.0	+9.8	21.6	34.5	49.0	58.1	70.4	89.8	110.8	132.9	−51
isothiocyanate	C_2H_3NS	−34.7	−8.3	+5.4	20.4	38.2	47.5	59.3	77.5	97.8	119.0	35.5
undecyl ketone (2-tridecanone)	$C_{13}H_{26}O$	86.8	117.0	131.8	147.8	165.7	176.6	191.5	214.0	238.3	262.5	28.5
isovalerate	$C_6H_{12}O_2$	−19.2	+2.9	14.0	26.4	39.8	48.2	59.8	77.3	96.7	116.7	
Monovinylacetylene (butenyne)	C_4H_4	−93.2	−77.7	−70.0	−61.3	−51.7	−45.3	−37.1	−24.1	−10.1	+5.3	
Myrcene	$C_{10}H_{16}$	14.5	40.0	53.2	67.0	82.6	92.6	106.0	126.0	148.3	171.5	
Myristaldehyde	$C_{14}H_{28}O$	99.0	132.0	148.3	166.2	186.0	198.3	214.5	240.4	267.9	297.8	23.5
Myristic acid (tetradecanoic acid)	$C_{14}H_{28}O_2$	142.0	174.1	190.8	207.6	223.5	237.2	250.5	272.3	294.6	318.0	57.5
Naphthalene	$C_{10}H_8$	52.6	74.2	85.8	101.7	119.3	130.2	145.5	167.7	193.2	217.9	80.2
1-Naphthoic acid	$C_{11}H_8O_2$	156.0	184.0	196.8	211.2	225.0	234.5	245.8	263.5	281.4	300.0	160.5
2-Naphthoic acid	$C_{11}H_8O_2$	160.8	189.7	202.8	216.9	231.5	241.3	252.7	270.3	289.5	308.5	184
1-Naphthol	$C_{10}H_8O$	94.0	125.5	142.0	158.0	177.8	190.0	206.0	229.6	255.8	282.5	96
2-Naphthol	$C_{10}H_8O$	......	128.6	145.5	161.8	181.7	193.7	209.8	234.0	260.6	288.0	122.5
1-Naphthylamine	$C_{10}H_9N$	104.3	137.7	153.8	171.6	191.5	203.8	220.0	244.9	272.2	300.8	50

TABLE 3-8 Vapor Pressures of Organic Compounds, up to 1 atm *(Continued)*

Compound: Name	Formula	Pressure, mm. Hg: 1	5	10	20	40	60	100	200	400	760	Melting point, °C.
		Temperature, °C.										
2-Naphthylamine	$C_{10}H_9N$	108.0	141.6	157.6	175.8	195.7	208.1	224.3	249.7	277.4	306.1	111.5
Nicotine	$C_{10}H_{14}N_2$	61.8	91.8	107.2	123.7	142.1	154.7	169.5	193.8	219.8	247.3	
2-Nitroaniline	$C_6H_6N_2O_2$	104.0	135.7	150.4	167.7	186.0	197.8	213.0	236.3	260.0	284.5	71.5
3-Nitroaniline	$C_6H_6N_2O_2$	119.3	151.5	167.8	185.5	204.2	216.5	232.1	255.3	280.2	305.7	114
4-Nitroaniline	$C_6H_6N_2O_2$	142.4	177.6	194.4	213.2	234.2	245.9	261.8	284.5	310.2	336.0	146.5
2-Nitrobenzaldehyde	$C_7H_5NO_3$	85.8	117.7	133.4	150.0	168.8	180.7	196.2	220.0	246.8	273.5	40.9
3-Nitrobenzaldehyde	$C_7H_5NO_3$	96.2	127.4	142.8	159.0	177.7	189.5	204.3	227.4	252.1	278.3	58
Nitrobenzene	$C_6H_5NO_2$	44.4	71.6	84.9	99.3	115.4	125.8	139.9	161.2	185.8	210.6	+5.7
Nitroethane	$C_2H_5NO_2$	−21.0	+1.5	12.5	24.8	38.0	46.5	57.8	74.8	94.0	114.0	−90
Nitroglycerin	$C_3H_5N_3O_9$	127	167	188	210	235	251	......	......	......	......	11
Nitromethane	CH_3NO_2	−29.0	−7.9	+2.8	14.1	27.5	35.5	46.6	63.5	82.0	101.2	−29
2-Nitrophenol	$C_6H_5NO_3$	49.3	76.8	90.4	105.8	122.1	132.6	146.4	167.6	191.0	214.5	45
2-Nitrophenyl acetate	$C_8H_7NO_4$	100.0	128.0	142.0	155.8	172.8	181.7	194.1	213.0	233.5	253.0	
1-Nitropropane	$C_3H_7NO_2$	−9.6	+13.5	25.3	37.9	51.8	60.5	72.3	90.2	110.6	131.6	−108
2-Nitropropane	$C_3H_7NO_2$	−18.8	+4.1	15.8	28.2	41.8	50.3	62.0	80.0	99.8	120.3	−93
2-Nitrotoluene	$C_7H_7NO_2$	50.0	79.1	93.8	109.6	126.3	137.6	151.5	173.7	197.7	222.3	−4.1
3-Nitrotoluene	$C_7H_7NO_2$	50.2	81.0	96.0	112.8	130.7	142.5	156.9	180.3	206.8	231.9	15.5
4-Nitrotoluene	$C_7H_7NO_2$	53.7	85.0	100.5	117.7	136.0	147.9	163.0	186.7	212.5	238.3	51.9
4-Nitro-1,3-xylene (4-nitro-m-xylene)	$C_8H_9NO_2$	65.6	95.0	109.8	125.8	143.3	153.8	168.5	191.7	217.5	244.0	+2
Nonacosane	$C_{29}H_{60}$	234.2	269.8	286.4	303.6	323.2	334.8	350.0	373.2	397.2	421.8	63.8
Nonadecane	$C_{19}H_{40}$	133.2	166.3	183.5	200.8	220.0	232.8	248.0	271.8	299.8	330.0	32
n-Nonane	C_9H_{20}	+1.4	25.8	38.0	51.2	66.0	75.5	88.1	107.5	128.2	150.8	−53.7
1-Nonanol	$C_9H_{20}O$	59.5	86.1	99.7	113.8	129.0	139.0	151.3	170.5	192.1	213.5	−5
2-Nonanone	$C_9H_{18}O$	32.1	59.0	72.3	87.2	103.4	113.8	127.4	148.2	171.2	195.0	−19
Octacosane	$C_{28}H_{58}$	226.5	260.3	277.4	295.4	314.2	326.8	341.8	364.8	388.9	412.5	61.6
Octadecane	$C_{18}H_{38}$	119.6	152.1	169.6	187.5	207.4	219.7	236.0	260.6	288.0	317.0	28
n-Octane	C_8H_{18}	−14.0	+8.3	19.2	31.5	45.1	53.8	65.7	83.6	104.0	125.6	−56.8
n-Octanol (1-octanol)	$C_8H_{18}O$	54.0	76.5	88.3	101.0	115.2	123.8	135.2	152.0	173.8	195.2	−15.4
2-Octanone	$C_8H_{16}O$	23.6	48.4	60.9	74.3	89.8	99.0	111.7	130.4	151.0	172.9	−16
n-Octyl acrylate	$C_{11}H_{20}O_2$	58.5	87.7	102.0	117.8	135.6	145.6	159.1	180.2	204.0	227.0	
iodide (1-Iodooctane)	$C_8H_{17}I$	45.8	74.8	90.0	105.9	123.8	135.4	150.0	173.3	199.3	225.5	−45.9
Oleic acid	$C_{18}H_{34}O_2$	176.5	208.5	223.0	240.0	257.2	269.8	286.0	309.8	334.7	360.0	14
Palmitaldehyde	$C_{16}H_{32}O$	121.6	154.6	171.8	190.0	210.0	222.6	239.5	264.1	292.3	321.0	34
Palmitic acid	$C_{16}H_{32}O_2$	153.6	188.1	205.8	223.8	244.4	256.0	271.5	298.7	326.0	353.8	64.0
Palmitonitrile	$C_{16}H_{31}N$	134.3	168.3	185.8	204.2	223.8	236.6	251.5	277.1	304.5	332.0	31
Pelargonic acid	$C_9H_{18}O_2$	108.2	126.0	137.4	149.8	163.7	172.3	184.4	203.1	227.5	253.5	12.5
Pentachlorobenzene	C_6HCl_5	98.6	129.7	144.3	160.0	178.5	190.1	205.5	227.0	251.6	276.0	85.5
Pentachloroethane	C_2HCl_5	+1.0	27.2	39.8	53.9	69.9	80.0	93.5	114.0	137.2	160.5	−22
Pentachloroethylbenzene	$C_8H_5Cl_5$	96.2	130.0	148.0	166.0	186.2	199.0	216.0	241.8	269.3	299.0	
Pentachlorophenol	C_6HCl_5O	......	......	......	192.2	211.2	223.4	239.6	261.8	285.0	309.3	188.5
Pentacosane	$C_{25}H_{52}$	194.2	230.0	248.2	266.1	285.6	298.4	314.0	339.0	365.4	390.3	53.3
Pentadecane	$C_{15}H_{32}$	91.6	121.0	135.4	150.2	167.7	178.4	194.0	216.1	242.8	270.5	10
1,3-Pentadiene	O_5H_8	−71.8	−53.8	−45.0	−34.8	−23.4	−16.5	−6.7	+8.0	24.7	42.1	
1,4-Pentadiene	C_5H_8	−83.5	−66.2	−57.1	−47.7	−37.0	−30.0	−20.6	−6.7	+8.3	26.1	
Pentaethylbenzene	$C_{16}H_{26}$	86.0	120.0	135.8	152.4	171.9	184.2	200.0	224.1	250.2	277.0	
Pentaethylchlorobenzene	$C_{16}H_{25}Cl$	90.0	123.8	140.7	158.1	178.2	191.0	208.0	230.3	257.2	285.0	
n-Pentane	C_5H_{12}	−76.6	−62.5	−50.1	−40.2	−29.2	−22.2	−12.6	+1.9	18.5	36.1	−129.7
iso-Pentane (2-methylbutane)	C_5H_{12}	−82.9	−65.8	−57.0	−47.3	−36.5	−29.6	−20.2	−5.9	+10.5	27.8	−159.7
neo-Pentane (2,2-dimethylpropane)	C_5H_{12}	−102.0	−85.4	−76.7	−67.2	−56.1	−49.0	−39.1	−23.7	−7.1	+9.5	−16.6
2,3,4-Pentanetriol	$C_5H_{12}O_3$	155.0	189.3	204.5	220.5	239.6	249.8	263.5	284.5	307.0	327.2	
1-Pentene	C_5H_{10}	−80.4	−63.3	−54.5	−46.0	−34.1	−27.1	−17.7	−3.4	+12.8	30.1	
α-Phellandrene	$C_{10}H_{16}$	20.0	45.7	58.0	72.1	87.8	97.6	110.6	130.6	152.0	175.0	
Phenanthrene	$C_{14}H_{10}$	118.2	154.3	173.0	193.7	215.8	229.9	249.0	277.1	308.0	340.2	99.5
Phenethyl alcohol (phenyl cellosolve)	$C_8H_{10}O_2$	58.2	85.9	100.0	114.8	130.5	141.2	154.0	175.0	197.5	219.5	
2-Phenetidine	$C_8H_{11}NO$	67.0	94.7	108.6	123.7	139.9	149.8	163.5	184.0	207.0	228.0	
Phenol	C_6H_6O	40.1	62.5	73.8	86.0	100.1	108.4	121.4	139.0	160.0	181.9	40.6
2-Phenoxyethanol	$C_8H_{10}O_2$	78.0	106.6	121.2	136.0	152.2	163.2	176.5	197.6	221.0	245.3	11.6
2-Phenoxyethyl acetate	$C_{10}H_{12}O_3$	82.6	113.5	128.0	144.5	162.3	174.0	189.2	211.3	235.0	259.7	−6.7
Phenyl acetate	$C_8H_8O_2$	38.2	64.8	78.0	92.3	108.1	118.1	131.6	151.2	173.5	195.9	
Phenylacetic acid	$C_8H_8O_2$	97.0	127.0	141.3	156.0	173.6	184.5	198.2	219.5	243.0	265.5	76.5
Phenylacetonitrile	C_8H_7N	60.0	89.0	103.5	119.4	136.3	147.7	161.8	184.2	208.5	233.5	−23.8
Phenylacetyl chloride	C_8H_7ClO	48.0	75.3	89.0	103.6	119.8	129.8	143.5	163.8	186.0	210.0	
Phenyl benzoate	$C_{13}H_{10}O_2$	106.8	141.5	157.8	177.0	197.6	210.8	227.8	254.0	283.5	314.0	70.5
4-Phenyl-3-buten-2-one	$C_{10}H_{10}O$	81.7	112.2	127.4	143.8	161.3	172.6	187.8	211.0	235.4	261.0	41.5
Phenyl isocyanate	C_7H_5NO	10.6	36.0	48.5	62.5	77.7	87.7	100.6	120.8	142.7	165.6	
isocyanide	C_7H_5N	12.0	37.0	49.7	63.4	78.3	88.0	'01.0	120.8	142.3	165.0	
Phenylcyclohexane	$C_{12}H_{16}$	67.5	96.5	111.3	126.4	144.0	154.2	169.3	191.3	214.6	240.0	+7.5
Phenyl dichlorophosphate	$C_6H_5Cl_2O_2P$	66.7	95.9	110.0	125.9	143.4	153.6	168.0	189.8	213.0	239.5	
m-Phenylene diamine (1,3-phenylenediamine)	$C_6H_8N_2$	99.8	131.2	147.0	163.8	182.5	194.0	209.9	233.0	259.0	285.5	62.8
Phenylglyoxal	$C_8H_6O_2$	......	75.0	87.8	100.7	115.5	124.2	136.2	153.8	173.5	193.5	73
Phenylhydrazine	$C_6H_8N_2$	71.8	101.6	115.8	131.5	148.2	158.7	173.5	195.4	218.2	243.5	19.5
N-Phenyliminodiethanol	$C_{10}H_{15}NO_2$	145.0	179.2	195.8	213.4	233.0	245.3	260.6	284.5	311.3	337.8	
1-Phenyl-1,3-pentanedione	$C_{11}H_{12}O_2$	98.0	128.5	144.0	159.9	178.0	189.8	204.5	226.7	251.2	276.5	
2-Phenylphenol	$C_{12}H_{10}O$	100.0	131.6	146.2	163.3	180.3	192.2	205.9	227.9	251.8	275.0	56.5
4-Phenylphenol	$C_{12}H_{10}O$	......	......	176.2	193.8	213.0	225.3	240.9	263.2	285.5	308.0	164.5
3-Phenyl-1-propanol	$C_9H_{12}O$	74.7	102.4	116.0	131.2	147.4	156.8	170.3	191.2	212.8	235.0	
Phenyl isothiocyanate	C_7H_5NS	47.2	75.6	89.8	115.5	122.5	133.3	147.7	169.6	194.0	218.5	−21.0
Phorone	$C_9H_{14}O$	42.0	68.3	81.5	95.6	111.3	121.4	134.0	153.5	175.3	197.2	28
iso-Phorone	$C_9H_{14}O$	38.0	66.7	81.2	96.8	114.5	125.6	140.6	163.3	188.7	215.2	
Phosgene (carbonyl chloride)	CCl_2O	−92.9	−77.0	−69.3	−60.3	−50.3	−44.0	−35.6	−22.3	−7.6	+8.3	−104
Phthalic anhydride	$C_8H_4O_3$	96.5	121.3	134.0	151.7	172.0	185.3	202.3	228.0	256.8	284.5	130.8
Phthalide	$C_8H_6O_2$	95.5	127.7	144.0	161.3	181.0	193.5	210.0	234.5	261.8	290.0	73
Phthaloyl chloride	$C_8H_4Cl_2O_2$	86.3	118.3	134.2	151.0	170.0	182.2	197.8	222.0	248.3	275.8	88.5
2-Picoline	C_6H_7N	−11.1	+12.6	24.4	37.4	51.2	59.9	71.4	89.0	108.4	128.8	−70
Pimelic acid	$C_7H_{12}O_4$	163.4	196.2	212.0	229.3	247.0	258.2	272.0	294.5	318.5	342.1	103
α-Pinene	$C_{10}H_{16}$	−1.0	+24.6	37.3	51.4	66.8	76.8	90.1	110.2	132.3	155.0	−55
β-Pinene	$C_{10}H_{16}$	+4.2	30.0	42.3	58.1	71.5	81.2	94.0	114.1	136.1	158.3	
Piperidine	$C_5H_{11}N$	......	−7.0	+3.9	15.8	29.2	37.7	49.0	66.2	85.7	106.0	−9

TABLE 3-8 Vapor Pressures of Organic Compounds, up to 1 atm *(Continued)*

Compound		Pressure, mm. Hg										Melting point, °C.
Name	Formula	1	5	10	20	40	60	100	200	400	760	
		Temperature, °C.										
Piperonal	$C_8H_6O_3$	87.0	117.4	132.0	148.0	165.7	177.0	191.7	214.3	238.5	263.0	37
Propane	C_3H_8	−128.9	−115.4	−108.5	−100.9	−92.4	−87.0	−79.6	−68.4	−55.6	−42.1	−187.1
Propenylbenzene	C_9H_{10}	17.5	43.8	57.0	71.5	87.7	97.8	111.7	132.0	154.7	179.0	−30.1
Propionamide	C_3H_7NO	65.0	91.0	105.0	119.0	134.8	144.3	156.0	174.2	194.0	213.0	79
Propionic acid	$C_3H_6O_2$	4.6	28.0	39.7	52.0	65.8	74.1	85.8	102.5	122.0	141.1	−22
anhydride	$C_6H_{10}O_3$	20.6	45.3	57.7	70.4	85.6	94.5	107.2	127.8	146.0	167.0	−45
Propionitrile	C_3H_5N	−35.0	−13.6	−3.0	+8.8	22.0	30.1	41.4	58.2	77.7	97.1	−91.9
Propiophenone	$C_9H_{10}O$	50.0	77.9	92.2	107.6	124.3	135.0	149.3	170.2	194.2	218.0	21
n-Propyl acetate	$C_5H_{10}O_2$	−26.7	−5.4	+5.0	16.0	28.8	37.0	47.8	64.0	82.0	101.8	−92.5
iso-Propyl acetate	$C_5H_{10}O_2$	−38.3	−17.4	−7.2	+4.2	17.0	25.1	35.7	51.7	69.8	89.0	
n-Propyl alcohol (1-propanol)	C_3H_8O	−15.0	+5.0	14.7	25.3	36.4	43.5	52.8	66.8	82.0	97.8	−127
iso-Propyl alcohol (2-propanol)	C_3H_8O	−26.1	−7.0	+2.4	12.7	23.8	30.5	39.5	53.0	67.8	82.5	−85.8
n-Propylamine	C_3H_9N	−64.4	−46.3	−37.2	−27.1	−16.0	−9.0	+0.5	15.0	31.5	48.5	−83
Propylbenzene	C_9H_{12}	6.3	31.3	43.4	56.8	71.6	81.1	94.0	113.5	135.7	159.2	−99.5
Propyl benzoate	$C_{10}H_{12}O_2$	54.6	83.8	98.0	114.3	131.8	143.3	157.4	180.1	205.2	231.0	−51.6
n-Propyl bromide (1-bromopropane)	C_3H_7Br	−53.0	−33.4	−23.3	−12.4	−0.3	+7.5	18.0	34.0	52.0	71.0	−109.9
iso-Propyl bromide (2-bromopropane)	C_3H_7Br	−61.8	−42.5	−32.8	−22.0	−10.1	−2.5	+8.0	23.8	41.5	60.0	−89.0
n-Propyl n-butyrate	$C_7H_{14}O_2$	−1.6	+22.1	34.0	47.0	61.5	70 3	82.6	101.0	121.7	142.7	−95.2
isobutyrate	$C_7H_{14}O_2$	−6.2	+16.8	28.3	40.6	54.3	63.0	73.9	91.8	112.0	133.9	
iso-Propyl isobutyrate	$C_7H_{14}O_2$	−16.3	+5.8	17.0	29.0	42.4	51.4	62.3	80.2	100.0	120.5	
Propyl carbamate	$C_4H_9NO_2$	52.4	77.6	90.0	103.2	117.7	126.5	138.3	155.8	175.8	195.0	
n-Propyl chloride (1-chloropropane)	C_3H_7Cl	−68.3	−50.0	−41.0	−31.0	−19.5	−12.1	−2.5	+12.2	29.4	46.4	−122.8
iso-Propyl chloride (2-chloropropane)	C_3H_7Cl	−78.8	−61.1	−52.0	−42.0	−31.0	−23.5	−13.7	+1.3	18.1	36.5	−117
iso-Propyl chloroacetate	$C_5H_9ClO_2$	+3.8	28.1	40.2	53.9	68.7	78.0	90.3	108.8	128.0	148.6	
Propyl chloroglyoxylate	$C_5H_7ClO_3$	9.7	32.3	43.5	55.6	68.8	77.2	88.0	104.7	123.0	150.0	
Propylene	C_3H_6	−131.9	−120.7	−112.1	−104.7	−96.5	−91.3	−84.1	−73.3	−60.9	−47.7	−185
Propylene glycol (1,2-Propanediol)	$C_3H_8O_2$	45.5	70.8	83.2	96.4	111.2	119.9	132.0	149.7	168.1	188.2	
Propylene oxide	C_3H_6O	−75.0	−57.8	−49.0	−39.3	−28.4	−21.3	−12.0	+2.1	17.8	34.5	−112.1
n-Propyl formate	$C_4H_8O_2$	−43.0	−22.7	−12.6	−1.7	+10.8	18.8	29.5	45.3	62.6	81.3	−92.9
iso-Propyl formate	$C_4H_8O_2$	−52.0	−32.7	−22.7	−12.1	−0.2	+7.5	17.8	33.6	50.5	68.3	
4,4′-iso-Propylidenebisphenol	$C_{15}H_{16}O_2$	193.0	224.2	240.8	255.5	273.0	282.9	297.0	317.5	339.0	360.5	
n-Propyl iodide (1-iodopropane)	C_3H_7I	−36.0	−13.5	−2.4	+10.0	23.6	32.1	43.8	61.8	81.8	102.5	−98.8
iso-Propyl iodide (2-iodopropane)	C_3H_7I	−43.3	−22.1	−11.7	0.0	+13.2	21.6	32.8	50.0	69.5	89.5	−90
n-Propyl levulinate	$C_8H_{14}O_3$	59.7	86.3	99.9	114.0	130.1	140.6	154.0	175.6	198.0	221.2	
iso-Propyl levulinate	$C_8H_{14}O_3$	48.0	74.5	88.0	102.4	118.1	127.8	141.8	161.6	185.2	208.2	
Propyl mercaptan (1-propanethiol)	C_3H_8S	−56.0	−36.3	−26.3	−15.4	−3.2	+4.6	15.3	31.5	49.2	67.4	−112
2-iso-Propylnaphthalene	$C_{13}H_{14}$	76.0	107.9	123.4	140.3	159.0	171.4	187.6	211.8	238.5	266.0	
iso-Propyl β-naphthyl ketone (2-isobutyronaphthone)	$C_{14}H_{14}O$	133.2	165.4	181.0	197.7	215.6	227.0	242.3	264.0	288.2	313.0	
2-iso-Propylphenol	$C_9H_{12}O$	56.6	83.8	97.0	111.7	127.5	137.7	150.3	170.1	192.6	214.5	15.5
3-iso-Propylphenol	$C_9H_{12}O$	62.0	90.3	104.1	119.8	136.2	146.6	160.2	182.0	205.0	228.0	26
4-iso-Propylphenol	$C_9H_{12}O$	67.0	94.7	108.0	123.4	139.8	149.7	163.3	184.0	206.1	228.2	61
Propyl propionate	$C_6H_{12}O_2$	−14.2	+8.0	19.4	31.6	45.0	53.8	65.2	82.7	102.0	122.4	−76
4-iso-Propylstyrene	$C_{11}H_{14}$	34.7	62.3	76.0	91.2	108.0	118.4	132.8	153.9	178.0	202.5	
Propyl isovalerate	$C_8H_{16}O_2$	+8.0	32.8	45.1	58.0	72.8	82.3	95.0	113.9	135.0	155.9	
Pulegone	$C_{10}H_{16}O$	58.3	82.5	94.0	106.8	121.7	130.2	143.1	162.5	189.8	221.0	
Pyridine	C_5H_5N	−18.9	+2.5	13.2	24.8	38.0	46.8	57.8	75.0	95.6	115.4	−42
Pyrocatechol	$C_6H_6O_2$	......	104.0	118.3	134.0	150.6	161.7	176.0	197.7	221.5	245.5	105
Pyrocaltechol diacetate (1,2-phenylene diacetate)	$C_{10}H_{10}O_4$	98.0	129.8	145.7	161.8	179.8	191.6	206.5	228.7	253.3	278.0	
Pyrogallol	$C_6H_6O_3$	......	151.7	167.7	185.3	204.2	216.3	232.0	255.3	281.5	309.0	133
Pyrotartaric anhydride	$C_5H_6O_3$	69.7	99.7	114.2	130.0	147.8	158.6	173.8	196.1	221.0	247.4	
Pyruvic acid	$C_3H_4O_3$	21.4	45.8	57.9	70.8	85.3	94.1	106.5	124.7	144.7	165.0	13.6
Quinoline	C_9H_7N	59.7	89.6	103.8	119.8	136.7	148.1	163.2	186.2	212.3	237.7	−15
iso-Quinoline	C_9H_7N	63.5	92.7	107.8	123.7	141.6	152.0	167.6	190.0	214.5	240.5	24.6
Resorcinol	$C_6H_6O_2$	108.4	138.0	152.1	168.0	185.3	195.8	209.8	230.8	253.4	276.5	110.7
Safrole	$C_{10}H_{10}O_2$	63.8	93.0	107.6	123.0	140.1	150.3	165.1	186.2	210.0	233.0	11.2
Salicylaldehyde	$C_7H_6O_2$	33.0	60.1	73.8	88.7	105.2	115.7	129.4	150.0	173.7	196.5	−7
Salicylic acid	$C_7H_6O_3$	113.7	136.0	146.2	156.8	172.2	182.0	193.4	210.0	230.5	256.0	159
Sebacic acid	$C_{10}H_{18}O_4$	183.0	215.7	232.0	250.0	268.2	279.8	294.5	313.2	332.8	352.3	134.5
Selenophene	C_4H_4Se	−39.0	−16.0	−4.0	+9.1	24.1	33.8	47.0	66.7	89.8	114.3	
Skatole	C_9H_9N	95.0	124.2	139.6	154.3	171.9	183.6	197.4	218.8	242.5	266.2	95
Stearaldehyde	$C_{18}H_{36}O$	140.0	174.6	192.1	210.6	230.8	244.2	260.0	285.0	313.8	342.5	63.5
Stearic acid	$C_{18}H_{36}O_2$	173.7	209.0	225.0	243.4	263.3	275.5	291.0	316.5	343.0	370.0	69.3
Stearyl alcohol (1-octadecanol)	$C_{18}H_{38}O$	150.3	185.6	202.0	220.0	240.4	252.7	269.4	293.5	320.3	349.5	58.5
Styrene	C_8H_8	−7.0	+18.0	30.8	44.6	59.8	69.5	82.0	101.3	122.5	145.2	−30.6
Styrene dibromide [(1,2-dibromoethyl) benzene]	$C_8H_8Br_2$	86.0	115.6	129.8	145.2	161.8	172.2	186.3	207.8	230.0	254.0	
Suberic acid	$C_8H_{14}O_4$	172.8	205.5	219.5	238.2	254.6	265.4	279.8	300.5	322.8	345.5	142
Succinic anhydride	$C_4H_4O_3$	92.0	115.0	128.2	145.3	163.0	174.0	189.0	212.0	237.0	261.0	119.6
Succinimide	$C_4H_5NO_2$	115.0	143.2	157.0	174.0	192.0	203.0	217.4	240.0	263.5	287.5	125.5
Succinyl chloride	$C_4H_4Cl_2O_2$	39.0	65.0	78.0	91.8	107.5	117.2	130.0	149.3	170.0	192.5	17
α-Terpineol	$C_{10}H_{18}O$	52.8	80.4	94.3	109.8	126.0	136.3	150.1	171.2	194.3	217.5	35
Terpenoline	$C_{10}H_{16}$	32.3	58.0	70.6	84.8	100.0	109.8	122.7	142.0	163.5	185.0	
1,1,1,2-Tetrabromoethane	$C_2H_2Br_4$	58.0	83.3	95.7	108.5	123.2	132.0	144.0	161.5	181.0	200.0	
1,1,2,2-Tetrabromoethane	$C_2H_2Br_4$	65.0	95.5	110.0	126.0	144.0	155.1	170.0	192.5	217.5	243.5	
Tetraisobutylene	$C_{16}H_{32}$	63.8	93.7	108.5	124.5	142.2	152.6	167.5	190.0	214.6	240.0	
Tetracosane	$C_{24}H_{50}$	183.8	219.6	237.6	255.3	276.3	288.4	305.2	330.5	358.0	386.4	51.1
1,2,3,4-Tetrachlorobenzene	$C_6H_2Cl_4$	68.5	99.6	114.7	131.2	149.2	160.0	175.7	198.0	225.5	254.0	46.5
1,2,3,5-Tetrachlorobenzene	$C_6H_2Cl_4$	58.2	89.0	104.1	121.6	140.0	152.0	168.0	193.7	220.0	246.0	54.5
1,2,4,5-Tetrachlorobenzene	$C_6H_2Cl_4$	......	......	......	......	146.0	157.7	173.5	196.0	220.5	245.0	139
1,1,2,2-Tetrachloro-1,2-difluoroethane	$C_2Cl_4F_2$	−37.5	−16.0	−5.0	+6.7	19.8	28.1	38.6	55.0	73.1	92.0	26.5
1,1,1,2-Tetrachloroethane	$C_2H_2Cl_4$	−16.3	+7.4	19.3	32.1	46.7	56.0	68.0	87.2	108.2	130.5	−68.7
1,1,2,2-Tetrachloroethane	$C_2H_2Cl_4$	−3.8	+20.7	33.0	46.2	60.8	70.0	83.2	102.2	124.0	145.9	−36
1,2,3,5-Tetrachloro-4-ethylbenzene	$C_8H_6Cl_4$	77.0	110.0	126.0	143.7	162.1	175.0	191.6	215.3	243.0	270.0	
Tetrachloroethylene	C_2Cl_4	−20.6	+2.4	13.8	26.3	40.1	49.2	61.3	79.8	100.0	120.8	−19.0
2,3,4,6-Tetrachlorophenol	$C_6H_2Cl_4O$	100.0	130.3	145.3	161.0	179.1	190.0	205.2	227.2	250.4	275.0	69.5
3,4,5,6-Tetrachloro-1,2-xylene	$C_8H_6Cl_4$	94.4	125.0	140.3	156.0	174.2	185.8	200.5	223.0	248.3	273.5	
Tetradecane	$C_{14}H_{30}$	76.4	106.0	120.7	135.6	152.7	164.0	178.5	201.8	226.8	252.5	5.5

TABLE 3-8 Vapor Pressures of Organic Compounds, up to 1 atm *(Continued)*

Compound		Pressure, mm. Hg										Melting point, °C.
Name	Formula	1	5	10	20	40	60	100	200	400	760	
		Temperature, °C.										
Tetradecylamine	$C_{14}H_{31}N$	102.6	135.8	152.0	170.0	189.0	200.2	215.7	239.8	264.6	291.2	
Tetradecyltrimethylsilane	$C_{17}H_{38}Si$	120.0	150.7	166.2	183.5	201.5	213.3	227.8	250.0	275.0	300.0	
Tetraethoxysilane	$C_8H_{20}O_4Si$	16.0	40.3	52.6	65.8	81.1	90.7	103.6	123.5	146.2	168.5	
1,2,3,4-Tetraethylbenzene	$C_{14}H_{22}$	65.7	96.2	111.6	127.7	145.8	156.7	172.4	196.0	221.4	248.0	11.6
Tetraethylene glycol	$C_8H_{18}O_5$	153.9	183.7	197.1	212.3	228.0	237.8	250.0	268.4	288.0	307.8	
Tetraethylene glycol chlorohydrin	$C_8H_{17}ClO_4$	110.1	141.8	156.1	172.6	190.0	200.5	214.7	236.5	258.2	281.5	
Tetraethyllead	$C_8H_{20}Pb$	38.4	63.6	74.8	88.0	102.4	111.7	123.8	142.0	161.8	183.0	−136
Tetraethylsilane	$C_8H_{20}Si$	−1.0	+23.9	36.3	50.0	65.3	74.8	88.0	108.0	130.2	153.0	
Tetralin	$C_{10}H_{12}$	38.0	65.3	79.0	93.8	110.4	121.3	135.3	157.2	181.8	207.2	−31.0
1,2,3,4-Tetramethylbenzene	$C_{10}H_{14}$	42.6	68.7	81.8	95.8	111.5	121.8	135.7	155.7	180.0	204.4	−6.2
1,2,3,5-Tetramethylbenzene	$C_{10}H_{14}$	40.6	65.8	77.8	91.0	105.8	115.4	128.3	149.9	173.7	197.9	−24.0
1,2,4,5-Tetramethylbenzene	$C_{10}H_{14}$	45.0	65.0	74.6	88.0	104.2	114.8	128.1	149.5	172.1	195.9	79.5
2,2,3,3-Tetramethylbutane	C_8H_{18}	−17.4	+3.2	13.5	24.6	36.8	44.5	54.8	70.2	87.4	106.3	−102.2
Tetramethylene dibromide (1,4-dibromobutane)	$C_4H_8Br_2$	32.0	58.8	72.4	87.6	104.0	115.1	128.7	149.8	173.8	197.5	−20
Tetramethyllead	$C_4H_{12}Pb$	−29.0	−6.8	+4.4	16.6	30.3	39.2	50.8	68.8	89.0	110.0	−27.5
Tetramethyltin	$C_4H_{12}Sn$	−51.3	−31.0	−20.6	−9.3	+3.5	11.7	22.8	39.8	58.5	78.0	
Tetrapropylene glycol monoisopropyl ether	$C_{15}H_{32}O_5$	116.6	147.8	163.0	179.8	197.7	209.0	223.3	245.0	268.3	292.7	
Thioacetic acid (mercaptoacetic acid)	$C_2H_4O_2S$	60.0	87.7	101.5	115.8	131.8	142.0	154.0	......	......	......	−16.5
Thiodiglycol (2,2′-thiodiethanol)	$C_4H_{10}O_2S$	42.0	96.0	128.0	165.0	210.0	240.5	285				
Thiophene	C_4H_4S	−40.7	−20.8	−10.9	0.0	+12.5	20.1	30.5	46.5	64.7	84.4	−38.3
Thiophenol (benzenethiol)	C_6H_6S	18.6	43.7	56.0	69.7	84.2	93.9	106.6	125.8	146.7	168.0	
α-Thujone	$C_{10}H_{16}O$	38.3	65.7	79.3	93.7	110.0	120.2	134.0	154.2	177.8	201.0	
Thymol	$C_{10}H_{14}O$	64.3	92.8	107.4	122.6	139.8	149.8	164.1	185.5	209.2	231.8	51.5
Tiglaldehyde	C_5H_8O	−25.0	−1.6	+10.0	23.2	37.0	45.8	57.7	75.4	95.5	116.4	
Tiglic acid	$C_5H_8O_2$	52.0	77.8	90.2	103.8	119.0	127.8	140.5	158.0	179.2	198.5	64.5
Tiglonitrile	C_5H_7N	−25.5	−2.4	+9.2	22.1	36.7	46.0	58.2	77.8	99.7	122.0	
Toluene	C_7H_8	−26.7	−4.4	+6.4	18.4	31.8	40.3	51.9	69.5	89.5	110.6	−95.0
Toluene-2,4-diamine	$C_7H_{10}N_2$	106.5	137.2	151.7	167.9	185.7	196.2	211.5	232.8	256.0	280.0	99
2-Toluic nitrile (2-tolunitrile)	C_8H_7N	36.7	64.0	77.9	93.0	110.0	120.8	135.0	156.0	180.0	205.2	−13
4-Toluic nitrile (4-tolunitrile)	C_8H_7N	42.5	71.3	85.8	101.7	109.5	130.0	145.2	167.3	193.0	217.6	29.5
2-Toluidine	C_7H_9N	44.0	69.3	81.4	95.1	110.0	119.8	133.0	153.0	176.2	199.7	−16.3
3-Toluidine	C_7H_9N	41.0	68.0	82.0	96.7	113.5	123.8	136.7	157.6	180.6	203.3	−31.5
4-Toluidine	C_7H_9N	42.0	68.2	81.8	95.8	111.5	121.5	133.7	154.0	176.9	200.4	44.5
2-Tolyl isocyanide	C_8H_7N	25.2	51.0	64.0	78.2	94.0	104.0	117.7	137.8	159.9	183.5	
4-Tolylhydrazine	$C_7H_{10}N_2$	82.4	110.0	123.8	138.6	154.1	165.0	178.0	198.0	219.5	242.0	65.5
Tribromoacetaldehyde	C_2HBr_3O	18.5	45.0	58.0	72.1	87.8	97.5	110.2	130.0	151.6	174.0	
1,1,2-Tribromobutane	$C_4H_7Br_3$	45.0	73.5	87.8	103.2	120.2	131.6	146.0	167.8	192.0	216.2	
1,2,2-Tribromobutane	$C_4H_7Br_3$	41.0	69.0	83.2	98.6	116.0	127.0	141.8	163.5	188.0	213.8	
2,2,3-Tribromobutane	$C_4H_7Br_3$	38.2	66.0	79.8	94.6	111.8	122.2	136.3	157.8	182.2	206.5	
1,1,2-Tribromoethane	$C_2H_3Br_3$	32.6	58.0	70.6	84.2	100.0	110.0	123.5	143.5	165.4	188.4	−26
1,2,3-Tribromopropane	$C_3H_5Br_3$	47.5	75.8	90.0	105.8	122.8	134.0	148.0	170.0	195.0	220.0	16.5
Triisobutylamine	$C_{12}H_{27}N$	32.3	57.4	69.8	83.0	97.8	107.3	119.7	138.0	157.8	179.0	−22
Triisobutylene	$C_{12}H_{24}$	18.0	44.0	56.5	70.0	86.7	96.7	110.0	130.2	153.0	179.0	
2,4,6-Tritertbutylphenol	$C_{18}H_{30}O$	95.2	126.1	142.0	158.0	177.4	188.0	203.0	226.2	250.6	276.3	
Trichloroacetic acid	$C_2HCl_3O_2$	51.0	76.0	88.2	101.8	116.3	125.9	137.8	155.4	175.2	195.6	57
Trichloroacetic anhydride	$C_4Cl_6O_3$	56.2	85.3	99.6	114.3	131.2	141.8	155.2	176.2	199.8	223.0	
Trichloroacetyl bromide	C_2BrCl_3O	−7.4	+16.7	29.3	42.1	57.2	66.7	79.5	98.4	120.2	143.0	
2,4,6-Trichloroaniline	$C_6H_4Cl_3N$	134.0	157.8	170.0	182.6	195.8	204.5	214.6	229.8	246.4	262.0	78
1,2,3-Trichlorobenzene	$C_6H_3Cl_3$	40.0	70.0	85.6	101.8	119.8	131.5	146.0	168.2	193.5	218.5	52.5
1,2,4-Trichlorobenzene	$C_6H_3Cl_3$	38.4	67.3	81.7	97.2	114.8	125.7	140.0	162.0	187.7	213.0	17
1,3,5-Trichlorobenzene	$C_6H_3Cl_3$	......	63.8	78.0	93.7	110.8	121.8	136.0	157.7	183.0	208.4	63.5
1,2,3-Trichlorobutane	$C_4H_7Cl_3$	+0.5	27.2	40.0	55.0	71.5	82.0	96.2	118.0	143.0	169.0	
1,1,1-Trichloroethane	$C_2H_3Cl_3$	−52.0	−32.0	−21.9	−10.8	+1.6	9.5	20.0	36.2	54.6	74.1	−30.6
1,1,2 Trichloroethane	$C_2H_3Cl_3$	−24.0	−2.0	+8.3	21.6	35.2	44.0	55.7	73.3	93.0	113.9	−36.7
Trichloroethylene	C_2HCl_3	−43.8	−22.8	−12.4	−1.0	+11.9	20.0	31.4	48.0	67.0	86.7	−73
Trichlorofluoromethane	CCl_3F	−84.3	−67.6	−59.0	−49.7	−39.0	−32.3	−23.0	−9.1	+6.8	23.7	
2,4,5-Trichlorophenol	$C_6H_3Cl_3O$	72.0	102.1	117.3	134.0	151.5	162.5	178.0	201.5	226.5	251.8	62
2,4,6-Trichlorophenol	$C_6H_3Cl_3O$	76.5	105.9	120.2	135.8	152.2	163.5	177.8	199.0	222.5	246.0	68.5
Tri-2-chlorophenylthiophosphate	$C_{18}H_{12}Cl_3O_3PS$	188.2	217.2	231.2	246.7	261.7	271.5	283.8	302.8	322.0	341.3	
1,1,1-Trichloropropane	$C_3H_5Cl_3$	−28.8	−7.0	+4.2	16.2	29.9	38.3	50.0	67.7	87.5	108.2	−77.7
1,2,3-Trichloropropane	$C_3H_5Cl_3$	+9.0	33.7	46.0	59.3	74.0	83.6	96.1	115.6	137.0	158.0	−14.7
1,1,2-Trichloro-1,2,2-trifluoroethane	$C_2Cl_3F_3$	−68.0	−49.4	−40.3	−30.0	−18.5	−11.2	−1.7	+13.5	30.2	47.6	−35
Tricosane	$C_{23}H_{48}$	170.0	206.3	223.0	242.0	261.3	273.8	289.8	313.5	339.8	366.5	47.7
Tridecane	$C_{13}H_{28}$	59.4	98.3	104.0	120.2	137.7	148.2	162.5	185.0	209.4	234.0	−6.2
Tridecanoic acid	$C_{13}H_{26}O_2$	137.8	166.3	181.0	195.8	212.4	222.0	236.0	255.2	276.5	299.0	41
Triethoxymethylsilane	$C_7H_{18}O_3Si$	−1.5	+22.8	34.6	47.2	61.7	70.4	82.7	101.0	121.8	143.5	
Triethoxyphenylsilane	$C_{12}H_{20}O_3Si$	71.0	98.8	112.6	127.2	143.5	153.2	167.5	188.0	210.5	233.5	
1,2,4-Triethylbenzene	$C_{12}H_{18}$	46.0	74.2	88.5	104.0	121.7	132.2	146.8	168.3	193.7	218.0	
1,3,4-Triethylbenzene	$C_{12}H_{18}$	47.9	76.0	90.2	105.8	122.6	133.4	147.7	168.3	193.2	217.5	
Triethylborine	$C_6H_{15}B$	......	......	−148.0	−140.6	−131.4	−125.2	−116.0	−101.0	−81.0	−56.2	
Triethyl camphoronate	$C_{15}H_{26}O_6$	......	150.2	166.0	183.6	201.8	213.5	228.6	250.8	276.0	301.0	135
citrate	$C_{12}H_{20}O_7$	107.0	138.7	144.0	171.1	190.4	202.5	217.8	242.2	267.5	294.0	
Triethyleneglycol	$C_6H_{14}O_4$	114.0	144.0	158.1	174.0	191.3	201.5	214.6	235.2	256.6	278.3	
Triethylheptylsilane	$C_{13}H_{30}Si$	70.0	99.8	114.6	130.3	148.0	158.2	174.0	196.0	221.0	247.0	
Triethyloctylsilane	$C_{14}H_{32}Si$	73.7	104.8	120.6	137.7	155.7	168.0	184.3	208.0	235.0	262.0	
Triethyl orthoformate	$C_7H_{16}O_3$	+5.5	29.2	40.5	53.4	67.5	76.0	88.0	106.0	125.7	146.0	
phosphate	$C_6H_{15}O_4P$	39.6	67.8	82.1	97.8	115.7	126.3	141.6	163.7	187.0	211.0	
Triethylthallium	$C_6H_{15}Tl$	+9.3	37.6	51.7	67.7	85.4	95.7	112.1	136.0	163.5	192.1	−63.0
Trifluorophenylsilane	$C_6H_5F_3Si$	−31.0	−9.7	+0.8	12.3	25.4	33.2	44.2	60.1	78.7	98.3	
Trimethallyl phosphate	$C_{12}H_{21}PO_4$	93.7	131.0	149.8	169.8	192.0	207.0	225.7	255.0	288.5	324.0	
2,3,5-Trimethylacetophenone	$C_{11}H_{14}O$	79.0	108.0	122.3	137.5	154.2	165.7	179.7	201.3	224.3	247.5	
Trimethylamine	C_3H_9N	−97.1	−81.7	−73.8	−65.0	−55.2	−48.8	−40.3	−27.0	−12.5	+2.9	−117.1
2,4,5-Trimethylaniline	$C_9H_{13}N$	68.4	95.9	109.0	123.7	139.8	149.5	162.0	182.3	203.7	234.5	67
1,2,3-Trimethylbenzene	C_9H_{12}	16.8	42.9	55.9	69.9	85.4	95.3	108.8	129.0	152.0	176.1	−25.5
1,2,4-Trimethylbenzene	C_9H_{12}	13.6	38.3	50.7	64.5	79.8	89.5	102.8	122.7	145.4	169.2	−44.1
1,3,5-Trimethylbenzene	C_9H_{12}	9.6	34.7	47.4	61.0	76.1	85.8	98.9	118.6	141.0	164.7	−44.8
2,2,3-Trimethylbutane	C_7H_{16}	......	......	−18.8	−7.5	+5.2	13.3	24.4	41.2	60.4	80.9	−25.0
Trimethyl citrate	$C_9H_{14}O_7$	106.2	146.2	160.4	177.2	194.2	205.5	219.6	241.3	264.2	287.0	78.5

TABLE 3-8 Vapor Pressures of Organic Compounds, up to 1 atm *(Concluded)*

Compound: Name	Formula	1	5	10	20	40	60	100	200	400	760	Melting point, °C.
		Pressure, mm. Hg / Temperature, °C.										
Trimethyleneglycol (1,3-propanediol)	$C_3H_8O_2$	59.4	87.2	100.6	115.5	131.0	141.1	153.4	172.8	193.8	214.2	
1,2,4-Trimethyl-5-ethylbenzene	$C_{11}H_{16}$	43.7	71.2	84.6	99.7	106.0	126.3	140.3	160.3	184.5	208.1	
1,3,5-Trimethyl-2-ethylbenzene	$C_{11}H_{16}$	38.8	67.0	80.5	96.0	113.2	123.8	137.9	158.4	183.5	208.0	
2,2,3-Trimethylpentane	C_8H_{18}	−29.0	−7.1	+3.9	16.0	29.5	38.1	49.9	67.8	88.2	109.8	−112.3
2,2,4-Trimethylpentane	C_8H_{18}	−36.5	−15.0	−4.3	+7.5	20.7	29.1	40.7	58.1	78.0	99.2	−107.3
2,3,3-Trimethylpentane	C_8H_{18}	−25.8	−3.9	+6.9	19.2	33.0	41.8	53.8	72.0	92.7	114.8	−101.5
2,3,4-Trimethylpentane	C_8H_{18}	−26.3	−4.1	+7.1	19.3	32.9	41.6	53.4	71.3	91.8	113.5	−109.2
2,2,4-Trimethyl-3-pentanone	$C_8H_{16}O$	14.7	36.0	46.4	57.6	69.8	77.3	87.6	102.2	118.4	135.0	
Trimethyl phosphate	$C_3H_9O_4P$	26.0	53.7	67.8	83.0	100.0	110.0	124.0	145.0	167.8	192.7	
2,4,5-Trimethylstyrene	$C_{11}H_{14}$	48.1	77.0	91.6	107.1	124.2	135.5	149.8	171.8	196.1	221.2	
2,4,6-Trimethylstyrene	$C_{11}H_{14}$	37.5	65.7	79.7	94.8	111.8	122.3	136.8	157.8	182.3	207.0	
Trimethylsuccinic anhydride	$C_7H_{10}O_3$	53.5	82.6	97.4	113.8	131.0	142.2	156.5	179.8	205.5	231.0	
Triphenylmethane	$C_{19}H_{16}$	169.7	188.4	197.0	206.8	215.5	221.2	228.4	239.7	249.8	259.2	93.4
Triphenylphosphate	$C_{18}H_{15}O_4P$	193.5	230.4	249.8	269.7	290.3	305.2	322.5	349.8	379.2	413.5	49.4
Tripropyleneglycol	$C_9H_{20}O_4$	96.0	125.7	140.5	155.8	173.7	184.6	199.0	220.2	244.3	267.2	
Tripropyleneglycol monobutyl ether	$C_{13}H_{28}O_4$	101.5	131.6	147.0	161.8	179.8	190.2	204.4	224.4	247.0	269.5	
Tripropyleneglycol monoisopropyl ether	$C_{12}H_{26}O_4$	82.4	112.4	127.3	143.7	161.4	173.2	187.8	209.7	232.8	256.6	
Tritolyl phosphate	$C_{21}H_{21}O_4P$	154.6	184.2	198.0	213.2	229.7	239.8	252.2	271.8	292.7	313.0	
Undecane	$C_{11}H_{24}$	32.7	59.7	73.9	85.6	104.4	115.2	128.1	149.3	171.9	195.8	−25.6
Undecanoic acid	$C_{11}H_{22}O_2$	101.4	133.1	149.0	166.0	185.6	197.2	212.5	237.8	262.8	290.0	29.5
10-Undecenoic acid	$C_{11}H_{20}O_2$	114.0	142.8	156.3	172.0	188.7	199.5	213.5	232.8	254.0	275.0	24.5
Undecan-2-ol	$C_{11}H_{24}O$	71.1	99.0	112.8	127.5	143.7	153.7	167.2	187.7	209.8	232.0	
n-Valeric acid	$C_5H_{10}O_2$	42.2	67.7	79.8	93.1	107.8	116.6	128.3	146.0	165.0	184.4	−34.5
iso-Valeric acid	$C_5H_{10}O_2$	34.5	59.6	71.3	84.0	98.0	107.3	118.9	136.2	155.2	175.1	−37.6
γ-Valerolactone	$C_5H_8O_2$	37.5	65.8	79.8	95.2	101.9	122.4	136.5	157.7	182.3	207.5	
Valeronitrile	C_5H_9N	−6.0	+18.1	30.0	43.3	57.8	66.9	78.6	97.7	118.7	140.8	
Vanillin	$C_8H_8O_3$	107.0	138.4	154.0	170.5	188.7	199.8	214.5	237.3	260.0	285.0	81.5
Vinyl acetate	$C_4H_6O_2$	−48.0	−28.0	−18.0	−7.0	+5.3	13.0	23.3	38.4	55.5	72.5	
2-Vinylanisole	$C_9H_{10}O$	41.9	68.0	81.0	94.7	110.0	119.8	132.3	151.0	172.1	194.0	
3-Vinylanisole	$C_9H_{10}O$	43.4	69.9	83.0	97.2	112.5	122.3	135.3	154.0	175.8	197.5	
4-Vinylanisole	$C_9H_{10}O$	45.2	72.0	85.7	100.0	116.0	126.1	139.7	159.0	182.0	204.5	
Vinyl chloride (1-chloroethylene)	C_2H_3Cl	−105.6	−90.8	−83.7	−75.7	−66.8	−61.1	−53.2	−41.3	−28.0	−13.8	−153.7
cyanide (acrylonitrile)	C_3H_3N	−51.0	−30.7	−20.3	−9.0	+3.8	11.8	22.8	38.7	58.3	78.5	−82
fluoride (1-fluoroethylene)	C_2H_3F	−149.3	−138.0	−132.2	−125.4	−118.0	−113.0	−106.2	−95.4	−84.0	−72.2	−160.5
Vinylidene chloride (1,1-dichloroethene)	$C_2H_2Cl_2$	−77.2	−60.0	−51.2	−41.7	−31.1	−24.0	−15.0	−1.0	+14.8	31.7	−122.5
4-Vinylphenetole	$C_{10}H_{12}O$	64.0	91.7	105.6	120.3	136.3	146.4	159.8	180.0	202.8	225.0	
2-Xenyl dichlorophosphate	$C_{12}H_9Cl_2PO$	138.2	171.1	187.0	205.0	223.8	236.0	251.5	275.3	301.5	328.5	
2,4-Xyaldehyde	$C_9H_{10}O$	59.0	85.9	99.0	114.0	129.7	139.8	152.2	172.3	194.1	215.5	75
2-Xylene (2-xylene)	C_8H_{10}	−3.8	+20.2	32.1	45.1	59.5	68.8	81.3	100.2	121.7	144.4	−25.2
3-Xylene (3-xylene)	C_8H_{10}	−6.9	+16.8	28.3	41.1	55.3	64.4	76.8	95.5	116.7	139.1	−47.9
4-Xylene (4-xylene)	C_8H_{10}	−8.1	+15.5	27.3	40.1	54.4	63.5	75.9	94.6	115.9	138.3	+13.3
2,4-Xylidine	$C_8H_{11}N$	52.6	79.8	93.0	107.6	123.8	133.7	146.8	166.4	188.3	211.5	
2,6-Xylidine	$C_8H_{11}N$	44.0	72.6	87.0	102.7	120.2	131.5	146.0	168.0	193.7	217.9	

TABLE 3-9 Vapor Pressures of Organic Compounds, above 1 atm*

Compound: Name	Formula	1	2	5	10	20	30	40	50	60	Critical point: t_c, °C.	Critical point: P_c, atm.
		Pressure, atm. / Temperature, °C.										
Acetic acid	$C_2H_4O_2$	118.1	143.5	180.3	214.0	252.0	276.5	297.0	312.5	......	321.6	57.2
anhydride	$C_4H_6O_3$	139.6	162.0	194.0	221.5	253.0	272.8	288.5	......	......	296	46
Acetone	C_3H_6O	56.5	78.6	113.0	144.5	181.0	205.0	214.5	......	......	235.0	47.0
Acetylene	C_2H_2	−84.0	−71.6	−50.2	−32.7	−10.0	+4.8	16.8	26.8	34.8	36.0	62.0
Allene (propadiene)	C_3H_4	−35.0	−18.4	+8.0	33.2	64.5	85.5	103.5	118.0	......	120.7	51.8
Aniline	C_6H_7N	184.4	212.8	254.8	292.7	342.0	375.5	400.0	422.4	......	426	52.4
Benzene	C_6H_6	80.1	103.8	142.5	178.8	221.5	249.5	272.3	290.3	......	290.5	50.1
Bromobenzene	C_6H_5Br	156.2	186.2	232.5	274.5	327.0	359.8	387.5	......	......	397	44.6
1,3-Butadiene	C_4H_6	−4.5	+15.3	47.0	76.0	114.0	139.8	158.0	......	......	161.8	42.6
iso-Butane (2-methylpropane)	C_4H_{10}	−11.7	+7.5	39.0	66.8	99.5	120.5	......	......	......	134.0	37.0
n-Butane	C_4H_{10}	−0.5	+18.8	50.0	79.5	116.0	140.6	......	......	......	152.8	36.0
iso-Butyl alcohol (2-methylpropanol-1)	$C_4H_{10}O$	108.0	127.3	156.2	182.0	212.5	232.0	251.0	......	......	265	48
n-Butyl alcohol (1-butanol)	$C_4H_{10}O$	117.5	139.8	172.5	203.0	237.0	259.0	277.0	......	......	287	48.4
sec-Butyl alcohol (2-butanol)	$C_4H_{10}O$	99.5	118.2	147.5	172.0	204.0	230.0	251.0	......	......	265	48
tert-Butyl alcohol (trimethyl carbinol)	$C_4H_{10}O$	82.9	102.0	130.0	154.2	184.5	207.0	222.5	......	......	235	49
iso-Butyl formate	$C_5H_{10}C_2$	98.2	121.8	157.8	192.4	234.0	261.0	......	......	......	278.0	38.0
Butyric acid	$C_4H_8O_2$	163.5	188.3	225.0	257.0	295.0	319.0	338.0	352.0	......	355	52.0
iso-Butyric acid	$C_4H_8O_2$	154.5	179.8	217.0	250.0	289.0	315.0	336.0	......	......	336	40.0
Carbon dioxide	CO_2	−78.2	−69.1	−56.7	−39.5	−18.9	−5.3	+5.9	14.9	22.4	31.1	73.0
disulfide	CS_2	46.5	69.1	104.8	136.3	175.5	201.5	222.8	240.0	256.0	273.0	72.9
monoxide	CO	−191.3	−183.5	−170.7	−161.0	−149.7	−141.9	......	......	......	−138.7	34.6
tetrachloride	CCl_4	76.7	102.0	141.7	178.0	222.0	251.2	276.0	......	......	283.1	45.0
Chlorobenzene	C_6H_5Cl	132.2	160.2	205.0	245.3	292.8	324.4	349.8	......	......	359.2	44.6
Chlorodifluoromethane	$CHClF_2$	−40.8	−24.7	+0.3	24.0	52.0	70.3	85.3	......	......	96	48.7
Chloroform (trichloromethane)	$CHCl_3$	61.3	83.9	120.0	152.3	191.8	216.5	237.5	254.0	......	260	54.9
1-Chloro-1,2,2-trifluoroethylene	C_2ClF_3	−27.9	−11.1	+15.5	40.0	71.1	91.9	......	......	......	107.0	39.0
Chlorotrifluoromethane	$CClF_3$	−81.2	−66.7	−42.7	−18.5	+12.0	34.8	52.8	......	......	53	40.3
Cyanogen	C_2N_2	−21.0	−4.4	+21.4	44.6	72.6	91.6	106.5	118.2	......	126.6	58.2
Cyclohexane	C_6H_{12}	80.7	106.0	146.4	184.0	228.4	257.5	......	......	......	279.9	39.8
1,2-Dibromoethane	$C_2H_4Br_2$	131.5	157.7	200.0	237.0	269.0	286.0	295.0	300.0	304.5	309.8	70.6
Dichlorodifluoromethane	CCl_2F_2	−29.8	−12.2	+16.1	42.4	74.0	95.6	......	......	......	111.5	39.6
1,1-Dichloroethane	$C_2H_4Cl_2$	57.3	80.2	117.3	150.3	192.7	220.0	243.0	261.5	......	261.5	50.0
1,2-Dichloroethane	$C_2H_4Cl_2$	83.7	108.1	147.8	183.5	226.5	254.0	272.0	285.0	......	288.4	53.0
cis-1,2-Dichloroethylene	$C_2H_2Cl_2$	59.0	82.1	119.3	152.3	194.0	221.5	244.5	260.0	......	271.0	57.9
trans-1,2-Dichloroethylene	$C_2H_2Cl_2$	47.8	69.8	104.0	135.7	174.0	199.8	220.0	236.5	......	243.3	54.5
Dichlorofluoromethane	$CHCl_2F$	8.9	28.4	59.0	87.0	121.2	144.0	162.6	177.5	......	178.5	51.0
1,2-Dichloro-1,1,2,2-tetrafluoroethane	$C_2Cl_2F_4$	3.5	22.8	54.0	82.3	117.5	140.9	......	......	......	145.7	32.3
Diethylamine	$C_4H_{11}N$	55.5	77.8	113.0	145.3	184.5	210.0	......	......	......	223.3	36.6
Diethyl ether	$C_4H_{10}O$	34.6	56.0	90.0	122.0	159.0	183.3	......	......	......	193.8	35.5
sulfide	$C_4H_{10}S$	88.0	112.0	153.8	190.2	234.0	263.0	......	......	......	283.8	39.1

*Compiled from the extended tables published by D. R. Stull in *Ind. Eng. Chem.*, 39, 517 (1947). For data on gasoline and aircraft fuels see Hibbard, NACA Res. Mem. E56I21, 1956 (declassified 1958). Extensive data for aqueous solutions of ethylene glycol, diethylene glycol, triethylene glycol, and propylene glycol from −20 to 300°F are contained in *Glycols*, a Union Carbide Corporation publication. See also Fig. 3-3.

TABLE 3-9 Vapor Pressures of Organic Compounds, above 1 atm *(Concluded)*

Compound: Name	Formula	Pressure, atm. (Temperature, °C.): 1	2	5	10	20	30	40	50	60	Critical point: t_c, °C.	P_c, atm.
Dimethylamine	C_2H_7N	7.4	25.0	53.9	80.0	111.7	132.2	149.8	162.6	......	164.5	52.4
2,3-Dimethylbutane	C_6H_{14}	58.0	82.0	120.3	155.7	198.7	225.5	......	......	......	227.4	30.7
Dimethyl ether	C_2H_6O	−23.7	−6.4	+20.8	45.5	75.7	96.0	112.1	125.2	......	126.9	52.0
oxalate	$C_4H_6O_4$	163.3	189.6	228.7	......	......	......	......	......	......	260	9.5
sulfide	C_2H_6S	36.0	57.8	92.3	124.5	163.8	188.5	209.0	224.5	......	229.9	54.6
n-Dodecane	$C_{12}H_{26}$	216.2	249.2	300.0	345.8	......	......	......	......	......	385	17.5
Ethane	C_2H_6	−88.6	−75.0	−52.8	−32.0	−6.4	+10.0	23.6	......	......	32.3	48.2
Ethyl acetate	$C_4H_8O_2$	77.1	100.6	136.6	169.7	209.5	235.0	......	......	......	250.1	37.9
alcohol (ethanol)	C_2H_6O	78.4	97.5	126.0	151.8	183.0	203.0	218.0	230.0	242.0	243.5	63.1
Ethylamine	C_2H_7N	16.6	35.7	65.3	91.8	124.0	146.0	163.0	176.0	......	183.2	55.5
Ethyl benzene	C_8H_{10}	136.2	163.5	207.5	246.3	294.5	326.5	......	......	......	346.4	38.1
bromide	C_2H_5Br	38.4	60.2	95.0	126.8	164.3	188.0	206.5	220.0	229.5	230.8	61.5
chloride	C_2H_5Cl	12.3	32.5	64.0	92.6	127.3	149.5	167.0	180.5	......	187.2	52.0
fluoride	C_2H_5F	−32.0	−16.7	+7.7	30.2	57.5	75.7	90.0	......	......	102.2	49.6
formate	$C_3H_6O_2$	54.3	76.0	110.5	142.2	180.0	205.0	225.0	......	......	235.3	46.8
isobutyrate	$C_6H_{12}O_2$	110.1	135.5	174.2	210.0	253.0	280.0	......	......	......	280.0	30.0
mercaptan (ethanethiol)	C_2H_6S	35.0	56.6	90.7	121.9	159.5	184.3	204.7	220.0	......	225.5	54.2
methyl ether	C_3H_8O	7.5	26.5	56.4	84.0	108.0	141.4	160.0	......	......	164.7	43.4
propionate	$C_5H_{10}O_2$	99.1	123.8	162.7	197.8	240.0	264.5	......	......	......	272.8	33.2
propyl ether	$C_5H_{12}O$	61.7	85.3	123.1	156.2	197.2	223.0	......	......	......	227.4	32.1
Ethylene	C_2H_4	−103.7	−90.8	−71.1	−52.8	−29.1	−14.2	−1.5	+8.9	......	9.6	50.7
Fluorobenzene	C_6H_5F	84.7	109.9	148.5	184.4	227.6	257.0	279.3	......	......	286.5	44.7
n-Heptane	C_7H_{16}	98.4	124.8	165.7	202.8	247.5	......	......	......	......	266.8	26.9
n-Hexane	C_6H_{14}	68.7	93.0	131.7	166.6	209.4	......	......	......	......	234.8	29.6
Hydrogen cyanide (hydrocyanic acid)	CHN	25.9	45.8	75.8	102.7	135.0	153.8	169.9	183.5	......	183.5	50.0
Iodobenzene	C_6H_5I	188.6	220.0	270.0	315.7	371.5	406.0	437.2	......	......	448	44.7
Methane	CH_4	−161.5	−152.3	−138.3	−124.8	−108.5	−96.3	−86.3	......	......	−82.1	45.8
Methyl acetate	$C_3H_6O_2$	57.8	79.5	113.1	144.2	181.0	205.0	225.0	......	......	233.7	46.3
acetylene (propyne)	C_3H_4	−23.3	−7.1	+19.5	43.8	74.0	94.0	111.5	125.0	......	128	52.8
alcohol	CH_4O	64.7	84.0	112.5	138.0	167.8	186.5	203.5	214.0	224.0	240.0	78.7
Methylamine	CH_5N	−6.3	+10.1	36.0	59.5	87.8	106.3	121.8	133.7	144.6	156.9	73.6
Methyl bromide	CH_3Br	3.6	23.3	54.8	84.0	121.7	147.5	170.2	190.0	......	194	51.6
butyrate	$C_5H_{10}O_2$	102.3	127.5	166.7	203.0	244.5	272.0	......	......	......	281.2	34.2
chloride	CH_3Cl	−24.0	−6.4	+22.0	47.3	77.3	97.5	113.8	126.0	137.5	143.8	65.8
fluoride	CH_3F	−78.2	−64.5	−42.0	−21.0	+2.6	15.5	26.5	36.0	43.5	44.9	62.0
formate	$C_2H_4O_2$	32.0	51.9	83.5	112.0	147.2	169.7	188.5	213.0	......	214.0	59.1
iodide	CH_3I	42.4	65.5	101.8	138.0	176.5	206.0	228.5	248.0	......	255	54.6
isobutyrate	$C_5H_{10}O_2$	92.6	116.7	155.2	190.2	232.0	259.5	......	......	......	267.5	33.9
mercaptan (methanethiol)	CH_4S	6.8	26.1	55.9	83.4	117.5	140.0	157.7	172.0	185.0	196.8	71.4
propionate	$C_4H_8O_2$	79.8	103.0	139.8	172.6	212.5	239.0	......	......	......	257.4	39.3
n-Octane	C_8H_{18}	125.6	152.7	196.2	235.8	281.4	......	......	......	......	296.2	24.7
iso Pentane (2-methylbutane)	C_5H_{12}	27.8	48.8	82.8	114.5	154.0	180.3	......	......	......	187.8	32.8
n-Pentane	C_5H_{12}	36.1	58.0	92.4	124.7	164.3	191.3	......	......	......	197.2	33.0
neo-Pentane (2,2-dimethylpropane)	C_5H_{12}	+9.5	29.5	61.1	90.7	127.6	152.5	......	......	......	159.0	33.0
Phenol	C_6H_6O	181.9	208.0	248.2	283.8	328.7	358.0	382.1	400.0	418.7	419	60.5
Phosgene (carbonyl chloride)	CCl_2O	8.3	27.3	57.2	85.0	119.0	141.8	159.8	174.0	......	181.7	56.0
Propane	C_3H_8	−42.1	−25.6	+1.4	26.9	58.1	78.7	94.8	......	......	96.8	42.0
Propionic acid	$C_3H_6O_2$	141.1	160.0	186.0	203.5	220.0	228.0	233.0	238.0	......	239.5	53.0
Propyl acetate	$C_5H_{10}O_2$	101.8	126.8	165.7	200.5	242.8	269.0	......	......	......	276.2	33.2
iso-Propyl alcohol (2-propanol)	C_3H_8O	82.5	101.3	130.2	155.7	186.0	205.0	220.2	232.0	......	235	53
n-Propyl alcohol (1-propanol)	C_3H_8O	97.8	117.0	149.0	177.0	210.8	232.3	250.0	......	......	263.7	49.9
Propylamine	C_3H_9N	48.5	69.8	102.8	133.4	170.0	194.3	214.5	......	......	223.8	46.8
Propyl formate	$C_4H_8O_2$	81.3	104.3	142.0	176.4	217.5	245.0	......	......	......	264.8	39.5
Propylene	C_3H_6	−47.7	−31.4	−4.8	+19.8	49.5	70.0	85.0	......	......	91.4	45.4
Tetramethylsilane	$C_4H_{12}Si$	27.0	48.0	82.0	113.0	152.0	178.0	......	......	......	185	33
Toluene	C_7H_8	110.6	136.5	178.0	215.8	262.5	292.8	319.0	......	......	320.6	41.6
Trichlorofluoromethane	CCl_3F	23.7	44.1	77.3	108.2	146.7	172.0	194.0	......	......	198.0	43.2
1,1,2-Trichloro-1,2,2-trifluoroethane	$C_2Cl_3F_3$	47.6	70.0	105.5	138.0	177.7	205.0	......	......	......	214.1	33.7

VAPOR PRESSURES OF SOLUTIONS

UNITS CONVERSIONS

For this subsection, the following units conversions are applicable:

°F = 9⁄5 °C + 32.

To convert millimeters of mercury to pounds-force per square inch, multiply by 0.01934.

To convert cubic feet to cubic meters, multiply by 0.02832.

To convert bars to pounds-force per square inch, multiply by 14.504.

To convert bars to kilopascals, multiply by 1×10^2.

TABLE 3-10 Partial Pressures of Water over Aqueous Solutions of HCl*

$\log_{10} pmm = A - B/T$, which, however, agrees only approximately with the table. The table is more nearly correct.

Partial pressure of H_2O, mmHg, °C

% HCl	A	B	0°	5°	10°	15°	20°	25°	30°	35°	40°	45°	50°	60°	70°	80°	90°	100°	110°
6	8.99156	2282	4.18	6.04	8.45	11.7	15.9	21.8	29.1	39.4	50.6	66.2	86.0	139	220	333	492	715	
10	8.99864	2295	3.84	5.52	7.70	10.7	14.6	20.0	26.8	35.5	47.0	61.5	80.0	130	204	310	463	677	960
14	8.97075	2300	3.39	4.91	6.95	9.65	13.1	18.0	24.1	31.9	42.1	55.3	72.0	116	185	273	425	625	892
18	8.98014	2323	2.87	4.21	5.92	8.26	11.3	15.4	20.6	27.5	36.4	47.9	62.5	102	162	248	374	550	783
20	8.97877	2334	2.62	3.83	5.40	7.50	10.3	14.1	19.0	25.1	33.3	43.6	57.0	93.5	150	230	345	510	729
22	9.02708	2363	2.33	3.40	4.82	6.75	9.30	12.6	17.1	22.8	30.2	39.8	52.0	85.6	138	211	317	467	670
24	8.96022	2356	2.05	3.04	4.31	6.03	8.30	11.4	15.4	20.4	27.1	35.7	46.7	77.0	124	194	290	426	611
26	9.01511	2390	1.76	2.60	3.71	5.21	7.21	9.95	13.5	18.0	24.0	31.7	41.5	69.0	112	173	261	387	555
28	8.97611	2395	1.50	2.24	3.21	4.54	6.32	8.75	11.8	15.8	21.1	27.9	36.5	60.7	99.0	154	234	349	499
30	9.00117	2422	1.26	1.90	2.73	3.88	5.41	7.52	10.2	13.7	18.4	24.3	32.0	53.5	87.5	136	207	310	444
32	9.03317	2453	1.04	1.57	2.27	3.25	4.55	6.37	8.70	11.7	15.7	21.0	27.7	46.5	76.5	120	184	275	396
34	9.07143	2487	0.85	1.29	1.87	2.70	3.81	5.35	7.32	9.95	13.5	18.1	24.0	40.5	66.5	104	161	243	355
36	9.11815	2526	0.68	1.03	1.50	2.19	3.10	4.41	6.08	8.33	11.4	15.4	20.4	34.8	57.0	90.0	140	212	311
38	9.20783	2579	0.53	0.81	1.20	1.75	2.51	3.60	5.03	6.92	9.52	13.0	17.4	29.6	49.1	77.5	120	182	266
40	9.33923	2647	0.41	0.63	0.94	1.37	2.00	2.88	4.09	5.68	7.85	10.7	14.5	25.0	42.1	67.3	105	158	230
42	9.44953	2709	0.31	0.48	0.72	1.06	1.56	2.30	3.28	4.60	6.45	8.90	12.1	21.2	35.8	57.2	89.2	135	195

*Accuracy, ca. 2 percent for solutions of 15 to 30 percent HCl between 0 and 100°; for solutions of > 30 percent HCl the accuracy is ca. 5 percent at the lower temperatures and ca. 15 percent at the higher temperatures. Below 15 percent HCl, the accuracy is ca. 5 percent at the lower temperatures and higher strengths to ca. 15 to 20 percent at the lower strengths and perhaps 15 to 20 percent at the higher temperatures and lower strengths.

TABLE 3-11 Partial Pressures of HCl over Aqueous Solutions of HCl*

$\log_{10} pmm = A - B/T$, which, however, agrees only approximately with the table. The table is more nearly correct.

mmHg, °C

%HCl	A	B	0°	5°	10°	15°	20°	25°	30°	35°	40°	45°	50°	60°	70°	80°	90°	100°	110°
2	11.8037	4736	....	....	0.0000117	0.000023	0.000044	0.000084	0.000151	0.000275	0.00047	0.00083	0.00140	0.00380	0.0100	0.0245	0.058	0.132	0.280
4	11.6400	4471	0.000018	0.000036	.000069	.000131	.00024	.00044	.00077	.00134	.0023	.00385	.0064	.0165	.0405	.095	.21	.46	.93
6	11.2144	4202	.000066	.000125	.000234	.000425	.00076	.00131	.00225	.0038	.0062	.0102	.0163	.040	.094	.206	.44	.92	1.78
8	11.0406	4042	.000118	.000323	.000583	.00104	.00178	.0031	.00515	.0085	.0136	.022	.0344	.081	.183	.39	.82	1.64	3.10
10	10.9311	3908	.00042	.00075	.00134	.00232	.00395	.0067	.0111	.0178	.0282	.045	.069	.157	.35	.73	1.48	2.9	5.4
12	10.7900	3765	.00099	.00175	.00305	.0052	.0088	.0145	.0234	.037	.058	.091	.136	.305	.66	1.34	2.65	5.1	9.3
14	10.6954	3636	.0024	.00415	.0071	.0118	.0196	.0316	.050	.078	.121	.185	.275	.60	1.25	2.50	4.8	9.0	16.0
16	10.6261	3516	.0056	.0095	.016	.0265	.0428	.0685	.106	.163	.247	.375	.55	1.17	2.40	4.66	8.8	16.1	28
18	10.4957	3376	.0135	.0225	.037	.060	.095	.148	.228	.345	.515	.77	1.11	2.3	4.55	8.6	15.7	28	48
20	10.3833	3245	.0316	.052	.084	.132	.205	.32	.48	.72	1.06	1.55	2.21	4.4	8.5	15.6	28.1	49	83
22	10.3172	3125	.0734	.119	.187	.294	.45	.68	1.02	1.50	2.18	3.14	4.42	8.6	16.3	29.3	52	90	146
24	10.2185	2995	.175	.277	.43	.66	1.00	1.49	2.17	3.14	4.5	6.4	8.9	16.9	31.0	54.5	94	157	253
26	10.1303	2870	.41	.64	.98	1.47	2.17	3.20	4.56	6.50	9.2	12.7	17.5	32.5	58.5	100	169	276	436
28	10.0115	2732	1.0	1.52	2.27	3.36	4.90	7.05	9.90	13.8	19.1	26.4	35.7	64	112	188	309	493	760
30	9.8763	2593	2.4	3.57	5.23	7.60	10.6	15.1	21.0	28.6	39.4	53	71	124	208	340	542	845	
32	9.7523	2457	5.7	8.3	11.8	16.8	23.5	32.5	44.5	60.0	81	107	141	238	390	623	970		
34	9.6061	2316	13.1	18.8	26.4	36.8	50.5	68.5	92	122	161	211	273	450	720				
36	9.5262	2229	29.0	41.0	56.4	78	105.5	142	188	246	322	416	535	860					
38	9.4670	2094	63.0	87.0	117	158	210	277	360	465	598	758	955						
40	9.2156	1939	130	176	233	307	399	515	627	830									
42	8.9925	1800	253	332	430	560	709	900											
44	8.8621	1681	510	655	840														
46	....	....	940																

*Accuracy, ca. 2 percent for solutions of 15 to 30 percent HCl between 0 and 100°; for solutions of > 30 percent HCl the accuracy is ca. 5 percent at the lower temperatures and ca. 15 percent at the higher temperatures. Below 15 percent HCl, the accuracy is ca. 5 percent at the lower temperatures and higher strengths to ca. 15 to 20 percent at the lower strengths and perhaps 15 to 20 percent at the higher temperatures and lower strengths.

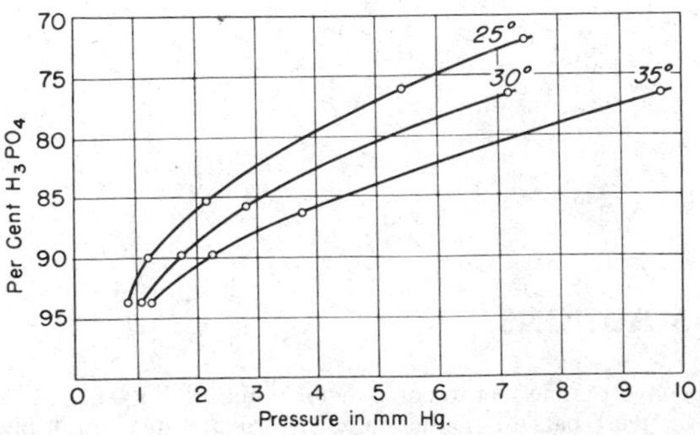

FIG. 3-1 Vapor pressures of H_3PO_4 aqueous; partial pressure of H_2O vapor. *(Courtesy of Victor Chemical Works, Stauffer Chemical Company; measurements by W. H. Woodstock.)*

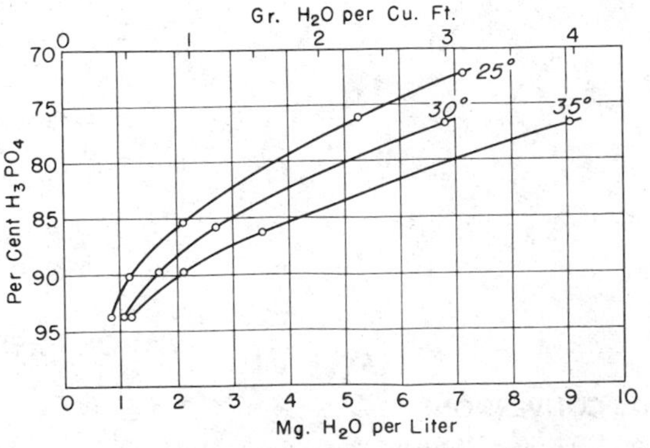

FIG. 3-2 Vapor pressures of H_3PO_4 aqueous; weight of H_2O in saturated air. *(Courtesy of Victor Chemical Works, Stauffer Chemical Company; measurements by W. H. Woodstock.)*

TABLE 3-12 Partial Pressures of H_2O and SO_2 over Aqueous Solutions of Sulfur Dioxide*

Partial pressures of H_2O and SO_2, mmHg, °C

Grams SO_2 per 100 g. water	10°C.		20°C.		30°C.		40°C.		50°C.		60°C.		70°C.		80°C.		90°C.		100°C.		110°C.		120°C.		130°C.	
	H_2O	SO_2	H_2O	SO_2	H_2O	SO_2	H_2O	SO_2	H_2O	SO_2	H_2O	SO_2	H_2O	SO_2	H_2O	SO_2	H_2O	SO_2	H_2O	SO_2	H_2O	SO_2	H_2O	SO_2	H_2O	SO_2
0.0	9.2	...	17.5	...	31.8	...	55.3	...	92.5	...	149.5	...	234	...	355	...	526	...	760	...	1074	....	1488	...	2026	
0.5	9.2	21	17.5	29	31.7	42	55.2	60	92.3	83	149.2	111	234	144	354	182	525	225	758	274	1072	326	1486	377	2024	420
1.0	9.2	42	17.4	59	31.7	85	55.1	120	92.2	164	149.0	217	233	281	354	356	524	445	757	548	1071	661	1484	775	2022	879
1.5	9.2	64	17.4	90	31.6	129	55.0	181	92.0	247	148.8	328	233	426	353	543	523	684	756	850	1070	1032				
2.0	9.1	86	17.4	123	31.6	176	55.0	245	91.9	333	148.6	444	233	581	353	746	523	940								
2.5	9.1	108	17.4	157	31.5	224	54.9	311	91.8	421	148.3	562	232	739	352	956										
3.0	9.1	130	17.3	191	31.5	273	54.7	378	91.6	511	148.1	682	232	897												
3.5	9.1	153	17.3	227	31.5	324	54.7	447	91.5	603	147.9	804														
4.0	9.1	176	17.3	264	31.4	376	54.6	518	91.4	698																
4.5	9.1	199	17.3	300	31.4	428	54.5	588	91.2	793																
5.0	9.1	223	17.2	338	31.3	482	54.4	661																		
5.5	9.0	247	17.2	375	31.3	536	54.4	733																		
6.0	9.0	271	17.2	411	31.2	588	54.3	804																		
6.5	9.0	295	17.2	448	31.2	642																				
7.0	9.0	320	17.1	486	31.1	698																				
7.5	9.0	345	17.1	524	31.1	752																				
8.0	9.0	370	17.1	562	31.0	806																				
8.5	9.0	395	17.0	600																						
9.0	9.0	421	17.0	638																						
9.5	8.9	447	17.0	676																						
10.0	8.9	473	17.0	714																						
10.5	8.9	499	17.0	751																						
11.0	8.9	526	16.9	789																						
11.5	8.9	553																								
12.0	8.9	580																								
12.5	8.9	608																								
13.0	8.8	635																								
13.5	8.8	662																								
14.0	8.8	689																								
14.5	8.8	716																								
15.0	8.8	743																								
15.5	8.8	771																								
16.0	8.8	799																								

* From "International Critical Tables," vol. 3, p. 302, McGraw-Hill.

TABLE 3-13 Water Partial Pressure, bar, over Aqueous Sulfuric Acid Solutions*

°C	Weight percent, H_2SO_4									
	10.0	20.0	30.0	40.0	50.0	60.0	70.0	75.0	80.0	85.0
0	.582E−02	.534E−02	.448E−02	.326E−02	.193E−02	.836E−03	.207E−03	.747E−04	.197E−04	.343E−05
10	.117E−01	.107E−01	.909E−02	.670E−02	.405E−02	.180E−02	.467E−03	.175E−03	.490E−04	.952E−05
20	.223E−01	.205E−01	.174E−01	.130E−01	.802E−02	.367E−02	.995E−03	.388E−03	.115E−04	.245E−04
30	.404E−01	.373E−01	.319E−01	.241E−01	.151E−01	.710E−02	.201E−02	.811E−03	.253E−03	.589E−04
40	.703E−01	.649E−01	.558E−01	.427E−01	.272E−01	.131E−01	.387E−02	.162E−02	.531E−03	.133E−03
50	.117	.109	.939E−01	.725E−01	.470E−01	.232E−01	.715E−02	.809E−02	.106E−02	.286E−03
60	.189	.175	.152	.119	.782E−01	.395E−01	.127E−01	.565E−02	.204E−02	.584E−03
70	.296	.275	.239	.188	.126	.651E−01	.217E−01	.997E−02	.376E−02	.114E−02
80	.449	.417	.365	.290	.196	.104	.360E−01	.170E−01	.668E−02	.213E−02
90	.664	.617	.542	.434	.298	.161	.578E−01	.281E−01	.115E−01	.383E−02
100	.957	.891	.786	.634	.441	.244	.905E−01	.452E−01	.192E−01	.666E−02
110	1.349	1.258	1.113	.904	.638	.360	.138	.708E−01	.312E−01	.112E−01
120	1.863	1.740	1.544	1.264	.903	.519	.206	.108	.493E−01	.183E−01
130	2.524	2.361	2.101	1.732	1.253	.734	.301	.162	.760E−01	.291E−01
140	3.361	3.149	2.810	2.333	1.708	1.020	.431	.236	.115	.451E−01
150	4.404	4.132	3.697	3.090	2.289	1.392	.605	.339	.170	.682E−01
160	5.685	5.342	4.793	4.031	3.021	1.870	.837	.478	.246	.101
170	7.236	6.810	6.127	5.185	3.930	2.475	1.138	.662	.350	.147
180	9.093	8.571	7.731	6.584	5.045	3.233	1.525	.902	.489	.208
190	11.289	10.658	9.640	8.259	6.397	4.169	2.017	1.212	.673	.291
200	13.861	13.107	11.887	10.245	8.020	5.312	2.632	1.606	.913	.401
210	16.841	15.951	14.505	12.576	9.948	6.696	3.395	2.101	1.220	.542
220	20.264	19.225	17.529	15.287	12.217	8.354	4.331	2.714	1.609	.724
230	24.160	22.960	20.992	18.414	14.864	10.322	5.466	3.467	2.096	.952
240	28.561	27.188	24.927	21.992	17.929	12.641	6.831	4.381	2.699	1.237
250	33.494	31.939	29.364	26.056	21.452	15.351	8.458	5.480	3.435	1.587
260	38.984	37.240	34.334	30.642	25.472	18.496	10.382	6.788	4.326	2.012
270	45.055	43.116	39.865	35.784	30.030	22.121	12.640	8.333	5.395	2.525
280	51.726	49.590	45.984	41.514	35.168	26.274	15.269	10.142	6.663	3.136
290	59.015	56.681	52.715	47.865	40.926	31.003	18.311	12.242	8.155	3.857
300	66.934	64.407	60.081	54.868	47.346	36.360	21.808	14.665	9.897	4.701
310	75.495	72.781	68.100	62.553	54.470	42.395	25.804	17.438	11.912	5.680
320	84.705	81.816	76.792	70.947	62.337	49.164	30.343	20.591	14.227	6.806
330	94.567	91.518	86.172	80.077	70.988	56.721	35.473	24.153	16.867	8.093
340	105.083	101.894	96.252	89.969	80.463	65.123	41.240	28.154	19.855	9.551
350	116.251	112.946	107.043	100.646	90.802	74.426	47.692	32.622	23.217	11.193

*Vermeulen, Dong, Robinson, Nguyen, and Gmitro, AIChE meeting, Anaheim, Calif., 1982; and private communication from Prof. Theordore Vermeulen, Chemical Engineering Dept., University of Calfornia, Berkeley.

TABLE 3-13 Water Partial Pressure, bar, over Aqueous Sulfuric Acid Solutions *(Concluded)*

°C	Weight percent, H_2SO_4									
	90.0	92.0	94.0	96.0	97.0	98.0	98.5	99.0	99.5	100.0
0	.518E−06	.242E−06	.107E−06	.401E−07	.218E−07	.980E−08	.569E−08	.268E−08	.775E−09	.196E−09
10	.159E−05	.762E−06	.344E−06	.130E−06	.713E−07	.323E−07	.188E−07	.888E−08	.258E−08	.655E−09
20	.448E−05	.220E−05	.101E−05	.390E−06	.215E−06	.978E−07	.572E−07	.271E−07	.789E−08	.201E−08
30	.117E−04	.587E−05	.275E−05	.108E−05	.598E−06	.275E−06	.161E−06	.766E−07	.224E−07	.575E−08
40	.285E−04	.146E−04	.696E−05	.278E−05	.155E−05	.720E−06	.424E−06	.202E−06	.595E−07	.153E−07
50	.652E−04	.341E−04	.166E−04	.672E−05	.379E−05	.177E−05	.105E−05	.503E−06	.149E−06	.384E−07
60	.141E−03	.754E−04	.372E−04	.154E−04	.875E−05	.413E−05	.245E−05	.118E−05	.350E−06	.910E−07
70	.290E−03	.158E−03	.795E−04	.334E−04	.192E−04	.912E−05	.544E−05	.263E−05	.784E−06	.205E−06
80	.569E−03	.316E−03	.162E−03	.691E−04	.400E−04	.192E−04	.115E−04	.559E−05	.168E−05	.439E−06
90	.107E−02	.606E−03	.315E−03	.137E−03	.801E−04	.388E−04	.234E−04	.114E−04	.343E−05	.903E−06
100	.194E−02	.112E−02	.590E−03	.261E−03	.154E−03	.752E−04	.455E−04	.223E−04	.674E−05	.178E−05
110	.338E−02	.198E−02	.107E−02	.479E−03	.285E−03	.141E−03	.855E−04	.420E−04	.128E−04	.339E−05
120	.571E−02	.341E−02	.186E−02	.851E−03	.511E−03	.254E−03	.155E−03	.766E−04	.233E−04	.623E−05
130	.938E−02	.569E−02	.315E−02	.146E−02	.886E−03	.445E−03	.273E−03	.135E−03	.414E−04	.111E−04
140	.150E−01	.923E−02	.519E−02	.245E−02	.149E−02	.757E−03	.467E−03	.232E−03	.711E−04	.191E−04
150	.233E−01	.146E−01	.832E−02	.399E−02	.245E−02	.125E−02	.776E−03	.387E−03	.119E−03	.321E−04
160	.354E−01	.225E−01	.130E−01	.633E−02	.393E−02	.202E−02	.126E−02	.629E−03	.194E−03	.526E−04
170	.526E−01	.340E−01	.199E−01	.983E−02	.614E−02	.319E−02	.199E−02	.999E−03	.309E−03	.840E−04
180	.766E−01	.502E−01	.298E−01	.149E−01	.941E−02	.492E−02	.309E−02	.155E−02	.482E−03	.131E−03
190	.110	.729E−01	.438E−01	.222E−01	.141E−01	.744E−02	.469E−02	.236E−02	.735E−03	.201E−03
200	.154	.104	.631E−01	.325E−01	.208E−01	.110E−01	.698E−02	.352E−02	.110E−02	.300E−03
210	.213	.146	.894E−01	.467E−01	.300E−01	.161E−01	.102E−01	.516E−02	.161E−02	.442E−03
220	.290	.201	.125	.660E−01	.427E−01	.230E−01	.147E−01	.743E−02	.232E−02	.638E−03
230	.389	.273	.171	.918E−01	.598E−01	.325E−01	.208E−01	.105E−01	.329E−02	.906E−03
240	.514	.366	.232	.126	.825E−01	.451E−01	.290E−01	.147E−01	.460E−02	.127E−02
250	.673	.485	.310	.170	.112	.618E−01	.398E−01	.202E−01	.633E−02	.174E−02
260	.870	.635	.409	.227	.151	.835E−01	.540E−01	.274E−01	.858E−02	.237E−02
270	1.112	.822	.534	.300	.200	.111	.723E−01	.366E−01	.115E−01	.317E−02
280	1.407	1.052	.689	.391	.263	.147	.957E−01	.485E−01	.152E−01	.420E−02
290	1.763	1.335	.880	.505	.341	.192	.125	.634E−01	.199E−01	.548E−02
300	2.190	1.676	1.112	.646	.437	.248	.162	.820E−01	.257E−01	.708E−02
310	2.696	2.088	1.394	.817	.556	.316	.208	.105	.328E−01	.905E−02
320	3.292	2.578	1.732	1.025	.701	.400	.264	.133	.415E−01	.114E−01
330	3.990	3.159	2.133	1.274	.875	.502	.331	.167	.520E−01	.143E−01
340	4.801	3.843	2.608	1.571	1.083	.624	.413	.208	.646E−01	.178E−01
350	5.738	4.641	3.164	1.922	1.331	.770	.511	.256	.795E−01	.218E−01

TABLE 3-14*a* Sulfur Trioxide Partial Pressure, bar, over Aqueous Sulfuric Acid Solutions*

°C	Weight percent, H_2SO_4									
	10.0	20.0	30.0	40.0	50.0	60.0	70.0	75.0	80.0	85.0
0	.644E−29	.103E−27	.205E−26	.688E−25	.368E−23	.341E−21	.784E−19	.174E−17	.531E−16	.229E−14
10	.149E−27	.223E−26	.395E−25	.113E−23	.522E−22	.415E−20	.796E−18	.158E−16	.417E−15	.141E−13
20	.278E−26	.394E−25	.626E−24	.156E−22	.621E−21	.426E−19	.685E−17	.121E−15	.280E−14	.767E−13
30	.426E−25	.577E−24	.832E−23	.181E−21	.630E−20	.376E−18	.509E−16	.808E−15	.164E−13	.371E−12
40	.549E−24	.714E−23	.941E−22	.181E−20	.555E−19	.288E−17	.331E−15	.473E−14	.851E−13	.162E−11
50	.602E−23	.757E−22	.921E−21	.158E−19	.429E−18	.195E−16	.191E−14	.246E−13	.395E−12	.643E−11
60	.573E−22	.699E−21	.789E−20	.122E−18	.294E−17	.118E−15	.985E−14	.116E−12	.165E−11	.234E−10
70	.477E−21	.567E−20	.599E−19	.843E−18	.181E−16	.643E−15	.461E−13	.492E−12	.634E−11	.791E−10
80	.352E−20	.410E−19	.408E−18	.524E−17	.101E−15	.319E−14	.197E−12	.192E−11	.223E−10	.249E−09
90	.233E−19	.266E−18	.250E−17	.296E−16	.516E−15	.145E−13	.775E−12	.693E−11	.731E−10	.734E−09
100	.139E−18	.157E−17	.140E−16	.153E−15	.242E−14	.606E−13	.283E−11	.232E−10	.223E−09	.204E−08
110	.756E−18	.844E−17	.719E−16	.730E−15	.105E−13	.236E−12	.961E−11	.729E−10	.641E−09	.538E−08
120	.377E−17	.418E−16	.340E−15	.323E−14	.424E−13	.858E−12	.307E−10	.215E−09	.174E−08	.135E−07
130	.174E−16	.191E−15	.150E−14	.133E−13	.160E−12	.293E−11	.922E−10	.601E−09	.446E−08	.324E−07
140	.743E−16	.815E−15	.615E−14	.517E−13	.569E−12	.943E−11	.262E−09	.159E−08	.109E−07	.745E−07
150	.297E−15	.325E−14	.237E−13	.188E−12	.191E−11	.287E−10	.710E−09	.403E−08	.256E−07	.165E−06
160	.111E−14	.122E−13	.862E−13	.649E−12	.608E−11	.833E−10	.183E−08	.974E−08	.575E−07	.351E−06
170	.393E−14	.430E−13	.296E−12	.212E−11	.184E−10	.231E−09	.453E−08	.226E−07	.125E−06	.725E−06
180	.131E−13	.144E−12	.967E−12	.622E−11	.532E−10	.610E−09	.107E−07	.505E−07	.260E−06	.145E−05
190	.415E−13	.458E−12	.301E−11	.197E−10	.147E−09	.155E−08	.246E−07	.109E−06	.527E−06	.282E−05
200	.125E−12	.139E−11	.893E−11	.561E−10	.391E−09	.379E−08	.542E−07	.228E−06	.103E−05	.534E−05
210	.362E−12	.404E−11	.254E−10	.154E−09	.100E−08	.894E−08	.116E−06	.462E−06	.198E−05	.986E−05
220	.100E−11	.112E−10	.695E−10	.405E−09	.246E−08	.204E−07	.240E−06	.911E−06	.368E−05	.178E−04
230	.265E−11	.301E−10	.183E−09	.103E−08	.587E−08	.450E−07	.482E−06	.175E−05	.668E−05	.314E−04
240	.678E−11	.777E−10	.465E−09	.253E−08	.135E−07	.965E−07	.944E−06	.328E−05	.119E−04	.543E−04
250	.167E−10	.193E−09	.114E−08	.602E−08	.303E−07	.201E−06	.180E−05	.600E−05	.206E−04	.923E−04
260	.399E−10	.466E−09	.272E−08	.139E−07	.660E−07	.408E−06	.336E−05	.108E−04	.352E−04	.154E−03
270	.920E−10	.109E−08	.628E−08	.312E−07	.140E−06	.807E−06	.612E−05	.189E−04	.590E−04	.253E−03
280	.206E−09	.247E−08	.141E−07	.683E−07	.288E−06	.156E−05	.109E−04	.326E−04	.973E−04	.408E−03
290	.449E−09	.545E−08	.308E−07	.145E−06	.580E−06	.295E−05	.191E−04	.553E−04	.158E−03	.649E−03

*Vermeulen, Dong, Robinson, Nguyen, and Gmitro, AIChE meeting, Anaheim, Calif., 1982; and private communication from Prof. Theodore Vermeulen, Chemical Engineering Dept., University of California, Berkeley.

TABLE 3-14*a* Sulfur Trioxide Partial Pressure, bar, over Aqueous Sulfuric Acid Solutions *(Concluded)*

	Weight percent, H_2SO_4									
°C	10.0	20.0	30.0	40.0	50.0	60.0	70.0	75.0	80.0	85.0
300	.953E−09	.117E−07	.657E−07	.302E−06	.114E−05	.546E−05	.329E−04	.921E−04	.253E−03	.102E−02
310	.197E−08	.245E−07	.136E−06	.614E−06	.220E−05	.990E−05	.556E−04	.151E−03	.398E−03	.158E−02
320	.397E−08	.502E−07	.277E−06	.122E−05	.414E−05	.176E−04	.923E−04	.245E−03	.621E−03	.242E−02
330	.782E−08	.100E−06	.551E−06	.237E−05	.766E−05	.308E−04	.151E−03	.391E−03	.956E−03	.367E−02
340	.151E−07	.196E−06	.107E−05	.452E−05	.139E−04	.529E−04	.243E−03	.617E−03	.145E−02	.550E−02
350	.285E−07	.376E−06	.204E−05	.846E−05	.246E−04	.893E−04	.387E−03	.963E−03	.219E−02	.815E−02

	Weight percent, H_2SO_4									
°C	90.0	92.0	94.0	96.0	97.0	98.0	98.5	99.0	99.5	100.0
0	.671E−13	.216E−12	.677E−12	.240E−11	.500E−11	.124E−10	.224E−10	.502E−10	.182E−09	.755E−09
10	.345E−12	.107E−11	.326E−11	.114E−10	.234E−10	.578E−10	.104E−09	.232E−09	.839E−09	.347E−08
20	.159E−11	.475E−11	.141E−10	.482E−10	.986E−10	.241E−09	.433E−09	.961E−09	.346E−08	.142E−07
30	.664E−11	.192E−10	.557E−10	.186E−09	.376E−09	.911E−09	.163E−08	.360E−08	.129E−07	.528E−07
40	.254E−10	.709E−10	.201E−09	.655E−09	.131E−08	.315E−08	.562E−08	.123E−07	.440E−07	.179E−06
50	.897E−10	.242E−09	.669E−09	.214E−08	.424E−08	.101E−07	.179E−07	.391E−07	.139E−06	.560E−06
60	.294E−09	.771E−09	.207E−08	.647E−08	.127E−07	.299E−07	.528E−07	.115E−06	.405E−06	.163E−05
70	.904E−09	.230E−08	.602E−08	.184E−07	.357E−07	.833E−07	.146E−06	.316E−06	.111E−05	.444E−05
80	.261E−08	.643E−08	.165E−07	.492E−07	.946E−07	.218E−06	.381E−06	.820E−06	.286E−05	.114E−04
90	.712E−08	.171E−07	.426E−07	.124E−06	.237E−06	.541E−06	.940E−06	.201E−05	.698E−05	.276E−04
100	.184E−07	.430E−07	.105E−06	.300E−06	.565E−06	.127E−05	.220E−05	.470E−05	.162E−04	.638E−04
110	.456E−07	.103E−06	.247E−06	.689E−06	.128E−05	.287E−05	.494E−05	.105E−04	.359E−04	.141E−03
120	.108E−06	.238E−06	.555E−06	.152E−05	.280E−05	.619E−05	.106E−04	.224E−04	.764E−04	.298E−03
130	.244E−06	.526E−06	.120E−05	.321E−05	.586E−05	.128E−04	.219E−04	.459E−04	.156E−03	.606E−03
140	.533E−06	.112E−05	.250E−05	.656E−05	.118E−04	.257E−04	.435E−04	.910E−04	.308E−03	.119E−02
150	.112E−05	.230E−05	.504E−05	.129E−04	.231E−04	.497E−04	.837E−04	.174E−03	.588E−03	.226E−02
160	.229E−05	.459E−05	.983E−05	.247E−04	.438E−04	.932E−04	.156E−03	.324E−03	.109E−02	.416E−02
170	.453E−05	.886E−05	.186E−04	.459E−04	.806E−04	.170E−03	.283E−03	.586E−03	.196E−02	.746E−02
180	.870E−05	.166E−04	.343E−04	.829E−04	.144E−03	.301E−03	.499E−03	.103E−02	.343E−02	.130E−01
190	.163E−04	.304E−04	.615E−04	.146E−03	.252E−03	.520E−03	.859E−03	.177E−02	.587E−02	.222E−01
200	.297E−04	.543E−04	.108E−03	.251E−03	.429E−03	.878E−03	.144E−02	.296E−02	.981E−02	.370E−01
210	.528E−04	.946E−04	.185E−03	.422E−03	.714E−03	.145E−02	.237E−02	.486E−02	.161E−01	.603E−01
220	.919E−04	.161E−03	.309E−03	.694E−03	.117E−02	.235E−02	.383E−02	.781E−02	.258E−01	.965E−01
230	.157E−03	.269E−03	.508E−03	.112E−02	.187E−02	.373E−02	.605E−02	.123E−01	.405E−01	.152
240	.261E−03	.441E−03	.819E−03	.178E−02	.293E−02	.582E−02	.939E−02	.191E−01	.627E−01	.234
250	.428E−03	.708E−03	.130E−02	.276E−02	.453E−02	.891E−02	.143E−01	.291E−01	.955E−01	.356
260	.690E−03	.112E−02	.202E−02	.423E−02	.688E−02	.134E−01	.215E−01	.437E−01	.143	.532
270	.109E−02	.174E−02	.309E−02	.638E−02	.103E−01	.200E−01	.319E−01	.646E−01	.212	.786
280	.170E−02	.266E−02	.466E−02	.948E−02	.152E−01	.293E−01	.465E−01	.943E−01	.309	1.144
290	.261E−02	.401E−02	.694E−02	.139E−01	.221E−01	.423E−01	.670E−01	.136	.444	1.646
300	.395E−02	.595E−02	.102E−01	.201E−01	.318E−01	.604E−01	.953E−01	.193	.632	2.339
310	.589E−02	.873E−02	.148E−01	.287E−01	.451E−01	.852E−01	.134	.272	.889	3.289
320	.868E−02	.126E−01	.211E−01	.405E−01	.632E−01	.119	.186	.378	1.236	4.575
330	.126E−01	.181E−01	.299E−01	.565E−01	.877E−01	.164	.256	.520	1.703	6.303
340	.181E−01	.255E−01	.418E−01	.780E−01	.120	.224	.348	.708	2.323	8.603
350	.258E−01	.357E−01	.578E−01	.107	.164	.303	.470	.956	3.142	11.640

TABLE 3-14*b* Sulfuric Acid Partial Pressure, bar, over Aqueous Sulfuric Acid*

	Weight Percent, H_2SO_4									
°C	10.0	20.0	30.0	40.0	50.0	60.0	70.0	75.0	80.0	85.0
0	.576E−21	.843E−20	.141E−18	.344E−17	.109E−15	.438E−14	.249E−12	.200E−11	.161E−10	.121E−09
10	.634E−20	.874E−19	.131E−17	.276E−16	.769E−15	.273E−13	.135E−11	.101E−10	.743E−10	.490E−09
20	.588E−19	.769E−18	.104E−16	.193E−15	.474E−14	.149E−12	.649E−11	.447E−10	.305E−09	.179E−08
30	.468E−18	.584E−17	.721E−16	.119E−14	.259E−13	.725E−12	.278E−10	.178E−09	.113E−08	.594E−08
40	.324E−17	.389E−16	.441E−15	.649E−14	.127E−12	.317E−11	.108E−09	.643E−09	.379E−08	.181E−07
50	.197E−16	.229E−15	.241E−14	.320E−13	.562E−12	.126E−10	.380E−09	.212E−08	.117E−07	.513E−07
60	.107E−15	.121E−14	.119E−13	.144E−12	.228E−11	.462E−10	.124E−08	.646E−08	.334E−07	.135E−06
70	.526E−15	.581E−14	.535E−13	.592E−12	.851E−11	.156E−09	.373E−08	.183E−07	.888E−07	.336E−06
80	.235E−14	.254E−13	.221E−12	.225E−11	.295E−10	.492E−09	.105E−07	.485E−07	.222E−06	.786E−06
90	.960E−14	.102E−12	.844E−12	.798E−11	.956E−10	.145E−08	.279E−07	.121E−06	.522E−06	.175E−05
100	.353E−13	.381E−12	.300E−11	.264E−10	.291E−09	.402E−08	.698E−07	.287E−06	.117E−05	.371E−05
110	.127E−12	.132E−11	.997E−11	.824E−10	.835E−09	.106E−07	.166E−06	.644E−06	.249E−05	.752E−05
120	.418E−12	.432E−11	.312E−10	.243E−09	.227E−08	.264E−07	.375E−06	.138E−05	.508E−05	.147E−04
130	.129E−11	.132E−10	.924E−10	.678E−09	.589E−08	.631E−07	.814E−06	.285E−05	.995E−05	.277E−04
140	.375E−11	.385E−10	.259E−09	.181E−08	.146E−07	.144E−06	.169E−05	.565E−05	.188E−04	.503E−04
150	.103E−10	.106E−09	.694E−09	.460E−08	.346E−07	.316E−06	.340E−05	.108E−04	.343E−04	.889E−04
160	.272E−10	.279E−09	.178E−08	.112E−07	.789E−07	.670E−06	.659E−05	.200E−04	.608E−04	.152E−03
170	.682E−10	.702E−09	.436E−08	.264E−07	.174E−06	.137E−05	.124E−04	.359E−04	.104E−03	.255E−03
180	.164E−09	.170E−08	.103E−07	.599E−07	.369E−06	.271E−05	.225E−04	.627E−04	.175E−03	.416E−03
190	.378E−09	.394E−08	.234E−07	.131E−06	.760E−06	.521E−05	.400E−04	.107E−03	.286E−03	.663E−03
200	.842E−09	.883E−08	.514E−07	.278E−06	.152E−05	.975E−05	.691E−04	.177E−03	.457E−03	.104E−02
210	.181E−08	.191E−07	.109E−06	.573E−06	.295E−05	.178E−04	.117E−03	.288E−03	.715E−03	.159E−02
220	.376E−08	.401E−07	.226E−06	.115E−05	.559E−05	.316E−04	.193E−03	.459E−03	.110E−02	.239E−02
230	.758E−08	.817E−07	.455E−06	.224E−05	.103E−04	.549E−04	.311E−03	.717E−03	.166E−02	.354E−02
240	.148E−07	.162E−06	.889E−06	.427E−05	.186E−04	.935E−04	.494E−03	.110E−02	.245E−02	.515E−02
250	.283E−07	.312E−06	.170E−05	.793E−05	.329E−04	.156E−03	.770E−03	.166E−02	.358E−02	.740E−02
260	.526E−07	.588E−06	.316E−05	.144E−04	.569E−04	.255E−03	.118E−02	.247E−02	.516E−02	.105E−01
270	.954E−07	.108E−05	.577E−05	.257E−04	.965E−04	.411E−03	.178E−02	.362E−02	.733E−02	.147E−01
280	.169E−06	.194E−05	.103E−04	.450E−04	.161E−03	.650E−03	.265E−02	.524E−02	.103E−01	.203E−01
290	.294E−06	.342E−05	.180E−04	.771E−04	.263E−03	.101E−02	.389E−02	.750E−02	.143E−01	.278E−01
300	.500E−06	.591E−05	.309E−04	.130E−03	.424E−03	.156E−02	.563E−02	.106E−01	.196E−01	.376E−01
310	.834E−06	.100E−04	.522E−04	.215E−03	.672E−03	.236E−02	.805E−02	.148E−01	.266E−01	.504E−01
320	.137E−05	.167E−04	.865E−04	.352E−03	.105E−02	.352E−02	.114E−01	.205E−01	.359E−01	.670E−01
330	.220E−05	.273E−04	.141E−03	.565E−03	.162E−02	.519E−02	.159E−01	.281E−01	.480E−01	.883E−01
340	.349E−05	.440E−04	.227E−03	.895E−03	.246E−02	.757E−02	.221E−01	.382E−01	.636E−01	.116
350	.544E−05	.698E−04	.360E−03	.140E−02	.369E−02	.109E−01	.303E−01	.516E−01	.836E−01	.150

	Weight percent, H_2SO_4									
°C	90.0	92.0	94.0	96.0	97.0	98.0	98.5	99.0	99.5	100.0
0	.534E−09	.803E−09	.112E−08	.148E−08	.167E−08	.187E−08	.196E−08	.206E−08	.217E−08	.228E−08
10	.200E−08	.296E−08	.409E−08	.540E−08	.609E−08	.679E−08	.714E−08	.750E−08	.788E−08	.827E−08
20	.677E−08	.993F−08	.136E−07	.179E−07	.201E−07	.224E−07	.236E−07	.247E−07	.260E−07	.273E−07
30	.211E−07	.306E−07	.415E−07	.543E−07	.611E−07	.680E−07	.714E−07	.749E−07	.786E−07	.824E−07
40	.607E−07	.870E−07	.117E−06	.153E−06	.171E−06	.191E−06	.200E−06	.210E−06	.220E−06	.230E−06
50	.163E−06	.231E−06	.309E−06	.400E−06	.449E−06	.498E−06	.523E−06	.548E−06	.574E−06	.600E−06
60	.411E−06	.575E−06	.765E−06	.985E−06	.110E−05	.122E−05	.128E−05	.134E−05	.140E−05	.147E−05
70	.976E−06	.135E−05	.179E−05	.229E−05	.256E−05	.283E−05	.297E−05	.310E−05	.325E−05	.339E−05
80	.220E−05	.302E−05	.396E−05	.504E−05	.562E−05	.622E−05	.652E−05	.681E−05	.712E−05	.743E−05
90	.473E−05	.642E−05	.835E−05	.106E−04	.118E−04	.130E−04	.136E−04	.143E−04	.149E−04	.155E−04
100	.973E−05	.131E−04	.169E−04	.213E−04	.237E−04	.261E−04	.274E−04	.285E−04	.298E−04	.310E−04
110	.192E−04	.256E−04	.328E−04	.412E−04	.457E−04	.503E−04	.527E−04	.549E−04	.572E−04	.595E−04
120	.366E−04	.482E−04	.614E−04	.767E−04	.849E−04	.935E−04	.977E−04	.102E−03	.106E−03	.110E−03
130	.672E−04	.879E−04	.111E−03	.138E−03	.153E−03	.168E−03	.175E−03	.182E−03	.190E−03	.197E−03
140	.120E−03	.155E−03	.195E−03	.241E−03	.266E−03	.292E−03	.304E−03	.316E−03	.329E−03	.341E−03
150	.207E−03	.266E−03	.332E−03	.408E−03	.449E−03	.493E−03	.514E−03	.534E−03	.554E−03	.574E−03
160	.348E−03	.444E−03	.550E−03	.673E−03	.740E−03	.810E−03	.844E−03	.876E−03	.909E−03	.941E−03
170	.572E−03	.723E−03	.889E−03	.108E−02	.119E−02	.130E−02	.135E−02	.140E−02	.145E−02	.150E−02
180	.917E−03	.115E−02	.140E−02	.170E−02	.186E−02	.204E−02	.212E−02	.220E−02	.227E−02	.235E−02
190	.144E−02	.179E−02	.217E−02	.262E−02	.286E−02	.312E−02	.325E−02	.336E−02	.348E−02	.359E−02
200	.221E−02	.273E−02	.329E−02	.395E−02	.431E−02	.470E−02	.488E−02	.505E−02	.522E−02	.538E−02
210	.333E−02	.408E−02	.490E−02	.585E−02	.637E−02	.693E−02	.720E−02	.744E−02	.768E−02	.791E−02
220	.494E−02	.601E−02	.715E−02	.850E−02	.924E−02	.100E−01	.104E−01	.108E−01	.111E−01	.114E−01
230	.719E−02	.869E−02	.103E−01	.122E−01	.132E−01	.143E−01	.149E−01	.153E−01	.158E−01	.162E−01
240	.103E−01	.124E−01	.146E−01	.171E−01	.186E−01	.201E−01	.209E−01	.215E−01	.221E−01	.227E−01
250	.146E−01	.174E−01	.203E−01	.238E−01	.257E−01	.278E−01	.289E−01	.297E−01	.305E−01	.314E−01
260	.203E−01	.240E−01	.279E−01	.326E−01	.352E−01	.380E−01	.394E−01	.405E−01	.416E−01	.427E−01
270	.279E−01	.329E−01	.380E−01	.441E−01	.475E−01	.513E−01	.531E−01	.545E−01	.560E−01	.574E−01
280	.380E−01	.444E−01	.510E−01	.589E−01	.633E−01	.683E−01	.706E−01	.725E−01	.744E−01	.762E−01
290	.510E−01	.592E−01	.676E−01	.778E−01	.835E−01	.900E−01	.930E−01	.954E−01	.978E−01	.100
300	.678E−01	.782E−01	.888E−01	.102	.109	.117	.121	.124	.127	.130
310	.892E−01	.102	.115	.132	.141	.151	.156	.160	.164	.167
320	.116	.132	.149	.169	.180	.193	.199	.204	.209	.213
330	.150	.170	.190	.214	.228	.245	.252	.258	.263	.269
340	.192	.216	.240	.270	.287	.307	.317	.323	.330	.336
350	.243	.272	.301	.337	.358	.383	.394	.402	.410	.417

*Vermeulen, Dong, Robinson, Nguyen, and Gmitro, AIChE meeting, Anaheim, Calif., 1982; and private communication from Prof. Theodore Vermeulen, Chemical Engineering Dept., University of California, Berkeley.

TABLE 3-15 Total Pressure, bar, of Aqueous Sulfuric Acid Solutions*

	Weight percent, H_2SO_4									
°C	10.0	20.0	30.0	40.0	50.0	60.0	70.0	75.0	80.0	85.0
0	.582E−02	.534E−02	.448E−02	.326E−02	.193E−02	.836E−03	.207E−03	.747E−04	.197E−04	.343E−05
10	.117E−01	.107E−01	.909E−02	.670E−02	.405E−02	.180E−02	.467E−03	.175E−03	.490E−04	.952E−05
20	.223E−01	.205E−01	.174E−01	.130E−01	.802E−02	.367E−02	.995E−03	.388E−03	.115E−03	.245E−04
30	.404E−01	.373E−01	.319E−01	.241E−01	.151E−01	.710E−02	.201E−02	.811E−03	.253E−03	.589E−04
40	.703E−01	.649E−01	.558E−01	.427E−01	.272E−01	.131E−01	.387E−02	.162E−02	.531E−03	.134E−03
50	.117	.109	.939E−01	.725E−01	.470E−01	.232E−01	.715E−02	.309E−02	.106E−02	.286E−03
60	.189	.175	.152	.119	.782E−01	.395E−01	.127E−01	.565E−02	.204E−02	.584E−03
70	.296	.275	.239	.188	.126	.651E−01	.217E−01	.997E−01	.376E−02	.114E−02
80	.449	.417	.365	.290	.196	.104	.360E−01	.170E−01	.668E−02	.213E−02
90	.664	.617	.542	.434	.298	.161	.578E−01	.281E−01	.115E−01	.383E−02
100	.957	.891	.786	.634	.441	.244	.905E−01	.452E−01	.192E−01	.666E−02
110	1.349	1.258	1.113	.904	.638	.360	.138	.708E−01	.312E−01	.112E−01
120	1.863	1.740	1.544	1.264	.903	.519	.206	.108	.493E−01	.183E−01
130	2.524	2.361	2.101	1.732	1.253	.734	.301	.162	.760E−01	.291E−01
140	3.361	3.149	2.810	2.333	1.708	1.020	.431	.236	.115	.451E−01
150	4.404	4.132	3.697	3.090	2.289	1.392	.605	.339	.170	.683E−01
160	5.685	5.342	4.793	4.031	3.021	1.870	.837	.478	.246	.101
170	7.236	6.810	6.127	5.185	3.930	2.475	1.138	.662	.350	.147
180	9.093	8.571	7.731	6.584	5.045	3.233	1.525	.902	.489	.209
190	11.289	10.658	9.640	8.259	6.397	4.169	2.017	1.212	.673	.292
200	13.861	13.107	11.887	10.245	8.020	5.312	2.633	1.606	.913	.402
210	16.841	15.951	14.505	12.576	9.948	6.696	3.396	2.101	1.221	.544
220	20.264	19.225	17.529	15.287	12.217	8.354	4.331	2.715	1.610	.726
230	24.160	22.960	20.992	18.414	14.864	10.322	5.466	3.468	2.098	.956
240	28.561	27.188	24.927	21.992	17.929	12.641	6.832	4.382	2.701	1.242
250	33.494	31.939	29.364	26.056	21.452	15.351	8.459	5.481	3.439	1.594
260	38.984	37.240	34.334	30.642	25.472	18.496	10.384	6.791	4.332	2.023
270	45.055	43.116	39.865	35.784	30.030	22.122	12.642	8.337	5.402	2.540
280	51.726	49.590	45.984	41.514	35.168	26.275	15.272	10.147	6.673	3.157
290	59.015	56.681	52.715	47.866	40.926	31.004	18.315	12.250	8.170	3.886
300	66.934	64.407	60.081	54.869	47.347	36.361	21.814	14.675	9.916	4.740
310	75.495	72.781	68.101	62.553	54.470	42.398	25.812	17.453	11.939	5.732
320	84.705	81.816	76.792	70.947	62.338	49.168	30.355	20.611	14.264	6.876
330	94.567	91.518	86.172	80.078	70.990	56.727	35.489	24.182	16.916	8.185
340	105.083	101.894	96.252	89.970	80.466	65.130	41.262	28.193	19.920	9.672
350	116.251	112.947	107.043	100.647	90.806	74.437	47.723	32.674	23.303	11.351

	Weight percent, H_2SO_4									
°C	90.0	92.0	94.0	96.0	97.0	98.0	98.5	99.0	99.5	100.0
0	.518E−06	.243E−06	.109E−06	.416E−07	.235E−07	.117E−07	.768E−08	.479E−08	.313E−08	.323E−08
10	.159E−05	.765E−06	.348E−06	.136E−06	.774E−07	.391E−07	.261E−07	.166E−07	.113E−07	.124E−07
20	.449E−05	.221E−05	.102E−05	.407E−06	.235E−06	.121E−06	.812E−07	.528E−07	.373E−07	.435E−07
30	.117E−04	.590E−05	.279E−05	.113E−05	.659E−06	.344E−06	.234E−06	.155E−06	.114E−06	.141E−06
40	.385E−04	.147E−04	.708E−05	.293E−05	.173E−05	.914E−06	.630E−06	.425E−06	.323E−06	.425E−06
50	.653E−04	.344E−04	.169E−04	.712E−05	.425E−05	.228E−05	.159E−05	.109E−05	.861E−06	.120E−05
60	.141E−03	.759E−04	.380E−04	.164E−04	.987E−05	.538E−05	.379E−05	.264E−05	.216E−05	.319E−05
70	.291E−03	.159E−03	.813E−04	.357E−04	.218E−04	.120E−04	.856E−05	.605E−05	.514E−05	.804E−05
80	.571E−03	.319E−03	.166E−03	.742E−04	.458E−04	.257E−04	.184E−04	.132E−04	.117E−04	.193E−04
90	.107E−02	.612E−03	.324E−03	.148E−03	.921E−04	.524E−04	.390E−04	.277E−04	.253E−04	.441E−04
100	.195E−02	.113E−02	.607E−03	.283E−03	.178E−03	.103E−03	.751E−04	.555E−04	.527E−04	.966E−04
110	.340E−02	.201E−02	.110E−02	.521E−03	.332E−03	.194E−03	.143E−03	.107E−03	.106E−03	.204E−03
120	.575E−02	.346E−02	.192E−02	.929E−03	.598E−03	.354E−03	.263E−03	.201E−03	.206E−03	.414E−03
130	.944E−02	.578E−02	.327E−02	.161E−02	.104E−02	.626E−03	.470E−03	.363E−03	.387E−03	.814E−03
140	.151E−01	.939E−02	.539E−02	.270E−02	.177E−02	.107E−02	.815E−03	.639E−03	.708E−03	.155E−02
150	.235E−01	.149E−01	.866E−02	.441E−02	.293E−02	.180E−02	.137E−02	.109E−02	.126E−02	.287E−02
160	.357E−01	.230E−01	.136E−01	.703E−02	.471E−02	.293E−02	.226E−02	.183E−02	.219E−02	.516E−02
170	.532E−01	.347E−01	.208E−01	.110E−01	.741E−02	.466E−02	.363E−02	.299E−02	.372E−02	.905E−02
180	.775E−01	.514E−01	.312E−01	.167E−01	.114E−01	.726E−02	.571E−02	.478E−02	.619E−02	.155E−01
190	.111	.747E−01	.460E−01	.250E−01	.172E−01	.111E−01	.880E−02	.749E−02	.101E−01	.260E−01
200	.156	.107	.665E−01	.367E−01	.255E−01	.166E−01	.133E−01	.115E−01	.161E−01	.427E−01
210	.216	.150	.944E−01	.530E−01	.371E−01	.245E−01	.198E−01	.175E−01	.253E−01	.687E−01
220	.295	.207	.132	.752E−01	.531E−01	.354E−01	.289E−01	.260E−01	.392E−01	.109
230	.396	.282	.182	.105	.749E−01	.505E−01	.417E−01	.382E−01	.596E−01	.169
240	.525	.379	.247	.145	.104	.710E−01	.592E−01	.553E−01	.895E−01	.258
250	.688	.503	.331	.197	.143	.985E−01	.830E−01	.790E−01	.132	.389
260	.881	.660	.439	.264	.193	.135	.115	.112	.193	.577
270	1.141	.856	.575	.351	.258	.153	.157	.156	.279	.846
280	1.447	1.099	.744	.460	.341	.245	.213	.215	.398	1.225
290	1.817	1.398	.954	.597	.446	.324	.285	.295	.562	1.751
300	2.261	1.761	1.211	.767	.578	.425	.379	.399	.785	2.476
310	2.791	2.199	1.524	.977	.742	.553	.498	.536	1.085	3.465
320	3.417	2.723	1.901	1.234	.944	.713	.649	.714	1.486	4.800
330	4.153	3.347	2.353	1.545	1.191	.911	.840	.944	2.018	6.586
340	5.011	4.084	2.889	1.919	1.491	1.156	1.078	1.239	2.718	8.957
350	6.006	4.949	3.523	2.366	1.852	1.456	1.374	1.614	3.631	12.079

*Vermeulen, Dong, Robinson, Nguyen, and Gmitro, AIChE meeting, Anaheim, Calif., 1982; and private communication from Prof. Theodore Vermeulen, Chemical Engineering Dept., University of California, Berkeley.

TABLE 3-16 Partial Pressures of HNO_3 and H_2O over Aqueous Solutions of HNO_3

mmHg

Percentages are weight % HNO_3 in solution.

°C.	20%		25%		30%		35%		40%		45%		50%	
	HNO_3	H_2O	HNO_3	H_2O	HNO_3	H_2O	HNO_3	H_2O	HNO_3	H_2O	HNO_3	H_2O	HNO_3	H_2O
0	....	4.1	....	3.8	....	3.6	.....	3.3	.....	3.0	.....	2.6	.....	2.1
5	....	5.7	....	5.4	....	5.0	.....	4.6	.....	4.2	.....	3.6	.....	3.0
10	....	8.0	....	7.6	....	7.1	.....	6.5	.....	5.8	.....	5.0	0.12	4.2
15	.....	10.9	....	10.3	....	9.7	.....	8.9	.....	8.0	0.10	6.9	.18	5.8
20	....	15.2	....	14.2	....	13.2	.....	12.0	.....	10.8	.15	9.4	.27	7.9
25	....	20.6	....	19.2	....	17.8	.....	16.2	0.12	14.6	.23	12.7	.39	10.7
30	....	27.6	....	25.7	....	23.8	0.09	21.7	.17	19.5	.33	16.9	.56	14.4
35	....	36.5	....	33.8	....	31.1	.13	28.3	.25	25.5	.48	22.3	.80	19.0
40	....	47.5	....	44	0.11	41	.20	37.7	.36	33.5	.68	29.3	1.13	25.0
45	....	62	0.09	57.5	.17	53	.28	48	.52	43	.96	38.0	1.57	32.5
50	....	80	.13	75	.25	69	.42	63	.75	56	1.35	49.5	2.18	42.5
55	0.09	100	.18	94	.35	87	.59	79	1.04	71	1.83	62.5	2.95	54
60	.13	128	.28	121	.51	113	.85	102	1.48	90	2.54	80	4.05	70
65	.19	162	.40	151	.71	140	1.18	127	2.05	114	3.47	100	5.46	88
70	.27	200	.54	187	1.00	174	1.63	159	2.80	143	4.65	126	7.25	110
75	.38	250	.77	234	1.38	217	2.26	198	3.80	178	6.20	158	9.6	138
80	.53	307	1.05	287	1.87	267	3.07	243	5.10	218	8.15	195	12.5	170
85	.74	378	1.44	352	2.53	325	4.15	297	6.83	268	10.7	240	16.3	211
90	1.01	458	1.95	426	3.38	393	5.50	359	9.0	325	13.7	292	20.9	258
95	1.37	555	2.62	517	4.53	478	7.32	436	11.7	394	17.8	355	26.8	315
100	1.87	675	3.50	628	6.05	580	9.7	530	15.5	480	23.0	430	34.2	383
105	2.50	800	4.65	745	7.90	690	12.7	631	20.0	573	29.2	520	43.0	463
110	....	.....	....	.....	....	.....	16.5	755	25.7	688	37.0	625	54.5	560
115	....	.....	....	.....	....	.....	.....	.....	32.5	810	46	740	67	665
120	....	.....	....	.....	....	.....	.....	.....	.....	.....	.....	.....	84	785

°C.	55%		60%		65%		70%		80%		90%		100%
	HNO_3	H_2O	HNO_3	H_2O	HNO_3	H_2O	HNO_3	H_2O	HNO_3	H_2O	HNO_3	H_2O	HNO_3
0	.....	1.8	0.19	1.5	0.41	1.3	0.79	1.1	2	.....	5.5	....	11
5	0.14	2.5	.28	2.1	.60	1.8	1.12	1.6	3	.....	8	....	15
10	.21	3.5	.41	3.0	.86	2.6	1.58	2.2	4	1.2	11	....	22
15	.31	4.9	.59	4.1	1.21	3.5	2.18	3.0	6	1.7	15	....	30
20	.45	6.7	.84	5.6	1.68	4.9	3.00	4.1	8	2.4	20	....	42
25	.66	9.1	1.21	7.7	2.32	6.6	4.10	5.5	10.5	3.2	27	1	57
30	.93	12.2	1.66	10.3	3.17	8.8	5.50	7.4	14	4	36	1.3	77
35	1.30	16.1	2.28	13.6	4.26	11.6	7.30	9.8	18.5	5.5	47	1.8	102
40	1.82	21.3	3.10	18.1	5.70	15.5	9.65	12.8	24.5	7	62	2.4	133
45	2.50	28.0	4.20	23.7	7.55	20.0	12.6	16.7	32	9.5	80	3	170
50	3.41	36.3	5.68	31	10.0	26.0	16.5	21.8	41	12	103	4	215
55	4.54	46	7.45	39	12.8	33.0	21.0	27.3	52	15	127	5	262
60	6.15	60	9.9	51	16.8	43.0	27.1	35.3	67	20	157	6.5	320
65	8.18	76	13.0	64	21.7	54.5	34.5	44.5	85	25	192	8	385
70	10.7	95	16.8	81	27.5	68	43.3	56	106	31	232	10	460
75	13.9	120	21.8	102	35.0	86	54.5	70	130	38	282	13	540
80	18.0	148	27.5	126	43.5	106	67.5	86	158	48	338	16	625
85	23.0	182	34.8	156	54.5	131	83	107	192	60	405	20	720
90	29.4	223	43.7	192	67.5	160	103	130	230	73	480	24	820
95	37.3	272	55.0	233	83.5	195	125	158	278	89	570	29	
100	47	331	69.5	285	103	238	152	192	330	108	675	35	
105	58.5	400	84.5	345	124	288	183	231	392	129	790	42	
110	73	485	103	417	152	345	221	270	465	155			
115	90	575	126	495	181	410	262	330	545	185			
120	110	685	156	590	218	490	312	393	640	219			
125	......	.....	187	700	260	580	372	469					

TABLE 3-17 Partial Pressures of H_2O and HBr over Aqueous Solutions of HBr at 20 to 55°C

mmHg

% HBr	20°C.		25°C.		50°C.		55°C.	
	HBr	H_2O	HBr	H_2O	HBr	H_2O	HBr	H_2O
32	.....	...	0.0016					
34	.....	...	.0022					
36	.....	...	.0033					
38	.....	...	.0061					
40	.....	...	.011					
42	.....	...	.023					
44	.....	...	.048					
46	.....	...	.10					
48	0.09	6.2	.13	8.2	1.3	30.2	2.0	38
50	.23	4.5	.37	6.1	3.2	24.3	4.6	31
52	.71	3.3	1.1	4.5	7.2	19.3	10.2	25
54	2.2	2.4	3.2	3.3	17	16.0	23.0	21
56	6.8	1.7	9.3	2.4	40	13.3	51	18
58	21	1.3	27	1.9	91	10.4	115	14
60	.....	...	....	...	....	....	260	11.4

TABLE 3-18 Partial Pressures of HI over Aqueous Solutions of HI at 25°C

mmHg

%HI......	4	46	48	50	52	54	56
p_{HI}......	0.00064	0.0010	0.0022	0.0050	0.013	0.035	0.10

TABLE 3-19 Vapor Pressures of the System: Water–Sulfuric Acid–Nitric Acid

For these data reference must be made to the graphs of *International Critical Tables*, vol. 3, pp. 306–308.

TABLE 3-20 Total Vapor Pressures of Aqueous Solutions of CH_3COOH

Percentages of weight % acetic acid in the solution
mmHg

°C.	25%	50%	75%
20	16.3	15.7	15.3
25	22.1	21.4	20.8
30	29.6	28.8	27.8
35	39.4	38.3	36.6
40	51.7	50.2	48.1
45	67.0	65.0	62.0
50	87.2	85.0	80.1
55	110	107	102
60	141	138	130
65	178	172	162
70	223	216	203
75	277	269	251
80	342	331	310
85	419	407	376
90	510	497	458
95	618	602	550
100	743	725	666

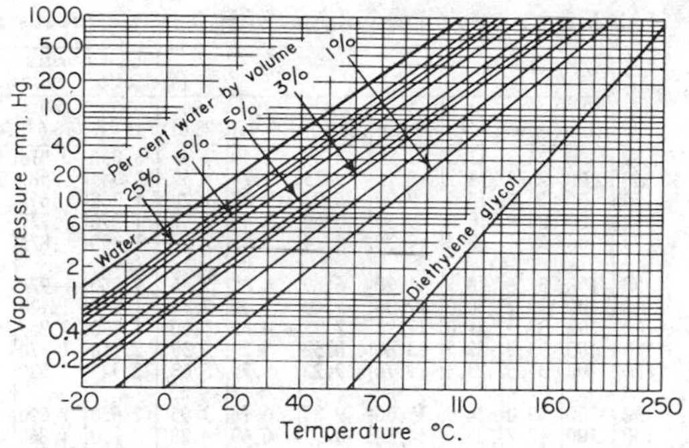

FIG. 3-3 Vapor pressure of aqueous diethylene glycol solutions. *(Courtesy of Carbide and Carbon Chemicals Corp.)*

TABLE 3-21 Partial Pressures of H_2O over Aqueous Solutions of NH_3*

Pressures are in pounds per square inch absolute

t, °F.	Molal concentration of ammonia in the solutions in percentages (Weight concentration of ammonia in the solution in percentages)																			
	0 (0)	5 (4.74)	10 (9.50)	15 (14.29)	20 (19.10)	25 (23.94)	30 (28.81)	35 (33.71)	40 (38.64)	45 (43.59)	50 (48.57)	55 (53.58)	60 (58.62)	65 (63.69)	70 (68.79)	75 (73.91)	80 (79.07)	85 (84.26)	90 (89.47)	95 (94.72)
32	0.09	0.084	0.079	0.074	0.070	0.065	0.060	0.056	0.051	0.047	0.042	0.038	0.034	0.030	0.025	0.021	0.017	0.013	0.008	0.004
40	.12	.115	.108	.101	.095	.089	.083	.076	.070	.064	.058	.052	.046	.040	.035	.029	.023	.015	.012	.006
50	.18	.17	.16	.15	.14	.13	.12	.11	.10	.094	.085	.076	.068	.059	.051	.042	.034	.025	.017	.008
60	.26	.24	.23	.21	.20	.19	.17	.16	.15	.13	.12	.11	.097	.085	.073	.061	.049	.037	.024	.012
70	.36	.34	.32	.30	.28	.26	.25	.23	.21	.19	.17	.15	.14	.12	.10	.086	.069	.052	034	.017
80	.51	.48	.45	.42	.40	.37	.34	.32	.29	.27	.24	.22	.19	.17	.14	.12	.096	.072	.048	.024
90	.70	.66	.63	.58	.55	.51	.47	.44	.40	.37	.33	.30	.26	.23	.20	.16	.13	.10	.066	.033
100	.95	.90	.85	.79	.74	.69	.64	.59	.55	.50	.45	.41	.36	.31	.27	.22	.18	.13	.090	045
110	1.27	1.20	1.14	1.07	1.00	.93	.86	.80	.73	.67	.60	.54	.48	.42	.36	.30	.24	.18	.120	.061
120	1.69	1.60	1.51	1.42	1.33	1.24	1 15	1.06	.97	.89	.80	.72	.64	.56	.48	.40	.32	.24	.160	.081
130	2.22	2.10	1.98	1.86	1.74	1.62	1.51	1.39	1.28	1.17	1.05	.95	.84	.74	.63	.53	.42	.32	.210	.100
140	2.89	2.73	2.57	2.42	2.26	2.11	1.96	1.81	1.66	1.52	1.37	1.23	1.10	.96	.82	.69	.55	.41	.270	.140
150	3.72	3.51	3.31	3.11	2.91	2.72	2.52	2.33	2.14	1.95	1.76	1.59	1.41	1.24	1.06	.88	.71	.53	.350	.180
160	4.74	4.48	4.22	3.97	3.71	3.46	3.22	2.97	2.73	2.49	2.25	2.02	1.80	1.58	1.35	1.12	.90	.67	.450	.220
170	5.99	5.66	5.34	5.02	4.70	4.38	4.07	3.75	3.45	3.15	2.84	2.56	2.28	1.99	1.71	1.42	1.13	1.85	.570	.300
180	7.51	7.10	6.69	6.30	5.89	5.49	5.10	4.71	4.33	3.94	3.57	3.21	2.85	2.50	2.14	1.77	1.42	1.06		
190	9.34	8.83	8.32	7.82	7.32	6.83	6.34	5.86	5.38	4.91	4.44	3.99	3.55	3.10	2.65					
200	11.53	10.90	10.27	9.65	9.04	8.43	7.83	7.23	6.64	6.06	5.48	4.93	4.38	3.81						
210	14.12	13.35	12.58	11.82	11.07	10.32	9.59	8.86	8.13	7.42	6.71	6.04	5.34							
220	17.19	16.25	15.32	14.39	13.48	12.57	11.67	10.78	9.90	9.03	8.17	7.31								
230	20.78	19.64	18.51	17.40	16.29	15.19	14.11	13.03	11.97	10.91	9.87									
240	24.97	23.60	22.25	20.91	19.58	18.26	16.95	15.66	14.38	13.12	11.86									
250	29.83	28.20	26.58	25.00	23.39	21.82	20.25	18.71	17.18	15.67										

* Wilson, *Univ. Ill., Eng. Expt. Sta. Bull.* 146.

TABLE 3-22 Mole Percentages of H_2O over Aqueous Solutions of NH_3*

t, °F.	Molal concentration of ammonia in the solutions in percentages (Weight concentration of ammonia in the solutions in percentages)																			
	0 (0)	5 (4.74)	10 (9.50)	15 (14.29)	20 (19.10)	25 (23.94)	30 (28.81)	35 (33.71)	40 (38.64)	45 (43.59)	50 (48.57)	55 (53.58)	60 (58.62)	65 (63.69)	70 (68.79)	75 (73.91)	80 (79.07)	85 (84.26)	90 (89.47)	95 (94.72)	100 (100.00)
32	100	24.3	13.2	7.63	4.43	2.50	1.43	0.856	0.514	0.335	0.216	0.151	0.109	0.0816	0.0585	0.0457	0.0345	0.0249	0.0146	0.00689	0.00
40	100	25.3	14.1	8.15	4.73	2.74	1.59	.943	.581	.372	.248	.172	.124	.0914	.0706	.0533	.0395	.0243	.0185	.00879	
50	100	26.6	15.2	9.09	5.24	3.03	1.78	1.060	.652	.434	.290	.202	.148	.1095	.0838	.0630	.0477	.0332	.0215	.00959	
60	100	27.9	16.2	9.50	5.69	3.42	1.97	1.210	.777	.481	.331	.238	.172	.1290	.0986	.0754	.0566	.0406	.0251	.01125	
70	100	29.1	17.4	10.30	6.14	3.65	2.27	1.390	.873	.569	.383	.266	.205	.1510	.112	.0882	.0656	.0474	.0296	.0135	
80	100	31.6	18.5	11.20	6.89	4.08	2.45	1.550	.978	.659	.444	.323	.230	.1750	.130	.103	.0772	.0528	.0351	.0167	
90	100	32.7	20.0	12.00	7.40	4.47	2.73	1.730	1.100	.742	.505	.366	.267	.2020	.157	.115	.0884	.0647	.0408	.0194	
100	100	34.4	21.0	12.90	7.92	4.85	3.00	1.890	1.250	.834	.574	.420	.307	.2290	.179	.135	.104	.0714	.0473	.0226	
110	100	35.9	22.2	13.80	8.59	5.29	3.30	2.110	1.370	.932	.644	.466	.347	.2640	.208	.157	.118	.0846	.0540	.0262	
120	100	37.5	23.4	14.70	9.22	5.75	3.63	2.320	1.520	1.044	.714	.529	.395	.3020	.233	.180	.135	.0970	.0619	.0300	
130	100	39.0	24.5	15.60	9.85	6.18	3.95	2.550	1.690	1.160	.811	.596	.444	.3430	.263	.205	.154	.1117	.0703	.0339	
140	100	40.7	25.8	16.50	10.50	6.69	4.28	2.790	1.860	1.286	.906	.663	.501	.3840	.297	.232	.175	.124	.0786	.0385	
150	100	42.3	27.1	17.50	11.20	7.19	4.63	3.080	2.040	1.410	1.004	.741	.558	.4320	.334	.257	.197	.140	.0892	.0439	
160	100	44.1	28.3	18.40	11.90	7.69	5.01	3.300	2.230	1.550	1.110	.818	.617	.4800	.372	.287	.218	.154	.1005	.0499	
170	100	45.6	29.6	19.40	12.70	8.22	5.38	3.580	2.430	1.700	1.220	.904	.689	.5300	.414	.320	.242	.174	.112	.0567	
180	100	47.3	30.9	20.40	13.40	8.76	5.78	3.870	2.640	1.850	1.340	.994	.756	.5860	.456	.352	.268	.192			
190	100	48.7	32.2	21.40	14.10	9.31	6.18	4.160	2.860	2.020	1.460	1.087	.830	.6420	.501						
200	100	50.4	33.4	22.30	14.90	9.88	6.59	4.470	3.080	2.190	1.580	1.187	.907	.7010							
210	100	52.1	34.7	23.40	15.70	10.45	7.03	4.780	3.310	2.360	1.720	1.272	.983								
220	100	53.7	36.1	24.40	16.40	11.05	7.48	5.100	3.560	2.540	1.860	1.390									
230	100	55.2	37.3	25.40	17.30	11.63	7.91	5.440	3.810	2.730	2.000										
240	100	56.8	38.6	26.50	18.00	12.24	8.36	5.780	4.060	2.920	2.150										
250	100	58.4	39.8	27.50	18.80	12.88	8.82	6.120	4.340	3.120											

* Wilson, *Univ. Ill., Eng. Expt. Sta. Bull.* 146.

TABLE 3-23 Partial Pressures of NH_3 over Aqueous Solutions of NH_3*

Pressures are in pounds per square inch absolute

t, °F.	Molal concentration of ammonia in the solutions in percentages (Weight concentration of ammonia in the solutions in percentages)																		
	5 (4.74)	10 (9.50)	15 (14.29)	20 (19.10)	25 (23.94)	30 (28.81)	35 (33.71)	40 (38.64)	45 (43.59)	50 (48.57)	55 (53.58)	60 (58.62)	65 (63.69)	70 (68.79)	75 (73.91)	80 (79.07)	85 (84.26)	90 (89.47)	95 (94.72)
32	0.26	0.52	0.90	1.51	2.67	4.27	6.54	8.93	14.13	19.36	25.12	31.13	36.74	42.69	45.92	49.26	52.13	54.89	58.01
40	.33	.66	1.14	1.92	3.16	5.13	7.98	11.98	17.14	23.33	30.15	37.15	43.69	49.56	54.40	58.31	61.62	64.77	68.31
50	.47	.89	1.50	2.53	4.16	6.63	10.24	15.24	21.56	29.17	37.46	45.86	53.79	60.82	66.63	71.26	75.22	79.05	83.40
60	.62	1.19	2.00	3.21	5.36	8.48	13.06	19.15	26.92	36.14	46.12	56.22	65.81	73.99	80.90	86.44	91.04	95.67	100.65
70	.83	1.52	2.60	4.28	6.87	10.76	16.33	23.84	33.20	44.25	56.29	68.32	79.42	89.26	97.42	104.01	109.55	114.83	120.61
80	1.04	1.98	3.34	5.45	8.69	13.52	20.29	29.40	40.69	53.84	67.97	82.36	95.52	107.06	116.42	124.20	130.57	136.35	143.70
90	1.36	2.52	4.25	6.88	10.89	16.76	25.04	35.94	49.45	64.99	81.61	98.35	113.79	127.22	138.18	147.02	154.46	161.74	169.73
100	1.72	3.20	5.34	8.60	13.53	20.68	30.57	43.57	59.49	77.85	97.27	116.81	134.70	150.23	162.94	173.22	181.97	190.13	199.17
110	2.14	4.00	6.65	10.64	16.65	25.21	37.01	52.43	71.20	92.59	115.16	137.62	158.42	176.18	190.85	203.02	212.71	222.22	232.79
120	2.67	4.95	8.21	13.09	20.30	30.54	44.56	62.62	84.44	109.40	135.48	161.44	185.14	205.81	222.28	236.05	247.14	258.24	270.02
130	3.28	6.09	10.05	15.93	24.58	36.74	53.16	74.27	99.69	128.45	158.45	188.16	215.14	238.70	257.87	272.88	286.08	298.46	311.80
140	3.97	7.41	12.21	19.23	29.43	43.77	62.97	87.53	116.72	149.93	184.17	218.18	248.70	275.33	297.12	314.45	328.99	342.93	358.46
150	4.78	8.92	14.70	23.09	35.09	51.91	74.28	102.51	136.15	173.64	212.91	251.24	286.00	316.24	340.82	360.39	376.57	392.45	409.62
160	5.68	10.70	17.57	27.45	41.56	61.03	86.91	119.37	157.71	200.45	244.98	288.38	327.82	361.75	389.08	411.30	429.73	447.35	466.38
170	6.75	12.67	20.85	32.41	48.89	71.48	101.09	138.30	181.95	230.36	280.54	329.42	373.61	411.59	442.28	466.67	487.85	507.63	528.50
180	7.90	14.96	24.56	38.13	57.19	83.07	116.97	159.37	208.66	263.43	319.89	374.25	424.10	466.26	500.63	528.08	551.24		
190	9.23	17.55	28.78	44.49	66.49	96.22	134.89	182.72	238.39	299.86	363.11	424.15	479.40	526.15					
200	10.70	20.45	33.49	51.58	76.90	110.85	154.58	208.56	270.94	340.02	410.17	478.62	539.79						
210	12.26	23.68	38.76	59.65	88.48	126.83	176.24	236.97	307.08	383.99	462.36	537.56							
220	14.02	27.15	44.61	68.43	101.24	144.74	200.46	268.30	346.07	431.43	518.19								
230	15.95	31.09	51.06	78.14	115.45	164.17	226.67	302.53	389.29	483.53									
240	17.92	35.40	58.00	89.02	130.94	185.79	255.26	339.72	435.78	540.44									
250	20.12	40.09	65.74	100.69	147.66	209.37	286.89	380.42	486.73										

* Wilson, *Univ. Ill., Eng. Expt. Sta. Bull.* 146.

TABLE 3-24 Total Vapor Pressures of Aqueous Solutions of NH_3*

Pressures are in pounds per square inch absolute

t,°F.	Molal concentration of ammonia in the solutions in percentages (Weight concentration of ammonia in the solutions in percentages)																				
	0 (0)	5 (4.74)	10 (9.50)	15 (14.29)	20 (19.10)	25 (23.94)	30 (28.81)	35 (33.71)	40 (38.64)	45 (43.59)	50 (48.57)	55 (53.58)	60 (58.62)	65 (63.69)	70 (68.79)	75 (73.91)	80 (79.07)	85 (84.26)	90 (89.47)	95 (94.72)	100 (100.00)
32	0.09	0.34	0.60	0.97	1.58	2.60	4.20	6.54	9.93	14.18	19.40	25.16	31.16	36.77	42.72	45.94	49.28	52.14	54.90	58.01	62.29
40	.12	.45	.77	1.24	2.01	3.25	5.21	8.06	12.05	17.20	23.39	30.20	37.20	43.73	49.60	54.43	58.33	61.64	64.78	68.32	73.32
50	.18	.64	1.05	1.65	2.67	4.29	6.75	10.35	15.34	21.65	29.26	37.54	45.93	53.85	60.87	66.67	71.29	75.25	79.07	83.41	89.19
60	.26	.86	1.42	2.21	3.51	5.55	8.65	13.22	19.30	27.05	36.26	46.23	56.32	65.90	74.06	80.96	86.49	91.08	95.69	100.66	107.6
70	.36	1.17	1.84	2.90	4.56	7.13	11.01	16.56	24.05	33.39	44.42	56.44	68.46	79.54	89.36	97.51	104.08	109.60	114.86	120.63	128.8
80	.51	1.52	2.43	3.76	5.85	9.06	13.86	20.61	29.69	40.96	54.08	68.19	82.55	95.69	107.20	116.54	124.30	130.64	136.40	143.72	153.0
90	.70	2.02	3.15	4.83	7.43	11.40	17.23	25.48	36.34	49.82	65.32	81.91	98.61	114.02	127.42	138.34	147.15	154.56	161.81	169.76	180.6
100	.95	2.62	4.05	6.13	9.34	14.22	21.32	31.16	44.12	59.99	78.30	97.68	117.17	135.01	150.50	163.16	173.40	182.10	190.22	199.22	211.9
110	1.27	3.34	5.14	7.72	11.64	17.58	26.07	37.81	53.16	71.87	93.19	115.7	138.10	158.84	176.54	191.15	203.26	212.89	222.34	232.85	247.0
120	1.69	4.27	6.46	9.63	14.42	21.54	31.69	45.62	63.59	85.33	110.2	136.2	162.08	185.70	206.29	222.68	236.37	247.38	258.40	270.1	286.4
130	2.22	5.38	8.07	11.91	17.67	26.20	38.25	54.55	75.55	100.86	129.5	159.	189.00	215.88	239.33	258.40	273.3	286.4	298.67	311.9	330.3
140	2.89	6.70	9.98	14.63	21.49	31.54	45.73	64.78	89.19	118.24	151.3	185.4	219.28	249.66	276.15	297.81	315.0	329.4	343.2	358.6	379.1
150	3.72	8.29	12.23	17.81	26.00	37.81	54.43	76.61	104.65	138.1	175.4	214.5	252.65	287.24	317.3	341.7	361.1	377.1	392.8	409.8	432.2
160	4.74	10.16	14.92	21.54	31.16	45.02	64.25	89.88	122.10	160.2	202.7	247.0	290.18	329.4	363.1	390.2	412.2	430.4	447.8	466.6	492.8
170	5.99	12.41	18.01	25.87	37.11	53.27	75.55	104.84	141.75	185.1	233.2	283.1	331.7	375.6	413.3	443.7	467.8	488.7	508.2	528.8	558.4
180	7.51	15.00	21.65	30.86	44.02	62.68	88.17	121.68	163.7	212.6	267.0	323.1	377.1	426.6	468.4	502.4	529.5	552.3			
190	9.34	18.06	25.87	36.60	51.81	73.32	102.56	140.75	188.1	243.3	304.3	367.1	427.7	492.5	528.8						
200	11.53	21.60	30.72	43.14	60.62	85.33	118.68	161.81	215.2	277.0	345.5	415.1	483.0	543.6							
210	14.12	25.61	36.26	50.58	70.72	98.80	136.42	185.10	245.1	314.5	390.7	468.4	542.9								
220	17.19	30.27	42.47	59.00	81.91	113.81	156.41	211.24	278.2	355.1	439.6	525.5									
230	20.78	35.59	49.60	68.46	94.43	130.64	178.28	239.70	314.5	400.2	493.4										
240	24.97	41.52	57.65	78.91	108.60	149.20	202.74	270.92	354.1	448.9	552.3										
250	29.83	48.32	66.67	90.74	124.08	169.48	229.62	305.60	397.6	502.4											

* Wilson, *Univ. Ill., Eng. Expt. Sta. Bull.* 146.

TABLE 3-25 Partial Pressures of H_2O over Aqueous Solutions of Sodium Carbonate

mmHg

t, °C.	$\%Na_2CO_3$						
	0	5	10	15	20	25	30
0	4.5	4.5					
10	9.2	9.0	8.8				
20	17.5	17.2	16.8	16.3			
30	31.8	31.2	30.4	29.6	28.8	27.8	26.4
40	55.3	54.2	53.0	57.6	50.2	48.4	46.1
50	92.5	90.7	88.7	86.5	84.1	81.2	77.5
60	149.5	146.5	143.5	139.9	136.1	131.6	125.7
70	239.8	235	230.5	225	219	211.5	202.5
80	355.5	348	342	334	325	315	301
90	526.0	516	506	494	482	467	447
100	760.0	746	731	715	697	676	648

TABLE 3-26 Partial Pressures of H_2O and CH_3OH over Aqueous Solutions of Methyl Alcohol*

Mole fraction CH_3OH	39.9°C.		Mole fraction CH_3OH	59.4°C.	
	P_{H_2O}, mm. Hg	P_{CH_3OH}, mm. Hg		P_{H_2O}, mm. Hg	P_{CH_3OH}, mm. Hg
0	54.7	0	0	145.4	0
14.99	39.2	66.1	22.17	106.9	210.1
17.85	38.5	75.5	27.40	102.2	240.2
21.07	37.2	85.2	33.24	96.6	272.1
27.31	35.8	100.6	39.80	91.7	301.9
31.06	34.9	108.8	47.08	84.8	335.6
40.1	32.8	127.7	55.5	76.9	373.7
47.0	31.5	141.6	69.2	57.8	439.4
55.8	27.3	158.4	78.5	43.8	486.6
68.9	20.7	186.6	85.9	30.1	526.9
86.0	10.1	225.2	100.0	0	609.3
100.0	0	260.7			

* "International Critical Tables," vol. 3, p. 290, McGraw-Hill.

TABLE 3-27 Partial Pressures of H_2O over Aqueous Solutions of Sodium Hydroxide

mmHg

Conc. g. NaOH/100 g. H_2O	Temperature, °C.											
	0	20	40	60	80	100	120	160	200	250	300	350
0	4.6	17.5	55.3	149.5	355.5	760.0	1,489	4,633	11,647	29,771	64,200	123,600
5	4.4	16.9	53.2	143.5	341.5	730.0	1,430	4,450	11,200	28,600	61,800	118,900
10	4.2	16.0	50.6	137.0	325.5	697.0	1,365	4,260	10,750	27,500	59,300	114,100
20	3.6	13.9	44.2	120.5	288.5	621.0	1,225	3,860	9,800	25,300	54,700	105,400
30	2.9	11.3	36.6	101.0	246.0	537.0	1,070	3,460	8,950	23,300	50,800	98,000
40	2.2	8.7	28.7	81.0	202.0	450.0	920	3,090	8,150	21,500	47,200	91,600
50	...	6.3	20.7	62.5	160.5	368.0	770	2,690	7,400	19,900	44,100	85,800
60	...	4.4	15.5	47.0	124.0	294.0	635	2,340	6,750	18,400	41,200	80,700
70	...	3.0	10.9	34.5	94.0	231.0	515	2,030	6,100	17,100	38,700	76,000
80	...	2.0	7.6	24.5	70.5	179.0	415	1,740	5,500	15,800	36,300	71,900
90	...	1.3	5.2	17.5	53.0	138.0	330	1,490	5,000	14,700	34,200	68,100
100	...	0.9	3.6	12.5	38.5	105.0	262	1,300	4,500	13,650	32,200	64,600
120	...	....	1.7	6.3	20.5	61.0	164	915	3,650	11,800	28,800	58,600
140	...	....	....	3.0	11.0	35.5	102	765	2,980	10,300	25,900	53,400
160	...	....	....	1.5	6.0	20.5	63	470	2,430	8,960	23,300	49,000
180	...	....	....	.....	3.5	12.0	40	340	1,980	7,830	21,200	45,100
200	...	....	....	.....	2.0	7.0	25	245	1,620	6,870	19,200	41,800
250	...	....	....	.....	0.5	2.0	8	110	985	5,000	15,400	35,000
300	...	....	....	.....	0.1	0.5	2.7	50	610	3,690	12,500	29,800
350	...	....	....	.....	.....	.....	0.9	23	380	2,750	10,300	25,700
400	...	....	....	.....	.....	.....	.....	11	240	2,080	8,600	22,400
500	...	....	....	.....	.....	.....	.....	.....	100	1,210	6,100	17,500
700	...	....	....	.....	.....	.....	.....	.....	.....	440	3,300	11,500
1000	...	....	....	.....	.....	.....	.....	.....	.....	.....	1,470	6,800
2000	...	....	....	.....	.....	.....	.....	.....	.....	.....	150	1,760
4000	...	....	....	.....	.....	.....	.....	.....	.....	.....	.....	120
8000	...	....	....	.....	.....	.....	.....	.....	.....	.....	.....	7

WATER-VAPOR CONTENT OF GASES

CHARTS FOR GASES AT HIGH PRESSURES

The accompanying figures are useful in determining the water-vapor content of gases at high pressure in contact with liquid water. Data for air are given in Fig. 3-4. Figure 3-5 shows the water-vapor content of hydrogen and nitrogen in contact with liquid water at high pressures. For additional experimental values of the water content of compressed nitrogen in contact with water at 100, 200, and 300 atm and up to 230°C, see Saddington and Krase, *J. Am. Chem. Soc.*, **56**, 360 (1934). Results to 100°C are shown in Fig. 3-8, and comparisons with Bartlett's values at 50°C are included. Figure 3-6 shows the water-vapor content of compressed gases in contact with liquid water. Figure 3-7 shows the volume percentage of water vapor in gases expanded from high-pressure contact with liquid water.

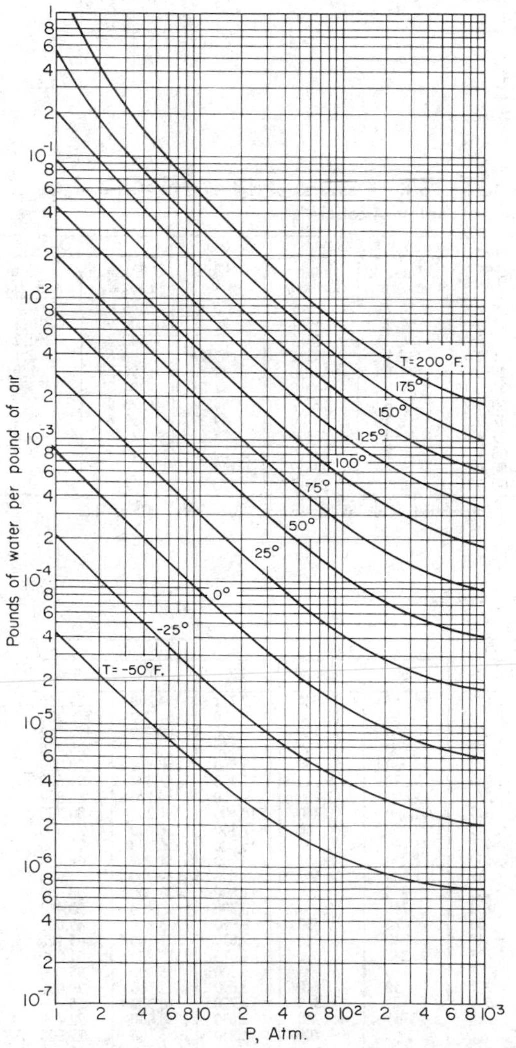

FIG. 3-4 Water content of air. °C = (°F − 32) × 5/9. (*Landsbaum, Dadds, and Stutzman. Reprinted from vol. 47, p. 102, January 1955, issue of* Ind. Eng. Chem. *Copyright 1955 by the American Chemical Society and reproduced by permission of the copyright owner.*)

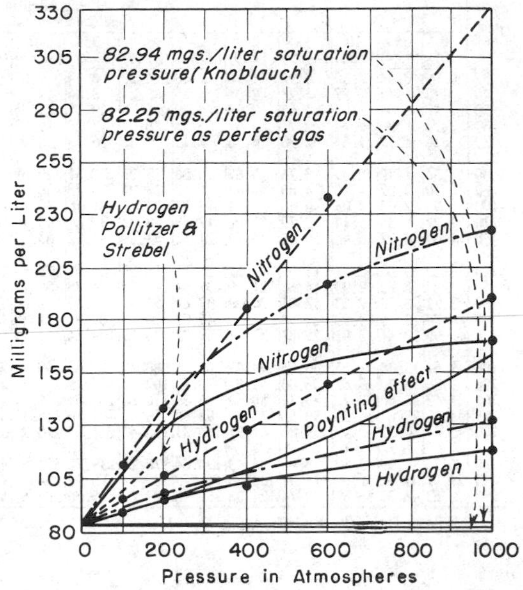

FIG. 3-5 Water-vapor content of hydrogen and nitrogen in contact with liquid water at high pressures at 50°C, ----, calculated to perfect-gas volume; ———, calculated to actual volume; -·-·-·, calculated to free space. °F = 9/5 °C + 32; to convert milligrams per liter to pounds per cubic foot, multiply by 6.243×10^{-5}. [*Bartlett,* J. Am. Chem. Soc., *49, 65 (1927).*]

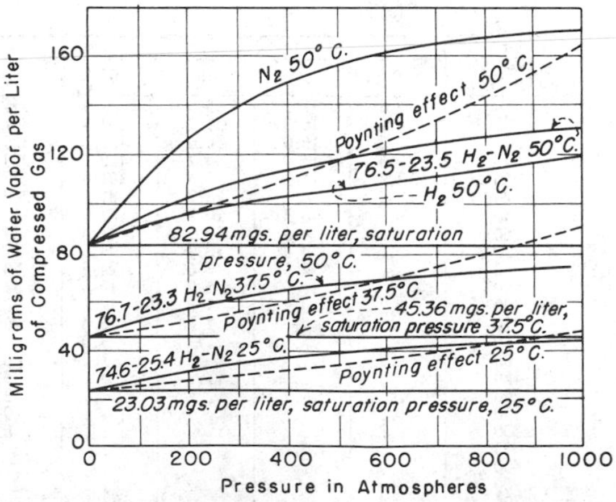

FIG. 3-6 Water-vapor content of N_2–H_2 mixtures in contact with liquid water at 25.0, 37.5, and 50.0°C. °F = 9/5 °C + 32; to convert milligrams per liter to pounds per cubic foot, multiply by 6.243×10^{-5}. [*Bartlett,* J. Am. Chem. Soc., *49, 65 (1927).*]

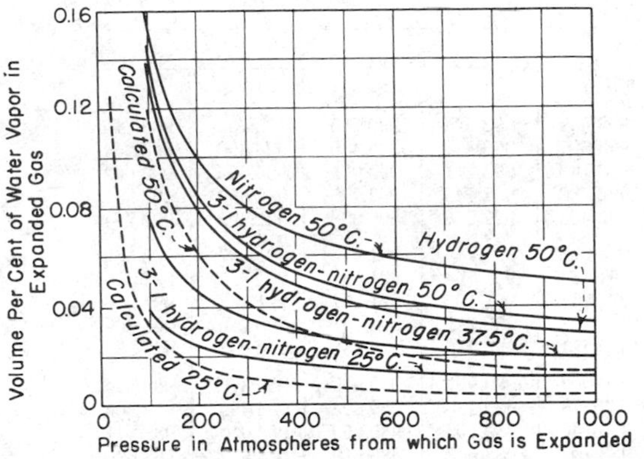

FIG. 3-7 Volume percentage of water vapor in N_2–H_2 mixture expanded from high-pressure contact with liquid water at 25.0, 37.5, and 50.0°C. °F = % °C + 32. [*Bartlett,* J. Am. Chem. Soc., *49, 65 (1927).*]

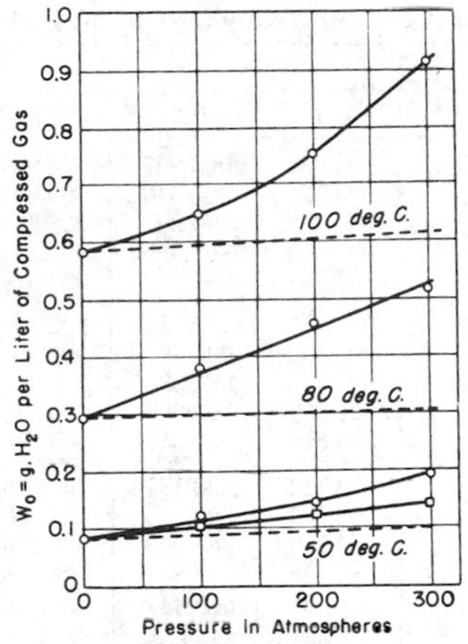

FIG. 3-8 Effect of pressure on the water-vapor content of compressed N_2 gas; ———, Poynting relation; □, Bartlett; ○, experimental. °F = % °C + 32; to convert grams per liter to pounds per cubic foot, multiply by 6.243×10^{-2}. [*Saddington and Krase,* J. Am. Chem. Soc., *56, 360 (1934).*]

DENSITIES OF PURE SUBSTANCES

UNITS CONVERSIONS

For this subsection, the following units conversions are applicable:
°F = % °C + 32.

To convert kilograms per cubic meter to pounds per cubic foot, multiply by 0.06243.

TABLE 3-28 Density (kg/m³) of Water from 0 to 100°C*

	ρ, kg/m³									
t, °C	0.0	0.1	0.2	0.3	0.4	0.5	0.6	0.7	0.8	0.9
0	999.839	999.846	999.852	999.859	999.865	999.871	999.877	999.882	999.888	999.893
1	999.898	999.903	999.908	999.913	999.917	999.921	999.925	999.929	999.933	999.936
2	999.940	999.943	999.946	999.949	999.952	999.954	999.956	999.959	999.961	999.962
3	999.964	999.966	999.967	999.968	999.969	999.970	999.971	999.971	999.972	999.972
4	999.972	999.972	999.972	999.971	999.971	999.970	999.969	999.968	999.967	999.965
5	999.964	999.962	999.960	999.958	999.956	999.954	999.951	999.949	999.946	999.943
6	999.940	999.937	999.934	999.930	999.926	999.923	999.919	999.915	999.910	999.906
7	999.901	999.897	999.892	999.887	999.882	999.877	999.871	999.866	999.860	999.854
8	999.848	999.842	999.836	999.829	999.823	999.816	999.809	999.802	999.795	999.788
9	999,781	999.773	999.765	999.758	999.750	999.742	999.734	999.725	999.717	999.708
10	999.699	999.691	999.682	999.672	999.663	999.654	999.644	999.635	999.625	999.615
11	999.605	999.595	999.584	999.574	999.563	999.553	999.542	999.531	999.520	999.509
12	999.497	999.486	999.474	999.462	999.451	999.439	999.426	999.414	999.402	999.389
13	999.377	999.364	999.351	999.338	999.325	999.312	999.299	999.285	999.272	999.258
14	999.244	999.230	999.216	999.202	999.188	999.173	999.159	999.144	999.129	999.114
15	999.099	999.084	999.069	999.054	999.038	999.022	999.007	998.991	998.975	998.958
16	998.943	998.926	998.910	999.894	998.877	998.860	998.843	998.826	998.809	998.792
17	998.775	998.757	998.740	998.722	998.704	998.686	998.668	998.650	998.632	998.614
18	998.595	998.577	998.558	998.539	998.520	998.502	998.482	998.463	998.444	998.425
19	998.405	998.385	998.366	998.346	998.326	998.306	998.286	998.265	998.245	998.224
20	998.204	998.183	998.162	998.141	998.120	998.099	998.078	998.057	998.035	998.014
21	997.992	997.971	997.949	997.927	997.905	997.883	997.860	997.838	997.816	997.793
22	997.770	997.747	997.725	997.702	997.679	997.656	997.632	997.609	997.585	997.562
23	997.538	997.515	997.491	997.467	997.443	997.419	997.394	997.370	997.345	997.321
24	997.296	997.272	997.247	997.222	997.197	997.172	997.146	997.121	997.096	997.070
25	997.045	997.019	996.993	996.967	996.941	996.915	996.889	996.863	996.836	996.810
26	996.783	996.757	996.730	996.703	996.676	996.649	996.622	996.595	996.568	996.540
27	996.513	996.485	996.458	996.430	996.402	996.374	996.346	996.318	996.290	996.262
28	996.233	996.205	996.176	996.148	996.119	996.090	996.061	996.032	996.003	995.974
29	995.945	995.915	995.886	995.856	995.827	995.797	995.767	995.737	995.707	995.677
30	995.647	995.617	995.586	995.556	995.526	995.495	995.464	995.433	995.403	995.372
31	995.341	995.310	995.278	995.247	995.216	995.184	995.153	995.121	995.090	995.058
32	995.026	994.997	994.962	994.930	994.898	994.865	994.833	994.801	994.768	994.735
33	994.703	994.670	994.637	994.604	994.571	994.538	994.505	994.472	994.438	994.405
34	994.371	994.338	994.304	994.270	994.236	994.202	994.168	994.134	994.100	994.066

*From "Water: Density at Atmospheric Pressure and Temperatures from 0 to 100°C," *Tables of Standard Handbook Data*, Standartov, Moscow, 1978. To conserve space, only a few tables of density values are given. The reader is reminded that density values may be found as the reciprocal of the specific volume values tabulated in the "Thermodynamic Properties: Tables" subsection.

TABLE 3-28 Density (kg/m³) of Water from 0 to 100°C *(Concluded)*

	ρ, kg/m³									
t, °C	0.0	0.1	0.2	0.3	0.4	0.5	0.6	0.7	0.8	0.9
35	994.032	993.997	993.963	993.928	993.893	993.859	993.824	993.789	993.754	993.719
36	993.684	993.648	993.613	993.578	993.543	993.507	993.471	993.436	993.400	993.364
37	993.328	993.292	993.256	993.220	993.184	993.148	993.111	993.075	993.038	993.002
38	992.965	992.928	992.891	992.855	992.818	992.780	992.743	992.706	992.669	992.631
39	992.594	992.557	992.519	992.481	992.444	992.406	992.368	992.330	992.292	992.254
40	992.215	992.177	992.139	992.100	992.062	992.023	991.985	991.946	991.907	992.868
41	991.830	991.791	991.751	991.712	992.673	991.634	991.594	991.555	991.515	991.476
42	991.436	991.396	991.357	991.317	991.277	991.237	991.197	991.157	991.116	991.076
43	991.036	990.995	990.955	990.914	990.873	990.833	990.792	990.751	990.710	990.669
44	990.628	990.587	990.546	990.504	990.463	990.421	990.380	990.338	990.297	990.255
45	990.213	990.171	990.129	990.087	990.045	990.003	989.961	989.919	989.876	989.834
46	989.792	989.749	989.706	989.664	989.621	989.578	989.535	989.492	989.449	989.406
47	989.363	989.320	989.276	989.233	989.190	989.146	989.103	989.059	989.015	988.971
48	988.928	988.884	988.840	988.796	988.752	988.707	988.663	988.619	988.574	988.530
49	988.485	988.441	988.396	988.352	988.307	988.262	988.217	988.172	988.127	988.082
50	988.037	987.992	987.946	987.901	987.844	987.810	987.764	987.719	987.673	987.627
51	987.581	987.536	987.490	987.444	987.398	987.351	987.305	987.259	987.213	987.166
52	987.120	987.073	987.027	986.980	986.933	986.886	986.840	986.793	986.746	986.699
53	986.652	986.604	986.557	986.510	986.463	986.415	986.368	986.320	986.272	986.225
54	986.177	986.129	986.081	986.033	985.985	985.937	985.889	985.841	985.793	985.745
55	985.696	985.648	985.599	985.551	985.502	985.454	985.405	985.356	985.307	985.258
56	985.219	985.160	985.111	985.062	985.013	984.963	984.914	984.865	984.815	984.766
57	984.716	984.666	984.617	984.567	984.517	984.467	984.417	984.367	984.317	984.267
58	984.217	984.167	984.116	984.066	984.016	983.965	983.914	983.864	983.813	983.762
59	983.712	983.661	983.610	983.559	983.508	983.457	983.406	983.354	983.303	983.252
60	983.200	983.149	983.097	983.046	982.994	982.943	982.891	982.839	982.787	982.735
61	982.683	982.631	982.579	982.527	982.475	982.422	982.370	982.318	982.265	982.213
62	982.160	982.108	982.055	982.002	981.949	981.897	981.844	981.791	981.738	981.685
63	981.631	981.578	981.525	981.472	981.418	981.365	981.311	981.258	981.204	981.151
64	981.097	981.043	980.989	980.935	980.881	980.827	980.773	980.719	980.665	980.611
65	980.557	980.502	980.443	980.393	980.339	980.284	980.230	980.175	980.120	980.065
66	980.011	979.956	979.901	979.846	979.791	979.736	979.680	979.625	979.570	979.515
67	979.459	979.403	979.348	979.293	979.237	979.181	979.126	979.070	979.014	978.958
68	978.902	978.846	978.790	978.734	978.678	978.621	978.565	978.509	978.452	978.396
69	978.339	978.283	978.226	978.170	978.113	978.056	977.999	977.942	977.885	977.828
70	977.771	977.714	977.657	977.600	977.543	977.485	977.428	977.370	977.313	977.255
71	977.198	977.140	977.082	977.025	976.967	976.909	976.851	976.793	976.735	976.677
72	976.619	976.561	976.503	976.444	976.386	976.327	976.269	976.211	976.152	976.093
73	976.035	975.976	975.917	975.858	975.800	975.741	975.682	975.623	975.564	975.504
74	975.445	975.386	975.327	975.267	975.208	975.148	975.089	975.029	974.970	974.910
75	974.850	974.791	974.731	974.671	974.611	974.551	974.491	974.431	974.371	974.311
76	974.250	974.190	974.130	974.069	974.009	973.948	973.888	973.827	973.767	973.706
77	973.645	973.584	973.524	973.463	973.402	973.341	973.280	973.218	973.157	973.096
78	973.025	972.974	972.912	972.851	972.789	972.728	972.666	972.605	972.543	972.481
79	972.419	972.358	972.296	972.234	972.172	972.110	972.048	971.986	971.923	971.861
80	971.799	971.737	971.674	971.612	971.549	971.487	971.424	971.361	971.299	971.236
81	971.173	971.110	971.048	970.985	970.922	970.859	970.796	970.732	970.669	970.606
82	970.543	970.479	970.416	970.353	970.289	970.226	970.162	970.098	970.035	969.971
83	969.907	969.843	969.772	969.715	969.652	969.587	969.523	969.459	969.395	969.331
84	969.267	969.202	969.138	969.073	969.009	968.944	968.880	968.815	968.751	968.686
85	968.621	968.556	968.491	968.427	968.362	968.297	969.232	968.166	968.101	968.036
86	967.971	967.906	967.840	967.775	967.709	967.641	967.578	967.513	967.447	967.381
87	967.316	967.250	967.184	967.118	967.052	966.986	966.920	966.854	966.788	966.722
88	966.656	966.589	966.523	966.457	966.390	966.324	966.257	966.191	966.124	966.057
89	965.991	965.924	965.857	965.790	965.723	965.656	965.589	965.522	965.455	965.388
90	965.321	965.254	965.187	965.119	965.052	964.984	964.917	964.849	964.782	964.714
91	964.647	954.579	964.511	964.443	964.376	964.308	964.240	964.172	964.104	964.036
92	963.967	963.899	963.831	963.763	963.694	963.626	963.558	963.489	963.421	963.352
93	963.284	963.215	963.146	963.077	963.009	962.940	962.871	962.802	962.733	962.664
94	962.595	962.526	962.457	962.387	962.318	962.249	962.180	962.110	962.041	961.971
95	961.902	961.832	961.762	961.693	961.693	961.553	961.483	961.414	961.344	961.274
96	961.204	961.134	961.064	960.993	960.923	960.853	960,783	960.712	960.642	960.572
97	960.501	960.431	960.360	960.289	960.219	960.148	960.077	960.006	959.936	959.865
98	959.794	959.723	959.652	959.581	959.510	959.438	959.367	959.296	959.225	959.153
99	959.082	959.010	958.939	958.867	958.796	958.724	958.653	958.581	958.509	958.431
100	958.365									

TABLE 3-29 Density of Mercury from 0 to 350°C*

t, °C	Density, kg/m³									
	0	1	2	3	4	5	6	7	8	9
0	13595.08	13592.61	13590.14	13587.68	13585.21	13582.75	13580.29	13577.82	13575.36	13572.90
10	13570.44	13567.98	13565.52	13563.06	13560.60	13558.14	13555.69	13553.23	13550.78	13548.32
20	13545.87	13543.41	13540.96	13538.51	13536.06	13533.61	13531.16	13528.71	13526.26	13523.81
30	13521.36	13518.91	13516.47	13514.02	13511.58	13509.13	13506.69	13504.25	13501.80	13499.36
40	13496.92	13494.48	13492.04	13489.60	13487.16	13484.72	13482.29	13479.85	13477.41	13474.98
50	13472.54	13470.11	13467.67	13465.24	13462.81	13460.38	13457.94	13455.51	13453.08	13450.65
60	13448.22	13445.80	13443.37	13440.94	13438.51	13436.09	13433.66	13431.23	13428.81	13426.39
70	13423.96	13421.54	13419.12	13416.69	13414.27	13411.85	13409.43	13407.01	13404.59	13402.17
80	13399.75	13397.34	13394.92	13392.50	13390.08	13387.67	13385.25	13382.84	13380.42	13378.01
90	13375.59	13373.18	13370.77	13368.36	13365.94	13363.53	13361.12	13358.71	13356.30	13353.89
100	13351.5	13349.1	13346.7	13344.3	13341.9	13339.4	13337.0	13334.6	13332.2	13329.8
110	13327.4	13325.0	13322.6	13320.2	13317.8	13315.4	13313.0	13310.6	13308.2	13305.8
120	13303.4	13301.0	13298.6	13296.2	13293.8	13291.4	13288.9	13286.6	13284.2	13281.8
130	13279.4	13277.0	13274.6	13272.2	13269.8	13267.4	13265.0	13262.6	13260.2	13257.8
140	13255.4	13253.0	13250.6	13248.2	13245.8	13243.4	13241.0	13238.7	13236.3	13233.9
150	13231.5	13229.1	13226.7	13224.3	13221.9	13219.5	13217.1	13214.7	13212.4	13210.0
160	13207.6	13205.2	13202.8	13200.4	13198.0	13195.6	13193.2	13190.8	13188.5	13186.1
170	13183.7	13181.3	13178.9	13176.5	13174.1	13171.7	13169.4	13167.0	13164.6	13162.2
180	13159.8	13157.4	13155.0	13152.6	13150.3	13147.9	13145.5	13143.1	13140.7	13138.3
190	13136.0	13133.6	13131.2	13128.3	13126.4	13124.0	13121.7	13119.3	13116.9	13114.5
200	13112.1	13109.7	13107.4	13105.0	13102.6	13100.2	13097.8	13095.4	13093.1	13090.7
210	13088.3	13085.9	13083.5	13081.1	13078.8	13076.4	13074.0	13071.6	13069.2	13066.8
220	13064.5	13062.1	13059.7	13057.3	13054.9	13052.6	13050.2	13047.8	13045.4	13043.0
230	13040.6	13038.3	13035.9	13033.5	13031.1	13028.7	13026.4	13024.0	13021.6	13019.2
240	13016.8	13014.5	13012.1	13009.7	13007.3	13004.9	13002.5	13000.2	12997.8	12995.4
250	12993.0	12990.6	12988.3	12985.9	12983.5	12981.1	12978.7	12976.3	12974.0	12971.6
260	12969.2	12966.8	12964.4	12962.0	12959.7	12957.3	12954.9	12952.5	12950.1	12947.7
270	12945.4	12943.0	12940.6	12938.2	12935.8	12933.4	12931.1	12928.7	12926.3	12923.9
280	12921.5	12919.1	12916.7	12914.4	12912.0	12909.6	12907.2	12904.8	12902.4	12900.0
290	12897.7	12895.3	12892.9	12890.5	12888.1	12885.7	12883.3	12880.9	12878.5	12876.2
300	12873.8	12871.4	12869.0	12866.6	12864.2	12861.8	12859.4	12857.0	12854.6	12852.2
310	12849.9	12847.5	12845.1	12842.7	12840.3	12837.9	12835.5	12833.1	12830.7	12828.3
320	12825.9	12823.5	12821.1	12818.7	12816.3	12813.9	12811.5	12809.1	12806.7	12804.3
330	12801.9	12799.5	12797.1	12794.7	12792.3	12789.9	12787.5	12785.1	12782.7	12780.2
340	12777.8	12775.4	12773.0	12770.6	12768.2	12765.8	12763.4	12761.0	12758.6	12756.1
350	12753.7									

*From "Mercury—Density and Thermal Expansion at Atmospheric Pressure and Temperatures from 0 to 350°C," *Tables of Standard Handbook Data,* Standartov, Moscow, 1978. The density values obtainable from those cited for the specific volume of the saturated liquid in the "Thermodynamic Properties" subsection show minor differences. No attempt was made to adjust either set.

TABLE 3-30 Densities of Gases at Standard Conditions (0°C, 1 atm)*

Gas	Formula	Mol. wt.	Density G./l.	Density Lb./cu. ft.
Acetylene	C_2H_2	26.02	1.1708	0.0732
Air	....	....	1.2928	0.0808
Ammonia	NH_3	17.03	0.7708	0.0482
Argon	A	39.91	1.7828	0.1114
Bromine	Br_2	159.83	7.1388	0.4460
Butane	C_4H_{10}	58.08	2.5985	0.1623
Carbon dioxide	CO_2	44.00	1.9768	0.1235
Carbon monoxide	CO	28.00	1.2501	0.0781
Carbon oxychloride	$COCl_2$	98.91	4.5313	0.2830
Carbon oxysulfide	COS	60.06	2.7201	0.1700
Chlorine	Cl_2	70.91	3.2204	0.2011
Chlorine monoxide	Cl_2O	86.91	3.8874	0.2428
Cyanogen	C_2N_2	52.02	2.3348	0.1459
Ethane	C_2H_6	30.05	1.3567	0.0848
Ethyl chloride	C_2H_5Cl	64.50	2.8700	0.1793
Ethylene	C_2H_4	28.03	1.2644	0.0783
Fluorine	F_2	38.00	1.6354	0.1022
Helium	He	4.00	0.1769	0.0111
Hydrogen	H_2	2.016	0.0898	0.0056
Hydrogen chloride	HCl	36.47	1.6394	0.1024
Hydrogen fluoride	HF	20.01	0.9218	0.0576
Hydrogen iodide	HI	127.94	5.7245	0.3576
Hydrogen selenide	H_2Se	81.22	3.6134	0.2258
Hydrogen sulfide	H_2S	34.08	1.5392	0.0961
Hydrogen telluride	H_2Te	129.52	5.8034	0.3625
Krypton	Kr	82.90	3.6431	0.2275
Methane	CH_4	16.03	0.7167	0.0448
Methyl chloride	CH_3Cl	50.48	2.3044	0.1440
Neon	Ne	20.40	0.8713	0.0544
Nitric oxide	NO	30.01	1.3401	0.0837
Nitrogen	N_2	28.02	1.2507	0.0782
Nitrous oxide	N_2O	44.02	1.9781	0.1235
Nitrosyl chloride	NOCl	65.47	2.9864	0.1865
Oxygen	O_2	32.00	1.4289	0.0892
Phosphine	PH_4	34.05	1.5293	0.0955
Silicon fluoride	SiF_2	104.06	4.6541	0.2907
Sulfur dioxide	SO_2	64.06	2.9268	0.1828
Xenon	X	130.20	5.7168	0.3570

*For other tables see Atack, "Handbook of Chemical Data," pp. 90, 91, Reinhold, New York, 1957; "Smithsonian Physical Tables," 9th ed., Table 255, 1954; "American Institute of Physics Handbook," p. 2–200, McGraw-Hill, New York, 1957.

NOTE: For ideal-gas behavior, density may be calculated as MP/RT.

DENSITIES OF AQUEOUS INORGANIC SOLUTIONS

UNITS AND UNITS CONVERSIONS

Densities are given in grams per cubic centimeter. To convert to pounds per cubic foot, multiply by 62.43. °F = % °C + 32.

ADDITIONAL REFERENCES

For more detailed data on densities see *International Critical Tables:* tabular index, vol. 3, p. 1; abrasives, vol. 2, p. 87; air, moist, vol. 1, p. 71; building stones, vol. 2, p. 52; clays, vol. 2, p. 56; coals, vol. 2, p. 135; compounds, vol. 1, pp. 106, 176, 313, 341; elements, vol. 1, pp. 102, 340; fibers, vol. 2, p. 237; gases and vapors, vol. 3, pp. 3, 345; glass, vol. 2, p. 93; liquids and vitreous solids, vol. 3, p. 22; vol. 1, pp. 102, 340; vol. 2, pp. 456, 463; vol. 3, pp. 20, 35; liquid coolants and saturated vapors are available from WADC-TR-59-598, 1959; plastics are collected in the *Handbook of Chemistry and Physics*, Chemical Rubber Publishing Co.; solid helium, neon, argon, fluorine, and methane data are given by Johnson (ed.), WADD-TR-60-56, 1960; temperatures of maximum solubility, vol. 3, p. 107; metals, vol. 2, p. 463; oils, fats, and waxes, vol. 2, p. 201; orthobaric, vol. 3, pp. 202, 228, 237, 244; petroleums, vol. 2, pp. 137, 144; plastics, vol. 2, p. 296; porcelains, vol. 2, pp. 68, 75; refrigerating brines, vol. 2, p. 327; rubber, vol. 2, pp. 255, 259; soaps, vol. 5, p. 447; metallic solid solutions, vol. 2, p. 358; solids, vol. 3, pp. 43, 45; vol. 2, p. 456; vol. 3, p. 21; solutions and mixtures, vol. 3, pp. 17, 51, 95, 104, 107, 111, 125, 130; woods, vol. 2, p. 1. Also see the *Handbook of Chemistry and Physics*, Chemical Rubber Publishing Co., 40th ed., etc.

TABLE 3-31 Aluminum Sulfate [$Al_2(SO_4)_3$]

%	d_4^{15}	%	d_4^{15}
1	1.0093	16	1.1770
2	1.0195	20	1.2272
4	1.0404	24	1.2803
8	1.0837	26	1.3079
12	1.1293		

TABLE 3-32 Ammonia (NH_3)

%	−15°C.	−10°C.	−5°C.	0°C.	5°C.	10°C.	20°C.	25°C.	%	d_4^{15}
1	.....	0.9943	0.9954	0.9959	0.9958	0.9955	0.9939	0.993	32	0.889
2	.....	.9906	.9915	.9919	.9917	.9913	.9895	.988	36	.877
4	.....	.9834	.9840	.9842	.9837	.9832	.9811	.980	40	.865
8	0.970	.9701	.9701	.9695	.9686	.9677	.9651	.964	45	.849
12	.958	.9576	.9571	.9561	.9548	.9534	.9501	.948	50	.832
16	.947	.9461	.9450	.9435	.9420	.9402	.9362	.934	60	.796
20	.....	.9353	.9335	.9316	.9296	.9275	.9229		70	.755
24	.....	.9249	.9226	.9202	.9179	.9155	.9101		80	.711
28	.....	.9150	.9122	.9094	.9067	.9040	.8980		90	.665
30	.....	.9101	.9070	.9040	.9012	.8983	.8920		100	.618

TABLE 3-33 Ammonium Acetate* (CH_3COONH_4)

%	d_4^{25}
1	0.9992
2	1.0013
4	1.0055
8	1.0136
12	1.0216
16	1.0294
20	1.0368
24	1.0439
28	1.0507
30	1.0540
35	1.0618
40	1.0691
45	1.0760

* For data at 16°C. for 3(1)52 per cent see Atack "Handbook of Chemical Data," p. 33, Reinhold, New York, 1957.

TABLE 3-34 Ammonium Bichromate [$(NH_4)_2CR_2O_7$]

%	d_4^{12}
1	1.0051
2	1.0108
4	1.0223
8	1.0463
12	1.0715
16	1.0981
20	1.1263

TABLE 3-35 Ammonium Chloride (NH_4Cl)

%	0°C.	10°C.	20°C.	30°C.	50°C.	80°C.	100°C.
1	1.0033	1.0029	1.0013	0.9987	0.9910	0.9749	0.9617
2	1.0067	1.0062	1.0045	1.0018	.9940	.9780	.9651
4	1.0135	1.0126	1.0107	1.0077	.9999	.9842	.9718
8	1.0266	1.0251	1.0227	1.0195	1.0116	.9963	.9849
12	1.0391	1.0370	1.0344	1.0310	1.0231	1.0081	.9975
16	1.0510	1.0485	1.0457	1.0422	1.0343	1.0198	1.0096
20	1.0625	1.0596	1.0567	1.0532	1.0454	1.0312	1.0213
24	1.0736	1.0705	1.0674	1.0641	1.0564	1.0426	1.0327

TABLE 3-36 Ammonium Chromate [$(NH_4)_2CrO_4$]

%	°C.	d_4^t
3.80	20	1.0219
10.52	13	1.0627
19.75	13.7	1.1189
28.04	19.6	1.1707

TABLE 3-37 Ammonium Nitrate (NH_4NO_3)

%	0°C.	10°C.	25°C.	40°C.	60°C.	80°C.
1.0	1.0043	1.0039	1.0011	0.9961	0.9870	0.9755
2.0	1.0088	1.0082	1.0051	1.0000	.9908	.9793
4.0	1.0178	1.0168	1.0132	1.0079	.9985	.9869
8.0	1.0358	1.0340	1.0297	1.0238	1.0142	1.0024
12.0	1.0539	1.0515	1.0464	1.0400	1.0301	1.0181
16.0	1.0721	1.0691	1.0633	1.0565	1.0462	1.0342
20.0	1.0905	1.0870	1.0806	1.0734	1.0627	1.0506
24.0	1.1090	1.1051	1.0982	1.0907	1.0796	1.0673
28.0	1.1277	1.1234	1.1161	1.1082	1.0968	1.0844
30.0	1.1371	1.1327	1.1252	1.1171	1.1055	1.0931
40.0	1.1862	1.1810	1.1727	1.1640	1.1515	1.1385
50.0	1.2380	1.2320	1.2229	1.2136	1.2006	1.1868

TABLE 3-38 Ammonium Sulfate [$(NH_4)_2SO_4$]

%	0°C.	20°C.	40°C.	80°C.	100°C.
1	1.0061	1.0041	0.9980	0.9777	0.9644
2	1.0124	1.0101	1.0039	.9836	.9705
4	1.0248	1.0220	1.0155	.9953	.9826
8	1.0495	1.0456	1.0387	1.0187	1.0066
12	1.0740	1.0691	1.0619	1.0421	1.0303
16	1.0980	1.0924	1.0849	1.0653	1.0539
20	1.1215	1.1154	1.1077	1.0883	1.0772
24	1.1448	1.1383	1.1304	1.1111	1.1003
28	1.1677	1.1609	1.1529	1.1338	1.1232
35	1.2072	1.2800	1.1919	1.1731	1.1629
40	1.2350	1.2277	1.2196	1.2011	1.1910
50	1.2899	1.2825	1.2745	1.2568	1.2466

TABLE 3-39 Arsenic Acid ($H_3A_3O_4$)

%	d_4^{15}	%	d_4^{15}
1	1.0057	20	1.1447
2	1.0124	30	1.2331
6	1.0398	40	1.3370
10	1.0681	50	1.4602
16	1.1128	60	1.6070
		70	1.7811

TABLE 3-40 Barium Chloride ($BaCl_2$)

%	0°C.	20°C.	40°C.	60°C.	80°C.	100°C.
2	1.0181	1.0159	1.0096	1.0004	0.9890	0.9755
4	1.0368	1.0341	1.0275	1.0181	1.0066	.9931
8	1.0760	1.0721	1.0648	1.0551	1.0434	1.0299
12	1.1178	1.1128	1.1047	1.0948	1.0827	1.0692
16	1.1627	1.1564	1.1478	1.1373	1.1249	1.1113
20	1.2105	1.2031	1.1938	1.1828	1.1702	1.1563
24	......	1.2531	1.2430	1.2316	1.2186	1.2045
26	......	1.2793	1.2688	1.2571	1.2440	1.2298

TABLE 3-41 Cadmium Nitrate [$Cd(NO_3)_2$]

%	d_4^{18}	%	d_4^{18}
2	1.0154	20	1.1904
4	1.0326	25	1.2488
8	1.0683	30	1.3124
12	1.1061	40	1.4590
16	1.1468	50	1.6356

TABLE 3-42 Calcium Chloride ($CaCl_2$)

%	−5°C.	0°C.	20°C.	30°C.	40°C.	60°C.	80°C.	100°C.	120°C.*	140°C.
2	......	1.0171	1.0148	1.0120	1.0084	0.9994	0.9881	0.9748	0.9596	0.9428
4	......	1.0346	1.0316	1.0286	1.0249	1.0158	1.0046	.9915	.9765	.9601
8	1.0708	1.0703	1.0659	1.0626	1.0586	1.0492	1.0382	1.0257	1.0111	.9954
12	1.1083	1.1072	1.1015	1.0978	1.0937	1.0840	1.0730	1.0610	1.0466	1.0317
16	1.1471	1.1454	1.1386	1.1345	1.1301	1.1202	1.1092	1.0973	1.0835	1.0691
20	1.1874	1.1853	1.1775	1.1730	1.1684	1.1581	1.1471	1.1352	1.1219	1.1080
25	......	1.2376	1.2284	1.2236	1.2186	1.2079	1.1965	1.1846		
30	......	1.2922	1.2816	1.2764	1.2709	1.2597	1.2478	1.2359		
35	......	......	1.3373	1.3316	1.3255	1.3137	1.3013	1.2893		
40	......	......	1.3957	1.3895	1.3826	1.3700	1.3571	1.3450		

* Corrected to atmospheric pressure.

TABLE 3-43 Calcium Hydroxide [$Ca(OH)_2$]

%	d_4^{15}	d_4^{25}
0.05	0.99979	0.99773
.10	1.00044	.99838
.15	1.00110	.99904

TABLE 3-44 Calcium Hypochlorite* ($CaOCl_2$)

% total salt	d_4^{15}
2	1.0169
4	1.0345
6	1.0520
8	1.0697
10	1.0876
12	1.1060

* $CaOCl_2$ = 89.15%;
$CaCl_2$ = 7.31%;
$Ca(ClO_3)_2$ = 0.26%;
$Ca(OH)_2$ = 2.92%.

TABLE 3-45 Calcium Nitrate [$Ca(NO_3)_2$]

%	6°C.	18°C.	25°C.	30°C.
2*	1.0157	1.0137	1.0120	1.0105
4	1.0316	1.0291	1.0272	1.0256
8	1.0641	1.0608	1.0585	1.0565
12	1.0979	1.0937	1.0911	1.0887
16	1.1330	1.1279	1.1250	1.1224
20	1.1694	1.1636	1.1602	1.1575
25	1.2168	1.2106	1.2065	1.2032
30	......	1.260		
35	......	1.311		
40	......	1.365		
45	......	1.422		
68*	......	1.747	1.741	1.736

* Supercooled tetrahydrate (m.p. 41.4°C.).

TABLE 3-46 Chromic Acid (CrO_3)

%	d_4^{15}	%	d_4^{15}
1	1.006	20	1.163
2	1.014	26	1.220
6	1.045	30	1.260
10	1.076	40	1.371
16	1.127	50	1.505
		60	1.663

TABLE 3-47 Chromium Chloride ($CrCl_3$)

%	d_4^{18}		
	Violet	Green	Equilibrium mixture of violet and green
1	1.0076	1.0071	1.0075
2	1.0166	1.0157	1.0165
4	1.0349	1.0332	1.0347
8	1.0724	1.0691	1.0722
12	1.1114	1.1065	1.1111
14	1.1316		

TABLE 3-48 Copper Nitrate [$Cu(NO_3)_2$]

%	d_4^{20}	%	d_4^{20}
1	1.007	12	1.107
2	1.015	16	1.147
4	1.032	20	1.189
8	1.069	25	1.248

TABLE 3-49 Copper Sulfate ($CuSO_4$)

%	0°C.	20°C.	40°C
1	1.0104	1.0086	1.0024
4	1.0429	1.0401	1.0332
8	1.0887	1.084	1.0764
12	1.1379	1.1308	1.1222
16	......	1.180	
18	......	1.206	

TABLE 3-50 Cuprous Chloride (Cu_2Cl_2)

%	0°C.	20°C.	40°C.
1	1.0095	1.0072	1.002
4	1.0387	1.036	1.0305
8	1.0788	1.0754	1.0682
12	1.1208	1.1165	1.107
16	1.1653	1.1595	1.151
20	1.2121	1.2052	1.1953

TABLE 3-51 Ferric Chloride ($FeCl_3$)

%	0°C.	10°C.	20°C.	30°C.
1	1.0086	1.0084	1.0068	1.0040
2	1.0174	1.0168	1.0152	1.0122
4	1.0347	1.0341	1.0324	1.0292
8	1.0703	1.0692	1.0669	1.0636
12	1.1088	1.107[illegible]	1.1040	1.1006
16	1.1475	1.1449	1.1418	1.1386
20	1.1870	1.1847	1.1820	1.1786
25	1.2400	1.2380	1.2340	1.2290
30	1.2970	1.2950	1.2910	1.2850
35	1.3605	1.3580	1.3530	1.3475
40	1.4280	1.4235	1.4175	1.4115
45	......	1.4920	1.4850	
50	......	1.5610	1.5510	

TABLE 3-52 Ferric Sulfate [$Fe_2(SO_4)_3$]

%	$d_4^{17.5}$
1	1.0072
2	1.0157
4	1.0327
8	1.0670
12	1.1028
16	1.1409
20	1.1811
30	1.3073
40	1.4487
50	1.6127
60	1.7983

TABLE 3-53 Ferric Nitrate [$Fe(NO_3)_3$]

%	d_4^{18}
1	1.0065
2	1.0144
4	1.0304
8	1.0636
12	1.0989
16	1.1359
20	1.1748
25	1.2281

TABLE 3-54 Ferrous Sulfate ($FeSO_4$)

%	15°C.	18°C.	20°C.
0.2	......	1.00068	1.0002
0.4	......	1.00275	1.0022
0.8	......	1.00645	1.0062
1.0	1.0090	1.0085	1.0082
4.0	1.0380	1.0375	
8.0	1.0790	1.0785	
12.0	1.1235	1.1220	
16.0	1.1690	1.1675	
20.0	1.2150	1.2135	

TABLE 3-55 Hydrogen Bromide (HBr)

%	d_4^{4}	d_4^{10}	d_4^{25}
1.0	1.0073	1.0068	1.0041
2.0	1.0146	1.0139	1.0111
4.0	1.0295	1.0285	1.0255
6.0	1.0448	1.0435	1.0402
8.0	1.0604	1.0589	1.0552
10.0	1.0764	1.0747	1.0707
12.0	1.0928	1.0910	1.0867
14.0	1.1097	1.1078	1.1032
16.0	1.1272	1.1251	1.1202
18.0	1.1453	1.1430	1.1377
20.0	1.1640	1.1615	1.1557
22.0	1.1832	1.1806	1.1743
24.0	1.2030	1.2003	1.1935
26.0	1.2235	1.2206	1.2134
28.0	1.2446	1.2415	1.2340
30.0	1.2663	1.2630	1.2552
40.0	1.3877	1.3838	1.3736
50.0	1.5305	1.5257	1.5127
60.0	1.6950	1.6892	1.6731
65.0	1.7854	1.7792	1.7613

TABLE 3-56 Hydrogen Cyanide (HCN)

%	d_4^{15}
1	0.998
2	.996
4	.993
8	.984
12	.971
16	.956
82	.752
90	.724
100	.691

TABLE 3-57 Hydrogen Chloride (HCl)

%	−5°C.	0°C.	10°C.	20°C.	40°C.	60°C.	80°C.	100°C.
1	1.0048	1.0052	1.0048	1.0032	0.9970	0.9881	0.9768	0.9636
2	1.0104	1.0106	1.0100	1.0082	1.0019	.9930	0.9819	.9688
4	1.0213	1.0213	1.0202	1.0181	1.0116	1.0026	0.9919	.9791
6	1.0321	1.0319	1.0303	1.0279	1.0211	1.0121	1.0016	.9892
8	1.0428	1.0423	1.0403	1.0376	1.0305	1.0215	1.0111	.9992
10	1.0536	1.0528	1.0504	1.0474	1.0400	1.0310	1.0206	1.0090
12	1.0645	1.0634	1.0607	1.0574	1.0497	1.0406	1.0302	1.0188
14	1.0754	1.0741	1.0711	1.0675	1.0594	1.0502	1.0398	1.0286
16	1.0864	1.0849	1.0815	1.0776	1.0692	1.0598	1.0494	1.0383
18	1.0975	1.0958	1.0920	1.0878	1.0790	1.0694	1.0590	1.0479
20	1.1087	1.1067	1.1025	1.0980	1.0888	1.0790	1.0685	1.0574
22	1.1200	1.1177	1.1131	1.1083	1.0986	1.0886	1.0780	1.0668
24	1.1314	1.1287	1.1238	1.1187	1.1085	1.0982	1.0874	1.0761
26	1.1426	1.1396	1.1344	1.1290	1.1183	1.1076	1.0967	1.0853
28	1.1537	1.1505	1.1449	1.1392	1.1280	1.1169	1.1058	1.0942
30	1.1648	1.1613	1.1553	1.1493	1.1376	1.1260	1.1149	1.1030
32	......	......	......	1.1593				
34	......	......	......	1.1691				
36	......	......	......	1.1789				
38	......	......	......	1.1885				
40	......	......	......	1.1980				

TABLE 3-58 Hydrogen Fluoride (HF)

%	d_4^{20}	d_4^{0}
5	1.020	1.017
10	1.040	1.035
20	1.080	1.070
30	1.119	1.101
40	1.159	1.130
50	1.198	1.155
60	1.235	
70	1.258	
80	1.259	
90	1.178	
95	1.089	
100	1.0005	

TABLE 3-59 Hydrogen Peroxide (H_2O_2)

%	d_4^{18}	%	d_4^{18}
1	1.0022	26	1.0959
2	1.0058	28	1.1040
4	1.0131	30	1.1122
6	1.0204	35	1.1327
8	1.0277	40	1.1536
10	1.0351	45	1.1749
12	1.0425	50	1.1966
14	1.0499	55	1.2188
16	1.0574	60	1.2416
18	1.0649	70	1.2897
20	1.0725	80	1.3406
22	1.0802	90	1.3931
24	1.0880	100	1.4465

TABLE 3-60 Hydrofluosilic Acid (H_2SiF_6)

%	$d_4^{17.5}$	%	$d_4^{17.5}$
1	1.0080	16	1.1373
2	1.0161	20	1.1748
4	1.0324	25	1.2235
8	1.0661	30	1.2742
12	1.1011	34	1.3162

TABLE 3-61 Magnesium Chloride ($MgCl_2$)

%	0°C.	20°C.	40°C.	60°C.	80°C.	100°C.
2	1.0168	1.0146	1.0084	0.9995	0.9883	0.9753
4	1.0338	1.0311	1.0248	1.0159	1.0050	.9923
8	1.0683	1.0646	1.0580	1.0493	1.0388	1.0269
12	1.1035	1.0989	1.0921	1.0836	1.0735	1.0622
16	1.1395	1.1342	1.1272	1.1188	1.1092	1.0984
20	1.1764	1.1706	1.1635	1.1552	1.1460	1.1359
25	1.2246	1.2184	1.2111	1.2031	1.1942	1.1847
30	1.2754	1.2688	1.2614	1.2535	1.2451	1.2360

TABLE 3-62 Magnesium Sulfate ($MgSO_4$)

%	0°C.	20°C.	30°C.	40°C.	50°C.	60°C.	80°C.
2	1.0210	1.0186	1.0158	1.0123	1.0081	1.0032	0.9916
4	1.0423	1.0392	1.0362	1.0326	1.0283	1.0234	1.0118
8	1.0858	1.0816	1.0782	1.0743	1.0700	1.0650	1.0534
12	1.1309	1.1256	1.1220	1.1179	1.1135	1.1083	1.0968
16	1.1777	1.1717	1.1679	1.1637	1.1592		
20	1.2264	1.2198	1.2159	1.2117	1.2072		
26	1.3032	1.2961	1.2922	1.2879	1.2836		

TABLE 3-63 Nickel Chloride ($NiCl_2$)

%	d_4^{18}
1	1.0082
2	1.0179
4	1.0375
8	1.0785
12	1.1217
16	1.1674
20	1.2163
30	1.353

TABLE 3-64 Nickel Nitrate [$Ni(NO_3)_2$]

%	d_4^{20}
1	1.0065
2	1.0150
4	1.0325
8	1.0688
12	1.1070
16	1.1480
20	1.191
30	1.311
35	1.377

TABLE 3-65 Nickel Sulfate ($NiSO_4$)

%	d_4^{18}
1	1.0091
2	1.0198
4	1.0415
8	1.0852
12	1.1325
16	1.1825
18	1.2090

TABLE 3-66 Nitric Acid (HNO_3)

%	0°C.	5°C.	10°C.	15°C.	20°C.	25°C.	30°C.	40°C.	50°C.	60°C.	80°C.	100°C.
1	1.0058	1.00572	1.00534	1.00464	1.00364	1.00241	1.0009	0.9973	0.9931	0.9882	0.9767	0.9632
2	1.0117	1.01149	1.01099	1.01018	1.00909	1.00778	1.0061	1.0025	.9982	.9932	.9816	.9681
3	1.0176	1.01730	1.01668	1.01576	1.01457	1.01318	1.0114	1.0077	1.0033	.9982	.9865	.9730
4	1.0236	1.02315	1.02240	1.02137	1 02008	1.01861	1.0168	1.0129	1.0084	1.0033	.9915	.9779
5	1.0296	1.02904	1.02816	1.02702	1.02563	1.02408	1.0222	1.0182	1.0136	1.0084	.9965	.9829
6	1.0357	1.03497	1.03397	1.03272	1.03122	1.02958	1.0277	1.0235	1.0188	1.0136	1.0015	.9879
7	1.0418	1.0410	1.0399	1.0385	1.0369	1.0352	1.0333	1.0289	1.0241	1.0188	1.0066	.9929
8	1.0480	1.0471	1.0458	1.0443	1.0427	1.0409	1.0389	1.0344	1.0295	1.0241	1.0117	.9980
9	1.0543	1.0532	1.0518	1.0502	1.0485	1.0466	1.0446	1.0399	1.0349	1.0294	1.0169	1.0032
10	1.0606	1.0594	1.0578	1.0561	1.0543	1.0523	1.0503	1.0455	1.0403	1.0347	1.0221	1.0083
11	1.0669	1.0656	1.0639	1.0621	1.0602	1.0581	1.0560	1.0511	1.0458	1.0401	1.0273	1.0134
12	1.0733	1.0718	1.0700	1.0681	1.0661	1.0640	1.0618	1.0567	1.0513	1.0455	1.0326	1.0186
13	1.0797	1.0781	1.0762	1.0742	1.0721	1.0699	1.0676	1.0624	1.0568	1.0509	1.0379	1.0238
14	1.0862	1.0845	1.0824	1.0803	1.0781	1.0758	1.0735	1.0681	1.0624	1.0564	1.0432	1.0289
15	1.0927	1.0909	1.0887	1.0865	1.0842	1.0818	1.0794	1.0739	1.0680	1.0619	1.0485	1.0341
16	1.0992	1.0973	1.0950	1.0927	1.0903	1.0879	1.0854	1.0797	1.0737	1.0675	1.0538	1.0393
17	1.1057	1.1038	1.1014	1.0989	1.0964	1.0940	1.0914	1.0855	1.0794	1.0731	1.0592	1.0444
18	1.1123	1.1103	1.1078	1.1052	1.1026	1.1001	1.0974	1.0913	1.0851	1.0787	1.0646	1.0496
19	1.1189	1.1168	1.1142	1.1115	1.1088	1.1062	1.1034	1.0972	1.0908	1.0843	1.0700	1.0547
20	1.1255	1.1234	1.1206	1.1178	1.1150	1.1123	1.1094	1.1031	1.0966	1.0899	1.0754	1.0598
21	1.1322	1.1300	1.1271	1.1242	1.1213	1.1185	1.1155	1.1090	1.1024	1.0956	1.0808	1.0650
22	1.1389	1.1366	1.1336	1.1306	1.1276	1.1247	1.1217	1.1150	1.1083	1.1013	1.0862	1.0701
23	1.1457	1.1433	1.1402	1.1371	1.1340	1.1310	1.1280	1.1210	1.1142	1.1070	1.0917	1.0753
24	1.1525	1.1501	1.1469	1.1437	1.1404	1.1374	1.1343	1.1271	1.1201	1.1127	1.0972	1.0805
25	1.1594	1.1569	1.1536	1.1503	1.1469	1.1438	1.1406	1.1332	1.1260	1.1185	1.1027	1.0857
26	1.1663	1.1638	1.1603	1.1569	1.1534	1.1502	1.1469	1.1394	1.1320	1.1244	1.1083	1.0910
27	1.1733	1.1707	1.1670	1.1635	1.1600	1.1566	1.1533	1.1456	1.1381	1.1303	1.1139	1.0963
28	1.1803	1.1777	1.1738	1.1702	1.1666	1.1631	1.1597	1.1519	1.1442	1.1362	1.1195	1.1016
29	1.1874	1.1847	1.1807	1.1770	1.1733	1.1697	1.1662	1.1582	1.1503	1.1422	1.1251	1.1069
30	1.1945	1.1917	1.1876	1.1838	1.1800	1.1763	1.1727	1.1645	1.1564	1.1482	1.1307	1.1122
31	1.2016	1.1988	1.1945	1.1906	1.1867	1.1829	1.1792	1.1708	1.1625	1.1542	1.1363	1.1175
32	1.2088	1.2059	1.2014	1.1974	1.1934	1.1896	1.1857	1.1772	1.1687	1.1602	1.1419	1.1228
33	1.2160	1.2131	1.2084	1.2043	1.2002	1.1963	1.1922	1.1836	1.1749	1.1662	1.1476	1.1281
34	1.2233	1.2203	1.2155	1.2113	1.2071	1.2030	1.1988	1.1901	1.1812	1.1723	1.1533	1.1335
35	1.2306	1.2275	1.2227	1.2183	1.2140	1.2098	1.2055	1.1966	1.1876	1.1784	1.1591	1.1390
36	1.2375	1.2344	1.2294	1.2249	1.2205	1.2163	1.2119	1.2028	1.1936	1.1842	1.1645	1.1440
37	1.2444	1.2412	1.2361	1.2315	1.2270	1.2227	1.2182	1.2089	1.1995	1.1899	1.1699	1.1490
38	1.2513	1.2479	1.2428	1.2381	1.2335	1.2291	1.2245	1.2150	1.2054	1.1956	1.1752	1.1540
39	1.2581	1.2546	1.2494	1.2446	1.2399	1.2354	1.2308	1.2210	1.2112	1.2013	1.1805	1.1589
40	1.2649	1.2613	1.2560	1.2511	1.2463	1.2417	1.2370	1.2270	1.2170	1.2069	1.1858	1.1638
41	1.2717	1.2680	1.2626	1.2576	1.2527	1.2480	1.2432	1.2330	1.2229	1.2126	1.1911	1.1687
42	1.2786	1.2747	1.2692	1.2641	1.2591	1.2543	1.2494	1.2390	1.2287	1.2182	1.1963	1.1735
43	1.2854	1.2814	1.2758	1.2706	1.2655	1.2606	1.2556	1.2450	1.2345	1.2238	1.2015	1.1783
44	1.2922	1.2880	1.2824	1.2771	1.2719	1.2669	1.2618	1.2510	1.2403	1.2294	1.2067	1.1831
45	1.2990	1.2947	1.2890	1.2836	1.2783	1.2732	1.2680	1.2570	1.2461	1.2350	1.2119	1.1879
46	1.3058	1.3014	1.2955	1.2901	1.2847	1.2795	1.2742	1.2630	1.2519	1.2406	1.2171	1.1927
47	1.3126	1.3080	1.3021	1.2966	1.2911	1.2858	1.2804	1.2690	1.2577	1.2462	1.2223	1.1976
48	1.3194	1.3147	1.3087	1.3031	1.2975	1.2921	1.2867	1.2750	1.2635	1.2518	1.2275	1.2024
49	1.3263	1.3214	1.3153	1.3096	1.3040	1.2984	1.2929	1.2811	1.2693	1.2575	1.2328	1.2073
50	1.3327	1.3277	1.3215	1.3157	1.3100	1.3043	1.2987	1.2867	1.2748	1.2628	1.2377	1.2118
51	1.3391	1.3339	1.3277	1.3218	1.3160	1.3102	1.3045	1.2923	1.2802	1.2680	1.2425	1.2163
52	1.3454	1.3401	1.3338	1.3278	1.3219	1.3160	1.3102	1.2978	1.2856	1.2731	1.2473	1.2208
53	1.3517	1.3462	1.3399	1.3338	1.3278	1.3218	1.3159	1.3033	1.2909	1.2782	1.2521	1.2252
54	1.3579	1.3523	1.3459	1.3397	1.3336	1.3275	1.3215	1.3087	1.2961	1.2833	1.2568	1.2296
55	1.3640	1.3583	1.3518	1.3455	1.3393	1.3331	1.3270	1.3141	1.3013	1.2883	1.2615	1.2339
56	1.3700	1.3642	1.3576	1.3512	1.3449	1.3386	1.3324	1.3194	1.3064	1.2932	1.2661	1.2382
57	1.3759	1.3700	1.3634	1.3569	1.3505	1.3441	1.3377	1.3246	1.3114	1.2981	1.2706	1.2424
58	1.3818	1.3757	1.3691	1.3625	1.3560	1.3495	1.3430	1.3298	1.3164	1.3029	1.2751	1.2466
59	1.3875	1.3813	1.3747	1.3680	1.3614	1.3548	1.3482	1.3348	1.3213	1.3077	1.2795	1.2507
60	1.3931	1.3868	1.3801	1.3734	1.3667	1.3600	1.3533	1.3398	1.3261	1.3124	1.2839	1.2547
61	1.3986	1.3922	1.3855	1.3787	1.3719	1.3651	1.3583	1.3447	1.3308	1.3169	1.2881	1.2587
62	1.4039	1.3975	1.3907	1.3838	1.3769	1.3700	1.3632	1.3494	1.3354	1.3213	1.2922	1.2625
63	1.4091	1.4027	1.3958	1.3888	1.3818	1.3748	1.3679	1.3540	1.3398	1.3255	1.2962	1.2661
64	......	1.4078	1.4007	1.3936	1.3866	1.3795	1.3725					
65	......	1.4128	1.4055	1.3984	1.3913	1.3841	1.3770					
66	......	1.4177	1.4103	1.4031	1.3959	1.3887	1.3814					
67	......	1.4224	1.4150	1.4077	1.4004	1.3932	1.3857					
68	......	1.4271	1.4196	1.4122	1.4048	1.3976	1.3900					
69	......	1.4317	1.4241	1.4166	1.4091	1.4019	1.3942					
70	......	1.4362	1.4285	1.4210	1.4134	1.4061	1.3983					
71	......	1.4406	1.4328	1.4252	1.4176	1.4102	1.4023					
72	......	1.4449	1.4371	1.4294	1.4218	1.4142	1.4063					
73	......	1.4491	1.4413	1.4335	1.4258	1.4182	1.4103					
74	......	1.4532	1.4454	1.4376	1.4298	1.4221	1.4142					

TABLE 3-66 Nitric Acid (HNO_3) *(Concluded)*

%	0°C.	5°C.	10°C.	15°C.	20°C.	25°C.	30°C.	40°C.	50°C.	60°C.	80°C.	100°C.
75	......	1.4573	1.4494	1.4415	1.4337	1.4259	1.4180					
76	......	1.4613	1.4533	1.4454	1.4375	1.4296	1.4217					
77	......	1.4652	1.4572	1.4492	1.4413	1.4333	1.4253					
78	......	1.4690	1.4610	1.4529	1.4450	1.4369	1.4288					
79	......	1.4727	1.4647	1.4565	1.4486	1.4404	1.4323					
80	......	1.4764	1.4683	1.4601	1.4521	1.4439	1.4357					
81	......	1.4800	1.4718	1.4636	1.4555	1.4473	1.4391					
82	......	1.4835	1.4753	1.4670	1.4589	1.4507	1.4424					
83	......	1.4869	1.4787	1.4704	1.4622	1.4540	1.4456					
84	......	1.4903	1.4820	1.4737	1.4655	1.4572	1.4487					
85	......	1.4936	1.4852	1.4769	1.4686	1.4603	1.4518					
86	......	1.4968	1.4883	1.4799	1.4716	1.4633	1.4548					
87	......	1.4999	1.4913	1.4829	1.4745	1.4662	1.4577					
88	......	1.5029	1.4942	1.4858	1.4773	1.4690	1.4605					
89	......	1.5058	1.4970	1.4885	1.4800	1.4716	1.4631					
90	......	1.5085	1.4997	1.4911	1.4826	1.4741	1.4656					
91	......	1.5111	1.5023	1.4936	1.4850	1.4766	1.4681					
92	......	1.5136	1.5048	1.4960	1.4873	1.4789	1.4704					
93	......	1.5156	1.5068	1.4979	1.4892	1.4807	1.4722					
94	......	1.5177	1.5088	1.4999	1.4912	1.4826	1.4741					
95	......	1.5198	1.5109	1.5019	1.4932	1.4846	1.4761					
96	......	1.5220	1.5130	1.5040	1.4952	1.4867	1.4781					
97	......	1.5244	1.5152	1.5062	1.4974	1.4889	1.4802					
98	......	1.5278	1.5187	1.5096	1.5008	1.4922	1.4835					
99	......	1.5327	1.5235	1.5144	1.5056	1.4969	1.4881					
100	......	1.5402	1.5310	1.5217	1.5129	1.5040	1.4952					

TABLE 3-67 Perchloric Acid ($HClO_4$)

%	d_4^{15}	d_4^{20}	d_4^{25}	d_4^{50}	%	d_4^{15}	d_4^{20}	d_4^{50}
1	1.0050	......	1.0020	0.9933	28	1.1900	1.1851	1.1645
2	1.0109	......	1.0070	0.9986	30	1.2067	1.2013	1.1800
4	1.0228	......	1.0169	0.9906	32	1.2239	1.2183	1.1960
6	1.0348	......	1.0270	1.0205	34	1.2418	1.2359	1.2130
8	1.0471	......	1.0372	1.0320	36	1.2603	1.2542	1.2310
10	1.0597	......	1.0475	1.0440	38	1.2794	1.2732	1.2490
12	1.0726	......	......	1.0560	40	1.2991	1.2927	1.2680
14	1.0589	......	......	1.0680	45	1.3521	1.3450	1.3180
16	1.0995	......	......	1.0810	50	1.4103	1.4018	1.3730
18	1.1135	......	......	1.0940	55	1.4733	1.4636	1.4320
20	1.1279	......	......	1.1070	60	1.5389	1.5298	1.4950
22	1.1428	......	......	1.1205	65	1.6059	1.5986	1.5620
24	1.1581	......	......	1.1345	70	1.6736	1.6680	1.6290
26	1.1738	1.1697	......	1.1490				

TABLE 3-68 Phosphoric Acid (H_3PO_4)

°C.	2%	6%	14%	20%	26%	35%	50%	75%	100%
0	1.0113	1.0339	1.0811	1.1192					
10	1.0109	1.0330	1.0792	1.1167	1.1567	1.221	1.341		
20	1.0092	1.0309	1.0764	1.1134	1.1529	1.216	1.335	1.579	1.870
30	1.0065	1.0279	1.0728	1.1094	1.1484	1.211	1.329	1.572	1.862
40	1.0029	1.0241	1.0685	1.1048					

TABLE 3-69 Potassium Bicarbonate ($KHCO_3$)

°C.	1%	2%	4%	6%	8%	10%
0	1.0066	1.0134	1.0270			
10	1.0064	1.0132	1.0268			
15	1.0058	1.0125	1.0260	1.0396	1.0534	1.0674
20	1.0049	1.0117	1.0252			
30	1.0024	1.0092	1.0228			
40	0.9990	1.0058	1.0195			
50	.9949	1.0017	1.0154			
60	.9901	0.9969	1.0106			
80	.9786	.9855	0.9993			
100	.9653	.9722	.9860			

TABLE 3-70 Potassium Bromide (KBr)

%	d_4^{20}
1	1.0054
2	1.0127
6	1.0426
12	1.0903
20	1.1601
30	1.2593
40	1.3746

TABLE 3-71 Potassium Carbonate (K_2CO_3)

%	0°C.	10°C.	20°C.	40°C.	60°C.	80°C.	100°C.
1	1.0094	1.0089	1.0072	1.0010	0.9919	0.9803	0.9670
2	1.0189	1.0182	1.0163	1.0098	1.0005	.9889	.9756
4	1.0381	1.0369	1.0345	1.0276	1.0180	1.0063	.9931
8	1.0768	1.0746	1.0715	1.0640	1.0538	1.0418	1.0291
12	1.1160	1.1131	1.1096	1.1013	1.0906	1.0786	1.0663
16	1.1562	1.1530	1.1490	1.1399	1.1290	1.1170	1.1049
20	1.1977	1.1941	1.1898	1.1801	1.1690	1.1570	1.1451
24	1.2405	1.2366	1.2320	1.2219	1.2106	1.1986	1.1869
28	1.2846	1.2804	1.2756	1.2652	1.2538	1.2418	1.2301
30	1.3071	1.3028	1.2979	1.2873	1.2759	1.2640	1.2522
35	1.3646	1.3600	1.3548	1.3440	1.3324	1.3206	1.3089
40	1.4244	1.4195	1.4141	1.4029	1.3913	1.3795	1.3678
45	1.4867	1.4815	1.4759	1.4644	1.4528	1.4408	1.4290
50	1.5517	1.5462	1.5404	1.5285	1.5169	1.5048	1.4928

TABLE 3-72 Potassium Chromate (K_2CrO_4)

%	d_4^{15}	d_4^{18}
1	1.0073	1.0066
2	1.0155	1.0147
4	1.0321	1.0311
8	1.0659	1.0647
12	1.1009	1.0999
16	......	1.1366
20	......	1.1748
24	......	1.2147
28	......	1.2566
30	......	1.2784

TABLE 3-73 Potassium Chlorate ($KClO_3$)

°C.	1%	2%	3%	4%
0	1.0061	1.0124	1.0189	1.0256
10	1.0059	1.0122	1.0187	1.0254
20	1.0045	1.0109	1.0174	1.0241
30	1.0020	1.0085	1.0151	1.0218
40	0.9986	1.0051	1.0116	1.0183
60	.9895	0.9959	1.0024	1.0091
80	.9781	.9845	0.9910	0.9977
100	.9646	.9709	.9774	.9840

TABLE 3-74 Potassium Chloride (KCl)

%	0°C.	20°C.	25°C.	40°C.	60°C.	80°C.	100°C.
1.0	1.00661	1.00462	1.00342	0.99847	0.9894	0.9780	0.9646
2.0	1.01335	1.01103	1.00977	1.00471	.9956	.9842	.9708
4.0	1.02690	1.02391	1.02255	1.01727	1.0080	.9966	.9834
8.0	1.05431	1.05003	1.04847	1.04278	1.0333	1.0219	1.0088
12.0	1.08222	1.07679	1.07506	1.06897	1.0592	1.0478	1.0350
16.0	1.11068	1.10434	1.10245	1.09600	1.0861	1.0746	1.0619
20.0	1.13973	1.13280	1.13072	1.12399	1.1138	1.1024	1.0897
24.0	....	1.16226	1.15995	1.15299	1.1425	1.1311	1.1185
28.0	....	....	....	1.18304	1.1723	1.1609	1.1483

%	110°C.	120°C.	130°C.	140°C.
3.79	0.9733	0.9663	0.9583	0.9502
7.45	.9978	.9899	.9827	.9745
13.62	1.0388	1.0313	1.0238	1.0159

TABLE 3-75 Potassium Chrome Alum [$K_2Cr_2(SO_4)_4$]

%	d_4^{15}
1	1.007
2	1.016
6	1.052
10	1.089
14	1.129
20	1.193
30	1.315
40	1.456
50	1.615

TABLE 3-76 Potassium Hydroxide (KOH)

%	d_4^{15}
1.0	1.0083
2.0	1.0175
4.0	1.0359
6.0	1.0544
8.0	1.0730
10.0	1.0918
15.0	1.1396
20.0	1.1884
25.0	1.2387
30.0	1.2905
35.0	1.3440
40.0	1.3991
45.0	1.4558
50.0	1.5143
51.7	1.5355 (sat'd. soln.)

TABLE 3-77 Potassium Nitrate (KNO_3)

%	0°C.	10°C.	20°C.	40°C.	60°C.	80°C.	100°C.
1	1.00654	1.00615	1.00447	0.99825	0.9890	0.9776	0.9641
2	1.01326	1.01262	1.01075	1.00430	.9949	.9834	.9699
4	1.02677	1.02566	1.02344	1.01652	1.0068	.9951	.9816
8	1.05419	1.05226	1.04940	1.04152	1.0313	1.0192	1.0056
12	1.08221	1.07963	1.07620	1.06740	1.0567	1.0442	1.0304
16	....	....	1.10392	1.09432	1.0831	1.0703	1.0562
20	....	....	1.13261	1.12240	1.1106	1.0974	1.0831
24	....	....	1.16233	1.15175	1.1391	1.1256	1.1110

TABLE 3-78 Potassium Dichromate ($K_2Cr_2O_7$)

%	d_4^{20}
1	1.0052
2	1.0122
4	1.0264
6	1.0408
8	1.0554
10	1.0703

TABLE 3-79 Potassium Sulfate (K_2SO_4)

%	d_4^{20}
1	1.0063
2	1.0145
4	1.0310
6	1.0477
8	1.0646
10	1.0817

TABLE 3-80 Potassium Sulfite (K_2SO_3)

%	d_4^{15}
1	1.0073
2	1.0155
4	1.0322
8	1.0667
12	1.1026
16	1.1402
20	1.1793
24	1.2197
26	1.2404

TABLE 3-81 Sodium Acetate ($NaC_2H_3O_2$)

%	d_4^{20}
1	1.0033
2	1.0084
4	1.0186
8	1.0392
12	1.0598
18	1.0807
20	1.1021
26	1.1351
28	1.1462

TABLE 3-82 Sodium Arsenate (Na_3AsO_4)

%	d_4^{17}
1	1.0097
2	1.0207
4	1.0431
8	1.0892
10	1.1130
12	1.1373

TABLE 3-83 Sodium Bichromate ($Na_2Cr_2O_7$)

%	d_4^{15}
1	1.006
2	1.013
4	1.027
8	1.056
12	1.084
16	1.112
20	1.140
24	1.166
28	1.193
30	1.207
35	1.244
40	1.279
45	1.312
50	1.342

TABLE 3-84 Sodium Bromide (NaBr)

%	d_4^{17}
1	1.0060
2	1.0139
4	1.0298
8	1.0631
10	1.0803
12	1.0981
20	1.1745
30	1.2841
40	1.4138

TABLE 3-85 Sodium Formate (HCOONa)

%	d_4^{25}
1	1.003
2	1.009
4	1.022
8	1.048
12	1.074
16	1.100
20	1.127
24	1.155
28	1.184
30	1.199
35	1.236
40	1.274

TABLE 3-86 Sodium Carbonate (Na_2CO_3)

%	0°C.	10°C.	20°C.	30°C.	40°C.	60°C.	80°C.	100°C.
1	1.0109	1.0103	1.0086	1.0058	1.0022	0.9929	0.9814	0.9683
2	1.0219	1.0210	1.0190	1.0159	1.0122	1.0027	.9910	.9782
4	1.0439	1.0423	1.0398	1.0363	1.0323	1.0223	1.0105	.9980
8	1.0878	1.0850	1.0816	1.0775	1.0732	1.0625	1.0503	1.0380
12	1.1319	1.1284	1.1244	1.1200	1.1150	1.1039	1.0914	1.0787
14	1.1543	1.1506	1.1463	1.1417	1.1365	1.1251	1.1125	1.0996
16	......	......	......	1.1636				
18	......	......	......	1.1859				
20	......	......	......	1.2086				
24	......	......	......	1.2552				
28	......	......	......	1.3031				
30	......	......	......	1.3274				

TABLE 3-87 Sodium Chlorate ($NaClO_3$)

%	d_4^{18}	%	d_4^{18}
1	1.0053	18	1.1288
2	1.0121	20	1.1449
4	1.0258	22	1.1614
6	1.0397	24	1.1782
8	1.0538	26	1.1953
10	1.0681	28	1.2128
12	1.0827	30	1.2307
14	1.0977	32	1.2491
16	1.1131	34	1.2680

TABLE 3-88 Sodium Chloride (NaCl)

%	0°C.	10°C.	25°C.	40°C.	60°C.	80°C.	100°C.
1	1.00747	1.00707	1.00409	0.99908	0.9900	0.9785	0.9651
2	1.01509	1.01442	1.01112	1.00593	.9967	.9852	.9719
4	1.03038	1.02920	1.02530	1.01977	1.0103	.9988	.9855
8	1.06121	1.05907	1.05412	1.04798	1.0381	1.0264	1.0134
12	1.09244	1.08946	1.08365	1.07699	1.0667	1.0549	1.0420
16	1.12419	1.12056	1.11401	1.10688	1.0962	1.0842	1.0713
20	1.15663	1.15254	1.14533	1.13774	1.1268	1.1146	1.1017
24	1.18999	1.18557	1.17776	1.16971	1.1584	1.1463	1.1331
26	1.20709	1.20254	1.19443	1.18614	1.1747	1.1626	1.1492

TABLE 3-89 Sodium Chromate (Na_2CrO_4)

%	d_4^{18}
1	1.0074
2	1.0164
4	1.0344
8	1.0718
12	1.1110
16	1.1518
20	1.1942
24	1.2383
26	1.2611

TABLE 3-90 Sodium Hydroxide (NaOH)

%	0°C.	15°C.	20°C.	40°C.	60°C.	80°C.	100°C.
1	1.0124	1.01065	1.0095	1.0033	0.9941	0.9824	0.9693
2	1.0244	1.02198	1.0207	1.0139	1.0045	.9929	.9797
4	1.0482	1.04441	1.0428	1.0352	1.0254	1.0139	1.0009
8	1.0943	1.08887	1.0869	1.0780	1.0676	1.0560	1.0432
12	1.1399	1.13327	1.1309	1.1210	1.1101	1.0983	1.0855
16	1.1849	1.17761	1.1751	1.1645	1.1531	1.1408	1.1277
20	1.2296	1.22183	1.2191	1.2079	1.1960	1.1833	1.1700
24	1.2741	1.26582	1.2629	1.2512	1.2388	1.2259	1.2124
28	1.3182	1.3094	1.3064	1.2942	1.2814	1.2682	1.2546
32	1.3614	1.3520	1.3490	1.3362	1.3232	1.3097	1.2960
36	1.4030	1.3933	1.3900	1.3768	1.3634	1.3498	1.3360
40	1.4435	1.4334	1.4300	1.4164	1.4027	1.3889	1.3750
44	1.4825	1.4720	1.4685	1.4545	1.4405	1.4266	1.4127
48	1.5210	1.5102	1.5065	1.4922	1.4781	1.4641	1.4503
50	1.5400	1.5290	1.5253	1.5109	1.4967	1.4827	1.4690

TABLE 3-91 Sodium Nitrate ($NaNO_3$)

%	0°C.	20°C.	40°C.	60°C.	80°C.	100°C.
1	1.0071	1.0049	0.9986	0.9894	0.9779	0.9644
2	1.0144	1.0117	1.0050	.9956	.9840	.9704
4	1.0290	1.0254	1.0180	1.0082	.9964	.9826
8	1.0587	1.0532	1.0447	1.0340	1.0218	1.0078
12	1.0891	1.0819	1.0724	1.0609	1.0481	1.0340
16	1.1203	1.1118	1.1013	1.0892	1.0757	1.0614
20	1.1526	1.1429	1.1314	1.1187	1.1048	1.0901
24	1.1860	1.1752	1.1629	1.1496	1.1351	1.1200
28	1.2204	1.2085	1.1955	1.1816	1.1667	1.1513
30	1.2380	1.2256	1.2122	1.1980	1.1830	1.1674
35	1.2834	1.2701	1.2560	1.2413	1.2258	1.2100
40	1.3316	1.3175	1.3027	1.2875	1.2715	1.2555
45	......	1.3683	1.3528	1.3371	1.3206	1.3044

TABLE 3-92 Sodium Nitrite ($NaNO_2$)

%	d_4^{15}
1	1.0058
2	1.0125
4	1.0260
8	1.0535
12	1.0816
16	1.1103
20	1.1394

TABLE 3-93 Sodium Silicate

	Concentration, %												
	1	2	4	8	10	14	20	24	30	36	40	45	50
	Density$_4^{20}$												
$Na_2O/3.9SiO_2$	1.006	1.014	1.030	1.063	1.080	1.116	1.172	1.211	1.275				
$Na_2O/3.36SiO_2$	1.006	1.014	1.030	1.065	1.083	1.120	1.179	1.222	1.290	1.365			
$Na_2O/2.40SiO_2$	1.007	1.016	1.034	1.071	1.090	1.130							
$Na_2O/2.44SiO_2$	.....	.....	.....	.....	.....	.....	.....	.....	1.309	1.387	1.445		
$Na_2O/2.06SiO_2$	1.007	1.016	1.035	1.073	1.093	1.134	1.200	1.247	1.321	1.397	1.450	1.520	1.594
$Na_2O/1.69SiO_2$	1.007	1.017	1.036	1.077	1.098	1.141	1.210	1.259	1.337	1.424			

TABLE 3-94 Sodium Sulfate (Na_2SO_4)

%	0°C.	20°C.	30°C.	40°C.	60°C.	80°C.	100°C.
1	1.0094	1.0073	1.0046	1.0010	0.9919	0.9805	0.9671
2	1.0189	1.0164	1.0135	1.0098	1.0007	.9892	.9758
4	1.0381	1.0348	1.0315	1.0276	1.0184	1.0068	.9934
8	1.0773	1.0724	1.0682	1.0639	1.0544	1.0426	1.0292
12	1.1174	1.1109	1.1062	1.1015	1.0915	1.0795	1.0661
16	1.1585	1.1506	1.1456	1.1406	1.1299	1.1176	1.1042
20	1.2008	1.1915	1.1865	1.1813	1.1696	1.1569	
24	1.2443	1.2336	1.2292	1.2237			

TABLE 3-95 Sodium Sulfide (Na_2S)

%	d_4^{18}
1	1.0098
2	1.0211
4	1.0440
8	1.0907
12	1.1388
16	1.1885
18	1.2140

TABLE 3-96 Sodum Sulfite (Na_2SO_3)

%	d_4^{19}
1	1.0078
2	1.0172
4	1.0363
8	1.0751
12	1.1146
16	1.1549
18	1.1755

TABLE 3-97 Sodium Thiosulfate ($Na_2S_2O_3$)

%	d_4^{20}
1	1.0065
2	1.0148
4	1.0315
8	1.0654
12	1.1003
16	1.1365
20	1.1740
24	1.2128
28	1.2532
30	1.2739
35	1.3273
40	1.3827

TABLE 3-98 Sodium Thiosulfate Pentahydrate ($Na_2S_2O_3 \cdot 5H_2O$)

%	d_4^{19}
1	1.0052
2	1.0105
4	1.0211
8	1.0423
12	1.0639
16	1.0863
20	1.1087
24	1.1322
28	1.1558
30	1.1676
40	1.2297
50	1.2954

TABLE 3-99 Stannic Chloride ($SnCl_4$)

%	d_4^{15}
1	1.007
2	1.015
4	1.031
8	1.064
12	1.099
16	1.135
20	1.173
24	1.212
28	1.255
30	1.278
35	1.337
40	1.403
45	1.475
50	1.555
55	1.644
60	1.742
65	1.851
70	1.971

TABLE 3-100 Stannous Chloride ($SnCl_2$)

%	d_4^{15}
1	1.0068
2	1.0146
4	1.0306
8	1.0638
12	1.0986
16	1.1353
20	1.1743
24	1.2159
28	1.2603
30	1.2837
35	1.3461
40	1.4145
45	1.4897
50	1.5729
55	1.6656
60	1.7695
65	1.8865

TABLE 3-101 Sulfuric Acid (H_2SO_4)

%	0°C.	10°C.	15°C.	20°C.	25°C.	30°C.	40°C.	50°C.	60°C.	80°C.	100°C.
1	1.0074	1.0068	1.0060	1.0051	1.0038	1.0022	0.9986	0.9944	0.9895	0.9779	0.9645
2	1.0147	1.0138	1.0129	1.0118	1.0104	1.0087	1.0050	1.0006	.9956	.9839	.9705
3	1.0219	1.0206	1.0197	1.0184	1.0169	1.0152	1.0113	1.0067	1.0017	.9900	.9766
4	1.0291	1.0275	1.0264	1.0250	1.0234	1.0216	1.0176	1.0129	1.0078	.9961	.9827
5	1.0364	1.0344	1.0332	1.0317	1.0300	1.0281	1.0240	1.0192	1.0140	1.0022	.9888
6	1.0437	1.0414	1.0400	1.0385	1.0367	1.0347	1.0305	1.0256	1.0203	1.0084	.9950
7	1.0511	1.0485	1.0469	1.0453	1.0434	1.0414	1.0371	1.0321	1.0266	1.0146	1.0013
8	1.0585	1.0556	1.0539	1.0522	1.0502	1.0481	1.0437	1.0386	1.0330	1.0209	1.0076
9	1.0660	1.0628	1.0610	1.0591	1.0571	1.0549	1.0503	1.0451	1.0395	1.0273	1.0140
10	1.0735	1.0700	1.0681	1.0661	1.0640	1.0617	1.0570	1.0517	1.0460	1.0338	1.0204
11	1.0810	1.0773	1.0753	1.0731	1.0710	1.0686	1.0637	1.0584	1.0526	1.0403	1.0269
12	1.0886	1.0846	1.0825	1.0802	1.0780	1.0756	1.0705	1.0651	1.0593	1.0469	1.0335
13	1.0962	1.0920	1.0898	1.0874	1.0851	1.0826	1.0774	1.0719	1.0661	1.0536	1.0402
14	1.1039	1.0994	1.0971	1.0947	1.0922	1.0897	1.0844	1.0788	1.0729	1.0603	1.0469
15	1.1116	1.1069	1.1045	1.1020	1.0994	1.0968	1.0914	1.0857	1.0798	1.0671	1.0537
16	1.1194	1.1145	1.1120	1.1094	1.1067	1.1040	1.0985	1.0927	1.0868	1.0740	1.0605
17	1.1272	1.1221	1.1195	1.1168	1.1141	1.1113	1.1057	1.0998	1.0938	1.0809	1.0674
18	1.1351	1.1298	1.1271	1.1243	1.1215	1.1187	1.1129	1.1070	1.1009	1.0879	1.0744
19	1.1430	1.1375	1.1347	1.1318	1.1290	1.1261	1.1202	1.1142	1.1081	1.0950	1.0814
20	1.1510	1.1453	1.1424	1.1394	1.1365	1.1335	1.1275	1.1215	1.1153	1.1021	1.0885
21	1.1590	1.1531	1.1501	1.1471	1.1441	1.1410	1.1349	1.1288	1.1226	1.1093	1.0957
22	1.1670	1.1609	1.1579	1.1548	1.1517	1.1486	1.1424	1.1362	1.1299	1.1166	1.1029
23	1.1751	1.1688	1.1657	1.1626	1.1594	1.1563	1.1500	1.1437	1.1373	1.1239	1.1102
24	1.1832	1.1768	1.1736	1.1704	1.1672	1.1640	1.1576	1.1512	1.1448	1.1313	1.1176
25	1.1914	1.1848	1.1816	1.1783	1.1750	1.1718	1.1653	1.1588	1.1523	1.1388	1.1250
26	1.1996	1.1929	1.1896	1.1862	1.1829	1.1796	1.1730	1.1665	1.1599	1.1463	1.1325
27	1.2078	1.2010	1.1976	1.1942	1.1909	1.1875	1.1808	1.1742	1.1676	1.1539	1.1400
28	1.2160	1.2091	1.2057	1.2023	1.1989	1.1955	1.1887	1.1820	1.1753	1.1616	1.1476
29	1.2243	1.2173	1.2138	1.2104	1.2069	1.2035	1.1966	1.1898	1.1831	1.1693	1.1553
30	1.2326	1.2255	1.2220	1.2185	1.2150	1.2115	1.2046	1.1977	1.1909	1.1771	1.1630
31	1.2409	1.2338	1.2302	1.2267	1.2232	1.2196	1.2126	1.2057	1.1988	1.1849	1.1708
32	1.2493	1.2421	1.2385	1.2349	1.2314	1.2278	1.2207	1.2137	1.2068	1.1928	1.1787
33	1.2577	1.2504	1.2468	1.2432	1.2396	1.2360	1.2289	1.2218	1.2148	1.2008	1.1866
34	1.2661	1.2588	1.2552	1.2515	1.2479	1.2443	1.2371	1.2300	1.2229	1.2088	1.1946
35	1.2746	1.2672	1.2636	1.2599	1.2563	1.2526	1.2454	1.2383	1.2311	1.2169	1.2027
36	1.2831	1.2757	1.2720	1.2684	1.2647	1.2610	1.2538	1.2466	1.2394	1.2251	1.2109
37	1.2917	1.2843	1.2805	1.2769	1.2732	1.2695	1.2622	1.2550	1.2477	1.2334	1.2192
38	1.3004	1.2929	1.2891	1.2855	1.2818	1.2780	1.2707	1.2635	1.2561	1.2418	1.2276
39	1.3091	1.3016	1.2978	1.2941	1.2904	1.2866	1.2793	1.2720	1.2646	1.2503	1.2361
40	1.3179	1.3103	1.3065	1.3028	1.2991	1.2953	1.2880	1.2806	1.2732	1.2589	1.2446
41	1.3268	1.3191	1.3153	1.3116	1.3079	1.3041	1.2967	1.2893	1.2819	1.2675	1.2532
42	1.3357	1.3280	1.3242	1.3205	1.3167	1.3129	1.3055	1.2981	1.2907	1.2762	1.2619
43	1.3447	1.3370	1.3332	1.3294	1.3256	1.3218	1.3144	1.3070	1.2996	1.2850	1.2707
44	1.3538	1.3461	1.3423	1.3384	1.3346	1.3308	1.3234	1.3160	1.3086	1.2939	1.2796
45	1.3630	1.3553	1.3515	1.3476	1.3437	1.3399	1.3325	1.3251	1.3177	1.3029	1.2886
46	1.3724	1.3646	1.3608	1.3569	1.3530	1.3492	1.3417	1.3343	1.3269	1.3120	1.2976
47	1.3819	1.3740	1.3702	1.3663	1.3624	1.3586	1.3510	1.3435	1.3362	1.3212	1.3067
48	1.3915	1.3835	1.3797	1.3758	1.3719	1.3680	1.3604	1.3528	1.3455	1.3305	1.3159
49	1.4012	1.3931	1.3893	1.3854	1.3814	1.3775	1.3699	1.3623	1.3549	1.3399	1.3253
50	1.4110	1.4029	1.3990	1.3951	1.3911	1.3872	1.3795	1.3719	1.3644	1.3494	1.3348
51	1.4209	1.4128	1.4088	1.4049	1.4009	1.3970	1.3893	1.3816	1.3740	1.3590	1.3444
52	1.4310	1.4228	1.4188	1.4148	1.4109	1.4069	1.3991	1.3914	1.3837	1.3687	1.3540
53	1.4412	1.4329	1.4289	1.4248	1.4209	1.4169	1.4091	1.4013	1.3936	1.3785	1.3637
54	1.4515	1.4431	1.4391	1.4350	1.4310	1.4270	1.4191	1.4113	1.4036	1.3884	1.3735
55	1.4619	1.4535	1.4494	1.4453	1.4412	1.4372	1.4293	1.4214	1.4137	1.3984	1.3834
56	1.4724	1.4640	1.4598	1.4557	1.4516	1.4475	1.4396	1.4317	1.4239	1.4085	1.3934
57	1.4830	1.4746	1.4703	1.4662	1.4621	1.4580	1.4500	1.4420	1.4342	1.4187	1.4035
58	1.4937	1.4852	1.4809	1.4768	1.4726	1.4685	1.4604	1.4524	1.4446	1.4290	1.4137
59	1.5045	1.4959	1.4916	1.4875	1.4832	1.4791	1.4709	1.4629	1.4551	1.4393	1.4240
60	1.5154	1.5067	1.5024	1.4983	1.4940	1.4898	1.4816	1.4735	1.4656	1.4497	1.4344
61	1.5264	1.5177	1.5133	1.5091	1.5048	1.5006	1.4923	1.4842	1.4762	1.4602	1.4449
62	1.5375	1.5287	1.5243	1.5200	1.5157	1.5115	1.5031	1.4950	1.4869	1.4708	1.4554
63	1.5487	1.5398	1.5354	1.5310	1.5267	1.5225	1.5140	1.5058	1.4977	1.4815	1.4660
64	1.5600	1.5510	1.5465	1.5421	1.5378	1.5335	1.5250	1.5167	1.5086	1.4923	1.4766
65	1.5714	1.5623	1.5578	1.5533	1.5490	1.5446	1.5361	1.5277	1.5195	1.5031	1.4873
66	1.5828	1.5736	1.5691	1.5646	1.5602	1.5558	1.5472	1.5388	1.5305	1.5140	1.4981
67	1.5943	1.5850	1.5805	1.5760	1.5715	1.5671	1.5584	1.5499	1.5416	1.5249	1.5089
68	1.6059	1.5965	1.5920	1.5874	1.5829	1.5785	1.5697	1.5611	1.5528	1.5359	1.5198
69	1.6176	1.6081	1.6035	1.5989	1.5944	1.5899	1.5811	1.5724	1.5640	1.5470	1.5307
70	1.6293	1.6198	1.6151	1.6105	1.6059	1.6014	1.5925	1.5838	1.5753	1.5582	1.5417
71	1.6411	1.6315	1.6268	1.6221	1.6175	1.6130	1.6040	1.5952	1.5867	1.5694	1.5527
72	1.6529	1.6433	1.6385	1.6338	1.6292	1.6246	1.6155	1.6067	1.5981	1.5806	1.5637
73	1.6648	1.6551	1.6503	1.6456	1.6409	1.6363	1.6271	1.6182	1.6095	1.5919	1.5747
74	1.6768	1.6670	1.6622	1.6574	1.6526	1.6480	1.6387	1.6297	1.6209	1.6031	1.5857
75	1.6888	1.6789	1.6740	1.6692	1.6644	1.6597	1.6503	1.6412	1.6322	1.6142	1.5966
76	1.7008	1.6908	1.6858	1.6810	1.6761	1.6713	1.6619	1.6526	1.6435	1.6252	1.6074
77	1.7128	1.7026	1.6976	1.6927	1.6878	1.6829	1.6734	1.6640	1.6547	1.6361	1.6181
78	1.7247	1.7144	1.7093	1.7043	1.6994	1.6944	1.6847	1.6751	1.6657	1.6469	1.6286
79	1.7365	1.7261	1.7209	1.7158	1.7108	1.7058	1.6959	1.6862	1.6766	1.6575	1.6390

TABLE 3-101 Sulfuric Acid (H_2SO_4) *(Concluded)*

%	0°C.	10°C.	15°C.	20°C.	25°C.	30°C.	40°C.	50°C.	60°C.	80°C.	100°C.
80	1.7482	1.7376	1.7323	1.7272	1.7221	1.7170	1.7069	1.6971	1.6873	1.6680	1.6493
81	1.7597	1.7489	1.7435	1.7383	1.7331	1.7279	1.7177	1.7077	1.6978	1.6782	1.6594
82	1.7709	1.7599	1.7544	1.7491	1.7437	1.7385	1.7281	1.7180	1.7080	1.6882	1.6692
83	1.7815	1.7704	1.7649	1.7594	1.7540	1.7487	1.7382	1.7279	1.7179	1.6979	1.6787
84	1.7916	1.7804	1.7748	1.7693	1.7639	1.7585	1.7479	1.7375	1.7274	1.7072	1.6878
85	1.8009	1.7897	1.7841	1.7786	1.7732	1.7678	1.7571	1.7466	1.7364	1.7161	1.6966
86	1.8095	1.7983	1.7927	1.7872	1.7818	1.7763	1.7657	1.7552	1.7449	1.7245	1.7050
87	1.8173	1.8061	1.8006	1.7951	1.7897	1.7842	1.7736	1.7632	1.7529	1.7324	1.7129
88	1.8243	1.8132	1.8077	1.8022	1.7968	1.7914	1.7809	1.7705	1.7602	1.7397	1.7202
89	1.8306	1.8195	1.8141	1.8087	1.8033	1.7979	1.7874	1.7770	1.7669	1.7464	1.7269
90	1.8361	1.8252	1.8198	1.8144	1.8091	1.8038	1.7933	1.7829	1.7729	1.7525	1.7331
91	1.8410	1.8302	1.8248	1.8195	1.8142	1.8090	1.7986	1.7883	1.7783	1.7581	1.7388
92	1.8453	1.8346	1.8293	1.8240	1.8188	1.8136	1.8033	1.7932	1.7832	1.7633	1.7439
93	1.8490	1.8384	1.8331	1.8279	1.8227	1.8176	1.8074	1.7974	1.7876	1.7681	1.7485
94	1.8520	1.8415	1.8363	1.8312	1.8260	1.8210	1.8109	1.8011	1.7914		
95	1.8544	1.8439	1.8388	1.8337	1.8286	1.8236	1.8137	1.8040	1.7944		
96	1.8560	1.8457	1.8406	1.8355	1.8305	1.8255	1.8157	1.8060	1.7965		
97	1.8569	1.8466	1.8414	1.8364	1.8314	1.8264	1.8166	1.8071	1.7977		
98	1.8567	1.8463	1.8411	1.8361	1.8310	1.8261	1.8163	1.8068	1.7976		
99	1.8551	1.8445	1.8393	1.8342	1.8292	1.8242	1.8145	1.8050	1.7958		
100	1.8517	1.8409	1.8357	1.8305	1.8255	1.8205	1.8107	1.8013	1.7922		

%	$d_4^{5.96}$	%	$d_4^{13.00}$	$d_4^{18.00}$
0.005	1.000 0140	0.05	0.999 810	0.999 028
.01	1.000 0576	.1	1.000 185	.999 400
.02	1.000 1434	.2	1.000 912	1.000 119
.03	1.000 2276	.3	1.001 623	1.000 820
.04	1.000 3104	.4	1.002 326	1.001 512
.05	1.000 3920	.5	1.003 023	1.002 197
.06	1.000 4726	.6	1.003 716	1.002 877
.07	1.000 5523	.8	1.005 090	1.004 227
.08	1.000 6313	1.0	1.006 452	1.005 570
.09	1.000 7098	1.2	1.007 807	1.006 909
.10	1.000 7880	1.4	1.009 159	1.008 247
.15	1.001 1732	1.6	1.010 510	1.009 583
.20	1.001 5514	1.8	1.011 860	1.010 918
.25	1.001 9254	2.0	1.013 209	1.012 252
.30	1.002 2961	2.2	1.014 557	1.013 586
.35	1.002 6639	2.4	1.015 904	1.014 919
.40	1.003 0292			
.45	1.003 3923			
.50	1.003 7534			

TABLE 3-102 Zinc Bromide ($ZnBr_2$)

%	0°C.	20°C.	40°C.	60°C.	80°C.	100°C.
2	1.0188	1.0167	1.0102	1.0008	0.9890	0.9751
4	1.0381	1.0354	1.0285	1.0187	1.0065	0.9921
8	1.0777	1.0738	1.0660	1.0554	1.0422	1.0270
12	1.1186	1.1135	1.1046	1.0932	1.0789	1.0629
16	1.1609	1.1544	1.1445	1.1320	1.1169	1.1000
20	1.2043	1.1965	1.1855	1.1720	1.1560	1.1382
30	1.3288	1.3170	1.3030	1.2868	1.2688	1.2489
40	1.477	1.462	1.445	1.427	1.406	1.385
50	1.661	1.643	1.623	1.602	1.579	1.555
60	1.891	1.869	1.845	1.822	1.797	1.771
65	2 026	2 002	1.976	1.951	1.924	1.898

TABLE 3-103 Zinc Chloride ($ZnCl_2$)

%	0°C.	20°C.	40°C.	60°C.	80°C.	100°C.
2	1.0192	1.0167	1.0099	1.0003	0.9882	0.9739
4	1.0384	1.0350	1.0274	1.0172	1.0044	.9894
8	1.0769	1.0715	1.0624	1.0508	1.0369	1.0211
12	1.1159	1.1085	1.0980	1.0853	1.0704	1.0541
16	1.1558	1.1468	1.1350	1.1212	1.1055	1.0888
20	1.1970	1.1866	1.1736	1.1590	1.1428	1.1255
30	1.3062	1.2928	1.2778	1.2614	1.2438	1.2252
40	1.4329	1.4173	1.4003	1.3824	1.3637	1.3441
50	1.5860	1.5681	1.5495	1.5300	1.5097	1.4892
60	......	1.749				
70	......	1.962				

TABLE 3-104 Zinc Nitrate [$Zn(NO_3)_2$]

%	18°C.	%	18°C.
2	1.0154	18	1.1652
4	1.0322	20	1.1865
6	1.0496	25	1.2427
8	1.0675	30	1.3029
10	1.0859	35	1.3678
12	1.1048	40	1.4378
14	1.1244	45	1.5134
16	1.1445	50	1.5944

TABLE 3-105 Zinc Sulfate ($ZnSO_4$)

%	20°C.
2	1.019
4	1.0403
6	1.0620
8	1.0842
10	1.1071
12	1.1308
14	1.1553
16	1.1806

DENSITIES OF AQUEOUS ORGANIC SOLUTIONS*

UNITS AND UNITS CONVERSIONS

Unless otherwise noted, densities are given in grams per cubic centimeter. To convert to pounds per cubic foot, multiply by 62.43. °F = $\frac{9}{5}$ °C + 32.

From *International Critical Tables*, vol. 3, pp. 115–129. All compositions are in weight percent in vacuo. All density values are d_4^t = g/mL in vacuo.

TABLE 3-106 Formic Acid (HCOOH)

%	0°C.	15°C.	20°C.	30°C.	%	0°C.	15°C.	20°C.	30°C.	%	0°C.	15°C.	20°C.	30°C.	%	0°C.	15°C.	20°C.	30°C.
0	0.9999	0.9991	0.9982	0.9957	25	1.0706	1.0627	1.0609	1.0540	50	1.1349	1.1225	1.1207	1.1098	75	1.1953	1.1794	1.1769	1.1636
1	1.0028	1.0019	1.0019	0.9980	26	1.0733	1.0652	1.0633	1.0564	51	1.1374	1.1248	1.1223	1.1120	76	1.1976	1.1816	1.1785	1.1656
2	1.0059	1.0045	1.0044	1.0004	27	1.0760	1.0678	1.0656	1.0587	52	1.1399	1.1271	1.1244	1.1142	77	1.1999	1.1837	1.1801	1.1676
3	1.0090	1.0072	1.0070	1.0028	28	1.0787	1.0702	1.0681	1.0609	53	1.1424	1.1294	1.1269	1.1164	78	1.2021	1.1859	1.1818	1.1697
4	1.0120	1.0100	1.0093	1.0053	29	1.0813	1.0726	1.0705	1.0632	54	1.1448	1.1318	1.1295	1.1186	79	1.2043	1.1881	1.1837	1.1717
5	1.0150	1.0124	1.0115	1.0075	30	1.0839	1.0750	1.0729	1.0654	55	1.1472	1.1341	1.1320	1.1208	80	1.2065	1.1902	1.1806	1.1737
6	1.0179	1.0151	1.0141	1.0101	31	1.0866	1.0774	1.0753	1.0676	56	1.1497	1.1365	1.1342	1.1230	81	1.2088	1.1924	1.1876	1.1758
7	1.0207	1.0177	1.0170	1.0125	32	1.0891	1.0798	1.0777	1.0699	57	1.1523	1.1388	1.1361	1.1253	82	1.2110	1.1944	1.1896	1.1778
8	1.0237	1.0204	1.0196	1.0149	33	1.0916	1.0821	1.0800	1.0721	58	1.1548	1.1411	1.1381	1.1274	83	1.2132	1.1965	1.1914	1.1798
9	1.0266	1.0230	1.0221	1.0173	34	1.0941	1.0844	1.0823	1.0743	59	1.1573	1.1434	1.1401	1.1295	84	1.2154	1.1985	1.1929	1.1817
10	1.0295	1.0256	1.0246	1.0197	35	1.0966	1.0867	1.0847	1.0766	60	1.1597	1.1458	1.1424	1.1317	85	1.2176	1.2005	1.1953	1.1837
11	1.0324	1.0281	1.0271	1.0221	36	1.0993	1.0892	1.0871	1.0788	61	1.1621	1.1481	1.1448	1.1338	86	1.2196	1.2025	1.1976	1.1856
12	1.0351	1.0306	1.0296	1.0244	37	1.1018	1.0916	1.0895	1.0810	62	1.1645	1.1504	1.1473	1.1360	87	1.2217	1.2045	1.1994	1.1875
13	1.0379	1.0330	1.0321	1.0267	38	1.1043	1.0940	1.0919	1.0832	63	1.1669	1.1526	1.1493	1.1382	88	1.2237	1.2064	1.2012	1.1893
14	1.0407	1.0355	1.0345	1.0290	39	1.1069	1.0964	1.0940	1.0854	64	1.1694	1.1549	1.1517	1.1403	89	1.2258	1.2084	1.2028	1.1910
15	1.0435	1.0380	1.0370	1.0313	40	1.1095	1.0988	1.0963	1.0876	65	1.1718	1.1572	1.1543	1.1425	90	1.2278	1.2102	1.2044	1.1927
16	1.0463	1.0405	1.0393	1.0336	41	1.1122	1.1012	1.0990	1.0898	66	1.1742	1.1595	1.1565	1.1446	91	1.2297	1.2121	1.2059	1.1945
17	1.0491	1.0430	1.0417	1.0358	42	1.1148	1.1036	1.1015	1.0920	67	1.1766	1.1618	1.1584	1.1467	92	1.2316	1.2139	1.2078	1.1961
18	1.0518	1.0455	1.0441	1.0381	43	1.1174	1.1060	1.1038	1.0943	68	1.1790	1.1640	1.1604	1.1489	93	1.2335	1.2157	1.2099	1.1978
19	1.0545	1.0480	1.0464	1.0404	44	1.1199	1.1084	1.1062	1.0965	69	1.1813	1.1663	1.1628	1.1510	94	1.2354	1.2174	1.2117	1.1994
20	1.0571	1.0505	1.0488	1.0427	45	1.1224	1.1109	1.1085	1.0987	70	1.1835	1.1685	1.1655	1.1531	95	1.2372	1.2191	1.2140	1.2008
21	1.0598	1.0532	1.0512	1.0451	46	1.1249	1.1133	1.1108	1.1009	71	1.1858	1.1707	1.1677	1.1552	96	1.2390	1.2208	1.2158	1.2022
22	1.0625	1.0556	1.0537	1.0473	47	1.1274	1.1156	1.1130	1.1031	72	1.1882	1.1729	1.1702	1.1573	97	1.2408	1.2224	1.2170	1.2036
23	1.0652	1.0580	1.0561	1.0496	48	1.1299	1.1179	1.1157	1.1053	73	1.1906	1.1751	1.1728	1.1595	98	1.2425	1.2240	1.2183	1.2048
24	1.0679	1.0604	1.0585	1.0518	49	1.1324	1.1202	1.1185	1.1076	74	1.1929	1.1773	1.1752	1.1615	99	1.2441	1.2257	1.2202	1.2061
															100	1.2456	1.2273	1.2212	1.2073

TABLE 3-107 Acetic Acid (CH_3COOH)

%	0°C.	10°C.	15°C.	20°C.	25°C.	30°C.	40°C.	%	0°C.	10°C.	15°C.	20°C.	25°C.	30°C.	40°C.
0	0.9999	0.9997	0.9991	0.9982	0.9971	0.9957	0.9922	40	1.0621	1.0557	1.0522	1.0488	1.0450	1.0416	1.0338
1	1.0016	1.0013	1.0006	.9996	.9987	.9971	.9934	41	1.0633	1.0568	1.0532	1.0498	1.0460	1.0425	1.0346
2	1.0033	1.0029	1.0021	1.0012	1.0000	.9984	.9946	42	1.0644	1.0578	1.0542	1.0507	1.0469	1.0433	1.0353
3	1.0051	1.0044	1.0036	1.0025	1.0013	.9997	.9958	43	1.0656	1.0588	1.0551	1.0516	1.0477	1.0441	1.0361
4	1.0070	1.0060	1.0051	1.0040	1.0027	1.0011	.9970	44	1.0667	1.0598	1.0561	1.0525	1.0486	1.0449	1.0368
5	1.0088	1.0076	1.0066	1.0055	1.0041	1.0024	.9982	45	1.0679	1.0608	1.0570	1.0534	1.0495	1.0456	1.0375
6	1.0106	1.0092	1.0081	1.0069	1.0055	1.0037	.9994	46	1.0689	1.0618	1.0579	1.0542	1.0503	1.0464	1.0382
7	1.0124	1.0108	1.0096	1.0083	1.0068	1.0050	1.0006	47	1.0699	1.0627	1.0588	1.0551	1.0511	1.0471	1.0389
8	1.0142	1.0124	1.0111	1.0097	1.0081	1.0063	1.0018	48	1.0709	1.0636	1.0597	1.0559	1.0518	1.0479	1.0395
9	1.0159	1.0140	1.0126	1.0111	1.0094	1.0076	1.0030	49	1.0720	1.0645	1.0605	1.0567	1.0526	1.0486	1.0402
10	1.0177	1.0156	1.0141	1.0125	1.0107	1.0089	1.0042	50	1.0729	1.0654	1.0613	1.0575	1.0534	1.0492	1.0408
11	1.0194	1.0171	1.0155	1.0139	1.0120	1.0102	1.0054	51	1.0738	1.0663	1.0622	1.0582	1.0542	1.0499	1.0414
12	1.0211	1.0187	1.0170	1.0154	1.0133	1.0115	1.0065	52	1.0748	1.0671	1.0629	1.0590	1.0549	1.0506	1.0421
13	1.0228	1.0202	1.0184	1.0168	1.0146	1.0127	1.0077	53	1.0757	1.0679	1.0637	1.0597	1.0555	1.0512	1.0427
14	1.0245	1.0217	1.0199	1.0182	1.0159	1.0139	1.0088	54	1.0765	1.0687	1.0644	1.0604	1.0562	1.0518	1.0432
15	1.0262	1.0232	1.0213	1.0195	1.0172	1.0151	1.0099	55	1.0774	1.0694	1.0651	1.0611	1.0568	1.0525	1.0438
16	1.0278	1.0247	1.0227	1.0209	1.0185	1.0163	1.0110	56	1.0782	1.0701	1.0658	1.0618	1.0574	1.0531	1.0443
17	1.0295	1.0262	1.0241	1.0223	1.0198	1.0175	1.0121	57	1.0790	1.0708	1.0665	1.0624	1.0580	1.0536	1.0448
18	1.0311	1.0276	1.0255	1.0236	1.0210	1.0187	1.0132	58	1.0798	1.0715	1.0672	1.0631	1.0586	1.0542	1.0453
19	1.0327	1.0291	1.0269	1.0250	1.0223	1.0198	1.0142	59	1.0805	1.0722	1.0678	1.0637	1.0592	1.0547	1.0458
20	1.0343	1.0305	1.0283	1.0263	1.0235	1.0210	1.0153	60	1.0813	1.0728	1.0684	1.0642	1.0597	1.0552	1.0462
21	1.0358	1.0319	1.0297	1.0276	1.0248	1.0222	1.0164	61	1.0820	1.0734	1.0690	1.0648	1.0602	1.0557	1.0466
22	1.0374	1.0333	1.0310	1.0288	1.0260	1.0233	1.0174	62	1.0826	1.0740	1.0696	1.0653	1.0607	1.0562	1.0470
23	1.0389	1.0347	1.0323	1.0301	1.0272	1.0244	1.0185	63	1.0833	1.0746	1.0701	1.0658	1.0612	1.0566	1.0473
24	1.0404	1.0361	1.0336	1.0313	1.0283	1.0256	1.0195	64	1.0838	1.0752	1.0706	1.0662	1.0616	1.0571	1.0477
25	1.0419	1.0375	1.0349	1.0326	1.0295	1.0267	1.0205	65	1.0844	1.0757	1.0711	1.0666	1.0621	1.0575	1.0480
26	1.0434	1.0388	1.0362	1.0338	1.0307	1.0278	1.0215	66	1.0850	1.0762	1.0716	1.0671	1.0624	1.0578	1.0483
27	1.0449	1.0401	1.0374	1.0349	1.0318	1.0289	1.0225	67	1.0856	1.0767	1.0720	1.0675	1.0628	1.0582	1.0486
28	1.0463	1.0414	1.0386	1.0361	1.0329	1.0299	1.0234	68	1.0860	1.0771	1.0725	1.0678	1.0631	1.0585	1.0489
29	1.0477	1.0427	1.0399	1.0372	1.0340	1.0310	1.0244	69	1.0865	1.0775	1.0729	1.0682	1.0634	1.0588	1.0491
30	1.0491	1.0440	1.0411	1.0384	1.0350	1.0320	1.0253	70	1.0869	1.0779	1.0732	1.0685	1.0637	1.0590	1.0493
31	1.0505	1.0453	1.0423	1.0395	1.0361	1.0330	1.0262	71	1.0874	1.0783	1.0736	1.0687	1.0640	1.0592	1.0495
32	1.0519	1.0465	1.0435	1.0406	1.0372	1.0341	1.0272	72	1.0877	1.0786	1.0738	1.0690	1.0642	1.0594	1.0496
33	1.0532	1.0477	1.0446	1.0417	1.0382	1.0351	1.0281	73	1.0881	1.0789	1.0741	1.0693	1.0644	1.0595	1.0497
34	1.0545	1.0489	1.0458	1.0428	1.0392	1.0361	1.0289	74	1.0884	1.0792	1.0743	1.0694	1.0645	1.0596	1.0498
35	1.0558	1.0501	1.0469	1.0438	1.0402	1.0371	1.0298	75	1.0887	1.0794	1.0745	1.0696	1.0647	1.0597	1.0499
36	1.0571	1.0513	1.0480	1.0449	1.0412	1.0380	1.0306	76	1.0889	1.0796	1.0746	1.0698	1.0648	1.0598	1.0499
37	1.0584	1.0524	1.0491	1.0459	1.0422	1.0390	1.0314	77	1.0891	1.0797	1.0747	1.0699	1.0648	1.0598	1.0499
38	1.0596	1.0535	1.0501	1.0469	1.0432	1.0399	1.0322	78	1.0893	1.0798	1.0747	1.0700	1.0648	1.0598	1.0498
39	1.0608	1.0546	1.0512	1.0479	1.0441	1.0408	1.0330	79	1.0894	1.0798	1.0747	1.0700	1.0648	1.0597	1.0497

*For gasoline and aircraft fuels see Hibbard, NACA Res. Mem. E56I21 (declassified 1958).

TABLE 3-107 Acetic Acid (CH_3COOH) *(Concluded)*

%	0°C.	10°C.	15°C.	20°C.	25°C.	30°C.	40°C.	%	0°C.	10°C.	15°C.	20°C.	25°C.	30°C.	40°C.
80	1.0895	1.0798	1.0747	1.0700	1.0647	1.0596	1.0495	90	1.0865	1.0766	1.0708	1.0661	1.0605	1.0549	1.0445
81	1.0895	1.0797	1.0745	1.0699	1.0646	1.0594	1.0493	91	1.0857	1.0758	1.0700	1.0652	1.0597	1.0541	1.0436
82	1.0895	1.0796	1.0743	1.0698	1.0644	1.0592	1.0490	92	1.0848	1.0749	1.0690	1.0643	1.0587	1.0530	1.0426
83	1.0895	1.0795	1.0741	1.0696	1.0642	1.0589	1.0487	93	1.0838	1.0739	1.0680	1.0632	1.0577	1.0518	1.0414
84	1.0893	1.0793	1.0738	1.0693	1.0638	1.0585	1.0483	94	1.0826	1.0727	1.0667	1.0619	1.0564	1.0506	1.0401
85	1.0891	1.0790	1.0735	1.0689	1.0635	1.0582	1.0479	95	1.0813	1.0714	1.0652	1.0605	1.0551	1.0491	1.0386
86	1.0887	1.0787	1.0731	1.0685	1.0630	1.0576	1.0473	96	1.0798	......	1.0632	1.0588	1.0535	1.0473	1.0368
87	1.0883	1.0783	1.0726	1.0680	1.0626	1.0571	1.0467	97	1.0780	......	1.0611	1.0570	1.0516	1.0454	1.0348
88	1.0877	1.0778	1.0721	1.0675	1.0620	1.0564	1.0460	98	1.0759	......	1.0590	1.0549	1.0495	1.0431	1.0325
89	1.0872	1.0773	1.0715	1.0668	1.0613	1.0557	1.0453	99	1.0730	......	1.0567	1.0524	1.0468	1.0407	1.0299
								100	1.0697	.....	1.0545	1.0498	1.0440	1.0380	1.0271

TABLE 3-108 Oxalic Acid ($H_2C_2O_4$)

%	$d_4^{17.5}$	%	$d_4^{17.5}$
1	1.0035	8	1.0280
2	1.0070	10	1.0350
4	1.0140	12	1.0420

TABLE 3-109 Methyl Alcohol (CH_3OH)*

%	0°C.	10°C.	15.56°C.	20°C.	15°C.	%	0°C.	10°C.	15.56°C	20°C.	15°C.	%	0°C.	10°C.	15.56°C.	20°C.	15°C.
0	0.9999	0.9997	0.9990	0.9982	0.99913	35	0.9534	0.9484	0.9456	0.9433	0.94570	70	0.8869	0.8794	0.8748	0.8715	0.87507
1	.9981	.9980	.9973	.9965	.99727	36	.9520	.9469	.9440	.9416	.94404	71	.8847	.8770	.8726	.8690	.87271
2	.9963	.9962	.9955	.9948	.99543	37	.9505	.9453	.9422	.9398	.94237	72	.8824	.8747	.8702	.8665	.87033
3	.9946	.9945	.9938	.9931	.99370	38	.9490	.9437	.9405	.9381	.94067	73	.8801	.8724	.8678	.8641	.86792
4	.9930	.9929	.9921	.9914	.99198	39	.9475	.9420	.9387	.9363	.93894	74	.8778	.8699	.8653	.8616	.86546
5	.9914	.9912	.9904	.9896	.99029	40	.9459	.9403	.9369	.9345	.93720	75	.8754	.8676	.8629	.8592	.86300
6	.9899	.9896	.9889	.9880	.98864	41	.9443	.9387	.9351	.9327	.93543	76	.8729	.8651	.8604	.8567	.86051
7	.9884	.9881	.9872	.9863	.98701	42	.9427	.9370	.9333	.9309	.93365	77	.8705	.8626	.8579	.8542	.85801
8	.9870	.9865	.9857	.9847	.98547	43	.9411	.9352	.9315	.9290	.93185	78	.8680	.8602	.8554	.8518	.85551
9	.9856	.9849	.9841	.9831	.98394	44	.9395	.9334	.9297	.9272	.93001	79	.8657	.8577	.8529	.8494	.85300
10	.9842	.9834	.9826	.9815	.98241	45	.9377	.9316	.9279	.9252	.92815	80	.8634	.8551	.8503	.8469	.85048
11	.9829	.9820	.9811	.9799	.98093	46	.9360	.9298	.9261	.9234	.92627	81	.8610	.8527	.8478	.8446	.84794
12	.9816	.9805	.9796	.9784	.97945	47	.9342	.9279	.9242	.9214	.92436	82	.8585	.8501	.8452	.8420	.84536
13	.9804	.9791	.9781	.9768	.97802	48	.9324	.9260	.9223	.9196	.92242	83	.8560	.8475	.8426	.8394	.84274
14	.9792	.9778	.9766	.9754	.97660	49	.9306	.9240	.9204	.9176	.92048	84	.8535	.8449	.8400	.8366	.84009
15	.9780	.9764	.9752	.9740	.97518	50	.9287	.9221	.9185	.9156	.91852	85	.8510	.8422	.8374	.8340	.83742
16	.9769	.9751	.9738	.9725	.97377	51	.9269	.9202	.9166	.9135	.91653	86	.8483	.8394	.8347	.8314	.83475
17	.9758	.9739	.9723	.9710	.97237	52	.9250	.9182	.9146	.9114	.91451	87	.8456	.8367	.8320	.8286	.83207
18	.9747	.9726	.9709	.9696	.97096	53	.9230	.9162	.9126	.9094	.91248	88	.8428	.8340	.8294	.8258	.82937
19	.9736	.9713	.9695	.9681	.96955	54	.9211	.9142	.9106	.9073	.91044	89	.8400	.8314	.8267	.8230	.82667
20	.9725	.9700	.9680	.9666	.96814	55	.9191	.9122	.9086	.9052	.90839	90	.8374	.8287	.8239	.8202	.82396
21	.9714	.9687	.9666	.9651	.96673	56	.9172	.9101	.9065	.9032	.90631	91	.8347	.8261	.8212	.8174	.82124
22	.9702	.9673	.9652	.9636	.96533	57	.9151	.9080	.9045	.9010	.90421	92	.8320	.8234	.8185	.8146	.81849
23	.9690	.9660	.9638	.9622	.96392	58	.9131	.9060	.9024	.8988	.90210	93	.8293	.8208	.8157	.8118	.81568
24	.9678	.9646	.9624	.9607	.96251	59	.9111	.9039	.9002	.8968	.89996	94	.8266	.8180	.8129	.8090	.81285
25	.9666	.9632	.9609	.9592	.96108	60	.9090	.9018	.8980	.8946	.89781	95	.8240	.8152	.8101	.8062	.80999
26	.9654	.9618	.9595	.9576	.95963	61	.9068	.8998	.8958	.8924	.89563	96	.8212	.8124	.8073	.8034	.80713
27	.9642	.9604	.9580	.9562	.95817	62	.9046	.8977	.8936	.8902	.89341	97	.8186	.8096	.8045	.8005	.80428
28	.9629	.9590	.9565	.9546	.95668	63	.9024	.8955	.8913	.8879	.89117	98	.8158	.8068	.8016	.7976	.80143
29	.9616	.9575	.9550	.9531	.95518	64	.9002	.8933	.8890	.8856	.88890	99	.8130	.8040	.7987	.7948	.79859
30	.9604	.9560	.9535	.9515	.95366	65	.8980	.8911	.8867	.8834	.88662	100	.8102	.8009	.7959	.7917	.79577
31	.9590	.9546	.9521	.9499	.95213	66	.8958	.8888	.8844	.8811	.88433						
32	.9576	.9531	.9505	.9483	.95056	67	.8935	.8865	.8820	.8787	.88203						
33	.9563	.9516	.9489	.9466	.94896	68	.8913	.8842	.8797	.8763	.87971						
34	.9549	.9500	.9473	.9450	.94734	69	.8891	.8818	.8771	.8738	.87739						

* It should be noted that the values for 100 per cent do not agree with some data available elsewhere, *e.g.*, "American Institute of Physics Handbook," McGraw-Hill, New York, 1957. Also, see Atack, "Handbook of Chemical Data," Reinhold, New York, 1957.

TABLE 3-110 Ethyl Alcohol (C_2H_5OH)*

%	10°C.	15°C.	20°C.	25°C.	30°C.	35°C.	40°C.	%	10°C.	15°C.	20°C.	25°C.	30°C.	35°C.	40°C.
0	0.99973	0.99913	0.99823	0.99708	0.99568	0.99406	0.99225	50	0.92126	0.91776	0.91384	0.90985	0.90580	0.90168	0.89750
1	785	725	636	520	379	217	034	51	.91943	555	160	760	353	.89940	519
2	602	542	453	336	194	031	.98846	52	723	333	.90936	534	125	710	288
3	426	365	275	157	014	.98849	663	53	502	110	711	307	.89896	479	056
4	258	195	103	.98984	.98839	672	485	54	279	.90885	485	079	667	248	.88823
5	098	032	.98938	817	670	501	311	55	055	659	258	.89850	437	016	589
6	.98946	.98877	780	656	507	335	142	56	.90831	433	031	621	206	.88784	356
7	801	729	627	500	347	172	.97975	57	607	207	.89803	392	.88975	552	122
8	660	584	478	346	189	009	808	58	381	.89980	574	162	744	319	.87888
9	524	442	331	193	031	.97846	641	59	154	752	344	.88931	512	085	653
10	393	304	187	043	.97875	685	475	60	.89927	523	113	699	278	.87851	417
11	267	171	047	.97897	723	527	312	61	698	293	.88882	446	044	615	180
12	145	041	.97910	753	573	371	150	62	468	062	650	233	.87809	379	.86943
13	026	.97914	775	611	424	216	.96989	63	237	.88830	417	.87998	574	142	705
14	.97911	790	643	472	278	063	829	64	006	597	183	763	337	.86905	466
15	800	669	514	334	133	.96911	670	65	.88774	364	.87948	527	100	667	227
16	692	552	387	199	.96990	760	512	66	541	130	713	291	.86863	429	.85987
17	583	433	259	062	844	607	352	67	308	.87895	477	054	625	190	747
18	473	313	129	.96923	697	452	189	68	074	660	241	.86817	387	.85950	407
19	363	191	.96997	782	547	294	023	69	.87839	424	004	579	148	710	266
20	252	068	864	639	395	134	.95856	70	602	187	.86766	340	.85908	470	025
21	139	.96944	729	495	242	.95973	687	71	365	.86949	527	100	667	228	.84783
22	024	818	592	348	087	809	516	72	127	710	287	.85859	426	.84986	540
23	.96907	689	453	199	.95929	643	343	73	.86888	470	047	618	184	743	297
24	787	558	312	048	769	476	168	74	648	229	.85806	376	.84941	500	053
25	665	424	168	.95895	607	306	.94991	75	408	.85988	564	134	698	257	.83809
26	539	287	020	738	442	133	810	76	168	747	322	.84891	455	013	564
27	406	144	.95867	576	272	.94955	625	77	.85927	505	079	647	211	.83768	319
28	268	.95996	710	410	098	774	438	78	685	262	.84835	403	.83966	523	074
29	125	844	548	241	.94922	590	248	79	442	018	590	158	720	277	.82827
30	.95977	686	382	067	741	403	055	80	197	.84772	344	.83911	473	029	578
31	823	524	212	.94890	557	214	.93860	81	.84950	525	096	664	224	.82780	329
32	665	357	038	709	370	021	662	82	702	277	.83848	415	.82974	530	079
33	502	186	.94860	525	180	.93825	461	83	453	028	599	164	724	279	.81828
34	334	011	679	337	.93986	626	257	84	203	.83777	348	.82913	473	027	576
35	162	.94832	494	146	790	425	051	85	.83951	525	095	660	220	.81774	322
36	.94986	650	306	.93952	591	221	.92843	86	697	271	.82840	405	.81965	519	067
37	805	464	114	756	390	016	634	87	441	014	583	148	708	262	.80811
38	620	273	.93919	556	186	.92808	422	88	181	.82754	323	.81888	448	003	552
39	431	079	720	353	.92979	597	208	89	.82919	492	062	626	186	.80742	291
40	238	.93882	518	148	770	385	.91992	90	654	227	.81797	362	.80922	478	028
41	042	682	314	.92940	558	170	774	91	386	.81959	529	094	655	211	.79761
42	.93842	478	107	729	344	.91952	554	92	114	688	257	.80823	384	.79941	491
43	639	271	.92897	516	128	733	332	93	.81839	413	.80983	549	111	669	220
44	433	062	685	301	.91910	513	108	94	561	134	705	272	.79835	393	.78947
45	226	.92852	472	085	692	291	.90884	95	278	.80852	424	.79991	555	114	670
46	017	640	257	.91868	472	069	660	96	.80991	566	138	706	271	.78831	388
47	.92806	426	041	649	250	.90845	434	97	698	274	.79846	415	.78981	542	100
48	593	211	.91823	429	028	621	207	98	399	.79975	547	117	684	247	.77806
49	379	.91995	604	208	.90805	396	.89979	99	094	670	243	.78814	382	.77946	507
								100	.79784	360	.78934	506	075	641	203

* For data from −78° to 78°C., see "American Institute of Physics Handbook," p. 2-142, Table 2N-5, McGraw-Hill, New York, 1957.

TABLE 3-111 Densities of Mixtures of C_2H_5OH and H_2O at 20°C

g/mL

% alcohol by weight	Tenths of % 0	1	2	3	4	5	6	7	8	9
0	0.99823	804	785	766	748	729	710	692	673	655
1	636	618	599	581	562	544	525	507	489	471
2	453	435	417	399	381	363	345	327	310	292
3	275	257	240	222	205	188	171	154	137	120
4	103	087	070	053	037	020	003	*987	*971	*954
5	.98938	922	906	890	874	859	843	827	811	796
6	780	765	749	734	718	703	688	673	658	642
7	627	612	597	582	567	553	538	523	508	493
8	478	463	449	434	419	404	389	374	360	345
9	331	316	301	287	273	258	244	229	215	201
10	187	172	158	144	130	117	103	089	075	061
11	047	033	019	006	*992	*978	*964	*951	*937	*923
12	.97910	896	883	869	855	842	828	815	801	788
13	775	761	748	735	722	709	696	683	670	657
14	643	630	617	604	591	578	565	552	539	526
15	514	501	488	475	462	450	438	425	412	400
16	387	374	361	349	336	323	310	297	284	272
17	259	246	233	220	207	194	181	168	155	142
18	129	116	103	089	076	063	050	037	024	010
19	.96997	984	971	957	944	931	917	904	891	877
20	864	850	837	823	810	796	783	769	756	742
21	729	716	702	688	675	661	647	634	620	606
22	592	578	564	551	537	523	509	495	481	467
23	453	439	425	411	396	382	368	354	340	326
24	312	297	283	269	254	240	225	211	196	182
25	168	153	139	124	109	094	080	065	050	035
26	020	005	*990	*975	*959	*944	*929	*914	*898	*883
27	.95867	851	836	820	805	789	773	757	742	726
28	710	694	678	662	646	630	613	597	581	565
29	548	532	516	499	483	466	450	433	416	400
30	382	365	349	332	315	298	281	264	247	230
31	212	195	178	161	143	126	108	091	074	056
32	038	020	003	*985	*967	*950	*932	*914	*896	*878
33	.94860	842	824	806	788	770	752	734	715	697
34	679	660	642	624	605	587	568	550	531	512
35	494	475	456	438	419	400	382	363	344	325
36	306	287	268	249	230	211	192	172	153	134
37	114	095	075	056	036	017	*997	*978	*958	*939
38	.93919	899	879	859	840	820	800	780	760	740
39	720	700	680	660	640	620	599	579	559	539
40	518	498	478	458	437	417	396	376	356	335
41	314	294	273	253	232	212	191	170	149	129
42	107	086	065	044	023	002	*981	*960	*939	*918
43	.92897	876	855	834	812	791	770	749	728	707
44	685	664	642	621	600	579	557	536	515	493
45	472	450	429	408	386	365	343	322	300	279
46	257	236	214	193	171	150	128	106	085	063
47	041	019	*997	*976	*954	*932	*910	*889	*867	*845
48	.91823	801	780	758	736	714	692	670	648	626
49	604	582	560	538	516	494	472	450	428	406

% alcohol by weight	Tenths of % 0	1	2	3	4	5	6	7	8	9
50	0.91384	361	339	317	295	272	250	228	206	183
51	160	138	116	093	071	049	026	004	*981	*959
52	.90936	914	891	869	846	824	801	779	756	734
53	711	689	666	644	621	598	576	553	531	508
54	485	463	440	417	395	372	349	327	304	281
55	258	236	213	190	167	145	122	099	076	054
56	031	008	*985	*962	*939	*917	*894	*871	*848	*825
57	.89803	780	757	734	711	688	665	643	620	597
58	574	551	528	505	482	459	436	413	390	367
59	344	321	298	275	252	229	206	183	160	137
60	113	090	067	044	021	*998	*975	*951	*928	*905
61	.88882	859	836	812	789	766	743	720	696	673
62	650	626	603	580	557	533	510	487	463	440
63	417	393	370	347	323	300	277	253	230	206
64	183	160	136	113	089	066	042	019	*995	*972
65	.87948	925	901	878	854	831	807	784	760	737
66	713	689	666	642	619	595	572	548	524	501
67	477	454	430	406	383	359	336	312	288	265
68	241	218	194	170	147	123	099	075	052	028
69	004	*981	*957	*933	*909	*885	*862	*838	*814	*790
70	.86766	742	718	694	671	647	623	599	575	551
71	527	503	479	455	431	407	383	339	335	311
72	287	263	239	215	191	167	143	119	095	071
73	047	022	*998	*974	*950	*926	*902	*878	*854	*830
74	.85806	781	757	733	709	685	661	636	612	588
75	564	540	515	491	467	443	419	394	370	346
76	322	297	273	249	225	200	176	152	128	103
77	079	055	031	006	*982	*958	*933	*909	*884	*860
78	.84835	811	787	762	738	713	689	664	640	615
79	590	566	541	517	492	467	443	418	393	369
80	344	319	294	270	245	220	196	171	146	121
81	096	072	047	022	*997	*972	*947	*923	*898	*873
82	.83848	823	798	773	748	723	698	674	649	624
83	599	574	549	523	498	473	448	423	398	373
84	348	323	297	272	247	222	196	171	146	120
85	095	070	044	019	*994	*968	*943	*917	*892	*866
86	.82840	815	789	763	738	712	686	660	635	609
87	583	557	531	505	479	453	427	401	375	349
88	323	297	271	245	219	193	167	140	114	088
89	062	035	009	*983	*956	*930	*903	*877	*850	*824
90	.81797	770	744	717	690	664	637	610	583	556
91	529	502	475	448	421	394	366	339	312	285
92	257	230	203	175	148	120	093	066	038	010
93	.80983	955	928	900	872	844	817	789	761	733
94	705	677	649	621	593	565	537	509	480	452
95	424	395	367	338	310	281	253	224	195	166
96	138	109	080	051	022	*993	*963	*934	*905	*875
97	.79846	816	787	757	727	698	668	638	608	578
98	547	517	487	456	426	396	365	335	305	274
99	243	213	182	151	120	089	059	028	*997	*966
100	.78934									

* Indicates change in the first two decimal places.

TABLE 3-112 Specific Gravity (60°/60°F [(15.56°/15.56°C)] of Mixtures by (Volume) of C_2H_5OH and H_2O

% alcohol by volume at 60°F.	Tenths of %: 0	1	2	3	4	5	6	7	8	9
0	1.00000	*985	*970	*955	*940	*925	*910	*895	*880	865
1	0.99850	835	820	806	791	776	761	747	732	717
2	703	688	674	659	645	630	616	602	587	573
3	559	545	531	516	502	488	474	460	446	432
4	419	405	391	378	364	350	336	323	309	296
5	282	269	255	242	228	215	202	189	176	163
6	150	137	124	111	098	085	073	060	047	035
7	022	009	*997	*984	*972	*960	*947	*935	*923	*911
8	.98899	887	875	863	851	838	826	814	803	791
9	779	767	755	743	731	720	708	696	684	672
10	661	649	637	625	614	602	590	579	567	556
11	544	532	521	509	498	487	475	464	452	441
12	430	419	408	396	385	374	363	352	341	330
13	319	308	297	286	275	264	254	243	232	221
14	210	200	190	179	168	157	147	136	125	115
15	104	093	083	072	062	051	040	030	019	009
16	.97998	988	977	967	956	946	936	925	915	905
17	895	885	875	864	854	844	834	824	814	804
18	794	784	774	764	754	744	734	724	714	704
19	694	684	674	664	654	645	635	625	615	605
20	596	586	576	566	556	546	536	526	516	506
21	496	486	476	466	456	446	436	425	415	405
22	395	385	375	365	354	344	334	324	313	303
23	293	283	272	262	252	241	231	221	210	200
24	189	179	168	158	147	137	126	116	105	095
25	084	073	063	052	042	031	020	010	*999	*988
26	.96978	967	957	946	935	924	914	903	892	881
27	870	859	848	837	826	815	804	793	782	771
28	760	749	738	727	715	704	693	682	671	659
29	648	637	625	614	603	591	580	568	557	546
30	534	522	511	499	488	476	464	453	441	429
31	418	406	394	382	370	358	346	334	321	309
32	296	284	271	259	246	234	221	209	196	183
33	170	157	144	132	119	106	093	080	067	054
34	041	028	015	002	*988	*975	*962	*948	*935	*921
35	.95908	894	881	867	854	840	826	812	798	784
36	770	756	742	728	714	700	685	671	657	643
37	628	614	599	585	570	556	541	526	512	497
38	482	467	452	437	423	408	393	378	362	347
39	332	317	302	286	271	256	240	225	209	194
40	178	162	147	131	115	100	084	068	052	036
41	020	004	*988	*972	*956	*940	*923	*907	*891	*875
42	.94858	842	825	809	792	776	759	743	726	710
43	693	676	660	643	626	609	592	575	558	541
44	524	507	490	473	455	438	421	403	386	369
45	351	334	316	298	281	263	245	228	210	192
46	174	156	138	120	102	084	066	048	030	011
47	.93993	975	956	938	920	901	883	864	845	827
48	808	789	771	752	733	714	695	676	657	638
49	619	600	581	562	543	523	504	485	465	446

% alcohol by volume at 60°F.	Tenths of %: 0	1	2	3	4	5	6	7	8	9
50	0.93426	407	387	368	348	328	309	289	270	250
51	230	210	190	171	151	131	111	091	071	051
52	031	011	*991	*971	*951	*931	*911	*890	*870	*850
53	.92830	810	789	769	749	728	708	688	667	647
54	626	605	585	564	544	523	502	482	461	440
55	419	398	377	357	336	315	294	273	252	231
56	210	189	168	147	126	105	084	062	041	020
57	.91999	978	956	935	914	892	871	849	827	806
58	784	762	741	719	697	675	653	631	610	588
59	565	543	521	499	477	455	433	410	388	366
60	344	322	299	277	255	232	210	188	165	143
61	120	097	075	052	030	007	*984	*962	*939	*916
62	.90893	870	847	825	802	779	756	733	710	687
63	664	641	618	595	572	549	526	503	480	457
64	434	411	388	365	341	318	295	272	249	225
65	202	179	155	132	108	085	061	038	014	*991
66	.89967	943	920	896	872	848	825	801	777	753
67	729	705	681	657	633	609	585	561	537	513
68	489	465	441	416	392	368	343	319	295	270
69	245	220	196	171	147	122	098	073	048	024
70	.88999	974	950	925	900	875	850	825	801	776
71	751	725	700	675	650	625	600	574	549	524
72	499	474	448	423	397	372	346	321	296	270
73	244	218	193	167	141	116	090	064	039	013
74	.87987	961	935	910	884	858	832	806	780	754
75	728	702	676	650	623	597	571	545	518	492
76	465	439	412	386	359	332	306	279	252	226
77	199	172	145	118	092	065	038	011	*984	*957
78	.86929	902	875	847	820	793	766	738	711	684
79	656	629	601	574	546	518	491	463	435	408
80	380	352	324	296	269	241	213	185	157	129
81	100	072	044	015	*987	*959	*931	*902	*874	*846
82	.85817	789	760	732	703	674	646	617	588	560
83	531	502	473	444	415	386	357	328	299	270
84	240	211	181	152	122	093	063	033	004	*974
85	.84944	914	884	854	824	794	764	734	703	673
86	642	612	581	551	520	490	459	428	398	367
87	336	305	274	243	212	181	150	119	088	056
88	025	*994	*962	*930	*899	*867	*835	*803	*771	*739
89	.83707	675	643	610	578	545	513	480	447	415
90	382	349	315	282	249	216	183	150	116	083
91	049	015	*981	*947	*913	*879	*845	*810	*776	*741
92	.82705	670	635	600	565	529	494	458	423	387
93	351	315	279	243	206	170	133	096	059	022
94	.81984	947	909	871	834	796	757	719	681	642
95	603	564	525	486	446	407	367	327	287	247
96	206	165	125	084	042	001	*960	*918	*876	*834
97	.80792	750	707	664	620	577	533	489	445	401
98	356	311	265	219	173	127	080	033	*985	*937
99	.79889	841	792	743	693	643	593	543	492	441
100	389									

* Indicates change in first two decimal places.

TABLE 3-113 *n*-Propyl Alcohol (C_3H_7OH)

%	0°C.	15°C.	30°C.	%	0°C.	15°C.	30°C.	%	0°C.	15°C.	30°C.	%	0°C.	15°C.	30°C.	%	0°C.	15°C.	30°C.
0	0.9999	0.9991	0.9957	20	0.9789	0.9723	0.9643	40	0.9430	0.9331	0.9226	60	0.9033	0.8922	0.8807	80	0.8634	0.8516	0.8394
1	.9982	.9974	.9940	21	.9776	.9705	.9622	41	.9411	.9310	.9205	61	.9013	.8902	.8786	81	.8614	.8496	.8373
2	.9967	.9960	.9924	22	.9763	.9688	.9602	42	.9391	.9290	.9184	62	.8994	.8882	.8766	82	.8594	.8475	.8352
3	.9952	.9944	.9908	23	.9748	.9670	.9583	43	.9371	.9269	.9164	63	.8974	.8861	.8745	83	.8574	.8454	.8332
4	.9939	.9929	.9893	24	.9733	.9651	.9563	44	.9352	.9248	.9143	64	.8954	.8841	.8724	84	.8554	.8434	.8311
5	.9926	.9915	.9877	25	.9717	.9633	.9543	45	.9332	.9228	.9122	65	.8934	.8820	.8703	85	.8534	.8413	.8290
6	.9914	.9902	.9862	26	.9700	.9614	.9522	46	.9311	.9207	.9100	66	.8913	.8800	.8682	86	.8513	.8393	.8269
7	.9904	.9890	.9848	27	.9682	.9594	.9501	47	.9291	.9186	.9079	67	.8894	.8779	.8662	87	.8492	.8372	.8248
8	.9894	.9877	.9834	28	.9664	.9576	.9481	48	.9272	.9165	.9057	68	.8874	.8759	.8641	88	.8471	.8351	.8227
9	.9883	.9864	.9819	29	.9646	.9556	.9460	49	.9252	.9145	.9036	69	.8854	.8739	.8620	89	.8450	.8330	.8206
10	.9874	.9852	.9804	30	.9627	.9535	.9439	50	.9232	.9124	.9015	70	.8835	.8719	.8600	90	.8429	.8308	.8185
11	.9865	.9840	.9790	31	.9608	.9516	.9418	51	.9213	.9104	.8994	71	.8815	.8700	.8580	91	.8408	.8287	.8164
12	.9857	.9828	.9775	32	.9589	.9495	.9396	52	.9192	.9084	.8973	72	.8795	.8680	.8559	92	.8387	.8266	.8142
13	.9849	.9817	.9760	33	.9570	.9474	.9375	53	.9173	.9064	.8952	73	.8776	.8659	.8539	93	.8364	.8244	.8120
14	.9841	.9806	.9746	34	.9550	.9454	.9354	54	.9153	.9044	.8931	74	.8756	.8639	.8518	94	.8342	.8221	.8098
15	.9833	.9793	.9730	35	.9530	.9434	.9333	55	.9132	.9023	.8911	75	.8736	.8618	.8497	95	.8320	.8199	.8077
16	.9825	.9780	.9714	36	.9511	.9413	.9312	56	.9112	.9003	.8890	76	.8716	.8598	.8477	96	.8296	.8176	.8054
17	.9817	.9768	.9698	37	.9491	.9392	.9289	57	.9093	.8983	.8869	77	.8695	.8577	.8456	97	.8272	.8153	.8031
18	.9808	.9752	.9680	38	.9471	.9372	.9269	58	.9073	.8963	.8849	78	.8675	.8556	.8435	98	.8248	.8128	.8008
19	.9800	.9739	.9661	39	.9450	.9351	.9247	59	.9053	.8942	.8828	79	.8655	.8536	.8414	99	.8222	.8104	.7984
																100	.8194	.8077	.7958

TABLE 3-114 Isopropyl Alcohol (C_3H_7OH)

%	0°C.	15°C.*	15°C.*	20°C.	30°C
0	0.9999	0.9991	0.99913	0.9982	0.9957
1	.9980	.9973	.9972	.9962	.9939
2	.9962	.9956	.9954	.9944	.9921
3	.9946	.9938	.9936	.9926	.9904
4	.9930	.9922	.9920	.9909	.9887
5	.9916	.9906	.9904	.9893	.9871
6	.9902	.9892	.9890	.9877	.9855
7	.9890	.9878	.9875	.9862	.9839
8	.9878	.9864	.9862	.9847	.9824
9	.9866	.9851	.9849	.9833	.9809
10	.9856	.9838	.98362	.9820	.9794
11	.9846	.9826	.9824	.9808	.9778
12	.9838	.9813	.9812	.9797	.9764
13	.9829	.9802	.9800	.9876	.9750
14	.9821	.9790	.9788	.9776	.9735
15	.9814	.9779	.9777	.9765	.9720
16	.9806	.9768	.9765	.9754	.9705
17	.9799	.9756	.9753	.9743	.9690
18	.9792	.9745	.9741	.9731	.9675
19	.9784	.9730	.9728	.9717	.9658
20	.9777	.9719	.97158	.9703	.9642
21	.9768	.9704	.9703	.9688	.9624
22	.9759	.9690	.9689	.9669	.9606
23	.9749	.9675	.9674	.9651	.9587
24	.9739	.9660	.9659	.9634	.9569
25	.9727	.9643	.9642	.9615	.9549
26	.9714	.9626	.9624	.9597	.9529
27	.9699	.9608	.9605	.9577	.9509
28	.9684	.9590	.9586	.9558	.9488
29	.9669	.9570	.9568	.9540	.9467
30	.9652	.9551	.95493	.9520	.9446
31	.9634	......	.9530	.9500	.9426
32	.9615	......	.9510	.9481	.9405
33	.9596	......	.9489	.9460	.9383
34	.9577	......	.9468	.9440	.9361

%	0°C.	15°C.*	15°C.*	20°C.	30°C.
35	0.9557	......	0.9446	0.9419	0.9338
36	.9536	......	.9424	.9399	.9315
37	.9514	......	.9401	.9377	.9292
38	.9493	......	.9379	.9355	.9269
39	.9472	......	.9356	.9333	.9246
40	.9450	......	.93333	.9310	.9224
41	.9428	......	.9311	.9287	.9201
42	.9406	......	.9288	.9264	.9177
43	.9384	......	.9266	.9239	.9154
44	.9361	......	.9243	.9215	.9130
45	.9338	......	.9220	.9191	.9106
46	.9315	......	.9197	.9165	.9082
47	.9292	......	.9174	.9141	.9059
48	.9270	......	.9150	.9117	.9036
49	.9247	......	.9127	.9093	.9013
50	.9224	......	.91043	.9069	.8990
51	.9201	......	.9081	.9044	.8966
52	.9178	......	.9058	.9020	.8943
53	.9155	......	.9035	.8996	.8919
54	.9132	......	.9011	.8971	.8895
55	.9109	......	.8988	.8946	.8871
56	.9086	......	.8964	.8921	.8847
57	.9063	......	.8940	.8896	.8823
58	.9040	......	.8917	.8874	.8800
59	.9017	......	.8893	.8850	.8777
60	.8994	......	.88690	.8825	.8752
61	.8970	......	.8845	.8800	.8728
62	.8947	0.8829	.8821	.8776	.8704
63	.8924	.8805	.8798	.8751	.8680
64	.8901	.8781	.8775	.8727	.8656
65	.8878	.8757	.8752	.8702	.8631
66	.8854	.8733	.8728	.8679	.8607
67	.8831	.8710	.8705	.8656	.8583
68	.8807	.8686	.8682	.8632	.8559
69	.8784	.8662	.8658	.8609	.8535

%	0°C.	15°C.*	15°C.*	20°C.	30°C.
70	0.8761	0.8639	0.86346	0.8584	0.8511
71	.8738	.8615	.8611	.8560	.8487
72	.8714	.8592	.8588	.8537	.8464
73	.8691	.8568	.8564	.8513	.8440
74	.8668	.8545	.8541	.8489	.8416
75	.8644	.8521	.8517	.8464	.8392
76	.8621	.8497	.8493	.8439	.8368
77	.8598	.8474	.8470	.8415	.8344
78	.8575	.8450	.8446	.8391	.8321
79	.8551	.8426	.8422	.8366	.8297
80	.8528	.8403	.83979	.8342	.8273
81	.8503	.8379	.8374	.8317	.8248
82	.8479	.8355	.8350	.8292	.8224
83	.8456	.8331	.8326	.8268	.8200
84	.8432	.8307	.8302	.8243	.8175
85	.8408	.8282	.8278	.8219	.8151
86	.8384	.8259	.8254	.8194	.8127
87	.8360	.8234	.8229	.8169	.8201
88	.8336	.8209	.8205	.8145	.8078
89	.8311	.8184	.8180	.8120	.8053
90	.8287	.8161	.81553	.8096	.8029
91	.8262	.8136	.8130	.8072	.8004
92	.8237	.8110	.8104	.8047	.7979
93	.8212	.8085	.8079	.8023	.7954
94	.8186	.8060	.8052	.7998	.7929
95	.8160	.8034	.8026	.7973	.7904
96	.8133	.8008	.7999	.7949	.7878
97	.8106	.7981	.7972	.7925	.7852
98	.8078	.7954	.7945	.7901	.7826
99	.8048	.7926	.7918	.7877	.7799
100	.8016	.7896	.78913	.7854	.7770

* Two different observers; see "International Critical Tables," vol. 3, p. 120.

TABLE 3-115 Glycerol*

Glycerol, %	Density 15°C.	15.5°C.	20°C.	25°C.	30°C.
100	1.26415	1.26381	1.26108	1.15802	1.25495
99	1.26160	1.26125	1.25850	1.25545	1.25235
98	1.25900	1.25865	1.25590	1.25290	1.24975
97	1.25645	1.25610	1.25335	1.25030	1.24710
96	1.25385	1.25350	1.25080	1.24770	1.24450
95	1.25130	1.25095	1.24825	1.24515	1.24190
94	1.24865	1.24830	1.24560	1.24250	1.23930
93	1.24600	1.24565	1.24300	1.23985	1.23670
92	1.24340	1.24305	1.24035	1.23725	1.23410
91	1.24075	1.24040	1.23770	1.23460	1.23150
90	1.23810	1.23775	1.23510	1.23200	1.22890
89	1.23545	1.23510	1.23245	1.22935	1.22625
88	1.23280	1.23245	1.22975	1.22665	1.22360
87	1.23015	1.22980	1.22710	1.22400	1.22095
86	1.22750	1.22710	1.22445	1.22135	1.21830
85	1.22485	1.22445	1.22180	1.21870	1.21565
84	1.22220	1.22180	1.21915	1.21605	1.21300
83	1.21955	1.21915	1.21650	1.21340	1.21035
82	1.21690	1.21650	1.21380	1.21075	1.20770
81	1.21425	1.21385	1.21115	1.20810	1.20505
80	1.21160	1.21120	1.20850	1.20545	1.20240
79	1.20885	1.20845	1.20575	1.20275	1.19970
78	1.20610	1.20570	1.20305	1.20005	1.19705
77	1.20335	1.20300	1.20030	1.19735	1.19435
76	1.20060	1.20025	1.19760	1.19465	1.19170
75	1.19785	1.19750	1.19485	1.19195	1.18900
74	1.19510	1.19480	1.19215	1.18925	1.18635
73	1.19235	1.19205	1.18940	1.18650	1.18365
72	1.18965	1.18930	1.18670	1.18380	1.18100
71	1.18690	1.18655	1.18395	1.18110	1.17830
70	1.18415	1.18385	1.18125	1.17840	1.17565
69	1.18135	1.18105	1.17850	1.17565	1.17290
68	1.17860	1.17830	1.17575	1.17295	1.17020
67	1.17585	1.17555	1.17300	1.17020	1.16745
66	1.17305	1.17275	1.17025	1.16745	1.16470

Glycerol, %	Density 15°C.	15.5°C.	20°C.	25°C.	30°C.
65	1.17030	1.17000	1.16750	1.16475	1.16195
64	1.16755	1.16725	1.16475	1.16200	1.15925
63	1.16480	1.16445	1.16205	1.15925	1.15650
62	1.16200	1.16170	1.15930	1.15655	1.15375
61	1.15925	1.15895	1.15655	1.15380	1.15100
60	1.15650	1.15615	1.15380	1.15105	1.14830
59	1.15370	1.15340	1.15105	1.14835	1.14555
58	1.15095	1.15065	1.14830	1.14560	1.14285
57	1.14815	1.14785	1.14555	1.14285	1.14010
56	1.14535	1.14510	1.14280	1.14015	1.13740
55	1.14260	1.14230	1.14005	1.13740	1.13470
54	1.13980	1.13955	1.13730	1.13465	1.13195
53	1.13705	1.13680	1.13455	1.13195	1.12925
52	1.13425	1.13400	1.13180	1.12920	1.12650
51	1.13150	1.13125	1.12905	1.12650	1.12380
50	1.12870	1.12845	1.12630	1.12375	1.12110
49	1.12600	1.12575	1.12360	1.12110	1.11845
48	1.12325	1.12305	1.12090	1.11840	1.11580
47	1.12055	1.12030	1.11820	1.11575	1.11320
46	1.11780	1.11760	1.11550	1.11310	1.11055
45	1.11510	1.11490	1.11280	1.11040	1.10795
44	1.11235	1.11215	1.11010	1.10775	1.10530
43	1.10960	1.10945	1.10740	1.10510	1.10265
42	1.10690	1.10670	1.10470	1.10240	1.10005
41	1.10415	1.10400	1.10200	1.09975	1.09740
40	1.10145	1.10130	1.09930	1.09710	1.09475
39	1.09875	1.09860	1.09665	1.09445	1.09215
38	1.09605	1.09590	1.09400	1.09180	1.08955
37	1.09340	1.09320	1.09135	1.08915	1.08690
36	1.09070	1.09050	1.08865	1.08655	1.08430
35	1.08800	1.08780	1.08600	1.08390	1.08165
34	1.08530	1.08515	1.08335	1.08125	1.07905
33	1.08265	1.08245	1.08070	1.07860	1.07645
32	1.07995	1.07975	1.07800	1.07600	1.07380
31	1.07725	1.07705	1.07535	1.07335	1.07120

Glycerol, %	Density 15°C.	15.5°C.	20°C.	25°C.	30°C.
30	1.07455	1.07435	1.07270	1.07070	1.06855
29	1.07195	1.07175	1.07010	1.06815	1.06605
28	1.06935	1.06915	1.06755	1.06560	1.06355
27	1.06670	1.06655	1.06495	1.06305	1.06105
26	1.06410	1.06390	1.06240	1.06055	1.05855
25	1.06150	1.06130	1.05980	1.05800	1.05605
24	1.05885	1.05870	1.05720	1.05545	1.05350
23	1.05625	1.05610	1.05465	1.05290	1.05100
22	1.05365	1.05350	1.05205	1.05035	1.04850
21	1.05100	1.05090	1.04950	1.04780	1.04600
20	1.04840	1.04825	1.04690	1.04525	1.04350
19	1.04590	1.04575	1.04440	1.04280	1.04105
18	1.04335	1.04325	1.04195	1.04035	1.03860
17	1.04085	1.04075	1.03945	1.03790	1.03615
16	1.03835	1.03825	1.03695	1.03545	1.03370
15	1.03580	1.03570	1.03450	1.03300	1.03130
14	1.03330	1.03320	1.03200	1.03055	1.02885
13	1.03080	1.03070	1.02955	1.02805	1.02640
12	1.02830	1.02820	1.02705	1.02560	1.02395
11	1.02575	1.02565	1.02455	1.02315	1.02150
10	1.02325	1.02315	1.02210	1.02070	1.01905
9	1.02085	1.02075	1.01970	1.01835	1.01670
8	1.01840	1.01835	1.01730	1.01600	1.01440
7	1.01600	1.01590	1.01495	1.01360	1.01205
6	1.01360	1.01350	1.01255	1.01125	1.00970
5	1.01120	1.01110	1.01015	1.00890	1.00735
4	1.00875	1.00870	1.00780	1.00655	1.00505
3	1.00635	1.00630	1.00540	1.00415	1.00270
2	1.00395	1.00385	1.00300	1.00180	1.00035
1	1.00155	1.00145	1.00060	0.99945	0.99800
0	0.99913	0.99905	0.99823	0.99708	0.99568

* Bosart and Snoddy, *Ind. Eng. Chem.*, **20**, 1378 (1928).

TABLE 3-116 Hydrazine (N_2H_4)

%	d_4^{15}	%	d_4^{15}
1	1.0002	30	1.0305
2	1.0013	40	1.038
4	1.0034	50	1.044
8	1.0077	60	1.047
12	1.0121	70	1.046
16	1.0164	80	1.040
20	1.0207	90	1.030
24	1.0248	100	1.011
28	1.0286		

TABLE 3-117 Densities of Aqueous Solutions of Miscellaneous Organic Compounds*

d (resp., d_w, d_s) = density of the solution (resp., water; resp., the pure liquid solute) in g/mL. p_s (resp., p_w) = weight % of solute (resp., water) in the solution. "Range" = range of applicability of the equation.

Section A $d = d_w + Ap_s + Bp_s^2 + Cp_s^3$

Name	Formula	t, °C.	Range, p_s	A	B	C
Acetaldehyde	C_2H_4O	18	0– 30	$+0.0_3255$	-0.0_616	
Acetamide	C_2H_5NO	15	0– 6	$+0.0_3639$	$+0.0_4171$	
Acetone	C_3H_6O	0	0–100	-0.0_3856	-0.0_5449	-0.0_7588
		4	0–100	-0.0_37648	-0.0_41193	$+0.0_8272$
		15	0–100	-0.0_21009	-0.0_59682	-0.0_8624
		20	0–100	-0.0_21233	-0.0_53529	-0.0_75327
		25	0–100	-0.0_21171	-0.0_5904	-0.0_856
Acetonitrile	C_2H_3N	15	0– 16	-0.0_21175	-0.0_42024	
Allyl alcohol	C_3H_6O	0	0– 89	-0.0_33729	-0.0_41232	$+0.0_72984$
Benzenepentacarboxylic acid	$C_{11}H_6O_{10}$	25	0– 0.6	$+0.0_25615$	-0.0_2117	
Butyl alcohol (n-)	$C_4H_{10}O$	20	0– 7.9	-0.0_21651	$+0.0_4285$	
Butyric acid (n-)	$C_4H_8O_2$	18	0– 10	$+0.0_3414$	$+0.0_4131$	
		25	0– 62	$+0.0_35135$	-0.0_4166	$+0.0_611$
Chloral hydrate	$C_2H_3Cl_3O_2$	0	0– 70	$+0.0_24489$	$+0.0_42802$	-0.0_71291
		15	0– 78	$+0.0_24455$	$+0.0_42198$	$+0.0_74366$
		30	0– 90	$+0.0_24401$	$+0.0_41887$	$+0.0_76549$
Chloroacetic acid	$C_2H_3ClO_2$	20	0– 32	$+0.0_23648$	$+0.0_5302$	
		25	0– 86	$+0.0_23602$	$+0.0_5552$	$+0.0_722$
Citric acid (hydrate)	$C_6H_8O_7 + H_2O$	18	0– 50	$+0.0_23824$	$+0.0_41141$	$+0.0_717$
Dichloroacetic acid	$C_2H_2Cl_2O_2$	20	0– 30	$+0.0_24427$	$+0.0_5537$	$+0.0_77534$
		25	0– 97	$+0.0_24427$	$+0.0_5537$	$+0.0_77534$
Diethylamine hydrochloride	$C_4H_{12}ClN$	21	0– 36	$+0.0_334$	$+0.0_676$	
Ethylamine hydrochloride	C_2H_8ClN	21	0– 65	$+0.0_21193$	-0.0_5307	-0.0_747
Ethylene glycol	$C_2H_6O_2$	0	0–100	$+0.0_21483$	$+0.0_52992$	-0.0_75248
		15	0– 6	$+0.0_2133$	-0.0_5108	
Ethyl ether	$C_4H_{10}O$	20	0– 5	-0.0_2221	$+0.0_448$	
		25	0– 4.5	-0.0_2221	$+0.0_435$	
tartrate	$C_8H_{14}O_6$	15	0– 95	$+0.0_22367$	$+0.0_5358$	-0.0_76005
Formaldehyde	CH_2O	15	0– 40	$+0.0_22518$	-0.0_6658	$+0.0_6542$
Formamide	CH_3NO	25	22– 96	$+0.0_21217$	$+0.0_53199$	-0.0_72529
Furfural	$C_5H_4O_2$	20	0– 8	$+0.0_21827$	$+0.0_5366$	
		25	0– 8	$+0.0_21664$	$+0.0_421$	
Isoamyl alcohol	$C_5H_{12}O$	20	0– 2.5	$+0.0_2155$	$+0.0_43$	
Isobutyl alcohol	$C_4H_{10}O$	15	0– 8	-0.0_2146	$+0.0_56$	
		20	0– 8	-0.0_2169	$+0.0_438$	
Isobutyric acid	$C_4H_8O_2$	15	0– 9	$+0.0_352$		
		18	0– 9	$+0.0_345$		
		25	0– 12	$+0.0_337$		
Isovaleric acid	$C_5H_{10}O_2$	25	0– 5	$+0.0_3253$	-0.0_4282	
Lactic acid	C_3H_6O	25	0– 9	$+0.0_2231$	$+0.0_5186$	
Maleic acid	$C_4H_4O_4$	25	0– 40	$+0.0_234$	$+0.0_575$	
Malic acid	$C_4H_6O_5$	20	0– 40	$+0.0_23933$	$+0.0_5957$	
		25	0– 40	$+0.0_23736$	$+0.0_4175$	
Malonic acid	$C_3H_4O_4$	20	0– 40	$+0.0_2389$	$+0.0_41066$	
Methyl acetate	$C_3H_6O_2$	20	0– 20	$+0.0_340$	-0.0_574	
glucoside (α-)	$C_7H_{14}O_6$	0	26– 51	$+0.0_23336$	$+0.0_5996$	$+0.01544$
		30	26– 51	$+0.0_23151$	$+0.0_5975$	$+0.0_8978$
Nicotine	$C_{10}H_{14}N_2$	20	0– 60	$+0.0_3642$	$+0.0_5454$	-0.0_7687
Nitrophenol (p-)	$C_6H_5NO_3$	15	0– 1.5	$+0.0_23216$	-0.0_455	
Oxalic acid	$C_2H_2O_4$	0	0– 4	$+0.0_25898$	-0.0_33185	$+0.0_441$
		15	0– 4	$+0.0_2494$	-0.0_68	
		17.5	0– 9	$+0.0_2494$	-0.0_68	
		20	0– 4	$+0.0_25264$	-0.0_31996	$+0.0_4254$
		25	0– 4	$+0.0_25108$	-0.0_31607	$+0.0_4208$
Phenol	C_6H_6O	15	0– 5	$+0.0_2111$	-0.0_4283	
		80	0– 65	$+0.0_3462$	-0.0_686	
Phenylglycolic acid	$C_8H_8O_3$	25	0– 11	$+0.0_2207$	$+0.0_423$	
Picoline (α-)	C_6H_7N	25	0– 70	-0.0_4386	-0.0_51405	-0.0_74167
(β-)	C_6H_7N	25	0– 60	-0.0_4683	-0.0_513	
Propionic acid	$C_3H_6O_2$	18	0– 10	$+0.0_395$	-0.0_4172	
		25	0– 40	$+0.0_39245$	-0.0_699	$+0.0_7361$
Pyridine	C_5H_5N	25	0– 60	$+0.0_3229$	-0.0_5204	-0.0_828
Resorcinol	$C_6H_6O_2$	18	0– 52	$+0.0_2201$	$+0.0_5519$	-0.0_819
Succinic acid	$C_4H_6O_4$	25	0– 5.5	$+0.0_2304$		
Tartaric acid (d., l., or dl.)	$C_4H_6O_6$	15	0– 15	$+0.0_24482$	$+0.0_4185$	
		17.5	0– 50	$+0.0_24455$	$+0.0_4185$	
		20	0– 50	$+0.0_24432$	$+0.0_41837$	
		30	0– 50	$+0.0_24335$	$+0.0_4185$	
		40	0– 50	$+0.0_24265$	$+0.0_4185$	
		50	0– 50	$+0.0_24205$	$+0.0_4185$	
		60	0– 50	$+0.0_24155$	$+0.0_4185$	

* From "International Critical Tables," vol. 3, pp. 111–114.

TABLE 3-117 Densities of Aqueous Solutions of Miscellaneous Organic Compounds *(Concluded)*

Name	Formula	t, °C.	Range, p_s	A	B	C
Tetraethyl ammonium chloride	$C_8H_{20}ClN$	21	0– 63	$+0.0_3 1884$	$+0.0_5 6$	$+0.0_7 122$
Thiourea	CH_4N_2S	15	0– 7	$+0.0_2 2995$	$+0.0_5 374$	
Trichloroacetic acid	$C_2HCl_3O_2$	12.5	0– 61	$+0.0_2 499$	$+0.0_4 153$	
		20	10– 30	$+0.0_2 5053$	$+0.0_4 1387$	
		25	0– 94	$+0.0_2 5051$	$+0.0_5 6119$	$+0.0_6 1038$
Triethylamine hydrochloride	$C_6H_{16}ClN$	21	0– 54	$+0.0_4 6$	$+0.0_5 558$	$-0.0_8 69$
Trimethyl carbinol	$C_4H_{10}O$	20	0–100	$-0.0_2 117$	$-0.0_4 1908$	$+0.0_7 957$
		25	0–100	$-0.0_2 1286$	$-0.0_4 176$	$+0.0_7 887$
Urea	CH_4N_2O	14.8	0– 12	$+0.0_2 3213$	$-0.0_4 4802$	$+0.0_5 1216$
		18	0– 51	$+0.0_2 2718$	$+0.0_5 1552$	$+0.0_7 2573$
		20	0– 35	$+0.0_2 2702$	$+0.0_5 3712$	$-0.0_7 2285$
		25	0– 10	$+0.0_2 2728$	$-0.0_4 1817$	$+0.0_5 1379$
Urethane	$C_3H_7NO_2$	20	0– 56	$+0.0_2 1278$	$-0.0_5 245$	$-0.0_7 3437$
Valeric acid (n-)	$C_5H_{10}O_2$	25	0– 3	$+0.0_3 34$	$-0.0_4 27$	

Section B $d = d_s + Ap_w + Bp_w^2 + Cp_w^3$

Name	Formula	d_s	t, °C.	Range, p_w	A	B	C
Butyl alcohol (n-)	$C_4H_{10}O$	0.8097	20	0–20	$+0.0_2 2103$	$-0.0_4 113$	
Butyric acid (n-)	$C_4H_8O_2$	0.9534	25	0–38	$+0.0_2 1854$	$-0.0_4 2314$	
Ethyl ether	$C_4H_{10}O$	0.7077	25	0– 1.1	$+0.0_2 34$	$+0.0_3 36$	
Isobutyl alcohol	$C_4H_{10}O$	0.8170	0	0–14	$+0.0_2 2437$	$-0.0_4 285$	
		0.8055	15	0–16	$+0.0_2 224$	$-0.0_4 129$	
Isobutyric acid	$C_4H_8O_2$	0.9425	26	0–80	$+0.0_2 1808$	$-0.0_4 2358$	$+0.0_6 1253$
Nicotine	$C_{10}H_{14}N_2$	1.0093	20	0–40	$+0.0_2 199$	$-0.0_4 331$	$+0.0_7 315$
Picoline (α-)	C_6H_7N	0.9404	25	0–30	$+0.0_2 2715$	$-0.0_4 393$	
(β-)	C_6H_7N	0.9515	25	0–40	$+0.0_2 1925$	$-0.0_4 352$	$+0.0_6 25$
Pyridine	C_5H_5N	0.9776	25	0–40	$+0.0_2 1157$	$-0.0_5 536$	$-0.0_6 2$
Trimethyl carbinol	$C_4H_{10}O$	0.7856	20	0–20	$+0.0_2 2287$	$+0.0_5 275$	

Section C $d_t = d_o + At + Bt^2$

Name	Formula	p_s	d_0	Range, °C.	A	B
Allyl alcohol	C_3H_6O	76.60	0.9122	0–45	$-0.0_3 8$	$-0.0_5 27$
Butyl alcohol (n-)	$C_4H_{10}O$	80.95	0.8614	0–43	$-0.0_3 7292$	$-0.0_6 75$
Chloral hydrate	$C_2H_3Cl_3O_2$	2.00	1.0094	7–80	$-0.0_4 2597$	$-0.0_5 4313$
		10.00	1.0476	7–80	$-0.0_4 7955$	$-0.0_5 4253$
Ethyl tartrate	$C_7H_{14}O_6$	5.00	1.0150	15–80	$-0.0_3 2103$	$-0.0_5 2544$
		10.00	1.0270	15–80	$-0.0_3 2116$	$-0.0_6 2929$
		25.00	1.0665	15–80	$-0.0_3 401$	$-0.0_5 23$
Furfural	$C_5H_4O_2$	4.62	1.0125	22–74	$-0.0_3 232$	$-0.0_5 254$
		5.69	1.0140	22–74	$-0.0_3 221$	$-0.0_5 268$
		6.56	1.0155	22–74	$-0.0_3 211$	$-0.0_5 290$
Pyridine	C_5H_5N	9.34	1.0055	11–73	$-0.0_3 171$	$-0.0_5 3615$
		21.20	1.0115	14–73	$-0.0_3 378$	$-0.0_5 248$
		29.50	1.0145	12–72	$-0.0_3 463$	$-0.0_5 235$
		40.40	1.0182	9–74	$-0.0_3 605$	$-0.0_5 167$

DENSITIES OF MISCELLANEOUS MATERIALS

TABLE 3-118 Approximate Specific Gravities and Densities of Miscellaneous Solids and Liquids*

Water at 4°C and normal atmospheric pressure taken as unity. For more detailed data on any material, see the section dealing with the properties of that material.

Substance	Sp. gr.	Aver. weight lb./cu. ft.
Metals, Alloys, Ores		
Aluminum, cast-hammered	2.55–2.80	165
bronze	7.7	481
Brass, cast-rolled	8.4–8.7	534
Bronze, 7.9 to 14% Sn	7.4–8.9	509
phosphor	8.88	554
Copper, cast-rolled	8.8–8.95	556
ore, pyrites	4.1–4.3	262
German silver	8.58	536
Gold, cast-hammered	19.25–19.35	1205
coin (U.S.)	17.18–17.2	1073
Iridium	21.78–22.42	1383
Iron, gray cast	7.03–7.13	442
cast, pig	7.2	450
wrought	7.6–7.9	485
spiegeleisen	7.5	468
ferro-silicon	6.7–7.3	437
ore, hematite	5.2	325
ore, limonite	3.6–4.0	237
ore, magnetite	4.9–5.2	315
slag	2.5–3.0	172
Lead	11.34	710
ore, galena	7.3–7.6	465
Manganese	7.42	475
ore, pyrolusite	3.7–4.6	259
Mercury	13.6	849
Monel metal, rolled	8.97	555
Nickel	8.9	537
Platinum, cast-hammered	21.5	1330
Silver, cast-hammered	10.4–10.6	656
Steel, cold-drawn	7.83	489
machine	7.80	487
tool	7.70–7.73	481
Tin, cast-hammered	7.2–7.5	459
cassiterite	6.4–7.0	418
Tungsten	19.22	1200
Zinc, cast-rolled	6.9–7.2	440
blende	3.9–4.2	253
Various Solids		
Cereals, oats, bulk	0.51	26
barley, bulk	0.62	39
corn, rye, bulk	0.73	45
wheat, bulk	0.77	48
Cork	0.22–0.26	15
Cotton, flax, hemp	1.47–1.50	93
Fats	0.90–0.97	58
Flour, loose	0.40–0.50	28
pressed	0.70–0.80	47
Glass, common	2.40–2.80	162
plate or crown	2.45–2.72	161
crystal	2.90–3.00	184
flint	3.2–4.7	247
Hay and straw, bales	0.32	20
Leather	0.86–1.02	59
Paper	0.70–1.15	58
Potatoes, piled	0.67	44
Rubber, caoutchouc	0.92–0.96	59
goods	1.0–2.0	94
Salt, granulated, piled	0.77	48
Saltpeter	1.07	67
Starch	1.53	96
Sulfur	1.93–2.07	125
Wool	1.32	82

Substance	Sp. gr.	Aver. weight lb./cu. ft.
Timber, Air-dry		
Apple	0.66–0.74	44
Ash, black	0.55	34
white	0.64–0.71	42
Birch, sweet, yellow	0.71–0.72	44
Cedar, white, red	0.35	22
Cherry, wild red	0.43	27
Chestnut	0.48	30
Cypress	0.45–0.48	29
Elm, white	0.56	35
Fir, Douglas	0.48–0.55	32
balsam	0.40	25
Hemlock	0.45–0.50	29
Hickory	0.74–0.80	48
Locust	0.67–0.77	45
Mahogany	0.56–0.85	44
Maple, sugar	0.68	43
white	0.53	33
Oak, chestnut	0.74	46
live	0.87	54
red, black	0.64–0.71	42
white	0.77	48
Pine, Norway	0.55	34
Oregon	0.51	32
red	0.48	30
Southern	0.61–0.67	38–42
white	0.43	27
Poplar	0.43	27
Redwood, California	0.42	26
Spruce, white, red	0.45	28
Teak, African	0.99	62
Indian	0.66–0.88	48
Walnut, black	0.59	37
Willow	0.42–0.50	28
Various Liquids		
Alcohol, ethyl (100%)	0.789	49
methyl (100%)	0.796	50
Acid, muriatic, 40%	1.20	75
nitric, 91%	1.50	94
sulfuric, 87%	1.80	112
Chloroform	1.500	95
Ether	0.736	46
Lye, soda, 66%	1.70	106
Oils, vegetable	0.91–0.94	58
mineral, lubricants	0.88–0.94	57
Turpentine	0.861–0.867	54
Water, 4°C. max. density	1.0	62.428
100°C	0.9584	59.830
ice	0.88–0.92	56
snow, fresh fallen	0.125	8
sea water	1.02–1.03	64
Ashlar Masonry		
Bluestone	2.3–2.6	153
Granite, syenite, gneiss	2.4–2.7	159
Limestone	2.1–2.8	153
Marble	2.4–2.8	162
Sandstone	2.0–2.6	143
Rubble Masonry		
Bluestone	2.2–2.5	147
Granite, syenite, gneiss	2.3–2.6	153
Limestone	2.0–2.7	147
Marble	2.3–2.7	156
Sandstone	1.9–2.5	137

Substance	Sp. gr.	Aver. weight lb./cu. ft.
Dry Rubble Masonry		
Granite, syenite, gneiss	1.9–2.3	130
Limestone, marble	1.9–2.1	125
Sandstone, bluestone	1.8–1.9	110
Brick Masonry		
Hard brick	1.8–2.3	128
Medium brick	1.6–2.0	112
Soft brick	1.4–1.9	103
Sand-lime brick	1.4–2.2	112
Concrete Masonry		
Cement, stone, sand	2.2–2.4	144
slag, etc	1.9–2.3	130
cinder, etc	1.5–1.7	100
Various Building Materials		
Ashes, cinders	0.64–0.72	40–45
Cement, Portland, loose	1.5	94
Lime, gypsum, loose	0.85–1.00	53–64
Mortar, lime, set	1.4–1.9	103
Portland cement	2.08–2.25	94–135
Portland cement	3.1–3.2	196
Slags, bank slag	1.1–1.2	67–72
bank screenings	1.5–1.9	98–117
machine slag	1.5	96
slag sand	0.8–0.9	49–55
Earth, etc., Excavated		
Clay, dry	1.0	63
damp plastic	1.76	110
and gravel, dry	1.6	100
Earth, dry, loose	1.2	76
dry, packed	1.5	95
moist, loose	1.3	78
moist, packed	1.6	96
mud, flowing	1.7	108
mud, packed	1.8	115
Riprap, limestone	1.3–1.4	80–85
Riprap, sandstone	1.4	90
Riprap, shale	1.7	105
Sand, gravel, dry, loose	1.4–1.7	90–105
gravel, dry, packed	1.6–1.9	100–120
gravel, wet	1.89–2.16	126
Excavations in Water		
Clay	1.28	80
River mud	1.44	90
Sand or gravel	0.96	60
and clay	1.00	65
Soil	1.12	70
Stone riprap	1.00	65
Minerals		
Asbestos	2.1–2.8	153
Barytes	4.50	281
Basalt	2.7–3.2	184
Bauxite	2.55	159
Bluestone	2.5–2.6	159
Borax	1.7–1.8	109
Chalk	1.8–2.8	143
Clay, marl	1.8–2.6	137
Dolomite	2.9	181
Feldspar, orthoclase	2.5–2.7	162
Gneiss	2.7–2.9	175
Granite	2.6–2.7	165
Greenstone, trap	2.8–3.2	187
Gypsum, alabaster	2.3–2.8	159
Hornblende	3.0	187
Limestone	2.1–2.86	155
Marble	2.6–2.86	170
Magnesite	3.0	187
Phosphate rock, apatite	3.2	200
Porphyry	2.6–2.9	172

* From Marks, "Mechanical Engineers' Handbook," McGraw-Hill.

TABLE 3-118 Approximate Specific Gravities and Densities of Miscellaneous Solids and Liquids *(Concluded)*

Substance	Sp. gr.	Aver. weight lb./cu. ft.	Substance	Sp. gr.	Aver. weight lb./cu. ft.	Substance	Sp. gr.	Aver. weight lb./cu. ft.
Minerals (*Cont'd*)			**Bituminous Substances**			**Bituminous Substances (*Cont'd*)**		
Pumice, natural	0.37–0.90	40	Asphaltum	1.1–1.5	81	Petroleum	0.87	54
Quartz, flint	2.5–2.8	165	Coal, anthracite	1.4–1.8	97	refined (kerosene)	0.78–0.82	50
Sandstone	2.0–2.6	143	bituminous	1.2–1.5	84	benzine	0.73–0.75	46
Serpentine	2.7–2.8	171	lignite	1.1–1.4	78	gasoline	0.70–0.75	45
Shale, slate	2.6–2.9	172	peat, turf, dry	0.65–0.85	47	Pitch	1.07–1.15	69
						Tar, bituminous	1.20	75
Soapstone, talc	2.6–2.8	169	charcoal, pine	0.28–0.44	23			
Syenite	2.6–2.7	165	charcoal, oak	0.47–0.57	33	**Coal and Coke, Piled**		
			coke	1.0–1.4	75	Coal, anthracite	0.75–0.93	47–58
Stone, Quarried, Piled			Graphite	1.64–2.7	135	bituminous, lignite	0.64–0.87	40–54
Basalt, granite, gneiss	1.5	96	Paraffin	0.87–0.91	56	peat, turf	0.32–0.42	20–26
Greenstone, hornblende	1.7	107				charcoal	0.16–0.23	10–14
Limestone, marble, quartz	1.5	95				coke	0.37–0.51	23–32
Sandstone	1.3	82						
Shale	1.5	92						

NOTE: To convert pounds per cubic foot to kilograms per cubic meter, multiply by 16.02. °F = 9/5 °C + 32.

TABLE 3-119 Density (kg/m^3) of Selected Elements as a Function of Temperature

Temperature, K*	Al	Be†	Cr	Cu	Au	Ir	Fe	Pb	Mo	Ni	Pt	Ag	Zn†
	Element symbol												
50	2736	3650	7160	9019	19,490	22,600	7910	11,570	10,260	8960	21,570	10,620	7280
100	2732	3640	7155	9009	19,460	22,580	7900	11,520	10,260	8950	21,550	10,600	7260
150	2726	3630	7150	8992	19,420	22,560	7890	11,470	10,250	8940	21,530	10,575	7230
200	2719	3620	7145	8973	19,380	22,540	7880	11,430	10,250	8930	21,500	10,550	7200
250	2710	3610	7140	8951	19,340	22,520	7870	11,380	10,250	8910	21,470	10,520	7170
300	2701	3600	7135	8930	19,300	22,500	7860	11,330	10,240	8900	21,450	10,490	7135
400	2681	3580	7120	8885	19,210	22,450	7830	11,230	10,220	8860	21,380	10,430	7070
500	2661	3555	7110	8837	19,130	22,410	7800	11,130	10,210	8820	21,330	10,360	7000
600	2639	3530	7080	8787	19,040	22,360	7760	11,010	10,190	8780	21,270	10,300	6935
800	2591	....	7040	8686	18,860	22,250	7690	10,430	10,160	8690	21,140	10,160	6430
1000	2365	....	7000	8568	18,660	22,140	7650	10,190	10,120	8610	21,010	10,010	6260
1200	2305	....	6945	8458	18,440	22,030	7620	9,940	10,080	8510	20,870	9,850	
1400	2255	....	6890	7920	17,230	21,920	7520	.....	10,040	8410	20,720	9,170	
1600	....	....	6760	7750	16,950	21,790	7420	.....	10,000	8320	20,570	8,980	
1800	....	....	6700	7600	.....	21,660	7320	.....	9,950	7690	20,400		
2000	....	....	....	7460	.....	21,510	7030	.....	9,900	7450	20,220		

NOTE: Above the horizontal line the condensed phase is solid; below it, it is liquid.
*°R = 9/5 K.
†Polycrystalline form tabulated. Similar tables for an additional 45 elements appear in the *Handbook of Heat Transfer*, 2d ed., McGraw-Hill, New York, 1984.

SOLUBILITIES

UNITS CONVERSIONS

For this subsection, the following units conversions are applicable:

°F = 9/5 °C + 32.

To convert cubic centimeters to cubic feet, multiply by 3.532×10^{-5}.

To convert millimeters of mercury to pounds-force per square inch, multiply by 0.01934.

To convert grams per liter to pounds per cubic foot, multiply by 6.243×10^{-2}.

TABLE 3-120 Solubilities of Inorganic Compounds in Water at Various Temperatures*

This table shows the amount of anhydrous substance which is soluble in 100 g of water at the temperature in degrees Celsius as indicated; when the name is followed by †, the value is expressed in grams of substance in 100 cm^3 of saturated solution. Solid phase gives the hydrated form in equilibrium with the saturated solution.

	Substance	Formula	Solid phase	0°C.	10°C.	20°C.	30°C.	40°C.	50°C.	60°C.	70°C.	80°C.	90°C.	100°C.	
1	Aluminum chloride	$AlCl_3$	$6H_2O$			69.86$^{15°}$									1
2	sulfate	$Al_2(SO_4)_3$	$18H_2O$	31.2	33.5	36.4	40.4	46.1	52.2	59.2	66.1	73.0	80.8	89.0	2
3	Ammonium aluminum sulfate	$(NH_4)_2Al_2(SO_4)_4$	$24H_2O$	2.1	4.99	7.74	10.94	14.88	20.10	26.70				109.7$^{95°}$	3
4	bicarbonate	NH_4HCO_3		11.9	15.8	21	27								4
5	bromide	NH_4Br		60.6	68	75.5	83.2	91.1	99.2	107.8	116.8	126	135.6	145.6	5
6	chloride	NH_4Cl		29.4	33.3	37.2	41.4	45.8	50.4	55.2	60.2	65.6	71.3	77.3	6
7	chloroplatinate	$(NH_4)_2PtCl_6$			0.7									1.25	7
8	chromate	$(NH_4)_2CrO_4$					40.4								8
9	chromium sulfate	$(NH_4)_2Cr_2(SO_4)_4$	$24H_2O$			10.78$^{25°}$									9
10	dichromate	$(NH_4)_2Cr_2O_7$					47.17								10
11	dihydrogen phosphite	$NH_4H_2PO_3$		171		190$^{14.5°}$	260$^{31°}$								11
12	hydrogen phosphate	$(NH_4)_2HPO_4$				131^{15}									12
13	iodide	NH_4I		154.2	163.2	172.3	181.4	190.5	199.6	208.9	218.7	228.8		250.3	13
14	magnesium phosphate	NH_4MgPO_4	$6H_2O$	0.023		0.052		0.036	0.030	0.040	0.016	0.019			14
15	manganese phosphate	NH_4MnPO_4	$7H_2O$			0		0		0	0.005	0.007			15
16	nitrate	NH_4NO_3		118.3		192	241.8	297.0	344.0	421.0	499.0	580.0	740.0	871.0	16
17	oxalate	$(NH_4)_2C_2O_4$	$1H_2O$	2.2	3.1	4.4	5.9	8.0	10.3						17
18	perchlorate†	NH_4ClO_4†		11.56		20.85		30.58		39.05		48.19		57.01	18
19	persulfate	$(NH_4)_2S_2O_8$		58.2											19
20	sulfate	$(NH_4)_2SO_4$		70.6	73.0	75.4	78.0	81.0		88.0		95.3		103.3	20
21	thiocyanate	NH_4CNS		119.8	144	170	207.7								21
22	vanadate (meta)	NH_4VO_3				0.48	0.84	1.32	1.78		3.05				22
23	Antimonious fluoride	SbF_3		384.7		444.7	563.6								23
24	sulfide	Sb_2S_3				0.000175$^{18°}$									24
25	Arsenic oxide	As_2O_5		59.5	62.1	65.8	69.5	71.2		73.0		75.1		76.7	25
26	Arsenious sulfide	As_2S_3		5.17 × 10^{-5} at 18°											26
27	Barium acetate	$Ba(C_2H_3O_2)_2$	$3H_2O$	59	63	71									27
28	acetate	$Ba(C_2H_3O_2)_2$	$1H_2O$				75	79	77	74	74			75	28
29	carbonate	$BaCO_3$			0.0016$^{8°}$	0.0022$^{18°}$	0.0024 at 24.2°								29
30	chlorate	$Ba(ClO_3)_2$	$1H_2O$	20.34	26.95	33.80	41.70	49.61		66.81		84.84		104.9	30
31	chloride	$BaCl_2$	$2H_2O$	31.6	33.3	35.7	38.2	40.7	43.6	46.4	49.4	52.4		58.8	31
32	chromate	$BaCrO_4$		0.0002	0.00028	0.00037	0.00046								32
33	hydroxide	$Ba(OH)_2$	$8H_2O$	1.67	2.48	3.89	5.59	8.22	13.12	20.94		101.4			33
34	iodide	BaI_2	$6H_2O$	170.2	185.7	203.1	219.6								34
35	iodide	BaI_2	$2H_2O$					231.9		247.3		261.0		271.7	35
36	nitrate	$Ba(NO_3)_2$		5.0	7.0	9.2	11.6	14.2	17.1	20.3		27.0		34.2	36
37	nitrite	$Ba(NO_2)_2$	$1H_2O$			67.5						205.8		300	37
38	oxalate	BaC_2O_4			0.0016$^{8°}$	0.0022$^{18°}$	0.0024 at 24.2°								38
39	perchlorate	$Ba(ClO_4)_2$	$3H_2O$	205.8		289.1		358.7	426.3		495.2		562.3		39
40	sulfate	$BaSO_4$		1.15 × 10^{-4}	2.0 × 10^{-4}	2.4 × 10^{-4}	2.85 × 10^{-4}								40
41	Beryllium sulfate	$BeSO_4$	$6H_2O$				52		60.67						41
42	sulfate	$BeSO_4$	$4H_2O$				43.78	46.74			62		83	100	42
43	sulfate	$BeSO_4$	$2H_2O$									84.76	98	110	43
44	Boric acid	H_3BO_3		2.66	3.57	5.04	6.60	8.72	11.54	14.81	16.73	23.75	30.38	40.25	44
45	Boron oxide	B_2O_3		1.1	1.5	2.2		4.0		6.2		9.5		15.7	45
46	Bromine	Br_2		4.22	3.4	3.20	3.13								46
47	Cadmium chloride	$CdCl_2$	$4H_2O$	97.59	125.1										47
48	chloride	$CdCl_2$	$2\frac{1}{2}H_2O$	90.01			132.1								48
49	chloride	$CdCl_2$	$1H_2O$		135.1	134.5		135.3		136.5		140.4		147.0	49
50	cyanide	$Cd(CN)_2$				1.7$^{15°}$									50
51	hydroxide	$Cd(OH)_2$					2.6 × 10^{-4} at 25°								51
52	sulfate	$CdSO_4$		76.48	76.00	76.60		78.54		83.68			63.13	60.77	52
53	Calcium acetate	$Ca(C_2H_3O_2)_2$	$2H_2O$	37.4	36.0	34.7	33.8	33.2		32.7		33.5			53
54	acetate	$Ca(C_2H_3O_2)_2$	$1H_2O$										31.1	29.7	54

*By N. A. Lange. Abridged from "Table of Solubilities of Inorganic Compounds in Water at Various Temperatures" in Lange, *Handbook of Chemistry*, 10th ed., McGraw-Hill, New York, 1961. For tables of the solubility of gases in water at various temperatures, Atack (*Handbook of Chemical Data*, Reinhold, New York, 1957) gives values at closer temperature intervals, usually 1 or 5°C, than are tabulated here. For materials marked by ‡, additional data are given in tables subsequent to this one. For the solubility of various hydrocarbons in water at high pressures see *J. Chem. Eng. Data*, 4, 212 (1959).

TABLE 3-120 Solubilities of Inorganic Compounds in Water at Various Temperatures *(Continued)*

	Substance	Formula	Solid phase	0°C.	10°C.	20°C.	30°C.	40°C.	50°C.	60°C.	70°C.	80°C.	90°C.	100°C.	
1	Calcium bicarbonate	$Ca(HCO_3)_2$		16.15		16.60		17.05		17.50		17.95		18.40	1
2	chloride	$CaCl_2$	$6H_2O$	59.5	65.0	74.5	102								2
3	chloride	$CaCl_2$	$2H_2O$							136.8	141.7	147.0	152.7	159	3
4	fluoride	CaF_2				0.0016^18°	0.0017^26°								4
5	hydroxide	$Ca(OH)_2$		0.185	0.176	0.165	0.153	0.141	0.128	0.116	0.106	0.094	0.085	0.077	5
6	nitrate	$Ca(NO_3)_2$	$4H_2O$	102.0	115.3	129.3	152.6	195.9							6
7	nitrate	$Ca(NO_3)_2$	$3H_2O$					237.5	281.5						7
8	nitrate	$Ca(NO_3)_2$										358.7		363.6	8
9	nitrite	$Ca(NO_2)_2$	$4H_2O$	62.07		76.68									9
10	nitrite	$Ca(NO_2)_2$	$2H_2O$							132.6	151.9		244.8		10
11	oxalate	CaC_2O_4			6.7 × 10⁻⁴ at 13°	6.8 × 10⁻⁴ at 25°	9.5 × 10⁻⁴ at 50°	14 × 10⁻⁴ at 95°							11
12	sulfate	$CaSO_4$	$2H_2O$	0.1759	0.1928		0.2090	0.2097		0.2047	0.1966			0.1619	12
13	Carbon dioxide, 760 mm ‡	CO_2		0.3346	0.2318	0.1688	0.1257	0.0973	0.0761	0.0576				0	13
14	monoxide, 760 mm ‡	CO		0.0044	0.0035	0.0028	0.0024	0.0021	0.0018	0.0015	0.0013	0.0010	0.0006	0	14
15	Cesium chloride	CsCl		161.4	174.7	186.5	197.3	208.0	218.5	229.7	239.5	250.0	260.1	270.5	15
16	nitrate	$CsNO_3$		9.33	14.9	23.0	33.9	47.2	64.4	83.8	107.0	134.0	163.0	197.0	16
17	sulfate	Cs_2SO_4		167.1	173.1	178.7	184.1	189.9	194.9	199.9	205.0	210.3	214.9	220.3	17
18	Chlorine, 760 mm ‡	Cl_2		1.46	0.980	0.716	0.562	0.451	0.386	0.324	0.274	0.219	0.125	0	18
19	Chromic anhydride	CrO_3		164.9				174.0	182.1				217.5	206.8	19
20	Cupric chloride	$CuCl_2$	$2H_2O$	70.7	73.76	77.0	80.34	83.8	87.44	91.2		99.2		107.9	20
21	nitrate	$Cu(NO_3)_2$	$6H_2O$	81.8	95.28	125.1									21
22	nitrate	$Cu(NO_3)_2$	$3H_2O$					159.8		178.8		207.8			22
23	sulfate	$CuSO_4$	$5H_2O$	14.3	17.4	20.7	25	28.5	33.3	40		55		75.4	23
24	sulfide	CuS				3.3 × 10⁻⁵ at 18°									24
25	Cuprous chloride	CuCl				1.52^25°									25
26	Ferric chloride	$FeCl_3$		74.4	81.9	91.8			315.1			525.8		535.7	26
27	Ferrous chloride	$FeCl_2$	$4H_2O$		64.5		73.0	77.3	82.5	88.7		100			27
28	chloride	$FeCl_2$											105.3	105.8	28
29	nitrate	$Fe(NO_3)_2$	$6H_2O$	71.02		83.8				165.6					29
30	sulfate	$FeSO_4$	$7H_2O$	15.65	20.51	26.5	32.9	40.2	48.6						30
31	sulfate	$FeSO_4$	$1H_2O$								50.9	43.6	37.3		31
32	Hydrobromic acid, 760 mm	HBr		221.2	210.3	198			171.5					130	32
33	Hydrochloric acid, 760 mm	HCl		82.3			67.3	63.3	59.6	56.1					33
34	Iodine	I_2				0.029	0.04	0.056	0.078						34
35	Lead acetate	$Pb(C_2H_3O_2)_2$	$3H_2O$				55.04^25°								35
36	bromide	$PbBr_2$		0.4554		0.85	1.15	1.53	1.94	2.36		3.34		4.75	36
37	carbonate	$PbCO_3$				0.00011									37
38	chloride	$PbCl_2$		0.6728		0.99	1.20	1.45	1.70	1.98		2.62		3.34	38
39	chromate	$PbCrO_4$				7 × 10⁻⁶									39
40	fluoride	PbF_2			0.060	0.064	0.068								40
41	nitrate	$Pb(NO_3)_2$		38.8	48.3	56.5	66	75	85	95		115		38.8	41
42	sulfate	$PbSO_4$		0.0028	0.0035	0.0041	0.0049	0.0056							42
43	Magnesium bromide	$MgBr_2$	$6H_2O$	91.0	94.5	96.5	99.2	101.6	104.1	107.5		113.7		120.2	43
44	chloride	$MgCl_2$	$6H_2O$	52.8	53.5	54.5		57.5		61.0		66.0		73.0	44
45	hydroxide	$Mg(OH)_2$				0.0009^18°									45
46	nitrate	$Mg(NO_3)_2$	$6H_2O$	66.55				84.74					137.0		46
47	sulfate	$MgSO_4$	$7H_2O$		30.9	35.5	40.8	45.6							47
48	sulfate	$MgSO_4$	$6H_2O$	40.8	42.2	44.5	45.3		50.4	53.5	59.5	64.2	69.0	74.0	48
49	sulfate	$MgSO_4$	$1H_2O$									62.9		68.3	49
50	Manganous sulfate	$MnSO_4$	$7H_2O$	53.23	60.01										50
51	sulfate	$MnSO_4$	$5H_2O$		59.5	62.9	67.76								51
52	sulfate	$MnSO_4$	$4H_2O$			64.5	66.44	68.8	72.6						52
53	sulfate	$MnSO_4$	$1H_2O$						58.17	55.0	52.0	48.0	42.5	34.0	53
54	Mercurous chloride	HgCl		0.00014		0.0002		0.0007							54
55	Molybdic oxide	MoO_3	$2H_2O$			0.138	0.264	0.476	0.687	1.206	2.055	2.106			55
56	Nickel chloride	$NiCl_2$	$6H_2O$	53.9	59.5	64.2	68.9	73.3	78.3	82.2	85.2			87.6	56
57	nitrate	$Ni(NO_3)_2$	$6H_2O$	79.58		96.31		122.2							57
58	nitrate	$Ni(NO_3)_2$	$3H_2O$							163.1	169.1		235.1		58
59	sulfate	$NiSO_4$	$7H_2O$	27.22	32		42.46								59
60	sulfate	$NiSO_4$	$6H_2O$						50.15	54.80	59.44	63.17		76.7	60
61	Nitric oxide, 760 mm	NO		0.00984	0.00757	0.00618	0.00517	0.00440	0.00376	0.00324	0.00267	0.00199	0.00114	0	61
62	Nitrous oxide	N_2O			0.1705	0.1211									62

1	Potassium acetate	$KC_2H_3O_2$	$1\frac{1}{2}H_2O$	216.7	233.9	255.6	283.8	323.3	...	...	...	...	...	...	1
2	acetate	$KC_2H_3O_2$	$\frac{1}{2}H_2O$	...	...	...	...	...	337.3	350	364.8	380.1	396.3	...	2
3	alum	$K_2SO_4.Al_2(SO_4)_3$	$24H_2O$	3.0	4.0	5.9	8.39	11.70	17.00	24.75	40.0	71.0	109.0	...	3
4	bicarbonate	$KHCO_3$	...	22.4	27.7	33.2	39.1	45.4	...	60.0	...	...	...	...	4
5	bisulfate	$KHSO_4$	...	36.3	...	51.4	...	67.3	...	...	...	...	...	121.6	5
6	bitartrate	$KHC_4H_4O_6$	...	0.32	0.40	0.53	0.90	1.32	1.83	2.46	...	4.6	...	6.95	6
7	carbonate	K_2CO_3	$2H_2O$	105.5	108	110.5	113.7	116.9	121.2	126.8	133.1	139.8	147.5	155.7	7
8	chlorate	$KClO_3$	...	3.3	5	7.4	10.5	14	19.3	24.5	...	38.5	...	57	8
9	chloride	KCl	...	27.6	31.0	34.0	37.0	40.0	42.6	45.5	48.3	51.1	54.0	56.7	9
10	chromate	K_2CrO_4	...	58.2	60.0	61.7	63.4	65.2	66.8	68.6	70.4	72.1	73.9	75.6	10
11	dichromate	$K_2Cr_2O_7$	...	5	7	12	20	26	34	43	52	61	70	80	11
12	ferricyanide	$K_3Fe(CN)_6$	...	31	36	43	50	60	...	66	...	...	...	82.6^{104}	12
13	hydroxide	KOH	$2H_2O$	97	103	112	126	...	...	...	...	...	...	...	13
14	hydroxide	KOH	$1H_2O$	...	...	...	...	...	140	...	...	...	...	178	14
15	nitrate	KNO_3	...	13.3	20.9	31.6	45.8	63.9	85.5	110.0	138	169	202	246	15
16	nitrite	KNO_2	...	278.8	...	298.4	...	334.9	...	...	...	...	...	412.8	16
17	perchlorate	$KClO_4$	...	0.75	1.05	1.80	2.6	4.4	6.5	9	11.8	14.8	18	21.8	17
18	permanganate	$KMnO_4$	...	2.83	4.4	6.4	9.0	12.56	16.89	22.2	...	...	...	...	18
19	persulfate†	$K_2S_2O_8$†	†	1.62	2.60	4.49	7.19	9.89	...	...	...	...	...	...	19
20	sulfate	K_2SO_4	...	7.35	9.22	11.11	12.97	14.76	16.50	18.17	19.75	21.4	22.8	24.1	20
21	thiocyanate	$KCNS$	...	177.0	...	217.5	...	...	...	...	...	...	...	...	21
22	Silver cyanide	$AgCN$	...	...	...	2.2×10^{-5}	...	...	...	...	...	...	...	...	22
23	nitrate	$AgNO_3$	...	122	170	222	300	376	455	525	...	669	...	952	23
24	sulfate	Ag_2SO_4	...	0.573	0.695	0.796	0.888	0.979	1.08	1.15	1.22	1.30	1.36	1.41	24
25	Sodium acetate	$NaC_2H_3O_2$	$3H_2O$	36.3	40.8	46.5	54.5	65.5	83	139	...	...	...	...	25
26	acetate	$NaC_2H_3O_2$	...	119	121	123.5	126	129.5	134	139.5	146	153	161	170	26
27	bicarbonate	$NaHCO_3$	...	6.9	8.15	9.6	11.1	12.7	14.45	16.4	...	...	...	...	27
28	carbonate	Na_2CO_3	$10H_2O$	7	12.5	21.5	38.8	...	...	...	...	...	...	...	28
29	carbonate	Na_2CO_3	$1H_2O$	...	...	...	50.5	48.5	...	46.4	...	45.8	...	45.5	29
30	chlorate	$NaClO_3$	...	79	89	101	113	126	140	155	172	189	...	230	30
31	chloride	$NaCl$	...	35.7	35.8	36.0	36.3	36.6	37.0	37.3	37.8	38.4	39.0	39.8	31
32	chromate	Na_2CrO_4	$10H_2O$	31.70	50.17	88.7	...	...	...	...	...	...	...	...	32
33	chromate	Na_2CrO_4	$4H_2O$	...	...	...	88.7	95.96	104	114.6	...	...	...	...	33
34	chromate	Na_2CrO_4	...	...	...	...	...	...	...	...	123.0	124.8	...	125.9	34
35	dichromate	$Na_2Cr_2O_7$	$2H_2O$	163.0	...	177.8	...	...	244.8	...	316.7	376.2	...	...	35
36	dichromate	$Na_2Cr_2O_7$	...	...	...	...	...	...	...	...	...	...	...	426.3	36
37	dihydrogen phosphate	NaH_2PO_4	$2H_2O$	57.9	69.9	85.2	106.5	138.2	...	...	...	...	...	...	37
38	dihydrogen phosphate	NaH_2PO_4	$1H_2O$	...	...	...	...	...	158.6	...	...	...	...	...	38
39	dihydrogen phosphate	NaH_2PO_4	...	...	...	...	...	...	...	179.3	190.3	207.3	225.3	246.6	39
40	hydrogen arsenate	Na_2HAsO_4	$12H_2O$	7.3	15.5	26.5	37	47	...	65	...	85	...	...	40
41	hydrogen phosphate	Na_2HPO_4	$12H_2O$	1.67	3.6	7.7	20.8	...	...	...	...	...	...	...	41
42	hydrogen phosphate	Na_2HPO_4	$7H_2O$	...	...	...	...	51.8	...	...	...	...	...	...	42
43	hydrogen phosphate	Na_2HPO_4	$2H_2O$	...	...	...	...	...	80.2	82.9	88.1	92.4	102.9	...	43
44	hydrogen phosphate	Na_2HPO_4	...	...	...	...	...	...	...	...	...	...	...	102.2	44
45	hydroxide	$NaOH$	$4H_2O$	42	...	...	...	...	...	...	...	...	...	...	45
46	hydroxide	$NaOH$	$3\frac{1}{2}H_2O$	...	51.5	...	...	...	...	...	...	...	...	...	46
47	hydroxide	$NaOH$	$1H_2O$	...	...	109	119	129	145	174	...	...	...	...	47
48	hydroxide	$NaOH$	...	...	...	...	...	...	...	...	...	...	313	347	48
49	nitrate	$NaNO_3$	...	73	80	88	96	104	114	124	...	148	...	180	49
50	nitrite	$NaNO_2$	...	72.1	78.0	84.5	91.6	98.4	104.1	...	...	132.6	...	163.2	50
51	oxalate	$Na_2C_2O_4$	...	...	...	3.7	...	...	...	...	...	...	...	6.33	51
52	phosphate, tri-	Na_3PO_4	$12H_2O$	1.5	4.1	11	20	31	43	55	...	81	...	108	52
53	pyrophosphate	$Na_4P_2O_7$	$10H_2O$	3.16	3.95	6.23	9.95	13.50	17.45	21.83	...	30.04	...	40.26	53
54	sulfate	Na_2SO_4	$10H_2O$	5.0	9.0	19.4	40.8	...	...	...	...	...	...	...	54
55	sulfate	Na_2SO_4	$7H_2O$	19.5	30	44	...	...	...	...	...	...	...	...	55
56	sulfate	Na_2SO_4	...	...	...	...	...	48.8	46.7	45.3	...	43.7	...	42.5	56
57	sulfide	Na_2S	$9H_2O$	...	15.42	18.8	22.5	28.5	...	...	...	...	...	...	57
58	sulfide	Na_2S	$5\frac{1}{2}H_2O$	...	...	...	...	...	39.82	42.69	45.73	51.40	59.23	...	58
59	sulfide	Na_2S	$6H_2O$	...	...	...	...	...	36.4	39.1	43.31	49.14	57.28	...	59
60	sulfite	Na_2SO_3	$7H_2O$	13.9	20	26.9	36	...	...	...	...	...	...	...	60
61	sulfite	Na_2SO_3	...	...	...	...	...	28	28.2	28.8	...	28.3	...	...	61
62	tetraborate	$Na_2B_4O_7$	$10H_2O$	1.3	1.6	2.7	3.9	...	10.5	20.3	...	...	...	...	62
63	tetraborate	$Na_2B_4O_7$	$5H_2O$	...	...	...	...	...	...	...	24.4	31.5	41	52.5	63
64	vanadate (meta)	$NaVO_3$	$2H_2O$	...	...	$15.3^{25°}$	...	30.2	...	68.4	...	...	...	...	64

TABLE 3-120 Solubilities of Inorganic Compounds in Water at Various Temperatures *(Concluded)*

1	Sodium vanadate (meta)	$NaVO_3$	......	....	....	21.10[25°]	....	26.23	....	32.97	36.9	38.8[75°]	....	....	1
2	Stannous chloride	$SnCl_2$	......	83.9	....	269.8[15°]	....	....	....	....	....	....	....	....	2
3	sulfate	$SnSO_4$	......	....	....	19	....	....	....	....	....	....	....	18	3
4	Strontium acetate	$Sr(C_2H_3O_2)_2$	$4H_2O$	36.9	43.61	....	....	....	....	....	....	....	....	....	4
5	acetate	$Sr(C_2H_3O_2)_2$	$\frac{1}{2}H_2O$	....	42.95	41.6	39.5	....	37.35	....	36.24	36.10	....	36.4	5
6	chloride	$SrCl_2$	$6H_2O$	43.5	47.7	52.9	58.7	65.3	72.4	81.8	....	....	....	....	6
7	chloride	$SrCl_2$	$2H_2O$	....	....	....	....	....	....	....	85.9	90.5	....	100.8	7
8	nitrate	$Sr(NO_3)_2$	$1H_2O$	52.7	....	64.0	....	....	83.8	97.2	....	....	130.4	139	8
9	nitrate	$Sr(NO_3)_2$	$4H_2O$	40.1	....	70.5	....	....	....	....	....	....	....	....	9
10	nitrate	$Sr(NO_3)_2$	......	....	....	....	88.6	90.1	....	93.8	96	98	100	....	10
11	sulfate	$SrSO_4$	......	0.0113	....	0.0114	0.0114	....	....	....	....	....	....	....	11
12	Sulfur dioxide, 760 mm ‡	SO_2	......	22.83	16.21	11.29	7.81	5.41	4.5	....	....	....	....	....	12
13	Thallium sulfate	Tl_2SO_4	......	2.70	3.70	4.87	6.16	....	9.21	10.92	12.74	14.61	16.53	18.45	13
14	Thorium sulfate	$Th(SO_4)_2$	$9H_2O$	0.74	0.98	1.38	1.995	2.998	5.22	....	....	....	....	....	14
15	sulfate	$Th(SO_4)_2$	$8H_2O$	1.0	1.25	1.62	....	....	....	....	....	....	....	....	15
16	sulfate	$Th(SO_4)_2$	$6H_2O$	1.50	....	1.90	2.45	....	....	6.64	....	....	....	....	16
17	sulfate	$Th(SO_4)_2$	$4H_2O$	....	....	....	....	4.04	2.54	1.63	1.09	....	....	....	17
18	Zinc chlorate	$ZnClO_3$	$6H_2O$	145.0	152.5	....	....	....	....	....	....	....	....	....	18
19	chlorate	$ZnClO_3$	$4H_2O$	....	....	200.3	209.2	223.2	273.1	....	....	....	....	....	19
20	nitrate	$Zn(NO_3)_2$	$6H_2O$	94.78	....	118.3	....	....	....	....	....	....	....	....	20
21	nitrate	$Zn(NO_3)_2$	$3H_2O$	....	....	....	....	206.9	....	....	....	....	....	....	21
22	sulfate	$ZnSO_4$	$7H_2O$	41.9	47	54.4	....	....	....	....	....	....	....	....	22
23	sulfate	$ZnSO_4$	$6H_2O$	....	....	....	....	70.1	76.8	....	....	....	....	....	23
24	sulfate	$ZnSO_4$	$1H_2O$	....	....	....	....	....	....	....	....	86.6	83.7	80.8	24

The H in solubility tables (3-121 to 3-144) is the proportionality constant for the expression of Henry's law, $p = Hx$, where x = mole fraction of the solute in the liquid phase; p = partial pressure of the solute in the gas phase, expressed in atmospheres; and H = a proportionality constant and is in units of atmospheres of solute pressure in the gas phase per unit concentration of the solute in the liquid phase. (The unit of concentration of the solute in the liquid phase is moles solute per mole solution.)

TABLE 3-121 Acetylene (C_2H_2)

t, °C.	0	5	10	15	20	25	30
$10^{-3} \times H$*	0.72	0.84	0.96	1.08	1.21	1.33	1.46

"International Critical Tables," vol. 3, p. 260, McGraw-Hill, 1928.

TABLE 3-122 Air

t, °C.	0	5	10	15	20	25	30	35
$10^{-4} \times H$*	4.32	4.88	5.49	6.07	6.64	7.20	7.71	8.23
t, °C.	40	45	50	60	70	80	90	100
$10^{-4} \times H$*	8.70	9.11	9.46	10.1	10.5	10.7	10.8	10.7

"International Critical Tables," vol. 3, p. 257.

*H is calculated from the absorption coefficients of O_2 and N_2, taking into consideration the correction for constant argon content.

TABLE 3-123 Ammonia (NH_3)

Weight NH_3 per 100 weights H_2O	Partial pressure of NH_3, mm. Hg							
	0°C.	10°C.	20°C.	25°C.	30°C.	40°C.	50°C.	60°C.
100	947							
90	785							
80	636	987	1450	....	.....	3300		
70	500	780	1170	....	.....	2760		
60	380	600	945	....	.....	2130		
50	275	439	686	....	.....	1520		
40	190	301	470	....	719	1065		
30	119	190	298	....	454	692		
25	89.5	144	227	....	352	534	825	
20	64	103.5	166	....	260	395	596	834
15	42.7	70.1	114	....	179	273	405	583
10	25.1	41.8	69.6	....	110	167	247	361
7.5	17.7	29.9	50.0	....	79.7	120	179	261
5	11.2	19.1	31.7	....	51.0	76.5	115	165
4	.....	16.1	24.9	....	40.1	60.8	91.1	129.2
3	.....	11.3	18.2	23.5	29.6	45	67.1	94.3
2.5	.....	.....	15.0	19.4	24.4	(37.6)*	(55.7)	77.0
2	.....	.....	12.0	15.3	19.3	(30.0)	(44.5)	61.0
1.6	.....	.....	......	12.0	15.3	(24.1)	(35.5)	48.7
1.2	.....	.....	......	9.1	11.5	(18.3)	(26.7)	36.3
1.0	.....	.....	......	7.4	.....	(15.4)	(22.2)	30.2
0.5	.....	.....	......	3.4				

* Extrapolated values.

TABLE 3-124 Ammonia (NH_3)

Weight NH_3 per 100 weights H_2O	0.105	0.244	0.32	0.38	0.576	0.751	1.02
Partial pressure NH_3, mm. Hg, at 25°C.	0.791	1.83	2.41	2.89	4.41	5.80	7.96
Weight NH_3 per 100 weights H_2O	1.31	1.53	1.71	1.98	2.11	2.58	2.75
Partial pressure NH_3, mm. Hg, at 25°C.	10.31	11.91	13.46	15.75	16.94	20.86	22.38

"Landolt-Börnstein Physikalische-chemische Tabellen," Eg. I, p. 303, 1927. Phase-equilibrium data for the binary system NH_3-H_2O are given by Clifford and Hunter, *J. Phys. Chem.*, **37**, 101 (1933).

TABLE 3-125 Carbon Dioxide (CO_2)

Total pressure, atm.	Weight of CO_2 per 100 weights of H_2O*								
	12°C.	18°C.	25°C.	31.04°C.	35°C.	40°C.	50°C.	75°C.	100°C.
25	....	3.86	....	2.80	2.56	2.30	1.92	1.35	1.06
50	7.03	6.33	5.38	4.77	4.39	4.02	3.41	2.49	2.01
75	7.18	6.69	6.17	5.80	5.51	5.10	4.45	3.37	2.82
100	7.27	6.72	6.28	5.97	5.76	5.50	5.07	4.07	3.49
150	7.59	7.07	....	6.25	6.03	5.81	5.47	4.86	4.49
200	....	....	....	6.48	6.29	6.28	5.76	5.27	5.08
300	7.86	7.35	....	....	....	....	6.20	5.83	5.84
400	8.12	7.77	7.54	7.27	7.06	6.89	6.58	6.30	6.40
500	....	....	....	7.65	7.51	7.26	....	....	....
700	....	....	....	....	....	....	7.58	7.43	7.61

* In the original, concentration is expressed in cubic centimeters of CO_2 reduced to 0°C. and 1 atm.) dissolved in 1 g. of water.

TABLE 3-126 Carbon Monoxide (CO)

Partial pressure of CO, mm. Hg	$10^{-4} \times H$	
	17.7°C.	19.0°C.
900	4.77	4.88
2000	4.77	4.91
3000	4.77	4.93
4000	4.78	4.95
5000	4.80	4.97
6000	4.82	4.98
7000	4.86	5.02
8000	4.88	5.08

"International Critical Tables," vol. 3, p. 260.

TABLE 3-127 Carbonyl Sulfide (COS)

t, °C.	0	5	10	15	20	25	30
$10^{-3} \times H$	0.92	1.17	1.48	1.82	2.19	2.59	3.04

"International Critical Tables," vol. 3, p. 261.

TABLE 3-128 Chlorine (Cl_2)

Partial pressure of Cl_2, mm. Hg	Solubility, g. of Cl_2 per liter					
	0°C.	10°C.	20°C.	30°C.	40°C.	50°C.
5	0.488	0.451	0.438	0.424	0.412	0.398
10	.679	.603	.575	.553	.532	.512
30	1.221	1.024	.937	.873	.821	.781
50	1.717	1.354	1.210	1.106	1.025	.962
100	2.79	2.08	1.773	1.573	1.424	1.313
150	3.81	2.73	2.27	1.966	1.754	1.599
200	4.78	3.35	2.74	2.34	2.05	1.856
250	5.71	3.95	3.19	2.69	2.34	2.09
300	.....	4.54	3.63	3.03	2.61	2.31
350	.....	5.13	4.06	3.35	2.86	2.53
400	.....	5.71	4.48	3.69	3.11	2.74
450	.....	6.26	4.88	3.98	3.36	2.94
500	.....	6.85	5.29	4.30	3.61	3.14
550	.....	7.39	5.71	4.60	3.84	3.33
600	.....	7.97	6.12	4.91	4.08	3.52
650	.....	8.52	6.52	5.21	4.32	3.71
700	.....	9.09	6.90	5.50	4.54	3.89
750	.....	9.65	7.29	5.80	4.77	4.07
800	.....	10.21	7.69	6.[illegible]3	4.99	4.27
900	.....	......	8.46	6.68	5.44	4.62
1000	.....	......	9.27	7.27	5.89	4.97
1200	$Cl_2.8H_2O$ separates		10.84	8.42	6.81	5.67
1500	.....	......	13.23	10.14	8.05	6.70
2000	.....	......	17.07	13.02	10.22	8.38
2500	.....	......	21.0	15.84	12.32	10.03
3000	.....	......	......	18.73	14.47	11.70
3500	.....	......	......	21.7	16.62	13.38
4000	.....	......	......	24.7	18.84	15.04
4500	.....	......	......	27.7	20.7	16.75
5000	.....	......	......	30.8	23.3	18.46

Partial pressure of Cl_2, mm. Hg	Solubility, g. of Cl_2 per liter					
	60°C.	70°C.	80°C.	90°C.	100°C.	110°C.
5	0.383	0.369	0.351	0.339	0.326	0.316
10	.492	.470	.447	.431	.415	.402
30	.743	.704	.671	.642	.627	.598
50	.912	.863	.815	.781	.747	.722
100	1.228	1.149	1.085	1.034	.987	.950
150	1.482	1.382	1.294	1.227	1.174	1.137
200	1.706	1.580	1.479	1.396	1.333	1.276
250	1.914	1.764	1.642	1.553	1.480	1.413
300	2.10	1.932	1.793	1.700	1.610	1.542
350	2.28	2.10	1.940	1.831	1.736	1.661
400	2.47	2.25	2.08	1.965	1.854	1.773
450	2.64	2.41	2.22	2.09	1.972	1.880
500	2.80	2.55	2.35	2.21	2.08	1.986
550	2.97	2.69	2.47	2.32	2.19	2.09
600	3.13	2.83	2.59	2.43	2.29	2.19
650	3.29	2.97	2.72	2.55	2.41	2.28
700	3.44	3.10	2.84	2.66	2.50	2.37
750	3.59	3.23	2.96	2.76	2.60	2.47
800	3.75	3.37	3.08	2.87	2.69	2.56
900	4.04	3.63	3.30	3.08	2.89	2.74
1000	4.36	3.88	3.53	3.28	3.07	2.91
1200	4.92	4.37	3.95	3.67	3.43	3.25
1500	5.76	5.09	4.58	4.23	3.95	3.74
2000	7.14	6.26	5.63	5.17	4.78	4.49
2500	8.48	7.40	6.61	6.05	5.59	5.25
3000	9.83	8.52	7.54	6.92	6.38	5.97
3500	11.22	9.65	8.53	7.79	7.16	6.72
4000	12.54	10.76	9.52	8.65	7.94	7.42
4500	13.88	11.91	10.46	9.49	8.72	8.13
5000	15.26	13.01	11.42	10.35	9.48	8.84

TABLE 3-129 Chlorine Dioxide (ClO_2)

Vol. % of ClO_2 in gas phase	Weight of ClO_2, grams per liter of solution						
	0°C.	5°C.	10°C.	15°C.	20°C.	30°C.	40°C.
1	2.00	1.50	1.25	1.00	0.90	0.60	0.46
3	6.00	4.7	3.85	3.20	2.70	1.95	1.30
5	10.0	7.8	6.30	5.25	4.30	3.20	2.25
7	14.0	10.9	8.95	7.35	6.15	4.40	3.20
10	20.0	15.5	12.8	10.5	8.80	6.30	4.50
11	.....	17.0	14.0	11.7	9.70	7.00	5.00
12	.....	18.6	15.3	12.8	10.55	7.50	5.45
13	.....	20.3	16.6	13.8	11.5	8.20	5.85
14	.....	.....	18.0	14.9	12.3	8.80	6.35
15	.....	.....	19.2	16.0	13.2	9.50	6.80
16	.....	.....	20.3	17.0	14.2	10.1	7.20

Ishi, *Chem. Eng. (Japan)*, **22**, 153 (1958).

TABLE 3-130 Ethane (C_2H_6)

t, °C.	0	5	10	15	20	25	30	35
$10^{-4} \times H$	1.26	1.55	1.89	2.26	2.63	3.02	3.42	3.83

t, °C.	40	45	50	60	70	80	90	100
$10^{-4} \times H$	4.23	4.63	5.00	5.65	6.23	6.61	6.87	6.92

"International Critical Tables," vol. 3, p. 261.

TABLE 3-131 Ethylene (C_2H_4)

t, °C.	0	5	10	15	20	25	30
$10^{-3} \times H$	5.52	6.53	7.68	8.95	10.2	11.4	12.7

"International Critical Tables," vol. 3, p. 260.

TABLE 3-132 Helium (He)

t, °C.	0	10	20	30	40	50
$10^{-4} \times H$	12.9	12.6	12.5	12.4	12.1	11.5

See also Pray, Schweickert, and Minnich, *Ind. Eng. Chem.*, **44**, 1146 (1952).

TABLE 3-133 Hydrogen (H_2)

t, °C.	0	5	10	15	20	25	30	35
$10^{-4} \times H$	5.79	6.08	6.36	6.61	6.83	7.07	7.29	7.42

t, °C.	40	45	50	60	70	80	90	100
$10^{-4} \times H$	7.51	7.60	7.65	7.65	7.61	7.55	7.51	7.45

"International Critical Tables," vol. 3, p. 256.
See also Pray, Schweickert, and Minnich, *Ind. Eng. Chem.*, **44**, 1146 (1952)

TABLE 3-134 Hydrogen (H_2)

Partial pressure H_2, mm. Hg	$10^{-4} \times H$	
	19.5°C.	23°C.
900	7.42	....
1100	....	7.75
2000	7.42	7.76
3000	7.43	7.77
4000	7.47	7.81
5000	7.56	7.89
6000	7.70	8.00
7000	7.87	8.16
8200	....	8.41
8250	8.17	....

"International Critical Tables," vol. 3, p. 256.

TABLE 3-135 Hydrogen Chloride (HCl)

Weights of HCl per 100 weights of H_2O	Partial pressure of HCl, mm. Hg			
	0°C.	10°C.	20°C.	30°C.
78.6	510	840		
66.7	130	233	399	627
56.3	29.0	56.4	105.5	188
47.0	5.7	11.8	23.5	44.5
38.9	1.0	2.27	4.90	9.90
31.6	0.175	0.43	1.00	2.17
25.0	.0316	.084	0.205	0.48
19.05	.0056	.016	.0428	.106
13.64	.00099	.00305	.0088	.0234
8.70	.000118	.000583	.00178	.00515
4.17	.000018	.000069	.00024	.00077
2.04	....	.0000117	.000044	.000151

Weights of HCl per 100 weights of H_2O	Partial pressure of HCl, mm. Hg		
	50°C.	80°C.	110°C.
78.6			
66.7			
56.3	535		
47.0	141	623	
38.9	35.7	188	760
31.6	8.9	54.5	253
25.0	2.21	15.6	83
19.05	0.55	4.66	28
13.64	.136	1.34	9.3
8.70	.0344	0.39	3.10
4.17	.0064	.095	0.93
2.04	.00140	.0245	.280

Enthalpy and phase-equilibrium data for the binary system $HCl-H_2O$ are given by Van Nuys, *Trans. Am. Inst. Chem. Engrs.*, **39**, 663 (1943).

TABLE 3-136 Hydrogen Sulfide (H_2S)

t, °C.	0	5	10	15	20	25	30	35
$10^{-4} \times H$	2.68	3.15	3.67	4.23	4.83	5.45	6.09	6.76
t, °C.	40	45	50	60	70	80	90	100
$10^{-4} \times H$	7.45	8.14	8.84	10.3	11.9	13.5	14.4	14.8

"International Critical Tables," vol. 3, p. 259.

TABLE 3-137 Methane (CH_4)

t, °C.	0	5	10	15	20	25	30	35
$10^{-4} \times H$	2.24	2.59	2.97	3.37	3.76	4.13	4.49	4.86
t, °C.	40	45	50	60	70	80	90	100
$10^{-4} \times H$	5.20	5.51	5.77	6.26	6.66	6.82	6.92	7.01

"International Critical Tables," vol. 3, p. 260.

TABLE 3-138 Nitrogen (N_2)*

t, °C.	0	5	10	15	20	25	30	35
$10^{-4} \times H$	5.29	5.97	6.68	7.38	8.04	8.65	9.24	9.85
t, °C.	40	45	50	60	70	80	90	100
$10^{-4} \times H$	10.4	10.9	11.3	12.0	12.5	12.6	12.6	12.6

"International Critical Tables," vol. 3, p. 256. See also Pray, Schweickert, and Minnich, *Ind. Eng. Chem.*, **44**, 1146 (1952).
* Atmospheric nitrogen = 98.815 vol. % N_2 + 1.185 vol. % A.

TABLE 3-139 Nitrogen (N_2)

Partial pressure of N_2, mm. Hg	$10^{-4} \times H$	
	19.4°C.	24.9°C.
900	8.24	9.08
2000	8.32	9.15
3000	8.41	9.25
4000	8.49	9.38
5000	8.59	9.49
6000	8.74	9.62
7000	8.86	9.75
8100	9.04	
8200	....	9.91

See also Goodman and Krase [*Ind. Eng. Chem.*, **23**, 401(1931)] for values up to 169°C. and 300 atm.

TABLE 3-140 Oxygen (O_2)

t, °C.	0	5	10	15	20	25	30	35
$10^{-4} \times H$	2.55	2.91	3.27	3.64	4.01	4.38	4.75	5.07
t, °C.	40	45	50	60	70	80	90	100
$10^{-4} \times H$	5.35	5.63	5.88	6.29	6.63	6.87	6.99	7.01

"International Critical Tables," vol. 3, p. 257. Pray, Schweickert, and Minnich [*Ind. Eng. Chem.*, **44**, 1146 (1952)] give $H = 4.46 \times 10^{-4}$ at 25°C. and other values up to 343°C.

TABLE 3-141 Oxygen (O_2)

Partial pressure of O_2, mm. Hg	$10^{-4} \times H$	
	23°C.	25.9°C.
800	....	4.79
900	4.58	
2000	4.59	4.80
3000	4.60	4.83
4000	4.68	4.88
5000	4.73	4.92
6000	4.80	4.98
7000	4.88	5.05
8150	4.98	
8200	....	5.16

"International Critical Tables," vol. 3, p. 257. See also *Trans. Am. Soc. Mech. Engrs.*, **76**, 69 (1954) for solubility of O_2 for 100°F. $< T <$ 650°F., 300 $< P <$ 2000 lb./sq. in.

TABLE 3-142 Ozone (O_3)

t, °C.	0	5	10	15	20	25	30	35	40	50
$10^{-3} \times H$	1.94	2.18	2.48	2.88	3.76	4.57	5.98	8.18	12.0	27.4

"International Critical Tables," vol. 3, p. 257.

TABLE 3-143 Propylene (C_3H_6)

t, °C.	2	6	10	14	18
$10^{-3} \times H$	3.04	3.84	4.46	5.06	5.69

"International Critical Tables," vol. 3, p. 260.

TABLE 3-144 Sulfur Dioxide (SO_2)

Weight of SO_2 per 100 weights of H_2O	Partial pressure of SO_2, mm. Hg							
	0°C.	7°C.	10°C.	15°C.	20°C.	30°C.	40°C.	50°C.
20	646	657						
15	474	637	726					
10	308	417	474	567	698			
7.5	228	307	349	419	517	688		
5.0	148	198	226	270	336	452	665	
2.5	69	92	105	127	161	216	322	458
1.5	38	51	59	71	92	125	186	266
1.0	23.3	31	37	44	59	79	121	172
0.7	15.2	20.6	23.6	28.0	39.0	52	87	116
.5	9.9	13.5	15.6	19.3	26.0	36	57	82.0
.3	5.1	6.9	7.9	10.0	14.1	19.7		
.2	2.8	3.7	4.6	5.7	8.5	11.8	.....	31.0
.15	1.9	2.6	3.1	3.8	5.8	8.1	12.9	20.0
.10	1.2	1.5	1.75	2.2	3.2	4.7	7.5	12.0
.05	0.6	0.7	0.75	0.8	1.2	1.7	2.8	4.7
.02	.25	.3	.3	.3	0.5	0.6	0.8	1.3

THERMAL EXPANSION

UNITS CONVERSIONS

For this subsection, the following units conversion is applicable:
°F = 9/5 °C + 32.

ADDITIONAL REFERENCES

The tables given under this subject are reprinted by permission from the *Smithsonian Tables*. For more detailed data on thermal expansion see *International Critical Tables:* tabular index, vol. 3, p. 1; abrasives, vol. 2, p. 87; alloys, vol. 2, p. 463; building stones, vol. 2, p. 54; carbons, vol. 2, p. 303; elements, vol. 1, p. 102; enamels, vol. 2, p. 115; glass, vol. 2, p. 93; metals, vol. 2, p. 459; petroleums, vol. 2, p. 145; porcelains, vol. 2, pp. 70, 78; refractory materials, vol. 2, p. 83; solid insulators, vol. 2, p. 310.

THERMAL EXPANSION OF GASES

No tables of the coefficients of thermal expansion of gases are given in this edition. The coefficient at constant pressure, $1/v(\partial v/\partial T)_p$, for an ideal gas is merely the reciprocal of the absolute temperature. For a real gas or liquid, both it and the coefficient at constant volume, $1/p(\partial p/\partial T)_v$, should be calculated either from the equation of state or from tabulated PVT data.

TABLE 3-145 Linear Expansion of the Solid Elements*

C is the true expansion coefficient at the given temperature; *M* is the mean coefficient between given temperatures; where one temperature is given, the true coefficient at that temperature is indicated; α and β are coefficients in formula $l_t = l_0(1 + \alpha t + \beta t^2)$; l_0 is length at 0°C. (unless otherwise indicated, when, if x is the reference temperature, $l_t = l_x[1 + \alpha(t - t_x) + \beta(t - t_x)^2]$; l_t is length at t°C.).

Element	Temp., °C.	$C \times 10^4$	Temp. range, °C.	$M \times 10^4$	Temp. range, °C.	$\alpha \times 10^4$	$\beta \times 10^6$
Aluminum	20	0.224	100	0.235	0, 500	0.22	0.009
Aluminum	300	0.284	500	0.311			
Antimony	20	0.136‖	20	0.080⊥			
Arsenic	20	0.05					
Bismuth	20	0.014‖	20	0.103⊥			
Cadmium	0	0.54‖	−180, −140	0.59‖	20, 100	0.526‖	
Cadmium	0	0.20⊥	−180, −140	0.117⊥	20, 100	0.214⊥	
Carbon, diamond	50	0.012					
graphite	50	0.06					
Chromium	....	....	20, 100	0.068	20, 500	0.086	
Cobalt	20	0.123	....	....	6, 121	0.121	0.0064
Copper	20	0.162	100	0.166	0, 625	0.161	0.0040
Copper	200	0.170	300	0.175			
Gold	20	0.140	17, 100	0.143	0, 520	0.142	0.0022
Gold	....	....	−191, 17	0.132			
Indium	40	0.417					
Iodine	....	....	−190, 17	0.837			
Iridium	20	0.065	....	....	0, 80	0.0636	0.0032
Iridium	....	....	....	....	1070, 1720	0.0679	0.0011
Iron, soft	40	0.1210	0, 100	0.11			
cast	20	0.118	....	....	0, 750	0.1158	0.0053
wrought	20	0.119	....	....	0, 750	0.1170	0.0053
steel	20	0.114	....	....	0, 750	0.1118	0.0053
Lead (99.9)	....	....	20, 100	0.291	100, 240	0.269	0.011
	100	0.291	20, 200	0.300			
	280	0.343					
Magnesium	20	0.254	−100, + 20	0.240	+ 20, 500	0.2480	0.0096
			20, 100	0.260			
Manganese	20	0.233	0, 100	0.228			
			−190, 0	0.159	20, 300	0.216	0.0121
Molybdenum†	20	0.053	0, 100	0.052	−142, 19	0.0515	0.0057
			25, 100	0.049	19, +305	0.0501	0.0014
			25, 500	0.055			
Nickel	20	0.126	0, 100	0.130	−190, + 20	0.1308	0.0166
					+ 20, +300	0.1236	0.0066
					500, 1000	0.1346	0.0033
Osmium	40	0.066					
Palladium	20	0.1173	....	....	−190, +100	0.1152	0.00517
					0, 1000	0.1167	0.0022
Platinum	20	0.0887	....	....	−190, −100	0.0875	0.00314
	20	0.0893	....	....	0, + 80	0.0890	0.00121
					0, 1000	0.0887	0.00132
Potassium	....	....	0, 50	0.83			
Rhodium	40	0.0850	6, 21	0.0876	− 75, −112	0.0746	
Ruthenium	40	0.0963					
Selenium	0	0.439	0, 100	0.660			

TABLE 3-145 Linear Expansion of the Solid Elements* *(Concluded)*

Element	Temp., °C.	$C \times 10^4$	Temp. range, °C.	$M \times 10^4$	Temp. range, °C.	$\alpha \times 10^4$	$\beta \times 10^6$
Silicon	40	0.0763	− 3, + 18	0.0249	− 75, − 67	0.0182	
Silver	20	0.1846	0, 100	0.197	0, 875	0.1827	0.00479
	20	0.195	....	....	20, 500	0.1939	0.00295
Sodium	....	....	−190, −17	0.622	0, 50	0.72	
Steel, 36.4Ni	....	....	20, 260	0.031	260, 500	0.144	
			20, 340	0.055	340, 500	0.136	
Tantalum†	20	0.065	− 78, 0	0.059	20, 400	0.0646	0.0009
			0, 100	0.0655			
Tellurium	20	0.016‖	20	0.272⊥			
Thallium	40	0.302					
Tin	20	0.214	....	....	8, 95	0.2033	0.0263
	20	0.305‖	20	0.154⊥			
Tungsten†	27	0.0444	0, 100	0.045	−105, +502	0.0428	0.00058
Zinc	20‡	0.643‖	−140, −100	0.656‖	+ 0, 400	0.354	0.010
	20‡	0.125⊥	+ 20, 100	0.639‖			
	20	0.358	+ 20, 100	0.141⊥			

* "Smithsonian Tables." For more complete tabulations see Table 142, "Smithsonian Physical Tables," 9th ed., 1954; "Handbook of Chemistry and Physics," 40th ed., pp. 2239–2245, Chemical Rubber Publishing Co.; Goldsmith, and Waterman, WADC-TR-58-476, 1959; Johnson (ed.), WADD-TR-60-56, 1960, etc.

† Molybdenum, 300° to 2500°C.; $l_t = l_{300}[1 + 5.00 \times 10^{-6}(t - 300) + 10.5 \times 10^{-10}(t - 300)^2]$
Tantalum, 300° to 2800°C.; $l_t = l_{300}[1 + 6.60 \times 10^{-6}(t - 300) + 5.2 \times 10^{-10}(t - 300)^2]$
Tungsten, 300° to 2700°C.; $l_t = l_{300}[1 + 4.44 \times 10^{-6}(t - 300) + 4.5 \times 10^{-10}(t - 300)^2]$
Beryllium, 20° to 100°C.; 12.3×10^{-6} per °C.
Columbium, 0° to 100°C.; 7.2×10^{-6} per °C.
Tantalum, 20° to 100°C.; 6.6×10^{-6} per °C.

‡ Two errors in the data of zinc have been corrected. These values were taken from Grüneisen and Goens, *Z. Physik.*, **29**, 141 (1924).

TABLE 3-146 Linear Expansion of Miscellaneous Substances*

The coefficient of cubical expansion may be taken as three times the linear coefficient. t is the temperature or range of temperature, C the coefficient of expansion.

Substance	t°C	$C \times 10^4$
Amber	0–30	0.50
	0–09	0.61
Bakelite, bleached	20–60	0.22
Brass:		
Cast	0–100	0.1875
Wire	0–100	0.1930
Wire	0–100	0.1783 to 0.193
71.5 Cu + 27.7 Zn + 0.3 Sn + 0.5 Pb	40	0.1859
71 Cu + 29 Zn	0–100	0.1906
Bronze:		
3 Cu + 1 Sn	16.6–100	0.1844
3 Cu + 1 Sn	16.6–350	0.2116
3 Cu + 1 Sn	16.6–957	0.1737
86.3 Cu + 9.7 Sn + 4 Zn	40	0.1782
97.6 Cu + 2.2 Sn + 0.2 P, hard	0–80	0.1713
97.6 Cu + 2.2 Sn + 0.2 P, soft	0–80	0.1708
Caoutchouc	....	0.657 to 0.686
Caoutchouc	16.7–25.3	0.770
Celluloid	20–70	1.00
Constantan	4–29	0.1523
Duralumin, 94Al	20–100	0.23
	20–300	0.25
Ebonite	25.3–35.4	0.842
Fluorspar. CaF_2	0–100	0.1950
German silver	0–100	0.1836
Gold-platinum, 2 Au + 1 Pt	0–100	0.1523
Gold-copper, 2 Au + 1 Cu	0–100	0.1552
Glass:		
Tube	0–100	0.0833
Tube	0–100	0.0828
Plate	0–100	0.0891
Crown (mean)	0–100	0.0897
Crown (mean)	50–60	0.0954
Flint	50–60	0.0788
Jena thermometer 16III normal	0–100	0.081

Substance	t°C	$C \times 10^4$
Jena thermometer 59III	0–100	0.058
Jena thermometer 59III	−191 to +16	0.424
Gutta percha	20	1.983
Ice	−20 to −1	0.51
Iceland spar:		
Parallel to axis	0–80	0.2631
Perpendicular to axis	0–80	0.0544
Lead tin (solder) 2 Pb + 1 Sn	0–100	0.2508
Limestone	25–100	0.09
Magnalium	12–39	0.238
Manganin	....	0.181
Marble	15–100	0.117
Monel metal	25–100	0.14
	25–600	0.16
Paraffin	0–16	1.0662
Paraffin	16–38	1.3030
Paraffin	38–49	4.7707
Platinum-iridium, 10 Pt + 1 Ir	40	0.0884
Platinum-silver, 1 Pt + 2 Ag	0–100	0.1523
Porcelain	20–790	0.0413
Porcelain Bayeux	1000–1400	0.0553
Quartz:		
Parallel to axis	0–80	0.0797
Parallel to axis	−190 to +16	0.0521
Perpend. to axis	0–80	0.1337
Quartz glass	−190 to +16	−0.0026
Quartz glass	16 to 500	0.0057
Quartz glass	16 to 1000	0.0058
Rock salt	40	0.4040
Rubber, hard	0	0.691
Rubber, hard	−160	0.300
Speculum metal	0–100	0.1933
Steel, 0.14 C, 34.5 Ni	25–100	0.037
	25–600	0.136

Substance	t°C	$C \times 10^4$
Topaz:		
Parallel to lesser horizontal axis	0–100	0.0832
Parallel to greater horizontal axis	0–100	0.0836
Parallel to vertical axis	0–100	0.0472
Tourmaline:		
Parallel to longitudinal axis	0–100	0.0937
Parallel to horizontal axis	0–100	0.0773
Type metal	16.6–254	0.1952
Vulcanite	0–18	0.6360
Wedgwood ware	0–100	0.0890
Wood:		
Parallel to fiber:		
Ash	0–100	0.0951
Beech	2.34	0.0257
Chestnut	2.34	0.0649
Elm	2.34	0.0565
Mahogany	2.34	0.0361
Maple	2.34	0.0638
Oak	2.34	0.0492
Pine	2.34	0.0541
Walnut	2.34	0.0658
Across the fiber:		
Beech	2.34	0.614
Chestnut	2.34	0.325
Elm	2.34	0.443
Mahogany	2.34	0.404
Maple	2.34	0.484
Oak	2.34	0.544
Pine	2.34	0.341
Walnut	2.34	0.484
Wax white	10–26	2.300
Wax white	26–31	3.120
Wax white	31–43	4.860
Wax white	43–57	15.227

* "Smithsonian Tables." For a more complete tabulation see Tables 143, 144, "Smithsonian Physical Tables," 9th ed., 1954, also reprinted in "American Institute of Physics Handbook," McGraw-Hill, New York, 1957; "Handbook of Chemistry and Physics," 40th ed., pp. 2239–2245, Chemical Rubber Publishing Co. For data on many solids prior to 1926 see Gruneisen, "Handbuch der Physik," vol. 10, pp. 1–52, 1926, translation available as *N.A.S.A.* RE 2-18-59W, 1959. For eight plastic solids below 300°K. see Scott, "Cryogenic Engineering," p. 331, Van Nostrand, Princeton, N.J., 1959. For 11 other materials to 300°K. see Scott, *loc. cit.*, p. 333. For quartz and silica see Cook, *Brit. J. Appl. Phys.*, **7**, 285 (1956).

TABLE 3-147 Cubical Expansion of Liquids*

If V_0 is the volume at 0°, then at t° the expansion formula is $V_t = V_0(1 + \alpha t + \beta t^2 + \gamma t^3)$. The table gives values of α, β, and γ and of C, the true coefficient of cubical expansion at 20° for some liquids and solutions. Δt is the temperature range of the observation. Values for the coefficient of cubical expansion of liquids can be derived from the tables of specific volumes of the saturated liquid given as a function of temperature later in this section.

Liquid	Range	$\alpha \times 10^3$	$\beta \times 10^6$	$\gamma \times 10^8$	$C \times 10^3$ at 20°
Acetic acid	16–107	1.0630	0.12636	1.0876	1.071
Acetone	0–54	1.3240	3.8090	− 0.87983	1.487
Alcohol:					
Amyl	−15–80	0.9001	0.6573	1.18458	0.902
Ethyl, 30% by volume	18–39	0.2928	10.790	−11.87	
Ethyl, 50% by volume	0–39	0.7450	1.85	0.730	
Ethyl, 99.3% by volume	27–46	1.012	2.20	....	1.12
Ethyl, 500 atm. pressure	0–40	0.866			
Ethyl, 3000 atm. pressure	0–40	0.524			
Methyl	0–61	1.1342	1.3635	0.8741	1.199
Benzene	11–81	1.17626	1.27776	0.80648	1.237
Bromine	0–59	1.06218	1.87714	− 0.30854	1.132
Calcium chloride:					
5.8% solution	18–25	0.07878	4.2742	....	0.250
40.9% solution	17–24	0.42383	0.8571	....	0.458
Carbon disulfide	−34–60	1.13980	1.37065	1.91225	1.218
500 atm. pressure	0–50	0.940			
3000 atm. pressure	0–50	0.581			
Carbon tetrachloride	0–76	1.18384	0.89881	1.35135	1.236
Chloroform	0–63	1.10715	4.66473	− 1.74328	1.273
Ether	−15–38	1.51324	2.35918	4.00512	1.656
Glycerin	....	0.4853	0.4895	....	0.505
Hydrochloric acid, 33.2% solution	0–33	0.4460	0.215	....	0.455
Mercury	0–100	0.18182	0.0078	....	0.18186
Olive oil	....	0.6821	1.1405	− 0.539	0.721
Pentane	0–33	1.4646	3.09319	1.6084	1.608
Potassium chloride, 24.3% solution	16–25	0.2695	2.080	....	0.353
Phenol	36–157	0.8340	0.10732	0.4446	1.090
Petroleum, 0.8467 density	24–120	0.8994	1.396	....	0.955
Sodium chloride, 20.6% solution	0–29	0.3640	1.237	....	0.414
Sodium sulfate, 24% solution	11–40	0.3599	1.258	....	0.410
Sulfuric acid:					
10.9% solution	0–30	0.2835	2.580	....	0.387
100.0%	0–30	0.5758	−0.432	....	0.558
Turpentine	− 9–106	0.9003	1.9595	− 0.44998	0.973
Water	0–33	−0.06427	8.5053	− 6.7900	0.207

* "Smithsonian Tables," Table 269. For a detailed discussion of mercury data see Cook, *Brit. J. Appl. Phys.*, **7**, 285 (1956). For data on nitrogen and argon see Johnson (ed.), WADD-TR-60-56, 1960.

Bromoform[1] 7.7 − 50°C.
$V_t = 0.34204[1 + 0.00090411(t - 7.7) + 0.0000006766(t - 7.7)^2]$
0.34204 in the specific volume of bromoform at 7.7°C.
Glycerin[2] −62 to 0°C.
$V_t = V_0(1 + 4.83 \times 10^{-4}t - 0.49 \times 10^{-6}t^2)$
0 − 80°C.
$V_t = V_0(1 + 4.83 \times 10^{-4}t + 0.49 \times 10^{-6}t^2)$
Mercury[3] 0 − 300°C.
$V_t - V_0[1 + 10^{-8}(18153.8t + 0.7548t^2 + 0.001533t^2 + 0.00000536t^4)]$

[1] Sherman and Sherman, *J. Am. Chem. Soc.*, **50**, 1119 (1928). (An obvious error in their equation has been corrected.)
[2] Samsoen, *Ann. phys.*, (10) **9**, 91 (1928).
[3] Harlow, *Phil. Mag.*, (7) **7**, 674 (1929).

TABLE 3-148 Cubical Expansion of Solids*

If v_2 and v_1 are the volumes at t_2 and t_1, respectively, then $v_2 = v_1(1 + C\Delta t)$, C being the coefficient of cubical expansion and Δt the temperature interval. Where only a single temperature is stated, C represents the true coefficient of cubical expansion at that temperature.

Substance	t or Δt	$C \times 10^4$
Antimony	0–100	0.3167
Beryl	0–100	0.0105
Bismuth	0–100	0.3948
Copper†	0–100	0.4998
Diamond	40	0.0354
Emerald	40	0.0168
Galena	0–100	0.558
Glass, common tube	0–100	0.276
hard	0–100	0.214
Jena, borosilicate 59 III	20–100	0.156
pure silica	0–80	0.0129
Gold	0–100	0.4411
Ice	−20 to −1	1.1250
Iron	0–100	0.3550
Lead†	0–100	0.8399
Paraffin	20	5.88
Platinum	0–100	0.265
Porcelain, Berlin	20	0.0814
chloride	0–100	1.094
nitrate	0–100	1.967
sulfate	20	1.0754
Quartz	0–100	0.3840
Rock salt	50–60	1.2120
Rubber	20	4.87
Silver	0–100	0.5831
Sodium	20	2.13
Stearic acid	33.8–45.4	8.1
Sulfur, native	13.2–50.3	2.23
Tin	0–100	0.6889
Zinc†	0–100	0.8928

* "Smithsonian Tables," Table 268.
† See additional data below.

Aluminum[1] 100 − 530°C.
$V = V_0(1 + 2.16 \times 10^{-5}t + 0.95 \times 10^{-8}t^2)$
Cadmium[1] 130 − 270°C.
$V = V_0(1 + 8.04 \times 10^{-5}t + 5.9 \times 10^{-8}t^2)$
Copper[1] 110 − 300°C.
$V = V_0(1 + 1.62 \times 10^{-5}t + 0.20 \times 10^{-8}t^2)$
Colophony[2] 0 − 34°C.
$V = V_0(1 + 2.21 \times 10^{-4}t + 0.31 \times 10^{-6}t^2)$
34 − 150°C.
$V = V_{34}[1 + 7.40 \times 10^{-4}(t - 34) + 5.91 \times 10^{-6}(t - 34)^2]$
Lead[1] 100 − 280°C.
$V = V_0(1 + 1.60 \times 10^{-5}t + 3.2 \times 10^{-8}t^2)$
Shellac[2] 0 − 46°C.
$V = V_0(1 + 2.73 \times 10^{-4}t + 0.39 \times 10^{-6}t^2)$
46 − 100°C.
$V = V_{46}[1 + 13.10 \times 10^{-4}(t - 46) + 0.62 \times 10^{-6}(t - 46)^2]$
Silica (vitreous)[3] 0 − 300°C.
$V_t = V_0[1 + 10^{-8}(93.6t + 0.7776t^2 - 0.003315t^2 + 0.000005244t^4)$
Sugar (cane, amorphous)[2] 0 − 67°C.
$V_t = V_0(1 + 2.34 \times 10^{-4}t + 0.14 \times 10^{-6}t^2)$
67 − 160°C.
$V_t = V_{67}[1 + 5.02 \times 10^{-4}(t - 67) + 0.43 \times 10^{-6}(t - 67)^2]$
Zinc[1] 120 − 360°C.
$V_t = V_0(1 + 8.50 \times 10^{-5}t + 3.9 \times 10^{-8}t^2)$

[1] Uffelmann, *Phil. Mag.*, (7) **10**, 633 (1930).
[2] Samsoen, *Ann. phys.*, (10) **9**, 83 (1928).
[3] Harlow, *Phil. Mag.*, (7) **7**, 674 (1929).

JOULE-THOMSON EFFECT

UNITS CONVERSIONS

For this subsection the following units conversions are applicable:

To convert the Joule-Thomson coefficient, μ, in degrees Celsius per atmosphere to degrees Fahrenheit per atmosphere, multiply by 1.8.

°F = ⁹⁄₅ °C + 32; °R = ⁹⁄₅ K.

To convert bars to pounds-force per square inch, multiply by 14.504; to convert bars to kilopascals, multiply by 1×10^2.

ADDITIONAL REFERENCES

TABLE 3-149 References Available for the Joule-Thomson Coefficient

Gas	Pressure range, atm.				Temp. range, °C.			Unclassified
	0–10	10–50	50–200	>200	<0	0–300	>300	
Air	12, 15, 16 19, 35	12, 15, 19 35	15, 19, 35	....	19, 35	12, 15, 16 19, 35	....	3, 4, 18
Ammonia	28	....	....	....	....	28	....	2, 3
Argon	39	39	39	....	39	39		
Benzene	31	31	31	....	....	31	31	
Butane	26	26	....	....	....	26		
Carbon dioxide	7, 8, 28 37	7, 8, 37	7, 8, 37	....	7, 8, 37	7, 8, 9, 10 37		
Carbon monoxide	17	17	....	....	17	17		
Deuterium	....	22, 24, 25 1*	1,* 22, 24 25	....	1,* 22, 24, 25			
Dowtherm A	46	46	....	....	....	46	46	
Ethane	45	45	....	....	....	45		
Ethylene	....	....	....	....	....	9, 10		
Helium	1, 38	1, 38	38	....	1, 38	38	....	48
Hydrogen	24, 30	22, 24, 25 30	24, 30	....	22, 24, 25 30	24		
Methane	....	6	6	....	....	6		
Mixtures	....	....	....	....	....	9, 11		
Natural gas	....	....	33	33	33	33		
Nitrogen	13, 28, 40	13, 40	13, 40	13	13, 40	9, 10, 13 28, 40	13	19
Nitrous oxide	....	....	....	....	....	9, 10		
Pentane	26, 34, 44	34	34	....	....	26, 34, 44		
Propane	41	43	....	....	....	43		
Steam	28, 29, 42	29, 42, 47	42, 47	....	....	28, 29, 42 45	29, 42, 47	29, 47

*See also 14 (generalized chart); 18 (review, to 1919); 20–22; 23 (review, to 1948); 27 (review, to 1905); 32, 36, 41, 50.

References: 1. Baehr, Z. *Elektrochem.*, **60,** 515 (1956). 2. Beattie, *J. Math. Phys.*, **9,** 11 (1930). 3. Beattie, *Phys. Rev.*, **35,** 643 (1930). 4. Bradley and Hale, *Phys. Rev.*, **29,** 258 (1909). 5. Brown and Dean, *Bur. Stand. J. Res.*, **60,** 161 (1958). 6. Budenholzer, Sage, et al., *Ind. Eng. Chem.*, **29,** 658 (1937). 7. Burnett, *Phys. Rev.*, **22,** 590 (1923). 8. Burnett, Univ. Wisconsin Bull. 9(6), 1926. 9. Charnley, Ph.D. thesis, University of Manchester, 1952. 10. Charnley, Isles, et al., *Proc. R. Soc. (London)*, **A217,** 133 (1953). 11. Charnley, Rowlinson, et al., *Proc. R. Soc. (London)*, **A230,** 354 (1955). 12. Dalton, *Commun. Phys. Lab. Univ. Leiden*, no. 109c, 1909. 13. Deming and Deming, *Phys. Rev.*, **48,** 448 (1935). 14. Edmister, *Pet. Refiner*, **28,** 128 (1949). 15. Eucken, Clusius, et al., *Z. Tech. Phys.*, **13,** 267 (1932). 16. Eumorfopoulos and Rai, *Phil. Mag.*, **7,** 961 (1926). 17. Huang, Lin, et al., *Z. Phys.*, **100,** 594 (1936). 18. Hoxton, *Phys. Rev.*, **13,** 438 (1919). 19. Ishkin and Kaganev, *J. Tech. Phys. U.S.S.R.*, **26,** 2323 (1956). 20. Isles, Ph.D. thesis, Leeds University. 21. Jenkin and Pye, *Phil. Trans. R. Soc. (London)*, **A213,** 67 (1914); **A215,** 353 (1915). 22. Johnston, *J. Am. Chem. Soc.*, **68,** 2362 (1946). 23. Johnston, *Trans. Am. Soc. Mech. Eng.*, **70,** 651 (1948). 24. Johnston, Bezman, et al., *J. Am. Chem. Soc.*, **68,** 2367 (1946). 25. Johnston, Swanson, et al., *J. Am. Chem. Soc.*, **68,** 2373 (1946). 26. Kennedy, Sage, et al., *Ind. Eng. Chem.*, **28,** 718 (1936). 27. Kester, *Phys. Rev.*, **21,** 260 (1905). 28. Keyes and Collins, *Proc. Nat. Acad. Sci.*, **18,** 328 (1932). 29. Kleinschmidt, *Mech. Eng.*, **45,** 165 (1923); **48,** 155 (1926). 30. Koeppe, *Kältetechnik*, **8,** 275 (1956). 31. Lindsay and Brown, *Ind. Eng. Chem.*, **27,** *817* (1935). 32. Noell, dissertation, Munich, 1914, *Forschungsdienst*, 184, p. 1, 1916. 33. Palienko, *Tr. Inst. Ispol' z. Gaza, Akad. Nauk Ukr. SSR*, no. 4, p. 87, 1956. 34. Pattee and Brown, *Ind. Eng. Chem.*, **26,** 511 (1934). 35. Roebuck, *Proc. Am. Acad. Arts Sci.*, **60,** 537 (1925); **64,** 287 (1930). 36. Roebuck, see 49 below. 37. Roebuck and Murrell, *Phys. Rev.*, **55,** 240 (1939). 38. Roebuck and Osterberg, *Phys. Rev.*, **37,** 110 (1931); **43,** 60 (1933). 39. Roebuck and Osterberg, *Phys. Rev.*, **46,** 785 (1934). 40. Roebuck and Osterberg, *Phys. Rev.*, **48,** 450 (1935). 41. Roebuck, Murrell, et al., *J. Am. Chem. Soc.*, **64,** 400 (1942). 42. Sage, unpublished data, California Institute of Technology, 1959. 43. Sage and Lacy, *Ind. Eng. Chem.*, **27,** 1484 (1934). 44. Sage, Kennedy, et al., *Ind. Eng. Chem.*, **28,** 601 (1936). 45. Sage, Webster, et al., *Ind. Eng. Chem.*, **29,** 658 (1937). 46. Ullock, Gaffert, et al., *Trans. Am. Inst. Chem. Eng.*, **32,** 73 (1936). 47. Yang, *Ind. Eng. Chem.*, **45,** 786 (1953). 48. Zelmanov, *J. Phys. U.S.S.R.*, **3,** 43 (1940). 49. Roebuck, recalculated data. 50. Michels et al., van der Waals laboratory publications. Gunn, Cheuh, and Prausnitz, *Cryogenics*, **6,** 324 (1966), review equations relating the inversion temperatures and pressures. The ability of various equations of state to relate these was also discussed by Miller, *Ind. Eng. Chem. Fundam.*, **9,** 585 (1970); and Juris and Wenzel, *Am. Inst. Chem. Eng. J.*, **18,** 684 (1972). Perhaps the most detailed review is that of Hendricks, Peller, and Baron, NASA Tech. Note D 6807, 1972.

TABLE 3-150 Approximate Inversion-Curve Locus in Reduced Coordinates ($T_r = T/T_c$; $P_r = P/P_c$)*

P_r	0	0.5	1	1.5	2	2.5	3	4
T_{rL}	0.782	0.800	0.818	0.838	0.859	0.880	0.903	0.953
T_{rU}	4.984	4.916	4.847	4.777	4.706	4.633	4.550	4.401
P_r	5	6	7	8	9	10	11	11.79
T_{rL}	1.1	1.08	1.16	1.25	1.35	1.50	1.73	2.24
T_{rU}	4.23	4.06	3.88	3.68	3.45	3.18	2.86	2.24

*Calculated from the best three-constant equation recommended by Miller, *Ind. Eng. Chem. Fundam.*, **9**, 585 (1970). T_{rL} refers to the lower curve and T_{rU} to the upper curve.

TABLE 3-151a Joule-Thomson Data for Air*

P, atm.	Temp., °C. −150	−100	−75	−50	−25	0	25	50	75	100	150	200	250
1	....	0.5895	0.4795	0.3910	0.3225	0.2745	0.2320	0.1956	0.1614	0.1355	0.0961	0.0645	0.0409
20	....	.5700	.4555	.3690	.3010	.2580	.2173	.1830	.1508	.1258	.0883	.0580	.0356
60	0.0450	.4820	.3835	.3195	.2610	.2200	.1852	.1571	.1293	.1062	.0732	.0453	.0254
100	.0185	.2775	.2880	.2505	.2130	.1820	.1550	.1310	.1087	.0884	.0600	.0343	.0165
140	− .0070	.1360	.1855	.1825	.1650	.1450	.1249	.1070	.0889	.0726	.0482	.0250	.0092
180	− .0255	.0655	.1136	.1270	.1240	.1100	.0959	.0829	.0707	.0580	.0376	.0174	.0027
200	− .0330	.0440	.0855	.1065	.1090	.0950							

* Free of water and CO_2. Extracted from Table 261, "Smithsonian Physical Tables," 9th rev. ed., Washington, D.C., 1954. These data are corrected from earlier publications. μ in °C./atm.

TABLE 3-151b Approximate Inversion-Curve Locus for Air

P, bar	0	25	50	75	100	125	150	175	200	225
T_L, K	(112)*	114	117	120	124	128	132	137	143	149
T_U, K	653	641	629	617	606	594	582	568	555	541
P, bar	250	275	300	325	350	375	400	425	432	
T_L, K	156	164	173	184	197	212	230	265	300	
T_U, K	526	509	491	470	445	417	386	345	300	

*Hypothetical low-pressure limit.

TABLE 3-152a Joule-Thomson Data for Argon*

Temp., °C.	Pressure, atm. 1	20	60	100	140	180	200
−150	1.812	......	−0.0025	−0.0277	−0.0403	−0.0595	−0.0640
−125	1.112	1.102	.1250	.0415	.0090	−.0100	− .0165
−100	0.8605	0.8485	.6900	.2820	.1137	.0560	.0395
−75	.7100	.6895	.5910	.4225	.2480	.1537	.1215
−50	.5960	.5720	.4963	.3970	.2840	.2037	.1860
−25	.5045	.4805	.4210	.3460	.2763	.2140	.1950
0	.4307	.4080	.3600	.3010	.2505	.2050	.1883
25	.3720	.3490	.3077	.2628	.2213	.1890	.1745
50	.3220	.3015	.2650	.2297	.1947	.1700	.1580
75	.2695	.2557	.2285	.1993	.1710	.1505	.1415
100	.2413	.2277	.1975	.1715	.1490	.1320	.1255
125	.2105	.1980	.1707	.1480	.1300	.1153	.1100
150	.1845	.1720	.1485	.1285	.1123	.0998	.0945
200	.1377	.1280	.1102	.0950	.0823	.0715	.0675
250	.0980	.0910	.0785	.0665	.0555	.0485	.0468
300	.0643	.0607	.0530	.0445	.0370	.0370	.0276

* Extracted from Table 263, "Smithsonian Physical Tables," 9th rev. ed., Washington, D.C., 1954. These data are corrected from an earlier publication. μ in °C./atm.

TABLE 3-152*b* Approximate Inversion-Curve Locus for Argon

P, bar	0	25	50	75	100	125	150	175	200	225
T_L, K	94	97	101	105	109	113	118	123	128	134
T_U, K	765	755	744	736	726	716	705	694	683	671
P, bar	250	275	300	325	350	375	400	425	450	475
T_L, K	141	148	158	170	183	201	222	248	288	375
T_U, K	657	643	627	610	591	569	544	515	478	375

TABLE 3-153*a* Joule-Thomson Data for Carbon Dioxide*

Temp., °C.	Pressure, atm.							
	1	20	60	73	100	140	180	200
−75	......	−0.0200	−0.0200	−0.0232	−0.0228	−0.0240	−0.0250	−0.0290
−50	2.4130	− .0140	− .0150	− .0165	− .0160	− .0183	− .0228	− .0248
0	1.2900	1.4020	.0370	.0310	.0215	.0115	.0085	.0045
50	0.8950	.8950	.8800	.8225	.5570	.1720	.1025	.0930
100	.6490	.6375	.6080	.5920	.5405	.4320	.3000	.2555
125	.5600	.5450	.5160	.5068	.4750	.4130	.3230	.2915
150	.4890	.4695	.4430	.4380	.4155	.3760	.3102	.2910
200	.3770	.3575	.3400	.3325	.3150	.2890	.2600	.2455
250	.3075	.2885	.2625	.2565	.2420	.2235	.2045	.1975
300	.2650	.2425	.2080	.2002	.1872	.1700	.1540	.1505

* Extracted from Table 266, "Smithsonian Physical Tables," 9th rev. ed., Washington, D.C., 1954. These data are corrected from an earlier publication. μ in °C./atm.

TABLE 3-153*b* Approximate Inversion-Curve Locus for Carbon Dioxide*

P, bar	50	100	150	200	250	300	350	400	450
T_L, K	243	251	258	266	272	283	293	302	312
T_U, K	1290	1261	1233	1205	1175	1146	1111	1076	1045
P, bar	500	550	600	650	700	750	800	850	884
T_L, K	325	338	351	365	383	403	441	496	608
T_U, K	1015	983	950	914	878	840	796	739	608

*Interpolated from Vukalovich and Altunin's interpolation of data of Price, *Ind. Eng. Chem.*, 47, 1691 (1955). T_L = lower inversion temperature, and T_U = upper inversion temperature.

TABLE 3-154 Joule-Thomson Data for Helium*

T, K	160	180	200	220	240	260	280	300
μ	−0.0574	−0.0587	−0.0594	−0.0601	−0.0608	−0.0614	−0.0619	−0.0625
T, K	320	340	360	380	400	420	440	460
μ	−0.0629	−0.0634	−0.0637	−0.0640	−0.0643	−0.0645	−0.0645	−0.0643
T, K	480	500	520	540	560	580	600	
μ	−0.0640	−0.0636	−0.0630	−0.0622	−0.0611	−0.0587	−0.0540	

*Interpolated and converted from data in Table 262, *Smithsonian Physical Tables*, 9th rev. ed., Washington, 1954. These data are corrected from those in an earlier publication. μ is in °C/atm. Below about 200 atm little change in the coefficient with pressure occurs.

TABLE 3-155 Joule-Thomson Data for Nitrogen*

Temp., °C.	Pressure, atm. 1	20	33.5	60	100	140	180	200
−150	1.2659	1.1246	0.1704	0.0601	0.0202	−0.0056	−0.0211	−0.0284
−125	0.8557	0.7948	.7025	.4940	.1314	.0498	.0167	.0032
−100	.6490	.5958	.5494	.4506	.2754	.1373	.0765	.0587
−75	.5033	.4671	.4318	.3712	.2682	.1735	.1026	.0800
−50	.3968	.3734	.3467	.3059	.2332	.1676	.1120	.0906
−25	.3224	.3013	.2854	.2528	.2001	.1506	.1101	.0932
0	.2656	.2494	.2377	.2088	.1679	.1316	.1015	.0891
25	.2217	.2060	.1961	.1729	.1400	.1105	.0874	.0779
50	.1855	.1709	.1621	.1449	.1164	.0915	.0732	.0666
75	.1555	.1421	.1336	.1191	.0941	.0740	.0583	.0543
100	.1292	.1173	.1100	.0975	.0768	.0582	.0462	.0419
125	.1070	.0973	.0904	.0786	.0621	.0459	.0347	.0326
150	.0868	.0776	.0734	.0628	.0482	.0348	.0248	.0228
200	.0558	.0472	.0430	.0372	.0262	.0168	.0094	.0070
250	.0331	.0256	.0230	.0160	.0071	.0009	−.0037	−.0058
300	.0140	.0096	.0050	−.0013	−.0075	−.0129	−.0160	−.0171

* Extracted from Table 264, "Smithsonian Physical Tables," 9th rev. ed., Washington, D.C., 1954. These data are corrected from an earlier publication. μ in °C./atm.

TABLE 3-156 Approximate Inversion-Curve Locus for Deuterium

P, bar	0	25	50	75	100	125	150	175	194
T_L, K	(31)°	34	38	43	49	56	65	77	108
T_U, K	216	202	189	178	168	157	146	131	108

°Hypothetical low-pressure limit.

TABLE 3-157 Approximate Inversion-Curve Locus for Normal Hydrogen

P, bar	0	25	50	75	100	125	150	164
T_L, K	(28)°	32	38	44	52	61	73	92
T_U, K	202	193	183	171	157	141	119	92

°Hypothetical low-pressure limit.

TABLE 3-158 Approximate Inversion-Curve Locus for Methane

P, bar	25	50	75	100	125	150	175	200	225	250	275	300
T_L, K	. . .	161	166	172	176	182	189	195	202	209	217	225
P, bar	325	350	375	400	425	450	475	500	525	534		
T_L, K	234	243	254	265	277	292	309	331	365	400		
T_U, K	. . .	. . .	. . .	. . .	. . .	. . .	505	474	437	400		

TABLE 3-159 Approximate Inversion-Curve Locus for Ethane

P, bar	0	25	50	75	100	125	150	175	200	225
T_L, K	. . .	249	255	262	269	275	282	290	297	306
P, bar	250	275	300	325	350	375	400	425	450	475
T_L, K	315	325	335	345	357	370	383	398	415	432
P, bar	500	525	550	575	600					
T_L, K	453	477	505	545	626					

TABLE 3-160 Approximate Inversion-Curve Locus for Propane

P, bar	0	25	50	75	100	125	150	175	200	225	250	275
T_L, K	(296)°	303	311	318	327	336	345	355	365	374	389	403
P, bar	300	325	350	375	400	425	450	475	500	525	541	
T_L, K	418	435	452	473	495	521	551	586	628	686	780	

°Hypothetical low-pressure limit.

CRITICAL CONSTANTS

ADDITIONAL REFERENCES

Other data and estimation techniques for the elements are contained in Gates and Thodos, *Am. Inst. Chem. Eng. J.*, **6**, 50–54 (1960); and Ohse and von Tippelskirch, *High Temperatures—High Pressures*, **9**, 367–385 (1977). For inorganic substances see Mathews, *Chem. Rev.*, **72**, 71–100 (1972); for organics see Kudchaker, Alani, and Zwolinski, *Chem. Rev.*, **68**, 659–735 (1968); and for fluorocarbons see *Advances in Fluorine Chemistry*, app. B, Butterworth, Washington, 1963, pp. 173–175.

TABLE 3-161 Critical Constants of Elements and Inorganic and Organic Compounds

For additional values of the critical temperature and pressue, see Table 3-6, p. 3-45, and Table 3-9, p. 3-61, on vapor pressures above 1 atm.

Name	Formula	t_c, °C.	P_c, atm.	d_c, g./cc.
Acetaldehyde	C_2H_4O	188.0		
Acetic acid	$C_2H_4O_2$	321.6	57.2	0.351
anhydride	$C_4H_6O_3$	296.0	46.0	
Acetone	C_3H_6O	235.0	47.0	0.268
Acetonitrile	C_2H_3N	274.7	47.7	0.240
Acetylene	C_2H_2	36.0	62.0	0.231
Air	.....	−140.7	37.2	0.35 (0.31)
Allyl alcohol	C_3H_6O	272.0		
Allylene	C_3H_4	128.0		
Allyl ethyl ether	$C_5H_{10}O$	245		
sulfide	$C_6H_{10}S$	380		
Ammonia	NH_3	132.4	111.5	0.235
Amyl alcohol (t-)	$C_5H_{12}O$	272		
Aniline	C_6H_7N	426	52.4	
Anisole	C_7H_8O	369	41.3	
Argon	A	−122	48.0	0.531
Arsenic	As	803	342.0	
Benzene	C_6H_6	288.5	47.7	0.304
Benzonitrile	C_7H_5N	426.0	41.6	
Boron tribromide	BBr_3	300	....	0.90
Bromine	Br_2	311	102	0.848
Bromobenzene	C_6H_5Br	397	44.6	0.486
Butadiene-1,3	C_4H_6	152	42.7	0.245
Butane (n-)	C_4H_{10}	153	36.0	
Butyl acetate (n-)	$C_6H_{12}O_2$	306.0		
alcohol (n-)	$C_4H_{10}O$	287	48.4	
alcohol (s-)	$C_4H_{10}O$	265		
alcohol (t-)	$C_4H_{10}O$	235		
Butyric acid (n-)	$C_4H_8O_2$	355	....	0.302
Butyronitrile	C_4H_7N	309	37.4	
Capronitrile	$C_6H_{11}N$	349.0	32.2	
Carbon dioxide	CO_2	31.1	73.0	0.460
disulfide	CS_2	273.0	76.0	0.441
monoxide	CO	−139.0	35.0	0.311
oxysulfide	COS	105.0	61.0	
tetrachloride	CCl_4	283.1	45.0	0.558
Chlorine	Cl_2	144.0	76.1	0.573
Chlorobenzene	C_6H_5Cl	359.0	44.6	0.365
Chloroform	$CHCl_3$	263.0	....	0.516
Cresol (o-)	$CH_3.C_6H_4OH$	422.0	49.4	
(m-)	$CH_3.C_6H_4OH$	432.0	45.0	
(p-)	$CH_3.C_6H_4OH$	426.0	50.8	
Cyanogen	$(CN)_2$	128.0	59.0	
Cyclohexane	C_6H_{12}	281.0	40.4	0.270
Deuterium	D_2	−234.4	17.4	
Dichlordifluormethane	CCl_2F_2	111.5	39.56	0.555
Diethylamine	$(C_2H_5)_2NH$	223.5	36.2	0.246
Di-isobutyl	$(CH_3)_2CH(CH_2)_2CH(CH_3)_2$	277.0	24.5	0.237
isopropyl	C_6H_{14}	227.4	30.6	0.241
Dimethylamine	$(CH_3)_2NH$	164.6	51.7	
Dimethyl aniline	$C_6H_5N(CH_3)_2$	415.0	35.8	
toluidine (o-)	$C_8H_{11}N$	395.0	30.8	
Dipropylamine	$(C_3H_7)_2NH$	277.0	31.0	
Ethane	C_2H_6	32.1	48.8	0.21
Ethyl acetate	$CH_3COOC_2H_5$	250.1	37.8	0.308
alcohol	C_2H_5OH	243.1	63.1	0.2755
allyl ether	$C_2H_5OCH_2CHCH_2$	245.0		
amine	$C_2H_5NH_2$	183.2	55.5	
bromide	C_2H_5Br	231.0	....	0.513
butyrate	$C_3H_7COOC_2H_5$	293.0	30.0	0.276
caprylate	$C_7H_{15}COOC_2H_5$	386.0		
chloride	C_2H_5Cl	187.2	52.0	0.33
chloroformate	$ClCOOC_2H_5$	<235.0		
crotonate	$C_6H_{10}O_2$	326.0		
disulfide	$(C_2H_5)_2S_2$	369.0		
Ethylene	C_2H_4	9.7	50.5	0.22
chloride	$C_2H_4Cl_2$	290	....	0.45
oxide	C_2H_4O	192.0		
Fluorobenzene	C_6H_5F	286.0	44.6	0.354
Fluorine	F	−155.0	25.0	
Germanium tetrachloride	$GeCl_4$	277.0	38.0	
Helium	He	−267.9	2.26	0.0693
Heptane (n-)	C_7H_{16}	266.8	26.8	0.234
Heptyl alcohol (n-)	$C_7H_{15}OH$	365.0		
Hexane (n-)	C_6H_{14}	234.8	29.5	0.234
Hydrazine	N_2H_4	380.0	145.0	
Hydrogen	H_2	−239.9	12.8	0.0310
bromide	HBr	90.0	84.0	
chloride	HCl	51.4	81.6	0.42
cyanide	HCN	183.5	53.2	0.20
fluoride	HF	230.2		
iodide	HI	151.0	82.0	
selenide	H_2Se	138.0	88.0	
sulfide	H_2S	100.4	88.9	2.86
Iodine	I_2	553.0		
Iodobenzene	C_6H_5I	448.0	44.6	0.581
Isoamyl acetate	$CH_3COOC_5H_{11}$	326.0		
alcohol	$C_5H_{11}OH$	307.0		
butyrate	$C_3H_7COOC_5H_{11}$	346.0		
formate	$HCOOC_5H_{11}$	303.0	34.0	0.282
mercaptan	$C_5H_{11}SH$	321.0		
propionate	$C_2H_5COOC_5H_{11}$	338.0		
sulfide	$(C_5H_{11})_2S$	391.0		
Isobutane	C_4H_{10}	134.0	37.0	
Isobutyl acetate	$CH_3COOC_4H_9$	288.0	31.0	0.281
alcohol	C_4H_9OH	265.0	48.0	
butyrate	$C_3H_7COOC_4H_9$	338.0		
formate	$HCOOC_4H_9$	278.0	38.0	0.288
isobutyrate	$C_3H_7COOC_4H_9$	329.0		
isovalerate	$C_9H_{18}O_2$	348.0		
propionate	$C_7H_{14}O_2$	319.0		
Isobutyric acid	C_3H_7COOH	336.0	....	0.304
Isopentane	C_5H_{12}	187.8	32.8	0.234
Isopropyl alcohol	C_3H_7OH	235.0	53.0	
Isovaleric acid	C_4H_9OH	361.0		
Krypton	Kr	−63.8	54.3	1.10
Mercury	Hg	>1550	>200	4–5
Methane	CH_4	−82.5	45.8	0.162
Methyl acetate	CH_3COOCH_3	233.7	46.3	0.325
Methylal	$CH_2(OCH_3)_2$	224.0		
Methyl alcohol	CH_3OH	240.0	78.7	0.272
amine	CH_3NH_2	156.9	73.6	
aniline	$C_6H_5NHCH_3$	429.0	51.3	
butyrate	$C_3H_7COOCH_3$	281.3	34.2	0.300
chloride	CH_3Cl	143.1	65.8	0.37
ether	$(CH_3)_2O$	126.9	52.0	0.271
ethyl ether	$CH_3OC_2H_5$	164.7	43.4	0.270
sulfide	$CH_3SC_2H_5$	260.0	42.0	
fluoride	CH_3F	44.9	62.0	
formate	$HCOOCH_3$	214.0	59.15	0.349
isobutyrate	$C_3H_7COOCH_3$	267.55	33.7	0.301
mercaptan	CH_3SH	196.8	71.4	0.323
oxalate	$(CH_3O_2C)_2$	260.0	9.48	
propionate	$C_2H_5COOCH_3$	257.4	39.3	0.312
sulfide	$(CH_3)_2S$	229.9	54.6	0.306
valerate	$C_6H_{12}O_2$	294.0(d)	32.0	0.279
Methyl diethyl ether	$C_5H_{12}O_2$	254.0		
Neon	Ne	−228.7	25.9	0.484
Niton	Nt	+104.5	62.5	
Nitric oxide	NO	−94.0	65.0	0.52
Nitrogen	N_2	−147.1	33.5	0.3110
tetroxide	N_2O_4	158.0	100	1.785
Nitrous oxide	N_2O	36.5	71.7	0.45
Octane (n-)	C_8H_{18}	296.0	24.6	0.234

NOTE: °F = ⅘ °C + 32. To convert grams per cubic centimeter to pounds-mass per cubic foot, multiply by 62.43.

TABLE 3-161 Critical Constants of Elements and Inorganic and Organic Compounds *(Concluded)*

Name	Formula	t_c, °C.	P_c, atm.	d_c, g./cc.
Ethyl ether	$(C_2H_5)_2O$	194.6	35.5	0.2625
formate	$HCOOC_2H_5$	235.3	46.65	0.323
isobutyrate	$(CH_3)_2CHCOOC_2H_5$	280.0	30.0	0.276
isovalerate	$C_7H_{14}O_2$	315.0		
mercaptan	C_2H_5SH	225.5	54.2	0.301
methyl ether	$CH_3C_2H_5O$	164.7	43.4	0.270
sulfide	$C_2H_5CH_3S$	260.0	42.0	
nonylate	$C_{11}H_{22}O_2$	400.0		
propionate	$C_2H_5COOC_2H_5$	272.9	33.0	0.2965
propyl ether	$C_2H_5C_3H_7O$	227.4	32.1	0.258
sulfide	$(C_2H_5)_2S$	283.8	39.1	0.279
valerate	$C_7H_{14}O_2$	297.0		
Propionic acid	C_2H_5COOH	339.5	53.0	0.315
Propyl alcohol (n-)	C_3H_7OH	263.7	49.95	0.273
acetate	$CH_3COOC_3H_7$	276.2	32.9	0.296
Propylamine	$C_3H_7NH_2$	223.8	46.3	
Propyl butyrate	$C_3H_7COOC_3H_7$	327.0		
chloride (n-)	C_3H_7Cl	230.0	45.2	
formate	$HCOOC_3H_7$	264.85	40.1	0.309
Propylene	C_3H_6	92.3	45.0	0.233
Propyl ethyl ether	$C_3H_7OC_2H_5$	227.4	32.1	0.258
isobutyrate	$C_3H_7COOC_3H_7$	316.0		
isovalerate	$C_8H_{16}O_2$	336.0		
Propionitrile	C_3H_5N	291.2	41.3	0.241
Propyl propionate	$C_2H_5COOC_3H_7$	305.0		
Pyridine	C_6H_5N	344.0	60.0	
Quinoline	C_9H_7N	520.0		
Radon	Rn	104.0	62.0	
Sodium	Na	2546	343	
Octyl alcohol (n-)	$C_8H_{17}OH$	385.0		
(s-)	$C_8H_{17}OH$	364.0		
Oxygen	O_2	−118.8	49.7	0.430
Paraldehyde	$(C_2H_4O)_3$	290.0		
Pentane (n-)	C_5H_{12}	197.2	33.0	0.232
Phenetole	$C_6H_5OC_2H_5$	374.0	33.8	
Phenol	C_6H_5OH	419.0	60.5	
Phosgene	$COCl_2$	182.0	56.0	0.52
Phosphine	PH	51.0	64.0	0.30
Phosphonium chloride	PH_4Cl	49.0	73.0	0.226
Propane (n-)	C_3H_8	96.8	42.0	0.220
Silicon tetrafluoride	SiF_4	−1.5	50.0	
tetrahydride	SiH_4	−3.5	48.0	
Stannic tetrachloride	$SnCl_4$	318.7	37.0	0.742
Steam	H_2O	374.0	217.7	0.4
Sulfur	S	1040		
dioxide	SO_2	157.2	77.7	0.52
trioxide	SO_3	218.3	83.6	0.630
Thiophene	C_4H_4S	317.0	48.0	
Thymol	$CH_3C_6H_3(OH)C_3H_7$	425.0		
Toluene	$C_6H_5CH_3$	320.6	41.6	0.292
Tolunitrile	$CH_3C_6H_4CN$	450.0		
Triethylamine	$(C_2H_5)_3N$	262.0	30.0	0.251
Trimethylamine	$(CH_3)_3N$	161.0	41.0	
Tritium	....	−229.5	20.8	
Valeric acid (n-)	$C_4H_{10}O_2$	379.0		
Water	H_2O	374.15	218.4	0.323
Xenon	Xe	16.6	58.2	1.155

COMPRESSIBILITIES

INTRODUCTION

The increasing ranges of pressure and temperature of interest to technology for an ever-increasing number of substances would necessitate additional tables in this subsection as well as in the subsection "Thermodynamic Properties." Space restrictions preclude this. Hence, in the present revision, an attempt was made to update the fluid-compressibility tables for selected fluids and to omit tables for other fluids. The reader is thus referred to the fourth edition for tables on miscellaneous gases at 0°C, acetylene, ammonia, ethane, ethylene, hydrogen-nitrogen mixtures, and methyl chloride. The reader is also reminded that compressibilities can be calculated from the pressure–volume (or density)–temperature tables of the subsection "Thermodynamic Properties."

UNITS CONVERSIONS

For this subsection, the following units conversions are applicable:

°R = $\frac{9}{5}$ K.

To convert bars to pounds-force per cubic inch, multiply by 14.504.

To convert bars to kilopascals, multiply by 1×10^2.

TABLE 3-162 Compressibility Factor of Air*

Temp., °K.	Pressure, bars 1	5	10	20	40	60	80	100	150	200	250	300	400	500
75	0.0052	0.0260	0.0519	0.1036	0.2063	0.3082	0.4094	0.5099	0.7581	1.0025				
80		0.0250	0.0499	0.0995	0.1981	0.2958	0.3927	0.4887	0.7258	0.9588	1.1931	1.4139		
90	0.9764	0.0236	0.0471	0.0940	0.1866	0.2781	0.3686	0.4581	0.6779	0.8929	1.1098	1.3110	1.7161	2.1105
100	0.9797	0.8872	0.0453	0.0900	0.1782	0.2635	0.3498	0.4337	0.6386	0.8377	1.0395	1.2227	1.5937	1.9536
120	0.9880	0.9373	0.8660	0.6730	0.1778	0.2557	0.3371	0.4132	0.5964	0.7720	0.9530	1.1076	1.5091	1.7366
140	0.9927	0.9614	0.9205	0.8297	0.5856	0.3313	0.3737	0.4340	0.5909	0.7699	0.9114	1.0393	1.3202	1.5903
160	0.9951	0.9748	0.9489	0.8954	0.7803	0.6603	0.5696	0.5489	0.6340	0.7564	0.8840	1.0105	1.2585	1.4970
180	0.9967	0.9832	0.9660	0.9314	0.8625	0.7977	0.7432	0.7084	0.7180	0.7986	0.9000	1.0068	1.2232	1.4361
200	0.9978	0.9886	0.9767	0.9539	0.9100	0.8701	0.8374	0.8142	0.8061	0.8549	0.9311	1.0185	1.2054	1.3944
250	0.9992	0.9957	0.9911	0.9822	0.9671	0.9549	0.9463	0.9411	0.9450	0.9713	1.0152	1.0702	1.1990	1.3392
300	0.9999	0.9987	0.9974	0.9950	0.9917	0.9901	0.9903	0.9930	1.0074	1.0326	1.0669	1.1089	1.2073	1.3163
350	1.0000	1.0002	1.0004	1.0014	1.0038	1.0075	1.0121	1.0183	1.0377	1.0635	1.0947	1.1303	1.2116	1.3015
400	1.0002	1.0012	1.0025	1.0046	1.0100	1.0159	1.0229	1.0312	1.0533	1.0795	1.1087	1.1411	1.2117	1.2890
450	1.0003	1.0016	1.0034	1.0063	1.0133	1.0210	1.0287	1.0374	1.0614	1.0913	1.1183	1.1463	1.2090	1.2778
500	1.0003	1.0020	1.0034	1.0074	1.0151	1.0234	1.0323	1.0410	1.0650	1.0913	1.1183	1.1463	1.2051	1.2667
600	1.0004	1.0022	1.0039	1.0081	1.0164	1.0253	1.0340	1.0434	1.0678	1.0920	1.1172	1.1427	1.1947	1.2475
800	1.0004	1.0020	1.0038	1.0077	1.0157	1.0240	1.0321	1.0408	1.0621	1.0844	1.1061	1.1283	1.1720	1.2150
1000	1.0004	1.0018	1.0037	1.0068	1.0142	1.0215	1.0290	1.0365	1.0556	1.0744	1.0948	1.1131	1.1515	1.1889

*Calculated from values of pressure, volume (or density), and temperature in Vasserman, Kazavchinskii, and Rabinovich, "Thermophysical Properties of Air and Air Components," Moscow, Nauka, 1966, and NBS-NSF Trans. TT 70-50095, 1971; and Vasserman and Rabinovich, "Thermophysical Properties of Liquid Air and Its Components," Moscow, 1968, and NBS-NSF Trans. 69-55092, 1970.

TABLE 3-163 Compressibility Factors for Nitrogen*

Temp., °K.	Pressure, bars 1	5	10	20	40	60	80	100	200	300	400	500
70	0.0057	0.0287	0.0573	0.1143	0.2277	0.3400	0.4516	0.5623	1.1044	1.6308	Solid	Solid
80	0.9593	0.0264	0.0528	0.1053	0.2093	0.3122	0.4140	0.5148	1.0061	1.4797	1.9396	2.3879
90	0.9722	0.0251	0.0500	0.0996	0.1975	0.2938	0.3888	0.4826	0.9362	1.3700	1.7890	2.1962
100	0.9798	0.8910	0.0487	0.0966	0.1905	0.2823	0.3720	0.4605	0.8840	1.2852	1.6707	2.0441
120	0.9883	0.9397	0.8732	0.7059	0.1975	0.2822	0.3641	0.4438	0.8188	1.1684	1.5015	1.8223
140	0.9927	0.9635	0.9253	0.8433	0.6376	0.4251	0.4278	0.4799	0.7942	1.0996	1.3920	1.6726
160	0.9952	0.9766	0.9529	0.9042	0.8031	0.7017	0.6304	0.6134	0.8107	1.0708	1.3275	1.5762
180	0.9967	0.9846	0.9690	0.9381	0.8782	0.8125	0.7784	0.7530	0.8550	1.0669	1.2893	1.5105
200	0.9978	0.9897	0.9791	0.9592	0.9212	0.8882	0.8621	0.8455	0.9067	1.0760	1.2683	1.4631
250	0.9992	0.9960	0.9924	0.9857	0.9741	0.9655	0.9604	0.9589	1.0048	1.1143	1.2501	1.3962
300	0.9998	0.9990	0.9983	0.9971	0.9964	0.9973	1.0000	1.0052	1.0559	1.1422	1.2480	1.3629
350	1.0001	1.0007	1.0011	1.0029	1.0069	1.0125	1.0189	1.0271	1.0810	1.1560	1.2445	1.3405
400	1.0002	1.0011	1.0024	1.0057	1.0125	1.0199	1.0283	1.0377	1.0926	1.1609	1.2382	1.3216
450	1.0003	1.0018	1.0033	1.0073	1.0153	1.0238	1.0332	1.0430	1.0973	1.1606	1.2303	1.3043
500	1.0004	1.0020	1.0040	1.0081	1.0167	1.0257	1.0350	1.0451	1.0984	1.1575	1.2213	1.2881
600	1.0004	1.0021	1.0040	1.0084	1.0173	1.0263	1.0355	1.0450	1.0951	1.1540	1.2028	1.2657
800	1.0004	1.0017	1.0036	1.0074	1.0157	1.0237	1.0320	1.0402	1.0832	1.1264	1.1701	1.2140
1000	1.0003	1.0015	1.0034	1.0067	1.0136	1.0205	1.0275	1.0347	1.0714	1.1078	1.1449	1.1814

*Computed from pressure-volume-temperature tables in the Vasserman monographs referenced under Table 3-162.

TABLE 3-164 Compressibility Factors for Oxygen*

Temp., °K.	Pressure, bars 1	5	10	20	40	60	80	100	200	300	400	500
75	0.0043	0.0213	0.0425	0.0849	0.1693	0.2533	0.3368	0.4200	0.8301	1.2322	1.6278	2.0175
80	0.0041	0.0203	0.0406	0.0811	0.1616	0.2418	0.3214	0.4007	0.7912	1.1738	1.5495	1.9196
90	0.0038	0.0188	0.0376	0.0750	0.1494	0.2233	0.2966	0.3696	0.7281	1.0780	1.4211	1.7580
100	0.9757	0.0177	0.0354	0.0705	0.1404	0.2096	0.2783	0.3464	0.6798	1.0040	1.3206	1.6309
120	0.9855	0.9246	0.8367	0.0660	0.1302	0.1935	0.2558	0.3173	0.6148	0.8999	1.1762	1.4456
140	0.9911	0.9535	0.9034	0.7852	0.1334	0.1940	0.2527	0.3099	0.5815	0.8374	1.0832	1.3214
160	0.9939	0.9697	0.9379	0.8689	0.6991	0.3725	0.2969	0.3378	0.5766	0.8058	1.0249	1.2364
180	0.9960	0.9793	0.9579	0.9134	0.8167	0.7696	0.5954	0.5106	0.6043	0.8025	0.9990	1.1888
200	0.9970	0.9853	0.9705	0.9399	0.8768	0.8140	0.7534	0.6997	0.6720	0.8204	0.9907	1.1623
250	0.9987	0.9938	0.9870	0.9736	0.9477	0.9237	0.9030	0.8858	0.8563	0.9172	1.0222	1.1431
300	0.9994	0.9968	0.9941	0.9884	0.9771	0.9676	0.9597	0.9542	0.9560	0.9972	1.0689	1.1572
350	0.9998	0.9990	0.9979	0.9961	0.9919	0.9890	0.9870	0.9870	1.0049	1.0451	1.1023	1.1722
400	1.0000	1.0000	1.0000	1.0000	1.0003	1.0011	1.0022	1.0045	1.0305	1.0718	1.1227	1.1816
450	1.0002	1.0007	1.0015	1.0024	1.0048	1.0074	1.0106	1.0152	1.0445	1.0859	1.1334	1.1859
500	1.0002	1.0011	1.0022	1.0038	1.0075	1.0115	1.0161	1.0207	1.0523	1.0927	1.1380	1.1866
600	1.0003	1.0014	1.0024	1.0052	1.0102	1.0153	1.0207	1.0266	1.0582	1.0961	1.1374	1.1803
800	1.0003	1.0014	1.0026	1.0055	1.0109	1.0164	1.0219	1.0271	1.0565	1.0888	1.1231	1.1582
1000	1.0003	1.0013	1.0026	1.0053	1.0101	1.0149	1.0198	1.0253	1.0507	1.0783	1.1072	1.1369

*Calculated from pressure-volume-temperature tables in the Vasserman monographs listed under Table 3-162.

TABLE 3-165 Compressibility Factors of Normal Hydrogen*

Temperature, K	Pressure, bar											
	1	10	20	40	60	80	100	200	400	600	800	1000
20	0.0169	0.1680	0.3302	0.6430	0.9434	1.2346	1.5166	2.844				
40	0.9848	0.8340	0.6311	0.5240	0.6627	0.8118	0.9590	1.650	2.878	3.993	5.034	6.019
60	0.9955	0.9562	0.9169	0.8608	0.8498	0.8832	0.9432	1.347	2.158	2.902	3.598	4.263
80	0.9986	0.9776	0.9763	0.9655	0.9676	0.9842	1.0138	1.257	1.834	2.389	2.907	3.404
100	0.9998	0.9979	0.9976	1.0022	1.0133	1.0280	1.0528	1.225	1.659	2.095	2.512	2.902
200	1.0007	1.0066	1.0134	1.0275	1.0422	1.0575	1.0734	1.163	1.355	1.555	1.753	1.936
300	1.0005	1.0059	1.0117	1.0236	1.0357	1.0479	1.0603	1.124	1.253	1.383	1.510	1.636
400	1.0004	1.0048	1.0096	1.0192	1.0289	1.0386	1.0484	1.098	1.196	1.293	1.388	1.481
500	1.0004	1.0040	1.0080	1.0160	1.0240	1.0320	1.0400	1.080	1.159	1.236	1.311	1.385
600	1.0003	1.0034	1.0068	1.0136	1.0204	1.0272	1.0340	1.068	1.133	1.197	1.259	1.320
800	1.0002	1.0026	1.0052	1.0104	1.0156	1.0208	1.0259	1.051	1.100	1.147	1.193	1.237
1000	1.0002	1.0021	1.0042	1.0084	1.0126	1.0168	1.0209	1.041	1.080	1.117	1.153	1.187
2000	1.0009	1.0013	1.0023	1.0044	1.0065	1.0086	1.0107	1.021	1.040	1.057	1.073	1.088

*Calculated from PVT tables of McCarty, Hord, and Roder, NBS Monogr. 168, 1981.

TABLE 3-166 Compressibility Factors for Carbon Dioxide*

Temp., °C.	Pressure, bar											
	1	5	10	20	40	60	80	100	200	300	400	500
0	0.9933	0.9658	0.9294	0.8496								
50	0.9964	0.9805	0.9607	0.9195	0.8300	0.7264	0.5981	0.4239				
100	0.9977	0.9883	0.9764	0.9524	0.9034	0.8533	0.8022	0.7514	0.5891	0.6420		
150	0.9985	0.9927	0.9853	0.9705	0.9416	0.9131	0.8854	0.8590	0.7651	0.7623	0.8235	0.9098
200	0.9991	0.9953	0.9908	0.9818	0.9640	0.9473	0.9313	0.9170	0.8649	0.8619	0.8995	0.9621
250	0.9994	0.9971	0.9943	0.9886	0.9783	0.9684	0.9593	0.9511	0.9253	0.9294	0.9508	1.0096
300	0.9996	0.9982	0.9967	0.9936	0.9875	0.9822	0.9773	0.9733	0.9640	0.9746	1.0030	1.0464
350	0.9998	0.9991	0.9983	0.9964	0.9938	0.9914	0.9896	0.9882	0.9895	1.0053	1.0340	1.0734
400	0.9999	0.9997	0.9994	0.9989	0.9982	0.9979	0.9979	0.9984	1.0073	1.0266	1.0559	1.0928
450	1.0000	1.0000	1.0003	1.0005	1.0013	1.0023	1.0038	1.0056	1.0070	1.0412	1.0709	1.1067
500	1.0000	1.0004	1.0008	1.0015	1.0035	1.0056	1.0079	1.0107	1.0282	1.0522	1.0820	1.1165
600	1.0000	1.0007	1.0013	1.0030	1.0062	1.0093	1.0129	1.0168	1.0386	1.0648	1.0948	1.1277
700	1.0003	1.0010	1.0017	1.0036	1.0073	1.0161	1.0155	1.0198	1.0436	1.0707	1.1000	1.1318
800	1.0002	1.0009	1.0019	1.0040	1.0082	1.0122	1.0168	1.0212	1.0458	1.0731	1.1016	1.1324
900	1.0002	1.0009	1.0020	1.0041	1.0083	1.0128	1.0171	1.0221	1.0463	1.0726	1.1012	1.1303
1000	1.0002	1.0009	1.0021	1.0042	1.0084	1.0128	1.0172	1.0218	1.0460	1.0725	1.0725	1.1274

*Calculated from density-pressure-temperature data in Vukalovitch and Altunin, "Thermophysical Properties of Carbon Dioxide," Atomizdat, Moscow, 1965, and Collet's, London, 1968, translation.

TABLE 3-167 Compressibility Factors for Carbon Monoxide*

Temp., °K.	Pressure, atm. 1	4	7	10	40	70	100
200	0.9973	0.9893	0.9813	0.9734			
250	0.9989	0.9957	0.9926	0.9896	0.9632		
300	0.9997	0.9987	0.9977	0.9968	0.9907	0.9896	0.9935
350	1.0000	1.0002	1.0003	1.0005	1.0042	1.0112	1.0216
400	1.0002	1.0010	1.0017	1.0025	1.0042	1.0112	1.0216
450	1.0003	1.0014	1.0025	1.0035	1.0152	1.0285	1.0433
500	1.0004	1.0016	1.0029	1.0041	1.0172	1.0314	1.0469
600	1.0005	1.0018	1.0032	1.0045	1.0186	1.0332	1.0485
700	1.0005	1.0018	1.0032	1.0045	1.0183	1.0325	1.0470
800	1.0004	1.0017	1.0030	1.0044	1.0175	1.0309	1.0445
900	1.0004	1.0017	1.0029	1.0041	1.0166	1.0291	1.0418
1000	1.0004	1.0016	1.0027	1.0039	1.0156	1.0273	1.0391
1500	1.0003	1.0012	1.0021	1.0029	1.0115	1.0200	1.0286
2000	1.0002	1.0009	1.0016	1.0022	1.0088	1.0155	1.0221
2500	1.0002	1.0007	1.0013	1.0018	1.0071	1.0124	1.0178
3000	1.0002	1.0006	1.0010	1.0015	1.0059	1.0104	1.0148

*From Hilsenrath *et al.*, *N.B.S. Circ.* 564, 1955. Some of the above values have been rounded to four decimal places. Values at 10°K. increments below 1000°K. and at 50°K. increments for higher temperatures appear in the original, also for pressures below atmospheric.

TABLE 3-168 Compressibility Factors of Neon*

Temperature, K	Pressure, bar 1	5	10	20	40	60	80	100	200	300	400	500
50	0.9913	0.9472	0.9083	0.8013	0.3810	0.4398	0.4984	0.5850	0.9864	1.3659	1.7289	2.0794
100	0.9993	0.9970	0.9949	0.9913	0.9854	0.9245	0.9864	0.9930	1.0796	1.2197	1.3796	1.5473
150	1.0002	1.0017	1.0036	1.0078	1.0162	1.0262	1.0375	1.0497	1.1236	1.2131	1.3113	1.4150
200	1.0003	1.0023	1.0049	1.0100	1.0204	1.0318	1.0427	1.0551	1.1191	1.1909	1.2655	1.3422
250	1.0001	1.0022	1.0045	1.0097	1.0198	1.0295	1.0403	1.0502	1.1057	1.1633	1.2223	1.2822
300	1.0000	1.0020	1.0041	1.0091	1.0181	1.0277	1.0369	1.0469	1.0961	1.1476	1.1997	1.2520
400	1.0000	1.0017	1.0036	1.0074	1.0151	1.0216	1.0301	1.0376	1.0771	1.1172	1.1575	1.1981
500	1.0000	1.0014	1.0029	1.0058	1.0124	1.0188	1.0252	1.0316	1.0641	1.0963	1.1291	1.1621
600	1.0000	1.0012	1.0024	1.0049	1.0107	1.0160	1.0214	1.0267	1.0542	1.0814	1.1091	1.1369
800	1.0000	1.0009	1.0018	1.0043	1.0081	1.0123	1.0163	1.0206	1.0413	1.0622	1.0829	1.1039
1000	1.0000	1.0007	1.0014	1.0034	1.0068	1.0098	1.0132	1.0165	1.0330	1.0500	1.0670	1.0836

*Calculated from PVT values tabulated in Rabinovich (ed.), *Thermophysical Properties of Neon, Argon, Krypton and Xenon*, Standards Press, Moscow, 1976.

TABLE 3-169 Compressibility Factors of Argon*

Temperature, K	Pressure, bar 1	5	10	20	40	60	80	100	200	300	400	500
100	0.9773	0.0183	0.0366	0.0729	0.1449	0.2162	0.2867	0.3567	0.6975	1.0267	1.3470	1.6932
150	0.9932	0.9647	0.9273	0.8447	0.6101	0.2249	0.2781	0.3324	0.5934	0.8387	1.0732	1.2995
200	0.9972	0.9857	0.9713	0.9419	0.8810	0.8208	0.7624	0.7121	0.6870	0.8360	1.0051	1.1982
250	0.9988	0.9935	0.9869	0.9741	0.9494	0.9263	0.9056	0.8877	0.8590	0.9207	1.0262	1.1479
300	0.9995	0.9969	0.9941	0.9884	0.9777	0.9686	0.9611	0.9552	0.9533	0.9950	1.0673	1.1786
400	1.0001	0.9997	0.9998	0.9999	1.0004	1.0018	1.0031	1.0056	1.0280	1.0656	1.1157	1.1976
500	1.0002	1.0007	1.0012	1.0034	1.0071	1.0113	1.0154	1.0205	1.0501	1.0874	1.1301	1.1997
600	1.0003	1.0012	1.0025	1.0046	1.0094	1.0143	1.0198	1.0250	1.0553	1.0904	1.1291	1.1933
800	1.0003	1.0012	1.0023	1.0050	1.0102	1.0151	1.0205	1.0258	1.0532	1.0830	1.1147	1.1707
1000	1.0002	1.0013	1.0022	1.0050	1.0096	1.0142	1.0193	1.0239	1.0484	1.0736	1.0999	1.1497

*Calculated from PVT values tabulated in Rabinovich (ed.), *Thermophysical Properties of Neon, Argon, Krypton and Xenon*, Standard. Press, Moscow, 1976.

TABLE 3-170 Compressibility Factors of Krypton*

Temperature, K	Pressure, bar											
	1	5	10	20	40	60	80	100	200	300	400	500
150	0.9837	0.9155	0.0310	0.0618	0.1227	0.1829	0.2423	0.3012	0.5875	0.8636	1.1315	1.3932
200	0.9933	0.9648	0.9278	0.8459	0.6039	0.1870	0.2393	0.2903	0.5313	0.7568	0.9730	1.1820
250	0.9966	0.9841	0.9635	0.9265	0.8468	0.7605	0.6680	0.5810	0.5785	0.7461	0.9197	1.0891
300	0.9982	0.9899	0.9800	0.9595	0.9197	0.8807	0.8437	0.8097	0.7337	0.7954	0.9302	1.0627
350	0.9989	0.9949	0.9897	0.9793	0.9522	0.9415	0.9250	0.9110	0.8774	0.8992	0.9799	1.0664
400	0.9993	0.9967	0.9933	0.9867	0.9746	0.9635	0.9539	0.9459	0.9323	0.9570	1.0150	1.0910
450	0.9998	0.9985	0.9969	0.9939	0.9886	0.9838	0.9800	0.9774	0.9663	1.0011	1.0543	1.1142
500	0.9998	0.9992	0.9984	0.9970	0.9942	0.9921	0.9910	0.9906	1.0019	1.0311	1.0732	1.1258
600	1.0000	1.0003	1.0005	1.0012	1.0025	1.0043	1.0064	1.0091	1.0301	1.0618	1.1000	1.1431
800	1.0002	1.0010	1.0020	1.0041	1.0079	1.0122	1.0170	1.0214	1.0475	1.0779	1.1112	1.1147
1000	1.0002	1.0013	1.0023	1.0045	1.0091	1.0135	1.0184	1.0230	1.0486	1.0767	1.1063	1.1369

*Calculated from PVT values tabulated in Rabinovich (ed.), *Thermophysical Properties of Neon, Argon, Krypton and Xenon,* Standards Press, Moscow, 1976.

TABLE 3-171 Compressibility Factors of Xenon*

Temperature, K	Pressure, bar											
	1	5	10	20	40	60	80	100	200	300	400	500
200	0.9831	0.9088	0.0293	0.0584	0.1162	0.1733	0.2300	0.2861	0.5601	0.8253	1.0833	1.3356
250	0.9911	0.9545	0.9052	0.7887	0.1114	0.1642	0.2158	0.2663	0.5074	0.7355	0.9546	1.1670
300	0.9949	0.9736	0.9465	0.8885	0.7517	0.5492	0.2794	0.3016	0.5021	0.6997	0.8886	1.0707
350	0.9967	0.9834	0.9669	0.9322	0.8473	0.7840	0.7039	0.6249	0.5645	0.7124	0.8706	1.0269
400	0.9977	0.9892	0.9183	0.9562	0.9128	0.8696	0.8278	0.7888	0.6916	0.7642	0.8850	1.0148
450	0.9989	0.9928	0.9856	0.9714	0.9429	0.9163	0.8911	0.8679	0.7335	0.8331	0.9187	1.0224
500	0.9982	0.9951	0.9902	0.9810	0.9623	0.9452	0.9293	0.9156	0.8774	0.8953	0.9572	1.0412
600	0.9996	0.9979	0.9957	0.9917	0.9841	0.9772	0.9715	0.9667	0.9596	0.9791	1.0211	1.0799
800	1.0000	0.9998	1.0002	1.0004	1.0012	1.0020	1.0034	1.0054	1.0213	1.0476	1.0818	1.1222
1000	1.0000	1.0004	1.0015	1.0031	1.0144	1.0101	1.0133	1.0172	1.0394	1.0669	1.0979	1.1331

*Calculated from PVT values tabulated in Rabinovich (ed.), *Thermophysical Properties of Neon, Argon, Krypton and Xenon,* Standards Press, Moscow, 1976.

TABLE 3-172 Compressibility Factors of Methane (R50)*

Temperature, K	Pressure, bar											
	1	5	10	20	40	60	80	100	200	300	400	500
150	0.9854	0.9225	0.8275	0.0714	0.1411	0.2093	0.2763	0.3423	0.6599	0.9623	1.2537	1.5363
200	0.9936	0.9676	0.9339	0.8599	0.6784	0.3559	0.3172	0.3618	0.6141	0.8568	1.0887	1.3122
250	0.9965	0.9838	0.9680	0.9352	0.8682	0.8020	0.7386	0.6854	0.6899	0.8554	1.0359	1.2155
300	0.9983	0.9915	0.9830	0.9667	0.9343	0.9047	0.8783	0.8556	0.8280	0.9154	1.0432	1.1829
350	0.9991	0.9954	0.9911	0.9825	0.9662	0.9520	0.9401	0.9306	0.9227	0.9800	1.0723	1.1804
400	0.9995	0.9977	0.9953	0.9912	0.9835	0.9772	0.9726	0.9696	0.9779	1.0245	1.0986	1.1859
450	0.9997	0.9989	0.9979	0.9963	0.9935	0.9917	0.9911	0.9916	1.0098	1.0528	1.1152	1.1899
500	0.9999	0.9997	0.9995	0.9995	0.9996	1.0005	1.0022	1.0048	1.0285	1.0699	1.1248	1.1899
600	1.0000	1.0009	1.0020	1.0039	1.0081	1.0125	1.0171	1.0217	1.0540	1.0969	1.1470	1.2019
800	1.0003	1.0017	1.0034	1.0068	1.0130	1.0197	1.0263	1.0330	1.0678	1.1068	1.1496	1.1951
1000	1.0004	1.0014	1.0035	1.0071	1.0141	1.0207	1.0274	1.0342	1.0678	1.1033	1.1400	1.1790

*Calculated from PVT values tabulated in Goodwin, NBS Tech. Note 653, 1974, for temperatures up to 500 K, and from PVT values tabulated in Zhuravlev, *Thermophysical Properties of Gaseous and Liquid Methane,* Standartov, Moscow, 1969, and NBS-NSF transl. TT 70-50097, 1970.

TABLE 3-173 Compressibility Factors for Steam*

Pressure, lb./sq. in. abs.	Temp., °F.																		
	400	600	800	1000	1200	1400	1600	1800	2000	2200	2400	2600	2800	3000	3200	3400	3600	3800	4000
10	0.9965	0.9989	0.9992	0.9995	0.9999	0.9999	0.9999	1.0000	1.0000	1.0000	1.0001	1.0006	1.0012	1.0024	1.0053	1.0084	1.0145	1.0211	1.0332
15	0.9943	0.9972	0.9986	0.9993	0.9997	0.9998	0.9999	0.9999	1.0000	1.0000	1.0001	1.0004	1.0012	1.0022	1.0042	1.0072	1.0124	1.0188	1.0295
20	0.9930	0.9970	0.9981	0.9991	0.9995	0.9996	0.9998	0.9999	1.0000	1.0000	1.0001	1.0003	1.0011	1.0020	1.0036	1.0065	1.0112	1.0173	1.0269
40	0.9861	0.9940	0.9967	0.9981	0.9990	0.9994	0.9996	0.9998	0.9999	0.9999	1.0001	1.0003	1.0010	1.0018	1.0028	1.0054	1.0090	1.0139	1.0214
60	0.9788	0.9910	0.9951	0.9973	0.9984	0.9991	0.9994	0.9997	0.9999	0.9999	1.0001	1.0003	1.0009	1.0018	1.0024	1.0048	1.0080	1.0120	1.0186
80	0.9714	0.9878	0.9935	0.9963	0.9979	0.9987	0.9992	0.9996	0.9998	0.9999	1.0001	1.0003	1.0008	1.0016	1.0023	1.0044	1.0073	1.0108	1.0170
100	0.9469	0.9848	0.9919	0.9954	0.9974	0.9985	0.9990	0.9995	0.9998	0.9999	1.0001	1.0004	1.0007	1.0015	1.0022	1.0042	1.0067	1.0099	1.0157
150	0.9435	0.9770	0.9879	0.9931	0.9960	9.9976	0.9985	0.9993	0.9997	0.9998	1.0001	1.0004	1.0006	1.0014	1.0021	1.0039	1.0059	1.0087	1.0137
200	0.9216	0.9690	0.9839	0.9908	0.9947	0.9968	0.9980	0.9991	0.9996	0.9998	1.0001	1.0005	1.0007	1.0015	1.0021	1.0037	1.0055	1.0080	1.0126
400		0.9356	0.9675	0.9817	0.9893	0.9935	0.9960	0.9982	0.9992	0.9998	1.0002	1.0007	1.0011	1.0017	1.0023	1.0033	1.0049	1.0070	1.0105
600		0.8989	0.9509	0.9725	0.9839	0.9904	0.9942	0.9973	0.9988	0.9997	1.0002	1.0008	1.0014	1.0019	1.0026	1.0034	1.0048	1.0066	1.0097
800		0.8586	0.9336	0.9633	0.9790	0.9872	0.9925	0.9964	0.9985	0.9996	1.0003	1.0010	1.0016	1.0022	1.0029	1.0036	1.0049	1.0065	1.0094
1,000		0.8138	0.9162	0.9540	0.9733	0.9841	0.9905	0.9955	0.9981	0.9994	1.0004	1.0012	1.0019	1.0025	1.0032	1.0039	1.0052	1.0066	1.0092
1,500		0.6702	0.8695	0.9305	0.9600	0.9764	0.9859	0.9932	0.9971	0.9992	1.0007	1.0017	1.0026	1.0033	1.0040	1.0048	1.0059	1.0072	1.0096
2,000			0.8188	0.9067	0.9468	0.9687	0.9813	0.9900	0.9958	0.9990	1.0010	1.0023	1.0034	1.0042	1.0049	1.0058	1.0068	1.0082	1.0104
4,000			0.5608	0.8060	0.8942	0.9392	0.9647	0.9836	0.9930	0.9989	1.0024	1.0050	1.0069	1.0082	1.0093	1.0106	1.0118	1.0132	1.0149
6,000				0.7042	0.8442	0.9121	0.9497	0.9771	0.9907	0.9991	1.0048	1.0081	1.0110	1.0128	1.0139	1.0152	1.0165	1.0179	1.0195
8,000				0.6185	0.8003	0.8883	0.9371	0.9714	0.9895	1.0004	1.0075	1.0118	1.0152	1.0172	1.0188	1.0204	1.0216	1.0229	1.0242
10,000				0.5699	0.7657	0.8693	0.9274	0.9668	0.9890	1.0025	1.0105	1.0158	1.0196	1.0220	1.0240	1.0258	1.0271	1.0284	1.0298

*Calculated by P. E. Liley from various steam tables for the lower temperatures and from Paper B-11 by P. H. Kesselman and Yu. I. Blank, 7th. Int. Conf. Properties of Steam, Tokyo, 1968, for the higher temperatures.

TABLE 3-174 Compressibility Factors of Water Substance*

Temperature, K	Pressure, bar																			
	1	5	10	15	20	25	30	40	50	60	80	100	150	200	250	300	400	500	600	800	1000
400	0.990	0.003	0.006	0.009	0.012	0.014	0.017	0.023	0.029	0.035	0.046	0.058	0.086	0.114	0.143	0.171	0.227	0.282	0.336	0.445	0.552
450	0.993	0.003	0.006	0.009	0.012	0.014	0.016	0.022	0.027	0.033	0.043	0.054	0.080	0.107	0.134	0.159	0.206	0.255	0.304	0.402	0.498
500	0.996	0.980	0.958	0.930	0.901	0.878	0.016	0.021	0.026	0.031	0.042	0.052	0.077	0.102	0.127	0.152	0.201	0.249	0.297	0.390	0.482
550	0.997	0.985	0.969	0.956	0.939	0.922	0.904	0.865	0.822	0.773	0.042	0.052	0.077	0.102	0.126	0.150	0.181	0.198	0.289	0.378	0.464
600	0.998	0.990	0.979	0.970	0.961	0.948	0.935	0.910	0.885	0.858	0.798	0.726	0.082	0.107	0.131	0.155	0.201	0.246	0.290	0.375	0.457
650	0.999	0.992	0.984	0.977	0.968	0.959	0.958	0.937	0.919	0.902	0.864	0.824	0.702	0.514	0.177	0.183	0.221	0.260	0.303	0.383	0.460
700	1.000	0.994	0.988	0.984	0.976	0.967	0.966	0.952	0.941	0.929	0.900	0.876	0.800	0.716	0.618	0.503	0.326	0.316	0.340	0.406	0.476
750	1.000	0.996	0.991	0.988	0.981	0.975	0.971	0.961	0.955	0.945	0.927	0.907	0.856	0.801	0.743	0.682	0.557	0.465	0.435	0.456	0.509
800	1.000	0.997	0.993	0.991	0.985	0.982	0.976	0.970	0.966	0.957	0.945	0.929	0.892	0.853	0.813	0.773	0.693	0.620	0.568	0.538	0.561
850	1.000	0.997	0.995	0.992	0.989	0.984	0.981	0.977	0.973	0.967	0.957	0.946	0.917	0.889	0.860	0.831	0.775	0.715	0.679	0.631	0.629
900	1.000	0.998	0.997	0.993	0.992	0.989	0.986	0.982	0.979	0.974	0.965	0.958	0.936	0.915	0.893	0.872	0.830	0.792	0.760	0.714	0.700
950	1.000	0.998	0.997	0.994	0.994	0.993	0.991	0.985	0.983	0.980	0.973	0.967	0.950	0.933	0.916	0.901	0.867	0.839	0.816	0.780	0.761
1000	1.000	0.999	0.998	0.995	0.995	0.994	0.993	0.990	0.987	0.985	0.978	0.973	0.960	0.948	0.935	0.923	0.900	0.878	0.859	0.831	0.816
1200	1.000	1.000	0.999	0.998	0.998	0.997	0.997	0.995	0.994	0.994	0.992	0.990	0.986	0.982	0.975	0.968	0.961	0.957	0.949	0.942	0.937
1400	1.000	1.000	1.000	1.000	1.000	1.000	1.000	0.999	0.998	0.998	0.998	0.997	0.996	0.995	0.995	0.994	0.993	0.992	0.994	0.996	0.998
1600	1.000	1.000	1.000	1.000	1.000	1.000	1.000	1.000	1.000	1.000	1.000	1.000	1.001	1.002	1.002	1.004	1.006	1.009	1.012	1.015	1.020
1800	1.001	1.001	1.001	1.000	1.000	1.000	1.000	1.000	1.000	1.001	1.002	1.003	1.003	1.004	1.005	1.008	1.011	1.014	1.017	1.021	1.031
2000	1.003	1.002	1.002	1.002	1.002	1.002	1.002	1.002	1.002	1.003	1.003	1.004	1.004	1.006	1.008	1.011	1.014	1.018	1.021	1.032	1.043

*Calculated by P. E. Liley from various steam tables for the lower temperatures and from Pap. B-11 by P. H. Kesselman and Yu. I. Blank, 7th Internal Conference on the Properties of Steam, Tokyo, 1968, for the higher temperatures.

TABLE 3-175 Compressibilities of Liquids*

At the constant temperature t, the compressibility $\beta = (1/\bar{V}_0)(dV/dP)$. In general as P increases, β decreases rapidly at first and then slowly; the change of β with t is large at low pressures but very small at pressures above 1000 to 2000 megabars. 1 megabar = 0.987 atm. = 10^6 dynes/sq. cm. based upon the older usage, 1 bar = 1 dyne/sq. cm. The use of "bar" as a pressure unit is not encouraged.

Substance	Temp., °C.	Pressure, megabars	Compressibility per megabar $\beta \times 10^6$	Substance	Temp., °C.	Pressure, megabars	Compressibility per megabar $\beta \times 10^6$	Substance	Temp., °C.	Pressure, megabars	Compressibility per megabar $\beta \times 10^6$
Acetone	14	23	111	Ethyl acetate	20	400	75	Methyl alcohol	15	23	103
Acetone	20	500	61	alcohol	14	23	100	alcohol	20	200	95
Acetone	20	1,000	52	alcohol	20	500	63	alcohol	20	400	80
Acetone	40	12,000	9	alcohol	20	1,000	54	alcohol	20	500	65
Amyl alcohol	14	23	88	alcohol	20	12,000	8	alcohol	20	1,000	54
alcohol, iso	20	200	84	bromide	20	200	100	alcohol	20	12,000	8
alcohol, iso	20	400	70	bromide	20	400	82	Nitric acid	0	17	32
alcohol, n	20	500	61	bromide	20	500	70	Oils:			
alcohol, n	20	1,000	46	bromide	20	1,000	54	Almond	15	5	53
alcohol, n	20	12,000	8	bromide	20	12,000	8	Castor	15	5	46
alcohol, n	40	12,000	8	chloride	15	23	151	Linseed	15	5	51
Benzene	17	5	89	chloride	20	500	102	Olive	15	5	55
Benzene	20	200	77	chloride	20	1,000	66	Rapeseed	20	.....	59
Benzene	20	400	67	chloride	20	12,000	8	Phosphorus trichloride	10	250	71
Bromine	20	200	56	ether	25	23	188	trichloride	20	500	63
Bromine	20	400	51	ether	20	500	84	trichloride	20	1,000	47
Butyl alcohol, iso	18	8	97	ether	20	1,000	61	trichloride	20	12,000	8
alcohol, iso	20	200	81	ether	20	12,000	10	Propyl alcohol (n)	20	200	77
alcohol, iso	20	400	64	iodide	20	200	81	alcohol (n)	20	400	67
alcohol, iso	20	500	56	iodide	20	400	69	alcohol (n?)	20	500	65
alcohol, iso	20	1,000	46	iodide	20	500	64	alcohol (n?)	20	1,000	47
alcohol, iso	20	12,000	8	iodide	20	1,000	50	alcohol (n?)	20	12,000	7
Carbon bisulfide	16	21	86	iodide	20	12,000	8	Toluene	20	200	74
bisulfide	20	500	57	Gallium	30	300	3.97	Toluene	20	400	64
bisulfide	20	1,000	48	Glycerol	15	5	22	Turpentine	20	.....	74
bisulfide	20	12,000	6	Hexane	20	200	117	Water	20	13	49
tetrachloride	20	200	86	Hexane	20	400	91	Water	20	200	43
tetrachloride	20	400	73	Kerosene	20	500	55	Water	20	400	41
Chloroform	20	200	83	Kerosene	20	1,000	45	Water	20	500	39
Chloroform	20	400	70	Kerosene	20	12,000	8	Water	40	500	38
Dichloroethylsulfide	32	1,000	34	Mercury	20	300	3.95	Water	40	1,000	33
Dichloroethylsulfide	32	2,000	24	Mercury	22	500	3.97	Water	40	12,000	9
Ethyl acetate	13	23	103	Mercury	22	1,000	3.91	Xylene, meta	20	200	69
acetate	20	200	90	Mercury	22	12,000	2.37	meta	20	400	60

*"Smithsonian Tables," Table 106.

Scott ("Cryogenic Engineering," Van Nostrand, Princeton, N.J., 1959) gives data for liquid nitrogen (p. 283), oxygen (p. 276), and hydrogen (p. 303). For a convenient index to the high-pressure work of Bridgman see "American Institute of Physics Handbook," p. 2-163, McGraw-Hill, New York, 1957.

TABLE 3-176 Compressibilities of Solids

Many data on the compressibility of solids obtained prior to 1926 are contained in Gruneisen, *Handbuch der Physik*, vol. 10, Springer, Berlin, 1926, pp. 1–52; also available as translation, NASA RE 2-18-59W, 1959. See also Tables 271, 273, 276, 278, and other material in *Smithsonian Physical Tables*, 9th ed., 1954. For a review of high-pressure work to 1946 see Bridgman, *Rev. Mod. Phys.*, 18, 1 (1946).

LATENT HEATS

UNITS CONVERSIONS

For this subsection, the following units conversions are applicable:

°F = ⁹⁄₅ °C + 32.

To convert calories per gram-mole to British thermal units per pound-mole, multiply by 1.799; to convert calories per gram to British thermal units per pound, multiply by 1.799.

To convert millimeters of mercury to pounds-force per square inch, multiply by 1.934×10^{-2}.

TABLE 3-177 Heats of Fusion and Vaporization of the Elements and Inorganic Compounds*

Unless stated otherwise, the values have been taken from the compilations by K. K. Kelley on "Heats of Fusion of Inorganic Compounds," U.S. Bur. Mines Bull. 393 (1936), and "The Free Energies of Vaporization and Vapor Pressures of Inorganic Substances," U.S. Bur. Mines Bull. 383 (1935).

Substance	M.p., °C.	Heat of fusion,[a,b] cal./mole	B.p. at 1 atm., °C.	Heat of vaporization,[a,b] cal./mole
Aluminum:				
Al	660.0	2,550	2057	61,020
Al_2Br_6	97.5	5,420	256.4	10,920
Al_2Cl_6	192.5	16,960	180.2[c]	26,750[c]
$AlF_3.3NaF$	1000	16,380		
Al_2I_6	191.0	7,960	385.5	15,360
Al_2O_3	2045	(26,000)	3000	
Antimony:				
Sb	630.5	4,770	1440	46,670
$SbBr_3$	97	3,510		
$SbCl_3$	73.4	3,030	219	10,360
$SbCl_5$	4	2,400	172[d]	11,570
Sb_4O_6	655	(27,000)	1425	17,820
Sb_4S_6	546	11,200		
Argon:				
A	−189.3	290	−185.8	1,590
Arsenic:				
As	814	(6,620)	610[c]	31,000[c]
$AsBr_3$	31	2,810		
$AsCl_3$	− 16	2,420	122	7,570
AsF_5	− 80.7	2,800	− 52.8	4,980
As_4O_6	313	8,000	457.2	14,300
Barium:				
Ba	704	(1,400)[e]	1638	35,670
$BaBr_2$	847	6,000		
$BaCl_2$	960	5,370		
BaF_2	1287	3,000		
$Ba(NO_3)_2$	595	(5,980)		
$Ba_3(PO_4)_2$	1730	18,600		
$BaSO_4$	1350	9,700		
Beryllium:				
Be	1280	2,500[e]		
Bismuth:				
Bi	271.3	2,505	1420	
$BiBr_3$	....	....	461	18,020
$BiCl_3$	224	2,600	441	17,350
Bi_2O_3	817	6,800		
Bi_2S_5	747	8,900		
Boron:				
BBr_3	....	....	91.3	7,300
BCl_3	....	....	12.5	5,680
BF_3	−128	480	−100.9	4,620
B_2H_6	−165.5	....	− 92.4	3,685
B_3H_{10}	−119.8	....	16	6,470
B_5H_9	− 46.9	....	58	7,700
B_5H_{11}	....	....	67	8,500
$B_{10}H_{14}$	99.7	7,800	f	11,600
B_2H_5Br	−104	....	16	6,230
$B_3N_3H_6$	− 58	....	50.4	7,670
Bromine:				
Br_2	− 7.2	2,580	58.0	7,420
BrF_5	− 61.3	1,355	40.4	7,470
Cadmium:				
Cd	320.9	1,460	765	23,870
$CdBr_2$	568	(5,000)		
$CdCl_2$	568	5,300	967	29,860
CdF_2	1110	(5,400)		
CdI_2	387	3,660	796	25,400
CdO	....	....	1559[c]	53,820[c]
$CdSO_4$	1000	4,790		
Calcium:				
Ca	851	2,230	1487	36,580
$CaBr_2$	730	4,180		
$CaCO_3$	1282	(12,700)		
$CaCl_2$	782	6,100		
CaF_2	1392	4,100		
$Ca(NO_3)_2$	561	5,120		
CaO	2707	(12,240)		
$CaO.Al_2O_3.2SiO_2$	1550	29,400		
$CaO.MgO.2SiO_2$	1392	(18,200)		
$CaO.SiO_2$	1512	13,400		
$CaSO_4$	1297	6,700		
Carbon:				
C (graphite)	3600	11,000[e]		
CBr_4	90	1,050		
CCl_4	− 24.0	644	77	7,280
CF_4	....	....	−127.9	3,110
CH_4	−182.5	224	−161.4	2,040
C_2N_2	− 27.8	1,938[u]	− 21.1	5,576[u]
CNBr	52	....	....	11,010[c]
CNCl	− 5	2,240	13	6,300
Carbon (*Cont.*):				
CNF	....	....	− 72.8	5,780[c]
CNI	....	....	141	13,980[c]
CO	−205.0	200	−191.5	1,444
CO_2	− 57.5	1,900	− 78.4[c]	6,030[c,?]
COS	−138.8	1,129[k]	− 50.2	4,423[k]
$COCl_2$	....	....	8.0	5,990
CS_2	−112.0	1,049[l]		
Cerium:				
Ce	775	2,120		
Cesium:				
Cs	28.4	500	690	16,320
CsBr	....	....	1300	35,990
CsCl	642	3,600	1300	35,690
CsF	715	(2,450)	1251	34,330
CsI	....	....	1280	35,930
$CsNO_3$	407	3,250		
Chlorine:				
Cl_2	−101.0	1,531[m]	− 34.1	4,878[m]
ClF	....	....	−101	
ClF_3	....	....	11.3	5,890
Cl_2O	....	....	2.0	6,280
ClO_2	....	....	10.9	7,100
Cl_2O_7	....	....	79	8,480
Chromium:				
Cr	1550	3,930	2475	
CrO_2Cl_2	....	....	117	8,250
Cobalt:				
Co	1490	3,660		
$CoCl_2$	727	7,390	1050	27,170
Copper:				
Cu	1083.0	3,110	2595	72,810
Cu_2Br_2	....	....	1355	16,310
Cu_2Cl_2	430	4,890	1490	11,920
CuI	....	....	1336	15,940
$Cu_2(CN)_2$	473	(5,400)		
Cu_2O	1230	(13,400)		
CuO	1447	2,820		
Cu_2S	1127	5,500		
Fluorine:				
F_2	−223	....	−188.2	1,640
F_2O	....	....	−144.8	2,650
Gallium:				
Ga	29.8	1,336	2071	
Germanium:				
Ge	959	(8,300)		
GeH_4	−165	....	− 89.1	3,580
Ge_2H_6	−109	....	31.4	5,900
Ge_3H_8	−105.6	....	110.6	7,550
$GeHCl_3$	− 71	....	75[g]	8,000
$GeBr_4$	26.1	....	189	8,560
$GeCl_4$	− 49.5	....	84	7,030
$Ge(CH_3)_4$	− 88	....	44	6,460
Gold:				
Au	1063.0	3,030	2966	81,800
Helium:				
He	−271.4	....	−268.4	22
Hydrogen:				
H_2	−259.2	28	−252.7	216
HBr	− 86.9	575	− 66.7	4,210
HCl	−114.2	476	− 85.0	3,860
HCN	− 13.2	2,009[i]	25.7	6,027[i]
HF	− 83.0	1,094	33.3	7,460
$(HF)_6$	....	....	51.2	5,020
HI	− 50.8	686		
H_2O	0.0	1,436	100.0	9,729[h,q]
$H^2{}_2O (= D_2O)$	3.8	1,501[s]	101.4	9,945[r,q]
H_2O_2	− 2	2,520[c]	158	10,270
HNO_3	− 47	600		
H_3PO_2	17.4	2,310		
H_3PO_3	74	3,070		
H_3PO_4	42.4	2,520		
$H_4P_2O_6$	55	8,300		
H_2S	− 85.5	568[t]	− 60.3	4,463[t]
H_2S_2	− 87.6	1,805		
H_2SO_4	10.5	2,360		
H_2Se	....	....	− 41.3	4,880
H_2SeO_4	58	3,450		
H_2Te	− 48.9	1,670	− 2.2	5,650
Indium:				
In	156.4	781		

*See also subsection "Thermodynamic Properties."

TABLE 3-177 Heats of Fusion and Vaporization of the Elements and Inorganic Compounds *(Continued)*

Substance	M.p., °C.	Heat of fusion,[a,b] cal./mole	B.p. at 1 atm., °C.	Heat of vaporization,[a,b] cal./mole
Iodine:				
I_2	113.0	3,650	183	10,390
$ICl(\alpha)$	17.2	2,660		
$ICl(\beta)$	13.9	2,270		
IF_7	......	......	4[c]	7,460[c]
Iron:				
Fe	1530	3,560	2735	84,600
$FeCl_2$	677	7,800	1026	30,210
Fe_2Cl_6	304	20,590	319	12,040
$Fe(CO)_5$	− 21	3,250	105	9,000
FeO	1380	(7,700)		
FeS	1195	5,000		
Krypton:				
Kr	−157	360[e]	152.9	2,310[e]
Lead:				
Pb	327.4	1,224	1744	42,060
$PbBr_2$	488	4,290	914	27,700
$PbCl_2$	498	5,650	954	29,600
PbF_2	824	1,860	1293	38,300
PbI_2	412	5,970	872	24,850
$PbMoO_4$	1065	(25,800)		
PbO	890	2,820	1472	51,310
PbS	1114	4,150	1281	(50,000)
$PbSO_4$	1087	9,600		
$PbWO_4$	1123	(15,200)		
Lithium:				
Li	179	1,100	1372	32,250
$LiBO_2$	845	(5,570)		
LiBr	552	2,900	1310	35,420
LiCl	614	3,200	1382	35,960
LiF	847	(2,360)	1681	50,970
LiI	440	(1,420)	1171	40,770
LiOH	462	2,480		
Li_2MoO_4	705	4,200		
$LiNO_3$				
Li_2SiO_3	1177	7,210		
Li_4SiO_4	1249	7,430		
Li_2SO_4	857	3,040		
Li_2WO_4	742	(6,700)		
Magnesium:				
Mg	650	2,160	1107	32,520
$MgBr_2$	711	8,300		
$MgCl_2$	712	8,100	1418	32,690
MgF_2	1221	5,900		
MgO	2642	18,500		
$Mg_3(PO_4)_2$	1184	(11,300)		
$MgSiO_3$	1524	14,700		
$MgSO_4$	1127	3,500		
$MgZn_2$	589	(8,270)		
Manganese:				
Mn	1220	3,450	2152	55,150
$MnCl_2$	650	7,340	1190	29,630
$MnSiO_3$	1274	(8,200)		
$MnTiO_3$	1404	(7,960)		
Mercury:				
Hg	− 38.9	557	361	13,980
$HgBr_2$	241	3,960	319	14,080
$HgCl_2$	277	4,150	304	14,080
HgI_2	250	4,500	354	14,260
$HgSO_4$	850	(1,440)		
Molybdenum:				
Mo	2622	(6,660)	(4800)	(128,000)
MoF_6	17	2,500	36	6,000
MoO_3	745	(2,500)	1151	
Neon:				
Ne	−248.5	77	−246.0	440[e]
Nickel:				
Ni	1455	4,200	2730	87,300
$NiCl_2$	......	......	987[c]	48,360[c]
$Ni(CO)_4$	......	......	42.5	7,000
Ni_2S	645	(2,980)		
Ni_3S_2	790	5,800		
Nitrogen:				
N_2	−210.0	172	−195.8	1,336
NF_3	......	......	−129.0	3,000
NH_3	− 77.7	1,352[n]	− 33.4	5,581[n]
NH_4CNS	146	(4,700)		
NH_4NO_3	169.6	1,460		
N_2O	− 90.8	1,563	− 88.5	3,950
NO	−163.6	550	−151.7	3,307
N_2O_4	− 13	5,540	30	7,040
N_2O_5	......	......	32.4	13,800[c]
NOCl	......	......	− 6.4	6,140
Osmium:				
OsF_8	......	......	47.4	6,840
OsO_4 (yellow)	56	4,060	130	9,450
OsO_4 (white)	42	2,340		
Oxygen				
O_2	−218.9	106	−183.0	1,629
O_3	......	......	−111	2,880
Palladium:				
Pd	1554	4,120		
Phosphorus:				
P_4 (yellow)	44.2	615	280	12,520
P_4 (violet)	......	......	417[c]	25,600[c]
P_4 (black)	......	......	453[c]	33,100
PCl_3	......	......	74.2	7,280
PH_3	−133.8	270[c]	− 87.7	3,489[c]
P_4O_6	23.8	3,360	174	10,380
$P_4O_{10}(\alpha)$	569	17,080	591	20,670
$P_4O_{10}(\beta)$	......	......	358[c]	
$POCl_3$	1.1	3,110	105.1	8,380
P_2S_3	......	......	508	
Platinum:				
Pt	1773.5	4,700	(4400)	(107,000)
Potassium:				
K	63.5	574	776	18,920
KBO_2	947	(5,700)		
KBr	742	5,000	1383	37,060
KCl	770	6,410	1407	38,840
KCN	623	(3,500)		
KCNS	179	2,250		
K_2CO_3	897	7,800		
K_2CrO_4	984	6,920		
$K_2Cr_2O_7$	398	8,770		
KF	857	6,500		
KI	682	4,100	1324	34,690
K_2MoO_4	922	(4,000)		
KNO_3	338	2,840		
KOH	360	(2,000)	1327	30,850
KPO_3	817	2,110		
K_3PO_4	1340	8,900		
$K_4P_2O_7$	1092	14,000		
K_2SO_4	1074	8,100		
K_2TiO_3	810	(10,600)		
K_2WO_4	927	(4,400)		
Praseodymium:				
Pr	932	2,700		
Radon:				
Rn	−71	......	− 61.8	4,010
Rhenium:				
Re	(3000)			
Re_2O_7	296	15,340	362.4	18,060
Re_2O_8	147	3,800		
Rubidium:				
Rb	39.1	525	679	18,110
RbBr	677	3,700	1352	37,120
RbCl	717	4,400	1381	36,920
RbF	833	4,130	1408	39,510
RbI	638	2,990	1304	35,960
$RbNO_3$	305	1,340		
Selenium:				
Se_2	217	1,220	753	25,490
Se_6	......	......	736	20,600
SeF_6	......	......	− 45.8[c]	6,350[c]
SeO_2	......	......	317[c]	20,900
$SeOCl_2$	10	1,010	168	
Silicon:				
Si	1427	9,470	2290	
$SiCl_4$	− 67.6	1,845	56.8	6,860
Si_2Cl_6	− 1	......	139	
Si_3Cl_8	......	......	211.4	12,340
$(SiCl_3)_2O$	− 33	......	135.6	8,820
SiF_4	......	......	− 94.8[c]	6,130[c]
Si_2F_6	− 18.5	3,900	− 18.9[c]	10,400[c]
SiF_3Cl	−138	......	− 70.1	4,460
SiF_2Cl_2	−144	......	− 31.5	5,080
SiH_4	−185	......	−111.6	2,960
Si_2H_6	−132.5	......	− 14.3	5,110
Si_3H_8	−117	......	53.1	6,780
Si_4H_{10}	− 93.5	......	100	8,890
SiH_3Br	− 93.8	......	2.4	5,650
SiH_2Br_2	− 70.0	......	70.5	6,840
$SiHCl_3$	−126.5	......	31.8	6,360
$(SiH_3)_3N$	−105.6	......	48.7	6,850
$(SiH_3)_2O$	−144	......	− 15.4	5,350
SiO_2 (quartz)	1470	3,400	2230	
SiO_2 (cristobalite)	1700	2,100		
Silver:				
Ag	960.5	2,700	2212	60,720
AgBr	430	2,180		
AgCl	455	3,155	1564	42,520
AgCN	350	2,750		
AgI	557	2,250	1506	34,450
$AgNO_3$	209	2,755		
Ag_2S	842	3,360		
Ag_2SO_4	657	(4,300)		
Sodium:				
Na	97.7	630	914	23,120
$NaBO_2$	966	8,660		

TABLE 3-177 Heats of Fusion and Vaporization of the Elements and Inorganic Compounds *(Concluded)*

Substance	M.p., °C.	Heat of fusion,[a,b] cal./mole	B.p. at 1 atm., °C.	Heat of vaporization,[a,b] cal./mole
Sodium (*Cont.*):				
NaBr	747	6,140	1392	37,950
NaCl	800	7,220	1465	40,810
$NaClO_3$	255	5,290		
NaCN	562	(4,400)	1500	37,280
NaCNS	323	4,450		
Na_2CO_3	854	7,000		
NaF	992	7,000	1704	53,260
NaI	662	5,240		
Na_2MoO_4	687	3,600		
$NaNO_3$	310	3,760		
NaOH	322	2,000	1378	
$\frac{1}{2}Na_2O.\frac{1}{2}Al_2O_3.3SiO_2$	1107	13,150		
$NaPO_3$	988	(5,000)		
$Na_4P_2O_7$	970	(13,700)		
Na_2S	920	(1,200)		
Na_2SiO_3	1087	10,300		
$Na_2Si_2O_5$	884	8,460		
Na_2SO_4	884	5,830		
Na_2WO_4	702	5,800		
Strontium:				
Sr	757	2,190	1384	33,610
$SrBr_2$	643	4,780		
$SrCl_2$	872	4,100		
SrF_2	1400	4,260		
$Sr_3(PO_4)_2$	1770	18,500		
Sulfur:				
S (rhombic)	112.8	....	444.6	2,200
S (monoclinic)	119.2			
S_2Cl_2	....	....	138	8,720
SF_6	....	....	− 63.5[c]	5,600[c]
SO_2	− 75.5	1,769[p]	− 5.0	5,960[p]
$SO_3(\alpha)$	17	2,060	44.8	10,190
$SO_3(\beta)$	32.4	2,890		
$SO_3(\gamma)$	62.2	6,310		
$SOBr_2$	....	....	139.5	9,920
$SOCl_2$	....	....	75.4	7,600
SO_2Cl_2	....	....	69.2	7,760
Tellurium:				
Te	453	3,230	1090	
$TeCl_4$	....	....	392	16,830
TeF_6	....	....	− 38.6[c]	6,700[c]
Thallium:				
Tl	302.5	1,030	1457	38,810
TlBr	460	5,990	819	23,800
TlCl	427	4,260	807	24,420
Tl_2CO_3	273	4,400		
TlI	440	3,125	823	25,030
$TlNO_3$	207	2,290		
Tl_2S	449	3,000		
Tl_2SO_4	632	5,500		
Tin:				
Sn_4	231.8	1,720	2270	68,000
$SnBr_2$	232	(1,700)		
$SnBr_4$	30	3,000		
$SnCl_2$	247	3,050	623	20,740
$SnCl_4$	− 33.2	2,190	113	8,330
$Sn(CH_3)_4$	....	....	78.3	7,320
SnH_4	−149.8	....	− 52:3	4,420
SnI_4	143.5	(4,300)		
Titanium:				
$TiBr_4$	38.2	(2,060)		
$TiCl_4$	− 23	2,240	136	8,350
TiO_2	1825	(11,400)		
Tungsten:				
W	3390	(8,400)	(5900)	(176,000)
WF_6	− 0.4	1,800	17.3	6,350
Uranium:				
UF_6	....	....	55.1[c]	9,990[c]
Xenon:				
Xe	−111.5	740	−108.0	3,110
Zinc:				
Zn	419.5	1,595	907	27,430
$ZnCl_2$	283	(5,500)	732	28,710
$Zn(C_2H_5)_2$	....	....	118	8,960
ZnO	1975	4,470		
ZnS	1645	(9,000)		
Zirconium:				
$ZrBr_4$	....	....	357[c]	25,800[c]
$ZrCl_4$	....	....	311[c]	25,290[c]
ZrI_4	....	....	431[c]	29,030[c]
ZrO_2	2715	20,800		

[a] Values in parentheses are uncertain.
[b] For the freezing point or the normal boiling point unless otherwise stated.
[c] Sublimation.
[d] Decomposes at about 75°C.; value obtained by extrapolation.
[e] Bichowsky and Rossini, "Thermochemistry of the Chemical Substances," Reinhold, New York (1936).
[f] Decomposes before the normal boiling point is reached.
[g] Decomposes at about 40°C.; value obtained by extrapolation.
[h] See also pp. 3-206ff. on steam table.
[i] Giauque and Ruehrwein, *J. Am. Chem. Soc.*, **61**, 2626 (1939).
[j] Giauque and Egan, *J. Chem. Phys.*, **5**, 45 (1937).
[k] Kemp and Giauque, *J. Am. Chem. Soc.*, **59**, 79 (1937).
[l] Brown and Manov, *J. Am. Chem. Soc.*, **59**, 500 (1937).
[m] Giauque and Powell, *J. Am. Chem. Soc.* **61**, 1970 (1939).
[n] Overstreet and Giauque, *J. Am. Chem. Soc.*, **59**, 254 (1937).
[o] Stephenson and Giauque, *J. Chem. Phys.*, **5**, 149 (1937).
[p] Giauque and Stephenson, *J. Am. Chem. Soc.*, **60**, 1389 (1938).
[q] Osborne, Stimson, and Ginnings, *Bur. Standards J. Research*, **23**, 197, 261 (1939).
[r] Miles and Menzies, *J. Am. Chem. Soc.*, **58**, 1067 (1936).
[s] Long and Kemp, *J. Am. Chem. Soc.*, **58**, 1829 (1936).
[t] Giauque and Blue, *J. Am. Chem. Soc.*, **58**, 831 (1936).
[u] Ruehrwein and Giauque, *J. Am. Chem. Soc.*, **61**, 2940 (1939).

TABLE 3-178 Heats of Fusion of Organic Compounds

The values for the hydrocarbons are from the tables of the American Petroleum Institute Research Project 44 at the National Bureau of Standards, with some from Parks and Huffman, *Ind. Eng. Chem.*, 23, 1138 (1931).

The values for the nonhydrocarbon compounds were recalculated from data in *International Critical Tables*, vol. 5.

Hydrocarbon compounds	Formula	M.p., °C.	Heat of fusion, cal./g.
Paraffins:			
Methane	CH_4	−182.48	14.03
Ethane	C_2H_6	−183.23	22.712
Propane	C_3H_8	−187.65	19.100
n-Butane	C_4H_{10}	−138.33	19.167
2-Methylpropane	C_4H_{10}	−159.60	18.668
n-Pentane	C_5H_{12}	−129.723	27.874
2-Methylbutane	C_5H_{12}	−159.890	17.076
2,2-Dimethylpropane	C_5H_{12}	− 16.6	10.786
n-Hexane	C_6H_{14}	− 95.320	36.138
2-Methylpentane	C_6H_{14}	−153.680	17.407
2,2-Dimethylbutane	C_6H_{14}	− 99.73	1.607
2,3-Dimethylbutane	C_6H_{14}	−128.41	2.251
n-Heptane	C_7H_{16}	− 90.595	33.513
2-Methylhexane	C_7H_{16}	−118.270	21.158
3-Ethylpentane	C_7H_{16}	−118.593	22.555
2,2-Dimethylpentane	C_7H_{16}	−123.790	13.982
2,4-Dimethylpentane	C_7H_{16}	−119.230	15.968
3,3-Dimethylpentane	C_7H_{16}	−134.46	16.856
2,2,3-Trimethylbutane	C_7H_{16}	− 24.96	5.250
n-Octane	C_8H_{18}	− 56.798	43.169
2-Methylheptane	C_8H_{18}	−109.04	21.458
3-Methylpentane	C_8H_{18}	−120.50	23.795
4-Methylheptane	C_8H_{18}	−120.955	22.692
2,2-Dimethylhexane	C_8H_{18}	−121.18	24.226
2,5-Dimethylhexane	C_8H_{18}	− 91.200	26.903
3,3-Dimethylhexane	C_8H_{18}	−126.10	14.9
2-Methyl-3-ethylpentane	C_8H_{18}	−114.960	23.690
3-Methyl-3-ethylpentane	C_8H_{18}	− 90.870	22.657
2,2,3-Trimethylpentane	C_8H_{18}	−112.27	18.061
2,2,4-Trimethylpentane	C_8H_{18}	−107.365	19.278
2,3,3-Trimethylpentane	C_8H_{18}	−100.70	3.204
2,3,4-Trimethylpentane	C_8H_{18}	−109.210	19.392
2,2,3,3-Tetramethylbutane	C_8H_{18}	+100.69	14.900
n-Nonane	C_9H_{20}	− 53.9	41.2
n-Decane	$C_{10}H_{22}$	− 30.0	48.3
n-Undecane	$C_{11}H_{24}$	− 25.9	34.1
n-Dodecane	$C_{12}H_{26}$	− 9.6	51.3
Eicosane	$C_{20}H_{42}$	+ 36.4	52.0
Pentacosane	$C_{25}H_{52}$	+ 53.3	53.6
Tritriacontane	$C_{33}H_{68}$	+ 71.1	54.0
Aromatics:			
Benzene	C_6H_6	+ 5.533	30.100
Methylbenzene (Toluene)	C_7H_8	− 94.991	17.171
Ethylbenzene	C_8H_{10}	− 94.950	20.629
o-Xylene	C_8H_{10}	− 25.187	30.614
m-Xylene	C_8H_{10}	− 47.872	26.045
p-Xylene	C_8H_{10}	+ 13.263	38.526
n-Propylbenzene	C_9H_{12}	− 99.500	16.97
Isopropylbenzene	C_9H_{12}	− 96.028	19.22
1-Methyl-2-ethylbenzene	C_9H_{12}	− 80.833	21.13
Aromatics—(*Cont.*):			
1-Methyl-3-ethylbenzene	C_9H_{12}	− 95.55	15.14
1-Methyl-4-ethylbenzene	C_9H_{12}	− 62.350	25.29
1,2,3-Trimethylbenzene	C_9H_{12}	− 25.375	16.64
1,2,4-Trimethylbenzene	C_9H_{12}	− 43.80	24.54
1,3,5-Trimethylbenzene	C_9H_{12}	− 44.720	18.97
Naphthalene	$C_{10}H_8$	+ 80.0	36.0
Camphene	$C_{10}H_{12}$	+ 51	57
Durene	$C_{10}H_{14}$	+ 79.3	37.4
Isodurene	$C_{10}H_{14}$	− 24.0	23.0
Prehnitene	$C_{10}H_{14}$	− 7.7	20.0
p-Cymene	$C_{10}H_{14}$	− 68.9	17.1
n-Butyl benzene	$C_{10}H_{14}$	− 88.5	19.5
tert-Butyl benzene	$C_{10}H_{14}$	− 58.1	14.9
β-Methyl naphthalene	$C_{11}H_{10}$	+ 34.1	20.1
Diphenyl	$C_{12}H_{10}$	+ 68.6	28.8
Hexamethyl benzene	$C_{12}H_{18}$	+165.5	30.4
Diphenyl methane	$C_{13}H_{12}$	+ 25.2	26.4
Anthracene	$C_{14}H_{10}$	+216.5	38.7
Phenanthrene	$C_{14}H_{10}$	+ 96.3	25.0
Tolane	$C_{14}H_{10}$	+ 60	28.7
Stilbene	$C_{14}H_{12}$	+124	40.0
Dibenzil	$C_{14}H_{14}$	+ 51.4	30.7
Triphenyl methane	$C_{19}H_{16}$	+ 92.1	21.1
Alkyl cyclohexanes:			
Cyclohexane	C_6H_{12}	+ 6.67	7.569
Methylcyclohexane	C_7H_{14}	−126.58	16.429
Alkyl cyclopentanes:			
Cyclopentane	C_5H_{10}	− 93.80	2.068
Methylcyclopentane	C_6H_{12}	−142.445	19.68
Ethylcyclopentane	C_7H_{14}	−138.435	11.10
1,1-Dimethylcyclopentane	C_7H_{14}	− 69.73	3.36
cis-1,2-Dimethylcyclopentane	C_7H_{14}	− 53.85	3.87
trans-1,2-Dimethylcyclopentane	C_7H_{14}	−117.57	15.68
trans-1,3-Dimethylcyclopentane	C_7H_{14}	−133.680	17.93
Monoolefins:			
Ethene (Ethylene)	C_2H_4	−169.15	28.547
Propene (Propylene)	C_3H_6	−185.25	17.054
1-Butene	C_4H_8	−185.35	16.393
cis-2-Butene	C_4H_8	−138.91	31.135
trans-2-Butene	C_4H_8	−105.55	41.564
2-Methylpropene (isobutene)	C_4H_8	−140.35	25.265
1-Pentene	C_5H_{10}	−165.27	16.82
cis-2-pentene	C_5H_{10}	−151.363	24.239
trans-2-pentene	C_5H_{10}	−140.235	26.536
2-Methyl-1-butene	C_5H_{10}	−137.560	26.879
3-Methyl-1-butene	C_5H_{10}	−168.500	18.009
2-Methyl-2-butene	C_5H_{10}	−133.780	25.738
Acetylenes:			
Acetylene	C_2H_2	− 81.5	23.04
2-Butyne (dimethylacetylene)	C_4H_6	−132.23	40.808

Non-hydrocarbon compounds	Formula	M.p., °C.	Heat of fusion, cal./g.
Acetic acid	$C_2H_4O_2$	16.7	46.68
Acetone	C_3H_6O	−95.5	23.42
Acrylic acid	$C_3H_4O_2$	12.3	37.03
Allo-cinnamic acid	$C_9H_8O_2$	68	27.35
Aminobenzoic acid (o-)	$C_7H_7NO_2$	145	35.48
(m-)	$C_7H_7NO_2$	179.5	38.03
(p-)	$C_7H_7NO_2$	188.5	36.46
Amyl alcohol	$C_5H_{12}O$	−78.9	26.65
Anethole	$C_{10}H_{12}O$	22.5	25.80
Aniline	$C_6H_5NH_2$	−6.3	27.09
Anthraquinone	$C_{14}H_8O_2$	284.8	37.48
Apiol	$C_{12}H_{14}O_4$	29.5	25.80
Azobenzene	$C_{12}H_{10}N_2$	67.1	28.91
Azoxybenzene	$C_{12}H_{10}N_2O$	36	21.62
Benzil	$C_{14}H_{10}O_2$	95.2	22.15
Benzoic acid	$C_7H_6O_2$	122.45	33.90
Benzophenone	$C_{13}H_{10}O$	47.85	23.53
Benzylaniline	$C_{13}H_{13}N$	32.37	21.86
Bromocamphor	$C_{10}H_{15}BrO$	78	41.57
Bromochlorbenzene (o-)	C_6H_4BrCl	−12.6	15.41
(m-)	C_6H_4BrCl	−21.2	15.29
(p-)	C_6H_4BrCl	64.6	23.41
Bromoiodobenzene (o-)	C_6H_4BrI	21	12.18
(m-)	C_6H_4BrI	9.3	10.27
(p-)	C_6H_4BrI	90.1	16.60
Bromol hydrate	$C_2H_3Br_3O_2$	46	16.90
Bromophenol (p-)	C_6H_5BrO	63.5	20.50
Bromotoluene (p-)	C_7H_7Br	28	20.86
Butyl alcohol (n-)	$C_4H_{10}O$	−89.2	29.93
(t-)	$C_4H_{10}O$	25.4	21.88
Butyric acid (n-)	$C_4H_8O_2$	−5.7	30.04
Capric acid (n-)	$C_{10}H_{20}O_2$	31.99	38.87
Caprylic acid (n-)	$C_8H_{16}O_2$	16.3	35.40
Carbazole	$C_{12}H_9N$	243	42.05
Carbon tetrachloride	CCl_4	−22.8	41.57
Carvoxime (d-)	$C_{10}H_{15}NO$	71.5	23.29
(l-)	$C_{10}H_{15}NO$	71	23.41
(dl-)	$C_{10}H_{15}NO$	91	24.61
Cetyl alcohol	$C_{16}H_{34}O$	49.27	33.80
Chloracetic acid (α-)	$C_2H_3ClO_2$	61.2	31.06
(β-)	$C_2H_3ClO_2$	56	35.12
Chloral alcoholate	$C_4H_7Cl_3O_2$	9	24.03
hydrate	$C_2H_3Cl_3O_2$	47.4	33.18
Chloroaniline (p-)	C_6H_6ClN	71	37.15
Chlorobenzoic acid (o-)	$C_7H_5ClO_2$	140.2	39.30
(m-)	$C_7H_5ClO_2$	154.25	36.41
(p-)	$C_7H_5ClO_2$	239.7	49.21
Chloronitrobenzene (m-)	$C_6H_4ClNO_2$	44.4	29.38
(p-)	$C_6H_4ClNO_2$	83.5	31.51
Cinnamic acid	$C_9H_8O_2$	133	36.50
anhydride	$C_{18}H_{14}O_3$	48	28.14
Cresol (p-)	C_7H_8O	34.6	26.28
Crotonic acid (α-)	$C_4H_6O_2$	72	25.32
(cis-)	$C_4H_6O_2$	71.2	34.90
Cyanamide	CH_2N_2	44	49.81
Cyclohexanol	$C_6H_{12}O$	25.46	4.19

TABLE 3-178 Heats of Fusion of Organic Compounds *(Concluded)*

Non-hydrocarbon compounds	Formula	M.p., °C.	Heat of fusion, cal./g.
Dibromobenzene (o-)	$C_6H_4Br_2$	1.8	12.78
(m-)	$C_6H_4Br_2$	−6.9	13.38
(p-)	$C_6H_4Br_2$	86	20.55
Dibromophenol (2, 4-)	$C_6H_4Br_2O$	12	13.97
Dichloroacetic acid	$C_2H_2Cl_2O_2$	−4(?)	14.21
Dichlorobenzene (o-)	$C_6H_4Cl_2$	−16.7	21.02
(m-)	$C_6H_4Cl_2$	−24.8	20.55
(p-)	$C_6H_4Cl_2$	53.13	29.67
Dihydroxybenzene (o-)	$C_6H_6O_2$	104.3	49.40
(m-)	$C_6H_6O_2$	109.65	46.20
(p-)	$C_6H_6O_2$	172.3	58.77
Di-iodobenzene (o-)	$C_6H_4I_2$	23.4	10.15
(m-)	$C_6H_4I_2$	34.2	11.54
(p-)	$C_6H_4I_2$	129	16.20
Dimethyl tartrate (dl-)	$C_6H_{10}O_6$	87	35.12
(d-)	$C_6H_{10}O_6$	49	21.50
pyrone	$C_7H_8O_2$	132	56.14
Dinitrobenzene (o-)	$C_6H_4N_2O_4$	116.93	32.25
(m-)	$C_6H_4N_2O_4$	89.7	24.70
(p-)	$C_6H_4N_2O_4$	173.5	39.99
Dinitrotoluene (2, 4-)	$C_7H_6N_2O_4$	70.14	26.40
Dioxane	$C_4H_8O_2$	11.0	34.85
Diphenyl amine	$C_{12}H_{11}N$	52.98	25.23
Elaidic acid	$C_{18}H_{34}O_2$	44.4	52.08
Ethyl acetate	$C_4H_8O_2$	83.8	28.43
alcohol	C_2H_6O	−114.4	25.76
Ethylene dibromide	$C_2H_4Br_2$	10.012	13.52
Ethyl ether	$C_4H_{10}O$	−116.3	23.54
Formic acid	CH_2O_2	8.40	58.89
Glutaric acid	$C_6H_8O_4$	97.5	37.39
Glycerol	$C_3H_8O_3$	18.07	47.49
Glycol, ethylene	$C_2H_6O_2$	−11.5	43.26
Hydrazo benzene	$C_{12}H_{12}N_2$	134	22.89
Hydrocinnamic acid	$C_9H_{10}O_2$	48	28.14
Hydroxyacetanilide	$C_8H_9NO_2$	91.3	33.59
Iodotoluene (p-)	C_7H_7I	34	18.75
Isopropyl alcohol	C_3H_8O	−88.5	21.08
ether	$C_6H_{14}O$	−86.8	25.79
Lauric acid (n-)	$C_{12}H_{24}O_2$	43.22	43.72
Levulinic acid	$C_5H_8O_3$	33	18.97
Menthol (l-) (α)	$C_{10}H_{20}O$	43.5	18.63
Methyl alcohol	CH_4O	−97.8	23.7
Myristic acid	$C_{14}H_{28}O_2$	53.86	47.49
Methyl cinnamate	$C_{10}H_{10}O_2$	36	26.53
fumarate	$C_6H_8O_4$	102	57.93
oxalate	$C_4H_6O_4$	54.35	42.64
phenylpropiolate	$C_{10}H_8O_2$	18	22.86
succinate	$C_6H_{10}O_4$	19.5	35.72

Non-hydrocarbon compounds	Formula	M.p., °C.	Heat of fusion, cal./g.
Naphthol (α-)	$C_{10}H_8O$	95.0	38.94
(β-)	$C_{10}H_8O$	120.6	31.30
Naphthylamine (α-)	$C_{10}H_9N$	50	22.34
Nitroaniline (o-)	$C_6H_6N_2O_2$	71.2	27.88
(m-)	$C_6H_6N_2O_2$	114.0	40.97
(p-)	$C_6H_6N_2O_2$	147.3	36.46
Nitrobenzene	$C_6H_5NO_2$	5.85	22.52
Nitrobenzoic acid (o-)	$C_7H_5NO_4$	145.8	40.06
(m-)	$C_7H_5NO_4$	141.1	27.59
(p-)	$C_7H_5NO_4$	239.2	52.80
Nitronaphthalene	$C_{10}H_7NO_2$	56.7	25.44
Nitrophenol (o-)	$C_6H_5NO_3$	45.13	26.76
Palmitic acid	$C_{16}H_{32}O_2$	61.82	39.18
Paraldehyde	$C_6H_{12}O_3$	10.5	25.02
Pelargic acid (n-) (β-)	$C_9H_{18}O_2$	....	39.04
Pelargonic acid (n-) (α-)	$C_9H_{18}O_2$	12.35	30.63
Phenol	C_6H_6O	40.92	29.03
Phenylacetic acid	$C_8H_8O_2$	76.7	25.44
Phenylhydrazine	$C_6H_8N_2$	19.6	36.31
Propyl ether (n)	$C_6H_{14}O$	−126.1	20.66
Quinone	$C_6H_4O_2$	115.7	40.85
Stearic acid	$C_{18}H_{30}O_2$	68.82	47.54
Succinic anhydride	$C_4H_4O_3$	119	48.74
Succinonitrile	$C_4H_4N_2$	54.5	11.71
Tetrachloroxylene (o-)	$C_8H_6Cl_4$	86	21.02
(p-)	$C_8H_6Cl_4$	95	22.10
Thiophene	C_4H_4S	−39.4	14.11
Thiosinamine	$C_4H_8N_2S$	77	33.45
Thymol	$C_{10}H_{14}O$	51.5	27.47
Toluic acid (o-)	$C_8H_8O_2$	103.7	35.40
(m-)	$C_8H_8O_2$	108.75	27.59
(p-)	$C_8H_8O_2$	179.6	39.90
Toluidine (p-)	C_7H_9N	43.3	39.90
Tribromophenol (2, 4, 6-)	$C_6H_3Br_3O$	93	13.38
Trichloroacetic acid	$C_2HCl_3O_2$	57.5	8.60
Trinitroglycerol	$C_3H_5N_3O_9$	12.3	23.02
Trinitrotoluene (2, 4, 6-)	$C_7H_5N_3O_6$	80.83	22.34
Tristearin	$C_{57}H_{110}O_6$	70.8, 54.5	45.63
Undecylic acid (α-) (n-)	$C_{11}H_{22}O_2$	28.25	32.20
(β-) (n-)	$C_{11}H_{22}O_2$	....	42.91
Urethane	$C_3H_7NO_2$	48.7	40.85
Veratrol	$C_8H_{10}O_2$	22.5	27.45
Xylene dibromide (o-)	$C_8H_8Br_2$	95	24.25
(m-)	$C_8H_8Br_2$	77	21.45
dichloride (o-)	$C_8H_8Cl_2$	55	29.03
(m-)	$C_8H_8Cl_2$	34	26.64
(p-)	$C_8H_8Cl_2$	100	32.73

TABLE 3-179 Heats of Vaporization of Organic Compounds

The values for the hydrocarbons are from the tables of the American Petroleum Institute Research Project 44 at the National Bureau of Standards. The values for the hydrocarbon compounds were recalculated from data in *International Critical Tables*, vol. 5.

Hydrocarbon compounds	Formula	Temperature, °C.	ΔHv, cal./g.
Paraffins:			
Methane	CH_4	−161.6	121.87
Ethane	C_2H_6	−88.9	116.87
Propane	C_3H_8	25	81.76
		−42.1	101.76
n-Butane	C_4H_{10}	25	86.63
		−0.50	92.09
2-Methylpropane (isobutane)	C_4H_{10}	25	78.63
		−11.72	87.56
n-Pentane	C_5H_{12}	25	87.54
		36.08	85.38
2-Methylbutane (isopentane)	C_5H_{12}	25	81.47
		27.86	80.97
2,2-Dimethylpropane (neopentane)	C_5H_{12}	25	72.15
		9.45	75.37
n-Hexane	C_6H_{14}	25	87.50
		68.74	80.48
2-Methylpentane	C_6H_{14}	25	82.83
		60.27	76.89
3-Methylpentane	C_6H_{14}	25	83.96
		63.28	78.42
2,2-Dimethylbutane	C_6H_{14}	25	76.79
		49.74	73.75
2,3-Dimethylbutane	C_6H_{14}	25	80.77
		57.99	76.53
n-Heptane	C_7H_{16}	25	87.18
		98.43	76.45
2-Methylhexane	C_7H_{16}	25	83.02
		90.05	73.4
3-Methylhexane	C_7H_{16}	25	83.68
		91.95	74.1
3-Ethylpentane	C_7H_{16}	25	84.02
		93.47	74.3
2,2-Dimethylpentane	C_7H_{16}	25	77.36
		79.20	69.7
2,3-Dimethylpentane	C_7H_{16}	25	81.68
		89.79	72.9
2,4-Dimethylpentane	C_7H_{16}	25	78.44
		80.51	70.9
3,3-Dimethylpentane	C_7H_{16}	25	78.76
		86.06	70.6
2,2,3-Trimethylbutane	C_7H_{16}	25	76.42
		80.88	69.3
n-Octane	C_8H_{18}	25	86.80
		125.66	73.19
2-Methylheptane	C_8H_{18}	25	83.02
		117.64	70.3
3-Methylheptane	C_8H_{18}	25	83.35
		118.92	71.3
4-Methylheptane	C_8H_{18}	25	83.01
		117.71	70.91
3-Ethylhexane	C_8H_{18}	25	82.95
		118.53	71.7
2,2-Dimethylhexane	C_8H_{18}	25	78.02
		106.84	67.7
2,3-Dimethylhexane	C_8H_{18}	25	81.17
		115.60	70.2
2,4-Dimethylhexane	C_8H_{18}	25	79.02
		109.43	68.5
2,5-Dimethylhexane	C_8H_{18}	25	79.21
		109.10	68.6
3,3-Dimethylhexane	C_8H_{18}	25	78.54
		111.97	68.5
3,4-Dimethylhexane	C_8H_{18}	25	81.55
		117.72	70.2
2-Methyl-3-ethylpentane	C_8H_{18}	25	80.60
		115.65	69.7
3-Methyl-3-ethylpentane	C_8H_{18}	25	79.49
		118.26	69.3
2,2,3-Trimethylpentane	C_8H_{18}	25	77.24
		109.84	67.3
2,2,4-Trimethylpentane	C_8H_{18}	25	73.50
		99.24	64.87
2,3,3-Trimethylpentane	C_8H_{18}	25	77.87
		114.76	68.1
2,3,4-Trimethylpentane	C_8H_{18}	25	78.90
		113.47	68.37
2,2,3,3-Tetramethylbutane	C_8H_{18}	106.30	66.2
Alkyl benzenes:			
Benzene	C_6H_6	25	103.57
		80.10	94.14
Methylbenzene (toluene)	C_7H_8	25	98.55
		110.62	86.8
Ethylbenzene	C_8H_{10}	25	95.11
		136.19	81.0
1,2-Dimethylbenzene (o-xylene)	C_8H_{10}	25	97.79
		144.42	82.9
1,3-Dimethylbenzene (m-xylene)	C_8H_{10}	25	96.03
		139.10	82.0
1,4-Dimethylbenzene (p-xylene)	C_8H_{10}	25	95.40
		138.35	81.2
n-Propylbenzene	C_9H_{12}	25	91.93
		159.22	76.0
Isopropylbenzene	C_9H_{12}	25	89.77
		152.40	74.6
1-Methyl-2-ethylbenzene	C_9H_{12}	25	94.9
		165.15	77.3
1-Methyl-3-ethylbenzene	C_9H_{12}	25	93.3
		161.30	76.6
1-Methyl-4-ethylbenzene	C_9H_{12}	25	92.7
		162.05	76.4
1,2,3-Trimethylbenzene	C_9H_{12}	25	97.56
		176.15	79.6
1,2,4-Trimethylbenzene (pseudocumene)	C_9H_{12}	25	95.33
		169.25	78.0
1,3,5-Trimethylbenzene (mesitylene)	C_9H_{12}	25	94.40
		164.70	77.6
Alkyl cyclopentanes:			
Cyclopentane	C_5H_{10}	25	97.1
		49.26	93.1
Methylcyclopentane	C_6H_{12}	25	89.83
		71.81	83.2
Ethylcyclopentane	C_7H_{14}	25	88.6
		103.45	78.3
1,1-Dimethylcyclopentane	C_7H_{14}	25	82.5
		87.5	74.6
cis-1,2-Dimethylcyclopentane	C_7H_{14}	25	86.4
		99.3	77.0
trans-1,2-Dimethylcyclopentane	C_7H_{14}	25	83.9
		91.9	75.5
trans-1,3-Dimethylcyclopentane	C_7H_{14}	25	83.6
		90.8	75.3
Alkyl cyclohexanes:			
Cyclohexane	C_6H_{12}	25	93.81
		80.74	85.6
Methylcyclohexane	C_7H_{14}	25	86.07
		100.94	76.9
Ethylcyclohexane	C_8H_{16}	25	86.21
		131.79	73.7
1,1-Dimethylcyclohexane	C_8H_{16}	25	80.9
		119.50	70.7
cis-1,2-Dimethylcyclohexane	C_8H_{16}	25	84.59
		129.73	72.9
trans-1,2-Dimethylcyclohexane	C_8H_{16}	25	81.70
		123.42	71.1
cis-1,3-Dimethylcyclohexane	C_8H_{16}	25	83.49
		124.45	72.1
trans-1,3-Dimethylcyclohexane	C_8H_{16}	25	81.42
		120.09	70.9
cis-1,4-Dimethylcyclohexane	C_8H_{16}	25	83.13
		124.32	71.9
trans-1,4-Dimethylcyclohexane	C_8H_{16}	25	80.67
		119.35	70.4
Monoolefins:			
Ethene (ethylene)	C_2H_4	−103.71	115.39
Propene (propylene)	C_3H_6	−47.70	104.62
1-Butene	C_4H_8	25	86.8
		−6.25	93.36
cis-2-Butene	C_4H_8	25	94.5
		3.72	99.46
trans-2-Butene	C_4H_8	25	91.8
		0.88	96.94
2-Methylpropene (isobutene)	C_4H_8	25	87.7
		−6.90	94.22

Non-hydrocarbon compounds	Formula	Temperature, °C.	ΔHv, cal./g.
Acetal	$C_6H_{14}O_2$	102.9	66.18
Acetaldehyde	C_2H_4O	21	136.17
Acetic acid	$C_2H_4O_2$	118.3	96.75
		140	94.37
		220	81.23
		321	0
anhydride	$C_4H_6O_3$	137	92.2
Acetone	C_3H_6O	0	134.74
		20	131.87
		40	128.05
		60	123.51
		80	118.26
		100	112.76
		235	0

TABLE 3-179 Heats of Vaporization of Organic Compounds *(Concluded)*

Non-hydrocarbon compounds	Formula	Temperature, °C.	ΔH_v, cal./g.
Acetonitrile	C_2H_3N	80	173.68
Acetophenone	C_8H_8O	203.7	77.16
Acetyl chloride	C_2H_3ClO	51	78.84
Air	....	....	51.0
Allyl alcohol	C_3H_6O	96	163.41
Amyl alcohol (n-)	$C_5H_{11}OH$	131	120.17
alcohol (t-)	$C_5H_{11}OH$	102	105.83
amine (n-)	$C_5H_{13}N$	95	98.67
bromide (n-)	$C_5H_{11}Br$	129	48.26
ether (n-)	$C_{10}H_{22}O$	170	69.52
iodide (n-)	$C_5H_{11}I$	155	47.54
methyl ketone (n-)	$C_7H_{14}O$	149.2	82.66
Amylene	C_5H_{10}	12.5	75.01
Anethole (p-)	$C_{10}H_{12}O$	232	71.43
Aniline	C_6H_7N	183	103.68
Benzaldehyde	C_7H_6O	179	86.48
Benzonitrile	C_7H_5N	189	87.68
Benzyl alcohol	C_7H_8O	204.3	112.28
Butyl acetate (n-)	$C_6H_{12}O_2$	124	73.82
alcohol (n-)	$C_4H_{10}O$	116.8	141.26
alcohol (s-)	$C_4H_{10}O$	98.1	134.38
alcohol (t-)	$C_4H_{10}O$	83	130.44
formate	$C_4H_{10}O_2$	105.1	86.74
methyl ketone (n-)	$C_6H_{12}O$	127	82.42
propionate (n-)	$C_7H_{14}O_2$	144.9	71.74
Butyric acid (n-)	$C_4H_8O_2$	163.5	113.96
Butyronitrile (n-)	C_4H_7N	117.4	114.91
Bromobenzene	C_6H_5Br	155.9	57.60
Capronitrile	$C_6H_{11}N$	156	88.15
Carbon disulfide	CS_2	0	89.35
		46.25	84.09
		100	75.49
		140	67.37
tetrachloride	CCl_4	0	52.06
		76.75	46.42
		200	32.73
Carvacrol	$C_{10}H_{14}O$	237	68.09
Chloral	C_2HCl_3O	....	53.99
hydrate	$C_2H_3Cl_3O_2$	96	131.87
Chlorobenzene	C_6H_5Cl	130.6	77.59
Chloroethyl alcohol (2-)	C_2H_5ClO	126.5	122.94
acetate (β-)	$C_4H_7ClO_2$	141.5	80.75
Chloroform	$CHCl_3$	0	64.74
		40	60.92
		61.5	59.01
		100	55.19
		260	0
Chlorotoluene (o-)	C_7H_7Cl	158.1	72.63
(p-)	C_7H_7Cl	160.4	73.13
Cresol (m-)	C_7H_8O	202	100.58
Cyanogen	$(CN)_2$	0	102.97
chloride	$CNCl$	13	134.98
Cyclohexanol	$C_6H_{12}O$	161.1	108.22
Cycohexyl chloride	$C_6H_{11}Cl$	142.0	74.78
Dichloroacetic acid	$C_2H_2Cl_2O_2$	194.4	77.16
Dichlorodifluormethane	CCl_2F_2	−29.8	40.40
Diethylamine	$C_4H_{11}N$	58	91.02
carbonate	$C_5H_{10}O_3$	126	73.10
ketone	$C_5H_{10}O$	101	90.78
oxalate	$C_6H_{10}O_4$	185	67.61
Di-isobutylamine	$C_8H_{19}N$	134	65.70
Dimethyl aniline	$C_8H_{11}N$	193	80.75
carbonate	$C_3H_6O_3$	90	88.15
Dipropyl ketone	$C_7H_{14}O$	143.5	75.73
Dipropylamine (n-)	$C_6H_{15}N$	108	75.73
Ethyl acetate	$C_4H_8O_2$	0.0	102.01
alcohol	C_2H_6O	78.3	204.26
Ethylamine	C_2H_7N	15	145.97
Ethyl benzoate	$C_9H_{10}O_2$	213	64.50
bromide	C_2H_5Br	38.4	59.92
butyrate (n-)	$C_6H_{12}O_2$	118.9	74.68
caprylate	$C_{10}H_{20}O_2$	207	60.44
chloride	C_2H_5Cl	4.7	92.93
		15.0	92.45
		20.0	92.22
		25.0	91.98
Ethylene bromide	$C_2H_4Br_2$	130.8	46.23
chloride	$C_2H_4Cl_2$	0	85.29
		82.3	77.33
glycol	$C_2H_6O_2$	197	191.12
oxide	C_2H_4O	13	138.56
Ethyl ether	$C_4H_{10}O$	34.6	83.85
formate	$C_3H_6O_2$	53.3	97.18
iodide	C_2H_5I	71.2	45.61
Ethylidine chloride	$C_2H_4Cl_2$	0.0	76.69
		60	67.13
Ethyl isobutyl ether	$C_6H_{14}O$	79.0	74.78
isobutyrate	$C_6H_{12}O_2$	109.2	72.05
isovalerate	$C_7H_{14}O_2$	144	67.85
methyl ketone	C_4H_8O	78.2	105.93
methyl ketoxime	C_4H_9NO	182	115.87

Non-hydrocarbon compounds	Formula	Temperature, °C.	ΔH_v, cal./g.
Ethyl nonylate	$C_{11}H_{22}O_2$	227	58.05
propionate	$C_5H_{10}O_2$	97.6	80.08
propyl ether	$C_5H_{12}O$	60	82.66
valerate (n-)	$C_7H_{14}O_2$	98	77.16
Formic acid	CH_2O_2	101	119.93
Furane	C_4H_4O	31	95.32
Furfural	$C_5H_4O_2$	160.5	107.51
Heptyl alcohol (n-)	$C_7H_{16}O$	176	104.88
Hexylmethyl ketone	$C_8H_{16}O$	173	74.06
Hydrogen cyanide	HCN	20	210.23
Isoamyl acetate	$C_7H_{14}O_2$	143.6	69.04
alcohol	$C_5H_{12}O$	130.2	119.78
butyrate (n-)	$C_9H_{18}O_2$	169	61.88
formate	$C_6H_{12}O_2$	123	73.58
isobutyrate	$C_9H_{18}O_2$	168	57.57
propionate	$C_8H_{16}O_2$	161	65.22
valerate (n-)	$C_{10}H_{20}O_2$	187	56.14
Isobutyl acetate	$C_6H_{12}O_2$	115.3	73.75
alcohol	$C_4H_{10}O$	106.9	138.08
butyrate (n-)	$C_8H_{16}O_2$	157	64.50
formate	$C_5H_{10}O_2$	97	78.50
isovalerate	$C_9H_{18}O_2$	169	60.44
isobutyrate	$C_8H_{16}O_2$	148	63.31
propionate	$C_7H_{14}O_2$	137	65.94
valerate (n-)	$C_9H_{18}O_2$	169	57.81
Isobutyric acid	$C_4H_8O_2$	154	111.57
Isopropyl alcohol	C_3H_8O	82.3	159.35
methyl ketone	$C_5H_{10}O$	92	89.83
Isovaleric acid	$C_5H_{10}O_2$	176.3	101.05
Limonene	$C_{10}H_{16}$	165	69.52
Mesityl oxide	$C_6H_{10}O$	128	85.77
Methyl acetate	$C_3H_6O_2$	0.0	113.96
		56.3	98.09
Methylal	$C_3H_8O_2$	42	89.83
Methyl alcohol	CH_4O	0	284.29
		64.7	262.79
		100	241.29
		160	193.51
		200	148.12
		220	109.89
		240	0
amyl ketone (n-)	$C_7H_{14}O$	149.2	82.66
aniline	C_7H_9N	194	95.56
butyl ketone (n-)	$C_6H_{12}O$	127	82.42
butyrate (n-)	$C_5H_{10}O_2$	102.6	79.79
chloride	CH_3Cl	−23.8	102.25
		+15.0	96.04
		20.0	95.32
		25.0	94.60
ethyl ketone	C_4H_8O	78.2	105.93
ethyl ketoxime	C_4H_9NO	182	115.87
formate	$C_2H_4O_2$	31.3	112.35
hexyl ketone	$C_8H_{16}O$	173	74.06
iodide	CH_3I	42	45.87
isobutyrate	$C_5H_{10}O_2$	91.1	78.12
isopropyl ketone	$C_5H_{10}O$	92	89.83
isovalerate	$C_6H_{12}O_2$	116	72.39
phenyl ether	C_7H_8O	153	81.46
propionate	$C_4H_8O_2$	79.0	87.56
valerate (n-)	$C_6H_{12}O_2$	116	70.00
Naphthalene	$C_{10}H_8$	218	75.49
Nitrobenzene	$C_6H_5NO_2$	210	79.08
Nitromethane	CH_3NO_2	99.9	134.98
Octyl alcohol (n-)	$C_8H_{18}O$	196	97.47
alcohol (dl-) (sec-)	$C_8H_{18}O$	180	94.37
Phenyl methyl ether	C_7H_8O	153	81.46
Picoline (α-)	C_6H_7N	129	90.78
Piperidine	$C_5H_{11}N$	106	89.35
Propionic acid	$C_3H_6O_2$	139.3	98.81
Propionitrile	C_3H_5N	97	134.26
Propyl acetate (n-)	$C_5H_{10}O_2$	100.4	80.27
alcohol (n-)	C_3H_8O	97.2	164.36
butyrate (n-)	$C_7H_{14}O_2$	143.6	68.33
formate (n-)	$C_4H_8O_2$	80.0	88.13
isobutyrate (n-)	$C_7H_{14}O_2$	134	63.79
isovalerate (n-)	$C_8H_{16}O_2$	156	64.50
propionate (n-)	$C_6H_{12}O_2$	120.6	73.15
Pyridine	C_5H_5N	114.1	107.36
Salicylaldehyde	$C_7H_6O_2$	196	74.78
Tetrachloroethane (1,1,2,2-)	$C_2H_2Cl_4$	145	55.07
Tetrachloroethylene	C_2Cl_4	120.7	50.05
Toluidine (o-)	C_7H_9N	198	95.08
Trichloroethylene	C_2HCl_3	85.7	57.24
Valeronitrile (n-)	C_5H_9N	129	96.28

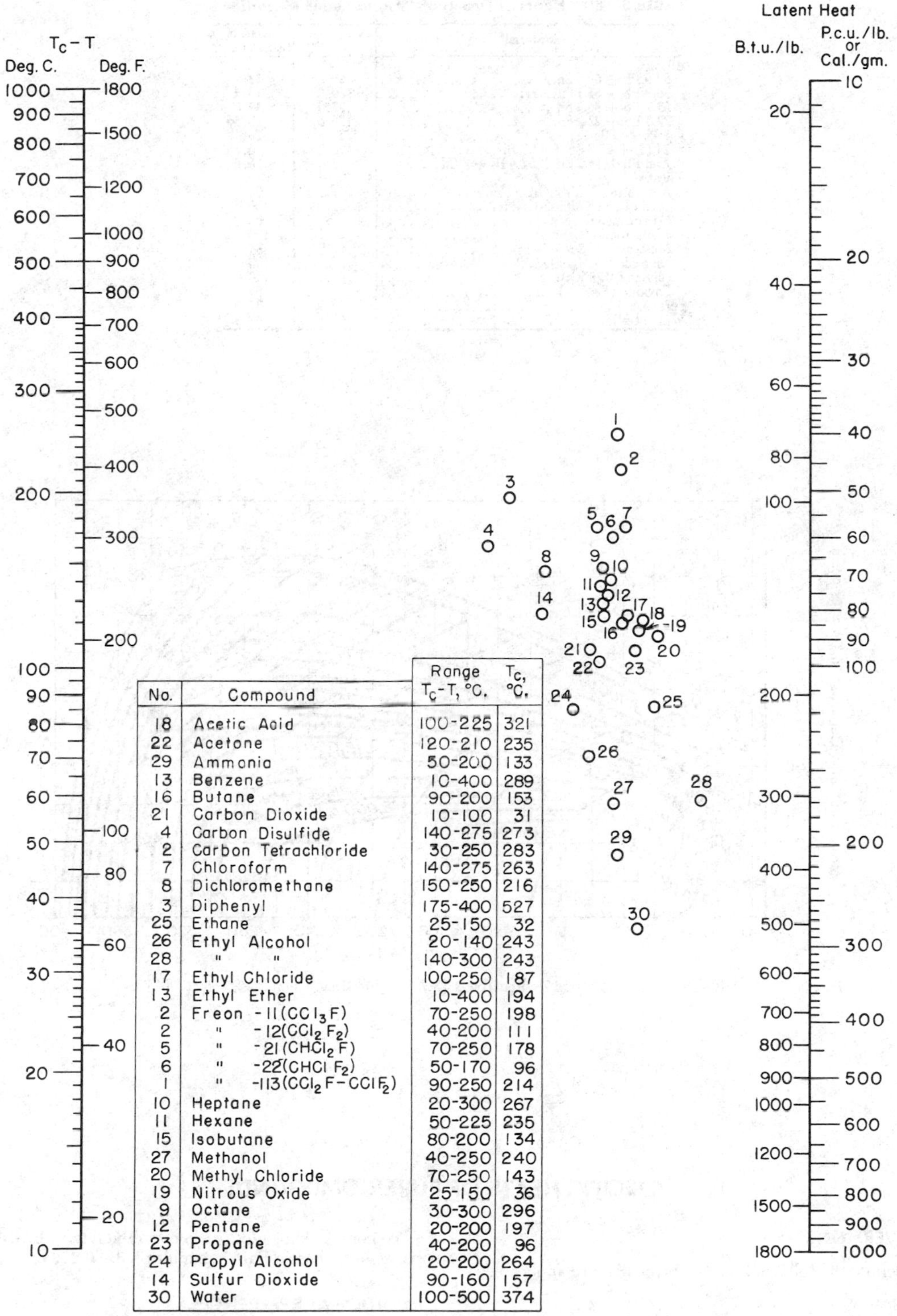

No.	Compound	Range T_c-T, °C.	T_c, °C.
18	Acetic Acid	100-225	321
22	Acetone	120-210	235
29	Ammonia	50-200	133
13	Benzene	10-400	289
16	Butane	90-200	153
21	Carbon Dioxide	10-100	31
4	Carbon Disulfide	140-275	273
2	Carbon Tetrachloride	30-250	283
7	Chloroform	140-275	263
8	Dichloromethane	150-250	216
3	Diphenyl	175-400	527
25	Ethane	25-150	32
26	Ethyl Alcohol	20-140	243
28	" "	140-300	243
17	Ethyl Chloride	100-250	187
13	Ethyl Ether	10-400	194
2	Freon -11(CCl_3F)	70-250	198
2	" -12(CCl_2F_2)	40-200	111
5	" -21($CHCl_2F$)	70-250	178
6	" -22($CHClF_2$)	50-170	96
1	" -113($CCl_2F-CClF_2$)	90-250	214
10	Heptane	20-300	267
11	Hexane	50-225	235
15	Isobutane	80-200	134
27	Methanol	40-250	240
20	Methyl Chloride	70-250	143
19	Nitrous Oxide	25-150	36
9	Octane	30-300	296
12	Pentane	20-200	197
23	Propane	40-200	96
24	Propyl Alcohol	20-200	264
14	Sulfur Dioxide	90-160	157
30	Water	100-500	374

FIG. 3-9 Latent heat of vaporization. (*Chilton, Colburn, and Vernon, personal communication. Based mainly on data from* International Critical Tables.)

TABLE 3-180 Heats of Fusion of Miscellaneous Materials

Material	M.p., °C.	Heat of fusion, cal./g.
Alloys		
30.5 Pb + 69.5 Sn	183	17
36.9 Pb + 63.1 Sn	179	15.5
63.7 Pb + 36.3 Sn	177.5	11.6
77.8 Pb + 22.2 Sn	176.5	9.54
1 Pb + 9 Sn	236	28
24 Pb + 27.3 Sn + 48.7 Bi	98.8	6.85
25.8 Pb + 14.7 Sn + 52.4 Bi + 7 Cd	75.5	8.4
Silicates		
Anorthite ($CaAl_2Si_2O_8$)	.....	100
Orthoclase ($KAlSi_2O_8$)	.....	100
Microcline ($KAlSi_3O_8$)	.....	83
Wollastonite ($CaSiO_3$)	.....	100
Malacolite ($Ca_3MgSi_4O_{12}$)	.....	94
Diopside ($CaMgSi_2O_6$)	.....	100
Olivine (Mg_2SiO_4)	.....	130
Fayalite (Fe_2SiO_4)	.....	85
Spermaceti	43.9	37.0
Wax (bees')	61.8	42.3

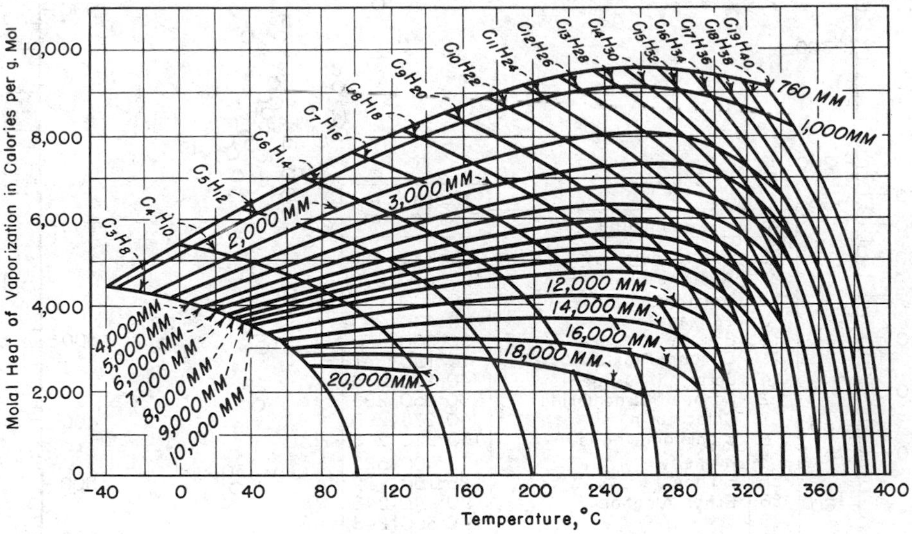

FIG. 3-10 Molal heats of vaporization of hydrocarbons. [*Schultz*, Ind. Eng. Chem., *22, 785 (1930).*]

SPECIFIC HEATS OF PURE COMPOUNDS

UNITS CONVERSIONS

For this subsection, the following units conversions are applicable:

°F = ⅝ °C + 32.
°F = 1.8 K.

To convert calories per gram-kelvin to British thermal units per pound–degree Rankine, multiply by 1.0; to convert calories per gram-mole-kelvin to British thermal units per pound–mole–degree Rankine, multiply by 1.0.

To convert kilojoules per kilogram-kelvin to British thermal units per pound–degree Rankine, multiply by 0.2388.

ADDITIONAL REFERENCES

Additional data are contained in the subsection "Thermodynamic Properties." Data on water are also contained in that subsection. Additional tables for water are referred to Eng. Sci. Data Item 68008, 251 Regent Street, London, England, which contains about 5000 values from 1 to 1000 bar, 0 to 1500°C.

TABLE 3-181 Heat Capacities of the Elements and Inorganic Compounds*

Substance	State†	Heat capacity at constant pressure (T = °K.; 0°C. = 273.1°K.), cal./deg. mol	Range of temperature, °K.	Uncertainty, %
Aluminum:[1]				
Al	*c*	$4.80 + 0.00322T$	273– 931	1
	l	7.00	931–1273	5
$AlBr_3$	*c*	$18.74 + 0.01866T$	273– 370	3
	l	29.5	370– 407	5
$AlCl_3$	*c*	$13.25 + 0.02800T$	273– 465	3
	l	31.2	465– 504	3
$AlCl_3.6H_2O$	*c*	76	288– 327	?
AlF_3	*c*	19.3	288– 326	5
$AlF_3.3\frac{1}{2}H_2O$	*c*	50.5	288– 326	?
$AlF_3.3NaF$	*c*	$38.63 + 0.04760T - 449200/T^2$	273–1273	2
	l	142	1273–1373	?
AlI_3	*c*	$16.88 + 0.02266T$	273– 464	3
	l	28.8	464– 480	5
Al_2O_3	*c*	$22.08 + 0.008971T - 522500/T^2$	273–1973	3
$Al_2O_3.SiO_2$	*c*, sillimanite	$40.79 + 0.004763T - 992800/T^2$	273–1573	3
	c, disthene	$41.81 + 0.005283T - 1211000/T^2$	273–1673	2
	c, andalusite	$43.96 + 0.001923T - 1086000/T^2$	273–1573	3
$3Al_2O_3.2SiO_2$	*c*, mullite	$59.65 + 0.0670T$	273– 576	5
$4Al_2O_3.3SiO_2$	*c*	$113.2 + 0.0652T$	273– 575	3
$Al_2(SO_4)_3$	*c*	63.5	273– 373	?
$Al_2(SO_4)_3.18H_2O$	*c*	235	288– 325	?
Antimony:				
Sb	*c*	$5.51 + 0.00178T$	273– 903	2
	l	7.15	903–1273	5
$SbBr_3$	*c*	$17.2 + 0.0293T$	273– 370	?
$SbCl_3$	*c*	$10.3 + 0.0511T$	273– 346	?
Sb_2O_3	*c*	$19.1 + 0.0171T$	273– 929	?
Sb_2O_4	*c*	$22.6 + 0.0162T$	273–1198	?
Sb_2S_3	*c*	$24.2 + 0.0132T$	273– 821	?
Argon:[2]				
A	*g*	4.97	All	0
Arsenic:				
As	*c*	$5.17 + 0.00234T$	273–1168	5
$AsCl_3$	*l*	31.9	286– 371	?
As_2O_3	*c*	$8.37 + 0.0486T$	273– 548	?
As_2S_3	*c*	25.8	293– 373	?
Barium:				
$BaCl_2$	*c*	$17.0 + 0.00334T$	273–1198	?
$BaCl_2.H_2O$	*c*	28.2	273– 307	?
$BaCl_2.2H_2O$	*c*	37.3	273– 307	?
$Ba(ClO_3)_2.H_2O$	*c*	51	289– 320	?
$BaCO_3$	*c*, α	$17.26 + 0.0131T$	273–1083	5
	c, β	30.0	1083–1255	15
$BaMoO_4$	*c*	34	273– 297	?
$Ba(NO_3)_2$	*c*	39.8	285– 371	?
$BaSO_4$	*c*	$21.35 + 0.0141T$	273–1323	5
Beryllium:[3,4]				
Be	*c*	$4.698 + 0.001555T - 121000/T^2$	273–1173	1
BeO	*c*	$8.69 + 0.00365T - 313000/T^2$	273–1175	5
$BeO.Al_2O_3$	*c*	25.4	273– 373	?
$BeSO_4$	*c*	20.8	273– 373	?
Bismuth:[4]				
Bi	*c*	$5.38 + 0.00260T$	273– 544	3
	l	7.60	544–1273	3
Bi_2O_3	*c*	$23.27 + 0.01105T$	273– 777	2
Bi_2S_3	*c*	30.4	284– 372	?
Boron:				
B	*c*	$1.54 + 0.00440T$	273–1174	5
B_2O_3	*gls*	$5.14 + 0.0320T$	273– 513	3
	gls	30.4	513– 623	3
BN	*c*	$1.61 + 0.00400T$	273–1173	5
Bromine:				
Br_2	*g*	9.00	300–2000	5
Cadmium:				
Cd	*c*	$5.46 + 0.002466T$	273– 594	1
	l	7.13	594– 973	5
CdO	*c*	$9.65 + 0.00208T$	273–2086	?
CdS	*c*	$12.9 + 0.00090T$	273–1273	?
$CdSO_4.\frac{8}{3}H_2O$	*c*	51.3	293	?
Calcium:				
Ca	*c*	$5.31 + 0.00333T$	273– 673	2
	c	$6.29 + 0.00140T$	673– 873	2
$CaCl_2$	*c*	$16.9 + 0.00386T$	273–1055	?
$CaCO_3$	*c*	$19.68 + 0.01189T - 307600/T^2$	273–1033	3
CaF_2	*c*	$14.7 + 0.00380T$	273–1651	?
$CaMg(CO_3)_2$	*c*	40.1	299– 372	?
$CaMoO_4$	*c*	33	273– 297	?
CaO	*c*	$10.00 + 0.00484T - 108000/T^2$	273–1173	2
$Ca(OH)_2$	*c*	21.4	276– 373	?
$CaO.Al_2O_3.2SiO_2$	*c*, anorthite	$63.13 + 0.01500T - 1537000/T^2$	273–1673	1
	gls	$67.41 + 0.01048T - 1874000/T^2$	273– 973	1

*From Kelley, U.S. Bur. Mines Bull. 371, 1934. For a revision see Kelley, U.S. Bur. Mines Bull. 477, 1948. Data for many elements and compounds are given by Johnson (ed.), WADD-TR-60-56, 1960, for cryogenic temperatures. Tabulated data for gases can be obtained from many of the references cited in the "Thermodynamic Properties" subsection and other tables in this section. Thinh, Duran, et al., *Hydrocarbon Process.*, **50**, 98 (January 1971), review previous equation fits and give newer fits for 408 hydrocarbons and related compounds. Later publications include Duran, Thinh, et al., *Hydrocarbon Process.*, **55**, 153 (August 1976); Thompson, *J. Chem. Eng. Data*, **22**(4), 431 (1977); and Passut and Danner, *Ind. Eng. Chem. Process Des. Dev.*, **11**, 543 (1972); 13, 193 (1974).

†The symbols in this column have the following meaning; *c*, crystal; *l*, liquid; *g*, gas; *gls*, glass.

TABLE 3-181 Heat Capacities of the Elements and Inorganic Compounds *(Continued)*

Substance	State†	Heat capacity at constant pressure (T = °K.; 0°C. = 273.1°K.), cal./deg. mol	Range of temperature, °K.	Uncertainty, %
Calcium—(*Cont.*):				
$CaO.MgO.2SiO_2$	*c*, diopside	$54.46 + 0.005746T - 1500000/T^2$	273–1573	1
	gls	$51.68 + 0.009724T - 1308000/T^2$	273– 973	1
$CaO.SiO_2$	*c*, wollastonite	$27.95 + 0.002056T - 745600/T^2$	273–1573	1
	c, pseudowollastonite	$25.48 + 0.004132T - 488100/T^2$	273–1673	1
	gls	$23.16 + 0.009672T - 487100/T^2$	273– 973	1
CaP_2O_6	*c*	39.5	287– 371	?
$CaSO_4$	*c*	$18.52 + 0.02197T - 156800/T^2$	273–1373	5
$CaSO_4.2H_2O$	*c*	46.8	282– 373	?
$CaWO_4$	*c*	27.9	292– 322	?
Carbon:[5]				
C	*c*, graphite	$2.673 + 0.002617T - 116900/T^2$	273–1373	2
	c, diamond	$2.162 + 0.003059T - 130300/T^2$	273–1313	3
CH_4	*g*	$5.34 + 0.0115T$	273–1200	2
CO[6]	*g*	$6.60 + 0.00120T$	273–2500	1½
CO_2	*g*	$10.34 + 0.00274T - 195500/T^2$	273–1200	1½
CS_2	*l*	18.4	293	?
Cerium:				
Ce	*c*	$5.88 + 0.00123T$	273– 908	?
CeO_2	*c*	15.1	273– 373	?
$Ce_2(MoO_4)_3$	*c*	96	273– 297	?
$Ce_2(SO_4)_3$	*c*	66.4	273– 373	?
$Ce_2(SO_4)_3.5H_2O$	*c*	131.6	273– 319	?
Cesium:				
Cs	*c*	$1.96 + 0.0182T$	273– 301	3
	l	8.00	302	3
	g	4.97	All	0
CsBr	*c*	$12.6 + 0.00259T$	273– 909	?
CsCl	*c*	$11.7 + 0.00309T$	273– 752	?
CsF	*c*	$11.3 + 0.00285T$	273– 957	?
CsI	*c*	$11.6 + 0.00268T$	273– 894	?
Chlorine:				
Cl_2	*g*	$8.28 + 0.00056T$	273–2000	1½
Chromium:[4]				
Cr	*c*	$4.84 + 0.00295T$	273–1823	5
	l	9.70	1823–1923	10
$CrCl_2$	*c*	23	286– 319	?
Cr_2O_3	*c*	$26.0 + 0.00400T$	273–2263	?
CrSb	*c*	$12.3 + 0.00120T$	273–1383	?
$CrSb_2$	*c*	$19.2 + 0.00184T$	273– 949	?
$Cr_2(SO_4)_3$	*c*	67.4	273– 373	?
Cobalt:[4]				
Co	*c*	$5.12 + 0.00333T$	273–1763	5
	l	8.40	1763–1873	5
$CoAs_2.CoS_2$	*c*	32.9	283– 373	?
CoSb	*c*	$11.7 + 0.00156T$	273–1464	?
Co_2Sn	*c*	$15.83 + 0.00950T$	273– 903	2
CoS	*c*	$10.6 + 0.00251T$	273–1373	?
$CoSo_4.7H_2O$	*c*	96	286– 303	?
Copper:[7]				
Cu	*c*	$5.44 + 0.001462T$	273–1357	1
	l	7.50	1357–1573	3
CuAl	*c*	$9.88 + 0.00500T$	273– 733	2
$CuAl_2$	*c*	$16.78 + 0.00366T$	273– 773	2
Cu_3Al	*c*	$19.61 + 0.01054T$	273– 775	2
CuI	*c*	$12.1 + 0.00286T$	273– 675	?
CuI_2	*c*	20.1	274– 328	?
CuO	*c*	$10.87 + 0.003576T - 150600/T^2$	273– 810	2
$CuO.SiO_2.H_2O$	*c*	29	293– 323	?
CuS	*c*	$10.6 + 0.00264T$	273–1273	?
Cu_2S	*c*, α	$9.38 + 0.0312T$	273– 376	3
	c, β	20.9	376–1173	2
CuS.FeS	*c*	24	292– 321	?
Cu_2Sb	*c*	$13.73 + 0.01350T$	273– 573	2
Cu_2Sb	*c*	$21.79 + 0.00900T$	273– 693	2
Cu_2Se	*c*, α	20.85	273– 383	5
	c, β	20.35	383– 488	5
Cu_3Si	*c*	$21.3 + 0.00587T$	273–1135	?
$CuSO_4$	*c*	24.1	282	?
$CuSO_4.H_2O$	*c*	31.3	282	?
$CuSO_4.3H_2O$	*c*	49.0	282	?
$CuSO_4.5H_2O$	*c*	67.2	282	?
Fluorine:[8]				
F_2	*g*	$6.50 + 0.00100T$	300–3000	5
Gallium:				
Ga_2O_3	*c*	$18.2 + 0.0252T$	273– 923	?
$Ga_2(SO_4)_3$	*c*	62.4	273– 373	?
Germanium:[4]				
Ge	*c*			
Gold:				
Au	*c*	$5.61 + 0.00144T$	273–1336	2
	l	7.00	1336–1573	5
$AuSb_2$	*c*, α	$17.12 + 0.00465T$	273– 628	1
	c, βγ	$11.47 + 0.01756T$	628– 713	?
Helium:[9]				
He	*g*	4.97	All	0
Hydrogen:[10]				
H	*g*	4.97	All	0
H_2	*g*	$6.62 + 0.00081T$	273–2500	2
HBr	*g*	$6.80 + 0.00084T$	273–2000	2

TABLE 3-181 Heat Capacities of the Elements and Inorganic Compounds *(Continued)*

Substance	State†	Heat capacity at constant pressure (T = °K.; 0°C. = 273.1°K.), cal./deg. mol	Range of tempera-ture, °K.	Uncer-tainty, %
Hydrogen—(*Cont.*):				
HCl	*g*	$6.70 + 0.00084T$	273–2000	1½
HI	*g*	$6.93 + 0.00083T$	273–2000	2
H_2O	*l*	See Table 3-174		
	g	$8.22 + 0.00015T + 0.00000134T^2$	300–2500	?
H_2S	*g*	$7.20 + 0.00360T$	300– 600	8
$H_2S_2O_7$	*c*	27	281	?
	l	58	308	?
Indium:				
In	*c*			
Iodine:				
I_2	*g*	9.00	300–2000	5
Iridium:				
Ir	*c*	$5.50 + 0.00148T$	273–1873	1
Iron:[4]				
Fe	*c*, α	$4.13 + 0.00638T$	273–1041	3
	c, β	$6.12 + 0.00336T$	1041–1179	3
	c, γ	8.40	1179–1674	5
	c, δ	10.0	1674–1803	5
	l	8.15	1803–1873	5
$FeAs_2$	*c*	17.8	283– 373	?
Fe_3C	*c*	$25.17 + 0.00223T$	273–1173	10
$FeCO_3$	*c*	22.7	293– 368	?
FeO	*c*	$12.62 + 0.001492T - 76200/T^2$	273–1173	2
Fe_2O_3	*c*	$24.72 + 0.01604T - 423400/T^2$	273–1097	2
Fe_3O_4	*c*	$41.17 + 0.01882T - 979500/T^2$	273–1065	2
$Fe_2O_3.3H_2O$	*c*	47.8	286– 373	?
FeS	*c*, α	$2.03 + 0.0390T$	273– 411	5
	c, β	$12.05 + 0.00273T$	411–1468	3
FeS_2	*c*	$10.7 + 0.01336T$	273– 773	?
FeSi	*c*	$10.54 + 0.00458T$	273– 903	2
Fe_2SiO_4	*c*	$33.57 + 0.01907T - 879700/T^2$	273–1161	2
$FeSO_4$	*c*	22	293– 373	?
$Fe_2(SO_4)_3$	*c*	66.2	273– 373	?
$FeSO_4.4H_2O$	*c*	63.6	282	?
$FeSO_4.7H_2O$	*c*	96	291– 319	?
Krypton:				
Kr	*g*	4.97	All	0
Lanthanum:				
La	*c*	$5.91 + 0.00100T$	273–1009	?
La_2O_3	*c*	$22.6 + 0.00544T$	273–2273	?
$La_2(MoO_4)_3$	*c*	86	273– 307	?
$La_2(SO_4)_3$	*c*	66.9	273– 373	?
$La_2(SO_4)_3.9H_2O$	*c*	152	273– 319	?
Lead:[4]				
Pb	*c*	$5.77 + 0.00202T$	273– 600	2
	l	6.8	600–1273	5
$Pb_3(AsO_4)_2$	*c*	65.5	286– 370	?
PbB_2O_4	*c*	26.5	288– 371	?
PbB_4O_7	*c*	41.4	289– 371	?
$PbBr_2$	*c*	$18.13 + 0.00310T$	273– 761	2
	l	27.4	761– 860	10
$PbCl_2$	*c*	$15.88 + 0.00835T$	273– 771	2
	l	27.2	771– 851	10
$2PbCl_2.NH_4Cl$	*c*	53.1	293	?
$PbCO_3$	*c*	21.1	286– 320	?
$PbCrO_4$	*c*	29.1	292– 323	?
PbF_2	*c*	$16.5 + 0.00412T$	273–1091	?
PbI_2	*c*	$18.66 + 0.00293T$	273– 648	2
	l	32.3	648– 776	20
$PbMoO_4$	*c*	30.4	292– 322	?
$Pb(NO_3)_2$	*c*	36.4	286– 320	?
PbO	*c*	$10.33 + 0.00318T$	273– 544	2
PbO_2	*c*	$12.7 + 0.00780T$	273– ?	?
$Pb_2P_2O_7$	*c*	48.3	284– 371	?
PbS	*c*	$10.63 + 0.00401T$	273– 873	3
$PbSO_4$	*c*	26.4	293– 372	?
PbS_2O_3	*c*	29	293– 373	?
$PbWO_4$	*c*	35	273– 297	?
Lithium:				
Li	*c*	$0.68 + 0.0180T$	273– 459	10
	g	4.97	All	0
LiBr	*c*	$11.5 + 0.00302T$	273– 825	?
$LiBr.H_2O$	*c*	22.6	278– 318	?
LiCl	*c*	$11.0 + 0.00339T$	273– 887	?
$LiCl.H_2O$	*c*	23.6	279– 360	?
LiF	*c*	$8.20 + 0.00520T$	273–1117	?
LiI	*c*	$12.5 + 0.00208T$	273– 723	?
$LiI.H_2O$	*c*	23.6	277– 359	?
$LiI.2H_2O$	*c*	32.9	277– 345	?
$LiI.3H_2O$	*c*	43.2	277– 347	?
$LiNO_3$	*c*	$9.17 + 0.0360T$	273– 523	5
	l	26.8	523– 575	5
Magnesium:[4]				
Mg	*c*	$6.20 + 0.00133T - 67800/T^2$	273– 923	1
	l	7.4	923–1048	10
MgAg	*c*	$10.58 + 0.00412T$	273– 905	2
Mg_4Al_3	*c*	$34.4 + 0.0198T$	273– 736	?
MgAu	*c*	$11.3 + 0.00189T$	273–1433	?
Mg_2Au	*c*	$16.2 + 0.00451T$	273–1073	?

TABLE 3-181 Heat Capacities of the Elements and Inorganic Compounds *(Continued)*

Substance	State†	Heat capacity at constant pressure (T = °K.; 0°C. = 273.1°K.), cal./deg. mol	Range of tempera-ture, °K.	Uncer-tainty, %
Magnesium—(*Cont.*):				
Mg_3Au	*c*	21.2 + 0.00614*T*	273–1103	?
$MgCl_2$	*c*	17.3 + 0.00377*T*	273– 991	?
$MgCl_2.6H_2O$	*c*	77.1	292– 342	?
$MgCO_3$	*c*	16.9	290	?
$MgCu_2$	*c*	14.96 + 0.00776*T*	273– 903	3
Mg_2Cu	*c*	15.5 + 0.00652*T*	273– 843	?
$MgNi_2$	*c*	15.87 + 0.00692*T*	273– 903	2
MgO	*c*	10.86 + 0.001197*T* − 208700/T^2	273–2073	2
$MgO.Al_2O_3$	*c*	28	288– 319	?
$MgO.SiO_2$	*c*, amphibole	25.60 + 0.004380*T* − 674200/T^2	273–1373	1
	c, pyroxene	23.35 + 0.008062*T* − 558800/T^2	273– 773	1
	gls	23.30 + 0.007734*T* − 542000/T^2	273– 973	1
$6MgO.MgCl_2.8B_2O_3$	*c*, α	58.7 + 0.408*T*	273– 538	5
	c, β	107.2 + 0.2876*T*	538– 623	5
$Mg(OH)_2$	*c*	18.2	292– 323	?
Mg_3Sb_2	*c*	28.2 + 0.00560*T*	273–1234	?
Mg_2Si	*c*	15.4 + 0.00415*T*	273–1343	?
$MgSO_4$	*c*	26.7	296– 372	?
$MgSO_4.H_2O$	*c*	33	282	?
$MgSO_4.6H_2O$	*c*	80	282	?
$MgSO_4.7H_2O$	*c*	89	291– 319	?
Manganese:				
Mn	*c*, α	3.76 + 0.00747*T*	273–1108	5
	c, β	5.06 + 0.00395*T*	1108–1317	5
	c, γ	4.80 + 0.00422*T*	1317–1493	5
	l	11.0	1493–1673	10
$MnCl_2$	*c*	16.2 + 0.00520*T*	273– 923	?
$MnCO_3$	*c*	7.79 + 0.0421*T* + 0.0000090T^2	273– 773	?
MnO	*c*	7.43 + 0.01038*T* − 0.00000362T^2	273–1923	?
Mn_2O_3	*c*	10.33 + 0.0530*T* − 0.0000257T^2	273–1173	?
Mn_3O_4	*c*	19.25 + 0.0538*T* − 0.0000209T^2	273–1773	?
MnO_2	*c*	1.92 + 0.0471*T* − 0.0000297T^2	273– 773	?
$Mn_2O_3.H_2O$	*c*	31	291– 322	?
MnS	*c*	10.21 + 0.00656*T* − 0.00000242T^2	273–1883	?
$MnSO_4$	*c*	27.5	293– 373	?
$MnSO_4.5H_2O$	*c*	78	290– 319	?
Mercury:[11]				
Hg	*l*	6.61	273– 630	1
	g	4.97	All	0
Hg_2	*g*	9.00	300–2000	5
HgCl	*c*	11.05 + 0.00370*T*	273– 798	?
$HgCl_2$	*c*	15.3 + 0.0103*T*	273– 553	?
$Hg(CN)_2$	*c*	25	285– 319	?
HgI	*c*	11.4 + 0.00461*T*	273– 563	?
HgI_2	*c*, α	17.4 + 0.004001*T*	273– 403	3
	c, β	20.2	403– 523	3
HgO	*c*	11.5	278– 371	?
HgS	*c*	10.9 + 0.00365*T*	273– 853	?
Hg_2SO_4	*c*	31.0	273– 307	?
Molybdenum:				
Mo	*c*	5.69 + 0.00188*T* − 50300/T^2	273–1773	5
MoO_3	*c*	15.1 + 0.0121*T*	273–1068	?
MoS_2	*c*	19.7 + 0.00315*T*	273– 729	?
Neon:[12]				
Ne	*g*	4.97	All	0
Nickel:[4]				
Ni	*c*, α	4.26 + 0.00640*T*	273– 626	2
	c, β	6.99 + 0.000905*T*	626–1725	5
	l	8.55	1725–1903	10
NiO	*c*	11.3 + 0.00215*T*	273–1273	?
NiS	*c*	9.25 + 0.00640*T*	273– 597	3
Ni_2Si	*c*	15.8 + 0.00329*T*	273–1582	?
NiSi	*c*	10.0 + 0.00312*T*	273–1273	?
Ni_3Sn	*c*	20.78 + 0.0102*T*	273– 904	2
$NiSO_4$	*c*	33.4	293– 373	?
$NiSO_4.6H_2O$	*c*	82	291– 325	?
NiTe	*c*	11.00 + 0.00433*T*	273– 700	2
Nitrogen:[13]				
N_2	*g*	6.50 + 0.00100*T*	300–3000	3
NH_3	*g*	6.70 + 0.00630*T*	300– 800	1½
NH_4Br	*c*	22.8	274– 328	?
NH_4Cl	*c*, α	9.80 + 0.0368*T*	273– 457	5
	c, β	5.0 + 0.0340*T*	457– 523	5
NH_4I	*c*	17.8	273– 328	?
NH_4NO_3	*c*	31.8	273– 293	?
$(NH_4)_2SO_4$	*c*	51.6	275– 328	?
NO	*g*	8.05 + 0.000233*T* − 156300/T^2	300–5000	2
Osmium:				
Os	*c*	5.686 + 0.000875*T*	273–1877	1
Oxygen:[14]				
O_2	*g*	8.27 + 0.000258*T* − 187700/T^2	300–5000	1
Palladium:				
Pd	*c*	5.41 + 0.00184*T*	273–1822	2
Phosphorus:				
P	*c*, yellow	5.50	273– 317	5
	c, red	0.21 + 0.0180*T*	273– 472	10
	l	6.6	317– 373	10
PCl_3	*l*	28.7	284– 371	?
P_4O_{10}	*c*	15.72 + 0.1092*T*	273– 631	2
	g	73.6	631–1371	3

TABLE 3-181 Heat Capacities of the Elements and Inorganic Compounds *(Continued)*

Substance	State†	Heat capacity at constant pressure (T = °K.; 0°C. = 273.1°K.), cal./deg. mol	Range of tempera-ture, °K.	Uncer-tainty, %
Platinum: [4]				
Pt	*c*	$5.92 + 0.00116T$	273–1873	1
Potassium:				
K	*c*	$5.24 + 0.00555T$	273– 336	5
	l	7.7	336– 373	5
	g	4.97	All	0
K_2	*g*	9.00	300–2000	5
$KAsO_3$	*c*	25.3	290– 372	?
KBO_2	*c*	$12.6 + 0.0126T$	273–1220	?
$K_2B_4O_7$	*c*	51.3	290– 372	?
KBr	*c*	$11.49 + 0.00360T$	273– 543	2
KCl	*c*	$10.93 + 0.00376T$	273–1043	2
$KClO_3$	*c*	25.7	289– 371	?
$KClO_4$	*c*	26.3	287– 318	?
$2KCl.CuCl_2.2H_2O$	*c*	63	292– 323	?
$2KCl.PtCl_4$	*c*	55	286– 319	?
$2KCl.SnCl_4$	*c*	54.5	292– 323	?
$2KCl.ZnCl_2$	*c*	43.4	279– 319	?
$2KCN.Zn(CN)_2$	*c*	57.4	277– 319	?
K_2CO_3	*c*	29.9	296– 372	?
K_2CrO_4	*c*	35.9	289– 371	?
$K_2Cr_2O_7$	*c*	$42.80 + 0.0410T$	273– 671	5
	l	96.9	671– 757	5
KF	*c*	$10.8 + 0.00284T$	273–1129	?
$K_4Fe(CN)_6$	*c*	80.1	273– 319	?
$K_4Fe(CN)_6.3H_2O$	*c*	114.5	273– 310	?
KH_2AsO_4	*c*	32	289– 319	?
KH_2PO_4	*c*	28.3	290– 320	?
$KHSO_4$	*c*	30	292– 324	?
$KMnO_4$	*c*	28	287– 318	?
KNO_3	*c*	$6.42 + 0.0530T$	273– 401	10
	c	28.8	401– 611	5
	l	29.5	611– 683	10
$K_2O.Al_2O_3.3SiO_2$	*c*, orthoclase	$69.26 + 0.00821T - 2331000/T^2$	273–1373	1½
	gls, orthoclase	$69.81 + 0.01053 - 2403000/T^2$	273–1373	1½
	c, microcline	$65.65 + 0.01102T - 1748000/T^2$	273–1373	1½
	gls, microcline	$64.83 + 0.01438T - 1641000/T^2$	273–1373	1½
$K_4P_2O_7$	*c*	63.1	290– 371	?
K_2SO_4	*c*	33.1	287– 371	?
$K_2S_2O_8$	*c*	37	293– 373	?
$K_2SO_4.Al_2(SO_4)_3.24H_2O$	*c*	352	292– 322	?
$K_2SO_4.Cr_2(SO_4)_3.24H_2O$	*c*	324	292– 324	?
$K_2SO_4.MgSO_4.6H_2O$	*c*	106	292– 323	?
$K_2SO_4.NiSO_4.6H_2O$	*c*	107	289– 319	?
$K_2SO_4.ZnSO_4.6H_2O$	*c*	120	293– 317	?
Prometheum:				
Pr	*c*			
Radon:				
Rn	*g*	4.97	All	0
Rhenium:				
Re	*c*	$6.30 + 0.00053T$	273–2273	?
Rhodium:				
Rh	*c*	$5.40 + 0.00219T$	273–1877	2
Rubidium:				
Rb	*c*	$3.27 + 0.0131T$	273– 312	2
	l	7.85	312– 373	5
RbBr	*c*	$11.6 + 0.00255T$	273– 954	?
RbCl	*c*	$11.5 + 0.00249T$	273– 987	?
Rb_2CO_3	*c*	28.4	291– 320	?
RbF	*c*	$11.3 + 0.00256T$	273–1048	?
RbI	*c*	$11.6 + 0.00263T$	273– 913	?
Scandium:				
Sc_2O_3	*c*	21.1	273– 373	?
$Sc_2(SO_4)_3$	*c*	62.0	273– 373	?
Selenium:				
Se	*c*	$4.53 + 0.00550T$	273– 490	2
	l	8.35	490– 570	3
Silicon:				
Si	*c*	$5.74 + 0.000617T - 101000/T^2$	273–1174	2
SiC	*c*	$8.89 + 0.00291T - 284000/T^2$	273–1629	2
$SiCl_4$	*l*	32.4	293– 373	?
SiO_2	*c*, quartz, α	$10.87 + 0.008712T - 241200/T^2$	273– 848	1
	c, quartz, β	$10.95 + 0.00550T$	848–1873	3½
	c, cristobalite, α	$3.65 + 0.0240T$	273– 523	2½
	c, cristobalite, β	$17.09 + 0.000454T - 897200/T^2$	523–1973	2
	gls	$12.80 + 0.00447T - 302000/T^2$	273–1973	3½
Silver: [4]				
Ag	*c*	$5.60 + 0.00150T$	273–1234	1
	l	8.2	1234–1573	3
Ag_3Al	*c*	$22.56 + 0.00570T$	273– 902	2
Ag_2Al	*c*	$16.85 + 0.00450T$	273– 903	2
$AgAl_{12}$	*c*	$58.62 + 0.0575T$	273– 768	5
AgBr	*c*	$8.58 + 0.0141T$	273– 703	6
	l	14.9	703– 836	5
AgCl	*c*	$9.60 + 0.00929T$	273– 728	2
	l	14.05	728– 806	5
AgCNO	*c*	18.7	273– 353	?
AgI	*c*, α	$8.58 + 0.0141T$	273– 423	6
$AgNO_3$	*c*, α	$18.83 + 0.0160T$	273– 433	2
	c, β	25.7	433– 482	5
	l	30.2	482– 541	5

TABLE 3-181 Heat Capacities of the Elements and Inorganic Compounds *(Continued)*

Substance	State†	Heat capacity at constant pressure (T = °K.; 0°C. = 273.1°K.), cal./deg. mol	Range of temperature, °K.	Uncertainty, %
Silver—(*Cont.*):				
Ag_3PO_4	*c*	37.5	293– 325	?
Ag_2S	*c*, α	18.8	273– 448	5
	c, β	21.8	448– 597	5
Ag_3Sb	*c*	$19.53 + 0.0160T$	273– 694	5
Ag_2Se	*c*, α	20.2	273– 406	5
	c, β	20.4	406– 460	5
Sodium:[15]				
Na	*c*	$5.01 + 0.00536T$	273– 371	1½
	l	7.50	371– 451	2
	g	4.97	All	0
$NaBO_2$	*c*	$10.4 + 0.0199T$	273–1239	?
$Na_2B_4O_7$	*c*	47.9	289– 371	?
$Na_2B_4O_7.10H_2O$	*c*	147	292– 323	?
NaBr	*c*	$11.74 + 0.00233T$	273– 543	2
NaCl	*c*	$10.79 + 0.00420T$	273–1074	2
	l	15.9	1073–1205	3
$NaClO_3$	*c*	$9.48 + 0.0468T$	273– 528	3
	l	31.8	528– 572	5
NaCNO	*c*	13.1	273– 353	?
Na_2CO_3	*c*	28.9	288– 371	?
NaF	*c*	$10.4 + 0.00289T$	273–1261	?
$Na_2HPO_4.7H_2O$	*c*	86.6	275– 307	?
$Na_2HPO_4.12H_2O$	*c*	133.4	275– 307	?
NaI	*c*	$12.5 + 0.00162T$	273– 936	?
$NaNO_3$	*c*	$4.56 + 0.0580T$	273– 583	5
	l	37.2	583– 703	10
$Na_2O.Al_2O_3.3SiO_2$	*c*, albite	$63.78 + 0.01171T - 1678000/T^2$	273–1373	1
	gls	$61.25 + 0.01768T - 1545000/T^2$	273–1173	1
$NaPO_3$	*c*	22.1	290– 319	?
$Na_4P_2O_7$	*c*	60.7	290– 371	?
Na_2SO_4	*c*	32.8	289– 371	?
$Na_2S_2O_3$	*c*	34.9	273– 307	?
$Na_2S_2O_3.5H_2O$	*c*	86.2	273– 307	?
Sodium-potassium alloys[15]	*l*			
Strontium:				
$SrBr_2$	*c*	$18.1 + 0.00311T$	273– 923	?
$SrBr_2.H_2O$	*c*	28.9	277– 370	?
$SrBr_2.6H_2O$	*c*	82.1	276– 327	?
$SrCl_2$	*c*	$18.2 + 0.00244T$	273–1143	?
$SrCl_2.H_2O$	*c*	28.7	276– 365	?
$SrCl_2.2H_2O$	*c*	38.3	277– 366	?
$SrCO_3$	*c*	21.8	281– 371	?
SrI_2	*c*	$18.6 + 0.00304T$	273– 783	?
$SrI_2.H_2O$	*c*	28.5	276– 363	?
$SrI_2.2H_2O$	*c*	39.1	275– 336	?
$SrI_2.6H_2O$	*c*	84.9	275– 333	?
$SrMoO_4$	*c*	37	273– 297	?
$Sr(NO_3)_2$	*c*	38.3	290– 320	?
$SrSO_4$	*c*	26.2	293– 369	?
Sulfur:[16]				
S	*c*, rhombic	$3.63 + 0.00640T$	273– 368	3
	c, monoclinic	$4.38 + 0.00440T$	368– 392	3
S_2	*g*	$8.58 + 0.00030T$	300–2500	5
S_2Cl_2	*l*	27.5	273– 332	?
SO_2	*g*	$7.70 + 0.00530T - 0.00000083T^2$	300–2500	2½
Tantalum:				
Ta	*c*	$5.91 + 0.00099T$	273–1173	2
Tellurium:				
Te	*c*	$5.19 + 0.00250T$	273– 600	3
Thallium:				
Tl	*c*, α	$5.32 + 0.00385T$	273– 500	1
	c, β	8.12	500– 576	1
	l	7.12	576– 773	3
TlBr	*c*	$12.53 + 0.00100T$	273– 733	10
	l	16.0	733– 800	10
TlCl	*c*	$12.56 + 0.00088T$	273– 700	5
	l	14.2	700– 803	10
Thorium:				
Th	*c*	6.40	273– 373	?
ThO_2	*c*	$14.6 + 0.00507T$	273–1273	?
$Th(SO_4)_2$	*c*	41.2	273– 373	?
Tin:‡				
Sn	*c*	$5.05 + 0.00480T$	273– 504	2
	l	6.6	504–1273	10
SnAu	*c*	$11.79 + 0.00233T$	273– 581	1
$SnCl_2$	*c*	$16.2 + 0.00926T$	273– 520	?
$SnCl_4$	*l*	38.4	286– 371	?
SnO	*c*	$9.40 + 0.00362T$	273–1273	?
SnO_2	*c*	$13.94 + 0.00565T - 252000/T^2$	273–1373	?
SnPt	*c*	$11.49 + 0.00190T$	273–1318	1
SnS	*c*	$12.1 + 0.00165T$	273–1153	?
SnS_2	*c*	$20.5 + 0.00400T$	273– 873	?
Titanium:				
Ti	*c*	$8.91 + 0.00114T - 433000/T^2$	273– 713	3
$TiCl_4$	*l*	35.7	285– 372	?
TiO_2	*c*	$11.81 + 0.00754T - 41900/T^2$	273– 713	3
Tungsten:				
W	*c*	$5.65 + 0.00866$	273–2073	1
WO_3	*c*	$16.0 + 0.00774T$	273–1550	?

TABLE 3-181 Heat Capacities of the Elements and Inorganic Compounds *(Concluded)*

Substance	State†	Heat capacity at constant pressure (T = °K.; 0°C. = 273.1°K.), cal./deg. mol	Range of temperature, °K.	Uncertainty, %
Uranium:				
U	c	6.64	273–372	?
U_8O_8	c	59.8	276–314	?
Vanadium:				
V	c	$5.57 + 0.00097T$	273–1993	?
Xenon:				
Xe	g	4.97	All	0
Zinc:[4]				
Zn	c	$5.25 + 0.00270T$	273–692	1
	l	$7.59 + 0.00055T$	692–1122	3
$ZnCl_2$	c	$15.9 + 0.00800T$	273–638	?
ZnO	c	$11.40 + 0.00145T - 182400/T^2$	273–1573	1
ZnS	c	$12.81 + 0.00095T - 194600/T^2$	273–1173	5
ZnSb	c	$11.5 + 0.00313T$	273–810	?
$ZnSO_4$	c	28	293–373	?
$ZnSO_4.H_2O$	c	34.7	282	?
$ZnSO_4.6H_2O$	c	80.8	282	?
$ZnSO_4.7H_2O$	c	100.2	273–307	?
Zirconium:				
ZrO_2	c	$11.62 + 0.01046T - 177700/T^2$	273–1673	5
$ZrO_2.SiO_2$	c	26.7	297–372	?

[1]See also Table 3-182. Data to 298°K are also given by Scott, *Cryogenic Engineering*, Van Nostrand, Princeton, N.J., 1959.
[2]For liquid and gas data see Johnson (ed.), WADD-TR-60-56, 1960.
[3]Stalder, NACA Tech. Note 4141, 1957 (Fig. 5), gives data from 400 to 2600°R.
[4]See also Table 3-182.
[5]For data from 400 to 5500°R see Stalder, NACA Tech. Note 4141, 1975 (Fig. 4).
[6]For *s*, *l*, *g* data see Johnson (ed.), WADD-TR-60-56, 1960.
[7]For data from 400 to 2350°R see Stalder, NACA Tech. Note 4141, 1957.
[8]For *s*, *l*, *g* data see Johnson (ed.), WADD-TR-60-56, 1960.
[9]For *l*, *g* data see Johnson (ed.), WADD-TR-60-56, 1960.
[10]For *s*, *l*, *g* data see Johnson (ed.), WADD-TR-60-56, 1960.
[11]See also Table 3-182; Douglas, Ball, et al., *Bur. Stand. J. Res.*, **46**, 334 (1951); Busey and Giaque, *J. Am. Chem. Soc.*, **75**, 806 (1953); Sheldon, ASME Pap. 49-A-30, 1949.
[12]For *s*, *l*, *g* data see Johnson (ed.), WADD-TR-56-60, 1960.
[13]For *s*, *l*, *g* data see Johnson (ed.), WADD-TR-56-60, 1960.
[14]For *s*, *l*, *g* data see Johnson (ed.), WADD-TR-56-60, 1960.
Ozone: For liquid see Brabets and Waterman, *J. Chem. Phys.*, **28**, 1212, 1958.
[15]For data on liquid Na-K alloys to 1500°F and for liquid Na to 1460°F see Lubarsky and Kaufman, NACA Rep. 1270, 1956.
[16]See also Evans and Wagman, *Bur. Stand. J. Res.* **49**, 141 (1952); Gratch, OTS PB 124957, 1950; Guthrie, Scott, et al., *J. Am. Chem. Soc.*, **76**, 1488 (1954).

TABLE 3-182 Specific Heat [kJ/(kg·K)] of Selected Elements

Symbol	Temperature, K														
	4	6	8	10	20	40	60	80	100	200	250	300	400	600	800
Al	0.00026	0.00050	0.00088	0.00140	0.0089	0.0775	0.214	0.357	0.481	0.797	0.859	0.902	0.949	1.042	1.134
Be	0.00008	......	......	0.00028	0.0014	......	......	......	0.195	1.109	1.537	1.840	2.191	2.605	2.823
Bi	0.00054	0.00220	0.00541	0.01040	0.0340	0.0729	0.092	0.102	0.109	0.120	0.121	0.122	0.123	0.142	0.136
Cr	0.00016	0.00029	0.00050	0.00081	0.0021	0.0107	0.059	0.127	0.190	0.382	0.424	0.450	0.501	0.565	0.611
Co	0.00036	0.00059	0.00085	0.00121	0.0048	0.0404	0.110	0.184	0.234	0.376	0.406	0.426	0.451	0.509	0.543
Cu	0.00011	0.00024	0.00048	0.00086	0.0076	0.059	0.137	0.203	0.254	0.357	0.377	0.386	0.396	0.431	0.448
Ge	......	......	0.00037	0.00081	0.0129	0.0619	0.108	0.153	0.192	0.286	0.305	0.323	0.343	0.364	0.377
Au	0.00018	0.00047	0.00126	0.00255	0.0163	0.0569	0.084	0.100	0.109	0.124	0.127	0.129	0.131	0.136	0.141
Ir	......	......	......	0.00032	0.0021	......	......	......	0.090	0.122	0.128	0.131	0.133	0.140	0.146
Fe	0.00038	0.00061	0.00090	0.00127	0.0039	0.0276	0.086	0.154	0.216	0.384	0.422	0.450	0.491	0.555	0.692
Pb	0.00075	0.00242	0.00747	0.01350	0.0531	0.0944	0.108	0.114	0.118	0.125	0.127	1.129	0.132	0.142	
Mg	0.00034	0.00080	0.00155	0.00172	0.0148	0.138	0.336	0.513	0.648	0.929	0.985	1.005	1.082	1.177	1.263
Hg	0.00417	0.01420	0.01820	0.02250	0.0515	0.0895	0.107	0.116	0.121	0.136	0.141	0.139	0.136	0.135	0.104
Mo	0.00011	0.00019	0.00032	0.00050	0.0029	0.0236	0.061	0.105	0.140	0.223	0.241	0.248	0.261	0.280	0.292
Ni	0.00054	0.00086	0.00121	0.00178	0.0058	0.0380	0.103	0.173	0.232	0.383	0.416	0.444	0.490	0.590	0.530
Pt	0.00019	0.00028	0.00067	0.00112	0.0077	0.0382	0.069	0.088	0.101	0.127	0.132	0.134	0.136	0.140	0.146
Ag	0.00016	0.00035	0.00093	0.00186	0.0159	0.0778	0.133	0.166	0.187	0.225	0.232	0.236	0.240	0.251	0.264
Sn	0.00024	0.00127	0.00423	0.00776	0.0400	0.108	0.149	0.173	0.189	0.214	0.220	0.222	0.245	0.257	0.257
Zn	0.00011	0.00029	0.00096	0.00250	0.0269	0.123	0.205	0.258	0.295	0.366	0.380	0.389	0.404	0.435	0.479

TABLE 3-183 Specific Heats of Organic Liquids*

From *International Critical Tables*, vol. 5, pp. 107–113, and a few data from other sources

Compound	Formula	Temperature, °C.	Sp. ht., cal./g. °C.	Compound	Formula	Temperature, °C.	Sp. ht., cal./g. °C.
Acetal	$C_6H_{14}O_2$	0	0.467	Bromochlorobenzene (o-)	C_6H_4BrCl	0	.215
		19–99	.520	(m-)	C_6H_4BrCl	0	.212
Acetic acid*	$C_2H_4O_2$	26–95	.522	Bromoiodobenzene (o-)	C_6H_4BrI	0	.153
Acetone*	C_3H_6O	3–22.6	.514			5–100	.160
		0	.506			3.2–64.6	.157
		24.2–49.4	.538			1.8–34	.157
Acetonitrile	C_2H_3N	21–76	.541	(m-)	C_6H_4BrI	0	.152
Acetophenone	C_8H_8O	20–196	.450			5–100	.158
Acetyl chloride	C_2H_3ClO	0	.339			3.2–64.5	.156
Allyl acetate	$C_5H_8O_2$	0	.430			1.7–34.1	.154
alcohol	C_3H_6O	0	.386			1.7–36.2	.149
		21–96	.665	Bromophenol	C_6H_5BrO	18–77	.316
benzoate	$C_{10}H_{10}O$	20	.388	Butane (n-)	C_4H_{10}	0	.549
butyrate	$C_7H_{12}O_2$	20	.451	Butyl alcohol (n-)	$C_4H_{10}O$	21–115	.687
chloride	C_3H_5Cl	0	.313			30	.582
chloroacetate	$C_5H_7ClO_2$	20	.396			−76.2	.443
dichloroacetate	$C_5H_6Cl_2O_2$	20	.332			−33.3	.453
isobutyrate	$C_7H_{12}O_2$	20	.448			2.3	.526
propionate	$C_6H_{10}O_2$	20	.451			19.2	.563
trichloroacetate	$C_5H_5Cl_3O_2$	20	.288	chloride (n-)	C_4H_9Cl	20	.451
valerate	$C_8H_{14}O$	20	.451	formate (n-)	$C_5H_{10}O_2$	20	.459
Aminobenzoic acid (o-)	$C_7H_7NO_2$	M. P.	.435	propionate	$C_7H_{14}O_2$	20	.459
(m-)	$C_7H_7NO_2$	M. P.	.435	valerate	$C_9H_{18}O_2$	20	.459
(p-)	$C_7H_7NO_2$	M. P.	.444	Butyric acid (n-)	$C_4H_8O_2$	0	.444
Amyl alcohol (d-primary)	$C_5H_{12}O$	22–125	.711			40	.501
(t-)	$C_5H_{12}O$	20–99	.753			20–100	.515
Amylene	C_5H_{10}	0	.282	Butyronitrile (n-)	C_4H_7N	21–113	.547
Anethole	$C_{10}H_{12}O$	23–233	.511				
		22.48	.551	Caproic acid	$C_6H_{12}O_2$	29–105	.531
		24.59	.564	Capronitrile	$C_6H_{11}N$	18–156	.541
		25.23	.612	Carbon tetrachloride*	CCl_4	0	.198
Aniline	C_6H_7N	8–82	.512			20	.201
						30	.200
Benzaldehyde	C_7H_6O	22–172	.428	Carvacrol	$C_{10}H_{14}O$	24–233	.575
Benzene*	C_6H_6	6–60	.419	Chloral	C_2HCl_3O	17–53	.250
		10	.340	hydrate	$C_2H_3Cl_3O_2$	55–88	.470
		65	.482	Chlorobenzene*	C_6H_5Cl	0	.273
Benzonitrile	C_7H_5N	22–186	.441			10	.298
Benzophenone (β-)	$C_{13}H_{10}O$	3–40	.382			20	.308
		0	.346	Chlorobenzoic acid (o-)	$C_7H_5ClO_2$	0	0.390
Benzyl alcohol	C_7H_8O	20–100	.511	(m-)	$C_7H_5ClO_2$	0	.265
		22–200	.540	(p-)	$C_7H_5ClO_2$	M. P.	.545
chloride	C_7H_7Cl	0	.323	Chloroform*	$CHCl_3$	0	.232
ethylene	C_9H_{10}	0	.393			15	.226
Bromobenzene	C_6H_5Br	0	.215			30	.234
		20–100	.231	Chlorophenol	C_6H_5ClO	0–20	.399
		16.9–65	.239	Chlorotoluene	C_7H_7Cl	0	.315

*See Fig. 3-11 for the specific heats of a number of substances as a function of the temperature. For additional data on the specific heat of liquid organic compounds see pp. 2278–2282 of *Handbook of Chemistry and Physics*, 40th ed., Chemical Rubber Publishing Co. Extensive data on ethylene glycol, diethylene glycol, triethylene glycol, and propylene glycol solutions are contained in Union Carbide Corporation publication F4763F, 1958.

TABLE 3-183 Specific Heats of Organic Liquids *(Continued)*

Compound	Formula	Temperature, °C.	Sp. ht., cal./g. °C.
Cresol (o-)	C_7H_8O	0–20	.497
(m-)	C_7H_8O	21–197	.551
		0–20	.477
Cresyl methyl ether (p-)	$C_8H_{10}O$	0	.404
Crotonic acid	$C_4H_6O_2$	71.4	.500
Cyclohexanol	$C_6H_{12}O$	15–18	.416
Cyclohexanone	$C_6H_{10}O$	15–18	.431
o-Cymene	$C_{10}H_{14}$	0	.398
Decahydronaphthalene (cis-)	$C_{10}H_{18}$	15–18	.393
Decane	$C_{10}H_{22}$ b.p. = 159	21–154	.588
	$C_{10}H_{22}$ b.p. = 162	0–50	.493
	$C_{10}H_{22}$ b.p. = 172	0–50	.500
Decylene (γ-)	$C_{10}H_{20}$	0–50	.467
Diallyl oxalate	$C_8H_{10}O_4$	20	.424
succinate	$C_{10}H_{14}O_4$	20	.450
Diamylene	$C_{10}H_{20}$	20–130	.543
Dibromobenzene (o-)	$C_6H_4Br_2$	0	.179
(m-)	$C_6H_4Br_2$	0	.175
Dibutyl oxalate	$C_{10}H_{18}O_4$	20	.439
Dichlorodifluormethane	CCl_2F_2	−43	.21
Dichloroacetic acid	$C_2H_2Cl_2O_2$	21–106	.349
		21–196	.348
Dichlorobenzene (o-)	$C_6H_4Cl_2$	0	.269
(m-)	$C_6H_4Cl_2$	0	.269
(p-)	$C_6H_4Cl_2$	53–99	.297
Diethylamine	$C_4H_{11}N$	22.5	.516
Diethyl carbonate	$C_5H_{10}O_3$	0	.245
		20–100	.462
ether (see Ether)	$C_4H_{10}O$	20.2–123	.473
ketone	$C_5H_{10}O$	20–98.5	.555
malate	$C_8H_{14}O_5$	24–186	.473
malonate	$C_7H_{12}O_4$	20	.431
oxalate	$C_6H_{10}O_4$	20	.431
succinate	$C_8H_{14}O_4$	20	.450
Dihydronaphthalene	$C_{10}H_{10}$	18–28	.345
Di-iodobenzene (m-)	$C_6H_4I_2$	34.2–99.6	.139
Di-isoamyl	$C_{10}H_{22}$	21.5–155	.588
oxalate	$C_{12}H_{22}O_4$	20	.447
Di-isobutylamine	C_8H_9N	22–130	.569
Dimethyl aniline	$C_8H_{11}N$	0–20	.416
		0	.403
naphthalene (β-)	$C_{12}H_{12}$	0	.392
pyrone	$C_7H_8O_2$	166	.547
Dinitrobenzene (m-)	$C_6H_4N_2O_4$	M. P.	.404
Diphenylamine	$C_{12}H_{11}N$	54	.437
		56	.441
		66	.480
Dipropyl ketone	$C_7H_{14}O$	20–140	.550
malonate	$C_9H_{16}O_4$	20	.431
oxalate (n-)	$C_8H_{14}O_4$	20	.431
succinate	$C_{10}H_{18}O_4$	20	.450
Dodecane	$C_{12}H_{26}$	14–20	.505
		0–50	.498
Dodecylene	$C_{12}H_{24}$	0–50	.455
Ether*	$C_4H_{10}O$	−100	.511
		−50	.515
		−5	.525
		0	.521
		+30	.545
		80	.687
		120	.800
		140	.819
		180	1.037
Ethyl acetate*	$C_4H_8O_2$	20	0.457
		20	.476
acetoacetate	$C_6H_{10}O_3$	0	.428
		20–100	.475
alcohol*	C_2H_6O (100%)	−20	.505
		0–98	.680
benzene	C_8H_{10}	0	.392
		30	.407
benzoate	$C_9H_{10}O_2$	20	.387
bromide*	C_2H_5Br	−100	0.194
		−20	.206
		5–10	.216
		10–15	.213
		15–20	.214
butyrate	$C_6H_{12}O_2$	20	.457
chloride	C_2H_5Cl	−28 to +4	.426
		0	.367
chloroacetate	$C_4H_7ClO_2$	9–138	.416
		20	.397
cresyl ether (p-)	$C_9H_{12}O$	0	.427
dichloroacetate	$C_4H_6Cl_2O_2$	20	.328
ether	$C_4H_{10}O$	0	.521
formate	$C_3H_6O_2$	14–49	.508
		−20 to +14	.454
Ethyl (*Cont.*):			
iodide*	C_2H_5I	−30	.156
		0	.161
		60	.171
isobutyrate	$C_6H_{12}O_2$	20	.457
propionate	$C_5H_{10}O_2$	20	.457
silicate	$C_8H_{20}SiO_4$	15–98	.424
sulfide	$C_4H_{10}S$	0	.468
		5–10	.470
		10–15	.473
		20–70	.477
trichloroacetate	$C_4H_5Cl_3O_2$	10–81	.294
		9–139	.305
		20	.284
valerate	$C_7H_{14}O_2$	20	.457
Ethylene bromide	$C_2H_4Br_2$	8–95	.182
		13–106	.175
		20	.173
chloride	$C_2H_4Cl_2$	−30	.278
		+20	.299
		30	.304
		50	.313
		60	.318
dichloroacetate	$C_6H_6Cl_4O_4$	0	.321
glycol*	$C_2H_6O_2$	−11.1	.535
		0	.542
		+ 2.5	.550
		5.1	.554
		14.9	.569
		19.9	.573
Formamide	CH_3NO	19	.549
Formic acid	CH_2O_2	0	.436
		15.5	.509
		20–100	.524
Furfural	$C_5H_4O_2$	0	.367
		20–100	.416
Glycerol*	$C_3H_8O_3$	15–50	.576
Glycol (ethylene)*	$C_2H_6O_2$	(see ethylene	glycol)
Heptaldehyde	$C_7H_{14}O$	0	.364
Heptane (n-)*	C_7H_{16}	0–50	.507
		20	.490
		30	.518
Heptylene	C_7H_{14}	0–50	.486
Heptylic acid	$C_7H_{14}O_2$	9	.556
Hexadecane (n-)	$C_{16}H_{34}$	0–50	.496
Hexadiene (1,5-)	C_6H_{10}	0	.405
Hexahydrocresol (o-)	$C_7H_{14}O$	15–18	.416
(m-)	$C_7H_{14}O$	15–18	.420
(p-)	$C_7H_{14}O$	15–18	.421
Hexane (n-)	C_6H_{14}	0–50	.527
		20–100	.600
Hexylene	C_6H_{12}	0–50	.504
Isoamyl acetate	$C_7H_{14}O_2$	20	.459
alcohol	$C_5H_{12}O$	0	.502
		20	.535
		30	.570
		47.9	.662
		10–117	.693
		21–130	.695
		75.5	.688
amine	$C_5H_{13}N$	22–91	.614
butyrate	$C_9H_{18}O_2$	20	.459
formate	$C_6H_{12}O_2$	16–65	.509
isobutyrate	$C_9H_{18}O_2$	20	.459
propionate	$C_8H_{16}O_2$	20	.459
succinate	$C_{14}H_{26}O_4$	0	.449
valerate	$C_{10}H_{20}O_2$	20	.459
Isobutane	C_4H_{10}	0	.549
Isobutyl acetate	$C_6H_{12}O_2$	20	.459
alcohol	$C_4H_{10}O$	21–109	.716
		30	.603
butyrate	$C_8H_{16}O_2$	20	0.459
succinate	$C_{12}H_{22}O_4$	0	.442
Isobutyric acid	$C_4H_8O_2$	20	.450
Isoheptane	C_7H_{16}	0–50	.501
Isopentane	C_5H_{12}	0	.512
		8	.527
Isovaleric acid	$C_5H_{10}O_2$	20	.463
		23–93	.590
Lauric acid	$C_{12}H_{24}O_2$	40–100	0.572
		57	.515
Mesitylene	C_9H_{12}	0	.393
Mesityl oxide	$C_6H_{10}O$	21–128	.521
Methyl acetate	$C_3H_6O_2$	15	.468
Methylal	$C_3H_8O_2$	15–41	.521

TABLE 3-183 Specific Heats of Organic Liquids *(Concluded)*

Compound	Formula	Temperature, °C.	Sp. ht., cal./g. °C.
Methyl alcohol*	CH_4O	5–10	.590
		15–20	.601
aniline	C_7H_9N	20–197	.512
benzoate	$C_8H_8O_2$	0	.363
butyl ketone	$C_6H_{12}O$	21–127	.553
butyrate (n-)	$C_5H_{10}O_2$	20	.459
chloroacetate	$C_3H_5ClO_2$	20	.382
cyclohexanone (o-)	$C_7H_{12}O$	15–18	.436
(m-)	$C_7H_{12}O$	15–18	.441
(p-)	$C_7H_{12}O$	15–18	.441
dichloroacetate	$C_3H_4Cl_2O_2$	20	.311
ethylketone	C_4H_8O	20–78	.549
ethylketoxime	C_4H_9NO	21.8–151.5	.650
formate	$C_2H_4O_2$	13–29	.516
hexyl ketone	$C_8H_{16}O$	22–168	.552
isobutyl ketone	$C_6H_{12}O$	20	.459
isopropyl ketone	$C_5H_{10}O$	20–91	.525
propionate	$C_4H_8O_2$	20	.459
trichloroacetate	$C_3H_3Cl_3O_2$	20	.267
valerate	$C_6H_{12}O_2$	20	.459
Methylene chloride	CH_2Cl_2	15–40	.288
Myristic acid	$C_{14}H_{28}O_2$	56–100	.539
Naphthalene	$C_{10}H_8$	87.5	.402
Naphthylamine (α-)	$C_{10}H_9N$	53.2	.475
		94.2	.476
Nitrobenzene	$C_6H_5NO_2$	10	.358
		30	.339
		50	.330
		70	.330
		90	.343
		120	.394
Nitrobenzoic acid (p-)	$C_7H_5NO_4$	M. P.	.449
Nitromethane	CH_3NO_2	17	.412
Nitronaphthalene (α-)	$C_{10}H_7NO_2$	58.6	.365
		61.4	.378
		94.3	.390
Nonane	C_9H_{20}	0–50	.503
Nonylene	C_9H_{18}	0–50	.485
Octane (n-)*	C_8H_{18}	0–50	.505
		20–123	.578
Octylene	C_8H_{16}	0–50	.486
Palmitic acid	$C_{16}H_{32}O_2$	65–104	.653
Paraldehyde	$C_6H_{12}O_3$	0	.436
Pentadecane	$C_{15}H_{32}$	0–50	.497
Pentadecylene	$C_{15}H_{30}$	0–50	.471
Phenetole	$C_8H_{10}O$	20	.446
Phenyl methyl ether	C_7H_8O	0	.405
		20–152	.483
Picoline (α-)	C_6H_7N	22–124	.434
Piperidine	$C_5H_{11}N$	20–98	.523
Propane	C_3H_8	0	.576
Propionaldehyde	C_3H_6O	0	.522
Propionic acid	$C_3H_6O_2$	0	.444
		20–137	.560
Propionitrile	C_3H_5N	0	.508
		19–95	.538
Propyl acetate (n-)	$C_5H_{10}O_2$	20	.459
benzene	C_9H_{12}	0	.400
benzoate	$C_{10}H_{12}O_2$	20	.398
butyrate	$C_7H_{14}O_2$	20	.459
Propyl (*Cont.*):			
chloroacetate	$C_5H_9ClO_2$	20	.414
dichloroacetate	$C_5H_8Cl_2O_2$	20	.341
formate (n-)	$C_4H_8O_2$	20	.459
isobutyrate	$C_7H_{14}O_2$	20	.459
phenyl ether	$C_9H_{12}O$	0	.429
propionate	$C_6H_{12}O_2$	20	.459
trichloroacetate	$C_5H_7Cl_3O_2$	20	.297
valerate	$C_8H_{16}O_2$	20	.459
Pseudocumene	C_9H_{12}	20	.414
Pyridine	C_5H_5N	20	.405
		21–108	.431
		0–20	.395
Quinoline	C_9H_7N	0–20	.352
Salicylaldehyde	$C_7H_6O_2$	18	.382
Salol	$C_{13}H_{10}O_3$	44.1	.391
Stearic acid	$C_{18}H_{36}O_2$	75–137	.550
Tetrachloroethane	$C_2H_2Cl_4$	20	.268
Tetrachloroethylene	C_2Cl_4	20	.216
		24	.211
Tetradecane	$C_{14}H_{30}$	0–50	0.497
Thymol (m-)	$C_{10}H_{14}O$	50	.566
Toluene*	C_7H_8	10	.364
		85	.534
		12–99	.440
Toluidine (o-)	C_7H_9N	0	.454
		22–195	.598
		40.5	.498
(p-)	C_7H_9N	43	.524
		58	.634
		94	.533
Trichloroethane	$C_2H_3Cl_3$	20	0.266
Trichloroethylene	C_2HCl_3	20	.223
Tridecane	$C_{13}H_{28}$	0–50	.499
Tridecylene	$C_{13}H_{26}$	0–50	.457
Trinitrotoluene (2,4,6-)	$C_7H_5N_3O_6$	....	.335
Undecane	$C_{11}H_{24}$	0–50	.501
Undecylene	$C_{11}H_{22}$	0–50	.482
Valeronitrile	C_5H_9N	23–121	.520
Xylene (o-)*	C_8H_{10}	30	.411
		39	.450
(m-)*	C_8H_{10}	0	.383
		9–40	.400
		16–35	.387
		30	.401
(p-)*	C_8H_{10}	0	.383
		30	.397
		40.8	.428
dibromide (o-)	$C_8H_8Br_2$	15–40	.183
(m-)	$C_8H_8Br_2$	15–40	.184
(p-)	$C_8H_8Br_2$	15–40	.180
dichloride (o-)	$C_8H_8Cl_2$	15–40	.283
(m-)	$C_8H_8Cl_2$	15–40	.295
(p-)	$C_8H_8Cl_2$	15–40	.282
tetrachloride (o-)	$C_8H_6Cl_4$	15–40	.240
(p-)	$C_8H_6Cl_4$	15–40	.242
Xylyl ethyl ether (2,4-)	$C_{10}H_{14}O$	0	.417

Specific heat = P.c.u. / (lb.) (deg. C.) = B.t.u. / (lb.) (deg. F.) = calories / (gm.) (deg. C.)

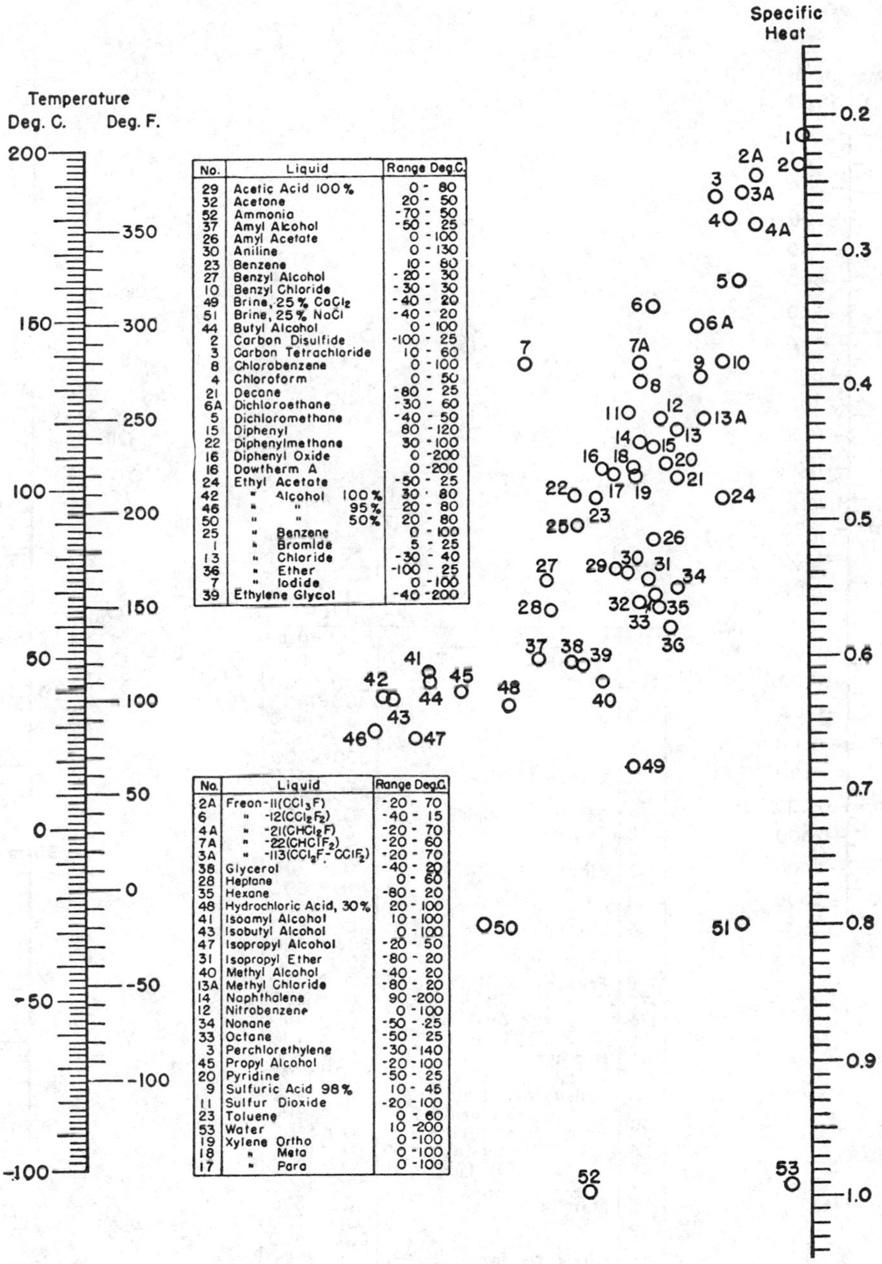

No.	Liquid	Range Deg.C.
29	Acetic Acid 100%	0 - 80
32	Acetone	20 - 50
52	Ammonia	-70 - 50
37	Amyl Alcohol	-50 - 25
26	Amyl Acetate	0 - 100
30	Aniline	0 - 130
23	Benzene	10 - 80
27	Benzyl Alcohol	-20 - 30
10	Benzyl Chloride	-30 - 30
49	Brine, 25% $CaCl_2$	-40 - 20
51	Brine, 25% NaCl	-40 - 20
44	Butyl Alcohol	0 - 100
2	Carbon Disulfide	-100 - 25
3	Carbon Tetrachloride	10 - 60
8	Chlorobenzene	0 - 100
4	Chloroform	0 - 50
21	Decane	-80 - 25
6A	Dichloroethane	-30 - 60
5	Dichloromethane	-40 - 50
15	Diphenyl	80 - 120
22	Diphenylmethane	30 - 100
16	Diphenyl Oxide	0 - 200
16	Dowtherm A	0 - 200
24	Ethyl Acetate	-50 - 25
42	" Alcohol 100%	30 - 80
46	" " 95%	20 - 80
50	" " 50%	20 - 80
25	" Benzene	0 - 100
1	" Bromide	5 - 25
13	" Chloride	-30 - 40
36	" Ether	-100 - 25
7	" Iodide	0 - 100
39	Ethylene Glycol	-40 - 200

No.	Liquid	Range Deg.C.
2A	Freon-11(CCl_3F)	-20 - 70
6	" -12(CCl_2F_2)	-40 - 15
4A	" -21($CHCl_2F$)	-20 - 70
7A	" -22($CHClF_2$)	-20 - 60
3A	" -113(CCl_2F-$CClF_2$)	-20 - 70
38	Glycerol	-40 - 20
28	Heptane	0 - 60
35	Hexane	-80 - 20
48	Hydrochloric Acid, 30%	20 - 100
41	Isoamyl Alcohol	10 - 100
43	Isobutyl Alcohol	0 - 100
47	Isopropyl Alcohol	-20 - 50
31	Isopropyl Ether	-80 - 20
40	Methyl Alcohol	-40 - 20
13A	Methyl Chloride	-80 - 20
14	Naphthalene	90 - 200
12	Nitrobenzene	0 - 100
34	Nonane	-50 - 25
33	Octane	-50 - 25
3	Perchlorethylene	-30 - 140
45	Propyl Alcohol	-20 - 100
20	Pyridine	-50 - 25
9	Sulfuric Acid 98%	10 - 45
11	Sulfur Dioxide	-20 - 100
23	Toluene	0 - 60
53	Water	10 - 200
19	Xylene Ortho	0 - 100
18	" Meta	0 - 100
17	" Para	0 - 100

FIG. 3-11 Specific heats of liquids. (*Chilton, Colburn, and Vernon, personal communication. Based on data from* International Critical Tables.)

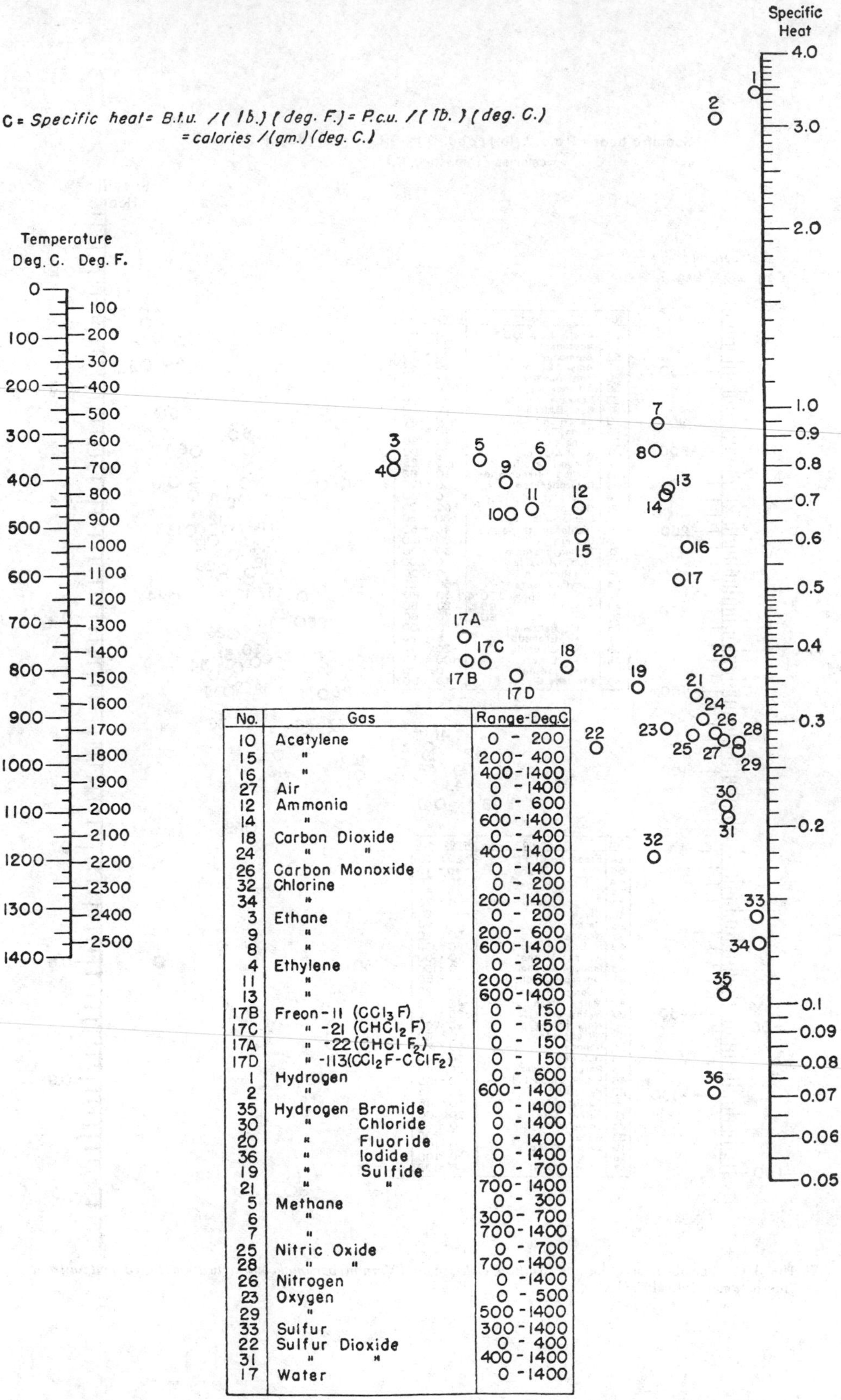

No.	Gas	Range-DegC
10	Acetylene	0 - 200
15	"	200 - 400
16	"	400 - 1400
27	Air	0 - 1400
12	Ammonia	0 - 600
14	"	600 - 1400
18	Carbon Dioxide	0 - 400
24	" "	400 - 1400
26	Carbon Monoxide	0 - 1400
32	Chlorine	0 - 200
34	"	200 - 1400
3	Ethane	0 - 200
9	"	200 - 600
8	"	600 - 1400
4	Ethylene	0 - 200
11	"	200 - 600
13	"	600 - 1400
17B	Freon-11 (CCl_3F)	0 - 150
17C	" -21 ($CHCl_2F$)	0 - 150
17A	" -22 ($CHClF_2$)	0 - 150
17D	" -113 (CCl_2F-$CClF_2$)	0 - 150
1	Hydrogen	0 - 600
2	"	600 - 1400
35	Hydrogen Bromide	0 - 1400
30	" Chloride	0 - 1400
20	" Fluoride	0 - 1400
36	" Iodide	0 - 1400
19	" Sulfide	0 - 700
21	" "	700 - 1400
5	Methane	0 - 300
6	"	300 - 700
7	"	700 - 1400
25	Nitric Oxide	0 - 700
28	" "	700 - 1400
26	Nitrogen	0 - 1400
23	Oxygen	0 - 500
29	"	500 - 1400
33	Sulfur	300 - 1400
22	Sulfur Dioxide	0 - 400
31	" "	400 - 1400
17	Water	0 - 1400

FIG. 3-12 Specific heats (C_p) of gases at 1-atm pressure.

TABLE 3-184 Specific Heats of Organic Solids

Recalculated from *International Critical Tables*, vol. 5, pp. 101–105

Compound	Formula	Temperature, °C.	Sp. ht., cal./g. °C.
Acetic acid	$C_2H_4O_2$	−200 to +25	0.330 + 0.00080*t*
Acetone	C_3H_6O	−210 to −80	0.540 + 0.0156*t*
Aminobenzoic acid (o-)	$C_7H_7NO_2$	85 to m.p.	0.254 + 0.00136*t*
(m-)	$C_7H_7NO_2$	120 to m.p.	0.253 + 0.00122*t*
(p-)	$C_7H_7NO_2$	128 to m.p.	0.287 + 0.00088*t*
Aniline	C_6H_7N	....	0.741
Anthracene	$C_{14}H_{10}$	50	0.308
		100	0.350
		150	0.382
Anthraquinone	$C_{14}H_8O_2$	0 to 270	0.258 + 0.00069*t*
Apiol	$C_{12}H_{14}O_4$	10	0.299
Azobenzene	$C_{12}H_{10}N_2$	28	0.330
Benzene	C_6H_6	−250	0.0399
		−225	0.0908
		−200	0.124
		−150	0.170
		−100	0.227
		− 50	0.299
		0	0.375
Benzoic acid	$C_7H_6O_2$	20 to m.p.	0.287 + 0.00050*t*
Benzophenone	$C_{13}H_{10}O$	−150	0.115
		−100	0.172
		− 50	0.220
		0	0.275
		+ 20	0.303
Betol	$C_{17}H_{12}O_3$	−150	0.129
		−100	0.167
		0	0.248
		+ 50	0.308
Bromoiodobenzene (o-)	C_6H_4BrI	− 50 to 0	0.143 + 0.00025*t*
(m-)	C_6H_4BrI	− 75 to −15	0.143
(p-)	C_6H_4BrI	− 40 to 50	0.116 + 0.00032*t*
Bromonaphthalene (β-)	$C_{10}H_7Br$	41	0.260
Bromophenol	C_6H_5BrO	32	0.263
Camphene	$C_{10}H_{16}$	35	0.380
Capric acid	$C_{10}H_{20}O_2$	8	0.695
Caprylic acid	$C_8H_{16}O_2$	− 2	0.628
Carbon tetrachloride	CCl_4	−240	0.013
		−200	0.081
		−160	0.131
		−120	0.162
		− 80	0.182
		− 40	0.201
Cerotic acid	$C_{27}H_{54}O_2$	15	0.387
Chloral alcoholate	$C_4H_7Cl_3O_2$	78	0.509
hydrate	$C_2H_3Cl_3O_2$	32	0.213
Chloroacetic acid	$C_2H_3ClO_2$	60	0.363
Chlorobenzoic acid (o-)	$C_7H_5ClO_2$	80 to m.p.	0.228 + 0.00084*t*
(m-)	$C_7H_5ClO_2$	94 to m.p.	0.232 + 0.00073*t*
(p-)	$C_7H_5ClO_2$	180 to m.p.	0.242 + 0.00055*t*
Chlorobromobenzene (o-)	C_6H_4BrCl	− 34	0.192
(m-)	C_6H_4BrCl	− 52	0.150
(p-)	C_6H_4BrCl	− 40	0.150
Crotonic acid	$C_4H_6O_2$	38 to 70	0.520 + 0.00020*t*
Cyamelide	$C_3H_3N_3O_3$	40	0.263
Cyanamide	CH_2N_2	20	0.547
Cyanuric acid	$C_3H_3N_3O_3$	40	0.318
Dextrin	$(C_6H_{10}O_5)_x$	0 to 90	0.291 + 0.00096*t*
Dextrose	$C_6H_{12}O_6$	−250	0.016
		−200	0.077
		−100	0.160
		0	0.277
		20	0.300
Dibenzyl	$C_{14}H_{14}$	28	0.363
Dibromobenzene (o-)	$C_6H_4Br_2$	− 36	0.248
(m-)	$C_6H_4Br_2$	− 25	0.134
(p-)	$C_6H_4Br_2$	− 50 to +50	0.139 + 0.00038*t*
Dichloroacetic acid	$C_2H_2Cl_2O_2$	....	0.406
Dichlorobenzene (o-)	$C_6H_4Cl_2$	− 48.5	0.185
(m-)	$C_6H_4Cl_2$	− 52	0.186
(p-)	$C_6H_4Cl_2$	− 50 to +53	0.219 + 0.0021*t*
Dicyandiamide	$C_2H_4N_4$	0 to 204	0.456
Dihydroxybenzene (o-)	$C_6H_6O_2$	−163 to m.p.	0.278 + 0.00098*t*
(m-)	$C_6H_6O_2$	−160 to m.p.	0.269 + 0.00118*t*
(p-)	$C_6H_6O_2$	−250	0.025
		−240	0.038
		−220	0.061
		−200	0.081
		−150 to m.p.	0.268 + 0.00093*t*
Di-iodobenzene (o-)	$C_6H_4I_2$	− 50 to +15	0.109 + 0.00026*t*
(m-)	$C_6H_4I_2$	− 52 to −42	0.100 + 0.00026*t*
(p-)	$C_6H_4I_2$	− 50 to +80	0.101 + 0.00026*t*
Dimethyl oxalate	$C_4H_6O_4$	10 to 50	0.212 + 0.0044*t*
Dimethylpyrene	$C_7H_8O_2$	50	0.368
Dinitrobenzene (o-)	$C_6H_4N_2O_4$	−160 to m.p.	0.252 + 0.00083*t*
(m-)	$C_6H_4N_2O_4$	−160 to m.p.	0.248 + 0.00077*t*
(p-)	$C_6H_4N_2O_4$	119 to m.p.	0.259 + 0.00057*t*
Diphenyl	$C_{12}H_{10}$	40	0.385

TABLE 3-184 Specific Heats of Organic Solids *(Continued)*

Compound	Formula	Temperature, °C.	Sp. ht., cal./g. °C.
Diphenylamine	$C_{12}H_{11}N$	26	0.337
Dulcitol	$C_6H_{14}O_6$	20	0.282
Erythritol	$C_4H_{10}O_4$	60	0.351
Ethyl alcohol	C_2H_6O (crystalline)	−190	0.232
		−180	0.248
		−160	0.282
		−140	0.318
		−130	0.376
	(vitreous)	−190	0.260
		−180	0.296
		−175	0.380
		−170	0.399
Ethylene glycol	$C_2H_6O_2$	−190 to −40	0.366 + 0.00110*t*
Formic acid	CH_2O_2	− 22	0.387
		0	0.430
Glutaric acid	$C_5H_8O_4$	20	0.299
Glycerol	$C_3H_8O_3$	−265	0.009
		−260	0.022
		−250	0.047
		−220	0.085
		−200	0.115
		−100	0.217
		0	0.330
Hexachloroethane	C_2Cl_6	25	0.174
Hexadecane	$C_{16}H_{34}$	....	0.495
Hydroxyacetanilide	$C_8H_9NO_2$	41 to m.p.	0.249 + 0.00154*t*
Iodobenzene	C_6H_5I	40	0.191
Isopropyl alcohol	C_3H_8O	−200 to −160	0.051 + 0.00165*t*
Lactose	$C_{12}H_{22}O_{11}$	20	0.287
	$C_{12}H_{22}O_{11}.H_2O$	20	0.299
Lauric acid	$C_{12}H_{24}O_2$	− 30 to +40	0.430 + 0.000027*t*
Levoglucosane	$C_6H_{10}O_5$	40	0.607
Levulose	$C_6H_{12}O_6$	20	0.275
Malonic acid	$C_3H_4O_4$	20	0.275
Maltose	$C_{12}H_{22}O_{11}$	20	0.320
Mannitol	$C_6H_{14}O_6$	0 to 100	0.313 + 0.00025*t*
Melamine	$C_3H_6N_6$	40	0.351
Myristic acid	$C_{14}H_{28}O_2$	0 to 35	0.381 + 0.00545*t*
Naphthalene	$C_{10}H_8$	−130 to m.p.	0.281 + 0.00111*t*
Naphthol (α-)	$C_{10}H_8O$	50 to m.p.	0.240 + 0.00147*t*
(β-)	$C_{10}H_8O$	61 to m.p.	0.252 + 0.00128*t*
Naphthylamine (α-)	$C_{10}H_9N$	0 to 50	0.270 + 0.0031*t*
Nitroaniline (o-)	$C_6H_6N_2O_2$	−160 to m.p.	0.269 + 0.000920*t*
(m-)	$C_6H_6N_2O_2$	−160 to m.p.	0.275 + 0.000946*t*
(p-)	$C_6H_6N_2O_2$	−160 to m.p.	0.276 + 0.001000*t*
Nitrobenzoic acid (o-)	$C_7H_5NO_4$	−163 to m.p.	0.256 + 0.00085*t*
(m-)	$C_7H_5NO_4$	66 to m.p.	0.258 + 0.00091*t*
(p-)	$C_7H_5NO_4$	−160 to m.p.	0.247 + 0.00077*t*
Nitronaphthalene	$C_{10}H_7NO_2$	0 to 55	0.236 + 0.00215*t*
Oxalic acid	$C_2H_2O_4$	−200 to +50	0.259 + 0.00076*t*
	$C_2H_2O_4.2H_2O$	−200	0.117
		−100	0.239
		0	0.338
		+ 50	0.385
		100	0.416
Palmitic acid	$C_{16}H_{32}O_2$	−180	0.167
		−140	0.208
		−100	0.251
		− 50	0.306
		0	0.382
		+ 20	0.430
Phenol	C_6H_6O	14 to 26	0.561
Phthalic acid	$C_8H_6O_4$	20	0.232
Picric acid	$C_6H_3N_3O_7$	−100	0.165
		0	0.240
		+ 50	0.263
		100	0.297
		120	0.332
Propionic acid	$C_3H_6O_2$	− 33	0.726
Propyl alcohol (n-)	C_3H_8O	−200	0.170
		−175	0.363
		−150	0.471
		−130	0.497
Pyrotartaric acid	$C_6H_8O_4$	20	0.301
Quinhydrone	$C_{12}H_{10}O_4$	−250	0.017
		−225	0.061
		−200	0.098
		−100	0.191
		0	0.256

TABLE 3-184 Specific Heats of Organic Solids *(Concluded)*

Compound	Formula	Temperature, °C.	Sp. ht., cal./g. °C.
Quinone	$C_6H_4O_2$	−250	0.031
		−225	0.082
		−200	0.113
		−150 to m.p.	0.282 + 0.00083*t*
Salol	$C_{13}H_{10}O_3$	32	0.289
Stearic acid	$C_{18}H_{36}O_2$	15	0.399
Succinic acid	$C_4H_6O_4$	0 to 160	0.248 + 0.00153*t*
Sucrose	$C_{12}H_{22}O_{11}$	20	0.299
Sugar (cane)	$C_{12}H_{22}O_{11}$	22 to 51	0.301
Tartaric acid	$C_4H_6O_6$	36	0.287
Tartaric acid	$C_4H_6O_6.H_2O$	−150	0.112
		−100	0.170
		− 50	0.231
		0	0.308
		+ 50	0.366
Tetrachloroethylene	C_2Cl_4	− 40 to 0	0.198 + 0.00018*t*
Tetryl	$C_7H_5N_5O_8$	−100	0.182
		− 50	0.199
		0	0.212
		+100	0.236
1 Tetryl + 1 picric acid	$C_{13}H_8N_8O_{15}$	−100 to +100	0.253 + 0.00072*t*
1 Tetryl + 2 TNT	$C_{21}H_{15}N_{11}O_{20}$	−100	0.172
		0	0.280
		+ 50	0.325
Thymol	$C_{10}H_{14}O$	0 to 49	0.315 + 0.0031*t*
Toluic acid (o-)	$C_8H_8O_2$	54 to m.p.	0.277 + 0.00120*t*
(m-)	$C_8H_8O_2$	54 to m.p.	0.239 + 0.00195*t*
(p-)	$C_8H_8O_2$	130 to m.p.	0.271 + 0.00106*t*
Toluidine (p-)	C_7H_9N	0	0.337
		20	0.387
		40	0.440
Trichloroacetic acid	$C_2HCl_3O_2$	solid	0.459
Trimethyl carbinol	$C_4H_{10}O$	− 4	0.559
Trinitrotoluene	$C_7H_5N_3O_6$	−100	0.170
		− 50	0.253
		0	0.311
		+100	0.385
Trinitroxylene	$C_8H_7N_3O_6$	−185 to +23	0.241
		20 to 50	0.423
Triphenylmethane	$C_{19}H_{16}$	0 to 91	0.189 + 0.0027*t*
Urea	CH_4N_2O	20	0.320

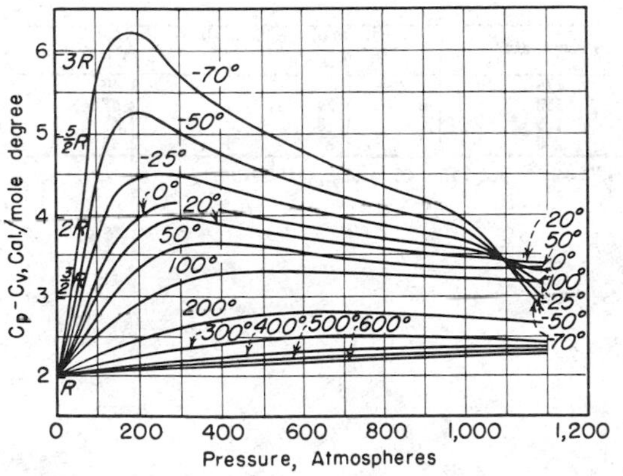

FIG. 3-13 The variation of $C_p - C_v$ for nitrogen with pressure at various temperatures. [*Deming and Shupe,* Phys. Rev., *37, 638 (1931).*]

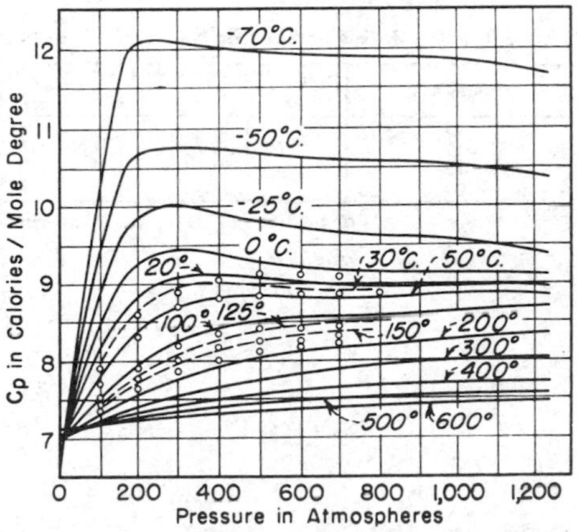

FIG. 3-14 The variation of heat capacity for nitrogen with pressure at various temperatures. [*Deming and Shupe,* Phys. Rev., *37, 638 (1931).*]

TABLE 3-185 C_p/C_v: Ratios of Specific Heats of Gases at 1-atm Pressure*

Compound	Formula	Temperature, °C.	Ratio of specific heats, $(\gamma) = C_p/C_v$
Acetaldehyde	C_2H_4O	30	1.14
Acetic acid	$C_2H_4O_2$	136	1.15
Acetylene	C_2H_2	15	1.26
		−71	1.31
Air	....	925	1.36
		17	1.403
		−78	1.408
		−118	1.415
Ammonia	NH_3	15	1.310
Argon	A	15	1.668
		−180	1.76 (?)
		0–100	1.67
Benzene	C_6H_6	90	1.10
Bromine	Br_2	20–350	1.32
Carbon dioxide	CO_2	15	1.304
		−75	1.37
disulfide	CS_2	100	1.21
monoxide	CO	15	1.404
		−180	1.41
Chlorine	Cl_2	15	1.355
Chloroform	$CHCl_3$	100	1.15
Cyanogen	$(CN)_2$	15	1.256
Cyclohexane	C_6H_{12}	80	1.08
Dichlorodifluormethane	CCl_2F_2	25	1.139
Ethane	C_2H_6	100	1.19
		15	1.22
		−82	1.28
Ethyl alcohol	C_2H_6O	90	1.13
ether	$C_4H_{10}O$	35	1.08
		80	1.086
Ethylene	C_2H_4	100	1.18
		15	1.255
		−91	1.35
Helium	He	−180	1.660
Hexane (n-)	C_6H_{14}	80	1.08
Hydrogen	H_2	15	1.410
		−76	1.453
		−181	1.597
bromide	HBr	20	1.42
chloride	HCl	15	1.41
		100	1.40
cyanide	HCN	65	1.31
		140	1.28
		210	1.24
Hydrogen (*Cont.*):			
iodide	HI	20–100	1.40
sulfide	H_2S	15	1.32
		−45	1.30
		−57	1.29
Iodine	I_2	185	1.30
Isobutane	C_4H_{10}	15	1.11
Krypton	Kr	19	1.68
Mercury	Hg	360	1.67
Methane	CH_4	600	1.113
		300	1.16
		15	1.31
		−80	1.34
		−115	1.41
Methyl acetate	$C_3H_6O_2$	15	1.14
alcohol	CH_4O	77	1.203
ether	C_2H_6O	6–30	1.11
Methylal	$C_3H_8O_2$	13	1.06
		40	1.09
Neon	Ne	19	1.64
Nitric oxide	NO	15	1.400
		−45	1.39
		−80	1.38
Nitrogen	N_2	15	1.404
		−181	1.47
Nitrous oxide	N_2O	100	1.28
		15	1.303
		−30	1.31
		−70	1.34
Oxygen	O_2	15	1.401
		−76	1.415
		−181	1.45
Pentane (n-)	C_5H_{12}	86	1.086
Phosphorus	P	300	1.17
Potassium	K	850	1.77
Sodium	Na	750–920	1.68
Sulfur dioxide	SO_2	15	1.29
Xenon	Xe	19	1.66

* From "International Critical Tables," vol. 5, pp. 80–82.

TABLE 3-186 Ratios of Specific Heats of Air at High Pressures

Pressure, atm.	Ratio of specific heats, $(\gamma) = C_p/C_v$		Pressure, atm.	Ratio of specific heats, $(\gamma) = C_p/C_v$	
	0°C.	−79.4°C.		0°C.	−79.4°C.
25	1.47	1.57	125	1.69	2.40
50	1.53	1.77	150	1.74	2.47
75	1.59	2.00	175	1.78	2.41
100	1.65	2.20	200	1.83	2.33

For additional data, see "International Critical Tables," vol. 5, pp. 115–116, and pp. 122–125.

SPECIFIC HEATS OF AQUEOUS SOLUTIONS

UNITS CONVERSIONS

For this subsection the following units conversions are applicable:

$^\circ F = \frac{9}{5}\,^\circ C + 32$.

To convert calories per gram-degree Celsius to British thermal units per pound-degree Fahrenheit, multiply by 1.0.

TABLE 3-187 Hydrochloric Acid

Mole % HCl	Specific heat, cal./g. °C. 0°C.	10°C.	20°C.	40°C.	60°C.
0.0	1.00				
9.09	0.72	0.72	0.74	0.75	0.78
16.7	.61	.605	.631	.645	.67
20.0	.58	.575	.591	.615	.638
25.9	.55	.....	.....	.....	.61

TABLE 3-188 Sulfuric Acid*

$\%H_2SO_4$	C_p at 20°C., cal./g. °C.	$\%H_2SO_4$,	C_p at 20°C., cal./g. °C.
0.34	0.9968	35.25	0.7238
0.68	.9937	37.69	.7023
1.34	.9877	40.49	.6770
2.65	.9762	43.75	.6476
3.50	.9688	47.57	.6153
5.16	.9549	52.13	.5801
9.82	.9177	57.65	.5420
15.36	.8767	64.47	.5012
21.40	.8339	73.13	.4628
22.27	.8275	77.91	.4518
23.22	.8205	81.33	.4481
24.25	.8127	82.49	.4467
25.39	.8041	84.48	.4408
26.63	.7945	85.48	.4346
28.00	.7837	89.36	.4016
29.52	.7717	91.81	.3787
30.34	.7647	94.82	.3554
31.20	.7579	97.44	.3404
33.11	.7422	100.00	.3352

*Vinal and Craig, *Bur. Standards J. Research*, **24**, 475 (1940).

TABLE 3-189 Nitric Acid

$\%HNO_3$ by Weight	Specific Heat at 20°C., Cal./g. °C.
0	1.000
10	0.900
20	.810
30	.730
40	.675
50	.650
60	.640
70	.615
80	.575
90	.515

TABLE 3-190 Phosphoric Acid*

$\%H_3PO_4$	C_p at 21.3°C. cal./g. °C.	$\%H_3PO_4$	C_p at 21.3°C. cal./g. °C.
2.50	0.9903	50.00	0.6350
3.80	.9970	52.19	.6220
5.33	.9669	53.72	.6113
8.81	.9389	56.04	.5972
10.27	.9293	58.06	.5831
14.39	.8958	60.23	.5704
16.23	.8796	62.10	.5603
19.99	.8489	64.14	.5460
22.10	.8300	66.13	.5349
24.56	.8125	68.14	.5242
25.98	.8004	69.97	.5157
28.15	.7856	69.50	.5160
29.96	.7735	71.88	.5046
32.09	.7590	73.71	.4940
33.95	.7432	75.79	.4847
36.26	.7270	77.69	.4786
38.10	.7160	79.54	.4680
40.10	.7024	80.00	.4686
42.08	.6877	82.00	.4593
44.11	.6748	84.00	.4500
46.22	.6607	85.98	.4419
48.16	.6475	88.01	.4359
49.79	.6370	89.72	.4206

Z. physik. Chem., A167, **42** (1933).

ADDITIONAL REFERENCES

For additional data see *International Critical Tables*, vol. 5, pp. 115–116, 122–125.

TABLE 3-191 Acetic Acid (at 38°C)

Mole % acetic acid	0	6.98	30.9	54.5	100
Cal./g. °C.	1.0	0.911	0.73	0.631	0.535

TABLE 3-192 Sodium Hydroxide (at 20°C)

Mole % NaOH	0	0.5	1.0	9.09	16.7	28.6	37.5
Cal./g. °C.	1.0	0.985	0.97	0.835	0.80	0.784	0.782

TABLE 3-193 Potassium Hydroxide (at 19°C)

Mole % KOH	0	0.497	1.64	4.76	9.09
Cal./g. °C.	1.0	0.975	0.93	0.814	0.75

TABLE 3-194 Ammonia

Mole % NH_3	Specific heat, cal./g. °C. 2.4°C.	20.6°C.	41°C.	61°C.
0	1.01	1.0	0.995	1.0
10.5	0.98	0.995	1.06	1.02
20.9	.96	.99	1.03	
31.2	.956	1.0		
41.4	.985			

TABLE 3-195 Sodium Carbonate*

$\%Na_2CO_3$ by weight	Temperature, °C. 17.6	30.0	76.6	98.0
0.000	0.9992	0.9986	1.0098	1.0084
1.498	.9807			
2.000	......	.9786		
2.901	.9597			
4.000	......	.9594		
5.000	.9428	......	0.9761	
6.000	......	.9392		
8.000	.9183			
10.000	.9086	......	.9452	
13.790	.8924			
13.840	......	.8881		
20.000	......	.8631	.8936	
25.000	......	......	.8615	0.8911

* *J. Chem. Soc.*, pp. 3062–3079 (1931).

TABLE 3-196 Sodium Chloride

Mole % NaCl	Specific heat, cal./g. °C. 6°C.	20°C.	33°C.	57°C.
0.249	.....	0.99		
.99	0.96	.97	0.97	
2.44	.91	.915	.915	0.923
9.09	.805	.81	.81	.82

TABLE 3-197 Potassium Chloride

Mole % KCl	Specific heat, cal./g. °C. 6°C.	20°C.	33°C.	40°C.
0.99	0.945	0.947	0.947	0.947
3.85	.828	.831	.835	.837
5.66	.77	.775	.778	.775
7.41	.....	.727		

TABLE 3-198 Zinc Sulfate

Composition	Temperature	Sp. ht., cal./g. °C.
$ZnSO_4 + 50H_2O$	20° to 52°C.	0.842
$ZnSO_4 + 200H_2O$	20° to 52°C.	.952

TABLE 3-199 Copper Sulfate

Composition	Temperature	Sp. ht., cal./g. °C.
$CuSO_4 + 50H_2O$	12° to 15°C.	0.848
$CuSO_4 + 200H_2O$	12° to 14°C.	.951
$CuSO_4 + 400H_2O$	13° to 17°C.	.975

TABLE 3-200 Methyl Alcohol

Mole % CH_3OH	Specific heat, cal./g. °C.		
	5°C.	20°C.	40°C.
5.88	1.02	1.0	0.995
12.3	0.975	0.982	.98
27.3	.877	.917	.92
45.8	.776	.811	.83
69.6	.681	.708	.726
100	.576	.60	.617

TABLE 3-201 Ethyl Alcohol

Mole % C_2H_5OH	Specific heat, cal./g. °C.		
	3°C.	23°C.	41°C.
4.16	1.05	1.02	1.02
11.5	1.02	1.03	1.03
37.0	0.805	0.86	0.875
61.0	.67	.727	.748
100.0	.54	.577	.621

TABLE 3-202 Normal Propyl Alcohol

Mole % C_3H_7OH	Specific heat, cal./g. °C.		
	5°C.	20°C.	40°C.
1.55	1.03	1.02	1.01
5.03	1.07	1.06	1.03
11.4	1.035	1.032	0.99
23.1	0.877	0.90	.91
41.2	.75	.78	.815
73.0	.612	.645	.708
100.0	.534	.57	.621

TABLE 3-203 Glycerol

Mole % $C_3H_5(OH)_3$	Specific heat, cal./g. °C.	
	15°C.	32°C.
2.12	0.961	0.960
4.66	.929	.924
11.5	.851	.841
22.7	.765	.758
43.9	.67	.672
100.0	.555	.576

NOTE.—For the specific heats of non-aqueous solutions, see "International Critical Tables," vol. 5, pp. 116, 125.

TABLE 3-204 Aniline (at 20°C)

Mol % aniline	100	95	90.5	82.3	75.2
Cal./g. °C.	0.497	0.52	0.53	0.56	0.581

SPECIFIC HEATS OF MISCELLANEOUS MATERIALS

TABLE 3-205 Specific Heats of Miscellaneous Liquids and Solids

Material	Specific Heat, cal./g. °C.
Alumina	0.2 (100°C.); 0.274 (1500°C.)
Alundum	0.186 (100°C.)
Asbestos	0.25
Asphalt	0.22
Bakelite	0.3 to 0.4
Brickwork	About 0.2
Carbon	0.168 (26° to 76°C.)
	0.314 (40° to 892°C.)
	0.387 (56° to 1450°C.)
(gas retort)	0.204
(See under Graphite)	
Cellulose	0.32
Cement, Portland Clinker	0.186
Charcoal (wood)	0.242
Chrome brick	0.17
Clay	0.224
Coal	0.26 to 0.37
tar oils	0.34 (15° to 90°C.)
Coal tars	0.35 (40°C.); 0.45 (200°C.)
Coke	0.265 (21° to 400°C.)
	0.359 (21° to 800°C.)
	0.403 (21° to 1300°C.)
Concrete	0.156 (70° to 312°F.); 0.219 (72° to 1472°F.)
Cryolite	0.253 (16° to 55°C.)
Diamond	0.147
Fireclay brick	0.198 (100°C.); 0.298 (1500°C.)
Fluorspar	0.21 (30°C.)
Gasoline	0.53
Glass (crown)	0.16 to 0.20
(flint)	0.117
(pyrex)	0.20
(silicate)	0.188 to 0.204 (0 to 100°C.)
	0.24 to 0.26 (0 to 700°C.)
wool	0.157
Granite	0.20 (20° to 100°C.)
Graphite	0.165 (26° to 76°C.); 0.390 (56° to 1450°C.)
Gypsum	0.259 (16° to 46°C.)
Kerosene	0.47
Limestone	0.217
Litharge	0.055
Magnesia	0.234 (100°C.); 0.188 (1500°C.)
Magnesite brick	0.222 (100°C.); 0.195 (1500°C.)
Marble	0.21 (18°C.)
Pyrites (copper)	0.131 (19° to 50°C.)
(iron)	0.136 (15° to 98°C.)
Quartz	0.17 (0°C.); 0.28 (350°C.,
Sand	0.191
Silica	0.316
Steel	0.12
Stone	About 0.2
Turpentine	0.42 (18°C.)
Wood (oak)	0.570
Most woods vary between	0.45 and 0.65

Oils (animal, vegetable, mineral oils)

$$C_p[\text{cal}/(g\cdot°C)] = A/\sqrt{d_4^{15}} + B(t - 15)$$

where d = density, g/cm^3.

°F = % °C + 32; to convert calories per gram-degree Celsius to British thermal units per pound-degree Fahrenheit, multiply by 1.0; to convert grams per cubic centimeter to pounds per cubic foot, multiply by 62.43.

Oils	A	B
Castor	0.500	0.0007
Citron	(0.438 at 54°C.)	
Fatty drying	0.440	0.0007
non-drying	0.450	0.0007
semidrying	0.445	0.0007
oils (except castor)	0.450	0.0007
Naphthene base	0.405	0.0009
Olive	(0.47 at 7°C.)	
Paraffin base	0.425	0.0009
Petroleum oils	0.415	0.0009

Porcelain	Average specific heat between 20°C. and			
	100°C.	300°C.	500°C.	1100°C.
Fired Berlin	0.189	0.203	0.222	0.337
Green Berlin	.185	.197	.228	
Fired Berlin (glaze)	.179	.189	.199	.245
Green Berlin (glaze)	.170	.183	.208	
Fired earthenware	.186	.203	.223	.324
Green earthenware	.181	.192	.215	

Material	Specific Heat, cal./g. °C.
Pyrex glass	0.20
Pyroxylin plastics	0.34 to 0.38
Rubber (vulcanized)	0.415
Silica brick	0.202 (100°C.); 0.195 (1500°C.)
Silicon carbide brick	0.202 (100°C.)
Silk	0.33
Stoneware (common)	0.185 to 0.191 (20° to 100°C.)
Wool	0.325
Zirconium oxide	0.11 (100°C.); 0.179 (1500°C.)

HEATS AND FREE ENERGIES OF FORMATION

TABLE 3-206 Heats and Free Energies of Formation of Inorganic and Organic Compounds

The values given in the following table for the heats and free energies of formation of inorganic compounds are derived from (*a*) Bichowsky and Rossini, "Thermochemistry of the Chemical Substances," Reinhold, New York, 1936; (*b*) Latimer, "Oxidation States of the Elements and Their Potentials in Aqueous Solution," Prentice-Hall, New York, 1938; (*c*) the tables of the American Petroleum Institute Research Project 44 at the National Bureau of Standards; and (*d*) the tables of Selected Values of Chemical Thermodynamic Properties of the National Bureau of Standards. The reader is referred to the preceding books and tables for additional details as to methods of calculation, standard states, etc.

The organic compounds in the following table are all given under the element carbon. The values for the non-hydrocarbons are largely from E. I. du Pont de Nemours & Co., Ammonia Department, Chemical Division, Experimental Station; and the values for the hydrocarbons are from the tables of the American Petroleum Institute Research Project 44 at the National Bureau of Standards.*

Compound	State†	Heat of formation‡§ ΔH (formation) at 25°C., kcal./mole	Free energy of formation‖¶ ΔF (formation) at 25°C., kcal./mole
Aluminum:			
Al	c	0.00	0.00
$AlBr_3$	c	−123.4	
	aq	−209.5	−189.2
Al_4C_3	c	−30.8	−29.0
$AlCl_3$	c	−163.8	
	aq, 600	−243.9	−209.5
AlF_3	c	−329	
	aq	−360.8	−312.6
AlI_3	c	−72.8	
	aq	−163.4	−152.5
AlN	c	−57.7	−50.4
$Al(NH_4)(SO_4)_2$	c	−561.19	−486.17
$Al(NH_4)(SO_4)_2.12H_2O$	c	−1419.36	−1179.26
$Al(NO_3)_3.6H_2O$	c	−680.89	−526.32
$Al(NO_3)_3.9H_2O$	c	−897.59	
Al_2O_3	c, corundum	−399.09	−376.87
$Al(OH)_3$	c	−304.8	−272.9
$Al_2O_3.SiO_2$	c, sillimanite	−648.7	
$Al_2O_3.SiO_2$	c, disthene	−642.4	
$Al_2O_3.SiO_2$	c, andalusite	−642.0	
$3Al_2O_3.2SiO_2$	c, mullite	−1874	
Al_2S_3	c	−121.6	
$Al_2(SO_4)_3$	c	−820.99	−739.53
	aq	−893.9	−759.3
$Al_2(SO_4)_3.6H_2O$	c	−1268.15	−1103.39
$Al_2(SO_4)_3.18H_2O$	c	−2120	
Antimony:			
Sb	c	0.00	0.00
$SbBr_3$	c	−59.9	
$SbCl_3$	c	−91.3	−77.8
$SbCl_5$	l	−104.8	
SbF_3	c	−216.6	
SbI_3	c	−22.8	
Sb_2O_3	c, I, orthorhombic	−165.4	−146.0
	c, II, octahedral	−166.6	
Sb_2O_4	c	−213.0	−186.6
Sb_2O_5	c	−230.0	−196.1
Sb_2S_3	c, black	−38.2	−36.9
Arsenic:			
As	c	0.00	0.00
$AsBr_3$	c	−45.9	
$AsCl_3$	l	−80.2	−70.5
AsF_3	l	−223.76	−212.27
AsH_3	g	43.6	37.7
AsI_3	c	−13.6	
As_2O_3	c	−154.1	−134.8
As_2O_5	c	−217.9	−183.9
As_2S_3	c	−20	−20
	amorphous	−34.76	
Barium:			
Ba	c	0.00	0.00
$BaBr_2$	c	−180.38	
	aq, 400	−185.67	−183.0
$BaCl_2$	c	−205.25	
	aq, 300	−207.92	−196.5
$Ba(ClO_3)_2$	c	−176.6	
	aq, 1600	−170.0	−134.4
$Ba(ClO_4)_2$	c	−210.2	
	aq, 800	...	−155.3
$Ba(CN)_2$	c	−48	
$Ba(CNO)_2$	c	−212.1	
	aq	...	−180.7
$BaCN_2$	c	−63.6	
$BaCO_3$	c, witherite	−284.2	−271.4
$BaCrO_4$	c	−342.2	
BaF_2	c	−287.9	
	aq, 1600	−284.6	−265.3
BaH_2	c	−40.8	−31.5
$Ba(HCO_3)_2$	aq	−459	−414.4
BaI_2	c	−144.6	
	aq, 400	−155.17	−158.52
$Ba(IO_3)_2$	c	−264.5	
	aq	−237.50	−198.35
$BaMoO_4$	c	−370	
Ba_3N_2	c	−90.7	
$Ba(NO_2)_2$	c	−184.5	
	aq	−179.05	−150.75
Barium (*Cont.*):			
$Ba(NO_3)_2$	c	−236.99	−189.94
	aq, 600	−227.74	
BaO	c	−133.0	
$Ba(OH)_2$	c	−225.9	
	aq, 400	−237.76	−209.02
$BaO.SiO_2$	c	−363	
$Ba_3(PO_4)_2$	c	−992	
$BaPtCl_6$	c	−284.9	
BaS	c	−111.2	
$BaSO_3$	c	−282.5	
$BaSO_4$	c	−340.2	−313.4
$BaWO_4$	c	−402	
Beryllium:			
Be	c	0.00	0.00
$BeBr_2$	c	−79.4	
	aq	−142	−127.9
$BeCl_2$	c	−112.6	
	aq	−163.9	−141.4
BeI_2	c	−39.4	
	aq	−112	−103.4
Be_3N_2	c	−134.5	−122.4
BeO	c	−145.3	−138.3
$Be(OH)_2$	c	−215.6	
BeS	c	−56.1	
$BeSO_4$	c	−281	
	aq	...	−254.8
Bismuth:			
Bi	c	0.00	0.00
$BiCl_3$	c	−90.5	−76.4
	aq	−101.6	
BiI_3	c	−24	
	aq	−27	
BiO	c	−49.5	−43.2
Bi_2O_3	c	−137.1	−117.9
$Bi(OH)_3$	c	−171.1	
Bi_2S_3	c	−43.9	−39.1
$Bi_2(SO_4)_3$	c	−607.1	
Boron:			
B	c	0.00	0.00
BBr_3	l	−52.7	
	g	−44.6	−50.9
BCl_3	g	−94.5	−90.8
BF_3	g	−265.2	−261.0
B_2H_6	g	7.5	19.9
BN	c	−32.1	−27.2
B_2O_3	c	−302.0	−282.9
	gls	−297.6	−280.3
$B(OH)_3$	c	−260.0	−229.4
B_2S_3	c	−56.6	
Bromine:			
Br_2	l	0.00	0.00
	g	7.47	0.931
BrCl	g	3.06	−0.63
Cadmium:			
Cd	c	0.00	0.00
$CdBr_2$	c	−75.8	−70.7
	aq, 400	−76.6	−67.6
$CdCl_2$	c	−92.149	−81.889
	aq, 400	−96.44	−81.2
$Cd(CN)_2$	c	36.2	
$CdCO_3$	c	−178.2	−163.2
CdI_2	c	−48.40	
	aq, 400	−47.46	−43.22
Cd_3N_2	c	39.8	
$Cd(NO_3)_2$	aq, 400	−115.67	−71.05
CdO	c	−62.35	−55.28
$Cd(OH)_2$	c	−135.0	−113.7
CdS	c	−34.5	−33.6
$CdSO_4$	c	−222.23	
	aq, 400	−232.635	−194.65
Calcium:			
Ca	c	0.00	0.00
$CaBr_2$	c	−162.20	
	aq, 400	−187.19	−181.86
CaC_2	c	−14.8	−16.0
$CaCl_2$	c	−190.6	−179.8
	aq	−209.15	−195.36

*For footnotes see end of table.

TABLE 3-206 Heats and Free Energies of Formation of Inorganic and Organic Compounds *(Continued)*

Compound	State†	Heat of formation‡§ ΔH (formation) at 25°C., kcal./mole	Free energy of formation‖¶ ΔF (formation) at 25°C., kcal./mole
Calcium (*Cont.*):			
$CaCN_2$	c	−85	
$Ca(CN)_2$	c	−43.3	
	aq	....	−54.0
$CaCO_3$	c, calcite	−289.5	−270.8
	c, aragonite	−289.54	−270.57
$CaCO_3.MgCO_3$	c	−558.8	
CaC_2O_4	c	−332.2	
$Ca(C_2H_3O_2)_2$	c	−356.3	
	aq	−364.1	−311.3
CaF_2	c	−290.2	
	aq	−286.5	−264.1
CaH_2	c	−46	−35.7
CaI_2	c	−128.49	
	aq, 400	−156.63	−157.37
Ca_3N_2	c	−103.2	−88.2
$Ca(NO_3)_2$	c	−224.05	−177.38
	aq, 400	−228.29	
$Ca(NO_3)_2.2H_2O$	c	−367.95	−293.57
$Ca(NO_3)_2.3H_2O$	c	−439.05	−351.58
$Ca(NO_3)_2.4H_2O$	c	−509.43	−409.32
CaO	c	−151.7	−144.3
$Ca(OH)_2$	c	−235.58	−213.9
	aq, 800	−239.2	−207.9
$CaO.SiO_2$	c, II, wollastonite	−377.9	−357.5
	c, I, pseudowollastonite	−376.6	−356.6
CaS	c	−114.3	−113.1
$CaSO_4$	c, insoluble form	−338.73	−311.9
	c, soluble form α	−336.58	−309.8
	c, soluble form β	−335.52	−308.8
$CaSO_4.\frac{1}{2}H_2O$	c	−376.13	
$CaSO_4.2H_2O$	c	−479.33	−425.47
$CaWO_4$	c	−387	
Carbon:			
C	c, graphite	0.00	0.00
	c, diamond	0.453	0.685
CO	g	−26.416	−32.808
CO_2	g	−94.052	−94.260
CH_4 methane	g	−17.889	−12.140
C_2H_6 ethane	g	−20.236	−7.860
C_3H_8 propane	g	−24.820	−5.614
C_4H_{10} n-butane	g	−29.812	−3.754
C_4H_{10} isobutane	g	−31.452	−4.296
C_5H_{12} n-pentane	g	−35.00	−1.96
	l	−41.36	−2.21
C_5H_{12} 2-methylbutane	g	−36.92	−3.50
	l	−42.85	−3.59
C_5H_{12} 2,2-dimethylpropane	g	−39.67	−3.64
C_6H_{14} n-hexane	g	−39.96	0.05
	l	−47.52	−0.91
C_6H_{14} 2-methylpentane	g	−41.66	−0.96
	l	−48.82	−1.73
C_6H_{14} 3-methylpentane	g	−41.02	−0.29
	l	−48.28	−1.12
C_6H_{14} 2,2-dimethylbutane	g	−44.35	−2.35
	l	−51.00	−2.88
C_6H_{14} 2,3-dimethylbutane	g	−42.49	−0.73
	l	−49.48	−1.44
C_7H_{16} n-heptane	g	−44.89	2.09
	l	−53.63	0.42
C_7H_{16} 2-methylhexane	g	−46.60	0.98
	l	−54.93	−0.47
C_7H_{16} 3-methylhexane	g	−45.96	1.10
	l	−54.35	−0.39
C_7H_{16} 3-ethylpentane	g	−45.34	2.59
	l	−53.77	1.06
C_7H_{16} 2,2-dimethylpentane	g	−49.29	0.09
	l	−57.05	−1.08
C_7H_{16} 2,3-dimethylpentane	g	−47.62	0.16
	l	−55.81	−1.27
C_7H_{16} 2,4-dimethylpentane	g	−48.30	0.72
	l	−56.17	−0.49
C_7H_{16} 3,3-dimethylpentane	g	−48.17	0.63
	l	−56.07	−0.69
C_7H_{16} 2,2,3-trimethylbutane	g	−48.96	0.76
	l	−56.63	−0.43
C_8H_{18} n-octane	g	−49.82	4.14
	l	−59.74	1.77
C_8H_{18} 2-methylheptane	g	−51.50	3.06
	l	−60.98	0.92
C_8H_{18} 3-methylheptane	g	−50.82	3.29
	l	−60.34	1.12
C_8H_{18} 4-methylheptane	g	−50.69	4.00
	l	−60.17	1.86
C_8H_{18} 3-ethylhexane	g	−50.40	3.95
	l	−59.88	1.80
Carbon (*Cont.*):			
C_8H_{18} 2,2-dimethylhexane	g	−53.71	2.56
	l	−62.63	−0.72
C_8H_{18} 2,3-dimethylhexane	g	−51.13	4.23
	l	−60.40	2.17
C_8H_{18} 2,4-dimethylhexane	g	−52.44	2.80
	l	−61.47	0.89
C_8H_{18} 2,5-dimethylhexane	g	−53.21	2.50
	l	−62.26	0.59
C_8H_{18} 3,3-dimethylhexane	g	−52.61	3.17
	l	−61.58	1.23
C_8H_{18} 3,4-dimethylhexane	g	−50.91	4.97
	l	−60.23	2.86
C_8H_{18} 2-methyl-3-ethylpentane	g	−50.48	5.08
	l	−59.69	3.03
C_8H_{18} 3-methyl-3-ethylpentane	g	−51.38	4.76
	l	−60.46	2.69
C_8H_{18} 2,2,3-trimethylpentane	g	−52.61	4.09
	l	−61.44	2.22
C_8H_{18} 2,2,4-trimethylpentane	g	−53.57	3.13
	l	−61.97	1.51
C_8H_{18} 2,3,3,-trimethylpentane	g	−51.73	4.52
	l	−60.63	2.54
C_8H_{18} 2,3,4-trimethylpentane	g	−51.97	4.32
	l	−60.98	2.34
C_8H_{18} 2,2,3,3,-tetramethylbutane	g	−53.99	4.88
	c	−64.23	2.74
C_2H_4 ethylene	g	12.496	16.282
C_3H_6 propylene	g	4.879	14.964
C_4H_8 1-butene	g	0.280	17.217
C_4H_8 cis-2-butene	g	−1.362	16.007
C_4H_8 trans-2-butene	g	−2.405	15.323
C_4H_8 2-methyl-2-propene	g	−3.343	14.574
C_5H_{10} 1-pentene	g	−5.000	18.787
C_5H_{10} cis-2-pentene	g	−6.710	17.173
C_5H_{10} trans-2-pentene	g	−7.590	16.575
C_5H_{10} 2-methyl-1-butene	g	−8.680	15.509
C_5H_{10} 3-methyl-1-butene	g	−6.920	17.874
C_5H_{10} 2-methyl-2-butene	g	−10.170	14.267
C_2H_2 acetylene	g	54.194	50.000
C_3H_4 methylacetylene	g	44.319	46.313
C_4H_6 1-butyne	g	39.70	48.52
C_4H_6 2-butyne	g	35.374	44.725
C_5H_8 1-pentyne	g	34.50	50.17
C_5H_8 2-pentyne	g	30.80	46.41
C_5H_8 3-methyl-1-butyne	g	32.60	49.12
C_6H_6 benzene	g	19.820	30.989
C_7H_8 toluene	l	11.718	29.756
	g	11.950	29.228
	l	2.867	27.282
C_8H_{10} ethylbenzene	g	7.120	31.208
	l	−2.977	28.614
C_8H_{10} o-xylene	g	4.540	29.177
	l	−5.841	26.370
C_8H_{10} m-xylene	g	4.120	28.405
	l	−6.075	25.730
C_8H_{10} p-xylene	g	4.290	28.952
	l	−5.838	26.310
C_9H_{12} n-propylbenzene	g	1.870	32.810
	l	−9.178	29.600
C_9H_{12} isopropylbenzene	g	0.940	32.738
	l	−9.848	29.708
C_9H_{12} 1-methyl-2-ethylbenzene	g	0.290	31.323
	l	−11.110	27.973
C_9H_{12} 1-methyl-3-ethylbenzene	g	−0.460	30.217
	l	−11.670	26.977
C_9H_{12} 1-methyl-4-ethylbenzene	g	−0.780	30.281
	l	−11.920	27.041
C_9H_{12} 1,2,3-trimethylbenzene	g	−2.290	29.319
	l	−14.013	25.679
C_9H_{12} 1,2,4-trimethylbenzene	g	−3.330	27.912
	l	−14.785	24.462
C_9H_{12} 1,3,5-trimethylbenzene	g	−3.840	28.172
	l	−15.184	24.832
C_5H_{10} cyclopentane	g	−18.46	9.23
	l	−25.31	8.70
C_6H_{12} methylcyclopentane	g	−25.50	8.55
	l	−33.08	7.53
C_7H_{14} ethylcyclopentane	g	−30.38	10.59
	l	−39.09	8.84

TABLE 3-206 Heats and Free Energies of Formation of Inorganic and Organic Compounds *(Continued)*

Compound	State†	Heat of formation‡§ ΔH (formation) at 25°C., kcal./mole	Free energy of formation‖¶ ΔF (formation) at 25°C., kcal./mole
Carbon (*Cont.*):			
C_6H_{12} cyclohexane	g	−29.43	7.59
	l	−37.34	6.39
C_7H_{14} methylcyclohexane	g	−37.00	6.52
	l	−45.46	4.86
C_8H_{16} ethylcyclohexane	g	−41.06	9.38
	l	−50.73	6.96
CH_4O methanol	g	−48.08	−38.62
	l	−57.04	−39.80
C_2H_6O ethanol	g	−52.23	−40.23
	l	−66.35	−41.76
C_3H_8O *n*-propanol	g	−61.17	−38.83
	l	−71.87	−39.84
C_3H_8O isopropanol	g	−62.41	−38.20
	l	−74.32	−38.83
$C_4H_{10}O$ *n*-butanol	g	−67.81	−38.88
	l	−79.61	−40.37
$C_4H_{10}O$ isobutanol	g	−69.05	−38.25
	l	−81.06	−39.36
$C_2H_6O_2$ ethylene glycol	g	−92.53	−71.26
	l	−107.91	−76.44
$C_3H_8O_3$ glycerol	g		
	l	−159.16	−113.65
C_6H_6O phenol	g	−21.71	−6.26
	l	−37.80	−11.02
C_7H_8O cresol	g	....	−13.17
C_2H_4O ethylene oxide	g	−16.1	−6.94
C_2H_6O dimethyl ether	g	−43.06	−26.06
	l	−51.3	
$C_4H_{10}O$ diethyl ether	l	−65.2	−27.75
CH_2O formaldehyde	g	−28.29	−26.88
C_2H_4O acetaldehyde	g	−39.72	−31.46
C_3H_4O acrolein	g	−20.50	−15.57
	l	−27.97	−16.17
C_3H_6O propionaldehyde	g	−49.15	−33.96
C_4H_8O *n*-butyraldehyde	g	−52.40	−73.24
C_7H_6O benzaldehyde	g	−9.57	5.85
	l	−21.23	2.24
C_8H_8O *p*-toluic aldehyde	g	−17.78	4.09
	l	−29.79	0.97
C_2H_2O ketene	g	−14.78	−14.30
	l	−18.78	−13.32
C_3H_6O acetone	g	−51.79	−36.45
	l	−59.32	−37.16
$C_5H_{10}O$ diethylketone	l	−73.8	
CH_2O_2 formic acid	g	−86.67	−80.24
	l	−97.8	−82.7
½$(CH_2O_2)_2$ bimolecular formic acid	g	−93.85	−81.90
$C_2H_4O_2$ acetic acid	g	−104.72	−91.24
	l	−116.2	−93.56
$C_3H_6O_2$ propionic acid	g	−108.75	−88.27
	l	−121.7	−91.65
$C_2H_4O_3$ hydroxyacctic acid	l	−155.33	−125.57
$C_6H_{10}O_4$ adipic acid	g	−216.19	−163.96
	l	−235.51	−177.17
$C_2H_4O_2$ methyl formate	g	−84.69	−71.37
	l	−95.26	−71.53
$C_4H_6O_2$ methyl acrylate	g	−70.10	−56.78
	l	−82.76	−58.13
$C_4H_8O_2$ ethyl acetate	g	−102.02	−74.93
	l	−110.72	−76.11
$C_5H_{10}O_2$ ethyl propionate	g	−112.36	−77.37
	l	−122.16	−79.16
$C_4H_6O_3$ acetic anhydride	g	−148.82	−119.29
	l	−155.16	−121.75
$C_6H_{10}O_3$ propionic anhydride	g	−147.32	−109.78
	l	−161.53	−113.66
CS_2 carbon disulfide	g	28.11	16.13
COS carbonyl sulfide	g	−33.83	−40.85
C_2N_2 cyanogen	g	73.82	71.02
HCN hydrogen cyanide	g	31.1	27.94
	l	25.2	29.0
	aq, 100	25.2	26.8
C_2H_3N acetonitrile	g	19.81	
CH_5N methylamine	g	−6.7	6.6
C_2H_7N ethylamine	g	−12.24	10.01
C_3H_9N propylamine	g	−16.45	14.38
$C_4H_{11}N$ butylamine	g	−15.60	19.55
$C_6H_{13}N$ hexamethyleneimine	g	−14.37	31.52
	l	−24.90	28.84
CH_2N_2 cyanamide	l	11.18	24.30
	c	9.15	24.18
$C_6H_8N_2$ adiponitrile	g	33.34	61.43
	l	19.19	54.63
$C_6H_{16}N_2$ hexamethyienediamine	g	−30.57	28.91
CH_5N_3 guanidine	l	−27.48	7.34
	c	−30.68	6.33

Compound	State†	Heat of formation‡§ ΔH (formation) at 25°C., kcal./mole	Free energy of formation‖¶ ΔF (formation) at 25°C., kcal./mole
Carbon (*Cont.*):			
$C_3H_6N_6$ melamine	l	−19.33	40.80
CH_3NO formamide	g	−44.64	−36.60
C_2H_7NO ethanolamine	l	−62.52	27.50
CH_4N_2O urea	l	−77.55	−46.45
	c	−79.634	−47.118
Cerium:			
Ce	c	0.00	0.00
CeN	c	−78.2	−70.8
Cesium:			
Cs	c	0.00	0.00
CsBr	c	−97.64	
	aq, 500	−91.39	−94.86
CsCl	c	−106.31	
	aq, 400	−102.01	−101.61
Cs_2CO_3	c	−271.88	
CsF	c	−131.67	
	aq, 400	−140.48	−135.98
CsH	c	−12	−7.30
$CsHCO_3$	c	−230.6	
	aq, 2000	−226.6	−210.56
CsI	c	−83.91	
	aq, 400	−75.74	−82.61
$CsNH_2$	c	−28.2	
$CsNO_3$	c	−121.14	
	aq, 400	−111.54	−96.53
Cs_2O	c	−82.1	
CsOH	c	−100.2	
	aq, 200	−117.0	−107.87
Cs_2S	c	−87	
Cs_2SO_4	c	−344.86	
	aq	−340.12	−316.66
Chlorine:			
Cl_2	g	0.00	0.00
ClF	g	−25.7	
ClO	g	33	
ClO_2	g	24.7	29.5
ClO_3	g	37	
Cl_2O	g	18.20	22.40
Cl_2O_7	g	63	
Chromium:			
Cr	c	0.00	0.00
$CrBr_3$	aq	....	−122.7
Cr_3C_2	c	−21.008	−21.20
Cr_4C	c	−16.378	−16.74
$CrCl_2$	c	−103.1	−93.8
	aq	....	−102.1
CrF_2	c	−152	
CrF_3	c	−231	
CrI_2	c	−63.7	
	aq	....	−64.1
CrO_3	c	−139.3	
Cr_2O_3	c	−268.8	−249.3
$Cr_2(SO_4)_3$	aq	....	−626.3
Cobalt:			
Co	c	0.00	0.00
$CoBr_2$	c	−55.0	
	aq	−73.61	−61.96
Co_3C	c	9.49	7.08
$CoCl_2$	c	−76.9	−66.6
	aq, 400	−95.58	−75.46
$CoCO_3$	c	−172.39	−155.36
CoF_2	aq	−172.98	−144.2
CoI_2	c	−24.2	
	aq	−43.15	−37.4
$Co(NO_3)_2$	c	−102.8	
	aq	−114.9	−65.3
CoO	c	−57.5	
Co_3O_4	c	−196.5	
$Co(OH)_2$	c	−131.5	−108.9
$Co(OH)_3$	c	−177.0	−142.0
CoS	c	−22.3	−19.8
Co_2S_3	c	−40.0	
$CoSO_4$	c	−216.6	
	aq, 400	....	−188.9
Columbium:			
Cb	c	0.00	0.00
Cb_2O_5	c	−462.96	
Copper:			
Cu	c	0.00	0.00
CuBr	c	−26.7	−23.8
$CuBr_2$	c	−34.0	
	aq	−42.4	−33.25
CuCl	c	−31.4	−24.13
$CuCl_2$	c	−48.83	
	aq, 400	−64.7	
$CuClO_4$	aq	−28.3	1.34
$Cu(ClO_3)_2$	aq, 400	....	15.4
$Cu(ClO_4)_2$	aq	....	−5.5

TABLE 3-206 Heats and Free Energies of Formation of Inorganic and Organic Compounds *(Continued)*

Compound	State†	Heat of formation‡§ ΔH (formation) at 25°C., kcal./mole	Free energy of formation‖¶ ΔF (formation) at 25°C., kcal./mole
Copper (*Cont.*):			
CuI	c	−17.8	−16.66
CuI_2	c	−4.8	
	aq	−11.9	−8.76
Cu_3N	c	17.78	
$Cu(NO_3)_2$	c	−73.1	
	aq, 200	−83.6	−36.6
CuO	c	−38.5	−31.9
Cu_2O	c	−43.00	−38.13
$Cu(OH)_2$	c	−108.9	−85.5
CuS	c	−11.6	−11.69
Cu_2S	c	−18.97	−20.56
$CuSO_4$	c	−184.7	−158.3
	aq, 800	−200.78	−160.19
Cu_2SO_4	c	−179.6	
	aq	....	−152.0
Erbium:			
Er	c	0.00	0.00
$Er(OH)_3$	c	−326.8	
Fluorine:			
F_2	g	0.00	0.00
F_2O	g	5.5	9.7
Gallium:			
Ga	c	0.00	0.00
$GaBr_3$	c	−92.4	
$GaCl_3$	c	−125.4	
GaN	c	−26.2	
Ga_2O	c	−84.3	
Ga_2O_3	c	−259.9	
Germanium:			
Ge	c	0.00	0.00
Ge_3N_4	c	−15.7	
GeO_2	c	−128.6	
Gold:			
Au	c	0.00	0.00
AuBr	c	−3.4	
$AuBr_3$	c	−14.5	
	aq	−11.0	24.47
AuCl	c	−8.3	
$AuCl_3$	c	−28.3	
	aq	−32.96	4.21
AuI	c	0.2	−0.76
Au_2O_3	c	11.0	18.71
$Au(OH)_3$	c	−100.6	
Hafnium:			
Hf	c	0.00	0.00
HfO_2	c	−271.1	−258.2
Hydrogen:			
H_3AsO_3	aq	−175.6	−153.04
H_3AsO_4	c	−214.9	
	aq	−214.8	−183.93
HBr	g	−8.66	−12.72
	aq, 400	−28.80	−24.58
HBrO	aq	−25.4	−19.90
$HBrO_3$	aq	−11.51	5.00
HCl	g	−22.063	−22.778
	aq, 400	−39.85	−31.330
HCN	g	31.1	27.94
	aq, 100	24.2	26.55
HClO	aq, 400	−28.18	−19.11
$HClO_3$	aq	−23.4	−0.25
$HClO_4$	aq, 660	−31.4	−10.70
$HC_2H_3O_2$	l	−116.2	−93.56
	aq, 400	−116.74	−96.8
$H_2C_2O_4$	c	−196.7	
	aq, 300	−194.6	−165.64
HCOOH	l	−97.8	−82.7
	aq, 200	−98.0	−85.1
H_2CO_3	aq	−167.19	−149.0
HF	g	−64.2	−64.7
	aq, 200	−75.75	
HI	g	6.27	0.365
	aq, 400	−13.47	−12.35
HIO	aq	−38	−23.33
HIO_3	c	−56.77	
	aq	−54.8	−32.25
HN_3	g	70.3	78.50
HNO_3	g	−31.99	−17.57
	l	−41.35	−19.05
	aq, 400	−49.210	
$HNO_3.H_2O$	l	−112.91	−78.36
$HNO_3.3H_2O$	l	−252.15	−193.70
H_2O	g	−57.7979	−54.6351
	l	−68.3174	−56.6899
H_2O_2	l	−45.16	−28.23
	aq, 200	−45.80	−31.47
H_3PO_2	c	−145.5	
	aq	−145.6	−120.0
H_3PO_3	c	−232.2	
	aq	−232.2	−204.0
Hydrogen (*Cont.*):			
H_3PO_4	c	−306.2	
	aq, 400	−309.32	−270.0
H_2S	g	−4.77	−7.85
	aq, 2000	−9.38	
H_2S_2	l	−3.6	
H_2SO_3	aq, 200	−146.88	−128.54
H_2SO_4	l	−193.69	
	aq, 400	−212.03	
H_2Se	g	20.5	17.0
	aq	18.1	18.4
H_2SeO_3	c	−126.5	
	aq	−122.4	−101.36
H_2SeO_4	c	−130.23	
	aq, 400	−143.4	
H_2SiO_3	c	−267.8	−247.9
H_4SiO_4	c	−340.6	
H_2Te	g	36.9	33.1
H_2TeO_3	c	−145.0	−115.7
	aq	−145.0	
H_2TeO_4	aq	−165.6	
Indium:			
In	c	0.00	0.00
$InBr_3$	c	−97.2	
	aq	−112.9	−97.2
$InCl_3$	c	−128.5	
	aq	−145.6	−117.5
InI_3	c	−56.5	
	aq	−67.2	−60.5
InN	c	−4.8	
In_2O_3	c	−222.47	
Iodine:			
I_2	c	0.00	0.00
	g	14.88	4.63
IBr	g	10.05	1.24
ICl	g	4.20	−1.32
ICl_3	c	−21.8	−6.05
I_2O_5	c	−42.5	
Iridium:			
Ir	c	0.00	0.00
IrCl	c	−20.5	−16.9
$IrCl_2$	c	−40.6	−32.0
$IrCl_3$	c	−60.5	−46.5
IrF_6	l	−130	
IrO_2	c	−40.14	
Iron:			
Fe	c, α	0.00	0.00
$FeBr_2$	c	−57.15	
	aq, 540	−78.7	−69.47
$FeBr_3$	aq	−95.5	−76.26
Fe_3C	c	5.69	4.24
$Fe(CO)_5$	l	−187.6	
$FeCO_3$	c, siderite	−172.4	−154.8
$FeCl_2$	c	−81.9	−72.6
	aq	−100.0	−83.0
$FeCl_3$	c	−96.4	
	aq, 2000	−128.5	−96.5
FeF_2	aq, 1200	−177.2	−151.7
FeI_2	c	−24.2	
	aq	−47.7	−45
FeI_3	aq	−49.7	−39.5
Fe_4N	c	−2.55	0.862
$Fe(NO_3)_2$	aq	−118.9	−72.8
$Fe(NO_3)_3$	aq, 800	−156.5	−81.3
FeO	c	−64.62	−59.38
Fe_2O_3	c	−198.5	−179.1
Fe_3O_4	c	−266.9	−242.3
$Fe(OH)_2$	c	−135.9	−115.7
$Fe(OH)_3$	c	−197.3	−166.3
$FeO.SiO_2$	c	−273.5	
Fe_2P	c	−13	
FeSi	c	−19.0	
FeS	c	−22.64	−23.23
FeS_2	c, pyrites	−38.62	−35.93
	c, marcasite	−33.0	
$FeSO_4$	c	−221.3	−195.5
	aq, 400	−236.2	−196.4
$Fe_2(SO_4)_3$	aq, 400	−653.3	−533.4
$FeTiO_3$	c, ilmenite	−295.51	−277.06
Lanthanum:			
La	c	0.00	0.00
$LaCl_3$	c	−253.1	
	aq	−284.7	
La_3H_8	c	−160	
LaN	c	−72.0	−64.6
La_2O_3	c	−539	
LaS_2	c	−148.3	
La_2S_3	c	−351.4	
$La_2(SO_4)_3$	aq	−972	

TABLE 3-206 Heats and Free Energies of Formation of Inorganic and Organic Compounds *(Continued)*

Compound	State†	Heat of formation‡§ ΔH (formation) at 25°C., kcal./mole	Free energy of formation‖¶ ΔF (formation) at 25°C., kcal./mole
Lead:			
Pb	c	0.00	0.00
$PbBr_2$	c	−66.24	−62.06
	aq	−56.4	−54.97
$PbCO_3$	c, cerussite	−167.6	−150.0
$Pb(C_2H_3O_2)_2$	c	−232.6	
	aq, 400	−234.2	−184.40
PbC_2O_4	c	−205.3	
$PbCl_2$	c	−85.68	−75.04
	aq	−82.5	−68.47
PbF_2	c	−159.5	−148.1
PbI_2	c	−41.77	−41.47
$Pb(NO_3)_2$	c	−106.88	
	aq, 400	−99.46	−58.3
PbO	c, red	−51.72	−45.53
	c, yellow	−50.86	−43.88
PbO_2	c	−65.0	−52.0
Pb_3O_4	c	−172.4	−142.2
$Pb(OH)_2$	c	−123.0	−102.2
PbS	c	−22.38	−21.98
$PbSO_4$	c	−218.5	−192.9
Lithium:			
Li	c	0.00	0.00
LiBr	c	−83.75	
	aq, 400	−95.40	−95.28
$LiBrO_3$	aq	−77.9	−65.70
Li_2C_2	c	−13.0	
LiCN	aq	−31.4	−31.35
LiCNO	aq	−101.2	−94.12
$LiC_2H_3O_2$	aq	−183.9	−160.00
Li_2CO_3	c	−289.7	−269.8
	aq, 1900	−293.1	−267.58
LiCl	c	−97.63	
	aq, 278	−106.45	−102.03
$LiClO_3$	aq	−87.5	−70.95
$LiClO_4$	aq	−106.3	−81.4
LiF	c	−145.57	
	aq, 400	−144.85	−136.40
LiH	c	−22.9	
$LiHCO_3$	aq, 2000	−231.1	−210.98
LiI	c	−65.07	
	aq, 400	−80.09	−83.03
$LiIO_3$	aq	−121.3	−102.95
Li_3N	c	−47.45	−37.33
$LiNO_3$	c	−115.350	
	aq, 400	−115.88	−96.95
Li_2O	c	−142.3	
Li_2O_2	c	−151.9	−138.0
	aq	−159	
LiOH	c	−116.58	−106.44
	aq, 400	−121.47	−108.29
$LiOH.H_2O$	c	−188.92	
$Li_2O.SiO_2$	gls	−374	
Li_2Se	c	−84.9	
	aq	−95.5	−105.64
Li_2SO_4	c	−340.23	−314.66
	aq, 400	−347.02	
$Li_2SO_4.H_2O$	c	−411.57	−375.07
Magnesium:			
Mg	c	0.00	0.00
$Mg(AsO_4)_2$	c	−731.3	
	aq	−749	−630.14
$MgBr_2$	c	−123.9	
	aq, 400	−167.33	−156.94
$Mg(CN)_2$	aq	−39.7	−29.08
$MgCN_2$	c	−61	
$Mg(C_2H_3O_2)_2$	aq	−344.6	−286.38
$MgCO_3$	c	−261.7	−241.7
$MgCl_2$	c	−153.220	−143.77
	aq, 400	−189.76	
$MgCl_2.H_2O$	c	−230.970	−205.93
$MgCl_2.2H_2O$	c	−305.810	−267.20
$MgCl_2.4H_2O$	c	−453.820	−387.98
$MgCl_2.6H_2O$	c	−597.240	−505.45
MgF_2	c	−263.8	
MgI_2	c	−86.8	
	aq, 400	−136.79	−132.45
$MgMoO_4$	c	−329.9	
Mg_3N_2	c	−115.2	−100.8
$Mg(NO_3)_2$	c	−188.770	−140.66
	aq, 400	−209.927	−160.28
$Mg(NO_3)_2.2H_2C$	c	−336.625	
$Mg(NO_3)_2.6H_2O$	c	−624.48	−496.03
MgO	c	−143.84	−136.17
$MgO.SiO_2$	c	−347.5	−326.7
$Mg(OH)_2$	c, ppt.	−221.90	−200.17
	c, brucite	−223.9	−193.3
MgS	c	−84.2	
	aq	−108	
Magnesium (*Cont.*):			
$MgSO_4$	c	−304.94	−277.7
	aq, 400	−325.4	−283.88
MgTe	c	−25	
$MgWO_4$	c	−345.2	
Manganese:			
Mn	c, α	0.00	0.00
$MnBr_2$	c	−91	
	aq	−106	−97.8
Mn_3C	c	1.1	1.26
$Mn(C_2H_3O_2)_2$	c	−270.3	
	aq	−282.7	−227.2
$MnCO_3$	c	−211	−192.5
MnC_2O_4	c	−240.9	
$MnCl_2$	c	−112.0	−102.2
	aq, 400	−128.9	
MnF_2	aq, 1200	−206.1	−180.0
MnI_2	c	−49.8	
	aq	−76.2	−73.3
Mn_5N_2	c	−57.77	−46.49
$Mn(NO_3)_2$	c	−134.9	
	aq, 400	−148.0	−101.1
$Mn(NO_3)_2.6H_2O$	c	−557.07	−441.2
MnO	c	−92.04	−86.77
MnO_2	c	−124.58	−111.49
Mn_2O_3	c	−229.5	−209.9
Mn_3O_4	c	−331.65	−306.22
$MnO.SiO_2$	c	−301.3	−282.1
$Mn(OH)_2$	c	−163.4	−143.1
$Mn(OH)_3$	c	−221	−190
$Mn_3(PO_4)_2$	c	−736	
MnSe	c	−26.3	−27.5
MnS	c, green	−47.0	−48.0
$MnSO_4$	c	−254.18	−228.41
	aq, 400	−265.2	
$Mn_2(SO_4)_3$	c	−635	
	aq	−657	
Mercury:			
Hg	l	0.00	0.00
HgBr	g	23	18
$HgBr_2$	c	−40.68	−38.8
	aq	−38.4	−9.74
$Hg(C_2H_3O_2)_2$	c	−196.3	
	aq	−192.5	−139.2
$HgCl_2$	c	−53.4	−42.2
	aq	−50.3	−23.25
HgCl	g	19	14
Hg_2Cl_2	c	−63.13	
$Hg(CN)_2$	c	62.8	
	aq, 1110	66.25	
HgC_2O_4	c	−159.3	
HgH	g	57.1	52.25
HgI_2	c, red	−25.3	−24.0
HgI	g	33	23
Hg_2I_2	c	−28.88	−26.53
$Hg(NO_3)_2$	aq	−56.8	−13.09
$Hg_2(NO_3)_2$	aq	−58.5	−15.65
HgO	c, red	−21.6	−13.94
	c, yellow ppt.	−20.8	
Hg_2O	c	−21.6	−12.80
HgS	c, black	−10.7	−8.80
$HgSO_4$	c	−166.6	
Hg_2SO_4	c	−177.34	−149.12
Molybdenum:			
Mo	c	0.00	0.00
Mo_2C	c	4.36	2.91
Mo_2N	c	−8.3	
MoO_2	c	−130	−118.0
MoO_3	c	−180.39	−162.01
MoS_2	c	−56.27	−54.19
MoS_3	c	−61.48	−57.38
Nickel:			
Ni	c	0.00	0.00
$NiBr_2$	c	−53.4	
	aq	−72.6	−60.7
Ni_3C	c	9.2	8.88
$Ni(C_2H_3O_2)_2$	aq	−249.6	−190.1
$Ni(CN)_2$	aq	230.9	66.3
$NiCl_2$	c	−75.0	
	aq, 400	−94.34	−74.19
NiF_2	c	−157.5	
	aq	−171.6	−142.9
NiI_2	c	−22.4	
	aq	−42.0	−36.2
$Ni(NO_3)_2$	c	−101.5	
	aq, 200	−113.5	−64.0
NiO	c	−58.4	−51.7
$Ni(OH)_2$	c	−129.8	−105.6
$Ni(OH)_3$	c	−163.2	

TABLE 3-206 Heats and Free Energies of Formation of Inorganic and Organic Compounds *(Continued)*

Compound	State†	Heat of formation‡§ ΔH (formation) at 25°C., kcal./mole	Free energy of formation‖¶ ΔF (formation) at 25°C., kcal./mole
Nickel (*Cont.*):			
NiS	c	−20.4	
$NiSO_4$	c	−216	
	aq, 200	−231.3	−187.6
Nitrogen:			
N_2	g	0.00	0.00
NF_3	g	−27	
NH_3	g	−10.96	−3.903
	aq, 200	−19.27	
NH_4Br	c	−64.57	
	aq	−60.27	−43.54
$NH_4C_2H_3O_2$	c	−148.1	
	aq, 400	−148.58	−108.26
NH_4CN	c	−0.7	
	aq	3.6	20.4
NH_4CNS	c	−17.8	
	aq	−12.3	4.4
$(NH_4)_2CO_3$	aq	−223.4	−164.1
$(NH_4)_2C_2O_4$	c	−266.3	
	aq	−260.6	−196.2
NH_4Cl	c	−75.23	−48.59
	aq, 400	−71.20	
NH_4ClO_4	c	−69.4	
	aq	−63.2	−21.1
$(NH_4)_2CrO_4$	c	−276.9	
	aq	−271.3	−209.3
NH_4F	c	−111.6	
	aq	−110.2	−84.7
NH_4I	c	−48.43	
	aq	−44.97	−31.3
NH_4NO_3	c	−87.40	
	aq, 500	−80.89	
NH_4OH	aq	−87.59	
$(NH_4)_2S$	aq, 400	−55.21	−14.50
$(NH_4)_2SO_4$	c	−281.74	−215.06
	aq, 400	−279.33	−214.02
N_2H_4	l	12.06	
$N_2H_4.H_2O$	l	−57.96	
$N_2H_4.H_2SO_4$	c	−232.2	
N_2O	g	19.55	24.82
NO	g	21.600	20.719
NO_2	g	7.96	12.26
N_2O_4	g	2.23	23.41
N_2O_5	c	−10.0	
$NOBr$	l	11.6	19.26
$NOCl$	g	12.8	16.1
Osmium:			
Os	c	0.00	0.00
OsO_4	c	−93.6	−70.9
	g	−80.1	−68.1
Oxygen:			
O_2	g	0.00	0.00
O_3	g	33.88	38.86
Palladium:			
Pd	c	0.00	0.00
PdO	c	−20.40	
Phosphorus:			
P	c, white ("yellow")	0.00	0.00
	c, red ("violet")	−4.22	−1.80
P	g	150.35	141.88
P_2	g	33.82	24.60
P_4	g	13.2	5.89
PBr_3	l	−45	
PBr_5	c	−60.6	
PCl_3	g	−70.0	−65.2
	l	−76.8	−63.3
PCl_5	g	−91.0	−73.2
PH_3	g	2.21	−1.45
PI_3	c	−10.9	
P_2O_5	c	−360.0	
$POCl_3$	g	−138.4	−127.2
Platinum:			
Pt	c	0.00	0.00
$PtBr_4$	c	−40.6	
	aq	−50.7	
$PtCl_2$	c	−34	
$PtCl_4$	c	−62.6	
	aq	−82.3	
PtI_4	c	−18	
$Pt(OH)_2$	c	−87.5	−67.9
PtS	c	−20.18	−18.55
PtS_2	c	−26.64	−24.28
Potassium:			
K	c	0.00	0.00
K_3AsO_3	aq	−323.0	
K_3AsO_4	aq	−390.3	−355.7
KH_2AsO_4	c	−271.2	−236.7
KBr	c	−94.06	−90.8
	aq, 400	−89.19	−92.0
Potassium (*Cont.*):			
$KBrO_3$	c	−81.58	−60.30
	aq, 1667	−71.68	
$KC_2H_3O_2$	c	−173.80	
	aq, 400	−177.38	−156.73
KCl	c	−104.348	−97.76
	aq, 400	−100.164	−98.76
$KClO_3$	c	−93.5	−69.30
	aq, 400	−81.34	
$KClO_4$	c	−103.8	−72.86
	aq, 400	−101.14	
KCN	c	−28.1	
	aq, 400	−25.3	−28.08
$KCNO$	c	−99.6	
	aq	−94.5	−90.85
$KCNS$	c	−47.0	
	aq, 400	−41.07	−44.08
K_2CO_3	c	−274.01	
	aq, 400	−280.90	−264.04
$K_2C_2O_4$	c	−319.9	
	aq, 400	−315.5	−293.1
K_2CrO_4	c	−333.4	
	aq, 400	−328.2	−306.3
$K_2Cr_2O_7$	c	−488.5	
	aq, 400	−472.1	−440.9
KF	c	−134.50	
	aq, 180	−138.36	−133.13
$K_3Fe(CN)_6$	c	−48.4	
	aq	−34.5	
$K_4Fe(CN)_6$	c	−131.8	
	aq	−119.9	
KH	c	−10	−5.3
$KHCO_3$	c	−229.8	
	aq, 2000	−224.85	−207.71
KI	c	−78.88	−77.37
	aq, 500	−73.95	−79.76
KIO_3	c	−121.69	−101.87
	aq, 400	−115.18	−99.68
KIO_4	aq	−98.1	
$KMnO_4$	c	−192.9	−169.1
	aq, 400	−182.5	−168.0
K_2MoO_4	aq, 880	−364.2	−342.9
KNH_2	c	−28.25	
KNO_2	aq	−86.0	−75.9
KNO_3	c	−118.08	−94.29
	aq, 400	−109.79	−93.68
K_2O	c	−86.2	
$K_2O.Al_2O_3.SiO_2$	c, leucite	−1379.6	
	gls	−1368.2	
$K_2O.Al_2O_3.SiO_2$	c, adularia	−1784.5	
	c, microcline	−1784.5	
	gls	−1747	
KOH	c	−102.02	
	aq, 400	−114.96	−105.0
K_3PO_3	aq	−397.5	
K_3PO_4	aq	−478.7	−443.3
KH_2PO_4	c	−362.7	−326.1
K_2PtCl_4	c	−254.7	
	aq	−242.6	−226.5
K_2PtCl_6	c	−299.5	−263.6
	aq, 9400	−286.1	
K_2Se	c	−74.4	
	aq	−83.4	−99.10
K_2SeO_4	aq	−267.1	−240.0
K_2S	c	−121.5	
	aq, 400	−110.75	−111.44
K_2SO_3	c	−267.7	
	aq	−269.7	−251.3
K_2SO_4	c	−342.65	−314.62
	aq, 400	−336.48	−310.96
$K_2SO_4.Al_2(SO_4)_3$	c	−1178.38	−1068.48
$K_2SO_4.Al_2(SO_4)_3.24H_2O$	c	−2895.44	−2455.68
$K_2S_2O_6$	c	−418.62	
Rhenium:			
Re	c	0.00	0.00
ReF_6	g	−274	
Rhodium:			
Rh	c	0.00	0.00
RhO	c	−21.7	
Rh_2O	c	−22.7	
Rh_2O_3	c	−68.3	
Rubidium:			
Rb	c	0.00	0.00
$RbBr$	c	−95.82	
	g	−45.0	−52.50
	aq, 500	−90.54	−93.38
$RbCN$	aq	−25.9	
Rb_2CO_3	c	−273.22	
	aq, 220	−282.61	−263.78

TABLE 3-206 Heats and Free Energies of Formation of Inorganic and Organic Compounds *(Continued)*

Compound	State†	Heat of formation‡§ ΔH (formation) at 25°C., kcal./mole	Free energy of formation‖¶ ΔF (formation) at 25°C., kcal./mole
Rubidium (*Cont.*):			
$RbCl$	c	−105.06	−98.48
	g	−53.6	−57.9
	aq, ∞	−101.06	−100.13
RbF	c	−133.23	
	aq, 400	−139.31	−134.5
$RbHCO_3$	c	−230.01	
	aq, 2000	−225.59	−209.07
RbI	c	−81.04	
	g	−31.2	−40.5
	aq, 400	−74.57	−81.13
$RbNH_2$	c	−27.74	
$RbNO_3$	c	−119.22	
	aq, 400	−110.52	−95.05
Rb_2O	c	−82.9	
Rb_2O_2	c	−107	
$RbOH$	c	−101.3	
	aq, 200	−115.8	−106.39
Ruthenium:			
Ru	c	0.00	0.00
RuS_2	c	−46.99	−44.11
Selenium:			
Se	c, I, hexagonal	0.00	0.00
	c, II, red, monoclinic	0.2	
Se_2Cl_2	l	−22.06	−13.73
SeF_6	g	−246	−222
SeO_2	c	−56.33	
Silicon:			
Si	c	0.00	0.00
$SiBr_4$	l	−93.0	
SiC	c	−28	−27.4
$SiCl_4$	l	−150.0	−133.9
	g	−142.5	−133.0
SiF_4	g	−370	−360
SiH_4	g	−14.8	−9.4
SiI_4	c	−29.8	
Si_3N_4	c	−179.25	−154.74
SiO_2	c, cristobalite, 1600° form	−202.62	
	c, cristobalite, 1100° form	−202.46	
	c, quartz	−203.35	−190.4
	c, tridymite	−203.23	
Silver:			
Ag	c	0.00	0.00
$AgBr$	c	−23.90	−23.02
Ag_2C_2	c	84.5	
$AgC_2H_3O_2$	c	−95.9	
	aq	−91.7	−70.86
$AgCN$	c	33.8	38.70
Ag_2CO_3	c	−119.5	−103.0
$Ag_2C_2O_4$	c	−158.7	
$AgCl$	c	−30.11	−25.98
AgF	c	−48.7	
	aq, 400	−53.1	−47.26
AgI	c	−15.14	−16.17
$AgIO_3$	c	−42.02	−24.08
$AgNO_2$	c	−11.6	3.76
	aq	−2.9	9.99
$AgNO_3$	c	−29.4	−7.66
	aq, 6500	−24.02	−7.81
Ag_2O	c	−6.95	−2.23
Ag_2S	c	−5.5	−7.6
Ag_2SO_4	c	−170.1	−146.8
	aq	−165.8	−139.22
Sodium:			
Na	c	0.00	0.00
Na_3AsO_3	aq, 500	−314.61	
Na_3AsO_4	c	−366	
	aq, 500	−381.97	−341.17
$NaBr$	c	−86.72	
	aq, 400	−86.33	−87.17
$NaBrO$	aq	−78.9	
$NaBrO_3$	aq, 400	−68.89	−57.59
$NaC_2H_3O_2$	c	−170.45	
	aq, 400	−175.450	−152.31
$NaCN$	c	−22.47	
	aq, 200	−22.29	−23.24
$NaCNO$	c	−96.3	
	aq	−91.7	−86.00
$NaCNS$	c	−39.94	
	aq, 400	−38.23	−39.24
Na_2CO_3	c	−269.46	−249.55
	aq, 1000	−275.13	−251.36
$NaCO_2NH_2$	c	−142.17	
$Na_2C_2O_4$	c	−313.8	
	aq, 600	−309.92	−283.42
$NaCl$	c	−98.321	−91.894
	aq, 400	−97.324	−93.92
Sodium (*Cont.*):			
$NaClO_3$	c	−83.59	
	aq, 400	−78.42	−62.84
$NaClO_4$	c	−101.12	
	aq, 476	−97.66	−73.29
Na_2CrO_4	c	−319.8	
	aq, 800	−323.0	−296.58
$Na_2Cr_2O_7$	aq, 1200	−465.9	−431.18
NaF	c	−135.94	−129.0
	aq, 400	−135.711	−128.29
NaH	c	−14	−9.30
$NaHCO_3$	c	−226.0	−202.66
	aq	−222.1	−202.87
NaI	c	−69.28	
	aq, ∞	−71.10	−74.92
$NaIO_3$	aq, 400	−112.300	−94.84
Na_2MoO_4	c	−364	
	aq	−358.7	−333.18
$NaNO_2$	c	−86.6	
	aq	−83.1	−71.04
$NaNO_3$	c	−111.71	−87.62
	aq, 400	−106.880	−88.84
Na_2O	c	−99.45	−90.06
Na_2O_2	c	−119.2	−105.0
$Na_2O.SiO_2$	c	−383.91	−361.49
$Na_2O.Al_2O_3.3SiO_2$	c, natrolite	−1180	
$Na_2O.Al_2O_3.4SiO_2$	c	−1366	
$NaOH$	c	−101.96	−90.60
	aq, 400	−112.193	−100.18
Na_3PO_3	aq, 1000	−389.1	
Na_3PO_4	c	−457	
	aq, 400	−471.9	−428.74
Na_2PtCl_4	aq	−237.2	−216.78
Na_2PtCl_6	c	−272.1	
	aq	−280.9	
Na_2Se	c	−59.1	
	aq, 440	−78.1	−89.42
Na_2SeO_4	c	−254	
	aq, 800	−261.5	−230.30
Na_2S	c	−89.8	
	aq, 400	−105.17	−101.76
Na_2SO_3	c	−261.2	−240.14
	aq, 800	−264.1	−241.58
Na_2SO_4	c	−330.50	−302.38
	aq, 1100	−330.82	−301.28
$Na_2SO_4.10H_2O$	c	−1033.85	−870.52
Na_2WO_4	c	−391	
	aq	−381.5	−345.18
Strontium:			
Sr	c	0.00	0.00
$SrBr_2$	c	−171.0	
	aq, 400	−187.24	−182.36
$Sr(C_2H_3O_2)_2$	c	−358.0	
	aq	−364.4	−311.80
$Sr(CN)_2$	aq	−59.5	−54.50
$SrCO_3$	c	−290.9	−271.9
$SrCl_2$	c	−197.84	
	aq, 400	−209.20	−195.86
SrF_2	c	−289.0	
$Sr(HCO_3)_2$	aq	−459.1	−413.76
SrI_2	c	−136.1	
	aq, 400	−156.70	−157.87
Sr_3N_2	c	−91.4	−76.5
$Sr(NO_3)_2$	c	−233.2	
	aq, 400	−228.73	−185.70
SrO	c	−140.8	−133.7
$SrO.SiO_2$	gls	−364	
SrO_2	c	−153.3	−139.0
Sr_2O	c	−153.6	
$Sr(OH)_2$	c	−228.7	
	aq, 800	−239.4	−208.27
$Sr_3(PO_4)_2$	c	−980	
	aq	−985	−881.54
SrS	c	−113.1	
	aq	−120.4	−109.78
$SrSO_4$	c	−345.3	
	aq, 400	−345.0	−309.30
$SrWO_4$	c	−393	
Sulfur:			
S	c, rhombic	0.00	0.00
	c, monoclinic	−0.071	−0.023
	l, λ	0.257	0.072
	l, λμ equilibrium	....	0.071
	g	53.25	43.57
S_2	g	31.02	19.36
S_6	g	27.78	13.97
S_8	g	27.090	12.770
S_2Br_2	l	−4	
SCl_4	l	−13.7	

TABLE 3-206 Heats and Free Energies of Formation of Inorganic and Organic Compounds *(Concluded)*

Compound	State†	Heat of formation‡§ ΔH (formation) at 25°C., kcal./mole	Free energy of formation‖¶ ΔF (formation) at 25°C., kcal./mole
Sulfur (*Cont.*):			
S_2Cl_2	l	−14.2	−5.90
S_2Cl_4	l	−24.1	
SF_6	g	−262	−237
SO	g	19.02	12.75
SO_2	g	−70.94	−71.68
SO_3	g	−94.39	−88.59
	l	−103.03	−88.28
	c, α	−105.09	−88.22
	c, β	−105.92	−88.34
	c, γ	−109.34	−88.98
SO_2Cl_2	g	−82.04	−74.06
	l	−89.80	−75.06
Tantalum:			
Ta	c	0.00	0.00
TaN	c	−51.2	−45.11
Ta_2O_5	c	−486.0	−453.7
Tellurium:			
Te	c	0.00	0.00
$TeBr_4$	c	−49.3	
$TeCl_4$	c	−77.4	−57.4
TeF_6	g	−315	−292
TeO_2	c	−77.56	−64.66
Thallium:			
Tl	c	0.00	0.00
TlBr	c	−41.5	−39.43
	aq	−28.0	−32.34
TlCl	c	−49.37	−44.46
	aq	−38.4	−39.09
$TlCl_3$	c	−82.4	
	aq	−91.0	−44.25
TlF	aq	−77.6	−73.46
TlI	c	−31.1	−31.3
	aq	−12.7	−20.09
$TlNO_3$	c	−58.2	−36.32
	aq	−48.4	−34.01
Tl_2O	c	−43.18	
Tl_2O_3	c	−120	
TlOH	c	−57.44	−45.54
	aq	−53.9	−45.35
Tl_2S	c	−22	
Tl_2SO_4	c	−222.8	−197.79
	aq, 800	−214.1	−191.62
Thorium:			
Th	c	0.00	0.00
$ThBr_4$	c	−281.5	
	aq	−352.0	−295.31
ThC_2	c	−45.1	
$ThCl_4$	c	−335	
	aq	−392	−322.32
ThI_4	aq	−292.0	−246.33
Th_3N_4	c	−309.0	−282.3
ThO_2	c	−291.6	−280.1
$Th(OH)_4$	c, "soluble"	−336.1	
$Th(SO_4)_2$	c	−632	
	aq	−668.1	−549.2
Tin:			
Sn	c, II, tetragonal	0.00	0.00
	c, III, "gray," cubic	0.6	1.1
$SnBr_2$	c	−61.4	
	aq	−60.0	−55.43
$SnBr_4$	c	−94.8	
	aq	−110.6	−97.66
$SnCl_2$	c	−83.6	
	aq	−81.7	−68.94
$SnCl_4$	l	−127.3	−110.4
	aq	−157.6	−124.67
SnI_2	c	−38.9	
	aq	−33.3	−30.95
Tin (*Cont.*):			
SnO	c	−67.7	−60.75
SnO_2	c	−138.1	−123.6
$Sn(OH)_2$	c	−136.2	−115.95
$Sn(OH)_4$	c	−268.9	−226.00
SnS	c	−18.61	
Titanium:			
Ti	c	0.00	0.00
TiC	c	−110	−109.2
$TiCl_4$	l	−181.4	−165.5
TiN	c	−80.0	−73.17
TiO_2	c, III, rutil	−225.0	−211.9
	amorphous	−214.1	−201.4
Tungsten:			
W	c	0.00	0.00
WO_2	c	−130.5	−118.3
WO_3	c	−195.7	−177.3
WS_2	c	−84	
Uranium:			
U	c	0.00	0.00
UC_2	c	−29	
UCl_3	c	−213	
UCl_4	c	−251	
U_3N_4	c	−274	−249.6
UO_2	c	−256.6	−242.2
$UO_2(NO_3)_2.6H_2O$	c	−756.8	−617.8
UO_3	c	−291.6	
U_3O_8	c	−845.1	
Vanadium:			
V	c	0.00	0.00
VCl_2	c	−147	
VCl_3	l	−187	
VCl_4	l	−165	
VN	c	−41.43	−35.08
V_2O_2	c	−195	
V_2O_3	c	−296	−277
V_2O_4	c	−342	−316
V_2O_5	c	−373	−342
Zinc:			
Zn	c	0.00	0.00
ZnSb	c	−3.6	−3.88
$ZnBr_2$	c	−77.0	−72.9
	aq, 400	−93.6	
$Zn(C_2H_3O_2)_2$	c	−259.4	
	aq, 400	−269.4	−214.4
$Zn(CN)_2$	c	17.06	
$ZnCO_3$	c	−192.9	−173.5
$ZnCl_2$	c	−99.9	−88.8
	aq, 400	−115.44	
ZnF_2	aq	−192.9	−166.6
ZnI_2	c	−50.50	−49.93
	aq	−61.6	
$Zn(NO_3)_2$	aq, 400	−134.9	−87.7
ZnO	c, hexagonal	−83.36	−76.19
$ZnO.SiO_2$	c	−282.6	
$Zn(OH)_2$	c, rhombic	−153.66	
ZnS	c, wurtzite	−45.3	−44.2
$ZnSO_4$	c	−233.4	
	aq, 400	−252.12	−211.28
Zirconium:			
Zr	c	0.00	0.00
ZrC	c	−29.8	−34.6
$ZrCl_4$	c	−268.9	
ZrN	c	−82.5	−75.9
ZrO_2	c, monoclinic	−258.5	−244.6
$Zr(OH)_4$	c	−411.0	
$ZrO(OH)_2$	c	−337	−307.6

† The physical state is indicated as follows: *c*, crystal (solid); *l*, liquid; *g*, gas; *gls*, glass or solid supercooled liquid; *aq*, in aqueous solution. A number following the symbol *aq* applies only to the values of the heats of formation (not to those of free energies of formation); and indicates the number of moles of water per mole of solute; when no number is given, the solution is understood to be dilute. For the free energy of formation of a substance in aqueous solution, the concentration is always that of the hypothetical solution of unit molality.

‡ The increment in heat content, ΔH, in the reaction of forming the given substance from its elements in their standard states. When ΔH is negative, heat is evolved in the process, and, when positive, heat is absorbed.

§ The heat of solution in water of a given solid, liquid, or gaseous compound is given by the difference in the value for the heat of formation of the given compound in the solid, liquid, or gaseous state and its heat of formation in aqueous solution. The following two examples serve as an illustration of the procedure: (1) For NaCl(*c*) and NaCl(*aq*, $400H_2O$), the values of ΔH(formation) are, respectively, −98.321 and −97.324 kg.-cal. per mole. Subtraction of the first value from the second gives $\Delta H = 0.998$ kg.-cal. per mole for the reaction of dissolving crystalline sodium chloride in 400 moles of water. When this process occurs at a constant pressure of 1 atm., 0.998 kg.-cal. of energy are absorbed. (2) For HCl(*g*) and HCl(*aq*, $400H_2O$), the values for ΔH(formation) are, respectively, −22.06 and −39.85 kg.-cal. per mole. Subtraction of the first from the second gives $\Delta H = -17.79$ kg.-cal per mole for the reaction of dissolving gaseous hydrogen chloride in 400 moles of water. At a constant pressure of 1 atm. 17.79 kg.-cal. of energy are evolved in this process.

‖ The increment in the free energy, ΔF, in the reaction of forming the given substance in its standard state from its elements in their standard states. The standard states are: for a gas, fugacity (approximately equal to the pressure) of 1 atm.; for a pure liquid or solid, the substance at a pressure of 1 atm.; for a substance in aqueous solution, the hypothetical solution of unit molality, which has all the properties of the infinitely dilute solution except the property of concentration.

¶ The free energy of solution of a given substance from its normal standard state as a solid, liquid, or gas to the hypothetical one molal state in aqueous solution may be calculated in a manner similar to that described in footnote § for calculating the heat of solution.

NOTE: °F = 9/5 °C + 32; to convert kilocalories per gram-mole to British thermal units per pound-mole, multiply by 1.799×10^{-3}.

HEATS OF COMBUSTION

TABLE 3-207 Hydrogen, Carbon, Carbon Monoxide, and Hydrocarbons

Heats of combustion of additional compounds may be calculated from the heats of formation given in Table 3-206.

The following values are taken from the tables of the American Petroleum Institute Research Project 44 of the National Bureau of Standards on the Collection, Analysis, Calculation, and Compilation of Data on the Properties of Hydrocarbons.

Compound	Formula	State	Heat of combustion, $-\Delta Hc°$, at 25°C. and constant pressure, to form: H_2O (liq.) and CO_2 (gas)			H_2O (gas) and CO_2 (gas)		
			Kcal./mole	Cal./g.	B.t.u./lb.	Kcal./mole	Cal./g.	B.t.u./lb.
Hydrogen	H_2	gas	68.3174	33,887.6	60,957.7	57.7979	28,669.6	51,571.4
Carbon	C	solid, graph.	94.0518	7,831.1	14,086.8			
Carbon monoxide	CO	gas	67.6361	2,414.7	4,343.6			
Paraffins								
Methane	CH_4	gas	212.798	13,265.1	23,861	191.759	11,953.6	21,502
Ethane	C_2H_6	gas	372.820	12,399.2	22,304	341.261	11,349.6	20,416
Propane	C_3H_8	gas	530.605	12,033.5	21,646	488.527	11,079.2	19,929
Propane	C_3H_8	liq.*	526.782	11,946.8	21,490	484.704	10,992.5	19,774
n-Butane	C_4H_{10}	gas	687.982	11,837.3	21,293	635.384	10,932.3	19,665
n-Butane	C_4H_{10}	liq.*	682.844	11,748.9	21,134	630.246	10,843.9	19,506
2-Methylpropane (Isobutane)	C_4H_{10}	gas	686.342	11,809.1	21,242	633.744	10,904.1	19,614
2-Methylpropane (Isobutane)	C_4H_{10}	liq.*	681.625	11,727.9	21,096	629.027	10,822.9	19,468
n-Pentane	C_5H_{12}	gas	845.16	11,714.6	21,072	782.04	10,839.7	19,499
n-Pentane	C_5H_{12}	liq.	838.80	11,626.4	20,914	775.68	10,751.5	19,340
2-Methylbutane (Isopentane)	C_5H_{12}	gas	843.24	11,688.0	21,025	780.12	10,813.1	19,451
2-Methylbutane (Isopentane)	C_5H_{12}	liq.	837.31	11,605.8	20,877	774.19	10,730.9	19,303
2,2-Dimethylpropane (Neopentane)	C_5H_{12}	gas	840.49	11,649.8	20,956	777.37	10,775.0	19,382
2,2-Dimethylpropane (Neopentane)	C_5H_{12}	liq.	835.18	11,576.2	20,824	772.06	10,701.4	19,250
n Hexane	C_6H_{14}	gas	1,002.57	11,634.5	20,928	928.93	10,780.0	19,391
n-Hexane	C_6H_{14}	liq.	995.01	11,546.8	20,771	921.37	10,692.2	19,233
2-Methylpentane	C_6H_{14}	gas	1,000.87	11,614.8	20,893	927.23	10,760.2	19,356
2-Methylpentane	C_6H_{14}	liq.	993.71	11,531.7	20,743	920.07	10,677.1	19,206
3-Methylpentane	C_6H_{14}	gas	1,001.51	11,622.2	20,906	927.87	10,767.6	19,369
3-Methylpentane	C_6H_{14}	liq.	994.25	11,538.0	20,755	920.61	10,683.4	19,218
2,2-Dimethylbutane	C_6H_{14}	gas	998.17	11,583.5	20,837	924.53	10,728.9	19,299
2,2-Dimethylbutane	C_6H_{14}	liq.	991.52	11,506.3	20,698	917.88	10,651.7	19,161
2,3-Dimethylbutane	C_6H_{14}	gas	1,000.04	11,605.2	20,876	926.40	10,750.6	19,338
2,3-Dimethylbutane	C_6H_{14}	liq.	993.05	11,524.0	20,730	919.41	10,669.5	19,192
n-Heptane	C_7H_{16}	gas	1,160.01	11,577.2	20,825	1,075.85	10,737.2	19,314
n-Heptane	C_7H_{16}	liq.	1,151.27	11,489.9	20,668	1,067.11	10,650.0	19,157
2-Methylhexane	C_7H_{16}	gas	1,158.30	11,560.1	20,795	1,074.14	10,720.2	19,284
2-Methylhexane	C_7H_{16}	liq.	1,149.97	11,477.0	20,645	1,065.81	10,637.0	19,134
3-Methylhexane	C_7H_{16}	gas	1,158.94	11,566.5	20,806	1,074.78	10,726.6	19,295
3-Methylhexane	C_7H_{16}	liq.	1,150.55	11,482.8	20,655	1,066.39	10,642.8	19,145
3-Ethylpentane	C_7H_{16}	gas	1,159.56	11,572.7	20,817	1,075.40	10,732.7	19,306
3-Ethylpentane	C_7H_{16}	liq.	1,151.13	11,488.6	20,666	1,066.97	10,648.6	19,155
2,2-Dimethylpentane	C_7H_{16}	gas	1,155.61	11,533.3	20,746	1,071.45	10,693.3	19,235
2,2-Dimethylpentane	C_7H_{16}	liq.	1,147.85	11,455.8	20,607	1,063.69	10,615.9	19,096
2,3-Dimethylpentane	C_7H_{16}	gas	1,157.28	11,549.9	20,776	1,073.12	10,710.0	19,265
2,3-Dimethylpentane	C_7H_{16}	liq.	1,149.09	11,468.2	20,629	1,064.93	10,628.3	19,118
2,4-Dimethylpentane	C_7H_{16}	gas	1,156.60	11,543.1	20,764	1,072.44	10,703.2	19,253
2,4-Dimethylpentane	C_7H_{16}	liq.	1,148.73	11,464.6	20,623	1,064.57	10,624.7	19,112
3,3-Dimethylpentane	C_7H_{16}	gas	1,156.73	11,544.4	20,766	1,072.57	10,704.5	19,255
3,3-Dimethylpentane	C_7H_{16}	liq.	1,148.83	11,465.6	20,625	1,064.67	10,625.7	19,114
2,2,3-Trimethylbutane	C_7H_{16}	gas	1,155.94	11,536.6	20,752	1,071.78	10,696.6	19,241
2,2,3-Trimethylbutane	C_7H_{16}	liq.	1,148.27	11,460.0	20,614	1,064.11	10,620.1	19,104
n-Octane	C_8H_{18}	gas	1,317.45	11,533.9	20,747	1,222.77	10,705.0	19,256
n-Octane	C_8H_{18}	liq.	1,307.53	11,447.1	20,591	1,212.85	10,618.2	19,100
2-Methylheptane	C_8H_{18}	gas	1,315.76	11,519.1	20,721	1,221.08	10,690.2	19,230
2-Methylheptane	C_8H_{18}	liq.	1,306.28	11,436.1	20,572	1,211.60	10,607.2	19,080
3-Methylheptane	C_8H_{18}	gas	1,316.44	11,525.1	20,732	1,221.76	10,696.2	19,240
3-Methylheptane	C_8H_{18}	liq.	1,306.92	11,441.7	20,582	1,212.24	10,612.8	19,091
4-Methylheptane	C_8H_{18}	gas	1,316.57	11,526.2	20,734	1,221.89	10,697.3	19,243
4-Methylheptane	C_8H_{16}	liq.	1,307.09	11,443.2	20,584	1,212.41	10,614.3	19,093
3-Ethylhexane	C_8H_{18}	gas	1,316.87	11,528.8	20,738	1,222.19	10,699.9	19,247
3-Ethylhexane	C_8H_{18}	liq.	1,307.39	11,445.8	20,589	1,212.71	10,616.9	19,098
2,2-Dimethylhexane	C_8H_{18}	gas	1,313.56	11,499.9	20,686	1,218.88	10,671.0	19,195
2,2-Dimethylhexane	C_8H_{18}	liq.	1,304.64	11,421.8	20,546	1,209.96	10,592.9	19,055
2,3-Dimethylhexane	C_8H_{18}	gas	1,316.13	11,522.4	20,727	1,221.45	10,693.5	19,236
2,3-Dimethylhexane	C_8H_{18}	liq.	1,306.86	11,441.2	20,581	1,212.18	10,612.3	19,090
2,4-Dimethylhexane	C_8H_{18}	gas	1,314.83	11,511.0	20,706	1,220.15	10,682.1	19,215
2,4-Dimethylhexane	C_8H_{18}	liq.	1,305.80	11,431.9	20,564	1,211.12	10,603.0	19,073
2,5-Dimethylhexane	C_8H_{18}	gas	1,314.05	11,504.2	20,694	1,219.37	10,675.3	19,203
2,5-Dimethylhexane	C_8H_{18}	liq.	1,305.00	11,424.9	20,551	1,210.32	10,596.0	19,060
3,3-Dimethylhexane	C_8H_{18}	gas	1,314.65	11,509.4	20,703	1,219.97	10,680.5	19,212
3,3-Dimethylhexane	C_8H_{18}	liq.	1,305.68	11,430.9	20,562	1,211.00	10,602.0	19,071
3,4-Dimethylhexane	C_8H_{18}	gas	1,316.36	11,524.4	20,730	1,221.68	10,695.5	19,239
3,4-Dimethylhexane	C_8H_{18}	liq.	1,307.04	11,442.8	20,583	1,212.36	10,613.9	19,092
2-Methyl-3-ethylpentane	C_8H_{18}	gas	1,316.79	11,528.1	20,737	1,222.11	10,699.2	19,246
2-Methyl-3-ethylpentane	C_8H_{18}	liq.	1,307.58	11,447.5	20,592	1,212.90	10,618.6	19,101
3-Methyl-3-ethylpentane	C_8H_{18}	gas	1,315.88	11,520.2	20,723	1,221.20	10,691.3	19,232
3-Methyl-3-ethylpentane	C_8H_{18}	liq.	1,306.80	11,440.7	20,580	1,212.12	10,611.8	19,089
2,2,3-Trimethylpentane	C_8H_{18}	gas	1,314.66	11,509.5	20,703	1,219.98	10,680.6	19,212
2,2,3-Trimethylpentane	C_8H_{18}	liq.	1,305.83	11,432.2	20,564	1,211.15	10,603.3	19,073
2,2,4-Trimethylpentane	C_8H_{18}	gas	1,313.69	11,501.0	20,688	1,219.01	10,672.1	19,197
2,2,4-Trimethylpentane	C_8H_{18}	liq.	1,305.29	11,427.5	20,556	1,210.61	10,598.6	19,065
2,3,3-Trimethylpentane	C_8H_{18}	gas	1,315.54	11,517.2	20,717	1,220.86	10,688.3	19,226
2,3,3-Trimethylpentane	C_8H_{18}	liq.	1,306.64	11,439.3	20,577	1,211.96	10,610.4	19,086
2,3,4-Trimethylpentane	C_8H_{18}	gas	1,315.29	11,515.0	20,713	1,220.61	10,686.1	19,222
2,3,4-Trimethylpentane	C_8H_{18}	liq.	1,306.28	11,436.1	20,572	1,211.60	10,607.2	19,080
2,2,3,3-Tetramethylbutane	C_8H_{18}	gas	1,313.27	11,497.3	20,682	1,218.59	10,668.4	19,191

TABLE 3-207 Hydrogen, Carbon, Carbon Monoxide, and Hydrocarbons *(Continued)*

Compound	Formula	State	Heat of combustion, $-\Delta Hc°$, at 25°C. and constant pressure, to form: H_2O (liq.) and CO_2 (gas)			H_2O (gas) and CO_2 (gas)		
			Kcal./mole	Cal./g.	B.t.u./lb.	Kcal./mole	Cal./g.	B.t.u./lb.
2,2,3,3-Tetramethylbutane	C_8H_{18}	solid	1,303.03	11,407.7	20,520	1,208.35	10,578.8	19,029
n-Nonane	C_9H_{20}	gas	1,474.90	11,500.2	20,687	1,369.70	10,680.0	19,211
n-Nonane	C_9H_{20}	liq.	1,463.80	11,413.6	20,531	1,358.60	10,593.4	19,056
n-Decane	$C_{10}H_{22}$	gas	1,632.34	11,473.0	20,638	1,516.63	10,659.7	19,175
n-Decane	$C_{10}H_{22}$	liq.	1,620.06	11,386.7	20,483	1,504.35	10,573.4	19,020
n-Undecane	$C_{11}H_{24}$	gas	1,789.78	11,450.8	20,598	1,663.55	10,643.2	19,145
n-Undecane	$C_{11}H_{24}$	liq.	1,776.32	11,364.7	20,443	1,650.09	10,557.0	18,990
n-Dodecane	$C_{12}H_{26}$	gas	1,947.23	11,432.2	20,564	1,810.48	10,629.4	19,120
n-Dodecane	$C_{12}H_{26}$	liq.	1,932.59	11,346.3	20,410	1,795.84	10,543.4	18,966
n-Tridecane	$C_{13}H_{28}$	gas	2,104.67	11,416.5	20,536	1,957.40	10,617.6	19,099
n-Tridecane	$C_{13}H_{28}$	liq.	2,088.85	11,330.6	20,382	1,941.58	10,531.8	18,945
n-Tetradecane	$C_{14}H_{30}$	gas	2,262.11	11,402.9	20,512	2,104.32	10,607.5	19,081
n-Tetradecane	$C_{14}H_{30}$	liq.	2,245.11	11,317.2	20,358	2,087.32	10,521.8	18,927
n-Pentadecane	$C_{15}H_{32}$	gas	2,419.55	11,391.2	20,491	2,251.24	10,598.7	19,065
n-Pentadecane	$C_{15}H_{32}$	liq.	2,401.37	11,305.6	20,337	2,233.06	10,513.2	18,911
n-Hexadecane	$C_{16}H_{34}$	gas	2,577.00	11,380.9	20,472	2,398.17	10,591.1	19,052
n-Hexadecane	$C_{16}H_{34}$	liq.	2,557.64	11,295.4	20,318	2,378.81	10,505.6	18,898
n-Heptadecane	$C_{17}H_{36}$	gas	2,734.44	11,371.8	20,456	2,545.09	10,584.3	19,039
n-Heptadecane	$C_{17}H_{36}$	liq.	2,713.90	11,286.4	20,302	2,524.55	10,498.9	18,886
n-Octadecane	$C_{18}H_{38}$	gas	2,891.88	11,363.7	20,441	2,692.01	10,578.3	19,028
n-Octadecane	$C_{18}H_{38}$	liq.	2,870.16	11,278.4	20,288	2,670.29	10,493.0	18,875
n-Nonadecane	$C_{19}H_{40}$	gas	3,049.33	11,356.5	20,428	2,838.94	10,572.9	19,019
n-Nonadecane	$C_{19}H_{40}$	liq.	3,026.43	11,271.2	20,275	2,816.04	10,487.7	18,865
n-Eicosane	$C_{20}H_{42}$	gas	3,206.77	11,350.0	20,416	2,985.86	10,568.1	19,010
n-Eicosane	$C_{20}H_{42}$	liq.	3,182.69	11,264.7	20,263	2,961.78	10,482.8	18,857
	Alkyl benzenes							
Benzene	C_6H_6	gas	789.08	10,102.4	18,172	757.52	9,698.4	17,446
Benzene	C_6H_6	liq.	780.98	9,998.7	17,986	749.42	9,594.7	17,259
Methylbenzene (toluene)	C_7H_8	gas	943.58	10,241.4	18,422	901.50	9,784.7	17,601
Methylbenzene (toluene)	C_7H_8	liq.	934.50	10,142.8	18,245	892.42	9,686.1	17,424
Ethylbenzene	C_8H_{10}	gas	1,101.13	10,372.4	18,658	1,048.53	9,876.9	17,767
Ethylbenzene	C_8H_{10}	liq.	1,091.03	10,277.2	18,487	1,038.43	9,781.7	17,596
1,2-Dimethylbenzene (o-xylene)	C_8H_{10}	gas	1,098.54	10,348.0	18,614	1,045.94	9,852.5	17,723
1,2-Dimethylbenzene (o-xylene)	C_8H_{10}	liq.	1,088.16	10,250.2	18,438	1,035.56	9,754.7	17,547
1,3-Dimethylbenzene (m-xylene)	C_8H_{10}	gas	1,098.12	10,344.0	18,607	1,045.52	9,848.5	17,716
1,3-Dimethylbenzene (m-xylene)	C_8H_{10}	liq.	1,087.92	10,247.9	18,434	1,035.32	9,752.4	17,543
1,4-Dimethylbenzene (p-xylene)	C_8H_{10}	gas	1,098.29	10,345.6	18,610	1,045.69	9,850.1	17,719
1,4-Dimethylbenzene (p-xylene)	C_8H_{10}	liq.	1,088.16	10,250.2	18,438	1,035.56	9,754.7	17,547
n-Propylbenzene	C_9H_{12}	gas	1,258.24	10,469.1	18,832	1,195.12	9,943.9	17,887
n-Propylbenzene	C_9H_{12}	liq.	1,247.19	10,377.2	18,667	1,184.07	9,852.0	17,722
Isopropylbenzene (cumene)	C_9H_{12}	gas	1,257.31	10,461.4	18,818	1,194.19	9,936.2	17,873
Isopropylbenzene (cumene)	C_9H_{12}	liq.	1,246.52	10,371.6	18,657	1,183.40	9,846.4	17,712
1-Methyl-2-ethylbenzene	C_9H_{12}	gas	1,256.66	10,456.0	18,808	1,193.54	9,930.8	17,864
1-Methyl-2-ethylbenzene	C_9H_{12}	liq.	1,245.26	10,361.1	18,638	1,182.14	9,835.9	17,693
1-Methyl-3-ethylbenzene	C_9H_{12}	gas	1,255.92	10,449.8	18,797	1,192.80	9,924.6	17,853
1-Methyl-3-ethylbenzene	C_9H_{12}	liq.	1,244.71	10,356.5	18,630	1,181.59	9,831.3	17,685
1-Methyl-4-ethylbenzene	C_9H_{12}	gas	1,255.59	10,447.1	18,792	1,192.47	9,921.9	17,848
1-Methyl-4-ethylbenzene	C_9H_{12}	liq.	1,244.45	10,354.4	18,626	1,181.33	9,829.2	17,681
1,2,3-Trimethylbenzene (hemimellitene)	C_9H_{12}	gas	1,254.08	10,434.5	18,770	1,190.96	9,909.3	17,825
1,2,3-Trimethylbenzene (hemimellitene)	C_9H_{12}	liq.	1,242.36	10,337.0	18,594	1,179.24	9,811.8	17,650
1,2,4-Trimethylbenzene (pseudocumene)	C_9H_{12}	gas	1,253.04	10,425.8	18,754	1,189.92	9,900.7	17,809
1,2,4-Trimethylbenzene (pseudocumene)	C_9H_{12}	liq.	1,241.58	10,330.5	18,583	1,178.46	9,805.3	17,638
1,3,5-Trimethylbenzene (mesitylene)	C_9H_{12}	gas	1,252.53	10,421.6	18,747	1,189.41	9,896.4	17,802
1,3,5-Trimethylbenzene (mesitylene)	C_9H_{12}	liq.	1,241.19	10,327.2	18,577	1,178.07	9,802.1	17,632
n-Butylbenzene	$C_{10}H_{14}$	gas	1,415.44	10,546.3	18,971	1,341.80	9,997.6	17,984
n-Butylbenzene	$C_{10}H_{14}$	liq.	1,403.46	10,457.0	18,810	1,329.82	9,908.4	17,823
	Alkyl cyclopentanes							
Cyclopentane	C_5H_{10}	gas	793.39	11,313.1	20,350	740.79	10,563.1	19,001
Cyclopentane	C_5H_{10}	liq.	786.54	11,215.5	20,175	733.94	10,465.4	18,825
Methylcyclopentane	C_6H_{12}	gas	948.72	11,273.4	20,279	885.60	10,523.3	18,930
Methylcyclopentane	C_6H_{12}	liq.	941.14	11,183.3	20,117	878.02	10,433.2	18,768
Ethylcyclopentane	C_7H_{14}	gas	1,106.21	11,266.9	20,267	1,032.57	10,516.9	18,918
Ethylcyclopentane	C_7H_{14}	liq.	1,097.50	11,178.2	20,108	1,023.86	10,428.2	18,758
n-Propylcyclopentane	C_8H_{16}	gas	1,263.56	11,260.9	20,256	1,179.40	10,510.8	18,907
n-Propylcyclopentane	C_8H_{16}	liq.	1,253.74	11,173.4	20,099	1,169.58	10,423.3	18,750
n-Butylcyclopentane	C_9H_{18}	gas	1,421.10	11,257.7	20,250	1,326.42	10,507.6	18,901
n-Butylcyclopentane	C_9H_{18}	liq.	1,410.10	11,170.5	20,094	1,315.42	10,420.5	18,745
	Alkyl cyclohexanes							
Cyclohexane	C_6H_{12}	gas	944.79	11,226.7	20,195	881.67	10,476.7	18,846
Cyclohexane	C_6H_{12}	liq.	936.88	11,132.7	20,026	873.76	10,382.7	18,676
Methylcyclohexane	C_7H_{14}	gas	1,099.59	11,199.5	20,146	1,025.95	10,449.5	18,797
Methylcyclohexane	C_7H_{14}	liq.	1,091.13	11,113.3	19,991	1,017.49	10,363.3	18,642
Ethylcyclohexane	C_8H_{16}	gas	1,257.90	11,210.4	20,166	1,173.74	10,460.4	18,816
Ethylcyclohexane	C_8H_{16}	liq.	1,248.23	11,124.3	20,011	1,164.07	10,374.3	18,661
n-Propylcyclohexane	C_9H_{18}	gas	1,415.12	11,210.3	20,165	1,320.44	10,460.3	18,816
n-Propylcyclohexane	C_9H_{18}	liq.	1,404.34	11,124.9	20,012	1,309.66	10,374.9	18,663
n-Butylcyclohexane	$C_{10}H_{20}$	gas	1,572.74	11,213.0	20,170	1,467.54	10,463.0	18,821
n-Butylcyclohexane	$C_{10}H_{20}$	liq.	1,560.78	11,127.8	20,017	1,455.58	10,377.8	18,668
	Monoolefins							
Ethene (ethylene)	C_2H_4	gas	337.234	12,021.7	21,625	316.195	11,271.7	20,276
Propene (propylene)	C_3H_6	gas	491.987	11,692.3	21,032	460.428	10,942.3	19,683
1-Butene	C_4H_8	gas	649.757	11,581.3	20,833	607.679	10,831.3	19,484
cis-2-Butene	C_4H_8	gas	648.115	11,552.0	20,780	606.037	10,802.0	19,431
trans-2-Butene	C_4H_8	gas	647.072	11,533.4	20,747	604.994	10,783.4	19,397
2-Methylpropene (isobutene)	C_4H_8	gas	646.134	11,516.7	20,716	604.056	10,766.7	19,367
1-Pentene	C_5H_{10}	gas	806.85	11,505.1	20,696	754.25	10,755.1	19,346

TABLE 3-207 Hydrogen, Carbon, Carbon Monoxide, and Hydrocarbons *(Concluded)*

Compound	Formula	State	Heat of combustion, $-\Delta Hc°$, at 25°C. and constant pressure, to form: H_2O (liq.) and CO_2 (gas) Kcal./mole	Cal./g.	B.t.u./lb.	H_2O (gas) and CO_2 (gas) Kcal./mole	Cal./g.	B.t.u./lb.
cis-2-Pentene	C_5H_{10}	gas	805.34	11,483.5	20,657	752.74	10,733.5	19,308
trans-2-Pentene	C_5H_{10}	gas	804.26	11,468.1	20,629	751.66	10,718.1	19,280
2-Methyl-1-butene	C_5H_{10}	gas	803.17	11,452.6	20,601	750.57	10,702.6	19,252
3-Methyl-1-butene	C_5H_{10}	gas	804.93	11,477.7	20,646	752.33	10,727.7	19,297
2-Methyl-2-butene	C_5H_{10}	gas	801.68	11,431.3	20,563	749.08	10,681.3	19,214
Acetylenes								
Ethyne (acetylene)	C_2H_2	gas	310.615	11,930.2	21,460	300.096	11,526.2	20,734
Propyne (methylacetylene)	C_3H_4	gas	463.109	11,559.8	20,794	442.070	11,034.6	19,849
1-Butyne (ethylacetylene)	C_4H_6	gas	620.86	11,478.7	20,648	589.302	10,895.2	19,599
2-Butyne (dimethylacetylene)	C_4H_6	gas	616.533	11,398.7	20,504	584.974	10,815.2	19,455
1-Pentyne	C_5H_8	gas	778.03	11,422.5	20,547	735.95	10,804.7	19,436
2-Pentyne	C_5H_8	gas	774.33	11,368.2	20,449	732.25	10,750.4	19,338
3-Methyl-1-butyne	C_5H_8	gas	776.13	11,394.6	20,497	734.05	10,776.8	19,386

NOTE: °F = 9/5 °C + 32.
°Saturation pressure.

HEATS OF SOLUTION

TABLE 3-208 Heats of Solution of Inorganic Compounds in Water

Heat evolved, in kilogram-calories per gram formula weight, on solution in water at 18°C. Computed from data in Bichowsky and Rossini, *Thermochemistry of Chemical Substances*, Reinhold, New York, 1936.

Substance	Dilution*	Formula	Heat, kg.-cal./g.-mole
Aluminum bromide	aq	$AlBr_3$	+85.3
chloride	600	$AlCl_3$	+77.9
	600	$AlCl_3.6H_2O$	+13.2
fluoride	aq	AlF_3	+31
	aq	$AlF_3.\frac{1}{2}H_2O$	+19.0
	aq	$AlF_3.3\frac{1}{2}H_2O$	−1.7
iodide	aq	AlI_3	+89.0
sulfate	aq	$Al_2(SO_4)_3$	+126
	aq	$Al_2(SO_4)_3.6H_2O$	+56.2
	aq	$Al_2(SO_4)_3.18H_2O$	+6.7
Ammonium bromide	aq	NH_4Br	−4.45
chloride	∞	NH_4Cl	−3.82
chromate	aq	$(NH_4)_2CrO_4$	−5.82
dichromate	600	$(NH_4)_2Cr_2O_7$	−12.9
iodide	aq	NH_4I	−3.56
nitrate	∞	NH_4NO_3	−6.47
perborate	aq	$NH_4BO_3.H_2O$	−9.0
sulfate	∞	$(NH_4)_2SO_4$	−2.75
sulfate, acid	800	NH_4HSO_4	+0.56
sulfite	aq	$(NH_4)_2SO_3$	−1.2
	aq	$(NH_4)_2SO_3.H_2O$	−4.13
Antimony fluoride	aq	SbF_3	−1.7
iodide	aq	SbI_3	−0.8
Arsenic acid	aq	H_3AsO_4	−0.4
Barium bromate	∞	$Ba(BrO_3)_2.H_2O$	−15.9
bromide	∞	$BaBr_2$	+5.3
	∞	$BaBr_2.H_2O$	−0.8
	∞	$BaBr_2.2H_2O$	−3.87
chlorate	∞	$Ba(ClO_3)_2$	−6.7
	∞	$Ba(ClO_3)_2.H_2O$	−10.6
chloride	∞	$BaCl_2$	+2.4
	∞	$BaCl_2.H_2O$	−2.17
	∞	$BaCl_2.2H_2O$	−4.5
cyanide	aq	$Ba(CN)_2$	+1.5
	aq	$Ba(CN)_2.H_2O$	−2.4
	aq	$Ba(CN)_2.2H_2O$	−4.9
iodate	∞	$Ba(IO_3)_2$	−9.1
	∞	$Ba(IO_3)_2.H_2O$	−11.3
iodide	∞	BaI_2	+10.5
	∞	$BaI_2.H_2O$	+2.7
	∞	$BaI_2.2H_2O$	+0.14
	∞	$BaI_2.2\frac{1}{2}H_2O$	−0.58
	∞	$BaI_2.7H_2O$	−6.61
nitrate	∞	$Ba(NO_3)_2$	−10.2
perchlorate	∞	$Ba(ClO_4)_2$	−2.8
	∞	$Ba(ClO_4)_2.3H_2O$	−10.5
sulfide	∞	BaS	+7.2
Beryllium bromide	aq	$BeBr_2$	+62.6
chloride	aq	$BeCl_2$	+51.1
iodide	aq	BeI_2	+72.6
sulfate	aq	$BeSO_4$	+18.1
	aq	$BeSO_4.H_2O$	+13.5
	aq	$BeSO_4.2H_2O$	+7.9
	aq	$BeSO_4.4H_2O$	+1.1
Bismuth iodide	aq	BiI_3	+3

Substance	Dilution*	Formula	Heat, kg.-cal./g.-mole
Boric acid	aq	H_3BO_3	−5.4
Cadmium bromide	400	$CdBr_2$	+0.4
	400	$CdBr_2.4H_2O$	−7.3
chloride	400	$CdCl_2$	+3.1
	400	$CdCl_2.H_2O$	+0.6
	400	$CdCl_2.2\frac{1}{2}H_2O$	−3.00
nitrate	400	$Cd(NO_3)_2.H_2O$	+4.17
	400	$Cd(NO_3)_2.4H_2O$	−5.08
sulfate	400	$CdSO_4$	+10.69
	400	$CdSO_4.H_2O$	+6.05
	400	$CdSO_4.2\frac{2}{3}H_2O$	+2.51
Calcium acetate	∞	$Ca(C_2H_3O_2)_2$	+7.6
	∞	$Ca(C_2H_3O_2)_2.H_2O$	+6.5
bromide	∞	$CaBr_2$	+24.86
	∞	$CaBr_2.6H_2O$	−0.9
chloride	∞	$CaCl_2$	+4.9
	∞	$CaCl_2.H_2O$	+12.3
	∞	$CaCl_2.2H_2O$	+12.5
	∞	$CaCl_2.4H_2O$	+2.4
	∞	$CaCl_2.6H_2O$	−4.11
formate	400	$Ca(CHO_2)_2$	+0.7
iodide	∞	CaI_2	+28.0
	∞	$CaI_2.8H_2O$	+1.8
nitrate	∞	$Ca(NO_3)_2$	+4.1
	∞	$Ca(NO_3)_2.H_2O$	+0.7
	∞	$Ca(NO_3)_2.2H_2O$	−3.2
	∞	$Ca(NO_3)_2.3H_2O$	−4.2
	∞	$Ca(NO_3)_2.4H_2O$	−7.99
phosphate, mono-	aq	$Ca(H_2PO_4)_2.H_2O$	−0.6
dibasic	aq	$CaHPO_4.2H_2O$	−1
sulfate	∞	$CaSO_4$	+5.1
	∞	$CaSO_4.\frac{1}{2}H_2O$	+3.6
	∞	$CaSO_4.2H_2O$	−0.18
Chromous chloride	aq	$CrCl_2$	+18.6
		$CrCl_2.3H_2O$	+5.3
		$CrCl_2.4H_2O$	+2.0
iodide	aq	CrI_2	+5.7
Cobaltous bromide	aq	$CoBr_2$	+18.4
	aq	$CoBr_2.6H_2O$	−1.25
chloride	400	$CoCl_2$	+18.5
	400	$CoCl_2.2H_2O$	+9.8
	400	$CoCl_2.6H_2O$	−2.9
iodide	aq	CoI_2	+18.8
sulfate	400	$CoSO_4$	+15.0
	400	$CoSO_4.6H_2O$	−1.4
	400	$CoSO_4.7H_2O$	−3.6
Cupric acetate	aq	$Cu(C_2H_3O_2)_2$	+2.4
formate	aq	$Cu(CHO_2)_2$	+0.5
nitrate	200	$Cu(NO_3)_2$	+10.3
	200	$Cu(NO_3)_2.3H_2O$	−2.6
	200	$Cu(NO_3)_2.6H_2O$	−10.7
sulfate	800	$CuSO_4$	+15.9
		$CuSO_4.H_2O$	+9.3
		$CuSO_4.3H_2O$	+3.65
		$CuSO_4.5H_2O$	−2.85
Cuprous sulfate	aq	Cu_2SO_4	+11.6

* The numbers represent moles of water used to dissolve 1 g. formula weight of substance; ∞ means "infinite dilution"; and aq means "aqueous solution of unspecified dilution."

TABLE 3-208 Heats of Solution of Inorganic Compounds in Water *(Continued)*

Substance	Dilution*	Formula	Heat, kg.-cal./g.-mole
Ferric chloride	1000	$FeCl_3$	+31.7
	1000	$FeCl_3.2\frac{1}{2}H_2O$	+21.0
	1000	$FeCl_3.6H_2O$	+5.6
nitrate	800	$Fe(NO_3)_3.9H_2O$	−9.1
Ferrous bromide	aq	$FeBr_2$	+18.0
chloride	400	$FeCl_2$	+17.9
	400	$FeCl_2.2H_2O$	+8.7
	400	$FeCl_2.4H_2O$	+2.7
iodide	aq	FeI_2	+23.3
sulfate	400	$FeSO_4$	+14.7
	400	$FeSO_4.H_2O$	+7.35
	400	$FeSO_4.4H_2O$	+1.4
	400	$FeSO_4.7H_2O$	−4.4
Lead acetate	400	$Pb(C_2H_3O_2)_2$	+1.4
	400	$Pb(C_2H_3O_2)_2.3H_2O$	−5.9
bromide	aq	$PbBr_2$	−10.1
chloride	aq	$PbCl_2$	−3.4
formate	aq	$Pb(CHO_2)_2$	−6.9
nitrate	400	$Pb(NO_3)_2$	−7.61
Lithium bromide	∞	$LiBr$	+11.54
	∞	$LiBr.H_2O$	+5.30
	∞	$LiBr.2H_2O$	+2.05
	∞	$LiBr.3H_2O$	−1.59
chloride	∞	$LiCl$	+8.66
	∞	$LiCl.H_2O$	+4.45
	∞	$LiCl.2H_2O$	+1.07
	∞	$LiCl.3H_2O$	−1.98
fluoride	∞	LiF	−0.74
hydroxide	∞	$LiOH$	+4.74
	∞	$LiOH.\frac{1}{8}H_2O$	+4.39
	∞	$LiOH.H_2O$	+9.6
iodide	∞	LiI	+14.92
	∞	$LiI.\frac{1}{2}H_2O$	+10.08
	∞	$LiI.H_2O$	+6.93
	∞	$LiI.2H_2O$	+3.43
	∞	$LiI.3H_2O$	−0.17
nitrate	∞	$LiNO_3$	+0.466
	∞	$LiNO_3.3H_2O$	−7.87
sulfate	∞	Li_2SO_4	+6.71
	∞	$Li_2SO_4.H_2O$	+3.77
Magnesium bromide	∞	$MgBr_2$	+43.7
	∞	$MgBr_2.H_2O$	+35.9
	∞	$MgBr_2.6H_2O$	+19.8
chloride	∞	$MgCl_2$	+36.3
	∞	$MgCl_2.2H_2O$	+20.8
	∞	$MgCl_2.4H_2O$	+10.5
	∞	$MgCl_2.6H_2O$	+3.4
iodide	∞	MgI_2	+50.2
nitrate	∞	$Mg(NO_3)_2.6H_2O$	−3.7
phosphate	aq	$Mg_3(PO_4)_2$	+10.2
sulfate	∞	$MgSO_4$	+21.1
	∞	$MgSO_4.H_2O$	+14.0
	∞	$MgSO_4.2H_2O$	+11.7
	∞	$MgSO_4.4H_2O$	+4.9
	∞	$MgSO_4.6H_2O$	+0.55
	∞	$MgSO_4.7H_2O$	−3.18
sulfide	aq	MgS	+25.8
Manganic nitrate	400	$Mn(NO_3)_2$	+12.9
	400	$Mn(NO_3)_2.3H_2O$	−3.9
	400	$Mn(NO_3)_2.6H_2O$	−6.2
sulfate	aq	$Mn_2(SO_4)_3$	+22
Manganous acetate	aq	$Mn(C_2H_3O_2)_2$	+12.2
	aq	$Mn(C_2H_3O_2)_2.4H_2O$	+1.6
bromide	aq	$MnBr_2$	+15
	aq	$MnBr_2.H_2O$	+14.4
	aq	$MnBr_2.4H_2O$	+16.1
chloride	400	$MnCl_2$	+16.0
	400	$MnCl_2.2H_2O$	+8.2
	400	$MnCl_2.4H_2O$	+1.5
formate	aq	$Mn(CHO_2)_2$	+4.3
	aq	$Mn(CHO_2)_2.2H_2O$	−2.9
iodide	aq	MnI_2	+26.2
	aq	$MnI_2.H_2O$	+24.1
	aq	$MnI_2.2H_2O$	+22.7
	aq	$MnI_2.4H_2O$	+19.9
	aq	$MnI_2.6H_2O$	+21.2
sulfate	400	$MnSO_4$	+13.8
	400	$MnSO_4.H_2O$	+11.9
	400	$MnSO_4.7H_2O$	−1.7
Mercuric acetate	aq	$Hg(C_2H_3O_2)_2$	−4.0
bromide	aq	$HgBr_2$	−2.4
chloride	aq	$HgCl_2$	−3.3
nitrate	aq	$Hg(NO_3)_2.\frac{1}{2}H_2O$	−0.7
Mercurous nitrate	aq	$Hg_2(NO_3)_2.2H_2O$	−11.5
Nickel bromide	aq	$NiBr_2$	+19.0
	aq	$NiBr_2.3H_2O$	+0.2

Substance	Dilution*	Formula	Heat, kg.-cal./g.-mole
Nickel chloride	800	$NiCl_2$	+19.23
	800	$NiCl_2.2H_2O$	+10.4
	800	$NiCl_2.4H_2O$	+4.2
	800	$NiCl_2.6H_2O$	−1.15
iodide	aq	NiI_2	+19.4
nitrate	200	$Ni(NO_3)_2$	+11.8
	200	$Ni(NO_3)_2.6H_2O$	−7.5
sulfate	200	$NiSO_4$	+15.1
	200	$NiSO_4.7H_2O$	−4.2
Phosphoric acid, ortho-	400	H_3PO_4	+2.79
	400	$H_3PO_4.\frac{1}{2}H_2O$	−0.1
pyro-	aq	$H_4P_2O_7$	+25.9
	aq	$H_4P_2O_7.1\frac{1}{2}H_2O$	+4.65
Potassium acetate	∞	$KC_2H_3O_2$	+3.55
aluminum sulfate	600	$KAl(SO_4)_2$	+48.5
	600	$KAl(SO_4)_2.3H_2O$	+26.6
		$KAl(SO_4)_2.12H_2O$	−10.1
bicarbonate	2000	$KHCO_3$	−5.1
bromate	∞	$KBrO_3$	−10.13
bromide	∞	KBr	−5.13
carbonate	∞	K_2CO_3	+6.58
		$K_2CO_3.\frac{1}{2}H_2O$	+4.25
		$K_2CO_3.1\frac{1}{2}H_2O$	−0.43
chlorate	∞	$KClO_3$	−10.31
chloride	∞	KCl	−4.404
chromate	2185	K_2CrO_4	−4.9
chrome sulfate	600	$KCr(SO_4)_2$	+55
		$KCr(SO_4)_2.H_2O$	+42
		$KCr(SO_4)_2.2H_2O$	+33
		$KCr(SO_4)_2.6H_2O$	+7
		$KCr(SO_4)_2.12H_2O$	−9.5
cyanide	200	KCN	−3.0
dichromate	1600	$K_2Cr_2O_7$	−17.8
fluoride	∞	KF	+3.96
	∞	$KF.2H_2O$	−1.85
	∞	$KF.4H_2O$	−6.05
hydrosulfide	∞	KHS	+0.86
	∞	$KHS.\frac{1}{4}H_2O$	+1.21
hydroxide	∞	KOH	+12.91
	∞	$KOH.\frac{3}{4}H_2O$	+4.27
	∞	$KOH.H_2O$	+3.48
	∞	$KOH.7H_2O$	+0.86
iodate	∞	KIO_3	−6.93
iodide	∞	KI	−5.23
nitrate	∞	KNO_3	−8.633
oxalate	400	$K_2C_2O_4$	−4.6
		$K_2C_2O_4.H_2O$	−7.5
perchlorate	∞	$KClO_4$	−12.94
permanganate	400	$KMnO_4$	−10.4
phosphate, dihydrogen	aq	KH_2PO_4	+4.7
pyrosulfite	aq	$K_2S_2O_5$	−11.0
	aq	$K_2S_2O_5.\frac{1}{2}H_2O$	−10.22
sulfate	∞	K_2SO_4	−6.32
sulfate, acid	800	$KHSO_4$	−3.10
sulfide	∞	K_2S	−11.0
sulfite	aq	K_2SO_3	+1.8
	aq	$K_2SO_3.H_2O$	+1.37
thiocyanate	∞	$KCNS$	−6.08
thionate, di-	aq	$K_2S_2O_6$	−13.0
thiosulfate	∞	$K_2S_2O_3$	−4.5
Silver acetate	aq	$AgC_2H_3O_2$	−5.4
nitrate	200	$AgNO_3$	−4.4
Sodium acetate	∞	$NaC_2H_3O_2$	+4.085
	∞	$NaC_2H_3O_2.3H_2O$	−4.665
arsenate	500	Na_3AsO_4	+15.6
	500	$Na_3AsO_4.12H_2O$	−12.61
bicarbonate	1800	$NaHCO_3$	−4.1
borate, tetra-	900	$Na_2B_4O_7$	+10.0
	900	$Na_2B_4O_7.10H_2O$	−16.8
bromide	∞	$NaBr$	−0.58
	∞	$NaBr.2H_2O$	−4.57
carbonate	∞	Na_2CO_3	+5.57
	∞	$Na_2CO_3.H_2O$	+2.19
	∞	$Na_2CO_3.7H_2O$	−10.81
	∞	$Na_2CO_3.10H_2O$	−16.22
chlorate	∞	$NaClO_3$	−5.37
chloride	∞	$NaCl$	−1.164
chromate	800	Na_2CrO_4	+2.50
	800	$Na_2CrO_4.4H_2O$	−7.52
	800	$Na_2CrO_4.10H_2O$	−16.0
cyanide	200	$NaCN$	−0.37
	200	$NaCN.\frac{1}{2}H_2O$	−0.92
	200	$NaCN.2H_2O$	−4.41
fluoride	∞	NaF	−0.27
hydrosulfide	∞	$NaHS$	+4.62
	∞	$NaHS.2H_2O$	−1.49

TABLE 3-208 Heats of Solution of Inorganic Compounds in Water *(Concluded)*

Substance	Dilution*	Formula	Heat, kg.-cal./g.-mole
Sodium hydroxide	∞	$NaOH$	+10.18
	∞	$NaOH.\frac{1}{2}H_2O$	+8.17
	∞	$NaOH.\frac{2}{3}H_2O$	+7.08
	∞	$NaOH.\frac{3}{4}H_2O$	+6.48
	∞	$NaOH.H_2O$	+5.17
iodide	∞	NaI	+1.57
	∞	$NaI.2H_2O$	−3.89
metaphosphate	600	$NaPO_3$	+3.97
nitrate	∞	$NaNO_3$	−5.05
nitrite	aq	$NaNO_2$	−3.6
perchlorate	∞	$NaClO_4$	−4.15
phosphate, di-	1600	Na_2HPO_4	+5.21
tri-	1600	Na_3PO_4	+13
phosphate	1600	$Na_3PO_4.12H_2O$	−15.3
di-	1600	$Na_2HPO_4.2H_2O$	−0.82
	1600	$Na_2HPO_4.7H_2O$	−12.04
	1600	$Na_2HPO_4.12H_2O$	−23.18
phosphite, mono-	600	NaH_2PO_3	+0.90
	600	$NaH_2PO_3.2\frac{1}{2}H_2O$	−5.29
di-	800	Na_2HPO_3	+9.30
	800	$Na_2HPO_3.5H_2O$	−4.54
pyrophosphate	1600	$Na_4P_2O_7$	+11.9
	1600	$Na_4P_2O_7.10H_2O$	−11.7
di-	1200	$Na_2H_2P_2O_7$	−2.2
	1200	$Na_2H_2P_2O_7.6H_2O$	−14.0
sulfate	∞	Na_2SO_4	+0.28
	∞	$Na_2SO_4.10H_2O$	−18.74
sulfate, acid	800	$NaHSO_4$	+1.74
	800	$NaHSO_4.H_2O$	+0.15
sulfide	∞	Na_2S	+15.2
	∞	$Na_2S.4\frac{1}{2}H_2O$	+0.09
	∞	$Na_2S.5H_2O$	−6.54
	∞	$Na_2S.9H_2O$	−16.65
sulfite	∞	Na_2SO_3	+2.8
	∞	$Na_2SO_3.7H_2O$	−11.1
thiocyanate	∞	$NaONS$	−1.83
thionate, di-	aq	$Na_2S_2O_6$	−5.80
	aq	$Na_2S_2O_6.2H_2O$	−11.86

Substance	Dilution*	Formula	Heat, kg.-cal./g.-mole
Sodium thiosulfate	aq	$Na_2S_2O_3$	+2.0
	aq	$Na_2S_2O_3.5H_2O$	−11.30
Stannic bromide	aq	$SnBr_4$	+15.5
Stannous bromide	aq	$SnBr_2$	−1.6
iodide	aq	SnI_2	−5.8
Strontium acetate	∞	$Sr(C_2H_3O_2)_2$	+6.2
	∞	$Sr(C_2H_3O_2)_2.\frac{1}{2}H_2O$	+5.9
bromide	∞	$SrBr_2$	+16.4
	∞	$SrBr_2.H_2O$	+9.25
	∞	$SrBr_2.2H_2O$	+6.5
	∞	$SrBr_2.4H_2O$	+0.4
	∞	$SrBr_2.6H_2O$	−6.1
chloride	∞	$SrCl_2$	+11.54
	∞	$SrCl_2.H_2O$	+6.4
	∞	$SrCl_2.2H_2O$	+2.95
	∞	$SrCl_2.6H_2O$	−7.1
iodide	∞	SrI_2	+20.7
	∞	$SrI_2.H_2O$	+12.65
	∞	$SrI_2.2H_2O$	+10.4
	∞	$SrI_2.6H_2O$	−4.5
nitrate	∞	$Sr(NO_3)_2$	−4.8
	∞	$Sr(NO_3)_2.4H_2O$	−12.4
sulfate	∞	$SrSO_4$	+0.5
Sulfuric acid, pyro-	∞	$H_2S_2O_7$	−18.08
Zinc acetate	400	$Zn(C_2H_3O_2)_2$	+9.8
	400	$Zn(C_2H_3O_2)_2.H_2O$	+7.0
	400	$Zn(C_2H_3O_2)_2.2H_2O$	+3.9
bromide	400	$ZnBr_2$	+15.0
chloride	400	$ZnCl_2$	+15.72
iodide	aq	ZnI_2	+11.6
nitrate	400	$Zn(NO_3)_2.3H_2O$	−5
	400	$Zn(NO_3)_2.6H_2O$	−6.0
sulfate	400	$ZnSO_4$	+18.5
	400	$ZnSO_4.H_2O$	+10.0
	400	$ZnSO_4.6H_2O$	−0.8
	400	$ZnSO_4.7H_2O$	−4.3

NOTE: To convert kilocalories per gram-mole to British thermal units per pound-mole, multiply by 1.799×10^{-3}.

TABLE 3-209 Heats of Solution of Organic Compounds in Water (at Infinite Dilution and Approximately Room Temperature)

Recalculated and rearranged from *International Critical Tables*, vol. 5, pp. 148–150. (g·cal)/(g·mol) = Btu/(lb·mol) × 1.799.

Solute	Heat of Solution G.-cal./g.-mole Solute*
Acetic acid (solid), $C_2H_4O_2$	−2,251
Acetylacetone, $C_5H_8O_2$	−641
Acetylurea, $C_3H_6N_2O_2$	−6,812
Aconitic acid, $C_6H_6O_6$	−4,206
Ammonium benzoate, $C_7H_9NO_2$	−2,700
picrate	−8,700
succinate (n-)	−3,489
Aniline, hydrochloride, C_6H_8ClN	−2,732
Barium picrate	−4,708
Benzoic acid, $C_7H_6O_2$	−6,501
Camphoric acid, $C_{10}H_{16}O_4$	−502
Citric acid, $C_6H_8O_7$	−5,401
Dextrin, $C_{12}H_{20}O_{10}$	268
Fumaric acid, $C_4H_4O_4$	−5,903
Hexamethylenetetramine, $C_6H_{12}N_4$	4,780
Hydroxybenzamide (m-), $C_7H_7NO_2$	−4,161
(m-), (HCl)	−7,003
(o-), $C_7H_7NO_2$	−4,340
(p-)	−5,392
Hydroxybenzoic acid (o-), $C_7H_6O_3$	−6,350
(p-), $C_7H_6O_3$	−5,781
Hydroxybenzyl alcohol (o-), $C_7H_8O_2$	−3,203
Inulin, $C_{36}H_{62}O_{31}$	−96
Isosuccinic acid, $C_4H_6O_4$	−3,420
Itaconic acid, $C_5H_6O_4$	−5,922
Lactose, $C_{12}H_{22}O_{11}.H_2O$	−3,705
Lead picrate	−7,098
($2H_2O$)	−13,193
Magnesium picrate	14,699
($8H_2O$)	−15,894
Maleic acid, $C_4H_4O_4$	−4,441
Malic acid, $C_4H_6O_5$	−3,150
Malonic acid, $C_3H_4O_4$	−4,493
Mandelic acid, $C_8H_8O_3$	−3,090
Mannitol, $C_6H_{14}O_6$	−5,260
Menthol, $C_{10}H_{20}O$	0
Nicotine dihydrochloride, $C_{10}H_{16}Cl_2N_2$	6,561
Nitrobenzoic acid (m-), $C_7H_5NO_4$	−5,593
(o-), $C_7H_5NO_4$	−5,306
(p-), $C_7H_5NO_4$	−8,891
Nitrophenol (m-), $C_6H_5NO_3$	−5,210
(o-), $C_6H_5NO_3$	−6,310
(p-), $C_6H_5NO_3$	−4,493
Oxalic acid, $C_2H_2O_4$	−2,290
($2H_2O$)	−8,485
Phenol (solid), C_6H_6O	−2,605
Phthalic acid, $C_8H_6O_4$	−4,871
Picric acid, $C_6H_3N_3O_7$	−7,098
Piperic acid, $C_{12}H_{10}O_4$	−10,492
Piperonylic acid, $C_8H_6O_4$	−9,106
Potassium benzoate	−1,506
citrate	2,820
tartrate (n-) (0.5 H_2O)	−5,562
Pyrogallol, $C_6H_6O_3$	−3,705
Pyrotartaric acid	−5,019
Quinone	−3,991
Raffinose, $C_{18}H_{32}O_{16}$ ($5H_2O$)	−9,703
Resorcinol, $C_6H_6O_2$	−3,960
Silver malonate (n-)	−9,799
Sodium citrate (tri-)	5,270
picrate	−6,441
potassium tartrate	−1,817
($4H_2O$)	−12,342
succinate (n-)	2,390
($6H_2O$)	−10,994
tartrate (n-)	−1,121
($2H_2O$)	−5,882
Strontium picrate	7,887
($6H_2O$)	−14,412
Succinic acid, $C_4H_6O_4$	−6,405
Succinimide, $C_4H_5NO_2$	−4,302
Sucrose, $C_{12}H_{22}O_{11}$	−1,319
Tartaric acid (d-)	−3,451
Thiourea, CH_4N_2S	−5,330
Urea, CH_4N_2O	−3,609
acetate	−8,795
formate	−7,194
nitrate	−10,803
oxalate	−17,806
Vanillic acid	−5,160
Vanillin	−5,210
Zinc picrate	−11,496
($8H_2O$)	−15,894

* + denotes heat evolved, and − denotes heat absorbed. All values are positive unless otherwise noted. The data in the "International Critical Tables" were calculated by E. Anderson.

THERMODYNAMIC PROPERTIES

EXPLANATION OF TABLES

The following subsection presents information on the thermodynamic properties of a number of fluids. In some cases transport properties are also included.

Notation

c_p = specific heat
e = specific internal energy
h = enthalpy
k = thermal conductivity
p = pressure
s = specific entropy
t = temperature
T = absolute temperature
u = specific internal energy
μ = viscosity
v = specific volume
f = subscript denoting saturated liquid
g = subscript denoting saturated vapor

UNITS CONVERSIONS

For this subsection, the following units conversions are applicable:

c_p, specific heat: to convert kilojoules per kilogram-kelvin to British thermal units per pound–degree Fahrenheit, multiply by 0.23885.

e, internal energy: to convert kilojoules per kilogram to British thermal units per pound, multiply by 0.42992.

g, gravity acceleration: to convert meters per second squared to feet per second squared, multiply by 3.2808.

h, enthalpy: to convert kilojoules per kilogram to British thermal units per pound, multiply by 0.42992.

k, thermal conductivity: to convert watts per meter-kelvin to British thermal unit–feet per hour–square foot–degree Fahrenheit, multiply by 0.57779.

p, pressure: to convert bars to kilopascals, multiply by 1×10^2; to convert bars to pounds-force per square inch, multiply by 14.504; and to convert millimeters of mercury to pounds-force per square inch, multiply by 0.01934.

s, entropy: to convert kilojoules per kilogram-kelvin to British thermal units per pound–degree Rankine, multiply by 0.23885.

t, temperature: °F = $\frac{9}{5}$ °C + 32.

T, absolute temperature: °R = $\frac{9}{5}$ K.

u, internal energy: to convert kilojoules per kilogram to British thermal units per pound, multiply by 0.42992.

μ, viscosity: to convert pascal-seconds to pound-force–seconds per square foot, multiply by 0.020885; to convert pascal-seconds to c_p, multiply by 1000.

v, specific volume: to convert cubic meters per kilogram to cubic feet per pound, multiply by 16.018.

ρ, density: to convert kilograms per cubic meter to pounds per cubic foot, multiply by 0.062428.

ADDITIONAL REFERENCES

Bretsznajder, *Prediction of Transport and Other Physical Properties of Fluids,* Pergamon, New York, 1971. D'Ans and Lax, *Handbook for Chemists and Physicists* (in German), 3 vols., Springer-Verlag, Berlin. *Engineering Data Book,* Natural Gas Processors Suppliers Association, Tulsa, Okla. Ganic, Hartnett, and Rohsenow, *Handbook of Heat Transfer,* 2d ed., McGraw-Hill, New York, 1984. Gray, *American Institute of Physics Handbook,* 3d ed., McGraw-Hill, New York, 1972. Kay and Laby, *Tables of Physical and Chemical Constants,* Longman, London, various editions and dates. *Landolt-Börnstein Tables,* many volumes and dates, Springer-Verlag, Berlin. Lange, *Handbook of Chemistry,* McGraw-Hill, New York, various editions and dates. Partington, *Advanced Treatise on Physical Chemistry,* 5 vols., Longman, London, 1950. Raznjevic, *Handbook of Thermodynamic Tables and Charts,* McGraw-Hill, New York, 1976 and other editions. Reid, Sherwood, and Prausnitz, *The Properties of Gases and Liquids,* 3d ed., McGraw-Hill, New York, 1975. Reynolds, *Thermodynamic Properties in SI,* Department of Mechanical Engineering, Stanford University, 1979. Stephan and Lucas, *Viscosity of Dense Fluids,* Plenum, New York and London, 1979. *Selected Values of Properties of Chemical Compounds* and *Selected Values of the Properties of Hydrocarbons and Related Compounds,* Thermodynamics Research Center, Texas A&M University, College Station, looseleaf, intermittent publication. Touloukian and Ho, McGraw-Hill/CINDAS Data Series on Material Properties, in publication, 1983. Vargaftik, *Tables of the Thermophysical Properties of Gases and Liquids,* Wiley, New York, 1975. Vargaftik, Filippov, Tarzimanov, and Totskiy, *Thermal Conductivity of Liquids and Gases* (in Russian), Standartov, Moscow, 1978. Weast, *Handbook of Chemistry and Physics,* Chemical Rubber Co., Boca Raton, Fla., annually.

TABLE 3-210 Saturated Acetylene*

Temperature, K	Pressure, bar	v_{cond}, m³/kg	v_g, m³/kg	h_{cond}, kJ/kg	h_g, kJ/kg	s_{cond}, kJ/(kg·K)	s_g, kJ/(kg·K)
162.0	0.101	......	5.081	158	983	2.967	8.062
169.3	0.203	......	2.644	173	994	3.039	7.889
173.9	0.304	......	1.805	182	999	3.095	7.797
180.0	0.507	......	1.116	194	1007	3.161	7.672
184.3	0.709	......	0.810	203	1011	3.216	7.596
189.1	1.013	......	0.5780	214	1015	3.272	7.511
192.4^t	1.283	......	0.4617	221	1018	3.312	7.455
192.4^t	1.283	0.00164	0.4617	378	1018	4.127	7.455
200.9	2.027	0.00165	0.3011	411	1027	4.296	7.362
209.4	3.040	0.00169	0.2074	445	1035	4.461	7.280
221.5	5.066	0.00174	0.1264	493	1046	4.684	7.180
230.4	7.093	0.00179	0.0907	528	1052	4.837	7.111
240.7	10.13	0.00186	0.0635	565	1058	4.990	7.037
253.2	15.20	0.00195	0.0420	602	1061	5.133	6.947
263.0	20.27	0.00204	0.0309	628	1061	5.231	6.878
271.6	25.33	0.00213	0.0240	654	1060	5.326	6.822
278.9	30.40	0.00223	0.0193	680	1057	5.414	6.767
284.9	35.46	0.00232	0.0159	704	1051	5.494	6.716
290.4	40.53	0.00242	0.0133	727	1041	5.576	6.658
300.0	50.66	0.00270	0.0093	778	1017	5.737	6.534
307.8	60.80	0.00335	0.0061	850	968	5.965	6.351
308.7^c	62.47	0.00434	0.0043	908	908	6.158	6.158

*Values recalculated into SI units from those of Din, *Thermodynamic Functions of Gases,* vol. 2, Butterworth, London, 1956. Above the solid line the condensed phase is solid; below the line it is liquid. t = triple point; c = critical point.

TABLE 3-211 Saturated Air*

T, K	P_f, bar	P_g, bar	v_f, m^3/kg	v_g, m^3/kg	h_f, kJ/kg	h_g, kJ/kg	s_f, kJ/(kg·K)	s_g, kJ/(kg·K)	c_{pf}, kJ/(kg·K)	μ_f, 10^{-4} Pa·s	k_f, W/(m·K)
60	......	......	1.040.−3	5.55	−159.2	59.7	2.528	6.255	....	3.25	0.180
62	......	......	1.050.−3	3.73	−155.2	61.7	2.585	6.164	....	2.98	0.176
64	0.123	0.071	1.060.−3	2.57	−151.4	63.6	2.641	6.080	....	2.75	0.173
66	0.174	0.104	1.070.−3	1.82	−147.8	65.5	2.696	6.002	....	2.54	0.169
68	0.239	0.147	1.080.−3	1.313	−144.2	67.4	2.747	5.929	....	2.36	0.166
70	0.323	0.205	1.089.−3	0.968	−140.6	69.2	2.797	5.862	1.817	2.21	0.163
72	0.429	0.280	1.101.−3	0.728	−137.1	71.0	2.847	5.799	1.827	2.07	0.160
74	0.560	0.376	1.113.−3	0.556	−133.5	72.8	2.895	5.740	1.838	1.95	0.156
76	0.721	0.495	1.125.−3	0.431	−129.9	74.5	2.941	5.685	1.849	1.84	0.152
78	0.915	0.644	1.136.−3	0.339	−126.3	76.2	2.988	5.634	1.861	1.74	0.148
80	1.146	0.825	1.146.−3	0.270	−122.6	77.8	3.034	5.585	1.873	1.65	0.145
82	1.420	1.043	1.160.−3	0.217	−118.8	79.4	3.079	5.540	1.885	1.58	0.142
84	1.741	1.305	1.173.−3	0.177	−115.0	80.9	3.123	5.496	1.898	1.51	0.139
86	2.114	1.614	1.187.−3	0.145	−111.2	82.3	3.167	5.454	1.912	1.44	0.135
88	2.544	1.976	1.201.−3	0.120	−107.4	83.6	3.209	5.414	1.927	1.38	0.132
90	3.036	2.397	1.216.−3	0.1002	−103.5	84.8	3.251	5.376	1.944	1.32	0.128
92	3.596	2.884	1.231.−3	0.0843	− 99.5	85.9	3.293	5.340	1.962	1.27	0.125
94	4.229	3.441	1.247.−3	0.0713	− 95.5	87.0	3.335	5.304	1.982	1.23	0.121
96	4.940	4.075	1.265.−3	0.0607	− 91.5	87.9	3.376	5.270	2.003	1.18	0.117
98	5.736	4.792	1.283.−3	0.0520	− 87.5	88.7	3.416	5.236	2.027	1.14	0.114
100	6.621	5.599	1.302.−3	0.0447	− 83.3	89.3	3.456	5.204	2.053	1.10	0.110
105	9.265	8.056	1.355.−3	0.0312	− 72.8	90.2	3.553	5.124	2.137	1.02	0.102
110	12.59	11.22	1.418.−3	0.0222	− 61.9	90.1	3.649	5.045	2.264	0.95	0.093
115	16.68	15.21	1.495.−3	0.0159	− 50.3	88.4	3.747	4.964	2.477	0.87	0.084
120	21.61	20.14	1.596.−3	0.0115	− 37.5	84.8	3.850	4.877	2.916	0.75	0.076
125	27.43	26.14	1.757.−3	0.0081	− 22.0	78.2	3.969	4.776	4.585	0.42	0.067
130	34.16	33.32	2.075.−3	0.0054	0.4	66.1	4.136	4.644			
132.55^c	....	37.69	3.196.−3	0.0032	37.4	37.4	4.410	4.410	∞	...	∞

*Liquid properties extracted or converted from Vasserman and Rabinovich, *Thermophysical Properties of Liquid Air and Its Components*, Moscow, 1968, and NBS-NSF transl. TT 69-55092, 1970. Copyrighted material. Reproduced by permission. Vapor properties extracted or converted from Vasserman, Kazavchinskii, and Rabinovich, *Thermophysical Properties of Air and Its Components*, Nauka, Moscow, 1966, and NBS-NSF transl. TT 70-50095, 1971. Copyrighted material. Reproduced by permission. Note that on pages 150–151 of the TT 69-55092 publication certain values of TT 70-50095 were adjusted. As a complete retabulation was not given, the tables here are based upon the two separate publications, as indicated. See also Table 3-218 for the argon-oxygen-nitrogen equilibrium data. c = critical point. The notation 1.040.−3 signifies 1.040 × 10^{-3}.

TABLE 3-212 Thermophysical Properties of Compressed Air*

Pressure, bar		Temperature, K												
		80	90	100	120	140	160	180	200	220	240	260	280	300
1	v	....	0.251	0.281	0.340	0.399	0.457	0.515	0.537	0.631	0.688	0.746	0.803	0.861
	h	....	87.9	98.3	118.8	139.1	159.3	179.5	199.7	219.8	239.9	260.0	280.2	300.3
	s	Mix	5.650	5.759	5.946	6.103	6.238	6.357	6.463	6.559	6.647	6.727	6.802	6.871
	C_p	....	1.044	1.032	1.020	1.014	1.010	1.008	1.007	1.006	1.006	1.006	1.006	1.007
	μ	....	0.064	0.071	0.085	0.097	0.109	0.121	0.133	0.144	0.154	0.165	0.175	0.185
	k	....	0.0084	0.0093	0.0112	0.0129	0.0147	0.0164	0.0181	0.0198	0.0214	0.0231	0.0247	0.0263
5	v	0.00115	0.00122	0.0509	0.0646	0.0773	0.0895	0.102	0.114	0.125	0.137	0.149	0.160	0.172
	h	−122.3	−103.3	90.6	113.6	135.3	156.4	177.1	197.7	218.1	238.5	258.8	279.1	299.4
	s	3.031	3.250	5.246	5.455	5.623	5.763	5.885	5.994	6.092	6.180	6.262	6.337	6.406
	C_p	1.868	1.941	1.212	1.107	1.065	1.045	1.033	1.025	1.020	1.017	1.015	1.013	1.013
	μ	1.794	1.163	0.077	0.087	0.098	0.110	0.122	0.134	0.145	0.155	0.165	0.175	0.185
	k	0.146	0.128	0.0103	0.0119	0.0135	0.0151	0.0168	0.0185	0.0201	0.0217	0.0234	0.0250	0.0265
10	v	0.00115	0.00121	0.00130	0.0298	0.0370	0.0436	0.0499	0.0561	0.0621	0.0681	0.0741	0.0800	0.0859
	h	−122.0	−103.1	−83.2	106.2	130.2	152.5	174.1	195.2	216.1	236.7	257.3	277.8	298.3
	s	3.028	3.246	3.452	5.214	5.398	5.548	5.675	5.786	5.885	5.975	6.058	6.134	6.204
	C_p	1.863	1.932	2.041	1.270	1.146	1.093	1.065	1.049	1.038	1.031	1.026	1.023	1.201
	μ	1.816	1.177	0.838	0.089	0.101	0.112	0.124	0.135	0.146	0.156	0.166	0.176	0.186
	k	0.146	0.128	0.111	0.0126	0.0141	0.0157	0.0173	0.0189	0.0205	0.0221	0.0237	0.0253	0.0268
20	v	0.00114	0.00121	0.00129	0.0116	0.0167	0.0206	0.0241	0.0274	0.0306	0.0337	0.0368	0.0398	0.0428
	h	−121.3	−102.5	−82.9	85.2	118.5	144.3	167.7	190.1	211.9	233.2	254.3	275.2	296.0
	s	3.022	3.239	3.442	4.882	5.140	5.312	5.450	5.568	5.672	5.765	5.849	5.927	5.998
	C_p	1.853	1.916	2.010	2.237	1.390	1.215	1.141	1.101	1.076	1.061	1.050	1.042	1.037
	μ	1.859	1.205	0.857	0.098	0.106	0.116	0.127	0.137	0.148	0.158	0.168	0.178	0.187
	k	0.147	0.130	0.112	0.0152	0.0157	0.0169	0.0182	0.0197	0.0212	0.0228	0.0243	0.0258	0.0273
40	v	0.00114	0.00120	0.00128	0.00153	0.0058	0.0090	0.0114	0.0131	0.0148	0.0165	0.0182	0.0198	0.0214
	h	−120.0	−101.4	−82.2	−39.8	83.6	125.3	154.3	179.7	203.5	226.3	248.5	270.2	291.7
	s	3.011	3.225	3.424	3.807	4.745	5.025	5.196	5.330	5.444	5.543	5.632	5.712	5.786
	C_p	1.834	1.886	1.958	2.432	3.193	1.610	1.335	1.221	1.159	1.122	1.097	1.081	1.068
	μ	1.943	1.261	0.896	0.516	0.132	0.129	0.135	0.144	0.154	0.163	0.172	0.182	0.191
	k	0.149	0.132	0.115	0.0814	0.0460	0.0201	0.0206	0.0217	0.0229	0.0242	0.0256	0.0270	0.0284
60	v	0.00113	0.00119	0.00126	0.00147	0.00222	0.00505	0.00687	0.00833	0.00963	0.0108	0.0120	0.0131	0.0142
	h	−118.6	−100.3	−81.4	−40.8	22.8	90.0	132.6	163.9	191.1	216.1	240.0	263.1	285.6
	s	3.000	3.211	3.407	3.773	4.260	4.798	5.020	5.174	5.298	5.404	5.497	5.581	5.657
	C_p	1.818	1.860	1.915	2.205	4.808	2.338	1.594	1.361	1.249	1.186	1.146	1.119	1.100
	μ	2.028	1.318	0.936	0.559	0.277	0.153	0.149	0.154	0.161	0.169	0.178	0.186	0.195
	k	0.150	0.134	0.117	0.0861	0.0480	0.0360	0.0238	0.0240	0.0248	0.0258	0.0270	0.0283	0.0296
80	v	0.00113	0.00119	0.00126	0.00145	0.00188	0.00327	0.00480	0.00601	0.00706	0.00803	0.00894	0.00981	0.0107
	h	−117.2	−99.1	−80.4	−41.3	9.0	78.4	125.3	158.7	187.1	212.9	237.3	260.8	283.7
	s	2.989	3.198	3.391	3.745	4.138	4.597	4.875	5.051	5.186	5.299	5.396	5.484	5.562
	C_p	1.802	1.838	1.881	2.078	2.992	3.029	1.887	1.510	1.342	1.250	1.194	1.156	1.130
	μ	2.12	1.38	0.977	0.597	0.356	0.194	0.167	0.166	0.170	0.177	0.184	0.191	0.200
	k	0.152	0.134	0.120	0.0901	0.0599	0.0420	0.0278	0.0268	0.0269	0.0276	0.0286	0.0296	0.0308
100	v	0.00112	0.00118	0.00125	0.00142	0.00174	0.00252	0.00366	0.00467	0.00556	0.00637	0.00713	0.00785	0.00855
	h	−115.8	−97.8	−79.4	−41.3	3.9	61.7	111.8	148.8	179.4	206.7	232.2	256.4	279.9
	s	2.978	3.186	3.376	3.721	4.076	4.457	4.753	4.949	5.095	5.214	5.315	5.406	5.486
	C_p	1.789	1.818	1.852	1.992	2.506	2.874	2.114	1.650	1.431	1.311	1.239	1.191	1.158
	μ	2.21	1.44	1.02	0.631	0.405	0.249	0.193	0.181	0.181	0.185	0.191	0.198	0.205
	k	0.154	0.137	0.122	0.0936	0.0669	0.0500	0.0327	0.0299	0.0293	0.0295	0.0302	0.0311	0.0320
150	v	0.00111	0.00116	0.00122	0.00137	0.00158	0.00194	0.00247	0.00309	0.00369	0.00425	0.00478	0.00529	0.00578
	h	−112.2	−94.5	−76.6	−40.1	0.5	45.2	89.5	129.2	163.2	193.4	221.0	247.0	271.8
	s	2.954	3.157	3.342	3.673	3.988	4.287	4.548	4.757	4.919	5.051	5.161	5.257	5.343
	C_p	1.789	1.818	1.852	1.992	2.506	2.874	2.114	1.650	1.431	1.311	1.239	1.267	1.220
	μ	2.44	1.60	1.13	0.709	0.490	0.349	0.266	0.229	0.215	0.211	0.212	0.215	0.220
	k	0.157	1.142	0.127	0.101	0.0785	0.0588	0.0455	0.0389	0.0360	0.0348	0.0346	0.0349	0.0354
200	v	0.00110	0.00115	0.00120	0.00133	0.00150	0.00174	0.00206	0.00245	0.00287	0.00328	0.00368	0.00407	0.00446
	h	−108.5	−91.2	−73.6	−38.0	0.2	40.2	79.8	117.6	152.2	183.6	212.5	239.6	265.5
	s	2.930	3.130	3.312	3.634	3.931	4.198	4.432	4.631	4.796	4.932	5.048	5.149	5.238
	C_p	1.733	1.747	1.761	1.809	1.905	1.988	1.953	1.814	1.643	1.501	1.396	1.321	1.266
	μ	2.70	1.78	1.25	0.782	0.561	0.420	0.331	0.279	0.253	0.241	0.236	0.235	0.237
	k	0.161	0.146	0.132	0.107	0.0868	0.0691	0.0559	0.0476	0.0429	0.0405	0.0393	0.0389	0.0389
250	v	0.00109	0.00114	0.00119	0.00130	0.00144	0.00162	0.00186	0.00214	0.00244	0.00276	0.00307	0.00338	0.00368
	h	−104.8	−87.6	−70.3	−35.4	1.3	38.9	75.8	111.7	145.6	177.1	206.6	234.3	260.8
	s	2.909	3.106	3.285	3.601	3.886	4.138	4.355	4.544	4.706	4.843	4.961	5.064	5.155
	C_p	1.712	1.722	1.733	1.767	1.824	1.854	1.831	1.748	1.635	1.522	1.427	1.353	1.297
	μ	2.96	1.97	1.39	0.855	0.625	0.476	0.385	0.327	0.292	0.272	0.262	0.257	0.256
	k	0.165	0.150	0.137	0.113	0.0935	0.0769	0.0641	0.0552	0.0495	0.0460	0.0441	0.0430	0.0426

*For sources, units, and remarks, see Table 3-211. v = specific volume, m^3/kg; h = specific enthalpy, kJ/kg; s = specific entropy, kJ/kg·K); C_p = specific heat at constant pressure, kJ/(kg·K); μ = viscosity, 10^{-4} Pa·s; and k = thermal conductivity, W/(m·K).

Temperature, K												
350	400	450	500	600	800	1000	1200	1400	1600	1800	2000	2500
1.005	1.148	1.292	1.436	1.723	2.297	2.872	3.446	4.020	4.594	5.168	5.743	7.200
350.7	401.2	452.1	503.4	607.5	822.5	1046.8	1278	1515	1764	2017	2279	3011
7.026	7.161	7.282	7.389	7.579	7.888	8.138	8.349	8.531	8.695	8.844	8.983	9.308
1.009	1.014	0.021	1.030	1.051	1.099	1.141	1.175	1.207	1.248	1.286	1.337	1.665
0.208	0.230	0.251	0.270	0.306	0.370	0.424	0.473	0.527	0.584	0.637	0.689	0.818
0.0301	0.0336	0.0371	0.0404	0.0466	0.0577	0.0681	0.0783	0.0927	0.106	0.120	0.137	0.222
0.201	0.230	0.259	0.288	0.345	0.460	0.575	0.690	0.805	0.920	1.034	1.149	1.438
350.0	400.8	451.8	503.2	607.4	822.6	1046.9	1279	1516	1764	2017	2278	2981
6.563	6.698	6.818	6.927	7.116	7.426	7.676	7.887	8.069	8.233	8.382	8.520	8.832
1.014	1.017	1.024	1.032	1.053	1.100	1.142	1.175	1.208	1.248	1.285	1.326	1.516
0.208	0.230	0.251	0.270	0.306	0.370	0.425	0.473	0.527	0.584	0.637	0.689	0.818
0.0303	0.0338	0.0372	0.0405	0.0467	0.0578	0.0681	0.0783	0.0927	0.106	0.120	0.136	0.195
0.101	0.115	0.130	0.144	0.173	0.231	0.288	0.345	0.403	0.460	0.518	0.575	0.720
349.2	400.2	451.4	502.9	607.3	822.7	1047.2	1279	1516	1765	2018	2279	2974
6.361	6.497	6.618	6.727	6.917	7.226	7.477	7.688	7.870	8.034	8.183	8.321	8.630
1.019	1.021	1.027	1.034	1.055	1.100	1.142	1.175	1.208	1.248	1.284	1.324	1.481
0.209	0.231	0.252	0.271	0.306	0.370	0.425	0.473	0.527	0.584	0.637	0.689	0.817
0.0305	0.0340	0.0374	0.0407	0.0469	0.0579	0.0682	0.0784	0.0927	0.106	0.120	0.135	0.187
0.0503	0.0577	0.0650	0.0723	0.0868	0.116	0.145	0.173	0.202	0.231	0.260	0.288	0.360
347.7	399.1	450.7	502.4	607.2	823.0	1047.7	1280	1517	1766	2019	2279	2970
6.158	6.295	6.417	6.526	6.716	7.027	0.277	7.489	7.671	7.835	7.984	8.121	8.428
1.030	1.029	1.033	1.039	1.057	1.102	1.143	1.176	1.209	1.249	1.284	1.322	1.456
0.210	0.232	0.253	0.272	0.307	0.371	0.425	0.474	0.527	0.584	0.637	0.689	0.817
0.0309	0.0344	0.0377	0.0410	0.0471	0.0581	0.0685	0.0787	0.0928	0.106	0.120	0.135	0.181
0.0252	0.0290	0.0327	0.0364	0.0438	0.0583	0.0728	0.0872	0.102	0.116	0.130	0.145	0.181
344.6	397.0	449.2	501.5	606.9	823.7	1048.8	1281	1519	1768	2021	2281	2969
5.950	6.090	6.212	6.323	6.515	6.826	7.077	7.289	7.473	7.636	7.785	7.922	8.229
1.051	1.044	1.044	1.049	1.063	0.105	1.145	1.177	1.210	1.249	1.284	1.322	1.438
0.213	0.235	0.255	0.274	0.309	0.372	0.426	0.474	0.527	0.584	0.637	0.689	0.817
0.0318	0.0351	0.0384	0.0416	0.0476	0.0584	0.0687	0.0789	0.0928	0.106	0.120	0.135	0.177
0.0169	0.0194	0.0220	0.0245	0.0294	0.0392	0.0489	0.0585	0.0681	0.0776	0.0872	0.0968	0.1207
340.4	394.0	447.1	500.6	606.8	824.3	1050.0	1283	1521	1770	2023	2284	2969
5.824	5.967	6.091	6.202	6.396	6.708	6.960	7.172	7.355	7.520	7.669	7.806	8.112
1.072	1.059	1.055	1.057	1.069	1.108	1.147	1.178	1.210	1.249	1.286	1.322	1.430
0.217	0.237	0.257	0.275	0.310	0.373	0.427	0.475	0.527	0.584	0.637	0.689	0.817
0.0328	0.0359	0.0391	0.0422	0.0481	0.0588	0.0690	0.0790	0.0929	0.106	0.120	0.134	0.176
0.0127	0.0147	0.0166	0.0185	0.0223	0.0296	0.0369	0.0442	0.0513	0.0585	0.0657	0.0729	0.0908
339.0	393.1	446.5	499.8	606.7	825.1	1051.1	1284	1522	1772	2025	2285	2971
5.733	5.878	6.004	6.116	6.311	6.624	6.877	7.089	7.273	7.437	7.586	7.723	8.029
1.091	1.073	1.066	1.065	1.075	1.111	1.149	1.180	1.210	1.249	1.286	1.322	1.426
0.220	0.240	0.259	0.278	0.312	0.374	0.428	0.475	0.527	0.584	0.637	0.689	0.817
0.0337	0.0368	0.0398	0.0428	0.0486	0.0592	0.0693	0.0793	0.0929	0.106	0.120	0.134	0.175
0.0102	0.0118	0.0134	0.0149	0.0180	0.0239	0.0298	0.0356	0.0413	0.0470	0.0528	0.0584	0.0729
336.5	391.3	445.3	499.0	606.6	825.8	1052.4	1286	1524	1774	2027	2288	2972
5.661	5.807	5.935	6.048	6.244	6.559	6.812	7.024	7.208	7.373	7.522	7.659	7.964
1.110	1.087	1.076	1.073	1.080	1.114	1.151	1.181	1.211	1.250	1.288	1.323	1.423
0.224	0.243	0.262	0.280	0.314	0.375	0.429	0.477	0.527	0.584	0.637	0.689	0.817
0.0347	0.0376	0.0405	0.0434	0.0491	0.0595	0.0696	0.0795	0.0930	0.106	0.120	0.134	0.175
0.00695	0.00806	0.00914	0.0102	0.0123	0.0163	0.0202	0.0241	0.0279	0.0317	0.0356	0.0394	0.0490
330.9	387.5	442.9	497.5	606.6	827.8	1055.5	1290	1529	1779	2033	2294	2977
5.525	5.677	5.807	5.922	6.121	6.439	6.693	6.906	7.092	7.256	7.405	7.543	7.848
1.151	1.117	1.099	1.092	1.093	1.121	1.155	1.184	1.213	1.252	1.290	1.325	1.418
0.235	0.252	0.270	0.286	0.318	0.379	0.431	0.478	0.527	0.584	0.637	0.689	
0.0374	0.0398	0.0424	0.0451	0.0504	0.0605	0.0703	0.0801	0.0932	0.106	0.120	0.133	
0.00534	0.00620	0.00702	0.00783	0.00940	0.0125	0.0154	0.0184	0.0212	0.0241	0.0269	0.0298	0.0370
326.5	384.5	440.9	496.6	607.0	829.9	1058.7	1294	1533	1783	2038	2299	2982
5.426	5.581	5.715	5.831	6.033	6.353	6.608	6.822	7.009	7.173	7.323	7.460	7.765
1.184	1.141	1.119	1.108	1.104	1.128	1.160	1.187	1.214	1.254	1.292	1.326	1.415
0.248	0.262	0.278	0.293	0.324	0.382	0.434	0.481	0.528	0.585	0.638		
0.0400	0.0420	0.0423	0.0467	0.0517	0.0614	0.0711	0.0808	0.0934	0.106	0.120		
0.00440	0.00509	0.00576	0.00642	0.00770	0.0102	0.0126	0.0149	0.0172	0.0195	0.0218	0.0241	0.0298
323.2	382.3	439.6	496.0	607.6	832.2	1062.0	1298	1538	1789	2043	2304	2988
5.348	5.506	5.641	5.760	5.963	6.286	6.542	6.757	6.944	7.108	7.258	7.396	7.701
1.208	1.161	1.135	1.121	1.115	1.135	1.164	1.190	1.216	1.256	1.294	1.328	1.414
0.262	0.273	0.286	0.301	0.329	0.386	0.437	0.483	0.528	0.585			
0.0429	0.0443	0.0462	0.0484	0.0531	0.0624	0.0718	0.0814	0.0937	0.106			

TABLE 3-212 Thermophysical Properties of Compressed Air* *(Concluded)*

Pressure, bar		Temperature, K												
		80	90	100	120	140	160	180	200	220	240	260	280	300
300	v	0.00108	0.00112	0.00117	0.00127	0.00139	0.00155	0.00173	0.00195	0.00219	0.00243	0.00269	0.00294	0.00318
	h	−101.0	−84.0	−67.0	−32.4	3.1	39.2	74.5	109.0	142.0	173.2	202.7	230.8	257.7
	s	2.888	3.083	3.260	3.572	3.849	4.090	4.298	4.480	4.637	4.773	4.891	4.995	5.088
	C_p	1.694	1.703	1.713	1.740	1.769	1.777	1.751	1.689	1.607	1.518	1.438	1.370	1.316
	μ	3.24	2.18	1.53	0.932	0.687	0.529	0.433	0.370	0.329	0.303	0.288	0.280	0.276
	k	0.168	0.154	0.141	0.118	0.0996	0.0836	0.0710	0.0619	0.0555	0.0514	0.0487	0.0471	0.0462
400	v	...	0.00110	0.00114	0.00123	0.00133	0.00145	0.00158	0.00173	0.00189	0.00206	0.00224	0.00242	0.00260
	h	...	−76.6	−59.8	−25.9	8.3	42.4	75.8	108.5	140.1	170.5	199.7	227.8	254.8
	s	...	3.042	3.216	3.523	3.788	4.016	4.214	4.386	4.537	4.669	4.786	4.890	4.983
	C_p	...	1.674	1.686	1.704	1.702	1.685	1.654	1.607	1.550	1.490	1.431	1.378	1.331
	μ	...	2.63	1.86	1.10	0.802	0.631	0.500	0.446	0.397	0.364	0.341	0.325	0.316
	k	...	0.161	0.149	0.127	0.110	0.0946	0.0823	0.0729	0.0660	0.0610	0.0574	0.0550	0.0533
500	v	...	0.00109	0.00112	0.00120	0.00128	0.00138	0.00148	0.00160	0.00173	0.00186	0.00199	0.00213	0.00227
	h	...	−69.0	−52.3	−18.7	14.4	47.4	79.8	111.4	142.0	171.7	200.5	228.4	255.4
	s	...	3.005	3.177	3.482	3.743	3.966	4.151	4.317	4.463	4.593	4.708	4.811	4.905
	C_p	...	1.655	1.670	1.686	1.667	1.644	1.598	1.557	1.509	1.461	1.415	1.371	1.331
	μ	...	3.13	2.24	1.31	0.924	0.710	0.0560	0.512	0.459	0.420	0.391	0.370	0.356
	k	...	0.167	0.156	0.135	0.119	0.104	0.0916	0.0822	0.0749	0.0694	0.0653	0.0622	0.0599
600	v	...	...	...	...	...	...	...	0.00151	0.00161	0.00172	0.00183	0.00194	0.00205
	h	...	...	...	...	...	...	...	116.0	146.1	175.3	203.6	231.2	258.1
	s	...	...	...	...	...	...	...	2.263	4.406	4.533	4.646	4.749	4.842
	C_p	...	...	...	...	...	...	...	1.525	1.480	1.438	1.398	1.361	1.327
	μ	...	...	...	...	...	...	...	...	0.516	0.472	0.439	0.414	0.396
	k	...	...	...	...	...	...	...	0.0903	0.0828	0.0769	0.0724	0.0689	0.0662
800	v	...	...	...	...	...	...	...	...	0.00147	0.00155	0.00163	0.00171	0.00179
	h	...	...	...	...	...	...	...	...	157.4	185.9	213.7	240.3	267.3
	s	...	...	...	...	...	...	...	...	4.318	4.442	4.553	4.653	4.745
	C_p	...	...	...	...	...	...	...	...	1.445	1.406	1.372	1.342	1.314
	μ	...	...	...	...	...	...	...	...	...	...	0.529	0.497	0.473
	k	...	...	...	...	...	...	...	...	0.0964	0.0901	0.0850	0.0809	0.0776
1000	v	...	...	...	...	...	...	...	...	...	...	0.00151	0.00157	0.00163
	h	...	...	...	...	...	...	...	...	...	...	226.4	253.2	279.5
	s	...	...	...	...	...	...	...	...	...	...	4.482	4.582	4.672
	C_p	...	...	...	...	...	...	...	...	...	...	1.355	1.327	1.303
	μ	...	...	...	...	...	...	...	...	...	...	...	...	0.546
	k	...	...	...	...	...	...	...	...	...	...	0.0961	0.0916	0.0878

Temperature, K												
350	400	450	500	600	800	1000	1200	1400	1600	1800	2000	2500
0.00379	0.00437	0.00493	0.00548	0.00656	0.00864	0.0107	0.0126	0.0145	0.0164	0.0183	0.0202	0.0250
320.9	380.9	438.9	495.9	608.5	834.5	1065.3	1302	1542	1794	2049	2310	2993
5.283	5.443	5.580	5.700	5.906	6.230	6.488	6.703	6.891	7.056	7.206	7.344	7.648
1.226	1.176	1.148	1.133	1.124	1.140	1.168	1.193	1.217	1.257	1.298	1.330	1.413
0.276	0.284	0.296	0.308	0.335	0.390	0.440	0.485	0.529				
0.0457	0.0466	0.0481	0.0501	0.0544	0.0634	0.0726	0.0820	0.0940				
0.00304	0.00348	0.00390	0.00432	0.00514	0.00673	0.00826	0.00977	0.0111	0.0126	0.0140	0.0155	0.0190
319.1	380.0	439.0	496.8	611.0	839.4	1072.0	1310	1552	1804	2059	2321	3004
5.181	5.344	5.483	5.605	5.813	6.142	6.401	6.618	6.808	6.972	7.123	7.261	7.566
1.246	1.195	1.166	1.149	1.138	1.151	1.176	1.199	1.222	1.258	1.301	1.333	1.412
0.307	0.308	0.315	0.325	0.348	0.398	0.446	0.490					
0.0513	0.0512	0.0521	0.0535	0.0571	0.0653	0.0740	0.0832					
0.00262	0.00296	0.00330	0.00364	0.00430	0.00558	0.00683	0.00804	0.00911	0.0103	0.0114	0.0126	0.0154
319.9	381.3	440.8	499.1	614.3	844.6	1078.8	1318	1561	1814	2070	2332	3015
5.103	5.267	5.408	5.531	5.741	6.072	6.333	6.550	6.743	6.907	7.058	7.196	7.501
1.255	1.206	1.176	1.159	1.148	1.159	1.183	1.205	1.226	1.265	1.306	1.337	1.412
0.338	0.333	0.336	0.343	0.361	0.407	0.452	0.495					
0.0568	0.0557	0.0560	0.0569	0.0598	0.0672	0.0755	0.0844					
0.00234	0.00262	0.00290	0.00318	0.00374	0.00481	0.00586	0.00689	0.00776	0.00873	0.00970	0.0107	0.0130
322.6	384.2	444.0	502.6	618.5	850.1	1085.5	1326	1570	1824	2080	2343	3026
5.041	5.205	5.346	5.470	5.681	6.014	6.277	6.495	6.690	6.854	7.005	7.144	7.449
1.258	1.211	1.182	1.166	1.154	1.166	1.189	1.210	1.231	1.267	1.310	1.341	1.412
0.370	0.359	0.358	0.361	0.375	0.416	0.459	0.501					
0.0620	0.0602	0.0598	0.0603	0.0625	0.0691	0.0770	0.0857					
0.00200	0.00221	0.00242	0.00263	0.00304	0.00385	0.00465	0.00544	0.00608	0.00681	0.00754	0.00826	0.0101
331.6	393.8	453.4	512.3	625.8	862.0	1099.3	1341	1588	1844	2101	2365	3049
4.943	5.108	5.250	5.374	5.586	5.922	6.136	6.407	6.605	6.769	6.921	7.060	7.366
1.257	1.216	1.188	1.172	1.161	1.175	1.198	1.219	1.240	1.275	1.318	1.347	1.412
0.432	0.411	0.402	0.399	0.405	0.436	0.474	0.512					
0.0718	0.0688	0.0673	0.0669	0.0679	0.0730	0.0800	0.0881					
0.00180	0.00196	0.00213	0.00230	0.00262	0.00328	0.00392	0.00455	0.00507	0.00565	0.00624	0.00681	0.00825
343.4	405.1	465.3	524.4	641.2	875.1	1113.3	1356	1606	1863	2121	2386	3071
4.869	5.034	5.176	5.300	5.513	5.850	6.115	6.337	6.539	6.703	6.856	6.995	7.302
1.254	1.217	1.192	1.175	1.164	1.179	1.204	1.225	1.248	1.283	1.325	1.354	1.413
0.494	0.463	0.446	0.438	0.435	0.456	0.489	0.524					
0.0810	0.0768	0.0744	0.0733	0.0732	0.0768	0.0830	0.0906					

TABLE 3-213 Enthalpy and Psi Functions for Ideal-Gas Air*

T, K	h, kJ/kg	Ψ	T, K	h, kJ/kg	Ψ	T, K	h, kJ/kg	Ψ
200	200.0	−0.473	650	659.8	1.339	1200	1278	2.376
210	210.0	−0.400	660	670.5	1.364	1220	1301	2.406
220	220.0	−0.329	670	681.1	1.388	1240	1325	2.435
230	230.1	−0.262	680	691.8	1.412	1260	1349	2.463
240	240.1	−0.197	690	702.5	1.436	1280	1372	2.491
250	250.1	−0.135	700	713.3	1.459	1300	1396	2.519
260	260.1	−0.076	710	724.0	1.482	1320	1420	2.547
270	270.1	−0.018	720	734.8	1.505	1340	1444	2.574
280	280.1	0.037	730	745.6	1.528	1360	1467	2.601
290	290.2	0.090	740	756.4	1.550	1380	1491	2.627
300	300.2	0.142	750	767.3	1.572	1400	1515	2.653
310	310.3	0.191	760	778.2	1.594	1420	1539	2.679
320	320.3	0.240	770	789.1	1.615	1440	1563	2.705
330	330.4	0.286	780	800.0	1.637	1460	1587	2.730
340	340.4	0.332	790	811.0	1.658	1480	1612	2.755
350	350.5	0.376	800	821.9	1.679	1500	1636	2.779
360	360.6	0.419	810	832.9	1.699	1520	1660	2.803
370	370.7	0.461	820	844.0	1.720	1540	1684	2.827
380	380.8	0.502	830	855.0	1.740	1560	1709	2.851
390	390.9	0.541	840	866.1	1.760	1580	1738	2.875
400	401.0	0.580	850	877.2	1.780	1600	1758	2.898
410	411.2	0.618	860	888.3	1.800	1620	1782	2.921
420	421.3	0.655	870	899.4	1.819	1640	1806	2.944
430	431.5	0.691	880	910.6	1.838	1660	1831	2.966
440	441.7	0.727	890	921.8	1.857	1680	1855	2.988
450	451.8	0.761	900	933.0	1.876	1700	1880	3.010
460	462.1	0.795	910	944.2	1.895	1720	1905	3.032
470	472.3	0.829	920	955.4	1.914	1740	1929	3.054
480	482.5	0.861	930	966.7	1.932	1760	1954	3.075
490	492.8	0.893	940	978.0	1.950	1780	1979	3.096
500	503.1	0.925	950	989.3	1.969	1800	2003	3.117
510	513.4	0.956	960	1000.6	1.987	1820	2028	3.138
520	523.7	0.986	970	1011.9	2.004	1840	2053	3.158
530	534.0	1.016	980	1023.3	2.022	1860	2078	3.178
540	544.4	1.045	990	1034.7	2.039	1880	2102	3.198
550	554.8	1.074	1000	1046.1	2.057	1900	2127	3.218
560	565.2	1.102	1020	1068.9	2.091	1920	2152	3.238
570	575.6	1.130	1040	1091.9	2.125	1940	2177	3.258
580	586.1	1.158	1060	1114.9	2.158	1960	2202	3.277
590	596.5	1.185	1080	1138.0	2.190	1980	2227	3.296
600	607.0	1.211	1100	1161.1	2.223	2000	2252	3.215
610	617.5	1.238	1120	1184.3	2.254	2050	2315	3.362
620	628.1	1.264	1140	1207.6	2.285	2100	2377	3.408
630	638.6	1.289	1160	1230.9	2.316	2150	2440	3.453
640	649.2	1.314	1180	1254.3	2.346	2200	2504	3.496

*Values rounded off from Chappell and Cockshutt, Nat. Res. Counc. Can. Rep. NRC LR 759 (NRC No. 14300), 1974. This source tabulates values of seven thermodynamic functions at 1-K increments from 200 to 2200 K in SI units and at other increments for two other unit systems. An earlier report (NRC LR 381, 1963) gives a more detailed description of an earlier fitting from 200 to 1400 K. In the above table h = specific enthalpy, kJ/kg, and $\Psi_2 - \Psi_1 = \log_{10} (P_2/P_1)_s$ for an isentrope. In terms of the Keenan and Kaye function ϕ, $\Psi = (\log_{10} e/R) \cdot \phi$.

TABLE 3-214 Thermodynamic Properties of the International Standard Atmosphere*

Z, m	T, K	P, bar	ρ, kg/m³	g, m/s²	M	a, m/s	μ, Pa·s	k, W/(m·K)	λ, m	H, m
0	288.15	1.01325	1.2250	9.80665	28.964	340.29	1.79.−5	2.54.−5	6.63.−8	0
1,000	281.65	0.89876	1.1117	9.8036	28.964	336.43	1.76.−5	2.49.−5	7.31.−8	1,000
2,000	275.15	0.79501	1.0066	9.8005	28.964	332.53	1.73.−5	2.43.−5	8.07.−8	2,999
3,000	268.66	0.70121	0.90925	9.7974	28.964	328.58	1.69.−5	2.38.−5	8.94.−8	2,999
4,000	262.17	0.61660	0.81935	9.7943	28.964	324.59	1.66.−5	2.33.−5	9.92.−8	3,997
5,000	255.68	0.54048	0.73643	9.7912	28.964	320.55	1.63.−5	2.28.−5	1.10.−7	4,996
6,000	249.19	0.47217	0.66011	9.7882	28.964	316.45	1.59.−5	2.22.−5	1.23.−7	5,994
7,000	242.70	0.41105	0.59002	9.7851	28.964	312.31	1.56.−5	2.17.−5	1.38.−7	6,992
8,000	236.22	0.35651	0.52579	9.7820	28.964	308.11	1.53.−5	2.12.−5	1.55.−7	7,990
9,000	229.73	0.30800	0.46706	9.7789	28.964	303.85	1.49.−5	2.06.−5	1.74.−7	8,987
10,000	223.25	0.26499	0.41351	9.7759	28.964	299.53	1.46.−5	2.01.−5	1.97.−7	9,984
15,000	216.65	0.12111	0.19476	9.7605	28.964	295.07	1.42.−5	1.95.−5	4.17.−7	14,965
20,000	216.65	0.05529	0.08891	9.7452	28.964	295.07	1.42.−5	1.95.−5	9.14.−7	19,937
25,000	221.55	0.02549	0.04008	9.7300	28.964	298.39	1.45.−5	1.99.−5	2.03.−6	24,902
30,000	226.51	0.01197	0.01841	9.7147	28.964	301.71	1.48.−5	2.04.−5	4.42.−6	29,859
40,000	250.35	2.87.−3	4.00.−3	9.6844	28.964	317.19	1.60.−5	2.23.−5	2.03.−5	39,750
50,000	270.65	8.00.−4	1.03.−3	9.6542	28.964	329.80	1.70.−5	2.40.−5	7.91.−5	49,610
60,000	247.02	2.20.−4	3.10.−4	9.6241	28.964	315.07	1.58.−5	2.21.−5	2.62.−4	59,439
70,000	219.59	5.22.−5	8.28.−5	9.5942	28.964	297.06	1.44.−5	1.98.−5	9.81.−4	69,238
80,000	198.64	1.05.−5	1.85.−5	9.5644	28.964	282.54	1.32.−5	1.80.−5	4.40.−3	79,006
90,000	186.87	1.84.−6	3.43.−6	9.5348	28.95	.....	....	....	2.37.−2	88,744
100,000	195.08	3.20.−7	5.60.−7	9.5052	28.40	.....	....	....	0.142	98,451
150,000	634.39	4.54.−9	2.08.−9	9.3597	24.10	.....	....	....	33	146,542
200,000	854.56	8.47.−10	2.54.−10	9.2175	21.30	.....	....	....	240	193,899
250,000	941.33	2.48.−10	6.07.−11	9.0785	19.19	.....	....	....	890	240,540
300,000	976.01	8.77.−11	1.92.−11	8.9427	17.73	.....	....	....	2600	286,480
400,000	995.83	1.45.−11	2.80.−12	8.6799	15.98	.....	....	....	1.6.+4	376,320
500,000	999.24	3.02.−12	5.22.−13	8.4286	14.33	.....	....	....	7.7.+4	463,540
600,000	999.85	8.21.−13	1.14.−13	8.1880	11.51	.....	....	....	2.8.+5	548,252
800,000	999.99	1.70.−13	1.14.−14	7.7368	5.54	.....	....	....	1.4.+6	710,574
1,000,000	1000.00	7.51.−14	3.56.−15	7.3218	3.94	.....	....	....	3.1.+6	864,071

*Extracted from *U.S. Standard Atmosphere, 1976, National Oceanic and Atmospheric Administration*, National Aeronautics and Space Administration and the U.S. Air Force, Washington, 1976. Z = geometric altitude, T = temperature, P = pressure, g = acceleration of gravity, M = molecular weight, a = velocity of sound, μ = viscosity, k = thermal conductivity, λ = mean free path, and H = geopotential altitude. The notation 1.79.−5 signifies 1.79×10^{-5}.

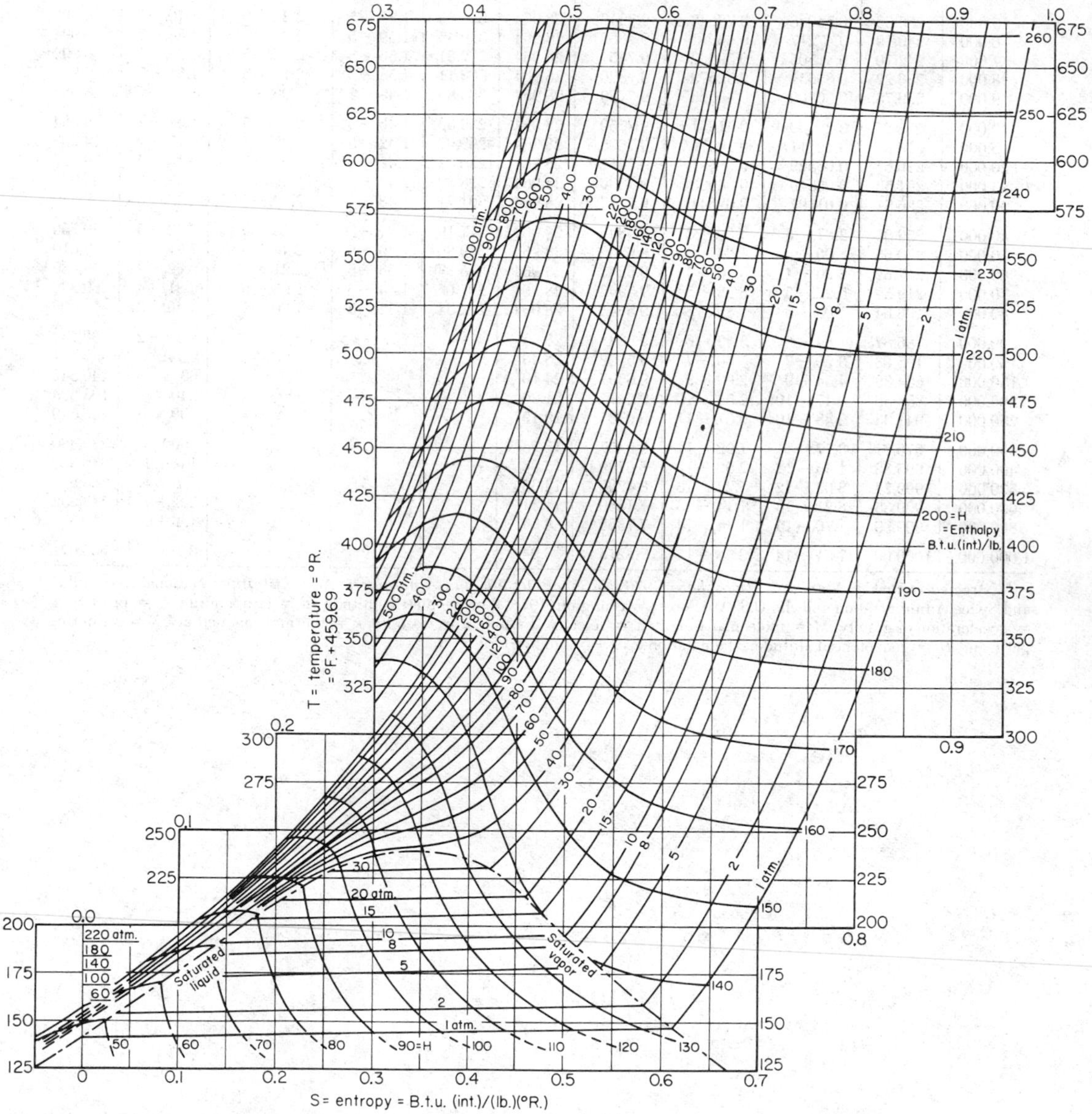

FIG. 3-15 Temperature-entropy diagram for air. [*Landsbaum, Dadds, Stevens, et al.,* Am. Inst. Chem. Eng. J., *1(3), 303 (1955). Reproduced by permission of the authors and of the editor, American Institute of Chemical Engineers.*]

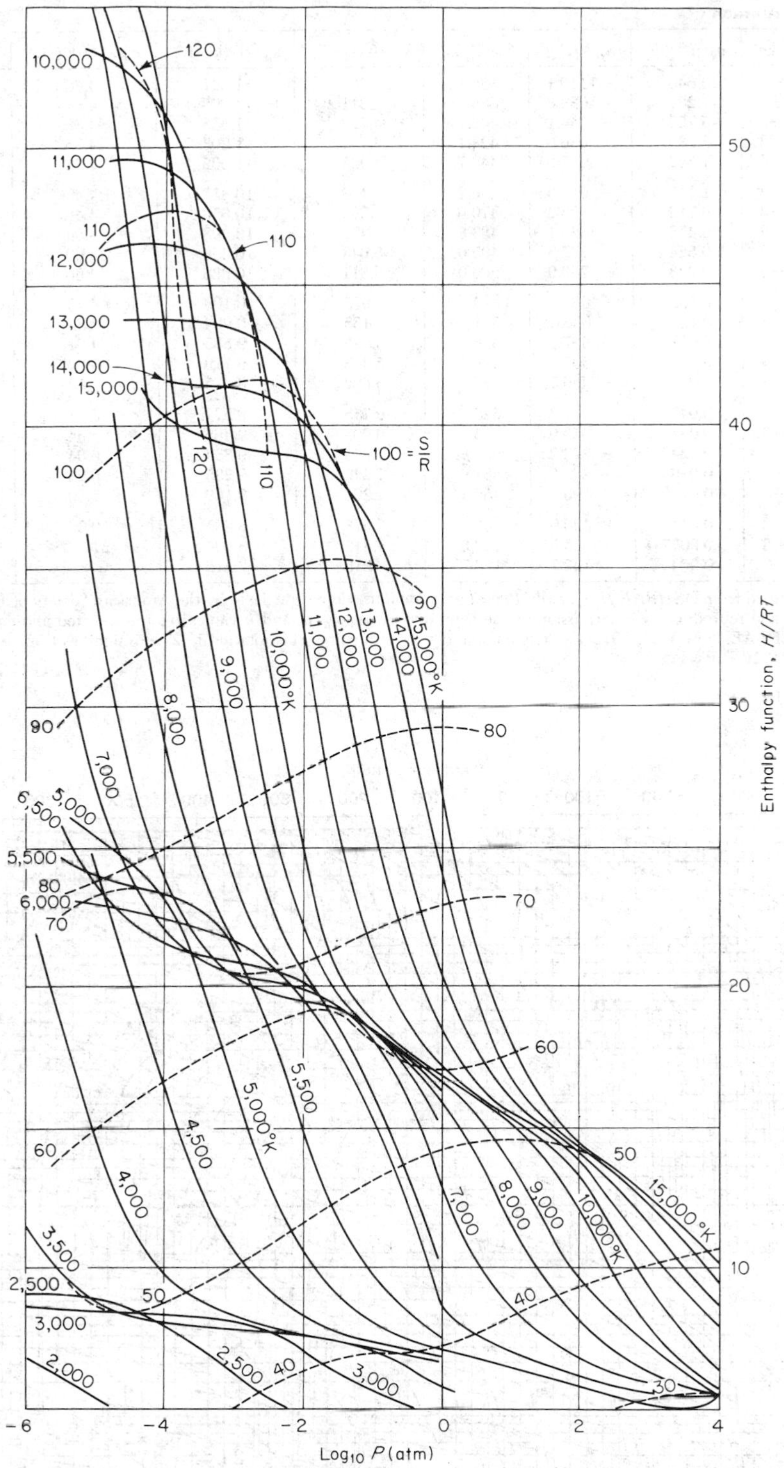

FIG. 3-16 Enthalpy–log-pressure ratio diagram for air. Drawn from tables in J. Hilsenrath and M. Klein, AEDC-TDR-63-161, 1963. A later report (AEDC-TDR-65-58, 1965) also contains equilibrium-composition tables. The low-temperature composition of 0.78084 N_2, 0.20946 O_2, 0.00934 Ar, 0.00033 CO_2, and 0.00003 Ne is equivalent to a molecular weight of 28.967. The gas content R was assumed to be 1.98726 cal/(mol·K), and the normal density ρ_N as 1.29313 $\times 10^{-3}$ g/cm.^{3} For temperatures from 10,000 to 10,000,000 K., F. R. Gilmore, DASA 1971-1, *Thermal Radiation Phenomena*, vol. I, 1967 (AD654054), has prepared similar tables. This source also contains tables for lower temperatures for many of the components.

TABLE 3-215 Saturated Ammonia*

T, K	P, bar	v_f, m³/kg	v_g, m³/kg	h_f, kJ/kg	h_g, kJ/kg	s_f, kJ/(kg·K)	s_g, kJ/(kg·K)	c_{pf}, kJ/(kg·K)	μ_f, 10^{-4} Pa·s	k_f, W/(m·K)
195.5^t	0.0608	1.327.−3	15.648	−1110.1	380.1	4.203	11.827	4.73	4.25	0.715
200	0.0865	1.372.−3	11.237	−1088.8	388.5	4.311	11.698	4.61	4.07	0.709
210	0.1775	1.394.−3	5.729	−1044.1	406.7	4.529	11.438	4.38	3.69	0.685
220	0.3381	1.417.−3	3.135	−1000.6	424.1	4.731	11.207	4.35	3.34	0.661
230	0.6044	1.442.−3	1.822	− 957.0	440.7	4.925	11.002	4.38	3.02	0.638
240	1.0226	1.468.−3	1.115	− 912.9	456.2	5.113	10.817	4.43	2.73	0.615
250	1.6496	1.495.−3	0.712	− 868.2	470.6	5.294	10.650	4.48	2.45	0.592
260	2.5529	1.524.−3	0.472	− 823.1	483.8	5.471	10.498	4.54	2.20	0.569
270	3.8100	1.551.−3	0.324	− 777.3	495.6	5.643	10.358	4.60	1.97	0.546
280	5.5077	1.589.−3	0.228	− 730.9	506.0	5.811	10.228	4.66	1.76	0.523
290	7.741	1.626.−3	0.165	− 683.8	514.7	5.975	10.108	4.73	1.58	0.500
300	10.61	1.666.−3	0.121	− 636.0	521.5	6.135	9.994	4.82	1.41	0.477
310	14.24	1.710.−3	0.091	− 587.2	526.1	6.293	9.885	4.91	1.26	0.454
320	18.72	1.760.−3	0.069	− 537.5	528.2	6.448	9.779	5.02	1.13	0.431
330	24.20	1.815.−3	0.053	− 486.7	527.5	6.602	9.675	5.17	1.02	0.408
340	30.79	1.878.−3	0.0410	− 434.3	523.3	6.755	9.571	5.37	0.92	0.385
350	38.64	1.952.−3	0.0319	− 380.0	515.1	6.908	9.465	5.64	0.83	0.361
360	47.90	2.039.−3	0.0249	− 323.2	501.8	7.063	9.354	6.04	0.75	0.337
370	58.74	2.148.−3	0.0194	− 262.6	481.9	7.222	9.235	6.68	0.69	0.313
380	71.35	2.291.−3	0.0149	− 196.5	452.7	7.391	9.100	7.80	0.61	0.286
390	85.98	2.499.−3	0.0113	− 120.9	408.1	7.578	8.935	10.3	0.50	0.254
400	103.0	2.882.−3	0.0077	− 23.5	329.0	7.813	8.694	21.	0.39	0.21
405.4^c	113.0	4.255.−3	0.0043	142.7	142.7	8.216	8.216	∞	0.25	∞

**P*, *v*, *h*, and *s* values condensed from *ASHRAE Handbook, 1981: Fundamentals.* Copyright 1981 by the American Society of Heating, Refrigerating and Air-Conditioning Engineers, Inc., and reproduced by permission of the copyright owner. c_p, μ, and k values are interpolated and converted from *Thermophysical Properties of Refrigerants*, ASHRAE, New York, 1976. t = triple point; c = critical point. The notation 1.327.−3 signifies 1.327×10^{-3}. At 195.5 K the viscosity of the saturated liquid is 4.25×10^{-4} Pa·s.

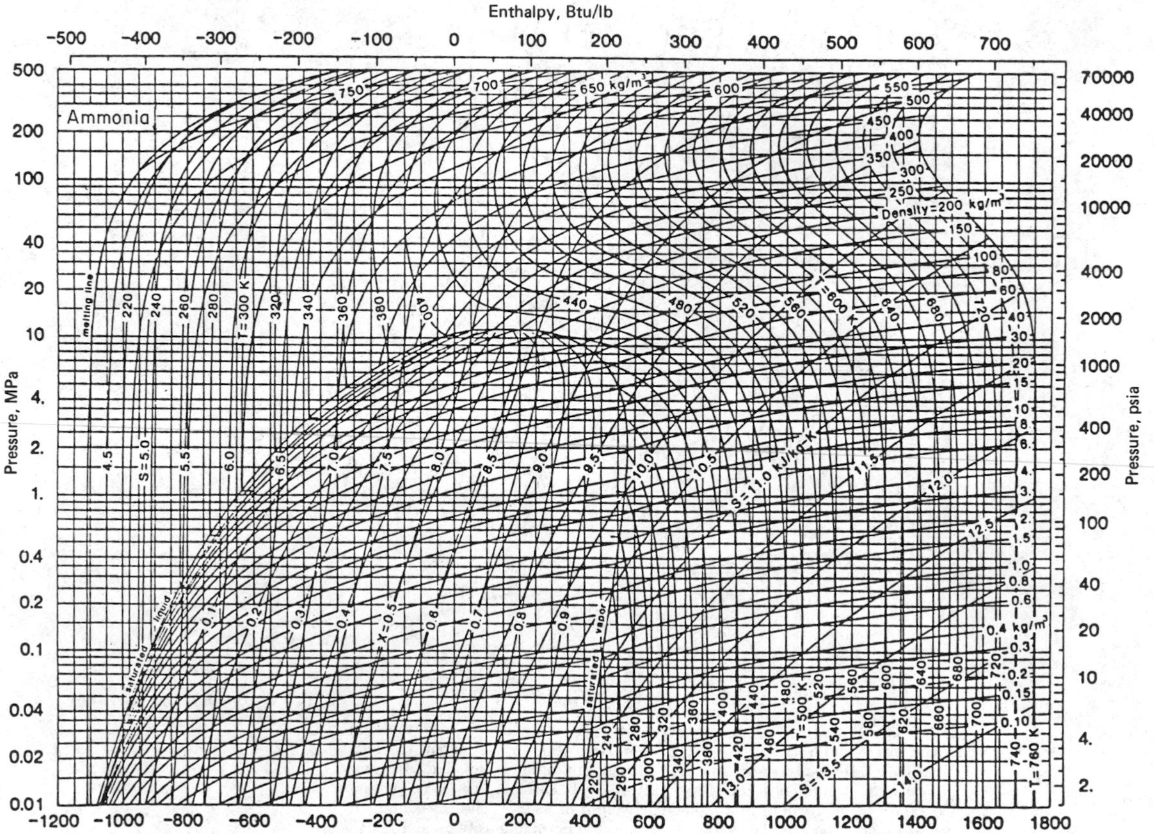

FIG. 3-17 Enthalpy–log-pressure diagram for ammonia. 1 MPa = 10 bar. *(Copyright 1981 by the American Society of Heating, Refrigerating and Air-Conditioning Engineers and reproduced by permission of the copyright owner.)*

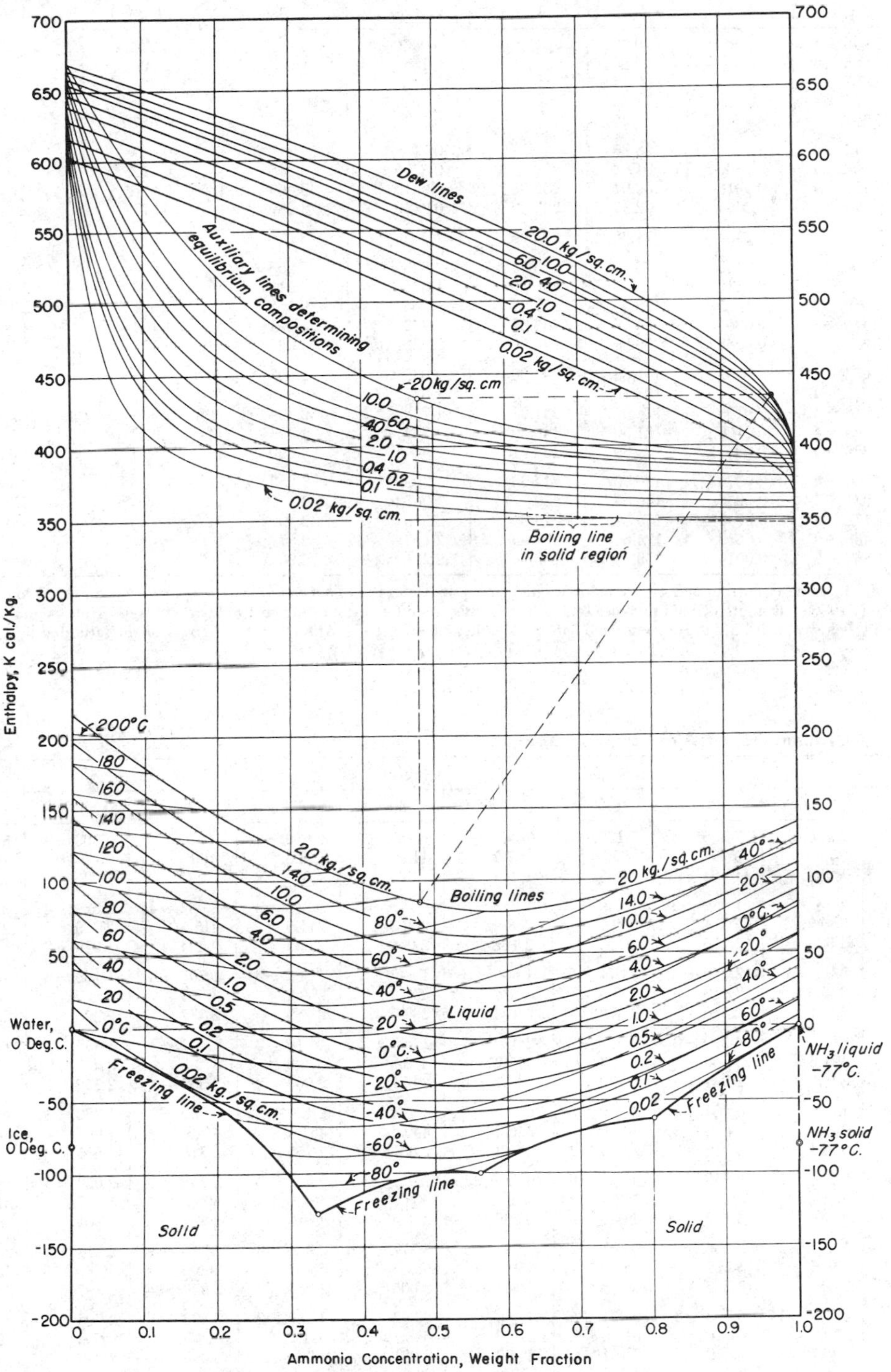

FIG. 3-18 Enthalpy-concentration diagram for aqueous ammonia. Reference states: enthalpies of liquid water at 0°C and liquid ammonia at −77° C are zero. NOTE: In order to determine equilibrium compositions, a vertical may be erected from any liquid composition on any boiling line and its intersection with the appropriate auxiliary line determined. A horizontal from this intersection will establish the equilibrium vapor composition on the appropriate dew line. An example at 48 percent ammonia and 20 kg/cm^2 is indicated (*Bosnjakovic*, Technische Thermodynamik, *T. Steinkopff, Leipzig, 1935*). A slightly improved diagram with melting curves added appears as Tafel IV of the second edition of this work, published by T. Steinkopff, Dresden and Leipzig, 1961.

TABLE 3-216 Saturated Argon (R740)*

T, K	P, bar	v_f, m³/kg	v_g, m³/kg	h_f, kJ/kg	h_g, kJ/kg	s_f, kJ/(kg·K)	s_g, kJ/(kg·K)	c_{pf}, kJ/(kg·K)	μ_f, 10^{-4} Pa·s	k_f, W/(m·K)
10	.....	5.646.−4	.....	0.20	.....	0.0266	.....	0.083		
20	.....	5.666.−4	.....	2.20	.....	0.1559	.....	0.306		
30	.....	5.707.−4	.....	6.12	.....	0.3129	.....	0.466		
40	.....	5.763.−4	.....	11.30	.....	0.4610	.....	0.560		
50	.....	5.831.−4	.....	17.26	.....	0.5937	.....	0.627		
60	.....	5.912.−4	.....	23.85	.....	0.7138	.....	0.687		
70	0.082	6.008.−4	2.1800	31.08	229.08	0.8250	3.415	0.752		
80	0.406	6.125.−4	0.3918	39.07	232.88	0.9316	3.364	0.836		
83.8^t	0.687	6.178.−4	0.2434	42.34	235.06	0.9720	3.280	0.877		
83.8^t	0.687	7.068.−4	0.2434	71.88	235.06	1.333	3.280	1.050	2.93	0.134
85	0.790	7.107.−4	0.2145	73.16	235.55	1.348	3.258	1.058	2.81	0.132
87.3	1.013	7.174.−4	0.1710	75.61	236.39	1.375	3.216	1.073	2.60	0.128
90	1.338	7.269.−4	0.1327	78.55	237.37	1.403	3.168	1.091	2.40	0.124
95	2.137	7.440.−4	0.0864	84.15	238.91	1.462	3.091	1.124	2.08	0.116
100	3.247	7.628.−4	0.0588	89.85	240.20	1.520	3.023	1.158	1.82	0.109
110	6.665	8.064.−4	0.0299	101.83	241.66	1.632	2.903	1.229	1.46	0.096
115	9.107	8.322.−4	0.0221	108.11	241.78	1.685	2.848	1.274	1.32	0.090
120	12.13	8.618.−4	0.0166	114.62	241.33	1.738	2.794	1.336	1.21	0.084
125	15.81	8.965.−4	0.0126	121.50	240.30	1.792	2.743	1.427	1.12	0.078
130	20.23	9.620.−4	0.0096	128.79	238.41	1.846	2.690	1.550	1.01	0.072
135	25.49	9.906.−4	0.0074	136.76	234.60	1.902	2.633	1.752	0.89	0.066
140	31.68	1.061.−3	0.0056	145.58	230.74	1.961	2.570	.....	0.75	0.060
145	38.93	1.172.−3	0.0041	155.73	223.09	2.026	2.490	.....	0.60	0.054
150	47.39	1.468.−3	0.0026	174.64	204.35	2.133	2.331	.....	0.45	
150.9	48.98	1.867.−3	0.0019	189.94	189.94	2.201	2.201	.....	0.28	∞

*Values extracted and in some cases rounded off from those cited in Rabinovich (ed.), *Thermophysical Properties of Neon, Argon, Krypton and Xenon,* Standards Press, Moscow, 1976. This source contains values for the compressed state for pressures up to 1000 bar, etc. t = triple point. Above the solid line the condensed phase is solid; below it, it is liquid. The notation 5.646.−4 signifies 5.646×10^{-4}. At 83.8 K the viscosity of the saturated liquid is 2.93×10^{-4} Pa·s = 0.000293 Ns/m².

TABLE 3-217 Thermodynamic Properties of Compressed Argon*

T, K		Pressure, bar										
		1	100	200	300	400	500	600	700	800	900	1000
100	v	0.2035	7.420.−4	7.255.−4	7.120.−4	7.006.−4	6.907.−4	6.819.−4	6.050.−4	6.009.−4	5.976.−4	5.935.−4
	h	243.4	93.6	97.9	102.5	107.2	112.0	116.8	91.1	96.1	101.0	106.0
	s	3.299	1.494	1.464	1.438	1.414	1.393	1.372	1.037	1.026	1.016	1.007
200	v	0.4151	2.96.−3	1.430.−5	1.159.−3	1.045.−3	9.778.−4	9.312.−4	8.962.−4	8.683.−4	8.454.−4	8.260.−4
	h	296.4	250.2	217.1	209.1	207.9	209.2	211.9	215.1	218.9	223.0	227.4
	s	3.667	2.538	2.276	2.173	2.112	2.068	2.033	2.004	1.979	1.957	1.936
300	v	0.6241	5.96.−3	2.976.−3	2.071.−3	1.666.−3	1.443.−3	1.304.−3	1.207.−3	1.136.−3	1.081.−3	1.037.−3
	h	348.6	330.9	316.3	306.6	301.4	299.3	299.2	300.5	302.7	305.6	310.0
	s	3.879	2.872	2.686	2.572	2.493	2.435	2.389	2.352	2.320	2.293	2.269
400	v	0.8326	8.37.−3	4.279.−3	2.957.−3	2.322.−3	1.955.−3	1.719.−3	1.557.−3	1.435.−3	1.344.−3	1.271.−3
	h	400.7	391.3	383.6	378.4	375.2	373.8	373.8	374.8	376.6	379.2	382.0
	s	4.028	3.048	2.881	2.780	2.707	2.651	2.603	2.565	2.533	2.505	3.480
500	v	1.0409	1.062.−2	5.464.−3	3.772.−3	2.940.−3	2.448.−3	2.124.−3	1.899.−3	1.730.−3	1.607.−3	1.506.−3
	h	452.8	447.7	444.3	442.0	440.9	440.6	441.4	422.9	444.7	447.1	449.9
	s	4.145	3.174	3.018	2.924	2.854	2.801	2.755	2.718	2.685	2.658	2.633
600	v	1.2489	1.280.−2	6.589.−3	4.539.−3	3.525.−3	2.922.−3	2.522.−3	2.238.−3	2.023.−3	1.866.−3	1.736.−3
	h	504.9	502.4	501.6	501.4	501.8	503.0	504.6	506.6	508.7	511.2	5.13.9
	s	4.240	3.274	3.122	3.031	2.966	2.914	2.870	2.834	2.801	2.774	2.750
700	v	1.4569	1.495.−2	7.686.−3	5.281.−3	4.088.−3	3.377.−3	2.906.−3	2.570.−3	2.317.−3	2.123.−3	1.966.−3
	h	556.9	556.5	556.9	558.0	559.8	561.8	564.2	566.9	569.6	527.5	575.3
	s	4.320	3.356	3.207	3.118	3.054	3.005	2.963	2.928	2.897	2.870	2.845
800	v	1.6659	1.708.−2	8.768.−3	6.011.−3	4.640.−3	3.822.−3	3.280.−3	2.893.−3	2.603.−3	2.376.−3	2.196.−3
	h	609.9	609.8	611.0	612.9	615.2	618.1	621.2	624.5	627.8	631.3	634.8
	s	4.389	3.427	3.279	3.191	3.129	3.081	3.039	3.005	2.975	2.948	2.924
900	v	1.8739	1.920.−2	9.841.−3	6.732.−3	5.183.−3	4.259.−3	3.646.−3	3.209.−3	2.881.−3	2.626.−3	2.423.−3
	h	661.0	662.7	664.6	667.2	670.1	673.3	676.8	680.7	684.4	688.3	692.3
	s	4.451	3.490	3.342	3.255	3.193	3.145	3.105	3.071	3.042	3.016	2.992
1000	v	2.0819	2.131.−2	1.091.−2	7.448.−3	5.723.−3	4.692.−3	4.008.−3	3.520.−3	3.156.−3	2.872.−3	2.645.−3
	h	713.1	715.4	717.9	720.9	724.3	727.8	731.5	735.6	739.8	744.1	748.5
	s	4.506	3.545	3.398	3.312	3.250	3.203	3.163	3.129	3.100	3.074	3.051

*Values extracted and in some cases rounded off from those cited in Rabinovich (ed.), *Thermophysical Properties of Neon, Argon, Krypton and Xenon,* Standards Press, Moscow, 1976. v = specific volume, m³/kg; h = specific enthalpy, kJ/kg; s = specific entropy, kJ/(kg·K). This source contains an exhaustive tabulation of values. The notation 7.420.−4 signifies 7.420×10^{-4}.

TABLE 3-218 Liquid-Vapor Equilibrium Data for the Argon-Nitrogen-Oxygen System*

Liquid mole fraction		Vapor mole fraction			Temperature, °R	Relative volatility			Pressure activity coefficient			Enthalpy, Btu (lb·mol)		Heat capacity, Btu/(lb·mol·°R)	
N_2/N_2+O_2	Ar	N_2	Ar	O_2		N_2/Ar	N_2/O_2	Ar/O_2	N_2	Ar	O_2	Liquid	Vapor	Liquid	Vapor
Pressure, 1 atm															
0.	0.	0.	0.	1.0000	162.4	2.575	4.010	1.557	1.118	1.165	0.999	−1841.	1093.	13.2	7.406
0.	0.01	0.	0.0154	0.9845	162.3	2.581	4.007	1.553	1.117	1.161	1.000	−1844.	1087.	13.1	7.374
0.	0.02	0.	0.0306	0.9694	162.2	2.586	4.004	1.548	1.115	1.158	1.000	−1847.	1082.	13.1	7.342
0.	0.03	0.	0.0456	0.9544	162.1	2.592	4.001	1.544	1.113	1.155	1.000	−1850.	1076.	13.1	7.311
0.	0.04	0.	0.0603	0.9397	162.0	2.597	3.998	1.540	1.112	1.151	1.001	−1852.	1071.	13.0	7.281
0.	0.05	0.	0.0748	0.9253	161.9	2.602	3.995	1.535	1.110	1.148	1.001	−1855.	1066.	13.0	7.251
0.	0.07	0.	0.1031	0.8970	161.7	2.613	3.989	1.526	1.107	1.142	1.002	−1860.	1056.	12.9	7.192
0.	0.10	0.	0.1439	0.8561	161.5	2.629	3.979	1.513	1.103	1.132	1.003	−1868.	1041.	12.9	7.107
0.	0.20	0.	0.2687	0.7313	160.7	2.682	3.941	1.469	1.091	1.104	1.010	−1893.	997.	12.6	6.847
0.	0.40	0.	0.4796	0.5204	159.4	2.786	3.852	1.382	1.076	1.058	1.034	−1938.	924.	11.9	6.406
0.	0.60	0.	0.6605	0.3395	158.5	2.888	3.746	1.297	1.075	1.026	1.072	−1978.	862.	11.3	6.026
0.	0.80	0.	0.8293	0.1707	157.7	2.991	3.632	1.214	1.087	1.008	1.127	−2015.	807.	10.7	5.669
0.	0.90	0.	0.9136	0.0865	157.5	3.042	3.572	1.174	1.099	1.003	1.162	−2032.	779.	10.4	5.491
0.10	0.	0.3135	0.	0.6865	157.7	2.621	4.111	1.568	1.103	1.168	1.012	−1834.	1060.	13.2	7.410
0.10	0.01	0.3095	0.0119	0.6786	157.6	2.626	4.106	1.563	1.102	1.164	1.012	−1837.	1057.	13.1	7.386
0.10	0.02	0.3056	0.0237	0.6707	157.6	2.631	4.100	1.558	1.100	1.161	1.012	−1839.	1053.	13.1	7.361
0.10	0.03	0.3017	0.0354	0.6630	157.6	2.636	4.095	1.554	1.099	1.157	1.013	−1842.	1049.	13.1	7.337
0.10	0.04	0.2978	0.0470	0.6553	157.6	2.641	4.090	1.549	1.098	1.154	1.013	−1844.	1045.	13.1	7.313
0.10	0.05	0.2939	0.0585	0.6476	157.5	2.645	4.085	1.544	1.096	1.151	1.013	−1846.	1042.	13.0	7.289
0.10	0.07	0.2863	0.0812	0.6325	157.5	2.655	4.074	1.534	1.094	1.144	1.014	−1851.	1034.	13.0	7.242
0.10	0.10	0.2752	0.1145	0.6103	157.4	2.669	4.058	1.520	1.090	1.135	1.015	−1858.	1024.	12.9	7.173
0.10	0.20	0.2399	0.2207	0.5394	157.2	2.717	4.003	1.473	1.080	1.106	1.022	−1882.	990.	12.5	6.951
0.10	0.40	0.1759	0.4170	0.4072	157.0	2.812	3.887	1.382	1.070	1.061	1.045	−1926.	928.	11.9	6.540
0.10	0.60	0.1169	0.6036	0.2795	156.9	2.906	3.766	1.296	1.072	1.029	1.082	−1969.	871.	11.3	6.147
0.10	0.80	0.0595	0.7933	0.1471	156.9	3.001	3.640	1.213	1.086	1.009	1.134	−2009.	813.	10.7	5.746
0.10	0.90	0.0303	0.8937	0.0762	157.1	3.048	3.576	1.173	1.099	1.004	1.166	−2029.	783.	10.4	5.534
0.20	0.	0.5095	0.	0.4905	154.0	2.641	4.155	1.573	1.085	1.171	1.026	−1814.	1035.	13.2	7.422
0.20	0.01	0.5042	0.0096	0.4861	154.0	2.646	4.149	1.568	1.084	1.168	1.026	−1816.	1032.	13.2	7.402
0.20	0.02	0.4990	0.0192	0.4818	154.0	2.651	4.143	1.563	1.083	1.164	1.026	−1819.	1029.	13.1	7.382
0.20	0.03	0.4938	0.0288	0.4775	154.1	2.655	4.137	1.558	1.082	1.161	1.027	−1821.	1027.	13.1	7.362
0.20	0.04	0.4886	0.0383	0.4731	154.1	2.660	4.131	1.553	1.081	1.158	1.027	−1824.	1024.	13.1	7.342
0.20	0.05	0.4834	0.0477	0.4688	154.1	2.665	4.125	1.548	1.080	1.154	1.027	−1826.	1021.	13.0	7.322
0.20	0.07	0.4732	0.0666	0.4602	154.1	2.674	4.112	1.538	1.078	1.148	1.028	−1831.	1016.	13.0	7.283
0.20	0.10	0.4580	0.0946	0.4474	154.2	2.688	4.094	1.523	1.075	1.139	1.030	−1839.	1008.	12.9	7.224
0.20	0.20	0.4083	0.1866	0.4051	154.3	2.735	4.032	1.474	1.068	1.110	1.036	−1863.	981.	12.6	7.031
0.20	0.40	0.3123	0.3680	0.3197	154.8	2.829	3.907	1.381	1.062	1.064	1.058	−1911.	930.	11.9	6.648
0.20	0.60	0.2162	0.5550	0.2288	155.4	2.921	3.779	1.294	1.068	1.032	1.093	−1958.	877.	11.3	6.252
0.20	0.80	0.1144	0.7602	0.1254	156.2	3.009	3.647	1.212	1.086	1.011	1.140	−2004.	819.	10.7	5.817
0.20	0.90	0.0593	0.8744	0.0663	156.7	3.052	3.580	1.173	1.099	1.006	1.169	−2026.	787.	10.4	5.574
0.40	0.	0.7333	0.	0.2667	148.7	2.629	4.124	1.569	1.050	1.187	1.065	−1748.	997.	13.3	7.452
0.40	0.01	0.7279	0.0070	0.2651	148.8	2.634	4.119	1.564	1.049	1.183	1.065	−1751.	996.	13.3	7.437
0.40	0.02	0.7226	0.0140	0.2635	148.8	2.640	4.114	1.558	1.049	1.179	1.065	−1754.	994.	13.2	7.422
0.40	0.03	0.7172	0.0210	0.2619	148.9	2.645	4.108	1.553	1.048	1.176	1.066	−1757.	992.	13.2	7.407
0.40	0.04	0.7118	0.0280	0.2602	148.9	2.650	4.103	1.548	1.048	1.172	1.066	−1760.	991.	13.2	7.392
0.40	0.05	0.7064	0.0350	0.2586	149.0	2.656	4.098	1.543	1.047	1.169	1.066	−1763.	989.	13.1	7.377
0.40	0.07	0.6956	0.0491	0.2553	149.1	2.667	4.087	1.533	1.047	1.162	1.066	−1770.	986.	13.1	7.347
0.40	0.10	0.6794	0.0703	0.2503	149.3	2.683	4.072	1.517	1.046	1.152	1.067	−1779.	981.	13.0	7.301
0.40	0.20	0.6244	0.1426	0.2331	149.9	2.737	4.018	1.468	1.044	1.122	1.071	−1810.	964.	12.6	7.145
0.40	0.40	0.5075	0.2977	0.1948	151.2	2.841	3.907	1.375	1.048	1.074	1.088	−1871.	928.	11.9	6.811
0.40	0.60	0.3743	0.4776	0.1482	152.8	2.939	3.788	1.289	1.062	1.039	1.116	−1932.	885.	11.3	6.423
0.40	0.80	0.2121	0.7010	0.0870	154.8	3.025	3.657	1.209	1.084	1.016	1.154	−1991.	829.	10.7	5.944
0.40	0.90	0.1141	0.8382	0.0477	155.9	3.063	3.586	1.171	1.099	1.008	1.177	−2020.	794.	10.4	5.651
0.60	0.	0.8569	0.	0.1431	144.9	2.575	3.993	1.551	1.024	1.218	1.126	−1663.	970.	13.6	7.483
0.60	0.01	0.8521	0.0056	0.1424	145.0	2.582	3.991	1.546	1.024	1.214	1.125	−1667.	969.	13.5	7.471
0.60	0.02	0.8472	0.0111	0.1416	145.1	2.589	3.988	1.541	1.024	1.210	1.125	−1672.	968.	13.5	7.459
0.60	0.03	0.8424	0.0167	0.1409	145.1	2.595	3.985	1.536	1.024	1.206	1.124	−1676.	967.	13.4	7.446
0.60	0.04	0.8375	0.0224	0.1402	145.2	2.602	3.983	1.531	1.024	1.202	1.124	−1680.	966.	13.4	7.434
0.60	0.05	0.8326	0.0280	0.1395	145.3	2.609	3.980	1.526	1.023	1.198	1.123	−1684.	965.	13.4	7.421
0.60	0.07	0.8227	0.0394	0.1380	145.5	2.622	3.975	1.516	1.023	1.190	1.122	−1692.	963.	13.3	7.396
0.60	0.10	0.8076	0.0566	0.1357	145.7	2.642	3.966	1.501	1.023	1.179	1.121	−1704.	960.	13.2	7.357
0.60	0.20	0.7557	0.1163	0.1280	146.5	2.707	3.937	1.454	1.025	1.145	1.120	−1744.	948.	12.8	7.224
0.60	0.40	0.6391	0.2507	0.1102	148.4	2.833	3.867	1.365	1.036	1.090	1.126	−1824.	923.	12.0	6.926
0.60	0.60	0.4938	0.4191	0.0871	150.6	2.945	3.777	1.283	1.056	1.049	1.143	−1903.	888.	11.3	6.556
0.60	0.80	0.2961	0.6500	0.0539	153.5	3.037	3.662	1.206	1.083	1.021	1.169	−1978.	837.	10.7	6.055
0.60	0.90	0.1647	0.8047	0.0306	155.2	3.071	3.592	1.170	1.098	1.010	1.185	−2014.	801.	10.4	5.723
0.80	0.	0.9384	0.	0.0616	142.0	2.501	3.811	1.524	1.013	1.273	1.214	−1570.	949.	14.0	7.514
0.80	0.01	0.9340	0.0047	0.0613	142.0	2.509	3.811	1.519	1.013	1.268	1.212	−1575.	948.	13.9	7.503
0.80	0.02	0.9296	0.0094	0.0610	142.1	2.517	3.812	1.515	1.013	1.263	1.210	−1580.	947.	13.9	7.492
0.80	0.03	0.9252	0.0142	0.0607	142.2	2.525	3.812	1.510	1.013	1.257	1.209	−1585.	947.	13.8	7.481
0.80	0.04	0.9207	0.0189	0.0604	142.3	2.533	3.813	1.506	1.013	1.258	1.207	−1590.	946.	13.8	7.470
0.80	0.05	0.9162	0.0237	0.0601	142.4	2.540	3.813	1.501	1.013	1.247	1.205	−1595.	945.	13.7	7.459
0.80	0.07	0.9071	0.0334	0.0595	142.6	2.556	3.814	1.492	1.012	1.238	1.202	−1606.	944.	13.6	7.437
0.80	0.10	0.8934	0.0481	0.0586	142.8	2.580	3.814	1.478	1.013	1.224	1.197	−1621.	942.	13.5	7.403
0.80	0.20	0.8452	0.0994	0.0554	143.8	2.658	3.814	1.435	1.015	1.181	1.185	−1671.	934.	13.0	7.284
0.80	0.40	0.7339	0.2177	0.0483	146.0	2.809	3.797	1.352	1.028	1.112	1.173	−1772.	916.	12.2	7.013
0.80	0.60	0.5869	0.3740	0.0391	148.7	2.943	3.751	1.275	1.051	1.062	1.173	−1870.	889.	11.4	6.662
0.80	0.80	0.3690	0.6059	0.0252	152.2	3.045	3.662	1.202	1.082	1.026	1.184	−1964.	843.	10.7	6.153
0.80	0.90	0.2117	0.7736	0.0147	154.5	3.078	3.595	1.168	1.098	1.013	1.193	−2007.	806.	10.3	5.790

*Calculated values, from Wilson, Silverberg, and Zellner, USAF Aero Propulsion Laboratory, Rep. APL TDR 64-64 (AD 603 151), 1964. Relative volatility = $\alpha_{i-j} = (y_i/x_i)(x_j/y_j)$, where x = liquid composition, y = vapor composition. Pressure activity coefficient = y_iP/x_ip^0, where p^0 = vapor pressure. These data were confirmed by the analyses of Bender, *Cryogenics*, **13**, 11 (1973); and by Elshayal and Lu, *J. Chem. Eng. Data*, **16**, 31 (1971).

TABLE 3-218 Liquid-Vapor Equilibrium Data for the Argon-Nitrogen-Oxygen System* *(Continued)*

Liquid mole fraction		Vapor mole fraction			Temperature, °R	Relative volatility			Pressure activity coefficient			Enthalpy, Btu (lb·mol)		Heat capacity, Btu/(lb·mol·°R)	
N_2/N_2+O_2	Ar	N_2	Ar	O_2		N_2/Ar	N_2/O_2	Ar/O_2	N_2	Ar	O_2	Liquid	Vapor	Liquid	Vapor
0.90	0.	0.9709	0.	0.0291	140.6	2.459	3.710	1.509	1.015	1.311	1.271	−1522.	939.	14.2	7.530
0.90	0.01	0.9667	0.0044	0.0289	140.7	2.468	3.712	1.504	1.014	1.305	1.268	−1527.	938.	14.2	7.519
0.90	0.02	0.9624	0.0088	0.0288	140.8	2.476	3.714	1.500	1.014	1.299	1.265	−1533.	938.	14.1	7.509
0.90	0.03	0.9581	0.0133	0.0286	140.9	2.485	3.716	1.496	1.013	1.293	1.263	−1538.	937.	14.1	7.498
0.90	0.04	0.9538	0.0177	0.0285	141.0	2.493	3.718	1.491	1.013	1.287	1.260	−1544.	937.	14.0	7.488
0.90	0.05	0.9494	0.0222	0.0284	141.1	2.502	3.720	1.487	1.013	1.281	1.257	−1550.	936.	14.0	7.477
0.90	0.07	0.9407	0.0312	0.0281	141.3	2.519	3.724	1.478	1.013	1.270	1.252	−1561.	935.	13.8	7.456
0.90	0.10	0.9274	0.0450	0.0276	141.6	2.545	3.729	1.465	1.012	1.254	1.245	−1578.	934.	13.7	7.423
0.90	0.20	0.8808	0.0931	0.0261	142.6	2.629	3.743	1.424	1.014	1.204	1.226	−1633.	928.	13.2	7.310
0.90	0.40	0.7723	0.2048	0.0229	144.9	2.793	3.755	1.344	1.026	1.126	1.200	−1745.	912.	12.3	7.050
0.90	0.60	0.6262	0.3552	0.0186	147.8	2.938	3.733	1.270	1.050	1.069	1.190	−1853.	889.	11.4	6.708
0.90	0.80	0.4018	0.5860	0.0122	151.7	3.048	3.660	1.201	1.081	1.029	1.192	−1956.	846.	10.7	6.197
0.90	0.90	0.2339	0.7589	0.0072	154.2	3.082	3.597	1.167	1.098	1.014	1.197	−2004.	809.	10.3	5.822
0.97	0.	0.9916	0.	0.0084	139.8	2.429	3.638	1.498	1.018	1.342	1.316	−1488.	933.	14.4	7.541
0.97	0.01	0.9874	0.0042	0.0084	139.9	2.438	3.641	1.494	1.018	1.335	1.313	−1494.	932.	14.4	7.531
0.97	0.02	0.9832	0.0085	0.0083	140.0	2.447	3.644	1.489	1.017	1.329	1.309	−1500.	932.	14.3	7.520
0.97	0.03	0.9790	0.0127	0.0083	140.1	2.456	3.647	1.485	1.017	1.322	1.306	−1505.	931.	14.2	7.510
0.97	0.04	0.9748	0.0170	0.0083	140.2	2.465	3.650	1.481	1.016	1.315	1.302	−1511.	931.	14.2	7.500
0.97	0.05	0.9705	0.0213	0.0082	140.3	2.474	3.653	1.477	1.016	1.309	1.299	−1517.	930.	14.1	7.489
0.97	0.07	0.9619	0.0300	0.0081	140.5	2.492	3.658	1.468	1.015	1.296	1.293	−1529.	929.	14.0	7.469
0.97	0.10	0.9488	0.0432	0.0080	140.8	2.519	3.667	1.456	1.014	1.278	1.284	−1547.	928.	13.9	7.437
0.97	0.20	0.9032	0.0893	0.0076	141.8	2.608	3.691	1.415	1.014	1.224	1.257	−1606.	923.	13.3	7.327
0.97	0.40	0.7965	0.1969	0.0066	144.2	2.780	3.722	1.339	1.025	1.137	1.220	−1725.	910.	12.3	7.073
0.97	0.60	0.6513	0.3433	0.0054	147.2	2.934	3.718	1.267	1.049	1.075	1.202	−1841.	889.	11.4	6.737
0.97	0.80	0.4236	0.5728	0.0036	151.3	3.050	3.658	1.199	1.081	1.031	1.198	−1951.	847.	10.7	6.226
0.97	0.90	0.2489	0.7489	0.0021	153.9	3.084	3.597	1.167	1.098	1.015	1.200	−2002.	810.	10.3	5.844
1.00	0.	1.0000	0.	0.	139.4	2.416	3.607	1.493	1.021	1.357	1.338	−1473.	930.	14.5	7.546
1.00	0.01	0.9959	0.0041	0.0000	139.5	2.425	3.611	1.489	1.020	1.350	1.334	−1479.	929.	14.4	7.535
1.00	0.02	0.9917	0.0083	0.0000	139.6	2.434	3.614	1.485	1.019	1.343	1.330	−1485.	929.	14.4	7.525
1.00	0.03	0.9875	0.0125	0.0000	139.7	2.443	3.617	1.480	1.019	1.336	1.326	−1491.	929.	14.3	7.515
1.00	0.04	0.9833	0.0167	0.0000	139.8	2.452	3.621	1.476	1.018	1.329	1.322	−1497.	928.	14.3	7.505
1.00	0.05	0.9791	0.0209	0.0000	139.9	2.462	3.624	1.472	1.018	1.322	1.318	−1503.	928.	14.2	7.495
1.00	0.07	0.9705	0.0295	0.0000	140.1	2.480	3.630	1.464	1.017	1.309	1.311	−1516.	927.	14.1	7.474
1.00	0.10	0.9576	0.0424	0.0000	140.4	2.507	3.640	1.452	1.016	1.290	1.301	−1534.	926.	13.9	7.443
1.00	0.20	0.9122	0.0878	0.0000	141.5	2.598	3.668	1.412	1.015	1.232	1.271	−1595.	921.	13.4	7.333
1.00	0.40	0.8063	0.1937	0.0000	143.9	2.774	3.708	1.337	1.025	1.142	1.230	−1716.	909.	12.4	7.082
1.00	0.60	0.6616	0.3384	0.	147.0	2.932	3.712	1.266	1.048	1.077	1.208	−1836.	888.	11.4	6.749
1.00	0.80	0.4327	0.5674	0.	151.1	3.050	3.657	1.199	1.080	1.032	1.201	−1949.	848.	10.7	6.838
1.00	0.90	0.2553	0.7447	0.	153.8	3.085	3.598	1.166	1.098	1.016	1.201	−2001.	811.	10.3	5.853
						Pressure, 4 atm									
0.	0.	0.	0.	1.0000	190.6	2.020	2.776	1.375	0.987	1.111	1.000	−1466.	1224.	13.4	8.122
0.	0.01	0.	0.0137	0.9863	190.5	2.023	2.775	1.372	0.986	1.109	1.001	−1470.	1218.	13.4	8.094
0.	0.02	0.	0.0272	0.9728	190.4	2.026	2.773	1.369	0.985	1.106	1.001	−1473.	1212.	13.4	8.065
0.	0.03	0.	0.0405	0.9595	190.3	2.030	2.772	1.366	0.985	1.104	1.002	−1477.	1207.	13.4	8.037
0.	0.04	0.	0.0537	0.9463	190.2	2.033	2.770	1.362	0.984	1.102	1.002	−1480.	1201.	13.4	8.010
0.	0.05	0.	0.0668	0.9332	190.1	2.037	2.768	1.359	0.983	1.100	1.003	−1484.	1196.	13.3	7.982
0.	0.07	0.	0.0924	0.9076	189.9	2.043	2.765	1.353	0.982	1.096	1.004	−1490.	1185.	13.3	7.929
0.	0.10	0.	0.1299	0.8701	189.6	2.053	2.759	1.344	0.980	1.089	1.005	−1500.	1169.	13.3	7.850
0.	0.20	0.	0.2471	0.7529	188.8	2.086	2.738	1.313	0.974	1.070	1.013	−1533.	1121.	13.1	7.605
0.	0.40	0.	0.4548	0.5452	187.5	2.148	2.688	1.251	0.967	1.038	1.034	−1593.	1037.	12.7	7.166
0.	0.60	0.	0.6410	0.3590	186.5	2.207	2.627	1.190	0.967	1.016	1.066	−1647.	964.	12.3	6.771
0.	0.80	0.	0.8190	0.1811	185.8	2.264	2.560	1.131	0.976	1.003	1.110	−1698.	896.	11.9	6.389
0.	0.90	0.	0.9084	0.0916	185.5	2.292	2.524	1.101	0.983	1.000	1.137	−1723.	862.	11.8	6.196
0.10	0.	0.2393	0.	0.7607	186.2	2.063	2.831	1.372	0.994	1.120	1.020	−1452.	1193.	13.7	8.150
0.10	0.01	0.2363	0.0116	0.7521	186.1	2.066	2.828	1.369	0.994	1.117	1.021	−1455.	1188.	13.6	8.126
0.10	0.02	0.2334	0.0230	0.7436	186.1	2.069	2.825	1.365	0.993	1.115	1.021	−1459.	1184.	13.6	8.102
0.10	0.03	0.2305	0.0344	0.7351	186.0	2.072	2.822	1.362	0.992	1.113	1.021	−1462.	1180.	13.6	8.078
0.10	0.04	0.2277	0.0457	0.7267	186.0	2.075	2.820	1.359	0.991	1.110	1.022	−1465.	1175.	13.6	8.054
0.10	0.05	0.2248	0.0569	0.7183	186.0	2.078	2.817	1.356	0.990	1.108	1.022	−1469.	1171.	13.5	8.031
0.10	0.07	0.2192	0.0792	0.7017	185.9	2.083	2.811	1.349	0.988	1.104	1.023	−1475.	1163.	13.5	7.984
0.10	0.10	0.2109	0.1120	0.6772	185.8	2.092	2.802	1.340	0.986	1.097	1.024	−1485.	1150.	13.4	7.915
0.10	0.20	0.1843	0.2174	0.5983	185.5	2.119	2.772	1.308	0.979	1.077	1.030	−1517.	1110.	13.2	7.691
0.10	0.40	0.1353	0.4150	0.4497	185.2	2.173	2.707	1.246	0.971	1.044	1.049	−1578.	1037.	12.8	7.269
0.10	0.60	0.0896	0.6045	0.3058	185.0	2.224	2.638	1.186	0.971	1.020	1.078	−1637.	968.	12.4	6.860
0.10	0.80	0.0452	0.7961	0.1588	185.1	2.273	2.564	1.128	0.978	1.006	1.117	−1693.	900.	12.0	6.444
0.10	0.90	0.0229	0.8957	0.0814	185.2	2.296	2.526	1.100	0.985	1.002	1.141	−1720.	865.	11.8	6.226
0.20	0.	0.4170	0.	0.5831	182.4	2.094	2.861	1.366	0.997	1.128	1.041	−1431.	1166.	13.8	8.189
0.20	0.01	0.4126	0.0099	0.5776	182.4	2.097	2.857	1.363	0.996	1.125	1.041	−1434.	1162.	13.8	8.167
0.20	0.02	0.4082	0.0198	0.5721	182.4	2.100	2.854	1.359	0.995	1.123	1.042	−1438.	1159.	13.8	8.146
0.20	0.03	0.4038	0.0297	0.5666	182.4	2.102	2.851	1.356	0.994	1.120	1.042	−1441.	1155.	13.8	8.125
0.20	0.04	0.3994	0.0395	0.5611	182.4	2.105	2.847	1.353	0.993	1.118	1.042	−1445.	1152.	13.7	8.104
0.20	0.05	0.3951	0.0493	0.5557	182.4	2.107	2.844	1.349	0.992	1.116	1.042	−1448.	1149.	13.7	8.083
0.20	0.07	0.3864	0.0688	0.5448	182.4	2.112	2.837	1.343	0.990	1.111	1.043	−1455.	1142.	13.7	8.041
0.20	0.10	0.3735	0.0979	0.5286	182.5	2.120	2.827	1.333	0.988	1.104	1.044	−1465.	1132.	13.6	7.978
0.20	0.20	0.3316	0.1932	0.4752	182.6	2.145	2.792	1.301	0.982	1.083	1.049	−1498.	1099.	13.3	7.771
0.20	0.40	0.2506	0.3808	0.3686	183.0	2.194	2.720	1.240	0.974	1.049	1.065	−1562.	1035.	12.9	7.363
0.20	0.60	0.1706	0.5714	0.2580	183.6	2.239	2.645	1.181	0.974	1.024	1.090	−1625.	971.	12.4	6.944
0.20	0.80	0.0883	0.7742	0.1376	184.3	2.281	2.568	1.126	0.980	1.008	1.124	−1687.	904.	12.0	6.497
0.20	0.90	0.0452	0.8834	0.0715	184.8	2.301	2.528	1.099	0.986	1.003	1.145	−1717.	867.	11.8	6.255
0.40	0.	0.6560	0.	0.3441	176.3	2.124	2.859	1.346	0.992	1.146	1.090	−1372.	1121.	14.1	8.277
0.40	0.01	0.6506	0.0077	0.3417	176.4	2.127	2.856	1.343	0.991	1.143	1.090	−1376.	1119.	14.0	8.260
0.40	0.02	0.6453	0.0155	0.3393	176.4	2.129	2.853	1.340	0.991	1.140	1.090	−1380.	1117.	14.0	8.242

TABLE 3-218 Liquid-Vapor Equilibrium Data for the Argon-Nitrogen-Oxygen System* *(Concluded)*

Liquid mole fraction		Vapor mole fraction			Temperature, °R	Relative volatility			Pressure activity coefficient			Enthalpy, Btu (lb·mol)		Heat capacity, Btu/(lb·mol·°R)	
N_2/N_2+O_2	Ar	N_2	Ar	O_2		N_2/Ar	N_2/O_2	Ar/O_2	N_2	Ar	O_2	Liquid	Vapor	Liquid	Vapor
0.40	0.03	0.6400	0.0232	0.3369	176.5	2.132	2.850	1.337	0.990	1.138	1.090	−1384.	1115.	14.0	8.224
0.40	0.04	0.6347	0.0310	0.3345	176.6	2.135	2.846	1.333	0.990	1.135	1.090	−1387.	1112.	13.9	8.206
0.40	0.05	0.6293	0.0387	0.3320	176.6	2.137	2.843	1.330	0.989	1.133	1.090	−1391.	1110.	13.9	8.188
0.40	0.07	0.6186	0.0543	0.3271	176.8	2.143	2.837	1.324	0.988	1.128	1.090	−1399.	1106.	13.9	8.153
0.40	0.10	0.6025	0.0778	0.3197	177.0	2.151	2.827	1.315	0.986	1.120	1.090	−1410.	1099.	13.8	8.098
0.40	0.20	0.5482	0.1575	0.2943	177.7	2.176	2.794	1.284	0.982	1.098	1.091	−1449.	1077.	13.5	7.915
0.40	0.40	0.4348	0.3259	0.2393	179.2	2.223	2.725	1.226	0.978	1.061	1.100	−1525.	1029.	13.0	7.527
0.40	0.60	0.3104	0.5141	0.1756	180.9	2.264	2.652	1.171	0.979	1.033	1.116	−1600.	975.	12.5	7.095
0.40	0.80	0.1684	0.7334	0.0982	182.9	2.297	2.573	1.120	0.985	1.013	1.139	−1674.	910.	12.0	6.597
0.40	0.90	0.0882	0.8595	0.0523	184.1	2.309	2.531	1.096	0.989	1.006	1.153	−1710.	872.	11.8	6.312
0.60	0.	0.8076	0.	0.1924	171.7	2.120	2.798	1.320	0.986	1.173	1.154	−1296.	1086.	14.2	8.369
0.60	0.01	0.8023	0.0064	0.1913	171.8	2.123	2.796	1.317	0.986	1.170	1.153	−1301.	1084.	14.2	8.354
0.60	0.02	0.7971	0.0127	0.1902	171.9	2.127	2.794	1.314	0.985	1.167	1.152	−1305.	1083.	14.1	8.338
0.60	0.03	0.7918	0.0192	0.1891	172.0	2.130	2.792	1.311	0.985	1.164	1.152	−1310.	1082.	14.1	8.322
0.60	0.04	0.7865	0.0256	0.1879	172.1	2.133	2.790	1.308	0.985	1.161	1.151	−1315.	1080.	14.1	8.306
0.60	0.05	0.7812	0.0321	0.1868	172.2	2.137	2.788	1.305	0.984	1.159	1.150	−1319.	1079.	14.0	8.289
0.60	0.07	0.7704	0.0451	0.1845	172.4	2.143	2.784	1.299	0.984	1.153	1.149	−1329.	1076.	14.0	8.257
0.60	0.10	0.7542	0.0649	0.1810	172.6	2.153	2.778	1.290	0.983	1.144	1.148	−1343.	1072.	13.9	8.208
0.60	0.20	0.6980	0.1332	0.1688	173.7	2.184	2.757	1.262	0.981	1.119	1.143	−1389.	1056.	13.6	8.038
0.60	0.40	0.5739	0.2848	0.1413	175.9	2.239	2.708	1.209	0.980	1.076	1.141	−1482.	1021.	13.0	7.666
0.60	0.60	0.4261	0.4667	0.1073	178.5	2.283	2.648	1.160	0.984	1.043	1.145	−1573.	976.	12.5	7.228
0.60	0.80	0.2413	0.6963	0.0625	181.6	2.311	2.576	1.115	0.989	1.019	1.155	−1661.	916.	12.0	6.690
0.60	0.90	0.1293	0.8367	0.0340	183.4	2.317	2.533	1.093	0.992	1.009	1.161	−1704.	876.	11.8	6.367
0.80	0.	0.9152	0.	0.0848	168.0	2.095	2.699	1.288	0.986	1.216	1.239	−1209.	1057.	14.3	8.462
0.80	0.01	0.9102	0.0055	0.0843	168.1	2.099	2.699	1.286	0.986	1.212	1.237	−1215.	1056.	14.2	8.447
0.80	0.02	0.9052	0.0110	0.0839	168.2	2.103	2.699	1.283	0.986	1.209	1.235	−1220.	1055.	14.2	8.432
0.80	0.03	0.9001	0.0165	0.0834	168.3	2.107	2.698	1.280	0.985	1.205	1.234	−1226.	1054.	14.2	8.417
0.80	0.04	0.8950	0.0221	0.0829	168.4	2.112	2.698	1.278	0.985	1.201	1.232	−1232.	1053.	14.1	8.402
0.80	0.05	0.8899	0.0277	0.0825	168.5	2.116	2.698	1.275	0.984	1.198	1.230	−1237.	1052.	14.1	8.387
0.80	0.07	0.8795	0.0390	0.0815	168.7	2.124	2.698	1.270	0.984	1.190	1.227	−1249.	1050.	14.0	8.356
0.80	0.10	0.8638	0.0562	0.0801	169.1	2.136	2.697	1.263	0.983	1.180	1.222	−1266.	1047.	14.0	8.309
0.80	0.20	0.8087	0.1162	0.0751	170.3	2.175	2.693	1.238	0.981	1.148	1.208	−1322.	1037.	13.7	8.148
0.80	0.40	0.6826	0.2535	0.0638	173.1	2.244	2.674	1.192	0.983	1.095	1.188	−1434.	1012.	13.1	7.786
0.80	0.60	0.5231	0.4274	0.0496	176.4	2.295	2.636	1.149	0.988	1.055	1.176	−1543.	976.	12.5	7.347
0.80	0.80	0.3077	0.6624	0.0299	180.3	2.323	2.576	1.109	0.993	1.024	1.171	−1647.	920.	12.0	6.777
0.80	0.90	0.1684	0.8150	0.0166	182.7	2.325	2.535	1.090	0.994	1.012	1.169	−1698	880.	11.8	6.420
0.90	0.	0.9596	0.	0.0404	166.3	2.077	2.641	1.271	0.990	1.245	1.292	−1163.	1044.	14.3	8.509
0.90	0.01	0.9547	0.0051	0.0402	166.4	2.081	2.641	1.269	0.990	1.241	1.289	−1169.	1043.	14.3	8.494
0.90	0.02	0.9497	0.0103	0.0399	166.5	2.086	2.642	1.267	0.989	1.236	1.287	−1175.	1042.	14.2	8.479
0.90	0.03	0.9448	0.0155	0.0397	166.7	2.091	2.643	1.264	0.989	1.232	1.284	−1181.	1042.	14.2	8.464
0.90	0.04	0.9397	0.0208	0.0395	166.8	2.095	2.644	1.262	0.988	1.228	1.281	−1188.	1041.	14.2	8.449
0.90	0.05	0.9347	0.0260	0.0393	166.9	2.100	2.645	1.260	0.988	1.224	1.279	−1194.	1040.	14.1	8.434
0.90	0.07	0.9245	0.0367	0.0388	167.2	2.109	2.646	1.255	0.987	1.215	1.274	−1206.	1039.	14.1	8.404
0.90	0.10	0.9090	0.0529	0.0381	167.5	2.122	2.649	1.248	0.986	1.203	1.267	−1225.	1036.	14.0	8.358
0.90	0.20	0.8546	0.1096	0.0358	168.8	2.166	2.653	1.225	0.984	1.167	1.246	−1287.	1028.	13.7	8.198
0.90	0.40	0.7287	0.2407	0.0305	171.8	2.242	2.651	1.182	0.985	1.107	1.215	−1409.	1007.	13.1	7.840
0.90	0.60	0.5659	0.4102	0.0239	175.3	2.299	2.627	1.143	0.990	1.062	1.193	−1527.	975.	12.5	7.400
0.90	0.80	0.3387	0.6467	0.0146	179.7	2.328	2.575	1.106	0.995	1.027	1.179	−1640.	922.	12.0	6.818
0.90	0.90	0.1873	0.8045	0.0082	182.4	2.329	2.536	1.089	0.995	1.013	1.173	−1694.	882.	11.8	6.446
0.97	0.	0.9882	0.	0.0118	165.2	2.062	2.598	1.260	0.995	1.269	1.334	−1130.	1035.	14.3	8.541
0.97	0.01	0.9834	0.0050	0.0117	165.3	2.067	2.599	1.257	0.994	1.264	1.330	−1136.	1035.	14.3	8.527
0.97	0.02	0.9784	0.0099	0.0116	165.5	2.072	2.601	1.255	0.994	1.259	1.327	−1143.	1034.	14.2	8.512
0.97	0.03	0.9735	0.0149	0.0116	165.6	2.077	2.602	1.253	0.993	1.254	1.324	−1149.	1033.	14.2	8.497
0.97	0.04	0.9685	0.0200	0.0115	165.7	2.082	2.604	1.251	0.992	1.250	1.321	−1156.	1033.	14.2	8.482
0.97	0.05	0.9635	0.0250	0.0114	165.8	2.087	2.605	1.248	0.992	1.245	1.317	−1163.	1032.	14.2	8.467
0.97	0.07	0.9534	0.0353	0.0113	166.1	2.097	2.608	1.244	0.991	1.236	1.311	−1176.	1031.	14.1	8.437
0.97	0.10	0.9380	0.0509	0.0111	166.5	2.111	2.612	1.237	0.989	1.222	1.302	−1195.	1029.	14.0	8.391
0.97	0.20	0.8840	0.1056	0.0104	167.9	2.158	2.624	1.216	0.987	1.182	1.276	−1261.	1022.	13.7	8.233
0.97	0.40	0.7584	0.2327	0.0089	170.9	2.240	2.633	1.176	0.986	1.116	1.235	−1390.	1003.	13.1	7.877
0.97	0.60	0.5940	0.3990	0.0070	174.6	2.302	2.620	1.138	0.991	1.067	1.206	−1516.	974.	12.5	7.436
0.97	0.80	0.3597	0.6361	0.0043	179.3	2.332	2.574	1.104	0.996	1.030	1.185	−1635.	923.	12.0	6.846
0.97	0.90	0.2003	0.7972	0.0024	182.1	2.331	2.537	1.088	0.996	1.014	1.176	−1692.	883.	11.8	6.464
1.00	0.	1.0000	0.	0.	164.7	2.057	2.580	1.254	0.997	1.280	1.352	−1115.	1032.	14.3	8.555
1.00	0.01	0.9951	0.0049	0.0000	164.9	2.061	2.581	1.252	0.997	1.275	1.349	−1122.	1031.	14.3	8.541
1.00	0.02	0.9902	0.0098	0.0000	165.0	2.066	2.583	1.250	0.996	1.270	1.346	−1129.	1031.	14.3	8.526
1.00	0.03	0.9853	0.0147	0.0000	165.1	2.071	2.584	1.248	0.995	1.265	1.342	−1136.	1030.	14.2	8.511
1.00	0.04	0.9803	0.0197	0.0000	165.3	2.076	2.586	1.246	0.995	1.260	1.338	−1142.	1029.	14.2	8.496
1.00	0.05	0.9753	0.0247	0.0000	165.4	2.081	2.588	1.243	0.994	1.255	1.335	−1149.	1029.	14.2	8.481
1.00	0.07	0.9653	0.0347	0.0000	165.7	2.091	2.591	1.239	0.993	1.245	1.328	−1163.	1028.	14.1	8.451
1.00	0.10	0.9499	0.0501	0.0000	166.1	2.106	2.596	1.233	0.991	1.231	1.319	−1183.	1026.	14.0	8.405
1.00	0.20	0.8960	0.1040	0.0000	167.4	2.154	2.610	1.212	0.988	1.189	1.289	−1250.	1019.	13.7	8.247
1.00	0.40	0.7706	0.2294	0.0000	170.6	2.239	2.626	1.173	0.987	1.120	1.244	−1382.	1002.	13.1	7.892
1.00	0.60	0.6055	0.3945	0.	174.4	2.303	2.617	1.136	0.992	1.069	1.211	−1511.	973.	12.5	7.451
1.00	0.80	0.3684	0.6316	0.	179.1	2.333	2.574	1.103	0.996	1.030	1.188	−1633.	923.	12.0	6.858
1.00	0.90	0.2058	0.7942	0.	182.0	2.332	2.537	1.088	0.996	1.015	1.177	−1691.	884.	11.8	6.471

TABLE 3-219 Saturated Benzene*

T, K	P, bar	v_f, m³/kg	v_g, m³/kg	h_f, kJ/kg	h_g, kJ/kg	s_f, kJ/(kg·K)	s_g, kJ/(kg·K)	c_{pf}, kJ/(kg·K)	μ_f, 10^{-4} Pa·s	k_f, W/(m·K)
290	0.0860	1.133.−6	3.569.−3	371.1	810.3	2.172	3.686	1.719	6.75	0.147
300	0.1382	1.147.−6	2.292.−3	388.3	820.4	2.229	3.670	1.746	5.80	0.144
310	0.2139	1.162.−6	1.525.−3	405.9	830.8	2.286	3.657	1.774	5.14	0.141
320	0.3206	1.176.−6	1.046.−3	423.8	841.5	2.344	3.650	1.804	4.52	0.138
330	0.4665	1.192.−6	7.379.−4	442.1	852.4	2.400	3.643	1.836	3.95	0.135
340	0.6615	1.207.−6	5.332.−4	460.8	863.6	2.455	3.641	1.868	3.55	0.132
350	0.9162	1.224.−6	3.938.−4	479.6	875.0	2.510	3.641	1.890	3.23	0.129
360	1.2419	1.241.−6	2.965.−4	498.7	886.7	2.564	3.642	1.920	2.99	0.126
370	1.6517	1.259.−6	2.233.−4	518.1	898.6	2.617	3.646	1.950	2.72	0.123
380	2.1588	1.277.−6	1.767.−4	537.7	910.6	2.669	3.651	1.989	2.46	0.120
390	2.7774	1.297.−6	1.393.−4	557.6	922.9	2.592	3.657	2.030	2.24	0.117
400	3.5228	1.318.−6	1.112.−4	577.9	935.2	2.644	3.665	2.070	2.05	0.114
410	4.4091	1.340.−6	8.972.−5	598.6	947.8	2.823	3.674	2.110	1.88	0.111
420	5.4540	1.363.−6	7.309.−5	619.7	960.4	2.873	3.684	2.160	1.73	0.107
430	6.6739	1.388.−6	6.003.−5	641.3	973.0	2.924	3.695	2.210	1.60	0.104
440	8.0861	1.415.−6	4.965.−5	663.5	985.6	2.974	3.706	2.260	1.48	0.101
450	9.7088	1.444.−6	4.131.−5	686.3	998.2	3.025	3.718	2.320	1.37	0.098
460	11.451	1.475.−6	3.455.−5	709.7	1010.7	3.075	3.730	2.380	1.28	0.095
470	13.660	1.510.−6	2.901.−5	733.8	1022.9	3.126	3.742	2.450	1.10	0.092
480	16.028	1.548.−6	2.441.−5	758.6	1034.9	3.179	3.753	2.519	1.12	0.089
490	18.685	1.591.−6	2.059.−5	784.3	1046.4	3.230	3.765	2.590	1.05	0.086
500	21.651	1.640.−6	1.736.−5	810.9	1057.3	3.284	3.777	2.670	0.98	0.083
510	24.952	1.697.−6	1.462.−5	838.5	1067.5	3.336	3.785	2.750	0.91	
520	28.613	1.765.−6	1.226.−5	867.2	1076.6	3.391	3.794	2.839	0.84	
530	32.669	1.849.−6	1.020.−5	897.2	1084.3	3.446	3.800	2.941	0.77	
540	37.161	2.126.−6	8.349.−6	928.8	1089.5	3.504	3.802	....	0.70	
550	42.144	2.258.−6	6.616.−6	963.2	1090.4	3.565	3.797	....	0.65	
560	47.696	2.512.−6	4.696.−6	1007.3	1077.6	3.642	3.769	....	0.60	
562.2	48.979	3.290.−6	3.290.−6	1043.0	1043.0	3.706	3.706			

*Converted from a tabulation by Counsell, Lawrenson, and Lees, Nat. Phys. Lab. Teddington (U.K.) Rep. Chem. 52, 1976. Another tabulation by Kesselman et al., in Vargaftik (ed.), *Tables on the Thermophysical Properties of Liquids and Gases*, Hemisphere, Washington and London, 1975, shows some differences. The notation 1.133.−6 signifies 1.133×10^{-6}.

TABLE 3-220 Saturated Bromine*

T, K	P, bar	v_f, m³/kg	v_g, m³/kg	h_f, kJ/kg	h_g, kJ/kg	s_f, kJ/(kg·K)	s_g, kJ/(kg·K)	c_{pf}, kJ/(kg·K)	μ_f, 10^{-4} Pa·s	k_f, W/(m·K)
260	0.042	3.106.−4	3.195	−147.2	51.8	0.903	1.669	0.486	13.4	0.131
280	0.124	3.168.−4	1.169	−138.9	56.2	0.933	1.629	0.479	11.5	0.127
300	0.310	3.232.−4	0.5002	−131.6	60.6	0.956	1.597	0.475	9.3	0.122
320	0.680	3.311.−4	0.2425	−124.2	64.8	0.978	1.570	0.473	7.8	0.118
340	1.330	3.385.−4	0.1309	−112.3	71.1	1.004	1.539	0.471	6.7	0.114
360	2.384	3.464.−4	0.0767	−108.6	73.1	1.026	1.531	0.470	5.7	0.109
380	4.010	3.550.−4	0.0477	−100.6	76.9	1.048	1.515	0.471	5.0	0.104
400	6.390	3.647.−4	0.0311	− 93.4	80.6	1.063	1.501	0.475	4.5	0.099
420	9.730	3.752.−4	0.0211	− 85.8	84.0	1.084	1.488	0.480	4.0	0.094
440	14.25	3.885.−4	0.0148	− 77.7	87.1	1.103	1.477	0.489	3.7	0.089
460	20.17	4.023.−4	0.0107	− 69.0	89.9	1.122	1.467	0.503	3.3	0.084
480	27.75	4.179.−4	0.00786	− 59.7	92.2	1.142	1.457	0.527	3.1	0.079
500	37.21	4.378.−4	0.00589	− 49.3	94.0	1.161	1.448	0.595	2.8	0.073
520	48.81	4.623.−4	0.00445	− 37.7	95.0	1.183	1.438	0.710	2.6	0.066
540	62.80	4.938.−4	0.00337	− 24.0	94.8	1.207	1.428	0.860	2.5	0.059
560	79.41	5.368.−4	0.00251	− 7.1	92.5	1.237	1.414	1.063	2.3	0.050
580	98.90	6.250.−4	0.00167	18.8	82.5	1.280	1.390	2.31	2.2	0.035
584.2[c]	103.4	8.475.−4	0.00085	64.8	64.8	1.356	1.356	∞	2.1	∞

*Reproduced or converted from a tabulation by Seshadri, Viswanath, and Kuloor, *Ind. J. Technol.*, **6**, 191–198 (1970). c = critical point.

TABLE 3-221 Saturated 1,3-Butadiene*

Temp., °F.	Abs. pressure, lb/sq. in.	Volume, cu. ft./lb.		Enthalpy, B.t.u./lb.		Entropy, B.t.u./(lb.)(°R)		Temp., °F.	Abs. pressure, lb/sq. in.	Volume, cu ft./lb.		Enthalpy, B.t.u./lb.		Entropy, B.t.u./(lb.)(°R)	
		Liquid	Vapor	Liquid	Vapor	Liquid	Vapor			Liquid	Vapor	Liquid	Vapor	Liquid	Vapor
t	p	v_f	v_g	h_f	h_g	s_f	s_g	t	p	v_f	v_g	h_f	h_g	s_f	s_g
−164.05	0.010	0.02097	5706.	122.61	341.8	0.5904	1.3317	10	10.728	0.02429	8.441	204.97	386.7	.8092	1.1962
−160	.013	.02104	4504.	124.44	342.7	.5973	1.3256	20	13.45	.02453	6.840	210.05	389.6	.8199	1.1942
−140	.045	.02136	1406.	133.55	347.3	.6267	1.2953	30	16.68	.02478	5.595	215.19	392.4	.8305	1.1925
−120	.130	.02170	516.5	142.72	352.0	.6546	1.2707	40	20.49	.02503	4.617	220.40	395.3	.8410	1.1910
−100	.329	.02205	216.7	151.96	356.9	.6810	1.2509	50	24.94	.02529	3.840	225.66	398.2	.8514	1.1899
− 90	.500	.02224	146.4	156.61	359.5	.6938	1.2425	60	30.11	.02557	3.218	231.00	401.1	.8617	1.1890
− 80	.740	.02242	101.44	161.29	362.0	.7062	1.2350	70	36.05	.02585	2.715	236.40	404.0	.8719	1.1883
− 70	1.071	.02261	71.88	165.99	364.7	.7184	1.2283	80	42.84	.02614	2.305	241.88	406.8	.8821	1.1878
− 60	1.516	.02280	52.00	170.72	367.3	.7304	1.2223	90	50.57	.02645	1.968	247.43	409.7	.8922	1.1874
− 50	2.103	.02300	38.33	175.49	370.0	.7422	1.2170	100	59.30	.02678	1.689	253.0	412.5	.9023	1.1872
− 40	2.867	.02320	28.75	180.29	372.7	.7538	1.2123	120	80.11	.02747	1.262	264.6	418.2	.9223	1.1873
− 30	3.841	.02341	21.91	185.14	375.5	.7652	1.2081	140	105.93	.02823	0.9576	276.4	423.6	.9422	1.1877
− 20	5.068	.02362	16.94	190.02	378.2	.7764	1.2045	160	137.4	.02909	.7362	288.6	428.9	.9620	1.1883
− 10	6.592	.02384	13.27	194.96	381.0	.7875	1.2013	180	175.4	.03007	.5715	301.3	433.9	.9817	1.1891
0	8.461	.02406	10.525	199.94	383.9	.7984	1.1985	200	220.5	.03121	.4465	315.	439.	1.001	1.190

* C. H. Meyers, C. S. Cragoe, and E. F. Mueller, *J. Research N. B. S.* **39**, 507 (1947).

TABLE 3-222 Saturated Normal Butane (R600)*

T, K	P, bar	v_f, m³/kg	v_g, m³/kg	h_f, kJ/kg	h_g, kJ/kg	s_f, kJ/(kg·K)	s_g, kJ/(kg·K)	c_{pf}, kJ/(kg·K)	μ_f, 10^{-4} Pa·s	k_f, W/(m·K)
134.9[t]	6.7.−6	1.360.−3	28630	0.00	494.21	2.3056	5.9702	1.946	15.8	0.181
140	1.7.−5	1.369.−3	11635	9.95	499.96	2.3778	5.8779	1.953	14.4	0.179
150	8.7.−5	1.387.−3	2470	29.44	511.39	2.5121	5.7251	1.970	12.0	0.175
160	3.5.−4	1.405.−3	654	49.10	523.13	2.6389	5.6016	1.985	9.94	0.171
170	1.17.−3	1.424.−3	207	68.94	535.16	2.7592	5.5017	2.001	8.26	0.167
180	3.37.−3	1.443.−3	76.4	88.97	547.48	2.8738	5.4211	2.018	6.87	0.163
190	8.53.−3	1.463.−3	31.8	109.22	560.07	2.9835	5.3564	2.035	5.71	0.160
200	1.94.−2	1.484.−3	14.7	129.71	572.93	3.0887	5.3048	2.055	4.83	0.156
210	4.05.−2	1.505.−3	7.39	150.45	586.06	3.1900	5.2643	2.077	4.15	0.152
220	7.81.−2	1.528.−3	4.00	171.49	599.42	3.2879	5.2331	2.101	3.61	0.148
230	0.1411	1.551.−3	2.31	192.83	613.02	3.3828	5.2097	2.128	3.18	0.144
240	0.2408	1.575.−3	1.40	214.50	626.83	3.4749	5.1929	2.158	2.83	0.140
250	0.3915	1.601.−3	0.893	236.52	640.82	3.5647	5.1818	2.192	2.55	0.136
260	0.6100	1.628.−3	0.592	258.92	654.97	3.6523	5.1755	2.231	2.31	0.132
270	0.9155	1.656.−3	0.406	281.72	669.24	3.7380	5.1732	2.274	2.10	0.128
280	1.3297	1.686.−3	0.286	309.94	683.60	3.8220	5.1744	2.323	1.93	0.124
290	1.8765	1.718.−3	0.207	328.62	697.99	3.9046	5.1783	2.377	1.77	0.120
300	2.5811	1.752.−3	0.1533	352.77	712.36	3.9860	5.1846	2.437	1.62	0.116
310	3.4706	1.790.−3	0.1156	377.46	726.67	4.0663	5.1928	2.503	1.47	0.113
320	4.5731	1.830.−3	0.0885	402.71	740.84	4.1458	5.2025	2.577	1.34	0.109
330	5.9179	1.874.−3	0.0687	428.61	754.80	4.2248	5.2132	2.657	1.21	0.105
340	7.5354	1.923.−3	0.0539	455.25	768.49	4.3035	5.2248	2.746	1.08	0.101
350	9.4573	1.978.−3	0.0427	482.74	781.79	4.3822	5.2367	2.842	0.97	0.097
360	11.72	2.041.−3	0.0340	511.22	794.60	4.4613	5.2485	2.947	0.87	0.093
370	14.35	2.114.−3	0.0272	540.88	806.72	4.5412	5.2597	3.062	0.78	0.089
380	17.40	2.200.−3	0.0218	571.94	817.86	4.6225	5.2696	3.20	0.69	0.085
390	20.90	2.307.−3	0.0174	604.76	827.56	4.7058	5.2771	3.34	0.62	0.081
400	24.92	2.447.−3	0.0138	639.85	834.95	4.7922	5.2800	3.50	0.55	0.077
410	29.54	2.652.−3	0.0106	678.30	838.10	4.8842	5.2740	3.69	0.49	0.074
420	34.86	3.048.−3	0.0075	723.89	830.34	4.9903	5.2437	3.84	0.44	0.072
425.2[c]	37.96	4.405.−3	0.0044	783.50	783.50	5.1290	5.1290	∞		∞

*Values rounded and reproduced or converted from Goodwin, NBSIR 79-1621, 1979. t = triple point; c = critical point. The notation 6.7.−6 signifies 6.7×10^{-6}.

TABLE 3-223 Superheated Normal Butane*

P, bar		Temperature, K 150	200	250	300	350	400	450	500	600	700
	v	0.00139	0.00148	0.00160	0.4106	0.4847	0.5575	0.6297	0.7013	0.8440	0.9861
1.013	*h*	29.6	129.8	236.6	718.9	810.7	913.1	1026.0	1149.0	1423	1730
	s	2.512	3.088	3.564	5.334	5.616	5.889	6.155	6.414	6.913	7.386
	v	0.00139	0.00148	0.00160	0.00175	0.0909	0.1078	0.1238	0.1393	0.1693	0.1988
5	*h*	30.0	130.2	237.0	352.9	798.5	904.3	1019.3	1143.7	1420	1728
	s	2.511	3.088	3.563	3.985	5.363	5.645	5.916	6.178	6.680	7.155
	v	0.00139	0.00148	0.00160	0.00175	0.00198	0.0502	0.0593	0.0677	0.0835	0.0987
10	*h*	30.6	130.8	237.4	353.3	482.7	891.9	1010.3	1136.8	1415	1725
	s	2.510	3.087	3.562	3.983	4.382	5.524	5.803	6.069	6.575	7.052
	v	0.00138	0.00148	0.00160	0.00174	0.00196	0.0205	0.0268	0.0318	0.0406	0.0487
20	*h*	31.7	131.8	238.4	354.0	482.6	860.0	990.1	1122.0	1406	1718
	s	2.509	3.085	3.560	3.980	4.376	5.364	5.670	5.948	6.464	6.945
	v	0.00138	0.00148	0.00159	0.00174	0.00195	0.00240	0.0156	0.0198	0.0263	0.0320
30	*h*	32.8	132.9	239.3	354.7	482.6	637.3	965.5	1105.9	1396	1711
	s	2.507	3.082	3.557	3.976	4.370	4.783	5.570	5.866	6.394	6.880
	v	0.00138	0.00148	0.00159	0.00173	0.00194	0.00234	0.0097	0.0137	0.0192	0.0237
40	*h*	33.9	134.0	240.3	355.4	482.7	633.6	932.2	1088.1	1387	1705
	s	2.505	3.080	3.555	3.973	4.365	4.768	5.468	5.797	6.341	6.832
	v	0.00138	0.00148	0.00159	0.00173	0.00193	0.00229	0.00549	0.0101	0.0149	0.0188
50	*h*	35.0	135.0	241.3	356.2	428.8	631.0	877.0	1068.2	1377	1699
	s	2.503	3.078	3.552	3.970	4.360	4.755	5.329	5.734	6.297	6.792
	v	0.00138	0.00148	0.00159	0.00172	0.00192	0.00255	0.00352	0.00764	0.0121	0.0155
60	*h*	36.2	136.1	242.3	356.9	483.1	629.1	825.1	1046.4	1367	1692
	s	2.501	3.076	3.550	3.967	4.355	4.745	5.204	5.673	6.258	6.759
	v	0.00138	0.00147	0.00158	0.00172	0.00190	0.00219	0.00286	0.00482	0.00868	0.0114
80	*h*	38.4	138.3	244.2	358.5	483.7	626.5	798.1	1001.5	1347	1680
	s	2.498	3.072	3.545	3.960	4.346	4.727	5.130	5.559	6.191	6.704
	v	0.00138	0.00147	0.00158	0.00171	0.00188	0.00214	0.00264	0.00368	0.00669	0.00901
100	*h*	40.6	140.4	246.2	360.1	484.5	624.9	787.9	971.3	1329	1668
	s	2.495	3.069	3.540	3.954	4.337	4.712	5.095	5.310	6.134	6.658
	v	0.00137	0.00146	0.00156	0.00167	0.00178	0.00200	0.00225	0.00258	0.00349	0.00460
200	*h*	51.9	151.3	257.9	368.8	490.3	624.4	773.3	933.7	1270	1623
	s	2.478	3.049	3.518	3.927	4.301	4.660	5.010	5.348	5.960	6.849
	v	0.00136	0.00145	0.00154	0.00164	0.00176	0.00191	0.00209	0.00231	0.00284	0.00345
300	*h*	63.2	162.2	266.7	378.3	498.0	629.2	773.4	928.4	1255	1603
	s	2.462	3.032	3.498	3.903	4.273	4.623	4.962	5.288	5.884	6.419
	v	0.00136	0.00144	0.00152	0.00162	0.00173	0.00185	0.00200	0.00217	0.00255	0.00298
400	*h*	74.5	173.3	277.4	388.2	506.8	636.2	778.0	930.2	1253	1600
	s	2.447	3.015	3.479	3.882	4.248	4.593	4.927	5.247	5.836	6.366
	v	0.00136	0.00143	0.00151	0.00160	0.00170	0.00181	0.00193	0.00207	0.00240	0.00272
500	*h*	85.8	184.4	288.1	398.4	516.3	644.5	784.8	935.3	1256	1599
	s	2.432	2.999	3.461	3.863	4.226	4.569	4.898	5.215	5.799	6.328

*Converted and rounded from tables of Goodwin, NBSIR 79-1621, 1979.

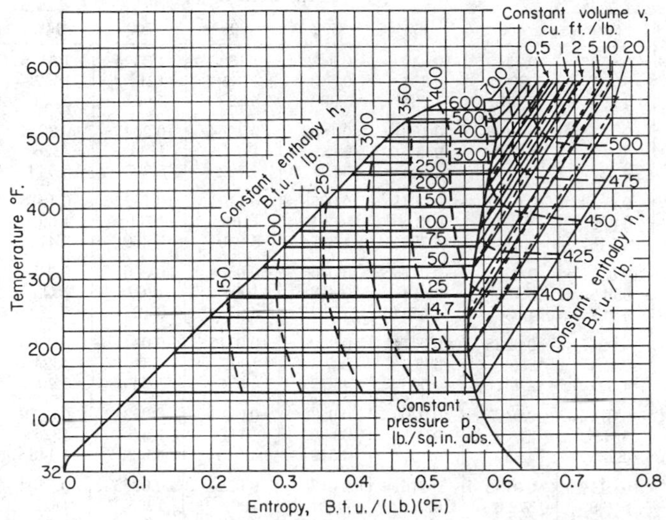

FIG. 3-19 Temperature-entropy diagram for *n*-butanol. (*By L. W. Shemilt*, Proc. J. Conf. Thermodynam. Transp. Prop. Fluids, *London, 1958.*)

TABLE 3-224 Saturated Carbon Dioxide*

T, K	P, bar	v_f, m³/kg	v_g, m³/kg	h_f, kJ/kg	h_g, kJ/kg	s_f, kJ/(kg·K)	s_g, kJ/(kg·K)	c_{pf}, kJ/(kg·K)	μ_f, 10^{-4} Pa·s	k_f, W/(m·K)
216.6	5.180	8.484.−4	0.0712	386.3	731.5	2.656	4.250	1.707	. . .	0.182
220	5.996	8.574.−4	0.0624	392.6	733.1	2.684	4.232	1.761	. . .	0.178
225	7.357	8.710.−4	0.0515	401.8	735.1	2.723	4.204	. . . .	. . .	0.171
230	8.935	8.856.−4	0.0428	411.1	736.7	2.763	4.178	1.879	1.64	0.164
235	10.75	9.011.−4	0.0357	420.5	737.9	2.802	4.152	. . . .	. . .	0.160
240	12.83	9.178.−4	0.0300	430.2	738.9	2.842	4.128	1.933	1.45	0.156
245	15.19	9.358.−4	0.0253	440.1	739.4	2.882	4.103	. . . .	. . .	0.148
250	17.86	9.554.−4	0.0214	450.3	739.6	2.923	4.079	1.992	1.28	0.140
255	20.85	9.768.−4	0.0182	460.8	739.4	2.964	4.056	. . . .	. . .	0.134
260	24.19	1.000.−3	0.0155	471.6	738.7	3.005	4.032	2.125	1.14	0.128
270	32.03	1.056.−3	0.0113	494.4	735.6	3.089	3.981	2.410	1.02	0.116
275	36.59	1.091.−3	0.0097	506.5	732.8	3.132	3.954	. . . .	. . .	0.109
280	41.60	1.130.−3	0.0082	519.2	729.1	3.176	3.925	2.887	0.91	0.102
290	53.15	1.241.−3	0.0058	547.6	716.9	3.271	3.854	3.724	0.79	0.088
300	67.10	1.470.−3	0.0037	585.4	690.2	3.393	3.742	. . . .	0.60	0.074
304.2[c]	73.83	2.145.−3	0.0021	636.6	636.6	3.558	3.558	∞	0.31	∞

°c = critical point. The notation 8.484.−4 signifies 8.484 × 10^{-4}.

TABLE 3-225 Superheated Carbon Dioxide*

P, bar		Temperature, K									
		300	350	400	450	500	600	700	800	900	1000
1	*v*	0.5639	0.6595	0.7543	0.8494	0.9439	1.1333	1.3324	1.5115	1.7005	1.8894
	h	809.3	853.1	899.1	947.1	997.0	1102	1212	1327	1445	1567
	s	4.860	4.996	5.118	5.231	5.337	5.527	5.697	5.850	5.990	6.120
5	*v*	0.1106	0.1304	0.1498	0.1691	0.1882	0.2264	0.2645	0.3024	0.3403	0.3782
	h	805.5	850.3	897.0	945.5	995.8	1101	1211	1326	1445	1567
	s	4.548	4.686	4.810	4.925	50.31	5.222	5.392	5.546	5.685	5.814
10	*v*	0.0539	0.0642	0.0742	0.0841	0.0938	0.1131	0.1322	0.1513	0.1703	0.1893
	h	800.7	846.9	894.4	943.5	994.1	1100	1211	1326	1445	1567
	s	4.405	4.548	4.674	4.790	4.897	5.089	5.260	5.414	5.555	5.683
20	*v*	0.0255	0.0311	0.0364	0.0416	0.0466	0.0564	0.0661	0.0757	0.0853	0.0948
	h	790.2	839.8	889.3	939.4	990.8	1098	1209	1325	1444	1567
	s	4.249	4.402	4.534	4.653	4.762	4.955	5.127	5.282	5.423	5.551
30	*v*	0.0159	0.0201	0.0238	0.0274	0.0309	0.0375	0.0441	0.0505	0.0570	0.0633
	h	778.5	832.4	883.8	935.2	987.3	1096	1208	1324	1444	1566
	s	4.144	4.341	4.447	4.569	4.679	4.876	5.049	5.204	5.346	5.474
40	*v*	0.0110	0.0146	0.0175	0.0203	0.0230	0.0281	0.0331	0.0379	0.0428	0.0476
	h	764.9	824.6	878.3	931.1	984.3	1094	1205	1323	1443	1566
	s	4.055	4.239	4.380	4.507	4.619	4.818	4.993	5.148	5.291	5.419
50	*v*	0.0080	0.0112	0.0138	0.0161	0.0183	0.0224	0.0265	0.0304	0.0343	0.0382
	h	748.2	816.3	872.6	926.9	981.1	1091	1205	1322	1443	1566
	s	3.968	4.179	4.330	4.457	4.572	4.773	4.948	5.104	5.247	5.377
60	*v*	0.0058	0.0090	0.0113	0.0133	0.0151	0.0187	0.0221	0.0254	0.0286	0.0318
	h	726.9	807.7	866.9	922.7	977.8	1089	1204	1321	1442	1565
	s	3.878	4.126	4.314	4.416	4.532	4.736	4.912	5.069	5.212	5.341
80	*v*	.....	0.0062	0.0081	0.0097	0.0112	0.0140	0.0166	0.0191	0.0216	0.0240
	h	.....	788.4	855.1	914.2	971.3	1085	1201	1320	1441	1565
	s	.....	4.029	4.208	4.347	4.468	4.675	4.854	5.011	5.155	5.286
100	*v*	.....	0.0045	0.0062	0.0076	0.0089	0.0111	0.0133	0.0153	0.0173	0.0193
	h	.....	766.2	843.0	905.7	964.9	1081	1198	1318	1440	1564
	s	.....	3.936	4.144	4.290	4.417	4.627	4.808	4.967	5.111	5.241
150	*v*	.....	0.0023	0.0038	0.0049	0.0058	0.0074	0.0089	0.0103	0.0117	0.0130
	h	.....	704.5	811.9	884.8	949.4	1072	1192	1314	1437	1562
	s	.....	3.716	4.005	4.177	4.313	4.536	4.722	4.884	5.030	5.162
200	*v*	.....	0.0017	0.0027	0.0035	0.0043	0.0056	0.0067	0.0078	0.0088	0.0099
	h	.....	670.0	783.2	865.2	934.9	1063	1186	1310	1435	1561
	s	.....	3.591	3.894	4.088	4.234	4.468	4.668	4.824	4.970	5.104
300	*v*	.....	.....	0.0018	0.0023	0.0029	0.0038	0.0046	0.0053	0.0060	0.0067
	h	.....	.....	745.3	834.0	910.6	1047	1176	1303	1431	1559
	s	.....	.....	3.747	3.956	4.118	4.367	4.573	4.743	4.886	5.021
400	*v*	.....	.....	0.0015	0.0018	0.0022	0.0029	0.0035	0.0041	0.0047	0.0052
	h	.....	.....	728.1	814.6	893.3	1035	1168	1298	1428	1558
	s	.....	.....	3.663	3.867	4.033	4.292	4.497	4.671	4.824	4.960
500	*v*	.....	.....	.....	0.0016	0.0018	0.0024	0.0029	0.0034	0.0038	0.0043
	h	.....	.....	.....	803.5	881.9	1027	1162	1294	1426	1557
	s	.....	.....	.....	3.805	3.970	4.234	4.443	4.620	4.774	4.913

*Interpolated and rounded from Vukalovich and Altunin, *Thermophysical Properties of Carbon Dioxide*, Atomizdat, Moscow, 1965; and Collett, England, 1968.

TABLE 3-226 Saturated Carbon Monoxide*

T, K	P, bar	v_f, m^3/kg	v_g, m^3/kg	h_f, kJ/kg	h_g, kJ/kg	s_f, kJ/(kg·K)	s_g, kJ/(kg·K)
81.62	1.01	1.268.−3	0.0666	150.25	365.30	3.005	5.640
83.36	1.52	1.295.−3	0.0631	158.56	368.07	3.104	5.559
88.25	2.03	1.317.−3	0.0606	165.00	370.00	3.178	5.501
96.16	4.05	1.385.−3	0.0547	182.76	374.21	3.368	5.359
101.51	6.08	1.440.−3	0.0513	195.0	375.98	3.489	5.271
105.69	9.12	1.489.−3	0.0318	204.8	376.6	3.580	5.206
109.17	10.13	1.535.−3	0.0253	213.2	376.6	3.656	5.152
116.08	15.20	1.651.−3	0.0163	231.0	374.5	3.807	5.043
121.48	20.27	1.778.−3	0.0116	246.3	370.2	3.918	4.948
125.97	25.33	1.936.−3	0.0085	261.2	363.6	4.041	4.854
129.84	30.40	2.168.−3	0.0063	277.6	313.15	4.161	4.747
132.91^c	34.96	3.337.−3	0.0033				

*Pressure and volume values converted, and enthalpy and entropy values reproduced, from Hust and Stewart, NBS Tech. Note 202, 1963. This source gives values at and above 72.373 K at closer pressure intervals. c = critical point. The notation 1.268.−3 signifies 1.268 × 10^{-3}.

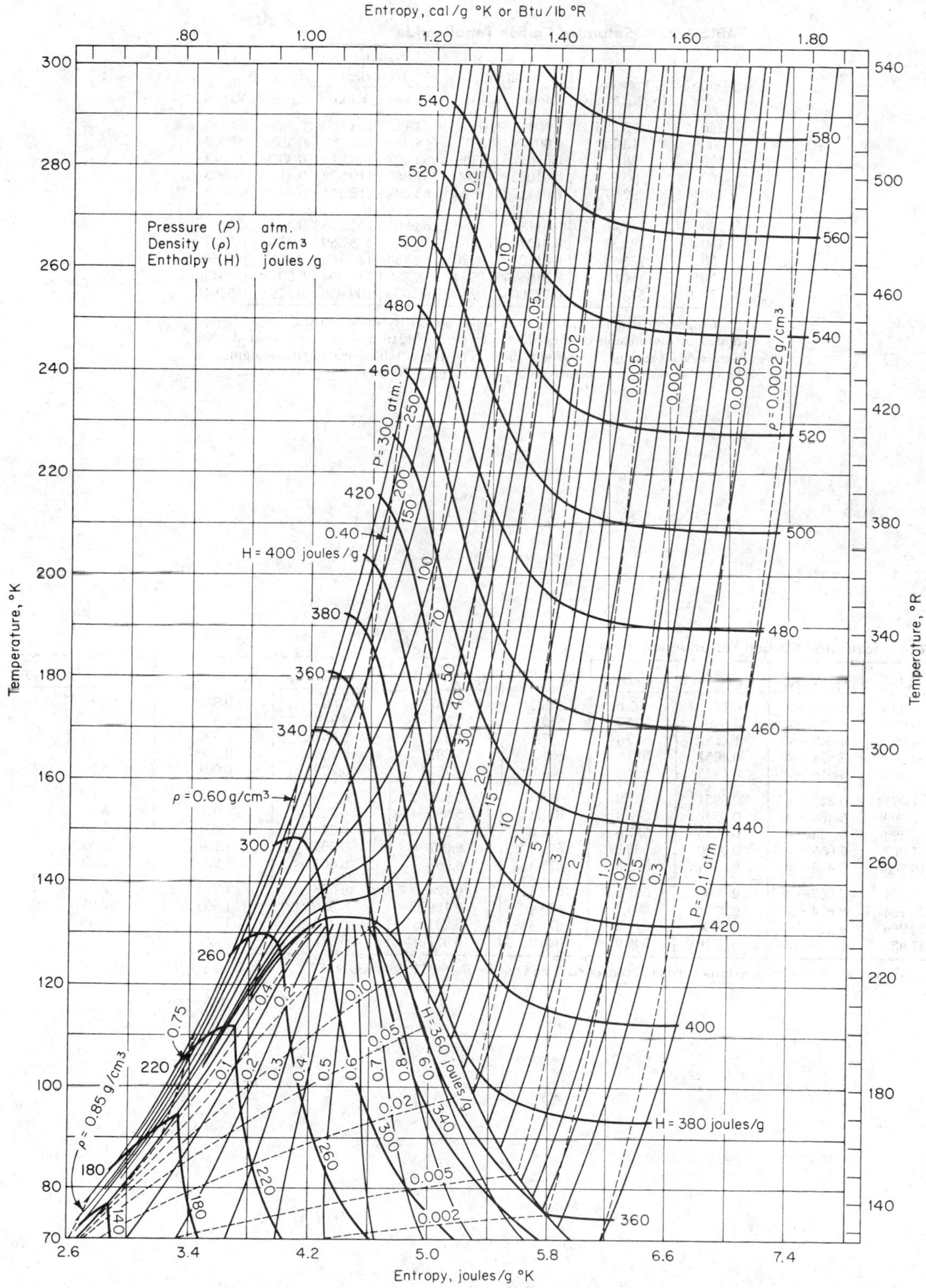

FIG. 3-20 Temperature-entropy diagram for carbon monoxide. Pressure P, in atmospheres; density ρ, in grams per cubic centimeter; enthalpy H, in joules per gram. *(From Hust and Stewart, NBS Tech. Note 202, 1963.)*

TABLE 3-227 Saturated Carbon Tetrafluoride*

Temp., °F.	Pressure, lb./sq. in. abs.	Volume, cu. ft./lb.		Enthalpy, B.t.u./lb.		Entropy, B.t.u./(lb.)(°F.)	
		Liquid	Vapor	Liquid	Vapor	Liquid	Vapor
−220	5.726	0.00943	5.023	44.112	105.150	0.3638	0.6185
−200	13.748	0.00978	2.231	48.394	107.083	0.3809	0.6069
−180	28.69	0.01017	1.1232	53.007	108.920	0.3979	0.5978
−160	53.67	0.01062	0.6209	57.952	110.592	0.4148	0.5905
−140	92.07	0.01117	0.3674	63.224	112.014	0.4316	0.5842
-120	147.49	0.01184	0.2280	68.801	113.075	0.4482	0.5785
-100	223.81	0.01273	0.1451	74.715	113.589	0.4646	0.5727
-80	325.74	0.01403	0.0920	81.154	113.163	0.4813	0.5656
-60	460.61	0.01658	0.0534	89.236	110.394	0.5011	0.5540
-50.19	543.16	0.02560	0.0256	100.636	100.636	0.5284	0.5284

*Condensed from Chari, Ph.D. Thesis, University of Michigan, 1960. Reproduced by permission of the University of Michigan and E. I. du Pont de Nemours & Company. Copyright 1961. The thesis tables are for more significant figures at 1°F. increments from −270° to −51°F.

TABLE 3-228 Saturated Carbon Tetrafluoride (R14)*

T, K	P, bar	v_f, m³/kg	v_g, m³/kg	h_f, kJ/kg	h_g, kJ/kg	s_f, kJ/(kg·K)	s_g, kJ/(kg·K)	c_{pf}, kJ/(kg·K)	μ_f, 10^{-4} Pa·s	k_f, W/(m·K)
100	0.0089	5.370.−4	10.77	495.8	648.4	5.487	7.003	0.887	. . .	0.136
110	0.0286	5.515.−4	3.648	502.7	652.9	5.556	6.919	0.887	. . .	0.128
120	0.0924	5.668.−4	1.228	510.4	657.1	5.624	6.847	0.890	. . .	0.119
130	0.2986	5.834.−4	0.4051	518.8	661.1	5.691	6.786	0.896	. . .	0.111
140	0.6901	6.018.−4	0.1855	527.7	664.8	5.757	6.736	0.904	3.56	0.104
150	1.4074	6.225.−4	0.0951	537.2	668.3	5.822	6.696	0.922	3.28	0.097
160	2.598	6.460.−4	0.0532	549.4	671.4	5.885	6.662	0.975	3.03	0.089
170	4.426	6.733.−4	0.0318	557.6	674.0	5.947	6.629	1.031	2.80	0.081
180	7.067	7.055.−4	0.0200	568.2	676.1	6.007	6.607	1.104	2.59	0.072
190	10.702	7.449.−4	0.0131	579.3	677.4	6.066	6.583	1.203	2.39	0.064
200	15.531	7.957.−4	0.0087	591.0	677.8	6.124	6.558	1.334	2.19	0.057
210	21.794	8.674.−4	0.0058	603.5	676.4	6.182	6.536	1.506	2.01	0.049
220	29.269	9.931.−4	0.0036	618.5	671.4	6.233	6.490	1.73	1.85	0.042
227.5[c]	37.45	1.598.−3	0.0016	646.9	646.9	6.371	6.371	∞	. . .	∞

*P, v, h, and s values interpolated, extrapolated, and converted from Oguchi, *Reito*, **52**, 869–889, 1977. c = critical point. The notation 5.370.−4 signifies 5.370×10^{-4}.

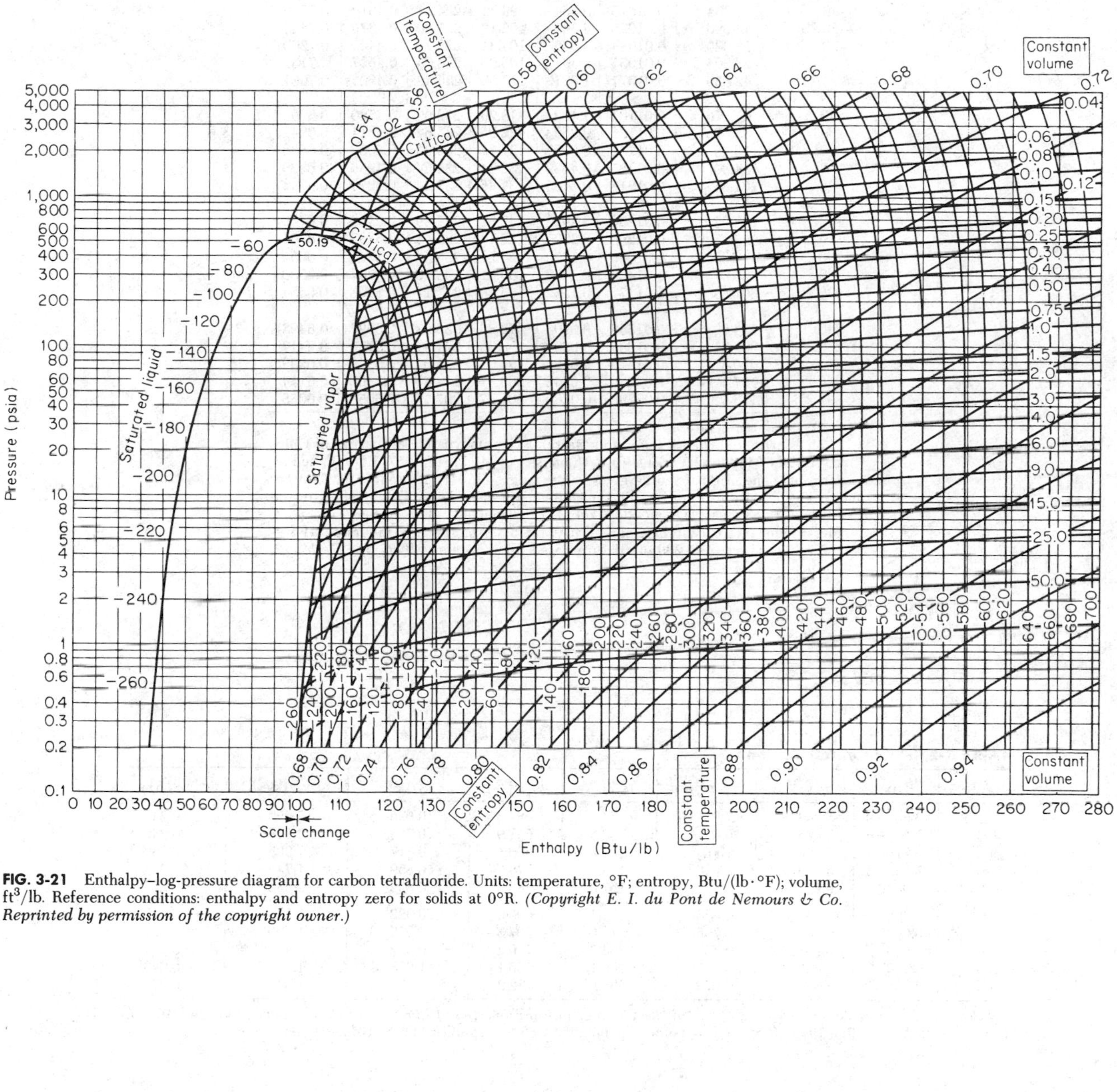

FIG. 3-21 Enthalpy–log-pressure diagram for carbon tetrafluoride. Units: temperature, °F; entropy, Btu/(lb · °F); volume, ft^3/lb. Reference conditions: enthalpy and entropy zero for solids at 0°R. *(Copyright E. I. du Pont de Nemours & Co. Reprinted by permission of the copyright owner.)*

TABLE 3-229 Chlorine*

Temp., °F.	Pressure, lb./sq. in. abs.	Volume, cu. ft./lb. Liquid	Volume, cu. ft./lb. Vapor	Enthalpy, B.t.u./lb. Liquid	Enthalpy, B.t.u./lb. Vapor	Entropy B.t.u./(lb.)(°F.) Liquid	Entropy B.t.u./(lb.)(°F.) Vapor
−40	11.234	0.01015	5.564	99.72	224.81	0.4318	0.7298
−29.29	14.696	0.01025	4.346	102.19	225.86	0.4376	0.7249
−20	18.330	0.01034	3.546	104.34	226.76	0.4425	0.7209
0	28.504	0.01055	2.361	108.96	228.63	0.4527	0.7131
20	42.539	0.01077	1.631	113.59	230.42	0.4625	0.7061
40	61.276	0.01101	1.163	118.26	232.09	0.4720	0.6998
60	85.606	0.01126	0.8503	122.96	233.64	0.4811	0.6941
80	116.46	0.01154	0.6357	127.71	235.03	0.4900	0.6888
100	154.80	0.01185	0.4839	132.54	236.24	0.4986	0.6839
120	201.65	0.01218	0.3739	137.46	237.24	0.5070	0.6792
140	258.03	0.01256	0.2923	142.51	237.99	0.5154	0.6746
160	325.05	0.01298	0.2305	147.74	238.45	0.5237	0.6701
180	403.87	0.01346	0.1829	153.21	238.54	0.5321	0.6655
200	495.67	0.01402	0.1453	159.01	238.16	0.5407	0.6606
220	601.83	0.01471	0.1152	165.29	237.17	0.5496	0.6554
240	723.69	0.01559	0.0903	172.26	235.28	0.5592	0.6493
260	862.81	0.01683	0.0690	180.38	231.91	0.5700	0.6417
280	1020.8	0.01910	0.0486	191.04	225.03	0.5840	0.6299
290	1107.5	0.02286	0.0342	200.81	214.87	0.5966	0.6154
291.21[c]	1118.4	0.02796	0.0280	207.77	207.77	0.6058	0.6058

[c] = critical temperature.

*Values extracted from Kapoor and Martin, "Thermodynamic Properties of Chlorine," Engineering Research Institute, University of Michigan, 1957. Reproduced by permission. The original source gives values from −130°F. at usually 10°F. increments and usually to more significant figures than cited here. Other tables in the "Chlorine Handbook," Diamond Alkali Co., Cleveland, Ohio, differ somewhat. Values in the metric system from −70°C. at 5°C. increments in Landolt-Börnstein, vol. IVa, pp. 238–239, 1967, are based upon Ziegler, *Chem.-Ing.-Tech.*, **22**, 229 (1950).

TABLE 3-230 Saturated Cesium*

T, K	P, bar	v_f, m³/kg	v_g, m³/kg	h_f, kJ/kg	h_g, kJ/kg	s_f, kJ/(kg·K)	s_g, kJ/(kg·K)	c_{pf}, kJ/(kg·K)
301.6[m]	2.66.−9	5.444.−4	7.01.+7	74.6	637.6	0.696	2.563	0.245
400	3.83.−6	5.615.−4	6.54.+4	98.5	651.9	0.765	2.148	0.240
500	3.11.−4	5.800.−4	1001	122.0	666.1	0.817	1.905	0.232
600	5.65.−3	5.999.−4	65.63	144.9	678.4	0.859	1.748	0.224
700	0.0440	6.215.−4	9.671	167.0	688.9	0.893	1.638	0.219
800	0.2029	6.443.−4	2.353	188.7	698.3	0.922	1.559	0.217
900	0.6620	6.689.−4	0.796	210.6	707.3	0.975	1.500	0.222
1000	1.693	6.954.−4	0.335	233.2	716.4	0.972	1.455	0.231
1200	6.790	7.628.−4	0.097	281.1	736.1	1.015	1.394	0.248
1500	27.6	8.84.−4	0.029	358.8	772.2	1.072	1.345	0.275

*Converted from tables in Vargaftik, *Tables of the Thermophysical Properties of Liquids and Gases*, Nauka, Moscow, 1972, and Hemisphere, Washington, 1975. m = melting point. The notation 2.66.−9 signifies 2.66×10^{9}.

TABLE 3-231 Saturated Decane*

T, K	P, bar	v_f, m³/kg	v_g, m³/kg	h_f, kJ/kg	h_g, kJ/kg	s_f, kJ/(kg·K)	s_g, kJ/(kg·K)	c_{pf}, kJ/(kg·K)	μ_f, 10^{-4} Pa·s	k_f, W/(m·K)
243.5[m]	0.00001	1.319.−3	20750	418.1	812.5	2.561	4.092	2.119	25.0	0.149
260	0.00006	1.334.−3	3300.	452.7	836.3	2.699	4.120	2.109	16.6	0.144
280	0.00042	1.356.−3	443.	495.3	866.9	2.856	4.158	2.155	11.3	0.139
300	0.00197	1.381.−3	88.74	539.0	899.2	3.007	4.200	2.217	8.2	0.134
320	0.00720	1.410.−3	22.73	584.0	933.2	3.153	4.246	2.286	6.5	0.129
340	0.02155	1.442.−3	8.883	631.1	968.9	3.303	4.296	....	5.2	0.124
360	0.05522	1.478.−3	3.763	680.1	1006.2	3.443	4.350	....	4.16	0.119
380	0.1248	1.515.−3	1.750	730.7	1045.0	3.581	4.408	....	3.52	0.116
400	0.2549	1.552.−3	0.892	782.0	1085.0	3.712	4.469	....	2.98	0.110
420	0.4789	1.591.−3	0.490	835.6	1126.2	3.842	4.534	....	2.54	
440	0.8387	1.632.−3	0.290	889.6	1168.4	3.968	4.602	....	2.23	
447.3	1.0133	1.650.−3	0.243	909.4	1184.0	4.014	4.627	....	2.09	
460	1.3852	1.682.−3	0.178	944.5	1211.4	4.089	4.670			
480	2.1745	1.735.−3	0.115	1002.6	1255.2	4.213	4.739			
500	3.2690	1.797.−3	0.0759	1062.7	1299.4	4.335	4.808			
520	4.733	1.868.−3	0.0525	1124.5	1344.4	4.456	4.879			
540	6.633	1.952.−3	0.0369	1190.1	1389.5	4.573	4.949			
560	9.062	2.067.−3	0.0248	1256.1	1432.2	4.698	5.011			
580	12.16	2.255.−3	0.0154	1318.5	1468.1	4.802	5.060			
600	16.12	2.588.−3	0.0093	1384.5	1495.6	4.913	5.098			
617.5[c]	20.97	4.238.−3	0.0042	1483.2	1483.2	5.073	5.073			

*Values converted from Das and Kuloor, *Ind. J. Technol.*, **5**, 75 (1967). m = melting point; c = critical point. The notation 1.319.−3 signifies 1.319×10^{-3}.

TABLE 3-232 Saturated Normal Deuterium*

T, K	P, bar	v_f, m³/kg	v_g, m³/kg	h_f, kJ/kg	h_g, kJ/kg	s_f, kJ/(kg·K)	s_g, kJ/(kg·K)
18.71	0.1709	0.005752	2.232	−161.1	158.6	4.54	21.62
19	0.1944	0.005771	1.988	−160.0	159.1	4.68	21.48
20	0.2944	0.005840	1.365	−152.8	163.9	4.97	20.81
21	0.4297	0.005914	0.968	−145.9	167.6	5.30	20.23
22	0.6072	0.005993	0.705	−138.7	170.6	5.63	19.69
23	0.8344	0.00608	0.5256	−131.4	173.0	5.95	19.18
24	1.1192	0.00617	0.3995	−123.8	174.6	6.26	18.70
25	1.4694	0.00627	0.3088	−116.1	175.5	6.57	18.23
26	1.8932	0.00638	0.2421	−108.2	175.7	6.87	17.79
27	2.3989	0.00650	0.1921	−100.2	175.1	7.16	17.36
28	2.995	0.00663	0.1540	−92.0	173.8	7.44	16.94
29	3.690	0.00678	0.1246	−83.6	171.7	7.72	16.52
30	4.493	0.00694	0.1015	−74.9	168.7	8.00	16.12
31	5.412	0.00713	0.0831	−65.9	165.0	8.27	15.72
32	6.457	0.00735	0.0683	−56.5	160.3	8.54	15.32
33	7.455	0.00761	0.0563	−46.4	154.7	8.83	14.92
34	8.962	0.00793	0.0465	−35.5	148.0	9.12	14.52
35	10.44	0.00834	0.0382	−23.2	140.0	9.45	14.11
36	12.09	0.00890	0.0311	−8.6	130.1	9.82	13.67
37	13.91	0.00976	0.0249	10.0	117.1	10.28	13.17
38	15.92	0.01158	0.0185	39.7	95.0	11.01	12.47
38.34[c]	16.65	0.01433	0.0143	69.2	69.2	11.76	11.76

*Condensed and converted from tables of Prydz, NBS Rep. 9276, 1967. c = critical point.

TABLE 2-233 Saturated Deuterium Oxide*

T, K	P, bar	v_f, m^3/kg	v_g, m^3/kg	h_f, kJ/kg	h_g, kJ/kg	s_f, kJ/(kg·K)	s_g, kJ/(kg·K)
277.0[t]	0.00668	9.047.−4	172.2	0.0	2320.9	0.000	8.380
278.2	0.00720	9.045.−4	160.4	5.0	2322.5	0.0188	8.351
283.2	0.01030	9.042.−4	114.1	25.9	2330.9	0.0920	8.233
288.2	0.01449	9.043.−4	82.48	46.9	2339.3	0.166	8.122
293.2	0.02011	9.047.−4	60.45	67.8	2347.6	0.239	8.016
298.2	0.02758	9.054.−4	44.88	88.7	2356.0	0.311	7.915
303.2	0.03730	9.063.−4	33.71	109.6	2364.0	0.382	7.818
308.2	0.04990	9.075.−4	25.59	130.5	2372.3	0.450	7.725
313.2	0.06598	9.091.−4	19.66	151.5	2380.7	0.518	7.637
318.2	0.08638	9.108.−4	15.24	172.4	2388.6	0.585	7.550
323.2	0.1120	9.127.−4	11.93	193.3	2396.6	0.650	7.468
333.2	0.1831	9.170.−4	7.52	234.7	2413.3	0.776	7.315
353.2	0.4439	9.274.−4	3.27	318.4	2445.1	1.020	7.042
373.2	0.9646	9.403.−4	1.58	402.0	2474.8	1.253	6.807
398.2	2.2427	9.599.−4	0.72	507.5	2509.6	1.527	6.555
423.2	4.653	9.835.−4	0.362	612.5	2541.8	1.781	6.341
448.2	8.806	1.012.−3	0.198	718.8	2569.4	2.020	6.149
473.2	15.46	1.044.−3	0.115	826.8	2585.7	2.256	5.973
498.2	25.52	1.082.−3	0.0704	938.5	2597.0	2.483	5.812
523.2	39.99	1.133.−3	0.0447	1055.2	2598.7	2.707	5.658
548.2	60.04	1.200.−3	0.0290	1177.4	2587.0	2.930	5.501
573.2	86.97	1.276.−3	0.0191	1306.7	2555.6	3.153	5.332
598.2	122.4	1.392.−3	0.0124	1445.6	2492.4	3.356	5.132
623.2	168.3	1.596.−3	0.0075	1607.1	2366.5	3.631	4.850
644.7[c]	218.4	2.950.−3	0.0030				

*Extracted or converted from values in Kazavchinskii, Kesselman, et al., *Thermophysical Properties of Heavy Water,* Moscow and Leningrad, 1963; NBS-NSF transl. 70-50094, 1971. t = triple point; c = critical point. The notation 9.047.−4 signifies 9.047×10^{-4}.

TABLE 3-234 Deuterium Oxide Gas at 1-kg/cm³ Pressure

T (K)	400	450	500	550	600	650	700	750
v, m^3/kg	1.676	1.895	2.112	2.322	2.535	2.747	2.960	3.172
h, kJ/kg	2525	2619	2712	2807	2904	3002	3102	3205
s, kJ/(kg·K)	6.931	7.151	7.349	7.529	7.697	7.855	8.003	8.153

TABLE 3-235 Saturated Diphenyl*

t,	P, bar	v_f, m^3/kg	v_g, m^3/kg	h_f, kJ/kg	h_g, kJ/kg	s_f, kJ/(kg·K)	s_g, kJ/(kg·K)	c_{pf}, kJ/(kg·K)	μ_f, 10^{-4} Pa·s	k_f, W/(m·K)
343	0.0010	1.010.−3	252.5	0.0	444.2	0.000	1.298	1.760	15.0	0.139
350	0.0016	1.014.−3	156.1	13.0	444.2	0.036	1.266	1.782	13.5	0.138
360	0.0029	1.021.−3	85.0	30.0	446.7	0.084	1.236	1.813	11.7	0.136
370	0.0049	1.030.−3	49.9	47.2	449.7	0.130	1.213	1.844	10.3	0.135
380	0.0064	1.037.−3	29.9	65.0	454.5	0.178	1.200	1.875	9.1	0.133
390	0.0129	1.046.−3	18.3	82.7	462.7	0.224	1.194	1.906	8.1	0.132
400	0.0200	1.054.−3	11.7	99.3	461.2	0.273	1.202	1.936	7.3	0.130
420	0.0432	1.072.−3	5.84	139.9	499.0	0.358	1.228	1.998	6.0	0.127
440	0.0879	1.092.−3	3.021	180.3	532.4	0.451	1.267	2.060	5.0	0.125
460	0.1694	1.112.−3	1.652	222.7	569.7	0.545	1.378	2.122	4.3	0.122
480	0.3112	1.132.−3	0.9594	267.6	611.6	0.652	1.367	2.184	3.7	0.119
500	0.5218	1.154.−3	0.4452	314.9	651.8	0.746	1.424	2.246	3.3	0.116
520	0.8375	1.177.−3	0.3652	361.5	687.8	0.824	1.477	2.308	2.7	0.113
540	1.290	1.204.−3	0.2261	404.5	723.8	0.915	1.529	2.370	2.4	0.110
560	1.941	1.230.−3	0.1447	457.2	762.7	1.032	1.582	2.432	2.2	0.107
580	2.818	1.258.−3	0.0977	522.3	801.7	1.125	1.635	2.494	1.90	0.105
600	3.926	1.291.−3	0.0685	563.7	842.4	1.223	1.688	2.556	1.71	0.102
620	5.408	1.326.−3	0.0504	630.4	886.4	1.316	1.740	2.618	1.54	0.099
640	7.328	1.366.−3	0.0381	689.1	930.9	1.375	1.748	2.680	1.39	0.096
660	9.572	1.412.−3	0.0301	745.9	977.1	1.457	1.791	2.741	1.24	0.093
680	12.05	1.465.−3	0.0236	802.8	1024.9	1.585	1.856	2.803	1.10	0.090
700	15.21	1.529.−3	0.0186	860.1	1073.1	1.663	1.951	2.865	0.97	0.087
720	19.14	1.56.−3	0.0147	917.5	1116.7	1.746	2.003	2.93		
740	23.93	1.70.−3	0.0113	975.2	1152.8	1.822	2.058	3.00		
760	28.71	1.95.−3	0.0085	1033.1	1182.5	1.901	2.099			
780	34.83	2.16.−3	0.0058	1091.2	1163.0	1.977	2.107			
800	42.46	3.18.−3	0.0032	1148.4	1148.4	2.047	2.047			

*Interpolated by P. E. Liley from the Landolt-Börnstein band IVa, p. 557, 1967 tables based on *Technical Data on Fuel*, British National Committee, World Energy Conference, London.

TABLE 3-236 Saturated Ethane (R170)*

T, K	P, bar	v_f, m^3/kg	v_g, m^3/kg	h_f, kJ/kg	h_g, kJ/kg	s_f, kJ/(kg·K)	s_g, kJ/(kg·K)	c_{pf}, kJ/(kg·K)	μ_f, 10^{-4} Pa·s	k_f, W/(m·K)
90.4^t	1.131.−5	1.534.−3	21945	176.8	769.4	2.560	9.113	2.260	14.19	0.215
100	1.110.−4	1.546	2484.5	198.7	782.4	2.790	8.627	2.274	9.37	0.208
110	7.467.−3	1.573	407.0	221.5	795.0	3.008	8.222	2.284	6.57	0.201
120	3.545.−3	1.615	93.61	244.4	807.2	3.207	7.897	2.292	4.89	0.194
130	1.291.−2	1.644	27.83	267.4	819.3	3.391	7.637	2.302	3.81	0.187
140	3.831.−2	1.675	10.08	290.5	831.4	3.562	7.426	2.316	3.07	0.180
150	9.672.−2	1.708	4.263	313.7	843.5	3.722	7.254	2.333	2.55	0.174
160	0.2146	1.743	2.039	337.2	855.6	3.873	7.113	2.355	2.17	0.167
170	0.4290	1.780	1.075	360.9	867.6	4.017	6.998	2.383	1.88	0.160
180	0.7874	1.819	0.6139	384.9	879.4	4.154	6.901	2.417	1.65	0.153
190	1.347	1.862	0.3738	409.3	890.8	4.285	6.819	2.458	1.47	0.147
200	2.174	1.908	0.2395	434.2	901.7	4.412	6.750	2.508	1.33	0.140
210	3.340	1.958	0.1602	459.7	911.9	4.535	6.689	2.568	1.21	0.133
220	4.922	2.014	0.1109	485.9	921.4	4.655	6.635	2.640	1.11	0.126
230	7.004	2.076	0.0789	512.8	929.6	4.773	6.585	2.730	1.03	0.119
240	9.670	2.148	0.0573	540.8	936.6	4.890	6.539	2.843	0.96	0.112
250	13.01	2.231	0.0423	569.9	941.9	5.006	6.493	2.991	0.82	0.106
260	17.12	2.330	0.0316	600.7	945.4	5.123	6.449	3.214	0.73	0.099
270	22.10	2.452	0.0237	633.6	946.4	5.233	6.392	3.511	0.64	0.092
280	28.06	2.613	0.0177	669.3	943.6	5.370	6.350	4.011	0.55	0.085
290	35.14	2.847	0.0129	709.8	934.7	5.502	6.278	5.089	0.44	0.078
300	43.54	3.295	0.0087	761.6	910.8	5.669	6.166	9.919	0.31	0.067
305.3^c	48.71	4.891	0.0048	841.2	841.2	5.919	5.919	∞		

*Values reproduced or converted from Goodwin, Roder, and Straty, NBS Tech. Note 684, 1976. t = triple point; c = critical point. The notation 1.131.−5 signifies 1.131×10^{-5}.

TABLE 3-237 Superheated Ethane*

P, bar		Temperature, K: 100	150	200	250	300	350	400	450	500	600	700
1.013	v	0.00156	0.00171	0.5310	0.6725	0.8118	0.9500	1.0877	1.2250	1.3622	1.6360	1.9096
	h	198.9	313.8	909.3	984.7	1068.3	1161.5	1265.3	1379.8	1504.6	1783	2097
	s	2.790	3.722	6.993	7.330	7.634	7.921	8.198	8.467	8.730	9.237	9.720
5	v	0.00156	0.00171	0.00191	0.1288	0.1595	0.1890	0.2178	0.2464	0.2747	0.3308	0.3867
	h	199.4	314.3	434.5	973.3	1060.3	1155.6	1260.7	1376.1	1501.5	1781	2096
	s	2.789	3.720	4.411	6.858	7.175	7.468	7.748	8.020	8.284	8.793	9.227
10	v	0.00156	0.00171	0.00190	0.0590	0.0765	0.0923	0.1073	0.1220	0.1365	0.1650	0.1933
	h	200.0	314.9	435.0	956.5	1050.0	1148.2	1255.0	1371.5	1497.9	1777	2094
	s	2.788	3.719	4.408	6.618	6.959	7.262	7.547	7.821	8.087	8.598	9.083
20	v	0.00156	0.00170	0.00190	0.00222	0.0346	0.0438	0.0521	0.0599	0.0674	0.0822	0.0966
	h	201.3	316.1	435.9	569.8	1026.1	1132.3	1243.3	1362.4	1490.5	1774	2090
	s	2.785	3.715	4.404	4.999	6.710	7.038	7.334	7.614	7.884	8.399	8.886
40	v	0.00155	0.00170	0.00189	0.00219	0.0118	0.0193	0.0244	0.0288	0.0329	0.0407	0.0482
	h	203.9	318.5	437.9	569.9	947.9	1096.2	1218.6	1343.8	1475.9	1764	2083
	s	2.780	3.709	4.394	4.982	6.309	6.770	7.097	7.391	7.670	8.194	8.686
60	v	0.00155	0.00170	0.00188	0.00217	0.00290	0.0109	0.0152	0.0185	0.0215	0.0270	0.0321
	h	206.5	321.0	439.8	570.3	738.1	1050.9	1192.0	1324.8	1461.2	1754	2077
	s	2.775	3.702	4.385	4.966	5.574	6.557	6.934	7.247	7.535	8.068	8.564
80	v	0.00155	0.00169	0.00188	0.00215	0.00273	0.00667	0.0106	0.0134	0.0158	0.0201	0.0459
	h	209.1	323.4	441.9	570.9	728.1	993.8	1163.6	1305.5	1446.7	1745	2070
	s	2.769	3.696	4.377	4.951	5.522	6.345	6.800	7.135	7.432	7.975	8.476
100	v	0.00155	0.00169	0.00187	0.00213	0.00263	0.00465	0.00791	0.0104	0.0124	0.0160	0.0193
	h	211.7	325.8	443.9	571.8	722.7	924.4	1134.7	1286.3	1432.4	1736	2064
	s	2.764	3.690	4.368	4.938	5.486	6.166	6.682	7.040	7.348	7.900	8.406
150	v	0.00155	0.00168	0.00185	0.00209	0.00247	0.00328	0.00488	0.00655	0.00805	0.0107	0.0130
	h	218.1	332.0	449.2	574.6	716.4	887.4	1075.2	1242.3	1399.3	1715	2050
	s	2.752	3.674	4.348	4.907	5.423	5.955	6.457	6.851	7.182	7.758	8.274
200	v	0.00154	0.00167	0.00184	0.00205	0.00237	0.00291	0.00383	0.00495	0.00605	0.00806	0.00986
	h	224.6	338.2	454.7	578.2	714.8	870.5	1041.7	1210.2	1327.3	1697	2038
	s	2.738	3.660	4.329	4.880	5.377	5.863	6.320	6.717	7.059	7.651	8.176
300	v	0.00153	0.00166	0.00181	0.00200	0.00225	0.00259	0.00307	0.00367	0.00433	0.00563	0.00686
	h	237.6	350.6	465.9	586.8	715.9	860.9	1014.9	1175.5	1338.7	1671	2019
	s	2.715	3.632	4.294	4.833	5.309	5.757	6.168	6.547	6.891	7.496	8.032
400	v	0.00153	0.00165	0.00179	0.00195	0.00216	0.00244	0.00276	0.00316	0.00361	0.00454	0.00545
	h	250.6	363.2	477.6	596.6	723.7	861.6	1008.3	62.5	1322.7	1657	2008
	s	2.692	3.605	4.262	4.793	5.257	5.688	6.080	6.443	6.780	7.388	7.930
500	v	0.00152	0.00163	0.00176	0.00192	0.00210	0.00232	0.00258	0.00288	0.00322	0.00392	0.00465
	h	263.5	375.8	489.5	607.1	732.0	866.5	1009.3	1159.5	1316.9	1650	2003
	s	2.670	3.580	4.234	4.758	5.213	5.634	6.015	6.369	6.00	7.306	7.851

*Converted and rounded off from the tables of Goodwin, Roder, and Straty, NBS Tech. Note 684, 1976. v = specific volume, m^3/kg; h = specific enthalpy, kJ/kg; s = specific entropy, kJ/(kg·K).

TABLE 3-238 Saturated Ethylamine ($C_2H_5NH_2$)*

Temp., °F. t	Abs. Pressure, lb./sq. in. p	Volume, cu. ft./lb. vapor v_g	Enthalpy, B.t.u./lb. Liquid h_f	Enthalpy, B.t.u./lb. Vapor h_g	Entropy, B.t.u./(lb.)(°R.) Liquid s_f	Entropy, B.t.u./(lb.)(°R.) Vapor s_g
− 58	0.335	270.8	− 7.82	284.78	−0.0263	0.7094
− 40	.740	134.5	0.00	290.93	.0000	.6931
− 22	1.408	72.87	10.93	296.84	.0253	.6788
− 4	2.546	41.71	23.01	303.05	.0498	.6664
+ 5	3.342	32.32	27.61	306.04	0.619	.6609
23	5.590	20.00	38.91	311.92	.0857	.6512
41	8.960	12.88	50.36	317.50	.1091	.6430
68	16.896	7.156	68.18	326.50	.1437	.6332
86	24.72	5.039	80.50	332.46	.1664	.6281
113	41.49	3.136	99.52	341.44	.1999	.6223

* "Refrigerating Data Book," 5th ed., American Society of Refrigerating Engineers, New York, 1942.

TABLE 3-239 Saturated Ethyl Chloride*

Temp., °F. t	Abs. pressure, lb./sq. in. p	Volume, cu. ft./lb. Liquid v_f	Volume, cu. ft./lb. Vapor v_g	Enthalpy, B.t.u./lb. Liquid h_f	Enthalpy, B.t.u./lb. Vapor h_g	Entropy, B.t.u./(lb.)(°R.) Liquid s_f	Entropy, B.t.u./(lb.)(°R.) Vapor s_g
−22	2.20	0.01657	34.4	−23.1	158.2	−0.0497	0.3642
−13	2.85	.01669	26.95	−19.2	160.7	− .0410	.3615
− 4	3.66	.01682	21.33	−15.4	163.1	− .0324	.3591
+ 5	4.65	.01695	17.06	−11.6	165.4	− .0241	.3566
14	5.85	.01708	13.77	− 7.7	167.8	− .0159	.3543
23	7.28	.01721	11.21	− 3.8	170.2	− .0079	.3523
32	8.99	.01735	9.21	0.0	172.5	.0000	.3506
41	11.01	.01749	7.62	+ 3.8	174.7	+ .0077	.3488
50	13.37	.01763	6.36	7.7	177.0	.0154	.3475
59	16.11	.01777	5.34	11.6	179.3	.0228	.3459
68	19.29	.01792	4.51	15.4	181.4	.0302	.3446
77	22.94	.01807	3.84	19.2	183.5	.0374	.3433
86	27.10	.01822	3.29	23.1	185.7	.0445	.3423
95	31.82	.01838	2.83	26.9	187.7	.0515	.3412
104	37.17	.01854	2.44	30.8	189.9	.0583	.3402
113	43.16	.01870	2.13	34.6	191.8	.0651	.3394
122	49.88	.01887	1.86	38.5	193.8	.0718	.3386
131	57.36	.01904	1.63	42.3	195.6	.0783	.3377

* *Am. Soc. Refrigerating Eng. Circ.* 9, 1926.

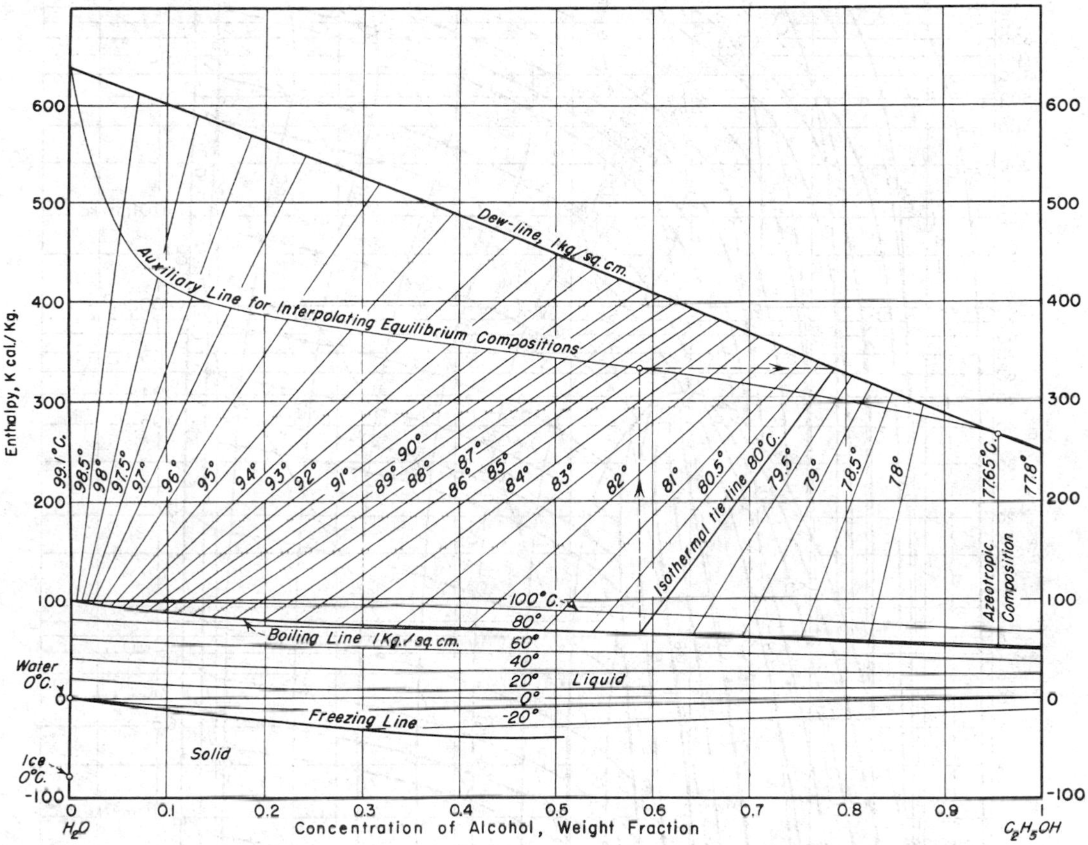

FIG. 3-22 Enthalpy-concentration diagram for aqueous ethyl alcohol. Reference states: enthalpies of liquid water and ethyl alcohol at 0°C are zero. NOTE: In order to interpolate equilibrium compositions, a vertical may be erected from any liquid composition on the boiling line and its intersection with the auxiliary line determined. A horizontal from this intersection will establish the equilibrium vapor composition on the dew line. (*Bosnjakovic,* Technische Thermodynamik, *T. Steinkopff, Leipzig, 1935.*)

TABLE 3-240 Saturated Ethylene (R1150)*

T, K	P, bar	v_f, m^3/kg	v_g, m^3/kg	h_f, kJ/kg	h_g, kJ/kg	s_f, kJ/(kg·K)	s_g, kJ/(kg·K)	c_{pf}, kJ/(kg·K)	μ_f, 10^{-4} Pa·s	k_f, W/(m·K)
104.0[t]	0.0012	1.417.−3	265	241	803	. . . .	. . . .	2.480	7.5	0.258
110	0.0032	1.449.−3	109	251	808	. . . .	. . . .	2.457	5.63	0.252
120	0.0141	1.500.−3	27.8	275	820	. . . .	. . . .	2.436	4.20	0.242
130	0.0469	1.553.−3	8.72	299	832	. . . .	. . . .	2.418	3.28	0.232
140	0.118	1.605.−3	3.611	324	844	. . . .	. . . .	2.405	2.65	0.222
150	0.269	1.659.−3	1.676	347	856	. . . .	. . . .	2.397	2.20	0.212
160	0.562	1.711.−3	0.8281	371.3	868.6	4.077	7.185	2.395	1.87	0.202
170	1.053	1.762.−3	0.4605	395.4	875.7	4.224	7.049	2.404	1.62	0.192
180	1.822	1.811.−3	0.2761	419.6	882.7	4.362	6.935	2.410	1.43	0.182
190	2.959	1.865.−3	0.1752	443.8	888.9	4.494	6.837	2.425	1.28	0.172
200	4.559	1.923.−3	0.1162	467.8	894.1	4.617	6.749	2.450	1.15	0.162
210	6.723	1.986.−3	0.0801	492.6	899.8	4.731	6.670	2.505	1.05	0.152
220	9.560	2.056.−3	0.0570	518.0	905.7	4.844	6.608	2.580	0.97	0.142
230	13.18	2.142.−3	0.0415	544.2	911.1	4.956	6.551	2.706	0.90	0.132
240	17.71	2.243.−3	0.0305	571.5	912.2	5.070	6.490	2.915	0.83	0.122
250	23.28	2.370.−3	0.0224	600.8	908.1	5.188	6.417	3.260	0.77	0.112
260	30.03	2.539.−3	0.0164	633.0	900.1	5.313	6.340	3.775	0.71	0.102
270	38.11	2.798.−3	0.0118	673.2	887.7	5.459	6.253	4.990		
275	42.71	2.997.−3	0.0096	700.3	874.2	5.551	6.194			
283.1[c]	50.97	4.739.−3	0.0047	799.1	799.1	5.903	5.903	∞	. . .	∞

*P, v, h, and s values extracted from Vashchenko, Voinov, et al., *Thermodynamic and Transport Properties of Ethylene and Propylene*, Standartov, Moscow, 1971; transl. NBSIR 75-763–COM-75-11276, NTIS, 1975. t = triple point; c = critical point.

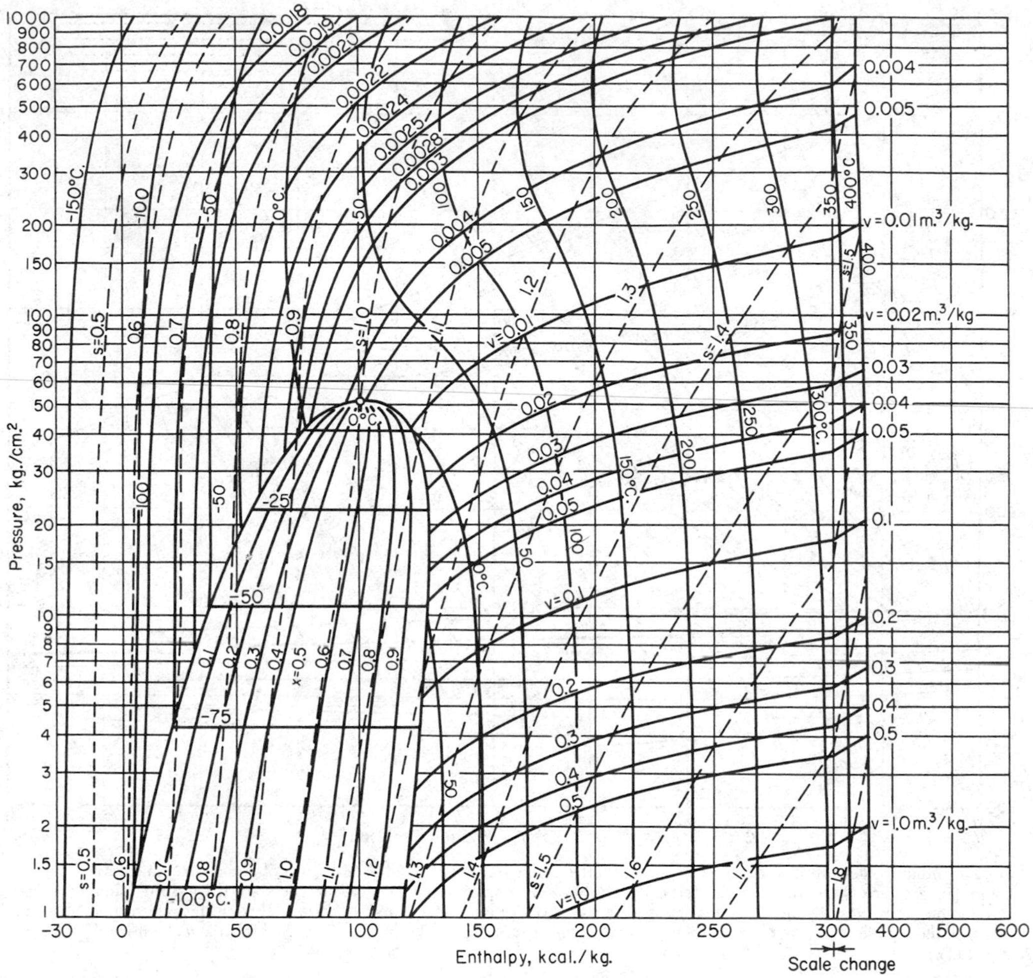

FIG. 3-23 Enthalpy-pressure diagram for ethylene. [*From Benzler and Koch,* Chem. Ing. Tech., *27, 71 (1955). Copyright Verlag Chemie GmbH, Weinheim, Bergstrasse, Germany.*]

TABLE 3-241 Saturated Fluorine*

T, K	P, bar	v_f, m³/kg	v_g, m³/kg	h_f, kJ/kg	h_g, kJ/kg	s_f, kJ/(kg·K)	s_g, kJ/(kg·K)	c_{pf}, kJ/(kg·K)	μ_f, 10^{-4} Pa·s	k_f, W/(m·K)
53.5[t]	0.0025	5.866.−4	46.2	−158.6	40.9	1.602	5.314	1.446	8.8	0.186
55	0.0041	5.898.−4	17.1	−153.5	42.0	1.642	5.235	1.442	8.0	0.184
60	0.0155	6.005.−4	8.46	−149.1	45.8	1.768	5.004	1.437	6.0	0.177
65	0.0477	6.119.−4	2.93	−141.8	49.6	1.885	4.816	1.442	4.7	0.170
70	0.1230	6.240.−4	1.24	−134.4	53.2	1.995	4.666	1.450	3.8	0.162
75	0.276	6.369.−4	0.583	−127.0	56.8	2.097	4.540	1.460	3.19	0.154
80	0.555	6.508.−4	0.309	−119.5	60.1	2.194	4.433	1.474	2.71	0.146
85	1.019	6.657.−4	0.176	−111.9	63.3	2.285	4.342	1.498	2.33	0.137
90	1.740	6.819.−4	0.108	−104.3	66.1	2.372	4.262	1.535	2.00	0.129
95	2.802	6.997.−4	0.069	− 96.5	68.6	2.455	4.191	1.555	1.76	0.120
100	4.280	7.193.−4	0.0466	− 88.6	70.7	2.535	4.127	1.585	1.53	0.112
105	6.280	7.412.−4	0.0323	− 80.5	72.4	2.612	4.068	1.630	1.36	0.103
110	8.885	7.659.−4	0.0231	− 72.2	73.6	2.688	4.012	1.692	1.21	0.095
115	12.20	7.948.−4	0.0168	− 63.6	74.1	2.763	3.959	1.782	1.08	0.087
120	16.33	8.283.−4	0.0125	− 54.5	73.9	2.837	3.906	1.888	0.96	0.080
125	21.37	8.696.−4	0.0093	− 44.9	72.7	2.912	3.864	2.05	0.86	0.073
130	27.48	9.223.−4	0.0069	− 34.5	70.2	2.989	3.795	2.33	0.74	0.066
135	34.72	9.963.−4	0.0051	− 22.7	65.6	3.073	3.727	2.90	0.63	0.070
140	43.47	1.119.−3	0.0036	− 8.4	56.9	3.170	3.636	3.64	0.49	0.105
144.3[c]	52.15	1.743.−3	0.0017	23.9	23.9	3.388	3.388	∞	. . .	∞

*Values reproduced or converted from Prydz and Straty, NBS Tech. Note 392, rev., September 1973. t = triple point; c = critical point. The notation 5.866.−4 signifies 5.866×10^{-4}.

TABLE 3-242 Fluorine Gas at Atmospheric Pressure*

T, K	84.95	90	100	120	140	160	180	200	220	240	260	280	300
v, m^3/kg	0.1776	0.1892	0.2118	0.2562	0.3002	0.3439	0.3874	0.4309	0.4744	0.5176	0.5610	0.6043	0.6476
h, kJ/kg	63.22	67.30	75.27	90.96	106.53	122.06	137.62	153.2	169.0	184.9	201.0	217.2	233.7
s, kJ/(kg·K)	4.342	4.390	4.474	4.616	4.737	4.840	4.932	5.014	5.090	5.158	5.221	5.282	5.340

*Extracted from Prydz and Straty, NBS Tech. Note 392, 1970. This source is recommended for other pressures and temperatures. Other information is contained in *J. Chem. Phys.*, **53**, 2359 (1970); and *J. Res. NBS*, **74A**, 499, 661, 747 (1970).

TABLE 3-243 Saturated Helium^3*

T, K	P, bar	v_f, m^3/kg	v_g, m^3/kg	h_f, kJ/kg	h_g, kJ/kg	s_f, kJ/(kg·K)	s_g, kJ/(kg·K)
1.0	0.0122	0.01222	1.72	−5.69	6.75	2.28	14.72
1.1	0.0182	0.01224	1.33	−5.49	7.40	2.65	14.34
1.2	0.0274	0.01227	1.02	−5.26	8.03	2.95	14.02
1.3	0.0370	0.01231	0.805	−5.01	8.65	3.20	13.70
1.4	0.0517	0.01236	0.649	−4.75	9.27	3.40	13.41
1.5	0.0659	0.01241	0.526	−4.47	9.88	3.60	13.13
1.6	0.0871	0.01247	0.437	−4.17	10.46	3.80	12.88
1.7	0.107	0.01254	0.363	−3.84	11.04	3.91	12.53
1.8	0.137	0.01262	0.308	−3.47	11.60	4.01	12.38
1.9	0.163	0.01271	0.260	−3.07	12.15	4.13	12.14
2.0	0.202	0.01282	0.222	−2.64	12.68	4.26	11.91
2.1	0.237	0.01294	0.189	−2.17	13.19	4.40	11.69
2.2	0.284	0.01308	0.164	−1.55	13.67	4.55	11.47
2.3	0.326	0.01324	0.142	−0.99	14.13	4.71	11.25
2.4	0.385	0.01343	0.124	−0.34	14.57	4.87	11.04
2.5	0.438	0.01365	0.109	0.36	14.98	5.03	10.84
2.6	0.508	0.01390	0.096	1.16	15.37	5.20	10.64
2.7	0.576	0.01419	0.085	2.01	15.89	5.38	10.41
2.8	0.653	0.01456	0.074	2.96	16.40	5.57	10.17
2.9	0.732	0.01497	0.064	4.01	16.37	5.77	9.92
3.0	0.803	0.01549	0.055	5.28	16.32	6.00	9.66
3.1	0.907	0.01614	0.047	6.70	16.20	6.24	9.34
3.2	1.023	0.01720	0.039	8.44	15.98	6.54	8.90
3.3	1.128	0.01902	0.028	10.66	14.50	6.96	8.35
3.32^c	1.165	0.02394	0.024	13.25	13.25	7.50	7.50

*Converted and smoothed from a tabulation of Gibbons and Nathan, USAF Rep. AFML-TR-67-175, 1967. c = critical point.

TABLE 3-244 Saturated Helium[a]*

T, K	P, bar	v_f, m³/kg	v_g, m³/kg	h_f, kJ/kg	h_g, kJ/kg	s_f, kJ/(kg·K)	s_g, kJ/(kg·K)	c_{pf}, kJ/(kg·K)	μ_f, 10^{-4} Pa·s	k_f, W/(m·K)
2.18$^\lambda$	0.0503	6.839.−3	0.8491	3.25	25.41	1.65	11.92	. . .	0.0360	0.0147
2.2	0.0533	6.842.−3	0.8099	3.28	25.51	1.67	11.85	3.16	0.0361	0.0148
2.3	0.0672	6.860.−3	0.6654	3.57	25.91	1.80	11.57	1.80	0.0367	0.0155
2.4	0.0834	6.882.−3	0.5540	3.82	26.30	1.90	11.32	2.25	0.0371	0.0160
2.5	0.1021	6.970.−3	0.4665	4.05	26.68	1.99	11.09	2.13	0.0365	0.0276
2.6	0.1235	6.936.−3	0.3967	4.27	27.04	2.07	10.87	2.10	0.0373	0.0168
2.7	0.1479	6.969.−3	0.3402	4.50	27.38	2.15	10.66	2.14	0.0373	0.0171
2.8	0.1753	7.004.−3	0.2941	4.73	27.71	2.23	10.47	2.22	0.0372	0.0174
2.9	0.2060	7.043.−3	0.2559	4.98	28.03	2.30	10.28	2.31	0.0370	0.0176
3.0	0.2402	7.085.−3	0.2241	5.23	28.33	2.38	10.11	2.42	0.0368	0.0178
3.1	0.2780	7.130.−3	0.1972	5.50	28.61	2.46	9.94	2.54	0.0365	0.0180
3.2	0.3198	7.180.−3	0.1745	5.78	28.87	2.54	9.77	2.67	0.0362	0.0182
3.3	0.3655	7.232.−3	0.1550	6.08	29.11	2.62	9.62	2.80	0.0358	0.0184
3.4	0.4154	7.290.−3	0.1382	6.39	29.32	2.71	9.46	2.95	0.0355	0.0186
3.5	0.4698	7.352.−3	0.1237	6.72	29.52	2.79	9.31	3.10	0.0351	0.0188
3.6	0.5289	7.419.−3	0.1110	7.07	29.69	2.88	9.17	3.28	0.0347	0.0190
3.7	0.5926	7.492.−3	0.0999	7.43	29.84	2.96	9.03	3.47	0.0342	0.0191
3.8	0.6614	7.571.−3	0.0901	7.82	29.96	3.05	8.88	3.68	0.0338	0.0192
3.9	0.7353	7.657.−3	0.0815	8.22	30.06	3.14	8.74	3.92	0.0333	0.0194
4.0	0.8147	7.752.−3	0.0702	8.65	30.12	3.23	8.60	4.19	0.0328	0.0195
4.1	0.8996	7.856.−3	0.0669	9.11	30.15	3.33	8.46	4.51	0.0323	0.0196
4.2	0.9901	7.971.−3	0.0606	9.59	30.14	3.43	8.32	4.88	0.0318	0.0196
4.3	1.0872	8.100.−3	0.0550	10.10	30.08	3.53	8.18	5.32	0.0313	0.0197
4.4	1.190	8.244.−3	0.0499	10.66	29.98	3.64	8.03	5.86	0.0307	0.0197
4.5	1.299	8.407.−3	0.0452	11.25	29.81	3.75	7.88	6.55	0.0301	0.0199
4.6	1.416	8.596.−3	0.0408	11.90	29.58	3.87	7.71	7.44	0.0295	0.0200
4.7	1.537	8.818.−3	0.0368	12.61	29.25	4.00	7.54	8.68	0.0289	0.0202
4.8	1.670	9.086.−3	0.0329	13.40	28.80	4.14	7.35	10.5	0.0282	0.0205
4.9	1.808	9.427.−3	0.0292	14.30	28.18	4.30	7.13	13.6	0.0274	0.0210
5.0	1.955	9.893.−3	0.0255	15.39	27.28	4.50	6.87	19.9	0.0264	0.0222
5.1	2.110	1.063.−2	0.0214	16.82	25.83	4.75	6.51	38.5	0.0251	0.0263
5.20^c	2.275	1.436.−2	0.0144	21.36	21.36	5.59	5.59	∞	. . . . .	∞

*Values extracted from McCarty, NBS Tech. Note 631, 1972. λ = lambda point; c = critical point.

TABLE 3-245 Superheated Helium*

P, bars		Temp., °C. 0	100	200	300	400	500	600	800	1000
1	v	5.677	7.754	9.831	11.908	13.985	16.063	18.140	22.294	26.448
	h	0.327	519.6	1039	1558	2078	2597	3116	4155	5193
	s	0.0116	1.620	2.853	3.849	4.684	5.403	6.035	7.106	7.993
5	v	1.138	1.553	1.968	2.384	2.799	3.215	3.630	4.461	5.291
	h	1.636	520.9	1040	1560	2079	2598	3117	4156	5194
	s	−3.343	−1.723	−0.490	0.506	1.341	2.060	2.692	3.763	4.650
10	v	0.5704	0.780	0.986	1.193	1.401	1.609	1.816	2.232	2.647
	h	3.272	522.5	1042	1561	2080	2600	3119	4157	5196
	s	−4.782	−3.162	−1.929	−0.934	−0.098	0.621	1.252	2.323	3.211
20	v	0.2867	0.3904	0.4942	0.5979	0.7017	0.9093	1.1169	1.3245	1.8435
	h	6.544	525.8	1045	1564	2083	2603	3122	4160	5199
	s	−6.221	−4.601	−3.368	−2.373	−1.537	−0.818	−0.187	0.884	1.771
50	v	0.1164	0.1579	0.1993	0.2408	0.2822	0.3257	0.3652	0.4481	0.5311
	h	16.360	535.5	1055	1574	2093	2612	3131	4169	5207
	s	−8.121	−6.501	−5.268	−4.273	−3.438	−2.719	−2.088	−1.017	−0.130
100	v	0.0597	0.0803	0.1010	0.1217	0.1424	0.1631	0.1838	0.2252	0.2666
	h	37.720	551.7	1071	1590	2108	2627	3146	4184	5222
	s	−9.555	−7.936	−6.703	−5.709	−4.874	−4.155	−3.524	−2.454	−1.567
150	v	0.0407	0.0545	0.0682	0.0820	0.0958	0.1095	0.1233	0.1509	0.1785
	h	49.080	567.9	1087	1605	2124	2643	3161	4199	5236
	s	−10.391	−8.773	−7.541	−6.546	−5.712	−4.994	−4.363	−3.293	−2.407
200	v	0.0312	0.0416	0.0518	0.0622	0.0725	0.0828	0.0931	0.1137	0.1344
	h	65.440	584.1	1103	1621	2140	2658	3176	4213	5250
	s	−10.983	−9.635	−8.134	−7.139	−6.306	−5.588	−4.957	−3.888	−3.002

*Extracted from Tsederberg, Popov, *et al.*, "Thermodynamic and Thermophysical Properties of Helium," Atomizdat, Moscow, 1969, and NBS-NSF TT 50096, 1971. Copyright material. Reproduced by permission. This source contains entries for many more temperatures and pressures than can be reproduced here. v = volume, cu. m./kg.; h = enthalpy, kJ./kg.; s = entropy, kJ./(kg.)(°K.).

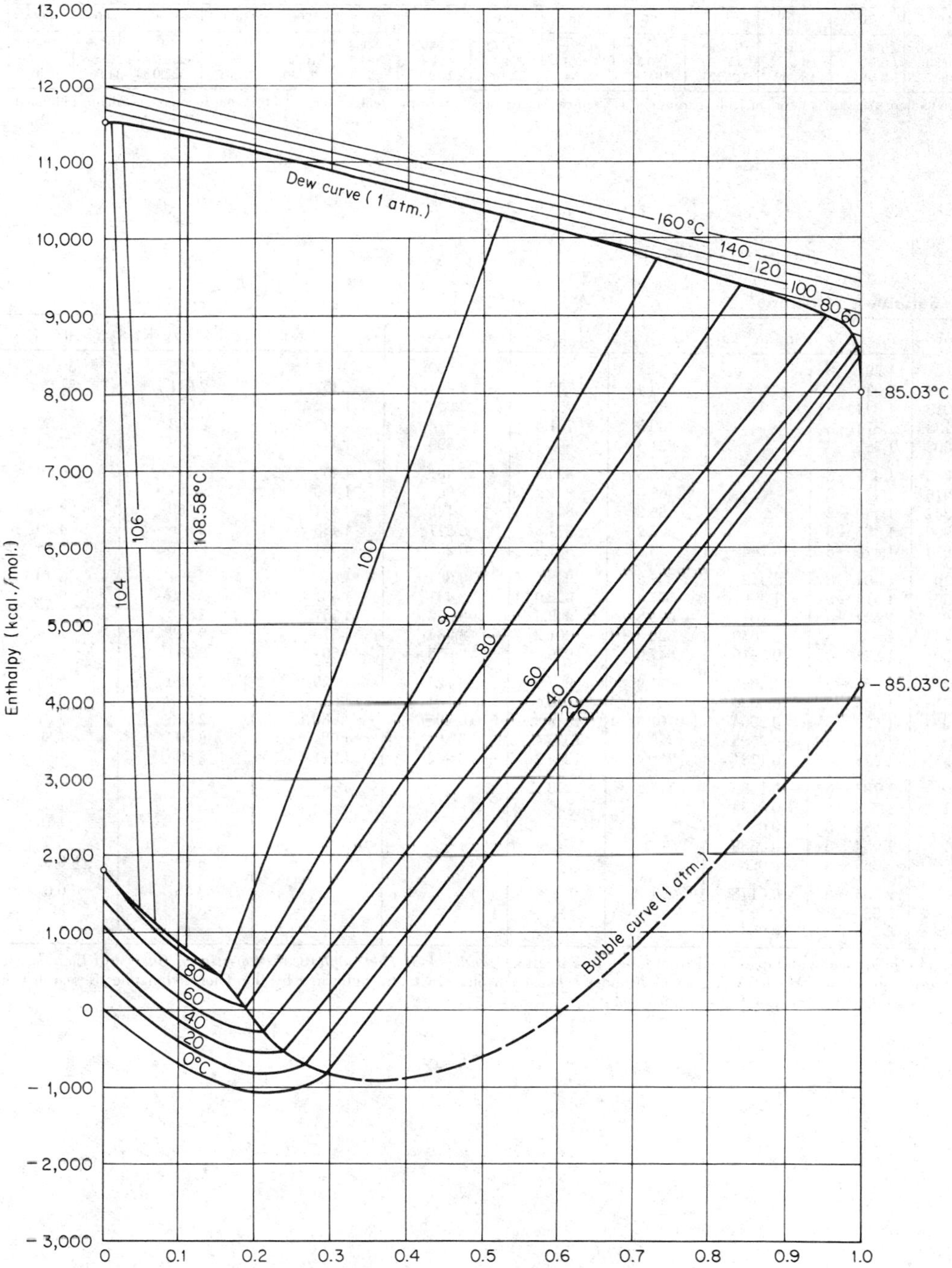

$$\psi = \frac{M_{HCl}}{M_{H_2O} + M_{HCl}} \quad [Mol_{HCl}/Mol]$$

FIG. 3-24 Enthalpy-concentration diagram for aqueous hydrogen chloride at 1 atm. Reference states: enthalpy of liquid water at 0°C is zero; enthalpy of pure saturated HCl vapor at 1 atm (−85.03°C) is 8000 kcal/mol. NOTE: It should be observed that the weight basis includes the vapor, which is particularly important in the two-phase region. Saturation values may be read at the ends of the tie lines. [*Van Nuys*, Trans. Am. Inst. Chem. Eng., *39, 663 (1943).*]

TABLE 3-246 Helium[4] Gas at Atmospheric Pressure*

T, K	4.224	5	10	20	30	40	50	75	100	200	300	400	500	600	800	1000
v, m³/kg	0.0591	0.0834	0.1612	0.4094	0.6161	0.8218	1.0273	1.5403	2.053	4.102	6.154	8.191	10.24	12.31	16.40	20.50
h, kJ/kg	30.30	36.18	64.91	117.95	170.24	222.4	274.4	404.4	534.2	1054	1573	2092	2612	3131	4170	5208
s, kJ/(kg·K)	8.327	9.614	13.369	17.321	19.442	20.94	22.10	24.21	25.71	29.30	31.41	32.90	34.06	35.01	36.50	37.66

From McCarty, NBS Rep. 9762, 1970. Reproduced by permission. The source contains values for further temperatures and for other functions, usually to additional significant figures.

TABLE 3-247 Saturated *n*-Heptane*

T, K	P, bar	v_f, m³/kg	v_g, m³/kg	h_f, kJ/kg	h_g, kJ/kg	s_f, kJ/(kg·K)	s_g, kJ/(kg·K)	c_{pf}, kJ/(kg·K)	μ_f, 10^{-4} Pa·s	k_f, W/(m·K)
182.6[t]	.....	1.292.−3	.....	284.1	.....	2.260	....	2.025	39.4	0.150
200	0.00002	1.316.−3	.....	319.4	722.6	2.441	4.457	2.011	21.0	0.148
220	0.00019	1.344.−3	.....	359.7	757.1	2.636	4.442	2.026	12.6	0.145
240	0.00133	1.374.−3	.....	400.5	791.4	2.814	4.443	2.063	8.52	0.142
250	0.00303	1.389.−3	.....	421.3	808.3	2.899	4.447	2.088	7.23	0.140
260	0.00635	1.405.−3	.....	442.3	824.9	2.981	4.453	2.117	6.52	0.137
270	0.01316	1.422.−3	.....	463.6	841.2	3.061	4.460	2.147	5.46	0.135
280	0.02347	1.440.−3	.....	485.2	857.8	3.140	4.471	2.180	4.83	0.132
290	0.03997	1.457.−3	.....	507.2	874.8	3.217	4.485	2.216	4.29	0.129
300	0.06674	1.475.−3	3.744	529.6	891.9	3.293	4.501	2.252	3.85	0.126
310	0.1070	1.494.−3	2.412	552.3	908.9	3.367	4.517	2.291	3.48	0.123
320	0.1656	1.514.−3	1.596	575.4	926.0	3.441	4.537	2.329	3.17	0.121
330	0.2461	1.534.−3	1.101	598.8	943.3	3.513	4.557	2.370	2.89	0.119
340	0.3614	1.555.−3	0.7650	622.8	961.2	3.584	4.579	2.412	2.66	0.116
350	0.5130	1.578.−3	0.5510	647.0	979.1	3.655	4.604	2.454	2.45	0.114
360	0.712	1.601.−3	0.4058	671.9	997.5	3.725	4.629	2.500	2.24	0.111
370	0.967	1.625.−3	0.3036	697.1	1016.1	3.794	4.656	2.548	2.04	0.109
371.6	1.013	1.629.−3	0.2904	701.9	1019.8	3.805	4.660	2.556	2.01	0.108
380	1.289	1.651.−3	0.2308	723.9	1035.4	3.864	4.684	2.60	1.86	0.107
390	1.689	1.678.−3	0.1781	750.4	1054.2	3.932	4.711	2.65	1.71	0.105
400	2.180	1.708.−3	0.1388	777.2	1073.2	4.000	4.740	2.70	1.58	0.103
420	3.471	1.775.−3	0.0734	....	.....	....	.....	2.81	1.35	0.099
440	5.268	1.853.−3	0.0576	....	.....	....	.....	2.93	1.15	0.095
460	7.691	1.954.−3	0.0389	....	.....	....	.....	3.05	0.97	0.091
480	10.92	2.065.−3	0.0265	....	.....	....	.....	3.19	0.82	0.087
500	15.10	2.235.−3	0.0178	....	.....	....	.....	3.38	0.67	0.080
520	20.43	2.52.−3	.....	....	.....	....	.....	3.7		
540.1[c]	27.35	4.3.−3	0.0043	....	.....	....				

*Values of P and v interpolated and converted from tables in Vargaftik, *Handbook of Thermophysical Properties of Gases and Liquids*, Hemisphere, Washington, and McGraw-Hill, New York, 1975. Values of h and s calculated from API tables published by the Thermodynamics Research Center, Texas A&M University, College Station. t = triple point; c = critical point.

TABLE 3-248a Saturated Normal Hydrogen*

T, K	P, bar	v_f, m³/kg	v_g, m³/kg	h_f, kJ/kg	h_g, kJ/kg	s_f, kJ/(kg·K)	s_g, kJ/(kg·K)	c_{pf}, kJ/(kg·K)	μ_f, 10^{-4} Pa·s	k_f, W/(m·K)
13.95[t]	0.072	0.01298	7.974	218.3	667.4	14.079	46.635	6.36	0.255	0.073
14	0.074	0.01301	7.205	219.6	669.3	14.173	46.301	6.47	0.248	0.075
15	0.127	0.01316	4.488	226.4	678.2	14.640	44.763	6.91	0.218	0.083
16	0.204	0.01332	2.954	233.8	686.7	15.104	43.418	7.36	0.194	0.089
17	0.314	0.01348	2.032	241.6	694.7	15.568	42.227	7.88	0.175	0.093
18	0.461	0.01366	1.449	249.9	702.1	16.032	41.158	8.42	0.159	0.095
19	0.654	0.01387	1.064	258.8	708.8	16.498	40.188	8.93	0.146	0.097
20	0.901	0.01407	0.8017	268.3	714.8	16.966	39.299	9.45	0.135	0.098
21	1.208	0.01430	0.6177	278.4	720.2	17.440	38.485	10.13	0.125	0.100
22	1.585	0.01455	0.4828	289.2	724.4	17.919	37.710	10.82	0.116	0.101
23	2.039	0.01483	0.3829	300.8	727.6	18.405	36.973	11.69	0.108	0.101
24	2.579	0.01515	0.3072	313.3	729.8	18.901	36.266	12.52	0.101	0.101
25	3.213	0.01551	0.2489	326.7	730.7	19.408	35.579	13.44	0.094	0.100
26	3.950	0.01592	0.2032	341.2	730.2	19.929	34.900	14.80	0.088	0.098
27	4.800	0.01639	0.1667	357.0	728.0	20.473	34.221	16.17	0.082	0.096
28	5.770	0.01696	0.1370	374.3	723.7	21.041	33.524	18.48	0.076	0.094
29	6.872	0.01765	0.1125	393.6	716.6	21.650	32.795	22.05	0.070	0.091
30	8.116	0.01854	0.0919	415.4	705.9	22.315	32.002	26.59	0.065	0.087
31	9.510	0.01977	0.0738	441.3	689.7	23.075	31.091	36.55	0.058	0.086
32	11.07	0.02174	0.0571	474.7	663.2	24.032	29.926	65.37	0.051	0.092
33.18[c]	13.13	0.03182	0.0318	565.4	565.4	26.680	26.680	∞	....	∞

*Values extracted and occasionally rounded off from McCarty, Hord, and Roder, NBS Monogr. 168, 1981. t = triple point; c = critical point.

TABLE 248*b* Compressed Normal Hydrogen*

Pressure, bar		Temperature (K) 15	20	30	40	50	60	80	100	150	200
0.1	*v*	6.076	8.176	12.333	16.473	20.606	24.736	32.991	41.244	61.870	82.495
	h	679.2	731.6	835.5	938.9	1042.3	1146	1356	1575	2172	2826
	s	46.02	49.04	53.25	56.23	58.53	60.43	63.45	65.89	70.68	74.46
1	*v*	0.0131	0.0141	1.196	1.625	2.046	2.463	3.295	4.123	6.190	8.254
	h	227.3	268.3	826.0	932.7	1037.9	1143	1354	1574	2172	2826
	s	14.62	16.96	43.56	46.63	48.98	50.89	53.93	56.38	61.17	64.96
5	*v*	0.0131	0.0140	0.2006	0.3039	0.3958	0.4839	0.6553	0.8238	1.241	1.655
	h	231.7	272.1	775.0	903.4	1017.6	1128	1345	1568	2170	2826
	s	14.57	16.88	35.80	39.52	42.07	44.07	47.20	49.68	54.66	58.31
10	*v*	0.0130	0.0138	0.0181	0.1376	0.1895	0.2366	0.3255	0.4116	0.6221	0.8303
	h	237.2	277.0	412.1	861.8	991.1	1109	1334	1560	2167	2826
	s	14.50	16.77	22.09	35.95	38.85	40.99	44.23	46.75	51.63	55.44
20	*v*	0.0129	0.0136	0.0167	0.0521	0.0866	0.1135	0.1611	0.2057	0.3129	0.4179
	h	248.2	286.9	406.5	752.0	934.7	1070	1312	1546	2163	2826
	s	14.37	16.58	21.33	31.07	35.19	37.67	41.15	43.76	48.71	52.55
40	*v*	.....	0.0133	0.0155	0.0216	0.0376	0.0533	0.0796	0.1033	0.1586	0.2119
	h	.....	307.3	413.5	589.3	823.5	997	1271	1521	2155	2826
	s	.....	16.26	20.50	25.49	30.73	33.91	37.87	40.65	45.75	49.64
60	*v*	.....	0.0130	0.0147	0.0182	0.0254	0.0351	0.0532	0.0697	0.1073	0.1433
	h	.....	328.0	427.2	570.1	757.0	940	1237	1499	2149	2828
	s	.....	15.98	19.95	24.03	28.19	31.54	35.82	38.76	43.99	47.92
80	*v*	.....	0.0127	0.0142	0.0167	0.0211	0.0273	0.0406	0.0531	0.0818	0.1090
	h	.....	348.9	443.5	572.3	732.8	905	1210	1482	2146	2831
	s	.....	15.74	19.53	23.21	26.78	29.93	34.34	37.37	42.72	46.69
100	*v*	.....	0.0125	0.0138	0.0158	0.0190	0.0233	0.0335	0.0434	0.0666	0.0885
	h	.....	369.8	461.1	581.5	727.4	888	1192	1469	2144	2835
	s	.....	15.53	19.19	22.63	25.88	28.80	33.19	36.28	41.73	45.73
200	*v*	.....	0.0117	0.0125	0.0136	0.0150	0.0167	0.0207	0.0253	0.0368	0.0480
	h	.....	474.4	556.1	658.7	776.9	908	1182	1458	2156	2869
	s	.....	14.71	17.99	20.93	23.56	25.94	29.88	32.97	38.59	42.72
400	*v*	.....	.....	0.0113	0.0119	0.0126	0.0134	0.0151	0.0171	0.0225	0.0279
	h	.....	.....	751.0	841.9	945.4	1059	1303	1560	2249	2973
	s	.....	.....	16.59	19.20	21.50	23.58	27.07	29.94	35.48	39.67
600	*v*	.....	.....	0.0106	0.0110	0.0115	0.0120	0.0131	0.0144	0.0178	0.0214
	h	.....	.....	941.5	1027	1124	1231	1463	1709	2385	3107
	s	.....	.....	15.68	18.14	20.29	22.24	25.57	28.31	33.74	37.92
800	*v*	.....	.....	.....	0.0104	0.0107	0.0111	0.0120	0.0130	0.0155	0.0181
	h	.....	.....	.....	1209	1302	1405	1628	1870	2535	3255
	s	.....	.....	.....	17.35	19.43	21.30	24.50	27.20	32.54	36.70
1000	*v*	.....	.....	.....	0.0099	0.0102	0.0106	0.0112	0.0120	0.0140	0.0160
	h	.....	.....	.....	1387	1478	1578	1796	2032	2692	3403
	s	.....	.....	.....	16.72	18.75	20.58	23.70	26.33	31.63	35.75

*Values extracted and sometimes rounded off from the tables of McCarty, Hord, and Roder, NBS Monogr. 168, 1981. This source contains an exhaustive tabulation of property values for both the normal and the para forms of hydrogen. v = specific volume, m^3/kg; h = specific enthalpy, kJ/kg; s = specific entropy, kJ/(kg·K).

Temperature, K										
250	300	350	400	450	500	600	700	800	900	1000
103.12	123.23	144.35	164.97	185.60	206.22	247.46	288.70	329.94	371.18	412.43
3517	4227	4945	5668	6393	7118	8571	10028	11493	12969	14458
77.53	80.13	82.34	84.27	85.98	87.51	90.15	92.40	94.36	96.10	97.66
10.32	12.38	14.44	16.50	18.57	20.63	24.75	28.88	33.00	37.13	41.25
3517	4227	4946	5669	6393	7118	8571	10029	11494	12969	14459
68.03	70.63	72.85	74.78	76.48	78.01	80.66	82.91	84.86	86.60	88.17
2.069	2.482	2.895	3.307	3.720	4.132	4.957	5.782	6.607	7.432	8.257
3518	4229	4948	5671	6396	7121	8574	10032	11497	12973	14462
61.39	63.99	66.21	68.14	69.84	71.37	74.02	76.27	78.23	79.96	81.53
1.038	1.245	1.451	1.658	1.864	2.070	2.483	2.896	3.308	3.720	4.133
3519	4231	4951	5674	6399	7125	8578	10036	11501	12977	14467
58.52	61.12	63.34	65.28	66.98	68.51	71.16	73.41	75.37	77.10	78.67
0.522	0.6259	0.7294	0.8328	0.9361	1.040	1.246	1.452	1.658	1.865	2.071
3522	4235	4956	5680	6406	7132	8586	10044	11509	12985	14475
55.65	58.26	60.48	62.41	64.12	65.65	68.30	70.55	72.51	74.24	75.81
0.2644	0.3166	0.3685	0.4204	0.4721	0.5238	0.6271	0.7303	0.8335	0.9366	1.040
3527	4244	4967	5692	6419	7146	8601	10059	11525	13002	14492
52.76	55.38	57.61	59.55	61.26	62.79	65.44	67.69	69.65	71.39	72.95
0.1786	0.2136	0.2483	0.2829	0.3174	0.3519	0.4209	0.4897	0.5585	0.6273	0.6961
3533	4253	4978	5705	6432	7160	8616	10075	11542	13018	14508
51.05	53.69	55.92	57.86	59.58	61.11	63.76	66.02	67.97	70.51	71.28
0.1357	0.1621	0.1882	0.2142	0.2401	0.2660	0.3177	0.3694	0.4120	0.4726	0.5242
3540	4263	4989	5718	6446	7174	8631	10091	11558	13035	14525
49.84	52.49	54.73	56.67	58.39	59.92	62.57	64.83	66.79	68.52	70.09
0.1099	0.1312	0.1521	0.1730	0.1937	0.2145	0.2559	0.2972	0.3385	0.3798	0.4211
3547	4273	5001	5731	6460	7189	8647	10107	11574	13051	14542
48.89	51.55	53.79	55.74	57.46	59.00	61.65	63.90	65.87	67.60	69.17
0.0588	0.0695	0.0801	0.0905	0.1001	0.1114	0.1321	0.1528	0.1734	0.1941	0.2147
3594	4329	5064	5798	6531	7263	8724	10187	11656	13134	14625
45.94	48.62	50.89	52.85	54.58	56.12	58.78	61.04	63.00	64.74	66.31
0.0334	0.0388	0.0441	0.0493	0.0545	0.0597	0.0701	0.0804	0.0908	0.1011	0.1114
3716	4458	5202	5943	6681	7416	8883	10349	11820	13300	14792
42.98	45.68	47.97	49.95	51.69	53.24	55.91	58.17	60.14	61.88	63.45
0.0249	0.0285	0.0321	0.0355	0.0390	0.0425	0.0494	0.0562	0.0631	0.0700	0.0768
3854	4600	5349	6095	6836	7574	9045	10513	11985	13466	14958
41.24	43.95	46.26	48.26	50.00	51.56	54.24	56.50	58.47	60.21	61.78
0.0207	0.0234	0.0260	0.0286	0.0312	0.0338	0.0390	0.0441	0.0492	0.0543	0.0594
4003	4748	5501	6249	6993	7734	9207	10677	12150	13631	15124
40.03	42.73	45.05	47.05	48.81	50.37	53.05	55.32	57.29	59.03	60.60
0.0181	0.0202	0.0223	0.0244	0.0265	0.0286	0.0327	0.0367	0.0408	0.0449	0.0490
4156	4898	5654	6405	7151	7893	9370	10842	12316	13797	15289
39.10	41.79	44.12	46.12	47.88	49.45	52.14	54.41	56.38	58.12	59.69

TABLE 3-249 Saturated *para*-Hydrogen*

T, K	P, bar	v_f, m³/kg	v_g, m³/kg	h_f, kJ/kg	h_g, kJ/kg	s_f, kJ/(kg·K)	s_g, kJ/(kg·K)	c_{pf} kJ/(kg·K)	μ_f, 10^{-4} Pa·s	k_f, W/(m·K)
13.8[t]	0.070	0.0130	7.97	−308.9	140.3	4.97	37.52	6.37	0.255	0.073
14	0.079	0.0130	7.20	−307.6	142.1	5.06	37.19	6.47	0.248	0.075
15	0.134	0.0132	4.49	−300.9	151.1	5.53	36.65	6.91	0.218	0.082
16	0.216	0.0133	2.96	−293.4	159.6	5.99	34.31	7.36	0.194	0.089
17	0.329	0.0135	2.03	−285.6	167.6	6.45	33.11	7.88	0.175	0.092
18	0.482	0.0137	1.45	−277.3	175.0	6.92	32.05	8.42	0.159	0.095
19	0.682	0.0139	1.07	−268.4	181.7	7.38	31.08	8.93	0.146	0.097
20	0.935	0.0141	0.802	−258.9	187.7	7.85	30.19	9.45	0.135	0.098
21	1.250	0.0143	0.618	−248.8	193.0	8.32	29.37	10.13	0.125	0.100
22	1.634	0.0146	0.483	−237.9	197.3	8.80	28.60	10.82	0.116	0.101
23	2.096	0.0148	0.383	−226.3	200.5	9.29	27.86	11.69	0.108	0.101
24	2.645	0.0152	0.307	−213.9	202.7	9.78	27.15	12.52	0.101	0.100
25	3.288	0.0155	0.249	−200.4	203.6	10.29	26.46	13.44	0.094	0.099
26	4.035	0.0159	0.203	−185.9	203.1	10.81	25.79	14.81	0.088	0.098
27	4.892	0.0164	0.167	−170.2	200.9	11.36	25.11	16.18	0.082	0.096
28	5.88	0.0170	0.137	−152.9	196.5	11.93	24.41	18.5	0.076	0.094
29	6.98	0.0177	0.113	−133.6	189.5	12.54	23.68	22.1	0.070	0.091
30	8.23	0.0185	0.092	−111.7	178.8	13.20	22.89	26.6	0.065	0.087
31	9.63	0.0198	0.074	− 85.8	162.6	13.96	21.98	36.6	0.058	0.088
32	11.20	0.0217	0.057	− 52.4	136.1	14.92	20.81	65.4	0.051	0.092
33[c]	12.93	0.0318	0.032	38.3	38.3	17.56	17.56	∞	....	∞

*Values extracted and occasionally rounded off from McCarty, Hord, and Roder, NBS *Monogr.* 168, 1981. t = triple point; c = critical point.

TABLE 3-250 Saturated Hydrogen Peroxide*

T, K	P, bar	v_f, m³/kg	v_g, m³/kg	h_f, kJ/kg	h_g, kJ/kg	s_f, kJ/(kg·K)	s_g, kJ/(kg·K)	c_{pf}, kJ/(kg·K)	μ_f, 10^{-4} Pa·s	k_f, W/(m·K)
273	0.0004	0.00068	1672	−5577	−4027	2.990	8.662	1.45	18.0	0.483
300	0.0031	0.00069	235	−5510	−3995	3.224	8.269	1.48	11.3	0.481
350	0.0564	0.00072	15.1	−5376	−3933	3.631	7.758	1.54	4.3	0.474
400	0.4521	0.00076	2.12	−5238	−3878	4.032	7.440	1.61	2.2	0.464
450	2.143	0.00081	0.487	−5091	−3820	4.346	7.172	1.68	1.3	0.453
500	7.126	0.00088	0.155	−4945	−3777	4.656	6.992	1.75	0.89	0.443
550	18.56	0.00095	0.0605	−4794	−3745	4.941	6.846	1.82	0.65	0.431
600	40.75	0.00107	0.0268	−4635	−3731	5.209	6.720	1.90	0.50	0.416
650	79.27	0.00125	0.0125	−4463	−3746	5.485	6.582			
700	141.7	0.00171	0.0048	−4195	−3860	5.682	6.339			
708.5[c]	155.3	0.00284	0.0028	−4012	−4012	5.732	5.732			

*Values reproduced or converted from a tabulation by Tsykalo and Tabachnikov in V. A. Rabinovich (ed.), *Thermophysical Properties of Gases and Liquids*, Standartov, Moscow, 1968; NBS-NSF transl. TT 69-55091, 1970. The reader may be reminded that very pure hydrogen peroxide is very difficult to obtain owing to its decomposition or instability. c = critical point.

TABLE 3-251 Saturated Isobutane (R600a)*

T, K	P, bar	v_f, m³/kg	v_g, m³/kg	h_f, kJ/kg	h_g, kJ/kg	s_f, kJ/(kg·K)	s_g, kJ/(kg·K)	c_{pf}, kJ/(kg·K)	μ_f, 10^{-4} Pa·s	k_f, W/(m·K)
113.6[t]	1.9.−7	1.349.−3	8.60.+6	0.0	485.3	1.863	6.136			
120	9.3.−7	1.360.−3	1.84.+6	11.0	491.1	1.957	5.957	1.78		
140	4.8.−5	1.396.−3	4210	46.0	510.1	2.226	5.541	1.87		0.163
160	8.2.−4	1.435.−3	278.2	82.1	530.8	2.467	5.272	1.93	. . .	0.158
180	0.0070	1.476.−3	36.66	119.5	533.0	2.688	5.097	1.99	9.46	0.149
200	0.0369	1.520.−3	7.723	158.5	576.7	2.893	4.984	2.05	6.06	0.142
220	0.1374	1.568.−3	2.265	199.0	601.5	3.086	4.916	2.12	4.21	0.134
240	0.3989	1.621.−3	0.8432	241.4	627.4	3.270	4.878	2.19	3.11	0.127
260	0.9600	1.680.−3	0.3738	285.8	654.2	3.446	4.863	2.28	2.40	0.120
270	1.4081	1.712.−3	0.2617	308.8	667.7	3.532	4.861	2.33	2.14	0.117
280	2.0020	1.746.−3	0.1882	332.3	681.3	3.617	4.863	2.39	1.93	0.113
290	2.7686	1.784.−3	0.1385	356.4	694.9	3.700	4.867	2.46	1.75	0.110
300	3.7365	1.824.−3	0.1040	381.1	708.4	3.783	4.874	2.53	1.59	0.106
310	4.934	1.868.−3	0.0794	406.4	721.7	3.865	4.882	2.61	1.46	0.102
320	6.392	1.916.−3	0.0614	432.4	734.8	3.946	4.891	2.70	1.35	0.099
330	8.140	1.971.−3	0.0481	459.2	747.7	4.028	4.902	2.81	1.25	0.095
340	10.21	2.032.−3	0.0380	486.9	760.0	4.109	4.912	2.92	1.15	0.092
350	12.64	2.103.−3	0.0301	515.7	771.8	4.191	4.923	3.04	1.05	0.088
360	15.46	2.187.−3	0.0240	545.6	782.7	4.273	4.932	3.17	0.95	0.083
370	18.72	2.289.−3	0.0190	577.1	792.3	4.357	4.939	3.31	0.85	0.080
380	22.48	2.420.−3	0.0150	610.6	799.8	4.444	4.942	3.45	0.75	0.076
390	26.82	2.604.−3	0.0115	647.1	803.7	4.536	4.937	3.62	0.63	0.071
400	31.86	2.920.−3	0.0083	689.6	799.6	4.639	4.915	3.85	0.51	0.065
408.0[c]	36.55	4.464.−3	0.0045	752.5	752.5	4.791	4.791	∞	. . .	∞

*Values reproduced or converted from Goodwin, NBSIR 79-1612, 1979. t = triple point; c = critical point. The notation 1.9.−7 signifies 1.9×10^{-7}.

TABLE 3-252 Saturated Krypton*

T, K	P, bar	v_f, m³/kg	v_g, m³/kg	h_f, kJ/kg	h_g, kJ/kg	s_f, kJ/(kg·K)	s_g, kJ/(kg·K)	c_{pf}, kJ/(kg·K)	μ_f, 10^{-4} Pa·s	k_f, W/(m·K)
10	. . . . .	3.235.−4	. . . . .	0.22	. . . .	0.0256	. . . .	0.070		
20	. . . . .	3.246.−4	. . . . .	1.59	. . . .	0.1141	. . . .	0.188		
30	. . . . .	3.265.−4	. . . . .	3.84	. . . .	0.2034	. . . .	0.247		
40	. . . . .	3.288.−4	. . . . .	6.49	. . . .	0.2791	. . . .	0.276		
50	. . . . .	3.313.−4	. . . . .	9.37	. . . .	0.3431	. . . .	0.295		
60	. . . . .	3.341.−4	. . . . .	12.40	. . . .	0.3982	. . . .	0.311		
70	. . . . .	3.372.−4	. . . . .	15.57	. . . .	0.4471	. . . .	0.327		
80	. . . . .	3.407.−4	. . . . .	18.97	. . . .	0.4925	. . . .	0.345		
90	. . . . .	3.446.−4	. . . . .	22.58	. . . .	0.5353	. . . .	0.366		
100	. . . . .	3.492.−4	. . . . .	26.42	. . . .	0.5765	. . . .	0.389		
110	. . . . .	3.544.−4	. . . . .	30.52	. . . .	0.6165	. . . .	0.414		
115.76	. . . . .	3.579.−4	. . . . .	33.18	. . . .	0.6390	. . . .	0.427		
115.76	0.732	4.090.−4	0.1529	52.78	161.8	0.8095	1.751	0.547		
119.76	1.013	4.143.−4	0.1136	54.99	162.6	0.8279	1.726	0.545		
120	1.032	4.146.−4	0.1116	55.09	162.6	0.8291	1.724	0.544	3.72	0.0900
130	2.112	4.284.−4	0.0578	60.55	164.1	0.8724	1.669	0.542	3.16	0.0828
140	3.878	4.440.−4	0.0330	66.02	165.3	0.9124	1.622	0.546	2.64	0.0756
150	6.552	4.619.−4	0.0201	71.58	166.1	0.9499	1.580	0.559	2.20	0.0688
160	10.37	4.831.−4	0.0130	77.34	166.4	0.9859	1.543	0.587	1.87	0.0625
170	15.57	5.091.−4	0.0086	83.48	166.0	1.022	1.507	0.641	1.54	0.0558
180	22.41	5.423.−4	0.0059	90.26	164.6	1.058	1.472	0.734	1.28	0.0494
190	31.20	5.882.−4	0.0040	98.19	161.8	1.098	1.433	0.905	1.05	0.0433
200	42.23	6.641.−4	0.0026	108.40	156.0	1.147	1.386	1.515	0.80	0.0348
209.39	54.96	1.098.−3	0.0011	133.90	133.9	1.262	1.262	∞	. . .	∞

*Values extracted and in some cases rounded off from those cited in Rabinovich (ed.), *Thermophysical Properties of Neon, Argon, Krypton and Xenon*, Standards Press, Moscow, 1976. This source contains values for the compressed state for pressures up to 1000 bar, etc. The notation 3.235.−4 signifies 3.235×10^{-4}.

TABLE 3-253 Compressed Krypton*

Temperature, K		Pressure, bar: 1	10	20	40	60	80	100	200	400	600	800	1000
100	*v*	3.49.−4	3.49.−4	3.49.−4	3.48.−4	3.48.−4	3.47.−4	3.47.−4	3.45.−4	3.42.−4	3.39.−4	3.36.−4	3.33.−4
	h	26.42	26.69	26.99	27.59	28.18	28.78	29.38	32.38	38.42	44.47	50.52	56.57
	s	0.5765	0.5760	0.5755	0.5745	0.5735	0.5724	0.5714	0.5667	0.5580	0.5503	0.5432	0.5366
200	*v*	0.1971	0.0184	8.39.−3	3.00.−3	6.19.−4	5.94.−4	5.76.−4	5.27.−4	4.83.−4	4.58.−4	4.41.−4	4.28.−4
	h	183.1	179.3	174.5	159.4	105.6	104.4	103.6	102.7	105.3	109.7	114.9	120.3
	s	1.859	1.618	1.533	1.405	1.129	1.116	1.106	1.073	1.037	1.013	0.993	0.977
300	*v*	0.2971	0.0292	0.0143	6.84.−3	4.37.−3	3.14.−3	2.41.−3	1.09.−3	6.92.−4	5.94.−4	5.44.−4	5.13.−4
	h	208.1	206.3	204.2	200.0	195.7	191.2	186.6	166.8	155.1	155.2	158.2	162.3
	s	1.961	1.728	1.654	1.575	1.525	1.485	1.451	1.333	1.239	1.196	1.169	1.149
400	*v*	0.3966	0.0394	0.0196	9.67.−3	6.37.−3	4.73.−3	3.75.−3	1.85.−3	1.01.−3	7.79.−4	6.76.−4	6.14.−4
	h	233.0	231.9	230.7	228.3	225.9	223.6	221.3	211.4	199.7	196.8	197.9	200.8
	s	2.032	1.802	1.730	1.657	1.612	1.579	1.552	1.463	1.368	1.317	1.284	1.259
500	*v*	0.4960	0.0495	0.0247	0.0123	8.20.−3	6.15.−3	4.91.−3	2.49.−3	1.33.−3	9.81.−4	8.22.−4	7.29.−4
	h	257.8	257.1	256.3	254.9	253.3	251.9	250.5	244.5	236.9	234.2	234.7	237.4
	s	2.088	1.858	1.788	1.716	1.673	1.642	1.617	1.537	1.451	1.400	1.365	1.340
600	*v*	0.5953	0.0596	0.0298	0.0149	9.96.−3	7.49.−3	6.01.−3	3.07.−3	1.64.−3	1.18.−3	9.67.−4	8.44.−4
	h	282.7	282.2	281.7	280.7	279.7	278.8	277.9	274.2	269.6	268.1	269.1	271.7
	s	2.133	1.904	1.834	1.763	1.721	1.691	1.667	1.591	1.511	1.462	1.428	1.403
700	*v*	0.6946	0.0696	0.0348	0.0175	0.0117	8.80.−3	7.07.−3	3.62.−3	1.93.−3	1.38.−3	1.11.−3	9.56.−4
	h	307.5	307.2	306.9	306.2	305.6	305.1	304.5	302.2	299.8	299.6	301.1	304.0
	s	2.171	1.942	1.873	1.803	1.761	1.732	1.708	1.634	1.557	1.511	1.478	1.453
800	*v*	0.7939	0.0795	0.0399	0.0200	0.0134	0.0101	8.11.−3	4.16.−3	2.21.−3	1.57.−3	1.25.−3	1.07.−3
	h	332.3	332.2	331.9	331.6	331.2	330.9	330.5	329.3	328.6	329.4	331.5	334.6
	s	2.204	1.975	1.906	1.837	1.795	1.766	1.743	1.671	1.596	1.551	1.518	1.494
900	*v*	0.8931	0.0895	0.0448	0.0225	0.0151	0.0114	9.13.−3	4.68.−3	2.48.−3	1.75.−3	1.39.−3	1.18.−3
	h	357.1	357.0	356.9	356.8	356.6	356.4	356.3	355.8	356.3	358.0	360.7	364.2
	s	2.233	2.005	1.936	1.866	1.825	1.796	1.773	1.702	1.628	1.584	1.553	1.528
1000	*v*	0.9924	0.0994	0.0498	0.0250	0.0168	0.0126	0.0102	5.20.−3	2.74.−3	1.93.−3	1.53.−3	1.29.−3
	h	381.9	381.9	381.9	381.8	381.8	381.8	381.8	381.9	383.4	385.9	389.0	392.8
	s	2.260	2.031	1.962	1.893	1.852	1.823	1.800	1.729	1.657	1.614	1.583	1.559

*Values extracted and in some cases rounded off from those cited in Rabinovich (ed), *Thermophysical Properties of Neon, Argon,, Krypton and Xenon*, Standards Press, Moscow, 1976. This source contains an exhaustive tabulation of values. v = specific volume, m^3/kg; h = specific enthalpy, kJ/kg; s = specific entropy, kJ/(kg·K). The notation 3.49.−4 signifies 3.49×10^{-4}.

TABLE 3-254 Saturated Lithium*

T, K	P, bar	v_f, m^3/kg	v_g, m^3/kg	h_f, kJ/kg	h_g, kJ/kg	s_f, kJ/(kg·K)	s_g, kJ/(kg·K)	c_{pf}, kJ/(kg·K)
453.7^m	1.78.−13	1.912.−3	....	1703	24259	6.776	56.492	4.30
500	8.21.−12	1.946.−3	....	1905	24390	7.199	52.169	4.34
600	4.18.−9	1.988.−3	....	2334	24674	7.983	45.216	4.23
700	3.51.−7	2.028.−3	2.40.+7	2697	24869	8.633	40.307	4.19
800	9.57.−6	2.070.−3	9.94.+5	3174	25162	9.192	36.678	4.17
900	1.24.−4	2.114.−3	8.55.+4	3590	25341	9.682	33.850	4.16
1000	9.60.−4	2.160.−3	1.22.+4	4006	25477	10.120	31.591	4.16
1200	0.0204	2.262.−3	669.3	4835	25654	10.876	28.225	4.14
1400	0.1794	2.370.−3	86.06	5668	25778	11.518	25.882	4.19
1500	0.4269	2.433.−3	38.17	6088	25845	11.808	24.979	4.20

*Converted from tables in Vargaftik, *Tables of the Thermophysical Properties of Liquids and Gases*, Nauka, Moscow, 1972, and Hemisphere, Washington, 1975. m = melting point. The notation 1.78.−13 signifies 1.78×10^{-13}.

TABLE 3-255 Saturated Mercury*

T, K	P, bar	$v_f \times 10^5$, m^3/kg	v_g, m^3/kg	h_f, kJ/kg	h_g, kJ/kg	h_{fg}, kJ/kg	s_f, kJ(kg·K)	s_g, kJ(kg·K)
203.15	$2.298 \cdot 10^{-11}$	7.26239	$3.665 \cdot 10^{9}$	33.131	342.637	309.506	0.32434	1.84787
213.15	$1.288 \cdot 10^{-10}$	7.27570	$6.862 \cdot 10^{8}$	34.567	343.674	309.107	0.33124	1.78142
223.15	$6.169 \cdot 10^{-10}$	7.28900	$1.499 \cdot 10^{8}$	35.997	344.710	308.713	0.33780	1.72123
233.15	$2.580 \cdot 10^{-9}$	7.30231	$3.746 \cdot 10^{7}$	37.422	345.746	308.324	0.34404	1.66647
243.15	$9.573 \cdot 10^{-9}$	7.31563	$1.053 \cdot 10^{7}$	38.842	346.782	307.940	0.35001	1.61647
253.15	$3.198 \cdot 10^{-8}$	7.32896	$3.281 \cdot 10^{6}$	40.258	347.819	307.561	0.35571	1.57065
263.15	$9.736 \cdot 10^{-8}$	7.34229	$1.120 \cdot 10^{6}$	41.668	348.855	307.187	0.36118	1.52852
273.15	$2.728 \cdot 10^{-7}$	7.35563	$4.150 \cdot 10^{5}$	43.074	349.891	306.817	0.36642	1.48967
283.15	$7.101 \cdot 10^{-7}$	7.36898	$1.653 \cdot 10^{5}$	44.476	350.927	306.451	0.37146	1.45375
293.15	$1.729 \cdot 10^{-6}$	7.38234	$7.026 \cdot 10^{4}$	45.874	351.964	306.090	0.37631	1.42045
303.15	$3.968 \cdot 10^{-6}$	7.39572	$3.167 \cdot 10^{4}$	47.268	353.000	305.732	0.38099	1.38951
313.15	$8.626 \cdot 10^{-6}$	7.40911	$1.505 \cdot 10^{4}$	48.659	354.036	305.377	0.38550	1.36068
323.15	$1.786 \cdot 10^{-5}$	7.42252	$7.501 \cdot 10^{3}$	50.046	355.072	305.026	0.38986	1.33378
333.15	$3.356 \cdot 10^{-5}$	7.43594	$3.905 \cdot 10^{3}$	51.430	356.108	304.678	0.39408	1.30862
343.15	$6.724 \cdot 10^{-5}$	7.44938	$2.115 \cdot 10^{3}$	52.810	357.145	304.335	0.39816	1.28505
353.15	$1.232 \cdot 10^{-4}$	7.46285	$1.188 \cdot 10^{3}$	54.188	358.181	303.993	0.40212	1.26292
363.15	$2.182 \cdot 10^{-4}$	7.47633	$6.899 \cdot 10^{2}$	55.563	359.217	303.654	0.40596	1.24213
373.15	$3.745 \cdot 10^{-4}$	7.48984	413.0	56.936	360.253	303.317	0.40969	1.22255
383.15	$6.247 \cdot 10^{-4}$	7.50337	254.2	58.306	361.289	302.983	0.41331	1.20408
393.15	$1.015 \cdot 10^{-3}$	7.51693	153.6	59.674	362.326	302.652	0.41684	1.18665
403.15	$1.608 \cdot 10^{-3}$	7.53052	103.9	61.039	363.362	302.323	0.42027	1.17017
413.15	$2.491 \cdot 10^{-3}$	7.55415	68.75	62.403	364.397	301.994	0.42361	1.15456
423.15	$3.778 \cdot 10^{-3}$	7.55780	46.43	63.765	365.433	301.668	0.42687	1.13978
433.15	$5.618 \cdot 10^{-3}$	7.57148	31.96	65.125	366.469	301.344	0.43004	1.12575
443.15	$8.204 \cdot 10^{-3}$	7.58520	22.39	66.484	367.504	301.020	0.43314	1.11242
453.15	$1.178 \cdot 10^{-2}$	7.59897	15.95	67.842	368.539	300.697	0.43617	1.09975
463.15	$1.664 \cdot 10^{-2}$	7.61277	11.54	69.198	369.574	300.376	0.43913	1.08768
473.15	$2.315 \cdot 10^{-2}$	7.62662	8.469	70.553	370.609	300.056	0.44203	1.07619
483.15	$3.177 \cdot 10^{-2}$	7.64051	6.301	71.908	371.642	299.734	0.44486	1.06524
493.15	$4.304 \cdot 10^{-2}$	7.65444	4.748	73.261	372.676	299.415	0.44763	1.05478
503.15	$5.758 \cdot 10^{-2}$	7.66843	3.621	74.614	373.708	299.094	0.45035	1.04479
513.15	$7.614 \cdot 10^{-2}$	7.68247	2.793	75.967	374.740	298.773	0.45301	1.03525
523.15	$9.959 \cdot 10^{-2}$	7.69656	2.176	77.319	375.771	298.452	0.45562	1.02611
533.15	0.12892	7.71071	1.7132	78.671	376.800	298.129	0.45818	1.01737
543.15	0.16527	7.72491	1.3613	80.023	377.829	297.806	0.46069	1.00899
553.15	0.20993	7.73918	1.0912	81.375	378.855	297.480	0.46316	1.00095
563.15	0.26435	7.75351	0.88213	82.728	379.880	297.152	0.46558	0.99324
573.15	0.33015	7.7679	0.71874	84.080	380.904	296.824	0.46796	0.98584
583.15	0.40910	7.7823	0.59002	85.434	381.925	296.491	0.47030	0.97893
593.15	0.50320	7.7969	0.48779	86.788	382.944	296.156	0.47260	0.97190
603.15	0.61460	7.8115	0.40600	88.143	383.960	295.817	0.47487	0.96532
613.15	0.74567	7.8262	0.34008	89.499	384.973	295.474	0.47709	0.95899
623.15	0.89896	7.8409	0.28660	90.856	385.984	295.128	0.47929	0.95289
633.15	1.0772	7.8558	0.24291	92.215	386.991	294.776	0.48145	0.94702
643.15	1.2834	7.8707	0.20702	93.575	387.994	294.419	0.48358	0.94135
653.15	1.5207	7.8858	0.17735	94.937	388.994	294.057	0.48568	0.93589
663.15	1.9725	7.9008	0.15269	96.300	389.989	293.689	0.48774	0.93061
673.15	2.1024	7.9160	0.13207	97.666	390.980	293.314	0.48978	0.92552
683.15	2.454	7.9313	0.11476	99.033	391.966	292.933	0.49180	0.92059
693.15	2.852	7.9467	0.10014	100.403	392.947	292.544	0.49378	0.91583
703.15	3.299	7.9622	0.08775	101.775	393.923	292.148	0.49574	0.91123
713.15	3.801	7.9778	0.07719	103.150	394.893	291.743	0.49768	0.90677
723.15	4.362	7.9935	0.06815	104.528	395.858	291.330	0.49959	0.90245
733.15	4.986	8.0094	0.06039	105.908	396.816	290.908	0.50148	0.89827
743.15	5.679	8.0252	0.05369	107.292	397.767	290.475	0.50335	0.89422
753.15	6.446	8.0413	0.04789	108.679	398.711	290.032	0.50519	0.89029
763.15	7.292	8.0574	0.04285	110.069	399.649	289.580	0.50702	0.88647
773.15	8.222	8.074	0.03846	111.463	400.579	289.116	0.50882	0.88277
783.15	9.242	8.090	0.03462	112.861	401.501	288.640	0.51061	0.87917
793.15	10.358	8.106	0.03124	114.262	402.415	288.153	0.51238	0.87568
803.15	11.576	8.123	0.02827	115.668	403.321	287.653	0.51412	0.87228
813.15	12.901	8.140	0.02565	117.078	404.218	287.140	0.51586	0.86898
823.15	14.340	8.157	0.02333	118.492	405.106	286.614	0.51757	0.86576
833.15	15.899	8.174	0.02126	119.911	405.985	286.074	0.51927	0.86263

*From Vukalovich, Ivanov, Fokin, and Yakovlev, *Thermophysical Properties of Mercury*, Standartov, Moscow, 1971. For the saturated liquid the specific volume at 203.15 K is 7.26239 × 10^{-5} m^3/kg, etc. All the tabular values for 203.15 K, 213.15 K, 223.15 K, and 233.15 K represents a metastable equilibrium between the subcooled liquid and the saturated vapor.

TABLE 3-255 Saturated Mercury* *(Concluded)*

T, K	P, bar	$v_f \times 10^5$, m^3/kg	v_g, m^3/kg	h_f, kJ/kg	h_g, kJ/kg	h_{fg}, kJ/kg	s_f, kJ(kg·K)	s_g, kJ(kg·K)
843.15	17.584	8.191	0.019426	121.335	406.855	285.520	0.52095	0.85959
853.15	19.403	8.209	0.017785	122.763	407.715	284.952	0.52262	0.85662
863.15	21.36	8.226	0.016317	124.197	408.565	284.368	0.52427	0.85372
873.15	23.46	8.244	0.015000	125.636	409.405	283.769	0.52591	0.85090
883.15	25.72	8.262	0.013815	127.080	410.235	283.155	0.52753	0.84815
893.15	28.14	8.280	0.012748	128.530	411.054	282.524	0.52914	0.84546
903.15	30.72	8.298	0.011784	129.986	411.861	281.875	0.53074	0.84284
913.15	33.47	8.316	0.010911	131.448	412.658	281.210	0.53232	0.84028
923.15	36.41	8.335	0.010120	132.915	413.444	280.529	0.53389	0.83777
933.15	39.53	8.353	0.009401	134.389	414.218	279.829	0.53545	0.83533
943.15	42.85	8.372	0.008746	135.869	414.980	279.111	0.53700	0.83294
953.15	46.36	8.391	0.008150	137.356	415.731	278.375	0.53854	0.83060
963.15	50.09	8.410	0.007604	138.850	416.469	277.619	0.54006	0.82831
973.15	54.03	8.430	0.007105	140.350	417.195	276.845	0.54158	0.82606
983.15	58.20	8.450	0.006648	141.858	417.909	276.051	0.54308	0.82387
993.15	62.59	8.468	0.006228	143.372	418.610	275.238	0.54458	0.82172
1003.15	67.22	8.488	0.005842	144.894	419.298	274.404	0.54607	0.81961
1013.15	72.10	8.508	0.005487	146.424	419.974	273.550	0.54754	0.81754
1023.15	77.22	8.529	0.005159	147.961	420.636	272.675	0.54901	0.81552
1033.15	82.60	8.550	0.004856	149.506	421.286	271.780	0.55047	0.81353
1043.15	88.25	8.570	0.004576	151.059	421.923	270.864	0.55192	0.81158
1053.15	94.17	8.590	0.004317	152.619	422.546	269.927	0.55336	0.80966
1063.15	100.37	8.612	0.004077	154.188	423.156	268.968	0.55479	0.80778
1073.15	106.85	8.632	0.003854	155.766	423.752	267.986	0.55621	0.80593

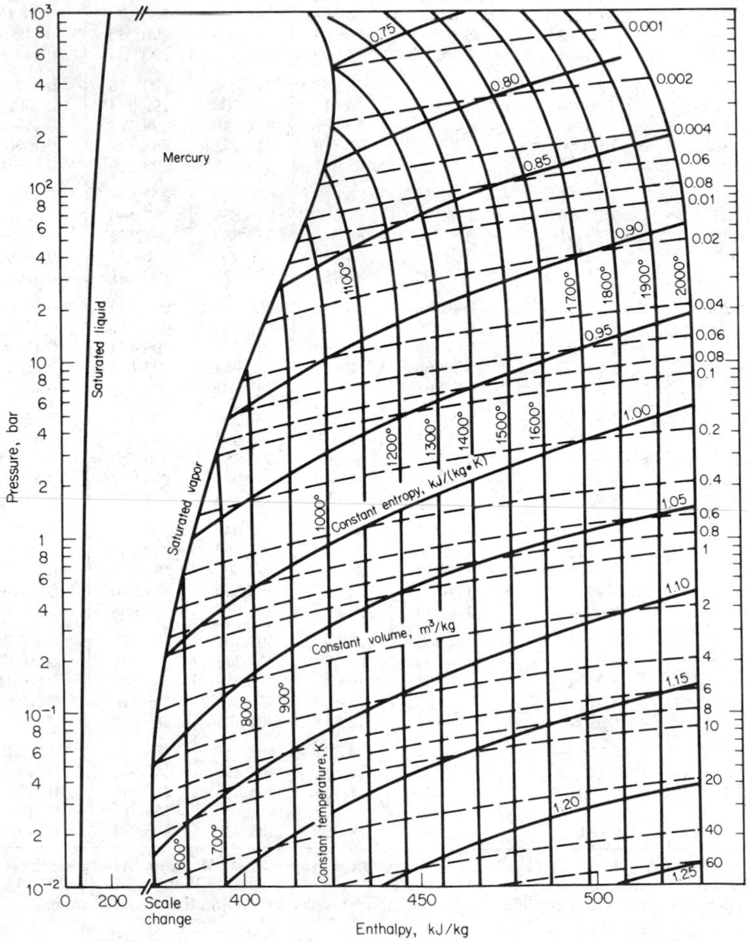

FIG. 3-25 Enthalpy–log-pressure diagram for mercury. *(Drawn from tabular data in footnote reference to Table 3-255.)*

TABLE 3-256 Saturated Methane*

T, K	P, bar	v_f, m^3/kg	v_g, m^3/kg	h_f, kJ/kg	h_g, kJ/kg	s_f, kJ/(kg·K)	s_g, kJ/(kg·K)	c_{pf}, kJ/(kg·K)	μ_f, 10^{-4} Pa·s	k_f, W/(m·K)
90.7[t]	0.117	2.215.−3	3.976	216.4	759.9	4.231	10.225	3.288	2.02	0.225
95	0.198	2.244.−3	2.463	232.5	769.0	4.406	10.034	3.318	1.71	0.215
100	0.345	2.278.−3	1.479	246.3	776.9	4.556	9.862	3.369	1.56	0.206
105	0.565	2.316.−3	0.940	263.2	785.7	4.719	9.710	3.425	1.33	0.197
110	0.884	2.353.−3	0.625	280.1	794.5	4.882	9.558	3.478	1.22	0.189
115	1.325	2.396.−3	0.430	297.7	802.5	5.035	9.436	3.525	1.09	0.181
120	1.919	2.438.−3	0.306	315.3	810.4	5.188	9.314	3.570	0.98	0.173
125	2.693	2.487.−3	0.223	333.5	817.3	5.332	9.062	3.620	0.89	0.165
130	3.681	2.536.−3	0.167	351.7	824.1	5.476	8.810	3.679	0.81	0.158
135	4.912	2.594.−3	0.127	370.6	829.5	5.614	8.871	3.755	0.73	0.150
140	6.422	2.652.−3	0.098	389.5	834.8	5.751	8.932	3.849	0.66	0.143
145	8.246	2.722.−3	0.077	409.5	844.4	5.885	8.891	3.965	0.61	0.136
150	10.41	2.792.−3	0.061	429.4	853.9	6.019	8.849	4.101	0.56	0.129
155	12.97	2.882.−3	0.049	450.8	848.5	6.151	8.725	4.27	0.51	0.122
160	15.94	2.971.−3	0.039	472.1	843.0	6.283	8.601	4.47	0.46	0.115
165	19.39	3.095.−3	0.032	495.4	840.0	6.417	8.513	4.75	0.42	0.108
170	23.81	3.218.−3	0.026	518.6	837.0	6.551	8.424	5.16	0.38	0.101
175	27.81	3.419.−3	0.020	545.8	827.6	6.697	8.315	5.89	0.34	0.094
180	32.86	3.619.−3	0.016	572.9	818.1	6.843	8.205	7.27	0.30	0.088
185	38.59	3.979.−3	0.012	605.4	797.7	7.017	8.049	11.1	0.25	0.085
190	45.20	4.900.−3	0.008	661.6	750.7	7.293	7.762	70.	0.19	0.090
190.6[c]	45.99	6.233.−3	0.006	704.4	704.4	7.516	7.516	∞	0.17	∞

*Values reproduced or converted from Goodwin, NBS Tech. Note 653, 1974. t = triple point; c = critical point. The notation 2.215.−3 signifies 2.215×10^{-3}.

TABLE 3-257 Superheated Methane*

P, bar		Temperature, K: 100	150	200	250	300	350	400	450	500
1	v	0.00228	0.7661	1.0299	1.2915	1.5521	1.8122	2.0719	2.3669	2.5911
	h	246.4	879.0	984.3	1090.4	1199.8	1314.8	1437.4	1568.8	1708.9
	s	4.555	10.152	10.757	11.230	11.629	11.983	12.310	12.618	12.914
5	v	0.00228	0.1434	0.2006	0.2549	0.3083	0.3611	0.4136	0.4657	0.5181
	h	247.0	865.0	976.1	1084.7	1195.5	1311.5	1434.7	1566.6	1706.9
	s	4.553	9.256	9.896	10.381	10.785	11.142	11.471	11.781	12.066
10	v	0.00227	0.0643	0.0968	0.1254	0.1528	0.1798	0.2063	0.2327	0.2590
	h	247.8	843.6	965.5	1077.9	1190.6	1307.9	1432.0	1564.1	1705.3
	s	4.549	8.797	9.501	10.002	10.414	10.775	11.106	11.417	11.715
20	v	0.00227	0.00277	0.0446	0.0606	0.0751	0.0891	0.1027	0.1162	0.1295
	h	249.4	429.8	941.9	1063.6	1180.7	1300.6	1426.5	1560.3	1702.1
	s	4.542	6.003	9.059	9.603	10.030	10.400	10.736	11.050	11.349
40	v	0.00226	0.00274	0.0176	0.0281	0.0363	0.0438	0.0510	0.0579	0.0648
	h	252.5	430.8	879.3	1032.9	1160.5	1286.0	1415.7	1552.1	1696.0
	s	4.528	5.973	8.465	9.155	9.621	10.008	10.354	10.674	10.978
60	v	0.00226	0.00271	0.00615	0.0173	0.0234	0.0287	0.0338	0.0386	0.0432
	h	255.7	432.2	734.0	999.8	1140.0	1271.7	1405.1	1544.2	1690.0
	s	4.515	5.946	7.623	8.847	9.359	9.765	10.121	10.440	10.756
80	v	0.00225	0.00268	0.00411	0.0119	0.0171	0.0213	0.0252	0.0289	0.0324
	h	258.9	433.8	660.5	964.4	1119.7	1257.7	1394.9	1536.6	1684.4
	s	4.502	5.920	7.209	8.590	9.158	9.584	9.951	10.283	10.595
100	v	0.00224	0.00266	0.00375	0.00888	0.0133	0.0169	0.0201	0.0231	0.0260
	h	262.1	435.5	644.5	928.5	1099.6	1244.2	1385.2	1529.4	1679.0
	s	4.489	5.897	7.090	8.364	8.991	9.437	9.814	10.153	10.469
150	v	0.00223	0.00261	0.00337	0.00555	0.00852	0.0111	0.0134	0.0155	0.0175
	h	270.2	440.7	630.2	860.0	1054.1	1213.1	1362.8	1513.0	1667.0
	s	4.458	5.843	6.930	7.953	8.664	9.155	9.555	9.907	10.233
200	v	0.00221	0.00256	0.00318	0.00447	0.00644	0.00837	0.0101	0.0118	0.0133
	h	278.3	446.5	626.5	825.0	1019.8	1187.2	1343.8	1498.9	1656.9
	s	4.429	5.796	6.829	7.719	8.426	8.944	9.362	9.727	10.060
300	v	0.00218	0.00249	0.00296	0.00369	0.00474	0.00593	0.00708	0.00818	0.00924
	h	294.7	459.6	629.2	804.4	982.9	1153.6	1316.8	1478.5	1642.2
	s	4.373	5.714	6.690	7.471	8.122	8.649	9.085	9.465	9.811
400	v	......	0.00244	0.00282	0.00336	0.00406	0.00486	0.00569	0.00560	0.00729
	h	......	473.8	637.7	802.4	970.1	1137.8	1303.0	1467.7	1634.7
	s	......	5.645	6.588	7.323	7.935	8.451	8.893	9.280	9.633
500	v	......	0.00239	0.00272	0.00315	0.00368	0.00428	0.00492	0.00555	0.00616
	h	......	488.8	648.9	807.7	969.0	1132.8	1297.8	1464.2	1633.2
	s	......	5.584	6.507	7.215	7.802	8.307	8.748	9.139	9.496

*Converted and rounded off from the tables of Goodwin, NBS Tech. Note 654, 1974. v = specific volume, m^3/kg; h = specific enthalpy, kJ/kg; s = specific entropy, kJ/(kg·K).

TABLE 3-258 Saturated Methanol*

T, K	P, bar	v_f, m^3/kg	v_g, m^3/kg	h_f, kJ/kg	h_g, kJ/kg	s_f, kJ/(kg·K)	s_g, kJ/(kg·K)	c_{pf}, kJ/(kg·K)
175.4[t]	1.887.−6	1.105.−3	241200	314.3	1624.3	2.766	10.233	2.177
180	3.909.−6	1.111.−3	119600	324.4	1629.8	2.822	10.075	2.181
190	1.693.−5	1.124.−3	29690	346.1	1642.2	2.940	9.761	2.190
200	6.274.−5	1.137.−3	8278	367.8	1654.5	3.051	9.485	2.201
210	2.035.−4	1.150.−3	2680	389.9	1666.9	3.169	9.240	2.215
220	5.881.−4	1.163.−3	970.7	412.5	1679.2	3.264	9.022	2.281
230	1.539.−3	1.176.−3	385.3	435.6	1691.8	3.366	8.829	2.251
240	3.687.−3	1.189.−3	168.7	459.1	1704.2	3.467	8.655	2.275
250	8.189.−3	1.203.−3	79.02	483.0	1716.6	3.664	8.499	2.303
260	1.700.−2	1.216.−3	39.49	507.1	1728.8	3.659	8.358	2.337
270	3.327.−2	1.230.−3	20.93	531.5	1740.7	3.751	8.230	2.375
280	6.208.−2	1.244.−3	11.62	556.1	1752.1	3.840	8.112	2.420
290	0.1094	1.259.−3	6.778	581.0	1763.2	3.928	8.005	2.471
300	0.1860	1.274.−3	4.095	606.2	1773.5	4.013	7.904	2.528
310	0.3043	1.290.−3	2.566	631.7	1783.1	4.097	7.811	2.59
320	0.4817	1.306.−3	1.661	657.5	1791.9	4.179	7.723	2.67
330	0.7395	1.323.−3	1.103	683.6	1799.8	4.259	7.640	2.75
337.5	1.0012	1.337.−3	0.8110	703.4	1804.8	4.310	7.581	2.81
340	1.1044	1.342.−3	0.7533	710.1	1806.5	4.338	7.562	2.83
350	1.6082	1.361.−3	0.5256	736.9	1812.4	4.415	7.488	2.91
360	2.288	1.381.−3	0.3752	764.3	1817.4	4.492	7.417	3.01
370	3.188	1.403.−3	0.2723	792.5	1821.6	4.569	7.350	3.11
380	4.357	1.426.−3	0.2015	820.2	1825.1	4.643	7.287	3.22
390	5.845	1.452.−3	0.1512	856.4	1828.1	4.736	7.227	3.34
400	7.703	1.480.−3	0.1155	888.8	1830.7	4.817	7.172	3.48
410	10.00	1.510.−3	0.0893	917.7	1832.8	4.888	7.120	3.63
420	12.83	1.543.−3	0.0698	947.7	1834.0	4.959	7.069	3.80
430	16.26	1.581.−3	0.0542	982.2	1834.1	5.039	7.020	4.05
440	20.40	1.624.−3	0.0425	1022.7	1832.4	5.130	6.971	4.33
450	25.30	1.674.−3	0.0333	1069.2	1828.4	5.233	6.920	4.68
460	31.08	1.733.−3	0.0263	1120.6	1820.8	5.344	6.866	5.06
470	37.80	1.806.−3	0.0202	1175.8	1807.5	5.460	6.804	5.29
480	45.61	1.896.−3	0.0156	1233.4	1785.6	5.578	6.728	5.54
490	54.66	2.021.−3	0.0115	1292.5	1751.4	5.696	6.633	5.77
500	65.17	2.214.−3	0.0084	1353.7	1704.3	5.816	6.516	6.1
510	77.43	2.69.−3	0.0059	1437.6	1649.4	5.975	6.391	
512.7[c]	81.03	3.64.−3	0.0036	1532	1532	6.170	6.170	∞

*Interpolated from Zubarev, Prusakov, and Sergeyeva, *Thermophysical Properties of Methanol—A Handbook*, Standartov, Moscow, 1973 (in Russian). v = specific volume, m^3/kg; h = specific enthalpy, kJ/kg; s = specific entropy, kJ/(kg·K); c_p = specific heat at constant pressure, kJ/(kg·K). The notation 1.887.−6 signifies 1.887×10^{-6}.

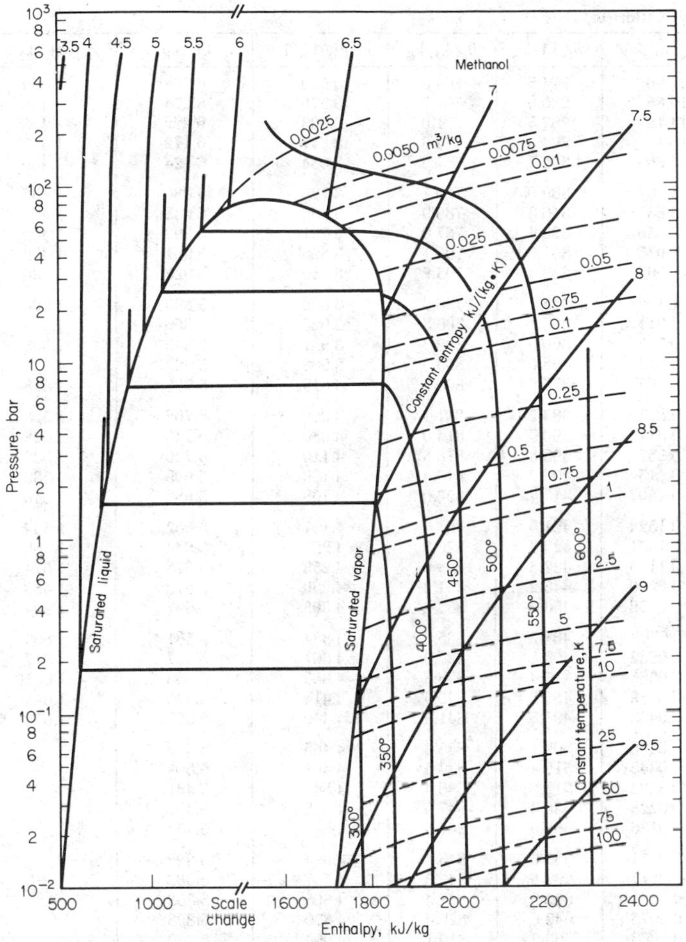

FIG. 3-26 Enthalpy–log-pressure diagram for methanol. (*Drawn* from tabular data in footnote reference to Table 3-258.)

TABLE 3-259 Saturated Methylamine (CH_3NH_2)*

Temp., °F. t	Abs. pressure, lb./sq. in. p	Volume, cu. ft./lb., vapor v_g	Enthalpy, B.t.u./lb. Liquid h_f	Enthalpy, B.t.u./lb. Vapor h_g	Entropy, B.t.u./lb. (°R.) Liquid s_f	Entropy, B.t.u./lb. (°R.) Vapor s_g	Temp., °F. t	Abs. pressure, lb./sq. in. p	Volume, cu. ft./lb., vapor v_g	Enthalpy, B.t.u./lb. Liquid h_f	Enthalpy, B.t.u./lb. Vapor h_g	Entropy, B.t.u./lb. (°R.) Liquid s_f	Entropy, B.t.u./lb. (°R.) Vapor s_g
− 58	1.322	98.93	−12.4	374.8	−0.0503	0.9135	23	15.99	10.01	45.6	404.3	.1015	.8443
− 40	2.532	56.72	0.0	381.7	.0000	.9092	41	24.49	6.796	59.2	410.1	.1292	.8300
− 22	4.551	32.56	12.7	388.4	.0294	.8877	68	43.52	3.985	80.2	418.8	.1698	.8115
− 4	7.78	19.84	25.7	395.0	.0586	.8689	86	61.53	2.962	94.5	424.4	.1963	.8008
+ 5	10.03	15.54	32.6	398.5	.0731	.8603	113	98.76	1.867	116.9	432.6	.2360	.7873

* "Refrigerating Data Book," 5th ed., American Society of Refrigerating Engineers, New York, 1942.

TABLE 3-260 Saturated Methyl Chloride*

T, K	P, bar	v_f, m³/kg	v_g, m³/kg	h_f, kJ/kg	h_g, kJ/kg	s_f, kJ/(kg·K)	s_g, kJ/(kg·K)	c_{pf}, kJ/(kg·K)	μ_f, 10^{-4} Pa·s	k_f, W/(m·K)
175	0.0117	8.84.−4	27.90	274.5	764.3	3.529	6.328	1.469		
180	0.0165	8.91.−4	19.85	280.9	767.7	3.570	6.274	1.472		
185	0.0233	8.97.−4	14.12	287.5	771.0	3.603	6.222	1.475		
190	0.0327	9.04.−4	10.12	294.5	774.3	3.647	6.172	1.477		
195	0.0462	9.10.−4	7.208	301.7	777.5	3.684	6.124	1.480		
200	0.0653	9.17.−4	5.137	309.0	780.7	3.722	6.080	1.483	4.44	0.241
205	0.0919	9.25.−4	3.835	316.3	783.9	3.756	6.038	1.486	4.27	0.236
210	0.1315	9.33.−4	2.656	323.7	787.0	3.791	5.998	1.489	4.11	0.232
215	0.181	9.40.−4	1.975	331.0	790.1	3.825	5.961	1.492	3.96	0.228
220	0.243	9.48.−4	1.505	338.4	793.2	3.859	5.928	1.496	3.82	0.224
225	0.319	9.56.−4	1.168	345.7	796.3	3.892	5.896	1.500	3.69	0.219
230	0.417	9.65.−4	0.911	353.1	799.3	3.925	5.866	1.504	3.57	0.215
235	0.539	9.73.−4	0.718	360.5	802.3	3.957	5.845	1.508	3.46	0.211
240	0.688	9.81.−4	0.572	368.0	805.3	3.988	5.822	1.513	3.35	0.207
245	0.866	9.89.−4	0.462	375.6	808.2	4.019	5.786	1.518	3.25	0.202
250	1.076	9.98.−4	0.377	383.2	811.1	4.050	5.762	1.523	3.16	0.198
255	1.328	10.08.−4	0.311	390.7	814.0	4.080	5.740	1.528	3.08	0.194
260	1.627	10.18.−4	0.257	398.3	816.8	4.110	5.720	1.533	3.00	0.190
265	1.970	10.27.−4	0.215	406.0	819.4	4.139	5.699	1.539	2.92	0.186
270	2.364	10.36.−4	0.1807	413.7	822.0	4.168	5.680	1.546	2.85	0.182
275	2.830	10.46.−4	0.1524	421.5	824.4	4.197	5.662	1.554	2.78	0.177
280	3.347	10.57.−4	0.1301	429.4	826.8	4.225	5.644	1.565	2.72	0.173
285	3.936	10.68.−4	0.1115	437.3	829.0	4.253	5.628	1.574	2.66	0.169
290	4.612	10.79.−4	0.0960	445.2	831.2	4.280	5.612	1.583	2.61	0.165
295	5.361	10.91.−4	0.0830	453.2	833.2	4.308	5.597	1.594	2.56	0.160
300	6.189	11.03.−4	0.0723	461.2	835.2	4.334	5.581	1.605	2.51	0.156
305	7.110	11.15.−4	0.0632	469.3	837.0	4.361	5.567	1.617	2.46	0.152
310	8.111	11.27.−4	0.0556	477.4	838.8	4.388	5.553	1.631	2.42	0.148
315	9.243	11.40.−4	0.0489	485.6	840.5	4.414	5.540	1.644	2.37	0.143
320	10.47	11.55.−4	0.0433	493.8	841.9	4.440	5.527	1.658	2.33	0.139
325	11.78	11.70.−4	0.0386	502.1	843.3	4.465	5.516	....	2.30	0.135
330	13.27	11.86.−4	0.0343	510.4	844.5	4.491	5.504	....	2.27	0.131
340	16.52	12.17.−4	0.0282	518.8	846.4	4.542	5.481	....	2.12	0.124
350	20.53	12.54.−4	0.0228	538.3	847.5	4.592	5.457	....	1.99	0.117
360	25.29	12.97.−4	0.0186	562.9	847.6	4.643	5.434	....	1.87	0.110
370	30.74	13.47.−4	0.0151	581.6	845.9	4.694	5.398	....	1.77	0.103
380	36.99	14.11.−4	0.0117	602.8	842.6	4.747	5.382	....	1.67	0.095
390	44.05	14.67.−4	0.0096	622.9	837.4	4.805	5.358	....	1.59	0.086
400	52.29	15.66.−4	0.0075	643.6	826.4	4.870	5.323	....	1.51	0.075
405	56.6	16.48.−4	0.0063	663.2	819.1	4.904	5.289			
410	61.5	17.97.−4	0.0052	677.3	807.1	4.954	5.256			
415	67.4	21.10.−4	0.0038	714.1	778.6	5.025	5.200			
416[c]	69.0	27.40.−4	0.0027	749.3	749.3	5.116	5.116			

*Interpolated by P. E. Liley from the Landolt-Börnstein band IVa, p. 677, 1967 tables by Steinle/Dienemann. c = critical point. The notation 8.84.−4 signifies 8.84×10^{-4}.

TABLE 3-261 Superheated Methyl Chloride*

v, volume, ft^3/lb; h, enthalpy, Btu/lb; s, entropy, Btu/(lb·°R)
Parenthetic figures under pressures are saturation temperatures.

Temp., °F.	Abs. pressure, 6 lb./sq. in. (−44.8°F.)			Temp., °F.	Abs. pressure, 10 lb./sq. in. (−26.1°F.)			Temp., °F.	Abs. pressure, 20 lb./sq. in. (2.5°F.)		
t	v	h	s	t	v	h	s	t	v	h	s
Sat.	14.45	189.96	0.4580	Sat.	8.993	192.64	0.4446	Sat.	4.710	196.58	0.4270
−40	14.62	190.77	.4599	−20	9.124	193.67	.4471	20	4.917	199.90	.4341
−20	15.36	194.27	.4681	0	9.567	197.32	.4552	40	5.146	203.75	.4420
0	16.09	197.84	.4760	20	10.01	201.04	.4630	60	5.373	207.66	.4496
20	16.82	201.48	.4838	40	10.45	204.78	.4707	80	5.599	211.65	.4572
40	17.55	205.19	.4914	60	10.89	208.62	.4782	100	5.823	215.69	.4645
60	18.27	209.01	.4989	80	11.33	212.53	.4856	120	6.046	219.80	.4717
80	18.99	212.88	.5061	100	11.77	216.50	.4928	140	6.268	223.99	.4788
100	19.71	216.82	.5133	120	12.21	220.54	.5000	160	6.489	228.24	.4858
120	20.42	220.84	.5204	140	12.65	224.67	.5069	180	6.709	232.56	.4927
140	21.14	224.94	.5274	160	13.08	228.86	.5138	200	6.929	236.94	.4994
160	21.86	229.11	.5342	180	13.52	233.13	.5206	220	7.147	241.40	.5061
180	22.57	233.36	.5410	200	13.95	237.47	.5273	240	7.365	245.96	.5127
200	23.29	237.69	.5476	220	14.38	241.89	.5339	260	7.583	250.55	.5192
220	24.00	242.09	.5542	240	14.81	246.42	.5045	280	7.801	255.23	.5256
240	24.71	246.60	.5607	260	15.24	250.98	.5469	300	8.019	259.96	.5319
260	25.42	251.15	.5672	280	15.67	255.66	.5532				

Temp., °F.	Abs. pressure, 50 lb./sq. in. (47.8°F.)			Temp., °F.	Abs. pressure, 100 lb./sq. in. (89.6°F.)			Temp., °F.	Abs. pressure, 200 lb./sq. in. (140.3°F.)		
t	v	h	s	t	v	h	s	t	v	h	s
Sat.	1.992	202.09	0.4043	Sat.	1.025	206.11	0.3872	Sat.	0.517	209.60	0.3702
60	2.054	204.65	.4094	100	1.055	208.58	.3920	160	.551	215.30	.3796
80	2.154	208.89	.4174	120	1.111	213.33	.4003	180	.582	220.87	.3886
100	2.252	213.18	.4252	140	1.165	218.15	.4084	200	.612	226.35	.3971
120	2.348	217.52	.4328	160	1.217	222.94	.4163	220	.641	231.75	.4052
140	2.443	221.88	.4402	180	1.268	227.71	.4239	240	.668	237.12	.4129
160	2.537	226.32	.4475	200	1.318	232.50	.4312	260	.695	242.39	.4204
180	2.630	230.79	.4546	220	1.367	237.32	.4384	280	.721	247.65	.4276
200	2.722	235.32	.4616	240	1.415	242.20	.4455	300	.747	252.93	.4346
220	2.813	239.90	.4684	260	1.463	247.06	.4523	320	.772	258.21	.4414

Temp., °F.	Abs. pressure, 50 lb./sq. in. (47.8°F.)			Temp., °F.	Abs. pressure, 100 lb./sq. in. (89.6°F.)			Temp., °F.	Abs. pressure, 200 lb./sq. in. (140.3°F.)		
t	v	h	s	t	v	h	s	t	v	h	s
240	2.903	244.58	.4752	280	1.511	251.96	.4591	340	.797	263.51	.4481
260	2.993	249.27	.4818	300	1.557	256.92	.4657	360	.822	268.84	.4547
280	3.083	254.02	.4884	320	1.603	261.93	.4722	380	.847	274.19	.4612
300	3.173	258.83	.4948	340	1.649	266.97	.4786	400	.870	279.59	.4676
320	3.261	263.71	.5011	360	1.695	272.07	.4849	420	.895	285.05	.4738
340	3.349	268.65	.5074	380	1.739	277.25	.4911	440	.918	290.52	.4800

* Tanner, Banning, and Matthewson, *Ind. Eng. Chem.*, **31**, 878 (1939). Copyright, 1939, E. I. du Pont de Nemours & Co., Inc.

TABLE 3-262 Saturated Methylene Chloride (CH_2Cl_2)*

Temp., °F.	Abs. pressure, lb./sq. in.	Volume, cu. ft./lb., vapor	Enthalpy, B.t.u./lb.		Entropy, B.t.u./(lb.)(°R.)	
			Liquid	Vapor	Liquid	Vapor
t	p	v_g	h_f	h_g	s_f	s_g
10	1.38	42.55	3.4	164.4	0.0072	0.3502
20	1.92	31.40	6.8	165.6	.0151	.3461
30	2.56	23.90	10.2	166.9	.0222	.3425
40	3.38	18.60	13.6	168.0	.0285	.3377
60	5.52	11.68	20.4	170.1	.0410	.3292
80	8.81	7.50	27.2	172.0	.0520	.3202
100	13.25	5.14	34.0	173.7	.0620	.3113
120	19.20	3.65	40.8	175.0	.0714	.3031
140	26.79	2.69	47.6	176.0	.0795	.2935

* "Refrigerating Data Book," 5th ed., American Society of Refrigerating Engineers, New York, 1942.

TABLE 3-263 Saturated Methyl Formate ($HCOOOCH_3$)*

Temp., °F.	Abs. pressure, lb./sq. in.	Volume, cu. ft./lb., vapor	Enthalpy, B.t.u./lb.		Entropy, B.t.u./(lb.)(°R.)	
			Liquid	Vapor	Liquid	Vapor
t	p	v_g	h_f	h_g	s_f	s_g
0	1.50	54.0	0	232.5	0.0000	0.5055
20	2.70	31.0	10.3	236.6	.0219	.4934
40	4.66	18.9	20.6	240.7	.0432	.4837
60	7.61	12.0	30.9	244.8	.0633	.4748
80	12.07	7.98	41.2	248.8	.0825	.4670
100	18.26	5.38	51.5	252.9	.1015	.4615
120	27.24	3.74	61.8	257.0	.1192	.4559
140	38.41	2.65	72.1	261.0	.1375	.4525

* "Refrigerating Data Book," 5th ed., American Society of Refrigerating Engineers, New York, 1942.

TABLE 3-264 Saturated Neon*

T, K	P, bar	v_f, m³/kg	v_g, m³/kg	h_f, kJ/kg	h_g, kJ/kg	s_f, kJ/(kg·K)	s_g, kJ/(kg·K)	c_{pf}, kJ/(kg·K)	μ_f, 10^{-4} Pa·s	k_f, W/(m·K)
10	. . . . .	6.654. −4	. . . . .	0.75	. . . . .	0.0992	. . . . .	0.278		
20	. . . . .	6.823. −4	. . . . .	6.78	. . . . .	0.4906	. . . . .	0.945		
24.6^m	. . . . .	6.696. −4	. . . . .	11.96	. . . . .	0.7257	. . . . .	1.345		
24.6^m	0.434	8.012. −4	0.2266	28.22	117.0	1.388	5.006	1.802	1.57	0.146
26	0.718	8.172. −4	0.1429	30.90	118.1	1.494	4.846	1.868	1.37	0.132
28	1.321	8.413. −4	0.0817	34.75	119.3	1.634	4.653	1.955	1.16	0.124
30	2.238	8.687. −4	0.0501	38.80	120.1	1.771	4.483	2.052	1.00	0.115
32	3.552	9.001. −4	0.0323	43.06	120.6	1.905	4.329	2.163	0.84	0.106
34	5.352	9.370. −4	0.0217	47.57	120.6	2.036	4.184	2.302	0.71	0.097
36	7.728	9.820. −4	0.0149	52.34	119.9	2.166	4.043	2.506	0.59	0.088
38	10.78	1.039. −3	0.0104	57.52	118.4	2.297	3.900	2.825	0.48	0.078
40	14.62	1.116. −3	0.0073	63.33	115.8	2.435	3.749	3.436	0.38	0.069
42	19.39	1.232. −3	0.0050	69.82	111.8	2.582	3.582	5.26	0.31	0.059
44	25.22	1.538. −3	0.0031	80.83	103.0	2.812	3.316	25.0	0.25	
44.4^c	26.53	2.070. −3	0.0021	92.50	92.5	3.062	3.062	∞	. . . . .	∞

*Values extracted and in some cases rounded off from those cited in Rabinovich (ed.), *Thermophysycal Properties of Neon, Argon, Krypton and Xenon*, Standards Press, Moscow, 1976. m = melting point; c = critical point. The notation 6.654. −4 signifies 6.654×10^{-4}. This source contains values for the compressed state up to 1000 bar, etc.

TABLE 3-265 Compressed Neon*

Temperature, K		Pressure, bar											
		1	10	20	40	60	80	100	200	400	600	800	1000
	v	0.4117	0.0410	0.0204	0.0102	6.76. −3	5.08. −3	4.09. −3	2.22. −3	1.42. −3	1.18. −3	1.06. −3	9.74. −4
100	h	195.4	194.0	192.4	189.4	186.6	184.0	181.6	174.1	173.3	180.0	189.2	199.2
	s	6.129	5.168	4.869	4.556	4.363	4.221	4.106	3.739	3.386	3.197	3.066	2.964
	v	0.8243	0.0828	0.0416	0.0210	0.0142	0.0107	8.69. −3	4.61. −3	2.61. −3	1.95. −3	1.63. −3	1.43. −3
200	h	298.5	298.4	298.4	298.2	298.2	298.2	298.3	299.4	304.6	312.4	321.8	332.1
	s	6.844	5.893	5.605	5.315	5.143	5.020	4.924	4.620	4.308	4.124	3.994	3.893
	v	1.236	0.1241	0.0624	0.0315	0.0212	0.0160	0.0129	6.77. −3	3.71. −3	2.69. −3	2.18. −3	1.87. −3
300	h	401.6	401.8	402.2	402.8	403.5	404.1	404.9	408.8	417.8	427.8	438.5	449.7
	s	7.262	6.312	6.026	5.739	5.570	5.450	5.357	5.065	4.769	4.593	4.469	4.372
	v	1.648	0.1654	0.0830	0.0418	0.0281	0.0212	0.0171	8.88. −3	4.77. −3	3.40. −3	2.72. −3	2.30. −3
400	h	504.6	505.0	505.4	506.4	507.4	508.3	509.3	514.4	525.3	536.7	548.4	560.2
	s	7.558	6.609	6.323	6.037	5.896	5.750	5.657	5.369	5.078	4.907	4.785	4.690
	v	2.060	0.2066	0.1036	0.0521	0.0350	0.0264	0.0213	0.0110	5.82. −3	4.10. −3	3.24. −3	2.73. −3
500	h	607.6	608.1	608.6	609.7	610.8	611.9	613.0	618.8	630.7	642.9	655.2	667.5
	s	7.788	6.839	6.553	6.267	6.100	5.981	5.889	5.601	5.313	5.144	5.023	4.929
	v	2.472	0.2478	0.1242	0.0625	0.0419	0.0316	0.0254	0.0130	6.85. −3	4.80. −3	3.77. −3	3.15. −3
600	h	710.6	711.1	711.7	712.9	714.1	715.3	716.5	722.5	735.0	747.8	760.5	773.2
	s	7.975	7.027	6.741	6.455	6.288	6.169	6.077	5.791	5.504	5.335	5.215	5.122
	v	2.884	0.2890	0.1449	0.0728	0.0487	0.0367	0.0295	0.0151	7.89. −3	5.49. −3	4.29. −3	3.57. −3
700	h	813.5	814.1	814.7	816.0	817.2	818.5	819.7	826.0	838.9	851.9	865.0	878.1
	s	8.134	7.186	6.900	6.614	6.447	6.328	6.236	5.950	5.664	5.496	5.376	5.284
	v	3.296	0.3302	0.1655	0.0831	0.0556	0.0419	0.0336	0.0172	8.92. −3	6.18. −3	4.81. −3	3.98. −3
800	h	916.5	917.1	917.7	919.0	920.3	921.6	922.9	929.3	942.4	955.7	969.0	982.2
	s	8.272	7.323	7.038	6.752	6.585	6.466	6.374	6.088	5.802	5.634	5.515	5.423
	v	3.708	0.3714	0.1861	0.0934	0.0625	0.0470	0.0378	0.0192	9.96. −3	6.87. −3	5.32. −3	4.40. −3
900	h	1020	1020	1021	1022	1023	1025	1026	1033	1046	1059	1073	1086
	s	8.393	7.444	7.159	6.873	6.706	6.588	6.496	6.210	5.924	5.756	5.637	5.545
	v	4.120	0.4126	0.2067	0.1037	0.0693	0.0522	0.0419	0.0213	0.0110	7.56. −3	5.84. −3	4.81. −3
1000	h	1123	1123	1124	1125	1126	1128	1129	1136	1149	1163	1176	1190
	s	8.502	7.553	7.267	6.982	6.815	6.696	6.604	6.318	6.032	5.856	5.746	5.654

*Values extracted and in some cases rounded off from those cited in Rabinovich (ed.), *Thermophysical Properties of Neon, Argon, Krypton and Xenon*, Standards Press, Moscow, 1976. This source contains an exhaustive tabulation of values. v = specific volume, m³/kg; h = specific enthalpy, kJ/kg; s = specific entropy, kJ/(kg·K). The notation 6.76. −3 signifies 6.76×10^{-3}.

TABLE 3-266 Nitric Oxide

For ideal-gas thermodynamic functions see Beckett and Haar, *Proc. J. Conf. Thermodynam. Transport Prop. Fluids*, Institution of Mechanical Engineers, London, 1958, p. 33; Fickett and Cowan, *AEC Rep.* LA1727, 1954. For properties at higher pressures see Opfell, Schlinger, et al., *Ind. Eng. Chem.*, **46**, 189 (1954) (−80° to 220°F, to 3000 psia); Kobe and Pennington, *Pet. Refiner*, **29**, 129 (1950) (H, C_p, etc., to 3500°).

TABLE 3-267 Saturated Nitrogen (R728)*

T, K	P, bar	v_f, m^3/kg	v_g, m^3/kg	h_f, kJ/kg	h_g, kJ/kg	s_f, kJ/(kg·K)	s_g, kJ/(kg·K)	c_{pf}, kJ/(kg·K)	μ_f, 10^{-4} Pa·s	k_f, W/(m·K)
63.15^t	0.1253	1.155	1477	−148.5	64.1	2.459	5.826	1.928	. . .	0.170
65	0.1743	1.165	1091	−144.9	65.8	2.516	5.757	1.930	2.74	0.160
70	0.3859	1.193	525.6	−135.2	70.5	2.657	5.595	1.937	2.17	0.151
75	0.7609	1.224	281.8	−125.4	74.9	2.789	5.460	1.948	1.77	0.141
77.35	1.0133	1.239	216.9	−120.8	76.8	2.849	5.404	1.955	1.60	0.136
80	1.369	1.258	164.0	−115.6	78.9	2.913	5.345	1.964	1.48	0.132
85	2.287	1.297	101.7	−105.7	82.3	3.032	5.244	1.989	1.27	0.123
90	3.600	1.340	66.28	−95.6	85.0	3.147	5.152	2.028	1.10	0.114
95	5.398	1.390	44.87	−85.2	86.8	3.256	5.067	2.086	0.97	0.105
100	7.775	1.447	31.26	−74.5	87.7	3.363	4.985	2.176	0.87	0.097
105	10.83	1.514	22.23	−63.8	87.4	3.469	4.904	2.319	0.79	0.088
110	14.67	1.597	15.98	−51.4	85.6	3.575	4.820	2.566	0.71	0.080
115	19.40	1.714	11.47	−38.1	81.8	3.687	4.729	3.063	0.60	0.071
120	25.15	1.892	8.031	−21.4	74.3	3.821	4.619	. . . .	0.48	0.063
125	32.05	2.324	5.016	5.1	57.2	4.024	4.444	. . . .	0.32	0.052
126.25^c	33.96	3.289	3.289	34.8	34.8	4.252	4.252	∞	. . .	∞

*Reproduced and converted from Vasserman and Rabinovich, *Thermophysical Properties of Liquid Air and Its Components*, Standartov, Moscow, 1968; and Israel Program for Scientific Translations, TT 69-55092, 1970. t = triple point; c = critical point.

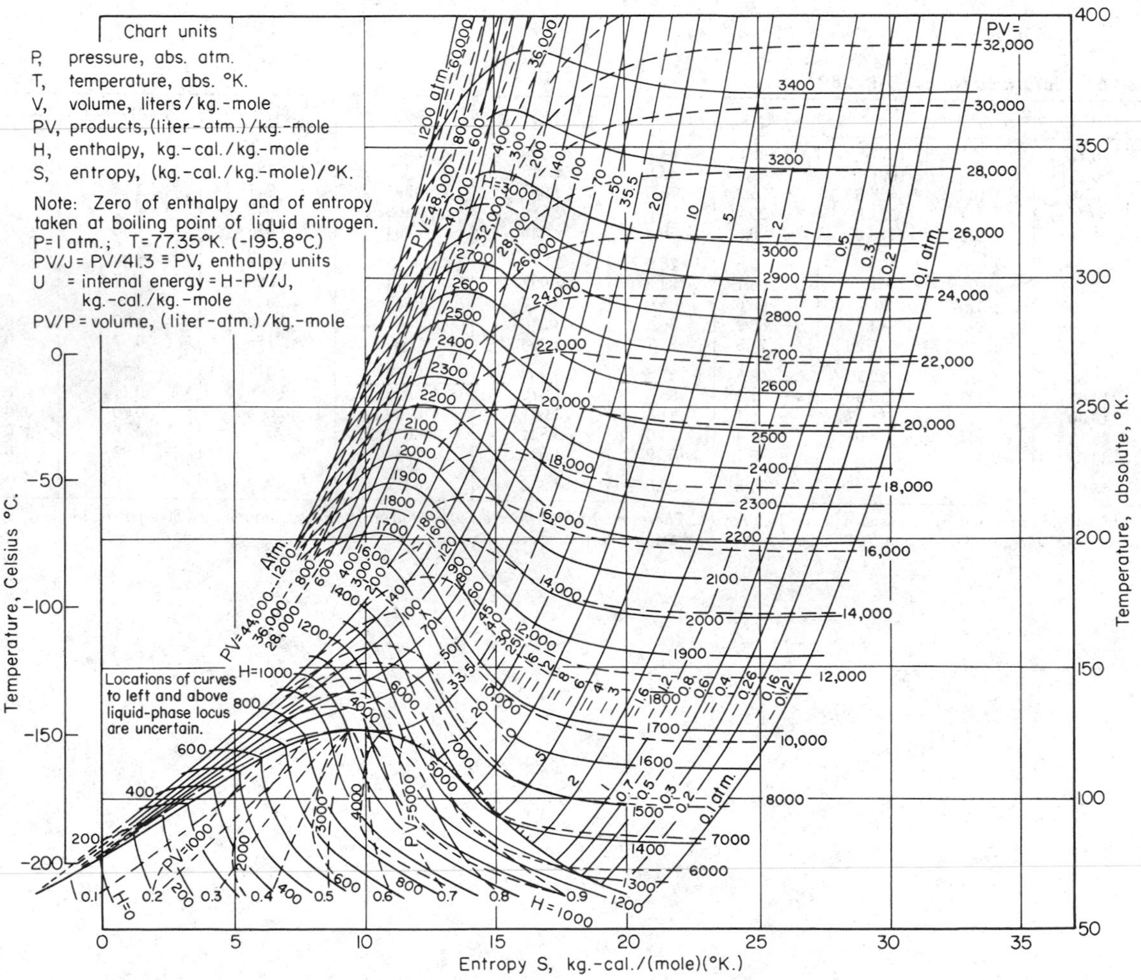

FIG. 3-27 Temperature-entropy diagram for nitrogen. Section of *T-S* diagram for nitrogen by E. S. Burnett, 1950. *(Reprinted from U.S. Bur. Mines Rep. Invest. 4729.)*

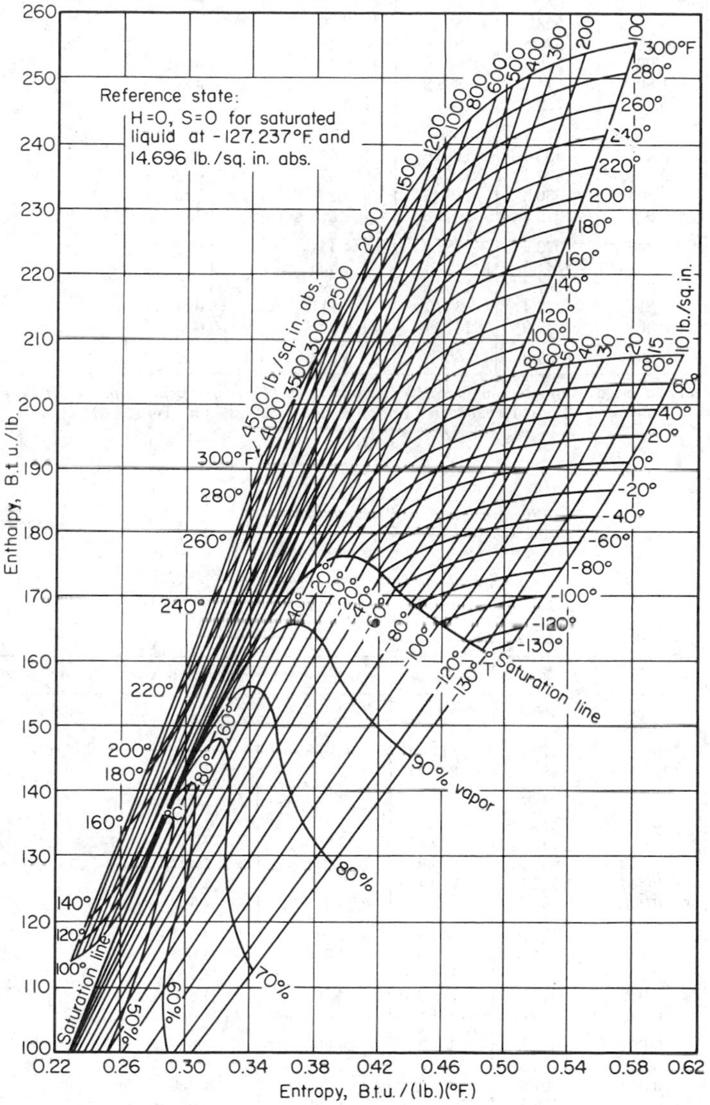

FIG. 3-28 Mollier diagram for nitrous oxide. *(Fig. 9, Univ. Texas Rep., Cont. DAI-23-072-ORD-685, June 1, 1956, by Couch and Kobe. Reproduced by permission.)* Some irregularity in the compressibility factors from 80 to 160 atm, 50 to 100°C exists *(Couch, private communication, 1967)*. See Couch et al., *J. Chem. Eng. Data*, **6**, 229, 233 (1961) for PVT data.

TABLE 3-268 Nonane*

T, K	P, bar	v_f, m^3/kg	v_g, m^3/kg	h_f, kJ/kg	h_g, kJ/kg	s_f, kJ/(kg·K)	s_g, kJ/(kg·K)	c_{pf}, kJ/(kg·K)	μ_f, 10^{-4} Pa·s	k_f, W/(m·K)
219.7^t	2.6.−6	....	.....	358.4	.....	2.424	....	2.07	33.5	0.150
220	2.7.−6	....	.....	359.2	.....	2.427	....	2.07	33.0	0.150
240	3.74.−5	....	.....	400.6	.....	2.607	....	2.08	17.9	0.145
260	2.97.−4	....	.....	442.2	828.7	2.774	4.210	2.10	12.1	0.140
280	1.61.−3	....	.....	484.8	859.4	2.932	4.243	2.16	8.7	0.134
300	6.40.−3	1.404.−3	30.35	528.6	891.7	3.083	4.282	2.22	6.53	0.129
320	0.0203	1.436.−3	10.19	573.8	925.6	3.229	4.324	2.30	5.13	0.123
340	0.0547	1.471.−3	4.00	622.0	961.1	3.370	4.368	...	4.16	0.118
360	0.1279	1.508.−3	1.80	671.3	998.2	3.511	4.419	...	3.44	0.112
380	0.2678	1.548.−3	0.894	722.5	1036.5	3.650	4.476	...	2.91	0.107
400	0.513	1.591.−3	0.485	776.7	1076.0	3.788	4.536	...	2.50	0.101
420	0.911	1.637.−3	0.286	833.3	1116.6	3.927	4.601	...	2.18	0.096
440	1.521	1.690.−3	0.161	890.2	1157.1	4.053	4.660	...	....	0.092
460	2.401	1.748.−3	0.104	950.3	1199.2	4.186	4.727	...	....	0.089
480	3.639	1.815.−3	0.069	1012.1	1241.3	4.316	4.794	...	....	0.085
500	5.309	1.895.−3	0.045	1076.2	1282.9	4.444	4.857	...	....	0.082
520	7.437	2.00.−3	0.030	1141.3	1324.5	4.569	4.921			
540	10.20	2.13.−3	0.021	1207.7	1363.8	4.691	3.980			
560	13.76	2.35.−3	0.013	1275.4	1338.7	4.811	5.029			
580	18.02	2.78.−3	0.008	1342.9	1318.1	4.927	5.056			
594.6^c	22.90	4.23.−3	0.004	1305.2	1305.2	5.032	5.032			

*Values of p and v interpolated and converted from tables in Vargaftik, *Handbook of Thermophysical Properties of Gases and Liquids*, Hemisphere, Washington, and McGraw-Hill, New York, 1975. Values of h and s calculated from API tables published by Texas A&M University, College Station. t = triple point; c = critical point.

TABLE 3-269 Octane*

T, K	P, bar	v_f, m^3/kg	v_g, m^3/kg	h_f, kJ/kg	h_g, kJ/kg	s_f, kJ/(kg·K)	s_g, kJ/(kg·K)	c_{pf}, kJ/(kg·K)	μ_f, 10^{-4} Pa·s	k_f, W/(m·K)
216.4^t	1.49.−5	....	......	365.9	.....	2.487	...	2.033	2.25	0.149
220	2.41.−5	....	......	373.2	.....	2.520	....	2.035	2.01	0.148
240	2.18.−4	1.353.−3	700	414.1	811.4	2.698	4.207	2.059	1.24	0.143
260	0.0014	1.368.−3	125	455.8	842.1	2.865	4.259	2.105	0.87	0.138
280	0.0061	1.384.−3	31.9	498.4	873.5	3.023	4.312	2.165	0.65	0.133
300	0.0207	1.420.−3	10.7	542.4	906.2	3.175	4.366	2.231	0.504	0.128
320	0.0575	1.457.−3	4.01	589.8	939.8	3.325	4.419	....	0.405	0.123
340	0.1384	1.495.−3	1.752	637.9	974.6	3.471	4.461	....	0.334	0.118
360	0.3000	1.536.−3	0.844	687.1	1010.4	3.611	4.509	....	0.282	0.112
380	0.5856	1.582.−3	0.448	737.7	1047.3	3.747	4.562	....	0.244	0.107
400	1.0507	1.632.−3	0.252	790.1	1084.8	3.881	4.617	....	0.200	0.102
420	1.758	1.685.−3	0.155	843.1	1123.6	4.010	4.677	....	0.167	0.099
440	2.797	1.747.−3	0.100	897.5	1162.5	4.137	4.740	....	0.143	0.095
460	4.246	1.818.−3	0.066	954.8	1202.0	4.264	4.802	....	0.121	0.091
480	6.201	1.904.−3	0.045	1013.5	1241.8	4.388	4.864	....	0.103	0.087
500	8.785	2.013.−3	0.031	1072.8	1281.2	4.508	4.924	....	0.086	0.083
520	12.15	2.16.−3	0.021	1136.0	1318.6	4.629	4.980	....	0.072	
540	16.46	2.37.−3	0.014	1201.5	1352.4	4.749	5.028	....	0.058	
560	21.98	2.81.−3	0.008	1276.7	1370.4	4.880	5.048	....	0.044	
568.8^c	24.97	4.26.−3	0.004	1331.7	1331.7	4.977	4.977			

*Values of p and v interpolated and converted from tables in Vargaftik, *Handbook of Thermophysical Properties of Gases and Liquids*, Hemisphere, Washington, and McGraw-Hill, New York, 1975. Values of h and s calculated from API tables published by Texas A&M University, College Station. t = triple point; c = critical point.

TABLE 3-270 Saturated Oxygen (R732)*

T, K	P, bar	v_f, m^3/kg	v_g, m^3/kg	h_f, kJ/kg	h_g, kJ/kg	s_f, kJ/(kg·K)	s_g, kJ/(kg·K)	c_{pf}, kJ/(kg·K)	μ_f, 10^{-4} Pa·s	k_f, W/(m·K)
54.35^t	0.0015	0.776	93980	−189.8	48.9	2.156	6.548			
55	0.0018	0.778	77920	−188.9	49.5	2.172	6.507			
60	0.0073	0.790	21240	−181.1	53.8	2.308	6.223			
65	0.0233	0.802	7200	−173.3	58.1	2.432	5.992			
70	0.0624	0.816	2894	−165.5	62.4	2.545	5.801			
75	0.1448	0.827	1330	−159.2	66.6	2.631	5.642	1.570	3.04	0.170
80	0.3003	0.845	680.7	−149.7	70.8	2.754	5.510	1.589	2.54	0.164
85	0.5677	0.862	379.7	−141.7	74.9	2.849	5.397	1.607	2.16	0.157
90	0.9943	0.880	227.1	−133.7	78.8	2.940	5.301	1.625	1.88	0.151
90.18	1.0133	0.881	223.2	−133.4	78.9	2.943	5.297	1.626	1.87	0.151
95	1.634	0.899	143.9	−125.4	82.4	3.045	5.216	1.645	1.66	0.144
100	2.547	0.920	95.46	−117.1	85.7	3.113	5.141	1.672	1.51	0.138
105	3.794	0.944	65.81	−108.6	88.5	3.196	5.073	1.706	1.34	0.131
110	5.443	0.970	46.81	− 99.9	90.8	3.276	5.009	1.752	1.20	0.125
115	7.559	0.998	34.15	− 90.0	92.6	3.354	4.950	1.814	1.07	0.118
120	10.21	1.031	25.42	− 81.6	93.6	3.432	4.892	1.896	0.97	0.111
125	13.48	1.070	19.21	− 71.8	93.9	3.510	4.836	2.004	0.86	0.103
130	17.44	1.116	14.67	− 61.5	93.3	3.588	4.779	2.148	0.78	0.096
135	22.19	1.170	11.25	− 50.6	91.6	3.667	4.720	2.341	0.70	0.088
140	27.82	1.237	8.612	− 38.9	88.4	3.748	4.657	2.629	0.60	0.080
145	34.45	1.332	6.499	− 25.9	82.9	3.833	4.583	3.141	0.52	0.072
150	42.23	1.487	4.705	− 10.8	73.1	3.928	4.487	3.935		
154.77^c	50.87	2.464	2.464	35.2	35.2	4.219	4.219	∞	. . .	∞

*Reproduced and converted from Vasserman and Rabinovich, *Thermophysical Properties of Liquid Air and Its Components*, Standartov, Moscow, 1968; and Israel Program for Scientific Translations, TT 69-55092, 1970. t = triple point; c = critical point.

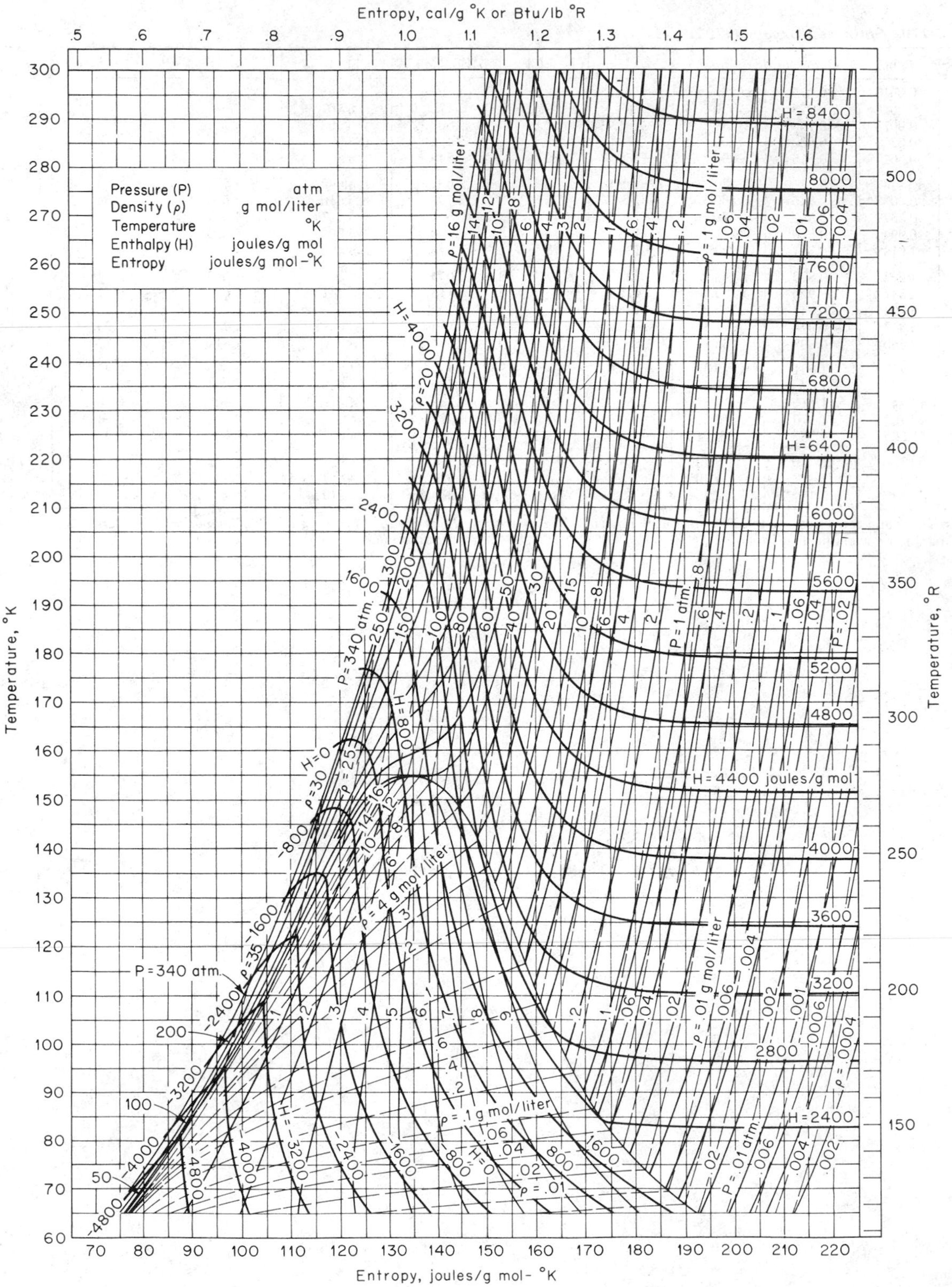

FIG. 3-29 Temperature-entropy chart for oxygen. Pressure P, atm; density ρ, (g·mol)/L; temperature, K; enthalpy H, J/(g·mol); entropy, J/(g·mol·K). *(NBS Chart D-56. Reproduced by permission.)*

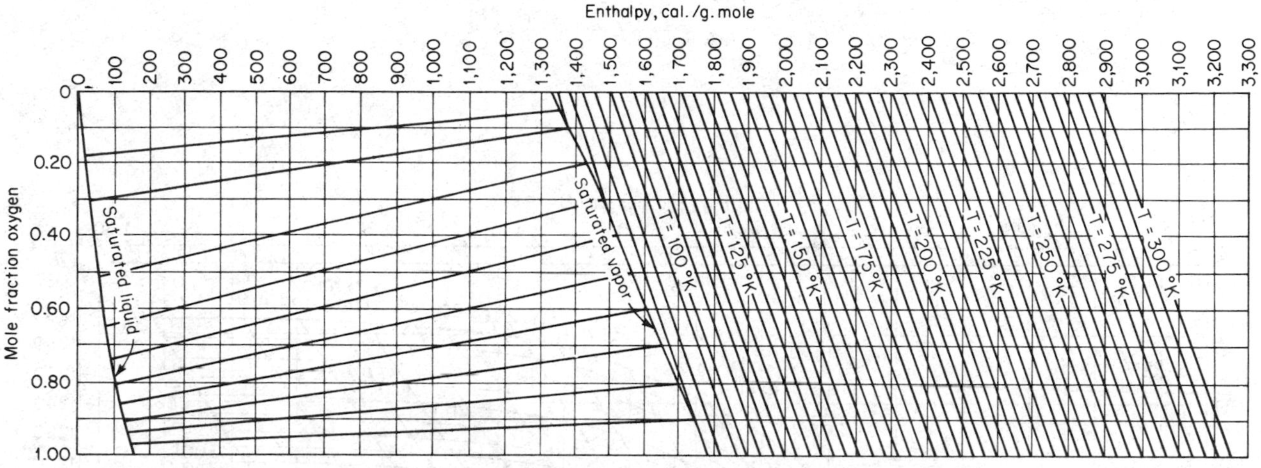

FIG. 3-30 Enthalpy-concentration diagram for oxygen-nitrogen mixture at 1 atm. Reference states: enthalpies of liquid oxygen and liquid nitrogen at the normal boiling point of nitrogen are zero. (*Dodge*, Chemical Engineering Thermodynamics, *McGraw-Hill, New York, 1944.*) Wilson, Silverberg, and Zellner, AFAPL TDR 64-64 (AD 603151), 1964, p. 314, present extensive vapor-liquid equilibrium data for the three-component system argon-nitrogen-oxygen as well as for binary systems including oxygen-nitrogen.

TABLE 3-271 Pentane

Canjar and Manning (*Thermodynamic Properties and Reduced Correlations for Gases*, Gulf, Houston, 1967) give extensive tables and an enthalpy–log-pressure diagram, based upon Brydon, Walen, and Canjar [*Chem. Eng. Prog. Symp. Ser.*, **49**, 7, 151–157 (1951)]. For isopentane, Arnold, Liou, and Eldridge [*J. Chem. Eng. Data*, **10**, 88 (1965)] used the Benedict-Webb-Rubin equation to generate information to 600°F and 60 atm. Das and Kuloor used the same equation in *Ind. J. Technol.*, **5**, 46 (1967) to calculate information up to 1500 K and 1000 atm.

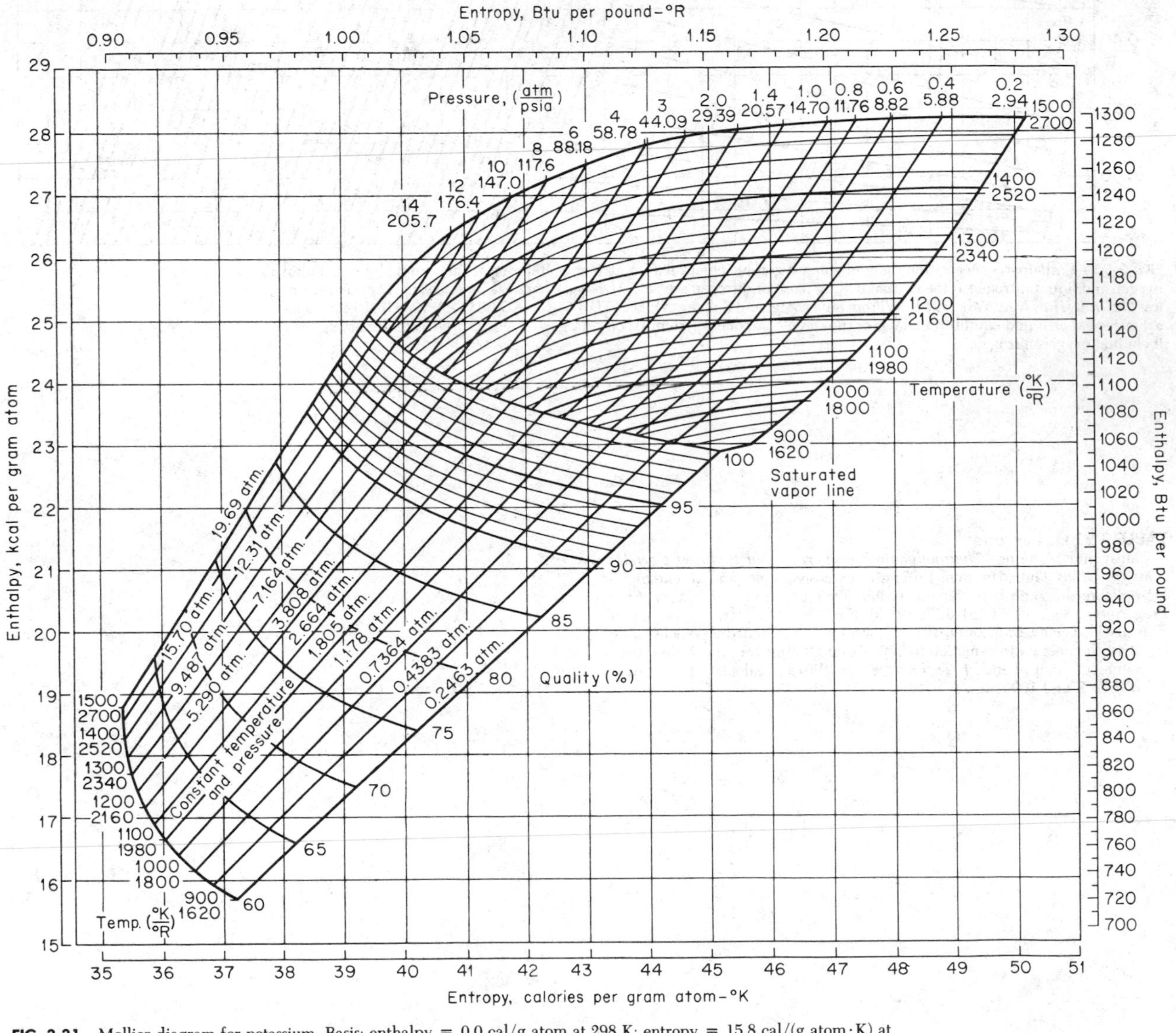

FIG. 3-31 Mollier diagram for potassium. Basis: enthalpy = 0.0 cal/g atom at 298 K; entropy = 15.8 cal/(g atom·K) at 298 K. *(Aerojet-General Rep. AGN8194, vol. 2, 1967. Reproduced by permission.)*

TABLE 3-272 Saturated Propane (R290)*

T, K	P, bar	v_f, m^3/kg	v_g, m^3/kg	h_f, kJ/kg	h_g, kJ/kg	s_f, kJ/(kg·K)	s_g, kJ/(kg·K)	c_{pf}, kJ/(kg·K)	μ_f, 10^{-4} Pa·s	k_f, W/(m·K)
85.5^t	3.0.−9	1.364.−3	5.37.+7	124.92	690.02	1.8738	8.3548	1.92		
90	1.5.−8	1.373.−3	1.12.+7	133.56	693.58	1.9723	8.0953	1.92		
100	3.2.−7	1.392.−3	5.85.+5	152.74	702.23	2.1743	7.6163	1.93		
110	3.9.−6	1.412.−3	53275	172.03	711.71	2.3581	7.2377	1.94		
120	3.1.−5	1.432.−3	7350	191.46	721.78	2.5271	6.9343	1.95		
130	1.8.−4	1.453.−3	1400	211.03	732.27	2.6838	6.6885	1.96		
140	7.7.−4	1.475.−3	344	230.77	743.07	2.8300	6.4881	1.98		
150	2.74.−3	1.497.−3	103	250.67	754.12	2.9674	6.3237	2.00	6.61	0.191
160	8.22.−3	1.521.−3	36.8	270.78	765.37	3.0971	6.1886	2.02	5.54	0.183
170	0.0214	1.545.−3	15.0	291.10	776.80	3.2202	6.0775	2.04	4.67	0.175
180	0.0495	1.570.−3	6.84	311.66	788.40	3.3377	5.9862	2.07	3.97	0.166
190	0.1035	1.597.−3	3.43	332.48	800.15	3.4503	5.9114	2.10	3.27	0.158
200	0.1993	1.625.−3	1.868	353.61	812.03	3.5586	5.8502	2.13	2.98	0.150
210	0.3574	1.654.−3	1.087	375.07	824.01	3.6631	5.8005	2.16	2.65	0.143
220	0.6031	1.686.−3	0.669	396.90	836.04	3.7645	5.7603	2.20	2.36	0.136
230	0.9661	1.719.−3	0.432	419.16	848.08	3.8631	5.7280	2.25	2.07	0.129
240	1.4800	1.754.−3	0.290	442.07	860.07	3.9605	5.7022	2.29	1.86	0.123
250	2.1819	1.792.−3	0.2020	465.58	871.94	4.0563	5.6817	2.34	1.69	0.117
260	3.1118	1.833.−3	0.1445	489.70	883.62	4.1505	5.6656	2.41	1.53	0.111
270	4.3120	1.878.−3	0.1059	514.45	895.02	4.2433	5.6528	2.48	1.40	0.106
280	5.8278	1.927.−3	0.0791	539.88	906.03	4.3349	5.6426	2.56	1.29	0.100
290	7.7063	1.982.−3	0.0600	566.06	916.54	4.4257	5.6343	2.65	1.19	0.096
300	9.9973	2.044.−3	0.0461	593.11	926.41	4.5160	5.6270	2.76	1.10	0.091
310	12.75	2.115.−3	0.0357	621.18	935.45	4.6062	5.6200	2.89	0.93	0.086
320	16.03	2.200.−3	0.0279	650.49	943.38	4.6971	5.6124	3.06	0.82	0.082
330	19.88	2.301.−3	0.0218	681.37	949.79	4.7896	5.6030	3.28	0.72	0.078
340	24.36	2.430.−3	0.0170	714.38	953.92	4.8850	5.5896	3.62	0.62	0.073
350	29.56	2.607.−3	0.0130	750.52	954.23	4.9861	5.5681	4.23	0.52	0.069
360	35.55	2.896.−3	0.0095	792.50	946.56	5.0997	5.5277	5.98	0.40	0.066
369.8^c	42.42	4.566.−3	0.0046	879.20	879.20	5.3300	5.3300	∞	0.29	∞

*Values converted and mostly rounded off from those of Goodwin, NBSIR 77-860, 1977. t = triple point; c = critical point. The notation 3.0.−9 signifies 3.0×10^{-9}.

TABLE 3-273 Saturated Potassium*

T, K	P, bar	v_f, m^3/kg	v_g, m^3/kg	h_f, kJ/kg	h_g, kJ/kg	s_f, kJ/(kg·K)	s_g, kJ/(kg·K)	c_{pf}, kJ/(kg·K)
336.4^m	1.37.−9	0.001208	....	93.8	2327	1.928	8.567	0.822
400	1.84.−7	0.001229	4.64.+6	145.5	2342	2.068	7.559	0.805
500	3.13.−5	0.001266	3.39.+4	225.1	2390	2.246	6.576	0.785
600	9.26.−4	0.001304	3164	302.7	2433	2.388	5.937	0.771
700	0.01022	0.001346	142.3	379.4	2468	2.506	5.490	0.762
800	0.06116	0.001389	26.75	455.5	2498	2.608	5.161	0.761
1000	0.7322	0.001488	2.691	609.7	2552	2.780	4.722	0.792
1200	3.913	0.001605	0.584	773.5	2610	2.929	4.459	0.846
1400	12.44	0.001742	0.207	948.0	2679	3.063	4.299	0.899
1500	20.0	0.001816	0.132	1040.0	2718	3.123	4.209	0.924

*Converted from tables in Vargaftik, *Tables of the Thermophysical Properties of Liquids and Gases*, Nauka, Moscow, 1972; and Hemisphere, Washington, 1975. m = melting point. The notation 1.37.−9 signifies 1.37×10^{-9}.

TABLE 3-274 Saturated Propylene (R 1270)*

T, K	P, bar	v_f, m³/kg	v_g, m³/kg	h_f, kJ/kg	h_g, kJ/kg	s_f, kJ/(kg·K)	s_g, kJ/(kg·K)	c_{pf}, kJ/(kg·K)	μ_f, 10^{-4} Pa·s	k_f, W/(m·K)
160	0.0123	1.453.−3	25.66	285.0	783.9	3.281	6.399	2.080	4.37	0.183
170	0.0310	1.477.−3	10.73	305.9	795.1	3.487	6.285	2.083	3.80	0.178
180	0.0696	1.504.−3	5.025	326.5	806.1	3.524	6.188	2.087	3.30	0.173
190	0.1426	1.531.−3	2.581	347.8	818.1	3.639	6.114	2.090	2.86	0.168
200	0.2638	1.559.−3	1.435	368.9	829.9	3.747	6.052	2.094	2.48	0.162
210	0.4731	1.588.−3	0.8528	390.1	842.0	3.851	6.003	2.102	2.16	0.157
220	0.7852	1.619.−3	0.5354	412.2	854.8	3.952	5.964	2.120	1.94	0.152
230	1.240	1.653.−3	0.3511	434.7	867.1	4.050	5.930	2.145	1.73	0.147
240	1.876	1.690.−3	0.2390	457.5	879.0	4.147	5.903	2.178	1.55	0.142
250	2.736	1.728.−3	0.1680	480.1	889.8	4.240	5.879	2.220	1.40	0.137
260	3.864	1.769.−3	0.1210	503.3	899.8	4.320	5.845	2.272	1.26	0.132
270	5.311	1.816.−3	0.0890	527.3	909.0	4.409	5.823	2.340	1.14	0.126
280	7.129	1.866.−3	0.0665	551.8	917.0	4.493	5.797	2.434	1.04	0.121
290	9.376	1.923.−3	0.0505	575.9	923.7	4.576	5.775	2.550	0.95	0.116
300	12.11	1.987.−3	0.0389	600.6	930.0	4.657	5.755	2.694	0.87	0.110
310	15.37	2.062.−3	0.0302	625.4	935.2	4.736	5.735	2.870	0.78	0.106
320	19.21	2.149.−3	0.0236	651.3	938.7	4.817	5.715	3.115	0.70	0.100
330	23.73	2.261.−3	0.0185	678.8	940.0	4.900	5.692	3.438	0.61	0.095
340	28.99	2.406.−3	0.0142	710.6	937.5	4.993	5.660	3.920	0.54	0.089
350	35.09	2.626.−3	0.0105	747.7	929.0	5.099	5.617	4.750	0.47	0.082
355	38.49	2.794.−3	0.0089	771.5	922.7	5.166	5.592	5.500	0.44	0.076
360	42.14	3.069.−3	0.0071	802.6	913.0	5.248	5.555	6.750	0.41	
365.1[c]	46.41	4.329.−3	0.0043	873.4	873.4	5.432	5.432	∞	. . .	∞

*P, v, h, and s values extracted from Vaschenko, Voinov, et al., *Thermodynamic and Transport Properties of Ethylene and Propylene*, Standartov, Moscow, 1971; transl. NBSIR 75-763–COM-75-11276, NTIS, 1975. c = critical point.

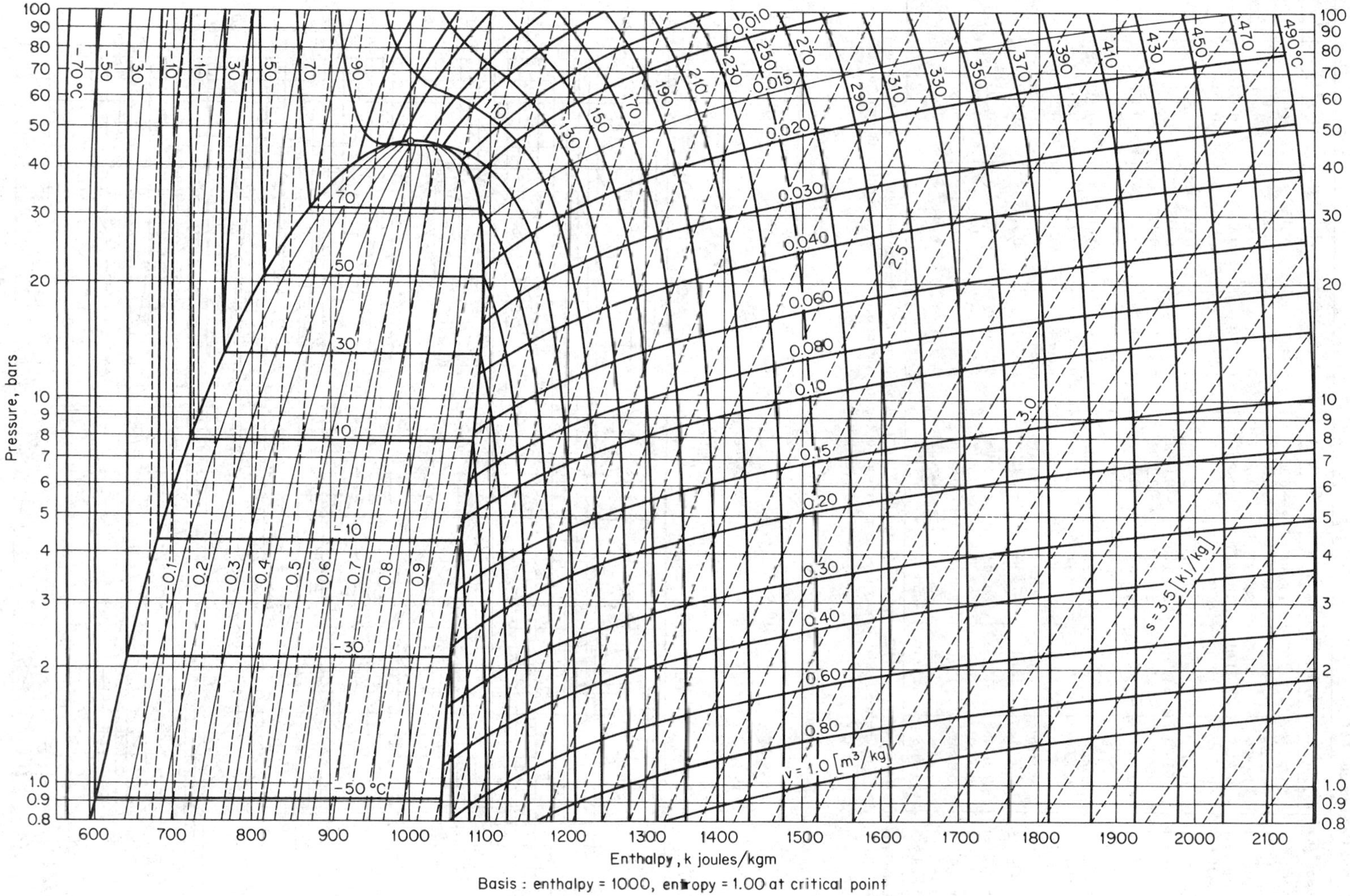

FIG. 3-32 Enthalpy–log-pressure diagram for propylene. 1 kJ/kg = 0.2388 cal/g = 0.4299 Btu/lb; 1 kJ/(kg·K) = 0.2388 cal/(g·K) = 0.2388 Btu/(lb·°R); 1 m^3/kg = 16.02 ft^3/lb. [*From Stephan and Scherer,* Chem. Ing. Tech., *33, 417 (1961). Copyright Verlag Chemie GmbH, Weinheim, Bergstrasse, Germany. Reproduced by permission.*]

TABLE 3-275 Saturated Refrigerant 11*

T, K	P, bar	v_f, m³/kg	v_g, m³/kg	h_f, kJ/kg	h_g, kJ/kg	s_f, kJ/(kg·K)	s_g, kJ/(kg·K)	c_{pf}, kJ/(kg·K)	μ_f, 10^{-4} Pa·s	k_f, W/(m·K)
200	0.0043	5.901.−4	28.06	−14.37	186.30	−0.0651	0.9431	0.815	1.674	0.115
220	0.0417	6.061.−4	6.272	−8.20	195.89	−0.0361	0.8925	0.828	1.142	0.110
240	0.0768	6.225.−4	1.882	4.97	205.85	0.0210	0.8581	0.842	0.831	0.104
260	0.2215	6.398.−4	0.703	21.01	216.06	0.0851	0.8353	0.856	0.635	0.098
270	0.3514	6.491.−4	0.458	29.53	221.23	0.1172	0.8272	0.863	0.563	0.095
280	0.5364	6.587.−4	0.309	38.25	226.40	0.1489	0.8209	0.870	0.504	0.093
290	0.7917	6.688.−4	0.216	47.10	231.58	0.1799	0.8160	0.878	0.454	0.090
300	1.1341	6.794.−4	0.154	56.06	236.73	0.2102	0.8124	0.887	0.413	0.087
310	1.5821	6.908.−4	0.113	65.10	241.83	0.2397	0.8099	0.897	0.377	0.084
320	2.1556	7.027.−4	0.0847	74.22	246.88	0.2686	0.8081	0.907	0.346	0.081
330	2.876	7.156.−4	0.0645	83.42	251.84	0.2967	0.8071	0.917	0.320	0.079
340	3.764	7.293.−4	0.0500	92.72	256.69	0.3243	0.8065	0.928	0.297	0.076
350	4.845	7.442.−4	0.0392	102.12	261.40	0.3513	0.8064	0.939	0.276	0.073
360	6.142	7.603.−4	0.0311	111.64	265.95	0.3778	0.8065	0.950	0.259	0.070
380	9.487	7.974.−4	0.0201	131.12	274.40	0.4298	0.8069	0.975	0.229	0.065
400	14.02	8.435.−4	0.0134	151.38	281.69	0.4808	0.8066	1.004	0.203	0.059
420	19.98	9.042.−4	0.0090	172.76	287.20	0.5317	0.8041	1.04	0.169	0.053
440	27.65	9.930.−4	0.0059	196.01	289.72	0.5840	0.7970	1.09	0.131	0.048
460	37.36	1.167.−3	0.0036	223.85	285.36	0.6435	0.7773	1.19	0.084	0.037
471.2[c]	44.09	1.799.−3	0.0018	258.70	258.70	0.7162	0.7162	∞	0.033	∞

*Values reproduced or converted from Table 1, p. 17.75, *ASHRAE Handbook, 1981: Fundamentals*, American Society of Heating, Refrigerating and Air-Conditioning Engineers, Atlanta, 1981. Copyright material. Reproduced by permission of the copyright owner. c = critical point. The notation 5.901.−4 signifies 5.901×10^{-4}.

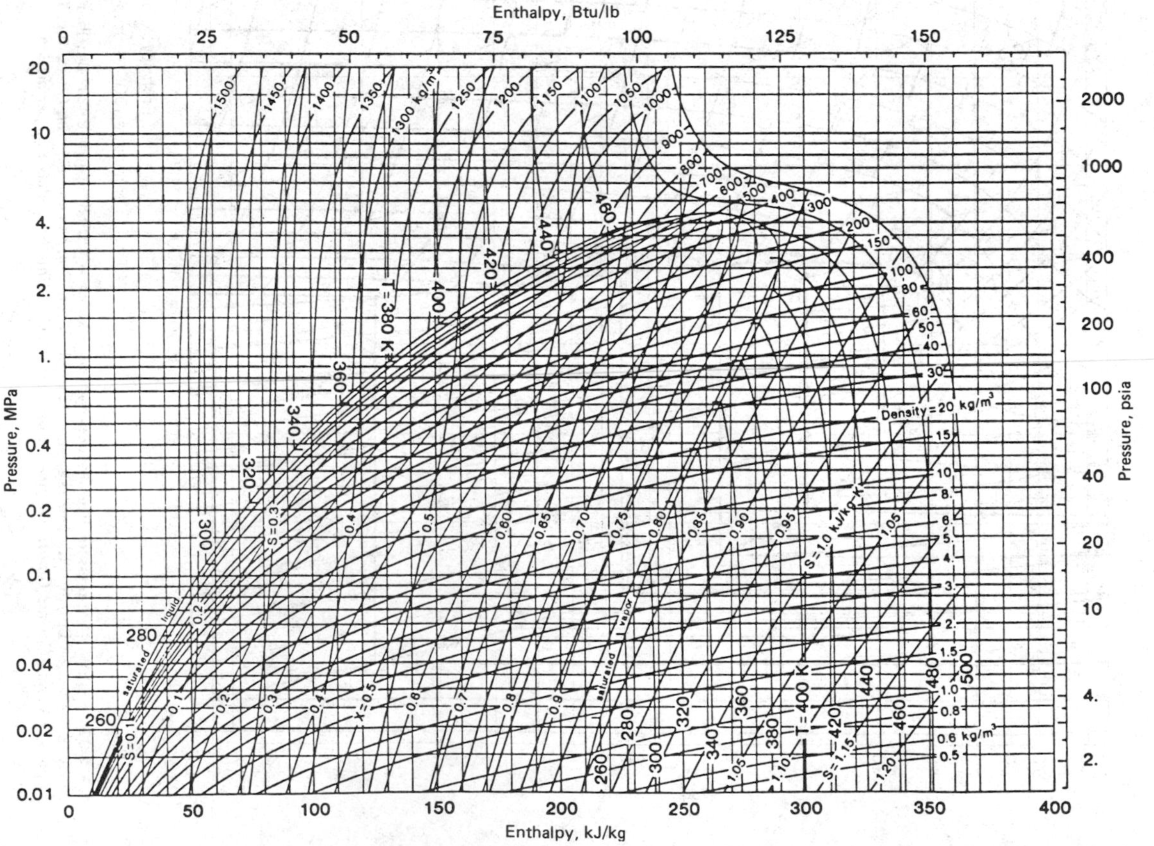

FIG. 3-33 Enthalpy–log-pressure diagram for Refrigerant 11. 1 MPa = 10 bar. *(Copyright 1981 by the American Society of Heating, Refrigerating and Air-Conditioning Engineers and reproduced by permission of the copyright owner.)*

TABLE 3-276 Saturated Refrigerant 12*

T, K	P, bar	v_f, m³/kg	v_g, m³/kg	h_f, kJ/kg	h_g, kJ/kg	s_f, kJ/(kg·K)	s_g, kJ/(kg·K)	c_{pf}, kJ/(kg·K)	μ_f, 10^{-4} Pa·s	k_f, W/(m·K)
150	0.00091	5.767.−4	179.12	294.6	496.0	3.492	4.835	0.808	18.9	0.123
160	0.00305	5.849.−4	36.05	302.3	500.2	3.543	4.780	0.817	15.1	0.119
170	0.00871	5.926.−4	13.40	310.3	504.5	3.591	4.734	0.827	12.1	0.116
180	0.02178	6.024.−4	5.666	318.3	508.9	3.637	4.696	0.836	9.69	0.113
190	0.04877	6.118.−4	2.665	326.5	513.5	3.681	4.665	0.845	7.94	0.109
200	0.0996	6.217.−4	1.370	334.8	518.1	3.724	4.640	0.855	6.64	0.105
210	0.1879	6.139.−4	0.7589	343.2	522.7	3.765	4.620	0.864	5.65	0.102
220	0.3317	6.431.−4	0.4476	351.8	527.4	3.805	4.603	0.873	4.88	0.098
230	0.5531	6.549.−4	0.2784	360.6	531.1	3.844	4.590	0.882	4.26	0.094
240	0.8781	6.675.−4	0.1811	369.5	536.8	3.881	4.579	0.891	3.77	0.090
250	1.3359	6.810.−4	0.1225	378.0	541.5	3.918	4.570	0.902	3.37	0.087
260	1.959	6.970.−4	0.08559	387.7	546.1	3.954	4.563	0.913	3.03	0.083
270	2.784	7.112.−4	0.06147	397.0	550.7	3.989	4.558	0.926	2.75	0.080
280	3.825	7.282.−4	0.04543	406.5	555.1	4.023	4.554	0.942	2.52	0.076
290	5.184	7.470.−4	0.03888	416.1	559.4	4.057	4.551	0.959	2.31	0.072
300	6.840	7.678.−4	0.02582	426.0	563.5	4.090	4.548	0.979	2.14	0.069
310	8.860	7.912.−4	0.01992	436.0	567.3	4.122	4.546	1.005	2.00	0.065
320	11.29	8.173.−4	0.01553	446.2	570.9	4.154	4.543	1.041	1.86	0.061
330	14.17	8.478.−4	0.01218	456.8	574.0	4.186	4.541	1.093	1.74	0.058
340	17.58	8.840.−4	0.00957	467.8	576.5	4.218	4.538	1.166	1.60	0.054
350	21.57	9.286.−4	0.00750	479.4	578.2	4.250	4.533	1.264	1.45	0.050
360	26.19	9.868.−4	0.00582	492.1	578.7	4.285	4.525	1.39	1.28	0.046
370	31.56	1.072.−3	0.00439	506.4	577.2	4.322	4.514	1.55	1.06	0.041
380	37.76	1.237.−3	0.00305	524.7	571.2	4.369	4.900	....	0.75	
385[c]	41.31	1.876.−3	0.00188	551.1	551.1	4.437	4.437	∞	0.31	∞

*P, v, h, and s data interpolated from Perelshteyn (ed.), *Tables and Diagrams of the Thermodynamic Properties of Refrigerants 12, 13, and 22*, Moscow, 1971. c_p, μ, and k data interpolated and converted from *Thermophysical Properties of Refrigerants*, American Society of Heating, Refrigerating and Air-Conditioning Engineers, New York, 1976. c = critical point. The notation 5.767.−4 signifies 5.767×10^{-4}.

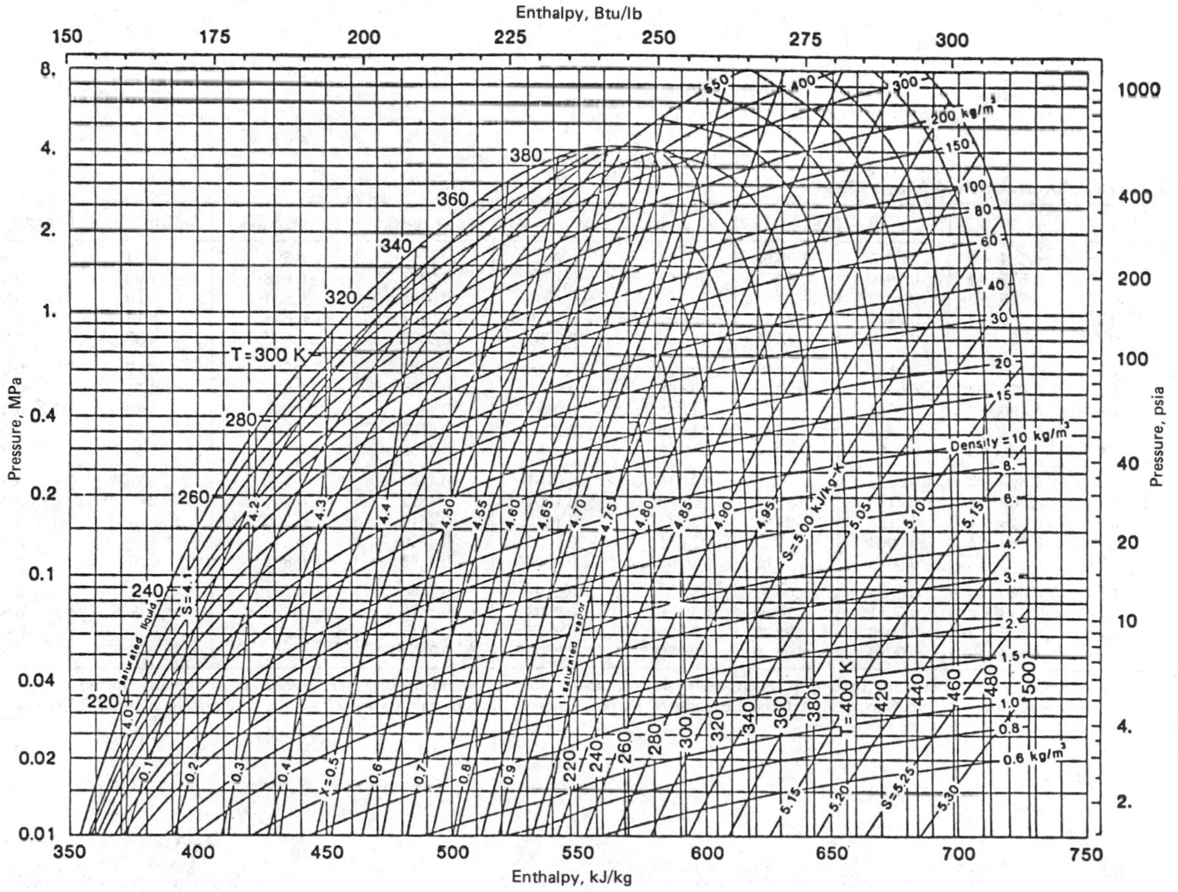

FIG. 3-34 Enthalpy–log-pressure diagram for Refrigerant 12. 1 MPa = 10 bar. *(Copyright 1981 by the American Society of Heating, Refrigerating and Air-Conditioning Engineers and reproduced by permission of the copyright owner.)*

TABLE 3-277 Saturated Refrigerant 13*

T, K	P, bar	v_f, m³/kg	v_g, m³/kg	h_f, kJ/kg	h_g, kJ/kg	s_f, kJ/(kg·K)	s_g, kJ/(kg·K)	c_{pf}, kJ/(kg·K)	μ_f, 10^{-4} Pa·s	k_f, W/(m·K)
91	3.817.−6	5.367.−4	19557	238.1	424.9	3.080	5.133			
100	3.418.−5	5.448.−4	2392	243.8	429.0	3.140	4.990			
110	2.563.−4	5.538.−4	347.0	251.1	433.1	3.205	4.860			
120	0.00137	5.635.−4	70.25	258.2	437.2	3.267	4.759			
130	0.00571	5.739.−4	18.15	265.8	441.3	3.327	4.677			
140	0.01895	5.850.−4	5.865	273.7	455.3	3.385	4.610			
150	0.05250	5.969.−4	2.2617	281.7	449.3	3.441	4.558	0.826	6.83	0.114
160	0.1258	6.095.−4	1.0019	290.0	453.5	3.494	4.516	0.845	5.60	0.109
170	0.2680	6.231.−4	0.4962	298.4	457.6	3.545	4.482	0.865	4.59	0.104
180	0.5186	6.380.−4	0.2689	307.1	461.8	3.594	4.454	0.884	3.83	0.099
190	0.9269	6.536.−4	0.1567	315.9	465.9	3.642	4.431	0.898	3.26	0.093
200	1.5507	6.709.−4	9.69.−2	325.0	469.9	3.688	4.413	0.910	2.82	0.088
210	2.456	6.899.−4	6.28.−2	334.3	473.8	3.732	4.397	0.924	2.48	0.083
220	3.712	7.110.−4	4.24.−2	343.8	477.5	3.777	4.385	0.943	2.20	0.078
230	5.396	7.346.−4	2.95.−2	353.6	481.0	3.820	4.374	0.972	1.97	0.072
240	7.589	7.615.−4	2.11.−2	363.5	484.1	3.862	4.364	1.014	1.79	0.067
250	10.37	7.928.−4	1.53.−2	373.9	486.1	3.903	4.355	1.072	1.63	0.062
260	13.85	8.302.−4	1.13.−2	384.7	489.1	3.944	4.346	1.151	1.50	0.057
270	18.13	8.769.−4	8.28.−3	396.2	490.5	3.986	4.336	1.255	1.34	0.051
280	23.32	9.320.−4	6.10.−3	408.8	490.6	4.029	4.323	1.386	1.14	0.045
290	29.57	1.035.−3	4.34.−3	423.6	488.3	4.080	4.303	1.549	0.87	0.038
300	37.05	1.284.−3	2.60.−3	445.3	477.5	4.151	4.257	1.75	0.52	
302.0[c]	38.70	1.808.−3	1.81.−3	463.1	463.1	4.209	4.209	∞	0.29	∞

*P, v, h, and s data interpolated from Perelshteyn (ed.), *Tables and Diagrams of the Thermodynamic Properties of Refrigerants 12, 13 and 22*, Moscow, 1971. c_p, μ, and k data interpolated and converted from *Thermophysical Properties of Refrigerants*, American Society of Heating, Refrigerating and Air-Conditioning Engineers, New York, 1976. c = critical point. The notation 3.817.−6 signifies 3.817×10^{-6}.

TABLE 3-278 Saturated Refrigerant 13B1*

T, K	P, bar	v_f, m³/kg	v_g, m³/kg	h_f, kJ/kg	h_g, kJ/kg	s_f, kJ/(kg·K)	s_g, kJ/(kg·K)	c_{pf}, kJ/(kg·K)	μ_f, 10^{-4} Pa·s	k_f, W/(m·K)
170	0.059	4.594.−4	1.6015	−40.90	90.95	−0.2033	0.5723	0.597	9.54	0.101
180	0.127	4.677.−4	0.7840	−34.75	94.37	−0.1682	0.5491	0.618	7.60	0.096
190	0.250	4.765.−4	0.4190	−28.51	97.83	−0.1345	0.5305	0.634	6.20	0.091
200	0.455	4.860.−4	0.2407	−22.17	101.32	−0.1020	0.5154	0.648	5.13	0.086
210	0.777	4.961.−4	0.1467	−15.68	104.82	−0.0704	0.5033	0.663	4.33	0.082
215.4	1.013	5.020.−4	0.1147	−12.09	106.70	−0.0536	0.4978	0.670	3.97	0.079
220	1.254	5.071.−4	0.0940	− 9.02	108.28	−0.0396	0.4936	0.676	3.71	0.077
230	1.933	5.190.−4	0.0628	− 2.19	111.68	−0.0094	0.4857	0.690	3.22	0.073
240	2.863	5.321.−4	0.0433	4.83	114.99	0.0202	0.4793	0.703	2.83	0.068
250	4.096	5.466.−4	0.0308	12.03	118.16	0.0494	0.4739	0.721	2.51	0.063
260	5.690	5.627.−4	0.0224	19.44	121.16	0.0781	0.4693	0.742	2.25	0.059
270	7.703	5.809.−4	0.0166	27.06	123.93	0.1064	0.4652	0.767	2.04	0.054
280	10.20	6.018.−4	0.0124	34.94	126.41	0.1345	0.4612	0.800	1.84	0.049
290	13.25	6.264.−4	0.0094	43.11	128.51	0.1625	0.4570	0.842	1.69	0.045
300	16.91	6.562.−4	0.0072	51.68	130.09	0.1908	0.4522	0.891	1.57	0.040
310	21.28	6.940.−4	0.0055	60.81	130.97	0.2197	0.4460	0.951	1.45	0.035
320	26.44	7.458.−4	0.0041	70.80	130.76	0.2503	0.4376	1.09	1.26	0.030
330	32.48	8.295.−4	0.0030	82.42	128.59	0.2845	0.4245	1.29	0.99	0.026
340.2[c]	39.64	1.344.−3	0.0013	108.70	108.70	0.3605	0.3605	∞	0.35	∞

*Values reproduced or converted from Table 4, p. 17.83, *ASHRAE Handbook, 1981: Fundamentals*, American Society of Heating, Refrigerating and Air-Conditioning Engineers, Atlanta, 1981. Copyright material. Reproduced by permission of the copyright owner. c = critical point. The notation 4.594.−4 signifies 4.594×10^{-4}.

TABLE 3-279 Saturated Refrigerant 21*

Temp., °F.	Pressure, lb./sq. in. abs.	Volume, cu. ft./lb.		Enthalpy, B.t.u./lb.		Entropy, B.t.u./(lb.)(°F.)	
		Liquid	Vapor	Liquid	Vapor	Liquid	Vapor
−40	1.358	0.01058	32.09	0.00	114.56	0.0000	0.2730
−20	2.578	0.01075	17.66	4.71	116.96	0.0109	0.2663
0	4.582	0.01093	10.35	9.44	119.37	0.0214	0.2606
5	5.243	0.01097	9.132	10.63	119.97	0.0240	0.2593
20	7.699	0.01112	8.085	11.81	120.57	0.0265	0.2581
40	12.32	0.01132	4.130	19.04	124.19	0.0414	0.2519
60	18.90	0.01153	2.773	23.98	126.60	0.0511	0.2486
80	27.96	0.01176	1.923	29.03	128.98	0.0606	0.2458
86	31.23	0.01183	1.733	30.56	129.68	0.0634	0.2450
100	40.04	0.01200	1.371	34.18	131.29	0.0699	0.2434
120	55.75	0.01226	1.001	39.46	133.53	0.0791	0.2414
140	75.72	0.01254	0.7457	44.86	135.66	0.0882	0.2396
160	100.6	0.01284	0.5646	50.43	137.69	0.0972	0.2381

TABLE 3-280 Saturated Refrigerant 22*

T, K	P, bar	v_f, m³/kg	v_g, m³/kg	h_f, kJ/kg	h_g, kJ/kg	s_f, kJ/(kg·K)	s_g, kJ/(kg·K)	c_{pf}, kJ/(kg·K)	μ_f, 10^{-4} Pa·s	k_f, W/(m·K)
150	0.0017	6.209.−4	83.40	268.2	547.3	3.355	5.215	1.059	....	0.161
160	0.0054	6.293.−4	28.20	278.2	552.1	3.430	5.141	1.058	....	0.156
170	0.0150	6.381.−4	10.85	288.3	557.0	3.494	5.075	1.057	0.770	0.151
180	0.0369	6.474.−4	4.673	298.7	561.9	3.551	5.013	1.058	0.647	0.146
190	0.0821	6.573.−4	2.225	308.6	566.8	3.605	4.963	1.060	0.554	0.141
200	0.1662	6.680.−4	1.145	318.8	571.6	3.657	4.921	1.065	0.481	0.136
210	0.3116	6.794.−4	0.6370	329.1	576.5	3.707	4.885	1.071	0.424	0.131
220	0.5470	6.917.−4	0.3772	339.7	581.2	3.756	4.854	1.080	0.378	0.126
230	0.9076	7.050.−4	0.2352	350.6	585.9	3.804	4.828	1.091	0.340	0.121
240	1.4346	7.195.−4	0.1532	361.7	590.5	3.852	4.805	1.105	0.309	0.117
250	2.174	7.351.−4	0.1037	373.0	594.9	3.898	4.785	1.122	0.282	0.112
260	3.177	7.523.−4	0.07237	384.5	599.0	3.942	4.768	1.143	0.260	0.107
270	4.497	7.733.−4	0.05187	396.3	603.0	3.986	4.752	1.169	0.241	0.102
280	6.192	7.923.−4	0.03803	408.2	606.6	4.029	4.738	1.193	0.225	0.097
290	8.324	8.158.−4	0.02838	420.4	610.0	4.071	4.725	1.220	0.211	0.092
300	10.956	8.426.−4	0.02148	432.7	612.8	5.113	4.713	1.257	0.198	0.087
310	14.17	8.734.−4	0.01643	445.5	615.1	4.153	4.701	1.305	0.186	0.082
320	18.02	9.096.−4	0.01265	458.6	616.7	4.194	4.688	1.372	0.176	0.077
330	22.61	9.535.−4	9.753.−3	472.4	617.3	4.235	4.674	1.460	0.167	0.072
340	28.03	1.010.−3	7.479.−3	487.2	616.5	4.278	4.658	1.573	0.151	0.067
350	34.41	1.086.−3	5.613.−3	503.7	613.3	4.324	4.637	1.718	0.130	0.062
360	41.86	1.212.−3	4.036.−3	523.7	605.5	4.378	4.605	1.897	0.106	
369.3[c]	49.89	2.015.−3	2.015.−3	570.0	570.0	4.501	4.501	∞		

*P, v, h, and s data interpolated from Perelshteyn (ed.), *Tables and Diagrams of the Thermodynamic Properties of Refrigerants 12, 13 and 22*, Moscow, 1971. c_p, μ, and k data interpolated and converted from *Thermophysical Properties of Refrigerants*, American Society of Heating, Refrigerating and Air-Conditioning Engineers, New York, 1976. c = critical point. The notation 6.209.−4 signifies 6.209×10^{-4}.

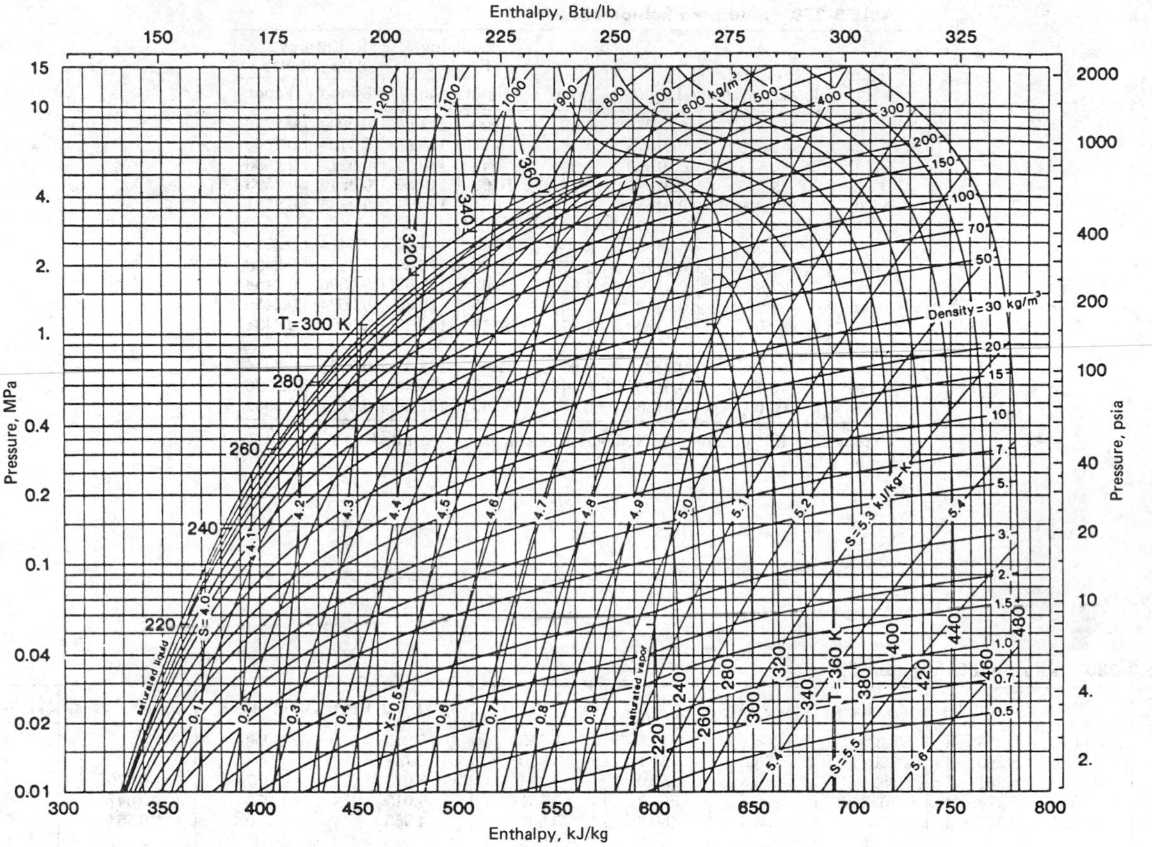

FIG. 3-35 Enthalpy–log-pressure diagram for Refrigerant 22. 1 MPa = 10 bar. *(Copyright 1981 by the American Society of Heating, Refrigerating and Air-Conditioning Engineers and reproduced by permission of the copyright owner.)*

TABLE 3-281 Saturated Refrigerant 23*

Temp., °F.	Pressure, lb./sq. in. abs.	Volume, cu. ft./lb.		Enthalpy, B.t.u./lb.		Entropy, B.t.u./(lb.)(°F.)	
		Liquid	Vapor	Liquid	Vapor	Liquid	Vapor
−190	0.622	0.01030	66.18	−44.44	71.64	−0.1301	0.3004
−180	1.068	0.01040	39.87	−41.28	72.83	−0.1185	0.2895
−160	2.774	0.01060	16.34	−35.39	75.17	−0.0982	0.2707
−140	6.280	0.01082	7.619	−29.82	77.42	−0.0802	0.2552
−120	12.75	0.01107	3.928	−24.32	79.53	−0.0636	0.2421
−100	23.71	0.01137	2.192	−18.65	81.45	−0.0474	0.2309
−80	40.99	0.01171	1.300	−12.74	83.16	−0.0315	0.2210
−60	66.67	0.01211	0.8165	−6.52	84.62	−0.0157	0.2123
−40	103.0	0.01258	0.5325	0.00	85.78	0.0000	0.2044
−20	152.5	0.01316	0.3583	6.78	86.60	0.0155	0.1970
0	217.6	0.01388	0.2463	13.80	86.95	0.0307	0.1899
20	301.6	0.01482	0.1710	21.13	86.65	0.0459	0.1825
40	408.2	0.01614	0.1176	29.12	85.25	0.0616	0.1739
60	543.0	0.01830	0.0768	38.89	81.48	0.0799	0.1618
70	623.4	0.02039	0.0581	45.63	77.18	0.0922	0.1518
78.7[c]	701.4	0.03051	0.0305	60.77	60.77	0.1198	0.1198

*Unpublished data of Allied Chemical Company, 1970. Reproduced by permission. [c] = critical temperature.

TABLE 3-282 Saturated Refrigerant 32*

Temp., °F.	Pressure, lb./sq. in. abs.	Volume cu. ft./lb.		Enthalpy, B.t.u./lb.		Entropy, B.t.u./(lb.)(°F.)	
		Liquid	Vapor	Liquid	Vapor	Liquid	Vapor
−150	0.510	0.0118	124.859	−39.77	143.75	−0.1094	0.4832
−140	0.828	0.0119	82.794	−36.36	145.31	−0.0986	0.4697
−120	1.984	0.0122	34.995	−29.44	148.37	−0.0776	0.4459
−100	4.257	0.0125	17.732	−22.33	151.33	−0.0572	0.4828
−80	8.338	0.0129	9.150	−15.05	154.13	−0.0376	0.4080
−60	15.134	0.0132	5.236	−7.62	156.73	−0.0125	0.3927
−40	25.760	0.0136	3.173	0.00	159.09	0.0000	0.3791
−20	41.534	0.0140	2.016	7.76	161.17	0.0180	0.3669
0	63.953	0.0144	1.331	15.66	162.92	0.0354	0.3558
20	94.684	0.0149	0.907	23.71	164.29	0.0524	0.3454
40	135.57	0.0155	0.634	31.96	165.21	0.0690	0.3356
60	188.61	0.0161	0.451	40.46	165.58	0.0854	0.3261
80	256.06	0.0168	0.324	49.31	165.26	0.1017	0.3165
100	340.41	0.0177	0.235	58.69	164.02	0.1183	0.3064
120	444.50	0.0189	0.169	68.93	161.44	0.1356	0.2952
140	571.61	0.0207	0.119	80.57	156.72	0.1545	0.2815
160	725.67	0.0240	0.078	94.77	147.60	0.1768	0.2621
173.12^c	849.62	0.0373	0.037	119.30	119.30	0.2148	0.2148

^{c} = critical temperature.

*Unpublished data from Specialty Chemicals Division, Allied Chemical Corporation. Reproduced by permission. The original tables give more properties and are at 1°F. increments. An enthalpy–log pressure diagram from 1 to 1000 lb./sq. in. abs., −120° to 600° F., was also prepared by this corporation.

TABLE 3-283 Saturated Refrigerant 113*

T, K	P, bar	v_f, m³/kg	v_g, m³/kg	h_f, kJ/kg	h_g, kJ/kg	s_f, kJ/(kg·K)	s_g, kJ/(kg·K)	c_{pf}, kJ/(kg·K)	μ_f, 10^{-4} Pa·s	k_f, W/(m·K)
240	0.0233	5.908.−4	4.548	5.70	171.97	0.0241	0.7169	0.845	17.9	0.087
250	0.0435	5.986.−4	2.537	14.19	178.06	0.0587	0.7142	0.877	14.8	0.084
260	0.0767	6.066.−4	1.492	22.83	184.22	0.0926	0.7134	0.895	12.3	0.083
270	0.1290	6.150.−4	0.9189	31.65	190.46	0.1259	0.7141	0.916	10.4	0.081
280	0.2076	6.237.−4	0.5893	40.63	196.75	0.1585	0.7161	0.933	8.9	0.079
290	0.3217	6.328.−4	0.3917	49.77	203.08	0.1906	0.7192	0.946	7.6	0.077
300	0.4817	6.422.−4	0.2687	59.07	209.44	0.2221	0.7233	0.958	6.6	0.075
310	0.6999	6.522.−4	0.1895	68.51	215.80	0.2530	0.7281	0.971	5.9	0.073
320	0.9897	6.626.−4	0.1370	78.09	222.17	0.2833	0.7336	0.983	5.2	0.071
330	1.3657	6.737.−4	0.1012	87.80	228.53	0.3131	0.7396	0.992	4.7	0.069
340	1.8347	6.854.−4	0.0762	97.64	234.86	0.3424	0.7460	1.000	4.2	0.066
350	2.4406	6.979.−4	0.0584	107.58	241.16	0.3711	0.7528	1.013	3.8	0.065
360	3.174	7.112.−4	0.0454	117.65	247.41	0.3993	0.7598	1.029	3.4	0.062
370	4.062	7.255.−4	0.0357	127.82	253.59	0.4270	0.7669	1.042	3.2	0.060
380	5.123	7.411.−4	0.0284	138.11	259.70	0.4542	0.7742	1.059	2.9	0.058
390	6.379	7.580.−4	0.0229	148.52	265.71	0.4810	0.7815	1.084	2.7	0.056
400	7.849	7.767.−4	0.0185	159.07	271.59	0.5075	0.7888	1.109	2.46	0.054
410	9.556	7.975.−4	0.0151	169.78	277.31	0.5336	0.7958	1.14	2.28	0.052
420	11.52	8.211.−4	0.0124	180.69	282.83	0.5595	0.8027	1.18	2.10	0.050
430	13.78	8.483.−4	0.0102	191.85	288.09	0.5853	0.8091	1.22	1.93	0.047
440	16.35	8.806.−4	0.0083	203.35	292.98	0.6112	0.8149	1.27	1.75	0.045
450	19.26	9.201.−4	0.0068	215.31	297.38	0.6375	0.8198	1.32	1.58	0.042
460	22.56	9.713.−4	0.0055	227.97	301.03	0.6645	0.8234	1.38	1.33	0.039
470	26.29	1.044.−3	0.0044	241.79	303.41	0.6933	0.8244	1.45	1.07	0.035
480	30.52	1.174.−3	0.0032	258.16	303.00	0.7264	0.8198	1.54	0.77	0.031
487.5^c	34.11	1.754.−3	0.0018	288.10	288.10	0.7828	0.7828	∞	0.30	∞

*Values reproduced or converted from Table 8, p. 17.91, *ASHRAE Handbook, 1981: Fundamentals*, American Society of Heating, Refrigerating and Air-Conditioning Engineers, Atlanta, 1981. Copyright material. Reproduced by permission of the copyright owner. c = critical point. The notation 5.908.−4 signifies 5.908×10^{-4}.

TABLE 3-284 Saturated Refrigerant 114*

T, K	P, bar	v_f, m³/kg	v_g, m³/kg	h_f, kJ/kg	h_g, kJ/kg	s_f, kJ/(kg·K)	s_g, kJ/(kg·K)	c_{pf}, kJ/(kg·K)	μ_f 10^{-4} Pa·s	k_f, W/(m·K)
190	0.0058	6.326.−4	15.823	−42.58	125.78	−0.2091	0.6794	0.765	23.9	0.093
200	0.0137	6.344.−4	7.094	−31.87	131.01	−0.1542	0.6648	0.787	18.2	0.090
210	0.029	6.366.−4	3.465	−21.48	136.41	−0.1035	0.6541	0.810	14.3	0.088
220	0.059	6.391.−4	1.822	−11.37	141.95	−0.0565	0.6466	0.831	11.5	0.085
230	0.109	6.421.−4	1.021	− 1.50	147.61	−0.0126	0.6419	0.854	9.4	0.082
240	0.190	6.457.−4	0.604	8.18	153.36	0.0286	0.6393	0.877	7.9	0.080
250	0.317	6.500.−4	0.375	17.74	159.18	0.0676	0.6387	0.900	6.61	0.077
260	0.505	6.554.−4	0.2431	27.22	165.05	0.1047	0.6396	0.923	5.66	0.075
270	0.773	6.619.−4	0.1633	36.71	170.95	0.1405	0.6418	0.946	4.96	0.072
280	1.143	6.700.−4	0.1132	46.27	176.85	0.1751	0.6452	0.967	4.30	0.069
290	1.636	6.799.−4	0.0807	55.95	182.75	0.2090	0.6494	0.991	3.80	0.067
300	2.279	6.918.−4	0.0590	65.79	188.61	0.2422	0.6543	1.015	3.35	0.064
310	3.096	7.060.−4	0.0440	75.79	194.44	0.2748	0.6598	1.038	3.02	0.061
320	4.116	7.224.−4	0.0334	85.92	200.19	0.3067	0.6657	1.062	2.69	0.059
330	5.366	7.412.−4	0.0257	96.16	205.84	0.3379	0.6719	1.087	2.48	0.056
340	6.877	7.624.−4	0.0201	106.49	211.37	0.3685	0.6781	1.111	2.27	0.054
350	8.683	7.863.−4	0.0158	116.96	216.71	0.3984	0.6843	1.136	2.07	0.051
360	10.82	8.135.−4	0.0125	127.63	221.82	0.4280	0.6903	1.160	1.91	0.048
370	13.32	8.453.−4	0.0099	138.60	226.57	0.4575	0.6957	1.185	1.76	0.045
380	16.24	8.836.−4	0.0079	149.99	230.84	0.4872	0.7002	1.210	1.59	0.042
390	19.62	9.324.−4	0.0062	162.01	234.36	0.5176	0.7032	1.236	1.39	0.038
400	23.52	1.001.−3	0.0048	175.03	236.61	0.5496	0.7036	1.261	1.17	0.034
410	28.00	1.118.−3	0.0035	190.13	236.20	0.5857	0.6980	1.5	0.87	0.030
419.0^c	32.61	1.795.−3	0.0018	219.90	219.90	0.6559	0.6559	∞	0.34	∞

*Values reproduced or converted from Table 9, p. 17.93, *ASHRAE Handbook, 1981: Fundamentals*, American Society of Heating, Refrigerating and Air-Conditioning Engineers, Atlanta, 1981. Copyright material. Reproduced by permission of the copyright owner. c = critical point. The notation 6.326.−4 signifies 6.326×10^{-4}.

TABLE 3-285 Saturated Refrigerant 115*

Temp., °F.	Pressure, lb./sq. in. abs.	Volume, cu. ft./lb.		Enthalpy, B.t.u./lb.		Entropy, B.t.u./(lb.)(°F.)	
		Liquid	Vapor	Liquid	Vapor	Liquid	Vapor
−100	2.327	0.00966	10.57	−13.07	45.83	−0.0335	0.1302
−80	4.573	0.00986	5.624	−8.78	48.39	−0.0219	0.1286
−60	8.306	0.01009	3.218	−4.43	50.96	−0.0108	0.1278
−40	14.13	0.01033	1.953	0.00	53.53	0.0000	0.1275
−20	22.74	0.01060	1.245	4.50	56.07	0.0104	0.1277
0	34.94	0.01090	0.8257	9.09	58.56	0.0206	0.1282
20	51.59	0.01123	0.5657	13.76	61.00	0.0305	0.1290
40	73.65	0.01161	0.3979	18.54	63.35	0.0401	0.1298
60	102.1	0.01204	0.2857	23.45	65.60	0.0496	0.1308
80	138.1	0.01255	0.2081	28.54	67.71	0.0591	0.1317
100	182.7	0.01316	0.1530	33.85	69.63	0.0686	0.1325
120	237.3	0.01393	0.1125	39.50	71.24	0.0782	0.1330
140	303.2	0.01496	0.0817	45.67	72.36	0.0884	0.1329
160	382.0	0.01664	0.0567	52.76	72.42	0.0996	0.1314
170	427.0	0.01838	0.0444	56.56	71.33	0.1055	0.1290
175.89^c	457.6	0.0261	0.0261	64.30	64.30	0.1175	0.1175

*Unpublished data of General Chemicals Division, Allied Chemical Company. Used by permission. ^{c} = critical temperature.

TABLE 3-286 Refrigerant 142b*

T, K	P, bar	v_f, m^3/kg	v_g, m^3/kg	h_f, kJ/kg	h_g, kJ/kg	s_f, kJ/(kg·K)	s_g, kJ/(kg·K)	c_{pf}, kJ/(kg·K)	μ_f, 10^{-4} Pa·s	k_f, W/(m·K)
200	0.0380	7.505.−4	4.337	−24.36	200.49	−0.1123	1.0119	...	....	0.123
210	0.0728	7.626.−4	2.374	−17.48	206.82	−0.0788	0.9893	1.15	....	0.118
220	0.1314	7.751.−4	1.373	−10.21	213.28	−0.0450	0.9708	1.17	....	0.114
230	0.2252	7.883.−4	0.833	− 2.52	219.82	−0.0109	0.9558	1.18	....	0.111
240	0.3691	8.019.−4	0.527	9.92	229.74	0.0414	0.9387	1.19	0.517	0.109
250	0.5815	8.164.−4	0.346	14.32	233.05	0.0592	0.9341	1.21	0.466	0.103
260	0.8846	8.317.−4	0.234	23.54	239.66	0.0952	0.9264	1.22	0.422	0.099
270	1.3046	8.480.−4	0.162	33.32	246.18	0.1320	0.9204	1.24	0.385	0.095
280	1.8714	8.653.−4	0.115	43.68	252.57	0.1695	0.9155	1.26	0.355	0.091
290	2.6184	8.843.−4	0.0838	54.60	258.77	0.2076	0.9116	1.28	0.329	0.088
300	3.583	9.047.−4	0.0619	66.07	264.69	0.2462	0.9082	1.30	0.305	0.084
310	4.803	9.273.−4	0.0464	78.07	270.26	0.2851	0.9051	1.32	0.285	0.080
320	6.324	9.525.−4	0.0353	90.55	275.40	0.3243	0.9020	1.34	0.267	0.075
330	8.187	9.810.−4	0.0271	103.45	280.01	0.3634	0.8985	...	0.241	0.072
340	10.44	1.014.−3	0.0210	116.71	283.99	0.4024	0.8943	...	0.216	0.068
350	13.13	1.052.−3	0.0164	130.30	287.23	0.4409	0.8893	...	0.192	0.064
360	16.30	1.099.−3	0.0129	144.18	289.61	0.4791	0.8831	...	....	0.060
370	20.01	1.157.−3	0.0102	158.45	291.01	0.5170	0.8753	...	....	0.056
380	24.29	1.235.−3	0.0080	173.45	291.22	0.5557	0.8656	...	....	0.052
390	29.20	1.348.−3	0.0062	190.16	289.77	0.5974	0.8528	...	....	0.048
400	34.78	1.541.−3	0.0046	212.57	284.04	0.6521	0.8307	...	....	0.044
410[c]	41.5	2.300.−3	0.0023	255.00	255.00	.....	....	...	....	∞

*Values reproduced and converted from Table 10, p. 17.95, *ASHRAE Handbook, 1981: Fundamentals*, American Society of Heating, Refrigerating and Air-Conditioning Engineers, Atlanta, 1981. Copyright material. Reproduced by permission of the copyright owner. c = critical point. The notation 7.505.−4 signifies 7.505×10^{-4}.

TABLE 3-287 Saturated Refrigerant 152a*

T, K	P, bar	v_f, m^3/kg	v_g, m^3/kg	h_f, kJ/kg	h_g, kJ/kg	s_f, kJ/(kg·K)	s_g, kJ/(kg·K)	c_{pf}, kJ/(kg·K)	μ_f, 10^{-4} Pa·s	k_f, W(m·K)
170	0.0078	8.683.−4	27.425	−56.60	273.86	−0.2801	1.6638	0.667	10.8	0.134
180	0.0184	8.811.−4	12.297	−48.92	281.29	−0.2363	1.5983	0.742	8.87	0.130
190	0.0397	8.944.−4	6.014	−40.86	288.92	−0.1927	1.5430	0.815	7.27	0.126
200	0.0792	9.084.−4	3.165	−32.34	296.74	−0.1490	1.4964	0.886	5.96	0.122
210	0.1480	9.234.−4	1.772	−23.31	304.70	−0.1050	1.4570	0.956	4.98	0.118
220	0.2614	9.391.−4	1.047	−13.68	312.78	−0.0603	1.4236	1.027	4.23	0.114
230	0.4393	9.558.−4	0.647	− 3.39	320.91	−0.0146	1.3954	1.098	3.65	0.110
240	0.7071	9.737.−4	0.416	7.64	329.06	0.0322	1.3714	1.188	3.18	0.106
250	1.0955	9.929.−4	0.2769	19.48	337.14	0.0804	1.3510	1.247	2.81	0.102
260	1.6411	1.013.−3	0.1897	32.19	345.10	0.1300	1.3335	1.326	2.50	0.098
270	2.3860	1.036.−3	0.1333	45.81	352.84	0.1811	1.3182	1.410	2.25	0.095
280	3.3775	1.061.−3	0.0957	60.37	360.27	0.2337	1.3047	1.499	2.03	0.091
290	4.6677	1.088.−3	0.0700	75.90	367.28	0.2877	1.2924	1.593	1.85	0.087
300	6.3132	1.118.−3	0.0520	92.41	373.75	0.3430	1.2808	1.695	1.70	0.083
310	8.3740	1.152.−3	0.0391	109.91	379.52	0.3996	1.2693	1.804	1.57	0.079
320	10.91	1.191.−3	0.0298	128.40	384.41	0.4574	1.2574	1.921	1.45	0.075
330	14.00	1.237.−3	0.0228	147.94	388.20	0.5163	1.2444	2.048	1.35	0.071
340	17.69	1.292.−3	0.0175	168.62	390.54	0.5767	1.2294	2.185	1.26	0.067
350	22.06	1.362.−3	0.0135	190.74	390.91	0.6391	1.2110	2.333	1.18	0.064
360	27.17	1.453.−3	0.0102	215.18	388.19	0.7059	1.1865			
370	33.10	1587.−3	0.0072	245.98	377.49	0.7878	1.1432			

*Values reproduced and converted from Table 11, p. 17.97, *ASHRAE Handbook, 1981: Fundamentals*, American Society of Heating, Refrigerating and Air-Conditioning Engineers, Atlanta, 1981. Copyright material. Reproduced by permission of the copyright owner. The notation 8.683.−4 signifies 8.683×10^{-4}.

TABLE 3-288 Saturated Refrigerant 216*

Temp., °F.	Pressure, lb./sq. in. abs.	Volume, cu. ft./lb.		Enthalpy, B.t.u./lb.		Entropy, B.t.u./(lb.)(°F.)	
		Liquid	Vapor	Liquid	Vapor	Liquid	Vapor
−40	0.339	0.00927	59.957	0.000	62.415	0.0000	0.1487
−20	0.713	0.00942	29.749	4.778	65.276	0.0111	0.1487
0	1.382	0.00958	15.986	9.541	68.208	0.0217	0.1493
20	2.497	0.00974	9.184	14.298	71.199	0.0318	0.1504
40	4.247	0.00992	5.582	19.056	74.239	0.0415	0.1520
60	6.862	0.01010	3.558	23.821	77.319	0.0509	0.1538
80	10.612	0.01030	2.361	28.598	80.429	0.0599	0.1559
100	15.797	0.01050	1.6215	33.391	83.559	0.0686	0.1582
120	22.753	0.01073	1.1462	38.205	86.701	0.0770	0.1607
140	31.845	0.01097	0.8304	43.049	89.845	0.0852	0.1632
160	43.468	0.01124	0.6142	47.930	92.981	0.0931	0.1658
180	58.046	0.01153	0.4623	52.861	96.099	0.1009	0.1685
200	76.033	0.01186	0.3529	57.857	99.186	0.1085	0.1712
220	97.913	0.01223	0.2725	62.939	102.225	0.1161	0.1739
240	124.21	0.01266	0.2121	68.132	105.196	0.1235	0.1765
260	155.50	0.01317	0.1660	73.474	108.066	0.1309	0.1790
280	192.40	0.01378	0.1300	79.015	110.789	0.1384	0.1813
300	235.63	0.01458	0.1013	84.835	113.282	0.1460	0.1834
320	286.03	0.01570	0.0776	91.089	115.373	0.1539	0.1851
340	344.81	0.01764	0.0565	98.234	116.538	0.1628	0.1856
355.98^c	399.45	0.02771	0.0277	110.248	110.248	0.1773	0.1773

*From published data, Chemicals Division, Union Carbide Corporation. Used by permission. The paper describing these data is by Shank, *ASHRAE J.*, 7, 94–101 (1965).
^{c} = critical temperature.

TABLE 3-289 Saturated Refrigerant 245*

T, K	P, bar	v_f, m^3/kg	v_g, m^3/kg	h_f, kJ/kg	h_g, kJ/kg	s_f, kJ/(kg·K)	s_g, kJ/(kg·K)
172	0.0034	6.46.−4	31.49	−63.4	133.8	−0.3131	0.8327
180	0.0076	6.57.−4	14.63	−55.9	138.7	−0.2707	0.8099
190	0.0190	6.70.−4	6.20	−46.2	145.1	−0.2182	0.7885
200	0.0425	6.83.−4	2.91	−36.0	151.7	−0.1666	0.7725
210	0.0870	6.97.−4	1.48	−25.7	158.5	−0.1157	0.7612
220	0.1654	7.11.−4	0.822	−14.8	165.4	−0.0654	0.7539
230	0.2946	7.25.−4	0.475	− 3.6	172.5	−0.0156	0.7500
240	0.4958	7.40.−4	0.292	8.0	179.6	0.0337	0.7487
250	0.7946	7.55.−4	0.192	19.9	186.8	0.0824	0.7497
260	1.2204	7.72.−4	0.125	32.3	194.0	0.1305	0.7525
270	1.806	7.89.−4	0.0862	44.9	201.1	0.1781	0.7567
280	2.584	8.08.−4	0.0611	57.9	208.3	0.2249	0.7621
290	3.600	8.30.−4	0.0443	71.1	215.3	0.2711	0.7683
300	4.888	8.53.−4	0.0327	84.6	222.2	0.3161	0.7751
310	6.491	8.80.−4	0.0246	98.4	228.9	0.3614	0.7822
320	8.456	9.11.−4	0.0186	112.6	235.3	0.4057	0.7893
330	10.83	9.48.−4	0.0143	127.1	241.4	0.4497	0.7960
340	13.67	9.93.−4	0.0111	142.1	246.9	0.4937	0.8018
350	17.04	0.00105	0.0084	157.2	251.5	0.5382	0.8060
360	21.02	0.00113	0.0063	174.7	254.8	0.5844	0.8071
370	25.71	0.00125	0.0045	193.6	255.2	0.6349	0.8013
375	28.46	0.00137	0.0036	205.2	252.5	0.6649	0.7953
380.1^c	31.37	0.00204	0.0020	231.8	231.8	0.7341	0.7341

*Values converted from tables of Shank, *Thermodynamic Properties of UCON 245 Refrigerant*, Union Carbide Corporation, New York, 1966. See also Shank, *J. Chem. Eng. Data*, **12,** 474–480 (1967). c = critical point. The notation 6.46.−4 signifies 6.46 $\times 10^{-4}$.

TABLE 3-290 Refrigerant C 318*

T, K	P, bar	v_f, m^3/kg	v_g, m^3/kg	h_f, kJ/kg	h_g, kJ/kg	s_f, kJ/(kg·K)	s_g, kJ/(kg·K)	c_{pf}, kJ/(kg·K)	μ_f, 10^{-4} Pa·s	k_f, W/(m·K)
200	0.0216	5.507.−4	3.810	353.5	498.0	3.909	4.560			
210	0.0449	5.593.−4	1.931	361.0	500.1	3.947	4.564			
220	0.0875	5.683.−4	1.038	369.2	502.2	3.984	4.569			
230	0.1608	5.778.−4	0.588	377.6	504.4	4.022	4.574	0.98	11.7	0.088
240	0.2810	5.879.−4	0.349	386.4	510.9	4.060	4.578	1.00	9.55	0.085
250	0.466	5.988.−4	0.2166	395.6	517.4	4.097	4.584	1.02	7.90	0.082
260	0.741	6.106.−4	0.1401	405.2	524.0	4.133	4.592	1.03	6.63	0.078
270	1.133	6.234.−4	0.0938	415.1	530.7	4.172	4.599	1.05	5.64	0.075
280	1.672	6.375.−4	0.0647	425.8	537.3	4.210	4.609	1.07	4.85	0.071
290	2.392	6.529.−4	0.0458	436.2	543.9	4.247	4.618	1.09	4.22	0.068
300	3.325	6.694.−4	0.0332	447.3	550.4	4.284	4.626	1.12	3.70	0.065
310	4.522	6.893.−4	0.0245	458.7	556.9	4.322	4.638	1.15	3.20	0.061
320	6.007	7.115.−4	0.0184	470.5	563.3	4.359	4.648	1.18	2.94	0.058
330	7.826	7.365.−4	0.0139	482.7	569.4	4.396	4.659	1.23	2.66	0.054
340	10.018	7.666.−4	0.0106	495.2	575.4	4.433	4.669	1.27	2.33	0.051
350	12.632	8.034.−4	0.0082	508.1	581.0	4.469	4.678	1.32	2.00	0.048
360	15.71	8.508.−4	0.0062	521.5	585.8	4.507	4.685	1.39		
370	19.33	9.172.−4	0.0047	535.6	589.9	4.544	4.691			
380	23.59	1.031.−3	0.0033	551.4	591.5	4.585	4.691			
388.5^c	27.83	1.613.−3	0.0016	577.2	577.2	4.651	4.651			

*Values of P, v, h, and s interpolated, extrapolated, and converted from tables of Oguchi, *Reito*, **52**, 869–889 (1977). Values of c_p, μ, and k interpolated and converted from tables in *Thermophysical Properties of Refrigerants*, American Society of Heating, Refrigerating and Air-Conditioning Engineers, New York, 1976. c = critical point.

TABLE 3-291 Saturated Refrigerant 500*

T, K	P, bar	v_f, m^3/kg	v_g, m^3/kg	h_f, kJ/kg	h_g, kJ/kg	s_f, kJ/(kg·K)	s_g, kJ/(kg·K)	c_{pf}, kJ/(kg·K)	μ_f, 10^{-4} Pa·s	k_f, W/(m·K)
200	0.1219	6.966.−4	1.360	−29.56	185.87	−0.1363	0.9408	1.044	6.11	0.113
210	0.2258	7.090.−4	0.766	−21.03	191.25	−0.0948	0.9161	1.018	5.15	0.109
220	0.3936	7.222.−4	0.457	−12.17	196.63	−0.0536	0.8955	0.997	4.42	0.106
230	0.6511	7.361.−4	0.286	− 2.97	201.96	−0.0130	0.8782	0.987	3.85	0.102
240	1.0291	7.509.−4	0.187	6.58	207.23	0.0277	0.8638	0.987	3.42	0.098
250	1.5632	7.668.−4	0.1261	16.50	212.40	0.0680	0.8517	0.997	3.04	0.094
260	2.2932	7.839.−4	0.0879	26.78	217.45	0.1082	0.8415	1.017	2.74	0.090
270	3.2624	8.024.−4	0.0628	37.44	222.35	0.1481	0.8329	1.048	2.48	0.086
280	4.5172	8.226.−4	0.0459	48.48	227.06	0.1878	0.8257	1.089	2.26	0.082
290	6.1064	8.450.−4	0.0342	59.91	231.56	0.2275	0.8194	1.140	2.08	0.078
300	8.0809	8.699.−4	0.0259	71.76	235.79	0.2671	0.8139	1.201	1.92	0.074
310	10.49	8.981.−4	0.0198	84.05	239.69	0.3067	0.8088	1.273	1.77	0.070
320	13.40	9.306.−4	0.0154	96.83	243.19	0.3464	0.8038	1.355	1.63	0.066
330	16.86	9.690.−4	0.0119	110.17	246.14	0.3864	0.7985	1.447	1.48	0.062
340	20.93	1.016.−3	0.0093	124.20	248.36	0.4271	0.7922	1.550	1.34	0.058
350	25.70	1.077.−3	0.0072	139.18	249.47	0.4689	0.7841	1.663		
360	31.25	1.162.−3	0.0055	155.66	248.71	0.5135	0.7721	1.919		
370	37.72	1.307.−3	0.0040	175.59	244.26	0.5650	0.7509	2.07		
378.6^c	44.26	2.012.−3	0.0020	219.50	219.50	0.6729	0.6729	∞		

*Values reproduced and converted from Table 12, p. 17.99, *ASHRAE Handbook, 1981: Fundamentals*, American Society of Heating, Refrigerating and Air-Conditioning Engineers, Atlanta, 1981. Copyright material. Reproduced by permission of the copyright owner. c = critical point. The notation 6.966.−4 signifies 6.966×10^{-4}.

TABLE 3-292 Saturated Refrigerant 502*

T, K	P, bar	v_f, m³/kg	v_g, m³/kg	h_f, kJ/kg	h_g, kJ/kg	s_f, kJ/(kg·K)	s_g, kJ/(kg·K)	c_{pf}, kJ/(kg·K)	μ_f, 10^{-4} Pa·s	k_f, W/(m·K)
200	0.2274	6.381.−4	0.646	−29.04	153.34	−0.1337	0.7782	1.018	5.72	0.103
210	0.4098	6.507.−4	0.374	−20.83	158.42	−0.0937	0.7599	1.036	4.88	0.099
220	0.6965	6.640.−4	0.228	−12.15	163.49	−0.0534	0.7449	1.055	4.23	0.095
230	1.1251	6.783.−4	0.146	− 2.99	168.50	−0.0128	0.7328	1.075	3.71	0.091
240	1.7392	6.938.−4	0.0969	6.66	173.42	0.0280	0.7228	1.097	3.28	0.087
250	2.5867	7.105.−4	0.0665	16.78	178.20	0.0691	0.7148	1.120	2.94	0.083
260	3.7188	7.289.−4	0.0470	27.36	182.81	0.1102	0.7082	1.144	2.65	0.079
270	5.1893	7.492.−4	0.0340	38.36	187.21	0.1514	0.7027	1.170	2.41	0.075
280	7.0530	7.720.−4	0.0251	49.77	191.35	0.1923	0.6980	1.197	2.18	0.072
290	9.3660	7.979.−4	0.0188	61.55	195.16	0.2330	0.6937	1.225	1.99	0.068
300	12.19	8.280.−4	0.0143	73.68	198.56	0.2734	0.6896	1.254	1.79	0.064
310	15.57	8.637.−4	0.0109	86.17	201.43	0.3134	0.6852	1.285	1.59	0.060
320	19.60	9.081.−4	0.0084	99.06	203.57	0.3532	0.6798	1.317	1.40	0.056
330	24.35	9.666.−4	0.0064	112.53	204.62	0.3933	0.6723	1.351	1.23	0.052
340	29.95	1.053.−3	0.0048	127.13	203.71	0.4351	0.6604	1.386	1.07	0.048
350	36.62	1.220.−3	0.0033	145.44	197.82	0.4859	0.6355	1.422	0.93	0.044
355.3[c]	40.75	1.786.−3	0.0018	174.00	174.00	0.5634	0.5634			

*Values reproduced and converted from Table 13, p. 17.101, *ASHRAE Handbook, 1981: Fundamentals*, American Society of Heating, Refrigerating and Air-Conditioning Engineers, Atlanta, 1981. Copyright material. Reproduced by permission of the copyright owner. c = critical point. The notation 6.381.−4 signifies 6.381×10^{-4}.

TABLE 3-293 Saturated Refrigerant 503*

T, K	P, bar	v_f, m³/kg	v_g, m³/kg	h_f, kJ/kg	h_g, kJ/kg	s_f, kJ/(kg·K)	s_g, kJ/(kg·K)	c_{pf}, kJ/(kg·K)	μ_f, 10^{-4} Pa·s	k_f, W/(m·K)
150	0.0750	6.384.−4	1.894	−89.60	111.02	−0.4694	0.8681	0.482	6.12	0.128
160	0.1798	6.478.−4	0.837	−79.73	115.40	−0.4057	0.8139	0.554	5.05	0.123
170	0.3828	6.585.−4	0.414	−69.55	119.70	−0.3441	0.7691	0.620	4.16	0.116
180	0.7395	6.700.−4	0.224	−59.08	123.84	−0.2844	0.7318	0.682	3.43	0.111
190	1.3187	6.850.−4	0.130	−48.36	127.77	−0.2267	0.7003	0.747	2.94	0.105
200	2.1999	7.014.−4	0.0803	−37.45	131.45	−0.1710	0.6735	0.817	2.56	0.099
210	3.4713	7.204.−4	0.0520	−26.36	134.84	−0.1173	0.6503	0.896	2.25	0.094
220	5.2281	7.426.−4	0.0350	−15.10	137.87	−0.0656	0.6298	0.988	1.98	0.088
230	7.5713	7.687.−4	0.0242	− 3.65	140.49	−0.0155	0.6112	1.017	1.73	0.082
240	10.61	8.001.−4	0.0172	8.07	142.58	0.0334	0.5939	1.227	1.52	0.076
250	14.46	8.386.−4	0.0124	20.22	143.98	0.0817	0.5767	1.382	1.33	0.070
260	19.25	8.874.−4	0.0090	33.10	144.38	0.1305	0.5585	1.57	1.17	0.065
270	25.13	9.526.−4	0.0064	47.22	143.23	0.1816	0.5373	1.79	1.03	0.059
280	32.27	1.050.−3	0.0045	63.64	139.25	0.2384	0.5085	2.03	0.91	0.054
290	40.87	1.264.−3	0.0028	86.41	127.51	0.3131	0.4548	2.35		
292.6[c]	43.57	1.773.−3	0.0018	110.20	110.20	0.3864	0.3864	∞	. . .	∞

**P*, *v*, *h*, and *s* values reproduced and converted from Table 14, p. 17.103, *ASHRAE Handbook, 1981: Fundamentals*, American Society of Heating, Refrigerating and Air-Conditioning Engineers, Atlanta, 1981. Copyright material. Reproduced by permission of the copyright owner. c_p, μ, and k values interpolated and converted from *Thermophysical Properties of Refrigerants*, American Society of Heating, Refrigerating and Air-Conditioning Engineers, New York, 1976. c = critical point. The notation 6.384.−4 signifies 6.384×10^{-4}.

TABLE 3-294 Saturated Refrigerant 504*

Temp., °F.	Pressure, lb./sq. in. abs.	Volume, cu. ft./lb.		Enthalpy, B.t.u./lb.		Entropy, B.t.u./(lb.)(°F.)	
		Liquid	Vapor	Liquid	Vapor	Liquid	Vapor
−120	2.964	0.01095	15.31	−21.48	86.69	−0.0565	0.2609
−100	6.042	0.01119	7.874	−16.39	89.31	−0.0420	0.2519
−80	11.34	0.01146	4.372	−11.12	91.84	−0.0277	0.2435
−60	19.85	0.01175	2.585	−5.65	94.25	−0.0137	0.2362
−40	32.76	0.01206	1.609	0.00	96.50	0.0000	0.2299
−20	51.44	0.01242	1.045	5.85	98.58	0.0135	0.2244
0	77.41	0.01282	0.7029	11.91	100.45	0.0269	0.2195
20	112.3	0.01328	0.4859	18.22	102.09	0.0401	0.2150
40	158.0	0.01379	0.3431	24.81	103.44	0.0533	0.2107
60	216.2	0.01443	0.2458	31.78	104.41	0.0667	0.2065
80	289.2	0.01522	0.1773	39.25	104.85	0.0804	0.2020
100	379.1	0.01629	0.1274	47.43	104.49	0.0948	0.1968
120	488.3	0.01783	0.0893	56.78	102.72	0.1107	0.1899
140	618.1	0.02083	0.0578	69.97	97.70	0.1322	0.1784
150	692.2	0.02597	0.0394	76.96	89.76	0.1432	0.1642

*Unpublished data of Allied Chemical Company, 1970. Used by permission.

TABLE 3-295 Saturated Rubidium*

T, K	P, bar	v_f, m³/kg	v_g, m³/kg	h_f, kJ/kg	h_g, kJ/kg	s_f, kJ/(kg·K)	s_g, kJ/(kg·K)	c_{pf}, kJ/(kg·K)
312.7^m	2.46.−9	6.75.−4	.....	118.7	1036	0.998	3.932	0.379
400	1.69.−6	6.98.−4	2.3.+5	151.6	1057	1.091	3.855	0.375
500	1.73.−4	7.22.−4	2790	188.8	1078	1.174	2.958	0.369
600	0.0037	7.46.−4	156.6	225.4	1096	1.241	2.692	0.362
700	0.0317	7.73.−4	20.75	261.3	1111	1.296	2.511	0.357
800	0.1584	8.10.−4	4.662	296.8	1124	1.343	2.378	0.353
1000	1.467	8.65.−4	0.605	367.6	1150	1.422	2.205	0.360
1200	6.466	9.40.−4	0.159	440.1	1179	1.490	2.104	0.385
1400	18.6	1.03.−3						
1500	28.5	1.08.−3						

*Converted from tables in Vargaftik, *Tables of the Thermophysical Properties of Liquids and Gases*, Nauka, Moscow, 1972, and Hemisphere, Washington, 1975. m = melting point. The notation 2.46.−9 signifies 2.46×10^{-9}.

TABLE 3-296 Saturated Sodium*

T, K	P, bar	v_f, m³/kg	v_g, m³/kg	h_f, kJ/kg	h_g, kJ/kg	s_f, kJ/(kg·K)	s_g, kJ/(kg·K)	c_{pf}, kJ/(kg·K)
380^m	2.631.−10	1.081.−3	5.222.+9	500	5003	2.853	14.703	1.379
400	1.385.−9	1.086.−3	1.044.+9	527	5020	2.924	14.156	1.370
500	7.523.−7	1.114.−3	2.395.+6	662	5101	3.225	12.102	1.328
600	4.926.−5	1.145.−3	4.359.+4	793	5168	3.464	10.756	1.296
800	8.904.−3	1.211.−3	312.8	1048	5260	3.831	9.095	1.259
1000	0.1963	1.286.−3	17.08	1299	5322	4.111	8.134	1.259
1200	1.5037	1.372.−3	2.572	1554	5386	4.343	7.536	1.295
1400	6.2538	1.469.−3	0.691	1820	5453	4.548	7.142	1.367
1600	18.02	1.581.−3	0.259	2104	5504	4.737	6.862	1.475
1800	40.91	1.713.−3	0.118	2410	5540	4.916	6.654	1.566
2000	70.49	1.884.−3	6.08.−2	2734	5558	5.083	6.495	1.682
2200	137.9	2.123.−3	3.44.−2	3985	5550	5.244	6.365	1.892
2400	219.5	2.493.−3	2.04.−2	3480	5497	5.407	6.248	2.377
2600	327.0	3.230.−3	1.19.−2	3977	5342	5.594	6.119	4.573
2733^c	413.6	5.501.−3	5.50.−3	4773	4773			

*Reproduced or conveted from Padilla, Argonne Nat. Lab. Rep. ANL 8095, 1974. Some values missing in this publication were supplied in a private communication. m = melting point; c = critical point. The notation 2.631.−10 signifies 2.631×10^{-10}.

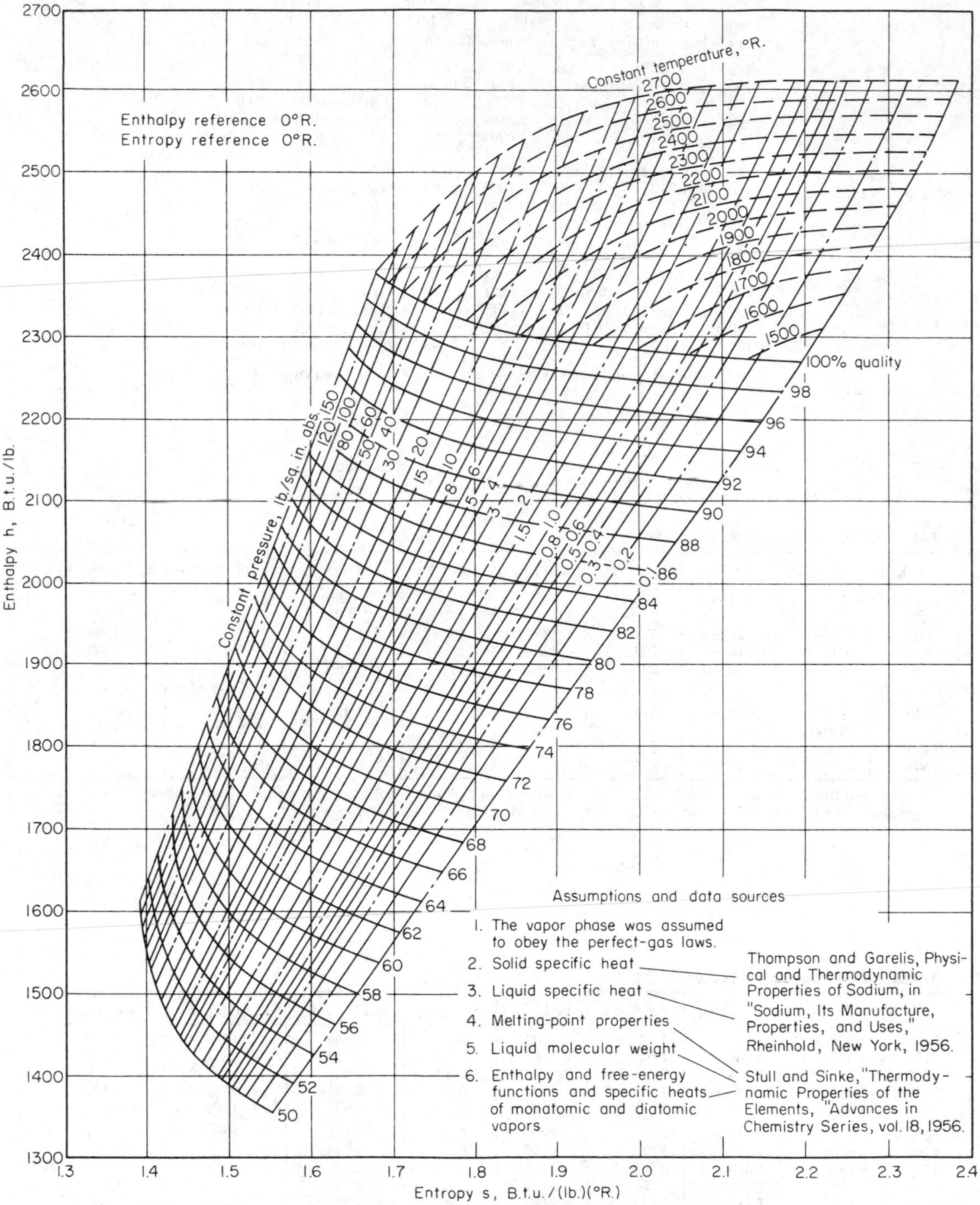

FIG. 3-36 Mollier diagram for sodium. Enthalpy reference, −0°R; entropy reference, −0°R. (*From Meisl and Shapiro,* Thermodynamic Properties of Alkali Metal Vapors and Mercury, *TIS Rep. R60FPD358, General Electric Co., Cincinnati, June 1960. This diagram is reproduced by permission.*)

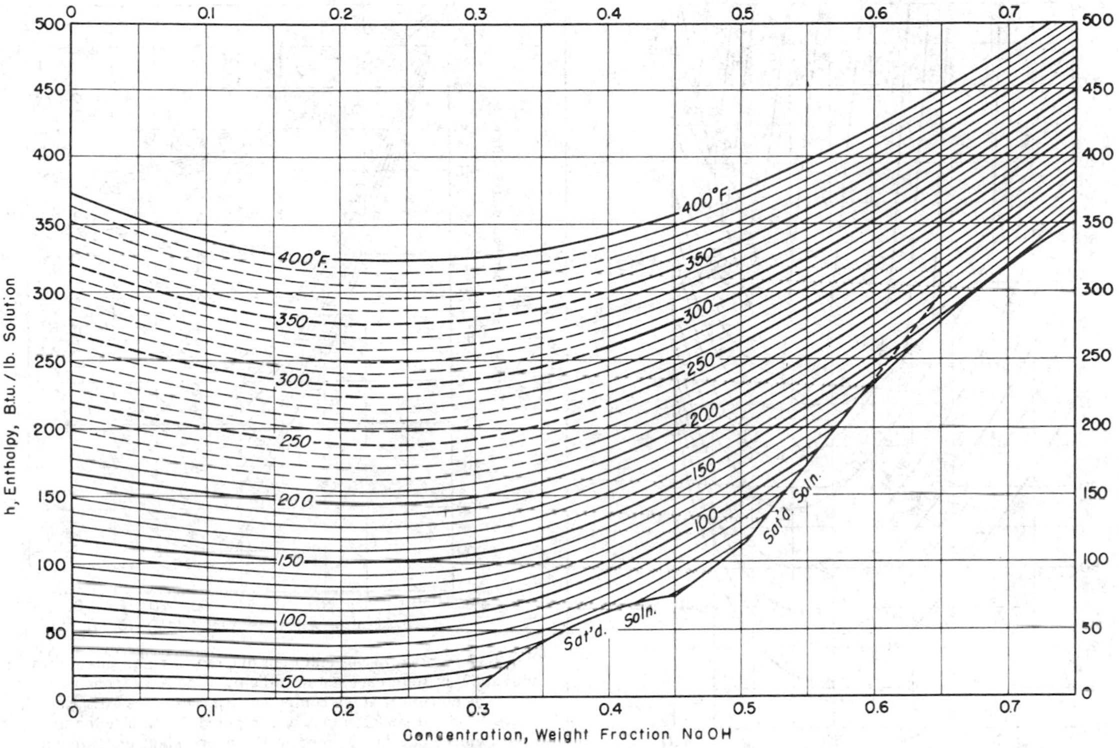

FIG. 3-37 Enthalpy-concentration diagram for aqueous sodium hydroxide at 1 atm. Reference states: enthalpy of liquid water at 32°F and vapor pressure is zero; partial molal enthalpy of infinitely dilute NaOH solution at 64°F and 1 atm is zero. [*McCabe*, Trans. Am. Inst. Chem. Eng., *31*, *129 (1935)*.]

TABLE 3-297 Saturated Sulfur Dioxide*

T, K	P, bar	v_f, m³/kg	v_g, m³/kg	h_f, kJ/kg	h_g, kJ/kg	s_f, kJ/(kg·K)	s_g, kJ/(kg·K)	c_{pf}, kJ/(kg·K)	μ_f, 10^{-4} Pa·s	k_f, W/(m·K)
200	0.02056	6.189.−4	12.602	7.4	433.3	0.033	2.212	1.280	12.3	
210	0.04569	6.284.−4	5.946	9.1	446.1	0.041	2.159	1.284	10.6	
220	0.09997	6.384.−4	2.876	28.6	453.8	0.123	2.075	1.288	8.37	
230	0.1844	6.488.−4	1.605	43.5	459.5	0.198	2.001	1.293	7.03	
240	0.3202	6.596.−4	0.9602	56.5	464.5	0.254	1.952	1.299	5.97	
250	0.5430	6.707.−4	0.5864	70.0	469.7	0.308	1.906	1.308	5.11	0.262
260	0.8778	6.819.−4	0.3745	85.1	474.5	0.363	1.865	1.317	4.39	0.243
270	1.3634	6.938.−4	0.2479	99.8	479.3	0.425	1.827	1.328	3.78	0.224
280	2.0402	7.057.−4	0.1699	114.8	484.3	0.473	1.793	1.343	3.30	0.206
290	2.9574	7.184.−4	0.1197	129.2	488.5	0.523	1.763	1.363	2.87	0.190
300	4.1675	7.312.−4	0.08647	143.1	492.5	0.568	1.732	1.389	2.51	0.174
310	5.7372	7.447.−4	0.06366	157.1	496.3	0.612	1.706	1.422	2.19	0.162
320	7.8226	7.590.−4	0.04707	170.1	498.9	0.649	1.678	1.459	1.91	0.151
330	10.301	7.847.−4	0.03572	183.0	501.2	0.690	1.654	1.499	1.67	0.139
340	13.229	8.066.−4	0.02792	196.0	502.5	0.731	1.633	1.546	1.46	0.128
350	16.759	8.303.−4	0.02209	211.2	502.9	0.781	1.614	1.603	1.27	0.117
360	21.01	8.571.−4	0.01755	223.7	503.1	0.817	1.593	1.68	1.11	0.108
370	26.01	8.877.−4	0.01399	239.9	502.9	0.862	1.573	1.75	0.96	0.098
380	31.92	9.236.−4	0.01110	257.9	502.7	0.910	1.555	1.84	0.84	0.089
390	38.76	9.671.−4	0.00877	277.7	500.7	0.962	1.534	1.97	0.73	0.081
400	46.67	1.023.−3	0.00685	300.2	496.7	1.020	1.511	2.12	0.63	0.072
410	55.80	1.098.−3	0.00559	326.2	489.5	1.083	1.481	....	0.53	0.064
420	66.19	1.235.−3	0.00387	355.6	474.1	1.155	1.436	....	0.44	0.055
425.1[c]	78.81	1.906.−3	0.00191	423.6	423.6	1.304	1.304			

*Values interpolated and converted from tables of Kang, McKetta, et al., Bur. Eng. Res. Repr. 59, University of Texas, Austin, 1961. See also *J. Chem. Eng. Data*, **6**, 220–227 (1961); and *Am. Inst. Chem. Eng. J.*, **7**, 418 (1961). c = critical point. The notation 6.189.−4 signifies 6.189 × 10^{-4}.

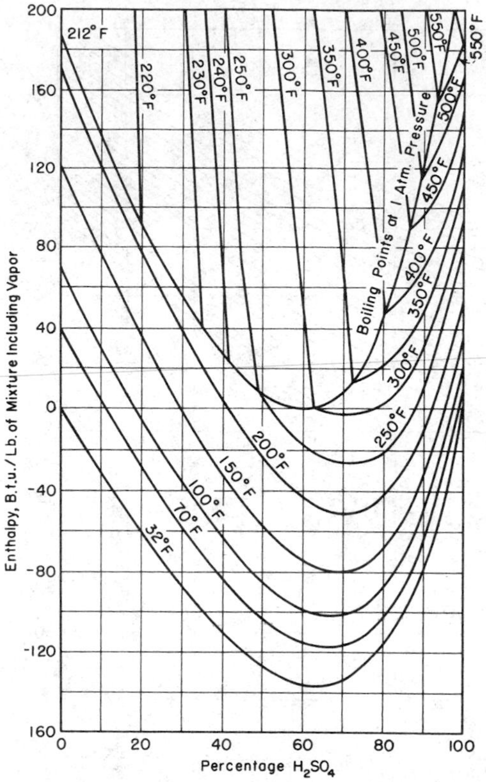

FIG. 3-38 Enthalpy-concentration diagram for aqueous sulfuric acid at 1 atm. Reference states: enthalpies of pure-liquid components at 32°F and vapor pressures are zero. NOTE: It should be observed that the weight basis includes the vapor, which is particularly important in the two-phase region. The upper ends of the tie lines in this region are assumed to be pure water. (*Hougen and Watson*, Chemical Process Principles, *part I, Wiley, New York, 1943.*)

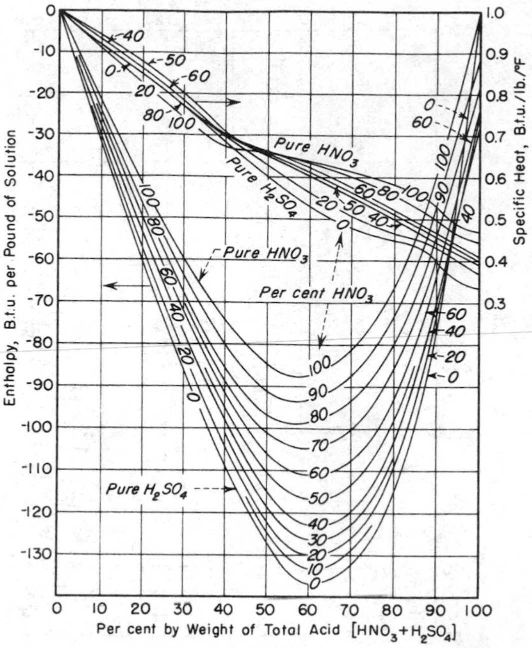

FIG. 3-39 Enthalpy-concentration diagram for aqueous sulfuric and nitric acids at 32°F. Reference states: enthalpy of pure components at 32°F is zero. NOTE: The percent HNO_3 is computed on a water-free basis. Enthalpies at temperatures other than 32°F may be computed by utilizing the specific-heat data given, which may be assumed to be independent of temperature as a first approximation. [*McKinley and Brown*, Chem. Metall. Eng., *49, 142 (1942).*]

TABLE 3-298 Saturated Trichloroethylene (C_2HCl_3)*

Temp., °F. t	Abs. pressure, lb./sq. in. p	Volume, cu. ft./lb. vapor v_g	Enthalpy, B.t.u./lb. Liquid h_f	Enthalpy, B.t.u./lb. Vapor h_g	Entropy, B.t.u./(lb.)(°R.) Liquid s_f	Entropy, B.t.u./(lb.)(°R.) Vapor s_g
0	0.150	261.3	0.0	112.60	0.0000	0.2451
10	.194	197.6	1.86	114.18	.0039	.2433
20	.252	153.3	3.90	116.00	.0081	.2429
30	.352	114.5	6.12	117.92	.0127	.2411
40	.492	85.8	8.42	119.83	.0178	.2412
50	.672	63.2	11.10	122.15	.0227	.2405
60	.900	47.53	13.80	124.43	.0278	.2404
70	1.20	36.00	16.80	127.02	.0324	.2407
80	1.56	28.00	19.84	129.62	.0374	.2414
90	1.98	22.46	23.22	132.54	.0428	.2422
100	2.45	18.54	26.60	135.41	.0480	.2428
110	3.15	14.62	29.92	138.22	.0530	.2432
120	3.98	11.73	33.60	141.38	.0581	.2440
130	5.05	9.20	37.70	144.90	.0632	.2450
140	6.31	7.70	42.00	148.57	.0680	.2459

* "Refrigerating Data Book," 5th ed., American Society of Refrigerating Engineers, New York, 1942.

TABLE 3-299 Saturated Toluene*

T, K	P, bar	v_f, m^3/kg	v_g, m^3/kg	h_f, kJ/kg	h_g, kJ/kg	s_f, kJ/(kg·K)	s_g, kJ/(kg·K)	c_{pf}, kJ/(kg·K)	μ_f, 10^{-4} Pa·s	k_f, W/(m·K)
270	0.0076	1.127.−6	0.0349	316.7	745.7	2.236	3.825	1.64	8.02	0.141
280	0.0139	1.138.−6	0.0191	333.0	756.1	2.295	3.806	1.66	6.96	0.138
290	0.0246	1.150.−6	0.0106	349.6	766.8	2.353	3.792	1.68	6.10	0.136
300	0.0418	1.162.−6	0.0082	366.5	777.8	2.410	3.782	1.71	5.41	0.133
310	0.0682	1.175.−6	0.0041	383.7	789.2	2.467	3.776	1.74	4.83	0.131
320	0.1072	1.188.−6	2.67.−3	401.3	800.9	2.522	3.771	1.78	4.34	0.128
330	0.1633	1.201.−6	1.80.−3	419.6	812.9	2.577	3.771	1.81	3.93	0.126
340	0.2416	1.215.−6	1.25.−3	437.4	825.2	2.632	3.772	1.84	3.58	0.124
350	0.3480	1.230.−6	8.91.−4	456.0	837.8	2.686	3.777	1.88	3.28	0.121
360	0.4894	1.245.−6	6.48.−4	475.1	850.7	2.739	3.783	1.92	3.01	0.119
370	0.6736	1.261.−6	4.81.−4	494.6	863.8	2.792	3.791	1.96	2.78	0.117
380	0.9090	1.277.−6	3.64.−4	514.4	877.2	2.846	3.801	2.01	2.56	0.114
390	1.2049	1.294.−6	2.79.−4	534.7	890.9	2.898	3.811	2.05	2.37	0.112
400	1.5713	1.312.−6	2.18.−4	555.4	904.8	2.950	3.824	2.09	2.19	0.110
420	2.5589	1.350.−6	1.37.−4	598.1	933.1	3.054	3.852	2.17	1.89	0.105
440	3.965	1.393.−6	9.00.−5	642.3	962.0	3.156	3.883	2.24	1.64	0.101
460	5.892	1.443.−6	6.11.−5	688.1	991.3	3.258	3.917	2.31	. . .	0.096
480	8.451	1.499.−6	4.26.−5	735.5	1021.1	3.358	3.953	2.38	. . .	0.091
500	11.76	1.567.−6	3.03.−5	784.4	1051.3	3.457	3.989	2.45	. . .	0.086
520	15.96	1.651.−6	2.19.−5	834.9	1081.4	3.554	4.027	2.53	. . .	0.082
540	21.99	1.761.−6	1.58.−5	887.3	1109.6	3.651	4.062	2.65	. . .	0.078
560	27.65	1.919.−6	1.13.−5	942.8	1132.1	3.750	4.088	2.82	. . .	0.074
580	35.56	2.213.−6	7.59.−6	1005.6	1142.3	3.857	4.093			
590	40.16	2.650.−6	5.28.−6	1050.2	1128.1	3.932	4.063			
591.8^c	41.04	3.432.−6	3.43.−6	1084.9	1084.9	3.989	3.989			

*Values converted and mostly rounded off from the tables of Counsell, Lawrenson, and Lees, Nat. Phys. Lab., Teddington (U.K.) Rep. Chem. 52, 1976. c = critical point. The notation 1.127.−6 signifies 1.127×10^{-6}.

TABLE 3-300 Saturated Solid/Vapor Water*

Temp., °F.	Pressure, lb./sq. in. abs.	Volume, cu. ft./lb.		Enthalpy, B.t.u./lb.		Entropy, B.t.u./(lb.)(°F.)	
		Solid	Vapor	Solid	Vapor	Solid	Vapor
−160	4.949. − 8	0.01722	3.607. + 9	−222.05	990.38	−0.4907	3.5549
−150	1.620. − 7	0.01723	1.139. + 9	−218.82	994.80	−0.4801	3.4387
−140	4.928. − 7	0.01724	3.864. + 8	−215.49	999.21	−0.4695	3.3301
−130	1.403. − 6	0.01725	1.400. + 8	−212.08	1003.63	−0.4590	3.2284
−120	3.757. − 6	0.01726	5.386. + 7	−208.58	1008.05	−0.4485	3.1330
−110	9.517. − 6	0.01728	2.189. + 7	−204.98	1012.47	−0.4381	3.0434
−100	2.291. − 5	0.01729	9.352. + 6	−201.28	1016.89	−0.4277	2.9591
−90	5.260. − 5	0.01730	4.186. + 6	−197.49	1021.31	−0.4173	2.8796
−80	1.157. − 4	0.01731	1.955. + 6	−193.60	1025.73	−0.4069	2.8045
−70	2.443. − 4	0.01732	9.501. + 5	−189.61	1030.15	−0.3965	2.7336
−60	4.972. − 4	0.01734	4.788. + 5	−185.52	1034.58	−0.3862	2.6664
−50	9.776. − 4	0.01735	2.496. + 5	−181.34	1039.00	−0.3758	2.6028
−45	1.354. − 3	0.01736	1.824. + 5	−179.21	1041.21	−0.3707	2.5723
−40	1.861. − 3	0.01737	1.343. + 5	−177.06	1043.42	−0.3655	2.5425
−35	2.540. − 3	0.01737	9.961. + 4	−174.88	1045.63	−0.3604	2.5135
−30	3.440. − 3	0.01738	7.441. + 4	−172.68	1047.84	−0.3552	2.4853
−25	4.627. − 3	0.01739	5.596. + 4	−170.46	1050.05	−0.3501	2.4577
−20	6.181. − 3	0.01739	4.237. + 4	−168.21	1052.26	−0.3449	2.4308
−15	8.204. − 3	0.01740	3.228. + 4	−165.94	1054.47	−0.3398	2.4046
−10	1.082. − 2	0.01741	2.475. + 4	−163.65	1056.67	−0.3347	2.3791
−5	1.419. − 2	0.01741	1.909. + 4	−161.33	1058.88	−0.3295	2.3541
0	1.849. − 2	0.01742	1.481. + 4	−158.98	1061.09	−0.3244	2.3297
5	2.396. − 2	0.01743	1.155. + 4	−156.61	1063.29	−0.3193	2.3039
10	3.087. − 2	0.01744	9.060. + 3	−154.22	1065.50	−0.3142	2.2827
15	3.957. − 2	0.01744	7.144. + 3	−151.80	1067.70	−0.3090	2.2600
16	4.156. − 2	0.01745	6.817. + 3	−151.32	1068.14	−0.3080	2.2555
18	4.581. − 2	0.01745	6.210. + 3	−150.34	1069.02	−0.3060	2.2466
20	5.045. − 2	0.01745	5.662. + 3	−149.36	1069.90	−0.3039	2.2378
22	5.552. − 2	0.01746	5.166. + 3	−148.38	1070.38	−0.3019	2.2291
24	6.105. − 2	0.01746	4.717. + 3	−147.39	1071.66	−0.2998	2.2205
26	6.708. − 2	0.01746	4.311. + 3	−146.40	1072.53	−0.2978	2.2119
28	7.365. − 2	0.01746	3.943. + 3	−145.40	1073.41	−0.2957	2.2034
30	8.080. − 2	0.01747	3.608. + 3	−144.40	1074.29	−0.2937	2.1950
31	8.461. − 2	0.01747	3.453. + 3	−143.90	1074.73	−0.2927	2.1908
32	8.858. − 2	0.01747	3.305. + 3	−143.40	1075.16	−0.2916	2.1867

*Condensed from *Fundamentals*, American Society of Heating, Refrigerating and Air-Conditioning Engineers, 1967 and 1972. Reproduced by permission. The validity of many standard reference tables has been critically reviewed by Jancso, Pupezin, and van Hook, *J. Phys. Chem.*, **74**, 2984 (1970). This source is recommended for further study. The notation 4.949.−8, 3.607.+9, etc., means 4.949×10^{-8}, 3.607×10^{9}, etc.

TABLE 3-301 Saturated Steam: Temperature Table*

Temp., °F.	Pressure, lb./sq. in. abs.	Volume, cu. ft./lb.		Enthalpy, B.t.u./lb.		Entropy, B.t.u./(lb.)(°F.)	
		Liquid	Vapor	Liquid	Vapor	Liquid	Vapor
32.018	0.08865	0.016022	3302.4	0.000	1075.5	0.0000	2.1872
35	0.09991	0.016020	2948.1	3.002	1076.8	0.0061	2.1767
40	0.12163	0.016019	2445.8	8.027	1079.0	0.0162	2.1594
45	0.14744	0.016020	2037.8	13.044	1081.2	0.0262	2.1426
50	0.17796	0.016023	1704.8	18.054	1083.4	0.0361	2.1262
55	0.21392	0.016027	1432.0	23.059	1085.6	0.0458	2.1102
60	0.25611	0.016033	1207.6	28.060	1087.7	0.0555	2.0946
65	0.30545	0.016041	1022.1	33.057	1089.9	0.0651	2.0794
70	0.36292	0.016050	868.4	38.052	1092.1	0.0745	2.0645
75	0.42964	0.016060	740.3	43.045	1094.3	0.0839	2.0500
80	0.50683	0.016072	633.3	48.037	1096.4	0.0932	2.0359
85	0.59583	0.016085	543.6	53.027	1098.6	0.1024	2.0221
90	0.69813	0.016099	468.1	58.018	1100.8	0.1115	2.0086
95	0.81534	0.016114	404.4	63.008	1102.9	0.1206	1.9954
100	0.94294	0.016130	350.4	67.999	1105.1	0.1295	1.9825
110	1.2750	0.016165	265.39	77.98	1109.3	0.1472	1.9577
120	1.6927	0.016204	203.26	87.97	1113.6	0.1646	1.9339
130	2.2230	0.016247	157.33	97.96	1117.8	0.1817	1.9112
140	2.8892	0.016293	122.98	107.89	1122.0	0.1985	1.8895
150	3.7184	0.016343	97.07	117.95	1126.1	0.2150	1.8686
160	4.7414	0.016395	77.27	127.96	1130.2	0.2313	1.8487
170	5.9926	0.016451	62.06	137.97	1134.2	0.2473	1.8295
180	7.5110	0.016510	50.225	148.00	1138.2	0.2631	1.8111
190	9.340	0.016572	40.957	158.04	1142.1	0.2787	1.7934
200	11.526	0.016637	33.639	168.09	1146.0	0.2940	1.7764
210	14.123	0.016705	27.816	178.15	1149.7	0.3091	1.7600
212	14.696	0.016719	26.799	180.17	1150.5	0.3121	1.7568
220	17.186	0.016775	23.148	188.23	1153.4	0.3241	1.7442
230	20.779	0.016849	19.381	198.33	1157.1	0.3388	1.7290
240	24.968	0.016926	16.321	208.45	1160.6	0.3533	1.7142
250	29.825	0.017066	13.819	218.59	1164.0	0.3677	1.7000
260	35.427	0.017089	11.762	228.76	1167.4	0.3819	1.6862
270	41.856	0.017175	10.060	238.95	1170.6	0.3960	1.6729
280	49.200	0.017264	8.644	249.17	1173.8	0.4098	1.6599
290	57.550	0.01736	7.4603	259.4	1167.8	0.4236	1.6473
300	67.005	0.01745	6.4658	269.7	1179.7	0.4372	1.6351
320	89.643	0.01766	4.9138	290.4	1185.2	0.4640	1.6116
340	117.992	0.01787	3.7878	311.3	1190.1	0.4902	1.5892
360	153.01	0.01811	2.9573	332.3	1194.4	0.5161	1.5678
380	195.73	0.01836	2.3353	353.6	1198.0	0.5416	1.5473
400	247.26	0.01864	1.8630	375.1	1201.0	0.5667	1.5274
420	308.78	0.01894	1.4997	396.9	1203.1	0.5915	1.5080
440	381.54	0.01926	1.2169	419.0	1204.4	0.6161	1.4890
460	466.87	0.01961	0.99424	441.5	1204.8	0.6405	1.4704
480	566.15	0.02000	0.81717	464.5	1204.1	0.6648	1.4518
500	680.86	0.02043	0.67492	487.9	1202.2	0.6890	1.4333
520	812.53	0.02091	0.55956	512.0	1199.0	0.7133	1.4146
540	962.79	0.02146	0.46513	536.8	1194.3	0.7378	1.3954
560	1133.38	0.02207	0.38714	562.4	1187.7	0.7625	1.3757
580	1326.17	0.02279	0.32216	589.1	1179.0	0.7876	1.3550
600	1543.2	0.02364	0.26747	617.1	1167.7	0.8134	1.3330
620	1786.9	0.02466	0.22081	646.9	1153.2	0.8403	1.3092
640	2059.9	0.02595	0.18021	679.1	1133.7	0.8686	1.2821
660	2365.7	0.02768	0.14431	714.9	1107.0	0.8995	1.2498
680	2708.6	0.03037	0.11117	758.5	1068.5	0.9365	1.2086
700	3094.3	0.03662	0.07519	825.2	991.7	0.9924	1.1359
702	3135.5	0.03824	0.06997	835.0	979.7	1.0006	1.1210
704	3177.2	0.04108	0.06300	854.2	956.2	1.0169	1.1046
705.47	3208.2	0.05078	0.05078	906.0	906.0	1.0612	1.0612

*Extracted and condensed from 1967 A.S.M.E. Steam Tables. Copyright reserved. Reproduced by permission.

TABLE 3-302 Saturated Steam

Temperature, K	Pressure, bar*	Volume, m³/kg		Enthalpy, kJ/kg		Entropy, kJ/(kg·K)		Specific heat, C_p, kJ/(kg·K)		Viscosity, Ns/m²		Thermal conductivity, W/(m·K)		Prandtl no.		Surface tension, N/m	Temperature, K
		Condensed†	Vapor	Condensed†	Vapor	Condensed†	Vapor	Condensed†	Vapor	Condensed†	Vapor	Condensed†	Vapor	Condensed†	Vapor	Condensed†	
150	6.30. −11	1.073. −3	9.55. +9	−539.6	2273	−2.187	16.54	1.155	...	...	...	3.73	...	...	...	...	150
160	7.72. −10	1.074. −3	9.62. +8	−525.7	2291	−2.106	15.49	1.233	...	...	...	3.52	...	...	...	...	160
170	7.29. −9	1.076. −3	1.08. +8	−511.7	2310	−2.026	14.57	1.311	...	...	...	3.34	...	...	...	...	160
180	5.38. −8	1.077. −3	1.55. +7	−497.8	2328	−1.947	13.76	1.389	...	...	...	3.18	...	...	...	...	180
190	3.23. −7	1.078. −3	2.72. +6	−483.8	2347	−1.868	13.03	1.467	...	...	...	3.04	...	...	...	...	190
200	1.62. −6	1.079. −3	5.69. +5	−467.5	2366	−1.789	12.38	1.545	...	...	...	2.91	...	...	...	...	200
210	7.01. −6	1.081. −3	1.39. +5	−451.2	2384	−1.711	11.79	1.623	...	...	...	2.79	...	...	...	...	210
220	2.65. −5	1.082. −3	3.83. +4	−435.0	2403	−1.633	11.20	1.701	...	...	...	2.69	...	...	...	...	220
230	8.91. −5	1.084. −3	1.18. +4	−416.3	2421	−1.555	10.79	1.779	...	...	...	2.59	...	...	...	...	230
240	3.72. −4	1.085. −3	4.07. +3	−400.1	2440	−1.478	10.35	1.857	...	...	...	2.50	...	...	...	...	240
250	7.59. −4	1.087. −3	1.52. +3	−381.5	2459	−1.400	9.954	1.935	...	...	...	2.42	...	...	...	...	250
255	1.23. −3	1.087. −3	956.4	−369.8	2468	−1.361	9.768	1.974	...	...	...	2.38	...	...	...	...	255
260	1.96. −3	1.088. −3	612.2	−360.5	2477	−1.323	9.590	2.013	...	...	...	2.35	...	...	...	...	260
265	3.06. −3	1.089. −3	400.4	−351.2	2486	−1.281	9.461	2.052	...	...	...	2.31	...	...	...	...	265
270	4.69. −3	1.090. −3	265.4	−339.6	2496	−1.296	9.255	2.091	...	...	...	2.27	...	...	...	...	270
273.15	6.11. −3	1.091. −3	206.3	−333.5	2502	−1.221	9.158	2.116	...	...	...	2.26	...	...	...	...	273.15
273.15	0.00611	1.000. −3	206.3	0.0	2502	0.000	9.158	4.217	1.854	1750. −6	8.02. −6	0.569	0.0182	12.99	0.815	0.0755	273.15
275	0.00697	1.000. −3	181.7	7.8	2505	0.028	9.109	4.211	1.855	1652. −6	8.09. −6	0.574	0.0183	12.22	0.817	0.0753	275
280	0.00990	1.000. −3	130.4	28.8	2514	0.104	8.980	4.198	1.858	1422. −6	8.29. −6	0.582	0.0186	10.26	0.825	0.0748	280
285	0.01387	1.000. −3	99.4	49.8	2523	0.178	8.857	4.189	1.861	1225. −6	8.49. −6	0.590	0.0189	8.81	0.833	0.0743	285
290	0.01917	1.001. −3	69.7	70.7	2532	0.251	8.740	4.184	1.864	1080. −6	8.69. −6	0.598	0.0193	7.56	0.841	0.0737	290
295	0.02617	1.002. −3	51.94	91.6	2541	0.323	8.627	4.181	1.868	959. −6	8.89. −6	0.606	0.0195	6.62	0.849	0.0727	295
300	0.03531	1.003. −3	39.13	112.5	2550	0.393	8.520	4.179	1.872	855. −6	9.09. −6	0.613	0.0196	5.83	0.857	0.0717	300
305	0.04712	1.005. −3	27.90	133.4	2559	0.462	8.417	4.178	1.877	769. −6	9.29. −6	0.620	0.0201	5.20	0.865	0.0709	305
310	0.06221	1.007. −3	22.93	154.3	2568	0.530	8.318	4.178	1.882	695. −6	9.49. −6	0.628	0.0204	4.62	0.873	0.0700	310
315	0.08132	1.009. −3	17.82	175.2	2577	0.597	8.224	4.179	1.888	631. −6	9.69. −6	0.634	0.0207	4.16	0.883	0.0692	315
320	0.1053	1.011. −3	13.98	196.1	2586	0.649	8.151	4.180	1.895	577. −6	9.89. −6	0.640	0.0210	3.77	0.894	0.0683	320
325	0.1351	1.013. −3	11.06	217.0	2595	0.727	8.046	4.182	1.903	528. −6	10.09. −6	0.645	0.0213	3.42	0.901	0.0675	325
330	0.1719	1.016. −3	8.82	237.9	2604	0.791	7.962	4.184	1.911	489. −6	10.29. −6	0.650	0.0217	3.15	0.908	0.0666	330
335	0.2167	1.018. −3	7.09	258.8	2613	0.854	7.881	4.186	1.920	453. −6	10.49. −6	0.655	0.0220	2.88	0.916	0.0658	335
340	0.2713	1.021. −3	5.74	279.8	2622	0.916	7.804	4.188	1.930	420. −6	10.69. −6	0.660	0.0223	2.66	0.925	0.0649	340
345	0.3372	1.024. −3	4.683	300.7	2630	0.977	7.729	4.191	1.941	389. −6	10.89. −6	0.665	0.0226	2.45	0.933	0.0641	345
350	0.4163	1.027. −3	3.846	321.7	2639	1.038	7.657	4.195	1.954	365. −6	11.09. −6	0.668	0.0230	2.29	0.942	0.0632	350
355	0.5100	1.030. −3	3.180	342.7	2647	1.097	7.588	4.199	1.968	343. −6	11.29. −6	0.671	0.0233	2.14	0.951	0.0623	355
360	0.6209	1.034. −3	2.645	363.7	2655	1.156	7.521	4.203	1.983	324. −6	11.49. −6	0.674	0.0237	2.02	0.960	0.0614	360
365	0.7514	1.038. −3	2.212	384.7	2663	1.214	7.456	4.209	1.999	306. −6	11.69. −6	0.677	0.0241	1.91	0.969	0.0605	365

370	0.9040	1.041.−3	1.861	405.8	2671	1.271	7.394	4.214	2.017	289.−6	11.89.−6	0.679	0.0245	1.80	0.978	0.0595	370
373.15	1.0133	1.044.−3	1.679	419.1	2676	1.307	7.356	4.217	2.029	279.−6	12.02.−6	0.680	0.0248	1.76	0.984	0.0589	373.15
375	1.0815	1.045.−3	1.574	426.8	2679	1.328	7.333	4.220	2.036	274.−6	12.09.−6	0.681	0.0249	1.70	0.987	0.0586	375
380	1.2869	1.049.−3	1.337	448.0	2687	1.384	7.275	4.226	2.057	260.−6	12.29.−6	0.683	0.0254	1.61	0.995	0.0576	380
385	1.5233	1.053.−3	1.142	469.2	2694	1.439	7.218	4.232	2.080	248.−6	12.49.−6	0.685	0.0258	1.53	1.004	0.0566	385
390	1.794	1.058.−3	0.980	490.4	2702	1.494	7.163	4.239	2.104	237.−6	12.69.−6	0.686	0.0263	1.47	1.013	0.0556	390
400	2.455	1.067.−3	0.731	532.9	2716	1.605	7.058	4.256	2.158	217.−6	13.05.−6	0.688	0.0272	1.34	1.033	0.0536	400
410	3.302	1.077.−3	0.553	575.6	2729	1.708	6.959	4.278	2.221	200.−6	13.42.−6	0.688	0.0282	1.24	1.054	0.0515	410
420	4370	1.088.−3	0.425	618.6	2742	1.810	6.865	4.302	2.291	185.−6	13.79.−6	0.688	0.0293	1.16	1.075	0.0494	420
430	5.699	1.099.−3	0.331	661.8	2753	1.911	6.775	4.331	2.369	173.−6	14.14.−6	0.685	0.0304	1.09	1.10	0.0472	430
440	7.333	1.110.−3	0.261	705.3	2764	2.011	6.689	4.36	2.46	162.−6	14.50.−6	0.682	0.0317	1.04	1.12	0.0451	440
450	9.319	1.123.−3	0.208	749.2	2773	2.109	6.607	4.40	2.56	152.−6	14.85.−6	0.678	0.0331	0.99	1.14	0.0429	450
460	11.71	1.137.−3	0.167	793.5	2782	2.205	6.528	4.44	2.68	143.−6	15.19.−6	0.673	0.0346	0.95	1.17	0.0407	460
470	14.55	1.152.−3	0.136	838.2	2789	2.301	6.451	4.48	2.79	136.−6	15.54.−6	0.667	0.0363	0.92	1.20	0.0385	470
480	17.90	1.167.−3	0.111	883.4	2795	2.395	6.377	4.53	2.94	129.−6	15.88.−6	0.660	0.0381	0.89	1.23	0.0362	480
490	21.83	1.184.−3	0.0922	929.1	2799	2.479	6.312	4.59	3.10	124.−6	16.23.−6	0.651	0.0401	0.87	1.25	0.0339	490
500	26.40	1.203.−3	0.0766	975.6	2801	2.581	6.233	4.66	3.27	118.−6	16.59.−6	0.642	0.0423	0.86	1.28	0.0316	500
510	31.66	1.222.−3	0.0631	1023	2802	2.673	6.163	4.74	3.47	113.−6	16.95.−6	0.631	0.0447	0.85	1.31	0.0293	510
520	37.70	1.244.−3	00525	1071	2801	2.765	6.093	4.84	3.70	108.−6	17.33.−6	0.621	0.0475	0.84	1.35	0.0269	520
530	44.58	1.268.−3	0.0445	1119	2798	2.856	6.023	4.95	3.96	104.−6	17.72.−6	0.608	0.0506	0.85	1.39	0.0245	530
540	52.38	1.294.−3	0.0375	1170	2792	2.948	5.953	5.08	4.27	101.−6	18.1.−6	0.594	0.0540	0.86	1.43	0.0221	540
550	61.19	1.323.−3	0.0317	1220	2784	3.039	5.882	5.24	4.64	97.−6	18.6.−6	0.580	0.0583	0.87	1.47	0.0197	550
560	71.08	1.355.−3	0.0269	1273	2772	3.132	5.808	5.43	5.09	94.−6	19.1.−6	0.563	0.0637	0.90	1.52	0.0173	560
570	82.16	1.392.−3	0.0228	1328	2757	3.225	5.733	5.68	5.67	91.−6	19.7.−6	0.548	0.0698	0.94	1.59	0.0150	570
580	94.51	1.433.−3	0.0193	1384	2737	3.321	5.654	6.00	6.40	88.−6	20.4.−6	0.528	0.0767	0.99	1.68	0.0128	580
590	108.3	1.482.−3	0.0163	1443	2717	3.419	5.569	6.41	7.35	84.−6	21.5.−6	0.513	0.0841	1.05	1.84	0.0105	590
600	123.5	1.541.−3	0.0137	1506	2682	3.520	5.480	7.00	8.75	81.−6	22.7.−6	0.497	0.0929	1.14	2.15	0.0084	600
610	137.3	1.612.−3	0.0115	1573	2641	3.627	5.318	7.85	11.1	77.−6	24.1.−6	0.467	0.103	1.30	2.60	0.0063	610
620	159.1	1.705.−3	0.0094	1647	2588	3.741	5.259	9.35	15.4	72.−6	25.9.−6	0.444	0.114	1.52	3.46	0.0045	620
625	169.1	1.778.−3	0.0085	1697	2555	3.805	5.191	10.6	18.3	70.−6	27.0.−6	0.430	0.121	1.65	4.20	0.0035	625
630	179.7	1.856.−3	0.0075	1734	2515	3.875	5.115	12.6	22.1	67.−6	28.0.−6	0.412	0.130	2.0	4.8	0.0026	630
635	190.9	1.935.−3	0.0066	1783	2466	3.950	5.025	16.4	27.6	64.−6	30.0.−6	0.392	0.141	2.7	6.0	0.0015	635
640	202.7	2.075.−3	0.0057	1841	2401	4.037	4.912	26	42	59.−6	32.0.−6	0.367	0.155	4.2	9.6	0.0008	640
645	215.2	2.351.−3	0.0045	1931	2292	4.223	4.732	90	....	54.−6	37.0.−6	0.331	0.178	12	26	0.0001	645
647.3‡	221.2	3.170.−3	0.0032	2107	2107	4.443	4.443	∞	∞	45.−6	45.0.−6	0.238	0.238	∞	∞	0.0000	647.3‡

*1 bar = 10^5 N/m^2.

†Above the solid line, the condensed phase is solid; below it is liquid.

‡Critical temperature.

NOTE: The notations 6.30.−11, 1.073.−3, 9.55.+9, etc. signify 6.30×10^{-11}, 1.073×10^{-3}, 955×10^9, etc.

TABLE 3-303 Thermodynamic Properties of Compressed Steam*

Temperature, K	Pressure, bar									
	0.1	0.5	1	5	10	20	40	60	80	100
v	16.12	1.027.−3	1.027.−3	1.027.−3	1.027.−3	1.026.−3	1.025.−3	1.024.−3	1.023.−3	1.023.−3
350 *h*	2644	321.7	231.8	322.1	322.5	323.3	324.9	326.4	328.1	329.7
s	8.327	1.037	1.037	1.037	1.037	1.036	1.035	1.034	1.032	1.031
v	18.44	3.67	1.827	1.067.−3	1.067.−3	1.066.−3	1.065.−3	1.064.−3	1.063.−3	1.061.−3
400 *h*	2739	2735	2730	533.1	533.4	534.1	535.4	536.8	538.2	539.6
s	8.581	7.831	7.502	1.601	1.600	1.599	1.597	1.595	1.593	1.592
v	20.75	4.14	2.063	0.410	1.124.−3	1.123.−3	1.121.−3	1.119.−3	1.118.−3	1.116.−3
450 *h*	2835	2833	2830	2804	749.0	749.8	750.8	751.9	753.0	754.1
s	8.811	8.061	7.736	6.949	2.110	2.107	2.105	2.102	2.099	2.097
v	23.07	4.61	2.298	0.452	0.221	0.104	1.201.−3	1.198.−3	1.196.−3	1.193.−3
500 *h*	2932	2931	2929	2912.4	2891.2	2839.4	975.9	976.3	976.8	977.3
s	9.012	8.261	7.944	7.177	6.823	6.422	2.578	2.575	2.571	2.567
v	27.7	5.53	2.76	0.548	0.271	0.133	0.0630	0.0396	0.0276	0.0201
600 *h*	3131	3130	3129	3120	3109	3087	3036	2976	2906	2820
s	9.374	8.630	8.309	7.560	7.223	6.875	6.590	6.224	5.997	5.775
v	32.3	6.46	3.23	0.643	0.319	0.158	0.0769	0.0500	0.0346	0.0283
700 *h*	3335	3335	3334	3328	3322	3307	3278	3247	3214	3179
s	9.692	8.946	8.625	7.877	7.550	7.215	6.864	6.644	6.431	6.334
v	36.9	7.38	3.69	0.736	0.367	0.182	0.0889	0.0589	0.0436	0.0343
800 *h*	3547	3546	3546	3542	3537	3526	3506	3485	3464	3442
s	9.971	9.228	8.908	8.161	7.837	7.507	7.151	6.965	6.809	6.685
v	41.5	8.31	4.15	0.829	0.414	0.206	0.102	0.0674	0.0501	0.0398
900 *h*	3765	3765	3764	3761	3757	3750	3737	3719	3704	3688
s	10.228	9.485	9.165	8.420	8.097	7.770	7.462	7.237	7.092	6.975
v	41.5	8.31	4.15	0.829	0.414	0.206	0.102	0.0674	0.0501	0.0398
1000 *h*	3990	3990	3990	3987	3984	3978	3967	3955	3944	3935
s	10.466	9.723	9.402	8.659	8.336	8.011	7.682	7.486	7.345	7.233
v	69.2	13.9	6.92	1.385	0.692	0.341	0.1730	0.1153	0.0865	0.0692
1500 *h*	5231	5228	5227	5225	5224	5221	5217	5212	5207	5203
s	11.47	10.77	10.40	9.66	9.34	9.015	8.693	8.503	8.368	8.262
v	93.0	18.6	9.26	1.850	0.925	0.462	0.231	0.1543	0.1157	0.0926
2000 *h*	6832	6734	6706	6662	6649	6639	6629	6623	6619	6616
s	12.38	11.58	11.25	10.48	10.15	9.828	9.503	9.313	9.178	9.073
v	123.7	24.0	11.90	2.35	1.171	0.583	0.291	0.1942	0.1457	0.1166
2500 *h*	10417	9330	9046	8621	8504	8413	8342	8307	8285	8269
s	13.95	12.73	12.28	11.35	10.80	10.62	10.26	10.06	9.920	9.810

*v = specific volume, m^3/kg; h = specific enthalpy, kJ/kg; s = specific entropy, kJ/(kg·K). The notation 1.027.−3 signifies 1.027×10^{-3}.

Pressure, bar										
150	200	250	300	400	500	600	700	800	900	1000
1.020.−3 333.7 1.028	1.018.−3 337.7 1.025	1.016.−3 341.7 1.022	1.014.−3 344.7 1.019	1.009.−3 353.8 1.013	1.005.−3 361.8 1.007	1.002.−3 369.7 1.001	9.977.−4 377.7 0.996	9.937.−4 385.7 0.991	9.900.−4 393.7 0.985	9.865.−4 401.7 0.979
1.059.−3 543.1 1.587	1.056.−3 546.5 1.583	1.053. 550.1 1.578	1.050.−3 553.5 1.574	1.045.−3 560.6 1.565	1.041.−3 567.8 1.557	1.035.−3 574.9 1.549	1.031.−3 582.1 1.541	1.027.−3 589.3 1.533	1.022.−3 596.5 1.526	1.018.−3 603.8 1.518
1.112.−3 756.8 2.558	1.108.−3 759.5 2.549	1.105.−3 762.3 2.541	1.101.−3 765.2 2.533	1.094.−3 771.0 2.517	1.088.−3 776.9 2.502	1.082.−3 783.0 2.488	1.076.−3 789.6 2.474	1.070.−3 795.3 2.461	1.065.−3 801.6 2.449	1.059.−3 807.9 2.437
1.187.−3 978.8 2.558	1.181.−3 980.3 2.549	1.175.−3 981.9 2.541	1.170.−3 983.7 2.533	1.160.−3 987.4 2.517	1.151.−3 991.5 2.502	1.142.−3 995.9 2.488	1.134.−3 1000.5 2.474	1.126.−3 1005.3 2.461	1.119.−3 1010.3 2.449	1.112.−3 1015.4 2.437
1.519.−3 1499 3.501	1.483.−3 1489 3.469	1.454.−3 1479 3.443	1.428.−3 1472 3.419	1.392.−3 1462 3.379	1.362.−3 1456 3.346	1.337.−3 1452 3.316	1.315.−3 1449 3.290	1.296.−3 1447 3.266	1.280.−3 1447 3.244	1.265.−3 1447 3.223
1.724.−2 3082 6.037	1.157.−2 2965 5.770	7.986.−3 2821 5.494	5.416.−3 2635 5.179	2.630.−3 2233 4.554	2.038.−3 2084 4.308	1.831.−3 2021 4.192	1.716.−3 1986 4.116	1.639.−3 1962 4.058	1.589.−3 1946 4.012	1.536.−3 1931 3.972
2.195.−2 3386 6.444	1.575.−2 3325 6.252	1.201.−2 3261 6.086	9.512.−3 3193 5.934	6.391.−3 3047 5.654	4.576.−3 2895 5.397	3.496.−3 2734 5.175	2.866.−3 2648 4.998	2.484.−3 2567 4.864	2.239.−3 2508 4.761	2.072.−3 2465 4.701
2.590.−2 3649 6.755	1.899.−2 3609 6.587	1.483.−2 3568 6.449	1.207.−2 3526 6.327	8.619.−3 3440 6.119	6.581.−3 3354 5.940	5.257.−3 3269 5.780	4.348.−3 3188 5.637	3.704.−3 3113 5.510	3.454.−3 3049 5.399	2.907.−3 2995 5.305
2.954.−2 3904 7.023	2.186.−2 3874 6.867	1.726.−2 3845 6.741	1.420.−2 3816 6.633	1.038.−2 3756 6.453	8.102.−3 3697 6.302	6.605.−3 3640 6.172	5.557.−3 3584 6.055	4.792.−3 3532 5.951	4.212.−3 3482 5.856	3.763.−3 3435 5.727
0.0461 5202 8.074	0.0346 5198 7.936	0.0277 5186 7.827	0.0231 5180 7.738	0.0173 5171 7.597	0.0139 5157 7.484	0.0116 5144 7.391	0.00993 5133 7.310	0.00871 5120 7.239	0.00776 5108 7.176	0.00700 5095 7.118
0.0619 6613 8.883	0.0465 6610 8.748	0.0372 6608 8.642	0.0311 6605 8.555	0.0234 6599 8.418	0.0188 6595 8.310	0.0157 6590 8.222	0.0135 6585 8.147	0.0119 6581 8.082	0.0106 6577 8.024	0.0096 6574 7.971
0.0778 8269 9.610	0.0584 8269 9.468	0.0468 8269 9.358	0.0391 8268 9.270	0.0294 8267 9.129	0.0236 8265 9.020	0.0197 8261 8.930	0.0170 2856 8.854	0.0149 8250 8.788	0.0133 8244 8.730	0.0120 8240 8.677

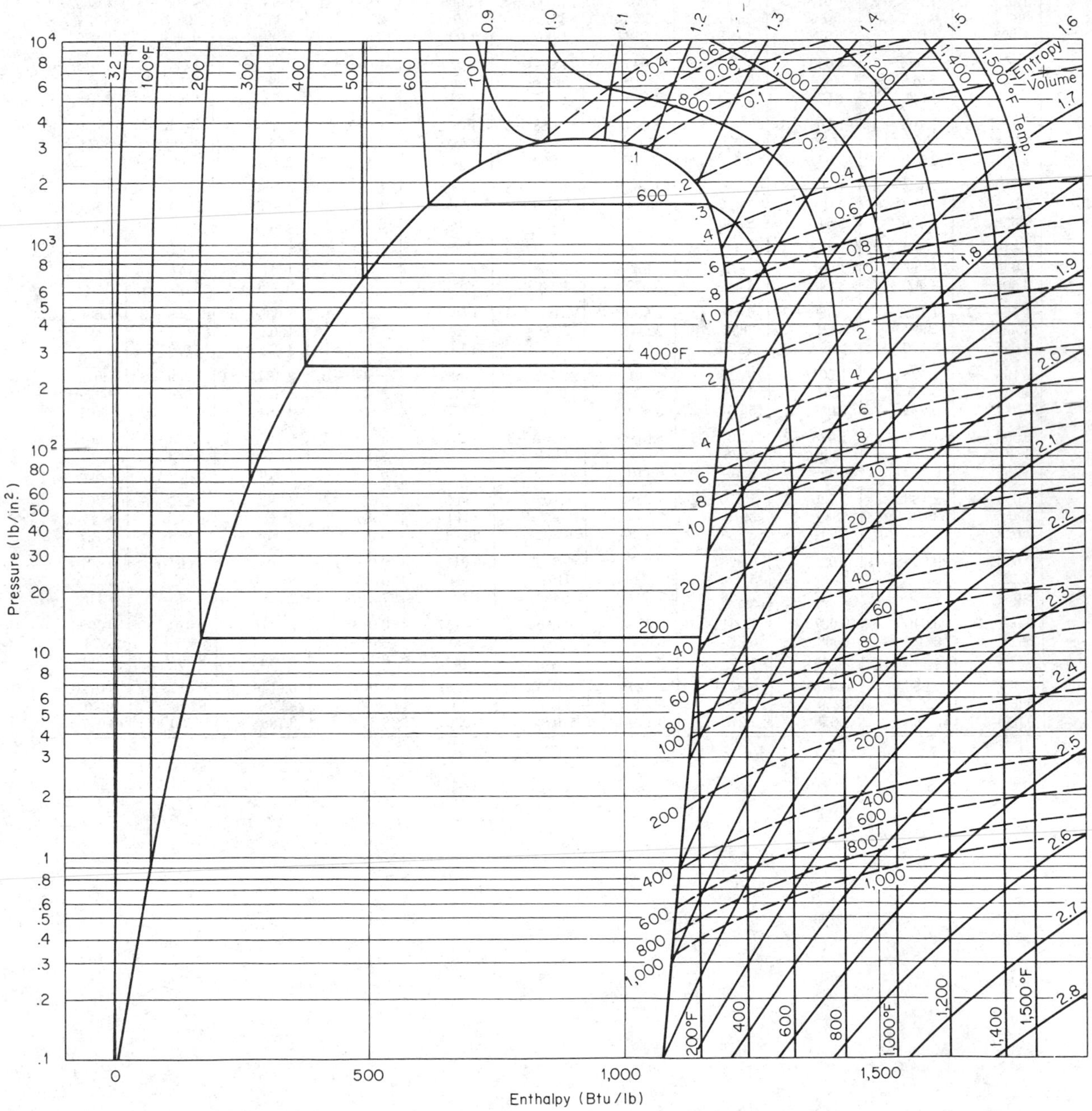

FIG. 3-40 Enthalpy–log-pressure diagram for water and steam. Volume, ft³/lb; entropy, Btu/(16·°R). *(Drawn from the 1967 ASME Steam Tables. Copyright 1967 by the American Society of Mechanical Engineers; reproduced by permission.)*

TABLE 3-304 Specific Heat and Other Thermophysical Properties of Water Substance*

Pressure, bar		Temperature, K: 300	350	400	450	500	600	700	800	900	1000	1200	1400	1600	1800	2000
1	μ	8.57.−4	3.70.−4	1.32.−5	1.52.−5	1.73.−5	2.15.−5	2.57.−5	2.98.−5	3.39.−5	3.78.−5	4.48.−5	5.06.−5	5.65.−5	6.19.−5	6.70.−5
	c_p	4.18	4.19	1.99	1.97	1.98	2.02	2.09	2.15	2.22	2.29	2.43	2.58	2.73	3.02	3.79
	k	0.614	0.668	0.0268	0.0311	0.0358	0.0464	0.0581	0.0710	0.0843	0.0981	0.13	0.16	0.21	0.33	0.57
	Pr	5.81	2.32	0.980	0.967	0.955	0.936	0.920	0.906	0.891	0.881	0.83	0.80	0.75	0.57	0.45
5	μ	8.57.−4	3.70.−4	2.17.−4	1.49.−5	1.72.−5	2.15.−5	2.57.−5	2.98.−5	3.39.−5	3.78.−5	4.45.−5	5.06.−5	5.65.−5	6.19.−5	6.70.−5
	c_p	4.18	4.19	4.26	2.21	2.10	2.07	2.11	2.16	2.23	2.29	2.43	2.58	2.73	2.98	3.40
	k	0.614	0.668	0.689	0.0335	0.0369	0.0469	0.0585	0.0713	0.0846	0.0984	0.13	0.16	0.20	0.28	0.43
	Pr	5.82	2.32	1.34	0.983	0.973	0.947	0.925	0.907	0.892	0.881	0.83	0.81	0.77	0.65	0.53
10	μ	8.57.−4	3.70.−4	2.17.−4	1.51.−4	1.71.−5	2.15.−5	2.58.−5	2.99.−5	3.39.−5	3.78.−5	4.45.−5	5.06.−5	5.65.−5	6.19.−5	6.70.−5
	c_p	4.18	4.19	4.25	4.39	2.29	2.13	2.13	2.18	2.24	2.30	2.44	2.58	2.73	2.95	3.29
	k	0.615	0.668	0.689	0.677	0.0380	0.0474	0.0590	0.0717	0.0851	0.0988	0.13	0.16	0.20	0.26	0.39
	Pr	5.82	2.32	1.34	0.981	1.028	0.963	0.931	0.908	0.892	0.881	0.84	0.82	0.78	0.70	0.57
20	μ	8.56.−4	3.71.−4	2.18.−4	1.51.−4	1.68.−5	2.15.−5	2.59.−5	3.00.−5	3.40.−5	3.79.−5	4.46.−5	5.06.−5	5.65.−5	6.19.−5	6.70.−5
	c_p	4.17	4.19	4.25	4.39	2.84	2.26	2.19	2.21	2.26	2.32	2.45	2.59	2.73	2.92	3.21
	k	0.616	0.669	0.689	0.679	0.0402	0.0485	0.0599	0.0726	0.0859	0.0996	0.13	0.16	0.20	0.25	0.36
	Pr	5.80	2.32	1.34	0.979	1.19	0.999	0.946	0.912	0.893	0.881	0.84	0.82	0.79	0.72	0.60
40	μ	8.55.−4	3.71.−4	2.18.−4	1.52.−4	1.19.−4	2.15.−5	2.61.−5	3.02.−5	3.42.−5	3.80.−5	4.47.−5	5.07.−5	5.65.−5	6.19.−5	6.70.−5
	c_p	4.17	4.19	4.25	4.38	4.65	2.60	2.32	2.28	2.30	2.34	2.46	2.59	2.73	2.90	3.14
	k	0.617	0.671	0.690	0.680	0.644	0.516	0.0620	0.0744	0.0877	0.101	0.13	0.16	0.19	0.24	0.33
	Pr	5.78	2.31	1.34	0.977	0.862	1.08	0.975	0.924	0.895	0.881	0.84	0.82	0.80	0.73	0.63
60	μ	8.54.−4	3.72.−4	2.19.−4	1.53.−4	1.20.−4	2.14.−5	2.63.−5	3.04.−5	3.43.−5	3.82.−5	4.48.−5	5.07.−5	5.66.−5	6.19.−5	6.70.−5
	c_p	4.16	4.18	4.24	4.37	4.63	3.11	2.47	2.35	2.34	2.37	2.48	2.60	2.73	2.89	3.11
	k	0.619	0.672	0.692	0.682	0.646	0.0561	0.0645	0.0764	0.0895	0.103	0.13	0.16	0.19	0.24	0.32
	Pr	5.74	2.31	1.34	0.976	0.859	1.19	1.008	0.984	0.899	0.879	0.84	0.82	0.81	0.74	0.65
80	μ	8.53.−4	3.72.−4	2.19.−4	1.53.−4	1.20.−4	2.14.−5	2.66.−5	3.06.−5	3.45.−5	3.83.−5	4.48.−5	5.08.−5	5.66.−5	6.19.−5	6.70.−5
	c_p	4.16	4.18	4.24	4.36	4.62	3.88	2.65	2.43	2.39	2.40	2.49	2.61	2.73	2.88	3.09
	k	0.620	0.674	0.693	0.684	0.648	0.0628	0.0672	0.0785	0.0914	0.105	0.13	0.16	0.10	0.24	0.31
	Pr	5.72	2.31	1.34	0.975	0.856	1.33	1.046	0.946	0.902	0.877	0.84	0.83	0.81	0.74	0.66
100	μ	8.52.−4	3.73.−4	2.20.−4	1.53.−4	1.21.−4	2.14.−5	2.69.−5	3.08.−5	3.47.−5	3.85.−5	4.49.−5	5.08.−5	5.66.−5	6.19.−5	6.70.−5
	c_p	4.15	4.17	4.23	4.35	4.60	5.22	2.85	2.52	2.44	2.44	2.50	2.62	2.73	2.88	3.08
	k	0.622	0.675	0.694	0.685	0.651	0.0730	0.0704	0.0807	0.0934	0.107	0.13	0.16	0.19	0.24	0.31
	Pr	5.69	2.31	1.34	0.975	0.853	1.74	1.088	0.960	0.905	0.876	0.84	0.83	0.81	0.74	0.67
150	μ	8.51.−4	3.74.−4	2.22.−4	1.56.−4	1.22.−4	8.22.−5	2.72.−5	3.12.−5	3.51.−5	3.89.−5	4.52.−5	5.09.−5	5.67.−5	6.19.−5	6.70.−5
	c_p	4.14	4.16	4.22	4.34	4.54	......	3.55	2.74	2.57	2.53	2.54	2.65	2.75	2.88	3.06
	k	0.624	0.678	0.699	0.693	0.657	0.520	0.079	0.086	0.098	0.110	0.14	0.16	0.19	0.23	
	Pr	5.64	2.30	1.34	0.974	0.842	......	1.22	0.994	0.916	0.891	0.84	0.83	0.82	0.76	
200	μ	8.50.−4	3.75.−4	2.24.−4	1.57.−4	1.23.−4	8.32.−5	2.80.−5	3.17.−5	3.54.−5	3.93.−5	4.54.−5	5.11.−5	5.67.−5		
	c_p	4.12	4.15	4.21	4.32	4.51	......	4.67	3.04	2.71	2.62	2.57	2.67	2.76	2.88	3.05
	k	0.626	0.681	0.702	0.697	0.661	0.525	0.095	0.095	0.104	0.113	0.14	0.16	0.19		
	Pr	5.59	2.29	1.34	0.974	0.833	......	1.38	1.014	0.925	0.903	0.84	0.83	0.82		
250	μ	8.49.−4	3.76.−4	2.26.−4	1.59.−4	1.23.−4	8.41.−5	2.89.−5	3.24.−5	3.59.−5	3.98.−5	4.56.−5	5.12.−5	5.68.−5		
	c_p	4.12	4.14	4.20	4.30	4.49	5.90	6.16	3.40	2.86	2.71	2.61	2.69	2.77	2.89	3.04
	k	0.627	0.683	0.705	0.701	0.672	0.537	0.112	0.103	0.110	0.119	0.136	0.16			
	Pr	5.57	2.28	1.34	0.974	0.826	0.924	1.590	1.070	0.940	0.910	0.85	0.84			
300	μ	8.49.−4	3.77.−4	2.28.−4	1.60.−4	1.24.−4	8.50.−5	3.7.−5	3.4.−5	3.64.−5	4.02.−5	4.59.−5	5.14.−5	5.68.−5		
	c_p	4.10	4.13	4.19	4.29	4.44	5.60	10.20	3.82	3.03	2.81	2.65	2.72	2.78	2.90	3.04
	k	0.629	0.685	0.708	0.704	0.675	0.548	0.173	0.113	0.113	0.123	0.14				
	Pr	5.53	2.27	1.34	0.973	0.820	0.859	2.18	1.149	0.976	0.917	0.87				
400	μ	8.49.−4	3.80.−4	2.30.−4	1.62.−4	1.26.−4	8.64.−5	5.3.−5	3.6.−5	3.8.−5	4.1.−5	4.6.−5	5.17.−5			
	c_p	4.08	4.12	4.16	4.26	4.42	5.31	13.20	4.86	3.39	3.01	2.70	2.77	2.81	2.91	3.04
	k	0.631	0.689	0.714	0.710	0.676	0.567	0.327	0.145	0.129	0.134	0.15				
	Pr	5.49	2.26	1.34	0.971	0.817	0.799	2.14	1.207	0.999	0.926					
500	μ	8.50.−4	3.82.−4	2.31.−4	1.64.−4	1.28.−4	8.83.−5	5.8.−5	4.0.−5	4.0.−5	4.2.−5	4.7.−5				
	c_p	4.06	4.10	4.15	4.23	4.38	5.08	8.44	5.70	3.90	3.21	2.77	2.81	2.84	2.92	3.04
	k	0.634	0.695	0.719	0.717	0.693	0.583	0.378	0.186	0.147	0.145					
	Pr	5.44	2.25	1.33	0.971	0.814	0.773	1.30	1.225	1.061	0.932					

TABLE 3-304 Specific Heat and Other Thermophysical Properties of Water Substance* *(Continued)*

Pressure, bar	Temperature, K 300	350	400	450	500	600	700	800	900	1000	1200	1400	1600	1800	2000
μ	8.51. −4	3.85. −4	2.32. −4	1.66. −4	1.30. −4	9.17. −5	6.5. −5	4.4. −5	4.2. −5	4.4. −5					
600 c_p	4.04	4.08	4.13	4.20	4.33	4.92	6.93	6.83	4.19	3.38	2.87	2.86	2.86	2.92	3.04
k	0.639	0.699	0.725	0.725	0.700	0.597	0.420	0.239	0.170	0.159					
Pr	5.38	2.24	1.32	0.970	0.812	0.755	1.073	1.175	1.035	0.935					
μ	8.52. −4	3.87. −4	2.33. −4	1.69. −4	1.33. −4	9.50. −5	6.9. −5	4.9. −5	4.5. −5	4.6. −5					
700 c_p	4.01	4.07	4.12	4.17	4.29	4.78	6.12	6.26	4.62	3.59	2.94	2.91	2.88	2.93	3.05
k	0.644	0.706	0.730	0.732	0.707	0.614	0.442	0.279	0.198	0.177					
Pr	5.33	2.23	1.32	0.970	0.810	0.739	1.047	1.098	1.010	0.935					
μ	8.53. −4	3.90. −4	2.34. −4	1.72. −4	1.36. −4	9.82. −5	7.3. −5	5.4. −5	4.8. −5	4.8. −5					
800 c_p	3.99	4.05	4.10	4.15	4.26	4.67	5.60	6.09	4.77	3.75	3.01	2.96	2.91	2.95	3.05
k	0.648	0.709	0.735	0.736	0.714	0.625	0.478	0.320	0.228	0.193					
Pr	5.28	2.23	1.31	0.970	0.808	0.725	0.855	1.028	1.003	0.933					
μ	8.54. −4	3.93. −4	2.35. −4	1.74. −4	1.38. −4	1.00. −4	7.6. −5	5.8. −5	5.1. −5	5.0. −5					
900 c_p	3.98	4.03	4.08	4.13	4.23	4.57	5.29	5.86	4.85	3.86	3.08	3.00	2.94	2.97	3.06
k	0.651	0.713	0.738	0.742	0.724	0.636	0.496	0.351	0.260	0.210					
Pr	5.23	2.22	1.30	0.969	0.806	0.712	0.810	0.968	0.950	0.919					
μ	8.56. −4	3.96. −4	2.36. −4	1.76. −4	1.40. −4	1.02. −4	7.9. −5	6.2. −5	5.4. −5	5.1. −5					
1000 c_p	3.97	4.02	4.06	4.11	4.20	4.47	5.08	5.51	4.88	3.96	3.16	3.05	2.97	2.98	3.07
k	0.653	0.717	0.743	0.747	0.731	0.650	0.516	0.372	0.288	0.228					
Pr	5.19	2.22	1.30	0.968	0.804	0.701	0.778	0.918	0.900	0.886					

*μ = viscosity, Ns/m^2; c_p = specific heat at constant pressure, kJ/(kg·K); k = thermal conductivity, W/(m·K); Pr = Prandtl number.

TABLE 3-305 Water and Steam of High Pressure*

Pressure, kB.		Temp., °C. 0	50	100	150	200	250	300	350	400
1	*v*	0.9567	0.9733	1.0002	1.0362	1.0821	1.1400	1.2140	1.3115	1.4444
	h	95	294	495	698	903	1113	1329	1555	1794
	s	−0.010	0.658	1.238	1.748	2.206	2.628	3.022	3.400	3.768
2	*v*	0.9244	0.9434	0.9678	0.9978	1.0338	1.0767	1.1276	1.1879	1.2595
	h	181	375	572	769	967	1167	1369	1575	1783
	s	−0.039	0.615	1.179	1.675	2.117	2.518	2.888	3.232	3.554
4	*v*	0.8785	0.8978	0.9190	0.9429	0.9695	0.9989	1.0316	1.0679	1.1082
	h	344	532	723	914	1105	1295	1485	1676	1867
	s	−0.111	0.530	1.081	1.561	1.986	2.368	2.716	3.035	3.330
6	*v*	0.8455	0.8643	0.8833	0.9031	0.9247	0.9482	0.9732	0.9999	1.0285
	h	507	682	873	1060	1246	1432	1617	1802	1985
	s	−0.127	0.450	0.998	1.469	1.886	2.259	2.597	2.906	3.189
8	*v*		0.8380	0.8558	0.8728	0.8908	0.9102	0.9308	0.9526	0.9754
	h		826	1020	1205	1389	1572	1754	1935	2115
	s		0.370	0.927	1.393	1.804	2.171	2.504	2.807	3.084
10	*v*		0.8165	0.8335	0.8486	0.8642	0.8806	0.8980	0.9163	0.9356
	h		966	1165	1350	1532	1712	1892	2071	2248
	s		0.290	0.863	1.328	1.735	2.098	2.426	2.725	2.999
20	*v*			0.7601	0.7705	0.7807	0.7909	0.8013	0.8120	0.8229
	h			1862	2052	2231	2408	2583	2755	2926
	s			0.605	1.084	1.485	1.840	2.159	2.448	2.711
50	*v*						0.6732	0.6791	0.6850	0.6911
	h						4341	4512	4681	4848
	s						1.388	1.699	1.981	2.239

*Extracted from Table 13 of Juza, *An Equation of State for Water and Steam: Steam Tables in the Critical Region and in the Range from 1000 to 100,000 Bars*, Academia, Prague, 1966. [Pressure in kilobars, volume in cm^3/g, enthalpy in J/g, entropy in J(g·K), temperature in °C.] The original tables give values for 0(25)500(5)1000°C, 1 to 100 kbar.

TABLE 3-306 Sulfur Hexafluoride

Mears, Rosenthal, and Sinka [*J. Phys. Chem.*, 73, 2254 (1969)] derive a fitting to the Martin-Hou equation of state, covering the range 25 to 130°C, 10 to 75 bar. Vapor pressure and virial and critical data are discussed. O'Hare (ANL Rep. 7315, 1968) reviews and tabulates data for 24 chalcogen fluorides, including SF, SF_2, SF_4, SF_5, SF_6, and other sulfur, selenium, tellurium, and oxygen fluorides.

TABLE 3-307 Saturated Xenon*

T, K	P, bar	v_f, m³/kg	v_g, m³/kg	h_f, kJ/kg	h_g, kJ/kg	s_f, kJ/(kg·K)	s_g, kJ/(kg·K)	c_{pf}, kJ/(kg·K)	μ_f, 10^{-4} Pa·s	k_f, W/(m·K)
10	.....	2.642. −4	.....	0.19	....	0.0236	....	0.058		
20	.....	2.650. −4	.....	1.21	....	0.0901	....	0.133		
30	.....	2.661. −4	.....	2.74	....	0.1510	....	0.164		
40	.....	2.675. −4	.....	4.47	....	0.2003	....	0.178		
50	.....	2.689. −4	.....	6.31	....	0.2410	....	0.186		
60	.....	2.704. −4	.....	8.21	....	0.2755	....	0.191		
80	.....	2.737. −4	.....	12.14	....	0.3319	....	0.202		
100	.....	2.776. −4	.....	16.30	....	0.3783	....	0.214		
120	.....	2.820. −4	.....	20.81	....	0.4197	....	0.231		
140	.....	2.874. −4	.....	25.67	....	0.4581	....	0.251		
160	.....	2.941. −4	.....	30.94	....	0.4946	....	0.270		
161.4^m	.....	2.946. −4	.....	31.30	....	0.4969	....	0.271		
161.4^m	0.816	3.372. −4	0.1219	48.98	145.5	0.6072	1.206	0.350		
170	1.336	3.439. −4	0.0776	52.01	146.5	0.6253	1.181	0.349	4.50	0.0707
180	2.218	3.523. −4	0.0487	55.52	147.5	0.6452	1.156	0.349	3.99	0.0663
190	3.480	3.615. −4	0.0321	59.04	148.3	0.6641	1.134	0.352	3.51	0.0622
200	5.212	3.715. −4	0.0220	62.61	148.9	0.6820	1.113	0.357	3.09	0.0582
210	7.504	3.828. −4	0.0156	66.25	149.2	0.6994	1.095	0.365	2.71	0.0542
220	10.45	3.955. −4	0.0113	70.00	149.4	0.7163	1.077	0.379	2.39	0.0506
230	14.16	4.100. −4	0.0084	73.91	149.2	0.7330	1.060	0.400	2.09	0.0468
240	18.72	4.271. −4	0.0063	78.05	148.5	0.7498	1.044	0.432	1.83	0.0429
250	24.25	4.476. −4	0.0047	82.54	147.5	0.7671	1.027	0.482	1.60	0.0393
260	30.87	4.730. −4	0.0036	87.52	145.7	0.7855	1.009	0.560	1.38	0.0355
270	38.69	5.079. −4	0.0027	93.30	142.8	0.8058	0.989	0.685	1.18	0.0313
280	47.86	5.689. −4	0.0019	100.6	138.0	0.8308	0.964	0.995	0.95	0.0275
289.7^c	58.21	9.091. −4	0.0009	120.0	120.0	0.8962	0.896	∞	...	∞

*Values extracted and in some cases rounded off from those cited in Rabinovich (ed.), *Thermophysical Properties of Neon, Argon, Krypton and Xenon*, Standards Press, Moscow, 1976. This source contains values for the compressed state for pressures up to 1000 bar, etc. m = melting point; c = critical point. The notation 2.642. −4 signifies 2.642×10^{-4}.

TABLE 3-308 Compressed Xenon*

T, K		Pressure, bar 1	100	200	300	400	500	600	700	800	900	1000
100	v	2.776. −4	2.764. −4	2.752. −4	2.742. −4	2.731. −4	2.721. −4	2.711. −4	2.702. −4	2.693. −4	2.684. −4	2.675. −4
	h	16.30	18.84	21.40	23.95	26.50	29.05	31.59	34.13	36.67	39.21	41.74
	s	0.3783	0.3762	0.3742	0.3723	0.3704	0.3686	0.3669	0.3652	0.3636	0.3621	0.3802
200	v	0.1245	3.623. −4	3.547. −4	3.484. −4	3.430. −4	3.383. −4	3.342. −4	3.304. −4	3.270. −4	3.240. −4	3.211. −4
	h	151.8	64.22	66.14	68.19	70.34	72.56	74.83	77.13	79.46	81.81	84.18
	s	1.228	0.6727	0.6643	0.6570	0.6505	0.6446	0.6391	0.6340	0.6292	0.6247	0.6204
300	v	0.1890	5.729. −4	4.769. −4	4.431. −4	4.220. −4	4.068. −4	3.955. −4	3.862. −4	3.783. −4	3.716. −4	3.657. −4
	h	168.0	106.4	101.6	101.3	102.0	103.3	104.9	106.7	108.5	110.6	112.8
	s	1.294	0.8401	0.8073	0.7908	0.7789	0.7691	0.7608	0.7540	0.7477	0.7424	0.7370
400	v	0.2527	1.998. −3	8.759. −4	6.452. −4	5.604. −4	5.141. −4	4.839. −4	4.622. −4	4.457. −4	4.325. −4	4.217. −4
	h	183.9	164.2	145.4	137.4	134.7	134.1	134.5	135.5	136.8	138.3	140.0
	s	1.340	1.012	0.9330	0.8945	0.8730	0.8581	0.8467	0.8373	0.8292	0.8220	0.8162
500	v	0.3163	2.899. −3	1.389. −3	9.449. −4	7.577. −4	6.593. −4	5.986. −4	5.570. −4	5.268. −4	5.038. −4	4.859. −4
	h	199.8	187.8	177.1	169.4	165.1	163.0	162.3	162.4	163.1	164.3	165.7
	s	1.375	1.065	1.004	0.9664	0.9409	0.9228	0.9088	0.8975	0.8881	0.8801	0.8731
600	v	0.3798	3.673. −3	1.823. −3	1.240. −3	9.699. −4	8.206. −4	7.273. −4	6.636. −4	6.172. −4	5.820. −4	5.545. −4
	h	215.7	207.4	200.3	194.8	191.1	188.9	187.9	187.6	188.0	188.8	189.9
	s	1.404	1.101	1.047	1.013	0.9885	0.9700	0.9555	0.9435	0.9334	0.9247	0.9172
700	v	0.4432	4.397. −3	2.217. −3	1.513. −3	1.175. −3	9.815. −4	8.583. −4	7.734. −4	7.115. −4	6.642. −4	6.268. −4
	h	231.5	225.6	220.6	216.7	213.8	212.2	211.3	211.1	211.3	212.0	213.1
	s	1.428	1.129	1.078	1.047	1.023	1.006	0.9916	0.9797	0.9695	0.9606	0.9528
800	v	0.5066	5.093. −3	2.587. −3	1.769. −3	1.370. −3	1.137. −3	9.870. −4	8.824. −4	8.057. −4	7.469. −4	7.005. −4
	h	247.4	243.0	239.5	236.7	234.8	233.6	233.0	232.9	233.3	234.0	235.0
	s	1.450	1.152	1.103	1.073	1.052	1.035	1.021	1.009	0.9988	0.9901	0.9823
900	v	0.5700	5.773. −3	2.944. −3	2.014. −3	1.557. −3	1.288. −3	1.112. −3	9.893. −4	8.989. −4	8.289. −4	7.737. −4
	h	263.2	260.1	257.5	255.7	254.4	253.6	253.4	253.7	254.2	254.9	256.1
	s	1.468	1.172	1.125	1.096	1.075	1.058	1.045	1.033	1.023	1.015	1.007
1000	v	0.6333	6.441. −3	3.291. −3	2.252. −3	1.738. −3	1.435. −3	1.235. −3	1.094. −3	9.899. −4	9.097. −4	8.461. −4
	h	279.1	276.8	275.1	273.9	273.2	272.9	273.0	273.4	274.1	275.1	276.2
	s	1.485	1.190	1.143	1.115	1.095	1.079	1.065	1.054	1.044	1.036	1.028

*Values extracted and in some cases rounded off from those cited in Rabinovich (ed.), *Thermophysical Properties of Neon, Argon, Krypton and Xenon*, Standards Press, Moscow, 1976. This source contains an exhaustive tabulation of values. v = specific volume, m³/kg; h = specific enthalpy, kJ/kg; s = specific entropy, kJ/(kg·K). The notation 2.776. −4 signifies 2.776×10^{-4}.

TRANSPORT PROPERTIES

INTRODUCTION

Extensive tables of the viscosity and thermal conductivity of air and of water or steam for various pressures and temperatures are given with the thermodynamic-property tables. The thermal conductivity and the viscosity for the saturated-liquid state are also tabulated for many fluids along with the thermodynamic-property tables earlier in this section.

UNITS CONVERSIONS

For this subsection the following units conversions are applicable:

Diffusivity: to convert square centimeters per second to square feet per hour, multiply by 3.8750; to convert square meters per second to square feet per hour, multiply by 38,750.

Pressure: to convert bars to pounds-force per square inch, multiply by 14.504.

Temperature: $^\circ F = \frac{9}{5}\,^\circ C + 32$; $^\circ R = \frac{9}{5}\,K$.

Thermal conductivity: to convert watts per meter-kelvin to British thermal unit–feet per hour–square foot–degree Fahrenheit, multiply by 0.57779; and to convert British thermal unit–feet per hour–square foot–degree Fahrenheit to watts per meter-kelvin, multiply by 1.7307.

Viscosity: to convert pascal-seconds to centipoises, multiply by 1000.

ADDITIONAL REFERENCES

An extensive coverage of the general pressure and temperature variation of thermal conductivity is given in the monograph by Vargaftik, Filippov, Tarzimanov, and Totskiy, *Thermal Conductivity of Liquids and Gases* (in Russian), Standartov, Moscow, 1978. For a similar work on viscosity, see Stephan and Lucas, *Viscosity of Dense Fluids*, Plenum, New York and London, 1979. Tables and polynomial fits for 38 refrigerants in both the gaseous and the liquid state are contained in *ASHRAE Thermophysical Properties of Refrigerants*, American Society of Heating, Refrigerating and Ventilating Engineers, New York (now in Atlanta), 1976. For polynomial fittings for various properties for the dilute gas, see Andrews and Biblarz, Rep. NPS67-81-001 (AD A096 388), Naval Postgraduate School, Monterey, Calif., 1981. The McGraw-Hill/CINDAS Data Series on Material Properties is scheduled to present a large compilation of values of many properties for many substances. For thermal conductivity, two other sources are Ho, Powell, and Liley. *J. Phys. Chem. Ref. Data*, **1**, 279–421 (1972), **3**, suppl. 1 (1974); and Childs, Ericks, and Powell, NBS Monogr. 131, 1973.

TABLE 3-309 Transport Properties of Selected Gases at Atmospheric Pressure*

	Thermal conductivity, W/(m·K) Temperature, K					Viscosity, 10^{-4} Pa·s Temperature, K					Prandtl number, dimensionless Temperature, K			
Substance	250	300	400	500	600	250	300	400	500	600	250	300	400	500
Acetone	0.0080	0.0115	0.0201	0.0310	.....	....	0.077	0.101	0.128	0.156				
Acetylene	0.0162	0.0213	0.0332	0.0452	0.0561	....	0.104	0.135	0.164					
Ammonia	0.0197	0.0246	0.0364	0.0506	0.0656	0.085	0.102	0.139	0.175	0.211	....	0.91	0.87	0.86
Argon	0.0152	0.0177	0.0223	0.0264	0.0301	0.195	0.229	0.289	0.343	0.390	0.669	0.668	0.666	0.663
Benzene	0.0077	0.0104	0.0195	0.0335	0.0524	....	0.076	0.101	0.127	0.154				
Bromine	0.0038	0.0048	0.0067	.....	.....	....	....	0.203	0.260	0.291				
Butane	0.0117	0.0160	0.0264	0.0377	.....	....	0.076	0.101	0.125	0.151	....	0.805	0.820	
CO_2	0.0129	0.0166	0.0244	0.0323	0.0403	0.126	0.150	0.196	0.239	0.278	0.793	0.778	0.752	0.734
CCl_4	0.0053	0.0067	0.0099	0.0126	.....	....	0.101	0.131	0.162	0.191				
Chlorine	0.0071	0.0089	0.0124	0.0156	0.0190	....	0.136	0.178	0.218	0.259				
Deuterium	0.122	0.141	0.176	.....	.....	0.111	0.126	0.153	0.178	0.201				
Ethane	0.0156	0.0218	0.0360	0.0516	0.0685	0.079	0.094	0.123	0.148	0.171	0.817	0.773	0.746	0.746
Ethylene	0.0152	0.0214	0.0342	0.0491	0.0653	0.087	0.103	0.135	0.162	0.187	0.812	0.796	0.769	0.750
Helium	0.134	0.150	0.180	0.211	0.247	0.176	0.199	0.243	0.284	0.322	0.671	0.668	0.663	0.661
Heptane	0.0082	0.0120	0.0214	0.0325	0.0447	....	....	0.080	0.099	1.116				
Hydrogen	0.156	0.182	0.221	0.256	0.291	0.080	0.090	0.109	0.126	0.143	0.71	0.71	0.71	0.71
Methane	0.0277	0.0343	0.0484	0.0671	0.0948	0.095	0.112	0.142	0.170	0.195	0.742	0.739	0.737	0.736
Nitrogen	0.0222	0.0260	0.0325	0.0386	0.0441	0.156	0.180	0.223	0.261	0.295	0.721	0.714	0.708	0.707
Oxygen	0.0225	0.0267	0.0343	0.0412	0.0480	0.179	0.207	0.258	0.306	0.348				
Pentane	0.0107	0.0152	0.0250	0.0362										
Propane	0.0129	0.0183	0.0295	0.0417	.....	0.069	0.082	0.108	0.131	....	0.810	0.774	0.788	0.826
Propylene	0.0114	0.0168	0.0226	0.0430	0.0580	0.073	0.087	0.115	0.141	....	0.860	0.797	0.762	
R 11	.....	0.0078	0.0119	.....	.....	0.094	0.110	0.144	....	....	....	0.814	0.761	
R 12	0.0072	0.0097	0.0151	0.0208	.....	0.108	0.126	0.162	....	....	0.827	0.781	0.745	0.708
R 13	0.0091	0.0121	0.0185	0.0248	.....	0.123	0.145	0.190	....	....	0.796	0.766	0.759	0.757
R 21	.....	0.0088	0.0135	0.0181	.....	0.100	0.115	0.154	....	....	....	0.779	0.773	
R 22	0.0080	0.0109	0.0170	0.0230	0.0290	0.109	0.129	0.168	....	....	0.820	0.771	0.760	
SO_2	0.0078	0.0096	0.0143	0.0200	0.0256	....	0.129	0.175	0.217	0.256				

*An approximate interpolation scheme is to plot the logarithm of the viscosity or the thermal conductivity versus the logarithm of the absolute temperature. At 250 K the viscosity of gaseous argon is to be read as 1.95×10^{-5} Pa·s = 0.0000195 Ns/m^2.

TABLE 3-310 Viscosities of Gases: Coordinates for Use with Fig. 3-41

Gas	X	Y	Gas	X	Y
Acetic acid	7.7	14.3	Hydrogen bromide	8.8	20.9
Butene	9.2	13.7	Hydrogen chloride	8.8	18.7
Butylene	8.9	13.0	Hydrogen cyanide	9.8	14.9
Carbon disulfide	8.0	16.0	Hydrogen iodide	9.0	21.3
Carbon monoxide	11.0	20.0	Hydrogen sulfide	8.6	18.0
Chlorine	9.0	18.4	Iodine	9.0	18.4
Chloroform	8.9	15.7	Mercury	5.3	22.9
Cyanogen	9.2	15.2	Methyl alcohol	8.5	15.6
Cyclohexane	9.2	12.0	Nitric oxide	10.9	20.5
Ethyl acetate	8.5	13.2	Nitrosyl chloride	8.0	17.6
Ethyl alcohol	9.2	14.2	Nitrous oxide	8.8	19.0
Ethyl chloride	8.5	15.6	Propyl alcohol	8.4	13.4
Ethyl ether	8.9	13.0	Toluene	8.6	12.4
Fluorine	7.3	23.8	2, 3, 3-Trimethylbutane	9.5	10.5
Freon-11	10.6	15.1	Xenon	9.3	23.0
Freon-113	11.3	14.0			
Helium	10.9	20.5			
Hexane	8.6	11.8			
$3H_2 + 1N_2$	11.2	17.2			

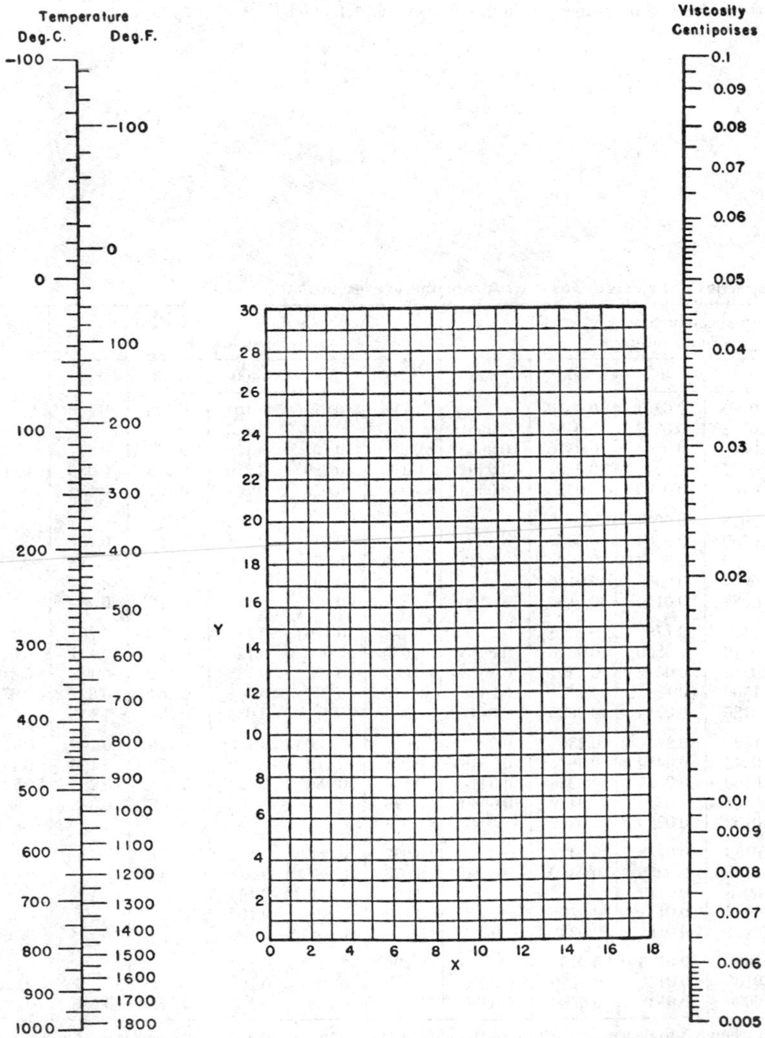

FIG. 3-41 Nomograph for viscosities of gases at 1 atm. For coordinates, see Table 3-310.

TABLE 3-311 Viscosities of Gases: Coordinates for Use with Fig. 3-42*

Gas	X	Y	$\mu \times 10^7$ p	Ref.
Acetic acid	7.0	14.6	825 (50°C)	1
Acetone	8.4	13.2	735	1
Acetylene	9.3	15.5	1017	1
Air	10.4	20.4	1812	1
Ammonia	8.4	16.0	1000	1
Amylene (β)	8.6	12.2	676	1
Argon	9.7	22.6	2215	1
Arsine	8.6	20.0	1576	1
Benzene	8.7	13.2	746	1
Bromine	8.8	19.4	1495	1
Butane (*n*)	8.6	13.2	735	1
Butane (iso)	8.6	13.2	744	1
Butyl acetate (iso)	5.7	16.3	778	1
Butylene (α)	8.4	13.5	761	1
Butylene (β)	8.7	13.1	746	1
Butylene (iso)	8.3	13.9	786	1
Butyl formate (iso)	6.6	16.0	840	1
Cadmium	7.8	22.5	5690 (500)	1
Carbon dioxide	8.9	19.1	1463	1
Carbon disulfide	8.5	15.8	990	1
Carbon monoxide	10.5	20.0	1749	1
Carbon oxysulfide	8.2	17.9	1220	1
Carbon tetrachloride	8.0	15.3	966	1
Chlorine	8.8	18.3	1335	1
Chloroform	8.8	15.7	1000	1
Cyanogen	8.2	16.2	1002	1
Cyclohexane	9.0	12.2	701	1
Cyclopropane	8.3	14.7	870	1
Deuterium	11.0	16.2	1240	1
Diethyl ether	8.8	12.7	730	1
Dimethyl ether	9.0	15.0	925	1
Diphenyl ether	8.6	10.4	610 (50)	1
Diphenyl methane	8.0	10.3	605 (50)	1
Ethane	9.0	14.5	915	1
Ethanol	8.2	14.5	835	1
Ethyl acetate	8.4	13.4	743	1
Ethyl chloride	8.5	15.6	978	1
Ethylene	9.5	15.2	1010	1
Ethyl propionate	12.0	12.4	890	1
Fluorine	7.3	23.8	2250	2
Freon-11	8.6	16.2	1298 (93)	3
Freon-12	9.0	17.4	1496 (93)	3
Freon-14	9.5	20.4	1716	5
Freon-21	9.0	16.7	1389 (93)	3
Freon-22	9.0	17.7	1554 (93)	3
Freon-113	11.0	14.0	1166 (93)	3
Freon-114	9.4	16.4	1364 (93)	3
Helium	11.3	20.8	1946	1
Heptane (*n*)	8.6	10.6	618 (50)	1
Hexane (*n*)	8.4	12.0	644	1
Hydrogen	11.3	12.4	880	1
Hydrogen-helium				1
10% H_2, 90% He	11.0	20.5	1780 (0)	
25% H_2, 75% He	11.0	19.4	1603 (0)	
40% H_2, 60% He	10.7	18.4	1431 (0)	
60% H_2, 40% He	10.8	16.7	1227 (0)	
81% H_2, 19% He	10.5	15.0	1016 (0)	
Hydrogen–sulfur dioxide				4
10% H_2, 90% SO_2	8.7	18.1	1259 (17)	
20% H_2, 80% SO_2	8.6	18.2	1277 (17)	
50% H_2, 50% SO_2	8.9	18.3	1332 (17)	
80% H_2, 20% SO_2	9.7	17.7	1306 (17)	
Hydrogen bromide	8.4	21.6	1843	1
Hydrogen chloride	8.5	19.2	1425	1
Hydrogen cyanide	7.1	14.5	737	1
Hydrogen iodide	8.5	21.5	1830	1
Hydrogen sulfide	8.4	18.0	1265	1
Iodine	8.7	18.7	1730 (100)	1
Krypton	9.4	24.0	2480	1
Mercury	7.4	24.9	4500 (200)	1
Mercuric bromide	8.5	19.0	2253	1
Mercuric chloride	7.7	18.7	2200 (200)	1
Mercuric iodide	8.4	18.0	2045 (200)	1
Mesitylene	9.5	10.2	660 (50)	1
Methane	9.5	15.8	1092	1
Methane (deuterated)	9.5	17.6	1290	1
Methanol	8.3	15.6	935	1
Methyl acetate	8.4	14.0	870 (50)	
Methyl acetylene	8.9	14.3	867	1
3-Methyl-1-butene	8.0	13.3	716	1
Methyl butyrate (iso)	6.6	15.8	824	1
Methyl bromide	8.1	18.7	1327	1
Methyl chloride	8.5	16.5	1062	1
3-Methylene-1-butene	8.0	13.3	716	1
Methylene chloride	8.5	15.8	989	1
Methyl formate	5.1	18.0	923	6
Neon	11.1	25.8	3113	1
Nitric oxide	10.4	20.8	1899	1
Nitrogen	10.6	20.0	1766	1
Nitrous oxide	9.0	19.0	1460	1
Nonane (*n*)	9.2	8.9	554 (50)	1
Octane (*n*)	8.8	9.8	586 (50)	1
Oxygen	10.2	21.6	2026	1
Pentane (*n*)	8.5	12.3	668	1
Pentane (iso)	8.9	12.1	685	1
Phosphene	8.8	17.0	1150	1
Propane	8.9	13.5	800	1
Propanol (*n*)	8.4	13.5	770	1
Propanol (iso)	8.4	13.6	774	1
Propyl acetate	8.0	14.3	797	1
Propylene	8.5	14.4	840	1
Pyridine	8.6	13.3	830 (50)	1
Silane	9.0	16.8	1148	1
Stannic chloride	9.1	16.0	1330 (100)	1
Stannic bromide	9.0	16.7	142 (100)	1
Sulfur dioxide	8.4	18.2	1250	1
Thiazole	10.0	14.4	958	1
Thiophene	8.3	14.2	901 (50)	1
Toluene	8.6	12.5	686	1
2,2,3-Trimethylbutane	10.0	10.4	691 (50)	1
Trimethylethane	8.0	13.0	686	1
Water	8.0	16.0	1250 (100)	1
Xenon	9.3	23.0	2255	
Zinc	8.0	22.0	5250 (500)	1

*Viscosity at 20°C unless otherwise indicated. From Beerman, *Meas. Control*, 154–157 (June 1982).

References:

1. I. F. Golubev, *Viscosity of Gases and Gas Mixtures*, Moscow 1959; transl. U.S. Department of Commerce, Clearinghouse for Federal Scientific and Technical Information, Springfield, Va., TT 70-50022, ISPT Cat. No. 5680, Table 4, Jerusalem 1970.
2. R. H. Perry and C. H. Chilton, *Chemical Engineers' Handbook*, 5th ed., McGraw-Hill, New York, 1973, pp. 3-210, 3-211.
3. Ibid., Table 3-282, p. 3-210.
4. By interpolation of data in Ref. 1.
5. *Thermophysical Properties of Refrigerants*, American Society of Heating, Refrigerating and Air-Conditioning Engineers, New York.
6. N. A. Lange, *Handbook of Chemistry*, 4th ed., Handbook Publishers, Sandusky, Ohio, 1941.

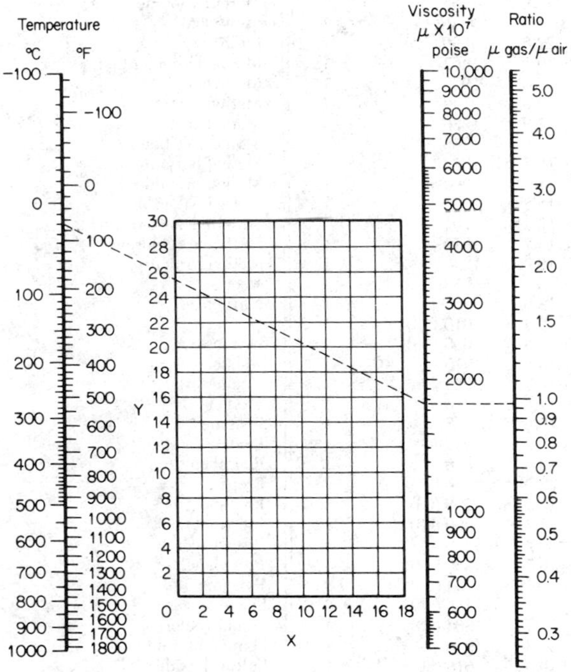

FIG. 3-42 Nomograph for determining (*a*) absolute viscosity of a gas as a function of temperature near ambient pressure and (*b*) relative viscosity of a gas compared with air. For coordinates see Table 3-311. To convert poises to pascal-seconds, multiply by 0.1. [*From Beerman,* Meas. Control, *154–157 (June 1982).*]

TABLE 3-312 Viscosities of Liquids: Coordinates for Use with Fig. 3-43

Liquid	X	Y
Acetaldehyde	15.2	4.8
Acetic acid, 100%	12.1	14.2
Acetic acid, 70%	9.5	17.0
Acetic anhydride	12.7	12.8
Acetone, 100%	14.5	7.2
Acetone, 35%	7.9	15.0
Acetonitrile	14.4	7.4
Acrylic acid	12.3	13.9
Allyl alcohol	10.2	14.3
Allyl bromide	14.4	9.6
Allyl iodide	14.0	11.7
Ammonia, 100%	12.6	2.0
Ammonia, 26%	10.1	13.9
Amyl acetate	11.8	12.5
Amyl alcohol	7.5	18.4
Aniline	8.1	18.7
Anisole	12.3	13.5
Arsenic trichloride	13.9	14.5
Benzene	12.5	10.9
Brine, $CaCl_2$, 25%	6.6	15.9
Brine, NaCl, 25%	10.2	16.6
Bromine	14.2	13.2
Bromotoluene	20.0	15.9
Butyl acetate	12.3	11.0
Butyl acrylate	11.5	12.6
Butyl alcohol	8.6	17.2
Butyric acid	12.1	15.3
Carbon dioxide	11.6	0.3
Carbon disulfide	16.1	7.5
Carbon tetrachloride	12.7	13.1
Chlorobenzene	12.3	12.4
Chloroform	14.4	10.2
Chlorosulfonic acid	11.2	18.1
Chlorotoluene, ortho	13.0	13.3
Chlorotoluene, meta	13.3	12.5
Chlorotoluene, para	13.3	12.5
Cresol, meta	2.5	20.8
Cyclohexanol	2.9	24.3
Cyclohexane	9.8	12.9
Dibromomethane	12.7	15.8
Dichloroethane	13.2	12.2
Dichloromethane	14.6	8.9
Diethyl ketone	13.5	9.2
Diethyl oxalate	11.0	16.4
Diethylene glycol	5.0	24.7
Diphenyl	12.0	18.3
Dipropyl ether	13.2	8.6
Dipropyl oxalate	10.3	17.7
Ethyl acetate	13.7	9.1
Ethyl acrylate	12.7	10.4
Ethyl alcohol, 100%	10.5	13.8
Ethyl alcohol, 95%	9.8	14.3
Ethyl alcohol, 40%	6.5	16.6
Ethyl benzene	13.2	11.5
Ethyl bromide	14.5	8.1
2-Ethyl butyl acrylate	11.2	14.0
Ethyl chloride	14.8	6.0
Ethyl ether	14.5	5.3
Ethyl formate	14.2	8.4
2-Ethyl hexyl acrylate	9.0	15.0
Ethyl iodide	14.7	10.3
Ethyl propionate	13.2	9.9
Ethyl propyl ether	14.0	7.0
Ethyl sulfide	13.8	8.9
Ethylene bromide	11.9	15.7
Ethylene chloride	12.7	12.2
Ethylene glycol	6.0	23.6
Ethylidene chloride	14.1	8.7
Fluorobenzene	13.7	10.4
Formic acid	10.7	15.8
Freon-11	14.4	9.0
Freon-12	16.8	5.6
Freon-21	15.7	7.5
Freon-22	17.2	4.7
Freon-113	12.5	11.4
Glycerol, 100%	2.0	30.0
Glycerol, 50%	6.9	19.6
Heptane	14.1	8.4
Hexane	14.7	7.0
Hydrochloric acid, 31.5%	13.0	16.6
Iodobenzene	12.8	15.9
Isobutyl alcohol	7.1	18.0
Isobutyric acid	12.2	14.4
Isopropyl alcohol	8.2	16.0
Isopropyl bromide	14.1	9.2
Isopropyl chloride	13.9	7.1
Isopropyl iodide	13.7	11.2
Kerosene	10.2	16.9
Linseed oil, raw	7.5	27.2
Mercury	18.4	16.4
Methanol, 100%	12.4	10.5
Methanol, 90%	12.3	11.8
Methanol, 40%	7.8	15.5
Methyl acetate	14.2	8.2
Methyl acrylate	13.0	9.5
Methyl *i*-butyrate	12.3	9.7
Methyl *n*-butyrate	13.2	10.3
Methyl chloride	15.0	3.8
Methyl ethyl ketone	13.9	8.6
Methyl formate	14.2	7.5
Methyl iodide	14.3	9.3
Methyl propionate	13.5	9.0
Methyl propyl ketone	14.3	9.5
Methyl sulfide	15.3	6.4
Napthalene	7.9	18.1
Nitric acid, 95%	12.8	13.8
Nitric acid, 60%	10.8	17.0
Nitrobenzene	10.6	16.2
Nitrogen dioxide	12.9	8.6
Nitrotoluene	11.0	17.0
Octane	13.7	10.0
Octyl alcohol	6.6	21.1
Pentachloroethane	10.9	17.3
Pentane	14.9	5.2
Phenol	6.9	20.8
Phosphorus tribromide	13.8	16.7
Phosphorus trichloride	16.2	10.9
Propionic acid	12.8	13.8
Propyl acetate	13.1	10.3
Propyl alcohol	9.1	16.5
Propyl bromide	14.5	9.6
Propyl chloride	14.4	7.5
Propyl formate	13.1	9.7
Propyl iodide	14.1	11.6
Sodium	16.4	13.9
Sodium hydroxide, 50%	3.2	25.8
Stannic chloride	13.5	12.8
Succinonitrile	10.1	20.8
Sulfur dioxide	15.2	7.1
Sulfuric acid, 110%	7.2	27.4
Sulfuric acid, 100%	8.0	25.1
Sulfuric acid, 98%	7.0	24.8
Sulfuric acid, 60%	10.2	21.3
Sulfuryl chloride	15.2	12.4
Tetrachloroethane	11.9	15.7
Thiophene	13.2	11.0
Titanium tetrachloride	14.4	12.3
Toluene	13.7	10.4
Trichloroethylene	14.8	10.5
Triethylene glycol	4.7	24.8
Turpentine	11.5	14.9
Vinyl acetate	14.0	8.8
Vinyl toluene	13.4	12.0
Water	10.2	13.0
Xylene, ortho	13.5	12.1
Xylene, meta	13.9	10.6
Xylene, para	13.9	10.9

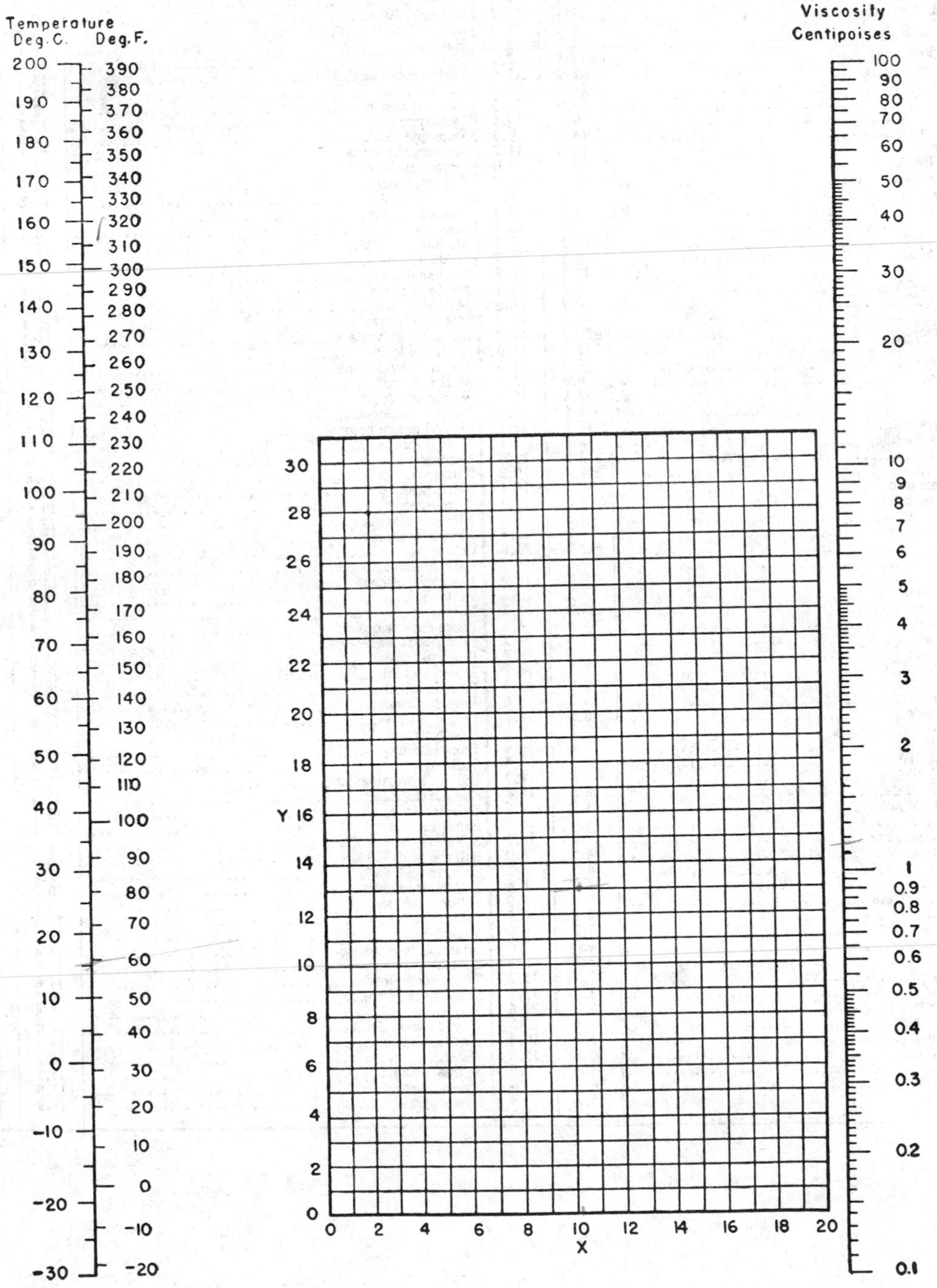

FIG. 3-43 Nomograph for viscosities of liquids at 1 atm. For coordinates see Table 3-312. To convert centipoises to pascal-seconds, multiply by 0.001.

TABLE 3-313 Thermal Conductivity of Liquids

$k = \mathrm{Btu/(h \cdot ft^2)(°F/ft)}$

A linear variation with temperature may be assumed. The extreme values given constitute also the tmperature limits over which the data are recommended.

Liquid	t, °F.	k
Acetic acid 100%[5]	68	0.099
50%[5]	68	.20
Acetone[4]	86	.102
	167	.095
Allyl alcohol[11]	77–86	.104
Ammonia[8]	5–86	.29
Ammonia, aqueous 26%[5]	68	.261
	140	.29
Amyl acetate[7]	50	.083
alcohol (n-)[6]	86	.094
	212	.089
(iso-)[12]	86	.088
	167	.087
Aniline[9]	32–68	.100
Benzene[12]	86	.092
	140	.087
Bromobenzene[12]	86	.074
	212	.070
Butyl acetate (n-)[11]	77–86	.085
alcohol (n-)[4]	86	.097
	167	.095
(iso-)[4]	50	.091
Calcium chloride brine 30%[5]	86	.32
15%[5]	86	.34
Carbon disulfide[4]	86	.093
	167	.088
tetrachloride[10]	32	.107
	154	.094
Chlorobenzene[12]	50	.083
Chloroform[10]	86	.080
Cymene (para-)[12]	86	.078
	140	.079
Decane (n-)[12]	86	.085
	140	.083
Dichlorodifluoromethane[12]	20	.057
	60	.053
	100	.048
	140	.043
	180	.038
Dichloroethane[10]	122	.082
Dichloromethane[10]	5	.111
	86	.096
Ethyl acetate[7]	68	.101
alcohol 100%[2]	68	.105
80%	68	.137
60%	68	.176
40%	68	.224
20%	68	.281
100%[2]	122	.087
benzene[12]	86	.086
	140	.082
bromide[4]	68	.070
ether[4]	86	.080
	167	.078
iodide[4,5]	104	.064
	167	.063
Ethylene glycol[7]	32	.153
Gasoline[6,12]	86	.078
Glycerol 100%[1]	68	.164
80%	68	.189
60%	68	.220
40%	68	.259
20%	68	.278
100%[1]	212	.164
Heptane (n-)[12]	86	.081
	140	.079

Liquid	t, °F.	k
Hexane (n-)[12]	86	0.080
	140	.078
Heptyl alcohol (n-)[6]	86	.094
	167	.091
Hexyl alcohol (n-)[6]	86	.093
	167	.090
Kerosene[4]	68	.086
	167	.081
Mercury[7]	82	4.83
Methyl alcohol 100%[2]	68	0.124
80%	68	.154
60%	68	.190
40%	68	.234
20%	68	.284
100%	122	.114
chloride[8,10]	5	.111
	86	.089
Nitrobenzene[12]	86	.095
	212	.088
Nitromethane[12]	86	.125
	140	.120
Nonane (n-)[12]	86	.084
	140	.082
Octane (n-)[12]	86	.083
	140	.081
Oils[5,12*]	86	.079
Oils, castor[9]	68	.104
	212	.100
Oils, olive[9]	68	.097
	212	.095
Paraldehyde[12]	86	.084
	212	.078
Pentane (n-)[12]	86	.078
	167	.074
Perchloroethylene[10]	122	.092
Petroleum ether[4]	86	.075
	167	.073
Propyl alcohol (n-)[6]	86	.099
	167	.095
alcohol (iso-)[12]	86	.091
	140	.090
Sodium	212	49
	410	46
Sodium chloride brine 25.0%[5]	86	0.33
12.5%[5]	86	.34
Sulfuric acid 90%[5]	86	.21
60	86	.25
30	86	.30
Sulfur dioxide[8]	5	.128
	86	.111
Toluene[4,12]	86	.086
	167	.084
β-Trichloroethane[10]	122	.077
Trichloroethylene[10]	122	.080
Turpentine[7]	59	.074
Vaseline[7]	59	.106
Water[13]	32	.343
	100	.363
	200	.393
	300	.395
	420	.376
	620	.275
Xylene (ortho-)[7]	68	.090
(meta-)[7]	68	.090

*Thermal conductivity data for a number of oils are available from Reference **12.** For many oils an average value of 0.079 may be used.

[1] Bates, *Ind. Eng. Chem.*, **28,** 494 (1936).
[2] Bates, Hazzard, and Palmer, *Ind. Eng. Chem.*, **10,** 314 (1938).
[3] Benning, A. F., private communication, 1940.
[4] Bridgman, *Proc. Am. Acad. Arts Sci.*, **59,** 141 (1923).
[5] Chilton and Genereaux, personal communication, 1939, based on data selected from the literature.
[6] Daniloff, *J. Am. Chem. Soc.*, **54,** 1328 (1932).
[7] "International Critical Tables," McGraw-Hill, New York, 1929.
[8] Kardos, Z. *Ver. deut. Ing.*, **77,** 1158 (1933); Z. *ges. Kälte-Ind.*, **41,** 1, 29 (1934).
[9] Kaye and Higgins, *Proc. Roy. Soc.* (*London*), **A117,** 459 (1928).
[10] DuPont Chlorinated Hydrocarbons, *Tech. Bull.*, Electrochemicals Dept., du Pont, Buffalo, N. Y., 1938.
[11] Shiba, *Sci. Papers Inst. Phys. Chem. Research* (*Tokyo*), **16,** 205 (1931).
[12] Smith, *Trans. Am. Soc. Mech. Engrs.*, **58,** 719 (1936).
[13] Timrot and Vargaftik, *J. Tech. Phys.* (*U.S.S.R.*), **10,** 1063 (1940).

NOTE: To convert British thermal unit–feet per hour–square foot–degree Fahrenheit to watts per meter-kelvin, multiply by 1.7307.

TABLE 3-314 Thermal Conductivity of Gases*

To obtain conductivity in watts per meter-kelvin, multiply by 10^{-2}.

Temp., °K.	Substance														
	Air	NH_3	Ar	CCl_4	CO_2	C_2H_6	He	H_2	Kr	CH_4	Ne	N_2	O_2	H_2O	Xe
100	0.93	. . .	0.66	. . .	. . .	. . .	7.2	6.7	. . .	1.08	2.19	0.96	0.93		
150	1.38	. . .	0.96	. . .	. . .	. . .	9.5	10.1	0.50	1.84	3.04	1.39	1.38		
200	1.80	1.53	1.26	. . .	0.94	. . .	11.5	13.1	0.68	2.17	3.62	1.83	1.83	. . .	0.38
250	2.21	1.96	1.52	. . .	1.30	. . .	13.4	15.7	0.80	2.75	4.29	2.22	2.26	. . .	0.48
300	2.62	2.47	1.77	0.69	1.66	2.15	15.1	18.3	1.00	3.42	4.89	2.59	2.66	61	0.58
350	3.00	3.04	2.00	0.85	2.04	2.84	16.6	20.4	1.13	4.00	5.46	2.93	2.98	67	0.66
400	3.38	3.70	2.22	1.01	2.43	3.56	18.4	22.5	1.26	4.93	6.01	3.27	3.30	2.66	0.74
450	3.73	4.40	2.44	1.16	2.83	4.36	20.1	24.7	1.38	5.79	6.53	3.59	3.63	3.10	0.82
500	4.07	5.25	2.66	1.30	3.25	. . .	21.8	26.6	1.51	6.68	7.03	3.89	4.12	3.58	0.90
600	4.69	6.70	3.07	1.44	4.07	. . .	25.0	30.5	1.75	8.52	7.97	4.46	4.73	4.63	1.05
700	5.24	. . .	3.41	1.58	4.81	. . .	27.8	34.2	1.98	10.46	8.86	4.98	5.28	5.81	1.20
800	5.73	. . .	3.74	. . .	5.51	. . .	30.4	37.8	2.21	. . .	9.71	5.48	5.89	7.08	1.35
900	6.20	. . .	4.06	. . .	6.18	. . .	33.0	41.2	2.42	. . .	10.53	5.97	6.49	8.41	1.49
1000	6.67	. . .	4.4	. . .	6.82	. . .	35.4	44.8	2.62	. . .	11.34	6.47	7.10	9.78	1.64
1200	7.63	. . .	4.9	. . .	8.0	. . .	40.5	52.8	2.98	. . .	12.16	7.6	8.2	. . .	1.95

*Condensed from Vargaftik et al., *Heat Conductivity of Gases and Liquids*, FTD-MT-24-133-71 (AD 736963), 1971. The Russian original appeared in 1970. This source contains 256 references.

TABLE 3-315 Viscosity of Sucrose Solutions*

Viscosity in centipoises

Temp., °C.	Percentage sucrose by weight			Temp., °C.	Percentage sucrose by weight		
	20	40	60		20	40	60
0	3.818	14.82		50	0.974	2.506	14.06
5	3.166	11.60		55	0.887	2.227	11.71
10	2.662	9.830	113.9	60	0.811	1.989	9.87
15	2.275	7.496	74.9	65	0.745	1.785	8.37
20	1.967	6.223	56.7	70	0.688	1.614	7.18
25	1.710	5.206	44.02	75	0.637	1.467	6.22
30	1.510	4.398	34.01	80	0.592	1.339	5.42
35	1.336	3.776	26.62	85	0.552	1.226	4.75
40	1.197	3.261	21.30	90	. . . . .	1.127	4.17
45	1.074	2.858	17.24	95	. . . . .	1.041	3.73

*"International Critical Tables," vol. 5, p. 23. Bingham and Jackson, *Bur. Standards Bull.* **14**, p. 59, 1919.

TABLE 3-316 Prandtl Number of Air*

Temperature, K	Pressure, bar											
	1	5	10	20	30	40	50	60	70	80	90	100
80	mix	2.31	2.32	2.35	2.37	2.40	2.42	2.45	2.48	2.51	2.54	2.57
90	0.796	1.76	1.77	1.78	1.79	1.81	1.82	1.83	1.85	1.87	1.89	1.91
100	0.786	0.872	1.54	1.53	1.53	1.53	1.53	1.53	1.53	1.54	1.54	1.55
120	0.773	0.813	0.89	1.44	1.65	1.54	1.48	1.43	1.40	1.38	1.36	1.34
140	0.763	0.782	0.82	0.94	1.20	1.59	2.14	2.43	2.07	1.78	1.62	1.52
160	0.754	0.765	0.78	0.84	0.92	1.03	1.13	1.25	1.37	1.65	1.83	1.72
180	0.745	0.754	0.763	0.792	0.830	0.876	0.932	1.00	1.07	1.14	1.20	1.25
200	0.738	0.743	0.749	0.766	0.788	0.812	0.841	0.87	0.90	0.95	0.97	1.00
240	0.724	0.727	0.729	0.737	0.746	0.756	0.767	0.78	0.80	0.81	0.81	0.82
280	0.710	0.711	0.713	0.717	0.721	0.726	0.731	0.737	0.742	0.75	0.75	0.76
300	0.705	0.707	0.708	0.712	0.715	0.717	0.721	0.725	0.728	0.732	0.737	0.742
350	0.699	0.699	0.699	0.701	0.703	0.705	0.707	0.709	0.711	0.712	0.714	0.716
400	0.694	0.694	0.694	0.695	0.696	0.697	0.698	0.699	0.700	0.701	0.703	0.704
450	0.691	0.691	0.691	0.691	0.692	0.692	0.693	0.693	0.694	0.695	0.695	0.696
500	0.689	0.689	0.689	0.689	0.689	0.690	0.690	0.690	0.690	0.691	0.691	0.691
600	0.690	0.690	0.690	0.689	0.689	0.689	0.689	0.689	0.689	0.690	0.690	0.690
700	0.696	0.696	0.695	0.695	0.695	0.695	0.695	0.695	0.695	0.695	0.695	0.695
800	0.705	0.704	0.704	0.704	0.704	0.703	0.703	0.703	0.703	0.702	0.702	0.702
900	0.709	0.709	0.708	0.708	0.708	0.708	0.708	0.708	0.708	0.708	0.708	0.708
1000	0.711	0.711	0.711	0.711	0.711	0.710	0.710	0.710	0.710	0.709	0.709	0.709

*Compiled by P. E. Liley from tables of specific heat at constant pressure, thermal conductivity, and viscosity given in SI units for integral kelvin temperatures and pressures in bars by Vasserman, *Thermophysical Properties of Air and Its Components* and *Thermophysical Properties of Liquid Air and Its Components*, Nauka, Moscow, and in translated form by the National Bureau of Standards, Washington. The number of significant figures given above reflects the similar numbers appearing for the constituent properties in the source references. While reasonable agreement occurs for atmospheric pressure with some other works, the fragmentary data available for the saturated, etc., states show large deviations.

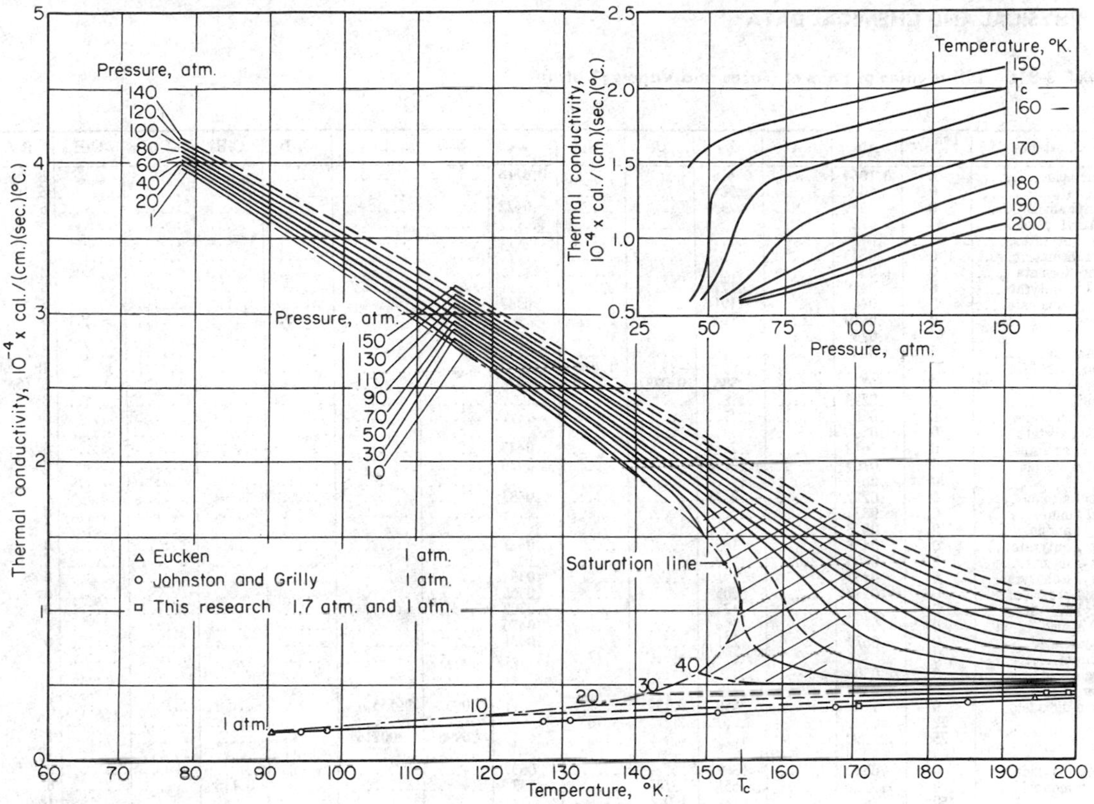

Fig. 3-44 Thermal conductivity of oxygen. To convert calories per centimeter-second-degree Celsius to British thermal unit-feet per hour-square foot-degree Fahrenheit, multiply by 241.75. [*From Ziebland and Burton,* Br. J. Appl. Phys., *6, 416 (1955). Crown copyright reserved. Reproduced by permission of the publishers and of the Controller, H. M. Stationery Office, London, England.*]

TABLE 3-317 Prandtl Number of Liquid Refrigerants*

Refrigerant	No.	180	200	220	240	260	280	300	320	340	360	380
		Temperature, K										
Trichlorofluoromethane	11		11.9	8.64	6.73	5.33	4.74	4.18				
Dichlorodifluoromethane	12	7.00	5.25	4.27	3.65	3.27	3.08	3.04	3.19	3.44	4.00	
Chlorotrifluoromethane	13		2.96	2.67	2.69	3.05	3.57		—	—	—	—
Bromotrifluoromethane	13B1	4.80	3.75	3.27	2.94	2.83	3.03	3.61	4.52		—	—
Dichlorofluoromethane	21			5.72	4.50	3.87	3.48	3.25	3.16	3.17		
Chlorodifluoromethane	22	4.68	3.76	3.23	2.93	2.79	2.77	2.87	3.18	3.54		—
Methyl chloride	40			2.53	2.42	2.40	2.45	2.60	2.85			
Trichlorotrifluoroethane	113	—	—	—					7.04	6.23	5.61	5.18
Dichlorotetrafluoroethane	114	25.7	15.13	11.18	8.59	6.94	5.77	5.06	4.78	4.82		
Chloropentafluoroethane	115	—	7.85	6.16	5.21	4.67	4.40	4.46	4.90		—	—
Ethane	170	2.55	2.29	2.22	2.40	2.70			—	—	—	—
Propane	290	5.28	4.46	3.88	3.44	3.16	3.02	3.16				
Octafluorocyclobutane	C318	—	—	—	11.2	8.74	7.35	6.37	5.87	5.96		—
Dichlorodifluoromethane/difluoroethane	500		5.78	4.23	3.40	3.13	3.01	3.13	3.35	3.72		—
Chlorodifluoromethane/chloropentafluoroethane	502		5.73	4.71	4.13	3.81					—	—
Trifluoromethane/chlorotrifluoromethane	503	2.10	2.09	2.24	2.43	2.89		—	—	—	—	—
Methylene fluoride/chloropentafluoroethane	504		4.90	3.60	3.04	2.79	2.69	2.85	3.30	—	—	—
Butane	600	8.35	6.19	5.20	4.44	3.83	3.44	3.22	3.07	3.02		
Isobutane (2-methyl propane)	600a		8.26	6.36	5.18	4.49	3.93	3.66	3.53	3.53	3.77	4.68
Ammonia	717	—			1.97	1.76	1.54	1.40	1.29	1.24	1.25	1.34
Water	718	—	—	—	—	—	10.3	5.69	3.65	2.60	1.99	1.59
Ethylene	1150	1.85	1.74	1.78	2.07	2.70	4.4	—	—	—	—	—
Propylene	1270	3.80	2.24	1.88	1.71	1.71	1.88	2.24	3.91	4.73		—

*Dashes indicate inaccessible states. Average uncertainty is about 20 percent. Values derived from formulations for thermal conductivity, specific heat at constant pressure, and viscosity contained in *Thermophysical Properties of Refrigerants*, American Society of Heating, Refrigerating and Air-Conditioning Engineers, New York, 1976. For further details see M. W. Johnson, M.S.M.E. thesis, Purdue University, West Lafayette, Ind., 1976.

TABLE 3-318 Diffusivities of Pairs of Gases and Vapors (1 atm)

D_v in cm²/s

Substance	Temp., °C.	Air	A	H_2	O_2	N_2	CO_2	N_2O	CH_4	C_2H_6	C_2H_4	n-C_4H_{10}	i-C_4H_{10}	Ref.
Acetic acid	0	0.1064		0.416			0.0716							8
Acetone	0	.109		.361										6, 16
n-Amyl alcohol	0	.0589		.235			.0422							8
sec-Amyl alcohol	30	.072												5
Amyl butyrate	0	.040												8
Amyl formate	0	.0543												8
i-Amyl formate	0	.058												8
Amyl isobutyrate	0	.0419		.171										8
Amyl propionate	0	.046		.1914			.0347							8
Aniline	0	.0610												8
	30	.075												5
Anthracene	0	.0421												8
Argon	20					0.194								18
Benzene	0	.077		.306	0.0797		.0528							8, 15
Benzidine	0	.0298												8
Benzyl chloride	0	.066												8
n-Butyl acetate	0	.058												8
i-Butyl acetate	0	.0612		.2364			.0425							8
n-Butyl alcohol	0	.0703		.2716			.0476							8
	30	.088												5
i-Butyl alcohol	0	.0727		.2771			.0483							8
Butyl amine	0	.0821												8
i-Butyl amine	0	.0853												8
i-Butyl butyrate	0	.0468		.185			.0327							8
i-Butyl formate	0	.0705												8
i-Butyl isobutyrate	0	.0457		.191			.0364							8
i-Butyl proprionate	0	.0529		.203			.0366							8
i-Butyl valerate	0	.0424		.173			.0308							8
Butyric acid	0	.067		.264			.0476							8
i-Butyric acid	0	.0679		.271			.0471							8
Cadmium	0					.17								13
Caproic acid	0	.050												8
i-Caproic acid	0	.0513												8
Carbon dioxide	0	.138		.550	.139			0.096	0.153					8
	20					.163								19
	25							.0996*	.00215†					1, 9
	500‡				.9									18
Carbon disulfide	0	.0892		.369			.063							8
Carbon monoxide	0			.651	.185		.137				0.116			8
	450‡				1.0									18
Carbon tetrachloride	0			.293	0.0636									16, 17
Chlorobenzene	30	.075												5
Chloroform	0	.091												6
Chloropicrin	25	.088												10
m-Chlorotoluene	0	.054												8
o-Chlorotoluene	0	.059												8
p-Chlorotoluene	0	.051												8
Cyanogen chloride	0	.111												10
Cyclohexane	15		0.0719	.319	.0744	.0760								3
	45	.086												6
n-Decane	90			.306		.0841								3
Diethylamine	0	.0884												8
2,3-Dimethyl butane	15		.0657	.301	.0753	.0751								3
Diphenyl	0	.0610												8
n-Dodecane	126			.308		.0813								3
Ethane	0			.459										8
Ethanol	0			.377			.0686							20
Ether (diethyl)	0	.0778		.298			.0546							7, 8
Ethyl acetate	0	.0715		.273			.0487							8
	30	.089												5
Ethyl alcohol	0	.102		.375			.0685							8
Ethyl benzene	0	.0658												8
Ethyl *n*-butyrate	0	.0579		.224			.0407							8
Ethyl *i*-butyrate	0	.0591		.229			.0413							8
Ethylene	0			.486										8
Ethyl formate	0	.0840		.337			.0573							8
Ethyl propionate	0	.068		.236			.0450							4, 8
Ethyl valerate	0	.0512		.205			.0367							8
Eugenol	0	.0377												8
Formic acid	0	.1308		.510			.0874							8
Helium	0		.641											8
	20					.705								19
n-Heptane	38								.066§					
n-Hexane	15		.0663	.290	.0753	.0757								3
Hexyl alcohol	0	.0499		.200			.0351							8
Hydrogen	0	.611			.697	.674	.550	.535	.625	0.459	0.486	0.272	0.277	8
	25						.646			.537	.726			2
	500				4.2									18
Hydrogen cyanide	0	.173												10
Hydrogen peroxide	60	.188												11
Iodine	0	.07				.070								8, 12, 14
Mercury	0	.112		.53		.13								8, 12, 13
Mesitylene	0	.056												8
Methane	500				1.1									18
Methyl acetate	0	.084		.333			.0567							8
Methyl alcohol	0	.132		.506			.0879							8
Methyl butyrate	0	.0633		.242			.0446							8
Methyl *i*-butyrate	0	.0639		.257			.0451							8
Methyl cyclopentane	15		.0731	.318	0.0742	.0758								3
Methyl formate	0	.0872												8
Methyl propionate	0	.0735		.295			.0528							8

TABLE 3-318 Diffusivities of Pairs of Gases and Vapors (1 atm.) (*Continued*)

Substance	Temp., °C.	Air	A	H_2	O_2	N_2	CO_2	N_2O	CH_4	C_2H_6	C_2H_4	n-C_4H_{10}	i-C_4H_{10}	Ref.
Methyl valerate	0	0.0569												8
Naphthalene	0	.0513												8
Nitrogen	0				0.181									8
	25						0.165			0.148	0.163	0.0960	0.0908	2
Nitrous oxide	0			0.535			.096							8
n-Octane	0	.0505												8
	30		0.0642	.271	.0705	0.0710								3
Oxygen	0	.178		.697		.181	.139							8
Phosgene	0	.095												10
Propionic acid	0	.0829		.330			.0588							8
Propyl acetate	0	.067												8
n-Propyl alcohol	0	.085		.315			.0577							8
i-Propyl alcohol	0	.0818												8
	30	.101												5
n-Propyl benzene	0	.0481												8
i-Propyl benzene	0	.0489												8
n-Propyl bromide	0	.085												8
i-Propyl bromide	0	.0902												8
Propyl butyrate	0	.0530		.206			.0364							8
Propyl formate	0	.0712		.281			.0490							8
n-Propyl iodide	0	.079												8
i-Propyl iodide	0	.0802												8
n-Propyl isobutyrate	0	.0549		.212			.0388							8
i-Propyl isobutyrate	0	.059												8
Propyl propionate	0	.057		.212			.0395							8
Propyl valerate	0	.0466		.189			.0341							8
Safrol	0	.0434												8
i-Safrol	0	.0455												8
Sulfur hexafluoride	25			.418										2
Toluene	0	.076	.071											4, 8
	30	.088												5
Trimethyl carbinol	0	.087												8
2,2,4-Trimethyl pentane	30		.0618	.288	.0688	.0705								3
2,2,3-Trimethyl heptane	90			.270		.0684								3
n-Valeric acid	0	.050												8
i-Valeric acid	0	.0544		.212			.0376							8
Water	0	.220		.75			.138							8, 20
	450				1.3									18

*320 mm. Hg.
†40 atm.
‡Also at other temperatures.
§Strong function of concentration.

References for Table 3-299

[1] Amdur, Irvine, Mason, and Ross, *J. Chem. Phys.*, **20**, 436 (1952).
[2] Boyd, Stein, Steingrimsson, and Rumpel, *J. Chem. Phys.*, **19**, 548 (1951).
[3] Cummings and Ubbelohde, *J. Chem. Soc.* (*London*), 1953, p. 3751.
[4] Fairbanks and Wilke, *Ind. Eng. Chem.*, **42**, 471 (1950).
[5] Gilliland, *Ind. Eng. Chem.*, **26**, 681 (1934).
[6] Gorynnova and Kuvskinskii, *Zhur. Tekh. Fiz.*, **18**, 1421 (1948).
[7] Hansen, Dissertation, Jena, 1907.
[8] "International Critical Tables," vol. 5, p. 62.
[9] Jeffries and Drickamer, *J. Chem. Phys.*, **22**, 436 (1954).
[10] Klotz and Miller, *J. Am. Chem. Soc.*, **69**, 2557 (1947).
[11] McMurtrie and Keyes, *J. Am. Chem. Soc.*, **70**, 3755 (1948).
[12] Mullaly and Jacques, *Phil. Mag.*, **48** (6), 1105 (1924).
[13] Spier, *Physica*, **6**, 453 (1939); **7**, 381 (1940).
[14] Topley and Whytlaw-Gray, *Phil. Mag.*, **4**, 873 (1927).
[15] Trautz and Ludwig, *Ann. Physik*, **5** (5), 887 (1930).
[16] Trautz and Muller, *Ann. Physik*, **22**, 353 (1935).
[17] Trautz and Ries, *Ann. Physik*, **8**, 163 (1931).
[18] Walker and Westenberg, *J. Chem. Phys.*, **32**, 436 (1960).
[19] Westenberg and Walker, *J. Chem. Phys.*, **26**, 1753 (1957).
[20] Winkelmann, *Wied. Ann.*, **22**, 1, 152 (1884); **23**, 203 (1884); **26**, 105 (1885); **33**, 445 (1888); **36**, 92 (1889).

In this table are a representative selection of diffusion coefficients. The subsection "Prediction and Correlation of Physical Properties" should be consulted for estimation techniques. As general references, the works by Hirschfelder, Curtiss, and Bird, *Molecular Theory of Gases and Liquids*, Wiley, New York, 1964; Chapman and Cowling, *The Mathematical Theory of Non-Uniform Gases*, Cambridge, New York, 1970; Reid and Sherwood, *The Properties of Gases and Liquids*, McGraw-Hill, New York, 1964; and Bretsznajder, *Prediction of Transport and Other Physical Properties of Fluids*, Pergamon, New York, 1971, may be found useful. The most exhaustive recent compilation for gases is by Mason and Marrero, *J. Phys. Chem. Ref. Data*, **1** (1972). Unfortunately, the Mason and Marrero work cites only equations and equation constants and not direct tabulations. For these, the Landolt-Börnstein series is suggested.

TABLE 3-319 Diffusivities in Liquids (25°C)

Dilute solutions and 1 atm unless otherwise noted; use $D_L\mu/T$ = constant to estimate effect of temperature; * indicates that reference gives effect of concentration.

Solute	Solvent	$D_L \times 10^5$, sq. cm./sec.	Estimated possible, error, ± %	Ref.
Acetal*	Ethanol	1.25	5	11
Acetamide*	Ethanol	0.68	5	11
Acetamide*	Water	1.19	3	11
Acetic acid	Acetone	3.31	...	4
Acetic acid	Benzene	2.11	...	1, 4
Acetic acid	Carbon tetrachloride	1.49	...	4
Acetic acid	Ethylene glycol	0.13	...	4
Acetic acid	Toluene	2.26	...	4
Acetic acid*	Water	1.24	3	11
Acetonitrile	Water	1.66	5	11
Acetylene	Water	1.78, 2.11	...	1, 24
Allyl alcohol*	Ethanol	1.06	5	11
Allyl alcohol	Water	1.19	6	11
Ammonia*	Water	1.7, 2.0, 2.3	...	1, 11
i-Amyl alcohol*	Ethanol	0.87	5	11
i-Amyl alcohol	Water	1.0	8	11, 25
Benzene	Carbon tetrachloride	1.53	...	7
Benzene (50 mole %)	*n*-Decane	1.72	...	26
Benzene (50 mole %)	2,4-Dimethyl pentane	2.49	...	26
Benzene (50 mole %)	*n*-Dodecane	1.40	...	26
Benzene (50 mole %)	*n*-Heptane	2.47	...	26
Benzene (50 mole %)	*n*-Hexadecane	0.96	...	26
Benzene (50 mole %)	*n*-Octadecane	0.86	...	26
Benzoic acid	Acetone	2.62	...	4
Benzoic acid	Benzene	1.38	...	4
Benzoic acid	Carbon tetrachloride	0.91	...	4
Benzoic acid	Ethylene glycol	0.043	...	4
Benzoic acid	Toluene	1.49	...	4
Bromine	Benzene	2.7	...	11
Bromine	Carbon disulfide	4.1	...	11
Bromine	Water	1.3	...	11
Bromobenzene	Benzene	2.30	...	25
Bromoform*	Acetone	2.90	...	11
Bromoform	*i*-Amyl alcohol	0.53	...	11
Bromoform	Ethanol	1.08	5	11
Bromoform*	Ethyl ether	3.62	...	11
Bromoform	Methanol	2.20	...	23
Bromoform	*n*-Propanol	0.94	...	11
n-Butanol	Water	0.96	5	1, 11, 18, 25
Caffeine	Water	0.63	6	11
Carbon dioxide	Ethanol	4.0	6	11
Carbon dioxide	Water	1.96	1	1, 3, 5, 20, 24, 28
Carbon disulfide (50 mole %, 200 atm.)	*n*-Butanol	3.57	...	14
Carbon disulfide (50 mole %, 200 atm.)	*i*-Butanol	2.42	...	14
Carbon disulfide (50 mole %, 218 atm.)	Chlorobenzene	3.00	...	14
Carbon disulfide (50 mole %, 200 atm.)	2,4-Dimethyl pentane	3.63	...	14
Carbon disulfide (50 mole %, 100 atm.)	*n*-Heptane	3.0	...	14
Carbon disulfide (50 mole %, 50 atm.)	Methyl cyclohexane	3.5	...	14
Carbon disulfide (50 mole %, 200 atm.)	*n*-Octane	3.10	...	14
Carbon disulfide (50 mole %)	Toluene	2.06	...	14
Carbon tetrachloride	Benzene	2.04	3	7, 9
Carbon tetrachloride*	Cyclohexane	1.49	2	9, 10*
Carbon tetrachloride	Decalin	0.776	2	9
Carbon tetrachloride	Dioxane	1.02	2	9
Carbon tetrachloride*	Ethanol	1.50	2	9, 10*
Carbon tetrachloride	*n*-Heptane	3.17	2	9
Carbon tetrachloride	Kerosene	0.961	2	9
Carbon tetrachloride	Methanol	2.30	2	9
Carbon tetrachloride	*i*-Octane	2.57	2	9
Carbon tetrachloride	Tetralin	0.735	2	9
Chloral*	Ethanol	0.68	5	11
Chloral hydrate	Water	0.77	7	11
Chlorine	Water	1.44	4	1, 28
Chlorobenzene	Benzene	2.66	...	25
Chloroform	Benzene	2.50	6	1, 25
Chloroform	Ethanol	1.38	3	11
Cinnamic acid	Acetone	2.41	...	4
Cinnamic acid	Benzene	1.12	...	4
Cinnamic acid	Carbon tetrachloride	0.76	...	4
Cinnamic acid	Toluene	2.41	...	4
1,1'-Dichloropropanol	Water	1.0	6	11
Dicyanodiamide*	Water	1.18	4	11
Diethyl ether	Benzene	2.73	...	25
Diethyl ether	Water	0.85	...	2
2,4-Dimethyl pentane (50 mole %)	*n*-Dodecane	1.44	...	26
2,4-Dimethyl pentane (50 mole %)	*n*-Hexadecane	0.88	...	26
Ethanol*	Water	1.28	4	1, 7, 9,* 11,* 22
Ethyl acetate	Ethyl benzoate	0.94	...	6
Ethylene dichloride	Benzene	2.8	...	1, 25
Formic acid	Acetone	3.77	...	4
Formic acid	Benzene	2.28	...	4
Formic acid	Carbon tetrachloride	1.89	...	4
Formic acid	Ethylene glycol	0.094	...	4
Formic acid	Toluene	2.65	...	
Formic acid	Water	1.37	10	11
Glucose	Water	0.69	6	11
Glycerol	*i*-Amyl alcohol	0.12	...	11

TABLE 3-319 Diffusivities in Liquids (25°C) (*Continued*)

Solute	Solvent	$D_L \times 10^5$, sq. cm./sec.	Estimated possible error, ± %	Ref.
Glycerol	Ethanol	0.56	...	11
Glycerol*	Water	0.94	6	1, 11*
n-Heptane (50 mole %)	*n*-Dodecane	1.58	...	26
n-Heptane (50 mole %)	*n*-Hexadecane	1.00	...	26
n-Heptane (50 mole %)	*n*-Octadecane	0.92	...	26
n-Heptane (50 mole %)	*n*-Tetradecane	1.29	...	26
Hexamethylene tetramine	Water	0.67	...	11
Hydrogen chloride*	Water	3.10	3	4, 11,* 12*
Hydrogen	Water	5.85 (4.4?)	...	1, 11, 24(?)
Hydrogen sulfide	Water	1.61	...	1
Hydroquinone*	Ethanol	0.53	5	11
Hydroquinone*	Water	0.88, 1.12	...	2, 11*
Iodine	Acetic acid	1.13	...	11
Iodine	Anisole	1.25	...	11
Iodine	Benzene	1.98	...	9, 19, 23
Iodine	Bromobenzene	1.25	10	4, 11, 19
Iodine	Carbon disulfide	3.2	...	11, 19, 23
Iodine	Carbon tetrachloride	1.45	8	9, 11, 19
Iodine	Chloroform	2.30	3	11, 23
Iodine	Cyclohexane	1.80	...	4
Iodine	Dioxane	1.07	...	9
Iodine*	Ethanol	1.30	...	4, 11*
Iodine	Ethyl acetate	2.2	...	11, 19
Iodine	Ethyl ether	3.61	...	11
Iodine	Ethylene bromide	0.93	...	11
Iodine	*n*-Heptane	3.4, 2.5	...	9, 11, 19
Iodine	*n*-Hexane	4.15	...	4, 9
Iodine	Mesitylene	1.49	...	9
Iodine	Methanol	1.74	...	19
Iodine	Methyl cyclohexane	2.1	...	4
Iodine	*n*-Octane	2.76	...	4
Iodine	Tetrabromoethane	2.0	...	11
Iodine	*n*-Tetradecane	0.96	...	4
Iodine	Toluene	2.1	...	11
Iodine	*m*-Xylene	1.82	...	9, 11
Iodobenzene	Ethanol	1.09	3	11
Lactose*	Water	0.49	5	11
Maltose*	Water	0.48	5	11
Mannitol*	Water	0.65	5	11
Methanol	Water	1.6	...	1, 7, 11
Micotine*	Water	0.60	8	11
Nitric acid*	Water	2.98	2	11
Nitrobenzene	Carbon tetrachloride	1.00	...	7
Nitrogen	Water	1.9	...	1, 24
Nitrous oxide	Water	1.8	...	1, 11
Oxalic acid*	Water	1.61	2	11
Oxygen	Glycerol*-water (106 poise)	0.24	...	13
Oxygen	Sucrose*-water (125 poise)	0.25	...	13
Oxygen	Water	2.5	20	1, 3, 15, 21, 24
Pentaerythritol*	Water	0.77	4	11
Phenol	*i*-Amyl alcohol	0.2	...	11
Phenol	Benzene	1.68	...	1
Phenol	Carbon disulfide	3.7	...	11
Phenol	Chloroform	2.0	...	11
Phenol	Ethanol	0.89	...	11
Phenol	Ethyl ether	3.9	...	11
n-Propanol	Water	1.1	...	1, 7, 11
Pyridine*	Ethanol	1.24	3	11
Pyridine	Water	0.76	7	11
Pyrogallol	Water	0.74	7	11
Raffinose*	Water	0.41	4	11
Resorcinol*	Ethanol	0.46	5	11
Resorcinol*	Water	0.87	4	11
Saccharose*	Water	0.49	4	11
Stearic acid*	Ethanol	0.65	5	11
Succinic acid*	Water	0.94	...	11
Sucrose	Water	0.56	6	2, 27
Sulfur dioxide	Water	1.7	...	15, 17
Sulfuric acid*	Water	1.97	3	11
Tartaric acid*	Water	0.80	10	11
1,1,2,2-Tetrabromoethane	1,1,2,2-Tetrachloroethane	0.61	4	11
Toluene	*n*-Decane	2.09	...	4
Toluene	*n*-Dodecane	1.38	...	4
Toluene	*n*-Heptane	3.72	...	4
Toluene	*n*-Hexane	4.21	...	4
Toluene	*n*-Tetradecane	1.02	...	4
Urea	Ethanol	0.73	...	11
Urea	Water	1.37	2	8, 11
Urethane	Water	1.06	...	11, 25
Water	Glycerol	0.021	...	16

References for Table 3-319

[1]Arnold, *J. Am. Chem. Soc.*, **52,** 3937 (1930).
[2]Calvet, *J. Chim. Phys.*, **44,** 47 (1947).
[3]Carlson, *J. Am. Chem. Soc.*, **33,** 1027 (1911).
[4]Chang and Wilke, *J. Phys. Chem.*, **59,** 592 (1955).
[5]Davidson and Cullen, *Trans. Inst. Chem. Eng.*, **35,** 51 (1957).
[6]Dummer, Z. *Anorg. Chem.*, **109,** 31 (1949).
[7]Gerlach, *Ann. Phys. (Leipzig)*, **10,** 437 (1931).
[8]Gosting and Akeley, *J. Am. Chem. Soc.*, **74,** 2058 (1952).
[9]Hammond and Stokes, *Trans. Faraday Soc.*, **49,** 890 (1953); **49,** 886 (1953).
[10]Hammond and Stokes, *Trans. Faraday Soc.*, **52,** 781 (1956).
[11]*International Critical Tables*, vol. 5, p. 63.
[12]James, Hollingshead, and Gordon, *J. Chem. Phys.*, **7,** 89 (1939); **7,** 836 (1939).
[13]Jordon, Ackermann, and Berger, J. *Am. Chem. Phys.*, **78,** 2979 (1956).
[14]Koeller and Drickamer, *J. Chem. Phys.*, **21,** 575 (1953.)
[15]Kolthoff and Miller, *J. Am. Chem. Soc.*, **63,** 1013 (1941).

TABLE 3-320 Thermal Conductivities of Some Building and Insulating Materials*

$k = \text{Btu}/(\text{h}\cdot\text{ft}^2)(°\text{F}/\text{ft})$

Material	Apparent density ρ, lb./cu. ft. at room temperature	t, °C.	k
Aerogel, silica, opacified	8.5	120	0.013
		290	.026
Asbestos-cement boards	120	20	.43
Asbestos sheets	55.5	51	.096
Asbestos slate	112	0	.087
	112	60	.114
Asbestos	29.3	−200	.043
	29.3	0	.090
	36	0	.087
	36	100	.111
	36	200	.120
	36	400	.129
	43.5	−200	.090
	43.5	0	.135
Aluminum foil (7 air spaces per 2.5 in.)	0.2	38	.025
		177	.038
Ashes, wood	......	0–100	.041
Asphalt	132	20	.43
Boiler scale (Note 1)			
Bricks:			
Alumina (92–99% Al_2O_3 by wt.) fused	......	427	1.8
Alumina (64–65% Al_2O_3 by wt.)	......	1315	2.7
(See also Bricks, fire clay)	115	800	0.62
	115	1100	.63
Building brick work	......	20	.4
Carbon	96.7	....	3.0
Chrome brick (32% Cr_2O_3 by wt.)	200	200	.67
	200	650	.85
	200	1315	1.0
Diatomaceous earth, natural, across strata (Note 2)	27.7	204	0.051
	27.7	871	.077
Diatomaceous, natural, parallel to strata (Note 2)	27.7	204	.081
	27.7	871	.106
Diatomaceous earth, molded and fired (Note 2)	38	204	.14
	38	871	.18
Diatomaceous earth and clay, molded and fired (Note 2)	42.3	204	.14
	42.3	871	.19
Diatomaceous earth, high burn, large pores (Note 3)	37	200	.13
	37	1000	.34
Fire clay (Missouri)	......	200	.58
		600	.85
		1000	.95
		1400	1.02
Kaolin insulating brick (Note 3)	27	500	0.15
	27	1150	.26
Kaolin insulating firebrick (Note 4)	19	200	.050
	19	760	.113
Magnesite (86.8% MgO, 6.3% Fe_2O_3, 3% CaO, 2.6% SiO_2 by wt.)	158	204	2.2
	158	650	1.6
	158	1200	1.1
Silicon carbide brick, recrystallized (Note 3)	129	600	10.7
	129	800	9.2
	129	1000	8.0
	129	1200	7.0
	129	1400	6.3
Calcium carbonate, natural	162	30	1.3
White marble	......	....	1.7
Chalk	96	....	0.4
Calcium sulfate (4H_2O), artificial	84.6	40	.22
plaster (artificial)	132	75	.43
(building)	77.9	25	.25
Cambric (varnished)	......	38	.091
Carbon, gas	......	0–100	2.0
Carbon stock	94	−184	0.55
	[illegible]	0	3.6
Cardboard, corrugated	......	....	0.037
Celluloid	87.3	30	.12
Charcoal flakes	11.9	80	.043
	15	80	.051
Clinker (granular)	......	0–700	.27
Coke, petroleum	......	100	3.4
		500	2.9
Coke, petroleum (20–100 mesh)	62	400	0.55
Coke (powdered)	......	0–100	.11
Concrete (cinder)	......	....	.20
(stone)	......	....	.54
(1:4 dry)	......	....	.44

Material	Apparent density ρ, lb./cu. ft. at room temperature	t, °C.	k
Cotton wool	5	30	0.024
Cork board	10	30	.025
Cork (regranulated)	8.1	30	.026
(ground)	9.4	30	.025
Diatomaceous earth powder, coarse (Note 2)	20.0	38	.036
	20.0	871	.082
fine (Note 2)	17.2	204	.040
	17.2	871	.074
molded pipe covering (Note 2)	26.0	204	.051
	26.0	871	.088
4 vol. calcined earth and 1 vol. cement, poured and fired (Note 2)	61.8	204	.16
	61.8	871	.23
Dolomite	167	50	1.0
Ebonite	......	....	0.10
Enamel, silicate	38	....	0.5–0.75
Felt, wool	20.6	30	0.03
Fiber insulating board	14.8	21	.028
Fiber, red	80.5	20	.27
(with binder, baked)	......	20–97	.097
Gas carbon	......	0–100	2.0
Glass	......	....	0.2–0.73
Borosilicate type	139	30–75	0.63
Window glass	......	....	0.3–0.61
Soda glass	......	....	0.3–0.44
Granite	......	....	1.0–2.3
Graphite, longitudinal	......	20	95.
powdered, through 100 mesh	30	40	0.104
Gypsum (molded and dry)	78	20	.25
Hair felt (perpendicular to fibers)	17	30	.021
Ice	57.5	0	1.3
Infusorial earth, see diatomaceous earth			
Kapok	0.88	20	0.020
Lampblack	10	40	.038
Lava	......	....	.49
Leather, sole	62.4	....	.092
Limestone (15.3 vol. % H_2O)	103	24	.54
Linen	......	30	.05
Magnesia (powdered)	49.7	47	.35
Magnesia (light carbonate)	13	21	0.034
Magnesium oxide (compressed)	49.9	20	.32
Marble	......	....	1.2–1.7
Mica (perpendicular to planes)	......	50	0.25
Mill shavings	......	....	0.033–0.05
Mineral wool	9.4	30	0.0225
	19.7	30	.024
Paper	......	....	.075
Paraffin wax	......	0	.14
Petroleum coke	......	100	3.4
	......	500	2.9
Porcelain	......	200	0.88
Portland cement, see concrete	......	90	.17
Pumice stone	......	21–66	.14
Pyroxylin plastics	......	....	.075
Rubber (hard)	74.8	0	.087
(para)	......	21	.109
(soft)	......	21	0.075–0.092
Sand (dry)	94.6	20	0.19
Sandstone	140	40	1.06
Sawdust	12	21	0.03
Scale (Note 1)			
Silk	6.3	....	.026
varnished	......	38	.096
Slag, blast furnace	......	24–127	.064
Slag wool	12	30	.022
Slate	......	94	.86
Snow	34.7	0	.27
Sulfur (monoclinic)	......	100	0.09–0.097
(rhombic)	......	21	0.16
Wall board, insulating type	14.8	21	.028
Wall board, stiff paste board	43	30	.04
Wood shavings	8.8	30	.034
Wood (across grain):			
Balsa	7–8	30	0.025–0.03
Oak	51.5	15	0.12
Maple	44.7	50	.11
Pine, white	34.0	15	.087
Teak	40.0	15	.10
White fir	28.1	60	.062
Wood (parallel to grain):			
Pine	34.4	21	.20
Wool, animal	6.9	30	.021

* Marks, "Mechanical Engineers' Handbook," 4th ed., McGraw-Hill, New York, 1941. "International Critical Tables," McGraw-Hill, 1929, and other sources. For additional data, see pp. 458–459.

Note 1: B. Kamp [*Z. tech. Physik*, **12**, 30 (1931)] shows the effect of increased porosity in decreasing thermal conductivity of boiler scale. Partridge [University of Michigan, *Eng. Research Bull.* 15, 1930] has published a 170-page treatise on Formation and Properties of Boiler Scale.

Note 2: Townshend and Williams, *Chem. & Met.*, **39**, 219 (1932).

Note 3: Norton, "Refractories," 2d ed., McGraw-Hill, New York, 1942.

Note 4: Norton, private communication.

TABLE 3-321 Thermal-Conductivity–Temperature Table for Metals*

Thermal conductivities tabulated in watts per meter-kelvin

Substance	Temperature, K 10	20	40	60	80	100	200	300	400	500	600	800	1000	1200	1400
Alumina	7	32	121	174	160	125	55	36	26	20	16	10	8	7	6
Aluminum	38,000	13,500	2,300	850	380	300	237	273	240	237	232	220	93	99	105
Antimony	470	230	110	80	60	48	32	26	22	20					
Beryllium oxide	47	196	810	1,400	1,650	1,490	480	272	196	146	111	70	47	33	25
Bismuth	240	100	45	31	24	22	18	16	14	12					
Boron	165	305	400	327	230	170	45	25	15	12					
Cadmium	900	250	150	120	110	110	105	104	101	99					
Chromium	400	570	450	250	180	158	111	90	87	85	81	71	65	62	61
Cobalt	250	450	380	250	190	160	120	100	85	70					
Constantan	4	9	16	18	19	20	23	25	27	30					
Copper	19,000	10,700	2,100	850	570	483	413	398	392	388	383	371	357	342	
Gallium	2,200	640	250	200	170	140	100	85							
Gold	2,800	1,500	520	380	350	345	327	315	312	309	304	292	278	262	
Graphite†	27	108	135	81	54	39	15	10	7	5	4	3	3	2	2
Graphite‡	81	420	1,630	2,980	4,290	4,980	3,250	2,000	1,460	1,140	930	680	530	440	370
Hastelloy	1	3	4	5	6	7	9	10	11	13					
Inconel	2	4	8	10	11	11	14	15							
Iridium	1,300	1,900	750	360	230	172	147	145	143	140					
Iron	710	1,000	560	270	170	132	94	80	69	61	55	43	33	28	31
Lead	175	57	43	42	41	40	37	35	34	33	31	19	22	24	26
Magnesium	1,200	1,300	620	290	190	169	159	156	153	151	149	146	84	98	112
Magnesium oxide	1,100	3,100	2,200	950	460	260	75	48	36	27	21	13	10	8	7
Manganese	2	2	4	5	5	6	7	8	9	9					
Manganin	2	4	9	11	13	13	17	22	28	34	40				
Mercury	54	40	35	33	33	32	32	8	10	11	12	13	14		
Molybdenum	150	280	350	250	210	179	143	138	134	130	126	118	112	105	100
Nickel	2,600	1,700	570	290	200	158	106	91	80	72	66	67	72	76	80
Nylon	0.04	0.10	0.17	0.20	0.23	0.25	0.28	0.30							
Palladium	1,200	610	160	100	88	80	78	78	78	80					
Platinum	1,200	490	130	92	82	79	75	73	72	72	72	73	78	78	81
PTFE§	0.94	1.43	1.94	2.1	2.15	2.16	2.20	2.25	2.3	2.5					
Pyrex	0.12	0.20	0.33	0.42	0.51	0.57	0.88	1.1	1.6	2.1					
Quartz	1,200	480	82	40	30										
Rhodium	2,900	3,900	1,000	370	250	190	160	150	145	140					
Rubber	......	.....	0.13	0.15	0.16	0.17	0.20	0.22	0.24	0.25					
Selenium (axis)	140	57	25	15	10	8	6	4	3	2					
Silica	......	.....	.....	....	....	....	....	1.34	1.52	1.70	1.87	2.22	2.60		
Silver	16,500	5,200	1,100	630	500	430	425	424	420	413	405	389	374	358	
Tantalum	108	146	88	68	62	59	58	57	58	58	59	59	60	61	62
Tellurium	300	93	29	17	13	11	6	4	3	3					
Tin	......	320	130	101	90	84	72	67	62	60					
Titanium	14	28	39	37	33	31	26	21	20	20	19				
Tungsten	......	.....	880	330	310	280	190	180	170	150	140				
Uranium	......	.....	....	20	22	23	26	28	30	32					
Zinc	......	.....	....	150	135	130	123	120	116	110	110				
Zirconium	100	110	59	42	38	34	25	23	22	21	21				

*Especially at low temperatures, the thermal conductivity can often be markedly reduced by even small traces of impurities. This table, for the highest-purity specimens available, should thus be used with caution in applications with commercial materials. From Perry, *Engineering Manual*, 3d ed., McGraw-Hill, New York, 1976.

†Parallel to basal plane.

‡Perpendicular to basal plane.

§Also known as Teflon, etc.

TABLE 3-322 Thermal Conductivity of Chromium Alloys*

k = Btu/(h·ft²)(°F/ft)

American Iron and Steel Institute Type No.	k at 212°F.	k at 932°F.
301, 302, 302B, 303, 304, 316†	9.4	12.4
308	8.8	12.5
309, 310	8.0	10.8
321, 347	9.3	12.8
403, 406, 410, 414, 416†	14.4	16.6
430, 430F†	15.1	15.2
442	12.5	14.2
501, 502†	21.2	19.5

*Table 3-322 is based on information from manufacturers.
†Shelton and Swanger (National Bureau of Standards), *Trans. Am. Soc. Steel Treat.*, **21**, 1061–1078 (1933).

TABLE 3-323 Thermal Conductivity of Some Alloys at High Temperature*

°R.	Thermal conductivity, B.t.u./(ft.)(hr.)(°R.)					
	Kovar	Advance	Monel	Hastelloy A	Inconel	Nichrome V
500	7.8	....	9.0	5.6	6.0	5.5
600	8.3	11.4	10.2	6.2	6.5	6.1
700	8.6	12.6	11.2	6.8	7.0	6.7
800	8.7	13.9	12.3	7.3	7.6	7.3
900	8.7	15.1	13.4	7.8	8.1	7.8
1000	8.9	16.4	14.4	8.4	8.6	8.4
1100	9.2	17.6	15.4	9.0	9.1	9.0
1200	9.5	18.8	16.5	9.5	9.7	9.5
1300	9.8	20.0	17.6	10.1	10.2	10.1
1400	10.2	21.2	18.7	10.7	10.8	10.7
1500	10.5	22.5	19.8	11.3	11.3	11.3
1600	10.8	23.8	20.8	11.8	11.8	11.9
1700	11.1	25.0	21.9	12.3	12.4	12.4
1800	11.3	26.2	23.0	12.9	13.0	13.0
1900	11.5	27.4	24.0	13.4	13.6	13.5
2000	11.8	28.7	25.1	14.0	14.0	14.1
2100	12.1	30.0	26.1	14.6	14.5	14.7
2200	12.3	....	27.2	15.1	15.0	15.3

* Silverman, *J. Metals*, **5**, 631 (1953). Copyright American Institute of Mining, Metallurgical and Petroleum Engineers, Inc.

TABLE 3-324 Thermal Conductivities of Some Materials for Refrigeration and Building Insulation*

k = Btu/(h·ft²)(°F/ft) at approximately room temperature

Material	Apparent density, lb./cu. ft. room temp.	k
Soft flexible materials in sheet form:		
Chemically treated wood fiber	2.2	0.023
Eel grass between paper	3.4–4.6	0.021–0.022
Felted cattle hair	11–13	0.022
Flax fibers between paper	4.9	.023
Hair and asbestos fibers, felted	7.8	.023
Insulating hair, and jute	6.1–6.3	0.022–0.023
Jute and asbestos fibers, felted	10.0	0.031
Loose materials:		
Cork, regranulated, fine particles	8–9	.025
Charcoal, 6 mesh	15.2	.031
Diatomaceous earth, powdered	10.6	.026
Glass wool, curled	4–10	.024
Gypsum in powdered form	26–34	0.043–0.05
Mineral wool, fibrous	6	0.0217
	10	.0225
	14	.0233
	18	.0242
Sawdust	12	.034
Wood shavings, from planer	8.8	.034
Semiflexible materials in sheet form:		
Flax fiber	13.0	.026
Semirigid materials in board form:		
Corkboard	7.0	.0225
	10.6	.025
Mineral wool, block, with binder	16.7	.031
Stiff fibrous materials in sheet form:		
Wood pulp	16.2–16.9	.028
Sugar-cane fiber	13.2–14.8	.028
Cellular gypsum	8	.029
	12	.037
	18	.049
	24	.064
	30	.083

* Abstracted from *U.S. Bur. Standards Letter Circ.* 227, Apr. 19, 1927.

TABLE 3-325 Thermal Conductivities of Insulating Materials at High Temperatures*

k = Btu/(h·ft²)(°F/ft)

Material	For temperatures, °F. up to	Mean temperatures, °F.									
		100	200	300	400	500	600	800	1000	1500	2000
Laminated asbestos felt (approx. 40 laminations per in.)	700	0.033	0.037	0.040	0.044	0.048					
Laminated asbestos felt (approx. 20 laminations per in.)	500	.045	.050	.055	.060	.065					
Corrugated asbestos (4 plies per in.)	300	.050	.058	.069							
85% magnesia (density, 13 lb./cu. ft.)	600	.034	.036	.038	.040						
Diatomaceous earth, asbestos and bonding material	1600	.045	.047	.049	.050	.053	0.055	0.060	0.065		
Diatomaceous earth brick	1600	.054	.056	.058	.060	.063	.065	.069	.073		
Diatomaceous earth brick	2000	.127	.130	.133	.137	.140	.143	.150	.158	0.176	
Diatomaceous earth brick	2500	.128	.131	.135	.139	.143	.148	.155	.163	.183	0.203
Diatomaceous earth powder (density, 18 lb./cu. ft.)	....	.039	.042	.044	.048	.051	.054	.061	.068		
Rock wool	....	.030	.034	.039	.044	.050	.057				

Asbestos cement, 1.2; 85% magnesia cement, 0.05; asbestos and rock wool cement, 0.075 approx.
* Marks, "Mechanical Engineers' Handbook," 4th ed., McGraw-Hill, New York, 1941.

TABLE 3-326 Thermal Conductivities of Insulating Materials at Moderate Temperatures (Nusselt)*

k = Btu/(h·ft²)(°F/ft)

Material	Weight, lb./cu. ft.	Temperatures, °F.						
		32	100	200	300	400	600	800
Asbestos	36.0	0.087	0.097	0.110	0.117	0.121	0.125	0.130
Burned infusorial earth for pipe coverings	12.5	.043	.046	.052	.057	.062	.073	.085
Insulating composition (loose)	25.0	.040	.046	.050	.053	.055		
Cotton	5.0	.032	.035	.039				
Silk hair	9.1	.026	.030	.034				
Silk	6.3	.025	.028	.034				
Wool	8.5	.022	.027	.033				
Pulverized cork	10.0	.021	.026	.032				
Infusorial earth (loose)	22.0	.035	.039	.045	.047	.050	.053	

* Marks, "Mechanical Engineers' Handbook," 4th ed., McGraw-Hill, New York, 1941.

TABLE 3-327 Thermal Conductivities of Insulating Materials at Low Temperatures (Gröber)*

k = Btu/(h·ft²)(°F/ft)
See also data in Sec. 12.

Material	Weight, lb./cu. ft.	Temperatures, °F.				
		32	−50	−100	−200	−300
Asbestos	44.0	0.135	0.132	0.130	0.125	0.100
Asbestos	29.0	.0894	.0860	.0820	.0720	.0545
Cotton	5.0	.0325	.0302	.0276	.0235	.0198
Silk	6.3	.0290	.0256	.0235	.0196	.0155

* Marks, "Mechanical Engineers' Handbook," 4th ed., McGraw-Hill, New York, 1941.

TABLE 3-328 Thermal Diffusivity (m^2/s) of Selected Elements*

Element	Temperature, K									
	20	40	60	80	100	200	400	600	800	1000
Aluminum	0.50	0.012	0.0014	4.4.−4	2.3.−4	1.1.−4	9.4.−5	8.4.−5	7.4.−5	6.6.−5
Beryllium	......	......	......	......	0.0036	1.5.−4	4.0.−5	2.6.−5	2.1.−5	1.7.−5
Chromium	0.038	0.0037	5.9.−4	2.0.−4	1.2.−4	4.1.−5	2.6.−5	2.0.−5	1.7.−5	1.4.−5
Copper	0.16	0.0040	6.9.−4	3.1.−4	2.2.−4	1.3.−4	1.1.−4	1.0.−4	9.0.−5	9.0.−5
Gold	0.005	4.5.−4	2.3.−4	1.8.−4	1.5.−4	1.3.−4	1.2.−4	1.2.−4	1.1.−4	9.8.−5
Iridium	0.046	......	......	......	8.4.−5	5.6.−5	4.8.−5	4.4.−5	4.1.−5	3.5.−5
Iron	0.043	3.2.−3	4.9.−4	1.6.−4	8.2.−5	3.1.−5	1.8.−5	1.3.−5	1.1.−5	1.0.−5
Lead	9.3.−5	3.9.−5	3.3.−5	3.1.−5	2.9.−5	2.6.−5	2.3.−5	2.0.−5	1.3.−5	1.5.−5
Molybdenum	0.0095	0.0014	4.0.−4	2.0.−4	1.3.−4	6.3.−5	5.1.−5	4.5.−5	4.2.−5	3.8.−5
Nickel	0.033	0.0017	3.1.−4	1.3.−4	8.0.−5	3.1.−5	1.9.−5	1.3.−5	1.4.−5	1.5.−5
Platinum	0.0029	1.6.−4	6.3.−5	4.3.−5	3.6.−5	2.7.−5	2.5.−5	2.5.−5	2.5.−5	2.5.−5
Silver	0.031	0.0013	4.5.−4	2.8.−4	2.3.−4	1.8.−4	1.7.−4	1.6.−4	1.5.−4	1.4.−4
Zinc	0.0046	3.1.−4	1.0.−4	7.0.−5	5.5.−5	4.7.−5	3.9.−5	3.4.−5	1.8.−5	2.2.−5

*Tables for up to 24 temperatures for 47 elements appear in the *Handbook of Heat Transfer*, 2d ed., McGraw-Hill, New York, 1984. The notation 3.2.−4 signifies 2.3×10^{-4}.

TABLE 3-329 Thermophysical Properties of Selected Nommetallic Solid Substances

Material	Density, kg/m^3	Emissivity	Specific heat, kJ/(kg·K)	Thermal conductivity, W/(m·K)	Thermal diffusivity, $m^2/s \times 10^6$
Alumina	3975	...	0.765	36.0	11.9
Asphalt	2110	...	0.920	0.06	0.03
Bakelite	1300	...	1.465	1.4	0.74
Beryllia	3000	0.82	1.030	270	88.
Brick	1925	0.93	0.835	0.72	0.45
Brick, fireclay	2640	0.93	0.960	1.0	0.39
Carbon, amorphous	1950	0.86	0.724	1.6	1.13
Clay	1460	0.91	0.880	1.3	1.01
Coal	1350	0.80	1.26	0.26	0.15
Cotton	80	...	1.30	0.06	0.58
Diamond	3500		0.509	2300.	1290.
Granite	2630	...	0.775	2.79	1.37
Hardboard	1000	...	1.38	0.15	0.11
Magnesite	3025	0.38	1.13	4.0	1.2
Magnesia	3635	0.72	0.943	48.	14.
Oak	770	0.90	2.38	0.18	0.10
Paper	930	0.83	1.34	0.011	0.01
Pine	525	0.84	2.75	0.12	0.54
Plaster board	800	0.91	...	0.17	
Plywood	540	....	1.22	0.12	0.18
Pyrex	2250	0.92	0.835	1.4	0.74
Rubber	1150	0.92	2.00	0.2	0.09
Rubber, foam	70	0.90	...	0.03	
Salt	....	0.34	0.854	7.1	
Sandstone	2150	0.59	0.745	2.9	1.8
Silica	....	0.79	0.743	1.3	
Sapphire	3975	0.48	0.765	46.	15.
Silicon carbide	3160	0.86	0.675	490.	230.
Soil	2050	0.38	1.84	0.52	0.14
Teflon	2200	0.92	0.35	0.26	0.34
Thoria	4160	0.28	0.71	14.	4.7
Urethane foam	70	...	1.05	0.03	0.36
Vermiculite	120	...	0.84	0.06	0.60

NOTE: Difficulties of accurately characterizing many of the specimens mean that many of the values presented here must be regarded as being of order of magnitude only. For some materials, actual measurement may be the only way to obtain data of the required accuracy. To convert kilograms per cubic meter to pounds per cubic foot, multiply by 0.062428; to convert kilojoules per kilogram-kelvin to British thermal units per pound–degree Fahrenheit, multiply by 0.23885.

PREDICTION AND CORRELATION OF PHYSICAL PROPERTIES

INTRODUCTION

In the absence of reliable experimental data, the methods presented here provide physical-property estimates which are sufficiently accurate for many engineering applications. These techniques have been selected on the basis of accuracy, generality, and in most cases simplicity and are divided into 11 categories: (1) pure-component constants: critical properties, acentric factor, and normal boiling temperature; (2) vapor density; (3) liquid density; (4) vapor pressure; (5) enthalpy of vaporization; (6) ideal-gas heat capacity (including heat and Gibbs energy of formation); (7) liquid heat capacity; (8) viscosity; (9) thermal conductivity; (10) diffusion coefficients; and (11) surface tension. Reliabiliy estimates, as well as illustrative examples, are given for most of the correlations. Symbols are listed under "Nomenclature"; applicable dimensional units are shown individually with each equation. Literature citations, indicated by superscript numbers, are at the end under "References."

UNITS

The correlations of physical properties in this section are in terms of particular sets of units and are not easily generalized to both SI and U.S. customary units. The specific units to be applied with the different correlations are defined at the points of use. For unit conversion factors the reader is referred to Sec. 1.

NOMENCLATURE

Symbol	Definition and equations
a	Activity; also constant in Eqs. (3-36), (3-37), (3-43), (3-45), (3-51), (3-76), and (3-146)
A	Constant in Eqs. (3-11), (3-48), (3-66), (3-67), (3-68), (3-79), (3-91), (3-100), (3-125), (3-126), and (3-142); also, interaction parameter in Eq. (3-117)
A°	Constant in Eq. (3-69)
A_0	Constant in Eq. (3-51)
b	Constant in Eqs. (3-37), (3-44), (3-45), (3-51), (3-76), and (3-146)
B	Constant in Eqs. (3-11), (3-27), (3-48), (3-66), (3-67), (3-68), (3-79), (3-91), (3-100), and (3-125)
$B^{(0)}$	Second virial coefficient parameter defined in Eqs. (3-31) and (3-33)
$B^{(1)}$	Second virial coefficient parameter defined in Eqs. (3-32) and (3-34)
$B^{(2)}$	Second virial coefficient parameter defined in Eqs. (3-36) and (3-37)
B°	Constant in Eq. (3-69)
B_0	Constant in Eq. (3-51)
c	Constant in Eqs. (3-51) and (3-76)
C	Constant in Eqs. (3-11), (3-27), (3-66), (3-67), (3-91), (3-100), and (3-125); also, interaction parameter in Eq. (3-122)
C°	Constant in Eq. (3-69)
C_i	Contribution of atomic group in Eq. (3-86); concentration of electrolyte in Eq. (3-131)
C_0	Constant in Eq. (3-51)
C_p	Heat capacity at constant pressure
C_v	Heat capacity at constant volume
CZ, C_1, C_2, C_3, C_4, C_5	Constants in Eq. (3-52)
d	Constant in Eq. (3-76)
$\mathcal{D}$	Diffusion coefficient
D	Constant in Eqs. (3-11), (3-27), (3-67), and (3-91)
D°	Constant in Eq. (3-69)
E	Constant in Eq. (3-11)
f	Function of
g	Defined in Eq. (3-119)
G	Interaction parameter in Eqs. (3-108) and (3-109)
H	Constant in Eq. (3-124)
k	Interaction parameter in Eq. (3-40)
K	Defined in Eq. (3-141)
ln	Denotes natural logarithm
$(\ln U)^0$	Defined in Eq. (3-59)
$(\ln U)^1$	Defined in Eq. (3-60)
log	Denotes common logarithm
m	Slope in Eq. (3-53)
M	Molecular weight
n	Number of components
n_i	Number of atomic groups of type i in Eq. (3-86)
N	Defined in Eq. (3-63); also, constant in Eq. (3-124)
P	Pressure
$[P]$	Parachor
p^{sat}	Vapor pressure
q	Constant in Eqs. (3-142) and (3-155)
Q	Defined in Eqs. (3-69) and (3-150)
r	Constant in Eq. (3-56); molecular radius in Eq. (3-137)
R	Universal gas constant
s	"Surface" fraction in Eq. (3-10)
S	Defined in Eqs. (3-121) and (3-122)
T	Temperature
V	Molar volume
w	Weight fraction of component in mixture
x	Mole fraction of component in liquid phase
y	Mole fraction of component in vapor phase
Z	Compressibility factor $= PV/RT$
$Z^{(0)}$, $Z^{(1)}$	Compressibility-factor parameters in Eq. (3-23)
Z_{RA}	Constant in Eq. (3-57)
	Greek symbols
α	Constant in Eq. (3-51); defined in Eq. (3-144)
α_c	Defined in Eq. (3-70)
α_1, α_2, α_3	Constants in Eq. (3-91)
β	Defined in Eq. (3-18); constant in Eq. (3-135)
β_1, β_2	Constants in Eq. (3-91)
γ	Constant in Eq. (3-51)
γ_1, γ_2, γ_3	Constants in Eq. (3-91)
Γ	Defined in Eq. (3-112)
δ_v	Critical-volume-interaction parameter defined in Eq. (3-12)
δ_1, δ_2, δ_3	Constants in Eq. (3-91)
Δ	Defined in Eq. (3-97)
Δ_p	Critical-pressure-group-contribution increment in Eq. (3-14)
Δ_T	Critical-temperature-group-contribution increment in Eq. (3-1)
Δ_v	Critical-volume-group-contribution increment in Eq. (3-7)
ΔG_f	Gibbs energy of formation
ΔH_f	Heat of formation
ΔH_v	Enthalpy of vaporization
ΔS°	Parameter in Eq. (3-124)
ΔS_f	Entropy of formation

	Greek symbols
ΔV_i	Critical-volume-group-contribution increment in Eq. (3-6)
η	Viscosity; also optical-isomer-correction term for B in Eq. (3-79)
$(\eta_L\,\xi)^{(0)}$	Defined in Eq. (3-104)
$(\eta_L\,\xi)^{(1)}$	Defined in Eq. (3-105)
θ	T_b/T_c in Eqs. (3-16) and (3-17)
λ	Thermal conductivity
μ_p	Dipole moment
μ_r	Function of dipole moment in Eq. (3-36)
ν	Interaction parameter defined in Eq. (3-11); diffusion-volume increment in Eq. (3-133)
ξ	Defined in Eqs. (3-84) and (3-106)
ξ'	Defined in Eq. (3-86)
π	Constant = 3.14159 . . .
ρ	Molar density
$\mathcal{P}$	Defined in Eq. (3-92)
σ	Surface tension; also, symmetry correction term for B in Eq. (3-79) and characteristic ion coefficient in Eq. (3-131)
Σ	Sum of
ϕ	"Volume" fraction defined in Eqs. (3-5) and (3-129); interaction parameter defined in Eqs. (3-88), (3-89), and (3-90); association factor in Eq. (3-139)
χ	Generalized parameter in Eq. (3-56)
ψ	Defined in Eqs. (3-70), (3-154), and (3-155)
ω	Acentric factor defined in Eq. (3-15)
Ω_a	Constant in Eq. (3-46)
Ω_b	Constant in Eq. (3-47)
	Superscripts
BP	At bubble point
0	At low pressure
sat	In the saturated state
	Subscripts
A,B	Components in a mixture
b	At normal boiling point
c	At critical point
G	Of the vapor
HI	Upper limit
i	Of the ith component
j	Of the jth component
k	Of the kth component
L	Of the liquid
LO	Lower limit
mix	Of a mixture
o	Of organic compound
r	Reduced quantity
w	Of water
1, 2, 3	Different levels of temperature or pressure or components in a mixture

PURE-COMPONENT CONSTANTS

Critical Temperature The critical temperature T_c of **organic** compounds is best estimated from Lydersen's correlation, Eq. (3-1).[137] Δ_T values are obtained from group contributions in Table 3-330. The normal boiling point T_b must be known. Both T_b and T_c are expressed as absolute temperatures, K (or °R).

$$T_c = \frac{T_b}{0.567 + \Sigma\,\Delta_T - (\Sigma\,\Delta_T)^2} \tag{3-1}$$

TABLE 3-330 Lydersen's Critical-Property Increments

	Δ_T	Δ_p	Δ_v
Nonring increments			
$-\mathrm{CH_3}$	0.020	0.227	55
$-\overset{\vert}{\mathrm{C}}\mathrm{H_2}$	0.020	0.227	55
$-\underset{\vert}{\overset{\vert}{\mathrm{C}}}\mathrm{H}$	0.012	0.210	51
$-\underset{\vert}{\overset{\vert}{\mathrm{C}}}-$	0.00	0.210	41
$=\mathrm{CH_2}$	0.018	0.198	45
$=\overset{\vert}{\mathrm{C}}\mathrm{H}$	0.018	0.198	45
$=\overset{\vert}{\mathrm{C}}-$	0.0	0.198	36
$=\mathrm{C}=$	0.0	0.198	36
$\equiv\mathrm{CH}$	0.005	0.153	(36)
$\equiv\mathrm{C}-$	0.005	0.153	(36)
Ring increments			
$-\mathrm{CH_2}-$	0.013	0.184	44.5
$-\underset{\vert}{\overset{\vert}{\mathrm{C}}}\mathrm{H}$	0.012	0.192	46
$-\underset{\vert}{\overset{\vert}{\mathrm{C}}}-$	(−0.007)	(0.154)	(31)
$=\overset{\vert}{\mathrm{C}}\mathrm{H}$	0.011	0.154	37
$=\overset{\vert}{\mathrm{C}}-$	0.011	0.154	36
$=\mathrm{C}=$	0.011	0.154	36
Halogen increments			
$-\mathrm{F}$	0.018	0.224	18
$-\mathrm{Cl}$	0.017	0.320	49
$-\mathrm{Br}$	0.010	(0.50)	(70)
$-\mathrm{I}$	0.012	(0.83)	(95)
Oxygen increments			
−OH (alcohols)	0.082	0.06	(18)
−OH (phenols)	0.031	(−0.02)	(3)
−O− (nonring)	0.021	0.16	20
−O− (ring)	(0.014)	(0.12)	(8)
$-\overset{\vert}{\mathrm{C}}=\mathrm{O}$ (nonring)	0.040	0.29	60
$-\overset{\vert}{\mathrm{C}}=\mathrm{O}$ (ring)	(0.033)	(0.2)	(50)

Errors in T_c are normally less than 2 percent except for high-molecular-weight compounds (>1000). The accuracy is, however, unknown for molecules with multipolar groups, e.g., glycols.

For **inorganic** compounds and nonmetallic elements, Gambill[65] suggests, as an approximation,

$$T_c = 1.63\,T_b \tag{3-2}$$

TABLE 3-330 Lydersen's Critical-Property Increments *(Continued)*

	Δ_T	Δ_p	Δ_v
Oxygen increments (*Cont.*)			
HC=O (aldehyde) (with one free bond on C)	0.048	0.33	73
−COOH (acid)	0.085	(0.4)	80
−COO− (ester)	0.047	0.47	80
=O (except for combinations above)	(0.02)	(0.12)	(11)
Nitrogen increments			
$-NH_2$	0.031	0.095	28
−NH (nonring) (with one further free bond on N)	0.031	0.135	(37)
−NH (ring) (with one further free bond on N)	(0.024)	(0.09)	(27)
−N− (nonring) (with one further free bond on N)	0.014	0.17	(42)
−N− (ring) (with one further free bond on N)	(0.007)	(0.13)	(32)
−CN	(0.060)	(0.36)	(80)
$-NO_2$	(0.055)	(0.42)	(78)
Sulfur increments			
−SH	0.015	0.27	55
−S− (nonring)	0.015	0.27	55
−S− (ring)	(0.008)	(0.24)	(45)
=S	(0.003)	(0.24)	(47)
Miscellaneous			
−Si− (with free bonds above and below)	0.03	(0.54)	
−B− (with one further free bond on B)	(0.03)		

NOTE: There are no increments for hydrogen. All bonds shown as free are connected with atoms other than hydrogen. Values in parentheses are based upon too few experimental values to be reliable. From vapor-pressure measurements and a calculational technique similar to that of Fishtine,[58] it has been suggested that the C−H ring increment (C with three further free bonds) common to two condensed saturated rings be given the value of $\Delta_T = 0.064$.

Deviations of 5 percent (average) and 11 percent (maximum) were found for 40 substances tested. For **metallic** elements,[75]

$$T_c = 0.4T_b^{1.254} \tag{3-3}$$

In Eq. (3-3), T_b and T_c are in kelvins. The error found using Eq. (3-3) is similar to that noted with Eq. (3-2).

To estimate the **true critical temperature** for **mixtures,** few reliable methods have been developed, and even for these best agreement is found when treating simple hydrocarbon mixtures (with, perhaps, small fractions of H_2S, CO_2, CO, and the inert gases). Li[125] suggests

$$T_{c,\text{mix}} = \sum_j \phi_j T_{cj} \tag{3-4}$$

where ϕ_j is a "volume" fraction based on critical volumes of the pure components,

$$\phi_j = (y_j V_{cj}) \bigg/ \left(\sum_i y_i V_{ci}\right) \tag{3-5}$$

In Eq. (3-5), y_j is the mole fraction of component j. A similar equation for $T_{c,\text{mix}}$ has been proposed by Chueh and Prausnitz,[34] using a "surface" fraction. Spencer et al.[224] found Eq. (3-4) to yield an average deviation of only 4 K when 135 binary hydrocarbon mixtures were tested.

Critical Volume The critical volume V_c of **organic** compounds is usually estimated by a group contribution method. Vetere[264] proposed Eq. (3-6).

$$V_c = 33.04 + \left(\sum_i M_i \times \Delta V_i\right)^{1.029} \tag{3-6}$$

M_i is the molecular weight of group i in Table 3-331; ΔV_i is the volume contribution of group i from this table. V_c is calculated in

TABLE 3-331 Vetere Group Contributions to Estimate Critical Volumes

Group	ΔV_i
Nonring	
In linear chain	
CH_3, CH_2, CH, C	3.360
In side chain	
CH_3, CH_2, CH, C	2.888
$=CH_2$, =CH, =C− (with free bonds on CH and C)	2.940
=C=	2.908
≡CH, ≡C−	2.648
Ring	
CH_2, CH, C	2.813
=CH, =C− (with free bonds on CH and C)	2.538
F	0.770
Cl	1.237
Br	0.899
I	0.702
−OH (alcohols)	0.704
−OH (phenols)	1.553
−O− (nonring)	1.075
−O− (ring)	0.790
−O− (epoxy)	−0.252
−C=O (nonring) (with one further free bond on C)	1.765
−C=O (ring) (with one further free bond on C)	1.500
HC=O (aldehyde) (with one free bond on C)	2.333
−COOH	1.652
−COO−	1.607
$-NH_2$	2.184
−NH (nonring) (with one further free bond on N)	2.333
−NH (ring) (with one further free bond on N)	1.736
−N− (nonring) (with one further free bond on N)	1.793
−N− (ring) (with one further free bond on N)	1.883
−CN	2.784
$-NO_2$	1.559
−SH	1.537
−S− (nonring)	0.591
−S− (ring)	0.911

units of cm^3/mol. A similar but somewhat less accurate method was suggested by Lydersen:[137]

$$V_c = 40 + \sum_i \Delta_v \tag{3-7}$$

with Δ_v values as given in Table 3-330. Again V_c has the units of cm^3/mol. Since so few reliable experimental values of V_c exist, the accuracy of Eq. (3-6) or Eq. (3-7) is not well known. Except for high-molecular-weight materials, errors of less than 5 to 7 percent would be expected.

If the molal volume of the liquid is available at the normal boiling point V_{Lb}, then the simple relation of Tyn and Calus[253] may be useful to estimate V_c:

$$V_c = 3.31\, V_{Lb}^{0.954} \tag{3-8}$$

This relation is preferable to group-contribution methods, but reliable values of V_{Lb} must be available. Both V_c and V_{Lb} are expressed in cm^3/mol.

Even fewer data are available to evaluate estimation methods for **true-mixture critical volumes.** For hydrocarbon mixtures (with, perhaps, CO_2 or H_2S), Prausnitz and his coworkers[34,210] suggest

$$V_{c,\text{mix}} = \sum_j s_j V_{cj} + \sum_i \sum_j s_i s_j \nu_{ij} \tag{3-9}$$

s_j is a "surface" fraction defined as

$$s_j = (y_j V_{cj}^{2/3}) \Big/ \left(\sum_k y_k V_{ck}^{2/3}\right) \tag{3-10}$$

with V_{cj} the critical volume of pure j and y_j the mole fraction of j. ν_{ij} is an interaction parameter such that $\nu_{ii} = 0$ and ν_{ij} $(i \neq j)$ is calculated from Eqs. (3-11) and (3-12) and Table 3-332.

$$\nu_{ij} = \frac{V_{ci} + V_{cj}}{2} (A + B\delta_v + C\delta_v^2 + D\delta_v^3 + E\delta_v^4) \tag{3-11}$$

$$\delta_v = \left| \frac{V_{ci}^{2/3} - V_{cj}^{2/3}}{V_{ci}^{2/3} + V_{cj}^{2/3}} \right| \tag{3-12}$$

Spencer et al.[225] evaluated Eq. (3-9) for 31 hydrocarbon-hydrocarbon and hydrocarbon-nonhydrocarbon binaries; average errors of 10.5 percent were found.

A somewhat simpler method developed to estimate $V_{c,\text{mix}}$ for binaries is described by Li,[126]

$$V_{c,\text{mix}} = y_1 V_{c1} + y_2 V_{c2} - \frac{y_1 y_2 (V_{c1} - V_{c2})^2}{(V_{c1} + V_{c2})} \tag{3-13}$$

Tests with 84 binary mixtures led to an average error of less than 5 percent.

Critical Presure The critical pressure P_c of **organic** compounds can be estimated from Eq. (3-14);[137] Δ_p contributions are given in Table 3-330.

$$P_c = \frac{M}{(0.34 + \Sigma\, \Delta_p)^2} \tag{3-14}$$

M is the molecular weight, and P_c is calculated in atm. Errors using Eq. (3-14) are usually less than 5 percent except for high-molecular-weight compounds or those with several polar groups; for these latter cases, the errors expected in using Eq. (3-14) are unknown.

The **true-mixture critical pressure** is often highly nonlinear in its dependence on mole fraction, and no reliable estimation method is available. Kreglewski and Kay[119] have used conformal-solution theory, while Chueh and Prausnitz[34] employ the Redlich-Kwong equation of state with estimated values of $T_{c,\text{mix}}$ and $V_{c,\text{mix}}$ to determine $P_{c,\text{mix}}$. Actually any equation of state believed to be applicable in the critical region could be so used. The detailed methods are described in Reid et al.[196]

Acentric Factor The acentric factor ω is often necessary in many correlation equations. It can be calculated from the definition[182,183]

$$\omega = -\log_{10} P_r^{\text{sat}} - 1.000 \tag{3-15}$$

with P_r^{sat} equal to the reduced vapor pressure at $T_r = T/T_c = 0.7$. If the necessary vapor-pressure data are not available, ω may be estimated by a correlation proposed by Lee and Kesler,[120]

$$\omega = \frac{-\ln P_c - 5.92714 + 6.09648\theta^{-1} + 1.28862 \ln \theta - 0.169347\theta^6}{15.2518 - 15.6875\theta^{-1} - 13.4721 \ln \theta + 0.43577\theta^6} \tag{3-16}$$

with P_c in atm and $\theta = T_b/T_c$.

Normal Boiling Temperature The normal boiling temperature T_b is usually available for materials which have been synthesized and studied. For the relatively rare case in which an approximate value of T_b must be estimated, Miller[158] has combined the group-contribution methods for T_c, P_c, and V_c with Eq. (3-8) and the Rackett[190] liquid-density relation to give

$$T_b = 0.012186\, \theta e^{\beta} \tag{3-17}$$

where T_b is in kelvins, θ is T_b/T_c and is estimated from Eq. (3-1) by using Table 3-330, while β is given by

$$\beta = \frac{[(1-\theta)^{2/7} - 0.048] \ln V_c + (1-\theta)^{2/7} \ln P_c + 1.255}{(1-\theta)^{2/7}} \tag{3-18}$$

with V_c in cm^3/mol and P_c in atm. When tested with data for 25 diverse compounds, the average error in T_b was about 10 percent. Note that only the structure of the compound and its molecular weight must be known.

The **critical** compressibility factor Z_c is obtained from T_c, P_c, and V_c as

$$Z_c = P_c V_c / R T_c \tag{3-19}$$

R is the universal gas constant.

VAPOR DENSITY

Since vapor density is a function of both temperature and pressure, it is correlated by equations of state, i.e., expressions which relate specific volume V, absolute temperature T, and absolute pressure P. The simplest equation of state is the perfect- or ideal-gas law:

$$PV = RT \tag{3-20}$$

where R = universal gas constant. This equation, which applies to both pure components and mixtures, is reasonably accurate up to a pressure of about 2 atm for most compounds and even higher for small, nonassociating molecules.

Example 1 At 25 psia and 300°F = 759.67°R, the specific volume of steam is reported to be 17.829 ft^3/lb.[156] Using Eq. (3-20) with R = 10.7314 (psia·ft^3)/(lb·mol·°R),

$$V = \frac{(10.7314)(759.67)}{25.0} = 326.093 \text{ ft}^3/(\text{lb·mol})$$

Table 3-332 Values of the Constants in Eq. (3-11)[9]

Binary	A	B	C	D	E
Aromatic-aromatic	0	0	0	0	0
Containing at least one cycloparaffin	0	0	0	0	0
Paraffin-aromatic	0.0753	−3.332	2.220	0	0
System with CO_2 or H_2S	−0.4957	17.1185	−168.56	587.05	−698.89
All other systems	0.1397	−2.9672	1.8337	−1.536	0

Using a molecular weight of water = 18.0152, $V = 326.093/18.0152 = 18.101$ ft^3/lb.

For higher pressures or for more accurate calculations, more complicated correlations have been developed. With a few exceptions, these efforts have fallen into four categories:

Corresponding States The nonideality of a gas can be expressed by the compressibility factor Z:

$$Z = PV/RT \tag{3-21}$$

Z = unity for a perfect gas. For pure components, the compressibility factor has been correlated with reduced temperature T_r and reduced pressure P_r:[170,19]

$$Z = f(T_r, P_r) \tag{3-22}$$

where $T_r = T/T_c$
$P_r = P/P_c$
T_c = absolute critical temperature
P_c = absolute critical pressure

Expected errors in using this approach are 4 to 6 percent[196] at pressures greater than 2 atm; for greater accuracy, a third correlating parameter is required.

The compressibility factor at the critical point Z_c has been often used as this third parameter,[138,98,53] but the one recommended here is the acentric factor in the form

$$Z = Z^{(0)} + \omega Z^{(1)} \tag{3-23}$$

where $Z^{(0)} = f(T_r, P_r)$
ω = acentric factor
$Z^{(1)} = f(T_r, P_r)$

Pitzer et al.[183] originally tabulated $Z^{(0)}$ and $Z^{(1)}$ as functions of T_r and P_r, and several modifications and extensions to wider ranges of T_r and P_r have been published.[208,120,132] A recent effort is that of Hsiao and Lu,[99] who used 15,000 data points (53 compounds) to determine their functions of $Z^{(0)}$ and $Z^{(1)}$. The parameter ranges are $0.2 \leq T_r \leq 5.0$, $0 \leq P_r \leq 12.0$, and $0 \leq \omega \leq 0.9065$. The resulting $Z^{(0)}$ and $Z^{(1)}$ are plotted in Figs. 3-45 and 3-46 respectively as functions of T_r and P_r. The authors report an average absolute deviation between calculated and experimental Z values of 0.51 percent. This correlation should not be used for quantum gases and highly polar compounds.

Example 2 The three required parameters for methyl chloride are[196] $T_c = 416.3$ K, $P_c = 65.9$ atm, and $\omega = 0.156$. At $T = 100°C = 373.15$ K, and $P = 10$ atm, $T_r = 373.15/416.3 = 0.896$, and $P_r = 10/65.9 = 0.152$. From

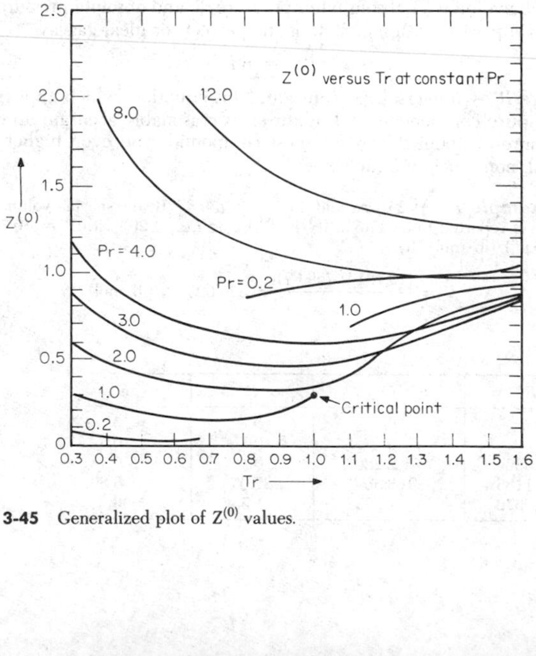

FIG. 3-45 Generalized plot of $Z^{(0)}$ values.

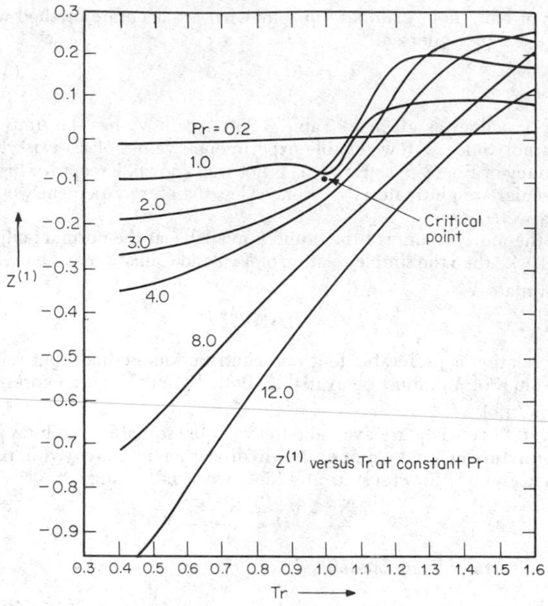

FIG. 3-46 Generalized plot of $Z^{(1)}$ values.

Fig. 3-45, $Z^{(0)} = 0.927$, and from Fig. 3-46, $Z^{(1)} = -0.031$. Substituting into Eq. (3-23),

$$Z = 0.927 + 0.156(-0.031) = 0.922$$

The reported experimental value is 0.9152.[100]

To use Eq. (3-23) for mixtures, equations are required which relate the T_c, P_c, and ω values for the mixture to the pure-component values and the mixture composition. The simplest expressions[196] which give acceptable results for an n-component mixture are

$$T_{c,\text{mix}} = \sum_{i=1}^{n} y_i T_{ci} \tag{3-24}$$[114]

$$P_{c,\text{mix}} = \frac{R\left(\sum_{i=1}^{n} y_i Z_{ci}\right) T_{c,\text{mix}}}{\sum_{i=1}^{n} y_i V_{ci}} \tag{3-25}$$[186]

$$\omega_{\text{mix}} = \sum_{i=1}^{n} y_i \omega_i \tag{3-26}$$[109]

where $T_{c,\text{mix}}$ = critical temperature of the mixture
$P_{c,\text{mix}}$ = critical pressure of the mixture
ω_{mix} = acentric factor of the mixture
T_{ci} = critical temperature of component i
Z_{ci} = critical compressibility factor of component i
V_{ci} = critical molar volume of component i
ω_i = acentric factor of component i
y_i = mole fraction of component i
R = gas constant in units consistent with those of $T_{c,\text{mix}}$ and V_{ci}

Virial Equation The virial equation is the only equation of state based on theory; it can be derived from statistical mechanics:

$$P = \frac{RT}{V} + \frac{RTB}{V^2} + \frac{RTC}{V^3} + \frac{RTD}{V^4} + \cdots \tag{3-27}$$

The parameters B, C, D, . . . are called the second, third, fourth, . . . virial coefficients and are functions of temperature and composition but not of pressure. The number of terms required for accurate predictions depends on the pressure: the higher the pressure, the more terms must be used. Although there is much information available on second virial coefficients, very little is known about the third coefficient, and essentially nothing about the fourth and higher coefficients. It is therefore common to truncate the virial equation after the second term:

$$P = RT/V + RTB/V^2 \tag{3-28}$$

When Eq. (3-21) is used in Eq. (3-28), an often more convenient form results:

$$Z = 1 + B/V \tag{3-29}$$

Because of the truncation, Eqs. (3-28) and (3-29) should not be used if the calculated V is less than twice the critical volume.

For pure components, second virial coefficients (B) have been critically evaluated and compiled by Dymond and Smith.[51] They can also be estimated by the correlation of Pitzer and Curl[184] for nonpolar compounds:

$$BP_c/RT_c = B^{(0)} + \omega B^{(1)} \tag{3-30}$$

where P_c = absolute critical pressure
T_c = absolute critical temperature
R = gas constant
ω = acentric factor

$B^{(0)}$ and $B^{(1)}$ are functions of reduced temperature ($T_r = T/T_c$) as modified by Tsonopoulos:[252]

$$B^{(0)} = 0.1445 - 0.330/T_r - 0.1385/T_r^2 - 0.0121/T_r^3 - 0.000607/T_r^8 \tag{3-31}$$

$$B^{(1)} = 0.0637 + 0.331/T_r^2 - 0.423/T_r^3 - 0.008/T_r^8 \tag{3-32}$$

Abbott[218] recommends the simpler equations

$$B^{(0)} = 0.083 - 0.422/T_r^{1.6} \tag{3-33}$$

$$B^{(1)} = 0.139 - 0.172/T_r^{4.2} \tag{3-34}$$

Example 3 The required properties to calculate the second virial coefficient for benzene at 380 K are[196] T_c = 562.1 K, P_c = 48.3 atm, and ω = 0.212. T_r = 380/562.1 = 0.6760. By using Eqs. (3-31) and (3-32),

$$B^{(0)} = 0.1445 - 0.330/0.6760 - 0.1385/(0.6760)^2 - 0.0121/(0.6760)^3 - 0.000607/(0.6760)^8 = -0.6998$$

$$B^{(1)} = 0.0637 + 0.331/(0.6760)^2 - 0.423/(0.6760)^3 - 0.008/(0.6760)^8 = -0.7647$$

Substituting into Eq. (3-30),

$$BP_c/RT_c = -0.6998 + (0.212)(-0.7647) = -0.8619$$

Using a value of R = 82.056 (atm·cm³)/(mol·K),

$$B = (-0.8619)(82.056)(562.1)/(48.3) = -823 \text{ cm}^3/\text{mol}$$

Using Eqs. (3-33) and (3-34),

$$B^{(0)} = 0.083 - 0.422/(0.6760)^{1.6} = -0.7066$$

$$B^{(1)} = 0.139 - 0.172/(0.6760)^{4.2} = -0.7517$$

Substituting into Eq. (3-30),

$$BP_c/RT_c = -0.7066 + (0.212)(-0.7517) = -0.8660$$

$$B = (-0.8660)(82.056)(562.1)/48.3 = -827 \text{ cm}^3/\text{mol}$$

The reported value is −810 ± 40 cm³/mol.[51]

For non-hydrogen-bonding polar compounds (e.g., ketones, ethers, aldehydes), Tsonopoulos[252] suggests that Eq. (3-30) be modified by the addition of another term:

$$BP_c/RT_c = B^{(0)} + \omega B^{(1)} + B^{(2)} \tag{3-35}$$

where

$$B^{(2)} = a/T_r^6 \tag{3-36}$$

a = $-2.410 \times 10^{-4}\mu_r - 4.308 \times 10^{-21}\mu_r^8$
μ_r = $10^5\mu_p^2 P_c/T_c^2$
μ_p = dipole moment, debyes
P_c = critical pressure, atm
T_c = critical temperature, K

Example 4 To calculate the second virial coefficient for diethyl ether at 340 K, the required properties are[196] T_c = 466.7 K, P_c = 35.9 atm, ω = 0.281, and μ_p = 1.3 debyes. T_r = 340/466.7 = 0.7285, and $\mu_r = 10^5(1.3)^2(35.9)/(466.7)^2$ = 27.855. Using Eqs. (3-31), (3-32), and (3-36),

$$B^{(0)} = 0.1445 - 0.330/0.7285 - 0.1385/(0.7285)^2 - 0.0121/(0.7285)^3 - 0.000607/(0.7285)^8 = -0.6084$$

$$B^{(1)} = 0.0637 + 0.331/(0.7285)^2 - 0.423/(0.7285)^3 - 0.008/(0.7285)^8 = -0.5075$$

$$B^{(2)} = [-2.140 \times 10^{-4}(27.855) - 4.308 \times 10^{-21}(27.855)^8]/(0.7285)^6 = -0.03988$$

Substituting into Eq. (3-35),

$$BP_c/RT_c = -0.6084 + (0.281)(-0.5075) + (-0.03988) = -0.7909$$

With R = 82.056 (atm·cm³)/(mol·K),

$$B = (-0.7909)(82.056)(466.7)/(35.9) = -844 \text{ cm}^3/\text{mol}$$

Dymond and Smith[51] select a value of − 880 ± 20 cm³/mol.

For hydrogen-bonding molecules, Tsonopoulos[252] recommends the use of Eq. (3-35) with

$$B^{(2)} = a/T_r^6 - b/T_r^8 \tag{3-37}$$

where a and b must be specified for each compound. He reports the following values:

	a	b
Methanol	0.0878	0.0560
Ethanol	0.0878	0.0572
n-Propanol	0.0878	0.0447
iso-Propanol	0.0878	0.0537
n-Butanol	0.0878	0.0367
sec-Butanol	0.0878	0.0487
iso-Butanol	0.0878	0.0481
tert-Butanol	0.0878	0.0508
Phenol	−0.0136	0.0
Water	0.0279	0.0229

A number of other methods have been proposed to calculate second virial coefficients for polar compounds, a few of which are Vetere,[196] Polak and Lu,[185] and Halm and Stiel.[83]

The second virial coefficient for mixtures of n components, B_{mix}, is obtained by the exact expression

$$B_{mix} = \sum_{i=1}^{n}\sum_{j=1}^{n} y_i y_j B_{ij} \tag{3-38}$$

where y_i = mole fraction of component i
B_{ii} = pure-component second virial coefficient of component i
B_{ij} = binary-interaction second virial coefficient for components i and j ($B_{ij} = B_{ji}$)

Thus, for a ternary mixture of components 1, 2, 3,

$$B_{mix} = y_1^2B_1 + y_2^2B_2 + y_3^2B_3 + 2y_1y_2B_{12} + 2y_1y_3B_{13} + 2y_2y_3B_{23} \tag{3-39}$$

The interaction coefficients B_{ij} may be calculated by using Eqs. (3-30) through (3-37) with the following combination rules:[252]

$$T_{cij} = (T_{ci}T_{cj})^{1/2}(1 - k_{ij}) \tag{3-40}$$

$$P_{cij} = \frac{4T_{cij}(P_{ci}V_{ci}/T_{ci} + P_{cj}V_{cj}/T_{cj})}{(V_{ci}^{1/3} + V_{cj}^{1/3})^3} \quad (3\text{-}41)$$

$$\omega_{ij} = \tfrac{1}{2}(\omega_i + \omega_j) \quad (3\text{-}42)$$

k_{ij} is a characteristic constant for each binary pair. For a first approximation, k_{ij} may be assumed to be zero. Lists of these constants for various pairs of compounds have been published.[252,33,96]

Equations (3-40), (3-41), and (3-42) are sufficient for nonpolar binaries. For polar-nonpolar binaries, B_{ij} is assumed to have no polar term, and so $B^{(2)}$ in Eq. (3-35) may be set to zero for these binary pairs. For polar-polar binaries, the a and b parameters in Eqs. (3-36) and (3-37) may be obtained by

$$a_{ij} = \tfrac{1}{2}(a_i + a_j) \quad (3\text{-}43)$$

$$b_{ij} = \tfrac{1}{2}(b_i + b_j) \quad (3\text{-}44)$$

Mention should also be made of the Hayden-O'Connell correlation[87] for the second virial coefficient. It includes the "chemical"-theory approach for associating compounds of Nothnagel et al.[171] and so is particularly appropriate for mixtures containing organic acids and other dimerizing molecules.

Two-Parameter Equations of State Beginning with the famous van der Waals equation,[254] a great number of two-constant equations of state have been proposed. The most widely used is the Redlich-Kwong:[192]

$$P = \frac{RT}{V - b} - \frac{a}{T^{0.5}V(V + b)} \quad (3\text{-}45)$$

where a and b are the constant parameters. Using the thermodynamic stability criteria[166] at the critical point, $(dP/dV)_{T_c} = 0$, and $(d^2P/dV^2)_{T_c} = 0$, in combination with Eq. (3-45), the values of a and b for pure components can be shown to be functions of the critical temperature T_c and critical pressure P_c:

$$a = \Omega_a R^2 T_c^{2.5}/P_c \quad (3\text{-}46)$$

$$b = \Omega_b R T_c/P_c \quad (3\text{-}47)$$

where $\Omega_a = [(9)(2^{1/3} - 1)]^{-1} = 0.42748\ldots$
$\Omega_b = (2^{1/3} - 1)/3 = 0.086640\ldots$

A more useful form for solving the Redlich-Kwong equation can be obtained by combining Eq. (3-21) with Eq. (3-45), which results in a cubic equation in Z:

$$Z^3 - Z^2 + (A - B^2 - B)Z - AB = 0 \quad (3\text{-}48)$$

where $A = aP/R^2T^{2.5}$ (3-49)
$B = bP/RT$ (3-50)

The largest real root of Eq. (3-48) is the correct value for the vapor. This equation is usually accurate to within 5 percent for vapor properties[196] except in the region of the critical point. It predicts a critical compressibility factor equal to ⅓; the value for most compounds is between 0.23 and 0.29.

For mixtures, Redlich and Kwong recommend the following expressions for a and b in Eqs. (3-45), (3-49), and (3-50):

$$a_{\text{mix}}^{1/2} = \sum_{i=1}^{n} y_i a_i^{1/2}$$

$$b_{\text{mix}} = \sum_{i=1}^{n} y_i b_i$$

where a_{mix} = a constant for the mixture
b_{mix} = b constant for the mixture
n = number of components in the mixture
y_i = mole fraction of component i
a_i = a constant calculated from Eq. (3-46) for component i
b_i = b constant calculated from Eq. (3-47) for component i

The success of the Redlich-Kwong correlation as a reliable two-parameter equation of state has resulted in a number of modifications. These usually have a and b as functions of temperature and/or the acentric factor and often involve an empirical binary-interaction parameter. Examples of modifications are Chueh and Prausnitz,[33] Zudkevitch and Joffe,[283] and Soave.[219]

Other two-term equations include the Martin[147] and Peng-Robinson.[180] Martin[148] has made an extensive critical review of all such equations.

Benedict-Webb-Rubin Equation One of the most popular equations of state is the Benedict-Webb-Rubin (BWR) equation.[8] Although it has eight constant parameters, both vapor and liquid properties may be calculated from it, and it is quite accurate for those compounds and their mixtures for which the constants have been determined. This equation was originally developed for gases and light hydrocarbons, but subsequent work has produced parameters for heavier compounds. However, it is generally limited to nonpolar molecules.

Expressed in terms of the molar density ρ, the BWR equation is

$$P = RT\rho + (B_0RT - A_0 - C_0/T^2)\rho^2 + (bRT - a)\rho^3 + a\alpha\rho^6 + (c\rho^3/T^2)(1 + \gamma\rho^2)\exp(-\gamma\rho^2) \quad (3\text{-}51)$$

where B_0, A_0, C_0, b, a, α, c, and γ are the eight constants. When solved for density, this equation can have up to six roots; the lowest value is the vapor quantity, and the highest is for the liquid.

It is recommended that C_0 be made a function of temperature, particularly at temperatures below and in the region of the normal boiling point.[8,113,177,220] This temperature dependence can be expressed as

$$C_0^{1/2} = CZ + C_1T + C_2T^2 + C_3T^3 + C_4T^4 + C_5T^5 \quad (3\text{-}52)$$

where CZ, C_1, C_2, C_3, C_4, C_5 are constant parameters. Equation (3-52) should be used between the temperature limits T_{LO} and T_{HI}, which are specific to each compound. For temperatures below T_{LO}, Eq. (3-52) should be linearly extrapolated with temperature; i.e.,

$$C_0^{1/2}(T) = C_0^{1/2}(T_{LO}) + m(T - T_{LO}) \quad (3\text{-}53)$$

where $m = [d(C_0^{1/2})]/dT$ at T_{LO}. The conventional value of C_0 (see Table 3-333) should be used at temperatures greater than T_{HI}; Eq. (3-52) should predict this value at T_{HI}.

A number of tabulations of BWR constants have been published.[8,113,177,220,15,37,128] Buck[28] has made a critical evaluation of these parameters; her recommendations for the constants in Eq. (3-51) are listed in Table 3-333 for 39 compounds. Those designated "vap." or "liq." should be used only for that phase; all others can be used for both phases. The selected parameters for use in Eq. (3-52) for 24 components are in Table 3-334, including T_{LO} and T_{HI}. These parameters are modifications of those from the listed references; they have been adjusted to be consistent with Eqs. (3-52) and (3-53).

Hydrogen is a special case. C_0 is temperature-independent, but b and γ are functions of temperature (in the same units as Table 3-333):

$$b = 3.3835927 \times 10^{-4} + 330.78829/RT \quad (3\text{-}54)$$

where $R = 8314.33\ (\text{Pa}\cdot\text{m}^3)/(\text{kmol}\cdot\text{K})$. For the temperature range T_1 to T_2,

$$\gamma = 3.2273131 \times 10^{-3} - 3.5776759 \times 10^{-6}(T_2 - T) \quad (3\text{-}55)$$

where $T_1 = 88.72$ K
$T_2 = 255.38$ K

For $T < T_1$, $\gamma = 2.6310576 \times 10^{-3}$.
For $T > T_2$, $\gamma = 3.2273131 \times 10^{-3}$.

For mixtures of n components, the eight constant parameters in Eq. (3-51) are functions of composition:

$$\chi_{\text{mix}} = \left(\sum_{i=1}^{n} y_i \chi_i^{1/r}\right)^r \quad (3\text{-}56)$$

where χ_{mix} = parameter for the mixture
y_i = mole fraction of component i
χ_i = pure-component parameter of component i

TABLE 3-333 BWR Constants: Eq. (3-51)*

Compound; Ref. no.	B_0	A_0	C_0	b	a	c	α	γ
Argon, vap. 37	.2228260D−01	83432.73	.1331537D+10	.2152890D−02	2921.787	.8088204D+08	.3558895D−04	.2338271D−02
Benzene 177	.5030059D−01	659632.3	.3475575D+12	.7663324D−01	564417.2	.1192082D+12	.7000384D−03	.2930112D−01
1,3-Butadiene 113	.9545566D−01	751863.7	.1053829D+12	.2800429D−01	141002.1	.2483083D+11	.7099623D−03	.2351617D−01
Isobutane 177	.1375469	1036866.	.8612441D+11	.4243716D−01	196343.6	.2898188D+11	.1074148D−02	.3400144D−01
n-Butane 177	.1243634	1021878.	.1006030D+12	.3999982D−01	190737.5	.3206134D+11	.1101390D−02	.3400144D−01
1-Butene 113	.1160274	917536.8	.9395851D+11	.3481707D−01	170440.0	.2785805D+11	.9109468D−03	.2959782D−01
cis-2-Butene 113	.1219749	995327.2	.1086163D+12	.3844685D−01	194288.7	.3384097D+11	.1057049D−02	.3268343D−01
Isobutylene 177	.1160274	907229.2	.9396083D+11	.3481706D−01	171523.7	.2785812D+11	.9109468D−03	.2959576D−01
Carbon dioxide 177	.4991091D−01	277379.8	.1404080D+11	.4124070D−02	13863.44	.1511650D+10	.8466750D−04	.5393795D−02
Carbon monoxide 37	.5454250D−01	135899.1	.8675538D+09	.2631580D−02	3713.561	.1053780D+09	.1350000D−03	.6000000D−02
n-Decane 177	.1727488	1729405.	.1333578D+13	.5076313	4190001.	.1056277D+13	.7370452D−02	.1380472
n-Decane, liq. 37	−6.231890	−3629.259	.1336477D+11	1.967010	.1267799D+08	.4488231D+09	.2144590D−02	.0
2,2-Dimethylbutane, vap. 37	.1921400	1199891.	.3404013D+12	.1400000	1024193.	.1771465D+12	.2189000D−02	.5650000D−01
2,3-Dimethylbutane, vap. 37	.1900000	1664770.	.2587233D+12	.7900000D−01	475781.7	.1149633D+12	.3594800D−02	.7500000D−01
2,2-Dimethylpentane, vap. 37	.8150500D−01	1236469.	.2241816D+12	.1772100	1194216.	.2288526D+12	.2764000D−02	.6799000D−01
2,2-Dimethylpropane 37	.1705300	1313527.	.1289867D+12	.6681200D−01	353674.9	.5532345D+11	.2000000D−02	.5000000D−01
Ethane 177	.6277381D−01	421081.0	.1819794D+11	.1112248D−01	34975.72	.3320337D+10	.2434045D−03	.1180048D−01
Ethylene 177	.5568449D−01	338397.5	.1328834D+11	.8600374D−02	26244.96	.2140129D+10	.1780112D−03	.9230404D−02
Helium 37	.2366100D−01	4150.475	16442.01	−.1972700D−06	−58.09874	−559.4153	−.7267300D−05	.7794200D−03
n-Heptane 177	.1990091	1775356.	.4808841D+12	.1519604	1050278.	.2502893D+12	.4356386D−02	.9000388D−01
Isohexane, vap. 37	.1729000	1512782.	.2887762D+12	.1215000	752702.9	.1418550D+12	.2350000D−02	.6200000D−01
n-Hexane 177	.1778167	1462927.	.3363485D+12	.1091355	721148.8	.1532908D+12	.2811034D−02	.6668780D−01
Hydrogen 177	.2084701D−01	15722.71	.3657087D+08	.3383593D−03	165.3975	736831.6	.1165677D−03	.3227313D−02
Hydrogen sulfide 177	.3484816D−01	314415.5	.1957241D+11	.4425055D−02	14691.79	.1895020D+10	.7032341D−04	.4555457D−02
Methane 177	.4260090D-01	187966.5	.2287011D+10	.3380183D-02	5005.791	.2578892D+09	.1243668D-03	.6000246D-2
Methane 37	.4546250D−01	182277.6	.3226006D+10	.2520330D−02	4409.664	.3635338D+09	.3300000D−03	.1050000D−01
Methyl Acetylene 113	.6946977D−01	517598.6	.6491471D+11	.1483247D−01	70643.74	.1113163D+11	.2736646D−03	.1245534D−01
3-Methylhexane, vap. 37	.9142300D−01	1449961.	.3198222D+12	.1432100	768590.7	.1342759D+12	.2815500D−02	.7446000D−01
3-Methylpentane, vap. 37	.8150500D−01	1236469.	.2241816D+12	.1122400	605072.4	.9682212D+11	.2250000D−02	.6289000D−01
Nitric oxide 37	.6045508D−01	222483.2	.1819362D+10	−.7531544D−02	−35546.99	−.1167642D+10	.1563696D−04	.2000000D−02
Nitrogen 177	.4579973D−01	120835.5	.5967371D+09	.1981615D−02	1509.846	.5553628D+08	.2915640D−03	.7500286D−02
Nitrogen, vap. 37	.4074260D−01	106760.3	.8165782D+09	.2327700D−02	2543.460	.7380614D+08	.1272000D−03	.5300000D−02
Nitrous oxide, vap. 37	.3945845D−01	257782.4	.1498999D+11	.5100164D−02	16996.43	.1615728D+10	.6555943D−04	.4812941D−02
n-Nonane 177	.1633034	1621681.	.1010965D+13	.3911164	3062233.	.7238104D+12	.5817363D−02	.1169047
n-Nonane, liq. 37	−2.320950	−1333.032	−.3246625D+12	.8564660	5589077.	.7921801D+11	.2328990D−02	.0
n-Octane 177	.1518030	1490560.	.7306480D+12	.2847121	2112782.	.4568033D+12	.4365266D−02	.9574765D−01
Oxygen 37	.3532850D−07	96345.08	.3307612D+10	.3588347D−02	16484.56	.1299734D+10	−3.927059	.3010000D−01
Isopentane 177	.1600566	1296603.	.1769537D+12	.6681485D−01	380622.8	.7042554D+11	.1700106D−02	.4630206D−01
n-Pentane 177	.1567541	1234134.	.2149415D+12	.6681485D−01	412907.5	.8351451D+11	.1810113D−02	.4750203D−01
1-Pentene 113	.1279253	1120050.	.1407172D+12	.4228867D−01	229299.3	.4598094D+11	.1219347D−02	.3594937D−01
Propane 177	.9731520D−01	696362.2	.5150139D+11	.2250097D−01	96032.18	.1307179D+11	.6072144D−03	.2200091D−01
Propylene 177	.8506621D−01	619345.7	.4450210D+11	.1870669D−01	78436.45	.1039777D+11	.4557252D−03	.1829077D−01
Sulfur dioxide, vap. 37	.2618200D−01	214853.6	.8043584D+11	.1465300D−01	85587.20	.1148519D+11	.7195500D−04	.5923600D−02

*Pressure = Pa; volume = $m^3/(k \cdot mol)$; temperature = K.
Some values are in double precision scientific notation; e.g., .7900000D-01 = 0.79×10^{-1}.

TABLE 3-334 BWR Constants: Eq. (3-52)

Compound	Ref.	T_{LO} / T_{HI}	CZ / C_0 at T_{HI}	C_1	C_2	C_3	C_4	C_5
Benzene	177	273.33	2949872.9	−37868.184	240.33684	−.75612542	.11811081D−02	−.73359957D−06
		373.16	.3475575D+12					
1,3-Butadiene	113	212.50	195054.89	932.12515	−2.3794609	.21395330D−02	.0	.0
		408.82	.1053829D+12					
Isobutane	177	163.30	2215799.1	−46903.055	447.81248	−2.1084100	.49127655D−02	−.45387795D−05
		269.27	.8612441D+11					
n-Butane	177	164.77	1219928.4	−21614.603	196.75588	−.87321324	.19092915D−02	−.16522697D−05
		294.27	.1006030D+12					
cis-2-Butene	113	217.79	123351.56	2173.3742	−8.6258725	.15314599D−01	−.10264811D−04	.0
		435.84	.1086162D+12					
Isobutylene	177	165.41	1279640.5	−23372.551	214.43769	−.96056110	.21194115D−02	−.18489505D−05
		294.27	.9396083D+11					
Carbon dioxide	177	185.19	2107495.3	−45523.556	410.90705	−1.8345281	.40608036D−02	−.35715229D−05
		249.72	.1404080D+11					
n-Decane	177	276.17	948004.86	1647.0679	−3.5402534	.22773707D−02	.0	.0
		592.19	.1333578D+13					
Ethane	177	88.72	111892.84	−115.58780	4.1372066	−.26136501D−01	.70916299D−04	−.77287552D−07
		235.94	.1819795D+11					
Ethylene	177	88.72	139802.24	−2198.2366	40.219210	−.31442506	.11474663D−02	−.16070436D−05
		205.38	.1328834D+11					
n-Heptane	177	187.16	536598.05	1925.1615	−16.788424	.80255078D−01	−.17616065D−03	.14121718D−06
		372.05	.4808841D+12					
n-Hexane	177	186.01	524832.90	−76.350471	3.2050984	−.17389375D−01	.46383657D−04	−.49742039D−07
		349.83	.3363485D+12					
Hydrogen	177	0.0	6047.6331	.0	.0	.0	.0	.0
		0.0	.3657387D+08					
Hydrogen sulfide	177	201.12	66654.740	598.04646	−1.5216043	.11833064D−02	.0	.0
		361.55	.1957241D+11					
Methane	177	88.72	63810.054	−1156.5851	22.430422	−.19660009	.83585195D−03	−.14082303D−05
		160.94	.2287011D+10					
Methyl acetylene	113	201.19	166289.46	605.35598	−1.2716348	.41330659D−03	.91581956D−06	.0
		395.94	.6491471D+11					
Nitrogen	177	0.0	24428.204	.0	.0	.0	.0	.0
		0.0	.5967371D+09					
n-Nonane	177	262.25	854581.17	1177.6457	−2.4124373	.14489840D−02	.0	.0
		601.34	.1010965D+13					
n-Octane	177	262.22	717592.42	1024.9385	−2.1607919	.13762442D−02	.0	.0
		568.53	.7306480D+12					
Isopentane	177	182.87	1828222.2	−31661.795	277.34169	−1.2012055	.25899463D−02	−.22291006D−05
		285.94	.1769536D+12					
n-Pentane	177	183.88	797193.52	−7288.3825	56.467595	−.20786825	.37376742D−03	−.26587363D−06
		349.83	.2149415D+12					
1-Pentene	113	232.39	−208376.38	6992.4807	−31.691218	.64512488D−01	−.49198343D−04	.0
		469.01	.1407171D+12					
Propane	177	88.72	183657.63	224.77276	−.27168015	.40766214D−02	−.24567358D−04	.36696788D−07
		283.16	.5150139D+11					
Propylene	177	88.72	201974.48	−710.66638	12.355683	−.82712738D−01	.26064327D−03	−.31886558D−06
		255.38	.4450210D+11					

°Units, C_0: Pa $[(K \cdot m^3)/(kmol)]^2$; temperature: K.

The values of r for each constant are

Constant χ	r	Constant χ	r
B_0	1	a	3
A_0	2	α	3
C_0	2	c	3
b	3	γ	2

There have been several efforts to generalize the BWR equation or modifications of it to apply to more types of compounds.[52,176,237,277] Two of the more widely used correlations of this type are those of Starling[228,229] and Lee and Kesler.[120] Starling has proposed an 11-constant equation in which the pure-component parameters are functions of the critical temperature, critical volume, acentric factor, and a binary interaction parameter. In the Lee-Kesler method, a modified BWR equation is used to calculate $Z^{(0)}$ and $Z^{(1)}$ in Eq. (3-23) by relating the compressibility factor of a real fluid to properties of a simple fluid (acentric factor = 0) and those of n-octane as a reference fluid. The authors report average errors of less than 2 percent for both the vapor and the liquid phases when comparing with primarily hydrocarbon data.

In recent years, a new approach has been tried in the development of equations of state for fluids: the hard-sphere equations. Examples of this technique include Carnahan-Starling[30] and Kreglewski et al.[118,117] Further effort along these lines has included perturbation theory, which relates properties of the system of interest to the known properties of a reference system. The perturbed-hard-sphere equation of Nakamura et al.[169] is a good example of this procedure, as is the Beret-Prausnitz perturbed-hard-chain equation[11] as modified by Donohue and Prausnitz.[46] This equation has been further extended by Gmehling et al.[67] to include chemical-dimerization equilibriums. These correlations show great promise in the calculation of both vapor and liquid properties for very nonideal, polar mixtures.

LIQUID DENSITY

For most pure compounds, at least one experimental saturated-liquid density is available.[196,284,285] When this is the case, the Rackett equation[190] as modified by Spencer and Danner[221] can be used to calculate saturated-liquid densities at absolute temperature T:

$$\rho^{sat} = P_c/RT_cZ_{RA}^{[1+(1-T_r)^{2/7}]} \tag{3-57}$$

where ρ^{sat} = saturated-liquid molar density
P_c = critical pressure
R = gas constant
T_c = absolute critical temperature
Z_{RA} = constant determined from experimental data
T_r = reduced temperature = T/T_c

Expected errors are less than 1 percent except for alcohols and acids, for which the errors are less than 2 percent.[223] Spencer and Adler[223] list values of Z_{RA} for 165 compounds. If Z_{RA} cannot be determined from experimental information, then the critical compressibility factor Z_c may be used in place of Z_{RA}; however, this will increase the possible error to 3 or 4 percent.[221]

Example 5 Spencer and Adler report the following values for benzene: P_c = 48.33 atm, T_c = 562.15 K, and Z_{RA} = 0.26967. For a temperature T of 423.15 K, T_r = 423.15/562.15 = 0.75274. Substituting in Eq. (3-57) with R = 82.05606 (atm·cm³)/(mol·K),

$$\rho^{sat} = 48.33/(82.05606)(562.15)(0.26967)^{[1+(1-0.75274)^{2/7}]}$$
$$= 0.0093592 \text{ mol/cm}^3$$

Since the molecular weight of benzene is 78.1134, ρ^{sat} = (0.0093592)(78.1134) = 0.7331 g/cm³. The reported value is 0.7295 g/cm³.[284]

Other methods requiring at least one experimental point have been documented.[196,85]

If no experimental data are available, then the correlation proposed by Bhirud[13] may be used for nonpolar compounds at absolute temperature T:

$$\ln (V^{sat}P_c)/RT_c) = (\ln U)^0 + \omega(\ln U)^1 \tag{3-58}$$

where V^{sat} = saturated-liquid molar volume
P_c = critical pressure
R = gas constant
T_c = absolute critical temperature
ω = acentric factor

$(\ln U)^0$ and $(\ln U)^1$ are functions of the reduced temperature $T_r = T/T_c$:

$$(\ln U)^0 = \ln T_r + 1.39644 - 24.076T_r + 102.615T_r^2 - 255.719T_r^3 + 355.805T_r^4 - 256.671T_r^5 + 75.1088T_r^6 \tag{3-59}$$

$$(\ln U)^1 = 13.4412 - 135.7437T_r + 533.380T_r^2 - 1091.453T_r^3 + 1231.43T_r^4 - 728.227T_r^5 + 176.737T_r^6 \tag{3-60}$$

Bhirud reports an average error of less than 1 percent when comparing this correlation against 752 data points for 24 compounds (mostly hydrocarbons).

Example 6 For n-hexane, the required properties[196] are P_c = 29.3 atm, T_c = 507.4 K, and ω = 0.296. At T = 423.15 K, T_r = 423.15/507.4 = 0.83396. Substituting in Eqs. (3-59) and 3-60),

$$(\ln U)^0 = \ln 0.83396 + 1.39644 - (24.076)(0.83396) + (102.615)(0.83396)^2 - (255.719)(0.83396)^3 + (355.805)(0.83396)^4 - (256.671)(0.83396)^5 + (75.1088)(0.83396)^6 = -1.98203$$

$$(\ln U)^1 = 13.4412 - (135.7437)(0.83396) + (533.380)(0.83396)^2 - (1091.453)(0.83396)^3 + (1231.43)(0.83396)^4 - (728.227)(0.83396)^5 + (176.737)(0.83396)^6 = -0.511916$$

Substituting in Eq. (3-58),

$$\ln (V^{sat}P_c/RT_c) = -1.98203 + (0.296)(-0.511916) = -2.13356$$

Rearranging,

$$V^{sat} = (RT_c/P_c)e^{-2.13356}$$

Using a value of R = 82.05606 (atm·cm³)/(mol·K),

$$V^{sat} = \frac{(82.05606)(507.4)}{(29.3)} e^{-2.13356} = 168.3 \text{ cm}^3\text{/mol}$$

The experimental density is 0.5217 g/cm³.[284] With a molecular weight of 86.1766, the corresponding molar volume is 86.1766/0.5217 = 165.2 cm³/mol.

Bhirud[12] has devised a similar expression for polar compounds, but it does require an experimental datum point.

If a reliable value of the critical volume is known, the liquid molar volume at the normal boiling point V_b may be estimated by the method of Tyn and Calus:[253]

$$V_b = 0.285\, V_c^{1.048} \tag{3-61}$$

where V_c = critical molar volume. For most compounds, this simple relation is accurate to within 3 percent.

Example 7 The critical volume of acetone is 209.0 cm³/mol.[196] Substituting in Eq. (3-61),

$$V_b = (0.285)(209.0)^{1.048} = 77.0 \text{ cm}^3\text{/mol}$$

The experimental value is 77.5 cm³/mol.[196]

For compressed liquids (i.e., at pressures greater than the vapor pressure), the Chueh and Prausnitz correlation[35] can be used to calculate the density ρ at absolute temperature T:

$$\rho = \rho^{sat}\left[1 + \frac{9Z_cN(P - P^{sat})}{P_c}\right]^{1/9} \tag{3-62}$$

where ρ^{sat} = saturated-liquid density
Z_c = critical compressibility factor
P = system pressure
P^{sat} = vapor pressure
P_c = critical pressure

N is a function of the acentric factor ω and the reduced temperature $T_r = T/T_c$, where T_c = absolute critical temperature:

$$N = [1.0 - 0.89\omega^{1/2}]\,[\exp (6.9547 - 76.2853\, T_r + 191.3060\, T_r^2 - 203.5472\, T_r^3 + 82.7631\, T_r^4)] \tag{3-63}$$

Example 8 For n-heptane at T = 423.15 K, ρ^{sat} = 0.5603 g/cm³,[284] and P^{sat} = 3.692 atm[284] = 3.741 bar, T_c = 540.2 K,[196] and so T_r = 423.15/540.2 = 0.7833. P_c = 27.0 atm[196] = 27.36 bar, Z_c = 0.263,[196] and ω = 0.351.[196] Substituting in Eq. (3-63),

$$N = [1.0 - (0.89)(0.351)^{1/2}]\,\{\exp [6.9547 - (76.2853)(0.7833) + (191.3060)(0.7833)^2 - (203.5472)(0.7833)^3 + (82.7631)(0.7833)^4]\}$$
$$= 0.05844$$

For P = 2000 bar in Eq. (3-62), we have

$$\rho = 0.5603\left[1 + \frac{(9)(0.263)(0.05844)(2000.0 - 3.741)}{27.36}\right]^{1/9} = 0.7320 \text{ g/cm}^3$$

The reported experimental value is 1/1.371 = 0.7294 g/cm³.[47]

Mixtures At moderate pressures, Amagat's law may be used for mixtures of similar compounds:

$$V_{mix} = \sum_{i=1}^{n} x_iV_i \tag{3-64}$$

where V_{mix} = liquid molar volume of the mixture
n = number of components
x_i = mole fraction of component i
V_i = liquid molar volume of pure component i at the temperature of the mixture

Note that the liquid molar volumes, *not* the densities, are being mole-fraction-averaged.

For a hydrocarbon mixture of n components, a modification of Eq. (3-57) may be used to calculate the bubble-point density.[222] The mixture values of the critical constants and Z_{RA} are obtained by mole-fraction averaging. At absolute temperature T, the expression for the bubble-point density ρ^{BP} is

$$\rho^{BP} = P_{c,\text{mix}}/RT_{c,\text{mix}}Z_{RA,\text{mix}}^{[1+(1-T_{r,\text{mix}})^{2/7}]} \quad (3\text{-}65)$$

where $P_{c,\text{mix}}$ = mixture critical pressure = $\sum_{i=1}^{n} x_i P_{ci}$

R = gas constant

$T_{c,\text{mix}}$ = mixture critical temperature = $\sum_{i=1}^{n} x_i T_{ci}$

$Z_{RA,\text{mix}} = \sum_{i=1}^{n} x_i Z_{RAi}$

$T_{r,\text{mix}}$ = reduced temperature of mixture = $T/T_{c,\text{mix}}$

x_i = mole fraction of component i

P_{ci} = critical pressure of pure component i

T_{ci} = critical temperature of component i

Z_{RAi} = constant determined from experimental data for pure component i

Example 9 The required pure-component properties for propane and n-butane are as follows:[223]

	Propane	n-Butane
P_c, atm	41.93	37.46
T_c, K	369.82	425.15
Z_{RA}	0.27664	0.27331

For a mixture of 30.85 mole percent propane at 367.95 K,

$$P_{c,\text{mix}} = (0.3085)(41.93) + (0.6915)(37.46) = 38.84 \text{ atm}$$

$$T_{c,\text{mix}} = (0.3085)(369.82) + (0.6915)(425.15) = 408.08 \text{ K}$$

$$Z_{RA,\text{mix}} = (0.3085)(0.27664) + (0.6915)(0.27331) = 0.27434$$

$$T_{r,\text{mix}} = 367.95/408.08 = 0.90166$$

Substituting in Eq. (3-65) by using a value of $R = 82.05606$ (atm·cm^3)/(mol·K),

$$\rho^{BP} = 38.84/(82.05606)(408.08)(0.27434)^{[1+(1-0.90166)^{2/7}]}$$
$$= 0.008235 \text{ mol/cm}^3$$

Using molecular weights of propane = 44.0962 and n-butane = 58.123, the mole average molecular weight is obtained as (0.3085)(44.0962) + (0.6915)(58.123) = 53.796. Thus,

$$\rho^{BP} = (0.008235)(53.796) = 0.4430 \text{ g/cm}^3$$

The experimental value is 0.4418 g/cm^3.[115] For dissimilar hydrocarbon and nonhydrocarbon mixtures, it is recommended that a more accurate critical-temperature mixing rule be used for $T_{r,\text{mix}}$.[222]

VAPOR PRESSURE

A widely used vapor-pressure correlation is the Antoine equation:[6]

$$\ln P^{\text{sat}} = A - B/(T + C) \quad (3\text{-}66)$$

where P^{sat} is vapor pressure and T is temperature. The constant parameters A, B, and C must be obtained by regressing experimental data;[18,111,235] however, these parameters have been tabulated for many substances.[18,172,196,284,285] The applicable temperature range is not large and in most cases corresponds to a pressure interval of about 10 to 1500 mmHg (0.0133 to 2.0 bar). This equation should *not* be used outside the stated temperature limits.

To extend the correlation interval, a more complicated expression such as the Riedel equation[200] is required:

$$\ln P^{\text{sat}} = A - B/T + C \ln T + DT^6 \quad (3\text{-}67)$$

When the constant parameters are determined by regressing experimental data, the range of this correlation extends to the critical point. A better fit of the data is often obtained when the exponent 6 is allowed to vary to other integer values.

If only two vapor-pressure points, (P_1^{sat}, T_1) and (P_2^{sat}, T_2), are known, then the Clapeyron equation may be used:

$$\ln P^{\text{sat}} = A - B/T \quad (3\text{-}68)$$

where $A = \ln P_1^{\text{sat}} + B/T_1$

$$B = \ln\left(\frac{P_2}{P_1}\right) \Big/ \left(\frac{1}{T_1} - \frac{1}{T_2}\right)$$

Equation (3-68) is a fairly good expression for approximating vapor pressure over small temperature ranges, but significant errors can result over large temperature intervals.

If the critical pressure P_c and the critical temperature T_c are known along with one other vapor-pressure point (such as the normal boiling point), then a reduced form of the Riedel equation may be used:

$$\ln P_r^{\text{sat}} = A^\circ - B^\circ/T_r + C^\circ \ln T_r + D^\circ T_r^6 \quad (3\text{-}69)$$

where P_r^{sat} = reduced vapor pressure = P^{sat}/P_c

T_r = reduced temperature = T/T_c

$A^\circ = -35Q$

$B^\circ = -36Q$

$C^\circ = 42Q + \alpha_c$

$D^\circ = -Q$

$Q = 0.0838(3.758 - \alpha_c)$

α_c can be determined by inserting the other known vapor-pressure point (P_1^{sat}, T_1) into Eq. (3-69) and solving for α_c:

$$\alpha_c = \frac{0.315\psi_1 - \ln P_{1r}^{\text{sat}}}{0.0838\psi_1 - \ln T_{1r}} \quad (3\text{-}70)$$

where $\psi_1 = -35 + 36/T_{1r} + 42 \ln T_{1r} - T_{1r}^6$

Example 10 Ambrose et al.[1] report the following data for acetone: P_c = 4700.0 kPa, T_c = 508.1 K, P_1^{sat} = 100.666 kPa, and T_1 = 329.026 K. Thus, T_{1r} = 329.026/508.1 = 0.64756, and P_{1r}^{sat} = 100.666/4700.0 = 0.021418. Substituting in Eq. (3-70),

$$\psi_1 = -35 + (36/0.64756) + 42 \ln 0.64756 - (0.64756)^6$$
$$= 2.2687$$
$$\alpha_c = \frac{(0.315)(2.2687) - \ln 0.021418}{(0.0838)(2.2687) - \ln 0.64756}$$
$$= 7.2970$$

Thus, the constant parameters in Eq. (3-69) are

$$Q = 0.0838(3.758 - 7.2970) = -0.29657$$
$$A^\circ = (-35)(-0.29657) = 10.380$$
$$B^\circ = (-36)(-0.29657) = 10.677$$
$$C^\circ = (42)(-0.29657) + 7.2970 = -5.1589$$
$$D^\circ = -(-0.29657) = 0.29657$$

Substituting in Eq. (3-69),

$$\ln P_r^{\text{sat}} = 10.380 - 10.677/T_r - 5.1589 \ln T_r + 0.29657 T_r^6$$

For a temperature T of 350.874 K, T_r = 350.874/508.1 = 0.69056, and the calculated reduced vapor pressure is given by

$$\ln P_r^{\text{sat}} = 10.380 - 10.677/0.69056 - 5.1589 \ln 0.69056$$
$$+ 0.29657(0.69056)^6$$
$$= -3.1391$$
$$P_r^{\text{sat}} = e^{-3.1391} = 0.043322$$
$$P^{\text{sat}} = (0.043322)(4700.0) = 203.61 \text{ kPa}$$

The experimental value is 201.571 kPa.[1]

There are also a number of other equations reported elsewhere[196,70,71] that can be applied with equivalent accuracy which require the critical and one other vapor pressure point.

ENTHALPY OF VAPORIZATION

The enthalpy of vaporization, ΔH_v, is related to vapor pressure by the thermodynamically exact Clausius-Clapeyron equation:

$$\Delta H_v = -R\,\Delta Z_v \frac{d \ln P^{\text{sat}}}{d(1/T)} \quad (3\text{-}71)$$

where R = gas constant
$\Delta Z_v = Z_G - Z_L$
Z_G = compressibility factor of saturated vapor
Z_L = compressibility factor of saturated liquid
P^{sat} = vapor pressure
T = absolute temperature

If accurate Z_G and Z_L data are available, excellent ΔH_v values can be obtained by differentiating one of the vapor-pressure correlations and using Eq. (3-71). If not, ΔZ_v can be estimated by Haggenmacher's equation:[82]

$$\Delta Z_v = \left(1 - \frac{P_r}{T_r^3}\right)^{1/2} \tag{3-72}$$

where P_r = reduced pressure = P/P_c
T_r = reduced temperature = T/T_c

However, Eq. (3-72) should be used only near or below the normal boiling point; even then, the accuracy of the resulting ΔH_v is significantly reduced.

Another reasonable technique is the corresponding-states approach suggested by Pitzer.[183] For a close approximation, an analytical representation[196] of this method for $0.6 < T_r \leq 1.0$ is

$$\Delta H_v/RT_c = 7.08(1 - T_r)^{0.354} + 10.95\omega(1 - T_r)^{0.456} \tag{3-73}$$

where R = gas constant
T_c = critical temperature (absolute)
T_r = reduced temperature = T/T_c
ω = acentric factor

Example 11 For propionaldehyde, the required properties[196] are T_c = 496.0 K and ω = 0.313. At temperature T = 302.69 K, T_r = 302.69/496.0 = 0.6103. Substituting into Eq. (3-73),

$$\Delta H_v/RT_c = (7.08)(1 - 0.6103)^{0.354} + (10.95)(0.313)(1 - 0.6103)^{0.456}$$
$$= 7.302$$

Using a value of 8.31433 J/(mol·K) for the gas constant,

$$\Delta H_v = (7.302)(8.31433)(496.0) = 30110 \text{ J/mol}$$

The reported experimental value is 29374 J/mol.[38]

The enthalpy of vaporization at the normal boiling point, ΔH_{vb}, can be estimated by an equation proposed by Riedel:[201]

$$\Delta H_{vb} = 1.093RT_c \left[T_{b_r} \frac{(\ln P_c - 1)}{0.930 - T_{b_r}} \right] \tag{3-74}$$

where R = gas constant
T_c = critical temperature (absolute)
T_{b_r} = reduced normal boiling-point temperature = T_b/T_c
P_c = critical pressure, atm

Example 12 The necessary properties for propionaldehyde[196] are T_b = 321.0 K, T_c = 496.0 K, and P_c = 47.0 atm. Thus, T_{b_r} = 321.0/496.0 = 0.6472. For R = 8.31433 J/(mol·K),

$$\Delta H_{vb} = (1.093)(8.31433)(496.0)\frac{0.6472(\ln 47.0 - 1)}{0.930 - 0.6472}$$
$$= 29400 \text{ J/mol}$$

The experimental value is 28297 J/mol.[38]

Other correlations of similar approach and accuracy may be found in Reid et al.[196]

The enthalpy of vaporization decreases with temperature and is zero at the critical point. If the value of an enthalpy of vaporization ΔH_{v1} is known at temperature T_1, this temperature dependency can be represented by the Watson relation[268] to calculate another enthalpy of vaporization ΔH_{v2} at any other temperature T_2:

$$\Delta H_{v2} = \Delta H_{v1} \left(\frac{1 - T_{r2}}{1 - T_{r1}}\right)^{0.38} \tag{3-75}$$

where T_r = reduced temperature T/T_c. Equation (3-75) will yield values satisfactory for engineering calculations between the normal boiling point and the critical point.

IDEAL-GAS HEAT CAPACITY

The ideal-gas heat capacity, C_p°, is a function of temperature but not of pressure ($P^\circ \to 0$). For mixtures, $C_{p,\text{mix}}^\circ$ is a mole-fraction average.

A number of analytical equations have been suggested to relate C_p° to temperature; the most common is a simple polynomial,

$$C_p^\circ = a + bT + cT^2 + dT^3 \tag{3-76}$$

Tables of C_p° as a function of temperature are given in the *JANAF Thermochemical Tables*[107] and in a compilation by Stull et al.[236] Constants for use in analytical relations as Eq. (3-76) are also available[101,179,247,249,280] for many compounds.

If C_p° values are in none of the tabulations noted here, there are available several estimation methods that require only the molecular structure. The methods of Thinh et al.[246] and Benson[175,211,214,230,9] are the most accurate but also the most complicated. They are reviewed and compared elsewhere.[196] The simple and reasonably accurate additive group-contribution method of Rihani and Doraiswamy makes use of the values shown in Table 3-335 to determine the constants in Eq. (3-76); in this case the temperature must be given in kelvins and C_p° is calculated in cal/(mol·K). Table 3-335 should not be used below about 270 K.

HEAT OF FORMATION

The standard heat of formation, ΔH_f°, in the ideal-gas state is required to determine heats of reaction. Domalski[45] has critically reviewed the literature and has recommended values for $\Delta H_{f\,298}^\circ$ for over 700 organic compounds; also Refs. 9, 10, 39, 107, and 236 are valuable sources of experimental values. Estimation techniques are also available.[175,211,214,230,9,196,60,61,248,3,98] A compromise between complexity and accuracy is found in the additive group-contribution scheme of Verma-Doraiswamy,[263] and values for the various atomic groups are shown in Table 3-335. The units for $\Delta H_{f\,298}^\circ$ are kcal/mol.

Only $\Delta H_{f\,298}^\circ$ may be found; for other temperatures,

$$\Delta H_{f\,T}^\circ = \Delta H_{f\,298}^\circ + \int_{298}^{T} C_p^\circ \, dT \tag{3-77}$$

Determination of C_p°, is discussed earlier.

GIBBS ENERGY

The standard Gibbs energy of formation, ΔG_f°, is of value to determine the extent of a chemical reaction. It may be found from standard heats and entropies of formation,

$$\Delta G_f^\circ = \Delta H_f^\circ - T\Delta S_f^\circ \tag{3-78}$$

or from standard reference compilations, e.g., Refs. 107 and 236. Estimation methods are not particularly accurate. Van Krevelen and Chermin[255,256] suggest

$$\Delta G_{f\,T}^\circ = A + BT \tag{3-79}$$

where A and B are found from the group contributions in Table 3-335; two temperature ranges are given, from 300 to 600 K and from 600 to 1500 K. In Eq. (3-79), T is in kelvins and ΔG_f° is in kilocalories per mole. In addition, one must correct for symmetry and optical isomers to obtain ΔG_f°. The symmetry correction is $R \ln \sigma$ and is added to the constant B. σ is the number of independent permutations of the entire molecule which appear identical to an observer. For example, for methane σ = 4, or for acetone σ = 2. If the compound has optical isomers, B must also be corrected by $-R \ln \eta$, where η is the number of such isomers.

With the use of Eq. (3-79) and Table 3-335, $\Delta G_{f\,T}^\circ$ can usually be estimated only within about ±5 kcal/mol.

TABLE 3-335 Group Contributions to Estimate C°_p, ΔH°_{f298}, and ΔG°_f*

Group	Heat capacity constants: a	$b \times 10^2$	$c \times 10^4$	$d \times 10^8$	ΔH°_f, 298 K	Gibbs energy constants, 300–600 K: A	$B \times 10^2$	600–1500 K: A	$B \times 10^2$
$-CH_3$	0.6087	2.1433	−0.0852	0.1135	10.25	−10.943	2.215	−12.310	2.436
$-CH_2-$	0.3945	2.1363	−0.1197	0.2596	−4.94	−5.193	2.430	−5.830	2.544
$-\overset{\vert}{C}H$ (–CH with three bonds)	−3.5232	3.4158	−0.2816	0.8015	−1.29	−0.705	2.910	−0.705	2.910
$-\overset{\vert}{\underset{\vert}{C}}-$	−5.8307	4.4541	−0.4208	1.263	0.62	1.958	3.735	4.385	3.350
H\C=CH₂ (H, one bond)	0.2773	3.4580	−0.1918	0.4130	15.02	13.737	1.655	12.465	1.762
\C=CH₂ (two bonds)	−0.4173	3.8857	−0.2783	0.7364	20.50	16.467	1.915	16.255	1.966
HC=CH (*cis*)	−3.1210	3.0860	−0.2359	0.5504	17.96	17.663	1.965	16.180	2.116
HC=CH (*trans*)	0.9377	2.9904	−0.1749	0.3918	17.83	17.187	1.915	15.815	2.062
>C=CH< (one H)	−1.4714	3.3842	−0.2371	0.6063	−20.10	20.217	2.295	19.584	2.354
>C=C<	0.4736	3.5183	−0.3150	0.9205	30.46	25.135	2.573	25.135	2.573
H\C=C=CH₂	2.2400	4.2896	−0.2566	0.5908	49.47	49.377	1.035	48.170	1.208
>C=C=CH₂	2.6308	4.1658	−0.2845	0.7277	51.30	51.084	1.474	51.084	1.474
HC=C=CH	−3.1249	6.6843	−0.5766	1.743	55.04	52.460	1.483	52.460	1.483
≡CH	.....	.....	.....	.....	27.10	27.048	−0.765	26.700	−0.704
≡C−	.....	.....	.....	.....	27.38	26.938	−0.525	26.555	−0.550
Conjugated alkene groups									
↔CH_2	.....	.....	.....	.....	(10.1)	5.437	0.675	4.500	0.832
↔CH< (H)	.....	.....	.....	.....	(12)	7.407	1.035	6.980	1.088
↔C<	.....	.....	.....	.....	.....	9.152	1.505	10.370	1.308
↘CH↙ (aromatic)	−1.4572	1.9147	−0.1233	0.2985	3.27	3.047	0.615	2.505	0.706
↘C−↙ (aromatic)	−1.3883	1.5159	−0.069	0.2659	5.55	4.675	1.150	5.010	0.988
↘C↔↙ (aromatic)	0.1219	1.2170	−0.0855	0.2122	4.48	3.513	0.568	3.998	0.485
Cycloparaffin-ring correction									
Three-membered	−3.5320	−0.0300	0.0747	−0.5514	24.13	23.458	−3.045	22.915	−2.966
Four-membered	−8.6550	1.0780	0.0425	0.0250	18.45	10.73	−2.65	10.60	−2.50
Five-membered									
Pentane	−12.2850	1.8609	−0.1037	0.2145	5.44	4.275	−2.350	2.665	−2.182
Pentene	−6.8813	0.7818	−0.0345	0.0591	....	−3.657	−2.395	−3.915	−2.250
Six-membered									
Hexane	−13.3923	2.1392	−0.0429	−0.1865	−0.76	−1.128	−1.635	−1.930	−1.504
Hexene	−8.0238	2.2239	−0.1915	0.5473	....	−9.102	−2.045	−8.810	−2.071
Branching in paraffins									
Side chain with two or more carbon atoms	.....	.....	.....	.....	0.80	1.31	0	1.31	0
Three adjacent $-\overset{\vert}{\underset{\vert}{C}}H$ groups	.....	.....	.....	.....	−1.2	−2.13	0	2.12	0

*Units: Cp, kcal/(mol·K); ΔH°_{f298}, kcal/mol; ΔG°_f, kcal/mol. To convert kilocalories per mole to British thermal units per pound-mole, multiply by 1800; to convert kilocalories per mole-kelvin to British thermal units per pound-mole-degree Fahrenheit, multiply by 1000."

TABLE 3-335 Group Contributions to Estimate C°_{pr}, ΔH°_{f298}, and ΔG°_f *(Continued)*

Group	Heat capacity constants				ΔH°_f, 298 K	Gibbs energy constants			
						300–600 K		600–1500 K	
	a	$b \times 10^2$	$c \times 10^4$	$d \times 10^8$		A	$B \times 10^2$	A	$B \times 10^2$
Adjacent $-\overset{\vert}{\underset{\vert}{C}}-$ and $-\overset{\vert}{\underset{\vert}{C}}H$ groups	.....	.....	.....	.....	0.6	1.80	0	1.80	0
Two adjacent $-\overset{\vert}{\underset{\vert}{C}}-$ groups	.....	.....	.....	.....	(5.4)	2.58	0	2.58	0
Branching in five-membered rings									
Single branching	.....	.....	.....	.....	0	−1.04	0	−1.69	0
Double branching									
1,1 position	.....	.....	.....	.....	0.30	−1.85	0	−1.19	−0.16
cis-1,2 position	.....	.....	.....	.....	0.70	−0.38	0	−0.38	0
trans-1,2 position	.....	.....	.....	.....	−1.10	−2.55	0	−0.945	−0.266
cis-1,3 position	.....	.....	.....	.....	−0.30	−1.20	0	−0.370	−0.166
trans-1,3 position	.....	.....	.....	.....	−0.90	−2.35	0	−0.800	−0.264
Branching in six-membered rings									
Single branching	.....	.....	.....	.....	0	−0.93	0	0.230	−0.192
Double branching									
1,1 position	.....	.....	.....	.....	2.44	0.835	−0.367	1.745	−0.556
cis-1,2 position	.....	.....	.....	.....	−0.20	−0.19	0	1.470	−0.276
trans-1,2 position	.....	.....	.....	.....	−2.69	−2.41	0	0.045	−0.398
cis-1,3 position	.....	.....	.....	.....	−2.98	−2.70	0	−1.647	−0.185
trans-1,3 position	.....	.....	.....	.....	−0.48	−1.60	0	0.260	−0.290
cis-1,4 position	.....	.....	.....	.....	−0.48	−1.11	0	−1.11	0
trans-1,4 position	.....	.....	.....	.....	−2.98	−2.80	0	−0.995	−0.245
Branching in aromatics									
Double branching									
1,2 position	.....	.....	.....	.....	0.94	1.02	0	1.02	0
1,3 position	.....	.....	.....	.....	0.38	−0.31	0	−0.31	0
1,4 position	.....	.....	.....	.....	0.58	0.93	0	0.93	0
Triple branching									
1,2,3 position	.....	.....	.....	.....	1.80	1.91	0	2.10	0
1,2,4 position	.....	.....	.....	.....	0.44	1.10	0	1.10	0
1,3,5 position	.....	.....	.....	.....	0.44	0	0	0	0
Oxygen-containing groups									
−OH (primary)	6.5128	−0.1347	0.0414	−0.1623	−41.2	−41.56	1.28	−41.56	1.28
−OH (secondary)	6.5128	−0.1347	0.0414	−0.1623	−43.8	−41.56	1.28	−41.56	1.28
−OH (tertiary)	6.5128	−0.1347	0.0414	−0.1623	−47.6	−41.56	1.28	−41.56	1.28
−OH (aromatic)	6.5128	−0.1347	0.0414	−0.1623	−45.1	−41.56	1.28	−41.56	1.28
−O−	2.8461	−0.0100	0.0454	−0.2728	−24.2	−15.79	−0.85		
−CHO−	3.5184	0.9437	0.0614	−0.6978	−29.71	−29.28	0.77	−30.15	0.83
⟩CO	1.0016	2.0763	−0.1636	0.4494	−31.48	−28.08	0.91	−28.08	0.91
−COOH	1.4055	3.4632	−0.2557	0.6886	−94.68	−98.39	2.86	−98.83	2.93
−COO−	2.7350	1.0751	0.0667	−0.9230	(−79.8)	−92.62	2.61	−92.62	2.61
O (ring)	−3.7344	1.3727	−0.1265	0.3789	−21.62	−18.37	0.80	−16.07	0.40
Nitrogen-containing groups									
−C≡N	4.5104	0.5461	0.0269	−0.3790	36.82	30.75	−0.72	30.75	−0.72
−N=C	5.0860	0.3492	0.0259	−0.2436	(44.4)	46.32	−0.89	46.32	−0.89
$-NO_2$	1.0898	2.6401	−0.1871	0.4750	−7.94	−9.0	3.70	−14.19	4.38
$-NH_2$ (aliphatic)	4.1783	0.7378	0.0679	−0.7310	3.21	2.82	2.71	−6.78	3.98
$-NH_2$ (aromatic)	4.1783	0.7378	0.0679	−0.7310	−1.27	2.82	2.71	−6.78	3.98
>NH (aliphatic)	−1.2530	2.1932	−0.1604	0.4237	13.47	12.93	3.16	12.93	3.16
>NH (aromatic)	−1.2530	2.1932	−0.1604	0.4237	8.50	12.93	3.16	12.93	3.16
>N− (aliphatic)	−3.4677	2.9433	−0.2673	0.7828	18.94	19.46	3.82	19.46	3.82
>N− (aromatic)	−3.4677	2.9433	−0.2673	0.7828	8.50	19.46	3.82	19.46	3.82
N (ring)	2.4458	0.3436	0.0171	−0.2719	....	11.32	1.11	12.26	0.96
Sulfur-containing groups									
−SH	2.5597	1.3347	−0.1189	0.3820	4.60	−10.68	1.07	−10.68	1.07
−S−	4.2256	0.1127	−0.0026	−0.0072	11.17	−3.32	1.42	−3.32	1.44
S (ring)	4.0824	−0.0301	0.0731	−0.6081	(7.8)	−0.97	0.51	−0.65	0.44
Halogen-containing groups									
−F	1.4382	0.3452	−0.0106	−0.0034	....	−45.10	0.20		
−Cl	3.0660	0.2122	−0.0128	0.0276	....	−8.25	0	−8.25	0
−Br	2.7605	0.4731	−0.0455	0.1420	....	−1.62	−0.26	−1.62	−0.26
−I	3.2651	0.4901	−0.0539	0.1782	....	7.80	0	7.80	0

TABLE 3-336 Group Contributions for Molar Liquid Heat Capacity at 20°C, cal/(g·mol·k)

Group	Value°
Alkane	
$-CH_3$	8.80
$-CH_2-$	7.26
$-\overset{\vert}{C}H-$	5.00
$-\overset{\vert}{\underset{\vert}{C}}-$	1.76
Olefin	
$=CH_2$	5.20
$=\overset{\vert}{C}-H$	5.10
$=\overset{\vert}{C}-$	3.80
Alkyne	
$-C\equiv H$	5.90
$-C\equiv$	5.90
In a ring	
$-\overset{\vert}{C}H-$	4.4
$-\overset{\vert}{C}=$ or $-\overset{\vert}{\underset{\vert}{C}}-$	2.9
$-CH=$	5.3
$-CH_2-$	6.2
Oxygen	
$-O-$	8.4
$>C=O$	12.66
$-\underset{\underset{H}{\vert}}{C}=O$	12.66
$-\overset{\overset{O}{\Vert}}{C}-OH$	19.1
$-\overset{\overset{O}{\Vert}}{C}O-$	14.5
$-CH_2OH$	17.5
$-\overset{\vert}{C}HOH$	18.2
$-\overset{\vert}{\underset{\vert}{C}}OH$	26.6
$-OH$	10.7
$-ONO_2$	28.5
Nitrogen	
$H-\overset{\overset{H}{\vert}}{N}-$	14.0
$-\overset{\overset{H}{\vert}}{N}-$	10.5
$-\overset{\vert}{N}-$	7.5
$-N=$ (in a ring)	4.5
$-C\equiv N$	13.9
Sulfur	
$-SH$	10.7
$-S-$	8.0
Halogen	
$-Cl$ (first or second on a carbon)	8.6
$-Cl$ (third or fourth on a carbon)	6.0
$-Br$	9.0
$-F$	4.0
$-I$	8.6
Hydrogen	
$H-$ (for formic acid, formates, hydrogen cyanide, etc.)	3.5

°Add 4.5 for any carbon group which fulfills the following criterion: a carbon group which is jointed by a single bond to a carbon group connected by a double or triple bond with a third carbon group. In some cases a carbon group fulfills this criterion in more ways than one. In these cases 4.5 should be added each time that the group fulfills the criterion.

Exceptions to the 4.5 addition rule:

1. No such extra 4.5 additions for $-CH_3$ groups.
2. For a $-CH_2-$ group fulfilling the 4.5-addition criterion add 2.5 instead of 4.5. However, when the $-CH_2-$ group fulfills the addition criterion in more ways than one, the addition should be 2.5 the first time and 4.5 for each subsequent addition.
3. No such extra addition for any carbon group in a ring.

This rule is illustrated for 1,4-pentadiene, i.e.,

$$C_{pL}\,(20°C) = 2(CH_2=) + 2\,(-CH=) + -CH_2- + \text{corrections noted in table}$$
$$= 2(5.20) + 2(5.10) + 7.26 + 2.5 + 4.5$$
$$= 34.9 \text{ cal/g·mol·k}$$

To convert calories per gram-mole-kelvin to British thermal units per pound-mole-degree Fahrenheit, multiply by 1.0.

LIQUID HEAT CAPACITY

The liquid heat capacity of **organic** compounds may be estimated at 20°C by the addition of group contributions in Table 3-336 *provided* the reduced temperature $[(20 + 273)/T_c(\mathrm{K})]$ is less than 0.75. Other, similar group-contribution methods are also available[110,213,133] to calculate C_{pL} at low reduced temperatures.

To determine heat capacities of organic liquids over wide temperature ranges, corresponding-states methods are normally used, and the *difference* between the heat capacities of the liquid and in the ideal-gas state is correlated with the acentric factor and the reduced temperature. An example is that ascribed to Sternling and Brown:[17]

$$\frac{C_{pL} - C_p^\circ}{R} = (0.5 + 2.2\omega)\,[3.67 + 11.64\,(1 - T_r)^4 + 0.634\,(1 - T_r)^{-1}] \qquad (3\text{-}80)$$

C_p° is the ideal-gas heat capacity at the same temperature used to evaluate C_{pL}; ω is the acentric factor, while R is the gas constant expressed in the same dimensions as C_{pL}. Equation (3-80) is usually accurate within 4 to 8 percent except at $T_r > 0.95$ or for highly polar compounds.

Other corresponding-states correlations are available[281,139,245] and are more accurate for polar liquids. All require, however, additional characterizing parameters such as the radius of gyration, association factors based upon second virial coefficients, etc. Most have been reviewed and illustrated elsewhere.[196]

There are, in addition, estimation methods based on tracing a thermodynamic cycle between the liquid and the ideal-gas state,[36,268,197,32] but these do not lead to rapid estimations of C_{pL}.

Miller et al.[160] present polynomial coefficients to allow one to calculate C_{pL} for many different compounds over a range in temperatures. For hydrocarbons, specific correlations of Hadden[80,81] and Luria and Benson[133] are quite accurate.

For **liquid mixtures** of **organic compounds,** a mole-fraction-average value of the heat capacities of the pure components is normally assumed. This procedure neglects any heat of mixing effects.

VISCOSITY

Gases Theoretical methods are available to calculate the viscosity of a gas at low pressures, but the intermolecular potential function applicable for the system must be known. Usually the Lennard-Jones 12-6 potential is assumed for nonpolar molecules and the Stockmayer potential for polar molecules.[22] The final result of the calculations is an expression for viscosity in terms of both an energy and a length parameter, each characteristic for a given molecule. The equations are presented in Ref. 196, and tabulated values of the energy and length parameters[242] are given for 75 compounds in App. C of Ref. 196.

Generally, however, viscosities of **pure** gases at **low pressures** are best estimated by corresponding-states methods. Thodos and coworkers[59,151,231,279] propose:

Nonpolar gases:

$$\eta\xi = 4.610T_r^{0.618} - 2.04e^{-0.449\,T_r} + 1.94e^{-4.058T_r} + 0.1 \qquad (3\text{-}81)$$

Polar gases, hydrogen-bonding, $T_r < 2.0$:

$$\eta\xi = (0.755T_r - 0.055)Z_c^{-5/4} \qquad (3\text{-}82)$$

Polar gases, non-hydrogen-bonding, $T_r < 2.5$:

$$\eta\xi = (1.90T_r - 0.29)^{4/5}Z_c^{-2/3} \qquad (3\text{-}83)$$

$$\text{where } \xi = T_c^{1/6}M^{-1/2}P_c^{-2/3} \qquad (3\text{-}84)$$

with η in μP, T_c in kelvins, and P_c in atmospheres. M is the molecular weight, and Z_c is the critical compressibility factor [Eq. (3-19)]. Equation (3-81) should not be used for H_2, He, or the diatomic halogen gases.

Reichenberg[193] suggested a somewhat different relation:

$$\eta\xi' = T_r[1 + 0.36\,T_r(T_r - 1)]^{-1/6} \qquad (3\text{-}85)$$

TABLE 3-337 Reichenberg's Values of C_i to Calculate the Gas-Viscosity Parameter ξ'

Group	Contribution C_i
$-CH_3$	9.04
$>CH_2$ (nonring)	6.47
$>CH-$ (nonring)	2.67
$>C<$ (nonring)	−1.53
$=CH_2$	7.68
$=CH-$ (nonring)	5.53
$>C=$ (nonring)	1.78
$\equiv CH$	7.41
$\equiv C-$ (nonring)	5.24
$>CH_2$ (ring)	6.91
$>CH-$ (ring)	1.16
$>C<$ (ring)	0.23
$=CH-$ (ring)	5.90
$>C=$ (ring)	3.59
$-F$	4.46
$-Cl$	10.06
$-Br$	12.83
$-OH$ (alcohols)	7.96
$>O$ (nonring)	3.59
$>C=O$ (nonring)	12.02
$-CHO$ (aldehydes)	14.02
$-COOH$ (acids)	18.65
$-COO-$ (esters) or HCOO (formates)	13.41
$-NH_2$	9.71
$>NH$ (nonring)	3.68
$=N-$ (ring)	4.97
$-CN$	18.13
$>S$ (ring)	8.86

where η is in μP and ξ' is defined as

$$\xi' = M^{-1/2} T_c^{-1} \Sigma n_i C_i \tag{3-86}$$

T_c is in kelvins, n_i is the number of atomic groups of type i, and C_i is the contribution of group i as found in Table 3-337.

Both the Thodos and the Reichenberg correlations usually will predict low-pressure gas viscosities to within 1 to 3 percent for nonpolar and 3 to 5 percent for polar materials.

The viscosity of a **gas mixture** at **low pressure** can be calculated from kinetic theory,[23,24,25,26,31,94] but the computations are relatively complicated. The rigorous solution may be approximated by a series expansion in which the first and most important term is

$$\eta_{\text{mix}} = \sum_{i=1}^{n} \frac{y_i \eta_i}{\sum_{j=1} y_j \phi_{ij}} \tag{3-87}$$

η_{mix} is the viscosity of a low-pressure gas containing n components with mole fractions y_i. η_i is the viscosity of pure i at the same temperature and at a low pressure. ϕ_{ij} is a parameter which may be estimated by Eqs. (3-88) and (3-89).[272]

$$\phi_{ij} = \frac{[1 + (\eta_i/\eta_j)^{1/2}(M_j/M_i)^{1/4}]^2}{[8(1 + M_i/M_j)]^{1/2}} \tag{3-88}$$

while

$$\phi_{ji} = (\eta_j/\eta_i)(M_i/M_j)\,\phi_{ij} \tag{3-89}$$

These relations were proposed by Wilke and have been extensively tested. Except in a few cases in which one of the mixture components was H_2 or He, errors seldom exceeded 3 to 4 percent.

Herning and Zipperer[92] suggest a simpler form for ϕ_{ij}, i.e.,

$$\phi_{ij} = (M_j/M_i)^{1/2} = \phi_{ji}^{-1} \tag{3-90}$$

Equation (3-90) is less accurate than Eq. (3-88) and should never be used if H_2 or He is present in the mixture.

Reichenberg[195] has examined all low-pressure gas-mixture viscosity relations and has suggested a modified form of Eq. (3-87) which retains higher-order terms. For mixtures containing H_2 or He, his relation does yield somewhat higher accuracy, but for most systems of interest to the chemical industry there is little change in overall accuracy.

For mixtures containing highly polar components, Brokaw[25,22] has proposed other ways to estimate ϕ_{ij}. Dean and Stiel[44] also suggest a corresponding-states estimation, but it is less accurate than the Wilke method [Eqs. (3-88) and (3-89)].

Sutton[240] has published a compendium of references to experimental gas-mixture viscosity data.

The viscosity of a gas near atmospheric pressure is quite insensitive to small changes in pressure, but, at *high pressure*, this is no longer true. Illustrated in Fig. 3-47 is a graph of the Reichenberg correlation[194] for the effect of pressure on gas viscosity as given in Eqs. (3-91) and (3-92).

$$\frac{\eta - \eta^\circ}{\eta^\circ\, \mathcal{P}} = \frac{A P_r^{1.5}}{B P_r + (1 + C P_r^D)^{-1}} \tag{3-91}$$

The constants A, B, C, and D are functions of the reduced temperature as follows:

$$A = \frac{\alpha_1}{T_r} \exp \alpha_2 T_r^{-\alpha_3} \qquad B = A(\beta_1 T_r - \beta_2)$$

$$C = \frac{\gamma_1}{T_r} \exp \gamma_2 T_r^{-\gamma_3} \qquad D = \frac{\delta_1}{T_r} \exp \delta_2 T_r^{-\delta_3}$$

$\alpha_1 = 1.9824 \times 10^{-3}$ $\quad \alpha_2 = 5.2683$ $\quad \alpha_3 = 0.5767$

$\beta_1 = 1.6552$ $\quad \beta_2 = 1.2760$

$\gamma_1 = 0.1319$ $\quad \gamma_2 = 3.7035$ $\quad \gamma_3 = 79.8678$

$\delta_1 = 2.9496$ $\quad \delta_2 = 2.9190$ $\quad \delta_3 = 16.6169$

$$\mathcal{P} = 1 - \frac{300(\mu_p)^2 P_c}{T_c^2} \tag{3-92}$$

with μ_p in debyes,* P_c in atmospheres, and T_c in kelvins. η° is the viscosity at atmospheric pressure and could be estimated separately with, for example, Eqs. (3-81) to (3-85). For nonpolar materials, $\mathcal{P} = 1$. Although this simple relation has not been extensively tested, it appears to yield errors of less than 10 percent even near the critical point.

Another way to estimate the viscosity of dense gases is to employ the empirical fact that the residual viscosity, $\eta - \eta^\circ$, is a function of reduced density ρ_r. As before, η is the viscosity at T and high pressure, and η° is the viscosity near atmospheric pressure but at the same temperature; ρ_r is the ratio of the gas density to the critical density. The most accurate and general residual correlations were suggested by Stiel and Thodos.[112,233]

Nonpolar gases:

$$[(\eta - \eta^\circ)\xi + 1]^{0.25} = 1.0230 + 0.23364\rho_r + 0.58533\rho_r^2 - 0.40758\rho_r^3 + 0.093324\rho_r^4 \tag{3-93}$$

*1 debye = $(\sqrt{10})(10^{-25})(\text{J}\cdot\text{m}^3)^{1/2}$.

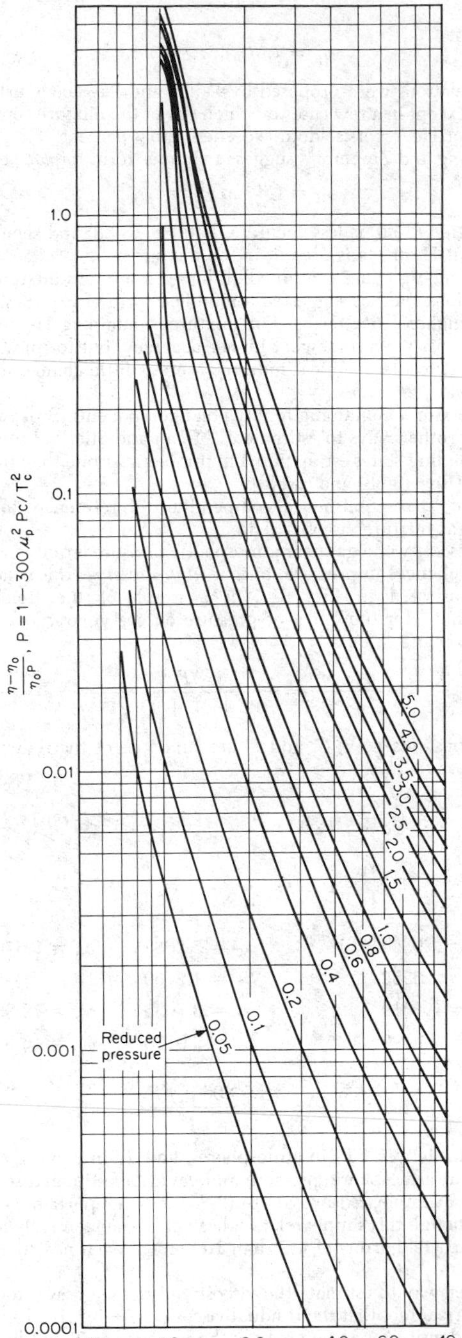

FIG. 3-47 Reichenberg correlation for dense-gas viscosities.

$\xi = T_c^{1/6}M^{-1/2}P_c^{-2/3}$, where T_c is in kelvins and P_c is in atmospheres and M is the molecular weight; η and η° are expressed in μP.

This relation is reported to be applicable in the range $0.1 \leq \rho_r < 3$.

Polar gases:

$$(\eta - \eta^\circ)\xi = 1.656\rho_r^{1.111} \qquad \rho_r \leq 0.1 \qquad (3\text{-}94)$$

$$(\eta - \eta^\circ)\xi = 0.0607(9.045\rho_r + 0.63)^{1.739} \qquad 0.1 \leq \rho_r < 0.9 \qquad (3\text{-}95)$$

$$\log \{4 - \log [(\eta - \eta^\circ)\xi]\} = 0.6439 - 0.1005\rho_r - \Delta \qquad 0.9 \leq \rho_r < 2.6 \qquad (3\text{-}96)$$

$$\text{where } \Delta = \begin{cases} 0 & 0.9 \leq \rho_r < 2.2 \\ (4.75)(10^{-4})(\rho_r^3 - 10.65)^2 & 2.2 \leq \rho_r < 2.6 \end{cases} \qquad (3\text{-}97)$$

The accuracy of Eqs. (3-93) through (3-97) is about 10 percent; the density of the high-pressure gas must be known or estimated from an equation of state.

None of the viscosity-estimation methods are applicable for gases at very low pressures when the molecular mean free path is in the same order as the characteristic dimension of the apparatus through which the gas is flowing.[27]

To estimate the viscosity of **gas mixtures** at **high pressure,** Fig. 3-47 [or Eqs. (3-91) and (3-92)] could be used if mixing rules were defined to calculate $T_{c,\text{mix}}$ and $P_{c,\text{mix}}$. If one or more of the components is polar, then the polar factor $\mathcal{P}$ would also have to be obtained. No rules presently exist to find $\mathcal{P}_{\text{mix}}$, but the modified Prausnitz and Gunn mixing rules[186] could be employed to determine $T_{c,\text{mix}}$ and $P_{c,\text{mix}}$.

$$T_{c,\text{mix}} = \sum_i y_i T_{ci}$$

$$Z_{c,\text{mix}} = \sum_i y_i Z_{ci} \qquad (3\text{-}98)$$

$$V_{c,\text{mix}} = \sum_i y_i V_{ci}$$

$$P_{c,\text{mix}} = \frac{Z_{c,\text{mix}} R T_{c,\text{mix}}}{V_{c,\text{mix}}}$$

where Z_c is the critical compressibility factor and y_i is the mole fraction of component i.

Dean and Stiel[44] propose a residual viscosity correlation similar to Eqs. (3-93) through (3-97):

$$(\eta_{\text{mix}} - \eta^\circ_{\text{mix}})\xi_{\text{mix}} = (1.08)[\exp(1.439\rho_{r,\text{mix}}) - \exp(-1.11\rho_{r,\text{mix}}^{1.858})] \qquad (3\text{-}99)$$

where η_{mix} = high-pressure mixture viscosity, μP
η°_{mix} = low-pressure mixture viscosity, μP
$\rho_{r,\text{mix}}$ = pseudo-reduced mixture density, $\rho_{\text{mix}}/\rho_{c,\text{mix}}$
ρ_{mix} = mixture density, (g·mol)/cm^3
$\rho_{c,\text{mix}}$ = pseudo-critical mixture density, (g·mol)/cm^3 = $P_{c,\text{mix}}/Z_{c,\text{mix}} RT_{c,\text{mix}}$
ξ_{mix} = $T_{c,\text{mix}}^{1/6} M_{\text{mix}}^{-1/2} P_{c,\text{mix}}^{-2/3}$

The mixture molecular weight M_{mix} is a mole-fraction average. The pseudo-critical-mixture parameters $Z_{c,\text{mix}}$, $T_{c,\text{mix}}$ and $P_{c,\text{mix}}$ are calculated from the modified Prausnitz and Gunn rules, Eq. (3-98).

Equation (3-99) is to be used only for nonpolar mixtures; it is said to be applicable both to gases at high pressure and to liquids at high temperature, but the accuracy for liquids with reduced densities greater than about 2 is expected to be poor. The equation has never been tested in any detail in the liquid region. When it was tested on nine gas mixtures at various densities, the average error found was 3.7 percent; most mixtures were composed of light hydrocarbons or hydrocarbons and inert gases. A graph of Eq. (3-99) is shown in Fig. 3-48, and for the simple systems shown the agreement is remarkable. A similar correlation has been proposed by Giddings.[68] In this case, different rules were chosen for determining the pseudo-critical constants. A good correlation was obtained for light-hydrocarbon mixtures; it was also found that the correlation could be improved if the mole-fraction molecular weight were employed as a third correlating parameter.

Liquids The viscosity of a liquid decreases with increasing temperature. At reduced temperatures below about $T_r \backsim 0.75$, one of the more accurate correlations is given by

$$\ln \eta_L = A + B/(T + C) \qquad (3\text{-}100)$$

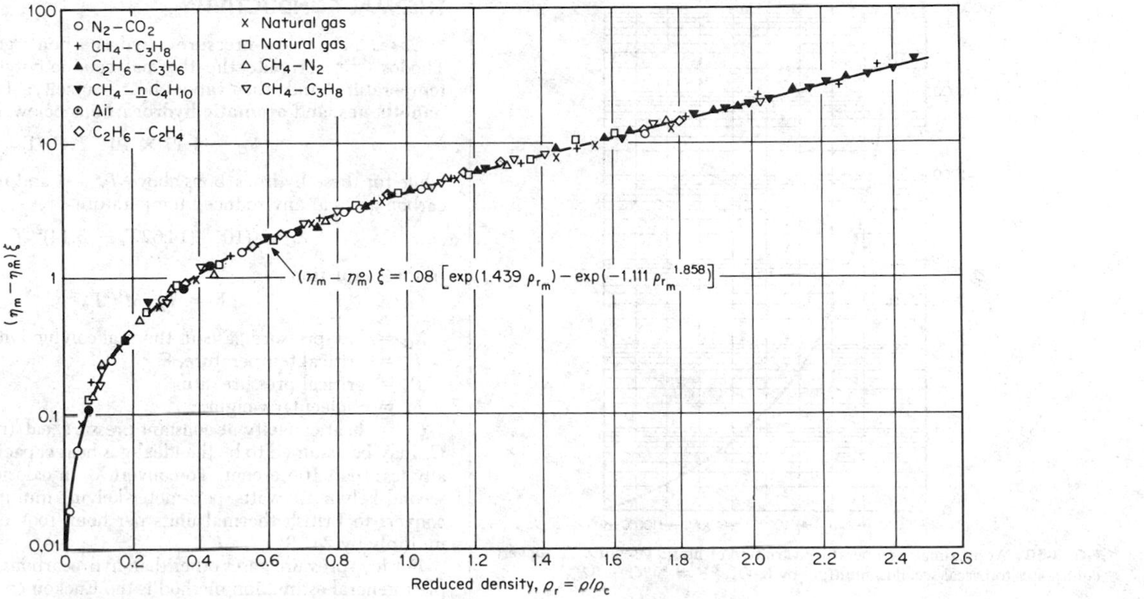

FIG. 3-48 Relationship between $(\eta - \eta^0)\xi$ and ρ_r for nonpolar mixtures.

The constants A, B, and C may be found from regressing experimental liquid-viscosity data. With little loss of accuracy, C may be approximated[69] as

$$C = 17.71 - 0.19\,T_b \tag{3-101}$$

where the temperature T and the normal boiling point T_b are expressed in kelvins. The constants A and B are often sensitive functions of the molecular structure. The simpler Andrade correlation[4,5] is often used in place of Eq. (3-100); i.e., if C is set equal to zero,

$$\ln \eta_L = A + B/T \tag{3-102}$$

Neither Eq. (3-100) nor Eq. (3-102) should be used if $T_r > 0.75$. For high-temperature liquids, Eq. (3-103) provides an approximate relation to estimate liquid viscosity:[123]

$$\eta_L\xi = (\eta_L\xi)^{(0)} + \omega(\eta_L\xi)^{(1)} \tag{3-103}$$

$$(\eta_L\xi)^{(0)} = 0.015174 - 0.02135T_r + 0.0075T_r^2 \tag{3-104}$$

$$(\eta_L\xi)^{(1)} = 0.042552 - 0.07674T_r + 0.0340T_r^2 \tag{3-105}$$

$$\xi = T_c^{1/6}M^{-1/2}P_c^{-2/3} \tag{3-106}$$

η_L is given in centipoises, and ω is the acentric factor. In Eq. (3-106), T_c is in kelvins, P_c in atmospheres, and M is the molecular weight. The correlation presented in Eq. (3-103) is shown in Fig. 3-49. It should be used only in the range $0.75 < T_r < 0.98$. At the critical point, $\eta_L\xi \backsim 0.00078$.

Equation (3-103) is similar to the corresponding-states gas-viscosity correlations described earlier; at high reduced temperatures, liquids become more "gaslike," and many property correlations are therefore based upon the law of corresponding states.

At low reduced temperatures, liquid structure plays a more specific role and no simple, reliable estimation equations are available to determine liquid viscosities. The available equations are reviewed elsewhere,[196] and a tabulation of parameters for use in the van Velzen et al. correlation[257,258] is available in Ref. 258; many of these parameters are reproduced in App. B of Ref. 196.

If one low-temperature liquid-viscosity datum point is available, viscosities at other temperatures may be approximated with Fig. 3-50.[124] To use the chart, locate the known value of viscosity on the ordinate and then extend the abscissa by the required number of degrees (Celsius) to locate the new viscosity. Figure 3-50 should be used only for organic liquids. The accuracy of such an extrapolation is generally ±20 percent.

The effect of **pressure** on the viscosity of **liquids** is not well understood. At reduced temperatures of 0.80 and above, Eqs. (3-93) through (3-97) provide a reasonable estimation if the applicable density ranges are not exceeded. At reduced temperatures below 0.80, no estimation method is available to predict the effect of pressure, but for molecules with simple molecular structure pressure effects are relatively insignificant up to about 50 atm.

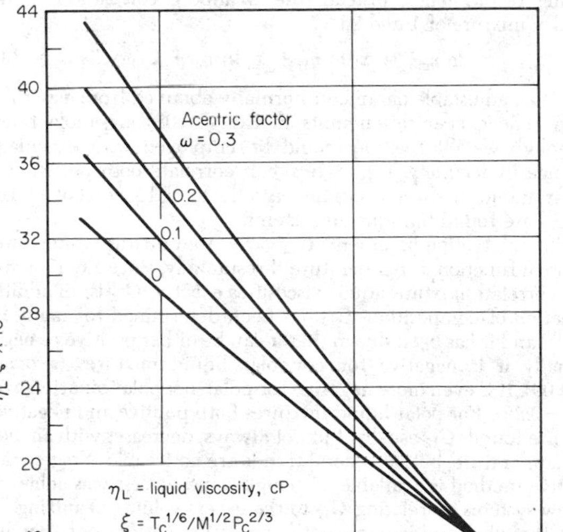

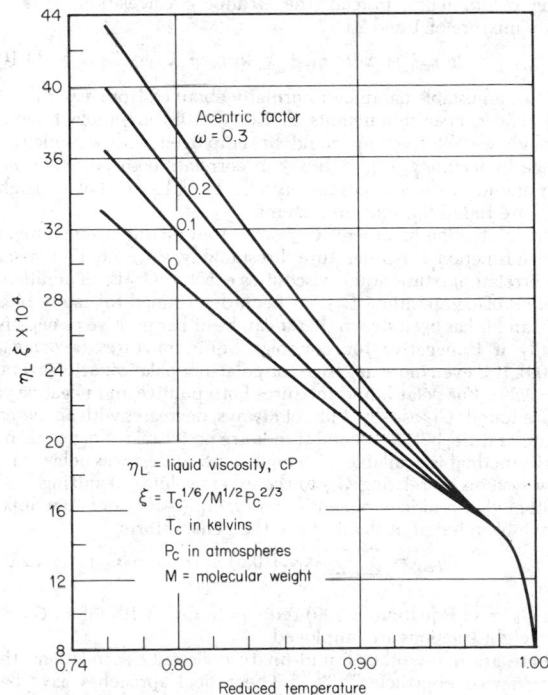

FIG. 3-49 Lestou-Stiel high-temperature liquid-viscosity correlation. To convert centipoises to pascal-seconds, multiply by 0.001; °R = 9⁄5 K.

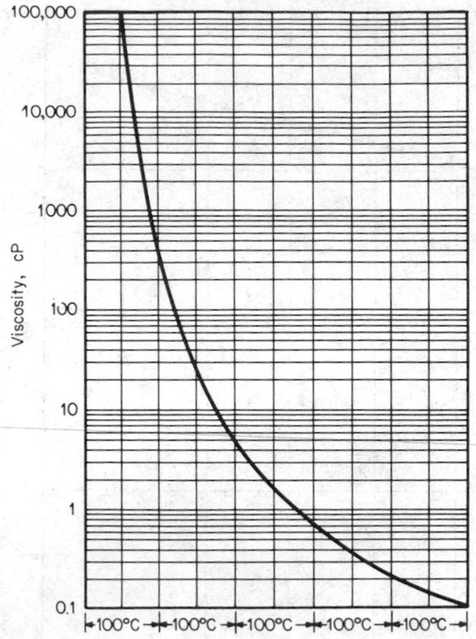

FIG. 3-50 Approximate temperature variation of liquid viscosity. To convert centipoises to pascal-seconds, multiply by 0.001; °F = $\frac{9}{5}$ °C + 32.

Equations to estimate the viscosities of **liquid mixtures** have been thoroughly studied by Irving.[102,103] For **nonpolar** mixtures, mixture viscosities can usually be predicted within ±5 to 10 percent by the simple correlation

$$\ln \eta_{\text{mix}} = \Sigma\, w_j \ln \eta_j \qquad (3\text{-}107)$$

where η_{mix} represents the mixture viscosity, η_j is the pure-component viscosity of j, and w_j is the *weight* fraction of j.

For liquid mixtures containing one or more polar constituents, including aqueous systems, Eq. (3-107) is often very inaccurate. Irving recommends, instead, the Grunberg correlation.[77,78] For a binary mixture of 1 and 2,

$$\ln \eta_{\text{mix}} = x_1 \ln \eta_1 + x_2 \ln \eta_2 + 2x_1x_2G_{12} \qquad (3\text{-}108)$$

G_{12} is an adjustable parameter normally obtained from experimental data. The concentration units in Eq. (3-108) are mole fractions, although weight fractions could be employed with essentially no change in accuracy. Eq. (3-108) can correlate both polar and nonpolar liquid-mixture viscosities usually to ±15 percent.[103] Higher errors are found for aqueous systems.

The interaction parameter G_{12} varies fom mixture to mixture and is also a function of temperature. Presumably, since Eq. (3-108) cannot correlate mixture liquid viscosities exactly, G_{12} is, in addition, a function of composition. G_{12} has been determined for many binaries,[103] and it has been shown that it can be either positive or negative. Usually it is negative for nonpolar liquid mixtures (average $\backsim$ −0.09). It is even more negative for polar-nonpolar binaries (average $\backsim$ −0.22). For polar liquid mixtures both positive and negative values are found. G_{12} usually, but not always, decreases with an increase in temperature, but no general trends are noticeable. No general predictive method is available, although some success was achieved for a few systems by relating G_{12} to the excess volume of mixing.[76]

Although a multicomponent form of Eq. (3-108) does not appear to have been tested, it should have the general form

$$\ln \eta_{\text{mix}} = \sum_i \sum_j x_i x_j \,[(0.5) \ln (\eta_i \eta_j) + G_{ij}] \qquad (3\text{-}109)$$

with $G_{ii} = 0$. Equation (3-109) reduces to Eq. (3-107) if all $G_{ij} = 0$ and weight fractions are employed.

There are many other liquid-mixture-viscosity correlations; they are reviewed elsewhere.[196,103,102] Theoretical aproaches have been summarized by Hanley.[86]

THERMAL CONDUCTIVITY

Gases For **low-pressure hydrocarbon** gases, Misic and Thodos[161,162] correlate the thermal conductivity λ_G with reduced temperature and the vapor heat capacity; i.e., for **methane, naphthenes,** and **aromatic hydrocarbons** below $T_r = 1.0$:

$$\lambda_G = 4.45 \times 10^{-6} T_r C_p/\Gamma \qquad (3\text{-}110)$$

while for these hydrocarbons above $T_r = 1$ and for all other hydrocarbon gases at any reduced temperature,

$$\lambda_G = (10^{-6})(14.52T_r - 5.14)^{2/3} C_p/\Gamma \qquad (3\text{-}111)$$

In these equations,

$$\Gamma = T_c^{1/6} M^{1/2} P_c^{-2/3} \qquad (3\text{-}112)$$

λ_G = low-pressure gaseous thermal conductivity, cal/(cm·s·K)
T_c = critical temperature, K
P_c = critical pressure, atm
M = molecular weight
C_p = heat capacity at constant pressure, cal/(mol·K)

C_p may be assumed to be the ideal-gas heat capacity. Errors are usually less than 10 percent. To convert λ_G in calories per centimeter-second-kelvin to watts per meter-kelvin, multiply by 418.68; to convert to British thermal units per hour–foot–degree Fahrenheit, multiply by 241.9.

For low-pressure gases other than hydrocarbons, probably the simplest general estimation method is the Eucken correlation

$$\lambda_G = (C_v + 4.47\,)(\eta/M) \qquad (3\text{-}113)$$

where η = low-pressure gaseous viscosity, P
C_v = heat capacity at constant volume, cal/(mol·K)

and the other terms are defined in Eqs. (3-110) and (3-111). η may be estimated by methods described earlier, while it is generally satisfactory to calculate C_v as $(C_p - R)$, where R = 1.986 cal/(mol·K). Errors encountered in using Eq. (3-113) normally are below 15 to 20 percent; more often than not, Eq. (3-113) underpredicts λ_G. A modified form of the Eucken relation[243] that attempts to account for internal energy transfer more realistically may be written as

$$\lambda_G = (1.32C_v + 3.52)(\eta/M) \qquad (3\text{-}114)$$

but this form usually *overpredicts* λ_G. Good results are quite often obtained by *averaging* the two estimated values of λ_G obtained from Eqs. (3-113) and (3-114).

Many other low-pressure gaseous thermal-conductivity estimation methods have been proposed; some are based on rigorous theory, while others are empirical and employ group contributions.[203,204] Most are reviewed in Ref. 196.

Thermal conductivities of low-pressure gases increase with temperature; over small temperature ranges $d\lambda/dT$ is almost constant and equal to 0.1 to 0.3 cal/(cm·s·K). Gases with larger values of λ_G usually have larger values of $d\lambda/dT$. Over quite wide temperature ranges, a more accurate way to predict temperature variations is

$$\lambda_{T2}/\lambda_{T1} = (T_2/T_1)^n \qquad (3\text{-}115)$$

where $n \backsim 1.8$.

The thermal conductivity of all **gases** increases with **pressure,** although from about 1 torr to 10 atm the effect of pressure is negligible. Below about 0.1 torr, λ_G is almost proportional to pressure. Very near the critical point, λ_G increases sharply to a maximum value.[212,79,144] Most correlations used to estimate the effect of pressure use Vargaftik's[259,260] suggestion that the difference between high-pressure thermal conductivity and the value at about 1 atm can be correlated with only the density of the gas. Stiel and Thodos[232] have suggested that, for nonpolar materials,

$$(\lambda_G - \lambda_G^0)\Gamma Z_c^5 = (14.0 \times 10^{-8})(e^{0.535\rho_r} - 1) \qquad \rho_r < 0.5$$
$$= (13.1 \times 10^{-8})(e^{0.67\rho_r} - 1.069) \qquad 0.5 \le \rho_r \le 2.0$$
$$= (2.976 \times 10^{-8})(e^{1.155\rho_r} + 2.016) \qquad 2.0 < \rho_r < 2.8$$
$$(3\text{-}116)$$

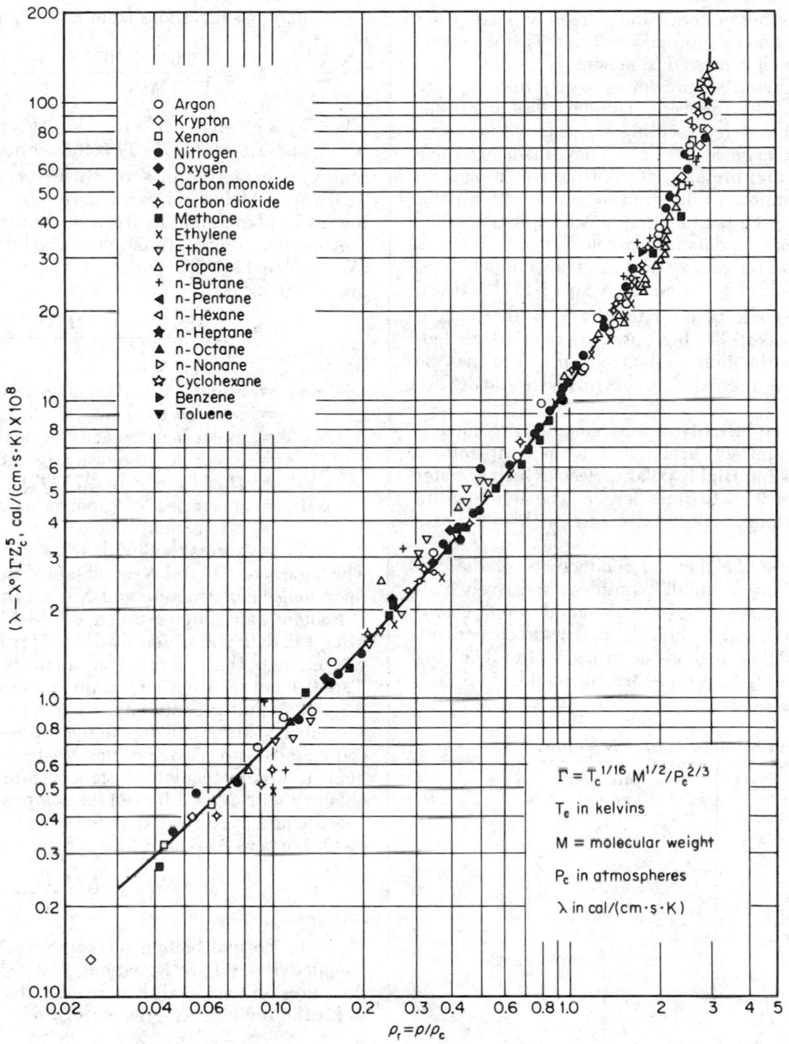

FIG. 3-51 Stiel and Thodos correlation for dense-gas thermal conductivities. To convert calories per centimeter-second-kelvin to British thermal units per foot–hour–degree Rankine, multiply by 241.9. °R = 9/5 K.

λ_G is the high-pressure thermal conductivity, whereas λ_G^0 is the low pressure (∼1-atm) value, both expressed in cal/(cm·s·K). Γ is defined in Eq. (3-112), Z_c is the critical compressibility factor, and ρ_r the reduced density, ρ/ρ_c.

A graph comparing predictions from these equations with available experimental data is given in Fig. 3-51. These correlations should not be used for polar substances or for hydrogen or helium; for such materials no general predictive estimation technique is available.

The thermal conductivity of **low-pressure gas mixtures** can be determined from the relation[267]

$$\lambda_{mix} = \sum_{i=1}^{n} \frac{y_i \lambda_i}{\sum_j y_j A_{ij}} \tag{3-117}$$

λ_{mix} is the low-pressure gas-mixture thermal conductivity, λ_i is the low-pressure thermal conductivity of pure i, while y_i is the mole fraction of i. The parameter A_{ij} can be estimated by several techniques.

For binaries of nonpolar or nonpolar-polar gases, A_{ij} can be equated to ϕ_{ij} [see Eq. (3-88)], or it may be calculated by the Mason-Saxena relation[149]

$$A_{ij} = \frac{[1 + (g_i/g_j)^{1/2}(M_i/M_j)^{1/4}]^2}{[8(1 + M_i/M_j)]^{1/2}} \tag{3-118}$$

with
$$\frac{g_i}{g_j} = \left(\frac{\Gamma_j}{\Gamma_i}\right)\left[\frac{\exp(0.0464T_{ri}) - \exp(-0.2412T_{ri})}{\exp(0.0464T_{rj}) - \exp(-0.2412T_{rj})}\right] \tag{3-119}$$

where M is the molecular weight and Γ is defined in Eq. (3-112).

For *both* polar and nonpolar gas mixtures, Lindsay and Bromley[129] suggest

$$A_{ij} = \frac{1}{4}\left\{1 + \left[\frac{\eta_i}{\eta_j}\left(\frac{M_j}{M_i}\right)^{3/4}\left(\frac{T + S_i}{T + S_j}\right)\right]^{1/2}\right\}^2\left(\frac{T + S_{ij}}{T + S_i}\right) \tag{3-120}$$

η_i is the pure, low-pressure viscosity of i, while

$$S_i = 1.5T_{b_i} \tag{3-121}$$

$$S_{ij} = S_{ji} = C(S_iS_j)^{1/2} \tag{3-122}$$

with $C \simeq 1$ except when either or both i and j are very polar. Then a lower value of $C \simeq 0.73$ is more appropriate. T and T_b (the normal boiling temperature) are both expressed in kelvins.

The estimation methods described previously to determine λ_{mix} are usually accurate to within 2 to 5 percent. The flexibility of Eq. (3-117) is such that it is capable of representing λ_{mix} data with either a maximum or a minimum. Gray et al.[72,73,74,141,142] have made extensive studies of its applicability and present criteria from which one can predict a priori if a maximum, a minimum, or an inflection point would be expected in a λ_{mix}-composition graph. A compilation of literature sources for available λ_{mix} data is available.[241]

For the effect of **pressure** on **gas-mixture** thermal conductivities, Eq. (3-116) may be used with λ°_{mix} found from Eq. (3-117). Critical properties for the mixture could be estimated by using the pseudocritical rules given in Eq. (3-98). The high-pressure gas-mixture density is required, and the estimation methods given before may be used. This technique has been tested,[196] and errors were usually less than about 10 percent.

Liquids The thermal conductivity of most common organic liquids ranges between 250 and 400 μcal/(cm·s·K) at temperatures below the normal boiling point. Highly polar molecules such as water and ammonia have values 2 to 3 times larger. The temperature dependence of λ_L is weak, and usually λ_L decreases with an increase in temperature.

Ho et al.[97] have reviewed the thermal conductivity of the elements, while Ewing et al.[54] and Gambill[66] consider respectively molten metals and molten-salt mixtures. Cryogenic liquids are discussed by Preston et al.[187] and Mo and Gubbins[165] and hydrocarbons by Mather et al.[153] The estimation methods described in the following paragraphs are applicable only for simple organic liquids.

TABLE 3-338 Robbins and Kingrea *H* and *N* Values

Functional group	Number of groups	H^*
Unbranched hydrocarbons		
Paraffins		0
Olefins		0
Rings		0
CH_3 branches	1	1
	2	2
	3	3
C_2H_5 branches	1	2
i-C_3H_7 branches	1	2
C_4H_9 branches	1	2
F substitutions	1	1
	2	2
Cl substitutions	1	1
	2	2
	3 or 4	3
Br substitutions	1	4
	2	6
I substitutions	1	5
OH substitutions	1 (iso)	1
	1 (normal)	−1
	2	0
	1 (tertiary)	5
Oxygen substitutions		
$-\overset{\vert}{C}=O$ (ketones, aldehydes)		0
$-\overset{O}{\overset{\Vert}{C}}-O-$ (acids, esters)		0
$-O-$ (ethers)		2
NH_2 substitutions	1	1

liquid density, g/cm³	N
< 1	1
> 1	0

*For compounds containing multiple functional groups, the H-factor contributions are additive.

Combining suggestions from Sato[143] and Riedel,[199]

$$\lambda_L = \left[\frac{2.64 \times 10^{-3}}{M^{1/2}}\right]\left[\frac{3 + 20(1 - T_r)^{2/3}}{3 + 20(1 - T_{b_r})^{2/3}}\right] \tag{3-123}$$

where λ_L is given in cal/(cm·s·K), M is the molecular weight, $T_r = T/T_c$, and $T_{b_r} = T_b/T_c$. T_b is the normal boiling point. All temperatures are expressed in K or °R. Errors noted in the use of Eq. (3-123) vary widely, and particularly poor results were obtained with low-molecular-weight hydrocarbons (error > 10–15 percent).

A more accurate estimation method is based on the suggestion by Weber[270] that λ_L is proportional to $C_p\rho^{4/3}$. As formulated by Robbins and Kingrea,[202]

$$\lambda_L = \frac{(88.0 - 4.94H)(10^{-3})}{\Delta S^{\circ}}\left(\frac{0.55}{T_r}\right)^N C_p\rho^{4/3} \tag{3-124}$$

where λ_L = liquid thermal conductivity, cal/(cm·s·K)
T_r = reduced temperature, T/T_c
C_p = molar heat capacity of the liquid, cal/(g·mol·K)
ρ = molar liquid density, (g·mol)/cm³
ΔS° = $H_{vb}/T_b + R \ln (273/T_b)$
ΔH_{vb} = molar heat of vaporization at the normal boiling point, cal/(g·mol)
T_b = normal boiling point, K

The parameters H and N are obtained from Table 3-338; H depends upon molecular structure and N upon the liquid density at 20°C.

Robbins and Kingrea tested Eq. (3-124) with 70 organic liquids, using 142 data points. Rarely did the errors exceed 10 percent. These workers state that the range of applicability is from a T_r of 0.4 to a T_r of 0.9, but their testing was invariably carried out only from about 0.5 to 0.7. Sulfur-containing compounds and inorganics cannot be treated. The abrupt change in the exponent N from zero to unity for compounds with mass densities greater or less than 1.0 g/cm³ is difficult to accept in many instances; often more reliable results are obtained with N = 1.0 even for compounds with $\rho > 1$ g/cm³.

Missenard[164] correlated λ_L for several homologous series with Eq. (3-125), where A, B, and C are shown in Table 3-339.

$$\lambda_L = A\left(\frac{CT_b - T}{CT_b - B}\right) \tag{3-125}$$

T_b is the normal boiling temperature, K. λ_L is the liquid thermal conductivity at T, K. Equation (3-125) should be used only for temperatures up to or slightly above T_b; the accuracy is about the same as for the Robbins-Kingrea correlation.

TABLE 3-339 Missenard Constants

Family	A, cal/(cm·s·K)	B, K	C
Saturated hydrocarbons	4.30×10^{-4}	114	2
Freons			
CCl_mF_n, $m + n = 4$	3.73×10^{-4}	56	2
$CHCl_3$ or CHF_3	3.63×10^{-4}	169	2
CH_2Cl_2 or CH_2F_2	3.51×10^{-4}	273	2
$CCl_mF_n - CCl_mF_n$, $m + n = 3$	2.75×10^{-4}	100	2
Halogen derivatives			
R−Cl	2.83×10^{-4}	290	2
R−Br	2.23×10^{-4}	327	2
R−I	1.89×10^{-4}	340	2
Esters			
R−COOCH	2.87×10^{-4}	373	2
R_1−COOC−R_2	2.68×10^{-4}	393	2
Ethers	2.56×10^{-4}	357	2
Ketones	2.68×10^{-4}	396	2
Benzene derivatives			
C_6H_5−R	2.32×10^{-4}	426	2
$C_6H_m-(CH_3)_n$, $m + n = 6$	2.30×10^{-4}	429	2
Alcohols			
Methanol	4.97×10^{-4}	273	2.5
Ethanol	4.11×10^{-4}	273	2.5
C_3OH, C_4OH, C_5OH	2.59×10^{-4}	484	2.5
C_6OH and higher	-3.70×10^{-4}	273	2.5
Acids	2.58×10^{-4}	526	2.5

TABLE 3-340 Values of *A* in Eq. (3-126)

	P_r					
T_r	1	5	10	50	100	200
0.8	0.036	0.038	0.038	0.038	0.038	0.038
0.7	0.018	0.025	0.027	0.031	0.032	0.032
0.6	0.015	0.020	0.022	0.024	0.025	0.025
0.5	0.012	0.0165	0.017	0.019	0.020	0.020

At moderate pressures, up to 30 to 40 atm, the effect of pressure on the thermal conductivity of liquids is usually neglected. At very high pressures, λ_L increases with an increase in pressure.[20] Lenoir[122] and Missenard[163] have presented general correlations to relate λ_L to pressure. The Missenard equation is

$$\lambda_L(P_r) = \lambda_L(\text{low pressure})(1 + AP_r^{0.7}) \tag{3-126}$$

where A is shown in Table 3-340 as a function of T_r and P_r. A graph of Eq. (3-126) is given in Fig. 3-52.

The thermal conductivity of **liquid mixtures** is usually less than would have been predicted with a mole- (or weight-) fraction average. Many correlating techniques have been suggested;[48,116,131,154,187,189,205,262,282,105,106,56,57,216,217,127] most are reviewed in Refs. 196 and 105. The Li[127] method is described as follows:

$$\lambda_{L\text{mix}} = \sum_i \sum_j \phi_i \phi_j \lambda_{ij} \tag{3-127}$$

where
$$\lambda_{ij} = 2(\lambda_{Li}^{-1} + \lambda_{Lj}^{-1})^{-1} \tag{3-128}$$

$$\phi_i = x_i V_i \Big/ \sum_k x_k V_k \tag{3-129}$$

x_i is the mole fraction of component i, and V_i is the molar volume of pure i. For a binary liquid of components 1 and 2,

$$\lambda_{L\text{mix}} = \phi_1^2 \lambda_{L1} + 2\phi_1\phi_2\lambda_{12} + \phi_2^2\lambda_{L2} \tag{3-130}$$

Errors in the use of Eq. (3-127) rarely exceed 3 to 4 percent for both aqueous and nonaqueous mixtures.

For dilute **ionic** solutions, the mixture thermal conductivity usually decreases with an increase in the concentration of the dissolved salts. To estimate the thermal conductivity of such mixtures, Jamieson and Tudhope[104] recommend the use of an equation proposed originally by Riedel[199] and tested by Vargaftik and Os'minin:[261]

$$\lambda_{\text{mix}}(20°\text{C}) = \lambda_{\text{H}_2\text{O}}(20°\text{C}) + \frac{1}{4.186}\sum_i \sigma_i C_i \tag{3-131}$$

where λ_{mix} = thermal conductivity of ionic solution at 20°C, cal/(cm·s·K)
$\lambda_{\text{H}_2\text{O}}$ = thermal conductivity of water at 20°C, cal/(cm·s·K)
C_i = concentration of the electrolyte, (g·mol)/L
σ_i = coefficient that is characteristic of each ion

Values of σ_i are shown in Table 3-341. To obtain λ_{mix} at other temperatures,

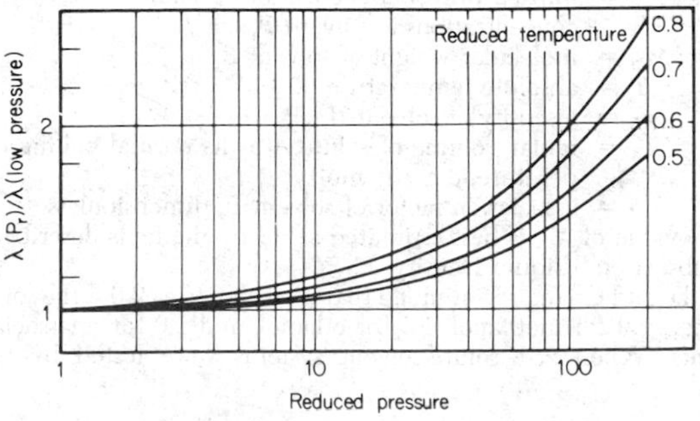

FIG. 3-52 Missenard correlation for liquid thermal conductivities at high pressures.

TABLE 3-341 Values of σ_i for Anions and Cations

Anion	$\sigma_i \times 10^5$	Cation	$\sigma_i \times 10^5$
OH^-	20.934	H^+	−9.071
F^-	2.0934	Li^+	−3.489
Cl^-	−5.466	Na^+	0.0000
Br^-	−17.445	K^+	−7.560
I^-	−27.447	NH_4^+	−11.63
NO_2^-	−4.652	Mg^{2+}	−9.304
NO_3^-	−6.978	Ca^{2+}	−0.5815
ClO_3^-	−14.189	Sr^{2+}	−3.954
ClO_4^-	−17.445	Ba^{2+}	−7.676
BrO_3^-	−14.189	Ag^+	−10.47
CO_3^{2-}	−7.560	Cu^{2+}	−16.28
SiO_3^{2-}	−9.300	Zn^{2+}	−16.28
SO_3^{2-}	−2.326	Pb^{2+}	−9.304
SO_4^{2-}	1.163	Co^{2+}	−11.63
$S_2O_3^{2-}$	8.141	Al^{3+}	−32.56
CrO_4^{2-}	−1.163	Th^{4+}	−43.61
$Cr_2O_7^{2-}$	15.93		
PO_4^{3-}	−20.93		
$Fe(CN)_6^{4-}$	18.61		
Acetate^-	−22.91		
Oxalate^{2-}	−3.489		

$$\lambda_{\text{mix}}(T) = \lambda_{\text{mix}}(20°\text{C})\frac{\lambda_{\text{H}_2\text{O}}(T)}{\lambda_{\text{H}_2\text{O}}(20°\text{C})} \tag{3-132}$$

Except for strong acids and alkalies at high concentrations, Eqs. (3-131) and (3-132) are usually accurate to within ±5 percent.

DIFFUSION COEFFICIENTS

Gases For pressures up to about 10 atm (or perhaps even higher), the diffusion coefficient for a **binary** mixture of gases A and B may be estimated from the Fuller, Schettler, and Giddings relation[62,63]

$$\mathcal{D}_{AB} = \frac{10^{-3}T^{1.75}[(M_A + M_B)/M_A M_B]^{1/2}}{P[(\Sigma\nu)_A^{1/3} + (\Sigma\nu)_B^{1/3}]^2} \tag{3-133}$$

where T is in kelvins, P is in atmospheres, and $\mathcal{D}_{AB}$ is in cm²/s. M_A and M_B are respectively the molecular weights of A and B. To determine $\Sigma\nu$ the values of the atomic diffusion volumes shown in Table 3-342 should be used. These values were determined by a regression analysis of 340 experimental diffusion-coefficient values of 153 binary systems. This correlation cannot distinguish between isomers. Nain and Ferron[168] found that the accuracy was poor in polar gas

TABLE 3-342 Atomic Diffusion Volumes for Use in Estimating $\mathcal{D}_{AB}$ by the Method of Fuller, Schettler, and Giddings*

Atomic and structural diffusion-volume increments v			
C	16.5	(Cl)	19.5
H	1.98	(S)	17.0
O	5.48	Aromatic ring	−20.2
(N)	5.69	Heterocyclic ring	−20.2
Diffusion volumes for simple molecules Σv			
H_2	7.07		
D_2	6.70	CO_2	26.9
He	2.88	N_2O	35.9
N_2	17.9	NH_3	14.9
O_2	16.6	H_2O	12.7
Air	20.1	(CCl_2F_2)	114.8
Ar	16.1	(SF_6)	69.7
Kr	22.8	(Cl_2)	37.7
(Xe)	37.9	(Br_2)	67.2
CO	18.9	(SO_2)	41.1

*Parentheses indicate that the value listed in based on only a few data points.

mixtures. A similar correction was suggested by Saksena and Saxena[206] for the more general case. Marrero and Mason[146] show that Eq. (3-133) is reliable from moderate to high temperatures but is often poor at low temperatures. Generally, however, Eq. (3-133) will predict $\mathcal{D}_{AB}$ within 5 to 10 percent. No compositon dependence is shown, as usually $\mathcal{D}_{AB}$ (for gax mixtures) is insensitive to this variable.

Theoretical predictive methods for $\mathcal{D}_{AB}$ are also available, but intermolecular distance and energy parameters must be known for both A and B. These methods and several empirical estimation techniques are reviewed in Ref. 3.

While Eq. (3-133) shows that $\mathcal{D}_{AB}$ is proportional to $T^{1.75}$, the power to which T is raised depends on the actual temperature.[40] The value of 1.75 represents an average and may not be appropriate if wide temperature ranges are to be considered.

As indicated earlier, at low to moderate pressures the diffusion coefficient for gases is inversely proportional to pressure or density. Most high-pressure diffusion experiments have been limited to the measurement of self-diffusion coefficients. Dawson et al.[43] carried out a particularly extensive study with methane from $0.8 < T_r < 1.9$, $0.3 < P_r < 7.4$ and were able to correlate their data with the following equation:

$$\frac{\mathcal{D}\rho}{(\mathcal{D}\rho)^\circ} = 1 + 0.053432\rho_r - 0.030182\rho_r^2 - 0.029725\rho_{zr}^3 \qquad (3\text{-}134)$$

where $\mathcal{D}$ = self-diffusion coefficient at T and ρ
ρ = density
$(\mathcal{D}\rho)^\circ$ = product of $\mathcal{D}$ and ρ evaluated at T but at low pressure
ρ_r = reduced density ρ/ρ_c

Equation (3-134) is plotted in Fig. 3-53. Below a value of $\rho_r = 1$ there is little effect of density, and $\mathcal{D}\rho \approx (\mathcal{D}\rho)^\circ$. This fact has been confirmed by other experimental self-diffusion data,[50,130,157,173,251] although there is a scatter of about ±10 percent. In this same range of reduced densities, Mathur and Thodos[152] have proposed that

$$\mathcal{D}\rho_r = \frac{10.7 \times 10^{-5} T_r}{\beta} \qquad (3\text{-}135)$$

where $\mathcal{D}$ = high-pressure self-diffusion coefficient, cm^2/s
ρ_r = reduced density ρ/ρ_c
T_r = reduced temperature T/T_c
β = $M^{1/2}P_c^{1/3}/T_c^{5/6}$, with P_c in atmospheres and T_c in kelvins

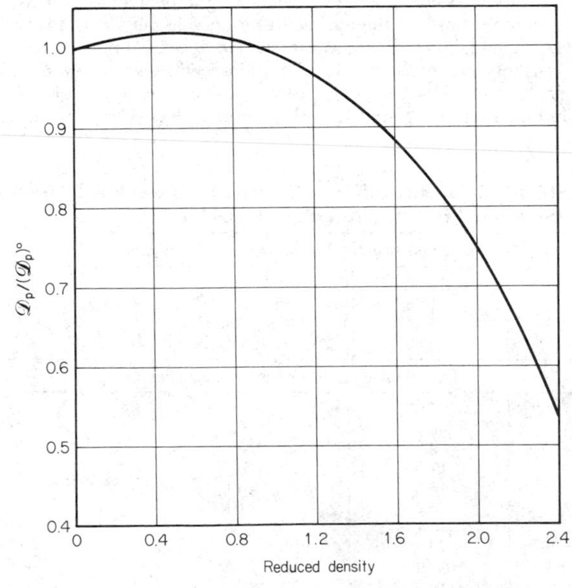

FIG. 3-53 Dawson-Khoury-Kobayashi correlation of self-diffusion with density.

Equation (3-135) would also show that $\mathcal{D}\rho/(\mathcal{D}\rho)^\circ = 1$ for $\rho_r < 1.0$, as indicated in Fig. 3-53. An alternative correlation leading to the same conclusion was proposed by Stiel and Thodos.[234]

In summary, for *self-diffusion* at high pressures, the product $\mathcal{D}\rho$ is essentially constant up to reduced densities of unity. At higher densities, the product $\mathcal{D}\rho$ decreases rapidly with further increases in density.

In contrast to this picture for self-diffusion, the few available data on binary diffusion indicate that $\rho\,\mathcal{D}_{AB}$ decreases with density even for pseudo-reduced densities° less than unity. The use of the density factor itself causes problems. At low densities, ρ, expressed in moles per volume, is independent of composition. However, at high densities the mixtures are usually not ideal, and ρ depends upon composition. The concomitant question of whether $\mathcal{D}_{AB}$ depends upon composition at high pressures has not yet been satisfactorily answered.

No estimation method to predict $\mathcal{D}_{AB}$ for high-pressure gases is yet available. If Eq. (3-134) or Eq. (3-135) is used, the calculated value of $\mathcal{D}_{AB}$ is too high.

Diffusion coefficients in multicomponent gas mixtures are difficult to calculate.[145] In the case of a dilute component i diffusing in a *homogeneous* mixture, where x_j is the mole fraction of j, then

$$\mathcal{D}_{i,\text{mix}} = \left[\sum_{\substack{j=1 \\ \neq i}}^{n} (x_j/\mathcal{D}_{ij})\right]^{-1} \qquad (3\text{-}136)$$

This simple relation (sometimes called Blanc's law[145,16]) was shown to apply to several ternary cases in which i was a trace component.[150] Deviations from Blanc's law are discussed by Sandler and Mason.[207]

The general theory of diffusion in multicomponent gas systems is covered by Hirschfelder, Curtiss, and Bird.[41,94] Approximate calculation methods were developed by Wilke.[273] The problem of diffusion in three-component gas systems has been generalized by Toor[250] and checked experimentally.[55,266,49]

Liquids Many liquid-diffusion-coefficient equations result from empirical modifications of the Stokes-Einstein equation, which predicts, for the diffusion of a large spherical molecule A through a dilute solvent B,[14,64]

$$\mathcal{D}_{AB} = RT/6\pi\eta_B r_A \qquad (3\text{-}137)$$

with η_B the viscosity of the solvent and r_A the radius of the spherical solute. Empiricism enters with the formulation

$$\mathcal{D}_{AB}^\circ = (T/\eta_B)f\,(\text{solute size}) \qquad (3\text{-}138)$$

where $\mathcal{D}_{AB}^\circ$ denotes that A is diffusing in essentially pure B. Thus, generally, for liquids $\mathcal{D}_{AB}^\circ \neq \mathcal{D}_{BA}^\circ$. The concentration dependence of $\mathcal{D}_{AB}$ is discussed later.

A widely used correlation for $\mathcal{D}_{AB}^\circ$ is the Wilke-Chang technique[274]

$$\mathcal{D}_{AB}^\circ = 7.4 \times 10^{-8} \frac{(\phi M_B)^{1/2} T}{\eta_B V_A^{0.6}} \qquad (3\text{-}139)$$

where D_{AB}° = mutual diffusion coefficient of solute A at very low concentrations in solvent B, cm^2/s
M_B = molecular weight of solvent B
T = absolute temperature, K
η_B = viscosity of solvent B, cP
V_A = molar volume of solute A at its normal boiling temperature, $cm^3/(g \cdot mol)$
ϕ = association factor of solvent B, dimensionless

The value of V_A is best estimated at T_b by methods described in the subsection "Liquid Density."

Wilke and Chang recommend that ϕ be chosen as 2.6 if the solvent is water, 1.9 for methanol, 1.5 for ethanol, and 1.0 for unassociated solvents. When 251 solute-solvent systems were tested by these

°Since true-mixture critical densities are not often known, a pseudo-critical density is used and defined as $\rho_c = V_{\text{mix},c}^{-1} = (\sum_j y_j V_{cj})^{-1}$.

authors, an average error of about 10 percent was noted. Hayduk and Laudie[91] reviewed diffusion-coefficient data for 87 different solutes in *water*. They reported that the Wilke-Chang correlation yielded an average error of 6.9 percent, although if the association factor for water is decreased from 2.6 to 2.26, the average error decreases to only 0.4 percent.

Olander[174] pointed out that whereas the Wilke-Chang correlation usually is satisfactory for most cases of an organic solute diffusing into water, the opposite situation of water diffusing into an organic solvent is not well predicted. He found the calculated value of $\mathcal{D}^{\circ}_{WB}$ to be about 2.3 times larger than the experimental value. He attributed this error to the fact that since water may diffuse as a tetramer, the value of V_w in Eq. (3-139) should be multiplied by 4; $V_w^{0.6}$ will then increase by a factor of 2.3. This explanation is disputed by Lusis.[134]

Many other modifications of the Wilke-Chang relation or Eq. (3-138) have been proposed.[275,215,93,276,2,88,121,29,89,136,209,191,227] However, for estimating diffusion coefficients in aqueous nonelectrolytes, it is the most accurate, particularly if the value of ϕ is reduced to 2.26 for organic solutes diffusing into water and, for water diffusing into organic solvents, if V_w is increased by a factor of 4.

In organic solutes diffusing into organic solvents, the Scheibel relation[209] is somewhat more accurate.

$$\mathcal{D}^{\circ}_{AB} = KT/\eta_B V_A^{1/3} \qquad (3\text{-}140)$$

where the symbols are defined in Eq. (3-139) and

$$K = (8.2 \times 10^{-8})\left[1 + \left(\frac{3V_B}{V_A}\right)^{2/3}\right] \qquad (3\text{-}141)$$

except that if benzene is the solvent (B) and $V_A < 2V_B$, then $K = 18.9 \times 10^{-8}$, and for other solvents, if $V_A < 2.5V_B$, $K = 17.5 \times 10^{-8}$.

Most of the estimation techniques introduced earlier in this subsection have assumed that $\mathcal{D}^{\circ}_{AB}$ varies inversely with the viscosity of the solvent. This inverse dependence originated from the Stokes-Einstein relation for a large (spherical) molecule diffusing through a continuum solvent (small molecules). If, however, the solvent is viscous, one may legitimately question whether this simple relationship is applicable. Davies et al.[42] found, for CO_2, that in various solvents $\mathcal{D}^{\circ}_{AB}\eta_B^{0.45} \approx$ constant for solvents ranging in viscosity from 1 to 27 cP, and these authors noted that in 1930 Arnold[7] had proposed an empirical estimation scheme in which $\mathcal{D}^{\circ}_{AB} \propto \eta^{-0.5}$. Hayduk and Cheng[90] investigated the effect of solvent viscosity more extensively and proposed that for nonaqueous systems

$$\mathcal{D}^{\circ}_{AB} = A\eta_B^q \qquad (3\text{-}142)$$

where the constants A and q are particular for a given solute. In Fig. 3-54 the diffusion coefficient of carbon tetrachloride is shown as a function of solvent viscosity. The applicability of Eq. (3-142) is clearly evident. Hayduk and Cheng list values of A and q for a few solutes. A rough correlation exists between A and q, indicating that solutes with lower diffusion coefficients vary in a more pronounced manner with solvent viscosity (or temperature). In Fig. 3-55 data for CO_2 diffusion in various solvents are shown. The solvent-viscosity range is reasonably large, and the correlation for organic solvents is not bad. For contrast, the data for water are also shown. These data fall well below the organic-solvent curve and have a slope close to -1. Hiss and Cussler[95] measured diffusion coefficients of n-hexane and naphthalene in hydrocarbons with viscosities ranging from 0.5 to 5000 cP and report that $\mathcal{D}^{\circ}_{AB} \propto \eta_B^{-2/3}$. These studies and others[64,135,269,89,167] show clearly that over wide temprature or solvent-viscosity ranges, simple empirical correlations, as presented earlier in this subsection, are inadequate. The diffusion coefficient does not decrease in proportion to an increase in solvent viscosity, but $\mathcal{D}^{\circ}_{AB} \propto \eta^q$, where q varies, usually from -0.5 to -1.

To determine the binary diffusion coefficient at finite concentrations, a correction is usually made to allow for the fact that the system may be nonideal in a thermodynamic sense. Also, to account for concentration variations, the relation of Vignes[265] is the most accurate. Thus, to estimate the liquid diffusion coefficient

$$\mathcal{D}_{AB} = \frac{(\mathcal{D}^{\circ}_{AB})^{x_B}(\mathcal{D}^{\circ}_{BA})^{x_A}}{\alpha} \qquad (3\text{-}143)$$

where $\mathcal{D}^{\circ}_{AB}$ and $\mathcal{D}^{\circ}_{BA}$ are the infinitely dilute diffusion coefficients estimated, for example, from Eq. (3-139) or Eq. (3-140), x_A and x_B are the mole fractions of A and B, and

$$\alpha = (\partial \ln a_A/\partial \ln x_A)_{T,P} = (\partial \ln a_B/\partial \ln x_B)_{T,P} \qquad (3\text{-}144)$$

where a is the activity.

Multicomponent-liquid diffusion coefficients are difficult to estimate, and the frame of reference from which to define diffusion rates must be clearly specified.[278] Perkins and Geankoplis[181] address the special case of the diffusion of a solute in a homogeneous mixed solvent. Equation (3-139) can be used if the association parameter is redefined as

$$\phi M = \sum_{\substack{j=1 \\ \neq A}}^{n} x_j \phi_j M_j \qquad (3\text{-}145)$$

where A is the diffusing solute, x_j the mole fraction of component j. ϕ_j is the association parameter of the A_j system, while M_j is the molecular weight of component j.

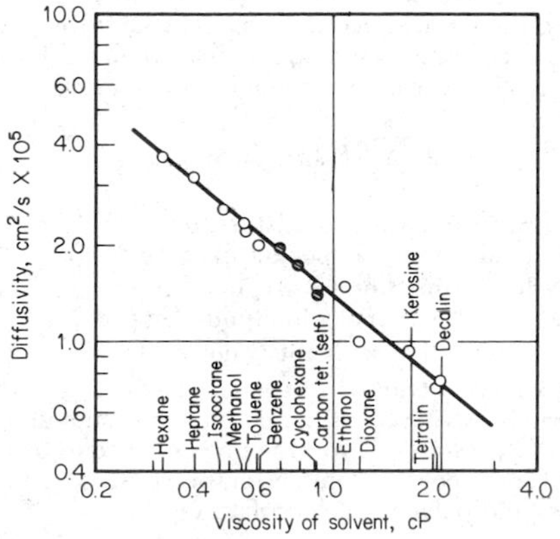

FIG. 3-54 Diffusivity of carbon tetrachloride. To convert centipoises to pascal-seconds, multiply by 0.001; to convert square centimeters per second to square feet per hour, multiply by 3.8750. [*From Hayduk and Cheng,* Chem. Eng. Sci., *26, 635 (1971).*]

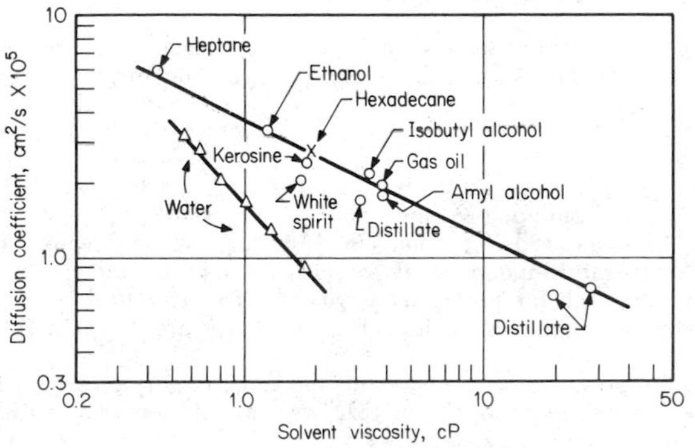

FIG. 3-55 Diffusion coefficient for carbon dioxide in various solvents; o, Davies, Ponter, and Craine, *Can. J. Chem. Eng.*, **45**, 372 (1967); x, Hayduk and Cheng, *Chem. Eng. Sci.*, **26**, 635 (1971); Δ, Himmelblau, *Chem. Rev.*, **64**, 527 (1964). To convert centipoises to pascal-seconds, multiply by 0.001; to convert square centimeters per second to square feet per hour, multiply by 3.8750.

SURFACE TENSION

Jasper[108] has made a critical evaluation of experimental surface tensions for approximately 2200 pure compounds. He correlates surface tension σ(dyn/cm) with temperature T(°C) over a specified temperature range as

$$\sigma = a - bT \quad (3\text{-}146)$$

a and b are listed for most of the substances. To obtain surface-tension values at a temperature higher than the upper temperature limit indicated by Jasper, the following expression may be used:

$$\sigma = d(1 - T_r)^e \quad (3\text{-}147)$$

where d and e are determined such that σ and $d\sigma/dT$ at the upper temperature limit T_{HI} have the same values calculated from both Eqs. (3-146) and (3-147):

$$e = \frac{b(T_c - T_{HI})}{a - bT_{HI}} \quad (3\text{-}148)$$

$$d = (a - bT_{HI})(1 - T_{HI_r})^{-e} \quad (3\text{-}149)$$

where T_c = critical temperature, °C
T_r = reduced temperature = $(T + 273.15)/(T_c + 273.15)$

Equation (3-147) correctly predicts that the surface tension becomes zero at the critical point.[198]

For non-hydrogen-bonded substances not found in Jasper,[108] use can be made of the corresponding-states approach of Brock and Bird[21] as modified by Miller:[159]

$$\sigma = P_c^{2/3}T_c^{1/3}Q(1 - T_r)^{11/9} \quad (3\text{-}150)$$

$$\text{where } Q = 0.1207\left(1 + \frac{T_{b_r}\ln P_c}{1 - T_{b_r}}\right) - 0.281$$

σ = surface tension, dyn/cm
P_c = critical pressure, atm
T_c = critical temperature, K
T_r = reduced temperature T/T_c
T_{b_r} = reduced normal boiling point temperature T_b/T_c

Errors are usually less than 5 percent.

Example 13 The required properties for ethyl mercaptan[196] are P_c = 54.2 atm, T_c = 499.0 K, and T_b = 308.2 K. For a temperature T of 303.15 K, T_{br} = 308.2/499.0 = 0.6176, and T_r = 303.15/499.0 = 0.6075. Substituting into Eq. (3-150),

$$Q = 0.1207\left[1 + \frac{0.6176 \ln 54.2}{1 - 0.6176}\right] - 0.281$$
$$= 0.6180$$
$$\sigma = (54.2)^{2/3}(499.0)^{1/3}(0.6180)(1 - 0.6075)^{11/9}$$
$$\sigma = 22.38 \text{ dyn/cm}$$

The reported experimental value is 22.68 dyn/cm.[108]

For hydrogen-bonded compounds, the approach originally suggested by Macleod[140] as further developed by Sugden[238,239] can be used:

$$\sigma^{1/4} = [P](\rho_L - \rho_G) \quad (3\text{-}151)$$

where σ = surface tension, dyn/cm
ρ_L = liquid density, mol/cm^3
ρ_G = vapor density, mol/cm^3

The temperature-independent parachor $[P]$ may be calculated by the additive scheme proposed by Quale.[188] The structural contributions for this method as modified by Reid et al.[196] are shown in Table 3-343.

At low pressures, where $\rho_L \gg \rho_V$, the vapor-density term may be neglected. Errors using Eq. (3-151) are normally less than 5 to 10 percent.

Example 14 For isobutyric acid, the liquid density is 0.968 g/cm^3 at 293.15 K.[196] With a molecular weight of 88.106, ρ_L = 0.968/88.106 = 0.01099 mol/cm^3. $[P]$ is determined from Table 3-343:

TABLE 3-343 Structural Contributions for Calculation of the Parachor

Carbon-hydrogen		Special groups	
C	9.0	=O(ketone)	
H	15.5	Three carbon atoms	22.3
CH_3-	55.5	Four carbon atoms	20.0
$-CH_2-$	40.0*	Five carbon atoms	18.5
$CH_3-CH(CH_3)-$	133.3	Six carbon atoms	17.3
$CH_3-CH_2-CH(CH_3)-$	171.9	-CHO	66
$CH_3-CH_2-CH_2-$		O (not noted above)	20
$CH(CH_3)-$	211.7	N (not noted above)	17.5
$CH_3-CH(CH_3)-CH_2-$	173.3	S	49.1
$CH_3-CH_2-CH(C_2H_5)$	209.5	P	40.5
$CH_3-C(CH_3)_2-$	170.4	F	26.1
$CH_3-CH_2-C(CH_3)_2-$	207.5	Cl	55.2
$CH_3-CH(CH_3)-$		Br	68.0
$CH(CH_3)-$	207.9	I	90.3
$CH_3-CH(CH_3)-C(CH_3)_2-$	243.5		
C_6H_5-	189.6	Ethylenic bond	
		Terminal	19.1
Special groups		2,3 position	17.7
$-COO-$	63.8	3,4 position	16.3
$-COOH$	73.8		
$-OH$	29.8	Triple bond	40.6
$-NH_2$	42.5	Ring closure	
$-O-$	20.0	Three-membered	12.5
$-NO_2$ (nitrite)	74	Four-membered	6.0
$-NO_3$ (nitrate)	93	Five-membered	3.0
$-CO(NH_2)$	91.7	Six-membered	0.8

*If $n > 12$ in $(-CH_2-)_n$, increase increment to 40.3.

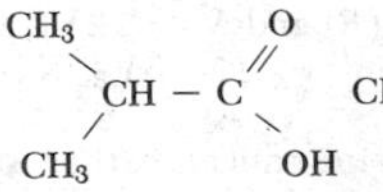

is made up of the two groups $CH_3-CH(CH_3)-$ and $-COOH$.

Therefore, $[P]$ = 133.3 + 73.8 = 207.1. With Eq. (3-151), neglecting the vapor density,

$$\sigma^{1/4} = (207.1)(0.01099) = 2.276$$
$$\sigma = 26.83 \text{ dyn/cm}$$

Jasper[108] quotes a value of 25.04 dyn/cm at 293.15 K.

Other methods for particular compound classes, e.g., alcohols and cryogenic liquids, may be found in Ref. 196.

The surface tension of a **liquid mixture** is not a simple function of the pure-component surface tensions because the composition of the mixture surface is not the same as that of the bulk. For **nonaqueous** solutions, the thermodynamics-derived correlation of Sprow and Prausnitz[226] results in errors of typically less than 2 to 3 percent. Unfortunately, use of this method requires activity-coefficient estimates for both the surface phase and the bulk liquid as well as pure-component surface tensions and liquid molar volumes. For less accurate values, Eq. (3-151) can be written for an n-component mixture:

$$\sigma_{mix}^{1/4} = \sum_{i=1}^{n} [P_i](\rho_{L_{mix}}x_i - \rho_{G_{mix}}y_i) \quad (3\text{-}152)$$

where σ_{mix} = surface tension of mixture, dyn/cm
$[P_i]$ = parachor of component i (Table 3-343)
ρ_{Lmix} = liquid-mixture density, mol/cm^3
x_i = mole fraction of i in liquid phase
ρ_{Gmix} = vapor-mixture density, mol/cm^3
y_i = mole fraction of i in vapor phase

Again, the term involving the vapor density and composition may be neglected at low pressures. Errors may be expected to be less than 5 to 10 percent for nonpolar systems; errors of 5 to 15 percent are likely for polar-polar and polar-nonpolar mixtures.

Example 15 For a mixture of 42.3 mole percent diethyl ether and 57.7 mole percent benzene, Hammick and Andrew[84] report a mixture density of 0.7996 g/cm^3 at 298.15 K. The molecular weight of diethyl ether is 74.1224; that of benzene is 78.1134. Therefore,

$$\rho_{L\mathrm{mix}} = 0.7996/[(0.423)(74.1224) + (0.577)(78.1134)]$$
$$= 0.01046 \text{ mol/cm}^3$$

From Table 3-343, diethyl ether consists of groups $2(CH_3-) + 2(-CH_2-) + (-O-)$. Thus, $[P]_{\mathrm{ether}} = 2(55.5) + 2(40.0) + 20.0 = 211.0$. Benzene consists of the groups C_6H_5- and H, resulting in $[P]_{\mathrm{benzene}} = 189.6 + 15.5 = 205.1$. Substituting in Eq. (3-152), neglecting the vapor terms:

$$\sigma_{\mathrm{mix}}^{1/4} = 0.01046[(0.423)(211.0) + (0.577)(205.1)]$$
$$= 2.171$$
$$\sigma_{\mathrm{mix}} = 22.21 \text{ dyn/cm}$$

The reported experimental value at 298.15 K is 21.81 dyn/cm.[84]

Surface tensions for **aqueous** solutions are more difficult to predict than those for nonaqueous mixtures because of the nonlinear dependence on mole fraction. Small concentrations of the organic material may significantly affect the mixture surface-tension value. For many binary organic-aqueous mixtures, the method of Tamura, Kurata, and Odani[244] may be used:

$$\sigma_{\mathrm{mix}}^{1/4} = \psi_w \sigma_w^{1/4} + \psi_o \sigma_o^{1/4} \tag{3-153}$$

where σ_{mix} = surface tension of mixture, dyn/cm
σ_w = surface tension of pure water, dyn/cm
σ_o = surface tension of pure organic component, dyn/cm

$$\psi_o = 1 - \psi_W \tag{3-154}$$

ψ_w is defined by the relation

$$\log_{10} \frac{(\psi_w)^q}{(1-\psi_w)} = \log_{10}\left[\frac{(x_w V_w)^q}{x_o V_o}(x_w V_w + x_o V_o)^{1-q}\right] + 0.441 \frac{q}{T}\left[\frac{\sigma_o V_o^{2/3}}{q} - \sigma_w V_w^{2/3}\right] \tag{3-155}$$

where x_w = bulk mole fraction of pure water
x_o = bulk mole fraction of pure organic component
V_w = molar volume of pure water, cm^3/mol
V_o = molar volume of pure organic component, cm^3/mol
q = constant depending upon type and size of organic constituent:

Materials	q	Example
Fatty acids, alcohols	Number of carbon atoms	Acetic acid, $q = 2$
Ketones	One less than the number of carbon atoms	Acetone, $q = 2$
Halogen derivatives of fatty acids	Number of carbons times ratio of molar volume of halogen derivative to parent fatty acid	Chloroacetic acid $q = 2\frac{V(\text{chloroacetic acid})}{V(\text{acetic acid})}$

Expected errors are less than 10 percent when q is less than 5 and within 20 percent when q is greater than 5. If the organic mole fraction is 0.01 or less, the simpler Szyszkowski equation as developed by Meissner and Michaels[155] may be used.

Example 16 This is an example of the use of Eq. (3-153) for a methanol-water mixture at 303.15 K when the methanol mole fraction is 0.122. From Jasper,[108] $\sigma_w = 71.40$ dyn/cm and $\sigma_o = 21.73$ dyn/cm. The specific volume of water is 1.0044 cm^3/g;[156] since the molecular weight of water is 18.0152, $V_w = (1.0044)(18.0152) = 18.09$ cm^3/mol. The density of methanol is reported to be 0.78196 g/cm^3;[271] with the molecular weight being 32.042, $V_o = 32.042/0.78196 = 40.98$ cm^3/mol. For methanol, $q = 1$. Therefore,

$$\log_{10}\frac{\psi_w}{1-\psi_w} = \log_{10}\left[\frac{(0.878)(18.09)}{(0.122)(40.98)}\right] + \frac{0.441}{303.15}[(21.73)(40.98)^{2/3} - (71.40)(18.09)^{2/3}]$$
$$= 0.502 - 0.340 = 0.162$$

Hence,

$$\psi_w/(1-\psi_w) = 10^{0.162} = 1.452$$
$$\psi_w = \frac{1.452}{2.452} = 0.592$$
$$\psi_o = 1 - 0.592 = 0.408$$

Substituting into Eq. (3-153),

$$\sigma_{\mathrm{mix}}^{1/4} = (0.592)(71.40)^{1/4} + (0.408)(21.73)^{1/4}$$
$$= 2.602$$
$$\sigma_{\mathrm{mix}} = 45.84 \text{ dyn/cm}$$

The reported experimental value is 46.1 dyn/cm.[244]

REFERENCES

1. Ambrose, D., C. H. S. Sprake, and R. Townsend, *J. Chem. Thermodyn.*, **6,** 693 (1974). 2. Amourdam, M. J., and G. S. Laddha, *J. Chem. Eng. Data,* **12,** 389 (1967). 3. Anderson, J. W., G. H. Beyer, and K. M. Watson, *Nat. Pet. News Tech. Sec.*, **36,** R476 (July 5, 1944). 4. Andrade, E. N. da C., *Nature*, **125,** 309 (1930). 5. Andrade, E. N. da C., *Phil. Mag.*, **17,** 497, 698 (1934). 6. Antoine, C., *Comptes Rendus*, 107, 681, 836 (1888). 7. Arnold, J. H., *J. Am. Chem. Soc.*, **52,** 3937 (1930). 8. Benedict, M., G. B. Webb, and L. C. Rubin, *J. Chem Phys.*, **8,** 334 (1940), **10,** 747 (1942); *Chem. Eng. Prog.*, **47,** 419 (1951). 9. Benson, S. W., *Thermochemical Kinetics*, Wiley, New York, 1968, chap. 2. 10. Benson, S. W., F. R. Cruickshank, D. M. Golden, G. R. Haugen, H. E. O'Neal, A. S. Rodgers, R. Shaw, and R. Walsh, *Chem. Rev.*, **69,** 279 (1969). 11. Beret, S., and J. M. Prausnitz, *Am. Inst. Chem. Eng. J.*, **21,** 1123 (1975). 12. Bhirud, V. L., *Am. Inst. Chem. Eng. J.*, **24,** 880 (1978). 13. Bhirud, V. L., *Am. Inst. Chem. Eng. J.*, **24,** 1127 (1978). 14. Bird, R. B., W. E. Stewart, and E. N. Lightfoot, *Transport Phenomena*, Wiley, New York, 1960, chap. 16. 15. Bishnoi, P. R., R. D. Miranda, and D. B. Robinson, *Hydrocarbon Process.*, **53**(11), 197 (1974). 16. Blanc, A., *J. Phys.*, **7,** 825 (1908). 17. Bondi, A., *Ind. Eng. Chem. Fundam.*, **5,** 443 (1966). 18. Boublik, T., V. Fried, and E. Hala, *The Vapor Pressures of Pure Substances*, Elsevier, New York, 1973. 19. Breedveld, G. J. F., and J. M. Prausnitz, *Am. Inst. Chem. Eng. J.*, **19,** 783 (1973). 20. Bridgman, P. W., *Proc. Am. Acad. Arts Sci.*, **59,** 154 (1923). 21. Brock, J. R., and R. B. Bird, *Am. Inst. Chem. Eng. J.*, **1,** 174 (1955). 22. Brokaw, R. S., *Ind. Eng. Chem. Process Des. Dev.*, **8,** 240 (1969). 23. Brokaw, R. S., NASA Tech. Note D-2502, November 1964. 24. Brokaw, R. S., *J. Chem. Phys.*, **42,** 1140 (1965). 25. Brokaw, R. S., NASA Tech. Note D-4496, April 1968. 26. Brokaw, R. S., R. A. Svehla, and C. E. Baker, NASA Tech. Note D-2580, January 1965. 27. Brown, G. P., A. Dinardo, G. K. Cheng, and T. K. Sherwood, *J. Appl. Phys.*, **17,** 802 (1946). 28. Buck, T. Królikowski, personal communication, 1979. 29. Caldwell, C. S., and A. L. Babb, *J. Phys. Chem.*, **60,** 14, 56 (1956). 30. Carnahan, N. F., and K. E. Starling, *J. Chem. Phys.*, **51,** 635 (1969); *Am. Inst. Chem. Eng. J.*, **18,** 1184 (1972). 31. Chapman, S., and T. G. Cowling, *The Mathematical Theory of Non-uniform Gases*, Cambridge, New York, 1939. 32. Chueh, P. L., and C. H. Deal, "Thermophysical Properties of Pure Chemical Compounds," paper presented at 65th annual AIChE meeting, New York, November 1972. 33. Chueh, P. L., and J. M. Prausnitz, *Ind. Eng. Chem. Fundam.*, **6,** 492 (1967). 34. Chueh, P. L., and J. M. Prausnitz, *Am. Inst. Chem. Eng. J.*, **13,** 1099 (1967). 35. Chueh, P. L., and J. M. Prausnitz, *Am. Inst. Chem. Eng. J.*, **15,** 471 (1969). 36. Chueh, C. F., and A. C. Swanson, *Chem. Eng. Prog.*, **69**(7), 83 (1973); *Can. J. Chem. Eng.*, **51,** 596 (1973). 37. Cooper, H. W., and J. C. Goldfrank, *Hydrocarbon Process.*, **46**(12), 141 (1967). 38. Counsell, J. F., and D. A. Lee, *J. Chem. Thermodyn.*, **4,** 915 (1972). 39. Cox, J. D., and G. Pilcher, *Thermochemistry of Organic and Organometallic Compounds*, Academic, London, 1970. 40. Cullinan, H. T., Jr., *Can. J. Chem. Eng.*, **49,** 130 (1971). 41. Curtiss, C. F., and J. O. Hirschfelder, *J. Chem. Phys.*, **17,** 550 (1949). 42. Davies, G. A., A. B. Ponter, and K. Craine, *Can. J. Chem. Eng.*, **45,** 372 (1967). 43. Dawson, R., F. Khoury, and R. Kobayashi, *Am. Inst. Chem. Eng. J.*, **16,** 725 (1970). 44. Dean, D. E., and L. I. Stiel, *Am. Inst. Chem. Eng. J.*, **11,** 526 (1965). 45. Domalski, E. S., *J. Phys. Chem. Ref. Data*, **1,** 221 (1972). 46. Donohue, M. D., and J. M. Prausnitz, *Am. Inst. Chem. Eng. J.*, **24,** 849 (1978). 47. Doolittle, A. K., *Chem. Eng. Prog. Symp. Ser.*, no. 44, 59, 1 (1963). 48. Dul'nev, G. N., and Y. P. Zarichnyak, *Thermophysical Properties of Substances and Materials*, 3d issue, Standards Press, Moscow, 1971, p. 103. 49. Duncan, J. B., and H. L. Toor, *Am. Inst. Chem. Eng. J.*, **8,** 38 (1962). 50. Durbin, L., and R. Kobayashi, *J. Chem. Phys.*, **37,** 1643 (1962). 51. Dymond, J. H., and E. B. Smith, *The Virial Coefficients of Gases*, Clarendon, Oxford, 1969. 52. Edmister, W. C., J. Vairogs, and A. J. Klekers, *Am. Inst. Chem. Eng. J.*, **14,** 479 (1968). 53. Edwards, M. N. B., and G. Thodos, *J. Chem. Eng. Data*, **19,** 14 (1974). 54. Ewing, C. T., B. E. Walker, J. A. Grand, and R. R. Miller, *Chem. Eng. Prog. Symp. Ser.*, **53**(20), 19 (1957). 55. Fairbanks, D. F., and C. R. Wilke, *Ind. Eng. Chem.*, **42,** 471 (1950). 56. Filippov, L. P., *Vestn. Mosk. Univ. Ser. Fiz. Mat. Estestv. Nauk*, (8), **10**(5), 67–69 (1955). 57. Filippov, L. P., and N. S. Novoselova, *Vestn. Mosk. Univ. Ser. Fiz. Mat. Estestv. Nauk*, (3),**10**(2), 37–40 (1955); *Chem. Abstr.*, **49,** 11366 (1955). 58. Fishtine, S. H., *Ind. Eng. Chem. Fundam.*, **2,** 149 (1963). 59. Flynn, L. W. and G. Thodos, *J. Chem. Eng. Data*, **6,** 457 (1961). 60. Franklin, J. L., *Ind. Eng. Chem.*, **41,** 1070 (1949). 61. Franklin, J. L., *J. Chem. Phys.*, **21,** 2029 (1953). 62. Fuller, E. N., and J. C. Giddings,

J. Gas Chromatogr., **3**, 222 (1965). 63. Fuller, E. N., P. D. Schettler, and J. C. Giddings, *Ind. Eng. Chem.*, **58**(5), 18 (1966). 64. Gainer, J. L., and A. B. Metzner, AIChE-IChemE Symp. Ser., no. 6, 1965, p. 74. 65. Gambill, W., unpublished. 66. Gambill, W. R., *Chem. Eng.*, **66**(16), 129 (1959). 67. Ghemling, J., D. D. Liu, and J. M. Prausnitz, *Chem. Eng. Sci.*, **34**, 951 (1979). 68. Giddings, J. D., Ph.D. thesis, Rice University, Houston, 1963. 69. Goletz, E., and D. Tassios, *Ind. Eng. Chem. Process Des. Dev.*, **16**, 75 (1977). 70. Gomez-Nieto, M., and G. Thodos, *Am. Inst. Chem. Eng. J.*, **23**, 904 (1977). 71. Gomez-Nieto, M., and G. Thodos, *Ind. Eng. Chem. Fundam.*, **17**, 45 (1978). 72. Gray, P., S. Holland, and A. O. S. Maczek, *Trans. Faraday Soc.*, **65**, 1032 (1969). 73. Gray, P., S. Holland, and A. O. S. Maczek, *Trans. Faraday Soc.*, **66**, 107 (1970). 74. Gray, P., and P. G. Wright, *Proc. R. Soc. (London)*, **A263**, 161 (1961). 75. Grosse, A. V., *J. Inorg. Chem.*, **22**, 23 (1961). 76. Grunberg, L., *An Equation for Predicting the Viscosity of Liquid Mixtures*, NEL Rep. No. 626, December 1976. 77. Grunberg, L., and A. H. Nissan, *Trans. Faraday Soc.*, **44**, 1013 (1948). 78. Grunberg, L., and A. H. Nissan, *Ind. Eng. Chem.*, **42**, 885 (1950). 79. Guildner, L. A., *Proc. Nat. Acad. Sci.*, **44**, 1149 (1958). 80. Hadden, S. T., *Hydrocarbon Process. Pet. Refiner*, **45**(7), 137 (1966). 81. Hadden, S. T., *J. Chem. Eng. Data*, **15**, 92 (1970). 82. Haggenmacher, J. E., *J. Am. Chem. Soc.*, **68**, 1633 (1946). 83. Halm, R. L., and L. I. Stiel, *Am. Inst. Chem. Eng. J.*, **17**, 259 (1971). 84. Hammick, D. L., and L. W. Andrew, *J. Chem. Soc.*, 754 (1929). 85. Hankinson, R. W., and G. H. Thompson, *Am. Inst. Chem. Eng. J.*, **25**, 653 (1979). 86. Hanley, H. J. M., *Cryogenics*, **16**, 643 (1976). 87. Hayden, J. G., and J. P. O'Connell, *Ind. Eng. Chem. Process Des. Dev.*, **14**, 209 (1975). 88. Hayduk, W., and W. D. Buckley, *Chem. Eng. Sci.*, **27**, 1997 (1972). 89. Hayduk, W., R. Casteñada, H. Bromfield, and R. R. Perras, *Am. Inst. Chem. Eng. J.*, **19**, 859 (1973). 90. Hayduk, W., and S. C. Cheng, *Chem. Eng. Sci.*, **26**, 635 (1971). 91. Hayduk, W., and H. Laudie, *Am. Inst. Chem. Eng. J.*, **20**, 611 (1974). 92. Herning, F., and L. Zipperer, *Gas Wasserfach*, **79**, 49 (1936). 93. Himmelblau, D. M., *Chem. Rev.*, **64**, 527 (1964). 94. Hirschfelder, J. O., C. F. Curtiss, and R. B. Bird, *Molecular Theory of Gases and Liquids*, Wiley, New York, 1954. 95. Hiss, T. G., and E. L. Cussler, *Am. Inst. Chem. Eng. J.*, **19**, 698 (1973). 96. Hiza, M. J., and A. G. Duncan, *Am. Inst. Chem. Eng. J.*, **16**, 733 (1970). 97. Ho, C. Y., R. W. Powell, and P. E. Liley, *J. Phys. Chem. Ref. Data*, **1**, 279 (1972). 98. Hougen, O. A., K. M. Watson and R. A. Ragatz, *Chemical Process Principles*, part II, Wiley, New York, 1959. 99. Hsiao, Y-J, and B. C-Y Lu, *Can. J. Chem. Eng.*, **57**, 102 (1979). 100. Hsu, C. C., and J. J. McKetta, *J. Chem. Eng. Data*, **9**, 45 (1964). 101. Huang, P. K., and T. E. Daubert, *Ind. Eng. Chem. Process Des. Dev.*, **13**, 193 (1974). 102. Irving, J. B., *Viscosities of Binary Liquid Mixtures: A Survey of Mixture Equations*, NEL Rep. No. 630, February 1977. 103. Irving, J. B., *Viscosities of Binary Liquid Mixtures: The Effectiveness of Mixture Equations*, NEL Rep. No. 631, February 1977. 104. Jamieson, D. T., and J. S. Tudhope, Nat. Eng. Lab. Glasgow Rep. No. 137, March 1964. 105. Jamieson, D. T., J. B. Irving, and J. S. Tudhope, *Liquid Thermal Conductivity: A Data Survey to 1973*, H. M. Stationery Office, Edinburgh, 1975. 106. Jamieson, D. T., and E. H. Hastings, in C. Y. Ho and R. E. Taylor (eds.), *Proceedings of the 8th Conference on Thermal Conductivity*, Plenum, New York, 1969, p. 631. 107. *JANAF Thermochemical Tables*, 2d ed., NSRDS-NBS 37, June 1971. 108. Jasper, J. J., *J. Phys. Chem. Ref. Data*, **1**, 841 (1972). 109. Joffe, J., *Ind. Eng. Chem. Fundam.*, **10**, 532 (1971). 110. Johnson, A. I., and C. J. Huang, *Can. J. Technol.*, **33**, 421 (1955). 111. Jordan, T. E., *Vapor Pressure of Organic Compounds*, Interscience, New York, 1954. 112. Jossi, J. A., L. I. Stiel, and G. Thodos, *Am. Inst. Chem. Eng. J.*, **8**, 59 (1962). 113. Kaufman, T. G., *Ind. Eng. Chem. Fundam.*, **7**, 115 (1968). 114. Kay, W. B., *Ind. Eng. Chem.*, **28**, 1014 (1936). 115. Kay, W. B., *J. Chem. Eng. Data*, **15**, 46 (1970). 116. Kerr, C. P., and J. Coates, paper, 67th national AIChE meeting, Atlanta, Feb. 15–18, 1970. 117. Kreglewski, A., and R. C. Wilhoit, *J. Phys. Chem.*, **78**, 1961 (1974). 118. Kreglewski, A., R. C. Wilhoit, and B. J. Zwolinski, *J. Chem. Eng. Data*, **18**, 432 (1973); *J. Phys. Chem.*, **77**, 2212 (1973). 119. Kreglewski, A., and W. B. Kay, *J. Phys. Chem.*, **73**, 3359 (1969). 120. Lee, B. I., and M. G. Kesler, *Am. Inst. Chem. Eng. J.*, **21**, 510 (1975). 121. Lees, F. P., and P. Sarram, *J. Chem. Eng. Data*, **16**, 41 (1971). 122. Lenoir, J. M., *Pet. Refiner*, **36**(8), 162 (1957). 123. Letsou, A., and L. I. Stiel, *Am. Inst. Chem. Eng. J.*, **19**, 409 (1973). 124. Lewis, W. K., and L. Squires, *Refiner Nat. Gasoline Manuf.*, **13**(12), 448 (1934). 125. Li, C. C., *Can. J. Chem. Eng.*, **19**, 709 (1971). 126. Li, C. C., *Can. J. Chem. Eng.*, **54**, 464 (1976). 127. Li, C. C., *Am. Inst. Chem. Eng. J.*, **22**, 927 (1976). 128. Lin, C.-J., and S. W. Hopke, *Chem. Eng. Prog. Symp. Ser.*, **70**(140), 37 (1974). 129. Lindsay, A. L., and L. A. Bromley, *Ind. Eng. Chem.*, **42**, 1508 (1950). 130. Lispicas, M., *J. Chem. Phys.*, **36**, 1235 (1962). 131. Losenicky, Z., *J. Phys. Chem.*, **72**, 4308 (1968). 132. Lu, B. C.-Y., C. Hsi, S.-D. Chang, and A. Tsang, *Am. Inst. Chem. Eng. J.*, **19**, 748 (1973). 133. Luria, M., and S. W. Benson, *J. Chem. Eng. Data*, **22**, 90 (1977). 134. Lusis, M. A., *Chem. Proc. Eng.*, **5**, 27 (May 1971). 135. Lusis, M. A., *Chem. Eng. Dev. (Bombay)*, 48 (January 1972); *Am. Inst. Chem. Eng. J.*, **20**, 207 (1974). 136. Lusis, M. A., and G. A. Ratcliff, *Am. Inst. Chem. Eng. J.*, **17**, 1492 (1971). 137. Lydersen, A. L., *Estimation of Critical Properties of Organic Compounds*, Univ. Wis. Coll. Eng., Eng. Exp. Stn. Rep. No. 3, Madison, April 1955. 138. Lydersen, A. L., R. A. Greenkorn, and O. A. Hougen, "Generalized Thermodynamic Properties of Pure Fluids," Univ. Wis. Coll. Eng., Eng. Exp. Stn. Rep. No. 4, Madison, October 1955. 139. Lyman, T. J., and R. P. Danner, *Am. Inst. Chem. Eng. J.*, **22**, 759 (1976). 140. Macleod, D. B., *Trans. Faraday Soc.*, **19**, 38 (1923). 141. Maczek, A. O. S., and P. Gray, *Trans. Faraday Soc.*, **65**, 1473 (1969). 142. Maczek, A. O. S., and P. Gray, *Trans. Faraday Soc.*, **66**, 127 (1970). 143. Maejima, T., private communication, 1973. 144. Mani, N., and J. E. S. Venart, *Adv. Cryog. Eng.*, **18**, 280 (1973). 145. Marrero, T. R., and E. A. Mason, *J. Phys. Chem. Ref. Data*, **1**, 3 (1972). 146. Marrero, T. R., and E. A. Mason, *Am. Inst. Chem. Eng. J.*, **19**, 498 (1973). 147. Martin, J. J., *Ind. Eng. Chem.*, **59**, 34 (1967). 148. Martin, J. J., *Ind. Eng. Chem. Fundam.*, **18**, 81 (1979). 149. Mason, E. A., and S. C. Saxena, *Phys. Fluids*, **1**, 361 (1958). 150. Mathur, G. P., and S. C. Saxena, *Ind. J. Pure Appl. Phys.*, **4**, 266 (1966). 151. Mathur, G. P., and G. Thodos, *Am. Inst. Chem. Eng. J.*, **9**, 596 (1963). 152. Mathur, G. P., and G. Thodos, *Am. Inst. Chem. Eng. J.*, **11**, 613 (1965). 153. Mathur, V. K., J. D. Singh, and W. M. Fitzgerald, *J. Chem. Eng. Data (Japan)*, **11**, 67 (1978). 154. McLaughlin, E., *Chem. Rev.*, **64**, 389 (1964). 155. Meissner, H. P., and A. S. Michaels, *Ind. Eng. Chem.*, **41**, 2782 (1949). 156. Meyer, C. A., et al., *ASME Steam Tables*, 3d ed., American Society of Mechanical Engineers, New York, 1977. 157. Mifflin, T. R., and C. O. Bennett, *J. Chem. Phys.*, **29**, 975 (1958). 158. Miller, C. O. M., private communication, April 1977. 159. Miller, D. G., *Ind. Eng. Chem. Fundam.*, **2**, 78 (1963). 160. Miller, J. W., Jr., G. R. Schorr, and C. L. Yaws, *Chem. Eng.*, **83**(23), 129 (1976). 161. Misic, D., and G. Thodos, *Am. Inst. Chem. Eng. J.*, **7**, 264 (1961). 162. Misic, D., and G. Thodos, *J. Chem. Eng. Data*, **9**, 540 (1963). 163. Missenard, A., *Rev. Gen. Thermodyn.*, **101**(5), 649 (1970). 164. Missenard, F. A., *Rev. Gen. Therm. Fr.*, **141**, 751 (1973). 165. Mo, K. C., and K. E. Gubbins, *Chem. Eng. Comm.*, **1**, 281 (1974). 166. Modell, M., and R. C. Reid, *Thermodynamics and Its Applications*, Prentice-Hall, Englewood Cliffs, N.J., 1974, sec. 7.5. 167. Moore, J. W., and R. M. Wellek, *J. Chem. Eng. Data*, **19**, 136 (1974). 168. Nain, V. P. S., and J. R. Ferron, *Ind. Eng. Chem. Fundam.*, **11**, 420 (1972). 169. Nakamura, R., G. J. F. Breedveld, and J. M. Prausnitz, *Ind. Eng. Chem. Process Des. Dev.*, **15**, 557 (1976). 170. Nelson, L. C., and E. F. Obert, *Trans. Am. Soc. Mech. Eng.*, **76**, 1057 (1954). 171. Nothnagel, K.-H., D. S. Abrams, and J. M. Prausnitz, *Ind. Eng. Chem. Process Des. Dev.*, **12**, 25 (1973). 172. Ohe, Shuzo, *Computer Aided Data Book of Vapor Pressure*, Data Book Publishing Co., Tokyo, 1976. 173. O'Hern, H. A., and J. J. Martin, *Ind. Eng. Chem.*, **47**, 2081 (1968). 174. Olander, D. R., *Am. Inst. Chem. Eng. J.*, **7**, 175 (1961). 175. O'Neal, H. E., and S. W. Benson, *J. Chem. Eng. Data*, **15**, 266 (1970). 176. Opfell, J. B., B. H. Sage, and K. S. Pitzer, *Ind. Eng. Chem.*, **48**, 2069 (1956). 177. Orye, R. V., *Ind. Eng. Chem. Process Des. Dev.*, **8**, 579 (1969). 178. Owens, E. J., and G. Thodos, *Am. Inst. Chem. Eng. J.*, **6**, 676 (1960). 179. Passut, C. A., and R. P. Danner, *Ind. Eng. Chem. Process Des. Dev.*, **11**, 543 (1972). 180. Peng, D. Y., and D. B. Robinson, *Ind. Eng. Chem. Fundam.*, **15**, 59 (1976). 181. Perkins, L. R., and C. J. Geankoplis, *Chem. Eng. Sci.*, **24**, 1035 (1969). 182. Pitzer, K. S., *J. Am. Chem. Soc.*, **77**, 3427 (1955). 183. Pitzer, K. S., D. Z. Lippmann, R. F. Curl, C. M. Huggins, and D. E. Peterson, *J. Am. Chem. Soc.*, **77**, 3433 (1955). 184. Pitzer, K. S., and R. F. Curl, *J. Am. Chem. Soc.*, **79**, 2369 (1957). 185. Polak, J., and B. C.-Y. Lu, *Can. J. Chem. Eng.*, **50**, 553 (1972). 186. Prausnitz, J. M., and R. D. Gunn, *Am Inst. Chem. Eng. J.*, **4**, 430, 494 (1958). 187. Preston, G. T., T. W. Chapman, and J. M. Prausnitz, *Cryogenics*, **7**(5), 274 (1967). 188. Quale, O. R., *Chem. Rev.*, **53**, 439 (1953). 189. Rabenovish, B. A., *Thermophysical Properties of Substances and Materials*, 3d ed., Standards Press, Moscow, 1971. 190. Rackett, H. G., *J. Chem. Eng. Data*, **15**, 514 (1970). 191. Reddy, K. A., and L. K. Doraiswamy, *Ind. Eng. Chem. Fundam.*, **6**, 77 (1967). 192. Redlich, O., and J. N. S. Kwong, *Chem. Rev.*, **44**, 233 (1949). 193. Reichenberg, D., *(a)* DCS Rep. 11, National Physical Laboratory, Teddington, England, August 1971; *(b) Am. Inst. Chem. Eng. J.*, **19**, 854 (1973); *(c)* **21**, 181 (1975). 194. Reichenberg, D., *The Viscosities of Pure Gases at High Pressures*, NPL Rep. Chem. 38, National Physical Laboratory, Teddington, England, August 1975. 195. Reichenberg, D., *New Simplified Methods for the Estimation of the Viscosities of Gas Mixtures at Moderate Pressures*, NPL Rep. Chem. 53, National Physical Laboratory, Teddington, England, May 1977. 196. Reid, R. C., J. M. Prausnitz, and T. K. Sherwood, *The Properties of Gases and Liquids*, 3d ed., McGraw-Hill, New York, 1977. 197. Reid, R. C., and J. E. Sobel, *Ind. Eng. Chem. Fundam.*, **4**, 328 (1965). 198. Rice, O. K., *J. Phys. Chem.*, **64**, 976 (1960). 199. Riedel, L., *Chem. Ing. Tech.*, **21**, 349 (1949); **23**, 59, 321,465 (1951). 200. Riedel, L., *Chem. Ing. Tech.*, **26**, 83 (1954). 201. Riedel, L., *Chem. Ing. Tech.*, **26**, 679 (1954). 202. Robbins, L. A., and C. L. Kingrea, *Hydrocarbon Process. Pet. Refiner*, **41**(5), 133 (1962); preprint of paper presented at the 27th midyear meeting, American Petroleum Institute, Refining Div., San Francisco, May 14, 1962. 203. Roy, D., and G. Thodos, *Ind. Eng. Chem. Fundam.*, **7**, 529 (1968). 204. Roy, D., and G. Thodos, *Ind. Eng. Chem. Fundam.*, **9**, 71 (1970). 205. Saksena, M. P., and Harminder, *Ind. Eng. Chem. Fundam.*, **13**,

245 (1974). 206. Saksena, M. P., and S. C. Saxena, *Int. J. Pure Appl. Phys.*, **4**, 109 (1966). 207. Sandler, S., and E. A. Mason, *J. Chem. Phys.*, **48**, 2873 (1968). 208. Satter, A., and J. M. Campbell, *Soc. Pet. Eng. J.*, 333 (December 1963). 209. Scheibel, E. G., *Ind. Eng. Chem.*, **46**, 2007 (1954). 210. Schick, L. M., and J. M. Prausnitz, *Am. Inst. Chem. Eng. J.*, **14**, 673 (1968). 211. Seaton, W. H., and E. Freedman, "Computer Implementation of a Second Order Additivity Method for the Estimation of Chemical Thermodynamic Data," paper presented at 65th annual meeting, AIChE, New York, November 1972. 212. Sengers, J. V., and P. H. Keyes, *Phys. Rev. Lett.*, **26**(2), 70 (1971). 213. Shaw, R., *J. Chem. Eng. Data*, **14**, 461 (1969). 214. Shaw, R., *J. Phys. Chem.*, **75**, 4047 (1971). 215. Shrier, A. L., *Chem. Eng. Sci.*, **22**, 1391 (1967). 216. Shroff, G. H., Ph.D. thesis, University of New Brunswick, Fredericton, 1968. 217. Shroff, G. H., in C. Y. Ho and R. E. Taylor (eds.), *Proceedings of the 8th Conference on Thermal Conductivity*, Purdue University, Oct. 7–10, 1968, Plenum, New York, 1969, p. 643. 218. Smith, J. M., and H. C. Van Ness, *Introduction to Chemical Engineering Thermodynamics*, 3d ed., McGraw-Hill, New York, 1975, p. 87. 219. Soave, G., *Chem. Eng. Sci.*, **27**, 1197 (1972). 220. Sood, S. K., and G. G. Haselden, *Am. Inst. Chem. Eng. J.*, **16**, 891 (1970). 221. Spencer, C. F., and R. P. Danner, *J. Chem. Eng. Data*, **17**, 236 (1972). 222. Spencer, C. F., and R. P. Danner, *J. Chem. Eng. Data*, **18**, 230 (1973). 223. Spencer, C. F., and S. B. Adler, *J. Chem. Eng. Data*, **23**, 82 (1978). 224. Spencer, C. F., T. E. Daubert, and R. P. Danner, *Am. Inst. Chem. Eng. J.*, **19**, 522 (1973). 225. Spencer, C. F., T. E. Daubert, and R. P. Danner, *Technical Data Book*, American Petroleum Institute, OP72, 539, Xerox University Microfilm, chap. 4, "Critical Properties." 226. Sprow, F. B., and J. M. Prausnitz, *Trans. Faraday Soc.*, **62**, 1105 (1966); *Can. J. Chem. Eng.*, **45**, 25 (1967). 227. Sridhar, T., and O. E. Potter, *Am. Inst. Chem. Eng. J.*, **23**, 590 (1977). 228. Starling, K. E., *Hydrocarbon Process.*, **50**(3), 101 (1971). 229. Starling, K. E., and M. S. Han, *Hydrocarbon Process.*, **51**(5), 129 (1972); **51**(6), 107 (1972). 230. Stein, S. E., D. M. Golden, and S. W. Benson, *J. Phys. Chem.*, **81**, 314 (1977). 231. Stiel, L. I., and G. Thodos, *Am. Inst. Chem. Eng. J.*, **7**, 611 (1961); **8**, 229 (1962); **10**, 266 (1964). 232. Stiel, L. I., and G. Thodos, *Am. Inst. Chem. Eng. J.*, **10**, 26 (1964). 233. Stiel, L. I., and G. Thodos, *Am. Inst. Chem. Eng. J.*, **10**, 275 (1964). 234. Stiel, L. I., and G. Thodos, *Can. J. Chem. Eng.*, **43**, 186 (1965). 235. Stull, D. R., *Ind. Eng. Chem.*, **39**, 517 (1947). 236. Stull, D. R., E. F. Westrum, and G. C. Sinke, *The Chemical Thermodynamics of Organic Compounds*, Wiley, New York, 1969. 237. Su, G. S., and D. S. Biswanath, *Am. Inst. Chem. Eng. J.*, **11**, 205 (1965). 238. Sugden, S., *J. Chem. Soc.*, 32 (1924). 239. Sugden, S., *J. Chem. Soc.*, 1177 (1924). 240. Sutton, J. R., *References to Experimental Data on Thermal Conductivity of Gas Mixtures*, NEL Rep. No. 612, May 1976. 241. Sutton, J. R., *References to Experimental Data on Viscosity of Gas Mixtures*, NEL Rep. No. 613, May 1976. 242. Svehla, R. A., *Estimated Viscosities and Thermal Conductivities at High Temperatures*, NASA-TRR-132, 1962. 243. Svehla, R. A., *Estimation Viscosities and Thermal Conductivities of Gases at High Temperatures*, NASA Tech. Rep. R-132, Lewis Research Center, Cleveland, 1962. 244. Tamura, M., M. Kurata, and H. Odani, *Bull. Chem. Soc. Japan*, **28**, 83 (1955). 245. Tarakad, R. R., and R. P. Danner, *Am. Inst. Chem. Eng. J.*, **23**, 944 (1977). 246. Thinh, T.-P., J.-L. Duran, and R. S. Ramalho, *Ind. Eng. Chem. Process Des. Dev.*, **10**, 576 (1971). 247. Thinh, T.-P., J.-L. Duran, R. S. Ramalho, and S. Kaliaguine, *Hydrocarbon Process.*, **50**(1), 98 (1971). 248. Thinh, T.-P., and T. K. Trong, *Can. J. Chem. Eng.*, **54**, 344 (1976). 249. Thompson, P. A., *J. Chem. Eng. Data*, **22**, 431 (1977). 250. Toor, H. L., *Am. Inst. Chem. Eng. J.*, **3**, 198 (1957). 251. Trappeniers, N. J., and P. H. Oosting, *Phys. Lett.*, **23**, 445 (1966). 252. Tsonopoulos, C., *Am. Inst. Chem. Eng. J.*, **20**, 263 (1974). 253. Tyn, M. T., and W. F. Calus, *Processing*, **21**(4), 16 (1975). 254. van der Waals, J. D., doctoral dissertation, Leiden, 1873. 255. van Krevelen, D. W., and H. A. G. Chermin, *Chem. Eng. Sci.*, **1**, 66 (1951). 256. van Krevelen, D. W., and H. A. G. Chermin, *Chem. Eng. Sci.*, **1**, 238 (1952). 257. van Velzen, D., R. L. Cardozo, and H. Langenkamp, *Ind. Eng. Chem. Fundam.*, **11**, 20 (1972). 258. van Velzen, D., R. L. Cardozo, and H. Langenkamp, "Liquid Viscosity and Chemical Constitution of Organic Compounds: A New Correlation and a Compilation of Literature Data," *Euratom*, 4735e, Joint Nuclear Research Centre, Ispra Establishment, Italy, 1972. 259. Vargaftik, N. B., "Thermal Conductivities of Compressed Gases and Steam at High Pressure," *Izv. Vses. Teplotekh. Inst.*, Nov. 7, 1951; private communication, Prof. N. V. TSederberg, Moscow Energetics Institute. 260. Vargaftik, N. B., *Proc. J. Conf. Thermodyn. Transport Prop. Fluids*, London, July 1957, Institution of Mechanical Engineers, London, 1958, p. 142. 261. Vargaftik, N. B., and Y. P. Os'minin, *Teploenergetika*, 3(7), 11 (1956). 262. Venart, J. E. S., and C. Krishnamurthy, *Proc. 7th Conf. Thermal Conductivity*, NBS Spec. Publ. 302, November 1967, p. 659. 263. Verma, K. K., and L. K. Doraiswamy, *Ind. Eng. Chem. Fundam.*, **4**, 389 (1965). 264. Vetere, A., private communication, December 1973; February 1976. 265. Vignes, A., *Ind. Eng. Chem. Fundam.*, **5**, 189 (1966). 266. Walker, R. E., N. De Haas, and A. A. Westenberg, *J. Chem. Phys.*, **32**, 1314 (1960). 267. Wassiljewa, A., *Phys. Z.*, **5**, 737 (1904). 268. Watson, K. M., *Ind. Eng. Chem.*, **35**, 398 (1943). 269. Way, P., Ph.D. thesis, Massachusetts Institute of Technology, Cambridge, Mass., 1971. 270. Weber, H. F., *Wiedmann's Ann., Ann. Phys. Chem.*, **10**, 103 (1880). 271. Wilhoit, R. C., and B. J. Zwolinski, *J. Phys. Chem. Ref. Data* **2**, suppl. 1, 1973. 272. Wilke, C. R., *J. Chem. Phys.*, **18**, 517 (1950). 273. Wilke, C. R., *Chem. Eng. Prog.*, **46**, 95 (1950). 274. Wilke, C. R., and P. Chang, *Am. Inst. Chem. Eng. J.*, **1**, 264 (1955). 275. Wise, D. L., and G. Houghton, *Chem. Eng. Sci.*, **21**, 999 (1966). 276. Witherspoon, P. A., and L. Bonoli, *Ind. Eng. Chem. Fundam.*, **8**, 589 (1969). 277. Yamada, T., *Am. Inst. Chem. Eng. J.*, **19**, 286 (1973). 278. Yon, C. M., and H. L. Toor, *Ind. Eng. Chem. Fundam.*, **7**, 319 (1968). 279. Yoon, P., and G. Thodos, *Am. Inst. Chem. Eng. J.*, **16**, 300 (1970). 280. Yuan, S. C., and Y. I. Mok, *Hydocarbon Process.*, **47**(3), 133 (1968); **47**(7), 153 (1968). 281. Yuan, T.-F., and L. I. Stiel, *Ind. Eng. Chem. Fundam.*, **9**, 393 (1970). 282. Zimmerling, W., and M. Ratzsch, *Chem. Tech.*, **28**(2), 84 (1976). 283. Zudkevitch, D., and J. Joffe, *Am. Inst. Chem. Eng. J.*, **16**, 112 (1970). 284. Zwolinski, B. J., et al., *Selected Values of Properties of Hydrocarbons and Related Compounds*, API Research Proj. 44, College Station, Tex. 285. Zwolinski, B. J., et al., *Selected Values of Properties of Chemical Compounds*, Thermodynamics Research Center, Texas A&M University, College Station.

Section 4

Reaction Kinetics, Reactor Design, and Thermodynamics

Kuang-Hui Lin, Ph.D., *Research Staff, Chemical Technology Division, Oak Ridge National Laboratory; Member, American Institute of Chemical Engineers. (Reaction Kinetics, Reactor Design)*

Hendrick C. Van Ness, D.Eng., P.E., *Department of Chemical and Environmental Engineering, Rensselaer Polytechnic Institute; Fellow, American Institute of Chemical Engineers; Member, American Chemical Society, American Society for Engineering Education. (Thermodynamics)*

Michael M. Abbott, Ph.D., *Department of Chemical and Environmental Engineering, Rensselaer Polytechnic Institute; Member, American Institute of Chemical Engineers. (Thermodynamics)*

REACTION KINETICS AND REACTOR DESIGN

THERMODYNAMICS

Nomenclature and Units

Symbol	Definition	SI units	U.S. customary units
A	Frequency factor	Variable	Variable
A_w	Heat-transfer area	m^2	ft^2
$\overline{A}$	Cross-section area of reactor tube	m^2	ft^2
a	Interfacial area between two phases	m^2/m^3 of two phases	ft^2/ft^3 of two phases
a	Wall-surface area per unit reactor volume	m^2/m^3	ft^2/ft^3
a_m	External surface area of catalyst per unit mass	m^2/kg	ft^2/lb
Bi	Biot number	Dimensionless	Dimensionless
C	Concentration of reacting component	mol/m^3	mol/ft^3
C_p	Heat capacity	$J/(kg \cdot K)$	$Btu/(lb \cdot °F)$
D	Diffusivity	m^2/s	ft^2/s
D_e	Effective diffusivity	m^2/s	ft^2/s
D_i	Fluid-phase diffusivity	m^2/s	ft^2/s
D_p	Particle diameter	m	ft
$\overline{D}$	Reactor diameter	m	ft
E	Activation energy	J/mol	cal/mol
F	Feed flow rate	mol/s	mol/h
F'	Volumetric feed rate of solid	m^3/s	ft^3/h
f_v	Volume fraction of dispersed liquid	Dimensionless	Dimensionless
G	Superficial mass flow rate	$kg/(s \cdot m^2)$	$lb/(h \cdot ft^2)$
H	Enthalpy	J/mol	Btu/mol
ΔH	Heat of reaction	J/mol	Btu/mol
h	Convective heat-transfer coefficient	$J/(s \cdot m^2 \cdot K)$	$Btu/(h \cdot ft^2 \cdot °F)$
h_g	Heat-transfer coefficient per unit external surface area of catalyst	$J/(s \cdot m^2 \cdot K)$	$Btu/(h \cdot ft^2 \cdot °F)$
j_D	Colburn's mass-transfer factor	Dimensionless	Dimensionless
K	Chemical-equilibrium constant	Variable	Variable
K	Distribution coefficient between two dissimilar phases	Variable	Variable
k	Specific reaction rate or rate constant	Variable	Variable
k'	Mass-transfer coefficient for a phase per unit volume of combined phases (2)	$[mol/(m^2 \cdot s)]/(mol/m^3)$	$[mol/(ft^2 \cdot h)]/(mol/ft^3)$
k_m	Mass-transfer coefficient per unit surface area of solid	m^2/s	ft^2/h
L	Length of reactor	m	ft
ℓ	Characteristic pore length of catalyst	m	ft
M_m	Mean molecular weight	kg/kmol	$lb/(lb \cdot mol)$
N	Moles of reacting components	mol	$lb \cdot mol$
N	Agitator-shaft speed	s^{-1}	min^{-1}
P	Total pressure	N/m^2	lbf/in^2 or atm
P	Power dissipated in mixing	J/s	$(ft \cdot lbf)/s$
Pr	Prandtl number	Dimensionless	Dimensionless
ΔP	Pressure drop	N/m^2	lbf/in^2 or atm
p	Partial pressure	N/m^2	lbf/in^2 or atm
Q	Heat generated by reaction	J/mol	Btu/mol
Q	Heat-transmission number	Dimensionless	Dimensionless
q_G	Rate of heat generation	J/s	Btu/h
q_p	Heat-generation potential	Dimensionless	Dimensionless
q_R	Rate of heat removal	J/s	Btu/h
R	Gas constant	$J/(mol \cdot K)$	$Btu/(lb \cdot mol \cdot °R)$
R	Outside radius of solid particle	m	ft
R	Recycle ratio	Dimensionless	Dimensionless
$\overline{R}$	Reactor radius	m	ft
Re	Reynolds number	Dimensionless	Dimensionless
r	Rate of reaction	$mol/(s \cdot m^3$ of reacting mixture)	$lb \cdot mol/(h \cdot ft^3$ of reaction mixture)
		$mol/(s \cdot m^2$ of surface area)	$lb \cdot mol/(h \cdot ft^2$ of surface area)
		$mol/(s \cdot g$ of solid)	$lb \cdot mol/(h \cdot lb$ of solid)
r	Radical position within a spherical particle	m	ft
$\overline{r}$	Average pore radius of catalyst	m	ft
r_c	Radius of unreacted solid core	m	ft
S	Space velocity	s^{-1}	h^{-1}
Sc	Schmidt number	Dimensionless	Dimensionless
S_g	Surface area of catalyst	m^2/g catalyst	m^2/g catalyst
St	Stanton number	Dimensionless	Dimensionless
T	Temperature	K	K or °R
t	Temperature	K	°C or °F
U	Overall heat-transfer coefficient	$J/(s \cdot m^2 \cdot K)$	$Btu/(s \cdot ft^2 \cdot °F)$
V	Volume of reacting fluid	m^3	ft^3
$\overline{V}$	Volume fraction of each of two immiscible phases	Dimensionless	Dimensionless
V_g	Catalyst pore volume	m^3/g of catalyst	m^3/g or ft^3/lb of catalyst
v	Molal volume of gas phase	m^3/mol	$ft^3/(lb \cdot mol)$
v	Volume of dispersed phase	m^3	ft^3
v_0	Volumetric flow rate of feed	m^3/s	ft^3/h
W	Mass of solid reactant or catalyst	g	lb
x	Fractional conversion	Dimensionless	Dimensionless
	Greek symbols		
α	Arrhenius number	Dimensionless	Dimensionless
β	Heat-of-reaction parameter	Dimensionless	Dimensionless
$\overline{\beta}$	Probability of poisoning	Dimensionless	Dimensionless
γ	Fractional change in volume of reacting mixture	Dimensionless	Dimensionless

Nomenclature and Units *(Continued)*

Symbol	Definition	SI units	U.S. customary units
ϵ	Void fraction of fluid-solid bed	Dimensionless	Dimensionless
ϵ	Emissivity or absorptivity	Dimensionless	Dimensionless
ϵ	Residence-time distribution (RTD); function defined by Eq. (4-71)	Dimensionless	Dimensionless
ε_p	Porosity of solid particle	Dimensionless	Dimensionless
η	Effectiveness factor	Dimensionless	Dimensionless
Θ	Time required for complete conversion of solid reactant	s	h
Θ	Space time	s	h
θ	Residence time	s	h
θ_i	Fraction of active centers occupied by component i	Dimensionless	Dimensionless
λ	Thermal conductivity	J/[(s·m²·K)/m]	Btu/[(h·ft²·°F)/ft]
$\bar{\lambda}$	Thermal conductivity of catalyst particle	J/[(s·m²·K)/m]	Btu/[(h·ft²·°F)/ft]
μ	Viscosity	g/(m·s)	lb/(ft·s) or g/(cm·s)(P)
ρ	Density	g/m³	lb/ft³
σ	Interfacial tension	N/m	dyn/cm
τ	Dimensionless time	Dimensionless	Dimensionless
τ	Thickness of solid-product layer	m	in
ϕ	Shape factor	Dimensionless	Dimensionless
ϕ_i	Flux of reacting component through surface of solid reactant	mol/(m²·s)	lb·mol/(ft²·h)

REACTION KINETICS AND REACTOR DESIGN

General References: Bamford and Tipper (eds.), *Comprehensive Chemical Kinetics*, vols. 1–22, Elsevier, Amsterdam, 1969–1980. Carberry, *Chemical and Catalytic Reaction Engineering*, McGraw-Hill, New York, 1976. Churchill, *The Interpretation and Use of Rate Data: The Rate Concept*, McGraw-Hill, New York, 1974. Cramer and Watkins, *Chemical Engineering Practice*, vol. 8: *Chemical Kinetics*, Butterworth, London, 1965. Froment and Bischoff, *Chemical Reactor Analysis and Design*, Wiley, New York, 1979. Gates, Katzer, and Schuit, *Chemistry of Catalytic Processes*, McGraw-Hill, New York, 1979. Hougen and Watson, *Chemical Process Principles*, part 3: *Kinetics and Catalysis*, Wiley, New York, 1947. Lapidus and Amundson (eds.), *Chemical Reactor Theory—A Review*, Prentice-Hall, Englewood Cliffs, N.J., 1977. Levenspiel, *Chemical Reaction Engineering*, 2d ed., Wiley, New York, 1972. Lewis (ed.), *Techniques of Chemistry*, vol. VI: *Investigation of Rates and Mechanism of Reactions*, part I—*General Considerations and Reactions at Conventional Rates*, Wiley, New York, 1974. March, *Advanced Organic Chemistry—Reactions, Mechanisms and Structure*, McGraw-Hill, New York, 1977. Rase, *Chemical Reactor Design for Process Plants*, vol. I: *Principles and Techniques*; vol. II: *Case Studies and Design Data*, Wiley, New York, 1977. Smith, *Chemical Engineering Kinetics*, 2d ed., McGraw-Hill, New York, 1970. Walas, *Reaction Kinetics for Chemical Engineers*, McGraw-Hill, New York, 1959.

Introduction Some of the tasks unique to the chemical engineering profession are the design and operation of chemical reactors for converting specific feed material (or reactants) into certain marketable products. To accomplish these tasks, chemical engineers must select the reactor type among many design alternatives, determine the required reactor size, and specify operating conditions. They must have knowledge of the rates of chemical reactions involved, the maximum conversion obtainable, the nature of physical processes interacting with the chemical reactions, and the parameters which influence the preceding.

In many cases, the rates of physical processes (e.g., mass transfer and heat transfer) involved in commonly used chemical reactors can be estimated adequately from the properties of substances participating in the reaction, the flow characteristics, the configuration of the reaction vessel, and so forth. In contrast, the chemical-process rate data for most industrially important reactions cannot be estimated reliably from theories and must be determined experimentally.

Although the chemical reactor is the heart of a chemical plant, one must not lose sight of critical functions performed by systems of supporting equipment in accomplishing the overall goal of the plant, which is to yield products most efficiently at the lowest possible costs. These supporting systems may include, for example, the reactor feed-preparation system (e.g., mixing, preheating, etc.) as well as the separation and postreaction treatment systems for intermediate as well as final products. The importance of these supporting systems cannot be overemphasized because they frequently exert significant influence over the yields and quality of the products. Thus, the characteristics and limitations of these systems which affect the reactor performance should be well understood to ensure successful design and operation of the plant.

FUNDAMENTALS OF CHEMICAL REACTION SYSTEMS

Principles of Reaction Kinetics

Basic Terminology The amount of a selected component A being converted or produced per unit time per unit quantity of a reference variable y in a chemically reacting system is defined as the **rate of reaction** r_A.

$$r_A = \frac{1}{y}\frac{dN_A}{d\theta} \tag{4-1a}$$

By definition, r_A is negative if A refers to a reactant, while it is positive if A is a reaction product. Although molal units are generally used as the amount N_A of the component being followed, mass units are also accepted. Other variables related to concentration are also employed in defining the rate. Examples are radioactivity, an optical property, and pressure. The reference variable y in homogeneous fluid reactions is usually the volume of the reacting fluid V or of the reactor V_R. In dealing with heterogeneous sytems, the mass of the solid W, as well as the surface of the solid S, may be taken as y for solid-fluid reactions, while use of the interfacial area for y is also encountered in fluid-fluid reactions. When the volume of the reacting fluid system remains constant, Eq. (4-1a) is simplified to

$$r_A = dC_A/d\theta \tag{4-1b}$$

where C_A is the concentration of component A and θ is the time.

A general **rate expression** for component i may be written as

$$r_i = f(C,T,P,m) \tag{4-2}$$

Here C represents concentrations of the reacting components, T the temperature, and P the total pressure; m includes parameters other than C, T, and P which have to be taken into consideration, especially in heterogeneous reactions (for example, parameters having influences on the mass- and energy-transport characteristics of the system; to be discussed later). In terms of concentrations of the reacting components $A, B, C, \ldots, E$, the rate expression for a simple irreversible reaction becomes

$$-r_A = kC_A^a C_B^b C_C^c \cdots C_E^e \tag{4-2a}$$

The proportionality constant k in Eq. (4-2a) is termed the **specific reaction rate** or simply the **rate constant.** k is markedly dependent upon temperature. Pressure and the presence of catalysts also affect k. The units and value of k vary with the specific component to which k refers, the order of reaction, and the units of C. The **order of reaction** is the sum of the exponents in Eq. (4-2a) determined empirically and need not be an integer. The order of reaction may also refer to the individual reacting components, e.g., ath order with respect to component A, bth order with respect to B, etc.

The reaction order agrees with the **molecularity** (i.e., the number of molecules actually taking part in the reaction) only for an **elementary reaction.** In this case the rate expression can be derived directly from the stoichiometric equation since it describes the true reaction mechanism. For example, the stoichiometric equation of an elementary reaction $aA + bB \rightarrow dD$ gives a rate expression: $-r_A = kC_A^a C_B^b$. Here the values of a and b must be positive integers. By contrast, there is, in general, no direct relation between the order of reaction and the reaction stoichiometry for a **nonelementary** reaction [e.g., see Eq. (4-15)].

The temperature dependency of the rate expression [Eq. (4-2)] is usually represented by the rate constant through the Arrhenius equation:

$$k = A \cdot e^{-E/RT} \tag{4-3}$$

In this equation A is called the **frequency factor** and has the same units as k. E is the **activation energy** and was considered by Arrhenius as the amount of energy in excess of the average energy level which the reactants must have in order for the reaction to proceed. Although the activation energy is not influenced by temperature in the moderate temperature range usually encountered, some exceptions have been reported. Other factors known to influence E include pressure and the presence of a catalyst.

Complex Reactions Most reactions important in industrial processes are quite complex in nature because their reaction mechanisms are considerably different from the stoichiometric equations. In such cases the reaction mechanism can frequently be determined by trial and error by postulating that the overall reaction takes place via two or more elementary reaction steps. The resulting overall rate expression is then compared with the experimental data to examine the closeness of agreement between the two. This procedure is repeated until a desired degree of accuracy is obtained.

The reaction steps in a complex reaction may proceed reversibly, concurrently, and/or consecutively. In any case, the net rate of the overall reaction may be considered to be the sum of the rates of all the individual reaction steps based on a selected component.

Reversible reactions. This type of reaction refers to the case in which conversion of reactants to products is far from complete at equilibrium and the reverse reaction becomes important. Thus, for the following set of reversible reactions each of which is elementary,

$$A + B \underset{k_R}{\overset{k_F}{\rightleftharpoons}} D + E \tag{4-4}$$

the net rate of disappearance of component A may be described by

$$-r_A = k_F C_A C_B - k_R C_D C_E \tag{4-5}$$

If the reacting system is an ideal solution (i.e., activities are proportional to concentrations), the following relation is valid at equilibrium:

$$\frac{C_D C_E}{C_A C_B} = \frac{k_F}{k_R} = K = \text{equilibrium constant} \tag{4-6}$$

Consecutive (or series) reactions. A simple case can be illustrated by

$$A \xrightarrow{k1} B \xrightarrow{k2} D \tag{4-7}$$

If an elementary reaction is assumed for each step,

$$-r_A = k_1 C_A \tag{4-8}$$

$$r_D = k_2 C_B \tag{4-9}$$

and

$$r_B = k_1 C_A - k_2 C_B \tag{4-10}$$

Shown in Fig. 4-1 is the concentration-time profile for the type of consecutive reactions represented by Eq. (4-7), with the assumption that k_1 and k_2 are of the same order of magnitude. The shapes of these curves vary with the type of reactor and the relative values of the rate constants as well as the orders of reaction.

Parallel (side or simultaneous) reactions. These are processes involving one or more reactants undergoing reactions according to more than one scheme. A simple example is

$$A \xrightarrow{k1} B \qquad A \xrightarrow{k2} D$$

with rate expressions under a constant-volume condition,

$$-r_A = -dC_A/d\theta = k_1 C_A^m + k_2 C_A^n \tag{4-11}$$

$$r_B = dC_B/d\theta = k_1 C_A^m \tag{4-12}$$

$$r_D = dC_D/d\theta = k_2 C_A^n \tag{4-13}$$

The relative production rate is obtained from Eqs. (4-12) and (4-13) as

$$\frac{dC_B}{dC_D} = \frac{k_1}{k_2} C_A^{m-n} \tag{4-14}$$

Equation (4-14) indicates that formation of B is favored by a high concentration of A when $m - n > 0$ while the reverse is true when $m - n < 0$. The concentration of A necessary to obtain a favorable product can be adjusted partially by selecting an appropriate type of reactor. Another example of parallel reactions,

$$A + B \xrightarrow{k1} D \qquad A + B \xrightarrow{k2} E$$

may be treated by a method similar to the one just discussed.

The rate equations under constant-volume conditions for simple cases of reversible, consecutive, and parallel reactions [e.g., Eqs. (4-5), (4-8) through (4-10), (4-11) through (4-13) with $m = n$] can be readily solved by analytical means to obtain rate constants, concentration-time relationships, etc. (see Table 4-1). For other cases (e.g., variable volume, nonelementary reaction steps, etc.), the analytical solution may be so complicated or even impossible that graphical and numerical methods are utilized.

Mixed reactions. This type implies combinations of reversible, consecutive, and/or parallel reactions. One of the common mixed reactions is the *chain reaction*, which can be illustrated by the well-known synthesis of HBr from hydrogen and bromine:

1:	$Br_2 \rightleftarrows Br\cdot + Br\cdot$	Initiation and termination
2:	$Br\cdot + H_2 \rightarrow HBr + H\cdot$	Propagation
3:	$H\cdot + Br_2 \rightarrow HBr + Br\cdot$	Propagation
4:	$H\cdot + HBr \rightarrow Br\cdot + H_2$	Propagation
5:	$H_2 + Br_2 \rightarrow 2HBr$	Overall stoichiometric reaction

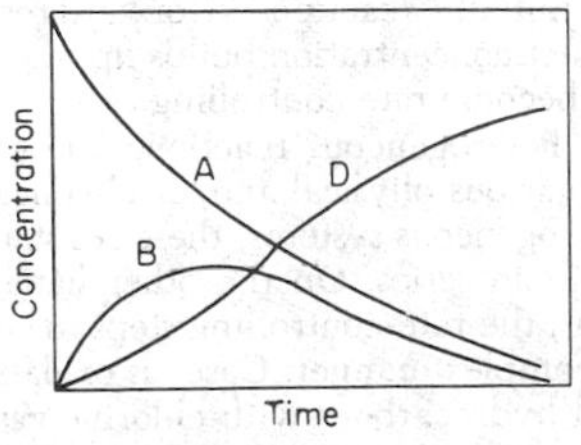

FIG. 4-1 Concentration-time profile for a consecutive reaction: $A \rightarrow B \rightarrow D$.

TABLE 4-1 Rate Equations and Examples of Simple Reactions

Order	Rate expression†	Integrated rate equation (constant volume)	Example of reaction [order with respect to]*
0	$-r_A = k$	$k\theta = C_{Ao} - C_A$	Pt-catalyzed decomposition of NH_3 (A-2); pyrolysis of Fe-sulfates (A-3)
½	$-r_A = kC_A^{1/2}$	$k\theta = 2(C_{Ao}^{1/2} - C_A^{1/2})$	Pyrolysis of ethyl nitrate (A-4); catalytic dehydrogenation of cyclohexane (A-5)
1	$-r_A = kC_A$	$k\theta = \ln \frac{C_{Ao}}{C_A}$	Pyrolysis of Ca-oxalate (A-6); thermal decomposition of H_2O_2 (A-7); pyrolysis of ethane (A-8)
2	$-r_A = kC_A^2$	$k\theta = \frac{1}{C_A} - \frac{1}{C_{Ao}}$	Uncatalyzed thermal polymerization of ethylene (A-9); ethane formation from methyl radicals (A-10); basic hydrolysis of difluoroamine (A-11)
2	$-r_A = kC_AC_B$	$k\theta = \frac{1}{\Delta_{BA}^0} \ln \frac{C_{Ac}(C_A + \Delta_{BA}^0)}{C_A C_{Bo}}$ $C_{Ao} \neq C_{Bo}$	Nitration of *p*-nitrotoluene in $H_2SO_4-HNO_3$ (A-12); formation of urethane, [isocyanates, alcohols] (A-13); condensation of urea with aldehyde (A-14)
3	$-r_A = kC_A^3$	$k\theta = \frac{1}{2}\left[\left(\frac{1}{C_A}\right)^2 - \left(\frac{1}{C_{Ao}}\right)^2\right]$	Decomposition of NH_4NO_2 in solution (A-15); reaction of active nitrogen (A-16)
3	$-r_A = kC_AC_BC_D$	$k\theta = \frac{1}{\Delta_{BA}^0 \cdot \Delta_{DA}^0} \ln \frac{C_{Ao}}{C_A} + \frac{1}{\Delta_{BD}^0 \cdot \Delta_{BA}^0} \ln \frac{C_{Bo}}{C_A + \Delta_{BA}^0} + \frac{1}{\Delta_{DB}^0 \cdot \Delta_{DA}^0} \ln \frac{C_{Do}}{C_A + \Delta_{DA}^0}$ $C_{Ao} \neq C_{Bo} \neq C_{Do}$	Reaction of Ag(II) and dithionate, [H^+, Ag, $S_2O_6^{2-}$] (A-17); addition of HI to cyclohexene, [C_6H_{10}, HI, I_2] (A-18)
3	$-r_A = kC_AC_B^2$	$k\theta = \frac{1}{(C_{Ao} + \Delta_{AB}^0)^2} \ln \frac{\widehat{C}_{Ao}\widehat{C}_B}{C_{Bo}C_A} + \frac{C_{Bo} - C_B}{(C_{Ao} + \Delta_{AB}^0)C_{Bo}C_B}$ $C_{Bo}/C_{Ao} \neq 2$	Reaction of thioacetamide [A] and ammonia [B] (A-19); reaction of propylene [B] and peracetic acid [A] in N_2 or Ar (A-20)

*See list of references at end of subsection. †$r_A = dC_A/d\theta$ for constant-volume case. NOTE: The integrated rate equation depends on both reaction stoichiometry and kinetics. The overall stoichiometry has been assumed equal to reaction orders for individual reactants. C_o's are initial concentrations. $\Delta_{BA}^0 = C_{Bo} - C_{Ao}$, $\Delta_{DA}^0 = C_{Do} - C_{Ao}$, $\Delta_{BD}^0 = C_{Bo} - C_{Do}$, $\Delta_{DB}^0 = C_{Do} - C_{Bo}$, $\Delta_{AB}^0 = C_{Ao} - C_{Bo}$.

The rate equation derived from this mechanism is

$$r_{HBr} = \frac{k_1 C_{H_2} C_{Br_2}^{1/2}}{1 + k_2 C_{HBr}/C_{Br_2}} \qquad (4\text{-}15)$$

Actually, k_1 and k_2 are different combinations of two or more rate constants resulting from the individual elementary reaction steps shown earlier. Reference A-1 discusses this reaction mechanism in detail. This case is also a good example of a nonelementary reaction since the rate expression obviously bears no direct relation to the overall reaction stoichiometry.

Additional examples of chain reactions include formation of phosgene from CO and Cl_2, reactions of a number of hydrocarbons under the influence of heat or ionizing radiation, and chain-growth polymerization. The existence of free radicals as the intermediates in chain reactions has been substantiated through the use of high-precision instruments such as the high-resolution spectroscope.

Rate-Controlling Steps In dealing with reaction kinetics, one process step may account for the major factor in determining the overall rate of reaction and is termed the *rate-controlling step*. The rate-controlling step may be present in homogeneous complex reactions as well as in heterogeneous reactions.

For example, in the reaction steps involved in the gas-phase formation of HBr, mentioned previously [see Eq. (4-15)], the rate-controlling steps are, initially, reactions 1 (forward reaction) and 2. However, as the product concentration builds up, the reverse reaction of 1 and reaction 4 become rate-controlling.

In the case of heterogeneous reactions, the rate-controlling step may result from various physical and/or chemical resistances. With noncatalytic heterogeneous systems, these resistances are frequently considered to exist in series. On the other hand, in heterogeneous catalytic reactions, the rate-controlling steps may be related to each other in a more complex manner. Gaseous oxidation of metals, chlorination of liquid hydrocarbons with chlorine gas, and reduction of ores to produce metals are examples of noncatalytic heterogeneous reactions in which the rate-controlling steps may vary according to the reaction conditions. Typical examples of catalytic heterogeneous reactions can be found in the catalytic cracking of petroleum fractions.

A more detailed discussion on rate-controlling steps appears under "Analyses of Reaction Kinetic Data: Identification of Rate-Controlling Steps."

Summary of Rate Equations for Simple and Complex Reactions Presented in Table 4-1 are rate expressions, integrated rate equations, and typical examples of simple-order reactions under constant-volume conditions. Experimental kinetic data may be analyzed through the use of either the differential or the integrated form of rate equations listed in the table to determine the order of reaction (for details, see "Analyses of Reaction Kinetic Data"). Information similar to that in Table 4-1 for typical complex reactions is summarized in Table 4-2.

No attempt has been made to cite all known models of reactions. Only those major types of reactions which can be illustrated by actual examples are selected. Examples of reactions are included to show the usefulness of various reaction models and schemes. However, the validity of each model as applied to a particular example depends on the specific reaction conditions; hence the original data in the reference should be evaluated by the user of the table to determine the applicability of the model.

Noncatalytic Heterogeneous Reaction Systems In a heterogeneous reaction system, the overall rate expression becomes complicated because of interaction between physical and chemical processes. This complication is introduced by the requirement that reactants in one phase have to be transported to the other phase containing other reactants where the reactions take place. The physical and chemical rate processes in the uncatalyzed heterogeneous system can be considered to take place in series and/or in parallel. Hence, the overall rate expression may be formulated by combining terms for the various process steps involved. The physical rate expressions

TABLE 4-2 Rate Equations and Examples for Typical Complex Reactions

Reaction scheme	Rate expression	Integrated rate equation or suggested method	Example of reaction*
$A \xrightarrow{k1} B$ $A \xrightarrow{k2} D$	$r_B = k_1C_A,\ r_D = k_2C_A$ $-r_A = r_B + r_D$	$(k_1 + k_2)\theta = -\ln (C_A/C_{Ao})$	Liquid-phase pyrolysis of α-pinene (A-21)
$A \xrightarrow{k1} D$ $B \xrightarrow{k2} D$	$r_D = k_1C_A + k_2C_B$	$C_D = (C_{Ao} + C_{Bo}) - (C_{Ao}e^{-k1\theta} + C_{Bo}e^{-k2\theta})$ $C_{Do} = 0$	Hydrolysis of mixed tertiary aliphatic chlorides (A-22)
$A \xrightarrow{k1} D + E$ $A + B \xrightarrow{k2} D + F$	$-r_A = k_1C_A + k_2C_AC_B$ $-r_B = k_2C_AC_B$	$k_2\theta = \dfrac{1}{k_1/k_2 + (C_{Bo} - C_{Ao})} \ln \left(\dfrac{C_{Ao}}{k_1/k_2 + C_{Bo}} \cdot \dfrac{k_1/k_2 + C_B}{C_A}\right)$	Hydrolysis of primary butenyl chloride [A], $OH^- \equiv [B]$ (A-23); dehydrochlorination of dichloroethane [A] in the presence of chlorine [B] (A-24)
$A \xrightarrow{k1} B \xrightarrow{k2} D$	$-r_A = k_1C_A,\ r_D = k_2C_B$ $r_B = k_1C_A - k_2C_B$	$C_B = [C_{Ao}k_1/(k_2 - k_1)](e^{-k1\theta} - e^{-k2\theta})$ $C_D = C_{Ao}[1 + (k_2e^{-k1\theta} - k_1e^{-k2\theta})/(k_1 - k_2)]$ $C_{Bo} = C_{Do} = 0$	Radioactive decay; hydration of acetaldehyde [A] (A-25)
$A \xrightarrow{k1} B \xrightarrow{k2}$ $D \xrightarrow{k3} E \xrightarrow{k4} \cdots$	$-r_A = k_1C_A,\ r_B = 0,$ $r_D = 0,\ r_E = 0, \ldots$	$C_A = C_{Ao}e^{-k1\theta},\ C_B = (k_1/k_2)C_A$ $C_D = (k_1/k_3)C_A,\ C_E = (k_1/k_4)C_A$	Radioactive decay with steady production of daughter isotopes
$A + B \xrightarrow{k1} D + E$ $A + D \xrightarrow{k2} F + E$	$-r_A = k_1C_AC_B + k_2C_AC_D$ $r_D = k_1C_AC_B - k_2C_AC_D$ $-r_B = k_1C_AC_B$ $r_F = k_2C_AC_D$	For detailed discussion on various methods of obtaining integrated rate equations, see Ref. A-1	See Ref. A-1; also reaction between oxygen atoms [A] and chlorine [B] (A-26)
$A \underset{k2}{\overset{k1}{\rightleftharpoons}} B$	$-r_A = k_1C_A - k_2C_B$	$(k_1 + k_2)\theta = \ln [(C_{Ao} - C_{Ae})/(C_A - C_{Ae})]$ $C_{Bo} = 0$, C_{Ae} = equilibrium concentration of &A	Acid-catalyzed hydrolysis of lactone (A-27)
$A \underset{k2}{\overset{k1}{\rightleftharpoons}} B + D$	$-r_A = k_1C_A - k_2C_BC_D$	$k_1\theta = \dfrac{C_{Ao} - C_{Ae}}{C_{Ao} + C_{Ae}} \ln \dfrac{C_{Ao}^2 - C_{Ae}C_A}{(C_A - C_{Ae})C_{Ao}}$	Decomposition of ethanol [A] in the presence of activated alumina (A-28)
$A + B \underset{k2}{\overset{k1}{\rightleftharpoons}} D + E$	$-r_A = k_1C_AC_B - k_2C_DC_E$	$C_{BO} = C_{DO}$ $k_1\theta = \dfrac{x_e}{2C_{Ao}(1 - x_e)} \ln \dfrac{x_e - (2x_e - 1)x}{x_e - x}$ $C_{Do} = C_{Eo} = 0,\ C_{Ao} = C_{Bo}$	Esterification of ethanol [A] and acetic acid [B] with HCl as catalyst (A-29)
$2A \underset{k2}{\overset{k1}{\rightleftharpoons}} D + E$	$-r_A = k_1C_A^2 - k_2C_DC_E$	$x = (C_{Ao} - C_A)/C_{Ao}$, x_e = value of x at equilibrium	Vapor-phase decomposition of HI [A] to H_2 [D] and I_2 [E] (A-30)

*See list of references at end of subsection.

vary not only with the types of phase present but also with the system boundary conditions as well as overall process conditions, including contacting patterns between phases.

Mechanism of Uncatalyzed Heterogeneous Reactions This type of reaction may consist of the following steps:

a. Diffusion of reactants from the bulk of the first phase to the interface between the two phases. If an additional layer of solid product and inert material (e.g., ash in solids) is present at the interface, the reactants would have to overcome the resistance of this layer before they could reach the surface of the second phase, where other reactants are present.

b. Diffusion of reactants from the interface to the bulk of the second phase.

c. Chemical reaction between the reactants from phase 1 and those in phase 2.

d. Diffusion of products within the second phase and/or out of phase 2 into the bulk of phase 1.

Whether these four steps proceed successively or simultaneously and which one of the steps controls the overall reaction rate depend upon the phases involved and the specific reaction concerned, as well as on process conditions.

Factors Affecting Heterogeneous Reactions Since interaction of chemical and physical processes is involved in heterogeneous reactions, the overall rate is under the influence of factors that affect both type of processes. These may include:

1. Mass-transfer factors, e.g., diffusion characteristics of fluid phases.
2. Contact patterns of phases; e.g., each phase may be in one of the two ideal flow patterns, i.e., plug or back-mix flow. There are a number of possible combinations of contacting patterns.
3. Fluid dynamic factors, e.g., mass velocity, degree of turbulence, etc.
4. Interfacial surface area.
5. Geometry of the reaction vessel.
6. Chemical kinetic factors, i.e., activation energy, concentrations of reactants, etc.
7. Temperature and pressure.

Some of these factors are not completely independent and may interact with one another. For example, factor 3 would exert influences on factors 1 and 4 in the case of a fluid-fluid reaction, and it is also related to factors 2 and 5.

The overall rate equation would have to take these factors into account. However, because of the features characteristic of a specific heterogeneous system, the form of the equations and the design procedures are quite different between the fluid-fluid and fluid-solid systems. Nevertheless, in combining rate equations for various process steps in either system, the equations must be first expressed in equivalent forms. Thus, when the mass-transfer rate is defined in terms of the mass flux (mass per unit time per unit surface), the rate of chemical reaction should also be based on unit area instead of unit volume.

Examples of Industrial Heterogeneous Reactions Shown in Table 4-3 are some of the industrially important noncatalytic heterogeneous reactions. Numerous technical data are available for gas-liquid reactions as the result of extensive investigations of this type of reaction. On the other hand, studies of solid-solid reactions are rather limited.

Mathematical Model for Fluid-Solid Reactions The general case of a fluid-solid reaction of the type $aA(g \text{ or } l) + bB(s) \rightarrow$ products under isothermal condition may be represented by differential material balances in terms of fluid reactant A and solid reactant B,

TABLE 4-3 Industrial Noncatalytic Heterogeneous Reaction*

Gas-solid
1. Action of chlorine on uranium oxide to recover volatile uranium chloride
2. Removal of iron oxide impurity from titanium oxide by volatilization by action of chlorine
3. Combustion and gasification of coal
4. Manufacture of hydrogen by action of steam on iron
5. Manufacture of blue gas by action of steam on carbon
6. Calcium cyanamide by action of atmospheric nitrogen on calcium carbide
7. Burning of iron sulfide ores with air
8. Nitriding of steel

Liquid-solid
9. Ion exchange
10. Acetylene by action of water on calcium carbide
11. Cyaniding of steel
12. Hydration of lime
13. Action of liquid sulfuric acid on solid sodium chloride or on phosphate rock or on sodium nitrate
14. Leaching of uranium ores with sulfuric acid

Gas-liquid
15. Sodium thiosulfate by action of sulfur dioxide on aqueous sodium carbonate and sodium sulfide
16. Sodium nitrite by action of nitric oxide and oxygen on aqueous sodium carbonate
17. Sodium hypochlorite by action of chlorine on aqueous sodium hydroxide
18. Ammonium nitrate by action of ammonia on aqueous nitric acid
19. Nitric acid by absorption of nitric oxide in water
20. Recovery of iodine by action of sulfur dioxide on aqueous sodium iodate
21. Hydrogenation of vegetable oils with gaseous hydrogen
22. Desulfurization of gases by scrubbing with aqueous ethanolamines

Liquid-liquid
23. Caustic soda by reaction of sodium amalgam and water
24. Nitration of organic compounds with aqueous nitric acid
25. Formation of soaps by action of aqueous alkalies on fats or fatty acids
26. Sulfur removal from petroleum fractions by aqueous ethanolamines
27. Treating of petroleum products with sulfuric acid

Solid-solid
28. Manufacture of cement
29. Boron carbide from boron oxide and carbon
30. Calcium silicate from lime and silica
31. Calcium carbide by reaction of lime and carbon
32. Leblanc soda ash

Gas-liquid-solid
33. Hydrogenation or liquefaction of coal in oil slurry

*Adapted from Walas, *Reaction Kinetics for Chemical Engineers*, McGraw-Hill, New York, 1959.

respectively, as follows:

$$\frac{1}{r^2}\frac{\partial}{\partial r}\left(D_e r^2 \frac{\partial \overline{C}_A}{\partial r}\right) - r_A\rho_p = \frac{\partial}{\partial \theta}(\varepsilon_p \overline{C}_A) \qquad (4\text{-}16)$$

$$-r_B\rho_p = \partial C_B/\partial\theta \qquad (4\text{-}17)$$

where r is the radial position within a spherical particle, D_e and $\overline{C}_A$ represent the effective diffusivity and the molar concentration respectively of A in the ash layer, ρ_p and ε_p are the density and porosity respectively of the particle, and C_B is the molar concentration of B.

General analytical solution (integration) of the pair of partial differential equations is not possible. Analytical solution of such equations [(4-16) and (4-17)] for certain cases may be feasible, however, when simplifying assumptions are applied. Thus, the pseudo-steady-state approximation (B-1, B-2, B-3) would eliminate the transient term $\partial/\partial\theta\ (\varepsilon_p\overline{C}_A)$ in Eq. (4-16). For fluid-solid reactions with first order in the fluid concentration and with constant diffusivity and solid porosity, Eqs. (4-16) and (4-17) may be reduced to a single partial differential equation by transforming $\overline{C}_A$ and C_B (B-4, B-5). Furthermore, this equation can then be transformed into an ordinary differential equation under the pseudo-steady-state condition, which may be solved analytically. It should be pointed out that the pseudo-steady-state approximation is justified for gas-solid reactions, but it may not be applicable to liquid-solid reactions (e.g., ion exchange).

Equations (4-16) and (4-17) may also be solved by the numerical method (B-6), for example, for fluid-solid reactions with the following rate equations:

$$r_A\rho_p = ak\overline{C}_A^m C_B^n \qquad (4\text{-}18)$$

and

$$r_B\rho_p = k\overline{C}_A^m C_B^n \qquad (4\text{-}19)$$

where a is the moles of fluid A reacting with 1 mol of solid reactant B. The results thus obtained are presented in a graphical form.

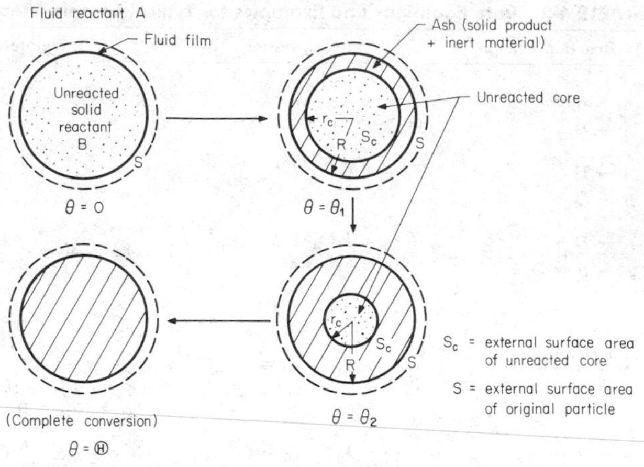

FIG. 4-2 Schematics of an unreacted-core model for fixed-sized particles.

Of various mathematical models that have been proposed to describe fluid-solid reactions (mostly for gas-solid reactions), a simplified version called the *unreacted-core* (or *shell-progressive*) model has been shown to represent the actual behavior closely in most cases based on experimental evidences. The model considers fluid reactant to initiate the reaction at the outer layer of the solid particle (reactant B), converting this layer into fluid and/or solid products and inert material ("ash"). The reaction then proceeds successively inward, constantly reducing the size of the core of unreacted material. Two different cases may be considered for this model. The first assumes that the continuous formation of solid product and inert material (ash) without flaking off would maintain the particle size unchanged (Fig. 4-2). In the second, the particle size changes with the progress of reaction owing to the formation of gaseous products, flaking off of the solids, etc. Additional assumptions for both of these cases include (1) spherical particles, (2) an irreversible reaction of the type $aA(g$ or $l) + bB(s) \rightarrow$ products (fluid and/or solid), with ath order in A and zero order in B, and (3) isothermal condition.

Case 1. *Fixed-size particles.* One of three process steps may control the overall reaction rate: namely, diffusion through the gas film, diffusion through the ash (converted solids and inert material), or chemical reaction. An analysis for the case in which diffusion through the ash controls the overall rate is outlined here. Under this condition the instantaneous rate of reaction in terms of A becomes equal to the diffusion rate of A through the ash layer; that is,

$$-r_A = -(dN_A/d\theta) = 4\pi r^2\phi_A = \text{constant} \qquad (4\text{-}20)$$

where ϕ_A is the flux of A through a spherical surface of radius r in the ash layer and may be represented by Fick's law:

$$\phi_A = D_e(dC_A/dr) \qquad (4\text{-}21)$$

Here D_e is the effective diffusivity of A through the ash layer.

Also

$$\frac{dN_A}{d\theta} = \frac{a}{b}\frac{dN_B}{d\theta}$$

and

$$dN_B = 4\pi r_c^2\, dr_c\rho_B$$

where r_c refers to the radius of the unreacted core, and ρ_B is its density in mol/m^3 (mol/ft^3). Combining these two expressions with Eqs. (4-20) and (4-21) and integrating gives

$$\frac{\theta}{\Theta_2} = 1 - 3\left(\frac{r_c}{R}\right)^2 + 2\left(\frac{r_c}{R}\right)^3 = 1 - 3(1 - x_B)^{2/3} + 2(1 - x_B) \qquad (4\text{-}22)$$

In this equation Θ_2 is the time required for a complete conversion of B, R is either the outside radius of the fixed-size particle or the initial radius of the shrinking particle (case 2), and x_B is the fractional conversion of B at time θ.

The integrated rate equations under other conditions are summarized in Table 4-4. Examples of fluid-solid reactions which approximate the case of fixed-particle sizes are (1) ion exchange, (2) production of calcium cyanamide by nitrogenation of calcium carbide, (3) production of metallic oxides by roasting of sulfide ores in air, (4) nitriding of steel, and (5) oxidative regeneration of coke-laden catalyst.

TABLE 4-4 Rate Equations for Fluid-Solid Reactions

(Unreacted-core model)

$aA(\text{fluid}) + bB(\text{s}) \rightarrow \text{products}$

Case	Controlling step	Rate equation
Fixed-size particles	Diffusion through gas film	$\dfrac{\theta}{\Theta_1} = 1 - \left(\dfrac{r_c}{R}\right)^3 = x_B$ $\Theta_1 = \dfrac{a\rho_B R}{3bk_m C_{A1}}$
	Diffusion through ash	$\dfrac{\theta}{\Theta_2} = 1 - 3\left(\dfrac{r_c}{R}\right)^2 + 2\left(\dfrac{r_c}{R}\right)^3$ $= 1 - 3(1 - x_B)^{2/3} + 2(1 - x_B)$ $\Theta_2 = \dfrac{a\rho_B R^2}{6bD_e C_{A1}}$
	Chemical reaction (*a*th order with respect to A)	$\dfrac{\theta}{\Theta_3} = 1 - \dfrac{r_c}{R}$ $= 1 - (1 - x_B)^{1/3}$ $\Theta_3 = \dfrac{a\rho_B R}{bk_s C_{A1}^a}$
Particle size varies; no ash formation	Diffusion through gas film°	For small particle (Stokes regime): $\dfrac{\theta}{\Theta} = 1 - (1 - x_B)^{2/3}$ $\Theta = \dfrac{\rho_B y R^2}{2bDC_{A1}}$ For large particle: $\dfrac{\theta}{\Theta} = 1 - (1 - x_B)^{1/2}$ $\Theta = KR^{3/2}/C_{A1}$
	Chemical reaction (*a*th order with respect to A)	$\dfrac{\theta}{\Theta_3} = 1 - \dfrac{r_c}{R}$ $= 1 - (1 - x_B)^{1/3}$ $\Theta_3 = \dfrac{a\rho_B R}{bk_s C_{A1}^a}$

Θ's are times required for complete conversion for individual controlling steps.

k_m and k_s are mass-transfer coefficient and reaction rate constant respectively, based on unit surface in m/s (ft/h) for k_m, and also for k_s when $a = 1$ (first-order reaction).

D_e is effective diffusivity of A in a porous structure in m^2/s (ft^2/h). C_{A1} is concentration of A in main fluid stream in mol/m^3 (mol/ft^3).

°Adapted from Ref. B-10; y is mole fraction of A in fluid, and K is a constant.

Case 2. Variable particle size. If no ash (converted solids and inert material) layer covers the unreacted core as the reaction proceeds, the particle would continue to shrink with time. Since no ash layer is present, there could be only two controlling steps, gas-film diffusion or chemical reaction. The rate equation in case 1 is directly applicable when chemical reaction controls the overall rate (Table 4-4). When diffusion through the gas film controls, the situation becomes more complex owing to the changing particle size in turn changing the film resistance. Generally, a semiempirical approach is used to develop a rate equation for this condition, and it frequently results in a fairly complex relationship. The rate equations presented in Table 4-4 have been derived on the basis of an empirical mass-transfer equation (B-7).

Further details on the derivation of rate equations based on the unreacted-core model are found in references such as Carberry (B-8), Froment and Bischoff (B-9), and Levenspiel (B-10).

Concurrent action of controlling steps. The analysis of fluid-solid reactions in terms of the simplified unreacted-core model just discussed assumes that only one control step prevails in the overall process. In practice, however, all control steps considered may play their roles simultaneously under certain conditions, and the relative importance of individual steps may vary with progress of the reaction. Thus, the control step associated with diffusion through the ash layer may be insignificant in the early stage of the process but should become more and more dominant as the thickness of the layer increases with the extent of the reaction.

Under isothermal conditions, a more general rate expression for the gas-solid system representing all the control steps (or resistances) may be derived from Eqs. (4-16) and (4-17) on the basis of the pseudo-steady-state approximation and by introducing suitable boundary conditions (B-2, B-11). A general rate expression in terms of the overall time required for a desired degree of conversion may also be obtained by summing up the expressions for individual controlling steps listed in Table 4-4 since these steps are present in series.

Particles of nonspherical geometry. A detailed discussion pertaining to development of rate expressions for solids of the flat-plate geometry was presented by Carberry and Gorring (B-12). The resulting rate equation for the flat-plate particle

$$x_B = -\alpha + \sqrt{\alpha^2 + \delta^2} \tag{4-23}$$

has been shown to also represent the rate expression for the spherical particles up to $x_B \doteq 0.3$. In Eq. (4-23),

$$\alpha = 3\left(\frac{1}{\text{Da}} + \frac{1}{\text{Bi}}\right) \quad \text{and} \quad \delta^2 = \frac{18\theta b D_e C_{A1}}{\rho_B R^2}$$

where Da = Damköhler number = $k_s R/D_e$
Bi = Biot number = $k_m R/D_e$
C_{A1} = concentration of A in main fluid stream, mol/m^3 (mol/ft^3)

The so-called grain model developed by Sohn and Szekely (B-13) considers the solid particle to comprise a matrix of very small grains between which the fluid reactant can readily diffuse through the pores. The individual grains are assumed to behave according to the unreacted-core model. An approximate solution for the model equations incorporating geometric factors for grains and particles is applicable to slabs and cylinders as well as to spheres (B-14). A similar model was utilized by Pigford and Sliger (B-15) in an analysis of experimental data involving an SO_2 gas-limestone reaction.

Nonisothermal condition. In certain fluid-solid reaction systems, significant temperature gradients exist within the solid particle owing to, for example, a rapid reaction taking place at the ash–unreacted-core interface. Experimental data acquired under such nonisothermal condition have been analyzed for cases involving hydrofluorination of UO_2 (B-16) and the burning of coke from fireclay particles (B-17). The rate expressions developed in the preceding subsections are not directly applicable to such cases since they are based on isothermal condition.

Derivation of rate equations for nonisothermal fluid-solid (spherical) reactions would require a mass-continuity equation

$$\frac{1}{r^2}\frac{\partial}{\partial r}\left(D_e r^2 \frac{\partial \overline{C}_A}{\partial r}\right) = 0 \tag{4-24}$$

and a heat-balance equation

$$\frac{1}{r^2}\frac{\partial}{\partial r}\left(\lambda_e r^2 \frac{\partial T}{\partial r}\right) = 0 \tag{4-25}$$

where λ_e is the effective thermal conductivity in a solid particle, kJ/(m·s·K) [Btu/(ft·h·°F)]. Both equations assume the pseudo-steady-state approximation. For the reaction of first order in A and zero order in solid, the solutions (B-8, B-18) of Eqs. (4-24) and (4-25) in terms of the concentration of A and temperature at the ash–unreacted-core interface, C_{Ac} and T_c respectively, assume the forms

$$\frac{C_{Ac}}{C_{A1}} = \frac{1}{1 + D_a\,\Delta x_B\,[(\Delta x_B/Bi) + (1 - \Delta x_B)]} \tag{4-26}$$

where $\Delta x_B = (1 - x_B)^{1/3} = r_c/R$

or

$$\frac{dN_A}{d\theta} = \frac{4\pi r_c^2 k_s C_{A1}}{1 + D_a\,\Delta x_B\,[(\Delta x_B/Bi) + (1 - \Delta x_B)]} \tag{4-27}$$

and

$$\frac{T_c}{T_1} = 1 + D_a\,\beta\,\frac{C_{AC}}{C_{A1}}\,\Delta x_B\,[(\Delta x_B/Bi_h) + (1 - \Delta x_B)] \tag{4-28}$$

where $\beta = -\Delta H C_{A1} D_e/\lambda_e T_1$
$-\Delta H$ = heat of reaction, kJ/mol [Btu/(lb·mol)]
T_1 = temperature in bulk fluid, K (°R)
Bi_h = thermal Biot number

Criteria for applicability of unreacted-core (Shell-progressive) model. The unreacted-core model has been shown to offer simplified analysis of most noncatalytic gas-solid reactions and to afford a good approximation of the reaction-system behavior. There are several known cases that deviate significantly from this model. In general, these reaction systems are associated with highly porous solid reactant and products, resulting in either a uniform reaction zone (chemical-reaction-controlled) or a diffused zone of reaction. Useful criteria for determining applicability of the unreacted-core model have been suggested by Carberry (B-8). In essence the criteria specify that:

1. $Da \geq 4 \times 10^4/LS$, where LS is the ratio of total (BET) area to geometric (external) area of the solid particle.
2. Validity of the pseudo-steady-state approximation which effectively implies that only gas-solid reaction systems are amenable to analysis by the unreacted-core model.
3. The solid phase in the reaction system for any LS values should consist of a virtually impervious unreacted core and a porous product layer.

Rate Equations from Experimental Data Rate equations derived from experimental data for fluid-solid reactions frequently appear in the published literature. For example, the overall rate of reaction in the reduction of magnetite, Fe_3O_4, by hydrogen has been shown (B-19) to follow the expression

$$r = k_1 \frac{(p_{H_2} - p_{H_2O})/K}{1 + k_2 p_{H_2} + k_3 p_{H_2O}}$$

where k's are rate constants for different steps, p's are partial pressures, and $K = p_{H_2O}/p_{H_2}$. The rate of fluorination of uranium oxide or plutonium oxide is expressed by (B-20):

$$k\theta = 1 - (1 - x)^{1/3}$$

This equation is similar to that for the chemical-reaction-controlled case shown in Table 4-4. The reduction of $Eu_2(SO_4)_3$ by CO proceeds in two stages, both of which can be described by (B-21) equations of the form

$$x = 1 - e^{k\theta n}$$

Here x refers to the fractional conversion, and n is a constant.

The rate of dissolution of uranium dioxide in a sulfuric acid solution at low concentration ($C_{H^+} < 0.003M$) under 100 or 200 lbf/in² oxygen (P_0) is controlled by C_{H^+}. However, the controlling factor is shifted to P_0 as C_{H^+} is increased. The reaction rate is represented by (B-22):

$$r = (k_1 k_2 A P_0 C_{H^+})/(k_1 P_0 + k_2 C_{H^+})$$

where A is the surface area of UO_2.

Formation of polysulfides $(NH_4)_2S_x$ by a two-step mechanism is involved in the leaching of sulfur from ores with ammonium sulfide. The first step is controlled by both chemical reaction and diffusion through the ore layer, while diffusion is the rate-determining factor in the second step. The kinetics of the process follow (B-23):

$$1 - (1 - x)^{1/3} = 0.00460\theta u^{1/3}\,\Delta C/(r_0 x_0 \rho)$$

for the first step and

$$1 - (1 - x)^{1/3} = 0.031\left(\frac{\theta\,\Delta C}{r_0 x_0 \rho}\right)^{1/2}$$

for the second step. In these equations, x is the degree of leaching after time θ; u is the linear flow rate of solvent; and ρ, x_0, and r_0 are the density, the initial sulfur content, and the initial particle radius respectively of ore.

$$\Delta C = \frac{(C_n - C_0) - (C_n - C)}{\ln[(C_n - C_0)/(C_n - C)]}$$

where C_n is the saturated sulfur concentration in the solvent, and C_0 and C are sulfur concentrations in the solvent entering and leaving the apparatus respectively.

Mathematical Model for Liquid-Liquid Reactions There are two cases that are relatively simple to analyze. Both cases assume that an elementary second-order reaction: $A + B \rightarrow$ products takes place in two partially miscible liquid phases 1 and 2. Under this condition, the overall reaction rate is governed by (*a*) the rates of diffusion to the interface, (*b*) the reaction at the interface, and (*c*) separate homogeneous reactions in each of the two phases.

Case 1. *Chemical-reaction-controlling.* When agitation is highly efficient, the dispersed phase would be divided into very fine droplets, producing a large surface area for mass transfer. Hence, an equilibrium distribution of the components between the phases may be approached and be described by the distribution coefficient. The overall rate r_A is then controlled by the chemical reaction and is the sum of the rates in the individual phases, r_{A1} and r_{A2}:

$$-r_A = -(r_{A1} + r_{A2}) = \frac{1}{V_1 + V_2}(k_1 C_{A1} C_{B1} V_1 + k_2 C_{A2} C_{B2} V_2)$$

Introducing the distribution coefficients,

$$K_A = C_{A1}/C_{A2} \qquad \text{and} \qquad K_B = C_{B1}/C_{B2}$$

$$-r_A = \left(k_1 \overline{V}_1 + \frac{k_2 \overline{V}_2}{K_A K_B}\right) C_{A1} C_{B1} \tag{4-29}$$

where $\overline{V}_1$ and $\overline{V}_2$ are the volume fractions of the individual phases. The rate constants k_1 and k_2 in the liquid-liquid system are different and, in general, vary with the concentration as well as with the temperature.

Approximation to examples of case 1 can be found in the nitration of benzene or toluene with a mixture of nitric and sulfuric acids (B-24, B-25).

Case 2. *Mass-transfer-controlling.* If phase equilibrium has not been achieved and the reaction is fast compared with the rate of mass transfer, there would be two liquid films on either side of the interface between the phases which offer resistance to mass transfer (two-film model). Thus, the reaction occurs as the reactants A and B diffuse through the films into the bulk of the other phase.

Assume that A is the major component in phase 1 while phase 2 is predominantly B and that phase equilibrium is attained at the interface. Then, the rate of mass transfer of A from phase 1 to phase 2 would be equal to the rate of disappearance of A by chemical reaction in phase 2, where C_{A2} drops to a steady-state value C^0_{A2}. Hence, the rates of disappearance of A by chemical reaction:

In phase 1,

$$-r_{A1} V_1 = k_1 C_{A1} C^0_{B1} V_1 \tag{4-30}$$

In phase 2,

$$-r_{A2} V_2 = k_2 C^0_{A2} C_{B2} V_2$$

and by mass transfer to phase 2,

$$-\frac{dn_{A1}}{d\theta} = k_2' a\left(\frac{C_{A1}}{K_A} - C^0_{A2}\right)(V_1 + V_2) \tag{4-31}$$

In these equations, C^0_{B1} is the steady-state value of C_{B1} in phase 1, k'_2 the mass-transfer coefficient for phase 2 per unit volume of both phases in [mol/(m^2·s)]/(mol/m^3 driving force), or [mol/(ft^2·h)]/(mol/ft^3 driving force), and a the interfacial area in m^2 per unit volume of both phases. Since $-dn_{A1}/d\theta = -r_{A2}V_2$, from Eqs. (4-30) and (4-31),

$$C^0_{A2} = \frac{k'_2 a(C_{A1}/K_A)(V_1 + V_2)}{k_2 V_2 C_{B2} + k'_2 a(V_1 + V_2)} \tag{4-32}$$

Similarly,

$$C^0_{B2} = \frac{k'_1 a(C_{B2}/K_B)(V_1 + V_2)}{k_1 V_1 C_{A1} + k'_1 a(V_1 + V_2)} \tag{4-33}$$

The values of k'_1a and k'_2a are strongly dependent on the physical configuration of the contact system (see Secs. 14 and 21), and use of an empirical relation is suggested in obtaining these values for scale-up purposes.

An example of liquid-liquid reaction in which no phase equilibrium exists and both mass transfer and chemical reaction are important is the hydrolysis of fat to produce glycerin and fatty acid (B-26).

Gas-Liquid Reactions When gaseous reactants are absorbed by a liquid, and a chemical reaction takes place, both chemical reaction and mass transfer may control the overall reaction rate. For analysis of this type of system, Sec. 14 should be consulted. Books by Astarita (B-27), Danckwerts (B-28), and Sherwood, Pigford, and Wilke (B-29) should also be helpful. Treatments similar to these references are likewise applicable to the liquid-liquid system if chemical reaction is involved in only one of two partially miscible phases and the overall rate is governed by both mass transfer and chemical reaction.

Solid-State Reactions The solid-state reactions discussed in the following paragraphs refer to decompositions and interaction of solids, and are generally more complex than those of other systems discussed previously. There have been few systematic analyses of chemical reactions in the solid state. Fairly comprehensive treatments of such reactions are presented in two recent publications by Bamford and Tipper (B-30) and Schmalzried (B-31) respectively.

General characteristics. Nature of atomic bonding, crystalline structure, and crystal-lattice imperfections are among the important factors that influence the reactivity of solids. The imperfections include, for example, point defects, nonstoichiometry, impurity inclusions, dislocations, edges, etc., all of which may lead to a reduction of lattice-stabilizing forces, which, in turn, would tend to increase the reactivity of solids. Detailed discussion on the subject of chemistry and physics of the solid state is beyond the scope of this section and should be referred to other sources such as Refs. B-30 and B-31.

On the basis of various experimental evidence, the generally accepted (B-30) kinetic principles for the interpretation of solid-state reaction data are:

1. The rate of solid-state reaction is proportional to the total effective area of the reactant-product interface.
2. Under isothermal condition, the rate of interface advance through an isotropic solid reactant is constant.
3. The temperature effect on the rate constant of most solid-state reactions may be described by the Arrhenius equation.
4. The overall rate of the solid-state reaction may be controlled by the diffusion rates of reactants through the product layers, which may function as a barrier between the reactants.

Items 1 and 2 are valid only when there is no preferential removal of a constituent and no melting of solid reactants and when the product layer offers no appreciable resistance for interactions between solid reactants. Other factors known to influence the kinetic behavior of solid-state reactions include particle-size distribution of reactants, pretreatment of reactants, presence of impurities, and melting of reactants.

The general characteristics of solid-state reactions (including decomposition and solid-solid reactions) based on experimental observations are summarized in Table 4-5, which illustrates several

TABLE 4-5 Summary Characteristics of Solid-State Reactions*

Type of chemical reaction	Possible reaction paths	Potential control steps
Decomposition of a single solid reactant	*a.* Reactant and/or product nucleation and growth	Interfacial phenomena or geometry
	b. Reaction at immobile surface	Interface or diffusion
Solid-solid reaction	*a.* Reactant and/or product nucleation and growth *b.* Barrier-product-layer formation *c.* Reaction at immobile surface	Reaction generally controlled by mobility of reactants and/or diffusion of reactant through the product barrier layer

*Adapted from Ref. B-30.

possible reaction paths and potential control steps for individual paths. The control steps, however, may shift as the reaction progresses. Furthermore, the reaction mechanism is affected by various factors mentioned earlier. For example, dehydration of $CaSO_4 \cdot 2H_2O$ involves the interconversion of different forms, including dihydrate, hemihydrates (α and β pseudomorphs), and anhydrous salts (hexagonal and orthorhombic) as follows (B-32):

$$CaSO_4 \cdot 2H_2O \rightarrow CaSO_4 \text{ (hexagonal)} \rightarrow CaSO_4 \text{ (orthorhombic)}$$

$$\nwarrow\searrow \qquad \swarrow$$

$$CaSO_4 \cdot \tfrac{1}{2} H_2O \ (\alpha \text{ and } \beta)$$

Experimental rate data and microscopic observations confirmed the complexity in the kinetic behavior of this system. The rate and mechanism of the reactions are quite sensitive to process conditions. Thus, dehydration of $CaSO_4 \cdot 2H_2O$ in the temperature range of 353 to 383 K was controlled by both nucleation and phase boundary, whereas diffusion mechanism prevailed in the range of 383 to 425 K. Likewise, the mechanism of dehydration of $CaSO_4 \cdot \tfrac{1}{2} H_2O$ was governed by reaction conditions as well as by its crystalline structural forms (i.e., α or β).

Rate equations for solid-state reactions. Presented in Table 4-6 are selected rate equations derived from various theories of solid-state reactions under isothermal conditions. Also shown are examples of reactions for individual rate equations. Since the reaction mechanism could be influenced by process conditions and other factors (e.g., particle size, impurity, pretreatment, etc.), the original references should be consulted to determine the specific conditions under which individual rate equations are applicable to these reactions.

Further examples of solid-state reactions. Kinetic equations for the thermal decomposition of solids have been classified (B-49) into the following four types, based on published experimental data and theories:

$$dx/d\theta = kx^a(1 - x)^b$$
$$dx/d\theta = k(1 - x)^b$$
$$dx/d\theta = kx^a$$
$$x = k\theta^n$$

where x is the degree of dissociation or transformation; and a, b, and n are constants.

The equation of the form $k\theta = 1 - (1 - x)^{1/3}$ has been shown to describe the dehydration of $LiCl \cdot H_2O$, and the thermal decomposition of NH_4VO_3, CuS, FeS_2, and $LiBr \cdot NH_3$.

The isothermal decomposition of $CaCO_3$, calcite, and magnesite in vacuo has been reported (B-50) to follow $dv/d\theta = 4\pi k(\gamma - \tau)\phi$. Here v is the volume of undecomposed carbonates, γ represents the initial spherical equivalent radius of the carbonate particles, τ is the thickness of the product layer at time θ, and ϕ is the shape factor. Another investigator (B-51) employed two empirical equations:

$$t = A + B\theta + C\theta^2$$

and

$$1 - (1 - x)^{1/3} = \alpha\theta + \beta\theta^2 + \delta\,\theta^3$$

TABLE 4-6 Selected Rate Equations for Isothermal Solid-State Reactions

Rate equation*	Reaction type	Example of reaction
$x^{1/n} = k\theta$	Acceleratory rate	Decomposition of Ni-formate [$n = 2$; $0.03 < x < 0.5$] (B-33)
		Decomposition of BeH_2 [$n = \sim 6$; $0.1 < x < 0.35$] (B-34)
$\ln x = k\theta$		Vacuum decomposition of Ag-oxalate (B-35)
$-\ln (1 - x)^{1/n} = k\theta$ $n = 2, 3$ and 4	Sigmoid rate	Dehydration of $NiSO_4 \cdot 6H_2O$ [$n = 2$] (B-36); decomposition of Ca-, Sr- and Ba- formates [$n = 4$] (B-37)
$\ln [x/(1 - x)] = k\theta$		Decomposition of $Th(HCO_2)_4$ (B-38) and $U(HCO_2)_4$ (B-39)
	Diffusion-controlled	
$x^2 = k\theta$	One-dimensional	$Ca_3(PO_4)_2 + CO(NH_2)_2 \cdot HNO_3$[$-180 + 200$ mesh] (B-40)
$(1 - x) \ln (1 - x) + x = k\theta$	Two-dimensional	$CaCO_3 + MoO_3$ (B-41)
$[1 - (1 - x)^{1/3}]^2 = k\theta$ $[1 - (2x/3)] - (1 - x)^{2/3} = k\theta$	Three-dimensional	$Li_2CO_3 + Fe_2O_3$ (B-42)
$1 - (1 - x)^{1/2} = k\theta$	Contracting area	$UO_2 + SiC$ [>1650 K] (B-43); decomposition of $Ag_2S_2O_6$ (B-44)
$1 - (1 - x)^{1/3} = k\theta$	Contracting volume	Dehydration of $LiSO_4 \cdot H_2O$ (B-45); decomposition of dolomite (B-46)
$-\ln (1 - x) = k\theta$	First order	$Ca_3(PO_4)_2 + 5C$ (B-47); decomposition of α-PbN_6 (B-48)
$(1 - x)^{-1} = k\theta$	Second order	
$(1 - x)^{-2} = k\theta$	Third order	

*Adapted from Ref. B-30. See list of references at end of subsection. x = fraction of limiting reactant converted; k = rate constant. θ = reaction time corrected for incubation period, if any.

to express decomposition of a $CaCO_3$ cube in an isothermal vacuum or nitrogen atmosphere. In these equations, t is the temperature in the cube undergoing decomposition at time θ, x is the fraction decomposed, and A, B, C, α, β, and δ are constants characteristic of the given cube.

According to Hulbert (B-52), the decomposition of $MgSO_4$ at 920 to 1080°C is satisfactorily described by the nuclei growth equation $(k\theta)^m = -\ln (1 - x)$ with $m = 1.10$. The rate is influenced by the sample mass as well as the pelletization pressure. The solid-solid reaction between $BaCO_3$ and ZnO assumes the form (B-53) $k\theta = -\ln (1 - x)^{2/3}$

Catalysis in Homogeneous and Heterogeneous Systems Catalysis generally refers to chemical processes in which chemical reaction rates are subjected to the influence of substances which may or may not change chemically during the reaction. Such a substance is termed a *catalyst*. Basically a catalyst is considered to form an intermediate compound with some of the reactants which, in turn, interacts with other reactants to form desired products and to regenerate the catalyst. In this way, a catalyst enables a reaction to proceed at a faster rate by a mechanism that requires a lower activation energy than that for the uncatalyzed one. A catalyst may also modify the multipathed mechanism of a reaction system to achieve a desired product distribution.

Catalysis is customarily classified into two major categories, namely, homogeneous and heterogeneous catalytic reactions. In the former the catalyst forms a homogeneous phase with the reaction mixture, whereas in the latter the catalyst exists as a phase distinct from the reaction mixture.

Basic Characteristics of Catalysis These may be summarized as follows:

1. In a catalytic reaction, the catalyst is unchanged at the conclusion of the reaction, but it may participate in the intermediate steps so as to increase the reaction rate.
2. When more than one mechanism is feasible for a reaction, a catalyst may exhibit selectivity, favoring one mechanism over others. This situation usually results in a product distribution different from those obtained by other mechanisms. Proper application of catalyst selectivity would enable the desired reaction to be accelerated while the undesirable reactions are retarded.
3. The rate of reaction is generally proportional to the catalyst concentration. For a solid-catalyzed reaction, the surface area of the catalyst and the concentration of the so-called active centers or catalyst sites (locations of high chemical activity on the surface) become important.
4. In a reversible reaction the catalyst accelerates the reverse reaction as well as the forward one. Thus, the equilibrium composition of the reacting system would be the same as that of the uncatalyzed system.
5. In an autocatalytic reaction, in which one of the reaction products functions as a catalyst, a small amount of the product must be present to initiate the reaction. A plot of conversion versus time for this type of reaction results in a characteristic S-shaped curve.

Homogeneous Catalysis Homogeneous catalytic reactions have been observed to take place either in the gas phase or in the liquid phase. A great number of catalyses in this category have been found to follow rate equations which are first-order with respect to the catalyst concentration.

Gas-phase catalysis. A well-known example of homogeneous catalysis in the gas phase is the oxidation of SO_2 to SO_3, catalyzed by nitric oxide in the lead-chamber sulfuric acid process. The presence of nitric oxide promotes the normally very slow oxidation process by the following mechanism:

$$2NO + O_2 \rightarrow 2NO_2$$

$$SO_2 + NO_2 \rightarrow SO_3 + NO$$

Additional examples can be found in the molecular iodine-catalyzed pyrolytic decomposition of such organic compounds as acetaldehyde, formaldehyde, methyl alcohol, ethylene oxide, and several aliphatic ethers (C-1). Dehydration of tertiary alcohols catalyzed by hydrogen halides also proceeds in the gas phase (C-2).

Liquid-phase catalysis. Most homogeneous catalysis occurs in the liquid phase. Acid-base catalysis is the type of liquid-phase catalytic reaction that has been most extensively studied. Examples of important organic reactions the rates of which are controlled by acid-base catalysis include:

1. Inversion of sugars
2. Hydrolysis of esters and amides
3. Halogenation of acetone and nitroparaffins
4. Mutarotation of glucose
5. Esterification of alcohols
6. Enolization of aldehyde and ketones

For general acid- or general base-catalyzed reactions, the relationship between catalyst effectiveness (in terms of the rate constant) and the strength of the acid or base (ionization constant) has been frequently expressed by the Brönsted equation

$$k = CK^a$$

where k represents the rate constant (or catalytic constant), K is the ionization constant of either acid or base, and C and a are empirical constants governed by the type of reaction, the temperature, and the solvent. The observed values of a lie in the range 0.3 to 0.9. The Brönsted equation has been found to hold rather well for many acid- or base-catalyzed reactions.

According to the general concept of acids and bases proposed by Brönsted and Lowry, undissociated acids and bases are just as effective catalysts as are H^+ or OH^- ions. Thus, acid catalysis involves transfer of H^+ to the reactant being catalyzed, whereas acceptance of H^+ by the base prevails in the case of base catalysis. This implies that the reactant functions as a base in acid catalysis while it serves as an acid in base catalysis.

Further details on the subject of acid or base catalysis can be found in such books as those by Bell (C-3), Frost and Pearson (A-1), Hammett (C-4), and Emmett, Sabatier, and Reid (C-5).

In addition to acid or base catalysis, other types of liquid-phase catalytic reactions are known. They include reactions catalyzed by various metal ions many of which form a stable complex of the chelate type (C-6 to C-10). Additional discussion of subjects and examples pertaining to homogeneous catalysis is found in two recent publications (C-11 and C-12).

Heterogeneous Catalysis The catalytic reaction system under this classification involves two or more of gas, liquid, and solid phases. The catalyst is present in a phase different from those of the reactants. The solid-catalyzed fluid-phase reactions are by far the most important and commonly encountered types in industrial processes. Examples of solid catalysis can be found in a great number of processes producing inorganic as well as organic chemicals, such as mineral acids (HCl, H_2SO_4, HNO_3, etc.), ammonia, methanol, petrochemicals, and synthetic high polymers. Especially notable is production of petrochemicals in which solid catalysts play key roles in various reactions, including cracking, alkylation, polymerization, isomerization, aromatization, and dehydrogenation.

The fundamental technical data base on heterogeneous catalysis may be obtained from publications listed below:

1. *Advances in Catalysis*, Academic Press, Inc., New York (books published approximately on an annual basis)
2. American Chemical Society reprints for Divisions of Petroleum Chemistry, Polymer Chemistry, and Fuel Chemistry
3. *Chemical Engineering Progress*
4. *Industrial and Engineering Chemistry—Process Design and Development; Product Research and Development*
5. *Journal of Applied Polymer Science*
6. *Journal of Polymer Science*
7. *Journal of Catalysis*
8. *Kinetics and Catalysis*, translation from Russian (*Kinetika i Kataliz*)

Most of the catalysts and processes involved have been patented, and are summarized in the *United States Patent Gazette*. Foreign patents can usually be found in *Chemical Abstracts*. Recent literature in the field of catalysis has been reviewed by several authors (C-13).

Mechanism of Solid-Catalyzed Reactions The generally accepted hypothesis concerning the steps involved in solid-catalyzed fluid-phase reactions is outlined as follows:

1. Diffusion of reactants from the bulk of the fluid phase to the exterior surface of the catalyst and into the catalyst pores
2. Adsorption (chemisorption) of reactants onto the exterior and pore surfaces (active centers)
3. Reaction of the adsorbed reactants on the surfaces to form products
4. Desorption of the products to the fluid phase near the surfaces
5. Diffusion of the products through the pores and from the exterior surfaces into the bulk-fluid phase

Also to be considered are different deactivation mechanisms that lead to loss of catalytic efficiency, including deactivation by fouling, poisoning, and elevated temperatures.

The relative significance of the five catalytic-process steps in controlling the overall rate of reaction depends on widely different conditions. Normally, only a few of these steps govern the overall reaction rate, and other steps may be either neglected or combined. The fact that these steps do not necessarily proceed in series or in parallel frequently makes it impossible to combine them by simple means. In this case an empirical approach may satisfactorily describe the experimental data.

Factors Affecting Catalytic Reactions Each of the five catalytic-process steps is influenced by one or more factors, including the following:

1. Fluid dynamic factors, e.g., mass velocity
2. Catalyst properties, e.g., particle size, porosity and pore dimensions, and surface characteristics
3. Diffusion characteristics of fluid reactants and products
4. Activation energy requirements for adsorption and desorption of fluid reactants and products
5. Activation energy of the surface reaction
6. Thermal factors, e.g., temperature, heat-transport characteristics

Thus, for example, steps 1 and 5 are mainly governed by factors 1, 2, and 3; while steps 2 and 4 are determined by factors 2, 4, and 6. Factors 2, 5, and 6 influence step 3.

The methods for identifying steps controlling the overall reaction rate and the ways in which various factors may be incorporated in the reactor design are discussed in the subsections "Identification of Rate-Controlling Steps" and "Reactor Design."

Examples of Industrially Important Catalytic Processes Selected heterogeneous catalytic processes that find important commercial applications are listed in Table 4-7. In addition to the major reaction step involved in individual processes, the typical catalyst used and known poisons for the catalyst are also shown in the table. Further details on industrial catalytic processes are found in books by Thomas (C-14) and by Gates, Katzer and Schuit (C-15) and in periodicals such as *Hydrocarbon Processing*.

Summary of Rate Equations for Solid-Catalyzed Reactions Table 4-8 lists selected catalytic reaction schemes and a rate equation for each catalytic step assumed to be controlling the overall reaction rate. The significance of the mechanisms and rate equations can be illustrated by the following example. Consider the reaction

$$A + B \rightleftharpoons R$$

1. An adsorption–surface-reaction–desorption mechanism is assumed (where *s* represents a catalyst site or an active center)

(1) $A + s \rightleftharpoons As$ (2) $B + s \rightleftharpoons Bs$	Adsorption of *A* and *B* molecules.
(3) $As + B \rightleftharpoons Rs$	Adsorbed *A* reacts with *B* in the fluid. This assumes that there is no reaction of *A* with adsorbed *B*. Adsorbed *B* serves only to block catalyst sites.
(4) $Rs \rightleftharpoons R + s$	Desorption of product *R*.

2. For each of these reactions it is possible to formulate a rate expression involving both forward and reverse reactions. If the reaction occurs at near-equilibrium conditions, the concentrations are related by an equilibrium constant.

$$(1)\ r_1 = k_{1f}C_A(1 - \theta_A - \theta_B - \theta_R) - k_{1r}\theta_A$$

$$K_1 = \frac{\theta_A}{C_A(1 - \theta_A - \theta_B - \theta_R)}$$

$$(2)\ r_2 = k_{2f}C_B(1 - \theta_A - \theta_B - \theta_R) - k_{2r}\theta_B$$

$$K_2 = \frac{\theta_B}{C_B(1 - \theta_A - \theta_B - \theta_R)}$$

$$(3)\ r_3 = k_{3f}(\theta_A)C_B - k_{3r}\theta_R$$

$$K_3 = \theta_R/\theta_A C_B$$

$$(4)\ r_4 = k_{4f}\theta_R - k_{4r}C_R(1 - \theta_A - \theta_B - \theta_R)$$

$$K_4 = \frac{C_R(1 - \theta_A - \theta_B - \theta_R)}{\theta_R}$$

In these equations K_1, K_2, K_3, and K_4 are equilibrium constants for the individual steps. Note that K_4 is written for a desorption step and is the reciprocal of the usual adsorption equilibrium constant. θ_A, θ_B, and θ_R represent fractions of total catalyst sites (or active centers) occupied by A, *B*, and *R* respectively. k_f's are constants in the rate equation for adsorption, $r_f = k_{if}C_i(1 - \Sigma\theta_i)$, and include a conversion factor to allow for the total number of active sites per unit

TABLE 4-7 Example of Industrial Heterogeneous Catalytic Processes*

Process or product	Major reaction step	Catalytic type	Catalyst poisons
Ammonia	$N_2 + 3H_2 \rightleftharpoons 2NH_3$	FeO/Fe_2O_3 promoted by Al_2O_3 and K_2O	Moisture, CO, CO_2, O_2, compounds of S, P, and As
Aniline	(*a*) $C_6H_5OH + NH_3 \rightarrow C_6H_5NH_2 + CO_2 + H_2O$	Ni powder, Al_2O_3	Groups VA and VIA elements
	(*b*) $C_6H_5NO_2 + 3H_2 \rightarrow C_6H_5NH_2 + H_2O$	Raney-Ni or -Cu, Cu-chromite	
Butadiene	(*a*) $C_4H_{10} \rightleftharpoons C_4H_6 + 2H_2$	$Ca_8Ni(PO_4)_6$ Cr_2O_3 on Al_2O_3	Halides, O_2, S, P, Si
	(*b*) $C_4H_8 + ½O_2 \rightarrow C_4H_6 + H_2O$	Bi-molybdate	
	(*c*) $C_4H_8 \rightleftharpoons C_4H_6 + H_2$	$Fe_2O_3 + Cr_2O_3 + K_2O$	
Ethanol	$C_2H_4 + H_2O + C_2H_5OH$	H_3PO_4 on Kisselguhr	NH_3, O_2, S, organic base
Ethylene oxide	$C_2H_4 + ½O_2 \rightarrow C_2H_4O$	Ag-oxide on refractory oxide	Compounds of S
Formaldehyde	$CH_3OH + ½O_2 \rightarrow HCHO + H_2O$	Ag on Al_2O_3 Ag needles $FeO_3 + MoO_3$	Cl_2, S compounds
Methanol	$CO + 2H_2 \rightarrow CH_3OH$ (*a*) High pressure (*b*) Low pressure	 $ZnO + Cr_2O_3$ CuO	 S compounds, Fe, Ni S compounds
Nitric acid	$4NH_3 + 5O_2 \rightarrow 4NO + 6H_2O$ $2NO + O_2 \rightarrow 2NO_2$ $3NO_2 + H_2O \rightarrow 2HNO_3 + NO$	Pt on Rh	Compounds of As and Cl_2
Polyethylene	$nC_2H_4 \rightarrow (-CH_2-CH_2-CH_2-)_m$ Ziegler process	Al-alkyl-Ti tetrachloride precipitate	Moisture, alcohols, O_2, SO_2, COS, CO_2, CO
Styrene	$C_6H_5C_2H_5 \rightarrow C_6H_5CH \cdot CH_2 + H_2$	(*a*) $Fe_2O_3 + K_2O + Cr_2O_3$ (*b*) $Fe_2O_3 + K_2CO_3 + Cr_2O_3 + V_2O_5$	Halides, S compounds, O, P, Si
Sulfuric acid	$SO_2 + ½O_2 \rightarrow SO_3$ $SO_3 + H_2O \rightarrow H_2SO_4$	$V_2O_5 + K_2O$ on Kieselguhr	Halides, As, Te
Cracking, alkylation, and isomerization of petroleum fraction	Acid-catalyzed reactions	Synthetic aluminosilicate, $AlCl_3$ H_3PO_4	Organometallic compounds, organic bases
Desulfurization, denitrogenation, and deoxygenation	Interaction of H_2 with large hydrocarbon molecules to remove S, N_2, and O_2 from their structures	$(NiO + MoO_3)$ $(CoO + MoO_3)$ or $(NiO + WO_3)$ on alumina	H_2S, CO, CO_2, heavy hydrocarbon deposits, compounds of Na, As, Pb

*Adapted from Ref. D-1.

weight of catalyst. k_f is dependent on temperature and on the catalyst and its form. k_r's are equivalent constants for the desorption rate equations:

$$r_r = k_{ir}\theta_i$$

3. One of these steps is now assumed to be rate-controlling and the others to occur at near equilibrium. Let us assume, for now, that step 3, the surface reaction between adsorbed A and B in the fluid, is rate-controlling. The appropriate rate expression would then be that given for r_3; however, it involves the fraction of total catalyst sites occupied by A and by $R(\theta_A$ and $\theta_R)$. Available experimental data will be in terms of C_A, C_B, and C_R. Making use of the assumption that steps 1, 2, and 4 occur at near equilibrium will allow determination of θ_A and θ_R in terms of known quantities:

$$\theta_A = K_1C_A(1 - \theta_A - \theta_B - \theta_R)$$

$$\theta_R = \frac{C_R(1 - \theta_A - \theta_B - \theta_R)}{K_4}$$

and $\quad K_3 = k_{3f}/k_{3r} \quad$ or $\quad k_{3r} = k_{3f}/K_3$

Thus,

$$r_3 = k_{3f}K_1C_AC_B(1 - \theta_A - \theta_B - \theta_R) - \frac{k_{3f}C_R}{K_3}\frac{(1 - \theta_A - \theta_B - \theta_R)}{K_4}$$

$$\text{or} \quad r_3 = k_{3f}(1 - \theta_A - \theta_B - \theta_R)\left(K_1C_AC_B - \frac{C_R}{K_3K_4}\right) \qquad (4\text{-}34)$$

The fraction of total catalyst sites vacant has been defined as $1 - \theta_A - \theta_B - \theta_R = \theta_v$. Therefore,

$$1 - K_1C_A\theta_v - K_2C_B\theta_v - (C_R/K_4)\theta_v = \theta_v \qquad (4\text{-}35)$$

Solving for $\theta_v = 1 - \theta_A - \theta_B - \theta_R$ in Eq. (4-35) and substituting in Eq. (4-34) yield with simplification:

$$r_3 = \frac{k_{3f}K_1(C_AC_B - C_R/K_3K_4K_1)}{K_1C_A + K_2C_B + C_R/K_4 + 1} \qquad (4\text{-}36)$$

The product $K_1K_3K_4$ is equivalent to the overall equilibrium constant for all the productive steps in the assumed mechanism. Equation (4-36) may be rewritten as

$$r_3 = \frac{k(C_AC_B - C_R/K)}{K_AC_A + K_BC_B + K_RC_R + 1}$$

where $k = k_{3f}K_1$, $K = K_1K_3K_4$, $K_A = K_1$, $K_B = K_2$, and $K_R = 1/K_4$. Derived rate expressions of this type for several different postulated reaction mechanisms are presented in Table 4-8.

4. Experimental rate data are used to evaluate the constants k_{3f}, K_1, K_2, and K_4 in Eq. (4-36); bulk-fluid-phase concentrations can be used only if there are no diffusional resistances.

As can be seen in the example presented, the rate equations listed in Table 4-8 are based on a simplified version of surface phenomena.

Table 4-8. Mechanisms and Their Corresponding Rate Equations for Solid-Catalyzed Reactions

Chemical equation	Catalytic steps	Rate equation*
$A \rightleftharpoons R$	$A + s \rightleftharpoons As$	$r = \dfrac{k(C_A - C_R/K)}{1 + K_R C_R}$
	$As \rightleftharpoons Rs$	$r = \dfrac{k(C_A - C_R/K)}{1 + K_A C_A + K_R C_R}$
	$Rs \rightleftharpoons R + s$	$r = \dfrac{k(C_A - C_R/K)}{1 + K_A C_A}$
$A \rightleftharpoons R$	$2A + s \rightleftharpoons A_2s$	$r = \dfrac{k(C_A^2 - C_R^2/K^2)}{1 + K_R C_R + K'_R C_R^2}$
	$A_2s + s \rightleftharpoons 2As$	$r = \dfrac{k(C_A^2 - C_R^2/K^2)}{(1 + K_R C_R + K_A C_A^2)^2}$
	$As \rightleftharpoons Rs$	$r = \dfrac{k(C_A - C_R/K)}{1 + K_A C_A^2 + K_A' C_A + K_R C_R}$
	$Rs \rightleftharpoons R + s$	$r = \dfrac{k(C_A - C_R/K)}{1 + K_A C_A^2 + K_A' C_A}$
$A \rightleftharpoons R$	$A + 2s \rightleftharpoons 2A_{1/2}s$	$r = \dfrac{k(C_A - C_R/K)}{(1 + \sqrt{K_R C_R} + K_R' C_R)^2}$
	$2A_{1/2}s \rightleftharpoons Rs + s$	$r = \dfrac{k(C_A - C_R/K)}{(1 + \sqrt{K_A C_A} + K_R C_R)^2}$
	$Rs \rightleftharpoons R + S$	$r = \dfrac{k(C_A - C_R/K)}{1 + \sqrt{K_A C_A} + K_A' C_A}$
$A \rightleftharpoons R + S$	$A + s \rightleftharpoons As$	$r = \dfrac{k(C_A - C_R C_S/K)}{1 + K_{RS} C_R C_S + K_R C_R + K_S C_S}$
	$As + s \rightleftharpoons Rs + Ss$	$r = \dfrac{k(C_A - C_R C_S/K)}{(1 + K_A C_A + K_R C_R + K_S C_S)^2}$
	$Rs \rightleftharpoons R + s$ $Ss \rightleftharpoons S + s$	$r = \dfrac{k(C_A - C_R C_S/K)}{C_S(1 + K_A C_A + (K_{AS} C_A/C_S) + K_S C_S)}$
$A \rightleftharpoons R + S$	$A + s \rightleftharpoons As$	$r = \dfrac{k(C_A - C_R C_S/K)}{1 + K_R C_R + K_{RS} C_R C_S}$
	$As \rightleftharpoons Rs + S$	$r = \dfrac{k(C_A - C_R C_S/K)}{1 + K_A C_A + K_R C_R}$
	$Rs \rightleftharpoons R + s$	$r = \dfrac{k(C_A - C_R C_S/K)}{C_S(1 + K_A C_A + K_{AS} C_A/C_S)}$
$A + B \rightleftharpoons R$	$A + s \rightleftharpoons As$	$r = \dfrac{k(C_A - C_R/K C_B)}{1 + (K_{RB} C_R/C_B) + K_B C_B + K_R C_R}$
	$B + s \rightleftharpoons Bs$	$r = \dfrac{k(C_B - C_R/K C_A)}{1 + K_A C_A + (K_{RA} C_R/C_A) + K_R C_R}$
	$As + Bs \rightleftharpoons Rs + s$	$r = \dfrac{k(C_A C_B - C_R/K)}{(1 + K_A C_A + K_B C_B + K_R C_R)^2}$
	$Rs \rightleftharpoons R + s$	$r = \dfrac{k(C_A C_B - C_R/K)}{1 + K_A C_A + K_B C_B + K_{AB} C_A C_B}$
$A + B \rightleftharpoons R + S$	$A + s \rightleftharpoons As$	$r = \dfrac{k(C_A - C_R C_S/K C_B)}{1 + (K_{RS} C_R C_S/C_B)^{1/2} + K_B C_B + K_R C_R + K_S C_S}$
	$B + s \rightleftharpoons Bs$	$r = \dfrac{k(C_B - C_R C_S/K C_A)}{1 + (K_{RS} C_R C_S/C_A) + K_A C_A + K_R C_R + K_S C_S}$
	$As + Bs \rightleftharpoons Rs + Ss$	$r = \dfrac{k(C_A C_B - C_R C_S/K)}{(1 + K_A C_A + K_B C_B + K_R C_R + K_S C_S)^2}$
	$Rs \rightleftharpoons R + s$ $Ss \rightleftharpoons S + s$	$r = \dfrac{k[(C_A C_B/C_S) - C_R/K]}{1 + K_A C_A + K_B C_B + K_S C_S + K_{AB} C_A C_B/C_S}$
$A + B \rightleftharpoons R + S$	$A + 2s \rightleftharpoons 2A_{1/2}s$	$r = \dfrac{k(C_A - C_R C_S/K C_B)}{[1 + (K_{RS} C_R C_S/C_B)^{1/2} + K_B C_B + K_R C_R + K_S C_S]^2}$
	$B + s \rightleftharpoons Bs$	$r = \dfrac{k(C_B - C_R C_S/K C_A)}{1 + \sqrt{K_A C_A} + (K_{RS} C_R C_S/C_A) + K_R C_R + K_S C_S}$
	$2A_{1/2}s + Bs \rightleftharpoons Rs + Ss + s$	$r = \dfrac{k(C_A C_B - C_R C_S/K)}{(1 + \sqrt{K_A C_A} + K_B C_B + K_R C_R + K_S C_S)^3}$
	$Rs \rightleftharpoons R + s$	$r = \dfrac{k(C_A C_B/C_S - C_R/K)}{1 + K_A \sqrt{C_A} + K_B C_B + (K_{AB} C_A C_B/C_S) + K_S C_S}$
	$Ss \rightleftharpoons S + s$	$r = \dfrac{k(C_A C_B/C_R - C_S/K)}{1 + \sqrt{K_A C_A} + K_B C_B + K_R C_R + K_{AB} C_A C_B/C_R}$
$A + B \rightleftharpoons R + S$	$B + s \rightleftharpoons Bs$	$r = \dfrac{k(C_B - C_S C_R/K C_A)}{1 + K_R C_R + K_{RS} C_R C_S/C_A}$
	$A + Bs \rightleftharpoons Rs + S$	$r = \dfrac{k(C_A C_B - C_R C_S/K)}{1 + K_R C_R + K_B C_B}$
	$Rs \rightleftharpoons R + s$	$r = \dfrac{k[(C_A C_B/C_S) - C_R/K]}{1 + (K_{AB} C_A C_B/C_S) + K_B C_B}$

$K_{AB...}$ = combined equilibrium constants.
K = over-all equilibrium constant for the chemical equation.
k = constant.
r = rate of product formation
*The rate equation is opposite the catalytic step assumed to be rate-controlling.

Assumptions have been made that all active centers (or catalyst sites) function identically and that each molecule is adsorbed regardless of interaction between molecules of the same and different kinds. There is experimental evidence that catalytic activity is governed by a surface distribution of active centers which is not necessarily uniform. Likewise, as active centers are progressively occupied, interaction among the adsorbed molecules takes place, and the activation energy may increase gradually.

EXPERIMENTAL TECHNIQUES FOR KINETIC-DATA ACQUISITION

The principles of reaction kinetics have been presented in the preceding subsection, leading to rate expressions for various types of reactions in the homogeneous system as well as in the heterogeneous noncatalytic and catalytic systems. This subsection is concerned with experimental methods for acquiring the data to develop rate equations and to determine numerical rate parameters for specific reaction systems. The rate expressions, together with the values of rate parameters thus determined, will provide the data base for the design of reactor systems.

Since the mechanisms of reactions involved in most commercial processes are fairly complex and since time and costs for development of the design data base are limited, it is imperative that selection of experimental methods and equipment be carried out carefully and expeditiously. There are two areas to consider in the selection. One deals with the method and equipment (reactor) for conducting the reactions, and the other with those for monitoring the progress of those reactions. There are no rigid criteria for selection of the experimental method and equipment since these depend primarily upon the specific reactions involved. Some aspects of these issues follow.

Methods and Experimental Reactor Types The method of approach used in acquiring kinetic data generally dictates the type of experimental reactor. The experimental methods frequently utilized in generating the design data base may be divided into three types outlined here.

1. *Development of rate equation through verification of assumed mechanism.* This method initially assumes a mechanism based on the preliminary chemical data (e.g., analysis of the reaction mixture), which consists of several elementary reaction steps involving the formation of intermediate compounds, the sum of which represents an overall reaction. The experimental kinetic data (including the intermediate and final product distribution) are acquired to determine whether the data confirm the proposed mechanism and the corresponding rate equations. If the result should prove unsatisfactory, alternative mechanisms are assumed and tested until satisfactory agreement is achieved. More detailed discussion of kinetic-data analysis is presented in the subsection "Analyses of Reaction Kinetic Data." The chemical rate equation thus developed must be combined with the physical rate (heat- and mass-transport) equations derived from the same reaction system to provide the usable design data base. Some degree of extrapolation is possible with the kinetic data derived by this method.

2. *Empirical method with scale-down reactor.* An experimental reactor which is of the same type as that to be used in the large-scale plant but is of reduced scale is employed to secure the kinetic data. The purpose is to derive rate equations from experiments that simulate the process conditions of the plant to be designed. The detailed reaction mechanism involved is not considered. Instead, emphasis is placed on development of rate expressions that can be utilized to design the full-scale reactor system by the scale-up technique. The rate equations thus derived may be described in terms of process variables (e.g., temperature, pressure, feed-reactant composition, etc.) as well as parameters associated with physical rate processes (e.g., mass-transfer coefficient), with empirically determined reaction order and pseudo-rate constants. This method of approach generally permits the derivation of reliable rate equations without requiring an excessive amount of time when the reaction system is highly complex, involving extensive interactions between the chemical and physical rate processes. Numerous examples of such a reaction system are observed in coal-conversion processes. The results from this method, however, cannot be dependably extrapolated beyond the ranges of the process conditions tested.

3. *Empirical-statistical method.* This method develops empirical equations for the reaction system on the basis of statistical analysis of the kinetic data obtained from statistically designed experiments (e.g., factorial, fractional-factorial, and Box-Wilson; see Sec. 2: "Box-Wilson Experimental Designs" for details), using the scaled-down reactor as in the preceding method. A limited number of process variables with selected levels of variation are imposed in the experiments to generate sufficient data to develop statistically significant correlations that may express the rates of reaction and product yields in terms of these variables. The equations obtained in this way bear no direct relation to the reaction mechanism, nor do they describe the physical rate process involved.

Various types of experimental reactors for acquiring kinetic data by the methods outlined are described by Rase (D-1). Specially designed small-scale "microreactors" are suitable for method 1 when the reaction system can be operated under chemical-reaction-controlled conditions. The scaled-down (model) reactors are commonly used to expedite acquisition of the design data by methods 2 and 3 for subsequent scale-up to a full-scale reactor. Problems associated with small reactors, which frequently are caused by low flow and low holdup and reactor activity, are also discussed by Rase along with recommendations for mitigating the impact of these problems. A guide in tabular form for selection of a suitable experimental reactor is presented by Weekman (D-2). Selection is based on several criteria, including the reactor's facility in sampling and analysis and in maintaining isothermality, its ability to determine accurate residence-contact time and catalyst selectivity, and the severity of reactor construction problems.

Analytical and Monitoring Techniques and Equipment Selection of techniques and equipment for monitoring the progress of a reaction system is as important as selection of the experimental reactor type. Chemical analyses are commonly used in following reactions, mainly by determining concentrations of reactants and products in samples taken at predetermined locations and time intervals. Although the criteria for selection of monitoring techniques usually vary with the type of reaction involved, factors frequently considered are (1) compatibility with the reactor, (2) sample quantity requirement, (3) complexity, (4) accuracy, and (5) time required per analysis. The preferred technique would be one that does not require stopping the reaction and/or separate sampling of the reaction mixture and that can determine concentrations (or properties related to concentrations) directly in the mixture without interfering with the reaction itself. Some so-called online analytical techniques could satisfy these conditions.

There are types of reactions which require "unconventional" monitoring techiques, for example, fast reactions, reactions that produce unstable intermediates, and reactions in which one or more components are present in very low concentrations, all of which are difficult or impossible to follow by conventional chemical analyses. Among these unconventional techniques are nuclear-magnetic-resonance (NMR) spectroscopy, electron-spin-resonance (ESR) spectroscopy, electrochemical technique (e.g., polarography), and radioactive-tracer technique. These techniques and others have been gaining wider use.

Application of these techniques often requires the use of an elaborate setup including an electronic computer system and other sophisticated accessories which sometimes are considered a drawback. Advantages include (1) a minimum of time required for data acquisition, which may be automatically recorded; (2) continuous monitoring of reactions with practically no interference with the reactions; (3) capability to follow fast reactions; and (4) capability of some techniques to identify reaction intermediates which would help establish a plausible reaction mechanism. Presented in the following paragraphs is an outline of selected unconventional techniques. Detailed discussion of these techniques is beyond the scope of this section, and more extensive references such as those by Willard, Merritt, and Dean (D-3) and by Ewing (D-4) should be consulted. Useful information can be found also in a multivolume publication entitled

Techniques of Chemistry, which contains a comprehensive presentation of experimental techniques applicable to studies of reaction kinetics and mechanisms. Volume VI of this publication deals with methods for investigating rates and mechanisms of reactions (D-5).

Radioactive-Tracer Technique This technique makes use of a selected radioisotope which enables a given atom to be followed through a process of chemical or physical change. The radioactivity of the isotope as detected by the counting equipment makes it possible to determine even a trace quantity of a substance which cannot be measured by ordinary methods.

The first step in the use of radioactive tracers is to select the isotope and its chemical form for the particular reaction to be studied. Bases for selection of the isotope include its half-life, type of radiation, specific activity, radiation-energy level, and chemical and/or physical behavior. The chemical form of the isotope is governed by such factors as solubility and chemical reaction characteristics. The type of counting equipment is determined by the isotope and its chemical form.

Details of the radioactive-tracer technique can be found in a book by Kohl, Zentner, and Lukens (D-6) and an article by Campbell and Thompson (D-7).

Applications. Using ^{15}N as tracer over an iron catalyst at a total pressure of 73.3 kPa (550 torr) and temperatures of 305 to 340°C, it has been shown (D-8) that the chemisorbed N was in equilibrium with gaseous NH_3 but not with nitrogen. The rate-controlling steps were the N chemisorption and desorption, respectively, for the synthesis and decomposition of NH_3. A tracer technique employing ^{14}C has been utilized in following pyrolysis reactions of toluene (D-9) by determining the distributions of ^{14}C in the gaseous products and in the deposited carbon.

An apparatus capable of following chemical reactions occurring in the droplets of salt solutions has been reported (D-10). The basis of the technique is measurement of the change in radioactivity of the droplets and/or the effluent gases which have been prelabeled with a suitable radioisotope. Zinc-65 or ^{35}S was added to the reacting mixture as a tracer in the investigation of heterogeneous reactions between ZnO and tetramethylthiuram disulfide in xylene and rubber (D-11). ZnO was separated from a sample of the reacting mixture, its radioactivity determined, and the rate constant calculated. A technique and an apparatus using ^{131}I have been described in the investigation of iodination kinetics of lead and silver (D-12). Also discussed were the advantages of the technique over a quartz balance or autoradiography.

Nuclear-Magnetic-Resonance (NMR) Spectroscopy The basis of this technique is the resonance interaction between a high-frequency field and the nuclei of a compound placed in an external magnetic field. The nuclei of some isotopes having a spin quantum number of ½ possess a magnetic moment produced by the spin of the nucleus. Examples are ^{1}H, ^{11}B, ^{19}F, ^{31}P, ^{35}Cl, and ^{79}Br. Isotopes with both the number of neutrons and the number of protons being even (e.g., ^{12}C, ^{16}O, and ^{32}S) do not have any nuclear magnetic moment and cannot be detected by this technique. When electromagnetic radiation of a right frequency (the resonance frequency) is passed through the substance containing isotopes of the former category, it is absorbed by the nuclei which shift from a lower to a higher energy level. The NMR spectrum of the compound is then obtained by plotting the extent of absorption at different frequencies against the frequency. The position in the spectrum characterizes the structure of the group. Fine-structure information obtained with higher resolution furnishes knowledge about the environment of the atoms.

Applications. NMR data have been correlated with reactivity in the hydrolysis of alkyl vinyl esters in a series of aqueous organic solvents (D-13). In a study of kinetics of the spontaneous rearrangement of ionol quinobromide, the NMR spectra were utilized to establish first-order kinetics (D-14). The values of activation energy and frequency factor were also calculated. The application of high-resolution NMR to the study of equilibrium distribution of various molecular sizes and shapes (organic and inorganic) has been discussed by Van Wazer and Moedritzer (D-15). On the basis of the NMR results, the increase in the methoxy proton signal intensity of the ester in the esterification of acetic anhydride with methanol has been interpreted in terms of second-order kinetics (D-16).

Electron-Spin-Resonance (ESR) Spectroscopy This method makes use of the resonance phenomena resulting from a magnetic moment produced by the spin of an unpaired electron when present in free radicals. EPR (electron paramagnetic resonance) is another name for ESR. The principle of the ESR technique is similar to but more complicated than that of NMR. In the ESR spectrum, the number of observable absorption lines is equivalent to one plus twice the magnetic quantum number of the nucleus associated with the unpaired electron. Free radicals with very low concentrations of unpaired electrons can now be detected by the ESR. Examples are free radicals in polymers from free-radical-catalyzed polymerizations as well as those in materials irradiated by high-energy particles.

Applications. The ESR technique has been used to determine rates of exchange of Cu^{2+} and Zn^{2+} with the cation-exchange resins (polystyrene sulfonic acid) (D-17). The rearrangements in the coordination of Cr^{5+} and V^{4+} ions in polymerization catalysts in contact with various gases were followed by the same technique (D-18). Additional applications have been reported in the reaction between oxygen atoms and SO_2 at 299 ± 2 K under the total pressure of 93.3 to 400 Pa (0.7 to 3 torr) (D-19) and in the production of free radicals in the decomposition of hydrazine derivatives (D-20).

Absorption Spectrophotometry This technique measures the absorption intensity of electromagnetic radiation by a substance at various wavelengths. Three regions of spectrum are utilized in the measurement, namely, the ultraviolet and visible regions [200 to 800 nm (200 to 800 mμ)] and the infrared region (2.5 to 15 μm). In the infrared region, the magnitude of radiation energy would only initiate vibrations within the molecule. However, the energy level in the visible and ultraviolet regions is so high that electronic transitions may take place. The technique yields very specific information on the structure and identification of complex molecules. The measurement in the infrared region is less suitable for quantitative determinations than that in the ultraviolet and visible region.

Applications. Ultraviolet spectrophotometry has detected a sharp drop in the reaction rate between $KMnO_4$ and H_2O_2 with an increasing H_2O_2 concentration (D-21). This phenomenon was attributed to formation of $Mn(H_2O_2)_n^{2+}$, and a rate equation was derived from the results in terms of the MnO_4^- concentration. The technique has also been utilized in investigations of a number of other subjects including the free-radical mechanism of phenol conversion (D-22) and the hydrolysis of isopropyl phenylcarbamate in alkaline media (D-23).

Extensive applications of spectrophotometry in the infrared region have been made in studies of various surface characteristics of catalysts. An infrared spectrophotometer was used to monitor continuously the concentration of chemisorbed CO in the oxidation of CO over platinum catalysts (D-24). The results obtained by the same technique in the study of active centers of aluminosilicate catalyst and gamma alumina have indicated that both catalysts have electron acceptor centers (D-25). The H-bonded interaction between surface hydroxyl groups and the adsorbate carbonyl groups was confirmed by means of the infrared spectral data in the adsorption of acetone and acetaldehyde on silica (D-26).

Differential Thermal Analysis (DTA) The principle of this method is that the heat effect associated with a reaction is related to the rate of the reaction. Thus, measurement is made of the change in the liberation or absorption of heat by the test sample as a function of the temperature difference between the sample and a reference compound. Both the sample and the reference compound are subjected to gradual heating or cooling at predetermined rates. This is a versatile and relatively simple technique in characterizing various compounds.

Applications. A flowing-gas DTA method has produced three endothermic peaks in the DTA curves for the dehydration of $CuSO_4 \cdot 5H_2O$. Two exothermic peaks were obtained in the oxidation kinetics study of UO_2 by the same method (D-27). An application of DTA was reported (D-28) in the catalytic hydrogenation of benzene and diphenyl amine at high temperatures and pressures [~9.653 MPa (1400 lbf/in^2)]. Determination of kinetic parameters has also been carried out by means of the DTA technique for styrene polymerization (D-29) and for pyrolysis of polystyrene, polypropylene, and polyethylene (D-30).

Gas Chromatography The basis of chromatography is the separation of components of a sample owing to their differences in solubility or in adsorption in a stationary bed of material (either liquid or solid). When the sample (moving phase) is a gas, the technique is termed either gas-solid or gas-liquid chromatography, depending on whether the stationary phase is a solid or a liquid. Of the three methods of separation (elution, frontal, and displacement), the elution method is of practical importance. In this method, a sample is introduced into the carrier gas as a vapor which flows through the chromatographic system. Upon separation by the stationary phase, the sample components travel through the system at different speeds, entering a detector where individual components are identified.

Applications. Kinetic data for a side-chain chlorination of *o*-cresol have been obtained from the rate curves produced through the gas-chromatography technique (D-31). The same technique has been used in a study of coal pyrolysis in which the formation of methane and CO was found to follow a pseudo-first-order rate (D-32). By using a chromatography column as a reactor, a solid reactant (stationary phase) was subjected to pulses of a gaseous reactant, the course of reaction being followed simultaneously by chromatography (D-33).

Other Experimental Techniques Additional techniques frequently employed in reaction kinetics studies include mass spectrometry, polarography, the stopped-flow method, ultrasonics, the temperature-jump method, and thermogravimetric analysis. For details on how these techniques have been applied, refer to the original papers cited in the references D-34 and D-35.

ANALYSES OF REACTION KINETIC DATA

Testing of Mechanism Selection and design of an industrial reactor for a specific reaction is strongly dependent on knowledge of the reaction mechanism, which does not necessarily agree with the stoichiometric equation. Although there is no simple standard method for testing the reaction mechanism, a trial-and-error approach as illustrated below is generally employed.

Steps Followed in Establishing Reaction Mechanism

1. Postulate a simple mechanism with a corresponding stoichiometry.
2. If the stoichiometry appears to indicate the reaction to be one-step and elementary, proceed to acquire kinetic data and analyze them according to the integral method (see subsection "Derivation of Rate Equation").
3. For nonelementary reactions, assume that the overall reaction consists of several elementary reaction steps with formation of intermediate compounds.
4. Formulate a rate expression for each of the elementary steps, and sum up the individual rate expressions to describe the overall rate.
5. If the resulting rate expression agrees with the experimental kinetic data, the assumed mechanism is acceptable. Otherwise, continue to assume alternative mechanisms, and repeat step 4 until a desired degree of agreement is obtained.
6. Frequently it may be simpler to use a purely empirical approach to correlate the data (see "Derivation of Rate Equation").

Reaction Schemes and Type of Intermediates In postulating a reaction mechanism, consideration is given to a reaction scheme and types of intermediates based on the chemical characteristics of the compounds taking part in the reactions. Generally accepted are the chain and nonchain mechanisms. Types of possible intermediates produced in these mechanisms include transition complexes, highly reactive molecules, ions, and free radicals. Chain reactions are mostly carried by either free radicals or ions. The first three types of intermediates are normally involved in the nonchain reactions.

An example of a chain-reaction mechanism involving free radicals is hydrogen atom addition to propylene, toluene, or xylene (E-1):

$$H_2 \rightleftharpoons 2H\cdot \qquad \text{Initiation and termination}$$

$$H\cdot + RCH_3 \rightleftharpoons RH + CH_3\cdot \qquad \text{Propagation}$$

$$CH_3\cdot + H_2 \rightleftharpoons CH_4 + H\cdot \qquad \text{Propagation}$$

The following rate equation was derived from the above mechanism:

$$-r_{RCH_3} = K_1^{1/2} k_2 C_{RCH_3} C_{H_2}^{1/2}$$

which is reported to be valid over a wide range of reaction conditions. Additional examples of chain-reaction mechanisms and subjects relevant to free-radical chain reactions are found in Ref. E-2.

Formation of synthetic high polymers is industrially the most important example in which the reaction proceeds by a chain mechanism carried by free radicals and ions. A book by Lenz (E-3) is a good source for details on polymerization reactions. Reference E-4 contains a collection of papers pertaining to polymerization kinetics.

A nonchain mechanism involving a transition-type intermediate has been shown (E-5) to hold in the formation of chlorate in sodium hypochlorite solutions:

$$ClO^- + H^+ \underset{k_2}{\overset{k_1}{\rightleftharpoons}} HClO$$

$$2HClO \underset{k_4}{\overset{k_3}{\rightleftharpoons}} H_2Cl_2O_2$$

$$ClO^- + H_2Cl_2O_2 \xrightarrow{k_5} ClO_3^- + 2H^+ + 2Cl^-$$

If one assumes that the concentration of the hypothetical transition complex $H_2Cl_2O_2$ is constant and much lower than that of HClO, the rate of disappearance of ClO^- becomes

$$-r = ac^2C/(1 + bC)$$

where $a = 3k_3k_5/k_4$, $b = k_5/k_4$, c is the sum of concentration of HClO and $H_2Cl_2O_2$, and C is the ClO^- concentration.

The acid-catalyzed oxidation of isopropyl alcohol (E-6) follows a mechanism that produces ionic intermediates:

$$ROH_2^+ + H_2O_2 \rightarrow ROH + H_3O_2^+$$

$$H_3O_2^+ \rightarrow HO^+ + H_2O$$

$$HO^+ + ROH \rightarrow H^+ + HO\cdot + RO\cdot$$

$$HO^+ + ROH \text{ (or } H_2O_2) \rightarrow H^+ + \text{various organic compounds}$$

Molecular intermediates are generally very reactive, having very short lives, but they can be detected by special techniques.

Examples of Reaction-Mechanism Testing Various methods used in establishing reaction mechanisms and rate equations are found in the original papers cited in the preceding paragraph. In addition, books by Hougen and Watson (E-7), Frost and Pearson (A-1), and Levenspiel (B-10) may be referred to for detailed discussions on the same subject. An article by Kittrell and Hunter (E-8) describing statistical methods for chemical reaction modeling is another good reference.

Derivation of Rate Equation When the mechanism of a reaction appears to be relatively simple, the rate expression may be derived from experimental kinetic data by the integral or the differential method discussed later (isothermal condition). The acquisition of experimental data may be carried out by means of either a batch or a flow reactor. The former is generally used in homogeneous reactions because of its simplicity and versatility, whereas heterogeneous reactions are usually studied in a flow reactor.

Kinetic data indicating the extent of reaction at constant temperature are obtained by following the variation of the concentration of a selected component with time (batch reactor) or with location (flow reactor). Physical properties and other variables which bear a certain relation to the progress of the reaction can also be utilized (see discussion under "Analytical and Monitoring Techniques and Equipment").

The treatment of batch kinetic data varies, depending on whether the data have been obtained under constant-volume or variable-volume conditions. In a homogeneous constant-volume reaction system, the reaction rate is defined simply by Eq. (4-1*b*):

$$r_A = dC_A/d\theta \tag{4-1b}$$

On the other hand, when the volume V of the reacting mixture changes with the progress of the reaction, the form of the reaction rate expression in terms of concentration becomes somewhat involved. It can be shown from Eq. (4-1*a*) that

$$-r_A = \frac{C_A}{V}\frac{dV}{d\theta} + \frac{dC_A}{d\theta} \tag{4-37}$$

implying that variations of both the volume and the concentration with time are to be determined experimentally. Thus, it is obvious that integration of the simplest-order rate expression would be rather complicated under the variable-volume condition.

The difficulty in treating kinetic data under the variable-volume condition can be alleviated by following the approach discussed by Levenspiel (B-10). The technique assumes the volume V to be in linear relation with conversion and uses the fractional conversion x_A as the variable instead of C_A. Thus,

$$V = V_0(1 + \gamma_A x_A) \tag{4-38}$$

where V_0 is the initial volume of the reaction mixture and γ_A is defined as the fractional change in V between no conversion and complete conversion.

$$\gamma_A = (V_{x_A=1} - V_{x_A=0})/V_{x_A=0} \tag{4-39}$$

For example, the value of γ_A for a gaseous reaction $2A \rightarrow 3B$ under the isothermal condition is

$$\gamma_A = (3 - 2)/2 = 0.5$$

With the same reaction starting with one volume of inert gas (I) per two volumes of A,

$$2A + I \rightarrow 3B + I \qquad \gamma_A = (4 - 3)/3 = ⅓$$

It is to be noted that $V_{x_A=1}$ and $V_{x_A=0}$ cannot be computed from the stoichiometry for liquid-phase reactions.

By using the parameter γ_A, the definition of the reaction rate Eq. (4-1*a*) becomes, for the variable-volume case,

$$-r_A = -\frac{1}{V}\frac{dN_A}{d\theta} = -\frac{1}{V_0(1 + \gamma_A x_A)}\frac{d}{d\theta}N_{Ao}(1 - x_A)$$

or

$$-r_A = \frac{C_{Ao}}{1 + \gamma_A x_A}\frac{dx_A}{d\theta} \tag{4-40}$$

Integral Method (Constant Volume) This method of data analysis starts with the assumption of a simple reaction model, such as one of those listed in Table 4-1. For example, consider an elementary second-order reaction model,

$$2A \rightarrow \text{products}$$

If the isothermal condition is valid, the rate expression becomes

$$-r_A = -dC_A/d\theta = kC_A^2 \tag{4-41}$$

and the integrated rate equation listed can be used directly. Thus,

$$k\theta = 1/C_A - 1/C_{Ao} \tag{4-42}$$

There are two ways of utilizing this equation to test for the proposed model. The first is to compute the values of k at various experimental values of C_A and θ. Constancy of k values thus calculated confirms that the assumed model (second-order kinetics in this case) is correct. Otherwise, a different reaction model is indicated. An alternative is a graphical method of plotting $1/C_A - 1/C_{Ao}$ versus time θ. The slope of this plot being the value of k, a fairly straight line would imply the correctness of the assumed model.

Other reactions of simple orders shown in Table 4-1 may be analyzed by using a technique similar to the one described. The integral method may be used in analyzing a set of kinetic data which are somewhat scattered but may follow a relatively simple mechanism.

Kinetic-data analysis for the types of complex reactions summarized in Table 4-2 is more complex than for those in Table 4-1. However, the same approach can still be tried. As an example, consider the first-order reversible reaction $A \underset{k_2}{\overset{k_1}{\rightleftharpoons}} B$. This mechanism is considered valid if a plot of the data in the form of ln $[(C_{Ao} - C_{Ae})/(C_A - C_{Ae})]$ versus θ yields a straight line. The slope is then equal to $k_1 + k_2$, which combined with the equilibrium constant $K = k_1/k_2$ gives the values of k_1 and k_2.

Integral Method (Variable Volume) The basis of this method is the integrated rate equation containing the parameter γ_A defined by Eq. (4-39). Even with this simplification, integration of the rate equation is more complex than in the constant-volume case for the same reaction model.

The general form of the integrated rate equation is derived from Eq. (4-40),

$$\theta = C_{Ao}\int_0^{x_A}\frac{dx_A}{(1 + \gamma_A x_A)(-r_A)} \tag{4-43}$$

where $-r_A$ is the rate expression for the reaction model to be tested. For a first-order reaction in terms of the rate of disappearance of component A,

$$-r_A = kC_A = k\frac{N_A}{V} = k\frac{N_{Ao}}{V_0}\frac{1 - x_A}{1 + \gamma_A x_A} = kC_{Ao}\frac{1 - x_A}{1 + \gamma_A x_A}$$

Substituting this expression into Eq. (4-43),

$$k\theta = -\ln(1 - x_A) \tag{4-44}$$

Similarly, for a second-order reaction,

$$kC_{Ao}\theta = \gamma_A \ln(1 - x_A) + (1 + \gamma_A)\frac{x_A}{1 - x_A} \tag{4-45}$$

and for a zero-order reaction,

$$k\theta = (C_{Ao}/\gamma_A)\ln(1 + \gamma_A x_A) \tag{4-46}$$

From here on, the steps in testing the proposed reaction model are the same as those of the constant-volume case.

Differential Method This technique employs the differential rate expression directly in evaluating the assumed reaction model. The method is useful in dealing with complex mechanisms or when integration of the rate expression is difficult. However, data used in this method have to be more precise and extensive than those for the integral method.

The general principle of the differential method in treating a rate expression of nth order, for example, is to rearrange (under constant-volume condition) the equation

$$-r_A = -dC_A/d\theta = kC_A^n \tag{4-47}$$

into

$$\log(-dC_A/d\theta) = \log k + n \log C_A \tag{4-48}$$

The values of the derivative $dC_A/d\theta$ can be evaluated from the experimental data by means of a graphical or numerical method. Finite differences $\Delta C_A/\Delta\theta$ with small intervals $\Delta\theta$ may likewise be used to approximate the values of $dC_A/d\theta$. A plot of $dC_A/d\theta$ versus C_A on log-log coordinates would yield a straight line if the assumed reaction model is of the nth order. The values of k and n can then be determined from the resulting plot.

This method may be applicable to a reaction of the type $A + B \rightarrow$ products when initially equimolal concentrations of A and B are used, since

$$-r_A = kC_A^a C_B^b = kC_A^{a+b} = kC_A^n \tag{4-49}$$

Once this reaction mechanism has been confirmed and the values of n and k have been evaluated, the order with respect to individual reactants a and b can be determined as follows:

$$-r_A = kC_A^a C_B^{n-a} = kC_B^n(C_A/C_B)^a \tag{4-50}$$

or

$$\log(-r_A/kC_B^n) = a \log(C_A/C_B) \tag{4-51}$$

The order a is found by plotting Eq. (4-51) on log-log coordinates, and b is calculated as the difference between n and a.

If this method fails to fit the experimental kinetic data within normal experimental-error limits, probably a complex mechanism is indicated. In this case, partial information about the mechanism may be obtained by carrying out the reaction to a very limited extent (i.e., very low conversion of the reactant). There are two methods based on this principle. For example, assume that the reaction mechanism

$$A + B \underset{k_2}{\overset{k_1}{\rightleftharpoons}} D + E \qquad D + A \overset{k_3}{\rightarrow} \text{product}$$

is represented by the rate expression

$$-r_A = -\frac{dC_A}{d\theta} = \frac{k_1k_3C_A^2C_B + k_3C_D}{k_3C_A + k_2C_E} \tag{4-52}$$

Method 1. *Initial rate method.* With different initial concentrations of reactants, the rates of reaction are determined when the reaction has progressed to a limited extent such that the product concentrations C_D and C_E may be ignored. Then, from Eq. (4-52),

$$-r_A = k_1C_AC_B \tag{4-53}$$

Confirmation of this rate expression and evaluation of k_1 may be made following the steps described by Eqs. (4-49) through (4-51). As can be seen from this example, the initial rate method is useful in the preliminary screening of various mechanisms. However, establishment of a complete mechanism cannot be accomplished by this method alone.

Method 2. This is a modification of method 1. Again, referring to the same example, the first set of initial rate data is obtained with only reactants A and B present. Then, another series of runs is made, starting with compounds D and E alone, to acquire the second set of initial rate data for the reverse reaction. The first set of data would give Eq. (4-53) while the second set corresponds to

$$-r_D = (k_3/k_2)(C_D/C_E) \tag{4-54}$$

since C_A and C_B are extremely low with the second set of initial rate measurements. Examples of application of the initial rate method in kinetic studies on industrial processes may be found in Ref. E-9.

The preceding discussion is based on the constant-volume case. However, the same technique is applicable to reactions under the variable-volume condition. The only modification required is the use of the rate expressed by Eq. (4-40) (instead of $dC_A/d\theta$) if Eq. (4-38) is valid.

Empirical Approach When the reaction mechanism is difficult to determine, an empirical approach may be employed. One technique using this approach includes a purely mathematical manipulation of the "curve-fitting" method.

This method first determines a proper form of the empirical equation, e.g., the simple linear form ($y = a + bx$), the logarithmic form ($y = c + ax^n$), the semilogarithmic form ($y = ae^{bx}$), the reciprocal form [$y = x/(a + bx)$], etc. The form of the empirical equation is determined by a trial-and-error technique which frequently involves plotting the kinetic data on different types of graph papers in an attempt to produce a straight line.

The next step is to evaluate the constants in the empirical equation that fit the data. This is done either graphically or analytically by the method of averages or the method of least squares. Reference E-10 presents a comprehensive treatment of the subject on development of empirical correlations from multifactor data, including statistical analysis of errors and goodness of fit. Extensive use of computer programs to facilitate the curve-fitting procedure is discussed.

Further details on the testing of mechanisms and derivation of rate equations can be found in books by Frost and Pearson (A-1), Levenspiel (B-10), Walas (E-11), Smith (A-29), Hougen and Watson (E-7), and Hill (E-12).

Evaluation of Rate Constant and Arrhenius Parameters

Rate Constant For simple reactions (e.g., Table 4-1) the calculation of the rate constant is fairly straightforward once the rate expression has been confirmed by one of the methods discussed earlier. The rate constant k can be determined by taking an average of values calculated at individual experimental points, using either an integrated form or a differential form of the rate equation. The graphical method can likewise be used to determine the rate constant as the slope of the straight line when the experimental data are plotted according to the integrated rate equation listed in Table 4-1. If the plot is made on the log-log coordinates based on a differential rate equation such as Eq. (4-48), the value of k is obtained from the plot.

In complex reactions, since there is more than one rate constant, no simple general method is available for evaluating all the rate constants, and each reaction has to be treated by a special method. For example, in reversible reactions (see Table 4-2), the forward and reverse rate constants can be determined by a technique similar to the one for simple reactions if the equilibrium constant is incorporated in the rate equation [see example in last paragraph of subsection "Integral Method (Constant Volume)"].

Arrhenius Parameters These parameters include the activation energy E and the frequency factor A which characterize the Arrhenius equation

$$k = Ae^{-E/RT} \tag{4-3}$$

Although Eq. (4-3) is about 90 years old, it predicts the effect of temperature on the rate constant (for simple reactions) so accurately that it still finds wide application in a great number of reaction kinetics problems.

More elaborate expressions for improving the Arrhenius equation have been derived from various theories such as the collision theory and the transition-state theory. Generally these assume the form

$$k = k_0T^me^{-E/RT} \tag{4-55}$$

The value of m ranges from 0 to 3 or 4. For most reactions the effect of the exponential term is so much greater than that of the T^m term that Eq. (4-55) essentially reduces to Eq. (4-3).

Arrhenius parameters E and A are evaluated from the measurements of rate constants at several different temperatures. There are two methods commonly used. One is to plot Eq. (4-3) as in Fig. 4-3 and calculate the value of E from the slope of the straight line. This is based on the following equation derived from Eq. (4-3):

$$\ln k = \ln A - E/RT \tag{4-56}$$

The value of A is then computed from Eq. (4-3). A significant deviation from a straight-line plot frequently implies that the reaction is complex.

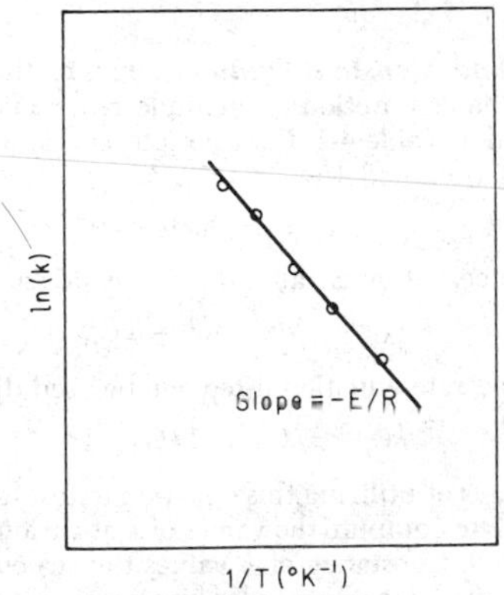

FIG. 4-3 Plot of Arrhenius equation.

The second method calculates E directly from adjacent pairs of k and T by simultaneously solving two sets of the Arrhenius equation

$$k_1 = Ae^{-E/RT_1} \quad \text{and} \quad k_2 = Ae^{-E/RT_2}$$

From these two equations,

$$E = \frac{R \ln (k_2/k_1)}{1/T_1 - 1/T_2} \tag{4-57}$$

A is computed from E values as before. More detailed discussion on the methods for estimation of rate parameters is presented by Benson (E-13).

Identification of Rate-Controlling Steps As outlined previously (see discussion under "Principles of Reaction Kinetics"), rate-controlling steps can be physical or chemical in nature, and are found in both homogeneous complex reactions and heterogeneous reactions.

Homogeneous Complex Reactions The presence of a rate-controlling step in this type of reaction can be identified through kinetic evidence that may be obtained by following the formation and disappearance of intermediates as well as the end products. Various experimental techniques have been employed in accomplishing this task, including ordinary "wet" chemical analysis, radioactive-tracer techniques, resonance method, absorption spectrophotometry, and others (see subsection "Analytical and Monitoring Techniques and Equipment"). The selection of a suitable technique depends on the types of reaction and intermediate. For example, the ESR technique is capable of measuring the concentration of free radicals. It also identifies the structure of free radicals. References given in the discussion under the previously mentioned subsection and under "Testing of Mechanism" should be consulted for details on how the rate-controlling steps may be identified using different techniques.

Heterogeneous Reactions It has been shown that heterogeneous reactions in general exhibit more complex behavior than homogeneous reactions because of the interaction between the physical and the chemical processes. The situation is even more complex with fluid-solid catalytic reactions. Accordingly, identification of the rate-controlling step is more complicated than with homogeneous reactions.

Consider a simple fluid-solid reaction of the type

$$aA(\text{fluid}) + bB(\text{solid}) \rightarrow \text{products}$$

The reaction is of the first order with respect to fluid A, and no ash is formed on the solid surface. Thus, the rate of chemical reaction r_A based on unit surface area of B is

$$-r_A = k_s C_{As} \tag{4-58}$$

and the rate of mass transfer of A through the fluid film by diffusion to the surface of B is

$$-r_d = k_m(C_{Am} - C_{As}) \tag{4-59}$$

where C_{As} and C_{Am} are the concentrations of A at the interface and in the bulk fluid respectively, k_s is the reaction rate constant based on unit surface area, and k_m is the mass-transfer coefficient. Under steady-state condition $-r_A = -r_d$, from which

$$C_{As} = \frac{k_m}{k_m + k_s} C_{Am}$$

This expression and either of the preceding two equations gives

$$-r_A = -r_d = -\bar{r} = \frac{k_s k_m}{k_s + k_m} C_{Am} \tag{4-60}$$

Equation (4-60) shows that when $k_s \gg k_m$,

$$-\bar{r} = k_m C_{Am}$$

which implies that mass transfer is the controlling step. On the other hand, if $k_m \gg k_s$,

$$-\bar{r} = k_s C_{Am}$$

which means that the chemical reaction is controlling the overall process.

Many industrially important heterogeneous reactions are more complex than in the example just discussed. The simplified qualitative guidelines generally used in setting up an experimental program and in evaluating the results may be summarized as follows.

1. *Chemical-reaction-controlling.* (*a*) A series of experimental runs at different temperatures frequently reveals whether a chemical reaction is controlling or not. This is based on the observation that most chemical reactions are considerably more sensitive to temperature change than physical processes. (*b*) The rate of fluid flow and particle sizes (of solid in fluid-solid reactions and of the dispersed phase in fluid-fluid reactions) usually have a much less pronounced effect on the chemical reaction than on mass transfer.

2. *Mass-transfer-controlling.* This phenomenon is generally identified by experiments using various particle sizes and flow rates of fluid and/or different degrees of agitation. In fluid-solid reactions, when a tightly adherent ash layer (solid product plus inert) is formed on the solid, the resistance of the ash layer usually predominates over the fluid-film resistance. However, a significant increase in the rate of the overall process with an increasing fluid velocity would indicate that mass transfer through the fluid film is controlling.

3. *Change in rate-controlling steps.* The steps that are rate-controlling can change in the course of the reaction, and they can overlap one another at certain stages. This situation could be due to changes in the composition of the reacting mixture as well as in particle size with the progress of the reaction.

Rate-controlling steps in solid-catalyzed reactions have been mentioned briefly (see subsection "Heterogeneous Catalysis" and the subsequent discussion). Comprehensive treatments of the subject are found in articles by Corrigan (E-9), Hougen and Watson (E-7), and Levenspiel (B-10).

SCALE-UP METHODS

It has been generally accepted that the design of a commercial-scale chemical reactor, which is the heart of a chemical plant, cannot be accomplished by a purely theoretical approach alone. To start with, at least laboratory and/or pilot-plant data on the reactions involved must be available. A satisfactory scale-up procedure may require a stepwise empirical approach in which the size of the reactor is increased successively, with a desired commercial size as the goal. This approach would result in cost and time before a commercial-scale plant could be constructed.

Basically, the rate of a given chemical reaction is independent of the size and structure of a reactor. The chemical reaction rate, however, is influenced by physical processes, e.g., mass transfer and heat transfer, which are usually controlled by the size and structure of the reactor. Thus, a chemical reaction is indirectly affected by the reactor type and scale in a manner that is usually unpredictable on a quantitative basis. These effects may result in a change of the overall rate of the same reaction, in a different product distribution, or both.

To deal with a complex relation between the overall rate of reaction and the type and size of the reactor by the conventional empirical method of scale-up would be time-consuming and expensive. Consequently, a number of alternative semiempirical methods have been proposed to alleviate this problem. Among these are the method of chemical similitude (dimensional similitude applied to chemical reactions) and the method of mathematical modeling, which is facilitated by the availability of high-speed computers. These methods still require the basic kinetic data from bench-scale and/or pilot-plant scale studies as well as practical experience.

Chemical Similitude This technique was originally developed by Damköhler (F-1) and expanded by several others including Bosworth (F-2). Walas (E-11) has presented a condensed version of Bosworth's treatment on chemical similitude.

The bases of this method are several dimensionless groups derived from the application of the laws of conservation of mass, momentum, and energy to the chemical reaction system. To illustrate the principles of the method, the following simplifying assumptions have been made by Walas:

1. The reaction system under consideration is a gas-phase first-order reaction of the type $A \rightarrow 2B$.
2. The analysis of the laws of conservation is based on a unit volume of a circular cylindrical reactor of radius $\bar{R}$ and length L.
3. Steady-state condition, i.e., no accumulation of mass, momentum, and energy with respect to time, prevails.

By introducing these assumptions into the partial differential equations (F-2) describing the conservation of mass, momentum, and energy, seven dimensionless groups shown in Table 4-9 are derived. The values of these groups for a model reactor and a prototype reactor must be the same to satisfy the condition of similitude.

TABLE 4-9 Dimensionless Groups in Chemical Reaction*

Homogeneous reaction	Heterogeneous reaction	Name of group
(a) $\frac{rL}{uC}$	(h) $\frac{r}{SC}$	Damköhler
(b) $\frac{r\overline{R}^2}{DC}$	(i) $\frac{rD_p^2}{DC}$	
(c) $\frac{u}{rL\,\Delta V}$	(i) $\frac{S}{r\,\Delta V}$	
(d) $\frac{\overline{R}u\rho}{\mu}$	(k) $\frac{D_pSL\rho}{\mu}$	Reynolds
(e) $\frac{C_pu\overline{R}^2\rho}{\lambda L}$	(l) $\frac{c_pS\rho\overline{R}^2}{\lambda + \frac{4}{3}D_p\rho\sigma T^3}$	Peclet
(f) $\frac{Qr\overline{R}^2}{\lambda\,\Delta T}$	(m) $\frac{Qr\overline{R}^2}{(\lambda + \frac{4}{3}D_p\rho\sigma T^3)\Delta T}$	
(g) $\frac{Qr}{\Delta T(4\alpha\epsilon\sigma T^3 - C_p\rho r\,\Delta V)}$	(n) $\frac{Qr}{C_p\rho r\,\Delta V\,\Delta T}$	

*From Walas, *Reaction Kinetics for Chemical Engineers*, McGraw-Hill, New York, 1959.

All nomenclature in Table 4-9 is consistent with the prior discussion. C is concentration in mol/m^3 (mol/ft^3); L is reactor length in m (ft); u is velocity in m/s (ft/s); r is the rate of reaction expressed as mol/(s·m^3) [mol/(s·ft^3)]; $\overline{R}$ is the reactor radius in m (ft); D is diffusivity in m^2/s (ft^2/s); ΔV is the volume change per unit disappearance of the reactant for which r is written in m^3/mol (ft^3/mol); ρ is the density in kg/m^3 (lb/ft^3); μ is the viscosity in N/(m·s) [lbf/(ft·s)]; C_p is the heat capacity in J/(kg·K) [Btu/(lb·°F)]; λ is the thermal conductivity in J/(s·m^2)(K/m) [Btu/(s·ft^2)(°F/ft)]; ΔT is the temperature in excess of the wall temperature in K; Q is the heat generated by reaction in J/mol (Btu/mol) reacted; D_p is the particle diameter in m (ft); S is the space velocity with units of s^{-1}; a is the wall surface area per unit of reactor volume in m^{-1} (ft^{-1}); σ is a proportionality constant of 57.75 kJ/(s·m^2·K^4) [0.484 Btu/(s·ft^2·°F^4); and ϵ is the emissivity or absorptivity of the reactor wall and is dimensionless.

In Table 4-9 groups (a) and (h) represent the ratio of the average time required for the reaction mixture to flow through the reactor to the mean life of a reacting component. Group (b) may be interpreted as the ratio of the average time needed for reacting molecules to diffuse across the reactor to the mean life of a reacting component. Group (i) has a meaning similar to that of (b) for a heterogeneous reactor. The term $\frac{4}{3}D_p\sigma T^3$ is the ratio of the contribution by thermal radiation to the effective thermal conductivity.

Reliable scale-up of a chemical reactor by means of dimensional similitude is usually limited to the relatively simple reaction system, e.g., first-order homogeneous reactions. In general, it is impractical, if not impossible, to have complete dimensional similitude in a chemical system. This is mainly attributed to the impossibility of realizing a condition such that the overall rate of a chemical reaction in reactors of widely different scales can be subjected to the influence of physical parameters to the same extent. This situation may be possible only in a case involving a fast reaction so that the overall rate is controlled by the physical rate process (i.e., mass transfer, heat transfer, etc.). However, in this case the significance of chemical similitude (of dimensionless groups containing reaction-rate parameters) diminishes, and dynamic and/or thermodynamic similitude (of groups containing force and/or thermal properties) plays the key role. At the other extreme, when the chemical reaction offers the greatest resistance in the overall rate, chemical similitude (i.e., of the Damköhler's group in particular) becomes important.

Partial dimensional similitude is also possible under special conditions listed in Table 4-10 for which some of the dimensionless groups may be omitted. It is to be noted that the relative importance of the dimensionless groups shown in Table 4-9 can change with operating conditions. This is especially true with respect to temperature.

TABLE 4-10 Special Cases Requiring Limited Similitude*

	Dimensionless groups of Table 4-9 that may be ignored	
Condition	Homogeneous	Heterogeneous
Batch, or low-flow, rate	a, c, d, e	h, j, k, l
Small diffusional resistance	b	i
Constant volume	c, g	j, n
Small radiation heat transfer	g	n
Adiabatic, with small heat of reaction	f, g	m, n

*From Walas, *Reaction Kinetics for Chemical Engineers*, McGraw-Hill, New York, 1959.

As complete dimensional similitude is nearly impossible, the net result of requiring chemical similitude is to fix automatically the reactor geometry and other physical parameters without complete satisfaction of other aspects of dimensional similitude. Rase (D-1) and Carberry (B-8) present comprehensive discussion on the subject of scale-up by dimensional similitude for homogeneous and heterogeneous reactors. Limitations of the technique as well as methods for simplification are described.

Mathematical Modeling One of the shortcomings of the method of dimensional similitude is that it reveals no direct quantitative information on the detailed mechanisms for the various rate processes concerned. Accordingly, extrapolation of the results from dimensional similitude is usually not recommended.

This shortcoming may be overcome by utilizing basic laws of physical and chemical rate processes to describe mathematically the operation of the reactor. The resulting mathematical model generally consists of a set of differential equations that are frequently so complex that they are impossible to solve by analytical methods. However, the solution of the mathematical model derived can often be accomplished by simulation of the model with analog or digital computers, or both. The success of this approach is governed not only by the completeness and reliability of the mathematical model but also by the amount and accuracy of the basic experimental data available.

Steps to Follow in Scale-Up by Mathematical Modeling There is no rigid procedure for scale-up of a chemical reactor by mathematical modeling. The steps to follow would vary from one specific case to another. However, an example of the stepwise approach to the reactor scale-up by modeling can be found in an article by Boreskov and Slin'ko (F-3). Additional examples for specific cases are found in such publications as those by Carberry (B-8) and Rase (F-4).

Consider the scale-up problem dealing with a catalytic reactor having a fixed bed of solid catalysts. The sequential steps applicable to this case are summarized in the following paragraphs.

Formulation of mathematical expressions. Various processes involved in fixed-bed catalytic reactions are represented by mathematical descriptions derived from basic laws of physical and chemical rate processes.

Since the size, pore structure, and inner surfaces of the catalyst pellets play important roles in the reaction, a process model for a single catalyst pellet is first derived. This may be done through analysis of laboratory experimental data for various particle sizes in combination with pore-structure analysis, if known. Selection of the optimum pore structure, size, and shape of the catalyst pellet is then made, taking into consideration such factors as the cost and activity of catalyst and the hydraulic resistance.

The mathematical description of the process for a layer of catalyst pellets is next formulated, based on the process model for one catalyst pellet incorporating corrections for nonuniformity of temperature and concentrations of the reaction mixture throughout the layer.

Process optimization. In this step mathematical methods of optimization (see Sec. 2) are employed to determine theoretical optimum conditions for the entire catalyst bed for a number of key parameters such as temperature, pressure, and composition. Such optimization methods frequently deal with conversion and product distribution. However, limitations from the technical viewpoint, e.g., the temperature limit, the explosive limits for the composition of the reaction mixture, etc., should also be taken into account. Reference should also be made to review articles (D-34, F-5) for various methods of optimization as applied to different chemical processes and reactor-design problems.

Preliminary selection of reactor types. In general, the optimum conditions established in the preceding step cannot be met exactly in real reactors. Therefore, the selection of reactor type would be made so as to approximate the optimum conditions as closely as possible. For this purpose, mathematical models of the process in several different types of reactors are derived. The optimum condition for selected parameters (e.g., temperature profile) is then compared with those calculated from the mathematical expressions for different reactors, and the type of reactor that most closely approaches the optimum would be selected.

For example, in the oxidation of ethylene to produce ethylene oxide, the optimum temperature programming would be to increase the temperature in accordance with the degree of conversion. This condition is imposed to minimize the complete combustion of ethylene, which requires a high activation energy. One type of fixed-bed catalytic reactor that can approximate this temperature program would be the one with stagewise catalyst beds having internal heat exchangers between the stages. It is to be noted that the operating conditions of the reactor should be determined to achieve the maximum profit since optimum operating conditions for the reactor may conflict with overall plant economics.

Process stability and parametric sensitivity. Stability of a chemical reactor for an exothermic process refers to a condition for which the rate of heat removal is equal to or greater than that of heat generated by the reaction and for which both temperature and concentration have no tendency to oscillate with increasing amplitudes. Stability is one of the important factors in the selection of a reactor and may be determined qualitatively from the mathematical model of the reactor (F-3).

Parametric sensitivity has been discussed by Bilous and Amundson (F-6) for the condition of a chemical reactor under which thermal behavior is very sensitive to small changes in process parameters. For example, a slight variation in the feed concentration or the reactor inlet temperature could cause a significant change in the temperature profile in the reactor. The parametric sensitivity of the reactor may be determined through the mathematical description of the process.

Further information on stability and parametric sensitivity of chemical reactors can be found in the references (B-9, D-34, F-5, F-6, F-7, F-8).

Final selection of reactor system. At this stage, technical as well as economic considerations are incorporated with the results from the preceding steps in order to make a decision on the reactor system as to the size of the experimental reactor and its operating conditions. The data from the experimental reactor would be used to make appropriate corrections for the mathematical model derived in the preceding steps. At this point, it may be desirable to go back to the previous steps for revision of earlier results.

To investigate the nonideality of the actual reactor system under the influence of physical-rate processes, some simulation method (e.g., the hydraulic analog) that excludes chemical-reaction effects could be employed. The reaction kinetic expression of the model (which can be made independent of the change in scale) in combination with correction for any nonideality can now be used to predict behavior of a large-scale reactor system.

Dimensional Similitude Combined with Mathematical Modeling This technique has been developed in an attempt to combine some of the advantages of both dimensional similitude and mathematical modeling. It is particularly useful when the complexity of the reactor system and the processes concerned makes a complete mathematical description of the system impossible. Thus, through this method some of the critical parameters for scale-up can be specified, and it may be possible to characterize the underlying rate processes quantitatively.

Examples of Scale-Up in Practice The published literature on the application of scale-up techniques is in general rather limited in comparison with literature on fundamental studies. The inherently complex nature of chemical rate processes has made scale-up problems of chemical reactors difficult. Published examples of scale-up in chemical processes and reactors in recent years have been largely based on the method of simulation by mathematical modeling. Summarized in the following paragraphs are some of the scale-up examples. Several case studies dealing with industrially important reactions presented by Rase (F-4) include examples to illustrate application of different scale-up methods.

Scale-Up by Mathematical Modeling Popularity of this technique has been stimulated by the availability of high-speed digital and analog computers.

Example 1 Isoprene Process Scale-up of the isoprene process from a bench-scale to a commercial-scale plant has been discussed by Garmon, Morrow, and Anhorn (F-9). The process consists of four key steps: (*a*) dimerization of propylene to 2-methyl-1-pentene, (*b*) isomerization of the latter to 2-methyl-2-pentene, (*c*) pyrolysis of 2-methyl-2-pentene to isoprene and methane, and (*d*) superfractionation to recover isoprene.

Mathematical modeling was applied to the initial step, dimerization, in a two-stage tubular reactor. A set of mathematical expressions was derived to describe (*a*) the temperature profile of the reactor system, (*b*) the net heat flow of process fluid, (*c*) the rate of dimer production, (*d*) the fluid density in the reactor, and (*e*) the reaction rate and conversion. Simulation by means of a digital computer under various operating conditions produced data that compare quite favorably with commercial-plant data. A 17,000-fold scale-up based on the bench-scale dimerization reactor was accomplished by this approach.

The isomerization step involved no mathematical modeling. Instead, a bench-scale experimental program was conducted in a heterogeneous catalytic reactor with a few selected catalysts in order to investigate the catalyst life, critical ratios of diameter to length for catalyst beds, and on-stream cycle times. Scale-up factors of 1000 to 7000 were achieved.

In the pyrolysis step, equipment of two different scales, including a small isothermal unit and a radiant furnace, was employed in bench-scale studies. The results from the latter established the most desirable firing patterns and the optimum product distribution at maximum isoprene yields.

The final step is the recovery of isoprene by fractionation, for which fairly accurate vapor-liquid equilibrium data were obtained for the hydrocarbons involved. The actual plant separation columns represented a 10,000-fold scale-up from a 7.62×10^{-2}-m (3-in) packed column.

Example 2 Applications of mathematical modeling to scale-up problems presented by Schoenemann (F-10) have dealt with three industrially important complex processes. Commercial-scale plants designed with this technique gave results which were in satisfactory agreement with those predicted. The balance sheets (mass, energy, and momentum), the prior decision on the accuracy required in both the theoretical calculations and experimental programs, and the preliminary estimates in design calculations based on the experimental data are the three important principles emphasized by the author.

a. Hexogen synthesis. Hexogen, $(CH_2)_3N_3(NO_2)_3$, is an explosive product formed by reaction between hexamethylenetetramine and a large excess of HNO_3 in the presence of ammonium nitrate. To retard one of the undesirable reactions leading to formation of N_2O, the reaction temperature was set at 75°C on the basis of analysis of Arrhenius plots. The temperature of the reaction mixture is critical because a slight increase in temperature would accelerate the highly exothermic reaction forming N_2O. To understand the thermal behavior of an industrial reactor, the heat balance was studied by employing a continuous agitated reactor with efficient back mixing. The required cooling surface of the commercial-scale reactor was then calculated. The heat balance, together with mathematical expressions and experimental data for other processes, has made possible scale-up of a commercial-scale plant producing 200 t per month directly from a bench-scale system.

b. Production of furfural from xylose. The process involves selective hydrolysis of wood, which is so complex that a highly simplified model has been used to represent the kinetics of the overall reaction. Consideration of the reaction mechanism consistent with the experimental data resulted in the following expression:

$$\text{Xylose} \xrightarrow{k_1} \text{intermediate} \xrightarrow{k'_1} \text{furfural} \xrightarrow{k_3} \text{resins}$$

$$\text{intermediate} \searrow \text{condensation} \qquad \text{furfural} \xrightarrow{k_2} \text{product}$$

A set of rate equations was derived for this mechanism by assuming that the intermediate concentration was independent of time. These equations indicated that controlling the reaction parameters alone would not give a satisfactory yield on a commercial-scale operation. Preliminary calculations indicated that the rate of the two undesirable reactions would be reduced by means of a continuous stagewise liquid-liquid extraction of furfural with Tetralin. A simplified rate expression was then derived to describe the formation of furfural. Process optimization was carried out by varying volumes and temperatures. A yield of 72 percent was obtained at the optimum condition. The accuracy of all the calculated results has been confirmed by experimental data from the pilot plant.

c. *Ethylene polymerization under high pressure.* The polymerization of ethylene is generally conducted in the temperature range from 160 to 300°C under pressures ranging from ~122 to 253 MPa (1200 to 2500 atm), using a small amount (~0.01 percent) of oxygen as the initiator. Both agitated reactors and tubular reactors are used, and both processes are characterized by a low conversion (8 to 15 percent).

The scale-up problem was concerned with designing reactors for producing 24,000 t of polyethylene per year. Pilot-plant data available included the temperature profiles of the reaction mixture and of the cooling water and the total conversion. One of the critical factors is control of the large amount of heat [3.76 MJ/kg (900 kcal/kg) of polyethylene] evolved. The reaction is quite sensitive to temperature and pressure as well as to the initial oxygen concentration.

To correlate the reaction rate with temperature, pressure, and degree of conversion, a simple radical chain mechanism was first postulated. Rate expressions were derived for the three individual steps, i.e., initiation, propagation, and termination. The overall reaction rate was developed from the three equations. Also derived was the rate of oxygen consumption. Three sets of curves showing the effect of key variables (heat removal, oxygen concentration, and pressure) on the progress of the reaction were obtained through computation by using the rate expressions so derived and estimated values of rate constants.

The next step was to confirm and improve the postulated rate equations by a bench-scale experimental investigation by means of a system that consisted of a rocking autoclave, a pressure multiplier, and accessories. The improved kinetic model resulted in the differential equations which were integrated numerically along the length of the reactor to predict the polymerization behavior in a commercial-scale reactor.

A commercial-scale plant with an annual capacity of 24,000 t of polyethylene was constructed, based on this scale-up technique without the need for a "semi-industrial" stage.

Example 3 Continuous Agitated-Tank Reactor System The fundamentals of scale-up for continuous agitated-tank reactor systems have been presented by Weber (F-11). The method utilizes pilot-plant batch data, and deals with both a scale-up problem and a problem of conversion from a batch to a continuous reactor system. The problems taken into consideration included reaction kinetics, economics, short circuiting, and heat transfer. Another paper (F-12) considered cases in which the pilot-plant stage may be omitted. Optimum operating conditions were determined through a limited economic analysis based on results computed from the mathematical model. The system stability was analyzed, and the design of controllers for the system was discussed.

Statistical Method of Scale-Up A scale-up method which is a modification of existing statistical techniques has been described by Novak, Lynn, and Harrington (F-13). The method designs the experimental program on the basis of information content and provides for a sequential program of investigation. The basic assumption is the presence of scale-up differences between separate stages of development. The method would direct the operating conditions toward the optimum area. A batch polymerization reaction was used as the example.

Scale-Up by Combined Method The technique of using dimensional similitude in combination with mathematical modeling was employed (F-14) in the design of a pilot plant and in evaluating the results to provide the basis for scale-up to a commercial-scale plant. A differential equation describing the material balance around a section of the system was first derived, and the equation made dimensionless by appropriate substitutions. Scale-up criteria were then established by evaluating the dimensionless groups. A mathematical model was next developed, based on the kinetics of the reaction, describing the effect of process variables on the conversion, yield, catalyst activity, etc. Kinetic parameters were determined by means of both analog and digital computers. A reaction of the type: $A \underset{k_{-1}}{\overset{k_1}{\rightleftharpoons}} B \underset{k_{-2}}{\overset{k_2}{\rightleftharpoons}} C$ was used as an example. The computer simulation also covered such operations as catalyst regeneration, product separation, and recycle, for the purpose of developing a realistic commercial-scale plant.

Scale-Up by Conventional Method Development of a production-scale fluidized-bed calciner [1.22-m (48-in) diameter] for the conversion of aqueous radioactive wastes to a solid form resulted from pilot-plant studies of three different scales [0.152-, 0.305-, and 0.61-m (6-, 12-, and 24-in) diameters] (F-15). Operating data from the pilot plants and the production plant were compared with respect to particle-size control, in-bed heat transfer, product properties, and fluidization characteristics. Most of the pilot-plant data were applicable to the production unit.

Another conventional scale-up techique has been illustrated (F-16) in the design of fluidized-droplet reactor systems that are used in such reactions as the alkylation and isomerization of hydrocarbons catalyzed by H_2SO_4 or HF. Three geometrically similar reaction vessels [0.19-, 0.457-, and 1.02-m (7.5-, 18-, and 40-in) diameters] were employed in the scale-up studies. An insight into the behavior of a commercial-scale system was achieved by conducting the catalyst holdup experiments at higher superficial velocities than those in the 1.02-m vessel.

A 1976 publication (F-17) describes the method of approach used in scale-up of a steam pyrolysis reactor (furnace) from the bench-scale unit to a full-scale commercial reactor having a nominal ethylene capacity of 25,000 t per year. The reactor produces olefins (with emphasis on ethylene) from a variety of feedstocks ranging from light naphthas to heavy vacuum gas oils. Of several process variables that affect yields and the product distribution, residence time (0.01 to 0.1 s) and reaction temperature (760 to 954°C) were found to be most critical, and these two were used as the key variables in the scale-up procedure. This represents a scale-up factor of ~10^2 from the bench-scale to the pilot plant and ~10^3 from the pilot plant to the commercial facility.

REACTOR DESIGN: BASIC PRINCIPLES AND DATA

Since all chemical processes are centered in the chemical reactor, one of the most important factors in determining overall process economics is the design of the reactor. Unlike equipment for mass- or heat-transport processes, there is no straightforward method for designing equipment to carry out a chemical reaction. This implies that the reactors must be designed to fulfill diverse requirements of specific reaction systems, as discussed later.

At the start of the design work, the following information is presumably available: (1) the reaction type (simple or complex), (2) the need for catalyst, (3) phases involved, (4) the mode of temperature and pressure control (isothermal, adiabatic, or other; the need for pressure, vacuum, etc.), and (5) the production capacity. In addition, basic data required include (1) the chemical rate expressions and variation of rate parameters with temperature, pressure, etc., (2) the heat- and mass-transfer characteristics, (3) physical and chemical (e.g., heat of reaction, equilibrium constant, etc.) properties of all components taking part in the reaction, and (4) corrosion-erosion characteristics of and any potential hazard associated with the reaction system. When this information and these data become available, the preliminary selection and sizing of the reactor can proceed.

Table 4-11 has been prepared to facilitate location of information needed in reactor design, most of which is discussed in detail in the paragraphs to follow. Sources of chemical rate data for industrially important reactions are found in Tables 4-18 through 4-20. For information pertaining to catalytic processes, refer to the material presented under "Reactors for Solid-Catalyzed Reactions."

Reactor Types Chemical reactors employed in various industrial processes come in a variety of designs which do not always bear any

TABLE 4-11 Quick Index to Reactor Design

Reactor type	Stirred tank (jacketed heat transfer)	Stirred tank (coiled heat transfer)	Tubular	Tower	Fluidized bed
Type of operation and reaction	Batch (small scale), semibatch, or continuous (large scale); homogeneous liquid phase, may have solids suspension (catalyst or reactant); high degree of back mixing		Single tube (small scale), or multitubular (large scale); may have packed catalyst beds; low degree of back mixing; homogeneous or heterogeneous	Continuous heterogeneous reactions, may have baffles or solids packing (catalyst, inert, or reactant)	Solid-fluid reactions, catalytic or noncatalytic
Reactor capacity equation*	Batch: Eq. (4-61*a*); Table 4-12 Continuous (back mix): Eq. (4-62); Table 4-12 Multiple-reactor system (back mix): Eqs. (4-63), (4-64)		Single plug flow: Eq. (4-62); Table 4-12 Multiple-reactor system (plug flow): Eqs. (4-65), (4-66)	Gas-liquid and liquid-liquid system: see Sec. 14, 15, 21 Solid-fluid system: Eq. (4-62)	Eq. (4-62); Table 4-12 (as a back-mix reactor)
Mass-transfer data			Packed-bed reactor: Eqs. (4-105), (4-106), (4-107); Figs. 4-17, 4-18	Gas-liquid: Secs. 14 and 18 Liquid-liquid: Eq. (4-103); 15, 21 Fluid-solid (packed bed): Eqs. (4-105), (4-106), (4-107); Figs. 4-17, 4-18	Eqs. (4-105), (4-106), (4-107); Figs. 4-17, 4-18
Heat-transfer data	Eq. (4-110); Sec. 11	Eq. (4-110); Sec. 11	Homogeneous flow: Sec. 10, 11 Heterogeneous flow: Eqs. (4-111), (4-112); Fig. 4-19; Sec. 10	Packed bed: Eqs. (4-111), (4-112); Fig. 4-19; Sec.10	Fig. 4-19; Eqs. (4-111), (4-112); Sec. 10
Fluid dynamic data	Sec. 19	Sec. 19	Empty tube: Sec. 5 Packed beds: Eq. (4-93) and Sec. 5	Packed tower: Sec. 18	References I-4; K-5–K-9

*Based on the assumption of ideal reactors. For cases with serious deviations from ideal behavior, see subsection "Characteristic Behavior of Reactors" and the references given therein.

specific relation to the reaction type or the type of operation. Frequently, equipment of several different designs is used for the same reaction.

In general, chemical reactors have been broadly classified in two ways, one according to the type of operation and the other according to design features. The former classification is mainly for homogeneous reactions and divides the reactors into batch, continuous, or semicontinuous types. A brief description of these types follows.

Batch Reactor This type takes in all the reactants at the beginning and processes them according to a predetermined course of reaction during which no material is fed into or removed from the reactor. Usually it is in a form of tank with or without agitation and is used primarily in a small-scale production. Most of the basic kinetic data for reactor design are obtained from this type.

Continuous Reactor Reactants are introduced and products withdrawn simultaneously in a continuous manner in this type of reactor. It may assume the shape of a tank, a tubular structure, or a tower and finds extensive applications in large-scale plants for the purpose of reducing the operating cost and facilitating control of product quality.

Semicontinuous Reactor In this category belong reactors that do not fit either of the preceding two types. In one case some of the reactants are charged at the beginning, and the remaining are fed continuously as the reaction progresses. Another type is similar to a batch reactor except that one or more of the products is removed continuously.

Chemical reactors have also been classified according to design features as follows:

Tank Reactor This is probably the most common type of reactor in use in the chemical industry. In most instances it is equipped with some means of agitation (e.g., stirring, rocking, or shaking) as well as provisions for heat transfer (e.g., jacket, external and internal heat exchangers). This type can accommodate either batch (Fig. 4-4*a*) or continuous operation (Fig. 4-4*b*) over wide ranges of temperatures and pressures. With the exception of very viscous liquids, a close approximation to perfect mixing (back mixing) can be achieved in a stirred-tank reactor. In a continuous operation, several stirred-tank reactors may be connected in series (Fig. 4-4*c*). Also in use in continuous operation is a single reaction vessel divided into a number of compartments, each of which is equivalent to a stirred-tank reactor (Fig. 4-4*d* and *e*). One of the tank mixer-reactors of special design is the helical ribbon mixer (Fig. 19-16), which is used in polymerization reactions. An example of the type of reactor shown in Fig. 4-4*d* in commercial use is the Kellogg cascade alkylator (Fig. 21-82).

Tubular Reactor This type of reactor is constructed of either a single continuous tube (Fig. 4-5*b*) or several tubes in parallel (Fig. 4-5*a*). The reactants enter at one end of the reactor, and the products leave from the other end, with a continuous variation in the composition of the reacting mixture in between. Heat transfer to or from the reactor may be accomplished by means of a jacket or a shell-and-tube design (Fig. 4-5*a*). The reactor tubes may be packed with catalyst pellets or inert solids. The tubular reactor finds application in cases in which back mixing of the reaction mixture in the flow direction is undesirable. Large-scale gaseous reactions such as the cracking of hydrocarbons, the conversion of air to NO, and the oxidation of NO to NO_2 are examples of the application of tubular reactors.

Tower Reactor A vertical cylindrical structure with a large height-to-diameter ratio characterizes this type of reactor. It may have baffles or solid packing (reactant, catalyst, or inert) or may be simply an empty tower, and is employed in continuous processes involving heterogeneous reactions. Examples are the lime kiln and gas-absorption units for gas-liquid reactions (Sec. 14) including packed towers, plate towers, and spray towers. (See Fig. 21-97).

Fluidized-Bed Reactor This is a vertical cylindrical vessel containing fine solid particles that are either catalysts or reactants. The fluid reactant stream is introduced at the bottom of the reactor at a rate such that solids are floated in the fluid stream without being carried out of the system. Under this condition the entire bed of particles behaves like a boiling liquid, which tends to equalize the composition of the reaction mixture and temperature throughout the bed

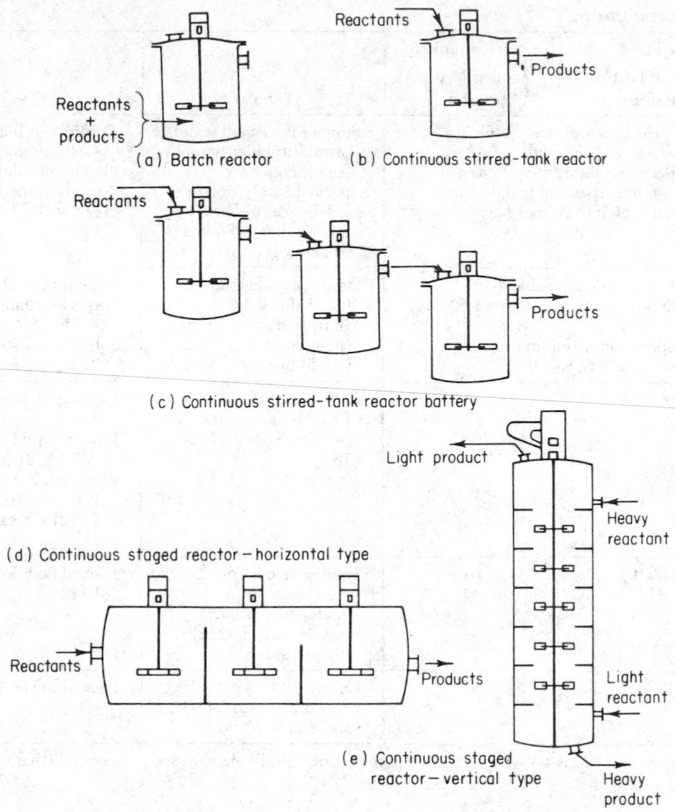

FIG. 4-4 Stirred-tank reactors.

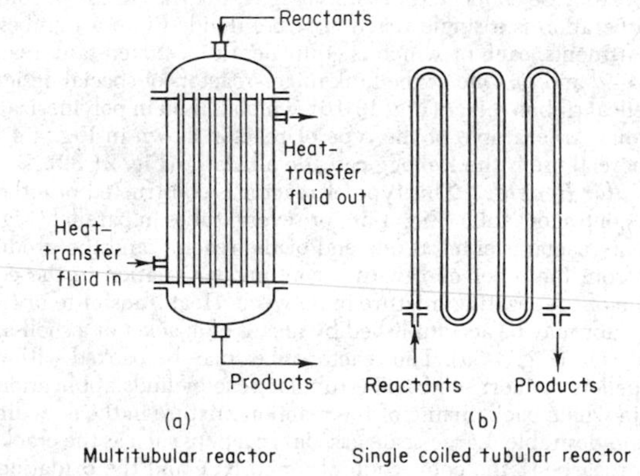

FIG. 4-5 Tubular reactors.

(i.e., some degree of back mixing is obtained). This is generally considered one of the distinct advantages of the fluidized-bed reactor. The attrition of catalyst with entrainment of the resulting fines is one of the disadvantages. This type has been applied extensively to solid-fluid reactions, such as the catalytic cracking of petroleum hydrocarbons, the conversion of uranium oxides to uranium fluorides, the reduction of some mineral ores, and the gasification of coal. A detailed discussion of the design features of fluidized-bed systems appears in Sec. 20, where illustrations of several industrial fluidized-bed reactors are also presented (e.g., Figs. 20-76, 20-92, and 20-95).

Slurry-Phase Reactor This type of reactor is characterized by a vertical column containing fine catalyst particles slurried with a liquid medium (e.g., oil), which may be one of the reactants. The gas reactant bubbling through the slurry dissolves into the liquid medium, where catalyzed reactions take place. This technique facilitates control of the temperature because of the high heat capacity and favorable heat-transfer characteristics of the liquid. Examples of production-scale slurry reactors are found in the hydrocracking of residual fuel oils (G-1 to G-3). The ebullating-bed reactor (commercial name, H-Coal reactor) developed by Hydrocarbon Research, Inc. (G-4), may be classified as this type. It is being used in the development of the coal-liquefaction process. The reaction system consists of coal and catalyst particles, oil, and hydrogen gas.

Basic Design Equations Major considerations in the design of a reactor system from the standpoint of process economics include the production capacity of the reactor and its capability in giving a desired product composition. The latter consideration is especially important in a process dealing with multiple reactions. To be able to select and size a reactor, some common parameters that permit a systematic comparison of various types of reactors are required.

The design equations are generally based on three types of ideal reactors. The ideal reactors can serve as the standards for comparison not only between themselves but also with actual reactors to determine the type and extent of nonideality of the latter. In this way, appropriate correction factors may be derived and applied to the design equation and data base for the ideal reactors. The first type of ideal reactor is the **batch reactor** discussed previously (see "Reactor Types"). All the reactants are introduced into the batch reactor, and no product is discharged until the process is concluded. This represents an unsteady-state process in regard to the composition of the reaction mixture. An assumption of complete mixing implies a homogeneous composition in the reactor at any instant.

The **plug-flow** reactor is the second type, which assumes complete mixing in the radial direction but allows for no diffusion in the flow direction (i.e., no back mixing). As a result, the velocity, temperature, and composition profiles are flat over any cross-sectional area perpendicular to the flow, but the composition varies along the flow path. The plug-flow reactor is the type depicted in Fig. 4-5.

The third type of ideal reactor is the **back-mix** (or stirred-tank) reactor with a steady-state continuous flow of feed and product streams. The feed entering the reactor immediately assumes a final uniform composition throughout the reactor because of perfect mixing. Accordingly, the exit stream would have the same composition as that within the reactor. When several back-mix reactors are connected in series as in Fig. 4-4*c* or *d*, the concentration of the reaction mixture in each reactor is uniform, but it changes from one reactor to the next. Thus, it can be shown that as the number of reactors is increased, the behavior of the back-mix reactor system should approach that of the plug-flow reactor.

Reactor Capacity The parameter that has been commonly used as a measure of the reactor capacity is either the **mean residence time** $\bar{\theta}$ or the **space time** Θ. The mean residence time is the average of time periods during which individual portions of the reaction mixture stay in the reactor and is mathematically described by a general form

$$\bar{\theta} = -\int_{N_{Ao}}^{N_{Af}} \frac{dN_A}{V(-r_A)} \tag{4-61}$$

where N_{Ao} and N_{Af} are moles of reactant A entering and leaving the reactor respectively. In the case of a flow reactor, V is the reactor volume actually occupied by the reaction mixture, and $-r_A$ represents the rate of disappearance of A. Equation (4-61) results from the material balance around the reactor and is applicable to the batch and the plug-flow reactors defined earlier. In terms of the fractional conversion x_A, Eq. (4-61) becomes

$$\bar{\theta} = N_{Ao} \int_0^{x_{Af}} \frac{dx_A}{V(-r_A)} \tag{4-61a}$$

Both Eqs. (4-61) and (4-61*a*) deal with a general case in which the volume V occupied by the reaction mixture may change with the progress of reaction. For a batch reactor, $\bar{\theta}$ may be considered as the average time required to obtain a desired conversion of a reactant for a given size of batch.

In the case of a back-mix reactor the mean residence time $\bar{\theta}$ can be represented by

$$\bar{\theta} = \frac{N_{Ao} x_A}{V(-r_A)} \tag{4-61b}$$

If V varies linearly with the fractional conversion x_A as in Eq. (4-38), then Eq. (4-61*b*) can be written as

$$\bar{\theta} = \frac{N_{Ao} x_A}{V_0(1 + \gamma_A x_A)(-r_A)} = \frac{C_{Ao} x_A}{(1 + \gamma_A x_A)(-r_A)} \tag{4-61c}$$

where V_0 is the volume of the reaction mixture under feed conditions. Similarly, when the relation in Eq. (4-38) holds for the plug-flow reactor, Eq. (4-61*a*) assumes the form

$$\bar{\theta} = C_{Ao} \int_0^{x_{Af}} \frac{dx_A}{(1 + \gamma_A x_A)(-r_A)} \tag{4-61d}$$

Under steady-flow conditions with a constant fluid density, $\bar{\theta}$ for a continuous-flow reactor may be computed from $\bar{\theta} = V/v$; here v is the volumetric flow rate of the reaction mixture.

In real continuous reactors behavior of the reaction mixture may deviate considerably from the ideality assumed for the back-mix and the plug-flow reactors respectively (see subsection "Characteristic Behavior of Reactors" for details). Consequently, the mean residence time $\bar{\theta}$ has little significance in such reactors since individual portions of the reacting fluid may have widely different values of $\bar{\theta}$ owing to nonideal flow through the reactor.

On the other hand, the **space time** Θ is defined as the time elapsed in processing one reactor volume of feed at specified conditions. That is, a space time of 1.5 h implies that 1.5 h would be required to process one reactor volume of feed at known conditions. The reciprocal of the space time is termed the **space velocity** S. The mathematical expression is

$$\Theta = \frac{V}{F/\rho} = \frac{V}{F_{Ao}/C_{Ao}} = \frac{V}{v_0} = \frac{1}{S} \tag{4-62}$$

where F and F_{Ao} are the flow rates, in moles per unit time, of the total feed and of reactant A in the feed, respectively; ρ is the molal density, in moles per unit volume, of the feed; C_{Ao} is the concentration of A in the feed; and v_0 is the volumetric flow rate of feed at entering conditions.

Thus, it is seen that $\bar{\theta}$ and Θ are in general different unless the fluid density remains constant throughout the reactor. The mean residence time $\bar{\theta}$ is used primarily to indicate the capacity of a batch reactor, whereas the volumetric feed capacity of a continuous-flow reactor is represented by the space time Θ or the space velocity S. As expressed by Eq. (4-61), calculation of $\bar{\theta}$ requires knowledge of the rate expression as well as the variation in V with progress of the reaction. By contrast, Θ as defined by Eq. (4-62) is determined from the data for the feed under the specified conditions.

Listed in Table 4-12 are mathematical expressions for the mean residence time and the space time as a measure of capacity for various ideal reactors. These equations have been derived on the basis of material balances and rate expressions and cover both general and special cases, including that of an isothermal nth-order reaction.

Sizing of a reactor using the equations in Table 4-12 may be illustrated by a general example for a plug-flow reactor. The space time Θ can be estimated from the laboratory kinetic data through the expression

$$\Theta = C_{Ao} \int \frac{dx_A}{-r_A}$$

Then from the definition of Θ [Eq. (4-62)],

$$V = (F_{Ao}/C_{Ao})\Theta = v_0\Theta$$

Hence, the required capacity of the reactor is computed from the specified feed rate and feed concentration of component A and the space time.

Material and Energy Balance Equations representing the material and energy balances are also required in the design of a chemical reactor system. A material balance describes the rates of chemical transformation of various components in terms of a specific rate expression, the feed flow rate, and the reactor volume. Thus the composition of the reaction mixture can be computed from the material balance. The rate of heat liberation or absorption expressed as a function of the reaction rate and various thermal properties of the reaction system is obtained from an energy balance. This information is then used in the design of the heat-transfer equipment for the reactor system.

As an example, material- and energy-balance equations for different types of reactors and various operations are summarized in Table 4-13, based on a complex reaction system of the following stoichiometry:

$$A + B \rightleftharpoons E$$
$$B + E \rightleftharpoons S$$

The *net* forward rates are r_1 and r_2 respectively for the first and second reactions, but no specific rate expressions (i.e., reaction mechanism) are assigned. These equations assume that no products E and S are present in the feed stream. Also presented in Table 4-13 are the applications, advantages, and disadvantages of various reactor types. The nomenclature applicable to Table 4-13 follows.

A_w = heat-transfer area, m^2 (ft^2)
C_{Ao}, C_{Bo} = feed concentration of components A and B, mol/m^3 (mol/ft^3)
C_{pi} = molar heat capacity of ith component, J/(mol·K) [Btu/(mol·°F)]
C_{io} = feed concentration of component i, mol/m^3 (mol/ft^3)

TABLE 4-12 Basic Design Equations: Capacity of Single Ideal Reactors in Terms of $\bar{\theta}$ and Θ*

Type of reactor	General design equation	Design equation with $V = V_0(1 + \gamma_A x_A)$†	Design equation for isothermal nth-order reaction with $V = V_0(1 + \gamma_A x_A)$†
Batch: $V =$ const.	$\bar{\theta} = C_{Ao} \int \frac{dx_A}{-r_A}$	$= C_{Ao} \int \frac{dx_A}{-r_A}$	$= \frac{1}{kC_{Ao}^{n-1}} \int \frac{dx_A}{(1 - x_A)^n}$
Batch: $\pi =$ const.	$\bar{\theta} = N_{Ao} \int \frac{dx_A}{V(-r_A)}$	$= C_{Ao} \int \frac{dx_A}{(1 + \gamma_A x_A)(-r_A)}$	$= \frac{1}{kC_{Ao}^{n-1}} \int \frac{(1 + \gamma_A x_A)^{n-1}}{(1 - x_A)^n} dx_A$
Plug flow	$\Theta = \frac{1}{S} = C_{Ao} \int \frac{dx_A}{-r_A}$	$= C_{Ao} \int \frac{dx_A}{-r_A}$	$= \frac{1}{kC_{Ao}^{n-1}} \int \frac{(1 + \gamma_A x_A)^n}{(1 - x_A)^n} dx_A$
Back-mix reactor	$\Theta = \frac{1}{S} = \frac{C_{Ao} x_A}{-r_A}$	$= \frac{C_{Ao} x_A}{-r_A}$	$= \frac{1}{kC_{Ao}^{n-1}} \frac{x_A(1 + \gamma_A x_A)^n}{(1 - x_A)^n}$

$x_A = (C_{Ao} - C_A)/C_{Ao}$ and $dx_A = -dC_A/C_{Ao}$ only when $\gamma_A = 0$.
π = total pressure.
*Adapted from Levenspiel, *Chemical Reaction Engineering*, Wiley, New York, 1962.
†See Eq. (4-38).

$\bar{D}$ = reactor diameter, m (ft)
F = feed rate, mol/s
H_{io} = enthalpy of ith component in feed stream, J/mol (Btu/mol)
H_i = enthalpy of ith component, J/mol (Btu/mol)
$\Delta H_1, \Delta H_2$ = heat of reaction for reactions 1 and 2, respectively, J/mol (Btu/mol)
L = length along a tubular reactor, m (ft)
N_i = number of moles of component i
r = reaction rate, mol/(s·m^3) [mol/(s·ft^3)]
r' = reaction rate, mol/(s·kg of catalyst) [mol/(s·lb of catalyst)]
T = temperature of reacting mixture, K (°F)
T_0 = reference temperature, K (°F)
T_w = temperature of heat source or sink to or from which the reaction mixture is transferring heat, K (°F)
U = overall heat-transfer coefficient, J/(s·m^2·K) [Btu/(s·ft^2·°F)]
V_0 = initial volume of reaction mixture, m^3 (ft^3)
V = reactor volume, m^3 (ft^3)
v = reaction-mixture velocity, m/s (ft/s)
$x_{A \cdots S}$ = mole fraction of component $A \cdots S$
ρ = molal density of reacting mixture at any point in tube, mol/m^3 (mol/ft^3)
ρ_c = catalyst density (as packed), kg/m^3 (lb/ft^3)
ρ_F = molar density of feed, mol/m^3 (mol/ft^3)
θ = time, s

Design Equations for Multiple-Reactor System When the reactor system consists of several reactors, the equations shown in Tables 4-12 and 4-13 are applicable to any single reactor in the system. The overall performance of the system depends not only on the reactor type but also on the manner in which the reactors are connected (i.e., series, parallel, or a combination of the two). In general, when a parallel arrangement of reactors exists in the system, the optimum system efficiency can be approached by adjusting the flow so that the compositions of the exit streams from the individual parallel branches before merging would be approximately the same.

Back-mix reactors. For a system having m equal-sized back-mix reactors in series in which a reaction of a simple order is taking place under constant-volume conditions, the following expression can be derived from a material balance for first-order reaction:

$$\Theta_{\text{overall}} = \frac{m}{k}\left[\left(\frac{C_{Ao}}{C_{Am}}\right)^{1/m} - 1\right] \tag{4-63}$$

where C_{Am} is the concentration of A in the discharge stream from the final reactor. For second-order reaction with an equal molal ratio of the reactants A and B,

$$C_{Am} = \frac{-1 + \sqrt[3]{-1 + \cdots + 2\sqrt{-1 + 2\sqrt{1 + 4k\Theta C_{Ao}}}}}{2k\Theta} \quad (m;\ m=2;\ m=1) \tag{4-64}$$

Here, Θ is the space time for each reactor. The solution of Eq. (4-64) would be formidable when the number of reactors m becomes large. However, numerical and graphical methods are available (G-5, B-10).

Plug-flow reactor. It can be readily shown that a number of plug-flow reactors with a total volume V_t arranged in series would result in the same degree of conversion as that from a single reactor of size V_t. The expressions for the overall space time Θ'_{overall} would be:

For first-order reaction:

$$\Theta'_{\text{overall}} = \frac{\ln (C_{Ao}/C_{Am})}{k} \tag{4-65}$$

For second-order reaction:

$$\Theta'_{\text{overall}} = \frac{C_{Ao}/C_{Am} - 1}{kC_{Ao}} \tag{4-66}$$

A system that consists of reactors in parallel may be treated in a manner similar to this, provided that the space times for the individual parallel branches are equal.

Characteristic Behavior of Reactors The design equations previously presented for the three basic types of reactors are based on the assumption that the reaction mixture behaves according to the ideal flow patterns characteristic of such reactors. That is, assumptions have been made of complete back mixing in a batch or a continuous stirred-tank reactor in one extreme and of plug flow in a tubular flow reactor in the other extreme. Mathematical models thus derived for the ideal reactors serve as a reference standard in the design of practical reactors.

Although practical reactors cannot completely fulfill these assumptions, many cases may be approximated by ideal reactor models without any serious error being introduced. On the other hand, there are a number of cases in which deviations from ideal behavior become so significant that an excessively large error would result if the design were based entirely on ideal models. Examples are the fixed-bed tubular catalytic reactor with wall cooling and the stirred-tank reactor handling a viscous reaction mixture. Such deviations can sometimes be taken care of by introducing correction factors into the ideal reactor models, but purely empirical methods of scale-up and design are also frequently employed.

TABLE 4-13 Basic Design Equations: Material and Energy Balances; General Features of Reactors

$A + B \rightleftharpoons E$: Net forward rate $= r_1$

$B + E \rightleftharpoons S$: Net forward rate $= r_2$

Reactor	Type of operation	Material balance	Energy balance°	Applications, advantages, and disadvantages
Batch	A and B charged initially; no product withdrawal during reaction.	$\frac{dN_E}{d\theta} = (r_1 - r_2)V$ $\frac{dN_S}{d\theta} = r_2V$ $N_A = C_{Ao}V - N_E - N_S$ $N_B = C_{Bo}V - N_E - 2N_S$	$\frac{dQ}{d\theta} = UA_w(T_w - T)$ $- (r_1\,\Delta H_1 + r_2\,\Delta H_2)V$	Appl.: small-scale or complicated process for expensive product Adv.: versatile; high product yield Disadv.: high operating cost
Semibatch	Only A is charged initially; B introduced continuously, and no product withdrawal during reaction. $V = V_0 + F\theta/\rho_F$	$\frac{dN_E}{d\theta} = (r_1 - r_2)V$ $\frac{dN_B}{d\theta} = C_{Bo}\frac{F}{\rho_F} - (r_1 + r_2)V$ $\frac{dN_S}{d\theta} = r_2V$ $N_A = C_{Ao}V_0 - N_E - N_S$	$\frac{dQ}{d\theta} = FC_{Bo}H_{Bo}/\rho_F$ $+ UA_w(T_w - T)$ $- (r_1\,\Delta H_1 + r_2\,\Delta H_2)V$ $A_w = 4V/D$ for jacketed heat-transfer area	Appl.: to regulate product distribution in homogeneous reaction Adv.: desired product distribution obtained by regulating concentration of a reactant; good temperature control. Disadv.: high operating cost
Semibatch	Only A is charged initially; B introduced continuously with constant product overflow; homogeneous liquid-phase reaction; V = constant	$\frac{dN_A}{d\theta} = -\frac{F}{\rho_F V}N_A - r_1V$ $\frac{dN_B}{d\theta} = -\frac{F}{\rho_F V}(N_B - C_{Bo}V)$ $-(r_1 + r_2)V$ $\frac{dN_E}{d\theta} = -\frac{FN_E}{\rho_F V} + (r_1 - r_2)V$ $\frac{dN_S}{d\theta} = -\frac{FN_S}{\rho_F V} + r_2V$	$\frac{dQ}{d\theta} = \frac{F}{\rho_F}(\Sigma C_{io}H_{io} - \frac{1}{V}\Sigma N_iH_i)$ $+ UA_w(T_w - T)$ $- (r_1\,\Delta H_1 + r_2\,\Delta H_2)V$	Same as above
Semibatch	Only A (liquid) is charged initially; B (gas) introduced continuously; no product withdrawal; V = constant	$\frac{dN_B}{d\theta} = C_{Bo}\frac{F}{\rho_F} - (r_1 + r_2)V$ $\frac{dN_E}{d\theta} = (r_1 - r_2)V$ $\frac{dN_S}{d\theta} = r_2V$ $N_A = C_{Ao}V - N_E - N_S$	$\frac{dQ}{d\theta} = (FC_{Bo}H_{Bo}/\rho_F)$ $+ UA_w(T_w - T)$ $- (r_1\,\Delta H_1 + r_2\,\Delta H_2)V$ $A_w = 4V/D$ for jacketed heat ransfer area	Same as above
Continuous back-mix flow	Continuous steady-state flow of feed and product streams; V = constant; complete back mixing in stirred-tank reactor	$\frac{dN_A}{d\theta} = 0 = (C_{Ao}V - N_A)\frac{F}{\rho_F V}$ $-r_1V$ $\frac{dN_B}{d\theta} = 0 = (C_{Bo}V - N_B)\frac{F}{\rho_F V}$ $-(r_1 + r_2)V$ $\frac{dN_E}{d\theta} = 0 = N_E\frac{F}{\rho_F V} + (r_1 - r_2)V$ $\frac{dN_S}{d\theta} = 0 = N_S\frac{F}{\rho_F V} + r_2V$	$\Sigma C_{io}H_{io} - \frac{1}{V}\Sigma N_iH_i)\frac{F}{\rho_F}$ $+ UA_w(T_w - T)$ $- (r_1\,\Delta H_1 + r_2\,\Delta H_2)V = 0$	Appl.: a large-scale production; normally several reactors used in series Adv.: lower operating cost; good control of product quality Disadv.: high capital cost
Continuous plug flow	Continuous steady-state flow of feed and product streams; complete radial mixing, no back mixing, i.e., no longitudinal mixing; packed catalyst-bed tubular reactor	$\frac{d(\rho v x_E)}{dL} = (r_1' - r_2')\rho_c$ $\frac{d(\rho v x_S)}{dL} = r_2'\rho_c$	$\frac{dQ'}{dL} = 4U(T_w - T)/\bar{D}$ $- (r_1'\,\Delta H_1 + r_2'\,\Delta H_2)\rho_c$ for jacketed heat-transfer area	Appl.: large-scale fluid reaction in which back mixing is undesirable Adv.: limited degree of back mixing Disadv.: high capital and operating costs, especially for slow reactions; catalyst replacement difficult

°For isothermal operation, $dQ/d\theta = dQ'/dL = 0$; for adiabatic operation, $T_w - T = 0$.
$Q = \Sigma N_iC_{pi}(T - T_0)$; $Q' = \rho v \Sigma x_iC_{pi}(T - T_0)$.

In addition to nonideal behavior, those reactor characteristics that govern the degree of conversion and the product composition should also be considered. They include reactor type and size, reaction temperature, and pressure. The influences of the last two parameters are especially complex in processes involving multiple homogeneous reactions as well as in nearly all heterogeneous reactions.

Effect of Reactor Types To analyze how reactor performance is influenced by reactor type, consider a plug-flow reactor and a back-mix reactor, both handling an irreversible nth-order reaction.

From the last two equations in the fourth column of Table 4-12 the following ratio can be formulated:

$$\frac{\eta_{\text{back mix}}}{\eta_{\text{plug}}} = \frac{(Z^n x_A)_{\text{back mix}}}{\left(\int_0^{x_A} Z^n \, dx_A\right)_{\text{plug}}} \tag{4-67}$$

where
$$\eta = \Theta C_{Ao}^{n-1} = VC_{Ao}^n/F_{Ao}$$
$$Z = (1 + \gamma_A x_A)/(1 - x_A)$$

The effect of the reactor type on the performance as represented by Eq. (4-67) is illustrated in Fig. 4-6 for reactions of the following types:

$$A \rightarrow \text{products} \qquad -r_A = kC_A^n \qquad n = 0 \text{ to } 3$$
$$A + B \rightarrow \text{products} \qquad -r_A = kC_AC_B \qquad \frac{C_{Ao}}{C_{Bo}} = 1$$
$$A + B + D \rightarrow \text{products} \qquad -r_A = kC_AC_BC_D \qquad C_{Ao}:C_{Bo}:C_{Do} = 1:1:1$$

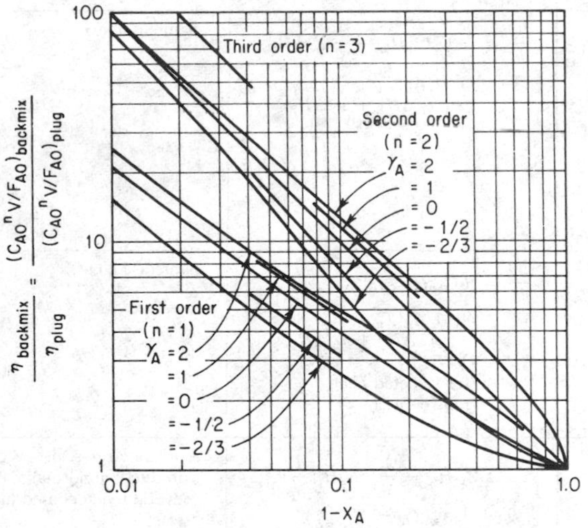

FIG. 4-6 Performance of plug-flow and back-mix reactors (single). (*Adapted from Levenspiel,* Chemical Reaction Engineering, *2d ed., Wiley, New York, 1972.*)

It is noted that when the same feed rate containing identical concentration of reactants is used in both types of reactors,

$$\frac{\eta_{\text{back mix}}}{\eta_{\text{plug}}} = \frac{V_{\text{back mix}}}{V_{\text{plug}}} = \frac{\Theta_{\text{back mix}}}{\Theta_{\text{plug}}}$$

Thus, Eq. (4-67) and Fig. 4-6 represent the reactor-volume ratio or the space-time ratio at various degrees of conversion.

According to Fig. 4-6, a larger volume is required for the back-mix reactor than for the plug-flow reactor at any given conversion. The volume ratio is strongly dependent on the order of reaction and the extent of conversion and to a lesser degree on the density variation as indicated by γ_A. The difference in the reactor-volume requirement diminishes as the required conversion is reduced.

An alternative way of comparing the performances of the two ideal-flow reactors is to compute the degree of conversion for both reactors of the same capacity. The result of such computation is shown in Fig. 4-7 (adapted from Ref. G-6) for an irreversible elementary reaction, $A + B \rightarrow R + S$, with an equimolar ratio of A to B in the feed.

Thermal Effects The process of chemical change is always accompanied by release or absorption of heat which influences not only the reaction rate but also the equilibrium yield and the product

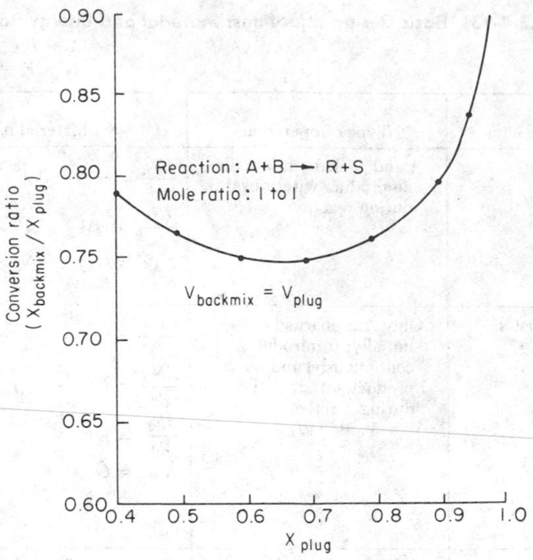

FIG. 4-7 Comparison of conversions between single plug-flow and back-mix reactors.

composition. The Arrhenius equation [Eq. (4-3)] predicts that the rate of a simple reaction as expressed by the rate constant k is favored by an increase in the reaction temperature. However, a continued increase in temperature does not always result in acceleration of the reaction rate. This phenomenon may be attributed to a number of causes as indicated by the general rate expression

$$r_i = f(C,T,P,m) \tag{4-2}$$

Consider the reaction rate of an exothermic irreversible reaction as a function of temperature and the reactant concentration alone described by

$$r_i = kf(C) \tag{4-2b}$$

Equation (4-2b) in combination with the Arrhenius equation indicates that the rate constant k would increase exponentially with temperature, thereby accelerating the reaction rate initially. However, depletion of the reactants with progress of the reaction counteracts the increase in the rate constant and causes the reaction rate to diminish even with a continued increase in reactor temperature. This behavior is depicted by Fig. 4-8, which shows the presence of a maximum rate for a batch reactor. A back-mix flow reactor exhibits a characteristic S-shaped curve. The exact shape of the curve depends on the mode of operations as well as on the reactor type and can be determined by plotting the energy balance for the system.

Behavior similar to that described has also been observed when the reaction mechanism changes with the temperature, e.g., a shift in the rate-controlling step from chemical reaction to diffusion. Examples for such behavior may be found in the gas-solid reactions (see subsection "Mathematical Model for Fluid-Solid Reactions" and Table 4-4). This change in the rate-controlling step can be explained by means of the example presented in Eqs. (4-58) through (4-60). In the low-temperature range, the overall process rate tends to be controlled by the chemical reaction. However, diffusion takes over as the rate-controlling step in the high-temperature range because the diffusion coefficient is generally much less sensitive to a temperature change than the rate constant.

In addition to the effects of the reactor temperature on the reaction rate, the process stability and parametric sensitivity should also be taken into account in the reactor design. These factors are especially important in tubular reactors. A brief discussion of these subjects has been presented previously (see "Steps to Follow in Scale-Up by Mathematical Modeling"). In view of the complex thermal characteristics of the various reactors previously discussed, it is obvious that an optimum temperature progression using the proper reactor

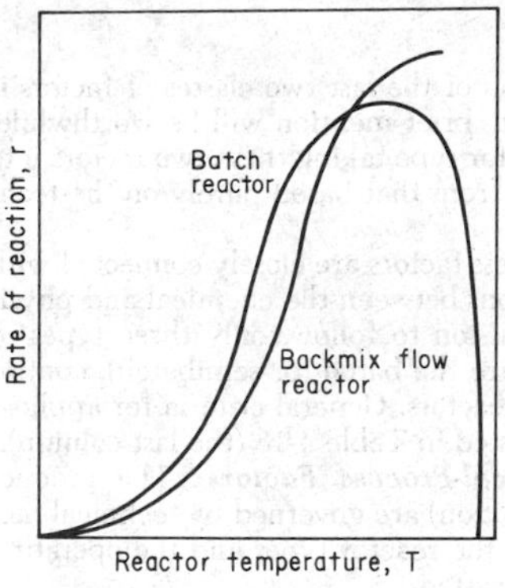

FIG. 4-8 Effect of reactor temperature on rate of reaction.

system is required to achieve a maximum rate of production. Detailed discussion of this subject is found in a number of references (G-7 through G-10).

Pressure Effects The effect of pressure is significant in a gas-phase reaction. Change in pressure also influences the activation energy of some reactions. The equilibrium conversion and product composition of gaseous reactions are affected by pressure when the number of moles varies with the progress of the reaction.

Consider a reversible gas-phase reaction:

$$aA + bB \underset{k_2}{\overset{k_1}{\rightleftharpoons}} rR + sS$$

It can be shown that the equilibrium constant K is related to a ratio K_y when the assumption of an ideal gas mixture is valid:

$$K = K_y \pi^{(r+s)-(a+b)} \tag{4-68}$$

where π is the total pressure and $K_y = y_R^r y_S^s / y_A^a y_B^b$. Here, y's are mole fractions. Equation (4-68) can be written as

$$y_R^r y_S^s / y_A^a y_B^b = K \pi^{(a+b)-(r+s)} \tag{4-68a}$$

which indicates that the mole-fraction ratio of products to reactants and the corresponding conversion at equilibrium are affected by the variation in the total pressure since K is independent of pressure. Thus, when the forward reaction accompanies an increase in the number of moles (i.e., $a + b < r + s$), the conversion is favored by low pressures toward the end of the reaction. The reverse is true when the number of moles decreases with reaction.

Because of the pressure effect just discussed, frequently it is desirable to conduct gas-phase reactions in tubular reactors in which a pressure profile is imposed along the reaction path so as to obtain a high conversion with a favorable product distribution. An example of the optimization of pressure in tubular reactors has been presented in a paper by van de Vusse and Voetter (G-11). An elementary reaction of the type $A \underset{k_2}{\overset{k_1}{\rightleftharpoons}} 2B$, with the rate equation

$$d\nu_A/d\theta = (-k_1 p_A + k_2 p_B^2)V \tag{4-69}$$

is first considered. With an ideal gas mixture assumed, this equation is converted to

$$\frac{1}{A\theta}\frac{d\nu_A}{dL} = -k_1 \frac{\nu_A \pi}{2\nu_0 - \nu_A} + 4k_2 \frac{(\nu_0 - \nu_A)^2 \pi^2}{(2\nu_0 - \nu_A)^2} = f[\pi(L), \nu_A] \tag{4-70}$$

In these equations, $\nu_A = p_A V = N_A RT$, ν_0 is the value of ν_A at $\theta = 0$, L is the length coordinate, and A is the cross-sectional area of the tube. The optimum pressure profile is determined from the function of Eq. (4-70) by calculating $\partial f/\partial \pi = 0$. Further discussion deals with concentration gradients to obtain a high yield with optimum selectivity for the reaction system:

$$A + B \rightarrow S \qquad A + A \rightarrow E$$

An experimental example of the pressure effects may be referred to a study by Siegel and Garti (G-12). The study dealt with catalytic hydrogenation of aromatic hydrocarbons, and the results demonstrate rather complex effects of pressure on the stereochemistry, rate, and product distribution.

Nonideal Flow Deviations from the ideality assumed in developing the basic reactor design equations are present in practical reactors, and the extent of the nonideality often varies considerably, depending on the scale and type of reactor. Important types of deviation from two ideal-flow reactors are outlined.

Deviation from plug-flow reactor. (1) Channeling of the reacting fluid through the catalyst packing and the presence of stagnant fluid pockets, (2) the presence of velocity and temperature gradients in the radial direction, and (3) diffusion in the direction of flow and back mixing as the result of fluid turbulence, thermal convective transport, and molecular diffusion.

Deviation from back-mix reactor. (1) Short circuiting and bypassing of the reacting fluid: i.e., certain portions of fluid may proceed directly from the feed inlet to the product discharge port without mixing with the contents of the reactor; (2) internal recycling of the fluid; and (3) the presence of stagnant fluid pockets.

Characterization of nonideal flow. The various types of behavior listed imply that different portions of the reacting fluid follow separate flow patterns through the reaction vessel, resulting in a wide distribution of residence times. Such deviations from ideality represent inefficiency in the reactor performance and cause reduction in production capacity. To alleviate this condition, there are various

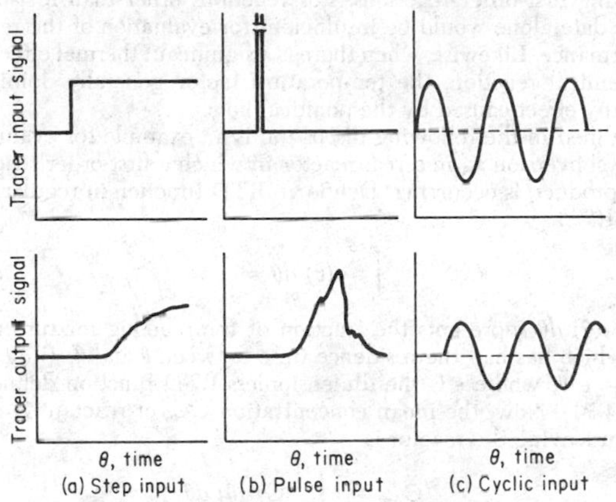

FIG. 4-9 Tracer output response curves for various types of tracer input.

methods often used to characterize the magnitude of deviation—the so-called stimulus-response techniques that utilize a tracer.

Three typical techniques of introducing a tracer into the reaction vessel are the step input, the pulse input, and the cyclic input. Figure 4-9 shows possible shapes of tracer signal-response curves determined at the exit of the vessel for different types of tracer input signals. Information thus obtained is applicable to the determination of conversion in a reactor, either directly or in combination with one of the several mathematical models proposed to account for the nonideal-flow condition (B-10, G-13, G-14).

Residence-time distribution (RTD). Information obtainable from the stimulus-response technique by using a tracer is the distribution of residence times for various elements of the reacting fluid.

For convenience in the data treatment, a dimensionless time τ defined as $\tau = \theta/\bar{\theta}$ is used in the mathematical analysis of the tracer data. Here θ is the elapsed time, and $\bar{\theta}$ refers to the mean residence time of the reacting mixture. Thus, τ is a measure of time in units of mean residence time. A dimensionless parameter ϵ defined by Eq. (4-71) is used as a measure of the RTD of different fluid elements at the exit of the reaction vessel:

$$\int_0^\infty \epsilon\, d\tau = 1 = \int_0^{\tau_0} \epsilon\, d\tau + \int_{\tau_0}^\infty \epsilon\, d\tau \tag{4-71}$$

where $\epsilon\, d\tau$ may be considered as the fraction of the reacting mixture at the reactor exit, the dimensionless residence time of which is between τ and $\tau + d\tau$. Also $\int_0^{\tau_0} \epsilon\, d\tau$ represents the fraction that has been in the reactor for a time shorter than τ_0, while $\int_{\tau_0}^\infty \epsilon\, d\tau$ is the fraction with the dimensionless residence time longer than τ_0. A plot of ϵ versus τ would yield a curve similar to the tracer output signal of Fig. 4-9*b* with the area under the curve being unity.

Tracer response curves. Two types of response curves are obtained from the tracer experiment. In both curves, a dimensionless tracer signal (or concentration) C/C_0 is plotted against the dimensionless time τ. C_0 is the input tracer concentration, whereas C is the tracer concentration measured at the exit. The response curve to a step tracer input is termed an **F curve** and is similar in appearance to the output signal of Fig. 4-9*a*. The second type, called a **C curve,** is the response to a pulse tracer injection, and has a shape similar to the output signal of Fig. 4-9*b*.

The method of obtaining the RTD from the tracer experiments is shown in the following relation:

$$\epsilon = C = (C/C_0)_{\text{pulse input}} \tag{4-72}$$

$$\epsilon = \frac{dF}{d\tau} = \frac{d}{d\tau}\left(\frac{C}{C_0}\right)_{\text{step input}} \tag{4-73}$$

Application of residence-time-distribution data. The RTD data resulting from the tracer experiment can be used to predict the performance of the reactor with good accuracy only for processes involving first-order reactions. For reactions other than first-order, tracer data alone would be insufficient for evaluation of the reactor performance. Likewise, when there is a significant thermal effect due to chemical reaction, the temperature factor generally dominates over any effect caused by the nonideal flow.

Outlined in the following discussion is an example for evaluation of the conversion x_A in a real reactor in which a first-order reaction, $A \rightarrow$ product, is occurring. Define an RTD function in regular time units, $\epsilon(\theta)$, as

$$\int_0^\infty \epsilon(\theta)\, d\theta = 1 \tag{4-74}$$

Thus $\epsilon(\theta)\, d\theta$ represents the fraction of the reacting mixture at the exit which has had the residence time between θ and $\theta + d\theta$. Also, $\epsilon(\theta) = \epsilon/\bar{\theta}$, where ϵ is the dimensionless RTD function defined by Eq. (4-71). Now, the mean concentration C_{Am} of reactant A in the stream leaving the reactor is

$$C_{Am} = \int_0^{\theta_1} C_A \epsilon(\theta)\, d\theta \tag{4-75}$$

where C_A is the concentration of A in the fraction of the reacting mixture at the exit with the residence time between θ and $\theta + d\theta$. θ_1 is the time after which the value of C_A becomes negligible. For a first-order reaction, $C_A = C_{Ao}e^{-k\theta}$. Therefore Eq. (4-75) in terms of the conversion x_A becomes

$$x_A = 1 - \frac{C_{Am}}{C_{Ao}} = 1 - \int_0^{\theta_1} e^{-k\theta}\epsilon(\theta)\, d\theta \tag{4-76}$$

The evaluation of $\int_0^{\theta_1} e^{-k\theta}\epsilon(\theta)\, d\theta$ can be made either graphically or numerically from the kinetic and tracer data (either **F** curve or **C** curve) through the relations described by Eqs. (4-72) and (4-73).

Further information on the subject of nonideal flow in practical chemical reactors can be obtained from a number of references by Danckwerts and others (G-15 to G-19, B-10).

Factors That Influence the Selection of Reactor Type At the time when a preliminary selection of the reactor type is made, the kinetics of the reaction as affected by key process parameters should have been established and the optimum operating conditions determined. In the design of an industrial reactor, the most important considerations in the selection of a reactor type are usually costs and profit. Accordingly, criteria for the selection would be such as to minimize costs and maximize profit. These criteria are determined by factors which may be classified in three different categories, namely, technical, economic, and social. The technical aspects refer to the chemical- and physical-process factors that control the product yield and quality, whereas the economic factors include the capital investment and the operating costs. The social factors include those that cannot be readily assigned direct monetary value, for example, safety and satisfaction of the operators and harmful effects of the reactor on the environment.

Although a detailed discussion of the last two classes of factors is beyond the scope of this section, brief mention will be worthwhile, as the final decision on the reactor type taking these two factors into consideration may be different from that based purely on the technical factors.

Chemical- and physical-process factors are closely connected with each other because of interactions between the chemical and physical rate processes. In the discussion to follow, only three types of reactors are considered. They are the batch (or semibatch), continuous stirred-tank, and tubular reactors. General criteria for application of these reactors are also listed in Table 4-13 (the last column).

Technical Factors: Chemical-Process Factors The product yield and its quality (or composition) are governed by technical factors which are characteristic of the reactor types and the operating conditions. Consider a reaction system:

$$aA + bB \rightarrow dD \qquad \text{(desired product)}$$
$$b'B + d'D \rightarrow sS \qquad \text{(undesirable)}$$

The **overall percent yield** Y of the desired product D based on the limiting reactant A may be defined as

$$Y = \frac{a}{d}\left(\frac{N_{Df} - N_{Do}}{N_{Ao} - N_{Af}}\right) \times 100 \tag{4-77}$$

where the subscripts f and o represent the reactor outlet (final) and inlet (initial) conditions respectively. The stoichiometric ratio a/d is included so that Y would have a value of 100 percent for complete conversion of A.

Chemical-process factors may include the reaction type (simple or complex), the rate of reaction, the production capacity, and the need for a catalyst. In a simple-order reaction the product distribution is fixed by the stoichiometry. A brief discussion on the preferred reactor type for some important reaction types follows.

Autocatalytic reaction. The acid-catalyzed hydrolysis of ester is an example of the autocatalytic reaction in which one of the products functions as a catalyst:

$$A + D \rightarrow B + D$$

where D is the catalyst. A small amount of D must be present to initiate the reaction.

In this type of reaction, the rate of reaction is initially low because C_D is small. With progress of the reaction the increase in C_D accelerates the rate to a maximum, after which the rate diminishes owing to a gradual reduction in C_A. To carry out this type of reaction in a continuous reactor system at a rate near the maximum (and to obtain a high yield), a preferred system would be the one that consists of a stirred-tank reactor followed by a tubular reactor. This is true because with efficient mixing the feed entering the stirred-tank reactor would immediately assume the same composition as the mixture in the tank, which can be adjusted to approach the maximum reaction rate. In the tubular reactor, the change in C_D from a high value to a low one is made gradual so as to sustain the reaction at a high rate for as long as feasible. An alternative arrangement may be to employ a stirred-tank reactor with provision for separating and recycling the reactant remaining in the product stream.

Polymerization. Reactions involved in the formation of polymers are one of the most industrially important examples of complex reactions involving a chain-transfer process. The product distribution is quite complicated, covering a wide range of chain lengths and molecular weights as a result of the influence of various chemical- and physical-process factors. The properties of the polymer product are controlled by the product distribution, which is the key factor in the selection of the reactor type.

Because of the highly viscous nature of the reacting mixture, the majority of polymerization reactions are conducted in a stirred-tank type of reactor, either continuously or batchwise. The decision is then between batch and continuous stirred-tank types and is governed by the requirements for production capacity and product distribution.

A batch reactor would be preferable for small-scale production. The chemical reaction system in a batch reactor is also characterized by the behavior in which all the reacting components tend to have a uniform residence time while the monomer concentration continues to decrease as the reaction progresses. The reverse is true of the continuous stirred-tank reactor. Accordingly, a batch reactor would yield a product having a narrower range of molecular-weight distribution than the continuous reactor. For slow polymerization reactions, a continuous system may require a battery of stirred-tank reactors to satisfy production requirements. Further details of examples on the selection of the reactor type in polymerization reactions are presented in Refs. G-20 and G-21. A number of publications pertaining to analysis and design of polymerization reactors are found in Refs. G-22 and G-23.

Consecutive reactions. The course of an elementary consecutive reaction of the type $A \xrightarrow{k1} B \xrightarrow{k2} D$ may be described by the concentration-time curves shown in Fig. 4-1. If B is the desired product, it can be seen that C_B goes through a maximum at some intermediate time θ_i. In a continuous stirred-tank reactor, different elements of the reacting fluid would have a variety of residence times owing to nonideal-flow conditions in the reactor, which makes it difficult to approach the maximum yield. Thus a batch reactor would be preferred for reactions of this type, as it is relatively simple to adjust the residence time of the entire reaction mixture to a value near the optimum θ_i. A tubular reactor with minimum back mixing would also produce a result comparable to that from a batch reactor.

Well-known examples of consecutive reactions include the oxidation of ethylene to produce ethylene oxide,

$$CH_2{=}CH_2 \underset{\text{air}}{\overset{\text{catalyst}}{\rightarrow}} \underset{\diagdown O \diagup}{CH_2{-}CH_2} \underset{\text{air}}{\overset{\text{catalyst}}{\rightarrow}} CO_2 + H_2O$$

the production of formaldehyde from methanol,

$$CH_3OH \underset{\text{air}}{\overset{\text{catalyst}}{\rightarrow}} HCHO + H_2O \underset{\text{air}}{\overset{\text{catalyst}}{\rightarrow}} CO_2 + H_2O$$

and radioactive decay,

Parallel reactions. Consider a pair of parallel reactions: $A \xrightarrow{k1} B$ and $A \xrightarrow{k2} D$, with rate expressions given by Eqs. (4-11) through (4-13). Assume that B is the desired product. The relative production rate is

$$dC_B/dC_D = (k_1/k_2)C_A^{m-n} \tag{4-14}$$

When $m > n$, the yield of B is favored by a high concentration of A. This condition can be achieved in a batch or tubular reactor in which the change in the reactant concentration is gradual and the average concentration of the reactant is higher than that in a continuous stirred-tank reactor. However, the yield in the latter type of reactor can be improved to some extent by using a battery of tanks in series.

Conversely, when $m < n$, a low-concentration of A would favor the yield of B. This is obviously satisfied best by the continuous-stirred-tank-reactor system. There are a few alternative methods. One is to use a low concentration of A in the feed, but this may not be economical. Another approach employs successive additions of small amounts of A to the reaction mixture at different stages of reaction, and is applicable to any of the three typical types of reactors under consideration.

References G-10, G-11 and G-24 should be consulted for further discussion on the selection of reactor types for parallel reactions.

Technical Factors: Physical-Process Factors Included in this category are the mode of heat transfer (isothermal, adiabatic, and others), the extent of back mixing, and the number and types of phases involved. These factors are generally so closely related to the chemical-process factors that they are not always separable.

Mode of heat transfer. To obtain a maximum yield from a reaction, a mode of heat transfer may have to be devised so that optimum temperature programming is realized. This temperature programming may be isothermal, adiabatic, or some predetermined temperature profile established from material and energy balances. A brief discussion of several typical cases is presented below.

In an exothermic reaction, when the amount of heat liberated is low to moderate so that the rate of reaction can be sustained despite the diminishing reactant concentration, a continuous stirred-tank reactor approaching adiabatic condition would be preferable. On the other hand, a tubular-flow-reactor system with external heat exchangers is desirable when an excessive amount of heat exerts an unfavorable effect on the equilibrium conversion. Either a tubular-flow-reactor or a continuous-stirred-tank-reactor system may be used with the injection of cold feed at various locations.

With endothermic reactions, as the temperature drops with the progress of reaction, both reaction rate and equilibrium conversion would diminish. Under the circumstances, the overall process may be controlled by the rate of heat transfer, and interstage heat exchangers would be required. Hence, the design of heat exchangers would probably become more important than the selection of the reactor type. However, a tubular flow reactor is usually desirable in this case.

Since sensitivity of a reaction to temperature is one of major factors in the selection of the reactor type as well as of the heat-transfer mode, it would be desirable to quantify this characteristic. Two parameters and a third which is the product of the first two have been defined and used (D-2) as the means of comparing different reaction systems in which a single reaction predominates. The first two are

Adiabatic factor=adiabatic temperature change for complete consumption of limiting reactant A

$$= \frac{C_{Ao}(-\Delta H_A)}{\rho_m C_{Pm}} \tag{4-78}$$

and

Temperature sensitivity
= fractional change in reaction rate with temperature

$$= \frac{d(-r_A)}{dT} \Big/ (-r_A) = \frac{E}{RT^2} \tag{4-79}$$

where ρ_m and C_{pm} are the molar density and heat capacity of total feed respectively. The product of these two parameters is a dimensionless group termed (D-2) the heat-generation potential q_p as in

$$q_p = \frac{C_{Ao}(-\Delta H_A)}{\rho_m C_{pm}} (E/RT_0^2) \tag{4-80}$$

Thus, the adiabatic factor is an indication of the extent of adiabatic temperature change if it is assumed that the reaction goes to completion, even though the actual extent of reaction may be limited by the equilibrium. Temperature sensitivity is a measure of the effect of temperature on the rate of heat generation or consumption. For a reaction system involving more than one reaction of equal dominance, these parameters can serve only as a means of qualitative comparison. A quantitative comparison will require knowledge of the relative rates of these reactions based on the established reactor model. Values of the adiabatic factor and heat-generation potential for various typical industrial reactions are listed in Ref. D-2.

Further considerations relevant to the heat-transfer mode should be given to the stability of reactor operations. To facilitate discussion on how reactor stability is influenced by the mode of heat transfer, consider a first-order irreversible, exothermic reaction represented by $A \rightarrow R$ in a continuous back-mix (or stirred-tank) reactor (CSTR). The rate of heat generation q_G is described by

$$q_G = kC_AV(-\Delta H_A) = kC_{Ao}(1 - x_A)V(-\Delta H_A) \tag{4-81}$$

From Eq. (4-61*b*) for a constant-volume first-order reaction,

$$\bar{\theta} = C_{Ao}x_A/kC_{Ao}(1 - x_A)$$

or

$$x_A = k\bar{\theta}/(1 + k\bar{\theta}) \tag{4-82}$$

$$C_{Ao}V = F_{Ao}\bar{\theta} \tag{4-83}$$

Combining Eqs. (4-81), (4-82), (4-83), and (4-3) gives

$$q_G = \frac{F_{A0}\bar{\theta}A(-\Delta H_A)}{\bar{\theta}A + e^{E/RT}} \tag{4-84}$$

The rate of heat removal q_R is the sum of the heat accompanying the reactor effluent stream and heat removed by the heat-transfer surface provided in the reactor:

$$q_R = C_pF(T - T_0) + UA_w(T - T_w)$$

or

$$q_R = (C_pF + UA_w)T - (C_pFT_0 + UA_wT_w) \tag{4-85}$$

Plots of Eqs. (4-84) and (4-85) are depicted as the solid-line curve q_G and broken lines q_R respectively in Fig. 4-10. The sigmoid shape of the heat-generation curve approaching a constant-q_G value reflects the characteristics of Eq. (4-84) since the exponential term diminishes rapidly in the high-temperature range. Linearity of the heat-removal curves results from the assumption that the molar heat capacity C_p and flow rate F of the feed stream and the overall heat-transfer coefficient U are insensitive to the temperature variation. Also, the heat-transfer area A_w, surface temperature T_w, and feed temperature T_0 are fixed for each q_R line. Thus, the temperature of the reaction system T is the only independent variable in Eq. (4-85). The intersections of the q_G curve with q_R lines, ℓ, m_1, m_2, m_3, and h represent steady states since $q_G = q_R$ at these points. The different q_R lines may be produced by varying T_0, U, and F. For example, when T_{01} is chosen and relatively high F or U or both are used, point ℓ corresponds to a negligible rate of reaction (and q_G). If the feed temperature is raised to T_{02} and either F or U or both are reduced somewhat, three intersections m_1, m_2, and m_3 are obtained; m_2, however, is unstable, as in this region the rate of change in q_G with respect to T is greater than that in q_R. Increasing the feed temperature further to T_{03} and slightly reducing F, for example, would result in the intersection h, which represents steady-state operation at very high conversion.

A similar approach may be used to develop q_G and q_R curves for complex reactions, although procedures are more complex. Under certain conditions there could be more than three intersections of the two curves. Likewise, the q_G-q_R relations for the plug-flow reactor are considerably more complex than the case discussed because both q_G and q_R as well as T vary along the length of the reactor. Further discussion on these cases is found in references such as B-8, B-9, B-10, D-1, and G-7.

Extent of back mixing. For most of the practical reactors, the extent of back mixing lies somewhere between a maximum for an ideal back-mix reactor and a minimum for an ideal plug-flow reactor. This behavior may be analyzed for a first-order isothermal reaction by considering either a system consisting of a number m of equal-sized CSTRs connected in series or a plug-flow reactor with effluent recycle. For the former case, material balances for individual reactors and successive substitution result in

$$\Theta_{bt} = \frac{m}{k}\left[\left(\frac{C_{Ao}}{C_{Am}}\right)^{1/m} - 1\right] \tag{4-63}$$

where Θ_{bt} is the overall space time for m CSTRs. As m increases, Eq. (4-63) approaches the corresponding equation for the plug-flow reactor

$$\Theta_P = \frac{1}{k}\ln\left(\frac{C_{Ao}}{C_{Am}}\right) \tag{4-65}$$

The effect of varying degrees of back mixing on the performance of the isothermal plug-flow reactor was evaluated by Gillespie and Carberry (G-25), who utilized the recycle ratio (R = quantity of effluent recycled/quantity of effluent discharged) as the mixing parameter. Space time for this case can be expressed by

$$\Theta_P = -\ln\left[\frac{\left(\frac{C_{Ao}}{C_{Am}}\right)(R+1)}{\left(\frac{C_{Ao}}{C_{Am}}\right)R + 1}\right] \tag{4-86}$$

Equation (4-86) approaches the expression for a single CSTR as R is increased. This can be shown by expanding the right-hand side of Eq. (4-86) by logarithmic series and solving for (C_{Ao}/C_{Am}). Mathematical models representing the effect of mixing on conversion in continuous reactors were developed by Weinstein and Alder (G-26) on the basis of micromixing and macromixing modes.

Types of phases involved. Many solid-catalyzed reactions are carried out in packed tubular flow reactors. Examples include ammonia synthesis, oxidation of SO_2, and cracking of petroleum hydrocarbons. However, a great number of variants of the basic reactors (i.e., tubular and stirred-tank) as well as reactors of special design are used. They include fluidized-bed reactors, slurry-phase reactors, kilns and hearth furnaces, and gas-absorption towers.

Economic Factors Since the complexity of the reactor does not necessarily bear any relation to the reactor type, no exact correlation is possible between costs and type of reactor. The costs of a reactor must therefore be evaluated for specific cases. Nevertheless, some qualitative guidelines are available in regard to capital and operating

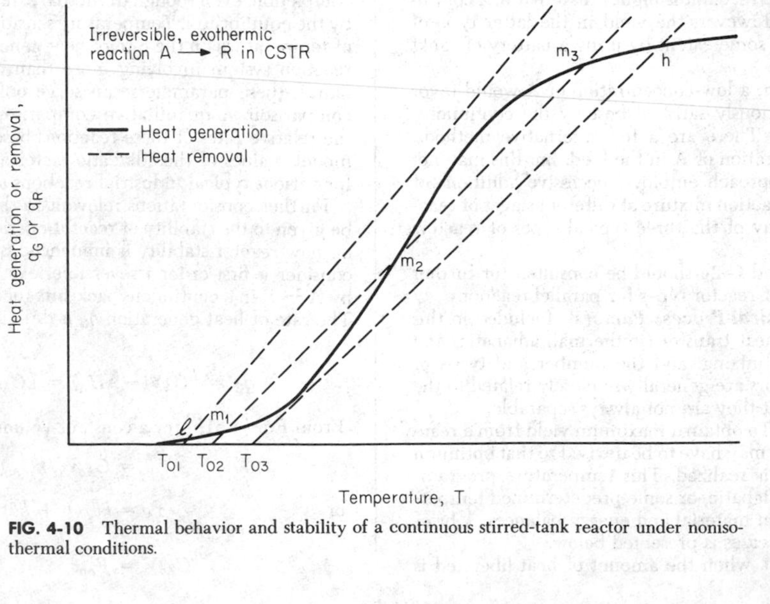

FIG. 4-10 Thermal behavior and stability of a continuous stirred-tank reactor under nonisothermal conditions.

costs for the three common reactor types, and those are listed in Table 4-13 (the last column).

Social Factors Factors in this category cannot be evaluated in terms of direct monetary values, but they can influence the decision on the reactor type. For example, one type of reactor may be safer to operate than others. Furthermore, the design feature of a reactor may facilitate the operation, thereby making it easier for the workers. Certain types of reactors may produce wastes that are difficult to treat and that tend to pollute the environment. These factors have been drawing more attention in recent years and are now playing an increasingly important role in plant design. In this connection, a book by Denbigh (G-27) is an interesting reference. Rase (D-1) also emphasized the critical needs for identifying and quantifying fire, physiological, and ecological hazards associated with the reaction itself as well as feed, product, and waste streams. Among the means for such quantification are applications of fundamentals of physical chemistry including thermodynamics in the assessment of explosion potential and limits of flammability. Compilations of hazard ratings of specific chemical processes and chemical substances are available in several references, e.g., Ref. G-28.

Noncatalytic Heterogeneous Reactors Discussion of this subject will be limited to reactors for fluid-solid reactions, with an emphasis on gas-solid reactions, since those for other types of heterogeneous reaction vessels (i.e., those for gas-liquid and liquid-liquid reactions) are presented in other sections. Examples of industrially important fluid-solid reactions are listed in Table 4-3. The primary considerations in the selection and design of a reactor for fluid-solid reactions may include (1) shape, particle-size distribution, and structure of solids; (2) physical properties of fluid and solid respectively, (3) kinetics and mechanism of reactions, and (4) fluid-solid flow and contact patterns.

Reactors for fluid-solid systems are characterized by many different design features. Typical examples are (1) fluidized-bed reactors (mixed solid flow; e.g., fluidized-bed iron-ore reduction); (2) fixed-bed reactors (fluid plug flow; e.g., ion-exchange column); (3) moving-bed reactors (both fluid and solids in plug flow; e.g., moving-bed oil-shale retort); and (4) rotary calciners-kilns (both fluid and solids in plug flow; e.g., reduction roasting of iron oxide ores). The flow patterns of fluid and solids indicated for the various types of reactors are approximations to ideal conditions. In many cases, however, flow and contact patterns of fluid and/or solids in industrial reactors are not readily identifiable in terms of specific ideal patterns.

Numerous theoretical and semiempirical models have been proposed to delineate behavior of fluid-solid systems in reactors with a final purpose of scale-up to commercial-scale reactors. While those models can sometimes approximate certain relatively simple systems, they are far from many real, industrially important reaction systems. This situation is primarily due to the sensitivity of some key design parameters to the reactor scale. The complexity in the reaction kinetics and mechanisms of fluid-solid systems also defies any simple analysis. Consequently, design of reactors in such cases is heavily dependent on past practical experiences accumulated in conjunction with experimental data bases of limited scopes. Thus, reactor scale-up based on theoretical and/or semiempirical models frequently fails mainly because the design data base has been obtained from equipment of much smaller scale than the required commercial size and under process conditions that are considerably different from those of actual operations. Despite these shortcomings, theoretical developments have contributed to understanding of the nature of reactors for fluid-solid reactions, and provided useful guidelines for design.

To illustrate the methods of approach in the analysis of fluid-solid reactors, simplified idealized systems considered by Levenspiel (B-10) are adapted for presentation in the following discussion. Five different cases are analyzed with the basic assumption that the fluid composition is uniform throughout the reactor. Also assumed are (1) ideal flow patterns (plug or mixed flow) for the solids, (2) applicability of the fixed-size-particle–unreacted-core model, and (3) control of the process by one type of mechanism (i.e., diffusion through gas film or ash layer or chemical reaction) at a time. The assumptions of the uniform fluid composition and mixed flow for solids represent good approximations in certain limited cases involving fluidized-bed reactors. In other cases, however, deviations from such conditions may be significant, especially with regard to the former assumption. More realistic analyses of such cases have been detailed by a number of researchers including Kunii and Levenspiel (H-1).

The results of analysis based on these simplified assumptions are summarized in Table 4-14. Notwithstanding the limitations imposed upon these correlations as discussed previously, they have been satisfactorily applied to design of reactors for some industrially important fluid-solid reactions. The methods of approach employed in the development of these correlations are illustrated below for two separate cases (cases 2 and 4).

Case 2. Plug flow of solids of different but unchanging particle sizes. For a fluid-solid reaction system consisting of a solid reactant B of mixed sizes ranging up to radius R_m and a fluid reactant A, the extent of the reaction may be described by

$$\begin{bmatrix}\text{Average}\\ \text{fractional}\\ \text{conversion of } B\end{bmatrix} = \bar{x} = 1 - \sum_{R_i=0}^{R_m}\left[1 - \begin{pmatrix}\text{fractional}\\ \text{conversion of } B \text{ of}\\ \text{size } R_i,\ x_B(R_i)\end{pmatrix}\right]\begin{bmatrix}\text{fraction of}\\ B \text{ of size } R_i\\ \text{in feed}\end{bmatrix} \tag{4-87a}$$

or

$$\bar{x}_B = 1 - \sum_{R_i=0}^{R_m} [1 - x_B(R_i)][F'(R_i)/F'] \tag{4-87b}$$

where F' represents the total volumetric or mass (if density of B does not change appreciably with reaction) flow of solid feed B, and $F'(R_i)$ is the volumetric or mass flow rate of B of size R_i in the feed. It is to be noted that Eq. (4-87b) is subject to the restriction $0 \leq x_B \leq 1$. This restriction takes into consideration a solid mixture that contains solids of small particle sizes which would be converted completely in less time than the solids residence time θ_p (for plug flow) in the reactor. An alternative form to Eq. (4-87b) is

$$\bar{x}_B = 1 - \sum_{R(\Theta)}^{R_m} [1 - x_B(R_i)][F'(R_i)/F'] \tag{4-88}$$

where $R(\Theta)$ refers to the size (radius) of the largest particle that would be converted completely in the reactor.

Thus, the ratio $F'(R_i)/F'$ is computed for different particle sizes by knowing the solid feed condition (i.e., the flow rate and particle-size distribution). The nature of the controlling step may be confirmed by a series of runs with solids of several different particle sizes to determine, for example, the relation between θ (the time required to obtain a specified fractional conversion) and the particle size R_i and by comparing the results with equations listed in Table 4-4. Once the controlling step has been identified, the expression for $x_B(R_i)$ is obtained from Table 4-4 and suitable experimental values (e.g., times required for complete conversion of solids of different particle sizes and reactor residence time) substituted to calculate the average fractional conversion $\bar{x}_B$.

Case 4. Mixed flow of solids of different but unchanging particle sizes with no solid entrainment. This case may be represented by a fluidized-bed reactor behaving as a mixed-flow reactor in which the conditions (e.g., size distribution) in the reactor are identical to those in the effluent stream. This implies that the mean residence time of solids of size R_i, $\bar{\theta}(R_i)$, is the same as that of all sizes in the reactor, $\bar{\theta}$. The residence times of individual solid particles, however, are not necessarily the same. The average fractional conversion of solids B of particle size R_i, $\bar{x}_B(R_i)$, is expressed as

$$\bar{x}_B(R_i) = 1 - \sum_{\theta=0}^{\infty}\left[1 - \begin{pmatrix}\text{fractional conversion of } B\\ \text{particles of size } R_i \text{ and ages}\\ \text{between } \theta \text{ and } \theta + d\theta\end{pmatrix}\right]\begin{bmatrix}\text{fraction of solids}\\ \text{effluent with size } R_i\\ \text{and reactor residence}\\ \text{time between } \theta \text{ and}\\ \theta + d\theta\end{bmatrix} \tag{4-89a}$$

TABLE 4-14 Solid Conversion–Time Correlations for Reactions in Gas-Solid Reactors*

Case	Particle-size distribution	Controlling step	Solid conversion–time correlations
		Plug flow of solids and uniform gas composition	
1	Uniform size		Rate equations listed in Table 4-4 are applicable.
2	Mixture of different but unchanging sizes		$\bar{x}_B = 1 - \sum_{R_i=0}^{Rm} [1 - x_B(R_i)] \frac{F'(R_i)}{F'} \quad 0 \le x_B \le 1$ or $\bar{x}_B = 1 - \sum_{R(\Theta)}^{Rm} [1 - x_B(R_i)] \frac{F'(R_i)}{F'}$
		Mixed flow of solids and uniform gas composition (fluidized bed)	
3	Uniform size	(a) Diffusion through gas film	$\bar{x}_B = (1 - e^{-\gamma})/\gamma;\ \gamma = \bar{\Theta}/\bar{\theta}$ or $\bar{x}_B = 1 - \frac{1}{2!}\gamma + \frac{1}{3!}\gamma^2 - \frac{1}{4!}\gamma^3 + \cdots$
		(b) Diffusion through ash	$\bar{x}_B = 1 - \frac{1}{5}\gamma + \frac{19}{420}\gamma^2 - \frac{41}{4620}\gamma^3 + \cdots$
		(c) Chemical reaction	$\bar{x}_B = 3/\gamma - 6/\gamma^2 + 6(1 - e^{-\gamma})/\gamma^3$ or $\bar{x}_B = 1 - \frac{1}{4}\gamma + \frac{1}{20}\gamma^2 - \frac{1}{120}\gamma^3 + \cdots$
4	Mixture of different but unchanging sizes	(a) Diffusion through gas film	$\bar{\bar{x}}_B = 1 - \sum^{Rm} (\frac{1}{2!}\gamma_i - \frac{1}{3!}\gamma_i^2 + \cdots) \frac{F'(R_i)}{F'};\ \gamma_i = \frac{\bar{\Theta}(R_i)}{\bar{\theta}}$
		(b) Diffusion through ash	$\bar{\bar{x}}_B = 1 - \sum^{Rm} (\frac{1}{5}\gamma_i - \frac{19}{420}\gamma_i^2 + \cdots) \frac{F'(R_i)}{F'}$
		(c) Chemical reaction	$\bar{\bar{x}}_B = 1 - \sum^{Rm} (\frac{1}{4}\gamma_i - \frac{1}{20}\gamma_i^2 + \cdots) \frac{F'(R_i)}{F'}$
		Mixed flow of solids and uniform gas composition (fluidized bed)	
5	Mixture of different but unchanging sizes; with solid entrainment	(a) Diffusion through gas film	$\bar{\bar{x}}_B = 1 - \sum^{Rm} (\frac{1}{2!}\bar{\gamma}_i - \frac{1}{3!}\bar{\gamma}_i^2 + \cdots) \frac{F_0'(R_i)}{F_0'};\ \bar{\gamma}_i = \frac{\bar{\Theta}(R_i)}{\bar{\theta}(R_i)}$
		(b) Diffusion through ash	$\bar{\bar{x}}_B = 1 - \sum^{Rm} (\frac{1}{5}\bar{\gamma}_i - \frac{19}{420}\bar{\gamma}_i^2 + \cdots) \frac{F_0'(R_i)}{F_0'}$
		(c) Chemical reaction	$\bar{\bar{x}}_B = 1 - \sum^{Rm} (\frac{1}{4}\bar{\gamma}_i - \frac{1}{20}\bar{\gamma}_i^2 + \cdots) \frac{F_0'(R_i)}{F_0'}$

*For reaction A(gas) + bB(solid) → products. Adapted from Ref. B-10.
$\bar{x}_B$, $\bar{\bar{x}}_B$: Mean fractional conversion of solid reactant.
Rm: Maximum solid particle size in feed.
R_i: Different particle sizes; (R_i) means that the variable preceding is a function of R_i.
F': Volumetric feed rate of solid.
F_0': Volumetric feed rate of solids for case 5.
$\bar{\theta}$: Mean residence time of solids in reactor.
$\bar{\Theta}$: Time required for complete conversion of single solid particle.

or

$$\bar{x}_B(R_i) = 1 - \int_0^{\Theta(R_i)} [1 - x_B(R_i)]\epsilon \, d\theta \tag{4-89b}$$

In this equation, ϵ is the exit age distribution or residence-time distribution (RTD) function defined previously [e.g., refer to Eq. (4-71)].

Since for mixed flow characterized by a fluidized bed

$$\epsilon = (e^{-\theta/\bar{\theta}})/\bar{\theta} \tag{4-90}$$

Eq. (4-89) becomes

$$\bar{x}_B(R_i) = 1 - \int_0^{\Theta(R_i)} [1 - x_B(R_i)] \left(\frac{e^{-\theta/\bar{\theta}}}{\bar{\theta}}\right) d\theta \tag{4-91}$$

The upper limit for the integral in Eq. (4-91) implies that it excludes particles of size R_i which remain in the reactor longer than $\Theta(R_i)$, the time required for complete conversion, since otherwise $x_B(R_i) > 1$. On the basis of an approach similar to Eq. (4-87a), the overall average fractional conversion for all particle sizes involved $\bar{\bar{x}}_B$ is described by

$$\bar{\bar{x}}_B = 1 - \sum_{R_i=0}^{Rm} [1 - \bar{x}_B(R_i)][F'(R_i)/F'] \tag{4-92}$$

Specific expressions for reaction systems that are subject to different controlling mechanisms may be derived from Eq. (4-92) by introducing respective $\bar{x}_B$ expressions from Table 4-4. The results thus obtained are listed in Table 4-14 (case 4).

Reactors for Solid-Catalyzed Reactions Solid-catalyzed fluid-phase reactions probably represent the most complex reaction system, and they are the most important and commonly encountered type in industrial processes. In view of such complexity and importance, this subsection presents a brief discussion of key factors to be considered in the design of reactors for solid-catalyzed processes. Among these factors are those that pertain to fluid dynamics (e.g., pressure drop across the reactor), mass transfer (external diffusion, pore diffusion, etc.), and characteristics of catalysts (e.g., porosity, internal and external surface areas, and susceptibility to poisons). Before proceeding with discussion of these factors, a concise description of basic catalyst constituents is presented to assist in understanding catalyst characteristics and functions and to facilitate subsequent discussion. Additional details relevant to catalyst constituents and preparations are found in references such as I-1 and I-2.

Basic Catalyst Constituents There are three major constituents involved in the preparation of a typical catalyst: the support (or carrier), the active catalyst agent, and the promoters which are often used.

Support or carrier. Materials frequently used as catalyst supports are porous solids with high total surface areas (external and internal) which would provide high concentrations of active sites per unit weight of catalyst. Such materials include activated aluminas, silicas, and aluminosilicates. The total surface area available from such a material varies not only with its crystalline structure but also with the method of catalyst preparation. The ranges of internal surface areas for selected catalysts are shown later in Table 4-16. The high surface area, however, is not always an advantage for catalytic reactions. In certain reactions, the fine pore structures of the high-area catalyst support tend to impede pore diffusion and intraparticle heat transfer, which may result in unfavorable product distribution and/or in sintering of the catalyst particles. The catalyst support may enhance the function of the catalyst agent.

Active catalyst agent. This is the constituent primarily accountable for the catalytic function and includes metals, semiconductors, and insulators. In general, the active catalyst agent must be prepared by one or more of the several chemical-processing steps, such as precipitation, leaching, thermal decomposition, and thermal fusion. For catalysts that require the support, the agent thus processed is deposited on the support by spraying or soaking, which is followed by drying, calcination, and, if necessary, activation through methods such as reduction or oxidation.

Promoter. Compounds that are added to enhance physical or chemical function of the catalyst are termed "promoters." They may be incorporated into the catalyst during any step in the chemical processing of catalyst constituents. In some cases, promoters are added during the course of reaction. A typical example of the physical promoter is Al_2O_3, which is added to the iron catalyst for the ammonia synthesis for the purpose of preventing the growth of the active agent (iron crystals). The chemical promoter generally augments the activity of the catalyst agent, but certain promoters are added to retard undesirable side reactions. Thus, for example, a minute quantity of ethylene chloride added as a chemical promoter to the ethylene oxidation catalyst enhances the product selectivity for ethylene oxide by suppressing production of CO_2.

Fluid Dynamic Factors Equations for evaluating pressure drop of a reacting fluid through packed beds have been derived on the basis of an approach similar to that for empty pipes. The mechanical-energy balance is generally used with an assumption of negligible static head. However, the size (in terms of equivalent diameter D_p) and shape of particles and the external void fraction of the bed ϵ must be taken into account in addition to the parameters usually considered for empty pipes. An equation for both laminar and turbulent flow has been developed by Ergun (I-3),

$$\frac{\Delta P}{L} = \left[\frac{150(1-\epsilon)\mu}{D_p} + 1.75G\right]\frac{1-\epsilon}{\epsilon^3}\frac{G}{D_p \rho g_c} \qquad (4\text{-}93)$$

where ΔP is the pressure drop across a bed of depth L, μ and ρ are the viscosity and density of the reacting fluid respectively, and G is the superficial mass flow rate, kg/(s·m²) [lb/(h·ft²)]. D_p is defined as $D_p = 6v_p/a_p$, where v_p and a_p are the volume and external surface area of a single particle. For packed beds of mixed particles, the surface mean equivalent diameter $\overline{D}_P$ is used in place of D_P. Equation (4-93) is based on the assumption that the packed bed consists of particles of similar size and shape throughout, with uniform packing, negligible wall effect, and no channeling.

Equation (4-93) may also be used as a good approximation for a fluidized-bed reactor up to the point of minimum fluidizing conditions. Beyond that range, however, the pressure-drop (ΔP) correlation and other fluid dynamic factors are considerably more complex than for the packed-bed reactor. For example, among the parameters that influence the ΔP value are the behavior of the fluidized bed (e.g., smooth fluidization, slugging, or channeling) and the particle-size distribution as well as the gas flow rate. The first two parameters are more closely interrelated than the third parameter. Thus, after reaching a peak ΔP at the point of minimum fluidization, the ΔP of a smoothly fluidizing bed would drop to a value approximately corresponding to the static pressure of the bed and remain nearly constant with further increase in the gas flow rate until the entrainment point is reached (curve *a*, Fig. 4-11). In contrast, a slugging bed (curve *b*) would display a wide fluctuation in the ΔP beyond the point of minimum fluidization, while a channeling bed (curve *c*) might exhibit the ΔP far below the bed static pressure. References such as H-1, I-4, and I-5 present the specific data base and detailed discussion for fluidized-bed reactors. Some aspects of the design of fluidized-bed systems are also discussed in Sec. 20.

Catalyst Porosity Catalysts used in industrial processes are usually porous pellets. The porous structure affords a considerable amount of the effective internal surface area per unit reactor volume. Because of the complex mechanism (see subsection "Mechanism of Solid-Catalyzed Reactions") involved in catalysis, the structure and internal surface area of the catalyst pores play extremely important roles in determining various key parameters. These parameters include the catalytic activity, permeability, diffusivity, thermal conductivity, effectiveness factor, and mechanical strength.

Several methods are available for characterizing catalyst pore structure. The BET method (I-6) and the method by Ritter and Drake (I-7) have been used as standards, the former for measurement of the internal surface area and the latter for determination of pore diameter and pore-size distribution. An improved method by Cranston and Inkley (I-8) evaluates the average pore size, pore-size distribution, internal surface area, and internal void volume based on the physical adsorption or desorption isotherms of nitrogen.

The importance of the catalyst surface area and pore structure becomes obvious when one considers the fact that solid-catalyzed

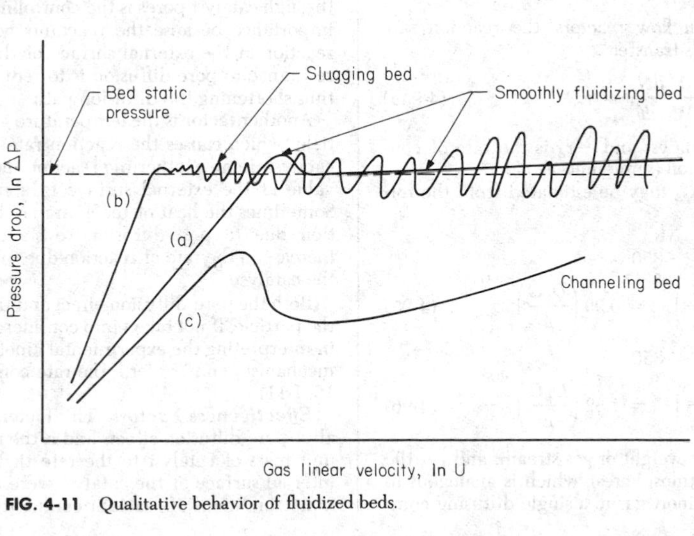

FIG. 4-11 Qualitative behavior of fluidized beds.

reactions proceed on the catalyst surface by means of adsorption and desorption processes and diffusion through the pores. However, the degree of importance varies, depending on the controlling step in the overall process. Thus, when surface chemical reaction is controlling, diffusion to the external surface and through the pores may be so fast that an isothermal reaction rate per unit catalyst weight will be essentially proportional to the total surface area, and the catalyst porosity becomes very important (because the internal surface is generally much larger than the external surface). In contrast, if pore diffusion is the controlling step, reaction on the external surface tends to predominate, and the porosity becomes of less importance since the internal surface area would not be utilized efficiently.

External Diffusion When the overall rate of reaction is subject to an appreciable influence of mass transfer of the gas phase to the external surface of catalysts, the mass-transfer rate can be described by

$$r'_m = k_{gA} a_m (p_{Ab} - p_{As}) \tag{4-94}$$

where r'_m = mass-transfer rate of component A, mol/(s·kg of catalyst) [(lb·mol)/(h·lb of catalyst)]
k_{gA} = mass-transfer coefficient for component A, mol/(s·Pa·m^2)[(lb·mol)/(h·atm·ft^2)]
a_m = external surface area of catalyst per unit mass, m^2/kg (ft^2/lb); $a_m = a/\rho_b$; a is obtained from Table 4-15
ρ_b = bulk density of catalyst bed, kg/m^3 (lb/ft^3)
p_{Ab} = partial pressure of diffusing component A in bulk stream, Pa (atm)
p_{As} = partial pressure of diffusing component A at solid surface, Pa (atm)

TABLE 4-15 External Surfaces Area per Volume of Catalyst Bed a in m^2/m^3 (ft^2/ft^3) of bed; $a_m = a/\rho_b$*

		Values of a			
	Mesh-size opening	% void space			
D_p, in.	(Tyler)	26	30	40	50
0.5	....	107	101	86.4	72
0.4	2	133	126	108	90
0.3	2.5	178	168	144	120
0.2	3.5	266	252	216	180
0.1	7-8	533	504	432	360
0.05	12-14	1066	1008	864	720
0.01	60	5330	5040	4320	3600

*To be used in Eq. (4-94).

Under steady-state conditions in flow reactors, the reaction rate becomes equal to the rate of mass transfer

$$-r'_A = -\frac{1}{W}\frac{dN_A}{d\theta} = r'_m \tag{4-94a}$$

where W is the total mass of catalyst and $-r'_A$ is the rate of consumption of A by chemical reaction (same units as r'_m).

The mass-transfer coefficient k_{gA} may be estimated from the following equations (E-7, p. 985):

For Reynolds number = $D_pG/\mu > 350$:

$$j_D = \frac{k_{gA} M_m p_{fA}}{G}\left(\frac{\mu}{\rho D_{Am}}\right)^{2/3} = 0.99\left(\frac{D_pG}{\mu}\right)^{-0.41} \tag{4-95}$$

For Reynolds number = $D_pG/\mu < 350$:

$$j_D = \frac{k_{gA} M_m p_{fA}}{G}\left(\frac{\mu}{\rho D_{Am}}\right)^{2/3} = 1.82\left(\frac{D_pG}{\mu}\right)^{-0.51} \tag{4-96}$$

where M_m is the mean molecular weight of gas stream, and p_{fA} the pressure film factor in pascals (atmospheres), which is analogous to the mean partial pressure of the inert gas in a single diffusing component in a stagnant gas. For a general gas-phase reaction $aA + bB \rightarrow rR + sS$, p_{fA} is defined as

$$p_{fA} = \pi - p_A(a + b - r - s)/a \tag{4-97}$$

D_{Am} is the mean diffusivity of component A. Other symbols have been defined previously [see Eqs. (4-93) and (4-94)].

With the mass-transfer coefficient known, the pressure differential $p_{Ab} - p_{As}$ can be evaluated through Eq. (4-94). The importance of external diffusion is indicated by the value of $p_{Ab} - p_{As}$, which may be ignored if $(p_{Ab} - p_{As})/p_{Ab}$ is less than a few percent. The value of $p_{Ab} - p_{As}$ can also be obtained from a chart (Fig. 4-12).

Evaluation of Partial Pressure and Temperature at Surface of Catalyst Particle The values of partial pressures and temperatures for any component i at the surface of catalyst particles can be estimated from charts (Figs. 4-12 and 4-13) prepared by Yoshida and Ramaswami and published by Hougen (I-9). To obtain the partial pressure drop for component i in terms of $\Delta y_i = \Delta p_i/\pi$ from Fig. 4-12, the following parameters are required:

Reynolds number: $\mathrm{Re} = G/a\phi\mu$
Schmidt number: $\mathrm{Sc} = (\mu/\rho D_{Am})_{gas}$
Mole fraction of pressure factor: $y_{fi} = p_{fi}/\pi$
Rate number: $R = r'_A/a_m\phi G_M$

where ϕ is the shape factor (being 1.0 for spheres, 0.91 for cylinders, and 0.90 for irregular grains) and G_M the molal mass velocity for the flowing gas stream. Similarly, the temperature drop Δt can be estimated from Fig. 4-13 from Re, heat transmission number Q, and Prandtl number $\mathrm{Pr} = C_p\mu/k_f$. Here k_f and C_p are the thermal conductivity and heat capacity of the fluid, respectively, and

$$Q = r_{mA}\,\Delta H_A/a_m\phi C_p G_M$$

where ΔH_A is the heat of reaction per mole of A at temperature T, and r_{mA} is the rate of mass transfer defined by

$$r_{mA} = k_{gA} a_m \phi (p_{Ab} - p_{As}) \tag{4-94b}$$

Likewise, the rate of heat transfer q_{mA} is described by

$$q_{mA} = r_{mA}\,\Delta H_A = h_g a_m \phi (t_b - t_s) \tag{4-98}$$

In this equation h_g refers to the heat-transfer coefficient per unit of external surface area of catalyst particle, and t_b and t_s are the bulk-fluid temperature and the catalyst surface temperature respectively.

The values of Δp and Δt thus obtained would facilitate the establishment of more reliable reaction models for solid-catalyzed gas-phase reactions from reaction-rate data.

Pore Diffusion and Internal-Temperature Gradient It was mentioned earlier (see "Catalyst Porosity") that when diffusion through catalyst pores is the controlling step, porosity is of secondary importance because the reactants would be consumed mainly by reaction at the external surface of the catalyst. One way to reduce resistance to pore diffusion is to reduce the particle size of catalyst, thus shortening the diffusion path.

Another factor is the temperature gradient within the catalyst particle, which causes the reaction rate to vary appreciably inside the catalyst. In an exothermic reaction the temperature increases from a value at the external surface to a maximum value at the center. Sometimes the heat of reaction may be so large that the rate reduction due to pore-diffusion resistance may be overcome by the increase in the rate of reaction due to a temperature gradient within the catalyst.

Both the pore-diffusion effect and the temperature gradient within the particle, if not taken into consideration, may lead to serious errors in interpreting the experimental kinetic data in terms of the reaction mechanism and "order," the rate constant and activation energy (I-10, I-11).

Effectiveness Factor This factor is useful in characterizing catalyst-pore-diffusion effects and is the ratio of actual reaction rate per unit mass of catalyst to the rate that would result if the complete internal surface of the catalyst were available for the reaction. The same conditions of reactant composition, temperature, and pressure

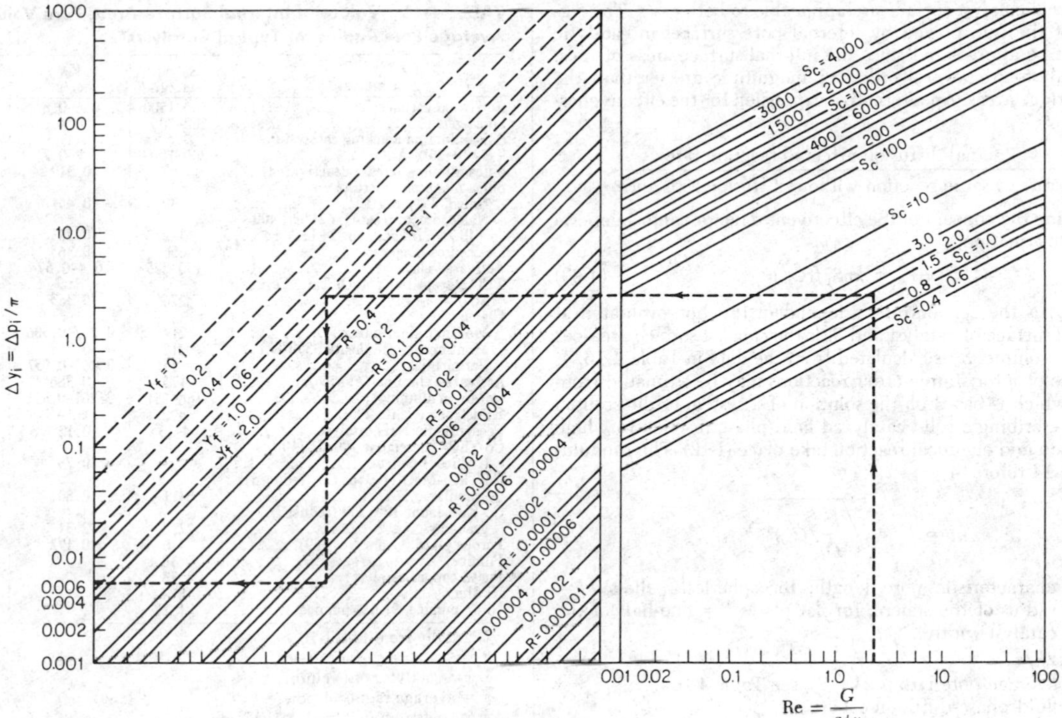

FIG. 4-12 Evaluation of partial-pressure gradients between a flowing fluid and the exterior surface of catalyst particles in a packed bed. [*Hougen*, Ind. Eng. Chem, *53(7), 509–528 (1961).*]

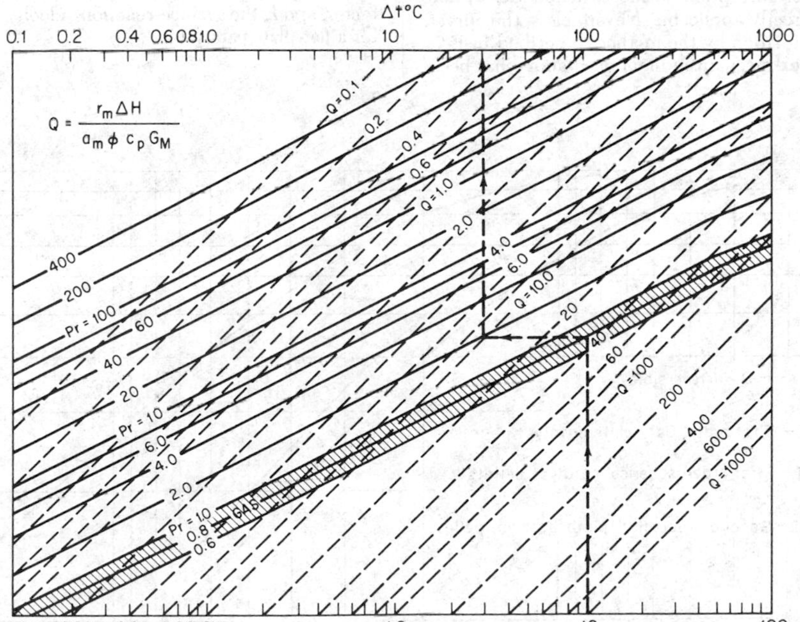

FIG. 4-13 Evaluation of temperature drop between a flowing fluid and the exterior surface of catalyst particles in a packed bed. (See Fig. 4-12.)

at the external catalyst surface are applicable to both cases. The significance of the role played by internal pore surfaces in catalytic activity can be appreciated since the internal surface areas of most porous catalysts are several orders of magnitude greater than the external surface areas. An alternative expression for the effectiveness factor η is

$$\eta = \frac{\text{actual diffusion-affected reaction rate}}{\text{rate of same reaction without diffusion resistance}}$$

Introducing the concept of the effectiveness factor η into a general reaction rate equation,

$$-r_A = k\eta S_v f(C_{io}) \qquad (4\text{-}99)$$

where $f(C_{io})$ is the concentration-dependent function evaluated at the external surface of catalyst and S_v is the catalyst surface area per unit reactor volume. S_v is calculated from the data in Table 4-16.

The value of η for simple-order reactions may be estimated from Fig. 4-14, which is based on the solution of a system of differential equations describing a solid-catalyzed fluid-phase reaction in which both diffusion and chemical reaction take place (I-12). The modulus m in Fig. 4-14 refers to

$$m = l\sqrt{\frac{k'}{D_i}(C_{io})^{n-1}}$$

where l = characteristic pore length; for spherical pellets l = radius of the sphere; for flat plates l = one-half of the catalyst width

$k' = 2k/\bar{r}$

$\bar{r}$ = average pore radius $2V_g/S_g$ (see Table 4-16)

D_i = fluid-phase-diffusion coefficient

C_{io} = concentration of reactant at external catalyst surface

n = true order of the chemical reaction

V_g = pore volume/g

S_g = surface area/g

The values of S_g, V_g, and r are obtained from Table 4-16. An isothermal condition with negligible adsorption effect has been assumed in Fig. 4-14. For a heterogeneous reaction system with a complex mechanism involving the adsorption–surface-reaction–desorption process, Fig. 4-14 is not directly applicable. Nevertheless, the curves may be used as an approximation by the methods described in E-7, I-10, and I-11. The complex mechanism mentioned above has been considered by Chu and Hougen (I-9) for a first-order solid-catalyzed reaction expressed by the rate equation

$$-r_A = \frac{\eta' k_s K_A p_A}{1 + K_A p_A} \qquad (4\text{-}99a)$$

where K_A is the adsorption equilibrium constant, η' the effectiveness factor, and k_s the surface-reaction velocity constant. The modulus m' for a flat-plate catalyst is

$$m' = l\sqrt{k_s v/D_e}$$

TABLE 4-16 Values of Internal Surface Area, Pore Volume, and Average Pore Radius for Typical Catalysts*

Catalyst	S_g, sq. m./g.	V_g, cc./g.	$\bar{r} = 2V_g/S_g$, A.
Activated carbons	500–1500	0.6–0.8	10–20
Silica gels	200–700	0.4	15–100
Silica-alumina cracking catalysts ~10–20% Al_2O_3	200–700	0.2–0.7	15–150
Silica-alumina (steam deactivated)	67	0.519	155
Silica-magnesia microsphere:			
Nalco, 25% MgO	630	0.451	14.3
Nalco, steam treated, 621°C., 400 lb./sq. in. gage for 24 hr.	322	0.283	17.6
Da-5 silica-magnesia	656	0.365	11.1
Activated clays	150–225	0.4–0.52	~100
TCC clay pellets (MgO, CaO, Fe_2O_3, SO_4) = ~10%	276	0.363	26.3
Clays:			
Montmorillonite (raw)	214	0.297–0.306	~28
Montmorillonite (heated 550°C.)	212	0.268	25.2
Vermiculite	35	0.063–0.057	~314
Activated alumina (Alorico)	175	0.388	45
CoMo on alumina	168–251	0.261–0.331	20–40
Kieselguhr (Celite 296)	4.2	1.14	11,000
Fe–synthetic NH_3 catalyst	4–13	0.12	200–1000
Co-ThO_2-Kieselguhr 100:18:100 (reduced) pellets	42.3	0.73	345
Co-ThO_2-MgO (100:6:12) (reduced) granular	84.1	0.80	190
Co-Kieselguhr 100:200 (reduced) granular	22.8	2.31	2030
Porous plate (Coors No. 760)	1.6	0.172	2150
Pumice	0.38		
Fused copper catalyst	0.23		
Ni film	8.4		
Ni on pumice, 91.8% pumice	1.27		

*Partially from Ref. I-10.

S_g = catalyst surface area.

V_g = catalyst pore volume.

r = average radius of pore.

Å = angstrom unit = 1×10^{-8} cm.

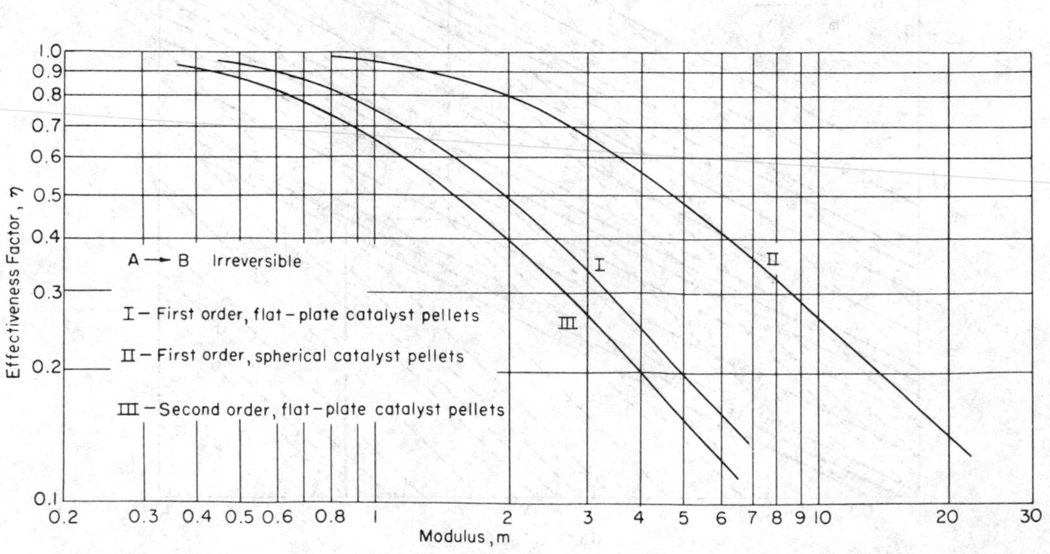

FIG. 4-14 Effectiveness factor for reactions of simple order.

where v is the molal volume of the gas phase, l represents one-half of the width of a flat-plate catalyst, and D_e the effective diffusivity. Figure 4-15 shows the relation between η' and m' for various values of $\zeta = 1(K_A\pi)$.

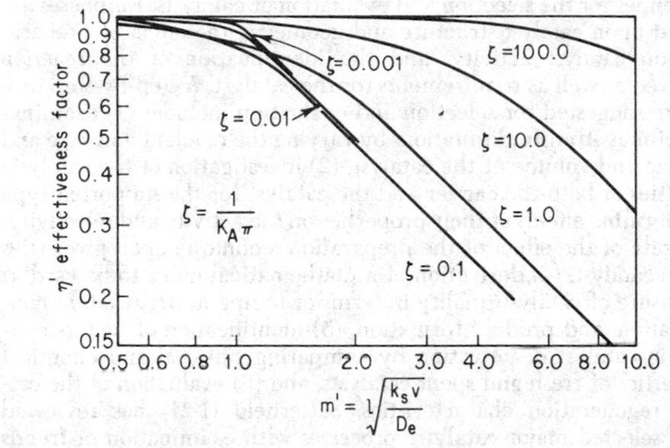

FIG. 4-15 Effectiveness factor for reaction model: $-r_A = (\eta' k_s K_A p_A)/(1 + K_A p_A)$. (See Fig. 4-12.)

Parameters affecting effectiveness factor. The effects of temperature gradients within the catalyst particles on the effectiveness factor have been investigated by Schilson (I-13), and by Mingle and Smith (I-14). The latter authors also demonstrated the effects of pore structure and thermal conductivity of catalyst particles on internal temperature and concentration gradients.

Variation of the effectiveness factor with the modulus as shown in Figs. 4-14 and 4-15 may be analyzed by dividing the curves into three regions. Hougen (I-9) considered Region 1 as the portion for $m < 1$, Region 2 as that for $1 < m < 10$, and Region 3 for $m > 10$ on the basis of the curve for a first-order reaction similar to that in Fig. 4-14. Region 1 is characterized by the value of η being almost unity, indicating a negligible diffusion resistance; Region 2, by the distinct curvature in the η versus m curve; and Region 3, by pore diffusion being the controlling step. In Region 1 the product selectivity for consecutive-type reactions will be subjected to the influence of the catalyst particle size as well as to nonselective poisoning. Except for Region 1, the effectiveness factor generally exhibits a downward trend with increasing temperature, as a result of the exponential rate of increase of the modulus with temperature.

The effectiveness factor is almost unaffected by the shape of the catalyst particle [spheres, flat plates, and cylinders (I-15)], but it is always lower for a spherical particle than for a flat plate of the same characteristic dimension (I-15, I-16) ($l = R/3$ where l is the half thickness of the plate and R the radius of the sphere).

In general, the effectiveness factor for a solid-catalyzed, nonisothermal reaction that accounts for the interphase (fluid–external-catalyst surface) and intraphase (fluid–internal-catalyst surface) concentration (or partial pressure) and temperature gradients may be expressed as a function of several dimensionless parameters as in (I-22, B-8)

$$\eta = f(m, \alpha, \beta, \mathrm{Bi}_m, \mathrm{Bi}_h) \tag{4-100}$$

where m is the Thiele modulus defined previously, α is the Arrhenius number E/RT_0, β is termed a heat-of-reaction parameter, $(-\Delta H)D_eC_{A0}/\bar{\lambda}T_0)$, Bi_m refers to the mass Biot number $k_{gA}\ell/D_e$ whereas Bi_h represents the thermal Biot number $h\ell/\bar{\lambda}$, T_0 is the bulk-fluid temperature, $\bar{\lambda}$ is the thermal conductivity of catalyst particle, and h is the convective heat-transfer coefficient. Weisz and Hicks (I-22) solved both the mass and the energy-balance equations numerically to obtain plots of η values as a function of m, α, and β for first-order reactions. The η values for the case involving CO oxidation over the supported Pt catalyst were computed by Carberry (B-8) on the basis of the rate equation

$$r_A = \frac{kC_A}{(1 + KC_A)^2} \tag{4-101}$$

where C_A represents the CO concentration and K is the equilibrium constant for the reaction. Reference B-8 also presents comprehensive discussion on the effectiveness factor from various fundamental viewpoints.

Poisoning Effects on Catalyst Any substance that appreciably diminishes the catalytic reaction rate is termed a "poison." The poisoning effects of such substances take place primarily by absorption on the catalyst surface. Reactants and products are sometimes considered as poisons (I-16). The manners in which poisons exert an effect include (1) masking the active centers, (2) changing the catalyst selectivity, (3) catalysis of undesirable side reactions (e.g., dehydrogenation of petroleum stocks by a small amount of nickel), (4) physical blocking of the pores and/or covering of active sites with inert deposits such as carbon, and (5) changing the catalyst structure.

Table 4-17 lists important metal catalysts which are readily poisoned due to a strong adsorptive bond between the catalyst and the poison.

TABLE 4-17 Catalysts Poisoned*

Fe (26)†	Co (27)	Ni (28)	Cu (29)
Ru (44)	Rh (45)	Pd (46)	Ag (47)
Os (76)	Ir (77)	Pt (78)	Au (79)

*From Maxted, *Advances in Catalysis*, vol. III, Academic, New York, 1951.
†The number in parentheses represents the atomic number of the metal.

Maxted (I-17) classified the common poisons for these catalysts into three categories: (1) compounds of elements in group VB or VIB of the periodic series, (2) compounds of catalytically toxic metals (Cu^+, Cu^{++}, Zn^{++}), and (3) multiple-bond molecules (e.g., CO_2, C_2H_4).

Determination of poisoned fraction. The methods used in determining the poisoned fraction may be illustrated by an example (I-18) in which $Pb(O \cdot CH_3CO)_2$ was the poison. The catalytic activity of poisoned Pt black was determined in the decomposition of H_2O_2 and oxidation of ethanol. The activity of the poisoned area A was correlated with that of the nonpoisoned area A_0 by an empirical equation,

$$A/A_0 = 1 - \delta C = 1 - \bar{\beta} g_a/Z \tag{4-102}$$

where δ represents the specific poisoning per unit concentration of poison C, $\bar{\beta}$ the probability of poisoning, g_a the amount of lead per gram of Pt black, and Z the number of active centers. The nature and structure of the adsorbed species on the catalyst surfaces can be determined by a technique such as the electron-spin-resonance method (ESR).

Effects of poisons on catalytic activity. The relation between catalytic activity and poison concentration varies, depending on the catalyst structure and the pellet modulus $m = l\sqrt{(k'/D_i)C_{io}^{n-1}}$ as defined previously. This relation is illustrated in Fig. 4-16. Catalytic activity diminishes linearly with an increasing fraction poisoned for a nonporous catalyst (curve A) having uniform distribution of poisons, whereas that for a porous catalyst (curve B) under the same condition decreases more gradually at first. When pore diffusion is the controlling factor (curves C and D with high m values), the reduction in the activity is quite drastic.

Generalized charts for estimation of the effectiveness factor with poisoning effects have been developed from the kinetic expressions of the Langmuir-Hinshelwood type by Roberts and Satterfield (I-16). A second-order surface reaction with significant poison effects (by either reactants or products) was considered.

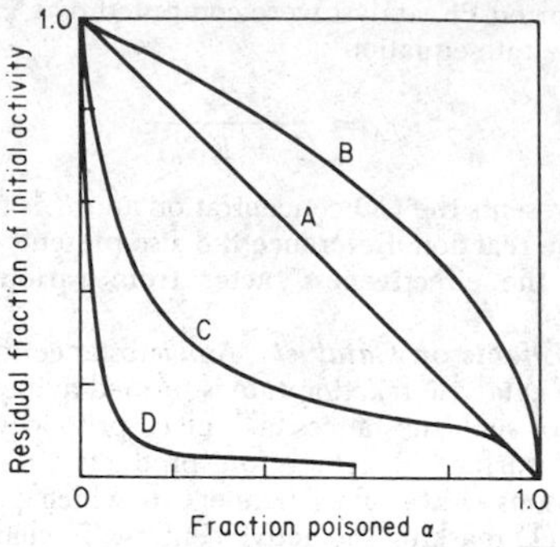

FIG. 4-16 Poisoning effect on catalyst activity. A = nonporous catalyst; B = uniform adsorption of poison; C = preferential adsorption near surface, m = 10; D = preferential adsorption near surface, m = 100. (*Wheeler,* Advances in Catalysis, *vol. 3, Academic, New York, 1951.*)

Selection of Catalyst The task of selecting a catalyst for a proposed chemical process is still dependent mainly on the empirical approach, even though numerous theories on catalysts and catalysis have been developed. Because of the complexity encountered in solid-catalyzed reactions, theoretical developments have been fragmentary and based largely on a number of simplifying assumptions which are often invalid in large commercial-scale operations. However, the use of a more rigorous approach frequently cannot be justified. Thus, applicability of the result from a purely theoretical approach to the selection of a suitable catalyst is rather limited.

In spite of this situation, theories of catalysis have been useful as an approximation to practical cases and as a qualitative guideline to facilitate understanding of experimental data. Consequently, in the selection of a catalyst, theory must be supported by the results from experimental investigation and evaluation of the catalysts as well as by experience. As an example, the theoretical development for a process may predict that pore diffusion is the controlling step and that the catalyst must have large pores and/or small particle sizes in order to achieve a high production rate and a favorable product selectivity. Considerations based on actual experience and experimental data would indicate that the mechanical strength of catalyst particles, in general, declines with increasing pore sizes. Likewise, reduction in particle size would result in increasing the pressure drop. Therefore, establishment of optimum values of pore and particle size must be based on a balance between the properties, process characteristics, and cost of the catalyst and the operating costs.

One of the difficulties in the selection of catalysts is that there is no absolute standard that can be used as a measure of activity for the enormous variety of catalysts available with diverse characteristics. Frequently, for purposes of comparison, some relative numerical values are assigned to various catalysts as activities based on reaction-rate measurements under specified conditions (initial reactant composition, temperature, pressure, etc.). In some instances a desirable catalyst may show a poorer performance than a less desirable one under conditions far removed from the optimum, and this may lead to an erroneous conclusion. In view of this situation, a systematic and consistent approach must be taken in evaluating and selecting catalysts.

Guides to systematic selection of catalysts. Realizing the fragmentary nature of work on catalyst selection, a number of investigators and specialists have been attempting to systematize and consolidate available experimental data. In an article by Hougen (I-9), current developments on the engineering aspects of solid catalysts were analyzed in detail. A book (I-19) published by the Academy of Sciences of the U.S.S.R. is a consolidation of contributions from more than 50 specialists in the field of catalysis and starts with a consideration of general problems of the scientific selection of catalysts including structural, energy, chemical, and electronic factors. These are followed by a discussion of kinetic factors affecting the catalyst selection, analyzing the relationship between the kinetic factors and the optimum catalyst and between the composition of the active center and the kinetics of heterogeneous catalysis. The need to account for nonhomogeneity of active centers in the selection of catalysts is emphasized. Finally, some rules governing the selection of oxide and metallic catalysts are presented, and attempts to clarify scientific principles in the search for new catalysts are described.

The concise paper by DeMaio and Naglieri (I-20) presents useful guidelines for the selection and evaluation of catalysts. Emphases are placed upon catalyst structure and geometry, the effect of preparation on catalyst activity, and the identification of the reaction involved as well as requirements for the catalyst. A step-by-step procedure suggested for selection and evaluation includes (1) examination of mass-transfer limitations by varying the reactant flow rate and the size and volume of the catalyst, (2) investigation of the catalytic activities of both the carrier and the catalyst for the supported type to determine effects of their properties on the activity and selectivity, (3) study of the effect of the preparation technique upon properties of the catalyst, (4) derivation of a mathematical index to be used as a measure of catalyst quality in terms of temperature, reactant concentration, and product formation, (5) identification of factors controlling catalyst deactivation by comparing physical and chemical properties of fresh and spent catalysts, and (6) evaluation of the catalyst regeneration characteristics. Satterfield (I-21) has reviewed some selected major catalytic processes with examination of trends in catalytic chemical processing.

Chemical Rate Data Listed in Tables 4-18 and 4-19 are references from which chemical rate data for selected chemical reactions may be obtained. In these tables the types of reaction and rate data are summarized, together with reaction conditions. Table 4-20 shows typical values of Arrhenius parameters for a number of important elementary reactions. Selected general references containing the chemical rate data base for pertinent reactions are summarized and tabulated in Table 4-21. Except for the last one (J-153), they are concerned primarily with the gas-phase reactions. These references present a critical evaluation of existing kinetic data and recommend reliable values of rate parameters whenever appropriate.

Physical Rate Data: Mass Transfer The problem of mass transfer becomes important in heterogeneous chemical reactions (catalyzed and uncatalyzed) because reactants are present in different phases and have to be transported to and/or across the interface before reactions can take place. Accordingly, the overall rate of reaction is subject to the influence of factors that affect the mass-transfer rate between phases. Outlined in this subsection are empirical equations and charts correlating the rate of mass transfer with various key parameters.

Gas-Liquid System A comprehensive discussion of the design of industrially important gas-absorption processes with accompanying chemical reaction is given in Sec. 14. Section 18 contains similar information for liquid-in-gas and gas-in-liquid dispersions.

Liquid-Liquid System For processes in which chemical reaction takes place in a partially or completely immiscible liquid-liquid system, the information presented in Secs. 15 and 21 should be helpful. In addition, some recent results outlined in the following paragraphs are useful.

The difference in density $\Delta\rho$ between the dispersed and the continuous phases is among the critical parameters that influence the rate of mass transfer in liquid-liquid as well as gas-liquid systems. This rate is generally favored by large $\Delta\rho$ (K-1). Since $\Delta\rho$ in the liquid-liquid system is usually small, one of the methods for promoting the mass transfer–chemical reaction for such a system is to provide large surface areas to facilitate intimate contact between the phases by means of agitation. Thus, reactors for liquid-liquid systems are mostly agitated vessels of different types (see Sec. 21).

It should be pointed out that the mass-transfer data for the liquid-liquid system could be subject to erroneous interpretation unless the effects of certain less obvious parameters are recognized. For example, very low concentrations of surface-active agents which may be present either in the feed or in the reaction-product stream could have dramatic effects on the interphase transport behavior (D-1). The nature of turbulence, which is not well understood, as well as agitator characteristics would also exert appreciable influence beyond certain ranges of power consumption. Therefore, reliable

TABLE 4-18 Chemical Rate Data for Inorganic Reactions*

Major reactants	Types of reaction and data	References
Ammonia	Synthesis in the presence of various catalysts; 250–500°C; rate-controlling steps, *M*, rate equations	J-69, J-70, J-71, J-72
Ammonia	Decomposition on various catalysts; 425–630°C; *M*, rate-controlling steps, rate equations	J-73, J-74, J-75
Ammonium nitrate	Thermal decomposition in fused salt solution; effect of added salt, rate curves, *M*	J-76
Ammonium perchlorate	Thermal decomposition; 200–280°C; *M*, *k*, AP, rate equation	J-1
Ammonium sulfite	Effect of Se compounds on oxidation of; 20°C, broad ranges of acidity and concentration; *M*, rate curves	J-2
Calcium carbide	Formation in fused phase; 1855–2000°C, *M*, rate curve	J-77
Calcium carbonate	Thermal decomposition in vacuum or nitrogen; 650–760°C; rate equation, *M*, *k*, AP	C-13, J-78
Calcium chloride	Thermal decomposition, 720–1200°C; rate and degree of decomposition	J-3
Calcium cyanamide	Thermal dissociation, ≤1300°C; *M*	J-4
Calcium hydroxide	Carbonation by CO_2, effects of agitation, concentration, time, temperature; 25–30°C; *M*, *k*	J-5
Calcium hydroxide	Preparation and thermal decomposition, 470–640°C; *k*, AP	J-6
Carbon	Reaction with water vapor; ~900°C; *M*, rate equation, *k*	J-7
Carbon dioxide	Reaction with coke; 815–1038°C; rate equation, *k*, *M*	J-79
Carbon dioxide	Reaction with hydrogen; 400–1050°C; rate equation, *k*, *M*	J-80
Carbon dioxide	Catalytic hydrogenation; rate data, contact potential difference, *M*	J-81
Carbon disulfide, caustic soda	Reaction in homogeneous system, 22–27°C; *M*, *k*	J-8
Carbon monoxide	Formation of CO_2 and C, 740–860°C; *k*, AP	J-9
Carbon monoxide	Water-gas conversion, 11.7 atm, conversion, yield	J-10
Carbon monoxide	Gas-phase reactions with Cl_2, NO_2; 381–473°C; *M*, *k*, AP	J-11
Carbon monoxide	Catalytic oxidation; 190–310°C; rate equation, catalytic activity, *M*	J-83, J-84
Carbon monoxide	Catalytic hydrogenation; rate equation, AP, overall heat-transfer coefficient	J-82
Iron (II)	Oxidation by aqueous Cl, 30°C; rate, *k*, AP	J-12
Lead sulfide	Effect of SO_2 on oxidation of; 500–1000°C; *M*, rate equation, industrial application	J-13
Magnetite	Reduction by hydrogen; 300–570°C; rate equation, *k*, *M*	J-85
Nickel and zinc oxides	Reduction by H_2; 350–500°C; *M*, *k*, AP, rate equation	J-14
Nitric acid	Catalytic oxidation of ammonia; space velocity, degree of conversion	J-86
Nitric acid	Optimization in commercial production; optimum diameter and plate spacing for absorption column	J-87
Nitrogen oxide	Reaction with water vapor; 102–425°C; rate of HNO_3 formation, composition of feed and product	J-15
Potassium carbonate	Reaction with SiO_2; *k*, AP, 630–865°C	J-16
Pyrite	Thermal decomposition in argon gas; ~400°C; rate equation, degree of conversion, *M*, AP	J-88
Pyritic material	Natural oxidation (sulfide-to-sulfate reaction); <45°C; rate data, *M*	J-89
Sodium and potassium carbonates	Thermal decomposition in drum furnace; 110–140°C (Na_2CO_3), 155–175°C (K_2CO_3); *M*, *k*, AP, rate equation	J-17
Sodium nitrate	Thermal decomposition; 570–670°C; effect of argon and air, *M*	J-98
Sodium sulfite	Oxidation by O_2; *k*	J-18
Sulfur dioxide	Oxidation using various catalysts; 400–600°C; rate equation, *M*, *k*, AP	J-90, J-91, J-92, J-93
Sulfur dioxide	Oxidation at high pressure (V_2O_5 catalyst); 342–400°C, 1.54–10.18 atm, rate equation, *k*, *M*	J-94
Sulfur dioxide	Oxidation in the presence of NO_2 and NO_3; 400–1050°C; rate equation, *k*, *M*, AP	J-95
Sulfur	Oxidation in aqueous-phase; rate equation, *k*, *M*, AP	J-96
Thiourea	Synthesis from $CaCN_2$ and $(NH_4)_2S$, 17–80°C; *k*, AP	J-19
Titanium dioxide	Chlorination in molten salts containing $FeCl_3$; *M*, *k*	J-20
Uranium or plutonium compound	Fluorination; rate equation, degree of completion, *k*, *M*	J-97
Zinc sulfide, C	Sublimation and reaction with C; rates of sublimation and reaction, AP	J-21

*k = rate constant, AP = Arrhenius parameters, and M = mechanism.

design and scale-up of the reactor for the liquid-liquid system have to be based primarily on data from the pilot-plant reactor with the scale and process conditions as close as possible to those of the full-scale plant.

The following empirical equation for the liquid-phase mass-transfer coefficient k_L, developed by Calderbank and Moo-Young (K-1) for the liquid-in-liquid dispersion system, may be used as a good approximation:

$$k_L \left(\frac{\mu}{\rho D_m}\right)^{2/3} = 0.13 \left[\left(\frac{P}{v}\right)\frac{\mu}{\rho^2}\right]^{1/4} \qquad (4\text{-}103)$$

where D_m refers to the molecular diffusivity of solute in solvent, P is the power dissipated, v is the volume of the dispersed phase, and μ and ρ are the viscosity and density, respectively, of the continuous phase.

The mean diameter of the dispersed liquid bubbles D_B can be estimated by K-1 and K-2:

$$D_B = 2.24 \left[\frac{\sigma^{0.6}}{(P/v)^{0.4}\rho^{0.2}}\right] f_v^{0.5} \left(\frac{\mu'}{\mu}\right)^{0.25} \qquad (4\text{-}104)$$

where f_v and μ' are the volume fraction and viscosity respectively of the dispersed liquid and σ is the interfacial tension. Equation (4-104) was derived from experimental data for a wide range of liquids with σ = 36 to 55 dyn/cm.

The importance of coalescence in determining the rate of some reactions and the effects of surface-active impurities and mass transfer on coalescence have been studied by several investigators (K-3, K-4).

Fluid-Solid System There have been numerous studies on mass and heat transfer in fluid-solid systems as indicated by the number of publications in the literature, including several books and sym-

TABLE 4-19 Chemical Rate Data for Organic Reactions*

Major reactants	Types of reactions and data	References
Acetaldehyde	Pyrolysis; *M*, *k*	J-22
Acetaldehyde	Reaction with atomic hydrogen in a discharge-flow system; 300 K, *M*, *k*	J-23
Acetylene	Dimerization in aqueous Cu_2Cl_2 solution; *M*, *k*	J-24
Acrylonitrile	Polymerization using various initiators; rate of reaction, rate equation, *M*, molecular-weight distribution	J-99, J-100
Acrylonitrile	Copolymerization with ethylene–sulfonic acid; rate of reaction, *k*, *M*, polymer composition, reactivity ratio	J-101
Aliphatic glycols	Esterification by acrylic and methacrylic acids; *M*, *k*	J-25
Aniline	Hydrogenation in vapor phase (catalytic); 160–210°C; rate equation, *M*	J-102
Aniline, acetic acid	Acetylation; conversion, *k*, AP	J-26
Benzaldehyde	Liquid-phase oxidation in benzene in presence of olefins; *M*, *k*	J-27
Benzaldehyde	Formation by nitric acid oxidation of benzyl ether in aqueous dioxane; rate equation, *k*, *M*, AP	J-103
Benzene	Catalytic oxidation with air on various catalysts; 180–230°C, 380–500°C, 35 atm, 1 atm; rate equation, *k*, *M*, AP	J-104, J-105
Benzene	Catalytic reaction with propylene; 150–200°C; rate equation, *k*, *M*	J-106
Benzene	Catalytic hydrogenation to cyclohexane; 140–220°C; rate equation, *k*, *M*, adsorption-equilibrium constant	J-107
Benzene	Bromination in aqueous $HClO_4$, 30°C; *k*, AP	J-28
Butadiene	Cyclotrimerization over catalyst, 40–60°C; rate equation, *k*, AP	J-29
m-Chloroaniline	Diazotization, 0–25°C, *M*, *k*, and rate equation, AP	J-30
Chloroethane	Thermal decomposition, 630–715°C, *k*, AP	J-31
Coal	Isothermal pyrolysis; 400–670°C, *M*, *k*, AP, gas-formation curve	J-32, J-108, J-109
Coal	Hydrogenation under high pressure; rate equation, *k*, *M*	J-110
Coal	Oxidation; ∼150°C, rate curves and equation, *M*, swelling index, calorific value change	J-111
Coke	Gasification by CO_2; 1000°C; rate data as function of partial pressures of CO and CO_2	J-112
Coke	Reactivity and adsorption properties; 950°C, *k*, specific surface	J-33
Cumene peroxide	Reaction with diethylamine in aqueous-solution; 60–80°C, *M*, *k*, AP	J-34
Cyclohexadienes	Gas-phase reactions with nitric oxide; 306–359°C, 3–7 torr and 64–436 torr; *M*, *k*, AP	J-35
Cyclohexane	Catalytic dehydrogenation; rate data, catalytic activity, chemisorption data	J-114
Cyclohexene	Catalytic dehydrogenation; 27.8–41.3°C, kinetic curve, *M*, AP	J-113
Cyclohexene	Reaction with 2,4-dinitrobenzene-sulfenyl bromide in C_6H_6 and in $CHCl_3$; *M*, *k*	J-36
Cyclopentene	Pyrolysis by static method; 416–532°C, initial pressure 12–30 torr, 5–600 min, *M*, *k*, AP	J-37
Ethane	Thermal decomposition, 522–610°C, 100–700 mmHg	J-38, A-8
Ethanol	Formation of ether over γ alumina; 150–195°C, 30–300 torr; rate equation, absorption coefficients, *M*	J-115
Ethanol	Oxidation in concentrated $NA_2Cr_2O_7$ solution; room temperature; rate equation, *k*, *M*	J-116
Ethanol	Catalytic conversion to butadiene; rate data, *M*	J-117
Ethyl acetate	Alkaline hydrolysis; *k*, AP	J-39
Ethyl acetate	Hydrolysis in aqueous H_2SO_4; 25°C, *M*, *k*, activity coefficient	J-40
Ethyl nitrate	Pyrolysis; 242–260°C, *M*, *k*, AP	J-41
Ethyl nitrite	Pyrolysis as affected by propylene; *M*, *k*, yield and composition of product	J-42
Ethylene, ethane, propane	Pyrolysis; 820–1125°C, *M*, *k*, AP	J-43
Ethylene	Chlorination, 320–380°C, *k*, AP	J-44
Ethylene	Catalytic oxychlorination; 260–300°C, catalytic activity, conversion, *M*	J-118
Ethylene	Emulsion polymerization; rate data, *M*	J-119
Ethylene	Polymerization; 160–300°C, 1200–2500 atm, rate equaion, *M*, scale-up data	F-7
Ethylene dichloride	Formation from gaseous ethylene and chlorine; rate equation in terms of mass-transfer coefficient, *M*	J-120
Ethylene oxide, water	Reaction at various concentrations; *k*, AP	J-45
Fluoromethane	Pyrolysis in flow system; 850–1100°C, *M*, *k*, AP	J-46
Formaldehyde	Condensation with olefins by thermal and catalytic paths; ∼200°C, product distribution, *M*	J-121
Formaldehyde	Gas-phase polymerization, 0–60°C, 16–216 torr; rate data, *M*, AP	J-122
Hydrazobenzene	Noncatalyzed thermal reaction; 140–220°C, *M*, *k*, product composition	J-47
Isobutane	Pyrolysis, initiation of chains; 824 K, *k*, AP	J-48
Isobutene, formaldehyde	Condensation to produce isoprene; 0.5°C, *M*, *k*, Henry coefficient	J-49
Isocyanates, alcohols	Formation of urethane, 15.5–31.5°C, *k*, AP	J-50
Isoprene	Polymerization in benzene; rate equations, *M*	J-123
Methane	Catalytic reaction with steam; 430–630°C, rate equation, conversion, product distribution, catalyst poisoning	J-124, J-125
Methane	Catalytic oxidation; 350–450°C, rate equation, *k*, *M*, AP	J-126
Methanol	Catalytic oxidation; 200–400°C, rate equation, *k*, *M*, AP, catalyst selectivity	J-127, J-128, J-129
Methanol	Dehydrogenation (catalytic); 260–350°C, 140–250 torr; rate equation, *k*, AP, *M*	J-130
Methanol	Nitration by various nitrating mixtures; *k*, reaction orders, conductance, and colorimetric data	J-51
Methyl methacrylate	Radical polymerization with various initiators, 25°C, 50°C, yield, rate equation, AP, *M*	J-131, J-132, J-133
Nitroacetanilide	Hydrolysis in aqueous H_2SO_4; 25–90°C, *M*, *k*, AP	J-52

TABLE 4-19 Chemical Rate Data for Organic Reactions* *(Continued)*

Major reactants	Types of reaction and data	References
Nitroalkanes	Acid-catalyzed hydrolysis; M, k, AP	J-53
Nitrobenzene	Hydrogenation to aniline and H_2; rate data, M	J-134
p-Nitrotoluene, o- and p-nitrochlorobenzene	Nitration in H_2SO_4-HNO_3 mixture; M, k, AP, activity coefficient	J-54
2,4-Dinitroacetanilide	Hydrolysis in aqueous H_2SO_4; 25°C, M, k, AP, activity coefficient	J-55
n-Pentane	Thermal cracking; 540°C, 50–180 torr, M, k, AP, product composition	J-56
Phenol, m-cresol	Oxidation by O_2 in aqueous medium; 25–80°C, pH 9.5–13; M, k, AP, yield	J-57
Phthalic anhydride, methanol	Esterification in batch reactor; 55–67°C, k, AP	J-58
Propane	Polymerization in methylene chloride; −15° ~ −50°C, rate equation, k, M	J-135
Propane	Vapor-phase reaction with NO_2; 200–350°C, M, k, AP	J-59
Propane butane	Thermal cracking; 475–570°C, 40–240 torr, contact time 1–10 min, M, k, AP, yield, product composition	J-60
Propionic acid	Thermal decomposition, 496–580°C, k, AP	J-61
Propylene	Diffusion polymerization; reaction rate data, monomer diffusion time, catalyst size	J-136
Propylene	Catalytic oxidation to acrolein; rate of partial reduction, k, AP	J-137
Propylene	Polymerization; 25–26°C, rate equation, k, AP, M	J-138
Propylene, peracetic acid	Reaction in N_2 or Ar; 20–60°C, 50 atm, M, k, AP, equilibrium constant	J-62
Styrene	Formation by thermal decomposition of ethylbenzene in flow reactor, 570–650°C, rate equation, k, AP, M	J-139
Styrene	Radical polymerization; reaction rate data, particle sizes, M	J-140
Styrene	Polymerization or copolymerization; 25°C, 30°C, rate equation, conversion, k, M, AP	J-141, J-142, J-143
Styrene	Epoxidation using a mixture of nitriles and H_2O_2; rate equation, k, M	J-144, J-145
Tetrachlorobenzene	Kinetics of formation of tetrachloraniline; k, AP, 230–275°C	J-63
Tetranitromethane	Gas-phase pyrolysis, 170–223.2°C, k, AP	J-64
Triphenylsilane	Pyrolysis, 425–525°C, k, M	J-65
Urea	Condensation with aldehydes, 24.2°C, k, catalytic parameters	J-66
Urea, acetaldehyde	Reaction in water; ~20°C, M, k, pH, AP	J-67
Vinyl chloride	Heterogeneous bulk polymerization; degree of conversion, rate equation, k, M	J-146
Vinyl chloride	Emulsion polymerization; rate equation, M	J-147
Xylene, pseudodocumene	Chloromethylation; 45–95°C, M, k, AP, conversion, product composition	J-68

*k = rate constant, AP = Arrhenius parameters, and M = mechanism.

posium proceedings on fluidization (e.g., I-4, K-5 to K-9). The correlations developed in these studies, however, are not always comparable because of the difference in the experimental methods and sometimes because of incomplete measurements of pertinent variables. Some of the convenient correlations to use in estimating the mass-transfer data have been developed by Chu, Kalil, and Wetteroth (K-10). The correlations shown here are for both gas- and liquid-solid systems covering both fixed- and fluidized-bed cases and over wide ranges of Schmidt number ($Sc = \mu_f/\rho_f D_m$ = 0.6 to 1400) and modified Reynolds number ($\overline{Re}$ = ~1 to 10,000). They are applicable to spherical and right cylindrical particles.

$$\text{For } 0 < \overline{Re} < 30; \quad j_D = 5.7(\overline{Re})^{-0.78} \tag{4-105}$$

$$\text{For } 30 < \overline{Re} < 10{,}000: \quad j_D = 177(\overline{Re})^{-0.44} \tag{4-106}$$

where

$$\overline{Re} = \text{modified Reynolds number} = \frac{D_p u \rho_f}{\mu_f(1-\epsilon)}$$

$$j_D = \text{Colburn's mass-transfer factor} = \frac{k_f M p}{G}\left(\frac{\mu_f}{\rho_f D_m}\right)^{2/3}$$

In these expressions u is the superficial fluid velocity; ρ_f, μ_f, k_f, G, and M are the density, viscosity, mass-transfer coefficient, mass velocity, and molecular weight respectively of fluid; ϵ is the void fraction, and p is similar to p_{fA} in Eq. (4-96).

For irregularly shaped particles over the $\overline{Re}/\phi'$ range from 20 to 10,000,

$$j_D = 2.05\phi'(\overline{Re}/\phi')^{-0.468} \tag{4-107}$$

where ϕ' refers to an empirical shape factor that is the measure of the ratio of effective specific areas of nonspherical to spherical particles. The values of ϕ' are 0.80 for flakes (8 to 10 or 14 to 18 mesh) and 0.65 for both Raschig rings [0.0127 to 0.0508 m (½ to 2 in)] and Berl saddles [0.0064 to 0.0127 m (¼ to ½ in)] respectively. Figure 4-17 is the basis for Eqs. (4-105) and (4-106), while Eq. (4-107) is derived from Fig. 4-18. The maximum errors associated with these correlations are estimated (K-10) to be approximately 20 percent.

Other investigators (e.g., Beek in Ref. K-8) correlated the mass-transfer data in terms of the product of Stanton number, $St = k_f\epsilon/u$, and $Sc^{2/3}$ as a function of $\overline{Re}$. Replotting of the data of Chu et al. (K-10) and those from other sources indicated that this type of plot is not suitable for covering the entire range from fixed bed to lean fluidized bed. Further replotting of the same set of data in terms of $St \cdot Sc^{2/3}$ versus $Re = D_p u \rho_f/\mu_f$ brought the data points for the fixed and fluidized beds close together and resulted in the following correlations (K-8):

$$\text{For } 5 < Re < 500: \quad St \cdot Sc^{2/3} = (0.81 \pm 0.05)\,Re^{-0.5} \tag{4-108}$$

$$\text{For } 50 < Re < 2000: \quad St \cdot Sc^{2/3} = (0.6 \pm 0.1)\,Re^{-0.43} \tag{4-109}$$

These correlations have been recommended (K-8) as the "most accurate and reliable" representation of the published data for most practical ranges of Sc and Re.

Much of the published mass-transfer data is based on measurements in very shallow beds and thus leads to questions as to how closely the results could be reproduced in large-scale systems (K-7). Kunii and Levenspiel (K-7) developed correlations based on the bubbling-bed model.

Physical Rate Data: Heat Transfer This subsection discusses heat-transfer data which are required in the energy-balance calculations (e.g., see Table 4-13) for reactor design. Reference should be made to Secs. 10 and 11, where prediction methods and empirical values may be found for specific physical situations. In developing energy-balance equations of the type listed in Table 4-13, assumptions were made that the temperature difference between the reacting mixture and the heating agent remains constant and uniform

TABLE 4-20 Typical Values of Arrhenius Parameters*

Reaction system	Activation energy, kcal./g.-mole	Frequency factor
First-order gaseous decompositions:		*sec.*$^{-1}$
Nitrogen tetroxide	13.9	8.0×10^{14}
Ethyl chlorocarbonate	29.1	9.2×10^{8}
Ethyl peroxide	31.5	5.1×10^{14}
Ethylidene dibutyrate	33.0	1.8×10^{10}
Acetic anhydride	34.5	1.0×10^{12}
Ethyl nitrite	37.7	1.4×10^{14}
tert-Butyl chloride	41.4	2.5×10^{12}
Methyl iodide	43.0	3.9×10^{12}
Paracetaldehyde	44.2	1.3×10^{15}
Ethylidene dichloride	49.5	1.2×10^{12}
Nitromethane	50.6	4.1×10^{13}
Azomethane	52.5	3.5×10^{16}
Propylene oxide	58.0	1.2×10^{14}
Dimethylethylacetic acid	60.0	3.3×10^{13}
1-Butene	63.0	5.0×10^{12}
Trimethylacetic acid	65.5	4.8×10^{14}
p-Xylene	76.2	5.0×10^{13}
Toluene	77.5	2.0×10^{13}
Second-order gaseous reactions between stable molecules:		*ml./mole-sec.*
$NO + O_3 \rightarrow NO_2 + O_2$	2.5	8×10^{11}
Cyclopentadiene	14.9	8.5×10^{7}
Cyclopentadiene + crotonaldehyde	15.2	1×10^{9}
Isoprene + acrolein	18.7	1.0×10^{9}
Butadiene + acrolein	19.7	1.5×10^{9}
Butadiene + crotonaldehyde	22.0	9.0×10^{8}
Isobutylene + HBr	22.5	1.6×10^{10}
$2NOCl \rightarrow 2NO + Cl_2$	24.0	9×10^{12}
1,3-Butadiene	25.3	4.7×10^{10}
1,3-Pentadiene	26.0	3.5×10^{10}
Isobutylene + HCl	28.8	1.0×10^{11}
Ethylene	37.7	7.1×10^{10}
Propylene	38.0	1.6×10^{10}
$H_2 + I_2 \rightarrow 2HI$	40.0	1×10^{14}
Isobutylene	43.0	2.0×10^{12}
Ethylene + H_2	43.2	4.0×10^{13}
$2HI \rightarrow H_2 + I_2$	44.0	6×10^{13}
Second-order reactions involving atoms or radicals:		
$H + HBr \rightarrow H_2 + Br$	1.2	
$H + Br_2 \rightarrow HBr + Br$	1.2	
$CH_3 + i\text{-}C_4H_{10}$	7.6	
$CH_3 + n\text{-}C_4H_{10}$	8.3	
$CH_3 + C_2H_6$	10.4	
Third-order gaseous reactions:		*ml.*^{2}*/mole*^{2}*-sec.*
$2NO + O_2 \rightarrow 2NO_2$	0 or negative	8×10^{9}
$2NO + Br_2 \rightarrow 2NOBr$	~4	$\sim 10^{11}$
$2NO + Cl_2 \rightarrow 2NOCl$	~4	$\sim 10^{9}$
$I + I + He \rightarrow I_2 + He$	0	0.34×10^{16}
$I + I + Ar \rightarrow I_2 + Ar$	0	0.72×10^{16}
$I + I + H_2 \rightarrow I_2 + H_2$	0	0.95×10^{16}
$I + I + CO_2 \rightarrow I_2 + CO_2$	0	2.7×10^{16}

*Adapted from Frost and Pearson, *Kinetics and Mechanism*, 2d ed., Wiley, New York, 1961.

TABLE 4-21 Chemical Rate Data Base: Selected General References

Reaction systems	Types of reactions and data	References
H_2-O_2 system of atoms, radicals, and molecules	High-temperature homogeneous gas-phase reactions; thermodynamic data, rate parameters with valid ranges, rate curves, tables of rate constants, experimental methods	J-148
H_2-N_2O_2 system of atoms, radicals, and molecules	Same types as above	J-149
Selected atoms, radicals, alkenes, alkynes, and aromatics	Gas-phase addition reactions of atoms and radicals; thermodynamic data, rate parameters with experimental conditions, experimental methods	J-150
Selected atoms, radicals, and molecules including hydrocarbons	Gas-phase exchange reactions, addition and recombination reactions, radical decompositions; rate parameters for modified Arrhenius equation with valid temperature ranges, methods	J-151
Selected molecules and free radicals of hydrocarbons	Gas-phase unimolecular, thermally induced, homogeneous reactions; thermodynamic data, rate parameters with experimental conditions and methods, reaction mechanisms	J-152
Selected inorganic molecules and hydrocarbons	Liquid-phase monomolecular, bimolecular, and trimolecular reactions; decomposition, substitution, and addition of radicals to molecules, radical recombination and disproportionation, ion-molecule reactions; rate parameters with valid temperature ranges, methods	J-153

throughout the reactor for a stirred-tank-type reactor (batch, semibatch, or continuous). Similarly, a flat temperature profile over any cross section normal to the direction of flow was assumed for a tubular reactor. Deviations from these ideal conditions are present in practical reactors as discussed earlier (see "Characteristic Behavior of Reactors"). The nonuniform temperature in a stirred-tank-type reactor could be due to either poor mixing or the relatively high viscosity of the reacting mixture. Likewise the deviation from the plug-flow assumption contributes to the development of temperature gradients in the radial direction as well as in the direction of flow. Such conditions may be treated by using an approach similar to the one for nonideal flow conditions. At present, information is sparse on the relationship between agitator design and the nonuniformity of temperatures to allow analysis of such problems in a stirred-tank reactor by an analytical method. The deviations in the case of a tubular flow reactor with packed solids have been studied by a number of investigators using the concept of effective thermal conductivity (e.g., see Refs. A-29, K-11, and K-12).

Stirred-Tank Reactors Heat-transfer behavior in this type of reactor (STR) is governed by the design of the reactor in terms of the vessel configuration, the type of heating surface, the agitator type and its arrangement in the vessel, and so forth. As a result, the heat-transfer correlation for STR, which is largely empirical in nature, varies with the specific design of STR. Attempts have been made to define a standard STR that could be used as the basis in the analysis to develop a "standard" empirical correlation (e.g., see Ref. K-13). Modified parameters are introduced into such a standard correlation for use in design of other "nonstandard" STRs.

Most of the empirical correlations (kettle-side coefficient) developed by different investigators for the jacket- and coil-heated STRs assume the following form:

$$\frac{h_k d}{\lambda} = \alpha \left(\frac{P^2 N\rho}{\mu}\right)^{\beta} \left(\frac{C_p}{\lambda}\right)^{\gamma} \left(\frac{\mu}{\mu_s}\right)^{\delta} \tag{4-110}$$

where d = diameter of kettle, m (ft)
h_k = kettle-side film coefficient at inside kettle surface or at outside surface of coil, J/(s·m^2·K) [Btu/(h·ft^2·°F)]
μ_s = viscosity of kettle mixture at temperature of kettle or coil surface, Pa·s [lb/(ft·h)]
μ = viscosity at bulk temperature of kettle mixture, Pa·s [lb/(ft·h)]
P = length of agitator, m (ft)
N = agitator shaft speed, revolutions per unit time
ρ = density of kettle mixture at bulk temperature, kg/m^3 (lb/ft^3)
C_p = specific heat of kettle fluid at bulk temperature, J/(kg·K) [Btu/(lb·°F)]
λ = thermal conductivity of kettle mixture at bulk temperature, J/(s·m·K) [Btu/(h·ft·°F)]

The values of the constant α and of exponents β, γ, and δ in Eq. (4-110) vary with the Reynolds-number range as well as the STR design as follows: α = 0.36–1.4, β = 0.5–0.75, γ = ~⅓, and δ = 0.14–0.24. Additional heat-transfer data base for STR, including values of these parameters for specific cases, is available in Sec. 10. The same type of data base and other types of information pertinent to STR are also found in Ref. D-1.

Jacket-side coefficient. Data for this case are limited. However, this coefficient is often high compared with the overall coefficient of heat transfer. Some empirical values are given in Sec. 10.

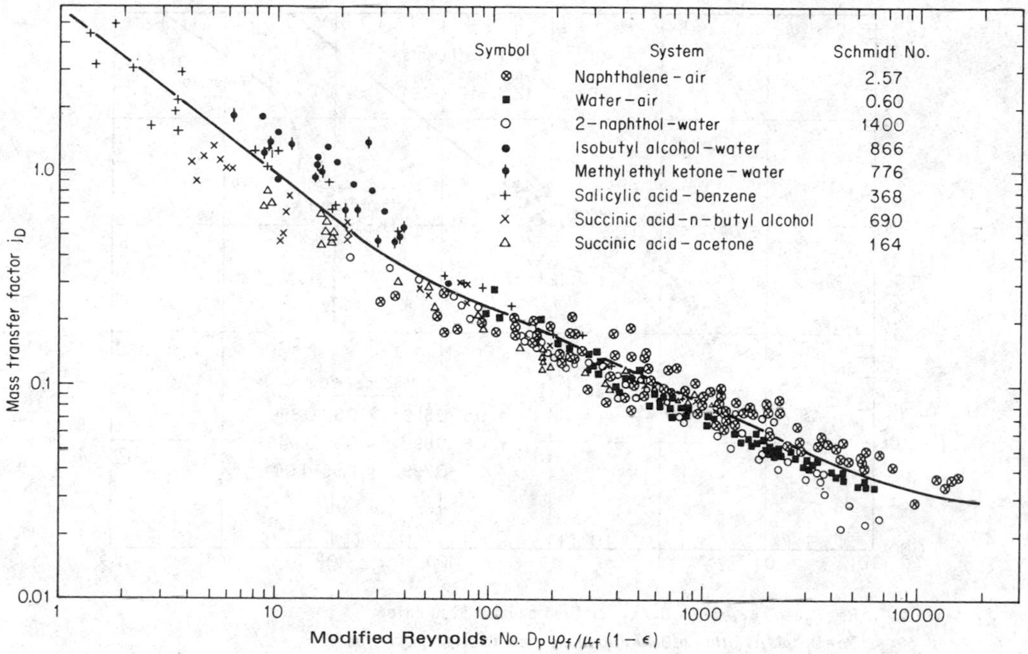

FIG. 4-17 Generalized mass-transfer correlation for fixed and fluidized beds of regularly shaped particles. [*Chu, Kalil, and Wetteroth,* Chem. Eng. Prog., *49, 141 (1953).*]

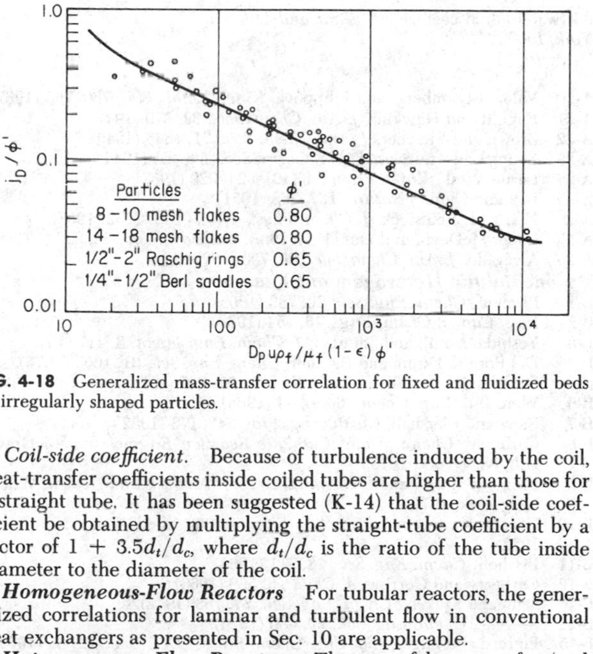

FIG. 4-18 Generalized mass-transfer correlation for fixed and fluidized beds of irregularly shaped particles.

Coil-side coefficient. Because of turbulence induced by the coil, heat-transfer coefficients inside coiled tubes are higher than those for a straight tube. It has been suggested (K-14) that the coil-side coefficient be obtained by multiplying the straight-tube coefficient by a factor of $1 + 3.5d_t/d_c$, where d_t/d_c is the ratio of the tube inside diameter to the diameter of the coil.

Homogeneous-Flow Reactors For tubular reactors, the generalized correlations for laminar and turbulent flow in conventional heat exchangers as presented in Sec. 10 are applicable.

Heterogeneous-Flow Reactors The rate of heat transfer (and therefore the temperature profile) in the fluid-solid reactor is one of the critical factors that affect the reaction rate, conversion, and product distribution. Therefore, the data base for heat transfer between the reactor wall and either fluid or bed of solids is required in the design of a suitable reactor that has provisions for removing or adding an estimated amount of heat to maintain the desired temperature profile in the reactor. As in the case of mass transfer in the fluid-solid system, the rate of heat transfer is subject to the influence of numerous variables, including properties of fluid and solid, superficial velocity, void fractions of the bed, and configuration of the reactor. Complex interactions among these variables as well as the difference in experimental methods thus lead to a number of different forms of empirical correlations developed by various investigators.

Wall-to-fluid coefficient. An integrated plot based on selected data appears in Fig. 4-19 (I-4), covering both fixed and fluidized beds.

Wall-to-bed coefficient. Selected empirical heat-transfer correlations for fixed- and fluidized-bed reactors are presented in Sec. 10. Additional correlations are found in publications such as Refs. I-4 and K-5 to K-9. Semiempirical correlations based on physical models are useful not only for design purposes but also for understanding possible heat-transfer mechanisms. Two of such correlations listed under "References" were developed by Mickley and Fairbanks (K-15) and Ziegler et al. (K-16) respectively.

$$\frac{hD_p}{\lambda_g} = \frac{1.13\,D_p}{\lambda_g}[(1-\epsilon_{mf})(1-f_b)n_w\rho_s C_{ps}\lambda_e]^{1/2} \quad (4\text{-}111)$$

$$\frac{hD_p}{\lambda_g} = 7.2 \Big/ \left(1+\frac{6\lambda_g\bar{\theta}}{\rho_s C_{ps} D_p^2}\right)^2 \quad (4\text{-}112)$$

where h = wall-to-bed heat-transfer coefficient, J/(s·m²·K) [Btu/(h·ft²·°F)]

λ_g = thermal conductivity of fluid, J/(s·m·K) [Btu/(h·ft·°F)]

ϵ_{mf} = void fraction at minimum fluidizing conditions

ρ_s = density of solids, kg/m³ (lb/ft³)

C_{ps} = specific heat of solids, J/(kg·K) [Btu/(lb·°F)]

f_b = fraction of time that an immersed surface is exposed to gas bubbles

n_w = bubble frequency at wall

λ_e = effective thermal conductivity of bed at minimum fluidizing conditions, J/(s·m·K) [Btu/(h·ft·°F)]

$\bar{\theta}$ = mean residence time of solids, s

The values of f_b, n_w, and λ_e may be obtained from the original reference (K-15).

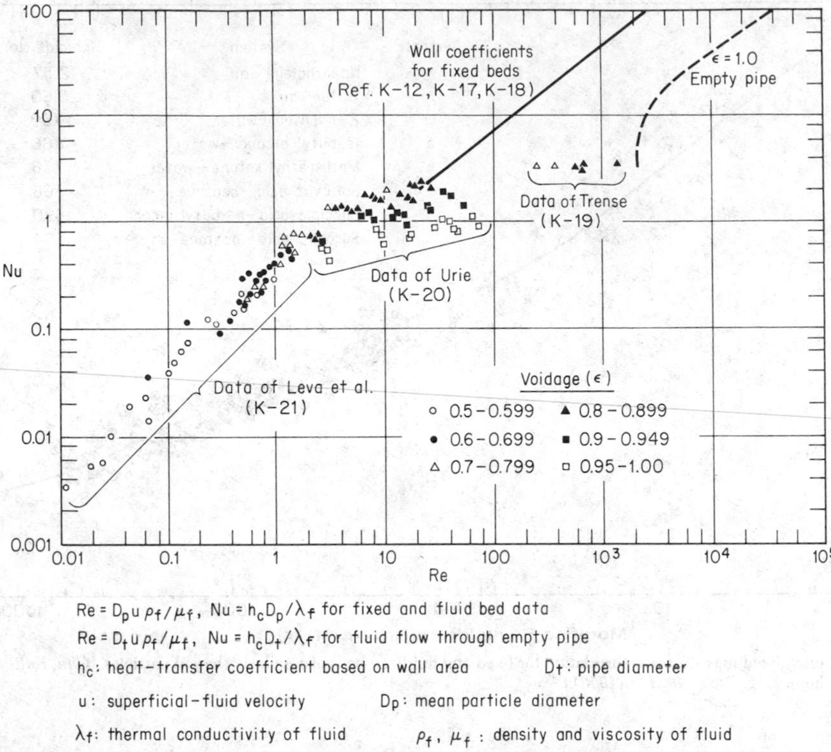

FIG. 4-19 Integrated plot of fluid-bed heat-transfer data (wall-to-fluid coefficient). (*Zenz and Othmer,* Fluidization and Fluid Particle Systems, *Reinhold, New York, 1960.*)

Recent studies (K-22 and K-23) indicated that an increase in pressure of gas-fluidized beds favors the wall-to-bed heat-transfer coefficients. The coefficient for the fine powders (<0.1 mm), however, did not change appreciably with pressure (K-22). The pressure was varied from 0.1 to 2.5 MPa (~1 to 25 atm) in one study (K-22), while it ranged from ~0.6 to 8.1 MPa (~6 to 80 atm) in the other (K-23).

REFERENCES

Fundamentals

Homogeneous Reactions

A-1 Frost and Pearson, *Kinetics and Mechanism*, 2d ed., Wiley, New York, 1961.
A-2 Hinshelwood and Burk, *J. Chem. Soc.*, **127**, 1051, 1114 (1925).
A-3 Nemecek and Pakarek, *Chem. Prum.*, **15**(3), 132–137 (1965).
A-4 Houser and Lee, *J. Phys. Chem.*, **71**(11), 3422–3426 (1967).
A-5 Andreev and Kiperman, *Kinet. Katal.*, **6**(5), 869–877 (1965).
A-6 Freeberg, Hartmen, Hisatsune, and Schempf, *J. Phys. Chem.*, **71**(2), 397–402 (1967).
A-7 Keating and Rozner, *J. Phys. Chem.*, **69**(10), 3658–3660 (1965).
A-8 Bartlit and Bliss, *Am. Inst. Chem. Eng. J.*, **11**(3), 562–572 (1965).
A-9 Laird, Morrell, and Seed, *Discuss. Faraday Soc.*, **22**, 126 (1956).
A-10 Gomer and Kistiakowsky, *J. Chem. Phys.*, **19**(1), 85 (1951).
A-11 Craig and Ward, *J. Am. Chem. Soc.*, **88**(19), 4526–4528 (1966).
A-12 Vinnik, Grabovskaya, and Arzamaskova, *Zh. Fiz. Khim.*, **41**(5), 1102–1107 (1967).
A-13 Pronina et al., *Kinet. Katal.*, **7**(3), 439–448 (1966).
A-14 Ogata, Kawasaki, and Okumura, *Tetrahedron*, **22**(6), 1731–1739 (1966).
A-15 Kolarov, Popyankov, and Angelov, *Monatsh. Chem.*, **96**(3), 949–958 (1965).
A-16 Smith, U.S. AEC Rep. UCRL-11763, 1965.
A-17 Veith, Guthals, and Viste, *Inorg. Chem.*, **6**(4), 667–669 (1967).
A-18 Dvorko and Shilov, *Kinet. Katal.*, **5**(6), 996–999 (1964).
A-19 Peters and Salajeghen, *Anal. Chem.*, **38**(13), 1824–1828 (1966).
A-20 Valov, Blyumberg, and Filippova, *Kinet. Katal.*, 8(4), 760–765 (1967).
A-21 Fuguitt and Hawkins, *J. Am. Chem. Soc.*, **69**, 319 (1947).
A-22 Brown and Fletcher, *J. Am. Chem. Soc.*, **71**, 1845 (1949).
A-23 Young and Andrews, *J. Am. Chem. Soc.*, **66**, 421 (1944).
A-24 Lerner et al., *Khim. Prom.*, **43**(12), 924–929 (1967).
A-25 Bell and Clunie, *Nature*, **167**, 363 (1951).
A-26 Niki and Weinstock, *J. Chem. Phys.*, **47**(9), 3249–3252 (1967).
A-27 Long, McDevit, and Dunkle, *J. Phys. Colloid Chem.*, **55**, 829 (1951).
A-28 Alvarado, *J. Am. Chem. Soc.*, **50**, 790 (1928).

Noncatalytic Heterogeneous Reactions

B-1 Bischoff, *Chem. Eng. Sci.*, **20**, 783 (1965).
B-2 Luss, *Can. J. Chem. Eng.*, **46**, 154 (1968).
B-3 Yoshida, Kunii, and Shimizu, *J. Chem. Eng. Japan*, **8**, 417 (1975).
B-4 Del Borghi, Dunn, and Bischoff, *Chem. Eng. Sci.*, **31**, 1065 (1976).
B-5 Dudokovic, *Am. Inst. Chem. Eng. J.*, **22**, 945 (1976).
B-6 Wen, *Ind. Eng. Chem.*, **60**(9), 34 (1968).
B-7 Ranz and Marshall, *Chem. Eng. Prog.*, **48**, 173 (1952).
B-8 Carberry, *Chemical and Catalytic Reaction Engineering*, McGraw-Hill, New York, 1976.
B-9 Froment and Bischoff, *Chemical Reactor Analysis and Design*, Wiley, New York, 1979.
B-10 Levenspiel, *Chemical Reaction Engineering*, 2d ed, Wiley, New York, 1972.
B-11 Bischoff, *Chem. Eng. Sci.*, **18**, 711 (1963).
B-12 Carberry and Gorring, *J. Catal.*, **5**, 529 (1966).
B-13 Sohn and Szekely, *Chem. Eng. Sci.*, **27**, 763 (1972).
B-14 ———, *Chem. Eng. Sci.*, **29**, 630 (1974).
B-15 Pigford and Sliger, *Ind. Eng. Chem. Process Des. Dev.*, **12**, 85 (1973).
B-16 Costa and Smith, *Am. Inst. Chem. Eng. J.*, **17**, 947 (1971).
B-17 Wang and Wen, *Am. Inst. Chem. Eng. J.*, **18**, 1231 (1972).
B-18 Luss and Amundson, *Am. Inst. Chem. Eng. J.*, **15**, 194 (1969).
B-19 Endom, Hedden, and Lehmann, *5th Int. Symp. React. Solids, Munich*, 1964, 632–646 (1965).
B-20 Vandenbussche, *Comm. Energ. At. (France)*, Rapp. No. 2859, 1966.
B-21 Sargutskii and Serebrennikov, *Zh. Neorg. Khim.*, **11**(1), 33 (1966).
B-22 Habashi and Thurston, *Energ. Nucl.*, **14**(4), 238–244 (1967).
B-23 Bretsnajder and Piskorski, *Chem. Stosow., Ser. A*, **11**(1), 3–24 (1967).
B-24 Lewis and Suen, *Ind. Eng. Chem., Industr.*, **32**, 1095 (1940).

B-25 McKinley and White, *Trans. Am. Inst. Chem. Eng.*, **40**, 143–175 (1944).
B-26 Jeffreys, Jenson, and Miles, *Trans. Inst. Chem. Eng. (London)*, **39**, 389 (1961).
B-27 Astarita, *Mass Transfer with Chemical Reaction*, Elsevier, Amsterdam, 1967.
B-28 Danckwerts, *Gas Liquid Reactions*, McGraw-Hill, New York, 1970.
B-29 Sherwood, Pigford, and Wilke, *Mass Transfer*, McGraw-Hill, New York, 1975.
B-30 Bamford and Tipper (eds.), *Comprehensive Chemical Kinetics*, vol. 22: *Reactions in the Solid State*, Elsevier, Amsterdam, 1980.
B-31 Schmalzried, *Solid State Reactions*, Verlag Chemie/Academic, Weinheim, Germany, 1974.
B-32 Ball and Norwood, *J. Chem. Soc. A*, 1633 (1969); 1476 (1970).
B-33 McGinn, Wheeler, and Galwey, *Trans. Faraday Soc.*, **66**, 1809 (1970).
B-34 Schneider and Nigh, *Combust. Flame*, **15**, 223 (1970).
B-35 Finch, Jacobs, and Thompkins, *J. Chem. Soc.*, 2053 (1954).
B-36 Thomas and Renshaw, *J. Chem. Soc. A*, 2749, 2753, 2756 (1969).
B-37 Canning and Hughes, *Thermochim. Acta*, **6**, 399 (1973).
B-38 Mentzen, *Ann. Chim.*, **3**, 367 (1968).
B-39 Buttress and Hughes, *J. Chem. Soc. A*, 1272 (1968).
B-40 Kutty and Murthy, *Indian J. Technol.*, **12**, 447 (1974).
B-41 Budnikov and Ginstling, *Principles of Solid State Chemistry*, trans. by Shaw, Maclaren, London, 1968.
B-42 Johnson and Gallagher, *J. Am. Ceram. Soc.*, **59**, 171 (1976).
B-43 Allen, Crofts, and Swan, *Reactivity of Solids: Proc. 7th Int. Symp.*, Chapman & Hall, London, 1977, p. 630.
B-44 Pavlyuchenko et al., *Vestsi Akad. Navuk BSSR, Ser. Khim. Navuk* (4), 105 (1966).
B-45 Bertrand, Lallemant, and Watelle-Marion, *J. Inorg. Nucl. Chem.*, **36**, 1303 (1974).
B-46 Britton, Gregg, and Winsor, *Trans. Faraday Soc.*, **48**, 63, 70 (1952).
B-47 Hussein, Kolta, Saba, and El Roudi, *Thermochim. Acta*, **10**, 177 (1974).
B-48 Jach, *Trans. Faraday Soc.*, **59**, 947 (1963).
B-49 Prodan and Pavlyachenko, *Geterogennyl Khim. Reaktsii, Inst. Obshch. Neorgan. Khim. Akad. Nauk BSSR*, **1965**, 20–43.
B-50 Muraishi, *Yamagata Daigaku Kiyo, Shizen Kagaku*, **6**(2), 187–196 (1964).
B-51 Auffredic, *C. R. Acad. Sci. (Paris), Ser. C*, **263**(19), 1093–1096 (1966).
B-52 Hulbert, *Mater. Sci. Eng.*, **2**(5), 262–268 (1968).
B-53 Hulbert and Klawitter, *J. Am. Ceram. Soc.*, **50**(9), 484–488 (1967).

Catalysis

C-1 Rollefson and Faull, *J. Am. Chem. Soc.*, **59**, 625 (1937).
C-2 Macoll and Stinson, *Proc. Chem. Soc.*, 80 (1958).
C-3 Bell, *Acid-Base Catalysis*, Oxford, Oxford, 1941.
C-4 Hammett, *Physical Organic Chemistry*, McGraw-Hill, New York, 1940.
C-5 Emmett, Sabatier, and Reid, *Catalysis Then and Now*, Franklin Pub. Co., Philadelphia, 1965.
C-6 Basolo and Pearson, *Mechanism of Inorganic Reactions*, Wiley, New York, 1958.
C-7 Bender, *Chem. Rev.*, **60**, 53 (1960).
C-8 Min'kov et al., *Kinet. Katal.*, **7**(4), 632–639 (1966).
C-9 Keier, *Probl. Kinet. Katal., Akad. Nauk SSSR*, **11**, 200–206 (1966).
C-10 Krause and Mennenga, *J. Prakt. Chem.*, **32**(5–6), 283–290 (1966).
C-11 Tsutsui and Ugo (eds.), *Fundamental Research in Homogeneous Catalysis*, Plenum, New York, 1977.
C-12 Parshall, *Homogeneous Catalysis*, Wiley, New York, 1980.
C-13 Kemball et al., *Catalysis: A Specialist Periodical Report*, vol. 1, Chemical Society, London, 1977.
C-14 Thomas, *Catalytic Processes and Proven Catalysts*, Academic, New York, 1970.
C-15 Gates, Katzer, and Schuit, *Chemistry of Catalytic Processes*, McGraw-Hill, New York, 1978.

Experimental Techiques

D-1 Rase, *Chemical Reactor Design for Process Plants*, vol. 1: *Principles and Techniques*, Wiley, New York, 1977.
D-2 Weekman, *Am. Inst. Chem. Eng. J.*, **20**, 833 (1974).
D-3 Willard, Merritt, and Dean, *Instrumental Methods of Analysis*, 4th ed., Van Nostrand, Princeton, N.J., 1965.
D-4 Ewing, *Instrumental Methods of Chemical Analysis*, 3d ed., McGraw-Hill, New York, 1969.
D-5 Hammes (ed.), *Techniques of Chemistry*, vol. VI: *Investigation of Rates and Mechanisms of Reactions (Two Parts)*, Wiley, New York, 1974.
D-6 Kohl, Zentner, and Lukens, *Radioisotope Applications Engineering*, Van Nostrand, Princeton, N.J., 1961.
D-7 Campbell and Thompson, "Radioisotopes in Studies of Chemisorption and Catalysis," in Cadenhead, Danielli, and Rosenberg (eds.), *Progress in Surface and Membrane Science*, Academic, New York, 1975, pp. 163–221.
D-8 Tanaka, *J. Res. Inst. Catal., Hokkaido Univ.*, **13**(2), 119–150 (1965).
D-9 Takeuchi, Sakaguchi, and Togashi, *Bull. Chem. Soc. Japan*, **39**(7), 1437–1439 (1966).
D-10 Foa et al., *J. Sci. Instrum.*, **44**(11), 936–938 (1967).
D-11 Shizuka and Azami, *Nippon Gomu Kyokaishi*, **39**(12), 999–1005 (1966).
D-12 Sugier, *Nukleonika*, **12**(9), 723–728 (1967).
D-13 Ledwith and Woods, *J. Chem. Soc., Ser. B*, **1966**(8), 753–757.
D-14 Brokskii et al., *Dokl. Akad. Nauk SSSR*, **172**(1), 122–125 (1967).
D-15 Van Wazer and Moedritzer, *Exch. React. Proc. Symp., Upton, N.Y.*, 1965, 23–26.
D-16 Schmid, Sofer, and Mayerboeck, *Monatsh. Chem.*, **99**(2), 463–468 (1968).
D-17 Fujita, *Japan. J. Appl. Phys.*, **5**(8), 701–710 (1966).
D-18 Van Reijen and Cosse, *Dicuss. Faraday Soc.*, **1966**(41), 277–289.
D-19 Mulcahy, Steven, and Ward, *Phys. Chem.*, **71**(7), 2124–2131 (1967).
D-20 Maruyama, Otsuki, and Iwao, *J. Org. Chem.*, **32**(1), 82–86 (1967).
D-21 Casado and Lizaso-Lamsfus, *An. R. Soc. Esp. Fis. Quim., Ser. B*, **63**(7–8), 739–748 (1967).
D-22 Vetchinkina et al., *Zh. Fiz. Khim.*, **40**(4), 762–765 (1966).
D-23 Briquet and Dondeyne, *Agriculture (Louvain)*, **13**(3), 529–554 (1965).
D-24 Heyne and Tompkins, *Proc. R. Soc. (London), Ser. A*, **292**(1431), 460–478 (1966).
D-25 Ignat'eva and Khalikova, *Zh. Prikl. Spektrosk.*, **5**(5), 642–647 (1966).
D-26 Young and Sheppard, *J. Catal.*, **7**(3), 223–233 (1967).
D-27 Ishii, Furumai, and Takeya, *Kogyo Kagaku Zasshi*, **70**(10), 1652–1656 (1967).
D-28 Ishii, Yahata, and Takeya, *Kagaku Kogaku*, **31**(9), 896–901 (1967).
D-29 Merzhanov, Abramov, and Abramova, *Zh. Fiz. Khim.*, **41**(k), 179–184 (1967).
D-30 Reich, *J. Appl. Polym. Sci.*, **11**(2), 161–170 (1967).
D-31 Bonath, Foertsch, and Saemann, *Chem. Ing. Tech.*, **38**(7), 739–742 (1966).
D-32 Berkowitz and Mullin, *Am. Chem. Soc., Div. Fuel Chem., Repr.* **10**(2), C-100–C-120 (1966).
D-33 Giordano, Bossi, and Paratella, *Chem. Eng. Sci.*, **21**(8), 621–630 (1966).
D-34 Lin, *Ind. Eng. Chem.*, **60**(5), 61–82 (1968).
D-35 Ibid., **61**(3), 42–66 (1969).

Analyses of Reaction Kinetic Data

E-1 Benson and Shaw, *J. Chem. Phys.*, **47**(10), 4052–4055 (1967).
E-2 Kochi (ed.), *Free Radicals*, vol. II, Wiley, New York, 1973.
E-3 Lenz, *Organic Chemistry of Synthetic High Polymers*, Interscience, New York, 1967.
E-4 Platzer (ed.), *Polymerization Kinetics and Technology*, American Chemical Society, Washington, 1973.
E-5 Yokoyama and Takayasu, *Kogyo Kagaku Zasshi*, **70**(10), 1619–1624 (1967).
E-6 Solyanikov and Denisov, *Dokl. Akad. Nauk SSSR*, **173**(5), 1106–1109 (1967).
E-7 Hougen and Watson, *Chemical Process Principles*, part III, Wiley, New York, 1947.
E-8 Kittrell and Hunter, *Statistical Methodology for Chemical Reaction Modeling*, AIChE Today ser., American Institute of Chemical Engineers, New York, 1970.
E-9 Albright and Crynes (eds.), *Industrial and Laboratory Pyrolyses*, American Chemical Society, Washington, 1976.
E-10 Daniel and Wood, *Fitting Equations to Data*, Wiley, New York, 1980.
E-11 Walas, *Reaction Kinetics for Chemical Engineers*, McGraw-Hill, New York, 1959.
E-12 Hill, *An Introduction to Chemical Engineering Kinetics and Reactor Design*, Wiley, New York, 1977.
E-13 Benson, *Thermochemical Kinetics*, 2d ed., Wiley, New York, 1976.

Scale-Up Methods

F-1 Damköhler, in Eucken and Jacob (eds.), *Der Chemie-Ingenieur*, vol. III, Edwards, Ann Arbor, Mich., 1937, part 1, p. 454.
F-2 Bosworth, *Transport Properties in Applied Chemistry*, Wiley, New York, 1956, pp. 303–325.
F-3 Boreskov and Slin'ko, *Theoretical Foundation of Chemical Engineering* (trans. from Russian), **1**(1), 3–12 (1967).
F-4 Rase, *Chemical Reactor Design for Process Plants*, vol. II: *Case Studies and Design Data*, Wiley, New York, 1977.
F-5 Denn, *Ind. Eng. Chem.*, **61**(2), 46–50 (1969).
F-6 Bilous and Amundson, *Am. Inst. Chem. Eng. J.*, **2**, 117 (1965).
F-7 Froment, *Ind. Eng. Chem.*, **59**(2), 18 (1967).
F-8 Carberry and White, *Ind. Eng. Chem.*, **61**(7), 27 (1969).
F-9 Garmon, Morrow, and Anhorn, *Chem. Eng. Prog.*, **61**(6), 57–61 (1965).

F-10 Schoenemann, *Genie Chim.*, **91**(6), 161–176 (1964).
F-11 Weber, *Chem. Eng. Prog.*, **49**(1), 26–34 (1953).
F-12 Saslow and Stevens, *Chem. Process Eng.*, **45**(3), 115–120 (1964).
F-13 Novak, Lynn, and Harrington, *Chem. Eng. Prog.*, **58**(2), 55–59 (1962).
F-14 Hamilton, Johnston, and Petersen, ibid., 51–54 (1962).
F-15 Cooper, Black, and Amberson, ibid., **61**(7), 89–96 (1965).
F-16 Malloy and Taylor, ibid., 101–105 (1965).
F-17 Leftin, Newsome, and Wolff, "Pyrolysis of Naphtha and of Kerosene in the Kellogg Millisecond Furnace," in Albright and Crynes (eds.), *Industrial and Laboratory Pyrolyses*, ACS Symp. Ser. 32, American Chemcal Society, Washington, 1976, pp. 373–391.

Reactor Design

Reactor Types, Characteristics, and Selection

G-1 von Urban, *Erdol Kohle*, **8**, 780 (1955).
G-2 Birthler et al., ibid., **12**, 71 (1959).
G-3 Zalai and Jancso, *Oil Gas J.*, **60**(12), 130 (1962).
G-4 Abrams, "The H-Coal Pilot Plant: History, Description, and Present Status," in Veziroglu (ed.), *Alternative Energy Sources*, vol. 7, Hemisphere Pub. Corp., Washington, 1977, p. 3265.
G-5 Eldridge and Piret, *Chem. Eng. Prog.*, **46**, 290 (1950).
G-6 Corrigan and Young, *Chem. Eng.*, 211 (December 1955).
G-7 Denbigh, *Chem. Eng. Sci.*, **8**, 125–132 (1958).
G-8 van Heerden, ibid., 133–145 (1958).
G-9 Horn, ibid., **14**, 77–89 (1961).
G-10 Aris, *The Optimal Design of Chemical Reactors*, Academic, New York, 1961.
G-11 van de Vusse and Voetter, *Chem. Eng. Sci.*, **14**, 90–100 (1961).
G-12 Siegel and Garti, "The Effect of Pressure on the Catalytic Hydrogenation of Aromatic Hydrocarbons on Rhodium," in Smith (ed.), *Catalysis in Organic Syntheses 1977*, Academic, New York, 1977, pp. 9–23.
G-13 Turner, *Br. Chem. Eng.*, **9**(6), 376–383 (1964).
G-14 Bischoff and McCracken, *Ind. Eng. Chem.*, **58**(7), 18–31 (1966).
G-15 Danckwerts, ibid., **2**, 1 (1958).
G-16 Ibid., **8**, 93 (1958).
G-17 Ibid., **9**, 78 (1958).
G-18 Zwietering, ibid., **11**, 1 (1959).
G-19 Ng and Rippin, *Third European Symposium on Chemical Reaction Engineering*, Pergamon, New York, 1964.
G-20 Denbigh, *J. Appl. Chem.*, **1**, 227 (1951).
G-21 Wall, Delbecq, and Florin, *J. Polym. Sci.*, **9**, 177 (1952).
G-22 Bouton and Chappelear (eds.), *Continuous Polymerization Reactors*, AIChE Symp. ser., vol. 72, no. 160, American Institute of Chemical Engineers, New York, 1976.
G-23 Platzer (ed.), *Polymerization Kinetics and Technology*, Adv. in Chemistry Ser. 128, American Chemical Society, Washington, 1973.
G-24 Denbigh, *Trans. Faraday Soc.*, **40**, 352 (1944).
G-25 Gillespie and Carberry, *Ind. Eng. Chem. Fundam.*, **5**(2), 164–171 (1966).
G-26 Weinstein and Alder, *Chem. Eng. Sci.*, **22**(1), 65–75 (1967).
G-27 Denbigh, *Science, Industry and Social Policy*, Oliver & Boyd, London, 1963.
G-28 Shabica, *Chem. Eng. Prog.*, **59**(9), 57 (1963).

Noncatalytic Heterogeneous Reactors

H-1 Kunii and Levenspiel, *Fluidization Engineering*, Wiley, New York, 1969.
H-2 Yagi and Kunii, *Chem. Eng. Sci.*, **16**, 364, 372, 380 (1961).

Reactors for Solid-Catalyzed Reactions

I-1 Ciapetta, Helm, and Baral, in Collier (ed.), *Catalysis in Practice*, Reinhold, New York, 1957, pp. 49–70.
I-2 Thomas, *Catalytic Processes and Proven Catalysts*, Academic, New York, 1970.
I-3 Ergun, *Chem. Eng. Prog.*, **48**(2), 89–94 (1952).
I-4 Zenz and Othmer, *Fluidization and Fluid-Particle Systems*, Reinhold, New York, 1960.
I-5 Miyauchi, Furusaki, Morooka, and Ikeda, "Transport Phenomena and Reaction in Fluidized Catalyst Beds," in *Advances in Chemical Engineering*, vol. 11, Academic, New York, 1981, pp. 275–448.
I-6 Emmett, *Advances in Catalysis*, vol. 1, Academic, New York, 1948, p. 650.
I-7 Ritter and Drake, *Ind. Eng. Chem., Anal. Ed.*, **17**, 787 (1945).
I-8 Cranston and Inkley, *Advances in Catalysis*, vol. 9, Academic, New York, 1957, pp. 143–154.
I-9 Hougen, *Ind. Eng. Chem.*, **53**(7), 509–528 (1961).
I-10 Wheeler, *Advances in Catalysis*, vol. 3, Academic, New York, 1951.
I-11 Weisz and Prater, ibid., vol. 6, 1954.
I-12 Thiele, *Ind. Eng. Chem.*, **31**, 917 (1939).
I-13 Schilson, Ph.D. thesis, University of Minnesota, Minneapolis, December 1957.
I-14 Mingle and Smith, *AIChE Conv., New Orleans*, March 1961.
I-15 Aris, *Chem. Eng. Sci.*, **6**, 262 (1957).
I-16 Roberts and Satterfield, *Ind. Eng. Chem. Fundam.*, **5**(3), 317–325 (1966).
I-17 Maxted, *Advances in Catalysis*, vol. 3, Academic, New York, 1951, p. 129.
I-18 Gorokhova et al., *Zh. Fiz. Khim.*, **39**(5), 1206–1210 (1965).
I-19 Balandin et al. (eds.), *Scientific Selection of Catalysts*, Academy of Science of U.S.S.R., transl. by Israel Program for Scientific Translations, Jerusalem, 1968.
I-20 DeMaio and Naglieri, *Chem. Eng.*, **75**, 127–132 (1968).
I-21 Satterfield, *Ind. Eng. Chem.*, **61**(6), 4–10 (1969).
I-22 Weisz and Hicks, *Chem. Eng. Sci.*, **17**, 265 (1962).

Chemical Rate Data

J-1 Davies, Jacobs, and Russell-Jones, *Trans. Faraday Soc.*, **63**(7), 1737–1748 (1967).
J-2 Zelionkaite, Janickis, and Pazarauskas, *Liet. TSR Mokslu Akad. Darb., Ser. B*, **1966**(2), 3–9.
J-3 Kuzhakhmetov and Knatyshenko, *Zh. Prikl. Khim.*, **39**(6), 1266–1271 (1966).
J-4 Schroeder and Kolaczkowski, *Chem. Stosow., Ser. A*, **10**(1), 19–33 (1966).
J-5 Morris and Woodburn, *S. Afr. Chem. Process.*, **2**(4), CP115–119 (1967).
J-6 Dave and Chopra, *J. Am. Ceram. Soc.*, **49**(10), 575 (1966).
J-7 Temkin, Cherednik, and Apel'baum, *Kinket. Katal.*, **9**(1), 95–103 (1968).
J-8 Hovenkamp, *Faserforsch. Textiltech.*, **17**(3), 100–105 (1966).
J-9 Pursley, Matula, and Witzell, *J. Phys. Chem.*, **70**(12), 3768–3770 (1966).
J-10 Giona and Toselii, *Ingegnere (Milan)*, **40**(12), 1124–1128 (1966).
J-11 Thomas and Woodman, *Trans. Faraday Soc.*, **63**(11), 2728–2736 (1967).
J-12 Crabtree and Schaefer, *Inorg. Chem.*, **5**(8), 1348–1351 (1966).
J-13 Gaivoronskii and Polyvyannyi, *Vestn. Akad. Nauk Kaz. SSR*, **23**(10), 11–19 (1967).
J-14 Pospisil and Cabicar, *Collect. Czech. Chem. Commun.*, **32**(11), 3832–3841 (1967).
J-15 Petrov and Kirillov, *Izv. Vyssh. Uchebn. Zaved. Khim. Khim. Tekhnol.*, **10**(4), 428–433 (1967).
J-16 Natveev and Frenkel, *Zh. Vses. Khim. Ova.*, **11**(4), 464–465 (1966).
J-17 Ishkin and Dubil, *Zh. Prikl. Khim.*, **41**(1), 52–58 (1968).
J-18 Barron and O'Hern, *Chem. Eng. Sci.*, **21**(5), 397–404 (1966).
J-19 Ganz and Revzin, *Zh. Prikl. Khim.*, **39**(4), 937–939 (1966).
J-20 Bezukladnikov, *Zh. Prikl. Khim.*, **40**(2), 291–296 (1967).
J-21 Tsygoda, Ponomarev, and Shkuridin, *Izv. Vyssh. Uchebn. Zaved. Tsvetn. Metall.*, **9**(2), 39–42 (1966).
J-22 Laidler and Liu, *Proc. R. Soc. (London), Ser. A*, **297**(1450), 365–375 (1967).
J-23 McKnight, Niki, and Weinstock, *J. Chem. Phys.*, **47**(12), 5219–5225 (1967).
J-24 Temkin, Tikhonov, Flid, and Galeev, *Kinet. Katal.*, **8**(6), 1236–1239 (1967).
J-25 Gareev and Men'shutin, *Kinet. Katal.*, **8**(6), 1369–1371 (1967).
J-26 Shah and Hussain, *Indian J. Technol.*, **4**(10), 287–289 (1966).
J-27 Ikawa, Tomizawa, and Yanagihara, *Can. J. Chem.*, **45**(16), 1900–1902 (1967).
J-28 Gallily, Schmidt, and Bernstein, *Chem. Eng. Sci.*, **22**(1), 35–42 (1967).
J-29 Zakharkin and Aklimedov, *Zh. Org. Khim.*, **2**(9), 1557–1561 (1966).
J-30 Aboul-Seoud, *Bull. Soc. Chim. Belg.*, **75**(5–6), 249–259 (1966).
J-31 Sypyak, Moin, and Shevchuk, *Dokl. Akad. Nauk SSSR*, **170**(4), 893–896 (1966).
J-32 Luther, Bergmann, and Sreenivasan, *Chem. Ing. Tech.*, **40**(7), 317–323 (1968).
J-33 Voloshin, Semisalov, Baskina, and Neserenko, *Koks Khim.*, **1967**(10), 21–23.
J-34 Grigoryan, Meliksetyan, and Beileryan, *Arm. Khim. Zh.*, **20**(5), 333–337 (1967).
J-35 Shaw, Cruickshank, and Benson, *J. Phys. Chem.*, **71**(13), 4538–4543 (1967).
J-36 Campbell and Hogg, *J. Chem. Soc., Ser. B*, **1967**(9), 889–892.
J-37 Tanji, Uchiyama, Amano, and Tokuhisa, *Kogyo Kagaku Zasshi*, **70**(3), 307–311 (1967).
J-38 Kudryavtseva and Vedeneev, *Kinet. Katal.*, **7**(2), 208–213 (1966).
J-39 Tsujikawa and Inoue, *Bull. Chem. Soc. Japan*, **39**(9), 1837–1842 (1966).
J-40 Vinnik and Librovich, *Zh. Fiz. Khim.*, **41**(8), 2013–2018 (1967).
J-41 Houser and Lee, *J. Phys. Chem.*, **71**(11), 3422–3426 (1967).
J-42 Marta and Seres, *Acta Univ. Szeged., Acta Phys. Chem.*, **12**(3–4), 113–116 (1966).
J-43 Nazarov and Torban, *Inf. Soobshch. Gos. Nauchno Issled. Proektn. Inst. Azotn. Prom. Prod. Org. Sint.*, **17**, 53–76 (1966).
J-44 Subbotin, Antonov, and Etlis, *Kinet. Katal.*, **7**(2), 202–207 (1966).

J-45 Miki, Ito, Ouchi, Moriya, and Tsuchiya, *Yukagoku*, **15**(6), 257–262 (1966).
J-46 Politanskii and Shevchuk, *Kinet. Katal.*, **8**(1), 12–17 (1967).
J-47 Hashimoto, Sunamoto, Shinkai, and Nakajo, *Kogyo Kagaku Zasshi*, **70**(10), 1705–1708 (1967).
J-48 Konar, Purnell, and Quinn, *J. Chem. Soc., Ser. A*, **1967**(10), 1543–1545.
J-49 Hellen, Jumbroso, and Coussemant, *Rev. Inst. Fr. Pet. Ann. Combust. Liq.*, **22**(5), 807–825 (1967).
J-50 Pronina, Spirin, Blagonravova, Aref'eva, and Gantmakher, *Kinet. Katal.*, **7**(3), 439–448 (1966).
J-51 Genich and Kustova, *Izv. Akad. Nauk SSSR, Ser. Khim.*, **1967**(4), 752–757.
J-52 Vinnik, Medvetskaya, Andreeva, and Tiger, *Zh. Fiz. Khim.*, **41**(1), 252–260 (1967).
J-53 Cundall and Locke, *J. Chem. Soc., Ser. B*, **1968**(2), 98–103.
J-54 Vinnik, Grabovskaya, and Arzamaskova, *Zh. Fiz. Khim.*, **41**(5), 1102–1107 (1967).
J-55 Vinnik and Medvestskaya, *Zh. Fiz. Khim.*, **41**(7), 1775–1782 (1967).
J-56 Kosyreva, Studentsov, and Stepukhovich, *Neftekhimiya*, **7**(4), 569–574 (1967).
J-57 Kirso, Kuiv, and Gubergrits, *Zh. Prikl. Khim.*, **40**(7), 1583–1589 (1967).
J-58 Hussain and Kamath, *Indian J. Technol.*, **5**(10), 321–324 (1967).
J-59 Ballod, Molchanova, and Shtern, *Neftekhimiya*, **7**(1), 115–123 (1967).
J-60 Karnaukhova, Stepukhovich, and Zolkin, ibid., (5), 738–745 (1967).
J-61 Blake and Hole, *J. Chem. Soc., Ser. B*, **1966**(6), 577–579.
J-62 Valov, Blyumberg, and Filippova, *Kinet. Katal.*, **8**(4), 760–765 (1967).
J-63 Shein and Kozorez, *Zh. Org. Khim.*, **2**(6), 1073–1075 (1966).
J-64 Sullivan and Axworthy, *J. Phys. Chem.*, **70**(10), 3366–3368 (1966).
J-65 Coutant and Levy, *Am. Chem. Soc., Div. Pet. Chem. Prepr.*, **11**(3), 291–296 (1966).
J-66 Ogata, Kawasaki, and Okumura, *Tetrahedron*, **22**(6), 1731–1739 (1966).
J-67 Ninagawa, Saiki, Hosono, and Okada, *Nippon Kagaku Zasshi*, **88**(2), 206–211 (1967).
J-68 Shein, Mironov, and Farberov, *Zh. Prikl. Khim.*, **40**(9), 2006–2014 (1967).
J-69 Tanaka, Yamamoto, and Matsuyama, *Proc. 3d Int. Congr. Catal., Amsterdam*, **1964**(1), 676–687 (1965).
J-70 Tamaru, ibid., 664–675 (1965).
J-71 Brill, *J. Polymer Sci., Part C*, no. **12**, 353–362 (1966).
J-72 Chesnokova, Gorbunov, Lachinov, Muravskaya, and Erdedi, *Kinet. Katal.*, **6**(2), 338–342 (1965).
J-73 Robertson and Willhoft, *Trans. Faraday Soc.*, **63**(2), 476–487.
J-74 Takazawa, *Shokubai (Tokyo)*, **8**(5), 390–401 (1966).
J-75 Vladov and Dinkov, *J. Catal.*, **5**(3), 412–418 (1966).
J-76 Barclay and Crewe, *J. Appl. Chem. (London)*, **17**(1), 21–26 (1967).
J-77 Torikai, *Kogyo Kagaku Zasshi*, **68**(1), 174–179 (1965).
J-78 Proks, *Chem. Zvesti*, **20**(10), 697–715 (1966).
J-79 Hottel, Williams, and Wu, *Am. Chem. Soc., Div. Fuel Chem., Prepr.*, **10**(3), 58–71 (1966).
J-80 Tingey, *J. Phys. Chem.*, **70**(5), 1406–1412 (1966).
J-81 Kozub, Rusov, and Vlasenko, *Kinet. Katal.*, **6**(3), 556–558 (1965).
J-82 Karn, Shultz, and Anderson, *Ind. Eng. Chem., Process Des. Dev.*, **4**(3), 266–270 (1965).
J-83 Najbar, Kuchynka, and Klier, *Collect. Czech. Chem. Commun.*, **31**(3), 959–969 (1966).
J-84 Biclanski, Dziembaj, and Slocrynski, *Bull. Acad. Pol. Sci. Ser. Sci. Chim.*, **14**(8), 659–672 (1966).
J-85 Endom, Hedden, and Lehmann, *Int. Symp. React. Solids, 5th Munich*, **1964**, 632–646; disc. 646–647 (1965).
J-86 Boeyum, Norw. Patent 107,384 (Sept. 4, 1965) (appl. Feb. 15, 1964).
J-87 Hellmer, *DECHEMA-Monogr.*, **55**(955–975), 127–141 (1964) (publ. 1965).
J-88 Zhukovzkii, Montilo, and Babadzhan, *Tr. Ural. Nauchno Issled. Proektn. Inst. Mednoi Prom.*, no. 8, 387–390 (1965).
J-89 Morth and Smith, *Am. Chem. Soc., Div. Fuel Chem., Prepr.*, **10**(1), 83–92 (1966).
J-90 Mars and Maessen, *Proc. 3d Int. Congr. Catal., Amsterdam*, **1964**(1), 266–282 (1965).
J-91 Kaneko, Odanaka, *J. Res. Inst. Catal., Hokkaido Univ.*, **13**(1), 29–43 (1965).
J-92 Ishii, *Kagaku Kogaku*, **29**(10), 779–780 (1965).
J-93 Brusset and Luquet, *Chim. Ind., Genie Chim.*, **96**(3), 557–564 (1966).
J-95 Cullis, Henson, and Trimm, *Proc. R. Soc. (London), Ser. A*, **295**(1440), 72–83 (1966).
J-96 Habashi and Bauer, *Ind. Eng. Chem. Fundam.*, **5**(4), 469–471 (1966).
J-97 Vandenbussche, *Comm. Energ. At. (France), Rapp. no. 28559*, 1966.
J-98 Bond and Jacobs, *J. Chem. Soc., Ser. A*, **1966**(9), 1265–1268.
J-99 Kiuchi and Watanabe, *Kogyo Kagaku Zasshi*, **66**(3), 370–373 (1963).
J-100 Kiuchi and Watanabe, *Kobunshi Kagaku*, **22**(245), 557–565 (1965).
J-101 Alfrey and Pfeifer, *J. Polym. Sci., Part A-1*, **4**(10), 2447–2460 (1966).
J-102 Hagiwara, Kojima, and Echigoya, *Bull. Chem. Soc. Japan*, **39**(8), 1800–1806 (1966).
J-103 Ogata and Sawaki, *J. Am. Chem. Soc.*, **88**(24), 5832–5837 (1966).
J-104 Germain, Gaschka, and Mayeux, *Bull. Soc. Chim. Fr.*, **1965**(5), 1445–1453.
J-105 Metelitsa and Denisov, *Zh. Fiz. Khim.*, **40**(9), 2162–2167 (1966).
J-106 Ogino and Kawakami, *Kogyo Kagaku Zasshi*, **68**(1), 45–49 (1965).
J-107 Nagata, Hashimato, Taniyama, Nishida, and Iwane, *Kagaku Kogaku*, **27**(8), 558–566 (1963).
J-108 Kertamus and Hill, *Am. Chem. Soc., Div. Fuel Chem., Prepr.* **8**(3), 89–96 (1964).
J-109 Mullin and Berkowitz, *Fuel*, **47**(1), 63–77 (1968).
J-110 Ishii, Maekawa, and Takeda, *Kagaku Kogaku*, **29**(12), 988–995 (1965).
J-111 Schuett, *Erdöl Kohle*, **19**(1), 32–38 (1966).
J-112 Heuchamps, *Bull. Soc. Chim. Fr.*, **1965**(10), 2955–2958.
J-113 Shimulis, *Kinet. Katal.*, **7**(3), 498–507 (1966).
J-114 Cusumano, Dembinski, and Sinfelt, *J. Catal.*, **5**(3), 471–475 (1966).
J-115 Knoezinger and Reiss, *Z. Phys. Chem.*, **154**(3–4), 136–149 (1967).
J-116 Bretsznajder and Marcinkowski, *Bull. Acad. Pol. Sci. Ser. Chim.*, **14**(11–12), 865–870 (1963).
J-117 Bhattacharyya and Sanyal, *J. Cataly.*, **7**(2), 152–158 (1967).
J-118 Todo, Kurita, and Hagiwara, *Kogyo Kagaku Zasshi*, **69**(8), 1463–1466 (1966).
J-119 Stryker, Mantell, and Helin, *J. Appl. Polym. Sci.*, **11**(1), 1–22 (1967).
J-120 Balasubramanian, Rihani, and Doraiswamy, *Ind. Eng. Chem. Fundam.*, **5**(2), 184–188 (1966).
J-121 Agami and Prevost, *C. R. Acad. Sci. (Paris), Ser. C*, **263**(2), 153–156 (1966).
J-122 Boyles and Toby, *J. Polym. Sci., Part B*, **4**(6), 411–415 (1966).
J-123 Cramond, Lawry, and Urwin, *Eur. Polym. J.*, **2**(2). 107–114 (1966).
J-124 Semenova and Fedosev, *Tr. Mosk. Khim. Tekhnol. Inst.*, no. **48**, 161–167 (1965).
J-125 Morita and Inoue, *Kogyo Kagaku Zasshi*, **68**(4), 659–663 (1965).
J-126 Germain et al., *Bull. Soc. Chim. Fr.*, **1965**(11), 3158–3162.
J-127 Wencke and Heise, *Monatsber. Dtsch. Akad. Wiss. Berlin*, **7**(12), 887–895 (1965).
J-128 Bliznakon, Jiru, and Klissurski, *Collect. Czech. Chem. Commun.*, **31**(7), 2995–2997 (1966).
J-129 Klisurski, *C. R. Acad Bulg. Sci.*, **19**(12), 1159–1162 (1966).
J-130 Kallo, *Z. Phys. Chem. (Frankfurt)*, **50**(3–4), 152–156 (1966).
J-131 Bamford and Hargreaves, *Proc. R. Soc., Ser. A*, **297**(1451), 425–439 (1967).
J-132 Sugimura and Minoura, *J. Polym. Sci., Part A-1*, **4**(11), 2735–2746 (1966).
J-133 Bamford, Fildes, and Maltmar, *Trans. Faraday Soc.*, **62**(9), 2544–2552 (1960).
J-134 Acres and Bond, *Platinum Met. Rev.*, **10**(4), 122–127 (1966).
J-135 Szell and Eastham, *J. Chem. Soc., Phys. Org.*, **1966**(1), 30–33.
J-136 Begley, *J. Polym. Sci., Part A-1*, **4**(2), 319–336 (1966).
J-137 Sachtler and De Boer, *Proc. 3d Int. Congr. Catal., Amsterdam*, **1964**(1), 263–265, disc. 252–263 (publ. 1965).
J-138 Soga and Keii, *J. Polym. Sci., Part A-1*, **4**(10), 2429–2439 (1966).
J-139 Hausmann and King, *Ind. Eng. Chem. Fundam.*, **5**(3), 295–301 (1966).
J-140 Williams and Bobalek, *Am. Chem. Soc., Div. Polym. Chem., Prepr.*, **5**(2), 688–693 (1984).
J-141 Anand, Deshpande, and Kapur, *J. Polym. Sci., Part A-1*, **5**(3), 665–673 (1967).
J-142 Potter, Johnson, Metz, and Bretton, *J. Polym. Sci., Part A-1*, **4**(2), 4-19–4-30 (1966).
J-143 Ureta, Smid, and Szware, ibid., (9), 2219–2229 (1966).
J-144 Suzuki, *Nippon Kagaku Zasshi*, **86**(12), 1318–1321 (1965).
J-145 Ogata and Sawaki, *Bull Chem. Soc. Japan*, **38**(2), 194–199 (1965).
J-146 Talamini and Giampietro, *J. Polym. Sci., Part A-2*, **4**(3), 535–537 (1966).
J-147 Ugelstad, Mork, Dahl, and Rangnes, *Am. Chem. Soc., Div. Polym. Chem., Prepr.*, **7**(2), 628–640 (1966).
J-148 Baulch, Drysdale, Horne, and Lloyd, *Evaluated Kinetic Data for High Temperature Reactions*, vol. 1: *Homogeneous Gas Phase Reactions of the H_2–O_2 System*, Butterworth, London, 1972.
J-149 Baulch, Drysdale, and Horne, *Evaluated Kinetic Data for High Temperature Reactions*, vol. 2: *Homogeneous Gas Phase Reactions of the H_2–N_2–O_2 System*, CRC Press, Cleveland, 1973.
J-150 Kerr and Personage, *Evaluated Kinetic Data on Gas Phase Addition Reactions: Reaction of Atoms and Radicals with Alkenes, Alkynes and Aromatic Compounds*, CRC Press, Cleveland, 1972.
J-151 Kondratiev, *Rate Constants of Gas Phase Reactions*, transl. by Holtschlag, ed. by Frostrom, COM-72-10014, National Bureau of Standards, Washington, 1972.
J-152 Benson and O'Neal, *Kinetic Data on Gas Phase Unimolecular Reac-*

tions, NSRDS-NBS21, National Bureau of Standards, Washington, February 1970.
J-153 Denisov, *Liquid Phase Reaction Rate Constants*, transl. by Johnston, IFI/Plenum, New York, 1974.

Physical Rate Data

K-1 Calderbank and Moo-Young, *Chem. Eng. Sci.*, **16**, 39 (1961).
K-2 Vermeulen, Williams, and Langlois, *Chem. Eng. Prog.*, **51**, 85F (1955).
K-3 Madden and Damerell, *Am. Inst. Chem. Eng. J.*, **8**, 233 (1962).
K-4 Groothuis and Zuiderweg, *Chem. Eng. Sci.*, **12**, 288 (1960).
K-5 Leva, *Fluidization*, McGraw-Hill, New York, 1959.
K-6 Zabrodsky, *Hydrodynamics and Heat Transfer in Fluidized Beds*, M.I.T., Cambridge, Mass., 1966.
K-7 Kunii and Levenspiel, *Fluidization Engineering*, Wiley, New York, 1969.
K-8 Davidson and Harrison (eds.), *Fluidization*, Academic, New York, 1971.
K-9 Grace and Matsen, *Fluidization*, Plenum, New York, 1980.
K-10 Chu, Kalil, and Wetteroth, *Chem. Eng. Prog.*, **49**, 141 (1953).
K-11 Yagi and Kunii, *Chem. Eng. (Japan)*, **18**, 576 (1954).
K-12 Yagi and Wakao, *Am. Inst. Chem. Eng. J.*, **5**, 79 (1959).
K-13 Chapman, Dallenbach, and Holland, *Trans. Inst. Chem. Eng. (London)*, **42**, T398 (1964).
K-14 McAdams, *Heat Transmission*, 3d ed., McGraw-Hill, New York, 1954.
K-15 Mickley and Fairbanks, *Am. Inst. Chem. Eng. J.*, **1**, 374 (1955).
K-16 Ziegler, Koppel, and Brazelton, *Ind. Eng. Chem. Fundam.*, **3**, 94, 324 (1964).
K-17 Plautz and Johnstone, ibid., **1**, 193 (1955).
K-18 Felix and Neill, *Amer. Inst. Chem. Eng. Rep. Heat Transfer Symp.* (December 1951).
K-19 Trense, Ph.D. thesis, Northwestern University, Evanston, Ill., 1954.
K-20 Urie, M.S. thesis, Massachusetts Institute of Technology, Cambridge, Mass., 1948.
K-21 Leva, Weintraub, and Grummer, *Chem. Eng. Prog.*, **45**, 563 (1949).
K-22 Xavier, King, Davidson, and Harrison, "Surface-Bed Heat Transfer in a Fluidized Bed at High Pressure," in Ref. K-9.
K-23 Borodulya, Gansha, and Podberezsky, "Heat Transfer in a Fluidized Bed at High Pressure," in Ref. K-9.

THERMODYNAMICS

GENERAL REFERENCES: Abbott and Van Ness, *Schaum's Outline of Theory and Problems of Thermodynamics*, McGraw-Hill, New York, 1972. Balzhizer, Samuels, and Eliassen, *Chemical Engineering Thermodynamics*, Prentice-Hall, Englewood Cliffs, N.J., 1972. Callen, *Thermodynamics*, Wiley, New York, 1960. Denbigh, *Principles of Chemical Equilibrium*, 3d ed, Cambridge, London, 1971. Hildebrand, Prausnitz, and Scott, *Regular and Related Solutions*, Van Nostrand Reinhold, New York, 1970. Hildebrand and Scott, *The Solubility of Nonelectrolytes*, 3d ed., Dover, New York, 1964. Lewis, Randall, Pitzer, and Brewer, *Thermodynamics*, 2d ed., McGraw-Hill, New York, 1961. McGlashan, *Chemical Thermodynamics*, Academic, New York, 1979. Modell and Reid, *Thermodynamics and Its Applications*, Prentice-Hall, Englewood Cliffs, N.J., 1974. Münster, *Classical Thermodynamics*, Wiley-Interscience, London, 1970. Prausnitz, *Molecular Thermodynamics of Fluid-Phase Equilibria*, Prentice-Hall, Englewood Cliffs, N.J., 1969. Prausnitz, Anderson, Grens, Eckert, Hsieh, and O'Connell, *Computer Calculations for Multicomponent Vapor-Liquid and Liquid-Liquid Equilibria*, Prentice-Hall, Englewood Cliffs, N.J., 1980. Prigogine and Defay, *Chemical Thermodynamics*, Longmans, London, 1954. Reid, Prausnitz, and Sherwood, *The Properties of Gases and Liquids*, 3d ed., McGraw-Hill, New York, 1977. Sandler, *Chemical and Engineering Thermodynamics*, Wiley, New York, 1977. Smith and Van Ness, *Introduction to Chemical Engineering Thermodynamics*, 3d ed., McGraw-Hill, New York, 1975. Van Ness and Abbott, *Classical Thermodynamics of Nonelectrolyte Solutions: With Applications to Phase Equilibria*, McGraw-Hill, New York, 1982.

INTRODUCTION

Thermodynamics is the branch of science that embodies the principles of energy transformation in macroscopic systems. The general restrictions which experience has shown to apply to all such transformations are known as the laws of thermodynamics. These laws are primitive; they cannot be derived from anything more basic.

The first law of thermodynamics states that energy is conserved, that although it can be altered in form and transferred from one place to another, the total quantity remains constant. Thus the first law of thermodynamics depends on the concept of energy, but conversely energy is an *essential* thermodynamic function because it allows the first law to be formulated. This coupling is characteristic of the primitive concepts of thermodynamics.

The words *system* and *surroundings* are similarly coupled. A system is taken to be any object, any quantity of matter, any region, etc., selected for study and set apart (mentally) from everything else, which is then called the surroundings. The imaginary envelope which encloses the system and separates it from its surroundings is called the *boundary* of the system.

This boundary may be imagined to have special properties which serve either (1) to *isolate* the system from its surroundings, or (2) to provide for *interaction* in specific ways between system and surroundings. An *isolated system* can exchange neither matter nor energy with its surroundings. If a system is not isolated, its boundaries may be imagined to permit either matter or energy or both to be exchanged with its surroundings. If the exchange of matter is allowed, the system is said to be *open;* if only energy and not matter may be exchanged, the system is *closed* (but not isolated), and its mass is constant.

When a system is isolated, it cannot be affected by its surroundings. Nevertheless, changes may occur within the system that are detectable with measuring instruments such as thermometers, pressure gauges, etc. However, such changes cannot continue indefinitely, and the system must eventually reach a final static condition of *internal equilibrium.*

For a closed system which interacts with its surroundings, a final static condition such that the system is not only internally at equilibrium but also in *external equilibrium* with its surroundings may also eventually be reached.

The concept of equilibrium is central in thermodynamics, for associated with the equilibrium condition of a system is the concept of *state.* A system has an identifiable, reproducible state when all its *properties* are fixed. The concepts of *state* and *property* are again coupled. One can equally well say that the properties of a system are fixed by its state. Certain properties are detected with measuring instruments such as thermometers and pressure gauges. The existence of other properties, such as *internal energy,* is recognized much more indirectly. The number of properties that must be set at given values in order to fix the state of a system depends on the system and must be determined from experience.

When a system is displaced from an equilibrium state, it undergoes a *process* during which its properties change until a new equilibrium state is reached. During such a process the system may be caused to interact with its surroundings so as to interchange energy in the forms of **heat** and **work** and so to produce in the system or surroundings changes considered desirable for one reason or another. A process that proceeds so that the system is never displaced more than differentially from an equilibrium state is said to be **reversible,** because such a process may be reversed to trace in the opposite direction the same path initially taken without requiring any additional work above that produced by the forward process.

The basis of thermodynamics rests upon experience and experiment. A number of resultant postulates are set forth as follows:

Postulate 1 *There exists a form of energy, known as* **internal energy** *U, which for systems in equilibrium states is an intrinsic property of the system, functionally related to the measurable coordinates which characterize the system.*

Postulate 2 *The* **total** *energy of any system and its surroundings is conserved.* **(First law of thermodynamics)**

Internal energy is quite distinct from kinetic and potential energy, which are external forms of energy. In applications of the first law of thermodynamics all forms of energy must be considered, including the internal energy. It is therefore clear that Postulate 2 depends on Postulate 1. For an isolated system the first law requires that its energy be constant. For a closed (but not isolated) system the first law requires that energy changes of the system be exactly compensated by energy changes in the surroundings. For such systems energy is exchanged between a system and its surroundings in two forms: heat and work.

Heat is energy crossing the system boundary under the influence of a temperature difference or gradient. A quantity of heat Q represents an amount of energy in transit between a system and its surroundings and is not a property of the system. The usual convention with respect to signs requires that numerical values of Q be taken as positive when heat is added to the system and as negative when heat leaves the system.

Work is again energy in transit between a system and its surroundings, but resulting from the displacement of an external force acting on the system. Like heat, it is not a property of the system. The sign convention chosen here requires that numerical values of work W be taken as positive when work is done by the system and as negative when work is done on the system. Thus W has the opposite sense of Q.

When applied to closed systems (constant mass) for which the only form of energy that changes is the internal energy, the first law of thermodynamics is expressed mathematically as

$$dU = dQ - dW \tag{4-113}$$

Note that dQ and dW are not exact differentials and that Q and W are not properties of the system or functions of the thermodynamic coordinates that characterize the system. On the other hand, dU represents a differential change in U, a property of the system. The differential **quantities** dQ and dW represent energy exchanges between the system and its surroundings and serve in the equation to account for the energy change of the surroundings. Integration of Eq. (4-113) gives for a finite process

$$\Delta U = Q - W \tag{4-114}$$

where Δ signifies the difference between the final and initial values of U.

Postulate 3 *There exists a property called* entropy S, *which for systems in equilibrium states is an intrinsic property of the system, functionally related to the measurable coordinates which characterize the system. For reversible processes changes in this property may be calculated by the equation*

$$dS = dQ_{rev}/T \tag{4-115}$$

where T is the absolute temperature of the system.

Postulate 4 *The entropy change of any system and its surroundings,* **considered together,** *resulting from any real process is positive and approaches a limiting value of zero for any process that approaches reversibility.* **(Second law of thermodynamics)**

In the same way that the first law of thermodynamics cannot be formulated without the prior recognition of internal energy as a property, so also the second law can have no complete and quantitative expression without a prior assertion of the existence of entropy as a property.

The second law requires that the entropy of an isolated system must either increase or in the limit, where the system has reached an equilibrium state, remain constant. For a closed (but not isolated) system it requires that any entropy decrease in either the system or its surroundings be more than compensated by an entropy increase in the other part, or that in the limit, where the process is reversible, the total entropy of system plus its surroundings remain constant.

The fundamental thermodynamic properties that arise in connection with the first and second laws of thermodynamics are internal energy and entropy. These properties together with the two laws for which they are essential apply to all types of systems. However, different types of systems are characterized by different sets of measurable coordinates or variables. The type of system most commonly encountered in chemical engineering applications is one for which the primary characteristic measurable variables are pressure, volume, temperature, and composition, not all of which are necessarily independent. Such systems are made up of fluids, liquid or gas, and are called PVT systems.

For closed systems of this kind the work of a reversible process may always be calculated from

$$dW_{rev} = PdV \tag{4-116}$$

where P is the absolute pressure and V is volume. This equation follows directly from the definition of mechanical work.

Postulate 5 *The macroscopic properties of homogeneous PVT systems in equilibrium states can be expressed as functions of pressure, temperature, and composition only.*

This postulate imposes an idealization and is the basis for all subsequent equations relating the properties of PVT systems for which the postulate is true. The PVT system serves as a satisfactory model in an enormous number of practical applications. In accepting this model one assumes that the effects of fields (e.g., electric, magnetic, or gravitational) are negligible and that surface and viscous shear effects are unimportant.

Temperature, pressure, and composition are considered here primarily as **conditions** imposed upon or exhibited by the system, and the functional dependence of the thermodynamic properties on these conditions is determined by experiment. This is quite direct for the volume, which can be measured, and leads immediately to the conclusion that there exists an **equation of state** relating volume to pressure, temperature, and composition for any particular homogeneous PVT system. Such equations of state find extensive use in the applications of thermodynamics.

Determination of the functional dependence of internal energy and entropy on pressure, temperature, and composition is wholly indirect and is realized through the network of equations to be developed presently. The basis for this development has now been established. All else results from definition and deduction.

Thermodynamics solves problems in terms of abstract quantities such as internal energy and entropy. The initial step in the solution of any problem in applied thermodynamics, once the problem has been defined, is to translate it into the terminology of thermodynamic variables so that the laws of thermodynamics may be imposed. The reverse process constitutes the final step, for ultimately results must be expressed in quantities having physical reality.

THERMODYNAMIC VARIABLES, DEFINITIONS, AND RELATIONSHIPS

Consider a single-phase closed system in which there are no chemical reactions. Under these restrictions the composition is fixed. If such a system undergoes a differential, reversible process, then by Eq. (4-113)

$$dU = dQ_{rev} - dW_{rev}$$

Substitution for dQ_{rev} and dW_{rev} by Eqs. (4-115) and (4-116) gives

$$dU = T\,dS - PdV$$

Although derived for a reversible process, this equation relates properties only and is valid for any change between equilibrium states in a closed system.

The quantities U, S, and V are **extensive** properties, dependent on the size of the system. On the other hand, T and P are **intensive,** independent of system size. Since the present treatment deals primarily with open systems, we employ a nomenclature which recognizes only intensive properties and which requires explicit inclusion of the mass of the system in all equations. If we henceforth define U, S, and V as the **molar** internal energy, entropy, and volume, then the preceding equation is written

$$d(nU) = T\,d(nS) - P\,d(nV) \tag{4-117}$$

where n is the number of moles of fluid contained in the system and is constant for the special case of a closed system. Note that

$$n = n_1 + n_2 + n_3 + \cdots = \Sigma n_i$$

where i is an index identifying the chemical species present. Alternatively U, S, and V can be taken as **specific** (unit mass) properties, in which case m replaces n.

Equation (4-117) shows that for the single-phase nonreacting closed system specified,

$$nU = u(nS, nV)$$

Thus

$$d(nU) = \left[\frac{\partial(nU)}{\partial(nS)}\right]_{nV,n} d(nS) + \left[\frac{\partial(nU)}{\partial(nV)}\right]_{nS,n} d(nV)$$

where the subscript n indicates that all mole numbers n_i (and hence n) are constant. Comparison with Eq. (4-117) shows that

$$\left[\frac{\partial(nU)}{\partial(nS)}\right]_{nV,n} = T \qquad (4\text{-}118)$$

and

$$\left[\frac{\partial(nU)}{\partial(nV)}\right]_{nS,n} = -P \qquad (4\text{-}119)$$

Consider now an **open** system consisting of a single phase. It is assumed that

$$nU = u(nS, nV, n_1, n_2, n_3, \ldots)$$

Then

$$d(nU) = \left[\frac{\partial(nU)}{\partial(nS)}\right]_{nV,n} d(nS) + \left[\frac{\partial(nU)}{\partial(nV)}\right]_{nS,n} d(nV) + \sum\left(\left[\frac{\partial(nU)}{\partial n_i}\right]_{nV,nS,nj} dn_i\right)$$

where the summation is over all species present in the system and subscript n_j indicates that all mole numbers are held constant except the ith. By definition, let

$$\mu_i = \left[\frac{\partial(nU)}{\partial n_i}\right]_{nV,nS,nj}$$

Together with Eqs. (4-118) and (4-119) this allows elimination of all the partial differential coefficients from the preceding equation:

$$d(nU) = T\,d(nS) - P\,d(nV) + \Sigma(\mu_i\,dn_i) \qquad (4\text{-}120)$$

Equation (4-120) is the fundamental equation interrelating the primary thermodynamic variables for single-phase PVT systems, and all other equations connecting the properties of such systems are derived from it. The quantity μ_i is called the **chemical potential** of component i, and it plays a vital role in the thermodynamics of phase and chemical equilibrium.

The most direct procedure for the derivation of additional property relations is as follows. Since $n_i = x_i n$, where x_i is the mole fraction of component i, Eq. (4-120) may be rewritten as

$$d(nU) - T\,d(nS) + P\,d(nV) - \Sigma[\mu_i\,d(x_i n)] = 0$$

Expansion of the differentials and collection of like terms give

$$n[dU - T\,dS + P\,dV - \Sigma(\mu_i\,dx_i)] + dn[U - TS + PV - \Sigma(x_i\mu_i)] = 0$$

Since n and dn are independent and arbitrary, the terms in brackets must separately be zero. Then

$$dU = T\,dS - P\,dV + \Sigma(\mu_i\,dx_i) \qquad (4\text{-}121)$$

$$U = TS - PV + \Sigma(x_i\mu_i) \qquad (4\text{-}122)$$

Equations (4-120) and (4-121) are similar, but there is an important difference between them. Equation (4-120) applies to a system of n mol where n may vary; whereas Eq. (4-121) applies to a system in which n is unity and invariant. Thus Eq. (4-121) is subject to the restraint that $\Sigma x_i = 1$ or that $\Sigma\,dx_i = 0$. In this equation the x_i's cannot be treated as though they were all independent variables. The n_i's in Eq. (4-120) are not subject to any such restraint.

Equation (4-120) dictates the possible combinations of terms that may be defined as additional primary functions. Provided that the summation $\Sigma x_i\mu_i$ is treated as a single term, there are just eight possible distinct combinations, and these are shown in Table 4-22. Other thermodynamic properties are related to these and arise by arbitrary definition. They are called auxiliary functions, and a number of them will be introduced later.

TABLE 4-22

Primary grouping	Symbol	Name	Alternative grouping
U	U	Internal energy	$TS - PV + \Sigma(x_i\mu_i)$
$U + PV$	H	Enthalpy	$TS + \Sigma(x_i\mu_i)$
$U - TS$	A	Helmholtz function	$-PV + \Sigma(x_i\mu_i)$
$U + PV - TS$	G	Gibbs function	$\Sigma(x_i\mu_i)$
$U - \Sigma(x_i\mu_i)$	X	Unnamed	$TS - PV$
$U + PV - \Sigma(x_i\mu_i)$	Y	Unnamed	TS
$U - TS - \Sigma(x_i\mu_i)$	Z	Unnamed	$-PV$
$U + PV - TS - \Sigma(x_i\mu_i)$		"Zero function"	

From Table 4-22 general expressions may be written for H, A, G, etc., in accord with their definitions. For example,

$$H = U + PV \quad \text{or} \quad nH = nU + P(nV)$$

Thus

$$d(nH) = d(nU) + P\,d(nV) + (nV)\,dP$$

Substitution for $d(nU)$ by Eq. (4-120) gives a general expression for the total differential $d(nH)$. The total differentials of the other properties are obtained similarly. The equations which result appear in Table 4-23. Each equation expresses a property (nU), (nH), etc., as a function of a particular set of independent variables; these are the *canonical variables* for the property. A similar set of equations can be developed from Eq. (4-121). This set results from the set in Table 4-23 by imposition of the restraints that $n = 1$ and $n_i = x_i$. The two sets are related exactly as Eq. (4-120) is related to Eq. (4-121). The equations written for $n = 1$ are of course less general than those of Table 4-23. Furthermore, the interdependence of the x_i's precludes those mathematical operations which depend on independence of the variables.

TABLE 4-23

$d(nU) = T\,d(nS) - P\,d(nV) + \Sigma(\mu_i\,dn_i)$	(4-120)
$d(nH) = T\,d(nS) + (nV)\,dP + \Sigma(\mu_i\,dn_i)$	(4-123)
$d(nA) = -(nS)\,dT - P\,d(nV) + \Sigma(\mu_i\,dn_i)$	(4-124)
$d(nG) = -(nS)\,dT + (nV)\,dP + \Sigma(\mu_i\,dn_i)$	(4-125)
$d(nX) = T\,d(nS) - P\,d(nV) - \Sigma(n_i\,d\mu_i)$	(4-126)
$d(nY) = T\,d(nS) + (nV)\,dP - \Sigma(n_i\,d\mu_i)$	(4-127)
$d(nZ) = -(nS)\,dT - P\,d(nV) - \Sigma(n_i\,d\mu_i)$	(4-128)
$0 = -(nS)\,dT + (nV)\,dP - \Sigma(n_i\,d\mu_i)$	(4-129)

Constant-Composition Systems For 1 mol of a homogeneous fluid of constant composition Eqs. (4-120) and (4-123) to (4-125) simplify to

$$dU = T\,dS - P\,dV \qquad (4\text{-}130)$$

$$dH = T\,dS + V\,dP \qquad (4\text{-}131)$$

$$dA = -P\,dV - S\,dT \qquad (4\text{-}132)$$

$$dG = V\,dP - S\,dT \qquad (4\text{-}133)$$

From these it is seen that

$$T = \left(\frac{\partial U}{\partial S}\right)_V = \left(\frac{\partial H}{\partial S}\right)_P \qquad (4\text{-}134)$$

$$-P = \left(\frac{\partial U}{\partial V}\right)_S = \left(\frac{\partial A}{\partial V}\right)_T \qquad (4\text{-}135)$$

$$V = \left(\frac{\partial H}{\partial P}\right)_S = \left(\frac{\partial G}{\partial P}\right)_T \qquad (4\text{-}136)$$

$$-S = \left(\frac{\partial A}{\partial T}\right)_V = \left(\frac{\partial G}{\partial T}\right)_P \qquad (4\text{-}137)$$

In addition, the common Maxwell equations result from application of the reciprocity relationship for exact differentials:

$$\left(\frac{\partial T}{\partial V}\right)_S = -\left(\frac{\partial P}{\partial S}\right)_V \tag{4-138}$$

$$\left(\frac{\partial T}{\partial P}\right)_S = \left(\frac{\partial V}{\partial S}\right)_P \tag{4-139}$$

$$\left(\frac{\partial P}{\partial T}\right)_V = \left(\frac{\partial S}{\partial V}\right)_T \tag{4-140}$$

$$\left(\frac{\partial V}{\partial T}\right)_P = -\left(\frac{\partial S}{\partial P}\right)_T \tag{4-141}$$

In all these equations it is understood that the partial derivatives are taken with composition held constant.

Enthalpy and Entropy as Functions of T and P At constant composition the molar thermodynamic properties are functions of temperature and pressure (Postulate 5). Thus,

$$dH = \left(\frac{\partial H}{\partial T}\right)_P dT + \left(\frac{\partial H}{\partial P}\right)_T dP \tag{4-142}$$

$$dS = \left(\frac{\partial S}{\partial T}\right)_P dT + \left(\frac{\partial S}{\partial P}\right)_T dP \tag{4-143}$$

The obvious next step is to eliminate the partial differential coefficients in favor of measurable quantities.

For this purpose the heat capacity at constant pressure is defined as

$$C_P = \left(\frac{\partial H}{\partial T}\right)_P \tag{4-144}$$

It is the property of the material and a function of temperature, pressure, and composition.

Equation (4-131) may first be divided by dT and restricted to constant pressure, and then be divided by dP and restricted to constant temperature, yielding the two equations

$$\left(\frac{\partial H}{\partial T}\right)_P = T\left(\frac{\partial S}{\partial T}\right)_P$$

$$\left(\frac{\partial H}{\partial P}\right)_T = T\left(\frac{\partial S}{\partial P}\right)_T + V$$

In view of Eq. (4-144), the first of these becomes

$$\left(\frac{\partial S}{\partial T}\right)_P = \frac{C_P}{T} \tag{4-145}$$

and in view of Eq. (4-141), the second equation becomes

$$\left(\frac{\partial H}{\partial P}\right)_T = V - T\left(\frac{\partial V}{\partial T}\right)_P \tag{4-146}$$

Combination of Eqs. (4-142), (4-144), and (4-146) gives

$$dH = C_P\,dT + \left[V - T\left(\frac{\partial V}{\partial T}\right)_P\right] dP \tag{4-147}$$

and combination of Eqs. (4-143), (4-145), and (4-141) gives

$$dS = \frac{C_P}{T}\,dT - \left(\frac{\partial V}{\partial T}\right)_P dP \tag{4-148}$$

Equations (4-147) and (4-148) are general equations expressing the enthalpy and entropy of homogeneous fluids *at constant composition* as functions of T and P. The coefficients of dT and dP are expressed in terms of measurable quantities.

Internal Energy and Entropy as Functions of T and V It is frequently more convenient to take T and V as independent variables rather than T and P. Because V is related to T and P through an equation of state, this is clearly permissible. In this case it is best to work with the internal energy and entropy, for which

$$dU = \left(\frac{\partial U}{\partial T}\right)_V dT + \left(\frac{\partial U}{\partial V}\right)_T dV \tag{4-149}$$

$$dS = \left(\frac{\partial S}{\partial T}\right)_V dT + \left(\frac{\partial S}{\partial V}\right)_T dV \tag{4-150}$$

The procedure now is analogous to that of the preceding section.

Define the **heat capacity at constant volume** by

$$C_V = \left(\frac{\partial U}{\partial T}\right)_V \tag{4-151}$$

It is a property of the material and a function of temperature, pressure, and composition.

Two relations follow immediately from Eq. (4-130):

$$\left(\frac{\partial U}{\partial T}\right)_V = T\left(\frac{\partial S}{\partial T}\right)_V$$

$$\left(\frac{\partial U}{\partial V}\right)_T = T\left(\frac{\partial S}{\partial V}\right)_T - P$$

As a result of Eq. (4-151) the first of these becomes

$$\left(\frac{\partial S}{\partial T}\right)_V = \frac{C_V}{T} \tag{4-152}$$

and as a result of Eq. (4-140) the second becomes

$$\left(\frac{\partial U}{\partial V}\right)_T = T\left(\frac{\partial P}{\partial T}\right)_V - P \tag{4-153}$$

Combination of Eqs. (4-149), (4-151), and (4-153) gives

$$dU = C_V\,dT + \left[T\left(\frac{\partial P}{\partial T}\right)_V - P\right] dV \tag{4-154}$$

and combination of Eqs. (4-150), (4-140), and (4-152) gives

$$dS = \frac{C_V}{T}\,dT + \left(\frac{\partial P}{\partial T}\right)_V dV \tag{4-155}$$

Equations (4-154) and (4-155) are general equations expressing the internal energy and entropy of homogeneous fluids *at constant composition* as functions of temperature and molar volume. The coefficients of dT and dV are expressed in terms of measurable quantities.

Heat-Capacity Relationships In Eqs. (4-147) and (4-154) both dH and dU are exact differentials, and application of the reciprocity relation leads to

$$\left(\frac{\partial C_P}{\partial P}\right)_T = -T\left(\frac{\partial^2 V}{\partial T^2}\right)_P \tag{4-156}$$

$$\left(\frac{\partial C_V}{\partial V}\right)_T = T\left(\frac{\partial^2 P}{\partial T^2}\right)_V \tag{4-157}$$

Thus the pressure or volume dependence of the heat capacities may be determined from PVT data. The temperature dependence of the heat capacities is, however, determined empirically and is often expressed by equations such as

$$C_P = \alpha + \beta T + \gamma T^2$$

Equations (4-148) and (4-155) both provide expressions for dS, which must be equal for the same change of state. Equating them and solving for dT gives

$$dT = \frac{T}{C_P - C_V}\left(\frac{\partial V}{\partial T}\right)_P dP + \frac{T}{C_P - C_V}\left(\frac{\partial P}{\partial T}\right)_V dV$$

However, at constant composition $T = T(P, V)$. Thus,

$$dT = \left(\frac{\partial T}{\partial P}\right)_V dP + \left(\frac{\partial T}{\partial V}\right)_P dV$$

Equating coefficients of either dP or dV in these two expressions for dT gives

$$C_P - C_V = T\left(\frac{\partial V}{\partial T}\right)_P \left(\frac{\partial P}{\partial T}\right)_V \tag{4-158}$$

Thus the *difference* between the two heat capacities may be determined from PVT data.

The *ratio* of these heat capacities is obtained by division of Eq. (4-145) by Eq. (4-152)

$$\frac{C_P}{C_V} = \frac{(\partial S/\partial T)_P}{(\partial S/\partial T)_V}$$

or

$$\frac{C_P}{C_V} = \frac{(\partial S/\partial V)_P(\partial V/\partial T)_P}{(\partial S/\partial P)_V(\partial P/\partial T)_V}$$

Replacement of each of the four partial derivatives through the appropriate Maxwell relation gives finally

$$\gamma = \frac{C_P}{C_V} = \left(\frac{\partial V}{\partial P}\right)_T \left(\frac{\partial P}{\partial V}\right)_S \qquad (4\text{-}159)$$

where γ is the symbol commonly used to represent the heat-capacity ratio.

The Ideal Gas The simplest equation of state is that for an ideal gas,

$$PV = RT$$

where R is a universal constant, values for which are given in Table 1-9. The following partial derivatives are obtained from the ideal-gas equation:

$$\left(\frac{\partial P}{\partial T}\right)_V = \frac{R}{V} = \frac{P}{T}$$

$$\left(\frac{\partial^2 P}{\partial T^2}\right)_V = 0$$

$$\left(\frac{\partial V}{\partial T}\right)_P = \frac{R}{P} = \frac{V}{T}$$

$$\left(\frac{\partial^2 V}{\partial T^2}\right)_P = 0$$

$$\left(\frac{\partial P}{\partial V}\right)_T = -\frac{P}{V}$$

The general equations for constant-composition fluids derived in the preceding subsections reduce to very simple forms when the relations for an ideal gas are substituted into them:

$$\left(\frac{\partial S}{\partial P}\right)_T = -\frac{R}{P} \qquad (4\text{-}141\ ideal)$$

$$\left(\frac{\partial H}{\partial P}\right)_T = 0 \qquad (4\text{-}146\ ideal)$$

$$\left(\frac{\partial S}{\partial V}\right)_T = \frac{R}{V} \qquad (4\text{-}140\ ideal)$$

$$\left(\frac{\partial U}{\partial V}\right)_T = 0 \qquad (4\text{-}153\ ideal)$$

$$dH = C_P\, dT \qquad (4\text{-}147\ ideal)$$

$$dS = (C_P/T)\, dT - R/P\, dP \qquad (4\text{-}148\ ideal)$$

$$dU = C_V\, dT \qquad (4\text{-}154\ ideal)$$

$$dS = (C_V/T)\, dT + (R/V)\, dV \qquad (4\text{-}155\ ideal)$$

$$\left(\frac{\partial C_p}{\partial P}\right)_T = 0 \qquad (4\text{-}156\ ideal)$$

$$\left(\frac{\partial C_V}{\partial V}\right)_T = 0 \qquad (4\text{-}157\ ideal)$$

$$C_P - C_V = R \qquad (4\text{-}158\ ideal)$$

$$\gamma = \frac{C_P}{C_V} = -\left(\frac{\partial \ln P}{\partial \ln V}\right)_S \qquad (4\text{-}159\ ideal)$$

It is clear from these equations that for an ideal gas H, U, C_P, and C_V are functions of temperature only and are independent of P and V. The entropy of an ideal gas, however, is a function of both T and P or of both T and V.

Variable-Composition Systems The composition of a system may vary because the system is open or because of chemical reactions even in a closed system. The equations developed here apply regardless of the cause of composition changes.

Partial Molar Properties The general homogeneous PVT system may contain any number of chemical species. The symbol M is taken to represent a molar thermodynamic property in general, where M may stand in turn for U, H, S, etc. The total system property is then nM, where $n = \Sigma n_i$ is the total moles of mixture in the system. One would expect the mixture property M to be related to the properties M_i of the pure chemical species which make up the mixture. However, no generally valid relationship is known, and the connection must be established experimentally for any particular system.

The constituents of a mixture of a solution do not have any identifiable separate thermodynamic properties. It is convenient, however, to think of the mixture property as being apportioned among the mixture constituents in some appropriate way. Once an apportioning procedure has been adopted, the resulting property values may quite logically be treated as though they were properties of the individual constituents as they exist in solution.

For a homogeneous PVT system *at constant temperature and pressure Postulate* 5 requires that

$$nM = f(n_1, n_2, n_3, \ldots)$$

Moreover, it is known from experiment that the total mixture property nM is extensive and is in fact a homogeneous function of the first degree in the mole numbers of the constituents. (That is, doubling, tripling, etc., of all the n_i's will double, triple, etc., nM). It follows from Euler's theorem on homogeneous functions that

$$nM = \sum \left(n_i \left[\frac{\partial(nM)}{\partial n_i}\right]_{T,P,n_j}\right)$$

For convenience define

$$\left[\frac{\partial(nM)}{\partial n_i}\right]_{T,P,n_j} = \bar{M}_i \qquad (4\text{-}160)$$

where $\bar{M}_i$ is called the **partial molar property** of i in solution. Substitution of Eq. (4-160) into the preceding equations gives

$$nM = \Sigma(n_i\bar{M}_i) \qquad (4\text{-}161)$$

The $\bar{M}_i$'s are homogeneous functions of zero degree in the mole numbers and are therefore intensive thermodynamic properties. Equation (4-160) is the formula that determines how the mixture property is apportioned among the mixture constituents. If the resulting molar properties $\bar{M}_i$ are treated as properties of the constituents in solution, then Eq. (4-161) shows that this apportioning procedure leads to the simple result that the total mixture property is the sum of the properties attributed to the constituents. Division of Eq. (4-161) by n puts it on a mole basis:

$$M = \Sigma(x_i\bar{M}_i) \qquad (4\text{-}162)$$

The equations developed for partial properties apply as well on a unit mass basis. In this case m replaces n, and the x_i's become mass fractions. As a result of Eq. (4-160) and the defining equations for H, A, and G it is readily shown that

$$\bar{H}_i = \bar{U}_i + P\bar{V}_i$$

$$\bar{A}_i = \bar{U}_i - T\bar{S}_i$$

$$\bar{G}_i = \bar{H}_i - T\bar{S}_i$$

Gibbs-Duhem Equation Equation (4-161) is perfectly general for any homogeneous PVT system in an equilibrium state. Changes in nM resulting from alteration of P, T, or the n_i's are given by the *total* differential of nM,

$$d(nM) = \Sigma(n_i\, d\bar{M}_i) + \Sigma(\bar{M}_i\, dn_i) \qquad (4\text{-}163)$$

Since the general functional relationship for nM is

$$nM = f(T,P,n_1,n_2,n_3,\ldots)$$

it is also true in general that

$$d(nM) = \left[\frac{\partial(nM)}{\partial T}\right]_{P,n} dT + \left[\frac{\partial(nM)}{\partial P}\right]_{T,n} dP + \Sigma(\overline{M}_i\, dn_i)$$

or

$$d(nM) = n\left(\frac{\partial M}{\partial T}\right)_{P,x} dT + n\left(\frac{\partial M}{\partial P}\right)_{T,x} dP + \Sigma(\overline{M}_i\, dn_i) \qquad (4\text{-}164)$$

where the subscript x indicates that all mole fractions are held constant.

Comparison of Eqs. (4-163) and (4-164) shows that they can both be generally true only if

$$n\left(\frac{\partial M}{\partial T}\right)_{P,x} dT + n\left(\frac{\partial M}{\partial P}\right)_{T,x} dP - \Sigma(n_i\, d\overline{M}_i) = 0$$

or

$$\left(\frac{\partial M}{\partial T}\right)_{P,x} dT + \left(\frac{\partial M}{\partial P}\right)_{T,x} dP - \Sigma(x_i\, d\overline{M}_i) = 0 \qquad (4\text{-}165)$$

Equation (4-165) is the general form of the Gibbs-Duhem equation and is valid for any thermodynamic property M in a homogeneous phase. For example, if M is taken to be the enthalpy H, then combination of Eqs. (4-144) and (4-146) with Eqs. (4-164) and (4-165) yields the general equations

$$d(nH) = nC_P\, dT + n\left[V - T\left(\frac{\partial V}{\partial T}\right)_{P,x}\right] dP + \Sigma(\overline{H}_i\, dn_i) \qquad (4\text{-}166)$$

$$C_P\, dT + \left[V - T\left(\frac{\partial V}{\partial T}\right)_{P,x}\right] dP - \Sigma(x_i\, d\overline{H}_i) = 0 \qquad (4\text{-}167)$$

Similar equations are readily derived when M takes on other identities.

At constant T and P, Eq. (4-165) becomes

$$\Sigma(x_i\, d\overline{M}_i) = 0 \text{ (constant } T \text{ and } P) \qquad (4\text{-}168)$$

Partial Molar Gibbs Function From Eqs. (4-120) and (4-123) to (4-125) of Table 4-23 it is seen that the chemical potential is related to each of the properties U, H, A, and G:

$$\mu_i = \left[\frac{\partial(nU)}{\partial n_i}\right]_{nS,nV,nj} = \left[\frac{\partial(nH)}{\partial n_i}\right]_{nS,P,nj} = \left[\frac{\partial(nA)}{\partial n_i}\right]_{nV,T,nj} = \left[\frac{\partial(nG)}{\partial n_i}\right]_{T,P,nj} \qquad (4\text{-}169)$$

It is clear by reference to Eq. (4-160) that the last member of this set is $\overline{G}_i$. Thus the chemical potential is identical with the partial molar Gibbs function, or

$$\mu_i = \overline{G}_i$$

From Table 4-22,

$$G = \Sigma(x_i\mu_i) = \Sigma(x_i\overline{G}_i)$$

and therefore

$$nG = \Sigma(n_i\overline{G}_i)$$

These are evidently special cases of Eqs. (4-162) and (4-161). In addition, Eqs. (4-125) and (4-129) with $\overline{G}_i$ substituted for μ_i become special cases of Eqs. (4-164) and (4-165):

$$d(nG) = -(nS)\, dT + (nV)\, dP + \Sigma(\overline{G}_i\, dn_i) \qquad (4\text{-}170)$$

and

$$-(nS)\, dT + (nV)\, dP - \Sigma(n_i\, d\overline{G}_i) = 0$$

or

$$-S\, dT + V\, dP - \Sigma(x_i\, d\overline{G}_i) = 0 \qquad (4\text{-}171)$$

This equation (4-171) is the most commonly encountered form of the Gibbs-Duhem equation.

The reciprocity relation for an exact differential may be applied systematically to the equations of Table 4-23 (but not so generally to equations restricted to $n = 1$). A number of equations result, among which are the Maxwell equations, already derived, and the two following particularly useful relations, which result from Eq. (4-125):

$$\left(\frac{\partial \mu_i}{\partial P}\right)_{T,n} = \left[\frac{\partial(nV)}{\partial n_i}\right]_{T,P,nj} = \overline{V}_i \qquad (4\text{-}172)$$

$$\left(\frac{\partial \mu_i}{\partial T}\right)_{P,n} = -\left[\frac{\partial(nS)}{\partial n_i}\right]_{T,P,nj} = -\overline{S}_i \qquad (4\text{-}173)$$

In a *constant-composition solution*, $\mu_i = \mu(T,P)$. Thus,

$$d\mu_i \equiv d\overline{G}_i = \left(\frac{\partial \mu_i}{\partial T}\right)_{P,n} dT + \left(\frac{\partial \mu_i}{\partial P}\right)_{T,n} dP$$

or

$$d\overline{G}_i = -\overline{S}_i\, dT + \overline{V}_i\, dP \qquad (4\text{-}174)$$

Comparison with the equation for 1 mol of a constant-composition solution as obtained from Eq. (4-125),

$$dG = -S\, dT + V\, dP$$

provides an example of the parallelism that exists between the equations for constant-composition solutions and the components in a constant-composition solution. This parallelism exists whenever the solution properties in the parent equation are related linearly (in the algebraic sense). Thus, in view of Eqs. (4-120), (4-123), and (4-124), we may write

$$d\overline{U}_i = T\, d\overline{S}_i - P\, d\overline{V}_i \qquad (4\text{-}175)$$

$$d\overline{H}_i = T\, d\overline{S}_i + \overline{V}_i\, dP \qquad (4\text{-}176)$$

$$d\overline{A}_i = -\overline{S}_i\, dT - P\, d\overline{V}_i \qquad (4\text{-}177)$$

Note that these relations hold only for the components of a *constant-composition* solution.

The Gibbs function can be nondimensionalized by dividing it by RT. Not only is G/RT dimensionless, but it is also a useful thermodynamic property and a function of temperature, pressure, and composition. The function standing in relation to G/RT as a partial molar property is $\overline{G}_i/RT$ or μ_i/RT. Thus Eq. (4-162) applied to this function becomes

$$\frac{G}{RT} = \sum\left(x_i\frac{\mu_i}{RT}\right)$$

In addition, Eqs. (4-164) and (4-165) may be written for these functions once the partial differential coefficients have been expressed in terms of measurable quantities. By definition $G = H - TS$. Thus

$$G/RT = H/RT - S/R$$

Differentiation gives

$$\left[\frac{\partial(G/RT)}{\partial T}\right]_{P,x} = \frac{1}{RT}\left(\frac{\partial H}{\partial T}\right)_{P,x} - \frac{H}{RT^2} - \frac{1}{R}\left(\frac{\partial S}{\partial T}\right)_{P,x}$$

Substitution for $(\partial H/\partial T)_{P,x}$ by Eq. (4-144) and for $(\partial S/\partial T)_{P,x}$ by Eq. (4-145) reduces this to

$$\left[\frac{\partial(G/RT)}{\partial T}\right]_{P,x} = -\frac{H}{RT^2}$$

Similarly,

$$\left[\frac{\partial(G/RT)}{\partial P}\right]_{T,x} = \frac{1}{RT}\left(\frac{\partial H}{\partial P}\right)_{T,x} - \frac{1}{R}\left(\frac{\partial S}{\partial P}\right)_{T,x}$$

Upon substitution for the two partial derivatives on the right by Eqs. (4-146) and (4-141), this reduces to

$$\left[\frac{\partial(G/RT)}{\partial P}\right]_{T,x} = \frac{V}{RT}$$

Equations (4-164) and (4-165) now take on the particular forms

$$d\left(\frac{nG}{RT}\right) = -\frac{nH}{RT^2}\,dT + \frac{nV}{RT}\,dP + \sum\left(\frac{\mu_i}{RT}\,dn_i\right) \qquad (4\text{-}178)$$

and

$$-\frac{H}{RT^2}\,dT + \frac{V}{RT}\,dP = \sum\left[x_i d\left(\frac{\mu_i}{RT}\right)\right] \qquad (4\text{-}179)$$

Equations (4-178) and (4-179) are alternatives to Eqs. (4-125) and (4-129). The utility of such general equations is that they represent most concisely a considerable amount of information. They are easily reduced to specialized cases and provide required partial derivatives and reciprocal relations by visual inspection. For example, applied to a constant-composition solution or to a pure material Eq. (4-178) becomes (on a molar basis)

$$d\left(\frac{G}{RT}\right) = -\frac{H}{RT^2} + \frac{V}{RT}\,dP \qquad (4\text{-}178a)$$

Parallel with this, an equation may be written for the partial properties in a constant-composition solution:

$$d\left(\frac{\overline{G}_i}{RT}\right) = -\frac{\overline{H}_i}{RT^2}\,dT + \frac{\overline{V}_i}{RT}\,dP \qquad (4\text{-}178b)$$

From these it immediately follows that

$$\left[\frac{\partial(G/RT)}{\partial T}\right]_{P,x} = -\frac{H}{RT^2} \qquad (4\text{-}180)$$

$$\left[\frac{\partial(\overline{G}_i/RT)}{\partial T}\right]_{P,x} = -\frac{\overline{H}_i}{RT^2} \qquad (4\text{-}181)$$

$$\left[\frac{\partial(G/RT)}{\partial P}\right]_{T,x} = \frac{V}{RT} \qquad (4\text{-}182)$$

$$\left[\frac{\partial(\overline{G}_i/RT)}{\partial P}\right]_{T,x} = \frac{\overline{V}_i}{RT} \qquad (4\text{-}183)$$

Equations (4-180) and (4-181) are called *Gibbs-Helmholtz equations.*

Table 4-24 shows a set of general equations having to do with the Gibbs function and certain related functions. The first two rows of this table have already been developed. An immediate purpose of the remainder of this theoretical treatment is to develop the latter sets of equations. The reason for emphasis on equations related to the Gibbs function is that the natural variables for this function are temperature, pressure, and the mole numbers, all measurable quantities with respect to real systems.

Auxiliary Thermodynamic Functions The auxiliary functions in common use arise by definition, and their use is simply a matter of convenience.

Compressibility Factor Z This quantity is defined by the equation

$$PV = ZRT$$

from which

$$Z = \frac{PV}{RT} = \frac{V}{RT/P} = \frac{V}{V'}$$

where V' is the molar volume as given by the ideal-gas law, $PV' = RT$. This is the simplest equation of state for a PVT system, and its use provides convenient "base" values, such as V', for the various thermodynamic properties. Similarly, H', S', and G' are the molar enthalpy, entropy, and Gibbs function that a PVT system would have if the ideal-gas law were the correct equation of state.

Generalized correlations of the compressibility factor as a function of reduced temperature and reduced pressure are discussed in Sec. 3.

Residual Functions Several classes of quantities may be defined which represent the difference between a property that would be obtained were the ideal-gas equation valid and the actual property. Thus, designating by M a molar property of a homogeneous fluid,

TABLE 4-24

General equations for open system based on Eq. (4-164)	General Gibbs-Duhem equation based on Eq. (4-165)
$d(nG) = -nS\,dT + nV\,dP + \Sigma(\mu_i\,dn_i)$ (4-125)	$-S\,dT + V\,dP = \Sigma(x_i\,d\mu_i)$ (4-129)
$d\left(\frac{nG}{RT}\right) = -\frac{nH}{RT^2}\,dT + \frac{nV}{RT}\,dP + \sum\left(\frac{\mu_i}{RT}\,dn_i\right)$ (4-178)	$-\frac{H}{RT^2}\,dT + \frac{V}{RT} = \sum\left(x_i\,d\frac{\mu_i}{RT}\right)$ (4-179)
$d(n\ln f) = \frac{n\,\Delta H'}{RT^2}\,dT + \frac{nV}{RT}\,dP + \sum\left(\ln\frac{\hat{f}_i}{x_i}\,dn_i\right)$ (4-197)	$\frac{\Delta H'}{RT^2}\,dT + \frac{V}{RT}\,dP = \Sigma\left(x_i\,d\ln\frac{\hat{f}_i}{x_i}\right)$ (4-198); $= \Sigma(x_i\,d\ln\hat{f}_i)$ (4-199)
$d(n\ln\phi) = \frac{n\,\Delta H'}{RT^2}\,dT - \frac{n\,\Delta V'}{RT}\,dP + \Sigma(\ln\hat{\phi}_i\,dn_i)$ (4-200)	$\frac{\Delta H'}{RT^2}\,dT - \frac{\Delta V'}{RT}\,dP = \Sigma(x_i\,d\ln\hat{\phi}_i)$ (4-202)
$d\left(\frac{nG^E}{RT}\right) = -\frac{n\,\Delta H}{RT^2}\,dT + \frac{n\,\Delta V}{RT}\,dP + \Sigma(\ln\gamma_i\,dn_i)$ (4-239)	$-\frac{\Delta H}{RT^2}\,dT + \frac{\Delta V}{RT}\,dP = \Sigma(x_i\,d\ln\gamma_i)$ (4-240)
These equations may be written for 1 mol of solution by setting $n = 1$ and $n_i = x_i$. In such equations the x_i's are subject to the restraint that $\Sigma x_i = 1$.	These equations may be multiplied directly by n.
Equation for 1 mol of pure material or constant-composition solution	**Equation for partial molar properties in a constant-composition solution**
$dG = -S\,dT + V\,dP$ (4-125a)	$d\overline{G}_i = d\mu_i = -\overline{S}_i\,dT + \overline{V}_i\,dP$ (4-125b)
$d\left(\frac{G}{RT}\right) = -\frac{H}{RT^2}\,dT + \frac{V}{RT}\,dP$ (4-178a)	$d\left(\frac{\mu_i}{RT}\right) = -\frac{\overline{H}_i}{RT^2}\,dT + \frac{\overline{V}_i}{RT}\,dP$ (4-178b)
$d\ln f = \frac{\Delta H'}{RT^2}\,dT + \frac{V}{RT}\,dP$ (4-197a)	$d\ln\hat{f}_i$, $d\ln\frac{\hat{f}_i}{x_i}$ $= \frac{H_i' - \overline{H}_i}{RT^2}\,dT + \frac{\overline{V}_i}{RT}\,dP$ (4-197b)
$d\ln\phi = \frac{\Delta H'}{RT^2}\,dT - \frac{\Delta V'}{RT}\,dP$ (4-200a)	$d\ln\hat{\phi}_i = \frac{H_i' - \overline{H}_i}{RT^2}\,dT - \frac{V_i' - \overline{V}_i}{RT}\,dP$ (4-200b)
$d\left(\frac{G^E}{RT}\right) = -\frac{\Delta H}{RT^2}\,dT + \frac{\Delta V}{RT}\,dP$ (4-239a)	$d\ln\gamma_i = d\left(\frac{\overline{G}_i^E}{RT}\right) = -\frac{\Delta\overline{H}_i}{RT^2}\,dT + \frac{\Delta\overline{V}_i}{RT}\,dP$ (4-239b)

we define

$$\Delta M' \equiv M'(T, P) - M(T, P) \qquad (4\text{-}184)$$

$$\Delta M'' \equiv M'(T, P^\circ) - M(T, P) \qquad (4\text{-}185)$$

$$\Delta M''' \equiv M'(T, V) - M(T, V) \qquad (4\text{-}186)$$

Here T, P, and V refer to the real substance, and P° is a fixed reference pressure (e.g., 1 bar or 1 atm). Although the three quantities $\Delta M'$, $\Delta M''$, and $\Delta M'''$ are related, they generally have different numerical values because the bases for the comparisons are different. We restrict ourselves here to the quantities $\Delta M'$ defined by Eq. (4-184); we call them *residual functions*. Residual functions (and also the quantities $\Delta M''$ and $\Delta M'''$) depend on interactions *between* molecules and not on characteristics of the individual molecules. Since the ideal-gas model presumes the absence of molecular interactions, deviations from ideality are measured by the residual functions. Examples of residual functions are

Residual volume:	$\Delta V' \equiv V' - V$
Residual enthalpy:	$\Delta H' \equiv H' - H$
Residual entropy:	$\Delta S' \equiv S' - S$
Residual Gibbs function:	$\Delta G' \equiv G' - G$

Fugacity Fugacity is defined in direct relation to the Gibbs function; there are two separate fundamental definitions. For a constant-composition mixture the fugacity f is defined so as to satisfy the equations

$$dG = RT\, d \ln f \text{ (constant } T, x) \qquad (4\text{-}187)$$

$$\lim_{P\to 0} \frac{f}{P} = 1 \qquad (4\text{-}188)$$

For species i in solution, the corresponding equations which define the fugacity $\hat{f}_i$ are

$$d\overline{G}_i = RT\, d \ln \hat{f}_i \text{ (constant } T) \qquad (4\text{-}189)$$

$$\lim_{P\to 0} \frac{\hat{f}_i}{x_i P} = 1 \qquad (4\text{-}190)$$

Equations (4-187) and (4-188) apply also to pure species i, a special case of a constant-composition mixture. For this case, these equations are usually written with subscript i affixed to G and f. Integration of Eq. (4-189) *at constant T* allows for both pressure and composition changes and in effect merely provides a change of variable:

$$\Delta \overline{G}_i = RT\, \Delta \ln \hat{f}_i \text{ (constant } T)$$

It can be shown that when the ideal-gas law is an appropriate equation of state, the fugacities become equal to pressures:

f becomes equal to P, the mixture pressure.
f_i becomes equal to P, the pressure on pure i.
$\hat{f}_i$ becomes equal to $x_i P$, the partial pressure of i in a gas mixture of mole fraction x_i in i.

The fugacity $\hat{f}_i$ of a component in solution is *not* a partial molar property with respect to f. However, there is a relation between $\hat{f}_i$ and f, which is shown as follows. By Eq. (4-187) for a mixture,

$$dG = RT\, d \ln f$$

Integration at constant T and composition from P° to P gives

$$G - G^\circ = RT \ln f - RT \ln f^\circ$$

If $P^\circ \to 0$, then by Eq. (4-188) $f^\circ = P^\circ$, and

$$G - G^\circ = RT \ln f - RT \ln P^\circ$$

For n mole,

$$nG - nG^\circ = nRT \ln f - nRT \ln P^\circ$$

Differentiation of this general equation with respect to n_i at constant T, P, and n_j gives

$$\left[\frac{\partial(nG)}{\partial n_i}\right]_{T,P,n_j} - \left[\frac{\partial(nG^\circ)}{\partial n_i}\right]_{T,P,n_j} = RT\left[\frac{\partial(n \ln f)}{\partial n_i}\right]_{T,P,n_j} - RT \ln P^\circ$$

or

$$\overline{G}_i - \overline{G}_i^\circ = RT\left[\frac{\partial(n \ln f)}{\partial n_i}\right]_{T,P,n_j} - RT \ln P^\circ \qquad (4\text{-}191)$$

For component i in a solution, Eq. (4-189) gives

$$d\overline{G}_i = RT\, d \ln \hat{f}_i$$

Integration at constant T and composition from P° to P gives

$$\overline{G}_i - \overline{G}_i^\circ = RT \ln \hat{f}_i - RT \ln \hat{f}_i^\circ$$

If $P^\circ \to 0$, then by Eq. (4-190) $\hat{f}_i^\circ = x_i P^\circ$. Thus

$$\overline{G}_i - \overline{G}_i^\circ = RT \ln \frac{\hat{f}_i}{x_i} - RT \ln P^\circ \qquad (4\text{-}192)$$

Comparison of Eqs. (4-191) and (4-192) shows that

$$\left[\frac{\partial(n \ln f)}{\partial n_i}\right]_{T,P,n_j} = \ln \frac{\hat{f}_i}{x_i} \qquad (4\text{-}193)$$

Since this is exactly the equation that defines a partial molar property, as seen from Eq. (4-160), it is clear that $\ln(\hat{f}_i/x_i)$ is related to $\ln f$ as a partial molar property. Equation (4-162) now provides the relation

$$\ln f = \sum \left(x_i \ln \frac{\hat{f}_i}{x_i}\right) \qquad (4\text{-}194)$$

Clearly $\ln f$ is a thermodynamic property that may be substituted for M in Eqs. (4-164) and (4-165), provided $\overline{M}$ is taken to be $\ln(\hat{f}_i/x_i)$. It remains only to determine the partial differential coefficients in Eqs. (4-164) and (4-165).

Equation (4-187) may be integrated for the hypothetical change from the ideal-gas state to the real state at constant T, P, and x:

$$G - G' = RT \ln f - RT \ln f'$$
$$= RT \ln f - RT \ln P$$

Thus

$$\ln f = G/RT - G'/RT + \ln P$$

Differentiation with respect to T at constant P and x gives

$$\left(\frac{\partial \ln f}{\partial T}\right)_{P,x} = \left[\frac{\partial(G/RT)}{\partial T}\right]_{P,x} - \left[\frac{\partial(G'/RT)}{\partial T}\right]_{P,x}$$

By Eq. (4-180), the Gibbs-Helmholtz equation, this becomes

$$\left(\frac{\partial \ln f}{\partial T}\right)_{P,x} = -\frac{H}{RT^2} + \frac{H'}{RT^2} = \frac{\Delta H'}{RT^2} \qquad (4\text{-}195)$$

By Eq. (4-187) and by Eq. (4-125) restricted to constant T and x and written for 1 mol,

$$dG = RT\, d \ln f = V\, dP \text{ (constant } T \text{ and } x)$$

Thus

$$\left(\frac{\partial \ln f}{\partial P}\right)_{T,x} = \frac{V}{RT} \qquad (4\text{-}196)$$

Equations (4-164) and (4-165) may now be specialized through use of Eqs. (4-195) and (4-196) to give

$$d(n \ln f) = \frac{n\,\Delta H'}{RT^2}\, dT + \frac{nV}{RT}\, dP + \sum \left(\ln \frac{\hat{f}_i}{x_i}\, dn_i\right) \qquad (4\text{-}197)$$

and

$$\frac{\Delta H'}{RT^2}\, dT + \frac{V}{RT}\, dP - \sum \left(x_i d \ln \frac{\hat{f}_i}{x_i}\right) = 0 \qquad (4\text{-}198)$$

Since

$$\Sigma(x_i\, d \ln x_i) = 0$$

Eq. (4-199) may also be written as

$$\frac{\Delta H'}{RT^2}\, dT + \frac{V}{RT}\, dP - \Sigma(x_i\, d \ln \hat{f}_i) = 0 \qquad (4\text{-}199)$$

These general equations are included in Table 4-24.

The fugacity of a component in solution $\hat{f}_i$ is related to its mole fraction x_i. Clearly $\hat{f}_i = f_i$ when $x_i = 1$, and presumably $\hat{f}_i = 0$ when $x_i = 0$. The simplest possible relation of $\hat{f}_i$ to x_i at constant T and P between these limits is a direct proportionality,

$$\hat{f}_i = x_i f_i$$

This is called the Lewis-Randall rule and is valid for certain **ideal solutions** to be discussed later. In general, deviations from this rule are observed. A typical plot of $\hat{f}_1$ versus x_1 for a binary system at constant T and P is shown in Fig. 4-20. When component 1 is present at high dilution, a tangent drawn to the end of the curve at $x_1 = 0$ must obviously represent the curve to a good approximation for some finite distance. Thus the equation $\hat{f}_1 = k_1 x_1$ must be valid as $x_1 \to 0$. This is Henry's law in its most general form, and k is Henry's-law constant for component 1.

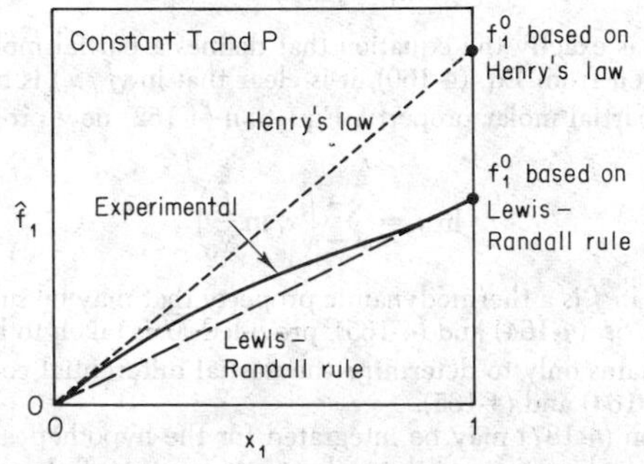

FIG. 4-20 Fugacity-composition relations for a binary mixture, showing standard-state fugacities based on Henry's law and on the Lewis-Randall rule.

Since $\ln (\hat{f}_i/x_i)$ is a partial molar property with respect to $\ln f$, the Gibbs-Duhem equation (4-198) at constant T and P for this property in a binary solution becomes

$$x_1 d \ln (\hat{f}_1/x_1) + x_2 d \ln (\hat{f}_2/x_2) = 0$$

Where Henry's law holds for component 1, $\hat{f}_1/x_1 = k_1$ and $d \ln (\hat{f}_1/x_1) = \ln k_1 = 0$. Therefore,

$$d \ln (\hat{f}_2/x_2) = 0 \qquad \text{or} \qquad \ln (\hat{f}_2/x_2) = K$$

When $x_2 = 1$, $\hat{f}_2 = f_2$; therefore, $K = \ln f_2$, and $\hat{f}_2 = x_2 f_2$, which is the Lewis-Randall rule for component 2. Similarly, when $\hat{f}_2 = k_2 x_2$, then $\hat{f}_1 = x_1 f_1$. This is the reason why the curve in Fig. 4-20 is drawn tangent to the straight line at $x_1 = 1$.

Fugacity Coefficients It is often convenient to deal with the ratio of fugacity to pressure instead of with the fugacity itself, and this ratio is called the fugacity coefficient ϕ. There are three such quantities:

For a mixture: $\phi = f/P$
For a pure material: $\phi_i = f_i/P$
For a constituent in solution: $\hat{\phi}_i = \hat{f}_i/x_i P$

For an ideal gas and for a real gas as $P \to 0$, all three fugacity coefficients are unity.

Equation (4-197) can be transformed through use of the following identity:

$$d(n \ln P) = n\, d \ln P + \ln P\, dn$$

Since $n = \Sigma n_i$,

$$dn = \Sigma\, dn_i$$

and

$$\ln P\, dn = \Sigma(\ln P\, dn_i)$$

Therefore,

$$d(n \ln P) = n(dP/P) + \Sigma(\ln P\, dn_i)$$

Subtraction of this equation from Eq. (4-197) gives

$$d\left(n \ln \frac{f}{P}\right) = \frac{n\,\Delta H'}{RT^2} dT + \frac{n}{RT}\left(V - \frac{RT}{P}\right) dP + \sum\left(\ln \frac{\hat{f}_i}{x_i P} dn_i\right)$$

The quantity $V - RT/P$ is $V - V'$, the negative of the residual volume $\Delta V'$. Thus,

$$d(n \ln \phi_i) = \frac{n\,\Delta H'}{RT^2} dT - \frac{n\,\Delta V'}{RT} dP + \Sigma(\ln \hat{\phi}_i\, dn_i) \qquad (4\text{-}200)$$

It is seen by comparison of Eqs. (4-200) and (4-164) that $\ln \hat{\phi}_i$ is related to $\ln \phi$ as a partial molar property. Thus, by Eq. (4-162),

$$\ln \phi = \Sigma(x_i \ln \hat{\phi}_i) \qquad (4\text{-}201)$$

and the appropriate Gibbs-Duhem equation follows from Eq. (4-165):

$$\frac{\Delta H'}{RT^2} dT - \frac{\Delta V'}{RT} dP - \Sigma(x_i\, d \ln \hat{\phi}_i) = 0 \qquad (4\text{-}202)$$

Equations (4-200) and (4-202) are included in Table 4-24.

SOLUTION THERMODYNAMICS

Property Changes of Mixing If M represents a molar thermodynamic property of a homogeneous fluid mixture, then ΔM is defined by

$$\Delta M = M - \Sigma(x_i M_i^\circ) \qquad (4\text{-}203)$$

where ΔM is called the property change of mixing, and M_i° is the molar property of pure i at the temperature of the mixture and in some *standard state* of specified pressure and composition. The standard state for a constituent is chosen for convenience and may be different for different constituents. The obvious standard state is that of pure i in its actual stable state at the mixture pressure. However, the stable state for a particular pure constituent may be a different kind of phase (liquid or gas) from that of the mixture at the same T and P. For saturated phases this is the rule rather than the exception for at least one constituent. In this case it is common to adopt as the standard state the hypothetical state of pure i at the mixture T and P *and in the same physical state as the mixture.* The value of M_i° must then be determined for this unstable or hypothetical state. The difficulty of doing this with certainty has led to the use of an alternative standard state based on Henry's law. Standard states will be discussed in detail later.

As an example take the volume of a liquid mixture as the property considered, and assume that all the pure components exist as stable liquids at the mixture T and P. Then $V_i^\circ = V_i$ and

$$\Delta V = V - \Sigma(x_i V_i)$$

Here ΔV is the volume increase or decrease observed relative to the total volume of unmixed liquids when 1 mol of mixture is formed at constant T and P.

The property changes of mixing are thermodynamic properties in their own right and are functions of temperature, pressure, and composition. Their use requires a clear statement of the standard states referred to.

Since by Eq. (4-162),

$$M = \Sigma(x_i \overline{M}_i)$$

Equation (4-203) can also be written

$$\Delta M = \Sigma(x_i \overline{M}_i) - \Sigma(x_i M_i^\circ) \equiv \Sigma[x_i(\overline{M}_i - M_i^\circ)]$$

or

$$\Delta M = \Sigma(x_i \overline{\Delta M}_i) \qquad (4\text{-}204)$$

where by definition

$$\overline{\Delta M}_i \equiv \overline{M}_i - M_i^\circ \qquad (4\text{-}205)$$

This quantity represents the property change of i as a result of a change of state of i from a pure material in some specified standard state to a constituent in solution at the same T. It is also a partial molar property with respect to ΔM and is a function of T, P, and x.

It is therefore possible to write very general equations for ΔM analogous to Eqs. (4-164) and (4-165).

$$d(n\,\Delta M) = n\left(\frac{\partial \Delta M}{\partial T}\right)_{P,x} dT + n\left(\frac{\partial \Delta M}{\partial P}\right)_{T,x} dP + \Sigma(\overline{\Delta M}_i\, dn_i) \qquad (4\text{-}206)$$

and
$$\left(\frac{\partial \Delta M}{\partial T}\right)_{P,x} dT + \left(\frac{\partial \Delta M}{\partial P}\right)_{T,x} dP - \Sigma(x_i\, d\overline{\Delta M}_i) = 0 \qquad (4\text{-}207)$$

Ideal Solutions and Standard States An ideal solution is by definition one for which the fugacity of each component in solution is given by

$$\hat{f}_i^{id} = x_i f_i^\circ \qquad (4\text{-}208)$$

at all pressures, temperatures, and compositions. As the name implies, this is an idealization in the same sense as is the concept of an ideal gas and is useful in much the same way. The equations that apply to ideal solutions are very simple, as shown later, and provide "base" values to which the properties of real solutions may be referred.

Equation (4-208) shows that $\hat{f}_i$ for an ideal solution is directly proportional to x_i. The proportionality constant is f_i°, the standard-state fugacity of pure i at the solution temperature, which clearly depends on the choice of standard state. Equation (4-208) implies that, for a given standard state and for fixed T and P, $\hat{f}_i$ depends only on x_i, independent of all other x's. Thus for an ideal solution a plot of $\hat{f}_i$ versus x_i is a straight line starting at the origin and terminating at $\hat{f}_i = f_i^\circ$ when $x_i = 1$. Clearly a different straight line results for each different choice of f_i°. Only two choices of the standard state have proved useful with any generality, and they are based on the behavior of a component in a real solution at the two concentration extremes, $x_i \to 0$ and $x_i \to 1$, where Henry's law and the Lewis-Randall rule, respectively, become valid, as discussed earlier. The two standard-state fugacities which result are shown in Fig. 4-20, together with their relation to a typical $\hat{f}_i$ versus x_1 plot for a binary system. It is presumed that the phase (liquid or gas) under consideration is stable throughout the entire composition range for the given T and P.

The straight dashed line labeled *Henry's law* represents an "ideal" behavior that is in fact observed only as $x_1 \to 0$. Extension of Henry's law to cover the entire composition range requires the postulation of a state of pure 1 having the value of f_1° shown uppermost on the figure. Since pure 1 does not actually exist in this state, the state is fictitious or hypothetical. However, the extrapolation necessary to establish f_1° is clearly defined. This value of f_1° depends on the nature of the other component.

The other straight line labeled *Lewis-Randall rule* in Fig. 4-20 represents a different "ideal," and is actually observed only as $x_1 \to 1$. For this case f_1° is the fugacity of pure 1 as it actually exists, namely f_1. The value of f_1 obviously does not depend on the nature of the other component.

Since both standard states are related to observed solution behavior at fixed T and P, the standard-state pressure for each is clearly the solution pressure.

Since the standard state based on the Lewis-Randall rule appears to be a real state of the pure material, the natural question is why any other standard state should ever be considered. The answer is that in the preceding discussion it was presumed that the phase considered was stable as a liquid or as a gas throughout the entire composition range. Unfortunately this is not necessarily so. There is always a range of conditions of P and T for which the full curve representing actual behavior for a given phase cannot be determined because the phase becomes unstable in some composition range. When the right-hand end of the experimental curve of Fig. 4-20 is inaccessible, the value of f_1° based on the Lewis-Randall rule must be determined by some sort of extrapolation. There is often no guide as to how to perform this extrapolation, and the resulting uncertainty in the standard-state value makes the use of this standard state undesirable. In this case the standard state based on Henry's law often proves useful because it is well defined even though it is a fictitious state and component-dependent.

The choice of standard states (which are sometimes advantageously chosen differently for different constituents of the same solution) is left open in the following development of the equations for ideal solutions. Of course, if a real solution were actually ideal, the three lines of Fig. 4-20 would coincide.

For either of the standard states discussed, logarithmic differentiation of Eq. (4-208) with respect to pressure gives

$$\left(\frac{\partial \ln \hat{f}_i^{id}}{\partial P}\right)_{T,x} - \left(\frac{\partial \ln f_i^\circ}{\partial P}\right)_T = 0$$

However, Eqs. (4-197*a*) and (4-197*b*) provide the relationships

$$\left(\frac{\partial \ln \hat{f}_i^{id}}{\partial P}\right)_{T,x} = \frac{\overline{V}_i^{id}}{RT} \qquad \text{and} \qquad \left(\frac{\partial \ln f_i^\circ}{\partial P}\right)_T = \frac{V_i^\circ}{RT}$$

Combinations of the above equations shows that

$$\overline{V}_i^{id} - V_i^\circ = 0$$

or
$$\overline{V}_i^{id} = V_i^\circ \qquad \text{and} \qquad \overline{\Delta V}_i^{id} = 0 \qquad (4\text{-}209)$$

It follows from Eq. (4-204) that

$$\Delta V^{id} = 0 \qquad (4\text{-}210)$$

and from Eq. (4-203) that

$$V^{id} = \Sigma(x_i V_i^\circ) \qquad (4\text{-}211)$$

Similarly, logarithmic differentiation of Eq. (4-208) with respect to temperature gives

$$\left(\frac{\partial \ln \hat{f}_i^{id}}{\partial T}\right)_{P,x} - \left(\frac{\partial \ln f_i^\circ}{\partial T}\right)_P = 0$$

From Eqs. (4-197*a*) and (4-197*b*)

$$\left(\frac{\partial \ln \hat{f}_i^{id}}{\partial T}\right)_{P,x} = \frac{H_i' - \overline{H}_i^{id}}{RT^2}$$

and
$$\left(\frac{\partial \ln f_i^\circ}{\partial T}\right)_P = \frac{H_i' - H_i^\circ}{RT^2}$$

Combination of the above equations leads to

$$\overline{H}_i^{id} - H_i^\circ = 0$$

or
$$\overline{H}_i^{id} = H_i^\circ \qquad \text{and} \qquad \overline{\Delta H}_i^{id} = 0 \qquad (4\text{-}212)$$

Also
$$\Delta H^{id} = 0 \qquad (4\text{-}213)$$

and
$$H^{id} = \Sigma(x_i H_i^\circ) \qquad (4\text{-}214)$$

Completely analogous equations are readily derived for the internal energy and the heat capacities.

If Eq. (4-189) is integrated for the change of state of i from that of a pure material in a standard state to that of a constituent in solution at the same T, then

$$\overline{G}_i - G_i^\circ = RT \ln \hat{f}_i / f_i^\circ \qquad (4\text{-}215)$$

In view of Eqs. (4-205) and (4-208) this becomes for an ideal solution

$$\overline{\Delta G}_i^{id} = RT \ln x_i \qquad (4\text{-}216)$$

From Eq. (4-204),

$$\Delta G^{id} = RT\Sigma(x_i \ln x_i) \qquad (4\text{-}217)$$

and by Eq. (4-203),

$$G^{id} = \Sigma(x_i G_i^\circ) + RT\Sigma(x_i \ln x_i) \qquad (4\text{-}218)$$

Use of the defining equation for the Gibbs function provides the relation

$$\overline{\Delta G_i^{id}} = \overline{\Delta H_i^{id}} - T\overline{\Delta S_i^{id}}$$

From this and Eqs. (4-212) and (4-216),

$$\overline{\Delta S_i^{id}} = -R \ln x_i \tag{4-219}$$

and by Eqs. (4-204) and (4-203),

$$\Delta S^{id} = -R\Sigma(x_i \ln x_i) \tag{4-220}$$

$$S^{id} = \Sigma(x_i S_i^\circ) - R\Sigma(x_i \ln x_i) \tag{4-221}$$

Excess Properties of Solutions *An excess property* is defined as the difference between an actual property and the property that would be calculated at the same conditions of temperature, pressure, and composition by the equations for an ideal solution. Thus by definition

$$M^E = M - M^{id} \tag{4-222}$$

$$\Delta M^E = \Delta M - \Delta M^{id} \tag{4-223}$$

where M^E is called the excess solution property and ΔM^E is the excess property change of mixing. Actually, they are identical, which is shown by substitution for both ΔM and ΔM^{id} by Eq. (4-203). In addition, there are the partial molar excess properties,

$$\overline{M_i^E} = \overline{M_i} - \overline{M_i^{id}} \tag{4-224}$$

$$\overline{\Delta M_i^E} = \overline{\Delta M_i} - \overline{\Delta M_i^{id}} \tag{4-225}$$

Again these two quantities are identical, and which one to use is a matter of personal preference. Use of M^E and $\overline{M_i^E}$ places emphasis on the solution property itself, whereas use of ΔM^E and $\overline{\Delta M_i^E}$ draws attention to the mixing by which the solution was formed. The partial molar relation which the latter properties bear to the former allows several general equations to be written analogous to Eqs. (4-204), (4-206), and (4-207) or to Eqs. (4-162), (4-164), and (4-165)

$$M^E = \Sigma(x_i \overline{M_i^E}) \tag{4-226}$$

$$d(nM^E) = n\left(\frac{\partial M^E}{\partial T}\right)_{P,x} dT + n\left(\frac{\partial M^E}{\partial P}\right)_{T,x} dP + \Sigma(\overline{M_i^E}\, dn_i) \tag{4-227}$$

$$\left(\frac{\partial M^E}{\partial T}\right)_{P,x} dT + \left(\frac{\partial M^E}{\partial P}\right)_{T,x} dP - \Sigma(x_i\, d\overline{M_i^E}) = 0 \tag{4-228}$$

Since ΔV^{id}, ΔH^{id}, ΔU^{id}, and ΔC_P^{id} are all zero, the excess properties are identical with the property changes of mixing themselves for these thermodynamic functions. For properties that incorporate the entropy this is not true, and the excess properties represent additional thermodynamic functions. Moreover, their use is advantageous because such properties provide the smallest and most sensitive measure of mixture properties in relation to pure-component (standard-state) properties.

The most useful of these properties is the excess Gibbs function. Combining Eqs. (4-222) and (4-217) gives

$$G^E \equiv \Delta G^E = \Delta G - \Delta G^{id} = \Delta G - RT\Sigma(x_i \ln x_i) \tag{4-229}$$

or

$$G^E = G - \Sigma(x_i G_i^\circ) - RT\Sigma(x_i \ln x_i) \tag{4-230}$$

Division of this last equation by RT puts it into dimensionless form,

$$\frac{G^E}{RT} = \frac{G}{RT} - \sum\left(x_i \frac{G_i^\circ}{RT}\right) - \Sigma(x_i \ln x_i) \tag{4-231}$$

Since G^E/RT is an excess thermodynamic property and a function of P, T, and $\underline{x}$, it may be substituted for M^E in Eqs. (4-226) to (4-228), where $\overline{M_i^E}$ becomes $\overline{G_i^E}/RT$. Equations (4-227) and (4-228) require evaluation of the partial differential coefficients of dT and dP in terms of measurable properties. This is done by differentiation of Eq. (4-231),

$$\left[\frac{\partial(G^E/RT)}{\partial T}\right]_{P,x} = \left[\frac{\partial(G/RT)}{\partial T}\right]_{P,x} - \sum\left(x_i\left[\frac{\partial(G_i^\circ/RT)}{\partial T}\right]_P\right)$$

and

$$\left[\frac{\partial(G^E/RT)}{\partial P}\right]_{T,x} = \left[\frac{\partial(G/RT)}{\partial P}\right]_{T,x} - \sum\left(x_i\left[\frac{\partial(G_i^\circ/RT)}{\partial P}\right]_T\right)$$

The partial derivatives on the right-hand sides of these equations are given by Eqs. (4-180) and (4-182). Thus

$$\left[\frac{\partial(G^E/RT)}{\partial T}\right]_{P,x} = -\frac{H}{RT^2} - \sum\left[x_i\left(-\frac{H_i^\circ}{RT^2}\right)\right] = -\frac{[H - \Sigma(x_i H_i^\circ)]}{RT^2} = -\frac{\Delta H}{RT^2}$$

and

$$\left[\frac{\partial(G^E/RT)}{\partial P}\right]_{T,x} = \frac{V}{RT} - \sum\left[x_i\frac{V_i^\circ}{RT}\right] = \frac{V - \Sigma(x_i V_i^\circ)}{RT} = \frac{\Delta V}{RT}$$

Equations (4-226) to (4-228) are now particularized as follows:

$$\frac{G^E}{RT} = \sum\left[x_i\frac{\overline{G_i^E}}{RT}\right] \tag{4-232}$$

$$d\left(\frac{nG^E}{RT}\right) = -\frac{n\,\Delta H}{RT^2} dT + \frac{n\,\Delta V}{RT} dP + \sum\left(\frac{\overline{G_i^E}}{RT} dn_i\right) \tag{4-233}$$

and

$$-\frac{\Delta H}{RT^2} dT + \frac{\Delta V}{RT} dP - \sum\left[x_i d\left(\frac{\overline{G_i^E}}{RT}\right)\right] = 0 \tag{4-234}$$

Equations (4-233) and (4-234) are based on standard states at the T and P of the system.

An auxiliary thermodynamic function known as the activity coefficient is defined for a constituent in solution by the equation

$$\gamma_i = \hat{f}_i/x_i f_i^\circ \tag{4-235}$$

where f_i° is a standard-state fugacity. Although other standard states are possible, the two already discussed are entirely adequate for the present treatment, and no others will be considered here.

The activity coefficient is directly related to the Gibbs function. Equation (4-235) may be written as

$$\ln \gamma_i = \ln(\hat{f}_i/f_i^\circ) - \ln x_i \tag{4-236}$$

Equations (4-215) and (4-216) are now used to substitute for the first and second terms, respectively, on the right side of Eq. (4-236):

$$\ln \gamma_i = \frac{\overline{G_i} - G_i^\circ}{RT} - \frac{\overline{G_i^{id}} - G_i^\circ}{RT} = \frac{\overline{G_i} - \overline{G_i^{id}}}{RT}$$

or

$$\ln \gamma_i = \overline{G_i^E}/RT \tag{4-237}$$

This equation shows that $\ln \gamma_i$ stands in relation to G^E/RT (or $\Delta G^E/RT$) as a partial molar property. This is made all the more evident by rewriting Eqs. (4-232) to (4-234) with the substitution indicated by Eq. (4-237).

$$G^E/RT = \Sigma(x_i \ln \gamma_i) \tag{4-238}$$

$$d\left(\frac{nG^E}{RT}\right) = -\frac{n\,\Delta H}{RT^2} dT + \frac{n\,\Delta V}{RT} dP + \Sigma(\ln \gamma_i\, dn_i) \tag{4-239}$$

$$-\frac{\Delta H}{RT^2} dT + \frac{\Delta V}{RT} dP - \Sigma(x_i\, d \ln \gamma_i) = 0 \tag{4-240}$$

Equations (4-239) and (4-240) are included in Table 4-24. In each equation the standard state *for a given constituent* must, of course, be the same throughout. In addition, the standard state must be at the sytem T and P.

The defining equations for the primary thermodynamic functions have their obvious analogs in terms of property changes of mixing and excess properties. For example, since

$$G = H - TS$$

then also
$$\Delta G = \Delta H - T\,\Delta S$$

and
$$G^E = H^E - TS^E$$

For the standard state based on the Lewis-Randall rule the standard-state properties M_i° become the properties of pure i at the mixture T and P and in the same physical state as the mixture. These properties are given the symbol M_i regardless of whether the state is real or hypothetical. Thus $M_i^\circ = M_i$, and hence $V_i^\circ = V_i$, $H_i^\circ = H_i$ $f_i^\circ = f_i$, etc., for the standard state based on the Lewis-Randall rule. For this standard state Eq. (4-235) for the activity coefficient becomes

$$\gamma_i = \hat{f}_i / x_i f_i$$

For a real solution in the limit as $x_i \to 1$, the Lewis-Randall rule becomes valid, and $\hat{f}_i = x_i f_i$. Thus,

$$\lim_{x_i \to 1} \gamma_i = 1$$

and this is true for each constituent of a real solution for which the standard state is taken so that $f_i^\circ = f_i$.

The selection of a standard state of i such that the activity coefficient of i becomes unity as x_i approaches a limiting value is said to "normalize" the activity coefficient. When all constituents are normalized with respect to the Lewis-Randall rule, such that each γ_i becomes unity when $x_i = 1$, the normalization is said to be symmetric.

For the alternative standard state based on Henry's law the standard-state properties M_i° become the properties of pure i at the mixture T and P but in a hypothetical state that would exist were Henry's law valid throughout the entire composition range of x_i from zero to unity. These values of M_i° can be related to the properties of component i in an infinitely dilute solution, i.e., to the partial molar properties of i as $x_i \to 0$. Equations (4-209) and (4-212) show that

$$V_i^\circ = \bar{V}_i^{id} \quad \text{and} \quad H_i^\circ = \bar{H}_i^{id}$$

Since any real solution becomes ideal with respect to Henry's law for a component at infinite dilution, it is clear that

$$V_i^\circ = \bar{V}_i^\infty \quad \text{and} \quad H_i^\circ = \bar{H}_i^\infty$$

where the superscript ∞ signifies the real state of infinite dilution of i in a specified solvent. In addition, reference to Fig. 4-20 shows that the slope of the Henry's-law line for component 1 of a binary system is given by

$$\lim_{k_1 \to 0} \frac{\hat{f}_1}{x_1} = \left(\frac{d\hat{f}_1}{dx_1}\right)_{x_1=0} = \frac{f_1^\circ}{1} = f_1^\circ$$

This slope is also Henry's-law constant k_i as expressed by the equation $\hat{f}_1 = k_1 x_1$. Thus for component 1 in a binary system $f_1^\circ = k_1$. Henry's constant k_1, as well as $\bar{V}_1^\infty$ and $\bar{H}_1^\infty$, depends on the nature of component 2 in a binary system. More generally, for a multicomponent system these quantities depend on the nature and composition of the solvent. For a specific solvent the standard-state properties $f_i^\circ = k_i$, $V_i^\circ = \bar{V}_i^\infty$, and $H_i^\circ = \bar{H}_i^\infty$ are experimentally accessible.

For this standard state the activity coefficient as given by Eq. (4-235) becomes

$$\gamma_i^* = \hat{f}_i / x_i k_i$$

where the asterisk on γ_i identifies an activity coefficient normalized with respect to Henry's law. This normalization requires that

$$\lim_{x_i \to 0} \gamma_i^* = 1$$

and follows from the fact that Henry's law $\hat{f}_i = k_i x_i$ becomes valid for any constituent in a real solution as $x_i \to 0$.

The standard state based on Henry's law is particularly useful for those constituents of a solution for which the standard state based on the Lewis-Randall rule cannot be precisely determined. Thus, when the Henry's-law standard state is used for one or more constituents of a solution, the Lewis-Randall-rule standard state is used for other constituents. When activity coefficients are normalized in two different ways for the same solution, the normalization is described as unsymmetric.

Henry's-law constants are functions of temperature and pressure, and this dependence is readily determined from Eq. (4-197a), Table 4-24, as written for the standard state:

$$d \ln f_i^\circ = \frac{H_i' - H_i^\circ}{RT^2}\,dT + \frac{V_i^\circ}{RT}\,dP$$

For this standard state $f_i^\circ = k_i$, $H_i^\circ = \bar{H}_i^\infty$, and $V_i^\circ = \bar{V}_i^\infty$. Therefore,

$$d \ln k_i = \frac{H_i' - \bar{H}_i^\infty}{RT^2}\,dT + \frac{\bar{V}_i^\infty}{RT}\,dP$$

whence
$$\left(\frac{\partial \ln k_i}{\partial T}\right)_P = \frac{H_i' - \bar{H}_i^\infty}{RT^2} \qquad (4\text{-}241)$$

and
$$\left(\frac{\partial \ln k_i}{\partial P}\right)_T = \frac{\bar{V}_i^\infty}{RT} \qquad (4\text{-}242)$$

Composition does not enter into these equations, because by its very nature k_i has significance only at $x_i = 0$. However, the values of $\bar{H}_i^\infty$ and $\bar{V}_i^\infty$ depend on the nature of the solvent and on its composition at $x_i = 0$.

The effect of different choices of standard states is best illustrated by example. Applied to the volume of a binary liquid mixture Eq. (4-203) becomes

$$V = x_1 V_1^\circ + x_2 V_2^\circ + \Delta V$$

There are four possible choices of the standard states discussed previously.

1. Base both standard states on the Lewis-Randall rule. Then

$$V = x_1 V_1 + x_2 V_2 + \Delta V$$

where V_1 and V_2 are the molar volumes of pure liquid 1 and pure liquid 2 at the mixture T and P.

2. Base both standard states on Henry's law. Then

$$V = x_1 \bar{V}_1^\infty + x_2 \bar{V}_2^\infty + \Delta V^*$$

where $\bar{V}_1^\infty$ and $\bar{V}_2^\infty$ are the partial molar volumes of the constituents when each is respectively at infinite dilution in the other at the mixture T and P. The asterisk is used to distinguish this volume change of mixing from that of choice 1.

3. Base the standard state of constituent 1 on Henry's law and that of constituent 2 on the Lewis-Randall rule. Then

$$V = x_1 \bar{V}_1^\infty + x_2 V_2 + \Delta V^{*1}$$

where the notation ΔV^{*1} indicates a volume change of mixing based on the Henry's-law standard state for constituent 1.

4. Base the standard state of constituent 1 on the Lewis-Randall rule and that of constituent 2 on Henry's law. Then

$$V = x_1 V_1 + x_2 \bar{V}_2^\infty + \Delta V^{*2}$$

The four volume changes of mixing arising from the different choices of standard states differ in general from one another. Nevertheless, the four equations are equivalent for purposes of calculating mixture volumes, provided that all quantities on the right-hand sides have been measured experimentally. However, if the volume changes of mixing are *not* known and *are to be neglected*, then use of the equation for which the volume change is most nearly zero should be advantageous.

Adequate experimental data are available for the system cyclohexane(1)-carbon tetrachloride(2) at 30°C to allow calculation of all four of the volume changes of mixing indicated previously as functions of x_1. The results of these calculations are shown in Fig. (4-21).

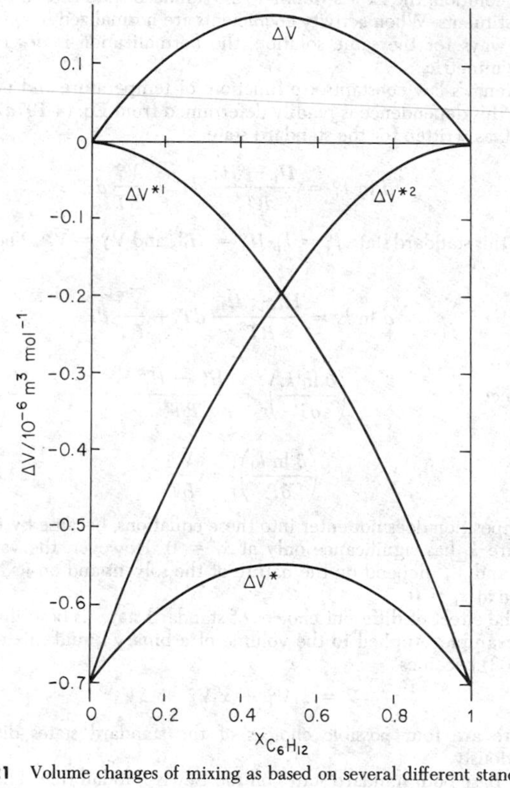

FIG. 4-21 Volume changes of mixing as based on several different standard states.

Comments on the four different cases follow.

1. The curve labeled ΔV represents the quantity that is directly measurable by experiment and is the volume change invariably employed when experimental data are available.

2. The curve labeled ΔV° illustrates the fact that the property changes of mixing need not vanish at the end points where x_1 becomes zero or unity. For this choice of standard states they clearly do not. There appears to be no advantage to the use of this volume change of mixing for any purpose at any composition.

3. The curve labeled $\Delta V^{\circ 1}$ goes to zero with zero slope at $x_1 = 0$, and it therefore exhibits very low values when x_1 is small. The reason is that at high dilution constituent 1 closely conforms to Henry's law and at the same time constituent 2 (being nearly pure) closely conforms to the Lewis-Randall rule, both in accord with the chosen standard states. If one is to neglect the volume change of mixing and to assume ideality for mixtures at small values of x_1, then this choice of standard states provides the idealization that most closely conforms to reality. [See Eq. (4-210).] The difficulty is that to take practical advantage of this circumstance one needs a value for $\overline{V}_1^{\infty}$ as shown by the equation for V that holds in this case. Determination of this quantity requires experimental measurements on the particular mixture considered at values of x_1 close to zero, and such data are rarely available unless the system has been thoroughly studied, in which case no approximation is called for. This problem accompanies every use of a standard state based on Henry's law.

4. The curve labeled $\Delta V^{\circ 2}$ is roughly the mirror image of that for $\Delta V^{\circ 1}$, and the same comments apply as in case 3 except that the constituents are interchanged.

Partial Molar Properties Equation (4-160), which defines a partial molar property, is of limited use for the calculation of numerical values from experimental data. The development of a working equation is based on the fact that *at constant T and P* a molar property M of a mixture is a function of $n - 1$ independent mole fractions,

$$M = m(x_1, x_2, \ldots, x_{i-1}, x_{i+1}, \ldots, x_n) \tag{4-243}$$

where x_i has been omitted as a dependent variable.

Expansion of Eq. (4-160) gives

$$\overline{M}_i = M(\partial n/\partial n_i)_{nj} + n(\partial M/\partial n_i)_{nj}$$

where the subscripts indicating constancy of T and P have been temporarily suppressed. If subscript k represents any constituent except i, then $n = \Sigma_k n_k + n_i$, and

$$(\partial n/\partial n_i)_{nj} = 1$$

Therefore

$$\overline{M}_i = M + n(\partial M/\partial n_i)_{nj} \tag{4-244}$$

By the application of the chain rule in conformity with Eq. (4-243),

$$\left(\frac{\partial M}{\partial n_i}\right)_{nj} = \sum_k \left[\left(\frac{\partial M}{\partial x_k}\right)\left(\frac{\partial x_k}{\partial n_i}\right)_{nj}\right] \tag{4-245}$$

where k runs over all constituents except the ith because x_i is not included in the functional relationship of Eq. (4-243). By definition, $x_k = n_k/n$, and therefore

$$\left(\frac{\partial x_k}{\partial n_i}\right)_{nj} = -\frac{n_k}{n^2}\left(\frac{\partial n}{\partial n_i}\right)_{nj} + \frac{1}{n}\left(\frac{\partial n_k}{\partial n_i}\right)_{nj}$$

However, $(\partial n/\partial n_i)_{nj} = 1$ and $(\partial n_k/\partial n_i)_{nj} = 0$, and therefore

$$(\partial x_k/\partial n_i)_{nj} = -\,n_k/n^2 = -\,x_k/n$$

Substitution into Eq. (4-245) gives

$$\left(\frac{\partial M}{\partial n_i}\right)_{nj} = -\frac{1}{n}\sum_k \left[x_k\left(\frac{\partial M}{\partial x_k}\right)\right]$$

Combination with Eq. (4-244) provides the result

$$\overline{M}_i = M - \sum_{k\neq i}\left[x_k\left(\frac{\partial M}{\partial x_k}\right)_{T,P,x_r}\right] \tag{4-246}$$

where the notation has been made explicit to show that k does not include i and that the partial derivatives are taken at constant T, P, and x_r, the subscript x_r indicating that all mole fractions are held constant *except x_i and the particular x_k of the derivative*.

Completely analogous equations exist for $\overline{\Delta M}_i$ and for $\overline{\Delta M}_i^E \equiv \overline{M}_i^E$.

Application of Eq. (4-246) to a binary system containing constituents 1 and 2 provides the equations

$$\overline{M}_1 = M + (1 - x_1)(dM/dx_1) \tag{4-247a}$$

$$\overline{M}_2 = M - x_1(dM/dx_1) \tag{4-247b}$$

where use has been made of the equation $x_1 + x_2 = 1$, and where it is implicit that the derivative is at constant T and P.

These equations lead directly to the self-evident graphical construction of Fig. 4-22, which shows the method of tangent intercepts for determination of $\overline{M}_1$ and $\overline{M}_2$. Figures 4-23 to 4-25 also employ this principle to show the interrelationships among the various types of solution properties that have been discussed.

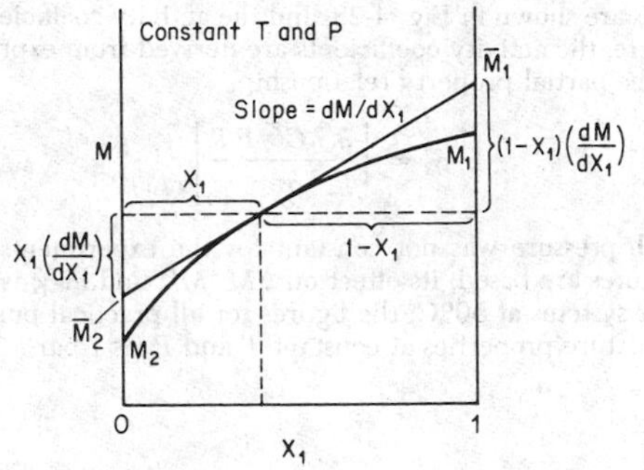

FIG. 4-22 Method of tangent intercepts.

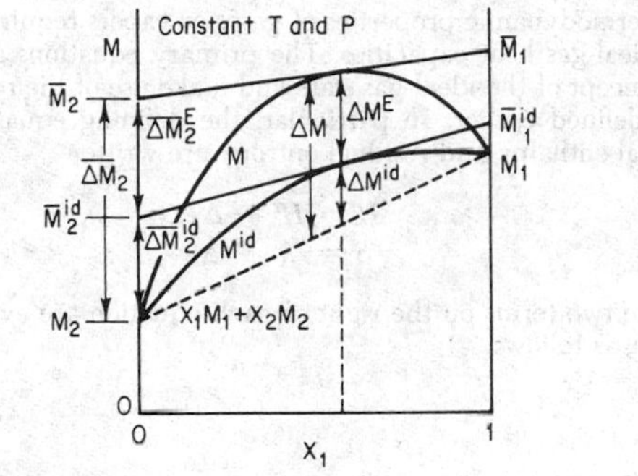

FIG. 4-23 Thermodynamic functions for binary solutions.

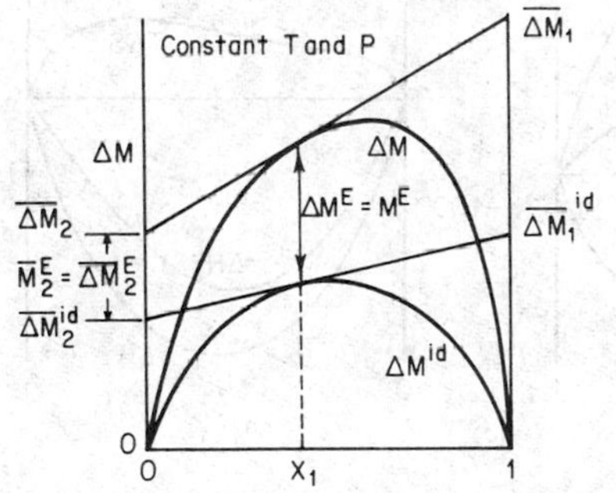

FIG. 4-24 Property changes of mixing for binary solutions.

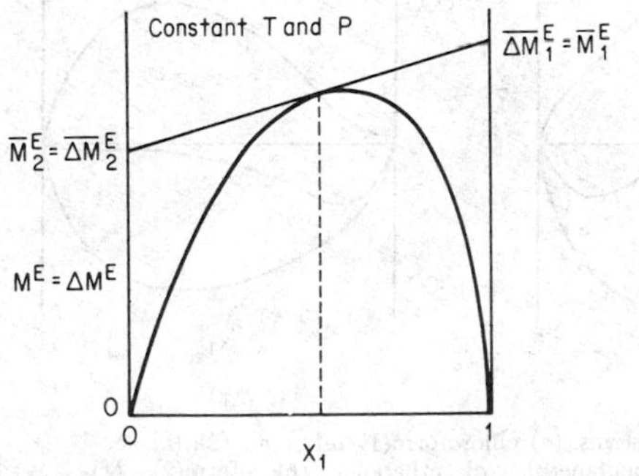

FIG. 4-25 Excess property changes of mixing for binary solutions.

Thermodynamic Behavior of Binary Liquid Solutions Since $M^E = \Delta M^E$, we have by Eq. (4-223) that

$$M^E = \Delta M - \Delta M^{id} \tag{4-248}$$

where (see previous discussions)

$$\Delta M^{id} = \begin{cases} 0 & (M = V, H, U, C_P, C_V) \\ -R\Sigma x_i \ln x_i & (M = S) \\ RT\Sigma x_i \ln x_i & (M = A, G) \end{cases}$$

These formulas for ΔM^{id} are valid whatever the choices of standard state for the various species. However, as demonstrated for ΔV by Fig. 4-21 and the accompanying discussion, the numerical values of the ΔM depend upon the choices of standard states; hence, by Eq. (4-248) so do the numerical values of the excess functions.

Property changes of mixing and excess functions find greatest application to the description of liquid mixtures at low reduced temperatures, i.e., at temperatures well below the critical temperature of each constituent species. At such temperatures, the liquid state is in principle *accessible* for each species, although (depending on the pressure) it may be metastable. Lewis-Randall standard states are then appropriate for all i, and thus we take

$$M_i^\circ = M_i \qquad \text{(all } i\text{)}$$

where M_i is molar property M for pure liquid i at the temperature and pressure of the solution. With this choice of standard state, all property changes of mixing and excess functions have the feature that

$$\lim_{x_i \to 1} \Delta M = \lim_{x_i \to 1} M^E = 0 \qquad \text{(any } i\text{)}$$

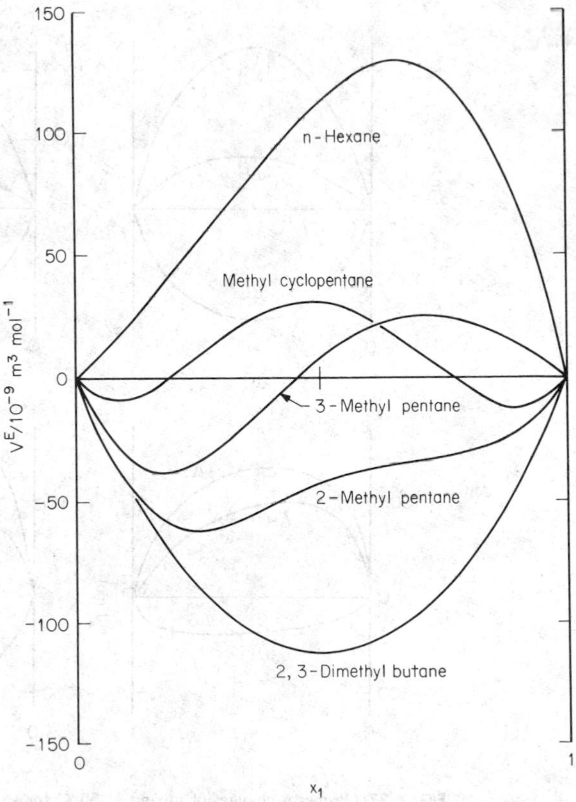

FIG. 4-26 Excess volumes at 25°C for liquid mixtures of cyclohexane(1) with some other C6 hydrocarbons.

Moreover, the activity coefficients are symmetrically normalized, and

$$\lim_{x_i \to 1} \gamma_i = 1 \qquad \text{(all } i\text{)}$$

The functions of interest to the chemical engineer are $\Delta V(=V^E)$, $\Delta H(=H^E)$, ΔS, S^E, ΔG, and G^E. The activity coefficient is also of special importance because of its usefulness in phase-equilibrium calculations, as shown later.

The behavior of *binary* liquid mixtures is advantageously displayed on plots of ΔM, M^E, and $\ln \gamma_i$ versus x_1 at constant T and P. The volume change of mixing $\Delta V(=V^E)$ is the most easily measured of all these quantities and is normally small. However, as illustrated by Fig. 4-26, it is capable of quite individualistic behavior, being very sensitive to effects of molecular size and shape and to differences in the nature and magnitude of intermolecular forces.

Of the other functions, the heat of mixing $\Delta H(=H^E)$ and the excess Gibbs function G^E are also experimentally accessible, ΔH by direct measurement and G^E (or γ_i) indirectly as a product of the reduction of phase-equilibrium data (usually for vapor-liquid equilibrium, discussed later). Given values for H^E and G^E, one may calculate the remaining property changes of mixing and excess functions, as summarized below:

$$\Delta G = G^E + RT\Sigma x_i \ln x_i \qquad (4\text{-}249)$$

$$S^E = (H^E - G^E)/T \qquad (4\text{-}250)$$

$$\Delta S = S^E - R\Sigma x_i \ln x_i \qquad (4\text{-}251)$$

Figure 4-27 displays plots of ΔH, ΔS, and ΔG, as functions of composition for six binary systems at 50°C. The corresponding excess functions are shown in Fig. 4-28, and the activity coefficients in Fig. 4-29. Here, the activity coefficients are derived from expressions for G^E via the partial-property relationship,

$$\ln \gamma_i = \left[\frac{\partial(nG^E/RT)}{\partial n_i}\right]_{T,P,n_j} \qquad (4\text{-}252)$$

Although pressure was not constant for the experiments on which these figures are based, its effect on ΔM, M^E, and $\ln \gamma_i$ is negligible for these systems at 50°C; the figures for all practical purposes represent mixture properties at constant T and P ($\approx$1 bar).

EVALUATION AND REPRESENTATION OF THERMODYNAMIC PROPERTIES

General The most satisfactory procedure for the calculation of the thermodynamic properties of gases or vapors requires PVT data and ideal-gas heat capacities. The primary equations are based on the concept of the ideal-gas state and make use of the residual functions defined earlier. In particular, the defining equations for the residual enthalpy and residual entropy are written

$$H = H' - \Delta H'$$

$$S = S' - \Delta S'$$

and the two terms on the right of each equation are evaluated separately, as follows.

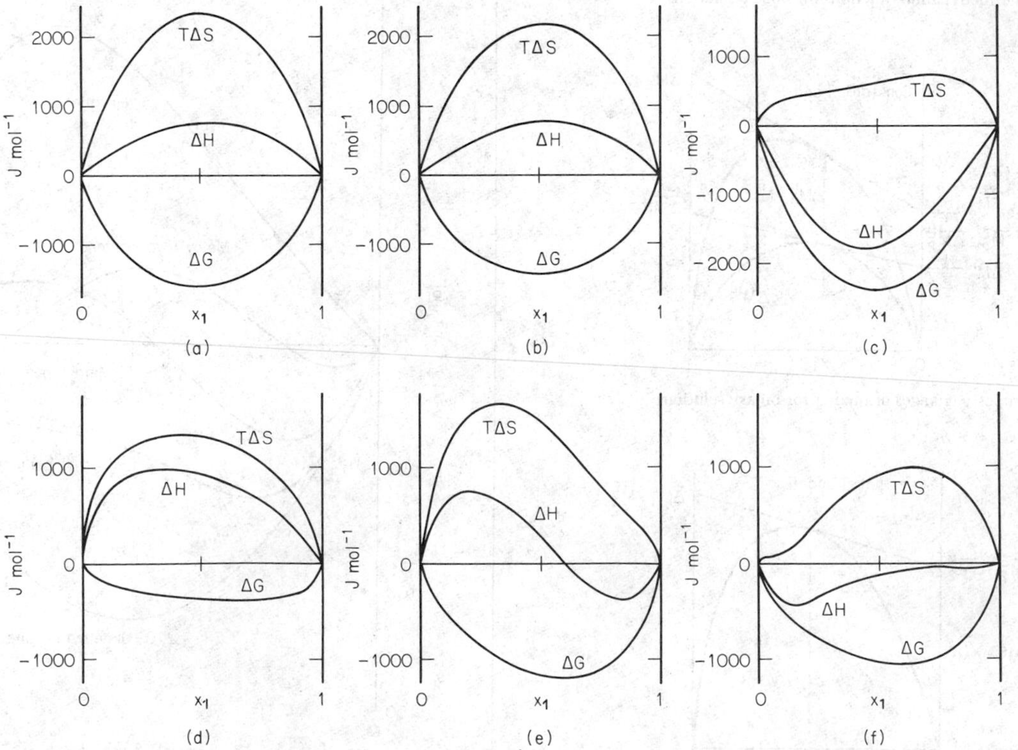

FIG. 4-27 Property changes of mixing at 50°C for six binary liquid systems: (*a*) chloroform(1)/n-heptane(2); (*b*) acetone(1)/methanol(2); (*c*) acetone(1)/chloroform(2); (*d*) ethanol(1)/n-heptane(2); (*e*) ethanol(1)/chloroform(2); (*f*) ethanol(1)/water(2).

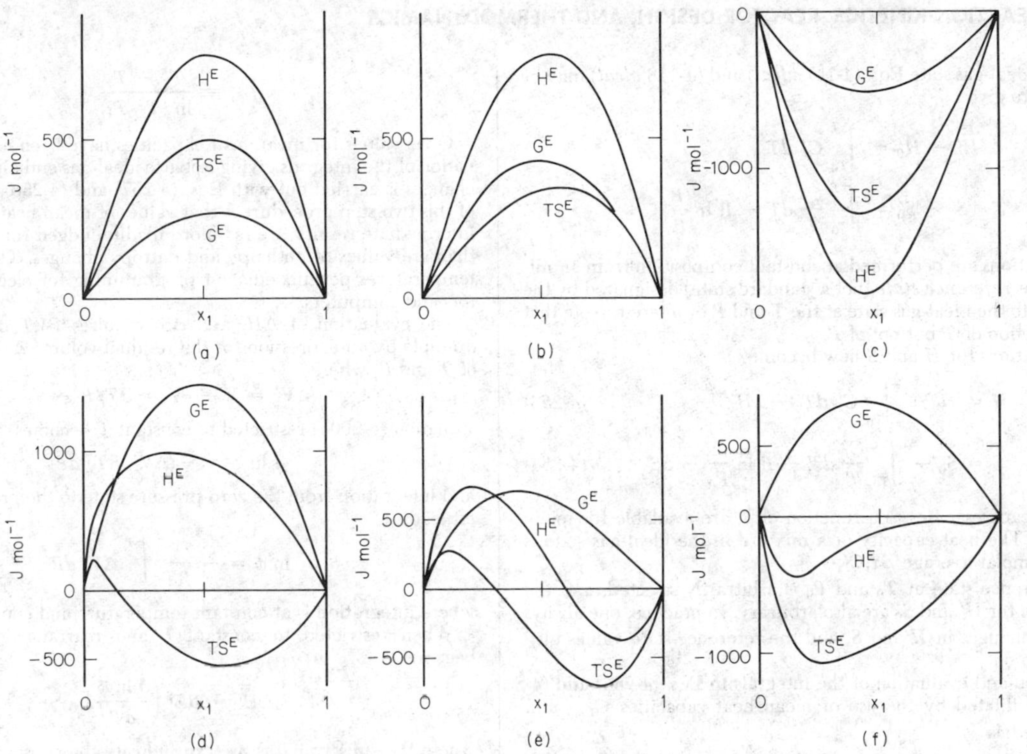

FIG. 4-28 Excess functions at 50°C for six binary liquid systems: (*a*) chloroform(1)/n-heptane(2); (*b*) acetone(1)/methanol(2); (*c*) acetone(1)/chloroform(2); (*d*) ethanol(1)/n-heptane(2); (*e*) ethanol(1)/chloroform(2); (*f*) ethanol(1)/water(2).

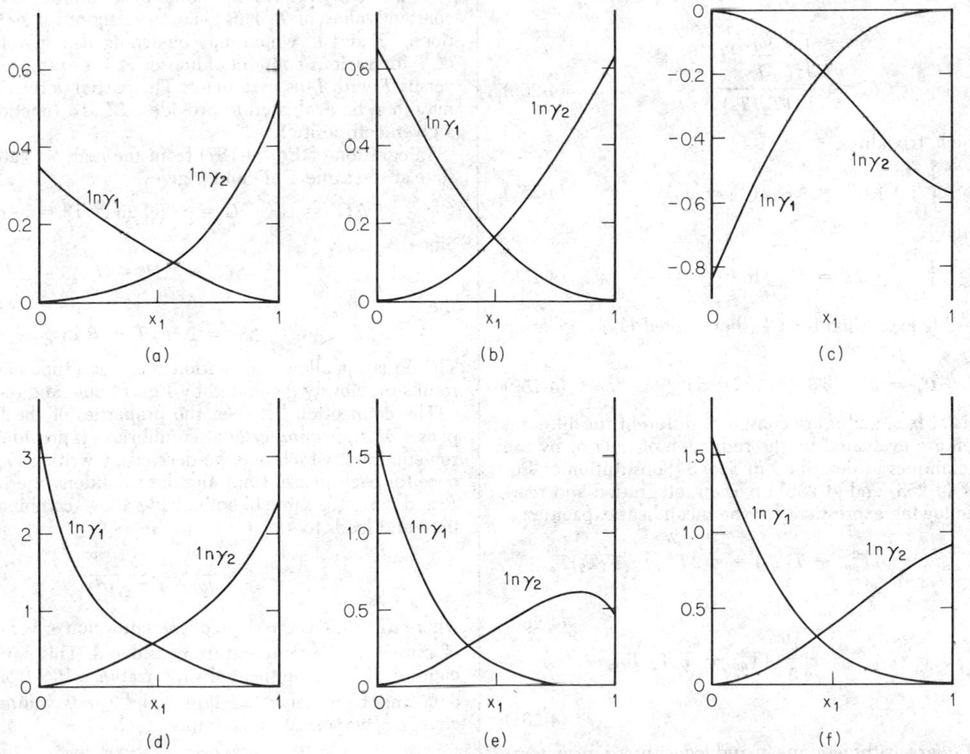

FIG. 4-29 Activity coefficients at 50°C in six binary liquid systems: (*a*) chloroform(1)/n-heptane(2); (*b*) acetone(1)/methanol(2); (*c*) acetone(1)/chloroform(2); (*d*) ethanol(1)/n-heptane(2); (*e*) ethanol(1)/chloroform(2); (*f*) ethanol(1)/water(2).

For the ideal-gas state Eqs. (4-147 *ideal*) and (4-148 *ideal*) may be integrated to give

$$H' - H'_0 = \int_{T_0}^{T} C'_P \, dT$$

$$S' - S'_0 = \int_{T_0}^{T} \frac{C'_P}{T} \, dT - R \ln \frac{P}{P_0}$$

The integrations are performed at constant composition from an initial ideal-gas *reference state* (not a standard state) designated by the subscript 0 to the ideal-gas state at the T and P of interest. Note that C'_P is a function of T but not of P.

The equations for H and S now become

$$H = H'_0 + \int_{T_0}^{T} C'_P \, dT - \Delta H' \tag{4-253}$$

$$S = S'_0 + \int_{T_0}^{T} \frac{C'_P}{T} \, dT - R \ln \frac{P}{P_0} - \Delta S' \tag{4-254}$$

Ideal-gas heat capacities as a function of T are available for many pure gases. The heat capacity of a mixture in the ideal-gas state is simply the molar average $\Sigma x_i C'_{P_i}$.

The reference state at T_0 and P_0 is arbitrarily selected, and the values taken for H'_0 and S'_0 are also arbitrary. In practice, one always deals with changes in H and S, and the reference-state values ultimately drop out.

Expression and evaluation of the integrals in Eqs. (4-253) and (4-254) are facilitated by the use of mean heat capacities $C_{P_{mh}}$ and $C_{P_{ms}}$, defined as

$$C'_{P_{mh}} \equiv \frac{\int_{T_1}^{T_2} C'_P \, dT}{T_2 - T_1} \tag{4-255}$$

and

$$C'_{P_{ms}} \equiv \frac{\int_{T_1}^{T_2} \frac{C'_P}{T} \, dT}{\ln (T_2/T_1)} \tag{4-256}$$

as a result of which, trivially,

$$\int_{T_1}^{T_2} C'_P \, dT = C'_{P_{mh}} (T_2 - T_1) \tag{4-257}$$

and

$$\int_{T_1}^{T_2} \frac{C'_P}{T} \, dT = C'_{P_{ms}} \ln (T_2/T_1) \tag{4-258}$$

A reasonably flexible expression for C'_P (there called C°_P) is given in Sec. 3:

$$C'_P = a + bT + cT^2 + dT^3 \tag{4-259}$$

Here, parameters a, b, c, and d are constants, different for different substances, which are evaluated by the reduction of data or by use of estimation techniques as described in Sec. 3. Substitution of Eq. (4-259) into Eqs. (4-255) and (4-256) gives on integration and rearrangement the following expressions for the mean heat capacities:

$$C'_{P_{mh}} = a + bT_{am} + \frac{c}{3}(4T_{am}^2 - T_1T_2) + d(2T_{am}^2 - T_1T_2)T_{am} \tag{4-260}$$

$$C'_{P_{ms}} = a + bT_{\ell m} + cT_{am}T_{\ell m} + \frac{d}{3}(4T_{am}^2 - T_1T_2)T_{\ell m} \tag{4-261}$$

Here, T_{am} and $T_{\ell m}$ are arithmetic-mean and logarithmic-mean temperatures, defined as

$$T_{am} \equiv (T_1 + T_2)/2 \tag{4-262}$$

$$T_{\ell m} \equiv \frac{T_2 - T_1}{\ln (T_2/T_1)} \tag{4-263}$$

Once values for mean heat capacities have been calculated, evaluation of the integrals giving isobaric ideal-gas enthalpy and entropy changes is carried out with Eqs. (4-257) and (4-258). An advantage of this two-step procedure is that values of mean heat capacities, the intermediate results, are far more readily judged for reasonableness than are values of enthalpy and entropy changes. The use of mean temperatures permits efficient programming for electronic calculators and computers.

The evaluation of $\Delta H'$ and $\Delta S'$ requires PVT data, given for example by an expression for the residual volume $\Delta V'$ as a function of T and P, where

$$\Delta V' = V' - V = RT/P - V$$

Equation (4-200*a*) restricted to constant T becomes

$$d \ln \phi = -(\Delta V'/RT) \, dP$$

and integration from the zero-pressure state to the pressure of interest leads to

$$\ln \phi = -\frac{1}{RT} \int_0^P \Delta V' \, dP \tag{4-264}$$

where integration is at constant temperature and composition.

When restricted to constant P and rearranged, Eq. (4-200*a*) becomes

$$\Delta H' = RT^2 \left(\frac{\partial \ln \phi}{\partial T} \right)_{P,y} \tag{4-265}$$

where the subscript y is used to indicate the constancy of all mole fractions in the gas mixture.

Experimental data for $\Delta V'$ as a function of P at constant T and y may be used to evaluate the integral of Eq. (4-264) either graphically or numerically for various values of P, and this is done for different constant values of T. This generates a set of values for $\ln \phi$ as a function of T and P, which may be arranged to give $\ln \phi$ as a function of T for various constant values of P: for example, as a plot of $\ln \phi$ versus T with P as parameter. The partial derivative of Eq. (4-265) may then be evaluated to provide $\Delta H'$ as a function of T and P for a given composition.

Integration of Eq. (4-187) from the real-gas state to the ideal-gas state at the same T, P, and y gives

$$\Delta G' = G' - G = -RT \ln (f/P) = -RT \ln \phi$$

Since

$$\Delta G' = \Delta H' - T \, \Delta S'$$

then

$$\Delta S' = \Delta H'/T + R \ln \phi \tag{4-266}$$

This equation allows calculation of $\Delta S'$ as a function of T and P from results previously generated by Eqs. (4-264) and (4-265).

The connection between the properties of the liquid and vapor phases *of a pure material* at equilibrium is provided by the Clapeyron equation, which may be derived by writing Eq. (4-197*a*) twice, once for each phase. Imposing the conditions of equilibrium that T, P, and f_i are the same in both phases allows combination of the equations and leads to

$$\frac{dP_i^{sat}}{dT} = \frac{\Delta H_i^{vap}}{T \, \Delta V_i^{vap}} \tag{4-267}$$

where dP_i^{sat}/dT is the slope of the saturation or vapor-pressure curve of pure i at the temperature considered. This equation allows calculation of the latent heat of vaporization ΔH_i^{vap} from vapor-pressure data and from saturated liquid and vapor volumes. The entropy change of vaporization is simply

$$\Delta S_i^{vap} = \Delta H_i^{vap}/T \tag{4-268}$$

Once one has saturated liquid enthalpies and entropies, the calculation of properties for pure compressed liquids is accomplished by direct integration of Eqs. (4-146) and (4-141):

$$H_i - H_i^{sat} = \int_{P_i^{sat}}^{P} \left[V_i - T\left(\frac{\partial V_i}{\partial T}\right)_P \right] dP \qquad (4\text{-}269)$$

$$S_i - S_i^{sat} = -\int_{P_i^{sat}}^{P} \left(\frac{\partial V_i}{\partial T}\right)_P dP \qquad (4\text{-}270)$$

In each case integration is at constant T. PVT data on pure liquids are required for these calculations.

The properties of liquid mixtures are usually determined from the properties of the pure constituents and experimental property-change-of-mixing data by the equation

$$H = \Sigma(x_i H_i) + \Delta H \qquad (4\text{-}271)$$

where the enthalpy change ΔH is commonly called the *heat of mixing* and is directly measurable. For the entropy

$$S = \Sigma(x_i S_i) + \Delta S^{id} + S^E$$

and in view of Eq. (4-220) this becomes

$$S = \Sigma(x_i S_i) - R\Sigma(x_i \ln x_i) + S^E \qquad (4\text{-}272)$$

Here S^E is not directly measurable but requires indirect calculation, usually from phase-equilibrium data.

Derived Properties from Equations of State

General Formulas The empirical representation of the PVT surface is treated in Sec. 3, where four methods are considered: corresponding-states correlations of the Pitzer type, the virial equation in density, two-parameter equations of state (as exemplified by the Redlich-Kwong equation), and the BWR equation of state. We treat here the computation of *derived properties* from such representations, first in general terms and then for each of the four cited cases.

PVT equations of state (or corresponding-states correlations) may be of two types: *volume-explicit*, in which case V (or Z) is given as a function of T, P, and composition, or *pressure-explicit*, in which case P (or Z) is given as a function of T,V, and composition. The working recipes for derived properties are different for the two cases; we develop them in the following paragraph.

Equation (4-264) is a general expression for $\ln \phi$. Noting that

$$\Delta V' = RT/P - V = -(RT/P)(Z - 1)$$

we can write it as

$$\ln \phi = \int_0^P (Z - 1)\frac{dP}{P} \qquad (4\text{-}273)$$

where the integration is performed at constant T and composition. The residual enthalpy is related to $\ln \phi$ by Eq. (4-265). Differentiating Eq. (4-273) with respect to T at constant P and composition, we thus find that

$$\Delta H' = RT^2 \int_0^P \left(\frac{\partial Z}{\partial T}\right)_{P,y} \frac{dP}{P} \qquad (4\text{-}274)$$

where again the integration is done at constant T and y. The residual entropy $\Delta S'$ is found by combination of Eqs. (4-273) and (4-274) according to Eq. (4-266).

In applications (e.g., to vapor-liquid equilibrium) it is usually the *component* fugacity coefficient $\hat{\phi}_i$ of species i in a *mixture* that is required. Given an expression for $\ln \phi$ as determined by Eq. (4-273), one finds the corresponding recipe for $\ln \hat{\phi}_i$ through the partial-property relationship.

$$\ln \hat{\phi}_i = \left[\frac{\partial(n \ln \phi)}{\partial n_i}\right]_{T,P,nj} \qquad (4\text{-}275)$$

One may either operate on the *result* of the integration of Eq. (4-273) according to Eq. (4-275) or apply Eq. (4-275) *directly* to Eq. (4-273), obtaining

$$\ln \hat{\phi}_i = \int_0^P (\bar{Z}_i - 1)\frac{dP}{P} \qquad (4\text{-}276)$$

where $\bar{Z}_i$ is the partial-compressibility factor, defined as

$$\bar{Z}_i \equiv \left[\frac{\partial(nZ)}{\partial n_i}\right]_{T,P,nj} \qquad (4\text{-}277)$$

Direct use of Eqs. (4-273) through (4-277) requires the availability of a *volume-explicit* equation of state or correlation. Many comprehensive equations of state are *pressure-explicit*, however, and alternative formulas must be used. To derive them, we proceed from Eq. (4-273). Since $PV = ZRT$, then

$$PdV + VdP = RTdZ \text{ (constant } T,y)$$

whence $\quad dP/P = dZ/Z - dV/V$ (constant T,y)

Substitution into Eq. (4-273) leads to

$$\ln \phi = Z - 1 - \ln Z - \int_\infty^V (Z - 1)\frac{dV}{V} \qquad (4\text{-}278a)$$

or, in terms of *molar density* ρ ($\equiv V^{-1}$),

$$\ln \phi = Z - 1 - \ln Z + \int_0^\rho (Z - 1)\frac{d\rho}{\rho} \qquad (4\text{-}278b)$$

The corresponding formulas for $\Delta H'$ are found most readily through Eq. (4-200a). Division of this equation by dT and restriction to constant V gives, on rearrangement,

$$\frac{\Delta H'}{RT^2} = \left(\frac{\partial \ln \phi}{\partial T}\right)_{V,y} + \left(\frac{1 - Z}{P}\right)\left(\frac{\partial P}{\partial T}\right)_{V,y}$$

Differentiation of Eq. (4-278a) provides the first term on the right, and differentiation of $P = ZRT/V$ provides the second. Substitution then leads to

$$\Delta H' = -RT^2 \int_\infty^V \left(\frac{\partial Z}{\partial T}\right)_{V,y} \frac{dV}{V} - RT(Z - 1) \qquad (4\text{-}279a)$$

or, in terms of molar density,

$$\Delta H' = RT^2 \int_0^\rho \left(\frac{\partial Z}{\partial T}\right)_{\rho,y} \frac{d\rho}{\rho} - RT(Z - 1) \qquad (4\text{-}279b)$$

As before, the residual entropy is found from $\ln \phi$ and $\Delta H'$ through Eq. (4-266).

With a pressure-explicit equation of state, one cannot determine $\ln \hat{\phi}_i$ by direct application of the partial-property definition, Eq. (4-273). We work instead through Eq. (4-200), which when divided by dn_i and restricted to constant T, nV, and n_j ($j \neq i$) leads to

$$\ln \hat{\phi}_i = \left[\frac{\partial(n \ln \phi)}{\partial n_i}\right]_{T,nV,nj} - \frac{n}{P}(Z - 1)\left(\frac{\partial P}{\partial n_i}\right)_{T,nV,nj}$$

But $P = (nZ)RT/nV$, and thus

$$\left(\frac{\partial P}{\partial n_i}\right)_{T,nV,nj} = \frac{P}{nZ}\left[\frac{\partial(nZ)}{\partial n_i}\right]_{T,nV,nj}$$

Combination of the last two equations gives

$$\ln \hat{\phi}_i = \left[\frac{\partial(n \ln \phi)}{\partial n_i}\right]_{T,nV,nj} - \left(\frac{Z - 1}{Z}\right)\left[\frac{\partial(nZ)}{\partial n_i}\right]_{T,nV,nj} \qquad (4\text{-}280a)$$

or, equivalently, if ρ is the favored volumetric variable,

$$\ln \hat{\phi}_i = \left[\frac{\partial(n \ln \phi)}{\partial n_i}\right]_{T,\rho/n,nj} - \left(\frac{Z - 1}{Z}\right)\left[\frac{\partial(nZ)}{\partial n_i}\right]_{T,\rho/n,nj} \qquad (4\text{-}280b)$$

Equations (4-280a and b) may either be applied to the results of the integrations in Eqs. (4-278a and b), or *directly* to Eqs. (4-278a and b). For the latter case, we obtain the following analogs of Eq. (4-276):

$$\ln \hat{\phi}_i = -\int_\infty^V \left\{\left[\frac{\partial(nZ)}{\partial n_i}\right]_{T,nV,nj} - 1\right\}\frac{dV}{V} - \ln Z \qquad (4\text{-}281a)$$

and

$$\ln \hat{\phi}_i = \int_0^\rho \left\{\left[\frac{\partial(nZ)}{\partial n_i}\right]_{T,\rho/n,nj} - 1\right\}\frac{d\rho}{\rho} - \ln Z \qquad (4\text{-}281b)$$

Pitzer's Corresponding-States Correlation In Sec. 3 is described a three-parameter corresponding-states correlation of the type first developed by Pitzer and coworkers [Pitzer et al., *J. Am. Chem. Soc.*, **77**, 3433 (1955)]:

$$Z = Z^{(0)} + \omega Z^{(1)} \tag{4-282}$$

Here, terms $Z^{(0)}$ and $Z^{(1)}$ are each functions of reduced temperature T_r and reduced pressure P_r, and ω is the acentric factor. The T_r and P_r dependencies of $Z^{(0)}$ and $Z^{(1)}$, as determined by Hsiao and Lu [*Can. J. Chem. Eng.*, **57**, 102 (1979)], are shown on Figs. 3-45 and 3-46. We illustrate here how this information can be used to determine correlations for the fugacity coefficient and for the dimensionless residual enthalpy $\Delta H'/RT_c$. A generalization of the procedure is given by Van Ness and Abbott (*Classical Thermodynamics of Nonelectrolyte Solutions: With Applications to Phase Equilibria*, McGraw-Hill, New York, 1982, secs. 4-3 and 4-4).

Eliminating P in favor of P_r, we may write Eq. (4-273) as

$$\ln \phi = \int_0^{P_r} (Z - 1)\frac{dP_r}{P_r} \tag{4-283}$$

where now the integration is carried out at constant composition and *reduced* temperature. Operating on Eq. (4-282) according to Eq. (4-283), we thus obtain that

$$\ln \phi = \ln \phi^{(0)} + \omega \ln \phi^{(1)} \tag{4-284}$$

where

$$\ln \phi^{(0)} = \int_0^{P_r} (Z^{(0)} - 1)\frac{dP_r}{P_r} \tag{4-285a}$$

and

$$\ln \phi^{(1)} = \int_0^{P_r} Z^{(1)}\frac{dP_r}{P_r} \tag{4-285b}$$

Note that the integrands of these two expressions are of different forms. Equation (4-285*a*) must be of the same functional form as Eq. (4-283) because $\phi^{(0)}$ represents (literally) the fugacity coefficient for substances with zero acentric factor (the "simple fluids" Ar, Kr, and Xe). The integral of Eq. (4-285*b*), which appears to be divergent, is not, because $Z^{(1)}$ approaches zero as $P_r \to 0$.

Division of Eq. (4-274) by RT_c gives, on elimination of T and P in favor of T_r and P_r,

$$\frac{\Delta H'}{RT_c} = T_r^2 \int_0^{P_r} \left(\frac{\partial Z}{\partial T_r}\right)_{P_r,y} \frac{dP_r}{P_r} \tag{4-286}$$

where again the integration is performed at constant T_r and y. Differentiating Eq. (4-282) and integrating according to Eq. (4-286), we

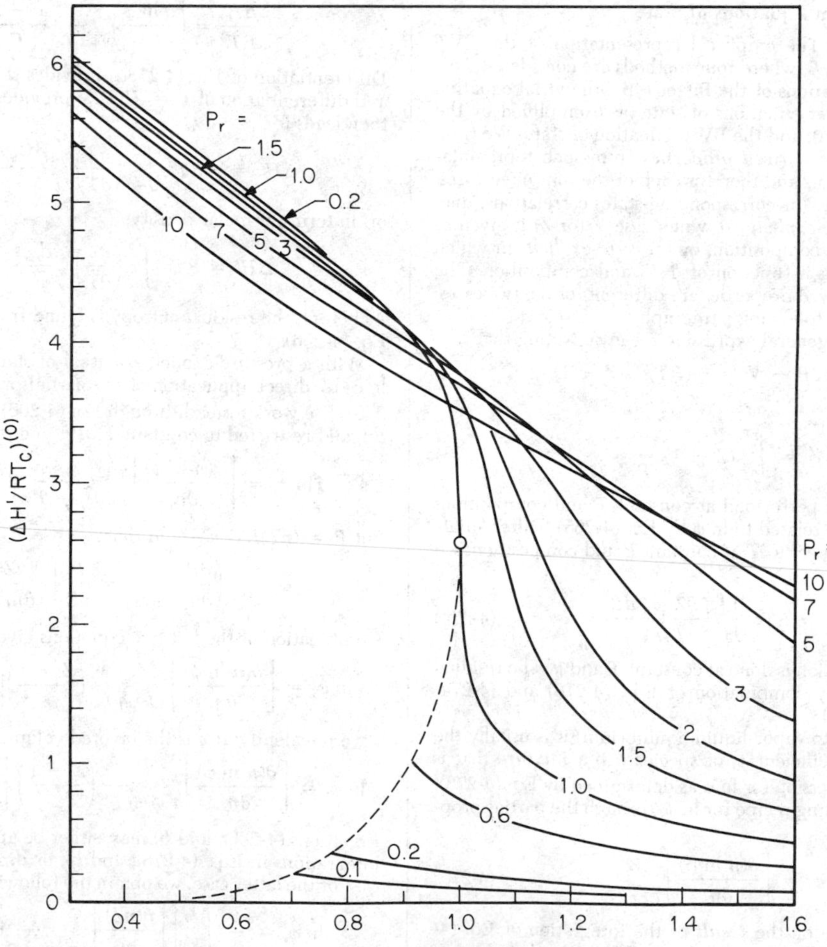

FIG. 4-30 Correlation of $(\Delta H'/RT_c)^{(0)}$, drawn from the tables of Lee and Kesler [Am. Inst. Chem. Eng. J., *21*, 510 (1975)].

find

$$\frac{\Delta H'}{RT_c} = \left(\frac{\Delta H'}{RT_c}\right)^{(0)} + \omega\left(\frac{\Delta H'}{RT_c}\right)^{(1)} \tag{4-287}$$

where

$$\left(\frac{\Delta H'}{RT_c}\right)^{(0)} = T_r^2 \int_0^{P_r} \left(\frac{\partial Z^{(0)}}{\partial T_r}\right)_{P_r,y} \frac{dP_r}{P_r} \tag{4-288a}$$

and

$$\left(\frac{\Delta H'}{RT_c}\right)^{(1)} = T_r^2 \int_0^{P_r} \left(\frac{\partial Z^{(1)}}{\partial T_r}\right)_{P_r,y} \frac{dP_r}{P_r} \tag{4-288b}$$

The Pitzer expression for the dimensionless residual entropy $\Delta S'/R$ is found from Eqs. (4-284), (4-285), (4-287), and (4-288) through the dimensionless analog of Eq. (4-266):

$$\Delta S'/R = (1/T_r)(\Delta H'/RT_c) + \ln \phi \tag{4-289}$$

Pitzer's original correlations for Z and the derived properties were determined graphically and presented in tabular form. Since then analytical refinements to the tables have been developed, with extended range and improved accuracy. One of the most comprehensive Pitzer-type correlations is that of Lee and Kesler [*Am. Inst. Chem. Eng. J.*, **21,** 510 (1975)]; these authors report an analytical correlation incorporating the BWR equation and in addition present tables for contributions to Z and the derived properties. The tables cover both the liquid and gas phases, and span the ranges $0.3 \leq T_r \leq 4.0$ and $0.01 \leq P_r \leq 10.0$. Shown on Figs. 4-30 and 4-31 are plots of values of $(\Delta H'/RT_c)^{(0)}$ and $(\Delta H'/RT_c)^{(1)}$ drawn from these tables, and on Figs. 4-32 through 4-35 the corresponding plots for the fugacity coefficient. Note that in Figs. 4-32 and 4-33 the ordinates are $-\ln \phi^{(0)}$ and $-\ln \phi^{(1)}$ with T_r the independent variable, whereas on Figs. 4-34 and 4-35 we plot isotherms, with $\phi^{(0)}$ and $\phi^{(1)}$ as the ordinates.

Although corresponding-states correlations are invariably based on data for pure materials, they may also be used for calculations of properties of mixtures and of partial molar properties. What is required is a set of recipes relating the corresponding-states parameters (T_c, P_c, and ω, in the case of Pitzer correlations) for a mixture to the pure-component values and to composition. Such recipes, examples of which are given by Eqs. 3-24, 3-25, and 3-26 of Sec. 3, define *pseudoparameters* for the mixture, so called because the values of T_c and P_c so defined do not necessarily correspond to the true-mixture critical properties and because ω for a mixture has no real physical significance.

We outline here a method for computing the constituent fugacity coefficient $\hat{\phi}_i$ for a species in a mixture described by Pitzer's correlation. The starting point is Eq. (4-275), which may be written in expanded form as

$$\ln \hat{\phi}_i = \ln \phi + n\left(\frac{\partial \ln \phi}{\partial n_i}\right)_{T,P,n_j} \tag{4-290}$$

Now $\ln \phi$ depends *explicitly* on T_r, P_r, and ω; the composition dependence is buried in the recipes for the pseudoparameters. Thus we have, by the chain rule, that

$$\left(\frac{\partial \ln \phi}{\partial n_i}\right)_{T,P,n_j} = \left(\frac{\partial \ln \phi}{\partial T_r}\right)_{P_r,\omega}\left(\frac{\partial T_r}{\partial T_c}\right)_{T,n_j}\left(\frac{\partial T_c}{\partial n_i}\right)_{n_j} + \left(\frac{\partial \ln \phi}{\partial P_r}\right)_{T_r,\omega}\left(\frac{\partial P_r}{\partial P_c}\right)_{P,n_j}\left(\frac{\partial P_c}{\partial n_i}\right)_{n_j} + \left(\frac{\partial \ln \phi}{\partial \omega}\right)_{T_r,P_r}\left(\frac{\partial \omega}{\partial n_i}\right)_{n_j} \tag{4-291}$$

But $T_r \equiv T/T_c$ and $P_r \equiv P/P_c$, and thus

$$\left(\frac{\partial T_r}{\partial n_i}\right)_{T,n_j} = -\frac{T}{T_c^2}\left(\frac{\partial T_c}{\partial n_i}\right)_{n_j} = -\frac{T_r}{T_c}\left(\frac{\partial T_c}{\partial n_i}\right)_{n_j} \tag{4-292}$$

$$\left(\frac{\partial P_r}{\partial n_i}\right)_{P,n_j} = -\frac{P}{P_c^2}\left(\frac{\partial P_c}{\partial n_i}\right)_{n_j} = -\frac{P_r}{P_c}\left(\frac{\partial n_c}{\partial n_i}\right)_{n_j} \tag{4-293}$$

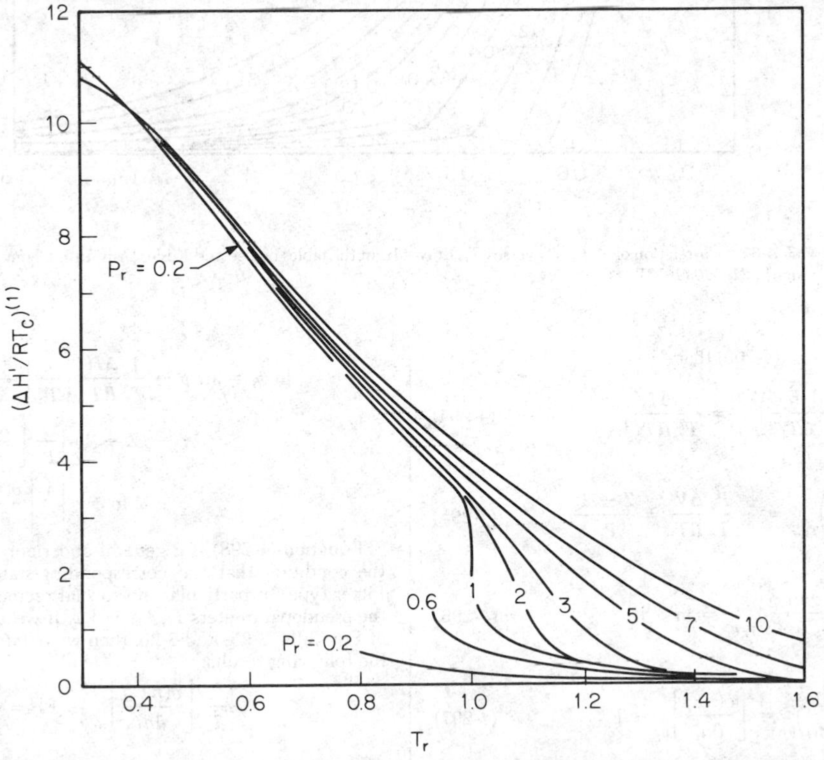

FIG. 4-31 Correlation of $(\Delta H'/RT_c)^{(1)}$, drawn from the tables of Lee and Kesler [Am. Inst. Chem. Eng. J., *21, 510 (1975)*].

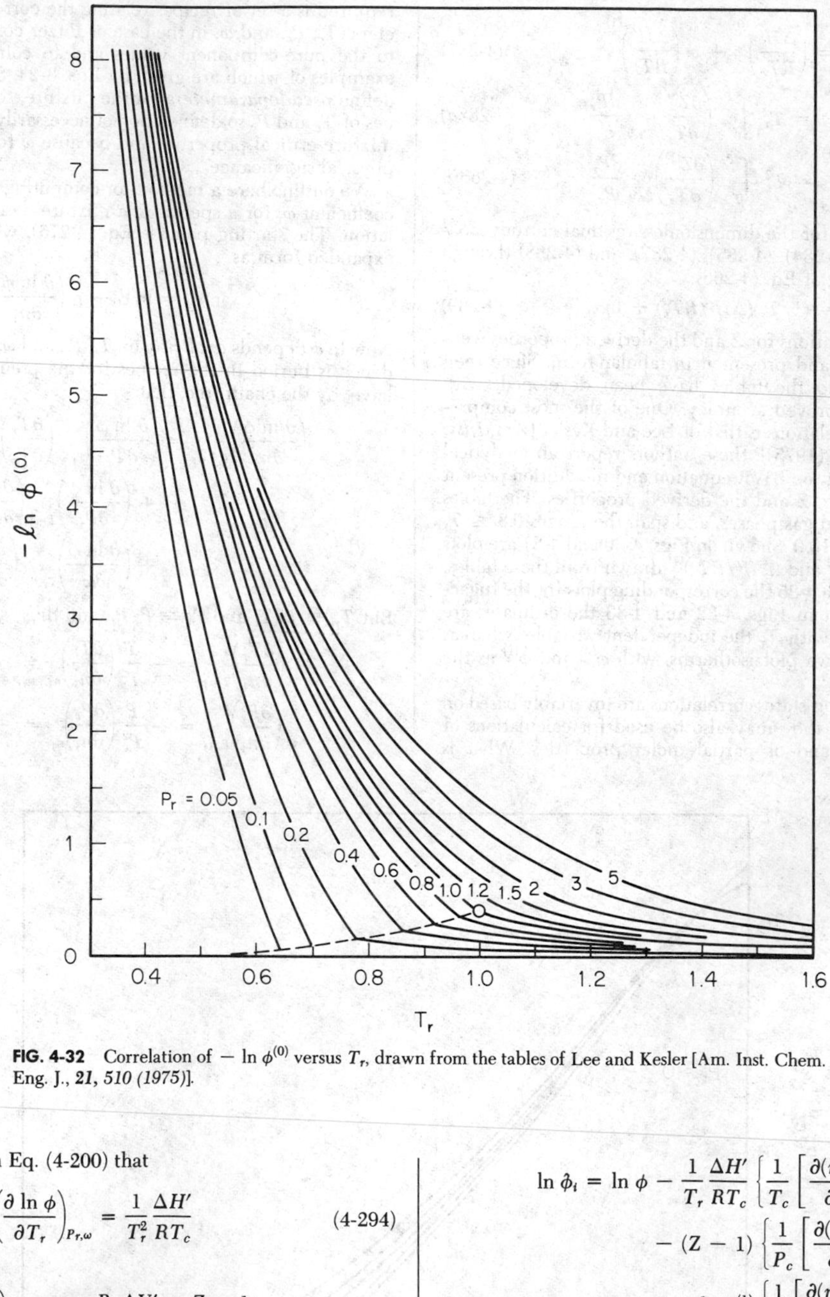

FIG. 4-32 Correlation of $-\ln \phi^{(0)}$ versus T_r, drawn from the tables of Lee and Kesler [Am. Inst. Chem. Eng. J., *21*, *510 (1975)*].

Moreover, we find from Eq. (4-200) that

$$\left(\frac{\partial \ln \phi}{\partial T_r}\right)_{P_r,\omega} = \frac{1}{T_r^2}\frac{\Delta H'}{RT_c} \tag{4-294}$$

and that

$$\left(\frac{\partial \ln \phi}{\partial P_r}\right)_{T_r,\omega} = -\frac{P_c}{T_c}\frac{\Delta V'}{RT_r} = \frac{Z-1}{P_r} \tag{4-295}$$

Also, by Eq. (4-284),

$$\left(\frac{\partial \ln \phi}{\partial \omega}\right)_{T_r,P_r} = \ln \phi^{(1)} \tag{4-296}$$

and, for any function F,

$$n\left(\frac{\partial F}{\partial n_i}\right)_{nj} = \left[\frac{\partial(nF)}{\partial n_i}\right]_{nj} - F \tag{4-297}$$

Letting $F = T_c$, P_c, and ω in Eq. (4-297) and combining the results with Eqs. (4-290) through (4-296), we obtain finally that

$$\begin{aligned}\ln \hat{\phi}_i = \ln \phi &- \frac{1}{T_r}\frac{\Delta H'}{RT_c}\left\{\frac{1}{T_c}\left[\frac{\partial(nT_c)}{\partial n_i}\right]_{nj} - 1\right\} \\ &- (Z-1)\left\{\frac{1}{P_c}\left[\frac{\partial(nP_c)}{\partial n_i}\right]_{nj} - 1\right\} \\ &+ \omega \ln \phi^{(1)}\left\{\frac{1}{\omega}\left[\frac{\partial(n\omega)}{\partial n_i}\right]_{nj} - 1\right\}\end{aligned} \tag{4-298}$$

Equation (4-298) is a general and rigorous result, subject only to the condition that the corresponding-states correlation be of the Pitzer type. In particular, no special recipes have been assumed for the pseudoparameters T_c, P_c, and ω. If we assume now the suitability of Eqs. 3-24, 3-25, and 3-26, then we obtain for the groups in braces the following results:

$$\left\{\frac{1}{T_c}\left[\frac{\partial(nT_c)}{\partial n_i}\right]_{nj} - 1\right\} = \frac{T_{ci} - T_c}{T_c} \tag{4-299a}$$

$$\left\{\frac{1}{P_c}\left[\frac{\partial(nP_c}{\partial n_i}\right]_{nj} - 1\right\} = \frac{T_{ci} - T_c}{T_c} + \frac{Z_{ci}}{\Sigma y_j Z_{cj}} - \frac{V_{ci}}{\Sigma y_j V_{cj}} \tag{4-299b}$$

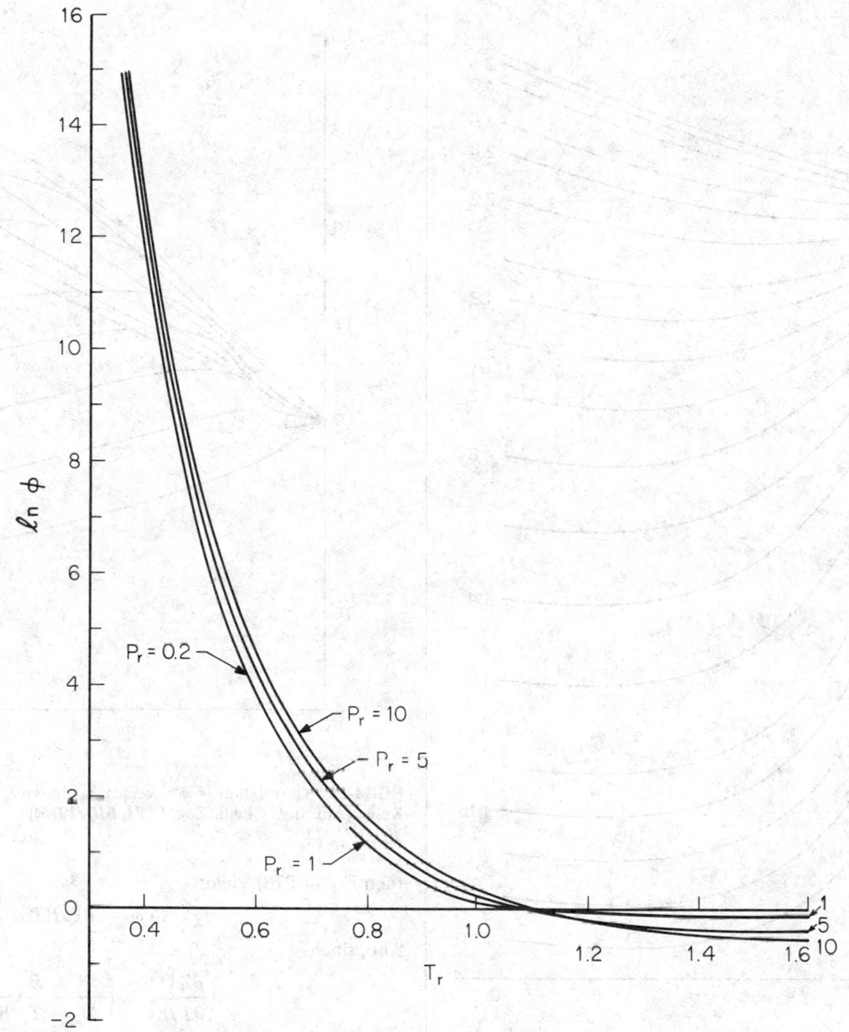

FIG. 4-33 Correlation of − ln $\phi^{(1)}$ versus T_r, drawn from the tables of Lee and Kesler [Am. Inst. Chem. Eng. J., *21*, 510 (1975)].

$$\left\{\frac{1}{\omega}\left[\frac{\partial(n\omega)}{\partial n_i}\right]_{n_j} - 1\right\} = \frac{\omega_i - \omega}{\omega} \qquad (4\text{-}299c)$$

Equations (4-298) and (4-299*a*, *b* and *c*), together with the pseudoparameter recipes and appropriate charts or graphs for Z, ln ϕ, and $\Delta H'/RT_c$, provide a basis for estimating component fugacities in mixtures from the Pitzer correlations. Note that all unsubscripted quantities in Eq. (4-298) are evaluated at T_r and P_r for the *mixture*, with the mixture T_c and P_c being given by the pseudocritical recipes.

Virial Equations of State The virial equation in *density* is described in Sec. 3. Formally, it is an infinite-series representation of the compressibility factor Z in powers of molar density ρ (or reciprocal molar volume V^{-1}) about the real-gas state at zero density (zero pressure):

$$Z = 1 + B\rho + C\rho^2 + D\rho^3 + \cdots \qquad (4\text{-}300)$$

The density-series virial coefficients B, C, D, . . . , depend on temperature and composition only. The composition dependencies are given by the exact recipes

$$B = \sum_i \sum_j y_i y_j B_{ij} \qquad (4\text{-}301)$$

$$C = \sum_i \sum_j \sum_k y_i y_j y_k C_{ijk} \qquad (4\text{-}302)$$

etc.

Although the virial equation itself is easily rationalized on empirical grounds, the "mixing rules" of Eqs. (4-301) and (4-302) follow rigorously from the methods of statistical mechanics.

An alternative expression to Eq. (4-300) may be proposed in which Z is expanded in powers of pressure about the real-gas state at zero pressure (zero density):

$$Z = 1 + B'P + C'P^2 + D'P^3 + \cdots \qquad (4\text{-}303)$$

Equation (4-303) is the virial equation in *pressure*, and B′, C′, D′, . . . , are the pressure-series virial coefficients; they, like the density-series coefficients, depend on T and y only. Moreover, the two sets of coefficients are related:

$$B' = B/RT \qquad (4\text{-}304a)$$

$$C' = (C - B^2)/(RT)^2 \qquad (4\text{-}304b)$$

etc.

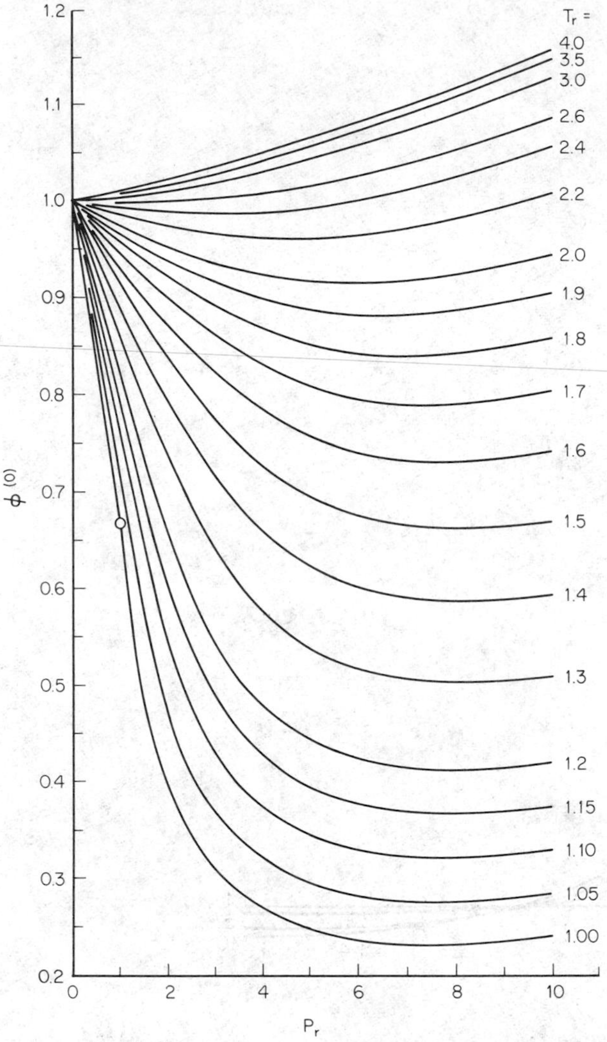

FIG. 4-34 Correlation of $\phi^{(0)}$ versus P_r, drawn from the tables of Lee and Kesler [Am. Inst. Chem. Eng. J., *21, 510 (1975)*].

Use of an *infinite* series for actual calculations is of course impossible, and so *truncations* of the virial equations are employed for numerical work. The degree of truncation is determined in part by the temperature and pressure for which calculations are to be done and also by the availability of the required virial coefficients. Data or correlations (see Sec. 3) are usually available for B, and sometimes for C [see, e.g., De Santis and Grande, *Am. Inst. Chem. Eng. J.*, **25**, 931 (1979)], but rarely for the higher coefficients, and thus one is normally restricted to truncations containing three terms or fewer. For work at low pressures, the two-term expression in pressure, with B' given by Eq. (4-304*a*), is usually preferred because of its simplicity:

$$Z = 1 + BP/RT \tag{4-305}$$

For more extreme conditions (but for densities not exceeding the critical value), the three-term truncation of Eq. (4-300) is used:

$$Z = 1 + B\rho + C\rho^2 \tag{4-306}$$

We may determine, corresponding to Eqs. (4-305) and (4-306), two sets of expressions for derived properties. We treat Eq. (4-305) first.

Equation (4-305) is explicit in *volume*, and thus we use Eqs. (4-273) through (4-275). Since

$$Z - 1 = BP/RT$$

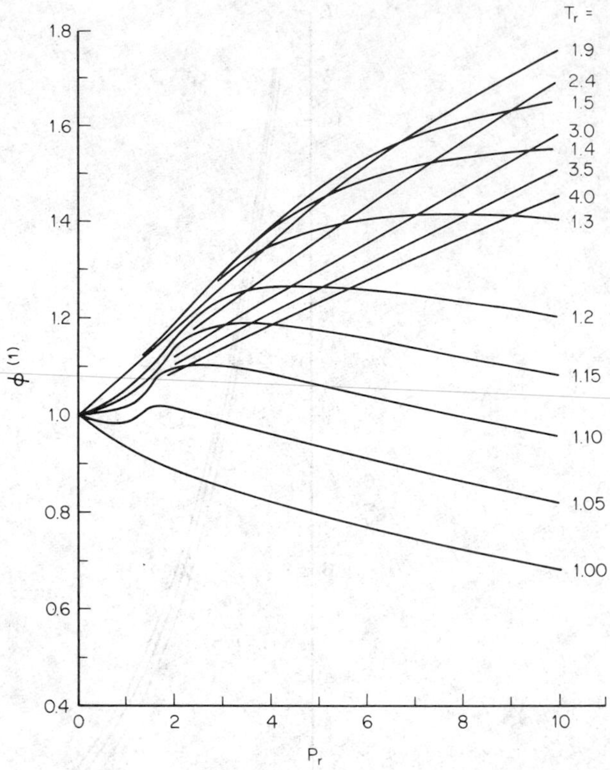

FIG. 4-35 Correlation of $\phi^{(1)}$ versus P_r, drawn from the tables of Lee and Kesler [Am. Inst. Chem. Eng. J., *21, 510 (1975)*].

then Eq. (4-273) yields

$$\ln \phi = BP/RT \tag{4-307}$$

Also, since

$$\left(\frac{\partial Z}{\partial T}\right)_{P,y} = \left(\frac{dB}{dT} - \frac{B}{T}\right)\frac{P}{RT}$$

then, by Eq. (4-274),

$$\Delta H' = \left(T\frac{dB}{dT} - B\right)P \tag{4-308}$$

The residual entropy is found by combination of Eqs. (4-307) and (4-308) via Eq. (4-266). We determine $\ln \hat{\phi}_i$ by application of Eq. (4-275) to Eq. (4-307), obtaining

$$\ln \hat{\phi}_i = \frac{P}{RT}\left[\frac{\partial(nB)}{\partial n_i}\right]_{T,n_j}$$

But, by Eq. (4-301),

$$nB = \frac{1}{n}\sum_k \sum_\ell n_k n_\ell B_{k\ell}$$

from which we find that

$$\left[\frac{\partial(nB)}{\partial n_i}\right]_{T,n_j} = 2\sum_k y_k B_{ki} - B \tag{4-309}$$

and thus

$$\ln \hat{\phi}_i = \left(2\sum_k y_k B_{ki} - B\right)\frac{P}{RT} \tag{4-310}$$

Equation (4-306) is explicit in *pressure*, and thus we use Eqs. (4-278) through (4-280). Since

$$Z - 1 = B\rho + C\rho^2$$

then Eq. (4-278*b*) yields

$$\ln \phi = 2B\rho + \frac{3}{2} C\rho^2 - \ln Z \tag{4-311}$$

Also, since

$$\left(\frac{\partial Z}{\partial T}\right)_{\rho,y} = \frac{dB}{dT}\rho + \frac{dC}{dT}\rho^2$$

then, by Eq. (4-279*b*),

$$\Delta H' = RT\left[\left(T\frac{dB}{dT} - B\right)\rho + \left(½T\frac{dC}{dT} - C\right)\rho^2\right] \tag{4-312}$$

The residual entropy is found by substitution of Eqs. (4-311) and (4-312) in Eq. (4-266). We determine $\ln \hat{\phi}_i$ by application of Eq. (4-281*b*). From Eq. (4-306) we find that

$$\left[\frac{\partial(nZ)}{\partial n_i}\right]_{T,\rho/n,nj} = 1 + \left\{B + \left[\frac{\partial(nB)}{\partial n_i}\right]_{T,nj}\right\}\rho + \left\{2C + \left[\frac{\partial(nC)}{\partial n_i}\right]_{T,nj}\right\}\rho^2$$

Substitution into Eq. (4-281*b*) gives, on integration,

$$\ln \hat{\phi}_i = \left\{B + \left[\frac{\partial(nB)}{\partial n_i}\right]_{T,nj}\right\}\rho + ½\left\{2C + \left[\frac{\partial(nC)}{\partial n_i}\right]_{T,nj}\right\}\rho^2 - \ln Z$$

Now the mole-number derivative of nB is given by Eq. (4-309); the corresponding derivative of nC is similarly found from Eq. (4-302) to be

$$\left[\frac{\partial(nC)}{\partial n_i}\right]_{T,nj} = 3\sum_k \sum_\ell y_k y_\ell C_{k\ell i} - 2C \tag{4-313}$$

Thus we obtain finally that

$$\ln \hat{\phi}_i = 2\rho \sum_k y_k B_{ki} + \tfrac{3}{2}\rho^2 \sum_k \sum_\ell y_k y_\ell C_{k\ell i} - \ln Z \tag{4-314}$$

In a process calculation, T and P, rather than T and ρ (or T and V) are usually the favored independent variables. Application of Eqs. (4-311), (4-312), and (4-314) therefore requires prior solution of Eq. (4-306) for Z or ρ. Since $Z \equiv P/\rho RT$, Eq. (4-306) may be written in the two equivalent forms

$$Z^3 - Z^2 - \left(\frac{BP}{RT}\right)Z - \frac{CP^2}{R^2T^2} = 0 \tag{4-315}$$

or

$$\rho^3 + \left(\frac{B}{C}\right)\rho^2 + \left(\frac{1}{C}\right)\rho - \frac{P}{CRT} = 0 \tag{4-316}$$

In the event that three real roots obtain for these equations, the largest Z (smallest ρ) is the value appropriate for the vapor phase. The smallest Z (largest ρ) is a liquid*like* root but has no quantitative significance, as the virial equations are suitable only for vapors and gases.

Redlich-Kwong Equation of State The Redlich-Kwong equation and its numerous variants are representative of the popular class of empirical cubic equations of state. The *original* Redlich-Kwong equation (the only one considered here) is given in Sec. 3 as

$$P = \frac{RT}{V - b} - \frac{a}{T^{1/2}V(V + b)} \tag{4-317}$$

where for a mixture parameters a and b are functions of composition:

$$a = (\Sigma y_i a_i^{1/2})^2 \tag{4-318}$$
$$b = \Sigma y_i b_i \tag{4-319}$$

Here the a_i and b_i are parameters for pure fluid i, constants for a particular substance but different for different substances.

Equation (4-317) is pressure-explicit and may be written in the alternative form

$$Z = \frac{V}{V - b} - \frac{a}{RT^{3/2}(V + b)} \tag{4-320}$$

Derived properties are therefore computed from Eqs. (4-278) through (4-280). From Eq. (4-278*a*) we find that

$$\ln \phi = Z - 1 - \ln\left(1 - \frac{b}{V}\right)Z - \frac{a}{bRT^{3/2}}\ln\left(1 + \frac{b}{V}\right) \tag{4-321}$$

and from Eq. (4-279*a*)

$$\Delta H' = \frac{3a}{2bT^{1/2}}\ln\left(1 + \frac{b}{V}\right) - RT(Z - 1) \tag{4-322}$$

The residual entropy is determined from Eqs. (4-321) and (4-322) with Eq. (4-266).

Computation of $\ln \hat{\phi}_i$ requires an expression for the mole-number derivative of nZ. This in turn requires expressions for the mole-number derivatives of the equation-of-state parameters a and b, found from the mixing rules of Eqs. (4-318) and (4-319) to be

$$\left[\frac{\partial(na)}{\partial n_i}\right]_{nj} = 2(aa_i)^{1/2} - a \tag{4-323}$$

and

$$\left[\frac{\partial(nb)}{\partial n_i}\right]_{nj} = b_i \tag{4-324}$$

Then we obtain from Eqs. (4-281*a*), (4-320), (4-323), and (4-324) that

$$\ln \hat{\phi}_i = \frac{b_i}{b}(Z - 1) - \ln\left(1 - \frac{b}{V}\right)Z + \frac{a}{bRT^{3/2}}\left[\frac{b_i}{b} - \frac{2(aa_i)^{1/2}}{a}\right]\ln\left(1 + \frac{b}{V}\right) \tag{4-325}$$

As with the three-term virial equation in density, use of the Redlich-Kwong equation for derived properties requires prior solution of the equation of state, Eq. (4-317) or Eq. (4-320), for Z or V.

Benedict-Webb-Rubin (BWR) Equation of State The BWR equation is given in Sec. 3. With Z as the dependent variable, it is

$$Z = 1 + \left(B_0 - \frac{A_0}{RT} - \frac{C_0}{RT^3}\right)\rho + \left(b - \frac{a}{RT}\right)\rho^2 + \frac{a\alpha}{RT}\rho^5 + \frac{c}{RT^3}\rho^2(1 + \gamma\rho^2)\exp(-\gamma\rho^2) \tag{4-326}$$

All eight parameters depend on composition; moreover, as noted in Sec. 3, parameters C_0, b, and γ are for some applications treated as functions of T. By Eq. (4-278*b*), the fugacity coefficient is given by

$$\ln \phi = 2\left(B_0 - \frac{A_0}{RT} - \frac{C_0}{RT^3}\right)\rho + \tfrac{3}{2}\left(b - \frac{a}{RT}\right)\rho^2 + \frac{6a\alpha}{5RT}\rho^5 + \frac{c}{2\gamma RT^3}[(2\gamma^2\rho^4 + \gamma\rho^2 - 2)\exp(-\gamma\rho^2) + 2] - \ln Z \tag{4-327}$$

Allowing for T dependence of C_0, b, and γ, we similarly find from Eq. (4-279*b*) that

$$\Delta H' = \left(-B_0RT + 2A_0 + \frac{4C_0}{T^2} - \frac{1}{T}\frac{dC_0}{dT}\right)\rho + ½\left(RT^2\frac{db}{dT} - 2bRT + 3a\right)\rho^2 - \frac{6a\alpha}{5}\rho^5 - \frac{c}{2\gamma T^2}[(2\gamma^2\rho^4 - \gamma\rho^2 - 6)\exp(-\gamma\rho^2) + 6] + \frac{c}{2\gamma^2 T}\frac{d\gamma}{dT}[(\gamma^2\rho^4 + 2\gamma\rho^2 + 2)\exp(-\gamma\rho^2) - 2] \tag{4-328}$$

The residual entropy is determined by Eqs. (4-266), (4-327), and (4-328).

Computation of $\ln \hat{\phi}_i$ is done via Eq. (4-281*b*). The result is

$$\ln \hat{\phi}_i \left\{ [B_0 + (\bar{B}_0)_i] - \frac{[A_0 + (\bar{A}_0)_i]}{RT} - \frac{[C_0 + (\bar{C}_0)_i]}{RT^3} \right\} \rho + \tfrac{1}{2}\left[(2b + \bar{b}_i) - \frac{(2a + \bar{a}_i)}{RT^2}\right]\rho^2 + \left(\frac{4a\alpha + a\bar{\alpha}_i + \alpha\bar{a}_i}{5RT}\right)\rho^5 + \frac{c}{2\gamma RT^3}\left\{\left[\left(1 + \frac{\bar{\gamma}_i}{\gamma}\right)\gamma^2\rho^4 + \left(2 - \frac{\bar{\gamma}_i}{\gamma} - \frac{\bar{c}_i}{c}\right)\gamma\rho^2 - 2\left(1 + \frac{\bar{c}_i}{c} - \frac{\bar{\gamma}_i}{\gamma}\right)\right]\exp(-\gamma\rho^2) + 2\left(1 + \frac{\bar{c}_i}{c} - \frac{\bar{\gamma}_i}{\gamma}\right)\right\} - \ln Z \qquad (4\text{-}329)$$

Here the quantities with overbars are "partial parameters" for species i, defined for arbitrary parameter π by

$$\bar{\pi}_i \equiv \left[\frac{\partial(n\pi)}{\partial n_i}\right]_{T,n_j} \qquad (4\text{-}330)$$

With mixing rules given as in Sec. 3, viz.,

$$\pi = \left(\sum_k y_k \pi_k^{1/r}\right)^r \qquad (4\text{-}331)$$

where r is a small integer (r = 1, 2, or 3), the recipe for $\bar{\pi}_i$ is

$$\bar{\pi}_i = \pi\left[r\left(\frac{\pi_i}{\pi}\right)^{1/r} - (r - 1)\right] \qquad (4\text{-}332)$$

For example, we have from Sec. 3 that $r = 3$ for $\pi \equiv c$; then

$$\bar{c}_i = c\left[3\left(\frac{c_i}{c}\right)^{1/3} - 2\right]$$

where c_i is the parameter for pure i and c is the parameter for the mixture, given by

$$c = \left(\sum_k y_k c_k^{1/3}\right)^3$$

Expressions for the Excess Gibbs Function In principle, equation-of-state procedures can be used for the calculation of liquid-phase as well as gas-phase properties, and much has been done in the development of *PVT* equations of state and corresponding-states correlations suitable for both phases. The available procedures are highly specific, however, finding major application to systems comprising nonpolar, nonassociating substances of relatively low molecular weight (i.e., cryogenic fluids and light hydrocarbons). For liquid mixtures exhibiting appreciable departure from ideal-solution behavior (e.g., most systems of the type shown in Figs. 4-27 through 4-29), the equation-of-state approach remains inappropriate, and instead use is made of the *excess functions,* described earlier.

The excess function of major importance for engineering calculations is the excess Gibbs function G^E, because its canonical variables are T, P, and x, the variables usually specified or sought in a design calculation. Knowing G^E as a function of T, P, and x, one can in principle compute from it not only activity coefficients (quantities required for equilibrium calculations) but also all other excess functions. Thus,

$$H^E = -RT^2\left[\frac{\partial(G^E/RT)}{\partial T}\right]_{P,x}$$

$$S^E = -\left(\frac{\partial G^E}{\partial T}\right)_{P,x}$$

$$V^E = \left(\frac{\partial G^E}{\partial P}\right)_{T,x}$$

and

$$\ln \gamma_i = \left[\frac{\partial(nG^E/RT)}{\partial n_i}\right]_{T,P,n_j}$$

In fact, as already noted in connection with Fig. 4-26, the excess volume for liquid mixtures is normally small, and the pressure dependence of G^E for liquids is often safely ignored. Thus engineering efforts at describing G^E center on representing its composition and temperature dependence.

Scores of expressions have been proposed for G^E, but only a handful find general use in the chemical industries. A major consideration is the number of adjustable parameters required per binary; since these parameters must be estimated from experiment or from the correlated results of experimental measurements, convenience dictates that the number be as small as possible. Practically, the optimum number of parameters is *two,* although precise representation of good solution data often requires three or four or more.

We consider here what seem to be the four most popular two-parameter expressions for G^E: the Margules equation, the van Laar equation, the Wilson equation, and the UNIQUAC equation. For conciseness, we represent the dimensionless excess Gibbs function G^E/RT by the symbol g:

$$g \equiv G^E/RT$$

The classical approach to the development of empirical binary expressions for G^E is to expand the function g/x_1x_2 as a power series in x_1, x_2, or some linear combination of the two. If only the first two terms are retained, there results on rearrangement the two-parameter Margules equation:

$$g \equiv G^E/RT = (A_{21}x_1 + A_{12}x_2)x_1x_2 \qquad (4\text{-}333)$$

Since $\ln \gamma_i$ is a partial molar property with respect to g, the corresponding expressions for the activity coefficients are most readily determined from Eqs. (4-247*a* and *b*), written for $M \equiv g$ as

$$\ln \gamma_1 = g + x_2\,(dg/dx_1) \qquad (4\text{-}334a)$$

$$\ln \gamma_2 = g - x_1\,(dg/dx_1) \qquad (4\text{-}334b)$$

Thus we find from Eq. (4-333) that

$$\ln \gamma_1 = x_2^2[A_{12} + 2(A_{21} - A_{12})x_1] \qquad (4\text{-}335a)$$

$$\ln \gamma_2 = x_1^2[A_{21} + 2(A_{12} - A_{21})x_2] \qquad (4\text{-}335b)$$

and that

$$\ln \gamma_1^\infty = A_{12}$$

$$\ln \gamma_2^\infty = A_{21}$$

where the γ_i^∞ are activity coefficients at infinite dilution, defined as

$$\gamma_i^\infty \equiv \lim_{x_i \to 0} \gamma_i$$

As an alternative to expanding g/x_1x_2 in powers of mole fraction, one may similarly develop a series representation of the reciprocal function x_1x_2/g. Retaining only the first two terms, we obtain on rearrangement the van Laar equation:

$$g = \frac{A'_{12}A'_{21}x_1x_2}{A'_{12}x_1 + A'_{21}x_2} \qquad (4\text{-}336)$$

The activity coefficients are given by

$$\ln \gamma_1 = \frac{A'_{12}A'^2_{21}x_2^2}{(A'_{12}x_1 + A'_{21}x_2)^2} \qquad (4\text{-}337a)$$

$$\ln \gamma_2 = \frac{A'_{21}A'^2_{12}x_1^2}{(A'_{12}x_1 + A'_{21}x_2)^2} \qquad (4\text{-}337b)$$

and thus

$$\ln \gamma_1^\infty = A'_{12}$$

$$\ln \gamma_2^\infty = A'_{21}$$

The Wilson equation [Wilson, *J. Am. Chem. Soc.,* **86,** 127 (1964)] was the first of the popular group of modern expressions for g called *local-composition equations.* Unlike the classical expressions, it is logarithmic in form:

$$g = -x_1 \ln(x_1 + \Lambda_{12}x_2) - x_2 \ln(x_2 + \Lambda_{21}x_1) \qquad (4\text{-}338)$$

The corresponding expressions for the $\ln \gamma_i$ are

$$\ln \gamma_1 = -\ln (x_1 + \Lambda_{12}x_2) + x_2\left(\frac{\Lambda_{12}}{x_1 + \Lambda_{12}x_2} - \frac{\Lambda_{21}}{x_2 + \Lambda_{21}x_1}\right) \qquad (4\text{-}339a)$$

$$\ln \gamma_2 = -\ln (x_2 + \Lambda_{21}x_1) + x_1\left(\frac{\Lambda_{21}}{x_2 + \Lambda_{21}x_1} - \frac{\Lambda_{12}}{x_1 + \Lambda_{12}x_2}\right) \qquad (4\text{-}339b)$$

whence

$$\ln \gamma_1^\infty = -\ln \Lambda_{12} + 1 - \Lambda_{21}$$
$$\ln \gamma_2^\infty = -\ln \Lambda_{21} + 1 - \Lambda_{12}$$

Finally, we cite the UNIQUAC equation [Abrams and Prausnitz, *Am. Inst. Chem. Eng. J.*, **21**, 116 (1975); Maurer and Prausnitz, *Fluid Phase Equilibria*, **2**, 91 (1978)], one of the more recent local-composition expressions. Here, g comprises two parts, a *combinatorial* term g(comb.) and a *residual* term g(resid.):

$$g = g(\text{comb.}) + g(\text{resid.}) \qquad (4\text{-}340)$$

Function g(comb.) contains pure-component parameters only, while g(resid.) incorporates two adjustable parameters to characterize each binary system:

$$g(\text{comb.}) = x_1 \ln (\Phi_1/x_1) + x_2 \ln (\Phi_2/x_2) + \tfrac{1}{2}z[q_1x_1 \ln (\theta_1/\Phi_1) + q_2x_2 \ln (\theta_2/\Phi_2)] \qquad (4\text{-}341)$$

and

$$g(\text{resid.}) = -x_1q_1 \ln (\theta_1 + \Gamma_{12}\theta_2) - x_1q_2 \ln (\theta_2 + \Gamma_{21}\theta_1) \qquad (4\text{-}342)$$

Here, Φ_i and θ_i are *segment* and *surface* fractions, defined as

$$\Phi_i \equiv x_ir_i/(x_1r_1 + x_2r_2) \qquad (4\text{-}343)$$

and

$$\theta_i \equiv x_iq_i/(x_1q_1 + x_2q_2) \qquad (4\text{-}344)$$

where r_i is a pure-component volume parameter and q_i a pure-component area parameter. Quantity z is a coordination number (taken as a constant for all mixtures, usually 10), and the Γ_{ij} are the adjustable parameters peculiar to each binary system. Expressions for the UNIQUAC activity coefficients are given by

$$\ln \gamma_1 = \ln (\Phi_1/x_1) + \tfrac{1}{2}zq_1 \ln (\theta_1/\Phi_1) + \Phi_2\left(\ell_1 - \frac{r_1}{r_2}\ell_2\right) - q_1 \ln (\theta_1 + \Gamma_{12}\theta_2) + \theta_2q_1\left(\frac{\Gamma_{12}}{\theta_1 + \Gamma_{12}\theta_2} - \frac{\Gamma_{21}}{\theta_2 + \Gamma_{21}\theta_1}\right)$$

and (4-345a)

$$\ln \gamma_2 = \ln (\Phi_2/x_2) + \tfrac{1}{2}zq_2 \ln (\theta_2/\Phi_2) + \Phi_1\left(\ell_2 - \frac{r_2}{r_1}\ell_1\right) - q_2 \ln (\theta_2 + \Gamma_{21}\theta_1) + \theta_1q_2\left(\frac{\Gamma_{21}}{\theta_2 + \Gamma_{21}\theta_1} - \frac{\Gamma_{12}}{\theta_1 + \Gamma_{12}\theta_2}\right) \qquad (4\text{-}345b)$$

where $\ell_1 \equiv \tfrac{1}{2}z(r_1 - q_1) - (r_1 - 1)$
and $\ell_2 \equiv \tfrac{1}{2}z(r_2 - q_2) - (r_2 - 1)$ (4-346)

Finally, we find from Eqs. (4-345*a* and *b*) that

$$\ln \gamma_1^\infty = \ln (r_1/r_2) + \tfrac{1}{2}zq_1 \ln \left(\frac{q_1r_2}{q_2r_1}\right) + \left(\ell_1 - \frac{r_1}{r_2}\ell_2\right) - q_1 \ln \Gamma_{12} + q_1(1 - \Gamma_{12}) \qquad (4\text{-}347a)$$

$$\ln \gamma_2^\infty = \ln (r_2/r_1) + \tfrac{1}{2}zq_2 \ln \left(\frac{q_2r_1}{q_1r_2}\right) + \left(\ell_2 - \frac{r_2}{r_1}\ell_1\right) - q_2 \ln \Gamma_{21} + q_2(1 - \Gamma_{21}) \qquad (4\text{-}347b)$$

The UNIQUAC equation is the most flexible of the four expressions in its ability to represent activity coefficients for binary systems. It is also the most complex, but this is of minor consequence for machine calculations. Of the remaining three expressions, the Margules equation is simplest, and comparative studies have shown it and the van Laar equation to be roughly equivalent for the correlation of vapor-liquid equilibriums for a wide variety of systems. However, the van Laar equation cannot represent "mixed" deviations from ideality (where G^E changes sign), nor can it reproduce interior extrema in the $\ln \gamma_i$ versus x_1 curves. On the other hand, the van Laar equation is superior to the Margules equation for asymmetric systems (those with significantly different values of γ_1^∞ and γ_2^∞) showing large positive deviations from ideality. The Wilson equation is best for systems of this type, but it cannot predict liquid-liquid phase splitting.

All four expressions for G^E can in principle be extended to multicomponent mixtures without the introduction of additional correlating parameters. However, the multicomponent versions of the Margules and van Laar equations are empirical and arbitrary [for examples, see Chien and Null, *Am. Inst. Chem. Eng. J.*, **18**, 1177 (1972)], and the quality of predictions for multicomponent systems can be substantially inferior to the goodness of fit obtained for the constituent binaries. The multicomponent extensions of the local-composition equations have a more rational basis and have been demonstrated to yield acceptable estimates of ternary VLE with parameters derived from binary VLE data only. For the Wilson equation, the extensions of Eqs. (4-338) and (4-339*a* and *b*) are

$$g = -\sum_i x_i \ln \left(\sum_j x_j\Lambda_{ij}\right) \qquad (4\text{-}348)$$

and

$$\ln \gamma_i = 1 - \ln \left(\sum_j x_j\Lambda_{ij}\right) - \sum_k \left(\frac{x_k\Lambda_{ki}}{\sum_j x_j\Lambda_{kj}}\right) \qquad (4\text{-}349)$$

where in both equations all summations extend over all species, with $\Lambda_{ij} = 1$ for $i = j$. For the UNIQUAC equation, we have

$$g(\text{comb.}) = \sum_i x_i \ln (\Phi_i/x_i) + \tfrac{1}{2}z \sum_i q_ix_i \ln (\theta_i/\Phi_i) \qquad (4\text{-}350)$$

and

$$g(\text{resid.}) = -\sum_i x_iq_i \ln \left(\sum_j x_j\Gamma_{ij}\right) \qquad (4\text{-}351)$$

whence

$$\ln \gamma_i = \ln (\Phi_i/x_i) + \tfrac{1}{2}zq_i \ln (\theta_i/\Phi_i) + \ell_i - \frac{\Phi_i}{x_i}\sum_j x_j\ell_j - q_i \ln \left(\sum_j \theta_j\Gamma_{ij}\right) + q_i - q_i \sum_k \left(\frac{\theta_k\Gamma_{ki}}{\sum_j x_j\Lambda_{kj}}\right) \qquad (4\text{-}352)$$

with

$$\Phi_i \equiv \frac{x_ir_i}{\sum_j x_jr_j} \qquad (4\text{-}353)$$

$$\theta_i \equiv \frac{x_iq_i}{\sum_j x_jq_j} \qquad (4\text{-}354)$$

and

$$\ell_j \equiv \tfrac{1}{2}x(r_j - q_j) - (r_j - 1) \qquad (4\text{-}355)$$

Again, all summations extend over all species, and $\Gamma_{ij} = 1$ for $i = j$.

No *generally* reliable recipes exist for the temperature dependence of the parameters in any of the expressions considered. The Wilson parameters Λ_{ij} and UNIQUAC parameters Γ_{ij} inherit a Boltzmann-type T dependence from the origins of the expressions for G^E, but computations of properties and phenomena sensitive to this dependence (e.g., heats of mixing and liquid-liquid solubility diagrams) generally yield results at best only in qualitative agreement with experiment. One must conclude that while two-*parameter* expressions may be able to correlate more or less satisfactorily the T and x dependence of G^E, no two-*constant* expressions have yet been proposed which do the job.

EQUILIBRIUM

Criteria The equations developed so far have been intended for application to PVT systems in equilibrium states. The criteria for thermal and mechanical equilibrium are well known and need not be discussed in detail here. They require uniformity of temperature and pressure throughout the system. The criteria for phase and chemical equilibriums are less obvious and are developed as follows.

Consider any closed PVT system, either homogeneous or heterogeneous, for which P and T are uniform throughout and which is in thermal and mechanical equilibrium with its surroundings. We assume, however, that the system is not initially at equilibrium with respect to mass transfer or with respect to chemical reaction. All changes in the system are irreversible and must necessarily bring the system closer to an equilibrium state. The first and second laws written for the entire system are

$$dU_{\text{system}} = dQ - dW$$

and

$$dS_{\text{system}} \geqq dQ/T$$

Combination of the two laws gives

$$dU_{\text{system}} + dW - T\,dS_{\text{system}} \leqq 0$$

Since mechanical equilibrium has been assumed,

$$dW = P\,dV_{\text{system}}$$

and

$$dU_{\text{system}} + P\,dV_{\text{system}} - T\,dS_{\text{system}} \leqq 0$$

The inequality applies to all incremental changes toward the equilibrium state, whereas the equality holds at the equilibrium state where any change is reversible.

Various constraints may be put on this expression to produce alternative criteria for the directions of irreversible processes and for the condition of equilibrium. For example, it follows immediately that

$$(dU_{\text{system}})_{S_{\text{system}},V_{\text{system}}} \leqq 0$$

Alternatively, other pairs of properties may be held constant. The most useful result comes from fixing T and P. If this is done, then

$$d(U_{\text{system}} + PV_{\text{system}} - TS_{\text{system}})_{T,P} \leqq 0$$

or

$$(dG_{\text{system}})_{T,P} \leqq 0$$

This expression shows that all irreversible processes occurring at constant T and P proceed in such a direction as to cause the Gibbs function of the system to decrease. Thus the equilibrium state of a closed system is that state having the minimum total Gibbs function attainable at the given T and P. At the equilibrium state differential variations may occur in the system at constant T and P without producing a change in G. This is the meaning of the equilibrium criterion

$$(dG_{\text{system}})_{T,P} = 0 \tag{4-356}$$

Consider now the application of this criterion to two phases in equilibrium in a closed system. Each phase taken separately is an *open* system, capable of exchanging mass with the other, and Eq. (4-125) may be written for each phase:

$$d(nG)' = -(nS)'\,dT + (nV)'\,dP + \Sigma(\mu_i'\,dn_i')$$

$$d(nG)'' = -(nS)''\,dT + (nV)''\,dP + \Sigma(\mu_i''\,dn_i'')$$

The primes and double primes denote the two phases. However, it has been assumed that T and P are uniform throughout the system. These two equations may be added to give dG_{system}, and if at the same time the equilibrium condition of Eq. (4-366) is imposed, the result is

$$(dG_{\text{system}})_{T,P} = \Sigma(\mu_i'\,dn_i') + \Sigma(\mu_i''\,dn_i'') = 0$$

However, the system is closed and without chemical reaction, and material balances require that

$$dn_i'' = -\,dn_i'$$

Therefore

$$\Sigma(\mu_i' - \mu_i'')\,dn_i' = 0$$

But the quantities dn_i' are independent and arbitrary, and it follows that

$$\mu_i' = \mu_i''$$

is a condition of phase equilibrium. This result is readily generalized to multiple phases by successive consideration of the phases by pairs. The general result is

$$\mu_i' = \mu_i'' = \mu_i''' = \cdots \tag{4-357}$$

By application of Eq. (4-189), Eq. (4-357) is readily transformed into the alternative criterion of phase equilibrium:

$$\hat{f}_i' = \hat{f}_i'' = \hat{f}_i''' = \cdots \tag{4-358}$$

These relations are fundamental to the prediction of multicomponent phase equilibriums.

For the case of equilibrium with respect to chemical reaction within a single-phase closed system, combination of Eqs. (4-125) and (4-356) leads immediately to

$$\Sigma(\mu_i\,dn_i) = 0 \tag{4-359}$$

For combined phase and chemical equilibrium the conditions of Eqs. (4-357) and (4-359) are superimposed.

The Phase Rule The *intensive* state of a PVT system is established when its temperature and pressure and the compositions of all phases are fixed. However, for equilibrium states these variables are not all independent, and fixing a limited number of them automatically establishes the others. This number of independent variables is given by the phase rule and is called the *number of degrees of freedom* of the system. It is this number of variables which may be arbitrarily specified and which must be so specified in order to fix the *intensive* state of a system at equilibrium. This number is the difference between the number of variables needed to characterize the system and the number of equations that may be written connecting these variables.

The following nomenclature is employed:

π = the number of phases
m = the number of chemical species
r = the number of independent chemical reactions
F = the number of degrees of freedom

The phase-rule variables are temperature, pressure, and $m - 1$ mole fractions in each phase. The number of these variables is $2 + (m - 1)\pi$. The masses of the phases are not phase-rule variables because they have nothing to do with the intensive state of the system. It should also be noted that the temperature and pressure have been assumed to be uniform throughout the system for an equilibrium state.

The equations that may be written connecting the phase-rule variables are:

1. For each species, $\mu_i' = \mu_i'' = \mu_i''' = \cdots = \mu_i^{\pi}$, giving $(\pi - 1)m$ phase-equilibrium equations.
2. For each independent chemical reaction $\Sigma_i \mu_i\,dn_i = 0$, giving r equations.

The total number of independent equations is therefore $(\pi - 1)m + r$. These equations in their fundamental forms relate chemical potentials, but these are functions of temperature, pressure, and compositions; therefore they represent relations connecting the phase-rule variables. Since F is the difference between the number of variables and the number of equations,

$$F = 2 + (m - 1)\pi - (\pi - 1)m - r$$

or

$$F = 2 - \pi + m - r \tag{4-360}$$

The number of independent chemical reactions can be determined systematically as follows:

1. Write *formation* reactions from the elements for each chemical compound present in the system.
2. Combine these reaction equations so as to eliminate from the set all elements not considered to be present as elements in the system. A systematic procedure is to select one equation and combine it with each of the other equations of the set so as to eliminate a particular element. This will usually reduce the set by one equation for each element eliminated, though it is possible that two or more elements may be simultaneously eliminated.

The set of r equations resulting is a complete set of independent reactions. More than one such set is often possible, but all sets number r and are equivalent.

Example 4

a. A system of two miscible nonreacting constituents in vapor-liquid equilibrium:

$$F = 2 - \pi + m - r = 2 - 2 + 2 - 0 = 2$$

The two degrees of freedom for this system may be satisfied by setting T and P, or T and y_1, or P and x_1, or x_1 and y_1, etc., at fixed values. Thus if one specifies that equilibrium be established at a particular T and P, then this state (if possible at all) can exist only at one liquid and one vapor composition. Once the two degrees of freedom have been used up, one is not free to specify anything else that would restrict the phase-rule variables. For example, one cannot *in addition* require that the system form an azeotrope (assuming this to be possible), for this requires $x_1 = y_1$, an equation not taken into account in the derivation of the phase rule. Thus the requirement that the system form an azeotrope imposes a special condition and reduces the number of degrees of freedom to one.

b. Gaseous system consisting of CO, CO_2, H_2, H_2O, and CH_4 in chemical equilibrium:

$$F = 2 - \pi + m - r = 2 - 1 + 5 - 2 = 4$$

The value of $r = 2$ is found by the method outlined. The formation reactions are

$$C + \tfrac{1}{2}O_2 \rightarrow CO$$
$$C + O_2 \rightarrow CO_2$$
$$H_2 + \tfrac{1}{2}O_2 \rightarrow H_2O$$
$$C + 2H_2 \rightarrow CH_4$$

Systematic elimination of C and O_2 from this set of chemical equations reduces the set to two. There are three possible pairs of equations that may result, depending on how the combination of equations is effected. Any *pair* of the following three equations represents a complete set of independent reactions, and all pairs are equivalent.

$$CH_4 + H_2O \rightleftharpoons CO + 3H_2$$
$$CO + H_2O \rightleftharpoons CO_2 + H_2$$
$$CH_4 + 2H_2O \rightleftharpoons CO_2 + 4H_2$$

The result, $F = 4$, means that one is free to specify, for example, T, P, and two mole fractions in an equilibrium mixture of these five chemical species, provided nothing else is arbitrarily set. Thus it is not possible simultaneously to require that the system be prepared from specified amounts of particular constituents.

Since the phase rule treats only the intensive state of a system, it applies to both closed and open systems. **Duhem's theorem,** on the other hand, is a rule relating to closed systems only: "For any closed system formed initially from given masses of prescribed chemical species, the equilibrium state is *completely determined* by any two properties of the system, provided only that these two properties are independently variable at the equilibrium state." The meaning of *completely determined* is that both the intensive and extensive states of the system are fixed: not only are T, P, and the phase compositions established, but so also are the masses of the phases.

Vapor-Liquid Equilibrium

General Vapor-liquid equilibrium relationships (as well as other interphase equilibrium relationships) are needed in the solution of many engineering problems. The required data for these relationships can be, and often are, directly measured by experiment. Such measurements are seldom easy, even for binary systems, and they become rapidly more difficult as the number of components increases. This is the incentive for the application of thermodynamics to the calculation of phase-equilibrium relationships. However, thermodynamics provides no more than a set of equations which interrelate the properties of materials, and without data of one kind or another thermodynamics is useless in the solution of practical problems. The level at which thermodynamics can profitably enter into the solution of a particular vapor-liquid-equilibrium problem depends on the nature of the system considered and on the kind of data available, whether from experiment or by prediction.

The general vapor-liquid-equilibrium problem involves a multicomponent system of m components for which the independent variables are T and P, $m - 1$ liquid mole fractions, and $m - 1$ vapor mole fractions. (Note that $\Sigma x_i = 1$ and $\Sigma y_i = 1$, where x_i and y_i represent liquid and vapor mole fractions respectively.) Thus there are $2m$ independent variables, and application of the phase rule shows that exactly m of these variables must be fixed in order to establish the intensive state of the system. This means that once m variables have been specified, the remaining m variables can be determined by simultaneous solution of the m equilibrium relations,

$$\hat{f}_i^L = \hat{f}_i^V \qquad (i = 1 \text{ to } m) \tag{4-361}$$

where the superscripts L and V denote the liquid and vapor phases, respectively.

In practice, one usually specifies either T or P *and* either the liquid-phase or the vapor-phase composition, thus fixing $1 + (m - 1) = m$ independent variables. The remaining m variables are then subject to calculation, provided that sufficient data are available to allow determination of all necessary thermodynamic properties.

Equation-of-State Approach The defining equation for the fugacity coefficient is written for each phase:

$$\text{Liquid:} \qquad \hat{f}_i^L = \hat{\phi}_i^L x_i P$$
$$\text{Vapor:} \qquad \hat{f}_i^V = \hat{\phi}_i^V y_i P$$

Equation (4-361) now becomes

$$x_i \hat{\phi}_i^L = y_i \hat{\phi}_i^V \qquad (i = 1 \text{ to } m) \tag{4-362}$$

This step introduces the compositions x_i and y_i into the equilibrium equations, but neither is explicit, because the $\hat{\phi}_i$'s are functions of composition. In addition, the $\hat{\phi}_i$'s are functions of T and P. Thus Eq. (4-362) represents m complex relationships connecting T, P, and the x_i's and y_i's. Even for computer solution of these equations it is essential that the $\hat{\phi}_i$'s be expressed analytically as functions of T, P, and composition. This requires an equation of state which represents the volumetric properties of both the liquid and the vapor phases throughout the range of temperatures, pressures, and compositions of interest. Given such an equation of state, Eq. (4-281) provides for the expression of the $\hat{\phi}_i$'s as functions of T, P, and composition.

Cubic equations of state (e.g., modifications of the Redlich-Kwong equation), as well as more comprehensive expressions (e.g., the BWR equation), are frequently used for this purpose, and, perhaps surprisingly, the simpler equations often perform as well as the complex ones. The appropriateness of the technique seems to be strongly influenced by the nature of the mixing rules, and the equation-of-state approach has therefore up to now proved generally satisfactory only for systems (e.g., those containing hydrocarbons and cryogenic fluids) exhibiting modest and well-behaved deviations from ideal-solution behavior in the liquid phase. Numerical examples of the procedures are given by Van Ness and Abbott (*Classical Thermodynamics of Nonelectrolyte Solutions: With Applications to Phase Equilibria*, McGraw-Hill, New York, 1982, Sec. 6–4).

Use of Activity Coefficients The primary difficulty in the method outlined in the preceding paragraphs is to describe adequately the volumetric behavior of liquid mixtures. For many vapor-liquid-equilibrium problems the pressure is low enough so that a relatively simple equation of state, such as the virial equation, may be used to provide the thermodynamic properties of the vapor phase. The liquid phase must be treated differently; moreover, it is not necessary to require a complete thermodynamic description of the liquid phase; only the excess Gibbs function is needed. Since the partial molar excess Gibbs function is directly related to the activity coefficient, the connection with fugacity is made through Eq. (4-235), which defines the activity coefficient. Thus

$$\hat{f}_i^L = \gamma_i x_i f_i^\circ$$

This equation is written for the liquid phase; for the vapor phase the same equation is used as before:

$$\hat{f}_i^V = \hat{\phi}_i y_i P$$

As a result of Eq. (4-361),

$$\gamma_i x_i f_i^\circ = \hat{\phi}_i y_i P \qquad (i = 1 \text{ to } m) \tag{4-363}$$

There is no need for the identifying superscripts L and V in this equation provided that it is kept in mind that γ_i and f_i° characterize the liquid phase and $\hat{\phi}_i$ the vapor phase.

For evaluation of $\hat{\phi}_i$ one relies as before on an equation of state. Examples of expressions for $\hat{\phi}_i$ which result are given by Eqs. (4-310), (4-314), and (4-325). The activity coefficient γ_i and the standard-state fugacity f_i° of a component in the liquid phase go together. The purpose of the activity coefficient is to relate the unknown fugacity $\hat{f}_i$ to its standard-state value f_i°. Clearly, the activity coefficient is useless unless the standard-state fugacity can be accurately evaluated. Although the standard state must always be at the temperature of the system, it is otherwise arbitrary. The common standard states are those discussed earlier, one based on the Lewis-Randall rule and the other on Henry's law, and no others will be considered here. In both cases f_i° is the fugacity of pure i at the temperature and pressure of the system, and in either case there are two equations which relate the activity coefficient to the excess Gibbs function. The first is Eq. (4-238),

$$G^E/RT = \Sigma(x_i \ln \gamma_i)$$

and the second follows from Eq. (4-246) because $\ln \gamma_i$ stands in relation to G^E/RT as a partial molar property.

$$\ln \gamma_i = \frac{G^E}{RT} - \sum_{k \neq i} x_k \left[\frac{\partial(G^E/RT)}{\partial x_k}\right]_{T,P,x_r} \tag{4-364}$$

where $r \neq i$ or k. Thus, for a given T, P, and composition, G^E can be determined from the activity coefficients, and conversely the activity coefficients can be determined if G^E is known as a function of composition. Thus what is required for the liquid phase is a relation between G^E and composition.

Equations in common use for this purpose have already been described. These presume that the standard states are states of the pure-liquid constituents at the T and P of the system and thus that $f_i^\circ = f_i$. By Eq. (4-197*a*) of Table 4-24

$$d \ln f_i = \frac{V_i^L}{RT}\, dP \text{ (constant } T)$$

Integration from the saturation (vapor) pressure of pure i to the system pressure leads to

$$f_i = f_i^{sat} \exp \int_{P_{sat}/i}^{P} \frac{V_i^L}{RT}\, dP$$

Substituting this expression for f_i° in Eq. (4-317) gives

$$\gamma_i x_i f_i^{sat} \exp \int_{P_{sat}/i}^{P} \frac{V_i^L}{RT}\, dP = \hat{\phi}_i y_i P$$

Since $f_i^{sat} = \phi_i^{sat} P_i^{sat}$, this equation may be written

$$y_i P \Phi_i = x_i \gamma_i P_i^{sat} \qquad (i = 1 \text{ to } m) \tag{4-365}$$

where
$$\Phi_i = \left(\frac{\hat{\phi}_i}{\phi_i^{sat}}\right) \exp\left[-\int_{P_{sat}/i}^{P} \frac{V_i^L}{RT}\, dP\right] \tag{4-366}$$

If V_i^L is assumed independent of pressure and $\hat{\phi}_i$ and ϕ_i^{sat} are based on the virial equation and evaluated by Eqs. (4-310) and (4-307), this reduces to

$$\Phi_i = \exp\left[\frac{P\bar{B}_i - P_i^{sat}B_{ii} - V_i^L(P - P_i^{sat})}{RT}\right] \tag{4-367}$$

where $\bar{B}_i$ is given by Eq. (4-309),

$$\bar{B}_i \equiv \left[\frac{\partial(nB)}{\partial n_i}\right]_{T,n_j} = 2\sum_k y_k B_{ki} - B \tag{4-368}$$

with B evaluated by Eq. (4-301).

The m equations represented by Eq. (4-365) in conjunction with Eq. (4-367) may be used for m unspecified phase-equilibrium variables. For a multicomponent system the calculation is formidable, but it is well suited to computer solution. The various types of problems encountered for nonelectrolyte systems at low to moderate pressures (well below the critical pressure) are discussed by Van Ness and Abbott (ibid., sec. 6–8) and treated in detail by Prausnitz et al. (*Computer Calculations for Multicomponent Vapor-Liquid and Liquid-Liquid Equilibria*, Prentice-Hall, Englewood Cliffs, N.J., 1980).

When an appropriate correlating equation for G^E is not available, reliable estimates of activity coefficients may often be obtained from one or the other of two group-contribution correlations. Both the ASOG (analytical-solution-of-groups) method and the UNIFAC (UNIQUAC functional-group-activity-coefficients) method are well developed and are fully described in monographs. (See Kojima and Tochigi, *Prediction of Vapor-Liquid Equilibria by the ASOG Method*, Elsevier, Amsterdam, 1979; and Fredenslund, Gmehling, and Rasmussen, *Vapor-Liquid Equilibria Using UNIFAC*, Elsevier, Amsterdam, 1977.) Revisions and extensions of the UNIFAC parameter tables are published periodically. [See Gmehling, Rasmussen, and Fredenslund, *Ind. Eng. Chem. Process Des. Dev.*, **20**, 331 (1981).]

Vapor-liquid-equilibrium calculations for systems at high pressure or at conditions near the critical region require special techniques. A useful review of this complex subject has been published by I. Wichterle [*Fluid Phase Equilibria*, **1**, 161, 225, 305; **2**, 59, 143 (1977–1978)].

Data Reduction Correlations for G^E and the activity coefficients are based on vapor-liquid-equilibrium data taken at low to moderate pressures. The ASOG and UNIFAC group-contribution methods depend for validity on parameters evaluated from a large base of such data. The process of finding a suitable thermodynamic description of vapor-liquid equilibrium from experimental measurements of equilibrium states is known as data reduction. We give here only a brief description of the treatment of data taken for *binary* systems under *isothermal* conditions. The full development of this topic is given by Van Ness and Abbott (ibid., sec. 6–7).

We presume the existence of an equation inherently capable of representing correct values of G^E for the liquid phase as a function of x_1:

$$g^+ = g(x_1; \alpha, \beta, \ldots) \tag{4-369}$$

where $g \equiv G^E/RT$ and $\alpha, \beta, \ldots$ represent adjustable parameters. The superscript $^+$ indicates a value predicted by Eq. (4-369) and is affixed to all predicted quantities to distinguish them from corresponding values determined directly from experimental data.

For isothermal, binary vapor-liquid-equilibrium data, a data point is fully described by measured values of x_1, P, y_1. From these, we calculate experimental values of γ_1 and γ_2 by Eq. (4-365),

$$\gamma_i = \frac{y_i P \Phi_i}{x_i P_i^{sat}} \qquad (i = 1, 2) \tag{4-370}$$

and experimental values of g by Eq. (4-238),

$$g = x_1 \ln \gamma_1 + x_2 \ln \gamma_2 \tag{4-371}$$

In Eq. (4-370), the factor Φ_i is evaluated by Eq. (4-367). The classical (precomputer) procedure is to find values of the parameters $\alpha, \beta, \ldots$ which minimize $\Sigma(\delta g)^2$ where

$$\delta g \equiv g^+ - g$$

and the summation $\Sigma(\delta g)^2$ is over all data points. This procedure is entirely satisfactory if the experimental values of γ_1 and γ_2 satisfy the Gibbs-Duhem equation, Eq. (4-240). However, if the data contain systematic error, this equation is not likely to be satisfied, and the data-reduction procedure yields a correlating equation that does not give the best possible reproduction of the experimental data.

Associated with each predicted value g^+ are predicted values of the activity coefficients:

$$\gamma_1^+ = \exp\left(g^+ + x_2 \frac{dg^+}{dx_1}\right) \qquad (4\text{-}372a)$$

and

$$\gamma_2^+ = \exp\left(g^+ - x_1 \frac{dg^+}{dx_1}\right) \qquad (4\text{-}372b)$$

These equations come from Eq. (4-364) applied to a binary system. The activity coefficients are related to predicted values of P^+ and y_i^+ by an equation analogous to Eq. (4-370):

$$\gamma_i^+ = (y_i^+ P^+ \Phi_i / x_i P_i^{sat}) \qquad (4\text{-}373)$$

Solving this equation for $y_i^+ P^+$, writing it for each species, and adding the two equations gives

$$P^+ = (x_1\gamma_1^+ P_1^{sat}/\Phi_1) + (x_2\gamma_2^+ P_2^{sat}/\Phi_2) \qquad (4\text{-}374)$$

By Eq. (4-373),

$$y_1^+ = (x_1\gamma_1^+ P_1^{sat}/P^+\Phi_1) \qquad (4\text{-}375)$$

These equations allow calculation of the *primary* residuals

$$\delta P \equiv P^+ - P \quad \text{and} \quad \delta y_1 \equiv y_1^+ - y_1$$

at the experimental values of x_1. Moreover, they provide alternative bases for data reduction. We can find values of the parameters α, β, . . . which minimize either $\Sigma(\delta P)^2$ or $\Sigma(\delta y_1)^2$. The former procedure is known as Barker's method [*Aust. J. Chem.*, **6**, 207 (1953)].

If the experimental activity coefficients satisfy the Gibbs-Duhem equation, then all procedures produce equivalent results, and all residuals (δg, δP, and δy_1) scatter about zero regardless of the procedure. The data are said to be thermodynamically consistent. If the experimental activity coefficients do not satisfy the Gibbs-Duhem equation, different procedures yield different results, and in each procedure one or more of the sets of residuals show systematic deviation from zero. The data are said to be thermodynamically inconsistent and contain systematic error. In this event, one may achieve a compromise fit of the data by minimizing the composite sum

$$S \equiv \Sigma\left(\frac{\delta P}{w_P}\right)^2 + \Sigma\left(\frac{\delta y_1}{w_y}\right)^2$$

where w_p and w_y are normalizing factors which make the residuals compatible for simultaneous treatment. This procedure provides a closer fit to the measured variables P and y than does minimization of $\Sigma(\delta g)^2$.

It is worth noting that the procedure based on minimization of Σ $(\delta P)^2$ does not require y_1 values and that the procedure based on minimization of $\Sigma(\delta y_1)^2$ does not require P values. Thus, the correlating parameters α, β, . . . can be evaluated from a data subset of x_1, P or x_1, y_1 measurements. It is in fact common practice now to measure just x_1, P data.

A compilation of the world's store of vapor-liquid-equilibrium data has been prepared by Gmehling and Onken (*Vapor-Liquid Equilibrium Data Collection*, DECHEMA Chemistry Data ser., vol. I, Frankfurt am Main, 1979–   ).

Determination of K Values Equilibrium ratios, or K values, are widely used in vapor-liquid-equilibrium calculations. A K value is simply the equilibrium ratio of the vapor to liquid composition:

$$K_i = y_i / x_i \qquad (4\text{-}376)$$

The various empirical correlations for K values that appear in the literature have little relation to thermodynamics. The proper thermodynamic equation for K values comes directly from Eq. (4-373):

$$K_i = \frac{y_i}{x_i} = \frac{f_i^\circ}{P}\frac{\gamma_i}{\hat{\phi}_i} \qquad (4\text{-}377)$$

In general, the equilibrium ratio or K value is a function of T, P, liquid composition, and vapor composition.

If the standard-state fugacity of the liquid is taken as f_i, the fugacity of pure liquid i at T and P, then

$$\frac{f_i^\circ}{P} = \frac{f_i^L}{P} = \phi_i^L$$

and Eq. (4-377) becomes

$$K_i = \frac{\phi_i^L}{\hat{\phi}_i}\gamma_i \qquad (4\text{-}378)$$

where $\hat{\phi}_i$ and γ_i refer to the vapor and liquid phases respectively. The well-known correlation for hydrocarbons of Chao and Seader [*Am. Inst. Chem. Eng. J.*, **7**, 598 (1961)] is based on this equation and is therefore on firm thermodynamic ground. The problem is always to get reasonable estimates for ϕ_i^L, $\hat{\phi}_i$, and γ_i. Chao and Seader used a generalized correlation based on Pitzer's work for ϕ_i^L, the generalized Redlich and Kwong equation for $\hat{\phi}_i$, and solubility parameters for γ_i. Improved estimation procedures appear in the literature [see, for example, Lee, Erbar, and Edmister, *Am. Inst. Chem. Eng. J.*, **19**, 349 (1973)]. In general, the Chao-Seader method and its variations are most successful when applied to mixtures of hydrocarbons and the common light gases.

Liquid-Liquid Equilibrium The general criteria of equilibrium, Eqs. (4-356), (4-357), and (4-358), apply as well to liquid-liquid as to vapor-liquid equilibrium. Denoting the equilibrium phases by the superscripts α and β, we write Eq. (4-358) for liquid-liquid equilibrium as

$$\hat{f}_i^\alpha = \hat{f}_i^\beta \qquad (i = 1 \text{ to } m)$$

Introduction of the activity coefficients makes this

$$x_i^\alpha\gamma_i^\alpha(f_i^\circ)^\alpha = x_i^\beta\gamma_i^\beta(f_i^\circ)^\beta \qquad (i = 1 \text{ to } m)$$

Since $(f_i^\circ)^\alpha$ and $(f_i^\circ)^\beta$ may both be taken as the fugacity of pure liquid i at the system T and P, this last equation becomes

$$x_i^\alpha\gamma_i^\alpha = x_i^\beta\gamma_i^\beta \; (i = 1 \text{ to } m)$$

Expressions for the activity coefficients may be obtained from various correlating equations for G^E, most notably the UNIQUAC equation. Furthermore, the UNIFAC method allows prediction of activity coefficients, and a special table of parameters has been developed for liquid-liquid-equilibrium calculations. [See Magnussen, Rasmussen, and Fredenslund, *Ind. Eng. Chem. Process Des. Dev.*, **20**, 331 (1981).]

A comprehensive treatment of liquid-liquid-equilibrium is given by Sørensen, Magnussen, Rasmussen, and Fredenslund [*Fluid Phase Equilibria*, **2**, 297 (1979); **3**, 47 (1979); **4**, 151 (1980)]. Liquid-liquid-equilibrium data are collected in a three-part set compiled by Sørensen and Arlt (*Liquid-Liquid Equilibrium Data Collection*, DECHEMA Chemistry Data ser., vol. V, Frankfurt am Main, 1979–1980, parts 1–3).

Chemical Equilibrium Chemical-equilibrium calculations have traditionally been made through the use of equilibrium constants, and although this is still a useful procedure for simple problems, it does not lend itself readily to the solution of complex equilibrium problems involving simultaneous reactions. Such calculations often are practical only if carried out with the aid of an automatic computing machine, and the calculational procedure used is dictated by its suitability to computer solution. The two types of problems are therefore treated separately in what follows.

Equilibrium Constants Consider a phase in which reaction occurs according to the equation

$$\nu_1 A_1 + \nu_2 A_2 \rightleftharpoons \nu_3 A_3 + \nu_4 A_4$$

where A_i represents a chemical species and ν_i represents the stoichiometric number. The criterion of chemical equilibrium as given by Eq. (4-359) may in this case be written

$$\mu_1\,dn_1 + \mu_2\,dn_2 + \mu_3\,dn_3 + \mu_4\,dn_4 = 0 \qquad (4\text{-}379)$$

Since the dn_i's result from chemical reaction only, they are related to the stoichiometric numbers. Specifically,

$$dn_2/dn_1 = \nu_2/\nu_1 \qquad dn_3/dn_1 = \nu_3/\nu_1 \qquad \text{etc.}$$

The stoichiometric numbers ν_i are by convention taken as positive for products and negative for reactants. The above equations may also be written

$$dn_1/\nu_1 = dn_2/\nu_2 = dn_3/\nu_3 = dn_4/\nu_4 = k\,d\epsilon \qquad (4\text{-}380)$$

where ϵ is the *extent of reaction* or *the reaction coordinate*, and appears by definition. The constant k is arbitrary.

The relations provided for the dn_i's by (4-380) may be substituted into (4-379) to give

$$\mu_1\nu_1\,k\,d\epsilon + \mu_2\nu_2 k\,d\epsilon + \mu_3\nu_3 k\,d\epsilon + \mu_4\nu_4 k\,d\epsilon = 0$$

from which

$$\Sigma(\nu_i\mu_i) = 0 \qquad (4\text{-}381)$$

The μ_i's may be eliminated from Eq. (4-381) by means of Eq. (4-215) written in the form

$$\mu_i = G_i^\circ + RT\ln\hat{a}_i$$

where by definition $\hat{a}_i = \hat{f}_i/f_i^\circ$, and is called an **activity.** The result of this substitution is

$$\Sigma[\nu_i(G_i^\circ + RT\ln\hat{a}_i)] = 0$$

or

$$\Sigma(\nu_i G_i^\circ) + RT\Sigma\ln\hat{a}_i^{\nu_i} = 0$$

or

$$\ln \Pi\hat{a}_i^{\nu_i} = \frac{-\Sigma(\nu_i G_i^\circ)}{RT}$$

The right-hand side of this equation is a function of temperature only for a given reaction and given standard states. It is therefore convenient to set it equal to ln K. This provides the equation

$$K = \Pi\hat{a}_i^{\nu_i} \qquad (4\text{-}382)$$

which defines K and in addition yields a second equation,

$$-RT\ln K = \Sigma(\nu_i G_i^\circ) \equiv \Delta G^\circ \qquad (4\text{-}383)$$

which relates K to the G_i° values of the standard states. Although K is called the **equilibrium constant,** it is a function of T. The quantity ΔG° appearing in Eq. (4-383) is merely the conventional way of indicating the quantity $\Sigma(\nu_i G_i^\circ)$. It is called the **standard Gibbs function change of reaction.**

The activities in Eq. (4-382) provide the connection between the *equilibrium* state of interest and the *standard* states of the constituents, for which data are presumed to be available. The standard states are arbitrary but must always be at the equilibrium temperature T. In the present treatment they are presumed to be states of the pure constituents. Although the standard states selected need not be the same for all constituents, it is essential that for a *particular* constituent the standard state represented by G_i° be the same state represented by f_i° upon which the activity $\hat{a}_i$ is based.

Equations (4-382) and (4-383) are very general and are often simplified by additional restriction of the standard states. Thus for the most important case of gas-phase reactions, the standard states are invariably taken as the ideal-gas states of the pure constituents at an absolute pressure of 1 atm. This means that f_i° = 1 atm for each constituent, and Eq. (4-382) may be written

$$K = \Pi(\hat{f}_i^{\nu_i}) \qquad (4\text{-}384)$$

Note that K does *not* have units and that the $\hat{f}_i$'s must be expressed in atmospheres. Although the atmosphere is not an SI unit, tables of thermochemical data converted to a basis consistent with this system are not yet in common use.

Even though ideal-gas *standard* states are used to define K, the nonidealities of the *equilibrium* state must be taken into account. This is accomplished by introduction of the fugacity coefficient,

$$\hat{\phi}_i = \hat{f}_i/y_i P$$

where y_i represents a gas-phase mole fraction and P is the absolute pressure in atmospheres. Substitution for $\hat{f}_i$ in Eq. (4-384) gives

$$K = (\Pi\hat{\phi}_i^{\nu_i})(\Pi y_i^{\nu_i})P^{\Sigma\nu_i} \qquad (4\text{-}385)$$

The procedure for solving Eqs. (4-383) and (4-385) to give a set of equilibrium y_i's depends on the availability of data for ΔG°. Given such data for a particular reaction at a given T, Eq. (4-383) is solved for K, and Eq. (4-385) is then used to determine the y_i's at the given P by a procedure to be illustrated later. One difficulty that must be overcome is that the $\hat{\phi}_i$'s depend on the y_i's, which are to be calculated. This complication disappears if the equilibrium mixture can be assumed an ideal gas; then each $\hat{\phi}_i$ is unity, and the term drops out. The problem is also resolved if the equilibrium mixture is taken to be an ideal solution. Then each $\hat{\phi}_i$ becomes ϕ_i, the fugacity coefficient of pure i at the mixture T and P. This quantity does not depend on composition and may be determined from experimental data, from a generalized correlation, or from an equation of state.

In the general case in which no simplifying assumptions are justified, an iterative procedure is indicated. As a first step one takes the $\hat{\phi}_i$'s equal to unity and determines a set of y_i's. This allows the $\hat{\phi}_i$'s to be calculated by means of an equation of state, for example, by Eq. (4-310), (4-314), or (4-325). A new set of y_i's is determined, and the process is continued to convergence.

For liquid-phase reactions the standard states are usually taken as the pure liquids at the equilibrium T and at an absolute pressure of 1 atm. Since this does *not* make f_i° = 1, Eq. (4-382) must be used for the equilibrium constant. It is now convenient to introduce the activity coefficient

$$\gamma_i = \hat{f}_i/x_i f_i$$

which gives for the activity

$$\hat{a}_i = \hat{f}_i/f_i^\circ = (\gamma_i f_i x_i/f_i^\circ)$$

where x_i is liquid-phase mole fraction. Both f_i and f_i° represent the fugacities of pure liquid i at temperature T, but at pressure P and at 1 atm, respectively. Since pressure often has little effect on the properties of liquids, the ratio f_i/f_i° is frequently taken as unity. When this is not acceptable, one can calculate this ratio by the equation

$$\ln\frac{f_i}{f_i^\circ} = \frac{1}{RT}\int_1^P V_i\,dP \cong \frac{V_i(P-1)}{RT}$$

When this pressure correction is negligible, then $\hat{a}_i = \gamma_i x_i$, and Eq. (4-382) becomes

$$K = (\Pi\gamma_i^{\nu_i})(\Pi x_i^{\nu_i}) \qquad (4\text{-}386)$$

Here the difficulty is to determine the γ_i's, which depend on the x_i's. This problem has not been solved for the general case. Two courses are open. The first is to assume that the equilibrium mixture is an ideal solution. Then γ_i = 1, and the term drops out. This gives

$$K = \Pi x_i^{\nu_i}$$

the so-called law of mass action. The only realistic alternative is recourse to an experimental determination of the equilibrium composition.

The reaction coordinate ϵ introduced by Eq. (4-380) provides a single variable to which the equilibrium mole fractions may be related. The arbitrary constant k in Eq. (4-380) may be set so as to normalize ϵ such that its limiting values are 0 and 1. Alternatively, it may be set equal to unity (or to any other value) in lieu of placing restrictions on the values that ϵ may assume. The simplest procedure is probably to let k = 1 and to take ϵ = 0 for the initial constitution of the system.

Example 5 Consider the reaction

$$CO + H_2O \rightleftharpoons CO_2 + H_2$$

Equation (4-380) with $k = 1$ becomes

$$\frac{dn_{CO}}{-1} = \frac{dn_{H_2O}}{-1} = \frac{dn_{CO_2}}{1} = \frac{dn_{H_2}}{1} = d\epsilon$$

Let the feed stream contain 3 mol CO, 1 mol H_2O, and 2 mol CO_2 for every mole of H_2 present. This initial constitution forms the *basis* for calculation, and for this nonequilibrium mixture ϵ is taken as 0. Integration of the above equation from the initial state to the equilibrium state gives

$$-\int_3^{n_{CO}} dn_{CO} = -\int_1^{n_{H_2O}} dn_{H_2O} = \int_2^{n_{CO_2}} dn_{CO_2}$$

$$= \int_1^{n_{H_2}} dn_{H_2} = \int_0^{\epsilon} d\epsilon$$

or

$$n_{CO} = 3 - \epsilon$$
$$n_{H_2O} = 1 - \epsilon$$
$$n_{CO_2} = 2 + \epsilon$$
$$n_{H_2} = 1 + \epsilon$$
$$\Sigma n_i = 7$$

Since $\Sigma n_i = 7$, $y_i = n_i/7$, and

$$\Pi y_i^{\nu_i} = \frac{\left(\frac{2+\epsilon}{7}\right)\left(\frac{1+\epsilon}{7}\right)}{\left(\frac{3-\epsilon}{7}\right)\left(\frac{1-\epsilon}{7}\right)} = \frac{(2+\epsilon)(1+\epsilon)}{(3-\epsilon)(1-\epsilon)}$$

At 1000 K, $\Delta G^\circ = -2680$ J for the reaction as written with the stoichiometric numbers representing gram-moles and for standard states as pure ideal gases at 1 atm. By Eq. (4-383),

$$\ln K = \frac{-\Delta G^\circ}{RT} = \frac{2680}{(8.314)(1000)} = 0.322$$

and

$$K = 1.38$$

If the equilibrium composition at 1 atm and 1000 K is required, then the assumption of ideal gases is appropriate, and Eq. (4-385) becomes

$$K = \Pi(y_i^{\nu_i})$$

from which

$$1.38 = \frac{(2+\epsilon)(1+\epsilon)}{(3-\epsilon)(1-\epsilon)}$$

giving finally

$$\epsilon = 0.258$$

Thus the equilibrium mixture contains

$n_{CO} =$	2.74 mol	$y_{CO} =$	0.391
$n_{H_2O} =$	0.74 mol	$y_{H_2O} =$	0.106
$n_{CO_2} =$	2.26 mol	$y_{CO_2} =$	0.323
$n_{H_2} =$	1.26 mol	$y_{H_2} =$	0.180
$\Sigma n_i =$	7.00 mol	$\Sigma y_i =$	1.000

The effect of temperature on the equilibrium constant is determined through Eq. (4-180), the Gibbs-Helmholtz equation. From this it follows that

$$\frac{d(\Delta G^\circ/RT)}{dT} = \frac{-\Delta H^\circ}{RT^2} \tag{4-387}$$

where ΔH° is called the standard heat of reaction and is the conventional symbol for $\Sigma(\nu_i H_i^\circ)$. Combination of Eqs. (4-383) and (4-387) gives

$$d \ln K/dT = \Delta H^\circ/RT^2 \tag{4-388}$$

For an endothermic reaction ΔH° is positive, and for an exothermic reaction it is negative. The dependence of ΔH° on temperature is given by

$$\frac{d(\Delta H^\circ)}{dT} = \Delta C_P^\circ \tag{4-389}$$

where $\Delta C_P^\circ \equiv \Sigma(\nu_i C_i^\circ)$

Integration of Eq. (4-389) requires that ΔC_P° be known as a function of T. Then

$$\Delta H^\circ = \Delta H_0 + \int \Delta C_P^\circ \, dT \tag{4-390}$$

where ΔH_0 is a constant of integration. Equation (4-390) expresses ΔH° as a function of T, and this allows integration of Eq. (4-390)

$$\ln K = \frac{1}{R}\int \frac{\Delta H^\circ}{T^2}\, dT + I \tag{4-391}$$

where I is another constant of integration.

In the more extensive compilations of data, these integrations have already been carried out, and values of ΔG° (and of ΔH°) are listed as a function of T. Reference is made in particular to *Selected Values of Properties of Chemical Compounds,* Thermodynamics Research Center, Texas A&M University, College Station, Texas. Kelley (U.S. Bur. Mines Bull. 584, 1960) has also tabulated much useful information.

Complex Chemical Equilibria When the equilibrium composition is determined by a number of simultaneous reactions, the computations required become complex and tedious but can be performed readily by automatic computers. The most direct scheme for doing this depends on direct minimization of the Gibbs function in accord with Eq. (4-356). The treatment here will be limited to gas-phase reactions in which the problem is to find the equilibrium composition for a given T and P and for a given initial feed to the system.

The total Gibbs function for the system is given by

$$nG = \Sigma(n_i \bar{G}_i)$$

and it is this quantity that is to be minimized with respect to the n_i's at constant T and P, subject to restraints imposed by material balances written for a closed system.

If $\bar{G}_i$ is eliminated by Eq. (4-215), then

$$nG = \Sigma(n_i G_i^\circ) + RT \sum \left(n_i \ln \frac{\hat{f}_i}{f_i^\circ}\right)$$

Since gas-phase reactions are being considered, $\hat{f}_i$ may be replaced by

$$\hat{f}_i = y_i P \hat{\phi}_i$$

Furthermore, the standard state is taken as the pure ideal-gas state at 1 atm for each constituent, and this makes each $f_i^\circ = 1$ atm. Because of this, the pressure P must be expressed in atmospheres. In addition, G_i° is arbitrarily set equal to zero for each chemical *element* in its standard state, and this gives

$$G_i^\circ = \Delta G_{f_i}^\circ$$

for each *compound,* where $\Delta G_{f_i}^\circ$ is the standard Gibbs function of formation of compound i from its constituent elements at temperature T. These substitutions yield

$$nG = \Sigma(n_i\, \Delta G_{f_i}^\circ) + (\Sigma n_i)RT \ln P + RT\Sigma(n_i \ln y_i) + RT\Sigma(n_i \ln \hat{\phi}_i)$$

The problem now is to find the set of n_i's which minimizes nG at constant T and P subject to the restraints of the material balances. The standard solution to this type of problem is through the method of Lagrange's undetermined multipliers. This requires that the restraints imposed by the material balances be incorporated in the expression for nG. The material-balance equations are developed as follows.

Let A_k be the total number of atomic weights of the kth element present in the system, as determined by its initial constitution. Let a_{ik} be the number of atoms of the kth element present in each molecule of chemical species i. Then, for each element k,

$$\sum_i (n_i a_{ik}) = A_k$$

which may also be written

$$\sum_i (n_i a_{ik}) - A_k = 0$$

This is multiplied by an undetermined constant λ_k, and the expression is then summed over all k, giving

$$\sum_k \left\{ \lambda_k \left[\sum_i (n_i a_{ik}) - A_k \right] \right\} = 0$$

Since this quantity is zero, it may be added to the right-hand side of the equation for nG:

$$nG = \sum_i (n_i \Delta G^\circ_{fi}) + \left(\sum_i n_i \right) RT \ln P + RT \sum_i n_i \ln \frac{n_i}{\sum_i n_i}$$

$$+ RT \sum_i (n_i \ln \hat{\phi}_i) + \sum_k \left\{ \lambda_k \left[\sum_i (n_i a_{ik}) - A_k \right] \right\}$$

where y_i has been replaced by $n_i/\Sigma_i n_i$. The final step is to differentiate partially with respect to n_i to form the derivatives $[\partial(nG)/\partial n_i]_{T,P,n_j}$, which are then set equal to zero. This leads to

$$\Delta G^\circ_{fi} + RT \ln P + RT \ln y_i + RT \ln \hat{\phi}_i + \sum_k (\lambda_k a_{ik}) = 0 \quad (4\text{-}392)$$

There are i such equations. In addition, there are k material-balance equations,

$$\sum_i (y_i a_{ik}) = A_k \Big/ \sum_i n_i \quad (4\text{-}393)$$

and, in addition,

$$\sum_i y_i = 1 \quad (4\text{-}394)$$

This provides a total of $i + k + 1$ equations.

The unknowns in these equations are the y_i's, of which there are i, the λ_k's, of which there are k, and $\Sigma_i n_i$, giving a total of $i + k + 1$ unknowns. Thus the set represented by Eqs. (4-392) to (4-394) can be solved for all unknowns.

However, Eq. (4-392) was derived on the presumption that the $\hat{\phi}_i$'s are known. If the phase is an ideal gas, then each $\hat{\phi}_i$ is indeed known and is unity, but for real gases each $\hat{\phi}_i$ is a function of the y_i's, which are to be determined. Thus an iterative procedure is indicated, which is initiated by setting $\hat{\phi}_i$'s equal to unity. Solution of the equations then provides a preliminary set of y_i's. For low pressures or high temperatures this result is usually quite adequate. When it is not, an equation of state is used, together with the calculated y_i's to give a new and more nearly correct set of $\hat{\phi}_i$'s for use in Eq. (4-393), and a new set of y_i's is determined. The process is repeated until successive iterations produce no significant change in the y_i's. All computations are done by computer, including calculation of the $\hat{\phi}_i$'s by equations such as Eqs. (4-310), (4-314), and (4-325).

It is important to note that in this procedure the question of what chemical reactions are involved never enters directly into any of the equations. However, the choice of a set of species is entirely equivalent to the choice of a set of independent reactions among the species. In any event a set of species or a set of independent reactions must be assumed, and different assumptions will, in general, produce different results.

Example 6 Assume that chemical equilibrium is established at 1 atm and 1000 K in a system containing the five chemical species CH_4, H_2O, CO, CO_2, and H_2. If the system is formed initially from 2 mol of CH_4 and 3 mol of H_2O, determine the equilibrium composition.

This system and the reactions which may take place in it were discussed earlier in connection with the phase rule.

The required values of A_k are determined from the initial feed, and the values of a_{ik} come directly from the chemical formulas of the species present:

	Element k		
	Carbon	Oxygen	Hydrogen
	A_k — no. of atomic wts. of k in system		
	$A_C = 2$	$A_O = 3$	$A_H = 14$
Species i	a_{ik} — atoms of k per molecule of i		
CH_4	$a_{CH_4,C} = 1$	$a_{CH_4,O} = 0$	$a_{CH_4,H} = 4$
H_2O	$a_{H_2O,C} = 0$	$a_{H_2O,O} = 1$	$a_{H_2O,H} = 2$
CO	$a_{CO,C} = 1$	$a_{CO,O} = 1$	$a_{CO,H} = 0$
CO_2	$a_{CO_2,C} = 1$	$a_{CO_2,O} = 2$	$a_{CO_2,H} = 0$
H_2	$a_{H_2,C} = 0$	$a_{H_2,O} = 0$	$a_{H_2,H} = 2$

At 1 atm and 1000 K the assumption of ideal gases should be justified, and the ln $\hat{\phi}_i$ term of Eq. (4-392) can be omitted. In addition ln P is zero and Eq. (4-392) becomes

$$\Delta G^\circ_{fi} + RT \ln y_i + \sum_k (\lambda_k a_{ik}) = 0$$

At 1000 K

$$\Delta G^\circ_{f\,CH_4} = 19.3 \text{ kJ/mol}$$
$$\Delta G^\circ_{f\,H_2O} = -192.6 \text{ kJ/mol}$$
$$\Delta G^\circ_{f\,CO} = -200.6 \text{ kJ/mol}$$
$$\Delta G^\circ_{f\,CO_2} = -395.9 \text{ kJ/mol}$$
$$\Delta G^\circ_{f\,H_2} = 0$$

The five equations resulting from Eq. (4-392) now become

$$CH_4\text{: } 19.3 + RT \ln y_{CH_4} + \lambda_C + 4\lambda_H = 0$$
$$H_2O\text{: } -192.6 + RT \ln y_{H_2O} + 2\lambda_H + \lambda_O = 0$$
$$CO\text{: } -200.6 + RT \ln y_{CO} + \lambda_C + \lambda_O = 0$$
$$CO_2\text{: } -395.9 + RT \ln y_{CO_2} + \lambda_C + 2\lambda_O = 0$$
$$H_2\text{: } RT \ln y_{H_2} + 2\lambda_H = 0$$

where $RT = 8.314$ kJ/mol.

From Eq. (4-393) there are three material balances:

$$C\text{: } y_{CH_4} + y_{CO} + y_{CO_2} = 2/\Sigma n_i$$
$$H\text{: } 4y_{CH_4} + 2y_{H_2O} + 2y_{H_2} = 14/\Sigma n_i$$
$$O\text{: } y_{H_2O} + y_{CO} + 2y_{CO_2} = 3/\Sigma n_i$$

In addition, Eq. (4-394) requires

$$y_{CH_4} + y_{H_2O} + y_{CO} + y_{CO_2} + y_{H_2} = 1$$

Simultaneous solution of these nine equations provides the following results:

$y_{CH_4} = 0.0199$			
$y_{H_2O} = 0.0995$	Σn_i	= 8.656
$y_{CO} = 0.1753$	λ_C/RT	= 0.797
$y_{CO_2} = 0.0359$	λ_O/RT	= 25.1
$y_{H_2} = 0.6694$	λ_H/RT	= 0.201
$\Sigma y_i = 1.0000$		

The λ_k's are of no real interest but are included to make the results complete.

OTHER KINDS OF SYSTEMS

All the material presented so far has dealt with *PVT* systems. Other kinds of systems, characterized by different sets of variables, are also subject to thermodynamic description and analysis. For example, the adsorbed layer of gas on a solid is characterized by the spreading pressure π, the surface area $\mathcal{A}$, and the temperature T. A network of equations for such a $\pi\mathcal{A}T$ system may be developed which is quite analogous to that for a *PVT* system. The treatment of gas-adsorbate equilibrium is also similar to that of vapor-liquid equilibrium. [See Van Ness, "Adsorption of Gases on Solids, *Ind. Eng. Chem. Fundam.*, 8, 464 (1969).]

Of the various kinds of systems other than the *PVT* system subject to thermodynamic treatment, one of the most important is the electrochemical cell, which is treated in some detail below.

ELECTROCHEMICAL CELLS

GENERAL REREFENCES: Alkire and Beck (eds.), *Tutorial Lectures in Electrochemical Engineering and Technology,* AIChE Symp. Ser. No. 204, vol. 77, American Institute of Chemical Engineers, New York, 1981. Harned and Owen, *The Physical Chemistry of Electrolyte Solutions,* 3d ed., Reinhold, New York, 1958. Lewis, Randall, Pitzer, and Brewer, *Thermodynamics,* 2d ed., McGraw-Hill, New York, 1961, chaps. 22–24. Liebhafsky and Cairns, *Fuel Cells and Fuel Batteries: A Guide to Their Research,* Wiley, New York, 1968. Mitchell (ed.), *Fuel Cells,* Academic, New York, 1963. Münster, *Classical Thermodynamics,* Wiley-Interscience, London, 1970, chap. XI. Newman, *Electrochemical Systems,* Prentice-Hall, Englewood Cliffs, N.J., 1973. Pytkowicz (ed.), *Activity Coefficients in Electrolyte Solutions,* vols. I and II, CRC Press, Boca Raton, Fla., 1979. Robinson and Stokes, *Electrolyte Solutions,* 2d rev. ed., Butterworth, London, 1965.

Introduction and Definitions When a chemical reaction can proceed spontaneously via an ionic mechanism, it is possible in principle to convert part of the energy change on reaction directly into electric energy without the intermediaries of a heat engine and generator. Conversely, some chemical reactions can be made to occur, via ionic mechanisms, by the addition of electric energy to a suitably contrived reactor system. Devices for accomplishing either of these ends are called **electrochemical cells,** or electrochemical reactors.

Electrochemical cells in which a chemical reaction is forced by added electric energy are called **electrolytic cells;** devices that produce electric energy via chemical reaction are called **galvanic cells.** Galvanic cells are customarily further classified according to their mode of operation: **fuel cells** are steady-state reactors to which reactants are continuously supplied and from which products are continuously withdrawn, while **primary cells** are unsteady-state devices containing fixed initial amounts of reactants. A third type of electrochemical cell, called a **secondary cell,** functions as a galvanic cell when in use but can be regenerated ("recharged") by reversing the cell reaction through the addition of electric energy. The familiar lead-acid automotive battery consists of series-connected secondary cells.

Central to the operation of any cell is the occurrence of ionic reactions which produce or consume electrons at isolated phases of the cell. These phases are called **electrodes** and must be good electronic conductors. In operation, a cell is connected to an external load or to an external voltage source, and electric charge is transferred by electrons between the electrodes through the external circuit. To complete the electric circuit through the cell, an additional mechanism must exist for internal charge transfer. This is provided by one or more **electrolytes,** which support charge transfer by ionic conduction. Electrolytes must be poor electronic conductors to prevent internal short circuiting of the cell.

The simplest electrochemical cell consists of at least three phases, including two electrodes and one or more electrolytes. The electrode at which an electron-producing ionic reaction occurs (e.g., $M \rightarrow M^+ + e^-$) is the **anode;** the electrode at which an electron-consuming reaction occurs (e.g., $N^+ + e^- \rightarrow N$) is called the **cathode.** The direction of electron flow in the external circuit is always from anode to cathode.

Electrical work can be done on or by charge moving in an electric field and is given in differential form by

$$dW_e = -\Delta\psi\, dq$$

where dq is a differential quantity of electricity transferred between two points between which the electrical potential difference is $\Delta\psi$. In applications of this equation to the passage of electricity through the external circuit of an electrochemical cell, the convention is adopted that the potential difference $\psi_{\text{cathode}} - \psi_{\text{anode}}$ is positive for a galvanic cell and negative for an electrolytic cell. This convention is consistent with the customary "positive" and "negative" designations for cell terminals: the cathode of a galvanic cell is *positive* (with respect to the anode), whereas the cathode of an electrolytic cell is *negative* (with respect to the anode). The difference $\psi_{\text{cathode}} - \psi_{\text{anode}}$ is the cell **voltage** and is here given the symbol $\mathcal{V}$. A definition of electrical work which provides a basis for the thermodynamic analysis of electrochemical cells is then

$$dW_e = -\mathcal{V}\, dq \tag{4-395}$$

where the definitions and conventions given in the preceding paragraphs further require that dq be negative.

The **overall cell reaction,** which must be known for a useful analysis of a cell, is the sum of the ionic reactions occurring in the cell. For example, the overall reaction for the discharge of a lead-acid secondary cell is

$$Pb + PbO_2 + 2H_2SO_4 \rightarrow 2PbSO_4 + 2H_2O$$

and the overall reaction of a hydrogen-oxygen fuel cell is

$$H_2 + \tfrac{1}{2}O_2 \rightarrow H_2O$$

The **charge** of an electrochemical cell can be regarded as an auxiliary thermodynamic function which provides a measure of the availability of chemical species participating in the overall cell reaction. It is given the symbol q and is related to the reaction coordinate ϵ by

$$dq = -\mathcal{N}\mathcal{F}\, d\epsilon \tag{4-396}$$

where $\mathcal{F}$ is the Faraday constant, and $\mathcal{N}$ is the number of moles of electrons produced at the anode per ν_i mol of component i produced (ν_i is the stoichiometric coefficient of i in the overall cell reaction). The definition of $d\epsilon$ is given by Eq. (4-380) with k set equal to unity. If dq, defined according to Eq. (4-396), is interpreted as the electronic charge transferred between the terminals via the external circuit per dn_i mol of i produced (in accordance with the definition $d\epsilon = dn_i/\nu_i$), then the conventions associated with Eq. (4-395) require that $d\epsilon$ always be taken as positive, that is, that dq be negative in *both* Eqs. (4-395) and (4-396). This is equivalent to requiring that the overall cell reaction be written in the *forward* direction, that is, in the direction in which it actually occurs during cell operation.

It is a matter of experience that a galvanic cell has associated with it an intrinsic voltage which is a characteristic of the cell and which is independent of any external circuit elements. This intrinsic voltage is called the **open-circuit voltage** of the cell and is here designated $\mathcal{V}_c$. A galvanic cell is conventionally represented as a pure voltage source of magnitude $\mathcal{V}_c$ associated with an internal resistance of magnitude $\mathcal{R}_c$. It is not $\mathcal{V}_c$, however, which in general determines the electrical work done by a cell, but rather the cell **terminal voltage,** which we represent by an unsubscripted $\mathcal{V}$. The terminal voltage, although it is a function of $\mathcal{V}_c$, also depends on the nature of the external circuit elements. This relationship is made evident by the simple circuit shown in Fig. 4-36.

Elements $\mathcal{R}_c$ and $\mathcal{V}_c$ represent the galvanic cell, for which points P and N are the positive and negative terminals, respectively. $\mathcal{V}_s$ is an external voltage source, and $\mathcal{R}_s$ is the equivalent resistance of the entire external circuit. Designating the current in the circuit by $\mathcal{I}$, a simple derivation gives

$$\mathcal{I} = (\mathcal{V}_s + \mathcal{V}_c)/(\mathcal{R}_s + \mathcal{R}_c) \tag{4-397a}$$

$$\mathcal{V} = (\mathcal{R}_s\mathcal{V}_c - \mathcal{R}_c\mathcal{V}_s)/(\mathcal{R}_s + \mathcal{R}_c) \tag{4-397b}$$

where $\mathcal{V} = \psi_P - \psi_N$.

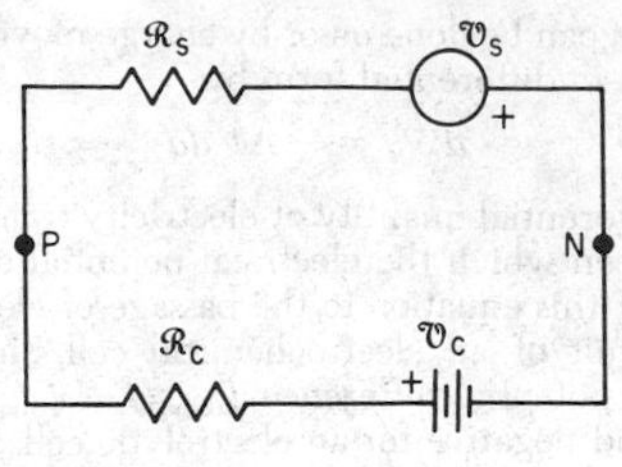

FIG. 4-36 Simple circuit containing a galvanic cell.

Equation (4-397*b*) shows how one could vary $\mathcal{V}$ by changing $\mathcal{V}_s$ and/or $\mathcal{R}_s$. One case is of particular interest. If the magnitude and polarity of $\mathcal{V}_s$ are adjusted so that $\mathcal{V}_s = -\mathcal{V}_c$, then the terminal voltage becomes equal to the open-circuit cell voltage, and, according to Eq. (4-397*b*), the current becomes zero. This suggests an analogy between a cell and a simple *PVT* system capable of performing mechanical work: when the open-circuit voltage of a cell ($\simeq$ internal pressure on a piston) is exactly balanced by an external voltage ($\simeq$ external pressure), a condition of zero current ($\simeq$ zero velocity or displacement) is obtained. The case for which $\mathcal{V}_s = -\mathcal{V}_c$ then clearly corresponds to a condition of external equilibrium. If at this condition no other changes tend to occur within the cell, then the system is also in internal equilibrium, and an infinitesimal displacement about this state constitutes a reversible process.

The characteristic cell voltage for conditions of both external and internal equilibrium is called the **electromotive force** (emf) and is given the symbol $\mathcal{E}$. As will be shown, the emf of a real cell is a well-defined quantity, but it is not in general equal to the open-circuit voltage because of the difficulty of attaining equilibrium conditions within a real cell. The reversible electrical work done by an electrochemical cell is then

$$dW_{e,\text{rev}} = -\mathcal{E}\, dq \tag{4-398}$$

Although this discussion has dealt with galvanic cells, Eq. (4-398) also applies to the reversible operation of an electrolytic cell. In fact, displacing an opposing external voltage $\mathcal{V}_s$ below or above the emf of a reversible cell causes the cell reaction to proceed in the forward or reverse direction, that is, makes the cell function either as a galvanic or as an electrolytic cell.

Thermodynamics of Reversible Cells If an electrochemical cell is considered a thermodynamic system, then the relationships among the variables characterizing such a system are obtained as a result of the first and second laws applied to the cell in reversible operation. Such a system may perform mechanical as well as electrical work, and the fundamental property relation is therefore

$$dU = T\, dS - P\, dV + \mathcal{E}\, dq \tag{4-399}$$

In this equation and henceforth unless otherwise noted, the thermodynamic properties refer to the system as a whole: that is, U is the *total* internal energy of the system, S is the *total* entropy, etc.

Insofar as the *properties* of the cell are concerned, an electrochemical cell may be considered a system of reacting chemical species characterized by three independent variables. The third variable, taken as q in Eq. (4-399), determines the extent of reaction. It could equally well be taken as ϵ, the reaction coordinate, and Eq. (4-399) would be replaced by

$$dU = T\, dS - P\, dV + (\partial U/\partial \epsilon)_{S,V}\, d\epsilon$$

However, it is the electrical nature of a cell that is of prime importance, and the identification of $\mathcal{E}$ as an electrical quantity conjugate to q in Eq. (4-399) dictates the use of this form of the fundamental property relation.

The enthalpy, Helmholtz function, and Gibbs function may be defined as for a *PVT* system:

$$H = U + PV \tag{4-400}$$

$$A = U - TS \tag{4-401}$$

$$G = U + PV - TS \tag{4-402}$$

Equations analogous to Eqs. (4-131) through (4-133) then follow directly:

$$dH = T\, dS + V\, dP + \mathcal{E}\, dq \tag{4-403}$$

$$dA = -P\, dV - S\, dT + \mathcal{E}\, dq \tag{4-404}$$

$$dG = V\, dP - S\, dT + \mathcal{E}\, dq \tag{4-405}$$

From Eqs. (4-399) and (4-403) to (4-405) one obtains, corresponding to Eqs. (4-134) through (4-137):

$$T = \left(\frac{\partial U}{\partial S}\right)_{V,q} = \left(\frac{\partial H}{\partial S}\right)_{P,q} \tag{4-406}$$

$$-P = \left(\frac{\partial U}{\partial V}\right)_{S,q} = \left(\frac{\partial A}{\partial V}\right)_{T,q} \tag{4-407}$$

$$V = \left(\frac{\partial H}{\partial P}\right)_{S,q} = \left(\frac{\partial G}{\partial P}\right)_{T,q} \tag{4-408}$$

$$-S = \left(\frac{\partial A}{\partial T}\right)_{V,q} = \left(\frac{\partial G}{\partial T}\right)_{P,q} \tag{4-409}$$

$$\mathcal{E} = \left(\frac{\partial U}{\partial q}\right)_{S,V} = \left(\frac{\partial H}{\partial q}\right)_{S,P} = \left(\frac{\partial A}{\partial q}\right)_{V,T} = \left(\frac{\partial G}{\partial q}\right)_{T,P} \tag{4-410}$$

In addition, a set of 12 Maxwell equations follows from the properties of exact differentials. Four of these are identical in form to Eqs. (4-138) through (4-141), where in addition q is understood to be held constant. Of the remaining eight, only two are of interest here. They are

$$\left(\frac{\partial \mathcal{E}}{\partial P}\right)_{T,q} = \left(\frac{\partial V}{\partial q}\right)_{T,P} \tag{4-411}$$

$$\left(\frac{\partial \mathcal{E}}{\partial T}\right)_{P,q} = -\left(\frac{\partial S}{\partial q}\right)_{T,P} \tag{4-412}$$

The important applications of these relationships are to connect $\mathcal{E}$ and its temperature and pressure derivatives with more accessible properties of the cell. This is accomplished through Eqs. (4-410) to (4-412), which all involve differential coefficients of the form $(\partial M/\partial q)_{T,P}$, where M is a total property of the system. These coefficients can be put in a more familiar form through the relationship between dq and $d\epsilon$ given by Eq. (4-396). Thus,

$$\left(\frac{\partial M}{\partial q}\right)_{T,P} = \left(\frac{\partial M}{\partial \epsilon}\right)_{T,P} \frac{d\epsilon}{dq}$$

or

$$\left(\frac{\partial M}{\partial q}\right)_{T,P} = -\frac{1}{\mathcal{N}\mathcal{F}}\left(\frac{\partial M}{\partial \epsilon}\right)_{T,P} \tag{4-413}$$

But $(\partial M/\partial \epsilon)_{T,P}$ is just the property change of the system resulting from the formation of ν_i mol of any component i at constant T and P and is given in terms of the reaction stoichiometry by

$$(\partial M/\partial \epsilon)_{T,P} = \Sigma \nu_i \overline{M}_i \equiv \Delta M_{T,P} \tag{4-414}$$

The summation in Eq. (4-414) is taken over the chemical species in physical states that exist within the equilibrated cell, so that in the most general case the summation is on the partial molar properties $\overline{M}_i$ for the species present in the various phases of the cell. This summation is conventionally abbreviated as $\Delta M_{T,P}$ and is called the **property change of reaction,** with conditions of constant temperature and pressure understood. It should be noted that Δ as used here is not a difference operator but rather stands for the differential operator $\partial/\partial\epsilon$.

Combining Eqs. (4-413) and (4-414),

$$\left(\frac{\partial M}{\partial q}\right)_{T,P} = -\frac{\Delta M_{T,P}}{\mathcal{N}\mathcal{F}} \tag{4-415}$$

from which Eqs. (4-410) through (4-412) become

$$\mathcal{E} = -\Delta G_{T,P}/\mathcal{N}\mathcal{F} \tag{4-416}$$

$$\left(\frac{\partial \mathcal{E}}{\partial P}\right)_{T,q} = -\frac{\Delta V_{T,P}}{\mathcal{N}\mathcal{F}} \tag{4-417}$$

$$\left(\frac{\partial \mathcal{E}}{\partial T}\right)_{P,q} = \frac{\Delta S_{T,P}}{\mathcal{N}\mathcal{F}} \tag{4-418}$$

According to Eq. (4-416), the emf is positive for a cell for which $\Delta G_{T,P}$ is negative. This corresponds to a cell for which the chemical reaction can proceed spontaneously, that is, to a galvanic cell. Conversely, $\mathcal{E}$ is negative for an electrolytic cell, for which the chemical reaction has a positive $\Delta G_{T,P}$.

A useful equation relating $\mathcal{E}$ and its temperature derivative to the enthalpy change on reaction follows from Eqs. (4-403), (4-413), (4-414), and (4-418). Since by Eq. (4-403),

$$\left(\frac{\partial H}{\partial q}\right)_{T,P} = T\left(\frac{\partial S}{\partial q}\right)_{T,P} + \mathcal{E}$$

then

$$\Delta H_{T,P} = \mathcal{N}\mathcal{F}\left[T\left(\frac{\partial \mathcal{E}}{\partial T}\right)_{P,q} - \mathcal{E}\right] \quad (4\text{-}419)$$

Equation (4-419) has found wide application in the use of reversible galvanic cells to establish enthalpy changes on reaction for systems that could be studied only with great difficulty (if at all) by calorimetric methods.

Equations giving the composition dependence of $\mathcal{E}$ follow directly from Eq. (4-416). Equation (4-215) allows transformation of the right side of this equation into variables commonly used in chemical equilibrium:

$$\mathcal{E} = -\frac{\Delta G^\circ}{\mathcal{N}\mathcal{F}} - \frac{RT}{\mathcal{N}\mathcal{F}} \ln \Pi \hat{a}_i^{\nu_i} \quad (4\text{-}420)$$

The first term on the right side of Eq. (4-420) is commonly designated $\mathcal{E}^\circ$ and is called the **standard** (state) **emf.** When all the species exist within the cell in their standard states, $\hat{a}_i = \hat{f}_i/f_i^\circ = 1$. and

$$\mathcal{E} = \mathcal{E}^\circ = -\Delta G^\circ/\mathcal{N}\mathcal{F}$$

Since $\mathcal{E}^\circ$ and f_i° are functions of temperature only, the variation of $\mathcal{E}$ with composition at constant T and P is, in difference form,

$$\Delta\mathcal{E}_{T,P} = \mathcal{E}(B) - \mathcal{E}(A) = -\frac{RT}{\mathcal{N}\mathcal{F}} \ln \Pi \left[\frac{\hat{f}_i(B)}{\hat{f}_i(A)}\right]^{\nu_i} \quad (4\text{-}421)$$

where the fugacities $\hat{f}_i(A)$ and $\hat{f}_i(B)$ are evaluated at states A and B of different composition.

Although Eqs. (4-416) through (4-421) are derived for a reversible electrochemical cell operating as a closed system at constant temperature and pressure, they also apply to reversible isothermal steady-flow cells (fuel cells) for which kinetic and gravitational potential energy effects can be ignored.

Example 7 A hydrogen-oxygen fuel cell operates at 1.013 bar (1 atm) and 25°C according to the overall reaction

$$H_2(g) + \tfrac{1}{2}O_2(g) \rightarrow H_2O(l)$$

for which $\mathcal{N} = 2$. The H_2, O_2, and H_2O exist within the cell and participate in the reaction as separate phases. Calculate the emf at 25°C and (*a*) 1 atm, (*b*) 5 atm, and (*c*) at 1 atm if air (79 mole percent N_2, 21 mole percent O_2) is used as the oxidant instead of pure oxygen.

a. For the given reaction and the stated conditions, $\Delta G_{T,P} = \Delta G^\circ = -237{,}190$ J. The value of $\mathcal{F}$ is 96,487 J V^{-1} mol^{-1}. Then, from Eq. (4-416),

$$\mathcal{E} = \frac{-237{,}190}{(2)(96{,}487)} = 1.229 \text{ V}$$

b. Equation (4-417) is applied in this case. This is a steady-flow system so that q is constant. Since the molar volume of liquid water is small (18.14 cm^3 mol^{-1}) and is insensitive to small changes in pressure, it may be taken as constant. For illustrative purposes H_2 and O_2 may be assumed to be ideal gases. Then

$$\Delta V_{T,P} = V_{H_2O}^{(l)} - \frac{RT}{P} - \tfrac{1}{2}\frac{RT}{P}$$

or

$$\Delta V_{T,P} = V_{H_2O}^{(l)} - \tfrac{3}{2}\frac{RT}{P}$$

Thus

$$d\mathcal{E}_T = \frac{1}{\mathcal{N}\mathcal{F}}(\tfrac{3}{2}RT\, d\ln P - V_{H_2O}^{(l)}\, dP)$$

or

$$\Delta\mathcal{E}_T = \frac{1}{\mathcal{N}\mathcal{F}}\left(\tfrac{3}{2}RT\int_1^5 d\ln P - V_{H_2O}^{(l)}\int_1^5 dP\right)$$

$$= \frac{1}{(2)(96{,}487)}\left[\frac{(3)(8.314)(298)}{(2)}(1.609) - \frac{(18.14)(4)}{(9.870)}\right]$$

$$= +0.031 \text{ V}$$

Thus

$$\mathcal{E} = 1.229 + 0.031 = 1.260 \text{ V}$$

c. Since N_2 does not participate in the reaction, $\nu_{N_2} = 0$, and the presence of N_2 may be ignored except as it affects the fugacity of O_2 in the oxidant phase. Also, the H_2 and H_2O phases are in the same states as in part *a.*, and so their fugacities are unchanged. Equation (4-421) is applied in this case:

$$\Delta\mathcal{E}_{T,P} = \frac{-\nu_{O_2}RT}{\mathcal{N}\mathcal{F}} \ln \frac{\hat{f}_{O_2}(B)}{\hat{f}_{O_2}(A)}$$

At one atmosphere, $\hat{f}_{O_2}$ may be approximated by the oxygen partial pressure. Thus.

$$\Delta\mathcal{E}_{T,P} = \frac{(8.314)(298)}{(2)(2)(96{,}487)} \ln \frac{(0.21)(1)}{(1)}$$

$$= -0.010 \text{ V}$$

and

$$\mathcal{E} = 1.229 - 0.010 = 1.219 \text{ V}$$

Standard Electrode Potentials In the selection of reactions in cell design, it is advantageous to have tabulations of emf's for as many cells as possible. A great reduction in the space required for such a summary is provided by the use of **standard electrode potentials.** An electrochemical cell may be considered to consist of two parts, called half cells, each of which is associated with one of the cell electrode reactions. The additive property of the Gibbs function then allows the whole-cell emf to be divided into two **electrode potentials,** ascribable respectively to the anodic and cathodic reactions. The number of tabulated potentials may be further reduced by presenting them at specified standard states; hence the name "standard" electrode potential.

Once a table of standard electrode potentials has been compiled, the standard emf's of new cells can be calculated by combination of the potentials for the appropriate half cells. (Ignored here are experimental problems involved in actually joining certain half-cell combinations.) Extrapolation of standard emf's to real cell conditions is then done with thermodynamic formulas presented earlier.

Although the *difference* between two electrode potentials has significance (i.e., as an emf), the magnitudes of individual potentials are quite arbitrary. Thus in working up a table of such data one must choose a **reference electrode** as a basis and assign an arbitrary potential to it. Most modern compilations are referred to a standard hydrogen electrode, which is arbitrarily given a potential of zero. The ionic reactions corresponding to electrode potentials may be tabulated as oxidation reactions (e.g., $M \rightarrow M^+ + e^-$) or as reduction reactions (e.g., $M^+ + e^- \rightarrow M$), or they may be written as they actually occur in a reversible cell, the other half cell of which contains a fixed reference electrode. The quantities so defined are usually called oxidation potentials, reduction potentials, and Gibbs electrode potentials respectively, and they will in general differ from one another in sign and/or magnitude. Each of these systems has its adherents, so that caution must be observed in using electrode-potential data extracted from the literature.

Real Galvanic Cells The emf of a galvanic cell is the *maximum* voltage theoretically obtainable under constant T, P (or isothermal steady-flow) conditions. Similarly, the absolute value of the emf for an electrolytic cell is the *minimum* applied voltage required to make the assumed overall reaction proceed at constant temperature and pressure. However, considerations of residence time and product yield and quality, rather than minimum voltage requirements, are usually of primary importance in electrolytic processing, so that thermodynamic considerations enter into electrolysis design calculations. The equations presented earlier therefore find their greatest application in galvanic-cell design for which, even though they are not generally applicable to terminal-voltage calculations, they nonetheless provide qualitative information on the effects of system variables on cell output and form the basis for efficiency calculations.

Although the cell terminal voltage $\mathcal{V}$ cannot be evaluated from thermodynamic considerations, a useful equation relating $\mathcal{E}$, $\mathcal{V}$, and the heat generated by the cell can be derived from the first and second laws. If the electrical work is given by Eq. (4-395), then

$$dU = dQ - P\,dV + \mathcal{V}\,dq \qquad (4\text{-}422)$$

Comparing Eqs. (4-398) and (4-422) for the same change in state, one obtains

$$dQ = T\,dS + (\mathcal{E} - \mathcal{V})\,dq$$

For a process at constant T and P, this becomes

$$dQ = \left[T\left(\frac{\partial S}{\partial q}\right)_{T,P} + (\mathcal{E} - \mathcal{V})\right] dq$$

which, if $\mathcal{E}$ and $\mathcal{V}$ are not functions of q (e.g., for a "saturated" cell), may be integrated to give

$$Q = T\,\Delta S_{T,P} - \mathcal{N}\mathcal{F}(\mathcal{E} - \mathcal{V}) \qquad (4\text{-}423)$$

Equation (4-323) applies both to closed systems at constant T and P and to isothermal steady-flow systems.

The $T\,\Delta S$ term in Eq. (4-323) is the reversible heat generation of the cell, because $\mathcal{V} = \mathcal{E}$ for a reversible cell and the second term goes to zero. If the cell is short- circuited, then the terminal voltage becomes zero, and Eq. (4-423) becomes

$$Q = \Delta H_{T,P}$$

that is, the entire enthalpy change on reaction is manifested as heat, the same as would obtain in an ordinary chemical reaction. The obvious interpretation of the second term on the right side of Eq. (4-423) is that it represents the irreversible heat generated by a real cell. Equation (4-423) is useful for translation of laboratory cell-performance data into the information required for heat-transfer calculations in cell design.

Cell-performance data are often presented in the form of a $\mathcal{V}$ versus $\mathcal{I}$ plot, where $\mathcal{I}$ is current or current density (current divided by the superficial electrode area perpendicular to the current within the cell). A typical performance curve for a fuel cell is shown in Fig. 4-37. The dashed horizontal line is the cell emf, and the intercept at $\mathcal{I} = 0$ is the open-circuit voltage $\mathcal{V}_c$. According to Eq. (4-423), the irreversible heat generation is proportional to the indicated difference $(\mathcal{E} - \mathcal{V})$. The S shape of the curve is characteristic of many cells and results from the cumulative effects of irreversibilities, which differ in their nature for different ranges of the current. An important effect at moderate currents is that due to the finite internal resistance of the cell; other effects which predominate at lower and higher currents reflect kinetic and mass-transfer limitations and are collectively called polarizations. Detailed discussion of these effects is beyond the scope of classical thermodynamics and can be found elsewhere.

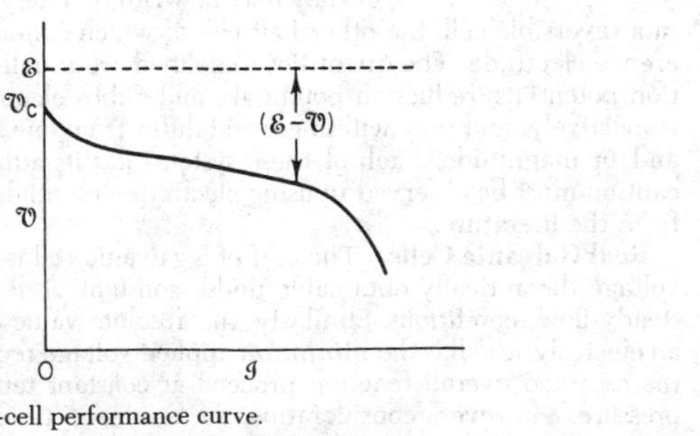

FIG. 4-37 Fuel-cell performance curve.

Thermodynamics of Electrolyte Solutions Calculation of the emf of a cell in which one or more of the reactants or products is an ionizable species in solution requires knowledge of the behavior of dissolved electrolyte systems. In addition, the feasibility of constructing essentially reversible galvanic cells has led to the widespread use of these cells as research tools for elucidating the physicochemical behavior of electrolyte solutions. Given here is a very brief account of some of the more important special features of the classical thermodynamics of electrolytes. (For more thorough treatments, see Denbigh, *Principles of Chemical Equilibrium*, 2d ed., Cambridge, London, 1966; and Lewis, Randall, Pitzer, and Brewer, *Thermodynamics*, McGraw-Hill, New York, 1961.)

In keeping with the rest of this treatment of thermodynamics, the present development employs mole fraction ("rational" scale) as the composition variable. Much of the chemical literature uses molality ("practical" scale), and one must use care in interpreting derived parameters reported elsewhere in terms of those defined in the following paragraphs. For concreteness, we deal with binary systems consisting of a **solvent** (subscript 0) in which an ionizable **solute** (subscript s) is dissolved. The solute-solvent distinction is generally meaningful for electrolyte solutions because one component is invariably a dissolved gas or solid which in practical applications rarely reaches high concentrations.

The existence of well-established limiting laws is essential for the useful application of thermodynamic principles to the estimation and correlation of properties of real systems. For binary electrolyte solutions, such laws have been found to exist for the dilute-solute region:

$$\lim_{\bar{x}_s \to 0}\left(\frac{\hat{f}_s}{K\bar{x}_s^{\nu}}\right) = 1 \qquad (4\text{-}424)$$

$$\lim_{\bar{x}_0 \to 1}\left(\frac{\hat{f}_0}{f_0\bar{x}_0}\right) = 1 \qquad (4\text{-}425)$$

Equations (4-424) and (4-425) may be thought of as modifications of Henry's law and the Lewis-Randall rule in which the exponent ν is an integer greater than unity whose interpretation will be discussed later. The variables $\bar{x}_s$ and $\bar{x}_0$ are not mole fractions as usually defined but are given by

$$\bar{x}_0 = \frac{x_0}{\nu + x_0(1 - \nu)} \qquad (4\text{-}426)$$

$$\bar{x}_s = \frac{\nu x_s}{1 + x_s(\nu - 1)} \qquad (4\text{-}427)$$

where x_0 and x_s are the usual ("nominal") mole fractions of solvent and solute respectively. Given the limiting laws for fugacity as expressed in Eqs. (4-424) and (4-425), the mole fractions defined in Eqs. (4-426) and (4-427) are not arbitrary but follow directly from the Gibbs-Duhem equation and the requirement that $\Sigma\bar{x}_i = 1$. The new mole fractions are called "true" mole fractions and are employed throughout this subsection.

The limiting fugacity laws suggest the definition of a number of special properties related to the chemical potential. Thus, for the *solute*,

$$\gamma_s = \hat{f}_s/K\bar{x}_s^{\nu} \qquad (4\text{-}428)$$

and

$$\hat{a}_s = \hat{f}_s/K \qquad (4\text{-}429)$$

so that

$$\lim_{\bar{x}_s \to 0} \gamma_s = 1 \qquad (4\text{-}430)$$

or

$$\lim_{\bar{x}_s \to 0}\left(\frac{\hat{a}_s}{\bar{x}_s^{\nu}}\right) = 1 \qquad (4\text{-}431)$$

Another set of definitions is often used for the solute. These are the **mean ionic activity coefficient** $\gamma_\pm$ and the **mean ionic activity** $\hat{a}_\pm$, which are related to the above quantities by

$$\gamma_\pm^{\nu} = \gamma_s \qquad (4\text{-}432)$$

and

$$\hat{a}_\pm^{\nu} = \hat{a}_s \qquad (4\text{-}433)$$

so that

$$\lim_{\bar{x}_s \to 0} \gamma_\pm = 1 \qquad (4\text{-}434)$$

and

$$\lim_{\bar{x}_s \to 0}\left(\frac{\hat{a}_\pm}{\bar{x}_s}\right) = 1 \qquad (4\text{-}435)$$

For the *solvent*, by definition

$$\gamma_0 = \hat{f}_0 / f_0 \bar{x}_0 \tag{4-436}$$

and

$$\hat{a}_0 = \hat{f}_0 / f_0 \tag{4-437}$$

so that

$$\lim_{\bar{x}_0 \to 1} \gamma_0 = 1 \tag{4-438}$$

or

$$\lim_{\bar{x}_0 \to 0} \left(\frac{\hat{a}_0}{\bar{x}_0} \right) = 1 \tag{4-439}$$

To get a larger numerical measure of the often very substantial solvent nonidealities at low solute concentrations, solvent behavior is sometimes characterized by an **osmotic coefficient** rather than by an activity coefficient. The osmotic coefficient g is defined by

$$\mu_0 = \mu_0^0 + gRT \ln \bar{x}_0 \tag{4-440}$$

Comparison of Eq. (4-440) with the alternative expression

$$\mu_0 = \mu_0^0 + RT \ln \gamma_0 \bar{x}_0$$

shows that g and γ_0 are related by

$$g = 1 + (\ln \gamma_0 / \ln \bar{x}_0) \tag{4-441}$$

and therefore

$$\lim_{\bar{x}_0 \to 1} g = 1 \tag{4-442}$$

So far no interpretation has been given to ν, and classical thermodynamics by itself cannot offer one. Experimentally it is found that, for a given solvent, ν can vary with the solute species, so that it is desirable to have some basis for fixing ν. This basis is provided by ionic theory, which postulates complete dissociation of the solute into its constituent ions at infinite dilution. Thus, if a molecule of solute contains ν_+ ions of positive valence and ν_- ions of negative valence; a total of $\nu_+ + \nu_-$ mol of ions results from the complete ionization of 1 mol of solute. Then, for a solution nominally containing n_0 mol of solvent and n_s mol of solute, the true number of moles will be $n_0 + (\nu_+ + \nu_-)n_s$ if the solute is completely ionized. The true mole fraction of the solvent is thus

$$\frac{x_0}{x_0 + (\nu_+ + \nu_-)(1 - x_0)}$$

which is identical to $\bar{x}_0$ defined by Eq. (4-426) if $\nu = \nu_+ + \nu_-$. This interpretation allows establishment of the correct limiting laws for a binary solution containing an ionizable solute simply from the chemical formula of the solute. For example, a dilute solution of $La_2(SO_2)_3$ in water contains 5 mol of ions per mole of solute, and one therefore defines

$$\gamma_s = \hat{f}_s / K \bar{x}_s^5$$
$$\hat{a}_s = \hat{f}_s / K$$
$$\gamma_0 = \hat{f}_0 / f_0 \bar{x}_0$$
$$\hat{a}_0 = \hat{f}_0 / f_0$$

where, as a result of Eqs. (4-426) and (4-427),

$$\bar{x}_0 = x_0 / (5 - 4x_0)$$
$$\bar{x}_s = 5x_s / (1 + 4x_s)$$

THERMODYNAMIC ANALYSIS OF PROCESSES

The object of a thermodynamic analysis of a real process is the determination of the efficiency of the process from the standpoint of energy utilization. Further, it is useful to calculate the influence of each irreversibility individually on the overall efficiency of the process.

The present treatment will be limited to consideration of steady-flow processes, for which the energy equation resulting from the first law of thermodynamics is

$$\Delta H + \Delta E_P + \Delta E_K = Q - W_s$$

or

$$W_s = Q - \Delta H - \Delta E_P - \Delta E_K \tag{4-443}$$

where H = enthalpy
E_P = potential energy
E_K = kinetic energy
Q = heat
W_s = shaft work

and Δ signifies a difference in values between outlet and inlet streams.

Figure 4-38 is a schematic representation of the general process considered, which may be simple or complex. We presume that the

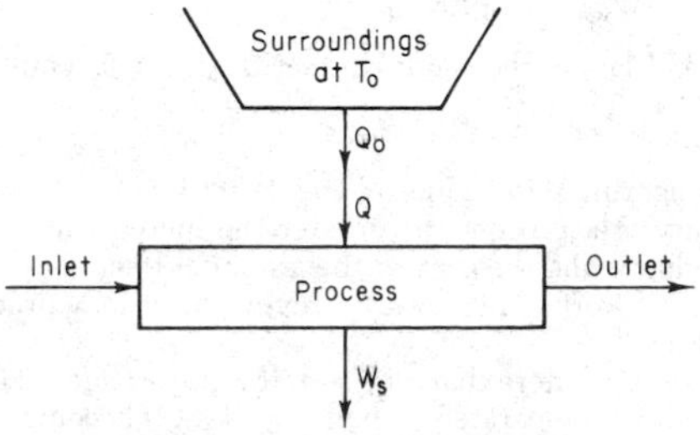

FIG. 4-38 Schematic diagram of a steady-flow process.

process exists in surroundings which constitute a heat reservoir at the constant temperature T_0. Heat exchange between process and surroundings causes entropy changes in the surroundings in the amount

$$\Delta S_0 = Q_0 / T_0$$

Since $Q_0 = -Q$, this equation may also be written as

$$Q = -T_0 \Delta S_0 \tag{4-444}$$

Combination of Eqs. (4-443) and (4-444) and rearrangement give

$$W_s = -T_0 \Delta S_0 - \Delta H - \Delta E_P - \Delta E_K \tag{4-445}$$

As it stands, this equation is of little use because the entropy change of the surroundings ΔS_0 is rarely known. However, for the special case of a completely reversible process, the second law of thermodynamics provides the equation

$$\Delta S + \Delta S_0 = 0$$

or

$$\Delta S_0 = -\Delta S \tag{4-446}$$

where ΔS is the entropy change of the flowing stream between inlet and outlet. Substitution of Eq. (4-446) into Eq. (4-445) provides

$$W_{ideal} = T_0 \Delta S - \Delta H - \Delta E_P - \Delta E_K \tag{4-447}$$

where the work is now represented by W_{ideal} so as to indicate clearly that it is the work associated with a completely reversible process for which the change of state is implied by the property changes ΔS, ΔH, ΔE_P, and ΔE_K. When these property changes are values taken for a real process, then Eq. (4-447) yields the work required to bring about *the same change of state* in a completely reversible process. It is the minimum work requirement or the maximum work obtainable, depending on whether the process requires or produces work. The stipulation of *complete* reversibility requires not only that the process be internally reversible but also that heat transfer between system and surroundings also be reversible. Such a process is taken as the standard or ideal against which to measure the efficiencies of real processes that accomplish the same change of state. Thus the thermodynamic efficiency η is given by

$$\eta(\text{work produced}) = W_s / W_{ideal} \tag{4-448a}$$

$$\eta(\text{work required}) = W_{ideal} / W_s \tag{4-448b}$$

The difference between the ideal work for a given change of state and the real work of a process that brings about the same change is called the lost work. Thus by definition,

$$W_{lost} = W_{ideal} - W_s \tag{4-449}$$

and is given as the difference between Eqs. (4-447) and (4-443), both written for the same change of state:

$$W_{lost} = T_0\,\Delta S - Q \tag{4-450}$$

Alternatively,

$$W_{lost} = T_0\,\Delta S + Q_0$$

Since $Q_0 = T_0\,\Delta S_0$

then $W_{lost} = T_0\,\Delta S + T_0\,\Delta S_0 = T_0(\Delta S + \Delta S_0)$

or

$$W_{lost} = T_0\,\Delta S_{total} \tag{4-451}$$

By the second law of thermodynamics, $\Delta S_{total} \geqq 0$. Thus,

$$W_{lost} \geqq 0 \tag{4-52}$$

The engineering significance of this result is clear. The greater the irreversibility of a process, the greater the increase in total entropy accompanying it and the greater the amount of energy that becomes unavailable as work. Thus every irreversibility in a process carries with it a price.

For processes of more than one step it is advantageous to calculate W_{lost} for each step separately. Then Eq. (4-449) becomes

$$\Sigma W_{lost} = W_{ideal} - W_s \tag{4-453}$$

For processes that require work this equation is written

$$W_s = W_{ideal} - \Sigma W_{lost} \tag{4-453a}$$

The terms on the right side represent an analysis of the actual work, showing the part ideally required to bring about the change of state and the parts that are required as a result of the irreversibilities in the various steps of the process.

For processes that produce work, Eq. (4-453) is written

$$W_{ideal} = W_s + \Sigma W_{lost} \tag{4-453b}$$

Here the terms on the right side represent an analysis of the ideal work showing the part actually produced and the parts that become unavailable because of irreversibilities in the various steps of the process.

Example 8 We wish to make a thermodynamic analysis of a simple Linde system for the separation of air into gaseous oxygen and nitrogen, as depicted in Fig. 4-39. Table 4-25 lists a set of operating conditions for the numbered points of the diagram. Heat leaks into the column of 147 J/mol of entering air and into the exchanger of 63 J/mol entering air have been assumed.

This problem will be worked on the basis of 1 mol of entering air, which is assumed to contain 79 mole percent N_2 and 21 percent O_2. A material balance on the nitrogen gives

$$0.79 = 0.9148x$$

or

$$x = 0.8636 \text{ mol of nitrogen product}$$

Therefore, the N_2 product stream contains 0.8636 mol, and the O_2 product stream contains 0.1364 mol.

Calculation of ideal work. If changes in kinetic and potential energies are neglected, Eq. (4-447) becomes

$$W_{ideal} = T_0\,\Delta S - \Delta H$$

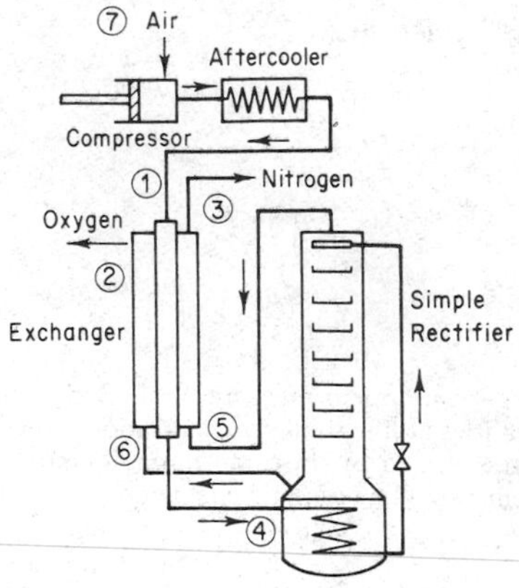

FIG. 4-39 Diagram of a simple gaseous oxygen process.

The surroundings temperature T_0 will be taken as 300 K. With the material quantities as calculated previously and the property values as given in Table 4-25, ΔH and ΔS for the overall process are determined as follows:

$$\Delta H = (13{,}460)(0.1364) + (12{,}074)(0.8636) - (12{,}407)(1)$$
$$= -144 \text{ J}$$
$$\Delta S = (118.48)(0.1364) + (114.34)(0.8636) - (117.35)(1)$$
$$= -2.445 \text{ J/K}$$

Therefore

$$W_{ideal} = (300)(-2.445) + 144$$
$$= -590 \text{ J}$$

Calculation of the actual work of compression. In order not to complicate this example, the work of compression will be calculated by means of the equation for an ideal gas in a three-stage reciprocating compressor with complete intercooling. This equation is based on isentropic compression in each stage, and it will be assumed that the work so calculated represents 80 percent of the actual work. This equation may be found in any number of standard textbooks on thermodynamics:

$$W_s = \frac{-n\gamma RT_1}{(0.8)(\gamma - 1)}\left[\left(\frac{P_2}{P_1}\right)^{(\gamma-1)/n\gamma} - 1\right]$$

where n = number of stages, here taken as 3
γ = ratio of heat capacities, here taken as 1.4
T_1 = initial absolute temperature, equal to 300 K
P_2/P_1 = overall pressure ratio, equal to 54.5
R = universal gas constant, equal to 8.314 J/(mol·K)

The efficiency factor of 0.8 has already been incorporated in the equation. Substitution of values gives

$$W_s = \frac{-(3)(1.4)(8.314)(300)}{(0.8)(0.4)}[(54.5)^{0.4/(3)(1.4)} - 1]$$
$$= 15{,}171 \text{ J}$$

The heat transferred to the surroundings during compression as a result of intercooling and aftercooling is determined from the first law:

$$Q = \Delta H + W_s = (12{,}046 - 12{,}407) - 15{,}171$$
$$= -15{,}532 \text{ J}$$

Calculation of lost work terms. The equation used is Eq. (4-450),

$$W_{lost} = T_0\,\Delta S - Q$$

and it remains only to evaluate ΔS and Q for the various steps of the process.

TABLE 4-25 States and Values of Properties for the Process of Fig. 4-39*

Point	P, bar	T, K	Composition	State	H, J/mol	S, J/(mol·K)
1	55.22	300	Air	Superheated	12,046	82.98
2	1.01	295	Pure O_2	Superheated	13,460	118.48
3	1.01	295	91.48% N_2	Superheated	12,074	114.34
4	55.22	147.2	Air	Superheated	5,850	52.08
5	1.01	79.4	91.48% N_2	Saturated vapor	5,773	75.82
6	1.01	90	Pure O_2	Saturated vapor	7,485	83.69
7	1.01	300	Air	Superheated	12,407	117.35

*Properties on the basis of Miller and Sullivan, U.S. Bur. Mines Tech. Pap. 424 (1928).

a. *Compression*

$$\Delta S = 82.98 - 117.35 = -34.37 \text{ J/K}$$
$$Q = -15{,}532 \text{ J}$$
$$W_{lost} = (300)(-34.37) + 15{,}532 = 5221 \text{ J}$$

b. *Heat exchanger*

$$\Delta S = (0.8636)(114.34 - 75.82) + (0.1364)(118.48 - 83.69) + (1)(52.08 - 82.98) = 7.11 \text{ J/K}$$
$$Q = 63 \text{ J}$$
$$W_{lost} = (300)(7.11) - 63 = 2070 \text{ J}$$

c. *Column*

$$\Delta S = (0.8636)(75.82) + (0.1364)(83.69) - (1)(52.08) = 24.81 \text{ J/K}$$
$$Q = 147 \text{ J}$$
$$W_{lost} = (300)(24.81) - 147 = 7296 \text{ J}$$

Summary. Since the process requires work, Eq. (4-453*a*) is appropriate for the thermodynamic analysis:

$$W_s = W_{ideal} - \Sigma W_{lost}$$

The various terms on the right appear as entries in the following summary of results:

	J	% of W_s
W_{ideal}	− 591	3.9
$-W_{lost}$		
Compression	−5221	34.4
Heat exchanger	−2070	13.6
Column	−7296	48.1
$W_s = W_{ideal} - \Sigma W_{lost}$	−15,178	100.0

The value of W_s determined by summing the individual terms should be the same as that calculated by the compressor formula (−15,171 J). The slight discrepancy is the result of the accumulation of round-off errors. The thermodynamic efficiency of the process is 3.9 percent, as given in the first row of the summary table. The largest lost-work term results from irreversibilities in the column.

Section 5*

Fluid and Particle Mechanics

Byron C. Sakiadis, Ph.D., *Senior Research Fellow, Engineering Technology Laboratory, E. I. du Pont de Nemours & Co.; Member, American Institute of Chemical Engineers.*

*The author acknowledges the contribution of the work of D. F. Boucher and G. E. Alves, editors of this section in the fifth edition of the *Handbook*.

PARTICLE DYNAMICS

Nomenclature and Units

In this listing, symbols used in the section are defined in a general way and appropriate SI and U.S. customary units are given. Specific definitions, as denoted by subscripts, are stated at the place of application in the section. Some specialized symbols used in the section are defined only at the place of application.

Symbol	Definition	SI units	U.S. customary units
A	Area	m²	ft²
a	Area	m²	ft²
a	Duct or channel width	m	ft
a	Coefficient, general		
B	Height	m	ft
b	Duct or channel height	m	ft
b	Coefficient, general		
C	Coefficient, general		
C	Conductance	m³/s	ft³/s
C_a	Capillary number	Dimensionless	Dimensionless
c_p	Constant-pressure specific heat	J/(kg·K)	Btu/(lb·°R)
c_v	Constant-volume specific heat	J/(kg·K)	Btu/(lb·°R)
D	Diameter	m	ft
d	Diameter	m	ft
E	Modulus of elasticity	N/m²	lbf/ft²
F	Force	N	lbf
F	Friction loss	(N·m)/kg	(ft·lbf)/lb
F	Correction factor	Dimensionless	Dimensionless
f	Frequency	Hz	1/s
f	Friction factor	Dimensionless	Dimensionless
G	Mass velocity	kg/(s·m²)	lb/(s·ft²)
g	Local acceleration due to gravity	m/s²	ft/s²
g_c	Dimensional constant	1.0 (kg·m)/(N·s²)	32.2 (lb·ft)/(lbf·s²)
h	Head of fluid, height	m	ft
i	Specific enthalpy	J/kg	Btu/lb
I	Electric current	A	A
J	Mechanical equivalent of heat	1.0 (N·m)/J	778 (ft·lbf)/Btu
K	Index, constant or flow parameter		
K	Fluid bulk modulus of elasticity	N/m²	lbf/ft²
k	Ratio of specific heats	Dimenionless	Dimensionless
L	Length	m	ft
M	Molecular weight	kg/mol	lb/mol
m	Mass	kg	lb
m	Thickness	m	ft
N	Number of data points or items	Dimensionless	Dimensionless
N	Frictional resistance	Dimensionless	Dimensionless
N_S	Strouhal number	Dimensionless	Dimensionless
N_{De}	Dean number	Dimensionless	Dimensionless
N_{Fr}	Froude number	Dimensionless	Dimensionless
N_{Re}	Reynolds number	Dimensionless	Dimensionless
N_{We}	Weber number	Dimensionless	Dimensionless
n	Pulsation frequency	Hz	1/s
n	Constant, general		
n	Number of items	Dimensionless	Dimensionless
p	Pressure	Pa	lbf/ft²
Q	Heat	J	Btu
Q	Volume	m³	ft³
q	Volume flow rate	m³/s	ft³/s
R	Gas constant	8314 J/(K·mol)	1545 (ft·lbf)/(mol·°R)
R	Radius	m	ft
R	Electrical resistance	Ω	Ω
R	Head reading	m	ft
r	Radius	m	ft
r	Pressure ratio	Dimensionless	Dimensionless
S	Specific surface area	m²/m³	ft²/ft³
S	Fluid head loss	Dimensionless	Dimensionless
S	Specific energy loss	m/s²	lbf/lb
S	Speed	m³/s	ft³/s
s	Specific entropy	J/(kg·K)	Btu/(lb·°R)
T	Temperature	K (°C)	°R (°F)
t	Time	s	s
u	Specific internal energy	J/kg	Btu/lb
u	Velocity	m/s	ft/s
V	Velocity	m/s	ft/s
v	Specific volume	m³/kg	ft³/lb
W	Work	N·m	lbf·ft
W	Weight	kg	lb
w	Weight flow rate	kg/s	lb/s
x	Weight fraction	Dimensionless	Dimensionless
x	Distance or length	m	ft
Y	Expansion factor	Dimensionless	Dimensionless
y	Distance or length	m	ft
Z	Vertical distance	m	ft
z	Gas-compressibility factor	Dimensionless	Dimensionless
z	Vertical distance	m	ft

Greek symbols

Symbol	Definition	SI units	U.S. customary units
α	Viscous-resistance coefficient	1/m²	1/ft²
α	Angle	°	°
β	Inertial-resistance coefficient	1/m	1/ft
β	Ratio of diameters	Dimensionless	Dimensionless
Γ	Liquid loading	kg/(s·m)	lb/(s·ft)
Γ	Pulsation intensity	Dimensionless	Dimensionless
δ	Thickness	m	ft
ε	Wall roughness	m	ft
ε	Voidage—fractional free volume	Dimensionless	Dimensionless
η	Viscosity, nonnewtonian fluids	Pa·s	lb/(ft·s)
θ	Angle	°	°
λ	Molecular mean free-path length	m	ft
μ	Viscosity	Pa·s	lb/(ft·s)
ν	Kinematic viscosity	m²/s	ft²/s
ρ	Density	kg/m³	lb/ft³
σ	Surface tension	N/m	lbf/ft
σ_c	Cavitation number	Dimensionless	Dimensionless
τ	Shear stress	N/m²	lbf/ft²
ϕ	Shape factor	Dimensionless	Dimensionless
ϕ	Angle	°	°
ψ	Sphericity	Dimensionless	Dimensionless

INTRODUCTION TO FLUID MECHANICS

General Reference: Streeter, *Handbook of Fluid Dynamics*, McGraw-Hill, New York, 1961.

NATURE OF FLUIDS

A **fluid** is a substance which undergoes continuous deformation when subjected to a shear stress. The resistance offered by a real fluid to such deformation is called fluid **viscosity.** For gases and for simple (low-molecular-weight) liquids, the viscosity is constant if static pressure and temperature are fixed. Such materials are called **newtonian.**

Let us consider two layers of fluid a distance y apart, as shown in Fig. 5-1, with the top layer moving parallel to the bottom layer at a velocity u relative to the bottom layer. With a newtonian fluid, to maintain this motion requires a force F, the magnitude of which is given by

$$F = \mu u A / g_c y \tag{5-1}$$

Expressed on a differential basis,

$$\tau = (\mu / g_c)(du/dy) \tag{5-2}$$

where μ = fluid viscosity, τ = shear stress, g_c = dimensional constant, and du/dy = velocity gradient. For definitions of F, A, u, and y, see Fig. 5-1. **Viscosity** can thus be considered to be a momentum conductivity analogous to thermal conductivity in conductive-heat transfer and to the diffusion coefficient in diffusive mass transfer.

An **ideal** or **perfect fluid** is a hypothetical gas or liquid which offers no resistance to shear and therefore has zero viscosity. The imaginary perfect fluid is not to be confused with a "perfect or ideal gas." In most flow problems, highly incorrect results are obtained if viscosity is neglected, although in the same problems the pressure-volume-temperature relations for perfect gases may often be safely used.

If the viscosity of a fluid is a function of the shear stress or equivalently of the shear rate ($= |du/dy|$), as well as of temperature and pressure, the fluid is called a **nonnewtonian fluid.** Nonnewtonian fluids are usually divided into three general classes: (1) those whose properties are independent of time under shear, (2) those whose properties are dependent upon duration of shear, and (3) those which exhibit many characteristics of a solid. See Bird, Armstrong, and Hassager, *Dynamics of Polymeric Liquids*, vol. 1: *Fluid Mechanics*, Wiley, New York, 1977; Metzner, "Flow of Non-Newtonian Fluids," in Streeter, op cit.; and Skelland, *Non-Newtonian Flow and Heat Transfer*, Wiley, New York, 1967.

1. *Time-independent.* The following three types of materials are in this class:

a. Bingham-plastic fluids are probably the simplest nonnewtonian fluids because they differ from newtonian fluids only in that their linear relationship between shear stress and shear rate does not go through the origin. This is illustrated by curve B in Fig. 5-2, which shows that a finite shear stress τ_y is required to initiate flow. For comparison, a newtonian fluid has a curve similar to curve A. Examples of fluids exhibiting Bingham-plastic behavior include water suspensions of rock [Wilhelm, Wroughton, and Loeffel, *Ind. Eng. Chem.*, **31,** 622–629 (1939)] or grains [Binder and Busher, *J. Appl. Mech.*, **13,** A101–A105 (1946)] and sewage sludge [Caldwell and Babbitt, *Ind. Eng. Chem.*, **33,** 249–256 (1941)].

b. Pseudoplastic materials include the majority of nonnewtonian fluids. Among them are polymeric solutions or melts and suspensions of paper pulp or pigments. The shape of the flow curve is shown as curve C in Fig. 5-2.

In general, the flow curve, over a range of shear rate, can be approximated by a straight line on a logarithmic plot, whereby

$$\tau = K(du/dy)|du/dy|^{n-1} \qquad \text{with } n < 1 \tag{5-3}$$

where τ = shear stress; du/dy = velocity gradient; K = consistency index, $N \cdot s^n/m^2$ [$(lbf \cdot s^n)/ft^2$]; and n = exponent, dimensionless. The fluid viscosity is given by

$$\eta = g_c K |du/dy|^{n-1} \tag{5-4}$$

where η = viscosity, Pa·s [lb/(ft·s)]; and g_c = dimensional constant. Note that the symbol μ for viscosity is reserved for newtonian fluids.

Fluids with a flow curve described by Eqs. (5-3) and (5-4) are referred to as **power-law nonnewtonian fluids.** Judicious choice of the numerical value of the material constants K and n can lead to accurate flow predictions even if these equations are applicable over only a limited range of shear rate. In general, the maximum shear stress (or shear rate) in the flow being considered must not exceed the maximum shear-stress (or shear-rate) limit of applicability of Eq. (5-3).

c. Dilatant materials exhibit rheological behavior opposite to that of pseudoplastics. Curve D (Fig. 5-2) shows the typical shape of the flow curve. As can be seen, apparent viscosity increases with increasing shear rate.

The flow behavior of dilatant materials over a limited range of shear rate can be represented with Eqs. (5-3) and (5-4) with $n > 1$. Some examples of dilatant materials are starch or mica suspensions in water, quicksand, and beach sand. Extensive discussions of dilatant suspensions, together with a listing of dilatant systems, are given by Bauer and Collins ("Thixotropy and Dilatancy," in Eirich, *Rheology*, vol. 4, Academic, New York, 1967); Green and Griskey [*Trans. Soc. Rheol*, **12**(1), 13–25, 27–37 (1968)]; and Griskey and Green [*Am. Inst. Chem. Eng. J.*, **17,** 725–728 (1971)].

2. *Time-dependent.* Included are those materials for which shear stress changes with duration of shear. Excluded are changes which might be produced through mechanical breaking or destruction of particles or molecular bonds.

a. Thixotropic fluids possess a structure the breakdown of which is a function of time under shear. As the structure breaks down with

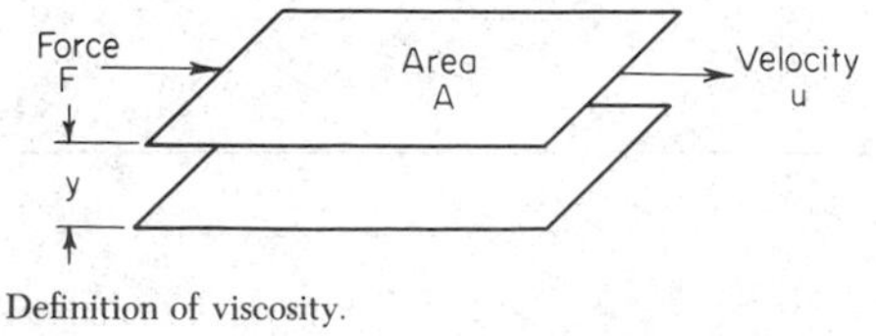

FIG. 5-1 Definition of viscosity.

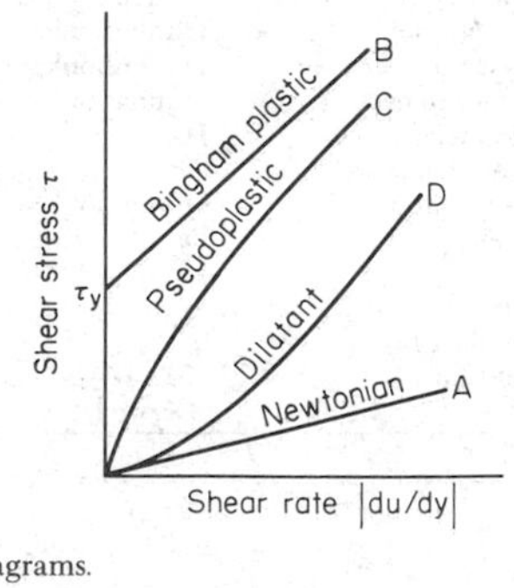

FIG. 5-2 Shear diagrams.

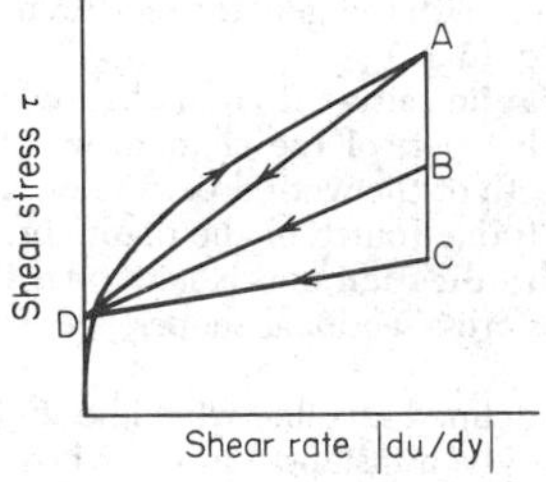

FIG. 5-3 Shear diagram for thixotropic fluid.

constant shear rate, shear stress decreases. This structure can rebuild itself if not prevented from doing so by externally applied forces. The shear diagram of a thixotropic fluid as obtained with a rotational viscometer is given in Fig. 5-3. The area within loop *DAD* is an indication of the amount of thixotropy. If the shear rate is held constant after point *A* has been reached on the up curve, the shear stress will decrease along path *AB* until point *C* is reached, beyond which no further breakdown can occur for that shear rate. If shear rate is then decreased, the down curve *CD* is followed. Any number of intermediate down curves, such as *BD*, are possible. Examples of such fluids are mayonnaise, drilling muds, paints, and inks. For a detailed discussion of thixotropic materials, together with a listing of thixotropic systems, see Bauer and Collins in Eirich, loc. cit.

b. Rheopectic materials will set up or build up, i.e., increase in apparent viscosity very rapidly upon being rhythmically shaken or tapped. Examples of these materials are bentonite sols, vanadium pentoxide sols, and gypsum suspensions in water. This phenomenon has been observed under constant shear rate (see Bauer and Collins in Eirich, loc. cit.).

3. *Viscoelastic fluids.* These fluids exhibit elastic recovery from deformations which occur during flow. Polymeric liquids comprise the largest group of fluids in this class. In the flow of these fluids, normal stresses (i.e., stresses perpendicular to direction of flow) in addition to the usual tangential stresses are built up. These normal stresses give rise to several unusual phenomena, for example, the "Weissenberg effect," in which the fluid has a tendency to climb up a shaft rotating in the fluid. For the steady-state unaccelerated flow of viscoelastic fluids, the equations developed for pseudoplastic fluids apply; the elastic properties are generally manifested as "end effects." For additional details on the flow behavior of viscoelastic fluids, see Bird, Armstrong, and Hassager, loc. cit.; Brodkey, *The Phenomena of Fluid Motions*, Addison-Wesley, Reading, Mass., 1967; Eirich, *Rheology*, vols. 1, 2, 3, 4, and 5, Academic, New York, 1956, 1958, 1960, 1967, and 1969 respectively; Frederickson, *Principles and Applications of Rheology*, Prentice-Hall, Englewood Cliffs, N.J., 1964; Lodge, *Elastic Liquids*, Academic, New York, 1964; McKelvey, *Polymer Processing*, Wiley, New York, 1962; Metzner, "Flow Behavior of Thermoplastics," in Bernhardt, *Processing of Thermoplastic Materials*, Reinhold, New York, 1959; and Middleman, *The Flow of High Polymers*, Interscience, New York, 1968.

The applications of nonnewtonian flow theory to operations in **polymer processing,** such as mixing, extrusion, calendering, fiber spinning, and sheet forming, are described in Bernhardt, *Processing of Thermoplastic Materials*, Reinhold, New York, 1959; McKelvey, loc. cit.; Middleman, *Fundamentals of Polymer Processing*, McGraw-Hill, New York, 1977; and Wilkinson, *Non-Newtonian Fluids*, Pergamon, New York, 1960.

The unit of viscosity (i.e., **absolute viscosity**) in the SI system is the **pascal-second** (Pa·s). One Pa·s equals 10 **poise** (P), 1000 **centipoise** (cP), or 0.672 lb/(ft·s). Conversion factors to other units are given in Sec. 1, Table 1-4 and 1-7.

Kinematic viscosity of a fluid of density ρ and viscosity μ or η is $\nu = \mu/\rho$ or η/ρ. A unit of kinematic viscosity called the **stoke** (St) equals 1 cm^2/s. **Fluidity** is the reciprocal of viscosity.

Special terminology commonly used to specify the viscosity of dilute **polymer solutions** is discussed in Billmeyer, *Textbook of Polymer Science*, 2d ed., Wiley, New York, 1971, pp. 84–89; and Eirich, *Rheology*, vol. 4, Academic, New York, 1967, chap. 9.

A wide variety of viscometers is available for the measurement of viscosity, such as capillary, rotational, orifice, falling-ball, and oscillatory types. They are described in Van Wazer, Lyons, Kim, and Colwell, *Viscosity and Flow Measurement*, Interscience, New York, 1963.

In several common commercial viscometers, kinematic viscosity is determined from the time of efflux (seconds) of a fixed volume of liquid through a standard capillary tube or orifice. In such instruments, the entrance and kinetic effects often constitute an important part of the resistance to flow. Consequently, the relation between time of efflux and kinematic viscosity is empirically determined.

The material constants *K* and *n* of **power-law nonnewtonian fluids** can be determined with rotational viscometers [see Krieger and Maron, *J. Appl. Phys.*, **25,** 72–75 (1954); and Metzner, "Non-Newtonian Technology," in Drew and Hoopes, *Advances in Chemical Engineering*, vol. I, Academic, New York, 1956]. Discussions of theories for various viscometers are given by Oka (Eirich, *Rheology*, vol. 3, Academic, New York, 1960, chap. 2), Van Wazer et al. (loc. cit.), and Wohl [*Chem. Eng.*, **75**(7), 99–104 (Mar. 25, 1968)]. Practical aspects of viscometry are given by Bowen [*Chem. Eng.*, **68**(17), 119; (18), 131 (1961)].

Detailed information on viscosity is presented in tables in Sec. 3. See also *International Critical Tables*, McGraw-Hill, New York, 1926–1933; Weast, *Handbook of Chemistry and Physics*, 59th ed., Chemical Rubber, Cleveland, 1978–1979, pp. F49–F61; and Dean, *Lange's Handbook of Chemistry*, 12th ed., McGraw-Hill, New York, 1978, pp. 10-99–10-116. Reasonable estimates of viscosity can often be made when no data are available. Methods available for estimation are summarized in Sec. 3.

TERMINOLOGY IN FLUID MECHANICS

A flow is said to be **steady** if it is invariant with time; i.e., the mass flow rate is constant, and all other quantities (temperature, pressure, velocity) are independent of time. Conversely, the flow is said to be **unsteady** if the mass flow rate and/or other quantities vary with time. Unsteady flow can result from control-valve action, from the action of reciprocating machinery, or from unstable two-phase flow.

A flow is said to be **accelerated** if it is unsteady or if the velocity varies in the general direction of flow. Many effects associated with nonnewtonian viscoelastic fluids generally occur in accelerated flow.

A stream is said to be **uniform** if the shape and size of its cross section are the same throughout the channel. A temperature or velocity is said to be uniform throughout a region when it has the same value at all parts of the region at a given instant.

The **mean mass velocity** *G* of a stream past a given cross section, taken perpendicularly to the general direction of flow through the apparatus, is the weight rate of flow divided by the area of the given cross section. Throughout a channel of uniform cross-sectional area, the mean mass velocity is constant unless there is an accumulation or depletion of material within the channel. When considering the flow through a bank of tubes or a bed of solids, the term **superficial mass velocity** is given to the quantity obtained on dividing the weight rate by the total cross-sectional area of the enclosing chamber (without subtracting that part of the cross section occupied by the obstructions).

The **mean linear velocity** *V* of a stream past any given cross section is commonly taken to be the quantity obtained when the corresponding mean mass velocity is divided by the average density at the given cross section. Unless the flow is isothermal, the term mean linear velocity cannot be interpreted unless the rule chosen for determination of the average density is stated definitely. Consequently it is preferable, when possible, to treat nonisothermal flow in terms of the mass velocity. **Superficial linear velocity** corresponds to the superficial mass velocity.

The **acoustic velocity,** or **velocity of sound,** in a fluid of large extent or contained in a rigid-walled vessel, is given by

$$V_a = \sqrt{g_c\left(\frac{\partial p}{\partial \rho}\right)_s} = \sqrt{g_c k\left(\frac{\partial p}{\partial \rho}\right)_T} = \sqrt{\frac{Kg_c}{\rho}} \qquad (5\text{-}5)$$

where g_c = dimensional constant; p = absolute pressure; ρ = fluid density; k = ratio of specific heats c_p/c_v, dimensionless; and K = fluid bulk modulus of elasticity. Subscript s denotes constancy of entropy; and subscript T, constancy of temperature. For perfect gases, $(\partial p/\partial \rho)_T = p/\rho = RT/M$, where T = absolute temperature, R = gas constant, and M = molecular weight. Consequently, for a perfect gas $V_a = \sqrt{g_c kRT/M}$. See Shapiro, *The Dynamics and Thermodynamics of Compressible Fluid Flow*, vol. I, Ronald, New York, 1953, pp. 45–48.

The ratio of the velocity of flow to the velocity of sound is called the **Mach number.** Flows at Mach number less than unity are called subsonic.

The **velocity head** $V^2/2g_c$ is the static head equivalent of the kinetic energy in a stream of uniform velocity V.

For definitions of thermodynamic terms such as internal energy, enthalpy, total heat, entropy, etc., see Sec. 4.

A **Reynolds number** N_{Re} is any of several dimensionless quantities of the form $LV\rho/\mu$ which are all proportional to the ratio of inertial force to viscous force in a flow system. Here L = a characteristic linear dimension of the flow channel, V = linear velocity, ρ = fluid density, and μ = fluid viscosity. The **critical** Reynolds number corresponds to the transition from turbulent flow to laminar flow as the velocity is reduced. Its value depends upon the channel geometry, being in the range of 2000 to 3000 for circular pipe (see Fig. 5-28 later in this section).

For **nonnewtonian fluids** transition from turbulent flow to laminar flow is predicted with the generalized Reynolds number defined subsequently by Eq. (5-77).

The **mean hydraulic radius** R_H of a channel is equal to the cross-sectional area of that part of the channel which is filled with fluid divided by the length of the wetted perimeter. The hydraulic radius of a circular pipe is one-fourth of the diameter; hence for a noncircular duct the **hydraulic diameter** is said to be 4 times the hydraulic radius. For various cross-sectional shapes, see Table 5-8 later in this section.

A **streamline** is defined as a line which lies in the direction of flow at every point at a given instant. **Laminar flow** is defined as a flow in which the streamlines remain distinct from one another over their entire length. The streamlines need not be straight or the flow steady as long as this criterion is fulfilled. This type of motion is also called **streamline flow** or **viscous flow.**

Laminar flow at $N_{Re} < 1$ is called **slow flow,** or creep flow. In such flows the inertial force relative to the viscous force can be neglected.

If the Reynolds number in a system exceeds the critical Reynolds number, the motion is generally found not to be laminar throughout the channel. Eddies generated in the initial zone of instability spread rapidly throughout the fluid, thereby producing a disruption of the entire flow pattern. The result is fluid turbulence superimposed upon the primary motion of translation, producing what is called **turbulent flow.** For additional information on turbulence and turbulent flow, see Hinze, *Turbulence*, 2d ed., McGraw-Hill, New York, 1975; Schlichting, *Boundary-Layer Theory*, 7th ed., McGraw-Hill, New York, 1979, pp. 555–779.

FLUID STATICS AND PRESSURE MEASUREMENT

GENERAL REFERENCES: *Fluid statics*. Baumeister, *Marks' Standard Handbook for Mechanical Engineers*, 8th ed., McGraw-Hill, New York, 1978. Hansen, *Fluid Mechanics*, Wiley, New York, 1967. Kaufman, *Fluid Mechanics*, McGraw-Hill, New York, 1963. Streeter and Wylie, *Fluid Mechanics*, 7th ed., McGraw-Hill, New York, 1979. Vennard and Street, *Elementary Fluid Mechanics*, 5th ed., Wiley, New York, 1975.

Pressure measurement, general. *ASME Power Test Code*, Part 2, PTC 19.2-1964. Considine, *Process Instruments and Controls Handbook*, 2d ed., McGraw-Hill, New York, 1974. Doolittle, *Mechanical Engineering Laboratory*, McGraw-Hill, New York, 1957. Jones, *Instrument Technology*, vol. 1, 3d ed., Butterworth, London, 1974. Sweeney, *Measurement Techniques in Mechanical Engineering*, Wiley, New York, 1953.

Pressure measurement, specific. For absolute pressures below 3.5 kPa (0.5 lbf/in^2) (about 1 inHg): Diels and Jaeckel, *Leybold Vacuum Handbook*, Pergamon, New York, 1966. Leck, *Pressure Measurement in Vacuum Systems*, 2d ed., Chapman & Hall, London, 1964. Lewin, *Fundamentals of Vacuum Science and Technology*, McGraw-Hill, New York, 1965, Steinherz, *Handbook of High-Vacuum Engineering*, Reinhold, New York, 1963. Van Atta, *Vacuum Science and Engineering*, McGraw-Hill, New York, 1965. For pressures over 140 MPa (about 20,000 lbf/in^2): Bridgman, *The Physics of High Pressures*, G. Bell, London, 1949. Tongue, *The Design and Construction of High-Pressure Chemical Plant*, 2d ed., Van Nostrand, Princeton, N.J., 1959.

DEFINITIONS

Fluid statics is concerned with the static properties and behavior of fluids. In the case of liquids, this subject is known as **hydrostatics;** in the case of gases it is called **pneumatics.**

A body of fluid in static equilibrium is being acted upon only by compressive forces. The intensity of this force, expressed in terms of force per unit area, is known as **static pressure.** It is normal to any surface on which it acts, and at any given point it has the same magnitude irrespective of the orientation of the surface. This is one way of stating **Pascal's law.** Another way is that the pressure at any point in a fluid at rest acts with equal intensity in all directions.

Gauge pressure is the difference between a given fluid pressure and that of the atmosphere. The readings of pressure gauges are commonly positive gauge pressures. A vacuum gauge may be used to show negative gauge pressures, i.e., fluid pressures less than atmospheric. **Absolute pressure** is the true total pressure and is equal to gauge pressure (taken with the proper sign) plus atmospheric pressure.

The term **static head** generally denotes the pressure in a fluid due to the head of fluid above the point in question. Its magnitude is given by the application of Newton's law (force = mass × acceleration). In the case of **liquids** (constant density), the static head p_h Pa (lbf/ft^2) is given by

$$p_h = h\rho g/g_c \qquad (5\text{-}6)$$

where h = head of liquid above the point, m (ft); ρ = liquid density; g = local acceleration due to gravity; and g_c = dimensional constant.

Dynamic pressure is the difference between **impact pressure** and **static head.** For moving fluids, a physical interpretation of the gauge readings cannot be given without specific knowledge of the position and orientation of the pressure tap relative to flow. The sum of static head and velocity head is called **total head** or **total pressure.** Specifications of pressure or piezometer taps are given in the subsection "Static Pressure."

LIQUID-COLUMN MANOMETERS

The **height,** or **head** [Eq. (5-6)], to which a fluid rises in an open vertical tube attached to an apparatus containing a liquid is a direct measure of the pressure at the point of attachment and is frequently used to show the level of liquids in tanks and vessels. This same principle can be applied with U-tube gauges (Fig. 5-4*a*) and equivalent devices (such as that shown in Fig. (5-4*b*) to measure pressure in terms of the head of a fluid other than the one under test. Most of these gauges may be used either as **open** or as **differential manometers.** The manometric fluid that constitutes the measured liquid column of these gauges may be any liquid immiscible with the fluid

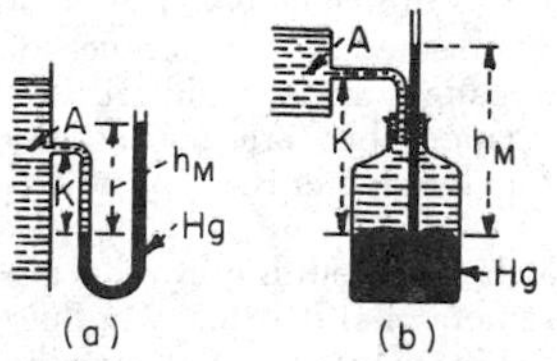

FIG. 5-4 Open manometers.

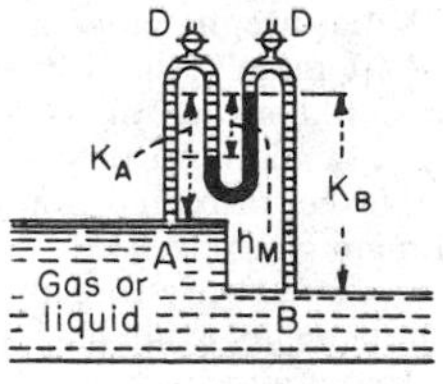

FIG. 5-5 Differential U tube.

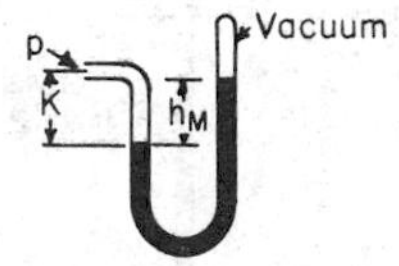

FIG. 5-6 Closed U tube.

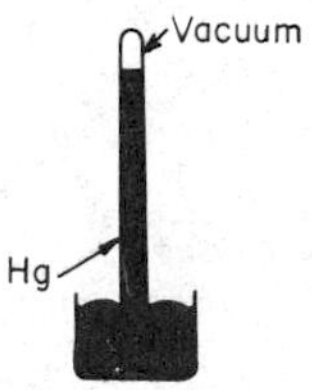

FIG. 5-7 Mercury barometer.

under pressure. For high vacuums or for high pressures and large pressure differences, the gauge liquid is a high-density liquid, generally mercury; for low pressures and small pressure differences, a low-density liquid (e.g., alcohol, water, or carbon tetrachloride) is used.

The **open U tube** (Fig. 5-4*a*) and the **open gauge** (Fig. 5-4*b*) each show a reading h_M m (ft) of manometric fluid. If the interface of the manometric fluid and the fluid of which the pressure is wanted is K m (ft) below the point of attachment, A, ρ_A is the density of the latter fluid at A, and ρ_M is that of the manometric fluid, then gauge pressure p_A Pa (lbf/ft^2) at A is

$$p_A = (h_M\rho_M - K\rho_A)(g/g_c) \qquad (5\text{-}7)^\circ$$

where g = local acceleration due to gravity and g_c = dimensional constant. The head H_A at A as meters (feet) of the fluid at that point is

$$h_A = h_M(\rho_M/\rho_A) - K \qquad (5\text{-}8)^\circ$$

When a gas pressure is measured, unless it is very high, ρ_A is so much smaller than ρ_M that the terms involving K in these formulas are negligible.

The **differential U tube** (Fig. 5-5) shows the pressure difference between taps A and B to be

$$p_A - p_B = [h_M(\rho_M - \rho_A) + K_A\rho_A - K_B\rho_B](g/g_c) \qquad (5\text{-}9)$$

where h_M is the difference in height of the manometric fluid in the U tube; K_A and K_B are the vertical distances of the upper surface of the manometric fluid above A and B respectively; ρ_A and ρ_B are the densities of the fluids at A and B respectively; and ρ_M is the density of the manometric fluid. If either pressure tap is above the higher level of manometric fluid, the corresponding K is taken to be negative. Valve D, which is kept closed when the gauge is in use, is used to vent off gas which may accumulate at these high points.

The **inverted differential U tube,** in which the manometric fluid may be a gas or a light liquid, can be used to measure liquid pressure differentials, especially for the flow of slurries where solids tend to settle out. Additional details on the use of this manometer can be obtained from Doolittle (op. cit., p. 18).

Closed U tubes (Fig. 5-6) using mercury as the manometric fluid serve to measure directly the absolute pressure p of a fluid, provided that the space between the closed end and the mercury is substantially a perfect vacuum.

The **mercury barometer** (Fig. 5-7) indicates directly the absolute pressure of the atmosphere in terms of height of the mercury column. Normal (standard) barometric pressure is 101.325 kPa by definition. Equivalents of this pressure in other units are 760 mm mercury (at 0°C), 29.921 inHg (at 0°C), 14.696 lbf/in^2, and 1 atm. For cases in which barometer readings, when expressed by the height of a mercury column, must be corrected to standard temperature (usually 0°C), appropriate temperature correction factors are given in ASME PTC, op. cit., pp. 23–26; and Weast, *Handbook of Chemistry and Physics*, 59th ed., Chemical Rubber, Cleveland, 1978–1979, pp. E39–E41.

°The line leading from the pressure tap to the gauge is assumed to be filled with fluid of the same density as that in the apparatus at the location of the pressure tap; if this is not the case, ρ_A is the density of the fluid actually filling the gauge line, and the value given for h_A must be multiplied by ρ_A/ρ, where ρ is the density of the fluid whose head is being measured.

Tube Size for Manometers To avoid capillary error, tube diameter should be sufficiently large and the manometric fluids of such densities that the effect of capillarity is negligible in comparison with the gauge reading. The effect of capillarity is practically negligible for tubes with inside diameters 12.7 mm (½ in) or larger (see ASME PTC, op. cit., p. 15). Small diameters are generally permissible for U tubes because the capillary displacement in one leg tends to cancel that in the other.

The capillary rise in a small vertical open tube of circular cross section dipping into a pool of liquid is given by

$$h = \frac{4\sigma g_c \cos\theta}{gD(\rho_1 - \rho_2)} \qquad (5\text{-}10)$$

Here σ = surface tension, D = inside diameter, ρ_1 and ρ_2 are the densities of the liquid and gas (or light liquid) respectively, g = local acceleration due to gravity, g_c = dimensional constant, and θ is the contact angle subtended by the heavier fluid. For most organic liquids and water, the contact angle θ is zero against glass, provided the glass is wet with a film of the liquid; for mercury against glass, θ = 140° (*International Critical Tables*, vol. IV, McGraw-Hill, New York, 1928, pp. 434–435). For further discussion of capillarity, see Schwartz, *Ind. Eng. Chem.*, **61**(1), 10–21 (1969).

MULTIPLYING GAUGES

To attain the requisite precision in measurement of small pressure differences by liquid-column manometers, means must often be devised to magnify the readings. Of the schemes that follow, the second and third may give tenfold multiplication; the fourth, as much as thirtyfold. In general, the greater the multiplication, the more elaborate must be the precautions in the use of the gauge if the gain in precision is not to be illusory.

1. *Change of manometric fluid.* In open manometers, choose a fluid of lower density. In differential manometers, choose a fluid such that the difference between its density and that of the fluid being measured is as small as possible.

2. *Inclined U tube (Fig. 5-8).* If the reading R m (ft) is taken as shown and R_0 m (ft) is the zero reading, by making the substitution $h_M = (R - R_0)\sin\theta$, the formulas of preceding paragraphs give ($p_A - p_B$) when the corresponding upright U tube is replaced by one inclined. For precise work, the gauge should be calibrated because of possible variations in tube diameter and slope.

3. *The draft gauge (Fig. 5-9).* Commonly used for low gas heads, this gauge has for one leg of the U a reservoir of much larger bore than the tubing that forms the inclined leg. Hence variations of level in the inclined tube produce little change in level in the reser-

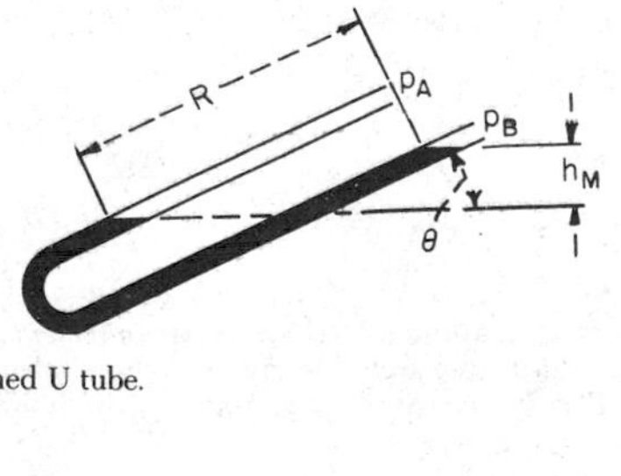

FIG. 5-8 Inclined U tube.

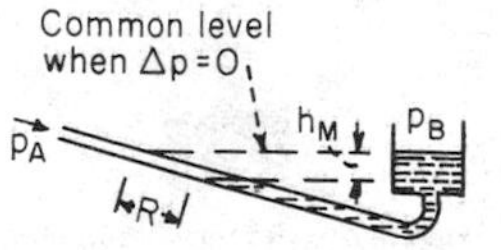

FIG. 5-9 Draft gauge.

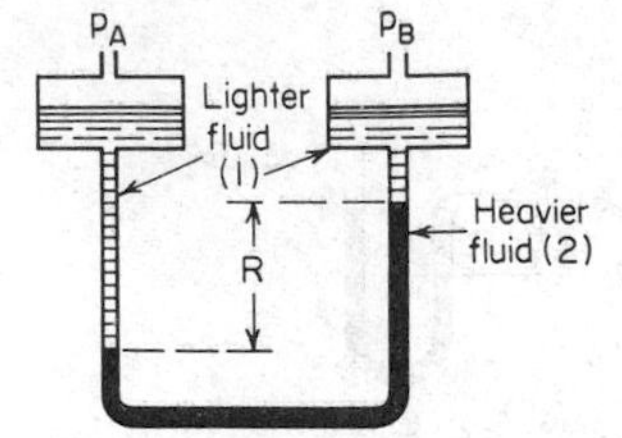

FIG. 5-10 Two-fluid U tube.

voir. Although h_M may be readily computed in terms of reading R and the dimensions of the tube, calibration of the gauge is preferable; often the changes of level in the reservoir are not negligible, and also variations in tube diameter may introduce serious error into the computation. Commercial gauges are often provided with a scale giving h_M directly in height of water column, provided a particular liquid (often not water) fills the tube; failure to appreciate that the scale is incorrect unless the gauge is filled with the specified liquid is a frequent source of error. If the scale reads correctly when the density of the gauge liquid is ρ_0, then the reading must be multiplied by ρ/ρ_0 if the density of the fluid actually in use is ρ.

4. *Two-fluid U tube (Fig. 5-10).* This is a highly sensitive device for measuring small gas heads. Let A be the cross-sectional area of each of the reservoirs and a that of the tube forming the U; let ρ_1 be the density of the lighter fluid and ρ_2 that of the heavier fluid; and if R is the reading and R_0 its value with zero pressure difference, then the pressure difference is

$$p_A - p_B = (R - R_0)\left(\rho_2 - \rho_1 + \frac{a}{A}\rho_1\right)\frac{g}{g_c} \tag{5-11}$$

where g = local acceleration due to gravity and g_c = dimensional constant.

When A/a is sufficiently large, the term $(a/A)\,\rho_1$ in Eq. (5-11) becomes negligible in comparison with the difference $(\rho_2 - \rho_1)$. However, this term should not be omitted without due consideration. In applying Eq. (5-11), the densities of the gauge liquids may not be taken from tables without the possibility of introducing serious error, for each liquid may dissolve appreciable quantities of the other. Before the gauge is filled, the liquids should be shaken together, and the actual densities of the two layers should be measured for the temperature at which the gauge is to be used. When high magnification is being sought, the U tube may have to be enclosed in a constant-temperature bath so that $(\rho_2 - \rho_1)$ may be accurately known. In general, if highest accuracy is desired, the gauge should be calibrated.

Several **micromanometers,** based on the liquid-column principle and possessing extreme precision and sensitivity, have been developed for measuring minute gas-pressure differences and for calibrating low-range gauges. Some of these micromanometers are available commercially. These micromanometers are free from errors due to capillarity and, aside from checking the micrometer scale, require no calibration. See Doolittle, op. cit., p. 21.

MECHANICAL PRESSURE GAUGES

The **Bourdon-tube gauge** indicates pressure by the amount of flection under internal pressure of an oval tube bent in an arc of a circle and closed at one end. These gauges are commercially available for all pressures below atmospheric and for pressures up to 700 MPa (about 100,000 lbf/in^2) above atmospheric. Details on Bourdon-type gauges are given by Harland [*Mach. Des.*, **40**(22), 69–74 (Sept. 19, 1968)].

A **diaphragm gauge** depends for its indication on the deflection of a diaphragm, usually metallic, when subjected to a difference of pressure between the two faces. These gauges are available for the same general purposes as Bourdon gauges but are not usually employed for high pressures. The aneroid barometer is a type of diaphragm gauge.

Small **pressure transducers with flush-mounted diaphragms** are commercially available for the measurement of either steady or fluctuating pressures up to 100 MPa (about 15,000 lbf/in^2). The metallic diaphragms are as small as 4.8 mm (3/16 in) in diameter. The transducer is mounted on the apparatus containing the fluid whose pressure is to be measured so that the diaphragm is flush with the inner surface of the apparatus. Deflection of the diaphragm is measured by unbonded strain gauges and recorded electrically.

With nonnewtonian fluids the pressure measured at the wall with non-flush-mounted pressure gauges may be in error (see subsection "Static Pressure").

Bourdon and diaphragm gauges that show both pressure and vacuum indications on the same dial are called **compound gauges.**

Conditions of Use Bourdon tubes should not be exposed to temperatures over about 65°C (about 150°F) unless the tubes are specifically designed for such operation. When the pressure of a hotter fluid is to be measured, some type of liquid seal should be used to keep the hot fluid from the tube. In using either a Bourdon or a diaphragm gauge to measure gas pressure, if the gauge is below the pressure tap of the apparatus so that liquid can collect in the lead, the gauge reading will be too high by an amount equal to the hydrostatic head of the accumulated liquid.

For measuring pressures of corrosive fluids, slurries, and similar process fluids which may foul Bourdon tubes, a **chemical gauge,** consisting of a Bourdon gauge equipped with an appropriate flexible diaphragm to seal off the process fluid, may be used. The combined volume of the tube and the connection between the diaphragm and the tube is filled with an inert liquid. These gauges are available commercially.

Further details on pressure-measuring devices are found in Sec. 22.

CALIBRATION OF GAUGES

Simple **liquid-column manometers** do not require calibration if they are so constructed as to minimize errors due to capillarity (see subsection "Liquid-Column Manometers"). If the scales used to measure the readings have been checked against a standard, the accuracy of the gauges depends solely upon the precision of determining the position of the liquid surfaces. Hence liquid-column manometers are primary standards used to calibrate other gauges.

For **high pressures** and, with commercial mechanical gauges, even for quite moderate pressures, a deadweight gauge (see ASME PTC, op. cit., pp. 36–41; Doolittle, op. cit., p. 33; Jones, op. cit., p. 43; Sweeney, op. cit., p. 104; and Tongue, op. cit., p. 29) is commonly used as the primary standard because it is safer and more convenient than use of manometers. When manometers are used as high-pressure standards, an extremely high mercury column may be avoided by connecting a number of the usual U tubes in series. Multiplying gauges are standardized by comparing them with a micromanometer. Procedure in the calibration of a gauge consists merely of connecting it, in parallel with a standard gauge, to a reservoir wherein constant pressure may be maintained. Readings of the unknown gauge are then made for various reservoir pressures as determined by the standard.

Calibration of **high-vacuum gauges** is described by Sellenger [*Vacuum*, **18**(12), 645–650 (1968)].

FLOW MEASUREMENT

GENERAL REFERENCES: Addison, *Hydraulic Measurements*, 2d ed., Wiley, New York, 1949. ASME Research Committee on Fluid Meters Report, *Fluid Meters—Their Theory and Application*, 6th ed., 1971. ASME Power Test Code, Part 5: "Measurement of Quantity of Materials," 1959. Considine, *Process Instruments and Controls Handbook*, 2d ed., McGraw-Hill, New York, 1974. Dean, *Aerodynamic Measurements*, M.I.T., Cambridge, Mass., 1953.

Ladenburg, Lewis, Pease, and Taylor, *Physical Measurements in Gas Dynamics and Combustion*, Princeton, Princeton, N.J., 1954. Ower and Pankhurst, *The Measurement of Air Flow*, 5th ed., Pergamon, Oxford, 1977. Spink, *Principles and Practice of Flowmeter Engineering*, 9th ed., Foxboro Co., Foxboro, Mass., 1967.

This subsection summarizes the techniques available for measuring static pressures, point velocities, and flow rates of flowing fluids. Coverage is generally limited to the primary, or actuating, elements. Secondary elements (e.g., pressure gauges or manometers) are, for a given primary device, more or less interchangeable. They are described elsewhere in this section and in Sec. 22.

STATIC PRESSURE

Local Static Pressure In a moving fluid, the local static pressure is equal to the pressure on a surface which moves with the fluid or to the normal pressure (for newtonian fluids) on a stationary surface which parallels the flow. The pressure on such a surface is measured by making a small hole perpendicular to the surface and connecting the opening to a pressure-sensing element (Fig. 5-11*a*). The hole is known as a piezometer opening or pressure tap.

Measurement of local static pressure is frequently difficult or impractical. If the channel is so small that introduction of any solid object disturbs the flow pattern and increases the velocity, there will be a reduction and redistribution of the static pressure. If the flow is in straight parallel lines, aside from the fluctuations of normal turbulence, the **flat disk** (Fig. 5-11*b*) and the **bent tube** (Fig. 5-11*c*) give satisfactory results when properly aligned with the stream. Slight misalignments can cause serious errors. Diameter of the disk should be 20 times its thickness and 40 times the static opening; the face must be flat and smooth, with the knife edges made by beveling the underside. The piezometer tube, such as that in Fig. 5-11*c*, should have openings with size and spacing as specified for a pitot tube (Fig. 5-13).

Readings given by open straight tubes (Fig. 5-11*d*) are too low. Readings of closed tubes oriented perpendicularly to the axis of the stream and provided with side openings (Fig. 5-11*e*) may be low by as much as two velocity heads.

Average Static Pressure In most cases, the object of a static-pressure measurement is to obtain a suitable average value for substitution in Bernoulli's theorem or in an equivalent flow formula. This can be done simply only when the flow is in straight lines parallel to the confining walls, such as in straight ducts at sufficient distance downstream from bends or other disturbances. For such streams, the sum of static head and gravitational potential head is the same at all points in a cross section taken perpendicularly to the axis of flow. Thus the exact location of a piezometer opening about the periphery of such a cross section is immaterial provided its elevation is known. However, in stating the static pressure, the custom is to give the value at the elevation corresponding to the centerline of the stream.

With flow in curved passages or with swirling flow, determination of a true average static pressure is, in general, impractical. In metering, straightening vanes are often placed upstream of the pressure tap to eliminate swirl.

Specifications for Piezometer Taps The size of a static opening should be small compared with the diameter of the pipe and yet large compared with the scale of surface irregularities. For reliable results, it is essential that (1) the surface in which the hole is made be substantially smooth and parallel to the flow for some distance on either side of the opening, and (2) the opening be flush with the surface and possess no "burr" or other irregularity around its edge. Rounding of the edge is often employed to ensure absence of a burr. Pressure readings will be high if the tap is inclined upstream, is rounded excessively on the upstream side, has a burr on the downstream side, or has an excessive countersink or recess. Pressure readings will be low if the tap is inclined downstream, is rounded excessively on the downstream side, has a burr on the upstream side, or protrudes into the flow stream. Errors resulting from these faults can be large.

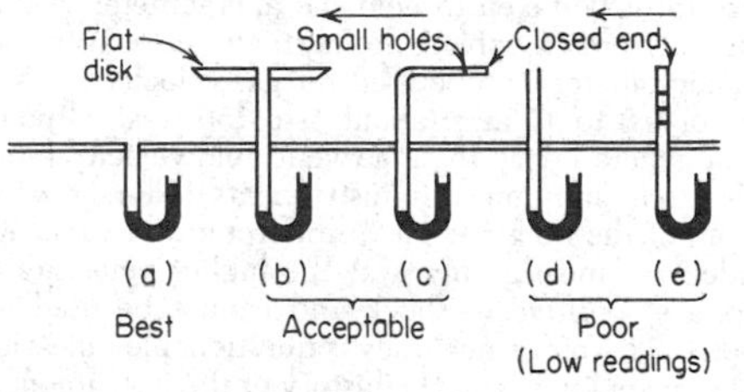

FIG. 5-11 Measurement of static pressure.

TABLE 5-1 Pressure-Tap Holes

Nominal inside pipe diameter, in	Maximum diameter of pressure tap, mm (in)	Radius of hole-edge rounding, mm (in)
1	3.18 (⅛)	<0.40 (1/64)
2	6.35 (¼)	0.40 (1/64)
3	9.53 (⅜)	0.40–0.79 (1/64–1/32)
4	12.7 (½)	0.79 (1/32)
8	12.7 (½)	0.79–1.59 (1/32–1/16)
16	19.1 (¾)	0.79–1.59 (1/32–1/16)

Recommendations for **pressure-tap dimensions** are summarized in Table 5-1. Data from several references were used in arriving at these composite values. The length of a pressure-tap opening prior to any enlargement in the tap channel should be at least two tap diameters, preferably three or more.

For information on prediction of static-hole error, see Shaw, *J. Fluid Mech.*, **7**, 550–564 (1960); Livesey, Jackson, and Southern, *Aircr. Eng.*, **34**, 43–47 (February 1962).

For nonnewtonian fluids, pressure readings with taps may also be low because of fluid-elasticity effects. This error can be largely eliminated by using flush-mounted diaphragms.

For information on the pressure-hole error for nonnewtonian fluids, see Han and Kim, *Trans. Soc. Rheol.*, **17**, 151–174 (1973); Novotny and Eckert, *Trans. Soc. Rheol.*, **17**, 227–241 (1973); and Higashitani and Lodge, *Trans. Soc. Rheol.*, **19**, 307–336 (1975).

A **piezometer ring** is a toroidal manifold into which are connected several sidewall static taps located around the perimeter of a common cross section. Its intent is to give an average pressure if differences in pressure other than those due to static head exist around the perimeter. However, there is generally no assurance that a true average is provided thereby. The principal advantage of the ring is that use of several holes in place of a single hole reduces the possibility of completely plugging the static openings.

Flush-mounted-diaphragm pressure-sensing elements are used primarily with fluids which degrade or decompose upon standing. They are discussed in greater detail in Sec. 22.

VELOCITY METERS

Pitot Tubes These tubes measure local or point velocities by measuring the difference between impact pressure and static pressure. The pitot tube shown in Fig. 5-12 consists of an impact tube

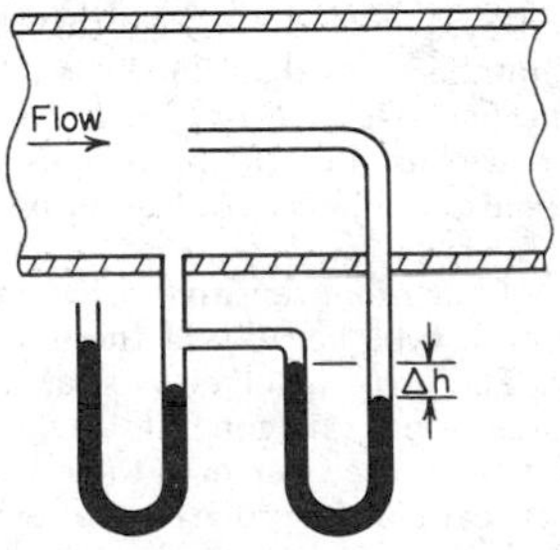

FIG. 5-12 Pitot tube with sidewall static tap.

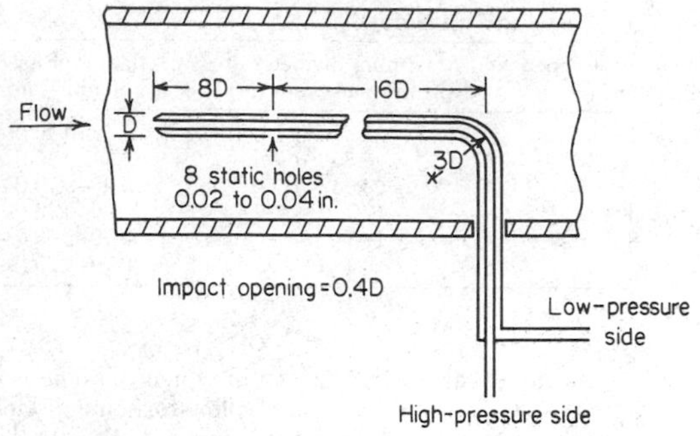

FIG. 5-13 Pitot-static tube.

whose opening faces directly into the stream to measure impact pressure, plus one or more sidewall taps to measure local static pressure. The combined pitot-static tube shown in Fig. 5-13 consists of a jacketed impact tube with one or more rows of holes, 0.51 to 1.02 mm (0.02 to 0.04 in) in diameter, in the jacket to measure the static pressure. Velocity V_0 m/s (ft/s) at the point where the tip is located is given by

$$V_0 = C\sqrt{2g_c\,\Delta h} = C\sqrt{2g_c(p_i - p_0)/\rho_0} \tag{5-12}$$

where C = coefficient, dimensionless; g_c = dimensional constant; Δh = differential pressure ($\Delta h_s g/g_c$) as shown in Fig. 5-12, expressed in (N·m)/kg [(ft·lbf)/lb or ft of fluid flowing]; Δh_s = differential height of static liquid column corresponding to Δh; g = local acceleration due to gravity; g_c = dimensional constant; p_i = impact pressure; p_0 = local static pressure; and ρ_0 = fluid density measured at pressure p_0 and the local temperature. With gases at velocities above 60 m/s (about 200 ft/s), compressibility becomes important, and the following equation should be used:

$$V_0 = C\sqrt{\frac{2g_c k}{k-1}\left(\frac{p_0}{\rho_0}\right)\left[\left(\frac{p_i}{p_0}\right)^{(k-1)/k} - 1\right]} \tag{5-13}$$

where k is the ratio of specific heat at constant pressure to that at constant volume. (See ASME Research Committee on Fluid Meters Report, op. cit., p. 105.) Coefficient C is usually close to 1.00 (±0.01) for simple pitot tubes (Fig. 5-12) and generally ranges between 0.98 and 1.00 for pitot-static tubes (Fig. 5-13).

There are certain **limitations** on the range of usefulness of pitot tubes. With gases, the differential is very small at low velocities; e.g., at 4.6 m/s (15.1 ft/s) the differential is only about 1.30 mm (0.051 in) of water (20°C) for air at 1 atm (20°C), which represents a lower limit for 1 percent error even when one uses a micromanometer with a precision of 0.0254 mm (0.001 in) of water. Equation (5-13) does not apply for Mach numbers greater than 0.7 because of the interference of shock waves. For supersonic flow, local Mach numbers can be calculated from a knowledge of the impact and true static pressures; see Ladenburg et al., op. cit., pp. 111–112.

With **liquids** at low velocities, the effect of the Reynolds number upon the coefficient is important. The coefficients are appreciably less than unity for Reynolds numbers less than 500 for pitot tubes and for Reynolds numbers less than 2300 for pitot-static tubes [see Folsom, *Trans. Am. Soc. Mech. Eng.*, **78,** 1447–1460 (1956)]. Reynolds numbers here are based on the probe outside diameter. Operation at low Reynolds numbers requires prior calibration of the probe.

The pitot-static tube is more sensitive to **yaw** or **angle of attack** than is the simple pitot tube because of the sensitivity of the static taps to orientation. The error involved is strongly dependent upon the exact probe dimensions. In general, angles greater than 10° should be avoided if the velocity error is to be 1 percent or less.

Disturbances upstream of the probe can cause large errors, in part because of the turbulence generated and its effect on the static-pressure measurement. A calming section of at least 50 pipe diameters is desirable. If this is not possible, the use of straightening vanes or a honeycomb is advisable.

The effect of **pulsating flow** on pitot-tube accuracy is treated by Ower et al., op. cit., pp. 310–312. For sinusoidal velocity fluctuations, the ratio of indicated velocity to actual mean velocity is given by the factor $\sqrt{1 + \lambda^2/2}$, where λ is the velocity excursion as a fraction of the mean velocity. Thus, the indicated velocity would be about 6 percent high for velocity fluctuations of ±50 percent, and pulsations greater than ±20 percent should be damped to avoid errors greater than 1 percent. The error increases as the frequency of flow oscillations approaches the natural frequency of the pitot tube and the density of the measuring fluid approaches the density of the process fluid [see Horlock and Daneshyar, *J. Mech. Eng. Sci.*, **15,** 144–152 (1973)].

Pressures substantially lower than true impact pressures are obtained with pitot tubes in turbulent flow of **dilute polymer solutions** [see Halliwell and Lewkowicz, *Phys. Fluids*, **18,** 1617–1625 (1975)].

Special Tubes A variety of special forms of the pitot tube have been evolved. Folsom (loc. cit.) gives a description of many of these special types together with a comprehensive bibliography. Included are the impact tube for **boundary-layer** measurements and **shielded total-pressure tubes.** The latter are insensitive to angle of attack up to 40°.

Chue [*Prog. Aerosp. Sci.*, **16,** 147–223 (1975)] reviews the use of the pitot tube and allied pressure probes for impact pressure, static pressure, dynamic pressure, flow direction and local velocity, skin friction, and flow measurements.

A reversed pitot tube, also known as a **pitometer,** has one pressure opening facing upstream and the other facing downstream. Coefficient C for this type is on the order of 0.85. This gives about a 40 percent increase in pressure differential as compared with standard pitot tubes and is an advantage at low velocities. There are commercially available very compact types of pitometers which require relatively small openings for their insertion into a duct.

The **pitot-venturi** flow element is capable of developing a pressure differential 5 to 10 times that of a standard pitot tube. This is accomplished by employing a pair of concentric venturi elements in place of the pitot probe. The low-pressure tap is connected to the throat of the inner venturi, which in turn discharges into the throat of the outer venturi. For a discussion of performance and application of this flow element, see Stoll, *Trans. Am. Soc. Mech. Eng.*, **73,** 963–969 (1951).

Anemometers An anemometer may be any instrument for measurement of gas velocity, e.g., a pitot tube, but usually the term refers to one of the following types.

The **vane anemometer** is a delicate revolution counter with jeweled bearings, actuated by a small windmill, usually 75 to 100 mm (about 3 to 4 in) in diameter, constructed of flat or slightly curved radially disposed vanes. Gas velocity is determined by using a stopwatch to find the time interval required to pass a given number of meters (feet) of gas as indicated by the counter. The velocity so obtained is inversely proportional to gas density. If the original calibration was carried out in a gas of density ρ_0 and the density of the gas stream being metered is ρ_1, the true gas velocity can be found as follows: From the calibration curve for the instrument, find $V_{t,0}$ corresponding to the quantity $V_m\sqrt{\rho_1/\rho_0}$, where V_m = measured velocity. Then the actual velocity $V_{t,1}$ is equal to $V_{t,0}\sqrt{\rho_0/\rho_1}$. In general, when working with air, the effects of atmospheric-density changes can be neglected for all velocities above 1.5 m/s (about 5 ft/s). In all cases, care must be taken to hold the anemometer well away from one's body or from any object not normally present in the stream.

Vane anemometers can be used for gas-velocity measurements in the range of 0.3 to 45 m/s (about 1 to 150 ft/s), although a given instrument generally has about a twentyfold velocity range. Bearing friction has to be minimized in instruments designed for accuracy at the low end of the range, while ample rotor and vane rigidity must be provided for measurements at the higher velocities. Vane anemometers are sensitive to shock and cannot be used in corrosive atmospheres. Therefore, accuracy is questionable unless a recent calibration has been made and the history of the instrument subsequent to calibration is known. For additional information, see Ower et al., op. cit., chap. VIII.

A **turbine flowmeter** consists of a straight flow tube containing a turbine which is free to rotate on a shaft supported by one or more bearings and located on the centerline of the tube. Means are provided for magnetic detection of the rotational speed, which is proportional to the volumetric flow rate. Its use is generally restricted to clean, noncorrosive fluids. Additional information on construction, operation, range, and accuracy can be obtained from Holzbock (*Instruments for Measurement and Control,* 2d ed., Reinhold, New York, 1962, pp. 155–162). For performance characteristics of these meters with liquids, see Shafer, *J. Basic Eng.*, **84,** 471–485 (December 1962); or May, *Chem. Eng.*, **78**(5), 105–108 (1971); and for the effect of density and Reynolds number when used in gas flowmetering, see Lee and Evans, *J. Basic Eng.*, **82,** 1043–1057 (December 1965).

The **current meter** is generally used for measuring velocities in open channels such as rivers and irrigation channels. There are two types, the cup meter and the propeller meter. The former is more widely used. It consists of six conical cups mounted on a vertical axis pivoted at the ends and free to rotate between the rigid arms of a U-shaped clevis to which a vaned tailpiece is attached. The wheel rotates because of the difference in drag for the two sides of the cup, and a signal proportional to the revolutions of the wheel is generated. The velocity is determined from the count over a period of time. The current meter is generally useful in the range of 0.15 to 4.5 m/s (about 0.5 to 15 ft/s) with an accuracy of ± 2 percent. For additional information see Creager and Justin, *Hydroelectric Handbook,* 2d ed., Wiley, New York, 1950, pp. 42–46.

The **hot-wire anemometer** consists essentially of an electrically heated fine wire (generally platinum) exposed to the gas stream whose velocity is being measured. An increase in fluid velocity, other things being equal, increases the rate of heat flow from the wire to the gas, thereby tending to cool the wire and alter its electrical resistance. In a constant-current anemometer, gas velocity is determined by measuring the resulting wire resistance; in the constant-resistance type, gas velocity is determined from the current required to maintain the wire temperature, and thus the resistance, constant. The difference in the two types is primarily in the electric circuits and instruments employed.

The hot-wire anemometer can, with suitable calibration, accurately measure velocities from about 0.15 m/s (0.5 ft/s) to supersonic velocities and detect velocity fluctuations with frequencies up to 200,000 Hz. Fairly rugged, inexpensive units can be built for the measurement of mean velocities in the range of 0.15 to 30 m/s (about 0.5 to 100 ft/s). More elaborate, compensated units are commercially available for use in unsteady flow and turbulence measurements. In calibrating a hot-wire anemometer, it is preferable to use the same gas, temperature, and pressure as will be encountered in the intended application. In this case the quantity $I^2R_w/\Delta t$ can be plotted against $\sqrt{V}$, where I = hot-wire current, R_w = hot-wire resistance, Δt = difference between the wire temperature and the gas bulk temperature, and V = mean local velocity. A procedure is given by Wasan and Baid [*Am. Inst. Chem. Eng. J.*, **17,** 729–731 (1971)] for use when it is impractical to calibrate with the same gas composition or conditions of temperature and pressure. Andrews, Bradley, and Hundy [*Int. J. Heat Mass Transfer,* **15,** 1765–1786 (1972)] give a calibration correlation for measurement of small gas velocities. The hot-wire anemometer is treated in considerable detail in Dean, op. cit., chap. VI; in Ladenburg et al., op. cit., art. F-2; by Grant and Kronauer, *Symposium on Measurement in Unsteady Flow,* American Society of Mechanical Engineers, New York, 1962, pp. 44–53; ASME Research Committee on Fluid Meters Report, op. cit., pp. 105–107; and by Compte-Bellot, *Ann. Rev. Fluid Mech.*, **8,** pp. 209–231 (1976).

The hot-wire anemometer can be modified for liquid measurements, although difficulties are encountered because of bubbles and dirt adhering to the wire. See Stevens, Borden, and Strausser, David Taylor Model Basin Rep. 953, December 1956; Middlebrook and Piret, *Ind. Eng. Chem.*, **42,** 1511–1513 (1950); and Piret et al., *Ind. Eng. Chem.*, **39,** 1098–1103 (1947).

The **hot-film anemometer** has been developed for applications in which use of the hot-wire anemometer presents problems. It consists of a platinum-film sensing element deposited on a glass substrate. Various geometries can be used. The most common involves a wedge with a 30° included angle at the end of a tapered rod. The wedge is commonly 1 mm (0.039 in) long and 0.2 mm (0.0079 in) wide on each face. Compared with the hot wire, it is less susceptible to fouling by bubbles or dirt when used in liquids, has greater mechanical strength when used with gases at high velocities and high temperatures, and can give a higher signal-to-noise ratio. For additional information see Ling and Hubbard, *J. Aeronaut. Sci.*, **23,** 890–891 (1956); and Ling, *J. Basic Eng.*, **82,** 629–634 (1960).

The **heated-thermocouple anemometer** measures gas velocity from the cooling effect of the gas stream flowing across the hot junctions of a thermopile supplied with constant electrical power input. Alternate junctions are maintained at ambient temperature, thus compensating for the effect of ambient temperature. For details see Bunker, *Proc. Instrum. Soc. Am.*, **9,** pap. 54-43-2 (1954).

A glass-coated bead **thermistor anemometer** can be used for the measurement of low fluid velocities, down to 0.001 m/s (0.003 ft/s) in air and 0.0002 m/s (0.0007 ft/s) in water [see Murphy and Sparks, *Ind. Eng. Chem. Fundam.*, **7,** 642–645 (1968)].

The **laser-Doppler anemometer** measures local fluid velocity from the change in frequency of radiation, between a stationary source and a receiver, due to scattering by particles along the wave path. A laser is commonly used as the source of incident illumination. The measurements are essentially independent of local temperature and pressure. This technique can be used in many different flow systems with transparent fluids containing particles whose velocity is actually measured. For a brief review of the laser-Doppler technique see Goldstein, *Appl. Mech. Rev.*, **27,** 753–760 (1974). For additional details see Durst, Melling, and Whitelaw, *Principles and Practice of Laser-Doppler Anemometry,* Academic, New York, 1976.

Traversing for Mean Velocity Mean velocity in a duct can be obtained by dividing the cross section into a number of equal areas, finding the local velocity at a representative point in each, and averaging the results. In the case of **rectangular passages,** the cross section is usually divided into small squares or rectangles and the velocity is found at the center of each. In **circular pipes,** the cross section is divided into several equal annular areas and a central circle. Readings of velocity are made at the intersections of a diameter and the set of circles which bisect the annuli and the central circle.

For an N-point traverse on a circular cross section, make readings on each side of the cross section at

$$100 \times \sqrt{(2n - 1)/N} \text{ percent} \qquad (n = 1, 2, 3 \text{ to } N/2)$$

of the pipe radius from the center. Traversing several diameters spaced at equal angles about the pipe is required if the velocity distribution is unsymmetrical. With a normal velocity distribution in a circular pipe, a 10-point traverse theoretically gives a mean velocity 0.3 percent high; a 20-point traverse, 0.1 percent high.

For normal velocity distribution in straight circular pipes at locations preceded by runs of at least 50 diameters without pipe fittings or other obstructions, the graph in Fig. 5-14 shows the ratio of mean velocity V to velocity at the center u_{max} plotted against the Reynolds number, where D = inside pipe diameter, ρ = fluid density, and μ = fluid viscosity, all in consistent units. Mean velocity is readily determined from this graph and a pitot reading at the center of the pipe if the quantity $Du_{max}\,\rho/\mu$ is less than 2000 or greater than 5000. The method is unreliable at intermediate values of the Reynolds number.

Methods for determining mean flow rate from probe measurements under nonideal conditions are described by Mandersloot, Hicks, and Langejan [*Chem. Eng. (London)*, no. 232, CE370-CE380 (1969)].

Flow Visualization A great many techniques have been developed for the visualization of velocity patterns, particularly for use in water-tunnel and wind-tunnel studies. In the case of **liquids,** the more common methods of revealing flow lines involve the use of dye traces, the addition of aluminum flake, plastic particles, and glass spheres, and the use of polarized light with a doubly refractive liquid

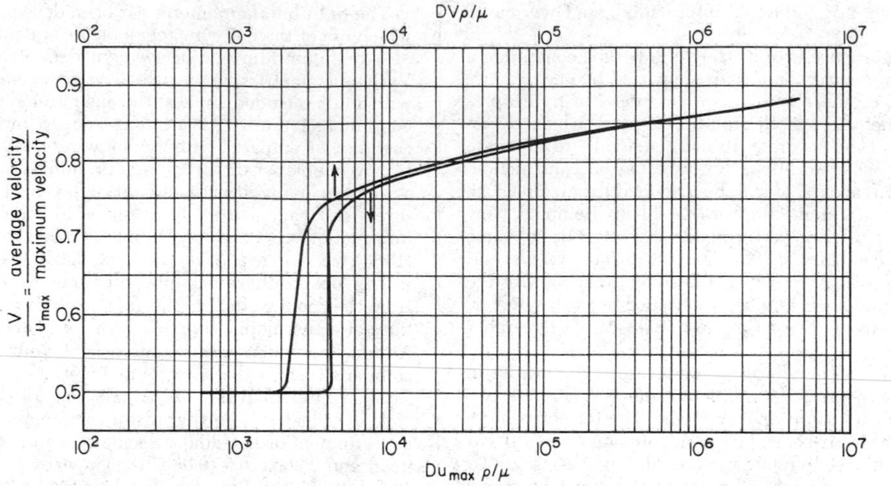

FIG. 5-14 Velocity ratio versus Reynolds number for smooth circular pipes. [*Based on data from Rothfus, Archer, Klimas, and Sikchi,* Am. Inst. Chem. Eng. J., *3, 208 (1957).*]

or suspension. For the last-named techniques, called *flow birefringence,* see Prados and Peebles, *Am. Inst. Chem. Eng. J.,* **5,** 225–234 (1959). The velocity pattern for laminar flow in a two-dimensional system can be quantitatively mapped by using an electrolytic-tank analog or a conductive-paper analog with a suitable combination of resistances, sources, and sinks. The hydrogen-bubble technique has been proposed for flow visualization and velocity field mapping in liquids. A fine wire, usually of the order of 0.013 to 0.05 mm (0.0005 to 0.002 in) in diameter, is employed as the negative electrode of a direct-current circuit in a water channel. Hydrogen bubbles, formed at the wire by periodic electrical pulses, are swept off by hydrodynamic forces and follow the flow. The bubbles are made visible by lighting at an oblique angle to the direction of view. For details see Schraub, Kline, Henry, Runstadler, and Little, *J. Basic. Eng.,* **87,** 429–444 (1965); or Davis and Fox, *J. Basic Eng.,* **89,** 771–781 (1967).

Thomas and Rice [*J. Appl. Mech.,* **40,** 321–325 (1973)] applied the hydrogen-bubble technique for velocity measurements in thin liquid films. Durelli and Norgard [*Exp. Mech.,* **12,** 169–177 (1972)] compare the flow birefringence and hydrogen-bubble techniques.

In the case of **gases,** flow lines can be revealed through the use of smoke traces or the addition of a lightweight powder such as balsa dust to the stream. One of the best smoke generators is the reaction of titanium tetrachloride with moisture in the air. A woodsmoke-generation system is described by Yu, Sparrow, and Eckert [*Int. J. Heat Mass Transfer,* **15,** 557–558 (1972)]. Tufts of wool or nylon attached at one end to a solid surface can be used to reveal flow phenomena in the vicinity of the surface. Optical methods commonly employed depend upon changes in the refractive index resulting from the presence of heated wires or secondary streams in the flow field or upon changes in density in the primary gas as a result of compressibility effects. The three common techniques are the shadowgraph, the schlieren, and the interferometer. All three theoretically can give quantitative information on the velocity profiles in a two-dimensional system, but in practice only the interferometer is commonly so used. The optical methods are described by Ladenburg et al. (op. cit., pp. 3–108). For additional information on other methods, see Goldstein, *Modern Developments in Fluid Dynamics,* vol. I, London, 1938, pp. 280–296.

The **water table** is frequently used to simulate two-dimensional compressible-flow phenomena in gases. It provides an effective, low-cost means for velocity and pressure-distribution studies or for flow visualization using either shadowgraph or schlieren techniques. In the water table, the wave velocity corresponds to the velocity of sound in the gas, streaming water flow corresponds to subsonic flow, shooting water flow corresponds to supersonic flow, and a hydraulic jump corresponds to a shock wave. From precise measurements of water depth, it is possible to calculate corresponding gas temperatures, pressures, and densities. For information on water-table design and operation see Orlin, Lindner, and Bitterly, *Application of the Analogy between Water Flow with a Free Surface and Two-Dimensional Compressible Gas Flow,* NACA Rep. 875, 1947, or Mathews, *The Design, Operation, and Uses of the Water Channel as an Instrument for the Investigation of Compressible-Flow Phenomena,* NACA Tech. Note 2008, 1950. Additional theoretical background can be obtained from Preiswerk, *Application of the Methods of Gas Dynamics to Water Flows with Free Surface,* part I: *Flows with No Energy Dissipation,* NACA Tech. Mem. 934, 1940; part II: *Flows with Momentum Discontinuities (Hydraulic Jumps),* NACA Tech. Mem. 935, 1940.

HEAD METERS

General Principles If a constriction is placed in a closed channel carrying a stream of fluid, there will be an increase in velocity, and hence an increase in kinetic energy, at the point of constriction. From an energy balance, as given by Bernoulli's theorem (see subsection "Energy Balance"), there must be a corresponding reduction in pressure. Rate of discharge from the constriction can be calculated by knowing this pressure reduction, the area available for flow at the constriction, the density of the fluid, and the coefficient of discharge C. The last-named is defined as the ratio of actual flow to the theoretical flow and makes allowance for stream contraction and frictional effects. The metering characteristics of commonly used head meters are reviewed and grouped by Halmi [*J. Fluids Eng.,* **95,** 127–141 (1973)].

Venturi Meters The standard Herschel-type venturi meter consists of a short length of straight tubing connected at either end to the pipe line by conical sections (see Fig. 5-15). Recommended proportions (*ASME PTC,* op. cit., p. 17) are entrance cone angle $\alpha_1 = 21 \pm 2°$, exit cone angle $\alpha_2 = 5$ to $15°$, throat length = one throat diameter, and upstream tap located 0.25 to 0.5 pipe diameter upstream of the entrance cone. The straight and conical sections should be joined by smooth curved surfaces for best results.

The practical working equation for weight rate of discharge, adopted by the ASME Research Committee on Fluid Meters for use with either gases or liquids, is

$$w = q_1\rho_1 = CYA_2\sqrt{\frac{2g_c(p_1 - p_2)\rho_1}{1 - \beta^4}} = KYA_2\sqrt{2g_c(p_1 - p_2)\rho_1} \quad (5\text{-}14)$$

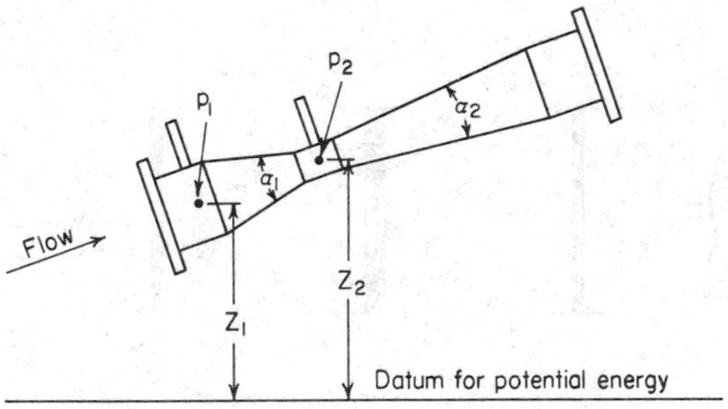

FIG. 5-15 Herschel-type venturi tube.

where A_2 = cross-sectional area of throat; C = coefficient of discharge, dimensionless; g_c = dimensional constant; $K = C/\sqrt{1-\beta^4}$, dimensionless; p_1, p_2 = pressure at upstream and downstream static pressure taps respectively; q_1 = volumetric rate of discharge measured at upstream pressure and temperature; w = weight rate of discharge; Y = expansion factor, dimensionless; β = ratio of throat diameter to pipe diameter, dimensionless; and ρ_1 = density at upstream pressure and temperature.

For the flow of **gases,** expansion factor Y, which allows for the change in gas density as it expands adiabatically from p_1 to p_2, is given by

$$Y = \sqrt{r^{2/k}\left(\frac{k}{k-1}\right)\left(\frac{1-r^{(k-1)/k}}{1-r}\right)\left(\frac{1-\beta^4}{1-\beta^4 r^{2/k}}\right)} \qquad (5\text{-}15)$$

for venturi meters and flow nozzles, where $r = p_2/p_1$ and k = specific heat ratio c_p/c_v. Values of Y computed from Eq. (5-15) are given in Fig. 5-16 as a function of r, k, and β.

For the flow of **liquids,** expansion factor Y is unity. The change in potential energy in the case of an inclined or vertical venturi meter must be allowed for. Equation (5-14) is accordingly modified to give

$$w = q_1\rho = CA_2\sqrt{\frac{[2g_c(p_1-p_2)+2g\rho(Z_1-Z_2)]\rho}{1-\beta^4}} \qquad (5\text{-}16)$$

where g = local acceleration due to gravity and Z_1, Z_2 = vertical heights above an arbitrary datum plane corresponding to the centerline pressure-reading locations for p_1 and p_2 respectively.

Value of the **discharge coefficient** C for a **Herschel-type venturi meter** depends upon the Reynolds number and to a minor extent upon the size of the venturi, increasing with diameter. A plot of C versus pipe Reynolds number is given in *ASME PTC*, op. cit., p. 19. A value of 0.984 can be used for pipe Reynolds numbers larger than 200,000.

Permanent pressure loss for a Herschel-type venturi tube depends upon diameter ratio β and discharge cone angle α_2. It ranges

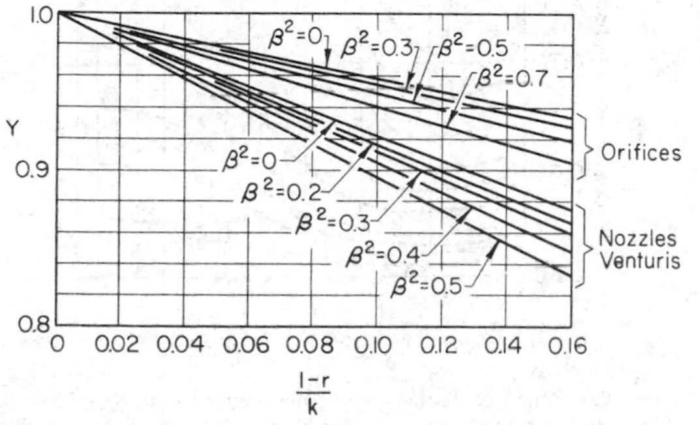

FIG. 5-16 Values of expansion factor Y for orifices, nozzles, and venturis.

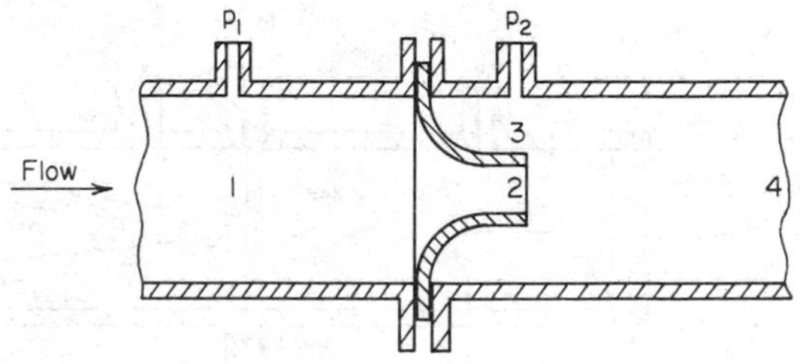

FIG. 5-17 Flow-nozzle assembly.

from 10 to 15 percent of the pressure differential $(p_1 - p_2)$ for small angles (5 to 7°) and from 10 to 30 percent for large angles (15°), with the larger losses occurring at low values of β (see *ASME PTC*, op. cit., p. 12). See Benedict, *J. Fluids Eng.*, **99,** 245–248 (1977), for a general equation for pressure loss for venturis installed in pipes or with plenum inlets.

For flow measurement of **steam and water mixtures** with a Herschel-type venturi in 2½-in- and 3-in-diameter pipes, see Collins and Gacesa, *J. Basic Eng.*, **93,** 11–21 (1971).

A variety of **short-tube** venturi meters are available commercially. They require less space for installation and are generally (although not always) characterized by a greater pressure loss than the corresponding Herschel-type venturi meter. Discharge coefficients vary widely for different types, and individual calibration is recommended if the manufacturer's calibration is not available. Results of tests on the Dall flow tube are given by Miner [*Trans. Am. Soc. Mech. Eng.*, **78,** 475–479 (1956)] and Dowdell [*Instrum. Control Syst.*, **33,** 1006–1009 (1960)]; and on the Gentile flow tube (also called Beth flow tube or Foster flow tube) by Hooper [*Trans. Am. Soc. Mech. Eng.*, **72,** 1099–1110 (1950)].

The use of a **multiventuri system** (in which an inner venturi discharges into the throat of an outer venturi) to increase both the differential pressure for a given flow rate and the signal-to-loss ratio is described by Klomp and Sovran [*J. Basic Eng.*, **94,** 39–45 (1972)].

Flow Nozzles A simple form of flow nozzle is shown in Fig. 5-17. It consists essentially of a short cylinder with a flared approach section. The approach cross section is preferably elliptical in shape but may be conical. Recommended contours for long-radius flow nozzles are given in *ASME PTC*, op. cit., p. 13. In general, the length of the straight portion of the throat is about one-half throat diameter, the upstream pressure tap is located about one pipe diameter from the nozzle inlet face, and the downstream pressure tap about one-half pipe diameter from the inlet face. For subsonic flow, the pressures at points 2 and 3 will be practically identical. If a conical inlet is preferred, the inlet and throat geometry specified for a Herschel-type venturi meter can be used, omitting the expansion section.

Rate of discharge through a flow nozzle for subcritical flow can be determined by the equations given for venturi meters, Eq. (5-14) for gases and Eq. (5-16) for liquids. The expansion factor Y for nozzles is the same as that for venturi meters [Eq. (5-15), Fig. 5-16]. The value of the discharge coefficient C depends primarily upon the pipe Reynolds number and to a lesser extent upon the diameter ratio β. Curves of recommended coefficients for long-radius flow nozzles with pressure taps located one pipe diameter upstream and one-half pipe diameter downstream of the inlet face of the nozzle are given in *ASME PTC*, op. cit., p. 15. In general, coefficients range from 0.95 at a pipe Reynolds number of 10,000 to 0.99 at 1,000,000.

The performance characteristics of pipe-wall-tap nozzles (Fig. 5-17) and throat-tap nozzles are reviewed by Wyler and Benedict [*J. Eng. Power*, **97,** 569–575 (1975)].

Permanent pressure loss across a subsonic flow nozzle is approximated by

$$p_1 - p_4 = \frac{1-\beta^2}{1+\beta^2}(p_1 - p_2) \qquad (5\text{-}17)$$

where p_1, p_2, p_4 = static pressures measured at the locations shown in Fig. 5-17; and β = ratio of nozzle throat diameter to pipe diameter, dimensionless. Equation (5-17) is based on a momentum bal-

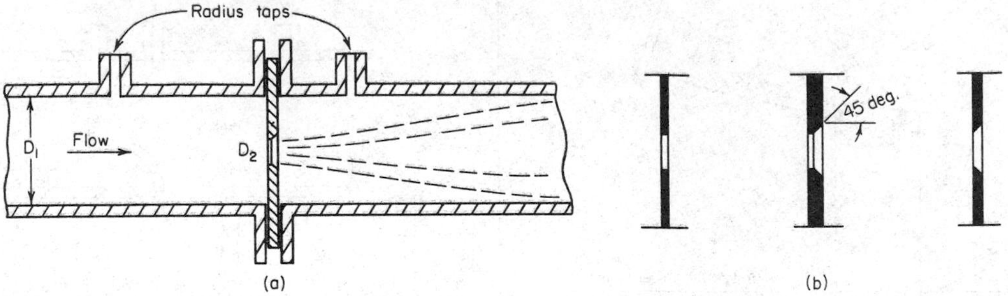

FIG. 5-18 Square-edged or sharp-edged orifices. The plate at the orifice opening must not be thicker than one-thirtieth of the pipe diameter, one-eighth of the orifice diameter, or one-fourth of the distance from the pipe wall to the edge of the opening. (*a*) Pipe-line orifice. (*b*) Types of plates.

ance assuming constant fluid density (see Lapple et al., *Fluid and Particle Mechanics*, University of Delaware, Newark, 1951, p. 13).

See Benedict, loc. cit., for a general equation for pressure loss for nozzles installed in pipes or with plenum inlets. Nozzles show higher loss than venturis. Permanent pressure loss for laminar flow depends on the Reynolds number in addition to β. For details, see Alvi, Sridharan, and Lakshamana Rao, *J. Fluids Eng.*, **100**, 299–307 (1978).

Critical Fow Nozzle For a given set of upstream conditions, the rate of discharge of a gas from a nozzle will increase for a decrease in the absolute pressure ratio p_2/p_1 until the linear velocity in the throat reaches that of sound in the gas at that location. The value of p_2/p_1 for which the acoustic velocity is just attained is called the critical pressure ratio r_c. The actual pressure in the throat will not fall below p_1r_c even if a much lower pressure exists downstream.

The **critical pressure ratio** r_c can be obtained from the following theoretical equation, which assumes a perfect gas and a frictionless nozzle:

$$r_c^{(1-k)/k} + \left(\frac{k-1}{2}\right)\beta^4 r_c^{2/k} = \frac{k+1}{2} \qquad (5\text{-}18)$$

This reduces, for $\beta \leq 0.2$, to

$$r_c = \left(\frac{2}{k+1}\right)^{k/(k-1)} \qquad (5\text{-}19)$$

where k = ratio of specific heats c_p/c_v and β = diameter ratio. A table of values of r_c as a function of k and β is given in the ASME Research Committee on Fluid Meters Report, op. cit., p. 68. For small values of β, r_c = 0.487 for k = 1.667, 0.528 for k = 1.40, 0.546 for k = 1.30, and 0.574 for k = 1.15.

Under **critical flow conditions,** only the upstream conditions p_1, v_1, and T_1 need be known to determine flow rate, which, for $\beta \leq$ 0.2, is given by

$$w_{max} = CA_2 \sqrt{g_c k \left(\frac{p_1}{v_1}\right)\left(\frac{2}{k+1}\right)^{(k+1)/(k-1)}} \qquad (5\text{-}20)$$

For a **perfect gas,** this corresponds to

$$w_{max} = CA_2 p_1 \sqrt{g_c k \left(\frac{M}{RT_1}\right)\left(\frac{2}{k+1}\right)^{(k+1)/(k-1)}} \qquad (5\text{-}21)$$

For **air,** Eq. (5-20) reduces to

$$w_{max} = C_1 C A_2 p_1/\sqrt{T_1} \qquad (5\text{-}22)$$

where A_2 = cross-sectional area of throat; C = coefficient of discharge, dimensionless; g_c = dimensional constant; k = ratio of specific heats, c_p/c_v; M = molecular weight; p_1 = pressure on upstream side of nozzle; R = gas constant; T_1 = absolute temperature on upstream side of nozzle; v_1 = specific volume on upstream side of nozzle; C_1 = dimensional constant, 0.0405 SI units (0.533 U.S. customary units); and w_{max} = maximum-weight flow rate.

Discharge coefficients for critical flow nozzles are, in general, the same as those for subsonic nozzles. See Grace and Lapple, *Trans. Am. Soc. Mech. Eng.*, **73**, 639–647 (1951); and Szaniszlo, *J. Eng. Power*, **97**, 521–526 (1975). Arnberg, Britton, and Seidl [*J. Fluids Eng.*, **96**, 111–123 (1974)] present discharge-coefficient correlations for circular-arc venturi meters at critical flow. For the calculation of the flow of natural gas through nozzles under critical-flow conditions, see Johnson, *J. Basic Eng.*, **92**, 580–589 (1970).

Orifice Meters A **square-edged** or **sharp-edged** orifice, as shown in Fig. 5-18, is a clean-cut square-edged hole with straight walls perpendicular to the flat upstream face of a thin plate placed crosswise of the channel. The stream issuing from such an orifice attains its minimum cross section (vena contracta) at a distance downstream of the orifice which varies with the ratio β of orifice to pipe diameter (see Fig. 5-19).

For a centered circular orifice in a pipe, the pressure differential is customarily measured between one of the following pressure-tap pairs. Except in the case of flange taps, all measurements of distance from the orifice are made from the upstream face of the plate.

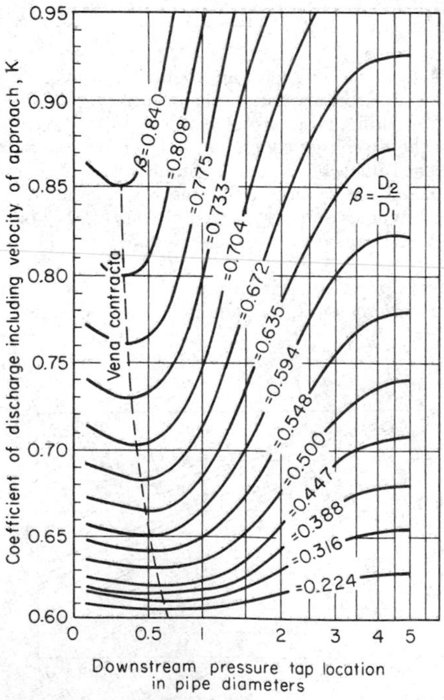

FIG. 5-19 Coefficient of discharge for square-edged circular orifices for N_{Re} > 30,000 with the upstream tap located between one and two pipe diameters from the orifice plate. [*Spitzglass,* Trans. Am. Soc. Mech. Eng., *44, 919 (1922).*]

1. *Corner taps.* Static holes drilled one in the upstream and one in the downstream flange, with the openings as close as possible to the orifice plate.

2. *Radius taps.* Static holes located one pipe diameter upstream and one-half pipe diameter downstream from the plate.

3. *Pipe taps.* Static holes located 2½ pipe diameters upstream and eight pipe diameters downstream from the plate.

4. *Flange taps.* Static holes located 25.4 mm (1 in) upstream and 25.4 mm (1 in) downstream from the plate.

5. *Vena-contracta taps.* The upstream static hole is one-half to two pipe diameters from the plate. The downstream tap is located at the position of minimum pressure (see Fig. 5-19).

Radius taps are best from a practical standpoint; the downstream pressure tap is located at about the mean position of the vena contracta, and the upstream tap is sufficiently far upstream to be unaffected by distortion of the flow in the immediate vicinity of the orifice (in practice, the upstream tap can be as much as two pipe diameters from the plate without affecting the results). Vena-contracta taps give the largest differential head for a given rate of flow but are inconvenient if the orifice size is changed from time to time. Corner taps offer the sometimes great advantage that the pressure taps can be built into the plate carrying the orifice. Thus the entire apparatus can be quickly inserted in a pipe line at any convenient flanged joint without having to drill holes in the pipe. Flange taps are similarly convenient, since by merely replacing standard flanges with special orifice flanges, suitable pressure taps are made available. Pipe taps give the lowest differential pressure, the value obtained being close to the permanent pressure loss.

Rate of discharge through an orifice meter is given by Eq. (5-12) for either liquids or gases. For the case of subsonic flow of a **gas** ($r_c < r < 1.0$), the expansion factor Y for orifices is approximated by

$$Y = 1 - [(1 - r)/k](0.41 + 0.35\beta^4) \qquad (5\text{-}23)$$

where r = ratio of downstream to upstream static pressure (p_2/p_1), k = ratio of specific heats (c_p/c_v), and β = diameter ratio. (See also Fig. 5-16.) Values of Y for supercritical flow of a gas ($r < r_c$) through orifices are given by Benedict [*J. Basic Eng.*, **93**, 121–137 (1971)]. For the case of **liquids,** expansion factor Y is unity, and Eq. (5-16) should be used, since it allows for any difference in elevation between the upstream and downstream taps.

Coefficient of discharge C for a given orifice type is a function of the Reynolds number N_{Re} (based on orifice diameter and velocity) and diameter ratio β. At Reynolds numbers greater than about 30,000, the coefficients are substantially constant. For square-edged or sharp-edged concentric circular orifices, the value will fall between 0.595 and 0.620 for vena-contracta or radius taps for β up to 0.8 and for flange taps for β up to 0.5. Figure 5-19 gives the coefficient of discharge K, including the velocity-of-approach factor ($1/\sqrt{1 - \beta^4}$), as a function of β and the location of the downstream tap. Precise values of K are given in *ASME PTC*, op. cit., pp. 20–39, for flange taps, radius taps, vena-contracta taps, and corner taps. Precise values of C are given in the ASME Research Committee on Fluid Meters Report, op. cit., pp. 202–207, for the first three types of taps.

The discharge coefficient of sharp-edged orifices was shown by Benedict, Wyler, and Brandt [*J. Eng. Power*, **97**, 576–582 (1975)] to increase with edge roundness. Typical as-purchased orifice plates may exhibit deviations on the order of 1 to 2 percent from ASME values of the discharge coefficient.

In the transition region (N_{Re} between 50 and 30,000), the coefficients are generally higher than the above values. Although calibration is generally advisable in this region, the curves given in Fig. 5-20 for corner and vena-contracta taps can be used as a guide. In the

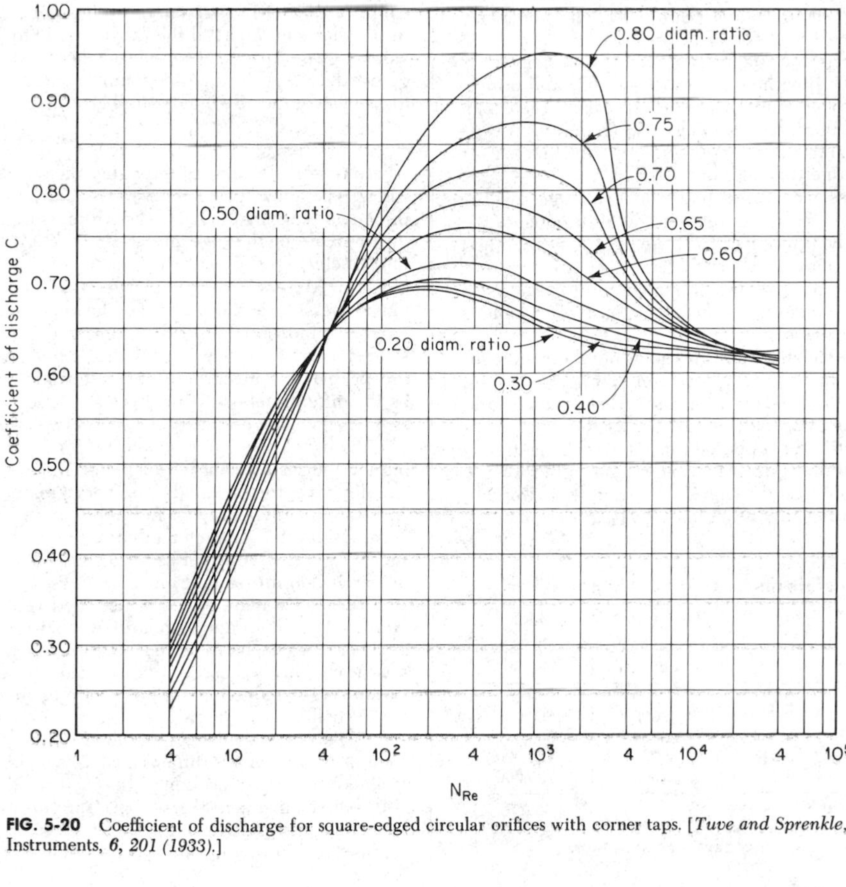

FIG. 5-20 Coefficient of discharge for square-edged circular orifices with corner taps. [*Tuve and Sprenkle, Instruments, 6, 201 (1933).*]

laminar-flow region ($N_{Re} < 50$), the coefficient C is proportional to $\sqrt{N_{Re}}$. For $1 < N_{Re} < 100$, Johansen [*Proc. R. Soc. (London)*, **A121**, 231–245 (1930)] presents discharge-coefficient data for sharp-edged orifices with corner taps. For $N_{Re} < 1$, Miller and Nemecek [ASME Paper 58-A-106 (1958)] present correlations giving coefficients for sharp-edged orifices and short-pipe orifices (L/D from 2 to 10). For short-pipe orifices (L/D from 1 to 4), Dickerson and Rice [*J. Basic Eng.*, **91**, 546–548 (1969)] give coefficients for the intermediate range ($27 < N_{Re} < 7000$). See also subsection "Contraction and Entrance Losses."

Permanent pressure loss across a concentric circular orifice with radius or vena-contracta taps can be approximated for turbulent flow by

$$(p_1 - p_4)/(p_1 - p_2) = 1 - \beta^2 \tag{5-24}$$

where p_1, p_2 = upstream and downstream pressure-tap readings respectively, p_4 = fully recovered pressure (four to eight pipe diameters downstream of the orifice), and β = diameter ratio. See *ASME PTC*, op. cit., Fig. 5.

See Benedict, *J. Fluids Eng.*, **99**, 245–248 (1977), for a general equation for pressure loss for orifices installed in pipes or with plenum inlets. Orifices show higher loss than nozzles or venturis. Permanent pressure loss for laminar flow depends on the Reynolds number in addition to β. See Alvi, Sridharan, and Lakshmana Rao, loc. cit., for details.

For the case of **critical flow** through a square- or sharp-edged concentric circular orifice (where $r \le r_c$, as discussed earlier in this subsection), use Eqs. (5-20), (5-21), and (5-22) as given for critical-flow nozzles. However, unlike nozzles, the flow through a sharp-edged orifice continues to increase as the downstream pressure drops below that corresponding to the critical pressure ratio r_c. This is due to an increase in the cross section of the vena contracta as the downstream pressure is reduced, giving a corresponding increase in the coefficient of discharge. At $r = r_c$, C is about 0.75, while at $r \cong 0$, C has increased to about 0.84. See Grace and Lapple, loc. cit.; and Benedict, *J. Basic Eng.*, **93**, 99–120 (1971).

Measurements by Harris and Magnall [*Trans. Inst. Chem. Eng. (London)*, **50**, 61–68 (1972)] with a venturi (β = 0.62) and orifices with radius taps (β = 0.60–0.75) indicate that the discharge coefficient for **nonnewtonian fluids**, in the range N'_{Re} (generalized Reynolds number) 3500 to 100,000, is approximately the same as for newtonian fluids at the same Reynolds number.

Quadrant-edge orifices have holes with rounded edges on the upstream side of the plate. The quadrant-edge radius is equal to the thickness of the plate at the orifice location. The advantages claimed for this type versus the square- or sharp-edged orifice are constant-discharge coefficients extending to lower Reynolds numbers and less possibility of significant changes in coefficient because of erosion or other damage to the inlet shape.

Values of discharge coefficient C and Reynolds numbers limit for constant C are presented in Table 5-2, based on Ramamoorthy and Seetharamiah [*J. Basic Eng.*, **88**, 9–13 (1966)] and Bogema and Monkmeyer (*J. Basic Eng.*, **82**, 729–734 (1960)]. At Reynolds numbers above those listed for the upper limits, the coefficients rise abruptly. As Reynolds numbers decrease below those listed for the lower limits, the coefficients pass through a hump and then drop off. According to Bogema, Spring, and Ramamoorthy [*J. Basic Eng.*, **84**, 415–418 (1962), the hump can be eliminated by placing a fine-mesh screen about three pipe diameters upstream of the orifice. This reduces the lower N_{Re} limit to about 500.

TABLE 5-2 Discharge Coefficients for Quadrant-Edge Orifices

β	C‡	K‡	Limiting N^*_{Re} for constant coefficient†	
			Lower	Upper
0.225	0.770	0.771	5,000	60,000
0.400	0.780	0.790	5,000	150,000
0.500	0.824	0.851	4,000	200,000
0.600	0.856	0.918	3,000	120,000
0.630	0.885	0.964	3,000	105,000

*Based on pipe diameter and velocity.
†For a precision of about ±0.5 per cent.
‡Can be used with corner taps, flange taps, or radius taps.

Permanent pressure loss across quadrant-edge orifices for turbulent flow is somewhat lower than given by Eq. (5-24). See Alvi, Sridharan, and Lakshmana Rao, loc. cit., for values of discharge coefficient and permanent pressure loss in laminar flow.

Segmental and **eccentric orifices** are frequently used for gas metering when there is a possibility that entrained liquids or solids would otherwise accumulate in front of a concentric circular orifice. This can be avoided if the opening is placed on the lower side of the pipe. For liquid flow with entrained gas, the opening is placed on the upper side. The pressure taps should be located on the opposite side of the pipe from the opening.

Coefficient C for a square-edged eccentric circular orifice (with opening tangent to pipe wall) varies from about 0.61 to 0.63 for β's from 0.3 to 0.5, respectively, and pipe Reynolds numbers > 10,000 for either vena-contracta or flange taps (where β = diameter ratio). For square-edged segmental orifices, the coefficient C falls generally between 0.63 and 0.64 for $0.3 \le \beta \le 0.5$ and pipe Reynolds numbers > 10,000, for vena-contracta or flange taps, where β = diameter ratio for an equivalent circular orifice = $\sqrt{\alpha}$ (α = ratio of orifice to pipe cross-sectional areas). Values of expansion factor Y are slightly higher than for concentric circular orifices, and the location of the vena contracta is moved farther downstream as compared with concentric circular orifices. For further details, see ASME Research Committee on Fluid Meters Report, op. cit., pp. 210–213.

For permanent pressure loss with segmental and eccentric orifices with laminar pipe flow see Lakshmana Rao and Sridharan, *Proc. Am. Soc. Civ. Eng., J. Hydraul. Div.*, **98** (HY 11), 2015–2034 (1972).

Annular orifices can also be used to advantage for gas metering when there is a possibility of entrained liquids or solids and for liquid metering with entrained gas present in small concentrations. Coefficient K was found by Bell and Bergelin [*Trans. Am. Soc. Mech. Eng.*, **79**, 593–601 (1957)] to range from about 0.63 to 0.67 for annulus Reynolds numbers in the range of 100 to 20,000 respectively for values of $2L/(D - d)$ less than 1 where L = thickness of orifice at outer edge, D = inside pipe diameter, and d = diameter of orifice disk. The annulus Reynolds number is defined as

$$N_{Re} = (D - d)(G/\mu) \tag{5-25}$$

where G = mass velocity through orifice opening and μ = fluid viscosity. The above coefficients were determined for β's (= d/D) in the range of 0.95 to 0.996 and with pressure taps located 19 mm (¾ in) upstream of the disk and 230 mm (9 in) downstream in a 5.25-in-diameter pipe.

Elbow Meters A pipe elbow can be used as a flowmeter for liquids if the differential centrifugal head generated between the inner and outer radii of the bend is measured by means of pressure taps located midway around the bend. Equation (5-16) can be used, except that the pressure-difference term $(p_1 - p_2)$ is now taken to be the differential centrifugal pressure and β is taken as zero if one assumes no change in cross section between the pipe and the bend. The discharge coefficient should preferably be determined by calibration, but as a guide it can be estimated within ±6 percent for circular pipe for Reynolds numbers greater than 10^5 from $C = 0.98\sqrt{R_c/2D}$, where R_c = radius of curvature of the centerline and D = inside pipe diameter in consistent units. See Murdock, Foltz, and Gregory, *J. Basic Eng.*, **86**, 498–506 (1964); or the ASME Research Committee on Fluid Meters Report, op. cit., pp. 75–77.

Accuracy Square-edged orifices and venturi tubes have been so extensively studied and standardized that reproducibilities within 1 to 2 percent can be expected between standard meters when new and clean. This is therefore the order of reliability to be had, if one assumes (1) accurate measurement of meter differential, (2) selection of the coefficient of discharge from recommended published literature, (3) accurate knowledge of fluid density, (4) accurate measurement of critical meter dimensions, (5) smooth upstream face of orifice, and (6) proper location of the meter with respect to other flow-disturbing elements in the system. Care must also be taken to avoid even slight corrosion or fouling during use.

Presence of **swirling flow** or an **abnormal velocity distribution**

TABLE 5-3 Locations of Orifices and Nozzles Relative to Pipe Fittings

Distances in pipe diameters, D_1

Type of fitting upstream	$\frac{D_2}{D_1}$	Distance, upstream fitting to orifice: Without straightening vanes	Distance, upstream fitting to orifice: With straightening vanes	Distance, vanes to orifice	Distance, nearest downstream fitting from orifice
Single 90° ell, tee, or cross used as ell	0.2	6			2
	0.4	6			
	0.6	9	9		
	0.8	20	12	8	4
2 short-radius 90° ells in form of S	0.2	7			2
	0.4	8	8		
	0.6	13	10	6	
	0.8	25	15	11	4
2 long- or short-radius 90° ells in perpendicular planes	0.2	15	9	5	2
	0.4	18	10	6	
	0.6	25	11	7	
	0.8	40	13	9	4
Contraction or enlargement	0.2	8	Vanes have no advantage		2
	0.4	9			
	0.6	10			
	0.8	15			4
Globe valve or stop check	0.2	9	9	5	2
	0.4	10	10	6	
	0.6	13	10	6	
	0.8	21	13	9	4
Gate valve, wide open, or plug cocks	0.2	6	Same as globe valve		2
	0.4	6			
	0.6	8			
	0.8	14			4

upstream of the metering element can cause serious metering error unless calibration in place is employed or sufficient straight pipe is inserted between the meter and the source of disturbance. Table 5-3 gives the minimum lengths of straight pipe required to avoid appreciable error due to the presence of certain fittings and valves either upstream or downstream of an orifice or nozzle. These values were extracted from plots presented by Sprenkle [*Trans. Am. Soc. Mech. Eng.*, **67**, 345–360 (1945)]. Table 5-3 also shows the reduction in spacing made possible by the use of straightening vanes between the fittings and the meter. Entirely adequate straightening vanes can be provided by fitting a bundle of thin-wall tubes within the pipe. The center-to-center distance between tubes should not exceed one-fourth of the pipe diameter, and the bundle length should be at least 8 times this distance.

The distances specified in Table 5-3 will be conservative if applied to venturi meters. For specific information on requirements for venturi meters, see a discussion by Pardoe appended to Sprenkle (op. cit.). Extensive data on the effect of installation on the coefficients of venturi meters are given elsewhere by Pardoe [*Trans. Am. Soc. Mech. Eng.*, **65**, 337–349 (1943)].

In the presence of **flow pulsations,** the indications of head meters such as orifices, nozzles, and venturis will often be undependable for several reasons. First, the measured pressure differential will tend to be high, since the pressure differential is proportional to the square of flow rate for a head meter, and the square root of the mean differential pressure is always greater than the mean of the square roots of the differential pressures. Second, there is a phase shift as the wave passes through the metering restriction which can affect the differential. Third, pulsations can be set up in the manometer leads themselves. Frequency of the pulsation also plays a part. At low frequencies, the meter reading can generally faithfully follow the flow pulsations, but at high frequencies it cannot. This is due to inertia of the fluid in the manometer leads or of the manometric fluid, whereupon the meter would give a reading intermediate between the maximum and minimum flows but having no readily predictable relation to the mean flow. Pressure transducers with flush-mounted diaphragms can be used together with high-speed recording equipment to provide accurate records of the pressure profiles at the upstream and downstream pressure taps, which can then be analyzed and translated into a mean flow rate.

The rather general practice of producing a steady differential reading by placing restrictions in the manometer leads can result in a reading which, under a fixed set of conditions, may be useful in control of an operation but which has no readily predictable relation to the actual average flow. If calibration is employed to compensate for the presence of pulsations, complete reproduction of operating conditions, including source of pulsations and waveform, is necessary to ensure reasonable accuracy.

According to Head [*Trans. Am. Soc. Mech. Eng.*, **78**, 1471–1479 (1956)], a pulsation-intensity limit of $\Gamma = 0.1$ is recommended as a practical pulsation threshold below which the performance of all types of flowmeters will differ negligibly from steady-flow performance (an error of less than 1 percent in flow due to pulsation). Γ is the peak-to-trough flow variation expressed as a fraction of the average flow rate. According to the ASME Research Committee on Fluid Meters Report (op. cit., pp. 34–35), the fractional metering error E for **liquid flow** through a head meter is given by

$$(1 + E)^2 = 1 + \Gamma^2/8 \qquad (5\text{-}26)$$

When the pulsation amplitude is such as to result in a greater-than-permissible metering error, consideration should be given to installation of a pulsation damper between the source of pulsations and the flowmeter. References to methods of pulsation-damper design are given in the subsection "Unsteady-State Behavior."

Pulsations are most likely to be encountered in discharge lines from reciprocating pumps or compressors and in lines supplying steam to reciprocating machinery. For **gas flow,** a combination involving a surge chamber and a constriction in the line can be used to damp out the pulsations to an acceptable level. The surge chamber is generally located as close to the pulsation source as possible, with the constriction between the surge chamber and the metering element. This arrangement can be used for either a suction or a discharge line. For such an arrangement, the metering error has been found to be a function of the Hodgson number N_H, which is defined as

$$N_H = Qn\,\Delta p_s/qp_s \qquad (5\text{-}27)$$

where Q = volume of surge chamber and pipe between metering element and pulsation source; n = pulsation frequency; ΔP_s = permanent pressure drop between metering element and surge chamber; q = average volume flow rate, based on gas density in the surge chamber; and p_s = pressure in surge chamber.

Herning and Schmid [*Z. Ver. Dtsch. Ing.*, **82**, 1107–1114 (1938)] presented charts for a simplex double-acting compressor for the prediction of metering error as a function of the Hodgson number and s, the ratio of piston discharge time to total time per stroke. Table 5-4*a* gives the minimum Hodgson numbers required to reduce the metering error to 1 percent as given by the charts (for specific heat ratios between 1.28 and 1.37). Schmid [*Z. Ver. Dtsch. Ing.*, **84**, 596–598 (1940)] presented similar charts for a duplex double-acting compressor and a triplex double-acting compressor for a specific heat ratio of 1.37. Table 5-4*b* gives the minimum Hodgson numbers cor-

TABLE 5-4*a* Minimum Hodgson Numbers

Simplex double-acting compressor

s	N_H	s	N_H
0.167	1.31	0.667	0.60
0.333	1.00	0.833	0.43
0.50	0.80	1.00	0.34

TABLE 5-4*b* Minimum Hodgson Numbers

Duplex double-acting compressor		Triplex double-acting compressor	
s	N_H	s	N_H
0.167	1.00	0.167	0.85
0.333	0.70	0.333	0.30
0.50	0.30	0.50	0.15
0.667	0.10	0.667	0.06
0.833	0.05	0.833	0.00
1.00	0.00	1.00	0.00

responding to a 1 percent metering error for these cases. The value of $Q \Delta p_s$ can be calculated from the appropriate Hodgson number, and appropriate values of Q and Δp_s selected so as to satisfy this minimum requirement.

AREA METERS

General Principles The underlying principle of an ideal area meter is the same as that of a head meter of the orifice type (see subsection "Orifice Meters"). The stream to be measured is throttled by a constriction, but instead of observing the variation with flow of the differential head across an orifice of fixed size, the constriction of an area meter is so arranged that its size is varied to accommodate the flow while the differential head is held constant.

A simple example of an area meter is a gate valve of the rising-stem type provided with static-pressure taps before and after the gate and a means for measuring the stem position. In most common types of area meters, the variation of the opening is automatically brought about by the motion of a weighted piston or float supported by the fluid. Two different cylinder- and piston-type area meters are described in the ASME Research Committee on Fluid Meters Report, op. cit., pp. 82–83.

Rotameters The rotameter, an example of which is shown in Fig. 5-21, has become one of the most popular flowmeters in the chemical-process industries. It consists essentially of a plummet, or "float," which is free to move up or down in a vertical, slightly tapered tube having its small end down. The fluid enters the lower end of the tube and causes the float to rise until the annular area between the float and the wall of the tube is such that the pressure drop across this constriction is just sufficient to support the float. Typically, the tapered tube is of glass and carries etched upon it a nearly linear scale on which the position of the float may be visually noted as an indication of the flow.

Interchangeable precision-bore glass tubes and metal metering tubes are available. Rotameters have proved satisfactory both for gases and for liquids at high and at low pressures. A single instrument can readily cover a tenfold range of flow, and by providing floats of different densities a two-hundredfold range is practicable. Rotameters are available with pneumatic, electric, and electronic transmitters for actuating remote recorders, integrators, and automatic flow controllers (see Considine, op. cit., pp. 4-35–4-36, and Sec. 22 of this *Handbook*).

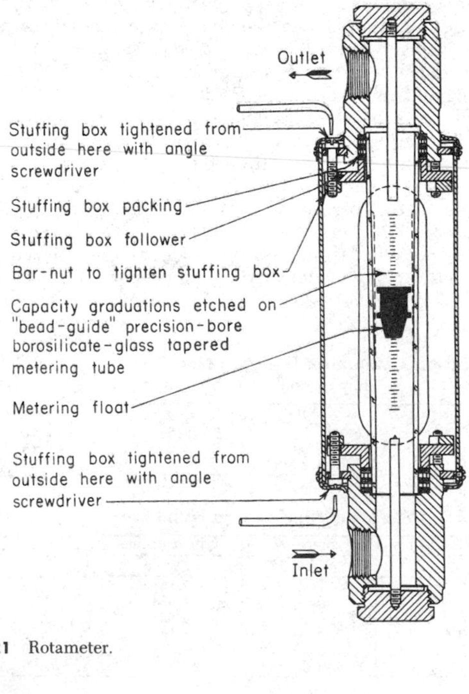

FIG. 5-21 Rotameter.

Rotameters require no straight runs of pipe before or after the point of installation. Pressure losses are substantially constant over the whole flow range. In experimental work, for greatest precision, a rotameter should be calibrated with the fluid which is to be metered. However, most modern rotameters are precision-made so that their performance closely corresponds to a master calibration plot for the type in question. Such a plot is supplied with the meter upon purchase.

According to Head [*Trans. Am. Soc. Mech. Eng.*, **76,** 851–862 (1954)], flow rate through a rotameter can be obtained from

$$w = q\rho = KD_f \sqrt{\frac{W_f(\rho_f - \rho)\rho}{\rho_f}} \tag{5-28}$$

and

$$K = \phi \left[\frac{D_t}{D_f}, \frac{\mu}{\sqrt{\dfrac{W_f(\rho_f - \rho)\rho}{\rho_f}}} \right] \tag{5-29}$$

where w = weight flow rate; q = volume flow rate; ρ = fluid density; K = flow parameter, $m^{1/2}/s$ ($ft^{1/2}/s$); D_f = float diameter at constriction; W_f = float weight; ρ_f = float density; D_t = tube diameter at point of constriction; and μ = fluid viscosity. The appropriate value of K is obtained from a composite correlation of K versus the parameters shown in Eq. (5-29) corresponding to the float shape being used. The relation of D_t to the rotameter reading is also required for the tube taper and size being used.

The ratio of flow rates for two different fluids A and B at the same rotameter reading is given by

$$\frac{w_A}{w_B} = \frac{K_A}{K_B} \sqrt{\frac{(\rho_f - \rho_A)\rho_A}{(\rho_f - \rho_B)\rho_B}} \tag{5-30}$$

A measure of self-compensation, with respect to weight rate of flow, for fluid-density changes can be introduced through the use of a float with a density twice that of the fluid being metered, in which case an increase of 10 percent in ρ will produce a decrease of only 0.5 percent in w for the same reading. The extent of immunity to changes in fluid viscosity depends upon the shape of the float.

According to Baird and Cheema [*Can. J. Chem. Eng.*, **47,** 226–232 (1969)], the presence of square-wave pulsations can cause a rotameter to overread by as much as 100 percent. The higher the pulsation frequency, the less the float oscillation, although the error can still be appreciable even when the frequency is high enough so that the float is virtually stationary. Use of a damping chamber between the pulsation source and the rotameter will reduce the error.

Additional information on rotameter theory is presented by Fischer [*Chem. Eng.*, **59**(6), 180–184 (1952)], Coleman [*Trans. Inst. Chem. Eng.*, **34,** 339–350 (1956)], and McCabe and Smith (*Unit Operations of Chemical Engineering*, 3d ed., McGraw-Hill, New York, 1976, pp. 215–218).

MASS FLOWMETERS

General Principles There are two main types of mass flowmeters: (1) the so-called true mass flowmeter, which responds directly to mass flow rate, and (2) the inferential mass flowmeter, which commonly measures volume flow rate and fluid density separately. A variety of types of true mass flowmeters have been developed, including the following: (*a*) the Magnus-effect mass flowmeter, (*b*) the axial-flow, transverse-momentum mass flowmeter, (*c*) the radial-flow, transverse-momentum mass flowmeter, (*d*) the gyroscopic transverse-momentum mass flowmeter, and (*e*) the thermal mass flowmeter. Type *b* is the basis for several commercial mass flowmeters, one version of which is briefly described here.

Axial-Flow Transverse-Momentum Mass Flowmeter This type is also referred to as an angular-momentum mass flowmeter. One embodiment of its principle involves the use of axial flow through a driven impeller and a turbine in series. The impeller imparts angular momentum to the fluid, which in turn causes a torque to be imparted to the turbine, which is restrained from rotating by a spring. The torque, which can be measured, is proportional to the rotational speed of the impeller and the mass flow rate.

Inferential Mass Flowmeters There are several types in this category, including the following:

1. *Head meters with density compensation.* Head meters such as orifices, venturis, or nozzles can be used with one of a variety of densitometers [e.g., based on (*a*) buoyant force on a float, (*b*) hydraulic coupling, (*c*) voltage output from a piezoelectric crystal, or (*d*) radiation absorption]. The signal from the head meter, which is proportional to ρV^2 (where ρ = fluid density and V = fluid velocity), is multiplied by ρ given by the densitometer. The square root of the product is proportional to the mass flow rate.

2. *Head meters with velocity compensation.* The signal from the head meter, which is proportional to ρV^2, is divided by the signal from a velocity meter to give a signal proportional to the mass flow rate.

3. *Velocity meters with density compensation.* The signal from the velocity meter (e.g., turbine meter, electromagnetic meter, or sonic velocity meter) is multiplied by the signal from a densitometer to give a signal proportional to the mass flow rate.

Additional information on mass-flowmeter principles can be obtained from Yeaple (*Hydraulic and Pneumatic Power and Control,* McGraw-Hill, New York, 1966, pp. 125–128), Halsell [*Instrum. Soc. Am. J.*, **7**, 49–62 (June 1960)], and Flanagan and Colman [*Control*, **7**, 242–245 (1963)]. Information on commercially available mass flowmeters is given in the latter two references.

WEIRS

Liquid flow in an open channel may be metered by means of a weir, which consists of a dam over which, or through a notch in which, the liquid flows. The terms "rectangular weir," "triangular weir," etc., generally refer to the shape of the notch in a notched weir. All weirs considered here have flat upstream faces that are perpendicular to the bed and walls of the channel.

Sharp-edged weirs have edges like those of square or sharp-edged orifices (see subsection "Orifice Meters"). Notched weirs are ordinarily sharp-edged. Weirs not in the sharp-edged class are, for the most part, those described as **broad-crested weirs.**

The head h_0 on a weir is the liquid-level height above the crest or base of the notch. The head must be measured sufficiently far upstream to avoid the drop in level occasioned by the overfall which begins at a distance about $2h_0$ upstream from the weir. Surface-level measurements should be made a distance of $3h_0$ or more upstream, preferably by using a stilling box equipped with a high-precision level gauge, e.g., a hook gauge or float gauge.

With sharp-edged weirs, the sheet of discharging liquid, called the "nappe," contracts as it leaves the opening and free discharge occurs. Rounding the upstream edge will reduce the contraction and increase the flow rate for a given head. A clinging nappe may result if the head is very small, if the edge is well rounded, or if air cannot flow in beneath the nappe. This, in turn, results in an increase in the discharge rate for a given head as compared with that for a free nappe. For further information on the effect of the nappe, see Gibson, *Hydraulics and Its Applications*, 5th ed., Constable, London, 1952; and Chow, *Open-Channel Hydraulics*, McGraw-Hill, New York, 1959.

Flow through a **rectangular weir** (Fig. 5-22) is given by

$$q = 0.415(L - 0.2h_0)h_0^{1.5}\sqrt{2g} \qquad (5\text{-}31)$$

where q = volume flow rate, L = crest length, h_0 = weir head, and g = local acceleration due to gravity. This is known as the modified Francis formula for a rectangular sharp-edged weir with two end corrections; it applies when the velocity-of-approach correction is small. The Francis formula agrees with experiments within 3 percent

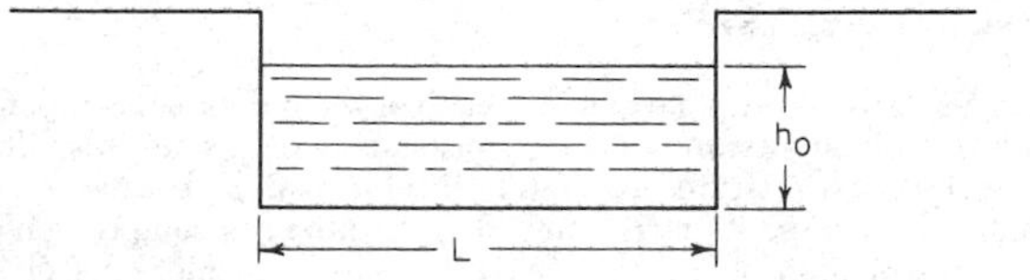

FIG. 5-22 Rectangular weir.

if (1) L is greater than $2h_0$, (2) velocity of approach is 0.6 m/s (2 ft/s) or less, (3) height of crest above bottom of channel is at least $3h_0$, and (4) h_0 is not less than 0.09 m (0.3 ft).

Narrow rectangular notches ($h_0 > L$) have been found to give about 93 percent of the discharge predicted by the Francis formula. Thus

$$q = 0.386Lh_0^{1.5}\sqrt{2g} \qquad (5\text{-}32)$$

In this case, no end corrections are applied even though the formula applies only for sharp-edged weirs. See Schoder and Dawson, *Hydraulics*, McGraw-Hill, New York, 1934, p. 175, for further details.

The **triangular-notch weir** has the advantage that a single notch can accommodate a wide range of flow rates, although this in turn reduces its accuracy. The discharge for sharp- or square-edged weirs is given by

$$q = (0.31h_0^{2.5}\sqrt{2g})/\tan\phi \qquad (5\text{-}33)$$

See Eq. (5-31) for nomenclature. Angle ϕ is illustrated in Fig. 5-23. Equations (5-31), (5-32), and (5-33) are applicable only to the flow of water. However, for the case of triangular-notch weirs Lenz [*Trans. Am. Soc. Civ. Eng.*, **108**, 759–802 (1943)] has presented correlations predicting the effect of viscosity over the range of 0.001 to 0.15 Pa·s (1 to 150 cP) and surface tension over the range of 0.03 to 0.07 N/m (30 to 70 dyn/cm). His equation predicts about an 8 percent increase in flow for a liquid of 0.1-Pa·s (100-cP) viscosity compared with water at 0.001 Pa·s (1 cP) and about a 1 percent increase for a liquid with one-half of the surface tension of water. For fluids of moderate viscosity, Ranga Raju and Asawa [*Proc. Am. Soc. Civ. Eng., J. Hydraul. Div.*, **103** (HY 10), 1227–1231 (1977)] find that the effect of viscosity and surface tension on the discharge flow rate for rectangular and triangular-notch (ϕ = 45°) weirs can be neglected when

$$(N_{Re})^{0.2}(N_{We})^{0.6} > 900 \qquad (5\text{-}34)$$

where N_{Re} (Reynolds number) = $\sqrt{gh_0^3}/\nu$, g = local acceleration due to gravity, h_0 = weir head, ν = kinematic viscosity; N_{We} (Weber number) = $\rho g h_0^2/g_c\sigma$, ρ = density, g_c = dimensional constant, and σ = surface tension.

For the flow of high-viscosity liquids over rectangular weirs, see Slocum, *Can. J. Chem. Eng.*, **42**, 196–200 (1964). His correlation is based on data for liquids with viscosities in the range of 2.5 to 500 Pa·s (25 to 5000 cP), in which range the discharge decreases markedly for a given head as viscosity is increased.

Information on other types of weirs can be obtained from Addison, op. cit.; Gibson, *Hydraulics and Its Applications*, 5th ed., Constable, London, 1952; Henderson, *Open Channel Flow*, Macmillan, New York, 1966; Linford, *Flow Measurement and Meters*, Spon, London, 1949; Lakshmana Rao, "Theory of Weirs," in *Advances in Hydroscience*, vol. 10, Academic, New York, 1975; and Urquhart, *Civil Engineering Handbook*, 4th ed., McGraw-Hill, New York, 1959.

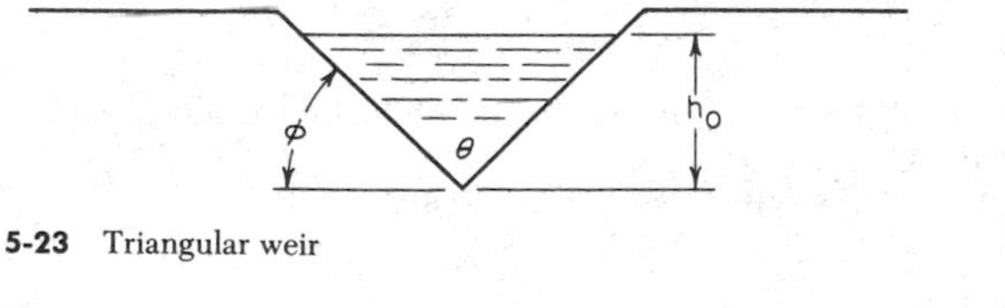

FIG. 5-23 Triangular weir

TWO-PHASE SYSTEMS

It is generally preferable to meter each of the individual components of a two-phase mixture separately prior to mixing, since it is difficult to meter such mixtures accurately. Problems arise because of fluctuations in composition with time and variations in composition over the cross section of the channel. Information on metering of such mixtures can be obtained from the following sources.

Gas-Solid Mixtures Carlson, Frazier, and Engdahl [*Trans. Am. Soc. Mech. Eng.*, **70,** 65–79 (1948)] describe the use of a **flow nozzle** and a **square-edged orifice** in series for the measurement of both the gas rate and the solids rate in the flow of a finely divided solid-in-gas mixture. The nozzle differential is sensitive to the flow of both phases, whereas the orifice differential is not influenced by the solids flow.

Farbar [*Trans. Am. Soc. Mech. Eng.*, **75,** 943–951 (1953)] describes how a **venturi meter** can be used to measure solids flow rate in a gas-solids mixture when the gas rate is held constant. Separate calibration curves (solids flow versus differential) are required for each gas rate of interest.

Cheng, Tung, and Soo [*J. Eng. Power*, **92,** 135–149 (1970)] describe the use of an **electrostatic probe** for measurement of solids flow in a gas-solids mixture.

Goldberg and Boothroyd [*Br. Chem. Eng.*, **14,** 1705–1708 (1969)] describe several types of solids-in-gas flowmeters and give an extensive bibliography.

Gas-Liquid Mixtures An empirical equation was developed by Murdock [*J. Basic Eng.*, **84,** 419–433 (1962)] for the measurement of gas-liquid mixtures using **sharp-edged orifice** plates with either radius, flange, or pipe taps.

An equation for use with **venturi meters** was given by Chisholm [*Br. Chem. Eng.*, **12,** 454–457 (1967)]. A procedure for determining steam quality via pressure-drop measurement with upflow through either venturi meters or sharp-edged orifice plates was given by Collins and Gacesa [*J. Basic Eng.*, **93,** 11–21 (1971)].

Liquid-Solid Mixtures Liptak [*Chem. Eng.*, **74**(4), 151–158 (1967)] discusses a variety of techniques that can be used for the measurement of solids-in-liquid suspensions or slurries. These include metering pumps, weigh tanks, magnetic flowmeter, ultrasonic flowmeter, gyroscope flowmeter, etc.

Shirato, Gotoh, Osasa, and Usami [*J. Chem. Eng. Japan*, **1,** 164–167 (January 1968)] present a method for determining the mass flow rate of suspended solids in a liquid stream wherein the liquid velocity is measured by an electromagnetic flowmeter and the flow of solids is calculated from the pressure drops across each of two vertical sections of pipe of different diameter through which the suspension flows in series.

FLUID DYNAMICS

GENERAL REFERENCES: Coulson and Richardson, *Chemical Engineering*, vol. 1, rev. 2d ed., Macmillan, New York, 1964; vol. 2, 2d ed., Pergamon, New York, 1968. Cremer and Davies, *Chemical Engineering Practice*, vol. 4 (1957), vol. 6 (1958), vol. 6 (1958), Academic, New York. Knudsen and Katz, *Fluid Dynamics and Heat Transfer*, McGraw-Hill, New York, 1958. Streeter, *Handbook of Fluid Dynamics*, McGraw-Hill, New York, 1961. Streeter, *Fluid Mechanics*, 5th ed., McGraw-Hill, New York, 1971. Vennard and Street, *Elementary Fluid Mechanics*, 5th ed., Wiley, New York, 1975. Whitaker, *Introduction to Fluid Mechanics*, Prentice-Hall, Englewood Cliffs, N.J., 1968.

ENERGY BALANCE

Total Energy Balance Consider a unit weight of fluid in a flow system, and let G = mass velocity, g = local acceleration due to gravity, g_c = dimensional constant, i = specific enthalpy, J = mechanical equivalent of heat, p = absolute static pressure, s = specific entropy, u = specific internal (or intrinsic) energy, V = linear velocity, v = specific volume, and Z = height above any arbitrary horizontal datum plane. Then the potential energy relative to the chosen reference level is Zg/g_c, the kinetic energy is $V^2/2g_c$, and the total energy of the unit weight of fluid is $(Ju + Zg/g_c + V^2/2g_c)$, all in units of specific energy J/kg (= N·m/kg) [(ft·lbf)/lb]. For steady flow of the fluid, there will be no accumulation or depletion of either fluid or energy within the system, and the total energy of the system can be altered only by adding or subtracting heat from the system or by external work on the system. Thus

$$\left(Ju_2 + \frac{Z_2 g}{g_c} + \frac{V_2^2}{2g_c}\right) - \left(Ju_1 + \frac{Z_1 g}{g_c} + \frac{V_1^2}{2g_c}\right) = JQ + W \qquad (5\text{-}35)$$

where subscripts 1 and 2 indicate conditions at inlet and outlet respectively; Q is the heat added, J/kg (Btu/lb), from sources external to the system; and W is the net external work, N·m/kg [(ft·lbf)/lb], done on the pound of fluid while in the apparatus. The term W may be subdivided as

$$W = p_1 v_1 - p_2 v_2 + W_e \qquad (5\text{-}36)$$

where W_e = work delivered by an external source, such as a blower or pump.

Combining Eqs. (5-35) and (5-36) gives

$$Ju_1 + \frac{Z_1 g}{g_c} + \frac{V_1^2}{2g_c} + p_1 v_1 + JQ + W_e = Ju_2 + \frac{Z_2 g}{g_c} + \frac{V_2^2}{2g_c} + p_2 v_2 \qquad (5\text{-}37)$$

This expression of the first law of thermodynamics is often known as the **overall energy balance** form of **Bernoulli's theorem.** No friction term occurs in this expression, since friction represents a conversion of mechanical energy into heat, with no change in the overall energy content of the system.

The kinetic-energy terms in the foregoing equations apply strictly only for a uniform velocity across a given cross section. For turbulent flow in circular pipes the term $V^2/2g_c$ is 3 to 8 percent too low, while for laminar flow in circular pipes the proper kinetic-energy term is V^2/g_c, which makes due allowance for the parabolic velocity distribution.

If $i[= u + (pv/J)$, J/kg (Btu/lb)] is the specific enthalpy (total heat content per unit weight), Eq. (5-37) takes a form convenient for use with steam or other fluids for which the thermal properties are tabulated or calculable:

$$(Z_1 - Z_2)(g/g_c) + J(i_1 - i_2) + JQ + W_e = (V_2^2 - V_1^2)/2g_c \qquad (5\text{-}38)$$

For adiabatic frictionless flow through a horizontal nozzle, Eq. (5-38) reduces to

$$J(i_1 - i_2) = \frac{G^2(v_2^2 - v_1^2)}{2g_c} \qquad (5\text{-}39)$$

For perfect gases [i.e., if pv = (constant)(T)],Eq. (5-39) becomes

$$J(i_1 - i_2) = J\int_{T_2}^{T_1} c_p\,dT = J(c_p)_{\text{avg}}(T_1 - T_2) = \frac{k}{k-1}(p_1 v_1 - p_2 v_2) \qquad (5\text{-}40)$$

where c_p = specific heat capacity at constant pressure; $(c_p)_{\text{avg}}$ = average specific heat capacity at constant pressure, for temperature interval T_2 to T_1; $k = c_p/c_v$, ratio of specific heat capacities; and T = absolute temperature.

Mechanical-Energy Balance The change in internal energy $J\,du$ of a pound of fluid may be expressed as $JT\,ds - p\,dv$. If other forms of energy, such as electric, are present and undergoing change, they should be included. The presence of friction renders the process irreversible, whence $JT\,ds = J\,dQ + dF$, where F = friction loss, (N·m)/kg [(ft·lbf)/lb]. With this consideration of friction loss and irreversibility, Eq. (5-37) becomes

$$\frac{Z_1 g}{g_c} + \frac{V_1^2}{2g_c} - \int_1^2 v\,dp - F + W_e = \frac{Z_2 g}{g_c} + \frac{V_2^2}{2g_c} \qquad (5\text{-}41)$$

Equation (5-41) is known as the **mechanical-energy** form of **Bernoulli's theorem.**

For **liquids,** the integral $\int_1^2 v\,dp$ becomes simply $(p_2 - p_1)v$ where v is substantially constant. For **gases,** the exact value of the integral depends upon the path followed by the expansion. For **isothermal flow** of a perfect gas

$$\int_1^2 v\,dp = -\frac{RT}{M}\ln\frac{v_2}{v_1} = -\frac{RT}{M}\ln\frac{p_1}{p_2} \qquad (5\text{-}42)$$

For **adiabatic flow,** see subsection "Compressible Flow."

System Pressure-Drop Evaluation There is a real distinction between pressure drop and friction loss. Pressure drop represents a conversion of pressure energy into any other form of energy, whereas friction loss represents a net loss in the total available work energy of the fluid. The two terms are related by Eq. (5-41).

There are two methods of evaluating the overall pressure drop for a system having a number of resistances in series. The first method involves calculation of the **pressure drop** for each individual resistance with due regard for algebraic sign and then summing up all such items for the entire system. The second method involves the calculation of **friction loss** for each individual resistance, summing up all such items and applying Eq. (5-41) to obtain the overall pressure drop. The pressure-drop summation should be used in systems involving branching lines. For the case of compressible flow with large pressure drops, see subsection "Compressible Flow."

MOMENTUM BALANCE

The **principle of conservation of momentum** states that the total momentum within a system remains constant during the exchange of momentum between two or more masses of the system. This is illustrated by the case of two masses m_M and m_N moving in the same direction, which impact and then travel together in the same direction with a common velocity V:

$$\frac{m_M V_M}{g_c} + \frac{m_N V_N}{g_c} = \frac{(m_M + m_N)V}{g_c} \qquad (5\text{-}43)$$

In this illustration there is a loss in kinetic energy through the exchange of momentum.

For flowing fluids in steady flow with uniform velocity, the **equation of momentum** is

$$F = w\,\Delta V/g_c = \rho q\,\Delta V/g_c \qquad (5\text{-}44)$$

where F = force acting on the fluid, w = weight rate of flow, ΔV = change in velocity, ρ = fluid density, q = volumetric flow rate, and g_c = dimensional constant. This equation is applied to several practical problems in the following subsections.

Jet Impact on a Plate For an open jet impinging upon a **stationary oblique flat plate** (Streeter, op. cit., 1971, pp. 158–159), as shown in Fig. (5-24), force F acting normal to the plate is given by

$$F = \rho q V/g_c \sin\theta \qquad (5\text{-}45)$$

This force is a vector quantity, and its components can be determined from vector addition. The division of flow is given by

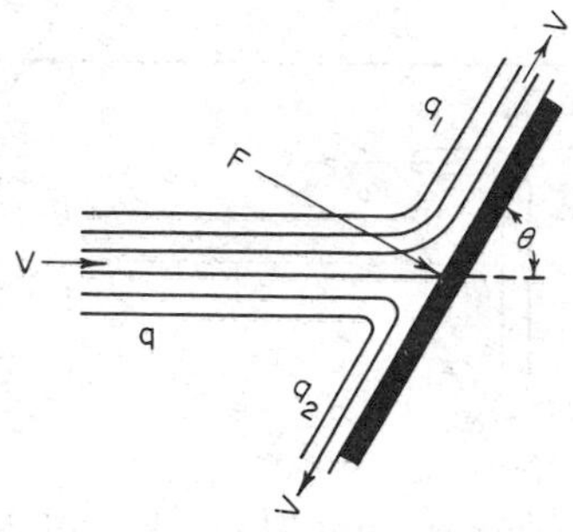

FIG. 5-24 Jet impact on stationary oblique flat plate.

$$q_1 = (q/2)(1 + \cos\theta) \qquad (5\text{-}46)$$

$$q_2 = (q/2)(1 - \cos\theta) \qquad (5\text{-}47)$$

where θ = angle of inclination of the plate and other symbols are as given before.

These equations assume that there is no loss of energy (thus the velocity must be unchanged) and that the plate is smooth (thus there is no tangential force).

For jet impact on a **stationary normal flat plate,** $\theta = 90°$ and

$$F = \rho q V/g_c \qquad (5\text{-}48)$$

$$q_1 = q_2 = q/2 \qquad (5\text{-}49)$$

Forces on Bends The forces exerted by a flowing fluid on a bend (Fig. 5-25) can be computed from a force balance on the bend (Streeter, op. cit., 1971, pp. 144–145; and Vennard and Street, op. cit., p. 228). From a balance of the forces due to pressure, momentum change, and weight along directions x and y, the forces F_x and F_y exerted **by the bend on the flowing fluid** are

$$F_x = p_1A_1 - p_2A_2\cos\theta + (\rho q/g_c)(V_1 - V_2\cos\theta) \qquad (5\text{-}50)$$

$$F_y = \frac{Wg}{g_c} + \left(p_2A_2 + \frac{\rho q V_2}{g_c}\right)\sin\theta \qquad (5\text{-}51)$$

where F_x = force in horizontal direction exerted by the bend on the flowing fluid; F_y = force in vertical direction exerted by the bend on the flowing fluid; p_1 = pressure, A_1 = area, and V_1 = velocity at the inlet to the bend; p_2 = pressure, A_2 = area, and V_2 = velocity at the outlet of the bend; q = fluid volume flow rate; ρ = fluid density; g = local acceleration due to gravity; g_c = dimensional constant; W = weight of fluid in the bend; and θ = angle subtended by the outlet.

The forces exerted **by the flowing fluid on the bend** are equal and opposite to those given by Eqs. (5-50) and (5-51).

Ejectors An ejector is a device in which the kinetic energy of one fluid (primary fluid) is used to pump another fluid (secondary fluid). The performance of this device can be computed by applica-

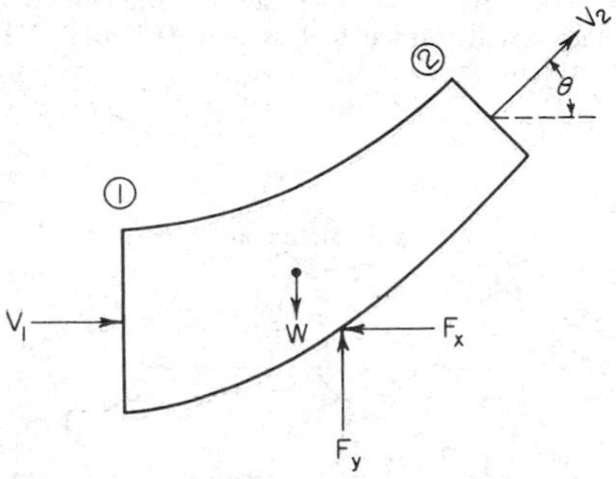

FIG. 5-25 Forces on a bend.

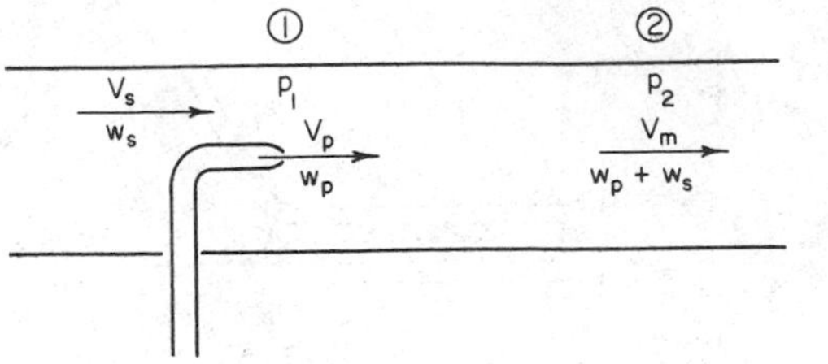

FIG. 5-26 Draft-tube ejector.

tion of the momentum equation. The method will be illustrated for a draft-tube ejector (Fig. 5-26). The primary, high-velocity fluid enters at 1 and mixes with the secondary, low-velocity fluid; mixing is assumed to be complete at 2. Through this mixing, a major portion of the momentum of the primary fluid is imparted to the secondary fluid, resulting in a static pressure at 2 greater than that at 1. This increase in static pressure is given by equating the pressure and momentum forces along the draft tube. For incompressible fluids, this increase $p_2 - p_1$ is given by

$$(p_2 - p_1)g_c A = w_p(V_p - V_m) + w_s(V_s - V_m) \quad (5\text{-}52)$$

where p_1 = pressure at plane 1, p_2 = pressure at plane 2, g_c = dimensional constant, A = cross-sectional area of the tube, w_p = weight rate of flow and V_p = velocity of the primary fluid, w_s = weight rate of flow and V_s = velocity of the secondary fluid, and V_m = velocity of the combined fluids.

Application of the momentum equation to ejectors of other types is discussed in Lapple et al., *Fluid and Particle Mechanics*, University of Delaware, Newark, Del., 1951, chap. 5.

Jet Behavior A **free jet,** upon leaving an outlet, will entrain the surrounding fluid and expand. The momentum of the jet is transferred to the surrounding fluid being entrained. There is some loss in momentum due to turbulence and to static-pressure gradients across the jet. A jet is considered free when its cross-sectional area is less than one-fifth of the total cross-sectional flow area of the region through which it is flowing [Elrod, *Heat. Piping Air Cond.*, **26**(3), 149–155 (1954)].

A **turbulent jet** is considered in this discussion to be a free jet whose jet Reynolds number is greater than 2000. Additional discussion on the relation between Reynolds number and turbulence in jets is given by Elrod (loc. cit.). A turbulent free jet (Fig. 5-27) has four flow regions [Tuve, *Heat. Piping Air Cond.*, **25**(1), 181–191 (1953)]:

1. Region of flow establishment—a short region whose length is about 5 nozzle diameters or slot heights (for a slot of infinite width). The fluid within the cone or core of the same length has a velocity about the same as the initial discharge velocity.
2. A transition region that extends to about 8 nozzle diameters, slightly less for slots.
3. Region of established flow—the principal region of the jet, extending to about 100 nozzle diameters or about 2000 slot heights.
4. A terminal region where the residual centerline or maximum velocity reduces rapidly within a short distance. For air jets, the residual velocity will reduce to less than 0.3 m/s (1 ft/s), usually regarded as still air.

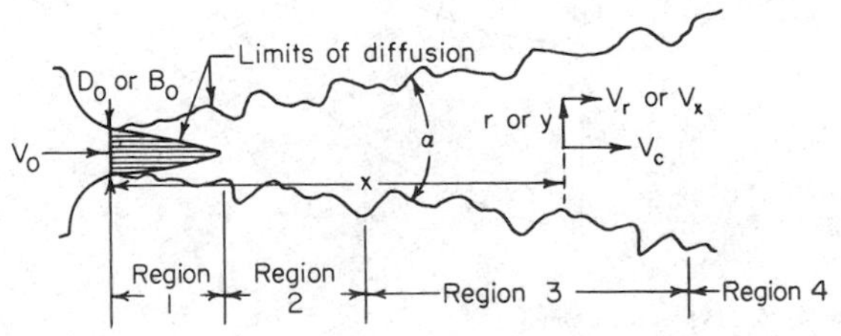

FIG. 5-27 Configuration of a turbulent free jet.

TABLE 5-5 Turbulent Free-Jet Characteristics
Where both jet fluid and entrained fluid are air

Rounded-inlet circular jet
Longitudinal distribution of velocity along jet center line*†
$\frac{V_c}{V_0} = K\frac{D_0}{x}$ for $7 < \frac{x}{D_0} < 100$
$K = 5$ for $V_0 = 2.5$ to 5.0 m/s
$K = 6.2$ for $V_0 = 10$ to 50 m/s
Radial distribution of longitudinal velocity†
$\log\left(\frac{V_c}{V_r}\right) = 40\left(\frac{r}{x}\right)^2$ for $7 < \frac{x}{D_0} < 100$
Jet angle*†
$\alpha \simeq 20°$ for $\frac{x}{D_0} < 100$
Entrainment of surrounding fluid‡
$\frac{q}{q_0} = 0.32\frac{x}{D_0}$ for $7 < \frac{x}{D_0} < 100$
Rounded-inlet, infinitely wide slot jet
Longitudinal distribution of velocity along jet centerline‡
$\frac{V_c}{V_0} = 2.28\left(\frac{B_0}{x}\right)^{0.5}$ for $5 < \frac{x}{B_0} < 2000$ and $V_0 = 12$ to 55 m/s
Transverse distribution of longitudinal velocity‡
$\log\left(\frac{V_c}{V_x}\right) = 18.4\left(\frac{y}{x}\right)^2$ for $5 < \frac{x}{B_0} < 2000$
Jet angle‡
α is slightly larger than that for a circular jet
Entrainment of surrounding fluid‡
$\frac{q}{q_0} = 0.62\left(\frac{x}{B_0}\right)^{0.5}$ for $5 < \frac{x}{B_0} < 2000$

*Nottage, Slaby, and Gojsza, *Heat. Piping Air Cond.*, **24**(1), 165–176 (1952).
†Tuve, *Heat. Piping Air Cond.*, **25**(1), 181–191 (1953).
‡Albertson, Dai, Jensen, and Rouse, *Trans. Am. Soc. Civ. Eng.*, **115**, 639–664 (1950), and Discussion, ibid., **115**, 665–697 (1950).

Table 5-5 gives characteristics of **rounded-inlet circular jets** and of **rounded-inlet infinitely wide slot jets** (aspect ratio or width-to-height ratio > 15). Information in the table is for a homogeneous air system under isothermal conditions, i.e., when both the jet fluid and the entrained fluid are air and of the same temperature.

Witze [*Am. Inst. Aeronaut. Astronaut. J.*, **12**, 417–418 (1974)] gives equations for the centerline velocity decay of different types of subsonic and supersonic circular free jets. Entrainment of surrounding fluid in the region of flow establishment is lower than in the region of established flow [see Hill, *J. Fluid Mech.*, **51**, 773–779 (1972)]. Data of Donald and Singer [*Trans. Inst. Chem. Eng. (London)*, **37**, 255–267 (1959)] indicate that jet angle and the coefficients in the equations given in Table 5-5 depend upon the fluids; for a water system, the jet angle for a circular jet is 14° and the entrainment ratio is about 70 percent of that for an air system. However, until more conclusive data are available, Table 5-5 can be used as a guide for other fluid systems. The following **nomenclature** is used: B_0 = slot height, D_0 = opening diameter, q = total jet flow rate at distance x, q_0 = initial jet flow rate, r = radius, V_c = jet centerline velocity, V_0 = initial jet velocity, V_r = longitudinal velocity at radius r, V_x = longitudinal velocity at distance y, x = distance from jet discharge, and y = distance from jet centerline. Characteristics of **rectangular jets** of various aspect ratios are given by Elrod (op. cit.).

Characteristics of **slot jets** discharging into a moving surrounding fluid are given by Weinstein, Osterle, and Forstall [*J. Appl. Mech.*, **23**, 437–443 (1956)] and Bradbury and Riley [*J. Fluid Mech.*, **27**, part 2, 381–394 (1967)]. Characteristics of **coaxial jets** are discussed by Forstall and Shapiro [*J. Appl. Mech.*, **17**, 399–408 (1950)], and those of **double concentric jets** are given by Chigier and Beer [*J. Basic Eng.*, **86**, 797–804 (1964)]. The behavior of **axisymmetric confined jets** is described by Barchilon and Curtet [*J. Basic Eng.*, **86**, 777–787 (1964)].

Density gradients will affect the rate of spread of a single-phase free jet. A jet of lower density than the surroundings spreads more rapidly than a jet of the same density as the surroundings, and conversely, a jet of higher density than the surroundings spreads less rapidly. Additional details are given by Keagy and Weller (*Proc. Heat Transfer Fluid Mech. Inst., ASME*, June 22–24, 1949, pp. 89–98) and Cleeves and Boelter [*Chem. Eng. Prog.*, **43**, 123–134 (1947)].

Laminar Jets Few experimental data exist on laminar jets [see Gutfinger and Shinnar, *Am. Inst. Chem. Eng. J.*, **10**, 631–639 (1964)]. Theoretical analysis for velocity distributions and entrainment ratios are available [see Schlichting, *Boundary-Layer Theory*, 7th ed., McGraw-Hill, New York, 1979, pp. 179–183, 230–234); and Morton, *Phys. Fluids*, **10**, 2120–2127 (1967)].

Theoretical analyses of jet flows for power-law **nonnewtonian fluids** are given by Vlachopoulos and Stournaras [*Am. Inst. Chem. Eng. J.*, **21**, 385–388 (1975)], Mitwally [*J. Fluids Eng.*, **100**, 363 (1978)], and Sridhar and Rankin [*J. Fluids Eng.*, **100**, 500 (1978)].

FLOW IN PIPES AND CHANNELS

Velocity Distribution: Circular Pipes For **laminar flow** in circular pipes, the velocity pattern is parabolic in shape with a maximum velocity at the center equal to 2 times the average velocity V. The local velocity u at any point in the cross section is given by

$$\frac{u}{V} = 2\left(1 - \frac{r^2}{r_w^2}\right) \qquad (5\text{-}53)$$

where r = radius at the point in question and r_w = radius of the pipe. See Knudsen and Katz, op. cit., p. 86.

The corresponding **distribution of residence time** for **laminar flow** is given by

$$F(\theta) = 1 - \frac{1}{4}\left(\frac{\theta_{avg}}{\theta}\right)^2 \qquad \text{for } \frac{\theta_{avg}}{\theta} < 2 \qquad (5\text{-}54)$$

where $F(\theta)$ = fraction of material in system for less than time θ and θ_{avg} = average residence time in system. See Danckwerts, *Chem. Eng. Sci.*, **2**, 1–12 (February 1953).

The residence-time distribution in long transfer lines can be favorably altered by insertion of one or more **flow inverters** in the transfer line. The function of each flow inverter is to divide the incoming fluid into at least two streams and to transpose the streams during flow through the inverter [see Boucher and Sakiadis, "Fluid Flow Inverter," U.S. Patent 3,128,794, 1964; and Nauman, *Am. Inst. Chem. Eng. J.*, **25**, 246–258 (1979)].

A theoretically derived distribution of residence time for laminar flow in **helical pipe coils** by Ruthven [*Chem. Eng. Sci.*, **26**, 1113–1121 (1971); *Chem. Eng. Sci.*, **33**, 628–629 (1978)] is given by

$$F(\theta) = 1 - \tfrac{1}{4}(\theta_{avg}/\theta)^{2.81} \qquad \text{for} \quad 0.50 < (\theta_{avg}/\theta) < 1.63 \qquad (5\text{-}55)$$

This relation was substantially confirmed by Trivedi and Vasudeva [*Chem. Eng. Sci.*, **29**, 2291–2295 (1974)] for $0.6 \leq N_{De} < 6$ and for $0.0036 < D/D_c < 0.097$, where $N_{De} = N_{Re}\sqrt{D/D_c}$ = Dean number, dimensionless; N_{Re} = Reynolds number, dimensionless; D = pipe diameter; and D_c = coil diameter. Measurements by Saxena and Nigam [*Chem. Eng. Sci.*, **34**, 425–426 (1979)] indicate that such a distribution will hold for $N_{De} > 1$. The residence-time distribution for helical coils is narrower than for straight circular pipes.

In the transition region (N_{Re} from 2000 to 3000), the velocity profile becomes more blunt and the ratio V/u_{max} increases (see Fig. 5-14). Velocity profile curves are given by Patel and Head [*J. Fluid Mech.*, **38**, part 1, 181–201 (1969)] for flow in smooth pipes in the Reynolds-number range of about 1500 to 10,000. At higher Reynolds numbers, the flow is generally fully turbulent, and the velocity profile in **smooth-wall pipes** is characterized by a laminar boundary layer ($y^+ < 5$), a turbulent core ($y^+ > 30$), and a buffer layer in between. The local velocity is given by the following relationships:

For the laminar boundary layer,

$$u^+ = y^+ \qquad \text{for} \quad y^+ < 5 \qquad (5\text{-}56a)$$

For the buffer layer,

$$u^+ = -3.05 + 5.00 \ln y^+ \qquad \text{for} \quad 5 < y^+ < 30 \qquad (5\text{-}56b)$$

For the turbulent core,

$$u^+ = 5.5 + 2.5 \ln y^+ \qquad \text{for} \quad y^+ > 30 \qquad (5\text{-}56c)$$

For **rough-wall pipes**, the local velocity in the turbulent core is given by

$$u^+ = 8.5 + 2.5 \ln y/\varepsilon \qquad \text{for} \quad y^+ > 30 \qquad (5\text{-}56d)$$

where $u^+ = u/u^\circ$; u = local velocity at distance y from the pipe wall; u° = friction velocity $(\sqrt{\tau_0 g_c/\rho})$; τ_0 = wall shear stress $(D\,\Delta p/\Delta L)$; g_c = dimensional constant; ρ = fluid density; Δp = pressure drop; D = inside pipe diameter; L = pipe length; $y^+ = yu^\circ\rho/\mu$, dimensionless; μ = fluid viscosity; and ε = height of wall roughness. For further details, see Knudsen and Katz, op. cit., pp. 154–169; and Cremer and Davies, op. cit., vol. 4, p. 401.

Equations describing the distribution of residence time for **turbulent flow** in pipes are given by Danckwerts, loc. cit.

Velocity Distribution: Other Shapes For velocity profiles under laminar- and turbulent-flow conditions in annuli, between infinite parallel planes, and in other noncircular cross sections, see Knudsen and Katz, op. cit.; Purday, *Mechanics of Viscous Flow*, chap. II, Dover, New York, 1949; Rouse, *Advanced Mechanics of Fluids*, Wiley, New York, 1959, p. 219; and Goldstein, *Modern Developments in Fluid Dynamics*, vol. 2, Oxford, London, 1938, pp. 359–360.

Analytically derived equations are presented by Straub, Silberman, and Nelson [*Trans. Am. Soc. Civ. Eng.*, **123**, 685–714 (1958)] for **laminar flow** through a variety of **open-channel** cross sections, including semicircular, rectangular, triangular, elliptical, and trapezoidal cross sections.

Experimentally determined velocity profiles are also presented by Straub et al. for **turbulent flow** in triangular troughs. Profiles for channels of various cross sections are given in O'Brien and Hickox, *Applied Fluid Mechanics*, McGraw-Hill, New York, 1937, pp. 268–270; and Chow, *Open-Channel Hydraulics*, McGraw-Hill, New York, 1959, pp. 24–29.

Residence-Time Distribution: Process Vessels An extensive treatment of distribution of residence time and of dispersion in a variety of typical process vessels is given by Levenspiel and Bischoff ("Patterns of Flow in Chemical Process Vessels," in Drew, Hoopes, and Vermeulen, *Advances in Chemical Engineering*, vol. 4, Academic, New York, 1963). The case of multiple stirred tanks in series is covered in detail by Stokes and Nauman [*Can. J. Chem. Eng.*, **48**, 723–725 (1970)]. Information on residence time and fluid mixing on commercial-scale sieve trays is given by Bell [*Am. Inst. Chem. Eng. J.*, **18**, 498–505 (1972)].

Incompressible Flow The flow can be considered to be incompressible if (1) the substance flowing is a liquid or (2) if it is a gas whose density changes within the system by no more than 10 percent. In this event, if the inlet density is employed, the resulting error in computed pressure drop will generally not exceed the uncertainty limits in the friction factor. In the event of larger changes in fluid density, e.g., gases with large pressure drops, the more exact methods described in the subsection "Compressible Flow" should be used.

General Formulas and Methods The problem of finding one of the three quantities—rate of discharge, size of channel, pressure or head loss—when the other two are given is solved by substituting the data of the problem in an appropriate form of the mechanical energy balance (see subsection "Energy Balance") after the term F, frictional loss of mechanical energy, has been evaluated. That part of F which arises from friction within the channel proper is considered later. The part due to fittings, bends, and the like, which often constitutes a major part of the friction, is discussed beginning with the subsection "Miscellaneous Pressure Losses."

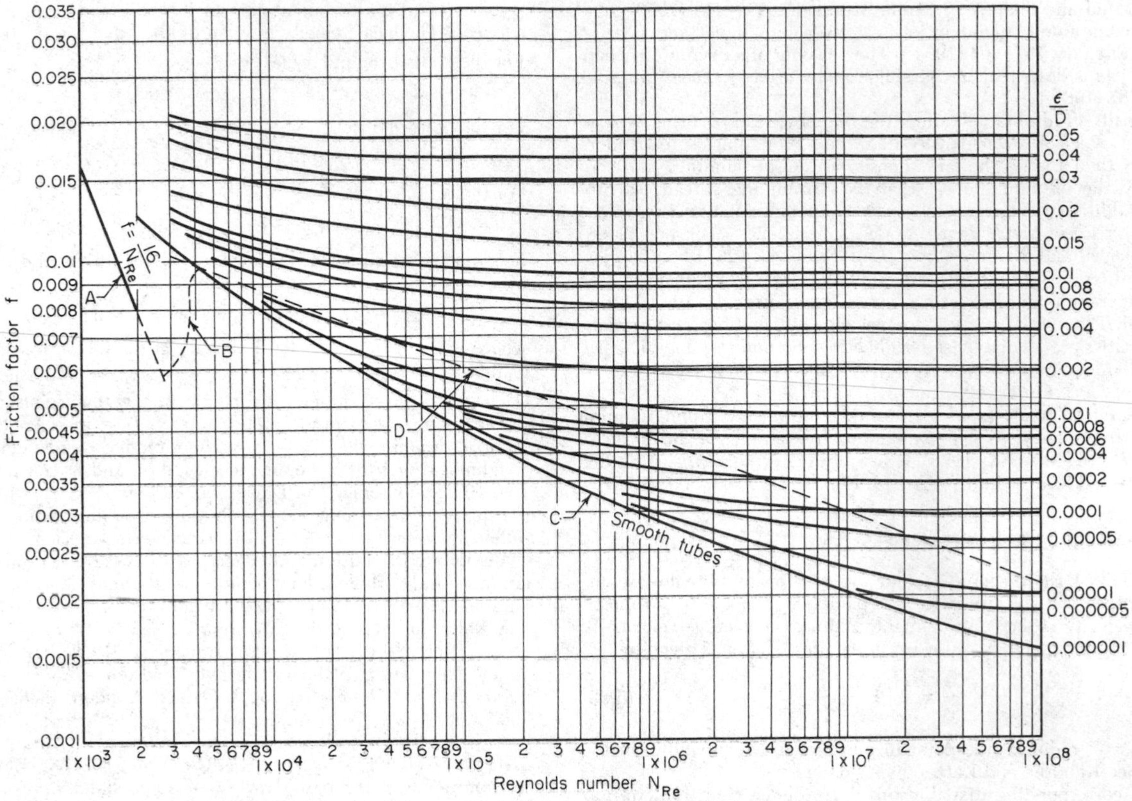

FIG. 5-28 Fanning friction factors. Reynolds number $N_{Re} = DV\rho/\mu$, where D = pipe diameter, V = velocity, ρ = fluid density, and μ = fluid viscosity. [*Based on Moody,* Trans. Am. Soc. Mech. Eng., *66, 671 (1944).*]

Circular Pipes The **Fanning,** or **Darcy, equation,** Eq. (5-57) for steady flow in uniform **circular pipes** running full of liquid under **isothermal conditions**

$$F = \left(\frac{4fL}{D}\right)\frac{V^2}{2g_c} = \left(\frac{4fL}{D}\right)h_v = \left(\frac{4fL}{D}\right)\frac{G^2}{2g_c\rho^2} = \frac{32fLw^2}{\pi^2\rho^2 g_c D^5} = \frac{32fLq^2}{\pi^2 g_c D^5} \qquad (5\text{-}57)$$

gives the friction loss F in specific energy, where D = duct diameter; L = duct length; ρ = fluid density; V = fluid velocity; h_v = velocity head ($V^2/2g_c$); G = mass velocity; w = weight rate of flow; q = volume rate of flow; g_c = dimensional constant; and f = **Fanning friction factor,** dimensionless. The pressure drop due to friction is $\Delta p = F\rho$.

The **Fanning friction factor** f is a function of the Reynolds number N_{Re} and the roughness of the channel inside surface ε. One widely used correlation [Moody, *Trans. Am. Soc. Mech. Eng.*, **66**, 671–684 (1944)], as shown in Fig. 5-28, is a plot of Fanning friction factor as a function of Reynolds number and relative roughness ε/D where ε = surface roughness, and D = pipe inside diameter. Values of ε for various materials are given in Table 5-6.

In turbulent flow, $N_{Re} > 3000$, the plot of Fig. 5-28 is a representation of the Colebrook equation [Colebrook, *J. Inst. Civ. Eng. (London)*, **11**, 133–156 (1938–1939)]

$$1/\sqrt{f} = -4 \log(\varepsilon/3.7D + 1.256/N_{Re}\sqrt{f}) \qquad (5\text{-}58)$$

For smooth pipes, $\varepsilon/D = 0$, Eq. (5-58) reduces to Prandtl's equation

$$1/\sqrt{f} = -4 \log(1.256/N_{Re}\sqrt{f}) \qquad (5\text{-}59)$$

Curve C in Fig. 5-28, whereas for very rough pipes it reduces to von Kármán's equation

$$1/\sqrt{f} = -4 \log(\varepsilon/3.7D) \qquad (5\text{-}60)$$

(see Schlichting, *Boundary-Layer Theory*, 7th ed., McGraw-Hill, New York, 1979, pp. 609–612, 620–623). When the flow rate is specified, the friction factor can be conveniently computed from the equation

$$1/\sqrt{f} = 3.60 \log(N_{Re}/7) \qquad (5\text{-}61)$$

[Churchill, *Chem. Eng.*, **84**(24), 91–92, 1977] or for $N_{Re} \leq 10^5$ by the Blasius equation

TABLE 5-6 Values of Surface Roughness for Various Materials*

Material	Surface roughness ε, mm
Drawn tubing (brass, lead, glass, and the like)	0.00152
Commercial steel or wrought iron	0.0457
Asphalted cast iron	0.122
Galvanized iron	0.152
Cast iron	0.259
Wood stove	0.183–0.914
Concrete	0.305–3.05
Riveted steel	0.914–9.14

*From Moody, *Trans. Am. Soc. Mech. Eng.*, **66**, 671–684 (1944); *Mech. Eng.*, **69**, 1005–1006 (1947). Additional values of ε for various types or conditions of concrete wrought-iron, welded steel, riveted steel, and corrugated-metal pipes are given in Brater and King, *Handbook of Hydraulics*, 6th ed., McGraw-Hill, New York, 1976, pp. 6-12–6-13. To convert millimeters to feet, multiply by 3.281×10^{-3}.

$$f = 0.0791/N_{Re}^{1/4} \tag{5-62}$$

(see Schlichting, loc. cit., pp. 596–600).

Curve B in Fig. 5-28 represents the transition region between laminar and turbulent flow.

Substitution of the equation for **Curve A,** Fig. 5-28, into Eq. (5-57) yields Poiseuille's law for laminar flow ($N_{Re} \leq 2000$); see Table 5-10 later. Note that the critical Reynolds number of 2000 does not represent the lower limit at which natural transition from laminar to turbulent flow occurs. Rather, it represents a limit at which turbulent flow in a long pipe can be maintained once it has been established. Care must be exercised when values of f are taken from the literature because the same name and symbol are sometimes used to denote various multiples of the f given by Fig. 5-28.

For **turbulent flow,** $N_{Re} > 3000$, the following equations, explicit in the unknowns, have been developed. For computation of **flow rate,**

$$q/D^2\sqrt{gDS_f} = -\pi/\sqrt{2}\log(\varepsilon/3.7D + 1.78\nu/D\sqrt{gDS_f}) \tag{5-63}$$

where S_f = height of fluid head loss per unit duct length ($\Delta p\, g_c/\rho g L$), dimensionless; ν = kinematic viscosity; g = local acceleration due to gravity; g_c = dimensional constant; and other symbols are as given earlier. Equation (5-63) is obtained by combining Eqs. (5-58) and (5-57) [Churchill, *Am. Inst. Chem. Eng. J.,* **19,** 375–376 (1973)]; and Swamee and Jain, *Proc. Am. Soc. Civ. Eng., J. Hydraul. Div.,* **102**(HY5), 657–664 (1976)]. For computation of **pressure drop** due to friction

$$\frac{D^5 g S_f}{q^2} = \frac{0.203}{\left[\log\left(\frac{\varepsilon}{3.7D} + \frac{5.74}{N_{Re}^{0.9}}\right)\right]^2} \tag{5-64}$$

with symbols as given earlier (Churchill, loc. cit.; and Swamee and Jain, loc. cit.). The error in using Eq. (5-64) is less than ±1 percent for the range $(5)(10^3) \leq N_{Re} \leq 10^8$, $10^{-6} \leq \varepsilon/D \leq 10^{-2}$ (Swamee and Jain, loc. cit.).

For computation of **duct diameter**

$$\frac{D^5 g S_f}{q^2} = 0.125\left[\left(\frac{\varepsilon^5 g S_f}{q^2}\right)^{0.25} + \left(\frac{\nu^5}{q^3 g S_f}\right)^{0.20}\right]^{0.20} \tag{5-65}$$

with symbols as given earlier. The error in using Eq. (5-65) is less than ±2 percent for the range $(3)(10^3 \leq N_{Re} \leq (3)(10^8)$, $(2)(10^{-6}) \leq \varepsilon/D \leq (2)(10^{-2})$ (see Swamee and Jain, loc. cit.). **Curve D** in Fig. 5-28, represented by the equation

$$f = 0.04/N_{Re}^{0.16} \tag{5-66}$$

is intended for use in problems of plant design when high accuracy is not required. It leads to conservative results for commercial pipe in turbulent flow [Genereaux, *Chem. Metall. Eng.,* **44,** 241–248 (1937)]. A **nomograph** to find pressure drop, taking into account pipe surface roughness, and a nomograph to determine flow rate or pipe size if either is unknown and if the pressure drop is known are given by Arnold [*Chem. Eng.,* **66**(11), 103–106 (1959)].

For **rough estimates** or **checks,** the velocity-head concept [Lapple, *Chem. Eng.,* **56**(5), 96–104 (1949)] can be applied to the first two forms of Eq. (5-57). The velocity head is $V^2/2g_c = h_v$, and the number of velocity-head losses in straight pipe is $4fL/D$. Typical values of h_v and L/D for one velocity-head loss are given in Table 5-7.

Noncircular Channels For **cross sections other than circular** of **ducts running full** or for **open channels** when the variation in depth is negligible, where the flow is in the **turbulent flow region,** the first three forms of the Fanning equation, Eq. (5-57), are applicable if D, wherever it occurs, is replaced by the hydraulic diameter (4 times the hydraulic radius, $4R_H$). Values of R_H for some common cross sections are given in Table 5-8. For flow in **annuli,** the friction factors based on the hydraulic diameter may be 5 to 10 percent greater than those for pipe; see Brighton and Jones, *J. Basic Eng.,* **86,** 835–842 (1964); Okiishi and Serovy, *J. Basic Eng.,* **89,** 823–836 (1967); and Lawn and Elliott, *J. Mech. Eng. Sci.,* **14,** 195–204 (1972).

TABLE 5-7 Approximate Values of Velocity Head and Pipe Length Equivalent to One Velocity-Head Loss*

Fluid	Fluid velocity, m/s	Velocity head, various units
Any fluid	3	4.5 (N·m)/kg
Water	3	4.5 kPa
	6	18.0 kPa
Air (40°C, 101.3 kPa)	15	13.0 mm water
	30	52.0 mm water

*Pipe length equivalent to one velocity-head loss

Fluid	L/D
Water	45 (f = 0.0055)
Air	55 (f = 0.0045)

To convert newton-meters per kilogram to feet of fluid flowing, multiply by 0.33455. To convert pascals to pounds-force per square inch, multiply by $(1.4504)(10^{-4})$.

For turbulent flow of mercury in annuli, see Dwyer, Hlavac, and Nimmo, *J. Fluids Eng.,* **98,** 113–116 (1976). The friction factor for turbulent flow in smooth **rectangular ducts of large aspect ratio** may be computed from

$$f = 0.0868/N_{Re}^{1/4} \tag{5-67}$$

for the range $(1.2)(10^4) < N_{Re} < (1.2)(10^6)$, where N_{Re} is based on the hydraulic diameter [Dean, *J. Fluids Eng.,* **100,** 215–223 (1978)]. In turbulent flow through straight channels of noncircular cross section there exist secondary flows which may augment the friction loss, by comparison with a pipe, at the same N_{Re}. For turbulent flow in

TABLE 5-8 Values of Hydraulic Radius R_H for Various Cross Sections

$$R_H = \frac{\text{area of stream cross section}}{\text{wetted perimeter}}; \text{ hydraulic diameter} = 4R_H$$

Shape of cross section	R_H
Pipes and ducts, running full	
Circle, diameter = D	$D/4$
Annulus, inner diameter = d; outer diameter = D	$(D - d)/4$
Square, side = D	$D/4$
Rectangle, sides a, b	$ab/[2(a + b)]$
Equilateral triangle, side = a	$a/4\sqrt{3}$
Ellipse, major axis = $2a$, minor axis = $2b$	$ab/[K(a + b)]$*
Open channels or partly filled ducts	
Rectangle, depth = y, width = b	$by/(b + 2y)$
Semicircle, free surface on a diameter D	$D/4$
Wide shallow stream on flat plate, depth = y	y
Triangular trough, ∠ = 90°, bisector vertical, depth = y, slant depth = d	$d/4 = y/2\sqrt{2}$
Trapezoid (depth = y, bottom width = b): Side slope 60° from horizontal	$y\left(\frac{b + y/\sqrt{3}}{b + 4y/\sqrt{3}}\right)$
Side slope 45°	$(yb + y^2)/(b + 2\sqrt{2}y)$
Film (thickness = t) on wall of vertical wetted wall tower of diameter = D	$t - t^2/D = t$ (approximate)

*Values of K. If $S = (a - b)/(a + b)$,

S =	0.1	0.2	0.3	0.4	0.5	0.6	0.7	0.8	0.9	1.0
K =	1.002	1.010	1.023	1.040	1.064	1.092	1.127	1.168	1.216	1.273

smooth **rectangular ducts,** the friction factor increases with increasing duct aspect ratio. Accurate friction factors can be obtained by using a modified Reynolds number

$$(N_{Re})_{mod} = F_1 N_{Re} \tag{5-68}$$

in place of N_{Re}, the Reynolds number based on the hydraulic diameter, where

$$F_1 \simeq 2/3 + 11/24\, b/a(2 - b/a) \tag{5-69}$$

and a/b is the duct aspect (width-to-height) ratio [Jones, *J. Fluids Eng.*, **98**, 173–181 (1976)]. The friction factors for turbulent flow through **elliptical channels,** 1.5:1 and 2:1 aspect ratio, are 8 and 13 percent respectively greater than those for pipe [Cain and Duffy, *Int. J. Mech. Sci.*, **13**, 451–459 (1971)]. The friction factors for turbulent flow through **triangular ducts** are smaller than those for pipe. See Carlson and Irvine, *J. Heat Transfer*, **83**, 441–444 (1961); and Aly, Trupp, and Gerrard [*J. Fluid Mech.*, **85**, 57–83 (1978)]. Rehme [*Int. J. Heat Mass Transfer*, **16**, 933–950 (1973)] and Malak, Hejna, and Schmid [*Int. J. Heat Mass Transfer*, **18**, 139–149 (1975)] developed equations for predicting friction factors for turbulent flow in **straight channels of complex shape** from shape factors derived for laminar flow in the same channels. Friction factors for **annuli** with various types of **inner fin tubes** and **annuli** which are **eccentric** are given by Knudsen and Katz, op. cit., pp. 193–205. Rehme [*Int. J. Heat Mass Transfer*, **15**, 2499–2517 (1972)] determined the friction loss in turbulent flow along **parallel rods in hexagonal arrangement,** in hexagonal channels. The friction factors, computed with the hydraulic diameter, were found to vary from about 0.6 of the pipe values, for maximum rod packing, to 1.1 of the pipe values for a rod spacing equal to the rod diameter.

Details of **transition from laminar to turbulent flow** vary with duct geometry. In rectangular ducts the Reynolds number, based on the hydraulic diameter, for laminar-turbulent-flow transition varies from 1900 to 2800 [Hanks and Ruo, *Ind. Eng. Chem. Fundam.*, **5**, 558–561 (1966)]. Laminar flow can persist to $N_{Re} \sim 14{,}000$ by reducing the background turbulence at the entrance to the duct; see Karnitz, Potter, and Smith, *J. Fluids Eng.*, **96**, 384–388 (1974); and Nishioka, Iida, and Ichikawa, *J. Fluid Mech.*, **72**, 731–751 (1975). In triangular ducts transition to turbulent flow begins at N_{Re} from 1600 to 1800; see Cope and Hanks, *Ind. Eng. Chem. Fundam.*, **11**, 106–117 (1972); and Bandopadhayay and Hinwood, *J. Fluid Mech.*, **59**, 775–783 (1973).

In the **laminar-flow region,** for **ducts running full,** the formulas given later in Table 5-10 should be used. These formulas do not include end corrections. The pressure drop $p_1 - p_2$ is that measured between static pressure taps L m (ft) apart in the wall of a continuous duct when sufficient distance is allowed between the inlet and the upstream pressure tap to ensure the existence of the normal velocity distribution at the latter point (see subsection "Contraction and Entrance Losses"). When short tubes are involved, these formulas will give highly incorrect results if the pressure drop is measured between terminal reservoirs without applying end corrections. For these corrections, see subsection "Enlargement and Exit Losses." Results of additional, largely analytical investigations of laminar flow in ducts were compiled by Shah and London (*Laminar Flow Forced Convection in Ducts*, Academic, New York, 1978).

For laminar and turbulent flow of gases through **cracks,** see Button, Grogan, Chivers, and Manning, *J. Fluids Eng.*, **100**, 453–458 (1978). Low-viscosity liquids flowing in 0.001-mm- (0.00004-in-) diameter or larger **microopenings,** at low N_{Re} and $L/D \geq 400$, generally follow Poiseuille's law, except at low-pressure drop in the smaller microopenings [Hedley, Olt, Holboke, and Wurstner, *J. Rheol.*, **22**, 91–112 (1978)].

Nonnewtonian Fluids The pressure drop due to friction for **laminar** flow of **nonnewtonian fluids** in circular pipes can be determined from a plot of $D\,\Delta p/4L$ versus $8V/D$. A typical flow curve on arithmetic coordinates is shown in Fig. 5-29. $D\,\Delta p/4L$ is the shear stress at the pipe wall τ_w, and $8V/D$ is uniquely related to shear rate at the wall, where D = pipe diameter, Δp = pressure drop, L = length of pipe, and V = velocity. The flow curve is determined in laboratory or small-scale tests with the particular nonnewtonian fluid.

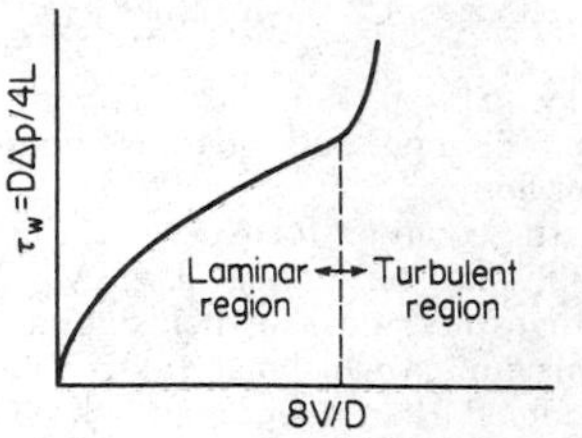

FIG. 5-29 General flow curve.

In the **laminar region** the curve (Fig. 5-29) is independent of pipe diameter for time-independent nonnewtonian fluids. If the fluid is time-dependent, a separate curve will be obtained for each pipe diameter and pipe length. For additional details see Alves, Boucher, and Pigford, *Chem. Eng. Prog.*, **48**, 385–393 (1952); Skelland, *Non-Newtonian Flow and Heat Transfer*, Wiley, New York, 1967, pp. 157–179.

For **Bingham plastics,** a typical curve is shown in Fig. 5-30. The theoretical flow equation [see Caldwell and Babbitt, *Ind. Eng. Chem.*, **33**, 249–256 (1941)] is

$$\frac{8V}{D} = \frac{g_c \tau_w}{\eta}\left[1 - \frac{4}{3}\frac{\tau_y}{\tau_w} + \frac{1}{3}\left(\frac{\tau_y}{\tau_w}\right)^4\right] \tag{5-70}$$

If $(\tau_y/\tau_w)^4$ is relatively small, i.e., at large shear stresses or pressure drops, Eq. (5-70) may be approximated by

$$\frac{8V}{D} = \frac{g_c}{\eta}\left(\tau_w - \frac{4}{3}\tau_y\right) \tag{5-71}$$

The error in omitting the last term in Eq. (5-70) is less than 2 percent for τ_y/τ_w less than 0.4. The apparent viscosity μ_a can be obtained from substituting Eq. (5-71) into Poiseuille's equation,

$$\mu_a = (g_c \tau_y D/6V) + \eta \tag{5-72}$$

In these equations, D = pipe inside diameter, g_c = dimensional constant, L = pipe length, Δp = static pressure drop, V = velocity, η = plastic viscosity, μ_a = apparent viscosity, τ_w = wall shear stress = $D\,\Delta p/4L$, and τ_y = yield stress.

For those fluids to which the **power law,** Eq. (5-3), applies,

$$D\,\Delta p/4L = K'(8V/D)^{n'} \tag{5-73}$$

where

$$K' = K\left(\frac{3n'+1}{4n'}\right)^{n'} \tag{5-74}$$

$$n' = \frac{d[\ln (D\,\Delta p/4L)]}{d[\ln (8V/D)]} = n \tag{5-75}$$

$$\left(-\frac{du}{dr}\right)_w = \left(\frac{3n'+1}{4n'}\right)\left(\frac{8V}{D}\right) \tag{5-76}$$

n' = power-law exponent, slope of the line from a plot of $D\,\Delta p/4L$ versus $8V/D$ on logarithmic coordinates; $(-du/dr)_w$ = shear rate at

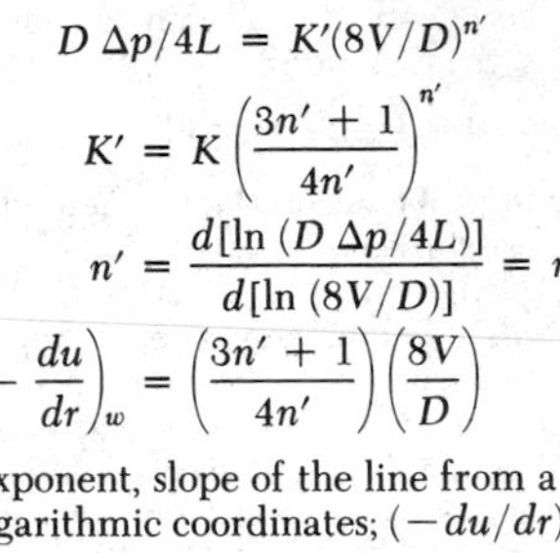

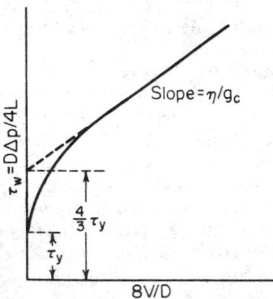

FIG. 5-30 Bingham-plastic flow curve.

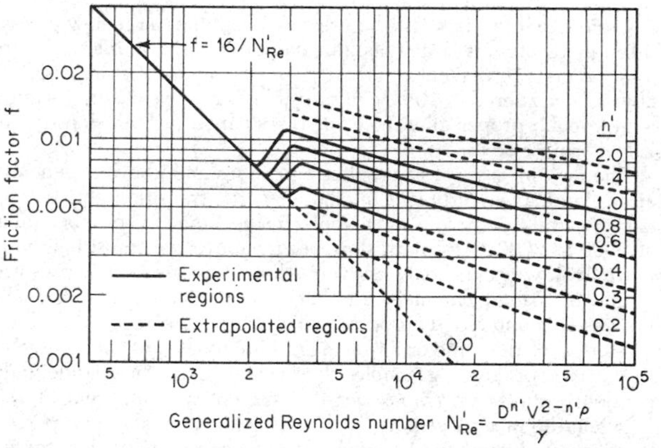

FIG. 5-31 Fanning friction factors for nonnewtonian flow. [*Dodge and Metzner,* Am. Inst. Chem. Eng. J., *5, 189 (1959).*]

wall, 1/s; K' = constant, $(\text{N}\cdot\text{s}^{n'})/\text{m}^2$ $[(\text{lbf}\cdot\text{s}^{n'})/\text{ft}^2]$; and other symbols are as defined after Eq. (5-72). For $n' = 1$, the fluid is newtonian; for $n' < 1$, pseudoplastic or Bingham-plastic; and for $n' > 1$, dilatant. K' is the value of $D\,\Delta p/4L$ for $8V/D = 1$. For newtonian fluids, $K' = \mu/g_c$.

The **limit of stable laminar flow** is usually taken as $N'_{\text{Re}} = 2100$, where

$$N'_{\text{Re}} = \text{generalized Reynolds number} = D^{n'}V^{2-n'}\rho/\gamma \qquad (5\text{-}77)$$

and

$$\gamma = g_c K' 8^{n'-1} \qquad (5\text{-}78)$$

or

$$N_{\text{Re}} = DV\rho/\mu_a = 2100 \qquad \text{for Bingham plastics} \qquad (5\text{-}79)$$

where ρ = density and other symbols are defined following Eq. (5-72). For additional details, see Skelland, op. cit., pp. 226–233.

In **turbulent flow,** the curve (Fig. 5-29) will break off. The point of break-off will differ for different pipe diameters. To predict the pressure drop for the turbulent flow of time-dependent nonnewtonian fluids, the Fanning friction factor is obtained from Fig. 5-31 by using the generalized Reynolds number. Further discussion of turbulent flow is given in Skelland, op. cit., pp. 180–239. For fluids that behave like Bingham plastics, $n' \simeq 1$ in turbulent flow. The pressure drop can then be predicted from the standard Fanning friction correlation for newtonian fluids, using for fluid viscosity the known plastic viscosity η [Barry, *J. Inst. Pet. (London)*, **57,** 74–85 (1971)].

A **reduction of friction loss** in **turbulent flow** of newtonian liquids can be achieved by the addition of soluble high-molecular-weight polymers in extremely low concentration. In these systems it is generally agreed that drag-reducing behavior is associated with the viscoelastic nature of the solutions effective in the shear region near the wall. A minimum molecular weight, characteristic of the polymer species, is necessary to initiate drag reduction at a specific flow rate. There is a critical concentration above which drag reduction will not occur [Kim, Little, and Ting, *J. Colloid Interface Sci.*, **47,** 530–535 (1974)]. Drag reduction in polymer solutions is reviewed by Hoyt [*J. Basic Eng.*, **94,** 258–285 (1972)], Little et al. [*Ind. Eng. Chem. Fundam.*, **14,** 283–296 (1975)], and Virk [*Am. Inst. Chem. Eng. J.*, **21,** 625–656 (1975)]. At maximum possible drag reduction in smooth pipes

$$1/\sqrt{f} = -19 \log (50.73/N_{\text{Re}}\sqrt{f}) \qquad (5\text{-}80)$$

or approximately

$$f = 0.58/N_{\text{Re}}^{0.58} \qquad (5\text{-}81)$$

for $4000 < N_{\text{Re}} < 40{,}000$, where f = Fanning friction factor, dimensionless; and N_{Re} = Reynolds number in pipe. The actual drag, below maximum, depends on the polymer system. For further details, see Virk, loc. cit.

For laminar flow of power-law nonnewtonian fluids in **rectangular channels** of large aspect ratio (Skelland, loc. cit.),

$$b\,\Delta p/2L = K(2n + 1/3n)^n(6q/b^2a)^n \qquad (5\text{-}82)$$

where b = channel height, a = channel width, and K,n = material constants [see Eqs. (5-74) and (5-75)]. For other rectangular channels see Schechter, *Am. Inst. Chem. Eng. J.*, **7,** 445–448 (1961); and Wheeler and Wissler, *Am. Inst. Chem. Eng. J.*, **11,** 207–212; 212–216 (1965). For **annular channels** (Bird, Armstrong, and Hassager, *Dynamics of Polymeric Liquids,* vol. 1: *Fluid Mechanics,* Wiley, New York, 1977),

$$\frac{D_0\,\Delta p}{4L} = K\left(\frac{2n+1}{2n}\right)^n \frac{1}{(1+\kappa)^n(1-\kappa)^{1+2n}}\left(\frac{32q}{\pi D_0^3}\right)^n \qquad (5\text{-}83)$$

$$\text{for} \quad n > 0.25 \quad \text{and} \quad \kappa > 0.5$$

where $\kappa = D_i/D_0$, D_0 = outer diameter of annulus, D_i = inner diameter of annulus, and K,n = material constants [see Eqs. (5-74) and (5-75)]. For further details, see Fredrickson and Bird, *Ind. Eng. Chem.*, **50,** 347–352 (1958), and *Ind. Eng. Chem. Fundam.*, **3,** 383 (1964); and Hanks and Larsen, *Ind. Eng. Chem. Fundam.*, **18,** 33–35 (1979). Measurements in annuli are presented by McEachern [*Am. Inst. Chem. Eng. J.*, **12,** 328–332 (1966)] and Tiu and Bhattacharyya [*Am. Inst. Chem. Eng. J.*, **20,** 1140–1144 (1974)]. For laminar flow of Bingham plastics in annuli, see Fredrickson and Bird, op. cit.; and Anshus, *Ind. Eng. Chem. Fundam.*, **13,** 162–164 (1974).

Rectilinear flow of nonnewtonian fluids, exhibiting normal stress effects in steady shear flow, in a **straight duct of arbitrary cross section** is not possible in general, and some secondary flow in the duct cross section is to be expected. However, the effect of the secondary flow on the flow rate may be small. Mitsuishi and Aoyagi [*Chem. Eng. Sci.*, **24,** 309–319 (1969)] discuss nonnewtonian flow in noncircular ducts.

Nonisothermal Flow In **nonisothermal flow of liquids,** f is sensibly increased if the liquid is being cooled and is decreased if the liquid is being heated. The available data, which are largely for oils, are approximated by first finding f as for isothermal flow of the liquid at the mainstream temperature and then dividing the result, in the case of cooling, by $(\mu_a/\mu_w)^{0.23}$ if in the laminar region or by $(\mu_a/\mu_w)^{0.11}$ if in the turbulent region; or, in the case of heating, by $(\mu_a/\mu_w)^{0.38}$ if in the laminar region or by $(\mu_a/\mu_w)^{0.17}$ if in the turbulent region. Here μ_a is the viscosity at the mainstream temperature, and μ_w is that at the wall temperature [see Sieder and Tate, *Ind. Eng. Chem.*, **28,** 1429–1435 (1936)]. In addition, $\rho = 1/v$ varies with temperature but, owing to incompressibility, not with p. Thus the following equation, Eq. (5-85) for pressure drop for ducts running full, such as heat-exchanger tubes, can be obtained by substituting the differential form of Eq. (5-57) into Eq. (5-41), dividing throughout by v, expressing ρ as a function of length because temperature is a function of length and integrating,

$$p_1 - p_2 = \frac{G^2(v_2 - v_1)}{g_c} + \int_0^L \left(\frac{fG^2v}{2g_cR_H} + \frac{g}{g_c}K\rho\right) dx \qquad (5\text{-}84)$$

$$p_1 - p_2 = \frac{G^2(v_2 - v_1)}{g_c} + \frac{fG^2v_{\text{avg}}L}{2g_cR_H} + \frac{g}{g_c}K\rho_{\text{avg}}L \qquad (5\text{-}85)$$

where p_1 = upstream static pressure; p_2 = downstream static pressure; G = mass velocity; v_1 = specific volume at upstream conditions; v_2 = specific volume at downstream conditions; g = local acceleration due to gravity; g_c = dimensional constant; f = Fanning friction factor, dimensionless; x = distance from inlet to a point downstream; L = duct length; v_{avg} = average specific volume; ρ_{avg} = average density; R_H = hydraulic radius; and $K = \sin\theta$ where θ = angle of inclination to horizontal. Note that $v_{\text{avg}} \neq 1/\rho_{\text{avg}}$ and both are averaged with respect to length, not with respect to p.

Equations and charts are presented by Christiansen and Gordon [*Am. Inst. Chem. Eng. J.*, **15,** 504–507 (1969)] for the prediction of friction loss in the laminar nonisothermal flow of either newtonian or nonnewtonian liquids when the liquids are being either heated or cooled in round tubes at constant wall temperature.

The flow of viscous liquids in small channels such as capillaries is often accompanied by large pressure drops. The conversion of work energy into heat via viscous shear can result in an appreciable tem-

TABLE 5-9 Average Values of *n* for Manning Formula,* Eq. (5-87)

Surface	n
Cast-iron pipe, fair condition	0.014
Riveted steel pipe	0.017
Vitrified sewer pipe	0.013
Concrete pipe	0.015
Wood-stave pipe	0.012
Planed-plank flume	0.012
Semicircular metal flumes, smooth	0.013
Semicircular metal flumes, corrugated	0.028
Canals and ditches:	
Earth, straight and uniform	0.023
Winding sluggish canals	0.025
Dredged earth channels	0.028
Natural-stream channels:	
Clean, straight bank, full stage	0.030
Winding, some pools and shoals	0.040
Same, but with stony sections	0.055
Sluggish reaches, very deep pools, rather weedy	0.070

*Brater and King, *Handbook of Hydraulics*, 6th ed., McGraw-Hill, New York, 1976, p. 7-22. For detailed information, see Chow, *Open-Channel Hydraulics*, McGraw-Hill, New York, 1959, pp. 110–123.

perature rise $\Delta T_a = \Delta p / c_v \rho J$, where ΔT_a = bulk temperature rise, Δp = pressure drop, c_v = fluid specific heat, ρ = fluid density, and J = work equivalent of heat. For example, it amounts to about 2.8°C (5°F) per 7 MPa (1000 lbf/in²) pressure drop for polymers and about 3.9°C (7°F) for hydrocarbon liquids. However, the increase in flow rate due to a reduction in viscosity is much greater than that corresponding to the bulk temperature rise, since a large part of the heat is generated near the channel walls, resulting in a flattening of the velocity profile. Compensation should generally be made for the heat effect if the Δp exceeds 1.4 MPa (200 lbf/in²) for adiabatic walls or 3.5 MPa (500 lbf/in²) for isothermal walls. For details, see Gerrard, Steidler, and Appeldoorn, *Ind. Eng. Chem. Fundam.*, **4**, 332–339 (1965), and **5**, 260–263 (1966).

Open Channels For **open channels,** the data are largely based on experiments with water in **turbulent flow,** and the results are usually given in terms of Chézy coefficients used in the Chézy formula [see Chow, *Open-Channel Hydraulics*, McGraw-Hill, New York, 1959, p. 93; Henderson, *Open Channel Flow*, Macmillan, New York, 1966, p. 91; and Streeter and Wylie, op. cit. (1979), p. 299]:

$$V = C\sqrt{R_H S} \tag{5-86}$$

where V = velocity; C = Chézy coefficient ($= \sqrt{2g_c/f}$), dimensionless $[(\text{lb}\cdot\text{ft})/(\text{lbf}\cdot\text{s}^2)]^{1/2}$; R_H = hydraulic radius; and S = loss in specific energy per unit channel length, or F/L [see Eq. (5-57)]. The Chézy coefficient is computed from the Manning formula:

$$C = C_1 (R_H)^{0.167}/n \tag{5-87}$$

where C_1 = dimensional constant, 0.320 in SI units (1.49 in U.S. customary units); and n = roughness factor given in Table 5-9. For the turbulent flow of **other newtonian fluids** in open channels, Eq. (5-57) and Table 5-8 should be used, or else C should be computed from the corresponding values of f. For the **laminar flow** of fluids in open channels, the equations given in Table 5-10 are recommended. In all cases capillary effects are negligible. For the effect of liquid contact angle on the flow rate for V troughs, see Ayyaswamy, Catton, and Edwards, *J. Appl. Mech.*, **41**, 332–336 (1974). For other cross sections, such as semicircular channels, elliptical channels, and trapezoidal channels, and for rough channels, see Straub, Silberman, and Nelson, *Trans. Am. Soc. Civ. Eng.*, **123**, 685–714 (1958). The lower critical Reynolds number, $N_{Re} = 4R_H V\rho/\mu$, for transition between laminar and turbulent flow in smooth channels is somewhat higher than that for smooth circular pipes, varying from 2000 to 4000, depending upon the cross-sectional shape [Straub et al., loc. cit.; Owen, *Trans. Am. Soc. Civ. Eng.*, **119**, 1157–1175 (1954)]. For the flow of **nonnewtonian fluids** in open channels, see Kozicki and Tiu, *Can. J. Chem. Eng.*, **45**, 127–134 (1967).

Compressible Flow If the pressure drop due to the flow of a gas through a system is large enough, compared with the inlet pressure, to occasion a 10 percent or greater decrease in gas density, then the flow is considered to be "compressible." In this event, formulas which make proper allowance for this change in both density and velocity must be used.

The flow of gases at atmospheric pressure and above is generally either laminar or turbulent in character, the transition between the two generally occurring somewhere in the Reynolds-number range of 2000 to 3000. **Turbulent flow** occurs above the transition, while **laminar flow** occurs below the transition. However, at low pressures or in very small channels, another flow phenomenon, **molecular flow,** is encountered. This occurs when the mean free path [see Eq. (5-109)] is of the same order of magnitude as the channel diameter. In this type of flow, a gas molecule migrates along the channel independently of other gas molecules. (See subsection "Molecular Flow.")

When the mean free path of the gas is less than about 65 percent but greater than about 1 percent of the channel diameter, the gas layer next to the channel wall assumes a certain velocity of slip. This is known as **slip flow** and may be taken to be a combination of laminar and molecular flows. Slip flow and molecular flow are most often encountered in high-vacuum technology. (See subsection "Slip Flow.")

Turbulent Flow The general formulas and methods in the subsection "Incompressible Flow" will apply only to a differential length of duct dx throughout which the density may be considered constant. Thus, in general, the Fanning equation, Eq. (5-57), must be used in differential form:

$$dF/dx = 4fV^2/2g_c D = fV^2/2g_c R_H = fG^2/2g_c\rho^2 R_H \tag{5-88}$$

Substitution of Eq. (5-88) in the differential form of the mechanical-energy balance [Eq. (5-41)] gives

$$v\,dp + \frac{V\,dV}{g_c} = -\left(\frac{fV^2}{2g_c R_H} + \frac{g}{g_c}\sin\theta\right)dx \tag{5-89}$$

where $v = 1/\rho$ = fluid specific volume and $(\sin\theta)\,dx = dz$ = vertical distance through which the fluid is raised when it moves distance dx along the pipe; W_e in Eq. (5-41) has been taken to be zero on the supposition that no pump is in the line. In a uniform duct, mass velocity $G = V/v$ is constant so that if p is known as a function of v only, $v\,dp$ can be written $\phi(v)\,dv$. Then, when $\sin\theta$ is constant, the variables in Eq. (5-89) are separable and integration is possible. The results in the following pages are obtained from Eq. (5-89) in the case of perfect gases ($pv = RT/M$).

Isothermal Flow in Horizontal Ducts Integration of Eq. (5-89) results in (Dodge, *Chemical Engineering Thermodynamics*, McGraw-Hill, New York, 1944, pp. 349–350)

$$p_1^2 - p_2^2 = \frac{fLG^2RT}{g_c R_H M}\left(1 + \frac{2R_H}{fL}\ln\frac{p_1}{p_2}\right) \tag{5-90}$$

In ducts of appreciable length, the last term in parentheses is generally negligible unless the pressure drop is very large. For example, if $L/D = 100$, $(p_1 - p_2)/p_1$ may be as great as 0.20 before the error in omitting the last term becomes as great as the uncertainty in the friction factor. When the last term is omitted, Eq. (5-90) may be written as

$$p_1 - p_2 = fLG^2/2g_c\rho_{avg}R_H \tag{5-91}$$

where ρ_{avg} = density at the average pressure $(p_1 + p_2)/2$. For a **round pipe,** Eq. (5-91) solved for weight rate of flow becomes

$$w = \frac{\pi}{8}\sqrt{\frac{(p_1^2 - p_2^2)g_c D^5 M}{fLRT}} \tag{5-92}$$

For a correction to Eq. (5-90) in the case of imperfect gases, see Madsen and Ramamoorthy, *J. Fluids Eng.*, **101**, 76–78 (1979).

In ordinary pipe lines, the flow is commonly more nearly adiabatic than truly isothermal.

Adiabatic Flow in Horizontal Ducts A convenient graphical method of integrating Eq. (5-89) for adiabatic flow in horizontal ducts [Lapple, *Trans. Am. Inst. Chem. Eng.*, **39**, 395–432 (1943);

TABLE 5-10 Laminar-Flow Formulas

Ducts running full

For liquids, let $N = \dfrac{\rho g_c}{\mu}\left(\dfrac{g}{g_c}\rho \sin\alpha + \dfrac{p_1 - p_2}{L}\right)$; for gases, let $N^a = \dfrac{g_c M}{2zRT\mu}\left(\dfrac{p_1^2 - p_2^2}{L}\right)$

Duct cross section	Theoretical equation for weight rate of flow
Circle,[b,c] diameter = D	$w = \pi D^4 N/128$ [for liquids this reduces to $p_1 - p_2 = 32\mu LV/g_c D^2$ (i.e., Poiseuille's law) if tube is horizontal]
Ellipse,[c] semiaxes = a,b	$w = \dfrac{\pi a^3 b^3}{a^2 + b^2}\left(\dfrac{N}{4}\right)$
Rectangle,[d] width = a, height = b	$w = ab^3 N/K$

where a/b =	1	3/2	2	3	4	5	10	∞
K =	28.45	20.43	17.49	15.19	14.24	13.73	12.81	12

Duct cross section	Theoretical equation for weight rate of flow
Broad parallel plates,[b,d] spacing = b; i.e., rectangle with $a/b = \infty$	$w = b^3 N/12$ per unit width
Annulus,[c] outer diameter = D_2, inner diameter = D_1	$w = \dfrac{\pi(D_2^2 - D_1^2)N}{128}\left[D_2^2 + D_1^2 - \dfrac{D_2^2 - D_1^2}{\ln (D_2/D_1)}\right]$
Isosceles triangle,[e] vertical < = $2\phi°$, bisector vertical, slant side = a	$w = a^4 N/K'$

where $K' = \dfrac{K(1 + \sin\phi)^2}{2(\sin\phi\cos\phi)^3}$

2ϕ =	10	20	30	40	50	60	70	80	90
K =	12.5	12.8	13.1	13.2	13.3	13.33	13.3	13.2	13.2

Duct cross section	Theoretical equation for weight rate of flow
Regular polygons,[f] area = A_c	$w = A_c^2 N/K$

where number of sides =	3	4	5	6	7	8	9	∞ (circle)
K =	34.64	28.45	26.77	26.07	25.81	25.54	25.42	8π

Open channels

Let $N = \rho^2 g \sin\alpha/\mu$, since necessarily $p_1 = p_2$; the following equations are valid only when the variation in depth is negligible

Channel cross section	Theoretical equation for weight rate of flow
Circle,[g] diameter = D, height of circular segment = H	$w = \pi D^4 N/128K$

where H/D =	0.1	0.2	0.3	0.4	0.5	0.6	0.7	0.8	0.9	1.0 (pipe)
K =	295	30.1	8.47	3.65	2.00	1.30	0.962	0.811	0.797	1

Channel cross section	Theoretical equation for weight rate of flow
Rectangle,[h] width = a, depth = $b/2$	$w = ab^3 N/2K$

where a/b =	1	2	3	4	5	10	∞ (broad stream)
K =	28.6	17.5	15.3	14.2	13.7	12.8	12

Channel cross section	Theoretical equation for weight rate of flow
Broad stream on flat plate,[h] depth = $b/2$	$w = b^3 N/24$ per unit breadth
V trough,[i] vertical ∠ = $2\phi°$, bisector vertical, slant depth = a	$w = a^4 N/K'$

where $K' = K/[2(\sin\phi\cos\phi)^3]$

2ϕ =	10	20	40	60	80	100	120
K =	12.14	12.45	13.20	13.83	14.18	14.18	13.83

Notation used in Table 5-10:

A_c = cross-sectional area
a,b,D,H = characteristic lengths
g = local acceleration due to gravity
g_c = dimensional constant
L = length of passage
M = molecular weight
p_1,p_2 = upstream and downstream static pressures
R = gas constant
T = absolute temperature
w = weight rate of flow
z = compressibility factor, dimensionless
α = angle between duct axis and horizontal
ρ = fluid density
μ = dynamic viscosity
2ϕ = central angle

[a]If the pressure drop is less than 10 percent of the downstream absolute pressure, the approximate expression $N = [\rho g_c(p_1 - p_2)]/\mu L$ may be used in case of gases.

[b]Dryden, Murnaghan, and Bateman, *Hydrodynamics*, Dover, New York, 1956, pp. 178, 184–185.

[c]Lamb, *Hydrodynamics*, 6th ed., Cambridge, New York, 1932, p. 587.

[d]Shah and London, *Laminar Flow Forced Convection in Ducts*, Academic, New York, 1978, p. 200.

[e]Sparrow, *Am. Inst. Chem. Eng. J.*, **8**, 599–604 (1962); see also Carlson and Irvine, *J. Heat Transfer*, **83**, 441–444 (1961).

[f]Cheng, *Proc. Third Int. Heat Transfer Conf.*, New York, **1**, 64–76 (1966); Shih, *Can. J. Chem. Eng.*, **45**, 285–294 (1967).

[g]Based on work of Sestak, *Can. J. Chem. Eng.*, **52**, 670–672 (1974); see also Straub et al., *Trans. Am. Soc. Civ. Eng.*, **123**, 685–714 (1958); Van Dromme and Hellinckx, *J. Heat Transfer*, **95**, 427–429 (1973).

[h]Owen, *Trans. Am. Soc. Civ. Eng.*, **119**, 1157–1175 (1954).

[i]Ayyaswamy, Catton, and Edwards, *J. Appl. Mech.*, **41**, 332–336 (1974); see also Straub et al., loc. cit.

Nomenclature for Perfect-Gas Equations and Charts

A = area
A_c = area for critical flow
D = diameter
f = Fanning friction factor, dimensionless
g_c = dimensional constant
$G = V\rho = V/v$ = mass velocity
G_c = critical or maximum mass velocity
$G^\circ = p_0 \sqrt{\dfrac{g_c M k}{R T_0}\left(\dfrac{2}{k+1}\right)^{(k+1)/(k-1)}}$
$k = c_p/c_v$, ratio of specific heat at constant pressure to that at constant volume, dimensionless
L = length
M = molecular weight
$N = fL/R_H$ = frictional resistance in velocity heads, dimensionless
p_c = absolute critical pressure
p_0 = absolute pressure in reservoir or large chamber
p_1, p_2 = absolute pressure at inlet and outlet respectively
p_3 = absolute pressure of surroundings
R = gas constant
R_H = hydraulic radius
$r_c = p_c/p_0$, critical pressure ratio, dimensionless
T = absolute temperature
V = velocity
V_c = critical velocity
v = specific volume
w = weight rate of flow
μ = viscosity
ρ = density

and Levenspiel, *Am. Inst. Chem. Eng. J.*, **23**, 402–403 (1977)] results by assuming that conditions of flow at the inlet arise from the adiabatic expansion of the gas through a frictionless nozzle leading from a chamber where the velocity is negligible. Such a chamber frequently is present in fact; departures of the real entrance from a perfect nozzle may be approximately allowed for by supposing the length of pipe to be increased. Data indicate that the friction factor is the same function of the Reynolds number for compressible flow as for incompressible flow (Shapiro, *The Dynamics and Thermodynamics of Compressible Fluid Flow*, vol. I, pp. 184–186, and vol. II, p. 1131, Ronald, New York, 1953 and 1954). For a given pipe diameter and mass flow rate, the friction factor depends upon the viscosity, which, in turn, depends upon temperature. Since in adiabatic compressible flow, Reynolds numbers are usually high (i.e., turbulent flow), variation of the friction factor due to temperature variations along the pipe length is small; thus a constant value of friction factor can be taken in the integrations.

Figure 5-32 shows the flow system and the usage of subscripts. For the system shown,

$$T_1/T_0 = (p_1/p_0)^{(k-1)/k} \tag{5-93}$$

$$\frac{p_0}{p_1} = \left[1 + \frac{G^2}{2g_c}\left(\frac{k-1}{k}\right)\frac{RT_1}{Mp_1^2}\right]^{k/(k-1)} \tag{5-94}$$

The charts of Fig. 5-33*a* and *b* show, for three values of k, ratio of specific heats, and for various values of frictional resistance N, number of velocity heads in the duct, the relation between p_3/p_0 and the ratio of mass velocity G in the duct to a parameter

$$G^\circ = p_0 \sqrt{\frac{g_c M k}{R T_0}\left(\frac{2}{k+1}\right)^{(k+1)/(k-1)}} \tag{5-95}$$

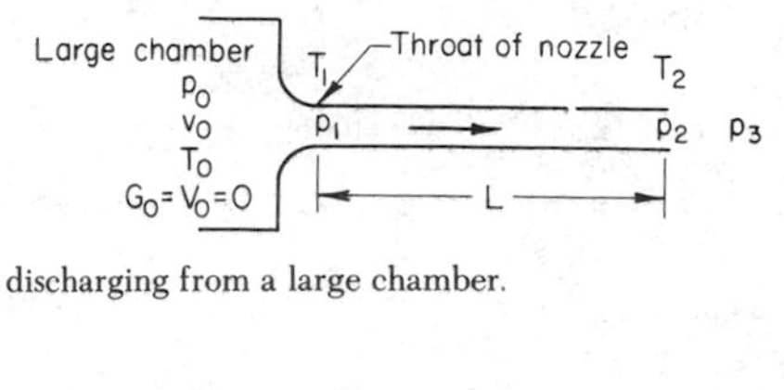

FIG. 5-32 Pipe discharging from a large chamber.

The quantity G° is the maximum mass velocity through a well-rounded nozzle in ideal isentropic flow. The pressure p_2 is equal to p_3 if, for a given N, the mass velocity is less than a certain maximum or critical value G_c; this is true in the unchoked-flow region of the charts. In this region the gas temperature T_2 at the pipe exit is given by

$$\frac{p_2}{p_0} = \sqrt{\frac{k-1}{2}\left(\frac{2}{k+1}\right)^{(k+1)/(k-1)}\left[\frac{(T_2/T_0)^2}{1-(T_2/T_0)}\right]}\;\frac{G}{G^\circ} \tag{5-96}$$

If p_3/p_0 is less than the value of p_2/p_0 corresponding to G_c, the flow is then independent of p_3 (see subsection "Critical Flow Nozzle"); this is true in the choking-flow region of the charts. In this region the pressure and temperature at the pipe exit are given by

$$p_2/p_0 = [2/(k+1)]^{k/(k-1)}(G_c/G^\circ) \tag{5-97}$$

$$T_2/T_0 = 2/(k+1) \tag{5-98}$$

Use of the charts is most easily understood by studying Example 1. When G is being sought, we assume $f = 0.0045$ to determine a trial value of G; then, using this approximate G to get a Reynolds number, we find f from Fig. 5-28 and repeat the calculation by using this revised estimate of f. When fittings are present in the pipe line, we increase the value of N calculated for the straight pipe by the number of velocity heads equivalent to the loss in the fittings (see Table 5-14). If, however, any cross section of a fitting is appreciably less than that of the pipe line, erroneous results may be obtained because the critical flow rate through the constriction may limit the capacity of the line; this occurs if the acoustic velocity (see subsection "Terminology in Fluid Mechanics") is approached at the constriction. For a sharp or abrupt inlet, the charts lead to approximately correct results if 0.5 is added to the value of N for the duct. In this case, however, the formulas given above for p_0/p_1 and T_1/T_0 are inapplicable because they apply only to a rounded entrance.

Example 1 It is desired to calculate the discharge rate of air to the atmosphere from a reservoir at 1-MPa gauge and 20°C through 10 m of straight 2-in Schedule 40 steel pipe (inside diameter = 5.25 cm) and three standard 90° elbows. The pipe inlet is abrupt.

Resistance	(L/D)	(N)
Inlet	. . .	0.50*
Straight pipe	190	3.43†
Elbows (three)	. . .	6.18

*Assumed an abrupt inlet.
†Calculated, assuming $f = 0.0045$.

From the conditions of the problem,
$T_0 = 293$ K
$p_0 = (1 + 0.101)(10^6) = (1.10)(10^6)$ Pa
$p_3 = (0.101)(10^6)$ Pa
$p_3/p_0 = 0.0918$
$M = 29$; $k = 1.4$

$$G^\circ = p_0 \sqrt{\frac{g_c M k}{R T_0}\left(\frac{2}{k+1}\right)^{(k+1)/(k-1)}} = (1.10)(10^6)$$
$$\sqrt{\frac{29 \times 1.4}{8314 \times 293}\left(\frac{2}{1.4+1}\right)^{(1.4+1)/(1.4-1)}}$$

$G^\circ = 2600$ kg/(s·m²)
Pipe cross section $= 0.785(5.25/100)^2 = 0.00216$ m²

(G/G°), from curve for total $N = 6.18$ and $p_3/p_0 = 0.0918$	0.47
G, kg/(s·m²)	1222
Discharge rate, kg/s	2.64
T_2/T_0, from Eq. (5-98), since flow is choked	0.833
T_2, K	244
Average gas temperature in pipe, °C	−5
p_2/p_0, from Eq. (5-97) ($G_c = G$), since flow is choked	0.25
p_2, MPa gauge	0.174
Viscosity μ at average gas temperature, Pa·s	1.71×10^{-5}
N_{Re} or (DG/μ)	3.75×10^6
f	0.0047

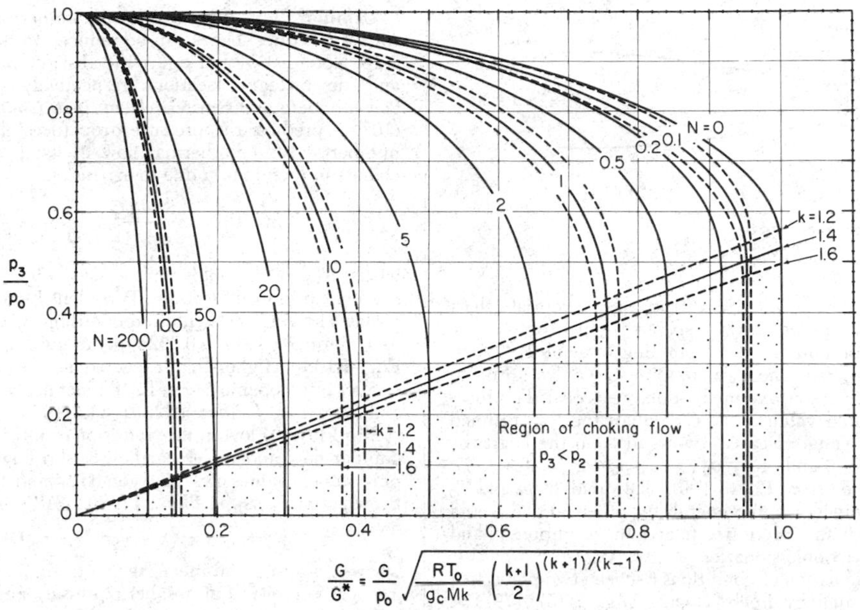

FIG. 5-33*a* Design chart for adiabatic flow of gases, useful for finding the allowable pipe length for given flow rate. [*Levenspiel*, Am. Inst. Chem. Eng. J., **23**, *402 (1977).*]

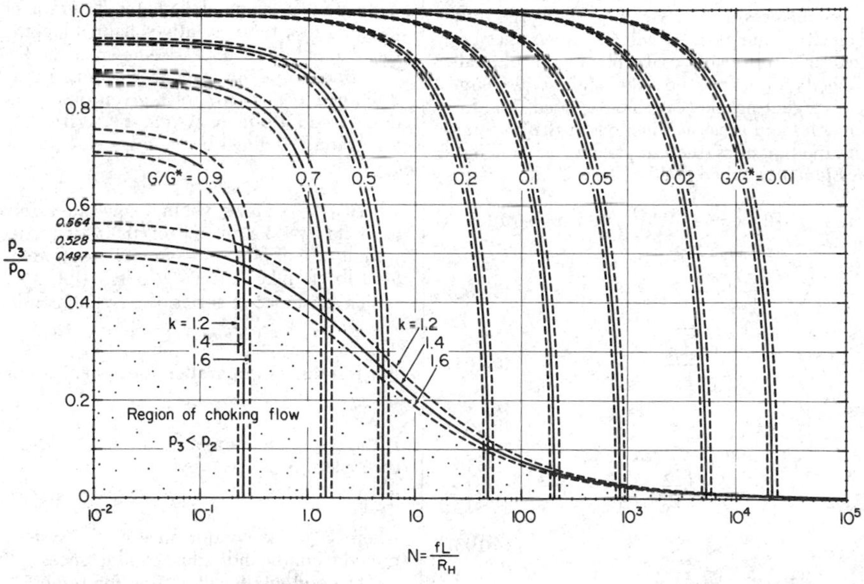

FIG. 5-33*b* Design chart for adiabatic flow of gases, useful for finding the discharge rate in a given piping system. [*Levenspiel*, Am. Inst. Chem. Eng. J., **23**, *402 (1977).*]

Since the value of f so obtained checks the assumed value reasonably well, it is not necessary in this case to repeat the determination of G.

Figure 5-33*a* and *b* can also be used to compute the **pressure drop between two locations along a pipe line.** For example, if the conditions at a given location and the flow rate are known and if the pressure at a location downstream is desired, let the known conditions be T_1 and p_1. Thus the ratios G/G_1°, p_2/p_1, and T_2/T_1 are referred to the known conditions and can be related to G/G°, p_1/p_0, and T_1/T_0 by

$$G/G_1^\circ = (G/G^\circ)[(\sqrt{T_1/T_0})/(p_1/p_0)] \tag{5-99}$$

$$p_2/p_1 = [(p_2/p_0)/(p_1/p_0)] \tag{5-100}$$

$$T_2/T_1 = [(T_2/T_0)/(T_1/T_0)] \tag{5-101}$$

where G_1° is a hypothetical quantity of no real physical significance defined by

$$G_1^\circ = p_1 \sqrt{\frac{g_c Mk}{RT_1}\left(\frac{2}{k+1}\right)^{(k+1)/(k-1)}} \tag{5-102}$$

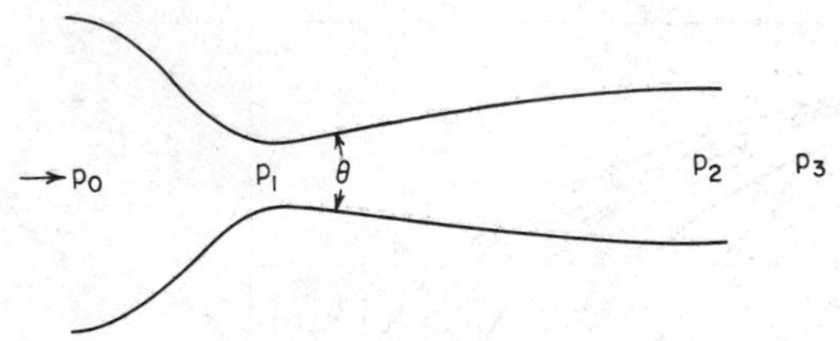

FIG. 5-34 Convergent-divergent nozzle.

First we assume a value of G/G°, and from the appropriate chart we read the corresponding value of p_1/p_0 ($= p_2/p_0$) for $N = 0$. From Eq. (5-96) we compute the corresponding value of T_1/T_0 ($= T_2/T_0$) for p_1/p_0 ($= p_2/p_0$). Then we compute G/G° from Eq. (5-99). Using this result as a new assumed value, we repeat the procedure until the computed value of G/G° equals the last assumed value. Using the final value of G/G°, we read from the chart the value of p_1/p_0 for $N = 0$ and of p_2/p_0 ($= p_3/p_0$) for $N = 4fL/D$; p_2 can then be computed from Eq. (5-100). The values of p_0 and T_0 that have been determined are, in general, hypothetical and would be the conditions in a chamber required to give the conditions p_1 and T_1 in the throat of a frictionless nozzle.

Charts for direct calculation of conditions between two points in a pipe have been presented by Loeb [*Chem. Eng.*, **76**(5), 179–184 (1969)] for known upstream or downstream conditions and by Powley [*Can. J. Chem. Eng.*, **36**, 241–245 (1958)] for known downstream conditions.

Flow through Convergent-Divergent Nozzles (Delaval Nozzle) For very high chamber pressures or very low discharge pressures, the discharge velocity from nozzles will be supersonic, i.e., greater than the critical or sonic velocity. To obtain the ultimate adiabatic expansion velocity of a gas, i.e., the highest discharge velocity, a convergent-divergent nozzle (Fig. 5-34) is used. The following relations for velocities and areas as functions of the pressures can be derived from thermodynamics (see Lapple, loc. cit.; and Sutton, *Rocket Propulsion Elements*, 2d ed., Wiley, New York, 1956, p. 53):

For $p_2/p_0 < r_c$ where $r_c = [2/(k+1)]^{k/(k-1)}$ [see Eq. (5-19)]

$$V_{1c} = \sqrt{g_c k p_c v_c} = \sqrt{\frac{2g_c k p_0 v_0}{k+1}} \tag{5-103}$$

$$G_{1c} = \left(\frac{2}{k+1}\right)^{(k+1)/2(k-1)} \sqrt{\frac{g_c k p_0}{v_0}} \tag{5-104}$$

$$p_{1c} = p_0 \left(\frac{2}{k+1}\right)^{k/(k-1)} \tag{5-105}$$

$$\left(\frac{A_{1c}}{A_2}\right)^2 = \left(\frac{2}{k-1}\right)\left(\frac{k+1}{2}\right)^{(k+1)/(k-1)}\left(\frac{p_2}{p_0}\right)^{2/k}\left[1-\left(\frac{p_2}{p_0}\right)^{(k-1)/k}\right] \tag{5-106}$$

$$\left(\frac{V_2}{V_{1c}}\right)^2 = \left(\frac{k+1}{k-1}\right)\left[1-\left(\frac{p_2}{p_0}\right)^{(k-1)/k}\right] \tag{5-107}$$

Only for $p_2 = p_3$ is the ultimate adiabatic expansion velocity realized in the nozzle. The expansion will be incomplete within the nozzle if $p_2 > p_3$, or a compression shock will result if $p_2 < p_3$.

The shape of the converging section is similar to that of a simple converging nozzle. The shape of the diverging section to obtain a uniform, parallel, shock-free supersonic stream of the discharge must be of a special shape; methods of design are given by Liepmann and Roshko (*Elements of Gasdynamics*, Wiley, New York, 1957, p. 284). If the nozzle is to be used as a thrust device, the diverging section can be a simple conical diverging section with a total angle $\theta = 30°$ (Sutton, loc. cit., p. 75). Note that, if $p_2/p_0 > [2/(k+1)]^{k/(k-1)}$, the discharge velocity will be subsonic; for such cases, a simple converging nozzle can be used.

Laminar Flow For gases under isothermal conditions, Eq. (5-91) for various duct shapes can be reduced to the forms given in Table 5-10. Note that p_1 and p_2 are the absolute pressures at the upstream and downstream locations respectively. Equation (5-108), by Schwartzberg and Gurevich [*Am. Inst. Chem. Eng. J.*, **16**, 762–766 (1970)], predicts the pressure drop for high-velocity (up to Mach number = 0.5) isothermal flow of gases with allowance for the change in momentum due to expansion.

$$p_1^2 - p_2^2 = \frac{8\mu RTG}{g_c MD}\left[\frac{8L}{D} + \frac{N_{Re}}{3}\ln\left(\frac{p_1}{p_2}\right)\right] \tag{5-108}$$

where N_{Re} = Reynolds number (see Fig. 5-28 for definition). See nomenclature in subsection "Turbulent Flow" for other symbols.

Gases flowing through **microopenings,** with diameters from 0.005 to 0.06 mm (0.0002 to 0.0024 in), at low N_{Re} and $L/D \geq 100$, obey Eq. (5-108). Higher-than-expected flow rates were obtained with smaller microopenings. For further details, see Hedley, Olt, Holboke, and Wurstner, *J. Rheol.*, **22**, 91–112 (1978).

Slip Flow Flow in the transition or intermediate range between laminar flow and molecular flow is slip flow. It exists for practical purposes for values of X between 0.014 and 1.0, where X is given by Brown et al. [*J. Appl. Phys.*, **17**, 802–813 (1946)] as

$$X = \sqrt{8/\pi}\,(\lambda/D) = (2\mu/p_m D)\sqrt{RT/Mg_c} \tag{5-109}$$

where λ = molecular mean free path length, D = channel diameter, μ = gas viscosity at atmospheric pressure and temperature T, p_m = arithmetic mean absolute pressure, R = gas constant, T = absolute gas temperature, M = molecular weight of gas, and g_c = dimensional constant. Thus for X = 0.014 (the approximate boundary between laminar flow and slip flow), the mean free path λ is about 1 percent of the channel diameter. The ratio of the molecular free path length λ to a representative channel length, which may be taken as the channel diameter D, is known as the Knudsen number.

In vacuum technology, it is common practice to refer to the **conductance,** or "speed," of a given pipe, pump, or system. The conductance of a pipe or system, m^3/s (ft^3/s), is the inverse of the resistance and is defined by

$$C = q'/\Delta p \tag{5-110}$$

where $q' = w/\rho'$ = volume flow rate referred to a pressure of 1 Pa (or 1 lbf/ft^2), $Pa \cdot m^3/s$ [$(lbf/ft^2)(ft^3/s)$]; Δp = pressure drop; w = weight rate of flow; $\rho' = M/RT$ = gas density at a pressure of 1 Pa (or 1 lbf/ft^2), $kg/(m^3 \cdot Pa)$ [$(lb/ft^3)/(lbf/ft^2)$].

For resistance in **series,** the system conductance is given by

$$1/C = 1/C_1 + 1/C_2 + 1/C_3 + \cdots \tag{5-111}$$

For resistances in **parallel,** the system conductance is given by

$$C = C_1 + C_2 + C_3 + \cdots \tag{5-112}$$

If the system contains a pump of speed S_p, m^3/s (ft^3/s), the speed of the system S_0 is given by

$$1/S_0 = 1/S_p + 1/C \tag{5-113}$$

where C is the conductance of the system [m^3/s (ft^3/s)] as determined from the individual conductances by Eqs. (5-111) and (5-112).

The **conductance** of a **pipe line** is most conveniently obtained for the slip-flow region by the method of Brown et al. (loc. cit.) which involves the use of Fig. 5-35. Conductance of the pipe line is first calculated as for laminar flow:

$$C_{lam} = q'_{lam}/\Delta p = g_c D^2 A p_m / 32\mu L \tag{5-114}$$

where C_{lam} = pipe-line conductance for laminar flow, m^3/s (ft^3/s); q'_{lam} = volume flow rate referred to a pressure of 1 Pa (or 1 lbf/ft^2), computed for laminar flow, $Pa \cdot (m^3/s)$ [$(lbf/ft^2)(ft^3/s)$]; Δp = pressure drop; A = cross-sectional area; L = pipe-line length; and other variables are as defined following Eq. (5-109). True conductance C of the pipe line is then obtained from

$$C = FC_{lam} \tag{5-115}$$

where correction factor F is obtained from Fig. 5-35 as a function of X.

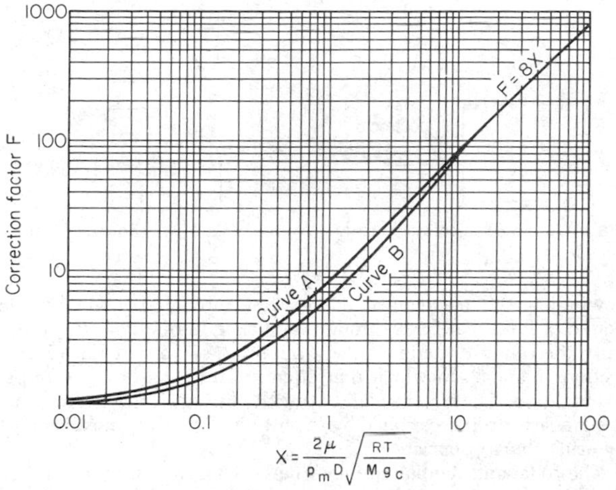

FIG. 5-35 Correction facor for Poiseuille's equation at low pressures. Curve *A*: experimental curve for glass capillaries and smooth metal tubes. [*From Brown et al.*, J. Appl. Phys., *17, 802 (1946)*.] Curve *B*: experimental curve for iron pipe. *(Riggle, Courtesy of E. I. du Pont de Nemours & Co.)*

For slip flow through **square channels,** see Milligan and Wilkerson, *J. Eng. Ind.*, **95,** 370–372 (1973).

For slip flow through **annuli** of various radius ratios, see Maegley and Berman, *Phys. Fluids*, **15,** 780–785 (1972).

Experimental data on the **conductance** of **sharp-edged orifices** in the slip-flow region is given by Knudsen [*Ann. Phys.*, **28,** 999–1016 (1909)] and Fujimoto and Kato [*Mem. Fac. Eng. Nagoya Univ.*, **22,** 262–275 (1970)].

For **elbows and tees,** few data are available. For design, conductance can be computed as for a straight pipe, using the path length through the fitting.

For additional information on slip flow, see Dushman and Lafferty, *Scientific Foundations of Vacuum Technique*, 2d ed., Wiley, New York, 1962, pp. 104–111; and Lawrance, *Vacuum Symp. Trans.*, Committee on Vacuum Techniques, Boston, 1954, pp. 55–62.

Molecular Flow It exists for practical purposes for values of X greater than 1.0, where X is defined by Eq. (5-109). This boundary between molecular flow and slip flow corresponds to a mean free path λ equal to about 65 percent of the channel diameter. The overall system conductance or speed for molecular flow is obtained from the individual conductances as indicated by Eqs. (5-111), (5-112), and (5-113).

For **circular pipe,** conductance can be obtained most conveniently by the method of Brown et al. (loc. cit.), which involves the use of Fig. 5-35. Conductance of the pipe line is first calculated as for laminar flow, using Eq. (5-114). Parameter X is computed, using Eq. (5-109), and the corresponding value of the correction factor F obtained from Fig. 5-35. The molecular-flow conductance is then computed as the product of F and the laminar-flow conductance, as given by Eq. (5-115). Note that the curve for glass and smooth metal pipe is a straight line in the molecular-flow region and is described by the equation $F = 8.0X$. For short pipes ($L/D < 100$), allowance should be made for entrance effects (see subsequent paragraph on short pipes).

For **rectangular channels,** the following method is given by Normand [*Ind. Eng. Chem.*, **40,** 783–787 (1948)]. The channel is treated as though it were a round pipe having an equivalent diameter D_e determined as follows:

(1) If $a/b < 3$
$$D_e = 2\sqrt{ab/\pi} \tag{5-116}$$

where a = major axis of rectangle and b = minor axis.

(2) If $a/b > 3$
$$D_e = \left(2.55K \frac{a^2b^2}{a+b}\right)^{1/3} \tag{5-117}$$

TABLE 5-11 Constants for Rectangular Channels

a/b	K	a/b	K
1.0	1.108	5	1.297
1.5	1.126	8	1.400
2.0	1.151	10	1.444
3.0	1.198		

TABLE 5-12 Constants for Circular Annulus

D_2/D_1	K	D_2/D_1	K
0	1.00	0.707	1.254
0.259	1.072	0.866	1.430
0.500	1.154	0.966	1.675

where K = constant obtained from Table 5-11. The value of D_e so obtained is then employed in place of D in Eqs. (5-109) and (5-114).

For **square channels,** see also Milligan and Wilkerson, loc. cit.

For a **circular annulus,** the following method is given by Guthrie and Wakerling (*Vacuum Equipment and Techniques*, McGraw-Hill, New York, 1949, pp. 37–38, 52–53). Conductance C, m^3/s (ft^3/s), is calculated as

$$C = C_1 K \frac{(D_1 - D_2)^2(D_1 + D_2)}{L} \sqrt{\frac{T}{M}} \tag{5-118}$$

where K = constant given in Table 5-12 as a function of D_2/D_1, dimensionless; M = molecular weight; T = absolute temperature, D_1 = outer diameter of annulus; D_2 = inner diameter of annulus; L = length of annulus; and C_1 = dimensional constant, 38.2 in SI units (93.3 in U.S. customary units). See also Maegley and Berman, loc. cit.

For an **orifice** (thin-walled or sharp-edged), the conductance C, m^3/s (ft^3/s), can be computed from

$$C = C_1\sqrt{T/M} \tag{5-119}$$

where T = absolute temperature; M = molecular weight; A = area of opening; and C_1 = dimensional constant, 36.5 in SI units (89.2 in U.S. customary units). See Guthrie and Wakerling, op. cit., pp. 19, 52–53. Fujimoto and Kato (loc. cit.) give experimental values of the conductance with air through sharp-edged orifices. They obtained molecular flow for $X > 0.16$, where X is defined by Eq. (5-109).

For a **short pipe** of circular cross section, the conductance as calculated for the orifice via Eq. (5-119) is multiplied by a correction factor K, which is a function of the length-to-diameter ratio. K can be approximated as follows (Kennard, *Kinetic Theory of Gases*, McGraw-Hill, New York, 1938, pp. 306–308):

For $0 \leq L/D \leq 0.75$

$$K = \frac{1}{1 + (L/D)} \tag{5-120}$$

For $L/D > 0.75$

$$K = \frac{1 + 0.8(L/D)}{1 + 1.90(L/D) + 0.6(L/D)^2} \tag{5-121}$$

where L = length of tube and D = diameter. More precise values of K can be obtained from Dushman and Lafferty (op. cit., pp. 94–95). For values of $L/D > 100$, the error due to neglecting the "end correction" (using the long-pipe formulas) will be less than 2 percent.

For a **slot seal,** with or without a sheet located in or passing through the seal, Yu and Sparrow [*J. Basic Eng.*, **70,** 405–410 (1970)] give a theoretically derived chart for predicting the mass flow of gas through the seal as a function of the ratio of seal plate thickness to gap opening.

For **elbows** and **tees,** conductance can be computed as for a straight pipe, using an equivalent length equal to the actual path length plus 1.33 pipe diameters. See Guthrie and Wakerling, op. cit., pp. 41–43.

Pump-down time θ for evacuating a vessel in the absence of in-leakage is given approximately by

$$\theta = (V_t/S_0) \ln [(p_1 - p_0)/(p_2 - p_0)] \tag{5-122}$$

where V_t = volume of vessel plus volume of piping between pump and vessel, S_0 = speed of system as given by Eq. (5-113), assumed independent of pressure; p_1 = initial vessel pressure; p_0 = lowest pump intake pressure attainable with the pump in question; and p_2 = final vessel pressure. For more precise calculations, see Dushman and Lafferty, op. cit., pp. 111–116.

The amount of inerts which have to be removed by a pumping system after the pump-down stage depends largely upon the in-leakage of air at the various fittings, connections, etc. A tabulation of average air in-leakages for various connections and pipe-line components is given by Jackson [*Chem. Eng. Prog.*, **44,** 347–352 (1948)]. The total of such leakages determines the size of pump required to maintain the desired vacuum.

For further details on molecular-flow phenomena, see Dushman and Lafferty, op. cit., pp. 87–104; Guthrie and Wakerling, op. cit., pp. 12–58; Lewin, *Fundamentals of Vacuum Science and Technology,* McGraw-Hill, New York, 1965, pp. 11–16; and Van Atta, *Vacuum Science and Engineering,* McGraw-Hill, New York, 1965, pp. 43–62.

Economic Pipe Diameter: Turbulent Flow In selecting the size of pipe to be used for a fluid-handling system, there is frequently a range of permissible diameters encompassing two or more standard sizes of pipe. In such cases the final selection should be made on an economic basis so that the last increment of investment reduces the operating cost enough to produce the required minimum return on investment. In the case of long cross-country pipe lines, of alloy pipe lines of appreciable length and complexity, or of pipe lines with control valves, detailed analyses of investment and operating costs should be made.

However, for pipe lines of the lengths usually encountered within chemical plants and petroleum refineries, use of one of the economic pipe-diameter charts for selection of the diameter is generally adequate. Such charts were presented by Genereaux [*Chem. Metall. Eng.*, **44**(5), 241–248 (1937)] and by Johnson and Maker [*Proc. Tenth Mid-Year Meet. Am. Pet. Inst.*, sec. III, 7–23 (1940)]. Both charts were based on the minimum cost of owning and operating the pipe line. Further details are given by Peters and Timmerhaus, *Plant Design and Economics for Chemical Engineers,* 3rd ed., McGraw-Hill, New York, 1980, chap. 10.

For the turbulent flow of **nonnewtonian fluids,** methods for computing the economic pipe diameter are outlined by Skelland (*Non-Newtonian Flow and Heat Transfer,* chap. 7, Wiley, New York, 1967).

Economic Pipe Diameter: Laminar Flow Pipe lines for the transport of high-viscosity liquids in chemical plants or petroleum refineries are seldom designed purely on the basis of economics. More often, the size is dictated by available pressure drop, by shear limitations, or by residence-time considerations.

An economic pipe-diameter chart for laminar flow was given by Sarchet and Colburn [*Ind. Eng. Chem.*, **32,** 1249–1252 (1940)]. An updated version was given by Peters and Timmerhaus (loc. cit., chap. 10). For the laminar flow of nonnewtonian fluids, methods for computing the economic pipe diameter are outlined by Skelland (op. cit., chap. 7).

Miscellaneous Pressure Losses Experimental determinations of the resistance of fittings and valves are ordinarily carried out by measuring the overall friction loss in a system made up of two or more lengths of straight pipe connected in series by a suitable number of identical fittings or valves. To obtain the loss due to the fittings or valves themselves, the friction loss in the straight pipe is subtracted from the overall or total friction loss. There are three distinct conventions for computing the length of the "straight pipe" in the test system: (1) the actual length of the centerline of the entire system is taken, (2) the lengths of the individual pieces of pipe that are actually straight are summed, or (3) the distances between the intersections of the extended centerlines of the successive straight pipes are added. The first convention, i.e., taking the **actual length of the centerline of the entire system,** is used here except as noted.

In the following compilation, F (N·m)/kg [(ft·lbf)/lb or ft of fluid flowing] is the mechanical energy lost through friction by each pound of fluid that flows through the fitting in question. F is identically the quantity designated by the same symbol in Eq. (5-41) if sections 1 and 2, to which that equation refers, are taken, respectively, immediately before and after the fitting. Hence, to calculate the pressure drop occasioned by the fitting, Eq. (5-41) must be solved by using the appropriate F.

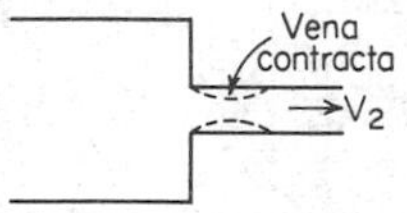

FIG. 5-36 Sudden contraction.

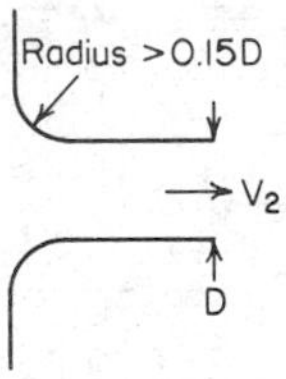

FIG. 5-37 Rounded entrance.

The following symbols are used here: A = cross-sectional area; D = inside diameter; g_c = dimensional constant; G = average mass velocity; K = coefficient expressed as number of velocity heads, dimensionless; L = length of pipe or duct; L_e = equivalent length of straight pipe; p = static pressure; V = average linear velocity; μ = viscosity; ρ = density. Subscripts are defined where they are used.

Equations and data apply to the flow of **incompressible fluids,** i.e., liquids, and gases and vapors for density changes less than 10 percent or for velocities less than 60 m/s (about 200 ft/s).

Contraction and Entrance Losses For a **sudden contraction** at a sharp-edged entrance to a pipe line or a sudden reduction in the cross-sectional area of a duct (Fig. 5-36), the loss of mechanical energy due to friction for **turbulent flow** is

$$F = K_c(V_2^2/2g_c) \tag{5-123}$$

where V_2 = average velocity in the smaller pipe; and K_c = coefficient, function of the ratio of the smaller cross-sectional area A_2 to the larger cross-sectional area A_1 and of the Reynolds numbers in both pipes, dimensionless. Values of K_c for turbulent flow (see Rouse, *Engineering Hydraulics,* Wiley, New York, 1950, p. 415) are given in Table 5-13. Slightly higher values are reported by Benedict, Carlucci, and Swetz [*J. Eng. Power,* **88,** 73–81 (1966)].

For a **trumpet-shaped** or **rounded entrance** (Fig. 5-37) with a radius of rounding greater than about 15 percent of the pipe diameter, the contraction coefficient K_c for **turbulent flow** is about 0.1 (see Vennard and Street, op. cit., pp. 420, 421). Rounding of the inlet prevents formation of the vena contracta (see Fig. 5-36), thereby resulting in a smaller loss.

For **laminar flow** the friction loss due to a sudden contraction in the pipe may be expressed in terms of an equivalent length L_e of straight discharge pipe. Following Holmes (dissertation, University of Delft, 1967), L_e is

$$L_e/D = a + bN_{Re} \tag{5-124}$$

where D = diameter of smaller pipe; N_{Re} = Reynolds number in smaller pipe; a = Couette coefficient, dimensionless; and b = Hagenbach coefficient, dimensionless. Coefficients a,b should be independent of the pipe cross-sectional areas for $A_2/A_1 < 0.2$. The theoretical value of a = 0.30 [Roscoe, *Philos. Mag.*, **40,** 338–351 (1949); Weissberg, *Phys. Fluids,* **5,** 1033–1036 (1962); Mills, *J. Mech. Eng. Sci.*, **10,** 133–140 (1968); and Boger, Gutpa, and Tanner, *J. Non-Newtonian Fluid Mech.*, **4,** 239–248 (1978)]. An experimental value of a = 0.31 (Holmes, loc. cit.). Experimental values, determined with tubes discharging in air, average a = 0.39 [Miller and Neme-

TABLE 5-13 Sudden Contraction-Loss Coefficient for Turbulent Flow

A_2/A_1	0	0.2	0.4	0.6	0.8	1.0
K_c	0.5	0.45	0.36	0.21	0.07	0

cek, ASME Pap. 58-A-106 (1958); Chong, Christiansen, and Baer, *J. Appl. Polym. Sci.*, **15,** 369–379 (1971); and Boger and Binnington, *Trans. Soc. Rheol.*, **21,** 515–534 (1977)]. The coefficient a so determined includes an exit loss which amounts to 0.12 (Boger, Gupta, and Tanner, loc. cit.). The theoretical value of b = 0.0391 [Shah, *J. Fluids Eng.*, **100,** 177–179 (1978)]. An experimental value of b = 0.0363 [Kaye and Rosen, *Am. Inst. Chem. Eng. J.*, **17,** 1269–1270 (1971)].

For laminar flow in **rectangular ducts** the friction loss due to a sudden contraction in the duct may be expressed in terms of an equivalent length L_e of straight discharge duct as

$$L_e/h = a' + b'N_{Re} \tag{5-125}$$

where h = height of smaller rectangle; N_{Re} = Reynolds number based on smaller-channel hydraulic diameter; a' = coefficient, dimensionless; and b' = coefficient, dimensionless. Coefficients a', b' should be independent of the duct cross-sectional areas when the smaller-duct cross section is negligibly small by comparison with the larger-duct cross section. The theoretical value of a', for rectangles with large width-to-height aspect ratio, is 0.42 (Roscoe, loc. cit.). For the same ducts, an experimental value of a' = 0.51 (Holmes, loc. cit.). A theoretical value of b' = 0.0281 (Shah, loc. cit.). For friction loss in laminar flow through rectangular ducts of other aspect ratios and concentric annular ducts at high Reynolds number, see Shah, loc. cit.

The cause of the friction loss is the work required to set up the parabolic velocity distribution for the laminar flow through the pipe. The hydrodynamic **entrance length** L_{hy} required for the centerline velocity to reach 99 percent of its fully developed value in **laminar flow,** starting from a uniform velocity profile at the pipe entrance, was computed by Atkinson, Brocklebank, Card, and Smith [*Am. Inst. Chem. Eng. J.*, **15,** 548–553 (1969)] to be

$$L_{hy}/D = 0.59 + 0.056N_{Re} \tag{5-126}$$

See also Langhaar, *J. Appl. Mech.*, **9,** A55–A58 (1942); Shapiro, Siegel, and Kline, *Proc. Second U.S. Nat. Cong. Appl. Mech.*, ASME, 1954, pp. 733–741; and Chen, *J. Fluids Eng.*, **95,** 153–158 (1973). For a Reynolds number DG/μ = 2000, the entrance effect would persist for about 110 diameters. An increased frictional resistance near the pipe inlet arises from the same cause in **turbulent flow** [Olson and Sparrow, *Am. Inst. Chem. Eng. J.*, **9,** 766–770 (1963); and Ross, *Trans. Am. Soc. Mech. Eng.*, **78,** 915–923 (1956)]. The effect is of less importance in the turbulent region than in the laminar region. The entrance length for turbulent flow in large steel pipes with bell-mouth entrance is $50L_{hy}/D$ [Wang and Tullis, *J. Fluids Eng.*, **92,** 62–68 (1974)]. However, the wall shear stress in the same steel pipes attains its fully developed value in only about $15L_{hy}/D$. The hydrodynamic entrance length in laminar flow for a **rectangular duct,** with very large width-to-height ratio, was computed by Atkinson, Brocklebank, Card, and Smith (loc. cit.) to be

$$L_{hy}/h = 0.625 + 0.022N_{Re} \tag{5-127}$$

where the Reynolds number is based on the hydraulic diameter. See also Chen, loc. cit. For turbulent-flow development in a square duct, see Ahmed and Brundrett, *Int. J. Heat Mass Transfer*, **14,** 365–375 (1971). For turbulent-flow development in annuli, see Sridhar, Nicol, and Padmanabha, *J. Appl. Mech.*, **37,** 25–28 (1970).

For circular tubes, the **pressure drop for laminar flow in the entrance length** following a rounded inlet in a horizontal pipe can be estimated from the curve given in Fig. 5-38. With an abrupt or square-edged inlet, for Reynolds numbers less than about 500 but probably greater than 50, Fig. 5-38 applies; for Reynolds numbers greater than about 500 the pressure drop will be greater because of the vena contracta and eddies which are found behind the inlet. For further data, see Kreith and Eisenstadt, *Trans. Am. Soc. Mech. Eng.*, **79,** 1070 (1957). For other theoretical work, see Chen, loc. cit.; and Shah, loc. cit. In Fig. 5-38, p_0 = static pressure in the reservoir, p_L = static pressure in the pipe at distance L; and L = distance from the beginning of the straight pipe, i.e., from exit of the inlet.

For the pressure drop for laminar flow in the entrance length of **rectangular ducts,** see Chen, loc. cit.; and Shah, loc. cit.

For laminar flow of **nonnewtonian (power-law) fluids** the friction loss due to a sudden contraction in the pipe is higher than for newtonian fluids. In slow flow ($N_{Re} < 1$), the equivalent length L_e of a sudden contraction ($A_2/A_1 < 0.2$) is about $0.30/n$ discharge-pipe diameter, where n is the fluid exponent (from Boger, Gupta, and Tanner, loc. cit.). Laminar flow of nonnewtonian **viscoelastic liquids** through circular channels with sudden contraction is characterized by the development of a vortex upstream of the contraction plane. In slow flow ($N_{Re} < 1$) the vortex size increases with increasing flow

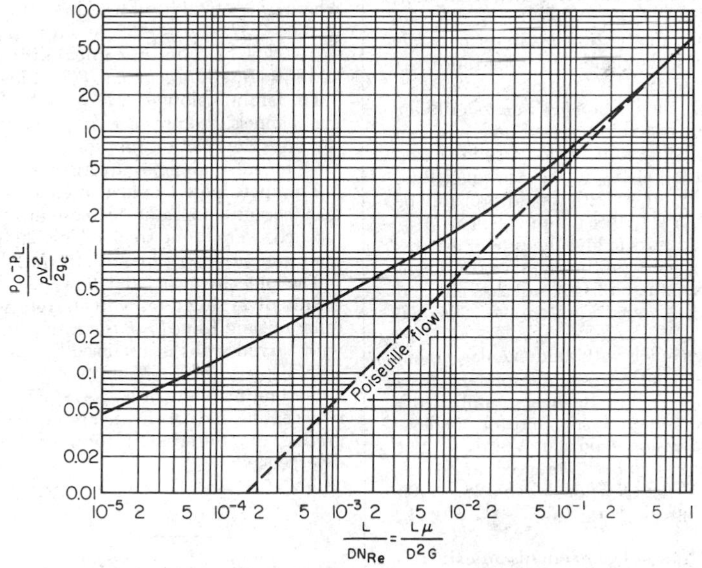

FIG. 5-38 Entrance effect in laminar flow. [*Based on Kreith and Eisenstadt,* Trans. Am. Soc. Mech. Eng., *79, 1070 (1957); and Shapiro, Siegel, and Kline,* Proc. Second Nat. Congr. Appl. Mech., *ASME, 1954, p. 733.*]

rate. At higher Reynolds number complex and distorted flow patterns are observed. For further details, see Cable and Boger, *Am. Inst. Chem. Eng. J.*, **24**, 869–879, 992–999 (1978), and **25**, 152–159 (1979). For flow at the inlet to an annular die, see Tan and Tiu, *J. Non-Newtonian Fluid Mech.*, **6**, 21–45 (1979). The pressure drop in flow of viscoelastic fluids through sharp-edged short (L/D from 1 to 10) orifices, discharging in air, is much higher than estimated from friction loss [see Bagley, *J. Appl. Phys.*, **28**, 624–627 (1957)].

For laminar slow flow ($N_{Re} < 1$) of fluids through **conical converging channels,** the pressure drop Δp due to friction between the ends of the converging section may be estimated by integrating the differential form of Eq. (5-73), provided the angle between the converging walls is small. For nonnewtonian (power-law) fluids

$$\Delta p/4(L/D)_e = K'(32q/\pi D_2^3)^n \tag{5-128}$$

$$\left(\frac{L}{D}\right)_e = \frac{1}{6n \tan(\alpha/2)}\left[1 - \left(\frac{D_2}{D_1}\right)^{3n}\right] \tag{5-129}$$

where D_1 = inlet diameter, D_2 = exit diameter, α = total angle between converging walls, q = volume flow rate, and K', n = material constants [see Eqs. (5-74) and (5-75)]. Measurements agree with predictions for channels with $\alpha < 11°$; see Kemblowski and Kiljanski, *Chem. Eng. J. (Lausanne)*, **9**, 141–151 (1975). The pressure drop in slow flow ($N_{Re} < 1$) through **two-dimensional converging channels** with **straight walls** may be estimated by integrating the differential form of Eq. (5-82), provided the angle between the converging walls is small.

Enlargement and Exit Losses For ducts of any cross section, the frictional loss for a **sudden enlargement** (Fig. 5-39) with **turbulent flow** is given by the simple Borda-Carnot equation,

$$F = \frac{(V_1 - V_2)^2}{2g_c} = \frac{V_1^2}{2g_c}\left(1 - \frac{A_1}{A_2}\right)^2 \tag{5-130}$$

where V_1 = velocity in smaller duct, V_2 = velocity in larger duct, A_1 = cross-sectional area of smaller duct, and A_2 = cross-sectional area of larger duct. Equation (5-130) is almost exact for water in turbulent flow (see Benedict et al., loc. cit.) and applies fairly well to the turbulent flow of gases and vapors with velocities less than 60 m/s (about 200 ft/s) [Kays, *Trans. Am. Soc. Mech. Eng.*, **72**, 1067–1074 (1950)]. For compressible subsonic and supercritical air flow, Benedict, Wyler, Dudek, and Gleed [*J. Eng. Power*, **98**, 327–334 (1976)] present tables for computing the friction loss in terms of the total pressure ratio across a sudden enlargement for area ratios A_1/A_2 from 0.1 to 0.5.

For slow **laminar flow** ($N_{Re} < 1$) the friction loss due to a sudden enlargement in the duct may be expressed in terms of an equivalent length of the smaller duct, which is the same as for a sudden contraction [see Eqs. (5-124) and (5-125), at $N_{Re} < 1$]. At higher Reynolds number the result obtained by Eq. (5-130) should probably be doubled. For a pipe with $A_1/A_2 < 0.1$, this is equivalent to b = 0.0313. At least, such a procedure appears to give the proper exit loss for the limiting case in which the velocity in the discharge or downstream reservoir is negligible [see Kreith and Eisenstadt, loc. cit.; and Willoughby and Kittle, *Ind. Eng. Chem. Fundam.*, **6**, 304–306 (1967)].

Flow of fluids through circular channels with sudden enlargement is characterized by the development of a vortex (secondary flow) downstream of the expansion plane. At low Reynolds number, the size of the vortex increases with expansion A_2/A_1, and for nonnewtonian (power-law) fluids with decreasing exponent. Vortex lengths up to nine upstream tube diameters have been observed, at $N_{Re} \sim 200$, $n > 0.65$ [see Macagno and Hung, *J. Fluid Mech.*, **28**, 43–64 (1967); and Halmos, Boger, and Cabelli, *Am. Inst. Chem. Eng. J.*, **21**, 540–549 (1975)].

Jetting of a liquid in slow flow ($N_{Re} < 1$) also entails **an exit friction loss,** which for pipes is equivalent to a pipe length of 0.12 diameter (see Boger, Gupta and Tanner, loc. cit.). For **nonnewtonian (power-law) liquids** the exit loss increases slightly with decreasing exponent. The cause of the friction loss is the rearrangement of the velocity profile near the pipe exit.

When nonnewtonian **viscoelastic liquids** are extruded through a die at a low Reynolds number, the extrudate expands to a diameter greater than the die diameter. The phenomenon, called **die swell,** is most pronounced with short dies [see Graessley, Glasscock, and Crawley, *Trans. Soc. Rheol.*, **14**, 519–544 (1970)]. For velocity-distribution measurements near the die exit, see Goulden and MacSporran, *J. Non-Newtonian Fluid Mech.*, **1**, 183–198 (1976); and Whipple and Hill, *Am. Inst. Chem. Eng. J.*, **24**, 664–671 (1978). The least swell measured with any liquid ($N_{Re} < 1$) is an extrudate-to-die-diameter ratio of about 1.10 [see Graessley, Glasscock, and Crawley, loc. cit.; and Batchelor, Berry, and Horsfall (*Polymer*, **14**, 297–299 (1973)]. For a theoretical computation of the die swell of newtonian liquids, see Allan, *Int. J. Numer. Methods Eng.*, **11**, 1621–1626 (1977). At higher flow rates the extrudate becomes distorted, reflecting the complex flow pattern at the die inlet (see Cable and Boger, loc. cit.). With molten polymers the phenomenon is called **melt fracture.** It occurs at a wall shear stress of about 10^5N/m^2 (14.5 lbf/in^2) [Tordella, *Rheol. Acta*, **1**(2–3), 216–221 (1958)].

If the transition from a small to a large duct of any cross-sectional shape is accomplished by a **uniformly diverging duct** (Fig. 5-40) with a straight axis, the total pressure change Δp between the ends of the diverging section may be estimated by integrating the differential form of Eq. (5-57), provided that the total angle α between the diverging walls is not greater than about 7°. Gibson (*Hydraulics and Its Applications*, 5th ed., Constable, London, 1952, p. 93) found for **turbulent flow,** based on his experiments with water flowing in conical diverging ducts of area ratios from 1:2.5 to 1:9, that the loss as a function of α decreases as α increases up to about 5 to 7° and then increases rapidly with increasing α. For α above 35°, the loss often exceeds considerably that accompanying a sudden expansion, reaching a maximum at around 60°; it then gradually decreases to that for a sudden expansion at α = 180°. A general and simple formulation for $\alpha > 7°$ cannot be given, since the measured losses are influenced by angle of divergence and ratio of expansion, the value of N_{Re}, nature of the initial velocity distribution, and length and cross-sectional shape of the downstream passage. As a guide, Gibson's results indicate that

$$\Delta h = K\frac{(V_1 - V_2)^2}{2g_c} \tag{5-131}$$

where Δh = total head loss, $K \cong 0.13$ for $\alpha \cong 5$ to 7°, $K = 0.0110\,\alpha^{1.22}$ for $7.5° < \alpha < 35°$, and α is in degrees.

The flow behavior in **conical diffusers** is discussed by McDonald and Fox [*Int. J. Mech. Sci.*, 8(2), 125–139 (1966)].

For **laminar slow flow** ($N_{Re} < 1$) of fluids through conical diverging channels, the pressure drop due to friction between the ends of the diverging section may be estimated from Eqs. (5-128) and (5-129), for small angle of divergence.

Trumpet-shaped enlargements for **turbulent flow** so designed that there is a constant decrease in velocity head per unit length of pipe axis were found to give from 20 to 60 percent less frictional loss than straight-taper pipes of the same length (Gibson, op. cit., p. 95).

The pressure drop and flow behavior in **two-dimensional uniformly diverging duct** with **straight walls** are discussed by Reneau, Johnston, and Kline [*J. Basic Eng.*, **89**, 141–150 (1967)], and the case with **curved walls** is discussed by Sagi and Johnston [ibid., 715–731 (1967)].

For **laminar slow flow** ($N_{Re} < 1$), the pressure drop due to friction between the ends of the diverging section may be estimated by integrating the differential form of Eq. (5-82) for small angle of divergence.

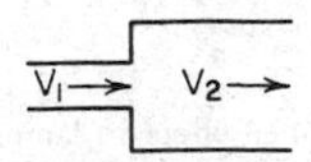

FIG. 5-39 Sudden enlargement.

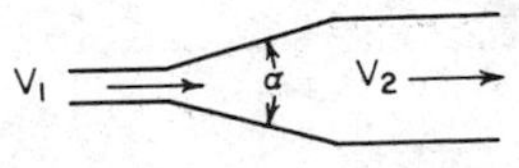

FIG. 5-40 Uniformly diverging duct.

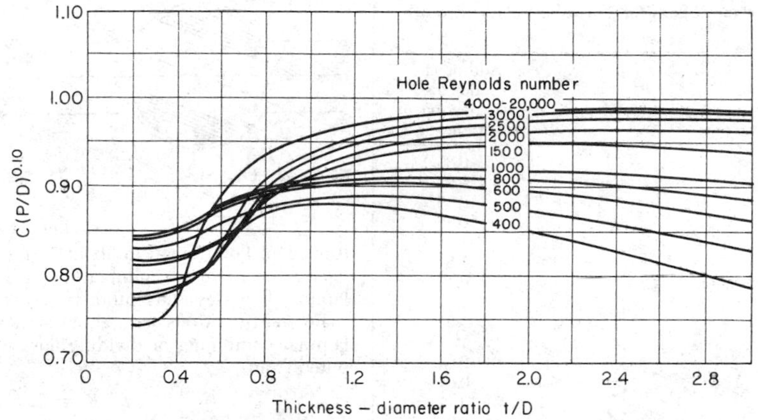

FIG. 5-41 Perforated-plate orifice coefficient versus hole Reynolds number and physical characteristics of plate. [*Smith and Van Winkle,* Am. Inst. Chem. Eng. J., *4, 266 (1958).*]

Orifices and Perforated Plates For a **concentric circular square-edged orifice** the overall frictional loss or permanent pressure drop for **turbulent flow** can be estimated by Eq. (5-24). If the flow is **laminar** in the upstream channel, data on short tubes (Kreith and Eisenstadt, loc. cit., p. 1074) indicate that the overall frictional loss F is equal to the orifice differential expressed in feet of fluid flowing. In slow laminar flow, with orifice $N_{Re} < 1$, the permanent pressure drop expressed in terms of an equivalent length L_e of straight pipe, of the same diameter as the orifice, is twice that obtained from Eq. (5-124), or 0.60 orifice diameter.

For a **rounded orifice** or **well-shaped nozzle** placed in a pipe line, the overall frictional loss for **turbulent flow** can be estimated by applying Eq. (5-17). For **laminar flow,** the overall frictional loss can be estimated from Fig. 5-38, taking the equivalent length L_e of a rounded inlet as about 0.2 orifice or nozzle exit diameter.

For **annular orifices,** the overall frictional loss for annulus Reynolds numbers in the range of 100 to 20,000 can be estimated by applying Eq. (5-14) together with the coefficient K given in the subsection "Orifice Meters."

For **perforated plates,** the permanent pressure drop can be computed from the correlation by Van Winkle et al. They correlated within ±5 percent the pressure-drop data of their own and of others for the flow of gases through **perforated plates with square-edged holes** on an equilateral triangular spacing for a **hole Reynolds-number range of 400 to 20,000 and P/D from 2 to 5.** The basic equation is

$$w = CA_fY\sqrt{\frac{2g_c\rho_1\,\Delta p}{1-(A_f/A_p)^2}} \tag{5-132}$$

where w = weight rate of flow; C = orifice coefficient described later, dimensionless; A_f = total free area of holes; A_p = total cross-sectional area of perforated plate; Y = expansion factor [Eq. (5-23)], dimensionless; g_c = dimensional constant; ρ_1 = fluid density at upstream pressure and temperature; and Δp = pressure drop across the plate. Orifice coefficient C as a function of Reynolds number and physical characteristics of the plate is given in Fig. 5-41 [Smith and Van Winkle, *Am. Inst. Chem. Eng. J.*, **4**, 266–268 (1958); see also Kolodzie and Van Winkle, *Am. Inst. Chem. Eng. J.*, **3**, 305–312 (1957)], where P = hole pitch (center-to-center distance); D = hole diameter; t = plate thickness; N_{Re} = Reynolds number based on hole diameter = $wD/A_f\mu$, dimensionless; and μ = fluid viscosity. If the inlet edge of the hole is slightly rounded, as may be the case for punched holes, the coefficient can be considerably higher than that given by Fig. 5-41. An indication of the effect of rounding can be seen from preceding information on entrance losses.

Fittings and Valves For **turbulent flow,** the additional frictional loss for fittings and valves can be allowed for by expressing the loss either as an equivalent length of straight pipe in pipe diameter, L_e/D, or as a number of velocity heads K lost in a pipe of the same size. K is defined as

$$K = \frac{\Delta F}{V^2/2g_c} \tag{5-133}$$

where ΔF = additional frictional loss (total frictional loss less frictional loss for centerline length of straight pipe), V = average fluid velocity, and g_c = dimensional constant. Quantities L_e/D and K are not entirely comparable, but both are accurate within the limits of available data or differences in details of available commercial fittings and valves. Theoretically K would be constant for all sizes of a given design of fitting or valve if all sizes were geometrically similar; however, geometric similarity is seldom achieved [De Craene, *Heat. Piping Air Cond.*, **27**(10), 90–95 (1955)]. Data indicate that resistance K tends to decrease with increasing fitting or valve size [De Craene, loc. cit.; and Pigott, *Trans. Am. Soc. Mech. Eng.*, **72,** 679–688 (1950)].

Representative values of K for many types of fittings and valves are given in Table 5-14. Approximate values of L_e/D can be obtained by multiplying K by 45 for liquids similar to water and by 55 for gases similar to air. Most of the values given are for standard screwed fittings and are probably accurate within ±30 percent. The difference in frictional loss between screwed, flanged, and welded ends has been found to be insignificant (De Craene, loc. cit.). Manufacturers and users of valves, especially control valves, have found it convenient to express valve capacity in terms of a flow coefficient C_v. This coefficient is related to K by

$$C_v = C_1d^2/\sqrt{K} \tag{5-134}$$

where C_v = valve-flow coefficient, gal/min of water at 60°F flowing under a valve pressure drop of 1 lbf/in²; d = inside diameter of valve, in; and C_1 = 29.9. Equivalent conditions in SI units are C_v = valve-flow coefficient, L/min of water at 15.5°C flowing under a valve pressure drop of 6.895 kPa; d = inside diameter of valve, cm; and C_1 = 17.54.

For **laminar flow,** data for the frictional loss of fittings and valves are meager [Beck and Miller, *J. Am. Soc. Nav. Eng.*, **56,** 62–83 (1944); Beck, ibid., **56,** 235–271, 366–388, 389–395 (1944); De Craene, loc. cit.; Karr and Schutz, *J. Am. Soc. Nav. Eng.*, **52,** 239–256 (1940); and Kittredge and Rowley, *Trans. Am. Soc. Mech. Eng.*, **79,** 1759–1766 (1957)]. The data of Kittredge and Rowley (loc. cit.) indicate that the additional frictional loss expressed as number of velocity heads K is constant for Reynolds numbers from over 2000

TABLE 5-14 Additional Frictional Loss for Turbulent Flow through Fittings and Valves[a]

Type of fitting or valve	Additional friction loss, equivalent no. of velocity heads, K
45° ell, standard[b,c,d,e,f]	0.35
45° ell, long radius[c]	0.2
90° ell, standard[b,c,e,f,g,h]	0.75
Long radius[b,c,d,e]	0.45
Square or miter[h]	1.3
180° bend, close return[b,c,e]	1.5
Tee, standard, along run, branch blanked off[e]	0.4
Used as ell, entering run[g,i]	1.0
Used as ell, entering branch[c,g,i]	1.0
Branching flow[i,j,k]	1[l]
Coupling[c,e]	0.04
Union[e]	0.04
Gate valve,[b,e,m] open	0.17
¾ open[n]	0.9
½ open[n]	4.5
¼ open[n]	24.0
Diaphragm valve,[o] open	2.3
¾ open[n]	2.6
½ open[n]	4.3
¼ open[n]	21.0
Globe valve,[e,m] bevel seat, open	6.0
½ open[n]	9.5
Composition seat, open	6.0
½ open[n]	8.5
Plug disk, open	9.0
¾ open[n]	13.0
½ open[n]	36.0
¼ open[n]	112.0
Angle valve,[b,e] open	2.0
Y or blowoff valve,[b,m] open	3.0
Plug cock[p] (Fig. 5-42) θ = 5°	0.05
10°	0.29
20°	1.56
40°	17.3
60°	206.0
Butterfly valve[p] (Fig. 5-43) θ = 5°	0.24
10°	0.52
20°	1.54
40°	10.8
60°	118.0
Check valve,[b,e,m] swing	2.0[q]
Disk	10.0[q]
Ball	70.0[q]
Foot valve[e]	15.0
Water meter,[h] disk	7.0[r]
Piston	15.0[r]
Rotary (star-shaped disk)	10.0[r]
Turbine-wheel	6.0[r]

[a]Lapple, *Chem. Eng.*, **56**(5), 96–104 (1949), general survey reference.
[b]"Flow of Fluids through Valves, Fittings, and Pipe," Tech. Pap. 410, Crane Co., 1969.
[c]Freeman, *Experiments upon the Flow of Water in Pipes and Pipe Fittings*, American Society of Mechanical Engineers, New York, 1941.
[d]Giesecke, *J. Am. Soc. Heat. Vent. Eng.*, **32**, 461 (1926).
[e]*Pipe Friction Manual*, 3d ed., Hydraulic Institute, New York, 1961.
[f]Ito, *J. Basic Eng.*, **82**, 131–143 (1960).
[g]Giesecke and Badgett, *Heat. Piping Air Cond.*, **4**(6), 443–447 (1932).
[h]Schoder and Dawson, *Hydraulics*, 2d ed., McGraw-Hill, New York, 1934, p. 213.
[i]Hoopes, Isakoff, Clarke, and Drew, *Chem. Eng. Prog.*, **44**, 691–696 (1948).
[j]Gilman, *Heat. Piping Air Cond.*, **27**(4), 141–147 (1955).
[k]McNown, *Proc. Am. Soc. Civ. Eng.*, **79**, Separate 258, 1–22 (1953); discussion, ibid., **80**, Separate 396, 19–45 (1954). For the effect of branch spacing on junction losses in dividing flow, see Hecker, Nystrom, and Qureshi, *Proc. Am. Soc. Civ. Eng., J. Hydraul. Div.*, **103**(HY3), 265–279 (1977).
[l]This is pressure drop (including friction loss) between run and branch, based on velocity in the mainstream before branching. Actual value depends on the flow split, ranging from 0.5 to 1.3 if mainstream enters run and from 0.7 to 1.5 if mainstream enters branch.
[m]Lansford, *Loss of Head in Flow of Fluids through Various Types of 1½-in. Valves*, Univ. Eng. Exp. Sta. Bull. Ser. 340, 1943.
[n]The fraction open is directly proportional to stem travel or turns of hand wheel. Flow direction through some types of valves has a small effect on pressure drop (see Freeman, op. cit.). For practical purposes this effect may be neglected.
[o]Streeter, *Prod. Eng.*, **18**(7), 89–91 (1947).
[p]Gibson, *Hydraulics and Its Applications*, 5th ed., Constable, London, 1952, p. 250.
[q]Values apply only when check valve is fully open, which is generally the case for velocities more than 1 m/s (about 3 ft/s) for water.
[r]Values should be regarded as approximate because there is much variation in equipment of the same type from different manufacturers.

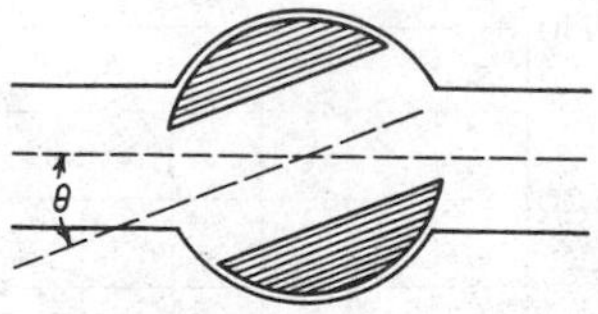

FIG. 5-42 Plug cock.

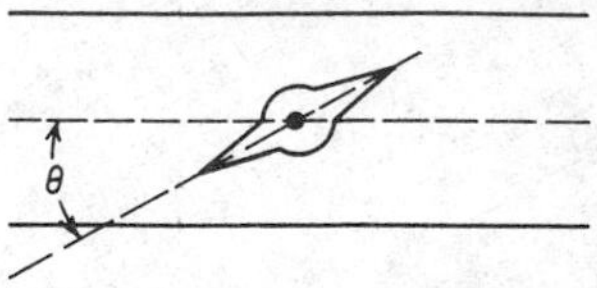

FIG. 5-43 Butterfly valve.

(turbulent flow) down to about 500; then K increases rapidly with decreasing Reynolds number. Typical values of K [Eq. (5-133) for laminar-flow Reynolds numbers N_{Re} are given in Table 5-15.

For **tee junctions** with equal area along the run and branch, in laminar combining or dividing flow, K [Eq. (5-133)] may be computed from

$$K = C/N_{Re} \qquad (5\text{-}135)$$

where C = coefficient given in Table 5-16, dimensionless [Jamison and Villemonte, *Proc. Am. Soc. Civ. Eng., J. Hydraul. Div.*, **97** (HY7), 1045–1063 (1971)]. In Eqs. (5-133) and Eq. (5-135), ΔF, V, and N_{Re} refer to the stream components prior to combination or division. The combined fitting loss may be computed from

$$\Delta F_m = \Delta F_1(V_1/V_m) + \Delta F_2(V_2/V_m) \qquad (5\text{-}136)$$

where ΔF_m = additional friction loss for combined stream, ΔF_1 = additional friction loss for flow along run, ΔF_2 = additional friction loss for flow in branch, V_1 = average fluid velocity of stream along run prior to junction in combining flow or after junction in dividing flow, V_2 = average fluid velocity of stream in branch prior to junction in combining flow or after junction in dividing flow; and V_m = average fluid velocity of combined stream.

Bends and Curved Pipe For **turbulent flow** in **smooth 90° bends** (Fig. 5-44*a*) and **segmental 90° bends** (Fig. 5-44*b*), total friction loss expressed as the ratio of equivalent length of straight pipe L_e to pipe diameter D as a function of the ratio of radius of curvature R to pipe diameter D, all in consistent units, is given in Fig. 5-45. The curve for smooth bends is based on various published data (see Fig. 5-45 for references) and represents most of the data with an uncertainty of probably ±25 percent. The curves for segmental bends are based on few data. For a 45° bend, total friction loss is about 65 percent of the loss for a 90° bend of a proportional number of segments, and similarly for a 180° bend, the loss is about 140 percent of that for a 90° bend, based on information presented by Conn, Colborne, and Brown [*Heat. Piping Air Cond.*, **25**(1), 201–205 (1953)], Ito (loc. cit.); Jorgensen (*Fan Engineering*, 7th ed., Buffalo Forge Co., Buffalo, 1970, p. 112); and Snyder (loc. cit.).

For flow through a **curved pipe** or **coil,** a secondary circulation of fluid called the double-eddy or Dean effect takes place in a plane at right angles to the main flow. Because of this circulation the friction loss in the curved pipe is greater than in an equal length of straight pipe. This circulation also stabilizes laminar flow, thus increasing the critical Reynolds number. The maximum Reynolds number, or **critical Reynolds number,** for laminar flow as a function of pipe diameter and coil diameter is given by Srinivasan, Nandapurkar, and Holland [*Chem. Eng. (London)*, No. 218, CE113–CE119 (May 1968)]:

$$(N_{Re})_{crit} = 2100(1 + 12\sqrt{D/D_c}) \qquad (5\text{-}137)$$

TABLE 5-15 Additional Frictional Loss for Laminar Flow through Fittings and Valves*

Type of fitting or valve	Additional frictional loss expressed as K			
	N_{Re} = 1000	500	100	50
90° ell, short radius	0.9	1.0	7.5	16
Gate valve	1.2	1.7	9.9	24
Globe valve, composition disk	11	12	20	30
Plug	12	14	19	27
Angle valve	8	8.5	11	19
Check valve, swing	4	4.5	17	55

*From curves by Kittredge and Rowley, *Trans. Am. Soc. Mech. Eng.*, **79**, 1759–1766 (1957).

TABLE 5-16 Additional Frictional Loss for Laminar Flow through Junctions

Type of flow	Additional frictional loss coefficient C	
	Along run	Branch
Combined, V_1/V_m = 1 to ¼	2100	
V_2/V_m = 0 to ¾		7300
Divided, V_1/V_m = 1, ¾	2100	
½	3650	
¼	6400	
V_2/V_m = 0 to ¾		7000

NOTE: V_1 = average velocity of stream along run; V_2 = average velocity of stream in branch; V_m = average velocity of combined stream.

for $10 < D_c/D < 250$, where $(N_{Re})_{crit} = (DG/\mu)_{crit}$ = critical Reynolds number, dimensionless; D = pipe diameter; D_c = coil diameter; G = mass velocity; and μ = fluid viscosity.

Total friction loss for **laminar flow** in **curved pipe** can be expressed in terms of an equivalent length L_e of straight pipe. Ratio of the equivalent to actual coil centerline length L_e/L is a function of the Dean number or $N_{Re}\sqrt{D/D_c}$ as shown in Fig. 5-46 (see also Srinivasan et al., loc. cit.). This curve is accurate to within ±5 percent. A summary of published theoretical work plus their theoretical analysis of pressure drop in laminar flow in a coiled tube is given by Larrain and Bonilla [*Trans. Soc. Rheol.*, **14**(2), 135–147 (1970)]. Theoretical analysis by Austin and Seader [*Am. Inst. Chem. Eng. J.*, **19**, 85–94 (1973)] indicates that, for $5 < D_c/D < 100$, a small second-order dependence upon the curvature ratio D_c/D exists beyond that shown by the correlation of Fig. 5-46.

The **friction loss** for **turbulent flow** can be computed from the Fanning equation, Eq. (5-57), where for industrial helices the friction factor f_c is given by the empirically determined equation (see Srinivasan et al., loc. cit.):

$$f_c = 0.08N_{Re}^{-0.25} + 0.01(D/D_c)^{0.5} \qquad (5\text{-}138)$$

Equation (5-138) is probably accurate within ±10 percent.

The pressure drop for flow in **spirals** is discussed by Srinivasan et al. (loc. cit.) and by Ali and Seshadri [*Ind. Eng. Chem. Process Des. Dev.*, **10**, 328–332 (1971)]. For friction loss in laminar flow through *semicircular ducts*, see Masliyah and Nandakumar, *Am. Inst. Chem. Eng. J.*, **25**, 478–487 (1979); for curved **channels of square section,** see Baylis, *J. Fluid Mech.*, **48**, 417–422 (1971); and for **curved rectangular channels,** see Cheng, Lin, and Ou, *J. Fluids Eng.*, **98**, 41–48 (1976).

For flow of **nonnewtonian (power-law) fluids** through coiled tubes, Mashelkar and Devarajan [*Trans. Inst. Chem. Eng. (London)*, **54**, 108–114 (1976)] propose the correlation

$$f_c = (9.07 - 9.44n + 4.37n^2)(D/D_c)^{0.5}N_{De}^{-0.768 + 0.122n} \qquad (5\text{-}139)$$

$$N_{De} = \frac{1}{8}\left(\frac{6n + 2}{n}\right)^n N'_{Re}\sqrt{\frac{D}{D_c}} \qquad (5\text{-}140)$$

where N_{De} = modified Dean number, dimensionless; N'_{Re} = generalized Reynolds number defined by Eq. (5-77); and n = material constant [see Eq. (5-75)]. This correlation was tested for the range N_{De} = 70 to 400, D/D_c = 0.01 to 0.135, and n = 0.35 to 1. See also Oliver and Asghar, *Trans. Inst. Chem. Eng. (London)*, **53**, 181–186 (1975).

Screens The flow through a screen can be considered as flow through a number of orifices or nozzles in parallel. Thus the pressure

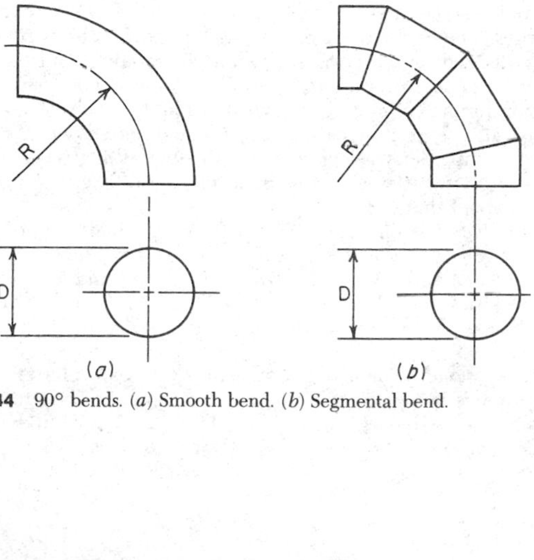

FIG. 5-44 90° bends. (*a*) Smooth bend. (*b*) Segmental bend.

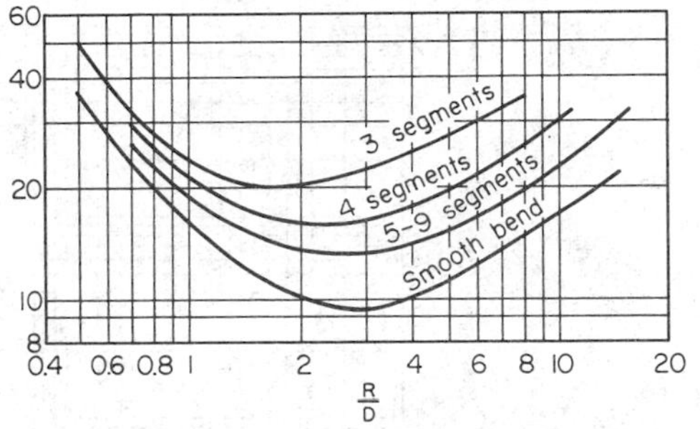

FIG. 5-45 Total friction loss in 90° bends. [*Smooth bend: based on information from Freeman,* Experiments upon the Flow of Water in Pipes and Pipe Fittings, *ASME, New York, 1941, p. 173; Ito,* J. Basic Eng., *82, 131 (1960); Locklin,* Trans. Am. Soc. Heat. Vent. Eng., *56, 479 (1950); Snyder,* Heat. Piping Air Cond., *7(1), 5 (1935). Segmental bends: from Locklin, loc. cit.*]

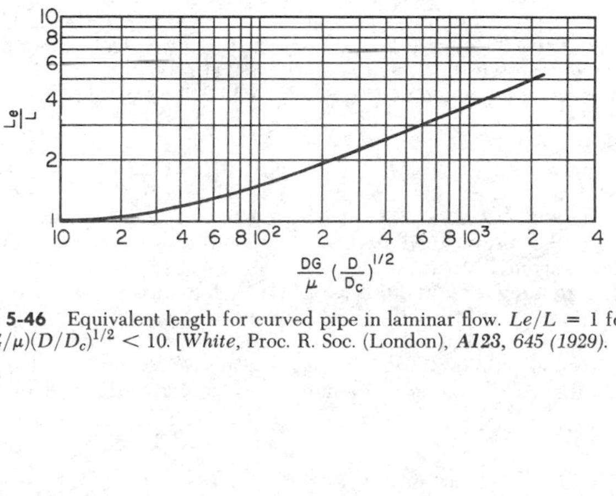

FIG. 5-46 Equivalent length for curved pipe in laminar flow. Le/L = 1 for $(DG/\mu)(D/D_c)^{1/2} < 10$. [*White,* Proc. R. Soc. (London), *A123, 645 (1929).*]

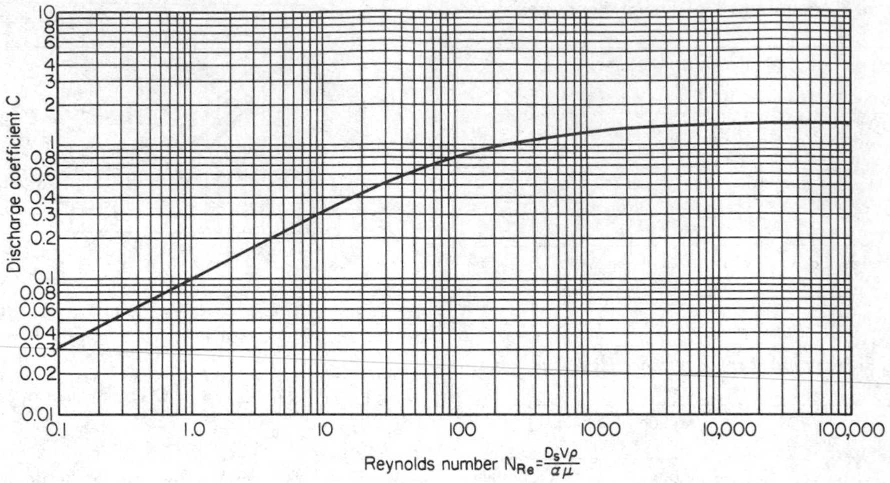

FIG. 5-47 Screen discharge coefficients, plain square-mesh screens. *(Courtesy of E. I. du Pont de Nemours & Co.)*

drop or head loss across a screen can be computed from an orifice-type equation. The resulting equation for head loss is

$$\Delta h = \left(\frac{n}{C^2}\right)\left(\frac{1 - \alpha^2}{\alpha^2}\right)\left(\frac{V^2}{2g_c}\right) \qquad (5\text{-}141)$$

where Δh = head loss, or F; n = number of screens in series, dimensionless; C = screen discharge coefficient, dimensionless; α = fractional free projected area of screen, dimensionless; V = superficial velocity ahead of screen; and g_c = dimensional constant. Experimental data (Grootenhuis, *Proc. Inst. Mech. Eng. (London)*, **A168**, 837–846 (1954)] indicate that for a series of screens the overall head loss is directly proportional to the number of screens in series, as given by Eq. (5-141), and is not affected by either the spacing between successive screens or by their orientation with respect to one another.

Screen discharge coefficient C is a function of screen Reynolds number, $N_{Re} = D_s V\rho/\alpha\mu$, where D_s = aperture width, ρ = fluid density, and μ = fluid viscosity. For **plain square-mesh screens,** α = 0.14 to 0.79, Lapple's plot of C versus N_{Re} is given in Fig. 5-47 (courtesy of E. I. du Pont de Nemours & Co.). This curve represents most of the data to within ±20 percent. In the laminar-flow region, $N_{Re} < 20$, the discharge coefficient can be computed from

$$C = 0.1\sqrt{N_{Re}} \qquad (5\text{-}142)$$

Coefficients greater than 1 probably indicate that the effective free area is larger than that of the projected area and that there is partial recovery of head due to the downstream rounding of the wires.

A correlation of overall frictional losses across plain square-mesh screens and sintered gauzes is given by Grootenhuis (loc. cit.). A correlation based on a packed-bed model for plain, twill, and "dutch" weaves is presented by Armour and Cannon [*Am. Inst. Chem. Eng. J.*, **14**, 415–420 (1968)]. For friction loss through monofilament fabrics see Pedersen, *Filtr. Sep.*, **11**, 586–589 (1974).

For screens inclined at angle θ to the flow direction use

$$V = V \cos\theta \qquad (5\text{-}143)$$

[Carothers and Baines, *J. Fluids Eng.*, **97**, 116–117 (1975)]. This modification to Eq. (5-141) applies for $N_{Re} > 500$, $C = 1.26$, $\alpha \le 0.97$, and $0 \le \theta \le 45$, for square-mesh screens and diamond-mesh netting. Screens inclined at an angle to the flow direction experience also a tangential stress.

For **nonnewtonian fluids** ($N_{Re} < 1$), the friction loss across a square-woven or full-twill-woven screen can be computed by considering the screen as a set of parallel tubes, each of diameter equal to the average minimal opening between adjacent wires, and length twice the wire diameter, without entrance effects [Carley and Smith, *Polym. Eng. Sci.*, **18**, 408–415 (1978)]. For screen stacks the friction loss of individual screens should be added.

Baffles For **segmental baffles,** such as tube-bundle baffles in heat exchangers, the overall friction loss for **turbulent flow** can be computed from

$$F = 2n_B(V_B^2/2g_c) = (n_B w^2/g_c\rho^2 A_B^2) \qquad (5\text{-}144)$$

where F = overall friction loss; n_B = number of baffles in series, dimensionless; V_B = fluid velocity through baffle opening (based on A_B); w = weight rate of flow; A_B = net free area of baffle opening; ρ = fluid density; and g_c = dimensional constant. Equation (5-144) is equivalent to treating the baffle opening as a segmental orifice with an overall discharge coefficient of about 0.7. No allowance is made for leakage, such as that between baffle and shell or between baffle and tubes.

TWO-PHASE FLOW

Liquids and Gases For **cocurrent flow with constant liquid-gas ratios,** considerable experimental and theoretical work has been done on prediction of pressure drop, volume fractions, and flow pattern for flow in pipes. A reliable general correlation has not as yet been developed, although correlations for specific flow systems have been published. Presented here are **guides** for the estimation of flow pattern, pressure drop, and volume fractions for flow in horizontal and vertical pipes.

In **horizontal pipe,** flow patterns have been reported in the literature and correlated empirically as functions of flow rates and flow properties. The boundaries between flow patterns, however, are not sharply defined, because the transitions are gradual and the mean boundaries depend upon interpretations of individual investigators and upon piping configurations and fluids under study. The following general types of **flow pattern** have been reported, where the values of the superficial velocities given are representative values for liquids with viscosities less than about 0.1 Pa·s (100 cP) and gases of densities about that of air:

1. *Bubble or froth flow.* This pattern, in which bubbles of gas are dispersed throughout the liquid, occurs for liquid superficial velocities from about 1.5 to 4.5 m/s (5 to 15 ft/s) and for gas superficial velocities from about 0.3 to 3 m/s (1 to 10 ft/s). (See subsection "Terminology in Fluid Mechanics" for definition of superficial velocity.)

2. *Plug flow.* This pattern, in which alternate plugs of liquid and gas move along the upper part of the pipe, occurs for liquid superficial velocities less than about 0.6 m/s (2 ft/s) and for gas superficial velocities less than about 0.9 m/s (3 ft/s).

3. *Stratified flow.* This pattern, in which the liquid flows along the bottom of the pipe and the gas flows over a smooth liquid-gas interface, occurs for liquid superficial velocities less than about 0.15 m/s (0.5 ft/s) and for gas superficial velocities from about 0.6 to 3 m/s (2 to 10 ft/s).

4. *Wavy flow.* This pattern is similar to stratified flow except that the interface has waves traveling in the direction of flow. This occurs for liquid superficial velocities less than about 0.3 m/s (1 ft/s) and for gas superficial velocities about 4.5 m/s (15 ft/s).

5. *Slug flow.* In this pattern a wave is picked up periodically by the rapidly moving gas to form a frothy slug which passes along the pipe at a greater velocity than the average liquid velocity. In this type of flow, slugs can cause severe and, in some cases, dangerous vibrations in equipment because of impact of the high-velocity slugs against such fittings as return bends.

6. *Annular flow.* In this pattern the liquid flows as a film around the pipe inside wall and the gas flows as a core. A portion of the liquid is entrained as a spray by the central gas core. This type of flow occurs for gas superficial velocities greater than about 6 m/s (20 ft/s). Determinations of entrainment are reported by Wicks and Dukler [*Am. Inst. Chem. Eng. J.*, **6**, 463–468 (1960)] and by Magiros and Dukler (in Lay and Malvern, *Developments in Mechanics*, vol. 1, Plenum, New York, 1961, pp. 532–553). Ishii and Grolmes [*Am. Inst. Chem. Eng. J.*, **21**, 308–318 (1975)] give inception criteria for droplet entrainment. Equations for estimating average droplet diameter are given by Tatterson, Dallman, and Hanratty [*Am. Inst. Chem. Eng. J.*, **23**, 68–76 (1977)].

7. *Spray or dispersed flow.* This pattern, in which nearly all the liquid is entrained as fine droplets by the gas, probably occurs for gas superficial velocities greater than 60 m/s (200 ft/s). The interaction between an air stream and a moving horizontal water surface is described by Hanratty and Engen [*Am. Inst. Chem. Eng. J.*, 3, 299–304 (1957)].

The flow pattern may be predicted from the correlation proposed by Baker [*Oil Gas J.*, 53(12), 185–190, 192, 195 (1954)] as shown in Fig. 5-48. In this figure,

$$\lambda = (\rho_G' \rho_L')^{1/2} \tag{5-145}$$

$$\psi = \frac{1}{\sigma'}\left[\frac{\mu_L'}{(\rho_L')^2}\right]^{1/3} \tag{5-146}$$

where G_G = gas mass velocity; G_L = liquid mass velocity; μ_L' = ratio of liquid viscosity to water viscosity, dimensionless; ρ_G' = ratio of gas density to air density, dimensionless; ρ_L' = ratio of liquid density to water density, dimensionless; σ' = ratio of liquid surface tension to water surface tension, dimensionless; and air and water properties are at 20° C (68°F) and 101.3 kPa (14.7 lbf/in²).

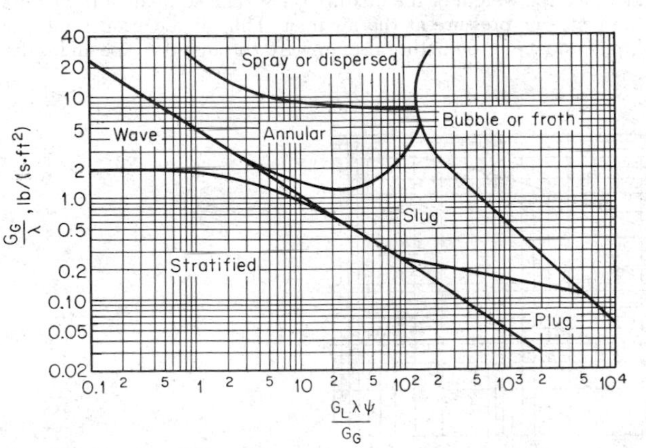

FIG. 5-48 Flow-pattern regions in cocurrent liquid-gas flow through horizontal pipes. To convert pounds per second per square foot to kilograms per second per square meter, multiply by 4.8824. [*From Baker,* Oil Gas J., *53(12), 185–190, 192, 195 (1954).*]

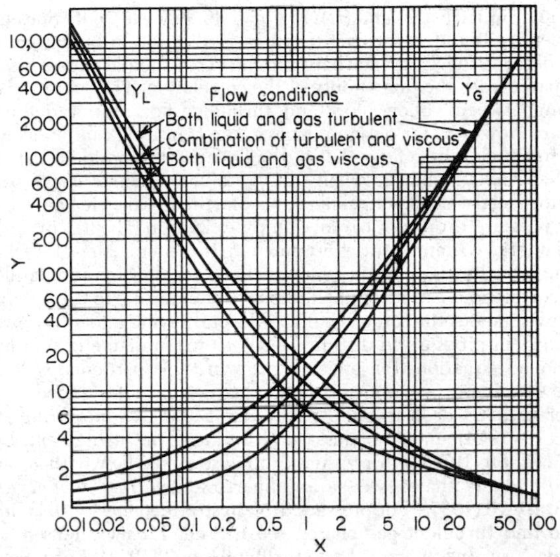

FIG. 5-49 Parameters for pressure drop in liquid-gas flow through horizontal pipes. [*Based on Lockhart and Martinelli,* Chem. Eng. Prog., *45, 39 (1949).*]

For additional details and references on flow pattern, see Anderson and Russell, *Chem. Eng.*, **72**(25), 135–144 (Dec. 6, 1965); Dukler and Wicks, in Acrivos, *Modern Chemical Engineering*, vol. I, Reinhold, New York, 1963, chap. 8; Scott, in Drew et al., *Advances in Chemical Engineering*, vol. 4, Academic, New York, 1963, pp. 199–277; and Aziz, Gregory, and Nicholson, *Can. J. Chem. Eng.*, **52**, 695–702 (1974). Taitel and Dukler [*Am. Inst. Chem. Eng. J.*, **22**, 47–55 (1976)] present a theoretically based flow-regime map for determining regime transitions which is in good agreement with the latest experimentally derived maps.

Two-phase pressure drop due to friction for cocurrent flow in **horizontal pipe** can be estimated by the semiempirical correlation of Lockhart and Martinelli [*Chem. Eng. Prog.*, **45**, 39–48 (1949)]. The basis of the correlation is that the two-phase pressure drop is equal to the single-phase pressure drop for either phase multiplied by a factor found to be a function of the single-phase pressure drops of the two phases:

$$(\Delta p/\Delta L)_{TP} = Y_L(\Delta p/\Delta L)_L \tag{5-147}$$

or

$$(\Delta p/\Delta L)_{TP} = Y_G(\Delta p/\Delta L)_G \tag{5-148}$$

where

$$Y_L = F_1(X) \quad \text{and} \quad Y_G = F_2(X) \tag{5-149}$$

$$X = \left[\frac{(\Delta p/\Delta L)_L}{(\Delta p/\Delta L)_G}\right]^{1/2} \tag{5-150}$$

Note that

$$Y_G = X^2 Y_L \tag{5-151}$$

The single-phase pressure-drop gradients $(\Delta p/\Delta L)_L$ and $(\Delta p/\Delta L)_G$ are computed from the Fanning equation, Eq. (5-57), by assuming that each phase is flowing alone in the pipe; that is, superficial velocities are used. The superficial velocities are based on the full cross-sectional area of the pipe:

$$V_L = q_L/A \quad \text{and} \quad V_G = q_G/A \tag{5-152}$$

where V_L = liquid-phase superficial velocity, V_G = gas-phase superficial velocity, q_L = liquid-phase volume flow rate, q_G = gas-phase volume flow rate, and A = pipe cross-sectional area.

Functions F_1 and F_2, Eq. (5-149), are shown as curves in Fig. 5-49. Separate curves are required for each flow regime, liquid in viscous* flow and gas in viscous flow (*vv*), liquid viscous-gas turbulent

*The term "viscous" is customarily used, rather than "laminar," in reference to two-phase flow.

(*vt*), and similarly (*tv*) and (*tt*). In Fig. 5-49, however, only one curve is given for liquid viscous-gas turbulent and liquid turbulent-gas viscous because the difference between the experimental curves is small compared with the uncertainty of the correlation. The transition criterion between viscous flow and turbulent flow is not definitely known. However, for design purposes, the transition can be taken as that for single-phase flow; that is, the flow can be considered viscous for $N_{Re} \leq 2000$ and turbulent for $N_{Re} > 2000$, where N_{Re} is based on superficial velocity. Lockhart and Martinelli (loc. cit.) correlated their pressure-drop data on flows in pipes 25 mm (about 1 in) or less in diameter within about ±50 percent. In general, the predictions are high for stratified, wavy, and slug flows and low for annular flow. The correlations can be applied to pipe diameters up to 0.1 m (about 4 in) with about the same accuracy. Several investigators have studied flows in pipes up to 0.25 m (about 10 in) in diameter and have developed equations for pressure drop in their particular systems; however, a better general correlation has not been developed.

For pressure drop in bubble flow, see Davis, *J. Fluids Eng.*, **96**, 173–179 (1974); and Kopalinsky and Bryant, *Am. Inst. Chem. Eng. J.*, **22**, 82–86 (1976). For pressure drop in stratified flow with laminar liquid and gas phases, see Yu and Sparrow, *Am. Inst. Chem. Eng. J.*, **13**, 10–16 (1967). For pressure drop in stratified flow with laminar liquid and turbulent gas phases, see Russell, Etchells, Jensen, and Arruda, *Am. Inst. Chem. Eng. J.*, **20**, 664–669 (1974). For pressure drop in stratified flow with turbulent liquid and gas phases, see Cheremisinoff and Davis, *Am. Inst. Chem. Eng. J.*, **25**, 48–56 (1979). For pressure drop in slug flow, see Heywood and Richardson, *Chem. Eng. Sci.*, **34**, 17–30 (1979). Mandhane, Gregory, and Aziz [*J. Pet. Technol.*, **29**, 1348–1358 (1977)] segregated a large amount of data into groups representing the different flow-pattern regions and evaluated the predictive ability of 16 different correlations, including 2 modified correlations. For highest accuracy they recommend a different correlation for each flow region. For other correlations and references to other correlations, see Dukler and Wicks in Acrivos, loc. cit.; Scott in Drew et al., loc. cit.; DeGance and Atherton, *Chem. Eng.*, **77**(15), 95–103 (1970); and Hsu and Graham, *Transport Processes in Boiling and Two-Phase Systems*, McGraw-Hill, New York, 1976.

Volume fraction or **holdup** of a phase for cocurrent flow in **horizontal pipe** can be predicted from the following correlation developed by Lockhart and Martinelli (loc. cit.):

$$R_L = F_3(X) \quad \text{and} \quad R_G = F_4(X) \tag{5-153}$$

where

$$R_L + R_G = 1 \tag{5-154}$$

R_L = fraction of pipe volume occupied by the liquid phase, dimensionless; R_G = fraction occupied by gas phase, dimensionless; and X = parameter defined by Eq. (5-150), dimensionless. Function F_3 as a curve of R_L versus X is given in Fig. 5-50. Lockart and Martinelli correlated data for pipe diameters of 25 mm (about 1 in) and less within ±50 percent of the curve shown. Indications are that liquid-volume fractions may be less than those predicted by Fig. 5-50 for liquid viscosities greater than 0.001 Pa·s (1 cP) [see Alves, *Chem. Eng. Prog.*, **59**, 449–456 (1954)] and greater than predicted for larger pipe diameters (see Baker, loc. cit.). Several investigators have developed equations for holdup in their particular systems, but a better general correlation has not been developed. For holdup in stratified flow with laminar liquid and gas phases, see Yu and Sparrow, loc. cit. For holdup in stratified flow with laminar liquid and turbulent gas phases, see Russell, Etchells, Jensen, and Arruda, loc. cit. For holdup in stratified flow with turbulent liquid and gas phases, see Cheremisinoff and Davis, loc. cit. Mandhane, Gregory, and Aziz [*J. Pet. Technol.*, **27**, 1017–1026 (1975)] segregated a large amount of data into groups representing the different flow-pattern regions and evaluated the predictive ability of 12 different correlations. No significant trend with pipe diameter was detected for any of the examined correlations. For highest accuracy they recommend a different correlation for each flow region.

A method of predicting pressure drop and holdup for **nonnewtonian fluids** in annular flow, with laminar liquid and turbulent gas phases, has been proposed by Eisenberg and Weinberger [*Am. Inst. Chem. Eng. J.*, **25**, 240–245 (1979)]. The method accounts for the shear-rate dependence of the liquid viscosity.

Pressure-drop data for a 1-in **inlet** or **feed tee**, with the liquid entering the run and the gas entering the branch, are given by Alves (loc. cit.). Pressure drop and division of two-phase **annular flow in a tee** are discussed by Fouda and Rhodes [*Trans. Inst. Chem. Eng. (London)*, **52**, 354–360 (1974)]. For **fittings** and **valves**, results by Chenoweth and Martin [*Pet. Refiner*, **34**(10), 151–155 (1955)] indicate that single-phase data can be used in their correlation for two-phase pressure drop. Smith, Murdock, and Applebaum [*J. Eng. Power*, **99**, 343–347 (1977)] evaluated existing correlations for two-phase flow of steam-water and other gas-liquid mixtures through sharp-edged **orifices** meeting American Society for Testing and Materials (ASTM) standards for flow measurement. The correlation of Murdock [*J. Basic Eng.*, **84**, 419–433 (1962)] may be used for these orifices. See also Collins and Gacesa, *J. Basic Eng.*, **93**, 11–21 (1971), for measurements with steam and water beyond the limits of this correlation.

For pressure drop and holdup in **inclined pipe** with upward or downward flow, see Beggs and Brill, *J. Pet. Technol.*, **25**, 607–617 (1973). Up to 10° from the horizontal upward pipe inclination has little effect on the holdup [Gregory, *Can. J. Chem. Eng.*, **53**, 384–388 (1975)].

For **upflow** in **vertical pipe,** considerable work has been done on gas lifts, a type of liquid pump. A gas lift consists simply of a vertical pipe, open at both ends, part of which is submerged below the surface of the liquid to be pumped. Compressed gas is admitted to the bottom of the pipe, forming a mixture of liquid and gas within the pipe. The gas reduces the average density of the mixture to a value at which the weight of the mixture is less than equivalent to the submergence or pressure at the air inlet. Thus at the proper rates of liquid and gas, the mixture rises upward through the pipe and is dis-

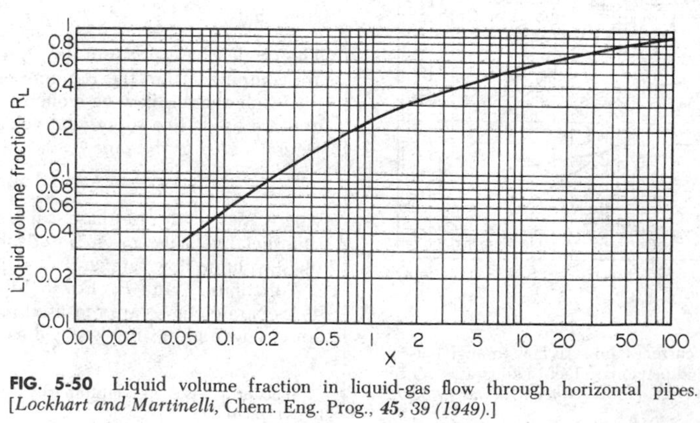

FIG. 5-50 Liquid volume fraction in liquid-gas flow through horizontal pipes. [*Lockhart and Martinelli,* Chem. Eng. Prog., *45, 39 (1949).*]

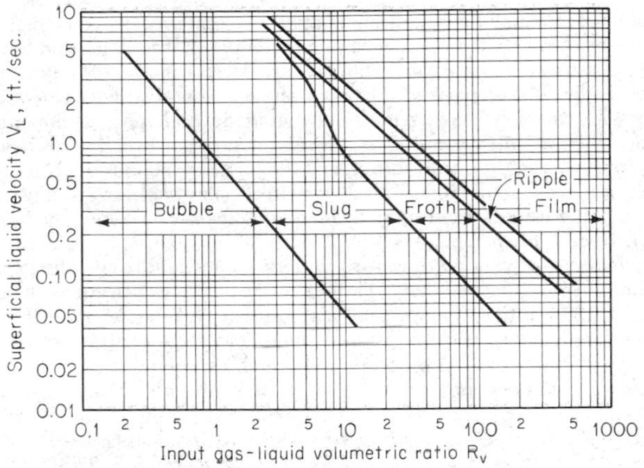

FIG. 5-51 Flow-pattern regions in cocurrent liquid-gas flow in upflow through vertical pipes. To convert feet per second to meters per second, multiply by 0.3048. [*Govier, Radford, and Dunn,* Can. J. Chem. Eng., *35, 58–70 (1957).*]

charged at the upper end. The submergence is the distance from the liquid surface to the air inlet. The lift is the distance from the liquid surface to the discharge. The submergence ratio is defined as

$$R_s = \frac{\text{submergence}}{\text{submergence} + \text{lift}} \qquad (5\text{-}155)$$

For **upflow** there have been observed the following general types of **flow pattern,** in which the values of the superficial velocities given are representative for liquids with viscosities less than about 0.01 Pa·s (100 cP) and gas densities about that of air:

1. *Bubble or aerated flow.* This pattern, in which the gas is dispersed as fine bubbles throughout the liquid, occurs for gas superficial velocities below about 0.6 m/s (2 ft/s).
2. *Piston, plug, or slug flow.* This pattern, in which the gas flows as large plugs, occurs for gas superficial velocities from about 0.6 to 9 m/s (2 to 30 ft/s).
3. *Froth flow.* In this pattern the gas bubbles mix with the liquid in a highly turbulent pattern.
4. *Ripple or wave flow.* In this pattern there is an upward-moving wavy layer of liquid on the wall.
5. *Annular or film flow.* This pattern, in which the liquid flows up the pipe as an annulus and the gas flows as a core, occurs for liquid superficial velocities less than about 0.6 m/s (2 ft/s) and for gas superficial velocities over about 9 m/s (30 ft/s). Data on entrainment are presented by Gill, Hewitt, and Lacey [*Chem. Eng. Sci.*, **20,** 71–88 (1965)]. Ishii and Grolmes (loc. cit.) give inception criteria for droplet entrainment.
6. *Mist flow.* In this pattern the liquid is carried as fine drops by the gas phase. Data indicate that this probably occurs for superficial gas velocities over about 21 m/s (70 ft/s) [Collier and Hewitt, *Trans. Inst. Chem. Eng. (London),* **39,** 127–136 (1961); and Yagi and Sasaki, *Chem. Eng. (Japan),* **17,** 216–223 (1953)].

The flow pattern may be predicted from the correlation proposed by Govier, Radford, and Dunn [*Can. J. Chem. Eng.,* **35,** 58–70 (1957)] as shown in Fig. 5-51. In this figure

V_L = superficial liquid velocity, ft/s

R_V = ratio of volumetric flow rates of gas to liquid entering pipe, dimensionless

For additional details or references on flow pattern, see Anderson and Russell, loc. cit.; Gosline, *Trans. Am. Inst. Mining Metall. Eng. (Pet. Dev. Technol.),* **118,** 56–70 (1936); Huntington et al., *Trans. Am. Inst. Mining Metall. Eng. (Pet. Dev. Technol.),* **136,** 79–90 (1940), and *Pet. Refiner,* 33(11), 208–211 (1954); and Oshinowo and Charles, *Can. J. Chem. Eng.,* **52,** 25–35 (1974).

During upflow, there is considerable slippage between the liquid and the gas; thus the **ratio of liquid to gas** within the pipe is greater than that at the inlet. Experimental data for the flow of water-air and kerosine-air mixtures in ½- and 2-in pipes are presented by Galegar, Stovall, and Huntington [*Pet. Refiner,* 33(11), 208–211 (1954)]; experimental data and a correlation for the flow of water-air mixtures in ½- to 2.5-in pipes are given by Govier and Short [*Can. J. Chem. Eng.,* **36,** 195–202 (1958)] and by Brown, Sullivan, and Govier [ibid., **38,** 62–66 (1960)]. A proposed correlation is presented by Hughmark [*Chem. Eng. Prog.,* **58**(4), 62–65 (1962)]. Some data on liquid entrainment are given by Anderson and Mantzouranis [*Chem. Eng. Sci.,* **12,** 233–242 (1960)]. Measurements of terminal bubble velocity in **slug flow** of air-water mixtures are given by Griffith and Wallis [*J. Heat Transfer,* **83,** 307–320 (1961)]; Nicklin, Wilkes, and Davidson [*Trans. Inst. Chem. Eng. (London),* **40,** 61–68 (1962)]; and Martin [*J. Fluids Eng.,* **98,** 715–722 (1976), and **99,** 263–264 (1977)].

For prediction of **pressure drop** in **upflow,** a proposed correlation together with references to other correlations is presented by Hughmark [*Ind. Eng. Chem. Fundam.,* **2,** 315–321 (1963)]. For pressure drop in bubble flow, see Davis, loc. cit. For prediction of pressure drop and film thickness in annular flow, when the amount of entrained liquid is known, see *DIMP (Design Institute Multiphase Processing),* vol. **3-1,** Am. Inst. Chem. Eng., 1977. A comparison of the predictive ability of available correlations is given by Vohra, Robinson, and Brill [*J. Pet. Technol.,* **26,** 829–832 (1974)]. The **performance of gas lifts** can be predicted by the theory presented by Stenning and Martin [*J. Eng. Power,* **90,** 106–110 (1968)] or from the data of Shaw (Texas Eng. Exp. Sta. Bull. 113, 1949).

For **upflow** in **helically coiled tubes,** the flow pattern, pressure drop, and holdup can be predicted by the correlations of Banerjee, Rhodes, and Scott [*Can. J. Chem. Eng.,* **47,** 445–453 (1969); and Akagawa, Sakaguchi, and Ueda [*Bull. JSME,* **14,** 564–571 (1971)].

A correlation for predicting **flow patterns** for **downflow** in vertical pipe has been proposed by Oshinowo and Charles (loc. cit.). Measurements of terminal bubble velocity in **slug flow** of air-water mixtures are given by Griffith and Wallis (loc. cit.), Nicklin, Wilkes, and Davidson (loc. cit.), and Martin (loc. cit.).

For **downflow** in vertical pipes, **pressure-drop** measurements for the **annular flow** of water-air mixtures in 1-in pipe and for the condensation of organic vapors inside a vertical tube condenser, 11.7-mm (about 0.46-in) inside diameter, have been tentatively correlated by Bergelin, Kegel, Carpenter, and Gazley (*Proc. Heat Transfer Fluid Mech. Inst.,* ASME, June 22–24, 1949, pp. 19–28) based on a modified Fanning friction factor as shown in Fig. 5-52. In this figure, f_G' = modified Fanning friction factor, dimensionless; $(N_{Re})_G$ = Reynolds number, dimensionless = $DV_G\rho_G/\mu_G$; D = inside diameter of pipe; V_G = gas superficial velocity = q_G/A; q_G = gas volume flow rate; A = cross-sectional area of pipe; ρ_G = gas density; μ_G = gas viscosity; σ_W/σ_L = ratio of surface tension of water at its boiling point to that of other liquid at its boiling point, dimensionless; ρ_L = liquid density; and Γ = liquid flow rate per unit pipe periphery. Viscosity of the liquid is omitted from this correlation because in the experimental investigations the viscosities of the liquids used did not vary greatly; thus the effect of viscosity could not be determined. Pressure drop is computed as follows, using the Fanning equation based on the gas phase (effect of the liquid appears as a "roughness" in the friction factor):

$$\Delta p = 4f_G' L\rho_G V_G^2/2g_c D \qquad (5\text{-}156)$$

where Δp = pressure drop, L = pipe length, and g_c = dimensional constant. Application of Eq. (5-156) to condensing flow in vertical condensers is described later in this subsection.

For prediction of pressure drop and film thickness when the amount of entrained liquid is known, see *DIMP,* vol. 3-1, loc. cit. Ishii and Grolmes (loc. cit.) give inception criteria for droplet entrainment.

For **downflow** in **helically coiled tubes,** some data on flow pattern, pressure drop, and holdup are presented by Casper [*Chem. Ing. Tech.,* **42,** 349–354 (1970)]. Gas-liquid **downflow** through symmetrical **Y junctions,** with stratified flow in the lines approaching the

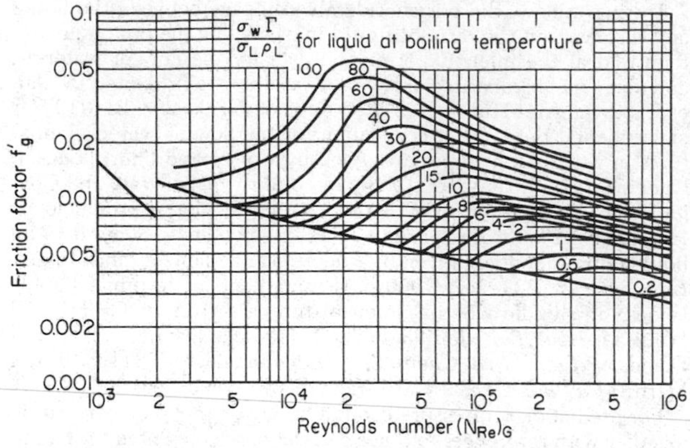

FIG. 5-52 Friction factors for condensing liquid-gas flow downward in vertical pipe. In this correlation Γ/ρ_L is in square feet per hour. To convert square feet per hour to square meters per second, multiply by 0.00155. (*Bergelin et al.,* Proc. Heat Transfer Fluid Mech. Inst., *ASME, 1949, p. 19.*)

junction, was investigated by Kubie and Gardner [*Chem. Eng. Sci.*, **33**, 319–329 (1978)]. They describe three flow regimes in the vertical line following the junction and give transition criteria.

Also involving **downflow** are **drain** and **overflow** pipes. The entrance to a drain pipe is flush with a horizontal surface; the entrance to an overflow pipe is above the horizontal surface, i.e., the pipe extends through and above the horizontal surface. When such pipes do not run full, considerable amounts of gas can be drawn down by the flowing liquid. The amount of gas entrained is a function of pipe diameter, pipe length, and liquid flow rate. Extensive data on air entrainment and head above the entrance as a function of water flow rate for pipe diameters from 43.9 to 148.3 mm (1.73 to 5.84 in) and lengths from about 1.22 to 5.18 m (4 to 17 ft) are reported by Kalinske (*Univ. Iowa Stud. Eng.*, Bull. 26, pp. 26–40, 1939–1940). For heads greater than the critical, the pipes will run full and no gas will be entrained. The critical head h for flow of water in drains and overflow pipes, as reported by Kalinske, is given as a function of pipe diameter D and length L in Fig. 5-53. From Kalinske's investigations, the height of protrusion of an overflow pipe apparently has no appreciable effect if the height of protrusion is greater than about one pipe diameter. For low heads, the liquid at the entrance to the pipe is generally swirling. The direction of swirl is theoretically counterclockwise (looking down) in the northern hemisphere. Practically, however, the direction of swirl will depend upon the approach conditions. For additional details, see Marris, *J. Appl. Mech.*, **34**, 11–15 (1967).

For conservative design of minimum liquid height for gas-liquid separators, McDuffle [*Am. Inst. Chem. Eng. J.*, **23**, 37–40 (1977)] recommends the relation

$$N_{Fr} \leq 1.6(h/D)^2 \tag{5-157}$$

where N_{Fr} = Froude number = $V_L \Big/ \sqrt{g\left(\frac{\rho_L - \rho_G}{\rho_L}\right) D}$, dimensionless; V_L = liquid velocity; g = acceleration due to gravity; ρ_L = liquid density; ρ_G = gas density; D = pipe inside diameter; and h = liquid height. Entrapped air bubbles, in liquids of viscosity less than 0.1 Pa·s (100 cP), can rise when

$$N_{Fr} \leq 0.31 \tag{5-158}$$

For additional information, see Simpson, *Chem. Eng.*, **75**(6), 192–214 (1968).

For **cocurrent flow** with **variable liquid-gas ratios,** a large amount of theoretical and experimental information is available on **flashing flow,** and a limited amount is available on **condensing flow. Flashing flow** occurs when a liquid which is initially saturated, i.e., at its boiling point, flows through a pipe. Several processes take place: (1) the pressure decreases because of friction, (2) the saturation temperature decreases because the pressure decreases, and (3) a portion of the liquid is vaporized. This two-phase flow occurs, with the ratio of the two phases continuously changing. Several types of flow are possible (see the beginning of this subsection); however, in this case, vapor is being produced throughout the liquid, which probably tends to produce bubble flow. The flow patterns observed for the flow of an evaporating refrigerant in a horizontal pipe are described by Zahn [*J. Heat Transfer,* **86**, 417–429 (1964)].

In the following **method for predicting flow conditions,** the principal assumptions are that (1) liquid and vapor flow with equal velocities, (2) elevation changes and heat losses are negligible, and (3) the mixture behaves as a compressible fluid. The basic relation between flow conditions and pipe geometry, which can be derived from the continuity and energy equations [see Benjamin and Miller, *Trans. Am. Soc. Mech. Eng.*, **64**, 657–669 (1942); Allen, ibid., **73**, 257–265 (1951); Bridge, *Heat. Piping Air Cond.*, **21**(5), 98–100 (1949); and Mikol, *Trans. Am. Soc. Refrig. Air Cond. Eng.*, **69**, 213–225 (1963)] is

$$v_m\, dp + \frac{G_m^2}{g_c}\left(v_m\, dv_m + \frac{4fv_m^2\, dL}{2D}\right) = 0 \tag{5-159}$$

Equation (5-159) can be integrated and solved graphically for the desired variable if the mathematical relation between mixture volume and pressure is known.

In general, Eq. (5-159) can be solved more conveniently by rewriting it and solving numerically:

$$\Delta L = -\frac{D}{2f}\left(\frac{g_c\, \Delta p}{G_m^2 \bar{v}_m} + \frac{\Delta v_m}{\bar{v}_m}\right) \tag{5-160}$$

where ΔL = increment of length; D = pipe diameter; f = Fanning friction factor, dimensionless; g_c = dimensional constant; G_m = mass velocity of mixture; Δp = increment of pressure (downstream minus upstream pressures); $\bar{v}_m$ = average specific volume of mixture over Δp; and Δv_m = incremental specific volume of mixture over Δp. The expansion is assumed to be an irreversible adiabatic (constant-enthalpy) process. The value of the friction factor f can be taken as 0.003, which is the average experimental value for the flow of a flashing mixture of water and steam determined by Benjamin and Miller (loc. cit.) and Bottomley [*Trans. North East Coast Inst. Eng. Shipbuild.*, **53**, 65–100 (1936)]. The value of the friction factor may be greater for flashing refrigerants in capillary tubes (see Mikol, loc. cit.).

In **numerical integration** of Eq. (5-160) for pipe length, the following general steps are taken: (1) mass velocity is known; (2) a pipe diameter is assumed; (3) increments of pressure are taken from the initial pressure to the final pressure; (4) average specific volume and the difference in specific volume of the mixture for each increment are obtained for a constant-enthalpy process, where the value of enthalpy is that of the saturated liquid at the initial pressure; (5) the increment of length for each increment of pressure taken is computed from Eq. (5-160), using the value of the friction factor given

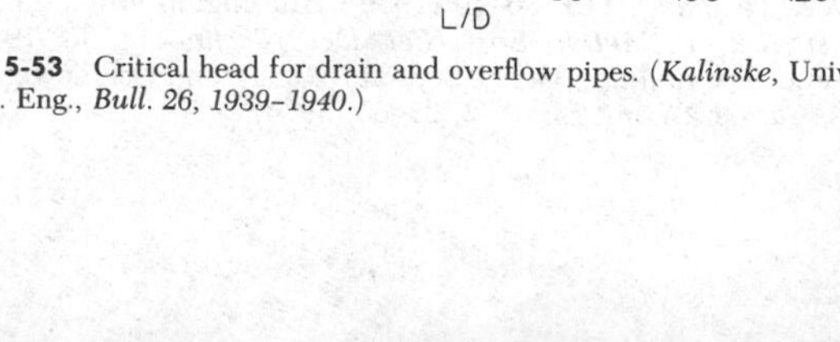

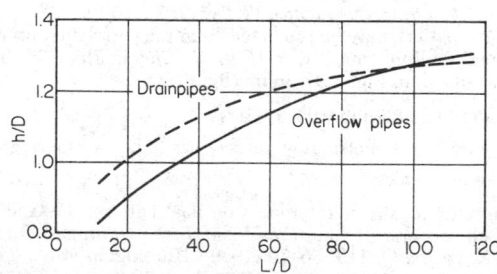

FIG. 5-53 Critical head for drain and overflow pipes. (*Kalinske,* Univ. Iowa Stud. Eng., *Bull. 26, 1939–1940.*)

above; and (6) total length of pipe is the sum of the length increments. Computations for the flow of water-steam mixtures in pipes are described by Bridge [*Heat. Piping Air Cond.*, **21**(4), 92–96 (1949)] and Keister [*Pet. Refiner*, **27**(11), 616–619 (1948)] and of refrigerants in capillary tubes by Mikol (loc. cit.). In this numerical integration, if the computed increment of length is negative, then the increment of pressure taken has been too large. A computed increment of length of zero is the condition for "critical flow" and corresponds to the final discharge pressure. When critical flow is indicated but the computed discharge pressure is larger than desired or the pipe length is shorter, a larger pipe diameter should be assumed and the integration repeated until the desired conditions are obtained. Solving Eq. (5-160) for pipe length when the other variables are known is comparatively straightforward, but solving for one of the other variables involves numerical integration on a trial-and-error basis. Charts for the maximum flow of water-steam mixtures in pipes, taking into account slip between the steam and the water, are presented by Moody [*J. Heat Transfer*, **88**, 285–295 (1966)].

The **critical-flow** condition in **flashing flow** is similar to that in the single-phase flow of gases through pipes. When the discharge pressure is gradually decreased, a pressure is reached at which further decrease in pressure will have no effect on pressure or flow rate within the pipe. The velocity at the discharge for these conditions is called the "critical velocity." For the single-phase flow of gases, this critical velocity is equal to the velocity of sound at the discharge pressure and temperature condition (see subsection "Terminology in Fluid Mechanics"). For the flow of flashing mixtures, the critical velocity may be much less than the velocity of sound in the vapor phase.

The **maximum** possible **rate of flow** of a flashing mixture through a pipe of constant cross section, if we assume that the mixture behaves as a compressible fluid, occurs when at the discharge end all the energy released by the differential drop in pressure is quantitatively converted into kinetic energy, i.e., frictionless flow [see Lapple et al., *Fluid and Particle Mechanics*, University of Delaware, Newark, 1951, p. 103; and Schweppe and Foust, *Chem. Eng. Prog.*, **49**, *Symp. Ser.* 5, 77–89 (1953)]:

$$(G_c)_m^2 = -g_c(dp/dv_m) \tag{5-161}$$

where $(G_c)_m$ = mixture critical mass velocity, g_c = dimensional constant, p = pressure, and v_m = specific volume of mixture. To account for slip between the vapor and the liquid, v_m is evaluated from the following equations [see Levy, *J. Heat Transfer*, **87**, 53–58 (1965)]:

$$v_m = \frac{v_G x^2}{\alpha} + \frac{v_L(1-x)^2}{1-\alpha} \tag{5-162}$$

$$\alpha = \left[1 + \frac{(1-x)(v_L/v_G)^{0.5}}{x}\right]^{-1} \tag{5-163}$$

where v_G = specific volume of vapor; v_L = specific volume of liquid; x = vapor weight fraction, dimensionless; and α = vapor volume fraction, dimensionless. The term dp/dv_m should be taken along a constant-entropy path for the conditions existing at the end of the pipe (for frictionless flow and reversible adiabatic expansion).

For the **critical flow** of **water-steam mixtures,** the maximum rate of flow and the fraction vaporized as functions of initial conditions and critical discharge pressures are given in Fig. 5-54 [Cruver and Moulton, *Am. Inst. Chem. Eng. J.*, **13**, 52–60 (1967)]. In Fig. 5-54, G_c = critical or maximum mass velocity; E = initial enthalpy + initial kinetic energy + heat added; p_c = absolute pressure inside tube at discharge end under critical flow conditions; and x = vapor weight fraction or fraction of liquid vaporized, dimensionless.

Flow of flashing mixtures of **water and steam** through **valves** is discussed by Allen [*Power Eng.*, **56**(5), 60–61, 102, 104, 106 (1952); **56**(6), 83–85 (1952); **57**(10), 94 (1953)]; through **orifices** and **nozzles** by Bailey [*Trans. Am. Soc. Mech. Eng.*, **73**, 1109–1116 (1951)], Burnell [*Engineering*, **164**, 572–576 (1947)], and Murdock [*J. Heat Transfer*, **84**, 419–433 (1962)]; flow of **carbon dioxide** liquid and vapor through orifices and nozzles by Hesson and Peek [*Am. Inst. Chem. Eng. J.*, **4**, 207–210 (1058)]; and flow of **refrigerant** liquid and vapor through orifices and nozzles by Min, Fauske, and Patrick [*Ind. Eng. Chem. Fundam.*, **5**, 50–55 (1966)] and Pasqua [*Refrig. Eng.*, **61**, 1084–1088, 1131 (1953)].

In flow of a **condensing vapor** in **horizontal straight circular pipe,** the pressure drop is the algebraic sum of the friction loss and the momentum changes. One general model [Soliman, Schuster, and Berenson, *J. Heat Transfer*, **90**, 267–276 (1968)] assumes that the

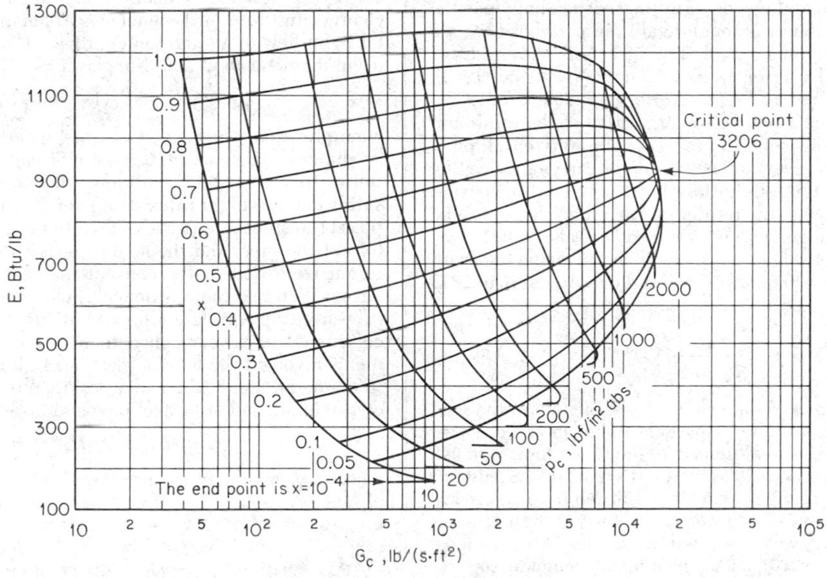

FIG. 5-54 Maximum discharge rates for water-steam mixtures inside pipes. To convert pounds per second per square foot to kilograms per second per square meter, multiply by 4.8824; to convert British thermal units per pound to joules per kilogram, multiply by 2326; and to convert pounds-force per square inch absolute to pascals absolute, multiply by 6895. [*Cruver and Moulton,* Am. Inst. Chem. Eng. J., *13, 52 (1967).*]

condensate flows as an annular ring with the vapor as the core. The Lockhart-Martinelli pressure-drop correlation, described earlier in this subsection, is recommended for predicting the friction loss. The momentum changes are estimated by the homogeneous-flow model [Andeen and Griffith, *J. Heat Transfer*, **90**, 211–222 (1968)], i.e., uniform velocity profile,

$$(\Delta p_a)_{1-2} = (G_m/g_c)(V_{m2} - V_{m1}) \tag{5-164}$$

where $(\Delta p_a)_{1-2}$ = pressure difference due to momentum changes between point 1 and point 2, G_m = mixture mass velocity, g_c = dimensional constant, V_{m2} = mixture average velocity at point 2, and V_{m1} = mixture average velocity at point 1. The pressure drop is computed for increments of pipe length commencing at the inlet from a knowledge of the condensation rate with length. The total pressure drop is the sum of the pressure drops for the increments.

For **condensing vapor flowing vertically downward** in a **circular pipe,** the pressure drop can be estimated by use of the differential form of Eq. (5-156):

$$-dp/dL = 4f'_G\rho_G V_G^2/2g_c D \tag{5-165}$$

From the heat-transfer characteristics the amount of vapor condensed as a function of pipe length can be computed. Then the pressure gradient $-dp/dL$ can be computed as a function of pipe length from Eq. (5-165) and Fig. 5-52. From a plot of pressure gradient versus pipe length, the pressure drop is given by integration of the area under the curve. Bergelin et al. (*Proc. Heat Transfer Fluid Mech. Inst.*, ASME, June 22–24, 1949, pp. 19–28) report considerable variation between experimental and predicted values; however, a more accurate method of prediction is not known. The overall pressure drop across an actual condenser includes the entrance and header losses.

Gases and Solids For the flow of gases and solids in a **horizontal pipe,** there are several possible modes of flow, depending upon the density of the solids, solids-to-gas weight-rate ratio (loading), and gas velocity. With low-density solids or low solids-to-gas ratios and high gas velocities, the solids may be fully suspended and fairly uniformly dispersed over the pipe cross section; with low solids-to-gas ratios and low gas velocities, the solids may bounce along the bottom of the pipe; with high solids-to-gas ratios and low gas velocities, the solids may settle to the bottom of the pipe and form dunes, with the particles moving from dune to dune, or form slugs, depending upon the nature of the particles. For additional details on modes of flow, see Coulson and Richardson, op. cit., vol. 2, p. 583; Korn, *Chem. Eng.*, **57**(3), 108–111 (1950); Patterson, *J. Eng. Power*, **81**, 43–54 (1959); and Wen and Simons, *Am. Inst. Chem. Eng. J.*, **5**, 263–267 (1959).

No single correlation is available for prediction of the **minimum carrying velocity** for all solid-to-gas ratios in **horizontal pipe.** Guides are available, however, for estimation of carrying velocities. For low solids-to-gas weight-rate ratios, such as are used in conventional pneumatic conveying, the minimum carrying velocity can be estimated by the following equation proposed by Dalla Valle [*Heat. Piping Air Cond.*, **4**, 639–641 (1932)], based on conveying tests with particles less than 8.1 mm (0.32 in) in size and density less than 2643 kg/m³ (165 lb/ft³), using air as the carrying gas:

$$V_{C,h} = C_1\left(\frac{\rho_s}{\rho_s + C_2}\right)D_s^{0.40} \tag{5-166}$$

where $V_{C,h}$ = minimum carrying velocity; ρ_s = density of the solid particles; D_s = diameter of largest particle to be conveyed; C_1 = dimensional constant, 132.4 in SI units (270 in U.S. customary units); and C_2 = dimensional constant, 998 in SI units (62.3 in U.S. customary units). Some data for single particles and mixed particle size are given by Zenz [*Ind. Eng. Chem. Fundam.*, **3**, 65–75 (1964)]. In practice, the actual conveying velocities used in systems with low solids-to-gas weight-rate ratios (<10), i.e., conventional pneumatic conveyors, are generally over 15 m/s (50 ft/s) [see Hudson, *Chem. Eng.*, **61**(4), 191–194 (1954); Jorgensen, *Fan Engineering*, 7th ed., Buffalo Forge Co., Buffalo, 1970, p. 486; Kunii and Levenspiel, *Fluidization Engineering*, Wiley, New York, 1969, chap. 12; and Zenz and Othmer, *Fluidization and Fluid-Particle Systems*, Reinhold, New York, 1960, pp. 220–222, 322]. For high solids-to-gas weight-rate ratios (>20), the actual gas velocities used are generally less than 7.5 m/s (25 ft/s) and are approximately equal to twice the actual solids velocities (Wen and Simons, loc. cit.).

Total **pressure drop** in **horizontal pipe** may be considered as the sum of the following individual pressure drops [Mehta, Smith, and Comings, *Ind. Eng. Chem.*, **49**, 986–992 (1957)]:

1. For acceleration of gas to the carrying velocity,

$$\Delta p_{a,G} = G_G V_G/2g_c \tag{5-167}$$

2. For acceleration of solid particles,

$$\Delta p_{a,s} = G_s V_s/g_c \tag{5-168}$$

3. For friction between gas and pipe wall,

$$\Delta p_{f,G} = 4f_G L\rho_{dG}V_G^2/2g_c D_t = 4f_G L G_G V_G/2g_c D_t \tag{5-169}$$

4. For combined friction between particles and pipe wall, between gas and particles, and between particles, if it is assumed that this friction can be expressed by a type of friction factor equation,

$$\Delta p_{f,s} = 4f_s L\rho_{ds}V_s^2/2g_c D_t = 4f_s L G_s V_s/2g_c D_t \tag{5-170}$$

Friction factor f_s can be related to the particle drag coefficient by a force balance on the particles in the pipe as follows:

$$4f_s = \frac{3\rho_G D_t C}{2\rho_s D_s}\left(\frac{V_G - V_s}{V_s}\right)^2 \tag{5-171}$$

In these equations, Δp = pressure drop; C = drag coefficient, dimensionless, obtained from Fig. 5-76, for N_{Re} (Reynolds number) $= D_s(V_G - V_s)\rho_G/\mu_G$; D_s = diameter of solid particle, D_t = diameter of tube or pipe; f_G = Fanning friction factor, dimensionless, obtained from Fig. 5-28; f_s = solids friction factor, dimensionless; $G_G = \rho_{dG}V_G$ = gas mass velocity; $G_s = \rho_{ds}V_s$ = solids mass velocity; g_c = dimensional constant; L = length of pipe; V_G = actual velocity of gas; V_s = actual velocity of solids; ρ_{dG} = dispersed gas density, weight of gas per unit volume of pipe; ρ_{ds} = dispersed solids density, weight of solids per unit volume of pipe; ρ_G = gas density; ρ_s = solids density; and μ_G = gas viscosity.

For **solids-to-gas weight-rate ratios less than 5 in horizontal pipes,** such as those usually employed in conventional pneumatic conveying systems, Hinkle (Ph.D. thesis, chemical engineering, Georgia Institute of Technology, Atlanta, 1953) found experimentally (for flow of air and solid particles 0.36 to 8.4 mm (0.014 to 0.33 in) in diameter in 2- and 3-in glass pipe) that

$$V_s = V'_G(1 - C_1 D_s^{0.3}\rho_s^{0.5}) \tag{5-172}$$

where for this case $V'_G = G_G/\rho_G$ = superficial gas velocity; and C_1 = dimensional constant, 0.0639 in SI units (0.179 in U.S. customary units). In such systems with low solids-to-gas ratios, $V_G \simeq V'_G$. For **solids-to-gas weight-rate ratios over 50,** such as those used in dense-phase transport, the particles tend to settle and move along the bottom of the pipe, and the flow pattern is therefore considerably different from that in the conventional dilute-phase pneumatic conveyor. Wen and Simons (loc. cit.) obtained the following empirical pressure-drop correlation for conditions in which the particles were considered to have reached their terminal velocity. Their experiments involved the flow of glass beads, less than 0.25 mm (0.01 in) in diameter, and coal powder, less than 0.75 mm (0.03 in) in diameter, with air in a ¼-in steel pipe and in ½- to 1-in glass pipes:

$$(\Delta p_f/L\rho_{ds})(D_t/D_s)^{0.25} = C_1 V_s^{0.45} \tag{5-173}$$

where Δp_f = sum of the pressure drops due to friction; L = length of pipe; $\rho_{ds} = G_s/V_s$ = dispersed solids density, weight of solids per unit volume of pipe; $G_s = w_s/A_t$ = superficial mass velocity of the solids; w_s = solids weight flow rate; A_t = cross-sectional area of pipe; D_t = diameter of pipe; D_s = diameter of solid particle; V_s = average actual velocity of solid particles, determined from Fig. 5-55; and C_1 = dimensional constant, 41.8 in SI units (2.5 in U.S. customary units). Average actual velocity of the air can be determined from the slippage between the air and the solids. Wen and Simons (loc. cit.) found that $V_a \simeq 2V_s$. Mass velocity of the air is given by

$$G_a = \rho_{da} V_a \quad (5\text{-}174)$$

where
$$\rho_{da} = (\rho_a/\rho_s)(\rho_s - \rho_{ds}) \quad (5\text{-}175)$$

and where G_a = mass velocity of air; V_a = average actual velocity of air; ρ_{da} = dispersed air density, weight of air per unit volume of pipe; ρ_a = air density; ρ_s = solids particle density; and ρ_{ds} = dispersed solids density, weight of solids per unit volume of pipe. The pressure-drop correlation, Eq. (5-173), is not dimensionless because only one gas, namely air, was used. Its uncertainty appears to be on the order of ± 35 percent. Additional investigation is needed to develop a general correlation.

For **upflow** of gases and solids in **vertical pipes**, the **minimum carrying velocity** for low solids-to-gas weight-rate ratios can be estimated by the following equation proposed by Dalla Valle (loc. cit.):

$$V_{C,v} = C_1 \left(\frac{\rho_s}{\rho_s + C_2}\right) D_s^{0.60} \quad (5\text{-}176)$$

where $V_{C,v}$ = minimum carrying velocity; ρ_s = density of the solid particles; D_s = diameter of largest particle to be conveyed; C_1 = dimensional constant, 566 in SI units (910 in U.S. customary units); and C_2 = dimensional constant, 998 in SI units (62.3 in U.S. customary units). Conveying velocities are given by Kunii and Levenspiel (loc. cit.). For high solids-to-gas weight-rate ratios, data on carrying velocities are not readily available.

Total pressure drop in vertical pipe may be considered as the sum of the following individual pressure drops [Hariu and Molstad, *Ind. Eng. Chem.*, **41,** 1148–1160 (1949); Mehta, Smith, and Comings, loc. cit.]:

1. For acceleration of gas to the carrying velocity, Eq. (5-167)
2. For acceleration of solid particles, Eq. (5-168)
3. For friction between gas and pipe wall, Eq. (5-169)
4. For combined friction between particles and pipe wall and the like, Eqs. (5-170) and (5-171)
5. For support of the column of gas,

$$\Delta p_{h,G} = G_G g L / V_G g_c \quad (5\text{-}177)$$

6. For support of solids,

$$\Delta p_{h,s} = G_s g L / V_s g_c \quad (5\text{-}178)$$

In Eqs. (5-177) and (5-178), $\Delta p_{h,G}$ = pressure drop to support column of gas; G_G = mass velocity of gas; V_G = actual velocity of gas in pipe; g = acceleration due to gravity; g_c = dimensional constant; L = column height, i.e., length of vertical pipe; $\Delta p_{h,s}$ = pressure drop to support solids; G_s = mass velocity of solids; and V_s = average actual velocity of solids in pipe. For low solids-to-gas weight-rate ratios, say, less than 5 to 10, $V_G - V_s$ is about equal to the free-fall terminal velocity of the particles, and V_G is about equal to V_G', the superficial gas velocity. Experimental data for specific systems are given in the immediately preceding references.

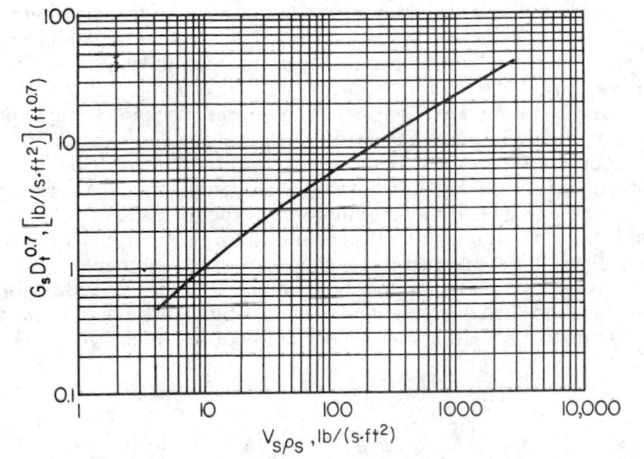

FIG. 5-55 Average velocity of solid particles. To convert pounds per second per square foot times foot$^{0.7}$ to kilograms per second per square meter times meter$^{0.7}$, multiply by 2.1254; to convert pounds per second per square foot to kilograms per second per square meter, multiply by 4.8824. [*Wen and Simons, Am. Inst. Chem. Eng. J.,* **5,** *263 (1959).*]

Yang [*Am. Inst. Chem. Eng. J.*, **24,** 548–552 (1978)] proposed a new correlation for computing the combined friction between particles and pipe wall.

For small particles (<0.06 mm diameter) in low solids-to-air ratios, the pressure drop for horizontal and vertical flows can be less than for the carrier air only. A reason for this drag reduction is the interaction of the particles with the fluid turbulence near the wall [Rossetti and Pfeffer, *Am. Inst. Chem. Eng. J.*, **18,** 31–39 (1972)].

For **dense-phase transport** in **vertical pipes,** information for small-diameter pipes is given by Sandy, Daubert, and Jones [*Chem. Eng. Prog.*, **66,** *Symp. Ser.* 105, 133–142 (1970)].

Choking imposes a limitation to the capacity of a pneumatic transport system. This occurs at a critical volume fraction of solid at which particles tend to agglomerate, and large unsteady pressure gradients are generated. Smith [*Chem. Eng. Sci.*, **33,** 745–749 (1978)] presents an equation for predicting choking occurrence.

Pneumatic conveying equipment is described in Sec. 7 and by Kraus (*Pneumatic Conveying of Bulk Materials,* Ronald, New York, 1968) and Stoess (*Pneumatic Conveying,* Interscience, New York, 1970).

The **flow of bulk solids through restrictions and bins** is discussed in symposium articles [*J. Eng. Ind.*, **91**(2), (1969)] and by Stepanoff (*Gravity Flow of Bulk Solids and Transportation of Solids in Suspension,* Wiley, New York, 1969).

Liquids and Solids Described here is the flow behavior of liquid-solids mixtures in pipes for cases in which the solids tend to settle out rapidly, i.e., "settling slurries." In these cases the particle size is generally larger than 0.25 mm (0.01 in). For cases in which the solids readily remain in suspension, as with solids of particle size less than 0.05 mm (0.002 in), the discussion given in the subsection "Nonnewtonian Flow" applies.

In **horizontal pipe,** gravity tends to stratify the mixture, the solids settling to the bottom. However, the solids can remain suspended in the liquid as a result of turbulence in the liquid due to high velocities or when the particles are less than about 3 mm (⅛ in). Settling of the solids has a pronounced effect upon pressure drop. At high velocities, the pressure gradient along the pipe is slightly greater than that for liquid alone flowing at the same velocity. With finely divided solids, the pressure gradient at high velocities may be less than that for the liquid alone because the solids interact with the turbulence [see Wilson, *Trans. Am. Soc. Civ. Eng.*, **107,** 1576–1594 (1942); and Bobkowicz and Gauvin, *Chem. Eng. Sci.*, **22,** 229–241 (1967)]. As velocity is reduced, the pressure gradient for the slurry will decrease, but it becomes increasingly greater than that for the liquid alone; a minimum pressure gradient is then reached, after which the pressure gradient increases with further reduction in velocity because of settling of the solids.

Transport velocity is a function of pipe diameter, particle size, particle density, solids concentration, and fluid properties. For particle sizes under 1 mm (0.04 in), Spells [*Trans. Inst. Chem. Eng. (London),* **33,** 79–84 (1955)] gives the following correlations on the flow of water suspensions of sands, ash, and lime, in **horizontal pipes** of 0.025- to 0.30-m (1- to 12-in) diameter:

1. Velocity to keep particles in suspension,

$$\frac{V_1^2}{gD_s}\frac{\rho_L}{(\rho_s - \rho_L)} = 0.0251\left(\frac{V_1 D_t \rho_m}{\mu_L}\right)^{0.775} \quad (5\text{-}179)$$

2. Velocity when pressure gradient for slurry becomes identical to that for a liquid with a density equal to the slurry and a viscosity the same as water,

$$\frac{V_2^2}{gD_s}\frac{\rho_L}{(\rho_s - \rho_L)} = 0.074\left(\frac{V_2 D_t \rho_m}{\mu_L}\right)^{0.775} \quad (5\text{-}180)$$

where, in Eqs. (5-179) and (5-180), g = acceleration due to gravity; D_s = particle diameter, such that 85 percent by weight of particles are smaller than D_s; ρ_s = particle density; ρ_L = liquid density; D_t = pipe diameter; ρ_m = slurry density; and μ_L = liquid viscosity.

For particle sizes greater than 2 mm (0.08 in), the velocity at which particles will begin to deposit is given by (see *The Transportation of Solids in Steel Pipelines*, Colorado School of Mines Research Foundation, Golden, Colo., 1963; and Stepanoff, *Pumps and Blowers—Two-Phase Flow*, Wiley, New York, 1965, p. 172)

$$V_c = 1.35 \left[\frac{2gD_t(\rho_s - \rho_L)}{\rho_L} \right]^{0.5} \quad (5\text{-}181)$$

where V_c = velocity at onset of deposition, g = acceleration due to gravity, D_t = pipe diameter, ρ_s = solid-particle density, and ρ_L = liquid density. A similar equation, with a coefficient dependent on solids concentration, has been proposed by Robinson and Graf [*Proc. Am. Soc. Civ. Eng., J. Hydraul. Div.*, **98**(HY7), 1221–1241 (1972)] for particle sizes less than 0.88 mm (0.035 in), in sand-water mixtures, with solids concentration less than 7 volume percent.

For large particles (diameter up to about one-third of the pipe diameter), Worster and Denny [*Proc. Inst. Mech. Eng. (London)*, **169**, 563–586 (1955)] give approximate values of the velocity for minimum pressure gradient for the flow of water suspensions of about 25 percent by volume of coal or gravel, as listed in Table 5-17. Additional velocity data for the flow of sand or similar suspensions are given by Condolios and Chapus [*Chem. Eng.*, **70**(14), 131–138 (July 8, 1963)]; Howard [*Trans. Am. Soc. Civ. Eng.*, **104**, 1334–1380 (1939)]; Newitt, Richardson, Abbott, and Turtle [*Trans. Inst. Chem. Eng. (London)*, **33**, 93–110 (1955)]; Smith [ibid., **33**, 85–92 (1955)]; and Wilson (loc. cit.). A general correlation for particles in the 0.04- to 2-mm range has been proposed by Hughmark [*Ind. Eng. Chem.*, **53**, 389 (1961)].

Although there are considerable data on **pressure drop** for flow of water suspensions of sand, gravel, coal, and the like in **horizontal** pipes, no single correlation has been found to be entirely satisfactory for all particle sizes, particle densities, concentrations, and pipe sizes. As a guide, the Durand et al. equation can be used [see *The Transportation of Solids in Steel Pipelines*, loc. cit.; Aude et al., *Chem. Eng.*, **78**(15), 74–90 (1971); and Govier and Aziz, *The Flow of Complex Mixtures in Pipes*, Van Nostrand Reinhold, New York, 1972, p. 686]:

$$\frac{i_m - i_L}{i_L} = 150c \left[\frac{D_t g(\rho_s - \rho_L)}{V^2 \rho_L \sqrt{C_D}} \right]^{3/2} \quad (5\text{-}182)$$

where i_m = pressure drop per unit pipe length for mixture = $(\Delta p_m/L)(\rho_m/\rho_L)$; i_L = pressure drop per unit pipe length for liquid flowing alone at velocity V; Δp_m = pressure drop for mixture; L = length of pipe; ρ_m = mixture density; ρ_L = liquid density; ρ_s = solid-particle density; D_t = pipe diameter; c = concentration as volume fraction of solids, dimensionless; g = acceleration due to gravity; V = mixture velocity; and C_D = particle drag coefficient, dimensionless (can be obtained by the method described under "Particle Dynamics: Applications").

TABLE 5-17 Velocity for Minimum Pressure Gradient, Water Suspensions of Coal or Gravel*

	Velocity, m/s (ft/s)†	
Pipe diameter, in	Coal‡	Gravel§
1	0.46 (1.5)	0.91 (3)
3	1.1 (3.5)	2.1 (7)
6	1.5 (5)	3.0 (10)
9	1.9 (6.3)	4.0 (13)
12	2.2 (7.3)	4.6 (15)
18	2.7 (8.8)	5.3 (17.5)

*Worster and Denny, *Proc. Inst. Mech. Eng. (London)*, **169**, 563–586 (1955).
†For concentrations about 25 percent by volume.
‡Specific gravity = 1.4.
§Specific gravity = 2.6.

Equation (5-182) is based on the flow of water suspensions of various solids of particle diameters of about 0.25 to 100 mm (0.01 to 4 in) with concentrations up to around 30 volume percent in pipe diameters from 0.038 to 0.71 m (1½ to 28 in). The equation is applicable for closely sized particles and for liquid velocities greater than the velocity at onset of deposition, i.e., over about 0.9 m/s (3 ft/s). For mixed particle sizes, the equation will be in error; the amount of error depends upon the amount of fines and the spread in particle size (see Smith, loc. cit.; and discussion by Durand and Condolios of the paper by Worster and Denny, loc. cit.). Turian and Yuan [*Am. Inst. Chem. Eng. J.*, **23**, 232–243 (1977); see also *Turian and Oroskar*, op. cit., **24**, 1144 (1978)] segregated a large body of data into four groups representing different flow regimes and developed more accurate empirical correlations for predicting the pressure drop in each flow regime.

Pressure-drop data for the flow of various **solids-water mixtures**, such as sands, gravels, and coal in water, are presented by Condolios and Chapus (loc. cit.), Howard (loc. cit.), Newitt et al. (loc. cit.), Smith (loc. cit.), Wilson (loc. cit.), and Worster and Denny (loc. cit.). Pressure-drop data for the flow of **paper stock** in pipes are given in the Data Section of *Standards of Hydraulic Institute*, Hydraulic Institute, 1965. The flow behavior of **fiber suspensions** is discussed by Bobkowicz and Gauvin [*Chem. Eng. Sci.*, **22**, 229–241 (1967)], Bugliarello and Daily [*TAPPI*, **44**, 881–893 (1961)], and Daily and Bugliarello [*TAPPI*, **44**, 497–512 (1961)].

In **upflow** in **vertical pipes** the concentration of solids will increase, and in **downflow** the concentration will decrease, because of the slip between the liquid and the solids. The **slip velocity**, i.e., the difference between liquid and solids velocities, can be taken as equal to the free-fall velocity of the solids in the liquid. The **pressure drop** for flows in which the nominal mixture velocity is greater than 4 or 5 times the free-fall velocity of the solids may be estimated by

$$i_m = i_L \pm \left[c\left(\frac{\rho_s}{\rho_L} - 1\right) + 1 \right] \quad (5\text{-}183)$$

where + is for upflow and − is for downflow. Nomenclature is the same as for Eq. (5-182). Additional details are given by Cloete, Miller, and Streat [*Trans. Inst. Chem. Eng. (London)*, **45**, T392–T400 (1967)] and Worster and Denny (loc. cit.).

FLUID DISTRIBUTION

Uniform fluid distribution is essential for efficient operation of chemical-processing equipment such as contactors and reactors, mixers, burners, heat exchangers, extrusion dies, and textile-spinning chimneys. To obtain optimum distribution, proper consideration must be given to flow behavior in the distributor, flow conditions upstream of the distributor, and flow conditions downstream of the distributor. Guides for the design of various types of fluid distributors are described later. These procedures take into account only the flow behavior in the distributor.

Perforated-Pipe Distributors The simple perforated pipe or sparger (Fig. 5-56), used in a wide variety of piping configurations, is a common type of distributor. As shown, the flow distribution is uniform; this is the case in which there is a proper balance between (1) kinetic energy and momentum force of the inlet stream, (2) friction losses along the length of pipe, and (3) pressure drop across the outlet holes. When inlet-stream kinetic energy and momentum force predominate, increasing amounts of fluid will be discharged as the flow travels toward the closed end; when friction losses along the pipe predominate, decreasing amounts of fluid will be discharged as the

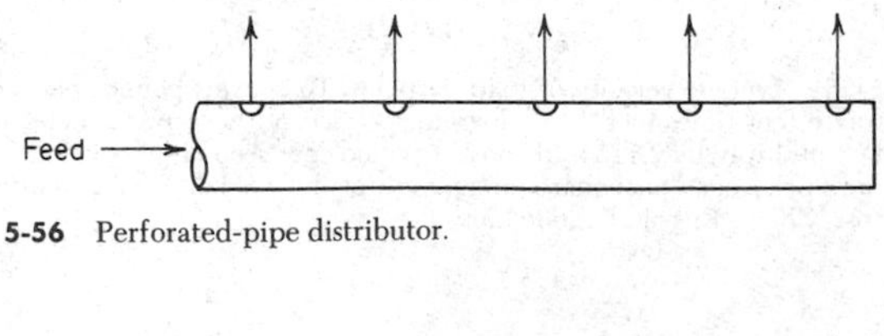

FIG. 5-56 Perforated-pipe distributor.

flow travels toward the closed end. When an upstream disturbance, such as produced by a bend, is superimposed upon a case of predominant inlet-stream kinetic energy and momentum force, the flow from the outlet holes near the distributor inlet and near the closed end can be greater than in the middle.

A **rule of thumb for design** of perforated-pipe distributors for turbulent flow such that the maldistribution is less than ±5 percent is this: the ratios of kinetic energy of the inlet stream to pressure drop across the outlet hole and of friction loss in the pipe to pressure drop across the outlet hole should be equal to or less than one-tenth [Senecal, *Ind. Eng. Chem.*, **49**, 993–997 (1957)]. From a knowledge of the velocity distribution at the inlet to the distributor, the kinetic energy can be computed from

$$\text{Kinetic specific energy} = \alpha V_i^2/2g_c \tag{5-184}$$

where V_i = average inlet velocity; g_c = dimensional constant; and α = correction factor to compensate for use of the average velocity, dimensionless. As a guide, α is 1.00 for plug flow (uniform velocity distribution), about 1.05 to 1.10 for turbulent flow in long straight ducts and 2 for laminar flow [Stoker, *Ind. Eng. Chem.*, **38**, 622–624 (1946)]. For other velocity distributions, the method of computing α is given by Rouse (*Engineering Hydraulics*, Wiley, New York, 1950, pp. 399–401).

Generally, an orifice coefficient of 0.60 to 0.63 is used; however, for perforated pipes the orifice coefficient is a function of hole size relative to pipe diameter and wall thickness, flow rate through the hole, flow rate in the pipe across the hole, and the like, and thus the value of the orifice coefficient could be considerably different from the values generally used (Senecal, loc. cit.). Additional experimental data are needed to define this function. In general, if the component of the hole outlet velocity normal to the pipe wall is larger than the velocity along the pipe, the effect of the pipe velocity on the orifice coefficient would appear to be small [Grobman, Dittrich, and Graves, *Trans. Am. Soc. Mech. Eng.*, **79**, 1601–1607 (1957)]. Bailey [*J. Mech. Eng. Sci.*, **17**, 338–347 (1975)] presents correlations for orifice coefficient, static-pressure regain coefficient, and angle of discharge for airflow in thin-walled pipe distributors. This information can be used to predict static-pressure and fluid-discharge distributions along uniformly perforated ducts.

The **pressure change** due to friction and to momentum recovery over the length of the perforated-pipe distributor can be shown by theory to be (Lapple et al., *Fluid and Particle Mechanics*, University of Delaware, Newark, 1951, p. 15)

$$\Delta h = \left(\frac{4fL}{3D} - 2\right)\frac{V_i^2}{2g_c} \tag{5-185}$$

where Δh = net loss in head between the inlet and closed end of the pipe; f = Fanning friction factor, dimensionless; L = pipe length; D = pipe diameter; V_i = average fluid velocity at inlet to pipe; and g_c = dimensional constant. Equation (5-185) applies to the case of uniform discharge at right angles to the pipe axis, with constant friction factor, in a pipe of constant cross section. Experimental investigations indicate that the factor 2 in Eq. (5-185), for momentum recovery, may be high. Measured values are closer to 1 [Van der Hegge Zijnen, *Appl. Sci. Res.*, **A3**, 144–162 (1951–1953); and Bailey, loc. cit.], which implies that the angle of fluid discharge is less than 90° with respect to the pipe axis. If a constant orifice coefficient is assumed, the percentage of maldistribution between the first and last outlets can be given by

$$\text{Percent maldistribution} = 100\left(1 - \sqrt{\frac{\Delta h_{01} - \Delta h}{\Delta h_{01}}}\right) \tag{5-186}$$

where Δh_{01} = difference in head across the first outlet.

Slot-Type Distributors These are generally used in sheeting dies for extrusion of films and coatings and in air knives for control of thickness of a material applied to a moving sheet.

A simple slotted pipe (Fig. 5-57) for turbulent-flow conditions can give severe maldistribution because this type of distributor does not readily give a discharge perpendicular to the slot [Koestel and Tuve, *Heat. Piping Air Cond.*, **20**(1), 153–157 (1948); and Koestel and Young, *Heat. Piping Air Cond.*, **23**(7), 111–115 (1951)]. The discharge angle will vary along the length of pipe. However, for slots in tapered ducts where the duct cross-sectional area decreases linearly to zero at the far end, the discharge angle will be constant along the length of duct (Koestel and Young, loc. cit.). One way to ensure an almost perpendicular discharge is to have the ratio of the area of the slot to the cross-sectional area of the pipe equal to or less than 0.1. As in the case of the perforated-pipe distributor, for good performance a proper balance has to be made of kinetic energy and momentum changes, friction and discharge losses, and upstream and downstream flow conditions. Another way to improve the discharge angle is to use turning vanes as described later.

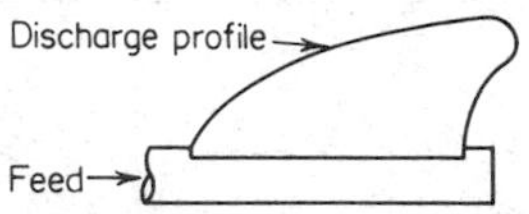

FIG. 5-57 Simple slotted-pipe distributor.

In practice, the following methods may be used to keep the diameter of the pipe to a minimum consistent with good performance (Senecal, loc. cit.):

1. Feed from both ends; this reduces the kinetic-energy term.
2. Modify cross-sectional design (Fig. 5-58); the slot is thus farther away from the influence of feed-stream velocity.
3. Increase pressure drop across the slot; this can be accomplished by lengthening the lips (see Fig. 5-58).
4. Use screens (see Fig. 5-58); screens upstream of the slot will increase the overall pressure drop.

Considerations to be taken into account when designing and using an air knife are discussed by Senecal (loc. cit.).

Design procedures for extrusion dies, when the flow is laminar as with highly viscous fluids, are presented by Bernhardt (*Processing of Thermoplastic Materials*, Reinhold, New York, 1959, pp. 248–281).

Turning Vanes In applications such as ventilation work, the discharge profile from slots (see Fig. 5-57) can be improved through the use of turning vanes. The tapered duct design is the most amenable for turning vanes because the discharge angle remains constant. One way of installing the vanes is shown in Fig. 5-59. The vanes should have a depth twice the spacing (*Heating, Ventilating, Air Conditioning Guide*, vol. 38, American Society of Heating, Refrigerating and Air-Conditioning Engineers, 1960, pp. 282–283) and a curvature of a circular arc which is tangent to the discharge angle θ of a slot

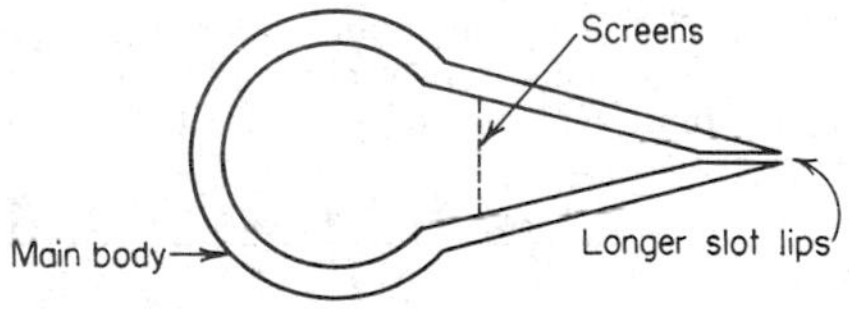

FIG. 5-58 Modified slot distributor.

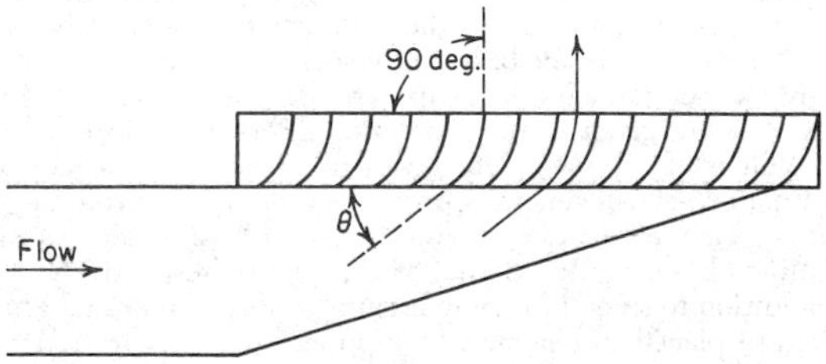

FIG. 5-59 Turning vanes in a slot distributor.

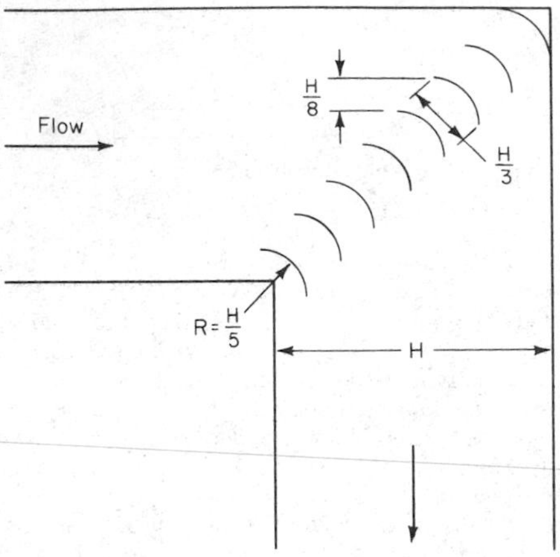

FIG. 5-60 Miter bend with vanes.

without vanes at the upstream end of the vanes and perpendicular to the slot at the downstream or discharge end (Koestel and Young, loc. cit.). Angle θ can be estimated from

$$\cot \theta = C_d A_s / A_d \qquad (5\text{-}187)$$

where A_s = slot area; A_d = duct cross-sectional area at upstream end; and C_d = discharge coefficient of slot, dimensionless.

Vanes may be used also to improve velocity distribution and to reduce friction loss in bends. For a miter bend with low-velocity flows, simple circular arcs (Fig. 5-60) can be used, and with high-velocity flows, vanes of special airfoil shapes are required. For additional details and references, see Ower and Pankhurst, *The Measurement of Air Flow*, Pergamon, New York, 1977, p. 102; Pankhurst and Holder, *Wind-Tunnel Technique*, Pitman, London, 1952, pp. 92–93; and Rouse, loc. cit., pp. 422–423. For a sweep bend, splitter vanes are used. These vanes are curved vanes extending from end to end of the bend and dividing the bend into several parallel channels. Additional details, together with a chart for determining the location of the splitters in the bend, are given by Jorgensen (*Fan Engineering*, 7th ed., Buffalo Forge Co., Buffalo, 1970, pp. 111, 117, 118).

Perforated Plates and Screens A nonuniform velocity profile of **turbulent flow** through channels or process equipment can be smoothed out to any desired degree by adding sufficient uniform resistance, such as perforated plates or screens across the flow channel.

By referring to Fig. 5-61, the amount of resistance required to smooth out a specific nonuniform velocity profile to a desired degree can be estimated from the following equations (Stoker, loc. cit.):

$$\frac{V_{2max}}{V_{avg}} = \sqrt{\frac{(V_{1max}/V_{avg})^2 + \alpha_2 - \alpha_1 + \alpha_2 K}{1 + K}} \qquad (5\text{-}188)$$

$$\Delta h = K(V_{avg}^2/2g_c) \qquad (5\text{-}189)$$

where V = velocity; $\Delta h = h_1 - h_2$, static head loss; α = correction factor to compensate for use of the average velocity, dimensionless; K = equivalent resistance of the uniform resistance expressed as number of velocity heads based on velocity V, dimensionless; g_c = dimensional constant; and subscripts are defined in Fig. 5-61. Typical values of α are given in the subsection "Perforated Pipe Distributors." Values of V_{max}/V_{avg} are given in Fig. 5-14. K as computed from Eq. (5-188) will generally be on the order of 10. From the value of Δh the characteristics of the perforated plates or screens can be determined by using Eq. (5-132) or Eq. (5-141) respectively.

In addition to smoothing flow nonuniformity, screens and **honeycombs** are placed in channels to suppress stream swirl and turbulence [Loehrke and Nagib, *J. Fluids Eng.*, **98**, 342–353 (1976)]. Further reduction in turbulence level and flow nonuniformity is obtained by contraction of the channel.

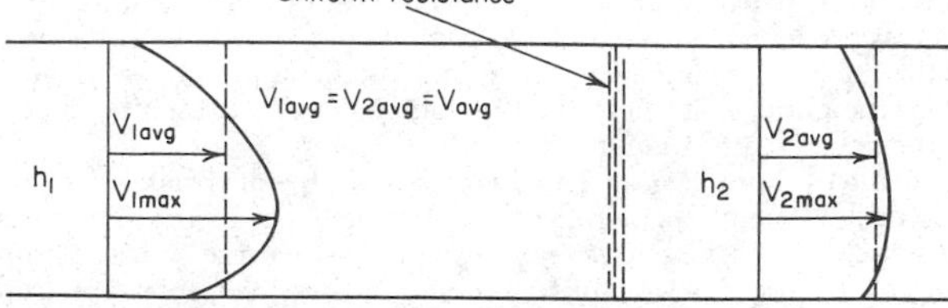

FIG. 5-61 Smoothing out a nonuniform profile in a channel.

Beds of Solids A suitable depth of solids can be employed as a fluid distributor. The depth should be such that the pressure drop across the bed is at least 10 velocity heads, based upon the superficial velocity through the bed. Methods of computing the pressure drop are given in the subsection "Fixed Beds of Granular Solids."

If a single liquid stream is fed to the top of a bed of solids with no gas flow, the flow will not become uniform until four or five bed diameters have been traversed [Baker, Chilton, and Vernon, *Trans. Am. Inst. Chem. Eng.*, **31**, 296–315 (1935)]. For the flow of a liquid downward through a bed of solids when the liquid was initially distributed uniformly, Akehata and Sato [*Chem. Eng. (Japan)*, **22**, 430–435 (1958)] found that the flow can become maldistributed in a distance of three to six bed diameters for either laminar or turbulent flow when the ratio of bed diameter to particle diameter is less than 15.

Velocity distribution of gas flow through beds of solids has been studied by Schwartz and Smith [*Ind. Eng. Chem.*, **45**, 1209–1218 (1953)]. They found that the velocity profile has a maximum value approximately one particle diameter from the wall, and the variation in the velocity profile was less than 20 percent for ratios of bed diameter to particle diameter of more than 30. These observations of velocity profile were confirmed by Calderbank and Pogorski [*Trans. Inst. Chem. Eng.*, **35**, 195–207 (1957)].

The spread of liquid trickling down a randomly packed bed and the wall flow in packed columns are discussed by Porter et al. [*Trans. Inst. Chem. Eng. (London)*, **46**, T69–T94 (1968)]. In the experiments with 0.05- to 0.3-m- (2- to 12-in-) diameter columns, packed with ½-in Raschig rings, the wall flow built up rapidly to an equilibrium value within a depth of 0.9 m (3 ft) of packing. The wall flow decreased as the ratio of column diameter to particle diameter increased, reaching an asymptotic value of about 15 percent of the total flow for ratios over about 25.

TUBE BANKS

It is not possible to obtain a fundamental correlation of data on pressure drop across tube banks by means of a single, simple friction-factor Reynolds-number curve. This is due to lack of geometric similarity among the large number of tube configurations and spacings encountered, two of which are illustrated in Fig. 5-62. Several investigators have made allowance for configuration and spacing by incorporating spacing correction factors in their friction-factor expressions or by use of multiple friction-factor plots. The best of these representations are described as follows.

Turbulent Region The correlation given by Grimison [*Trans. Am. Soc. Mech. Eng.*, **59**, 583–594 (1937)] is recommended for predicting pressure drop for turbulent flow ($N_{Re} \geq 2000$) across staggered or in-line tube banks for tube spacings [(a/D_t), (b/D_t)] ranging from 1.25 to 3.0. The pressure drop is given by

$$\Delta p = 4 f N_r \rho V_{max}^2 / 2g_c \qquad (5\text{-}190)$$

where f = friction factor, dimensionless; N_r = number of rows of tubes in the direction of flow, dimensionless; ρ = fluid density; V_{max} = fluid velocity through the minimum area available for flow; and g_c = dimensional constant.

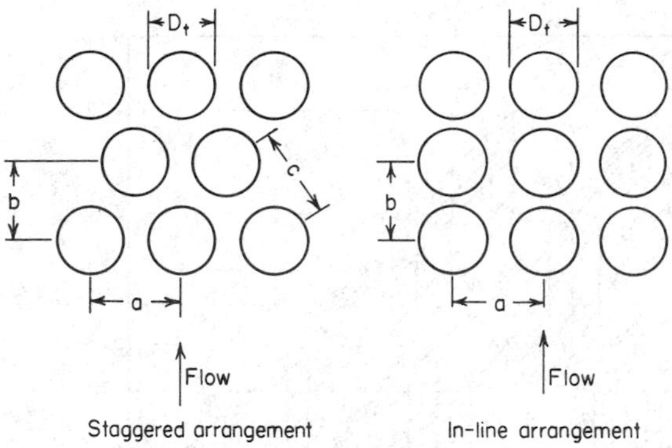

FIG. 5-62 Tube-bank configurations.

For banks of **staggered tubes,** the friction factor for isothermal flow is obtained from Fig. 5-63. Each "fence" (group of parametric curves) represents a particular Reynolds number defined as

$$N_{Re} = D_t V_{max} \rho / \mu \tag{5-191}$$

where D_t = tube outside diameter and μ = fluid viscosity. The numbers along each fence represent the transverse and inflow-direction spacings. The upper chart is for the case in which the minimum area for flow is in the transverse openings, while the lower chart is for the case in which the minimum area is in the diagonal openings. In the latter case, V_{max} must be based upon the area of the diagonal openings and N_r taken as the number of rows in the direction of flow less 1. A critical comparison of this method with all available data showed an average deviation from the data of the order of ±15 percent [Boucher and Lapple, *Chem. Eng. Prog.*, **44,** 117–134 (1948)]. For the case of tube spacings greater than three tube diameters, the correlation given by Gunter and Shaw [*Trans. Am. Soc. Mech. Eng.*, **67,** 643–660 (1945)] can be used as an approximation. As an **approximation,** the pressure drop can be taken as 0.72 velocity head (based on V_{max}) per row of tubes for tube spacings commonly encountered in practice (Lapple et al., *Fluid and Particle Mechanics,* University of Delaware, Newark, 1951, p. 40).

For banks of **in-line tubes,** the friction factor for isothermal flow is obtained from Fig. 5-64. Each fence represents a particular Reynolds number as defined in the preceding paragraph. Average deviation of this method from available data is on the order of ±15 percent. For tube spacings greater than three tube diameters, the charts given by Gram, Mackey, and Monroe [*Trans. Am. Soc. Mech. Eng.*, **80,** 25–35 (1958)] can be used. As an **approximation,** the pressure drop can be taken as 0.32 velocity head (based on V_{max}) per row of tubes (Lapple et al., op. cit., p. 40).

In the case of turbulent flow through **shallow banks** of tubes, the average friction factor per row will be somewhat higher than indi-

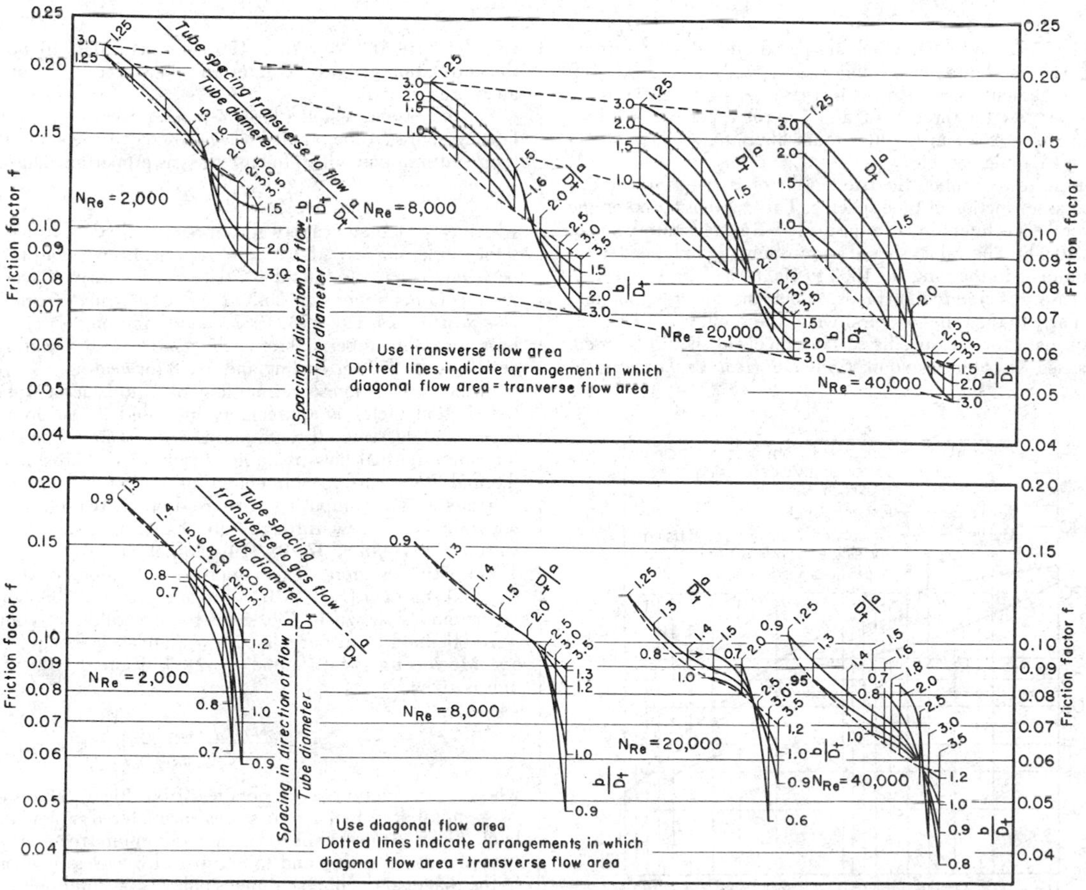

FIG. 5-63 Upper chart: friction factors for staggered tube banks with minimum fluid-flow area in transverse openings. Lower chart: friction factors for staggered tube banks with minimum fluid-flow area in diagonal openings. [*Grimison,* Trans. Am. Soc. Mech. Eng., *59, 583 (1937).*]

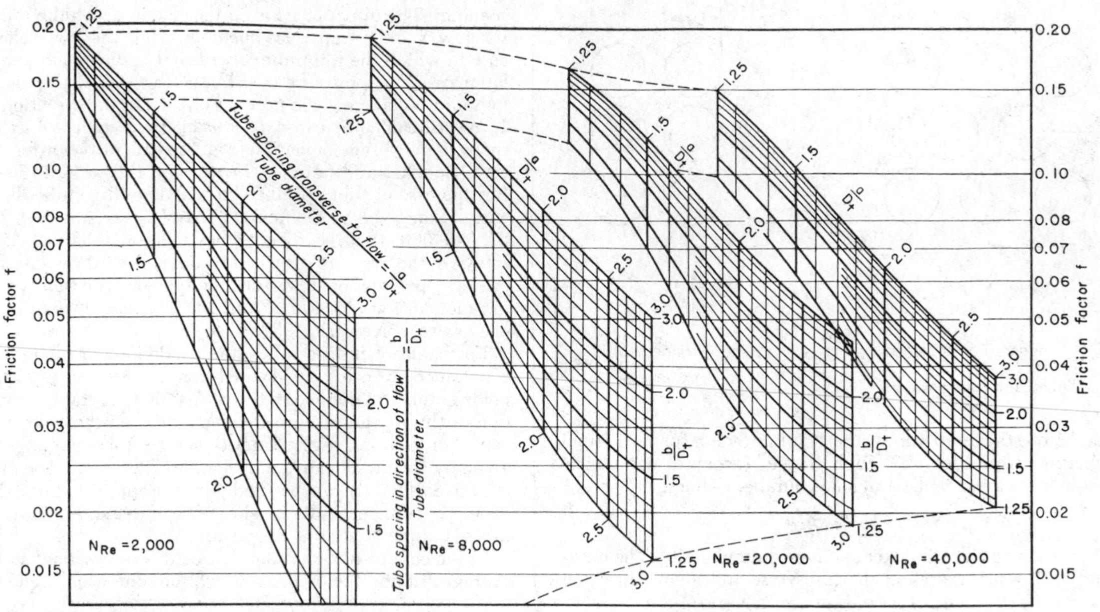

FIG. 5-64 Friction factors for in-line tube banks. [*Grimison*, Trans. Am. Soc. Mech. Eng., *59, 583 (1937).*]

cated by Figs. 5-63 and 5-64, which are based on banks 10 or more rows deep. The magnitude of this increase depends upon tube spacing and arrangement. A 30 percent increase per row for two rows, 15 percent per row for three rows, and 7 percent per row for four rows can be taken as about the maximum likely to be encountered (Boucher and Lapple, loc. cit.).

For a **single row** of tubes, the friction factor is given by curve *B* in Fig. 5-65 as a function of tube spacing. This curve is based on the data of several experimenters, all adjusted to a Reynolds number [Eq. (5-191)] of 10,000. The values should be substantially independent of Reynolds number in the range of 1000 to 100,000.

For the case of **extended surfaces,** which include fins mounted perpendicularly to the tubes or spiral-wound fins, pin fins, plate fins, etc., friction data for the specific surface involved should be used. For details, see Kays and London, *Compact Heat Exchangers,* 2d

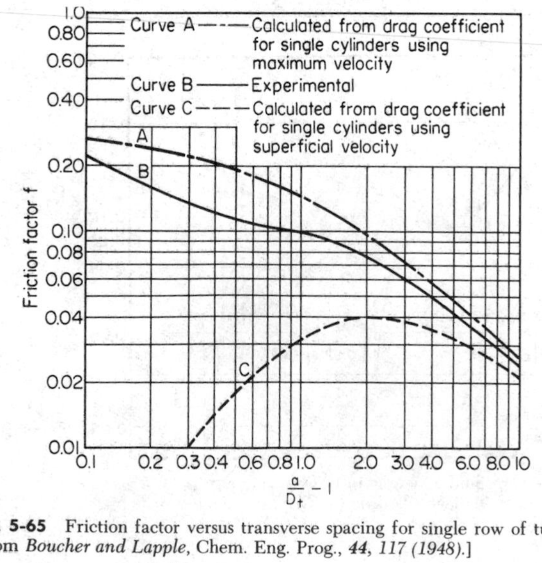

FIG. 5-65 Friction factor versus transverse spacing for single row of tubes. [From *Boucher and Lapple,* Chem. Eng. Prog., *44, 117 (1948).*]

ed., McGraw-Hill, New York, 1964. If specific data are not available, the correlation given by Gunter and Shaw (loc. cit.) can be used as an approximation.

For the case in which a large temperature change occurs in a gas flowing across a tube bundle, it is necessary to select a suitable mean temperature upon which to base the gas properties. This is given by

$$t_m = t_t + K\,\Delta t_m \tag{5-192}$$

where t_t = average tube-wall temperature, °C or °F; K = a constant, dimensionless; and Δt_m = log-mean temperature difference between the gas and the tubes, °C or °F. Values of K averaged from the recommendations of Chilton and Genereaux [*Trans. Am. Inst. Chem. Eng.,* **29,** 151–173 (1933)] and Grimison (loc. cit.) are as follows: for in-line tubes, 0.9 for cooling and −0.9 for heating; for staggered tubes, 0.75 for cooling and −0.8 for heating.

In the case of nonisothermal flow of **liquids** across tube bundles, the friction factor is appreciably increased if the liquid is being cooled and decreased if the liquid is being heated. The factors given for nonisothermal flow of liquids in pipes (see subsection "Nonisothermal Flow") are recommended for use in this case.

A method for computing pressure drop for two-phase, gas-liquid, horizontal cross-flow through tube banks is given by Diehl and Unruh [*Pet. Refiner,* **37**(10), 124–128 (1958)].

Transition Region This region extends approximately over the Reynolds-number [Eq. (5-191)] range of 200 to 2000. The friction-factor curves shown in Fig. 5-66 for five different configurations were obtained from Bergelin, Brown, and Doberstein [*Trans. Am. Soc. Mech. Eng.,* **74,** 953–960 (1952)]. Pressure drop for flow of liquids is given by

$$\Delta p = \frac{4 f_T N_r \rho V_{max}^2}{2 g_c}\left(\frac{\mu_s}{\mu_b}\right)^{0.14} \tag{5-193}$$

where f_T = friction factor obtained from Fig. 5-66, dimensionless; N_r = number of major restrictions encountered in flow through the bank (equal to number of rows when minimum free-flow area occurs in transverse openings and to number of rows less 1 when it occurs in the diagonal openings), dimensionless; ρ = fluid density; V_{max} = fluid velocity through minimum free-flow area; g_c = dimensional constant; μ_s = fluid viscosity at tube surface temperature; and μ_b = fluid viscosity at average bulk temperature. The use of Fig. 5-66 gives the friction factor within ±25 percent.

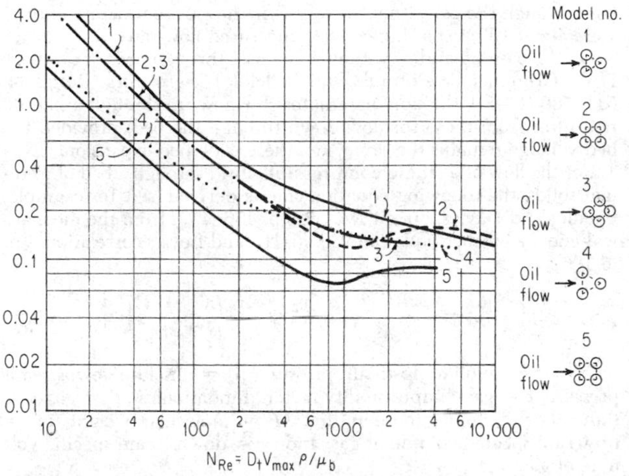

FIG. 5-66 Friction factors for transition-region flow across tube banks. (Pitch is the minimum center-to-center tube spacing.) [From *Bergelin, Brown, and Doberstein,* Trans. Am. Soc. Mech. Eng., *74, 953 (1952).*]

Model	Rows	D_t, in	Pitch/D_t
1	10	⅜	1.25
2	10	⅜	1.25
3	14	⅜	1.25
4	10	⅜	1.50
5	10	⅜	1.50

Laminar Region Bergelin, Colburn, and Hull (Univ. Delaware Eng. Exp. Sta. Bull. 2, 1950) recommend the following equations for pressure drop with laminar flow [$(N_{Re})_v < 100$] of liquids across banks of plain tubes with pitch ratios P/D_t of 1.25 and 1.50:

$$\Delta p = \frac{280 N_r}{(N_{Re})_v} \left(\frac{D_t}{P}\right)^{1.6} \left(\frac{\mu_s}{\mu_b}\right)^m \left(\frac{\rho V_{max}^2}{2g_c}\right) \tag{5-194}$$

$$m = 0.57/(N_{Re})_v^{0.25} \tag{5-195}$$

where $(N_{Re})_v = D_v V_{max} \rho/\mu$, dimensionless; D_v = volumetric hydraulic diameter [(4 × free-bundle volume)/(exposed surface area of tubes)]; P = pitch (= a for in-line arrangements, = a or c, whichever is smaller, for staggered arrangement); and other quantities are as defined following Eq. (5-193). Bergelin et al. show that pressure drop per row is independent of the number of rows in the bank with laminar flow. Equations (5-194) and (5-195) will predict the pressure drop within about ±25 percent.

The validity of extrapolating Eq. (5-194) to pitch ratios larger than 1.50 is not known. The correlation of Gunter and Shaw (loc. cit.) can be used as an approximation for such cases.

For the laminar flow of nonnewtonian solutions across tube banks, see Adams and Bell [*Chem. Eng. Prog.*, **64**, *Symp. Ser.* 82, 133–145 (1968)].

BEDS OF SOLIDS

Fixed Beds of Granular Solids Pressure-drop data on the flow of fluids through beds of granular solids are not readily correlated because of the variety of granular materials and of their packing arrangement. For the flow of a **single incompressible fluid** through an incompressible bed of granular solids, the pressure drop or other flow characteristics can be predicted from the correlation given by Leva [*Chem. Eng.*, **56**(5), 115–117 (1949), or *Fluidization*, McGraw-Hill, New York, 1959]. In this correlation,

$$\Delta p = \frac{2 f_m G^2 L (1-\epsilon)^{3-n}}{D_p g_c \rho \phi_s^{3-n} \epsilon^3} \tag{5-196}$$

where Δp = pressure drop; L = depth of bed; g_c = dimensional constant; D_p = average particle diameter, defined as the diameter of a sphere of the same volume as the particle; ϵ = voidage (fractional free volume), dimensionless; n = exponent, a function of the modified Reynolds number N'_{Re} given in Fig. 5-67, dimensionless; ϕ_s = shape factor of the solid, defined as the quotient of the area of a sphere equivalent to the volume of the particle divided by the

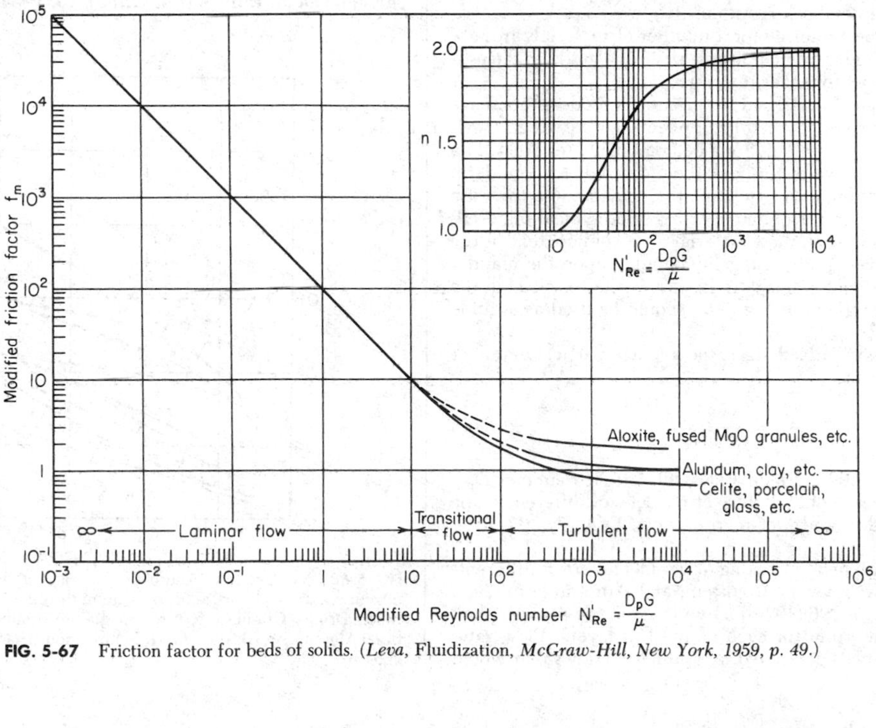

FIG. 5-67 Friction factor for beds of solids. (*Leva,* Fluidization, *McGraw-Hill, New York, 1959, p. 49.*)

TABLE 5-18 Shape Factors for Nonspherical Particles*

Material	Nature of grain	ϕ_s
Arnould's wire spirals	. . . .	0.2
Berl saddles	. . . .	.3
Coal dust, natural (up to 3/8 in.)	. . . .	.65
Coal dust, pulverized	. . . .	.73
Cork	. . . .	.69
Flue dust	Fused, spherical	.89
Flue dust	Fused, aggregates	.55
Fusain fibers	. . . .	.38
Glass, crushed	Jagged	.65
Mica flakes	. . . .	.28
Raschig rings	. . . .	.3
Sand:		
Average for various types	. . . .	.75
Flint sand	Jagged	.65
Flint sand	Jagged flakes	.43
Ottawa sand	Nearly spherical	.95
Sand	Rounded	.83
Sand	Angular	.73
Wilcox sand	Jagged	.60
Tungsten powder	. . . .	.89

*Values from Carman, *Trans. Inst. Chem. Eng. (London)*, **15**, 150–166 (1937), except value for Raschig rings from Brown et al., *Unit Operations*, Wiley, New York, 1950, p. 214.

actual surface of the particle, dimensionless; G = fluid superficial mass velocity based on empty chamber cross section; ρ = fluid density; and f_m = friction factor, a function of N'_{Re} given in Fig. 5-67. The modified Reynolds number N'_{Re} is defined as

$$N'_{Re} = D_p G/\mu \tag{5-197}$$

where μ = fluid viscosity.

In laminar flow ($N_{Re'} < 10$),

$$f_m = 100/N_{Re'} \tag{5-198}$$

For nonspherical particles,

$$D_p = \frac{6(1-\epsilon)}{\phi_s S} \tag{5-199}$$

where S = specific surface, or area of particle surface per unit volume of bed = $S_0(1-\epsilon)$, m^2/m^3 (ft^2/ft^3); and S_0 = area of particle surface per unit volume of solids, m^2/m^3 (ft^2/ft^3).

Values of the **shape factor** ϕ_s for a number of materials are tabulated in Table 5-18. This tabulation will serve as a guide in the estimation of shape factors for other materials.

The **actual voidage** ϵ of the bed in question is used in the above equations; thus no correction need be applied for wall effect. However, if the voidage is known only for a bed of a diameter other than the one in question, the voidage should be corrected. As a guide in estimating the correction, curves relating the voidage with the ratio of particular diameter to vessel diameter for some typical materials are given in Fig. 5-68. With most materials, the void content of the bed can be varied over quite a range, depending upon the manner and rate in which the material is introduced into the vessel. Thus the values of the voidage given in Fig. 5-68 should be used as a guide only.

For granular solids of **mixed sizes**, the average particle diameter D_p can be calculated as

$$\frac{1}{D_p} = \frac{x}{D_{p,x}} \tag{5-200}$$

where x = weight fraction, dimensionless, of particle diameter $D_{p,x}$. The average particle size of a mixture of particles of different shape is discussed by Standish and McGregor [*Chem. Eng. Sci.*, **33**, 618–619 (1978)].

In flow tests through beds consisting of parallel **layers** of different-size spheres, Sparrow, Beavers, Goldstein, and Bahrami [*Am. Inst. Chem. Eng. J.*, **22**, 194–196 (1976)] found that the pressure gradient was constant and the same for each of the bed layers. Flow rates through the layered beds computed by adding the flows that would pass through the constituent layers as if they were separate entities were about 10 percent higher than measured flow rates.

On the basis of studies with flow of gases through beds of sands by Leva, Grummer, Weintraub, and Pollchik [*Chem. Eng. Prog.*, **44**, 511–520 (1948)], the flow parameters for flow up through a bed do not differ from those for flow down through the bed, provided the bed voidage remains the same; i.e., there is no bed expansion.

For the flow of a **single compressible fluid** through a bed of granular solids, the following equation can be derived (see, for example, Cremer and Davies, op. cit., vol. 2, pp. 436–437) from the mechanical-energy-balance equation, Eq. (5-41), and Leva's correlation, Eq. (5-196):

$$p_1^2 - p_2^2 = \frac{2zRG^2T}{g_c M}\left[\ln\frac{v_2}{v_1} + \frac{2f_m L(1-\epsilon)^{3-n}}{\phi_s^{3-n}\epsilon^3 D_p}\right] \tag{5-201}$$

where p_1 = absolute upstream pressure; p_2 = absolute downstream pressure; z = gas compressibility factor, dimensionless; R = gas constant; T = absolute temperature; M = molecular weight; v_1 = upstream specific volume of gas; and v_2 = downstream specific volume of gas.

For an **approximation** of the pressure drop across beds of solids, the velocity-head concept can be employed. Equation (5-196) can be written as

$$\frac{\Delta p}{\rho} = \Delta h = \left[\frac{4f_m(1-\epsilon)^{3-n}}{\phi_s^{3-n}\epsilon^3}\right]\left(\frac{L}{D_p}\right)\left(\frac{V^2}{2g_c}\right) \tag{5-202}$$

where Δh = pressure-head loss, $V^2/2g_c$ = velocity head, and V = G/ρ = superficial fluid velocity. The pressure-head loss for a bed depth of one particle diameter (L/D_p = 1) for turbulent flow ($N'_{Re} > 100$) through a bed of spherical, or nearly so, particles can be shown from Eq. (5-202) and Figs. 5-67 and 5-68 to be

$$\Delta h \simeq 50(V^2/2g_c) \tag{5-203}$$

Other methods of correlation of pressure-drop data are given by Brown et al., *Unit Operations*, Wiley, New York, 1950, chap. 16; Coulson and Richardson, op. cit., vol. 2, chap. 1; Cremer and Davies, op. cit., vol. 2, pp. 376–421; Leva, *Fluidization*, McGraw-Hill, New York, 1959, chap. 3; and Zenz and Othmer, *Fluidization and Fluid-Particle Systems*, Reinhold, New York, 1960, chap. 5. For a review of single-phase flow through porous media and pore structure, see Dullien, *Chem. Eng. J. (Lausanne)*, **10**, 1–34 (1975).

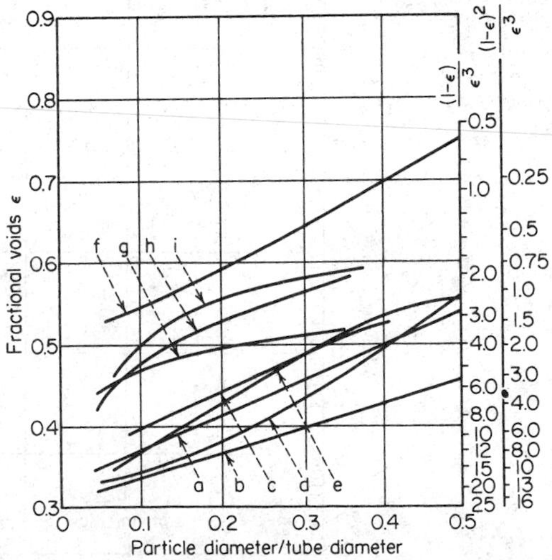

FIG. 5-68 Voidage in packed beds. Spherical: *a*, smooth, uniform; *b*, smooth, mixed; *c*, clay. Cylindrical: *d*, smooth, uniform; *e*, Alundum, uniform; *f*, clay Raschig rings. Granules: *g*, fused magnetite (synthetic ammonia catalyst); *h*, fused Alundum: *i*, Aloxite. (*Leva*, Fluidization, *McGraw-Hill, New York, 1959, p. 54.*)

The pressure drop in **laminar flow** of **nonnewtonian fluids** through beds of solids can be predicted from the correlation given by Christopher and Middleman [*Ind. Eng. Chem. Fundam.*, **4**, 422–426 (1965)]. In this correlation

$$\Delta p = \frac{150HLV^n(1-\varepsilon)^2}{D_p^2\phi_s^2\varepsilon^3} \tag{5-204}$$

$$H = \frac{K}{12}\left(9+\frac{3}{n}\right)^n\left[\frac{D_p^2\phi_s^2\varepsilon^4}{(1-\varepsilon)^2}\right]^{(1-n)/2} \tag{5-205}$$

where $V = G/\rho$ = superficial fluid velocity, K,n = power-law material constants, and all other symbols are as defined in Eq. (5-196). This correlation is supported by measurements of Christopher and Middleman (loc. cit.), Gregory and Griskey [*Am. Inst. Chem. Eng. J.*, **13**, 122–125 (1967)], Yu, Wen, and Bailie [*Can. J. Chem. Eng.*, **46**, 149–154 (1968)], Siskovic, Gregory, and Griskey [*Am. Inst. Chem. Eng. J.*, **17**, 281–285 (1971)], Kemblowski and Mertl [*Chem. Eng. Sci.*, **29**, 213–223 (1974)], and Kemblowski and Dziubinski [*Rheol. Acta*, **17**, 176–187 (1978)]. The measurements cover the range n = 0.50 to 1.60, and modified Reynolds number $N'_{Re} = 10^{-8}$ to 10, where

$$N'_{Re} = D_pV^{2-n}\rho/g_cH \tag{5-206}$$

For n = 1 (Newtonian fluid), Eq. (5-206) reduces to Eq. (5-197).

The maximum shear rate in the bed is

$$\left|\frac{du}{dr}\right|_{max} = \frac{3n+1}{4n}\left[\frac{12V(1-\varepsilon)}{D_p\phi_s\varepsilon^2}\right] \tag{5-207}$$

(Christopher and Middleman, loc. cit.), and the maximum shear stress is

$$\tau_{max} = \frac{\Delta_p\, D_p\varepsilon}{12.5L(1-\varepsilon)} \tag{5-208}$$

(Gregory and Griskey, loc. cit.). For n = 1 (newtonian fluid), $H = K = \mu$, fluid viscosity. In this limit Eqs. (5-204) and (5-205) reduce to Eq. (5-196) for the laminar region [see Eq. (5-198)] with a numerical constant of 25⁄6 instead of 2. For flow of nonnewtonian **viscoelastic fluids** through beds of granular solids, see Marshall and Metzner, *Ind. Eng. Chem. Fundam.*, **6**, 393–400 (1967); Siskovic, Gregory, and Griskey, loc. cit.; and Kemblowski and Dziubinski, loc. cit. See also Savins, *Ind. Eng., Chem.*, **61**(10), 18–47 (1969), for a review.

Tower Packings As in the case of beds of granular solids, pressure-drop data on the flow of fluids through beds of tower packings are not readily correlated. For the flow of a **single fluid** through a bed of tower packing (e.g., flow of gas over dry packing), pressure drop and other flow characteristics can be calculated by the methods described above for beds of granular solids (see also Sec. 18).

For the **countercurrent flow** of **liquids** and **gases**, the pressure drop is increased because of reduction in free volume by the liquid. A summary of data on **pressure drop** and **flooding** for several types of commercial packing is given in *Technical Data Related to Tower Packing*, Chemical Process Products Div., Norton Co., Akron, Ohio; and by Eckert, *Chem. Eng. Prog.*, **66**(3), 39–44 (1970). For countercurrent flow of liquids with a high-density difference, see Watson and McNeese, *Am. Inst. Chem. Eng. J.*, **19**, 230–237 (1973).

Some data and correlations on flow pattern, liquid distribution, liquid holdup, and pressure drop for **cocurrent flow** of **liquids and gases** through beds of tower packings and of granular solids are given by Reiss [*Ind. Eng. Chem. Process Des. Dev.*, **6**, 486–499 (1967)], Sweeney [*Am. Inst. Chem. Eng. J.*, **13**, 663–669 (1967)], Turpin and Huntington [*Am. Inst. Chem. Eng. J.*, **13**, 1196–1202 (1967)], and Specchia and Baldi [*Chem. Eng. Sci.*, **32**, 515–523 (1977)].

Fluidized Beds If the velocity of a fluid flowing up through a bed of granular solids is gradually increased, conditions will be met in which the fluid drag force, i.e., pressure drop × vessel cross-sectional area, will just equal the weight of the solids, and the particles will just begin to be in motion. This will be the onset of fluidization, or minimum fluidization. Since for most gas-solid systems, fluidization will begin for $N'_{Re} = D_pG_{mf}/\mu < 10$, values of mass velocity and voidage for the onset of fluidization can be related by (Leva, *Fluidization*, McGraw-Hill, New York, 1959, p. 63)

$$G_{mf} = \frac{0.005D_p^2g\rho_f(\rho_s-\rho_f)\phi_s^2\epsilon_{mf}^3}{\mu(1-\epsilon_{mf})} \tag{5-209}$$

where G_{mf} = fluid superficial mass velocity for minimum fluidization; D_p = particle diameter; g = local acceleration due to gravity; ρ_f = fluid density; ρ_s = solids density; ϕ_s = particle shape factor, dimensionless; ϵ_{mf} = voidage at minimum fluidization, dimensionless; and μ = fluid viscosity.

The **minimum voidage** can be determined by passing the fluid up through the bed and noting the bed height at incipient particle motion or minimum fluidization, and thus

$$\epsilon_{mf} = 1 - \frac{W}{L_{mf}A(\rho_s-\rho_f)} \tag{5-210}$$

where W = weight of solids in bed, L_{mf} = height of bed at minimum fluidization, and A = cross-sectional area of vessel. This and other methods of determining the minimum voidage are described by Leva (op. cit., pp. 20–21, 63). Typical values of minimum voidage as reported in the literature have been collected by Leva and are shown as a function of particle diameter in Fig. 5-69.

To improve the usefulness of Eq. (5-209), Leva and coworkers found (see Leva, op. cit., pp. 63–64) that variables ϕ_s and ϵ_{mf} could be related to N'_{Re}, thus reducing Eq. (5-209) to the following dimensional equation for $N'_{Re} < 5.0$:

$$G'_{mf} = C_1\frac{[\rho_f(\rho_s-\rho_f)]^{0.94}}{\mu^{0.88}}d_p^{1.82} \tag{5-211}$$

where G'_{mf} = fluid superficial mass velocity for minimum fluidization; d_p = particle diameter; ρ_f and ρ_s = density of fluid and solids respectively; μ = fluid viscosity; and C_1 = dimensional constant, 0.00930 in SI units (0.0284 in U.S. customary units). For beds of solids of mixed sizes, fluidization takes place over a range of fluid velocity [Vaid and Gupta, *Can. J. Chem. Eng.*, **56**, 292–296 (1978)].

The state of **bed expansion** can be predicted for both liquid-solid and gas-solid systems by the methods described in Leva, loc. cit., chap. 4 (see also Sec. 20).

Porous Media The flow of fluids through consolidated porous media is similar to that through beds of granular solids. However, because consolidated porous media consist of a highly complex net-

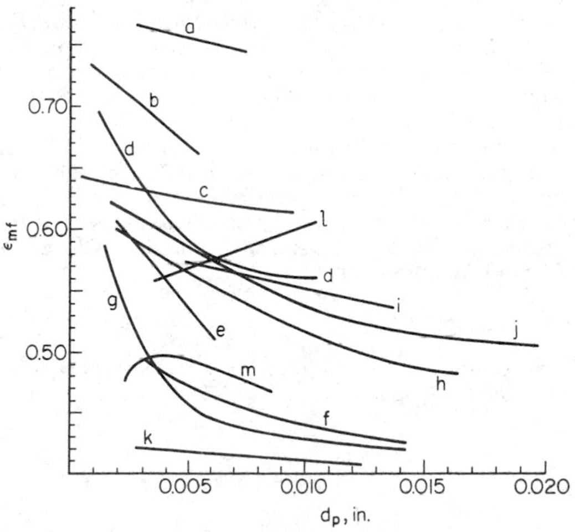

FIG. 5-69 Typical values of minimum voidage. (*a*) Soft brick; (*b*) absorption carbon; (*c*) broken Raschig rings; (*d*) coal and glass powder; (*e*) silicone carbide; (*f*) sand; (*g*) round sand, ϕ_s = 0.86; (*h*) sharp sand, ϕ_s = 0.67; (*i*) Fischer-Tropsch catalyst, ϕ_s = 0.58; (*j*) anthracite coal, ϕ_s = 0.63; (*k*) mixed round sand, ϕ_s = 0.86; (*l*) coke; (*m*) silicon carbide. To convert inches to meters, multiply by 0.0254. (*For references, see Leva,* Fluidization, *McGraw-Hill, New York, 1959, p. 21.*)

work of channels, description of the flow in terms of particle size or surface area, as for beds of granular solids, is difficult. For a review of single-phase flow through porous media and pore structure, see Dullien, loc. cit. In general, the function of pressure drop versus flow rate is similar in form to that for beds of granular solids; i.e., transition from laminar flow to turbulent flow is gradual. For this reason the function must include a viscous term and an inertial term, the form being as follows for the flow of an **incompressible fluid:**

$$(p_1 - p_2)/L = \alpha\mu V/g_c + \beta\rho V^2/g_c \qquad (5\text{-}212)$$

and for the isothermal flow of an ideal gas,

$$\frac{p_1^2 - p_2^2}{L} = \frac{2\alpha RT\mu G}{Mg_c} + \left(\beta + \frac{1}{L}\ln\frac{p_1}{p_2}\right)\left(\frac{2RTG^2}{Mg_c}\right) \qquad (5\text{-}213)$$

where p_1 = absolute upstream pressure; p_2 = absolute downstream pressure; L = thickness of the medium; V = superficial velocity of fluid (based on total cross section); G = superficial mass velocity of fluid; ρ = fluid density; μ = fluid viscosity; g_c = dimensional constant; M = molecular weight of gas; R = gas constant; T = absolute temperature; α = viscous resistance coefficient, $1/\text{m}^2$ ($1/\text{ft}^2$); and β = inertial resistance coefficient, 1/m (1/ft). For additional details, see Green and Duwez, *J. Appl. Mech.*, **18,** 39–45 (1951).

The isothermal-flow equation for gases may be used for **nonisothermal-flow** conditions by evaluating the fluid properties at the log-mean temperature [Koh, Dutton, Benson, and Fortini, *J. Heat Transfer*, **99,** 367–373 (1977)]. The general flow Eq. (5-212) is widely known as the **Forchheimer equation.**

For purely viscous flow, the second term (involving V^2) on the right-hand side of Eq. (5-212) becomes negligible, and the resulting equation is known as **Darcy's equation** of flow through porous media. The quantity $1/\alpha$ is referred to as the **permeability coefficient.** For further information on the Darcy equation, see Muskat, *Physical Principles of Oil Production*, McGraw-Hill, New York, 1949, chap. 3; Cremer and Davies, op. cit., vol. 2, pp. 406–414; and Rouse, *Engineering Hydraulics*, Wiley, New York, 1950, chap. V.

Values of α and β are determined experimentally for each type of porous medium. Data on pressure drop as a function of flow rate for various fluids are generally available from the manufacturer of such porous media as sintered metals. As a guide, values of α and β for specimens of sintered stainless steel, iron, and bronze are given by Green and Duwez (loc. cit.), and of sandstones by Cremer and Davies (op. cit., vol. 2, p. 417) and Firoozabadi and Katz [*J. Pet. Technol.*, **31,** 211–216 (1979)]. Pressure-drop data for flow of air and of liquids of various viscosities through specimens of sintered stainless steel and bronze are presented by Langhammer and Glick [*Prod. Eng.*, **24**(4), 179–182 (1953)]. For flow of gases through stainless-steel and copper porous-powder materials under isothermal and nonisothermal conditions, see Koh, Dutton, Benson, and Fortini, loc. cit.

Capillary diameters of porous media may be of the same order of magnitude as the mean free path of the molecules of the diffusing gas, even at several atmospheres pressure. For these cases, the slip-flow and molecular-flow equations (see subsections "Slip Flow" and "Molecular Flow") may be modified to apply to the flow of gases through porous media [see Monet and Vermeulen, *Chem. Eng. Prog.*, **55,** *Symp. Ser.* 25 (1959)].

FLOW AROUND OBJECTS

Vortex Shedding In the flow of fluids past objects and through orifices or similar restrictions, fluid vortices are shed periodically downstream from the object or other initiating element. Objects such as smokestacks, chemical-processing columns, suspended pipe lines, and electrical transmission lines can be subjected to damaging vibrations and forces due to the vortices, especially if the frequency of vortex shedding is close to the natural vibration frequency of the object. Also, such vortex shedding can produce sound, such as in the "Aeolian harp" or singing wires. For reviews of literature, see Krzywoblocki, *Appl. Mech. Rev.*, **6,** 393–397 (1953); and Marris, *J. Basic Eng.*, **86,** 185–196 (1964).

Development of the **vortex street,** commonly called "von Kármán vortex street," behind a cylindrical object in a flowing fluid is shown in Fig. 5-70 (see Rouse, op. cit., pp. 129–130). Velocity of the vortex street is given by

$$V_v = 0.86V \qquad (5\text{-}214)$$

where V_v = velocity of vortex street and V = free-stream velocity of the fluid.

Investigations have shown that the **frequency of vortex shedding** may be computed from the Strouhal number N_S, which in turn is a function of the Reynolds number N_{Re}. Over a wide range of Reynolds numbers, the Strouhal number is approximately constant (see Rouse, op. cit., pp. 129–130):

$$N_S = fD/V = 0.19 \qquad \text{for } 500 < N_{Re} < 10^5 \qquad (5\text{-}215)$$

where f = frequency; D = diameter of cylinder or effective width of object; V = free-stream velocity; N_{Re} = Reynolds number = $VD\rho/\mu$, dimensionless; ρ = fluid density; and μ = fluid viscosity. For $N_{Re} < 500$, the Strouhal number decreases with decreasing Reynolds number. Below N_{Re} = 40, the vortices are difficult to detect. For $N_{Re} > 10^5$, the Strouhal number increases rapidly with increasing Reynolds number. Above $N_{Re} = 4 \times 10^5$, the vortices are weak and aperiodic [Steidel, *J. Appl. Mech.*, **23,** 649–650 (1956)]. However, tall stacks have been observed to vibrate in winds with stack Reynolds numbers greater than 10^6 [see Den Hartog, *Proc. Nat. Acad. Sci.*, **40,** 155–157 (1954); Farquharson's discussion, pp. 1386–1387, of Ozker and Smith, *Trans. Am. Soc. Mech. Eng.*, **78,** 1381–1391 (1956); and Smith and McCarthy, *Mech. Eng.*, **87,** 38–41 (1965)]. Frequency measurements on a cylinder, 0.9 m (3 ft) in diameter, at $3.6 \times 10^5 < N_{Re} < 1.87 \times 10^7$ are discussed by Jones (*Am. Soc. Mech. Eng.*, Pap. 68-FE-36). Frequency of vibration of an object is equal to the frequency of vortex shedding if the frequency is not close to resonance; if it is 80 to 120 percent of resonance, the vortex shedding takes place at the resonant frequency of the object [Den Hartog, loc. cit.; see also Goldwag and Berry, *J. Eng. Power*, **90,** 213–217 (1968)].

An **alternating lateral force** F_K acting on the cylinder results from the vortex shedding. This force is perpendicular to the direction of flow. The maximum lateral force F_K is in the direction away from the last vortex, and for any object this force is given by (see Den Hartog, *Mechanical Vibrations*, 4th ed., McGraw-Hill, New York, 1956, pp. 305–309)

$$F_K = C_K A(\rho V^2/2g_c) \qquad (5\text{-}216)$$

where F_K = lateral force (sometimes called von Kármán force); C_K = von Kármán coefficient, dimensionless; A = projected area (perpendicular to the direction of flow); ρ = fluid density; V = free-stream velocity of fluid; and g_c = dimensional constant. C_K is dependent upon the shape of the object and upon flow characteristics. For a cylinder, C_K = 1.7 and A = diameter × length [see Rouse, op. cit., pp. 129–130; or Steinman, *Am. Sci.*, **42,** 397–438, 460 (1954)].

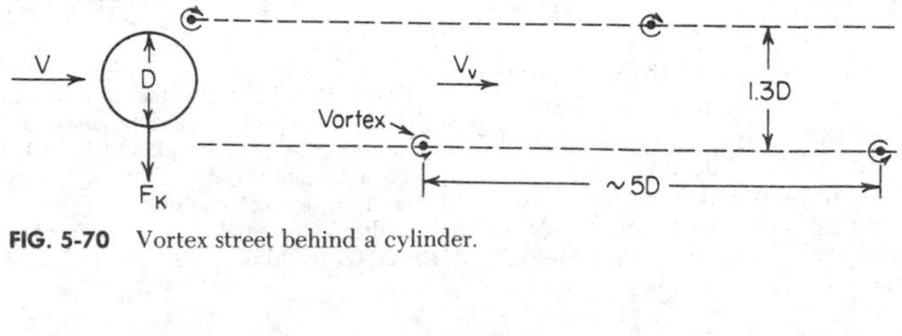

FIG. 5-70 Vortex street behind a cylinder.

Thus this force is about twice that due to fluid drag alone. If the cylinder is vibrating, the diameter term is replaced by an effective diameter which will never exceed twice the cylinder diameter (see Rouse, op. cit., pp. 129–130).

The following references pertain to discussions of vortex shedding in specific engineering structures: steel stacks (Ozker and Smith, loc. cit.; Smith and McCarthy, loc. cit.); chemical-processing columns [Freese, *J. Eng. Ind.*, **81,** 77–91 (1959)]; heat exchangers [Gainsboro, *Chem. Eng. Prog.*, **64**(3), 85–88 (1968); "Flow-Induced Vibration in Heat Exchangers," *Symp. Proc.*, ASME, New York, 1970]; suspended pipe lines [Baird, *Trans. Am. Soc. Mech. Eng.*, **77,** 797–804 (1955)]; and suspended cable (Steidel, loc. cit.).

Boundary-Layer Behavior When a fluid of low viscosity flows with an initial uniform velocity over a surface such as a flat plate, the velocity decreases until it is zero at the surface of the plate. This decrease in velocity takes place in a small layer of fluid called the "boundary layer." The flow in this layer may be laminar or turbulent; the transition can be estimated from the length Reynolds number.

Following are presented equations for **boundary-layer thickness** and **drag** for flow around finite flat plates (Prandtl, *Essentials of Fluid Dynamics*, Hafner, New York, 1952; Prandtl and Tietjens, *Applied Hydro- and Aero-Mechanics*, McGraw-Hill, New York, 1934; Schlichting, *Boundary-Layer Theory*, 7th ed., McGraw-Hill, New York, 1979) and continuous flat and cylindrical surfaces [Sakiadis, *Am. Inst. Chem. Eng. J.*, **7,** 26, 221, 467 (1961)]. Nomenclature common to these equations is: b = width of flat plate; D = total drag or force; g_c = dimensional constant; L = exposed length of surface; q = volume flow rate of fluid entrained; r = radius of continuous cylindrical surface; V = velocity of finite flat plate in stationary fluid, or free-stream velocity of fluid approaching stationary finite flat plate, or velocity of continuous surface; x = distance from leading edge or from orifice or slot to a point along the surface; $(N_{Re})_r = Vpr/\mu$ = Reynolds number based on cylinder radius, dimensionless; $(N_{Re})_x = V\rho x/\mu$ = Reynolds number based on axial length, dimensionless; $(N_{Re})_L$ = length Reynolds number where $x = L$, dimensionless; δ = boundary-layer thickness where local velocity $= 0.99V$; μ = fluid viscosity; and ρ = fluid density.

Finite Flat Plate For the case of the finite plate parallel with the fluid stream, the **critical length Reynolds number** at which the boundary layer becomes turbulent is (Schlichting, op. cit., p. 638)

$$(N_{Re})_x = V\rho x/\mu = 500{,}000 \qquad (5\text{-}217)$$

For a **laminar boundary layer,** the boundary-layer thickness along the plate is given by (Prandtl, op. cit.; Schlichting, op. cit.)

$$\delta = 5(x)(N_{Re})_x^{-0.5} \qquad (5\text{-}218)$$

Total drag on the plate, i.e., drag on both surfaces of the plate, is given by (Prandtl, op. cit.; Schlichting, op. cit.)

$$D = 1.328bLV^2(\rho/g_c)(N_{Re})_L^{-0.5} \qquad (5\text{-}219)$$

For **nonnewtonian (power-law) fluids** total drag on the plate is given by [Acrivos, Shah, and Petersen, *Am. Inst. Chem. Eng. J.*, **6,** 312–317 (1960); Hsu, ibid., **15,** 367–370 (1969)]

$$D = CbLV^2(\rho/g_c)(N_{Re})_L^{-1/(1+n)} \qquad (5\text{-}220)$$

n = 0.2	0.3	0.4	0.5	0.6	0.7	0.8	0.9	1
C = 2.075	1.958	1.838	1.727	1.627	1.538	1.460	1.390	1.328

where $(N_{Re})_L = \rho V^{2-n}L^n/g_cK$ = modified Reynolds number, dimensionless; and K,n = material constants [see Eq. (5-3)]. For a **turbulent boundary layer,** $(N_{Re})_L < 10^7$, the boundary-layer thickness is given by (Prandtl and Tietjens, op. cit.; Schlichting, op. cit.)

$$\delta = 0.37(x)(N_{Re})_x^{-0.2} \qquad (5\text{-}221)$$

and the total drag on both sides of the plate is given by (Prandtl and Tietjens, op. cit.; Schlichting, op. cit.)

$$D = 0.072bLV^2(\rho/g_c)(N_{Re})_L^{-0.2} \qquad (5\text{-}222)$$

For higher $(N_{Re})_x$, see Schlichting, op. cit. When the laminar boundary layer is a large part of the total length of the surface, the total drag can be computed by the following method (Schlichting, op. cit., p. 639): total drag for length L is computed by assuming a turbulent boundary layer from the leading edge; from this total drag the turbulent drag for length x_{crit} is subtracted; and to this total drag the laminar drag for the length x_{crit} is added.

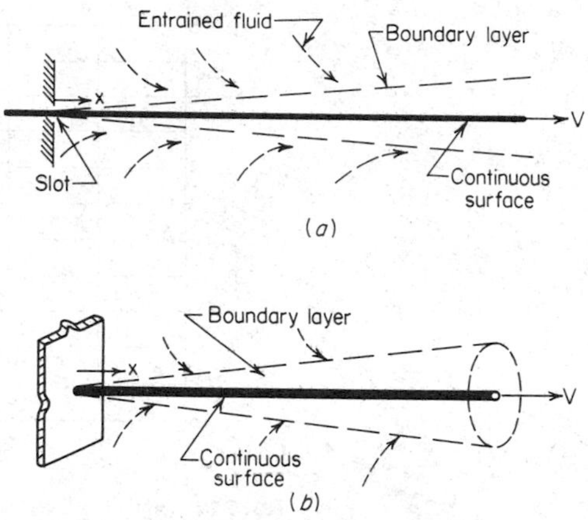

FIG. 5-71 Continuous surface. (*a*) Continuous flat surface. (*b*) Continuous cylindrical surface. [*Sakiadis*, Am. Inst. Chem. Eng. J., *7, 221, 467 (1961).*]

Finite Cylindrical Surface The characteristics of a **laminar boundary layer** along a finite cylindrical surface are given by Glauert and Lighthill [*Proc. R. Soc. (London),* **230A,** 188–203 (1955)], Jaffe and Okamura [Z. *Angew. Math. Phys.*, **19,** 564–574 (1968)], and Stewartson [*Q. Appl. Math.*, **13,** 113–122 (1955)]. For a **turbulent boundary layer** the total drag is given by

$$D = \bar{c}_f \pi r L(\rho/g_c)V^2 \qquad (5\text{-}223)$$

where $\bar{c}_f$ = average friction coefficient, dimensionless, given by [White, *J. Basic Eng.*, **94,** 200–206 (1972)]

$$\bar{c}_f = 0.0015 + [0.30 + 0.015(L/r)^{0.4}](N_{Re})_L^{-1/3} \qquad (5\text{-}224)$$

for $(N_{Re})_L = 10^6$ to 10^9, and $L/r < 10^6$.

Continuous Flat Surface This case (see Sakiadis, loc. cit.) is illustrated in Fig. 5-71*a*. The **critical length Reynolds number** at which the boundary layer becomes turbulent may be higher than given by Eq. (5-217) [see Tsou, Sparrow, and Kurtz, *J. Fluid Mech.*, **26,** 145–161 (1966)]. For a **laminar boundary layer,** its thickness is given by

$$\delta = 6.37(x)(N_{Re})_x^{-0.5} \qquad (5\text{-}225)$$

and the total drag on both sides of the surface is given by

$$D = 1.776bLV^2(\rho/g_c)(N_{Re})_L^{-0.5} \qquad (5\text{-}226)$$

Total quantity of fluid entrained or pumped by the surface is given by

$$q = 3.232bLV(N_{Re})_L^{-0.5} \qquad (5\text{-}227)$$

Tsou, Sparrow, and Goldstein [*Int. J. Heat Mass Transfer*, **10,** 219–235 (1967)] and Szeri, Yates, and Hai [*J. Lubr. Technol.*, **98,** 145–156 (1976)] experimentally verified the theoretical velocity field.

The laminar boundary layer on continuous flat surfaces moving through a **nonnewtonian (power-law) fluid** is discussed by Fox, Erickson, and Fan [*Am. Inst. Chem. Eng. J.*, **15,** 327–333 (1969)]. For a **turbulent boundary layer,** thickness is given by

$$\delta = 1.01(x)(N_{Re})_x^{-0.2} \qquad (5\text{-}228)$$

Total drag on both sides of the surface is given by

$$D = 0.056bLV^2(\rho/g_c)(N_{Re})_L^{-0.2} \qquad (5\text{-}229)$$

and total quantity of fluid entrained or pumped by the surface is given by

$$q = 0.252bLV(N_{Re})_L^{-0.2} \qquad (5\text{-}230)$$

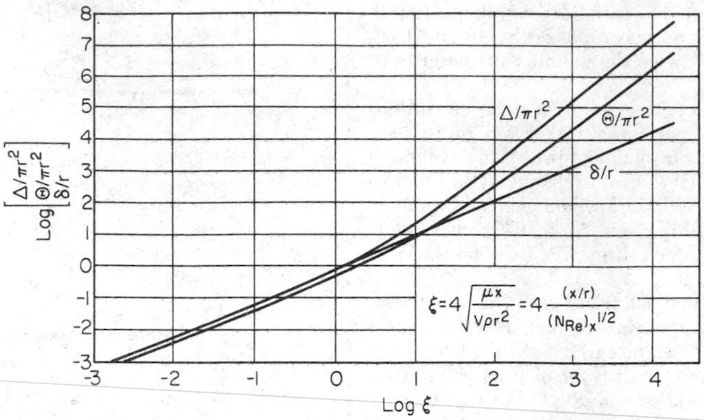

FIG. 5-72 Boundary-layer parameters for continuous cylindrical surfaces. [*Sakiadis, Am. Inst. Chem. J., 7, 467 (1961).*]

When the laminar boundary layer is a large part of the total length of the surface, the total drag can be computed by the method described for the finite flat plate.

An improved analysis of the turbulent boundary layer is presented by Tsou, Sparrow, and Goldstein (loc. cit.). Friction measurements by these investigators indicate that Eq. (5-229) underestimates the drag by about 15 percent.

Continuous Cylindrical Surface This case (see Sakiadis, loc. cit.) is illustrated in Fig. 5-71*b*. The **critical length Reynolds number** at which the boundary layer becomes turbulent is

$$(N_{Re})_{x,crit} = V\rho x/\mu = 200{,}000 \qquad (5\text{-}231)$$

(Barnett, in *Applied Polymer Symposia*, vol. 6, Interscience, New York, 1967, p. 53). For a **laminar boundary layer,** boundary-layer thickness δ can be determined from Fig. 5-72; total drag on the surface is given by

$$D = (\rho/g_c)V^2\Theta \qquad (5\text{-}232)$$

where Θ, called the "momentum area," is obtained from Fig. 5-72 for $x = L$; total quantity of fluid entrained or pumped by the surface is given by

$$q = V\,\Delta \qquad (5\text{-}233)$$

where Δ, called the "displacement area," is obtained from Fig. 5-72 for $x = L$. Further analysis of the laminar boundary layer is given by Crane [*Z. Angew. Math. Phys.*, **23**, 201–212 (1972)].

For a **turbulent boundary layer,** the drag can be estimated by Eqs. (5-223) and (5-224), developed for a finite cylindrical surface. Drag force measured on filaments by Kwon and Prevorsek [*J. Eng. Ind.*, **101**, 73–79 (1979)] is higher than predicted, possibly due to filament swaying.

The laminar boundary layer on a **deforming continuous** flat **surface,** with surface velocity varying linearly with axial distance from the slot, is discussed by Vleggaar [*Chem. Eng. Sci.*, **32**, 1517–1525 (1977)]. The corresponding case of a deforming continuous cylindrical surface is discussed by Crane [*Z. Angew. Math. Phys.*, **26**, 619–622 (1975)] and Vleggaar (loc. cit.).

ENTRAINMENT OF LIQUID FILMS

When a solid surface is withdrawn from a liquid pool in which it is submerged, a liquid film adheres to it. This is illustrated in Fig. 5-73. The thickness of film adhering to the surface depends on surface geometry and inclination, fluid properties, and speed of withdrawal. Tallmadge and Gutfinger [*Ind. Eng. Chem.*, **59**(1), 19–34 (1967)] review theory and early work.

Flat Plate The liquid-film thickness adhering to a flat plate withdrawn from a liquid pool, at low speed of withdrawal, is given by

$$h\left(\frac{\rho g}{g_c\sigma}\right)^{1/2} = \frac{0.944}{(1 - \cos\phi)^{1/2}}\,Ca^{2/3} \qquad (5\text{-}234)$$

where h = liquid-film thickness; ρ = liquid density; g = acceleration due to gravity; g_c = dimensional constant; σ = liquid surface tension; ϕ = angle of inclination with the horizontal (see Fig. 5-73); Ca = $\mu V/g_c\sigma$ = capillary number, dimensionless; μ = liquid viscosity; and V = speed of withdrawal. Equation (5-234) is valid for liquids other than water, and Ca < 0.03. For higher withdrawal speed, see Esmail and Hummel, *Am. Inst. Chem. Eng. J.*, **21**, 958–965 (1975); Spiers, Subbaraman, and Wilkinson, *Chem. Eng. Sci.*, **29**, 389–396 (1974); and White and Tallmadge [*Chem. Eng. Sci.*, **20**, 33–37 (1965). For higher withdrawal speed of an inclined flat surface, see Esmail and Hummel, *Chem. Eng. Sci.*, **34**, 125–129 (1979); Tallmadge, *Am. Inst. Chem. Eng. J.*, **17**, 243–246 (1971); and Tharmalingam and Wilkinson, *Chem. Eng. Sci.*, **33**, 1481–1487 (1978). For meniscus shape, see Esmail and Hummel, loc. cit., 1975; and Lee and Tallmadge, *Ind. Eng. Chem. Fundam.*, **15**, 258–266 (1976).

Cylindrical Surface The liquid-film thickness adhering to a vertical cylindrical surface withdrawn from a liquid pool, at low speed of withdrawal, is given by

$$h/R = 1.33Ca^{2/3}/(1 - 1.33Ca^{2/3}) \qquad (5\text{-}235)$$

where h = liquid-film thickness; R = radius of cylindrical surface; and Ca = capillary number, dimensionless. Equation (5-235) is valid for liquids other than water, Ca < 0.1, and $R(\rho g/g_c\sigma)^{1/2} < 0.085$; symbols are as defined for Eq. (5-234). For $R(\rho g/g_c\sigma)^{1/2} > 0.085$, see White and Tallmadge, *Am. Inst. Chem. Eng. J.*, **13**, 745–750 (1967). A cylindrical surface with $R(\rho g/g_c\sigma)^{1/2} > 3$ may be treated as a flat plate.

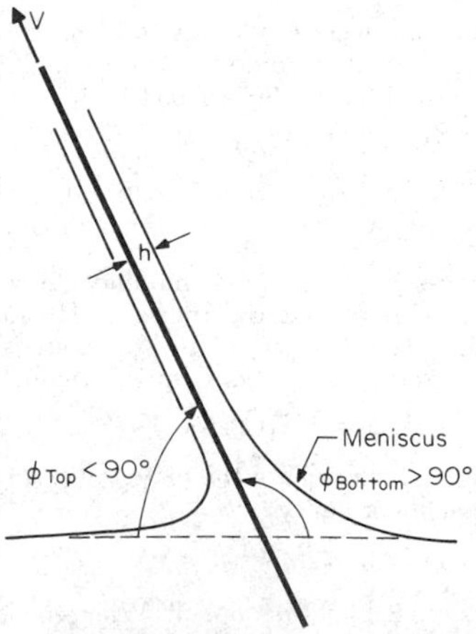

FIG. 5-73 Liquid entrainment by flat plate withdrawn from pool.

FALLING FILMS

Minimum Wetting Rate The minimum liquid rate required for complete wetting of a vertical surface is from about 0.03 to 0.3 kg/(s·m) [0.02 to 0.2 lb/(s·ft)] for water at room temperature. The exact rate depends upon the geometry and nature of the vertical surface, the conditioning of the surface, the surface tension of the liquid, and the mass transfer between the surrounding gas and the liquid. For additional details, see Morris and Jackson, *Absorption Towers*, Butterworth, London, 1953; Norman, *Absorption Distillation and Cooling Towers*, Wiley, New York, 1961; Ponter et al., *Int. J. Heat and Mass Transfer*, **10**, 349–359 (1967); *Trans. Inst. Chem. Eng. (London)*, **45**, 345–352 (1967); and Stainthorp and Allen, *Trans. Inst. Chem. Eng. (London)*, **43**, 85–91 (1965).

Laminar Flow Theoretical equations describing **laminar flow** of liquid films on **flat surfaces**, as presented by Cooper, Drew, and McAdams [*Ind. Eng. Chem.*, **26**, 428–431 (1934)] and by Fallah, Hunter, and Nash [*J. Soc. Chem. Ind.*, **53**, 369T–379T (1934)], are as follows:

$$m = \left[\frac{3\Gamma\mu}{g\rho_f(\rho_f - \rho_c)\sin\phi}\right]^{1/3} \tag{5-236}$$

$$u = \frac{g(\rho_f - \rho_c)\sin\phi}{\mu}\left(mx - \frac{x^2}{2}\right) \tag{5-237}$$

$$V = \frac{\Gamma}{m\rho_f} = \frac{g(\rho_f - \rho_c)m^2\sin\phi}{3\mu} \tag{5-238}$$

$$u_{max}/V = 1.50 \tag{5-239}$$

where m = film thickness; Γ = liquid loading per unit width of plate, kg/(s·m) [lb/(s·ft)]; μ = film liquid viscosity; g = acceleration due to gravity; ρ_f = density of film liquid; ρ_c = density of surrounding fluid; ϕ = angle of inclination with the horizontal; u = local film velocity; u_{max} = maximum velocity = velocity of the liquid surface; x = distance from plate, and V = average film velocity. These equations assume that no surface tractive force is present.

A number of investigators have shown that Eq. (5-236) accurately predicts the **film thickness** for laminar flow of low-viscosity liquids [<0.005 Pa·s (5 cP)] up to a critical Reynolds number ($4\Gamma/\mu$) generally found to be in the range of 1000 to 2000. However, Jackson [*Am. Inst. Chem. Eng. J.*, **1**, 231–240 (1955)] found that higher-viscosity liquids [0.01 to 0.02 Pa·s (10 to 20 cP)] gave film thicknesses appreciably below the predictions of Eq. (5-236) following the inception of surface wave motion. The latter was found to start at a Froude number ($N_{Fr} = V^2/gm$) of about 1.0. From this it follows that the volume flow rate at which waves start depends only upon the kinematic viscosity for the case of a vertical plate and a surrounding fluid of negligible density (i.e., a gas). Then $\Gamma/\rho_f = 3(\mu/\rho_f)$, which corresponds to $N_{Re} = 12$.

At Reynolds numbers of 25 or greater, **surface waves** will be present on the liquid film. The film surface velocity, however, will be within ±7 percent of that given by Eq. (5-239) [West and Cole, *Chem. Eng. Sci.*, **22**, 1388–1389 (1967)]. For the effect of surface tension in wavy film flow see Hoffman and Potts, *Ind. Eng. Chem. Fundam.*, **18**, 27–33 (1979).

Jackson (loc. cit.) has rederived Eqs. (5-236) through (5-239) for the case of films flowing on the inside wall of a **circular tube** for the case of a negligible surface tractive force. In this case, u_{max}/V is not a constant but varies from 1.5 up to 2.0 as film thickness increases from zero up to the pipe radius. Additional information on velocity distribution is given by Ho and Hummel [*Chem. Eng. Sci.*, **25**, 1225–1237 (1970)].

For laminar flow of nonnewtonian fluids down an inclined plane, see Bird, Armstrong, and Hassager, *Dynamics of Polymeric Liquids*, vol. 1: *Fluid Mechanics*, Wiley, New York, 1977, p. 215, 217; Astarita, Marrucci, and Palumbo, *Ind. Eng. Chem. Fundam.*, **3**, 333–339 (1964); and Cheng, *Ind. Eng. Chem. Fundam.*, **13**, 394–395 (1974).

Turbulent Flow An equation for estimating **film thickness** in the **turbulent-flow** regime ($N_{Re} > 2000$) was derived by Belkin et al. [*Am. Inst. Chem. Eng. J.*, **5**, 245–248 (1959)] by analogy with turbulent flow between parallel plates, as follows:

$$m = \frac{0.315\mu^{2/3}}{g^{1/3}\rho_f^{2/3}}(N_{Re}\sqrt{f})^{2/3} \tag{5-240}$$

From the friction data of Walker, Whan, and Rothfuss [*Am. Inst. Chem. Eng. J.*, **3**, 484–489 (1957)] for parallel plates,

$$f = 0.079(N_{Re})^{-0.25} \tag{5-241}$$

in the N_{Re} range of 3000 to 100,000. Equations (5-240) and (5-241) combined give

$$m = 0.304(\Gamma^{1.75}\mu^{0.25}/g\rho_f^2)^{1/3} \tag{5-242}$$

The derivation assumes negligible surface tractive force, a surrounding fluid of negligible density (i.e., a gas), and vertical orientation of the plate. This correlation was found to compare favorably with experimental data over the N_{Re} range of 3000 to 30,000. For a surface inclined at angle ϕ with the horizontal, substitute $g \sin\phi$ for g in Eq. (5-242).

Effect of Surface Traction If a drag is exerted on the surface of the film because of motion in the surrounding fluid, the film thickness will be reduced or increased, depending upon whether the drag is in parallel or counter, respectively, to the action of gravity. Thomas and Portalski [*Ind. Eng. Chem.*, **50**, 1081–1088 (1958)] and Dukler [*Chem. Eng. Prog.*, **55**(10), 62–67 (1959)] both employed the Nikuradse generalized velocity-distribution equations in developing procedures for computing film thickness and velocity distribution both with and without a surface tractive force. Dukler (loc. cit.) presented graphical solutions for the case of tractive force parallel with the action of gravity. Kosky [*Int. J. Heat Mass Transfer*, **14**, 1220–1224 (1971)] presented a generalized formulation which reduces to Eq. (5-242) for negligible surface traction and a surrounding fluid of negligible density. Generalized film-thickness data for falling-water films with cocurrent and countercurrent air flow in pipes, 13 to 20 mm (0.5 to 0.8 in) in diameter, are given by Zhivaikin [*Int. Chem. Eng.*, **2**, 337–341 (1962)].

Prediction of drainage rate from liquid films on flat and cylindrical surfaces is discussed by Tallmadge and Gutfinger [*Ind. Eng. Chem.*, **59**(11), 19–34, 1967)].

Flooding With countercurrent gas flow at high liquid rates or high gas rates, bridging of the liquid at either the bottom or the top of the tube can occur. This condition, known as flooding, is attended by a sudden large increase in pressure drop, by surging of the liquid in the tube, or by considerable entrainment of the liquid in the outgoing gas [further general description is given by Clift, Pritchard, and Nedderman, *Chem. Eng. Sci.*, **21**, 87–95 (1966)].

Gas mass velocities corresponding to liquid flooding at the bottom of the tubes are given in Figs. 5-74 and 5-75 (Holmes, courtesy of E. I. du Pont de Nemours & Co.). These figures apply for liquid mass velocities $G_L < 120$ kg/(s·m²) [25 lb/(s·ft²)] and for tubes having square tops. Figure 5-74 is for square-bottom tubes and Fig. 5-75 for slant-bottom tubes. The angle of the slant is 30° from the vertical, and if the gas approaches the tube from one side, the tapered end

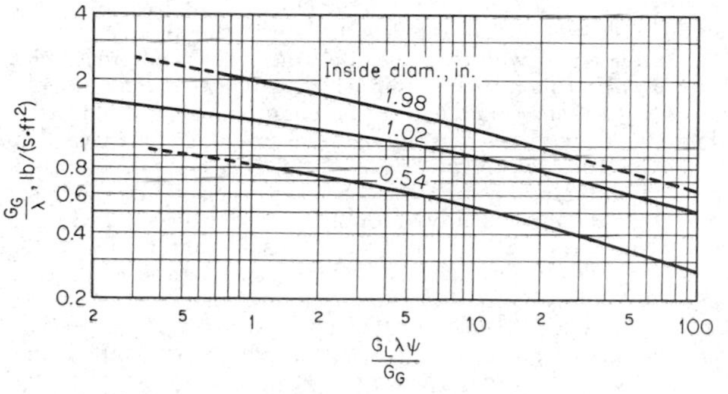

FIG. 5-74 Flooding in vertical tubes with square top and square bottom. To convert pounds per second per square foot to kilograms per second per square meter, multiply by 4.8824; to convert inches to millimeters, multiply by 25.4. *(Courtesy of E. I. du Pont de Nemours & Co.)*

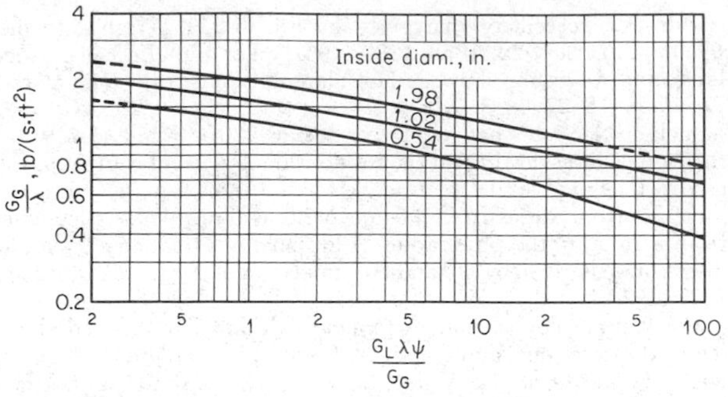

FIG. 5-75 Flooding in vertical tubes with square top and slant bottom. To convert pounds per second per square foot to kilograms per second per square meter, multiply by 4.8824; to convert inches to millimeters, multiply by 25.4. *(Courtesy of E. I. du Pont de Nemours & Co.)*

should be oriented so that the point faces the gas entrance. For $G_L > 120$ kg/(s·m²) [25 lb/(s·ft²)], flooding may occur at the top of the tubes. The nomenclature is G_G = superficial mass velocity of gas; G_L = superficial mass velocity of liquid; $\lambda = (\rho_G' \rho_L')^{1/2}$, dimensionless; $\psi = (1/\sigma')[\mu_L'/(\rho_L')^2]^{1/3}$, dimensionless; ρ_G' = ratio of gas density to air density; ρ_L' = ratio of liquid density to water density; σ' = ratio of liquid surface tension to water surface tension; and μ_L' = ratio of liquid viscosity to water viscosity; air and water properties are at 20°C (68°F) and 101.3 kPa (14.7 lbf/in²). For tube diameters less than 13 mm (½ in), bridging may occur along the tube because of the waves on the liquid surface (Kraybill, Ph.D. thesis, chemical engineering, University of Michigan, Ann Arbor, 1953). For large tubes, the indication is that the flooding gas velocity is independent of tube diameter; see Diehl and Koppany, *Chem. Eng. Prog.*, **65**, *Symp. Ser.* 92, 77–83 (1969); and Pushkina and Sorokin, *Heat Transfer—Soviet Research*, **1**(5) 56–64 (1969). Thus, for tubes of diameters larger than 50 mm (2 in), the curves for 1.98-in-diameter tubes (Figs. 5-74 and 5-75) should be used as a guide. Imura, Kusuda, and Funatsu [*Chem. Eng. Sci.*, **32**, 79–87 (1977)] compared a number of flooding correlations and proposed a new semiempirical correlation based on measurements with tubes ranging in diameter from 11.2 to 24.2 mm (0.44 to 0.95 in).

UNSTEADY-STATE BEHAVIOR

Water Hammer When a column of flowing fluid is suddenly stopped, a pounding of the line commonly known as **water hammer** is usually produced. For sudden flow stoppage, the pressure rise due to the deceleration of a truly incompressible fluid in a nonexpandable pipe would be infinite; the fluid in the line would behave as a "plug," and the pressure rise would be that corresponding to the inertia effects of this plug. Experience has shown that there is a finite maximum pressure rise because part of the kinetic energy of the moving fluid in the pipe is expended in stretching the pipe walls and compressing the fluid.

The equation for **maximum pressure** or **head rise** produced by a sudden flow change can be derived from Newton's second law, relating force to the rate of change of momentum, utilizing the velocity of the pressure waves which are set up owing to the inertia of the fluid in the line (see derivation by Moody, ASME-ASCE Symposium on Water Hammer, American Society of Mechanical Engineers, New York, 1933, pp. 25–28). The resulting equation is referred to as the Joukowsky or water-hammer equation:

$$h_{wh} = a(\Delta V)/g_c \qquad (5\text{-}243)$$

where

$$a = \sqrt{\frac{1}{\left[\frac{\rho}{g_c}\left(\frac{1}{K} + \frac{D}{bE}\right)\right]}} \qquad (5\text{-}244)$$

and h_{wh} = water-hammer head, N·m/kg [(ft·lbf)/lb or ft of fluid flowing]; a = velocity of wave propagation; ΔV = change in velocity; g_c = dimensional constant; ρ = fluid density; K = bulk modulus of elasticity of the fluid; D = pipe inside diameter; b = pipe-wall thickness; and E = modulus of elasticity of pipe-wall material. The maximum head rise given by Eq. (5-243) can also be developed if the flow is changed within the time that it takes the pressure wave to travel from the point of stoppage to the end of the pipe or to the location of total wave reflection and return; that is, within one period as given by

$$\tau = 2L/a \qquad (5\text{-}245)$$

where τ = pipe time period, L = length of pipe, and a = velocity of wave propagation.

For standard steel pipe, the value of the wave velocity a is about 1220 m/s (4000 ft/s).

If the time of flow stoppage is somewhat longer than one pipe period τ, the pressure rise will not be so great as that given by Eq. (5-243), since part of the direct pressure waves will be canceled by the reflected pressure waves. The actual pressure rise can be determined by use of the Allievi equations or charts (see Angus, *Hydraulics for Engineers*, 3d ed., Pitman, Toronto, 1943, pp. 283–284, 291–292; Kerr and Strowger, ASME-ASCE *Symposium on Water Hammer*, 1933, pp. 15–24; and Rich, *Hydraulic Transients*, McGraw-Hill, New York, 1951, pp. 24–27). Wood and Jones [*Proc. Am. Soc. Civ. Eng., J. Hydraul. Div.*, **99** (HY1), 167–178 (1973)] presented charts for more reliable estimates of water-hammer pressure for different valve-closure modes.

This anlaysis also applies to the pressure reduction for the reflected wave or on acceleration of flow. If the pressure reduction results in a static pressure at any point in the line below the vapor pressure of the fluid, the fluid in the line will separate or pull apart as the pressure wave passes that location. To keep the pipe from collapsing or bursting, provision must be made for protective devices such as relief valves to admit air when the fluid separates and to release air and some fluid when the fluid rejoins.

Additional details on water-hammer theory can be obtained from Angus, loc. cit., chap. XIV; ASME-ASCE *Symposium on Water Hammer*, 1933; Rich, loc. cit.; Wylie and Streeter, *Hydraulic Transients*, McGraw-Hill, New York, 1978.

Hydraulic Transients In the design and operation of most process pumping systems, consideration is given only to normal steady-state conditions; that is, based on continuous uninterrupted operations, the pumping system is specified for the given process flow rate and pressure. There are a few process pumping systems, however, in which sudden flow changes would cause damage to the pumping facilities or adversely affect the process.

Some of the potential problems to be considered in the design and operation of process pumping systems and methods of analyzing these systems are pointed out later.

In a complex system consisting of **several pumps in parallel** between a suction header and a discharge header, failure of power to one of the pumps could produce one of the following situations: (1) If there were no check valve in the failed pump discharge line, a considerable backflow could follow the failure, thus producing a further decrease in flow through the discharge header and producing a high reverse rotation of the pump which could cause damage to the motor. Experience indicates that a pump impeller could attain almost the design forward speed in the reverse direction in less than 1 s [Alves, *Am. Inst. Chem. Eng. J.*, **2**, 143–147 (1956)]. A flywheel installed on the pump shaft could be used to extend the pumping time while the individual system was being shut down. (2) If there were a check valve, the sudden backflow could slam the disk on the seat, possibly damaging the valve or setting up a dangerous pressure surge, or the disk could "hang up" for a time, allowing backflow to build up and then slam shut, thus setting up dangerous pressure surges. A flywheel on the pump shaft will extend the pumping time to permit a check valve to operate properly [for example, see *Power*, **84**(1), 57 (1940)].

Methods of analyzing pumping systems with possible hydraulic transients are described in the literature by several investigators, including Rich, loc. cit.; Parmakian, *Water Hammer Analysis*, Prentice-Hall, Englewood Cliffs, N.J., 1955; Streeter and Wylie, loc. cit.; Kittredge, *Trans. Am. Soc. Mech. Eng.*, **79**, 1307–1322 (1956); Knapp, *Trans. Am. Soc. Mech. Eng.*, **59**, 683–689 (1937); Angus, *Proc. Inst. Mech. Eng. (London)*, **136**, 245–331 (1937); and Alves, loc. cit. For the quick estimation of various hydraulic transients in pumping systems, charts are presented by Parmakian (loc. cit., pp. 87–91) and Kinno and Kennedy [*Proc. Am. Soc. Civ. Eng., J. Hydraul. Div.*, **91**(HY3), 247–270 (1965)].

The mechanism and effects of water hammer resulting from cavitating pumps are described by Carstens and Hagler [*Proc. Am. Soc. Civ. Eng., J. Hydraul. Div.*, **90**(HY6), 161–184 (1964)].

Pulsating Flow Flow pulsations in piping systems most often result from the presence of reciprocating machinery (compressors or pumps) in the system. Such pulsations generally adversely affect the performance of flowmeters and process-control elements and can cause vibration and ultimately equipment failure. It is important to recognize that the vibration and damage can result not only from the fundamental frequency of the pulse producer but also from higher harmonics of this frequency. The preferred solution is to minimize the problem by employing multipiston double-acting units. If this is not practical, a pulsation damper should be installed. See Buckley, *Techniques of Process Control*, Wiley, New York, 1964, chap. 16, for pulsation-damping theory.

Gas-Phase Pulsation Damping A general description of methods available for damping gas-phase flow pulsations is given by the M. W. Kellogg Co., *Design of Piping Systems*, 2d ed., Wiley, New York, 1956, pp. 279–283, 333–335. A tabulation of six different types of pulsation dampers and wave filters is given in Campbell, *Process Dynamics*, Wiley, New York, 1958, pp. 102–103, together with applicable formulas and attenuation characteristics. Included therein are the in-line surge chamber, closed-end resonator, low-pass filter, high-pass filter, bandpass filter, and band-elimination filter. Chilton and Handley [*Trans. Am. Soc. Mech. Eng.*, **74**, 931–943 (1952)] present charts for predicting the performance of a single-tank damper (in-line surge chamber) and a π-type filter (low-pass filter). Isakoff [*Ind. Eng. Chem.*, **47**, 413–421 (1955)] showed how a low-pass electrical filter could be used as an analog for a low-pass gas-pulsation damper, thereby facilitating design of the latter.

Liquid-Phase Pulsation Damping For liquid systems the custom is to employ gas-filled surge chambers attached to the pipe line on both the suction and the discharge sides of the pump and located as close to the pump as possible. Sizing of such surge chambers is discussed by Chilton and Handley [*Trans. Am. Soc. Mech. Eng.*, **77**, 225–230 (1955)]. Diaphragms or bellows are frequently used to separate the surge-chamber gas from the process liquid and thereby prevent gradual depletion of the gas. Drawings are presented by Dodge (in Yeaple, *Hydrostatic and Pneumatic Power and Control*, McGraw-Hill, New York, 1966, pp. 284–285) for six damper or accumulator designs. Equations for sizing such units are given by Greer Hydraulics, Bull. 500, 1957, and Robertshaw Controls Co., Cat. MA-11, 1967.

All the pulsation dampers referred to thus far are known as "passive" dampers, the functioning of which depends upon factors such as compressibility, mass, and viscous friction. If the pulsation frequency is so low that passive dampers become impractically large in size, then an "active" damper should be considered. Buckley (op. cit., pp. 142–144) gives the procedure for designing such an active damper.

Cavitation A practical definition of cavitation is the formation and collapse of vapor cavities in a flowing liquid. Such a vapor cavity can form anywhere in the flowing liquid where the local pressure is reduced to that of the liquid vapor pressure at the temperature of the flowing liquid. At these locations, some of the liquid vaporizes to form bubbles or cavities of vapor. Low-pressure zones can be produced by a local increase in velocity (in accordance with Bernoulli's equation; see subsection "Energy Balance") as in eddies or vortices, or over boundary contours; by rapid vibration of the boundary; by separation or parting of a liquid column owing to water hammer; or by an overall reduction in static pressure.

Collapse of the bubbles will begin when they are moved into regions where the local pressure is higher than the vapor pressure. Collapse of these cavities may produce objectionable noise and vibration, and extensive erosion or pitting of the boundary materials in the regions of bubble collapse. An exceedingly important effect of cavitation in liquid-handling facilities is the decrease in performance and efficiency of the equipment; for example, **valves** used as flow regulators [Tullis and Marschner, *Proc. Am. Soc. Civ. Eng., J. Hydraul. Div.*, **94**(HY1), 1–16 (1968); and Tullis, *Proc. Am. Soc. Civ. Eng., J. Hydraul. Div.*, **97** (HY12), 1931–1945 (1971)]; **pipe bends** [Kamiyama, *J. Basic Eng.*, **88**, 252–260 (1966)]; and **pumps** [Salemann, *J. Basic Eng.*, **81**, 167–180 (1959); and Spraker, *J. Eng. Power*, **87**, 309–318 (1965)].

In correlating equipment-performance data, a useful parameter is the dimensionless grouping called the cavitation number σ_c (or K):

$$\sigma_c = (p - p_v)/(\rho V^2/2g_c) \tag{5-246}$$

where p = static pressure (absolute) in undisturbed flow, p_v = liquid vapor pressure (absolute), ρ = liquid density, V = free-stream velocity of liquid, and g_c = dimensional constant. The cavitation number can be considered as the ratio of the net static pressure available to collapse the bubble to the dynamic pressure available to initiate the formation of the bubble. The value of the cavitation number for incipient cavitation $\sigma_{c,i}$ for a specific boundary condition or item of equipment is a characteristic of the geometry. Cavitation numbers for various head forms of cylinders, for disks, and for various hydrofoils are given by Holl and Wislicenus [*J. Basic Eng.*, **83**, 385–398 (1961)] and for various surface irregularities by Arndt and Ippen [*J. Basic Eng.*, **90**, 249–261 (1968)], Ball [*Proc. Am. Soc. Civ. Eng., J. Constr. Div.*, **89**(C02), 91–110 (1963)], and Holl [*J. Basic Eng.*, **82**, 169–183 (1960)]. As a guide only, for blunt forms the cavitation number is generally in the range of 1 to 2.5, and for somewhat streamlined forms the cavitation number is generally in the range of 0.2 to 0.5.

Cavitation conditions in sharp-edged **orifices** can be estimated from

$$\left(\frac{p_1 - p_v}{p_1 - p_B}\right)_{\text{crit}} = 2.6C_D^2 \tag{5-247}$$

where p_1 = total upstream pressure; p_v = liquid vapor pressure; p_B = back pressure; and C_D = orifice discharge coefficient, dimensionless. Equation (5-247) applies to circular orifices with a length-to-diameter ratio of 2:20 and to rectangular orifices with a length-to-channel-height ratio of 5.5:9 with a channel aspect ratio of 2:8. The onset of cavitation is affected by orifice-entrance sharpness. For further details, see Nurick, *J. Fluids Eng.*, **98**, 681–687 (1976).

Scaling up cavitation data obtained for a model must be done with caution. Investigations by Knapp [*Proc. Inst. Mech. Eng. (London)*, **166**, 150–163 (1952)] and Kermeen, McGraw, and Parkin [*Trans. Am. Soc. Mech. Eng.*, **77**, 533–541 (1955)] indicate that the cavitation number for incipient cavitation depends upon the free-stream velocity and the characteristic dimension or size. Additional details are given by Holl and Wislicenus (loc. cit.).

A discussion of cavitation, summarizing information on mechanism, cavitation numbers, modeling, and damage, is given by Eisenberg and Tulin ("Cavitation," sec. 12 of Streeter, *Handbook of Fluid Dynamics*, McGraw-Hill, New York, 1961) and by Knapp, Daily, and Hammit (*Cavitation*, McGraw-Hill, New York, 1970). Cavitation in viscoelastic media is discussed by Kaelble [*Trans. Soc. Rheol.*, **15**(2), 275–296 (1971)].

MODEL STUDIES

A **model** is a device or means which is so constituted that it can be used to predict accurately the performance of a "prototype." The **prototype** in turn is the full-scale physical system which is to be modeled. There are two general types of models:

1. Physically similar models which differ only in scale from the prototype
2. Physically dissimilar models such as mathematical models and electrical analogs

TABLE 5-19 Dimensionless Groupings and Their Significance

Name	Symbol	Formula	Special nomenclature*	Proportional to†	Where used
Bingham no.	N_{Bm}	$\tau_y g_c L/\eta V$	L = width of channel η = plastic viscosity τ_y = yield stress	Yield stress / Viscous stress	Flow of Bingham plastics
Blake no.	B	$V\rho/[\mu(1-\varepsilon)s]$	ε = voidage s = particle area/particle volume	Inertial force / Viscous force	Beds of solids
Bond no.	N_{Bo}	$(\rho-\rho')L^2 g/g_c\sigma$	L = diameter of droplet ρ = density of droplet ρ' = density of surrounding fluid σ = surface tension	Gravitational force / Surface-tension force	Atomization
Capillary no.	Ca	$\mu V/g_c\sigma$	σ = surface tension	Viscous force / Surface-tension force	Atomization; two-phase flow in beds of solids
Cauchy no.	N_C	$\rho V^2/g_c E_b$	E_b = bulk modulus of fluid	Inertial force / Compressibility force	Compressible flow
Cavitation no.	σ_c	$[(p-p_v)/\rho]/(V^2/2g_c)$	p = local absolute static pressure p_v = vapor pressure	Excess of local static head over vapor-pressure head / Velocity head	Cavitation
Dean no.	N_{De}	$(VL\rho/\mu)(L/2R)^{1/2}$	L = diameter of pipe R = radius of curvature	$N_{Re}\left(\frac{\text{centrifugal force}}{\text{inertial force}}\right)$	Flow in curved channels
Drag coefficient	C_d	$(\rho-\rho')Lg/\rho V^2$	L = characteristic dimension of object ρ = density of object ρ' = density of surrounding fluid	Gravitational force / Inertial force	Free-settling velocities
Elasticity no.	N_{El}	$\theta_r\mu/\rho L^2$	L = radius of pipe θ_r = relaxation time	Elastic force / Inertial force	Viscoelastic flow
Euler no.	N_{Eu}	$g_c(\Delta p_F/\rho)/V^2$	N = number of velocity heads	$\frac{\text{Friction head}}{2 \times \text{velocity head}} = \frac{N}{2}$	Fluid friction in conduits
Fanning friction factor	f	$g_c D(\Delta p_F/\rho)/2V^2 L$	$\Delta p_F/\rho$ = friction head D = characteristic diameter of cross section L = length of pipe $\Delta p_F/\rho$ = friction head	Shear stress at pipe wall expressed as number of velocity heads	Fluid friction in conduits
Froude no.	N_{Fr}	V^2/gL	L = characteristic dimension of system	Inertial force / Gravitational force	Wave and surface behavior
Hodgson no.	N_H	$V'f'\,\Delta p_F/\bar{q}p$	f' = frequency p = average static pressure Δp_F = pressure drop due to friction $\bar{q}$ = average volumetric flow rate V' = volume of system	Time constant of system / Period of pulsation	Pulsating gas flow
Ohnesorge no.	Z	$\mu/(\rho g_c L\sigma)^{1/2}$	L = characteristic dimension of system σ = surface tension	$\frac{\text{Viscous force}}{(\text{Inertial force} \times \text{surface-tension force})^{1/2}} = \frac{(N_{We})^{1/2}}{N_{Re}}$	Atomization
Pipe-line parameter	ρ_n	$aV_0/2g_c H$	a = water-hammer wave velocity H = static head V_0 = initial velocity	Maximum water-hammer pressure rise / 2 × static pressure	Water hammer (hydraulic transients)
Power no.	N_P	$Pg_c/L^5\rho n^3$	P = power to agitator L = characteristic dimension of agitator paddle n = rate of rotation	Drag force on paddle / Inertial force	Power consumption on agitated vessels
Prandtl velocity ratio	u^+	$u/(\tau_w g_c/\rho)^{1/2}$	u = local velocity τ_w = shear stress at wall	$\left(\frac{\text{Inertial force}}{\text{Wall shear force}}\right)^{1/2} = \frac{u}{V}\left(\frac{f}{2}\right)^{1/2}$	Turbulence
Reynolds no.	N_{Re}	$LV\rho/\mu$	L = characteristic dimension of the system	Inertial force / Viscous force	Dynamic similarity
Strouhal no.	N_S	$f'L/V$	f' = frequency L = characteristic dimension of obstacle	Reciprocal of vortex spacing expressed as number of obstacle diameters	Van Kármán vortex streets
Weber no.	N_{We}	$V^2\rho L/g_c\sigma$	L = characteristic dimension of system σ = surface tension	Inertial force / Surface-tension force	Bubble formation, breakup of liquid jet

*For units, see "General Nomenclature" in Table 5-20.
†See "Force Proportionalities," Table 5-20.

The principle of similarity must be observed in planning model studies. Four types of similarities are important in most chemical engineering studies, namely:

1. Geometric similarity (dimensional proportionality)
2. Mechanical similarity
   a. Static similarity (deformation proportionality)
   b. Kinematic similarity (time proportionality)
   c. Dynamic similarity (force proportionality)
3. Thermal similarity (temperature proportionality)
4. Chemical similarity (concentration proportionality)

Selection of dimensions and operating conditions so as to satisfy geometric, thermal, and chemical similarity requirements is usually relatively straightforward. Satisfaction of mechanical similarity requirements, however, generally involves proportionality control of certain critical groups of variables that are selected either by dimensional analysis or by inspection. The latter procedure, in particular, requires familiarity with known dimensionless numbers and their significance.

Table 5-19 gives a list of dimensionless numbers frequently encountered in fluid mechanics, together with their formulas, significance, area of use, and literature references. This is part of extensive tabulations given by Boucher and Alves [*Chem. Eng. Prog.*, **55**(9), 55–64 (1959); **59**(8), 75–83 (1963)]. More extensive tabulations are given by Catchpole and Fulford [*Ind. Eng. Chem.*, **58**(3), 46–60 (1966)] and Fulford and Catchpole [*Ind. Eng. Chem.*, **60**(3), 71–78 (1968)]. Table 5-20 gives a tabulation of force proportionalities.

Detailed treatments of the subject of model studies can be found in Johnstone and Thring, *Pilot Plants, Models, and Scale-Up Methods in Chemical Engineering*, McGraw-Hill, New York, 1957; Kline, *Similitude and Approximation Theory*, McGraw-Hill, New York, 1965; Langhaar, *Dimensional Analysis and Theory of Models*, Wiley, New York, 1951; and Murphy, *Similitude in Engineering*, Ronald, New York, 1950.

PARTICLE DYNAMICS

GENERAL REFERENCES: Brodkey, *The Phenomena of Fluid Motions*, Addison-Wesley, Reading, Mass., 1967. Brown et al., *Unit Operations*, Wiley, New York, 1950. Clift, Grace, and Weber, *Bubbles, Drops, and Particles*, Academic, New York, 1978. Davies, *Turbulence Phenomena*, Academic, New York, 1972. Knudsen and Katz, *Fluid Dynamics and Heat Transfer*, McGraw-Hill, New York, 1958. Lapple et al., *Fluid and Particle Mechanics*, University of Delaware, Newark, 1951. Levich, *Physicochemical Hydrodynamics*, Prentice-Hall, Englewood Cliffs, N.J., 1962. Orr, *Particulate Technology*, Macmillan, New York, 1966. Zenz and Othmer, *Fluidization and Fluid-Particle Systems*, Reinhold, New York, 1960.

Whenever relative motion exists between a particle and a surrounding fluid, the fluid will exert a drag upon the particle. The drag force on the particle is given by

$$F_d = CA_p\rho u^2/2g_c \qquad (5\text{-}248)$$

where F_d = drag force; C = drag coefficient, dimensionless; A_p = projected area particle in direction of motion; ρ = density of surrounding fluid; u = relative velocity between particle and fluid; and g_c = dimensional constant.

TABLE 5-20 Force Proportionalities

Buoyancy force $\propto L^3\rho\beta\,\Delta t g/g_c$
Centrifugal force $\propto \rho L^3V^2/g_cR$
Compressibility force $\propto E_bL^2$
Coriolis force $\propto 2(\rho/g_c)L^3nV \sin \alpha$
Elastic force $\propto \theta_r(\mu/g_c)V^2$
Gravitational force $\propto L^3(\rho - \rho')g/g_c$
Inertial force $\propto L^2\rho V^2/g_c$
Surface-tension force $\propto L\sigma$
Viscous force $\propto L\mu V/g_c$
Wall shear force $\propto \tau_wL^2$

General nomenclature

a = wave velocity, m/s (ft/s)
D = diameter, m (ft)
E_b = bulk modulus, N/m^2 (lbf/ft^2)
f' = frequency, Hz (1/s)
g = local acceleration due to gravity, m/s^2 (ft/s^2)
g_c = dimensional constant, 32.2 (lb·ft)/(lbf·s^2)
H = static head, (N·m)/kg [(ft·lbf)/lb]
L = characteristic dimension of system, m (ft)
N = number of velocity heads, dimensionless
n = rate of rotation, 1/s
P = power, (N·m)/s [(ft·lbf)/s]
$p, p_o, \bar{p}$ = pressure, Pa (lbf/ft^2)
Δp_F = pressure drop due to friction, Pa (lbf/ft^2)
q = volume flow rate, m^3/s (ft^3/s)
R = radius of curvature, m (ft)
s = particle of area per particle volume, 1/m (1/ft)
u = local velocity, m/s (ft/s)
V, V_0 = velocity, m/s (ft/s)
V' = volume, m^3 (ft^3)
α = angle, rad or °
β = coefficient of expansion, 1/°C (1/°F)
Δt = temperature difference, °C (°F)
ε = voidage, dimensionless
η = plastic viscosity, Pa·s [lb/(ft·s)]
θ_r = relaxation time, s
μ = viscosity, Pa·s [lb/(ft·s)]
ρ, ρ' = density, kg/m^3 (lb/ft^3)
σ = surface tension, N/m (lbf/ft)
τ_w = shear stress at wall, N/m^2 (lbf/ft^2)
τ_y = yield stress, N/m^2 (lbf/ft^2)

A particle falling under the action of gravity will accelerate until drag force just balances gravitational force, after which it will continue to fall at a constant velocity known as the **terminal** or **free-settling velocity** u_t, as given by

$$u_t = \sqrt{\frac{2gm_p(\rho_p - \rho)}{\rho\rho_pA_pC}} \qquad (5\text{-}249)$$

where g = local acceleration due to gravity; m_p = mass of particle, ρ_p = density of particle, and the rest of the symbols are as defined in the preceding paragraph. The drag coefficient C has been found to be a function of the shape of the particle and the Reynolds number $D_p\rho u/\mu$, where D_p = diameter of particle and μ = fluid viscosity.

The velocity of particles settling at higher Reynolds number fluctuates owing, among other factors, to vortex shedding. Oscillation is enhanced with increasing separation between the mass and geometric centers of the particle. Variations in mean velocity are less than 10 percent. As a result, the drag force on a particle fixed in space with fluid moving is somewhat lower than the drag force on a similar particle moving in stationary fluid.

The information presented hereafter on drag coefficients for rigid particles, gas bubbles in liquids, liquid drops in liquids, and liquid drops in gases applies in the absence of wall effects or concentration effects. These effects are generally not significant for container-to-particle-diameter ratios of 100 or more and for concentrations below 0.1 percent by volume. For the effects of lower ratios and higher concentrations, see later information under "Limitations."

Spherical Rigid Particles In slow flow ($N_{Re} < 0.1$) the velocity field about the sphere is symmetric. With increasing Reynolds number the field becomes progressively asymmetric. At about $N_{Re} = 20$ flow separation occurs at the rear stagnation point, and recirculation begins. At about $N_{Re} > 270$ vortices periodically form and move downstream (see Fig. 5-70). The frequency of vortex shedding reaches maximum at about $N_{Re} = 6000$. Marked changes in flow pattern occur at about $N_{Re} = 200{,}000$. For further details, see Clift, Grace, and Weber, op. cit., chap. 5. For the case of spherical particles, Eq. (5-249) becomes

$$u_t = \sqrt{\frac{4gD_p(\rho_p - \rho)}{3\rho C}} \qquad (5\text{-}250)$$

A plot of drag coefficient C versus N_{Re} is given as the solid curve in Fig. 5-76. For $N_{Re} < 0.1$,

$$C = 24/N_{Re} \qquad (5\text{-}251)$$

This corresponds to **Stokes' law,** which is usually written

$$F_d = 3\pi\mu uD_p/g_c \qquad (5\text{-}252)$$

The **terminal** settling velocity in the Stokes'-law region becomes

$$u_t = \frac{gD_p^2(\rho_p - \rho)}{18\mu} \qquad (5\text{-}253)$$

In the **intermediate region** ($N_{Re} < 1000$) the drag coefficient can be computed from

$$C = (24/N_{Re})(1 + 0.14N_{Re}^{0.70}) \qquad (5\text{-}254)$$

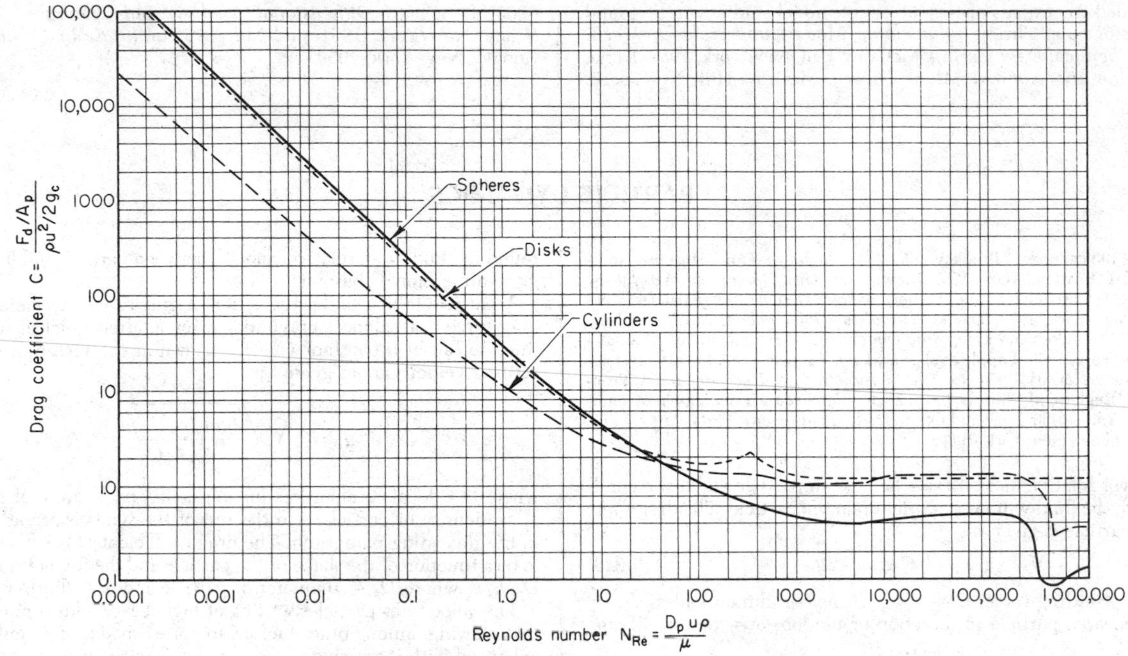

FIG. 5-76 Drag coefficients for spheres, disks, and cylinders. A_p = area of particle projected on a plane normal to direction of motion; C = overall drag coefficient, dimensionless; D_p = diameter of particle; F_d = drag or resistance to motion of body in fluid; g_c = dimensional constant; N_{Re} = Reynolds number, dimensionless; u = relative velocity between particle and main body of fluid; μ = fluid viscosity; and ρ = fluid density. [From *Lapple and Shepherd,* Ind. Eng. Chem., *32, 605 (1940).*]

within 6 percent. In the **Newton's-law** region, which covers the Reynolds-number range 1000 to 350,000, $C = 0.445$, within 13 percent. In this region, Eq. (5-249) becomes

$$u_t = 1.73\sqrt{gD_p(\rho_p - \rho)/\rho} \tag{5-255}$$

For $N_{Re} > 10^6$ the drag coefficient can be computed from

$$C = 0.19 - [(8)(10^4)/N_{Re}] \tag{5-256}$$

(Clift, Grace, and Weber, op. cit., p. 112).

These correlations for settling velocity are strictly applicable only in the case of newtonian fluids when the viscosity is independent of shear rate and duration of shear. Only limited information is available for particles settling in **nonnewtonian** fluids. A correlation is given by Dallon and Christiansen (Preprint 24C, *Symposium on Selected Papers*, part III, Sixty-first Annual Meeting of American Institute of Chemical Engineers, Los Angeles, Calif., Dec. 1–5, 1968) for spheres settling in pseudoplastic liquids; and by Ito and Kajiuchi [*J. Chem. Eng. Japan*, **2**(1), 19–24 (1969)] and Pazwash and Robertson [*J. Hydraul. Res.*, **13**, 35–55 (1975)] for spheres settling in Bingham plastics.

Nonspherical Rigid Particles The drag on a nonspherical particle depends upon its shape and its orientation with respect to the direction of motion. The orientation in free fall as a function of Reynolds number is given in Table 5-21. The drag coefficients for **disks** (flat side perpendicular to the direction of motion) and for **cylinders** (infinite length with axis perpendicular to the direction of motion) are given in Fig. 5-76 as a function of Reynolds number. The effect of length-to-diameter ratio for cylinders in the Newton's-law region is reported by Knudsen and Katz (op. cit., p. 301).

Pettyjohn and Christiansen [*Chem. Eng. Prog.*, **44**, 157–172 (1948)] present correlations which allow for the effect of particle shape on free-settling velocities for **isometric particles.** For $N_{Re} <$ 0.05, the terminal or free-settling velocity is given by

$$u_t = K_1 \frac{gD_s^2(\rho_p - \rho)}{18\mu} \tag{5-257}$$

$$K_1 = 0.843 \log (\psi/0.065) \tag{5-258}$$

where ψ = sphericity (surface area of a sphere having the same volume as the particle divided by the surface area of the particle), dimensionless; g = local acceleration due to gravity; D_s = "spherical" diameter (diameter of a sphere of equal volume); ρ_p = particle density; ρ = fluid density; and μ = fluid viscosity.

The terminal velocity of **axisymmetric particles in axial motion** can be computed from the theoretically derived correlation of Bowen and Masliyah [*Can. J. Chem. Eng.*, **51**, 8–15 (1973)]:

$$u_t = \frac{V'}{K_2} \frac{gD_s^2(\rho_p - \rho)}{18\mu} \tag{5-259}$$

$$K_2 = 0.244 + 1.035\Sigma - 0.712\Sigma^2 + 0.441\Sigma^3 \tag{5-260}$$

where V' = ratio of particle volume to volume of sphere with diameter D_s, dimensionless; Σ = ratio of surface area of particle to that

TABLE 5-21 Free-Fall Orientation of Particles*

Reynolds Number†	Orientation
0.1–5.5	All orientations are stable when there are three or more perpendicular axes of symmetry
5.5–200	Stable in position of maximum drag
200–500	Unpredictable. Disks and plates tend to wobble, while fuller bluff bodies tend to rotate
500–200,000	Rotation about axis of least inertia, frequently coupled with spiral translation

*From Becker, *Can. J. Chem. Eng.*, **37**, 85–91 (1959).
†Based on diameter of a sphere having the same surface area as the particle.

of a sphere with diameter D_s, dimensionless; D_s = diameter of sphere with perimeter equal to the maximum particle projected perimeter parallel to the direction of flow; and other symbols are as defined previously.

In the **Newton's-law** region, the terminal velocity is given by

$$u_t = \sqrt{\frac{4D_s(\rho_p - \rho)g}{3K_3\rho}} \qquad (5\text{-}261)$$

$$K_3 = 5.31 - 4.88\psi \qquad (5\text{-}262)$$

Symbols are as defined for Eqs. (5-257) and (5-258). Equations (5-258) and (5-262) are based on experiments on cube-octahedrons, octahedrons, cubes, and tetrahedrons for which the sphericity ψ ranges from 0.906 to 0.670 respectively. See also Clift, Grace, and Weber, op. cit., pp. 161–162.

For particles having sphericities less than 0.67, the correlations presented by Becker [*Can. J. Chem. Eng.*, **37**, 85–91 (1959)] should be employed. Reference to this paper is also recommended for the **intermediate-law region.** The settling characteristics of nonspherical particles are also discussed at some length by Zenz and Othmer (op. cit., chap. 6), by Brown et al. (op. cit., pp. 76–78), by Coulson and Richardson, *Chemical Engineering*, vol. 2, 2d ed., Pergamon, New York, 1968, pp. 148–152, and by Clift, Grace, and Weber (op. cit., chap. 6).

Drag coefficients for various two- and three-dimensional bodies are given by Hoerner, *Fluid-Dynamics Drag*, published by author, Midland Park, N.J., 1965, chap. III. Drag coefficients for flow normal to two-dimensional cylinders with a variety of cross sections are listed in Knudsen and Katz (op. cit., pp. 302–304). See also Clift, Grace, and Weber, op. cit., chaps. 4 and 6.

Gas Bubbles Fluid particles differ from solid particles in that internal circulation and particle deformation, both of which affect drag coefficient and terminal velocity, can occur. A large amount of information has been published on bubble dynamics in stationary low-viscosity liquid. However, much of it is conflicting, and no reliable generalized correlations are as yet available. It is now believed that many of the discrepancies in published data are due to the presence of surface-active agents which concentrate at the bubble surface and have a major effect on bubble shape and thus on terminal velocity, particularly in the intermediate size range.

Small bubbles [<3-mm (⅛-in) diameter] are approximately spherical and rise in straight lines in water; medium-sized bubbles [diameter 3 to 8 mm (⅛ to ⅓ in)] are flattened horizontally—generally ellipsoidal with the major axis in a horizontal plane—and rise with a rocking, oscillating, or spiral movement; while larger bubbles [diameter >8 mm (⅓ in)] are greatly deformed, assuming a mushroomlike shape, rise relatively straight, but are unstable and tend to break into smaller bubbles in water [see Peebles and Garber, *Chem. Eng. Prog.*, **49**, 88–97 (1953)]. For further information on bubble shape see Garner and Hammerton, *Chem. Eng. Sci.*, **3**, 1–11 (February 1954); Brodkey, op. cit., pp. 582–604; Levich, op. cit., chap. VII; Valentin, *Absorption in Gas-Liquid Dispersions*, Spon, London, 1967, pp. 14–23; Calderbank, *Chem. Eng. (London)*, no. 212, 209–233 (1967); and Clift, Grace, and Weber, op. cit., chap. 2.

The **drag-coefficient** curve for air bubbles rising in water as given by Haberman and Morton (David W. Taylor Model Basin Rep. 802, 1953) is shown in Fig. 5-77. The diameter of the bubble is taken to be the diameter of a sphere having the same volume as the bubble. This curve also serves as a good approximation for bubbles of other gases in water. Location of the drag-coefficient curve for a given system depends on the physical properties of the system, including liquid viscosity and interfacial tension. For liquids having properties appreciably different from water, see Haberman and Morton, loc. cit. For additional information on the effect of interfacial tension, see Harmathy, *Am. Inst. Chem. Eng. J.*, **6**, 281 (1960); Peebles and Garber, loc. cit.; Raymond and Zieminski, *Am. Inst. Chem. Eng. J.*, **17**, 57–65 (1971); and Clift, Grace, and Weber, op. cit., chap. 7.

For information on the motion of bubbles in **nonnewtonian liquids,** see Astarita and Apuzzo, *Am. Inst. Chem. Eng. J.*, **11**, 815–820 (1965)]; Calderbank, Johnson, and Loudon, *Chem. Eng. Sci.*, **25**, 235–256 (1970); and Acharya, Mashelkar, and Ulbrecht, *Chem. Eng. Sci.*, **32**, 863–872 (1977).

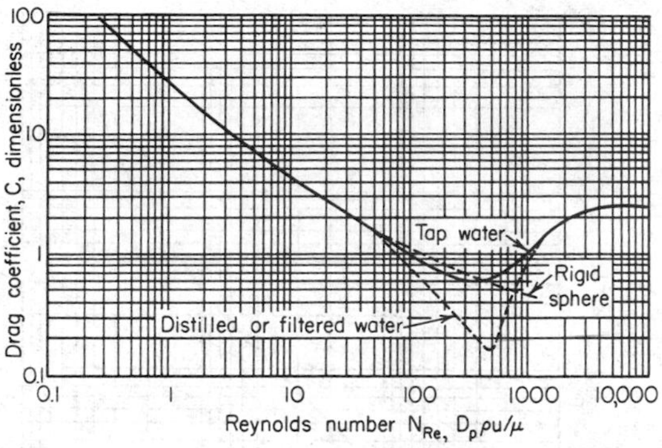

FIG. 5-77 Drag coefficient for air bubbles rising in water at room temperature. *(Haberman and Morton, David W. Taylor Model Basin Rep. 802, 1953.)*

Baker and Chao [*Am. Inst. Chem. Eng. J.*, **11**, 268–273 (1965)] studied bubble motion in an upward-flowing turbulent water stream. In general, the relative bubble terminal velocities were found to be slightly higher in the turbulent stream as compared with those in quiescent water.

For an exhaustive review of references on bubble phenomena for the period 1965–1970, see Gal-Or, Klinzing, and Tavlarides, *Ind. Eng. Chem.*, **61**, 21–34 (1969); and Tavlarides, Coulaloglou, Zeitlin, Klinzing, and Gal-Or, *Ind. Eng. Chem.*, **62**, 6–27 (1970).

Liquid Drops in Liquids Liquid drops will either rise or settle in a separate immiscible liquid medium, depending upon whether the drop density is less or greater, respectively, than the liquid-medium density. Very small drops behave like rigid spheres, and the terminal velocity can be approximated by use of the drag-coefficient curve for solid spheres for Reynolds numbers up to about 10 according to Warshay, Bogusz, Johnson, and Kintner [*Can. J. Chem. Eng.*, **37**, 29–36 (1959)]. The terminal velocity increases with drop size until drop oscillation sets in (in the Reynolds-number range of 50 to 500). The terminal velocity is greater than that given by the drag-coefficient curve for solids in the 10 to 500 Reynolds-number range because of internal circulation within the drop (see Davies, op. cit., chap. 8). As drop size is further increased, the terminal velocity decreases slightly and then levels off. For further information, see Klee and Treybal, *Am. Inst. Chem. Eng. J.*, **2**, 444–447 (1956); and Clift, Grace, and Weber, op. cit., chaps. 5 and 7. Drops of viscous liquids can be treated as rigid particles when the ratio of the dispersed phase viscosity to the continuous phase viscosity exceeds about 50.

For systems characterized by low liquid viscosities [~0.001 Pa·s (1 cP)] in the continuous phase and high interfacial tensions [0.025 to 0.045 N/m (25 to 45 dyn/cm)], the correlation of Hu and Kintner [*Am. Inst. Chem. Eng. J.*, **1**, 42–48 (1955)] as given in Fig. 5-78) is recommended. Here the dimensionless parameter $C(N_{We})(P^{0.15})$ is correlated with another dimensionless parameter $N_{Re}/P^{0.15}$, where C = drag coefficient, dimensionless; N_{We} = Weber number $(u_t^2 D_p \rho / \sigma_i g_c)$, dimensionless; u_t = relative terminal velocity between the drop and the liquid medium; D_p = drop diameter; ρ = density of medium; σ_i = interfacial tension; $P = (g_c^3 \rho^2 \sigma_i^3)/g\mu^4(\rho_d - \rho)$, dimensionless; g_c = dimensional constant; g = local acceleration due to gravity; μ = viscosity of liquid medium; ρ_d = drop density; and N_{Re} = Reynolds number $(D_p u_t \rho/\mu)$, dimensionless. A trial-and-error procedure is required to calculate the terminal velocity u_t corresponding to a given drop size D_p and known physical properties ρ, μ, σ_i, and ρ_p. First, select a value of C, and then calculate a trial value of u_t from Eq. (5-250). Then, calculate P, N_{Re}, and N_{We}. Determine a new value of C from the curve in Fig. 5-78. Repeat the procedure until two successive values of C check within the desired precision.

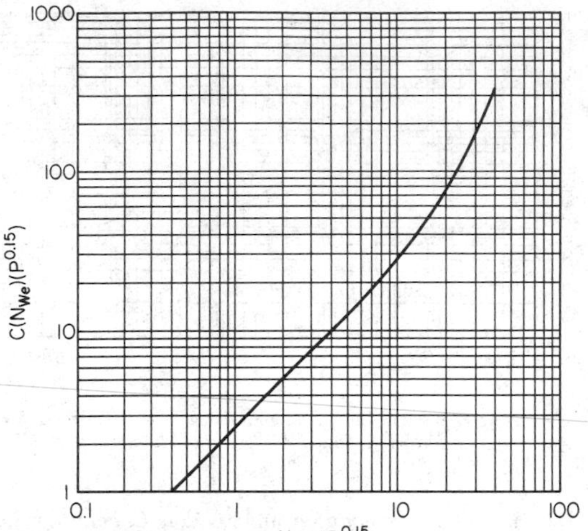

FIG. 5-78 Liquid drops in liquid media. [*From Hu and Kintner,* Am. Inst. Chem. Eng. J., *1, 42 (1955).*]

Klee and Treybal [*Am. Inst. Chem. Eng. J.*, **2,** 444–447 (1956)] have also presented correlations relating to the rise or fall of liquid drops in liquid media. Their correlations are based on data covering a wider range of interfacial tensions [0.0003 to 0.042 N/m (0.3 to 42 dyn/cm)] with continuous-phase viscosities around [0.001 Pa·s (1 cP)].

Thorsen, Stordalen, and Terjesen [*Chem. Eng. Sci.*, **23,** 413–426 (1968)] found that terminal velocities for drops of highly purified organic liquids falling through water greatly exceed those previously reported in the literature. The addition of small quantities of surface-active agents (10^{-3} g/L) gave velocities which checked the Hu and Kintner (loc. cit.) correlation. The latter should then be representative of conditions normally encountered in practice.

For drop velocities in **nonnewtonian liquids,** see Mhatre and Kinter, *Ind. Eng. Chem.*, **51,** 865–867 (1959)]; Marrucci, Apuzzo, and Astarita, *Am. Inst. Chem. Eng. J.*, **16,** 538–541 (1970); and Mohan et al., *Can. J. Chem. Eng.*, **50,** 37–40 (1972).

For an exhaustive review of references on drop phenomena for the period 1965–1970, see Gal-Or et al., loc. cit.; and Tavlarides et al., loc. cit.

Liquid Drops in Gases Liquid drops falling in gases appear to remain spherical and follow the same drag relationships as solid spherical particles up to a Reynolds number of about 100. Large drops will deform, with a resulting increase in drag, and in some cases will shatter. The largest water drop which will fall in air at its terminal velocity is about 8 mm (5/16 in) in diameter. The corresponding maximum velocity is about 9 m/s (30 ft/s).

Hughes and Gilliland [*Chem. Eng. Prog.*, **48,** 497–504 (1952)] correlated terminal-settling-velocity data for a variety of liquids in air on the basis of a drag-coefficient–Reynolds-number chart with lines of constant *Su* as given in Fig. 5-79, where Su = Suratman number = (1/Ohnesorge number)^2 = $g_c\sigma_i\rho D_p/\mu^2$; g_c = dimensional constant 32.2; σ_i = interfacial tension; ρ = gas density; D_p = drop diameter; and μ = gas viscosity. See also Clift, Grace, and Weber, op. cit., chap. 7.

For an exhaustive review of references on drop phenomena, see Gal-Or et al., loc. cit.; and Tavlarides et al., loc. cit.

Applications The use of a simple C versus N_{Re} plot for prediction of either **terminal velocity** or **particle diameter** involves trial and error, since both terms are involved in both the ordinate and the abscissa. However, the terms CN_{Re}^2 and C/N_{Re} do not include u_t or D_p respectively. Thus, to eliminate trial and error, values of these quantities can be calculated and plotted against each other or against N_{Re}. Then, depending upon whether u_t or D_p is unknown, the value of CN_{Re}^2 or C/N_{Re} respectively can be calculated, and the unknown obtained from the corresponding term. For spherical particles,

$$CN_{Re}^2 = \frac{4g\rho D_p^3(\rho_p - \rho)}{3\mu^2} \tag{5-263}$$

$$\frac{C}{N_{Re}} = \frac{4g\mu(\rho_p - \rho)}{3\rho^2 u_t^3} \tag{5-264}$$

Values of C, N_{Re}, CN_{Re}^2, and C/N_{Re} are given in Table 5-22 for spherical particles.

Figure 5-80 gives the terminal settling velocities of spherical particles of different densities settling in air and water at 70°F (about 21°C) under the action of gravity. These curves should be useful for ready reference in practical applications.

Limitations The relationships presented thus far have dealt with the motion of particles present in relatively dilute concentration and in bodies of fluid of relatively large cross section.

The terminal velocity of drops falling in line in air or in a liquid is greater than for an isolated drop at all Reynolds numbers [Arrowsmith and Foster, *Chem. Eng. J. (Lausanne)*, **5,** 243–250 (1973); and Zabel, Hanson, and Ingham, *Trans. Inst. Chem. Eng. (London)*, **51,** 162–164 (1973)]. In nonnewtonian viscoelastic liquids there is a critical separation distance between spheres falling in line. At distances less than the critical the spheres converge, whereas at distances greater than the critical they diverge [Riddle, Narvaez, and Bird, *J. Non-Newtonian Fluid Mech.*, **2,** 23–35 (1977)].

When the concentration increases, the particle-settling velocities decrease because of increases in both the apparent suspension viscosity and the density. This condition is encountered in sedimentation and is often referred to as **hindered settling.** The effect produces less than a 1 percent reduction in settling velocity for solids volumetric fractions below 0.1 percent. Maude and Whitmore [*Br. J. Appl. Phys.*, **9,** 477–482 (1958)] give the following relation for the effect of concentration on settling velocity for suspensions of spheres of uniform size

$$u_{ts} = u_t(1 - c)^n \tag{5-265}$$

where u_{ts} = terminal velocity in the suspension; u_t = terminal velocity of single sphere ($c = 0$); c = volume fraction of solid in the suspension, dimensionless; and n = function of Reynolds number ($D_pu_t\rho/\mu$) as given in Fig. 5-81 (where u_t is defined earlier and other terms are defined in the legend of Fig. 5-76), dimensionless. n = 4.65 for the Stokes'-law region ($N_{Re} < 0.3$) and 2.33 for the Newton's-law region ($N_{Re} > 1000$). See Fig. 5-81 for values for the intermediate region. Equation (5-265) applies approximately to the settling of particles of any size in a polydisperse system provided the volume fraction corresponding to all the particles is used in computing the terminal velocity [Richardson and Shabi, *Trans. Inst. Chem. Eng. (London)*, **38,** 33–42 (1960)]. The concentration effect is greater

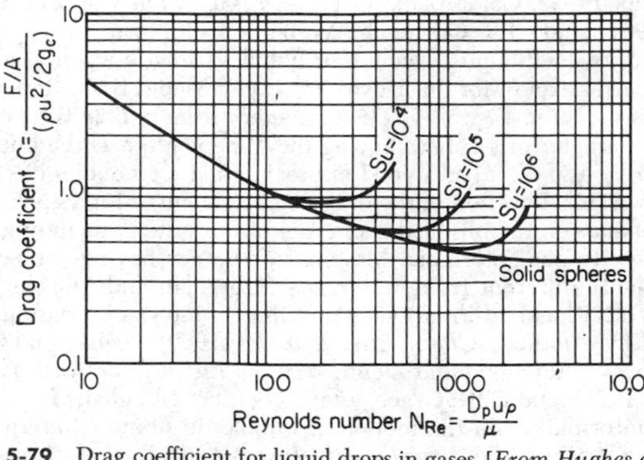

FIG. 5-79 Drag coefficient for liquid drops in gases. [*From Hughes and Gilliland,* Chem. Eng. Prog., *48, 497 (1952).*]

TABLE 5-22 Drag Coefficient and Related Functions for Spherical Particles*

N_{Re}†	C	CN_{Re}^2	C/N_{Re}
0.1	244	2.44	2440
0.2	124	4.96	620
0.3	83.8	7.54	279
0.5	51.5	12.9	103
0.7	37.6	18.4	53.8
1	27.2	27.2	27.2
2	14.8	59.0	7.38
3	10.5	94.7	3.51
5	7.03	176	1.41
7	5.48	268	0.782
10	4.26	426	0.426
20	2.72	$(1.09)(10^3)$	0.136
30	2.12	$(1.91)(10^3)$	0.0707
50	1.57	$(3.94)(10^3)$	0.0315
70	1.31	$(6.42)(10^3)$	0.0187
100	1.09	$(1.09)(10^4)$	0.0109
200	0.776	$(3.10)(10^4)$	$(3.88)(10^{-3})$
300	0.653	$(5.87)(10^4)$	$(2.18)(10^{-3})$
500	0.555	$(1.39)(10^5)$	$(1.11)(10^{-3})$
700	0.508	$(2.49)(10^5)$	$(7.26)(10^{-4})$
$(1)(10^3)$	0.471	$(4.71)(10^5)$	$(4.71)(10^{-4})$
$(2)(10^3)$	0.421	$(1.68)(10^6)$	$(2.11)(10^{-4})$
$(3)(10^3)$	0.400	$(3.60)(10^6)$	$(1.33)(10^{-4})$
$(5)(10^3)$	0.387	$(9.68)(10^6)$	$(7.75)(10^{-5})$
$(7)(10^3)$	0.390	$(1.91)(10^7)$	$(5.57)(10^{-5})$
$(1)(10^4)$	0.405	$(4.05)(10^7)$	$(4.05)(10^{-5})$
$(2)(10^4)$	0.442	$(1.77)(10^8)$	$(2.21)(10^{-5})$
$(3)(10^4)$	0.456	$(4.10)(10^8)$	$(1.52)(10^{-5})$
$(5)(10^4)$	0.474	$(1.19)(10^9)$	$(9.48)(10^{-6})$
$(7)(10^4)$	0.491	$(2.41)(10^9)$	$(7.02)(10^{-6})$
$(1)(10^5)$	0.502	$(5.02)(10^9)$	$(5.02)(10^{-6})$
$(2)(10^5)$	0.498	$(1.99)(10^{10})$	$(2.49)(10^{-6})$
$(3)(10^5)$	0.481	$(4.33)(10^{10})$	$(1.60)(10^{-6})$
$(3.5)(10^5)$	0.396	$(4.86)(10^{10})$	$(1.13)(10^{-6})$
$(3.75)(10^5)$	0.238	$(3.34)(10^{10})$	$(6.34)(10^{-7})$
$(4)(10^5)$	0.0891	$(1.43)(10^{10})$	$(2.23)(10^{-7})$
$(4.25)(10^5)$	0.0728	$(1.32)(10^{10})$	$(1.71)(10^{-7})$
$(4.5)(10^5)$	0.0753	$(1.53)(10^{10})$	$(1.67)(10^{-7})$
$(5)(10^5)$	0.0799	$(2.00)(10^{10})$	$(1.60)(10^{-7})$
$(7)(10^5)$	0.0945	$(4.63)(10^{10})$	$(1.35)(10^{-7})$
$(1)(10^6)$	0.110	$(1.10)(10^{11})$	$(1.10)(10^{-7})$
$(2)(10^6)$	0.150	$(6.00)(10^{11})$	$(7.50)(10^{-8})$
$(3)(10^6)$	0.163	$(1.47)(10^{12})$	$(5.44)(10^{-8})$
$(5)(10^6)$	0.174	$(4.35)(10^{12})$	$(3.48)(10^{-8})$
$(7)(10^6)$	0.179	$(8.75)(10^{12})$	$(2.55)(10^{-8})$
$(1)(10^7)$	0.182	$(1.82)(10^{13})$	$(1.82)(10^{-8})$

*Adopted from Clift, Grace, and Weber, op. cit., p. 112.
†For values of N_{Re} less than 0.1, $C = 24/N_{Re}$.

for nonspherical and angular particles than for spherical particles. For details, see Steinour, *Ind. Eng. Chem.*, **36**, 840–847 (1944).

When the diameter of the particle becomes appreciable with respect to the diameter of the container in which it is settling, the container walls will exert an additional retarding effect known as the **wall effect.** This can be allowed for in the case of **rigid spherical particles** by multiplying the terminal velocity as computed from Stokes's law by the factor k_w as given in Table 5-23. For values of $\beta < 0.05$, the Ladenburg correction as given by Eq. (5-266) can be used (see Zenz and Othmer, op. cit., pp. 208–209):

$$k_w = 1/(1 + 2.1\beta) \qquad (5\text{-}266)$$

where β = particle diameter divided by vessel diameter.

Sutterby [*Trans. Soc. Rheol.*, **17**, 559–573 (1973)] gives correction factors for wall and inertial effects for the region $N_{Re} < 3$, $\beta < 0.13$. See also Sutterby, ibid., **17**, 575–585 (1973), for end effects in settling in short tubes.

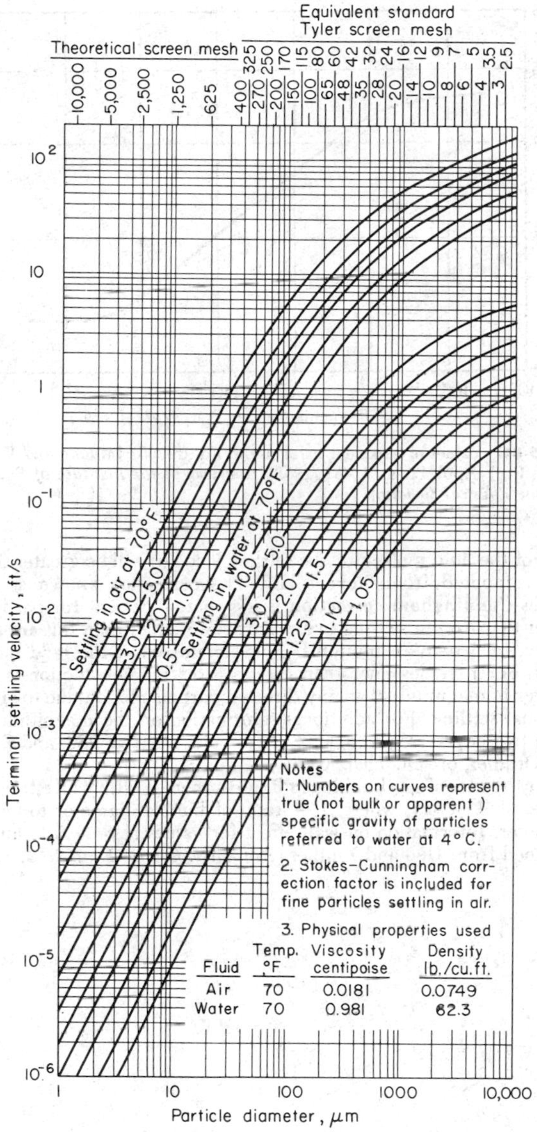

FIG. 5-80 Terminal velocities of spherical particles of different densities settling in air and water at 70°F under the action of gravity. To convert feet per second to meters per second, multiply by 0.3048. (*From Lapple et al.,* Fluid and Particle Mechanics, *University of Delaware, Newark, 1951, p. 292.*)

For the Newton's-law region, Harmathy (loc. cit.) developed the correction factor k_w' by which the computed terminal velocity for solid spheres should be multiplied to allow for the wall effect:

$$k_w' = \frac{1 - \beta^2}{\sqrt{1 + \beta^4}} \qquad (5\text{-}267)$$

The wall effect is independent of N_{Re} in the range of 100 to 10,000, $\beta < 0.6$ (Clift, Grace, and Weber, op. cit., pp. 226–227).

The wall effect and edge effects for cylinders at low N_{Re} are discussed by Stalnaker and Hussey [*Phys. Fluids*, **22**, 603–613 (1979)].

Stokes' law is subject to a lower limit in the case of rigid particles settling in a gas. When the particle size approaches the mean free

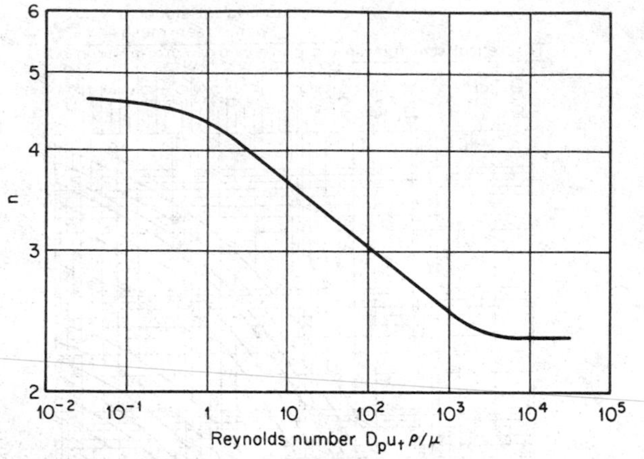

FIG. 5-81 Values of exponent n for use in Eq. (5-265). [*Maude and Whitmore,* Br. J. Appl. Phys., *9, 481 (1958). Courtesy of the Institute of Physics and the Physical Society.*]

path of the fluid molecules, the settling velocity will be greater than that computed from Stokes's law. The correction known as the **Stokes-Cunningham correction** is less than 1 percent for particles larger than 16 μm settling in air. Particles smaller than this are also subject to **Brownian motion** because of impact of the fluid molecules. For particles finer than 0.1 μm, this random motion is far greater in magnitude than any directed particle motion due to gravitational settling. For additional information on the magnitude of these two effects, see Lapple et al., op. cit., pp. 285–286; and Zenz and Othmer, op. cit., chap. 6.

For the case of **gas bubbles in liquids,** there is little wall effect for values of $\beta < 0.1$, where β = ratio of bubble diameter to vessel diameter. Information on wall effect for values of $\beta > 0.1$ can be obtained from Uno and Kintner [*Am. Inst. Chem. Eng. J.*, **2,** 420–424 (1956)], Maneri and Mendelson [*Chem. Eng. Prog.*, **64,** *Symp. Ser.*, 82, 72–80 (1968)], or Collins [*J. Fluid Mech.*, **28,** part 1, 97–112 (1967)].

TABLE 5-23 Wall Correction Factor for Rigid Spheres in Stokes' Law Region

β	k_w	β	k_w
0.0	1.000	0.4	0.279
0.05	0.885	0.5	0.170
0.1	0.792	0.6	0.0945
0.2	0.596	0.7	0.0468
0.3	0.422	0.8	0.0205

β = particle diameter divided by vessel diameter. (From Haberman and Sayre, *David W. Taylor Model Basin Report* 1143, 1958.)

For the case of **liquid drops in liquids,** information on wall effect can be obtained from Strom and Kintner [*Am. Inst. Chem. Eng. J.*, **4,** 153–156 (1958)], Harmathy (loc. cit.), or Haberman and Sayre (loc. cit., p. 34).

Particle Trajectories In the treatment up to this point, only one-dimensional steady-state motion has been considered, vertically up or down, under the action of gravity. Equations have been developed by Lapple and Shepherd [*Ind. Eng. Chem.*, **32,** 605–617 (1940)] for calculating the position-time histories for particles undergoing one-dimensional accelerated motion and two-dimensional motion.

In the Stokes'-law region, motion in a given direction is shown to be independent of motion in a direction perpendicular to the first, and the resulting trajectory is that of one motion superimposed upon the other. For other than the laminar regime, the motion of a particle in any direction will be influenced by velocity components in other directions. For example, the vertical terminal velocity of a particle will be lower the higher the horizontal velocity component. See also Brown et al., op. cit., pp. 79–83; Dalla Valle, *Micromeritics*, 2d ed., Pitman, New York, 1948, pp. 24–29; Zenz and Othmer, op. cit., pp. 216–220; Holland-Batt, *Trans. Inst. Chem. Eng. (London)*, **50,** 12–20 and 156–167 (1972); and Clift, Grace, and Weber, op. cit., chap. 11.

Section 6

Transport and Storage of Fluids

Raymond P. Genereaux, Ch.E., *Chemical Engineer (Retired), E. I. du Pont de Nemours & Co.; Fellow, American Institute of Chemical Engineers; Registered Professional Engineer (Delaware). (Section Editor)*

Charles J. B. Mitchell, B.S., *Principal Consultant (Retired), Engineering Service Division, E. I. du Pont de Nemours & Co.; Member, American Society of Mechanical Engineers, American Institute of Chemical Engineers. (Pumping of Liquids and Gases)*

C. Addison Hempstead, B.M.E., *Senior Design Consultant (Retired), E. I. du Pont de Nemours & Co.; Member, American Society of Mechanical Engineers, American Association for the Advancement of Science; Registered Professional Engineer (Ohio). (Process-Plant Piping)*

Bruce F. Curran, S.B., *Project Manager, Design Division, E. I. du Pont de Nemours & Co.; Registered Professional Engineer (Delaware). (Storage and Process Vessels)*

PUMPING OF LIQUIDS AND GASES

PROCESS-PLANT PIPING

STORAGE AND PROCESS VESSELS

Nomenclature and Units

In this listing, symbols used in Sec. 6 are defined in a general way with appropriate U.S. customary and SI units. Specific definitions denoted by subscripts are stated at the place of application. Some specialized symbols used in the section are defined at the place of application.

Symbol	Definition	SI units	U.S. customary units
A	Area	m^2	ft^2
A	Factor for determining minimum value of R_1		
C	Sum of mechanical allowances (thread or groove depth) plus corrosion or erosion allowances	mm	in
C	Cold-spring factor		
C	Constant		
C_1	Estimated self-spring or relaxation factor		
c_p	Specific heat at constant pressure	J/(kg·K)	Btu/(lb·°F)
c_v	Specific heat at constant volume	J/(kg·K)	Btu/(lb·°F)
d	Diameter; small-diameter truncated cone	cm or m	in
D, D_0	Outside diameter of pipe	mm	in
D	Cylinder diameter	m	ft
E	Quality factor		
E	Young's modulus of elasticity	MPa	kip/in² (ksi)
E_a	As-installed Young's modulus	MPa	kip/in² (ksi)
E_c	Casting quality factor		
E_j	Joint quality factor		
E_m	Minimum value of Young's modulus	MPa	kip/in² (ksi)
f	Stress-range reduction factor		
g	Acceleration of gravity	m/s^2	ft/s^2
H, h	Head (height of column of liquid)	m	ft
H	Depth of liquid	m	ft
h	Flexibility characteristic		
h	Height of truncated cone; depth of head	m	in
i	Stress-intensification factor		
i_i	In-plane stress-intensification factor		
i_o	Out-plane stress intensification factor		
k	Flexibility factor		
k	Adiabatic exponent c_p/c_v		
K_1	Constant in empirical flexibility equation		
L	Cylinder length	m	ft
L	Developed length of piping between anchors	m	ft
L	Dish radius	m	in
M_i, m_i	In-plane bending moment	N·mm	in·lbf
M_o	Out-plane bending moment	N·mm	in·lbf
M_t	Torsional moment	N·mm	in·lbf
NPSH	Net positive suction head	m	ft
n	Polytropic exponent		
N	Equivalent full temperature cycles		
p	Power	kW	hp
p	Pressure	Pa	lb/ft^2
P	Design gauge pressure	kPa	lbf/in^2
P_{ad}	Adiabatic power	kW	hp
q	Volume rate of flow	m^3/s	ft^3/s
Q	Volume rate of flow (liquids)	m^3/h	gal/min
Q	Volume rate of flow (gases)	m^3/h	ft^3/min (cfm)
R	Range of reaction forces or moments in flexibility analysis	N or N·mm	lbf or in·lbf
R	Cylinder radius	m	ft
R	Universal gas constant	J/(kg·K)	(ft·lbf)/(lbm·°R)
R_a	Estimated instantaneous reaction force or moment at installation temperature	N or N·mm	lbf or in·lbf
R_m	Estimated instantaneous maximum reaction force or moment at maximum or minimum metal temperature	N or N·mm	lbf or in·lbf
R_1	Effective radius of miter bend	mm	in
r_2	Mean radius of pipe using nominal wall thickness $\overline{T}$	mm	in
r_k	Knuckle radius	m	in
s	Specific gravity		
S	Basic allowable stress for metals, excluding factor E, or bolt design stress	MPa	kip/in² (ksi)
S_A	Allowable stress range for displacement stress	MPa	kip/in² (ksi)
S_E	Computed displacement-stress range	MPa	kip/in² (ksi)
S_L	Sum of longitudinal stresses	MPa	kip/in² (ksi)
S_T	Allowable stress at test temperature	MPa	kip/in² (ksi)
S_b	Resultant bending stress	MPa	kip/in² (ksi)
S_c	Basic allowable stress at minimum metal temperature expected	MPa	kip/in² (ksi)
S_h	Basic allowable stress at maximum metal temperature expected	MPa	kip/in² (ksi)

Symbol	Definition	SI units	U.S. customary units
S_t	Torsional stress	MPa	kip/in^2 (ksi)
T_s	Effective branch-wall thickness	mm	in
$\overline{T}$	Nominal wall thickness of pipe	mm	in
$\overline{T}_b$	Nominal branch-pipe wall thickness	mm	in
$\overline{T}_h$	Nominal header-pipe wall thickness	mm	in
t	Pressure design thickness	mm	in
t_m	Minimum required thickness, including mechanical, corrosion, and erosion allowances	mm	in
t_r	Pad or saddle thickness	mm	in
t	Head or shell radius	mm	in
U	Straight-line distance between anchors	m	ft

Symbol	Definition	SI units	U.S. customary units
V	Volume	m^3	ft^3
V	Velocity	m/s	ft/s
W	Work	J	ft·lb
w	Mass rate of flow	kg/s	lb/s
x	Value of expression $[(p_2/p_1)^{(k-1/k)} - 1]$		
y	Resultant of total displacement straings	mm	in
Z	Section modulus of pipe	mm^3	in^3
Z_e	Effective section modulus for branch	mm^3	in^3
	Greek symbols		
σ	Half-included angle	°	°
α, β, θ	Angles	°	°
ρ	Density	kg/m^3	lb/ft^3

PUMPING OF LIQUIDS AND GASES

PRINCIPLES OF LIQUID PUMPING

Means of Producing Fluid Flow Fluid pumping arises from the need to move fluids from one location to another through conduits or channels. A fluid is made to move through a conduit or channel by the transfer of energy. The means employed to cause fluids to flow are gravity, displacement, centrifugal force, electromagnetic force, transfer of momentum, mechanical impulse, and combinations of these six basic means. Aside from gravity, the means most commonly employed today is centrifugal force.

Displacement Discharge of a fluid from a vessel by partially or completely displacing its internal volume with a second fluid or by mechanical means is the principle upon which a great many fluid-transport devices operate. Included in this group are reciprocating-piston and diaphragm machines, rotary-vane and gear types, fluid piston compressors, acid eggs, and air lifts.

The large variety of displacement-type fluid-transport devices makes it difficult to list characteristics common to each. However, for most types it is correct to state that (1) they are adaptable to high-pressure operation, (2) the flow rate through the pump is variable (auxiliary damping systems may be employed to reduce the magnitude of pressure pulsation and flow variation), (3) mechanical considerations limit maximum throughputs, and (4) the devices are capable of efficient performance at extremely low-volume throughput rates.

Centrifugal Force Centrifugal force is applied by means of the centrifugal pump or compressor. Though the physical appearance of the many types of centrifugal pumps and compressors varies greatly, the basic function of each is the same, i.e., to produce kinetic energy by the action of centrifugal force and then to convert this energy into pressure by efficiently reducing the velocity of the flowing fluid.

In general, centrifugal fluid-transport devices have these characteristics: (1) discharge is relatively free of pulsation; (2) mechanical design lends itself to high throughputs, capacity limitations are rarely a problem; (3) the devices are capable of efficient performance over a wide range of pressures and capacities even at constant-speed operation; (4) discharge pressure is a function of fluid density; and (5) these are relatively small high-speed devices and less costly.

A device which combines the use of centrifugal force with mechanical impulse to produce an increase in pressure is the axial-flow compressor or pump. In this device the fluid travels roughly parallel to the shaft through a series of alternately rotating and stationary radial blades having airfoil cross sections. The fluid is accelerated in the axial direction by mechanical impulses from the rotating blades; concurrently, a positive-pressure gradient in the radial direction is established in each stage by centrifugal force. The net pressure rise per stage results from both effects.

Electromagnetic Force When the fluid is an electrical conductor, as is the case with molten metals, it is possible to impress an electromagnetic field around the fluid conduit in such a way that a driving force that will cause flow is created. Such pumps have been developed for the handling of heat-transfer liquids, especially for nuclear reactors.

Transfer of Momentum Deceleration of one fluid (motivating fluid) in order to transfer its momentum to a second fluid (pumped fluid) is a principle commonly used in the handling of corrosive materials, in pumping from inaccessible depths, or for evacuation. Jets and eductors are in this category.

Absence of moving parts and simplicity of construction have frequently justified the use of jets and eductors. However, they are relatively inefficient devices. When air or steam is the motivating fluid, operating costs may be several times the cost of alternative types of fluid-transport equipment. In addition, environmental considerations in today's chemical plants often inhibit their use.

Mechanical Impulse The principle of mechanical impulse when applied to fluids is usually combined with one of the other means of imparting motion. As mentioned earlier, this is the case in axial-flow compressors and pumps. The turbine or regenerative-type pump is another device which functions partially by mechanical impulse.

Measurement of Performance The amount of useful work that any fluid-transport device performs is the product of (1) the mass rate of fluid flow through it and (2) the total pressure differential measured immediately before and after the device, usually expressed in the height of column of fluid equivalent under adiabatic conditions. The first of these quantities is normally referred to as **capacity,** and the second is known as **head.**

Capacity This quantity is expressed in the following units. In SI units capacity is expressed in cubic meters per hour (m^3/h) for both liquids and gases. In U.S. customary units it is expressed in U.S. gallons per minute (gal/min) for liquids and in cubic feet per minute (ft^3/min) for gases. Since all these are volume units, the density or specific gravity must be used for conversion to mass rate of flow. When gases are being handled, capacity must be related to a pressure and a temperature, usually the conditions prevailing at the machine inlet. It is important to note that all heads and other terms in the following equations are expressed in height of column of liquid.

Total Dynamic Head The total dynamic head H of a pump is the total discharge head h_d minus the total suction head h_s.

Total Suction Head This is the reading h_{gs} of a gauge at the suction flange of a pump (corrected to the pump centerline*), plus the barometer reading and the velocity head h_{vs} at the point of gauge attachment:

$$h_s = h_{gs} + \text{atm} + h_{vs} \qquad (6\text{-}1)$$

If the gauge pressure at the suction flange is less than atmospheric, requiring use of a vacuum gauge, this reading is used for h_{gs} in Eq. (6-1) with a negative sign.

Before installation it is possible to estimate the total suction head as follows:

$$h_s = h_{ss} - h_{fs} \qquad (6\text{-}2)$$

where h_{ss} = static suction head and h_{fs} = suction friction head.

Static Suction Head The static suction head h_{ss} is the vertical distance measured from the free surface of the liquid source to the pump centerline plus the absolute pressure at the liquid surface.

Total Discharge Head The total discharge head h_d is the reading h_{gd} of a gauge at the discharge flange of a pump (corrected to the pump centerline*), plus the barometer reading and the velocity head h_{vd} at the point of gauge attachment:

$$h_d = h_{gd} + \text{atm} + h_{vd} \qquad (6\text{-}3)$$

Again, if the discharge gauge pressure is below atmospheric, the vacuum-gauge reading is used for h_{gd} in Eq. (6-3) with a negative sign.

Before installation it is possible to estimate the total discharge head from the static discharge head h_{sd} and the discharge friction head h_{fd} as follows:

$$h_d = h_{sd} + h_{fd} \qquad (6\text{-}4)$$

Static Discharge Head The static discharge head h_{sd} is the vertical distance measured from the free surface of the liquid in the receiver to the pump centerline,* plus the absolute pressure at the liquid surface. **Total static head** h_{ts} is the difference between discharge and suction static heads.

Velocity Since most liquids are practically incompressible, the relation between the quantity flowing past a given point in a given time and the velocity of flow is expressed as follows:

$$Q = Av \qquad (6\text{-}5)$$

This relationship in SI units is as follows:

$$v \text{ (for circular conduits)} = 3.54\, Q/d^2 \qquad (6\text{-}6)$$

where v = average velocity of flow, m/s; Q = quantity of flow, m^3/h; and d = inside diameter of conduit, cm.

This same relationship in U.S. customary units is

$$v \text{ (for circular conduits)} = 0.409\, Q/d^2 \qquad (6\text{-}7)$$

where v = average velocity of flow, ft/s; Q = quantity of flow, gal/min; and d = inside diameter of conduit, in.

Velocity Head This is the vertical distance by which a body must fall to acquire the velocity v.

$$h_v = v^2/2g \qquad (6\text{-}8)$$

Viscosity (See Sec. 5 for further information.) In flowing liquids the existence of internal friction or the internal resistance to relative motion of the fluid particles must be considered. This resistance is called viscosity. The viscosity of liquids usually decreases with rising temperature. Viscous liquids tend to increase the power required by a pump, to reduce pump efficiency, head, and capacity, and to increase friction in pipe lines.

Friction Head This is the pressure required to overcome the resistance to flow in pipe and fittings. It is dealt with in detail in Sec. 5.

*On vertical pumps, the correction should be made to the eye of the suction impeller.

Work Performed in Pumping To cause liquid to flow, work must be expended. A pump may raise the liquid to a higher elevation, force it into a vessel at higher pressure, provide the head to overcome pipe friction, or perform any combination of these. Regardless of the service required of a pump, all energy imparted to the liquid in performing this service must be accounted for; consistent units for all quantities must be employed in arriving at the work or power performed.

When arriving at the performance of a pump, it is customary to calculate its **power output,** which is the product of (1) the total dynamic head and (2) the mass of liquid pumped in a given time. In SI units power is expressed in kilowatts; horsepower is the conventional unit used in the United States.

In SI units,

$$\text{kW} = HQ\rho/3.670 \times 10^5 \qquad (6\text{-}9)$$

where kW is the pump power output, kW; H = total dynamic head, m (column of liquid); Q = capacity, m^3/h; and ρ = liquid density, kg/m^3.

When the total dynamic head H is expressed in pascals, then

$$\text{kW} = HQ/3.599 \times 10^6 \qquad (6\text{-}10)$$

In U.S. customary units,

$$\text{hp} = HQs/3.960 \times 10^3 \qquad (6\text{-}11)$$

where hp is the pump-power output, hp; H = total dynamic head, ft (column of liquid); Q = capacity, U.S. gal/min; and s = liquid specific gravity.

When the total dynamic head H is expressed in pounds-force per square inch, then

$$\text{hp} = HQ/1.714 \times 10^3 \qquad (6\text{-}12)$$

The **power input** to a pump is greater than the **power output** because of internal losses resulting from friction, leakage, etc. The **efficiency** of a pump is therefore defined as

$$\text{Pump efficiency} = (\text{power output})/(\text{power input}) \qquad (6\text{-}13)$$

Suction Limitations of a Pump Whenever the pressure in a liquid drops below the vapor pressure corresponding to its temperature, the liquid will vaporize. When this happens within an operating pump, the vapor bubbles will be carried along to a point of higher pressure, where they suddenly collapse. This phenomenon is known as **cavitation.** Cavitation in a pump should be avoided, as it is accompanied by metal removal, vibration, reduced flow, loss in efficiency, and noise. When the absolute suction pressure is low, cavitation may occur in the pump inlet and damage result in the pump suction and on the impeller vanes near the inlet edges. To avoid this phenomenon, it is necessary to maintain a **required net positive suction head** $(NPSH)_R$, which is the equivalent total head of liquid at the pump centerline less the vapor pressure p. Each pump manufacturer publishes curves relating $(NPSH)_R$ to capacity and speed for each pump.

When a pump installation is being designed, the **available net positive suction head** $(NPSH)_A$ must be equal to or greater than the $(NPSH)_R$ for the desired capacity. The $(NPSH)_A$ can be calculated as follows:

$$(\text{NPSH})_A = h_{ss} - h_{fs} - p \qquad (6\text{-}14)$$

If $(NPSH)_A$ is to be checked on an existing installation, it can be determined as follows:

$$(\text{NPSH})_A = \text{atm} + h_{gs} - p + h_{vs} \qquad (6\text{-}15)$$

Practically, the NPSH required for operation without cavitation and vibration in the pump is somewhat greater than the theoretical. The actual $(NPSH)_R$ depends on the characteristics of the liquid, the total head, the pump speed, the capacity, and impeller design. Any suction condition which reduces $(NPSH)_A$ below that required to prevent cavitation at the desired capacity will produce an unsatisfactory installation and can lead to mechanical difficulty.

Net-Positive-Suction-Head Curves for Centrifugal Hot-Water Pumps Figures 6-1 and 6-2 are typical of suction limitations of hot-liquid pumps. The NPSH required for different capacities and

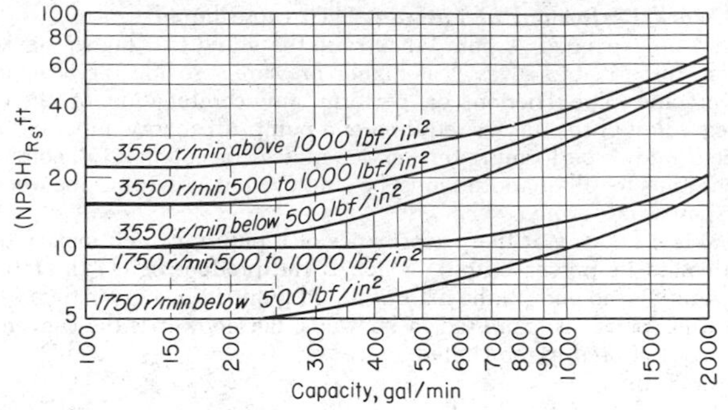

FIG. 6-1 Net positive suction head, centrifugal hot-water pumps, single suction; compiled from data by representative companies. The curves apply to water temperature up to 100°C (212°F). For temperatures above 100°C use temperature correction chart, Fig. 6-2. For speeds within ±25 percent of those shown, correct capacity according to r/min $\sqrt{\text{gal/min}}$ = constant. To convert gallons per minute to cubic meters per hour, multiply by 0.2271; to convert feet to meters, multiply by 0.3048; and to convert pounds-force per square inch to kilopascals, multiply by 6.895. *(By permission of Hydraulic Institute.)*

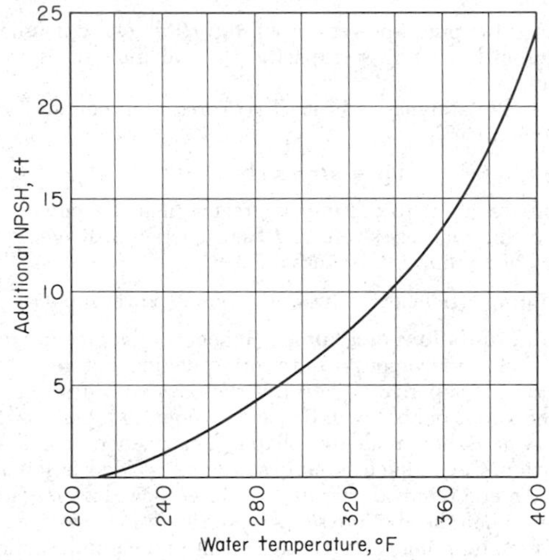

FIG. 6-2 Temperature correction chart, centrifugal hot-water pumps, single and double suction; additional suction head to be added to values given in Fig. 6-1. (°F − 32)5/9 = °C; to convert feet to meters, multiply by 0.3048. *(By permission of Hydraulic Institute.)*

speeds, including the additional NPSH required if the water temperature exceeds 100°C (212°F), is shown. The addition is necessitated by air entrainment.

For hot-liquid pumps taking suction from a source where the prevailing pressure is equivalent to the vapor pressure corresponding to its temperature, the NPSH available is the difference between the liquid level at the source and the pump centerline minus the entrance and friction losses in the suction piping.

Example 1 If a pump handles water at a temperature of 176.7°C (350°F) and a pressure of 1.034 MPa (150 lbf/in²) gauge or 1.136 MPa (164.7 lbf/in²) absolute at the suction nozzle, with 3.66-m/s (12-ft/s) velocity, what is the net positive suction head available $(\text{NPSH})_A$?

Solution

Vapor pressure of water at 176.7°C (350°F) = 0.928 MPa (134.6 lbf/in²)

Specific gravity of water at 176.7°C (350°F) = 0.89

Velocity head $v^2/2g$ = 0.68 m (2.22 ft)

g = 9.81 m/s² (32.17 ft/s²)

In SI units,

$$\text{NPSH}_A = (1.136 - 0.928)\ 102/0.89 + 0.68 = 24.52\ \text{m}$$

In U.S. customary units,

$$(\text{NPSH})_A = (164.7 - 134.6)\ 2.31/0.89 + 2.22 = 80.32\ \text{ft}$$

PUMP SELECTION

When selecting pumps for any service, it is necessary to know the liquid to be handled, the total dynamic head, the suction and discharge heads, and, in most cases, the temperature, viscosity, vapor pressure, and specific gravity. In the chemical industry, the task of pump selection is frequently further complicated by the presence of solids in the liquid and liquid corrosion characteristics requiring special materials of construction. Solids may accelerate erosion and corrosion, have a tendency to agglomerate, or require delicate handling to prevent undesirable degradation.

Range of Operation Because of the wide variety of pump types and the number of factors which determine the selection of any one type for a specific installation, the designer must first eliminate all but those types of reasonable possibility. Since range of operation is always an important consideration, Fig. 6-3 should be of assistance. The boundaries shown for each pump type are at best approximate, as unusual applications for which the best selection contradicts the chart will arise. In most cases, however, Fig. 6-3 will prove useful in limiting consideration to two or three types of pumps.

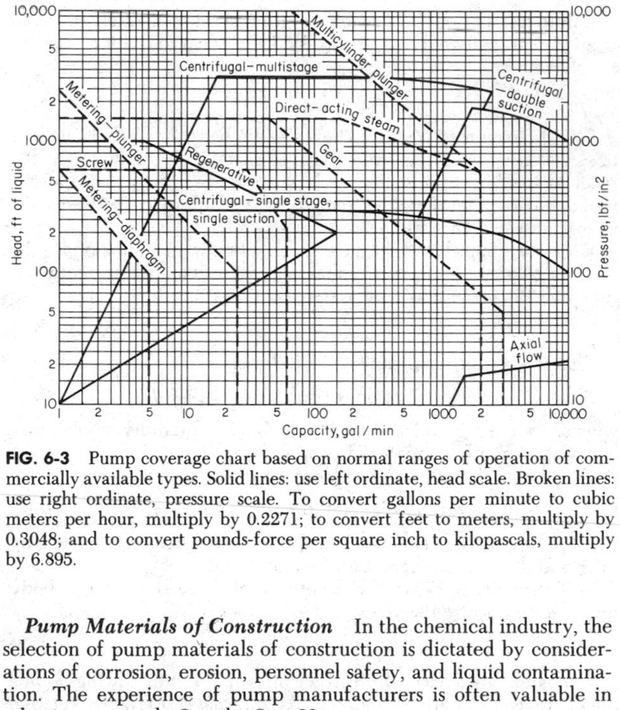

FIG. 6-3 Pump coverage chart based on normal ranges of operation of commercially available types. Solid lines: use left ordinate, head scale. Broken lines: use right ordinate, pressure scale. To convert gallons per minute to cubic meters per hour, multiply by 0.2271; to convert feet to meters, multiply by 0.3048; and to convert pounds-force per square inch to kilopascals, multiply by 6.895.

Pump Materials of Construction In the chemical industry, the selection of pump materials of construction is dictated by considerations of corrosion, erosion, personnel safety, and liquid contamination. The experience of pump manufacturers is often valuable in selecting materials. See also Sec. 23.

Presence of Solids When a pump is required to pump a liquid containing suspended solids, there are unique requirements which must be considered. Adequate clear-liquid hydraulic performance and the use of carefully selected materials of construction may not be all that is required for satisfactory pump selection. Dimensions of all internal passages are critical. Pockets and dead spots, areas where solids can accumulate, must be avoided. Close internal clearances are undesirable because of abrasion. Flushing connections for continuous or intermittent use should be provided.

For installations in which suspended solids must be handled with a minimum of solids breakage or degradation, such as pumps feeding filter presses, special attention is required; either a low-shear positive-displacement pump or a recessed-impeller centrifugal pump may be called for.

Ease of maintenance is of increasing importance in today's economy. Chemical pump installations that require annual maintenance costing 2 or 3 times the original investment are not uncommon. In most cases this expense is the result of improper selection.

Pump Costs Space does not permit presentation of adequate cost data on the many types of pumps discussed in this subsection. For two of the most common varieties, however, it is possible to give representative values. Figure 6-4 is a plot of costs for single-stage and two-stage pedestal-mounted centrifugal pumps, with a tabulation of factors to be applied when other than standard materials are used. Figure 6-5 may be employed to approximate costs for gear pumps of conventional design.

CENTRIFUGAL PUMPS

The centrifugal pump is the type most widely used in the chemical industry for transferring liquids of all types—raw materials, materials in manufacture, and finished products—as well as for general services of water supply, boiler feed, condenser circulation, condensate return, etc. These pumps are available through a vast range of sizes, in capacities from 0.5 m^3/h to 2×10^4 m^3/h (2 gal/min to 10^5 gal/min), and for discharge heads (pressures) from a few meters to approximately 48 MPa (7000 lbf/in^2). The size and type best suited to a particular application can be determined only by an engineering study of the problem.

The primary advantages of a centrifugal pump are simplicity, low first cost, uniform (nonpulsating) flow, small floor space, low maintenance expense, quiet operation, and adaptability for use with a motor or a turbine drive.

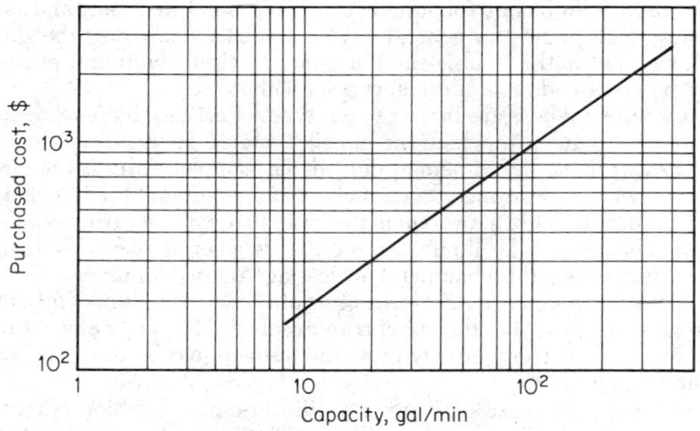

FIG. 6-5 Cost of positive-displacement gear pumps of ductile iron to handle nonviscous liquids at a maximum discharge pressure of 6.895×10^2 KPa (100 lbf/in^2) (first quarter of 1979). The cost includes the pump, drive, base plate, coupling, guard, standard mechanical seal, and relief valve. Refer to Fig. 6-4, "Material Factors," for the cost of pumps of other materials of construction. To convert gallons per minute to cubic meters per hour, multiply by 0.2271; to convert pounds-force per square inch to kilopascals, multiply by 6.895. *(Courtesy of E. I. du Pont de Nemours & Co.)*

A centrifugal pump, in its simplest form, consists of an impeller rotating within a casing. The **impeller** consists of a number of blades, either open or shrouded, mounted on a shaft that projects outside the casing. Its axis of rotation may be either horizontal or vertical, to suit the work to be done. **Closed-type,** or **shrouded,** impellers are generally the most efficient. **Open-** or **semiopen-type** impellers are used

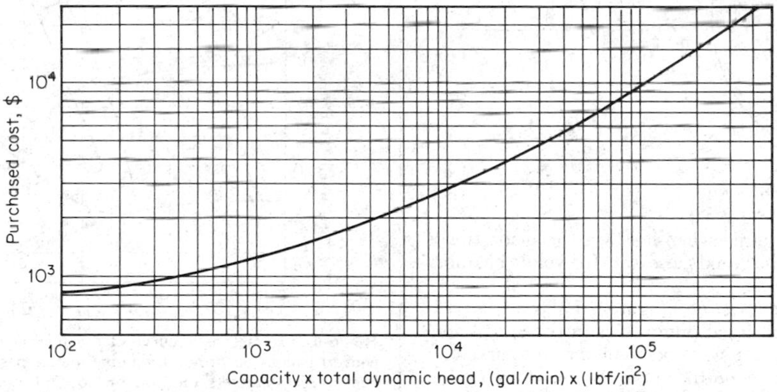

FIG. 6-4 Cost of general-purpose-single and two-stage-single-suction centrifugal pumps of ductile iron (first quarter of 1979). It includes the pump, drive, base plate, and coupling. For other materials of construction and for high suction pressures, multiply values from the chart by:

Material Factors*

Ductile iron	1.0	Durimet 20	1.7
304 stainless steel	1.4	Titanium	3.0
316 stainless steel	1.5	Hastelloy C	3.7
CD-4	1.6	Hastelloy B	4.1

Suction Pressure Factors*

Up to 150 lbf/in^2	1.0
150 to 500 lbf/in^2	1.6
500 to 1000 lbf/in^2	2.1

*Compound multipliers.

To convert gallons per minute to cubic meters per hour, multiply by 0.2271; to convert pounds-force per square inch to kilopascals, multiply by 6.895. *(Courtesy of E. I. du Pont de Nemours & Co.)*

for viscous liquids or for liquids containing solid materials and on many small pumps for general service. Impellers may be of the **single-suction** or the **double-suction** type—single if the liquid enters from one side, double if it enters from both sides.

Casings There are three general types of casings, but each consists of a chamber in which the impeller rotates, provided with inlet and exit for the liquid being pumped. The simplest form is the **circular casing,** consisting of an annular chamber around the impeller; no attempt is made to overcome the losses that will arise from eddies and shock when the liquid leaving the impeller at relatively high velocities enters this chamber. Such casings are seldom used.

Volute casings take the form of a spiral increasing uniformly in cross-sectional area as the outlet is approached. The volute efficiently converts the velocity energy imparted to the liquid by the impeller into pressure energy.

A third type of casing is used in **diffuser-type** or turbine pumps. In this type, **guide vanes** or **diffusers** are interposed between the impeller discharge and the casing chamber. Losses are kept to a minimum in a well-designed pump of this type, and improved efficiency is obtained over a wider range of capacities. This construction is often used in multistage high-head pumps.

Action of a Centrifugal Pump Briefly, the action of a centrifugal pump may be shown by Fig. 6-6. Power from an outside source is applied to shaft *A*, rotating the impeller *B* within the stationary casing *C*. The blades of the impeller in revolving produce a reduction in pressure at the entrance or eye of the impeller. This causes liquid to flow into the impeller from the suction pipe *D*. This liquid is forced outward along the blades at increasing tangential velocity.

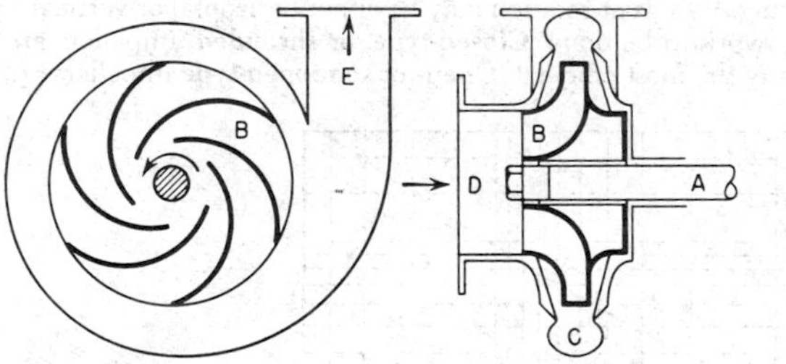

FIG. 6-6 A simple centrifugal pump.

The velocity head it has acquired when it leaves the blade tips is changed to pressure head as the liquid passes into the volute chamber and thence out the discharge *E*.

Centrifugal-Pump Characteristics Figure 6-7 shows a typical characteristic curve of a centrifugal pump. It is important to note that at any fixed speed the pump will operate along this curve and at no other points. For instance, on the curve shown, at 45.5 m^3/h (200 gal/min) the pump will generate 26.5-m (87-ft) head. If the head is increased to 30.48 m (100 ft), 27.25 m^3/h (120 gal/min) will be delivered. It is not possible to reduce the capacity to 27.25 m^3/h (120 gal/min) at 26.5-m (87-ft) head unless the discharge is throttled so that 30.48 m (100 ft) is actually generated within the pump. On pumps with variable-speed drivers such as steam turbines, it is possible to change the characteristic curve, as shown by Fig. 6-8.

It is important to remember that the head produced will be the same for any clean liquid of the same viscosity. The pressure rise, however, will vary in proportion to the specific gravity. Viscosities of less than 50 kPa·s (50 cP) do not affect the head materially.

Single-stage centrifugal pumps are available in capacities up to and over 1.136×10^4 m^3/h (50,000 gal/min) for heads (pressures) up to 488 m (1600 ft). They are available in a variety of designs for particular services (see Fig. 6-3).

Process Pumps This term is usually applied to single-stage pedestal-mounted units with single-suction overhung impellers and with a single packing box. These pumps are ruggedly designed for ease in dismantling and accessibility, with mechanical seals or packing arrangements, and are built especially to handle corrosive or otherwise difficult-to-handle liquids.

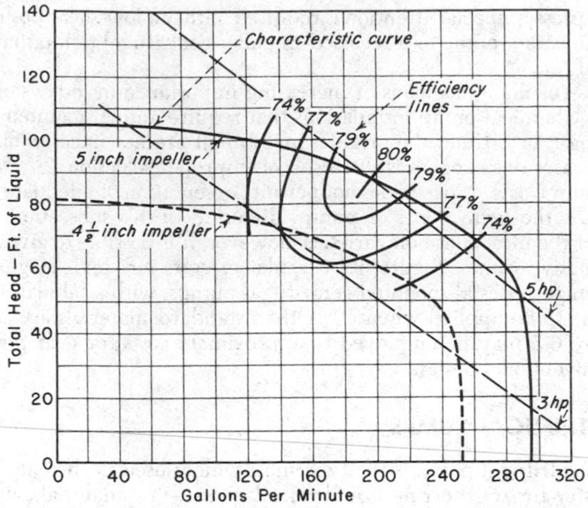

FIG. 6-7 Characteristic curve of a centrifugal pump operating at a constant speed of 3450 r/min. To convert gallons per minute to cubic meters per hour, multiply by 0.2271; to convert feet to meters, multiply by 0.3048; to convert horsepower to kilowatts, multiply by 0.746; and to convert inches to centimeters, multiply by 2.54.

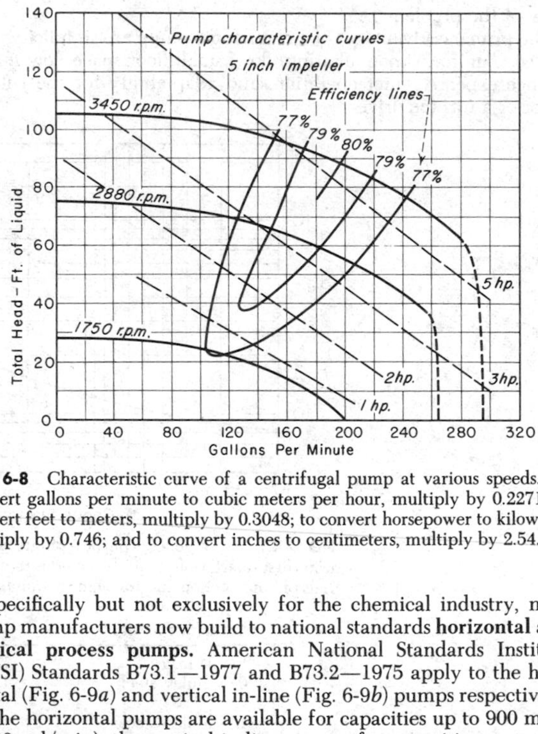

FIG. 6-8 Characteristic curve of a centrifugal pump at various speeds. To convert gallons per minute to cubic meters per hour, multiply by 0.2271; to convert feet to meters, multiply by 0.3048; to convert horsepower to kilowatts, multiply by 0.746; and to convert inches to centimeters, multiply by 2.54.

Specifically but not exclusively for the chemical industry, most pump manufacturers now build to national standards **horizontal and vertical process pumps.** American National Standards Institute (ANSI) Standards B73.1—1977 and B73.2—1975 apply to the horizontal (Fig. 6-9*a*) and vertical in-line (Fig. 6-9*b*) pumps respectively.

The horizontal pumps are available for capacities up to 900 m^3/h (4000 gal/min); the vertical in-line pumps, for capacities up to 320 m^3/h (1400 gal/min). Both horizontal and vertical in-line pumps are available for heads up to 120 m (400 ft). The intent of each ANSI specification is that pumps from all vendors for a given nominal capacity and total dynamic head at a given rotative speed shall be dimensionally interchangeable with respect to mounting, size, and location of suction and discharge nozzles, input shaft, base plate, and foundation bolts.

The vertical in-line pumps, although relatively new additions, are finding considerable use in chemical and petrochemical plants in the United States. An inspection of the two designs will make clear the relative advantages and disadvantages of each.

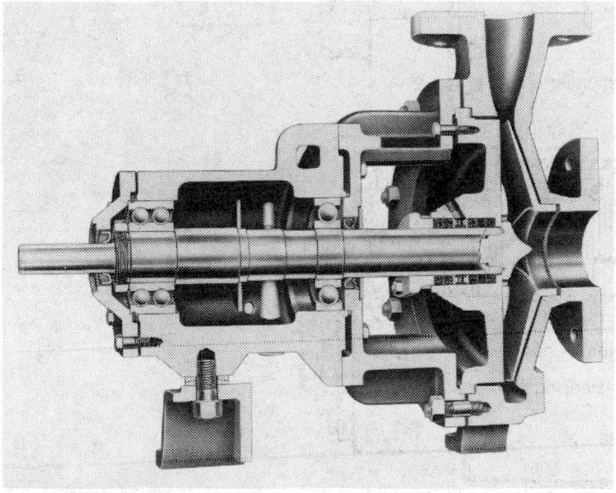

FIG. 6-9a Horizontal process pump conforming to American National Standards Institute (ANSI) Standard B73.1—1977.

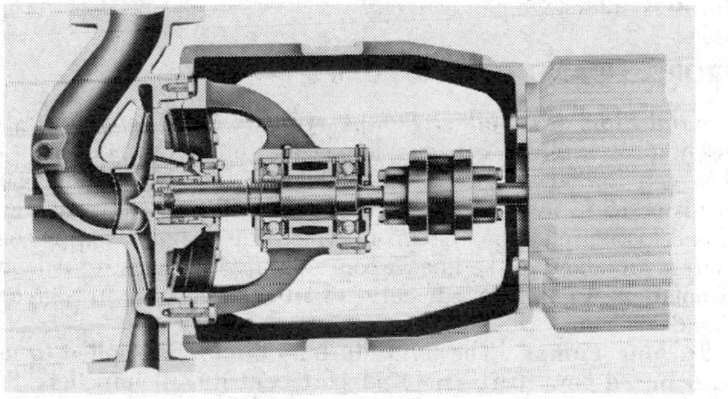

FIG. 6-9b Vertical in-line process pump conforming to ANSI Standard B73.2—1975. The pump shown is driven by a motor through flexible coupling. Not shown but also conforming to ANSI Standard B73.2 are vertical in-line pumps with rigid couplings and with no coupling (impeller-mounted on an extended motor shaft).

Chemical pumps are available in a variety of materials. Metal pumps are the most widely used. Although they may be obtained in iron, bronze, and iron with bronze fittings, an increasing number of pumps of ductile-iron, steel, and nickel alloys are being used. Pumps are also available in glass, glass-lined iron, carbon, rubber, rubber-lined metal, ceramics, and a variety of plastics, such units usually being employed for special purposes.

Sealing the Centrifugal Chemical Pump Although detailed treatment of **shaft seals** is presented in the subsection "Sealing of Rotating Shafts," it is appropriate to mention here the special problems of sealing centrifugal chemical pumps. Current practice demands that packing boxes be designed to accommodate both packing and mechanical seals. With either type of seal, one consideration is of paramount importance in chemical service: the liquid present at the sealing surfaces must be free of solids. Consequently, it is necessary to provide a secondary compatible liquid to flush the seal or packing whenever the process liquid is not absolutely clean.

The use of **packing** requires the continuous escape of liquid past the seal to minimize and to carry away the frictional heat developed. If the effluent is toxic or corrosive, quench glands or catch pans are usually employed. Although packing can be adjusted with the pump operating, leaking mechanical seals require shutting down the pump to correct the leak. Properly applied and maintained **mechanical seals** usually show no visible leakage. In general, owing to the more effective performance of mechanical seals, they have gained almost universal acceptance.

Double-Suction Single-Stage Pumps These pumps are used for general water-supply and circulating service and for chemical service when liquids that are noncorrosive to iron or bronze are being handled. They are available for capacities from about 5.7 m^3/h (25 gal/min) up to as high as 1.136×10^4 m^3/h (50,000 gal/min) and heads up to 304 m (1000 ft). Such units are available in iron, bronze, and iron with bronze fittings. Other materials increase the cost; when they are required, a standard chemical pump is usually more economical.

Close-Coupled Pumps (Fig. 6-10) Pumps equipped with a built-in electric motor or sometimes steam-turbine-driven (i.e., with pump impeller and driver on the same shaft) are known as close-coupled pumps. Such units are extremely compact and are suitable for a variety of services for which standard iron and bronze materials are satisfactory. They are available in capacities up to about 450 m^3/h (2000 gal/min) for heads up to about 73 m (240 ft). Two-stage units in the smaller sizes are available for heads to around 150 m (500 ft).

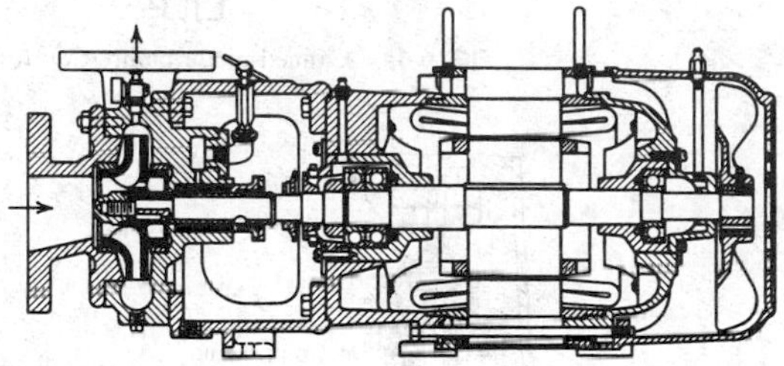

FIG. 6-10 Close-coupled pump.

Canned-Motor Pumps (Fig. 6-11) These pumps command considerable attention in the chemical industry. They are close-coupled units in which the cavity housing the motor rotor and the pump casing are interconnected. As a result, the motor bearings run in the process liquid and all seals are eliminated. Because the process liquid is the bearing lubricant, abrasive solids cannot be tolerated. Standard single-stage canned-motor pumps are available for flows up to 160 m^3/h (700 gal/min) and heads up to 76 m (250 ft). Two-stage units are available for heads up to 183 m (600 ft). Canned-motor pumps are being widely used for handling organic solvents, organic heat-transfer liquids, and light oils as well as many clean toxic or hazardous liquids or for installations in which leakage is an economic problem.

Vertical Pumps In the chemical industry, the term **vertical process pump** (Fig. 6-12) generally applies to a pump with a vertical shaft having a length from drive end to impeller of approximately 1 m (3.1 ft) minimum to 20 m (66 ft) or more. Vertical pumps are used as either **wet-pit pumps** (immersed) or **dry-pit pumps** (externally mounted) in conjunction with stationary or mobile tanks containing difficult-to-handle liquids. They have the following advantages: the liquid level is above the impeller, and the pump is thus self-priming; and the shaft seal is above the liquid level and is not wetted by the pumped liquid, which simplifies the sealing task. When no bottom connections are permitted on the tank (a safety consideration for highly corrosive or toxic liquid), the vertical wet-pit pump may be the only logical choice.

These pumps have the following disadvantages: intermediate or line bearings are generally required when the shaft length exceeds about 3 m (10 ft) in order to avoid shaft resonance problems; these bearings must be lubricated whenever the shaft is rotating. Since all wetted parts must be corrosion-resistant, low-cost materials may not be suitable for the shaft, column, etc. Maintenance is more costly since the pumps are larger and more difficult to handle.

For abrasive service, vertical cantilever designs requiring no line or foot bearings are available. Generally, these pumps are limited to about a 1-m (3.1-ft) maximum shaft length.

Sump Pumps These are small single-stage vertical pumps used to drain shallow pits or sumps. They are of the same general con-

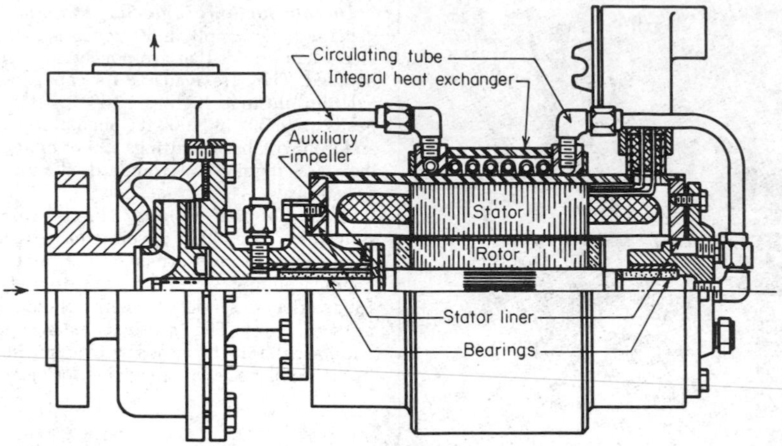

FIG. 6-11 Canned-motor pump *(Courtesy of Chempump Division, Crane Co.)*

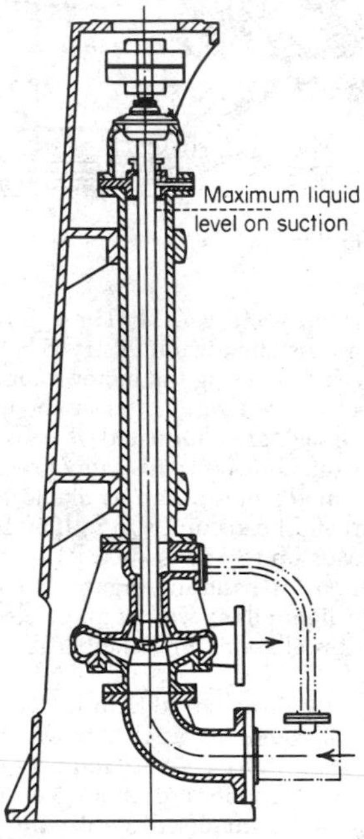

FIG. 6-12 Vertical process pump for dry-pit mounting. *(Courtesy of Lawrence Pumps, Inc.)*

struction as vertical process pumps but are not designed for severe operating conditions.

Multistage Centrifugal Pumps These pumps are used for services requiring heads (pressures) higher than can be generated by a single impeller. All impellers are in series, the liquid passing from one impeller to the next and finally to the pump discharge. The total head then is the summation of the heads of the individual impellers. Deep-well pumps, high-pressure water-supply pumps, boiler-feed pumps, fire pumps, and charge pumps for refinery processes are examples of multistage pumps required for various services.

Multistage pumps may be of the **volute type** (Fig. 6-13), with single- or double-suction impellers (Fig. 6-14), or of the **diffuser type** (Fig. 6-15). They may have horizontally split casings or, for extremely high pressures, 20 to 40 MPa (3000 to 6000 lbf/in^2), vertically split barrel-type exterior casings with inner casings containing diffusers, interstage passages, etc.

PROPELLER AND TURBINE PUMPS

Axial-Flow (Propeller) Pumps (Fig. 6-16) These pumps are essentially very-high-capacity low-head units. Normally they are designed for flows in excess of 450 m^3/h (2000 gal/min) against heads of 15 m (50 ft) or less. They are used to great advantage in closed-loop circulation systems in which the pump casing becomes merely an elbow in the line. A common installation is for calandria circulation. A characteristic curve of an axial-flow pump is given in Fig. 6-17.

Turbine Pumps The term "turbine pump" is applied to units with mixed-flow (part axial and part centrifugal) impellers. Such units are available in capacities from 20 m^3/h (100 gal/min) upward for heads up to about 30 m (100 ft) per stage. Turbine pumps are usually vertical.

A common form of turbine pump has the pump element mounted at the bottom of a column that serves as the discharge pipe (see Fig. 6-18). Such units are immersed in the liquid to be pumped and are commonly used for wells, condenser circulating water, large-volume drainage, etc. Another form of the pump has a shell surrounding the pumping element which is connected to the intake pipe. In this form, the pump is used on condensate service in power plants and for process work in oil refineries and elsewhere.

Regenerative Pumps Also referred to as turbine pumps because of the shape of the impeller, regenerative pumps employ a combination of mechanical impulse and centrifugal force to produce heads of several hundred meters (feet) at low volumes, usually less than 20 m^3/h (100 gal/min). The impeller, which rotates at high speed with small clearances, has many short radial passages milled on each side at the periphery. Similar channels are milled in the mating surfaces of the casing. Upon entering, the liquid is directed into the impeller passages and proceeds in a spiral pattern around the periphery, passing alternately from the impeller to the casing and receiving successive impulses as it does so. Figure 6-19 illustrates a typical performance-characteristic curve.

These pumps are particularly useful when low volumes of low-viscosity liquids must be handled at higher pressures than are normally available with centrifugal pumps. Close clearances limit their use to clean liquids. For very high heads, multistage units are available.

POSITIVE-DISPLACEMENT PUMPS

Whereas the total dynamic head developed by a centrifugal, mixed-flow, or axial-flow pump is uniquely determined for any given flow

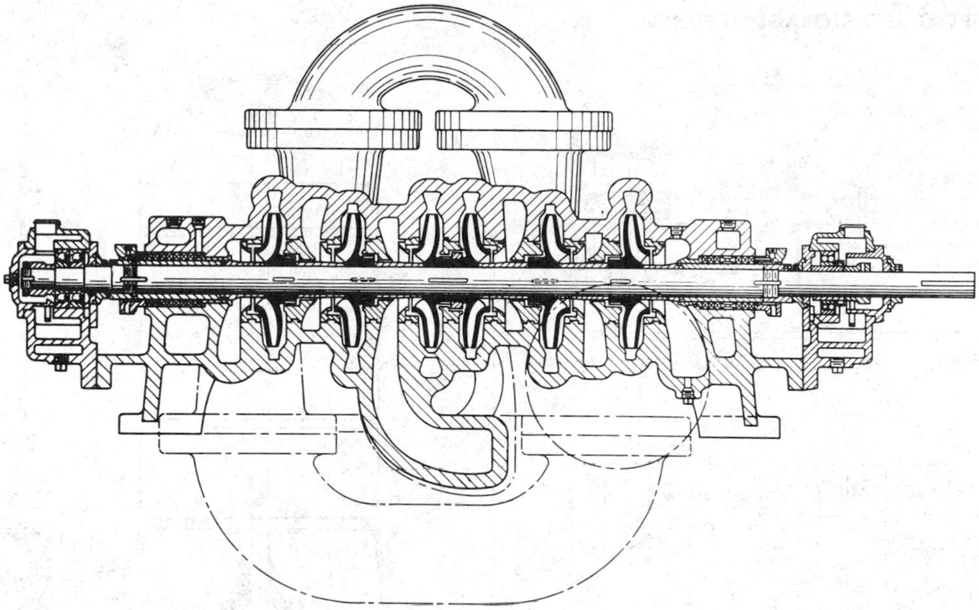

FIG. 6-13 Six-stage volute-type pump.

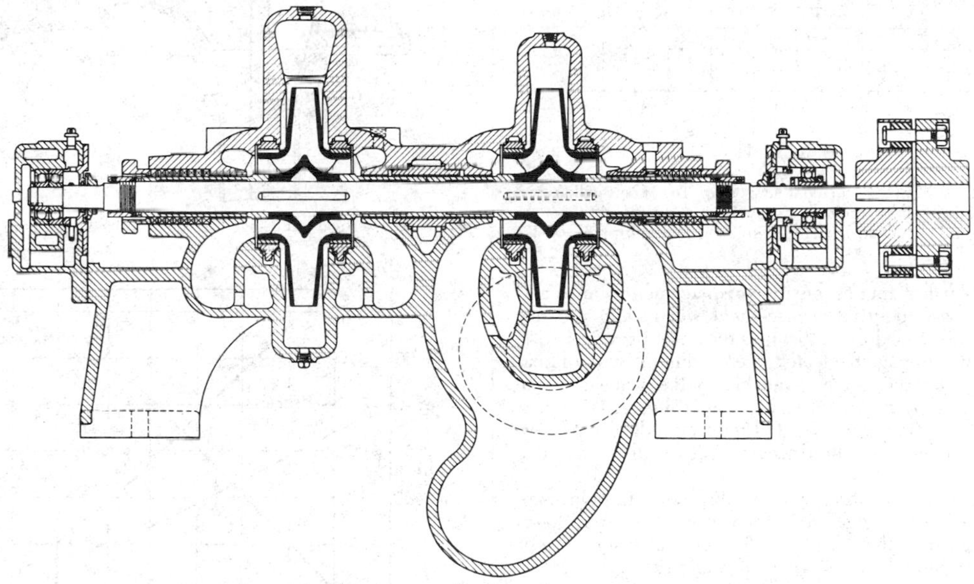

FIG. 6-14 Two-stage pump having double-suction impellers.

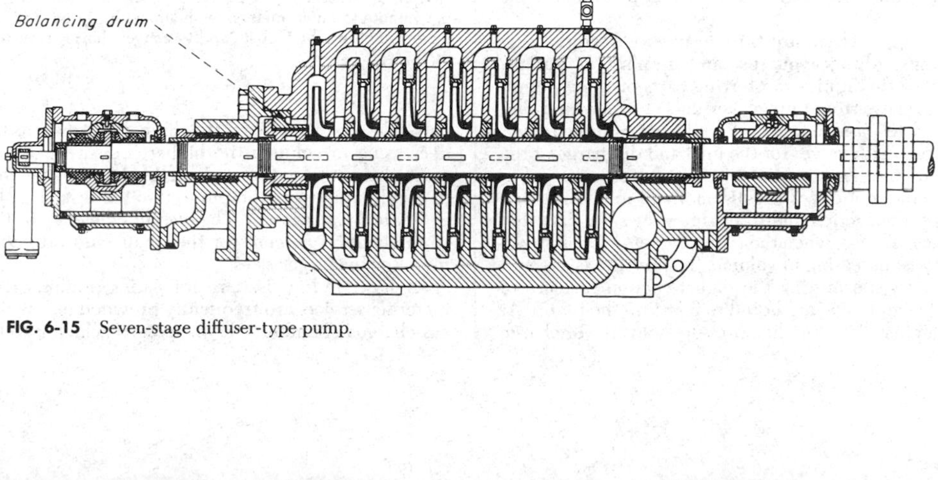

FIG. 6-15 Seven-stage diffuser-type pump.

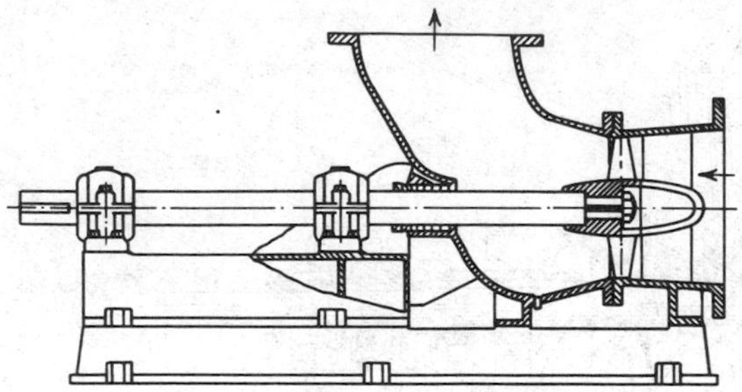

FIG. 6-16 Axial-flow elbow-type propeller pump. *(Courtesy of Lawrence Pumps, Inc.)*

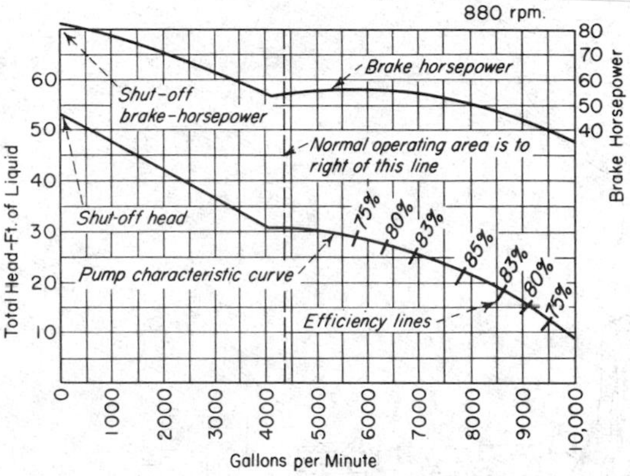

FIG. 6-17 Characteristic curve of an axial-flow pump. To convert gallons per minute to cubic meters per hour, multiply by 0.2271; to convert feet to meters, multiply by 0.3048; and to convert horsepower to kilowatts, multiply by 0.746.

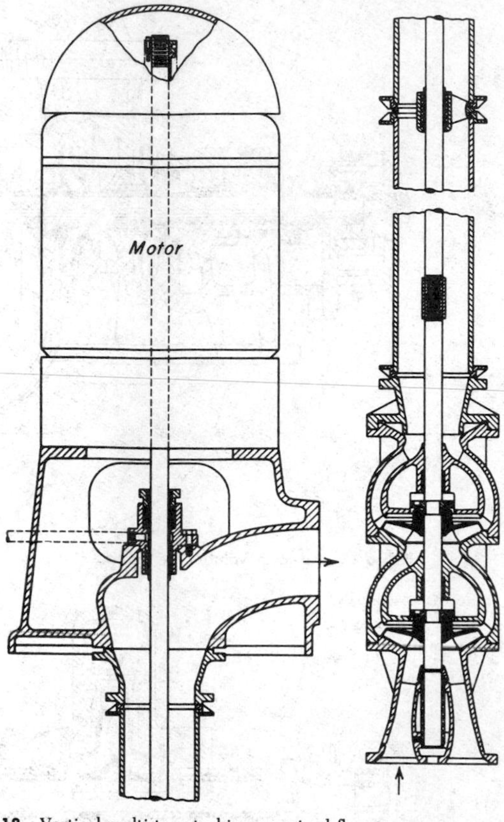

FIG. 6-18 Vertical multistage turbine, or mixed-flow, pump.

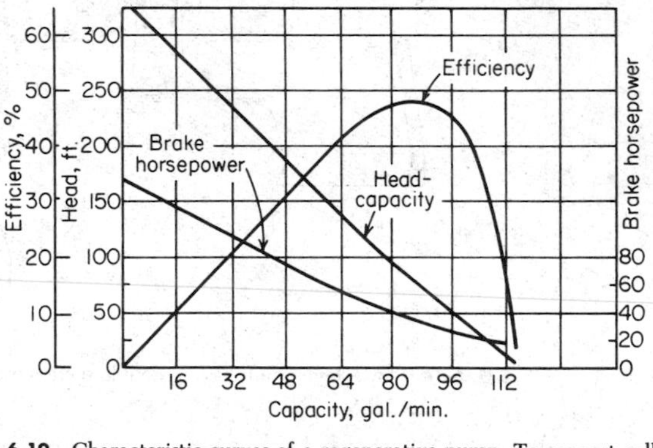

FIG. 6-19 Characteristic curves of a regenerative pump. To convert gallons per minute to cubic meters per hour, multiply by 0.2271; to convert feet to meters, multiply by 0.3048; and to convert horsepower to kilowatts, multiply by 0.746.

by the speed at which it rotates, **positive-displacement pumps** and those which approach positive displacement will ideally produce whatever head is impressed upon them by the system restrictions to flow. Actually, with slippage neglected, the maximum head attainable is determined by the power available in the drive and the strength of the pump parts. An automatic relief valve set to open at a safe pressure higher than the normal or maximum discharge pressure is generally required on the discharge side of all positive-displacement pumps.

In general, overall efficiencies of positive-displacement pumps are higher than those of centrifugal equipment because internal losses are minimized. On the other hand, the flexibility of each piece of equipment in handling a wide range of capacities is somewhat limited.

Positive-displacement pumps may be of either the **reciprocating** or the **rotary** type. In all positive-displacement pumps, a cavity or cavities are alternately filled and emptied of the pumped fluid by the action of the pump.

Reciprocating Pumps There are three classes of reciprocating pumps: **piston pumps, plunger pumps,** and **diaphragm pumps.** Basically, the action of the liquid-transferring parts of these pumps is the same, a cylindrical piston, plunger, or bucket or a round diaphragm being caused to pass or flex back and forth in a chamber. The device is equipped with valves for the inlet and discharge of the liquid being pumped, and the operation of these valves is related in a definite manner to the motions of the piston. In all modern-design reciprocating pumps, the suction and discharge valves are operated by pressure difference. That is, when the pump is on its suction stroke and the pump cavity is increasing in volume, the pressure is lowered within the pump cavity, permitting the higher suction pressure to open the suction valve and allowing liquid to flow into the pump. At the same time, the higher discharge-line pressure holds the discharge valve closed. Likewise on the discharge stroke, as the pump cavity is decreasing in volume, the higher pressure developed in the pump cavity holds the suction valve closed and opens the discharge valve to expel liquid from the pump into the discharge line.

Overall Efficiency The overall efficiency of these pumps varies from about 50 percent for the small pumps to about 90 percent or more for the larger sizes.

As shown in Fig. 6-20, reciprocating pumps, except when used for metering service, are frequently provided on the discharge side with gas-charged chambers, the purpose of which is to limit pressure pul-

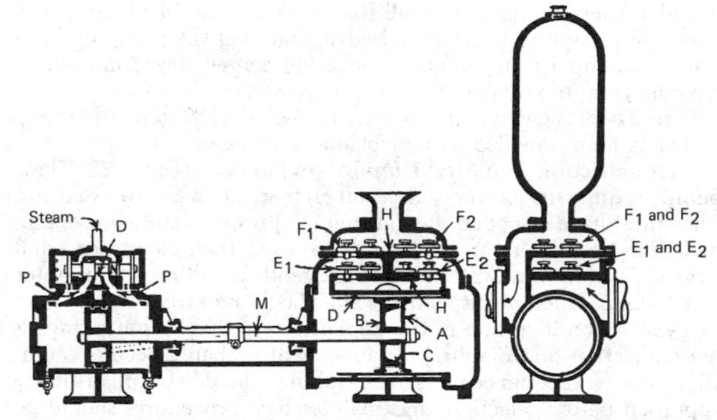

FIG. 6-20 Double-acting steam-driven reciprocating pump.

sation and to provide a more uniform flow in the discharge line. In many installations, surge chambers are required on the suction side as well. Piping layouts should be studied to determine the most effective size and location. If surge chambers are used, provision should be made to keep the chamber charged with gas. A surge chamber filled with liquid is of no value. A liquid-level gauge is desirable to permit checking the amount of gas in the chamber.

Reciprocating pumps may be of **single-cylinder** or **multicylinder** design. Multicylinder pumps have all cylinders in parallel for increased capacity. Piston-type pumps may be single-acting or double-acting; i.e., pumping may be accomplished from one or both ends of the piston. Plunger pumps are always single-acting. The following tabulation provides data on the flow variation of reciprocating pumps of various designs.

Number of cylinders	Single- or double-acting	Flow variation per stroke from mean, percent
Single	Single	+220 to − 100
Single	Double	+60 to − 100
Duplex	Single	+24.1 to − 100
Duplex	Double	+6.1 to − 21.5
Triplex	Single and double	+1.8 to − 16.9
Quintuplex	Single	+1.8 to − 5.2

Piston Pumps There are two ordinary types of piston pumps, simplex double-acting pumps and duplex double-acting pumps.

Simplex Double-Acting Pumps These pumps may be direct-acting (i.e., direct-connected to a steam cylinder) or power-driven (through a crank and flywheel from the crosshead of a steam engine). Figure 6-20 is a direct-acting pump, designed for use at pressures up to 0.690 MPa (100 lbf/in^2). In this figure, the piston consists of disks *A* and *B*, with packing rings *C* between them. A bronze liner for the water cylinder is shown at *D*. Suction valves are E_1 and E_2. Discharge valves are F_1 and F_2.

Duplex Double-Acting Pumps These pumps differ primarily from those of the simplex type in having two cylinders whose operation is coordinated. They may be direct-acting, steam-driven, or power-driven with crank and flywheel.

A duplex outside-end-packed **plunger pump** with pot valves, of the type used with hydraulic presses and for similar service, is shown in Fig. 6-21. In this drawing, plunger *A* is direct-connected to rod *B*, while plunger *C* is operated from the rod by means of yoke *D* and tie rods.

Plunger pumps differ from piston pumps in that they have one or more constant-diameter plungers reciprocating through packing glands and displacing liquid from cylinders in which there is considerable radial clearance. They are always single-acting, in the sense that only one end of the plunger is used in pumping the liquid.

Plunger pumps are available with one, two, three, four, five, or

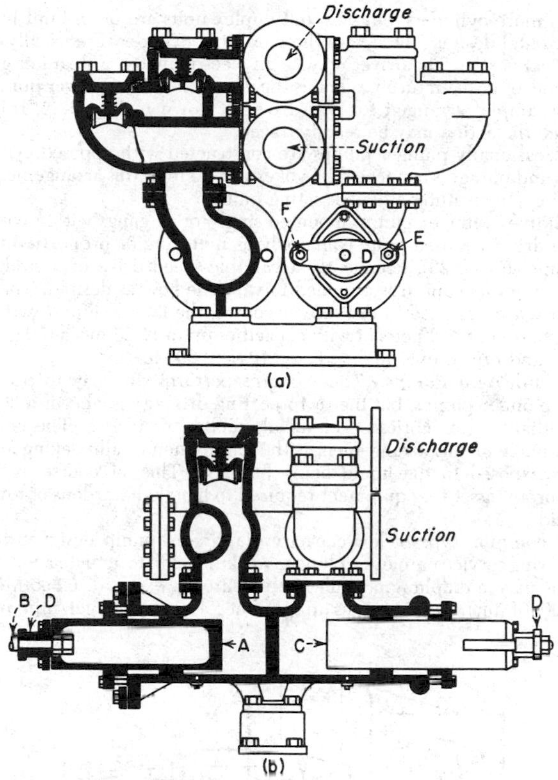

FIG. 6-21 Duplex single-acting plunger pump.

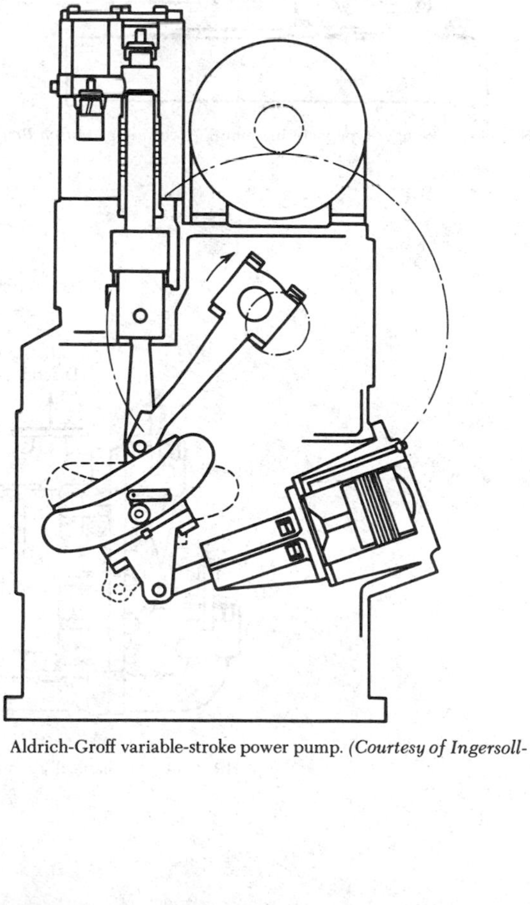

FIG. 6-22 Aldrich-Groff variable-stroke power pump. *(Courtesy of Ingersoll-Rand.)*

even more cylinders. Simplex and duplex units are often built in a horizontal design. Those with three or more cylinders are usually of vertical design. The driver may be an electric motor, a steam or gas engine, or a steam turbine. This is the common type of **power pump.** An example, arranged for belt drive, is shown in Fig. 6-22, from which the action may be readily traced.

Occasionally plunger pumps are constructed with opposed cylinders and plungers connected by yokes and tie rods; this arrangement, in effect, constitutes a double-acting unit.

Simplex plunger pumps mounted singly or in gangs with a common drive are used quite commonly as **metering** or **proportioning pumps** (Fig. 6-23). Frequently a variable-speed drive or a stroke-adjusting mechanism is provided to vary the flow as desired. These pumps are designed to measure or control the flow of liquid within a deviation of ± 2 percent with capacities up to 11.35 m^3/h (50 gal/min) and pressures as high as 68.9 MPa (10,000 lbf/in^2).

Diaphragm Pumps These pumps perform similarly to piston and plunger pumps, but the reciprocating driving member is a flexible diaphragm fabricated of metal, rubber, or plastic. The chief advantage of this arrangement is the elimination of all packing and seals exposed to the liquid being pumped. This, of course, is an important asset for equipment required to handle hazardous or toxic liquids.

A common type of low-capacity diaphragm pump designed for metering service employs a plunger working in oil to actuate a metallic or plastic diaphragm. Built for pressures in excess of 6.895 MPa (1000 lbf/in^2) with flow rates up to about 1.135 m^3/h (5 gal/min) per cylinder, such pumps possess all the characteristics of plunger-type metering pumps with the added advantage that the pumping head can be mounted in a remote (even a submerged) location entirely separate from the drive.

Figure 6-24 shows a high-capacity 22.7-m^3/h (100-gal/min) pump with actuation provided by a mechanical linkage.

Pneumatically Actuated Diaphragm Pumps (Fig. 6-25) These pumps require no power source other than plant compressed air. They must have a flooded suction, and the pressure is, of course, limited to the available air pressure. Because of their slow speed and large valves, they are well suited to the gentle handling of liquids for which degradation of suspended solids should be avoided.

A major consideration in the application of diaphragm pumps is the realization that diaphragm failure will probably occur eventually. The consequences of such failure should be realistically appraised before selection, and maintenance procedures should be established accordingly.

Rotary Pumps In rotary pumps the liquid is displaced by rotation of one or more members within a stationary housing. Because

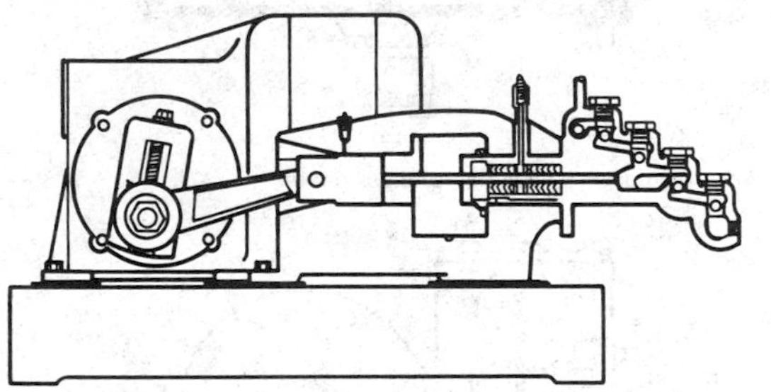

FIG. 6-23 Plunger-type metering pump. *(Courtesy of Milton Roy Co.)*

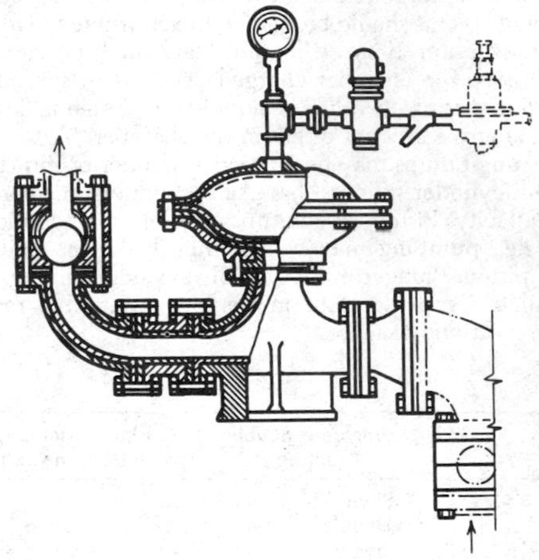

FIG. 6-25 Pneumatically actuated diaphragm pump for slurry service. *(Courtesy of Dorr-Oliver Inc.)*

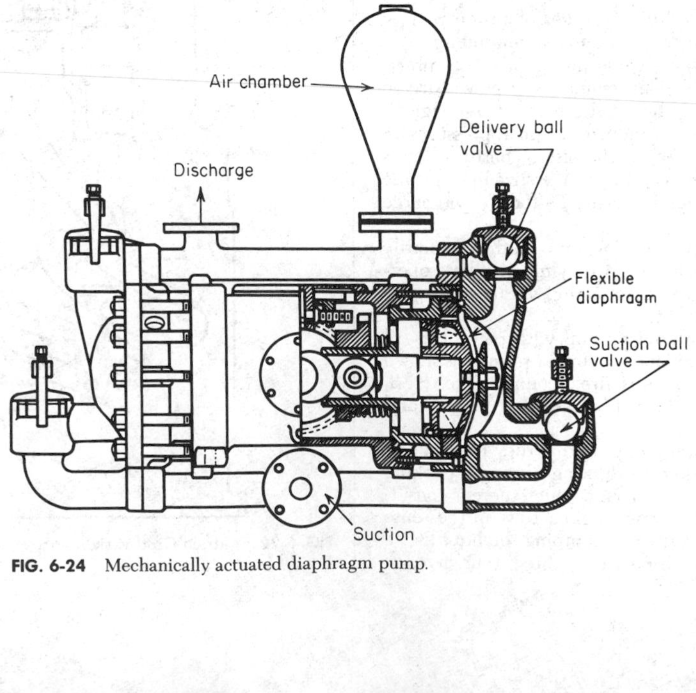

FIG. 6-24 Mechanically actuated diaphragm pump.

internal clearances, although minute, are a necessity in all but a few special types, capacity decreases somewhat with increasing pump differential pressure. Therefore, these pumps are not truly positive-displacement pumps. However, for many other reasons they are considered as such.

The selection of materials of construction for rotary pumps is critical. The materials must be corrosion-resistant, compatible when one part is running against another, and capable of some abrasion resistance.

Gear Pumps When two or more impellers are used in a rotary-pump casing, the impellers will take the form of toothed-gear wheels as in Fig. 6-26, of helical gears, or of lobed cams. In each case, these impellers rotate with extremely small clearance between them and between the surfaces of the impellers and the casing. In Fig. 6-26, the two toothed impellers rotate as indicated by the arrows; the suction connection is at the bottom. The pumped liquid flows into the spaces between the impeller teeth as these cavities pass the suction opening. The liquid is then carried around the casing to the discharge opening, where it is forced out of the impeller teeth mesh. The arrows indicate this flow of liquid.

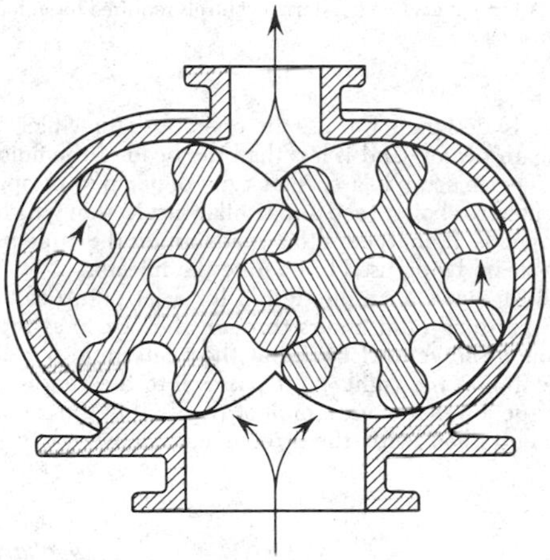

FIG. 6-26 Positive-displacement gear-type rotary pump.

Rotary pumps are available in two general classes, interior-bearing and exterior-bearing. The **interior-bearing type** is used for handling liquids of a lubricating nature, and the **exterior-bearing type** is used with nonlubricating liquids. The interior-bearing pump is lubricated by the liquid being pumped, and the exterior-bearing type is oil-lubricated.

The use of spur gears in gear pumps will produce in the discharge pulsations having a frequency equivalent to the number of teeth on both gears multiplied by the speed of rotation. The amplitude of these disturbances is a function of tooth design. The pulsations can be reduced markedly by the use of rotors with helical teeth. This in turn introduces end thrust, which can be eliminated by the use of double-helical or herringbone teeth.

Screw Pumps A modification of the helical gear pump is the screw pump. Figure 6-27 illustrates a two-rotor version in which the liquid is fed to either the center or the ends, depending upon the direction of rotation, and progresses axially in the cavities formed by the meshing threads or teeth. In three-rotor versions, the center rotor is the driving member while the other two are driven. Figure 6-28 shows still another arrangement, in which a metal rotor of unique design rotates without clearance in an elastomeric stationary sleeve.

Screw pumps, because of multiple dams that reduce slip, are well adapted for producing higher pressure rises, for example, 6.895 MPa (1000 lbf/in^2), especially when handling viscous liquids such as heavy oils. The all-metal pumps are generally subject to the same limitations on handling abrasive solids as conventional gear pumps. In addition, the wide bearing spans usually demand that the liquid have considerable lubricity to prevent metal-to-metal contact.

Among the liquids handled by rotary pumps are mineral oils, vegetable oils, animal oils, greases, glucose, viscose, molasses, paints, var-

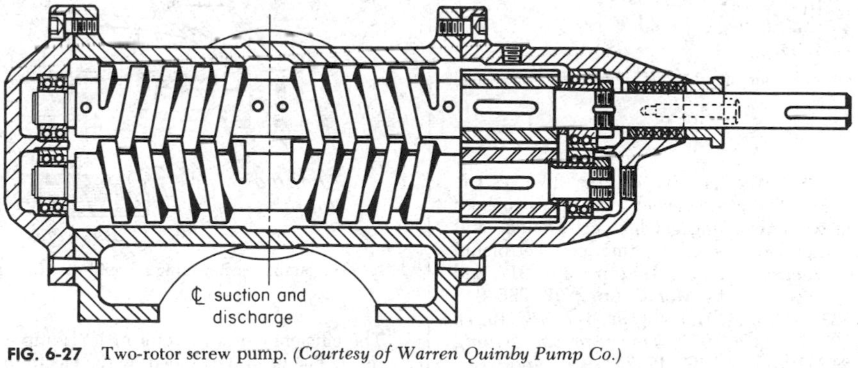

FIG. 6-27 Two-rotor screw pump. *(Courtesy of Warren Quimby Pump Co.)*

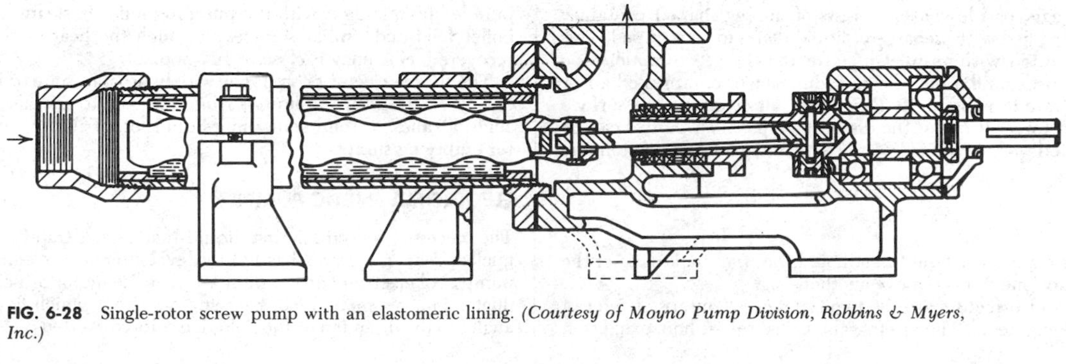

FIG. 6-28 Single-rotor screw pump with an elastomeric lining. *(Courtesy of Moyno Pump Division, Robbins & Myers, Inc.)*

nish, shellac, lacquers, alcohols, catsup, brine, mayonnaise, sizing, soap, tanning liquors, vinegar, and ink. Some screw-type units are specially designed for the gentle handling of large solids suspended in the liquid.

Fluid-Displacement Pumps In addition to pumps that depend on the mechanical action of pistons, plungers, or impellers to move the liquid, other devices for this purpose employ displacement by a secondary fluid. This group includes air lifts and acid eggs.

The **air lift** is a device for raising liquid by means of compressed air. In the past it was widely used for pumping wells, but it has been less widely used since the development of efficient centrifugal pumps. It operates by introducing compressed air into the liquid near the bottom of the well. The air-and-liquid mixture, being lighter than liquid alone, rises in the well casing. The advantage of this system of pumping lies in the fact that there are no moving parts in the well. The pumping equipment is an air compressor, which can be located on the surface.

A simplified sketch of an air lift for this purpose is shown in Fig. 6-29. Ingersoll-Rand has developed empirical information on air-lift performance which is available upon request.

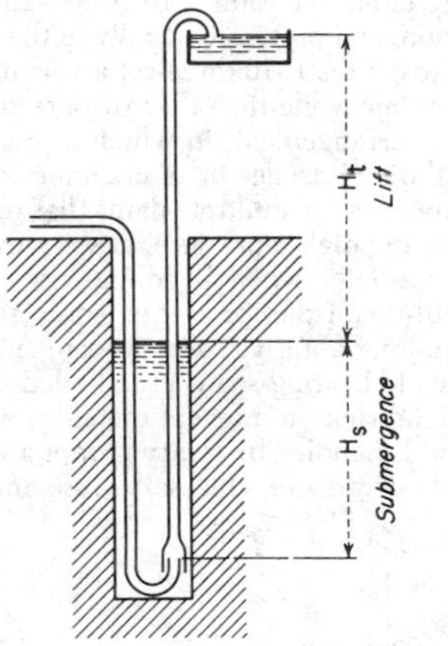

FIG. 6-29 Simplified sketch of an air lift, showing submergence and total head.

An important application of the gas-lift principle involves the extraction of oil from wells. There are several references to both practical and theoretical work involving gas lift performance and related problems. Recommended sources are American Petroleum Institute, *Drilling and Production Practices*, 1952, pp. 257–317, and 1939, p. 266; *Trans. Am. Soc. Mining Metall. Eng.*, **92**, 296–313 (1931), **103**, 170–186 (1933), **118**, 56–70 (1936), **192**, 317–326 (1951), **189**, 73–82 (1950), and **198**, 271–278 (1953); *Trans Am. Soc. Mining Metall.*, and *Pet. Eng.*, **213** (1958), and **207**, 17–24 (1956); and *Univ. Wisconsin Bull., Eng. Ser.*, **6**, no. 7 (1911, reprinted 1914).

An **acid egg**, or **blowcase**, consists of an egg-shaped container which can be filled with a charge of liquid that is to be pumped. This container is fitted with an inlet pipe for the charge, an outlet pipe for the discharge, and a pipe for the admission of compressed air or gas, as illustrated in Fig. 6-30. Pressure of air or gas on the surface of the liquid forces it out of the discharge pipe. Such pumps can be hand-operated or arranged for semiautomatic or automatic operation.

JET PUMPS

Jet pumps are a class of liquid-handling device that makes use of the momentum of one fluid to move another.

Ejectors and **injectors** are the two types of jet pumps of interest to chemical engineers. The ejector, also called the siphon, exhauster, or eductor, is designed for use in operations in which the head pumped against is low and is less than the head of the fluid used for pumping. The injector is a special type of jet pump, operated by steam and used for boiler feed and similar services, in which the fluid being pumped is discharged into a space under the same pressure as that of the steam being used to operate the injector.

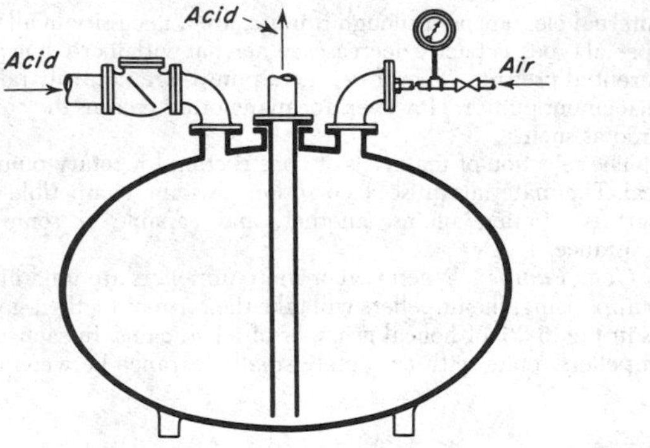

FIG. 6-30 A form of acid egg. External controls required for automatic operation are not shown.

Figure 6-31 shows a simple design for a jet pump of the ejector type. The pumping fluid enters through the nozzle at the left and passes through the venturi nozzle at the center and out of the discharge opening at the right. As it passes into the venturi nozzle, it develops a suction that causes some of the fluid in the suction chamber to be entrained with the stream and delivered through this discharge.

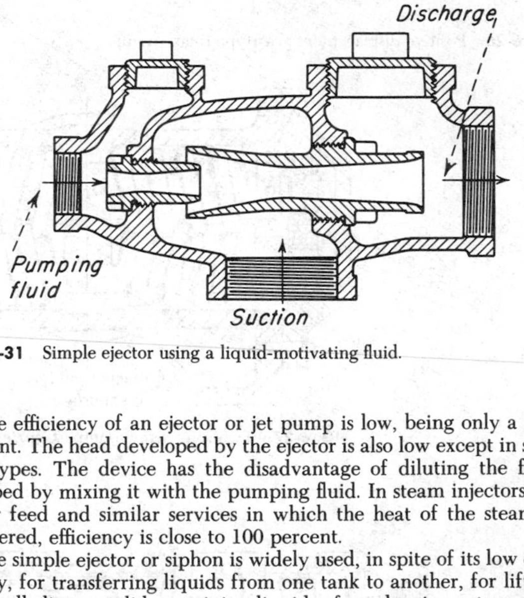

FIG. 6-31 Simple ejector using a liquid-motivating fluid.

The efficiency of an ejector or jet pump is low, being only a few percent. The head developed by the ejector is also low except in special types. The device has the disadvantage of diluting the fluid pumped by mixing it with the pumping fluid. In steam injectors for boiler feed and similar services in which the heat of the steam is recovered, efficiency is close to 100 percent.

The simple ejector or siphon is widely used, in spite of its low efficiency, for transferring liquids from one tank to another, for lifting acids, alkalies, or solid-containing liquids of an abrasive nature, and for emptying sumps.

ELECTROMAGNETIC PUMPS

The necessity of circulating liquid-metal heat-transfer media in nuclear-reactor systems has led to development of electromagnetic pumps. All electromagnetic pumps utilize the motor principle: a conductor in a magnetic field, carrying a current which flows at right angles to the direction of the field, has a force exerted on it, the force being mutually perpendicular to both the field and the current. In all electromagnetic pumps, the fluid is the conductor. This force, suitably directed in the fluid, manifests itself as a pressure if the fluid is suitably contained. The field and current can be produced in a number of different ways and the force utilized variously.

Both alternating- and direct-current units are available. While dc pumps (Fig. 6-32) are simpler, their high-current requirement is a definite limitation; ac pumps can readily obtain high currents by making use of transformers. Multipole induction ac pumps have been built in helical and linear configurations. Helical units are effective for relatively high heads and low flows, while linear induction pumps are best suited to large flows at moderate heads. Electromagnetic pumps are available for flow rates up to 2.271×10^3 m^3/h (10,000 gal/min), and pressures up to 2 MPa (300 lbf/in^2) are practical. Performance characteristics resemble those of centrifugal pumps.

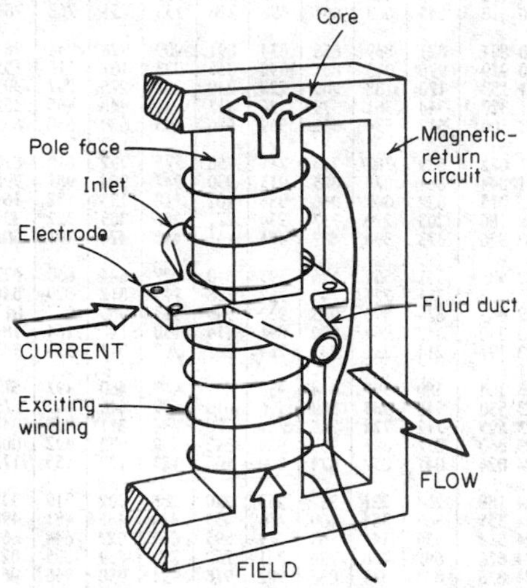

FIG. 6-32 Simplified diagram of a direct-current-operated electromagnetic pump.

COMPRESSION OF GASES

Theory of Compression In any continuous compression process the relation of absolute pressure p to volume V is expressed by the formula

$$pV^n = C = \text{constant} \tag{6-16}$$

The plot of pressure versus volume for each value of exponent n is known as the **polytropic** curve. Since the work W performed in proceeding from p_1 to p_2 along any polytropic curve (Fig. 6-33) is

$$W = \int_1^2 p\,dV \tag{6-17}$$

it follows that the amount of work required is dependent upon the polytropic curve involved and increases with increasing values of n. The path requiring the least amount of input work is $n = 1$, which is equivalent to **isothermal** compression. For **adiabatic** compression (i.e., no heat is being added or taken away during the process), $n = k$ = ratio of specific heat at constant pressure to that at constant volume.

Since most compressors operate along a polytropic path approaching the adiabatic, compressor calculations are generally based on the adiabatic curve.

Some formulas based upon the adiabatic equation and useful in compressor work are as follows:

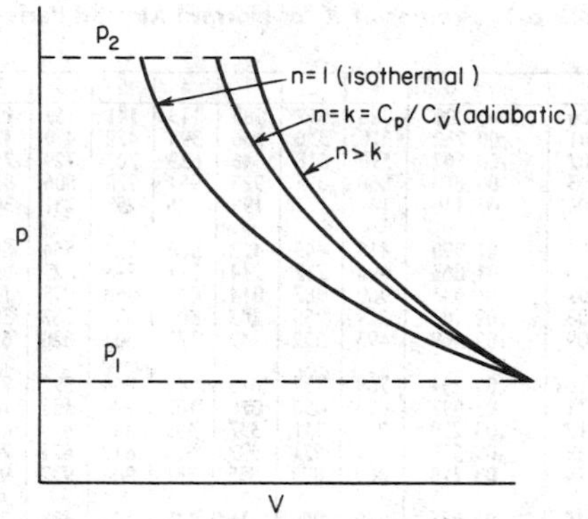

FIG. 6-33 Polytropic compression curves.

Pressure, volume, and **temperature** relations for perfect gases:

$$p_2/p_1 = (V_1/V_2)^k \tag{6-18}$$

$$T_2/T_1 = (V_1/V_2)^{k-1} \tag{6-19}$$

$$p_2/p_1 = (T_2/T_1)^{k/(k-1)} \tag{6-20}$$

Adiabatic Calculations Adiabatic head is expressed as follows:

In SI units,

$$H_{ad} = \frac{k}{k-1}\frac{RT_1}{9.806}\left[\left(\frac{p_2}{p_1}\right)^{(k-1)/k} - 1\right] \tag{6-21a}$$

where H_{ad} = adiabatic head, m; R = gas constant, J/(kg·K) = 8314/molecular weight; T_1 = inlet gas temperature, K; p_1 = absolute inlet pressure, kPa; and p_2 = absolute discharge pressure, kPa.

In U.S. customary units,

$$H_{ad} = \frac{k}{k-1}RT_1\left[\left(\frac{p_2}{p_1}\right)^{(k-1)/k} - 1\right] \tag{6-21b}$$

where H_{ad} = adiabatic head, ft; R = gas constant, (ft·lbf)/(lb·°R) = 1545/molecular weight; T_1 = inlet gas temperature, °R; p_1 = absolute inlet pressure, lbf/in^2; and p_2 = absolute discharge pressure, lbf/in^2.

The **work** expended on the gas during compression is equal to the product of the adiabatic head and the weight of gas handled. Therefore, the adiabatic power is as follows:

In SI units,

$$\text{kW}_{ad} = \frac{WH_{ad}}{10^3} = \frac{k}{k-1}\frac{WRT_1}{9806}\left[\left(\frac{p_2}{p_1}\right)^{(k-1)/k} - 1\right] \tag{6-22a}$$

$$\text{or}\quad \text{kW}_{ad} = 2.78 \times 10^{-4}\frac{k}{k-1}Q_1p_1\left[\left(\frac{p_2}{p_1}\right)^{(k-1)/k} - 1\right] \tag{6-23a}$$

where kW$_{ad}$ = power, kW; W = mass flow, kg/s × 9.806 N/kg; and Q_1 = volume rate of gas flow, m^3/h.

In U.S. customary units,

$$\text{hp}_{ad} = \frac{WH_{ad}}{550} = \frac{k}{k-1}\frac{WRT_1}{550}\left[\left(\frac{p_2}{p_1}\right)^{(k-1)/k} - 1\right] \tag{6-22b}$$

$$\text{or}\quad \text{hp}_{ad} = 4.36 \times 10^{-3}\frac{k}{k-1}Q_1p_1\left[\left(\frac{p_2}{p_1}\right)^{(k-1)/k} - 1\right] \tag{6-23b}$$

where hp$_{ad}$ = power, hp; W = mass flow, lb/s; and Q_1 = volume rate of gas flow, ft^3/min.

Adiabatic discharge temperature is

$$T_2 = T_1(p_2/p_1)^{(k-1)/k} \tag{6-24}$$

Air and a number of other gases have a value of k = 1.39 to 1.41. To simplify calculations for these gases, tables have been made of

TABLE 6-1 Values of X for Normal Air and Perfect Diatomic Gases*

$$X = r^{0.283} - 1$$

r	0	1	2	3	4	5	6	7	8	9	r	0	1	2	3	4	5	6	7	8	9
1.00	0.00 000	028	057	085	113	141	169	198	226	254	1.75	0.17 160	179	198	217	236	255	274	292	311	330
1.01	.00 282	310	338	366	394	422	450	478	506	534	1.76	.17 349	368	387	406	425	443	462	481	500	519
1.02	.00 562	590	618	646	673	701	729	757	785	812	1.77	.17 538	556	575	594	613	631	650	669	688	706
1.03	.00 840	868	895	923	951	978	006	034	061	089	1.78	.17 725	744	762	781	800	818	837	856	874	893
1.04	.01 116	144	171	199	226	253	281	308	336	363	1.79	.17 912	930	949	968	986	005	023	042	061	079
1.05	.01 390	418	445	472	500	527	554	581	608	636	1.80	.18 098	116	135	153	172	191	209	228	246	265
1.06	.01 663	690	717	744	771	798	825	852	879	906	1.81	.18 283	302	320	339	357	376	394	412	431	449
1.07	.01 933	960	987	014	041	068	095	122	148	175	1.82	.18 468	486	505	523	541	560	578	596	615	633
1.08	.02 202	229	255	282	309	336	362	389	416	442	1.83	.18 652	670	688	707	725	743	762	780	798	816
1.09	.02 469	495	522	549	575	602	628	655	681	708	1.84	.18 835	853	871	890	908	926	944	962	981	999
1.10	.02 734	760	787	813	840	866	892	919	945	971	1.85	.19 017	035	054	072	090	108	126	144	163	181
1.11	.02 997	024	050	076	102	129	155	181	207	233	1.86	.19 199	217	235	253	271	289	308	326	344	362
1.12	.03 259	285	311	337	363	389	415	441	467	493	1.87	.19 380	398	416	434	452	470	488	506	524	542
1.13	.03 519	545	571	597	623	649	675	700	726	752	1.88	.19 560	578	596	614	632	650	668	686	704	722
1.14	.03 778	804	829	855	881	906	932	958	983	009	1.89	.19 740	758	776	794	811	829	847	865	883	901
1.15	.04 035	060	086	111	137	162	188	213	239	264	1.90	.19 919	937	954	972	990	008	026	044	061	079
1.16	.04 290	315	341	366	391	417	442	467	493	518	1.91	.20 097	115	133	150	168	186	204	221	239	257
1.17	.04 543	569	594	619	644	670	695	720	745	770	1.92	.20 275	292	310	328	345	363	381	399	416	434
1.18	.04 796	821	846	871	896	921	946	971	996	021	1.93	.20 452	469	487	504	522	540	557	575	593	610
1.19	.05 046	071	096	121	146	171	196	221	245	270	1.94	.20 628	645	663	681	698	716	733	751	768	786
1.20	.05 295	320	345	370	394	419	444	469	493	518	1.95	.20 804	821	839	856	874	891	909	926	944	961
1.21	.05 543	567	592	617	641	666	691	715	740	764	1.96	.20 979	996	013	031	048	066	083	101	118	135
1.22	.05 789	813	838	862	887	911	936	960	985	009	1.97	.21 153	170	188	205	222	240	257	275	292	309
1.23	.06 034	058	082	107	131	155	180	204	228	253	1.98	.21 327	344	361	379	396	413	431	448	465	482
1.24	.06 277	301	325	350	374	398	422	446	470	495	1.99	.21 500	517	534	552	569	586	603	620	638	655
1.25	.06 519	543	567	591	615	639	663	687	711	735	2.00	.21 672	689	707	724	741	758	775	792	810	827
1.26	.06 759	783	807	831	855	879	903	927	951	974	2.01	.21 844	861	878	895	913	930	947	964	981	998
1.27	.06 998	022	046	070	094	117	141	165	189	212	2.02	.22 015	032	049	066	084	101	118	135	152	169
1.28	.07 236	260	283	307	331	354	378	402	425	449	2.03	.22 186	203	220	237	254	271	288	305	322	339
1.29	.07 472	496	520	543	567	590	614	637	661	684	2.04	.22 356	373	390	407	424	441	458	474	491	508
1.30	.07 708	731	754	778	801	825	848	871	895	918	2.05	.22 525	542	559	576	593	610	627	644	660	677
1.31	.07 941	965	988	011	035	058	081	104	128	151	2.06	.22 694	711	728	745	762	778	795	812	829	846
1.32	.08 174	197	220	243	267	290	313	336	359	382	2.07	.22 863	879	896	913	930	946	963	980	997	013
1.33	.08 405	428	451	474	497	520	543	566	589	612	2.08	.23 030	047	064	080	097	114	130	147	164	181
1.34	.08 635	658	681	704	727	750	773	795	818	841	2.09	.23 197	214	231	247	264	281	297	314	331	347
1.35	.08 864	887	910	932	955	978	001	023	046	069	2.10	.23 364	380	397	414	430	447	463	480	497	513
1.36	.09 092	114	137	160	182	205	228	250	273	295	2.11	.23 530	546	563	579	596	613	629	646	662	679
1.37	.09 318	341	363	386	408	431	453	476	498	521	2.12	.23 695	712	728	745	761	778	794	811	827	844
1.38	.09 543	566	588	611	633	655	678	700	723	745	2.13	.23 860	877	893	909	926	942	959	975	992	008
1.39	.09 767	790	812	834	857	879	901	923	946	968	2.14	.24 024	041	057	074	090	106	123	139	155	172
1.40	.09 990	012	035	057	079	101	123	145	168	190	2.15	.24 188	204	221	237	253	270	286	302	319	335
1.41	.10 212	234	256	278	300	322	344	366	389	411	2.16	.24 351	368	384	400	416	433	449	465	481	498
1.42	.10 433	455	477	499	521	542	564	586	608	630	2.17	.24 514	530	546	563	579	595	611	627	644	660
1.43	.10 652	674	696	718	740	761	783	805	827	849	2.18	.24 676	692	708	724	741	757	773	789	805	821
1.44	.10 871	892	914	936	958	979	001	023	045	066	2.19	.24 838	854	870	886	902	918	934	950	966	983
1.45	.11 088	110	131	153	175	196	218	239	261	283	2.20	.24 999	015	031	047	063	079	095	111	127	143
1.46	.11 304	326	347	369	390	412	433	455	476	498	2.21	.25 159	175	191	207	223	239	255	271	287	303
1.47	.11 520	541	562	584	605	627	648	669	691	712	2.22	.25 319	335	351	367	383	399	415	431	447	463
1.48	.11 734	755	776	798	819	840	862	883	904	925	2.23	.25 479	495	511	526	542	558	574	590	606	622
1.49	.11 947	968	989	010	032	053	074	095	116	138	2.24	.25 638	654	669	685	701	717	733	749	765	780
1.50	.12 159	180	201	222	243	264	286	307	328	349	2.25	.25 796	812	828	844	859	875	891	907	923	938
1.51	.12 370	391	412	433	454	475	496	517	538	559	2.26	.25 954	970	986	001	017	033	049	064	080	096
1.52	.12 580	601	622	643	664	685	706	726	747	768	2.27	.26 112	127	143	159	175	190	206	222	237	253
1.53	.12 789	810	831	852	872	893	914	935	956	977	2.28	.26 269	284	300	316	331	347	363	378	394	409
1.54	.12 997	018	039	060	080	101	122	142	163	184	2.29	.26 425	441	456	472	488	503	519	534	550	566
1.55	.13 205	225	246	266	287	308	328	349	370	390	2.30	.26 581	597	612	628	643	659	675	690	706	721
1.56	.13 411	431	452	472	493	513	534	554	575	595	2.31	.26 737	752	768	783	799	814	830	845	861	876
1.57	.13 616	636	657	677	698	718	739	759	780	800	2.32	.26 892	907	923	938	954	969	984	000	015	031
1.58	.13 820	841	861	881	902	922	942	963	983	003	2.33	.27 046	062	077	092	108	123	139	154	169	185
1.59	.14 024	044	064	085	105	125	145	165	186	206	2.34	.27 200	216	231	246	262	277	292	308	323	338
1.60	.14 226	246	267	287	307	327	347	367	387	408	2.35	.27 354	369	384	400	415	430	446	461	476	492
1.61	.14 428	448	468	488	508	528	548	568	588	608	2.36	.27 507	522	538	553	568	583	599	614	629	644
1.62	.14 628	648	668	688	708	728	748	768	788	808	2.37	.27 660	675	690	705	721	736	751	766	781	797
1.63	.14 828	848	868	888	908	928	948	968	988	007	2.38	.27 812	827	842	857	873	888	903	918	933	948
1.64	.15 027	047	067	087	107	126	146	166	186	206	2.39	.27 964	979	994	009	024	039	054	070	085	100
1.65	.15 225	245	265	284	304	324	344	363	383	403	2.40	.28 115	130	145	160	175	190	205	220	236	251
1.66	.15 423	442	462	481	501	521	540	560	580	599	2.41	.28 266	281	296	311	326	341	356	371	386	401
1.67	.15 619	638	658	678	697	717	736	756	775	795	2.42	.28 416	431	446	461	476	491	506	521	536	551
1.68	.15 814	834	853	873	892	912	931	951	970	990	2.43	.28 566	581	596	611	626	641	656	671	686	701
1.69	.16 009	028	048	067	087	106	125	145	164	184	2.44	.28 716	730	745	760	775	790	805	820	835	850
1.70	.16 203	222	242	261	280	299	319	338	357	377	2.45	.28 865	879	894	909	924	939	954	969	984	998
1.71	.16 396	415	434	454	473	492	511	531	550	569	2.46	.29 013	028	043	058	073	087	102	117	132	147
1.72	.16 588	607	626	646	665	684	703	722	741	760	2.47	.29 162	176	191	206	221	235	250	265	280	295
1.73	.16 780	799	818	837	856	875	894	913	932	951	2.48	.29 309	324	339	353	368	383	398	412	427	442
1.74	.16 970	989	008	027	046	065	084	103	122	141	2.49	.29 457	471	486	501	515	530	545	559	574	589

*Printed by permission of Compressed Air Data.

TABLE 6-1 Values of *X* for Normal Air and Perfect Diatomic Gases* *(Continued)*

r	0	1	2	3	4	5	6	7	8	9	*r*	0	1	2	3	4	5	6	7	8	9
2.50	0.29 604	618	633	647	662	677	691	706	721	735	2.75	0.33 147	161	174	188	202	215	229	243	256	270
2.51	.29 750	765	779	794	808	823	838	852	867	881	2.76	.33 284	297	311	325	338	352	366	379	393	407
2.52	.29 896	911	925	940	954	969	984	998	013	027	2.77	.33 420	434	448	461	475	488	502	516	529	543
2.53	.30 042	056	071	085	100	114	129	144	158	173	2.78	.33 556	570	584	597	611	624	638	651	665	679
2.54	.30 187	202	216	231	245	260	274	289	303	318	2.79	.33 692	706	719	733	746	760	773	787	801	814
2.55	.30 332	346	361	375	390	404	419	433	448	462	2.80	.33 828	841	855	868	882	895	909	922	936	949
2.56	.30 476	491	505	520	534	548	563	577	592	606	2.81	.33 963	976	990	003	017	030	044	057	070	084
2.57	.30 620	635	649	663	678	692	707	721	735	750	2.82	.34 097	111	124	138	151	165	178	191	205	218
2.58	.30 764	778	793	807	821	836	850	864	879	893	2.83	.34 232	245	259	272	285	299	312	326	339	352
2.59	.30 907	921	936	950	964	979	993	007	021	036	2.84	.34 366	379	393	406	419	433	446	459	473	486
2.60	.31 050	064	079	093	107	121	136	150	164	178	2.85	.34 500	513	526	540	553	566	580	593	606	620
2.61	.31 193	207	221	235	249	264	278	292	306	320	2.86	.34 633	646	660	673	686	700	713	726	739	753
2.62	.31 335	349	363	377	391	405	420	434	448	462	2.87	.34 766	779	793	806	819	832	846	859	872	886
2.63	.31 476	490	505	519	533	547	561	575	589	603	2.88	.34 899	912	925	939	952	965	978	991	005	018
2.64	.31 618	632	646	660	674	688	702	716	730	744	2.89	.35 031	044	058	071	084	097	110	124	137	150
2.65	.31 759	773	787	801	815	829	843	857	871	885	2.90	.35 163	176	190	203	216	229	242	255	269	282
2.66	.31 899	913	927	941	955	969	983	997	011	025	2.91	.35 295	308	321	334	347	361	374	387	400	413
2.67	.32 039	053	067	081	095	109	123	137	151	165	2.92	.35 426	439	452	466	479	492	505	518	531	544
2.68	.32 179	193	207	221	235	249	262	276	290	304	2.93	.35 557	570	584	597	610	623	636	649	662	675
2.69	.32 318	332	346	360	374	388	402	416	429	443	2.94	.35 688	701	714	727	740	753	767	780	793	806
2.70	.32 457	471	485	499	513	527	540	554	568	582	2.95	.35 819	832	845	858	871	884	897	910	923	936
2.71	.32 596	610	624	637	651	665	679	693	707	720	2.96	.35 949	962	975	988	001	014	027	040	053	066
2.72	.32 734	748	762	776	789	803	817	831	845	858	2.97	.36 079	092	105	118	131	144	157	169	182	195
2.73	.32 872	886	900	913	927	941	955	968	982	996	2.98	.36 208	221	234	247	260	273	286	299	312	324
2.74	.33 010	023	037	051	065	078	092	106	119	133	2.99	.36 337	350	363	376	389	402	415	428	440	453

r	0	1	2	3	4	5	6	7	8	9
3.0	0.3647	0.3659	0.3672	0.3685	0.3698	0.3711	0.3723	0.3736	0.3749	0.3761
3.1	.3774	.3786	.3799	.3811	.3824	.3836	.3849	.3861	.3874	.3886
3.2	.3898	.3911	.3923	.3935	.3947	.3959	.3971	.3984	.3996	.4008
3.3	.4020	.4032	.4044	.4056	.4068	.4080	.4091	.4103	.4115	.4127
3.4	.4139	.4150	.4162	.4174	.4186	.4197	.4209	.4220	.4232	.4244
3.5	.4255	.4267	.4278	.4290	.4301	.4313	.4324	.4335	.4347	.4358
3.6	.4369	.4380	.4392	.4403	.4414	.4425	.4437	.4448	.4459	.4470
3.7	.4481	.4492	.4503	.4514	.4525	.4536	.4547	.4558	.4569	.4580
3.8	.4591	.4602	.4612	.4623	.4634	.4645	.4656	.4666	.4677	.4688
3.9	.4698	.4709	.4720	.4730	.4741	.4752	.4762	.4773	.4783	.4794
4.0	.4804	.4815	.4825	.4835	.4846	.4856	.4867	.4877	.4887	.4898
4.1	.4908	.4918	.4928	.4939	.4949	.4959	.4970	.4980	.4990	.5000
4.2	.5010	.5020	.5030	.5040	.5050	.5060	.5070	.5080	.5090	.5100
4.3	.5110	.5120	.5130	.5140	.5150	.5160	.5170	.5179	.5189	.5199
4.4	.5209	.5219	.5228	.5238	.5248	.5258	.5267	.5277	.5287	.5296
4.5	.5306	.5316	.5325	.5335	.5344	.5354	.5363	.5373	.5382	.5392
4.6	.5401	.5411	.5420	.5430	.5439	.5449	.5458	.5467	.5477	.5486
4.7	.5495	.5505	.5514	.5523	.5533	.5542	.5551	.5560	.5570	.5579
4.8	.5588	.5597	.5606	.5616	.5625	.5634	.5643	.5652	.5661	.5670
4.9	.5679	.5688	.5697	.5706	.5715	.5724	.5733	.5742	.5751	.5760
5.0	.5769	.5778	.5787	.5796	.5805	.5814	.5822	.5831	.5840	.5849
5.1	.5858	.5867	.5875	.5884	.5893	.5902	.5910	.5919	.5928	.5936
5.2	.5945	.5954	.5962	.5971	.5980	.5988	.5997	.6006	.6014	.6023
5.3	.6031	.6040	.6048	.6057	.6065	.6074	.6082	.6091	.6099	.6108
5.4	.6116	.6125	.6133	.6142	.6150	.6159	.6167	.6175	.6184	.6192
5.5	.6200	.6209	.6217	.6225	.6234	.6242	.6250	.6258	.6267	.6275
5.6	.6283	.6291	.6300	.6308	.6316	.6324	.6332	.6340	.6349	.6357
5.7	.6365	.6373	.6381	.6389	.6397	.6405	.6413	.6421	.6430	.6438
5.8	.6446	.6454	.6462	.6470	.6478	.6486	.6494	.6502	.6509	.6517
5.9	.6525	.6533	.6541	.6549	.6557	.6565	.6573	.6581	.6588	.6596
6.0	.6604	.6612	.6620	.6628	.6635	.6643	.6651	.6659	.6666	.6674
6.1	.6682	.6690	.6697	.6705	.6713	.6721	.6729	.6736	.6744	.6752
6.2	.6759	.6767	.6774	.6782	.6789	.6797	.6805	.6812	.6820	.6827
6.3	.6835	.6843	.6850	.6858	.6865	.6873	.6880	.6888	.6895	.6903
6.4	.6910	.6918	.6925	.6933	.6940	.6948	.6955	.6963	.6970	.6978
6.5	.6985	.6992	.7000	.7007	.7014	.7021	.7028	.7036	.7043	.7050
6.6	.7058	.7065	.7073	.7080	.7087	.7095	.7102	.7110	.7117	.7124
6.7	.7131	.7138	.7145	.7153	.7160	.7167	.7174	.7181	.7189	.7196
6.8	.7203	.7210	.7217	.7224	.7232	.7239	.7246	.7253	.7260	.7267
6.9	.7274	.7281	.7288	.7295	.7302	.7309	.7316	.7323	.7330	.7338
7.0	.7345	.7352	.7359	.7366	.7373	.7380	.7386	.7393	.7400	.7407
7.1	.7414	.7421	.7428	.7435	.7442	.7449	.7456	.7463	.7470	.7477
7.2	.7483	.7490	.7497	.7504	.7511	.7518	.7524	.7531	.7538	.7545
7.3	.7552	.7559	.7565	.7572	.7579	.7586	.7592	.7599	.7606	.7613
7.4	.7620	.7626	.7633	.7640	.7646	.7653	.7660	.7666	.7673	.7680
7.5	.7687	.7693	.7700	.7706	.7713	.7720	.7726	.7733	.7740	.7746
7.6	.7753	.7760	.7766	.7773	.7779	.7786	.7792	.7799	.7806	.7812
7.7	.7819	.7825	.7832	.7838	.7845	.7851	.7858	.7864	.7871	.7877
7.8	.7884	.7890	.7897	.7903	.7910	.7916	.7923	.7929	.7936	.7942
7.9	.7949	.7955	.7961	.7968	.7974	.7981	.7987	.7993	.8000	.8006

*Printed by permission of Compressed Air Data.

TABLE 6-1 Values of X for Normal Air and Perfect Diatomic Gases* *(Continued)*

r	0	1	2	3	4	5	6	7	8	9
8.0	0.8013	0.8019	0.8025	0.8032	0.8038	0.8044	0.8051	0.8057	0.8063	0.8070
8.1	.8076	.8082	.8089	.8095	.8101	.8108	.8114	.8120	.8126	.8133
8.2	.8139	.8145	.8151	.8158	.8164	.8170	.8176	.8183	.8189	.8195
8.3	.8201	.8207	.8214	.8220	.8226	.8232	.8238	.8245	.8251	.8257
8.4	.8263	.8269	.8275	.8281	.8288	.8294	.8300	.8306	.8312	.8318
8.5	.8324	.8330	.8336	.8343	.8349	.8355	.8361	.8367	.8373	.8379
8.6	.8385	.8391	.8397	.8403	.8409	.8415	.8421	.8427	.8433	.8439
8.7	.8445	.8451	.8457	.8463	.8469	.8475	.8481	.8487	.8493	.8499
8.8	.8505	.8511	.8517	.8523	.8529	.8535	.8541	.8547	.8552	.8558
8.9	.8564	.8570	.8576	.8582	.8588	.8594	.8600	.8605	.8611	.8617
9.0	.8623	.8629	.8635	.8641	.8646	.8652	.8658	.8664	.8670	.8676
9.1	.8681	.8687	.8693	.8699	.8705	.8710	.8716	.8722	.8728	.8734
9.2	.8739	.8745	.8751	.8757	.8762	.8768	.8774	.8779	.8785	.8791
9.3	.8797	.8802	.8808	.8814	.8819	.8825	.8831	.8837	.8842	.8848
9.4	.8854	.8859	.8865	.8871	.8876	.8882	.8888	.8893	.8899	.8905
9.5	.8910	.8916	.8921	.8927	.8933	.8938	.8944	.8949	.8955	.8961
9.6	.8966	.8972	.8977	.8983	.8989	.8994	.9000	.9005	.9011	.9016
9.7	.9022	.9028	.9033	.9039	.9044	.9050	.9055	.9061	.9066	.9072
9.8	.9077	.9083	.9088	.9094	.9099	.9105	.9110	.9116	.9121	.9127
9.9	.9132	.9138	.9143	.9149	.9154	.9159	.9165	.9170	.9176	.9181
10.0	.9187	.9192	.9198	.9203	.9208	.9214	.9219	.9225	.9230	.9235
10.1	.9241	.9246	.9252	.9257	.9262	.9268	.9273	.9278	.9284	.9289
10.2	.9295	.9300	.9305	.9311	.9316	.9321	.9327	.9332	.9337	.9343
10.3	.9348	.9353	.9358	.9364	.9369	.9374	.9380	.9385	.9390	.9396
10.4	.9401	.9406	.9411	.9417	.9422	.9427	.9432	.9438	.9443	.9448
10.5	.9453	.9459	.9464	.9469	.9474	.9480	.9485	.9490	.9495	.9500
10.6	.9506	.9511	.9516	.9521	.9526	.9532	.9537	.9542	.9547	.9552
10.7	.9558	.9563	.9568	.9573	.9578	.9583	.9589	.9594	.9599	.9604
10.8	.9609	.9614	.9619	.9625	.9630	.9635	.9640	.9645	.9650	.9655
10.9	.9660	.9665	.9671	.9676	.9681	.9686	.9691	.9696	.9701	.9706
11.0	.9711	.9716	.9721	.9726	.9732	.9737	.9742	.9747	.9752	.9757
11.1	.9762	.9767	.9772	.9777	.9782	.9787	.9792	.9797	.9802	.9807
11.2	.9812	.9817	.9822	.9827	.9832	.9837	.9842	.9847	.9852	.9857
11.3	.9862	.9867	.9872	.9877	.9882	.9887	.9892	.9897	.9902	.9907
11.4	.9912	.9916	.9921	.9926	.9931	.9936	.9941	.9946	.9951	.9956
11.5	.9961	.9966	.9971	.9975	.9980	.9985	.9990	.9995	1.0000	1.0005
11.6	1.0010	1.0015	1.0019	1.0024	1.0029	1.0034	1.0039	1.0044	1.0049	1.0054
11.7	1.0058	1.0063	1.0068	1.0073	1.0078	1.0083	1.0087	1.0092	1.0097	1.0102
11.8	1.0107	1.0112	1.0116	1.0121	1.0126	1.0131	1.0136	1.0140	1.0145	1.0150
11.9	1.0155	1.0160	1.0164	1.0169	1.0174	1.0179	1.0184	1.0188	1.0193	1.0198
12.0	1.0203	1.0207	1.0212	1.0217	1.0222	1.0226	1.0231	1.0236	1.0241	1.0245

*Printed by permission of Compressed Air Data.

NOTE: Taken from Moss and Smith, Engineering Computations for Air and Gases, *Trans. Am. Soc. Mech. Engrs.*, vol. 52, 1930, paper APM-52-8. For nozzles $r = p_1/p_2$. For compressors and exhausters $r = p_2/p_1$.

r	x	r	x	r	x	r	x	r	x	r	x	r	x	r	x	r	x
12.5	1.0428	15.0	1.1520	17.5	1.2479	20.0	1.3345	22.5	1.4136	25.0	1.4867	27.5	1.5546	30.0	1.6183	32.5	1.6783
13.0	1.0666	15.5	1.1720	18.0	1.2659	20.5	1.3509	23.0	1.4287	25.5	1.5006	28.0	1.5678	30.5	1.6306	33.0	1.6899
13.5	1.0887	16.0	1.1916	18.5	1.2835	21.0	1.3669	23.5	1.4435	26.0	1.5144	28.5	1.5794	31.0	1.6434	33.5	1.7014
14.0	1.1103	16.5	1.2108	19.0	1.3008	21.5	1.3828	24.0	1.4581	26.5	1.5280	29.0	1.5933	31.5	1.6547	34.0	1.7127
14.5	1.1314	17.0	1.2295	19.5	1.3189	22.0	1.3983	24.5	1.4725	27.0	1.5414	29.5	1.6059	32.0	1.6666	34.5	1.7240

Values of X from 12.5 to 34.5 calculated by Ingersoll-Rand Co.

the bracketed expression $[(p_2/p_1)^{\frac{k-1}{k}} - 1]$ in these equations for a value of $k = 1.395$. These are known as X factors, and they are given in Table 6-1. By using X factors, the adiabatic formulas for $k = 1.395$ read as follows:

Adiabatic temperature, pressure, and volume relations:

$$V_1/V_2 = p_2/[(X + 1)p_1] \tag{6-25}$$

$$T_2/T_1 = X + 1 \tag{6-26}$$

$$T_2 - T_1 = T_1X = T_2[X/(X + 1)] \tag{6-27}$$

Adiabatic power:
In SI units,

$$\text{kW}_{ad} = 9.81 \times 10^{-4}Q_1p_1X \tag{6-28a}$$

In U.S. customary units,

$$\text{hp}_{ad} = 0.0154Q_1p_1X \tag{6-28b}$$

Adiabatic discharge temperature:

$$T_2 = T_1(X + 1) \tag{6-29}$$

To find the X factor X_G for a gas of any k value refer to Fig. 6-34. This figure gives values of X_G/X for gases having specific-heat ratios between 1.0 and 1.4. The factor X_G is then the product of X_G/X from Fig. 6-34 and the value X from Table 6-1 for desired compression ratio.

Adiabatic power for gases other than air:
In SI units,

$$\text{kW}_{ad} = 6.37 \times 10^{-4}Q_1p_1X/d \tag{6-30a}$$

In U.S. customary units,

$$\text{hp}_{ad} = 1 \times 10^{-2}Q_1p_1X/d \tag{6-30b}$$

where $d = 2.922\,(k - 1)/k$.

If the compression cycle approaches the isothermal condition, pV = constant, as is the case when several stages with intercoolers are used, a simple approximation of the power is obtained from the following formula:

In SI units,

$$\text{kW} = 2.78 \times 10^{-4}Q_1p_1 \ln p_2/p_1 \tag{6-31a}$$

In U.S. customary units,

$$\text{hp} = 4.4 \times 10^{-3}Q_1p_1 \ln p_2/p_1 \tag{6-31b}$$

For multistage compressors of N_s number of stages with adiabatic

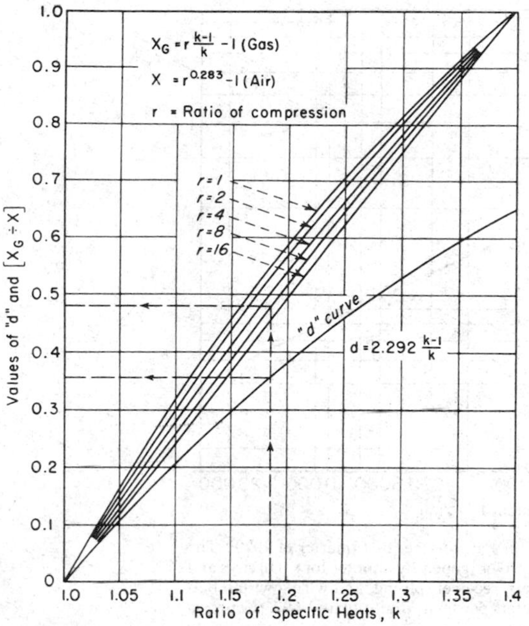

FIG. 6-34 Factors for use in adiabatic formula. Values of X to be used in finding X_G may be obtained from Table 6-1. *(By permission of Compressed Air Data.)*

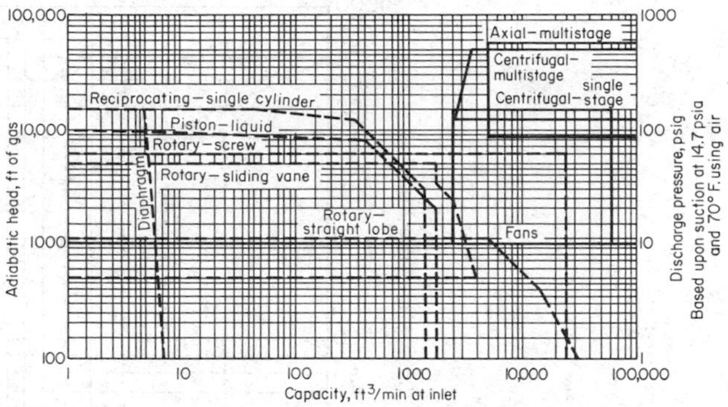

FIG. 6-35 Compressor coverage chart based on the normal range of operation of commercially available types shown. Solid lines: use left ordinate, head. Broken lines: use right ordinate, pressure. To convert cubic feet per minute to cubic meters per hour, multiply by 1.699; to convert feet to meters, multiply by 0.3048; and to convert pounds-force per square inch to kilopascals, multiply by 6.895; (°F − 32)⁵⁄₉ = °C.

compression in each stage, equal division of work between stages, and intercooling to the intake temperature, the following formulas are helpful:

In SI units,

$$kW_{ad} = \frac{6.37 \times 10^{-4} N_s Q_1 p_1}{d} \left(\sqrt[N_s]{X_G + 1} - 1\right) \qquad (6\text{-}32a)$$

In U.S. customary units,

$$hp_{ad} = \frac{1 \times 10^{-2} N_s Q_1 p_1}{d} \left(\sqrt[N_s]{X_G + 1} - 1\right) \qquad (6\text{-}32b)$$

$$T_2 = T_1 \sqrt[N_s]{X_G + 1} \qquad (6\text{-}33)$$

Compressor Selection To select the most satisfactory compression equipment, chemical engineers must consider a wide variety of types, each of which has peculiar advantages for particular applications. Among the major factors to be considered are flow rate, head or pressure, temperature limitations, method of sealing, method of lubrication, power consumption, serviceability, and cost. The coverage chart in Fig. 6-35 will assist in defining the range of performance of common types.

In chemical-plant service, additional problems must often be dealt with: gases may be highly corrosive or may carry abrasive solids in suspension; gases at elevated temperatures may create a potential explosion hazard, while air at the same temperatures may be handled quite normally; minute amounts of lubricating oil or water may contaminate the process gas and so may not be permissible; and for continuous-process use a high degree of equipment reliability is required, since frequent shutdowns for inspection or for maintenance cannot be tolerated.

Compressor Costs Figure 6-36 shows purchase and installation costs of reciprocating and centrifugal machines for 0.6895-MPa (100-lbf/in²) air service (first quarter of 1979).

FANS

Fans are used for low pressures, generally less than 3.447 kPa (0.5 lbf/in²). They are usually of the centrifugal or the axial-flow type. Both types are used for ventilating work, supplying draft to boilers and furnaces, moving large volumes of air or gas through ducts, supplying air for drying, conveying material suspended in the gas stream, removing fumes, etc.

Centrifugal Fans These fans are made in three general types: the straight-blade, or steel-plate, fan, the forward-curved-blade fan, and the backward-curved-blade fan.

Straight-blade fans (Fig. 6-37) have rotors of comparatively large diameter with 5 to 12 radial blades resembling paddle wheels. These fans operate at comparatively low speed. They are often used in exhaust work, particularly when wastes are carried in the air stream.

Forward-curved-blade fans (Fig. 6-38) are usually of the multiblade (20- to 64-blade) "sirocco" type. The rotors are of smaller diameter, and they operate at higher speeds than those of straight-blade units.

Backward-curved blade fans (Fig. 6-39) are of the multiblade (10- to 50-blade) type. Such fans have a wide range of usefulness.

Axial-flow fans are made in two general types, disk type and propeller type. **Disk-type fans** have plain or curved blades similar to those of an ordinary household fan. They are usually used for general circulation or exhaust work without ducts. **Propeller-type fans** (Fig. 6-40) have blades similar to aeronautical designs. Such fans may be two-staged.

Characteristic curves for the different types of fans are shown in Fig. 6-41.

The theory of operation of a centrifugal fan is much like that of a centrifugal pump, the pressure developed arising from two sources. These are centrifugal force due to the rotation of an enclosed volume of air or gas and to the velocity imparted to the air or gas by the blades and partly converted to pressure by the volute or scroll-shaped fan casing.

The centrifugal force developed by the rotor produces a compression of the air or gas which in fan engineering is called the *static pressure*. The amount of static pressure developed depends upon the ratio of the velocity of the air leaving the tips of the blades to the velocity of the air entering the fan at the heel of the blades. Therefore, the longer the blades, the greater the static pressure developed by the fan.

Operating efficiencies of fans range from 40 to 70 percent. Operating pressure is the sum of the static pressure and the velocity head of the air leaving the fan. In SI units the pressure is usually expressed in centimeters of water. The power output of the fan is expressed as follows:

$$kW = 2.72 \times 10^{-5} Qp \qquad (6\text{-}34a)$$

where kW is the fan power output, kW; Q is the fan volume, m³/h; and p is the fan-operating pressure, cm water column.

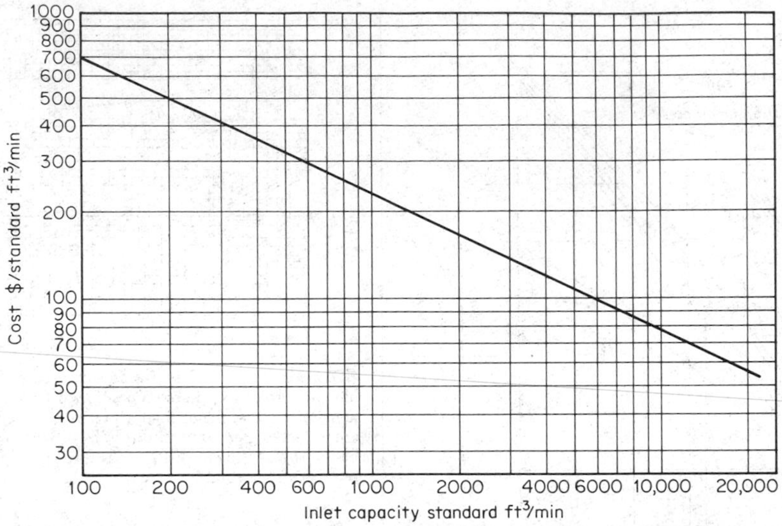

FIG. 6-36 Cost of compressors for 0.690-MPa (100-lbf/in^2) air service (first quarter of 1979). The cost shown is the installed cost and includes compressor, driver (generally a motor for small sizes and a steam turbine for larger sizes), foundation, dryers, filters, receiver, piping, and wiring within battery limits. To convert cubic feet per minute to cubic meters per hour, multiply by 1.699. *(Courtesy of E. I. du Pont de Nemours & Co.)*

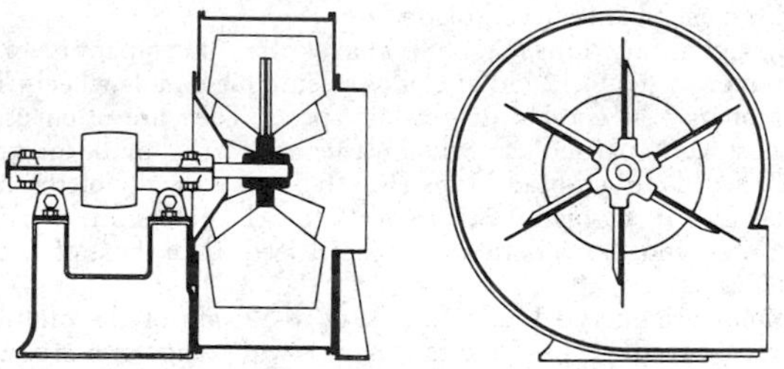

FIG. 6-37 Straight-blade, or steel-plate, fan.

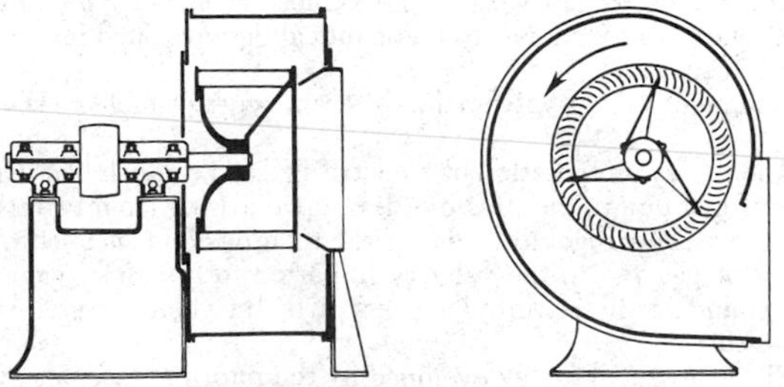

FIG. 6-38 Forward-curved-blade, or "sirocco"-type, fan.

In U.S. customary units,

$$\text{hp} = 1.57 \times 10^{-4} Qp \tag{6-34b}$$

where hp is the fan power output, hp; Q is the fan volume, ft^3/min; and p is the fan-operating pressure, inches water column.

$$\text{Efficiency} = \frac{\text{air power output}}{\text{shaft power input}} \tag{6-35}$$

Fan Performance The performance of a centrifugal fan varies with changes in conditions such as temperature, speed, and density of the gas being handled. It is important to keep this in mind in using the catalog data of various fan manufacturers, since such data are

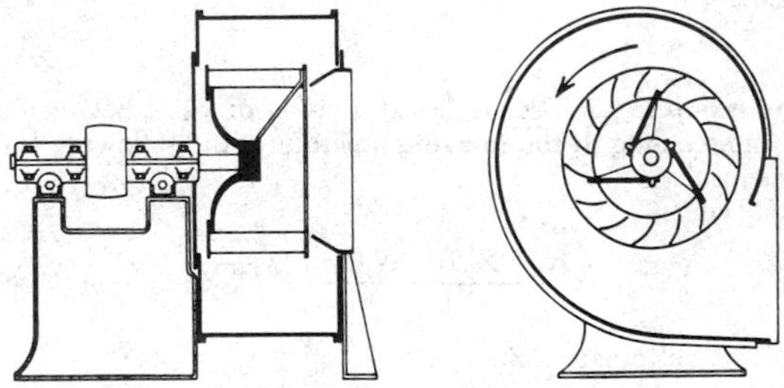

FIG. 6-39 Backward-curved-blade fan.

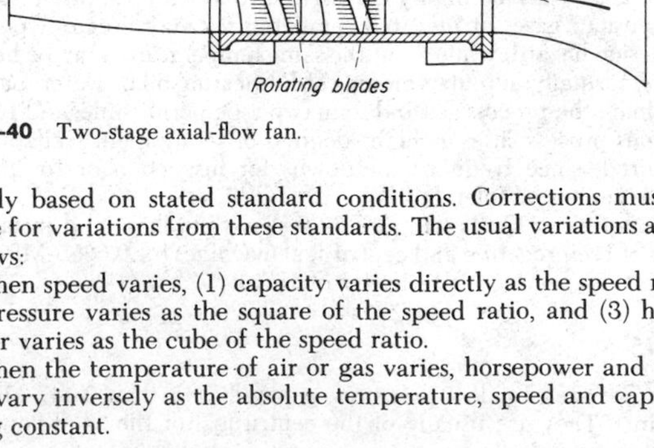

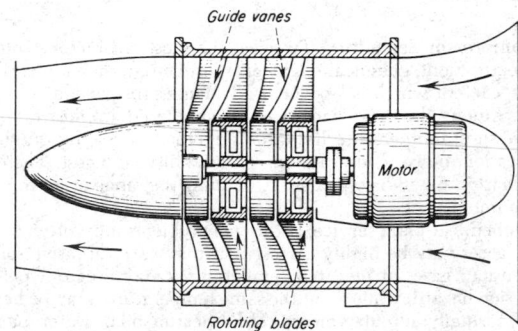

FIG. 6-40 Two-stage axial-flow fan.

usually based on stated standard conditions. Corrections must be made for variations from these standards. The usual variations are as follows:

When speed varies, (1) capacity varies directly as the speed ratio, (2) pressure varies as the square of the speed ratio, and (3) horsepower varies as the cube of the speed ratio.

When the temperature of air or gas varies, horsepower and pressure vary inversely as the absolute temperature, speed and capacity being constant.

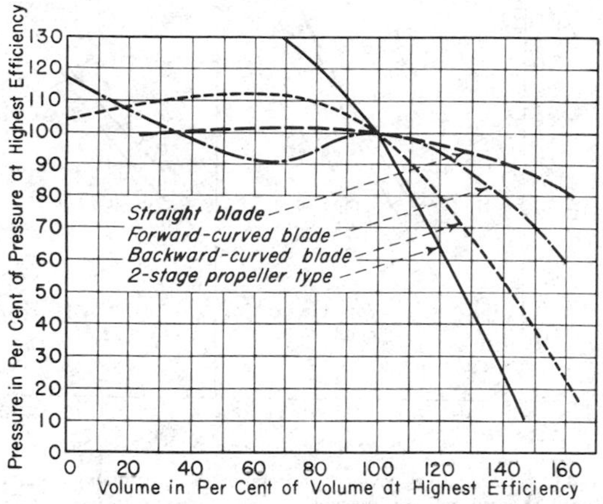

FIG. 6-41 Approximate characteristic curves of various types of fans.

When the density of air or gas varies, horsepower and pressure vary directly as the density, speed and capacity being constant.

Selection of Fans It is a common practice among fan manufacturers to publish in tabular form complete data showing capacities, pressures, speeds, and horsepowers of their fans under standard conditions of temperature and air density. These tables are of great use to heating and ventilating engineers and to others who specialize in fan engineering. Those who do not specialize along these lines, including chemical engineers, should not attempt to select fans from these tables. The proper course to follow is to put full data concerning the job to be done in the hands of fan manufacturers and allow them to specify the fan they are willing to guarantee to do the required work at the best obtainable economy. A comparison of several such proposals from manufacturers will indicate the best choice.

CENTRIFUGAL AND AXIAL COMPRESSORS

Centrifugal compressors, or turboblowers, are widely used to handle large volumes of gas at pressure rises from 3.447 kPa (0.5 lbf/in^2) up to several hundred kPa (lbf/in^2). The most important criterion, more important than pressure rise, is the pressure ratio, as pointed out later. For pressures below 3.447 kPa (0.5 lbf/in^2) one of the several types of fans is ordinarily selected.

Centrifugal compressors are widely used in the chemical industry for many services, e.g., for compressing process gas, supplying plant air, and conveying solid materials in suspension, as exhausters, and for ventilation, aeration, etc. They are used extensively in other industries, e.g., for supplying air or oxygen to blast furnaces in the iron and steel industry and as pipe-line boosters for natural gas.

Centrifugal compressors may be single or multistage within a single casing. In addition, several casings, either single or multistage, may be used in series. Intercoolers may be used between stages or between casings to cool the partially compressed gas to the initial or other desired temperature in order to minimize the power required for compression.

The principle of a centrifugal compressor is the same as that of a centrifugal pump. A major difference is that the air or gas handled in a centrifugal compressor or blower is compressible while the liquids handled in a pump are practically incompressible.

When selecting or sizing a centrifugal compressor, the combination of the most adverse conditions occurring simultaneously must be determined. These conditions are:

Lowest barometric pressure
Lowest intake pressure
Maximum intake temperature
Highest ratio of specific heats (k value)
Lowest specific gravity
Maximum intake volume
Maximum discharge pressure

Centrifugal compressor drivers are usually steam turbines, gas turbines, or electric motors. Speed-increasing gears may be used in conjunction with these drivers. Most centrifugal compressors operate at speeds above 60-Hz, two-pole motor speeds of 3600 r/min. Rotative speeds up to 50,000 r/min are not uncommon. For adequate performance aerodynamic design dictates the use of the higher speeds. The maximum impeller-tip speed is limited by the strength of the impeller material of construction.

In a centrifugal compressor as in a centrifugal pump, the head developed by the compressor is independent of the fluid handled. An examination of the formulas previously presented in this subsection shows that the pressure ratio is dependent upon the inlet temperature, molecular weight, and ratio of specific heats k. With air, having a molecular weight of 29, pressure ratios per stage when taking suction at room temperature are limited to about 1.4. With hydrogen, having a molecular weight of 2, pressure ratios are limited to about 1.025. For gases heavier than air such as carbon dioxide, pressure ratios considerably higher than 1.4 are attainable.

Because of rotor dynamics a single casing (Fig. 6-42) usually does not contain more than seven or eight stages. Two or more casings can be used in series. Intercoolers generally are used between stages or groups of stages to conserve power and limit the maximum temperature.

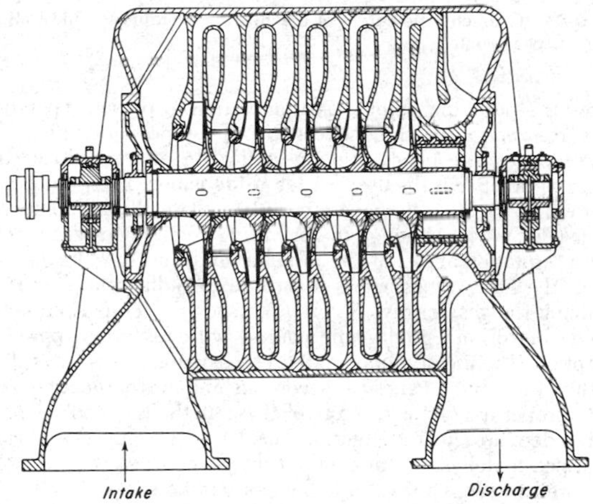

FIG. 6-42 Five-stage centrifugal compressor.

Typical **characteristic curves** of a multistage blower are shown in Fig. 6-43. From these curves it will be seen that a centrifugal compressor is essentially a **constant-pressure** machine and that power consumption is almost directly proportional to the volume delivered. For motor-driven blowers various devices such as hydraulic couplings, magnetic couplings, or wound-rotor motors may be used to obtain efficient operation at part loads or under adverse operating conditions, although the additional investment is rarely justified.

There is a minimum capacity for each blower, at every speed, below which operation becomes unstable. This instability is accompanied by a characteristic noise known as **pumping** or **surge.** The pumping limit is set largely by the impeller-discharge angle and normally lies in the neighborhood of 50 to 90 percent of the capacity at the best efficiency point. The primary cause of this behavior lies in the shape of the head-capacity curve, which after reaching a maximum begins to droop toward the zero-capacity point. When the capacity is reduced below this point, the pressure in the discharge pipe exceeds that produced by the blower and the flow tends to reverse momentarily. However, as soon as the flow is further reduced, the pressure in the discharge pipe drops and the blower begins to discharge into the pipe again. Such pulsations in pressure and capacity are magnified by the response of the compressible gas in the discharge system. Blowers should not be operated at volumes

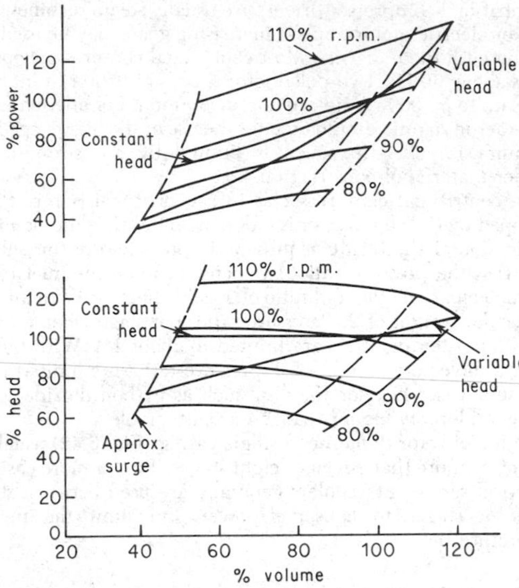

FIG. 6-43 Characteristic curves of a centrifugal compressor, illustrating the effect of speed control.

below the pumping point. Instrumentation to prevent operation in the surge zone is recommended to avoid compressor damage.

In addition to control of the operating range by speed variation, a common practice is the use of **inlet guide vanes.** These vanes can be adjusted to reduce the capacity and increase the stable operating range (Fig. 6-44). Although the primary role of guide vanes is to provide prerotation ahead of the impeller and thus to reduce entrance losses, the same vanes act as a throttle to reduce the flow rate by reducing the gas density. Figure 6-45 shows a comparison between speed control and guide-vane control with respect to power consumption. The use of a **blast gate** in the suction line is a less efficient method than guide vanes for accomplishing nearly the same result.

By proper use of one or more of these methods of control together with adequate instrumentation, centrifugal compressors can be equipped to deliver gas at constant discharge pressure, constant suction pressure, constant volume flow, or constant mass flow.

Advancement in the technology of high-speed rotating equip-

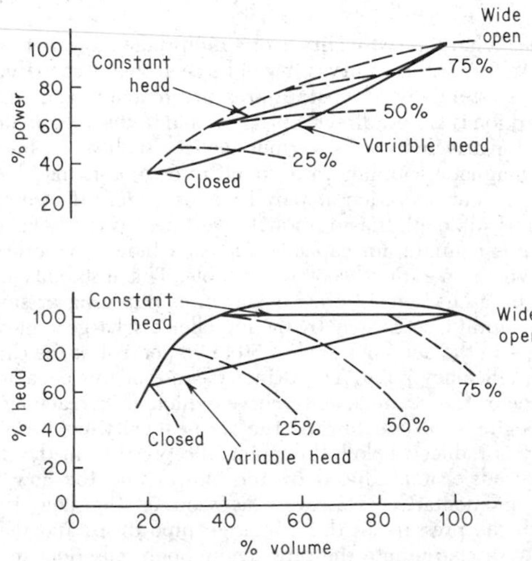

FIG. 6-44 Characteristic curves of a centrifugal compressor, illustrating effect of guide-vane control.

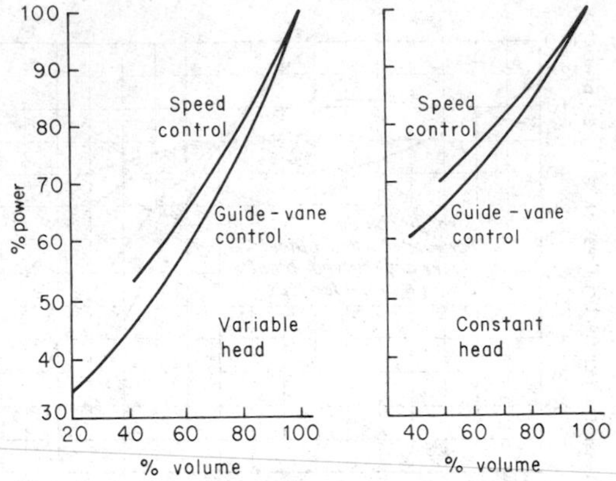

FIG. 6-45 Comparative power requirement of a centrifugal compressor; speed control versus guide-vane control.

ment, accelerated by work in the aircraft jet-engine field, has produced a family of **close-coupled gear-mounted compressors** for air and clean gases. These are well suited for 0.690-MPa (100-lbf/in^2) plant and instrument air systems with capacities available from 1×10^3 to 34×10^3 m^3/h (600 to 20,000 ft^3/min). Usually the driver is connected to a bull gear, which in turn drives one or more pinion gears. Single impellers, each with its own volute casing, are mounted on one or both ends of each pinion shaft. Intercooling is provided between stages. These machines are more compact and have a higher efficiency than multistage single-casing machines for the same service.

Stationary gas-turbine installations use almost exclusively the **axial-flow compressor.** Either of two designs, the **industrial type** or the **aircraft-derivative type,** is available. The former is characterized by a heavier, more rugged construction with all shafts supported on sleeve-type bearings; the latter, by aircraft-type construction and antifriction bearings.

In addition to its use with gas turbines, the axial-flow compressor is employed in the steel industry for blast-furnace blowers, in the chemical industry for large nitric acid plants, and for other large-air-volume or large-gas-volume uses. The air or gas stream is free from any lubricating oil or other contamination.

Figure 6-46 shows a typical axial-flow compressor. The rotating element consists of a single drum to which are attached several rows of decreasing-height blades having airfoil cross sections. Between each two rotating-blade rows is a stationary-blade row. All blade angles and areas are designed precisely for a given performance and **high efficiency.**

Pressure ratios per casing are comparable with those of centrifugal equipment, although flow rates are considerably higher for a given casing diameter because of the greater area of the flow path.

Because of the relatively steep head-capacity characteristic curve, the pumping point may be within 10 percent of the design flow, as illustrated in Fig. 6-47.

ROTARY BLOWERS AND COMPRESSORS

Rotary compressors, blowers, and vacuum pumps are machines of the positive-displacement type. Such units are essentially constant-volume machines with variable discharge pressure. The volume can be varied only by changing the speed or by bypassing or wasting some of the capacity of the machine. The discharge pressure will vary with the resistance on the discharge side of the system. A characteristic curve typical of the form produced by these rotary units is shown in Fig. 6-48. Rotary compressors are generally classified as of the straight-lobe type, screw type, sliding-vane type, and liquid-piston type.

Straight-Lobe Type This type is illustrated in Fig. 6-49. Such units are available for pressure differentials up to about 83 kPa (12

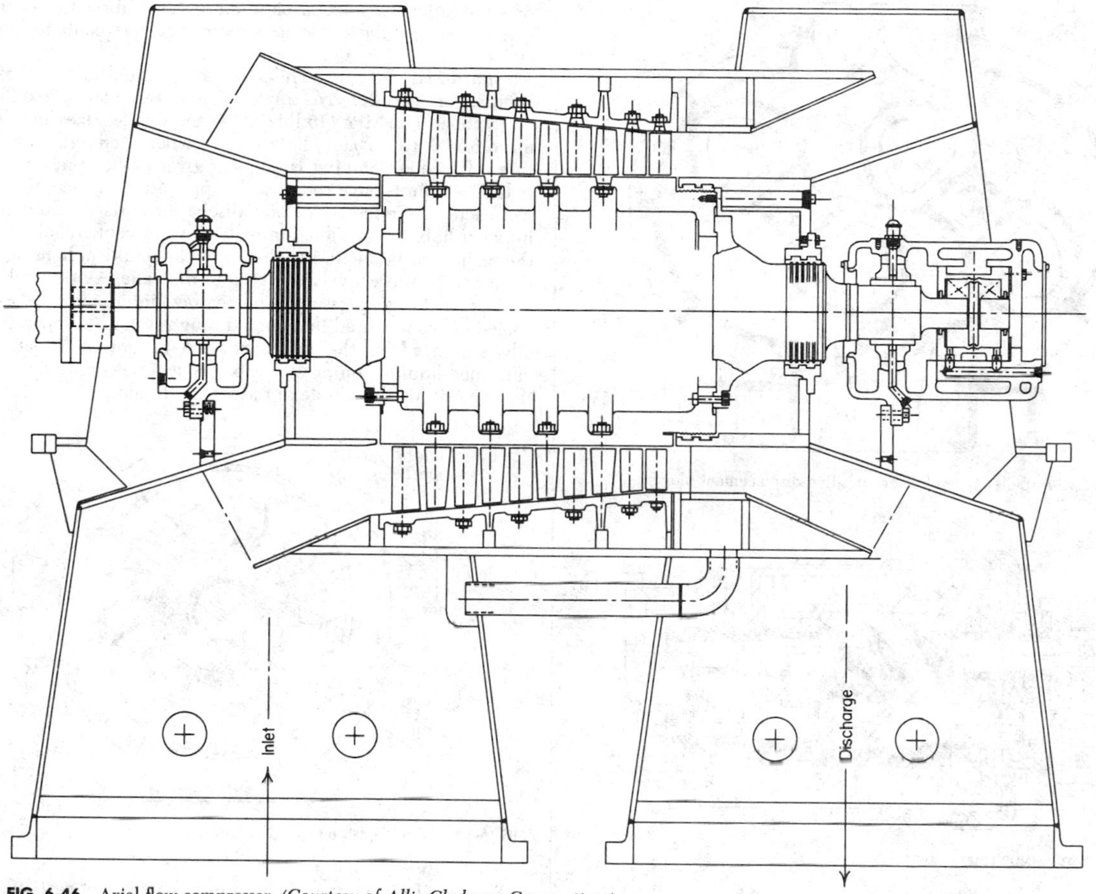

FIG. 6-46 Axial-flow compressor. *(Courtesy of Allis-Chalmers Corporation.)*

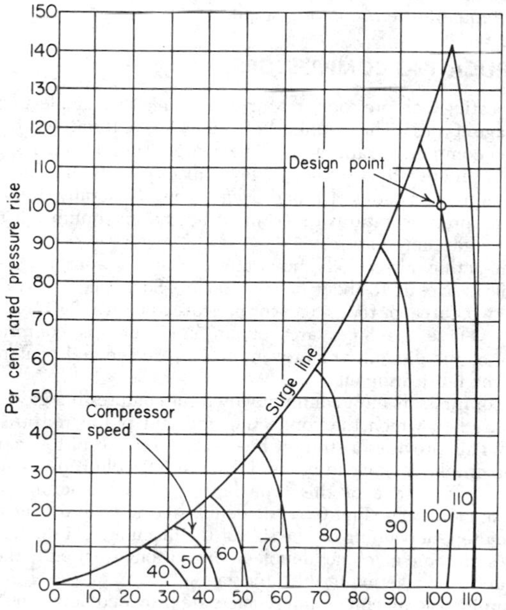

FIG. 6-47 Typical performance characteristics of an axial-flow compressor.

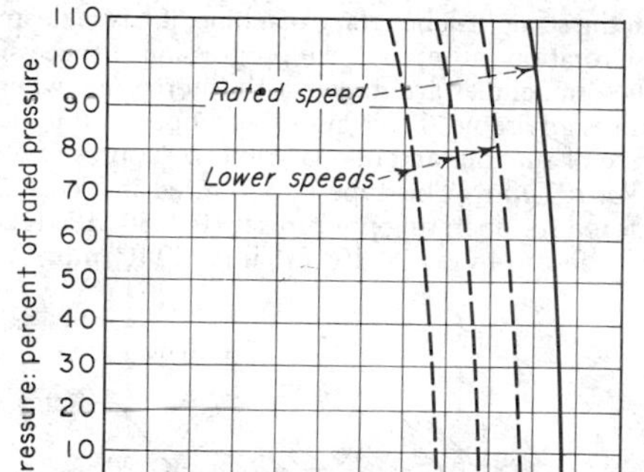

FIG. 6-48 Approximate performance curves for a rotary positive-displacement compressor. The safety valve in discharge line or bypass must be set to operate at a safe value determined by construction.

lbf/in^2) and capacities up to 2.549 × 10^4 m^3/h (15,000 ft^3/min). Sometimes multiple units are operated in series to produce higher pressures; individual-stage pressure differentials are limited by the shaft deflection, which must necessarily be kept small to maintain rotor and casing clearance.

Screw-Type This type of rotary compressor, as shown in Fig. 6-50, is capable of handling capacities up to about 4.248 × 10^4 m^3/h

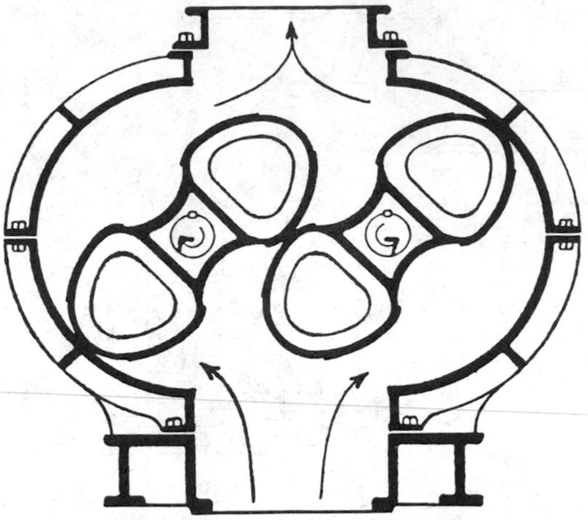

FIG. 6-49 Two-impeller type of rotary positive-displacement blower.

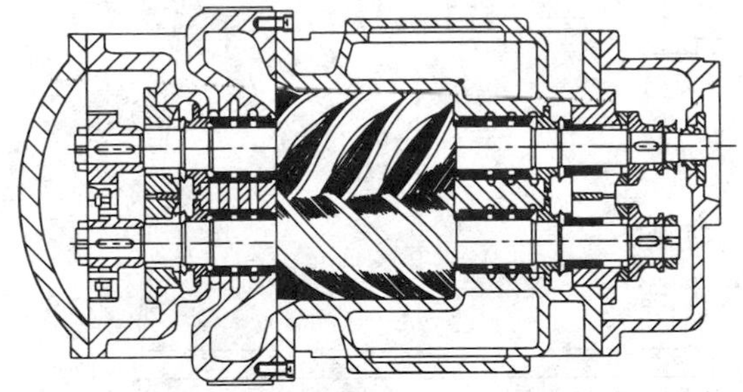

FIG. 6-50 Screw-type rotary compressor.

(25,000 ft^3/min) at pressure ratios of 4:1 and higher. Relatively small-diameter rotors allow rotative speeds of several thousand r/min. Unlike the straight-lobe rotary machine, it has male and female rotors whose rotation causes the axial progression of successive sealed cavities. These machines are staged with intercoolers when such an arrangement is advisable. Their high-speed operation usually necessitates the use of suction- and discharge-noise suppressors.

Sliding-Vane Type This type is illustrated in Fig. 6-51. These units are offered for operating pressures up to 0.86 MPa (125 lbf/in^2) and in capacities up to 3.4×10^3 m^3/h (2000 ft^3/min). Generally,

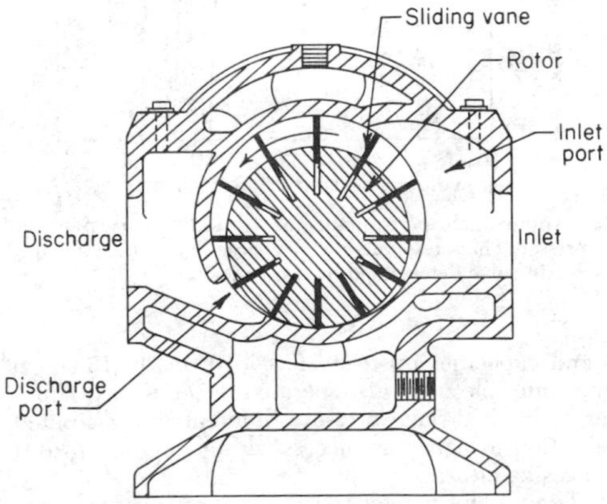

FIG. 6-51 Sliding-vane type of rotary compressor.

pressure ratios per stage are limited to 4:1. Lubrication of the vanes is required, and the air or gas stream therefore contains lubricating oil.

Liquid-Piston Type This type is illustrated in Fig. 6-52. These compressors are offered as single-stage units for pressure differentials up to about 0.52 MPa (75 lbf/in^2) in the smaller sizes and capacities up to 6.8×10^3 m^3/h (4000 ft^3/min) when used with a lower pressure differential. Staging is employed for higher pressure differentials. These units have found wide application as vacuum pumps on wet-vacuum service. Inlet and discharge ports are located in the impeller hub. As the vaned impeller rotates, centrifugal force drives the sealing liquid against the walls of the elliptical housing, causing the air to be successively drawn into the vane cavities and expelled against discharge pressure. The sealing liquid must be externally cooled unless it is used in a once-through system. A separator is usually employed in the discharge line to minimize carryover of entrained liquid. Compressor capacity can be considerably reduced if the gas is highly soluble in the sealing liquid.

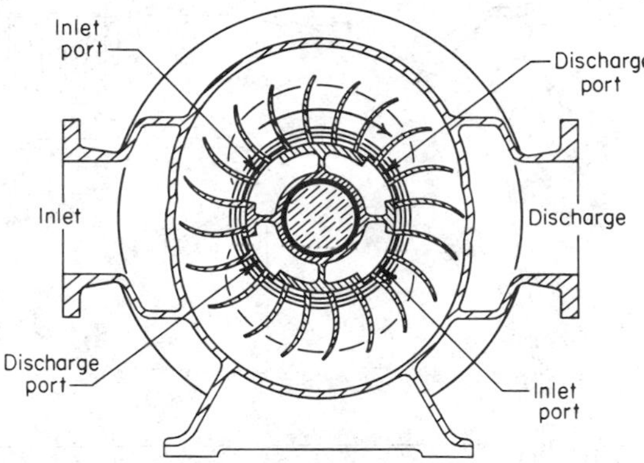

FIG. 6-52 Liquid-piston type of rotary compressor.

The liquid-piston type of compressor has been of particular advantage when hazardous gases are being handled. Because of the gas-liquid contact and because of the much greater liquid specific heat, the gas-temperature rise is very small.

RECIPROCATING COMPRESSORS

Reciprocating compressors are furnished in either single-stage or multistage types. The number of stages is determined by the required compressor ratio p_2/p_1. The compression ratio per stage is generally limited to 4, although low-capacity units are furnished with compression ratios of 8 and even higher. Generally, the maximum compression ratio is determined by the maximum allowable discharge-gas temperature.

Single-acting air-cooled and water-cooled air compressors are available in sizes up to about 75 kW (100 hp). Such units are available in one, two, three, or four stages for pressure as high as 24 MPa (3500 lbf/in^2). These machines are seldom used for gas compression because of the difficulty of preventing gas leakage and contamination of the lubricating oil.

The compressors most commonly used for compressing gases have a crosshead to which the connecting rod and piston rod are connected. This provides a straight-line motion for the piston rod and permits simple packing to be used. Figure 6-53 illustrates a simple single-stage machine of this type having a double-acting piston. Either single-acting (Fig. 6-54) or double-acting pistons (Fig. 6-55) may be used, depending on the size of the machine and the number of stages. In some machines double-acting pistons are used in the first stages and single-acting in the later stages.

On multistage machines, intercoolers are provided between stages. These heat exchangers remove the heat of compression from the gas and reduce its temperature to approximately the temperature exist-

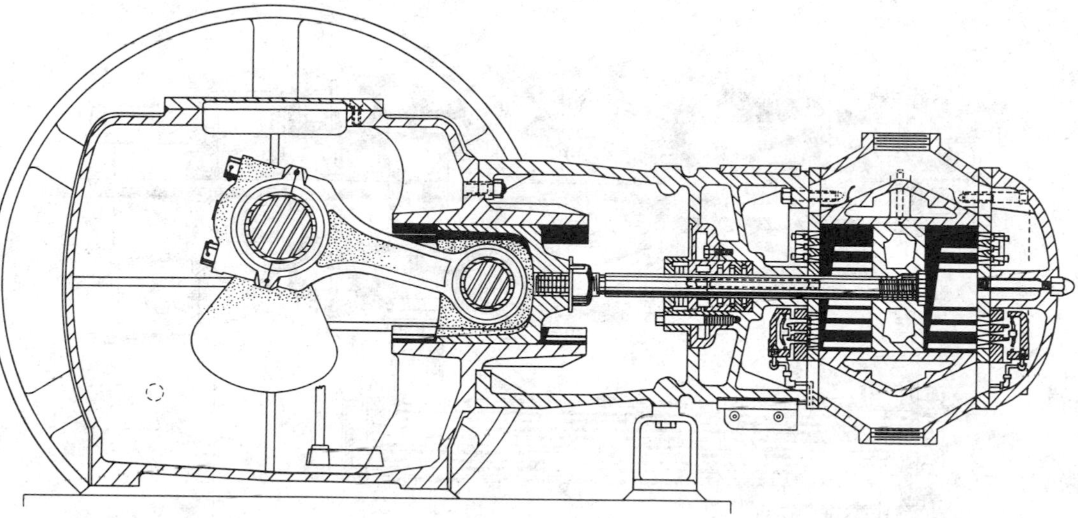

FIG. 6-53 Typical single-stage, double-acting water-cooled compressor.

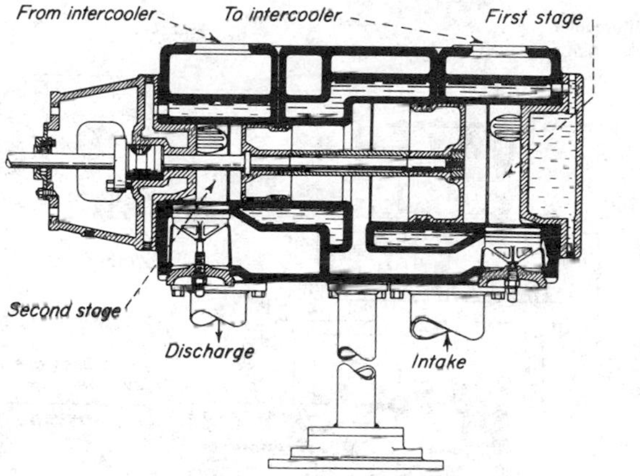

FIG. 6-54 Two-stage single-acting opposed piston in a single step-type cylinder.

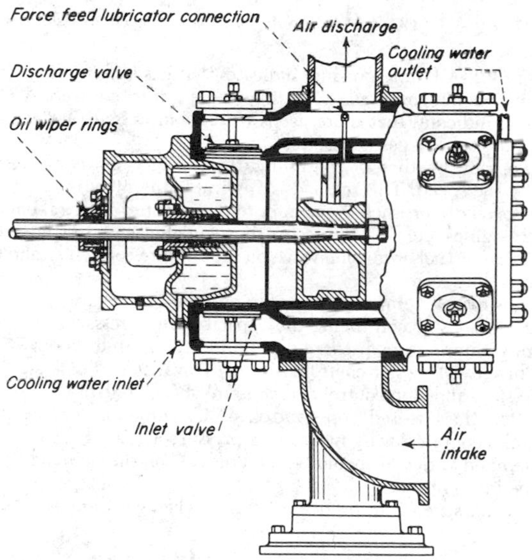

FIG. 6-55 Typical double-acting compressor piston and cylinder.

ing at the compressor intake. Such cooling reduces the volume of gas going to the high-pressure cylinders, reduces the power required for compression, and keeps the temperature within safe operating limits.

Figure 6-56 illustrates a two-stage compressor end such as might be used on the compressor illustrated in Fig. 6-53.

Compressors with horizontal cylinders such as illustrated in Figs. 6-53 to 6-56 are most commonly used because of their accessibility. However, machines are also built with vertical cylinders and other arrangements such as right-angle (one horizontal and one vertical cylinder) and V-angle.

Compressors up to around 75 kW (100 hp) usually have a single center-throw crank, as illustrated in Fig. 6-53. In larger sizes compressors are commonly of duplex construction with cranks on each end of the shaft (see Fig. 6-57). Some large synchronous motor-driven units are of four-corner construction; i.e., they are of double-duplex construction with two connecting rods from each of the two crank throws (see Fig. 6-58). Steam-driven compressors have one or more steam cylinders connected directly by piston rod or tie rods to the gas-cylinder piston or crosshead.

Valve Losses Above piston speeds of 2.5 m/s (500 ft/min), suction and discharge valve losses begin to exert significant effects on the actual internal compression ratio of most compressors, depending on the valve port area available. The obvious results are high temperature rise and higher power requirements than might be expected. These effects become more pronounced with higher-molecular-weight gases.

Control Devices In many installations the use of gas is intermittent, and some means of controlling the output of the compressor is therefore necessary. In other cases constant output is required despite variations in discharge pressure, and the control device must operate to maintain a constant compressor speed. Compressor capacity, speed, or pressure may be varied in accordance with requirements. The nature of the control device will depend on the function to be regulated. Regulation of pressure, volume, temperature, or some other factor determines the type of regulation required and the type of the compressor driver.

The most common control requirement is regulation of capacity. Many capacity controls, or unloading devices, as they are usually termed, are actuated by the pressure on the discharge side of the

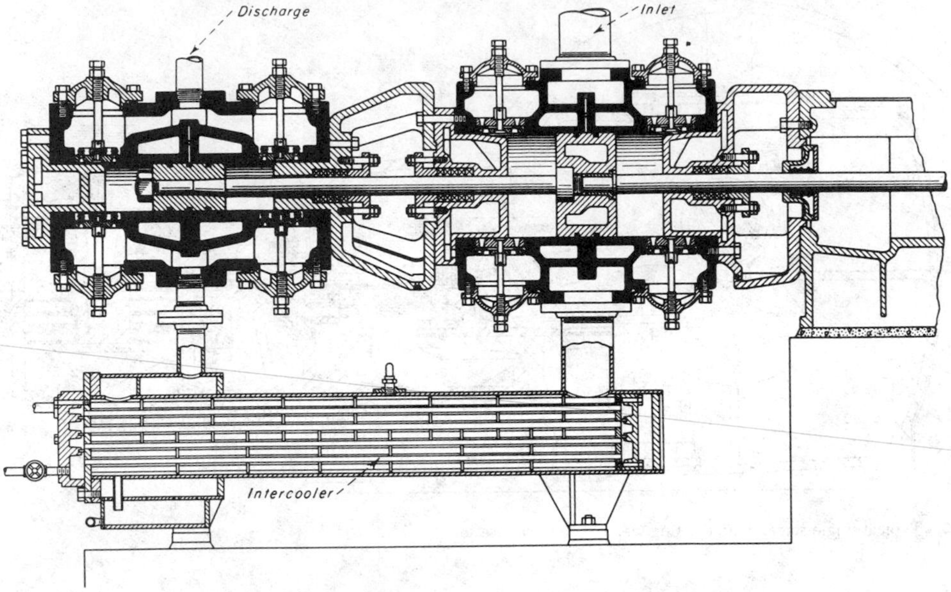

FIG. 6-56 Two-stage double-acting compressor cylinders with intercooler.

compressor. A falling pressure indicates that gas is being used faster than it is being compressed and that more gas is required. A rising pressure indicates that more gas is being compressed than is being used and that less gas is required.

An obvious method of controlling the capacity of a compressor is to vary the speed. This method is applicable to units driven by variable-speed drivers such as steam pistons, steam turbines, gas engines, diesel engines, etc. In these cases the regulator actuates the steam-admission or fuel-admission valve on the compressor driver and thus controls the speed.

Motor-driven compressors usually operate at constant speed, and other methods of controlling the capacity are necessary. On reciprocating compressors discharging into receivers, up to about 75 kW (100 hp), two types of control are usually available. These are automatic-start-and-stop control and constant-speed control.

Automatic-start-and-stop control, as its name implies, stops or starts the compressor by means of a pressure-actuated switch as the gas demand varies. It should be used only when the demand for gas will be intermittent.

Constant-speed control should be used when gas demand is fairly

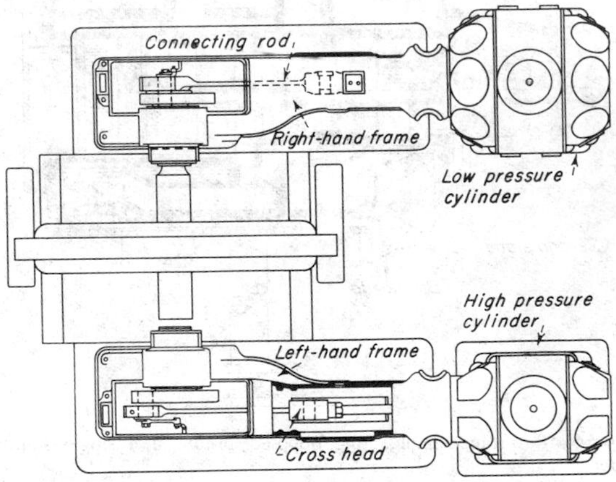

FIG. 6-57 Duplex two-stage compressor (plan view).

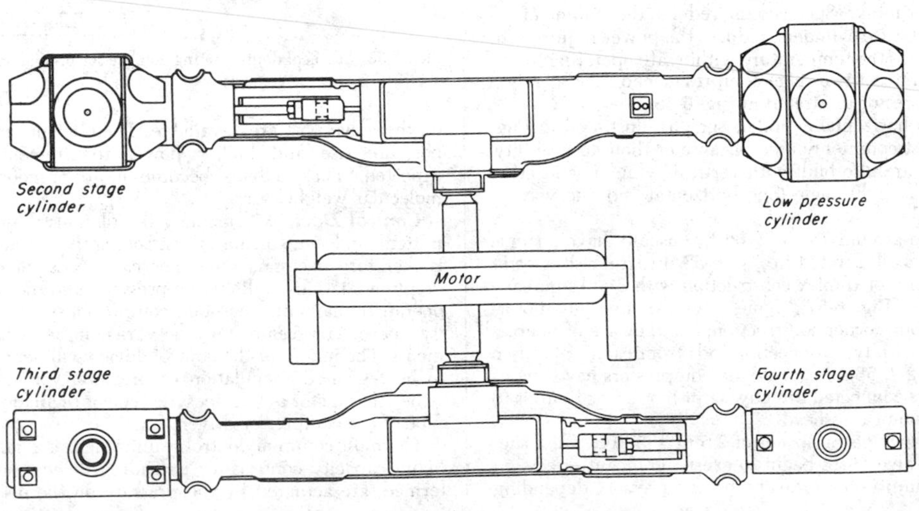

FIG. 6-58 Four-corner four-stage compressor (plan view).

constant. With this type of control, the compressor runs continuously but compresses only when gas is needed. Three methods of unloading the compressor with this type of control are in common use: (1) **closed suction unloaders,** (2) **open inlet-valve unloaders,** and (3) **clearance unloaders.** The closed suction unloader consists of a pressure-actuated valve which shuts off the compressor intake. Open inlet-valve unloaders (see Fig. 6-59) operate to hold the compressor inlet valves open and thereby prevent compression. Clearance unloaders (see Fig. 6-60) consist of pockets or small reservoirs which are opened when unloading is desired. The gas is compressed into them on the compression stroke and reexpands into the cylinder on the return stroke, thus preventing the compression of additional gas.

It is sometimes desirable to have a compressor equipped with both constant-speed and automatic-start-and-stop control. When this is done, a switch allows immediate selection of either type.

Motor-driven reciprocating compressors above about 75 kW (100 hp) in size are usually equipped with a step control. This is in reality a variation of constant-speed control in which unloading is accomplished in a series of steps, varying from full load down to no load. **Three-step control** (full load, one-half load, and no load) is usually accomplished with inlet-valve unloaders. **Five-step control** (full load, three-fourths load, one-half load, one-fourth load, and no load) is accomplished by means of clearance pockets (see Fig. 6-61). On some machines, inlet-valve and clearance-control unloading are used in combination.

Although such control devices are usually automatically operated, manual operation is satisfactory for some services. When manual operation is provided, it often consists of a valve or valves to open and close clearance pockets. In some cases, a movable cylinder head is provided for variable clearance in the cylinder (see Fig. 6-62).

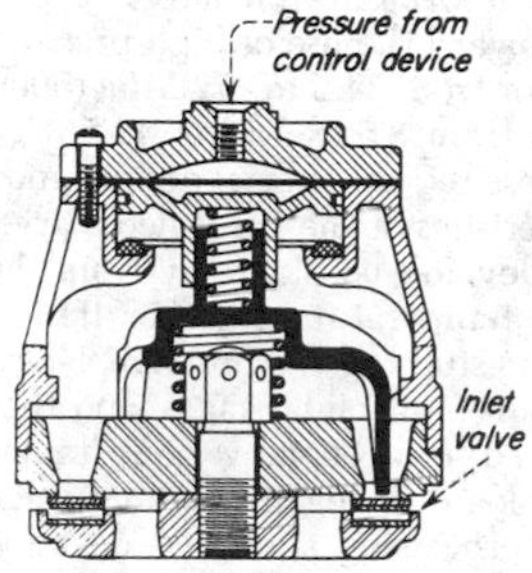

FIG. 6-59 Inlet-valve unloader.

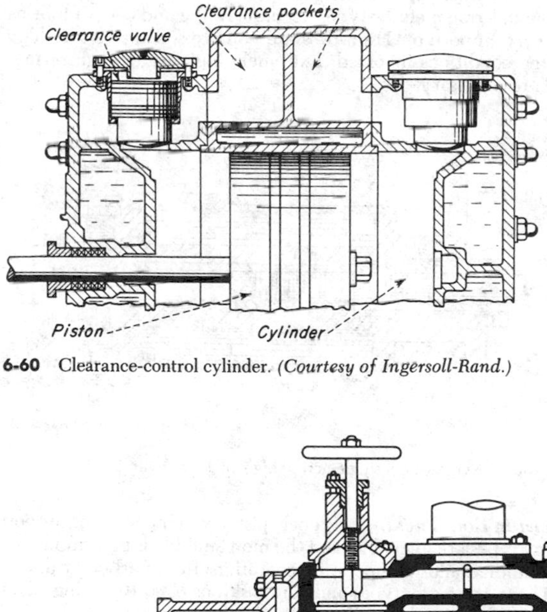

FIG. 6-60 Clearance-control cylinder. *(Courtesy of Ingersoll-Rand.)*

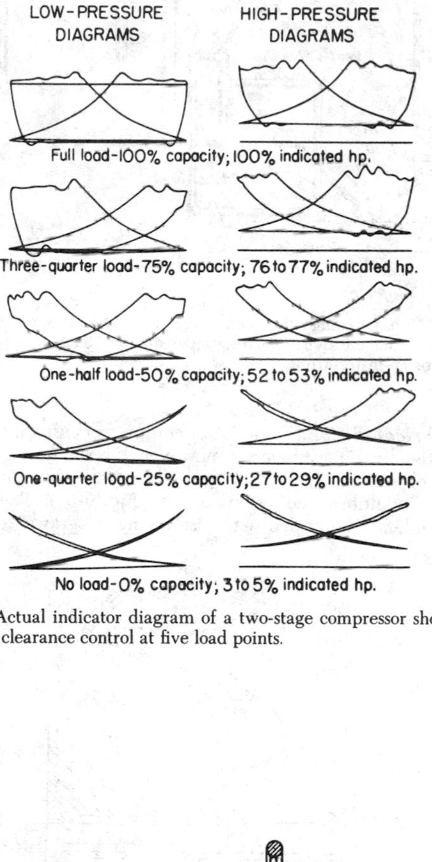

FIG. 6-61 Actual indicator diagram of a two-stage compressor showing the operation of clearance control at five load points.

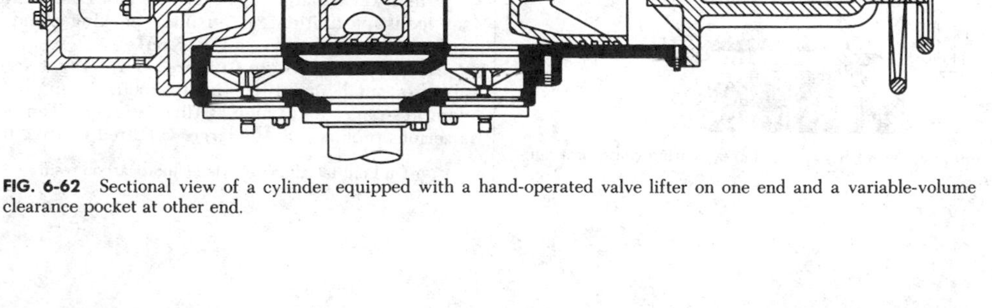

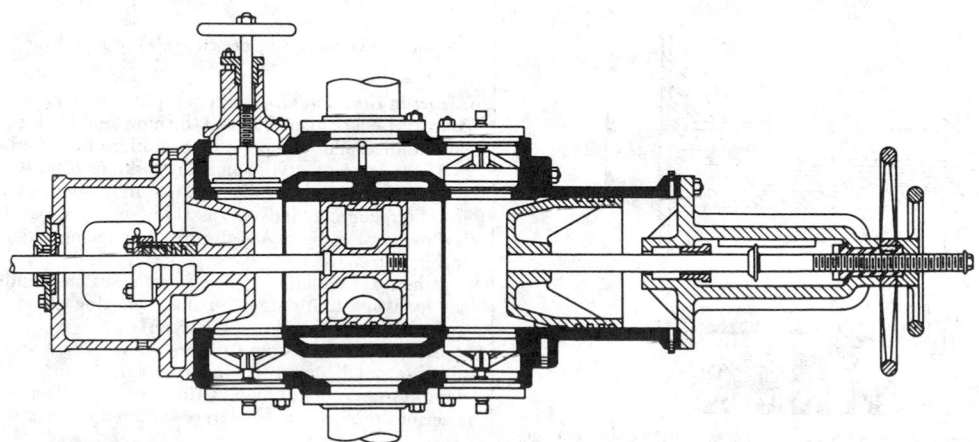

FIG. 6-62 Sectional view of a cylinder equipped with a hand-operated valve lifter on one end and a variable-volume clearance pocket at other end.

When no capacity control or unloading device is provided, it is necessary to provide bypasses between the inlet and discharge in order that the compressor can be started against no load (see Fig. 6-63).

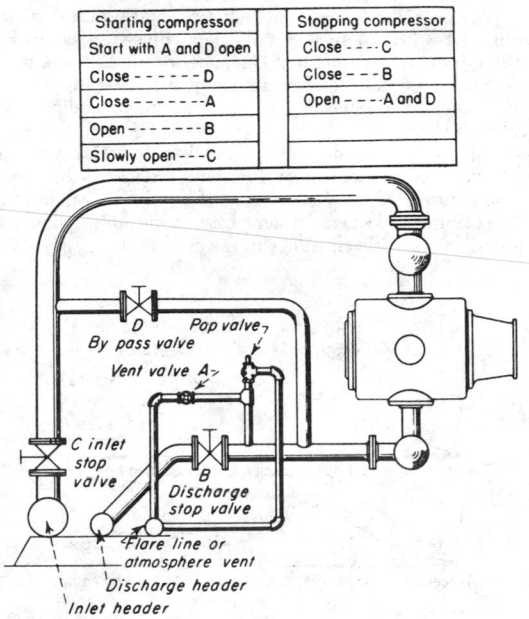

Starting compressor	Stopping compressor
Start with A and D open	Close - - - - C
Close - - - - - - D	Close - - - B
Close - - - - - - - A	Open - - - - A and D
Open - - - - - - - B	
Slowly open - - - C	

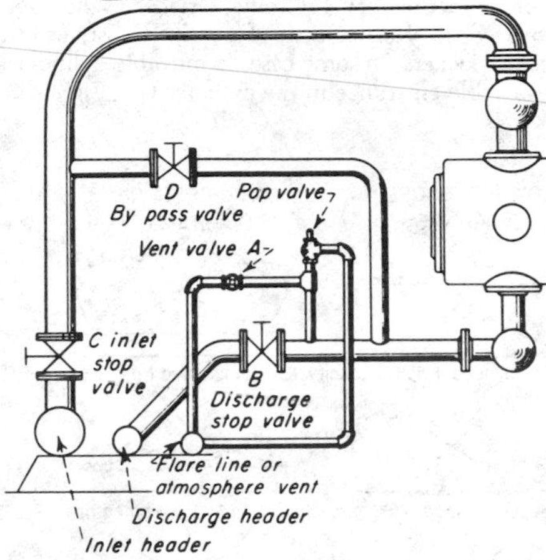

FIG. 6-63 Bypass arrangement for a single-stage compressor. On multistage machines each stage is bypassed in a similar manner. Such an arrangement is necessary for no-load starting.

Nonlubricated Cylinders Most compressors use oil to lubricate the cylinder. In some processes, however, the slightest oil contamination is objectionable. For such cases a number of manufacturers furnish a "nonlubricated" cylinder (see Fig. 6-64). The piston on these cylinders is equipped with piston rings of graphitic carbon or Teflon° as well as pads or rings of the same material to maintain proper clearance between the piston and the cylinder. Plastic packing of a type that requires no lubricant is used on the stuffing box. Although oil-wiper rings are used on the piston rod where it leaves the compressor frame, minute quantities of oil might conceivably enter the cylinder on the rod. If even such small amounts of oil are objectionable, an extended cylinder connecting piece can be furnished. This simply lengthens the piston rod enough so that no portion of the rod can alternately enter the frame and the cylinder.

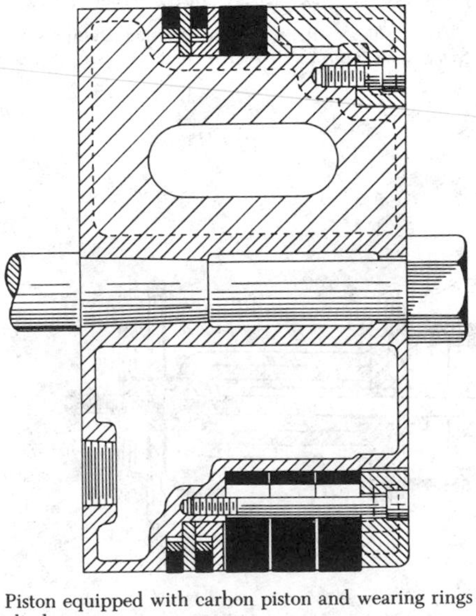

FIG. 6-64 Piston equipped with carbon piston and wearing rings for a nonlubricated cylinder.

In many cases, a small amount of gas leaking through the packing is objectionable. Special connecting pieces are furnished between the cylinder and the frame, which may be either single-compartment or double-compartment. These may be furnished gastight and vented back to the suction or filled with a sealing gas or fluid and held under a slight pressure.

High-Pressure Compressors There is a definite trend in the chemical industry toward the use of high-pressure compressors with discharge pressures of from 34.5 to 172 MPa (5000 to 25,000 lbf/in^2) and with capacities from 8.5×10^3 to 42.5×10^3 m^3/h (5000 to 25,000 ft^3/min). These require special design, and a complete knowledge of the characteristics of the gas is necessary.

The gas usually deviates considerably from the perfect-gas laws, and in many cases temperature or other limitations necessitate a thorough engineering study of the problem. These compressors usually have five, six, seven, or eight stages, and the cylinders must be properly proportioned to meet the various limitations involved and also to balance the load among the various stages. In many cases, scrubbing or other processing is carried on between stages. High-pressure cylinders are steel forgings with single-acting plungers (see Fig. 6-65). The compressors are usually designed so that the pressure load against the plunger is opposed by one or more single-acting pistons of the lower pressure stages. Piston-rod packing is usually of the segmental-ring metallic type. Accurate fitting and correct lubrication are very important. High-pressure compressor valves are designed for the conditions involved. Extremely high-grade engineering and skill are necessary.

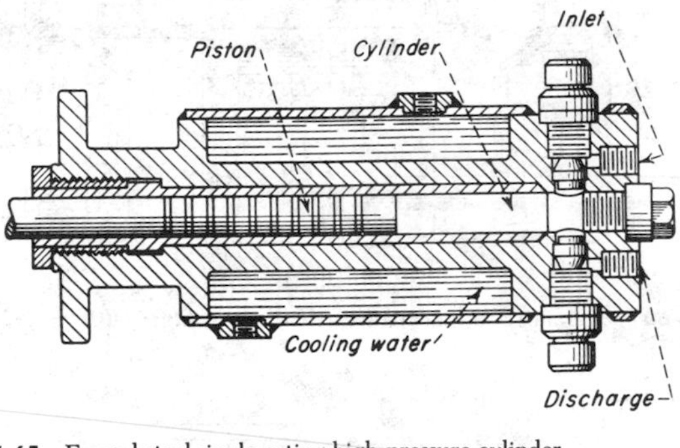

FIG. 6-65 Forged-steel single-acting high-pressure cylinder.

Piston-Rod Packing Proper piston-rod packing is important. Many types are available, and the most suitable is determined by the gas handled and the operating conditions for a particular unit.

There are many types and compositions of soft packing, semimetallic packing, and metallic packing. In many cases, metallic packing is to be recommended. A typical low-pressure packing arrangement is shown in Fig. 6-66. A high-pressure packing arrangement is shown in Fig. 6-67.

When wet, volatile, or hazardous gases are handled or when the service is intermittent, an auxiliary packing gland and soft packing are usually employed (see Fig. 6-68).

Metallic Diaphragm Compressors (Fig. 6-69) These are available for small quantities [up to about 17 m^3/h (10 ft^3/min)] for compression ratios as high as 10:1 per stage. Temperature rise is not a serious problem, as the large wall area relative to the gas volume

°® Du Pont tetrafluoroethylene fluorocarbon resin.

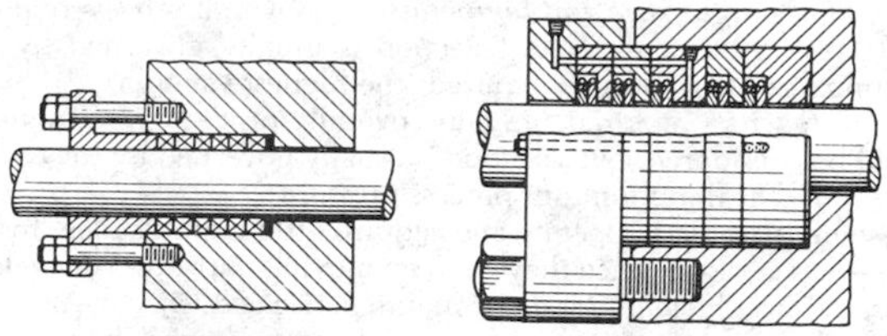

FIG. 6-66 Typical packing arrangements for low-pressure cylinders.

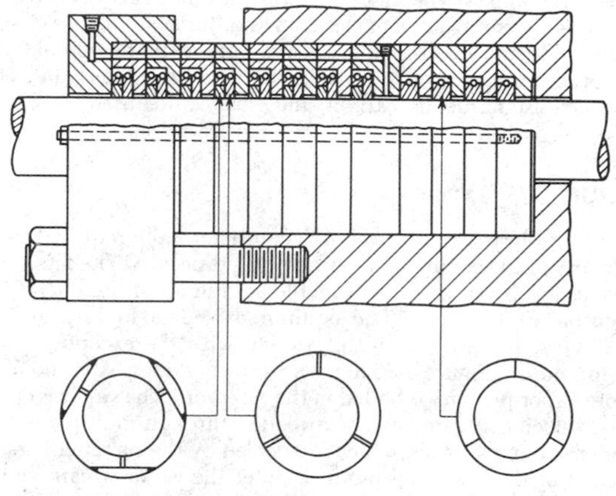

FIG. 6-67 Typical packing arrangement, using metallic packing, for high-pressure cylinders.

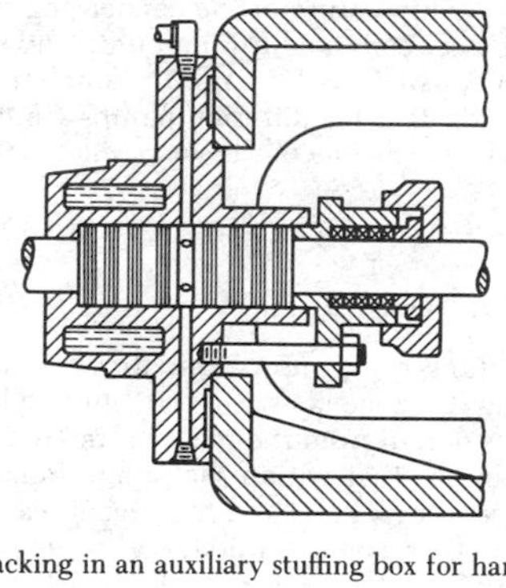

FIG. 6-68 Soft packing in an auxiliary stuffing box for handling gases.

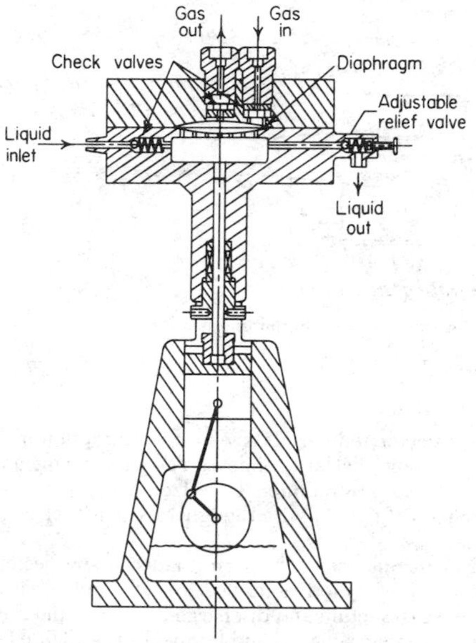

FIG. 6-69 High-pressure, low-capacity compressor having a hydraulically actuated diaphragm. *(Pressure Products Industries.)*

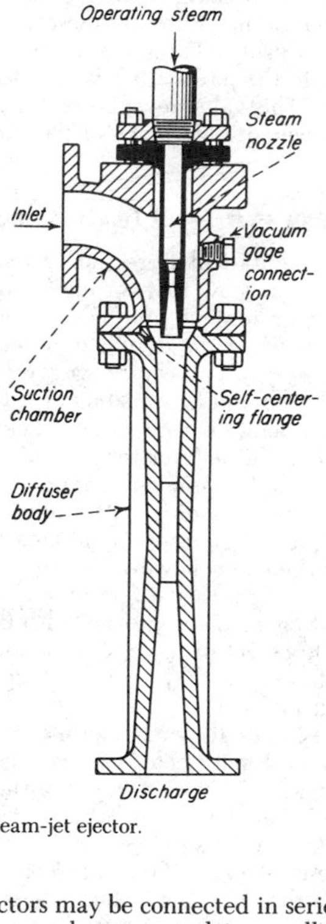

FIG. 6-70 Typical steam-jet ejector.

permits sufficient heat transfer to approach isothermal compression. These compressors possess the advantage of having no seals for the process gas. The diaphragm is actuated hydraulically by a plunger pump.

EJECTORS

An ejector is a simplified type of vacuum pump or compressor which has no pistons, valves, rotors, or other moving parts. Figure 6-70 illustrates a **steam-jet ejector.** It consists essentially of a steam nozzle which discharges a high-velocity jet across a suction chamber that is connected to the equipment to be evacuated. The gas is entrained by the steam and carried into a venturi-shaped diffuser which converts the velocity energy of the steam into pressure energy. Figure 6-71 shows a large-sized ejector, sometimes called a **booster ejector,** with multiple nozzles.

Two or more ejectors may be connected in series or stages. Also, a number of ejectors may be connected in parallel to handle larger quantities of gas or vapor.

Liquid- or air-cooled condensers are usually used between stages. Liquid-cooled condensers may be of either the direct-contact (barometric) or the surface type. By condensing vapor the load on the following stage is reduced, thus minimizing its size and reducing consumption of motive gas. Likewise, a **precondenser** installed

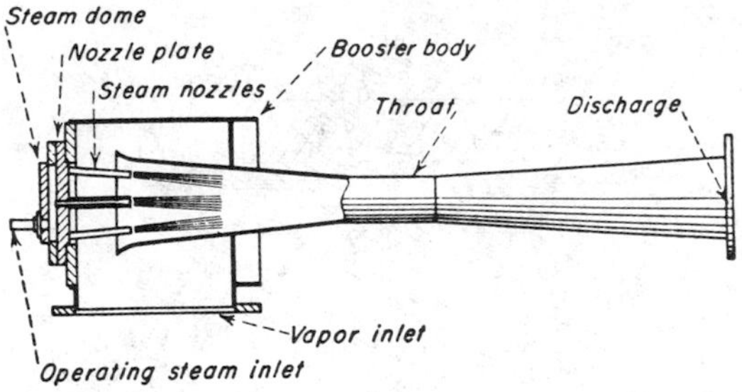

FIG. 6-71 Booster ejector with multiple steam nozzles.

ahead of an ejector reduces its size and consumption if the suction gas contains vapors that are condensable at the temperature condition available. An **aftercondenser** is frequently used to condense vapors from the final stage, although this does not affect ejector performance.

Ejector Performance The performance of any ejector is a function of the area of the motive-gas nozzle and venturi throat, pressure of the motive gas, suction and discharge pressures, and ratios of specific heats, molecular weights, and temperatures. Figure 6-72, based on the assumption of **constant-area mixing,** is useful in evaluating single-stage-ejector performance for compression ratios up to 10 and area ratios up to 100 (see Fig. 6-73 for notation).

For example,° assume that it is desired to evacuate air at 2.94 lbf/in^2 with a steam ejector discharging to 14.7 lbf/in^2 with available steam pressure of 100 lbf/in^2. Entering the chart at $p_{03}/p_{0b} = 5.0$, at $p_{0b}/p_{0a} = 2.94/100 = 0.0294$ the optimum area ratio is 12. Proceeding horizontally to the left, w_b/w_a is approximately 0.15 lb of air per 1 lb of steam. This value must be corrected for the temperature and molecular-weight differences of the two fluids by Eq. (6-36).

$$w/w_a = w_b/w_a\sqrt{T_{0a}M_b/T_{0b}M_a} \tag{6-36}$$

In addition, there are empirical correction factors which should be applied. Laboratory tests show that for ejectors with constant-area mixing the actual entrainment and compression ratios will be approximately 90 percent of the calculated values and even less at very small values of p_{0b}/p_{0a}. This compensates for ignoring wall friction in the mixing section and irreversibilities in the nozzle and diffuser. In theory, each point on a given design curve of Fig. 6-72 is associated with an optimum ejector for prevailing operating conditions. Adjacent points on the same curve represent theoretically different ejectors for the new conditions, the difference being that for each ratio of p_{0b}/p_{0a} there is an optimum area for the exit of the motive-gas nozzle. In practice, however, a segment of a given curve for constant A_2/A_t represents the performance of a single ejector satisfactorily for estimating purposes, provided that the suction pressure lies within 20 to 130 percent of the design suction pressure and the motive pressure within 80 to 120 percent of design motive pressure. Thus the curves can be used to select an optimum ejector for the design point and to estimate its performance at off-design conditions within the limits noted. Final ejector selection should, of course, be made with the assistance of a manufacturer of such equipment.

Uses of Ejectors For the operating range of steam-jet ejectors in vacuum applications, see the subsection "Vacuum Systems."

The choice of the most suitable type of ejector for a given application depends upon the following factors:

1. *Steam pressure.* Ejector selection should be based upon the minimum pressure in the supply line selected to serve the unit.
2. *Water temperature.* Selection is based on the maximum water temperature.
3. *Suction pressure and temperature.* Overall process requirements should be considered. Selection is usually governed by the minimum suction pressure required (the highest vacuum).
4. *Capacity required.* Again overall process requirements should be considered, but selection is usually governed by the capacity required at the minimum process pressure.

Ejectors are easy to operate and require little maintenance. Installation costs are low. Since they have no moving parts, they have long life, sustained efficiency, and low maintenance cost. Ejectors are suitable for handling practically any type of gas or vapor. They are also suitable for handling wet or dry mixtures or gases containing sticky or solid matter such as chaff or dust.

Ejectors are available in many materials of construction to suit process requirements. If the gases or vapors are not corrosive, the diffuser is usually constructed of cast iron and the steam nozzle of stainless steel. For more corrosive gases and vapors, many combinations of materials such as bronze, various stainless-steel alloys, and other corrosion-resistant metals, carbon, and glass can be used.

VACUUM SYSTEMS

Figure 6-74 illustrates the level of vacuum normally required to perform many of the common manufacturing processes. The attainment of various levels is related to available equipment in Fig. 6-75.

Vacuum Equipment The equipment shown in Fig. 6-75 has been discussed elsewhere in this section with the exception of the **diffusion pump.** Figure 6-76 depicts a typical design. A liquid of low absolute vapor pressure is boiled in the reservoir. The vapor is ejected at high velocity in a downward direction through multiple jets and is condensed on the walls, whicn are cooled by the surrounding coils. Molecules of the gas being pumped enter the vapor stream and are driven downward by collisions with the vapor molecules. The gas molecules are removed through the discharge line by a backing pump such as a rotary oil-sealed unit.

Diffusion pumps operate at very low pressures. The ultimate vacuum attainable depends somewhat upon the vapor pressure of the pump liquid at the temperature of the condensing surfaces. By providing a cold trap between the diffusion pump and the region being evacuated, pressures as low as 10^{-7} mmHg absolute are achieved in this manner. Liquids used for diffusion pumps are mercury and oils of low vapor pressure. Silicone oils have excellent characteristics for this service.

SEALING OF ROTATING SHAFTS

In rotary pumps, fans, compressors, agitators, etc., the shaft projects through the casing in a zone known as the "stuffing box" or "packing box." These terms derive from the fact that in order to separate two environments, namely, that within the pump, fan, compressor, agitator, etc., and the atmosphere, and to prevent leakage of one environment into the other, some material had to be stuffed or packed around the shaft where it passed through its casing. For many years soft packing was the material that was most often employed for this service.

In operation, rotating shafts in this equipment may be displaced both radially and axially. Small inaccuracies resulting from machining, manufacturing and assembly produce radial displacement, and differential thermal expansion causes axial displacement. Therefore the sealing device must be flexible. Another characteristic of the seal is compactness, which is dictated by design limitations of the equipment.

Packing A common type of rotating shaft seal consists of packing composed of fibers which are first woven, twisted, or braided into strands and then formed into coils, spirals, or rings. To ensure initial lubrication and to facilitate installation, the basic materials are often impregnated. Common materials are asbestos fabric, braided and twisted asbestos, rubber and duck, flax, jute, and metallic braids. The so-called plastic packings can be made up with varying amounts of fiber combined with a binder and lubricant for high-speed applica-

°All data are given in U.S. customary units since the charts are in these units. Conversion factors to SI units are given on the charts.

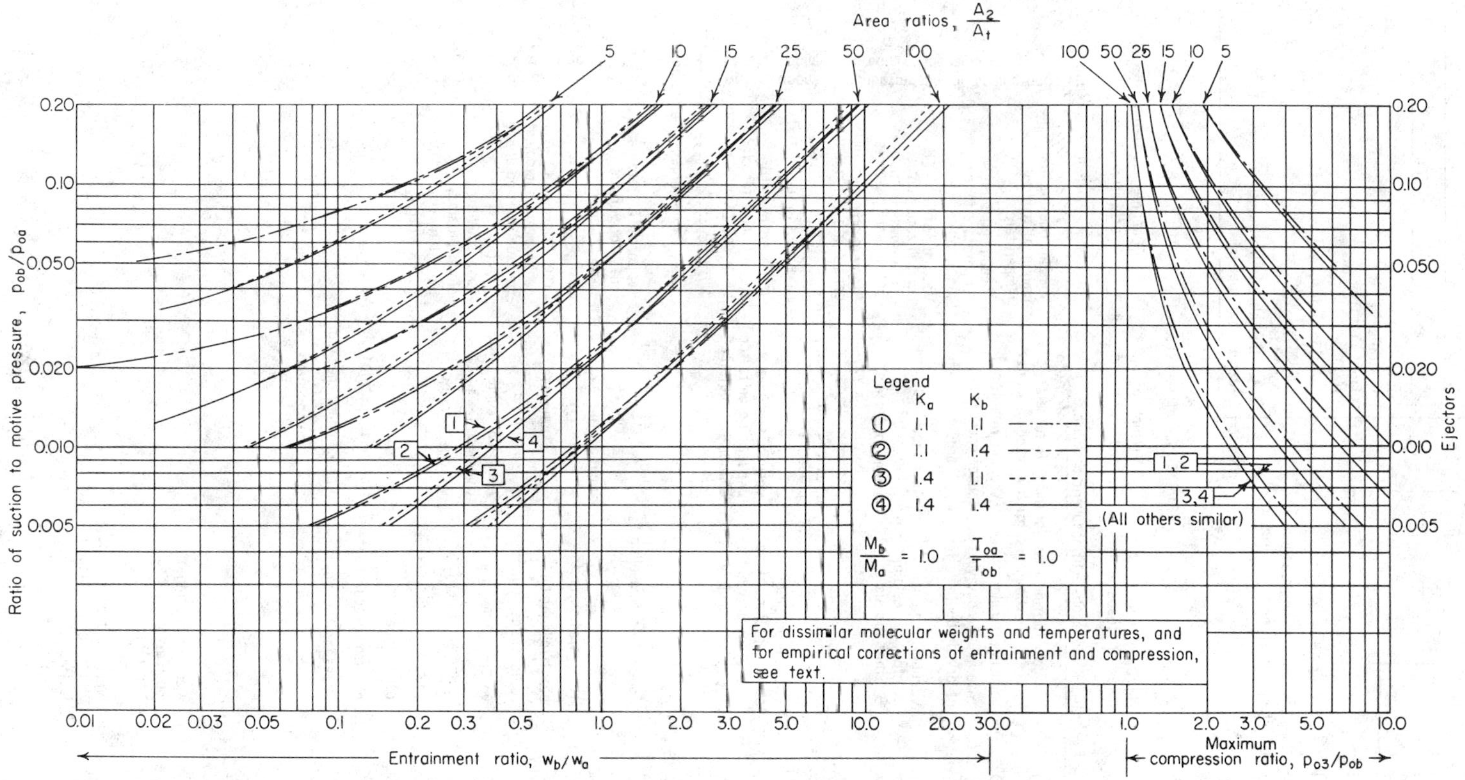

FIG. 6-72 Design curves for optimum single-stage ejectors. [*DeFrate and Hoerl,* Chem. Eng. Prog., **55**, *Symp. Ser. 21, 46 (1959).*]

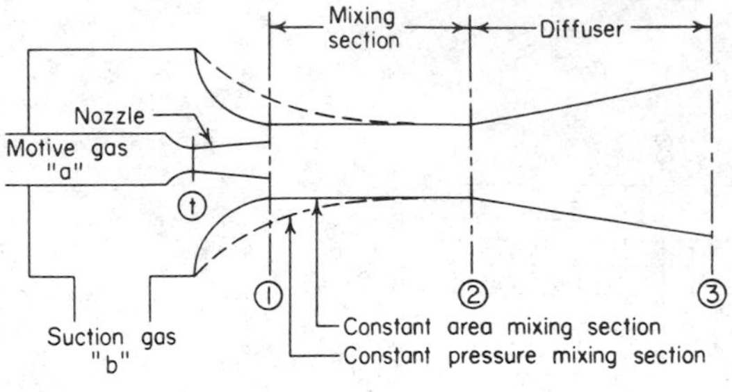

FIG. 6-73 Notation for Fig. 6-72.

tions. Maximum temperatures that base materials of packings withstand and still give good service are as follows:

	°C	°F
Flax	38	100
Cotton	93	200
Duck and rubber	149	300
Rubber	177	350
Metallic (lead-based)	218	425
Asbestos 1003	260	500
Asbestos 204	371	700
Metallic (aluminum-based)	552	1025
Metallic (copper-based)	829	1525

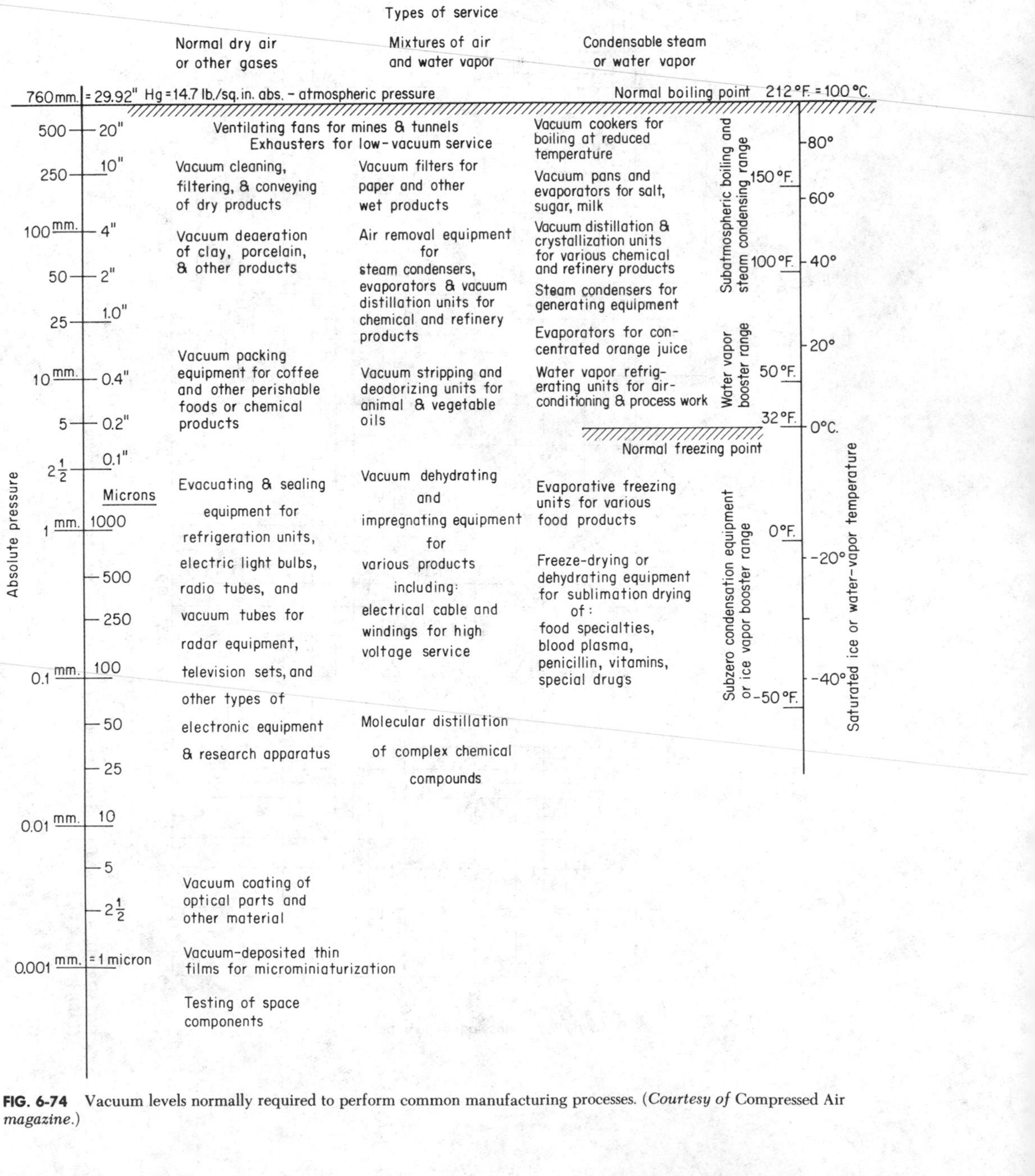

FIG. 6-74 Vacuum levels normally required to perform common manufacturing processes. (*Courtesy of* Compressed Air *magazine.*)

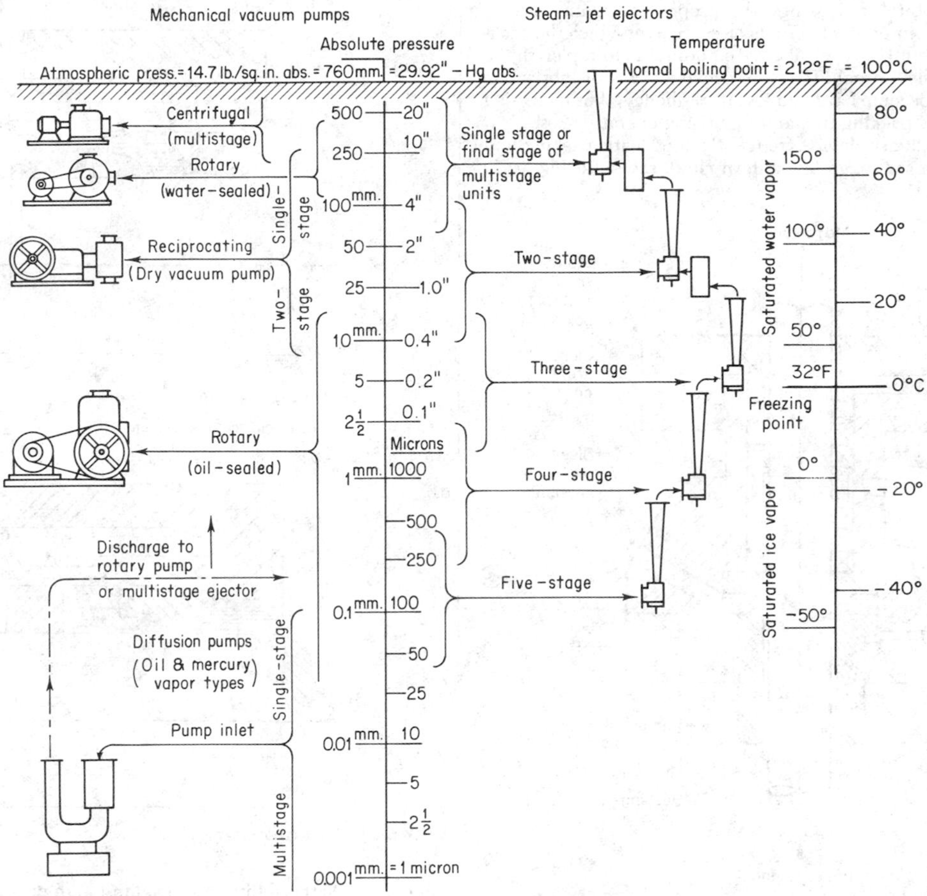

FIG. 6-75 Vacuum levels attainable with various types of equipment. (*Courtesy of* Compressed Air *magazine.*)

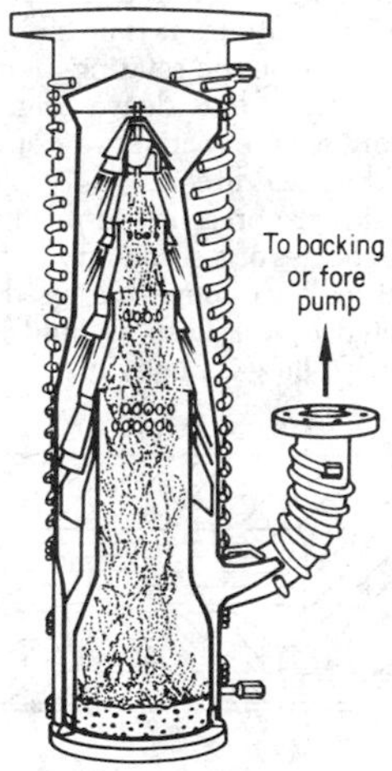

FIG. 6-76 Typical diffusion pump. (*Courtesy of* Compressed Air *magazine.*)

Packing may not provide a completely leak-free seal. With shaft surface speeds less than approximately 2.5 m/s (500 ft/min), the packing may be adjusted to seal completely. However, for higher speeds some leakage is required for lubrication, friction reduction, and cooling.

Application of Packing Coils and spirals are cut to form closed or nearly closed rings in the stuffing box. Clearance between ends should be sufficient to allow for fitting and possible expansion due to increased temperature or liquid absorption of the packing while in operation.

The correct form of the ring joint depends on materials and service requirements. Braided and flexible metallic packings usually have butt or square joints (Fig. 6-77*a*). With other packing material, service experience indicates that rings cut with bevel or skive joints (Fig. 6-77*b*) are more satisfactory. A slight advantage of the bevel joint over the butt joint is that the bevel permits a certain amount of sliding action, thus absorbing a portion of ring expansion.

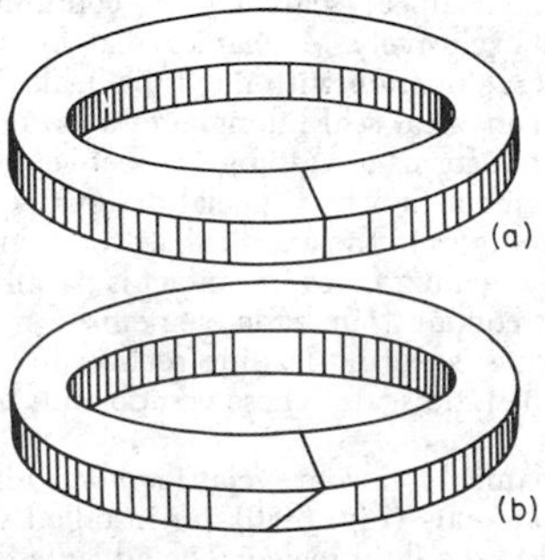

FIG. 6-77 Butt (*a*) and skive (*b*) joints for compression packing rings.

In the manufacture of packings, the proper grade and type of **lubricant** is usually impregnated for each service for which the packing is recommended. However, it may be desirable to replenish the lubricant during the normal life of the packing. Lack of lubrication causes packing to become hard and lose its resiliency, thus increasing friction, shortening packing life, and increasing operating costs.

An effective auxiliary device frequently used with packing and rotary shafts is the **seal cage** (or **lantern ring**), shown in Fig. 6-78.

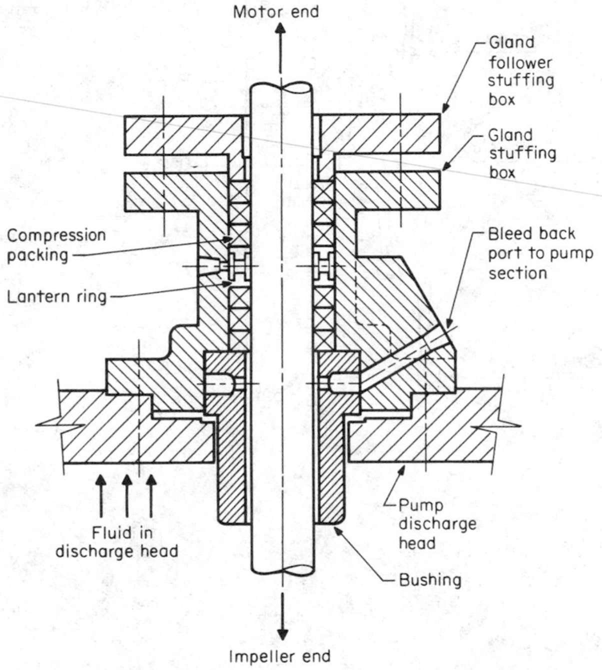

FIG. 6-78 Seal cage or lantern ring. *(Courtesy of Crane Packing Co.)*

The seal cage provides an annulus around the shaft for the introduction of a lubricant, oil, grease, etc. The seal cage is also used to introduce liquid for cooling, to prevent the entrance of atmospheric air, or to prevent the infiltration of abrasives from the process liquid.

The chief advantage of packing over other types of seals is the ease with which it can be adjusted or replaced. Most equipment is designed so that disassembly of major components is not required to remove or add packing rings. The major disadvantages of a packing-type seal are (1) short life, (2) requirement for frequent adjustment, and (3) need for some leakage to provide lubrication and cooling.

Mechanical Seals These are the most commonly used device for sealing against liquids when using rotating shafts. From the automobile-engine water pump to large high-horsepower boiler-feed-water pumps, mechanical seals are in continuous use and are employed almost exclusively. Mechanical seals are reliable, have long life, and, in general, operate with no visible leakage.

The term "mechanical seal" designates a prefabricated or packaged assembly that forms a running seal between flat precision-finished surfaces. Except in a few special designs, sealing surfaces are oriented at right angles to the axis of shaft rotation. The direction of forces holding the sealing faces in contact is parallel to the shaft. All mechanical seals contain four basic elements: a rotating seal ring, a stationary seal ring, a spring-loading section for maintaining seal-face contact, and static seals. These components are pointed out in Fig. 6-79.

Types Mechanical seals are classified broadly as internal or external. **Internal seals** (Fig. 6-80) are installed with all seal components exposed to the fluid sealed. The advantages of this arrangement are (1) the ability to seal against high pressure, since the hydrostatic force is normally in the same direction as the spring force; (2) protection of seal parts from external mechanical damage; and (3) reduction in the shaft length required.

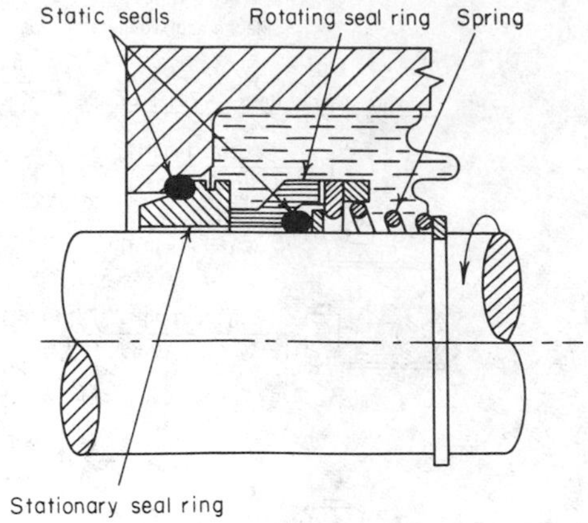

FIG. 6-79 Mechanical-seal components.

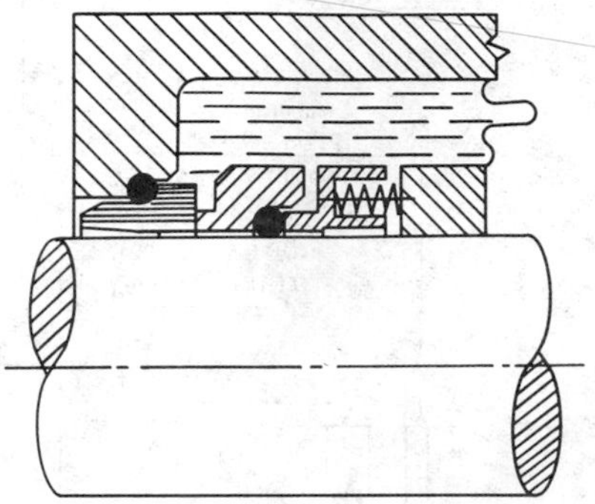

FIG. 6-80 Internal mechanical seal.

For high-pressure installations, it is possible to balance partially or fully the hydrostatic force on the rotating member of an internal seal by using a stepped shaft or shaft sleeve (Fig. 6-81). This method of relieving face pressure is an effective way of decreasing power consumption and extending seal life.

When abrasive solids are present and it is not permissible to introduce appreciable quantities of a secondary flushing fluid into the process, double internal seals are sometimes used (Fig. 6-82). Both sealing faces are protected by the flushing fluid injected between them even though the inward flow is negligible.

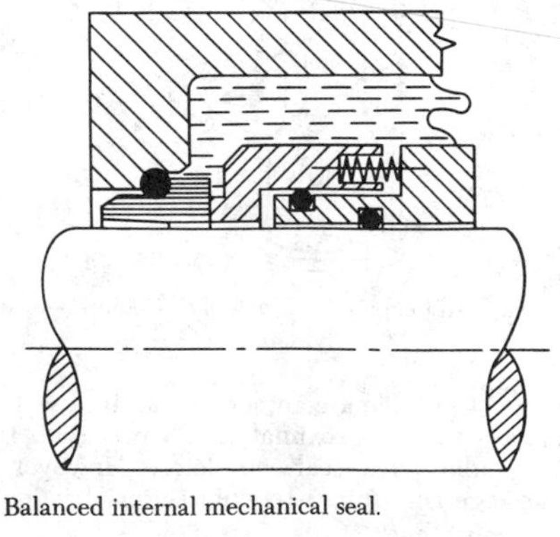

FIG. 6-81 Balanced internal mechanical seal.

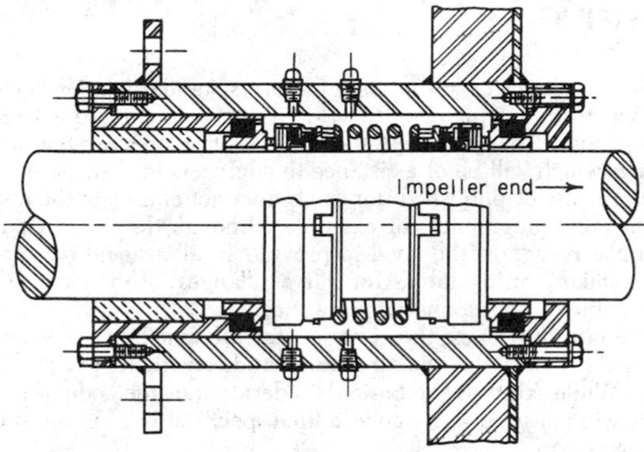

FIG. 6-82 Internal bellows-type double mechanical seal.

External seals (Fig. 6-83) are installed with all seal components protected from the process fluid. The advantages of this arrangement are that (1) fewer critical materials of construction are required, (2) installation and setting are somewhat simpler because of the exposed position of the parts, and (3) stuffing-box size is not a limiting factor. Hydraulic balancing is accomplished by proper proportioning of the seal face and secondary seal diameters.

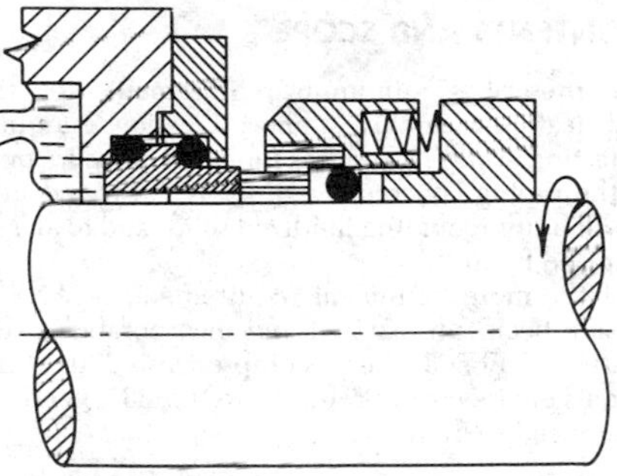

FIG. 6-83 External mechanical seal.

Throttle bushings (Fig. 6-84) are commonly used with single internal or external seals when solids are present in the fluid and the inflow of a flushing fluid is not objectionable. These close-clearance bushings are intended to serve as flow restrictions through which the maintenance of a small inward flow of flushing fluid prevents the entrance of a process fluid into the stuffing box.

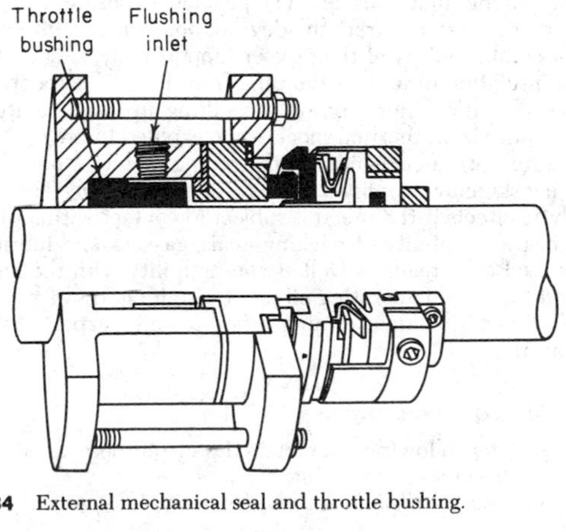

FIG. 6-84 External mechanical seal and throttle bushing.

Application Advantages of mechanical seals, as compared with conventional packed stuffing boxes, are reduced friction power loss, elimination of wear on shaft or shaft sleeve, negligible leakage over a long service life, and freedom from periodic maintenance. Mechanical seals are precision components and demand careful handling and installation.

Materials Springs and other metallic components are available in a wide variety of alloys and are usually selected on the basis of temperature and corrosion conditions. The use of a particular mechanical seal is frequently restricted by the temperature limitations of the organic materials used in the static seals. Most elastomers are limited to about 121°C (250°F). Teflon will withstand temperatures of 260°C (500°F) but softens appreciably above 204°C (400°F). Glass-filled Teflon is dimensionally stable up to 232 to 260°C (450 to 500°F).

One of the most common elements used for seal faces is carbon. Although compatible with most process media, carbon is affected by strong oxidizing agents, including fuming nitric acid, hydrogen chloride, and high-temperature air [above 316°C (600°F)]. Normal mating-face materials for carbon are tungsten or chromium carbide, hard steel, stainless steel, or one of the cast irons.

Other sealing-face combinations that have been satisfactory in corrosive service are carbide against carbide, ceramic against ceramic, ceramic against carbon, and carbon against glass. The ceramics have also been mated with the various hard-facing alloys. When selecting seal materials the possibility of galvanic corrosion must also be considered.

Gas Seals The design of face-type seals is much more critical for gas service than for liquids because of the comparatively low gas specific heat. A common gas seal is a labyrinth consisting of a number of circumferential knives or touch points arranged in series to provide successive expansion of the fluid (Fig. 6-85). As the differential pressure across any individual restriction is small, total leakage is minimized. When no external leakage of the process gas is permissible, a purge of some nontoxic gas is usually provided at an intermediate bleed point.

In some installations in which the presence of liquids in the gas stream is not undesirable, a liquid-buffered seal is used. This consists of a close-fitting bushing into which oil, water, etc. is injected. In refrigeration centrifugal compressors the bearings and the oil-buffered seals are located in a single enclosed housing.

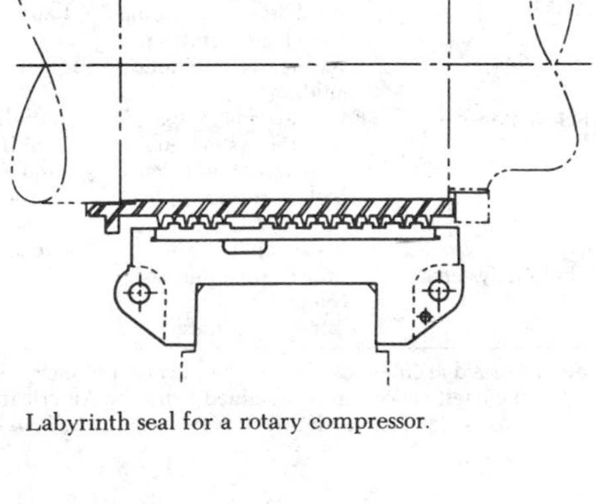

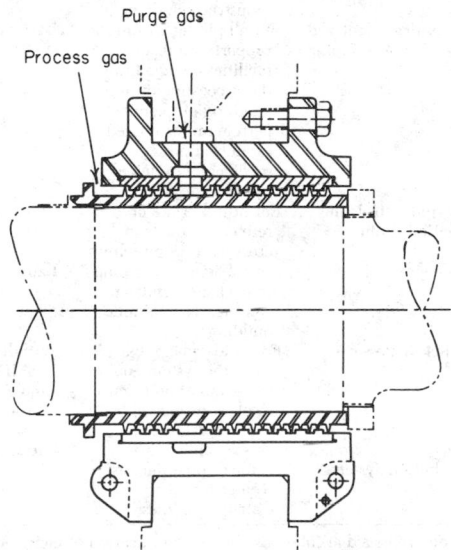

FIG. 6-85 Labyrinth seal for a rotary compressor.

PROCESS-PLANT PIPING

CODES AND STANDARDS

Units: Pipe and Tubing Sizes and Ratings In this subsection pipe and tubing sizes are generally quoted in units of inches. To convert inches to millimeters, multiply by 25.4. Ratings are given in pounds. To convert pounds to kilograms, multiply by 0.454.

Pressure-Piping Code The code for pressure piping (ANSI B31) consists of a number of sections which collectively constitute the code. Table 6-2 shows the status of the B31 code as of December 1980. The sections are published as separate documents for simplicity and convenience. The sections differ extensively.

The Chemical Plant and Petroleum Refinery Piping Code (ANSI B31.3) is a section of ANSI B31. It was derived from a merging of the code groups for chemical-plant (B31.6) and petroleum-refinery (B31.3) piping into a single committee. Some of the significant requirements of ANSI B31.3, Petroleum Refinery Piping (1980 edition), are summarized in the following presentation, which is aimed primarily at welded and seamless construction.

Where the word "code" is used in this subsection of the *Handbook* without other identification, it refers to the B31.3 section of ANSI B31. The code has been extensively quoted in this subsection of the *Handbook* with the permission of the publisher. The code is published by and copies are available from the American Society of Mechanical Engineers (ASME), 345 East 47th Street, New York, New York 10017. References to the ASME code are to the ASME Boiler and Pressure Vessel Code, also published by the American Society of Mechanical Engineers.

National Standards The American National Standards Institute (ANSI) and the American Petroleum Institute (API) have established dimensional standards for the most widely used piping components. Lists of these standards as well as specifications for pipe and fitting materials and testing methods of the American Society for Testing and Materials (ASTM), American Welding Society (AWS) specifications, and standards of the Manufacturers Standardization Society of the Valve and Fittings Industry (MSS) can be found in the ANSI B31 code sections. Many of these standards contain pressure-temperature ratings which will be of assistance to engineers in their design function. The use of published standards does not eliminate the need for engineering judgment. For example, although the code calculation formulas recognize the need to provide an allowance for corrosion, the standard rating tables for valves, flanges, fittings, etc., do not incorporate a corresponding allowance.

TABLE 6-2 Status of ANSI B31 Code for Pressure Piping

Standard number and designation	Scope and application	Remarks*
B31.1.0 Power Piping	For all piping in steam-generating stations	Latest issue: 1980
B31.2 Fuel Gas Piping	For fuel gas for steam-generating stations and industrial buildings	Latest issue: 1968
B31.3 Chemical Plant and Petroleum Refinery Piping	For all piping within the property limits of facilities engaged in the processing or handling of chemical, petroleum, or related products unless specifically excluded by the code	Latest issue: 1980
B31.4 Liquid Petroleum Transporation Piping Systems	For liquid crude or refined products in cross-country pipe lines	Latest issue: 1979
B31.5 Refrigeration Piping	For refrigeration piping in packaged units and commercial or public buildings	Latest issue: 1974
B31.7 Nuclear Power Piping	For fluids whose loss from the system could cause radiation hazard to plant personnel or the general public	Withdrawn; see ASME Boiler and Pressure Vessel Code, Sec. 3
B31.8 Gas Transmission and Distribution Systems	For gases in cross-country pipe lines as well as for city distribution lines	Latest issue: 1975

*Addenda are issued at intervals between publication of complete editions. Information on the latest issues can be obtained from the American Society of Mechanical Engineers, 345 East 47th Street, New York, N.Y. 10017.

The introduction to the code sets forth engineering requirements deemed necessary for the safe design and construction of piping systems. While safety is the basic consideration of the code, this factor alone will not necessarily govern final specifications for any pressure piping system.

Designers are cautioned that the code is not a design handbook and does not do away with the need for competent engineering judgment.

Governmental Regulations: OSHA Sections of the ANSI B31 code have been adopted with certain reservations or revisions by some state and local authorities as local codes.

The specific requirements for piping systems in certain services have been promulgated as Occupational Safety and Health Act (OSHA) regulations. These rules and regulations will presumably be revised and supplemented from time to time and may include specific requirements not contemplated in Sec. B31.3.

CODE CONTENTS AND SCOPE

The code prescribes minimum requirements for the materials, design, fabrication, assembly, support, erection, examination, inspection, and testing of piping systems subject to pressure or vacuum. The scope of the piping covered by B31.3 is illustrated in Fig. 6-86. It applies to all fluids including fluidized solids and to all services except as noted in the figure.

Some of the more significant requirements of ANSI B31.3 (1980 edition) have been summarized and incorporated in this section of the *Handbook*. For a more comprehensive treatment of code requirements engineers are referred to the B31.3 code and the standards referenced therein.

PIPE-SYSTEM MATERIALS

The selection of material to resist deterioration in service is outside the scope of the B31.3 code (see Sec. 23). Experience has, however, resulted in the following material considerations extracted from the code with the permission of the publisher, the American Society of Mechanical Engineers, New York.

General Considerations Considerations to be evaluated when selecting piping materials are (1) possible exposure to fire with respect to the loss of strength, degradation temperature, melting point, or combustibility of the pipe or support material; (2) ability of thermal insulation to protect the pipe from fire; (3) susceptibility of the pipe to brittle failure, possibly resulting in fragmentation hazards, or failure from thermal shock when exposed to fire or fire-fighting measures; (4) susceptibility of the piping material to crevice corrosion in stagnant confined areas (screwed joints) or adverse electrolytic effects if the metal is subject to contact with a dissimilar metal; (5) the suitability of packing, seals, gaskets, and lubricants or sealants used on threads as well as compatibility with the fluid handled; and (6) the refrigerating effect of a sudden loss of pressure on volatile fluids in determining the lowest expected service temperature.

Specific Material Precautions

Metals The following characteristics are to be evaluated when applying certain metals in piping:

1. *Irons: cast, malleable, and high silicon (14.5 percent).* Their

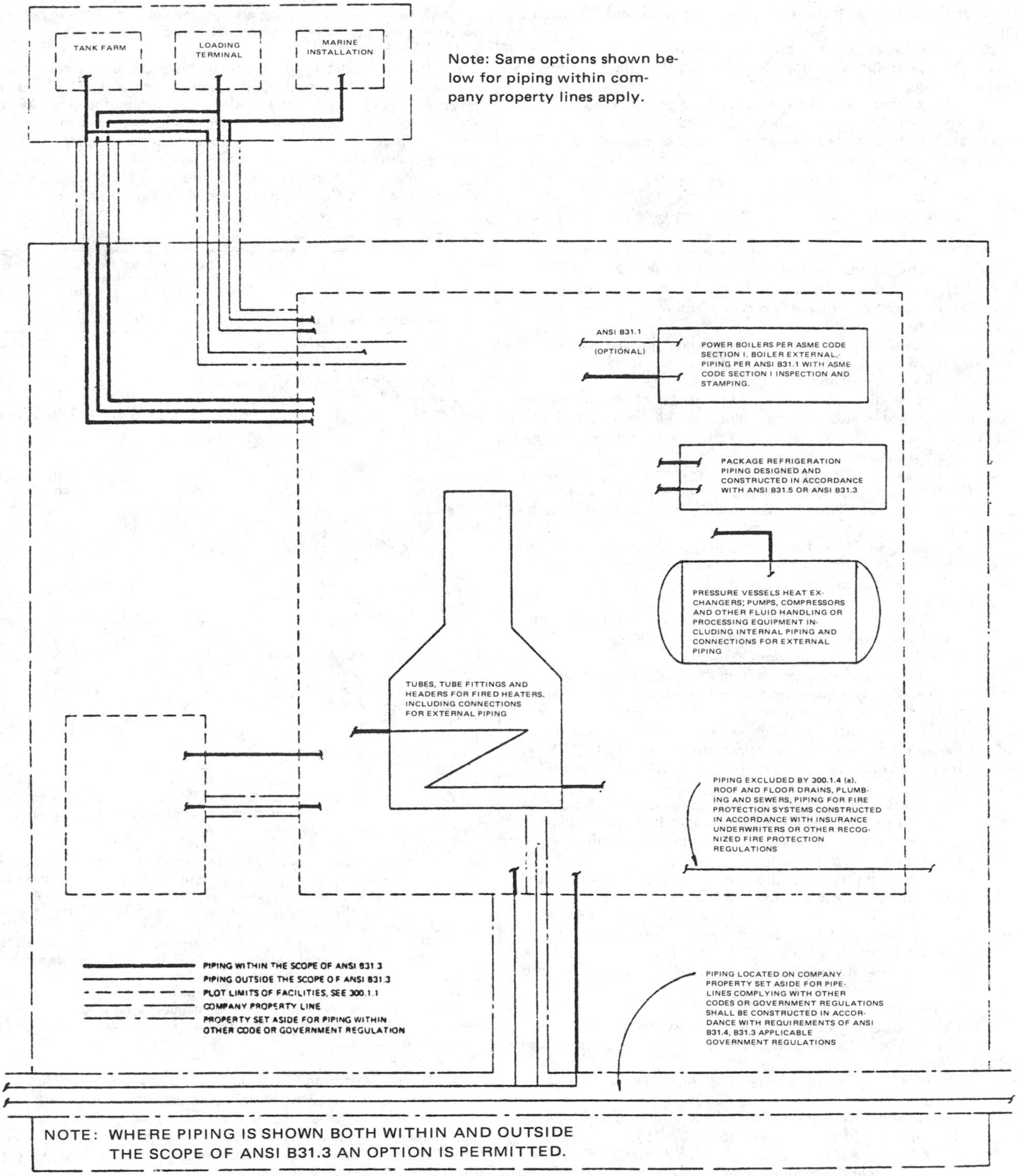

FIG. 6-86 Scope of piping covered by the Chemical Plant and Petroleum Refinery Piping Code, ANSI B31.3. (*From ASME, Chemical Plant and Petroleum Refinery Piping Code, ANSI B31.3—1980; reproduced with permission of the publisher, the American Society of Mechanical Engineers, New York.*)

lack of ductility and their sensitivity to thermal and mechanical shock.

2. *Carbon steel and low- and intermediate-alloy steels*

a. The possibility of embrittlement when handling alkaline or strong caustic fluids.

b. The possible conversion of carbides to graphite during long-time exposure to temperature above 427°C (800°F) of carbon steels, plain nickel steel, carbon-manganese steel, manganese-vanadium steel, and carbon-silicon steel.

c. The possible conversion of carbides to graphite during long-time exposure to temperatures above 468°C (875°F) of carbon-molybdenum steel, manganese-molybdenum-vanadium steel, and chromium-vanadium steel.

d. The advantages of silicon-killed carbon steel (0.1 percent silicon minimum) for temperatures above 480°C (900°F).

e. The possibility of hydrogen damage when piping material is exposed to hydrogen or to aqueous acid solutions under certain temperature-pressure conditions.

f. The possibility of deterioration when piping material is exposed to hydrogen sulfide.

3. *High-alloy (stainless) steels*

a. The possibility of stress-corrosion cracking of austenitic stainless steels exposed to media such as chlorides and other halides either internally or externally. The latter can result from improper selection or application of thermal insulation.

b. The susceptibility to intergranular corrosion of austenitic stainless steels after sufficient exposure to temperatures between 427 and 871°C (800 and 1600°F) unless stabilized or low-carbon grades are used.

c. The susceptibility to intercrystalline attack of austenitic stainless steels on contact with zinc or lead above their melting points or with many lead and zinc compounds at similarly elevated temperatures.

d. The brittleness of ferritic stainless steels at room temperature after service at temperatures above 370°C (700°F).

4. *Nickel and nickel-base alloys*

a. The susceptibility to grain boundary attack of nickel and nickel-base alloys not containing chromium when exposed to small quantities of sulfur at temperatures above 315°C (600°F).

b. The susceptibility to grain boundary attack of nickel-base alloys containing chromium at temperatures above 595°C (1100°F) under reducing conditions and above 760°C (1400°F) under oxidizing conditions.

c. The possibility of stress-corrosion cracking of nickel-copper alloy (70 Ni–30 Cu) in hydrofluoric acid vapor if the alloy is highly stressed or contains residual stresses from forming or welding.

5. *Aluminum and aluminum alloys*

a. The compatibility with aluminum of thread compounds used in aluminum threaded joints to prevent seizing and galling.

b. The possibility of corrosion from concrete, mortar, lime, plaster, or other alkaline materials used in buildings or other structures.

c. The susceptibility of alloys 5154, 5087, 5083, and 5456 to exfoliation or intergranular attack; and the upper temperature limit of 65°C (150°F) to avoid such deterioration.

6. *Copper and copper alloys*

a. The possibility of dezincification of brass alloys.

b. The susceptibility to stress-corrosion cracking of copper-based alloys.

c. The possibility of unstable acetylide formation when exposed to acetylene.

7. *Titanium and titanium alloys.* The possibility of deterioration of titanium and its alloys above 315°C (600°F).

8. *Zirconium and zirconium alloys.* The possibility of deterioration of zirconium and zirconium alloys above 315°C (600°F).

9. *Tantalum.* Above 300°C (570°F), the possibility of reactivity of tantalum with all gases except the inert gases. Below 300°C (570°F), the possibility of embrittlement of tantalum by nascent (monatomic) hydrogen (but not molecular hydrogen). Nascent hydrogen is produced by galvanic action or as a product of corrosion by certain chemicals.

Nonmetals The following are specific considerations to be evaluated when applying certain nonmetals in piping:

1. *Thermoplastics*

a. If thermoplastic piping is used above ground for compressed air or other compressed gases, special precautions should be observed. In determining the needed safeguarding for such services, the energetics and the specific failure mechanism need to be evaluated. Encasement of the plastic piping in shatter-resistant material may be considered.

b. Table 6-3 lists recommended minimum and maximum temperature limits for thermoplastic pipe materials.

TABLE 6-3 Temperature Limits for Thermoplastic Pipe*

	Recommended temperature limits			
	Minimum		Maximum	
Material (generic type)	°F	°C	°F	°C
Acrylonitrile-butadiene-styrene (ABS)	−30	−34	180	82
Cellulose acetate butyrate (CAB)	0	−18	140	60
Chlorinated polyether	0	−18	210	99
Polyacetal	0	−18	170	77
Polyethylene				
PE 1404	−30	−34	100	38
PE 2305	−30	−34	120	49
PE 2306	−30	−34	140	60
PE 3306	−30	−34	160	71
PE 3406	−30	−34	180	82
Polypropylene	30	−01	210	99
Poly (vinyl chloride)				
PVC 1120	0	−18	150	66
PVC 1220	0	−18	150	66
PVC 2110	0	−18	130	54
PVC 2112	0	−18	130	54
PVC 2116	0	−18	150	66
PVC 2120	0	−18	150	66
Chlorinated poly (vinyl chloride) (CPVC, 4120)	0	−18	210	99
Poly (vinylidene chloride)	40	4	160	71
Poly (vinylidene fluoride)	0	−18	275	135
Nylon	−30	−34	180	82
Polybutylene	0	−18	210	99
Poly (phenylene oxide) (POP 2125)	30	−01	210	99

*Extracted from the Chemical Plant and Petroleum Refinery Piping Code, ANSI B31.3—1980, with permission of the publisher, the American Society of Mechanical Engineers, New York.

These recommendations are for low-pressure applications with water and other fluids that do not significantly affect the properties of the particular thermoplastic. The upper temperature limits are reduced at higher pressures, depending on the combination of fluid and expected service life. Lower temperature limits are affected more by installation, environment, and safeguarding than by strength.

Because of low thermal conductivity, temperature gradients through the pipe wall may be substantial. Tabulated limits apply where more than half the wall thickness is at or above the stated temperature.

These recommendations apply only to products covered by ASTM standards listed in Appendix A, Table 3, of the code. Manufacturers should be consulted for temperature limits on the specific types and kinds of plastic not covered by those ASTM standards.

c. Table 6-4 lists minimum and maximum temperature limits for thermoplastic materials used as nonpressure retaining linings.

2. *Reinforced thermosetting resins.* Table 6-5 lists the normally accepted maximum temperature limits for reinforced-thermosetting-resin materials. The minimum recommended temperature is −29°C (−20°F) in all cases.

3. *Asbestos cement.* The normally accepted temperature limits for asbestos cement piping are −18°C (0°F) minimum and 93°C (200°F) maximum.

4. *Borosilicate glass and impregnated graphite.* Their lack of ductility and sensitivity to thermal and mechanical shock should be taken into account.

TABLE 6-4 Temperature Limits for Thermoplastics Used as Linings*

Material (generic type)	Minimum temperature °F	Minimum temperature °C	Maximum temperature °F	Maximum temperature °C
Poly (tetrafluorethylene)	−325	−198	500	260
Poly (fluorinated ethylene propylene)	−325	−198	400	204
Poly (vinylidene chloride)	0	−18	175	79
Poly (vinylidene fluoride)	0	−18	275	135
Polypropylene	0	−18	225	107
Poly (perfluoroalkoxy)	−325	−198	500	260

*Extracted from the Chemical Plant and Petroleum Refinery Piping Code, ANSI B31.3—1980, with permission of the publisher, the American Society of Mechanical Engineers.

Listed temperature limits apply to lining material only. Rules for establishing temperature limits for components being listed are covered elsewhere in this code.

These temperature limits are based on material tests and do not necessarily reflect evidence of successful use as piping-component linings at these temperatures. The designer should contact the manufacturer for specific applications, particularly as temperature limits are approached.

TABLE 6-5 Temperature Limits for Reinforced Thermosetting Resins*†

Material (generic type)	Maximum temperature °C	Maximum temperature °F
Epoxy, glass-fiber-reinforced	149	300
Polyester, glass-fiber-reinforced	93	200
Furan, glass-fiber-reinforced	93	200
Furan, carbon-reinforced	93	200

*Extracted from the Chemical Plant and Petroleum Refinery Piping Code, ANSI B31.3—1980, with permission of the publisher, the American Society of Mechanical Engineers, New York.

†Minimum recommended temperature of all materials is −29°C (−20°F).

METALLIC PIPE SYSTEMS: CARBON STEEL AND STAINLESS STEEL

The ferrous-metal piping systems comprising wrought carbon and alloy steels including stainless steels are the most widely used and the most completely covered by national standards.

Pipe and Tubing Pipe and tubing are divided into two main classes, seamless and welded. Seamless pipe, as a trade designation, refers to pipe made by forging a solid round, piercing it by simultaneously rotating and forcing it over a piercer point and further reducing it by rolling and drawing. However, seamless pipe and tubing are also produced by extrusion, casting into static or centrifugal molds, and by forging and boring. Seamless pipe has the same kilopascal (pounds-force per square inch) strength throughout the wall. Pierced seamless pipe frequently has the inside surface eccentric to the outside surface, resulting in nonuniform wall thickness.

Welded pipe is made from rolled strips formed into cylinders and seam-welded by various methods. The welds are credited with 60 to 100 percent of the strength of the pipe wall depending on welding and inspection procedures. Larger diameters and lower ratios of wall thickness to diameter can be obtained in welded pipe than can be obtained in seamless pipe (other than cast pipe). Uniform wall thickness is obtained. Hydrostatic testing does not reveal very short lengths of partially completed weld. This presents a possibility that small leaks may develop prematurely when corrosive fluids are being handled or the pipe is exposed to external corrosion. The weld must be taken into account in developing procedures for bending, flaring, and expanding the welded pipe.

Additional thickness and additional size and wall-thickness combinations are available as tubing. Two common classifications of tubing are "pressure" and "mechanical." Wall thickness (gauge) is specified as either "average wall" or "minimum wall." Minimum wall is more costly than average wall, and because of closer wall-thickness and diametral tolerance, both gauge systems make pressure tubing more costly than pipe. However, average-wall carbon steel electric-resistance-welded tubing, sizes 2⅜, 2⅞, 3½, and 4½ in outside diameter, produced from coiled strip on progressive forming rolls and electromagnetically rather than pressure-tested, competes vigorously with pipe.

Table 6-6 gives standard size and wall-thickness combinations together with capacity and weight.

Joints Pipe must be joined to pipe and to other components. Optimum design requires a minimum of assembly labor and provides the same resistance possessed by the pipe to (1) internal pressure as regards both rupture and leakage, (2) bending moments arising from spanning long distances between supports or from thermal expansion in piping containing offsets, (3) axial strain arising from internal pressure acting on changes in direction, blanks or closed valves, or thermal contraction in straight runs, and (4) rupture or leakage in event of fire.

However, joints in pipe buried in the soil, where the position of each length and component is fixed, need provide the same resistance as the pipe to internal pressure only; in event of earth settlement, the joints may be required to yield to resulting bending moments without leakage. Also, in piping subject to thermal expansion and contraction, some joints may be required to yield to resulting bending moments and axial strains without leakage.

The ideal pipe joint is free from changes in any dimension of the flow passage or direction of flow which would increase pressure drop or prevent complete drainage. It is free from crevices in which corrosion might be accelerated. It would require a minimum of labor to disassemble. Required frequency for disassembling the joint must be considered in making the selection. Generally speaking, joints which are easy to disassemble are deficient in one or more of the other requirements of the ideal joint.

Most joints involve modifications of the components being joined; those with the desired modifications can usually be purchased.

Welded Joints The most widely used joint in piping systems is the **butt-weld joint** (Fig. 6-87). In all ductile pipe metals which can be welded, pipe, elbows, tees, laterals, reducers, caps, valves, flanges, and V-clamp joints are available in all sizes and wall thicknesses with ends prepared for butt welding. Joint strength equal to the original pipe (except for work-hardened pipes which are annealed by the welding), unimpaired flow pattern, and generally unimpaired corrosion resistance more than compensate for the necessary careful alignment, skilled labor, and equipment required.

Plain-end pipe used for socket-weld joints (Fig. 6-88) is available in all sizes, but fittings and valves with socket-weld ends are limited to sizes 3 in and smaller, for which the extra cost of the socket is outweighed by much easier alignment and less skill needed in welding. The joint is not so resistant to bending stress as the butt-welded joint but is otherwise equal, except that for some fluids the crevice between the pipe and the socket may promote corrosion. ANSI B16.11—1973, Forged Steel Fittings, Socket-Welding and Threaded, requires that the wall thickness of the socket must be equal to or greater than 1.25 times the minimum pipe wall.

Branch Welds (Fig. 6-89) These welds eliminate the purchase of tees and require no more weld metal than tees. If the branch approaches the size of the run, careful end preparation of the branch pipe is required and the run pipe is weakened by the branch weld. See subsection "Pressure Design of Metallic Components: Wall Thickness" for rules for reinforcement. Reinforcing pads and fittings are commercially available. Use of the fittings facilitates visual inspection of the branch weld. See subsection "Welding, Brazing, or Soldering" for rules for welded joints.

Threaded Joints Pipe with **taper-pipe-thread** ends (Fig. 6-90), per ANSI B2.1, is available 12 in and smaller, subject to minimum-wall limitations. Fittings and valves with taper-pipe-thread ends are available in most pipe metals.

TABLE 6-6 Properties of Steel Pipe

Nominal pipe size, in	Outside diameter, in	Schedule no.	Wall thickness, in	Inside diameter, in	Cross-sectional area		Circumference, ft, or surface, ft^2/ft of length		Capacity at 1-ft/s velocity		Weight of plain-end pipe, lb/ft
					Metal, in^2	Flow, ft^2	Outside	Inside	U.S. gal/min	lb/h water	
⅛	0.405	10S	0.049	0.307	0.055	0.00051	0.106	0.0804	0.231	115.5	0.19
		40ST, 40S	.068	.269	.072	.00040	.106	.0705	.179	89.5	.24
		80XS, 80S	.095	.215	.093	.00025	.106	.0563	.113	56.5	.31
¼	0.540	10S	.065	.410	.097	.00092	.141	.107	.412	206.5	.33
		40ST, 40S	.088	.364	.125	.00072	.141	.095	.323	161.5	.42
		80XS, 80S	.119	.302	.157	.00050	.141	.079	.224	112.0	.54
⅜	0.675	10S	.065	.545	.125	.00162	.177	.143	.727	363.5	.42
		40ST, 40S	.091	.493	.167	.00133	.177	.129	.596	298.0	.57
		80XS, 80S	.126	.423	.217	.00098	.177	.111	.440	220.0	.74
½	0.840	5S	.065	.710	.158	.00275	.220	.186	1.234	617.0	.54
		10S	.083	.674	.197	.00248	.220	.176	1.112	556.0	.67
		40ST, 40S	.109	.622	.250	.00211	.220	.163	0.945	472.0	.85
		80XS, 80S	.147	.546	.320	.00163	.220	.143	0.730	365.0	1.09
		160	.188	.464	.385	.00117	.220	.122	0.527	263.5	1.31
		XX	.294	.252	.504	.00035	.220	.066	0.155	77.5	1.71
¾	1.050	5S	.065	.920	.201	.00461	.275	.241	2.072	1036.0	0.69
		10S	.083	.884	.252	.00426	.275	.231	1.903	951.5	0.86
		40ST, 40S	.113	.824	.333	.00371	.275	.216	1.665	832.5	1.13
		80XS, 80S	.154	.742	.433	.00300	.275	.194	1.345	672.5	1.47
		160	.219	.612	.572	.00204	.275	.160	0.917	458.5	1.94
		XX	.308	.434	.718	.00103	.275	.114	0.461	230.5	2.44
1	1.315	5S	.065	1.185	.255	.00768	.344	.310	3.449	1725	0.87
		10S	.109	1.097	.413	.00656	.344	.287	2.946	1473	1.40
		40ST, 40S	.133	1.049	.494	.00600	.344	.275	2.690	1345	1.68
		80XS, 80S	.179	0.957	.639	.00499	.344	.250	2.240	1120	2.17
		160	.250	0.815	.836	.00362	.344	.213	1.625	812.5	2.84
		XX	.358	0.599	1.076	.00196	.344	.157	0.878	439.0	3.66
1¼	1.660	5S	.065	1.530	0.326	.01277	.435	.401	5.73	2865	1.11
		10S	.109	1.442	0.531	.01134	.435	.378	5.09	2545	1.81
		40ST, 40S	.140	1.380	0.668	.01040	.435	.361	4.57	2285	2.27
		80XS, 80S	.191	1.278	0.881	.00891	.435	.335	3.99	1995	3.00
		160	.250	1.160	1.107	.00734	.435	.304	3.29	1645	3.76
		XX	.382	0.896	1.534	.00438	.435	.235	1.97	985	5.21
1½	1.900	5S	.065	1.770	0.375	.01709	.497	.463	7.67	3835	1.28
		10S	.109	1.682	0.614	.01543	.497	.440	6.94	3465	2.09
		40ST, 40S	.145	1.610	0.800	.01414	.497	.421	6.34	3170	2.72
		80XS, 80S	.200	1.500	1.069	.01225	.497	.393	5.49	2745	3.63
		160	.281	1.338	1.429	.00976	.497	.350	4.38	2190	4.86
		XX	.400	1.100	1.885	.00660	.497	.288	2.96	1480	6.41
2	2.375	5S	.065	2.245	0.472	.02749	.622	.588	12.34	6170	1.61
		10S	.109	2.157	0.776	.02538	.622	.565	11.39	5695	2.64
		40ST, 40S	.154	2.067	1.075	.02330	.622	.541	10.45	5225	3.65
		80ST, 80S	.218	1.939	1.477	.02050	.622	.508	9.20	4600	5.02
		160	.344	1.687	2.195	.01552	.622	.436	6.97	3485	7.46
		XX	.436	1.503	2.656	.01232	.622	.393	5.53	2765	9.03
2½	2.875	5S	.083	2.709	0.728	0.04003	.753	.709	17.97	8985	2.48
		10S	.120	2.635	1.039	.03787	.753	.690	17.00	8500	3.53
		40ST, 40S	.203	2.469	1.704	.03322	.753	.647	14.92	7460	5.79
		80XS, 80S	.276	2.323	2.254	.02942	.753	.608	13.20	6600	7.66
		160	.375	2.125	2.945	.02463	.753	.556	11.07	5535	10.01
		XX	.552	1.771	4.028	.01711	.753	.464	7.68	3840	13.69
3	3.500	5S	.083	3.334	0.891	.06063	.916	.873	27.21	13,605	3.03
		10S	.120	3.260	1.274	.05796	.916	.853	26.02	13,010	4.33
		40ST, 40S	.216	3.068	2.228	.05130	.916	.803	23.00	11,500	7.58
		80XS, 80S	.300	2.900	3.016	.04587	.916	.759	20.55	10,275	10.25
		160	.438	2.624	4.213	.03755	.916	.687	16.86	8430	14.32
		XX	.600	2.300	5.466	.02885	.916	.602	12.95	6475	18.58
3½	4.0	5S	.083	3.834	1.021	.08017	1.047	1.004	35.98	17.990	3.48
		10S	.120	3.760	1.463	.07711	1.047	0.984	34.61	17,305	4.97
		40ST, 40S	.226	3.548	2.680	.06870	1.047	0.929	30.80	15,400	9.11
		80XS, 80S	.318	3.364	3.678	.06170	1.047	0.881	27.70	13,850	12.50
4	4.5	5S	.083	4.334	1.152	.10245	1.178	1.135	46.0	23,000	3.92
		10S	.120	4.260	1.651	.09898	1.178	1.115	44.4	22,200	5.61
		40ST, 40S	.237	4.026	3.17	.08840	1.178	1.054	39.6	19,800	10.79
		80XS, 80S	.337	3.826	4.41	.07986	1.178	1.002	35.8	17,900	14.98

TABLE 6-6 Properties of Steel Pipe *(Continued)*

Nominal pipe size, in	Outside diameter, in	Schedule no.	Wall thickness, in	Inside diameter, in	Cross-sectional area: Metal, in^2	Cross-sectional area: Flow, ft^2	Circumference, ft, or surface, ft^2/ft of length: Outside	Circumference: Inside	Capacity at 1-ft/s velocity: U.S. gal/min	Capacity: lb/h water	Weight of plain-end pipe, lb/ft
		120	.438	3.624	5.58	.07170	1.178	0.949	32.2	16,100	19.00
		160	.531	3.438	6.62	.06647	1.178	0.900	28.9	14,450	22.51
		XX	.674	3.152	8.10	.05419	1.178	0.825	24.3	12,150	27.54
5	5.563	5S	.109	5.345	1.87	.1558	1.456	1.399	69.9	34,950	6.36
		10S	.134	5.295	2.29	.1529	1.456	1.386	68.6	34,300	7.77
		40ST, 40S	.258	5.047	4.30	.1390	1.456	1.321	62.3	31,150	14.62
		80XS, 80S	.375	4.813	6.11	.1263	1.456	1.260	57.7	28,850	20.78
		120	.500	4.563	7.95	.1136	1.456	1.195	51.0	25,500	27.04
		160	.625	4.313	9.70	.1015	1.456	1.129	45.5	22,750	32.96
		XX	.750	4.063	11.34	.0900	1.456	1.064	40.4	20,200	38.55
6	6.625	5S	0.109	6.407	2.23	0.2239	1.734	1.677	100.5	50,250	7.60
		10S	.134	6.357	2.73	.2204	1.734	1.664	98.9	49,450	9.29
		40ST, 40S	.280	6.065	5.58	.2006	1.734	1.588	90.0	45,000	18.97
		80XS, 80S	.432	5.761	8.40	.1810	1.734	1.508	81.1	40,550	28.57
		120	.562	5.501	10.70	.1650	1.734	1.440	73.9	36,950	36.39
		160	.719	5.187	13.34	.1467	1.734	1.358	65.9	32,950	45.34
		XX	.864	4.897	15.64	.1308	1.734	1.282	58.7	29,350	53.16
8	8.625	5S	.109	8.407	2.915	.3855	2.258	2.201	173.0	86,500	9.93
		10S	148	8.329	3.941	.3784	2.258	2.180	169.8	84,900	13.40
		20	.250	8.125	6.578	.3601	2.258	2.127	161.5	80,750	22.36
		30	.277	8.071	7.265	.3553	2.258	2.113	159.4	79,700	24.70
		40ST, 40S	.322	7.981	8.399	.3474	2,258	2.089	155.7	77,850	28.55
		60	.406	7.813	10.48	.3329	2.258	2.045	149.4	74,700	35.64
		80XS, 80S	.500	7.625	12.76	.3171	2.258	1.996	142.3	71,150	43.39
		100	.594	7.437	14.99	.3017	2.258	1.947	135.4	67,700	50.95
		120	.719	7.187	17.86	.2817	2.258	1.882	126.4	63,200	60.71
		140	.812	7.001	19.93	.2673	2.258	1.833	120.0	60,000	67.76
		XX	.875	6.875	21.30	.2578	2.258	1.800	115.7	57,850	72.42
		160	.906	6.813	21.97	.2532	2.258	1.784	113.5	56,750	74.69
10	10.75	5S	.134	10.482	4.47	.5993	2.814	2.744	269.0	134,500	15.19
		10S	.165	10.420	5.49	.5922	2.814	2.728	265.8	132,900	18.65
		20	.250	10.250	8.25	.5731	2.814	2.685	257.0	128,500	28.04
		30	.307	10.136	10.07	.5603	2.814	2.655	252.0	126,000	34.24
		40ST, 40S	.365	10.020	11.91	.5475	2.814	2.620	246.0	123,000	40.48
		80S, 60XS	.500	9.750	16.10	.5185	2.814	2.550	233.0	116,500	54.74
		80	.594	9.562	18.95	.4987	2.814	2.503	223.4	111,700	64.43
		100	.719	9.312	22.66	.4729	2.814	2.438	212.3	106,150	77.03
		120	.844	9.062	26.27	.4479	2.814	2.372	201.0	100,500	89.29
		140, XX	1.000	8.750	30.63	.4176	2.814	2.291	188.0	94,000	104.13
		160	1.125	8.500	34.02	.3941	2.814	2.225	177.0	88,500	115.64
12	12.75	5S	0.156	12.438	6.17	.8438	3.338	3.26	378.7	189,350	20.98
		10S	0.180	12.390	7.11	.8373	3.338	3.24	375.8	187,900	24.17
		20	0.250	12.250	9.82	.8185	3.338	3.21	367.0	183,500	33.38
		30	0.330	12.090	12.88	.7972	3.338	3.17	358.0	179,000	43.77
		ST, 40S	0.375	12.000	14.58	.7854	3.338	3.14	352.5	176,250	49.56
		40	0.406	11.938	15.74	.7773	3.338	3.13	349.0	174,500	53.52
		XS, 80S	0.500	11.750	19.24	.7530	3.338	3.08	338.0	169,000	65.42
		60	0.562	11.626	21.52	.7372	3.338	3.04	331.0	165,500	73.15
		80	0.688	11.374	26.07	.7056	3.338	2.98	316.7	158,350	88.63
		100	0.844	11.062	31.57	.6674	3.338	2.90	299.6	149,800	107.32
		120, XX	1.000	10.750	36.91	.6303	3.338	2.81	283.0	141,500	125.49
		140	1.125	10.500	41.09	.6013	3.338	2.75	270.0	135,000	139.67
		160	1.312	10.126	47.14	.5592	3.338	2.65	251.0	125,500	160.27
14	14	5S	0.156	13.688	6.78	1.0219	3.665	3.58	459	229,500	23.07
		10S	0.188	13.624	8.16	1.0125	3.665	3.57	454	227,000	27.73
		10	0.250	13.500	10.80	0.9940	3.665	3.53	446	223,000	36.71
		20	0.312	13.376	13.42	0.9750	3.665	3.50	438	219,000	45.61
		30, ST	0.375	13.250	16.05	0.9575	3.665	3.47	430	215,000	54.57
		40	0.438	13.124	18.66	0.9397	3.665	3.44	422	211,000	63.44
		XS	0.500	13.000	21.21	0.9218	3.665	3.40	414	207,000	72.09
		60	0.594	12.812	25.02	0.8957	3.665	3.35	402	201,000	85.05
		80	0.750	12.500	31.22	0.8522	3.665	3.27	382	191,000	106.13
		100	0.938	12.124	38.49	0.8017	3.665	3.17	360	180,000	130.85
		120	1.094	11.812	44.36	0.7610	3.665	3.09	342	171,000	150.79
		140	1.250	11.500	50.07	0.7213	3.665	3.01	324	162,000	170.21
		160	1.406	11.188	55.63	0.6827	3.665	2.93	306	153,000	189.11
16	16	5S	0.165	15.670	8.21	1.3393	4.189	4.10	601	300,500	27.90
		10S	0.188	15.624	9.34	1.3314	4.189	4.09	598	299,000	31.75
		10	0.250	15.500	12.37	1.3104	4.189	4.06	587	293,500	42.05

TABLE 6-6 Properties of Steel Pipe *(Continued)*

Nominal pipe size, in	Outside diameter, in	Schedule no.	Wall thickness, in	Inside diameter, in	Cross-sectional area		Circumference, ft, or surface, ft^2/ft of length		Capacity at 1-ft/s velocity		Weight of plain-end pipe, lb/ft
					Metal, in^2	Flow, ft^2	Outside	Inside	U.S. gal/min	lb/h water	
		20	0.312	15.376	15.38	1.2985	4.189	4.03	578	289,000	52.27
		30, ST	0.375	15.250	18.41	1.2680	4.189	3.99	568	284,000	62.58
		40, XS	0.500	15.000	24.35	1.2272	4.189	3.93	550	275,000	82.77
		60	0.656	14.688	31.62	1.1766	4.189	3.85	528	264,000	107.50
		80	0.844	14.312	40.19	1.1171	4.189	3.75	501	250,500	136.61
		100	1.031	13.938	48.48	1.0596	4.189	3.65	474	237,000	164.82
		120	1.219	13.562	56.61	1.0032	4.189	3.55	450	225,000	192.43
		140	1.438	13.124	65.79	0.9394	4.189	3.44	422	211,000	223.64
		160	1.594	12.812	72.14	0.8953	4.189	3.35	402	201,000	245.25
18	18	5S	0.165	17.670	9.25	1.7029	4.712	4.63	764	382,000	31.43
		10S	0.188	17.624	10.52	1.6941	4.712	4.61	760	379,400	35.76
		10	0.250	17.500	13.94	1.6703	4.712	4.58	750	375,000	47.39
		20	0.312	17.376	17.34	1.6468	4.712	4.55	739	369,500	58.94
		ST	0.375	17.250	20.76	1.6230	4.712	4.52	728	364,000	70.59
		30	0.438	17.124	24.16	1.5993	4.712	4.48	718	359,000	82.15
		XS	0.500	17.000	27.49	1.5763	4.712	4.45	707	353,500	93.45
		40	0.562	16.876	30.79	1.5533	4.712	4.42	697	348,500	104.67
		60	0.750	16.500	40.64	1.4849	4.712	4.32	666	333,000	138.17
		80	0.938	16.124	50.28	1.4180	4.712	4.22	636	318,000	170.92
		100	1.156	15.688	61.17	1.3423	4.712	4.11	602	301,000	207.96
		120	1.375	15.250	71.82	1.2684	4.712	3.99	569	284,500	244.14
		140	1.562	14.876	80.66	1.2070	4.712	3.89	540	270,000	274.22
		160	1.781	14.438	90.75	1.1370	4.712	3.78	510	255,000	308.50
20	20	5S	0.188	19.624	11.70	2.1004	5.236	5.14	943	471,500	39.78
		10S	.218	19.564	13.55	2.0878	5.236	5.12	937	467,500	46.06
		10	.250	19.500	15.51	2.0740	5.236	5.11	930	465,000	52.73
		20, ST	.375	19.250	23.12	2.0211	5.236	5.04	902	451,000	78.60
		30, XS	.500	19.000	30.63	1.9689	5.236	4.97	883	441,500	104.13
		40	.594	18.812	36.21	1.9302	5.236	4.92	866	433,000	123.11
		60	.812	18.376	48.95	1.8417	5.236	4.81	826	413,000	166.40
		80	1.031	17.938	61.44	1.7550	5.236	4.70	787	393,500	208.87
		100	1.281	17.438	75.33	1.6585	5.236	4.57	744	372,000	256.10
		120	1.500	17.000	87.18	1.5763	5.236	4.45	707	353,500	296.37
		140	1.750	16.500	100.3	1.4849	5.236	4.32	665	332,500	341.09
		160	1.969	16.062	111.5	1.4071	5.236	4.21	632	316,000	397.17
24	24	5S	0.218	23.564	16.29	3.0285	6.283	6.17	1359	679,500	55.37
		10, 10S	0.250	23.500	18.65	3.012	6.283	6.15	1350	675,000	63.41
		20, ST	0.375	23.250	27.83	2.948	6.283	6.09	1325	662,500	94.62
		XS	0.500	23.000	36.90	2.885	6.283	6.02	1295	642,500	125.49
		30	0.562	22.876	41.39	2.854	6.283	5.99	1281	640,500	140.68
		40	0.688	22.624	50.39	2.792	6.283	5.92	1253	626,500	171.29
		60	0.969	22.062	70.11	2.655	6.283	5.78	1192	596,000	238.35
		80	1.219	21.562	87.24	2.536	6.283	5.64	1138	569,000	296.58
		100	1.531	20.938	108.1	2.391	6.283	5.48	1073	536,500	367.39
		120	1.812	20.376	126.3	2.264	6.283	5.33	1016	508,000	429.39
		140	2.062	19.876	142.1	2.155	6.283	5.20	965	482,500	483.12
		160	2.344	19.312	159.5	2.034	6.283	5.06	913	456,500	542.13
30	30	5S	0.250	29.500	23.37	4.746	7.854	7.72	2130	1,065,000	79.43
		10, 10S	0.312	29.376	29.10	4.707	7.854	7.69	2110	1.055,000	98.93
		ST	0.375	29.259	34.90	4.666	7.854	7.66	2094	1,048,000	118.65
		20, XS	0.500	29.000	46.34	4.587	7.854	7.59	2055	1,027,500	157.53
		30	0.625	28.750	57.68	4.508	7.854	7.53	2020	1,010,000	196.08

5S, 10S, and 40S are extracted from Stainless Steel Pipe, ANSI B36.19—1976, with permission of the publisher, the American Society of Mechanical Engineers, New York. ST = standard wall, XS = extra strong wall, XX = double extra strong wall, and Schedules 10 through 160 are extracted from Wrought-Steel and Wrought-Iron Pipe, ANSI B36.10—1975, with permission of the same publisher. Decimal thicknesses for respective pipe sizes represent their nominal or average wall dimensions. Mill tolerances as high as ±12½ percent are permitted.

Plain-end pipe is produced by a square cut. Pipe is also shipped from the mills threaded, with a threaded coupling on one end, or with the ends beveled for welding, or grooved or sized for patented couplings. Weights per foot for threaded and coupled pipe are slightly greater because of the weight of the coupling, but it is not available larger than 12 in or lighter than Schedule 30 sizes 8 through 12 in, or Schedule 40 6 in and smaller.

To convert inches to millimeters, multiply by 25.4; to convert square inches to square millimeters, multiply by 645; to convert feet to meters, multiply by 0.3048; to convert square feet to square meters, multiply by 0.0929; to convert pounds per foot to kilograms per meter, multiply by 1.49; to convert gallons to cubic meters, multiply by 3.7854×10^{-3}; and to convert pounds to kilograms, multiply by 0.4536.

FIG. 6-87 Butt weld.

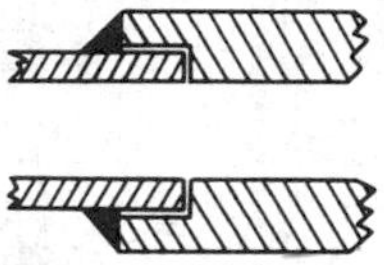

FIG. 6-88 Socket weld.

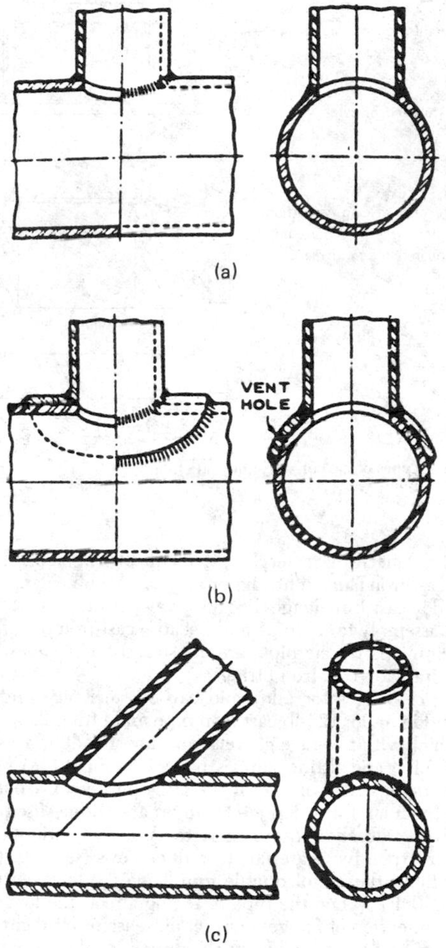

FIG. 6-89 Branch welds. (*a*) Without added reinforcement. (*b*) With added reinforcement. (*c*) Angular branch.

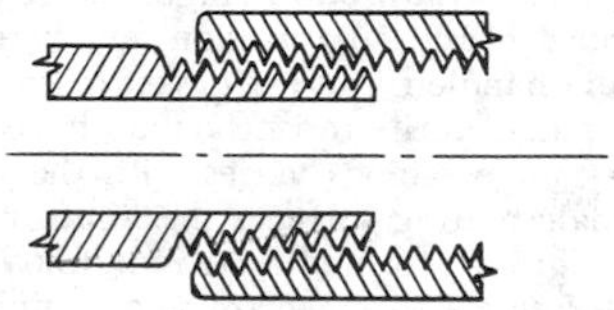

FIG. 6-90 Taper pipe thread.

Principal use of threaded joints is in sizes 2 in and smaller, in metals for which the most economically produced walls are thick enough to withstand considerable pressure and corrosion after reduction in thickness due to threading. For threaded joints over 2 in, assembly labor and size and cost of tools increase rapidly. Careful alignment, required at the start of assembly and during rotation of the components, as well as variation in length produced by diametral tolerances in the threads, severely limits preassembly of the components. Threading is not a precise machining operation, and filler materials known as "pipe dope" are necessary to block the spiral leakage path.

Threads notch the pipe and cause loss of strength and fatigue resistance. Enlargement and contraction of the flow passage at threaded joints creates turbulence; sometimes corrosion and erosion are aggravated at the point where the pipe has already been thinned by threading. The tendency of pipe wrenches to crush pipe and fittings limits the torque available for tightening threaded joints. For low-pressure systems, a slight rotation in the joint may be used to impart flexibility to the system, but this same rotation, unwanted, may cause leaks to develop in higher-pressure systems. In some metals, galling occurs when threaded joints are disassembled.

Straight Pipe Threads (Fig. 6-91) These are confined to lightweight couplings in sizes 2 in and smaller. Manufacturers of threaded pipe ship it with such couplings installed on one end of each pipe. The joint obtained is inferior to that obtained with taper threads. The code limits the joint shown in Fig. 6-91 to 1.0 MPa (150 lbf/in^2) gauge maximum, 182°C (360°F) maximum, and to nonflammable, nontoxic fluids.

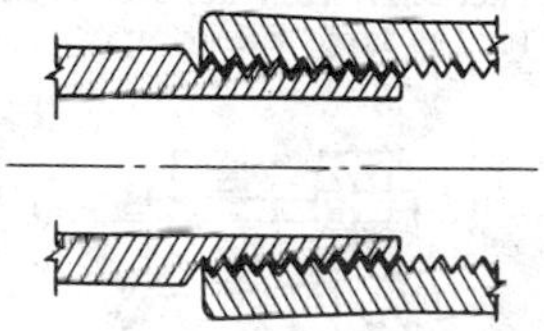

FIG. 6-91 Taper pipe to straight coupling thread.

When both components of a threaded joint are of weldable metal, the joint may be **seal-welded** as shown in Fig. 6-92. Seal welds may be used only to prevent leakage of threaded joints. They are not considered as contributing any strength to the joint. This type of joint is limited to new construction and is not suitable as a repair procedure, since pipe dope in the threads would interfere with welding. This method provides tight joints with a minimum of welding labor. When threaded joints used to join materials with widely different coefficients of thermal expansion are subject to temperature cycling, seal welding may be needed to prevent leakage.

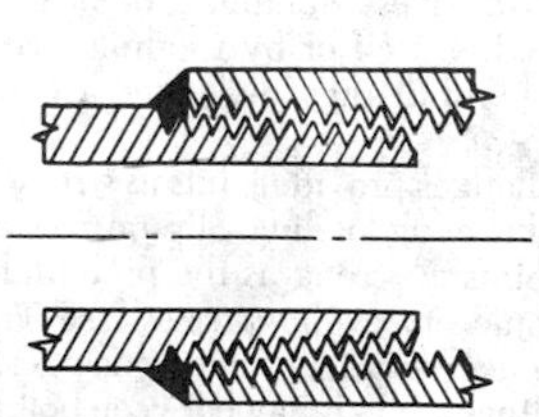

FIG. 6-92 Taper pipe thread seal-welded.

To assist in assembly and disassembly of both threaded and welded systems, **union joints** (Fig. 6-93) are used. They comprise metal-to-metal seats drawn together by a shouldered straight thread nut and are available both in couplings for joining two lengths of pipe and on the ends of some fittings. On threaded piping systems in which disassembly is not contemplated, union joints installed at intervals permit future further tightening of threaded joints. Tightening of heavy unions yields tight joints even if the pipe is slightly misaligned at the start of tightening.

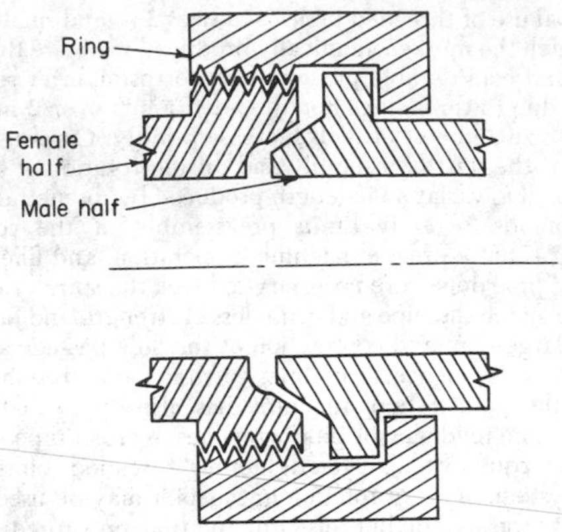

FIG. 6-93 Union.

Flanged Joints For sizes larger than 2 in when disassembly is contemplated, the flanged joint (Fig. 6-94) is the most widely used. Figure 6-95 and 6-96 illustrate the wide variety of types and facings available. Though flanged joints consume a large volume of metal, precise machining is required only on the facing. Flanged joints do not impose severe diametral tolerances on the pipe. Careful alignment prior to assembly of flat-face and raised-face flanges is not required, and the necessary wrenches are far smaller than those for screwed assembly for the same size of pipe.

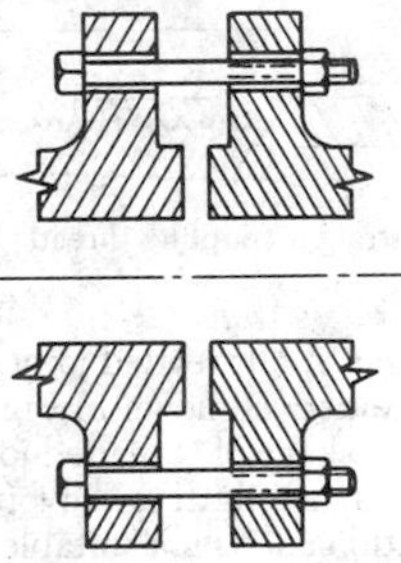

FIG. 6-94 Flanged joint.

Manufacturers offer **flanged-end pipe** in only a few metals. Otherwise, flanges are attached to pipe by various types of joints (Fig. 6-95). The lap joint involves a modification of the pipe which may be formed from the pipe itself or by welding a ring or a lap-joint stub end to it. **Flanged-end fittings** and valves are available in all sizes of most pipe metals.

Welding-neck flanges provide joints as strong as the pipe under all types of static and cycling loading. Slip-on, socket-weld, and lap-joint flanges provide joints as strong as the pipe under static loading but have lower resistance to cyclic stresses (see Table 6-42). Lap-joint flanges avoid the necessity of orienting flanges so that vertical and horizontal centerlines are halfway between bolt holes and permit orientation of the stems of flanged valves at any angle needed to provide clearance. The tolerance is $\frac{1}{8}$ in in the bolt holes; the necessity of making sure that the gasket does not protrude into the flow channel results in some disturbance of the flow pattern when flat-face and raised-face flanges are used. This can be eliminated by using welding-neck or socket-weld flanges with male-and-female or tongue-and-groove facings.

Dimensions of alloy and carbon steel and cast-iron pipe flanges with flat and raised faces are given in Tables 6-7 to 6-13 (see Fig. 6-96). The dimensions were extracted from Cast-Iron Pipe Flanges and Flanged Fittings, ANSI B16.1—1975, and Steel Pipe Flanges and Flanged Fittings, ANSI 16.5—1977, with the permission of the publisher, the American Society of Mechanical Engineers, New York. Against cast-iron flanged fittings or valves, steel pipe flanges are often preferred to cast-iron flanges because they permit welded rather than screwed assembly to the pipe and because cast-iron pipe flanges, not being reinforced by the pipe, are not so resistant to abuse as flanges cast integrally on cast-iron fittings.

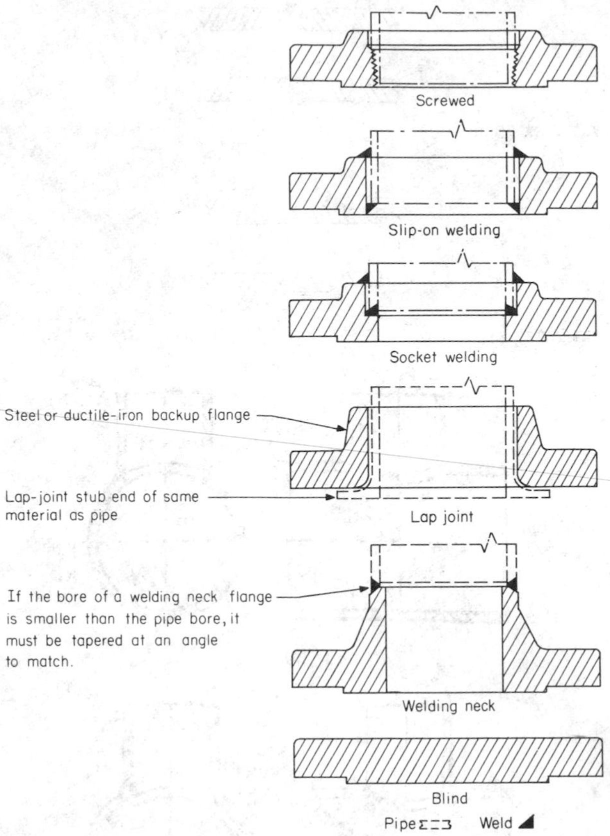

FIG. 6-95 Types of carbon and alloy steel flanges.

Facing of flanges for alloy and carbon steel pipe and fittings is shown in Fig. 6-96; 125-lb cast-iron pipe and fitting flanges have flat faces, which with full-face gaskets minimize bending stresses; 250-lb cast-iron pipe and fitting flanges have 1.5-mm ($\frac{1}{16}$-in) raised faces (wider than on steel flanges) for the same purpose. Carbon steel and ductile- (nodular-) iron lap-joint flanges are widely used as backup flanges with stub ends in piping systems of austenitic stainless steel and other expensive materials to reduce costs (see Fig. 6-95). The code prohibits the use of ductile-iron flanges at temperatures above 343°C (650°F). When the type of facing affects the length through the hub dimension of flanges, correct dimensions for commonly used facings can be determined from the dimensional data in Fig. 6-96.

Gaskets Gaskets must resist corrosion by the fluids handled. The more expensive male-and-female or tongue-and-groove facings may be required to seat hard gaskets adequately. With these facings the gasket generally cannot blow out. Flanged joints, by placing the gasket material under heavy compression and permitting only edge attack by the fluid handled, can use gasket materials which in other joints might not satisfactorily resist the fluid handled.

The finish of flange facings varies with the manufacturer. For raised-face or male mating surfaces the finish usually consists of a continuous spiral groove formed by a round-nosed tool or a V tool (serrated finish). Female surfaces are smooth-finished (i.e., without definite tool markings). Other finishes are concentric-grooved,

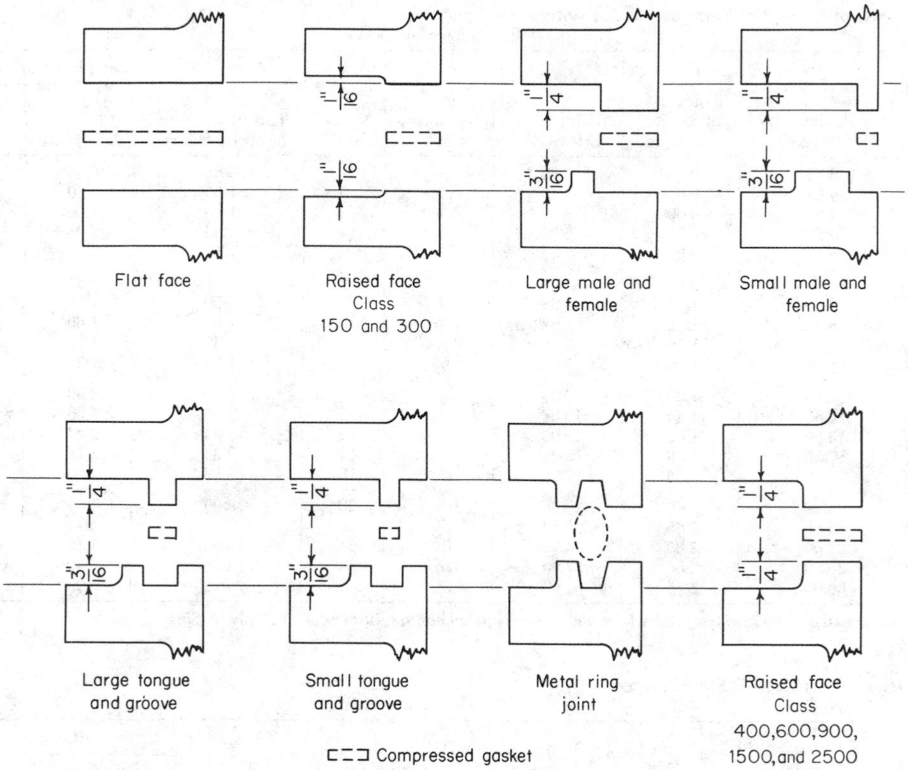

FIG. 6-96 Flange facings, illustrated on welding-neck flanges. (On small male-and-female facings the outside diameter of the male face is less than the outside diameter of the pipe, so this facing does not apply to screwed or slip-on flanges. A similar joint can be made with screwed flanges and threaded pipe by projecting the pipe through one flange and recessing it in the other. However, pipe thicker than Schedule 40 is required to avoid crushing gaskets.) To convert inches to millimeters, multiply by 25.4.

TABLE 6-7 Dimensions of Class 150-lb Flanges for Use with Steel Pipe*

All dimensions in inches

Nominal pipe size	Outside diameter of flange	Thickness of flange, minimum	Diameter of bolt circle	Diameter of bolts	No. of bolts	Length through hub: Threaded slip-on socket welding	Lap joint	Welding neck	ANSI B16.1, screwed (125-lb)
½	3.50	0.44	2.38	½	4	0.62	0.62	1.88	....
¾	3.88	0.50	2.75	½	4	0.62	0.62	2.06	....
1	4.25	0.56	3.12	½	4	0.69	0.69	2.19	0.69
1¼	4.62	0.62	3.50	½	4	0.81	0.81	2.25	0.81
1½	5.00	0.69	3.88	½	4	0.88	0.88	2.44	0.88
2	6.00	0.75	4.75	⅝	4	1.00	1.00	2.50	1.00
2½	7.00	0.88	5.50	⅝	4	1.12	1.12	2.75	1.12
3	7.50	0.94	6.00	⅝	4	1.19	1.19	2.75	1.19
3½	8.50	0.94	7.00	⅝	8	1.25	1.25	2.81	1.25
4	9.00	0.94	7.50	⅝	8	1.31	1.31	3.00	1.31
5	10.00	0.94	8.50	¾	8	1.44	1.44	3.50	1.44
6	11.00	1.00	9.50	¾	8	1.56	1.56	3.50	1.56
8	13.50	1.12	11.75	¾	8	1.75	1.75	4.00	1.75
10	16.00	1.19	14.25	⅞	12	1.94	1.94	4.00	1.94
12	19.00	1.25	17.00	⅞	12	2.19	2.19	4.50	2.19
14	21.00	1.38	18.75	1	12	2.25	3.12	5.00	2.25
16	23.50	1.44	21.25	1	16	2.50	3.44	5.00	2.50
18	25.00	1.56	22.75	1⅛	16	2.69	3.81	5.50	2.69
20	27.50	1.69	25.00	1⅛	20	2.88	4.06	5.69	2.88
24	32.00	1.88	29.50	1¼	20	3.25	4.38	6.00	3.25

*Dimensions from ANSI B16.5—1977, unless otherwise noted. To convert inches to millimeters, multiply by 25.4.

TABLE 6-8 Dimensions of Class 300 Flanges for Use with Steel Pipe*

All dimensions in inches

Nominal pipe size	Outside diameter of flange	Thickness of flange, minimum	Diameter of bolt circle	Diameter of bolts	No. of bolts	Length through hub: Threaded slip-on socket welding	Length through hub: Lap joint	Length through hub: Welding neck	Length through hub: ANSI B16.1, screwed (Class 250)
½	3.75	0.56	2.62	½	4	0.88	0.88	2.06	
¾	4.62	0.62	3.25	⅝	4	1.00	1.00	2.25	
1	4.88	0.69	3.50	⅝	4	1.06	1.06	2.44	0.88
1¼	5.25	0.75	3.88	⅝	4	1.06	1.06	2.56	1.00
1½	6.12	0.81	4.50	¾	4	1.19	1.19	2.69	1.12
2	6.50	0.88	5.00	⅝	8	1.31	1.31	2.75	1.25
2½	7.50	1.00	5.88	¾	8	1.50	1.50	3.00	1.43
3	8.25	1.12	6.62	¾	8	1.69	1.69	3.12	1.56
3½	9.00	1.19	7.25	¾	8	1.75	1.75	3.19	1.62
4	10.00	1.25	7.88	¾	8	1.88	1.88	3.38	1.75
5	11.00	1.38	9.25	¾	8	2.00	2.00	3.88	1.88
6	12.50	1.44	10.62	¾	12	2.06	2.06	3.88	1.94
8	15.00	1.62	13.00	⅞	12	2.44	2.44	4.38	2.19
10	17.50	1.88	15.25	1	16	2.62	3.75	4.62	2.38
12	20.50	2.00	17.75	1⅛	16	2.88	4.00	5.12	2.56
14	23.00	2.12	20.25	1⅛	20	3.00	4.38	5.62	2.69
16	25.50	2.25	22.50	1¼	20	3.25	4.75	5.75	2.88
18	28.00	2.38	24.75	1¼	24	3.50	5.12	6.25	
20	30.50	2.50	27.00	1¼	24	3.75	5.50	6.38	
24	36.00	2.75	32.00	1½	24	4.19	6.00	6.62	

*Dimensions from ANSI B16.5—1977, unless otherwise noted. To convert inches to millimeters, multiply by 25.4.

TABLE 6-9 Dimensions of Class 400 Steel Flanges*

All dimensions in inches

Nominal pipe size	Outside diameter of flange	Thickness of flange, minimum	Diameter of bolt circle	Diameter of bolts	No. of bolts	Length through hub: Threaded slip-on socket welding	Length through hub: Lap joint	Length through hub: Welding neck
½	Use Class 600 dimensions in these sizes.							
¾								
1								
1¼								
1½								
2								
2½								
3								
3½								
4	10.00	1.38	7.88	⅞	8	2.00	2.00	3.50
5	11.00	1.50	9.25	⅞	8	2.12	2.12	4.00
6	12.50	1.62	10.62	⅞	12	2.25	2.25	4.06
8	15.00	1.88	13.00	1	12	2.69	2.69	4.62
10	17.50	2.12	15.25	1¼	16	2.88	4.00	4.88
12	20.50	2.25	17.75	1¼	16	3.12	4.25	5.38
14	23.00	2.38	20.25	1¼	20	3.31	4.62	5.88
16	25.50	2.50	22.50	1⅜	20	3.69	5.00	6.00
18	28.00	2.62	24.75	1⅜	24	3.88	5.38	6.50
20	30.50	2.75	27.00	1½	24	4.00	5.75	6.62
24	36.00	3.00	32.00	1¾	24	4.50	6.25	6.88

*Dimensions from ANSI B16.5—1977. To convert inches to millimeters, multiply by 25.4.

lapped, or mirror (cold-water). The latter two are usually for application without gaskets.

In general, for 300-lb ANSI and lower-rated flanges compressed asbestos-sheet gaskets are used [400°C (750°F) maximum]. The metal-asbestos spiral-wound type is used for higher pressure and temperature services [593°C (1100°F) maximum], including services involving cyclic or difficultly contained fluids. The development of substitutes for asbestos in gaskets is being actively pursued because of the health hazard associated with asbestos. Metal-TFE and metal-graphite spiral-wound gaskets are available and may seal better than metal-asbestos gaskets. Spiral-wound gaskets are also used widely in high-pressure steam services. Spiral-wound gaskets should preferably be used with a smooth-flange finish.

The spiral-wound type furnished with a solid metallic ring on the outside to limit gasket compression provides protection against blowout when used with raised facing.

Metal-Ring-Joint Facing This is the most costly facing. The ring must be softer than the flange and is usually a softer grade of the same metal as the flange. It is used where other gasket materials are destroyed by the fluid being handled. In event of fire, it does not leak. Because the surfaces that the gasket contacts are below the flange face, it is the least likely facing to be damaged in handling.

TABLE 6-10 Dimensions of Class 600 Steel Flanges*

All dimensions in inches

Nominal pipe size	Outside diameter of flange	Thickness of flange, minimum	Diameter of bolt circle	Diameter of bolts	No. of bolts	Length through hub: Threaded slip-on socket welding	Length through hub: Lap joint	Length through hub: Welding neck
½	3.75	0.56	2.62	½	4	0.88	0.88	2.06
¾	4.62	0.62	3.25	⅝	4	1.00	1.00	2.25
1	4.88	0.69	3.50	⅝	4	1.06	1.06	2.44
1¼	5.25	0.81	3.88	⅝	4	1.12	1.12	2.62
1½	6.12	0.88	4.50	¾	4	1.25	1.25	2.75
2	6.50	1.00	5.00	⅝	8	1.44	1.44	2.88
2½	7.50	1.12	5.88	¾	8	1.62	1.62	3.12
3	8.25	1.25	6.62	¾	8	1.81	1.81	3.25
3½	9.00	1.38	7.25	⅞	8	1.94	1.94	3.38
4	10.75	1.50	8.50	⅞	8	2.12	2.12	4.00
5	13.00	1.75	10.50	1	8	2.38	2.38	4.50
6	14.00	1.88	11.50	1	12	2.62	2.62	4.62
8	16.50	2.19	13.75	1⅛	12	3.00	3.00	5.25
10	20.00	2.50	17.00	1¼	16	3.38	4.38	6.00
12	22.00	2.62	19.25	1¼	20	3.62	4.62	6.12
14	23.75	2.75	20.75	1¾	20	3.69	5.00	6.50
16	27.00	3.00	23.75	1½	20	4.19	5.50	7.00
18	29.25	3.25	25.75	1⅝	20	4.62	6.00	7.25
20	32.00	3.50	28.50	1⅝	24	5.00	6.50	7.50
24	37.00	4.00	33.00	1⅞	24	5.50	7.25	8.00

*Dimensions from ANSI B16.5—1977. To convert inches to millimeters, multiply by 25.4.

TABLE 6-11 Dimensions of Class 900 Steel Flanges*

All dimensions in inches

Nominal pipe size	Outside diamter of flange	Thickness of flange, minimum	Diameter of bolt circle	Diameter of bolts	No. of bolts	Length through hub: Threaded slip-on socket welding	Length through hub: Lap joint	Length through hub: Welding neck
½	Use Class 1500 dimensions in these sizes.							
¾								
1								
1¼								
1½								
2								
2½								
3	9.50	1.50	7.50	⅞	8	2.12	2.12	4.00
4	11.50	1.75	9.25	1⅛	8	2.75	2.75	4.50
5	13.75	2.00	11.00	1¼	8	3.12	3.12	5.00
6	15.00	2.19	12.50	1⅛	12	3.38	3.38	5.50
8	18.50	2.50	15.50	1⅜	12	4.00	4.50	6.38
10	21.50	2.75	18.50	1⅜	16	4.25	5.00	7.25
12	24.00	3.12	21.00	1⅜	20	4.62	5.62	7.88
14	25.25	3.38	22.00	1½	20	5.12	6.12	8.38
16	27.75	3.50	24.25	1⅝	20	5.25	6.50	8.50
18	31.00	4.00	27.00	1⅞	20	6.00	7.50	9.00
20	33.75	4.25	29.50	2	20	6.25	8.25	9.75
24	41.00	5.50	35.50	2½	20	8.00	10.50	11.50

*Dimensions from ANSI B16.5—1977. To convert inches to millimeters, multiply by 25.4.

Compared with raised or smooth faces, it is more difficult to disassemble because the flanges can be separated only in the axial direction.

Bolting Bolting requirements for ANSI flanged joints are presented in the code. For joining two steel flanges, by reference to ANSI B16.5, Steel Pipe Flanges and Flanged Fittings, the code requires alloy steel bolting, except that bolting for 150- and 300-lb flanges at 204°C (400°F) and lower may be made of ASTM A307 Grade B low-carbon externally threaded fasteners. The code limits this exception to −29°C (−20°F) minimum.

Steel 150-lb flanges may be bolted to cast-iron valves, fittings, or other cast-iron piping components having either Class 125 cast integral or screwed flanges. If such construction is used, it is preferred that the 1.5-mm (1/16-in) raised face on steel flanges be removed. If the raised face is removed and a flat-ring gasket extending to the inner edge of the bolt holes is used, the bolting shall not be stronger than carbon steel per ASTM A307 Grade B; if a full-face gasket is used, the bolting may be heat-treated carbon steel or alloy steel (ASTM A193). If the raised face of the steel flange is not removed, the bolting shall not be stronger than carbon steel ASTM A307 Grade B.

Steel 300-lb flanges may be bolted to cast-iron valves, fittings, or other cast-iron piping components having either Class 250 cast-iron integral or screwed flanges, without any change in the raised face on

TABLE 6-12 Dimensions of Class 1500 Steel Flanges*

All dimensions in inches

Nominal pipe size	Outside diameter of flange	Thickness of flange, minimum	Diameter of bolt circle	Diameter of bolts	No. of bolts	Length through hub: Threaded slip-on socket welding	Length through hub: Lap joint	Length through hub: Welding neck
½	4.75	0.88	3.25	¾	4	1.25	1.25	2.38
¾	5.12	1.00	3.50	¾	4	1.38	1.38	2.75
1	5.88	1.12	4.00	⅞	4	1.62	1.62	2.88
1¼	6.25	1.12	4.38	⅞	4	1.62	1.62	2.88
1½	7.00	1.25	4.88	1	4	1.75	1.75	3.25
2	8.50	1.50	6.50	⅞	8	2.25	2.25	4.00
2½	9.62	1.62	7.50	1	8	2.50	2.50	4.12
3	10.50	1.88	8.00	1⅛	8	2.88	2.88	4.62
4	12.25	2.12	9.50	1¼	8	3.56	3.56	4.88
5	14.75	2.88	11.50	1½	8	4.12	4.12	6.12
6	15.50	3.25	12.50	1⅜	12	4.69	4.69	6.75
8	19.00	3.62	15.50	1⅝	12	5.62	5.62	8.38
10	23.00	4.25	19.00	1⅞	12	6.25	7.00	10.00
12	26.50	4.88	22.50	2	16	7.12	8.62	11.12
14	29.50	5.25	25.00	2¼	16	....	9.50	11.75
16	32.50	5.75	27.75	2½	16	....	10.25	12.25
18	36.00	6.38	30.50	2¾	16	....	10.88	12.88
20	38.75	7.00	32.75	3	16	....	11.50	14.00
24	46.00	8.00	39.00	3½	16	....	13.00	16.00

*Dimensions from ANSI B15.5—1977. To convert inches to millimeters, multiply by 25.4.

TABLE 6-13 Dimensions of Class 2500 Steel Flanges*

All dimensions in inches

Nominal pipe size	Outside diameter of flange	Thickness of flange, minimum	Diameter of bolt circle	Diameter of bolts	No. of bolts	Length through hub: Threaded	Length through hub: Lap joint	Length through hub: Welding neck
½	5.25	1.19	3.50	¾	4	1.56	1.56	2.88
¾	5.50	1.25	3.75	¾	4	1.69	1.69	3.12
1	6.25	1.38	4.25	⅞	4	1.88	1.88	3.50
1¼	7.25	1.50	5.12	1	4	2.06	2.06	3.75
1½	8.00	1.75	5.75	1⅛	4	2.38	2.38	4.38
2	9.25	2.00	6.75	1	8	2.75	2.75	5.00
2½	10.50	2.25	7.75	1⅛	8	3.12	3.12	5.62
3	12.00	2.62	9.00	1¼	8	3.62	3.62	6.62
4	14.00	3.00	10.75	1½	8	4.25	4.25	7.50
5	16.50	3.62	12.75	1¾	8	5.12	5.12	9.00
6	19.00	4.25	14.50	2	8	6.00	6.00	10.75
8	21.75	5.00	17.25	2	12	7.00	7.00	12.50
10	26.50	6.50	21.25	2½	12	9.00	9.00	16.50
12	30.00	7.25	24.38	2¾	12	10.00	10.00	18.25

*Dimensions from ANSI B16.5—1977. To convert inches to millimeters, multiply by 25.4.

either flange. If such construction is used, the bolting shall not be stronger than carbon steel, ASTM A307 Grade B.

Cast-iron 25-lb and Class 125 integral or screwed companion flanges may be used with a full-face gasket or with a flat-ring gasket extending to the inner edge of the bolts. When a full-face gasket is used, the bolting may be of heat-treated carbon steel or alloy steel (ASTM A193). When a flat-ring gasket is used, the bolting shall not be stronger than carbon steel, per ASTM A307 Grade B.

When two Class 250 cast-iron integral or screwed companion flanges having 1.5-mm (1/16-in) raised faces are bolted together, the bolting shall not be stronger than carbon steel, per ASTM A307 Grade B.

Other Types of Piping Joints **Packed-gland joints** (Fig. 6-97) require no special end preparation of pipe but do require careful control of the diameter of the pipe. Thus the supplier of the pipe should be notified when packed-gland joints are to be used. Cast- and ductile-iron pipe, fittings, and valves are available with the bell cast on one or more ends. Glands, bolts, and gaskets are shipped with the pipe. Couplings equipped with packed glands at each end, known as Dresser couplings, are available in several metals. The joints can be assembled with small wrenches and unskilled labor, in limited space, and if necessary, under water.

Packed-gland joints are designed to take the same hoop stress as the pipe. They do not resist bending moments or axial forces tending to separate the joints but yield to them to an extent indicated by the vendor's allowable-angular-deflection and end movement specifications. Further angular or end movement produces leakage, but end movement can be limited by harnessing or bridling with a combination of rods and welded clips or clamps, or by anchoring to existing or new structures. The crevice between the bell and the spigot may promote corrosion. The joints are widely used in underground lines. They are not affected by limited earth settlement, and friction of the earth prevents end separation. When disassembly by moving pipe axially is not practical, packed-joint couplings which can be slid entirely onto one of the two lengths joined are available. However, the tendency of the packing to adhere to the pipe makes this difficult.

Poured joints (Fig. 6-98) require no special end preparation of the pipe or diametral control. They are used for brittle materials. Pipe,

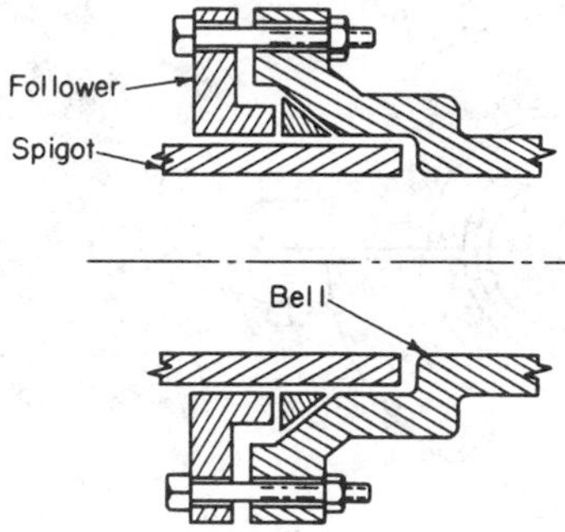

FIG. 6-97 Packed-gland joint.

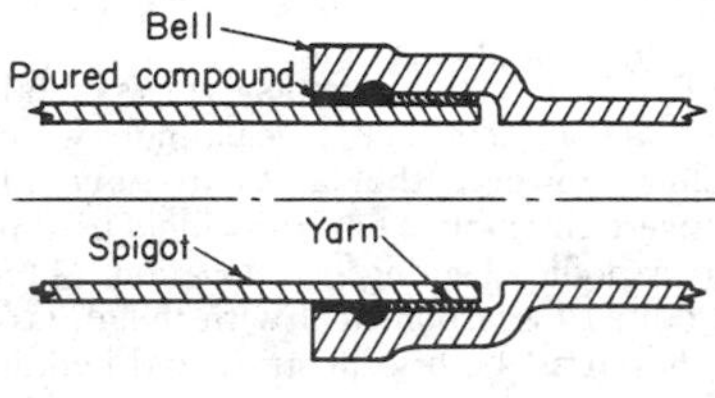

FIG. 6-98 Poured joint.

fittings, and valves are furnished with the bells cast on one or more ends. The pouring compound may be molten, or chemical-setting, or merely compacted; these choices are listed in descending order of ability to hold pressure. These joints cannot absorb angular or axial movement without leaking. Disassembly for maintenance is accomplished by cutting the pipe and reassembly by the use of a coupling with a bell at each end.

Push-on joints (Fig. 6-99) require diametral control of the end of the pipe. They are used for brittle materials. Pipe, fittings, and valves are furnished with the bells cast on one or more ends. Considerable force is required to push the spigot through the O ring; this is reduced by the extension on the O ring, which causes the friction of the pipe to elongate the cross section of the main portion of the O ring.

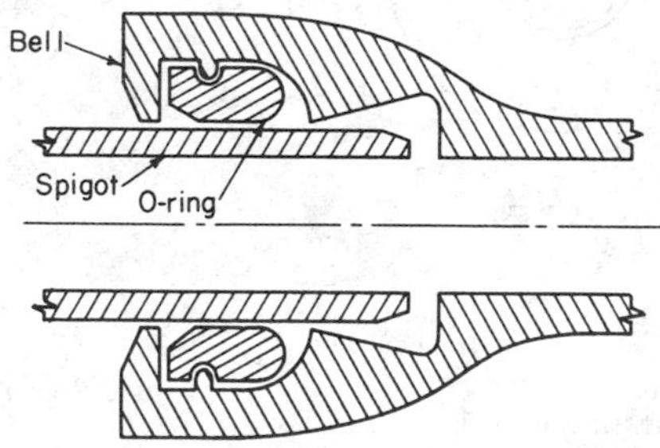

FIG. 6-99 Push-on joint.

Push-on joints do not resist bending moments or axial forces tending to separate the joints but yield to them to an extent limited by the vendor's allowable-angular-deflection and end-movement specifications. End movement can be limited by harnessing or bridling with a combination of rods and clamps, or by anchoring to existing or new structures. The joints are widely used on underground lines. They are not affected by limited earth settlement, and friction of the earth prevents end separation. A lubricant is used on the O ring during assembly. After this disappears, the O ring bonds somewhat to the spigot and disassembly is very difficult. Disassembly for maintenance is accomplished by cutting the pipe and reassembly by use of a coupling with a packed-gland joint on each end.

Expanded joints (Fig. 6-100) are confined to the smaller pipe sizes of ductile metals. A smooth finish is required on the outside of the pipe and on the faces of the ridges inside the bore. Pipe and bore must have the same coefficient of thermal expansion. Furthermore, it is essential that the pipe metal have a lower yield point than the metal into which it is expanded, except in cases in which the metal into which it is expanded is a thin cylinder temporarily backed by clamped heavy semicylindrical metal shells with a high yield point. An expanding tool is required, one for each size of pipe.

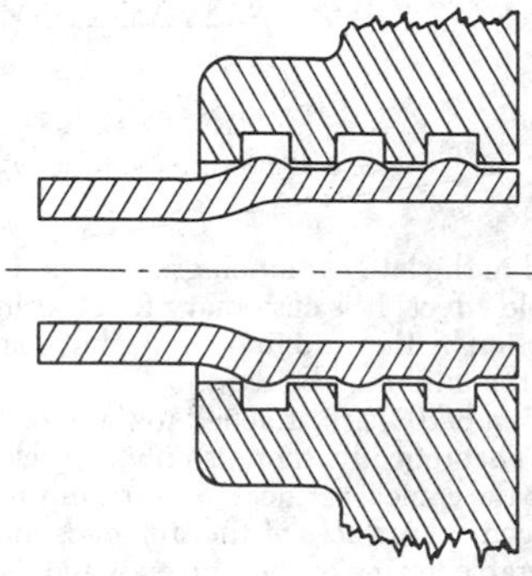

FIG. 6-100 Expanded joint.

After completion of the joint, it is difficult to determine whether the increase in the inside diameter of the pipe represents permanent stretch of the bore of the mating part or flow of metal into the grooves of the bore. An excess of the latter results in excessive thinning of the tube, while an insufficiency of the latter may cause the pipe to pull out of the bore under axial loading. In a variation, the expanded joint is combined with a flared joint to increase resistance to axial load. These joints are used to attach unions and Lovekin flanges to pipe. For alloy piping, composite Lovekin flanges in which the bore and raised face portion are made of the alloy, retained in the steel balance of the flange by an offset, are available.

Grooved joints (Fig. 6-101) are divided into two classes, cut grooves and rolled grooves. Rolled grooves are preferred because, compared with cut grooves, they are easier to form and reduce the metal wall less. However, they slightly reduce the flow area. They are limited to thin walls of ductile material, while cut grooves, because of their reduction of the pipe wall, are limited to thick walls. In the larger pipe sizes, some commonly used wall thicknesses are too thick for rolled grooves but too thin for cut grooves. The thinning of the walls impairs resistance to corrosion and erosion but not to internal pressure, because the thinned area is reinforced by the coupling.

Control of outside diameter is important. Permissible minus tolerance is limited, since it impairs the grip of the couplings. Plus tolerance makes it necessary to cut the cut grooves more deeply, increasing the thinning of the wall. Plus tolerance is not a problem with rolled grooves, since they are confined to walls thin enough so that the couplings can compress the pipe. Pipe is available from vendors already grooved and also with heavier-wall grooved ends welded on.

Grooved joints resist axial forces tending to separate the joints. Angular deflection, up to the limit specified by the vendor, may be used to absorb thermal expansion and to permit the piping to be laid on uneven ground. Compared with flanged joints, grooved joints will not pull misaligned pipe into alignment, and thus they require more support, but otherwise they require less labor for handling, assembly, and disassembly.

Gaskets are self-sealing against both internal and external pressure and are available in a wide variety of elastomers. However, successful performance of an elastomer as a flange gasket does not necessarily mean equally satisfactory performance in a grooved joint, since

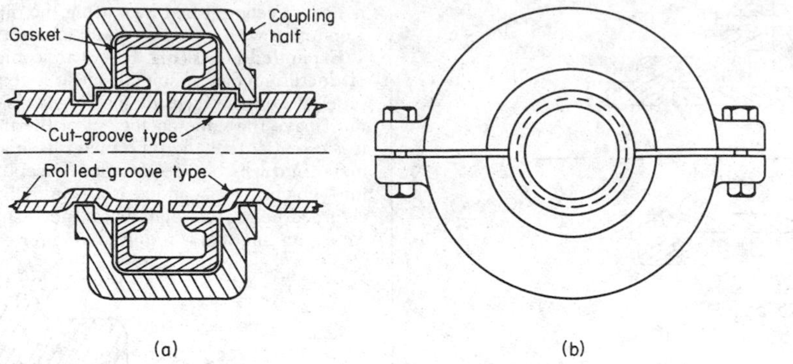

FIG. 6-101 Grooved joint. (*a*) Section. (*b*) End view.

exposure to the fluid in the latter is much greater and hardening has a greater unfavorable effect. It is customary to use couplings which are resistant to corrosion by the fluid in the pipe, but couplings which would contaminate the fluid may be used.

V-clamp joints (Fig. 6-102) are attached to the pipe by butt-weld or expanded joints. Theoretically, there is only one relative position of the parts in which the conical surfaces of the clamp are completely in contact with the conical surfaces of the stub ends. In actual practice, there is considerable flexing of the stub ends and the clamp; also complete contact is not required. This permits use of elastomeric gaskets as well as metal gaskets. Fittings are also available with integral conical shouldered ends.

Conical ends vary from machined forgings to roll-formed tubing, and clamps vary from machined forgings to bands to which several roll-formed channels are attached at their centers by spot welding. A hinge may be inserted in the band as a substitute for one of the draw bolts. Latches may also be substituted for draw bolts.

Compared with flanges, V-clamp joints use less metal, require less labor for assembly, and are less likely to leak under wide-range rapid temperature cycling. However, they are more susceptible to failure or damage from overtightening. They are widely used for high-alloy piping subject to periodic cleaning or relocation. Manufactured as forgings, they are used in carbon steel with metal gaskets for very high pressures. They resist both axial strain and bending moments. Each size of each type of joint is customarily rated by the vendor for both internal pressure and bending moment.

Seal-ring joints (Fig. 6-103) consist of hubs attached to the pipe, generally by welding. The joint is proprietary and sold under the registered trade name of Grayloc. The metal seal ring is in effect a self-energizing gasket. This joint is widely used in petrochemical plants for service at the higher pressures. Valves and other accessories are manufactured with Grayloc hub ends.

Pressure-seal joints (Fig. 6-104) are used for pressures of 4.4 MPa (600 lbf/in^2) and higher. They use less metal than flanged joints but

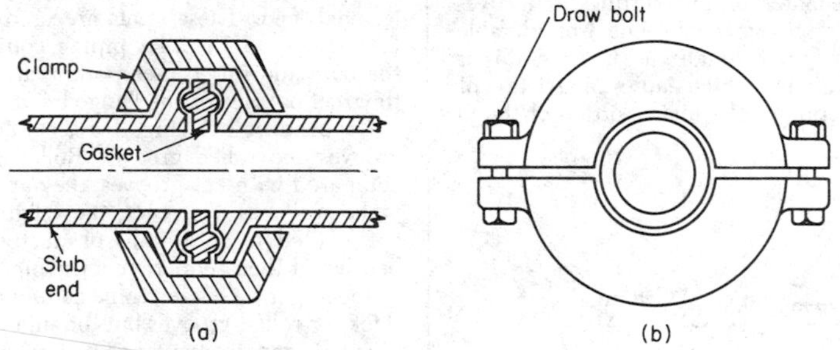

FIG. 6-102 V-clamp joint. (*a*) Section. (*b*) End view.

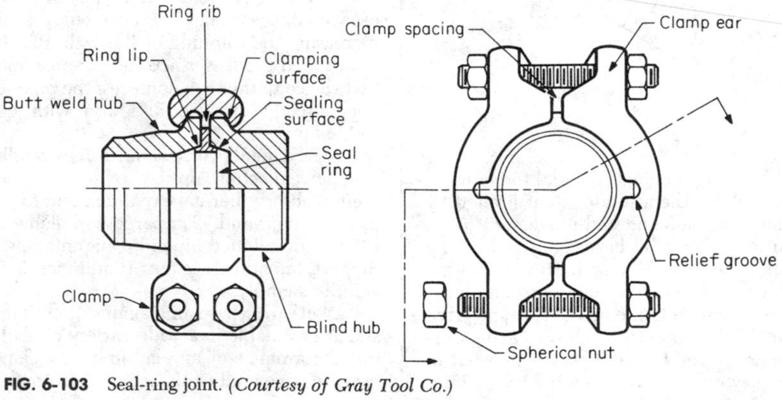

FIG. 6-103 Seal-ring joint. *(Courtesy of Gray Tool Co.)*

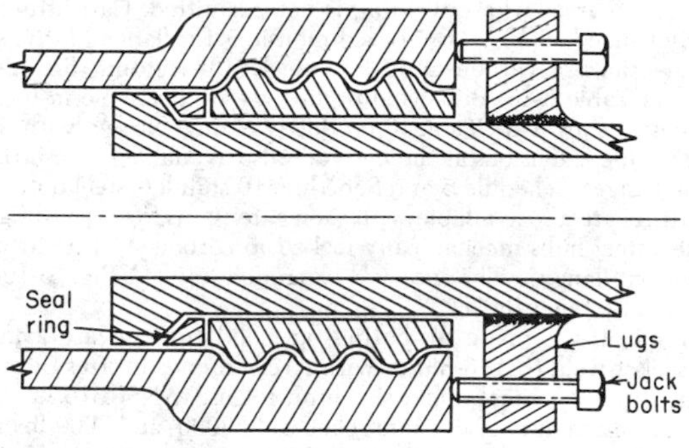

FIG. 6-104 Pressure-seal joint.

require much more machining of surfaces. There are several designs, in all of which increasing fluid pressure increases the force holding the sealing surfaces against each other. These joints are widely used as bonnet joints in carbon and alloy steel valves.

Tubing Joints **Flared-fitting joints** (see Fig. 6-105) are used for ductile tubing when the ratio of wall thickness to the diameter is small enough to permit flaring without cracking the inside surface. The tubing must have a smooth interior surface. The three-piece type avoids torsional strain on the tubing and minimizes vibration fatigue on the flared portion of the tubing. More labor is required for assembly, but the fitting is more resistant to temperature cycling than other tubing fittings and is unlikely to be damaged by overtightening, and its efficiency is not impaired by repeated assembly and disassembly. Size is limited because of the large number of machined surfaces. The nut and, in the three-piece type, the sleeve need not be of the same material as the tubing. For these fittings, less control of tubing diameter is required.

Compression-fitting joints (Fig. 6-106) are used for ductile tubing with thin walls. The outside of the tubing must be clean and smooth. Assembly consists only of inserting the tubing and tightening the nut. These are the least costly tubing fittings but are not resistant to vibration or temperature cycling.

Bite-type-fitting joints (Fig. 6-107) are used when the tubing has too high a ratio of wall thickness to diameter for flaring, when the tubing lacks sufficient ductility for flaring, and for low assembly-labor cost. The outside of the tubing must be clean and smooth. Assembly consists in merely inserting the tubing and tightening the nut. The sleeve must be considerably harder than the tubing yet still ductile enough to be diametrally compressed and must be as resistant to corrosion by the fluid handled as the tubing. The fittings are resistant to vibration but not to wide-range rapid temperature cycling. Compared with flared fittings, they are less suited for repeated assembly and disassembly, require closer diametral control of the tubing, and are more susceptible to damage from overtightening. They are widely used for oil-filled hydraulic systems at all pressures.

O-ring seal joints (Fig. 6-108) are also used for applications requiring heavy-wall tubing. The outside of the tubing must be clean and smooth. The joint may be assembled repeatedly, and as long as the tubing is not damaged, leaks can usually be corrected by replacing the O ring and the antiextrusion washer. This joint is used extensively in oil-filled hydraulic systems.

Soldered Joints (Fig. 6-109) These joints require precise control of the diameter of the pipe or tubing and of the cup in the fitting in order to cause the solder to draw into the clearance between the cup and the tubing by capillary action. Extrusion provides this diametral control, and the joints are most widely used in copper. A 50 percent lead, 50 percent tin solder is used for temperatures up to 93°C

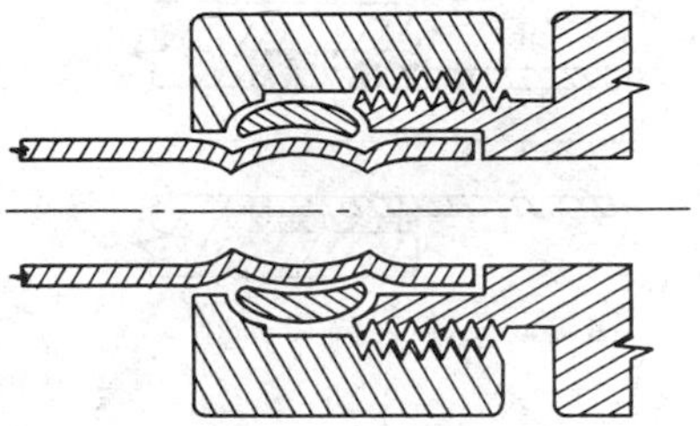

FIG. 6-106 Compression-fitting joint.

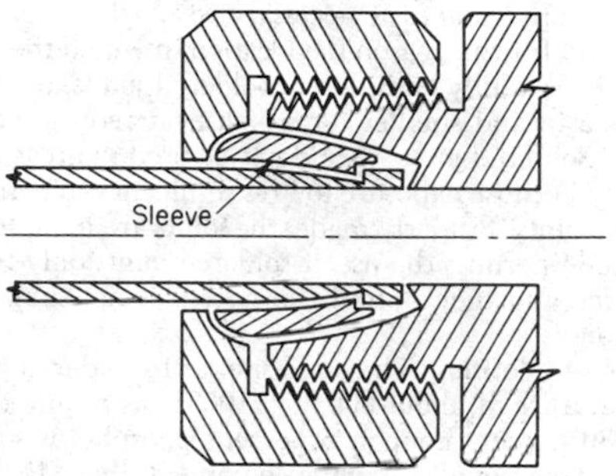

FIG. 6-107 Bite-type-fitting joint.

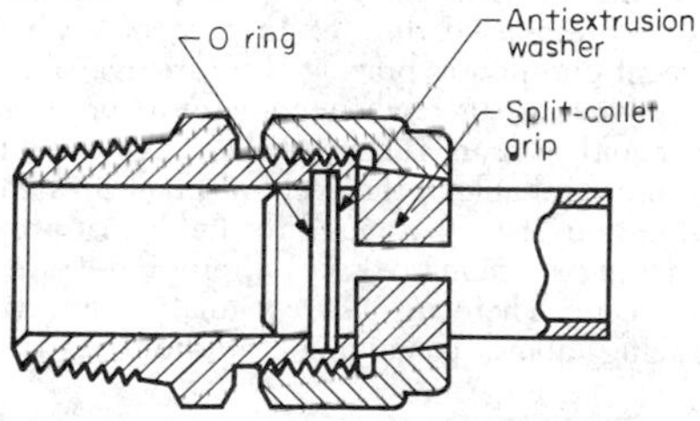

FIG. 6-108 O-ring seal joint. *(Courtesy of the Lenz Co.)*

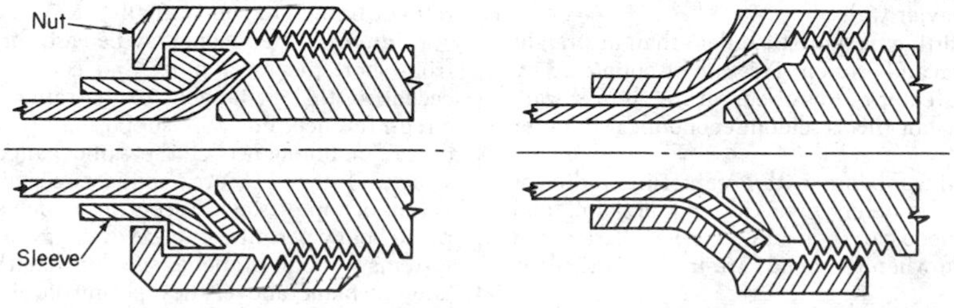

FIG. 6-105 Flared-fitting joint. *(a)* Three-piece. *(b)* Two-piece.

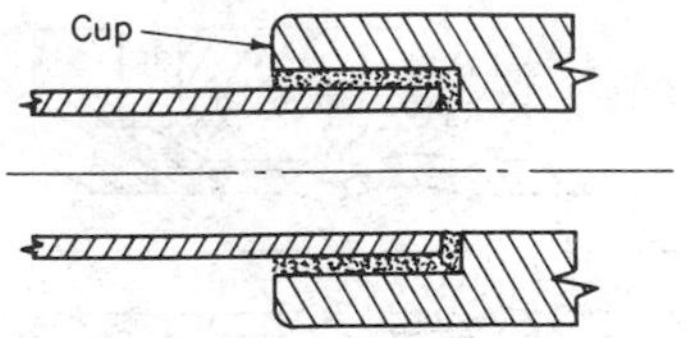

FIG. 6-109 Soldered, brazed, or cemented joint.

(200°F). Careful cleaning of the outside of the tubing and inside of the cup is required.

Heat for soldering is usually obtained from torches. The high conductivity of copper makes it necessary to use large flames for the larger sizes, and for this reason the location in which the joint will be made must be carefully considered. Soldered joints are most widely used in sizes 2 in and smaller for which heat requirements are less burdensome. Soldered joints should not be used in areas where plant fires are likely because exposure to fires results in rapid and complete failure of the joints. Properly made, the joints are completely impervious. The code permits the use of soldered joints only for Category D fluid service and then only if the system is not subject to severe cyclic condions.

Silver Brazed Joints These are similar to soldered joints except that a temperature of about 600°C (1100°F) is required. A 15 percent silver, 80 percent copper, 5 percent phosphorus solder is used for copper and copper alloys, while 45 percent silver, 15 percent copper, 16 percent zinc, 24 percent cadmium solders are used for copper, copper alloys, carbon steel, and alloy steel. Silver-brazed joints are used for temperatures up to 200°C (400°F). Cast-bronze fittings and valves with preinserted rings of 15 percent silver, 80 percent copper, 5 percent phosphorus brazing alloy are available.

Silver-brazed joints are used when temperature or the combination of temperature and pressure is beyond the range of soldered joints. They are also more reliable in the event of plant fires and are more resistant to vibration. If they are used for fluids that are flammable, toxic, or damaging to human tissue, appropriate safeguarding is required by the code. There are OSHA regulations governing the use of silver brazing alloys containing cadmium and other toxic materials.

Bends and Fittings Directional changes in piping systems require bends and elbow fittings. Bends may be made cold or hot. The outside wall is thinned by an amount that varies with the procedure used. Subsequent annealing is required for some materials. To prevent wrinkling and excessive flattening, sand packing is required for hot bending, and sand packing or flexible mandrels may be necessary for cold bending, depending on the ratios of the outside diameter of the pipe to the centerline radius of the bend and to the wall thickness of the pipe. For bends with a centerline radius of five nominal pipe diameters, internal support is not required when the wall thickness is at least 6 percent of the outside diameter of the pipe. Wrinkled bends are made by progressively heating the pipe only on the side which will be the inside of the bend.

Elbow fittings may be cast, forged, or hot- or cold-formed from short pieces of pipe or made by welding together pieces of miter-cut pipe. The thinning of pipe during the forming of elbows is compensated for by starting with heavier walls.

Flow in bends and elbow fittings is more turbulent than in straight pipe, thus increasing corrosion and erosion. This can be countered by selecting a component with greater radius of curvature, thicker wall, or smoother interior contour, but this is seldom economical in miter elbows.

Compared with elbow fittings, bends with a centerline radius of three or five nominal pipe diameters save the cost of joints and reduce pressure drop. Such bends are not suited for installation in a bank of pipes of unequal size when the bends are in the same plane as the bank.

Flanged fittings are used when pipe is likely to be dismantled for frequent cleaning or extensive revision, for lined piping systems, or for seasonal insertion of blanks as a substitute for valves. They are also used in areas where welding is not permitted. Cast fittings are usually flanged. Table 6-14 gives dimensions for flanged fittings.

Dimensions of carbon and alloy steel **butt-welding fittings** are shown in Table 6-15. Butt-welding fittings are available in the wall thicknesses shown in Table 6-6. Butt-welding elbows with short, straight pipe extensions at the ends are also available for insertion in slip-on flanges. Schedule 5 and Schedule 10 stainless-steel butt-welding fittings are also available with such extensions for expanding into stainless-steel hubs mechanically locked in carbon steel ANSI B16.5 dimension flanges. The use of expanded joints (Fig. 6-100) is restricted by the code.

Forged fittings made by boring out solid forgings are available with socket-weld (Fig. 6-88) or with screwed ends in sizes through 4 in, but 2 in is the usual upper size limit for use. ANSI B16.11—1973 gives minimum dimensions for socket-weld 3000- and 6000-lb classes and for screwed 2000-, 3000-, and 6000-lb classes. It also contains pressure-temperature ratings for the classes in various ferrous alloys. The use of socket-weld and screwed fittings is restricted by the code.

Steel forged fittings with **screwed** ends may be installed without pipe dope in the threads and seal-welded (Fig. 6-92) to secure bubble-tight joints with a minimum of welders' labor. They are not subject to deformation by pipe wrenches, and such couplings, bushings, and plugs are often used with the screwed fittings below.

ANSI B16.3—1977 gives dimensions of 150-lb **malleable-iron screwed fittings** through the 6-in size for 1.0 MPa (150 lbf/in^2) saturated steam and 2.1 MPa (300 lbf/in^2) at room temperature and for 300-lb malleable-iron screwed fittings through the 3-in size for 2.1 MPa (300 lbf/in^2) steam at 290°C (550°F) or 7.0 MPa (1000 lbf/in^2) at room temperature. These fittings are available with male threads or unions on one end for installation in confined spaces. Major use is in 150-lb elbows, tees, and reducers in sizes 2 in and smaller. They are less costly than forged fittings but cannot be seal-welded. The code does not permit the use of malleable iron in toxic service or in flammable service above either 150°C (300°F) or 2.76 MPa (400 lbf/in^2) gauge.

ANSI B16.4—1977 gives dimensions of 125-lb **cast-iron screwed fittings** through the 12-in size for 0.86 MPa (125 lb/in^2) saturated steam and 1.2 MPa (175 lbf/in^2) at 66°C (150°F) and of 250-lb cast-iron screwed fittings through the 12-in size for 1.72 MPa (250 lbf/in^2) saturated steam and for 2.76 MPa (400 lbf/in^2) at 66°C (150°F). The 125-lb fittings are made in regular 90° and 45° elbows, reducing elbows, regular and reducing tees, and crosses. The 250-lb fittings are made only in straight sizes. Major use is in 125-lb elbows, tees, and reducers in low-pressure noncritical service. The code does not permit the use of cast iron in toxic service or aboveground within process unit limits for flammable-fluid service above 150°C (300°F) or 1.0 MPa (150 lbf/in^2).

Tees Tees may be cast, forged, or hot- or cold-formed from short pieces of pipe. Though it is impossible to have the same flow simultaneously through all three end connections, it is not economical to produce or stock the great variety of tees which accurate sizing of end connections requires. It is customary to stock only tees with the two end (run) connections of the same size and the branch connection either of the same size as the run connections or one, two, or three sizes smaller. Adjacent reducers or reducing elbow fittings are used for other size reductions. Branch connections (see subsection "Joints") are often more economical than tees, particularly when the ratio of branch to run is small.

Reducers Reducers may be cast, forged, or hot- or cold-formed from short pieces of pipe. End connections may be concentric or eccentric, that is, tangent to the same plane at one point on their circumference. For pipe supported by hangers, concentric reducers permit maintenance of the same hanger length; for pipe laid on structural steel, eccentric reducers permit maintaining the same elevation of top of steel. Eccentric reducers with the common tangent plane below permit complete drainage of branched horizontal piping systems through branches smaller than the main. With the common tangent plane above, they permit liquid flow in horizontal lines to sweep the line free of gas or vapor.

Reducing elbow fittings permit change of direction and concentric size reduction in the same fitting.

TABLE 6-14 Dimensions of Flanged Fittings*

All dimensions in inches

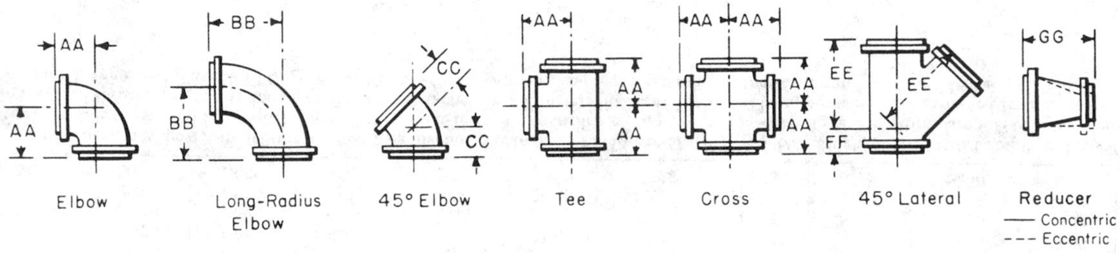

Nominal pipe size	ANSI B16.5, Class 150 ANSI B16.1, Class 125						ANSI B16.5, Class 300 ANSI B16.1, Class 250						ANSI B16.5, Class 400					ANSI B16.5 Class 600				
	AA	BB	CC	EE	FF	GG	AA	BB	CC	EE	FF	GG	AA	CC	EE	FF	GG	AA	CC	EE	FF	GG
½	....	....	....	....	....	....	....	....	....	....	....	....	Use Class 600 dimensions in these sizes					3.25	2.00	5.75	1.75	5.00
¾	....	....	....	....	....	....	....	....	....	....	....	....						3.75	2.50	6.75	2.00	5.00
1	3.50	5.00	1.75	5.75	1.75	4.50	4.00	5.00	2.25	6.50	2.00	4.50						4.25	2.50	7.25	2.25	5.00
1¼	3.75	5.50	2.00	6.25	1.75	4.50	4.25	5.50	2.50	7.25	2.25	4.50						4.50	2.75	8.00	2.50	5.00
1½	4.00	6.00	2.25	7.00	2.00	4.50	4.50	6.00	2.75	8.50	2.50	4.50						4.75	3.00	9.00	2.75	5.00
2	4.50	6.50	2.50	8.00	2.50	5.00	5.00	6.50	3.00	9.00	2.50	5.00						5.75	4.25	10.25	3.50	6.00
2½	5.00	7.00	3.00	9.50	2.50	5.50	5.50	7.00	3.50	10.50	2.50	5.50						6.50	4.50	11.50	3.50	6.75
3	5.50	7.75	3.00	10.00	3.00	6.00	6.00	7.75	3.50	11.00	3.00	6.00						7.00	5.00	12.75	4.00	7.25
3½	6.00	8.50	3.50	11.50	3.00	6.50	6.50	8.50	4.00	12.50	3.00	6.50						7.50	5.50	14.00	4.50	7.75
4	6.50	9.00	4.00	12.00	3.00	7.00	7.00	9.00	4.50	13.50	3.00	7.00	8.00	5.50	16.00	4.50	8.25	8.50	6.00	16.50	4.50	8.75
5	7.50	10.25	4.50	13.50	3.50	8.00	8.00	10.25	5.00	15.00	3.50	8.00	9.00	6.00	16.75	5.00	9.25	10.00	7.00	19.50	6.00	10.25
6	8.00	11.50	5.00	14.50	3.50	9.00	8.50	11.50	5.50	17.50	4.00	9.00	9.75	6.25	18.75	5.25	10.00	11.00	7.50	21.00	6.50	11.25
8	9.00	14.00	5.50	17.50	4.50	11.00	10.00	14.00	6.00	20.50	5.00	11.00	11.75	6.75	22.25	5.75	12.00	13.00	8.50	24.50	7.00	13.25
10	11.00	16.50	6.50	20.50	5.00	12.00	11.50	16.50	7.00	24.00	5.50	12.00	13.25	7.75	25.75	6.25	13.50	15.50	9.50	29.50	8.00	15.75
12	12.00	19.00	7.50	24.50	5.50	14.00	13.00	19.00	8.00	27.50	6.00	14.00	15.00	8.75	29.75	6.50	15.25	16.50	10.00	31.50	8.50	16.75
14	14.00	21.50	7.50	27.00	6.00	16.00	15.00	21.50	8.50	31.00	6.50	16.00	16.25	9.25	32.75	7.00	16.50	17.50	10.75	34.25	9.00	17.75
16	15.00	24.00	8.00	30.00	6.50	18.00	16.50	24.00	9.50	34.50	7.50	18.00	17.75	10.25	36.25	8.00	18.50	19.50	11.75	38.50	10.00	19.75
18	16.50	26.50	8.50	32.00	7.00	19.00	18.00	26.50	10.00	37.50	8.00	19.00	19.25	10.75	39.25	8.50	19.50	21.50	12.25	42.00	10.50	21.75
20	18.00	29.00	9.50	35.00	8.00	20.00	19.50	29.00	10.50	40.50	8.50	20.00	20.75	11.25	42.75	9.00	21.00	23.50	13.00	45.50	11.00	23.75
24	22.00	34.00	11.00	40.50	9.00	24.00	22.50	34.00	12.00	47.50	10.00	24.00	24.25	12.75	50.25	10.50	24.50	27.50	14.75	53.00	13.00	27.75

TABLE 6-14 Dimensions of Flanged Fittings *(Continued)*

All dimensions in inches

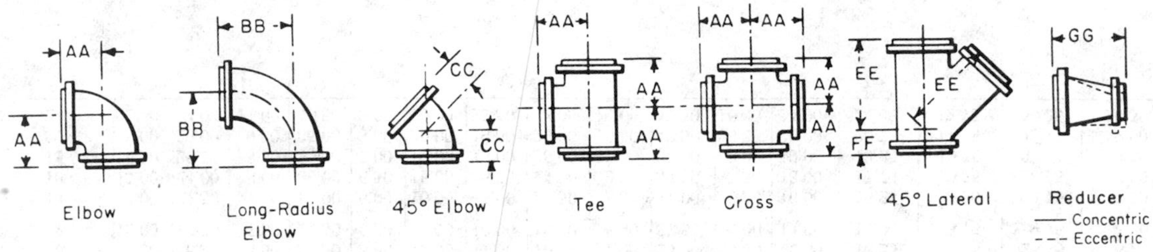

Nominal pipe size	ANSI B16.5, Class 900					ANSI B16.5, Class 1500					ANSI B16.5, Class 2500				
	AA	CC	EE	FF	GG	AA	CC	EE	FF	GG	AA	CC	EE	FF	GG
½	Use Class 1500 dimensions in these sizes					4.25	3.00	....	....	....	5.19				
¾						4.50	3.25	....	....	....	5.37				
1						5.00	3.50	9.00	2.50	5.00	6.06	4.00			
1¼						5.50	4.00	10.00	3.00	5.75	6.87	4.25			
1½						6.00	4.25	11.00	3.50	6.25	7.56	4.75			
2						7.25	4.75	13.25	4.00	7.25	8.87	5.75	15.25	5.25	9.50
2½						8.25	5.25	15.25	4.50	8.25	10.00	6.25	17.25	5.75	10.50
3	7.50	5.50	14.50	4.50	7.75	9.25	5.75	17.25	5.00	9.25	11.37	7.25	19.75	6.75	11.75
4	9.00	6.50	17.50	5.50	9.25	10.75	7.25	19.25	6.00	10.75	13.25	8.50	23.00	7.75	13.50
5	11.00	7.50	21.00	6.50	11.25	13.25	8.75	23.25	7.50	13.75	15.62	10.00	27.25	9.25	15.75
6	12.00	8.00	22.50	6.50	12.25	13.88	9.38	24.88	8.12	14.50	18.00	11.50	31.25	10.50	18.00
8	14.50	9.00	27.50	7.50	14.75	16.38	10.88	29.88	9.12	17.00	20.12	12.75	35.25	11.75	20.50
10	16.50	10.00	31.50	8.50	16.75	19.50	12.00	36.00	10.25	20.25	25.00	16.00	43.25	14.75	25.50
12	19.00	11.00	34.50	9.00	17.75	22.25	13.25	40.75	12.00	23.00	28.00	17.75	49.25	16.25	29.00
14	20.25	11.50	36.50	9.50	19.00	24.75	14.25	44.00	12.50	25.75					
16	22.25	12.50	40.75	10.50	21.00	27.25	16.25	48.25	14.75	28.25					
18	24.00	13.25	45.50	12.00	24.50	30.25	17.75	53.25	16.50	31.50					
20	26.00	14.50	50.25	13.00	26.50	32.75	18.75	57.75	17.75	34.00					
24	30.50	18.00	60.00	15.50	30.50	38.25	20.75	67.25	20.50	39.75					

°Outline drawings show ¼-in (6.5-mm) raised face machined onto flange, as for ANSI B16.5 400-lb and higher. ANSI B16.1 250-lb and ANSI B16.5 150- and 300-lb have 1/16-in (1.5-mm) raised face; ANSI B16.1 125-lb has no raised face. See Tables 6-7 through 6-13 for flange drillings. Dimensions for 400- and 600-lb fittings are identical for sizes ½ to 3½ in inclusive. Dimensions for 900- and 1500-lb fittings are identical for sizes ½ to 2½ in inclusive. To convert inches to millimeters, multiply by 25.4. The dimensions were extracted from Cast-Iron Pipe Flanges and Flanged Fittings, ANSI B16.1—1975, and Steel Pipe Flanges and Flanged Fittings, ANSI B16.5—1977, with permission of the publisher, the American Society of Mechanical Engineers, New York.

TABLE 6-15 Butt-Welding Fittings*

All dimensions in inches

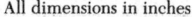

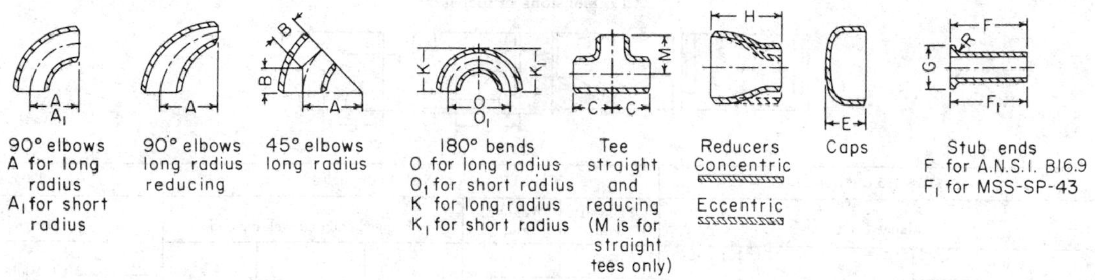

Pipe size	A	K	A1	K1	B	O	O1	M, C	H	E†	G	F	F1	R‡
½	1.50	1.88	....	....	0.62	3.00	....	1.00	....	1.00	1.38	3.00	2.00	0.12
¾ (1)	1.12	1.69	....	....	0.44	2.25	....	1.12	1.50	1.00	1.69	3.00	2.00	0.12
1	1.50	2.19	1.00	1.62	0.88	3.00	2.00	1.50	2.00	1.50	2.00	4.00	2.00	0.12
1¼	1.88	2.75	1.25	2.06	1.00	3.75	2.50	1.88	2.00	1.50	2.50	4.00	2.00	0.19
1½	2.25	3.25	1.50	2.44	1.12	4.50	3.00	2.25	2.50	1.50	2.88	4.00	2.00	0.25
2	3.00	4.19	2.00	3.19	1.38	6.00	4.00	2.50	3.00	1.50	3.62	6.00	2.50	0.31
2½	3.75	5.19	2.50	3.94	1.75	7.50	5.00	3.00	3.50	1.50	4.12	6.00	2.50	0.31
3	4.50	6.25	3.00	4.75	2.00	9.00	6.00	3.38	3.50	2.00	5.00	6.00	2.50	0.38
3½	5.25	7.25	3.50	5.50	2.25	10.50	7.00	3.75	4.00	2.50	5.50	6.00	3.00	0.38
4	6.00	8.25	4.00	6.25	2.50	12.00	8.00	4.12	4.00	2.50	6.19	6.00	3.00	0.44
5	7.50	10.31	5.00	7.75	3.12	15.00	10.00	4.88	5.00	3.00	7.31	8.00	3.00	0.44
6	9.00	12.31	6.00	9.31	3.75	18.00	12.00	5.62	5.50	3.50	8.50	8.00	3.50	0.50
8	12.00	16.31	8.00	12.31	5.00	24.00	16.00	7.00	6.00	4.00	10.62	8.00	4.00	0.50
10	15.00	20.38	10.00	15.38	6.25	30.00	20.00	8.50	7.00	5.00	12.75	10.00	5.00	0.50
12	18.00	24.38	12.00	18.38	7.50	36.00	24.00	10.00	8.00	6.00	15.00	10.00	6.00	0.50
14	21.00	28.00	14.00	21.00	8.75	42.00	28.00	11.00	13.00	6.50	16.25	12.00	6.00	0.50
16	24.00	32.00	16.00	24.00	10.00	48.00	32.00	12.00	14.00	7.00	18.50	12.00	6.00	0.50
18	27.00	36.00	18.00	27.00	11.25	54.00	36.00	13.50	15.00	8.00	21.00	12.00	6.00	0.50
20	30.00	40.00	20.00	30.00	12.50	60.00	40.00	15.00	20.00	9.00	23.00	12.00	6.00	0.50
24	36.00	48.00	24.00	36.00	15.00	72.00	48.00	17.00	20.00	10.50	27.25	12.00	6.00	0.50

*Extracted from Wrought-Steel Butt-Welding Fittings, ANSI B16.9—1978, and from Wrought-Steel Butt-Welding Short-Radius Elbows and Returns, ANSI B16.28—1978, with permission of the publisher, the American Society of Mechanical Engineers, New York. A and B dimensions of 1.50 and 0.75 in respectively may be furnished for NPS ¾ at the manufacturer's option. O and K dimensions may likewise be furnished in 2.00 in and 3.00 in respectively.

†For wall thicknesses greater than extra heavy, E is greater than shown here for sizes 2 in and larger.

‡For MSS SP-43 type B stub ends, which are designed to be backed up by slip-on flanges, R = ½₂ in for 4 in and smaller and ⅙₆ in for 6 through 12 in. To convert inches to millimeters multiply by 25.4.

Valves Valve bodies may be cast, forged, machined from bar stock, or fabricated from welded plate. Steel valves are available with screwed or socket-weld ends in the smaller sizes. Bronze and brass screwed-end valves are widely used for low-pressure service in steel systems. Table 6-16 gives contact-surface-of-face to contact-surface-of-face dimensions for flanged ferrous valves and end-to-end dimensions for butt-welding ferrous valves. Drilling of end flanges is shown in Tables 6-7 to 6-13. Bolt holes are located so that the stem is equidistant from the centerline of two bolt holes. Even if removal for maintenance is not anticipated, flanged valves are frequently used instead of butt-welding-end valves because they permit insertion of blanks for isolating sections of a loop piping system.

Ferrous valves are also available in nodular (ductile) iron, which has tensile strength and yield point approximately equal to cast carbon steel at temperatures of 343°C (650°F) and below and only slightly less elongation.

Valves serve not only to regulate the flow of fluids but also to isolate piping or equipment for maintenance without interrupting other connected units. Valve design should keep pressure, temperature changes, and strain from connected piping from distorting or misaligning the sealing surfaces. The sealing surfaces should be of such material and design that the valve will remain tight over a reasonable service period. The principal types are named, described, compared, and illustrated with line diagrams in subsequent subsections. In the line diagrams, the operating stem is shown in solid black, direction of flow by arrows on a thin solid line, and motion of valve parts by arrows on a dotted line. Moving parts are drawn with solid lines in the nearly closed position and with dotted lines in the fully open position. Packing is represented by an X in a square.

Gate Valves (Figs. 6-110) These valves are designed in two types. The wedge-shaped-gate, **inclined-seat** type is most commonly used. The wedge gate is usually solid but may be flexible (partly cut into halves by a plane at right angles to the pipe) or split (completely cleft by such a plane). Flexible and split wedges minimize galling of the sealing surfaces by distorting more easily to match angularly misaligned seats. In the double-disk **parallel-seat** type, an inclined-plane device mounted between the disks converts stem force to axial force, pressing the disks against the seats after the disks have been positioned for closing. This gate assembly distorts automatically to match both angular misalignment of the seats and longitudinal shrinkage of the valve body on cooling.

When shearing high-velocity flow of dense fluids, the gate assemblies shake violently, and for this service solid-wedge or flexible-wedge valves are preferred. When valve operation is manual, small bypass valves installed in parallel with the main valve may be used to eliminate the shake problem and to minimize manual effort in opening and closing the valves. Double-disk parallel-seat valves should be installed with the stem essentially vertical. All wedge gate valves are equipped with tongue-and-groove guides to keep the gate sealing surfaces from clattering on the seats and marring them during opening and closing. Depending on the velocity and density of the fluid stream being sheared, these guiding surfaces may be as cast, machined, or hard-surfaced and ground.

Gate valves may have nonrising stems, inside-screw rising stems,

TABLE 6-16 Dimensions of Valves*

All dimensions in inches

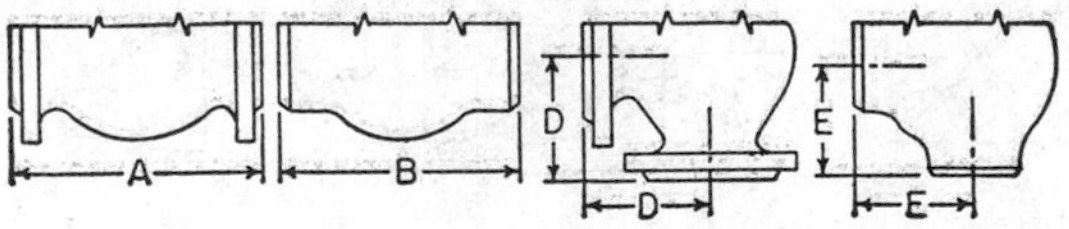

Nominal valve size	Class 125 cast iron					Class 150 steel, MSS-SP-42 through 12-in size					Class 250 cast iron		
	Flanged end					Flanged end	Welding end	Flanged end and welding end			Flanged end		
	Gate					Gate	Gate				Gate		
	Solid wedge A	Double disk A	Globe and lift check A	Angle and lift check D	Swing check A	Solid wedge and double disk A	Solid wedge and double disk B	Globe and lift check A and B	Angle and lift check D and E	Swing check A and B	Solid wedge and double disk A	Globe, lift check, and swing check A	Angle and lift check D
¼	....	....	....	....	....	4	4	4	2	4			
⅜	....	....	....	....	....	4	4	4	2	4			
½	....	....	....	....	....	4¼	4¼	4¼	2¼	4¼			
¾	....	....	....	....	....	4⅝	4⅝	4⅝	2½	4⅝			
1	....	....	....	....	....	5	5	5	2¾	5			
1¼	....	....	....	....	....	5½	5½	5½	3	5½			
1½	....	....	....	....	....	6½	6½	6½	3¼	6½			
2	7	7	8	4	8	7	8½	8	4	8	8½	10½	5¼
2½	7½	7½	8½	4¼	8½	7½	9½	8½	4¼	8½	9½	11½	5¾
3	8	8	9½	4¾	9½	8	11⅛	9½	4¾	9½	11⅛	12½	6¼
3½	8½	8½	....	....	....	†	....	....	....	....	12	14	7
4	9	9	11½	5¾	11½	9	12	11½	5¾	11½	15	15¾	7⅞
5	10	10	13	6½	13	10	15	14	7	13	15⅞	17½	8¾
6	10½	10½	14	7	14	10½	15⅞	16	8	14	16½	21	10½
8	11½	11½	19½	9¾	19½	11½	16½	19½	9¾	19½	18	24½	12¼
10	13	13	24½	12¼	24½	13	18	24½	12¼	24½	19¾	28	14
12	14	14	27½	13¾	27½	14	19¾	27½	13¾	27½	22½	†	
14	15	†	31	15½	31	15	22½	31	15½	31	24	†	
16	16	†	36	18	°	16	24	36	18	†	26		
18	17	†	....	....	°	17	26	....	....	†	28		
20	18	†	....	....	°	18	28	....	....	†			
24	20	†	....	....	°	20	32	....	....	†	31		

Nominal valve size	Class 300 steel				Class 400 steel				Class 600 steel						
	Flanged end and welding end				Flanged end and welding end				Flanged end and welding end						
	Gate				Gate				Gate					Angle and lift check	
	Solid wedge and double disk A and B	Globe and lift check A and B	Angle and lift check D and E	Swing check A and B	Solid wedge A and B	Double disk A and B	Globe, lift check, and swing check A and B	Angle and lift check D and E	Solid wedge A and B	Double disk A and B	Short pattern‡ B	Regular globe, regular lift check, swing check A and B	Short pattern‡ globe, short pattern lift check B	Regular D and E	Short pattern E
½	5½	6	3	....	6½	....	6½	3¼	6½	....	....	6½	....	3¼	
¾	6	7	3½	....	7½	....	7½	3¾	7½	....	....	7½	....	3¾	
1	6½§	8	4	8½	8½	8½	8½	4¼	8½	8½	5¼	8½	5¼	4¼	
1¼	7§	8½	4¼	9	9	9	9	4½	9	9	5¾	9	5¾	4½	
1½	7½	9	4½	9½	9½	9½	9½	4¾	9½	9½	6	9½	6	4¾	
2	8½	10½	5¼	10½	11½	11½	11½	5¾	11½	11½	7	11½	7	5¾	4¼
2½	9½	11½	5¾	11½	13	13	13	6½	13	13	8½	13	8½	6½	5
3	11⅛	12½	6¼	12½	14	14	14	7	14	14	10	14	10	7	6
4	12	14	7	14	16	16	16	8	17	17	12	17	12	8½	7
5	15	15¾	7⅞	15¾	18	18	18	9	20	20	15	20	15	10	8½
6	15⅞	17½	8¾	17½	19½	19½	19½	9¾	22	22	18	22	18	11	10
8	16½	22	11	21	23½	23½	23½	11¾	26	26	23	26	23	13	
10	18	24½	12¼	24½	26½	26½	26½	13¼	31	31	28	31	28	15½	
12	19¾	28	14	28	30	30	30	15	33	33	32	33	32	16½	
14	30	....	....	†	32½	30½	†	....	35	35	35	†			
16	33	....	....	†	35½	35½	†	....	39	39	39	†			
18	36	....	....	†	38½	38½	†	....	43	43	43	†			
20	39	....	....	†	41½	41½	†	....	47	47	47	†			
22	43	....	....	†	45	45	†	....	51	51	....	†			
24	45	....	....	†	48½	48½	†	....	55	55	55	†			

TABLE 6-16 Dimensions of Valves *(Continued)*

Nominal valve size	Class 900 steel, flanged end and welding end: Gate, Solid wedge A and B	Gate, Double disk A and B	Gate, Short pattern‡ B	Regular globe regular lift check, swing check A and B	Short pattern‡ globe, short pattern lift check B	Angle and lift check: Regular D and E	Angle and lift check: Short pattern E	Class 1500 steel, flanged end and welding end: Gate, Solid wedge A and B	Gate, Double disk A and B	Gate, Short pattern‡ B	Globe, lift check, swing check A and B	Angle and lift check D and E
¾	....	....	....	9	...	4½	....	....	....	....	9	4½
1	10	....	5½	10	...	5	....	10	....	5½	10	5
1¼	11	....	6½	11	...	5½	....	11	....	6½	11	5½
1½	12	....	7	12	...	6	....	12	....	7	12	6
2	14½	14½	8½	14½	...	7¼	....	14½	14½	8½	14½	7¼
2½	16½	16½	10	16½	...	8¼	....	16½	16½	10	16½	8¼
3	15	15	12	15	12	7½	6	18½	18½	12	18½	9¼
4	18	18	14	18	14	9	7	21½	21½	16	21½	10¾
5	22	22	17	22	17	11	8½	26½	26½	19	26½	13¼
6	24	24	20	24	20	12	10	27¾	27¾	22	27¾	13⅞
8	29	29	26	29	26	14½	13	32¾	32¾	28	32¾	16⅜
10	33	33	31	33	31	16½	15½	39	39	34	39	19½
12	38	38	36	38	36	19	18	44½	44½	39	44½	22¼
14	40½	40½	39	40½	39	20¼	19½	49½	49½	42	49½	24¾
16	44½	44½	43	....	...	....	....	54½	54½	47	....	....
18	48	48	†	....	...	....	....	60½	60½	53	....	....
20	52	52	†	....	...	....	....	65½	65½	58	....	....
24	61	61	†	....	...	....	....	76½	76½			

Nominal valve size	Class 2500 steel, flanged end and welding end: Gate, Solid wedge A and B	Gate, Double disk A and B	Gate, Short pattern‡ B	Globe, lift check, swing check A and B	Angle and lift check B
½	10⅜	....	....	10⅜	5 3/16
¾	10¾	....	....	10¾	5⅜
1	12⅛	....	7 5/16	12⅛	6 1/16
1¼	13¾	....	9⅛	13¾	6⅞
1½	15⅛	....	9⅛	15⅛	7 9/16
2	17¾	17¾	11	17¾	8⅞
2½	20	20	13	20	10
3	22¾	22¾	14½	22¾	11⅜
4	26½	26½	18	26½	13¼
5	31¼	31¼	21	31¼	15⅝
6	36	36	24	36	18
8	40¼	40¼	30	40¼	20⅛
10	50	50	36	50	25
12	56	56	41	56	28
14	....	....	44		
16	....	....	49		
18	....	....	55		

NOTE: Outline drawings for flanged valves shown ¼-in raised face machined onto flange, as for 400-lb cast-steel valves; 150- and 300-lb cast-steel valves and 250-lb cast-iron valves have 1/16-in raised faces; 125-lb cast-iron and 150-lb corrosion-resistant valves covered by MSS-SP-42 have no raised faces.

°Extracted from Face-to-Face and End-to-End Dimensions of Ferrous Valves, ANSI B16.10—1973, with permission of the publisher, the American Society of Mechanical Engineers, New York. To convert inches to millimeters, multiply by 25.4.

†Not shown in ANSI B16.10 but commercially available.

‡These dimensions apply to pressure-seal or flangeless bonnet valves only.

§Solid wedge only.

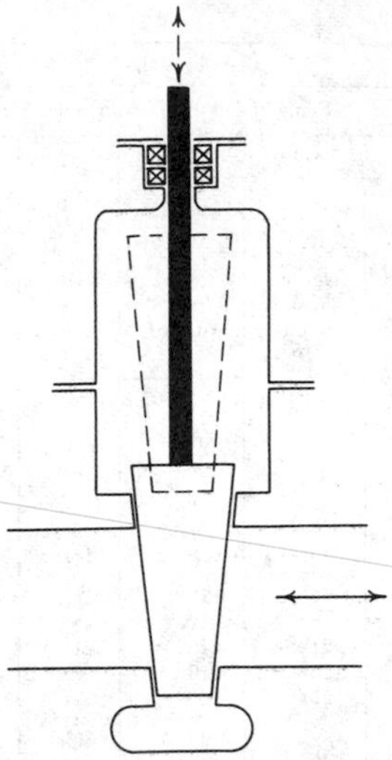

FIG. 6-110 Gate valve.

or outside-screw rising stems, listed in order of decreasing exposure of the stem threads to the fluid handled. Rising-stem valves require more space, but the position of the stem visually indicates the position of the gate. Indication is clearest on the outside-screw rising-stem valves, and on these the stem threads and thrust collars may be lubricated, reducing operating effort. The stem connection to the gate assembly prevents the stem from rotating.

Gate valves are used to minimize pressure drop in the open position and to stop the flow of fluid rather than to regulate it. The problem, when the valve is closed, of pressure buildup in the bonnet from cold liquids expanding or chemical action between fluid and bonnet should be solved by a relief valve or by notching the upstream seat ring.

Globe Valves (Fig. 6-111) These are designed as either inside-screw rising-stem or outside-screw rising-stem. Small valves generally are of the inside-screw type, while in larger sizes the outside-screw type is preferred. In most designs the disks are free to rotate on the stems; this prevents galling between the disk and the seat.

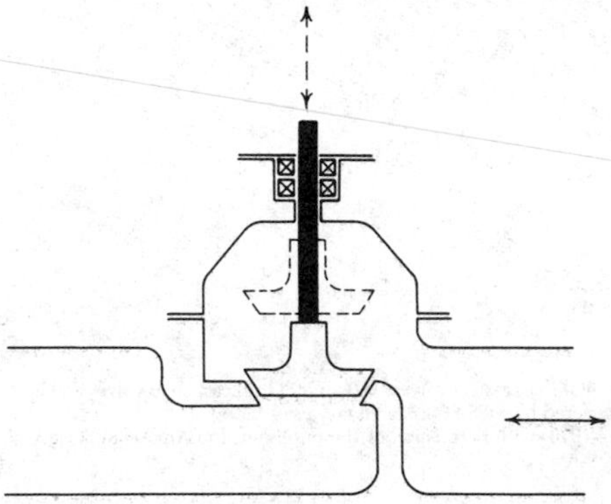

FIG. 6-111 Glove valve.

In the larger sizes, with conical seats, this swivel may permit enough misalignment to prevent proper sealing between the disk and the seat. When the valve is close to an elbow on the upstream side, the swivel also permits uneven distribution of the fluid to spin the disk on the stem. Guides above the disk, below the disk, or both are used to prevent misalignment and spinning. Misalignment can also be prevented by the use of spherical seats and designing the disk so that the pressure point of the stem on the disk is at the center of the sphere. In some designs, spinning and misalignment are prevented by rigidly attaching the disk to the stem, preventing rotation of the stem by lugs which ride along the yoke, and using a yoke bushing as in outside-screw-and-yoke gate valves.

Large globe valves should be installed with stems vertical. Globe valves are preferably installed with the higher-pressure side connected to the top of the disk. Exceptions occur (1) when blocked flow caused by separation of the disk from the stem would damage equipment or (2) when the valve is installed in seldom-used vertical drain lines in which accumulation of rust, scale, or sludge might prevent opening the valve.

Pressure drop through globe valves is much greater than that for gate valves. In Y-type globe valves, the stem and seat are at about 45° to the pipe instead of 90°. This reduces pressure drop but impairs alignment of seat and disk.

Globe valves in horizontal lines prevent complete drainage. Seat-wiper valves in which the disk may be rotated by a separate stem inside and concentric with the main stem are used to clear the seats of solid deposits.

Angle Valves (Fig. 6-112) These valves are similar to globe valves; the same bonnet, stem, and disk are used for both. They combine an elbow fitting and a globe valve into one component with a substantial saving in pressure drop. Flanged angle valves are easier to remove and replace than flanged globe valves.

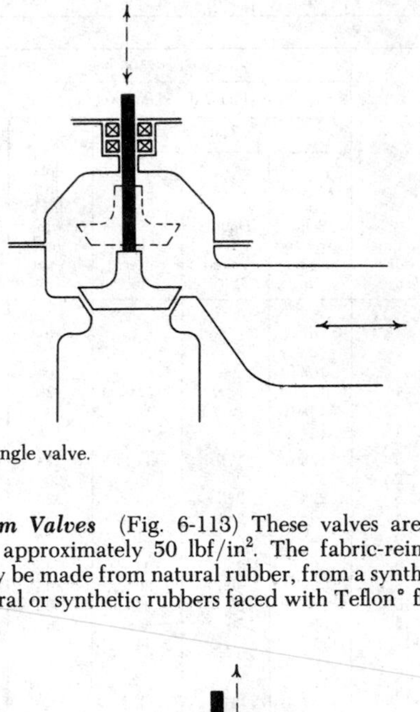

FIG. 6-112 Angle valve.

Diaphragm Valves (Fig. 6-113) These valves are limited to pressures of approximately 50 lbf/in^2. The fabric-reinforced diaphragms may be made from natural rubber, from a synthetic rubber, or from natural or synthetic rubbers faced with Teflon* fluorocarbon

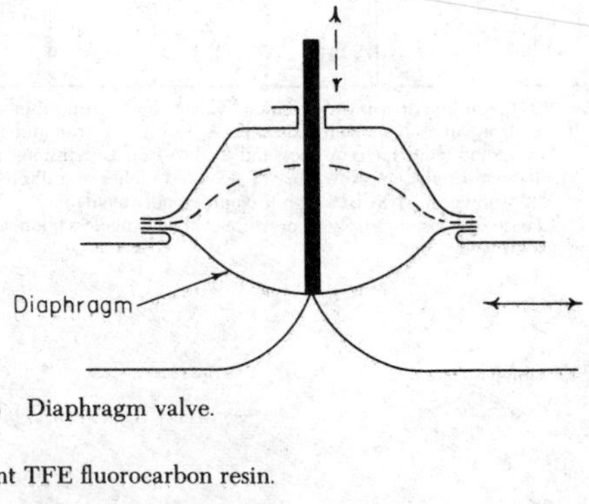

FIG. 6-113 Diaphragm valve.

*Du Pont TFE fluorocarbon resin.

resin. The simple shape of the body makes lining it economical. Elastomers have shorter lives as diaphragms than as linings because of flexing but still provide staisfactory service. Plastic bodies, which have low moduli of elasticity compared with metals, are practical in diaphragm valves since alignment and distortion are minor problems.

These valves are excellent for fluids containing suspended solids and can be installed in any position. Models in which the dam is very low, reducing pressure drop to a negligible quantity and permitting complete drainage in horizontal lines, are available. However, drainage can be obtained with any model simply by installing it with the stem horizontal. The only maintenance required is replacement of the diaphragm, which can be done very quickly without removing the valve from the line.

Plug Cocks (Fig. 6-114) These valves are limited to temperatures below 260°C (500°F) since differential expansion between the plug and the body results in seizure. The size and shape of the port divide these valves into different types. In order of increasing cost they are short venturi, reduced rectangular port; long venturi, reduced rectangular port; full rectangular port; and full round port.

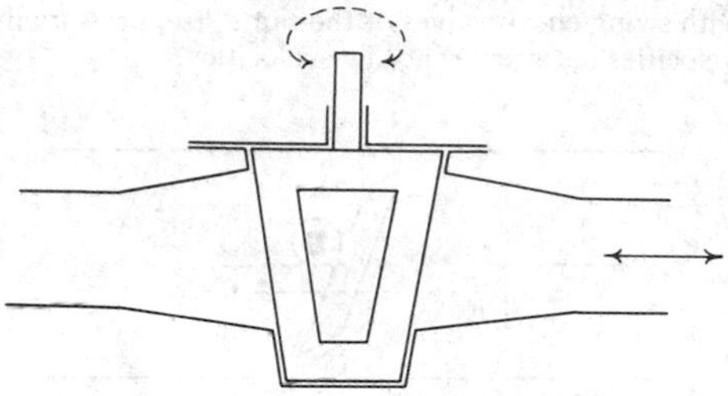

FIG. 6-114 Plug cock.

In lever-sealed plug cocks, tapered plugs are used. The plugs are raised by turning one lever, rotated by another lever, and reseated by the first lever. Lubricated plug cocks may use straight or tapered plugs. The tapered plugs may be raised slightly, to reduce turning effort, by injection of the lubricant, which also acts as a seal. Plastic is used in nonlubricated plug cocks as a body liner, a plug coating, or port seals in the body or on the plug.

In plug cocks other than lever-sealed plug cocks, the contact area between plug and body is large, and gearing is usually used in sizes 6 in and larger to minimize operating effort. There are several lever-sealed plug cocks incorporationg mechanisms which convert the rotary motion of a handwheel into sequenced motion of the two levers.

For lubricated plug cocks, the lubricant must have limited viscosity change over the range of operating temperature, must have low solubility in the fluid handled, and must be applied regularly. There must be no chemical reaction between the lubricant and the fluid which would harden or soften the lubricant or contaminate the fluid. For these reasons, lubricated plug cocks are most often used when there are a large number handling the same or closely related fluids at approximately the same temperature.

Lever-sealed plug cocks are used for throttling service. Because of the large contact area between plug and body, if a plug cock is operable, there is little likelihood of leakage when closed, and the handle position is a clearly visible indication of the valve position.

Ball Valves (Fig. 6-115 and 6-116) These valves are limited to temperatures that have little effect on their plastic seats. Since the sealing element is a ball, its alignment with the axis of the stem is not essential to tight shutoff. In free-ball valves the ball is free to move axially. Pressure differential across the valve forces the ball in the closed position against the downstream seat and the latter against the body. In fixed-ball valves, the ball rotates on stem extensions, with the bearings sealed with O rings. Plastic seats may be compressed or spring-loaded against the ball and the body by the assembly of the valves, or they may be forced against the ball by pressure across

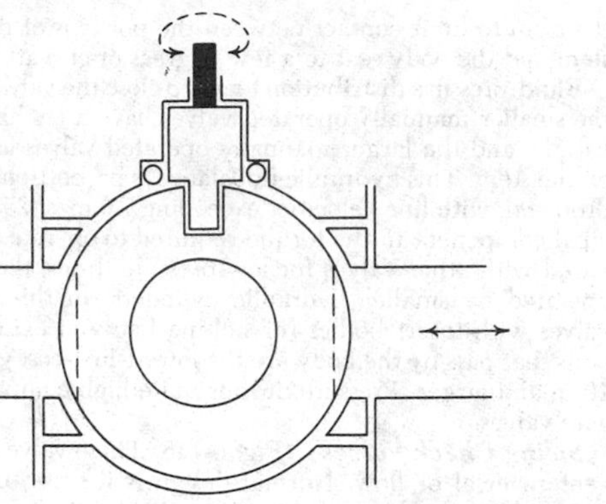

FIG. 6-115 Ball valve; free ball.

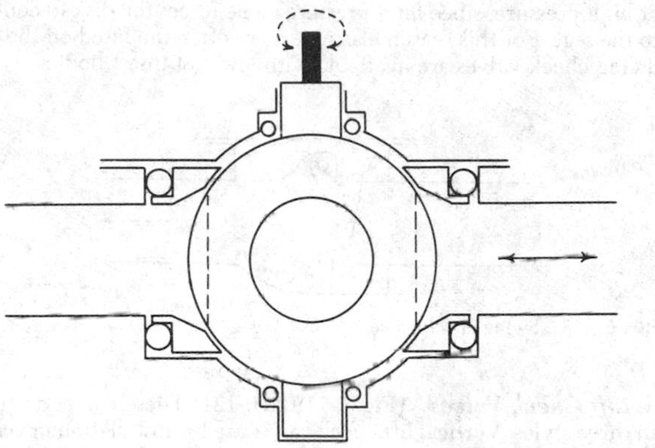

FIG. 6-116 Ball valve; fixed ball.

the valve acting against O rings which seal between the seat and the body.

Ball valves in which the ball and seats are inserted from above are known as top-entry ball valves. Replacement of seats is easiest in this type. The others are known as split-body valves. Some of these incorporate bolted assembly which permits their use as joints for assembly of the piping. Replacement of seats in this type is easiest when the body consists of three pieces with the ball and the seats contained in the middle piece.

For the larger sizes in high-pressure service, the fixed-ball type with O-ring seat seals requires less operating effort. However, these require two different plastic materials with resistance to the fluid and its temperature. Like plug cocks, ball valves may be either restricted-port or full-port, but the ports are always round and pressure drop is low.

Butterfly Valves (Fig. 6-117) These valves occupy less space in the line than any other valves. Relatively tight sealing without excessive operating torque and seat wear is accomplished by a variety of methods, such as resilient seats, piston rings on the disk, and inclining

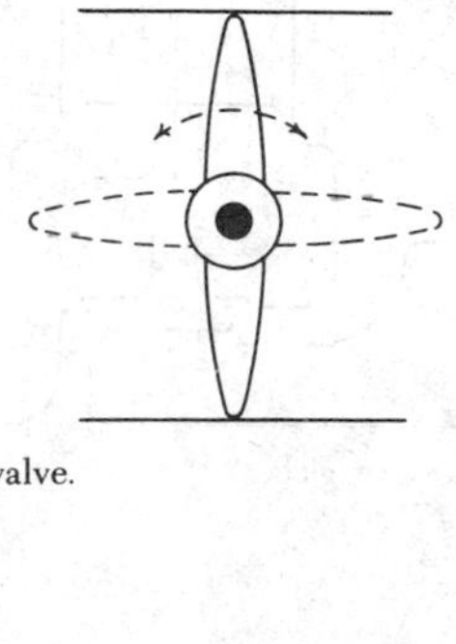

FIG. 6-117 Butterfly valve.

the stem to limit contact between the portions of disk closest to the stem and the body seat to a few degrees of curvature.

Fluid-pressure distribution tends to close the valve. For this reason, the smaller manually operated valves have a latching device on the handle, and the larger manually operated valves use worm gearing on the stem. This hydraulic unbalance is proportional to the pressure drop and, with line velocities exceeding 7.6 m/s (25 ft/s), is the principal component in the torque required to operate the valves. Compared with other valves for low-pressure drops, these valves can be operated by smaller hydraulic cylinders. In this service butterfly valves with insert bodies for bolting between existing flanges with bolts that pass by the body are the lowest-first-cost valve in pipe sizes 10 in and larger. Pressure drop is quite high compared with that of gate valves.

Swing Check Valves (Fig. 6-118) These valves are used to prevent reversal of flow. Normal design is for use only in horizontal lines, where the force of gravity on the disk is at a maximum at the start of closing and at a minimum at the end of closing. Unlike most other valves, check valves are more likely to leak at low pressure than at high pressure, since fluid pressure alone forces the disk to conform to the seat. For this reason elastomers are often mounted on the disk. Swing check valves are available with low-cost insert bodies.

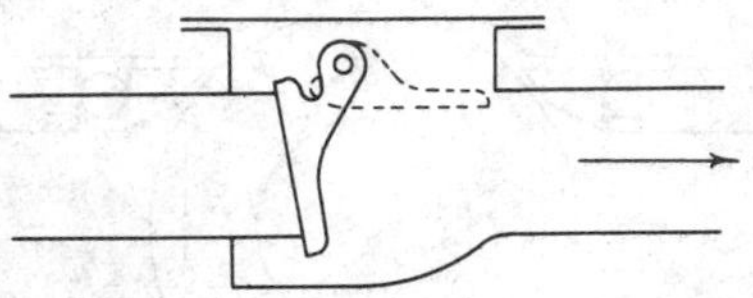

FIG. 6-118 Swing check valve.

Lift Check Valves (Fig. 6-119 to 6-121) These valves are made in three styles. Vertical lift check valves are for installation in vertical lines, where the flow is normally upward; globe check valves are for use in horizontal lines; angle check valves are for installation where a vertical line with upward flow turns horizontal. Globe and angle check valves normally incorporate an integral dashpot above the disk to slow the motion of the disk and reduce wear. In vertical lift check valves, this feature is found only in the larger sizes. Springs may be incorporated in the dashpots to speed closing, but this increases the pressure drop. Lift checks should not be used when the fluid contains suspended solids.

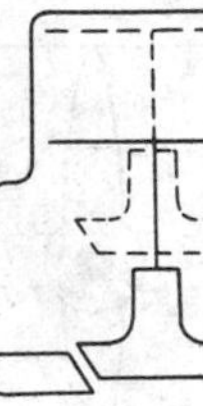

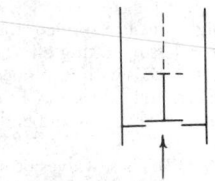

FIG. 6-119 Lift check valve, vertical.

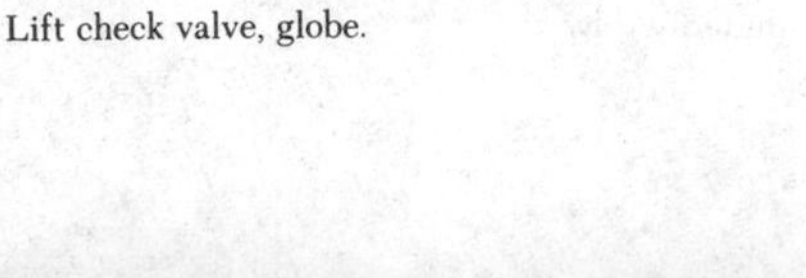

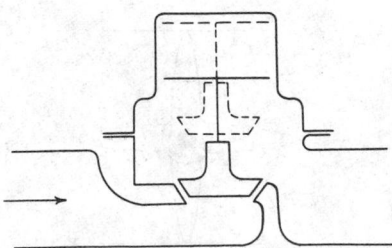

FIG. 6-120 Lift check valve, globe.

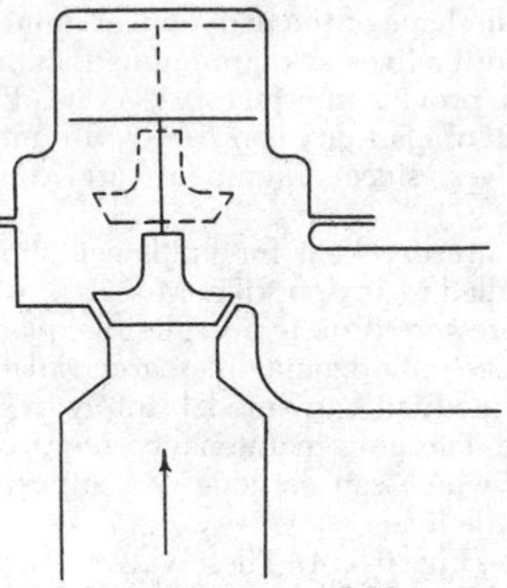

FIG. 6-121 Lift check valve, angle.

Tilting-Disk Check Valves (Fig. 6-122) These valves may be installed in a horizontal line or in lines in which the flow is vertically upward. The pivot point is located so that the distribution of pressure in the fluid handled speeds the closing but arrests slamming. Compared with swing check valves of the same size, pressure drop is less at low velocities but greater at high velocities.

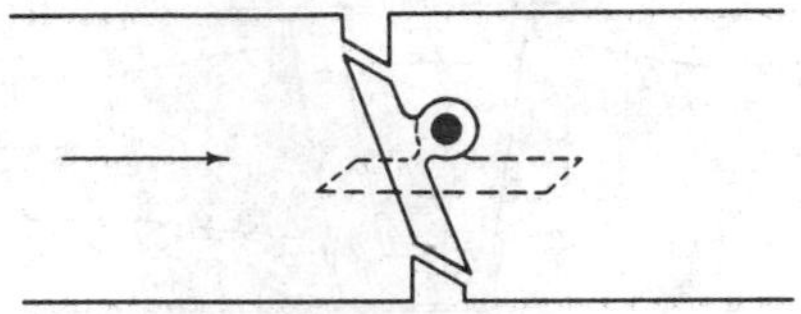

FIG. 6-122 Tilting-disk check valve.

Closure at the instant of reversal of flow is most nearly attained in these valves. This timing of closure is not the whole solution to noise and shock at check valves. For example, if cessation of pressure at the inlet of a valve produces flashing of the decelerating stream downstream from the valve or if stoppage of flow is caused by a sudden closure of a valve some distance downstream from the check valve and the stoppage is followed by returning water hammer, slower closure may be necessary. For these applications, tilting-disk check valves are equipped with external dashpots. They are also available with low-cost insert bodies.

Valve Trim Various alloys are available for valve parts such as seats, disks, and stems which must retain smooth finish for successful operation. The problem in seat materials is fivefold: (1) resistance to corrosion by the fluid handled and to oxidation at high temperatures, (2) resistance to erosion by suspended solids in the fluid, (3) prevention of galling (seizure at point of contact) by differences in material or hardness or both, (4) maintenance of high strength at high temperature, and (5) avoidance of distortion.

All valve trim materials have coefficients of thermal expansion which exceed those of cast or forged carbon steel by 24 to 45 percent and tend to cause distortion of seats and disks. To some extent leakage from this cause is prevented by closing the valve more tightly. Inserting a ring of high-temperature elastomer or plastic, either in or alongside the trim metal in the seat or disk, prevents leakage from this cause.

Cast Iron, Ductile Iron, and High-Silicon Iron

Cast Iron and Ductile Iron Cast iron and ductile iron provide more metal for less cost than steel in piping systems and are widely used in low-pressure services in which internal and external corrosion may cause a considerable loss of metal. They are widely used for underground water distribution. Cement lining is available at a nominal cost for handling water causing tuberculation.

Ductile iron has an elongation of 10 percent or more compared with essentially nil elongation for cast iron and has for all practical purposes supplanted cast iron as a cast piping material. It is usually centrifugally cast in rapidly revolving molds. This manufacturing method improves tensile strength and reduces porosity. Ductile-iron pipe is manufactured to ANSI A21.51—1976 and is available in nominal sizes from 3 through 54 in. Wall thicknesses are specified by seven standard thickness classes. Table 6-17 gives the outside diameter and standard thickness for various rated water working pressures for centrifugally cast ductile-iron pipe. The required wall thickness for underground installations increases with internal pressure, depth of laying, and weight of vehicles operating over the pipe. It is reduced by the degree to which the soil surrounding the pipe provides uniform support along the pipe and around the lower 180°. Tables are provided in ANSI A21.51 for determining wall-thickness-class recommendations for various installation conditions. The poured joint (Fig. 6-98) has been almost entirely superseded by the mechanical joint (Fig. 6-97) and the push-on joint (Fig. 6-99), which are better suited to wet trenches, bad weather, and unskilled labor and minimize strain on the pipe from ground settlement. Lengths vary between 5 and 6 m (between 18 and 20 ft), depending on the supplier. Stock fittings are designed for 1.72-MPa (250-lbf/in^2) cast iron or 2.41-MPa (350-lbf/in^2) ductile iron in sizes through 12 in and for 1.0- and 1.72-MPa (150- and 250-lbf/in^2) cast iron or 2.41-MPa (350-lbf/in^2) ductile iron in sizes 14 in and larger. Stock fittings include 22½° and 11¼° bends. Ductile-iron pipe is also supplied with flanges that match the dimensions of Class 125 flanges shown in ANSI B16.1 (see Table 6-7). These flanges are assembled to the pipe barrel by threaded joints.

High-Silicon Iron Duriron is a high-silicon iron containing approximately 14.5 percent silicon and 0.85 percent carbon. Durichlor is a special high-silicon iron containing appreciable amounts of molybdenum.

These alloys are available in the cast form only. Pipe and fittings are cast with the upset ends being joined by split flanges. Integrally cast flanged pipe is also available. Allowable working pressures cannot be stated in the manner customary for other types of pipe because of such variables as thermal shock, pulsating pressures, and the corrosive fluids being handled. Although rupture does not occur below 2.76-MPa (400-lbf/in^2) pressure in sizes up to and including 6 in, 0.3 MPa (50 lbf/in^2) is a normal recommendation, even though the pipe has been used for pressure considerably in excess of that figure.

Table 6-18 lists sizes 1 to 12 in, and larger sizes can be obtained. Bell-and-spigot pipe is produced in the weights and dimensions shown in Table 6-18; fittings are available. New hubless pipe utilizing TFE gaskets and stainless steel clamps to make a mechanical joint are now available (the trade name is Duriron MJ).

The coefficient of linear expansion of these alloys in the temperature range of 21 to 100°C (70 to 212°F) is 12.2×10^{-6}/°C (6.8×10^{-6}/°F), which is slightly above that of cast iron (National Bureau of Standards). Since these alloys have practically no elasticity, it is necessary to use expansion joints in relatively short pipe lines. Connections for flanged pipe, fittings, valves, and pumps are made to 125-lb American Standard drilling.

TABLE 6-18 High-Silicon Iron Pipe*

Size, inside diam., in.	Split flanged ends: Outside diam., in.	Wall thickness, in.	Standard† length, ft.	Weight per piece, lb.	Bell-and-spigot ends: Outside diam., in.	Wall thickness, in.	Standard† length, ft.	Weight per piece, lb.
1	1¾	⅜	3	15				
1½	2¼	⅜	3	18	2⅛	5/16	3	20
2	2¾	⅜	4	32	2⅝	5/16	4	30
2½	3¼	⅜	5	45				
3	3⅞	7/16	5	62·	3 11/16	11/32	5	68
4	4⅞	7/16	5	100	4⅝	5/16	5	89
6	7	½	5	180	6 11/16	11/32	5	133
8	9¼	⅝	6	265	9	½	5	232
10	11½	¾	6	433	11¼	⅝	5	341
12	14	1	6	694	13¼	⅝	5	463
15	....	....	...	...	16¾	⅞	5	680

*The Duriron Co.

†Laying lengths; lengths less than standard are available.

NOTE: To convert inches to millimeters, multiply by 25.4; to convert feet to meters, multiply by 0.3048; and to convert pounds to kilograms, multiply by 0.4536.

TABLE 6-17 Dimensions of Ductile-Iron Pipe*

Standard thickness for internal pressure†

		Rated water working pressure, lbf/in^2‡									
		150		200		250		300		350	
Pipe size, in.	Outside diameter, in	Thickness, in	Thickness class	Thickness, in	Thickness class	Thickness, in	Thickness class	Thickness, in	Thickness class	Thickness, in	Thickness class
3	3.96	0.25	51	0.25	51	0.25	51	0.25	51	0.25	51
4	4.80	0.26	51	0.26	51	0.26	51	0.26	51	0.26	51
6	6.90	0.25	50	0.25	50	0.25	50	0.25	50	0.25	50
8	9.05	0.27	50	0.27	50	0.27	50	0.27	50	0.27	50
10	11.10	0.29	50	0.29	50	0.29	50	0.29	50	0.29	50
12	13.20	0.31	50	0.31	50	0.31	50	0.31	50	0.31	50
14	15.30	0.33	50	0.33	50	0.33	50	0.33	50	0.33	50
16	17.40	0.34	50	0.34	50	0.34	50	0.34	50	0.34	50
18	19.50	0.35	50	0.35	50	0.35	50	0.35	50	0.35	50
20	21.60	0.36	50	0.36	50	0.36	50	0.36	50	0.39	51
24	25.80	0.38	50	0.38	50	0.38	50	0.41	51	0.44	52
30	32.00	0.39	50	0.39	50	0.43	51	0.47	52	0.51	53
36	38.30	0.43	50	0.43	50	0.48	51	0.53	52	0.58	53
42	44.50	0.47	50	0.47	50	0.53	51	0.59	52	0.65	53
48	50.80	0.51	50	0.51	50	0.58	51	0.65	52	0.72	53
54	57.10	0.57	50	0.57	50	0.65	51	0.73	52	0.81	53

*Extracted from the American Natonal Standard for Ductile-Iron Pipe, Centrifugally Cast in Metal Molds or Sand-Lined Molds, for Water or Other Liquids, ANSI A21.51—1976, with permission of the publisher, the American Society of Mechanical Engineers, New York.

†To convert from inches to millimeters, multiply by 25.4; to convert pounds-force per square inch to megapascals, multiply by 0.006895.

‡These pipe walls are adequate for the rated working pressure plus a surge allowance of 100 lbf/in^2. For the effect of laying conditions and depth of bury, see ANSI A21.51.

The use of high-silicon iron in flammable-fluid service or in Category M fluid service is prohibited by the code.

Nonferrous-Metal Piping Systems

Aluminum Seamless aluminum pipe and tube are produced by extrusion in essentially pure aluminum and in several alloys; 6-, 9-, and 12-m (20-, 30-, and 40-ft) lengths are available. Alloying and mill treatment improve physical properties, but welding reduces them. Essentially pure aluminum has an ultimate tensile strength of 65.5 MPa (9500 lbf/in^2) subject to a slight increase by mill treatment which is lost during welding. Alloy 6061, which contains 0.25 percent copper, 0.6 percent silicon, 1 percent magnesium, and 0.25 percent chromium, has an ultimate tensile strength of 124 MPa (18,000 lbf/in^2) in the annealed condition, 262 MPa (38,000 lbf/in^2), mill-treated as 6061-T6, and 165 MPa (24,000 lbf/in^2) at welded joints. Extensive use is made of alloy 1060, which is 99.6 percent pure aluminum, for hydrogen peroxide; of alloy 3003, which contains 1.2 percent manganese, for high-purity chemicals; and of alloys 6063 and 6061 for many other services. Alloy 6063 is the same as 6061 minus the chromium and has slightly lower mechanical properties.

Aluminum is not embrittled by low temperatures and is not subject to external corrosion when exposed to normal atmospheres. At 200°C (400°F) its strength is less than half that at room temperature. It is attacked by alkalies, by traces of copper, nickel, mercury, and other heavy-metal ions, and by prolonged contact with wet insulation. It suffers from galvanic corrosion when coupled to copper, nickel, or lead-base alloys but not when coupled to galvanized iron or austenitic stainless steel.

Aluminum pipe is stocked in 3003, 6061, and 6063 Schedule 40 through 10 in, Schedule 30 8 through 10 in, and standard-weight 12-in size. It is also stocked in 6063 as Schedule 5 through 6 in and Schedule 10 through 8 in (see Table 6-6).

Threaded **aluminum fittings** are seldom recommended for process piping. Wrought fittings with welding ends (see Table 6-15 for dimensions) and with grooved joint ends are available. Wrought 6061-T6 flanges with dimensions per Table 6-7 are also available. Cast flanges and flanged fittings, sand-cast as alloy B214, 3.8 percent magnesium alloy with 90-MPA (13,000-lbf/in^2) yield strength, or permanent mold cast as alloy 356-T6, 7 percent silicon, 0.3 percent magnesium alloy with 185-MPa (27,000-lbf/in^2) yield strength are available, but consideration must be given to the fact that the modulus of elasticity of aluminum is only slightly more than one-third that of ferrous alloys. See Table 6-14 for dimensions.

Aluminum-body diaphragm and ball valves are used extensively.

Copper and Copper Alloys Seamless copper, bronze, brass, copper-nickel-alloy, and copper-silicon-alloy pipe and tubing are produced by extrusion. Tubing is available in outside-diameter sizes from 1/16 to 16 in and in a range of wall thicknesses varying from 0.005 in for the smallest tubing to 0.75 in for the 16-in size. Tubing is usually specified by outside diameter and wall thickness. Table 11-2 in Sec. 11 gives dimensions of condenser and heat-exchanger tubing.

Seamless copper tubing is sold in water-tubing sizes (ASTM B88 and B306). These sizes are identified by a "standard" size designation dimensionally ⅛ in less than the nominal outside diameter. The tubing is also sold as outside-diameter copper tubing (ASTM B280).

Copper tubing is widely used in offices and laboratories for water, steam tracing, pneumatic control systems, compressed air, refrigeration, and inert-gas piping. Connections are made with flared-fitting joints (Fig. 6-105), compression-fitting joints (Fig. 6-106), bite-type-fitting joints (Fig. 6-107), and soldered or brazed joints (Fig. 6-109). Figure 6-109 is most economical for ¾-in size and larger. Ease of handling and bending favors the use of copper; it will usually survive a freeze-up without failure.

Copper water tubing ASTM B88 with dimensions and tolerances as given in Table 6-19 is available drawn or annealed in straight lengths of 6.1 m (20 ft) in types K, L, and M through 8-in size. Type K is available in 5.5-m (18-ft) lengths in 10-in size and 3.6-m (12-ft) lengths in 12-in size. Type L is available in 6.1-m (20-ft) lengths in 10-in size and 5.5-m (18-ft) lengths in 12-in size. Type M is available

TABLE 6-19 Copper Water Tubing—Types K, L, M (ASTM B88)*

Standard size, in	Nominal outside diameter, in	Average outside diameter tolerance, in†		Nominal wall thickness and tolerances, in						Theoretical weight, lb/ft		
				Type K		Type L		Type M				
		Annealed	Drawn	Wall thickness	Tolerance‡	Wall thickness	Tolerance‡	Wall thickness	Tolerance‡	Type K	Type L	Type M
¼	0.375	0.002	0.001	0.035	0.004	0.030	0.0035	§	§	0.145	0.126	§
⅜	0.500	0.0025	0.001	0.049	0.004	0.035	0.0035	0.025	0.0025	0.269	0.198	0.145
½	0.625	0.0025	0.001	0.049	0.004	0.040	0.0035	0.028	0.0025	0.344	0.285	0.204
⅝	0.750	0.0025	0.001	0.049	0.004	0.042	0.0035	§	§	0.418	0.362	§
¾	0.875	0.003	0.001	0.065	0.0045	0.045	0.004	0.032	0.003	0.641	0.455	0.328
1	1.125	0.0035	0.0015	0.065	0.0045	0.050	0.004	0.035	0.0035	0.839	0.655	0.465
1¼	1.375	0.004	0.0015	0.065	0.0045	0.055	0.0045	0.042	0.0035	1.04	0.884	0.682
1½	1.625	0.0045	0.002	0.072	0.005	0.060	0.0045	0.049	0.004	1.36	1.14	0.940
2	2.125	0.005	0.002	0.083	0.007	0.070	0.006	0.058	0.006	2.06	1.75	1.46
2½	2.625	0.005	0.002	0.095	0.007	0.080	0.006	0.065	0.006	2.93	2.48	2.03
3	3.125	0.005	0.002	0.109	0.007	0.090	0.007	0.072	0.006	4.00	3.33	2.68
3½	3.625	0.005	0.002	0.120	0.008	0.100	0.007	0.083	0.007	5.12	4.29	3.58
4	4.125	0.005	0.002	0.134	0.010	0.110	0.009	0.095	0.009	6.51	5.38	4.66
5	5.125	0.005	0.002	0.160	0.010	0.125	0.010	0.109	0.009	9.67	7.61	6.66
6	6.125	0.005	0.002	0.192	0.012	0.140	0.011	0.122	0.010	13.9	10.2	8.92
8	8.125	0.006	+0.002 −0.004	0.271	0.016	0.200	0.014	0.170	0.014	25.9	19.3	16.5
10	10.125	0.008	+0.002 −0.006	0.338	0.018	0.250	0.016	0.212	0.015	40.3	30.1	25.6
12	12.125	0.008	+0.002 −0.006	0.405	0.020	0.280	0.018	0.254	0.016	57.8	40.4	36.7

*Copyright American Society for Testing and Materials, 1916 Race Street, Philadelphia, Pa. 19103; reprinted/adapted with permission. To convert inches to millimeters, multiply by 25.4; to convert pounds per foot to kilograms per meter, multiply by 1.49.

†The average outside diameter of a tube is the average of the maximum and minimum outside diameter, as determined at any one cross section of the tube.

‡Maximum deviation at any one point.

§Indicates that the material is not generally available or that no tolerance has been established.

in 6.1-m (20-ft) lengths through 12-in size. All three types are available in 18.3-m (60-ft) or 30-m (100-ft) coils in sizes up to 1 in, in 18.3-m (60-ft) coils in 1¼- and 1½-in sizes, and in 12.2- or 13.7-m (40- or 45-ft) coils in 2-in size.

DWV tubing, ASTM B280, is available in 6-m (20-ft) straight lengths in the following size-wall combinations: 1¼ in, 0.040-in wall; 1½ in, 0.042-in wall; 2 in, 0.042-in wall; 3 in, 0.045-in wall; 4 in, 0.058-in wall; 5 in, 0.072-in wall; and 6 in, 0.083-in wall. DWV is available only in drawn temper. Outside-diameter copper tubing B280 is available in annealed or drawn temper, depending on size; it is used for refrigeration field service, automotive applications, and general service. Dimensions and tolerances are shown in Table 6-20. Drawn temper is available in 6.1-m (20-ft) straight lengths; annealed temper, in 15.2-m (50-ft) coils.

Too high a temperature or too long a heating period when silver-brazing ruins red-brass solder-joint fittings more quickly than wrought-copper fittings. The former are available in larger sizes. Yellow brass fails from dezincification in some waters.

Red-brass and bronze valves are available with female solder-joint ends for soldered copper-tubing piping systems.

Copper pipe is available per ASTM B42 with dimensions as in Table 6-21. Butt-welding fittings (Table 6-15) are available to fit copper pipe, as are screwed fittings per ANSI B16.15, but solder-end fittings of approximately the same dimensions as the screwed fittings and silver-brazing alloy comprise the usual method of assembly. Red-brass or bronze valves with ends identical to the fittings are available. Flanges and flanged fittings are seldom used, since soldered or silver-brazed joints can be melted apart and reassembled.

Threadless copper pipe, thinner than ASTM B42, is available with dimensions as in Table 6-22. Solder-end fittings similar to ANSI B16.15 screwed fittings and solder-end valves are used with this pipe.

Copper pipe is attacked by water originating in granite substrata, and for this reason red-brass pipe per ASTM B43 with red-brass screwed or solder-end fittings is sometimes used in its place.

70 percent copper, 30 percent nickel and **90 percent copper, 10 percent nickel** ASTM B466 are available as seamless pipe and welding fittings for handling brackish water in Schedule 10 and regular copper pipe thicknesses.

Copper-silicon alloy (96 percent cooper, 3 percent silicon, 1 percent manganese), per ASTM B315, is furnished as seamless pipe and welding fittings in Schedule 10 and regular and extra-strong copper pipe thicknesses. It is easier to weld than copper.

Lead and Lead-Lined Steel Pipe Lead and lead-lined steel pipe have been essentially eliminated as piping materials owing to health hazards in fabrication and installation and to environmental objections. Lead has been replaced by suitable plastic, reinforced plastic, plastic-lined steel, or high-alloy materials.

Magnesium Extruded magnesium tubing is available per ASTM B217—58 alloyed with aluminum, manganese, or zinc. Ultimate and yield strengths at 204°C (400°F) are about one-half those at room temperature. Outside-diameter range is ¼ through 8 in. Wall thickness ranges from a minimum of 0.028 in to a maximum of 0.031 in for the ¼-in diameter and from a minimum of 0.250 in to a maximum of 1.0 in for the 8-in diameter.

Nickel and Nickel Alloys A wide range of ferrous and nonferrous nickel and nickel-bearing alloys are available. They are usually

TABLE 6-20 Copper Outside-Diameter Tubing for Refrigeration Field Service and Automotive and General Service (ASTM B280)*

For mechanical or soldered fittings

Standard size, in	Outside diameter, in (mm)	Wall thickness, in (mm)	Weight, lb/ft (kg/m)	Tolerances† Average outside diameter, plus and minus, in (mm)‡	Tolerances† Wall thickness, plus and minus, in (mm)
			For coil		
⅛	0.125 (3.18)	0.030 (0.762)	0.0347 (0.0516)	0.002 (0.051)	0.003 (0.076)
3/16	0.187 (4.75)	0.030 (0.762)	0.0575 (0.0856)	0.002 (0.051)	0.0025 (0.064)
¼	0.250 (6.35)	0.030 (0.762)	0.0804 (0.120)	0.022 (0.051)	0.0025 (0.064)
5/16	0.312 (7.92)	0.032 (0.813)	0.109 (0.162)	0.002 (0.051)	0.0025 (0.064)
⅜	0.375 (9.52)	0.032 (0.813)	0.134 (0.199)	0.002 (0.051)	0.0025 (0.064)
½	0.500 (12.7)	0.032 (0.813)	0.182 (0.271)	0.002 (0.051)	0.0025 (0.064)
⅝	0.625 (15.9)	0.035 (0.889)	0.251 (0.373)	0.002 (0.051)	0.0030 (0.076)
¾	0.750 (19.1)	0.035 (0.889)	0.305 (0.454)	0.0025 (0.064)	0.0035 (0.089)
¾	0.750 (19.1)	0.042 (1.07)	0.362 (0.539)	0.0025 (0.064)	0.0035 (0.089)
⅞	0.875 (22.3)	0.045 (1.14)	0.455 (0.677)	0.003 (0.076)	0.004 (0.10)
1⅛	1.125 (28.6)	0.050 (1.27)	0.665 (0.975)	0.0035 (0.089)	0.004 (0.10)
1⅜	1.375 (34.9)	0.055 (1.40)	0.884 (1.32)	0.004 (0.10)	0.0045 (0.11)
1⅝	1.625 (41.3)	0.060 (1.52)	1.14 (1.70)	0.0045 (0.11)	0.0045 (0.11)
		For straight lengths (applicable to drawn-temper tube only)			
⅜	0.375 (9.52)	0.030 (0.762)	0.126 (0.187)	0.001 (0.025)	0.0035 (0.089)
½	0.500 (12.7)	0.035 (0.889)	0.198 (0.146)	0.001 (0.025)	0.0035 (0.089)
⅝	0.625 (15.9)	0.040 (1.02)	0.285 (0.424)	0.001 (0.025)	0.0035 (0.089)
¾	0.750 (19.1)	0.042 (1.07)	0.362 (0.539)	0.001 (0.025)	0.0035 (0.089)
⅞	0.875 (22.3)	0.045 (1.14)	0.455 (0.677)	0.001 (0.025)	0.004 (0.10)
1⅛	1.125 (28.6)	0.050 (1.27)	0.655 (0.975)	0.0015 (0.038)	0.004 (0.10)
1⅜	1.375 (34.9)	0.055 (1.40)	0.884 (1.32)	0.0015 (0.038)	0.0045 (0.11)
1⅝	1.625 (41.3)	0.060 (1.52)	1.14 (1.70)	0.002 (0.051)	0.0045 (0.11)
2⅛	2.125 (54.0)	0.070 (1.78)	1.75 (2.60)	0.002 (0.051)	0.006 (0.15)
2⅝	2.625 (66.7)	0.080 (2.03)	2.48 (3.69)	0.002 (0.051)	0.006 (0.15)
3⅛	3.125 (79.4)	0.090 (2.29)	3.33 (4.96)	0.022 (0.051)	0.007 (0.18)
3⅝	3.625 (92.1)	0.100 (2.54)	4.29 (6.38)	0.002 (0.051)	0.007 (0.18)
4⅛	4.125 (105)	0.110 (2.79)	5.38 (8.01)	0.002 (0.051)	0.009 (0.23)

*Copyright American Society for Testing and Materials, 1916 Race Street, Philadelphia, Pa. 19103; reprinted/adapted with permission.

†The tolerances listed represent the maximum deviation at any point.

‡The average outside diameter of a tube is the average of the maximum and minimum outside diameters as determined at any one cross section of the tube.

TABLE 6-21 Copper and Red-Brass Pipe (ASTM B42 and B43)*: Standard Dimensions, Weights, and Tolerances

Standard pipe size, in	Nominal outside diameter, in (mm)	Average outside diameter tolerances, in (mm), all minus†	Nominal wall thickness in (mm)	Tolerance in (mm)‡	Theoretical weight, lb/ft (kg/m)	Theoretical weight, lb/ft (kg/m)
			Regular pipe		Red brass	Copper
⅛	0.405 (10.3)	0.004 (0.10)	0.062 (1.57)	0.004 (0.10)	0.253 (0.376)	0.259 (0.385)
¼	0.540 (13.7)	0.004 (0.10)	0.082 (2.08)	0.005 (0.13)	0.447 (0.665)	0.457 (0.680)
⅜	0.675 (17.1)	0.005 (0.13)	0.090 (2.29)	0.005 (0.13)	0.627 (0.933)	0.641 (0.954)
½	0.840 (21.3)	0.005 (0.13)	0.107 (2.72)	0.006 (0.15)	0.934 (1.39)	0.955 (1.42)
¾	1.050 (26.7)	0.006 (0.15)	0.114 (2.90)	0.006 (0.15)	1.27 (1.89)	1.30 (1.93)
1	1.315 (33.4)	0.006 (0.15)	0.126 (3.20)	0.007 (0.18)	1.78 (2.65)	1.82 (2.71)
1¼	1.660 (42.2)	0.006 (0.15)	0.146 (3.71)	0.008 (0.20)	2.63 (3.91)	2.69 (4.00)
1½	1.900 (48.3)	0.006 (0.15)	0.150 (3.81)	0.008 (0.20)	3.13 (4.66)	3.20 (4.76)
2	2.375 (60.3)	0.008 (0.20)	0.156 (3.96)	0.009 (0.23)	4.12 (6.13)	4.22 (6.28)
2½	2.875 (73.0)	0.008 (0.20)	0.187 (4.75)	0.010 (0.25)	5.99 (8.91)	6.12 (9.11)
3	3.500 (88.9)	0.010 (0.25)	0.219 (5.56)	0.012 (0.30)	8.56 (12.7)	8.76 (13.0)
3½	4.000 (102)	0.010 (0.25)	0.250 (6.35)	0.013 (0.33)	11.2 (16.7)	11.4 (17.0)
4	4.500 (114)	0.012 (0.30)	0.250 (6.35)	0.014 (0.36)	12.7 (18.9)	12.9 (19.2)
5	5.562 (141)	0.014 (0.36)	0.250 (6.35)	0.014 (0.36)	15.8 (23.5)	16.2 (24.1)
6	6.625 (168)	0.016 (0.41)	0.250 (6.35)	0.014 (0.36)	19.0 (28.3)	19.4 (28.9)
8	8.625 (219)	0.020 (0.51)	0.312 (7.92)	0.022 (0.56)	30.9 (46.0)	31.6 (47.0)
10	10.750 (273)	0.022 (0.56)	0.365 (9.27)	0.030 (0.76)	45.2 (67.3)	46.2 (68.7)
12	12.750 (324)	0.024 (0.61)	0.375 (9.52)	0.030 (0.76)	55.3 (82.3)	56.5 (84.1)
			Extra strong pipe			
⅛	0.405 (10.3)	0.004 (0.10)	0.100 (2.54)	0.006 (0.15)	0.363 (0.540)	0.371 (0.552)
¼	0.540 (13.7)	0.004 (0.10)	0.123 (3.12)	0.007 (0.18)	0.611 (0.909)	0.625 (0.930)
⅜	0.675 (17.1)	0.005 (0.13)	0.127 (3.23)	0.007 (0.18)	0.829 (1.23)	0.847 (1.26)
½	0.840 (21.3)	0.005 (0.13)	0.149 (3.78)	0.008 (0.20)	1.23 (1.83)	1.25 (1.86)
¾	1.050 (26.7)	0.006 (0.15)	0.157 (3.99)	0.009 (0.23)	1.67 (2.48)	1.71 (2.54)
1	1.315 (33.4)	0.006 (0.15)	0.182 (4.62)	0.010 (0.25)	2.46 (3.66)	2.51 (3.73)
1¼	1.660 (42.2)	0.006 (0.15)	0.194 (4.93)	0.010 (0.25)	3.39 (5.04)	3.46 (5.15)
1½	1.900 (48.3)	0.006 (0.15)	0.203 (5.16)	0.011 (0.28)	4.10 (6.10)	4.19 (6.23)
2	2.375 (60.3)	0.008 (0.20)	0.221 (5.61)	0.012 (0.30)	5.67 (8.44)	5.80 (8.63)
2½	2.875 (73.0)	0.008 (0.20)	0.280 (7.11)	0.015 (0.38)	8.66 (12.9)	8.85 (13.2)
3	3.500 (88.9)	0.010 (0.25)	0.304 (7.72)	0.016 (0.41)	11.6 (17.3)	11.8 (17.6)
3½	4.000 (102)	0.010 (0.25)	0.321 (8.15)	0.017 (0.43)	14.1 (21.0)	14.4 (21.4)
4	4.500 (114)	0.012 (0.30)	0.341 (8.66)	0.018 (0.46)	16.9 (25.1)	17.3 (25.7)
5	5.562 (141)	0.014 (0.36)	0.375 (9.52)	0.019 (0.48)	23.2 (34.5)	23.7 (35.3)
6	6.625 (168)	0.016 (0.41)	0.437 (11.1)	0.027 (0.69)	32.2 (47.9)	32.9 (49.0)
8	8.625 (219)	0.020 (0.51)	0.500 (12.7)	0.035 (0.89)	48.4 (72.0)	49.5 (73.7)
10	10.750 (273)	0.022 (0.56)	0.500 (12.7)	0.040 (1.0)	61.1 (90.9)	62.4 (92.9)

*Copyright American Society for Testing and Materials, 1916 Race Street, Philadelphia, Pa. 19103; reprinted/adapted with permission. All tolerances are plus and minus except as otherwise indicated.

†The average outside diameter of a tube is the average of the maximum and minimum outside diameters as determined at any one cross section of the tube.

‡Maximum deviation at any one point.

selected because of their improved resistance to chemical attack or their superior resistance to the effects of high temperature. In general terms their cost and corrosion resistance are somewhat a function of their nickel content. The 300 Series stainless steels are the most generally used. Some other frequently used alloys are listed in Table 6-23 together with their nominal compositions. For metallurgical and corrosion resistance data, see Sec. 23.

Titanium Pipe per ASTM B337 is available welded or seamless via one of the following processes: extrusion, centrifugal casting, machining of bar stock, or powder compaction; Schedule 5S, 10S, 40S, and 80S, ⅛- through 24-in size. Extruded and drawn tubing per ASTM B338 is available from ¼-in outside diameter, 0.020- through 0.083-in wall, up through 3-in outside diameter. Cast welding fittings, flanges, and valves are also available. Titanium is used at temperatures up to 315°C (600°F). It is extremely notch-sensitive. Titanium alloys such as 6 Al-4V, with higher tensile strengths than straight titanium, are available. Unfortunately, they lack the corrosion resistance and weldability of the unalloyed material.

Zirconium (Tin 1.2 to 1.7 Percent) Tubing is available seamless ranging from ½- outside diameter by 0.030-in wall to 8-in outside diameter by 0.4-in wall, and welded up through 30-in outside diameter by ⅛-in wall. Cast valves and fittings are also available.

Flexible Metal Hose Deeply corrugated thin brass, bronze, Monel, aluminum, and steel tubes are covered with flexible braided-wire jackets to form flexible metal hose. Both tube and braid are brazed or welded to pipe-thread, union, or flanged ends. Failures are often the result of corrosion of the braided-wire jacket or of a poor jacket-to-fitting weld. Inside diameters range from ⅛ to 12 in. Maximum recommended temperature for bronze hose is approximately 230°C (450°F). Metal thickness is much less than for straight tube for the same pressure-temperature conditions; so accurate data on corrosion and erosion are required to make proper selection.

NONMETALLIC PIPE AND LINED PIPE SYSTEMS

Asbestos Cement Asbestos-cement pipe is seamless pipe made of silica and portland cement, compacted under heavy pressure, uniformly reinforced with asbestos fiber, and thoroughly cured. The interior surface is smooth, does not corrode, and does not tuberculate. Under normal conditions of operation, asbestos cement will handle

TABLE 6-22 Hard-Drawn Copper Threadless Pipe (ASTM B302)*

Standard pipe size, in	Nominal dimensions, in (mm)			Cross-sectional area of bore, in^2 (cm^2)	Nominal weight, lb/ft (kg/m)	Tolerances, in (mm)	
	Outside diameter	Inside diameter	Wall thickness			Average outside diameter, all minus†	Wall thickness, plus and minus
¼	0.540 (13.7)	0.410 (10.4)	0.065 (1.65)	0.132 (0.852)	0.376 (0.559)	0.004 (0.10)	0.0035 (0.089)
⅜	0.675 (17.1)	0.545 (13.8)	0.065 (1.65)	0.233 (1.50)	0.483 (0.719)	0.004 (0.10)	0.004 (0.10)
½	0.840 (21.3)	0.710 (18.0)	0.065 (1.65)	0.396 (2.55)	0.613 (0.912)	0.005 (0.13)	0.004 (0.10)
¾	1.050 (26.7)	0.920 (23.4)	0.065 (1.65)	0.665 (4.29)	0.780 (1.16)	0.005 (0.13)	0.004 (0.10
1	1.315 (33.4)	1.185 (30.1)	0.065 (1.65)	1.10 (7.10)	0.989 (1.47)	0.005 (0.13)	0.004 (0.10)
1¼	1.660 (42.2)	1.530 (38.9)	0.065 (1.65)	1.84 (11.9)	1.26 (1.87)	0.006 (0.15)	0.004 (0.10)
1½	1.900 (48.3)	1.770 (45.0)	0.065 (1.65)	2.46 (15.9)	1.45 (2.16)	0.006 (0.15)	0.004 (0.10)
2	2.375 (60.3)	2.245 (57.0)	0.065 (1.65)	3.96 (25.5)	1.83 (272)	0.007 (0.18)	0.006 (0.15)
2½	2.875 (73.0)	2.745 (69.7)	0.065 (1.65)	5.92 (38.2)	2.22 (3.30)	0.007 (0.18)	0.006 (0.15)
3	3.500 (88.9)	3.334 (84.7)	0.083 (2.11)	8.73 (56.3)	3.45 (513)	0.008 (0.20)	0.007 (0.18)
3½	4.000 (102)	3.810 (96.8)	0.095 (2.41)	11.4 (73.5)	4.52 (6.73)	0.008 (0.20)	0.007 (0.18)
4	4.500 (114)	4.286 (109)	0.107 (2.72)	14.4 (92.9)	5.72 (8.51)	0.010 (0.25)	0.009 (0.23)
5	5.562 (141)	5.298 (135)	0.132 (3.40)	22.0 (142)	8.73 (13.0)	0.012 (0.30)	0.010 (0.25)
6	6.625 (168)	6.309 (160)	0.158 (4.01)	31.3 (202)	12.4 (18.5)	0.014 (0.36)	0.010 (0.25)
8	8.625 (219)	8.215 (209)	0.205 (5.21)	53.0 (342)	21.0 (31.2)	0.018 (0.46)	0.014 (0.36)
10	10.750 (273)	10.238 (260)	0.256 (6.50)	82.3 (531)	32.7 (48.7)	0.018 (0.46)	0.016 (0.41)
12	12.750 (324)	12.124 (308)	0.313 (7.95)	115 (742)	47.4 (70.5)	0.018 (0.46)	0.020 (0.51)

*Copyright American Society for Testing and Materials, 1916 Race Street, Philadlphia, Pa. 19103; reprinted/adapted with permission.

†The average outside diameter of a tube is the average of the maximum and minimum outside diameters, as determined at any one cross section of the tube.

TABLE 6-23 Common Nickel and Nickel-Bearing Alloys

Common trade name or registered trademark	Code designation	Alloy no.	ASTM specification (pipe)	Nominal composition, %										
				Ni	Cr	Mo	Fe	C[a]	Si[a]	Mn	Cu	Cb	Co	W
Type 304 stainless steel		S30400	A312	9	19	. . .	70	0.08	. . .	2.0				
Type 316 stainless steel		S31600	A312	11	18	2.5	66.5	0.08	. . .	2.0				
Carpenter 20cb[b]	Ni-Cr-Fe-Mo-Cu-Cb stabilized	N08020	B464	33	20	2.5	38.5	0.06	. . .	2.0	3	1		
Incoloy 800[c]	Ni-Fe-Cr	N08800	B407	32.5	21		46	0.05	0.5	0.8	0.4			
Incoloy 825[c]	Ni-Fe-Cr-Mo-Cu	N08825	B423	42	21.5	3	30	0.03	0.2	0.5	2.2			
Hastelloy C-276[d]	Ni-Mo-Cr low carbon	N10276	B575[e]	54	15	16	5	0.02	0.08	1			2.5	4
Hastelloy B-2[d]	Ni-Mo	N10001	B333[e]	64	1	28	2	0.02	0.1	1				
Inconel 625[c]	Ni-Cr-Mo-Cb	N06625	B444	61	21.5	9	2.5	0.05	0.2	0.2		4		
Inconel 600[c]	Ni-Cr-Fe	N06600	B167	76	15.5	. . .	8	0.08	0.2	0.5	0.2			
Monel 400[c]	Ni-Cu	N04400	B165	66	. . . .	. . .	1.2	0.20	0.2	1	31.5			
Nickel 200[c]	Ni	N02200	B161	99+	. . . .	. . .	0.2	0.08						
Hastelloy G[d]	Ni-Cr-Fe-Mo-Cu	N06007	B622	42	22.2	6.5	19.5	0.05	1	1.5	2	2.2[f]	2.5[a]	1[a]

[a] Maximum.
[b] Registered trademark, Carpenter Technology Corp.
[c] Registered trademark, Huntington Alloys, Inc.
[d] Registered trademark, Cabot Corp.
[e] Plate.
[f] Cb + Ta.

solutions within a pH range of 4.5 to 14. It is a brittle material and undergoes expansion on wetting. There are stringent OSHA regulations pertaining to the fabrication and use of asbestos-containing materials. The most widely used joints are push-on joints. This pipe is used extensively for underground water systems, for paper-mill slurries and wastes, and for mine water. The push-on joints limit the temperature to 65°C (150°F). The light weight of the pipe minimizes handling labor, but careful handling is required to avoid damage. This pipe is available with an epoxy lining which increases its corrosion resistance.

Asbestos-cement fittings and valves are not available, but flanged fabricated-steel fittings lined with segments of asbestos-cement pipe and cement-lined cast-iron fittings with end bells for push-on joint to asbestos-cement pipe can be obtained. Adapters to regular cast-iron fittings are also available. When the pipe is installed aboveground, two guided supports per length of pipe are recommended, and when push-on joints are used, internal pressure thrusts at changes in direction, at reducers, at dead ends, and at valves must be resisted by braces. When poured flanges are used, expansion joints must be used also with braces to resist corresponding pressure thrust.

Pressure Pipe This pipe is made in three classes corresponding to working pressures of 0.7, 1.0, and 1.4 MPa (100, 150, and 200 lbf/in^2) (Table 6-24).

Gravity Sewer Pipe This pipe is made in five classes for varying depths of bury, trench dimension, soil, and vehicular loading (Table 6-25).

TABLE 6-24 Asbestos-Cement Pressure Pipe*

Nominal size	Length, ft.	Class 100† Inside diam., in.	Class 100† Wall, in.‡	Class 100† Wt., lb./ft.§	Class 150† Inside diam., in.	Class 150† Wall, in.‡	Class 150† Wt., lb./ft.§	Class 200† Inside diam., in.	Class 200† Wall, in.‡	Class 200† Wt., lb./ft.§
4	13	3.95	0.35	6.3	3.95	0.43	7.6	3.95	0.43	9.3
6	13	5.85	.42	10.6	5.85	.53	13.0	5.70	.60	15.4
8	13	7.85	.47	15.8	7.85	.63	19.9	7.60	.75	23.9
10	13	9.85	.52	21.8	10.00	.83	32.0	9.63	1.01	37.2
12	13	11.70	.64	29.7	12.00	.96	43.8	11.56	1.18	51.7
14	13	13.59	.74	38.9	14.00	1.11	58.5	13.59	1.31	69.0
16	13	15.50	.83	48.8	16.00	1.23	73.0	15.50	1.48	89.2

*Johns-Manville Co.
†Equivalent to working pressure, lb./sq. in.
‡Minimum thickness of machined end; balance of pipe is thicker.
§Pipe plus push-on joint coupling.

NOTE: To convert inches to millimeters, multiply by 25.4; to convert pounds per foot to kilograms per meter, multiply by 1.49; to convert pounds-force per square inch to megapascals, multiply by 0.00689; and to convert feet to meters, multiply by 0.3048.

TABLE 6-25 Asbestos-Cement Gravity Sewer Pipe*

Nominal size	Inside diam., in.	Class 1500† Wall, in.‡	Class 1500† Wt., lb./ft.	Class 2400† Wall, in.‡	Class 2400† Wt., lb./ft.	Class 3300† Wall, in.†	Class 3300† Wt., lb./ft.	Class 4000† Wall, in.‡	Class 4000† Wt., lb./ft.	Class 5000† Wall, in.‡	Class 5000† Wt., lb./ft.
6	6.00	0.46	8.5	0.49	9.5	0.57	11.1				
8	8.00	.51	12.6	.52	13.3	.61	15.6				
10	10.00	.56	17.6	.58	18.9	.68	22.0	0.75	24.3	0.85	27.6
12	12.05	.61	22.8	.63	24.3	.75	28.8	.82	31.5	0.93	35.8
14	14.05	....	....	.68	30.3	.81	35.8	.89	39.3	1.00	44.3
16	16.05	....	....	.73	37.0	.86	43.1	.95	47.6	1.07	53.7
18	18.05	....	....	.77	43.6	.91	50.9	1.01	56.5	1.13	63.3
20	20.05	....	....	.81	50.7	.96	59.2	1.06	65.4	1.19	73.6
24	24.05	....	....	.89	66.4	1.05	77.2	1.16	85.3	1.30	95.8
30	30.05	....	....	....	....	1.17	106.8	1.30	118.8	1.45	132.7
36	36.05	....	....	....	....	....	.....	1.42	155.0	1.59	173.8

Standard pipe length is 13 ft. except 6 in. Class 1500 is 10 ft. and 8 in. Class 1500 may also be 10 ft.
*Johns-Manville Co.
†Crushing strength per A.S.T.M. three-edge bearing method.
‡Thickness of wall of pipe excluding machined ends. Same coupling is used for all classes; it protects the machined ends from crushing loads.

NOTE: To convert inches to millimeters, multiply by 25.4; to convert pounds per foot to kilograms per meter, multiply by 1.49; and to convert feet to meters, multiply by 0.3048.

Impervious Graphite Impervious-graphite pipe, fittings, and valve bodies are made of electric-furnace graphite which, after extruding or molding, is rendered impervious by impregnation with synthetic resins. When impregnated with phenolic resin, it is resistant to most acids (including hydrofluoric), salts, and organic compounds. When impregnated with modified phenolic resin, it is resistant to strong alkalies and highly oxidizing materials. Ultimate tensile strength is low, 17.2 MPa (2500 lbf/in^2), and the material is brittle, but the modulus of elasticity is only 15,168 MPa (2.2×10^6 lbf/in^2). The material is highly resistant to thermal shock and is available with glass-cloth and resin armor for protection against physical abuse. Maximum continuous operating temperature is 170°C (340°F). Components are designed for operating pressure which increases from 0.3 MPa (50 lbf/in^2) at 170°C (340°F) to 0.5 MPa (75 lbf/in^2) at 21°C (70°F).

Table 6-26 lists standard sizes of pipe; ½-, ¾-, and ⅞-in sizes are heat-exchanger tubing, and standard fittings are not available for these sizes. Pipe is shipped threaded on request. National Form straight threads are used. Fittings made from the same material with the same thread form are available and include laps which can be screwed on the ends of pipe and stub ends which can be screwed into the fittings, both for the purpose of making flanged lap joints. All threaded joints are permanently bonded by special cements. Flanged joints use split cast-iron backup flanges which have 150-lb ANSI

TABLE 6-26 Standard Sizes of Impervious Graphite Pipe*

Nominal pipe size, in	Inside diameter, in	Outside diameter, in	Wall thickness, in	Maximum length, ft	Average weight, lb/ft	Inside cross-sectional area, ft^2	Circumference, ft, or surface, ft^2/ft of length: Inside	Circumference, ft, or surface, ft^2/ft of length: Outside
1	1	1½	¼	9	.74	.00545	.262	.393
1½	1½	2	¼	9	1.1	.01227	.393	.524
2	2	2¾	⅜	9	1.7	.0218	.524	.687
2½	2⅝	3	5/16	9	2.0	.0308	.622	.785
3	3	4	½	9	5.4	.0491	.785	1.047
4	4	5¼	⅝	9	8.1	.0873	1.047	1.374
6	6	7½	¾	9	15.6	.1965	1.571	1.964
8	8⅛	9 11/16	25/32	6	23.2	.360	2.127	2.536
10	10⅛	12 21/32	1 17/64	6	44.2	.559	2.650	3.313

NOTE: To convert inches to millimeters, multiply by 25.4; to convert feet to meters, multiply by 0.3048; to convert pounds per foot to kilograms per meter, multiply by 1.49; and to convert square feet to square meters, multiply by 0.0929.
*Courtesy Union Carbide Corporation, Carbon Products Division.

B16.5 bolting in sizes 6 in and smaller and 300-lb ANSI B16.5 bolting in sizes 8 in and larger. Asbestos sheet packing is used between the flange and the back of the lap to equalize bearing. Pipe can be sawed to length in the field and threaded with special tools. Synthetic elastomeric and Teflon gaskets are available. Diaphragm valves with impervious graphite bodies are available in sizes from 1 through 6 in. Maximum recommended support spacing is 2.7 m (9 ft), and valves should be supported independently.

Cement-Lined Steel Cement-lined steel pipe is made by lining steel pipe with special cement. Its use prevents pickup of iron by the fluid handled, corrosion of the metal by brackish water, and growth of tuberculation. Threaded pipe in sizes from ¾ to 4 in is stocked; however, cement-lined pipe in sizes smaller than 1½ in is not considered practical for common use.

The coefficients of expansion of iron and cement are nearly alike. Table 6-27 gives dimensions of cement-lined pipe.

TABLE 6-27 Cement-Lined Carbon-steel Pipe*

Standard pipe size, in.	Inside diam. after lining, in.	Thickness of lining, in.	Weight, per ft., lb.	Standard pipe size, in.	Inside diam. after lining, in.	Thickness of lining, in.	Weight per ft., lb.
3/4	0.70	0.06	1.3	3	2.70	0.13	8.3
1	.90	.07	1.9	4	3.60	.16	12.0
1 1/4	1.20	.08	2.5	6	5.40	.25	24.0
1 1/2	1.40	.09	3.0	8	7.40	.25	32.0
2	1.80	.10	4.1	10	9.40	.30	43.0
2 1/2	2.20	.10	6.6	12	11.40	.30	55.0

*To convert inches to millimeters, multiply by 25.4; to convert pounds per foot to kilograms per meter, multiply by 1.49.

Cement-lined carbon steel pipe larger than 4 in is shipped with flanged or welding ends. Welding does not damage the lining, which forms a slag protecting the weld. Shop cement lining of carbon steel pipe is covered by AWWA C205. Cement-lined carbon steel butt-welding fittings and flanged cast-iron fittings are available. AWWA C602 includes cement lining of both cast-iron and carbon steel water lines in place.

Chemical Ware Acidproof chemical-stoneware pipe and fittings withstand most acid, alkali, or other corrosives, the main exception being hydrofluoric acid. The range of sizes made with the bell-and-spigot joint and with plain butt ends is shown in Table 6-28.

TABLE 6-28 Chemical Stoneware: Bell-and-Spigot and Plain Butt-End Pipe*

Inside diam., in.	Outside diam., in.	Wall thickness, in.
1 1/2	2 1/4	3/8
2	2 3/4	3/8
3	4	1/2
4	5	1/2
5	6	1/2
6	7 1/4	5/8
8	9 1/2	3/4
10	11 3/4	7/8
12	13 3/4	7/8
14	15 3/4	7/8
15	17	1
16	18	1
18	20	1
20	22	1

Standard lengths up to 5 ft.
*Maurice A. Knight Co.
NOTE: To convert inches to millimeters, multiply by 25.4; to convert feet to meters, multiply by 0.3048.

Plain butt-end pipe is furnished with cemented-on flanges with ANSI B16.1 drilling or (for use in ventilating work in which the space is too limited for bell-and-spigot pipe) with a ring for joining with a steel band. Medium-pressure chemical-stoneware pipe armored with glass fiber reinforced with furan resin can be obtained. Flanges with ANSI B16.1 drilling bear against hubs formed from the armor.

Fittings and plug valves with ends to match the various types of pipe are available.

Vitrified-Clay Sewer Pipe This pipe is resistant to very dilute chemicals except hydrofluoric acid and is produced as standard-strength and extra-strength (ASTM C700). It is used for sewage, industrial waste, and storm water at atmospheric pressure. Elbows, Y branches, tees, reducers, and increasers are available. Assembly is by poured joints which allow for ample angular deflection. Joint compounds are of the hot-pour type or the cold mastic type; both adhere tightly to the scored clay surfaces but remain flexible enough to prevent leakage in the event of earth settlement. Pipe is also available with bituminous or plastic material die-cast on the outside of the spigot and the inside of the bell. The interfaces are a snug fit cemented by applying a solvent to them at the time of assembly.

TABLE 6-29 Vitrified-Clay Sewer Pipe*

Nominal size	Min. laying length, ft.	Min. outside diam. of barrel, in.	Min. wall thickness: Standard strength, in.	Min. wall thickness: Extra strength, in.
4	2	4 7/8	7/16	
6	2	7 1/16	1/2	9/16
8	2	9 1/4	9/16	3/4
10	2	11 1/2	11/16	7/8
12	2	13 3/4	13/16	1 1/16
16	3	17 3/16	15/16	1 3/8
18	3	20 5/8	1 1/8	1 3/4
21	3	24 1/8	1 5/16	2
24	3	27 1/2	1 1/2	2 1/4
27	3	31	1 11/16	2 1/2
30	3	34 3/8	1 7/8	2 3/4
33	3	37 5/8	2	3
36	3	40 3/4	2 1/16	3 1/4

*To convert inches to millimeters, multiply by 25.4.

Dimensions of pipe are given in Table 6-29. Choice between standard and extra strength is based on earth and vehicular loading.

Concrete Unreinforced-concrete sewer pipe is made with poured joint ends in sizes from 4 to 24 in conforming to ASTM C14. Reinforced-concrete culvert, storm-drain, and sewer pipe is made with poured joint or push-on joint ends conforming to ASTM C76 in five classes of reinforcement area and wall thickness in sizes from 12 through 108 in. Essentially the same pipe, except that it has push-on joint ends only, is available for water pressures up to 0.31 MPa (45 lbf/in^2) in sizes 12 through 96 in and lengths up through 5.6 m (16 ft) conforming to AWWA C302.

For higher water pressures, a steel cylinder approximately 1.6 mm (1/16 in) thick is embedded in the wall of the pipe, which prevents leakage through cracks, and to this there may be added prestressed circumferential reinforcing wire applied after the cylinder has been stiffened by cement lining. Such pipe is available in accordance with AWWA C300, sizes 20 through 96 in, for pressures 0.27 through 1.8 MPa (40 through 260 lbf/in^2), and in accordance with AWWA C301, sizes 16 through 96 in. Push-on joints are used. Pipe is also available with steel lugs welded to the reinforcing cages and projecting through the outside surface of the pipe for "bridling." This is known as "subaqueous pipe." Concrete fittings are also available. Concrete piping systems can be lined with special salt-glazed vitrified-clay liner plates, joined with a die-cast asphalt joint. Concrete pressure pipe is competitive with cement-lined ductile iron for underground plant water systems.

Glass Pipe and Tubing These are made from heat- and chemical-resistant borosilicate glass (e.g., Corning Glass Works No. 7740) ASTM C599. This glass is highly stable in acids and resists attack by alkalies in solutions in which pH is 8 or less. It is attacked by hydrofluoric acid and glacial phosphoric acid. Some important physical properties are:

Modulus of elasticity	9,750,000 lb/in^2 (67,224 MPa)
Specific gravity	2.23
Specific heat	0.20
Thermal conductivity at 75°F	8.1 Btu/(h·ft^2)(°F/in) [1.168 W/(m·K)]

Conical flanged glass pipe (Fig. 6-123) is made in the sizes shown in Table 6-30 and in lengths from 0.15 to 3 m (6 in to 10 ft). Maximum recommended working pressure is 0.3 MPa (50 lbf/in^2) through 3-in size, 0.24 MPa (35 lbf/in^2) for 4-in size, and 0.14 MPa (20 lbf/in^2) for 6-in size. Maximum sudden temperature differential is 93°C (200°F) through 3-in size, 80°C (175°F) for 4-in size, and 71°C (160°F) for 6-in size. Maximum operating temperature is 232°C (450°F). A complete line of fittings is available, and special parts are made to order. Thermal-expansion stresses should be completely relieved by tied Teflon corrugated expansion joints and offsets. Temperature rating may be limited by joint design and materials. Hangers should be padded to avoid scratching pipe, should fit loosely, and should be located 0.3 m (1 ft) from each end of each 3-m (10-ft) length.

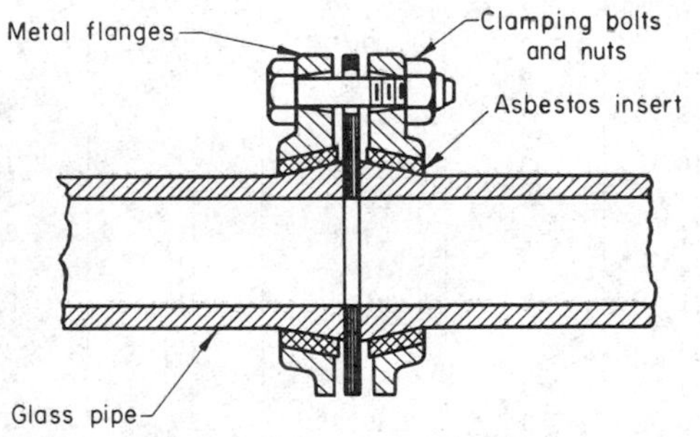

FIG. 6-123 Conical flanged joint.

Glass pipe can be furnished with an epoxy-resin coating reinforced with woven glass fiber to protect it from abuse. Equipped with special ball couplings, this may be used for 1-MPa (150-lbf/in^2) pressure.

For very low pressures, beaded-end pipe equipped with single-bolt band-type couplings is available.

Glass-Lined Steel Pipe This pipe is fully resistant to all acids except hydrofluoric and concentrated phosphoric acids at temperatures up to 121°C (250°F). It is also resistant to alkaline solutions at moderate temperatures. Glass-lined steel pipe can be used at temperatures up to 232°C (450°F) under some exposure conditions provided there are no excessive sudden temperature changes. The operating pressure rating of commonly available systems is 1 MPa (150 lbf/in^2). The glass lining is approximately 1.6 mm (1/16 in) thick. It is made by lining Schedule 40 steel pipe. Fittings are available in glass-lined cast iron, ductile iron, and steel. The fitting rating and recommended applications for fittings depend on the substrate material. Standard pipe sizes available are 1½ through 8 in. Larger-diameter pipe up to 48 in is available on a custom-order basis. A range of standard lengths is generally available from stock in 1½- through 4-in sizes. See Table 6-31 for dimensional data. Special Pfaudler-design steel split flanges drilled to ANSI Class 150 dimensions are used for assembly of the system.

TABLE 6-31 Glass-Lined Steel Pipe*

Size, in	Outside diameter, in	Approximate inside diameter, in	Range of standard lengths, in: Minimum†	Range of standard lengths, in: Maximum
1½	1.875	1.50	3½	120
2	2.375	1.95	4	120
3	3.500	2.95	4½	120
4	4.500	3.90	4½	120
6	6.625	5.95	5	120
8	8.625	7.85	5½	120

*From Pfaudler Company, division of Sybron Corp. To convert inches to millimeters, multiply by 25.4. Standard-length pipe spools are available in the following increments of length:

For lengths, in	Standard lengths available in length increments, in
3½–6	½
6–8	2
8–10	1
10–12	2
12–120	6

†Spacers are available in ½-in increments for making up lengths of less than the minimum spool length shown.

‡Spool lengths less than 120 in are available but are not standard.

Chemical-Porcelain Pipe Made of dense, nonporous material and fired at 1230°C (2250°F), chemical-porcelain pipe, fittings, and valves are inert to all acids except hydrofluoric but are not usually recommended for alkalies. Surfaces, except when ground for gasketing, are usually glazed for easy cleaning. Working pressures of 0.3 to 0.7 MPa (50 to 100 lbf/in^2) are recommended for valves and piping. Temperatures of 200°C (400°F) or more can be used, but sudden thermal shocks must be avoided.

TABLE 6-30 Glass Pipe and Tubing: Conical Flanged Joint*

Pipe size, in (mm)	Pipe outside diameter, in (mm)	Cone outside diameter, in (mm)	Wall thickness, in (mm)	Cone angle, °	Approximate weight per foot, lb (kg)
1 (25)	1 5/16 ± 0.016 (33 ± 0.4)	1 9/16 ± 0.016 (40 ± 0.4)	5/32 ± 0.016 (4.0 ± 0.4)	12	0.6 (0.27)
1½ (38)	1 27/32 ± 0.020 (47 ± 0.5)	2 1/8 ± 0.016 (54 ± 0.4)	11/64 ± 0.016 (4.4 ± 0.4)	12	1.0 (0.45)
2 (51)	2 11/32 ± 0.040 (60 ± 1.0)	2 5/8 ± 0.020 (67 ± 0.5)	11/64 ± 0.020 (4.4 ± 0.5)	12	1.13 (0.51)
3 (76)	3 13/32 ± 0.56) (87 ± 1.4)	3 25/32 ± 0.031 (96 ± 0.8)	13/64 ± 0.021 (5.2 ± 0.5)	12	2.0 (0.91)
4 (102)	4 17/32 ± 0.068 (115 ± 1.7)	5 23/64 ± 0.016 (136 ± 0.4)	17/64 ± 0.025 (6.7 ± 0.6)	21	3.4 (1.5)
6 (152)	6 21/32 ± 0.075 (169 ± 1.9)	7.553 ± 0.016 (192 ± 0.4)	5/16 ± 0.040 (7.9 ± 1.0)	21	6.3 (2.9)

*From Corning Glass Works. See Fig. 6-123.
NOTE: To convert feet to meters, multiply by 0.3048.

Cast-iron flanges (ANSI B16.1, 125-lb bolt spacing) are permanently attached to the porcelain with high strength acid-resistant cement. Flanged chemical-porcelain 90° and 45° elbows, tees, crosses, reducers, caps, and globe valves of the Y pattern are available. Armored chemical porcelain is furnished with 1.5- to 2.4-mm-(1/16- to 3/32-in-) thick woven glass cloth impregnated with and bonded to the porcelain by plastic cement. The armor is continuous end to end and runs under the flanges. It prevents abuse from cracking the porcelain and, if the porcelain is cracked, prevents rupture.

Fused Silica or Fused Quartz Containing 99.8 percent silicon dioxide, fused silica or fused quartz can be obtained as opaque or transparent pipe and tubing. The melting point is 1710°C (3100°F). Tensile strength is approximately 48 MPa (7000 lbf/in^2); specific gravity is about 2.2. The pipe and tubing can be used continuously at temperatures up to 1000°C (1830°F) and intermittently up to 1500°C (2730°F). The material's chief assets are noncontamination of most chemicals in high-temperature service, thermal-shock resistance, and high-temperature electrical insulating characteristics.

Transparent tubing is available in inside diameters from 1 to 125 mm in a range of wall thicknesses. Satin-surface tubing is available in inside diameters from 1/16 to 2 in, and sand-surface pipe and tubing are available in ½- to 24-in inside diameters and lengths up to 6 m (20 ft). Sand-surface pipe and tubing are obtainable in wall thicknesses varying from ⅛ to 1 in. Pipe and tubing sections in both opaque and transparent fused silica or fused quartz can be readily machine-ground to special tolerances for pressure joints or other purposes. Also, fused-silica piping and tubing can be reprocessed to meet special-design requirements. Manufacturers should be consulted for specific details.

Wood and Wood-Lined Steel Pipe Douglas fir, white pine, redwood, and cypress are the most common woods used for wood pipe. Wood-lined steel pipe is suitable for temperatures up to 82°C (180°F) and for pressures from 1.4 MPa (200 lbf/in^2) for the 4-in size, through 0.86 MPA (125 lbf/in^2) for the 10-in size, to 0.7 MPA (100 lbf/in^2) for sizes larger than 10 in. For fume stacks and similar uses, wood-stave pipe with rods on 0.3-m (1-ft) centers is most satisfactory because it permits periodic tightening. In recent years reinforced plastics have supplanted wood pipe in most applications.

Plastic-Lined and Rubber-Lined Steel Pipe Use of a variety of polymeric materials as liners for steel pipe rather than as piping systems solves problems which the relatively low tensile strength of the polymer at elevated temperature and high thermal expansion, compared with steel, would produce. The steel outer shell permits much wider spacing of supports, reliable flanged joints, and higher pressure and temperature in the piping. The size range is 1 through 12 in. The systems are flanged with 125-lb cast-iron, 150-lb ductile-iron, and 150- and 300-lb steel flanges. The linings are factory-installed in both pipe and fittings. Lengths are available up to 6 m (20 ft). Lined ball, diaphragm, and check valves and plug cocks are available.

One method of manufacture consists of inserting the liner into an oversize, approximately Schedule 40 steel tube and swaging the assembly to produce iron-pipe-size outside diameter, firmly engaging the liner which projects from both ends of the pipe. Flanges are then screwed onto the pipe, and the projecting liner is hot-flared over the flange faces nearly to the bolt holes. In another method, the liner is pushed into steel pipe having cold-flared laps backed up by flanges at the ends and then hot-flared over the faces of the laps. Pipe lengths made by either method may be shortened in the field and reflared with special procedures and tools. Square and tapered spacers are furnished to adjust for small discrepancies in assembly.

Saran Saran (Dow Chemical Co.) polyvinylidene chloride liners have excellent resistance to hydrochloric acid. Maximum temperature is 80°C (175°F).

Polypropylene Polypropylene liners (Hercules Incorporated) are used in sulfuric acid service. At 10 to 30 percent concentration the upper temperature limit is 93°C (200°F). In the range of 50 to 93 percent concentration, this drops from 66 to 24°C (from 150 to 75°F).

Kynar Kynar (Pennwalt Chemicals Corp.) vinylidene fluoride liners are used for many chemicals, including bromine and 50 percent hydrochloric acid.

TFE-, PFA-, and FEP-Lined Steel Pipe These are available in sizes from 1 through 12 in and in lengths through 6 m (20 ft). The liners are not affected by any concentration of acids, alkalies, or solvents, but vent holes or internal grooving is required in the steel pipe to release gases which permeate through the liners. Manufacturers should be consulted before use in vacuum service. Experience has determined that practical upper temperature limits are 204°C (400°F) for TFE (polytetrafluoroethylene) and PFA (perfluoroalkoxyl) and 149°C (300°F) for FEP (fluoroethylene polymer); 150-lb and 300-lb ductile-iron or steel flanged lined fittings and valves are used. The nonadhesive properties of the liner make it ideal for handling sticky or viscous substances. Thickness of the lining varies from 1.5 to 3.8 mm (60 to 150 mil), depending on pipe size. Only flanged joints are used.

Rubber-Lined Pipe This pipe is made in lengths up to 6 m (20 ft) with seamless, straight seam-welded and some types of spiral-welded pipe using various types of natural and synthetic adhering rubber. The type of rubber is selected to provide the most suitable lining for the specific service. In general, soft rubber is used for abrasion resistance, semihard for general service, and hard for the more severe service conditions. Multiple-ply lining and combinations of hard and soft rubber are available. The thickness of lining ranges from 3.2 to 6.4 mm (⅛ to ¼ in) depending on the service, the type of rubber, and the method of lining. Cast-steel, ductile-iron, and cast-iron flanged fittings are available rubber-lined. The fittings are usually purchased by the vendor since absence of porosity on the inner surface is essential. Pipe is flanged before rubber lining, and welding elbows and tees may be incorporated at one end of the length of pipe, subject to the conditions that the size of the pipe and the location of the fittings are such that the operator doing the lining can place a hand on any point on the interior surface of the fitting. Welds must be ground smooth on the inside, and a radius is required at the inner edge of the flange face.

The rubber lining is extended out over the face of flanges. With hard-rubber lining, a gasket is required. With soft-rubber lining, coating or a polyethylene sheet is required in place of a gasket to avoid bonding of the lining of one flange to the lining on the other and to permit disassembly of the flanged joint. Also, for pressures over 0.86 MPa (125 lbf/in^2), the tendency of soft-rubber linings to extrude out between the flanges may be prevented by terminating the lining inside the bolt holes and filling the balance of the space between the flange faces with a Masonite spacer of the proper thickness. Hard-rubber-lined gate, diaphragm, and swing check valves are available. In the gate valves, stem, wedge assembly, and seat rings, and in the check valves, hinge pin, flapper arm, disk, and seat ring must be made of metal resistant to the solution handled.

Plastic Pipe In contrast to other piping materials, plastic pipe is free from internal and external corrosion, is easily cut and joined, and does not cause galvanic corrosion when coupled to other materials. Allowable stresses and upper temperature limits are low. Normal operation is in the creep range. Fluids for which a plastic is not suited penetrate and soften it rather than dissolve surface layers. Coefficients of thermal expansion are high. The use of thermoplastic pipe in flammable service aboveground is prohibited by the code.

Support spacing must be much closer than for carbon steel. As temperature increases, the allowable stress for many plastic pipes decreases very rapidly, and heat from sunlight or adjacent hot uninsulated equipment has a marked effect. Successful economical underground use of plastic pipe does not necessarily indicate similar economies outdoors aboveground.

Plastic tubing is widely used for instrument air-signal connections.

Methods of joining include threaded joints with IPS dimensions, solvent-welded joints, heat-fused joints, and insert fittings. Schedules 40 and 80 (see Table 6-6) have been used as a source for standardized dimensions at joints. Some plastics are available in several grades with allowable stresses varying by a factor of 2 to 1. For the same plastic, ½-in Schedule 40 pipe of the strongest grade may have 4 times the allowable internal pressure of the weakest grade of a 2-in Schedule 40 pipe. For this reason, the plastic-pipe industry is shifting to standard dimension ratios (approximately the same ratio of diameter to wall thickness over a wide range of pipe sizes).

ASTM and the Plastics Pipe Institute, a division of the Society of the Plastics Industry, have established identifications for plastic pipe in which the first group of letters identifies the plastic, the two following numbers identify the grade of that plastic, and the last two numbers represent the design stress in the nearest lower (0.7-MPa (100-lbf/in^2) unit at 23°C (73.4°F).

Polyethylene Polyethylene (PE) pipe and tubing are available in sizes 42 in and smaller. They have excellent resistance at room temperature to salts, sodium and ammonium hydroxides, and sulfuric, nitric, and hydrochloric acids. Pipe and tubing are produced by extrusion from resins whose density varies with the manufacturing process. Physical properties and therefore wall thickness depend on the particular resin used. About 3 percent carbon black is added to provide resistance to ultraviolet light. Use of higher-density resin reduces splitting and pinholing in service and increases the strength of the material and the maximum service temperature.

ASTM D2104 covers PE pipe in sizes ½ through 6 in, with IPS Schedule 40 outside and inside diameters for insert-fitting joints. ASTM D2239 covers five standard dimension ratios of pipe diameter to wall thickness in sizes ½ through 6 in, with IPS Schedule 40 inside diameter for insert-fitting joints. ASTM D2447 covers sizes ½ through 12 in, with IPS Schedule 40 and 80 outside and inside diameters for use with heat-fusion socket-type and butt-type fittings. ASTM D3035 covers standard dimension ratios of pipe sizes from ½ through 6 in with IPS outside diameters. All these specifications cover five PE materials (see Table 6-3). Hydrostatic design stresses within the recommended temperature limits are given in Appendix A, Table 3, of the code. The hydrostatic design stress is the maximum tensile hoop stress due to internal hydrostatic water pressure that can be applied continuously with a high degree of certainty that failure of the pipe will not occur. Biaxially oriented polyethylene (PEO) pipe (ASTM D3287) has a higher hydrostatic design stress than PE pipe.

Polyethylene water piping is not damaged by freezing. Pipe and tubing 2 in and smaller are shipped in coils several hundred feet in length.

Clamped-insert joints (Fig. 6-124) are used for flexible plastic pipe up through the 2-in size. Friction between the pipe and the spud is developed both by forcing the spud into the pipe and by tightening the clamp. For the larger sizes, which have thicker walls, these methods cannot develop adequate friction. The joints also have high pressure drop. Stainless-steel bands are available. Inserts are available in nylon, polypropylene, and a variety of metals. A significant use for PE and PP pipe is the technique of rehabilitating deteriorated pipe lines by lining them with plastic pipe. Lining an existing pipe with plastic pipe has a large cost advantage over replacing the line, particularly if replacement of the old line would require excavation.

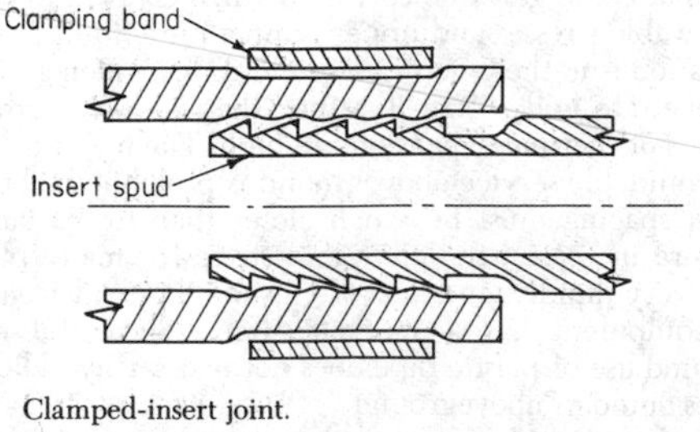

FIG. 6-124 Clamped-insert joint.

Polyvinyl chloride Polyvinyl chloride (PVC) and chlorinated polyvinyl chloride (CPVC) pipe and tubing are available in sizes 12 in and smaller for PVC and 4 in and smaller for CPVC. They have excellent resistance at room temperature to salts, ammonium hydroxide, and sulfuric, nitric, acetic, and hydrochloric acid but may be damaged by ketones, aromatics, and some chlorinated hydrocarbons.

Five PVC pipe materials having characteristic chemical resistance, impact strength, and hydrostatic design stresses are included in the group of ASTM pipe specifications pertaining to PVC. While all these materials have a −18°C (0°F) minimum-recommended-temperature limit (see Table 6-3), Code PVC-1120 and Code PVC-1220 materials become brittle at and below 4°C (40°F). On the other hand, Code PVC-2110, Code PVC-2112, and Code PVC-2216 materials have higher impact resistance but a lower hydrostatic design stress at elevated temperatures. Code PVC-2120 has the best combination of both properties. Allowable hydrostatic design stresses are given in Appendix A, Table 3, of the code, although no stresses are provided for temperatures above 38°C (100°F). The hydrostatic design stresses at 23°C (73.4°F) are 13.8 MPA (2000 lbf/in^2) for PVC-1120, PVC-1220, and PVC-2120, 11.0 MPa (1600 lbf/in^2) for PVC-2116 and CPVC-4116, 8.6 MPa (1250 lbf/in^2) for PVC-2116, and 6.9 MPa (1000 lbf/in^2) for PVC-2110. ASTM D1785 covers sizes from ⅛ through 12 in of PVC pipe in IPS Schedules 40, 80, and 120, except that Schedule 120 starts at ½ in and is not IPS for sizes from ½ through 3 in. ASTM D2241 covers the same size range but with IPS outside diameter and seven standard dimension ratios: 13.5, 17, 21, 26, 32.5, 41, and 64.

ASTM D2513 covers pipe in sizes from ½ through 12 in in both IPS outside diameter and plastic-tubing diameters from ¼ through 1¾ in with standard-dimension-ratio wall thicknesses. This product is intended for gas service. ASTM D2672 covers bell-end pipe in sizes from ⅛ through 8 in in IPS Schedule 40 and in IPS outside diameter and the same standard dimension ratios for wall thicknesses as in D2241. The pipe is intended to be joined by cementing. ASTM D2740 covers PVC-tubing diameters from ½ through 1¼ in with standard-dimension-ratio wall thicknesses.

Solvent-cemented joints (Fig. 6-109) are standard, but screwed joints are sometimes used with Schedule 80 pipe. Cemented joints must not be disturbed for 5 min and achieve full strength in 1 day. Because of the difference in thermal expansion, joints between PVC pipe and metal pipe should be flanged, using a PVC flange on the PVC pipe and a full-face gasket. Flanges are available with ANSI B16.5 150-lb drilling. Ball valves, Y-type globe valves, and diaphragm valves are available in PVC.

Polypropylene Polypropylene (PP) pipe and fittings have excellent resistance to most common organic and mineral acids and their salts, strong and weak alkalies, and many organic chemicals. They are available in sizes ½ through 6 in, in Schedules 40 and 80, but are not covered as such by ASTM specifications.

Reinforced-Thermosetting-Resin (RTR) Pipe Glass-reinforced epoxy resin has good resistance to nonoxidizing acids, alkalies, salt water, and corrosive gases. The glass reinforcement is many times stronger at room temperature than plastics, does not lose strength with increasing temperature, and reinforces the resin effectively up to 149°C (300°F). (See Table 6-5 for temperature limits.) The glass reinforcement is located near the outside wall, protected from the contents by a thick wall of resin and protected from the atmosphere by a thin wall of resin. Stock sizes are 2 through 12 in.

Pipe is supplied in 6- and 12-m (20- and 40-ft) lengths. It is more economical for long, straight runs than for systems containing numerous fittings. When the pipe is sawed to nonfactory lengths, it must be sawed very carefully to avoid cracking the interior plastic zone. A two-component cement may be used to bond lengths into socket couplings or flanges or cemented-joint fittings. Curing of the cement is temperature-sensitive; it sets to full strength in 45 min at 93°C (200°F), in 12 h at 38°C (100°F), and in 24 h at 10°C (50°F). Extensive use is made of shop-fabricated flanged preassemblies. Only flanged joints are used to bond to metallic piping systems. Compared with that of other plastics, the ratio of fitting cost to pipe cost is high. Cemented-joint fittings and flanged fittings are available. Flanged lined metallic valves are used.

RTR is more flexible than metallic pipe and consequently requires closer support spacing. While the recommended spacing varies among manufacturers and with the type of product, Table 6-32 gives typical hanger-spacing ranges. The pipe fabricator should be consulted for recommended hanger spacing on the specific pipe-wall construction being used.

Epoxy resin has a higher strength at elevated temperatures than polyester resins but is not as resistant to attack by some fluids. Some

TABLE 6-32 Typical Hanger-Spacing Ranges Recommended for Reinforced-Thermosetting-Resin Pipe

Nominal pipe size, in	2	3	4	6	8	10	12
Hanger-spacing range, ft	5–8	6–9	6–10	8–11	9–13	10–14	11–15

NOTE: Consult pipe manufacturer for recommended hanger spacing for the specific RTR pipe being used. Tabulated values are based on a specific gravity of 1.25 for the contents of the pipe. To convert feet to meters, multiply by 0.3048.

glass-reinforced epoxy-resin pipe is made with a polyester-resin liner. The coefficient of thermal expansion of glass-reinforced resin pipe is higher than that for carbon steel but much less than that for plastics.

Glass-reinforced polyester is the most widely used reinforced-resin system. A wide choice of polyester resins is available. The bisphenol resins resist strong acids as well as alkaline solutions. The size range is 2 through 12 in; the temperature range is shown in Table 6-5. Diameters are not standardized. Adhesive-cemented socket joints and hand-lay-up reinforced butt joints are used. For the latter, reinforcement consists of layers of glass cloth saturated with adhesive cement.

Haveg 41NA This is a proprietary thermoset plastic consisting of a phenol-formaldehyde resin and nonasbestos silicate fillers. It is furnished as pipe and fittings with several types of joints and is resistant to most acidic chemicals, especially hydrochloric acid. The standard joint uses split cast-iron flanges set in tapered grooves machined in the outside of the pipe. A facing and grooving tool is available. Standard lengths are 1.2 m (4 ft) in the ½- and ¾-in sizes and 3 m (10 ft) in all other sizes.

Flanges are drilled per ANSI B16.5, except that the bolt holes are smaller. Figure 6-125 shows pressure-temperature ratings for standard-wall pipe with standard joints. Pipe and fittings with cemented sleeve joints are also available for use when external corrosion might destroy cast-iron flanges. Y-type globe valves, diaphragm valves, and foot and check valves are available.

Haveg 61NA A proprietary nonasbestos silicate-filled furfuryl alcohol-formaldehyde resin pipe, Haveg 61NA is highly resistant to most acids and, with some reservations, to sodium hydroxide. It is also resistant to many hydrocarbons, halogenated organic compounds, and organic acids. Its pressure-temperature ratings are shown in Figure 6-125. A furfuryl resin pipe is also available from the Maurice A. Knight Co.

PIPING-SYSTEM DESIGN

Safeguarding Safeguarding may be defined as the provision of protective measures as required to ensure the safe operation of a proposed piping system. General considerations to be evaluated should include (1) the hazardous properties of the fluid, (2) the quantity of fluid which could be released by a piping failure, (3) the effect of a failure (such as possible loss of cooling water) on overall plant safety, (4) evaluation of effects on a reaction with the environment (i.e., possibility of a nearby source of ignition), (5) the probable extent of exposure of operating or maintenance personnel, and (6) the relative inherent safety of the piping by virtue of materials of construction, methods of joining, and history of service reliability.

Evaluation of safeguarding requirements might include engineered protection against possible failures such as thermal insulation, armor, guards, barricades, and damping for protection against severe vibration, water hammer, or cyclic operating conditions. Simple means to protect people and property such as shields for valve bonnets, flanged joints, and sight glasses should not be overlooked. The necessity for means to shut off or control flow in the event of a piping failure such as block valves or excess-flow valves should be examined.

Classification of Fluid Services The code applies to piping systems as illustrated in Fig. 6-86, but two categories of fluid services are segregated for special consideration as follows:

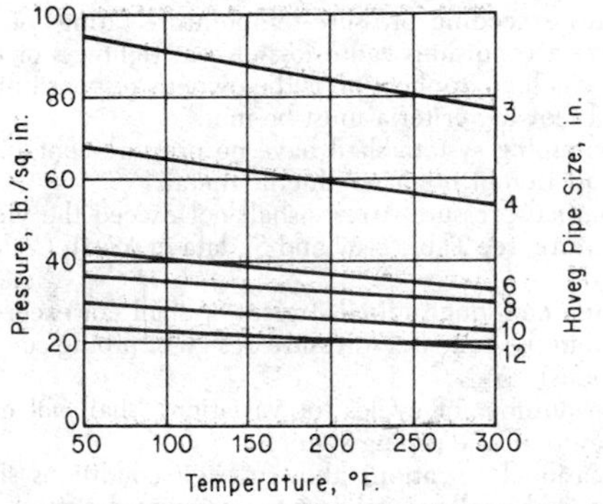

FIG. 6-125 Operating pressure-temperature ratings for Haveg 41NA and 61NA pipe and fittings. (°F − 32)% = °C; to convert pounds-force per square inch to kilopascals, multiply by 6.895; to convert inches to millimeters, multiply by 25.4.

Category D fluid service is defined as "a fluid service to which all the following apply: (1) the fluid handled is nonflammable and nontoxic; (2) the design gage pressure does not exceed 150 psi (1.0 MPa); and (3) the design temperature is between −20°F. (−29°C.) and 360°F. (182°C.)."

Category M fluid service is defined as "a fluid service in which a single exposure to a very small quantity of a toxic fluid, caused by leakage, can produce serious irreversible harm to persons on breathing or bodily contact, even when prompt restorative measures are taken."

The code assigns to the owner the responsibility for identifying those fluid services which are in Categories D and M. The design and fabrication requirements for Class M toxic-service piping are beyond the scope of this *Handbook*. See ANSI B31.3—1976, chap. VIII.

Design Conditions Definitions of the temperatures, pressures, and various forces applicable to the design of piping systems are as follows:

Design Pressure The design pressure of a piping system shall not be less than the pressure at the most severe condition of coincident pressure and temperature resulting in the greatest required component thickness or rating.

Design Temperature The design temperature is the material temperature representing the most severe condition of coincident pressure and temperature. For uninsulated metallic pipe with fluid below 38°C (100°F), the metal temperature is taken as the fluid temperature.

With fluid at or above 38°C (100°F) and without external insulation, the metal temperature is taken as a percentage of the fluid temperature unless a lower temperature is determined by test or calculation. For pipe, threaded and welding-end valves, fittings, and other components with a wall thickness comparable with that of the pipe, the percentage is 95 percent; for flanges and flanged valves and fittings, 90 percent; for lap-joint flanges, 85 percent; and for bolting, 80 percent.

With external insulation, the metal temperature is taken as the fluid temperature unless service data, tests, or calculations justify lower values. For internally insulated pipe, the design metal temperature shall be calculated or obtained from tests.

Ambient Influences If cooling results in a vacuum, the design must provide for external pressure or a vacuum breaker installed; also provision must be made for thermal expansion of contents trapped between or in closed valves. Nonmetallic or nonmetallic-lined pipe may require protection when ambient temperature exceeds design temperature.

Occasional variations of pressure or temperature, or both, above operating levels are characteristic of certain services. If the following criteria are met, such variations need not be considered in determining pressure-temperature design conditions. Otherwise, the most severe conditions of coincident pressure and temperature during the variation shall be used to determine design conditions. (Application of pressures exceeding pressure-temperature ratings of valves may under certain conditions cause loss of seat tightness or difficulty of operation. Such an application is the owner's responsibility.)

All the following criteria must be met:

1. The piping system shall have no pressure-containing components of cast iron or other nonductile metal.

2. Nominal pressure stresses shall not exceed the yield strength at temperature (see Table 6-37 and S_y data in ASME Code, Sec. VIII, Division 2).

3. Combined longitudinal stresses S_L shall not exceed the limits established in the code (see pressure design of piping components for S_L limitations).

4. The number of cycles (or variations) shall not exceed 7000 during the life of the piping system.

5. Occasional variations above design conditions shall remain within one of the following limits for pressure design:

a. When the variation lasts no more than 10 h at any one time and no more than 100 h per year, it is permissible to exceed the pressure rating or the allowable stress for pressure design at the temperature of the increased condition by not more than 33 percent.

b. When the variation lasts no more than 50 h at any one time and not more than 500 h per year, it is permissible to exceed the pressure rating or the allowable stress for pressure design at the temperature of the increased condition by not more than 20 percent.

Dynamic Effects Design must provide for impact (hydraulic shock, etc.), wind (exposed piping), earthquake (see ANSI A58.1), discharge reactions, and vibrations (of piping arrangement and support).

Weight considerations include (1) live loads (contents, ice, and snow), (2) dead loads (pipe, valves, insulation, etc.), and (3) test loads (test fluid).

Thermal-expansion and -contraction loads occur when a piping system is prevented from free thermal expansion or contraction as a result of anchors and restraints or undergoes large, rapid temperature changes or unequal temperature distribution because of an injection of cold liquid striking the wall of a pipe carrying hot gas.

Design Criteria: Metallic Pipe

The code uses three different approaches to design, as follows:

1. It provides for the use of dimensionally standardized components at their published pressure-temperature ratings.

2. It provides design formulas and maximum stresses.

3. It prohibits the use of materials, components, or assembly methods in certain conditions.

Components Having Specific Ratings These are listed in ANSI, API, and industry standards. These ratings are acceptable for design pressures and temperatures unless limited in the code. A list of component standards is given in Appendix E of the code. The following rating tables covering commonly used components have been extracted from the original document with permission of the publisher, the American Society of Mechanical Engineers, New York: Table 6-33 lists pressure-temperature ratings for flanges, flanged fittings, and flanged valves; and Table 6-34 lists hydrostatic-shell test pressures for flanges, flanged fittings, and flanged valves. Flanged joints, flanged valves in the open position, and flanged fittings may be subjected to system hydrostatic tests at a pressure not to exceed the hydrostatic-shell test pressure. Flanged valves in the closed position may be subjected to a system hydrostatic test at a pressure not to exceed 110 percent of the 100°F rating of the valve unless otherwise limited by the manufacturer.

Pressure-temperature ratings for soldered and brazed copper-tubing joints are given in Tables 6-35 and 6-36 respectively.

Components without Specific Ratings Components such as pipe and butt-welding fittings are generally furnished in nominal thicknesses. Fittings are rated for the same allowable pressures as pipe of the same nominal thickness and, along with pipe, are rated by the rules for pressure design and other provisions of the code.

Pressure Design of Metallic Components: Wall Thickness External-pressure stress evaluation of piping is the same as for pressure vessels. But an important difference exists when one is establishing design pressure and wall thickness for internal pressure as a result of the ASME Boiler and Pressure Vessel Code's requirement that the relief-valve setting be not higher than the design pressure. For vessels this means that the design is for a pressure 10 percent more or less above the intended maximum operating pressure to avoid popping or leakage from the valve during normal operation. However, on piping the design pressure and temperature are taken as the maximum intended operating pressure and coincident temperature combination which results in the maximum thickness. The temporary increased operating conditions listed under "Design Criteria" cover temporary operation at pressures that cause relief valves to leak or open fully. Allowable stresses for nearly 1000 materials are contained in the code. For convenience, the allowable stresses for commonly used materials have been extracted from the code and listed in Table 6-37.

For **straight metal pipe under internal pressure** the formula for minimum required wall thickness t_m is applicable for D_o/t ratios greater than 6. The more conservative Barlow and Lamé equations may also be used. Equation 6-37 includes a factor Y varying with material and temperature to account for the redistribution of circumferential stress which occurs under steady-state creep at high temperature and permits slightly lesser thickness at this range.

$$t_m = \frac{PD_o}{2(SE + PY)} + C \tag{6-37}$$

where (in consistent units)

P = design pressure
D_o = outside diameter of pipe
C = sum of allowances for corrosion, erosion, and any thread or groove depth. For threaded components the depth is h of ANSI B2.1, and for grooved components the depth is the depth removed (plus 1/64 in when no tolerance is specified).
SE = allowable stress (see Table 6-37)
S = basic allowable stress for materials, excluding casting, joint, or structural-grade quality factors
E = quality factor. The quality factor E is one or the product of more than one of the following quality factors: casting quality factor E_c, joint quality factor E_j (see Fig. 6-126), and structural-grade quality factor E_s of 0.92.
Y = coefficient having value in Table 6-38 for ductile ferrous materials, 0.4 for ductile nonferrous materials, and zero for brittle materials such as cast iron
t_m = minimum required thickness, in, to which manufacturing tolerance must be added when specifying pipe thickness on purchase orders. [Most ASTM specifications to which mill pipe is normally obtained permit minimum wall to be 12½ percent less than nominal. ASTM A155 for fusion-welded pipe permits minimum wall 0.25 mm (0.01 in) less than nominal plate thickness.] Pipe with t equal to or greater than $D/6$ or P/SE greater than 0.385 requires special consideration.

In addition to establishing the wall thickness for internal pressure, the stress values in Table 6-37 control other portions of the design. The **sum of the longitudinal stresses** S_L (in the corroded condition) due to internal pressure, weight of pipe and contents between supports, and other sustained loadings such as friction between a laid (not hung) long length of straight cold pipe and its supports when it is placed in service, shall not exceed the value of S_h. In this determination, for pipe with welded longitudinal seams, the longitudinal weld joint factor is disregarded. Also, when **thermal-expansion or contraction strains** are taken up primarily by bending or torsion, the local stresses so produced are limited to the following range designated as S_A:

$$S_A = f(1.25S_c + 0.25S_h) \tag{6-38}$$

where S_c = S from Table 6-37 at a minimum (cold) metal temperature normally expected during operation or shutdown (See Note 13, Table 6-37)
S_h = S from Table 6-37 at maximum (hot) metal temperature normally expected during operation or shutdown (See Note 13, Table 6-37)
f = stress-range reduction factor for total number of full temperature cycles over expected life (See Table 6-39)

TABLE 6-33 Pressure-Temperature Ratings for Flanges, Flanged Fittings, and Flanged Valves of Typical Materials,[4] lbf/in²

Material group	1.1	1.5	1.9	1.10	1.13	2.1	2.2	2.3	2.6	2.7
	Carbon									
Materials temperature, °F	Normal	C, ½Mo	1, 1¼Cr, ½Mo	2¼Cr, 1Mo	5Cr, ½Mo	Type 304	Type 316	Type 304L Type 316L	Type 309	Type 310
150-lb class										
−20 to 100	285	265	290			275	275	230	260	
200	260		260			235	240	195	230	
300	230		230			205	215	175	220	
400			200			180	195	160	200	
500			170			170		145	170	
600			140			140		140	140	
650			125			125		125	125	
700			110			110		110	110	
750			95			95		95	95	
800			80			80		80	80	
850			65			65		65	65	
900			50			50			50	
950			35			35			35	
1000			20			20			20	
300-lb class										
−20 to 100	740	695	750	750	750	720	720	600	670	
200	675	680	710	715	750	600	620	505	605	
300	655	655	675	675	730	530	560	455	570	
400	635	640	660	650	705	470	515	415	535	
500	600	620	640		665	435	480	380	505	
600	550		605			415	450	360	480	
650	535		590			410	445	350	465	
700	535		570			405	430	345	455	
750	505		530			400	425	335	445	
800	410		510		500	395	415	330	435	
850	270		485		440	390	405	320	425	
900	170		450		355	385	395		415	
950	105	280	380		260	375	385		385	
1000	50	165	225	270	190	325	365		335	350
1050			140	200	140	310	300		290	335
1100			95	115	105	260	325		225	290
1150			50	105	70	195	275		170	245
1200			35	55	45	155	205		130	205
1250						110	180		100	160
1300						85	140		80	120
1350						60	105		60	80
1400						50	75		45	55
1450						35	60		30	40
1500						25	40		25	25
400-lb class°										
−20 to 100	990	925	1000	1000	1000	960	960	800	895	
200	900	905	950	955	1000	800	825	675	805	
300	875	870	895	905	970	705	745	605	760	
400	845	855	880	865	940	630	685	550	710	
500	800	830	855		885	585	635	510	670	
600	730		805			555	600	480	635	
650	715		785			545	590	470	620	
700	710		755			540	575	460	610	
750	670		710			530	565	450	595	
800	550		675		665	525	555	440	580	
850	355		650		585	520	540	430	565	
900	230		600		470	510	525		555	
950	140	375	505		350	500	515		515	
1000	70	220	300	355	255	430	485		450	465
1050			185	265	190	410	480		390	445
1100			130	150	140	345	430		300	390
1150			70	140	90	260	365		230	330
1200			45	75	60	205	275		175	275
1250						145	245		135	215
1300						110	185		105	160
1350						85	140		80	105
1400						65	100		60	75
1450						45	80		40	50
1500						30	55		30	30

TABLE 6-33 Pressure-Temperature Ratings for Flanges, Flanged Fittings, and Flanged Valves of Typical Materials,[4] lbf/in² *(Continued)*

Material group	1.1	1.5	1.9	1.10	1.13	2.1	2.2	2.3	2.6	2.7
Materials	Carbon									
temperature, °F	Normal	C, ½Mo	1, 1¼Cr, ½Mo	2¼Cr, 1Mo	5Cr, ½Mo	Type 304	Type 316	Type 304L Type 316L	Type 309	Type 310
600-lb class°										
−20 to 100	1480	1390	1500	1500	1500	1440	1440	1200	1345	
200	1350	1360	1425	1430	1500	1200	1240	1015	1210	
300	1315	1305	1345	1355	1455	1055	1120	910	1140	
400	1270	1280	1315	1295	1410	940	1030	825	1065	
500	1200	1245	1285	1280	1330	875	955	765	1010	
600	1095		1210			830	905	720	955	
650	1075		1175			815	890	700	930	
700	1065		1135			805	865	685	910	
750	1010		1065			795	845	670	895	
800	825		1015		995	790	830	660	870	
850	535		975		880	780	810	645	850	
900	345		900		705	770	790		830	
950	205	560	755		520	750	775		775	
1000	105	330	445	535	385	645	725		670	700
1050			275	400	280	620	720		585	665
1100			190	225	205	515	645		445	585
1150			105	205	140	390	550		345	495
1200			70	110	90	310	410		260	410
1250						220	365		200	325
1300						165	275		160	240
1350						125	205		115	160
1400						90	150		90	110
1450						70	115		60	75
1500						50	85		50	50
900-lb class°										
−20 to 100	2220	2085	2250	2250	2250	2160	2160	1800	2015	
200	2025	2035	2135	2150	2250	1800	1860	1520	1815	
300	1970	1955	2020	2030	2185	1585	1680	1360	1705	
400	1900	1920	1975	1945	2115	1410	1540	1240	1600	
500	1795	1865	1925	1920	1995	1310	1435	1145	1510	
600	1640		1815			1245	1355	1080	1435	
650	1610		1765			1225	1330	1050	1395	
700	1600		1705			1210	1295	1030	1370	
750	1510		1595			1195	1270	1010	1340	
800	1235		1525		1490	1180	1245	985	1305	
850	805		1460		1315	1165	1215	965	1275	
900	515		1350		1060	1150	1180		1245	
950	310	845	1130		780	1125	1160		1160	
1000	155	495	670	805	575	965	1090		1010	1050
1050			410	595	420	925	1080		875	1000
1100			290	340	310	770	965		670	875
1150			155	310	205	585	825		515	740
1200			105	165	135	465	620		390	620
1250						330	545		300	485
1300						245	410		235	360
1350						185	310		175	235
1400						145	225		135	165
1450						105	175		95	115
1500						70	125		70	70
1500-lb class										
−20 to 100	3705	3470	3750	3750	3750	3600	3600	3000	3360	
200	3375	3395	3560	3580	3750	3000	3095	2530	3025	
300	3280	3260	3365	3385	3640	2640	2795	2270	2845	
400	3170	3200	3290	3240	3530	2350	2570	2065	2665	
500	2995	3105	3210	3200	3325	2185	2390	1910	2520	
600	2735		3025			2075	2255	1800	2390	
650	2685		2940			2040	2220	1750	2330	
700	2665		2840			2015	2160	1715	2280	
750	2520		2660			1990	2110	1680	2230	
800	2060		2540		2485	1970	2075	1645	2170	
850	1340		2435		2195	1945	2030	1610	2125	
900	860		2245		1765	1920	1970		2075	

TABLE 6-33 Pressure-Temperature Ratings for Flanges, Flanged Fittings, and Flanged Valves of Typical Materials,[4] lbf/in² (Continued)

Material group	1.1	1.5	1.9	1.10	1.13	2.1	2.2	2.3	2.6	2.7
Materials temperature, °F	Carbon Normal	Carbon C, ½Mo	1, 1¼Cr, ½Mo	2¼Cr, 1Mo	5Cr, ½Mo	Type 304	Type 316	Type 304L Type 316L	Type 309	Type 310
				600-lb class°						
950	515	1405	1885		1305	1870	1930			
1000	260	825	1115	1340	960	1610	1820		1680	1750
1050			685	995	705	1545	1800		1460	1665
1100			480	565	515	1285	1610		1115	1460
1150			260	515	345	980	1370		860	1235
1200			170	275	225	770	1030		650	1030
1250						550	910		495	805
1300						410	685		395	600
1350						310	515		290	395
1400						240	380		225	375
1450						170	290		155	190
1500						120	205		120	120
				2500-lb class						
−20 to 100	6170	5785	6250	6250	6250	6000	6000	5000	5600	
200	5625	5660	5930	5965	6250	5000	5160	4220	5040	
300	5470	5435	5605	5640	6070	4400	4660	3780	4740	
400	5280	5330	5485	5400	5880	3920	4280	3440	4440	
500	4990	5180	5350	5330	5540	3640	3980	3180	4200	
600	4560		5040			3460	3760	3000	3980	
650	4475		4905			3400	3700	2920	3880	
700	4440		4730			3360	3600	2860	3800	
750	4200		4430			3320	3520	2800	3720	
800	3430		4230		4145	3280	3460	2740	3620	
850	2230		4060		3660	3240	3380	2680	3540	
900	1430		3745		2945	3200	3280		3460	
950	860	2345	3145		2170	3120	3220		3220	
1000	430	1370	1860	2230	1600	2685	3030		2800	2915
1050			1145	1660	1170	2570	3000		2430	2770
1100			800	945	860	2145	2685		1860	2430
1150			430	860	570	1630	2285		1430	2060
1200			285	460	370	1285	1715		1085	1715
1250						915	1515		830	1345
1300						685	1145		660	1000
1350						515	860		485	660
1400						400	630		370	460
1450						285	485		260	315
1500						200	345		200	200

°For Group 1.1, do not use ASTM A181 Grade I or II materials.

NOTES:

1. Ratings shown apply to other material groups when column-dividing lines have been omitted.
2. Temperature notes for all material groups in Table 6-33.

Material group	Materials[3] (specification—grade)	See Notes
1.1	A105, A181-II, A216-WCB, A515-70	a, h
	A516-70	a, g
	A350-LF2, A537-C1.1	d
1.5	A182-F1, A204-A, A204-B, A217-WC1	b, h
	A352-LC1	d, m
1.9	A182-F11, A182-F12, A387-11, C1.2	m, c
	A217-WC6	j, m
1.10	A182-F22, A387-22, C1.2	c
	A217-WC9	m, j
1.13	A182-F5a, A217-C5	m
2.1	A182-F304, A182-F304H	n
	A240-304, A351-CF8	n, o
	A351-CF3	f
2.2	A182-F316, A182-F316H, A240-316	n, o
	A240-317, A351-CF8M	n, o
	A351-CF3M	g
2.3	A182-F304L, A240-304L	f
	A182-F316L, A240-316L	g
2.6	A240-309S, A351-CH8, A351-CH20	n, o
2.7	A182-F310, A240-310S	k, n
	A351-CK20	n

a. Permissible but not recommended for prolonged use above about 800°F.
b. Permissible but not recommended for prolonged use above about 850°F.

TABLE 6-33 Pressure-Temperature Ratings for Flanges, Flanged Fittings, and Flanged Valves of Typical Materials,[4] lbf/in² (Continued)

c. Permissible but not recommended for prolonged use above about 1100°F.
d. Not to be used over 650°F.
f. Not to be used over 800°F.
g. Not to be used over 850°F.
h. Not to be used over 1000°F.
i. Not to be used over 1050°F.
j. Not to be used over 1100°F.
k. For service temperatures 1050°F and above, assurance must be provided that grain size is not finer than ASTM No. 6.
l. Only killed steel shall be used above 850°F.
m. Use normalized and tempered material only.
n. At temperatures over 1000°F, use only when the carbon content is 0.04 percent or higher.
o. See ANSI B16.5 for heat treatment for service temperatures over 1000°F.
p. The ratings at −20 to 100°F given for the materials covered shall also apply at lower temperatures. The ratings for low-temperature service of the cast and forged materials listed in ASTM A352 and A350 shall be taken the same as the −20 to 100°F ratings for carbon steel.
q. Some of the materials listed in the rating tables undergo a decrease in impact resistance at temperatures lower than −20°F to such an extent as to be unable to resist safely shock loadings, sudden changes of stress, or high stress concentration.
3. See ANSI B16.5, Table 1A, for additional information and notes relating to specific materials.
4. Extracted from Steel Pipe Flanges and Flanged Fittings, ANSI B16.5 1977 and B16.34-1977, with permission of the publisher, the American Society of Mechanical Engineers, New York.
5. A product used under the jurisdiction of the ASME Boiler and Pressure Vessel Code and the ANSI Code for Pressure Piping B31.1 is subject to any limitation of those codes. This includes any maximum-temperature limitation for a material or a code rule governing the use of a material at a low temperature.
6. (°F − 32)5⁄9 = °C; to convert pounds-force per square inch to megapascals, multiply by 0.006895.

TABLE 6-34 Hydrostatic-Shell Test Pressures for Flanges, Flanged Fittings, and Flanged Valves of Typical Materials*

Material group no.	Shell test pressures by class, lbf/in² gauge						
	150	300	400	600	900	1500	2500
1.1	450	1125	1500	2225	3350	5575	9275
1.5	400	1050	1400	2100	3150	5225	8700
1.9	450	1125	1500	2250	3375	5625	9375
1.10	450	1125	1500	2250	3375	5625	9375
1.13	450	1125	1500	2250	3375	5625	9375
2.1	425	1100	1450	2175	3250	5400	9000
2.2	425	1100	1450	2175	3250	5400	9000
2.3	350	900	1200	1800	2700	4500	7500
2.6	400	1025	1350	2025	3025	5050	8400
2.7	400	1025	1350	2025	3025	5050	8400

*Extracted from Steel Pipe Flanges and Flanged Fittings, ANSI B16.5—1977, with permission of the publisher, the American Society of Mechanical Engineers, New York. Test temperature not to exceed 125°F. (°F − 32)5⁄9 = °C; to convert pounds-force per square inch to megapascals, multiply by 0.006895.

When the anticipated number of cycles is substantially less than 7000, useful information can be obtained from ASME Boiler and Pressure Vessel Code, Sec. III, "Nuclear Vessels."

However, if the sum of longitudinal stresses S_L enumerated is less than their stated limit S_h, the difference may be added to the term $0.25S_h$ in the equation limiting the stress range:

$$S_A = f[1.25(S_c + S_h) - S_L] \tag{6-39}$$

For flanges of nonstandard dimensions or for sizes beyond the scope of the approved standards, design shall be in accordance with the requirements of the ASME Boiler and Pressure Vessel Code, Sec. VIII, except that requirements for fabrication, assembly, inspection testing, and the pressure and temperature limits for materials of the Piping Code are to prevail. Countermoment flanges of flat face or otherwise providing a reaction outside the bolt circle are permitted if designed or tested in accordance with code requirements under pressure-containing components "not covered by standards and for which design formulas or procedures are not given."

In accordance with listed standards, **blind flanges** may be used at their pressure-temperature ratings. The minimum thickness of nonstandard blind flanges shall be the same as for a bolted flat cover, in

TABLE 6-35 Strength of Solder Joints*

Maximum recommended pressure-temperature ratings for solder joints made with copper tubing and wrought-copper and -bronze or cast-bronze solder-joint pressure fittings and using representive commercial solders

Joining material used in joints	Working temperatures, °F	Maximum working pressure, lbf/in²			
		⅛ to 1 in, inclusive†	1¼ to 2 in, inclusive†	2½ to 4 in, inclusive†	5 to 8 in, inclusive†
50-50 tin-lead solder‡	100	200	175	150	135
	150	150	125	100	90
	200	100	90	75	70
	250	85	75	50	45
95-5 tin-antimony solder	100	500	400	300	270
	150	400	350	275	250
	200	300	250	200	180
	250	200	175	150	135

NOTE: For extremely low working temperatures (in the 0 to −200°F range) it is recommended that a joining material melting at or above 1100°F be used. (Joining materials with melting points in excess of 800°F are defined as "brazing" alloys by the American Welding Society.) See Table 6-36.

*Extracted from ANSI B16.22—1973 with permission of the publisher, the American Society of Mechanical Engineers, New York. (°F − 32)5⁄9 = °C; to convert inches to millimeters, multiply by 25.4; to convert pounds-force per square inch to megapascals, multiply by 0.006895.

†Standard water-tubing sizes.

‡ASTM B32.66T Alloy Grade 50A.

TABLE 6-36 Strength of Silver-Brazed Joints*

Maximum recommended pressure-temperature ratings for brazed joints made with copper tubing and copper or copper-alloy fittings and using representative commercial brazing alloys

Outside-diameter size, in	lbf/in²			
	150°F (S = 5100 lbf/in²)	250°F (S = 4700 lbf/in²)	350°F (S = 4000 lbf/in²)	400°F (S = 3000 lbf/in²)
⅛	1790	1650	1400	1050
3/16	1190	1100	940	700
¼	890	825	700	525
5/16	840	780	660	500
⅜	780	720	615	460
½	680	625	530	400
⅝	615	565	480	360
¾	535	495	420	315
⅞	490	450	385	290
1⅛	420	390	330	250
1⅜	380	350	295	220
1⅝	350	320	275	205
2⅛	310	285	245	180
2⅝	286	265	225	170
3⅛	270	250	190	140
3⅝	260	240	200	150
4⅛	250	230	195	145
5⅛	225	210	180	135
6⅛	215	195	165	125

*Extracted from ANSI B16.41—January 1977 draft, with permission of the publisher, the American Society of Mechanical Engineers, New York. (°F − 32)5/9 = °C; to convert inches to millimeters, multiply by 25.4; to convert pounds-force per square inch to megapascals, multiply by 0.006875.

accordance with the rules of the ASME Boiler and Pressure Vessel Code, Sec. VIII.

Operational blanks shall be of the same thickness as blind flanges or may be calculated by the following formula (use consistent units):

$$t = d\sqrt{3P/16S} \qquad (6\text{-}40)$$

where d = inside diameter of gasket for raised- or flat (plain)-face flanges, or the gasket pitch diameter for retained gasketed flanges
P = internal design pressure or external design pressure
S = applicable allowable stress

Valves must comply with the applicable standards listed in Appendix E of the code and with the allowable pressure-temperature limits established thereby but not beyond the code-established service or materials limitations. Special valves must meet the same requirements as for countermoment flanges.

The code contains no specific rules for the design of **fittings** other than as branch openings. Ratings established by recognized standards are acceptable, however. ANSI Standard B16.5 for steel-flanged fittings incorporates a 1.5 shape factor and thus requires the entire fitting to be 50 percent heavier than a simple cylinder in order to provide reinforcement for openings and/or general shape. ANSI B16.9 for butt-welding fittings, on the other hand, requires only that the fittings be able to withstand the calculated bursting strength of the straight pipe with which they are to be used.

The thickness of **pipe bends** shall be determined as for straight pipe, provided the bending operation does not result in a difference between maximum and minimum diameters greater than 8 and 3 percent of the nominal outside diameter of the pipe for internal and external pressure respectively.

The maximum allowable internal pressure for multiple miter bends shall be the lesser value calculated from Eqs. (6-41) and (6-42). These equations are not applicable when θ exceeds 22.5°.

$$P = \frac{SEt}{r_2}\left(\frac{t}{t + 0.643 \tan\theta \sqrt{r_2 t}}\right) \qquad (6\text{-}41)$$

$$P = \frac{SEt}{r_2}\left(\frac{R_1 - r_2}{R_1 - 0.5r_2}\right) \qquad (6\text{-}42)$$

where nomenclature is the same as for straight pipe except as follows (see Fig. 6-127):

t = pressure design thickness
r_2 = mean radius of pipe
R_1 = effective radius of miter bend, defined as the shortest distance from the pipe centerline to the intersection of the planes of adjacent miter joints
θ = angle of miter cut, °
α = angle of change in direction at miter joint
= 2θ, °

For compliance with the code, the value of R_1 shall not be less than that given by Eq. (6-43).

$$R_1 = A/\tan\theta + D/2 \qquad (6\text{-}43)$$

where A has the following empirical values **(not valid in SI units);**

t, in	A
≤ 0.5	1.0
$0.5 < t < 0.88$	$2t$
≥ 0.88	$(2t/3) + 1.17$

Piping branch connections involve the same considerations as pressure-vessel nozzles. However, outlet size in proportion to piping header size is unavoidably much greater for piping. The current Piping Code rules for calculation of branch-connection reinforcement are similar to those of the ASME Boiler and Pressure Vessel Code, Sec. VIII, Division I (1980 edition) for a branch with axis at right angles to the header axis. If the branch connection makes an angle β with the header axis from 45 to 90°, the Piping Code requires that the area to be replaced be increased by dividing it by sin β. In such cases the half width of the reinforcing zone measured along the header axis is similarly increased, except that it may not exceed the outside diameter of the header. Some details of commonly used reinforced branch connections are given in Fig. 6-128.

The rules provide that a branch connection has adequate strength for pressure if a fitting (tee, lateral, or cross) is in accordance with an approved standard and is used within the pressure-temperature limitations or if the connection is made by welding a coupling or half coupling (wall thickness not less than the branch anywhere in reinforcement zone or less than extra heavy or 3000 lb) to the run and provided the ratio of branch to run diameters is not greater than one-fourth and that the branch is not greater than 2 in nominal diameter.

Dimensions of extra-heavy couplings are given in the *Steel Products Manual* published by the American Iron and Steel Institute. In ANSI B16.11—1966, 2000-lb couplings were superseded by 3000-lb couplings.

ANSI B31.3 states that the reinforcement area for resistance to external pressure is to be at least one-half of that required to resist internal pressure.

The code provides no guidance for analysis but requires that external and internal **attachments** be designed to avoid flattening of the pipe, excessive localized bending stresses, or harmful thermal gradients, with further emphasis on minimizing stress concentrations in cyclic service.

The code provides design requirements for **closures** which are flat, ellipsoidal, spherically dished, hemispherical, conical (without transition knuckles), conical convex to pressure, toriconical concave to pressure, and toriconical convex to pressure.

Openings in closures over 50 percent in diameter are designed as flanges in flat closures and as reducers in other closures. Openings of not over one-half of the diameter are to be reinforced as branch connections.

Thermal Expansion and Flexibility: Metallic Piping ANSI B31.3 requires that piping systems have sufficient flexibility to prevent thermal expansion or contraction or the movement of piping supports or terminals from causing (1) failure of piping supports from overstress or fatigue; (2) leakage at joints; or (3) detrimental stresses or distortions in piping or in connected equipment (pumps,

(Continued on page 6-87)

TABLE 6-37 Allowable Stresses in Tension for Materials (4, 13, 28)*

Specifications are ASTM unless otherwise indicated. Numbers in parentheses refer to notes at end of table.

Material	Specification	P no. (23)	Grade	Class	Factor, E	Minimum tensile strength, kip/in^2	Minimum yield strength, kip/in^2	Notes	Minimum temperature (18)	Minimum temperature to 100	200	300	400	500	600	650
Iron																
Centrifugally cast pipe																
	FS-WW-P421c	...	...	...	...	...	...	8, 10, 17	−20	6.0	6.0	6.0	6.0			
	AWWA C106	...	...	...	...	...	...	8, 10, 17	−20	6.0	6.0	6.0	6.0			
	AWWA C108	...	...	...	...	...	...	8, 10, 17	−20	6.0	6.0	6.0	6.0			
Carbon steel																
Seamless pipe and tubing																
	A53	1	A	Type S	...	48.0	30.0	1, 2	−20	16.0	16.0	16.0	16.0	16.0	14.8	14.5
	A53	1	B	Type S	...	60.0	35.0	1, 2	−20	20.0	20.0	20.0	20.0	18.9	17.3	17.0
	A106	1	A	...	...	48.0	30.0	2	−20	16.0	16.0	16.0	16.0	16.0	14.8	14.5
	A106	1	B	...	...	60.0	30.0	2	−20	20.0	20.0	20.0	20.0	18.9	17.3	17.0
	A106	1	C	...	...	70.0	40.0	2	−20	23.3	23.3	23.3	22.9	21.6	19.7	19.4
	A120	1	...	...	...	...	...	21	−20	12.0	11.4					
	A333	1	1	...	...	55.0	30.0	1, 2	−50	18.3	18.3	17.7	17.2	16.2	14.8	14.5
	A333	1	6	...	...	60.0	35.0	2	−50	20.0	20.0	20.0	20.0	18.9	17.3	17.0
	API 5L	1	A	...	...	48.0	30.0	1, 2	−20	16.0	16.0	16.0	16.0	16.0	14.8	14.5
	API5L	1	B	...	...	60.0	35.0	1, 2	−20	20.0	20.0	20.0	20.0	18.9	17.3	17.0
	API 5LX	SP2	X42	...	...	60.0	42.0	37, 38	−20	20.0	20.0	20.0	20.0			
	API 5LX	SP3	X46	...	...	63.0	46.0	37, 38	−20	21.0	21.0	21.0	21.0			
	API 5LX	SP3	X52	...	...	66.0	52.0	37, 38	−20	22.0	22.0	22.0	22.0			
	API 5LX	SP3	X52	...	...	72.0	52.0	37, 38	−20	24.0	24.0	24.0	24.0			
Electric-resistance-welded pipe																
	A53	1	A	Type E	0.85	48.0	30.0	1, 2	−20	13.6	13.6	13.6	13.6	13.6	12.6	12.3
	A53	1	B	Type E	0.85	60.0	35.0	1, 2	−20	17.0	17.0	17.0	17.0	16.1	14.7	14.5
	A120	1	...	...	0.85	...	...	21	−20	10.2	9.7					
	A135	1	A	...	0.85	48.0	30.0	1, 2	−20	13.6	13.6	13.6	13.6	13.6	12.6	12.3
	A135	1	B	...	0.85	60.0	35.0	1, 2	−20	17.0	17.0	17.0	17.0	16.1	14.7	14.5
	A333	1	1	...	0.85	55.0	30.0	1, 2	−50	15.6	15.6	15.0	14.6	13.8	12.6	12.3
	A333	1	6	...	0.85	60.0	35.0	2	−50	17.0	17.0	17.0	17.0	16.1	14.7	14.5
	A587	1	...	...	0.85	48.0	30.0	1, 2	−20	13.6	13.6	13.6	13.6	13.6	12.6	12.3
	API 5L	1	A25	I and II	0.85	45.0	25.0	1, 2	−20	12.8	12.8	12.3	11.8			
	API 5L	1	A	...	0.85	48.0	30.0	1, 2	−20	13.6	13.6	13.6	13.6	13.6	12.6	12.3
	API 5L	1	B	...	0.85	60.0	35.0	1, 2	−20	17.0	17.0	17.0	17.0	16.1	14.7	14.5
	API 5L	SP2	X42	...	0.85	60.0	42.0	37, 38	−20	17.0	17.0	17.0	17.0			
	API 5LX	SP3	X46	...	0.85	63.0	46.0	37, 38	−20	17.9	17.9	17.9	17.9			
	API 5LX	XP3	X52	...	0.85	66.0	52.0	37, 38	−20	18.7	18.7	18.7	18.7			
	API 5LX	SP3	X52	...	0.85	72.0	52.0	37, 38	−20	20.4	20.4	20.4	20.4			
Electric-fusion-welded pipe (straight seam)																
A570 GR A	A134	1	...	...	0.74	45.0	25.0	5, 21	−20	11.1	10.5	10.0				
A570 GR B	A134	1	...	...	0.74	49.0	30.0	5, 21	−20	12.1	11.4	10.9				
A570 GR C	A134	1	...	...	0.74	52.0	33.0	5, 21	−20	12.8	12.1	11.6				
A570 GR D	A134	1	...	...	0.74	55.0	40.0	5, 21	−20	13.6	12.8	12.2				
A570 GR E	A134	1	...	...	0.74	58.0	42.0	5, 21	−20	14.3	13.5	12.9				
Low- and intermediate-alloy steel																
Seamless pipe																
3½ Ni	A333	9B	3	...	...	65.0	35.0	...	−150	21.7	19.6	19.6	18.7	17.8	16.8	16.3
¾ Cr, ¾ Ni, Cu, Al	A333	4	4	...	...	60.0	35.0	...	−150	20.0	19.1	18.2	17.3	16.4	15.5	15.0
2¼ Ni	A333	9A	7	...	...	65.0	35.0	...	−100	21.7	19.6	19.6	18.7	17.6	16.8	16.3
9 Ni	A333	11A-SG1	8	...	...	100.0	75.0	40	−320	31.7	31.7					
C, ½ Mo	A335	3	P1	...	...	55.0	30.0	3	−20	18.3	18.3	17.5	16.9	16.3	15.7	15.4
5 Cr, ½ Mo	A335	5	P5	...	...	60.0	30.0	...	−20	20.0	18.1	17.4	17.2	17.1	16.8	16.6
1¼ Cr, ½ Mo	A335	4	P11	...	...	60.0	30.0	...	−20	20.0	18.7	18.0	17.5	17.2	16.7	16.2
2¼ Cr, 1 Mo	A335	5	P22	...	...	60.0	30.0	...	−20	20.0	18.5	18.0	17.9	17.9	17.9	17.9
Stainless steel																
Seamless pipe and tubing																
18Cr, 8Ni pipe	A312	8	TP304	...	...	75.0	30.0	7, 14, 16, 20	−425	20.0	20.0	20.0	18.7	17.5	16.4	16.2
18Cr, 8Ni pipe	A312	8	TP304H	...	...	75.0	30.0	16	−325	20.0	20.0	20.0	18.7	17.5	16.4	16.2
18Cr, 8Ni pipe	A312	8	TP304L	...	...	70.0	25.0	...	−425	16.7	16.7	16.7	15.8	14.8	14.0	13.7
25Cr, 20Ni pipe	A312	8	TP310	...	...	75.0	30.0	19, 24, 32	−325	20.0	20.0	20.0	20.0	20.0	19.2	18.8
25Cr, 20Ni pipe	A312	8	TP310	...	...	75.0	30.0	6, 19, 24, 32	−325	20.0	20.0	20.0	20.0	20.0	19.2	18.8
16Cr, 12Ni, 2Mo pipe	A312	8	TP316	...	...	75.0	30.0	14, 16	−325	20.0	20.0	20.0	19.3	17.9	17.0	16.7
16Cr, 12Ni, 2Mo pipe	A312	8	TP316H	...	...	75.0	30.0	16	−325	20.0	20.0	20.0	19.3	17.9	17.0	16.7
16Cr, 12Ni, 2Mo pipe	A312	8	TP316L	...	...	70.0	25.0	...	−325	16.7	16.7	16.7	15.5	14.4	13.5	13.2
18Cr, 10Ni, Cb pipe	A312	8	TP347	...	...	75.0	30.0	7, 14	−425	20.0	20.0	20.0	20.0	19.9	19.3	19.0
18Cr, 10Ni, Cb pipe	A312	8	TP347H	...	...	75.0	30.0	...	−325	20.0	20.0	20.0	20.0	19.9	19.3	19.0
Centrifugally cast pipe																
18Cr, 8Ni	A451	8	CPF8	...	0.90	70.0	30.0	14, 15, 16	−425	18.0	18.0	17.8	15.8	14.8	14.1	13.8
18Cr, 10Ni, 2Mo	A451	8	CPF8M	...	0.90	70.0	30.0	14, 15, 16	−425	18.0	18.0	18.0	17.5	16.3	15.4	15.0
18Cr, 10Ni, Cb	A451	8	CPF8C	...	0.90	70.0	30.0	7, 14, 15	−325	18.0	18.0	18.0	18.0	17.4	16.5	16.2
15Cr, 13Ni, 2Mo, Cb	A451	8	CPF10MC	...	0.90	70.0	30.0	7, 11, 14, 15	−325	18.0						
23Cr, 13Ni	A451	8	CPH8	...	0.90	65.0	28.0	11, 14, 15, 19	−325	16.8	16.8	16.8	16.8	16.8	16.2	15.7
23Cr, 13Ni	A451	8	CPH10 or CPH 20	...	0.90	70.0	30.0	9, 11, 14, 15, 19, 24	−325	18.0	18.0	18.0	18.0	18.0	17.3	16.9
25Cr, 20Ni	A451	8	CPK20	...	0.90	65.0	28.0	14, 15, 19, 24	−325	16.8	16.8	16.8	16.8	16.8	16.2	15.7
18Cr, 8Ni	A452	8	TP304H	...	0.85	75.0	30.0	15, 16	−325	17.0	17.0	17.0	15.9	14.8	14.0	13.8
16Cr, 12Ni, 2Mo	A452	8	TP316H	...	0.85	75.0	30.0	15, 16	−325	17.0	17.0	17.0	16.4	15.2	14.4	14.2
18Cr, 10Ni, Cb	A452	8	TP347H	...	0.85	75.0	30.0	15	−325	17.0	17.0	17.0	17.0	16.9	16.4	16.1

Metal temperature, °F (22)																
700	750	800	850	900	950	1000	1050	1100	1150	1200	1250	1300	1350	1400	1450	1500
14.4	10.7	9.3	7.9	6.5	4.5	2.5	1.6	1.0								
16.8	13.0	10.8	8.7	6.5	4.5	2.5	1.6	1.0								
14.4	10.7	9.3	7.9	6.5	4.5	2.5	1.6	1.0								
16.8	13.0	10.8	8.7	6.5	4.5	2.5	1.6	1.0								
19.2	14.8	12.0														
14.4	12.0	10.2	8.3	6.5	4.5	2.5	1.6	1.0								
16.8	13.0	10.8	8.7	6.5	4.5	2.5	1.6	1.0								
14.4	10.7	9.3	7.9	6.5	4.5	2.5	1.6	1.0								
16.8	13.0	10.8	8.7	6.5	4.5	2.5	1.6	1.0								
12.2	9.1	7.9	6.7	5.5	3.8	2.1	1.4	0.9								
14.0	11.0	9.2	7.4	5.5	3.8	2.1	1.4	0.9								
12.2	9.1	7.9	6.7	5.5	3.8	2.1	1.4	0.9								
14.0	11.0	9.2	7.4	5.5	3.8	2.1										
12.2	10.2	8.7	7.1	5.5	3.8	2.1	1.4	0.9								
14.0	11.0	9.2	7.4	5.5	3.8	2.1	1.4	0.9								
12.2	9.1	7.9	6.7	5.5	3.8	2.1	1.4	0.9								
14.0	11.0	9.2	7.4	5.5	3.8	2.1	1.4	0.9								
15.5	13.9	11.4	9.0	6.5	4.5	2.5	1.6	1.0								
15.5	13.9	11.4	9.0	6.5	4.5	2.5	1.6	1.0								
15.1	13.8	13.5	13.1	12.7	8.2	4.8										
16.3	13.2	12.8	12.1	10.9	8.0	5.8	4.2	2.9	2.0	1.3						
15.6	15.0	15.0	14.4	13.1	11.0	7.8	5.5	4.0	2.5	1.2						
17.9	17.9	15.2	14.5	12.8	11.0	7.8	5.8	4.2	3.0	2.0						
16.0	15.6	15.2	14.9	14.6	14.4	*13.8*	*12.2*	9.7	7.7	6.0	4.7	3.7	2.9	2.3	1.8	1.4
16.0	15.6	15.2	14.9	14.6	14.4	*13.8*	*12.2*	9.7	7.7	6.0	4.7	3.7	2.9	2.3	1.8	1.4
13.5	13.3	13.0	12.8	*11.9*	*9.9*	7.8	6.3	5.1	4.0	3.2	2.6	2.1	1.7	1.1	1.0	0.9
18.3	18.0	17.5	*14.6*	*13.9*	*12.5*	11.0	7.1	5.0	3.6	2.5	1.5	0.8	0.5	0.4	0.3	0.2
18.3	18.0	17.5	*14.6*	*13.9*	*12.5*	11.0	9.8	8.5	7.3	6.0	4.8	3.5	2.3	1.6	1.1	0.8
16.3	16.1	15.9	15.7	15.5	15.4	15.3	*14.5*	*12.4*	9.8	7.4	5.5	4.1	3.1	2.3	1.7	1.3
16.3	16.1	15.9	15.7	15.5	15.4	15.3	*14.5*	*12.4*	9.8	7.4	5.5	4.1	3.1	2.3	1.7	1.3
12.9	12.6	12.4	12.1	11.8	11.5	11.2	10.8	*10.2*	*8.8*	6.4	4.7	3.5	2.5	1.8	1.3	1.0
18.6	18.5	18.3	*15.4*	*14.9*	*14.8*	*14.0*	*12.1*	9.1	6.1	4.4	3.3	2.2	1.5	1.2	0.9	0.8
18.6	18.5	18.3	18.2	18.1	18.1	18.0	*17.1*	*14.2*	10.5	7.9	5.9	4.4	3.2	2.5	1.8	1.3
13.6	13.4	13.3	*11.6*	*11.4*	*11.1*	9.7	8.6	6.7	5.2	4.0	2.9	2.2	1.6	1.2	0.9	0.7
14.6	14.2	14.0	*13.2*	*13.0*	*12.6*	*11.8*	10.3	8.4	7.2	6.1	4.8	3.6	2.7	2.1	1.7	1.3
15.8	15.5	15.4	*12.6*	*12.5*	*12.3*	*12.1*	*11.7*	*9.7*	7.2	4.5	3.2	2.4	1.8	1.4	1.0	0.9
15.4	15.1	14.7	*11.5*	*11.2*	*10.6*	9.4	7.6	5.8	4.5	3.3	2.6	2.1	1.5	1.2	0.8	0.7
16.5	16.2	15.7	*12.2*	*12.0*	*11.2*	9.5	7.6	5.8	4.5	3.3	2.6	2.1	1.5	1.2	0.8	0.7
15.4	15.1	14.7	*11.5*	*11.2*	*10.7*	9.9	8.8	7.6	6.5	5.4	4.3	3.1	2.1	1.4	1.0	0.7
13.5	13.2	12.8	12.7	12.4	12.2	*11.7*	*10.3*	8.3	6.5	5.1	4.0	3.1	2.4	1.9	1.5	1.2
13.8	13.6	13.5	13.3	13.2	13.1	13.0	*12.3*	*10.5*	8.3	6.3	4.6	3.5	2.6	1.9	1.4	1.0
15.8	15.7	15.5	15.4	15.4	15.4	15.3	*14.5*	*12.1*	8.9	6.7	5.0	3.7	2.7	2.1	1.5	1.1

TABLE 6-37 Allowable Stresses in Tension for Materials (4, 13, 28)* *(Continued)*

Specifications are ASTM unless otherwise indicated. Numbers in parentheses refer to notes at end of table.

Material	Specification	P no. (23)	Grade	Class	Factor, E	Minimum tensile strength, kip/in²	Minimum yield strength, kip/in²	Notes	Minimum temperature (18)	Minimum temperature to 100	200	300	400	500	600	650
Electric-fusion-welded pipe and tubing																
18Cr, 8Ni pipe	A312	8	TP304	...	0.85	75.0	30.0	14, 16	−425	17.0	*17.0*	*17.0*	15.9	14.8	14.0	13.7
18Cr, 8Ni pipe	A312	8	TP304H	...	0.85	75.0	30.0	16	−325	17.0	*17.0*	*17.0*	15.9	14.8	14.0	13.7
18Cr, 8Ni pipe	A312	8	TP304L	...	0.85	70.0	25.0	...	−425	14.2	*14.2*	*14.2*	13.4	12.5	11.9	11.6
23Cr, 12Ni pipe	A312	8	TP309	...	0.85	75.0	30.0	19, 24, 32	−325	17.0	*17.0*	*17.0*	*17.0*	*17.0*	16.3	16.0
25Cr, 20Ni pipe	A312	8	TP310	...	0.85	75.0	30.0	19, 24, 32	−325	17.0	*17.0*	*17.0*	*17.0*	*17.0*	16.3	16.0
25Cr, 20Ni pipe	A312	8	TP310	...	0.85	75.0	30.0	6, 19, 24, 32	−325	17.0	*17.0*	*17.0*	*17.0*	*17.0*	16.3	16.0
16Cr, 12Ni, 2Mo pipe	A312	8	TP316	...	0.85	75.0	30.0	14, 16	−325	17.0	*17.0*	*17.0*	16.4	15.2	14.4	14.2
16Cr, 12Ni, 2Mo pipe	A312	8	TP316H	...	0.85	75.0	30.0	16	−325	17.0	*17.0*	*17.0*	16.4	15.2	14.4	14.2
16Cr, 12Ni, 2Mo pipe	A312	8	TP316L	...	0.85	70.0	25.0	...	−325	14.2	*14.2*	*14.2*	13.2	12.2	11.5	11.2
18Cr, 13Ni, 3Mo pipe	A312	8	TP317	...	0.85	75.0	30.0	14, 16	−325	17.0	*17.0*	*17.0*	16.4	15.2	14.4	14.2
18Cr, 10Ni, Ti pipe	A312	8	TP321	...	0.85	75.0	30.0	7, 14	−325	17.0	*17.0*	*17.0*	15.8	14.7	13.9	13.6
18Cr, 10Ni, Ti pipe	A312	8	TP321H	...	0.85	75.0	30.0	...	−325	17.0	*17.0*	*17.0*	15.8	14.7	13.9	13.6
18Cr, 10Ni, Cb pipe	A312	8	TP347	...	0.85	75.0	30.0	7, 14	−425	17.0	*17.0*	*17.0*	*17.0*	*16.9*	16.4	16.1
18Cr, 10Ni, Cb pipe	A312	8	TP347H	...	0.85	75.0	30.0	...	−325	17.0	*17.0*	*17.0*	*17.0*	*16.9*	16.4	16.1

Material	Specification	P no. (23), (30)	Temper	Class	Size range, in	Factor, E	Minimum tensile strength, kip/in²	Minimum yield strength, kip/in²	Notes	Minimum temperature (18)
Copper and copper alloy										
Seamless pipe and tubing										
Copper pipe	B42	31	Drawn	102, 120, 122	⅛–2, inclusive	...	45.0	40.0	9, 27	−325
Copper tubing	B88	31	Annealed	C10200, C12000, C12200,	...	...	30.0	9.0	9, 29	−325
Copper tubing	B88	31	Drawn	C10200, C12000, C12200,	...	...	36.0	30.0	9, 27, 29	−325
Cu, Ni 90/10	B466	34	Annealed	C70600	...	...	38.0	13.0	9	−325
Cu, Ni 70/30	B466	34	Annealed	C71500	...	...	50.0	18.0	9	−325

Material	Specification	P no. (23)	Grade	Class	Size range, in	Factor, E	Minimum tensile strength, kip/in²	Minimum yield strength, kip/in²	Notes	Minimum temperature	Minimum temperature to 100	200	300	400	500	600	650
Nickel and nickel alloy																	
Seamless pipe and tubing																	
Nickel	B161	41	200 (N02200)	Annealed	5 OD and under	...	55.0	15.0	...	−325	10.0	10.0	10.0	10.0	10.0	10.0	
Nickel	B161	41	200 (N02200)	Annealed	Over 5 OD	...	55.0	12.0	...	−325	8.0	8.0	8.0	8.0	8.0	8.0	
Low-C Ni	B161	41	201 (N02201)	Annealed	5 OD and under	...	50.0	12.0	...	−325	8.0	7.7	*7.5*	*7.5*	*7.5*	*7.5*	*7.5*
Low-C Ni	B161	41	201 (N02201)	Annealed	Over 5 OD	...	50.0	10.0	...	−325	6.7	*6.4*	*6.3*	*6.2*	*6.2*	*6.2*	*6.2*
Ni, Cu	B165	42	400 (N04400)	Annealed	5 OD and under	...	70.0	28.0	...	−325	18.7	16.4	15.4	14.8	14.8	14.8	14.8
Ni, Cu	B165	42	400 (N04400)	Annealed	Over 5 OD	...	70.0	25.0	...	−325	16.7	14.7	13.7	13.2	13.2	13.2	13.2
Ni, Cr, Fe	B167	43	600 (N06600)	Hot-finished or hot-finished annealed	5 OD and under	...	80.0	30.0	...	−325	20.0	*20.0*	*20.0*	*20.0*	*20.0*	*20.0*	*20.0*
Ni, Cr, Fe	B167	43	600 (N06600)	Hot-finished or hot-finished annealed	Over 5 OD	...	75.0	25.0	...	−325	16.7	*16.7*	*16.7*	*16.7*	*16.7*	*16.7*	*16.7*
Ni, Fe, Cr	B407	45	800 H (N08800)	Cold-drawn solution annealed or hot-finished	...	...	65.0	25.0	39	−325	16.7	*16.7*	*16.7*	*16.7*	*16.7*	*16.5*	*16.0*
Ni, Cr, Mo, Cb	B444	43	625 (N06625)	Annealed	...	...	120.0	60.0	42	−325	30.0	30.0	30.0	28.2	27.0	26.4	26.0
Welded pipe																	
Ni, Mo	B619	...	B (N10001)	Solution-annealed	...	0.85	100.0	45.0	...	−325	25.5	*25.5*	*25.5*	*25.5*	*25.5*	*25.5*	*25.0*
Ni, Mo	B619	44	B-2 (N10665)	Solution-annealed	...	0.85	110.0	51.0	...	−325	23.4	23.4	23.4	23.4	23.4	23.1	*23.1*
Ni, Mo, Cr	B619	44	C-4 (N06455)	Solution-annealed	...	0.85	100.0	40.0	...	−325	21.2	*21.2*	*21.2*	*21.2*	*21.0*	*20.7*	*20.5*
Ni, Mo, Cr	B619	44	C276 (N10276)	Solution-annealed	...	0.85	100.0	41.0	...	−325	23.2	*23.2*	*23.2*	*23.2*	*22.9*	*21.6*	*21.0*
Ni, Cr, Fe, Mo, Cu	B619	45	G1 (N06007)	Solution-annealed	...	0.85	90.0	35.0	...	−325	19.1	*19.1*	*19.1*	*18.6*	*18.3*	*17.9*	*17.8*
Ni, Cr, Mo, Fe	B619	...	X (N06002)	Solution-annealed	...	0.85	100.0	40.0	...	−325	22.6	*20.5*	*19.8*	*19.5*	*18.9*	*17.9*	*17.6*
Ni, Fe, Cr, Mo	B619	45	20-MOD (N08320)	Solution annealed	...	0.85	75.0	28.0	...	−325	15.9	*15.9*	*15.8*	*15.2*	*15.0*	*14.9*	*14.9*

Specification	P no.	Grade	Temper	Size range, in	Minimum tensile strength, kip/in²	Minimum yield strength, kip/in²	Notes	Minimum temperature, (18)	Metal temperature, °F (22): Minimum temperature, to 100	150	200	250	300	350	400
Aluminum alloy															
Seamless pipe and tubing															
B210	21	1060	0	0.018–0.500	8.5	2.5	2.6	−452	1.7	1.7	1.6	1.5	1.3	1.1	0.8
B210	21	3003	0	0.010–0.500	14.0	5.0	26	−452	3.3	3.3	3.3	3.1	2.4	1.8	1.4
B210	23	6061	T4	0.025–0.500	30.0	16.0	12, 26	−452	10.0	10.0	10.0	9.8	9.2	7.9	5.6
B210	23	6061	T4, T6 welded	...	24.0	...	35	−452	8.0	8.0	8.0	7.9	7.4	6.1	4.3

Metal temperature, °F (22)																
700	750	800	850	900	950	1000	1050	110	1150	1200	1250	1300	1350	1400	1450	1500
13.6	13.2	12.9	12.7	12.5	12.2	11.7	10.3	8.3	6.5	5.1	4.0	3.1	2.5	2.0	1.5	1.2
13.6	13.2	12.9	12.7	12.5	12.2	11.7	10.3	8.3	6.5	5.1	4.0	3.1	2.5	2.0	1.5	1.2
11.4	11.3	11.0	10.9	10.1	8.4	6.6	5.4	4.3	3.4	2.8	2.2	1.8	1.4	0.9	0.8	0.7
15.6	15.3	14.9	12.4	11.8	10.6	8.9	7.2	5.5	4.2	3.2	2.5	2.0	1.5	1.1	0.8	0.6
15.6	15.3	14.9	12.4	11.8	10.6	9.3	6.0	4.2	3.1	2.1	1.2	0.6	0.4	0.3	0.2	0.2
15.6	15.3	14.9	12.4	11.8	10.6	9.3	8.3	7.2	6.2	5.1	4.0	3.0	2.0	1.4	0.9	0.6
13.8	13.6	13.5	13.3	13.2	13.1	13.0	12.3	10.5	8.3	6.3	4.6	3.5	2.6	1.9	1.4	1.1
13.8	13.6	13.5	13.3	13.2	13.1	13.0	12.3	10.5	8.3	6.3	4.6	3.5	2.6	1.9	1.4	1.1
10.9	10.7	10.5	10.3	10.0	9.8	9.5	9.2	8.7	7.4	5.4	4.0	3.0	2.1	1.6	1.1	0.9
13.9	13.6	13.5	13.3	13.2	13.1	13.0	12.3	10.5	8.3	6.3	4.6	3.5	2.6	1.9	1.4	1.1
13.4	13.3	13.1	13.0	13.0	12.9	11.7	8.2	5.8	4.2	3.1	2.2	1.4	0.9	0.6	0.4	0.3
13.4	13.3	13.1	13.0	13.0	12.9	11.9	9.9	7.7	5.9	4.5	3.5	2.7	2.1	1.6	1.2	0.9
15.8	15.7	15.5	13.1	12.7	12.3	11.9	10.3	7.7	5.2	3.7	2.8	1.9	1.3	1.0	0.8	0.6
15.8	15.7	15.6	15.5	15.4	15.4	15.3	14.5	12.1	8.9	6.7	5.0	3.7	2.7	2.1	1.5	1.1

	Metal temperature, °F (22)											
Minimum temperature to 100	150	200	250	300	350	400	450	500	550	600	650	700
15.0	11.2	11.2	11.2	11.0	10.3	4.2						
6.0	6.0	5.9	5.8	5.0	3.8	2.5	1.5	0.8				
12.0	9.0	8.7	8.3	8.0	5.0	2.5	1.5	0.8				
8.7	8.3	8.1	8.0	7.8	7.7	7.5	7.3	7.2	7.0	6.0		
12.0	11.6	11.3	11.0	10.8	10.6	10.3	10.1	9.9	9.8	9.6	9.5	9.4

Metal temperature, °F (22)																
700	750	800	850	900	950	1000	1050	1100	1150	1200	1250	1300	1350	1400	1450	1500
7.4	7.3	7.2	5.8	4.5	3.7	3.0	2.4	2.0	1.5	1.2						
6.2	6.1	5.9	5.8	4.5	3.7	3.0	2.4	2.0	1.5	1.2						
14.8	14.6	14.2														
13.2	13.0	12.7														
20.0	20.0	20.0	19.6	16.0	10.6	7.0	4.5	3.0	2.2	2.0						
16.7	16.7	16.7	16.5	15.9	10.6	7.0	4.5	3.0	2.2	2.0						
15.7	15.4	15.3	15.1	14.8	14.6	14.4	13.7	13.5	11.2	8.4	6.9	5.4	4.5	3.6	3.0	2.5
26.0	26.0	26.0	26.0	26.0	26.0	26.0	26.0	26.0	21.0	13.2						
25.5	24.5	23.5														
23.0	22.9	22.8														
20.4	20.0	19.5														
20.4	20.0	19.5	19.2	18.9	18.7	18.5										
17.8	17.6	17.4	17.2	17.0	16.6	16.1										
17.3	17.0	16.8	16.7	16.7	16.3	15.8	15.3	14.9	12.3	9.6	8.0	6.5				
14.9	14.8	14.6														

TABLE 6-37 Allowable Stresses in Tension for Materials (4, 13, 28)* *(Continued)*

Design stresses for bolting materials

Material	Specification	Grade	Size range, in	Minimum tensile strength, kip/in^2	Minimum yield strength, kip/in^2	Notes	Minimum temperature, (18)	Minimum temperature, (to 100)	200	300	400	500	600	650
Carbon steel														
. . .	A307	B	. . .	60.0	. . .	22	−20	13.7	13.7	13.7	13.7	13.7		
. . .	A325	. . .	. . .	105.0	. . .	. . .	−20	19.3	19.3	19.3	19.3	19.3	19.3	19.3
. . .	A194	1, 2	. . .	. . .	. . .	25	−20							
. . .	A194	2H	. . .	. . .	. . .	25	−50							
Alloy steel														
Cr, Mo	A193	B7	2½ and under	125.0	105.0	33	−20	25.0	25.0	25.0	25.0	25.0	25.0	25.0
Cr, 0.2Mo	A193	B7M	2½ and under	100.0	80.0	. . .	−50	20.0	20.0	20.0	20.0	20.0	20.0	20.0
Cr, Mo, V	A193	B16	2½ and under	125.0	105.0	. . .	−20	25.0	25.0	25.0	25.0	25.0	25.0	25.0
C, Mo	A194	4	. . .	. . .	. . .	25								
Cr, Mo	A320	L7, L7A, L7B, L7C	2½ and under	125.0	105.0	31	−150	25.0	25.0	25.0	25.0	20.0	20.0	20.0
Stainless steel														
12 Cr	A193	B6	4 and under	110.0	85.0	19, 31	−20	21.2	21.2	21.2	21.2	21.2	21.2	21.2
304 solution-treated	A193	B8, Cl. 1	. . .	75.0	30.0	31, 32, 41	−325	18.8	15.6	14.0	12.9	12.1	11.4	11.2
316 solution treated	A193	B8M, Cl. 1	. . .	75.0	30.0	31, 32, 41	−325	18.8	16.1	14.6	13.3	12.5	11.8	11.5
304 strain-hardened	A193	B8, Cl. 2	Up to ¾	125.0	100.0	31, 32, 41	−325	25.0						
			¾ to 1	115.0	80.0	31, 32, 41	−325	20.0						
			Over 1 to 1¼	105.0	65.0	31, 32, 41	−325	16.2						
			Over 1¼ to 1½	100.0	50.0	31, 32, 41	−325	12.5						
316 strain-hardened	A193	B8M, Cl. 2	Up to ¾	110.0	95.0	31, 32, 41	−325	22.0	22.0	22.0	22.0	22.0	22.0	22.0
			¾ to 1	100.0	80.0	31, 32, 41	−325	20.0	20.0	20.0	20.0	20.0	20.0	20.0
			Over 1 to 1¼	95.0	65.0	31, 32, 41	−325	16.2	16.2	16.2	16.2	16.2	16.2	16.2
			Over 1¼ to 1½	90.0	50.0	31, 32, 41	−325	12.5	12.5	12.5	12.5	12.5	12.5	12.5
14 Cr, 24 Ni	A453	660A/B	. . .	130.0	85.0	19, 31	−20	21.3	20.7	20.5	20.4	20.3	20.2	20.2

Material	Specification	Grade	Temper	Size range, in	Minimum strength, kip/in^2	Minimum yield strength, kip/in^2	Notes	Minimum temperature, (18)
Aluminum and aluminum-base alloy								
	B211	2024	T4	0.500–4.500	62.0	42.0	34, 35	−325
	B211	6061	T6, T651	0.125–8.000	42.0	35.0	34, 35	−325
Copper and copper-base alloy								
Cu, Si	B98	C65500, C66100	Soft	. . .	52.0	15.0	43	−325
Cu, Si	B98	C65100	Bolt	Over ½ to 1	75.0	45.0	. . .	−325
Al, Bronze	B150	C64200	. . .	Over ½ to 1	85.0	45.0	. . .	−325
Al, Bronze	B150	C63000	. . .	½ to 1	100.0	50.0	. . .	−325
Al, Bronze	B150	C61400	. . .	Over ½ to 1	75.0	35.0	. . .	−325
Nickel and nickel-base alloy								
Nickel	B160	200 (N02200)	Cold-drawn	. . .	65.0	40.0	. . .	−325
Low C, Ni	B160	201 (N02100)	Annealed hot-finished	. . .	50.0	10.0	. . .	−325
Ni, Cu	B164	400 (N04400)	Hot-finished	All except hexagonal over 2½	80.0	40.0	. . .	−325
Ni, Cu	B164	400 (N04400)	Cold-drawn stress-relieved	. . .	84.0	50.0	36	−325
Ni, Cr, Fe	B166	600 (N06600)	Annealed	. . .	80.0	35.0	. . .	−325

NOTES:

Special note for the sixth edition: At this time, metric equivalents have not been provided for the allowable-stress tables of the piping code B31.3. They may be computed by the following relationships; (°F − 32) × ⅝ = °C; lbf/in^2 (stress) × 6.895 × 10^{-3} = MPa.

1. For temperatures above 480°C (900°F) consider the advantages of killed steel.
2. Conversion of carbides to graphite may occur after prolonged exposure to temperatures over 425°C (800°F).
3. Conversion of carbides to graphite may occur after prolonged exposure to temperatures over 468°C (875°F).
4. In shaded areas, allowable-stress values which are printed in *italics* exceed two-thirds of the expected yield strength at temperature. All other allowable-stress values in shaded areas are equal to 90 percent of expected yield strength at temperature. See ANSI B31.3.
5. A quality factor of 92 percent is included for structural grade.
6. The higher stress values at 566°C (1050°F) and above for this material shall be used only when the steel has an austenitic micrograin size No. 6 or less (coarser grain) as defined in ASTM E112. Otherwise the lower stress values shall be used.
7. For temperatures above 538°C (1000°F), these stress values may be used only if the material has been heat-treated at a temperature of 1090°C (2000°F) minimum.
8. There are restrictions in the code on the use of this material.
9. For use in code piping at the stated allowable stresses, the tensile and yield strengths listed in these tables must be verified by tensile tests at the mill; such tests shall be specified in the purchase order.
10. Pressure-temperature ratings of cast and forged parts as published in standards referenced in this code section may be used for parts meeting requirements of these standards. Allowable stresses for castings and forgings, where listed, are for use in the design of special components not furnished in accordance with such standards.
11. Certain forms of this material, as stated in Table 6-45, must be impact-tested to qualify for service below −29°C (−20°F). Alternatively, if provisions for impact testing are included in the material specification as supplementary requirements and are invoked, the material may be used down to the temperature at which the test was conducted in accordance with the specification.
12. For welded construction with work-hardened grades, use the stresses for annealed material; for welded construction with precipitation-hardened grades, use the special allowable stresses for welded construction given in the tables.
13. *SE* values shown in this table for welded pipe include the joint quality factor E_j for the longitudinal weld as required by Fig. 6-126 and, when applicable, the structural-grade quality factor E_S of 0.92. For some code computations, particularly with regard to expansion, flexibility, structural attachments, supports, and restraints, the longitudinal-joint quality factor E_j need not be considered. To determine the allowable stress *S* for use in code computations not utilizing the joint quality factor E_j divide the value *SE* shown in this table by the longitudinal-joint quality factor E_j tabulated in Fig. 6-126.
14. For temperatures above 38°C (100°F) these stress values apply only when the carbon content is 0.04 percent or higher.
15. Stress values shown include the casting quality factor shown in this table. Higher stress values can be used if special inspection is accomplished.
16. These unstabilized grades of stainless steel have an increasing tendency to intergranular carbide precipitation as the carbon content increases above 0.03 percent.
17. The allowable stress to be used for this gray-cast-iron material at its upper temperature limit of 232°C (450°F) is the same as that shown in the 204°C (400°F) column.

Metal temperature, °F (22)																
700	750	800	850	900	950	1000	1050	110	1150	1200	1250	1300	1350	1400	1450	1500
25.0	23.6	21.0	17.0	12.5	8.5	4.5										
20.0	20.0	18.5	16.2	12.5	8.5	4.5										
25.0	25.0	25.0	23.5	20.5	16.0	11.0	6.3	2.8								
20.0	20.0	20.0	16.2	12.5	8.5	4.5										
21.2	21.2 |	19.6	15.6	12.0												
11.0	10.8	10.5	10.3	10.1	9.9	9.7	9.5	8.8	7.7	6.0	4.7	3.7	2.9	2.3	1.8	1.4
11.3	11.0	10.9	10.8	10.7	10.7	10.6	10.5	10.3	9.3	7.4	5.4	4.1	3.0	2.2	1.7	1.2
22.0	22.0	22.0														
20.0	20.0	20.0														
16.2	16.2	16.2														
12.5	12.5	12.5														
20.1	20.0	19.9	19.9	19.9	19.8	19.8										

	Metal temperature, °F (22)																
Minimum tempera-ture, to 100	200	300	400	500	600	650	700	750	800	850	900	950	1000	1050	1100	1150	1200
10.5	10.5	10.4	4.5														
8.4	8.4	8.4	4.4														
10.0	10.0	10.0															
11.3	11.3	11.3															
21.3	21.3	21.3	20.8	12.6	9.9												
25.0	25.0	25.0	25.0	20.7	12.0	8.5	6.0										
18.8	18.8	18.8	18.4	16.1													
10.0	10.0	10.0	10.0	10.0	10.0												
6.7	6.4	6.3	6.2	6.2	6.2	6.2	6.2	6.0	5.9	5.8	4.8	3.7	3.0	2.4	2.0	1.5	1.2
20.0	20.0	20.0	20.0	20.0	20.0	20.0	19.2	18.5	14.5	8.5	4.0						
12.5	12.5	12.5	12.5	12.5													
20.0	20.0	20.0	20.0	20.0	20.0	19.8	19.6	19.4	19.1	18.7	16.0	10.6	7.0	4.5	3.0	2.2	2.0

18. The minimum temperature shown is that design minimum temperature for which the material is normally suitable without impact testing other than that required by the material specification. However, the use of a material at a design minimum temperature below −29°C (−20°F) is established by rules elsewhere in the code, including any necessary impact-test requirements.
19. These steels are intended for use at high temperatures; however, they may have low ductility and/or low impact properties at room temperature after being used above the temperature indicated by the single bar (|).
20. For pipe sizes NPS 8 and larger and for wall thicknesses of Schedule 140 or heavier, the minimum specification tensile strength is 483 MPa (70.0 kip/in^2).
21. There are restrictions on the use of this material in the text of the code.
22. A single bar (|) in these stress tables indicates that there are conditions other than stress which affect usage above or below the temperature as described in other referenced notes. A double bar (‖) after a tabled stress indicates that use of the material is prohibited above that temperature.
23. See ANSI B31.3 for a description of P-number groupings.
24. This material when used below −29°C (−20°F) requires impact testing if the carbon content is above 0.10 percent.
25. This is a product specification. No design stresses are necessary. Limitations on metal temperature for materials covered by this specification are:

	°C	°F
Grades 1 and 2	−29 to 480	−20 to 900
Grade 2H	−45 to 595	−50 to 1100
Grade 3	−29 to 595	−20 to 1100
Grade 4	−100 to 595	−150 to 1100
Grade 6	−29 to 425	−20 to 800
Grade 8FA (see Note 24)	−29 to 425	−20 to 800
Grades 8MA and 8TA	−198 to 815	−325 to 1500
Grades 8A and 8CA	−254 to 815	−425 to 1500

26. For use in code piping at the stated allowable stresses, the required minimum tensile and yield properties must be verified by tensile test at the mill. If such tests are not mandatory in the ASTM specification, they shall be specified in the purchase order.
27. After use above the temperature indicated by a single bar (|), use at a lower temperature shall be based on the stress values allowed for the annealed condition of the material.
28. The *SE* values in Table 6-37 are equal to the basic allowable stresses in tension *S* multiplied by a quality factor *E* (see subsection "Pressure Design of Metallic Components: Wall

TABLE 6-37 Allowable Stresses in Tension for Materials (4, 13, 28)* *(Continued)*

Thickness"). The design stress values for bolting materials are equal to the basic allowable stresses S. The stress values in shear shall be 0.80 times the allowable stresses in tension derived from tabulated values in Table 6-37 adjusted when applicable in accordance with Note 13. Stress values in bearing shall be twice those in shear.

29. Yield strengths listed are not included in ASTM specifications. The value shown is based on yield strengths of materials with similar characteristics.

30. The letter *a* indicates alloys which are not recommended for welding and which, if welded, must be individually qualified. The letter *b* indicates copper-base alloys which must be individually qualified.

31. These stress values are established from a consideration of strength only and will be satisfactory for average service. For bolted joints when freedom from leakage over a long period of time without retightening is required, lower stress values may be necessary as determined from the flexibility of the flange and bolts and corresponding relaxation properties.

32. For temperatures above 538°C (1000°F), these stress values apply only when the carbon content is 0.04 percent or higher.

33. For use at temperatures below −29 through −45°C (−20 through −50°F) this material must be quenched and tempered.

34. The stress values given for this material are not applicable when either welding or thermal cutting is employed.

35. For stress-relieved tempers (T351, T3510, T3511, T451, T4510, T4511, T651, T6510, T6511) stress values for material in the listed temper shall be used.

36. The maximum operating temperature is arbitrarily set at 260°C (500°F) because harder temper adversely affects design stress in the creep-rupture-temperature ranges.

37. Pipe produced to this specification is not intended for high-temperature service. The stress values apply to either nonexpanded or cold-expanded material in the as-rolled, normalized, or normalized and tempered condition.

38. Special P numbers SP-1, SP-2, and SP-3 of carbon steels are not included in P No. 1 because of a possible high-carbon–high-manganese combination which would require special consideration in qualification. Qualification of any high-carbon–high-manganese grade may be extended to other grades in its group.

39. Annealed at approximately 1150°C (2100°F).

40. If no welding is employed in the fabrication of piping from these materials, the allowable stress values may be increased to 230 MPa (33.3 kip/in^2).

41. For all design temperatures, the maximum hardness shall be Rockwell C35 immediately under the thread roots. The hardness shall be taken on a flat area at least 3 mm (⅛ in) across, prepared by removing threads. No more material than necessary shall be removed to prepare the area. Hardness determination shall be made at the same frequency as tensile tests.

42. The minimum tensile strength of the reduced section tensile specimen in accordance with QW-462.1 of ASME Code Sec. IX shall not be less than 758 MPa (110.0 kip/in^2).

43. Copper-silicon alloys are not always suitable when exposed to certain media and high temperature, particularly above 100°C (212°F). Users should satisfy themselves that the alloy selected is satisfactory for the service for which it is to be used.

*Table 6-37 and notes have been extracted from the Chemical Plant and Petroleum Refinery Piping Code, ANSI B31.3—1980, with permission of the publisher, the American Society of Mechanical Engineers, New York.

No.	Type of joint		Type of seam	Examination	Factor, E_j
1	Furnace butt weld, continuous		Straight	As required by listed specifications	0.60
2	Electric resistance weld		Straight or spiral	As required by listed specifications	0.85
3	Electric fusion weld				
	a Single butt weld (with or without filler metal)	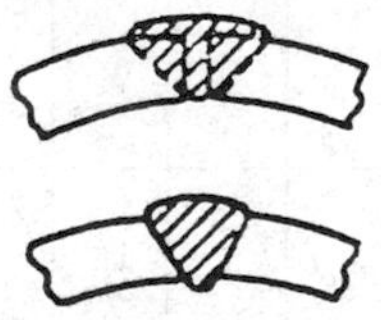	Straight or spiral	As required by listed specifications or this code Additionally spot-radiographed per ANSI B31.3, par. 336.6.1 Additionally 100 percent radiographed per ANSI B31.3, par. 336.4.5	0.80 0.90 1.00
	b Double butt weld (with or without filler metal)	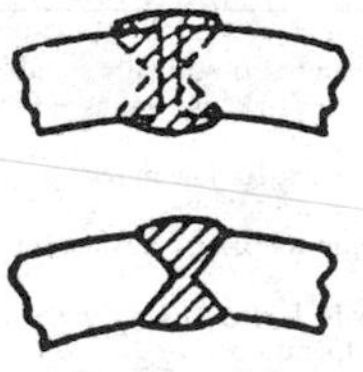	Straight or spiral (except as provided in 4*b*)	As required by listed specification or this code Additionally spot-radiographed per ANSI B31.3, par. 336.6.1 Additionally 100 percent radiographed per ANSI B31.3, par. 336.4.5	0.85 0.90 1.00
4	Per specific specifications				
	a ASTM A211	As permitted in specifications	Spiral	As required by specifications	0.75
	b Double submerged arc-welded pipe per API 5L or 5LX		Straight with one or two seams	As required by specifications, additionally examined by radiography for lengths of 200 mm (8 in) at each end	0.95

FIG. 6-126 Longitudinal and spiral-weld joint factor E_j. NOTE: It is not permitted to increase the joint quality factor by additional examination for joints 1, 2, and 4*a*. *(Extracted from ANSI B31.3—1980, with permission of the publisher, the American Society of Mechanical Engineers, New York.)*

TABLE 6-38 Values of Coefficient *Y* When *t* Is Less Than *D*/6*

Materials	Temperature, °C (°F) 485 (900) and lower	510 (950)	540 (1000)	560 (1050)	595 (1100)	620 (1150) and higher
Ferritic steels	0.4	0.5	0.7	0.7	0.7	0.7
Austenitic steels	0.4	0.4	0.4	0.4	0.5	0.7
Other ductile metals	0.4	0.4	0.4	0.4	0.4	0.4
Cast iron	0.0	. . .	. . .	. . .	. . .	. . .

*Extracted from ANSI B31.3—1980, with permission of the publisher, the American Society of Mechanical Engineers, New York.

TABLE 6-39 Stress-Range Reduction Factors *f**

Cycles, number	Factor, *f*
7000 and less	1.0
7000–14,000	0.9
14,000–22,000	0.8
22,000–45,000	0.7
45,000–100,000	0.6
Over 100,000	0.5

*Extracted from ANSI B31.3—1980, with permission of the publisher, the American Society of Mechanical Engineers, New York.

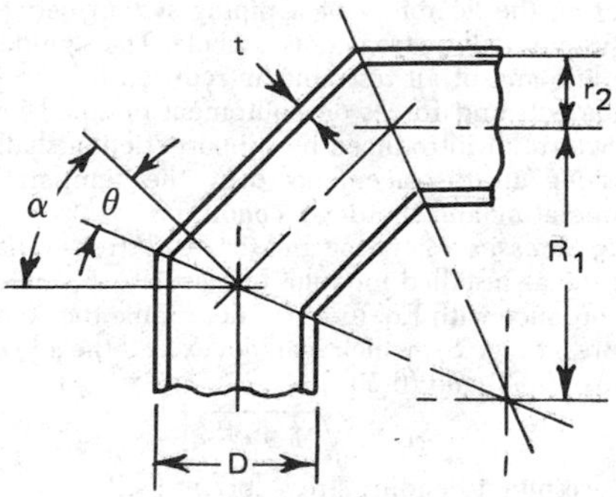

FIG. 6-127 Nomenclature for miter bends. *(Extracted from the Chemical Plant and Petroleum Refinery Code, ANSI B31.3—1976, with permission of the publisher, the American Society of Mechanical Engineers, New York.)*

turbines, or valves, for example), resulting from excessive thrusts or movements in the piping.

To assure that a system meets these requirements, the computed displacement–stress range S_E shall not exceed the allowable stress range S_A (Eqs. 6-38 and 6-39), the reaction forces R_m (Eq. 6-50) shall not be detrimental to supports or connected equipment, and movement of the piping shall be within any prescribed limits.

Displacement Strains Strains result from piping being displaced from its unrestrained position:

1. *Thermal displacements.* A piping system will undergo dimensional changes with any change in temperature. If it is constrained from free movement by terminals, guides, and anchors, it will be displaced from its unrestrained position.

2. *Reaction displacements.* If the restraints are not considered rigid and there is a predictable movement of the restraint under load, this may be treated as a compensating displacement.

3. *Externally imposed displacements.* Externally caused movement of restraints will impose displacements on the piping in addition to those related to thermal effects. Such movements may result from causes such as wind sway or temperature changes in connected equipment.

Total Displacement Strains Thermal displacements, reaction displacements, and externally imposed displacements all have equivalent effects on the piping system and must be considered together in determining total displacement strains in a piping system.

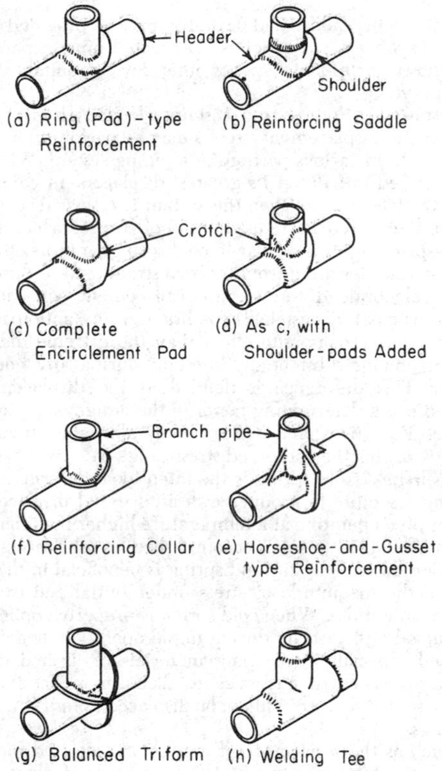

FIG. 6-128 Types of reinforcement for branch connections. *(From Kellogg,* Design of Piping Systems, *Wiley, New York, 1965.)*

Expansion strains may be taken up in three ways: by bending, by torsion, or by axial compression. In the first two cases maximum stress occurs at the extreme fibers of the cross section at the critical location. In the third case the entire cross-sectional area over the entire length is for practical purposes equally stressed.

Bending or torsional flexibility may be provided by bends, loops, or offsets; by corrugated pipe or expansion joints of the bellows type; or by other devices permitting rotational movement. These devices must be anchored or otherwise suitably connected to resist end forces from fluid pressure, frictional resistance to pipe movement, and other causes.

Axial flexibility may be provided by expansion joints of the slip-joint or bellows types, suitably anchored and guided to resist end forces from fluid pressure, frictional resistance to movement, and other causes.

Displacement Stresses Stresses may be considered proportional to the total displacement strain only if the strains are well distributed and not excessive at any point. The methods outlined here and in the code are applicable only to such a system. Poor distribution of strains (unbalanced systems) may result from:

1. Highly stressed small-size pipe runs in series with large and relatively stiff pipe runs

2. Local reduction in size or wall thickness or local use of a material having reduced yield strength (for example, girth welds of substantially lower strength than the base metal)

3. A line configuration in a system of uniform size in which expansion or contraction must be absorbed largely in a short offset from the major portion of the run

If unbalanced layouts cannot be avoided, appropriate analytical methods must be applied to assure adequate flexibility. If the designer determines that a piping system does not have adequate inherent flexibility, additional flexibility may be provided by adding bends, loops, offsets, swivel joints, corrugated pipe, expansion joints of the bellows or slip-joint type, or other devices. Suitable anchoring must be provided.

As contrasted with stress from sustained loads such as internal pressure or weight, displacement stresses may be permitted to cause limited overstrain in various portions of a piping system. When the system is operated initially at its greatest displacement condition, any yielding reduces stress. When the system is returned to its original condition, there occurs a redistribution of stresses which is referred to as self-springing. It is similar to cold springing in its effects.

While stresses resulting from thermal strain tend to diminish with time, the algebraic difference in displacement condition and in either the original (as-installed) condition or any anticipated condition with a greater opposite effect than the extreme displacement condition remains substantially constant during any one cycle of operation. This difference is defined as the displacement–stress range, and it is a determining factor in the design of piping for flexibility. See Eqs. (6-38) and (6-39) for the allowable stress range S_A and Eq. (6-45) for the computed stress range S_E.

Cold Spring Cold spring is the intentional deformation of piping during assembly to produce a desired initial displacement and stress. For pipe operating at a temperature higher than that at which it was installed, cold spring is accomplished by fabricating it slightly shorter than design length. Cold spring is beneficial in that it serves to balance the magnitude of stress under initial and extreme displacement conditions. When cold spring is properly applied, there is less likelihood of overstrain during initial operation; hence, it is recommended especially for piping materials of limited ductibility. There is also less deviation from as-installed dimensions during initial operation, so that hangers will not be displaced as far from their original settings.

Inasmuch as the service life of a system is affected more by the range of stress variation than by the magnitude of stress at a given time, no credit for cold spring is permitted in stress-range calculations. However, in calculating the thrusts and moments when actual reactions as well as their range of variations are significant, credit is given for cold spring.

Values of thermal-expansion coefficients to be used in determining total displacement strains for computing the stress range are determined from Table 6-40 as the algebraic difference between the value at design maximum temperature and that at the design minimum temperature for the thermal cycle under analysis.

Values for Reactions Values of thermal displacements to be used in determining total displacement strains for the computation of reactions on supports and connected equipment shall be determined as the algebraic difference between the value at design maximum (or minimum) temperature for the thermal cycle under analysis and the value at the temperature expected during installation.

The as-installed and maximum or minimum moduli of elasticity, E_a and E_m respectively, shall be taken as the values shown in Table 6-41.

Poisson's ratio may be taken as 0.3 at all temperatures for all metals.

The allowable stress range for displacement stresses S_A and permissible additive stresses shall be as specified in Eqs. (6-38) and (6-39) for systems primarily stressed in bending and/or torsion. For pipe or piping components containing longitudinal welds the basic allowable stress S may be used to determine S_A. (See Table 6-37, Note 13.)

Nominal thicknesses and outside diameters of pipe and fittings shall be used in flexibility calculations.

In the absence of more directly applicable data, the flexibility factor k and stress-intensification factor i shown in Table 6-42 may be used in flexibility calculations in Eq. 6-46. For piping components or attachments (such as valves, strainers, anchor rings, and bands) not covered in the table, suitable stress-intensification factors may be assumed by comparison of their significant geometry with that of the components shown.

Requirements for Analysis No formal analysis of adequate flexibility is required in systems which (1) are duplicates of successfully operating installations or replacements without significant change of systems with a satisfactory service record; (2) can readily be judged adequate by comparison with previously analyzed systems; or (3) are of uniform size, have no more than two points of fixation, have no intermediate restraints, and fall within the limitations of empirical Eq. (6-44):*

$$\frac{Dy}{(L - U)^2} \leq K_1 \tag{6-44}$$

where D = outside diameter of pipe, in (mm)
y = resultant of total displacement strains, in (mm), to be absorbed by the piping system
L = developed length of piping between anchors, ft (m)
U = anchor distance, straight line between anchors, ft (m)
K_1 = 0.03 for U.S. customary units listed
= 208.3 for SI units listed in parentheses

1. All systems not meeting these criteria shall be analyzed by simplified, approximate, or comprehensive methods of analysis appropriate for the specific case.
2. Approximate or simplified methods may be applied only if they are used in the range of configurations for which their adequacy has been demonstrated.
3. Acceptable comprehensive methods of analysis include analytical and chart methods which provide an evaluation of the forces, moments, and stresses caused by displacement strains.
4. Comprehensive analysis shall take into account stress-intensification factors for any component other than straight pipe. Credit may be taken for the extra flexibility of such a component.

In calculating the flexibility of a piping system between anchor points, the system shall be treated as a whole. The significance of all parts of the line and of all restraints introduced for the purpose of reducing moments and forces on equipment or small branch lines and also the restraint introduced by support friction shall be recognized. Consider all displacements over the temperature range defined by operating and shutdown conditions.

Flexibility Stresses Bending and torsional stresses shall be computed using the as-installed modulus of elasticity E_a and then combined in accordance with Eq. (6-45) to determine the computed displacement stress range S_E, which shall not exceed the allowable stress range S_A [Eqs. (6-38) and (6-39).]

$$S_E = \sqrt{S_b^2 + 4S_t^2} \tag{6-45}$$

where S_b = resultant bending stress, lbf/in² (MPa)
S_t = $M_t/2Z$ = torsional stress, lbf/in² (MPa)
M_t = torsional moment, in·lbf (N·mm)
Z = section modulus of pipe, in³ (mm³)

The resultant bending stresses S_b to be used in Eq. (6-45) for elbows and miter bends shall be calculated in accordance with Eq. (6-46), with moments as shown in Fig. 6-129.

$$S_b = \frac{\sqrt{(i_i M_i)^2 + (i_o M_o)^2}}{Z} \tag{6-46}$$

where S_b = resultant bending stress, lbf/in² (MPa)
i_i = in-plane stress-intensification factor from Table 6-42
i_o = out-plane stress-intensification factor from Table 6-42
M_i = in-plane bending moment, in·lbf (N·mm)
M_o = out-plane bending moment, in·lbf (N·mm)
Z = section modulus of pipe, in³ (mm³)

The resultant bending stresses S_b to be used in Eq. (6-45) for branch connections shall be calculated in accordance with Eqs. (6-47) and (6-48), with moments as shown in Fig. 6-130.

*WARNING: No general proof can be offered that this equation will yield accurate or consistently conservative results. It is not applicable to systems used under severe cyclic conditions. It should be used with caution in configurations such as unequal leg U bends ($L/U > 2.5$) or near-straight sawtooth runs, or for large thin-wall pipe ($i \geq 5$), or when extraneous displacements (not in the direction connecting anchor points) constitute a large part of the total displacement. There is no assurance that terminal reactions will be acceptably low even if a piping system falls within the limitations of Eq. (6-44).

TABLE 6-40 Thermal–Expansion Coefficients, U.S. Customary Units, for Metals*

Mean coefficient of linear thermal expansion between 70°F and indicated temperature, μin/(in·°F)

Tempera-ture, °F	Carbon steel, carbon-molybde-mum low-chromium (through 3 Cr Mo)	5 Cr Mo through 9 Cr Mo	Austenitic stainless steels, 18 Cr, 8 Ni	12 Cr 17 Cr 27 Cr	25 Cr, 20 Ni	Monel 67 Ni, 30 Cu	3½ Nickel	Alumi-num	Gray cast iron	Bronze	Brass	70 Cu, 30 Ni	Ni-Fe-Cr	Ni-Cr-Fe	Ductile iron
−325	5.00	4.70	8.15	4.30	....	5.55	4.76	9.90	....	8.40	8.20	6.65			
−300	5.07	4.77	8.21	4.36	....	5.72	4.90	10.04	....	8.45	8.24	6.76			
−275	5.14	4.84	8.28	4.41	....	5.89	5.01	10.18	....	8.50	8.29	6.86			
−250	5.21	4.91	8.34	4.47	....	6.06	5.15	10.33	....	8.55	8.33	6.97			
−225	5.28	4.93	8.41	4.53	....	6.23	5.30	10.47	....	8.60	8.37	7.08			
−200	5.35	5.05	8.47	4.59	....	6.40	5.45	10.61	....	8.65	8.41	7.19	....	....	4.65
−175	5.42	5.12	8.54	4.64	....	6.57	5.52	10.76	....	8.70	8.46	7.29	....	....	4.76
−150	5.50	5.20	8.60	4.70	....	6.75	5.59	10.90	....	8.75	8.50	7.40	....	....	4.87
−125	5.57	5.26	8.66	4.73	....	6.85	5.67	11.08	....	8.85	8.61	7.50	....	....	4.98
−100	5.65	5.32	8.75	4.85	....	6.95	5.78	11.25	....	8.95	8.73	7.60	....	....	5.10
−75	5.72	5.38	8.83	4.93	....	7.05	5.83	11.43	....	9.05	8.84	7.70	....	....	5.20
−50	5.80	5.45	8.90	5.00	....	7.15	5.88	11.60	....	9.15	8.95	7.80	....	....	5.30
−25	5.85	5.51	8.94	5.05	....	7.22	5.94	11.73	....	9.23	9.03	7.87	....	....	5.40
0	5.90	5.56	8.98	5.10	....	7.28	6.00	11.86	....	9.32	9.11	7.94	....	....	5.50
25	5.96	5.62	9.03	5.14	....	7.35	6.08	11.99	....	9.40	9.18	8.02	....	....	5.58
50	6.01	5.67	9.07	5.19	....	7.41	6.16	12.12	....	9.49	9.26	8.09	....		5.66
70	6.07	5.73	9.11	5.24	....	7.48	6.25	12.25	....	9.57	9.34	8.16	....	7.13	5.74
100	6.13	5.79	9.16	5.29	....	7.55	6.33	12.39	....	9.66	9.42	8.24	....	7.20	5.82
125	6.19	5.85	9.20	5.34	....	7.62	6.36	12.53	....	9.75	9.51	8.31	....	7.25	5.87
150	6.25	5.92	9.25	5.40	....	7.70	6.39	12.67	....	9.85	9.59	8.39	....	7.30	5.92
175	6.31	5.98	9.29	5.45		7.77	6.42	12.81		9.93	9.68	8.46		7.35	5.97
200	6.38	6.04	9.34	5.50	8.79	7.84	6.45	12.95	5.75	10.03	9.76	8.54	7.90	7.40	6.02
225	6.43	6.08	9.37	5.54	8.81	7.89	6.50	13.03	5.80	10.05	9.82	8.58	8.01	7.44	6.08
250	6.49	6.12	9.41	5.58	8.83	7.93	6.55	13.12	5.84	10.08	9.88	8.63	8.12	7.48	6.14
275	6.54	6.15	9.44	5.62	8.85	7.98	6.60	13.20	5.89	10.10	9.94	8.67	8.24	7.52	6.20
300	6.60	6.19	9.47	5.66	8.87	8.02	6.65	13.28	5.93	10.12	10.00	8.71	8.35	7.56	6.25
325	6.65	6.23	9.50	5.70	8.89	8.07	6.69	13.36	5.97	10.15	10.06	8.76	8.46	7.60	6.31
350	6.71	6.27	9.53	5.74	8.90	8.11	6.73	13.44	6.02	10.18	10.11	8.81	8.57	7.63	6.37
375	6.76	6.30	9.56	5.77	8.91	8.16	6.77	13.52	6.06	10.20	10.17	8.85	8.69	7.67	6.43

TABLE 6-40 Thermal-Expansion Coefficients, U.S. Customary Units, for Metals* *(Continued)*

Mean coefficient of linear thermal expansion between 70°F and indicated temperature, μin/(in·°F)

Temperature, °F	Carbon steel, carbon-molybdenum low-chromium (through 3 Cr Mo)	5 Cr Mo through 9 Cr Mo	Austenitic stainless steels, 18 Cr, 8 Ni	12 Cr 17 Cr 27 Cr	25 Cr, 20 Ni	Monel 67 Ni, 30 Cu	3½ Nickel	Aluminum	Gray cast iron	Bronze	Brass	70 Cu, 30 Ni	Ni-Fe-Cr	Ni-Cr-Fe	Ductile iron
400	6.82	6.34	9.59	5.81	8.92	8.20	6.80	13.60	6.10	10.23	10.23	8.90	8.80	7.70	6.48
425	6.87	6.38	9.62	5.85	8.92	8.25	6.83	13.68	6.15	10.25	10.29	....	8.82	7.72	6.57
450	6.92	6.42	9.65	5.89	8.92	8.30	6.86	13.75	6.19	10.28	10.35	....	8.85	7.75	6.66
475	6.97	6.46	9.67	5.92	8.92	8.35	6.89	13.83	6.24	10.30	10.41	....	8.87	7.77	6.75
500	7.02	6.50	9.70	5.96	8.93	8.40	6.93	13.90	6.28	10.32	10.47	....	8.90	7.80	6.85
525	7.07	6.54	9.73	6.00	8.93	8.45	6.97	13.98	6.33	10.35	10.53	....	8.92	7.82	6.88
550	7.12	6.58	9.76	6.05	8.93	8.49	7.01	14.05	6.38	10.38	10.58	....	8.95	7.35	6.92
575	7.17	6.62	9.79	6.09	8.93	8.54	7.04	14.13	6.42	10.41	10.64	....	8.97	7.88	6.95
600	7.23	6.66	9.82	6.13	8.94	8.58	7.08	14.20	6.47	10.44	10.69	....	9.00	7.90	6.98
625	7.28	6.70	9.85	6.17	8.94	8.63	7.12	....	6.52	10.46	10.75	....	9.02	7.92	7.02
650	7.33	6.73	9.87	6.20	8.95	8.68	7.16	....	6.56	10.48	10.81	....	9.05	7.95	7.04
675	7.38	6.77	9.90	6.23	8.95	8.73	7.19	....	6.61	10.50	10.86	....	9.07	7.98	7.08
700	7.44	6.80	9.92	6.26	8.96	8.78	7.22	....	6.65	10.52	10.92	....	9.10	8.00	7.11
725	7.49	6.84	9.95	6.29	8.96	8.83	7.25	....	6.70	10.55	10.98	....	9.12	8.02	7.14
750	7.54	6.88	9.99	6.33	8.96	8.87	7.29	....	6.74	10.57	11.04	....	9.15	8.05	7.18
775	7.59	6.92	10.02	6.36	8.96	8.92	7.31	....	6.79	10.60	11.10	....	9.17	8.08	7.22
800	7.65	6.96	10.05	6.39	8.97	8.96	7.34	....	6.83	10.62	11.16	....	9.20	8.10	7.25
825	7.70	7.00	10.08	6.42	8.97	9.01	7.37	....	6.87	10.65	11.22	....	9.22	....	7.27
850	7.75	7.03	10.11	6.46	8.98	9.06	7.40	....	6.92	10.67	11.28	....	9.25	....	7.31
875	7.79	7.07	10.13	6.49	8.99	9.11	7.43	....	6.96	10.70	11.34	....	9.27	....	7.34
900	7.84	7.10	10.16	6.52	9.00	9.16	7.45	....	7.00	10.72	11.40	....	9.30	....	7.37
925	7.87	7.13	10.19	6.55	9.05	9.21	7.47	....	7.05	10.74	11.46	....	9.32	....	7.41
950	7.91	7.16	10.23	6.58	9.10	9.25	7.49	....	7.10	10.76	11.52	....	9.35	....	7.44
975	7.94	7.19	10.26	6.60	9.15	9.30	7.52	....	7.14	10.78	11.57	....	9.37	....	7.47

1000	7.97	7.22	10.29	6.63	9.18	9.34	7.55	....	7.19	10.80	11.63	....	9.40	....	7.50
1025	8.01	7.25	10.32	6.65	9.20	9.39	....	....	....	10.83	11.69	....	9.42	....	
1050	8.05	7.27	10.34	6.68	9.22	9.43	....	....	....	10.85	11.74	....	9.45		
1075	8.08	7.30	10.37	6.70	9.24	9.48	....	....	....	10.88	11.80	....	9.47		
1100	8.12	7.32	10.39	6.72	9.25	9.52	....	....	....	10.90	11.85	....	9.50		
1125	8.14	7.34	10.41	6.74	9.29	9.57	....	....	....	10.93	11.91	....	9.52		
1150	8.16	7.37	10.44	6.75	9.33	9.61	....	....	....	10.95	11.97	....	9.55		
1175	8.17	7.39	10.46	6.77	9.36	9.66	....	....	....	10.98	12.03	....	9.57		
1200	8.19	7.41	10.48	6.78	9.39	9.70	....	....	....	11.00	12.09	....	9.60		
1225	8.21	7.43	10.50	6.80	9.43	9.75	....	....	....	....	....	....	9.64		
1250	8.24	7.45	10.51	6.82	9.47	9.79	....	....	....	....	....	....	9.68		
1275	8.26	7.47	10.53	6.83	9.50	9.84	....	....	....	....	....	....	9.71		
1300	8.28	7.49	10.54	6.85	9.53	9.88	....	....	....	....	....	....	9.75		
1325	8.30	7.51	10.56	6.86	9.53	9.92	....	....	....	....	....	....	9.79		
1350	8.32	7.52	10.57	6.88	9.54	9.96	....	....	....	....	....	....	9.83		
1375	8.34	7.54	10.59	6.89	9.55	10.00	....	....	....	....	....	....	9.86		
1400	8.36	7.55	10.60	6.90	9.56	10.04	....	....	....	....	....	....	9.90		
1425	....	....	10.64	....	....	....	....	....	....	....	....	....	9.94		
1450	....	....	10.68	....	....	....	....	....	....	....	....	....	9.98		
1475	....	....	10.72	....	....	....	....	....	....	....	....	....	10.01		
1500	....	....	10.77	....	....	....	....	....	....	....	....	....	10.05		

*Extracted from the Chemical Plant and Petroleum Refinery Piping Code, ANSI B31.3—1980, with permission of the publisher, the American Society of Mechanical Engineers, New York. These data are for information, and it is not implied that materials are suitable for all the temperatures shown. (°F−32) 5/9 = °C; to convert microinches per inch-degrees Fahrenheit to meters per meter-kelvins, multiply by 1.8.

TABLE 6-41 Modulus of Elasticity, U.S. Customary Units, for Metals*

Material	E = Modulus of elasticity, lbf/in² (multiply tabulated values by 10^6) — Temperature, °F																
	−325	−200	−100	70	200	300	400	500	600	700	800	900	1000	1100	1200	1300	1400	1500
Modulus of elasticity: ferrous materials																		
Carbon steels with carbon content 0.30 percent or less, 3½ Ni	30.0	29.5	29.0	27.9	27.7	27.4	27.0	26.4	25.7	24.8	23.4	18.5	15.4	13.0				
Carbon steels with carbon content above 0.30 percent	31.0	30.6	30.4	29.9	29.5	29.0	28.3	27.4	26.7	25.4	23.8	21.5	18.8	15.0	11.2			
Carbon-molybdenum steels, low-chromium steels through 3 Cr Mo	31.0	30.6	30.4	29.9	29.5	29.0	28.6	28.0	27.4	26.6	25.7	24.5	23.0	20.4	15.6			
Intermediate-chromium steels (5 Cr Mo through 9 Cr Mo)	29.4	28.5	28.1	27.4	27.1	26.8	26.4	26.0	25.4	24.9	24.2	23.5	22.8	21.9	20.8	19.5	18.1	
Austenitic steels (TP304, 310, 316, 321, 347)	30.4	29.9	29.4	28.3	27.7	27.1	26.6	26.1	25.4	24.8	24.1	23.4	22.7	22.0	21.3	20.7	19.3	17.9
Straight chromium steels (12 Cr, 17 Cr, 27 Cr)	30.8	30.3	29.8	29.2	28.7	28.3	27.7	27.0	26.0	24.8	23.1	21.1	18.6	15.6	12.2			
Gray cast iron	....	....	....	13.4	13.2	12.9	12.6	12.2	11.7	11.0	10.2							

Modulus of elasticity: nonferrous materials

Material	E = Modulus of elasticity, lbf/in² (multiply tabulated values by 10^6) — Temperature, °F															
	−325	−200	−100	70	100	200	300	400	500	600	700	800	900	1000	1100	1200
Monel (67 Ni, 30 Cu) and (66 Ni, 29 Cu—Al)	26.8	26.6	26.4	26.0	26.0	26.0	25.8	25.6	25.4	24.7	23.1	21.0	18.6	16.0	14.3	13.0
Copper-nickel (70 Cu, 30 Ni)	....	....	....	21.6	21.5	21.2	20.9	20.6	20.3	20.0	19.7	19.4				
Aluminum alloys	11.3	10.9	10.6	10.1	10.0	9.8	9.5	8.7	7.7							
Copper (99.98 percent Cu)	17.0	16.7	16.5	16.0	15.8	15.6	15.4	15.1	14.7	14.2	13.7					
Commercial brass (66 Cu, 34 Zn)	15.0	14.7	14.5	14.0	13.9	13.7	13.5	13.0	12.7	12.2	11.8					
Leaded tin bronze (88 Cu, 6 Sn, 1.5 Pb, 4.5 Zn)	14.2	13.8	13.5	13.0	12.9	12.7	12.4	12.0	11.7	11.3	10.9					

*Extracted from the Chemical Plant and Petroleum Refinery Piping Code, ANSI B 31.3—1980, with permission of the publisher, the American Society of Mechanical Engineers, New York. These data are for information, and it is not implied that materials are suitable for all the temperatures shown. (°F−32) 5/9 = °C; to convert pounds-force per square inch to megapascals, multiply by 0.006895.

For header (legs 1 and 2):

$$S_b = \frac{\sqrt{(i_i m_i)^2 + (i_o M_o)^2}}{Z} \tag{6-47}$$

For branch (leg 3):

$$S_b = \frac{\sqrt{(i_i m_i)^2 + (i_o M_o)^2}}{Z_e} \tag{6-48}$$

where S_b = resultant bending stress, lbf/in² (MPa)
Z_e = effective section modulus for branch, in³ (mm³)

$$Z_e = \pi\, r_2{}^2 T_s \tag{6-49}$$

r_2 = mean branch cross-sectional radius, in (mm)
T_s = effective branch wall thickness, in (mm) [lesser of $\overline{T}_h$ and $(i_o)(\overline{T}_b)$]
$\overline{T}_h$ = thickness of pipe matching run of tee or header exclusive of reinforcing elements, in (mm)
$\overline{T}_b$ = thickness of pipe matching branch, in (mm)
i_o = out-plane stress-intensification factor (Table 6-42)
i_i = in-plane stress-intensification factor (Table 6-42)

Allowable stress range S_A and permissible additive stresses shall be computed in accordance with Eqs. (6-38) and (6-39).

Required Weld Quality Assurance Any weld at which S_E exceeds 0.8 S_A for any portion of a piping system and the equivalent number of cycles N exceeds 7000 shall be fully examined in accordance with the requirements for severe cyclic service (presented later in this section).

Reactions: Metallic Piping Reaction forces and moments to be used in the design of restraints and supports and in evaluating the effects of piping displacements on connected equipment shall be based on the reaction range R for the extreme displacement conditions, considering the range previously defined for reactions and using E_a. The designer shall consider instantaneous maximum values of forces and moments in the original and extreme displacement conditions as well as the reaction range in making these evaluations.

Maximum Reactions for Simple Systems For two-anchor systems without intermediate restraints, the maximum instantaneous values of reaction forces and moments may be estimated from Eqs. (6-50) and (6-51).

For extreme displacement conditions, R_m. The temperature for this computation is the design maximum or design minimum temperature as previously defined for reactions, whichever produces the larger reaction:

$$R_m = R\left(1 - \frac{2C}{3}\right)\frac{E_m}{E_a} \tag{6-50}$$

where C = cold-spring factor varying from zero for no cold spring to 1.0 for 100 percent cold spring. (The factor 2/3 is based on experience, which shows that specified cold spring cannot be fully assured even with elaborate precautions.)
E_a = modulus of elasticity at installation temperature, lbf/in² (MPa)

TABLE 6-42 Flexibility Factor *k* and Stress-Intensification Factor *i**

Description	Flexibility factor k	Stress intensification factor[a,h] Out-plane, i_o	In-plane, i_i	Flexibility characteristic h	Sketch
Welding elbow[a,b,c,f,i] or pipe bend	$\frac{1.65}{h}$	$\frac{0.75}{h^{2/3}}$	$\frac{0.9}{h^{2/3}}$	$\frac{\bar{T}R_1}{(r_2)^2}$	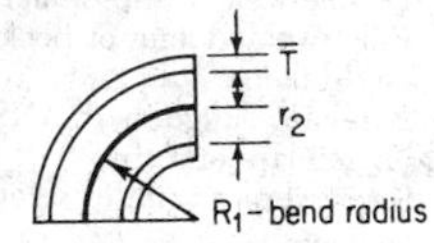
Closely spaced miter bend[a,b,c] $s < r_2\,(1 + \tan\theta)$	$\frac{1.52}{h^{5/6}}$	$\frac{0.9}{h^{2/3}}$	$\frac{0.9}{h^{2/3}}$	$\frac{\cot\theta}{2}\,\frac{\bar{T}_s}{(r_2)^2}$	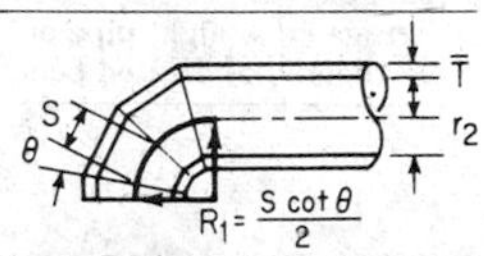
Single miter bend[a,b] or widely spaced miter bend $s \geq r_2\,(1+\tan\theta)$	$\frac{1.52}{h^{5/6}}$	$\frac{0.9}{h^{2/3}}$	$\frac{0.9}{h^{2/3}}$	$\frac{1+\cot\theta}{2}\,\frac{\bar{T}}{r_2}$	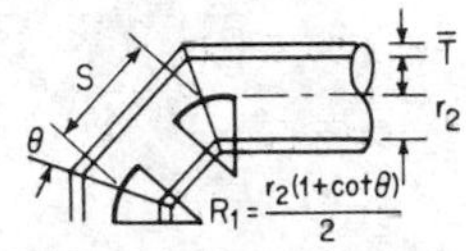
Welding tee[a,b,f] per ANSI B16.9 with $r_x > \frac{1}{8}\,D_b$ $T_c \geq 1.5\,\bar{T}$	1	$\frac{0.9}{h^{2/3}}$	$\frac{3}{4}\,i_o + \frac{1}{4}$	$4.4\,\frac{\bar{T}}{r_2}$	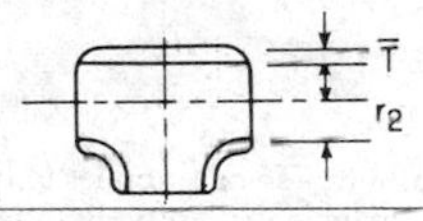
Reinforced fabricated[a,b,e] tee with pad or saddle	1	$\frac{0.9}{h^{2/3}}$	$\frac{3}{4}\,i_o + \frac{1}{4}$	$\frac{(\bar{T}+1/2\,t_r)^{5/2}}{\bar{T}^{3/2}\,r_2}$	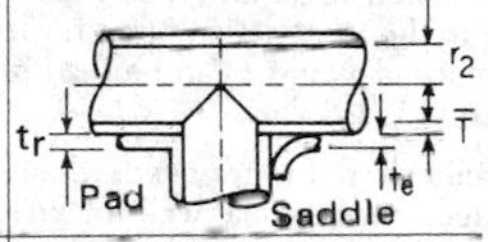
Unreinforced[a,b] fabricated tee	1	$\frac{0.9}{h^{2/3}}$	$\frac{3}{4}\,i_o + \frac{1}{4}$	$\frac{\bar{T}}{r_2}$	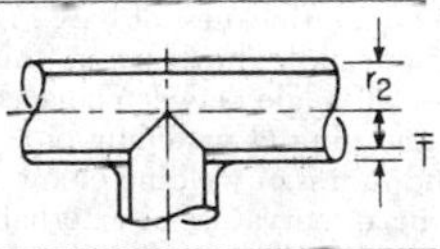
Extruded[a,b] welding tee $T_c < 1.5\,\bar{T}$	1	$\frac{0.9}{h^{2/3}}$	$\frac{3}{4}\,i_o + \frac{1}{4}$	$\left(1 + \frac{r_x}{r_2}\right)\frac{\bar{T}}{r_2}$	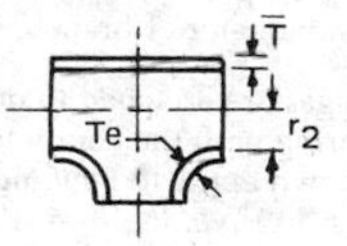
Welded-in[a,b] contour insert $r_x \geq \frac{1}{8}\,D_b$ $T_c \geq 1.5\,\bar{T}$	1	$\frac{0.9}{h^{2/3}}$	$\frac{3}{4}\,i_o + \frac{1}{4}$	$4.4\,\frac{\bar{T}}{r_2}$	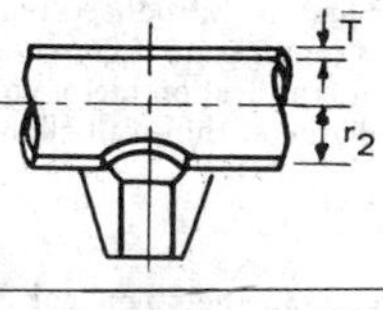
Branch[a,b,g] welded-on fitting (integrally reinforced)	1	$\frac{0.9}{h^{2/3}}$	$\frac{0.9}{h^{2/3}}$	$3.3\,\frac{\bar{T}}{r_2}$	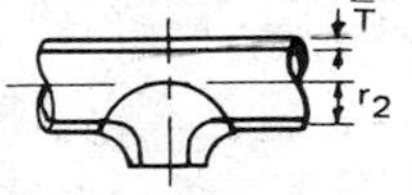

E_m = modulus of elasticity at design maximum or design minimum temperature, lbf/in² (MPa)

R = range of reaction forces or moments (derived from flexibility analysis) corresponding to the full displacement-stress range and based on E_a, lbf or in·lbf (N or N·mm)

R_m = estimated instantaneous maximum reaction force or moment at design maximum or design minimum temperature, lbf or in·lbf (N or N·mm)

For original condition, R_a. The temperature for this computation is the expected temperature at which the piping is to be assembled.

$$R_a = CR \quad \text{or} \quad C_1R, \text{ whichever is greater} \tag{6-51}$$

where nomenclature is as for Eq. (6-50) and

$C_1 = 1 - (S_h E_a / S_E E_m)$
= estimated self-spring or relaxation factor (use zero if value of C_1 is negative)

R_a = estimated instantaneous reaction force or moment at installation temperature, lbf or in·lbf (N or N·mm)

S_E = computed displacement–stress range, lbf/in² (MPa). See Eq. (6-45).

S_h = See Eq. (6-38).

TABLE 6-42 Flexibility Factor k and Stress-Intensification Factor i* *(Continued)*

Description	Flexibility factor k	Stress-intensification factor i
Butt-welded joint, reducer, or weld-neck flange	1	1.0
Double-welded slip-on flange	1	1.2
Fillet welded joint or pocket-weld flange	1	1.3
Lap-joint flange (with ANSI B16.9 lap-joint stub)	1	1.6
Screwed pipe joint or screwed flange	1	2.3
Corrugated straight pipe or corrugated or creased bend[d]	5	2.5

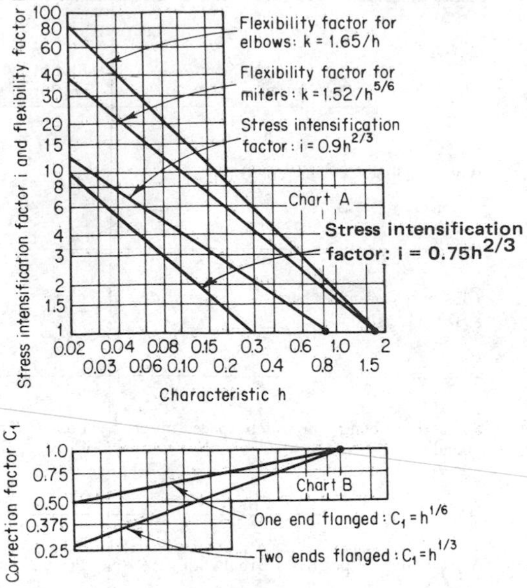

[a]The flexibility factor k applies to bending in any plane. The flexibility factors k and stress intensification factors i shall not be less than unity; factors for torsion equal unity. Both factors apply over the effective arc length (shown by heavy centerlines in the sketches) for curved and miter bends and to the intersection point for tees.

[b]The values of k and i can be read directly from Chart A by entering with the characteristic h computed from the formulas given above. Nomenclature is as follows:

$\bar{T}$ = for elbows and miter bends, the nominal wall thickness of the fitting, in (mm)
= for tees, the nominal wall thickness of the matching pipe, in
$\bar{T}_c$ = the crotch thickness of tees, in (mm)
t_r = pad or saddle thickness, in (mm)
θ = one-half angle between adjacent miter axes, °
r_2 = mean radius of matching pipe, in (mm)
R_1 = bend radius of welding elbow or pipe bend, in (mm)
r_x = radius of curvature of external contoured portion of outlet, in, measured in the plane containing the axes of the run and branch.
s = miter spacing at centerline, in (mm)
D_b = outside diameter of branch, in (mm)

[c]When flanges are attached to one or both ends, the values of k and i shall be corrected by the factors C_1, which can be read directly from Chart B, entering with the computed h.

[d]Factors shown apply to bending. Flexibility factor for torsion equals 0.9.

[e]When t_r is $> 1\frac{1}{2}\,\bar{T}$, use $h = 4(\bar{T}/r_2)$.

[f]Designers are cautioned that cast butt-welded fittings may have considerably heavier walls than that of the pipe with which they are used. Large errors may be introduced unless the effect of these greater thicknesses is considered.

[g]Designers must assure themselves that this fabrication has a pressure rating equivalent to that of straight pipe.

[h]A single intensification factor equal to $0.9/h^{2/3}$ may be used for both i_i and i_o if desired.

[i]In large-diameter thin-wall elbows and bends, pressure can significantly affect the magnitudes of k and i. To correct values from the table,

$$\text{Divide } k \text{ by } \left[1 + 6\left(\frac{P}{E_c}\right)\left(\frac{r_2}{t}\right)^{7/3}\left(\frac{R_1}{r_2}\right)^{1/3}\right] \qquad \text{Divide } i \text{ by } \left[1 + 3.25\left(\frac{P}{E_c}\right)\left(\frac{r_2}{t}\right)^{5/2}\left(\frac{R_1}{r_2}\right)^{2/3}\right]$$

*Extracted from the Chemical Plant and Petroleum Refinery Piping Code, ANSI B31.3—1980, with permission of the publisher, the American Society of Mechanical Engineers, New York.

Maximum Reactions for Complex Systems For multianchor systems and for two-anchor systems with intermediate restraints, Eqs. (6-50) and (6-51) are not applicable. Each case must be studied to estimate the location, nature, and extent of local overstrain and its effect on stress distribution and reactions.

Acceptable comprehensive methods of analysis are analytical, model-test, and chart methods, which evaluate for the entire piping system under consideration the forces, moments, and stresses caused by bending and torsion from a simultaneous consideration of terminal and intermediate restraints to thermal expansion and include all external movements transmitted under thermal change to the piping by its terminal and intermediate attachments. Correction factors, as provided by the details of these rules, must be applied for the stress intensification of curved pipe and branch connections and may be applied for the increased flexibility of such component parts.

Brock [in Crocker (ed.), *Piping Handbook*, 5th ed., McGraw-Hill, New York, 1967, sec. 4] provides further data on methods of analysis.

Expansion Joints All the foregoing applies to "stiff piping systems," i.e., systems without expansion joints (see detail 1 of Fig. 6-131). When space limitations, process requirements, or other considerations result in configurations of insufficient flexibility, capacity for deflection within allowable stress range limits may be increased successively by the use of one or more hinged bellows expansion joints, viz., semirigid (detail 2) and nonrigid (detail 3) systems, and expansion effects essentially eliminated by a free-movement joint (detail 4) system. Expansion joints for semirigid and nonrigid systems are

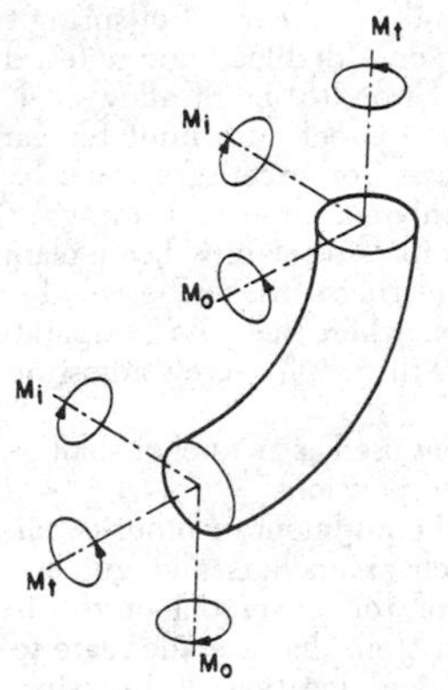

FIG. 6-129 Moments in bends. *(Extracted from the Chemical Plant and Petroleum Refinery Piping Code, ANSI B31.3—1976, with permission of the publisher, the American Society of Mechanical Engineers, New York.)*

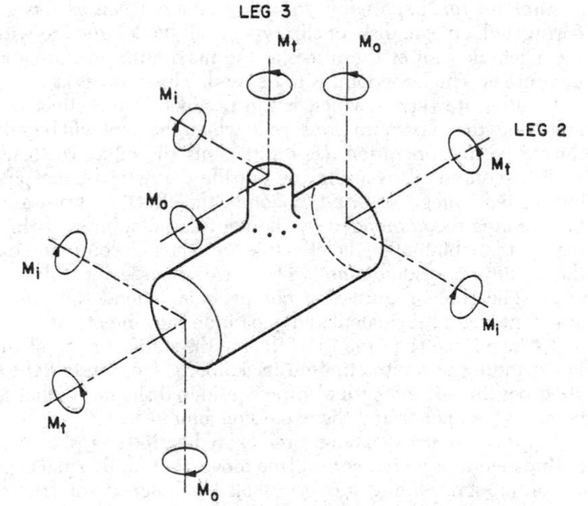

FIG. 6-130 Moments in branch connections. *(Extracted from the Chemical Plant and Petroleum Refinery Piping Code, ANSI B31.3—1976, with permission of the publisher, the American Society of Mechanical Engineers, New York.)*

restrained against longitudinal and lateral movement by the hinges with the expansion element under bending movement only and are known as "rotation" or "hinged" joints (see Fig. 6-132). Semirigid systems are limited to one plane; nonrigid systems require a minimum of three joints for two-dimensional and five joints for three-dimensional expansion movement.

Joints similar to that shown in Fig. 6-132, except with two pairs of hinge pins equally spaced around a gimbal ring, achieve similar results with a lesser number of joints.

Expansion joints for free-movement systems can be designed for axial or offset movement alone, or for combined axial and offset movements (see Fig. 6-133). For offset movement alone, the end load due to pressure and weight can be transferred across the joint by tie rods or structural members (see Fig. 6-134). For axial or combined movements, anchors must be provided to absorb the unbalanced pressure load and force bellows to deflect.

Commercial bellows elements are usually light-gauge [of the order of (0.05 to 0.10 in) thick] and are available in stainless and other alloy steels, copper, and other nonferrous materials. Multi-ply bellows, bellows with external reinforcing rings, and toroidal contour bellows are available for higher pressures. Since bellows elements are ordinarily rated for strain ranges which involve repetitive yielding, predictable

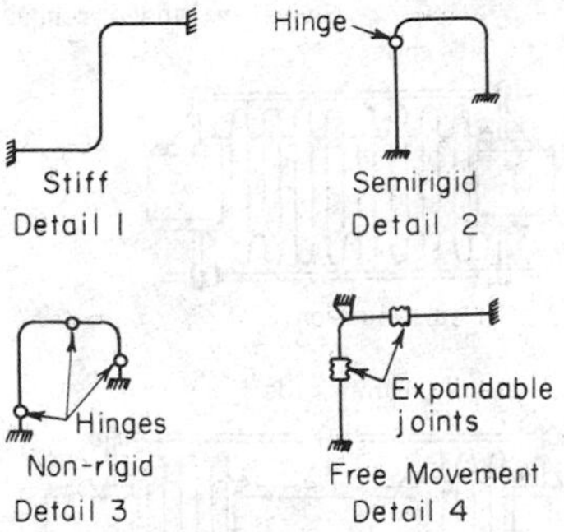

FIG. 6-131 Flexibility classification for piping systems. *(From Kellogg,* Design of Piping Systems, *Wiley, New York, 1965.)*

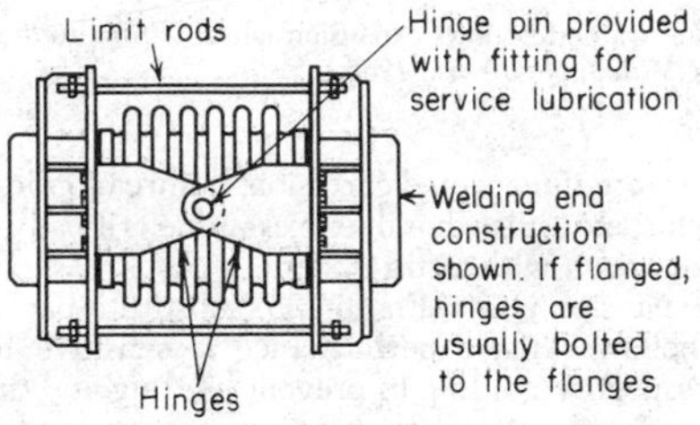

FIG. 6-132 Hinged expansion joint. (*From Kellogg,* Design of Piping Systems, *Wiley, New York, 1965.)*

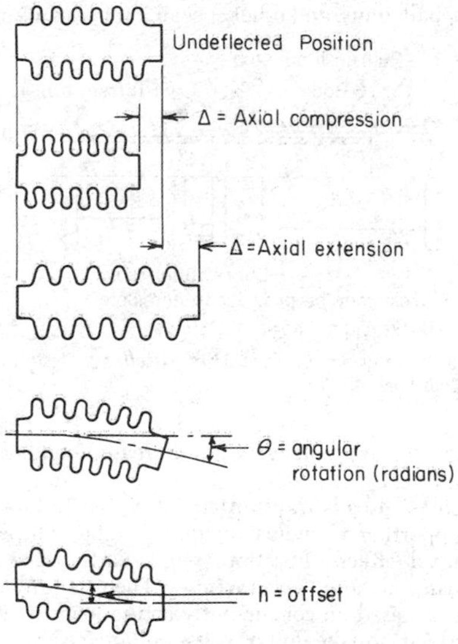

FIG. 6-133 Action of expansion bellows under various movements. *(From Kellogg,* Design of Piping Systems, *Wiley, New York, 1965.)*

performance is assured only by adequate fabrication controls and knowledge of the potential fatigue performance of each design. The attendant cold work can affect corrosion resistance and promote susceptibility to corrosion fatigue or stress corrosion; joints in a horizontal position cannot be drained and have frequently undergone pitting or cracking due to the presence of condensate during operation or offstream. For low-pressure essentially nonhazardous service, nonmetallic bellows of fabric-reinforced rubber or special materials are sometimes used. For corrosive service Teflon bellows may be used.

Because of the inherently greater susceptibility of expansion bel-

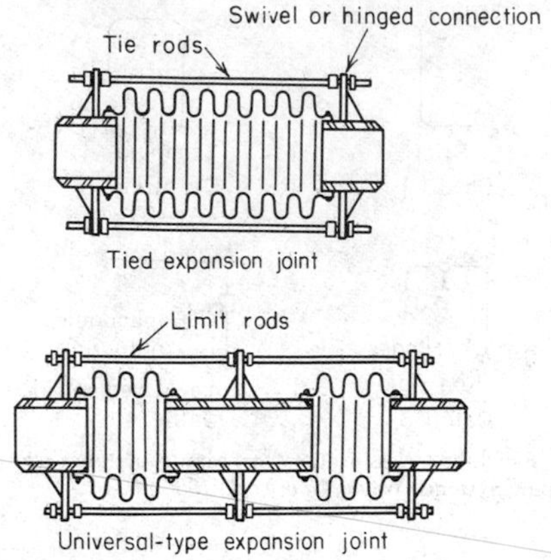

FIG. 6-134 Constrained-bellows expansion joints. (*From Kellogg,* Design of Piping Systems, *Wiley, New York, 1965.*)

lows to failure from unexpected corrosion, failure of guides to control joint movements, etc., it is advisable to examine critically their design choice in comparison with a stiff system.

Slip-type expansion joints (Fig. 6-135) substitute packing (ring or plastic) for bellows. Their performance is sensitive to adequate design with respect to guiding to prevent binding and the adequacy of stuffing boxes and attendant packing, sealant, and lubrication. Anchors must be provided for the unbalanced pressure force and for the friction forces to move the joint. The latter can be much higher than the elastic force required to deflect a bellows joint. Rotary packed joints, ball joints, and other special joints can absorb end load.

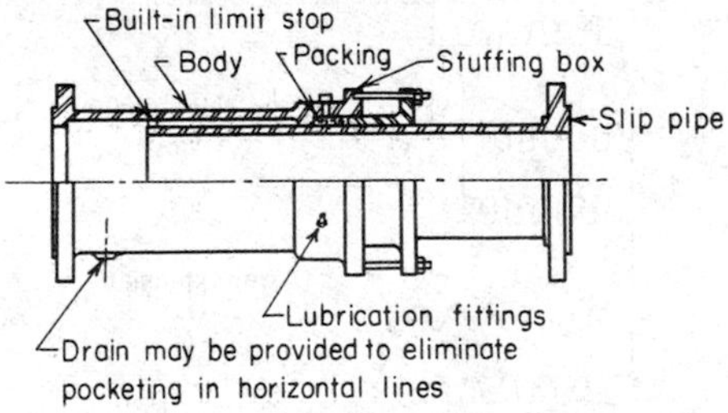

FIG. 6-135 Slip-type expansion joint. (*From Kellogg,* Design of Piping Systems, *Wiley, New York, 1965.*)

Corrugated pipe and corrugated and creased bends are also used to decrease stiffness.

Pipe Supports Loads transmitted by piping to attached equipment and supporting elements include weight, temperature- and pressure-induced effects, fibration, wind, earthquake, shock, and thermal expansion and contraction. The design of supports and restraints is based on concurrently acting loads (if it is assumed that wind and earthquake do not act simultaneously).

Resilient and constant-effort-type supports shall be designed for maximum loading conditions including test unless temporary supports are provided.

Though not specified in the code, supports for discharge piping from relief valves must be adequate to withstand the jet reaction produced by their discharge.

The code states further that pipe-supporting elements shall (1) avoid excessive interference with thermal expansion and contraction of pipe which is otherwise adequately flexible; (2) be such that they do not contribute to leakage at joints or excessive sag in piping requiring drainage; (3) be designed to prevent overstress, resonance, or disengagement due to variation of load with temperature; also, so that combined longitudinal stresses in the piping shall not exceed the code allowable limits; (4) be such that a complete release of the piping load will be prevented in the event of spring failure or misalignment, weight transfer, or added load due to test during erection; (5) be of steel or wrought iron; (6) be of alloy steel or protected from temperature when the temperature limit for carbon steel may be exceeded; (7) not be cast iron except for roller bases, rollers, anchor bases, etc., under mainly compression loading; (8) not be malleable or nodular iron except for pipe clamps, beam clamps, hanger flanges, clips, bases, and swivel rings; (9) not be wood except for supports mainly in compression when the pipe temperature is at or below ambient; and (10) have threads for screw adjustment which shall conform to ANSI B1.1.

A supporting element used as an anchor shall be designed to maintain an essentially fixed position.

To protect terminal equipment or other (weaker) portions of the system, restraints (such as anchors and guides) shall be provided where necessary to control movement or to direct expansion into those portions of the system that are adequate to absorb them. The design, arrangement, and location of restraints shall ensure that expansion–joint movements occur in the directions for which the joint is designed. In addition to the other thermal forces and moments, the effects of friction in other supports of the system shall be considered in the design of such anchors and guides.

Anchors for Expansion Joints Anchors (such as those of the corrugated, omega, disk, or slip type) shall be designed to withstand the algebraic sum of the forces at the maximum pressure and temperature at which the joint is to be used. These forces are:

1. Pressure thrust, which is the product of the effective thrust area times the maximum pressure to which the joint will be subjected during normal operation. (For slip joints the effective thrust area shall be computed by using the outside diameter of the pipe. For corrugated, omega, or disk-type joints, the effective thrust area shall be that area recommended by the joint manufacturer. If this information is unobtainable, the effective area shall be computed by using the maximum inside diameter of the expansion–joint bellows.)
2. The force required to compress or extend the joint in an amount equal to the calculated expansion movement.
3. The force required to overcome the static friction of the pipe in expanding or contracting on its supports, from installed to operating position. The length of pipe considered should be that located between the anchor and the expansion joint.

Support Fixtures Hanger rods may be pipe straps, chains, bars, or threaded rods which permit free movement for thermal expansion or contraction. Sliding supports shall be designed for friction and bearing loads. Brackets shall be designed to withstand movements due to friction in addition to other loads. Spring-type supports shall be designed for weight load at the point of attachment and to prevent misalignment, buckling, or eccentric loading of springs, and provided with stops to prevent spring overtravel. Compensating-type spring hangers are recommended for high-temperature and critical-service piping to make the supporting force uniform with appreciable movement. Counterweight supports shall have stops to limit travel. Hydraulic supports shall be provided with safety devices and stops to support load in the event of loss of pressure. Vibration dampers or sway braces may be used to limit vibration amplitude.

The code requires that the safe load for threaded hanger rods be based on the root area of the threads. This, however, assumes concentric loading. When hanger rods move to a nonvertical position so that the load is transferred from the rod to the supporting structure via the edge of one flat of the nut on the rod, it is necessary to consider the root area to be reduced by one-third. If a clamp is connected to a vertical line to support its weight, it is recommended that shear lugs be welded to the pipe, or that the clamp be located below a fitting or flange, to prevent slippage. Consideration shall be given to the localized stresses induced in the piping by the integral attachment. Typical pipe supports are shown in Fig. 6-136.

Much piping is supported from structures installed for other purposes. It is common practice to use beam formulas for tubular sections to determine stress, maximum deflection, and maximum slope of piping in **spans between supports.** When piping is supported from structures installed for that sole purpose and those structures rest on driven piles, detailed calculations are usually made to deter-

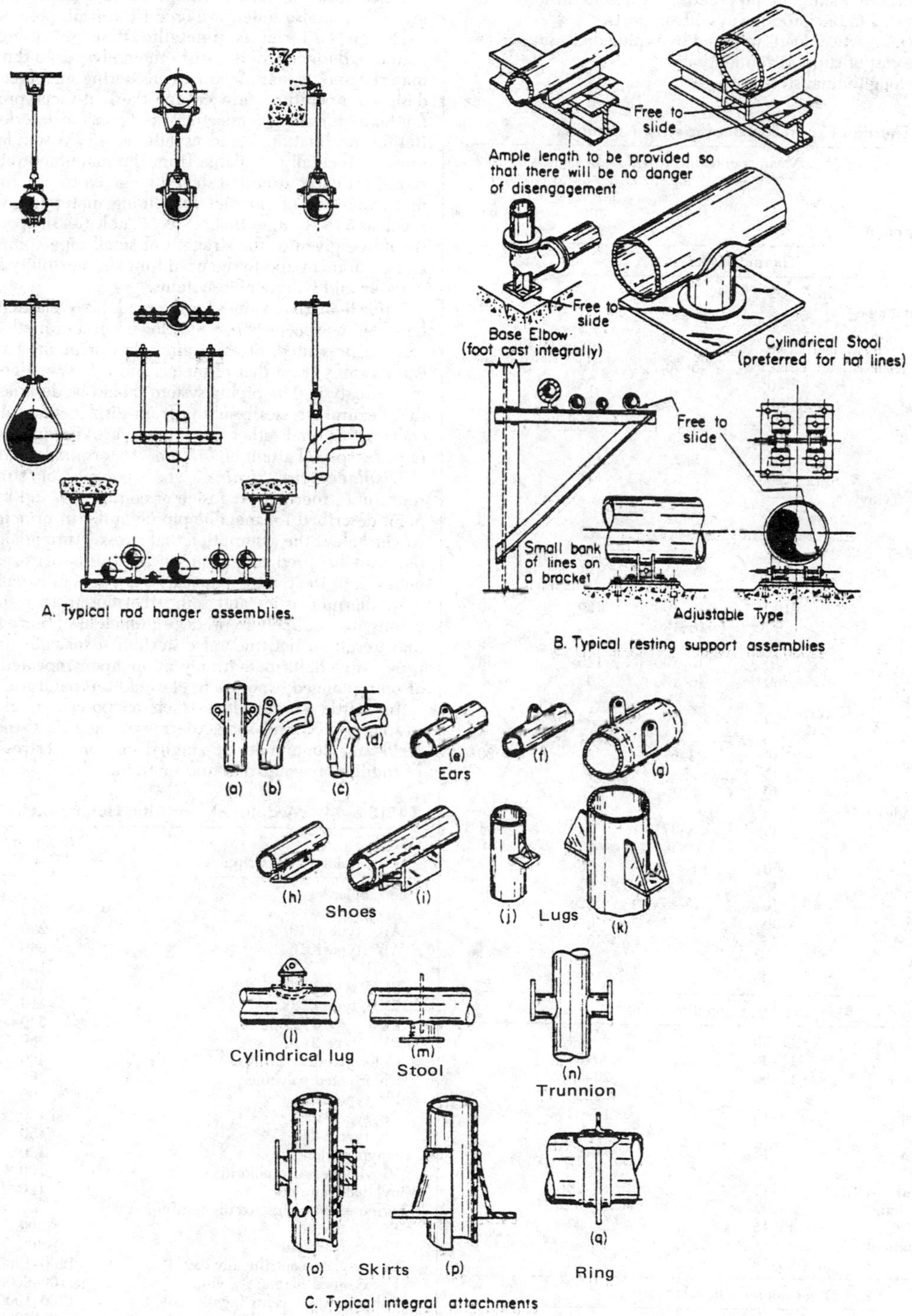

FIG. 6-136 Typical pipe supports and attachments. (*From Kellogg,* Design of Piping Systems, *Wiley, New York, 1965.*)

mine maximum permissible spans. Limits imposed on maximum slope to make the contents of the line drain to the lower end require calculations made on the weight per foot of the empty line. To avoid interference with other components, maximum deflection should be limited to 25.4 mm (1 in).

Pipe hangers are essentially frictionless but require taller pipe-support structures which cost more than structures on which pipe is laid. Devices that reduce friction between laid pipe subject to thermal movement and its supports are used to accomplish the following:

1. Reduce loads on anchors or on equipment acting as anchors.

2. Reduce the tendency of pipe acting as a column loaded by friction at supports to buckle sideways off supports.
3. Reduce nonvertical loads imposed by piping on its supports so as to minimize cost of support foundations.
4. Reduce longitudinal stress in pipe.

TABLE 6-43 Thermal Expansion Coefficients: Nonmetals*

Material description	Mean coefficients (divide table values by 10^6) in/(m·°F)	Range, °F	mm/mm, °C	Range, °C
	Thermoplastics			
Acetal AP2012	2	....	4	
Acrylonitrile-butadiene-styrene				
ABS 1208	60	....	108	
ABS 1210	55	45–55	99	8–12
ABS 1316	40	....	72	
ABS 2112	40	....	72	
Cellulose acetate butyrate				
CAB MH08	80	....	144	
CAB 5004	95	....	171	
Chlorinated poly (vinyl chloride)				
CPVC 4120	35	....	63	
Polybutylene PB 2110	72	....	130	
Polyether, chlorinated	45	....	81	
Polyethylene				
PE 1404	100	46–100	180	8–38
PE 2305	90	46–100	162	8–38
PE 2306	80	46–100	144	8–38
PE 3306	70	46–100	126	8–38
PE 3406	60	46–100	108	8–38
Polyphenylene POP 2125	30	....	54	
Polypropylene				
PP1110	48	33–67	86	0–20
PP1208	43	....	77	
PP2105	40	....	72	
Poly(vinyl chloride)				
PVC 1120	30	23–37	54	−5–+3
PVC 1220	35	34–40	63	1–4
PVC 2110	50	....	90	
PVC 2112	45	....	81	
PVC 2116	40	37–45	72	3–8
PVC 2120	30	....	54	
Vinylidine fluoride	85	....	153	
Vinylidine/vinyl chloride	100	....	180	
	Reinforced thermosetting resins			
Asbestos-phenolic	11–30	....	20–54	
Asbestos-epoxy	11–30	....	20–54	
Asbestos–polyester	11–30	....	20–54	
Glass-epoxy, centrifugal-cast	9–13	....	16–23	
Glass-polyester, centrifugal-cast	9–15	....	16–27	
Glass-polyester, filament-wound	9–11	....	16–20	
Glass-polyester, hand lay-up	12–15	....	22–27	
Glass-epoxy, filament-wound	9–13	....	16–23	
	Other nonmetallic materials			
Borosilicate glass	1.8	....	3	
Impregnated graphite	2.4	....	4	
Hard rubber (Buna N)	40	....	72	

*Extracted from the Chemical Plant and Petroleum Refinery Piping Code, ANSI B31.3—1980, with permission of the publisher, the American Society of Mechanical Engineers, New York. Individual compounds may vary from the values in the table by as much as 10 percent. Consult manufacturers for specific values for their products.

Linear bearing surfaces made of fluorinated hydrocarbons or of graphite and also rollers are used for this purpose.

Design Criteria: Nonmetallic Pipe In using a nonmetallic material, designers must satisfy themselves as to the adequacy of the material and its manufacture, considering such factors as strength at design temperature, impact- and thermal-shock properties, toxicity, methods of making connections, and possible deterioration in service. Rating information, based usually on ASTM standards or specifications, is generally available from the manufacturers of these materials. Particular attention should be given to provisions for the thermal expansion of nonmetallic piping materials, which may be as much as 5 to 10 times that of steel (Table 6-43). Special consideration should be given to the strength of small pipe connections to piping and equipment and to the need for extra flexibility at the junction of metallic and nonmetallic systems.

Table 6-44 gives values for the modulus of elasticity for nonmetals; however, no specific stress-limiting criteria or methods of stress analysis are presented. Stress-strain behavior of most nonmetals differs considerably from that of metals and is less well-defined for mathematic analysis. The piping system should be designed and laid out so that flexural stresses resulting from displacement due to expansion, contraction, and other movement are minimized. This concept requires special attention to supports, terminals, and other restraints.

Displacement Strains The concepts of strain imposed by restraint of thermal expansion or contraction and by external movement described for metallic piping apply in principle to nonmetals. Nevertheless, the assumption that stresses throughout the piping system can be predicted from these strains because of fully elastic behavior of the piping materials is not generally valid for nonmetals.

In thermoplastics and some thermosetting resins, displacement strains are not likely to produce immediate failure of the piping but may result in detrimental distortion. Especially in thermoplastics, progressive deformation may occur upon repeated thermal cycling or on prolonged exposure to elevated temperature.

In brittle nonmetallics (such as porcelain, glass, impregnated graphite, etc.) and some thermosetting resins, the materials show rigid behavior and develop high displacement stresses up to the point of sudden breakage due to overstrain.

TABLE 6-44 Modulus of Elasticity: Nonmetals*

Material description	*E*, kip/in² (73.4°F)	*E*, MPa (23°C)
Thermoplastics		
Acetal	410	2830
ABS, type 1210	250	1725
ABS, type 1316	340	2345
CAB	120	830
PVC, type 1120	420	2895
PVC, type 1220	410	2830
PVC, type 2110	340	2345
PVC, type 2116	380	2620
Chlorinated PVC	420	2895
Chlorinated polyether	160	1105
PE, type 2306	90	620
PE, type 3306	130	895
PE, type 3406	150	1035
Polypropylene	120	825
Vinylidene/vinyl chloride	100	690
Vinylidene fluoride	120	825
Thermosetting resins, axtally reinforced		
Epoxy-asbestos	1200	8280
Phenolic-asbestos	1200	8280
Epoxy-glass, centrifugally cast	1200–1900	8280–13100
Epoxy-glass, filament-wound	1100–2000	7580–13800
Polyester-glass, centrifugally cast	1200–1900	8280–13100
Polyester-glass, hand lay-up	800–1000	5510–6900
Other		
Borosilicate glass	9800	67,600
Impregnated graphite	2300	15900
Hard rubber (Buna N)	300	2070

*Extracted from the Chemical Plant and Petroleum Refinery Piping Code, ANSI B31.3—1980, with permission of the publisher, the American Society of Mechanical Engineers, New York.

Elastic Behavior The assumption that displacement strains will produce proportional stress over a sufficiently wide range to justify an elastic-stress analysis often is not valid for nonmetals. In brittle nonmetallic piping, strains initially will produce relatively large elastic stresses. The total displacement strain must be kept small, however, since overstrain results in failure rather than plastic deformation. In plastic and resin nonmetallic piping strains generally will produce stresses of the overstrained (plastic) type even at relatively low values of total displacement strain.

Further information on the design of thermoplastic piping can be found in the Plastics Pipe Institute's Technical Report TR-21.

FABRICATION, ASSEMBLY, AND ERECTION

Welding, Brazing, or Soldering Code requirements dealing with fabrication are more detailed for welding than for other methods of joining, since welding is used not only to join two pipes end to end but also to fabricate fittings which replace seamless fittings such as elbows and lap-joint stub ends. The code requirements for welding processes and operators are essentially the same as covered in the subsection on pressure vessels (i.e., qualification to Sec. IX of the ASME Boiler and Pressure Vessel Code) except that welding processes are not restricted, the material grouping (P number) must be in accordance with Appendix A, and welding positions are related to pipe position. The code also permits one fabricator to accept welders or welding operators qualified by another employer without requalification when welding pipe by the same or equivalent procedure. Procedure qualification may include a requirement for low-temperature toughness tests. See Table 6-45.

Filler metal is required to conform with the requirements of Sec. IX. Backing rings (of ferrous material), when used, shall be of weldable quality with sulfur limited to 0.05 percent. Backing rings of nonferrous and nonmetallic materials may be used provided they are proved satisfactory by procedure-qualification tests and provided their use has been approved by the designer.

The code requires internal alignment within the dimensional limits specified in the welding procedure and the engineering design without specific dimensional limitations. Internal trimming is permitted for correcting internal misalignment provided such trimming does not result in a finished wall thickness before welding of less than required minimum wall thickness t_m. When necessary, weld metal may be deposited on the inside or outside of the component to provide alignment or sufficient material for trimming.

Table 6-46 is a digest of code requirements for the quality of welds. The defects referred to are illustrated in Fig. 6-137.

The qualification of brazing procedures, brazers, or brazing operations is required in accordance with the requirements of Part QB, Sec. IX, ASME Code, except that for Category D fluid service at design temperatures not over 93°C (200°F). Such qualification is at the owner's option. The clearance between surfaces to be joined by brazing or soldering shall be no larger than is necessary to allow complete capillary distribution of the filler metal.

The only requirement for solderers is that they follow the procedure in the *Copper Tube Handbook* of the Copper Development Association.

Bending and Forming Pipe may be bent to any radius for which the bend-arc surface will be free of cracks and substantially free of buckles. The use of bends which are creased or corrugated is permitted. Bending may be by any hot or cold method permissible within the radii and material characteristics of the pipe being bent.

Postbend heat treatment may be required for bends in some materials; its necessity depends on the severity of the bend. The details of these requirements are spelled out in the code. Piping components may be formed by any suitable hot or cold pressing, rolling, forging, hammering, spinning, drawing, or other method. Thickness after forming shall not be less than required by design. Special rules cover the forming and pressure design verification of flared laps. Hot bending and hot forming shall be done within a temperature range consistent with material characteristics, end use, or postoperation heat treatment.

The development of fabrication facilities for bending pipe to the radius of commercial butt-welding long-radius elbows and forming flared metallic (Van Stone) laps on pipe are important techniques in reducing welded-piping costs. These techniques save both the cost of the ell or stub end and the welding operation required to attach the fitting to the pipe.

Preheating and Heat Treatment Preheating and postoperation heat treatment are used to avert or relieve the detrimental effects of the high temperature and severe thermal gradients inherent in the welding of metals. In addition, heat treatment may be needed to relieve residual stresses created during the bending or forming of metals. The code provisions shown in Tables 6-47 and 6-48 represent basic practices which are acceptable for most applications of welding, bending, and forming, but they are not necessarily suitable for all service conditions. The specification of more or less stringent preheating and heat-treating requirements is a function of those responsible for the engineering design.

Joining Nonmetallic Pipe Thermoplastic piping may be joined by a qualified hot-gas welding procedure, a qualified solvent-cement procedure, or by a qualified heat-fusion procedure. The general welding and heat-fusion procedures are described in ASTM D-2657 and solvent-cement procedures in ASTM D-2855. Two other techniques, for flared joints and elastomeric-sealed joints, are described in ASTM D-3140 and D-3139, respectively.

In joining reinforced thermosetting pipe it is particularly important that the pipe be cut without chipping or cracking it. It is also important to sand, file, or grind any mold-release agent from the surfaces to be cemented. Joints are built up layer by layer of adhesive-saturated reinforcement by following the manufacturer's recommended procedure. Application of adhesive to the surfaces to be joined and assembly of these surfaces shall produce a continuous bond and provide an adhesive seal to protect the reinforcement from attack by the contents of the pipe. Unfilled or unbonded areas of the joint are considered defects and must be repaired.

Assembly and Erection Flanged-joint faces shall be aligned to the design plane to within 1/16 in/ft (0.5 percent) maximum measured across any diameter, and flange bolt holes shall be aligned to within 3.2-mm (⅛-in) maximum offset. Flanged joints involving flanges with widely differing mechanical properties shall be assembled with extra care, and tightening to a predetermined torque is recommended.

The use of flat washers under bolt heads and nuts is a code requirement when assembling nonmetallic flanges. It is preferred that the bolts extend completely through their nuts; however, a lack of complete thread engagement not exceeding one thread is permitted by the code. In assembling nonmetallic lined joints consideration must be given to the need and means for maintaining electrical continuity when static sparking could occur. The assembly of cast-iron bell-and-spigot piping is covered in AWWA Standard C600.

Screwed joints which are intended to be seal-welded shall be made up without any thread compound.

EXAMINATION, INSPECTION, AND TESTING

Examination and Inspection The code differentiates between examination and inspection. "Examination" applies to quality-control functions performed by personnel of the piping manufacturer, fabricator, or erector. "Inspection" applies to functions performed for the owner by the authorized inspector.

The authorized inspector shall be designated by the owner and shall be the owner, an employee of the owner, an employee of an engineering or scientific organization, or an employee of a recognized insurance or inspection company acting as the owner's agent. The inspector shall not represent or be an employee of the piping erector, the manufacturer, or the fabricator unless the owner is also the erector, the manufacturer, or the fabricator.

The authorized inspector shall have a minimum of 10 years' experience in the design, fabrication, or inspection of industrial pressure piping. Each 20 percent of satisfactory work toward an engineering degree accredited by the Engineers' Council for Professional Development shall be considered equivalent to 1 year's experience, up to 5 years total.

It is the owner's responsibility, exercised through the authorized

TABLE 6-45 Requirements for Low-Temperature Toughness Tests*

Type of Material		Column A: At or above minimum temperature listed in Table 6-37 or Table 6-3		Column B: Below minimum temperatures listed in Table 6-37 or Table 6-3
Listed metallic materials	Ductile iron, malleable iron, Carbon steel, ASTM A36, ASTM A283	1. No additional requirements.	1. No additional requirements.	1. Shall not be used.
		Base metal	Deposited weld metal and heat-affected zone (See Note 1)	
	All other carbon steel, low-intermediate, and high-alloy steels, ferritic steels	2*a*. No additional requirements.	2*b*. When materials are fabricated or assembled by welding, the deposited weld metal and heat-affected zone shall be impact-tested if the design temperature is below −29°C (−20°F) unless conditions conform to Note 2.	2. Except when conditions conform to Note 2, the material shall be heat-treated to control its microstructure by a method appropriate to the material as outlined in the specification applicable to the product form and then impact-tested. (See Note 1). Deposited weld metal and heat-affected zone shall be impact-tested.
	Austenitic stainless steel	3*a*. If (1) the carbon content by analysis is greater than 0.10 percent or (2) the material is not in the solution-heat-treated condition, then impact testing is required for design temperatures below −29°C (−20°F). See Note 2.	3*b*. When materials are fabricated or assembled by welding, the deposited weld metal shall be impact-tested for design temperature below −29°C (−20°F) unless conditions conform to Note 2.	3. The material shall be impact-tested. See Note 2.
	Austenitic ductile iron, ASTM A571	4*a*. No additional requirements.	4*b*. Welding not permitted.	4. The material shall be impact-tested. This material shall not be used at design minimum temperatures lower than −196°C (−320°F). Welding is not permitted.
	Aluminum alloy, copper, copper alloy, nickel, nickel alloy, unalloyed titanium	5*a*. No additional requirements.	5*b*. No additional requirements except that when the composition of the filler metal is outside the range of composition for the base metal, testing shall be in accordance with column B, item 5.	5. Low-temperature tests such as tensile elongation and sharp-notch tensile strength (compared with unnotched tensile strength) shall have been conducted to provide assurance to the designer that the material and the deposited weld metal are suitable at the design minimum temperatures.
Listed nonmetallic materials		6. No additional requirements.	6. No additional requirements.	6. Below the recommended minimum temperatures, the designer shall have test results at or below the lowest expected service temperature which assure that the materials will have adequate toughness and are suitable at the design minimum temperatures.
Unlisted materials		Unlisted materials which conform to a published specification and are of composition, heat treatment, and product form comparable with those of listed materials shall be subject to the same requirements as the listed materials. All other unlisted materials conforming to a published specification shall be qualified as required by the applicable item in col. B.	Unlisted materials which conform to a published specification and are of composition, heat treatment, and product form comparable with those of listed materials shall be subject to the same requirements as the listed materials. All other unlisted materials conforming to a published specification shall be qualified as required by the applicable item in col. B.	Unlisted materials which conform to a published specification and are of composition, heat treatment, and product form comparable with those of listed materials shall be subject to the same requirements as the listed materials. All other unlisted materials conforming to a published specification shall be qualified as required by the applicable item in col. B.

*Extracted from the Chemical Plant and Petroleum Refinery Piping Code, ANSI B31.3—1980, with permission of the publisher, the American Society of Mechanical Engineers, New York.

NOTE: These toughness-test requirements are in addition to tests required by the material specification.

1. Any tests and associated acceptance criteria which are part of the welding-procedure qualification for filler materials and heat-affected zone need not be repeated.

2. Impact testing is not required if the design temperature is below −29°C (−20°F) but at or above −46°C (−50°F) and the maximum operating pressure of the fabricated or assembled components will not exceed 25 percent of the maximum allowable design pressure at ambient temperature and the combined longitudinal stress due to pressure, dead weight, and displacement strain (see Par. 319.2.1) does not exceed 41 MPa (6000 lbf/in^2).

TABLE 6-46 Limitations on Imperfections in Welds*

Imperfection†	When required examination is	Girth and miter-joint butt welds	Longitudinal butt welds‡	Fillet, socket, seal, and reinforcement attachment welds	Welded branch connections and fabricated laps
Cracks or lack of fusion	Any	None permitted	None permitted	None permitted	None permitted
Incomplete penetration	100% radiography	None permitted	None permitted	NA	None permitted
	Visual or random or spot radiography	A	None permitted	NA	A and H
Internal porosity	100% radiography	B	B	NA	B and H
	Random or spot radiography	C	C	NA	C and H
Slag inclusions or elongated defects	100% radiography	D	D	NA	D and H
	Random or spot radiography	E	E	NA	B and H
Undercutting	Any	Lesser of 1/32 in or $\overline{T}w/4$	None permitted	Lesser of 1/32 in or $\overline{T}w/4$	Lesser of 1/32 in or $\overline{T}w/4$
Surface porosity and exposed slag inclusion (3/16-in nominal wall thickness and less)		None permitted	None permitted	None permitted	None permitted
Concave root surface (suck-up)		F	F	NA	F and H
Weld reinforcement		G	G	G	G and H

NOTES:

NA: Not applicable.

A: The lesser of 1/32 in or 0.2 $\overline{T}w$ deep. The total length of such imperfections shall not exceed 1.5 in (38 mm) in any 6 in (150 mm) of weld length. (See Fig. 6-137).

B: An individual pocket of porosity shall not exceed the lesser of $\overline{T}w/3$ or 1/8 in in its greatest dimension. The total area of porosity projected radially through the weld shall not exceed an area equivalent to 3 times the area of a single maximum pocket allowable in any square inch (645 mm^2) of projected weld area.

C: An individual pocket of porosity shall not exceed the lesser of $\overline{T}w/2$ or 1/8 in in its greatest dimension. The total area of porosity projected radially through the weld shall not exceed an area equivalent to 3 times the area of a single maximum pocket allowable to any square inch (645 mm^2) of projected weld area.

D: The developed length of any single slag inclusion or elongated defect shall not exceed $\overline{T}w/3$. The total cumulative developed length of slag inclusions and/or elongated defects shall not exceed $\overline{T}w$ in any 12 $\overline{T}w$ length of weld. The width of a slag inclusion shall not exceed the lesser of 1/16 in or $\overline{T}w/3$.

E: The developed length of any single slag inclusion or elongated defect shall not exceed 2 $\overline{T}w$. The total cumulative developed length of slag inclusions and/or elongated defects shall not exceed 4 $\overline{T}w$ in any 6-in length of weld. The width of a slag inclusion shall not exceed the lesser of 1/8 in or $\overline{T}w/2$.

F: For single-sided welded joints, concavity of the root surface shall not reduce the total thickness of the joint, including reinforcement, to less than the thickness of the thinner of the components being jointed.

G: External weld reinforcement and internal weld protrusion shall be fused with and shall merge smoothly into the component surface. The thickness of external weld reinforcement and internal weld protrusion (when no backing ring is used) shall not exceed the following:

Wall thickness $\overline{T}w$, in	Weld reinforcement or protrusion, in, maximum
1/4 and under	1/16
over 1/4 through 1/2	1/8
over 1/2 through 1	5/32
over 1 (25.4 mm)	3/16

H: These requirements apply only to butt welds.

*Extracted from the Chemical Plant and Petroleum Refinery Piping Code, ANSI B31.3—1980, with permission of the publisher, the American Society of Mechanical Engineers, New York. To convert inches to millimeters, multiply by 25.4.

†See Fig. 6-137 for illustration of the defects.

‡This column applies to welds not made in accordance with a standard listed in Appendix A or Appendix E of the code.

inspector, to verify that all required examinations and testing have been completed and to inspect the piping to the extent necessary to be satisfied that it conforms to all applicable requirements of the code and the engineering design. This verification may include certifications and records pertaining to materials, components, heat treatment, examination and testing, and qualifications of operators and procedures. The authorized inspector may delegate the performance of inspection to a qualified person.

Inspection does not relieve the manufacturer, the fabricator, or the erector of responsibility for providing materials, components, and skill in accordance with requirements of the code and the engineering design, performing all required examinations, and preparing records of examinations and tests for the inspector's use.

Examination Methods The code establishes the types of examinations for evaluating various types of imperfections (see Table 6-49).

Personnel performing examinations other than visual shall be qualified in accordance with applicable portions of SNT TC-1A, *Recommended Practice for Nondestructive Testing Personnel Qualification and Certification*. Procedures shall be qualified as required in Par. T-150, Art. 1, Sec. V of the ASME Code. Limitations on imperfections shall be in accordance with the engineering design but shall at least meet the requirements of the code (see Tables 6-46 and 6-49) for the specific type of examination. Repairs shall be made as applicable.

Visual Examination This consists of observation of the portion of components, joints, and other piping elements that are or can be exposed to view before, during, or after manufacture, fabrication, assembly, erection, inspection, or testing. The examination includes verification of code and engineering design requirements for materials and components, dimensions, joint preparation, alignment, welding or joining, supports, assembly, and erection.

Visual examination shall be performed in accordance with Art. 9, Sec. V of the ASME Code.

Magnetic-Particle Examination This examination shall be performed in accordance with Art. 7, Sec. V of the ASME Code.

Liquid-Penetrant Examination This examination shall be performed in accordance with Art. 6, Sec. V of the ASME Code.

Radiographic Examination The following definitions apply to radiography required by the code or by the engineering design:

1. "Random radiography" applies only to girth butt welds. It is radiographic examination of the complete circumference of a specified percentage of the girth butt welds in a designated lot of piping.

2. "100 percent radiography" applies only to girth butt welds

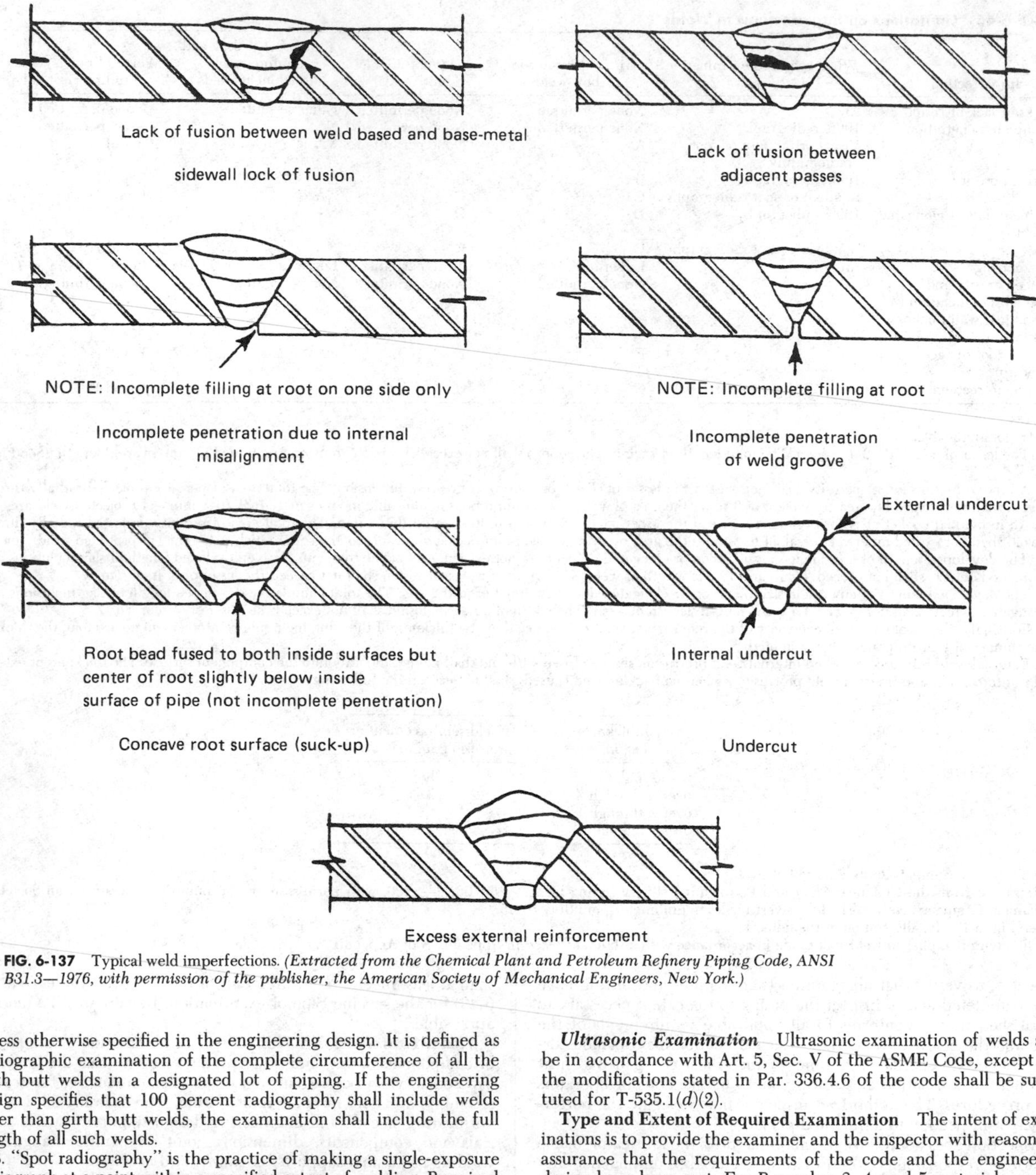

FIG. 6-137 Typical weld imperfections. *(Extracted from the Chemical Plant and Petroleum Refinery Piping Code, ANSI B31.3—1976, with permission of the publisher, the American Society of Mechanical Engineers, New York.)*

unless otherwise specified in the engineering design. It is defined as radiographic examination of the complete circumference of all the girth butt welds in a designated lot of piping. If the engineering design specifies that 100 percent radiography shall include welds other than girth butt welds, the examination shall include the full length of all such welds.

3. "Spot radiography" is the practice of making a single-exposure radiograph at a point within a specified extent of welding. Required coverage for a single spot radiograph is as follows:

a. For longitudinal welds, at least 150 mm (6 in) of weld length.

b. For girth, miter, and branch welds in piping 2½ in NPS and smaller, a single elliptical exposure which encompasses the entire weld circumference, and in piping larger than 2½ in NPS, at least 25 percent of the inside circumference or 150 mm (6 in), whichever is less.

Radiography of components other than castings and of welds shall be in accordance with Art. 2, Sec. V of the ASME Code. Limitations on imperfections in components other than castings and welds shall be as stated in Table 6-46 for the degree of radiography involved.

Ultrasonic Examination Ultrasonic examination of welds shall be in accordance with Art. 5, Sec. V of the ASME Code, except that the modifications stated in Par. 336.4.6 of the code shall be substituted for T-535.1(*d*)(2).

Type and Extent of Required Examination The intent of examinations is to provide the examiner and the inspector with reasonable assurance that the requirements of the code and the engineering design have been met. For P-number 3, 4, and 5 materials, examination shall be performed after any heat treatment has been completed.

Examination Normally Required Piping not covered by Category D fluid service or severe cyclic conditions shall be examined as follows or to any greater extent specified in the engineering design.

1. *Visual examination*

a. Sufficient materials and components, selected at random, to satisfy the examiner that they conform to specifications and are free from damage.

b. At least 5 percent of fabrication. For welds, each welder's or welding operator's work shall be represented, though not necessarily

TABLE 6-47 Preheat Temperatures*

Base-metal P number†	Weld-metal analysis A number‡	Base-material group	Nominal wall thickness		Minimum specified tensile strength, base metal		Minimum temperature			
							Required		Recommended	
			mm	in	MPa	kip/in^2	°C	°F	°C	°F
1	1	Carbon steel	< 25.4	< 1	≤ 490	≤ 71			10	50
			≥ 25.4	≥ 1	All	All			80	175
			All	All	> 490	> 71			80	175
3	2,11	Alloy steels Cr ½% maximum	< 12.7	< ½	≤ 490	≤ 71			10	50
			≥ 12.7	≥ ½	All	All			80	175
			All	All	≥ 490	> 71			80	175
4	3	Alloy steels Cr > ½% to 2%	All	All	All	All	150	300		
5	4,5	Alloy steels Cr 2¼% to 10%	All	All	All	All	175	350		
6	6	High-alloy steels: martensitic	All	All	All	All			150§	300§
7	7	High-alloy steels: ferritic	All	All	All	All			10	50
8	8,9	High-alloy steels: austenitic	All	All	All	All			10	50
9A, 9B, 9C	10	Nickel alloy steels	All	All	All	All			95	200
10A		Mn-V steel	All	All	All	All			80	175
10B		Cr-V steel	All	All	All	All			150	300
11A Group 1		9% Ni steel	All	All	All	All			10	50
P21–P52			All	All	All	All			10	50

*Extracted from the Chemical Plant and Petroleum Refinery Piping Code, ANSI B31.3—1980, with permission of the publisher, the American Society of Mechanical Engineers, New York.
†P number from ASME Code, Sec. IX, Table QW-422.
‡A number from ASME Code, Sec. IX, Table QW-442.
§Maximum interpass temperature 315°C (600°F).

each type of weld for each welder or welding operator. Limitations on imperfections shall be as stated in Table 6-46.

c. 100 percent of fabrication for longitudinal welds other than those in components made to material specifications recognized in the code. Limitations on imperfections are those of Table 6-46.

d. Random examination of the assembly of threaded, bolted, and other joints to satisfy the examiner that they conform to requirements.

e. Random examination during erection of piping, including checking of alignments, supports, and cold spring.

f. Examination of erected piping for evidence of damage that would require repair or replacement and for other evident deviations from the intent of the design.

2. *Other examination.* When piping is intended for service at temperatures above 186°C (366°F) or gauge pressures above 1.0 MPa (150 lbf/in^2) as designated in the engineering design, a minimum of 5 percent of circumferential butt welds shall be examined fully by random radiography or ultrasonic examination. The welds to be examined shall be selected to ensure that the work product of each individual welder or welding operator doing the production welding is included. They shall also be selected to maximize coverage of intersections with longitudinal joints. A minimum of 38 mm (1½ in) of the longitudinal welds shall be examined. In-process examination may be substituted for all or part of the radiographic or ultrasonic examination on a weld-for-weld basis if specified in the engineering design.

3. *In-process examination.* In-process examination comprises visual examination of the following as applicable:

a. Joint preparation and cleanliness
b. Preheating
c. Fit-up and internal alignment prior to welding
d. Weld position, electrode, and other variables specified by the welding procedure
e. Condition of the root pass after cleaning (external and, where accessible, internal), aided by liquid-penetrant or magnetic-particle examination when specified in the engineering design
f. Slag removal and weld condition between passes
g. Appearance of the finished weld

4. *Certification and records for components and materials.* The examiner shall be assured, by examination of certification, records, or other evidence, that the materials and components are of the specified grades and that they have received required heat treatment, examination, and testing. The examiner shall provide the inspector with a certification that all quality-control requirements of the code and of the engineering design have been met.

Category D Fluid-Service Piping This piping, as designated in the engineering design, shall be visually examined to the extent necessary to satisfy the examiner that components, materials, and work conform to the requirements of the code and the engineering design.

Piping Subject to Severe Cyclic Conditions Piping for other than Category D fluids to be used under severe cyclic conditions shall be examined as follows or to any greater extent specified in the engineering design.

1. *Visual examination*

a. All fabrication threaded, bolted, and other joints shall be examined.

b. All piping erection shall be examined to verify dimensions and alignment. Supports, guides, and points of cold spring shall be checked to assure that movement of the piping under all conditions of start-up, operation, and shutdown will be accommodated without binding or constraint.

2. *Other examination.* All circumferential butt welds and all fabricated branch connection welds comparable to Fig. 6-89 shall be examined by 100 percent radiography or (if specified in the engineering design) by ultrasonic examination. Limitations on imperfections are as specified in Table 6-46. The code also requires that a welding procedure which promotes a smooth, fully penetrated internal surface be employed and that the external surface of the completed weld be free of undercutting and finished to within 500 AARH. Socket welds and nonradiographed branch-connection welds shall be examined by magnetic-particle or liquid-penetrant methods.

Impact Testing Materials conforming to ASTM specifications listed in the code may generally be used at temperatures down to the lowest temperature listed for that material in the stress table without

TABLE 6-48 Requirements for Heat Treatment*

Base-metal P number†	Weld-metal analysis A number‡	Material group	Nominal wall thickness		Minimum specified tensile strength, base metal		Metal temperature range		h in, nominal wall§	Minimum time, h	Brinell hardness, maximum
			mm	in	MPa	kip/in²	°C	°F			
1	1	Carbon steel	≤ 19	≤ ¾	All	All	None	None	1	1	
			> 19	> ¾			595–650	1100–1200			
3	2, 11	Alloy steels	≤ 19	≤ ¾	≤ 490	≤ 71	None	None			
		Cr ½% max	> 19	> ¾	All	All	595–720	1100–1325	1	1	225
			All	All	> 490	> 71	595–720	1100–1325	1	1	225
4	3	Alloy steels	≤ 12.7	≤ ½	≤ 490	≤ 71	None	None			
		Cr > ½% to 2%	> 12.7	> ½	All	All	705–745	1300–1375	1	2	225
			All	All	> 490	> 71	705–745	1300–1375	1	2	225
5	4, 5	Alloy steels Cr 2¼% to 10%	≤ ½ and ≤ 3% Cr and ≤ 0.15% C		All	All	None	None			
			> ½ or > 3% Cr or > 0.15% C		All	All	705–760	1300–1400	1	2	241
		High-alloy steels:									
6	6	martensitic	All	All	All	All	730–790	1350–1450	1	2	241
		A240, Gr 429	All	All	All	All	620–660	1150–1225	1	2	241
		High-alloy steels:									
7	7	ferritic	All	All	All	All	None	None			
		High-alloy steels:									
8	8, 9	austenitic	All	All	All	All	None	None			
9A	10	Nickel alloy steels	≤ 19	≤ ¾	All	All	None	None			
9B			> 19	> ¾	All	All	595–635	1100–1175	½	1	
10A		Mn-V steel	≤ 19	≤ ¾	All	All	None	None	. . .	. . .	. . .
			> 19	> ¾	All	All	595–705	1100–1300	1	1	225
			All	All	> 490	> 71	595–705	1100–1300	1	1	225
10B		Cr-V steel	≤ 12.7	≤ ½	≤ 490	≤ 71	None	None	. . .	. . .	. . .
			> 12.7	> ½	All	All	595–730	1100–1350	1	1	225
			All	All	> 490	> 71	595–730	1100–1350	1	1	225
11A, Group 1		9% Ni steel	≤ 51	≤ 2	All	All	None	None			
			> 51	> 2	All	All	550–585	1025–1085	1	1	
								[Cooling rate > 150°C (300°F)/h to 315°C (600°F)]			

*Extracted from the Chemical Plant and Petroleum Refinery Piping Code, ANSI B31.3—1980, with permission of the publisher, the American Society of Mechanical Engineers, New York.

†P number from ASME Code, Sec. IX, Table QW-422, Special P numbers (SP-1, SP-2, SP-3) require special consideration in procedure qualification. The required thermal treatment shall be established by the engineering design and demonstrated by the procedure qualification.

‡A number from ASME Code, Sec. IX, Table QW-422.

§For SI equivalent, h/mm, divide h/in by 25.

additional testing. When welding or other operations are performed on these materials, additional low-temperature toughness tests may be required. The code requirements are listed in Table 6-45.

Pressure Testing Prior to initial operation, installed piping shall be pressure-tested to assure tightness except as permitted for Category D fluid service described later. The pressure test shall be maintained for a sufficient time to determine the presence of any leaks but not less than 10 min.

If repairs or additions are made following the pressure tests, the affected piping shall be retested except that, in the case of minor repairs or additions, the owner may waive retest requirements when precautionary measures are taken to assure sound construction.

When pressure tests are conducted at metal temperatures near the ductile-to-brittle transition temperature of the material, the possibility of brittle fracture shall be considered.

The test shall be hydrostatic, using water, with the following exceptions. If there is a possibility of damage due to freezing or if the operating fluid or piping material would be adversely affected by water, any other suitable liquid may be used. If a flammable liquid is used, its flash point shall not be less than 50°C (120°F), and consideration shall be given to the test environment.

The hydrostatic-test pressure at any point in the system shall be as follows:

1. Not less than 1½ times the design pressure.
2. For a design temperature above the test temperature, the min-

TABLE 6-49 Types of Examination for Evaluating Imperfections*

Type of imperfection	Type of examination: Visual	Liquid-penetrant or magnetic-particle	Ultrasonic or radiographic: Random	100%
Crack	X	X	X	X
Incomplete penetration	X		X	X
Lack of fusion	X		X	X
Weld undercutting	X			
Weld reinforcement	X			
Internal porosity			X	X
External porosity	X			
Internal slag inclusions			X	X
External slag inclusions	X			
Concave root surface	X		X	X

*Extracted from the Chemical Plant and Petroleum Refinery Piping Code, ANSI B31.3—1980, with permission of the publisher, the American Society of Mechanical Engineers, New York. For limitations on imperfections in welds see Table 6-46.

imum test pressure shall be as calculated by the following formula:

$$P_T = 1.5\ PS_T/S \qquad (6\text{-}52)$$

where PT = test hydrostatic gauge pressure, MPa (lbf/in^2)
P = internal design pressure, MPa (lbf/in^2)
S_T = allowable stress at test temperature, MPa (lbf/in^2)
S = allowable stress at design temperature, MPa (lbf/in^2)

If the test pressure as so defined would produce a stress in excess of the yield strength at test temperature, the test pressure may be reduced to the maximum pressure that will not exceed the yield strength at test temperature.

A preliminary air test at not more than 0.17-MPa (25-lbf/in^2) gauge pressure may be made prior to hydrostatic test in order to locate major leaks.

If hydrostatic testing is not considered practicable by the owner, a pneumatic test in accordance with the following procedure may be substituted, using air or another nonflammable gas.

If the piping is tested pneumatically, the test pressure shall be 110 percent of the design pressure. Pneumatic testing involves a hazard owing to the possible release of energy stored in compressed gas. Therefore, particular care must be taken to minimize the chance of brittle failure of metals and thermoplastics. The test temperature is important in this regard and must be considered when material is chosen in the original design. Any pneumatic test shall include a preliminary check at not more than 0.17-MPa (25-lbf/in^2) gauge pressure. The pressure shall be increased gradually in steps providing sufficient time to allow the piping to equalize strains during test and to check for leaks. If the test liquid in the system is subject to thermal expansion, precautions shall be taken to avoid excessive pressure.

At the owner's option, a piping system used only for Category D fluid service as defined in the subsection "Classification of Fluid Service" may be tested at the normal operating conditions of the system during or prior to initial operation by examining for leaks at every joint not previously tested. A preliminary check shall be made at not more than 0.17-MPa (25-lbf/in^2) gauge pressure when the contained fluid is a gas or a vapor. The pressure shall be increased gradually in steps providing sufficient time to allow the piping to equalize strains during testing and to check for leaks.

Tests alternative to those required by these provisions may be applied under certain conditions described in the code.

Piping required to have a sensitive leak test shall be tested by the gas- and bubble-formation testing method specified in Art. 10, Sec. V of the ASME Code or by another method demonstrated to have equal or greater sensitivity. The sensitivity of the test shall be at least (100 Pa·mL)/s [(10^3 atm·mL)/s] under test conditions. If a hydrostatic pressure test is used, it shall be carried out after the sensitive leak test.

Records shall be kept of each piping installation during the testing.

COMPARISON OF PIPING-SYSTEM COSTS

Piping may represent as much as 25 percent of the cost of a chemical-process plant. The installed cost of piping systems varies widely with the materials of construction and the complexity of the system. A study of piping costs shows that the most economical choice of material for a simple straight piping run may not be the most economical for a complex installation made up of many short runs involving numerous fittings and valves. The economics also depends heavily on the pipe size and fabrication techniques employed. Fabrication methods such as bending to standard long-radius-elbow dimensions and machine-flaring lap joints have a large effect on the cost of fabricating pipe from ductile materials suited to these techniques. Cost reductions of as high as 35 percent are quoted by some custom fabricators utilizing advanced techniques.

Figure 6-138 is based on data extracted from a comparison of the installed cost of piping systems of various materials published by the Dow Chemical Co. The chart shows the relative cost ratios for systems of various materials based on two installations, one consisting of 152 m (500 ft) of 2-in pipe in a complex piping arrangement and the other of 305 m (1000 ft) of 2-in pipe in a straight-run piping arrangement. Figure 6-138 is based on field-fabrication construction techniques using welding stubs, the method commonly used by contractors. A considerably different ranking would result from using other construction methods such as machine-formed lap joints and bends in place of welding elbows. Piping-cost experience shows that it is difficult to generalize and reflect accurate piping-cost comparisons. For an accurate comparison the cost for each type of material must be estimated individually on the basis of the actual fabrication and installation methods that will be used and the conditions anticipated for the proposed installation.

STORAGE AND PROCESS VESSELS

STORAGE OF LIQUIDS

Atmospheric Tanks The term "atmospheric tank" as used here applies to any tank that is designed to be used within plus or minus several hundred pascals (a few pounds per square foot) of atmospheric pressure. It may be either open to the atmosphere or enclosed. Minimum cost is usually obtained with a vertical cylindrical shape and a relatively flat bottom at ground level.

American Petroleum Institute (API) The institute has developed a series of atmospheric tank standards and specifications. Some of these are:

API Specification 12B, Bolted Production Tanks
API Specification 12D, Large Welded Production Tanks
API Specification 12F, Small Welded Production Tanks
API Standard 650, Steel Tanks for Oil Storage

American Water Works Association (AWWA) The association has many standards dealing with water handling and storage. A list of its publications is given in the *AWWA Handbook* (annually). AWWA D100, Standard for Steel Tanks—Standpipes, Reservoirs, and Elevated Tanks for Water Storage, contains rules for design and fabrication.

Although AWWA tanks are intended for water, they could be used for the storage of other liquids.

Underwriters Laboratories Inc. has published the following tank standards:

UL 58, Steel Underground Tanks for Flammable and Combustible Liquids
UL 142, Steel Aboveground Tanks for Flammable and Combustible Liquids

UL 58 covers horizontal steel tanks up to 190 m^3 (50,000 gal), with a maximum diameter of 3.66 m (12 ft), and a maximum length of six diameters. Thickness and a number of design and fabrication details are given. UL 142 covers horizontal steel tanks up to 190 m^3 (50,000 gal) (like UL 58), and vertical tanks up to 10.7-m (35-ft) height. Thickness and other details are given. The maximum diameter for a vertical tank is not specified.

The Underwriters Standards overlap API, but include tanks that are too small for API Standards. Underwriters Standards are, however, not as detailed as API and therefore put more responsibility on the designer. They do not specify grades of steel other than requiring weldability. Designers should also place their own limits on the diameter (or thickness) of vertical tanks. They can obtain guidance from API.

Posttensioned Concrete This material is frequently used for tanks to about 57,000 m^3 (15 $\times$ 10^6 gal), usually containing water. Their design is treated in detail by Creasy (*Prestressed Concrete Cylindrical Tanks*, Wiley, New York, 1961). For the most economical design of large open tanks at ground levels, he recommends limiting vertical height to 6 m (20 ft). Seepage can be a problem if unlined concrete is used with some liquids (e.g., gasoline).

Elevated Tanks These can supply a large flow when required,

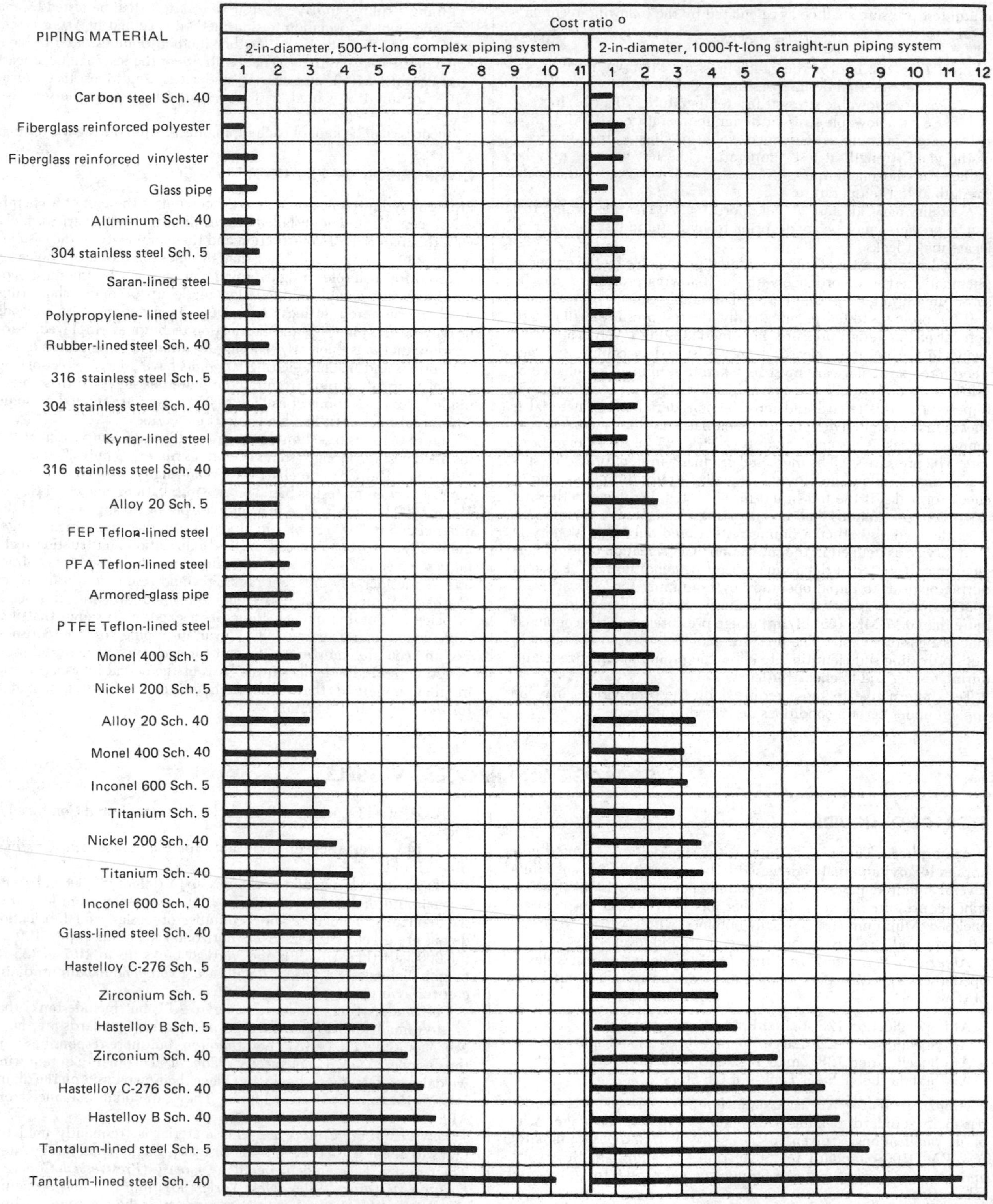

FIG. 6-138 Cost rankings and cost ratios for various piping materials. This figure is based on field-fabrication construction techniques using welding stubs, as this is the method most often employed by contractors. A considerably different ranking would result from using other construction methods, such as machined-formed lap joints, for the alloy pipe. °Cost ratio = (cost of listed item) ÷ (cost of Schedule 40 carbon steel piping system, field-fabricated by using welding stubs). (*Extracted with permission from* Installed Cost of Corrosion Resistant Piping, *copyright 1977, Dow Chemical Co.*)

but pump capacities need be only for average flow. Thus, they may save on pump and piping investment. They also provide flow after pump failure, an important consideration for fire systems.

Open Tanks These may be used to store materials that will not be harmed by water, weather, or atmospheric pollution. Otherwise, a roof, either fixed or floating, is required. **Fixed roofs** are usually either domed or coned. Large tanks have coned roofs with intermediate supports. Since negligible pressure is involved, snow and wind are the principal design loads. Local building codes often give required values.

Fixed-roof atmospheric tanks require **vents** to prevent pressure changes which would otherwise result from temperature changes and withdrawal or addition of liquid. API Standard 2000, Venting Atmospheric and Low Pressure Storage Tanks, gives practical rules for vent design. The principles of this standard can be applied to fluids other than petroleum products. Excessive losses of volatile liquids, particularly those with flash points below 38°C (100°F), may result from the use of open vents on fixed-roof tanks. Sometimes vents are manifolded and led to a vent tank, or the vapor may be extracted by a recovery system.

An effective way of preventing vent loss is to use one of the many types of **variable-volume tanks.** These are built under API Standard 650. They may have floating roofs of the double-deck or the single-deck type. There are lifter-roof types in which the roof either has a skirt moving up and down in an annular liquid seal or is connected to the tank shell by a flexible membrane. A fabric expansion chamber housed in a compartment on top of the tank roof also permits variation in volume.

Floating Roofs These must have a seal between the roof and the tank shell. If not protected by a fixed roof, they must have drains for the removal of water, and the tank shell must have a "wind girder" to avoid distortion. An industry has developed to retrofit existing tanks with floating roofs. Much detail on the various types of tank roofs is given in manufacturers' literature. Figure 6-139 shows types.

Pressure Tanks Vertical cylindrical tanks constructed with domed or coned roofs, which operate at pressures above several hundred pascals (a few pounds per square foot) but which are still relatively close to atmospheric pressure, can be built according to API Standard 650. The pressure force acting against the roof is transmitted to the shell, which may have sufficient weight to resist it. If not, the uplift will act on the tank bottom. The strength of the bottom, however, is limited, and if it is not sufficient, an anchor ring or a heavy foundation must be used. In the larger sizes uplift forces limit this style of tank to very low pressures.

As the size or the pressure goes up, curvature on all surfaces becomes necessary. Tanks in this category, up to and including a pressure of 103.4 kPa (15 lbf/in^2), can be built according to API Standard 620. Shapes used are spheres, ellipsoids, toroidal structures, and circular cylinders with torispherical, ellipsoidal, or hemispherical heads. The ASME Pressure Vessel Code (Sec. VIII of the ASME Boiler and Pressure Vessel Code), although not required below 103.4 kPa (15 lbf/in^2), is also useful for designing such tanks.

Tanks that could be subjected to vacuum should be provided with vacuum-breaking valves or be designed for vacuum (external pressure). The ASME Pressure Vessel Code contains design procedures.

Calculation of Tank Volume A tank may be a single geometrical element, such as a cylinder, a sphere, or an ellipsoid. It may also have a compound form, such as a cylinder with hemispherical ends or a combination of a toroid and a sphere. To determine the volume, each geometrical element usually must be calculated separately. Calculations for a full tank are usually simple, but calculations for partially filled tanks may be complicated.

To calculate the volume of a **partially filled horizontal cylinder** refer to Fig. 6-140. Calculate the angle α in degrees. Any units of length can be used, but they must be the same for H, R, and L. The liquid volume

$$V = LR^2\left(\frac{\alpha}{57.30} - \sin\alpha\cos\alpha\right) \tag{6-53}$$

This formula may be used for any depth of liquid between zero and the full tank, provided the algebraic signs are observed. If H is greater than R, $\sin\alpha\cos\alpha$ will be negative and thus will add numerically to $\alpha/57.30$. Table 6-50 gives liquid volume, for a partially filled horizontal cylinder, as a fraction of the total volume, for the dimensionless ratio H/D or $H/2R$.

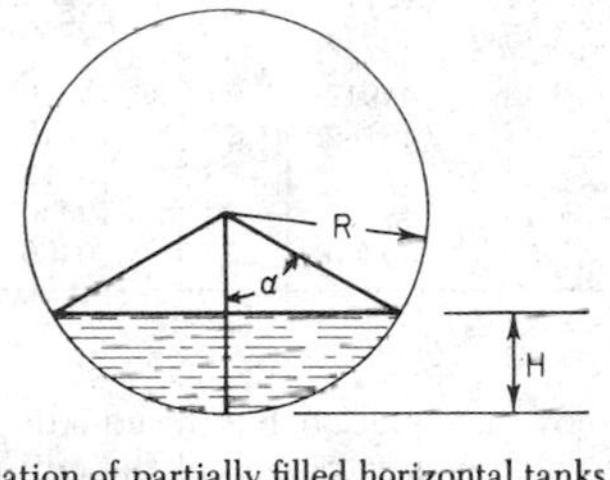

FIG. 6-140 Calculation of partially filled horizontal tanks. H = depth of liquid; R = radius; D = diameter; L = length; α = half of the included angle; and $\cos\alpha = 1 - H/R = 1 - 2H/D$.

The **volumes of heads** must be calculated separately and added to the volume of the cylindrical portion of the tank. The four types of heads most frequently used are the standard dished head,* torispherical or ASME head, ellipsoidal head, and hemispherical head. Dimensions and volumes for all four of these types are given in *Lukens Spun Heads*, Lukens Inc., Coatesville, Pennsylvania. Approximate volumes can also be calculated by the formulas in Table 6-51. Consistent units must be used in these formulas.

A partially filled horizontal tank requires the determination of the partial volume of the heads. The Lukens catalog gives approximate volumes for partially filled (axis horizontal) standard ASME and ellipsoidal heads. A formula for **partially filled heads**, by Doolittle [*Ind. Eng. Chem.*, **21**, 322–323 (1928)], is

$$V = 0.215\,H\,(3R - H) \tag{6-54}$$

where in consistent units V = volume, R = radius, and H = depth of liquid. Doolittle made some simplifying assumptions which affect

*The standard dished head does not comply with the ASME Pressure Vessel Code.

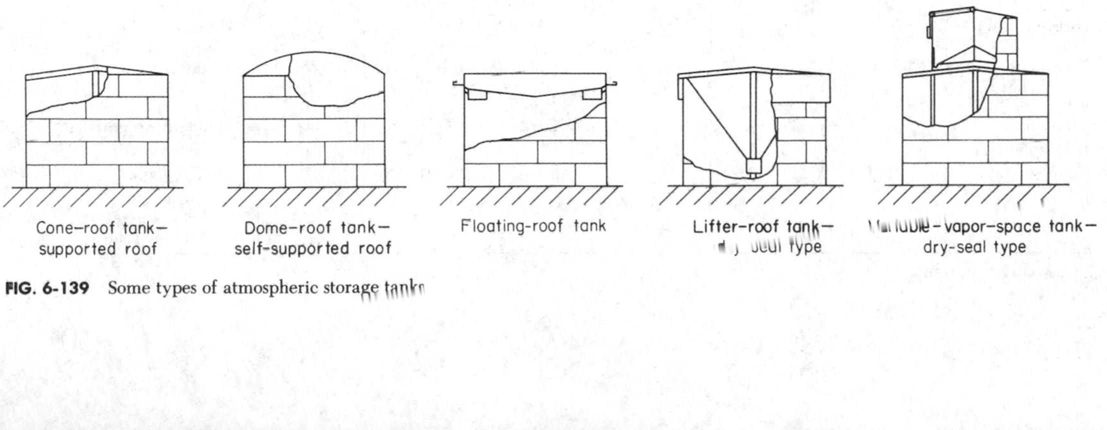

FIG. 6-139 Some types of atmospheric storage tanks

TABLE 6-50 Volume of Partially Filled Horizontal Cylinders

H/D	Fraction of volume	H/D	Fraction of volume	H/D	Fraction of volume	H/D	Fraction of volume
0.01	0.00169	0.26	0.20660	0.51	0.51273	0.76	0.81545
.02	.00477	.27	.21784	.52	.52546	.77	.82625
.03	.00874	.28	.22921	.53	.53818	.78	.83688
.04	.01342	.29	.24070	.54	.55088	.79	.84734
.05	.01869	.30	.25231	.55	.56356	.80	.85762
.06	.02450	.31	.26348	.56	.57621	.81	.86771
.07	.03077	.32	.27587	.57	.58884	.82	.87760
.08	.03748	.33	.28779	.58	.60142	.83	.88727
.09	.04458	.34	.29981	.59	.61397	.84	.89673
.10	.05204	.35	.31192	.60	.62647	.85	.90594
.11	.05985	.36	.32410	.61	.63892	.86	.91491
.12	.06797	.37	.33636	.62	.65131	.87	.92361
.13	.07639	.38	.34869	.63	.66364	.88	.93203
.14	.08509	.39	.36108	.64	.67590	.89	.94015
.15	.09406	.40	.37353	.65	.68808	.90	.94796
.16	.10327	.41	.38603	.66	.70019	.91	.95542
.17	.11273	.42	.39858	.67	.71221	.92	.96252
.18	.12240	.43	.41116	.68	.72413	.93	.96923
.19	.13229	.44	.42379	.69	.73652	.94	.97550
.20	.14238	.45	.43644	.70	.74769	.95	.98131
.21	.15266	.46	.44912	.71	.75930	.96	.98658
.22	.16312	.47	.46182	.72	.77079	.97	.99126
.23	.17375	.48	.47454	.73	.78216	.98	.99523
.24	.18455	.49	.48727	.74	.79340	.99	.99831
.25	.19550	.50	.50000	.75	.80450	1.00	1.00000

the volume given by the equation, but the equation is satisfactory for determining the volume as a fraction of the entire head. This fraction, calculated by Doolittle's formula, is given in Table 6-52 as a function of H/D_i (H is the depth of liquid, and D_i is the inside diameter). Table 6-52 can be used for standard dished, torispherical, ellipsoidal, and hemispherical heads with an error of less than 2 percent of the volume of the entire head. The error is zero when $H/D_i = 0$, 0.5, and 1.0. Table 6-52 cannot be used for conical heads.

When a tank volume cannot be calculated or when greater precision is required, **calibration** may be necessary. This is done by draining (or filling) the tank and measuring the volume of liquid. The measurement may be made by weighing, by a calibrated fluid meter, or by repeatedly filling small measuring tanks which have been calibrated by weight.

TABLE 6-52 Volume of Partially Filled Heads on Horizontal Tanks*

H/D_i	Fraction of volume	H/D_i	Fraction of volume	H/D_i	Fraction of volume	H/D_i	Fraction of volume
0.02	0.0012	0.28	0.1913	0.52	0.530	0.78	0.8761
.04	.0047	.30	.216	.54	.560	.80	.8960
.06	.0104	.32	.242	.56	.590	.82	.9145
.08	.0182	.34	.268	.58	.619	.84	.9314
.10	.0280	.36	.295	.60	.648	.86	.9467
.12	.0397	.38	.323	.62	.677	.88	.9603
.14	.0533	.40	.352	.64	.705	.90	.9720
.16	.0686	.42	.381	.66	.732	.92	.9818
.18	.0855	.44	.410	.68	.758	.94	.9896
.20	.1040	.46	.440	.70	.784	.96	.9953
.22	.1239	.48	.470	.72	.8087	.98	.9988
.24	.1451	.50	.500	.74	.8324	1.00	1.0000
.26	1676			.76	.8549		

*Based on Eq. (6-54).

Container Materials and Safety Storage tanks are made of almost any structural material. Steel and reinforced concrete are most widely used. Plastics and glass-reinforced plastics are used for tanks up to about 230 m^3 (60,000 gal). Resistance to corrosion, light weight, and lower cost are their advantages. Plastic and glass coatings are also applied to steel tanks. Aluminum and other nonferrous metals are used when their special properties are required. When expensive metals such as tantalum are required, they may be applied as tank linings or as clad metals.

Some grades of steel listed by API and AWWA Standards are of lower quality than is customarily used for pressure vessels. The stresses allowed by these standards are also higher than those allowed by the ASME Pressure Vessel Code. Small tanks containing nontoxic substances are not particularly hazardous and can tolerate a reduced factor of safety. Tanks containing highly toxic substances and very large tanks containing any substance can be hazardous. The designer must consider the magnitude of the hazard. The possibility of brittle behavior of ferrous metal should be taken into account in specifying materials (see subsection "Safety in Design").

Volume 1 of National Fire Codes (National Fire Protection Association, Quincy, Massachusetts) contains recommendations (Code 30) for venting, drainage, and dike construction of tanks for **flammable liquids.**

Container Insulation Tanks containing materials above atmospheric temperature may require insulation to reduce loss of heat.

TABLE 6-51 Volumes of Heads*

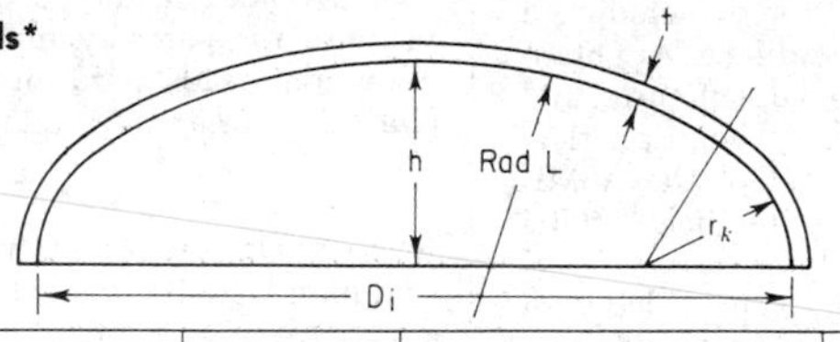

Type of head	Knuckle radius r_k	h	L	Volume	% Error	Remarks
Standard dished	Approx. $3t$	......	Approx. D_i	Approx. $0.050D_i^3 + 1.65tD_i^2$	±10	h varies with t
Torispherical or A.S.M.E.	$0.06L$	......	D_i	$0.0809D_i^3$	±0.1	r_k must be the larger of $0.06L$ and $3t$
Torispherical or A.S.M.E.	$3t$	......	D_i	Approx. $0.513hD_i^2$	±8	
Ellipsoidal	......	......	......	$\pi D_i^2 h/6$	0	
Ellipsoidal	......	$D_i/4$	......	$\pi D_i^3/24$	0	Standard proportions
Hemispherical	......	$D_i/2$	$D_i/2$	$\pi D_i^3/12$	0	
Conical	......	......	......	$\pi h(D_i^2 + D_i d + d^2)/12$	0	Truncated cone h = height d = diameter at small end

*Use consistent units.

Almost any of the commonly used insulating materials can be employed. Calcium silicate, glass fiber, mineral wool, cellular glass, and plastic foams are among those used. Tanks exposed to weather must have jackets or protective coatings, usually asphalt, to keep water out of the insulation.

Tanks with contents at lower than atmospheric temperature may require insulation to minimize heat absorption. The insulation must have a vapor barrier at the outside to prevent condensation of atmospheric moisture from reducing its effectiveness. An insulation not damaged by moisture is preferable. The insulation techniques presently used for refrigerated systems can be applied (see subsection "Low-Temperature and Cryogenic Storage").

Tank Supports Large vertical atmospheric steel tanks may be built on a base of about 150 cm (6 in) of sand, gravel, or crushed stone if the subsoil has adequate bearing strength. It can be level or slightly coned, depending on the shape of the tank bottom. The porous base provides drainage in case of leaks. A few feet beyond the tank perimeter the surface should drop about 1 m (3 ft) to assure proper drainage of the subsoil. API Standard 650, Appendix B, and API Standard 620, Appendix C, give recommendations for tank foundations.

The bearing pressure of the tank and contents must not exceed the **bearing strength** of the soil. Local building codes usually specify allowable soil loading. Some approximate bearing values are:

	kPa	Tons/ft^2
Soft clay (can be crumbled between fingers)	100	1
Dry fine sand	200	2
Dry fine sand with clay	300	3
Coarse sand	300	3
Dry hard clay (requires a pick to dig it)	350	3.5
Gravel	400	4
Rock	1000–4000	10–40

For high, heavy tanks, a foundation ring may be needed. Prestressed concrete tanks are sufficiently heavy to require foundation rings. Foundations must extend below the frost line. Some tanks that are not flat-bottomed may also be supported by soil if it is suitably graded and drained. When soil does not have adequate bearing strength, it may be excavated and backfilled with a suitable soil, or piles capped with a concrete mat may be required.

Spheres, spheroids, and toroids use steel or concrete saddles or are supported by columns. Some may rest directly on soil. Horizontal cylindrical tanks should have two rather than multiple saddles to avoid indeterminate load distribution. Small horizontal tanks are sometimes supported by legs. Most tanks must be designed to resist the reactions of the saddles or legs, and they may require reinforcing. Neglect of this can cause collapse. Tanks without stiffeners usually need to make contact with the saddles on at least 2.1 rad (120°) of their circumference. An elevated steel tank may have either a circle of steel columns or a large central steel standpipe. Concrete tanks usually have concrete columns. Tanks are often supported by buildings.

Pond and Underground Storage Low-cost liquid materials, if they will not be damaged by rain or atmospheric pollution, may be stored in **ponds.** A pond may be excavated or formed by damming a ravine. To prevent loss by seepage, the soil which will be submerged may require treatment to make it sufficiently impervious. This can also be accomplished by lining the pond with concrete, plastic film, or some other barrier. Prevention of seepage is especially necessary if the pond contains material that could contaminate present or future water supplies.

Underground Storage Investment in both storage facilities and land can often be reduced by underground storage. Porous media between impervious rocks are also used. Cavities can be formed in salt domes and beds by dissolving the salt and pumping it out. Geological formations suitable for some of these methods can be found in numerous locations. The most extensive application has been the storage of petroleum products, both liquid and gaseous, in the southwestern part of the United States. Chemicals have been handled in this way. Information on some installations is given in articles by Billue, Haight and Bernard, and Nixon [*Pet. Refiner,* **33,** 108–116 (1954)]. Another useful reference is *Relationships between Selected Physical Parameters and Cost Responses for the Deep-Well Disposal of Aqueous Industrial Wastes,* Technical Report to the U.S. Public Health Service, EHE 07-6801, CRWR28, by the Center for Research in Water Resources, University of Texas, Austin, August 1968. It contains an extensive bibliography.

Water is also stored underground when suitable formations are available. When an excess of surface water is available part of the time, the excess is treated, if required, and pumped into the ground to be retrieved when needed. Sometimes pumping is unnecessary, and it will seep into the ground.

Underground chambers are also constructed in frozen earth (see subsection "Low-Temperature and Cryogenic Storage"). Underground tunnel or tank storage is often the most practical way of storing hazardous or radioactive materials. A cover of 30 m (100 ft) of rock or dense earth can exert a pressure of about 690 kPa (100 lbf/in^2).

STORAGE OF GASES

Gas Holders Gas is sometimes stored in expandable gas holders of either the liquid-seal or dry-seal type. The liquid-seal holder is a familiar sight. It has a cylindrical container, closed at the top, and varies its volume by moving it up and down in an annular water-filled seal tank. The seal tank may be staged in several lifts (as many as five). Seal tanks have been built in sizes up to 280,000 m^3 (10 × 10^6 ft^3). The dry-seal holder has a rigid top attached to the sidewalls by a flexible fabric diaphragm which permits it to move up and down. It does not involve the weight and foundation costs of the liquid-seal holder. Additional information on gas holders can be found in *Gas Engineers Handbook,* Industrial Press, New York, 1966.

Solution of Gases in Liquids Certain gases will dissolve readily in liquids. In some cases in which the quantities are not large, this may be a practical storage procedure. Examples of gases that can be handled in this way are ammonia in water, acetylene in acetone, and hydrogen chloride in water. Whether or not this method is used depends mainly on whether the end use requires the anhydrous or the liquid state. Pressure may be either atmospheric or elevated. The solution of acetylene in acetone is also a safety feature because of the instability of acetylene.

Storage in Pressure Vessels, Bottles, and Pipe Lines The distinction between pressure vessels, bottles, and pipes is arbitrary. They can all be used for storing gases under pressure. A storage pressure vessel is usually a permanent installation. Storing a gas under pressure not only reduces its volume but also in many cases liquefies it at ambient temperature. Some gases in this category are carbon dioxide, several petroleum gases, chlorine, ammonia, sulfur dioxide, and some types of Freon. Pressure tanks are frequently installed underground.

Liquefied petroleum gas (LPG) is the subject of API Standard 2510, The Design and Construction of Liquefied Petroleum Gas Installations at Marine and Pipeline Terminals, Natural Gas Processing Plants, Refineries, and Tank Farms. This standard in turn refers to:

1. National Fire Protection Association (NFPA) Standard 58, Standard for the Storage and Handling of Liquefied Petroleum Gases
2. NFPA Standard 59, Standard for the Storage and Handling of Liquefied Petroleum Gases at Utility Gas Plants
3. NFPA Standard 59A, Standard for the Production, Storage, and Handling of Liquefied Natural Gas (LNG)

The API Standard gives considerable information on the construction and safety features of such installations. It also recommends minimum distances from property lines. The user may wish to obtain added safety by increasing these distances.

The term **bottle** is usually applied to a pressure vessel that is small enough to be conveniently portable. Bottles range from about 57 L (2 ft^3) down to CO_2 capsules of about 16.4 mL (1 in^3). Bottles are convenient for small quantities of many gases, including air, hydrogen, nitrogen, oxygen, argon, acetylene, Freon, and petroleum gas. Some are one-time-use disposable containers.

Pipe Lines A pipe line is not ordinarily a storage device. Pipes, however, have been buried in a series of connected parallel lines and used for storage. This avoids the necessity of providing foundations, and the earth protects the pipe from extremes of temperature. The economics of such an installation would be doubtful if it were designed to the same stresses as a pressure vessel. Storage is also obtained by increasing the pressure in operating pipe lines and thus using the pipe volume as a tank.

Low-Temperature and Cryogenic Storage This type is used for gases that liquefy under pressure at atmospheric temperature. In cryogenic storage the gas is at, or near to, atmospheric pressure and remains liquid because of low temperature. A system may also operate with a combination of pressure and reduced temperature. The term "cryogenic" usually refers to temperatures below −101°C (−150°F). Some gases, however, liquefy between −101°C and ambient temperatures. The principle is the same, but cryogenic temperatures create different problems with insulation and construction materials.

The liquefied gas must be maintained at or below its boiling point. Refrigeration can be used, but the usual practice is to cool by evaporation. The quantity of liquid evaporated is minimized by insulation. The vapor may be vented to the atmosphere (wasteful), it may be compressed and reliquefied, or it may be used.

At very low temperatures with liquid air and similar substances, the tank may have double walls with the interspace evacuated. The well-known Dewar flask is an example. Large tanks and even pipe lines are now built this way. An alternative is to use double walls without vacuum but with an insulating material in the interspace. Perlite and plastic foams are two insulating materials employed in this way. Sometimes both insulation and vacuum are used.

Materials Materials for liquefied-gas containers must be suitable for the temperatures, and they must not be brittle. Some carbon steels can be used down to −59°C (−75°F), and low-alloy steels to −101°C (−150°F) and sometimes −129°C (−200°F). Below these temperatures austenitic stainless steel (AISI 300 series) and aluminum are the principal materials.

Low temperatures involve problems of **differential thermal expansion.** With the outer wall at ambient temperature and the inner wall at the liquid boiling point, relative movement must be accommodated. Some systems for accomplishing this are patented. The Gaz Transport of France reduces dimensional change by using a thin inner liner of Invar. Another patented French system accommodates this change by means of the flexibility of thin metal which is creased. The creases run in two directions, and the form of the crossings of the creases is a feature of the system.

Low-temperature tanks may be installed underground to take advantage of the insulating value of the earth. Frozen-earth storage is also used. The frozen earth forms the tank. Some installations using this technique have been unsuccessful because of excessive heat absorption.

COST OF STORAGE FACILITIES

Contractors' bids offer the most reliable information on cost. Order-of-magnitude costs, however, may be required for preliminary studies. One way of estimating them is to obtain cost information from similar facilities and scale it to the proposed installation. Costs of steel storage tanks and vessels have been found to vary approximately as the 0.6 to 0.7 power of their weight [see Happel, *Chemical Process Economics*, Wiley, 1958, p. 267; also Williams, *Chem. Eng.*, **54**(12), 124 (1947)]. All estimates based on the costs of existing equipment must be corrected for changes in the price index from the date when the equipment was built. Considerable uncertainty is involved in adjusting data more than a few years old.

Figures 6-141, 6-142, and 6-143 are derived from a private communication from H. G. Garner. They give mid-1979 purchased costs for storage tanks. The prices for field-erected tanks are for multiple-tank installations erected by the contractor on foundations provided by the owner. Some cost information on tanks is given in various references cited in Sec. 25. Cost data vary considerably from one reference to another.

Prestressed (posttensioned) concrete tanks cost about 20 percent more than steel tanks of the same capacity. Once installed, however, concrete tanks require very little maintenance. A true comparison with steel would, therefore, require evaluating the maintenance cost of both types.

BULK TRANSPORT OF FLUIDS

Transportation is often an important part of product cost. Bulk transportation may provide significant savings. When there is a choice between two or more forms of transportation, the competition may result in rate reduction. Transportation is subject to considerable regulation, which will be discussed in some detail under specific headings.

Pipe Lines For quantities of fluid which an economic investigation indicates are sufficiently large and continuous to justify the investment, pipe lines are one of the lowest-cost means of transportation. They have been built up to 1.22 m (48 in) or more in diameter and about 3200 km (2000 mi) in length for oil, gas, and other products. Water is usually not transported more than 160 to 320 km (100 to 200 miles), but the conduits may be much greater than 1.22 m (48 in) in diameter. Open canals are also used for water transportation.

Petroleum pipe lines before 1969 were built to ASA (now ANSI) Standard B31.4 for liquids and Standard B31.8 for gas. These standards were seldom mandatory because few states adopted them. The U.S. Department of Transportation (DOT), which now has responsibility for pipe-line regulation, issued Title 49, Part 192—Transportation of Natural Gas and Other Gas by Pipeline: Minimum Safety Standards, and Part 195—Transportation of Liquids by Pipeline. These contain considerable material from B31.4 and B31.8. They allow generally higher stresses than the ASME Pressure Vessel Code would allow for steels of comparable strength. The enforcement of their regulations is presently left to the states and is therefore somewhat uncertain.

Pipe-line pumping stations usually range from 16 to 160 km (10 to 100 miles) apart, with maximum pressures up to 6900 kPa (1000 lbf/in^2) and velocities up to 3 m/s (10 ft/s) for liquid. Gas pipe lines have higher velocities and may have greater spacing of stations.

Tanks Tank cars (single and multiple tank), tank trucks, portable tanks, drums, barrels, carboys, and cans are used to transport fluids. Interstate transportation is regulated by the DOT. There are other regulating agencies—state, local, and private. Railroads make rules determining what they will accept, some states require compliance with DOT specifications on intrastate movements, and tunnel authorities as well as fire chiefs apply restrictions. Water shipments involve regulations of the U.S. Coast Guard. The American Bureau of Shipping sets rules for design and construction which are recognized by insurance underwriters.

The most pertinent **DOT regulations** (*Code of Federal Regulations*, Title 18, Parts 171–179 and 397) were published by R. M. Graziano (then agent and attorney for carriers and freight forwarders) in his tariff titled *Hazardous Materials Regulations of the Department of Transportation* (1978). New tariffs identified by number are issued at intervals, and interim revisions are sent out. Agents change at intervals.

Graziano's tariff lists many regulated (dangerous) commodities (Part 172, DOT regulations) for transportation. This includes those that are poisonous, flammable, oxidizing, corrosive, explosive, radioactive, and compressed gases. Part 178 covers specifications for all types of containers from carboys to large portable tanks and tank trucks. Part 179 deals with tank-car construction.

An Association of American Railroads (AAR) publication, *Specifications for Tank Cars*, covers many requirements beyond the DOT regulations.

Some additional details are given later. Because of frequent changes, it is always necessary to check the latest rules. The **shipper,** not the carrier, has the ultimate responsibility for shipping in the correct container.

Tank Cars These range in size from about 7.6 to 182 m^3 (2000

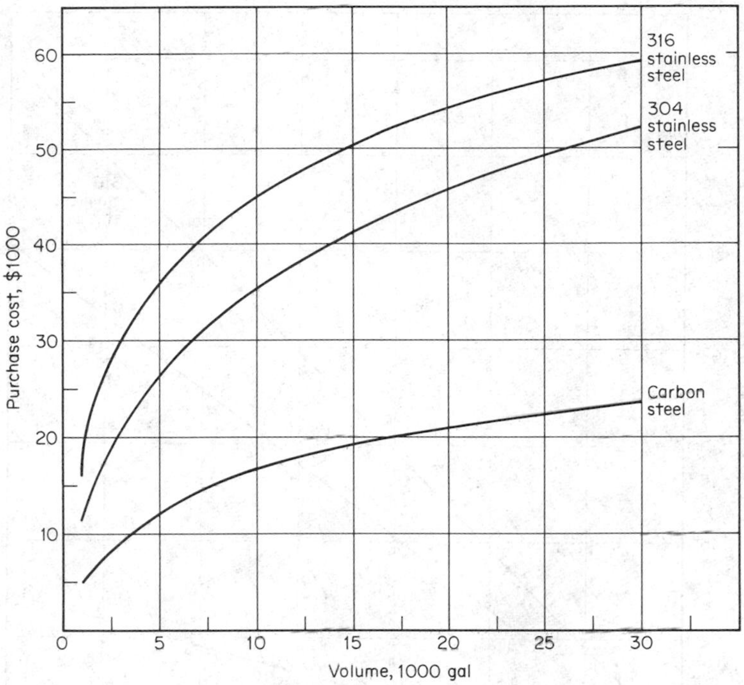

FIG. 6-141 Cost of shop-fabricated tanks in mid-1979 with ¼-in walls. Multiplying factors on carbon steel costs for other materials are: carbon steel, 1.0; rubber-lined carbon steel, 1.5; aluminum, 1.6; glass-lined carbon steel, 4.0; and fiber-reinforced plastic, 0.75 to 1.5. Multiplying factors on type 316 stainless-steel costs for other materials are: 316 stainless steel, 1.0; Monel, 1.5; Inconel, 1.5; nickel, 1.6; titanium, 3.0; and Hastelloy C, 3.5. Multiplying factors for wall thicknesses different from ¼ in are:

Thickness, in	Carbon steel	304 stainless steel	316 stainless steel
½	1.4	1.8	1.8
¾	2.1	2.5	2.6
1	2.7	3.3	3.5

To convert gallons to cubic meters, multiply by 3.785×10^{-3}.

to 48,000 gal), and a car may be single or multiunit. The DOT now limits them to 130 m^3 (34,500 gal) and 120,000 kg (263,000 lb) gross mass. Large cars usually result in lower investment per cubic meter and take lower shipping rates. Cars may be insulated to reduce heating or cooling of the contents. Certain liquefied gases may be carried in insulated cars; temperatures are maintained by evaporation (see subsection "Low-Temperature and Cryogenic Storage"). Cars may be heated by steam coils or by electricity. Some products are loaded hot, solidify in transport, and are melted for removal. Some low-temperature cargoes must be unloaded within a given time (usually 30 days) to prevent pressure buildup.

Tank cars are classified as pressure or general-purpose. Pressure cars have relief-valve settings of 517 kPa (75 lbf/in^2) and above. Those designated as general-purpose cars are, nevertheless, pressure vessels and may have relief valves or rupture disks. The DOT specification code number indicates the type of car. For instance, 105A500W indicates a pressure car with a test pressure of 3447 kPa (500 lbf/in^2) and a relief-valve setting of 2585 kPa (375 lbf/in^2). In most cases, loading and unloading valves, safety valves, and vent valves must be in a dome or an enclosure.

Companies shipping dangerous materials sometimes build tank cars with metal thicker than required by the specifications in order to reduce the possibility of leakage during a wreck or fire. The punching of couplers or rail ends into heads of tanks is a hazard.

Older tank cars have a center sill or beam running the entire length of the car. Most modern cars have no continuous sill, only short stub sills at each end. Cars with full sills have tanks anchored longitudinally at the center of the sill. The anchor is designed to be weaker than either the tank shell or the doubler plate between anchor and shell. Cars with stub sills have similar safeguards. Anchors and other parts are designed to meet AAR requirements.

The impact forces on car couplers put high stresses in sills, anchors, and doublers. This may start fatigue cracks in the shell, particularly at the corners of welded doubler plates. With brittle steel in cold weather, such cracks sometimes cause complete rupture of the tank. Large end radii on the doublers and tougher steels will reduce this hazard. Inspection of older cars can reveal cracks before failure.

A difference between tank cars and most pressure vessels is that tank cars are designed in terms of the theoretical ultimate or bursting strength of the tank. The test pressure is usually 40 percent of the bursting pressure (sometimes less). The safety valves are set at 75 percent of the test pressure. Thus, the maximum operating pressure is usually 30 percent of the bursting pressure. This gives a nominal factor of safety of 3.3, compared with 4.0 for Division 1 of the ASME Pressure Vessel Code.

The DOT rules require that pressure cars have relief valves designed to limit pressure to 82.5 percent (with certain exceptions) of test pressure (110 percent of maximum operating pressure) when

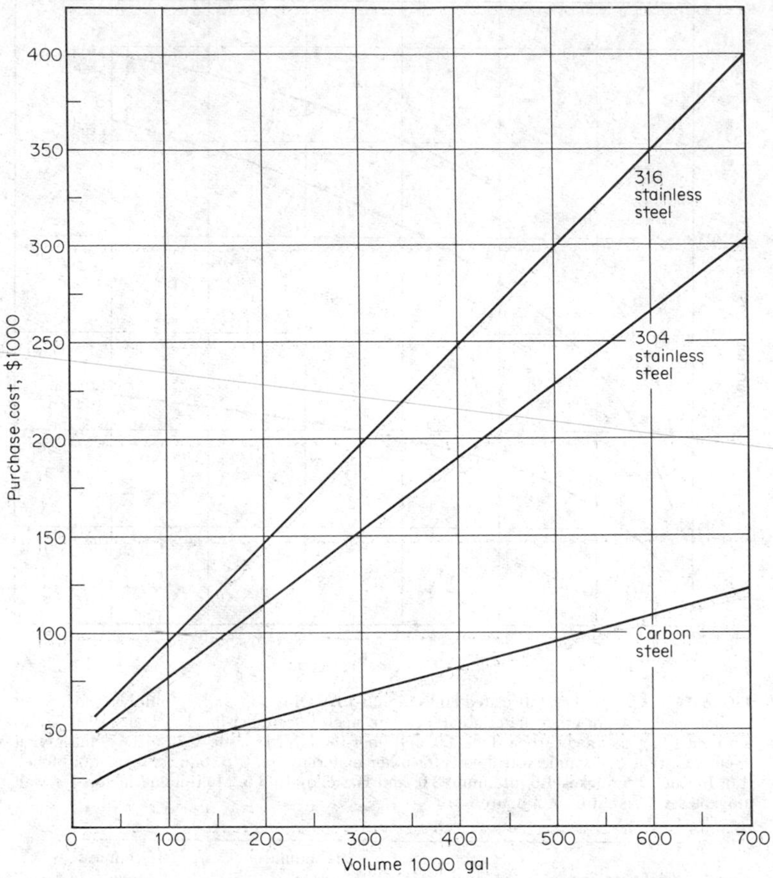

FIG. 6-142 Cost of small field-erected tanks in mid-1979, including stairs, platforms, and a normal complement of nozzles. The carbon steel curve is for API Standard 650 tanks, and the others are for stainless-steel tanks for atmospheric pressure with flat bottoms. The curves are for tanks purchased in quantities of three or more at a Gulf Coast site. Multiplying factors for other materials are: 316 stainless steel, 1.0; Monel, 1.5; Inconel, 1.5; nickel, 1.6; titanium, 3.0; and Hastelloy C, 3.5. Allowances should be added to the factored costs as follows: 10 percent for stiffener rings, 20 percent for API Standard 620, 15 percent for quantity of one tank, 10 percent for quantity of two tanks, 15 percent of steel cost for a congested working area, 50 percent of steel cost for an integral steel dike. To convert gallons to cubic meters, multiply by 3.785×10^{-3}.

exposed to fire. Appendix A of AAR Specifications deals with the flow capacity of relief devices. The formulas apply to cars in the upright position with the device discharging vapor. They may not protect the car adequately when it is overturned and the device is discharging liquid.

Appendix B of AAR Specifications deals with the certification of facilities. Fabrication, repairing, testing, and specialty work on tank cars must be done in certified facilities. The AAR certifies shops to build cars of certain materials, to do test work on cars, or to make certain repairs and alterations.

Tank Trucks These trucks may have single, compartmented, or multiple tanks. Many of their requirements are similar to those for tank cars, except that thinner shells are permitted in most cases. Trucks for nonhazardous materials are subject to few regulations other than the normal highway laws governing all motor vehicles. But trucks carrying hazardous materials must comply with DOT regulations, Parts 173, 177, 178, and 397. Maximum weight, axle loading, and length are governed by state highway regulations. Many states have limits in the vicinity of 31,750 kg (70,000 lb) total mass, 14,500 kg (32,000 lb) for tandem axles, and 18.3 m (60 ft) or less overall length. Some allow tandem trailers.

Truck cargo tanks (for dangerous materials) are built under Part 173 and Subpart J of Part 178, DOT regulations. This includes Specifications MC-306, MC-307, MC-312, and MC-331. MC-331 is required for compressed gas. Subpart J requires tanks for pressures above 345 kPa (50 lbf/in^2) in one case and 103 kPa (15 lbf/in^2) in another to be built according to the ASME Pressure Vessel Code. A particular issue of the code is specified.

Because of the demands of highway service, the DOT specifications have a number of requirements in addition to the ASME Code. These include design for impact forces and rollover protection for fittings.

Portable tanks, drums, or bottles are shipped by rail, ship, air, or truck. Portable tanks containing hazardous materials must conform to DOT regulations, Parts 173 and 178, Subpart H.

Some tanks are designed to be shipped by trailer and transferred to railcars or ships (see following discussion).

Marine Transportation Seagoing **tankers** are for high tonnage. The traditional tanker uses the ship structure as a tank. It is subdivided into a number of tanks by means of transverse bulkheads and a centerline bulkhead. More than one product can be carried. An elaborate piping system connects the tanks to a pumping plant which

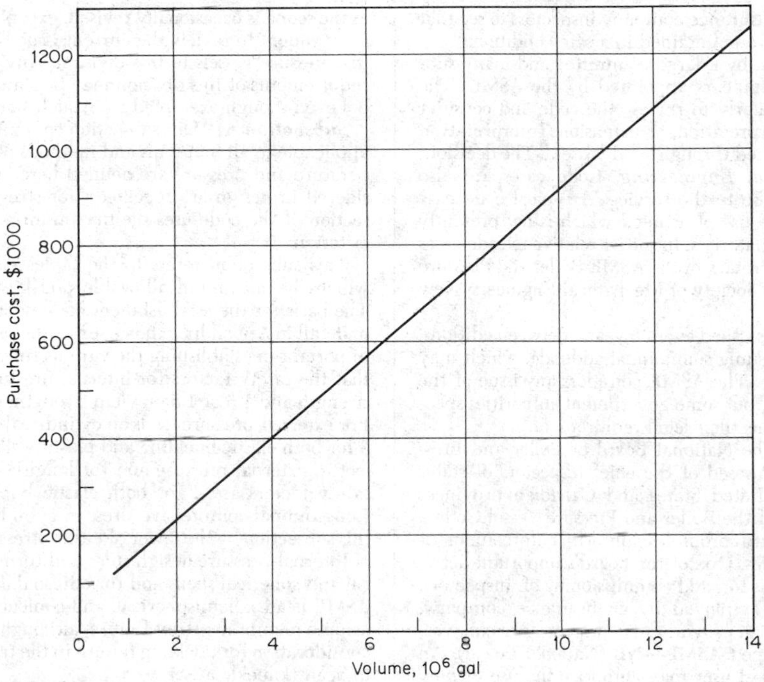

FIG. 6-143 Cost of large field-erected tanks in mid-1979, including stairs, platforms, and a normal complement of nozzles; curve for carbon steel API Standard 650 tanks in quantities of three or more at a Gulf Coast site. For type 304 stainless steel, multiply cost by 2.4; and for type 316 stainless steel, multiply cost by 3.1. Allowances should be added to the factored costs as follows: 10 percent for stiffener rings, 20 percent for API Standard 620, 15 percent for quantity of one tank, 10 percent for quantity of two tanks, and 15 percent of carbon steel cost for a congested working area. To convert gallons to cubic meters, multiply by 3.785×10^{-3}.

can discharge or transfer the cargo. Harbor and docking facilities appear to be the only limit to tanker size. The largest tanker size to date is about 500,000 deadweight tons. In the United States, tankers are built to specifications of the American Bureau of Shipping and the U.S. Coast Guard.

Low-temperature liquefied gases are shipped in special ships with insulation between the hull and an inner tank. Poisonous materials are shipped in separate tanks built into the ship. This prevents tank leakage from contaminating harbors. Separate tanks are also used to transport pressurized gases.

Barges are used on inland waterways. Popular sizes are up to 16 m (52½) wide by 76 m (250 ft) long, with 2.6 m (8½ ft) to 4.3 m (14 ft) draft. Cargo requirements and waterway limitations determine design. Use of barges of uniform size facilitates rafting them together.

Portable tanks may be stowed in the holds of conventional cargo ships or special container ships, or they may be fastened on deck.

Container ships have guides in the hold and on deck which hold boxlike containers or tanks. The tank is latched to a trailer chassis and hauled to shipside. A movable gantry, sometimes permanently installed on the ship, hoists the tank from the trailer and lowers it into the guides on the ship. This system achieves large savings in labor, but its application is sometimes limited by lack of agreement between ship operators and unions.

Portable tanks for regulated commodities in marine transportation must be designed and built under Coast Guard regulations (see discussion under "Pressure Vessels").

Materials of Construction for Bulk Transport Because of the more severe service, construction materials for transportation usually are more restricted than for storage. Most large pipe lines are constructed of steel conforming to API Specification 5L or 5LX. Most tanks (cars, etc.) are built of pressure-vessel steels or AAR specification steels, with a few of aluminum or stainless steel. Carbon steel tanks may be lined with rubber, plastic, nickel, glass, or other materials. In many cases this is practical and cheaper than using a stainless-steel tank. Other materials for tank construction may be proposed and used if approved by the appropriate authorities (AAR and DOT).

PRESSURE VESSELS

This discussion of pressure vessels is intended as an overview of the codes most frequently used for the design and construction of pressure vessels. Chemical engineers who design or specify pressure vessels should determine the federal and local laws relevant to the problem and then refer to the most recent issue of the pertinent code or standard before proceeding. Laws, codes, and standards are frequently changed.

A pressure vessel is a closed container of limited length (in contrast to the indefinite length of piping). Its smallest dimension is considerably larger than the connecting piping, and it is subject to pressures above 7 or 14 kPa (1 or 2 lbf/in^2). It is distinguished from a boiler, which in most cases is used to generate steam for use external to itself.

Code Administration The American Society of Mechanical Engineers has written the ASME Boiler and Pressure Vessel Code, which contains rules for the design, fabrication, and inspection of boilers and pressure vessels. The ASME Code is an American National Standard. Most states in the United States and all Canadian provinces have passed legislation which makes the ASME Code or certain parts of it their legal requirement. Only a few jurisdictions have adopted the code for all vessels. The others apply it to certain types of vessels or to boilers. States employ inspectors (usually under a chief boiler inspector) to enforce code provisions. The authorities also depend a great deal on insurance company inspectors to see that boilers and pressure vessels are maintained in a safe condition.

The ASME Code is written by a large committee and many subcommittees, composed of engineers appointed by the ASME. The Code Committee meets regularly to review the code and consider requests for its revision, interpretation, or extension. **Interpretation and extension** are accomplished through "code cases." The decisions are published in *Mechanical Engineering*. Code cases are also mailed to those who subscribe to the service. A typical code case might be the approval of the use of a metal which is not presently on the list of approved code materials. Inquiries relative to code cases should be addressed to the secretary of the ASME Boiler and Pressure Vessel Committee, American Society of Mechanical Engineers, New York.

A new edition of the code is issued every 3 years. Between editions, alterations are handled by issuing semiannual addenda, which may be purchased by subscription. The ASME considers any issue of the code to be adequate and safe, but some government authorities specify certain issues of the code as their legal requirement.

Inspection Authority The National Board of Boiler and Pressure Vessel Inspectors is composed of the chief inspectors of states and municipalities in the United States and Canadian provinces which have made any part of the Boiler and Pressure Vessel Code a legal requirement. This board promotes uniform enforcement of boiler and pressure-vessel rules. One of the board's important activities is providing examinations for, and commissioning of, inspectors. Inspectors so qualified and employed by an insurance company, state, municipality, or Canadian province may inspect a pressure vessel and permit it to be stamped ASME—NB (National Board). An inspector employed by a vessel user may authorize the use of only the ASME stamp. The ASME Code Committee authorizes fabricators to use the various ASME stamps. The stamps, however, may be applied to a vessel only with the approval of the inspector.

The ASME Boiler and Pressure Vessel Code consists of eleven sections as follows:

I. Power Boilers
II. Material Specifications (three parts)
III. Nuclear Power Plant Components
IV. Heating Boilers
V. Nondestructive Examination
VI. Recommended Rules for Care and Operation of Heating Boilers
VII. Recommended Rules for Care of Power Boilers
VIII. Pressure Vessels, Division 1
Pressure Vessels, Division 2, Alternative Rules
IX. Welding and Brazing Qualifications
X. Fiberglass-Reinforced Plastic Pressure Vessels
XI. Rules for Inservice Inspection of Nuclear Power Plant Components

Pressure vessels (as distinguished from boilers) are involved with Secs. II, III, V, VIII, IX, X, and XI. Section VIII, Division I, is the Pressure Vessel Code as it existed in the past (and will continue). Division 2 was brought out as a means of permitting higher design stresses while ensuring at least as great a degree of safety as in Division 1. These two divisions plus Secs. III and X will be discussed briefly here. They refer to Secs. II and IX.

ASME Code Section VIII, Division 1 Most pressure vessels used in the process industry in the United States are designed and constructed in accordance with Sec. VIII, Division 1. This division is divided into three subsections followed by appendixes.

Introduction The Introduction contains the scope of the division and defines the responsibilities of the user, the manufacturer, and the inspector. The scope defines pressure vessels as containers for the containment of pressure. It specifically excludes vessels having an internal pressure not exceeding 103 kPa (15 lbf/in^2) and further states that the rules are applicable for pressures not exceeding 20,670 kPa (3000 lbf/in^2). For higher pressures it is usually necessary to deviate from the rules in this division.

The scope covers many other less basic exclusions, and inasmuch as the scope is occasionally revised, except for the most obvious cases, it is prudent to review the current issue before specifying or designing pressure vessels to this division. Any vessel which meets all the requirements of this division may be stamped with the code *U* symbol even though exempted from such stamping.

Subsection A This subsection contains the general requirements applicable to all materials and methods of construction. Design temperature and pressure are defined here, and the loadings to be considered in design are specified. For stress failure and yielding, this section of the code uses the maximum-stress theory of failure as its criterion.

This subsection refers to the tables elsewhere in the division in which the maximum allowable tensile-stress values are tabulated. The basis for the establishment of these allowable stresses is defined in detail in Appendix P; however, as the safety factors used were very important in establishing the various rules of this division, it is noted that the safety factors for internal-pressure loads are 4 on ultimate strength and 1.6 or 1.5 on yield strength, depending on the material. For external-pressure loads on cylindrical shells, the safety factors are 3 for both elastic buckling and plastic collapse. For other shapes subject to external pressure and for longitudinal shell compression, the safety factors are 4 for both elastic buckling and plastic collapse. Longitudinal compressive stress in cylindrical elements is limited in this subsection by the lower of either stress failure or buckling failure.

Internal-pressure design rules and formulas are given for cylindrical and spherical shells and for ellipsoidal, torispherical (often called ASME heads), hemispherical, and conical heads. The formulas given assume membrane-stress failure, although the rules for heads include consideration for buckling failure in the transition area from cylinder to head (knuckle area).

Longitudinal joints in cylinders are more highly stressed than circumferential joints, and the code takes this fact into account. When forming heads, there is usually some thinning from the original plate thickness in the knuckle area, and it is prudent to specify the minimum allowable thickness at this point.

Unstayed flat heads and covers can be designed by very specific rules and formulas given in this subsection. The stresses caused by pressure on these members are bending stresses, and the formulas include an allowance for additional edge moments induced when the head, cover, or blind flange is attached by bolts. Rules are provided for quick-opening closures because of the risk of incomplete attachment or opening while the vessel is pressurized. Rules for braced and stayed surfaces are also provided.

External-pressure failure of shells can result from overstress at one extreme or from elastic instability at the other or at some intermediate loading. The code provides the solution for most shells by using a number of charts. One chart is used for cylinders where the shell diameter-to-thickness ratio and the length-to-diameter ratio are the variables. The rest of the charts depict curves relating the geometry of cylinders and spheres to allowable stress by curves which are determined from the modulus of elasticity, tangent modulus, and yield strength at temperatures for various materials or classes of materials. The text of this subsection explains how the allowable stress is determined from the charts for cylinders, spheres, and hemispherical, ellipsoidal, torispherical, and conical heads.

Frequently cost savings for cylindrical shells can result from reducing the effective length-to-diameter ratio and thereby reducing shell thickness. This can be accomplished by adding circumferential stiffeners to the shell. Rules are included for designing and locating the stiffeners.

Openings are always required in pressure-vessel shells and heads. Stress intensification is created by the existence of a hole in an otherwise symmetrical section. The code compensates for this by an area-replacement method. It takes a cross section through the opening, and it measures the area of the metal of the required shell that is removed and replaces it in the cross section by additional material (shell wall, nozzle wall, reinforcing plate, or weld) within certain distances of the opening centerline. These rules and formulas for calculation are included in Subsec. A.

When a cylindrical shell is drilled for the insertion of multiple tubes, the shell is significantly weakened and the code provides rules for tube-hole patterns and the reduction in strength that must be accommodated.

Fabrication tolerances are covered in this subsection. The tolerances permitted for shells for external pressure are much closer than those for internal pressure because the stability of the structure is dependent on the symmetry. Other paragraphs cover repair of defects during fabrication, material identification, heat treatment, and impact testing.

Inspection and testing requirements are covered in detail. Most vessels are required to be hydrostatic-tested (generally with water) at 1½ times the maximum allowable working pressure. Some enameled (glass-lined) vessels are permitted to be hydrostatic-tested at lower pressures. Pneumatic tests are permitted and are carried to at least 1¼ times the maximum allowable working pressure, and there is provision for proof testing when the strength of the vessel or any of its parts cannot be computed with satisfactory assurance of accuracy. Pneumatic or proof tests are rarely conducted.

Pressure-relief-device requirements are defined in Subsec. A. Set point and maximum pressure during relief are defined according to the service, the cause of overpressure, and the number of relief devices. Safety, safety relief, relief valves, rupture disk, breaking pin, and rules on tolerances for the relieving point are given.

Testing, certification, and installation rules for relieving devices are extensive. Every chemical engineer responsible for the design or operation of process units should become very familiar with these rules. The pressure-relief-device paragraphs are the only parts of Sec. VIII, Division 1, that are concerned with the installation and ongoing operation of the facility; all other rules apply only to the design and manufacture of the vessel.

Subsection B This subsection contains rules pertaining to the methods of fabrication of pressure vessels. Part UW is applicable to welded vessels. Service restrictions are defined. Lethal service is for "lethal substances," which are defined as poisonous gases or liquids of such a nature that a very small amount of the gas or the vapor of the liquid mixed or unmixed with air is dangerous to life when inhaled. It is stated that it is the user's responsibility to advise the designer or manufacturer if the service is lethal. All vessels in lethal service shall have all butt-welded joints fully radiographed, and when practical, joints shall be butt-welded. All vessels fabricated of carbon or low-alloy steel shall be postweld-heat-treated.

Low-temperature service is defined as being below −29°C (−20°F), and impact testing of many materials is required. The code is restrictive in the type of welding permitted.

Unfired steam boilers with design pressures exceeding 345 kPa (50 lbf/in^2) have restrictive rules on welded-joint design, and all butt joints require full radiography.

Pressure vessels subject to direct firing have special requirements relative to welded-joint design and postweld heat treatment.

This subsection includes rules governing welded-joint designs and the degree of radiography, with efficiencies for welded joints specified as functions of the quality of joint. These efficiencies are used in the formulas in Subsec. A for determining vessel thicknesses.

Details are provided for head-to-shell welds, tube sheet–to–shell welds, and nozzle–to–shell welds. Acceptable forms of welded staybolts and plug and slot welds for staying plates are given here.

Rules for the welded fabrication of pressure vessels cover welding processes, manufacturer's record keeping on welding procedures, welder qualification, cleaning, fit-up alignment tolerances, and repair of weld defects. Procedures for postweld heat treatment are detailed. Checking the procedures and welders and radiographic and ultrasonic examination of welded joints are covered.

Requirements for vessels fabricated by forging in Part UF include unique design requirements with particular concern for stress risers, fabrication, heat treatment, repair of defects, and inspection. Vessels fabricated by brazing are covered in Part UB. Brazed vessels cannot be used in lethal service, for unfired steam boilers, or for direct firing. Permitted brazing processes as well as testing of brazed joints for strength are covered. Fabrication and inspection rules are also included.

Subsection C This subsection contains requirements pertaining to classes of materials. Carbon and low-alloy steels are governed by Part UCS, nonferrous materials by Part UNF, high-alloy steels by Part UHA, and steels with tensile properties enhanced by heat treatment by Part UHT. Each of these parts includes tables of maximum allowable stress values for all code materials for a range of metal temperatures. These stress values include appropriate safety factors. Rules governing the application, fabrication, and heat treatment of the vessels are included in each part.

Part UHT also contains more stringent details for nozzle welding that are required for some of these high-strength materials. Part UCI has rules for cast-iron construction, Part UCL has rules for welded vessels of clad plate as lined vessels, and Part UCD has rules for ductile-iron pressure vessels.

A relatively recent addition to the code is Part ULW, which contains requirements for vessels fabricated by layered construction. This type of construction is most frequently used for high pressures, usually in excess of 13,800 kPa (2000 lbf/in^2).

There are several methods of layering in common use: (1) thick layers shrunk together; (2) thin layers, each wrapped over the other and the longitudinal seam welded by using the prior layer as backup; and (3) thin layers spirally wrapped. The code rules are written for either thick or thin layers. Rules and details are provided for all the usual welded joints and nozzle reinforcement. Supports for layered vessels require special consideration, in that only the outer layer could contribute to the support. For lethal service only the inner shell and inner heads need comply with the requirements in Subsec. B. Inasmuch as radiography would not be practical for inspection of many of the welds, extensive use is made of magnetic-particle and ultrasonic inspection. When radiography is required, the code warns the inspector that indications sufficient for rejection in single-wall vessels may be acceptable. Vent holes are specified through each layer down to the inner shell to prevent buildup of pressure between layers in the event of leakage at the inner shell.

Mandatory Appendixes These include a section on supplementary design formulas for shells not covered in Subsec. A. Formulas are given for thick shells, heads, and dished covers. Another appendix gives very specific rules, formulas, and charts for the design of bolted-flange connections. The nature of these rules is such that they are readily programmable for a digital computer, and most flanges now are designed by using computers. One appendix includes only the charts used for calculating shells for external pressure discussed previously. Jacketed vessels are covered in a separate appendix in which very specific rules are given, particularly for the attachment of the jacket to the inner shell. Other appendixes cover inspection and quality control.

Nonmandatory Appendixes These cover a number of subjects, primarily suggested good practices and other aids in understanding the code and in designing with the code. Several current nonmandatory appendixes will probably become mandatory.

Figure 6-144 illustrates a pressure vessel with the applicable code paragraphs noted for the various elements. Additional important paragraphs are referenced at the bottom of the figure.

ASME Code Section VIII, Division 2 Paragraph A-100e of Division 2 states: "In relation to the rules of Division 1 of Section VIII, these rules of Division 2 are more restrictive in the choice of materials which may be used but permit higher design stress intensity values to be employed in the range of temperatures over which the design stress intensity value is controlled by the ultimate strength or the yield strength; more precise design procedures are required and some common design details are prohibited; permissible fabrication procedures are specifically delineated and more complete testing and inspection are required." Most Division 2 vessels fabricated to date have been large or intended for high pressure and, therefore, expensive when the material and labor savings resulting from smaller safety factors have been greater than the additional engineering, administrative, and inspection costs.

The organization of Division 2 differs from that of Division 1.

Part A This part gives the scope of the division, establishes its jurisdiction, and sets forth the responsibilities of the user and the manufacturer. Of particular importance is the fact that no upper limitation in pressure is specified and that a user's design specification is required. The user or the user's agent shall provide require-

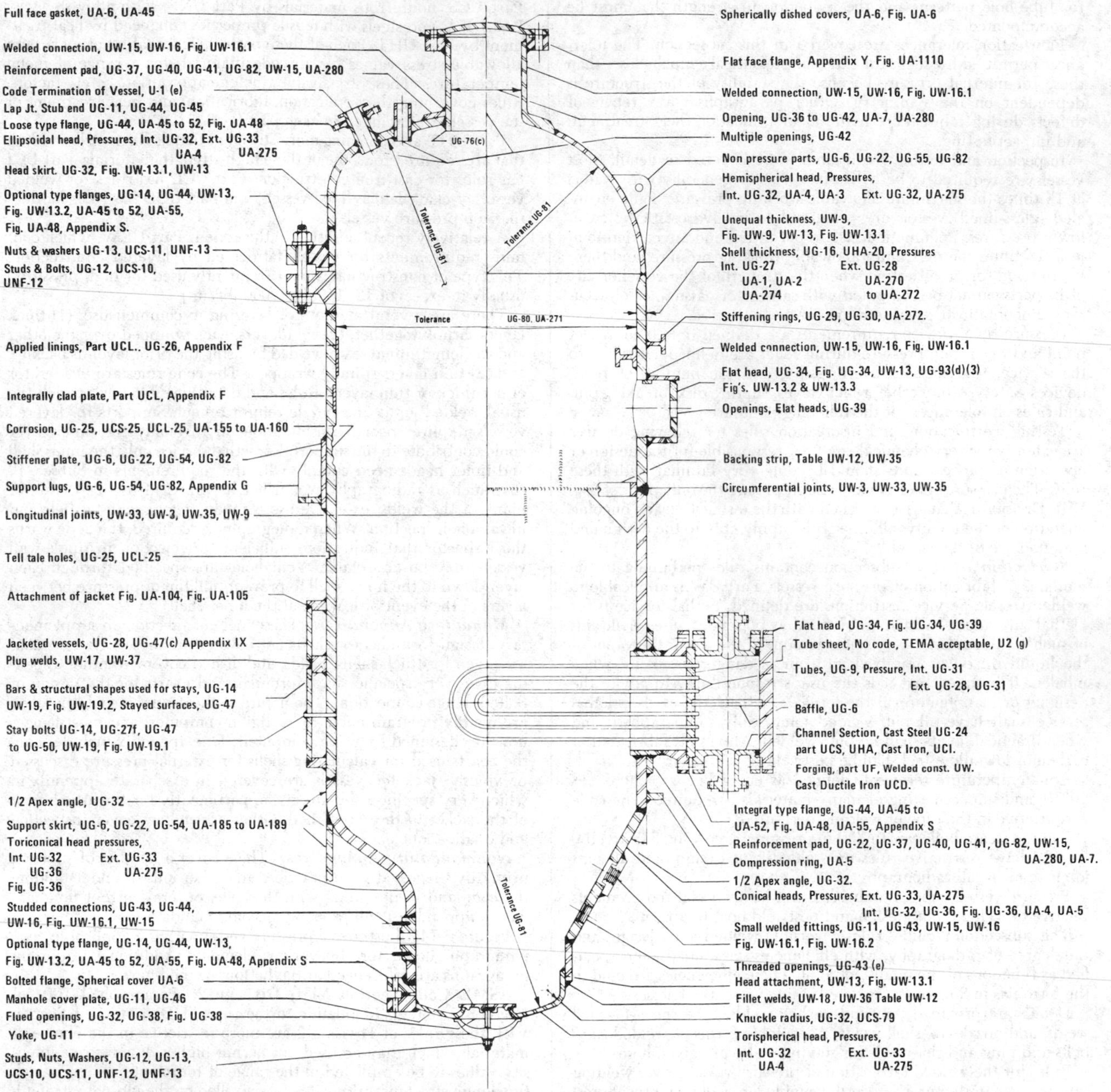

GENERAL NOTES

HEAT TREATMENT UG-85, UW-10, UW-40, UCS-56, TABLE UCS-56, UCS-79(d) UCS-85, UNF-56, UHA-32, UHA-105, & UCL-34
INSPECTION UG-90 THRU UG-97, U-1 (j)
JOINT EFFICIENCY UW-12, & TABLE UW-12
LETHAL SERVICE UW-2(a),UCD-2, & UCI-2
LOADINGS UG-22
LOW TEMPERATURE UG-84, UW-2(b), UCS-65, UCS-66, UCS-67, UNF-65, & UCL-27
MATERIALS UG-5 THRU UG-15, UG-18, UG-77, UCL-11 & UW-5 TABLES NF-1 & NF-2

PRESSURE, DESIGN UG-19, & UG-21
MAX. ALLOWABLE WORKING UG-98
TEMPERATURE, DESIGN UG-19, UG-20
PRESSURE VESSELS SUBJECT TO DIRECT FIRING UW-2(d), U-1(h)
RADIOGRAPHIC EXAM, UW-11, UW-51, UW-52, UCS-57, UNF-57, UHA-33, & UCL-35
SPOT EXAM OF WELDED JOINT UW-52
NO RADIOGRAPH UW-11(c)
RELIEF DEVICES UG-125 THROUGH UG-136, APP. XI
REPAIRS UG-78, UW-38, UW-40(d)
STRESS MAX. ALLOW., VALUE UG-23, UW-12(c), UNF-23, UHA-23, UCL-23

TEST, HYDROSTATIC, UG-99, UCI-99, UCL-52, & UA-60
PNEUMATIC UW-50 & UG-100
PROOF, UG-101
NON-DESTRUCTIVE, UG-103, UNF-58, & UHA-34
MAG. PART, UA-70 THRU UA-73
LIQ. PENE. UA-91 THRU UA-95
ULTRASONIC, UA-901 THRU UA-904
IMPACT, UG-84, UCS-66, UHA-51, NF-6
STAMPING & DATA, UG-115 THRU UG-120
UNFIRED STEAM BOILERS, UW-2(c), U-1(g)

FIG. 6-144 ASME Code Sec. VIII, Division 1: applicable paragraphs for design and construction details. *(Courtesy of Missouri Boiler and Tank Co.)*

ments for intended operating conditions in such detail as to constitute an adequate basis for selecting materials and designing, fabricating, and inspecting the vessel. The user's design specification shall include the method of supporting the vessel and any requirement for a fatigue analysis. If a fatigue analysis is required, the user must provide information in sufficient detail so that an analysis for cyclic operation can be made.

Part AM This part lists permitted individual construction materials, applicable specifications, special requirements, design stress-intensity values, and other property information. Of particular importance are the ultrasonic-test and toughness requirements. Among the properties for which data are included are thermal conductivity and diffusivity, coefficient of thermal expansion, modulus of elasticity, and yield strength. The design stress-intensity values include a safety factor of 3 on ultimate strength at temperature or 1.5 on yield strength at temperature.

Part AD This part contains requirements for the design of vessels. The rules of Division 2 are based on the maximum-shear theory of failure for stress failure and yielding. Higher stresses are permitted when wind or earthquake loads are considered. Any rules for determining the need for fatigue analysis are given here.

Rules for the design of shells of revolution under internal pressure differ from the Division 1 rules, particularly the rules for formed heads when plastic deformation in the knuckle area is the failure criterion. Shells of revolution for external pressure are determined on the same criterion, including safety factors, as in Division 1. Reinforcement for openings uses the same area-replacement method as Division 1; however, in many cases the reinforcement metal must be closer to the opening centerline.

The rest of the rules in Part AD for flat heads, bolted and studded connections, quick-actuating closures, and layered vessels essentially duplicate Division 1. The rules for support skirts are more definitive in Division 2.

Part AF This part contains requirements governing the fabrication of vessels and vessel parts.

Part AR This part contains rules for pressure-relieving devices.

Part AI This part contains requirements controlling inspection of vessel.

Part AT This part contains testing requirements and procedures.

Part AS This part contains requirements for stamping and certifying the vessel and vessel parts.

Appendixes Appendix 1 defines the basis used for defining stress-intensity values. Appendix 2 contains external-pressure charts, and Appendix 3 has the rules for bolted-flange connections; these two are exact duplicates of the equivalent appendixes in Division 1.

Appendix 4 gives definitions and rules for stress analysis for shells, flat and formed heads, and tube sheets, layered vessels, and nozzles including discontinuity stresses. Of particular importance are Table 4-120.1, "Classification of Stresses for Some Typical Cases," and Fig. 4-130.1, "Stress Categories and Limits of Stress Intensity." These are very useful in that they clarify a number of paragraphs and simplify stress analysis.

Appendix 5 contains rules and data for stress analysis for cyclic operation. Except in short-cycle batch processes, pressure vessels are usually subject to few cycles in their projected lifetime, and the endurance-limit data used in the machinery industries are not applicable. Curves are given for a broad spectrum of materials, covering a range from 10 to 1 million cycles with allowable stress values as high as 650,000 lbf/in^2. This low-cycle fatigue has been developed from strain-fatigue work in which stress values are obtained by multiplying the strains by the modulus of elasticity. Stresses of this magnitude cannot occur, but strains do. The curves given have a factor of safety of 2 on stress or 20 on cycles.

Appendix 6 contains requirements of experimental stress analysis, Appendix 8 has acceptance standards for radiographic examination, Appendix 9 covers nondestructive examination, Appendix 10 gives rules for capacity conversions for safety valves, and Appendix 18 details quality-control-system requirements.

The remaining appendixes are nonmandatory but useful to engineers working with the code.

General Considerations Most pressure vessels for the chemical-process industry will continue to be designed and built to the rules of Sec. VIII, Division 1. While the rules of Sec. VIII, Division 2, will frequently provide thinner elements, the cost of the engineering analysis, stress analysis and higher-quality construction, material control, and inspection required by these rules frequently exceeds the savings from the use of thinner walls.

Additional ASME Code Considerations

ASME Code Sec. III: Nuclear Power Plant Components This section of the code includes vessels, storage tanks, and concrete containment vessels as well as other nonvessel items.

ASME Code Sec. X: Fiberglass-Reinforced-Plastic Pressure Vessels This section is limited to four types of vessels: bag-molded and centrifugally cast, each limited to 1,000 kPa (150 lbf/in^2); filament-wound with cut filaments limited to 10,000 kPa (1500 lbf/in^2); and filament-wound with uncut filaments limited to 21,000 kPa (3000 lbf/in^2). Operating temperatures are limited to the range from +66°C (150°F) to −54°C (−65°F). Low modulus of elasticity and other property differences between metal and plastic required that many of the procedures in Sec. X be different from those in the sections governing metal vessels. The requirement that at least one vessel of a particular design and fabrication shall be tested to destruction has prevented this section from being widely used. The results from the combined fatigue and burst test must give the design pressure a safety factor of 6 to the burst pressure.

Safety in Design Designing a pressure vessel in accordance with the code will, under most circumstances, provide adequate safety. In the code's own words, however, the rules "cover minimum construction requirements for the design, fabrication, inspection, and certification of pressure vessels." The significant word is "minimum." The **ultimate responsibility** for safety rests with the user and the designer. They must decide whether anything beyond code requirements is necessary. The code cannot foresee and provide for all the unusual conditions to which a pressure vessel might be exposed. If it tried to do so, the majority of pressure vessels would be unnecessarily restricted. Some of the conditions that a vessel might encounter are unusually low temperatures, unusual thermal stresses, stress ratcheting caused by thermal cycling, vibration of tall vessels excited by von Kármán vortices caused by wind, very high pressures, runaway chemical reactions, repeated local overheating, explosions, exposure to fire, exposure to materials that rapidly attack the metal, containment of extremely toxic materials, and very large sizes of vessels. Large vessels, although they may contain nonhazardous materials, could, by their very size, create a serious hazard if they burst. The failure of the Boston molasses tank in 1919 killed 12 people. For pressure vessels which are outside code jurisdiction, there are sometimes special hazards in very-high-strength materials and plastics. There may be many others which the designers should recognize if they encounter them.

Metal fatigue, when it is present, is a serious hazard. Section VIII, Division 1, mentions rapidly fluctuating pressures. Division 2 and Sec. III do require a fatigue analysis. In extreme cases vessel contents may affect the fatigue strength (endurance limit) of the material. This is corrosion fatigue. Although most ASME Code materials are not particularly sensitive to corrosion fatigue, even they may suffer an endurance limit loss of 50 percent in some environments. High-strength heat-treated steels, on the other hand, are very sensitive to corrosion fatigue. It is not unusual to find some of these which lose 75 percent of their endurance in corrosive environments. In fact, in corrosion fatigue many steels do not have an endurance limit. The curve of stress versus cycles to failure (S/N curve) continues to slope downward regardless of the number of cycles.

Brittle fracture is probably the most insidious type of pressure-vessel failure. Without brittle fracture, a pressure vessel could be pressurized approximately to its ultimate strength before failure. With brittle behavior some vessels have failed well below their design pressures (which are about 25 percent of the theoretical bursting pressures). In order to reduce the possibility of brittle behavior, Division 2 and Sec. III require impact tests.

The subject of brittle fracture has been understood only since about 1950, and knowledge of some of its aspects is still inadequate. A notched or cracked plate of pressure-vessel steel, stressed at 66°C (150°F), would elongate and absorb considerable energy before breaking. It would have a ductile or plastic fracture. As the temperature is lowered, a point is reached at which the plate would fail in a brittle manner with a flat fracture surface and almost no elongation. The transition from ductile to brittle fracture actually takes place over a temperature range, but a point in this range is selected as the **transition temperature.** One of the ways of determining this temperature is the Charpy impact test (see ASTM Specification E-23). After the transition temperature has been determined by laboratory impact tests, it must be correlated with service experience on full-size plates. The literature on brittle fracture contains information on the relation of impact tests to service experience on some carbon steels.

A more precise but more elaborate method of dealing with the ductile-brittle transition is the **fracture-analysis diagram.** This uses a transition known as the **nil-ductility temperature** (NDT), which is determined by the drop-weight test (ASTM Standard E208) or the drop-weight tear test (ASTM Standard E436). The application of this diagram is explained in two papers by Pellini and Puzak (*Trans. Am. Soc. Mech. Eng.*, 429 (October 1964); Welding Res. Counc. Bull. 88, 1963.

Section VIII, Division 1, is rather lax with respect to brittle fracture. It allows the use of many steels down to −29°C (−20°F) without a check on toughness. Occasional brittle failures show that some vessels are operating below the nil-ductility temperature, i.e., the lower limit of ductility. Division 2 has resolved this problem by requiring impact tests in certain cases. Tougher grades of steel, such as the SA516 steels (in preference to SA515 steel), are available for a small price premium. Stress relief, steel made to fine-grain practice, and normalizing all reduce the hazard of brittle fracture.

Nondestructive testing of both the plate and the finished vessel is important to safety. In the analysis of fracture hazards, it is important to know the size of the flaws that may be present in the completed vessel. The four most widely used methods of examination are radiographic, magnetic-particle, liquid-penetrant, and ultrasonic.

Radiographic examination is either by **x-rays** or by **gamma radiation.** The former has greater penetrating power, but the latter is more portable. Few x-ray machines can penetrate beyond 300-mm (12-in) thickness.

Ultrasonic techniques use vibrations with a frequency between 0.5 and 20 MHz transmitted to the metal by a transducer. The instrument sends out a series of pulses. These show on a cathode-ray screen as they are sent out and again when they return after being reflected from the opposite side of the member. If there is a crack or an inclusion along the way, it will reflect part of the beam. The initial pulse and its reflection from the back of the member are separated on the screen by a distance which represents the thickness. The reflection from a flaw will fall between these signals and indicate its magnitude and position. Ultrasonic examination can be used for almost any thickness of material from a fraction of an inch to several feet. Its use is dependent upon the shape of the body because irregular surfaces may give confusing reflections. Ultrasonic transducers can transmit pulses normal to the surface or at an angle. Transducers transmitting pulses that are oblique to the surface can solve a number of special inspection problems.

Magnetic-particle examination is used only on magnetic materials. Magnetic flux is passed through the part in a path parallel to the surface. Fine magnetic particles, when dusted over the surface, will concentrate near the edges of a crack. The sensitivity of magnetic-particle examination is proportional to the sine of the angle between the direction of the magnetic flux and the direction of the crack. To be sure of picking up all cracks, it is necessary to probe the area in two directions.

Liquid-penetrant examination involves wetting the surface with a fluid which penetrates open cracks. After the excess liquid has been wiped off, the surface is coated with a material which will reveal any liquid that has penetrated the cracks. In some systems a colored dye will seep out of cracks and stain whitewash. Another system uses a penetrant that becomes fluorescent under ultraviolet light.

Each of these four popular methods has its advantages. Frequently, best results are obtained by using more than one method. Magnetic particles or liquid penetrants are effective on surface cracks. Radiography and ultrasonics are necessary for subsurface flaws. *No known method of nondestructive testing can guarantee the absence of flaws.* There are other less widely used methods of examination. Among these are eddy-current, electrical-resistance, acoustics, and thermal testing. *Nondestructive Testing Handbook* [Robert C. McMaster (ed.), Ronald, New York, 1959] gives information on many testing techniques.

The **eddy-current technique** involves an alternating-current coil along and close to the surface being examined. The electrical impedance of the coil is affected by flaws in the structure or changes in composition. Commercially, the principal use of eddy-current testing is for the examination of tubing. It could, however, be used for testing other things.

The **electrical-resistance method** involves passing an electric current through the structure and exploring the surface with voltage probes. Flaws, cracks, or inclusions will cause a disturbance in the voltage gradient on the surface. Railroads have used this method for many years to locate transverse cracks in rails.

The **hydrostatic test** is, in one sense, a method of examination of a vessel. It can reveal gross flaws, inadequate design, and flange leaks. Many believe that a hydrostatic test guarantees the safety of a vessel. This is not necessarily so. A vessel that has passed a hydrostatic test is probably safer than one that has not been tested. It can, however, still fail in service, even on the next application of pressure. Care in material selection, examination, and fabrication do more to guarantee vessel integrity than the hydrostatic test.

The ASME Codes recommend that hydrostatic tests be run at a temperature that is usually above the nil-ductility temperature of the material. This is, in effect, a pressure-temperature treatment of the vessel. When tested in the relatively ductile condition above the nil-ductility temperature, the material will yield at the tips of cracks and flaws and at points of high residual weld stress. This procedure will actually reduce the residual stresses and cause a redistribution at crack tips. The vessel will then be in a safer condition for subsequent operation. This procedure is sometimes referred to as **notch nullification.**

It is possible to design a hydrostatic test in such a way that it probably will be a proof test of the vessel. This usually requires, among other things, that the test be run at a temperature as low as and preferably lower than the minimum operating temperature of the vessel. Proof tests of this type are run on vessels built of ultrahigh-strength steel to operate at cryogenic temperatures.

Other Regulations and Standards Pressure vessels may come under many types of regulation, depending on where they are and what they contain. Although many states have adopted the ASME Boiler and Pressure Vessel Code, either in total or in part, any state or municipality may enact its own requirements. The federal government regulates some pressure vessels through the Department of Transportation, which includes the Coast Guard. If pressure vessels are shipped into foreign countries, they may face additional regulations.

Pressure vessels carried aboard United States–registered ships must conform to rules of the **U.S. Coast Guard.** Subchapter F of Title 46, *Code of Federal Regulations*, covers marine engineering. Of this, Parts 50 through 61 and 98 include pressure vessels. Many of the rules are similar to those in the ASME Code, but there are differences.

The **American Bureau of Shipping** (ABS) has rules that insurance underwriters require for the design and construction of pressure vessels which are a permanent part of a ship. Pressure cargo tanks may be permanently attached and come under these rules. Such tanks supported at several points are independent of the ship's structure and are distinguished from "integral cargo tanks" such as those in a tanker. ABS has pressure vessel rules in two of its publications. Most of them are in *Rules for Building and Classing Steel Vessels*.

Standards of Tubular Exchanger Manufacturers Association (TEMA) give recommendations for the construction of tubular heat exchangers. Although TEMA is not a regulatory body and there is no legal requirement for the use of its standards, they are widely accepted as a good basis for design. By specifying TEMA standards, one can obtain adequate equipment without having to write detailed specifications for each piece. TEMA gives formulas for the thickness of tube sheets. Such formulas are not in ASME Codes. (See further discussion of TEMA in Sec. 11.)

Vessels with Unusual Construction High pressures create design problems. The ASME Code Sec. VIII, Division 1, applies to vessels rated for pressures up to 20,670 kPa (3000 lbf/in^2). Division 2 is unlimited. At high pressures, special designs not necessarily in accordance with the code are sometimes used. At such pressures, a vessel designed for ordinary low-carbon-steel plate, particularly in large diameters, would become too thick for practical fabrication by ordinary methods. The alternatives are to make the vessel of high-strength plate, use a solid forging, or use multilayer construction.

High-strength steels with tensile strengths over 1380 MPa (200,000 lbf/in^2) are limited largely to applications for which weight is very important. Welding procedures are carefully controlled, and preheat is used. These materials are brittle at almost any temperature, and vessels must be designed to prevent brittle fracture. Flat spots and variations in curvature are avoided. Openings and changes in shape require appropriate design. The maximum permissible size of flaws is determined by fracture mechanics, and the method of examination must assure as much as possible that larger flaws are not present. All methods of nondestructive testing may be used. Such vessels require the most sophisticated techniques in design, fabrication, and operation.

Solid forgings are frequently used in construction for pressure vessels above 20,670 kPa (3000 lbf/in^2) and even lower. Almost any shell thickness can be obtained, but most of them range between 50 and 300 mm (2 and 12 in). The ASME Code lists forging materials with tensile strengths from 414 to 930 MPa (from 60,000 to 135,000 lbf/in^2). Brittle fracture is a possibility, and the hazard increases with thickness. Furthermore, some forging alloys have nil-ductility temperatures as high as 121°C (250°F). A forged vessel should have an NDT at least 17°C (30°F) below the design temperature. In operation, it should be slowly and uniformly heated at least to NDT before it is subjected to pressure. During construction, nondestructive testing should be used to detect dangerous cracks or flaws. Section VIII of the ASME Code, particularly Division 2, gives design and testing techniques.

As the size of a forged vessel increases, the sizes of ingot and handling equipment become larger. The cost may increase faster than the weight. The problems of getting sound material and avoiding brittle fracture also become more difficult. Some of these problems are avoided by use of **multilayer construction.** In this type of vessel, the heads and flanges are made of forgings, and the cylindrical portion is built up by a series of layers of thin material. The thickness of these layers may be between 3 and 50 mm (⅛ and 2 in), depending on the type of construction. There is an inner lining which may be different from the outer layers.

Although there are multilayer vessels as small as 380-mm (15-in) inside diameter and 2400 mm (8 ft) long, their principal advantage applies to the larger sizes. When properly made, a multilayer vessel is probably safer than a vessel with a solid wall. The layers of thin material are tougher and less susceptible to brittle fracture, have less probability of defects, and have the statistical advantage of a number of small elements instead of a single large one. The heads, flanges, and welds, of course, have the same hazards as other thick members. Proper attention is necessary to avoid cracks in these members.

There are several assembly techniques. One frequently used is to form successive layers in half cylinders and butt-weld them over the previous layers. In doing this, the welds are staggered so that they do not fall together. This type of construction usually uses plates from 6 to 12 mm (¼ to ½ in) thick. Another method is to weld each layer separately to form a cylinder and then shrink it over the previous layers. Layers up to about 50-mm (2-in) thickness are assembled in this way. A third method of fabrication is to wind the layers as a continuous sheet. This technique is used in Japan. The Wickel construction, fabricated in Germany, uses helical winding of interlocking metal strip. Each method has its advantages and disadvantages, and choice will depend upon circumstances.

Because of the possibility of voids between layers, it is preferable not to use multilayer vessels in applications where they will be subjected to fatigue. Inward thermal gradients (inside temperature lower than outside temperature) are also undesirable.

Articles on these vessels have been written by Fratcher [*Pet. Refiner,* **34**(11), 137 (1954)] and by Strelzoff, Pan, and Miller [*Chem. Eng.,* **75**(21), 143–150 (1968)].

Vessels for high-temperature service may be beyond the temperature limits of the stress tables in the ASME Codes. Section VIII, Division 1, makes provision for construction of pressure vessels up to 650°C (1200°F) for carbon and low-alloy steel and up to 815°C (1500°F) for stainless steels (300 series). If a vessel is required for temperatures above these values and above 103 kPa (15 lbf/in^2), it would be necessary, in a code state, to get permission from the state authorities to build it as a special project. Above 815°C (1500°F), even the 300 series stainless steels are weak, and creep rates increase rapidly. If the metal which resists the pressure operates at these temperatures, the vessel pressure and size will be limited. The vessel must also be expendable because its life will be short. Long exposure to high temperature may cause the metal to deteriorate and become brittle. Sometimes, however, economics favor this type of operation.

One way to circumvent the problem of low metal strength is to use a metal inner liner surrounded by insulating material, which in turn is confined by a pressure vessel. The liner, in some cases, may have perforations which will allow pressure to pass through the insulation and act on the outer shell, which is kept cool to obtain normal strength. The liner has no pressure differential acting on it and, therefore, does not need much strength. Ceramic linings are also useful for high-temperature work.

Lined vessels are used for many applications. Any type of lining can be used in an ASME Code vessel, provided it is compatible with the metal of the vessel and the contents. Glass, rubber, plastics, rare metals, and ceramics are a few types. The lining may be installed separately, or if a metal is used, it may be in the form of clad plate. The cladding on plate can sometimes be considered as a stress-carrying part of the vessel.

A **ceramic lining** when used with high temperature acts as an insulator so that the steel outer shell is at a moderate temperature while the temperature at the inside of the lining may be very high. Ceramic linings may be of unstressed brick, or prestressed brick, or cast in place. Cast ceramic linings or unstressed brick may develop cracks and are used when the contents of the vessel will not damage the outer shell. They are usually designed so that the high temperature at the inside will expand them sufficiently to make them tight in the outer (and cooler) shell. This, however, is not usually sufficient to prevent some penetration by the product.

Prestressed-brick linings can be used to protect the outer shell. In this case, the bricks are installed with a special thermosetting-resin mortar. After lining, the vessel is subjected to internal pressure and heat. This expands the steel vessel shell, and the mortar expands to take up the space. The pressure and temperature must be at least as high as the maximum that will be encountered in service. After the mortar has set, reduction of pressure and temperature will allow the vessel to contract, putting the brick in compression. The upper temperature limit for this construction is about 190°C (375°F). The installation of such linings is highly specialized work done by a few companies. Great care is usually exercised in operation to protect the vessel from exposure to unsymmetrical temperature gradients. Side nozzles and other unsymmetrical designs are avoided insofar as possible.

Concrete pressure vessels may be used in applications that require large sizes. Such vessels, if made of steel, would be too large and heavy to ship. Through the use of posttensioned (prestressed) concrete, the vessel is fabricated on the site. In this construction, the reinforcing steel is placed in tubes or plastic covers, which are cast into the concrete. Tension is applied to the steel after the concrete has acquired most of its strength.

Concrete nuclear reactor vessels, of the order of magnitude of 15-m (50-ft) inside diameter and length, have inner linings of steel which confine the pressure. After fabrication of the liner, the tubes for the cables or wires are put in place and the concrete is poured. High-strength reinforcing steel is used. Because there are thousands of reinforcing tendons in the concrete vessel, there is a statistical factor of safety. The failure of 1 or even 10 tendons would have little effect on the overall structure.

Plastic pressure vessels have the *advantages of chemical resistance* and light weight. Above 103 kPa (15 lbf/in^2), with certain exceptions, they must be designed according to the ASME Code section (see "Storage of Gases") and are confined to the three types of approved code construction. Below 103 kPa (15 lbf/in^2), any construction may be used. Even in this pressure range, however, the code should be used for guidance. Solid plastics, because of low strength and creep, can be used only for the lowest pressures and sizes. A stress of a few hundred pounds-force per square inch is the maximum for most plastics. To obtain higher strength, the filled plastics or filament-wound vessels, specified by the code, must be used. Solid-plastic parts, however, are often employed inside a steel shell, particularly for heat exchangers.

Graphite and ceramic vessels are used fully armored; that is, they are enclosed within metal pressure vessels. These materials are also used for boxlike vessels with backing plates on the sides. The plates are drawn together by tie bolts, thus putting the material in compression so that it can withstand low pressure.

Vessel Codes Other Than ASME Different design and construction rules are used in other countries. Chemical engineers concerned with pressure vessels outside the United States must become familiar with local pressure-vessel laws and regulations. *Boilers and Pressure Vessels*, an international survey of design and approval requirements published by the British Standards Institution, Maylands Avenue, Hemel Hempstead, Hertfordshire, England, in 1975, gives pertinent information for 76 political jurisdictions.

The British Code (British Standards) and the West German Code *(A. D. Merkblätter)* in addition to the ASME Code are most commonly permitted, although Netherlands, Sweden, and France also have codes. The major difference between the codes lies in factors of safety and in whether or not ultimate strength is considered. ASME Code, Sec. VIII, Division 1, vessels are generally heavier than vessels built to the other codes; however, the differences in allowable stress for a given material are less in the higher temperature (creep) range.

Engineers and metallurgists have developed alloys to comply economically with individual codes. In West Germany, where design stress is determined from yield strength and creep-rupture strength and no allowance is made for ultimate strength, steels which have a very high yield-strength-to-ultimate-strength ratio are used.

Other differences between codes include different bases for the design of reinforcement for openings and the design of flanges and heads. Some codes include rules for the design of heat-exchanger tube sheets, while others (ASME Code) do not. The Dutch Code *(Grondslagen)* includes very specific rules for calculation of wind loads, while the ASME Code leaves this entirely to the designer.

There are also significant differences in construction and inspection rules. Unless engineers make a detailed study of the individual codes and keep current, they will be well advised to make use of responsible experts for any of the codes.

Vessel Design and Construction The ASME Code lists a number of loads that must be considered in designing a pressure vessel. Among them are impact, weight of the vessel under operating and test conditions, superimposed loads from other equipment and piping, wind and earthquake loads, temperature-gradient stresses, and localized loadings from internal and external supports. In general, the code gives no values for these loads or methods for determining them, and no formulas are given for determining the stresses from these loads. Engineers must be knowledgeable in mechanics and strength of materials to solve these problems.

Some of the problems are treated by Brownell and Young, *Process Equipment Design*, Wiley, New York, 1959. ASME papers treat others, and a number of books published by the ASME are collections of papers on pressure-vessel design: *Pressure Vessels and Piping Design: Collected Papers, 1927–1959; Pressure Vessels and Piping Design and Analysis*, four volumes; and *International Conference: Pressure Vessel Technology*, published annually.

Throughout the year the Welding Research Council publishes bulletins which are final reports from projects sponsored by the council, important papers presented before engineering societies, and other reports of current interest which are not published in *Welding Research*. A large number of the published bulletins are pertinent for vessel designers.

Care of Pressure Vessels Protection against **excessive pressure** is largely taken care of by code requirements for relief devices. Exposure to fire is also covered by the code. The code, however, does not provide for the possibility of local overheating and weakening of a vessel in a fire. Insulation reduces the required relieving capacity and also reduces the possibility of local overheating.

A pressure-reducing valve in a line leading to a pressure vessel is not adequate protection against overpressure. Its failure will subject the vessel to full line pressure.

Vessels that have an operating cycle which involves the solidification and remelting of solids can develop excessive pressures. A solid plug of material may seal off one end of the vessel. If heat is applied at that end to cause melting, the expansion of the liquid can build up a high pressure and possibly result in yielding or rupture. Solidification in connecting piping can create similar problems.

Some vessels may be exposed to a runaway chemical reaction or even an explosion. This requires relief valves, rupture disks, or, in extreme cases, a barricade (the vessel is expendable). A vessel with a large rupture disk needs anchors designed for the jet thrust when the disk blows.

Vacuum must be considered. It is nearly always possible that the contents of a vessel might contract or condense sufficiently to subject it to an internal vacuum. If the vessel cannot withstand the vacuum, it must have vacuum-breaking valves.

Improper operation of a process may result in the vessel's **exceeding design temperature.** Proper control is the only solution to this problem. Maintenance procedures can also cause excessive temperatures. Sometimes the contents of a vessel may be burned out with torches. If the flame impinges on the vessel shell, overheating and damage may occur.

Excessively low temperature may involve the hazard of brittle fracture. A vessel that is out of use in cold weather could be at a subzero temperature and well below its nil-ductility temperature. In startup, the vessel should be warmed slowly and uniformly until it is above the NDT. A safe value is 38°C (100°F) for plate if the NDT is unknown. The vessel should not be pressurized until this temperature is exceeded. Even after the NDT has been passed, excessively rapid heating or cooling can cause high thermal stresses.

Corrosion is probably the greatest threat to vessel life. Partially filled vessels frequently have severe pitting at the liquid-vapor interface. Vessels usually do not have a corrosion allowance on the outside. Lack of protection against the weather or against the drip of corrosive chemicals can reduce vessel life. Insulation may contain damaging substances. Chlorides in insulating materials can cause cracking of stainless steels.

There are many ways in which a pressure vessel can suffer **mechanical damage.** The shells can be dented or even punctured, they can be dropped or have hoisting cables improperly attached, bolts can be broken, flanges are bent by excessive bolt tightening, gasket contact faces can be scratched and dented, rotating paddles can drag against the shell and cause wear, and a flange can be bolted up with a gasket half in the groove and half out. Most of these forms of damage can be prevented by care and common sense. If damage is repaired by straightening, as with a dented shell, it may be necessary to stress-relieve the repaired area. Some steels are susceptible to embrittlement by aging after severe straining. A safer procedure is to cut out the damaged area and replace it.

The National Board Inspection Code, published by the National Board of Boiler and Pressure Vessel Inspectors, Columbus, Ohio, is helpful. Any repair, however, is acceptable if it is made in accordance with the rules of the Pressure Vessel Code.

Pressure vessels should be **inspected periodically.** No rule can be given for the frequency of these inspections. Frequency depends on operating conditions. If the early inspections of a vessel indicate a low corrosion rate, intervals between inspections may be lengthened. Some vessels are inspected at 5-year intervals; others, as frequently as once a year. Measurement of corrosion is an important inspection

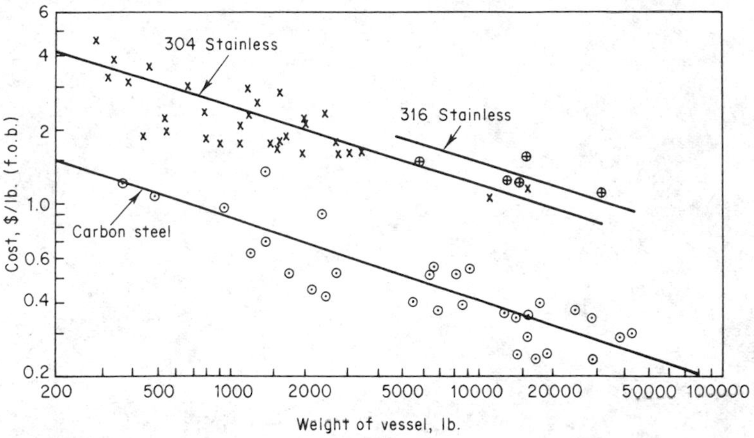

FIG. 6-145 Cost per pound of pressure vessels (1968). For carbon steel $C = 9.05\ W^{-0.34}$, for type 304 stainless steel $C = 25.6\ W^{-0.34}$, and for type 316 stainless steel $C = 34.2\ W^{-0.34}$, where C = FOB cost in dollars per pound and W = weight in pounds. To convert pounds to kilograms, multiply by 0.454.

item. One of the most convenient ways of measuring thickness (and corrosion) is to use an ultrasonic gauge. The location of the corrosion and whether it is uniform or localized in deep pits should be observed and reported. Cracks, any type of distortion, and leaks should be observed. Cracks are particularly dangerous because they can lead to sudden failure. Insulation is usually left in place during inspection of insulated vessels. If, however, severe external corrosion is suspected, the insulation should be removed. All forms of nondestructive testing are useful for examinations.

Care in **reassembling** the vessel is particularly important. Gaskets should be properly located, particularly if they are in grooves. Bolts should be tightened in proper sequence. In some critical cases and with large bolts, it is necessary to control bolt tightening by torque wrenches, micrometers, patented bolt-tightening devices, or heating bolts. After assembly, vessels are sometimes given a hydrostatic test.

Pressure-Vessel Cost and Weight The curves of Fig. 6-141 can be used for estimating cost (freight allowed) when a weight estimate is not available. The cost is based on some 1967 pressure-vessel costs. The prices are plotted as a function of vessel volume for average vessels 6.35 mm (¼ in) thick which are not of unusual design. Correction factors fo other thicknesses are given. Complicated vessels could cost considerably more. Guthrie [*Chem. Eng.*, **76**(6), 114–142 (1969)] also gives pressure-vessel cost data.

When vessels have complicated construction (large, heavy bolted connections, support skirts, etc.), it is preferable to estimate their weight and apply a unit cost in dollars per pound. Some data for vessels purchased in 1968 are plotted in Fig. 6-145. There is a variation of about 2 to 1 between the lowest and the highest costs. The unit FOB cost of carbon steel and type 304 stainless steel was found to vary as the −0.34 power of the weight. Stainless-steel vessels frequently include considerable carbon steel in the form of support skirts, brackets, legs, lap-joint flanges, bolts, etc. In calculating the equivalent weight of a stainless-steel vessel, each pound of carbon should be considered equivalent to 0.4 lb of stainless. See Tables 25-41 and 25-42 in Sec. 25 for appropriate cost indices for updating the data of Figs. 6-141 and 6-145.

Pressure-vessel weights are obtained by calculating the cylindrical shell and heads separately and then adding the weights of nozzles and attachments. Steel weighs 0.283 lb/in^3 and 40.7 lb/ft^3 for 1-in plate. Metal in heads can be approximated by calculating the area of the blank (disk) used for forming the head. The required diameter of blank can be calculated by multiplying the head outside diameter by the approximate factors given in Table 6-53. These factors make no allowance for the straight flange which is a cylindrical extension that is formed on the head. The blank diameter obtained from these factors must be increased by twice the length of straight flange, which is usually 1½ to 2 in but can be up to several inches in length. Manufacturers' catalogs give weights of heads.

Forming a head thins it in certain areas. To obtain the required minimum thickness of a head, it is necessary to use a plate that is initially thicker. Table 6-54 gives allowances for additional thickness.

Nozzles and flanges may add considerably to the weight of a vessel. Their weights can be obtained from manufacturers' catalogs (Taylor Forge Division of Gulf & Western Industries, Inc., Tube Turns Inc., Ladish Co., Lenape Forge, and others). Other parts such as skirts, legs, support brackets, and other details must be calculated.

TABLE 6-53 Factors for Estimating Diameters of Blanks for Formed Heads

	Ratio d/t	Blank diameter factor
A.S.M.E. head	Over 50	1.09
	30–50	1.11
	20–30	1.15
Ellipsoidal head	Over 20	1.24
	10–20	1.30
Hemispherical head	Over 30	1.60
	18–30	1.65
	10–18	1.70

d = head diameter
t = nominal minimum head thickness

TABLE 6-54 Extra Thickness Allowances for Formed Heads*

Minimum head thickness, in.	Extra thickness, in.		
	A.S.M.E. and Ellipsoidal		
	Head o.d. up to 150 in. incl.	Head o.d. over 150 in.	Hemispherical
Up to 0.99	1/16	1/8	3/16
1 to 1.99	1/8	1/8	3/8
2 to 2.99	1/4	1/4	5/8

*Lukens, Inc.

Section 7

Handling of Bulk Solids and Packaging of Solids and Liquids

Grantges J. Raymus, M.E., M.S.; *President, Raymus Associates, Incorporated, Packaging Consultants; Adjunct Professor and Assistant Director, Center for Packaging Engineering, Rutgers, The State University of New Jersey; formerly Manager of Packaging Engineering, Union Carbide Corporation; Registered Professional Engineer, California; Member, Packaging Institute, U.S.A.; ASME.*

TRANSPORTATION OF SOLIDS

PHYSICAL-DISTRIBUTION CONCEPT

Systems Approach **Physical distribution** is a term applied to a systems concept that comprises the entire spectrum of materials movement. The system begins with the storage and handling of raw materials and follows right on through the packaging and disposition of the finished product. The aim is the attainment of the **lowest overall cost** for the system as a whole, comprising the expenses borne by the manufacturer, transport carrier, warehouser, distributor, and customer. Even the manner in which the customer will handle the product is often taken into account.

Two main benefits accrue from a systems approach to materials handling and packaging. First, a trade-off of investment and operating costs is made possible; higher costs in some parts of a system become permissible in return for much lower costs in other parts. The net result is usually the lowest overall cost. If this is not the case, the reasons for incurring the higher costs can be identified and justified. The second benefit is that customers are not offended by ill-conceived packages, delivery vehicles, or product characteristics.

Mathematical modeling, using digital computers, aids in performing a systems-type analysis for either the entire system or parts of it. By means of integer or linear-programming techniques, optimum systems can be identified. The dynamic performance of these can then be determined by simulation techniques.

Determining the capacities of material-handling and packaging equipment is a primary consideration. Many **interacting variables** often are involved, such as an ever-changing or intermittent material-delivery rate, the capacity of intermediate storage and receiver bins, random stoppage or failure of equipment in the system, and the setup and cleanup time between product grades or blends. Variables frequently interact in such complex ways that conventional capacity analyses are impossible, especially if the interaction varies with time. Under such conditions the question of whether the system will deliver the required output can be answered only by simulation techniques.

Even when a total system analysis is unnecessary, the methodology of mathematical modeling is useful, because by considering each component of a system as a block of a flow sheet, the interrelationships become much clearer. Additional alternatives often become apparent, as does the need for more equipment-performance data.

Capacity Definitions In any analysis, the capacity per unit time of dynamic equipment (such as conveyors and bagging machines), as well as the rates at which they actually perform, must be defined more precisely and realistically than by a mere statement of kilograms or pounds per hour. Some useful definitions employed by the equipment industry are the following:

Instantaneous Rate This is a short-term rate when the equipment operates at the design rate or faster. Typical is the average weight handled over a short period of time, not exceeding 5 min.

Hourly Rate This intermediate rate takes into account equipment stoppages due principally to mechanical downtime rather than the equipment's idle time while it waits for action by other parts of the system.

Shift Rate A long-term rate, this reflects all causes of downtime, including idle time. Thus, the average per shift will vary, but by examining its range the practical capacity can be determined. Production time lost due to scheduling of the equipment affects the shift rate. On certain days the equipment has a shift rate close to the hourly rate, while on other days this rate is only half of the hourly rate. Examination of the reasons for this difference often identifies scheduled events as being responsible: the equipment was shut down for cleanup between product grades, product was unavailable for packaging because a bulk order had to be filled, or the product scheduled had a production rate which was half of the products normally made.

These capacity definitions are used to define responsibilities of both vendors and buyers. For instance, often a vendor is called in to examine a piece of equipment that does not perform at the "guaranteed" rate. Records of shift production are offered as proof. Yet the vendor then makes a test and shows that the guaranteed rate is met over a short interval. Who is correct? By defining rates the engineer responsible for the installation not only can avoid these situations but can obtain a better appreciation of potential plant situations.

CONVEYING OF BULK SOLIDS

CONVEYOR SELECTION

Selection of the correct conveyor for a specific bulk material in a specific situation is complicated by the large number of interrelated factors that must be considered. First, the alternatives among basic types must be weighed, and then the correct model and size must be chosen. Workability is the first criterion, but the degree of performance perfection that can be afforded must be established.

Because **standardized equipment designs** and complete engineering data are available for many common types of conveyors, their performance can be accurately predicted when they are used with materials having well-known conveying characteristics. However, even the best conveyors can perform disappointingly if material characteristics are unfavorable. It is often true that conveyor engineering is more of an art than a science; problems involving unusual materials or equipment should be approached with caution.

Many preengineered conveyor components can be purchased off the shelf; they are economical and easy to assemble, and they perform well on conventional applications (for which they are designed). However, it is advisable to check with the manufacturer to be sure that the application is proper.

Capacity requirement is a prime factor in conveyor selection. Belt conveyors, which can be manufactured in relatively large sizes to operate at high speeds, deliver large tonnages economically. On the other hand, screw conveyors become extremely cumbersome as they get larger and cannot be operated at high speeds without creating serious abrasion problems.

Length of travel is definitely limited for certain types of conveyors. With high-tensile-strength belting, the length limit on belt conveyors can be a matter of miles. Air conveyors are limited to 305 m (1000 ft); vibrating conveyors, to hundreds of meters or feet. In general, as length of travel increases, the choice among alternatives becomes narrower.

Lift can usually be handled most economically by vertical or inclined bucket elevators, but when lift and horizontal travel are combined, other conveyors should be considered. Conveyors that combine several directions of travel in a single unit are generally more expensive, but since they require only a single drive, this feature often compensates for the added base cost.

Material characteristics, both chemical and physical, should be considered, especially flowability. Abrasiveness, friability, and lump size are also important. Chemical effects (e.g., the effect of oil on rubber or of acids on metal) may dictate the structural materials out of which conveyor components are fabricated. Moisture or oxidation effects from exposure to the atmosphere may be harmful to the material being conveyed and require total enclosure of the conveyor or even an artificial atmosphere. Obviously, certain types of conveyors lend themselves to such special requirements better than others.

Processing requirements can be met by some conveyors with little or no change in design. For example, a continuous-flow conveyor may provide a desired cooling of the solids simply because it puts the conveyed material into direct contact with heat-conducting metals. Screen decks can be readily attached to vibrating conveyors for simple sizing and scalping operations, and special flights or casings on screw conveyors are available for a wide variety of processing operations such as mixing, dewatering, heating, and cooling.

Initial cost of a conveyor system is usually related to **life expectancy** as well as to the flow rate chosen. There is a great temptation to overdesign, which should be resisted. The first really long-distance belt conveyor was designed and fabricated to extremely high standards of quality. After 35 years it was still in operation with almost all its original components. Had this operation been planned for only a 10-year life, the conveyor system would have represented a bad case of overdesign. While there is a market for used conveyor equipment, it is extremely limited. Thus, it is important to gear conveyor quality to expected lifetime use.

Comparative costs for conveyor systems can be based only on studies of specific problems. For example, belt-conveyor idlers are available in a range of qualities that may make the best unit cost 3 times as much as the cheapest. Bearing quality, steel thickness, and diameter of rolls all affect cost, as does design for easy maintenance and repair. Therefore, it is necessary to make cost comparisons on the basis of a specific study for each conveyor application.

As a general guide to conveyor selection, Table 7-1 indicates **conveyor choices** on the basis of some common functions. Table 7-2 is designed to aid in **feeder selection** on the basis of the physical characteristics of the material to be handled. Table 7-3 is a coded listing of **material characteristics** to be used with Table 7-4, which describes the **conveying qualities** of some common materials. While these tables may serve as valuable guides, conveyor selection must be based on the **as-conveyed characteristics** of a material. For instance, if packing or aerating can occur in the conveyor, the machine's performance will not meet expectations if calculations are based on an average weight per cubic meter. Storage conditions, variations in ambient temperature and humidity, and discharge methods may all affect conveying characteristics. Such factors should be carefully considered before making a final conveyor selection.

TABLE 7-1 Conveyors for Bulk Materials*

Function	Conveyor Type
Conveying materials horizontally	Apron, belt, continuous flow, drag flight, screw, vibrating, bucket, pivoted bucket, air
Conveying materials up or down an incline	Apron, belt, continuous flow, flight, screw, skip hoist, air
Elevating materials	Bucket elevator, continuous flow, skip hoist, air
Handling materials over a combination horizontal and vertical path	Continuous flow, gravity-discharge bucket, pivoted bucket, air
Distributing materials to or collecting materials from bins, bunkers, etc.	Belt, flight, screw, continuous flow, gravity-discharge bucket, pivoted bucket, air
Removing materials from rail cars, trucks, etc.	Car dumper, grain-car unloader, car shaker, power shovel, air

*From FMC Corporation, Material Handling Systems Division.

TABLE 7-2 Feeders for Bulk Materials*

Material Characteristics	Feeder Type
Fine, free-flowing materials	Bar flight, belt, oscillating or vibrating, rotary vane, screw
Non-abrasive and granular materials, materials with some lumps	Apron, bar flight, belt, oscillating or vibrating, reciprocating, rotary plate, screw
Materials difficult to handle because of being hot, abrasive, lumpy, or stringy	Apron, bar flight, belt, oscillating or vibrating, reciprocating
Heavy, lumpy, or abrasive materials similar to pit-run stone and ore	Apron, oscillating or vibrating, reciprocating

*From FMC Corporation, Material Handling Systems Division.

TABLE 7-3 Classification System for Bulk Solids*

	Material characteristics	Class
Size	Very fine—< 149 μm (100 mesh)	A
	Fine—149 μm to 3.18 mm (100 mesh to ⅛ in)	B
	Granular—3.18 to 12.7 mm (⅛ to ½ in)	C
	Lumpy—containing lumps > 12.7 mm (½ in)	D
	Irregular—being fibrous, stringy, or the like	H
Flowability	Very free-flowing—angle of repose up to 30°	1
	Free-flowing—angle of repose 30 to 45°	2
	Sluggish—angle of repose 45° and up	3
Abrasiveness	Nonabrasive	6
	Mildly abrasive	7
	Very abrasive	8
Special characteristics	Contaminable, affecting use or salability	K
	Hygroscopic	L
	Highly corrosive	N
	Mildly corrosive	P
	Gives off dust or fumes harmful to life	R
	Contains explosive dust	S
	Degradable, affecting use or salability	T
	Very light and fluffy	W
	Interlocks or mats to resist digging	X
	Aerates and becomes fluid	Y
	Packs under pressure	Z

*From FMC Corporation, Material Handling Systems Division.
Example: A material which is granular, very free-flowing, mildly abrasive, and mildly corrosive would fall in classes C, 1, 7, and P, making its classification C17P.

To obtain a reliable **measurement of bulk density,** any wide-mouth vessel with a capacity of 1 ft^3 or more may be used. When such a determination must be made often, it is worthwhile to construct a test box from wood or light metal having a dimension of exactly 1 ft for length, width, and depth. The material to be weighed is poured into the test box to overfill it slightly. After the material has been leveled, the test box and its contents are weighed, and after adjustment has been made for the tare weight, the weight obtained is equivalent to the bulk density in the loose or flowing condition. If a loose density is to be determined, care should be exercised in filling the test box so as not to rap or vibrate the material. When a settled density is needed, the filling portion of the procedure is accompanied by rapping the walls of the box until no more material can be added. The density value obtained by this experiment (pounds per cubic foot) can be directly converted to SI units by multiplying by 16.02, giving density in kilograms per cubic meter.

Conveyor Drives Conveyor drives may account for from 10 to 30 percent of the total cost of the conveyor system, depending on specific job requirements. They may be of either fixed-speed or adjustable-speed type. **Fixed-speed drives** are used when the initially chosen conveyor speed does not require change during the course of normal operation. Simple sheave or sprocket changes suffice should minor speed alterations be needed. However, for major adjustments motor or speed-reducer changes are required. In any event, the conveyor must be shut down while the speed change is made. **Adjustable-speed drives** are designed for changing speed either manually or automatically while the conveyor is in operation, to meet variations in processing requirements.

The number of **speed reductions** is another way to classify conveyor drives. Most common of the speed-reduction methods is the

TABLE 7-4 Material Classes and Bulk Densities*

Material	Average bulk density, lb/ft³†	Class‡
Alum, lumpy	50–60	D26§
Alum, fine	45–50	B26§
Alumina	60	B28
Alumina gel	45	B27
Aluminum hydrate	18	C26
Ammonium chloride, crystalline	52	B26
Ammonium sulfate	45–58	§
Antimony powder		B27
Asbestos shred	20–25	H37WZ
Ashes, coal, dry, 3 in. and under	35–40	D37
Asphalt, crushed, ½ in. and under	45	C26
Bagasse	7–10	H36WXZ
Baking powder	41	A26
Bark, wood, refuse	10–20	H37X§
Bauxite, crushed, 3 in. and under	75–85	D28§
Bentonite, 100 mesh and under	50–60	A27Y§
Bicarbonate of soda	41	A26
Boneblack, 100 mesh and under	20–25	A27§
Bonechar, ⅛ in. and under	27–40	B27
Bonemeal	55–60	B27
Borate of lime		A26§
Borax, fine	53	B26
Boric acid, fine	55	B26
Calcium carbide	70–80	D27
Carbon black, pelletized	20–25	B16TZ§
Carbon black, powder	4–6	§
Casein	36	B27§
Cast-iron chips	130–200	C37
Cement, Portland	65–85	A27Y
Cement clinker	75–80	D28§
Chalk, lumpy	85–90	D37Z
Chalk, 100 mesh and under	70–75	A37YZ
Charcoal	18–25	D37T
Cinders, coal	40	D28§
Clay (see bentonite, fuller's earth, kaolin, and marl)		
Coal, anthracite	60	C27P
Coal, bituminous, mined, 50 mesh and under	50	B36P
Coal, bituminous, mined, sized	50	D26PT
Coal, bituminous, mined, slack, ½ in. and under	50	C36P
Coke, loose	23–32	D38TX§
Coke, petroleum, calcined	35–45	D28X
Coke breeze, ¼ in. and under	25–35	C38
Copper sulfate		D26
Cork, fine ground	12–15	B36WY
Cork, granulated	12–15	C36
Cryolite	110	D27
Cullet	80–120	D28§
Dicalcium phosphate	43	A36
Dolomite, lumpy	90–100	D27§
Ebonite, crushed, ½ in. and under	63–70	C26
Epsom salts	40–50	B26
Feldspar, ground, ⅛ in. and under	65–70	B27
Ferrous sulfate	50–75	C27
Flour, wheat	35–40	A36K§
Fluorspar	82	C37
Fly ash, dry	35–45	A18Y§
Fuller's earth, oil filter, burned	40	B28
Fuller's earth, oil filter, raw	35–40	B27
Fuller's earth, oil filter, spent	60–65	§
Glass batch	90–100	D28§
Glue, ground, ⅛ in. and under	40	B27
Graphite, flake	40	C26
Graphite, flour	28	A16Y
Gypsum, calcined, ½ in. and under	55–60	C27
Gypsum, calcined, powdered	60–80	A37
Gypsum, raw, 1 in. and under	90–100	D27
Ice, crushed	35–45	D16
Ilmenite	140	B28
Kaolin clay, 3 in. and under	163	D27
Lead arsenate	72	B36R
Lignite, air dried	45–55	D26
Lime, ground, ⅛ in. and under	60	B36Z
Lime, hydrated, ⅛ in. and under	40	B26YZ
Lime, hydrated, pulverized	32–40	A26YZ
Lime, pebble	53–56	D36
Limestone, agricultural, ⅛ in. and under	68	B27§
Limestone, crushed	85–90	D27§
Limestone dust	75	A37Y§
Magnesium chloride	33	C36
Manganese sulfate	70	C28
Marl	80	D27§
Mica, flakes	17–22	B17WY
Mica, ground	13–15	B27
Mica, pulverized	13–15	A27Y
Muriate of potash	77	B28
Naphthalene flakes	45	§
Oxalic acid crystals	60	B36L
Oyster shells, ground, ½ in. and under	53	C27
Oyster shells, whole		D27X
Phenol-formaldehyde molding powder	30–40	A36
Phosphate rock	75–85	D27§
Phosphate sand	90–100	B28
Phthalic anhydride flakes	30–35	C36XZ
Polyethylene pellets, high-density	35–45	C16K
Polyethylene pellets, low-density	28–40	C16K
Polypropylene pellets	35–50	C16K
Polystyrene cubes	35–40	C16K
Polyvinyl chloride pellets, compounds	35–55	C16K
Polyvinyl chloride resin, dispersion-type	12–18	A36KPY
Polyvinyl chloride resin, solvent, non-solvent, suspension types	20–35	A26KY
Potassium nitrate	76	C17P
Pumice, ⅛ in. and under	42–45	B38§
Salt, common dry, coarse	45–50	C37PL§
Salt, common dry, fine	70–80	B27PL§
Salt cake, dry, coarse	85	D27
Salt cake, dry, pulverized	65–85	B27
Saltpeter	80	B26S
Sand, bank, dry	90–110	B28
Sand, silica, dry	90–100	B18
Sawdust	10–13	§
Shale, crushed	85–90	C27
Shellac, powdered or granulated	31	B26K§
Silica gel	45	B28
Slag, furnace, granulated	60–65	C28
Slate, crushed, ½ in. and under	80–90	C27
Slate, ground, ⅛ in. and under	82	B27
Soap beads or granules		B26T
Soap chips	15–25	C26T§
Soap flakes	5–15	B26T§
Soap powder	20–25	B26§
Soapstone talc, fine	40–50	A37Z
Soda ash, heavy	55–65	B27
Soda ash, light	20–35	A27W
Sodium nitrate	70–80	§
Sodium sulfate (see salt cake)		
Starch	25–50	§
Steel chips, crushed	100–150	D38
Sugar, granulated	50–55	B26KT
Sugar, raw, cane, or beet	55–65	B36Z§
Sugar-beet pulp, dry	12–15	§
Sugar-beet pulp, wet	25–45	§
Sulfur, crushed, ½ in. and under	50–60	C26S§
Sulfur, lumpy, 3 in. and under	80–85	D26S§
Sulfur, powdered	50–60	B26SY§
Talcum powder	40–60	A27Y
Trisodium phosphate	60	B27
Vermiculite, expanded	16	C37W
Vermiculite ore	80	D27
Wood chips	10–30	H36WX§
Wood flour	16–36	§
Zinc oxide, heavy	30–35	A36Z§
Zinc oxide, light	10–15	A36WZ§

*Data supplied mostly by FMC Corporation, Material Handling Systems Division. To convert pounds per cubic foot to kilograms per cubic meter, multiply by 16.02.

† Weights of material, loose or slightly agitated. Weights are usually different when materials are settled or packed as in bins or containers.

‡ These classes represent observations under general conditions. Specific conditions may vary because of manufacturing processes and handling.

§ Class may vary considerably because of conditions.

two-step system, in which the motor is coupled to a speed reducer and the slow-speed shaft of the reducer is connected to the conveyor-drive shaft by a V belt or a roller chain. The second reduction not only permits the use of a simpler speed reducer but also allows a more flexible layout of the motor and reducer mounting plate. On many installations this eliminates the need for a specially designed drive mount.

Since it is good practice to maintain a selected inventory of spare parts for drives, economy can be achieved by **standardizing conveyor drives** throughout the plant. For example, intermediate speed reduction by means of V belts, sheaves or chains, and sprockets can frequently permit using the same speed-reducer size for several drives. Thus, it may be necessary to keep only one repair-stock speed reducer for a number of conveyors.

Conveyor Motors Motors for conveyor drives are generally three-phase, 60-Hz, 220-V units; 220/440-V; 550-V; four-wire, 208-V. Also common are 240- and 480-V ratings. Although many adjustable-speed drives use alternating-current induction motors, powered by ac alternators or ac-driven eddy-current clutches, there is a strong preference for direct-current motors when speed adjustments are required over a wide range at extremely accurate settings.

The **silicon-controlled rectifier** with a dc motor has become predominant in adjustable-speed drives for almost all commonly used conveyors when speed adjustment to process conditions is necessary. The low cost of this control device has influenced its use when speed synchronization among conveyors is required. This can also be done, of course, by changing sheave or sprocket ratios.

The **squirrel-cage motor** is most commonly used with belt conveyors and with drives up to 7.457 kW (10 hp); across-the-line starting is generally specified. Between 7.457 and 37.285 kW (10 and 50 hp), squirrel-cage motors are usually started by means of a manual reduced-voltage starter or a magnetic primary-resistance starter. Normal-torque motors are generally specified, with the assumption that if power is sufficient to drive the belt, sufficient starting torque can be developed. Motor selection for large conveyors should be based on a careful study, with particular emphasis on starting conditions (see also Sec. 24).

Auxiliary Equipment Elevating conveyors must be equipped with some form of **holdback** or **brake** to prevent reversal of travel and subsequent jamming when power is unexpectedly cut off. Ratchet and wedge roller-type holdbacks are commonly used. Solenoid brakes and spring clutches may also be employed.

Another problem with most conveyors is to cut out the driving force when a conveyor jams. **Torque-limiting devices** are often used, as are electrical controls which cut power to the drive motor. However, because of the high inertia of the motor rotor, it is sometimes desirable to eliminate the torque surge which may occur when the conveyor jams. A shear-pin hub is generally used in these cases, power being transmitted through a set of pins which are designed to shear at a fixed maximum torque. While equipment remains down until the pins can be replaced, there is an immediate disconnect between motor and conveyor which may prevent serious equipment damage. Special clutches are also used.

Unless a material discharges freely, **cleaners** are required on belt conveyors and may be helpful on others. Common types use a rotating brush, powered from the conveyor head-pulley shaft or independently, or a spring-mounted blade. The latter is applicable only at some point where the belt conveyor lies reasonably flat. Whenever cleaners are used, provision should be made for catching and chuting the material back into the main discharge stream or to a collecting container which can be periodically emptied.

Control of Conveyors Control has been enhanced considerably with the introduction of process-control computers and programmable controllers, which can be used to maintain rated capacities to close tolerances. This ability is especially useful if feed to the conveyor tends to be erratic. Through variable-speed drives, outputs can be adjusted automatically for changes in processing conditions. When the control devices are used in conjunction with strain-gauge or load-cell weight-sensing devices, actual discharge rates can be measured and employed in process calculations made by these devices, and output adjustments can be made automatically and accurately (see also Sec. 22).

SCREW CONVEYORS

The screw conveyor is one of the oldest and most versatile conveyor types. It consists of a helicoid flight (helix rolled from flat steel bar) or a sectional flight (individual sections blanked and formed into a helix from flat plate), mounted on a pipe or shaft and turning in a trough. Power to convey must be transmitted through the pipe or shaft and is limited by the allowable size of this member. Screw-conveyor capacities are generally limited to around 4.72 m^3/min (10,000 ft^3/h).

In addition to their conveying ability, screw conveyors can be adapted to a wide variety of **processing operations.** Almost any degree of mixing can be achieved with screw-conveyor flights cut, cut and folded, or replaced by a series of paddles. Use of ribbon flights allows sticky materials to be handled. Variable-pitch, tapered-flight, or stepped-flight units can give excellent control for feeder applications or on conveyors when precise control of the transport rate is required. Short-pitch screws are used for inclined and vertical conveying applications, and double-flight short-pitch units effectively deter flushing action. In addition to a wide variety of designs for components, screw conveyors may be fabricated in materials ranging from cast iron to stainless steel.

Use of hollow screws and pipes for circulating hot or cold fluids allows the screw conveyor to be used for heating, cooling, and drying operations. Jacketed casings may be used for the same purpose. It is relatively easy to seal a screw conveyor from the outside atmosphere so that it can operate outdoors without special protection. In fact, the conveyor can be completely sealed to operate in its own atmosphere at positive or negative pressure, and the casing can be insulated to maintain internal temperatures in areas of high or low ambient temperature. A further advantage is the fact that the casing can be designed with a drop bottom for easy cleaning to avoid contamination when different materials are to be run through the same system.

Since screw conveyors are usually made up of standard sections coupled together, special attention should be given to bending stresses in the couplings. Hanger bearings supporting the flights obstruct the flow of material when the trough is loaded above their level. Thus, with difficult materials, the load in the trough must be kept below this level, or special hanger bearings which minimize obstruction should be selected. Since screw conveyors operate at relatively low rotational speeds, the fact that the outer edge of the flight may be moving at a relatively high linear speed is often neglected. This may create a wear problem; if wear is too severe, it can be reduced by the use of hard-surfaced edges, detachable hardened flight segments, rubber covering, or high-carbon steels.

Power calculations for screw conveyors are well standardized. However, each manufacturer has grouped numerical constants in a different fashion and assigned slightly different values on the basis of individual design variations. Thus, in comparing screw-conveyor power requirements it is advisable to use a specific formula for specific equipment.

Required power is made up of two components, that necessary to drive the screw empty and that necessary to move the material. The first component is a function of conveyor length, speed of rotation, and friction in the conveyor bearings. The second is a function of the total weight of material conveyed per unit of time, conveyed length, and depth to which the trough is loaded. The latter power item is in turn a function of the internal friction and friction on metal of the conveyed material.

Table 7-5 indicates **screw-conveyor performance** on the basis of material classifications as listed in Table 7-4 and defined in Table 7-3. Table 7-6 gives a wide range of **capacities** and **power requirements** for various sizes of screws handling 801 kg/m^3 (50 lb/ft^3) of material of average conveyability. Within reasonable limits, values from Tables 7-5 and 7-6 can be interpolated for preliminary estimates and designs.

Typical **feed arrangements** are shown in Fig. 7-1. Plain spouts (Fig. 7-1*a*) may be used when the feed rate is fairly uniform and controlled by preceding equipment. The capacity of the conveyor should be well above the maximum rate of feed from either single or multiple feed points. The rotary cutoff valve (Fig. 7-1*b*) is an

TABLE 7-5 Screw-Conveyor Capacities and Loading Conditions*

Material class†	Screw diam., in	Max. lump size, in. 25% lumps	Max. lump size, in. 100% lumps	Capacity, cu. ft./hr.‡ At 1 r.p.m.	Capacity, cu. ft./hr.‡ At max. r.p.m.§	Approx. area occupied by material¶
A, B, C, D, and H 16, 26, 36	6	¾	½	2.27	375	
	9	1½	¾	8.0	1,200	
	12	2	1	19.3	2,700	
	14	2½	1¼	30.8	4,000	
	16	3	1½	46.6	5,600	
	18	3	2	66.1	7,600	
	20	3½	2	95.0	10,000	45%
A, B, C, D, and H 17, 27, 37	6	¾	½	1.5	75	
	9	1½	¾	5.6	280	
	12	2	1	13.3	665	
	14	2½	1¼	21.1	1,055	
	16	3	1½	31.4	1,570	
	18	3	2	45.4	2,270	
	20	3½	2	62.1	3,105	30%
A, B, C, D, and H 18, 28, 38	6	¾	½	0.75	25	
	9	1½	¾	2.8	90	
	12	2	1	6.7	200	
	14	2½	1¼	10.5	300	
	16	3	1½	15.7	425	
	18	3	2	22.7	590	
	20	3½	2	31.1	780	15%

*FMC Corporation, Material Handling Systems Division. To convert cubic feet per hour to cubic meters per hour, multiply by 0.02832; to convert screw diameter in inches to the nearest screw size in centimeters, multiply by 2.5. See Table 21-6 for the conversion of particle sizes from one measurement system to another.

†These classifications cover a broad list of materials that generally can be handled in a screw conveyor. Special consideration must be given to applications handling materials with the following characteristics:
Highly corrosive, Class N
Degradable, affecting use or salability, Class T
Interlocks or mats, Class X
Highly aerated or of fluid nature, Class Y

‡Capacity for horizontal conveyor uniformly fed. Volumetric capacity is based on material slightly agitated or fluffed. Material highly fluffed or aerated will decrease in weight and increase in volume.

§Maximum capacity for economical service.

¶Percentages higher than those indicated will result in excessive wear on hanger bearings and couplings.

enclosed dust-tight quick-acting valve for free-flowing materials. The rotary-vane feeder (Fig. 7-1*c*) delivers a uniform predetermined volume of material and may be driven from the screw or independently by constant- or variable-speed drive. Rack-and-pinion gates (Fig. 7-1*d*) are well suited to free-flowing materials in bins, hoppers, tanks, or silos and are also used as side inlet gates (Fig. 7-1*e*) for heavy or lumpy materials.

Typical **discharge arrangements** are shown in Fig. 7-2. Plain discharge openings (Fig. 7-2*a*) equipped with a discharge spout (Fig. 7-2*b*) are most common, although the open-end trough (Fig. 7-2*c*) is frequently used, as is the discharge-trough end (Fig. 7-2*e*). Open-bottom troughs (Fig. 7-2*g*) are often used for spreading material uniformly over a storage area. Flat-bottomed rack-and-pinion gates (Fig. 7-2*f*) allow selective discharge, as do hand slide gates (Fig. 7-2*d*). However, for perishable materials, the curved slide gate (Fig. 7-2*h*) eliminates the dead-storage pocket. Enclosed rack-and-pinion gates (Fig. 7-2*j*) give dust-tight operation, and rotary cutoff valves (Fig. 7-2*i*) allow quick shutoff and are readily adaptable to remote control. Air-cylinder-actuated gates have become more and more prominent because of the low investment required and the ease of connecting to automatic process-control centers.

BELT CONVEYORS

The belt conveyor is almost universal in application. It can travel for miles at speeds up to 5.08 m/s (1000 ft/min) and handle up to 4539 metric tons/h (5000 tons/h). It can also operate over short distances at speeds slow enough for manual picking, with a capacity of only a few kilograms per hour. However, it is not normally applicable to processing operations, except under unusual conditions.

Belt-conveyor slopes are limited to a maximum of about 30°, with those in the 18 to 20° range more common. Direction changes can occur only in the vertical plane of the belt path and must be carefully designed as vertical curves or relatively flat bends. Belt conveyors inside the plant may have higher initial cost than some other types of conveyors and, depending on idler design, may or may not require more maintenance. However, a belt conveyor given good routine maintenance can be expected to outlast almost any other type of conveyor. Thus, in terms of cost per ton handled, outstanding economy records have been established by belt conveyors.

Belt-conveyor design begins with a study of the material to be handled. Since weight per cubic meter or foot is an important factor, it should be accurately determined with the material in an as-handled condition. It is not wise to rely solely on published tables of weight per cubic meter or foot for various materials, since many processing operations will affect this by fluffing or compacting the material. Lump size is important, too. For a 600-mm (24-in) belt, uniform lump size can range up to about 102 mm (4 in). For each 152-mm

TABLE 7-6 Screw-Conveyor Data for 50-lb/ft³ Material and Pipe-Mounted Sectional Spiral Flights*

Capacity† Tons/h	Capacity† ft³/h	Diam. of flights, in	Diam. of pipe, in‡	Diam. of shafts, in	Hanger centers, ft	Max. size of lumps: All lumps	Max. size of lumps: Lumps 20 to 25%	Max. size of lumps: Lumps 10% or less	Speed, r/min	Max. torque capacity, in·lb	Feed section diam., in	hp at motor§ 15-ft. max. length	hp at motor§ 30-ft max. length	hp at motor§ 45-ft max. length	hp at motor§ 60-ft max. length	hp at motor§ 75-ft max. length	Max. hp capacity at speed listed	Cost, $¶
5	200	9	2½	2	10	¾	1½	2¼	40	7,600	6	0.43	0.85	1.27	1.69	2.11	4.8	2030
10	400	10	2½	2	10	¾	1½	2½	55	7,600	9	0.85	1.69	2.25	3.00	3.75	6.6	2250
15	600	10	2½	2	10	¾	1½	2½	80	7,600	9	1.27	2.25	3.38	3.94	4.93	9.6	2250
		12	2½	2	12	1	2	3	45	7,600	10	1.27	2.25	3.38	3.94	4.93	5.4	2655
		12	3½	3						16,400		1.27	2.25	3.38	3.94	4.93	11.7	2710
20	800	12	2½	2	12	1	2	3	60	7,600	10	1.69	3.00	3.94	4.87	5.63	7.2	2655
			3½	3						16,400		1.69	3.00	3.94	4.87	5.63	15.6	2710
25	1000	12	2½	2	12	1	2	3	75	7,600	10	2.12	3.75	4.93	5.63	6.55	9.0	2655
			3½	3						16,400		2.12	3.75	4.93	5.63	6.55	9.0	2710
		14	3½	3		1¼	2½	3½	45	16,400	12	2.12	3.75	4.93	5.63	6.55	11.7	3150
30	1200	14	3½	3	12	1¼	2½	3½	55	16,400	12	2.25	3.94	5.05	6.75	7.50	14.3	3150
35	1400	14	3½	3	12	1¼	2½	3½	65	16,400	12	2.62	4.58	5.90	7.00	8.75	16.9	3625
40	1600	16	3½	3	12	1½	3	4	50	16,400	14	3.00	4.50	6.75	8.00	10.00	13.0	3825

*Fairfield Engineering Co. data in U.S. customary system. To convert cubic feet per hour to cubic meters per hour, multiply by 0.02832; to convert tons per hour to metric tons per hour, multiply by 0.9078; and to convert screw size in inches to the nearest screw size in centimeters, multiply by 2.5.

†Capacities are based on screws carrying 31 percent of their cross section and, in the case of feed sections with half-pitch flights, based on 100 percent of their cross section.

‡Pipe sizes given are for ¼-in (6.35-mm) flights.

§Horsepowers listed are calculated for average conditions and are of the proper motor size with factors for length of conveyor, momentary overloads, etc., taken into consideration.

¶These are approximate 1980 prices for 30-ft (9-m) lengths, carbon steel construction, dustproof electrical equipment for common three-phase voltages. Drives are included in prices.

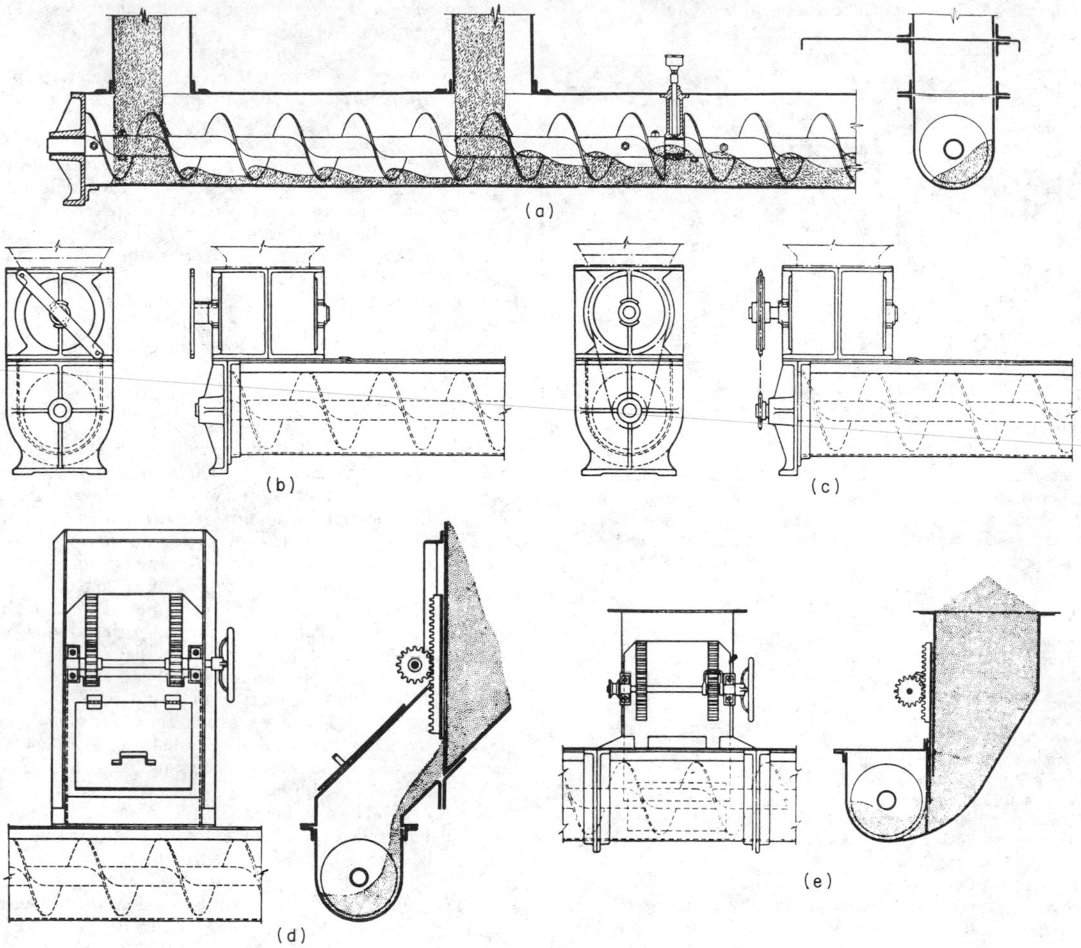

FIG. 7-1 Typical feed arrangments for screw conveyors. (*a*) Plain spouts or chutes. (*b*) Rotary cutoff valve. (*c*) Rotary-vane feeder. (*d*) Bin gate. (*e*) Side inlet gate. *(FMC Corporation, Material Handling Systems Division.)*

(6-in) increase in belt width, lump size can increase by about 51 mm (2 in). If material contains around 90 percent fines, lump size can be increased by around 50 percent. However, care should be taken to maintain uniform flow of material, with fine material reaching the belt first to protect it from impact damage. The larger the lump, the more danger of its falling off the belt or rolling back on inclines. With the belt running horizontally or sloping only slightly at the feed point, the problem of lumps falling off is minimized, especially if particular care is taken with feed-chute design.

Temperature and chemical activity of the conveyed material play important roles in **belt selection.** For example, natural rubber should be avoided with oily materials even when the material does not present an obviously oily surface. Special rubber, cotton, and asbestos-fiber belts are available to meet varying degrees of material temperature, and they should be used whenever high temperatures exist. Belts can be seriously and quickly damaged by high temperature, and the investment in what at first glance seems to be an extremely high-priced belt may prove most economical in the long run. There are many superperformance elastomers available for belt construction. These include neoprene, Teflon, Buna N rubber, and vinyls. Manufacturers are able to test products to be handled and often recommend several elastomer grades that will perform satisfactorily, each grade having a different first-cost–operating-life relation.

Moisture may create poor discharge conditions because of material sticking to the belt and to chutes, or it may even reduce capacity if it is present in enough quantity to give the material fluid properties. Even though abrasion may create problems with belt conveyors, these are easier to solve with properly designed belt systems than with most other conveyors.

In establishing belt-conveyor **tonnage requirements** it is important to work with peak rather than average loads. Only occasionally, because of intentional or accidental variations in production rates, are these two figures identical. The belt that runs empty half the time must carry twice the average load when it is working.

When a belt conveyor must **change direction,** it is often easier to use more than one conveyor. However, vertical curves can be designed and upward changes of direction accomplished with a pair of snub pulleys. If the belt pull is downward on the idlers, a simple flat pulley can be used for minor directional changes. In any case, using a single continuous belt eliminates the need for more than one drive. With a pair of snub pulleys, the carrying face of the belt is brought in contact with the pulley; hence special care must be taken to get a good discharge. When bending the belt over a flat pulley, belt speed must be slow enough to keep material from flying off the belt. In many situations the smooth curve, either concave or convex, is preferable. For a 61-cm (24-in) belt the minimum curve radius is about 61 m (200 ft), but for best operating conditions the curve should be carefully designed.

Operating conditions which affect belt-conveyor design include climate, surroundings, and hours of continuous service. Temperature and humidity extremes may dictate total enclosure of the belt; surroundings which involve such conditions as high temperature or cor-

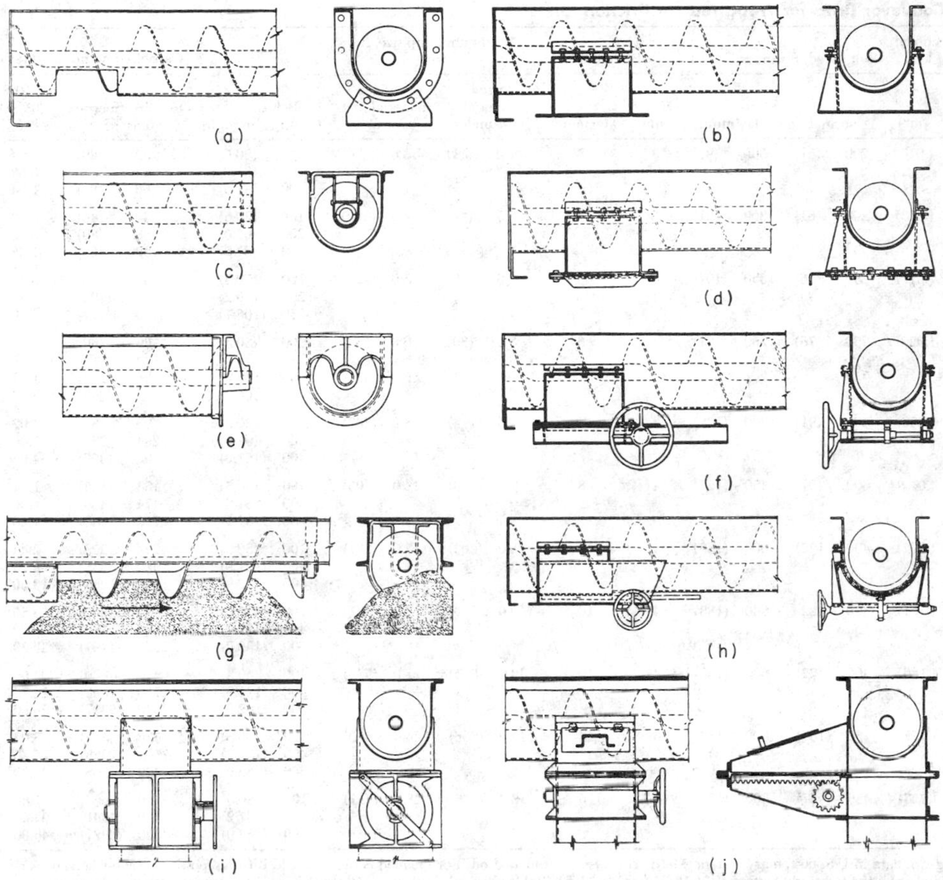

FIG. 7-2 Typical discharge arrangements for screw conveyors. (*a*) Plain discharge opening. (*b*) Discharge spout. (*c*) Open-end trough. (*d*) Hand slide gate. (*e*) Discharge trough end. (*f*) Rack-and-pinion flat side gate. (*g*) Open-bottom trough. (*h*) Rack-and-pinion curved slide gate. (*i*) Rotary cutoff valve. (*j*) Enclosed rack-and-pinion gate. *(FMC Corporation, Material Handling Systems Division.)*

rosive atmosphere can affect belt, machinery, and structure; and continuous service may require extremely high-quality components and even specially designed equipment for servicing while the belt is in operation. For example, idlers may be obtained with tilting stands which allow them to be tipped out of the way for service while the belt is running.

Belt width and speed are functions of bulk density of the material and lump size. Lowest first cost can often be obtained by using the narrowest possible belt for a given lump size and operating it at maximum speed. However, speed often may be limited by dusting, and sometimes it may be better economy to use a wider belt with fewer plies to combine the necessary tensile strength with good belt-troughing characteristics. Abrasiveness of the material can strongly affect speed and also lump size, for at higher speeds abrasive wear is increased and there is greater danger of lumps rolling off the belt. Ideally a belt should run with lump size, slope, and load of less than recommended maximums and with uniform feed introduced to the belt centrally as nearly as possible in the direction and speed of belt travel.

Power to drive a belt conveyor is made up of five components: power to drive the empty belt, to move the load against friction of the rotating parts, to raise or lower the load, to overcome inertia in putting material into motion, and to operate a belt-driven tripper if required. As with most other conveyor problems, it is advisable to work with formulas and constants from a specific manufacturer in making these calculations. For estimating purposes, typical data are given in Table 7-7.

Belt selection depends on power and development of the required tensile strength. Knowing drive-shaft power, belt tension can be calculated and a belt selected. However, since various combinations of width and ply thickness will develop the required strength, final selection is influenced by lump size, troughability of the belt, and ability of the belt to support the load between idlers. Thus it is necessary to use an empirical approach to arrive at a belt selection which meets all requirements.

Once final belt selection has been made, **idlers** and **return rolls** can also be selected. Figure 7-3 indicates the wide variety of belt supports for bulk-handling applications. Figure 7-3*a* and *b* consists of flat-belt arrangements of rollers or plate which allow material to be discharged by simple V-shaped plows. The flat plate-supported belt allows sidewalls to be erected to prevent dribble or to build up larger loads on the flat belt. As in Fig. 7-3*f*, larger capacity can also be achieved by troughing the plate. The 20° troughing idler with equal-length rolls (Fig. 7-3*c*) is the most common, with lighter materials adaptable to 45° idlers with short or long side rolls (Fig. 7-3*d* and *e*). Since the lighter materials do not require stiff belts for tensile strength, there is usually no problem with troughing.

With the proper idlers selected for size and service conditions, the most important step is to locate them properly. For long belts the tension varies considerably, and idlers should be spaced to hold belt sag to reasonable limits along the full length of travel. Too much belt sag can cause a significant power loss, but for most belts of ordinary length it is usually satisfactory to space idlers fairly closely at the feed point and then farther apart and uniformly for the rest of the conveyor length.

Loading and discharge points on belt conveyors need to accom-

TABLE 7-7 Belt-Conveyor Data for Troughed Antifriction Idlers*

Belt width, in (cm)	Cross-sectional area of load, ft² (m²)	Belt speed, ft/min (m/min), Normal	Belt speed, ft/min (m/min), Maximum	Belt plies, Minimum	Belt plies, Maximum	Maximum lump size, in (mm), Sized material, 80% under	Maximum lump size, in (mm), Unsized material, not over 20%	Belt speed, ft/min (m/min)	Capacity, tons/h (metric tons/h)	hp/10-ft (3.05-m) lift	hp/100-ft (30.48-m) centers	Add for tripper hp†
14 (35)	0.11 (.010)	200 (61)	300 (91)	3	5	2.0 (51)	3.0 (76)	100 (30.5)	32 (29)	0.34	0.44	
								200 (61.0)	64 (58)	0.68	0.68	2.0
								300 (91.5)	96 (87)	1.04	1.32	
16 (40)	0.14 (.013)	200 (61)	300 (91)	3	5	2.5 (64)	4.0 (102)	100 (30.5)	44 (40)	0.46	0.56	
								200 (61.0)	88 (80)	0.90	1.12	2.5
								300 (91.5)	132 (120)	1.36	1.68	
18 (45)	0.18 (.017)	250 (76)	350 (107)	4	6	3.0 (76)	5.0 (127)	100 (30.5)	54 (49)	0.58	0.70	
								250 (76.2)	134 (122)	1.42	1.76	3.0
								350 (106.7)	190 (172)	2.00	2.42	
20 (50)	0.22 (.020)	250 (76)	350 (107)	4	6	3.5 (89)	6.0 (152)	100 (30.5)	66 (60)	0.70	0.84	
								250 (76.2)	164 (148)	1.72	2.06	3.20
								350 (106.7)	230 (209)	2.44	2.90	
24 (60)	0.33 (.030)	300 (91)	400 (122)	4	7	4.5 (114)	8.0 (203)	100 (30.5)	98 (89)	1.02	1.02	
								300 (91.5)	294 (267)	3.06	3.04	3.5
								400 (121.9)	392 (356)	4.08	4.04	
30 (75)	0.53 (.049)	300 (91)	450 (137)	4	8	7.0 (178)	12.0 (305)	100 (30.5)	158 (143)	1.60	1.50	
								300 (91.5)	474 (430)	4.80	4.50	5.0
								450 (137.2)	710 (645)	7.20	6.74	
36 (90)	0.78 (.072)	400 (122)	600 (183)	4	9	8.0 (203)	15.0 (381)	100 (30.5)	230 (209)	2.44	1.59	
								400 (121.9)	920 (835)	9.74	6.36	7.0
								600 (182.9)	1380 (1253)	14.60	9.52	
42 (105)	1.09 (.101)	400 (122)	600 (183)	4	10	10.0 (254)	18.0 (457)	100 (30.5)	330 (300)	3.50	2.28	
								400 (121.9)	1320 (1198)	14.00	9.12	9.5
								600 (182.9)	1980 (1797)	23.20	13.68	
48 (120)	1.46 (.136)	400 (122)	600 (183)	4	12	12.0 (305)	21.0 (533)	100 (30.5)	440 (399)	4.66	3.04	
								400 (121.9)	1760 (1598)	18.70	12.14	12.8
								600 (182.9)	2640 (2397)	28.00	18.20	
54 (135)	1.90 (.177)	450 (137)	600 (183)	6	14	14.0 (356)	24.0 (610)	100 (30.5)	570 (517)	6.04	3.94	
								450 (137.2)	2564 (2328)	27.20	17.70	20.0
								600 (182.9)	3420 (3105)	36.20	23.60	
60 (150)	2.40 (.223)	450 (137)	600 (183)	6	16	16.0 (406)	28.0 (711)	100 (30.5)	720 (654)	7.64	4.98	
								450 (137.2)	3240 (2941)	34.40	22.40	23
								600 (182.9)	4320 (3921)	45.80	29.90	

*Fairfield Engineering Co. data in U.S. customary system. Metric conversion is rounded off. For inclined conveyors, add lift horsepower to center horsepower for total horsepower. For terminals multiply horsepower by the following factors: 0–50 ft (15.2 m), 1.20; 51–100 ft (30.5 m), 1.10; 101–150 ft (45.7 m), 1.05. For countershaft drives, multiply horsepower by 1.05 for each reduction (cut gears).

†Tripper horsepower is based on material bulk density of 100 lb/ft³ (1602 kg/m³) and a belt speed of 300 ft/min (91.4 m/min).

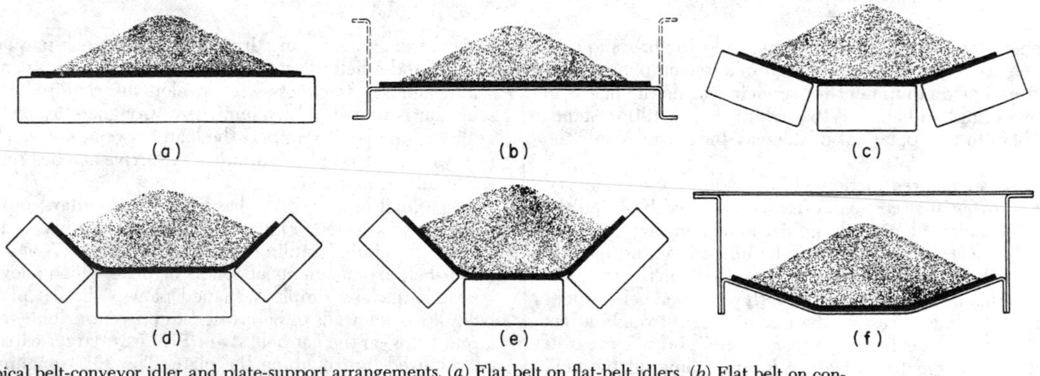

FIG. 7-3 Typical belt-conveyor idler and plate-support arrangements. (*a*) Flat belt on flat-belt idlers. (*b*) Flat belt on continuous plate. (*c*) Troughed belt on 20° idlers. (*d*) Troughed belt on 45° idlers with rolls of unequal length. (*e*) Troughed belt on 45° idlers with rolls of equal length. (*f*) Troughed belt on continuous plate. *(FMC Corporation, Material Handling Systems Division.)*

modate several factors. Figure 7-4*a* shows details for one type of rubber seal on a metal skirt plate. It is particularly important that material be loaded onto the belt in its center and in the direction of its travel, preferably with lumps falling on a layer of fine material. Fines can be delivered to the belt first by notching the feed chute or installing a screen section or grizzly bars. Figure 7-4*b* shows a heavy-duty loading-section design using not only rubber idler rolls but an additional short pad belt. Mass-flow bins and/or bin-flow-assisting devices are often used to minimize segregation of fines and to assure a uniform feed from a hopper onto a conveyor belt.

A clean discharge is vital to good belt life. On the return run the carrying side of the belt is in contact with the return rollers, and any material adhering to it is ground in or deposited on the roller. Extremely sticky material may require a belt-cleaning device in the

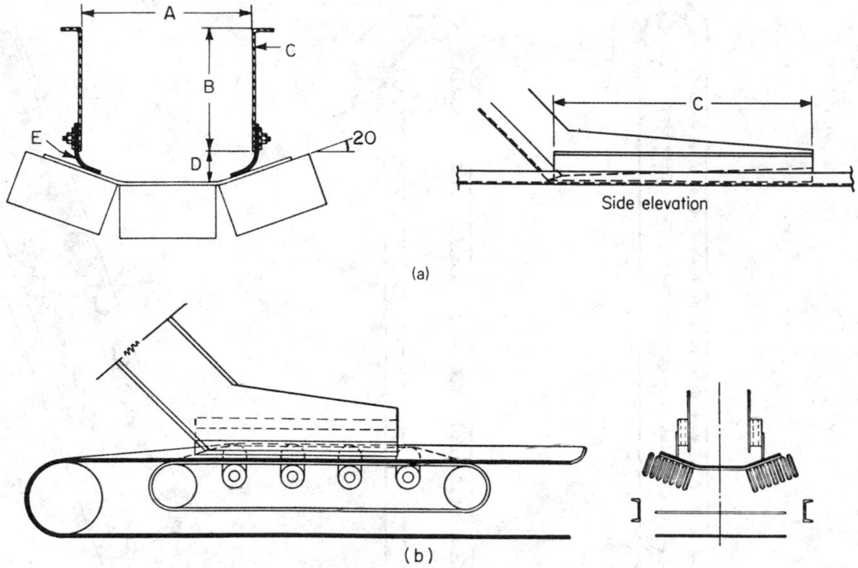

Skirt Plate and Seal Dimensions

Symbol	Name	Dimensions
A	Trough width	See table for (*a*).
B	Skirt depth	Minimum: 6 in (150 mm); maximum: 12 in (300 mm)
C	Skirt length	Minimum: 6 ft (1.8 m); maximum: 8 ft (2.4 m)
D	Skirt-to-belt clearance	See table for (*a*).
E	Seal specification	0.25-in- (6-mm-) thick live rubber, 6 in (150 mm) × *C*

Belt width, in (cm)	14 (35)	16 (40)	18 (45)	20 (50)	24 (60)	30 (75)	36 (90)	42 (110)	48 (125)	54 (140)	60 (155)
Trough width *A*, in (cm)	9 (23)	11 (28)	12 (30)	13 (33)	16 (41)	20 (51)	24 (61)	28 (71)	32 (81)	36 (91)	40 (102)
Skirt seal *D*, in (cm)	2.0 (5.1)	2.25 (5.7)	2.25 (5.7)	2.88 (7.3)	2.88 (7.3)	3.13 (8.0)	3.63 (9.2)	4.0 (10.1)	4.38 (11.1)	4.75 (12.1)	5.25(13.3)

FIG. 7-4 Belt-conveyor loading details. (*a*) Typical skirt-plate design and dimensions. (*b*) Pad belt and special roller-bearing idlers for heavy-duty loading. *(Stephens-Adamson Division, Allis-Chalmers Corporation.)*

form of a revolving brush, spring-mounted steel scrapers, rubber scraper blades, or sometimes a taut wire. When these devices are used, care should be taken that the dribble does not fall on the belt. Refer to the subsection "Storage and Weighing of Solids in Bulk," which deals with the criteria for bin design. For non-free-flowing materials, the combination of correct bin-discharge design and feeder-loading design is often a critical relation in which a slight error in either may produce a system in which the material will not flow at all.

BUCKET ELEVATORS

Bucket elevators are the simplest and most dependable units for making vertical lifts. They are available in a wide range of capacities and may operate entirely in the open or be totally enclosed. The trend is toward highly standardized units, but for special materials and high capacities it is wise to use specially engineered equipment. Main variations in quality are in casing thickness, bucket thickness, belt or chain quality, and drive equipment.

Spaced-Bucket Centrifugal-Discharge Elevators These elevators (Fig. 7-5*a*) are the most common. They are usually equipped with the style 1 or 2 buckets shown in Fig. 7-5*h*. Mounted on a belt or a chain, the buckets are spaced to prevent interference in loading or discharging. This type of elevator will handle almost any free-flowing fine or small-lump material such as grain, coal, or dry chemicals. Buckets are loaded partly by material flowing directly into them and partly by scooping material from the boot as shown in Fig. 7-5*e*. Speeds can be relatively high for fairly dense materials but must be lowered considerably for aerated or low-bulk-density materials [under 641 kg/m^3 (40 lb/ft^3)] to prevent fanning action.

Spaced-Bucket Positive-Discharge Elevators Elevators of this type (Fig. 7-5*b*) are essentially the same as centrifugal-discharge units except that the buckets are mounted on two strands of chain and are snubbed back under the head sprocket to invert them for positive discharge. These units are designed especially for materials which are sticky or tend to pack, and the slight impact of the chain seating on the snub sprocket combined with complete bucket inversion is generally sufficient to empty the buckets completely. In extreme cases, knockers may be used to hit the buckets at the discharge point to help free material. The speed of these units is relatively slow, and buckets must be larger or more closely spaced to reach capacity levels of the centrifugal style.

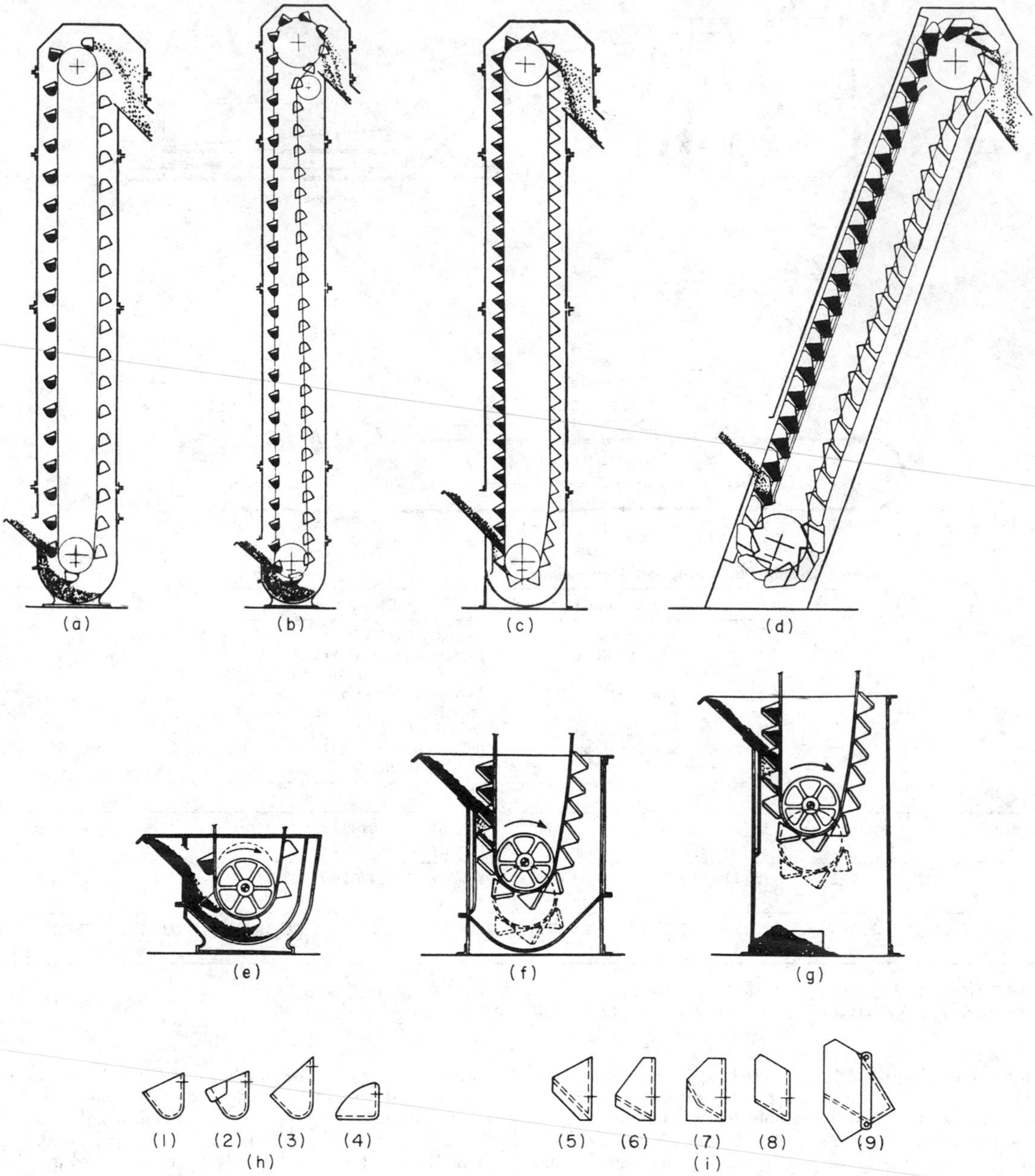

FIG. 7-5 Bucket-elevator types and bucket details. (*a*) Centrifugal-discharge spaced buckets. (*b*) Positive-discharge spaced buckets.(*c*) Continuous bucket. (*d*) Supercapacity continuous bucket. (*e*) Spaced buckets receive part of load direct and part by scooping from bottom. (*f*) Continuous: buckets are filled as they pass through loading leg, with feed spout above tail wheel. (*g*) Continuous: buckets in bottomless boot, with cleanout door. (*h*) Malleable-iron spaced buckets for centrifugal discharge. (*i*) Steel buckets for continuous-bucket elevators. *(Stephens-Adamson Division, Allis-Chalmers Corporation.)*

Continuous-Bucket Elevators These elevators (Fig. 7-5*c*) are generally used for larger-lump materials or for materials too difficult to handle with centrifugal-discharge units. Buckets are closely spaced, with the back of the preceding bucket serving as a discharge chute for the bucket which is dumping as it rounds the head pulley. Close bucket spacing reduces the speed at which the elevator must run to maintain capacities comparable with the spaced-bucket elevator. Gentle discharge prevents excessive degradation and makes this type of elevator effective for handling finely pulverized or aerated materials. Two boot styles and typical loading conditions are illustrated in Fig. 7-5*f* and *g*.

Supercapacity Continuous-Bucket Elevators Elevators of this type (Fig. 7-5*d*) are designed for high lifts and large-lump material. They handle high tonnages and are usually operated at an incline to improve loading and discharge conditions. Operating speeds are low, and because of the heavy loads the bucket-supporting chain is usually guided on the elevating and return runs.

Buckets for spaced-type elevators (Fig. 7-5*h*) are available in both malleable iron and steel in a variety of styles. Style 1 is standard, with style 2 identical except for a reinforced lip. Styles 3 and 4 are low-front designs for wet, stringy, or sticky materials which are difficult to discharge.

Continuous-type buckets (Fig. 7-5*i*) are generally back-mounted to chain or belt at close intervals. They are usually fabricated of steel. Style 5 is standard for normal materials, with style 6 a low-front type for better discharge of difficult materials. Style 7 buckets are used for additional capacity or large lumps, and style 8 for inclined crusher-type elevators. Style 9 buckets are designed for extremely high capacities and are usually side-mounted and hinged together.

Bucket-elevator horsepower can be calculated quite easily. For spaced buckets and digging boots it is equal to the desired capacity in tons per hour multiplied by the lift in feet and divided by 500. For continuous buckets with loading leg, the divisor is increased to 550. Both formulas include normal drive losses as well as loading pickup losses and are applicable for vertical and slightly inclined lifts. For estimating purposes, general bucket-elevator specifications are given for centrifugal units in Table 7-8 and for continuous units in Table 7-9.

V-Bucket Elevator-Conveyors These are still used for handling heavy materials, for coal, and, in light-duty designs, for lightweight free-flowing materials. Similar to the V-bucket type, but with buckets swinging freely on supporting shafts mounted between two strands of roller chain, is the pivoted-bucket conveyor. This type can be equipped with a fixed or movable tripper to dump buckets by overturning them. While considerably more expensive than the V-bucket conveyor, it eliminates the abrasion created by dragging material along in a trough and operates more smoothly at lower power per ton for heavy materials.

The most common chain conveyor is the bucket elevator already discussed, but there are a wide variety of special chain conveyors which are used so infrequently that they should be selected only on the specific recommendation of a qualified materials-handling engineer.

Skip Hoists These hoists, which operate on a batch rather than continuous principle, are not so widely used as in the past. However, for high lifts and extremely lumpy or hot materials, the skip hoist is still an economical and practical device.

Skip hoists may be designed to operate automatically or from a manual push-button station. They are usually classified as uncounterweighted, counterweighted, or balanced. Both the latter systems reduce operating-power requirements, and the balanced unit, using two buckets, can operate at twice the capacity of the others. Figure 7-6 illustrates these types as well as some of the common paths of travel which skip hoists may follow. Speed of operation is also a basis for skip-hoist classification, with multispeed motors required on high-speed operations to slow down bucket travel speed at loading and discharge points.

VIBRATING OR OSCILLATING CONVEYORS

Most vibrating conveyors are essentially directional-throw units which consist of a spring-supported horizontal pan vibrated by a direct-connected eccentric arm, rotating eccentric weights, an electromagnet, or a pneumatic or hydraulic cylinder. The motion imparted to the material particles may vary, but its purpose is to throw the material upward and forward so that it will travel along the conveyor path in a series of short hops.

The **capacity** of directional-throw vibrating conveyors is determined by the magnitude of trough displacement, frequency of this displacement, angle of throw, slope of trough, and ability of the material to receive and transmit through its mass the directional throw of the trough. The material itself is the most important factor. To be conveyed properly it should have a high friction factor on steel as well as a high internal friction factor so that conveying action is transmitted through its entire depth. Thus deep loads tend to move more slowly than thin ones. Material must also be dense enough to minimize the effect of air resistance on its trajectory, and it should not aerate. Tests have shown that granular materials handle better than pulverized materials and flat or irregular shapes better than spherical ones.

Classification of vibrating conveyors can probably best be based on drive characteristics as shown in Fig. 7-7. All these types transmit vibration to their supporting structures, but the direct, or positive, drive is the worst offender and should be mounted on a heavy supporting structure if it is not counterbalanced. Semipositive- and nonpositive-drive types reduce vibration effects because thrust is trans-

TABLE 7-8 Bucket-Elevator Specifications for Centrifugal-Discharge Buckets on Belt, Malleable-Iron, or Steel Buckets*

Size of bucket, in (mm), and bucket spacing, in (mm)†	Elevator centers, ft‡	Capacity, tons/h (metric tons/h)§	Size of lumps handled, in (mm)¶	Bucket speed, ft/min (m/min)	r/min, head shaft	hp required at head shaft	Additional hp/ft for intermediate lengths	Shaft diameter, in: Head	Shaft diameter, in: Tail	Diameter of pulleys, in: Head	Diameter of pulleys, in: Tail	Belt width, in
6 x 4 x 4¼ — 12	25	14 (12.7)	¾ (19.0)	225 (68.6)	43	1.0	0.02	1 15/16	1 11/16	20	14	7
	50	14 (12.7)	¾ (19.0)	225 (68.6)	43	1.6	0.02	1 15/16	1 11/16	20	14	7
(152 x 102 x 108) — (305)	75	14 (12.7)	¾ (19.0)	225 (68.6)	43	2.1	0.02	1 15/16	1 11/16	20	14	7
8 x 5 x 5½ — 14	25	27 (24.5)	1 (25.4)	225 (68.6)	43	1.6	0.04	1 15/16	1 11/16	20	14	9
	50	30 (27.2)	1 (25.4)	260 (79.2)	41	3.5	0.05	1 15/16	1 11/16	24	14	9
(203 x 127 x 140) — (356)	75	30 (27.2)	1 (25.4)	260 (79.2)	41	4.8	0.05	2 7/16	1 11/16	24	14	9
10 x 6 x 6¼ — 16	25	45 (40.8)	1¼ (32.0)	225 (68.6)	43	3.0	0.063	1 15/16	1 15/16	20	16	11
	50	52 (47.2)	1¼ (32.0)	260 (79.2)	41	5.2	0.07	2 7/16	1 15/16	24	16	11
(254 x 152 x 159) — (406)	75	52 (47.2)	1¼ (32.0)	260 (79.2)	41	7.2	0.07	2 15/16	1 15/16	24	16	11
12 x 7 x 7¼ — 18	25	75 (68.1)	1½ (38.1)	260 (79.2)	41	4.7	0.1	2 7/16	1 15/16	24	18	13
	50	84 (76.3)	1½ (38.1)	300 (91.4)	38	8.9	0.115	2 15/16	1 15/16	30	18	13
(305 x 178 x 184) — (457)	75	84 (76.3)	1½ (38.1)	300 (91.4)	38	11.7	0.115	3 7/16	2 7/16	30	18	13
14 x 7 x 7¼ — 18	25	100 (90.8)	1¾ (44.5)	300 (91.4)	38	7.3	0.14	2 15/16	2 7/16	30	18	15
	50	100 (90.8)	1¾ (44.5)	300 (91.4)	38	11.0	0.14	3 7/16	2 7/16	30	18	15
(355 x 179 x 184) — (457)	75	100 (90.8)	1¾ (44.5)	300 (91.4)	38	14.3	0.14	3 7/16	2 7/16	30	18	15
16 x 8 x 8½ — 18	25	150 (136.2)	2 (50.8)	300 (91.4)	38	8.5	0.165	2 15/16	2 7/16	30	20	18
	50	150 (136.2)	2 (50.8)	300 (91.4)	38	12.6	0.165	3 7/16	2 7/16	30	20	18
(406 x 203 x 216) — (457)	75	150 (136.2)	2 (50.8)	400 (121.9)	38	16.7	0.165	3 15/16	2 7/16	30	20	18

*From Stephens-Adamson Division, Allis-Chalmers Corporation.

†Bucket size given: width × projection × depth. Assumed bucket linear speed is 150 ft/min (45.7 m/min).

‡Elevator centers to nearest SI equivalent are 25 ft ≃8 m, 50 ft ≃15 m, and 75 ft ≃23 m.

§Capacities and horsepowers are given for materials having bulk densities of 100 lb/ft^3 (1602 kg/m^3). For other densities these will vary in direct proportion: a 50-lb/ft^3 material will reduce the capacity and horsepower required by 50 percent.

¶If the amount of lump product is less than 15 percent of the total, lump size may be twice that given.

TABLE 7-9 Bucket-Elevator Specifications for Continuous Buckets on Chain*

Size of bucket and bucket spacing, in (mm)†	Elevator centers, ft‡	Capacity, tons/h (metric tons/h)§	Size of lumps handled, in (mm)¶	r/min, head shaft	hp required at head shaft	Additional hp/ft for intermediate lengths	Shaft diameter, in: Head	Shaft diameter, in: Tail	Sprocket diameter, in: Head	Sprocket diameter, in: Tail
8 x 5½ x 7¾ — 8	25	35 (31.7)	1 (25.4)	28	1.8	0.06	1^{15}/16	1^{11}/16	20½	14
(203 x 140 x 197) — (203)	50	35 (31.7)	1 (25.4)	28	3.4	0.06	2 7/16	1^{11}/16	20½	14
	75	35 (31.7)	1 (25.4)	28	5.0	0.06	2^{15}/16	1^{11}/16	20½	14
10 x 7 x 11¾ — 12	25	60 (54.5)	1½ (38.1)	23	3.0	0.10	2 7/16	1^{15}/16	25	17½
(254 x 178 x 298) — (305)	50	60 (54.5)	1½ (38.1)	23	5.5	0.10	2 7/16	1^{15}/16	25	17½
	75	60 (54.5)	1½ (38.1)	23	8.0	0.10	2^{15}/16	1^{15}/16	25	17½
12 x 7 x 11¾ — 12	25	70 (63.5)	1½ (38.1)	23	3.5	0.12	2 7/16	1^{15}/16	25	17½
(305 x 178 x 298) — (305)	50	70 (63.5)	1½ (38.1)	23	6.5	0.12	2^{15}/16	1^{15}/16	25	17½
	75	70 (63.5)	1½ (38.1)	23	9.5	0.12	3 7/16	2 7/16	25	17½
14 x 7 x 11¾ — 12	25	80 (72.6)	1¾ (44.5)	23	4.0	0.14	2 7/16	2 7/16	25	17½
(356 × 178 × 298) — (305)	50	80 (72.6)	1¾ (44.5)	20	7.5	0.14	2^{15}/16	2 7/16	29	17½
	75	80 (72.6)	1¾ (44.5)	20	11	0.14	3 7/16	2 7/16	29	17½
14 x 8 x 11¾ — 12	25	100 (90.8)	2 (50.8)	20	5.0	0.17	2^{15}/16	2 7/16	29	17½
(356 x 203 x 298) — (305)	50	100 (90.8)	2 (50.8)	20	9.3	0.17	3 7/16	2 7/16	29	17½
	75	100 (90.8)	2 (50.8)	20	13.3	0.17	3^{15}/16	2 7/16	29	17½
16 x 8 x 11¾ — 12	25	115 (104.4)	2 (50.8)	20	6.0	0.20	2^{15}/16	2 7/16	29	17½
(406 x 203 x 298) — (305)	50	115 (104.4)	2 (50.8)	20	11	0.20	3^{15}/16	2 7/16	29	17½
	75	115 (104.4)	2 (50.8)	20	16	0.20	4 7/16	2 7/16	29	17½
18 x 8 x 11¾ — 12	25	130 (118.0)	2 (50.8)	20	7	0.22	2^{15}/16	2 7/16	29	17½
(406 x 203 x 298) — (305)	50	130 (118.0)	2 (50.8)	20	13	0.22	3^{15}/16	2 7/16	29	17½
	75	130 (118.0)	2 (50.8)	20	20	0.22	4 7/16	2 7/16	29	17½

*From Stephens-Adamson Division, Allis-Chalmers Corporation.
†Bucket size given: width × projection × depth. Assumed bucket linear speed is 150 ft/min (45.7 m/min).
‡Elevator centers to nearest SI equivalent are 25 ft ≃8 m, 50 ft ≃15 m, and 75 ft ≃23 m.
§Capacities and horsepowers are given for materials having bulk densities of 100 lb/ft^3 (1602 kg/m^3). For other densities these will vary in direct proportion: a 50-lb/ft^3 material will reduce the capacity and horsepower required by 50 percent.
¶If the total amount of lump product is less than 15 percent of the total, lump size may be twice that given.

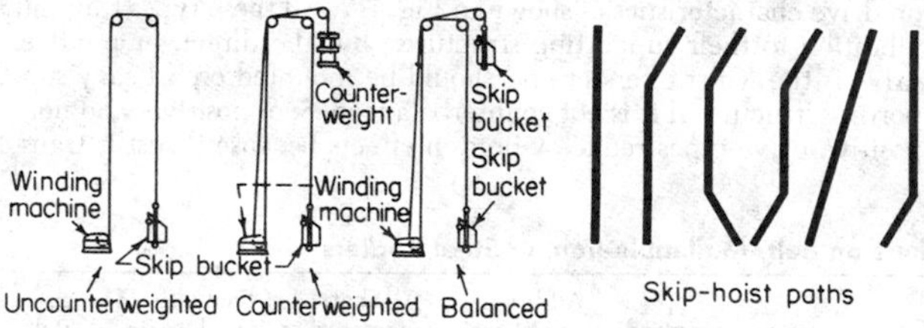

FIG. 7-6 Types of skip hoists and skip-hoist paths. *(Fairfield Engineering Co.)*

mitted over the entire support length rather than at a specific point. Regardless of drive type, care should be taken to mount the conveyor properly so that supporting structures will not be damaged. The frequency of vibration of the conveyor should in no case be at or near the natural frequency of the supporting structure.

Mechanical vibrating conveyors are designed to operate at specific frequencies and do not perform well at other frequencies without carefully designed alterations. Thus they are not adapted to frequent capacity changes except by varying the depth of material fed to the trough. Positive eccentric drives maintain their frequency and magnitude of stroke regardless of load, and serious drive damage can

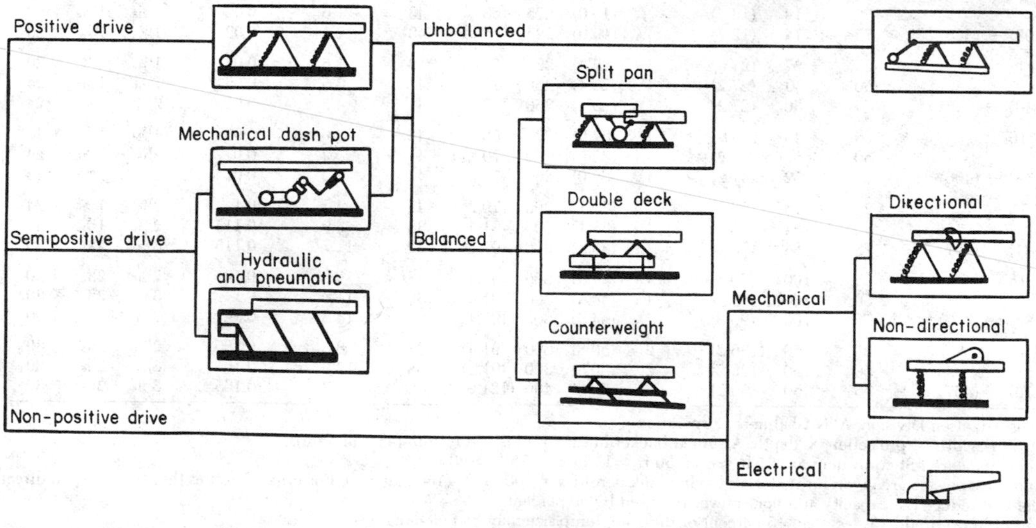

FIG. 7-7 Vibration-conveyor classification. (Modern Materials Handling.)

result from overloading. Rotating eccentric weights can also provide the motive force, and although they maintain a constant frequency, the magnitude of stroke is definitely affected by the load. Directional-throw mechanical vibrating conveyors are used primarily for conveying and do not usually perform well as feeders.

Electrical vibrating conveyors are characterized by the fact that there is no contact between the drive and the conveying medium. They operate on a pull-release cycle or a pull-push cycle, using direct current and pulsating electromagnets or alternating current combined with electromagnets and permanent magnets. While most electrical vibrating units are used as feeders, they also work well as conveyors. Most of them offer the advantage of capacity regulation through control of the electric current magnitude via rheostats. Figure 7-8 gives capacities as a function of pan size and power consumption.

Pneumatic and hydraulic vibrating conveyors have as their greatest asset elimination of explosion hazards. If pressurized air, water, or oil is available, they can be extremely practical since their drive design is relatively simple and pressure-control valves can be used to vary capacity either manually or automatically.

The **capacity** of vibrating conveyors is extremely broad, ranging from thousands of tons down to grams or ounces. Since so many variables affect their ability to convey, there is no simple formula for figuring capacity and power. Available data are generally the results of experiments and empirical equations, with most manufacturers providing selection charts for specific types of conveyors and materials. A typical leaf-spring unit is shown in Fig. 7-8, along with the graphical information required to select a standard unit. Conveyor lengths are limited to about 61 m (200 ft) with multiple drives and about 30.5 m (100 ft) with a single drive. There are many exceptions to these general limitations, and they should not preclude study of a specific problem when vibrating conveyors seem desirable.

Processing operations of many types can be carried out in vibrating conveyors because their simple conveying troughs can be modified quite easily. While tube and flat-pan troughs are most common, troughs can be provided in a wide variety of shapes and materials. Although conveying action is usually so gentle that abrasion problems do not arise, such problems can be easily solved when they do occur by the use of special materials or liners. Troughs are easily sealed to prevent contamination or for operation under positive or negative pressure. With screen or perforated deck plates, vibrating conveyors can dewater, rough-screen, scalp, or dry. Heating and cooling can also be handled by the use of air streams blowing over or through the material, infrared panels, resistance-heating panels, or contact with air- or water-cooled or heated trough casings. Special vibrating-conveyor designs are available for elevating at relatively steep slopes or up a spiral trough. There is probably no other conveyor so readily adaptable to the solution of processing problems.

CONTINUOUS-FLOW CONVEYORS

The principle of the continuous-flow conveyor is that when a surface is pulled transversely through a mass of granular, powdered, or small-lump material, it will pull along with it a cross section of material which is greater than the area of the surface itself. The conveying action of various designs of continuous-flow conveyors varies with the type of conveying flight but theoretically is not comparable with the action in a flight or drag conveyor. Flights vary from solid surfaces to skeleton designs, as shown in Fig. 7-9.

The continuous-flow conveyor is a totally enclosed unit which has a relatively high capacity per unit of cross-sectional area and can follow an irregular path in a single plane. These features make it extremely versatile. Figure 7-10 shows some typical arrangements and applications possible with these conveyors. Included is an example of the unit acting as a dewatering device (Fig. 7-10*c*).

These conveyors employ a chain-supported conveying element (some are cast integrally with the chain, which is designed with easily detachable knuckle joints). Thus the connecting element runs along the outside of the casing so that head and tail sections do not become excessively large because of projecting conveying elements. This means that the material feeding into the conveyor must fall past the chain element and travel in a reverse direction before passing into the actual conveying leg (see Fig. 7-10*a*). Since this affects the lump size that the conveyor can conveniently handle, the loop design (Fig. 7-10*c*) is sometimes used for better feeding conditions, or separate carrying runs and return runs are provided with inclined loading

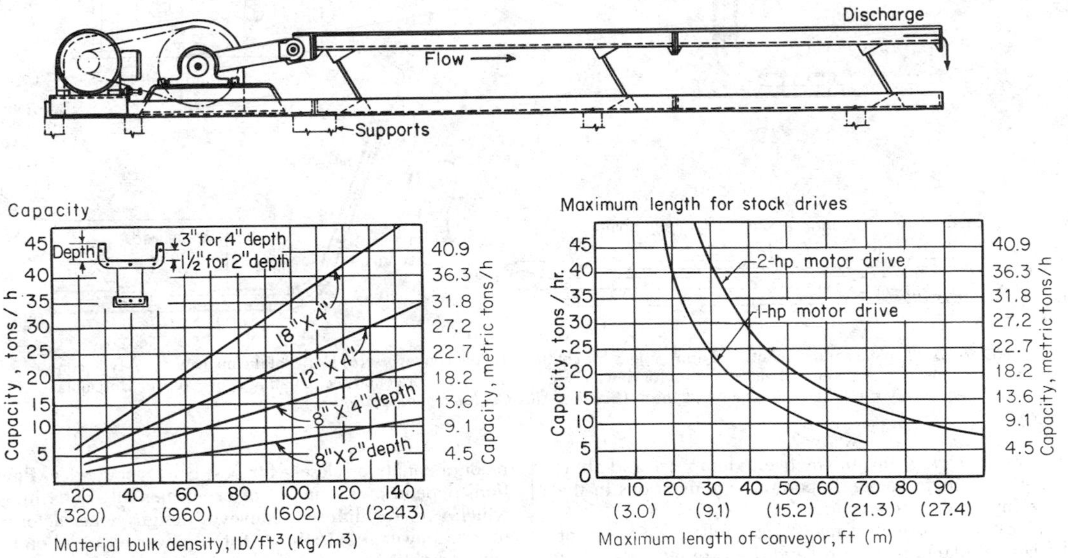

Pan width and stream depth: nearest metric equivalents

in	mm	in	mm
18 x 4	450 x 100	8 x 4	200 x 100
12 x 4	300 x 100	8 x 2	200 x 50

FIG. 7-8 Standardized leaf-spring mechanical oscillating conveyor with selection charts. Multiply pounds per cubic foot by 16.02 to get kilograms per cubic meeter; multiply feet by 0.3048 to get meters. *(FMC Corporation, Material Handling Systems Division.)*

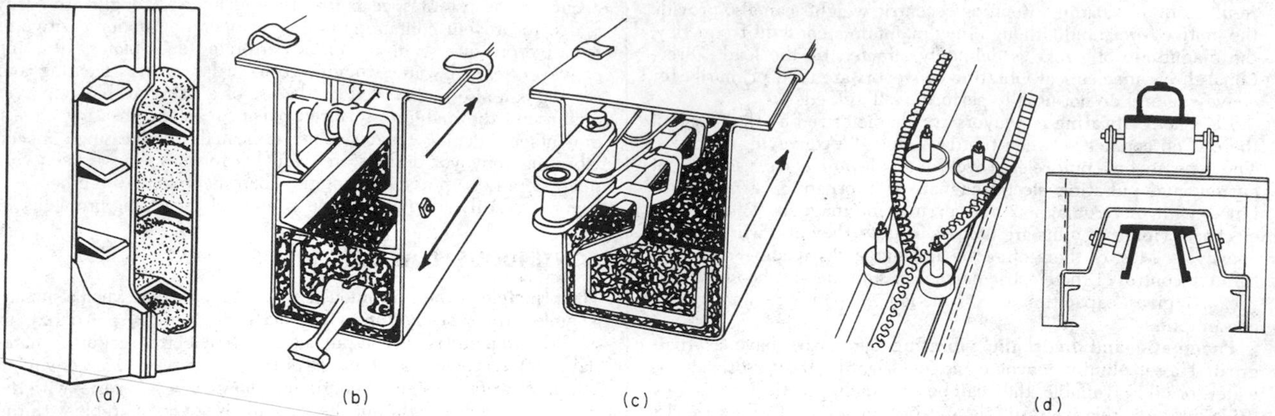

FIG. 7-9 Closed and open flights for continuous-flow conveyors. (*a*) and (*b*) Conveyor-elevator. (*c*) Horizontal conveyor with side-pull chain. (*d*) Detail of closed-belt conveyor; opening and closing rollers mesh and unmesh teeth in the same manner as a conventional clothing fastener. *(FMC Corporation, Material Handling Systems Division; Stephens-Adamson Division, Allis-Chalmers Corporation.)*

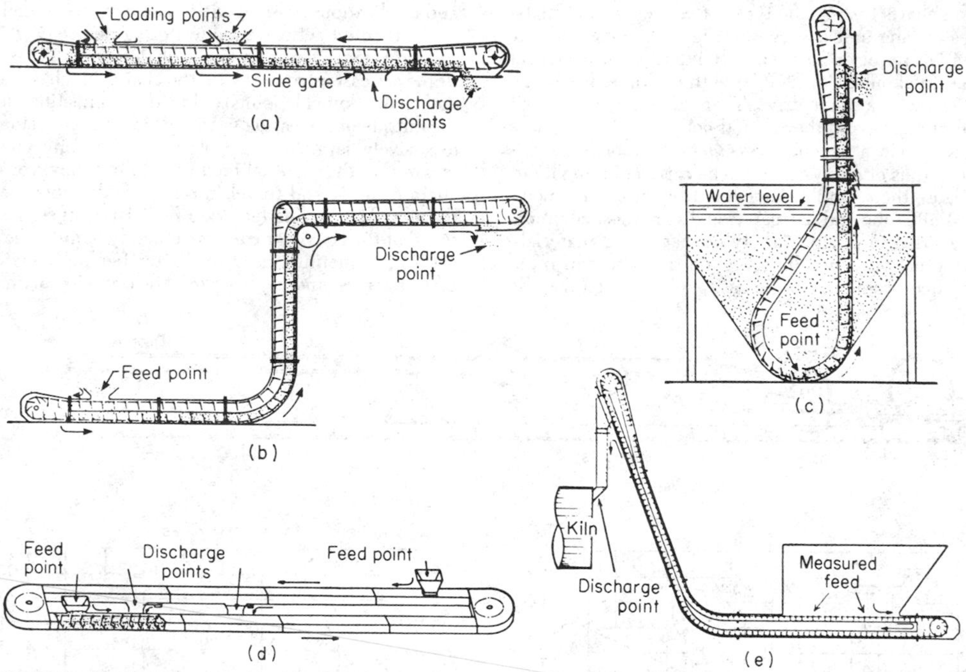

FIG. 7-10 Typical arrangements and applications for continuous-flow conveyors. (*a*) Horizontal conveyor. (*b*) Z-type conveyor-elevator. (*c*) Loop-feed elevator used for dewatering. (*d*) Side-pull horizontal recirculating conveyor. (*e*) Horizontal inclined conveyor-elevator. *(Stephens-Adamson Division, Allis-Chalmers Corporation.)*

chutes to the lower carrying run. In any event, lump size and abrasive characteristics of material are important considerations in the selection of continuous-flow conveyors.

The side-pull continuous-flow conveyor can follow a variety of paths in a horizontal plane, picking up and discharging material at many different points. Figure 7-9*c* is a detailed illustration of one type of conveying element, and Fig. 7-10*d* shows a typical arrangement with 180° turns. Triangular arrangements and rectangular layouts with 90° corners are also available.

The **capacity** of the continuous-flow conveyor is dependent on the particular design being considered. Limiting speeds are subject to considerable controversy. It is advisable to follow the manufacturer's recommendations closely for best conveyor service. **Power** calculations depend on a number of experimentally determined constants which vary for different conveyor designs. One factor contributing to total power requirements is the power required on bend corners where flights assume a radial position and tend to compress material which was fed between them when they were running in a parallel position. Noncompressible materials may require special clearances and feed conditions. Thus, while conveyor components have been well standardized, many materials will not convey well unless special design alterations are made.

Because of the fabrication required for casings and the precision fitting of conveying elements within it, the continuous-flow conveyor is normally an expensive unit. However, it occupies little space, needs little support because the casing forms a rigid box girder, may travel in several directions with only a single drive, is self-feeding, and can feed and discharge at several points. These factors may often compensate for what sometimes appears as a rather high cost per foot. Because it is adaptable to many processing operations, the continuous-flow conveyor is widely used in the chemical industry, in which there is a great deal of rehandling or requirements for many feed and discharge points. The conveyors can be designed for self-cleaning to allow different materials to be handled in the same unit without contamination.

Closed-Belt Conveyor This device, with zipperlike teeth which mesh to form a closed tube, is particularly adaptable to the problem of handling fragile materials which cannot be subjected to degradation. Since the belt is wrapped snugly around the material, it moves with the belt and is not subject to any form of internal movement except at feed and discharge. In addition, the belt can operate in many planes, with twists and turns to meet almost any layout condition within the fixed limit of curvature placed on the loaded belt. It can convey and elevate with only a single drive; multiple feed and discharge points are relatively easy to arrange.

The closed-belt conveyor is not readily adaptable to the handling of sticky materials, and special designs may be required for materials which are highly susceptible to aeration. Initial cost per foot is relatively high because of belting cost, but power requirements are low and with proper installation and maintenance belt life is good.

Since this type of conveyor is available in only one standard size, its capacity is determined by the belt speed and the fixed cross-sectional area. Tons-per-hour capacity is figured by multiplying the bulk density in pounds per cubic foot by the speed in feet per minute and a constant of 0.0021. Power requirements are quite low and figured in the same way as those for conventional belt conveyors.

Figure 7-9*d* illustrates a typical closed-belt-conveyor detail of the opening or closing mechanism and a cross section through a horizontal carrying-and-return run. Designs using two conventional conveyor belts have been developed to elevate material by pressing it between them, but their application is limited.

Flight Conveyors These devices are available in an almost infinite variety. Most flight-conveyor applications are open designs for rough conveying operations, but some are built with totally enclosed casings. Table 7-10 gives typical design and capacity information.

Apron Conveyors Probably the most common chain conveyors, these are available in a wide variety of designs for both horizontal and inclined travel. Their main application is the feeding of material at controlled rates, with lump sizes that are large enough to minimize dribble. The typical design is a series of pans mounted between two strands of roller chain, with pans overlapping to eliminate dribble, and often equipped with end plates for deeper loads. Pan design may vary according to material requirements. Figure 7-11 illustrates a typical apron-conveyor design, and Table 7-11 gives capacities for units with and without skirt plates. Apron-feeder applications range from fairly light-duty applications with light-gauge steel pans up to extremely heavy-duty applications requiring reinforced manganese steel pans with center supports. Table 7-11 values may be used in calculating capacities of other sizes, since this is a function of width of carrying surface, height of sides, speed, and bulk density. Apron-conveyor speeds are typically 0.25 to 0.38 m/s (50 to 75 ft/min). When these conveyors are used as feeders, velocities are kept in the 0.05- to 0.15-m/s (10- to 30-ft/min) range.

PNEUMATIC CONVEYORS

One of the most important material-handling techniques in the chemical industry is the movement of material suspended in a stream of air over horizontal and vertical distances ranging from a few to several hundred feet. Materials ranging from fine powders through 6.35-mm (¼-in) pellets and bulk densities of 16 to more than 3200 kg/m^3 (1 to more than 200 lb/ft^3) can be handled. A large, capable manufacturing industry supplies complete systems as well as components that users can incorporate into their own designs. Much engi-

TABLE 7-10 Flight-Conveyor Capacities*

	Maximum size of lumps		Capacity, tons/h (metric tons/h)† for various flight spacings, conveyor, horizontal, in (mm)			
Flight size and no. of strands, in (mm)	All lumps, (in) (mm)	10% lumps, in (mm)	18 (460)	24 (610)	36 (915)	Design type‡
10×4 (255 x 100—1	1½ (38)	3 (76)	32 (29)	25 (23)	16 (15)	1
12×5 (305 x 130)—1	1¾ (45)	3½ (89)	46 (42)	35 (32)	23 (21)	1
15×5 (380 x 130)—1	2 (51)	4 (102)	66 (60)	50 (45)	33 (30)	1
15×6 (380 x 155)—2	3½ (89)	7 (178)	87 (79)	67 (61)	44 (40)	2
16×8 (405 x 205)—2	4 (102)	8 (203)	110 (99)	82 (74)	55 (50)	2
18×8 (460 x 205)—2	5 (127)	9 (229)	124 (113)	93 (84)	62 (56)	2
20×10 (510 x 255)—2	6 (152)	10 (254)	... .	141 (128)	94 (85)	2
24×10 (610 x 255)—2	8 (203)	13 (305)	... .	176 (160)	116 (105)	2
30×10 (765 x 255)—2	10 (254)	14 (355)	... .	... .	250 (227)	2
12×5 (305 x 130)—1	1¾ (45)	3½ (89)	56 (51)	42 (38)	28 (25)	3
15×7 (380 x 180)—1	2½ (64)	4½ (114)	78 (71)	58 (53)	39 (35)	3
18×8 (460 x 205)—1	3 (76)	5 (127)	124 (113)	93 (84)	62 (56)	3
12×5 (305 x 130)—2	2 (51)	4 (102)	56 (51)	42 (38)	28 (25)	4
15×6 (380 x 155)—2	3 (76)	5 (127)	76 (69)	57 (52)	38 (34)	4
18×7 (460 x 180)—2	4 (102)	8 (203)	96 (87)	72 (65)	48 (44)	4
24×8 (610 x 205)—2	8 (203)	12 (305)	... .	124 (113)	83 (75)	4

*Data from Fairfield Engineering Co.

†Basis: 30-lb/ft^3 (480-kg/m^3) bulk density and conveyor velocity of 100 ft/min (30.5 m/min). For inclined conveyors capacities are reduced by factors given:

Slope off horizontal	Factor
15°	0.80
30°	0.55
45°	0.33

‡Type 1: malleable-iron conveyor flights; type 2: steel flights on roller chain; type 3: steel flights with wear shoes or rollers; type 4: steel flights on plain chain.

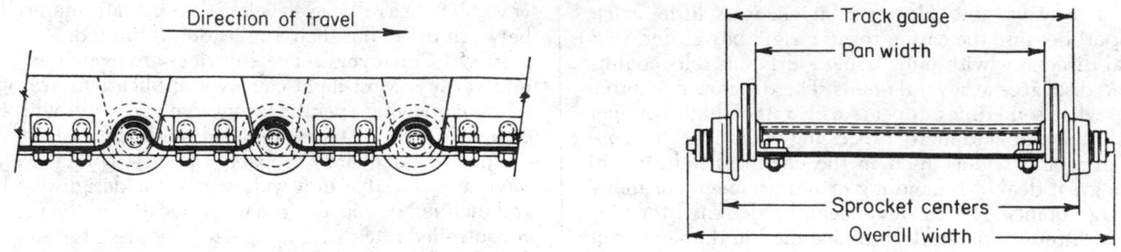

FIG. 7-11 Apron conveyors. *(Fairfield Engineering Co.)*

TABLE 7-11 Apron-Conveyor Capacities*

Capacity without skirts for various speeds and bulk densities for material depth of 4 in (102 mm) on pans

	50 ft/min (15.2 m/min)			100 ft/min (30.5 m/min)		
		tons/h (metric tons/h)			ton/h (metric tons/h)	
Apron width, in (mm)	ft^3/h (m^3/h)	50 lb/ft^3 (801 kg/m^3)	100 lb/ft^3 (1602 kg/m^3	ft^3/h (m^3/h)	50 lb/ft^3 (801 kg/m^3)	100 lb/ft^3 (1602 kg/m^3)
18 (460)	1125 (31.9)	28 (25)	56 (51)	2250 (63.7)	(56 (51)	112 (102)
24 (610)	1500 (42.5)	38 (34)	75 (68)	3000 (85.0)	75 (68)	150 (136)
30 (765)	1875 (53.2)	47 (43)	94 (85)	3750 (106.2)	94 (85)	188 (171)
36 (915)	2250 (63.7)	56 (51)	113 (102)	4500 (127.4)	113 (102)	226 (205)
42 (1070	2625 (74.3)	66 (60)	131 (119)	5250 (148.7)	131 (119)	262 (238)
48 (1220)	3000 (85.0)	75 (68)	150 (136)	6000 (170.0)	150 (136)	300 (272)
54 (1370)	3375 (95.6)	85 (77)	169 (153)	6750 (191.2)	169 (153)	338 (307)
60 (1525)	3750 (106.2)	94 (85)	188 (171)	7500 (212.4)	188 (171)	376 (341)

Capacities with skirts for various material depths, 0.75 loaded cross section and 10-ft/min (3-m/min) velocity

			Capacity, tons/h (metric tons/h); 50-lb/ft^3 (801-kg/m^3) bulk-density material; material depth on pans, in (mm)					
Pan width, in (mm)	Width between skirts, in (mm)	Max: lump size, in (mm)	4 (105)	8 (205)	12 (305)	18 (460)	21 (535)	24 (610)
18 (460)	16 (410)	3 (76)	5.0 (4.5)	10.0 (9.1)	15.0 (13.6)	22.5 (20.4)	26.3 (23.9)	30.0 (27.2)
24 (610)	22 (560)	4 (102)	6.9 (6.3)	13.7 (12.4)	20.6 (18.7)	31.0 (28.1)	36.1 (32.8)	41.2 (37.4)
30 (765)	28 (715)	6 (152)	8.8 (8.0)	17.5 (15.9)	26.2 (23.8)	39.3 (35.7)	45.9 (41.7)	52.5 (47.7)
36 (915)	34 (865)	8 (203)	10.7 (9.7)	21.3 (19.3)	32.0 (29.1)	48.0 (43.6)	56.0 (51.8)	64.0 (58.1)
42 (1070)	40 (1020)	10 (254)	12.5 (11.3)	25.0 (22.7)	37.5 (34.0)	56.3 (51.1)	65.7 (59.6)	75.0 (68.1)
48 (1220)	46 (1170)	12 (305)	14.4 (13.1)	28.8 (26.1)	43.2 (39.2)	64.8 (58.8)	75.6 (68.6)	86.3 (78.3)

*Data from Fairfield Engineering Co.

neering information is available from this industry in the form of brochures, data sheets, and nomographs.

The **capacity** of a pneumatic-conveying system depends on (1) product bulk density (and particle size and shape to some extent), (2) energy content of the conveying air over the entire system, (3) diameter of conveying line, and (4) equivalent length of conveying line.

Minimum capacity is achieved when the energy of the conveying air is just sufficient to move the product through the line without stoppage. To prevent such stoppage, it is good practice to provide an additional increment of air energy so that a factor of safety exists that allows for minor changes in product characteristics. An **optimum system** is one that repays, through operating economies, all design features above the minimum required, within the return-on-investment criteria set by the owner.

While successful and economical system designs can be devised by experienced process engineers, the competent technical aid available from equipment suppliers has led to a growing trend toward the purchase of complete systems, even on small jobs, rather than in-plant assembly from components on the basis of in-house designs. An idea of the change in **capital investment** for typical pneumatic-conveyor systems as a function of increasing transfer rates is given in Table 7-12.

Conveyor installations may be permanent or a combination of permanent and portable. The latter kind is often mounted on a bulk-delivery vehicle, which permits fast unloading into the customer's silo by the carrier without effort or equipment from the customer. Controls range from simple motor starters and hand-connected hoses to sophisticated, punched-card-directed, electropneumatic control systems.

Types of Systems Generally, pneumatic conveyors are classified according to five basic types: pressure, vacuum, combination pressure and vacuum, fluidizing, and the blow tank.

In **pressure systems** (Fig. 7-12*a*), material is dropped into an air stream (at above atmospheric pressure) by a rotary air-lock feeder. The velocity of the stream maintains the bulk material in suspension until it reaches the receiving vessel, where it is separated from the air by means of an air filter or cyclone separator.

Pressure systems are used for free-flowing materials of almost any particle size, up to 6.35-mm (¼-in) pellets, where flow rates over 151 kg/min (20,000 lb/h) are needed and where pressure loss through the system is about 305 mmHg (12 inHg). These systems are favored when one source must supply several receivers. Conveying air is usually supplied by positive-displacement blowers.

Vacuum systems (Fig. 7-12*b*) are characterized by material moving in an air stream of pressure less than ambient. The advantages of this type are that all the pumping energy is used to move the product and that material can be sucked into the conveyor line without the need of a rotary feeder or similar seal between the storage vessel and the conveyor. Material remains suspended in the air stream until it reaches a receiver. Here, a cyclone separator or filter (Fig. 7-12*c*) separates the material from the air, the air passing through the separator and into the suction side of the positive-displacement blower or some other power source.

Vacuum systems are typically used when flows do not exceed 6800

TABLE 7-12 Approximate Pneumatic-Conveyor Costs*
Product: Plastic pellets, ⅛-in (3.2-mm) cubes, 30-lb/ft^3 (481-kg/m^3) bulk density; equivalent length of system, 600 ft (183 m)

Flow rate, lb/h	(kg/h)	Conveyor pipe, inside diameter, in	(mm)	Power required, hp	Range of investment, $† Manual‡	Automatic§
10,000	(4,536)	4	(100)	25	26,000	46,000
25,000	(11,340)	6	(155)	60	50,000	75,000
50,000	(22,680)	6	(155)	125	87,000	112,000
100,000	(45,360)	8	(205)	200	175,000	200,000

*Courtesy of Whitlock, Inc.

†1980 costs. Equipment includes motor and blower package, cyclone receivers, railcar-unloading connections, high-level interlocks for stopping the motor and blower combination when the silos reach a full level, and all necessary piping. Installation is not included.

‡System includes a minimum control package, with most activities person-actuated, including the changing of feed lines to storage silos.

§System includes automatic actuation of most activities, with changing of feed lines to silos accomplished by diverter valves controlled automatically by silo-level controls.

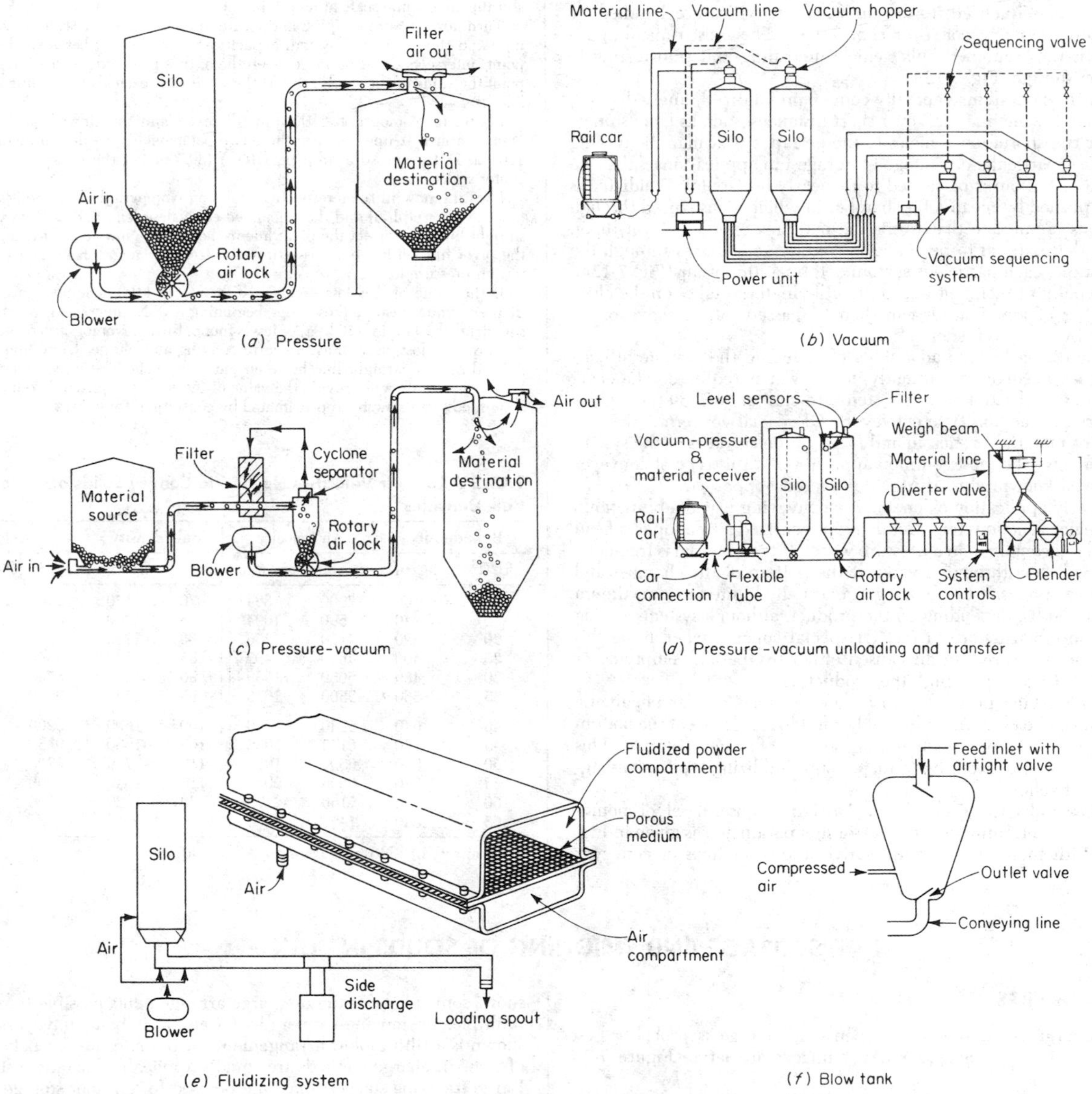

FIG. 7-12 Types of air-conveying systems. (*a*) Pressure. (*b*) Vacuum. (*c*) Pressure-vacuum. (*d*) Pressure-vacuum unloading and transfer. *(Whitlock, Inc.)* (*e*) Fluidizing system. *(Fuller Co.)* (*f*) Blow tank.

kg/h (15,000 lb/h), the equivalent conveyor length is less than 305 m (1000 ft), and several points are to be supplied from one source. They are widely used for finely divided materials. Of special interest are vacuum systems designed for flows under 7.6 kg/min (1000 lb/h), used to transfer materials short distances from storage bins or bulk containers to process units. This type of conveyor is widely used in plastics and other processing operations where the variety of conditions requires flexibility in choosing pickup devices, power sources, and receivers. Capital investment can be kept low, often in the range of $2000 to $7000.

Pressure-vacuum systems (Fig. 7-12*c*) combine the best of both the pressure and the vacuum methods. A vacuum is used to induce material into the conveyor and move it a short distance to a separator. Air passes through a filter and into the suction side of a positive-displacement blower. Material then is fed by a rotary feeder into the conveyor positive-pressure air stream, which comes from the blower discharge. Application can be very flexible, ranging from a central control station, with all interconnection activities electrically controlled and sequenced, to one in which activities are handled by manually changing conveyor connections. The most typical application is the combined bulk vehicle unloading and transferring to product storage (Fig. 7-12*d*).

Fluidizing systems generally convey prefluidized, finely divided, non-free-flowing materials over short distances, such as from storage bins or transportation vehicles to the entrance of a main conveying system. A particular advantage in storage-bin applications is that the bottom of the bin is permitted to be nearly horizontal. Fluidizing is accomplished by means of a chamber in which air is passed through a porous membrane that forms the bottom of the conveyor, upon which the material to be conveyed rests. As air passes through the membrane, each particle is surrounded by a film of air (Fig. 7-12*e*). At the point of incipient fluidization the material takes on the characteristics of free flow. It can then be passed into a conveyor air stream by a rotary feeder.

Prefluidizing has the advantage of reducing the volume of conveying air needed; consequently, less power is required. The characteristics of the rest of this system are similar to those of regular pressure- or vacuum-type conveyors. Of special concern is the tendency of material to stick to and build up on surfaces of the system components. The most common application of this type of conveyor is the well-known railroad Airslide covered hopper car.

An early application of pneumatic conveying was the **blow tank.** This device functions by introducing pressurized air on top of a head of material contained in a pressure vessel. If the material is free-flowing, it will flow through a valve at the bottom of the chamber and move through a short conveying line, usually limited to a maximum of 16 m (50 ft), depending on the product, although systems as long as 457 m (1500 ft) are in use. Of special concern when using this system are the surges of air caused either by the tank emptying or by the air breaking through the product.

The blow-tank principle can be used to feed regular pneumatic conveyors. Use of an Airslide or other fluidizing device at the bottom of the blow tank permits handling non-free-flowing materials. This principle is used extensively in pressure-fluidizing-type valve-bag-packing machines.

Nomographs for Preliminary Design A useful set of nomographs* for determining conveyor-design parameters is given in Fig. 7-13. With these charts, conservative approximations of conveyor size and power for given product bulk density, conveyor equivalent length, and required capacity can be obtained. Because pneumatic conveyors and their components are subject to continual improvements by a fast-changing supplier industry, manufacturers should be invited to submit alternative designs to that resulting from the use of the nomograph. Some large users of pneumatic conveyors have found it expedient to write computer programs for calculating system parameters.

To begin preliminary calculations, first determine the equivalent length of the system being considered. This length is the sum of the vertical and horizontal distances, plus an allowance for the pipe fittings used. Most common of these fittings are the long-radius 90° elbow pipe [equivalent length = 25 ft (7.6 m)] and the 45° elbow [equivalent length = 15 ft (4.6 m)].

The second step consists of choosing from Table 7-13 an initial air velocity that will move the product. An iterative procedure then begins by assuming a pipe diameter for the required capacity of the system.

Referring now to Nomograph 1, draw a straight line between the air-velocity and the pipe-diameter scales so that when the line is extended it will intersect the air-volume scale at a certain point.

Turn now to Nomograph 2 and locate in their respective scales the air volume and the calculated system capacity. A straight line between these two points intersects the scale in between them, thus providing at the intersection point the value of the solids ratio. If the solids ratio exceeds 15, assume a larger line size.

Locate in Nomograph 3 the pipe diameter and the air volume found in Nomograph 1. A line between these two points yields the design factor, or *P* 100 (30.5), the pressure drop per 100 ft (30.5 m), at the intersection of the center scale.

Locating now in their respective scales on Nomograph 4 the design factor (from Nomograph 3) and the calculated equivalent length, draw an extended straight line to intersect the pivot line in the center. Now connect this point in the pivot line with the solids-ratio scale (from Nomograph 2), and read the system pressure loss.

If the value of this loss exceeds 10 lb/in^2 (70 kPa), assume a larger pipe diameter and repeat all these steps, beginning with Nomograph 1. After a pressure drop of 10 lb/in^2 (70 kPa) or less is found, turn to Nomograph 5 and locate this pressure loss, as well as the corresponding air volume (from Nomograph 2), and draw a straight line between the two points. The intersection of the horsepower scale will provide the value of the power required. From this, the system cost can now be approximated by consulting Table 7-12.

TABLE 7-13 Air Velocities Needed to Convey Solids of Various Bulk Densities*

Bulk density		Air velocity		Bulk density		Air velocity	
lb/ft^3	kg/m^3	ft/min	m/min	lb/ft^3	kg/m^3	ft/min	m/min
10	160	2900	884	70	1120	7700	2347
15	240	3590	1094	75	1200	8000	2438
20	320	4120	1256	80	1280	8250	2515
25	400	4600	1402	85	1360	8500	2591
30	480	5050	1539	90	1440	8700	2652
35	560	5500	1676	95	1520	9000	2743
40	640	5840	1780	100	1600	9200	2804
45	720	6175	1882	105	1680	9450	2880
50	800	6500	1981	110	1760	9700	2957
55	880	6800	2072	115	1840	9900	3118
60	960	7150	2179	120	1920	10500	3200
65	1040	7450	2270				

*Courtesy of Flotronics Division, Allied Industries.

STORAGE AND WEIGHING OF SOLIDS IN BULK

STORAGE PILES

Discharge Arrangements Open-yard storage is probably best handled by belt conveyor when tonnages are large. Figure 7-14 shows some of the many discharge arrangements possible for single, multiple, or moving-tripper discharge from belt conveyors. Also shown is a tilting-plow arrangement for discharging flat belts. Most of these discharge methods are equally applicable for indoor storage. Large traveling stackers may also be used for outdoor storage. They may move along the length of a belt, forming a pile on one or both sides of the belt, or pivot about a fixed axis to form a circular pile.

*Nomographs prepared from data supplied by Flotronics Division, Allied Industries.

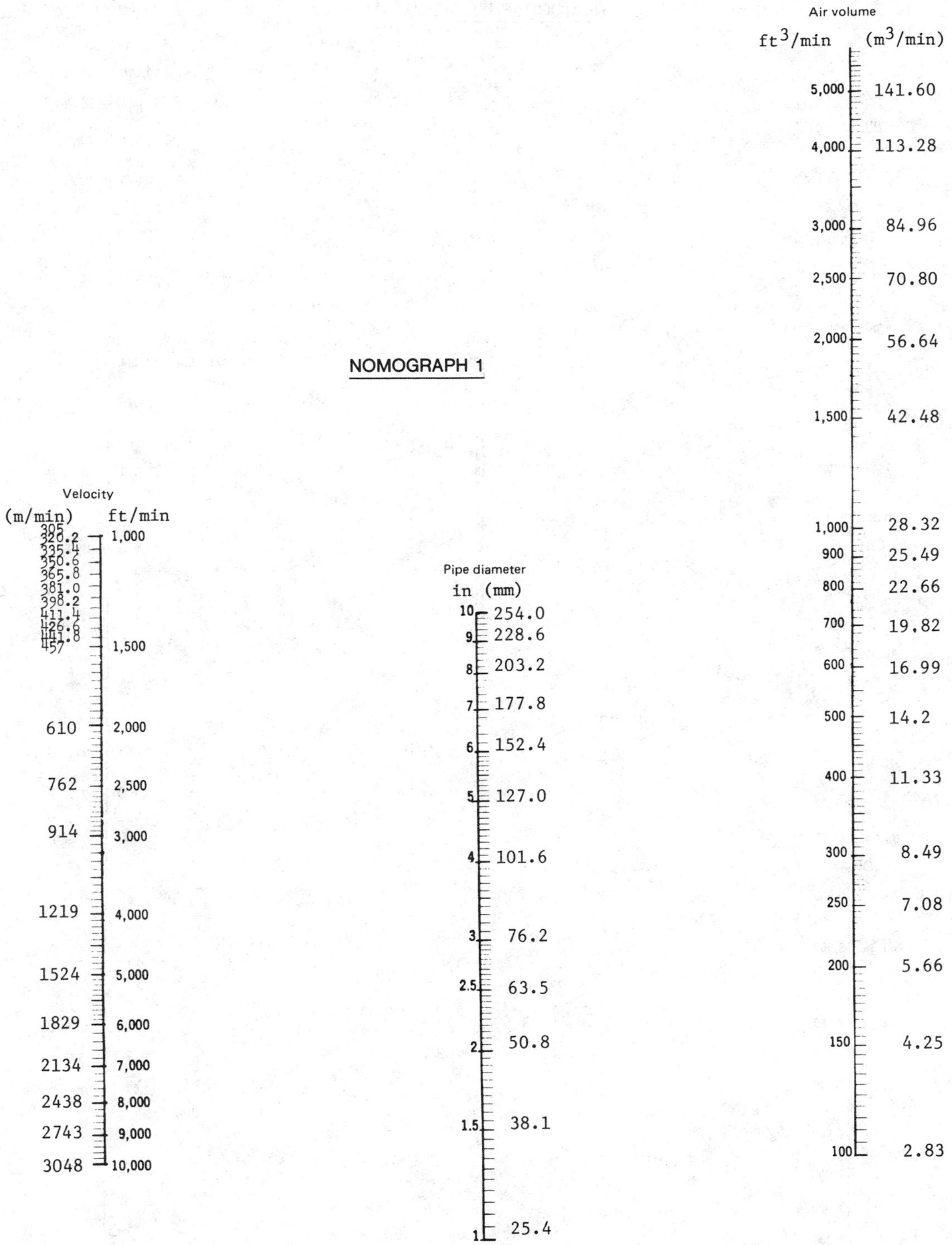

FIG. 7-13 Nomographs for determining conveyor-design parameters.

NOMOGRAPH 2

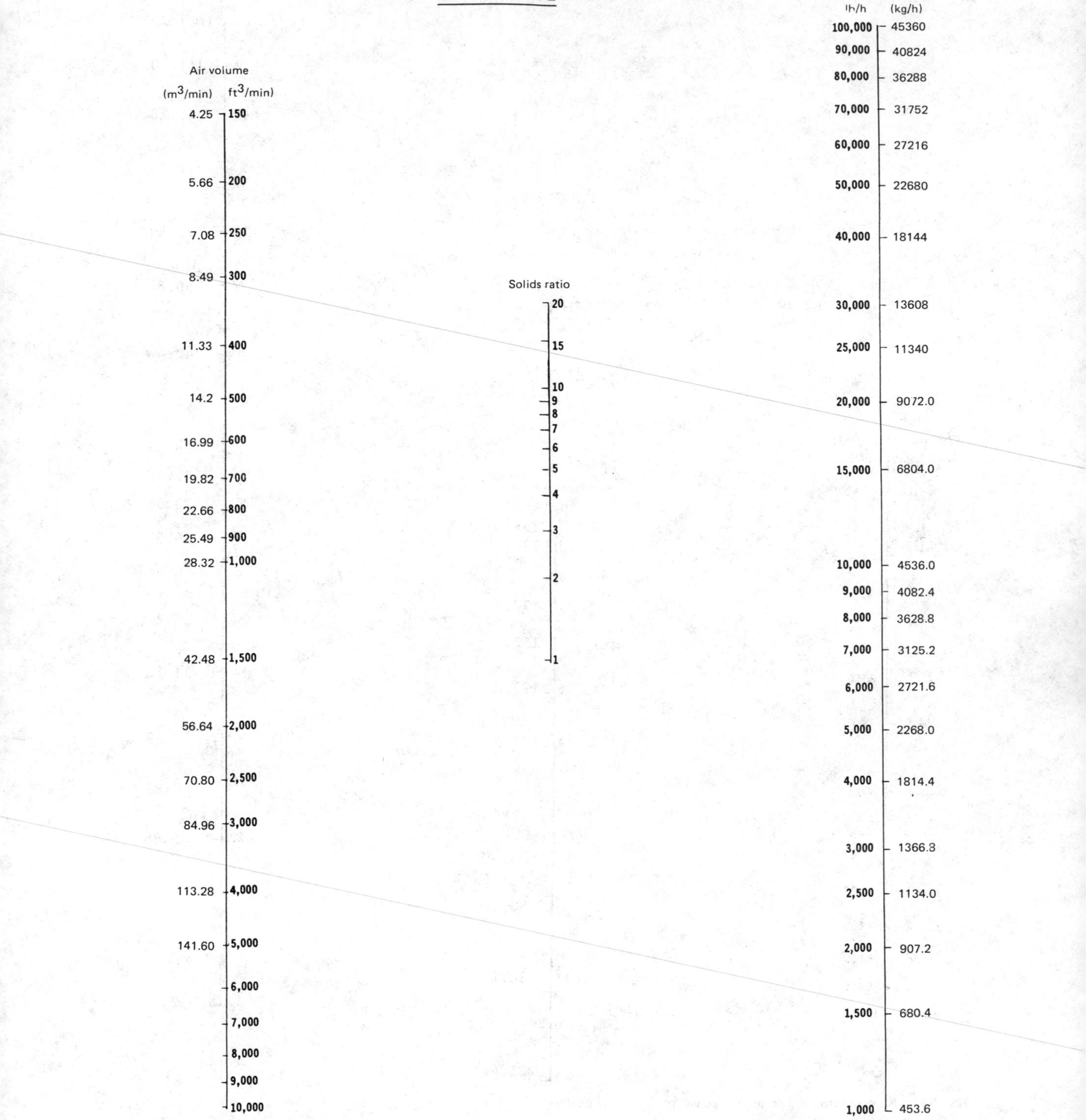

System capacity
lb/h (kg/h)
100,000 45360
90,000 40824
80,000 36288
70,000 31752
60,000 27216
50,000 22680
40,000 18144
30,000 13608
25,000 11340
20,000 9072.0
15,000 6804.0
10,000 4536.0
9,000 4082.4
8,000 3628.8
7,000 3125.2
6,000 2721.6
5,000 2268.0
4,000 1814.4
3,000 1366.3
2,500 1134.0
2,000 907.2
1,500 680.4
1,000 453.6
Air volume
(m3/min) ft3/min)
4.25 150
5.66 200
7.08 250
8.49 300
11.33 400
14.2 500
16.99 600
19.82 700
22.66 800
25.49 900
28.32 1,000
42.48 1,500
56.64 2,000
70.80 2,500
84.96 3,000
113.28 4,000
141.60 5,000
6,000
7,000
8,000
9,000
10,000
Solids ratio
20
15
10
9
8
7
6
5
4
3
2
1

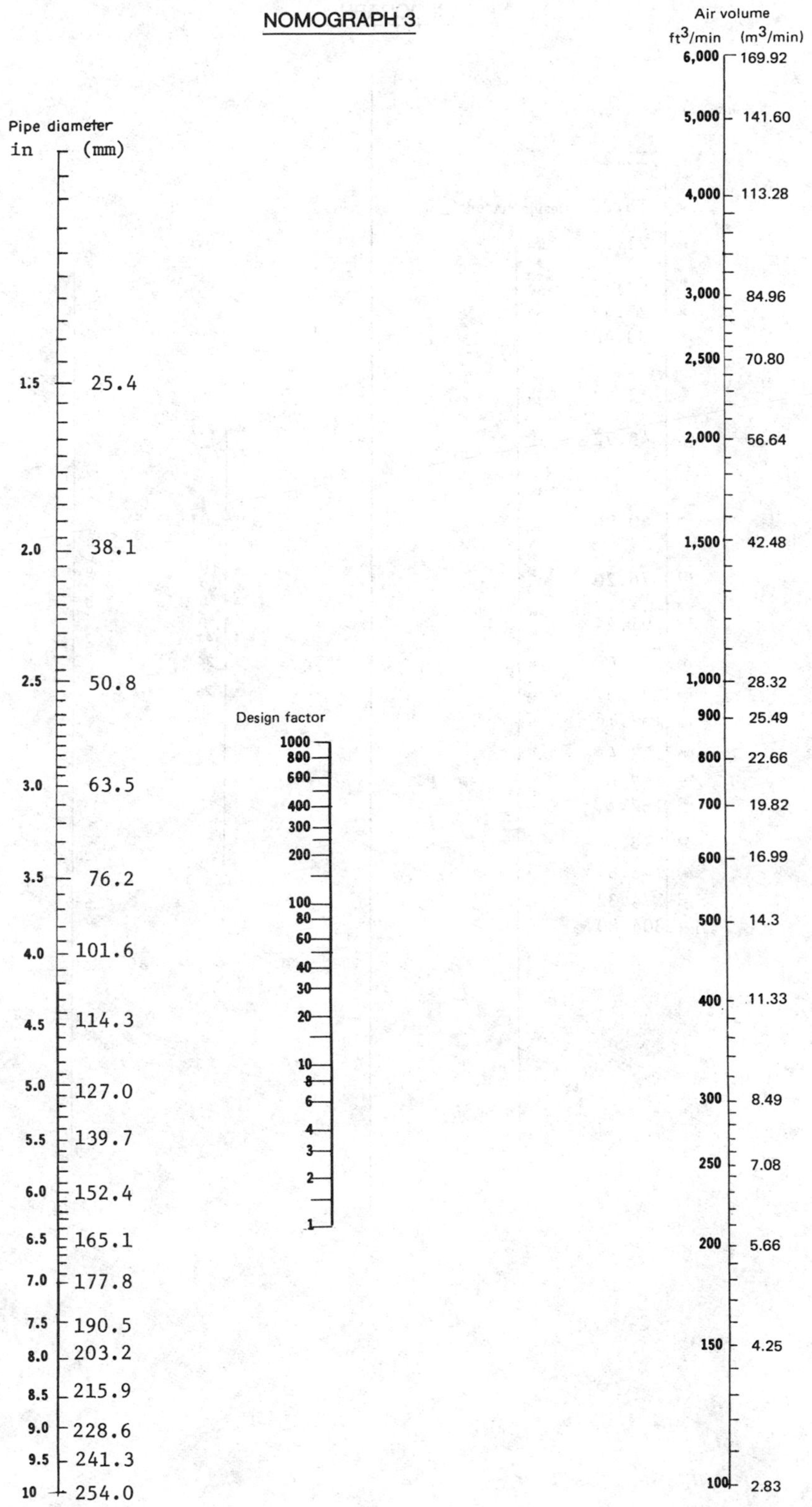

NOMOGRAPH 3
Pipe diameter
in (mm)
1.5 25.4
2.0 38.1
2.5 50.8
3.0 63.5
3.5 76.2
4.0 101.6
4.5 114.3
5.0 127.0
5.5 139.7
6.0 152.4
6.5 165.1
7.0 177.8
7.5 190.5
8.0 203.2
8.5 215.9
9.0 228.6
9.5 241.3
10 254.0
Design factor
1000
800
600
400
300
200
100
80
60
40
30
20
10
8
6
4
3
2
1
Air volume
ft3/min (m3/min)
6,000 169.92
5,000 141.60
4,000 113.28
3,000 84.96
2,500 70.80
2,000 56.64
1,500 42.48
1,000 28.32
900 25.49
800 22.66
700 19.82
600 16.99
500 14.3
400 11.33
300 8.49
250 7.08
200 5.66
150 4.25
100 2.83

NOMOGRAPH 4

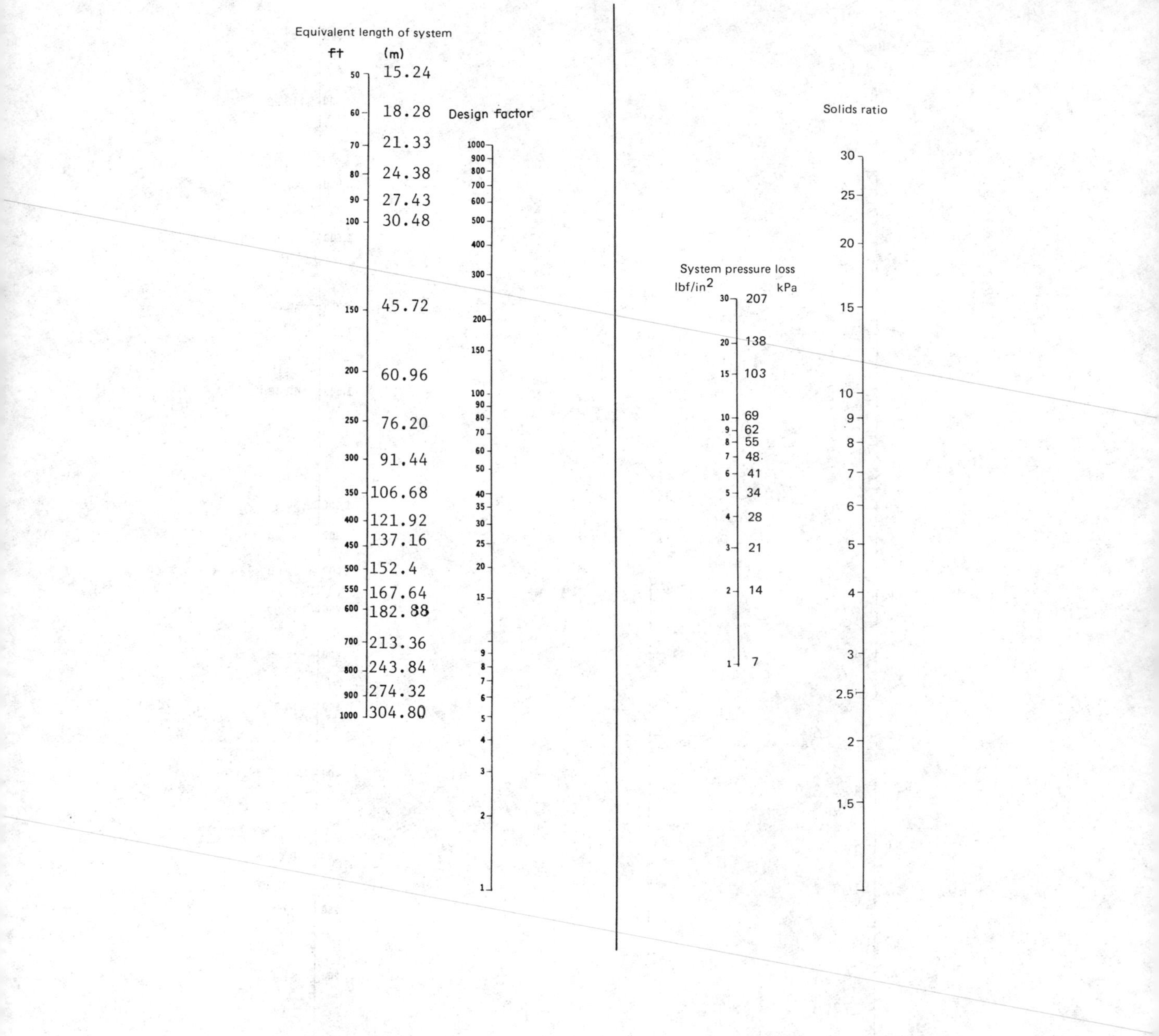

Equivalent length of system
ft (m)
50 15.24
60 18.28
70 21.33
80 24.38
90 27.43
100 30.48
150 45.72
200 60.96
250 76.20
300 91.44
350 106.68
400 121.92
450 137.16
500 152.4
550 167.64
600 182.88
700 213.36
800 243.84
900 274.32
1000 304.80
Design factor
1000 900 800 700 600 500 400 300 200 150 100 90 80 70 60 50 40 35 30 25 20 15 9 8 7 6 5 4 3 2 1
System pressure loss
lbf/in2 kPa
30 207
20 138
15 103
10 69
9 62
8 55
7 48
6 41
5 34
4 28
3 21
2 14
1 7
Solids ratio
30 25 20 15 10 9 8 7 6 5 4 3 2.5 2 1.5

NOMOGRAPH 5

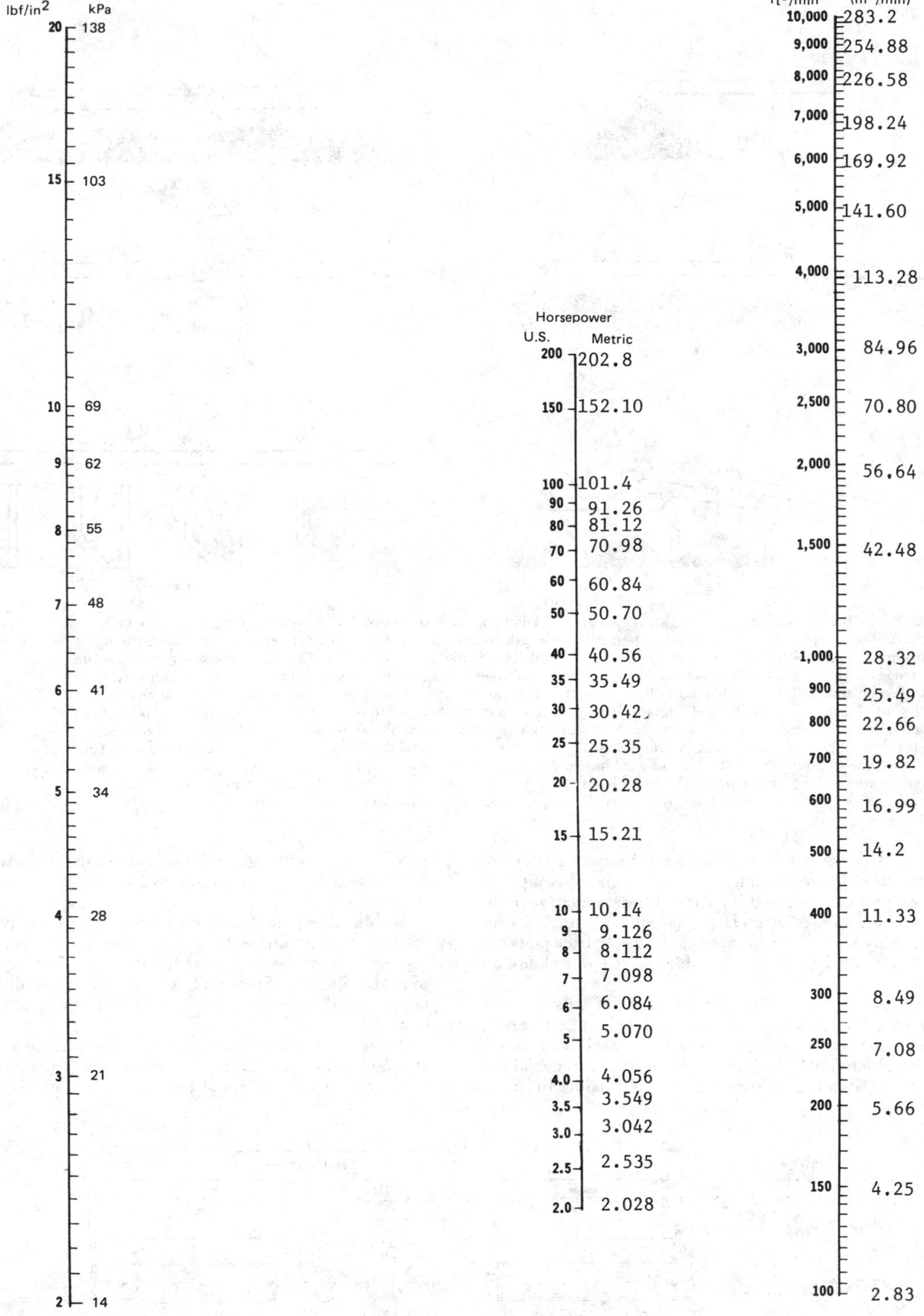
System pressure loss
lbf/in2 kPa
20 138
15 103
10 69
9 62
8 55
7 48
6 41
5 34
4 28
3 21
2 14
Horsepower
U.S. Metric
200 202.8
150 152.10
100 101.4
90 91.26
80 81.12
70 70.98
60 60.84
50 50.70
40 40.56
35 35.49
30 30.42
25 25.35
20 20.28
15 15.21
10 10.14
9 9.126
8 8.112
7 7.098
6 6.084
5 5.070
4.0 4.056
3.5 3.549
3.0 3.042
2.5 2.535
2.0 2.028
Air volume
ft3/min (m3/min)
10,000 283.2
9,000 254.88
8,000 226.58
7,000 198.24
6,000 169.92
5,000 141.60
4,000 113.28
3,000 84.96
2,500 70.80
2,000 56.64
1,500 42.48
1,000 28.32
900 25.49
800 22.66
700 19.82
600 16.99
500 14.2
400 11.33
300 8.49
250 7.08
200 5.66
150 4.25
100 2.83

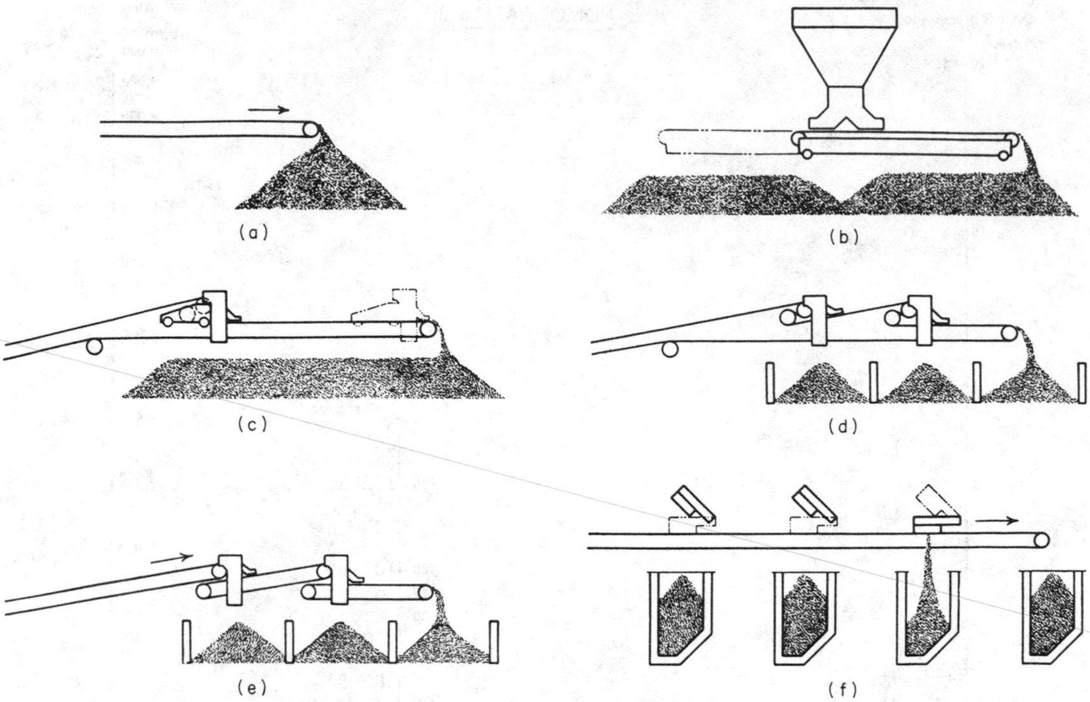

FIG. 7-14 Belt-conveyone discharge arrangements. (*a*) Discharge over an end pulley forms a conical pile at the end of the belt. (*b*) Discharge over either end pulley to distribute lengthwise by a reversible-shuttle conveyor. (*c*) Discharge through a traveling tripper, with or without a cross conveyor, to distribute material to one or both sides of the conveyor for the entire distance of tripper travel. Trippers can be propelled by a conveyor belt or by a separate motor. Motor-propelled trippers can also be automatically reversing to distribute material evenly or can be manually controlled to discharge at any desired point. (*d*) Discharge through fixed trippers, with or without a cross conveyor to one or both sides of the belt, to fixed bin openings or pile locations. This can also be done with multiple conveyors as shown in (*e*) or by stopping traveling trippers in the desired position. (*e*) Discharge from multiple conveyors through fixed discharge chutes, with or without a cross conveyor to one or both sides of the belt, to fixed bin openings or pile locations. (*f*) Discharge by hinged plows to one or more fixed locations along one or both sides of the conveyor. Plows may be adjusted to divide the discharge into several places simultaneously in the proportion desired. *(FMC Corporation, Material Handling Systems Division.)*

Reclaiming Underground-tunnel belts fed by special gates (Fig. 7-15) are often used for reclaiming, as is mobile shovel equipment. Cable-drag scrapers are also used for large outside storage areas and sometimes on inside storage when large, flat areas are used. A drag-scraper system may follow a single fixed cable line, or back posts may be provided to allow relocation of the cable line to cover almost any storage-space shape.

One development for handling large tonnages of bulk materials from storage is the **bucket-wheel reclaimer,** which consists of a series of buckets placed about the periphery of a large wheel that is carried by a fixed propulsion unit. The buckets empty onto a removal conveyor, usually of the belt type, which takes the product to further processing or handling. Bucket-wheel reclaimers capable of handling as little as 150 tons/h to as much as 20,000 tons/h (see Fig. 7-16) have been built.

Mobile equipment is often preferred to fixed types. Front-end loaders, scrapers, and bulldozers are used with increasing frequency, especially on projects of short duration or when capital investment must be limited. Front-end loaders are especially advantageous because of their ability to carry material as well as to plow or bulldoze it.

Angle of repose is the angle at which a material will rest on a pile. It is useful for determining the capacity of a bin or a pile. The angle of the cone that develops at the top of the pile when a bin is being

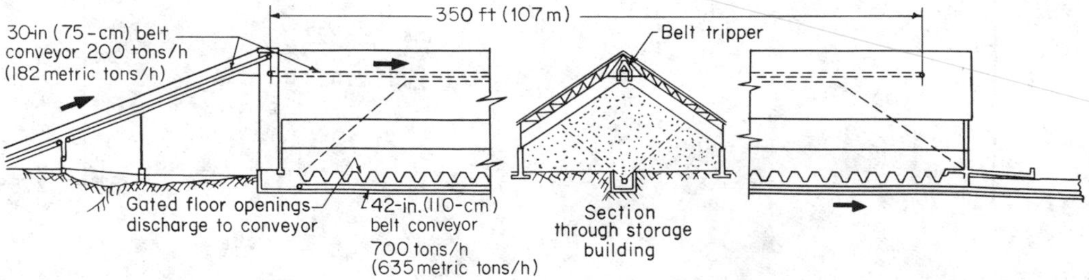

FIG. 7-15 Belt-conveyor storage and reclaiming in a flat-floor building. *(Stephens-Adamson Division, Allis-Chalmers Corporation.)*

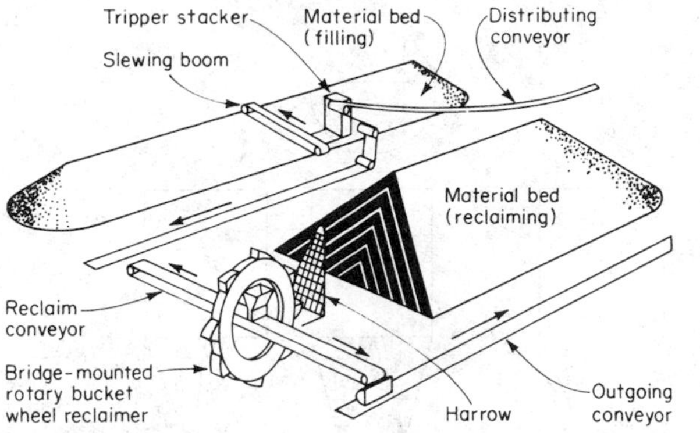

FIG. 7-16 Bucket-wheel reclaimer. Digging buckets mounted on wheel discharge on a belt conveyor for material transfer. (*Courtesy of* Mechanical Engineering.)

filled will be somewhat flatter than the angle of repose because of the effect of impact.

STORAGE BINS, SILOS, AND HOPPERS

Probably no section of the materials-handling and -storage art advanced as far in a decade (the 1960s) as did that of bin storage of bulk materials. Prior to this time, **storage-bin design** was a hit-or-miss empirical affair, in which success was assured only if the product was free-flowing. This was changed radically as a result of research led by Andrew W. Jenike. This work, which resulted in identifying the criteria that affect material flow in storage vessels, was first reported in Jenike's paper "Gravity Flow of Bulk Solids" (Bull. 108, University of Utah Engineering Experiment Station, October 1961). This paper set forth the equations defining bulk flow and the coefficients affecting flow.

Continuing experimentation verified these criteria, and in Bulletin 123 (November 1964) the subject was further defined by providing flow factors for a number of bin-hopper designs as well as specifications for determining experimentally the characteristics of bulk material affecting flow and storage. Along with the theory, Jenike produced a method of applying it, which includes equations and the physical measurement of material characteristics.

In what follows, a storage vessel is considered as consisting of a bin and a hopper. A bin is the upper section of the vessel and has vertical sides. The hopper, which has at least one sloping side, is the section between the bin and the outlet of the vessel.

Material-Flow Characteristics Two important definitions of the flow characteristics of a storage vessel are **mass flow,** which means that all the material in the vessel moves whenever any is withdrawn (Fig. 7-17), and **funnel flow,** which occurs when only a portion of the material flows (usually in a channel or rathole in the center of the system) when any material is withdrawn (Fig. 7-18). Some typical mass-flow designs are shown in Fig. 7-19.

Mass-flow bins feature the most sought-after characteristics of a storage vessel: unassisted flow whenever the bottom gate is opened. A funnel-flow bin may or may not flow but probably can be made to flow by some means.

Until Jenike developed the rationale for storage-vessel design, a common criterion was to measure the angle of repose, use this value as the hopper angle, and then fit the bin to whatever space was available. Too often, bins were designed from an architectural or structural-engineering viewpoint rather than from the role they were to play in a process. Economy of space is certainly one valid criterion in bin design, but others must be considered equally as well. Table 7-14 compares the principal characteristics of mass-flow and funnel-flow bins.

Although a mass-flow bin is obviously preferable to a funnel-flow vessel, the additional investment generally required must be justified.

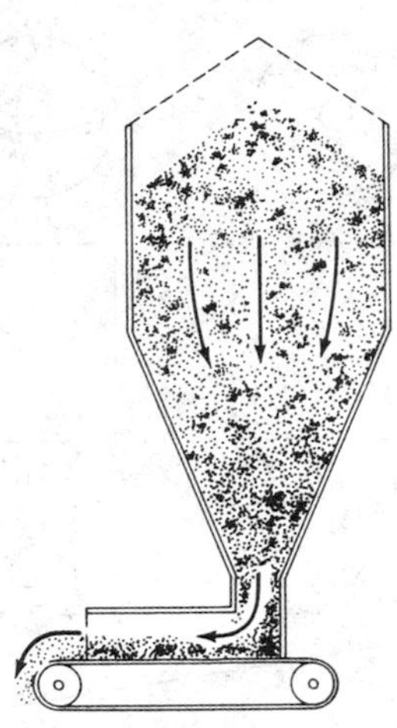

FIG. 7-17 Mass-flow bin. The material does not channel on discharge. (*Courtesy of* Chemical Engineering.)

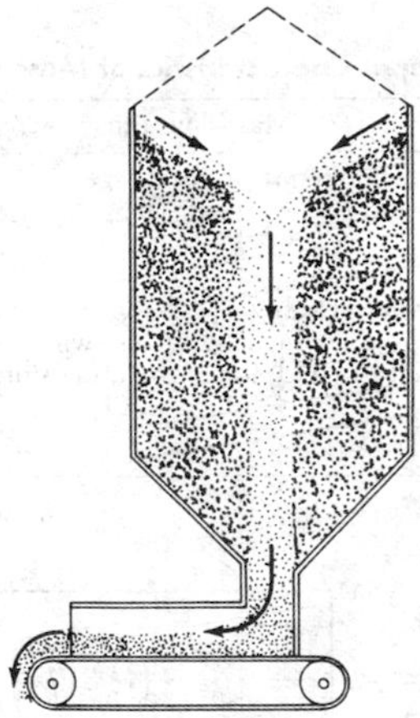

FIG. 7-18 Funnel-flow bin. The material segregates and develops ratholes. (*Courtesy of* Chemical Engineering.)

Often, this can be done by the reduced operating costs. But when installation space is limited, a compromise must be made, such as providing a special hopper design and sometimes even a feeder. Certainly, with mass-flow bins the feeder is not required for flow, but it might still be used for other reasons, such as conveying the material to the next process step.

Design Criteria Jenike's criteria permit an **engineering-economic analysis of storage** with about the same confidence level as in the rest of the process plant. His quantitative methods may be used to determine (1) whether the vessel will function with mass or funnel flow and (2) the outlet dimensions of the hopper so that product will flow. His methods also provide criteria for making engineering trade-offs between mass flow and funnel flow when product characteristics, space limitations, etc., dictate against design for mass flow.

The **relation between mass and funnel flows** for conical bins is shown in Fig. 7-20. The angle of kinematic friction ϕ', which is a measure of the friction coefficient between the solid and the material of construction used for the conical-shaped hopper, is measured with the "flow-factor tester." The degree of finish of the metal surface can have a large effect in determining whether the vessel will function in mass or funnel flow. Finer degrees of finish are being used more frequently, mostly because intuition has recommended this course. The kinematic angle of friction is also related to the degree of compression that the product undergoes in storage.

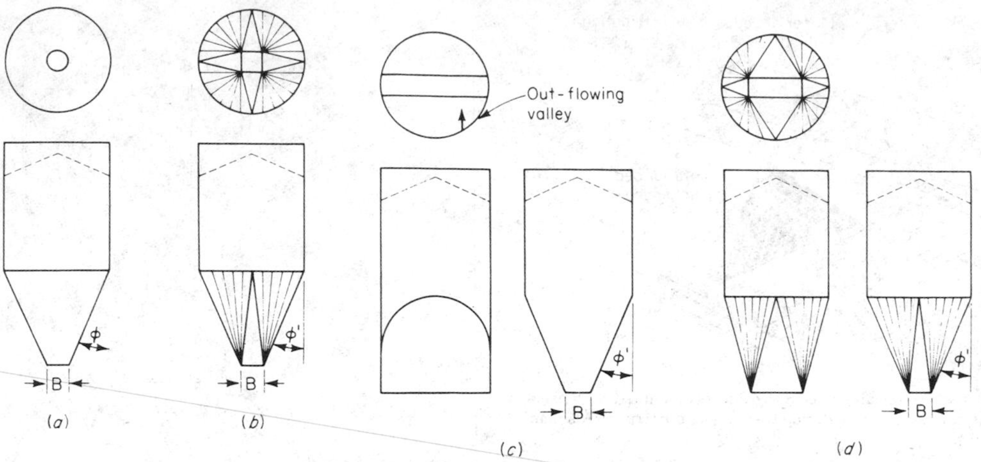

FIG. 7-19 Types of mass-flow bins. Type *c* is simple but has a valley. Although more difficult to make, type *d* has no valleys and is usually recommended. (*Courtesy of* Mechanical Engineering.)

TABLE 7-14 Principal Characteristics of Mass-Flow and Funnel-Flow Bins

Mass-flow Bins	Funnel-flow Bins
1. Particles segregate, but remit on discharge	1. Particles segregate and remain segregated
2. Powders deaerate and do not flood when the system discharges	2. First portion in is last one out
3. Flow is uniform	3. Product can remain in dead zones until complete cleanout of the system
4. Density of flow is constant	4. Product tends to bridge or arch, and then to rat-hole when discharging
5. Level indicators work reliably	5. Flow is erratic
6. Product does not remain in dead zones, where degradation can occur	6. Density can vary
7. Bin can be designed to yield non-segregating storage, or to function as a blender	7. Level indicators must be placed in critical positions so they will work properly
	8. Bins perform satisfactorily with free-flowing, large-particle solids

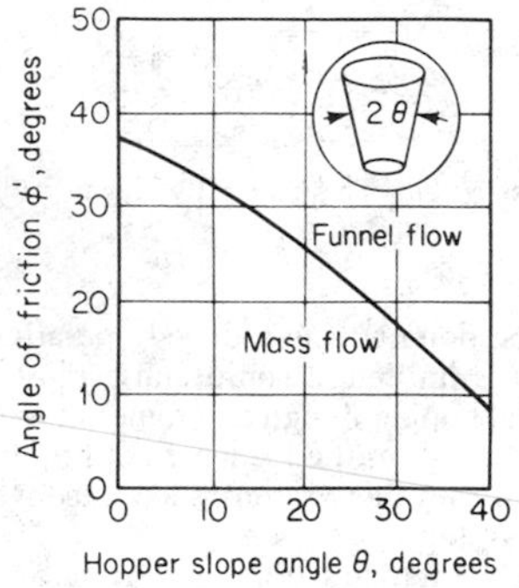

FIG. 7-20 Relation between mass and funnel flows for conical bins. (*Courtesy of* Chemical Engineering.)

Once a decision for mass or funnel flow has been made or a compromise made by including an expanded-flow bin, the hopper outlet and the type of feeder must be considered. Jenike's teaching on the flow through the bin opening is that materials that can be compacted (as opposed to being free-flowing) will be compacted because of storage-vessel shape and the packing characteristics of the product. When this happens, the material forms an **arch** that is capable of withstanding considerable stress.

Since the arch transfers the load to the hopper walls and in doing so applies so much pressure to them, the kinematic coefficient of friction ϕ' becomes great. The net result is that the "dome" or "bridge" that forms prevents any flow from the vessel. Force must then be applied to the arch so that it will collapse and flow will begin, even if erratically.

According to Jenike, when the strength of the arch f is exceeded by the internal stress s generated by a force applied above the dome, flow takes place. Summarizing:

When $f < s$, flow occurs.
When $f > s$, there is no flow.
When $f = s$, the critical point is reached.

To make a **flow analysis** when $f < s$, an element of material is observed as it moves through a storage vessel (Fig. 7-21). The pressure p on the element increases from zero at the entrance to a maximum value at the transition from the bin to the hopper. The pressure then decreases to zero linearly at the vertex of the hopper cone. The resultant strength f follows a similar pattern, though usually it has some value greater than zero. The stresses induced in the material in the hopper bottom by the weight of material above it are constant but decrease linearly to zero at the cone vertex. The f and s curves intersect at a point corresponding to the critical dimensions of the **bin opening** B.

Reducing this analysis to a technique for determining B, Jenike's method provides a practical way to measure and interpret the strength of a bulk solid as a function of consolidating pressure. To develop this relation, Jenike developed a **shear tester** that gives a flow function FF, which is a curve through a locus of points resulting from values of f and p obtained by the shear tester. This FF curve is plotted against a flow factor ff for the particular hopper being designed, as shown in Fig. 7-22.

The method makes use of the principle that a constant ratio of induced stress s in the stored contents to the consolidating pressure p exists. Thus, for any hopper design for which the ff curve is available, the shear-tester results can be plotted, and the point where $f = s$ is located. Since the distance at which this occurs above the hopper

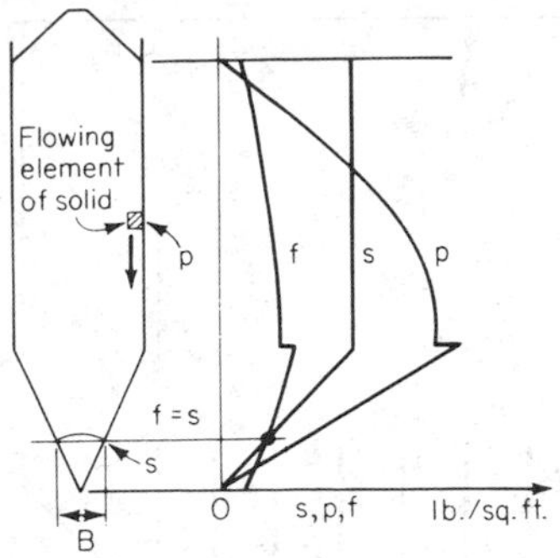

FIG. 7-21 Flow analysis is made by observing an element of material as it moves through the bin. (*Courtesy of* Chemical Engineering.)

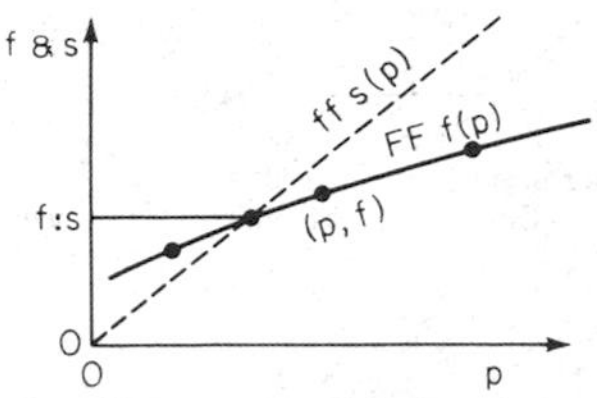

FIG. 7-22 Flow takes place only where *FF* lies below *ff*. (*Courtesy of* Chemical Engineering.)

vertex is also known, these values become the hopper dimensions at that point.

A useful approximation of *B* for a conical hopper is $B = 22f/\alpha$, where α is the bulk density of the stored product. The apparatus for determining the properties of solids has been developed and is offered for sale by the consulting firm of Jenike and Johansen, Winchester, Massachusetts, which also performs these tests on a contract basis. The flow-factor *FF* tester, a constant-rate-of-strain, direct-shear-type machine, gives the locus of points for the *FF* curve as well as ϕ', the kinematic coefficient of friction. A consolidating bench is used to prepare samples having different degrees of compaction for the flow-factor tester. This can be supplemented with the same bench enclosed in a controlled-temperature cabinet.

It is possible for some materials to produce an *FF* plot having no intersection with the *ff* curve. This indicates that a different hopper-bin design is needed or that the material cannot be made to flow. Figure 7-23 shows *FF* curves for several materials.

Jenike's method allows the chemical engineer to design bulk-storage vessels and to weigh cost versus performance with a high level of confidence that if the conditions in the real storage system are the same as those prevailing during the tests, the product will flow. It is up to the engineer, however, to establish the bounds of conditions that the product will encounter and to make appropriate tests. A product may not flow if its characteristics change, if radical temperature changes are encountered at the plant, or if moisture is left from an underdesigned dryer.

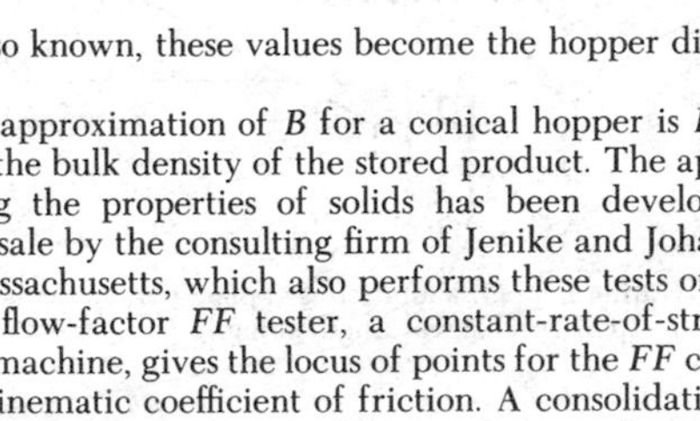

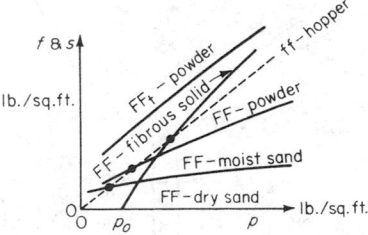

FIG. 7-23 *FF* curves for various materials. Multiply pounds-force per square foot by 0.0479 to get kilopascals. (*Courtesy of* Chemical Engineering.)

A further use of the Jenike method is its extension to the critical **structural design** of storage vessels. Because pressures can be calculated, it is possible to design for actual conditions rather than estimates. Also, flow-corrective devices may be designed by using his theory. References worth investigating on this topic are as follows:

Jenike, "Gravity Flow of Bulk Solids," Bull. 108, Utah Engineering Experiment Station, Salt Lake City, October 1961. Jenike, "Steady Gravity Flow of Frictional-Cohesive Solids in Converging Channels," *J. Appl. Mech.*, **31**, ser. E, 5–11 (March 1964). Jenike, "Why Bins Don't Flow," *Mech. Eng.*, May 1964, pp. 40–43. Jenike, "Storage and Flow of Solids," Bull. 123, Utah Engineering Experiment Station, Salt Lake City, November 1964. Jenike, "Gravity Flow of Frictional-Cohesive Solids—Convergence to Radial Stress Fields," *J. Appl. Mech.*, **32**, ser. E, 205–207 (March 1965). Jenike, "Quantitative Design of Mass-Flow Bins," *Powder Technol.*, January 1967, pp. 237–244. Jenike et al., "Flow Properties of Bulk Solids," *Proc. Am. Soc. Test. Mater.*, **60**, 1168–1181 (1960). Jenike and Johanson, "On the Theory of Bin Loads," Pap. 68-MH-8, American Society of Mechanical Engineers, Oct. 21–23, 1968. Jenike and Leser, "A Flow–No Flow Criterion in the Gravity Flow of Powders in Converging Channels," *Proc. Fourth Int. Congr. Rheology*, pt. 3, Brown University, Providence, R.I., Aug. 26–30, 1963, pp. 125–141. Johanson, "The Placement of Inserts to Correct Flow in Bins," *Powder Technol.*, January 1968, pp. 328–333. Johanson and Colijn, "New Design Criteria for Hoppers and Bins," *Iron Steel Eng.*, October 1964, pp. 85–104.

Specifying Bulk Materials for Best Flow Many flow problems can be eliminated at the source by rigid, accurate, and sensible specification of the physical characteristics of the material.

Particle size is one of the most common and controllable factors which affect the flowability of a given material. In general, it may be assumed that the larger the particle size and the freer the material is from fines, the more easily the material will flow. Specifications can dictate the desired particle size and uniformity of particle size for purchased raw materials. Stage grinding in the plant can reduce waste and improve flowability by producing a ground material with a minimum of fines, but this involves extra operations which may not be economically defensible.

Handling ease is often enhanced by **pelleting** the raw materials. The large particle size, uniformity of particle size, and hard, smooth surface of pellets all contribute to good flow.

Moisture content is another common and controllable flow factor. Most materials can safely absorb moisture up to a certain point; further addition of moisture can cause significant flow problems. Specifications can control the amount of moisture content present in purchased raw materials. Moisture content can be lowered in the plant by including a drying operation in the process line. The costs incurred in drying may be offset by more efficient flow, lower shipping cost, and control of deterioration losses.

Moisture control can also be effected by replacing the air in the material container or bin with a dry, stable gas—nitrogen, for example. This technique is also used to protect the material from certain types of deterioration, such as vitamin loss from food materials.

High temperatures can cause serious flow problems in some materials which contain glutens, sugars, or other soluble or low-melting-point components. These materials become sticky at high temperatures, and it may be necessary to install cooling equipment. As with drying equipment, a study should be made to determine if the additional cost of cooling can be offset by the savings effected by improved flow. Other possible advantages, such as the keeping qualities of the product at lower temperatures, should of course be considered.

Age appears to improve the flowability of certain materials. This is probably the result of particle-surface oxidation, more even moisture distribution, and the rounding of particle corners caused by handling.

Oil content does not materially decrease flowability. For example, the addition of oils and fats to animal-feed ingredients improves the quality of pellets made from these materials, making the pellet surfaces harder and enabling the pellets to resist attrition.

Gates (Fig. 7-24) are used to control flow from bins, hoppers, and processing equipment to feeders or directly to conveyors. They are

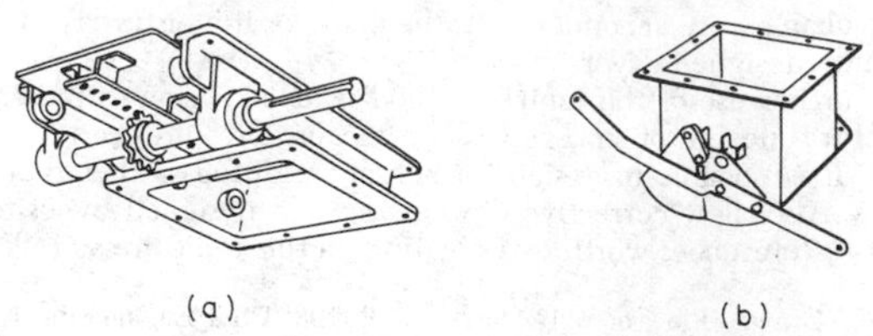

FIG. 7-24 (*a*) Rack-and-pinion gate. (*b*) Double-quadrant gate.

available in a wide range of styles, from the simple hand slide gate (which can frequently be very difficult to operate by hand) to the precision rack-and-pinion design, which is usually tightly sealed against dust and dribble. The rack-and-pinion gate operates manually with a minimum of effort and is easily adapted to electric, pneumatic, or hydraulic operation. The lever-operated quadrant gate is most often used when a quick-opening gate is desired. It is not designed to control the flow of material but rather to allow the free discharge of lumpy materials. There are hundreds of gate styles to select from, and when properly applied they can often eliminate the need for a more expensive feeder.

Solids-level controls are important for determining the level of materials in bins and hoppers and can also protect conveyors from damage due to jamming if placed in transfer and discharge chutes. They may simply activate an audio or visual warning signal, or they may be electrically tied into the conveying system to start or stop conveyors automatically. Many designs are available, ranging from expensive devices using radioactive isotopes down to simple paddles. The two designs shown in Fig. 7-25 depend on limit switches, with activation from a pendant cone on one and from a stainless-steel diaphragm on the other. In either case, the presence of material resting against the cone or diaphragm opens or closes the switch, activating a warning signal in the latter case and turning off power to the conveyor in the former.

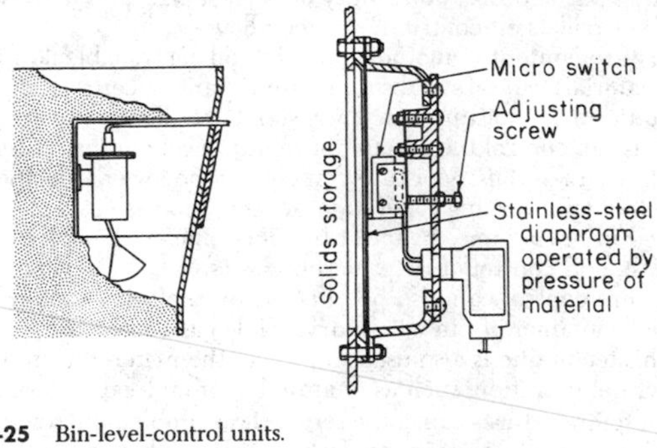

FIG. 7-25 Bin-level-control units.

FLOW-ASSISTING DEVICES AND FEEDERS

Often there are situations in which mass-flow bins cannot be installed for reasons such as space limitations and capacity requirements. Also, sometimes the product to be stored has an *FF* flow function that lies below the flow factor *ff*, bridging takes place, and unassisted mass flow is not possible. To handle these situations, a number of **flow assisters** are available, the most desirable of which use a feeder and a short mass-flow hopper to enlarge the flow channel of a funnel-flow bin. The choice of feeder or flow assister should always be made as part of the storage-vessel analysis. The resulting systems are then usually as effective as the mass-flow types.

Vibrating hoppers are one of the most important and versatile flow assisters. They are used to enlarge the storage-bin opening and to cause flow by breaking up material bridges. Figure 7-26 shows this type of feeder. Two basic types of vibrating hoppers are common: the gyrating kind, in which vibration is applied perpendicularly to

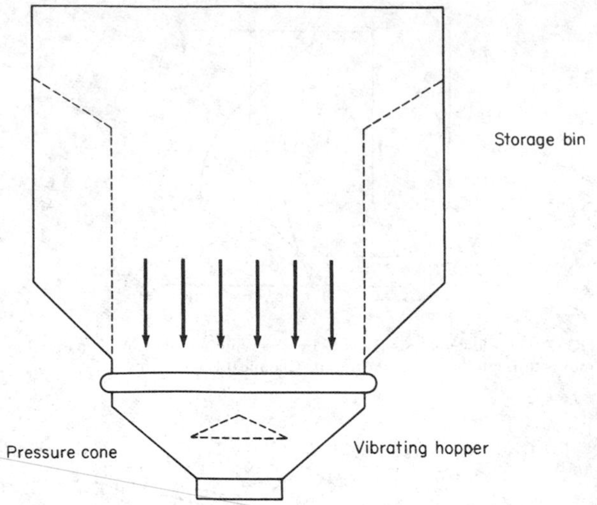

FIG. 7-26 Vibrating hopper. It enlarges the storage-bin opening and breaks up material bridging. (*Courtesy of* Mechanical Engineering.)

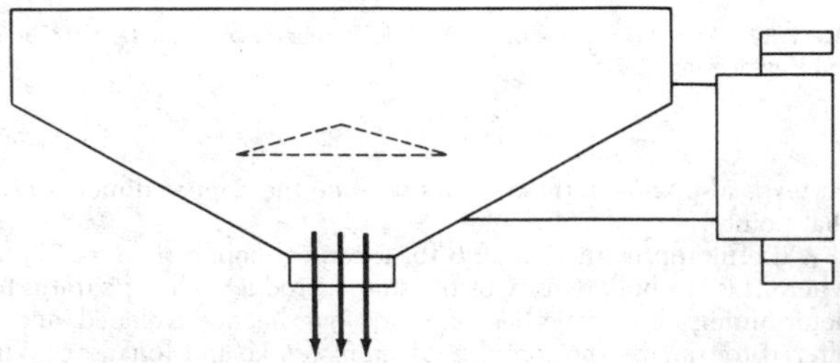

FIG. 7-27 Gyrating hopper. Vibration is applied perpendicularly to the flow channel. (*Courtesy of* Mechanical Engineering.)

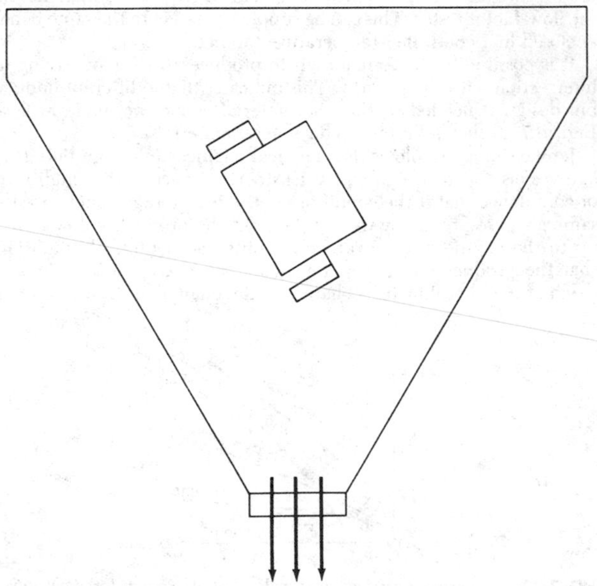

IG. 7-28 Whirlpool-type hopper. Two mounted motors provide lift-twist action to the material.

the flow channel (Fig. 7-27), and the whirlpool type, which by providing a combined twist and lift to the material, causes bridging to break (Fig. 7-28). One version of this type of flow assister is a bin that vibrates or oscillates in its entirety. Such bins are usually limited to a capacity of about 2.8 m^3 (100 ft^3).

Screw feeders are also used to assist in bin unloading and in producing uniform feed. Of importance here is the need for a variable-pitch screw to produce a uniform draw of material across the entire hopper opening (Fig. 7-29). For uniform flow to occur, the screw-feeder opening-to-diameter ratio should not exceed 6.

Belt or apron feeders can also be used to give uniform feed from a bin, but care must be taken that dead spots are not produced in the flow channel above the feeder belt (Fig. 7-30). The capacities of these feeders can be increased by tapering the outlet in the horizontal and vertical planes. To ensure the flow of non-free-flowing solids along the front bin wall, a sloping striker plate at the front of the hopper is necessary. Taper may be in one direction only. An apron feeder for large rock, for instance, would have bin skirts tight against the pan to prevent the rock from wedging between the hopper and feeder, and the taper would be in the horizontal plane only. For long slots, however, increasing slot width to provide taper becomes impractical.

Belts have been used successfully under slot openings as long as 30 m (100 ft), with a constant slot width of 205 mm (8 in). Provisions should be made for field adjustment of the space between the skirt and the belt to provide uniform flow along the entire length. Since the minimum distance between the skirt and the belt should allow the largest particle to pass under, very long belt feeders are limited to the finer solids.

The same principles apply to **table feeders.** The skirt is raised above the table in a spiral pattern to provide increased capacity in the direction of rotation (Fig. 7-31). The plow, located outside the bin, plows only the material that flows from under the skirt.

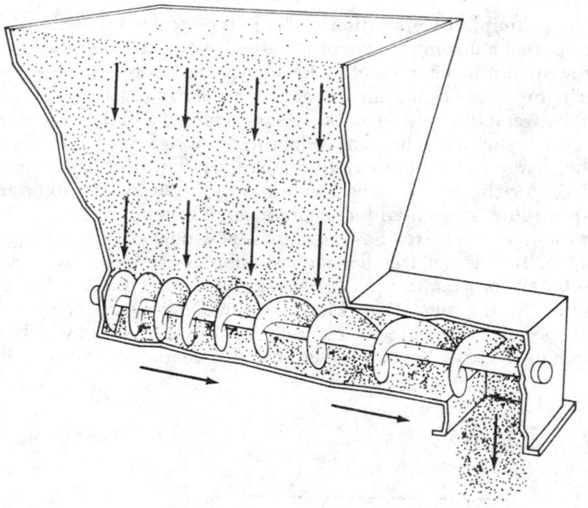

FIG. 7-29 Screw feeder. It needs a variable-pitch screw to produce a uniform draw of material. (*Courtesy of* Chemical Engineering.)

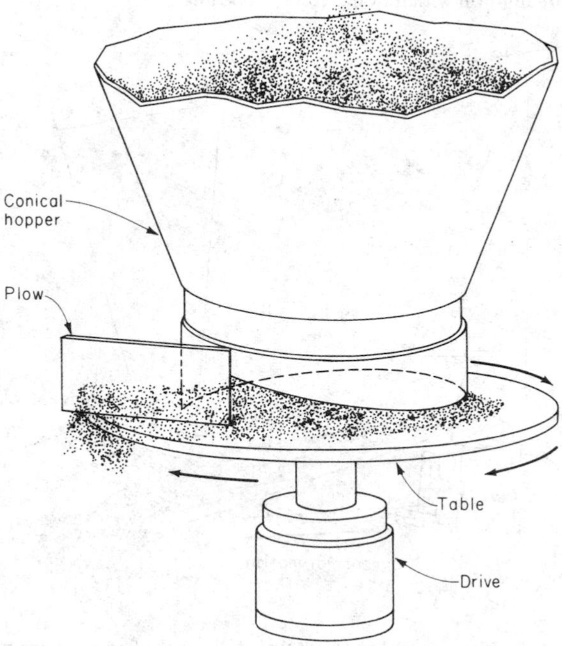

FIG. 7-31 Table feeder. The skirt is raised in a spiral pattern for increased capacity in the direction of rotation. (*Courtesy of* Chemical Engineering.)

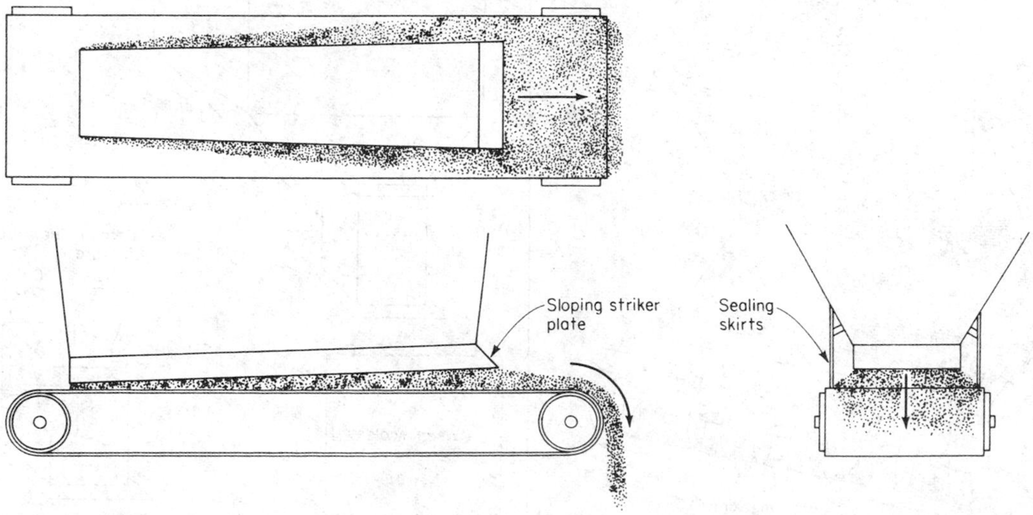

FIG. 7-30 Sloping striker plate in the belt of an apron feeder ensures the flow of non-free-flowing solids. (*Courtesy of* Chemical Engineering.)

Vibratory feeders also provide uniform flow along a slot opening of limited length (Fig. 7-32). Here also, the distance between the feeder pan and the hopper is increased in the feed direction. Slot length is limited by the motion of the feeder. Because in long slots the upward component of motion is not relieved by the front opening, solids tend to pack. This can cause flow problems with sticky solids as well as a large demand of power for free-flowing materials. To circumvent these difficulties, vibratory feeders and reciprocating-plate feeders are designed to feed across the slot. Although this kind of feeder may require several drives to accommodate extreme width, the drives are small because of the feeder's short length.

Star feeders with a collecting-screw conveyor (Fig. 7-33) provide highly uniform withdrawal along a slot opening. A vertical section of at least one outlet width should be added above the feeder to ensure uniform withdrawal across the opening.

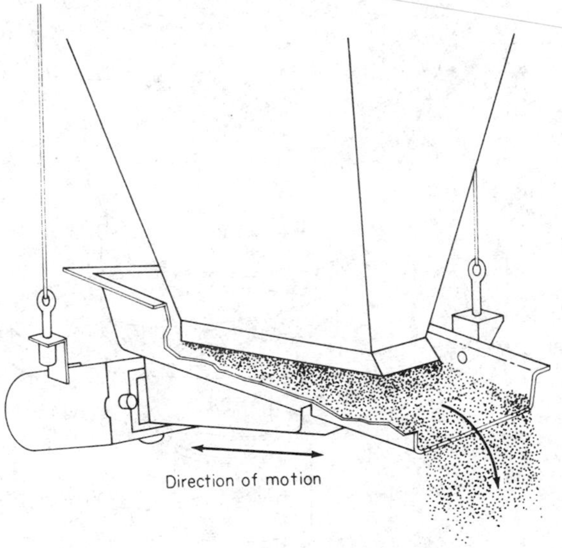

FIG. 7-32 Vibratory feeder. The distance between the feeder pan and the hopper is increased in the direction of feed. (*Courtesy of* Chemical Engineering.)

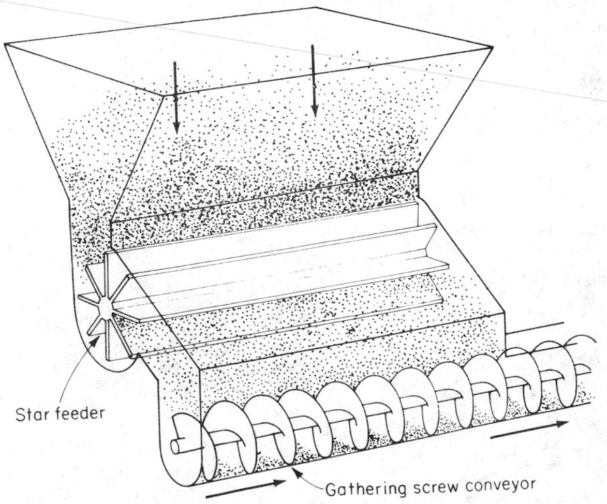

FIG. 7-33 Star feeder. The collecting screw ensures uniform withdrawal. (*Courtesy of* Chemical Engineering.)

Other methods of aiding bin unloading are rotating-arm units and air fluidizing pads.

WEIGHING OF BULK SOLIDS

Automatic weighing has largely replaced manual weighing in the chemical-process industries because of the advent of larger-capacity processes and the need to economize on labor. Also, the dependability of weighing equipment has increased markedly, and investment cost has decreased. Both batch and continuous weighing are used.

Batch Weighing In batch weighing, a given unit of weight is measured, and then the desired total weight is obtained through multiples of the given unit. Batching scales find use when small weighings are carried out either singly or a few in sequence.

Most batch scales involve a vessel mounted on a weigh beam, which is counterbalanced by a set of weights approximately equal to the desired weighing. A feed source mounted over the weigh vessel is activated or stopped by a signal generated by motion of the scale beam. Straight mechanical scale-control systems have largely been replaced by those having air or hydraulic-cylinder control of the feed source and weigh-vessel discharge. These are activated by electrical controls.

The **principle of operation** of batch-type scales is based on the concept that a flowing stream of material has constant density. If this is true, then if at some point in advance of the desired batch weight the stream is cut off, the amount of material flowing will remain constant between the time when the weight is sensed and the time when the flow is stopped. The total weight in the weigh vessel is the sum of the charge due to flow and the amount that flows during the cutoff period. For this reason, feed conditions to the scale are important. **Uniform flow** is essential for accurate batch weighings.

If the material is free-flowing, a mass-flow hopper (Fig. 7-34) can be used. If it is not free-flowing, an appropriate feeder such as a screw, belt, or vibratory feeder should be used. These feeders are described in the subsection "Flow-Assisting Devices and Feeders."

Of special interest in scale-control systems is the type in which the motion of the scale beam is sensed by a differential transformer or a

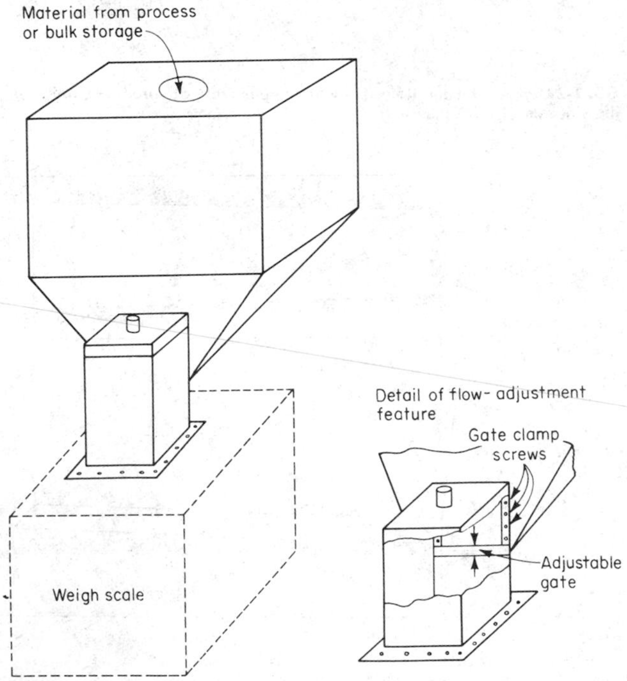

FIG. 7-34 Mass-flow hopper for free-flowing products, used with simultaneous fill-and-weigh and preweigh scales.

group of load cells. The output of such devices is proportional to the displacement of the scale beam, which in turn is proportional to the amount of material in the weigh bucket. This provides many benefits in addition to accurate weights. One of the most notable is the ability to use the output to indicate actual weight in the weigh vessels or, through a different calibration, to read variations from the desired weight.

Use of microprocessors allows this signal to be employed in a variety of useful ways: controlling weight, adjusting the scale to accommodate the slight changes in bulk density inherent in flowing bulk materials, and activating recording devices, printing heads, and label printers. These features were not possible with straight mechanical scales. Because the microprocessor can perform arithmetic operations and be programmed by using algebraic logic, many products previously considered difficult or impossible to weigh accurately can now be weighed with accuracies equal to those for free-flowing materials. This permits weighing to be done by **addition** or by **subtraction.** With the **additive**-type scale, material is **added** to the weigh vessel, which is being sensed by the scale-control system. With a **loss-of-weight scale,** material flows out of a vessel which is being continuously weighed.

When extreme accuracy of weighings is required, the feed to the weigh vessel is divided into two successive portions: a large **bulk charge,** followed by a final short **dribble feed,** which should have a flow rate of about 0.01 percent of the bulk rate.

A typical application of batch scales is the weighing of charges for packaging machines. Another is the weighing of given amounts of raw material and then dropping these into the next process unit, such as a mixer or an autoclave. Batch weigh scales are capable of a weighing accuracy of within ±0.1 percent when they are equipped with bulk and dribble controls.

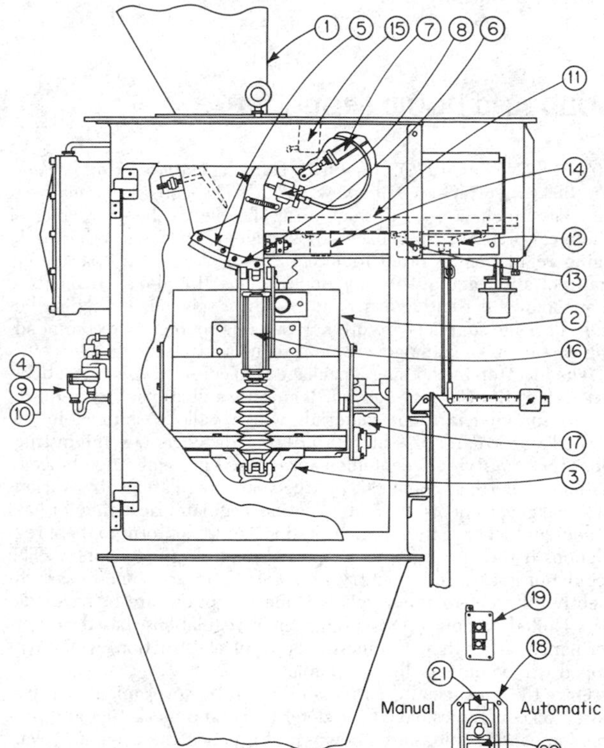

FIG. 7-35 Typical batch-type additive automatic scale. Components: (1) Bin. (2) Scale bucket. (3) Bucket gate. (4) Solenoid valve. (5) Bulk gate. (6) Dribble gate. (7) and (8) Air cylinders powered by solenoid-operated valves. (9) and (10) Solenoid-operated valves. (11) Scale beam. (12) Booster device. (13) and (14) Load cells. (15) Microswitches. (16) Air cylinder. (17) Microswitch. (18) Microprocessor. (19) Manual start switch. (*Courtesy of St. Regis Paper Co.*)

Additive Weigh Scale The sequence of operations involved in weighing a charge of material (Fig. 7-35) is as follows. A free-flowing product is available in the scale feed hopper (1). On depressing the manual start switch (19), the bulk gate (5) and the dribble gate (6) open. The product flows into the scale weigh bucket (2). Weight is sensed by strain gauges (13), (14), whose analog output is converted to a digital output by a circuit in the microprocessor (18), which reads the weight X times each second, depending on the sensivity needed. When a preset bulk weight (approximately 98 percent of the desired weight) is reached in the scale bucket (2), the microprocessor closes the bulk gate (5) and opens the dribble gate (6). Dribble feed commences, and when the desired weight is reached, the microprocessor causes the dribble gate to close, completing the weight measurement. The scale bucket gate automatically opens, discharging the product weighed to the next process stage. The microprocessor then displays the actual weight of the charge (20) and records, lists, prints (21), and signals any discrepancy in weight if this is outside the tolerance desired. It also can print a label for the batch if desired.

Loss-of-Weight Scale The sequence of operations in this scale (Fig. 7-36) is as follows: Depressing the initializing switch (1) causes the feeder (2) to fill the weigh hopper (3) until the level-control switch (5) opens, stopping the flow, closing the interlocking switch (5), and measuring and recording the initial weight W_0 in the weigh hopper (3). Depressing the start button (6) causes the feeder (4) and the bag packer to start simultaneously. The product is conveyed by the feeder (4) into the packer (7) and by the bag packer into the bag (not shown). The microprocessor (8) reads the analog-to-digital-converter signal (8), which is connected to the strain-gauge load cell (9), and subtracts weight W_i in the hopper (3) at time t_i from the initial weight W_0. The weight difference W_j is summed and recorded. When $W_j = W_s$, the desired weight, the microprocessor stops the feeder (4) and, X seconds later, the bag packer (7). The microprocessor then displays the value of weight W_f (10) and records, lists, and prints (11) any discrepancy between the desired weight and W_f,

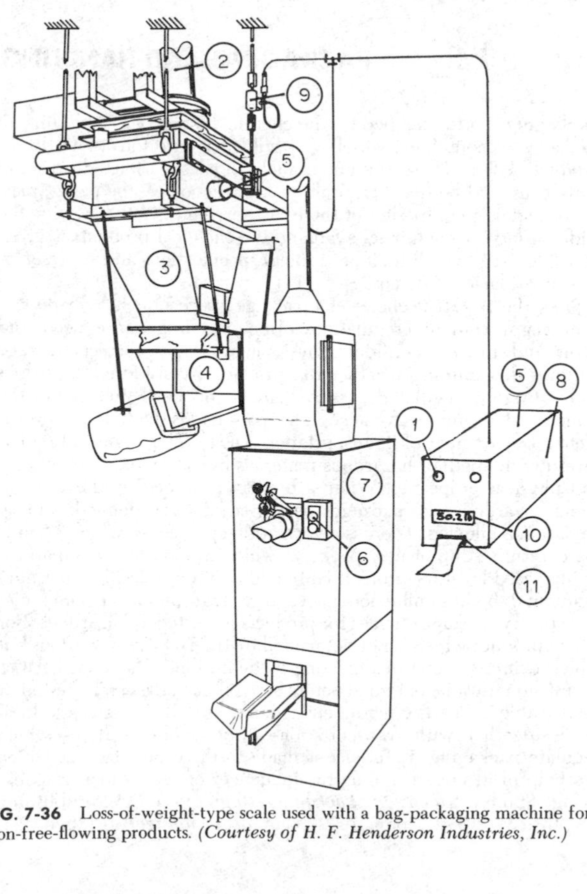

FIG. 7-36 Loss-of-weight-type scale used with a bag-packaging machine for non-free-flowing products. (*Courtesy of H. F. Henderson Industries, Inc.*)

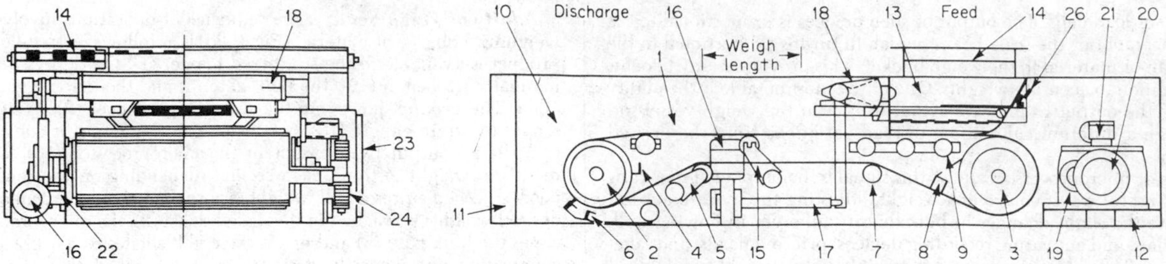

FIG. 7-37 Bulk continuous weigher. (1) Conveyor belt. (2) Head pulley. (3) Tail pulley. (4) Pulley scraper. (5) Spring-loaded take-up assembly. (6) Outer belt scraper (spring-loaded). (7) Belt tracking control. (8) Belt scraper (inner). (9) Pulley scraper. (10) Side channel. (11) Cross channel. (12) Cross channel. (13) Transition chute (optional). (14) Feed-cutoff gates. (15) Weigh idler(s). (16) Gate-screw drive motor. (17) Gate screw. (18) Manual gate-adjustment screw. (19) Tachogenerator (optional). (20) Variable-speed drive. (21) Speed adjustment for motor drive (20). (22) Coupling. (23) Side cover. (24) Weight-sensing elements. (25) Hopper storage (optional). (26) Adjustable heel plate (optional). *(Howe Richardson Company.)*

the weight actually obtained. The packaging system shown is designed for handling products which have very poor or erratic flow characteristics. In addition to its use with packaging equipment, the **loss-of-weight scale** can be employed for a wide variety of process applications.

Continuous Weighing This procedure involves a device that is sensitive both to the total amount of material flowing and to changes in the flow. The material is continuously brought over the weight-sensing elements of the continuous-weigh scale, which is capable of keeping track of the flow and its changes and eventually accounts for these when totaling them. Continuous-weighing scales use a section of a belt conveyor, over which the material to be weighed passes.

The belt is mounted on a weight-sensitive platform, typically equipped with load cells, which can detect minute changes in the weight of material passing over the belt. The load-cell output (which is usually a change in resistance proportional to weight) is integrated over short intervals and the condition of flow given. This may be a rate of flow or, at the end of a weight measurement, the total weight. Figure 7-37 shows a continuous weigher, sometimes referred to as a proportioner. Continuous-weighing scales are used mostly to feed materials to continuous processes at uniform, measured rates. They are capable of weighing within ± 1 percent error or even within 0.1 percent error under certain conditions.

All scales require continuous monitoring to assure that the desired set weight is maintained and does not drift off because of changes in product bulk-density or flow characteristics.

PACKAGING AND HANDLING OF SOLID AND LIQUID PRODUCTS

Packaging is often defined by the chemical industry as including all packages or containers which will hold 2 metric tons (4400 lb) of product or less. These containers include bags, cartons, drums and pails, cans, and bottles. Materials of construction can be paper, plastic, or metal or composites of these. A representative list showing the wide variety of containers available for chemical products is given in Table 7-15, which includes typical specifications and representative first-quarter 1980 costs.

Since the 1970s the choice of a package has become heavily dependent on governmental regulations, often preceding market needs and plant and process considerations. Choosing the correct package requires determining whether the product is considered hazardous and is thereby regulated by government. In the United States the principal determinant for acceptable packages is the federal government's Department of Transportation (DOT). This organization has carefully defined all hazardous materials by name and by chemical and physical properties. When a product is named by the DOT as being hazardous, the packages permitted are well defined, and no deviation is allowed. There is an exemption procedure which permits the careful testing of new packages which would allow comparable or increased levels of protection. In addition to the DOT, other governmental bodies influence packaging. Transportation companies also specify package choices for products not listed as hazardous. For a new product which is not yet named by the DOT as hazardous but which exhibits properties meeting its definitions of a hazard, DOT permission must be obtained before a given package can be tested or used. Table 7-16 lists government agencies at federal, state, and local levels together with transportation-company organizations which regulate packaging. Reference is made to three publications which are helpful in understanding the influences of regulations on packaging: Stanley Sacharow, *Packaging Regulation* (AVI Publishing Co., Inc., Westport, Conn., 1979); *Chemical Engineering* [**85**, no. 25, 103–108 (Nov. 6, 1978)]; Joseph F. Hanlon, *Handbook of Package Engineering* (McGraw-Hill, New York, 1971). The last-named work also has much useful information applicable to the packaging of chemical products. Specialized knowledge of DOT and other packaging regulations is often found in chemical-company packaging, transportation, and distribution departments. It is also available from management consulting firms and law firms specializing in distribution. Package suppliers are often a valuable source for recommending containers meeting government regulations.

The first step in choosing a package is to consult DOT regulations. Title 49 of the *Code of Fereral Regulations* gives package requirements for all hazardous materials which will be shipped in the United States. Parts 171 through 179 of Title 49 contain full instructions for identifying acceptable packages. Of growing importance to nations exporting chemical products are regulations by various national governments. Individual nations regulate packaging for hazardous and other materials, and products must conform to these regulations if they are to enter or pass through a given country. The regulations vary from country to country and are sometimes sufficiently different to cause delays while exceptions are being made. The United Nations (UN) is developing regulations (based on performance rather than specific materials of construction) which it is hoped will be universally acceptable.

Once the package alternatives permitted by government or transportation companies have been determined and marketing and production considerations are known, performance and economic evaluation must be made. This evaluation should consider packaging as part of a system. Not only must the package itself be considered, but so must factors which affect the package or are affected by it. If a choice of shipping in bulk form or in packages exists, cost comparisons must be made (Table 7-17).

Metric-system dimensions for packaging are not used extensively

TABLE 7-15 Typical 1980 Cost of Containers for Chemical Products

Container size and description	Unit cost, US$	Usable volume, ft^3	Usable volume, m^3
Metal drums[a,b,c]			
55 gal (208 L), steel, tight head, 18-20-18 gauge,[d] DOT-17E, new	$ 16.70	7.35	0.208
55 gal (208 L), steel, tight head, all 16 gauge, DOT-17C, new	21.72	7.35	0.208
55 gal (208 L), steel, open head, 18 gauge, Rule 40, new[e]	21.99	7.35	0.208
55 gal (208 L), steel, open head, 16-18-18 gauge, DOT-17H, new[e]	24.27	7.35	0.208
55 gal (208 L), steel, tight head or open head, 18 or 18-20-18 gauge, used, reconditioned	11.50	7.35	0.208
55 gal (208 L), Type 304 stainless steel, tight head, 16 gauge, DOT-5C (delivered)	230.00	7.35	0.208
30 gal (113 L), steel, tight head, 20 gauge, DOT-17E	12.60	4.00	0.113
30 gal (113 L), steel, open head, 20 gauge, Rule 40[e]	14.33	4.00	0.113
16 gal (61 L), steel, open head, lug cover, without fittings, 22 gauge (delivered)	7.20	2.14	0.061
55 gal (208 L), steel, dip-galvanized, tight head, 18 gauge, DOT-17E	44.00	7.35	0.208
56.1 gal (204 L), steel, with 40-mil PE insert, external fittings, 18-20-18 gauge, 55-gal usable volume, DOT-2SL37M (delivered), open head[e]	33.13	7.20	0.204
56.1 gal (204 L), DOT-25L37M, but tight head	33.13	7.20	0.204
55 gal (208 L), of blow-molded high-density PE, having steel lifting ring	22.25	7.20	0.204
Cans and pails[b]			
Pail, 5 gal (19 L), steel, tight head, 26 gauge, black steel, PE pour spout, unlined (delivered)	$ 2.72	0.67	0.019
Pail, 5 gal (19 L), 26 gauge, black steel, open head, unlined, lug cover, wire-bail handle.	2.20	0.67	0.019
Can, 1 gal (3.8 L), friction-wedge lid, handle (paint can)	1.01	0.1335	0.004
Can, 1 qt (0.95 L), friction-wedge lid (paint can)	.44	0.034	0.001
Can, 1 gal (3.8 L), oblong F style, handle, 1¼-in (32-mm) screw cap	1.12	0.1335	0.004
Can, 1 pt (0.48 L), oblong F style, 1-in (25-mm) screw cap	.37	0.0167	0.0005
Square can, 5 gal (20 L), blow-molded PE with 2¾-in cap (often called "5-gal squares")	2.47	0.67	0.019
Can, 1 pt (0.48 L), round-cone-top style, with 1-in (25-mm) cap	.48	0.0167	0.0005
Fiber drums[b,f]			
61 gal (231 L), 9 ply, 400-lb (181-kg) load limit, dry products only, Rule 403	$ 7.85	8.15	0.231
55 gal (208 L), 9 ply, 400-lb (181-kg) load limit, dry products only, Rule 403	7.21	7.35	0.208
47 gal (178 L), 9 ply, 400-lb (181-kg) load limit, dry products only Rule 403	6.94	6.28	0.179
41 gal (155 L), 9 ply, 400-lb (181-kg) load limit, dry products only, Rule 403	6.38	5.48	0.155
30 gal (113 L), 9 ply, 400-lb (181-kg) load limit, dry products only, Rule 403	5.65	4.00	0.113
30 gal (113 L), 7 ply, 225-lb (102-kg) load limit, dry products only, Rule 403	5.05	4.00	0.113
15 gal (56.8 L), 6 ply, 150-lb (68-kg) load limit, dry products only Rule 403	3.75	2.00	0.057
55 gal (208 L), 9 ply, PE barrier, 400-lb (181-kg) load limit, Rule 403	7.62	7.35	0.208
55 gal (208 L), 9 ply, PE-aluminum foil liner, 400-lb (181-kg) load limit, Rule 403	9.21	7.35	0.208
55 gal (208 L), 10 ply, blow-molded, 15-mil PE liquidtight liner, tight head, steel cover with 2-in and ¾-in NPT1 openings, 600-lb (272-kg) load limit, DOT-21C.21CP liquid products	16.50	7.35	0.208
30 gal (113 L), 9 ply, same as preceding except 450-lb (204-kg) load limit	12.74	4.00	0.113
30 gal (113 L), 8 ply, 300-lb (136-kg) load limit, removable fiber cover, no barrier	6.03	4.00	0.113
15 gal (56.8 L), 6 ply, same as preceding except 150-lb (68-kg) load limit	3.98	2.00	0.057
1 gal (3.8 L), 5 ply, same as preceding except 150-lb (68-kg) load limit	1.71	0.1335	0.004
55 gal (208 L), 9 ply, 400-lb (181-kg) load limit, semisquare removable fiber cover, Ro-Con style	7.05	7.35	0.208
45 gal (170 L)	6.41	6.01	0.170

Remarks

Black steel drums lined at extra cost:

Lining	No. coats	Cost/drum, $
Baked resin,	1	2.59
pigmented	2	3.19
Epoxy phenolic composite	2	3.35

Prices shown include ¾-in and 2-in fittings and are for unlined drums; add 50 cents per drum for delivery.

Approximate drums per carload

55-gal size = 360 drums
30-gal size = 592 drums
16-gal size = 1225 drums

U.S. standard gauge equivalents

Gauge	in	mm
16	0.0598	0.0152
18	0.0478	0.0121
20	0.0359	0.0091
22	0.0299	0.0076
24	0.0239	0.0061

Approximate drums per carload

61-gal size = 300 drums
55-gal size = 318 drums
47-gal size = 424 drums
41-gal size = 552 drums
30-gal size = 592 drums
15-gal size = 1272 drums
1-gal size = 17,365 drums

Drum type	Outside dimensions: Diameter, in	Diameter, cm	Height, in	Height, cm
55-gal lever top	21	53.3	40¾	103.5
55-gal lever top	23½	59.7	30¾	78.1
55 gal lever top	22	55.9	34¾	88.2
41-gal lever top	20½	45.1	30¼	76.8
30-gal lever top	19	48.3	26¼	66.7
6.28-ft^3 rectangular	17⅝°	44.8	37½	95.3
55-gal liquid	22	55.9	37½	95.3
30-gal liquid	19	48.3	28	71.1
55-gal fiber	20⅜	51.8	40¾	103.5
30-gal fiber	17⅜	44.1	30¾	78.1

°Side dimension, square.

TABLE 7-15 Typical 1980 Cost of Containers for Chemical Products *(Continued)*

Container size and description	Unit cost, US$	Usable volume ft³	Usable volume m³
Bags: multiwall paper and polyethylene film[g]			
Pasted valve bag, 20½- × 22-in face, 5½-in top and bottom (520 × 560 × 140 mm) with 1-mil free film, 2/50, 1/60 kraft, PE internal sleeve	$ 0.25	1.33	0.038
Sewn valve bag, 15 × 5½ × 30¼ in, 5½-in PE internal sleeve (380 × 140 × 770 × 140 mm) with 1-mil free film, 2/50, 1/60 kraft	0.29	1.33	0.038
Pasted valve bag, 18½ × 22¾ in, 3½-in top and bottom (470 × 580 × 90 mm), PE internal sleeve, 3/50 kraft	0.16	0.84	0.024
Sewn open-mouth bag, 20 × 4 × 30¾ in (510 × 100 × 780 mm), 3/50, 1/60 kraft	0.27	2.00	0.057
Sewn valve bag, 19 × 5 × 33½ in, 5½-in tuck-in sleeve (480 × 130 × 850 × 140 mm), 3/50, 1/60 kraft	0.31	2.00	0.057
Pasted valve bag, 24 × 25¼ in, 8-in top and bottom (610 × 640 × 200 mm), tuck-in sleeve, 3/50, 1/60 kraft	0.29	2.00	0.057
Pasted open-mouth baler bags, 22 × 24 in, 6-in bottom (560 × 610 × 150 mm), 1/130 kraft (or 2/70)	0.16	1.33	0.038
Flat tube open-mouth bag, 10-mil PE film, plain, 20½ × 34¼ in (520 × 870 mm)	0.41	1.33	0.038
Square-end valve bag, 20½- × 22-in face, 5½-in top and bottom (520 × 560 × 140 mm), 8-mil PE film, plain	0.28	1.33	0.038
Pinch-style open-mouth bag, 20 × 4 × 30¾ (510 × 100 × 780 mm), 1/10 PE 50, 2/50, 1/60 kraft, plain, no printing	0.35	2.00	0.057
Small bags, pouches, and folding boxes[h]			
Pouch, 8¾ × 16¾ in (220 × 425 mm), 2-ply PE film, 2-mil- (0.05-mm) thickness per ply	$ 0.068	0.12	0.0034
Bag, sugar-packet style, 6 × 2¾ × 16¾ in (150 × 70 × 425 mm), 2/40-lb basis weight, natural kraft paper	0.059	0.12	0.0034
Bag, pinch style, 8¾ × 3 × 21 in (220 × 75 × 530 mm), 2/40-lb basis weight, natural kraft paper	0.059	0.12	0.0034
Folding box, 5 × 1 × 8 in (125 × 25 × 200 mm), reverse-tuck design, 12-point kraft board with bleached white exterior	0.117	0.028	0.0008
Folding box, 9½ × 4½ × 15 in (240 × 115 × 380 mm), full-overlap top and bottom, 30-point chipboard with bleached white exterior	0.234	0.37	0.0105

For tuck-in sleeve, add $0.01/bag. Unit cost is for unprinted bag. For printing add the following up charges. U.S. dollars per 1000 bags:
1 side, 1 color, $ 7.50
1 side, 2 colors, $ 9.50
1 side, 3 colors, $12.50
2 sides, 1 color, $ 9.50
2 sides, 2 colors, $12.90
2 sides, 3 colors, $16.00

Polyethylene-film gauges

mil	Actual mm	Nearest mm
0.5	0.0127	0.01
1.0	0.0254	0.03
1.5	0.0381	0.04
1.75	0.0445	0.04
2.0	0.0508	0.05
8.0	0.2032	0.20

Multiwall kraft-paper basis-weight equivalents

U.S. customary, lb/3000 ft² ream	SI, g/m²
40	65
50	81
60	97

Permeability of common packaging films*

Type of film	Water-vapor transmission†	Gas permeability‡ O_2	N_2	CO_2	Water absorption
Cellophane, nitrocellulose-coated	0.3	1	1	13	High
Nylon	19	25	160	160	Medium
Polycarbonate	11	300	50	1000	Medium
Polyester, oriented	1.7	4	1	16	Low
Polyethylene, low-density	1.3	550	180	2900	Low
Polyethylene, high-density	0.3	600	70	4500	Low
Polypropylene	0.7	240	60	800	Low
Saran	0.2	14	12	4	Low

*From J. R. Hanlon, *Handbook of Package Engineering*, McGraw-Hill, New York, 1971.
†g loss, 24 h/(100 in²·mil), at 95°F, 90 percent relative humidity.
‡cc, 24 h/(100²·mil), at 77°F, 50 percent relative humidity; ASTM D1434.

Container size and description	Unit cost, US$	Usable volume ft³	Usable volume m³
Corrugated cartons and bulk boxes[h]			
Regular slotted carton (RSC), 24 × 16 × 6 in (610 × 405 × 150 mm), 275-lb-test double wall, stapled (stitched) joint	0.70	1.33	0.038
RSC, 16 × 6 × 24 in (405 × 150 × 610 mm), 275-lb-test double wall, stitched joint, end-opening style	0.59	1.33	0.038
Bag in box, RSC, 15 × 15 × 22 in (380 × 380 × 560 mm), 275-lb-test double wall, stitched liner, 600-lb-test double wall, 6-mil (0.15-mm) PE liner	2.15	2.86	0.081
Bulk box, 600/600 (test in lb for both pieces), laminated inner lining, approximately 41 × 34 × 36 in (1040 × 865 × 915 mm); includes special wood pallet and 8-mil (0.2-mm) blown low-density PE liner	18.00	5.00	0.142
Carboys, plastic drums, jars, and bottles[h]			
Carboy, 13½ gal (51 L), PE, blow-molded	25.90	1.35	0.038
Drum, PE, 15 gal (57 L), blow-molded, ICC-34 (DOT 34)[c]	29.40	2.00	0.057
Carboy, 15 gal (57 L), glass, nitric acid service, wooden crate	48.00	2.00	0.057
Jug, 1 gal (3.78 L), glass, with finger handle, plastic 38-mm cap, with corrugated reshipper carton	0.718	0.1335	0.004
Bottle, 1 qt (0.95 L), glass, Boston round, plastic 28-mm cap	0.258	0.034	0.001
Jar, 1 qt (0.95 L), glass, wide mouth, plastic 89-mm cap	0.370	0.034	0.001
Jar, 1 qt (0.95 L), glass, plastic 63-mm cap	0.212	0.034	0.001
Jar, 1 gal (3.78 L), PE, wide mouth, plastic 100-mm cap	0.709	0.1335	0.004
Bottle, 1 gal (3.78 L), round, PE, narrow neck, plastic 38-mm cap	0.448	0.1335	0.004
Bottle, 1 qt (0.95 L), PE, narrow neck, plastic 28-mm cap	0.189	0.034	0.001
Jar, 1 pt (0.47 L), PE, wide mouth, plastic 53-mm cap	0.195	0.017	0.0005

Cost US$[h]

	Expendable grade		Warehouse reusable grade	
	9-block type	Stringer type	9-block type	Stringer type
Pallets[h]				
40 × 48 in (1015 × 1220 mm)	5.47	5.38	8.24	7.59
35 × 42 in (890 × 1065 mm)	4.73	4.65	7.67	7.06
42 × 48 in (1065 × 1220 mm)	8.09	7.95	9.35	8.61
48 × 48 in (1220 × 1220 mm)	8.72	8.58	10.09	9.29
44 × 50 in (1115 × 1270 mm)	9.96	9.80	11.41	10.51

Wrap materials	US$/lb
Film, PE, Grade ADL, blown type	0.58
Film, PE, Grade ASF (shrinkable)	0.58
Film, polypropylene, shrinkable, yield before shrinkage = 31,100 in²/(lb·mil)	1.90
Paper, kraft, wrapping quality, 50-lb/ream basis-weight yield = 3000 ft²/ream	0.25
Film, PE, stretchable type for pallet wrap, 1.5 mil × 20 in (0.04 × 510 mm) wide	0.60

[a]Drum has 2-in and ¾-in national-pipe-thread (NPT) openings in head.
[b]Carload quantity price, FOB east-coast manufacturer's plant.
[c]DOT = U.S. Department of Transportation.
[d]Sequence of top, body, and bottom gauges. For example, 18-20-18 = 18-gauge top, 20-gauge body, and 18-gauge bottom.
[e]Removable head secured with bolted ring with screw draw-up.
[f]Drums are of plain fiber, have steel cover and bottom, and have lever-operated closing ring.
[g]Carload-quantity price, FOB buyer's plant.
[h]Truckload-quantity price, FOB east-coast buyer's plant.

Polyethylene-film* yield table

Thickness		Yield	
in	mm	in²/lb	m²/kg
0.001	0.025	30,000	19.4
0.0015	0.040	24,000	15.5
0.002	0.050	15,000	9.7
0.003	0.075	10,000	6.5
0.004	0.100	7,500	4.8
0.008	0.200	3,750	2.4
0.010	0.250	3,000	1.9

*Flat sheeting.

TABLE 7-16 Agency and Administrative Law

Title	Symbol	Regulate or affect
International		
United Nations	UN	Packages and labeling for products moving among member nations.
Intergovernmental Maritime Consultative Organization	IMCO	
Federal		
Department of the Treasury Federal Alcohol Administration Act	FAA (ATF)	Packages and labeling for alcohols, tobacco, firearms, and explosives.
Department of Transportation Transportation Safety Act, Title 49	DOT	Packaging and labeling for all hazardous materials shipped in interstate commerce.
U.S. Coast Guard, Title 46	USCG	Set packaging, labeling, blocking, and bracing for all freight moving by United States–registry ships on lakes, rivers, or oceans.
Department of Labor Occupational Safety and Health Act, Title 29	OSHA	Package-filling and -handling machinery, workplace design, warehouse practice, and acceptability of packages from workplace and warehouse viewpoint.
Department of Health and Human Resources		Packages, packaging machinery, and workplace from viewpoint of their effect on food and drug purity; package labeling and marking.
Food and Drug Administration Federal Food, Drug, and Cosmetic Act, Title 21	FDA	
Environmental Protection Agency, Title 40	EPA	Packaging facilities, packaging and labeling, package disposal, workplace refuse disposal, cleanup and disposal of spills.
Clean Air Act	CAA	
Clean Water Act	CWA	
Resource Conservation and Recovery Act	RCRA	
Federal Insecticide, Fungicide, and Rodenticide Act	FIFRA	
Toxic Substances Control Act	TSCA	
State		
Departments of Transportation Labor Environmental Protection Agriculture Others	Example: New Jersey Department of Transportation	Packaging, labeling, workplace, packaging machinery, etc., in *intrastate* commerce. Regulations generally parallel those of federal departments but frequently have important differences and additional requirements.
City		
Departments of Health Labor Fire Protection	City fire department; example: NYFD	Packages, packaging machinery, packaging facilities, and materials which are transported, stored, and handled in the city. These local laws are in addition to the requirements of state and federal law.
Industry associations		
Air Line Pilots Association	APA	Materials which can be carried on commercial aircraft piloted by Air Line Pilots Association members.
International Air Transport Association	IATA	Materials which can be carried on members' aircraft and packaging and labeling requirements for them.
Association of American Railroads Bureau of Explosives	AAR B of E	Packages and loading, blocking, and bracing for all hazardous products shipped by rail in the United States. Bureau standards are generally accepted by all railroads and are the basis for R. M. Graziano's Tariff No. 32, *Hazardous Materials Regulations of the Department of Transportation by Air, Rail, Highway, and Water*, Dec. 15, 1978, or latest edition.
Uniform Freight Classification Committee, Rules 40 and 41	UFC	Set packaging standards for all freight moving by *rail*.
American Bureau of Shipping National Cargo Bureau	ABS	Packaging, labeling, loading, blocking, and bracing for all freight moving by United States–registry ships on lakes, rivers, or oceans.
National Motor Freight Traffic Association National Motor Freight Classification	NMFC	Set packaging standards for all freight moving by highway.
National Fire Protection Association	NFPA	Packages, packaging facilities, and warehouse designs and operation.
Special carriers		
United Parcel Service	UPS	Packaging, labeling, and size and weight of small packages carrying hazardous materials. Requirements meet DOT standards. Quantities generally do not exceed 1 gal (3.785 L).
U.S. Postal Service	USPO	Materials which may be shipped and packaging, labeling, and size and weight of small packages handled by parcel post.

TABLE 7-17 Container Comparative Data[a]

Product: Thermoplastic pellets of 33-lb/ft^3 (528.7-kg/m^3) bulk density

Parameter	Domestic paper bag	Bulk corrugated-paper box	Intermediate bulk container	Bulk hopper truck	Ship container	Railroad hopper car	Intermediate bulk container
Construction	Multiwall pasted valve bag, 4-ply construction, with inner ply having an extrusion coating of low-density PE	Corrugated box, one-half RSC design, made of 600-lb burst-strength-test double wallboard, A-C flute laminated; includes cap, inner, and wood pallet	Flexible bag made of woven polypropylene of 2000 denier, 12 x 12 weave, with PE liner and nylon support straps	Welded-aluminum tank with pneumatic unloading pump; undercarriage equipped with pneumatic tires	Welded- or riveted-aluminum construction, with International Organization for Standardization (ISO) end castings for lifting by standard spreader	Welded-steel construction with plastic-coated interior, equipped with 100-ton trucks for 4-ft, 8½-in-gauge tracks	Rigid container made of welded-aluminum construction with butterfly discharge valve and fill port
Size	16 x 25 x 6.5 in (405 x 635 x 165 mm)	41 x 34 x 36 in (1040 x 865 x 915 mm)	53 in diameter x 53 in high (1350 x 1350 mm)	8 ft wide x 30 ft long (2.4 x 9.1 m)	8 x 8 x 35 ft (2.4 x 2.4 x 10.7 m)	5700 ft^3 (161.4 m^3)[b]	42 x 48 x 84 in (1070 x 1220 x 2135 mm)
Capacity, lb (kg)	55.1 (25)	900 (408)	2205 (1000)	42,000 (19,051)	50,000 (22,680)	180,000 (81,648)	2205 (1000)
Tare weight, lb (kg)	0.7 (0.32)	50 (22.7)	20 (9.1)	20,000 (9072)	4200 (1905)	100,000 (45,360)	255 (115.7)
Unit cost, US$	$0.32	$18.00	$40.00	$40,000	$8000	$45,000[c]	$1204
Lease cost, US$/month				$490[d]	....	$310[d]	$18[d]
Useful life	1 trip	1 trip	10 trips	15 years	5 years	25 years	10 years
Practical shipping radius, mi (km)	Any	Any	Any	250 (402) maximum	Any	300 (483) minimum	Any
Cost of typical system, US$/100 lb (45 kg)[e]	$2.89	$4.37	$2.86	$0.45	$0.45	$0.18	$2.61
Plant investment, US$ × 1000[b]	$50–100	$12–132	$15–140	$12–175	$25–175	$20–175	$15–140

[a]Based on 1980 prices.
[b]See Fig. 7-45 for dimensions.
[c]Mileage credit of US$0.24 per mile is paid to owners or lessees of this type of hopper car. To gain mileage credit car must be loaded.
[d]Lease cost is based on a period equal to the useful life.
[e]Includes cost of container, filling, handling, and storage for minimum volume of 1000 tons per month (908 metric tons per month) but does not include freight or amortization of filling equipment.
[f]Represents typical investment in filling and handling equipment. Actual amount depends on plant layout and nature of existing facilities.

in the United States, but initial steps are being taken to permit use of these dimensions. The subject is under intensive study by both the packaging-supply and the packaging-using industries. SI equivalents are usually available from package suppliers, but at present all ordering in the United States is done in the U.S. customary system. Table 7-18 gives the degree of expected metric conversion. In the United States suppliers are using millimeters as the principal metric measure. When a soft conversion is made, increments of 5 mm are used. A converted package dimension is rounded up or down to the nearest multiple of 5 mm. For example, a bag-face width of 16 in equals 406.4 mm, which would be rounded down to 405 mm.

For kraft paper, conversion will eventually be a hard conversion to two basis weights of 75 and 90 g/m^2, which will replace the 40-, 50-, and 60-lb basis weight currently used.

TABLE 7-18 Expected Metric Conversion of Packages for Chemical Products in the United States

Package type	Degree of conversion*
Bags (paper, plastic)	Hard
Boxes (paper)	Hard
Drums (fiber)	Hard
Drums (steel)	Soft
Pails (steel, plastic)	Soft
Cans (steel, fiber)	Hard
Bottles (glass, plastic)	Hard
Paper (kraft)	Hard

*Hard conversion: resized package dimensions to hold the nearest acceptable metric unit. Example: 50-lb (22.7-kg) multiwall paper bag changed to hold 25 kg (55.1 lb). Size can be limited by the maximum-size package available.

Soft conversion: the volume of the package is unchanged, and its metric equivalent is stated. Example: 55-gal drum equals 208 L.

LIQUID PACKAGING

Containers Containers for liquids consist principally of drums, pails, and cans made of steel or plastic and of bottles and vials made of plastic or glass. The chemical industry is often involved with all these containers, but the most frequently used packages for industrial chemicals are steel drums and pails. For exotic products, stainless-steel drums and pails are available. The most common types used are 208-L (55-gal) drums and 19-L (5-gal) pails.

Once the appropriate package has been determined by consulting governmental and carrier regulations, the type of material compatible with the product needs to be determined. A wide variety of coatings is available for lining carbon steel drums and pails. Suppliers are often able on the basis of experience to assist in determining a lining which will be compatible with the product. When prior information is unavailable, laboratory tests can determine compatibility. Laboratory tests are often desirable before field trials. In some instances, a product may not be compatible with metal. This circumstance has led to an important new container, the **all-plastic 208-L (55-gal) drum.** Made from blow-molded high-density polyethylene, this container is especially useful for products which might react with carbon steel or whose value does not warrant stainless steel. Special treatments are available to make the inner surface of the plastic drum impervious to penetration of many products. Sulfonation and fluorination are prominent among these processes.

Two basic designs of steel drums and pails exist: the tight-head and the open-head. Tight-head drums have both top and bottom members permanently fastened to the drum body. Open-head drums have only the bottom permanently attached. As the term "open-head" implies, the top of the drum does not have a permanently fixed cover; rather, a removable head is used. This head is designed so that a locking ring secures it to the drum body. Open-head drums and pails are usually employed for viscous products or for mixtures and slurries which are difficult to pump through lines 50 mm (2 in) or smaller. Tight-head drums and pails are used for low-viscosity products. No set rule can be given for the viscosities above which an open-head drum or pail must be used.

Closures for drums and pails need to be determined together with the gasket material to be used. Consideration must be given to compatibility with the product and to the vibration which the container will encounter during transportation. The torque required to produce closure integrity is thus a significant factor. The typical closure sizes used in United States practice for tight-head drums are a 2-in national pipe thread (NPT) and a ¾-in NPT in the top head. For open-head drums, market considerations determine whether or not these fittings are used.

Filling Line Among filling-line considerations are filling and weighing equipment, mechanical handling of empty and filled drums, loading of filled drums onto transportation vehicles, work-station design for the safe and efficient use of personnel, and conformance to Occupational Safety and Health Administration (OSHA) and other codes. A typical drum-filling line, capable of handling two drums per minute, is shown is Fig. 7-38.

Filling and Weighing of Drums This procedure is divided into two parts: delivery of the liquid to the drum and weighing out the desired amount. Pumping the liquid product through a series of delivery pipes to the drum-filling point should follow good practice whereby reasonable velocities and pressure losses are maintained. The terminal point of the filling line is a control valve which is activated by a signal from a weight-sensing unit or scale. Valves may be pneumatically, hydraulically, or electrically operated, their operation being actuated either by an electric or pneumatic system or manually. The filling nozzle may be either top-fill or bottom-fill. Top filling is usually employed for most products, especially viscous materials or slurries. Bottom filling is used for low-viscosity products, for those having flash points under 37.8°C (100°F), or for places where static electricity is a concern. Also, products which tend to foam are bottom-filled. With a bottom-fill installation sufficient headroom is needed to permit the filling nozzle to be withdrawn from the drum. With both types of filling nozzles, a provision must be made for collecting product which dribbles from the end of the nozzle after filling is complete.

Weighing The weighing apparatus can be as simple as a platform scale in which the operator shuts off the filling nozzle when the desired weight has been reached. Automatic weighing can employ as advanced a design as a strain-gauge load-cell system activating the flow-cutoff mechanism through a microprocessor or a minicomputer. The same principles of filling and weighing as were described in the subsection "Weighing of Bulk Solids" hold for liquids.

Work-Station Design Critical consideration must be given to the design of work stations so that filling operators work in a safe environment and are used productively. Methods design, time studies, and predetermined work-element data are helpful in determining the amount of work involved and the proper sequence of operations to permit good productivity. Of special importance (and often overlooked) is having the operator perform service functions on a drum while another drum is being filled. This is especially significant with an automatic system which does not require operator attention. The following activities can often be undertaken while a drum is being filled: removing closure plugs from the empty drum which will be filled next; replacing and tightening the closures in the drum which preceded the drum being filled; labeling and marking code numbers on the drum; and starting the filled, sealed, and labeled drum down the handling system leading to storage or transportation.

Safety Regulations Consideration must be given to safety regulations for the electrical grounding of the drum during filling, the handling of product vapors, the handling of possible inadvertent spills and splashes of product, and the design of the work station to conform to OSHA and state and local codes. Operators must be protected from contact with the product, and their physical movements must not be such as could cause potential injury. Work-station design benefits from consultation with governmental bodies and with equipment vendors and consultants.

Small Liquid Packages The packaging of small packages of liquids is a specialized field. High-speed bottle and can fillers typically are of volumetric rather than weigh design. Up to a size of 3.8 L (1 gal) volumetric fillers are used almost universally when the filling rate exceeds 10 containers per minute. Below this rate, filling is con-

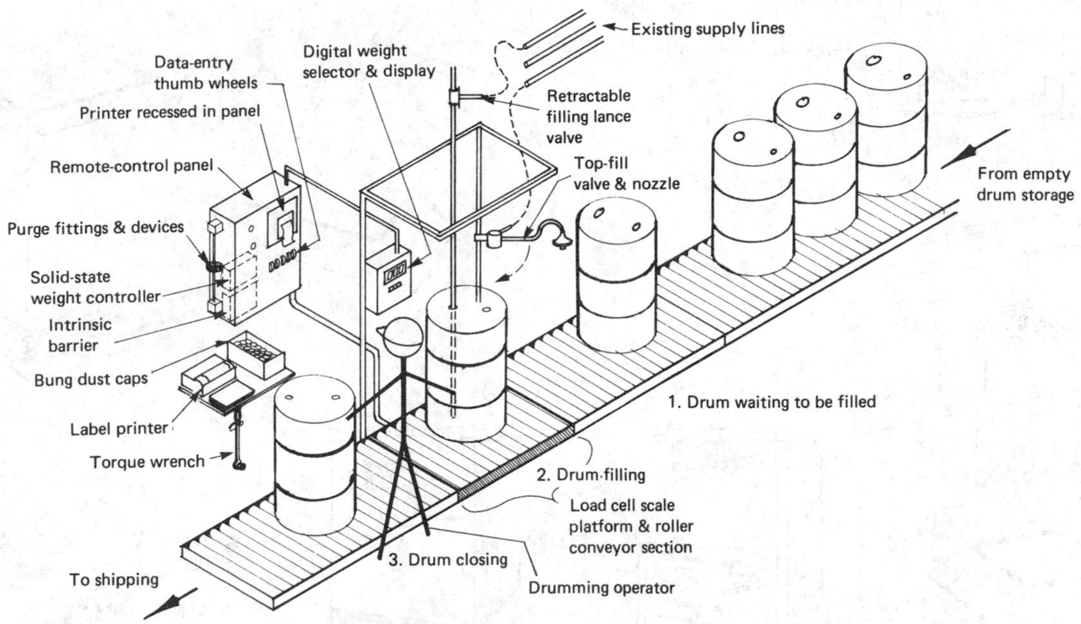

FIG. 7-38 Typical high-precision liquid filling and weighing system for packaging 208-L (55-gal) steel drums and similar smaller containers. *(Courtesy of H. F. Henderson Industries, Inc.)*

trolled by weight or even volumetrically by an operator activating manual controls.

SOLIDS PACKAGING

Containers for solids include bags, bulk boxes, cartons, and drums. While the intermediate flexible bulk container (IBC) has become an important package of world commerce, the most used package remains the mutliwall paper bag, supplemented by bags of similar design made of plastic film or plastic woven mesh.

Multiwall Paper Bags These bags (Fig. 7-39), made from plies of kraft paper or from combinations of kraft and special-purpose papers and plastics, are the most common packages for almost any pelleted or powdered material as well as for briquettes or bats of such solids as synthetic rubber, waxes, and insulation.

Empty bags are ordinarily shipped compressed (to obtain high load density) and on pallets, the most common of which measure 1220 by 1065 mm (48 by 42 in), 1220 by 1015 mm (48 by 40 in), and 1270 by 1115 mm (50 by 44 in). The number of empty bags per pallet varies with size, 1500 to 2000 being common. A typical filled pallet weighs about 907 kg (2000 lb). Pallet loads are often triple-tiered in warehouses.

Two bag designs are common: the **valve** and the **open-mouth** types. The **valve** bag has both ends closed during fabrication, filling being accomplished through a small opening (valve) in one corner of the bag. The open-mouth bag has one end closed at the factory and the other after filling.

Most **open-mouth bags** are closed by sewing, whereas adhesive is applied to the pinch type. The pinch bag has been the subject of intensive development by the bag industry, and as a result it has substantially displaced the sewn open-mouth bag. Reasons for this change include the ease, reliability, and repeatability of closing (sealing) equipment as well as the close control of the sealing adhesive applied by the bagmaker. The preapplied adhesive is activated by the closing machine in the user's plant. The positive closure of the pinch bag produces a container which completely seals in the product. Valve bags can also be sealed if they are provided with sealable sleeves. An advantage of valve bags over the open-mouth type is the availability of highly productive filling machines that not only require less labor than open-mouth filling equipment but also are capable of higher packing rates. In addition, the sealed valve bag permits the greatest pallet-load density. Bags may be fabricated from a variety of readily available flexible materials. In addition to kraft paper, barrier materials are available to prevent moisture or gases from entering or leaving the bag. These range in permeability from polyethylenes to aluminum foil. Table 7-15 presents some of the more prominent moisture barriers and their typical properties.

In the interest of standardization, the Packaging Institute, U.S.A. Chemical Packaging Committee (formerly the Chemical Manufacturers Association Packaging Committee) recommends four sizes of expendable four-way-entry pallets for bagged chemicals given in Table 7-19. The most common pattern consists of five bags on a 1220- by 1065-mm (48- by 42-in) pallet. This pallet size permits maximum trailer loading by placing the 1065-mm dimension across the van; maximum rail loadings are also possible if the 1220-mm side is placed across the car.

Sewn valve bags and sewn open-mouth bags are less important, except possibly for products with densities over 960 kg/m^3 (60 lb/ft^3) or for individual or small-lot shipments. These bag designs have the advantage of providing an easy grasp of the bag at the end of the sewing line without allowing fine powders to sift through the closure.

Valve bags usually rely on a labyrinth of paper or plastic film to seal off the valve. The automatic internal valve, while adequately protecting the contents of the bag, does allow a small amount of sifting of fine powders.

The starting point in bag-size determination is the weight or volume of product to be packaged and its bulk density (aerated and settled). Also to be considered are particle size, shape, and weight; degree of aeration at time of packaging; flow characteristics; temperature and relative humidity; type of handling system up to and including the filling machine; bag-closing method; bag style; and pallet size and pattern. Three sets of dimensions are needed: (1) **tube**—outside length and width of tube before bag closures are fabricated; (2) **finished face**—length, width, and thickness of bag after fabrication; (3) **filled face**—length, width, and thickness of bag after filling. Table 7-20 and Fig. 7-39 show these dimensions and their interrelations.

A first approximation of size can be determined from Fig. 7-40, which applies to sewn valve, sewn open-mouth, pinch-type open-mouth, and pasted valve bags. The resulting tube width and length

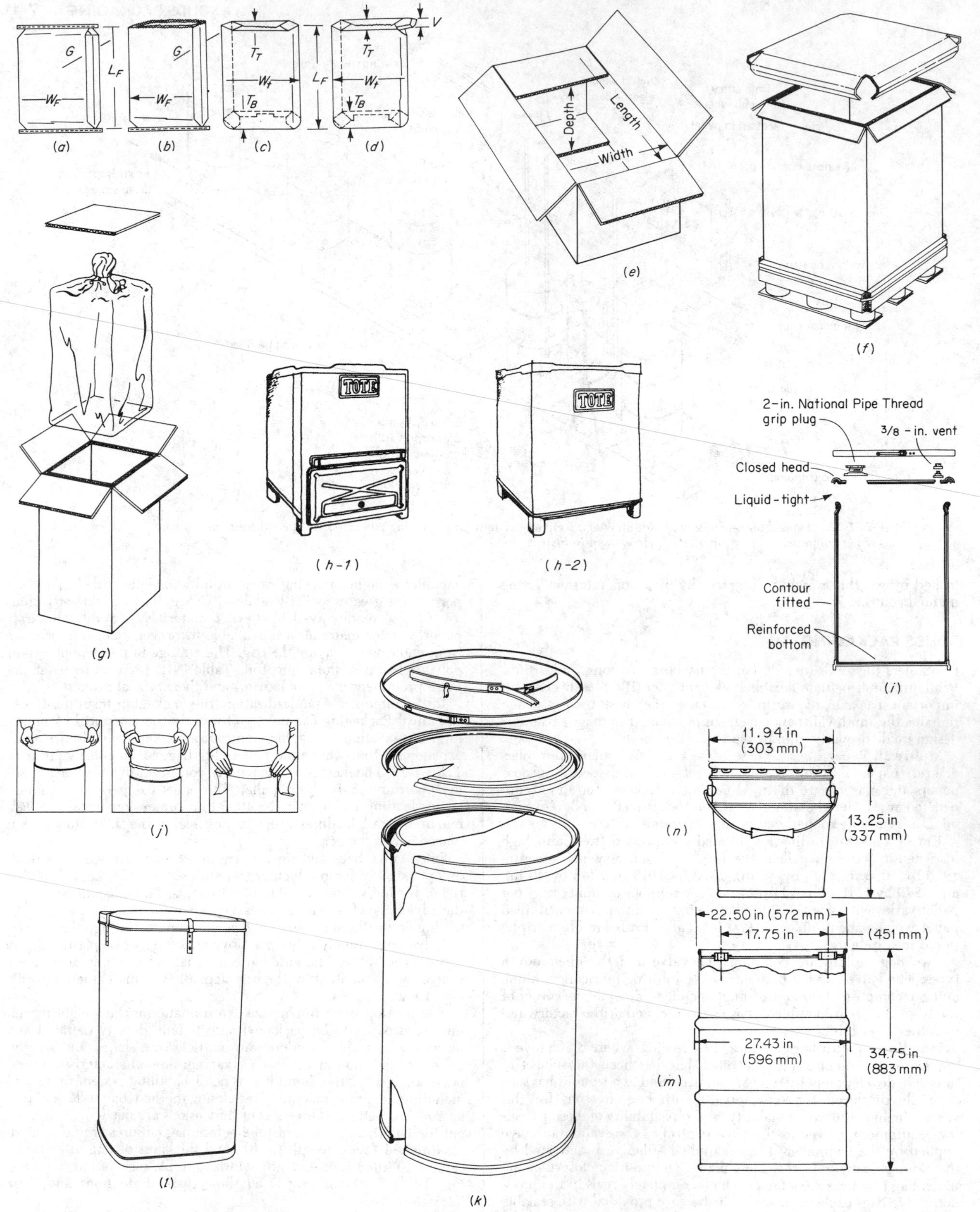

FIG. 7-39 Typical packages used for chemical products. (*a*) Sewn valve bag. (*b*) Sewn open-mouth bag; pinch-bottom-type open-mouth bag. (*c*) Pasted valve bag. (*d*) Pasted valve tuck-in-sleeve bag. (*e*) Principal (inside) dimensions of a regular slotted carton (RSC). (*f*) Bulk box of corrugated fiberboard for product weighing 450 kg (990 lb). (*g*) Bag in box. (*h*) Tote-bin rigid intermediate bulk container: 1—for solids, 2—for liquids. *(Courtesy of Hoover Universal, Inc., Materials Handling Division.)* (*i*) Liquid-type polyethylene drum with fiber overpack. (*j*) All-fiber drum with removal top. (*k*) Lever-locking fiber drum. (*l*) Ro-Con rectangular fiber drum; clip-mounted top. *(Courtesy of Greif Bros. Corp.)* (*m*) 208-L (55-gal) DOT-type 17E closed-head steel drum. (*n*) 19-L (5-gal) DOT-type 37A80 open-head universal steel pail with lug-type cover. [*(m) and (n) Courtesy of Eastern Steel Barrel Corp., Piscataway, N.J.*]

TABLE 7-19 Preferred Bag-Pallet Sizes*

Pallet size		Filled-bag face width		Filled-bag face length		Bags per tier (pattern)
in	mm	in	mm	in	mm	
48 × 40	1220 × 1015	16	405	24	610	5
48 × 42	1220 × 1065	16	405	24	610	5
52 × 44	1320 × 1115	17.25	440	26	660	5
52 × 36	1320 × 915	16	405	36	915	3

*From Packaging Institute, U.S.A. Chemical Packaging Committee (formerly Chemical Manufacturers Association Chemical Packaging Committee).

TABLE 7-20 Multiwall-Paper-Bag Dimensions

Bag type	Tube dimensions	Finished-face dimensions	Filled-face dimensions*	Valve dimensions
Sewn open-mouth	Width = $W_t = W_f + G_f$ Length = $L_t = L_f$	Width = $W_f = W_t - G_f$ Length = $L_f = L_t$ Gusset = G_f	Width = $W_F = W_f + \frac{1}{2}$ in. Length = $L_F = L_f - 0.67\ G_f$ Thickness = $G_F = G_f + \frac{1}{2}$ in.	
Sewn valve	Width = $W_t = W_f + G_f$ Length = $L_t = L_f$	Width = $W_f = W_t - G_f$ Length = $L_f = L_t$ Gusset = G_f	Width = $W_F = W_f + 1$ in. Length = $L_F = L_f - 0.67\ G_f$ Thickness = $G_F = G_f + 1$ in.	Width = $V = G_f \pm \frac{1}{2}$ in. †
Pasted valve	Width = $W_t = W_f$ Length = L_t	Width = $W_f = W_t$ Length = $L_f = L_t - (T_T + T_B)/2 - 1$ Thickness at top = T_T‡ Thickness at bottom = T_B‡	Width = $W_F = W_f - T_T + 1$ in. Length = $L_F = L_f - T_T + 1$ in. Thickness = $T_F = T_T + \frac{1}{2}$ in.	Width = $V = T_T \begin{cases} +0 \text{ in.} \\ -1 \text{ in.} \end{cases}$ §

Meaning of subscripts: B = bottom; f = finished-face; F = filled-face; t = tube; T = top.

*Formulas are based on conditions of bags after mechanical flattening.

† Valve dimension is flat width, which must not exceed $\pm\frac{1}{2}$ in. + G to maintain good closure. Circumference of valve = twice the width.

‡ T_T and T_B are usually equal; if they differ, use average. T = thickness.

§ Valve dimension is flat width. Valve width can be made less than top width without affecting closure properties.

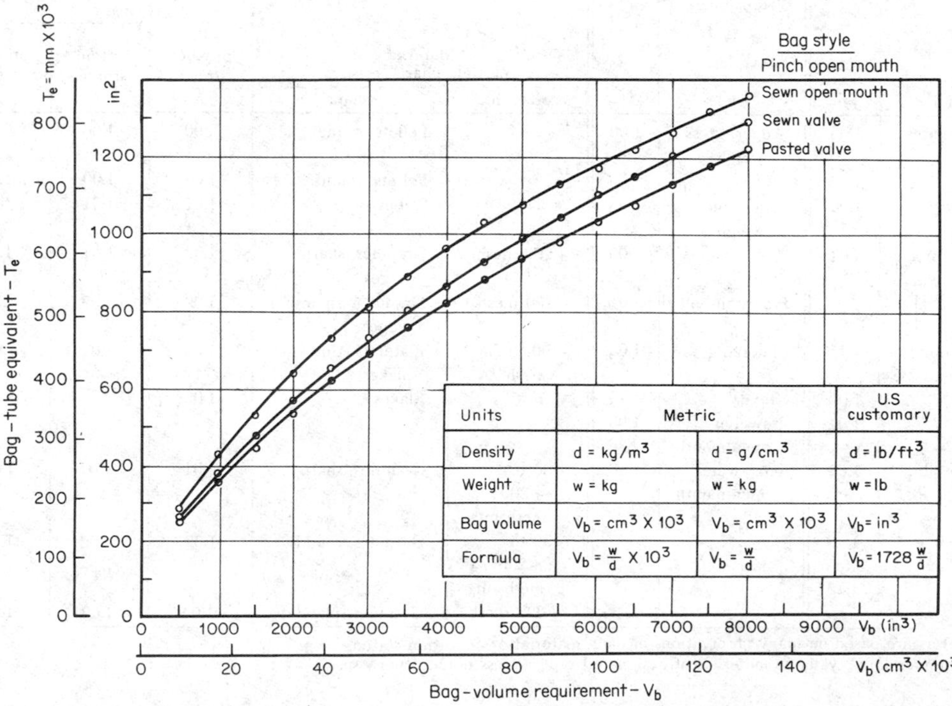

FIG. 7-40 Multiwall-bag-sizing graph. *(Raymus Associates, Incorporated.)*

can then be converted into finished and filled dimensions, and bag samples ordered for field verification. Changing bag size to accommodate different weights, density variations, pallet patterns, etc., becomes a simple matter through the use of the graph. Correction factors for particular situations such as special plies, type of filling machine, storage system, and product characteristics are given in Table 7-21.

To use the graph, given the weight of material to be packed, follow these steps: (1) obtain the settled and loose (or aerated) bulk densities of the product (a 1-ft^3 box serves this purpose well), and then calculate the average of the two densities; (2) calculate the bag-volume requirement from the relation V_b = [W lb (weight to be packed)·1728] d lb/ft^3 (average density); (3) multiply V_b by the product of the correction factors (Table 7-21), which reflect product, storage, and packaging conditions; (4) from Fig. 7-40 obtain the bag-tube equivalent T_e; and (5) using the corrected V_b, determine the bag size needed for palletizing.

Example 1 55.1 lb (25 kg) of plastic pellets having a bulk density of 38.5 lb/ft^3 (615 kg/m^3) are to be packaged in pasted valve bags constructed of three kraft plies and a free polyethylene (PE) 2-mil (0.05-mm) liner. Bags will be palletized in a 5-bag pattern, 40 bags per pallet, on 48- by 40-in (1220- by 1016-mm) pallets. Filled bags are permitted to overhang the pallet by 0.5 in (15 mm). Determine the proper bag size.

$$\text{Density} = 38.5 \text{ lb/ft}^3$$
$$V_b = (50/35) \times 1728 = 2470$$

Correction factor (from Table 7-21):

For barrier sheet of 2-mil polyethylene film	1.05
For filling machine, fluidizing type	1.02
For ⅛-in- (3.2-mm-) particle-size pellets	1.00
For storage and handling 24 h	1.00
Overall correction factor (product of above)	1.07

$$\text{Corrected } V_b = 2470 \times 1.07 = 2650$$
$$T_e \text{ (from Fig. 7-40)} = 640$$

For first approximation let $T_T = T_B = 6$ in, and $L_f = 24 - 1 = 23$ in.

$$\text{Since } L_f = L_t - (T_T + T_B)/2 - 1$$
$$L_t = 23 + 6 + 1 = 30 \text{ in}$$

and $T_e = W_t L_t = 640$

$$W_t = 640/L_t = 640/30 = 21.3 \text{ in}$$

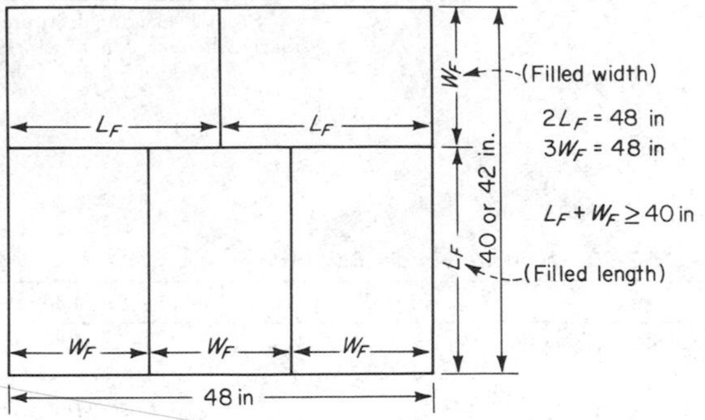

Since $W_F = W_f - T_T + 1$, and $W_t = W_f$, then

$$W_F = 21.3 - 6 + 1 = 16.3 \text{ in}$$

Checking for pallet-length conformity,

$$2L_F = 48 \text{ in, or } 2(24) = 48 \text{ in}$$
$$3W_F = 48 \text{ in, or } 3(16.3) = 48.9 \text{ in}$$

Checking for pallet-width conformity,

$$W_F + L_F \geq 40 \text{ in}$$

or

$$16.3 + 24 = 40.3 \text{ in}$$

TABLE 7-21 Bag-Volume-Equivalent (BVE) Correction Factors *f* for Specific Conditions*

NOTE: Refer to Table 21-6 for metric particle sizes.

Barrier and type of bag construction material		Filling-machine characteristic		Conveying, handling, and storage conditions					
				Material particle characteristic		Mechanical conveying		Pneumatic conveying	
						Off-product stream	From storage	Off-product stream	From storage
Type of material†	*f*	Type	*f*	Size, in.	Shape	*f*	*f*	*f*	*f*
Asphalt-kraft laminates	1.03	Auger, gross weigh	1.00	1/16	Pellets, round	1.00	1.00	1.00	1.00
				1/8, 3/16	Pellets, round	1.00	1.00	1.00	1.00
Polyethylene extrusion-coated on kraft	1.05	Auger, net weigh	1.03	1/8, 1/4	Cubes	1.01	1.01	1.01	1.01
Polyethylene, extrusion-coated on kraft	1.01	Belt	1.05	+200 mesh	Granules, sharp edges	1.02	1.01	1.03	1.02
with random partial perforations*		Preweigh-belt	1.01	+200 mesh	Granules, smooth edges	1.02	1.02	1.03	1.03
Polyethylene–aluminum foil–kraft laminate	1.15	Fluidizing	1.02	−50, +200 mesh	Platelets (tiny flakes)	1.03	1.02	1.05	1.03
Wax-coated kraft	1.09	Gravity	1.15	+1/4, −3/8	Flakes	1.02		1.03	
Glassine	1.04	Preweigh open-mouth	1.00						
Free polyethylene film 1 to 4 mils thickness	1.03	Gross-weigh open-mouth	1.10	−30 to +200 mesh mix	Granules, sharp	1.01	1 01	1.03	1.02
Polyethylene-film bag (extruded tubular film with heat-sealed ends)	1.10 to 1.15			−30 to +200 mesh mix	Granules, round	1.02	1.02	1.05	1.03
				−325 mesh	Granules, platelets	1.04	1.03	1.07	1.03

*Approximate factors are based on many observations of each material or condition stated.
† Applies to all commercially available materials of designated type, unless noted otherwise.

Summary:
Bag size: 21½ in (face width) × 23 in (face length) × 6 in top and bottom (545 × 585 × 150 mm).
Pallet size: Use 48 in × 40 in (1220 × 1015 mm).
Overall filled pallet dimensions: 48.9 × 40.3 in (1242 × 1024 mm).
This example can also be carried out in the SI system by using Fig. 7-40.

Liners At the time of setup of filling, many containers for bulk solids are lined with a polyethylene (PE)-film bag, the purpose being to prevent sifting of fine particles, retard moisture pickup or release, or prevent product contamination by the construction material of the container. Liner length should be sufficient to permit the top to be closed by heat sealing or wire tying. The film gauge (thickness) needed depends on the weight, bulk density, and particle roughness of the contents. Gauges of 0.05 to 0.25 mm (2 to 10 mils) are common. For ease of placing the liner in the carton, the gusseted type is preferred; and for convenience in handling, liners are usually made as continuous tubes, with heat seals and tear-off perforations at intervals equal to one bag length (see Table 7-15 for costs).

Form-Fill-Seal, Small Bags and Pouches, and Baler Bags Product weights from a few grams to 11 kg (25 lb) are often placed in packages made of plastic film, paper, or combinations of these. Groups of packages are then shipped in cartons or baler bags (see Table 7-15 for costs).

Form-fill-seal is a machine process that forms a tube from plastic-coated paper stock, heat-seals one end, fills the resultant bag, heat-seals the other end, and then cuts off the filled bag. This method has the advantage over filling small bags or pouches in that the cost of package fabrication is avoided until the package is actually needed; packaging labor is reduced to one attendant who can service a number of machines. Also, order lead time is shortened because standard, merchant plastic film or paper can be bought from local stock, often avoiding waits of 4 to 8 weeks for fabricated bags. Offsetting this is the higher investment in equipment and the service and maintenance problems associated with automatic equipment.

Small bags and pouches are made from one or more plies of paper or plastic film. The two main types of paper bags are the satchel-bottom and the pinch-bottom. Both types usually have a gusset that helps form a rectangular cross section (a useful trait when packing in cartons or baler bags). Although order lead time is longer than for form-fill-seal and operating labor is greater, capital and maintenance costs are smaller (Table 7-15), and equipment reliability is greater. These small packages require a master shipping container. Corrugated cartons are used extensively, as is the flexible baler bag.

Baler bags are pasted open-mouth bags with one or more plies and of either satchel-bottom or self-opening (gusseted) design. Pouches are loaded into the baler with their long axis parallel to that of the baler. Since the pouches must be tightly packed, mechanical-compression loading equipment is mandatory.

Rigid Intermediate Bulk Containers Rigid IBCs are made of metal or plastic suitable for the product and service intended. Sizes available range from 0.17 m^3 (6 ft^3) to 2.83 m^3 (100 ft^3). This type of container is intended for reuse and a useful life of up to 20 years. Important economic considerations are the cost of returning the empty container to the filling location and the cleaning, handling, and storage of it. Figure 7-39 illustrates a metal container. Table 7-17 gives economic information comparing this type of container with other containers of larger or smaller volume.

Flexible Intermediate Bulk Containers These containers are an important development of the 1970s. Made from woven polyolefins or other materials, flexible IBCs are available in a wide variety of volumes and can handle up to 1800 kg (4000 lb), depending on construction. This type of container can be equipped with a thermoplastic liner when it is necessary to protect the product against moisture or other contamination. Handling is accomplished by forklift truck or by hoist. Filling and weighing of flexible IBCs can be accomplished on specially designed weigh scales or volumetrically if the container is weighed at a remote location after filling and that weight is used as a basis for invoicing. Filling is carried out through a flexible port at the top of the container, while unloading is accomplished through a similar flexible member at the bottom. Table 7-22 gives dimensional and volumetric data. Figure 7-41 shows typical container designs and types of loading and discharge spouts.

TABLE 7-22 Flexible-Type Intermediate Bulk Containers: Dimension and Capacity Data (Variable Data)*

Height		Usable volume		Maximum bulk density	
mm	in	m^3	ft^3	kg/m^3	lb/ft^3
800	31.5	0.7	24.6	1445	90
915	36.0	0.8	28.1	1250	78
1040	41.0	0.9	31.7	1120	70
1145	45.0	1.0	35.3	995	62
1270	50.0	1.1	38.3	930	58
1385	54.5	1.2	42.4	835	52
1500	59.0	1.3	45.9	770	48
1600	63.0	1.4	49.4	720	45

*From Bonar Co., Ltd.

NOTE: Maximum weight, 1 metric ton (2205 lb); cross-section dimensions, 890 by 890 mm (35 by 35 in); tare weight, 3 kg (7 lb); material of construction, woven polypropylene body and polyester lifting straps.

Boxes **Bulk boxes** (Fig. 7-39) of corrugated kraft paper for dry bulk products fall into two broad categories: large, for 0.5- to 2-ton loads, and small, for loads of 23 to 68 kg (50 to 150 lb). Large boxes are used extensively for resin shipment; small ones, for certain regulated materials (such as caustic soda) and for low-bulk-density products that are assessed excessive freight rates if packed in drums.

A bulk box, sometimes called *bag in box,* consists of a box within a box plus other elements such as end pads, PE bag liners, and closing materials (tape, glue, staples). The double-wall corrugated kraft board consists of an outside liner, a corrugating medium, a center liner, another corrugating medium, and an inside liner; the single-wall board consists of an inner and outer liner with a corrugating medium center. The specifications for each depend on service requirements; 4100 kPa (600 lbf/in^2) burst strength is common for 454-kg (1000-lb) loads; 1900 kPa (275 lbf/in^2), for 68-kg (150-lb) loads. Materials of construction that resist high humidity and wetting with water are available.

Advantages of this container are its reclosing feature and its effi-

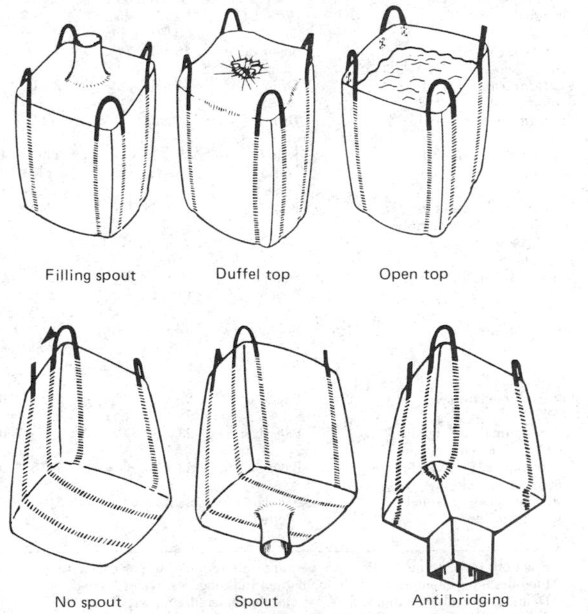

FIG. 7-41 Typical flexible-bulk-container designs and loading- and unloading-spout designs. *(Courtesy of Bonar Co., Ltd.)*

cient use of storage and shipping space. Disadvantages are the space required to store box components before assembly and the limited reuse market. The lead time for ordering made-to-order boxes ranges from 3 to 6 weeks. Filling equipment is similar to that for drums. Setting up the box can require two persons because of the unwieldiness of the components. Table 7-23 gives an idea of filling speeds for several types of filling arrangements and box styles.

Wire-bound wood boxes (typical loads, 1 to 2 tons) have limited use for chemical products. The box body, consisting of thin wooden slats held in place by steel wire twisted around each slat, is fastened to a solid-deck wood pallet. The top also consists of wire-bound wooden pieces. A PE liner protects the product and prevents it from falling through the slats. Disadvantages of the container are the labor needed for setup and the space required for knocked-down boxes. Since manufacturers are usually near sources of hardwood, shipping costs to users may be high; lead times of 3 to 4 weeks are common.

Folding boxes made from chipboard are used for consumer-size units [from a few grams or ounces to about 11 kg (25 lb)] of such products as insecticides, snow-melting compounds, salt, and food additives. PE liners are often included to protect the product from moisture or to prevent it from sifting through minute openings in the top and bottom folds. Order lead time is 6 to 8 weeks. Knocked-down folding boxes are dense and store efficiently when palletized. A typical pallet measures 760 by 915 mm (30 by 36 in) with the load 1220

TABLE 7-23 Performance Data for Packaging Systems

		Package detail					Product detail			
								Bulk density		
Type of filling and weighing machine	No. of filling spouts	Type	Size, in	Size, mm	Construction plies—basis weight	Closure	Material	lb/ft³	kg/m³	Particle size, U.S. standard§
Fluidizing, SFW‡	4	Pasted valve bag	20 x 25 5 top width	510 x 635 125	4-170 (= 4-ply, 170 lb)	Inner sleeve	PVC‡	38	609	−60 mesh
Centrifugal belt with PWS‡	2	Pasted valve bag	21 x 25—5¼ top	535 x 635 135	4-170 PE‡ barrier	Inner sleeve	PE‡	30	481	⅛-in pellets
PWS, open-mouth filler SMC‡	1	SOM‡ bag	16 x 5 x 30 5¼ top	405 x 125 x 760	4-170 PE barrier	Sewn, tape-bound	PE	30	481	⅛-in pellets
Fluidizing, SFW‡	2	Pasted valve bag	20 x 25—5¼ top	510 x 635 135	4-170	Tuck-in sleeve	PVC	36	577	−60 mesh
Fluidizing, SFW	3	Pasted valve bag	21 x 25	535 x 635	4-170	Tuck-in sleeve	PE	30	481	⅛-in pellets
Impeller, SFW	4	Pasted valve bag	18½ x 27½ (face)	470 x 700	3-170	Insert sleeve	Portland cement	94	1506	−325 mesh
Auger, SFW, net-weigh	1	Sewn valve bag	16 x 5 x 28	405 x 125 x 710	4-190	Tuck-in sleeve	PE	32	513	5/32-in cubes
Centrifugal belt, SFW	2	Sewn valve bag	15 x 5 x 36	380 x 125 x 915	3-150 PE barrier	Insert sleeve	Fertilizer, mixed	55	881	−10 to +100 mesh
PWS, open-mouth filler, SMC‡	1	SOM‡ bag	17 x 4 x 36	431 x 100 x 915	3-150 PE barrier	Sewn, no tape	Fertilizer, mixed	55	881	−10 to +100 mesh
Fluidizing, SFW	4	Pasted valve bag	18½ x 26 5¼ top	470 x 660 135	3-150 PE barrier	PE inner sleeve	Fertilizer, mixed	55	881	−10 to + 100 mesh
PWS, open-mouth, heat sealer	1	PE flat-tube bag	16 x 30½	405 x 775	10-mil PE	Heat-sealed	Fertilizer, mixed	55	881	−10 to +100 mesh
PWS, form-fill-seal	1	Forms, gusseted bag	16 x 5 x 30	405 x 125 x 760	6 mil PE	Heat-sealed	LDPE	30	480	⅛ in (3 mm)
SFW, liquid fill and weigh	1	Steel drum U.S. DOT 17 E	55 gal (208 L)—23.5 in dia. × 34.75 in high	596 x 883	18-ga (0.0428-in) ends, 20-ga (0.0324-in) body	2-in, ¾-in NPT bungs	Lacquer solvent		0.839	sp. gr.
Gravity, SFW	1	Sewn valve bag	16 x 5 x 28	405 x 125 x 710	4-190	Tuck-in sleeve	Polystyrene	32	513	5/32-in cubes
Platform scale, autofill cutoff, SFW	1	Drum	55 gal	208L	6-ply fiber (300 lb)	Lever-locked steel cover	PE master batch	30	481	⅛-in pellets
Platform scale, manual cutoff	1	Drum	55 gal	208L	6-ply fiber (300 lb)	Lever-locked steel cover	Cleaning compound	45	721	−20 to +80 mesh
Platform scale, autofill cutoff, SFW	1	Bulk box, 3-mil PE liner	15 x 15 x 24	380 x 380 x 610	Outer: 275-lb test, DW.‡ liner: 600-lb test, DW	Staples	Insecticide, technical grade	40	640	−200 mesh
Platform scale, autofill cutoff, SFW, automatic staple closer	1	Bulk box	41 x 34 x 36	1040 x 860 x 915	Inner, outer boxes: 600-lb test, DW kraft board	Staples	PE	30	481	⅛-in pellets
Vertical auger, SFW	1	Small bag Pouch	10 x 4 x 25 14 x 27	255 x 100 x 635 355 x 685	3-120 paper, 2- to 4-mil PE	Glued, heat-sealed	Insecticide powder	20	320	−325 mesh
Vertical auger, SFW	1	Folding box	6½ x 3½ x 9	165 x 90 x 230	12-point reprocessed board with 2-mil PE liner	Glued, tied PE liner	Sprayable insecticide powder	20	320	−10 μm
Form-fill pouch maker, PWS	2	Pouch	8½ x 15	215 x 380	1- to 3-mil PE film	Heat-seal, hot-wire cutoff	Detergent, spray-dried	39	625	−30 to +60 mesh
Baler, manual package in feed, mechanized closing	. . .	Baler bag	23 x 30	585 x 760	2-140	Glued	12, 5-lb (2.3 kg) bags herbicide	45	721	−325 mesh
Baler, manual package in feed, manual closing	. . .	Baler bag	23 x 30	585 x 760	2-140	Glued	12, 5-lb (2.3-kg) bags herbicide	45	721	−325 mesh
Corrugated case, manual package in feed, mechanized closing	. . .	Regular slotted carton	24 x 16 x 7	610 x 405 x 180	275 DW	Glued	12, 5-lb (2.3 kg) bags herbicide	45	721	−325 mesh

°Fractions indicate the portion of a person's time required to perform activity; these are additive to compute the number of people needed.

†Includes equipment and installation but not building or services needed.

‡Definition of abbreviations: SFW = simultaneous fill-and-weigh; PWS = preweigh scale; SMC = sewing-machine closer; DW = double wall; SOM = sewn open-mouth; PE = polyethylene; PVC = polyvinyl chloride.

§See Table 21-6 for metric equivalent of particle sizes given.

mm (48 in) high. Filling equipment, which can also be used for filling pouches, small bags, and glass jars, ranges from small, manually operated units to high-production, fully automatic units. Most common is the manually operated gross-weigher type.

Shipping cartons for liquids in cans and bottles, bulk solids in jars, pouches, and folding boxes, and briquetted items with or without individual packaging are usually made of corrugated kraft paper. Since the containers within the cartons can normally support vertically imposed loads, cartons are less sturdily built than bulk boxes. The most common styles are the regular slotted carton (RSC), the end-opening RSC, and the center, special-fuel overlap slotted container. End joints may be stapled, stitched, glued, or taped.

Specifications include dimensions of length, width, and depth, in that order (Fig. 7-39*e*). When boxes are set up and closed by automatic equipment, dimensional tolerances become critical. Cartons are shipped knocked down to the user from plants located in all industrial centers. Because order lead time is 4 to 6 weeks, inventories of empty boxes require considerable space.

Often, the size of items packaged in corrugated cartons either does not permit interlocking of layers of cartons, or leaves considerable void space between them. Since calculating by hand the best size of carton for maximum palletizing density requires considerable effort, computer programs are available for sale, lease, or use on a job basis. Examples are the CAPE-ARRANGE program (Danray Corp., Dal-

Weight of contents		Packaging rate, packages/min		Weight variation from average		Packaging personnel needed*				Approximate† investment (1981)		Package handling	
lb	kg	Avg	Instant	oz	gr	Package setup, supply	Filling-machine operators	Package closers	Palletizers, loaders, attendants	Filling machine	System	Package conveyorized	Automatic palletizing
50	22.7	12	17	4	114	1	1	0	2	62,000	246,000	Yes	No
50	22.7	16	22	0.5	14	0.5	1	0	0.5	86,000	310,000	Yes	Yes
50	22.7	8	12	0.5	14	1	1	1	2	25,000	112,000	Yes	No
50	22.7	6	8	4	114	1	1	0	1	44,000	185,000	Yes	No
50	22.7	16	24	3	85	1	1	0	1	54,000	310,000	Yes	Yes
94	42.7	22	28	8	227	1.5	1	0	0.5	62,000	369,000	Yes	Yes
50	22.7	1	2	3	85	0.25	0.25	0	0.5	15,000	29,000	No	No
80	36.4	12	16	8	227	1	1	0	2	37,000	185,000	Yes	No
80	36.4	16	22	4	114	1	1	2	2	25,000	135,000	Yes	No
80	36.4	18	24	16	455	1	1	0	2	67,000	197,000	Yes	No
50	22.7	18	24	4	114	1	1	1	2	37,000	160,000	Yes	No
50	25	8	12	1	28	1	0	0	0	600,000	1,840,000	Yes	Yes
385	175	2	3	6	170	0.25	0.5	0.5	1.75	65,000	100,000	Yes	No
50	22.7	0.2	0.4	16	455	0.25	0.5	0	0.25	2,000	7,000	No	No
250	113.6	1	4	2	57	1	0.5	0.5	1	15,000	62,000	Yes	No
300	136.4	0.5	1	0.5	14	0.25	0.25	0.25	0.25	4,000	12,000	Yes	No
100	45.5	0.5	1	4	114	1	1	1	1	25,000	49,000	No	No
900	409.1	0.33	0.50	8	227	2	0.75	0.25	. . .	25,000	123,000	Yes	
10	4.5	5	10	1	28	1	1	1	3	12,000	62,000	Yes	No
1.5	0.682	8	12	0.5	14	1	1	1	3	12,000	25,000	Yes	No
2.5	1.136	10	12	0.5	14	1	. . .	. . .	2	62,000	86,000	Yes	No
60	27.3	1.5	3	. . .	. . .	0.5	0.5	0.5	0.5	25,000	37,000	Yes	No
60	27.3	1.5	3	. . .	. . .	0.5	0.5	1	1	12,000	5,000	No	No
60	27.3	1	2	. . .	. . .	0.5	0.5	0	1	17,000	25,000	Yes	No

las, Texas), SPACE I (Physical Distribution Services, Marketing Publications, Inc., Washington, D.C.), and a program called CADES developed by the Rutgers University Center for Packaging Engineering, Piscataway, New Jersey.

Drums Drums (Fig. 7-39), made of either **steel** or **fiber,** rank next in importance after the multiwall paper bag. For dry solids or slurries, fiber drums predominate; for liquids, the steel drum. Steel drums of the open-head design are used for dry products when the product is hazardous, is to be stored outdoors, or is of a density that will cause reasonable weights to exceed the limits for fiber drums. Although only a few sizes are common, fiber drums can be made to order in almost any size and diameter-length combinations for volumes of 2 to 285 L (0.75 to 75 gal) and for weights ranging from 25 to 250 kg (60 to 550 lb; Table 7-15).

Advantages of the drum are protection of contents, ease of reclosure, and appreciable reuse-resale value. A serious limitation is the inefficient use of space because of the cylindrical shape, which results in high storage and transportation costs. To overcome this, a fiber drum with a square cross section (Ro-Con drum) and the bulk corrugated bag in box have been developed.

Fiber drums decorated with advertising cost from $0.50 to $2 each, depending on complexity and number of colors. The most common type is the multiple-ply kraft-paper body with a steel bottom and a reinforcing top hoop crimpled to the drum. A steel lid, secured by a locking ring tightened by a lever system, fits over the body. For **vapor protection,** barriers are incorporated among the plies, or liners are used as the first ply in contact with the product. Among barrier and liner materials are PE, aluminum and steel foil, polyesters, and silicones. When liquids are to be contained, blow-molded PE liners are used. Free-film PE liners inserted by the user yield a combination of barrier and liner properties at less cost than having the liners as part of the drum body.

Fiber drums are also made with a removable fiber top and a fiber bottom that is either removable or permanently fastened to the body. This drum has limited reuse, but it costs less than the lever-locked metal-top type. Filling equipment consists most commonly of an operator-controlled spout connected to a supply bin resting on a platform scale. Table 7-23 shows the labor productivity of several systems.

Steel drums are made from cold-rolled-steel sheet formed into a cylinder. The longitudinal seam is made by electric resistance welding. The rolling hoops are expanded into the body wall by a special hydraulic fixture, and the ends are crimped to the body to form a leakproof joint. Sealing compounds are often used to assure leakproof joints. The top head has openings to allow installation by resistance welding of the closures or bungs. These closures have U.S. standard-pipe-thread fittings, usually one 2-in and one ¾-in fitting, to allow connection to the loading and unloading equipment.

PACKAGING OPERATIONS

Dry-bulk-packaging operations are divided into two categories: weighing and filling a package that is itself the shipping container and weighing and filling small packages that are in turn placed in outer packages for shipment. The choice of equipment and the way in which it is combined into a system depend on such factors as the product and its chemical, physical, and rheological properties; the type of package to be filled; the total packaging output required; the instantaneous and average rates of filling; cost, attitude, and availability of labor; space available for equipment; storage, shipping, and transportation conditions; cost and availability of capital; seasonality of packaging activity; expected duration of the venture; sanitary, safety, packaging, and working conditions imposed by regulatory bodies; maintainability and reliability of equipment; changes expected in the product and in the demand for it; and nature of the product market (i.e., industrial, consumer, agricultural, or government).

Weighing There are two principal types of package-weighing and -filling equipment: **simultaneous fill-and weigh,** with which the material is weighed as it is poured into the container; and **preweigh,** with which the material is weighed prior to being poured into the package. The former applies mainly to valve bags, pouches, bulk boxes, and bags in boxes; the latter, to open-mouth bags, small bags, and cartons, to form-fill-seal, and, at times, to valve bags.

There is a further distinction between net weighers and gross weighers. **Net weighers** are defined by the ratio (0.3 to 0.5) of weight of charged material to weight of weighing vessel and associated parts. Preweigh scales are examples of net weighers. With **gross weighers,** of which simultaneous fill-and-weigh is an example, the ratio is usually greater than unity. Net weighers are accurate within ± 0.125 to ± 0.25 percent; gross weighers, from ± 0.5 to 1.0 percent. Maintaining certain scale-feed conditions is critical in obtaining accuracy and sustaining a given production rate; appropriate feeding devices and surge bins are of great importance. If desired, weight accuracy can be increased, at greater cost, by special modifications and accessories such as load cells and microprocessor controls, bulk and dribble devices, and feeders.

The **weight accuracy** of a dynamic weighing device is expressed as a plus or minus percentage deviation from a given *set weight*, which can only approximate the desired *actual weight*. The dynamic nature of weighing requires that the scale respond to changing static conditions as well as to a series of constant dynamic conditions. Minor variations in product density can cause the set weight to drift, the result being unacceptable packaged weight. Scale sensitivity is often suspected to be at fault, when in fact it is the set weight that has drifted. This is easily verified by check-weighing a series of weighings and determining their standard deviation.

Check Weighing Because of drifting set weight and the influence of federal and state legislation on allowable deviation from advertised weights, a major new phase of package filling and weighing is that of check weighing. This can be done manually with a platform scale and then following a simple statistical procedure and control chart. There are devices applicable to preweigh scales which perform and record a static weighing just prior to discharging to the filling machine. There are in-line check weighers that weigh each package and pass or reject it depending on its weight, and keep a log of the results. One development permits continuous automatic readjustment of the scale set weight by means of a process minicomputer that records each weighing and then, from a series of these, computes whether or not the set weight is drifting. An automatic adjustment is then made of the scale poise weight.

Filling and Weighing Equipment Of special interest in the selection of filling equipment from the wide variety available (Table 7-23) and combining it into the total system is the equipment's relation to instantaneous output, average output, and personnel. Methodizing, subdividing into work elements, and prediction of the time required for each job function by means of standardized data such as methods-time measurement (MTM) and general-purpose data (GPD) permit accurate identification of jobs and work content. Actual average output can thus be calculated and, from this, the instantaneous output.

Instantaneous rate, which is the rate that equipment manufacturers imply in their guarantees of performance, is defined as the number of packages produced per minute with the equipment operating under steady-state conditions. **Average rate,** the measure that the user needs to plan production and output commitments, can be defined as the arithmetic average (packages per minute) produced over a production shift (usually 8 h). Equipment reliability must be taken into consideration in determining average rates because malfunction downtime can have a significant effect on rate values. Also, the effect of production scheduling on equipment idle time and changeover from one product to another needs to be considered.

Valve-Bag-Filling Equipment Although multiwall paper bags and plastic bags can be filled by a wide variety of equipment, the simultaneous fill-and-weigh (gross-weigher) type predominates. Net-weigher-type equipment using a preweigh scale which discharges into a valve-bag filler is finding increased favor when greater weight accuracy is required. The most widely used category is the gross weigher of the pressure-fluidizing type, for which these parameters and ranges hold:

Parameter	Capability range*
Particle size	⅜-in (9.5-mm) pellets to submicrometer
Bulk density	0.5 to 200 lb/ft^3 (8 to 3200 kg/m^3)
Filling spout	1 to 4
Bagged weight	20 to 150 lb (10 to 70 kg)
Bagged volume	1 to 6 ft^3 (0.03 to 0.17 m^3)
Material of construction in contact with product	Carbon steel, stainless steel, plastic-coated steel, aluminum
Output capability	1 to 30 bags per minute
Bag-valve size	3 to 5½ in (75 to 140 mm)
Weight error: simultaneous fill-and-weigh scale	+2 to ±4 oz (60 to 120 g)
Weight error: preweigh scale	±1 oz (30 g)

*SI equivalents are rounded off.

Fluidizing Bag Fillers These fillers can meet any production requirements, ranging from pilot-plant scale through heavy-duty, conveyorized high-tonnage installations. A chamber is provided with an air pad at the bottom, adjacent to a filling spout. A column of product over this section, which is what causes flow, may be opened to the atmosphere or enclosed and pressurized. When the desired bag weight is reached, a system is activated by the integral weigh scale to close the valve through which material flows to the bag. Fluidizing and pressurizing air is best provided by a positive-displacement blower at 1.5 kW (3 hp) per filling spout.

Of special interest on multiple-spout conveyor-equipped fluidizers and on certain types of screw and belt filling machines is a combination operator's seat, bag rest, and tuck-in-sleeve work aid. This device places the operator in an optimum work position after filling, to allow easy and positive tucking of the sleeve. Extensive use of the PE film internal sleeve, however, has reduced the significance of the tuck-in-sleeve feature. Several types of heat-sealable valve-bag sleeves are available, as is equipment for closing them automatically. These are used when even slight leakage of product from an internal sleeve bag is unacceptable.

Bags can be automatically placed on valve-bag packers by means of an **automatic bag-placing device,** which consists of a magazine holding approximately 100 empty pasted valve bags and of a mechanism for removing the bag from the magazine, opening the valve and placing it on the filling spout, and initiating the filling-discharge cycle. The device's installed cost can be recovered in about 1 year's operation, based on typical wage rates paid in the United States for packaging-line labor.

Auger or Screw-Type Bag Fillers These fillers are usually applied to tuck-in-sleeve-type valve bags, for which production rates of one to two bags per minute and weight error limits of ±1 percent are required. Single-screw filling-spout designs (ordinarily of the net-weigh type) with simultaneous fill-and-weigh features are most common.

Gross-weight fillers need a feeding device such as a screw, vibrator, or belt, depending on the product. Particle size from 12.7-mm (½-in) pellets to 44-micrometer (325-mesh) powders can be handled, as can bulk densities ranging from 80 to 3200 kg/m^3 (5 to 200 lb/ft^3). Power requirements range from 373 W to 5.6 kW (0.5 to 7.5 hp). Weight accuracy is obtained by braking the motor to a rapid stop once the correct weight has been reached and the scale system has actuated the electrical or mechanical control system. Although fluidizing packers have diminished the importance of the screw type, the latter will always find application when space is a problem and investment must be low.

Centrifugal Belt-Type Packers This packer is used extensively on granular or pelleted products whose bulk densities range from 400 to 1600 kg/m^3 (25 to 100 lb/ft^3). Single-spout, simultaneous fill-and-weigh fillers, which consist basically of a short-belt conveyor, handle one to three bags per minute at weight accuracies within ±1 percent; the two-spout design is most common in high-speed conveyor-equipped installations, with which preweigh scales are used. Up to 30 bags per minute can be handled, with weight accuracy within ±0.1 percent or better.

Impeller-Type Fillers Used extensively for finely divided materials such as portland cement, plaster, lime, and talc, these fillers contain an impeller that turns in a casing (similar to a centrifugal pump) to move the product into the bag. Most impeller machines are installed with conveyors, although single-spout machines have been used when bag handling is done manually. Bulk densities are limited to 800 kg/m^3 (50 lb/ft^3) and higher. Portland-cement filling rates of up to thirty 43-kg (94-lb) bags per minute are possible with weight accuracies within about ±2 percent. Power requirements range from 3.7 to 7.4 kW (5 to 10 hp) per filling spout. Impeller fillers are being superseded by the fluidizing type because of the latter's better weight accuracy, cleanliness, and reduced investment and operating cost.

Gravity-Type Fillers These fillers are available in either the gross-weigher type or the net-weight type using a preweigh scale. Gross-weigher types are used in marginal operations for which investment must be limited and performance is not critical. Packing rates of 0.5 bag per minute and weight accuracies within ±5 percent are possible. Only free-flowing pellets and granules can be handled practically. The net-weight type utilizes a highly accurate preweigh scale which is placed 3 to 5 m (10 to 15 ft) over the bag-filling spout. Gravitational energy of the falling charge of product is used to force the product into the bag. Rates of up to six 25-kg (50-lb) bags per minute per scale-fill spout unit are possible. This type of equipment requires a free-flowing material and can handle the range of 250-micrometer through 4.8-mm (60-mesh through 4-mesh) pellets. Bulk densities as low as 400 kg/m^3 (25 lb/ft^3) can be handled.

Open-Mouth-Bag-Filling Equipment Two considerations in choosing this type of equipment (Table 7-23) and in deciding between open-mouth and valve-bag systems are the labor required for a given output and the capacity limitation of the closing system. With open-mouth bags, weighing and filling are usually done by a net-weight preweigh scale; gross weighers are sometimes used on low outputs. Operating principles and installation practice for automatic scales have been described earlier in this subsection.

Preweigh scales discharge to a chute system to which a bag is attached. The kinetic energy of the charge as it reaches the bottom permits the bag to stand without lateral support on a closing-machine conveyor. The filled bag is then dropped to a short-belt conveyor that passes the bag through a closing machine. Empty bags are held onto the chute system by hand or by a bag-clamp arrangement. These scales handle from 8 to 35 charges per minute. Weight accuracies are commensurate with product value and weight laws.

Bag Closures Conventional multiwall paper open-mouth bags are closed by sewing; the pinch-bottom type, by hot-melt adhesive. Three styles of sewn closure are used. The simplest and fastest consists of sewing with cotton or polyester thread, with needle and looper threads entwined in a chain-fashion stitch. This is adequate for low-cost products, for which sifting through the sewing is not objectionable. An improved method consists of adding a flat tape over the open mouth and sewing through it with the needle and looper threads. An additional thread, called filter cord, can be added between the needle thread and the tape to increase siftproofness, but this reduces closing rates.

Complete **siftproofness** can be had by the "tape-over-sewn" procedure, whereby the tape is glued onto the finished sewn closure by a device downstream from the sewing head. For siftproofness at high production rates, the pinch-style glued closure is used. The closing unit applies a 1.6-mm (0.0625-in) bead of thermoplastic adhesive to the staggered end of each ply and then folds these over and presses them to the bag face. Pressure and cooling are applied to the closure to set the adhesive prior to discharging the bag. An alternative type of pinch-bag closure has the adhesive preapplied to the open end by the bagmaker. After the bag has been filled, the closing machine reactivates the adhesive by heat prior to sealing.

PE-film bags are closed by heat-sealing together the face and back of the bag. The closing unit consists of a pair of belts that support the top of the bag and guide it through a heated section that fuses the face and back. This is followed by a cooling section.

Drum and Bulk-Box Filling This process consists of three operations: setting up, filling and weighing, and closing. Because setting up bulk boxes is cumbersome, a well-methodized workplace, equipped with work aids, is recommended. Weighing and filling can be done manually or automatically. There is enough similarity between the two ways for manual systems to be mechanized later.

The most common installation consists of a conveyor line with a platform scale at a central location. This scale may be a simple dial type, which the operator watches to stop flow. The first mechanization step is to add a cutoff switch to the scale. Filling rates of 5 to 10 kg/s (10 to 20 lb/s), with weight accuracies within ± 1 percent, are possible. Check weighing is easily accomplished by observing the net weight on the dial. A skilled worker can operate a manual system to within a few grams or ounces of the desired weight. Preweigh scales are occasionally used for free-flowing products, when the net weight is 100 kg (200 lb) or less or is a multiple of a weight that can be set on the scale and repeated to get the wanted weight. The main advantage of preweighing is higher accuracy.

Pouch, Small-Bag, and Carton Filling This process involves the two main operations of filling and closing. An additional step, in the case of cartons, is setting up and placing plastic pouch liners. Filling and weighing may involve either preweighing or simultaneous filling and weighing; the former is usually resorted to when high weight accuracy is required. With appropriate feeding devices, these fillers can be adapted for can and jar filling.

Plastic pouches are closed by heat sealing. Paper bags and pouches may be sewn by equipment similar to that for multiwall paper open-mouth bags but with appropriate conveyor and bag guides. A preferred method is adhesive closing by either a hot melt or a liquid that tacks rapidly. Adhesive-closing equipment can meet outputs ranging from a few bags to as many as 50 bags per minute. The rate needed determines whether the filling procedure is manual or automatic. For rates up to 20 packages per minute, manual operation is practical. Above this rate, some degree of mechanization becomes necessary. Economics dictates the choice between several manual lines and an automated one.

Form-Fill-Seal Premade small packages, when filled at high rates, present a problem because of the need to handle and store empty packages. To cope with this requirement the form-fill-seal type of packaging, which not only simplifies the supply problem but produces a superior package, has evolved. This method involves two main functions: a weigh cycle and a package make-fill cycle. Rates of up to 50 packages per minute are possible, with multiples of this rate on machines with two or more stations. Preweigh scales are of the same type used for small-package filling.

At present, form-fill-seal is limited to products having reasonably free-flowing particles with low dust concentrations. Because heat sealing to form the pouch has been largely responsible for the success of this system, thermoplastic films or other plastics and papers with a thermoplastic coating are required. The choice of form-fill-seal versus premade packages depends on economics but usually applies to materials that are nonseasonal.

Large-size form-fill-seal equipment for industrial packages has been introduced. Capable of packaging 25-kg (50-lb) bags, such units use PE sheeting or tubing in roll form. A bag is made just prior to being filled. An advantage of this system is lower labor and material costs, but this is offset by the increased complexity of the equipment. Rates of eight to twenty 25-kg (50-lb) bags are possible. Since preweigh scales are used, high accuracy can be attained.

Carton and Baler-Bag Loading, Wrapping, and Sealing Corrugated boxes may be used for shipping flexible or rigid small containers; baler bags, for flexible ones. Corrugated boxes are loaded manually, semiautomatically (manual loading of the carton set up by machine), or fully automatically. Manual setup and loading are practical for up to 3 cases per minute, semiautomatic up to 10, and automatic up to 40. Associated with each are conveyors that bring packages to the carton loader and remove filled cartons from the sealer.

Carton sealing is carried out automatically by adhesives, tape, or staples or manually by tape or staples. Carton closer-sealers have become so attractively priced that small operations can justify their use even when the balance of the line is manually operated. Case closer-sealers that take different package sizes in random order are available.

Baler bags can be manually loaded, but the preferred practice involves specifically designed compression units. Their use permits making an integral load in which all package parts share the forces imposed by shipping. A manually loaded compression unit handles 2 to 4 balers per minute; semiautomatic units, with a mechanical package feed and a manual baler-bag application unit, can handle 15 to 20. Baler bags are automatically closed with tape or adhesive, the latter being preferred especially for automated operations.

Wrapping, Bundling, and Shrink Packaging These techniques have limited applications for chemical products. Wrapping and bundling are substitutes for cartons and baler bags, their advantage being that the package is made from roll stock.

Shrink packaging is a significant development. The most important application in the chemical industry is in unitizing packages for palletized shipment. A cover of shrinkable PE film serves to bind a pallet load and permit it to absorb considerably higher transportation forces than it would if packed by any other method. Palletizing, adhesives, and strapping are eliminated, which offsets the cost of the shrink wrap. But it is the reduced damage in shipment that makes shrink wrap so economically attractive.

The hand-applied shroud of shrinkable PE usually consists of a premade bag large enough to envelop the load. Equipment to shrink the wrap ranges from small propane-fired hand-held units, which take about 5 min to shrink a 1-ton load, to fully automatic conveyor lines handling up to 30 pallets per minute.

Stretch-Wrap Packaging An important new development, this type of packaging is an alternative to shrink packaging. It consists of wrapping a pallet load of product with a thermoplastic film which is applied under tension and which envelops the sides of the entire load. Special machinery accomplishes this procedure correctly. An advantage of stretch wrap is that it does not require the precise fit and shrinkage properties of a shrink-wrap bag to ensure tight binding of the load to the pallet. As long as the stretch film meets tensile and elongation properties and the stretch-wrap machine is in proper adjustment, a satisfactory result will be obtained. Material cost for stretch wrap is 75 to 100 percent of that of shrink wrap. An economic advantage is that stretch wrap is less energy-intensive than shrink wrap by a factor of 100 or more. Labor requirements are also less, in that one operator can start the stretch-wrap process by attaching the film to the pallet and the machine completes the operation automatically. With shrink wrap, two operators are usually needed because of the unwieldiness of the shrink bag.

Package Marking and Labeling Label information may be divided into two classes, that fixed for each container and that which varies from package to package or from batch to batch. Examples of fixed information are name and address, net weight, name of product, and warnings about product hazards. Variable information includes batch, blend, or lot number, consecutive package number, coded information, and possibly date of manufacture. Export packages require outside-package dimensions and the gross, tare, and net weights in U.S. customary and SI units. When there is uncertainty as to label requirements, experienced legal advice should be sought. Fixed information is usually printed by the package maker. Variable information can be applied manually with rubber stamps or stencils or by automatic in-line marking equipment. A new system uses the digital computer for printing labels. Each label text is stored in the computer, which on command will print the required label by using a high-speed line printer. This system is expected to be used widely when many different labels are required and the text is subject to frequent modification.

Automatic reading of label information by remote electronic scanning devices has been the subject of development effort. Of interest is the reading of container serial numbers and contents as the container is being handled without having personnel perform this task. The objective is greater accuracy of identification at lower cost. For refillable containers, use of this remote-reading ability together with the computer for processing will aid in collecting rental costs and in container turnover.

PACKAGE HANDLING AND STORAGE

Warehouse Requirements Finished packages of the chemical industry are usually bags, drums, pails, or cartons (the last-named containing smaller units). Equipment for package handling and storage may be grouped into three main performance categories: (1) from packaging to pallet-unit loading, (2) from pallet-unit loading to storage or shipping, and (3) from storage to shipping.

The trend has been to the use of decentralized warehouses, with less and less finished product being stored at the producing plant. A typical plant inventory consists of 2 to 3 days of production, but at stocking points the inventory may amount to as much as 15 to 30 days. Although high inventory turnover is desirable, the variety of product grades and specifications often leads to longer storage times. Because of this, storage equipment and related conveyors are available to permit either a high velocity of product movement in and out of the plant warehouse or a virtually static one.

Mechanical handling of products in warehouses began with the forklift truck and pallet combination. Since warehouses had no moving equipment, pallet loads were set on the floor or placed on top of one another. Although this procedure is still practiced, loads now move into a storage-rack system, which permits storing pallet loads in vertical columns that make fuller use of the volume or height of the warehouse. Conveyors are also used to carry the pallet loads to storage, retrieve them, and send them to shipping. Forklift trucks are usually involved in all these movements.

Package-Handling Systems The **control of package-handling systems** may depend on simple motor starters, on interlocked relays with photocell control, or on computers. Solid-state controls are finding much application in the last two systems.

A second type of control required is that of the package or pallet itself as it is handled by conveyors and other equipment. This handling may consist of right-angle transfers in a vertical lift or of a set of restrainers on the sides of a belt conveyor.

System Analysis The choice of a specific handling system must take into consideration trade-offs that can be made among different types of equipment and between people-operated and -controlled equipment and automation. After a considerable trend to full automation, industry is beginning to recognize the unique ability of human beings in the handling of complex operations in which the decision process is too complicated for computer control. Another disadvantage of automation is the high cost of specialized maintenance required, which can cost annually between 5 and 10 percent of the original equipment cost and sometimes more. New technical skills are also often necessary.

Factors that enter into any economic analysis of handling-warehousing systems are (1) expected mechanical and economic life of the system; (2) annual maintenance cost; (3) capital requirements and expected return on investment; (4) building-construction cost and land value; (5) detailed analysis of each work position (to determine trade-offs of labor and equipment; expected future costs and availability of labor are important); (6) relation of system control and personnel used in system (trade-offs of people versus mechanical control); (7) type of information system (computerized or manual); and (8) expected changed in product, container, unit pallet loads, and customer preferences during the life of the system.

Forklift Trucks The backbone of most in-plant handling systems in the chemical industry is the forklift truck. Available in capacities ranging from 1 to 50 tons, the most commonly used are 1-, 1.5-, and 2-ton vehicles, with the 3-ton unit occasionally being used (Fig. 7-

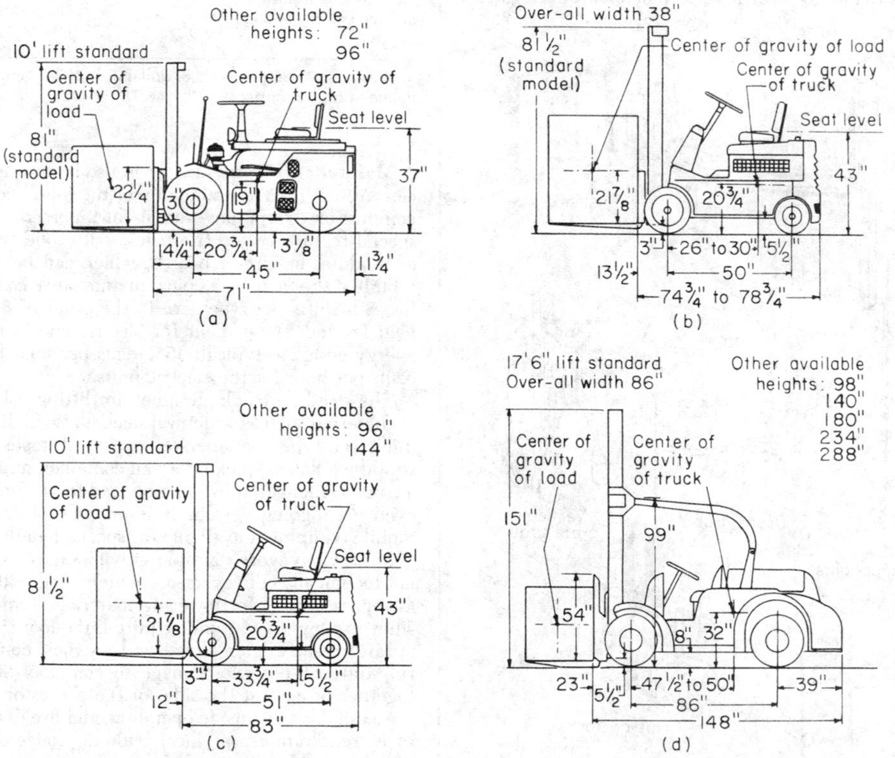

FIG. 7-42 Dimensions of representative forklift trucks. (*a*) 1000- to 2000-lb cpaacity. (*b*) 3000- to 4000-lb capacity. (*c*) 5000-lb capacity. (*d*) 10,000- to 12,000-lb capacity. Multiply pounds by 0.4536 to get kilograms; multiply inches by 0.0254 to get meters. *(Hyster Co.)*

42). The trucks are usually powered by internal-combustion engines that consume liquefied petroleum gas (LPG) or by electricity by means of storage batteries.

With internal-combustion engines, automatic transmissions are frequently used; these are easily justified when vehicles must make many moves during the day. Smooth as is the control afforded by automatic transmissions, it is nevertheless inferior to that provided by electric trucks, especially those with solid-state controls. Gasoline and diesel power are also used, but mostly for outdoor equipment and very-heavy-duty units.

The lift-truck industry is competitive, with innovations being introduced frequently. Competent sales and service are available at low cost from most manufacturers or their dealers. Application sales engineering (a very worthwhile service) is generally supplied at no cost.

The many **options available** for lift trucks fall into two classes: vehicle specialties, which include controls, transmissions, guards, etc.; and accessories, which are devices that handle specific types of loads (Fig. 7-43). Included in this second category are high-lift masts, up to 7 m (24 ft); handling attachments for circular products, such as drums and roll goods; attachments such as carton clamps; and the fork side-to-side shifting mechanism.

Worthy of particular notice among accessories is the **side shifter** that is used to move trucks horizontally, about 100 mm (4 in) from side to side. The modest cost of this feature is returned in a few months' operation through reduced handling time, maintenance, and product damage. The driver first positions the truck approximately in front of where the load is to be set down and then makes the final horizontal adjustment by means of the side shifter. Without this mechanism, two or three maneuverings of the truck are necessary, with the load never quite being placed in the ideal spot. Correct positioning is important for pallet loads, which should be placed as tightly together as possible.

Lift trucks are available to meet a variety of **clearance restrictions.** Noteworthy is narrow-aisle equipment. Another accessory worthy of consideration is the multilift mast, which permits lifting loads over 3.7 m (12 ft). Of special importance in specifying any mast is that it will clear the various door openings it must enter, which includes those of trucks, railcars, and buildings. To meet most conditions, the collapsed height of the mast must be 2235 mm (88 in). An ideal lift truck for chemical-plant distribution warehouses would have 2000-kg (4000-lb) capacity; electric (battery) propulsion; solid-state controls; power steering; Trilift mast, up to 4.9 m (16 ft) [2235 mm (88 in) collapsed]; side shifter; operator guard; solid tires (except for outside use); and adjustable forks.

Exceptions to the preceding requirements would apply where explosionproof equipment is needed; building ceiling heights are such that the standard 3.7-m (12-ft) lift is all that will ever be needed; and loads will never exceed 1 to 1.5 tons.

Capital investments in forklift equipment vary with specifications. Table 7-24 compares the cost of the electric-propulsion truck just described with an LPG-operated alternative. The operating cost is primarily for energy, with electric consumption being cheaper than liquid fuels.

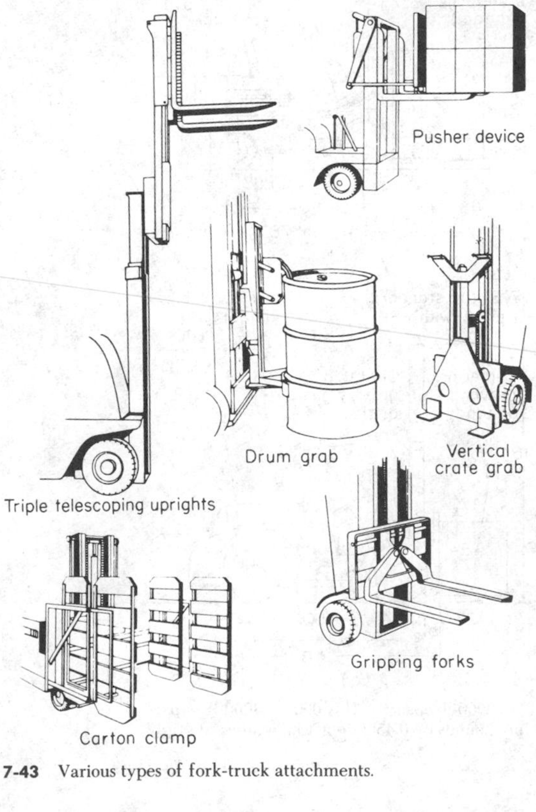

FIG. 7-43 Various types of fork-truck attachments.

TABLE 7-24 Initial Capital Investment Comparison between Liquefied Petroleum Gas and Electric Forklift Trucks of 2-Ton Capacity*

Item	Liquefied petroleum gas	Electric
Basic truck	$22,000	$26,000
Automatic transmission	700	....
Solid-state controls (standard)	....	....
Trilift mast, to 4800 mm (189 in)	2,500	2,500
Side shifter	2,500	2,500
Power steering (standard)	....	....
Solid tires (standard)	....	....
Storage battery and charger	....	9,000
Total	$27,700	$40,000

*Based on 1981 prices (U.S. dollars). Initial inventory of repair parts is not included in these prices.

Maintenance on gas trucks is also higher than with electric vehicles. About 5 percent annually of the initial cost applies to internal-combustion equipment, and about 2 percent annually to electric. A special feature on electric trucks with solid-state controls is the use of modules or circuit boards, which can be replaced as units and rebuilt at the factory. Typical maintenance costs for trucks operating five 8-h shifts per week are in the order of $1.75 per hour for gas vehicles and $1 per hour for electric ones. Under these conditions, energy costs are typically 5.2 cents per hour for gas trucks and 2.8 cents per hour for the electric units.

The **straddle truck,** designed for lifting bolsters with heavy loads or materials such as structural steel, is also finding application in handling van-type containers of packaged goods. For example, it can straddle a flatcar, pick off a van container, and deposit it directly on a truck-trailer rig. It can also be used for loading railroad flatcars and even oceangoing vessels. It is just one of many special pieces of mobile equipment available for special handling problems.

Slide Conveyors Simple gravity slides and spiral chutes, while not technically conveyors, are widely used with conveyor systems or as separate units for lowering materials from one floor to another. They are low in cost and require little floor space if slopes are held at fairly steep angles. However, they must be used only after a careful study of possible damage to containers from bumping either together or against the sides of the chutes or slides. Enclosed units are available for outside operation, and fire doors can be provided to meet requirements of local building codes. Multiple-blade chutes may be used for service to several floors, with separate inlet and outlet points. Blades may be lapped and riveted to eliminate the possibility of containers hanging up on exposed edges. Flight sections may also be flanged and bolted together.

Speed of containers sliding down a spiral may be controlled by the pitch of the spiral or by banking the outer or inner edge of the blade. Banking tends to throw the container to one side of the blade, thus varying its total travel distance. While usually fabricated of steel, blades may be specified in different materials, as required by specific applications.

Because of the steep pitch required, **slides** are limited in application. They are most commonly used to bridge the gap between roller-conveyor systems on two floors, because the roller conveyor can take the container off the slide rapidly and eliminate or reduce the chance for collisions. Slides may also be used when containers can be chuted from an upper floor to a manually loaded carrier. The use of several rollers at the feed point is recommended for easy delivery to the sloping section. If the drop is short and containers light, a roller cleanout will prevent backup of containers on the slide. The slope of gravity slides is a function of container weight, size, and friction characteristics and should be selected with care to be sure that containers do not move either too swiftly or not at all. Slides usually use flat steel sheet.

Gravity Wheel Conveyors These can be used as pusher units set horizontally or inclined for gravity flow. They are highly standardized and are usually sold in 1.5- or 3-m (5- or 10-ft) sections; special lengths are available at extra charge. Since wheel conveyors give what is essentially "point" support to containers, it is generally recommended that at least six wheels be located under the load at all times. Thus wheel arrangement is dictated by the smallest container that the line will handle. Only flat-bottomed containers can be handled on wheel conveyors, with the exception of fairly stiff-walled bags, which handle satisfactorily. This is due to the fact that the separate roller supports tend to pull the bag wall taut and flatten it out. Roller conveyors, on the contrary, tend to ripple the bag surface and prevent its movement. Wheel conveyors may also be specially designed for handling smooth-walled cylindrical shapes.

Wheels are available in a number of different designs, including variations in contour and material in contact with the container. Rubber or plastic tires are not uncommon. Through shafts may be used, with several wheels mounted on each shaft; stub bolts with a single wheel may be mounted to the side frame, or short shafts supported by bent bars may also be used. Wheel conveyors are generally used on lighter loads, and although manufacturers may offer widths up to 915 mm (36 in) or more, the smaller widths [up to about 457 mm (18 in)] are generally standard. Load ratings are generally given as the total uniform load which a standard section will support.

Since wheel units are relatively light, they have relatively low inertia, and loads may be started and stopped quite easily. In addition, wheel bearings are designed with loose tolerance to reduce starting friction. Metal plates or projecting hardwood slats are commonly used as stops on conveyor lines. Special hinged sections for passage of personnel through the conveyor line are available, and standard supports from floor or ceiling are recommended. Wheel-conveyor units are widely used for live storage, and special telescoping units are available for extension and retraction to meet variable conditions. Wheel conveyors are sometimes powered by a pressure belt or other methods but are most widely used as pusher or gravity lines. They are adaptable only for end discharge or side discharge by lifting, since the individual rollers tend to grip the container and prevent its sliding off the line at right angles to direction of travel. Grades for wheel lines may be figured at about two-thirds of the values shown for roller lines in Table 7-25. Care should be taken not to overload the conveyor sections since they will assume a concave shape and prevent movement.

Roller Conveyors Gravity rollers are considerably heavier than the wheels on wheel conveyors, and the weight is concentrated at a greater distance from the shaft centerline. Hence, roller conveyors have a greater inertia; they are harder to start and harder to stop, require more slope than wheel units, and on long runs tend to speed up containers at an accelerating rate. Typical **roller-conveyor grades** are shown in Table 7-25.

Spiral-roller units are usually equipped with tapered rollers to compensate for the difference in distance traveled by the inner and outer edges of the container. Tapered rollers are also used on curved sections of ordinary roller-conveyor lines.

Rollers are available in a wide **variety of constructions,** with tube ends either bored or formed to take the bearing insert. Bearings may be plain, with nylon rapidly becoming the most popular material for this type. Ball bearings are probably most common and are available with a variety of seals, or the bearing may be left unprotected. Lubrication fittings may be provided on a drilled shaft, or bearings may be prelubricated and sealed for life. Roller shafts are usually dead and may be cut from hexagonal stock to fit a similar opening in the side frame, or they may be round with ends milled flat to prevent turning. Rollers may be mounted in side frames in a variety of ways, above the side frames when containers are to be slid off the line or below when there is danger of the containers falling off.

Gravity roller conveyors can handle containers with protruding edges, i.e., steel drums, which is one of their advantages over wheel conveyors. However, they are not generally suitable for bags since the sides tend to sag between supports and prevent forward motion.

As with gravity wheel conveyors, roller units are highly standardized and auxiliary equipment is available for supporting the line

TABLE 7-25 Grades for Roller Conveyors

Package	Unlubricated ball bearings, in (mm)		Greased packed ball bearings, in (mm)	
	10 ft (3 m)	90° curve	10 ft (3 m)	90° curve
Corrugated cartons				
10–20 lb (5–10 kg)	6 (150)	5 (125)	NR	NR
20–50 lb (10–25 kg)	5 (125)	4 (100)	8 (200)	6 (150)
Steel drums, 55-gal (208-L)				
Empty	6 (150)	5 (125)	7 (175)	6 (150)
Full (450–600 lb) (200–275 kg)	5 (125)	4 (100)	6 (150)	5 (125)
Fiber drums				
150–300 lb (70–135 kg)	5 (125)	4 (100)	6 (150)	5 (125)
Multiwall paper bags				
50–100 lb (25-50 kg)	8 (200)	10 (250)	NR	NR

NOTE: NR = not a recommended practice. Grades are in total number of inches drop required in each 10-ft (3-m) section or 90° 2.5-ft (0.75-m) inside-radius curve. Grades required for roller conveyors vary somewhat, depending upon the size and spacing of the rolls used. The grades suggested are for average conditions with rolls of a size and capacity to suit the material handled. For level push lines, the average amount of push required to start the package from rest is about 3 percent of its weight. With heavy loads, a pitch of about ⅛ in/ft is recommended. This is not enough for the package to travel by gravity but will decrease the amount of push necessary. For wheel conveyors, use grade approximately two-thirds of that shown. SI equivalents are given to the nearest practical dimension.

from ceiling or floor. Many special rollers are available for retarding containers if speed becomes too great for safe handling. Switches, brakes, hinged sections, spurs, and frogs are also available.

Roller conveyors are quite frequently **powered,** the simplest method being use of a pressure belt in contact with the lower surface of the rolls. Shown in Fig. 7-44 is a special ripple belt with raised pads which is capable of starting up the load but does not build up excessive blocked pressure if the line fills up. Other similar drives are available, with varying degrees of control over the applied power. Most expensive of the powered roller units are those in which each roll is equipped with V-belt or chain drives. Pusher bars suspended from overhead chain conveyors may also be used to move containers along a roller line.

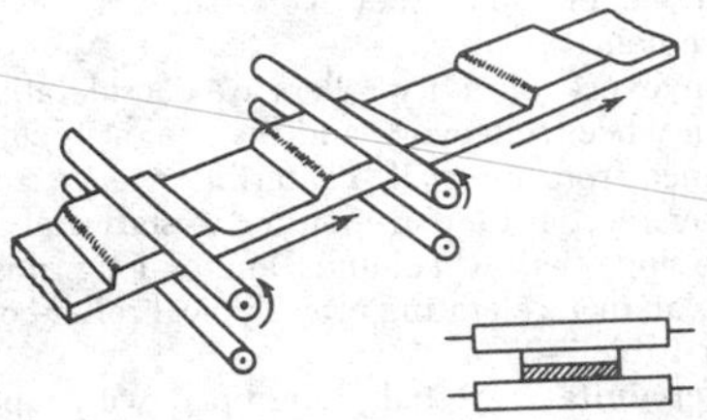

FIG. 7-44 Ripple belt for roller-conveyor drive.

One of the most important control devices on roller-conveyor lines is the escapement mechanism which allows containers to be released from a line individually. Powered escapement mechanisms are commonly available on highly mechanized systems. Their main function is to space out the containers so that they can be handled as discrete units.

Flat-Belt Conveyors These powered conveyors can lift containers up inclines. With the aid of special belt surfacing, grades may be quite steep. Belts also keep containers spaced out in exactly the way in which they are placed on the conveyor. However, because of the relatively high friction containers cannot be slid off belts by pushing devices.

Belt-conveyor designs use both roller and slider bed supports for the flat belt. The variety of designs available allows proper selection of flat belts for heavy or light loads and for various applications such as carton filling or emptying.

Chain Conveyors These devices for handling containers are available in either roller-chain designs or less costly types. There is a variety of **slat conveyors** that use both single and double strands of roller chain, as well as a slider type using cheaper chain. In general, slat chain conveyors are used only on loads which are too heavy for economical handling by belt, roller, or wheel units or which have odd shapes not suitable for roller or wheel units. They are particularly adaptable to pallet handling, as are simple open strands of chain with flat-surfaced attachments.

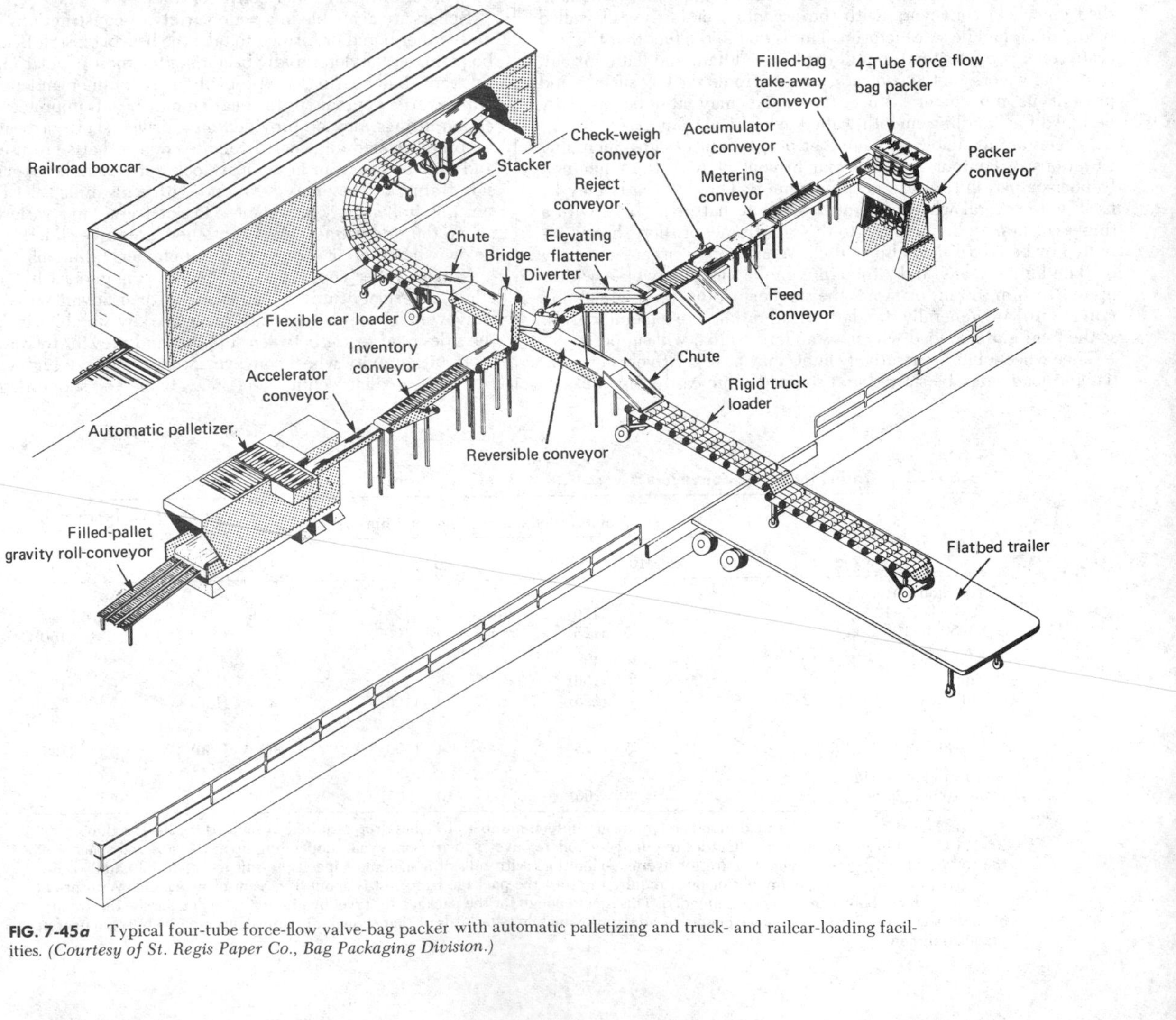

FIG. 7-45a Typical four-tube force-flow valve-bag packer with automatic palletizing and truck- and railcar-loading facilities. *(Courtesy of St. Regis Paper Co., Bag Packaging Division.)*

The most commonly used warehouse chain conveyor is the **tow chain.** Chain may be mounted overhead or in the floor, and trucks being towed can be designed for automatic detachment at a specific point. While the overhead chain is often used and is usually easy to support from structural members in the ceiling, the in-floor chain is probably most common. Automatic disengagement is possible should trucks encounter an obstruction or accidentally strike warehouse personnel. The two-chain conveyor is, of course, most economical when large tonnages are moved over a fixed path.

Chain-type **elevators,** such as arm and tray units, are commonly used for drums and barrels. Slight gravity runs at feed and discharge allow these units to roll on and off the conveyor easily and without special equipment.

Elevators Cable-type elevators are usually selected for heavy loads such as full pallets or large containers. They can be made fully automatic and are able to serve many floor levels. The use of properly designed elevator systems is often the only economical solution to multistory-plant problems.

Conveyor Accessories These may be divided into two groups, those which act on the container and those which are acted on by the container. In the first group are such items as deflectors, palletizers, pushers (powered by fluid, air, or mechanical linkage), upenders, sealers, staplers, and similar devices. In the second group are such items as electric eyes for counting or identification via printed or color codes, check weighers, mechanical counters, and other devices contributing to automatic conveyor-line operation.

Automatic Palletizers These machines receive packages from production by conveyor. The packages are then arranged in tiers, and the tiers are placed on pallets. The mechanism to accomplish this consists of package-handling conveyors, package-moving stops, rams, etc.; a package-tier-pattern assembly plate; an empty-pallet-handling conveyor and elevator; a filled-pallet-handling conveyor; and electrical regulators to control the tier-pattern formation. Automatic pallet loaders can handle 40 to 80 packages per minute, or one to two pallet loads. Capital investment is about $125,000 for the basic machine, not installed. Semiautomatic operator-directed palletizers capable of handling 10 to 20 packages per minute are available; they cost approximately $60,000, not installed.

Package-handling systems can be designed to handle almost any situation of package type, packaging machine, and warehousing-transportation requirement. Figure 7-45*a* shows a typical bag-handling system in which both palletized loads and loose bags are handled. A system of this type can handle thirty 25-kg (50-lb) bags per minute. Figure 7-45*b* shows a system using the pinch-bottom-type open-mouth bag. Such a system can handle eight 25-kg bags per minute. Performance data for both examples can be found in Table 7-23.

Storage of Packaged Items The inventory needed to support a given sales level is increasing in quantity as well as in the number of places where inventories are maintained to provide better service. Since a major portion of the chemical-process industries is located in urban centers where space is extremely valuable, efficient ways of storing packaged inventory have become very important. A similar situation exists when inventories are maintained at production plants. Here, space may be more readily available than in urban locations, but there is the question of whether to use the space for storage or for processing. These situations have led to the development of the storage-rack concept.

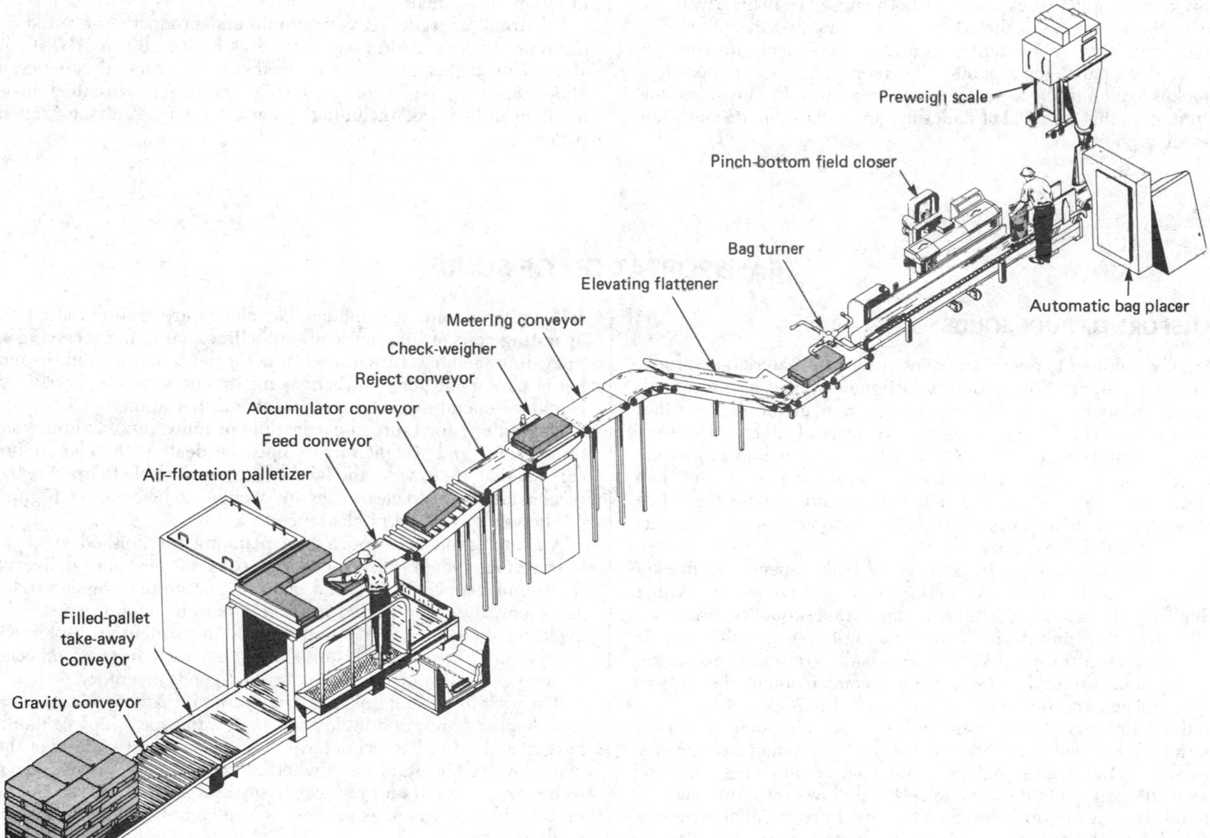

FIG. 7-45*b* Typical pinch-bottom system with automatic bag hanging and semiautomatic palletizing. *(Courtesy of St. Regis Paper Co., Bag Packaging Division.)*

Storage racks permit storing pallet loads of packages vertically as well as horizontally. Most pallet loads can be tiered two or three pallets high, with one resting on top of another (provided the packages are able to withstand the weight of the pallets above). Because the racks bear the pallet weight, stacks six to eight and even more pallets high are possible. Forklift trucks and stacker cranes are used to place and remove the pallets.

From an inventory-turnover point of view, four major rack-storage systems are possible: drive-in, drive-through, flow, and aisle.

Drive-in racks, which are practical up to a height of 10 m (30 ft), are serviced by forklift trucks. The inventory system required is last-in–first-out (LIFO), which many consider inefficient. Capital investment (installed) for a 5000-pallet rack system is about $50 per stored pallet, lift truck not included. Drive-in racks make good use of floor space, having a higher ratio of storage to aisle space than aisle racks.

A typical drive-in rack consists of a steel structure to support palletized goods at the pallet edge, with the center of the pallet unsupported. The space between pallet support members is sufficient to permit a lift truck to drive in to place or retrieve a load. These racks are usually made to accommodate 12 pallets, which are positioned from the service aisle to the end of the rack. Because of the rack, each pallet position has the ability to hold 6 to 8 pallets vertically.

In operation, the lift truck takes the first pallet load and drives to the end of the rack to set down the pallet. With the second pallet, the truck enters the rack with the pallet elevated to permit clearing the support member. This procedure is repeated until the rack is filled. Lift-truck productivity is low because the driver must possess agility and skill to manipulate pallets extended on the truck.

Drive-through racks are similar to the drive-in kind, differing mainly in having lift-truck access at both ends. The main advantage of drive-through racks is that they allow a first-in–first-out (FIFO) type of inventory management. Capital investment, installed, is about $50 per pallet for a 5000-pallet rack structure. In operation, the rack is loaded in the same way as a drive-in rack. The unloading is different, in that removal of pallets begins at the opposite end from the loading point.

Flow racks are similar to drive-through racks in that they are loaded from one end and unloaded from the opposite end. However, the truck does not enter the rack. Rather, each lane in the rack is equipped with a conveyor (roller, wheel, or belt, depending on pallet characteristics) which both supports the pallet and transports it (by gravity) from the entry point to the discharge end or to the nearest pallet. As a pallet is removed, the remaining ones flow to the removal point. This is a FIFO system of inventory.

The installed capital investment is about $210 per pallet for a 5000-pallet system. A characteristic of drive-in, drive-through, and flow racks is that, at any one point in time, only one product can occupy a given storage lane. Products are not mixed because of the complications that this practice presents in inventory management. In any event, there is seldom any need to mix products in the chemical industry because products are made in lots, blends, etc., and a storage lane is ordinarily designed to accommodate either a complete lot or some fraction of a lot. The result is that the total storage space available rarely is completely used. This is a problem that aisle racks overcome.

Aisle racks, used when there is a rapid turnover of inventory, permit a storage depth of only one or two pallet loads but offer the advantage of instant access to the stored item. Although this requires only a minimum of lift-truck time to store or retrieve a pallet, a high percentage of floor space must be devoted to aisles.

The inventory system needed is FIFO, which is desirable when inventories are subject to obsolescence or deterioration or when they consist of raw materials that fluctuate widely in value. The capital investment for a typical aisle-rack storage system having a 5000-pallet capacity (1 deep) is $35 per stored pallet (this does not include lift-truck investment).

Lift-truck operation is very simple and productive, even at 7-m (20-ft) elevations. Aisle racks can be as high as 30 m (100 ft), but above 7 m stacker cranes are favored over lift trucks because cranes allow servicing high storage at high rates. The installed investment in aisle racks, including a stacker crane, is about $280 per pallet.

TRANSPORTATION OF SOLIDS

TRANSPORT OF BULK SOLIDS

Originally confined to the shipment of crude raw materials and fuels, the term "transportation of bulk solids" now applies also to manufactured products, which often become raw materials for other industries. In recent years, increasing tonnages of highly processed, finished chemical products have moved to customers in large bulk units. A useful definition of a bulk shipment is any unit greater than 2000 kg (4000 lb) or 2 m^3 (70 ft^3). The containers available range from small portable hoppers of 2-m^3 (70-ft^3) capacity to railroad cars of 255-m^3 (9000-ft^3) capacity.

The choice of shipping in package or bulk depends on market requirements and economics. Products from different sources that tend to have the same characteristics (appearance, quality, price) are usually offered in bulk form. Those tending to be specialties, while sometimes offered in small bulk units, usually are sold in packages. Many products are sold in both ways. A comparison of the costs of typical package and bulk units is given in Table 7-15.

Bulk Containers These containers may be either open or closed. Generally, it is the effect of the weather on the product that governs the choice. High-value materials, such as certain ores, may be shipped in open containers, while relatively low-cost items, such as portland cement, require closed containers. Further influencing the choice of bulk containers is whether deliveries are made by truck, railroad, or water.

When customers maintain small inventories, **truck** delivery is often used, provided the location of the supply point is nearby, usually 550 km (300 mi) or less, and deliveries are frequent. If, however, a user maintains large inventories, deliveries are ordinarily made by **rail.** Other parameters influencing choice are transportation cost; operating costs of supplier loading facilities; customer receiving and unloading facilities; turnaround time for the container and the number of trips made per year (hence, investment write-off per trip); and container-operating cost, exclusive of transportation.

In planning for railroad-car loading or unloading facilities, many dimensional and weight factors must be dealt with. The common carriers that are to serve the facility are usually able to provide technical assistance as to clearances and weights to be handled. Figure 7-46 shows a typical set of clearances.

An interesting new concept in planning for finished goods and bulk storage (when rail is used principally for customer delivery) is the use of hopper cars instead of fixed storage bins. Since products are eventually to be loaded into cars, there is much to be saved by avoiding double handling and capital investment. A systemwide analysis often will show this to be the least costly method, especially if there is a policy of minimum finished-goods inventory.

The most important bulk containers are railroad hopper cars, highway hopper trucks, portable bulk bins, van-type (ship) containers, barges, and ships. Factors determining the suitability of any of these containers (after establishing whether open or closed containers are to be used) depend on product physical properties, the most important of which are ease of flow, corrosiveness, and sensitivity to contamination.

Railroad Hopper Cars Hopper cars follow three basic designs: (1) covered, with bottom unloading ports; (2) open, with bottom unloading ports; and (3) open, without unloading ports. Three types of unloading systems are used: gravity, pressure-differential, and fluidized. For the open-type car without unloading ports, clamshell

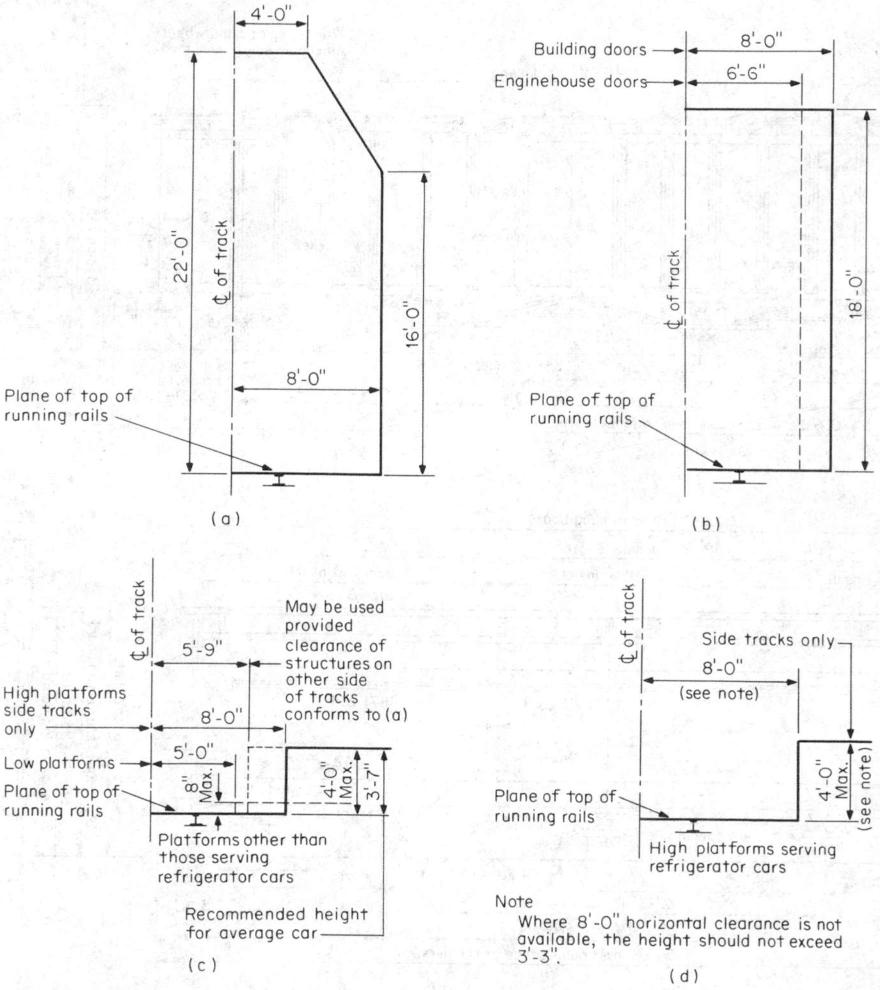

FIG. 7-46 Typical railroad clearances (United States practice). (*a*) Clearance diagram for structures other than platforms adjacent to industrial sidetracks. (*b*) Clearance diagram for building doors. (*c*) Clearance diagram for platforms. (*d*) Clearance diagram for platforms. All clearances are those recommended for new construction by the American Railway Engineering Association, Engineering Division, Association of American Railroads. Multiply inches by 0.0254 to get meters; multiply feet by 0.3048 to get meters.

buckets are often used. The car usually is loaded through ports located on the top of the car. Figure 7-47 shows examples of common types of covered hopper cars. Figure 7-48 shows fluidizing unloading ports.

Table 7-26 gives dimensions of hopper cars and other cars typically used in the chemical industry. Vacuum-pressure systems are used most frequently for unloading covered hopper cars. For certain free-flowing materials, in both covered and open-top hopper cars, shakeout devices are useful (Fig. 7-49).

Because of the railroad-car shortage that has persisted for many years, boxcars are often used for bulk materials. Lined with suitable materials to prevent contamination and with special bulkheads at each door, these cars are acceptable substitutes for covered hopper cars even though unloading is more difficult. Vacuum conveying wands are used to pick up the material, as are front-end-loader-type vehicles.

Loading of hopper cars and trucks can be done with most types of conveyors: air, belt, screw, etc. When an extremely full loading is required, centrifugal trimmers are frequently used. Available in a range of capacities, they can be engineered for any size of unit, up to a shiphold cargo (Fig. 7-50).

Track hoppers are needed for some boxcar and bottom-dump-car shipments. Since boxcars discharge to one side, fairly light construction can be used for the hoppers, which are located to one side of the tracks. However, for bottom-dump cars, the hoppers must be located on the centerline of the tracks. This requires heavy track girders over a hopper and feeder conveyor pit. Figure 7-51 shows a single hopper designed for use with a bucket elevator. Typical dimensions are given, but hopper depth must be set to give sufficient angle for material to flow well. Belts or reciprocating-plate feeders commonly carry the material to the bucket elevator.

Hopper Trucks These trucks are used to transport by highway a

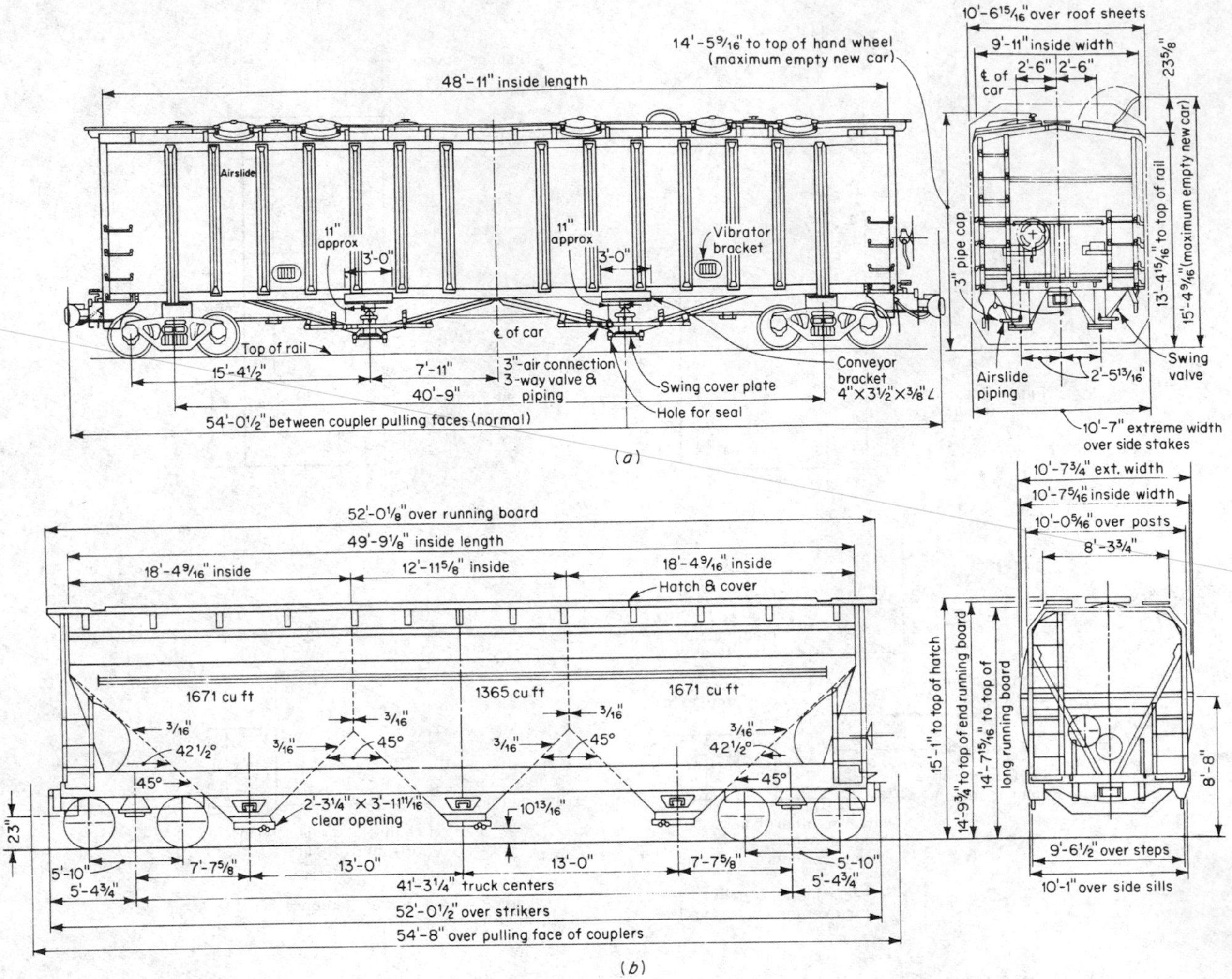

FIG. 7-47 Typical covered hopper cars. (*a*) 100-ton, 4180-ft^3 Airslide car; lightweight 69,600 lb; 193,400-lb load limit. *(Courtesy of Transportation Division, General American Transportation Corp.)* (*b*) 100-ton, 4700-ft^3 center-flow car; lightweight 62,800 lb; 200,200-lb load limit. *(Courtesy of Chessie System railroads.)* Data are given for United States railroads, which do not use SI dimensions. To convert to SI dimensions (millimeters), change dimensions shown to inches and multiply by 25.4. To convert volume to cubic meters, multiply by 0.02832. For weight, multiply by 0.4536 to obtain kilograms.

TABLE 7-26 Typical Railroad-Car Dimensions and Capacities*

Type of car	AAR* class	Nominal inside dimensions, Length	Width	Height	Nominal outside dimensions, Length	Width	Height	Volume, cu. ft. × 100	Weight, lb. × 1000
ACF center-flow hopper car	LO	....	....	....	39 ft. 8 in.	10 ft. 8 in.	14 ft. 10 in.	29.7	207
ACF center-flow hopper car	LO	....	....	....	54 ft. 8 in.	10 ft. 9 in.	15 ft. 1 in.	47.0	200
ACF center-flow hopper car	LO	....	....	....	59 ft. 2 in.	10 ft. 9 in.	15 ft. 1 in.	52.5	200
GATX	LO	....	....	....	42 ft. 0 in.	10 ft. 8 in.	14 ft. 4 in.	26.0	140
Airslide hopper car	LO	....	....	....	54 ft. 6 in.	10 ft. 7 in.	14 ft. 6 in.	41.8	192
Hopper car	HT	42 ft. 10 in.	9 ft. 8 in.	....	43 ft. 10 in.	10 ft. 6 in.	10 ft. 8 in.	27.5	157
Gondola car	GB	41 ft. 6 in.	9 ft. 4 in.	2 ft. 5 in.	42 ft. 9 in.	10 ft. 2 in	6 ft. 2 in.	9.6	100
Boxcar	XM	41 ft. 0 in.	9 ft. 6 in.	10 ft. 0 in.	42 ft. 8 in.	10 ft. 6 in.	14 ft. 7 in.	42.0	100
Boxcar	XM	50 ft. 7 in.	9 ft. 6 in.	10 ft. 8 in.	55 ft. 2 in.	10 ft. 6 in.	15 ft. 9 in.	51.2	100
Boxcar with DF† equipment	XL	50 ft. 6 in.	9 ft. 5 in.	10 ft. 6 in.	57 ft. 7 in.	10 ft. 6 in.	14 ft. 10 in.	49.5	100

*From Association of American Railroads. Data are given for United States railroads, which do not use SI dimensions. To convert to SI dimensions (millimeters), change dimensions shown to inches and multiply by 25.4. To convert volume to cubic meters, multiply by 0.02832. For weight, multiply by 0.4536 to obtain kilograms.

†Damage-free.

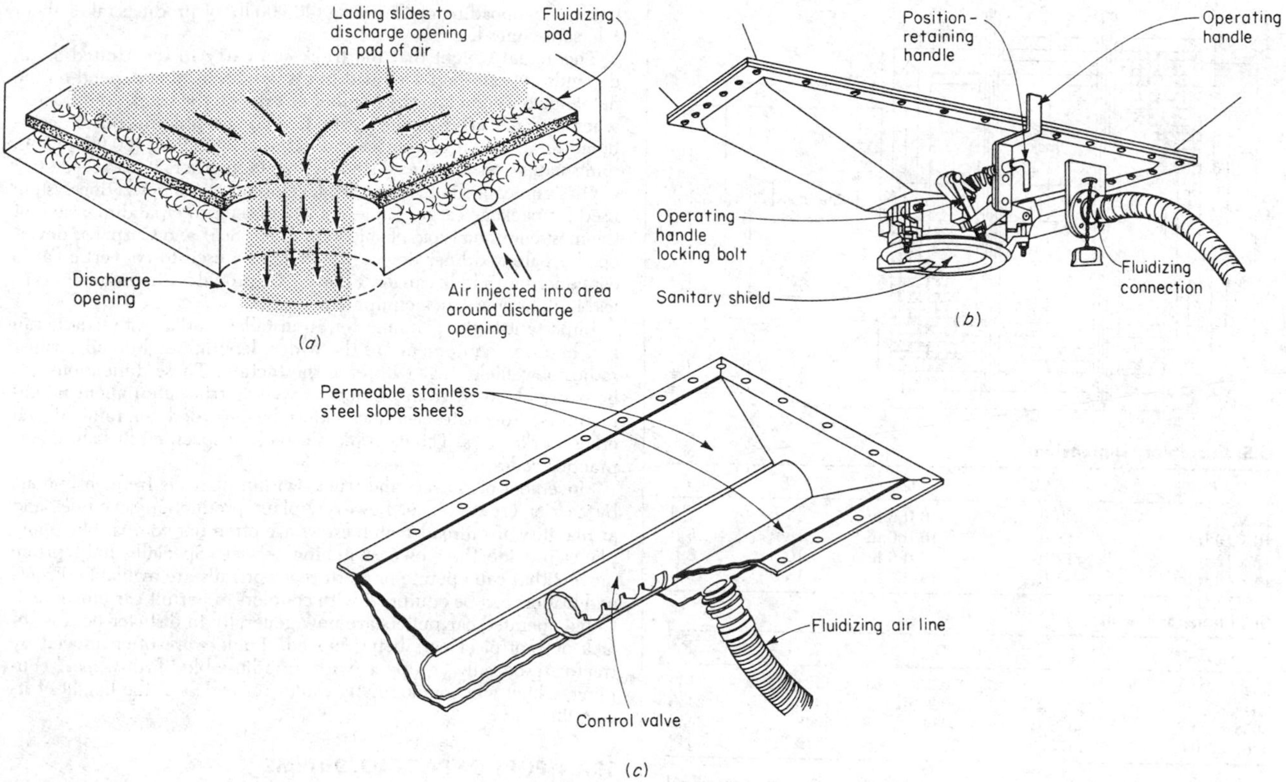

FIG. 7-48 Fluidizing outlets for hopper cars. (*a*) Air introduced through fluidizing pad makes powder flow toward opening. (*b*) ACF center-flow butterfly outlet controls discharge of fluidized bulk powders. (*c*) Another type of fluidizing butterfly outlet. *(Shippers Car Line Division, ACF Industries Incorporated.)*

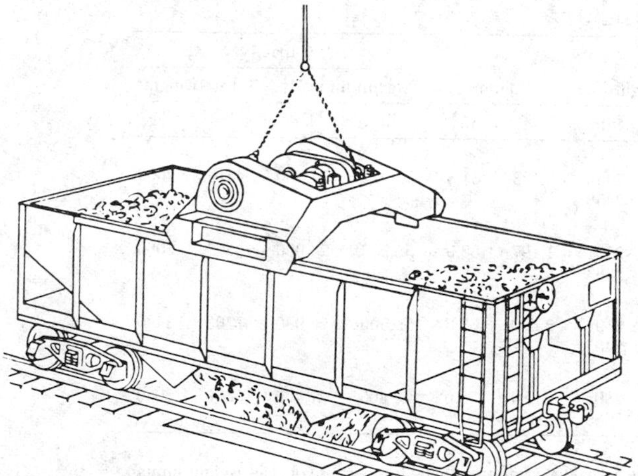

FIG. 7-49 Shaker-type car-unloaded device.

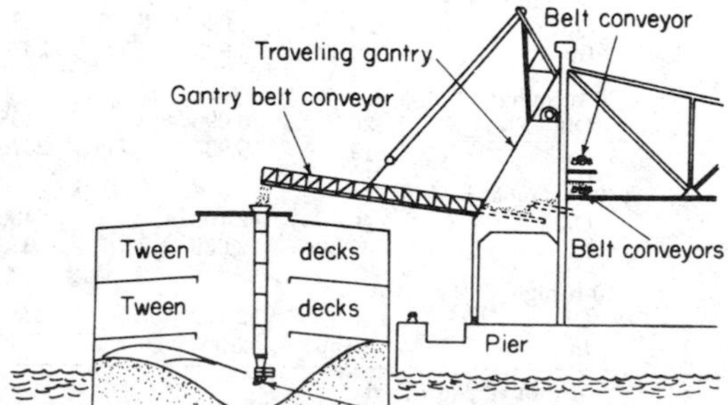

FIG. 7-50 Ship-loading system with trimmer and telescoping chute. *(Stephens-Adamson Division, Allis-Chalmers Corporation.)*

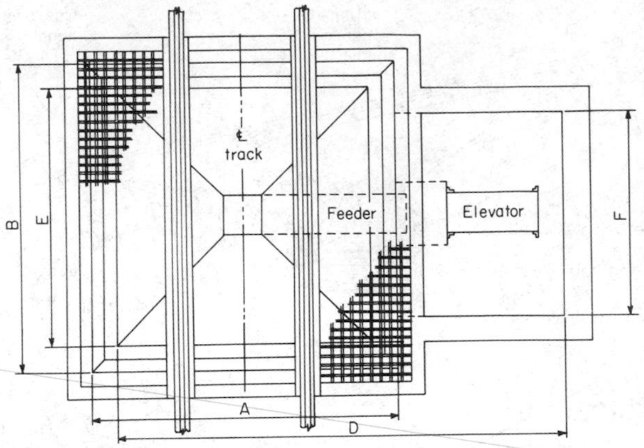

U.S. Customary Dimensions

A x B	C	D	E	F
10 x 7 ft	9 ft 6 in	16 ft 6 in	5 ft	7 ft
10 x 10 ft	11 ft	16 ft 6 in	8 ft	8 ft
12 x 12 ft	12 ft	17 ft 6 in	10 ft	8 ft
14 x 18 ft	15 ft	19 ft	15 ft	10 ft

SI Dimensions, mm

A x B	C	D	E	F
3048 x 2133	2895	5029	1524	2133
3048 x 3048	3353	5029	2438	2438
3658 x 3658	3658	5334	3048	2438
4267 x 5486	4572	5791	4572	3048

FIG. 7-51 Single-track hoppers for elevators. C = depth of pit for spaced bucket elevator. The depth must be increased for the continuous-bucket design. *(Fairfield Engineering Co.)*

wide variety of materials. Vehicle types range from the open-dumping kind to the closed type. Most common is the type that unloads by pressure differential into its own pneumatic-conveying system, which is temporarily connected to a storage silo. On this type of truck, the unloading of 18,100 kg (40,000 lb) of products takes about 1 h, sometimes less.

The actual weight that the truck can carry in the United States depends on state-highway load limits, which in turn depend on the net vehicle weight and the number of axles on the truck (and tractor, when a trailer arrangement is used). The accepted maximum combined total weight of vehicle and cargo is 36,200 kg (80,000 lb). In some states, this is reduced slightly, while in others it is exceeded.

Of significance is the rapidly developing containerization system used for package cargo. Table 7-27 gives the principal dimensions of the most common types of ship containers. SeaLand Corp. has developed a patented liner device which can be used to convert a cargo container into a bulk carrier. Figure 7-52 provides dimensions of typical bulk hopper-truck equipment.

Important in the planning for an installation that is to handle rail and highway equipment are the width, length, height, and turning radius of vehicles that will serve the facility. These dimensions can be easily obtained from carriers as well as from equipment manufacturers. Adequate clearances must be provided for railroad and other work crews. The clearances are often specified in state labor-practice codes.

Movement of railcars and trucks within plants is frequently done by carrier crews. Since, however, plant production schedules and availability of railroad switch crews are often not compatible, many plants provide their own switching service. Specially built prime movers that can operate on both roads or rails are available. Front-end loaders can be equipped with couplers to permit car movement. Cable-operated car pullers are now generally in disfavor because of lack of control of cars being moved. Trailers are often moved by tractors especially equipped with an adjustable "fifth-wheel" coupling, which will couple to any trailer regardless of the height of its coupling.

TRANSPORT OF PACKAGED ITEMS

Vehicle Choice Small units such as **bags, boxes, cartons, carboys, cans, and drums** are usually transported in closed van-type vehicles, which may range from small pickup and delivery vehicles of 1400-kg (3000-lb) capacity to trailers capable of holding 23,600 kg (52,000 lb). There has been a trend to higher and wider vehicles, but loading and unloading facilities should be designed to handle not only the newest and largest vehicles but also the older, smaller versions. Figure 7-53 shows typical trailers with their principal dimensions.

TABLE 7-27 Typical-Ship-Container Data*

	Length		Width		Height		Volume		Capacity: Maximum load†		Capacity: Tare weight	
Type	ft	mm	ft	mm	ft	mm	ft^3	m^3	lb	kg	lb	kg
20-ft standard‡												
Out	20	6,096	8	2438	8	2438	1,123	31.8	52,913	24,000	4410	2000
In	19.479	5,935	7.771	2370	7.406	2258						
20-ft high§												
Out	20	6,096	8	2438	8.5	2591	1,197	33.9	52,913	24,000	4585	2080
In	19.479	5,935	7.771	2370	7.813	2383						
40-ft standard												
Out	40	12,192	8	2438	8.5	2591	2,430	68.8	59,500	26,990	7700	3490
In	39.594	12,069	7.781	2373	7.896	2405						
40-ft high												
Out	40	12,192	8	2438	9.5	2895	2,684	76.0	60,400	27,400	6800	3080
In	39.563	12,059	7.688	2344	8.823	2689						

*From Hapag Lloyd.
†This is the maximum load that the container will safely carry. The actual load will usually be less because of road weight limits, which vary from country to country and among states and other political subdivisions. In planning, these limits need to be determined with governmental authorities.
‡Liquid tank containers having these outside dimensions and holding 5055 U.S. gal (19,140 L) are available.
§This type of container is also available for dry bulk cargo.

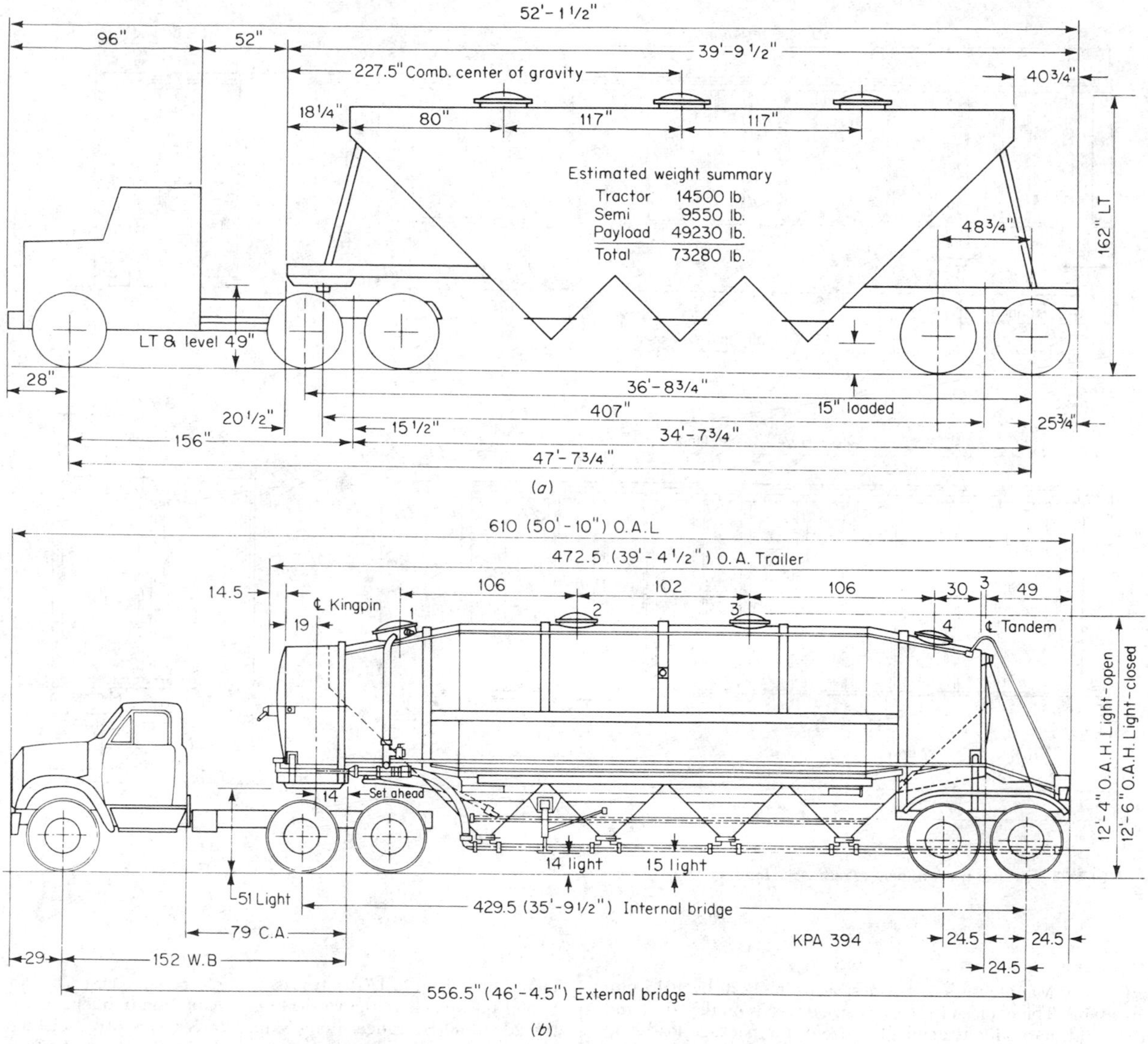

FIG. 7-52 Bulk hopper trucks. (*a*) Typical cement tractor-trailer truck. (*Trailmobile Inc.*) (*b*) Tractor trailer used for plastics. (*Butler Mfg. Co.*) To convert data to the SI system, change the dimensions shown to inches and multiply by 25.4. To convert volume to cubic meters, multiply cubic feet by 0.02832.

The use of closed railroad cars has declined somewhat in the handling of packaged chemical products in favor of trailers hauled piggyback fashion by the railroads. This trailer-on-flatcar approach combines the convenience and flexibility of trucks with the low cost and high speed offered by the railroads. Covered railroad cars for hauling packaged freight include not only standard boxcars but much special equipment offering heating, insulation, refrigeration, high volume for low-density products, and special protection for fragile items. Table 7-26 shows principal dimensions for some of this equipment. Of special note is the so-called damage-free (DF) equipment that provides bulkheads with the car. These form modules within the car to keep the freight from shifting during car movement, thus reducing damage.

Pallets These portable platforms, on which packaged materials can be handled and stored (Fig. 7-54 shows several designs), can be had in a variety of standard sizes and in almost any custom-made size. The dimensions, however, tend to be set by the transportation vehicle in which they will move. The older and most common 2235-mm (88-in) truck width and the 2743-mm (108-in) boxcar width have resulted in a "standard" pallet size of 1065 by 1220 mm (42 by 48 in), which fits two across in a truck (the 1065-mm side) and two across in a boxcar (the 1220-mm side), with adequate clearance for maneuvering the lift truck handling them.

There are several variations of this basic size, including the well-

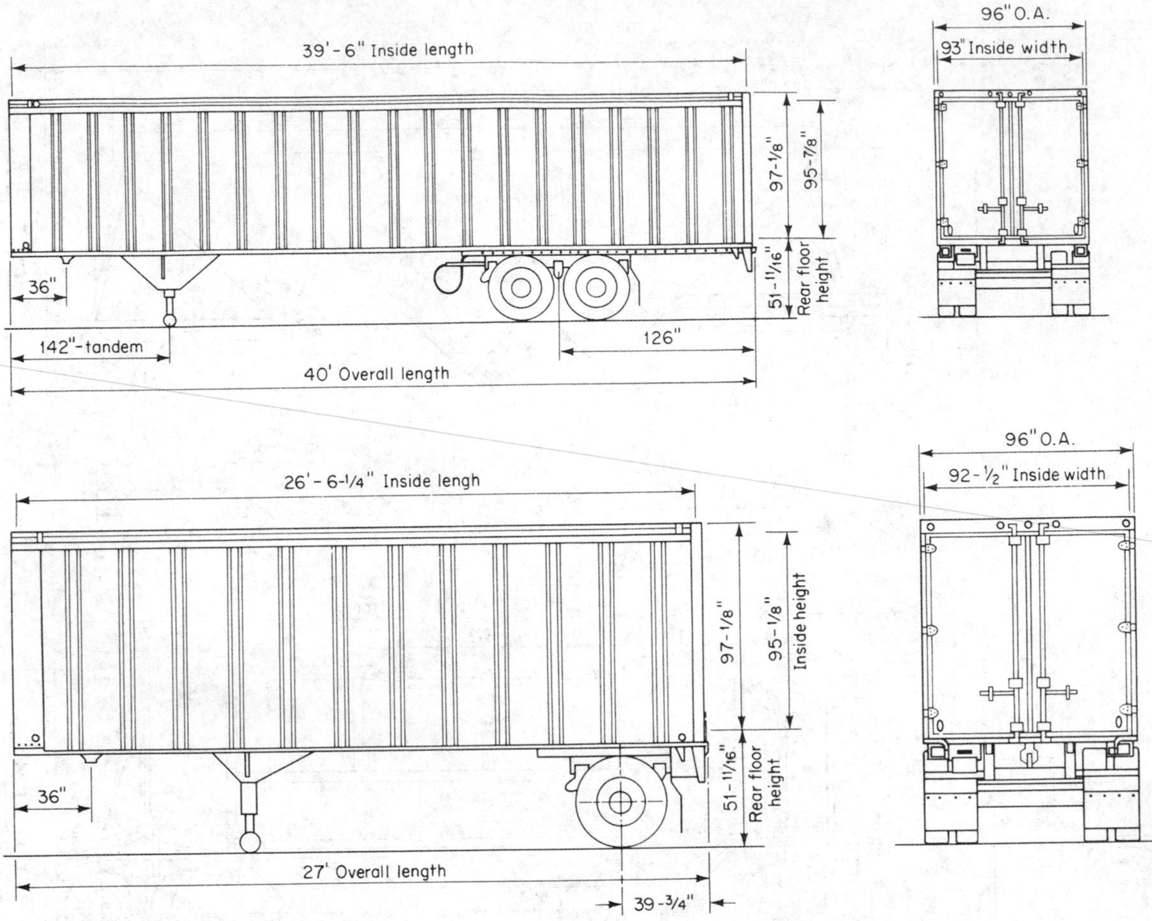

FIG. 7-53 Typical truck trailers. *(Trailmobile Inc.)*

used Grocery Manufacturers of America size of 1220 by 1015 mm (48 by 40 in). The choice of the exact size depends on the truck and boxcar width normally available, the size of the package load, and the customer's receiving and handling facilities. Ideally, the sum of the package dimensions should exactly fit the pallet, but in practice this is virtually impossible. The following rules of thumb are helpful:

For bags: exact pallet dimensions, or up to 13-mm (½-in) overhang on each side

For cartons: pallet dimensions or underhang by 13 mm on each dimension

For drums, cylinders, etc.: pallet dimensions or underhang by as much as 25 mm (1 in)

Pallet patterns to achieve these conditions are numerous. Figure 7-55 shows common patterns used in the chemical-process industries.

The traditional material for pallet construction has been hardwood such as oak, ash, and maple. Yellow pine is also often used. Nails and adhesives are used to join component pieces.

The growing shortage of hardwood has increased the cost of wooden pallets to a point at which plastic pallets and composites of wood, paper, and plastics are economically feasible. Much development work is being done on plastic-pallet design to handle typical loadings. Because of the cost of disposing of expendable pallets, returnable ones are often justified.

Blocking and Bracing of Packaged Loads All transportation vehicles impart significant forces to the packages they contain. Forces of up to 2 G are regularly encountered in rail and ocean shipments, and of up to 1 G in trucks. Some forces are caused by vibration of the vehicle in the vertical plane; vibration frequencies are in the 20- to 40-Hz range. Longitudinal forces caused by starting or braking are of similar magnitude under normal conditions, but severe coupling action or the starting of a long train, through slack action, can cause these forces to reach 6 to 8 G and sometimes higher. To protect packaged freight from damage restraining systems have been developed. There are **energy-absorbing blocking and bracing systems** which can absorb these forces with little damage to the packages. Readily available at low cost, these systems are economically justified by reducing or eliminating repackaging cost and by lowering product loss due to package failure. This subject is treated in detail by J. J. Dempsey in *Methods for Loading, Bracing and Blocking of Packaged Goods in Transportation Equipment* (E. I. du Pont de Nemours & Co., Applied Technology Division, Wilmington, Del. 19898), which may be purchased from Du Pont at nominal cost. **Simulation of transportation systems** on a laboratory scale is used to predict the effectiveness of freight-restraining systems before actual tryout. The effect of a system on controlling damage at various levels of impact and vibration can be determined quickly and at low cost. As a result, the risk of substantial product loss during initial trials of a new system is reduced significantly. This service is offered by several firms and by the Center for Packaging Engineering, Shock and Vibration Laboratory, Rutgers, The State University of New Jersey, Piscataway, New Jersey.

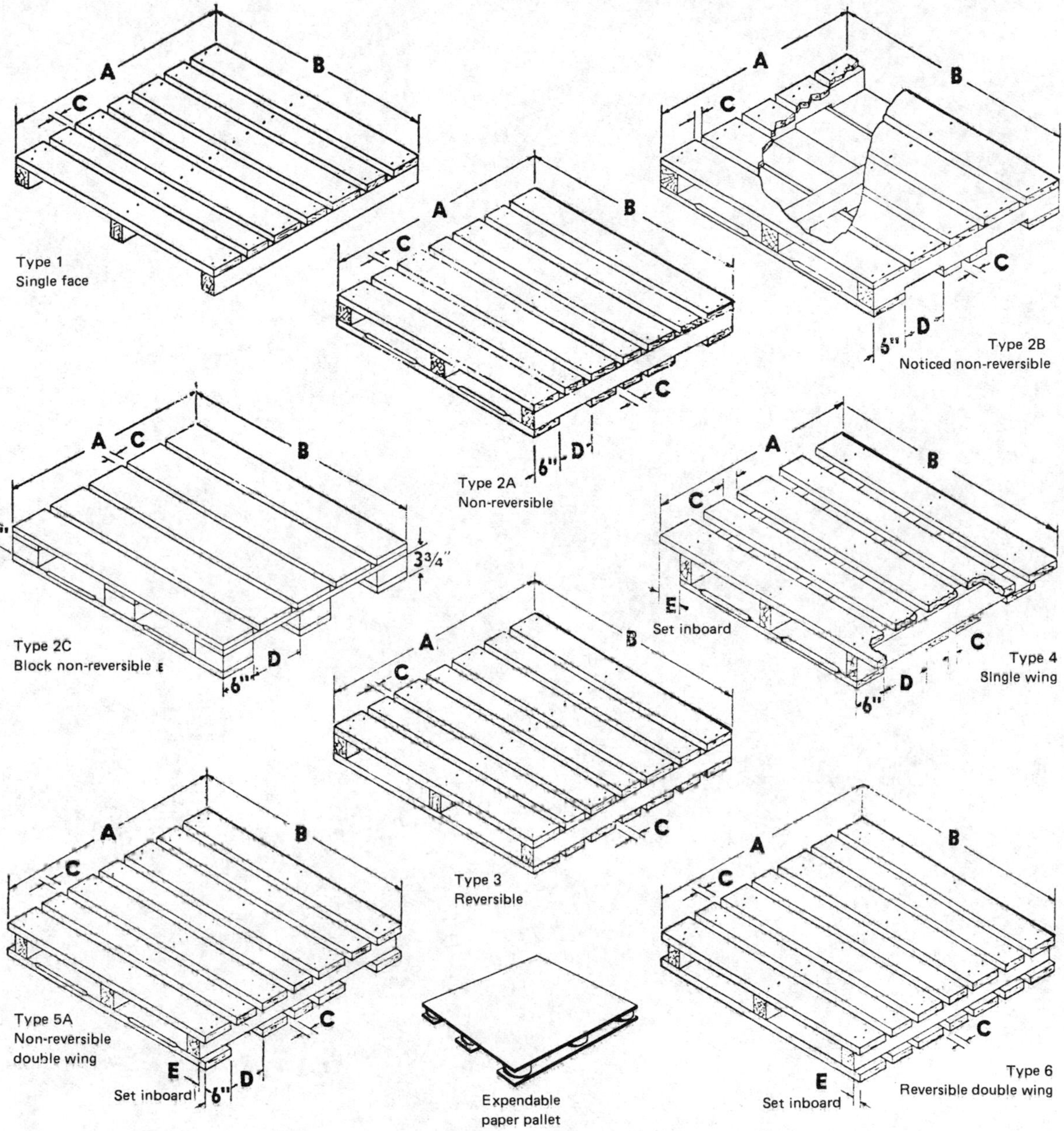

FIG. 7-54 Types of pallets. Pallets designated as types 1 through 6 are standard designs based on nomenclature and data of the National Wooden Pallet and Container Association, Washington, D.C., 20036. The expendable paper pallet is made of corrugated kraft paper.

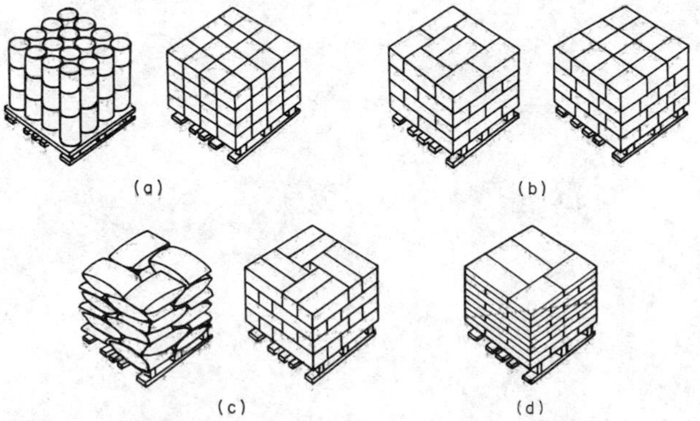

FIG. 7-55 Typical pallet patterns. (*a*) Block pattern is commonly used, although it is often unstable. It may be made more secure by encircling the top tier of containers with wire or strapping. (*b*) Brick pattern is the most commonly used. Containers are interlocked to make a relatively stable load by placing alternate tiers at a 90° position to each other. (*c*) Pinwheel pattern is used when the brick pattern is found to be unstable. Alternate tiers can be interlocked. (*d*) Three-to-two interlocking pattern is used extensively for bagged products. All pallet loads benefit from load-securing systems such as stretch wrap, shrink wrap, or palletizing adhesive.

Section 8

Size Reduction and Size Enlargement

Richard H. Snow, Ph.D., *Director, National Institute for Petroleum and Energy Research; Member, American Chemical Society, American Institute of Mining, Metallurgical and Petroleum Engineers, Sigma Xi; Fellow, American Institute of Chemical Engineers. (Section Editor)*

Brian H. Kaye, Ph.D, *Professor of Physics and Director of Institute for Fine Particles Research, Laurentian University; Member, American Institute of Physics, Sigma Xi. (Particle-Size Analysis)*

C. Edward Capes, Ph.D, *Senior Research Officer and Head, Chemical Engineering Section, National Research Council, Ottawa, Canada; Member, Canadian Society of Chemical Engineering, American Institute of Mining, Metallurgical and Petroleum Engineers; Fellow, Chemical Institute of Canada. (Size Enlargement)*

Guggilam C. Sresty, M.S., *Senior Engineer, IIT Research Institute; Member, American Institute of Chemical Engineers, American Institute of Mining, Metallurgical and Petroleum Engineers. (Crushing and Grinding Equipment)*

PARTICLE-SIZE ANALYSIS

PRINCIPLES OF SIZE REDUCTION

CRUSHING AND GRINDING EQUIPMENT

CRUSHING AND GRINDING PRACTICE

SIZE ENLARGEMENT

Nomenclature and Units

Symbol	Definition	SI units	U.S. customary units
A	Coefficient in double Schumann equation		
a	Constant		
$a_{k,k}$	Coefficient in mill equations		
$a_{k,n}$	Coefficient in mill equations		
B	Matrix of breakage function		
$\Delta B_{k,u}$	Breakage function		
b	Constant		
C	Constant		
C_s	Impact-crushing resistance	kWh/cm	(ft·lb)/in
D	Mill diameter	m	ft
D_b	Ball or rod diameter	cm	in
D_{mill}	Diameter of mill	m	ft
d	Differential		
d	Distance between rolls of crusher	cm	in
E	Work done in size reduction	kWh	hp·h
E	Energy input to mill	kW	hp
E_i	Bond work index	kWh/Mg	(hp·h)/ton
E_i	Work index of mill feed		
E_2	Net power input to laboratory mill	kW	hp
erf	Normal probability function		
F	As subscript, referring to feed stream		
F	Bonding force	kg/kg	lb/lb
g	Acceleration due to gravity	cm/s^2	ft/s^2
I	Unit matrix in mill equations		
i	Tensile strength of agglomerates	kg/cm^2	lb/in^2
K	Constant		
k	Parameter in size-distribution equations	cm	in
k	As subscript, referring to size of particles in mill and classifier parameters		
L	As subscript, referring to discharge from a mill or classifier		
L	Length of rolls	cm	in
L	Inside length of tumbling mill	m	ft
M	Mill matrix in mill equations		
m	Dimensionless parameter in size-distribution equations		
N	Mean-coordination number		
N_c	Critical speed of mill	r/min	r/min
ΔN	Incremental number of particles in size-distribution equation		
n	Dimensionless parameter in size-distribution equations		
n	Constant, general		
n_r	Percent critical speed of mill		
O	As subscript, referring inlet to stream		
P	As subscript, referring to product stream		
P_k	Fraction of particles coarser than a given sieve opening		
p	Number of short-time intervals in mill equations		
Q	Capacity of roll crusher	cm^3/min	ft^3/min
q	Total mass throughput of a mill	g/s	lb/s
g_c	Coarse-fraction mass flow rate	g/s	lb/s
q_F	Mass flow rate of fresh material to mill	g/s	lb/s
q_f	Fine-fractiom mass flow rate	g/s	lb/s
q_o	Feed mass flow rate	g/s	lb/s
q_p	Mass flow rate of classifier product	g/s	lb/s
q_R	Mass flow rate of classifier tailings	g/s	lb/s
q_R	Recycle mass flow rate to a mill	g/s	lb/s
R	Recycle		
R	Reid solution		
r	Dimensionless parameter in size-distribution equations		
S	Rate function	s^{-1}	s^{-1}
$\overline{S}$	Corrected rate function	s^{-1}	s^{-1}
S′	Matrix of rate function	Mg/kWh	ton/(hp·h)
$S_G(X)$	Grindability function	s^{-1}	s^{-1}
S_u	Grinding-rate function		
s	Parameter in size-distribution equations		
s	Peripheral speed of rolls	cm/min	in/min
t	Time	s	s
u	Settling velocity of particles	cm/s	ft/s
W	Vector of differential size distribution of a stream		
w_k	Weight fraction retained on each screen		
w_u	Weight fraction of upper-size particles		
w_t	Material holdup in mill	g	lb
X	Particle size or sieve size	cm	in
X'	Parameter in size-distribution equations	cm	in
ΔX_i	Particle-size interval	cm	in
X_i	Midpoint of particle-size interval, ΔX_i	cm	in
X_0	Constant, for classifier design		
X_f	Feed-particle size	cm	in
X_m	Mean size of increment in size-distribution equations	cm	in
X_p	Product-particle size	cm	in
X_p	Size of coarser feed to mill	cm	in
X_{25}	Particle size corresponding to 25 percent classifier-selectivity value	cm	in
X_{50}	Particle size corresponding to 50 percent classifier-selectivity value	cm	in
X_{75}	Particle size corresponding to 75 percent classifier-selectivity value	cm	in
ΔX_k	Difference between opening of successive screens	cm	in
x	Weight fraction of liquid		
Y	Cumulative fraction by weight undersize in size-distribution equations		
Y	Cumulative fraction by weight undersize or oversize in classifier equations		
Y	Fraction of particles between two sieve sizes		
ΔY	Incremental weight of particles in size-distribution equations	g	lb
ΔY_{ci}	Cumulative size-distribution intervals of coarse fractions	cm	in
ΔY_{fi}	Cumulative size-distribution intervals of fine fractions	cm	in
Z	Matrix of exponentials		

Nomenclature and Units (*Continued*)

Symbol	Definition	SI units	U.S. customary units
	Greek symbols		
β	Sharpness index of a classifier		
δ	Angle of contact	rad	°
ϵ	Volume fraction of void space		
Z	Residence time in the mill	s	s
η_x	Size-selectivity parameter		
μ	Viscosity of fluid	$(N \cdot S)/m^2$	P
ρ_f	Density of fluid	g/cm^3	lb/in^3
ρ_ℓ	Density of liquid	g/cm^3	lb/in^3
ρ_s	Density of solid	g/cm^3	lb/in^3
Σ	Summation		
σ	Standard deviation		
σ	Surface tension	N/cm	dyn/cm

PARTICLE-SIZE ANALYSIS

GENERAL REFERENCES: Beddow, *Particulate Science and Technology,* Chemical Publishing, New York, 1980. Cadle, *Particle Size,* Reinhold, New York, 1965. Chamot and Mason, *Handbook of Chemical Microscopy,* vol. 1, 3d ed., Wiley, New York, 1958. Herdan, *Small Particle Statistics,* 2d ed., Butterworth, London, 1960. Kaye, *Direct Characterization of Fineparticles,* Wiley, New York, 1981. Klug and Alexander, *X-Ray Diffraction Procedures for Polycrystalline and Amorphous Materials,* Wiley, New York, 1954. Orr and Dallavalle, *Fine Particle Measurement,* Macmillan, New York, 1959. *Particle Size Analysis,* symposium, Institution of Chemical Engineers, London, 1947. *Particle Size Analysis,* conference, Loughborough, England, 1966, publ. British Society of Analytical Chemists, London, 1967. Stevens, *Microphotography,* Wiley, New York, 1957. Van de Hulst, *Light Scattering by Small Particles,* Wiley, New York, 1957. Whitby and McFarlans, *Bibliography of Particle Size Analysis,* 1111 references, University of Minnesota, Minneapolis, 1959.

PARTICLE-SIZE DISTRIBUTION

Specifications for Particulates Feed to and finished products from size-reduction operations are defined in terms of the sizes involved. It is also well to know whether the ultimate individual particle is being measured, or, if any aggregation or agglomeration of particles exists, whether this has been created by the size-reduction operation.

The fullest description of a powder is given by its **particle-size distribution.** This can be plotted in terms of cumulative percent oversize or undersize in relation to the diameters of particles, or it can be plotted as a distribution of the amounts present in each unit of diameter against the several diameters. It is common to employ a weight basis for percentage, but there are some data in the literature in which frequency, or number of particles, is used. The basis of percentage, whether weight, frequency, or some less commonly used factor, should be specified, as should also be stated the diameter, its units, and preferably whether it is determined by sieve, settling velocity, or otherwise.

Figure 8-1 presents two sets of distributions, one cumulative and the other in unit intervals. The slopes of the 5-μm intervals of the cumulative curves are converted to percent per micrometer and plotted as a block, or histogram, from which smooth curves are derived. Powder *A* has a narrower, or tighter, size range for the bulk of its weight than powder *B.* Both materials have the same weights below and above the size marked by the arrow.

Simpler treatments of distribution are possible. In some cases the significant value is the top size. Since the 100 percent point is dubious, some amount, such as 95 or 98 percent, so specified, can be called top size. In other cases the percent on some sieve helps define coarseness. Merely stating that all or all but a small percentage passes a given sieve is inadequate for defining the true fineness of a material. Complete particle-size analysis to show distribution is essential for most comparisons and calculations.

Particle-Size Equations A number of equations have been proposed to correlate the quantity of a particulate material with its particle size to obtain a distribution relationship [Harris, *Trans. Am. Inst. Min. Metall. Pet. Eng.,* **241,** 343–357 (1968); Fagerholt, *Particle Size Distribution of Products Ground in Tube Mill,* Gads Forlag, Copenhagen, 1945]. In the literature it is often assumed that a powder must follow some distribution, such as the **Rosin-Rammler-Bennett** [Rosin and Rammler, *J. Inst. Fuel,* **7,** 29–36 (1933); Bennett, ibid., **10,** 22–39 (1936)]:

$$Y = 1 - [\exp - (X/X')^n] \tag{8-1}$$

or the **Gates-Gaudin-Schumann** distribution [Schumann, *Am. Inst. Min. Metall. Pet. Eng.,* Tech. Pap. 1189, *Min. Technol.* (1940)]:

$$Y = (X/k)^m \tag{8-2}$$

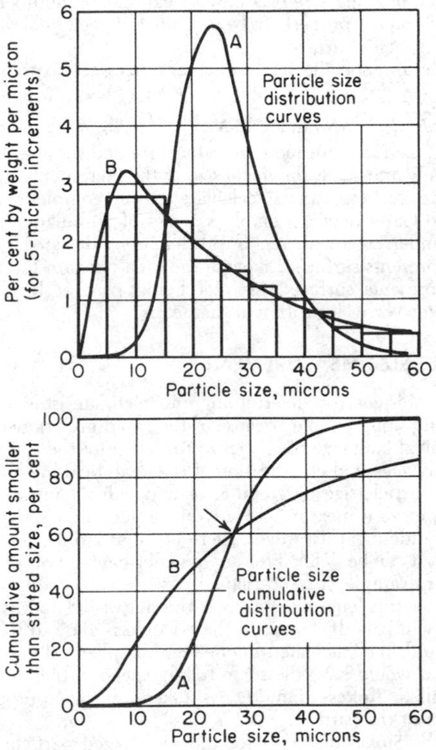

FIG. 8-1 Particle-size distribution curves for simple powders.

or the **logarithmic-probability** distribution [Hatch and Choate, *J. Franklin Inst.,* **207,** 369 (1929)]:

$$Y = \operatorname{erf}\left(\frac{\ln X/X'}{\sigma}\right) \tag{8-3}$$

or the **Gaudin-Meloy** distribution [Gaudin and Meloy, *Trans. Am. Inst. Min. Metall. Pet. Eng.,* **223,** 40–50 (1962)]:

$$Y = 1 - \left(1 - \frac{X}{X'}\right)^r \tag{8-4}$$

where Y = cumulative fraction by weight undersize; X = size; k, X' = parameters with dimension of size; m, n, r = dimensionless exponents; erf = normal probability function; and σ = standard-deviation parameter.

There is no fundamental reason why a particular powder must obey one of these empirical laws; forcing it to do so will result in error. Furthermore, it is difficult to tell whether the fit is good because *any* cumulative-size plot will give the appearance of a good fit; random numbers appear to fit the Rosin-Rammler curve 9 times out of 10 [Kaye, *Staub* (March 1964)]. Differential plots show size deviations more clearly [Kottler, *J. Franklin Inst.,* **250,** 399, 419, 499 (1950); **251,** 617 (1951)].

In special cases a few size-data points can be plotted and the rest of the curve assumed to follow trends previously established. This may be done in testing the application of a mill to a particular mineral when many previous runs have been made for similar materials and conditions.

Several of these laws are useful simply for curve-fitting purposes. The Rosin-Rammler can represent a distribution with a peak in the differential curve; the Gates-Gaudin-Schumann has the advantage of simplicity; and the Gaudin-Meloy has the advantage that it can fit a variety of curves found in practice. Size data may also be presented in tabular form, thus avoiding the problem of curve fitting.

Average particle size of a powder can have a number of values, depending on the property to be accentuated: weight or volume, surface, and specific surface.

Volume and specific surface average sizes are expressed respectively as $\sqrt[3]{\Sigma\ \Delta Y/\Sigma Y X_m^{-3}}$ and $\Sigma\ \Delta Y/\Sigma\ \Delta Y X_m^{-1}$ on a weight basis and as $\sqrt[3]{\Sigma\ \Delta N X_m^3/\Sigma\ \Delta N}$ and $\Sigma\ \Delta N X_m^3/\Sigma\ \Delta N X_m^2$ on a number basis, where ΔY and ΔN are incremental weight and number of particles respectively and X_m is the mean size of the increment.

Specific surface can be calculated from complete distribution data. The Gates diagram employs a plot of cumulative percent by weight undersize versus reciprocal diameter; the area beneath the curve represents surface. Likewise the area beneath the Roller diagram represents surface, this plot being percent by weight per micrometer versus logarithm of diameter.

PARTICLE-SIZE MEASUREMENT

Many techniques for determining the characteristics of powders have been published, but because of the rapid growth of the subject the technical language often lacks precision and there is not even a universally accepted classification of the analytical methods.

If the particle-size distribution of a powder composed of hard, smooth spheres is measured by any of the techniques, the measured values are identical. However, there are many different size distributions that can be defined for any powder having nonspherical particles. For example, if a rod-shaped particle is placed on a sieve, its diameter, not its length, determines the size of the aperture through which it will pass. If, however, the particle is allowed to settle in a viscous fluid, the calculated diameter of a sphere of the same substance that would have the same falling speed in the same fluid, or the so-called **Stokes diameter,** is taken as the appropriate size parameter of the particle.

Since the Stokes diameter for the rod-shaped particle will obviously differ from the true diameter of the rod, this difference represents added information concerning the particle shape. The fact that all measured distributions for spherical particles are identical represents not the ideal case but the degenerate simple case.

The term **shape** is used in two senses: (1) to denote the spatial configuration of an individual particle or (2) as a factor to correlate the mean particle sizes of a fine-particle system as measured by two methods based on different physical principles.

Sampling of Powders An essential prerequisite to accurate particle-size analysis is proper powder sampling. Kaye (op. cit., 1981) has reviewed the relative efficiency of various devices. An efficient sampling device in widespread use is the spinning riffler shown in Fig. 8-2. In this device a ring of containers rotates under the powder feed. If the powder flows for a long time compared with the period

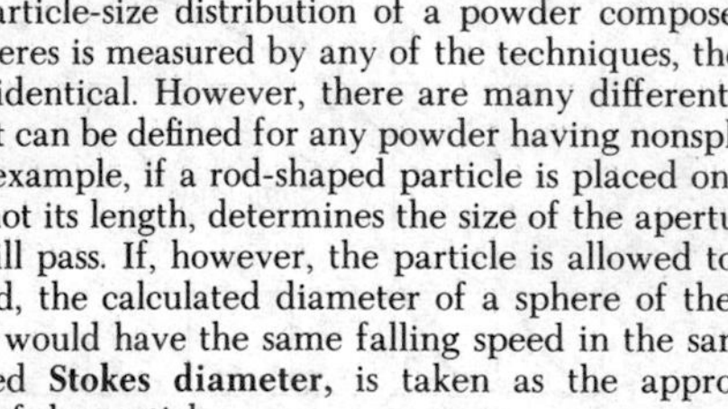

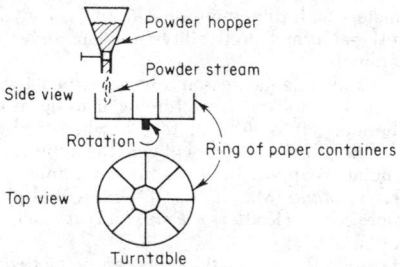

FIG. 8-2 Spinning-riffler sampling device.

of one revolution, the sample in each container will be made up of many small portions drawn from all points in the bulk. The speed of rotation must be kept low enough to avoid creating air currents that would blow the fines away from the cups. A spinning riffler with 20 cups divides a sample to 1/8000 of the original powder volume in only three passes.

Microscope Methods In microscope methods of size analysis, direct measurements are made on enlarged images of the particles. In the simplest technique, linear measurements of particles are made by using a scale placed on top of the particle image. For all shapes of particles except spherical, it is necessary to define precisely the measurement made on the profile. Some dimensions are defined in Fig. 8-3.

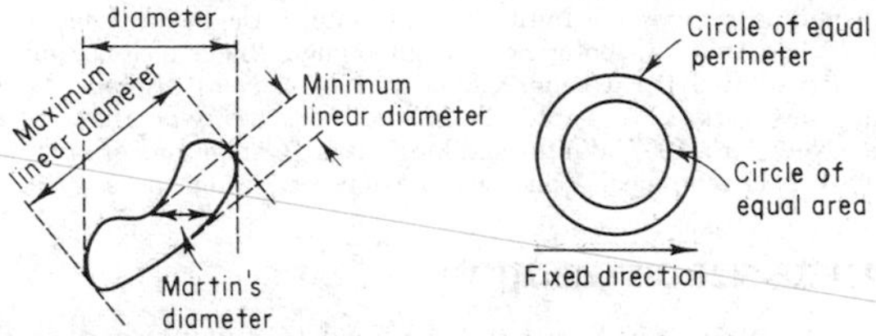

FIG. 8-3 Various dimensions of an irregular particle.

The term **diameter** can be properly used to describe any linear intercept through a profile of any shape.

The two major problems in microscope methods of analysis are the collection of sufficient data to ensure adequate precision in the derived parameters and the elimination of variables in the data due to operator performance. Statistical fluctuations in the occurrence of any given size profile (among the particle images) is a problem that becomes particularly acute if the size ratio of particles is more than 10:1 (Kaye, op. cit., 1981). A time-consuming feature is that the scale must be placed on top of the image and then oriented twice to measure maximum and minimum diameters. It is quicker to measure parameters that are independent of orientation, such as the diameters of the circles of equal area and equal perimeter, shown in Fig. 8-3. Data-gathering speed can also be increased by providing a set of comparison circles, but this method introduces operator bias.

The Zeiss Endter equipment for measuring circles of equal area (when photomicrographs are available) is effective in reducing the subjective element [Endter and Gebauer, *Optik,* **97,** 13 (1956)].

A second technique for speeding up information gathering is to measure so-called **statistical diameters** rather than physical diameters of the image profile. **Feret's diameter** (see Fig. 8-3) is the perpendicular projection, onto a fixed direction, of the tangents to the extremities of the particle profile. **Martin's diameter** is the line, parallel to the fixed direction, that divides the particle profile into two equal areas. Since the magnitude of the statistical diameters varies with orientation for a particular particle, these diameters have meaning only when a sufficient number of measurements is averaged. The problem in using these diameters is that frequently the technologist does not know what "sufficient" means in this context (Loughborough Conference, loc. cit.).

The falling prices of electrooptic devices and minicomputers led to the development of many automated devices for measuring the dimensions of fine-particle images. (For a list of suppliers see Kaye, op. cit., 1981.) The speed and sophistication of such devices made it possible to devise new methods for characterizing the shape of fine particles. It is convenient to split the new shape methods into two groups. The first group can be called **Fourier methods,** and the second group can be called **fractal methods.** In Fourier techniques one seeks to transform the shape characteristics of a fine particle into a signature waveform. Thus in the basic method employed by Beddow and coworkers the centroid of a profile is taken as a reference point. A vector is then rotated about this centroid with the tip of the vector touching the periphery. A plot of the magnitude of the vector against angular position is a wave-type function. This waveform is then subjected to Fourier analysis. It has been shown that the lower-frequency harmonics constituting the complex wave correspond to the gross external morphology of the fine particle whereas the higher frequencies correspond to the texture of the fine particle. (J. K. Beddow, *Particulate Science and Technology,* Chemical Publishing, New York, 1980.)

Fractal logic was introduced into fine-particle science by Kaye and coworkers, who have shown that the noneuclidean logic of Mandelbrot can be applied to a description of the ruggedness of a profile. It has been shown that a combination of the use of fractal dimensions and geometric shape factors such as aspect ratio (length divided by diameter) or elongation can be used to describe a population of fine particles of various shapes and that these can be related to the functional properties of the particles (Kaye, op. cit., 1981).

Sedimentation Methods In sedimentation methods of particle-size analysis, the Stokes diameter distribution of the powder is deduced from a study of concentration changes occurring within a settling suspension. The method is based on Stokes' law,

$$D = \sqrt{\frac{18\mu u \times 10^8}{(\rho_s - \rho_f)g}} \tag{8-5}$$

where μ is viscosity, P; u is velocity, cm/s; ρ_s is density of solid particle, g/cm^3; ρ_f is density of fluid; D is diameter of a sphere, μm; and g is acceleration due to gravity, cm/s^2.

The Stokes equation theoretically is valid only for a sphere. However, the difference between the volume of the irregular particle and the volume of the equivalent sphere does not represent an "error" but represents useful information on the shape of the particle. The more irregular the particle, the smaller the ratio of these volumes.

An experimental problem is to obtain adequate dispersion of the particles before sedimentation analysis. A good general rule is to disperse at least as severely as the powder will be dispersed in subsequent processing.

Equations to calculate size distributions from sedimentation data are usually based on the assumption that the particles fall freely in the suspension. At many concentrations used in sedimentation methods of analysis, interaction between the falling particles can cause the size of the particles to be overestimated.

In the **pipet method** (Fig. 8-4), concentration changes within the suspension are monitored by taking samples from the body of the suspension by using a pipet. The particle-size distribution is calculated from the measured concentration changes, the sample being assumed to be taken from region *D* in Fig. 8-4*a*.

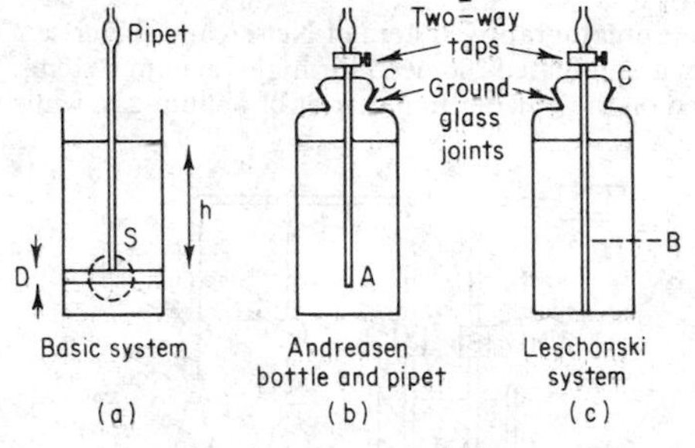

FIG. 8-4 Equipment used in the pipet method of size analysis.

The simplest equipment using this technique [Andreasen, *Kolloid-Z.,* **49,** 253 (1929)] is shown in Fig. 8-4*b*. When the suspension has begun to settle, a series of samples is withdrawn at specified time intervals. The main objections to this system are as follows. First, the withdrawal of a sample may affect the behavior of the settling suspension otherwise than by altering the distance of fall to the sampling zone. Second, the presence of the pipet in the system disturbs the settling behavior of the suspensions. The actual sampling zone is *S* in Fig. 8-4*a*, a sphere centered on the tip of the pipet. Particles below *A* are not replaced from above, the density becomes less in this region, and turbulent convection currents arise under the tip of the pipet. To overcome the density convection currents, Leschonski's apparatus (Fig. 8-4*c*) extends the pipet to the base of the vessel. Suspension is sucked into holes at *B*, arranged around the circumference of the stem. The sampling zone is a better approximation to the theoretical zone *D*.

The precision of a pipet method of analysis is influenced by the shape and volume of the pipet, the ratio of the diameter of the sampling sphere to the depth at which measurements are made, and the rate at which samples are withdrawn. These factors should be closely controlled. If the concentration of the samples is determined by drying the solids and weighing, care must be taken to allow for the weight of surface-active agents present.

In **hydrometer methods** of particle-size analysis small hydrometers are used to follow density gradients with a suspension. This method is widely used in soil-science studies. It is not of high accuracy but costs little and usually gives sufficient information for the purposes at hand (ASTM Spec. Pub. 234, 1959).

In **photosedimentation methods,** changes of concentration are measured by passing a beam of light through the suspension. Optical methods have high sensitivity, and only a small sample is required. At the low concentrations used, the particles are sufficiently far apart for free-fall conditions to apply. The resulting measurements are obtained as electric signals in a form suitable for automatic recording. Only if the particles are as small as the wavelength of light does a difficulty arise because the simple laws of geometric optics cannot be used to calculate the opacity of such a suspension.

Light-Diffraction Methods In laser techniques a suspension of the particles to be characterized is placed in a laser beam. The energy distribution in the complex diffraction pattern generated is analyzed by using a computer. Several assumptions are made in the transformation of the diffraction pattern into particle-size data, and the various companies marketing machines of this kind offer different interpretive programs, depending upon the type of distribution anticipated in the system being studied. Instruments of this kind include the Microtrac instrument manufactured by Leeds & Northrup Co., the CILAS granulometer, and the Malvern particle-size analyzer.

The laser has also generated a whole group of methods in which the velocity of moving particles is measured by examining the Doppler shift in the reflected light leaving the fine particles [Doyle, Thompson, and Stevenson (eds.), *Laser Velocimetry and Particle Sizing,* Hemisphere Publishing, Washington, 1979]. Doppler methods have been used to study the internal dynamics of cyclones (Stenhouse, "Particle Trajectories in Uniflow Cyclones," Second World Filtration Congress, 1979; Mazumder et al., "Realtime Measurement of Stack Emissions," application of SPART analyzer, in Doyle, Thompson, and Stevenson, op. cit.).

Sedimentation-Balance Methods In sedimentation-balance methods the weight of the sedimenting fine particles is measured as they accumulate on a balance pan hung within an initially homogeneous suspension. The techniques are relatively inexpensive but tend to be slow because of the time required for the smallest fine particles to settle out over a given column height.

Centrifugal Methods For the analysis of very fine particles, centrifugal sedimentation techniques are used. **Disk centrifuges** have an advantage over the tube type in that the particles settle radially, eliminating errors due to impingement on the walls of the tubes.

The equipment developed by Slater and Cohen (Fig. 8-5) is essentially a **centrifugal pipet** device. Size distribution is calculated from the measured solids concentrations of a series of samples withdrawn through the central drainage pillar at various time intervals. The calculations are complicated because of the different initial accelerations of particles starting at different distances from the center of rotation.

In the **ICI centrifugal sedimentometer,** the suspension under test is injected into clear fluid in the spinning disk through the entry port, and a layer of suspension is formed over the free surface of the liquid. Samples are extracted after given amounts of centrifuging by

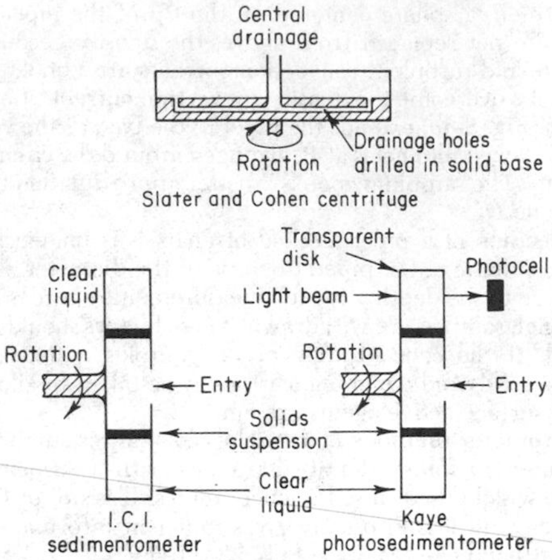

FIG. 8-5 Basic operation of a disk-centrifuge system.

placing a scoop through the entry port shown in Fig. 8-5. Stability problems of the liquid system can be critical with this instrument, since a relatively high solids concentration must be injected to obtain a sufficient sample.

The **centrifugal disk photosedimentometer,** although developed independently by Kaye and coworkers, is similar to the ICI device. Concentration changes are monitored by a light beam. This apparatus offers good resolution of information; after one analysis is complete, a second sample can be injected without stopping and cleaning the disk. Very low solids concentrations can be used, but the light-scattering properties of small particles make it difficult to interpret concentration changes. (For a fuller discussion of these instruments see Kaye, op. cit., 1981.)

Stream Methods In these techniques, the particles to be measured are examined individually in a stream of fluid. As the fluid passes through a sensing zone, the presence of particles is detected by the perturbation they cause. The **sensing zone** is variously monitored by using light beams, ultrasonic waves, and electrical resistance measurements. A widely used stream method is the Coulter counter [Kubitschek, *Research*, **13**, 129 (1960)]. In this instrument the size of the particle is deduced from the resistance change in a column of electrolyte as the particle passes through the column.

It is essential in stream methods to use very low particle concentrations because the signal received from two small particles is indistinguishable from that coming from a single larger particle. This is called a coincidence error.

Recently stream methods based upon the measurement of electrical resistance have been modified to allow the measurement of length, and the resolution and accuracy of the instruments have been improved by improving designs for the orifice through which the particles pass. The HIAC instrument is a widely used device for measuring the size of particles in the stream of fluid moving past a light beam.

Sieving Methods Sieving is probably the most frequently used and abused method of analysis because the equipment, analytical procedure, and basic concepts are deceptively simple. In sieving, the particles are presented to equal-sized apertures that constitute a series of go–no-go gauges. Sieve analysis presents three major difficulties: (1) In commercial sieves not all the apertures are identical although the highest-quality sieves minimize dimension deviations. (2) The sieving surfaces are easily damaged in use. (3) The particles must be efficiently presented to the aperture of the sieve.

In the past the lower limit for sieve analysis was usually set at 43 μm (325 mesh) because this was the practical limit for the manufacture of woven wire-mesh sieves. Also, at about this size the surface forces operating within the powder system made it very difficult to move the powders through the apertures.

More recently the lower limit on aperture size has been removed by the introduction of electroformed sieves. The deviations of the actual aperture size are of the order of 2 μm from the nominal size. The finer sizes (below 40 μm) have to be supported on coarser grids. With these finer sieves, traditional sieving methods are not possible. The Alpine air-swept sieve can be used. In this device a rotating jet below the sieving surface cleans the apertures. The return air helps to push the fines through the apertures.

In an analytical procedure called **felvation,** the particles are fluidized in a compartment below the sieving surface. The flow rate of fluid through the powder is gradually increased until fine particles are elutriated past the sieving surface. The flow rate of fluid is then further increased until particles just not able to pass through the sieve are forced against the apertures. Sieving is complete when the fluid above the sieving surface becomes clear [Kaye and Jackson, *Powder Technol.*, **1,** 43 (1967)].

Elutriation Methods In the simplest type of vertical-gravity elutriator the fluid tends to carry the particles up through a column, but this motion is opposed by gravity so that only particles smaller than a certain critical size are swept out. Terminal velocities are given in Sec. 5: "Particle Dynamics." However, the upward fluid flow is never uniform; hence the separation is very crude. For example, eddies occur. Also, since the velocity profile across the moving fluid is parabolic, larger particles can be supported in the middle of the column than at the sides. If gas is used, electrostatic charging of the particles can be a problem. An improved design that minimizes these problems is described by Leschoswki and Rumpf [*Powder Technol.*, **2**(3), 175–185 (1969)]. It uses a porous plate to establish a uniform-fluid-flow profile and a decreasing tube diameter to accelerate the transport of fines once they have been classified from the bed.

A *cyclone* is a centrifugal elutriator, though not usually so regarded (see Sec. 20: "Dust Collection Equipment"). An ingenious modification to the traditional cyclone is termed a cyclosizer [Kelsall and McAdam, *Trans. Inst. Chem. Eng. (London)*, **41,** 84 (1963)]. By inverting the cyclone and adding an apex chamber, fine particles that theoretically should be swept out of the cyclone are recycled, corresponding to successive passage of the particles through many similar cyclones.

Surface Area from Gas-Adsorption Measurements If the quantity of gas required to cover a powder with a complete monolayer of gas molecules is measured, the surface area can be calculated by using the cross-sectional area of the gas molecules. The surface area measured by gas-adsorption techniques is directly related to the surface area available in chemical reactions. Outgassing of the powder before analysis should be conducted very carefully to ensure reproducibility. Many theories on gaseous adsorption have been advanced, but measurements on powders are usually interpreted by using the BET theory [Brunauer, Emmett, and Teller, *J. Am. Chem. Soc.*, **60,** 309 (1938)].

A **gas chromatography** system of Nelsen and Eggertsen, shown in Fig. 8-6, has eliminated the need for high-vacuum systems. Nitrogen is adsorbed on the powder in a carrier of helium gas, while the pow-

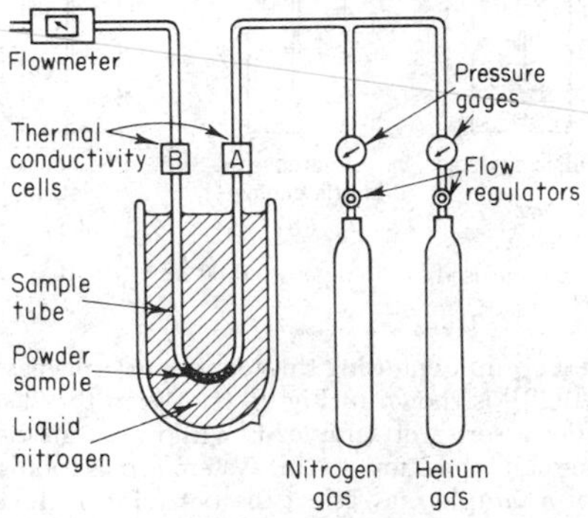

FIG. 8-6 Apparatus for measuring the surface area of a powder. Conductivity cells determine the percentage of nitrogen present in the gas stream as the stream enters and leaves the sample tube.

der is in a container surrounded by liquid nitrogen. When the coolant is removed, desorption occurs. The concentration of nitrogen is measured by a conductivity cell. Preliminary purging is important.

Permeability Techniques Size characteristics can be inferred from the resistance offered to the flow of a fluid through a pressed plug of powder. The Kozeny-Carman equation is widely used to calculate an average particle size or to estimate the surface area of the powder. Because of the very simple model which is postulated as a basis for the development of the Kozeny-Carman equation, there is really no reason why surface areas deduced from permeability methods should agree with surface areas measured by other analytical techniques. A source of confusion is that the values of the surface area calculated by using the Kozeny-Carman equation should theoretically be independent of the absolute porosity of the powder plug on which the measurements are made. In practice, the measured value is a function of the porosity, since as the pore structure collapses, the distribution of pore sizes changes and the effective tortuous path is different at each porosity. Empirically, for most powders the effect decreases after a certain limiting porosity is achieved, provided that the pressure applied in consolidating the particles does not crush or deform them. It is good practice to determine the relationship between the porosity and the measured surface area for each powder. Experimental procedures should then be standardized to use porosities in the range in which changes least affect the result.

A new type of permeability equipment in which the traditional rigid permeability cell is replaced by a rubber-membrane-lined cell has been described by Kaye and Legault [*Powder Technol.*, **23**, 179–186 (1979)]. The powder to be characterized is placed in this soft-walled cell and then compressed by using hydrostatic compression. This results in a more uniform plug and facilitates automation of the technique.

Online Procedures The growing trend toward automation in industry has resulted in many studies of very rapid procedures for generating size information so that feedback loops can be instituted as an integral part of size-reduction processes. Many of these techniques are modifications of more traditional methods. An excellent discussion of the various methods for on-stream characterization is to be found in a publication by the Engineering Foundation [Herbst and Sastry (eds.), *On-Stream Characterization and Control of Particulate Processes*, Engineering Foundation, New York, N.Y. 10017, 1978].

Davies [*Am. Lab.*, 97–110 (April 1978)] reviewed the various methods being used or tried. A Coulter counter with 16 channels can yield an analysis within 90 s. Various laser-diffraction devices such as the Microtac can be obtained in a form to sample and analyze rapidly flowing-liquid suspensions such as a taconite-ore stream. The Cyclosensor of Kelsall and Restarick automatically measures the amounts of particles above and below the cut size. Piezoelectric balances are being used to measure the sizes of dust particles in air as they impact on a sensor.

PRINCIPLES OF SIZE REDUCTION

GENERAL REFERENCES: Annual reviews of size reduction, *Ind. Eng. Chem.*, October or November issues, by Work from 1934 to 1965, by Work and Snow in 1966 and 1967, and by Snow in 1968, 1969, and 1970; and in *Powder Technol.*, **5**, 351 (1972), and **7** (1973); Snow and Luckie, **10**, 129 (1973), **13**, 33 (1976), **23**(1), 31 (1979). *Chemical Engineering Catalog*, Reinhold, New York, annually. Cremer-Davies, *Chemical Engineering Practice*, vol. 3: *Solid Systems*, Butterworth, London, and Academic, New York, 1957. *Crushing and Grinding: A Bibliography*, Chemical Publishing, New York, 1960. *European Symposia on Size Reduction:* 1st, Frankfurt, 1962, publ. 1962, Rumpf (ed.), Verlag Chemie, Düsseldorf; 2d, Amsterdam, 1966, publ. 1967, Rumpf and Pietsch (eds.), *DECHEMA-Monogr.*, **57;** 3d, Cannes, 1971, publ. 1972, Rumpf and Schönert (eds.), *DECHEMA-Monogr.*, **69.** Gaudin, *Principles of Mineral Dressing*, McGraw-Hill, New York, 1939. International Mineral Processing Congresses: *Recent Developments in Mineral Dressing*, London, 1952, publ. 1953, Institution of Mining and Metallurgy; *Progress in Mineral Dressing*, Stockholm, 1957, publ. London, 1960, Institution of Mining and Metallurgy; 6th, Cannes, 1962, publ. 1965, Roberts (ed.), Pergamon, New York; 7th, New York, 1964, publ. 1965, Arbiter (ed.), vol. 1: *Technical Papers*, vol. 2: *Milling Methods in the Americas*, Gordon and Breach, New York; 8th, Leningrad, 1968; 9th, Prague, 1970; 10th, London, 1973; 11th, Cagliari, 1975; 12th, São Paulo, 1977. Lowrison, *Crushing and Grinding*, CRC Press, Cleveland, 1974. *Pit and Quarry Handbook*, Pit & Quarry Publishing, Chicago, 1968. Richards and Locke, *Text Book of Ore Dressing*, 3d ed., McGraw-Hill, New York, 1940. Rose and Sullivan, *Ball, Tube and Rod Mills*, Chemical Publishing, New York, 1958. Snow, *Bibliography of Size Reduction*, vols. 1 to 9 (an update of the previous bibliography to 1973, including abstracts and index), U.S. Bur. Mines Rep. SO122069, available IIT Research Institute, Chicago, Ill. 60616. Stern, "Guide to Crushing and Grinding Practice," *Chem. Eng.*, **69**(25), 129 (1962). Taggart, *Elements of Ore Dressing*, McGraw-Hill, New York, 1951.

Since a large part of the literature is in the German language, availability of English translations is important. Translation numbers cited in this section refer to translations available through the National Translation Center, Crerar Library, Chicago, Ill. 60637. Also, volumes of selected papers in English translation are available from the Institute for Mechanical Processing Technology, Karlsruhe Technical University, Karlsruhe, Germany.

PROPERTIES OF SOLIDS

A single particle or lump has linear size, surface, hardness, and structure. The **linear size** may be the diameter for a sphere, an edge length for a cube, or some fictitious average linear dimension for an irregularly shaped lump. The **surface** is the exterior of most particles, although some have internal pore surface. The surface is readily calculated for cubes and spheres but must be estimated or measured for irregular shapes. **Hardness** is indicated by the conventional scratch criterion and can be measured by indentation. **Structure** may be homogeneous or heterogeneous.

A mixture of particles such as those in a powder may be defined in terms of particle-size distribution, surface, specific surface, and limiting particle size. **Particle-size distribution** is a function giving the proportionate amount of each size of individual particles in the powder. The **surface** is a summation of the individual grain surfaces, and **specific surface** is the surface of a unit of weight or volume. **Limiting particle size** is the size of the largest or smallest particles in the powder.

Grindability is a measure of the rate of grinding of material in a particular mill (discussed later).

Single-Particle Fracture More fundamental knowledge of the breaking action occurring within mills depends on developing knowledge of the mechanism of single-particle **fracture.** The early workers [Smekal, *Z. Ver. Dtsch. Ing. Beh. Verfahrenstech.*, no. 6, 159–165 (1938), NTC translation 70-14798; and Smekal *Z. Ver. Dtsch. Ing.*, **81**(46), 1321–1326 (1937), NTC translation 70-14799] investigated the breakage of cubes. This gives misleading results when cubes are crushed between platens because surface irregularities concentrate the load and give nonuniform load distribution. More meaningful measurements can be made with spheres, which approximate the shapes of particles broken in mills.

The force required to crush a single particle that is spherical near the contact regions is given by the equation of Hertz (Timoschenko and Goodier, *Theory of Elasticity*, 2d ed., McGraw-Hill, New York, 1951).

In an experimental and theoretical study on glass spheres Frank and Lawn [*Proc. R. Soc. (London)*, **A299**(1458), 291 (1967)] observed the repeated formation of ring cracks as increasing load was applied, causing the circle of contact to widen. Eventually a load is reached at which the crack deepens to form a cone crack, and at a sufficient load this propagates across the sphere to cause breakage into fragments. The authors' photographs show how the size of flaws that happen to be encountered at the edge of the circle of contact can result in a distribution of breakage strengths. Thus the mean value of breakage strength depends partly on intrinsic strength and partly on the extent of flaws present. From the measured breaking load and the Hertz theory one can calculate the apparent tensile strength σ_0, which is the maximum stress under the circle of contact normal to the direction of crack propagation. This tensile strength is the most appropriate one to use for breakage in mills, although the crushing strength of cubes still is often used as a rule of thumb. The propagation of cracks across spheres and disks has been recorded by high-speed spark cinematographs by Rumpf et al. (*Second European Symposium on Size Reduction*, op. cit., 1966, p. 57). They attempt to extend the Hertz theory deeper into the sphere although it is not valid far from the point of load application. The stress at points within the sphere far from the point of load application is given by the Bousinesque theory [Sternberg and Rosenthal, *J. Appl. Mech.*, (12), 413 (1952); and Hiramatsu and Oka, *Int. J. Rock Mech. Min. Sci.*, **3,** 89 (1966)].

Snow and Paulding (Heywood Memorial Conference, Loughborough University, England, September 1973) observed that when breakage occurs, the finest fragments arise near the circle of contact, where the stored internal stress is highest. They postulated that the fragment-size distribution could be calculated by assuming that the local fragment size is correlated with the locally stored stress energy just before fracture occurs. Calculated fragment-size distributions are roughly similar to those that they measured for glass spheres and various hard minerals as well as to distributions measured by Bergstrom and Sollenberger [*Trans. Am. Inst. Min. Metall. Pet. Eng.*, **220,** 373–379 (1961)]. From this it can be concluded that the wide distribution of fragment sizes from milling is inherent in the breakage process itself and that attempts to improve grinding efficiency by weakening the particles will result in coarser fragments which may require a further break to reach the desired size.

Different mills are designed to apply the force in different ways [Rumpf, *Chem. Ing. Tech.*, **31,** 323–327 (1959), NTC translation 61-12395]. The detailed prediction of grinding rates and product-size distribution from mills awaits the development of a simulation model based on the physics of fracture. An initial attempt is that of Buss and Shubert (*Third European Symposium on Size Reduction*, op. cit., 1972, p. 233), who assume that mill performance is given by the sum of breakage events which are similar to *single-particle breakage experiments* in the laboratory. A paper by Schönert [*Trans. Am. Inst. Min. Metall. Pet. Eng.*, **252**(1), 21 (1972)] summarizes single-particle-breakage data from numerous publications from the Technical University of Karlsruhe, Germany. Hildinger [*Freiberg. Forschungsh.*, **A480,** 19 (1970)] and Steier and Schönert (*Third European Symposium on Size Reduction*, op. cit., 1972, p. 135) report more experimental results on the probability of breakage of single particles by drop-weight experiments.

The laboratory breakage of particles in high-velocity impact simulates the action in hammer mills and jet mills. In the Karpinski-Tervo [*Trans. Am. Inst. Min. Metall. Pet. Eng.*, **229,** 126 (1964)] impact-testing machine, particles are dropped in front of a rotating paddle in vacuum and the fragments are collected. Results are given in Lyall and Tervo [*Proc. Fifth Can. Rock Mech. Symp.*, Toronto, 171 (1969)].

Grindability Grindability is the amount of product from a particular mill meeting a particular specification in a unit of grinding time e.g., tons per hour passing 200 mesh. The chief purpose of a study of grindability is to evaluate the size and type of mill needed to produce a specified tonnage and the power requirement for grinding. So many variables affect grindability that this concept can be used only as a rough guide to mill sizing; it says nothing about product-size distribution or type or size of mill. If a particular energy law is assumed, then the grinding behavior in various mills can be expressed as an energy coefficient or **work index** (discussed later). This more precise concept is limited by the inadequacies of these laws but often provides the only available information.

The technology based on grindability and energy considerations is being supplanted by computer simulation of milling circuits (see subsection "Simulation of Milling Circuits"), in which the gross concept of grindability is replaced by the **rate of breakage function** (sometimes called the selection function), which is the grindability of each particle size referred to the fraction of that size present.

Factors of hardness, elasticity, toughness, and cleavage are important in determining grindability. Grindability is related to modulus of elasticity and speed of sound in the material [Dahlhoff, *Chem. Ing. Tech.*, **39**(19), 1112–1116 (1967)].

The **hardness** of a mineral as measured by the **Mohs scale** is a criterion of its resistance to crushing [Fahrenwald, *Trans. Am. Inst. Min. Metall. Pet. Eng.*, **112,** 88 (1934)]. It is a fairly good indication of the abrasive character of the mineral, a factor that determines the wear on the grinding media. Arranged in increasing order or hardness, the Mohs scale is as follows: 1, talc; 2, gypsum; 3, calcite; 4, fluoride; 5, apatite; 6, feldspar; 7, quartz; 8, topaz; 9, corundum; and 10, diamond.

Materials of hardness 1 to 3 inclusive may be classed as soft; 4 to 7, as intermediate; and the others, as hard.

Soft Materials (1) Talc, dried filter-press cakes, soapstone, waxes, aggregated salt crystals; (2) gypsum, rock salt, crystalline salts in general, soft coal; (3) calcite, marble, soft limestone, barites, chalk, brimstone.

Intermediate Hardness (4) Fluorite, soft phosphate, magnesite, limestone; (5) apatite, hard phosphate, hard limestone, chromite, bauxite; (6) feldspar, ilmenite, orthoclase, hornblendes.

Hard Materials (7) Quartz, granite; (8) topaz; (9) corundum, sapphire, emery; (10) diamond.

A hardness classification of stone based on the **compressive strength** of 1-in cubes is as follows, for loadings in pounds-force per square inch: very soft, 10,000; soft, 15,000; medium, 20,000; hard, 25,000; very hard, 30,000.

Grindability Methods Laboratory experiments on single particles have been used to correlate grindability. In the past it has usually been assumed that the total energy applied could be related to the grindability whether the energy is applied in a single blow or by repeated dropping of a weight on the sample [Gross and Zimmerly, *Trans. Am. Inst. Min. Metall. Pet. Eng.*, **87,** 27, 35 (1930)]. In fact, the results depend on the way in which the force is applied (Axelson, Ph.D. thesis, University of Minnesota, 1949). In spite of this, the results of large mill tests can often be correlated within 25 to 50 percent by a simple test, such as the number of drops of a particular weight needed to reduce a given amount of feed to below a certain mesh size.

Two methods having particular application for **coal** are known as the ball-mill and Hardgrove methods. In the **ball-mill method,** the relative amounts of energy necessary to pulverize different coals are determined by placing a weighed sample of coal in a ball mill of a specified size and counting the number of revolutions required to grind the sample so that 80 percent of it will pass through a No. 200 sieve. The grindability index in percent is equal to the quotient of 50,000 divided by the average of the number of revolutions required by two tests (ASTM designation D-408).

In the **Hardgrove method** a prepared sample receives a definite amount of grinding energy in a miniature ball-ring pulverizer. The unknown sample is compared with a coal chosen as having 100 grindability. The Hardgrove grindability index $= 13 + 6.93W$, where W is the weight of material passing the No. 200 sieve (see ASTM designation D-409).

Chandler [*Bull. Br. Coal Util. Res. Assoc.*, **29**(10), 333; (11), 371 (1965)] finds no good correlation of grindability measured on 11 coals with roll crushing and attrition, and so these methods should be used with caution. The Bond grindability method is described in the subsection "Capacity and Power Consumption."

Manufacturers of various types of mills maintain laboratories in which grindability tests are made to determine the suitability of their machines. When grindability comparisons are made on small equipment of the manufacturers' own class, there is a basis for scale-up to commercial equipment. This is better than relying on a grindability index obtained in a ball mill to estimate the size and capacity of different types such as hammer or jet mills.

Mill Wear In general, hard materials, coarse particles, and fast motion are conducive to wear in mills. Wear may be resisted by using materials in the zones of mill wear which are harder than the material being ground.

Abrasion and erosion are different mechanisms that can predominate under different conditions [Maratray, *Rev. Ind. Miner.*, **52** (11), 713 (1970)]. Abrasion increases until the hardness of the abrasive is 1.5 times that of the metal. Pressure has more influence when both materials are soft; velocity has a complex influence. Wear is greater in wet than in dry grinding; this, however, is still unexplained. Harder materials of construction may also fail by breaking, so a compromise has to be adopted. This is illustrated in South African practice in ball-mill liners [French and Lissner, *J. S. Afr. Inst. Min. Metall.*, 71 (September 1968); 229 (December 1968); 475 (1969)]. Causing particles to grind each other helps materially. When mill movements are slow, resilient materials like rubber often withstand the wear of hard particles. Diehl and Griffiths [*Can. Min. J.*, **91,** 76 (June 1970)] summarize successful results with rubber liners in iron-ore mills. They show longer life and lower costs in all but the biggest mills. Rubber linings cause no change in capacity or grind. In the 1930s they failed tests mainly because no adequate method of attachment was found. Current methods use a rubber lifter bar that secures flat liner sheets.

The use of hard-surfacing techniques by welding and by inserts has contributed greatly to better maintenance and lower downtime [Lutes and Reid, *Chem. Eng.*, **63**(6), 243 (1956)]. Extensive data on the wear and cost of various types of steels in ore grinding have been reported [Norman and Loeb, *Trans. Am. Inst. Min. Metall. Pet. Eng.*, **183,** 330 (1949)].

Mill wear or abrasion becomes critical on high-peripheral-speed equipment, particularly high-speed close-clearance hammer mills. Pulverizing Machinery Co. has developed a reasonably reliable **abrasion test,** using a given weight (5 lb) of feed to a Bantam (small-scale model) Mikro-Pulverizer employing a standard rotor speed and drilled perforated screen (usually 0.027-in-diameter holes) and forged hard-faced hammers. The drilled perforations are examined under a microscope as being clean cuts before testing and are reexamined after test. The length of cutting along the screen surface is measured in micrometers on a calibrated eyepiece, and this figure is termed "abrasion index." Up to 20 μm is usually within economic limits and above 100 μm outside economic limits unless there are unusual aspects to the case.

An abrasion index in terms of kilowatthour input per pound of metal lost furnishes a useful indication. Rough values are quoted in Table 8-1.

Safety The explosion hazard of such nonmetallic materials as sulfur, starch, wood flour, cereal dust, dextrin, coal, pitch, hard rubber, and plastics is often not appreciated (Hartmann and Nagy, U.S. Bur. Mines Rep. Invest. 3751, 1944). Explosions and fires may be initiated by discharges of static electricity, sparks from flames, hot surfaces, and spontaneous combustion. Metal powders present a hazard because of their **flammability.** Their combustion is favored during grinding operations in which ball, hammer, or ring-roller mills are employed and during which a high grinding temperature may be reached.

Many finely divided metal powders in suspension in air are potential **explosion hazards,** and causes for ignition of such dust clouds are numerous [Hartmann and Greenwald, *Min. Metall.*, **26,** 331 (1945)]. Concentration of the dust in air and its particle size are important factors that determine explosibility. Below a lower limit of concentration, no explosion can result because the heat of combustion is insufficient to propagate it. Above a maximum limiting concentration, an explosion cannot be produced because insufficient oxygen is available. The finer the particles, the more easily is ignition accomplished and the more rapid is the rate of combustion. This is illustrated in Fig. 8-7.

Isolation of the mills, use of nonsparking materials of construction, and magnetic separators to remove foreign magnetic material from the feed are useful **precautions** (Hartman, Nagy, and Brown, U.S. Bur. Mines Rep. Invest. 3722, 1943). Stainless steel has less sparking tendency than ordinary steel or forgings.

Reduction of the oxygen content of air present in grinding systems is a means for preventing dust explosions in equipment (Brown, U.S. Dep. Agri. Tech. Bull. 74, 1928). Maintenance of oxygen content below 12 percent should be safe for most materials, but 8 percent is recommended for sulfur grinding. The use of **inert gas** has particular

TABLE 8-1 Abrasion Index Test Results*

Material	Number tested	Product diameter, μm	Work index E_i	Average abrasion index† A_i
Alnico	1	.....	...	0.3850
Alumina	6	15,500	...	0.6447
Asbestos cement pipe	1	13,330	...	0.0073
Cement clinker	2	12,100	10.9	0.0409
Cement raw material	4	.....	10.5	0.0372
Chrome ore	1	10,200	9.6	0.1200
Coke	1	.....	20.7	0.3095
Copper ore	12	12,900	11.2	0.0950
Coral rock	1	.....	...	0.0061
Diorite	1	.....	19.4	0.2303
Dolomite	5	.....	11.3	0.0160
Gneiss	1	.....	20.1	0.5360
Gold ore	2	.....	14.8	0.2000
Granite	11	15,200	14.4	0.3937
Gravel	2	.....	19.0	0.3051
Hematite	3	.....	8.6	0.0952
Iron ore (misc.)	4	.....	5.4	0.0770
Lead zinc ore	3	.....	8.3	0.1520
Limestone	19	13,000	12.1	0.0256
Magnesite	3	14,400	16.8	0.0750
Magnetite	2	.....	10.2	0.2517
Manganese ore	1	.....	17.2	0.1133
Nickel ore	2	.....	11.9	0.0215
Perlite	2	.....	...	0.0452
Pumice	1	.....	11.9	0.1187
Quartz	7	.....	12.8	0.1831
Quartzite	3	.....	12.2	0.6905
Rare earths	1	.....	...	0.0288
Rhyolite	2	13,200	...	0.4993
Schist-biotite	1	.....	23.5	0.1116
Shale	2	11,200	11.2	0.0060
Slag	1	.....	15.8	0.0179
Slate	1	.....	13.8	0.1423
Sulfur	1	.....	11.5	0.0001
Taconite	7	.....	16.2	0.6837
Trap rock	11	14,900	19.9	0.3860
Average		13,250	13.8	0.228

*Allis-Chalmers Corporation.

†Abrasion index is the fraction of a gram weight lost by the standard steel paddle in 1 h of beating 1600 g of ¾- by ½-in particles. The product averages 80 percent passing 13,250 μm.

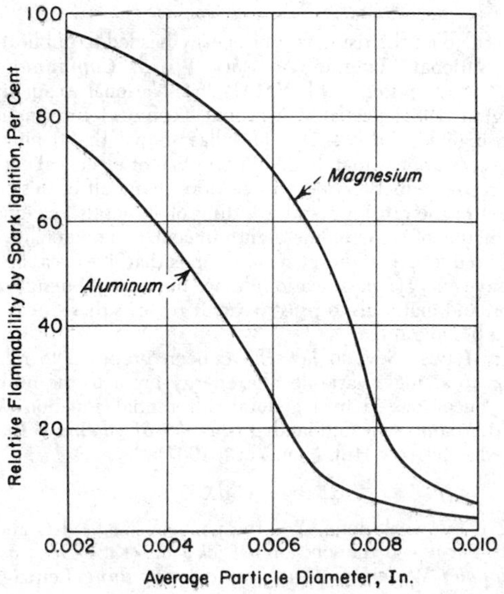

FIG. 8-7 Effect of fineness on the flammability of metal powders. *(Hartmann, Nagy, and Brown, U.S. Bur. Mines Rep. Invest. 3722, 1943.)*

adaptation to pulverizers equipped with air classification; flue gas can be used for this purpose, and it is mixed with the air normally present in a system (see subsection "Chemicals and Soaps" for sulfur grinding). Despite the protection afforded by the use of inert gas, equipment should be provided with explosion vents, and structures should be designed with venting in mind [Brown and Hanson, *Chem. Metall. Eng.*, **40,** 116 (1933)].

Hard rubber presents a fire hazard when reduced on steam-heated rolls (see subsection "Organic Polymers"). Its dust is explosive [Twiss and McGowan, *India Rubber J.*, **107,** 292 (1944)].

An annual publication, *National Fire Codes for the Prevention of Dust Explosions*, is available from the National Fire Protection Association, Quincy, Massachusetts, and should be of interest to those handling hazardous powders.

ATTAINABLE PRODUCT SIZE AND ENERGY REQUIRED

The fineness to which a material is ground has a marked effect on its production rate. Figure 8-8 is an example showing how the capacity decreases and the specific energy and cost increase as the product is ground finer.

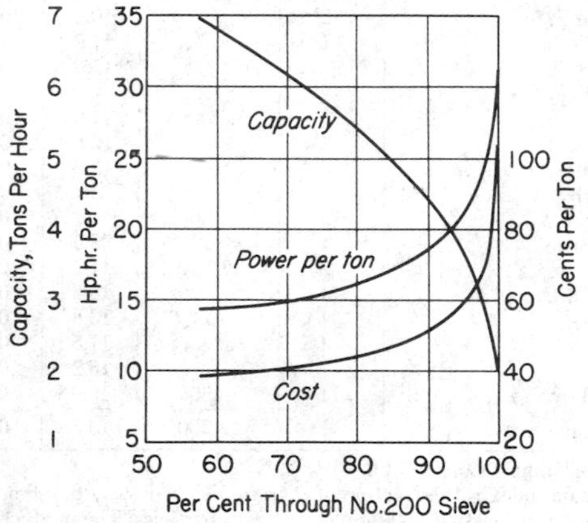

FIG. 8-8 Variation in capacity, power, and cost of grinding relative to fineness of product.

Concern about the rising cost of energy has led to publication of a report (National Materials Advisory Board, *Comminution and Energy Consumption*, Publ. NMAB-364, National Academy Press, Washington, 1981; available National Technical Information Service, Springfield, Va. 22151). This has shown that United States industries use approximately 32 billion kWh of electrical energy per annum in size-reduction operations. More than half of this energy is consumed in the crushing and grinding of minerals, one-quarter in the production of cement, one-eighth in coal, and one-eighth in agricultural products. The report recommends that five areas be considered to save energy: classification-device design, mill design, control, additives, and materials to resist wear. It reviews these areas with an extensive bibliography.

Energy Laws Several laws have been proposed to relate size reduction to a single variable, the energy input to the mill. These laws are encompassed in a general differential equation (Walker, Lewis, McAdams, and Gilliland, *Principles of Chemical Engineering*, 3d ed., McGraw-Hill, New York, 1937):

$$dE = -C\,dX/X^n \tag{8-6}$$

where E is the work done, X is the particle size, and C and n are constants. For $n = 1$ the solution is *Kick's law* (Kick, *Das Gasetz der propertionalen Widerstande und seine Anwendung*, Leipzig, 1885). The law can be written

$$E = C \log (X_F/X_P) \tag{8-7}$$

where X_F is the feed-particle size, X_P is the product size, and X_F/X_P is the reduction ratio. For $n > 1$ the solution is

$$E = \left(\frac{C}{n-1}\right)\left(\frac{1}{X_P^n - 1} - \frac{1}{X_F^n - 1}\right) \tag{8-8}$$

For $n = 2$ this becomes *Rittinger's law*, which states that the energy is proportional to the new surface produced (Rittinger, *Lehrbuch der Aufbereitungskunde*, Ernst and Korn, Berlin, 1867).

The *Bond law* corresponds to the case in which $n = 1.5$ [Bond, *Trans. Am. Inst. Min. Metall. Pet. Eng.*, **193,** 484 (1952)]:

$$E = 100\,E_i\left(\frac{1}{\sqrt{X_P}} - \frac{1}{\sqrt{X_F}}\right) \tag{8-9}$$

where E_i is the **Bond work index,** or work required to reduce a unit weight from a theoretical infinite size to 80 percent passing 100 μm. Extensive data on the work index have made this law useful for rough mill sizing. Summary data are given in Table 8-2.

The work index may be found experimentally from laboratory crushing and grinding tests or from commercial mill operations. Some rules of thumb for extrapolating the work index to conditions different from those measured are that for dry grinding the index must be increased by a factor of 1.34 over that measured in wet grinding; for open-circuit operations another factor of 1.34 is required over that measured in closed circuit; if the product size X_p is extrapolated below 70 μm, an additional correction factor is $(10.3 + X_p)/1.145X_p$. Also for a jaw or gyratory crusher the work index may be estimated from

$$E_i = 2.59C_s/\rho_s \tag{8-10}$$

where C_s = impact crushing resistance, (ft·lb)/in of thickness required to break; ρ_s = specific gravity; and E_i is expressed in kWh/ton.

None of the energy laws apply well in practice, and they have failed to yield a starting point for further development of understanding of milling. They are mainly of historical interest. Most of the early papers supporting one law or another were based on extrapolations of size distributions to finer sizes on the assumption of one or another size-distribution law. With present particle-size-analysis techniques applicable to the finest sizes, such confusion is no longer necessary. The relation of energy expenditure to the size distribution produced has been thoroughly examined [Arbiter and Bhrany, *Trans. Am. Inst. Min. Metall. Pet. Eng.*, **217,** 245–252 (1960); Harris, *Inst. Min. Metall. Trans.*, **75**(3), C37 (1966); Holmes, *Trans. Inst. Chem. Eng. (London)*, **35,** 125–141 (1957); and Kelleher, *Br. Chem. Eng.*, **4,** 467–477 (1959); **5,** 773–783 (1960)].

Grinding Efficiency The energy efficiency of a grinding operation is defined as the energy consumed compared with some ideal energy requirement.

The theoretical energy efficiency of grinding operations is 0.06 to 1 percent, based on values of the surface energy of quartz [Martin, *Trans. Inst. Chem. Eng. (London)*, **4,** 42 (1926); Gaudin, *Trans. Am. Inst. Min. Metall. Pet. Eng.*, **73,** 253 (1926)]. Uncertainty in these results is due to uncertainty in the theoretical surface energy.

A definitive monograph (Kuznetzov, *Surface Energy of Solids*, English translation, H. M. Stationery Office, London, 1957) established that most laboratory methods of measuring surface energy introduce large errors, but the cleavage method of Obreimov [Gilman, *J. Appl. Phys.*, **31,** 2208 (1960)] gave results for sodium chloride that agree with theoretical lattice calculations. Later studies by Raasch [*Int. J. Frac. Mech.*, **7**(9), 289 (1971)] and by Burns [*Philos. Mag.*, **25**(1), 131 (1972)] conclude that these measurements are valid when 50 percent corrections are added for the bending energy of the crystal. Kuznetzov ranks other materials by a relative wear test. His results substantiate the efficiencies given earlier. Attempts to measure efficiency of the grinding process by calorimetry involve errors that exceed the theoretical surface energy of the material being ground.

Practical energy efficiency is defined as the efficiency of technical grinding compared with that of laboratory crushing experiments. Practical efficiencies of 25 to 60 percent have been shown [Wilson,

TABLE 8-2 Average Work Indices for Various Materials*

Material	No. of tests	Specific gravity	Work index†
All materials tested	2088	—	13.81
Andesite	6	2.84	22.13
Barite	11	4.28	6.24
Basalt	10	2.89	20.41
Bauxite	11	2.38	9.45
Cement clinker	60	3.09	13.49
Cement raw material	87	2.67	10.57
Chrome ore	4	4.06	9.60
Clay	9	2.23	7.10
Clay, calcined	7	2.32	1.43
Coal	10	1.63	11.37
Coke	12	1.51	20.70
Coke, fluid petroleum	2	1.63	38.60
Coke, petroleum	2	1.78	73.80
Copper ore	308	3.02	13.13
Coral	5	2.70	10.16
Diorite	6	2.78	19.40
Dolomite	18	2.82	11.31
Emery	4	3.48	58.18
Feldspar	8	2.59	11.67
Ferrochrome	18	6.75	8.87
Ferromanganese	10	5.91	7.77
Ferrosilicon	15	4.91	12.83
Flint	5	2.65	26.16
Fluorspar	8	2.98	9.76
Gabbro	4	2.83	18.45
Galena	7	5.39	10.19
Garnet	3	3.30	12.37
Glass	5	2.58	3.08
Gneiss	3	2.71	20.13
Gold ore	209	2.86	14.83
Granite	74	2.68	14.39
Graphite	6	1.75	45.03
Gravel	42	2.70	25.17
Gypsum rock	5	2.69	8.16
Ilmenite	7	4.27	13.11
Iron ore	8	3.96	15.44
Hematite	79	3.76	12.68
Hematite—specular	74	3.29	15.40
Oolitic	6	3.32	11.33
Limanite	2	2.53	8.45
Magnetite	83	3.88	10.21
Taconite	66	3.52	14.87
Kyanite	4	3.23	18.87
Lead ore	22	3.44	11.40
Lead-zinc ore	27	3.37	11.35
Limestone	119	2.69	11.61
Limestone for cement	62	2.68	10.18
Manganese ore	15	3.74	12.46
Magnesite, dead burned	1	5.22	16.80
Mica	2	2.89	134.50
Molybdenum	6	2.70	12.97
Nickel ore	11	3.32	11.88
Oil shale	9	1.76	18.10
Phosphate fertilizer	3	2.65	13.03
Phosphate rock	27	2.66	10.13
Potash ore	8	2.37	8.88
Potash salt	3	2.18	8.23
Pumice	4	1.96	11.93
Pyrite ore	4	3.48	8.90
Pyrrhotite ore	3	4.04	9.57
Quartzite	16	2.71	12.18
Quartz	17	2.64	12.77
Rutile ore	5	2.84	12.12
Sandstone	8	2.68	11.53
Shale	13	2.58	16.40
Silica	7	2.71	13.53
Silica sand	17	2.65	16.46
Silicon carbide	7	2.73	26.17
Silver ore	6	2.72	17.30
Sinter	9	3.00	8.77
Slag	12	2.93	15.76
Slag, iron blast furnace	6	2.39	12.16
Slate	5	2.48	13.83
Sodium silicate	3	2.10	13.00
Spodumene ore	7	2.75	13.70
Syenite	3	2.73	14.90
Tile	3	2.59	15.53
Tin ore	9	3.94	10.81
Titanium ore	16	4.23	11.88
Trap rock	49	2.86	21.10
Uranium ore	20	2.70	17.93
Zinc ore	10	3.68	12.42

*Allis-Chalmers Corporation.

†Caution should be used in applying the average work index values listed here to specific installations, since individual variations between materials in any classification may be quite large.

Min. Technol., Tech. Publ. 810, 1937; and Bond and Maxson, *Trans. Am. Inst. Min. Metall. Pet. Eng.*, **134**, 296 (1939)].

An **energy coefficient** is sometimes based on Rittinger's law, i.e., new surface produced per unit of energy input. Usually time of grinding is the experimental variable, which is expressed indirectly as energy. The energy coefficient may also be expressed as tons per horsepower-hour passing a certain size. The value of this coefficient is between about 0.02 and 0.1 for wet ball-mill pulverizing hard to medium-hard minerals to No. 200 sieve size (74 μm).

The curves in Fig. 8-9 show decreasing production rate with increasing **moisture content.** (Occasionally, a small amount of water may be beneficial over complete dryness.) All three materials were being ground to 99.9 percent through a No. 200 sieve.

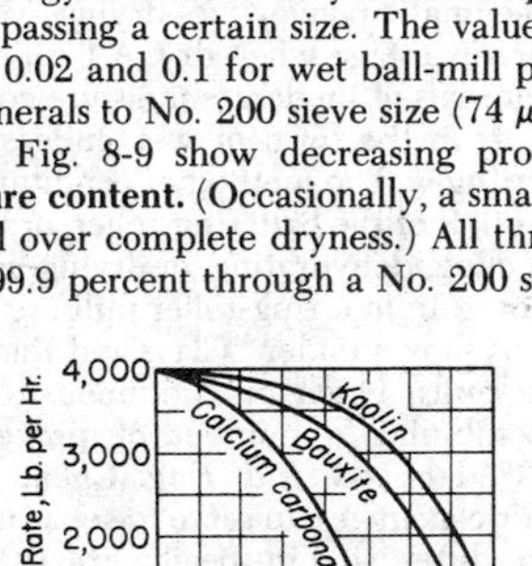

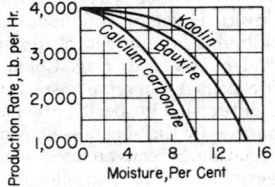

FIG. 8-9 Effect of moisture on the production rate of a pulverizer. [*Work*, Chem. Metall. Eng., *40, 306 (1933).*]

Dry versus Wet Grinding Ball mills have a large field of application for wet grinding in closed circuit with size classifiers. If the presence of liquid with the finished product is not objectionable or the feed is moist or wet, wet grinding generally is preferable to dry grinding. The net production in wet grinding at different meshes in the Bond grindability test varies from 145 to 200 percent of that in dry grinding [Maxson, Cadena, and Bond, *Trans. Am. Int. Min. Metall. Pet. Eng.*, **112**, 130–145, 161 (1934)]. In fine dry grinding, surface forces come into action to cause cushioning and ball coating with a less efficient use of energy. Other factors that influence choice are the performance of subsequent dry or wet classification steps, the cost of drying, and the capability of subsequent processing steps for handling a wet product.

It has long been thought that a **limiting size** is attainable. New technologies such as pressed ceramics and Xerox toners require finer sizes, and this again questions the existence of a limit. There are three theories for such a limit. Bradshaw [*J. Chem. Phys.*, **19**, 1057–1059 (1951)] thought that *reagglomeration* is responsible, especially in ball mills. Schönert and Steier [*Chem. Ing. Tech.*, **43**(13), 773 (1971)] suggest two other causes: *plastic deformation* and the difficulty of *stressing* fine particles to their breaking point. The latter stems from the Griffith crack theory, which requires that the particle have enough stored stress energy to allow a crack to propagate. A 10-μm glass particle requires 140 kPa/mm^2 tensile stress. Although both of these mechanisms can be limiting, recent experimental evidence indicates that plastic deformation can increase the resistance of even the most brittle materials on a fine scale. Rumpf and Schönert (*Third European Symposium on Size Reduction*, op. cit., 1972, p. 27) observed plastic deformation in crushing fine glass spheres. Schönert and Steier (loc. cit.) in electron-microscope photographs observed plastic deformation in crushing limestone particles as large as 3 to 4 μm and quartz particles of 2 to 3 μm. This deformation spreads a stress that would otherwise produce brittle fracture. Gane [*Philos. Mag.*, **25**(1), 25 (1972)] observed plastic deformation in magnesium oxide crystals 0.2 to 0.4 μm in size. The strengths average 180 kg/mm^2, which is 15 times the strength of large MgO crystals but one-tenth of the theoretical strength. Further proof is given by Weichert and Schönert [*J. Mech. Phys. Solids* (22), 127 (1974)], who analyze and measure the temperature rise at a propagating crack tip. They estimate that irreversible deformation occurs in a zone of radius about 30 A running along at the tip. The energy release causes temperatures as high as 1500 K above ambient temperature at the tip. This temperature explains plastic flow and even emitted light in some cases. Therefore, it is proved that plastic deformation can limit the grinding size attainable. Other means than size reduction must be found if particles much finer than 0.5 μm are wanted.

In practice it is found that finer size can be achieved by **wet grinding** than by **dry grinding.** In wet grinding by ball mills or vibratory mills with suitable surfactants, product sizes of 0.5 μm are attainable. In dry grinding the size is generally limited by ball coating (Bond and Agthe, *Min. Technol.*, AIME Tech. Publ. 1160, 1940) to about 15 μm. In dry grinding with hammer mills or ring-roller mills the limiting size is about 10 to 20 μm. Jet mills are generally limited to a product mean size of 15 μm, although dense particles can be ground to 5 μm because of the greater ratio of inertia to aerodynamic drag.

Dispersing Agents and Grinding Aids There is no doubt that grinding aids are helpful under some conditions. For example, surfactants make it possible to ball-mill magnesium in kerosine to 0.5-μm size [Fochtman, Bitten, and Katz, *Ind. Eng. Chem. Prod. Res. Dev.*, **2**, 212–216 (1963)]. Without surfactants the size attainable was 3 μm, and of course the rate of grinding was very slow at sizes below this. Also, the water in wet grinding may be considered to act as an additive.

Chemical agents that increase the rate of grinding are an attractive prospect since their cost is low. However, despite a voluminous literature on the subject, there is no accepted scientific method to choose such aids; there is not even agreement on the mechanisms by which they work.

In **wet grinding** there are several theories, which have been reviewed [Somasundaran and Lin, *Ind. Eng. Chem. Process Des. Dev.*, **11**(3), 321 (1972); Snow, annual reviews, op. cit., 1970–1974. See also Rose, *Ball and Tube Milling*, Constable, London, 1958, pp. 245–249]. The *Rehbinder theory* (Rehbinder, Schreiner, and Zhigach, *Hardness Reducers in Rock Drilling*, Moscow Academy of Science, 1944, transl. Council for Scientific and Industrial Research, Melbourne, Australia, 1948) holds that the hardness and strength of materials are decreased by adsorption of surface-active species because of reductions in surface energy. But the changes in drilling rate observed are too high to be caused by the slight changes in surface energy which surfactants could provide. According to Westwood [*J. Mater. Sci.*, **9**, 1871 (1974)], the additives may alter the near-surface structure and thereby influence the near-surface plastic flow and fracture behavior. These theories may be valid for drilling, but the evidence to apply them to milling is very tenuous. Another theory applicable to ball milling is that the additives decrease the state of flocculation of particles and make them more susceptible to impact. They may act either by electrostatic repulsion due to adsorbed ionic molecules or by steric-hindrance repulsion in which long-chain polymer molecules adsorbed on the particles act as spatial barriers to prevent the close approach of other particles. The electrostatic effect can be measured by measuring the zeta potential (Riddick, *Control of Colloid Stability through Zeta Potential*, Zeta-Meter Inc., 1968), but measurements of the effect of zeta potential on grinding have not been conclusive (Snow, loc. cit.). Steric effects are even more difficult to measure.

Additives can alter the rate of wet ball milling by changing the slurry viscosity or by altering the location of particles with respect to the balls. These effects are discussed under "Tumbling Mills." In conclusion, there is still no theoretical way to select the most effective additive. Empirical investigation, guided by the principles discussed earlier, is the only recourse. There are a number of commercially available grinding aids that may be tried. Also, a kit of 450 surfactants that can be used for systematic trials (Model SU-450, Chemservice Inc., West Chester, Pa. 19380) is available.

Numerous experimental studies lead to the conclusion that dry **grinding** is limited by ball coating and that additives function by reducing the tendency to coat (Bond and Agthe, op. cit.). Most materials coat if they are ground fine enough, and softer materials coat at larger sizes than hard materials. The presence of more than a few percent of soft gypsum promotes ball coating in cement-clinker grinding. The presence of a considerable amount of coarse particles above 35 mesh inhibits coating. Balls coat more readily as they become scratched. Small amounts of moisture may increase or decrease ball coating, and dry materials also coat.

Materials used as grinding aids include solids such as graphite, oleoresinous liquid materials, volatile solids, and vapors. The complex effects of vapors have been extensively studied [Goette and Ziegler, *Z. Ver. Dtsch. Ing.*, **98**, 373–376 (1956); and Locher and von Seebach, *Ind. Eng. Chem. Process Des. Dev.*, **11**(2), 190 (1972)], but water is the only vapor used in practice.

The most effective additive for dry grinding is fumed silica that has been treated with methyl silazane [Tulis, *J. Hazard. Mater.*, **4**, 3 (1980)].

SIZE REDUCTION COMBINED WITH OTHER OPERATIONS

Batch ball mills with low ball charges can be used in **dry mixing** or standardizing of dyes, pigments, colors, and insecticides to incorporate wetting agents and inert extenders (see also Sec. 21). Disk mills, hammer mills, and other high-speed disintegration equipment are useful for final intensive blending of insecticide compositions, earth colors, cosmetic powders, and a variety of other finely divided materials that tend to agglomerate in ribbon and conical blenders.

Mills with air-classification units may be equipped so that the **circulating air** can be **conditioned** by mixing with hot or cold air or gases introduced into the mill or by dehumidification to prepare the air for the grinding of hygroscopic materials. Liquid sprays or gases may be injected into the mill or air stream, for mixing with the material being pulverized to effect chemical reaction or surface treatment. Heat-sensitive materials with low softening temperatures are amenable to pulverizing if proper **temperature control** is exercised. Compositions containing fats and waxes are pulverized and blended readily if refrigerated air is introduced into their grinding systems (U.S. Patents 1,739,761 and 2,098,798; see also subsection "Organic Polymers").

The **drying** of materials while they are being pulverized or disintegrated is known variously as "flash" or "dispersion" drying; a generic term is "pneumatic conveying" drying. Many materials can be ground to better advantage when dry. A flash-drying system can be used for raw materials of moderate moisture content and also for precipitated products in the form of wet sludges or cakes coming from filters or centrifuges. The method of conditioning the air is the same whether the mill is of the ball, ring-roller, or hammer-mill type used for heating, cooling, dehydrating, or drying. Data for the grinding and drying of bauxite in a ring-roller mill are given in Table 8-3. A drying system is shown under "Clays and Kaolins," Fig. 8-60.

Ball and pebble mills, batch or continuous, offer considerable opportunity for combining a number of **processing steps** that include grinding [Underwood, *Ind. Eng. Chem.*, **30**, 905 (1938)]. Mills followed by air classifiers can serve to **separate components of mixtures** because of differences in specific gravity and particle size. The removal of impurities by this means is known as **cleaning, concentrating,** or **beneficiating.** Screens are used to separate coarse particles, not easily pulverized, from fine particles of the component that are pulverized readily. Grinding followed by *froth flotation* has become the beneficiation method most widely used for metallic ores

TABLE 8-3 Operating Data for Grinding and Drying of Bauxite in a Ring-Roller Mill

Initial moisture, %	9.75
Final moisture, %	0.75
Feed, lb./hr.	12,560
Product, lb./hr.	11,420
Moisture evaporated, lb.	1,140
Temperature of gases entering mill, °F.	700
Temperature of gases leaving mill, °F.	170
Temperature of feed, °F.	70
Temperature of material leaving mill, °F.	150
Oil consumed, gal.	14.3
Heating value of oil, B.t.u./gal.	142,000
Thermal efficiency, %	68.5
Total power for drying and pulverizing, hp.	105
Power for drying, hp.	10
Final product, % through No. 100 sieve	90

and also for nonmetallic minerals such as feldspar. Magnetic separation is the chief means used for upgrading taconite iron ore (see subsection "Ores and Minerals"). Magnetic separators frequently are employed to remove tramp magnetic solids from the feed to high-speed hammer and disk mills.

Most ores are heterogeneous, and the objective of grinding is to release the valuable mineral component so that it can be separated. Calculations based on a random-breakage model assuming no preferential breakage [Wiegel and Li, *Trans. Am. Inst. Min. Metall. Pet. Eng.*, **238**, 179–191 (1967)] agreed remarkably well with plant data on the efficiency of release of mineral grains. Figure 8-10 shows that as the ratio of abundance of mineral *A* to *B* becomes small, mineral *B* can be liberated by grinding to larger particle sizes such that the ratio of grain to particle size can be smaller. Many authors have assumed that breakage occurs preferentially along grain boundaries, but there is scant evidence for this. On the contrary, Gorski [*Bull. Acad. Pol. Sci. Ser. Sci. Tech.*, **20**(12), 929 (1972); CA 79, 20828k], from analysis of microscope sections, finds an intercrystalline character of comminution of dolomite regardless of the type of crusher used. In general, to separate mineral particles, the whole ore must be ground to a size substantially smaller than that of the grains.

Size Reduction Combined with Size Classification Grinding systems are batch or continuous in operation (Fig. 8-11). Most large-scale operations are continuous; batch ball or pebble mills are used only when small quantities are to be processed. Batch operation involves a high labor cost for charging and discharging the mill.

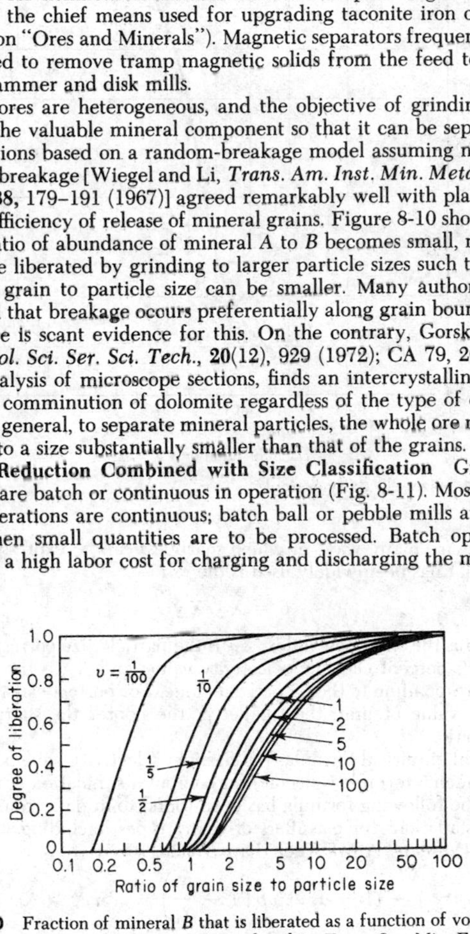

FIG. 8-10 Fraction of mineral *B* that is liberated as a function of volumetric abundance ratio V and size ratio. [*Wiegel and Li*, Trans. Soc. Min. Eng.–Am. Inst. Min. Metall. Pet. Eng., *238, 179 (1967).*]

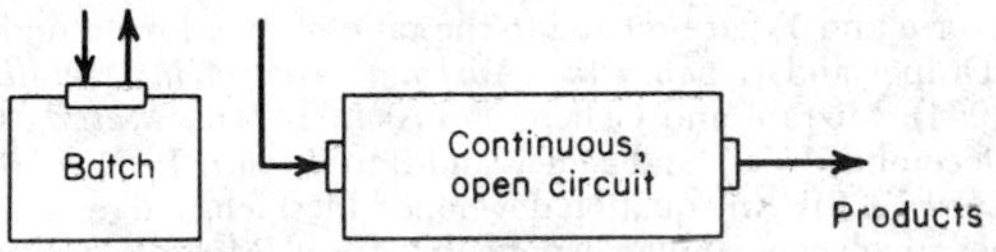

FIG. 8-11 Batch and continuous grinding systems.

Continuous operation is accomplished in open or closed circuit, as illustrated in Figs. 8-11 and 8-12. Most crushing and grinding equipment can be operated in **closed circuit** with size classifiers. **Operating economy** is the object of closed-circuit grinding to meet a limiting size specification. The idea is to remove the material from the mill before all of it is ground, separate the fine product in a classifier, and return the coarse for regrinding with the new feed to the mill. A mill with the fines removed in this way performs much more efficiently. Coarse material returned to a mill by a classifier is known as the **circulating load;** its rate may be from 1 to 10 times the production rate. The ability of the mill to transport material may limit the recycle rate; tube mills for use in such circuits may be designed with a smaller length-to-diameter ratio and hence a larger hydraulic gradient for more flow or with compartments separated by diaphragms with lifters.

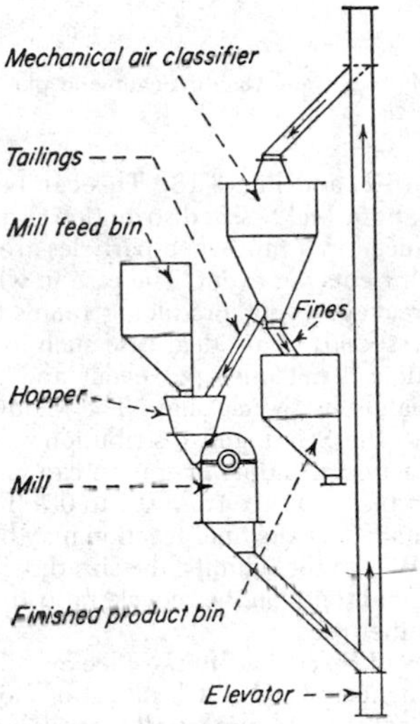

FIG. 8-12 Hammer mill in closed circuit with an air classifier.

There are many possible **arrangements** for connecting classifiers in closed circuit with grinding equipment; the most suitable depends upon the type and number of grinding units and the nature of the finished product. A simple arrangement of a hammer mill with air classifier is shown in Fig. 8-12.

In an arrangement used for a three-compartment mill with a single classifier, coarse tailings and fines from the classifier may be returned to any compartment, depending upon the amount of fines required in the finished product. Many other flow sheets may be designed for compartmented mills as well as for other units (see Fig. 8-61).

Internal size classification plays an essential role in the functioning of machines for dry grinding in the fine-size range; particles are retained in the grinding zone until they are as small as required in the finished product; then they are allowed to discharge.

By closed-circuit operation a product is obtained with more uniform size distribution than would be obtained by batch or continuous open-circuit operation to the same maximum limiting size; i.e., the size distribution is narrower. The product from the closed-circuit system will have a larger proportion of particles of the desired size. On the other hand, making a *product size within narrow limits* (such as between 20 and 40 μm) is often requested but usually is not possible regardless of the grinding circuit used. The reason is that particle breakage is a random process, both as to the probability of breakage of particles and as to the sizes of fragments produced from each breakage event. The narrowest size distribution ideally attainable is one that has a slope of 1.0 when plotted on Gates-Gaudin-Schumann

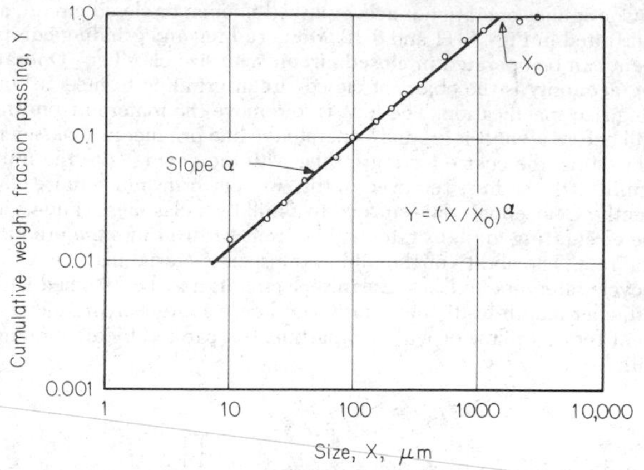

FIG. 8-13 Example of a Gates-Gaudin-Schumann plot of mill-product-size distribution.

coordinates [Eq. (8-2) and Fig. 8-13]. This can be demonstrated by examining the Gaudin-Meloy size distribution [Eq. (8-4)]. This is the distribution produced in a mill when particles are cut into pieces of random size, with r cuts per event. The case in which r is large corresponds to a breakage event producing many fines. The case in which r is 1 corresponds to an ideal case such as a knife cutter, in which each particle is cut once per event and the fragments are removed immediately by the classifier. The Meloy distribution with $r = 1$ reduces to the Schumann distribution with a slope of 1.0. Therefore, no practical grinding operation can have a slope greater than 1.0. Slopes typically range from 0.5 to 0.7. The specified product may still be made, but the finer fraction may have to be disposed of in some way. Within these limits, the size distribution of the classifier product depends both on the recycle ratio and on the sharpness of cut of the classifier used.

Characteristics of Size Classifiers (See Sec. 21: "Screening" on screening equipment and "Wet Classification" on wet classifiers.) Types of classifiers and commercially available equipment are described in the subsection "Particle-Size Classifiers Used with Grinding Mills." The American Institute of Chemical Engineers Equipment Testing Procedures Committee has published a procedure for particle-size classifiers (*Particle-Size Classifiers—A Guide to Performance Evaluation*, American Institute of Chemical Engineers, New York, 1980), including definitions which are followed here.

Three parameters define the performance of a classifier. These are *cut size, sharpness* of cut, and *capacity*. Cut size, X_{50}, is the size at which 50 percent of the material goes into the coarse product and 50 percent into the fine. (This should not be confused with the "cutoff size," a name sometimes given to the top size of the fine product.)

Size selectivity is the most thorough method of expressing classifier performance under a given set of operating conditions. Cut size and sharpness can be calculated from size-selectivity data. Size selectivity is defined by

$$\eta_X = \frac{\text{quantity of size } X \text{ entering coarse fraction}}{\text{quantity of size } X \text{ in feed}} \tag{8-11}$$

An equivalent mathematical expression is, on a *mass* basis,

$$\eta_X = \frac{q_c dY_c}{q_0 dY_0} = \frac{q_c dY_c}{q_c dY_c + q_f dY_f} \tag{8-12}$$

where Y_c is the cumulative percent by mass of coarse fraction less than particle size X, Y_f is the cumulative percent by mass of fine fraction less than particle size X, Y_0 is the cumulative percent by mass of feed less than particle size X, q_c is the coarse-fraction mass flow rate, q_f is the fine-fraction mass flow rate, and q_0 is the feed mass flow rate.

For purposes of calculating size selectivity from cumulative particle size distribution data, Eq. (8-12) can be expressed in incremental form as follows:

$$\eta_{X_i} = \frac{q_c\, \Delta Y_{ci}}{q_c\, \Delta Y_{ci} + q_f\, \Delta Y_{fi}} \tag{8-13}$$

where ΔY_{ci} and ΔY_{fi} are the cumulative size-distribution intervals of coarse and fine fractions associated with the size interval ΔX_i respectively. An interval representative size X_i is arbitrarily taken as the midpoint of ΔX_i.

See the American Institute of Chemical Engineers classifier test procedure for a sample calculation of classifier selectivity. This example is plotted in Fig. 8-14.

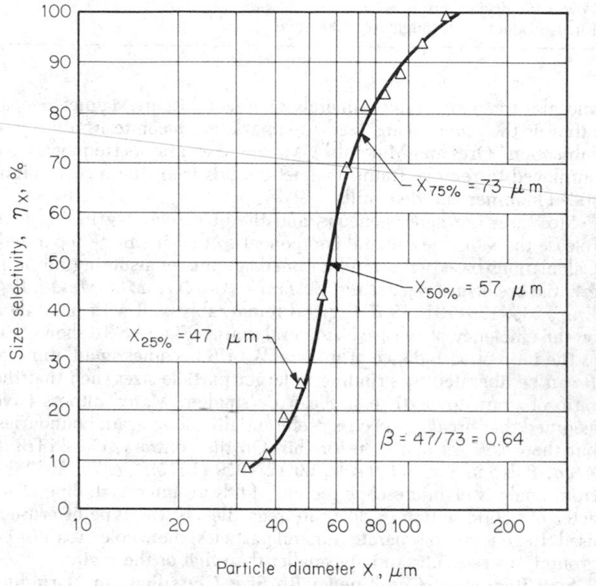

FIG. 8-14 Size-selectivity example.

There are many ways in which sharpness can be expressed. One index that has been widely used is the ratio

$$\beta = X_{25}/X_{75} \tag{8-14}$$

where β is the sharpness index, X_{75} is the particle size corresponding to the 75 percent classifier selectivity value, and X_{25} is the particle size corresponding to the 25 percent value. For perfect classification, β has a value of unit; the smaller β, the poorer the sharpness of classification.

Several empirical formulas for classifier selectivity have been proposed. Such a formula is needed for computer simulation of mill circuits. The following formula has been found to fit data from several field installations for classifiers of many types, including vibrating screens (Vaillant, AIME Tech. Pap. 67B26, 1967).

$$C_x = 1 - (1 - a) \exp b \left(1 - \frac{X}{X_0}\right) \quad \text{for } X > X_0$$
$$= a \quad \text{for } X \leq X_0 \tag{8-15}$$

where a, b, and X_0 are constants and X is the particle size. The agreement is especially good for wet classifying systems. For wet cyclones the factor a and Y_{50} are related to the ratio of overflow to underflow rates [Draper and Lynch, *Proc. Australas. Inst. Min. Metall.*, **209,** 109 (1964); Mizrahi and Cohen, *Trans. Inst. Min. Metall.*, C318–329 (December 1966); and Lynch and Rao, *Indian J. Tech.*, **6,** 106–114 (April 1968)]. An equation developed by stochastic reasoning for cyclones involves a similar exponential form [Molerus, *Chem Ing. Tech.*, **39**(13), 792–796 (1967)].

It has been suggested that the circulating load can be calculated by a material balance from size analyses of the feed, fine product, and coarse product of the classifier in a closed-circuit grinding system [Bond, *Rock Prod.*, **41,** 64 (January 1938)]. However, since size analyses are subject to error, it is better to use this information to check the size analyses (Vaillant, op. cit.). The appropriate equation is (Dahl, *Classifier Test Manual*, Portland Cem. Assoc. Bull. MRB-53, 1954)

$$\frac{q_R}{q_R + q_P} = R = \frac{Y_L(X) - Y_P(X)}{Y_R(X) - Y_P(X)} = \text{a constant for all } X \tag{8-16}$$

where q_R is tailings, q_P is classifier product, L is mill discharge, R is recycle, and Y is either the fraction of particles in a stream between two sieve sizes or the cumulative fraction retained or passing a sieve of size X.

SIMULATION OF MILLING CIRCUITS

The energy laws of Bond, Kick, and Rittinger relate to grinding from some average feed size to some product size but do not take into account the behavior of different sizes of particles in the mill. Computer simulation, based on population-balance models [Bass, *Z. Angew. Math. Phys.*, **5**(4), 283 (1954)], traces the breakage of each size of particle as a function of grinding time. Furthermore, the simulation models separate the breakage process into two aspects: a breakage rate and a mean fragment-size distribution. These are both functions of the size of particle being broken. They usually are not derived from knowledge of the physics of fracture but are empirical functions fitted to milling data. The following formulation is given in terms of a discrete representation of size distribution; there are comparable equations in integrodifferential form.

Batch Grinding Let w_k = the weight fraction of material retained on each screen of a nest of n screens; w_k is related to P_k, the fraction coarser than size X_k, by

$$w_k = (\partial P_k/\partial X_k)\,\Delta X_k \tag{8-17}$$

where ΔX_k is the difference between the openings of screens k and $k + 1$. The **grinding-rate function** S_u is the rate at which the material of upper size u is selected for breakage in an increment of time, relative to the amount of that size present:

$$dw_u/dt = -S_u w_u \tag{8-18}$$

The **breakage function** $\Delta B_{k,u}$ gives the size distribution of product breakage of size u into all smaller sizes k. Since some fragments from size u are large enough to remain in the range of size u, the term $\Delta B_{u,u}$ is not zero, and

$$\sum_{k=n}^{u} \Delta B_{k,u} = 1 \tag{8-19}$$

The differential equation of batch grinding is deduced from a balance on the material in the size range k. The rate of accumulation of material of size k equals the rate of production from all larger sizes minus the rate of breakage of material of size k:

$$\frac{dw_k}{dt} = \sum_{u=1}^{k} (w_u S_u(t)\,\Delta B_{k,u}) - S_k(t) w_k \tag{8-20}$$

In general, S_u is a function of all the milling variables. $\Delta B_{k,u}$ is also a function of breakage conditions. If it is assumed that these functions are constant, then relatively simple solutions of the grinding equation are possible, including an analytical solution [Reid, *Chem. Eng. Sci.*, **20**(11), 953–963 (1965)] and matrix solutions [Broadbent and Callcott, *J. Inst. Fuel*, **29,** 524–539 (1956); **30,** 18–25 (1967); and Meloy and Bergstrom, *7th Int. Min. Proc. Congr. Tech. Pap.*, 1964, pp. 19–31].

Solution of Batch-Mill Equations In general, the grinding equation can be solved by numerical methods—for example, the Euler technique (Austin and Gardner, *1st European Symposium on Size Reduction*, 1962) or the Runge-Kutta technique. The matrix method is a particularly convenient formulation of the Euler technique.

Reid's **analytical solution** is useful for calculating the product as a function of time t for a constant feed composition. It is

$$w_{Lk} = \sum_{n=1}^{k} a_{k,n} \exp(-\bar{S}_n\,\Delta t) \tag{8-21}$$

where the subscript L refers to the discharge of the mill, zero to the entrance, and $\bar{S}_n$ = 1 "corrected" rate function defined by $\bar{S}_n = (1 - \Delta B_{n,n})$ and B is then normalized with $\Delta B_{n,n} = 0$. The coefficients are

$$a_{k,k} = w_{0k} - \sum_{n=1}^{k-1} a_{k,n} \tag{8-22}$$

and

$$a_{k,n} = \sum_{u=n}^{k-1} \frac{S_u\,\Delta B_{k,u} a_{n,u}}{\bar{S}_k - \bar{S}_n} \tag{8-23}$$

The coefficients are evaluated in order since they depend on the coefficients already obtained for larger sizes.

The basic idea behind the **Euler method** is to set the change in w per increment of time as

$$\Delta w_k = (dw_k/dt)\,\Delta t \tag{8-24}$$

where the derivative is evaluated from Eq. (8-20). Equation (8-24) is applied repeatedly for a succession of small time intervals until the desired duration of milling is reached.

In the **matrix method** a modified rate function is defined, $S'_k = S_k\,\Delta t$ as the amount of grinding that occurs in some small time Δt. The result is

$$\mathbf{w}_L = (\mathbf{I} + \mathbf{S'B} - \mathbf{S'})\mathbf{w}_F = \mathbf{M}\mathbf{w}_F \tag{8-25}$$

where the quantities **w** are vectors, **S′** and **B** are the matrices of rate and breakage functions, and **I** is the unit matrix. This follows because the result obtained by multiplying these matrices is just the sum of products obtained from the Euler method. Equation (8-25) has a physical meaning. The unit matrix times $\mathbf{w}_F$ is simply the amount of feed that is not broken. $\mathbf{S'B}\mathbf{w}_F$ is the amount of feed that is selected and broken into the vector of products. $\mathbf{S'}\mathbf{w}_F$ is the amount of material that is broken out of its size range and hence must be subtracted from this element of the product. The entire term in parentheses can be considered as a mill matrix **M.** Thus the milling operation transforms the feed vector into the product vector. Meloy and Bergstrom (op. cit.) pointed out that when Eq. (8-25) is applied over a series of p short-time intervals, the result is

$$\mathbf{w}_L = \mathbf{M}^p\mathbf{w}_F \tag{8-26}$$

Matrix multiplication happens to be commutative in this special case. It is easy to raise a matrix to a power on a computer since three multiplications give the eighth power, etc. Therefore the matrix formulation is well adapted to computer use.

Continuous-Mill Simulation Batch-grinding experiments are the simplest type of experiments to produce data on grinding coefficients. But scale-up from batch to continuous mills must take into account the **residence-time distribution** in a continuous mill. This distribution is apparent if a tracer experiment is carried out. For this purpose background ore is fed continuously, and a pulse of tagged feed is introduced at time t_0. This tagged material appears in the effluent distributed over a period of time, as shown by a typical curve in Fig. 8-15. Because of this distribution some portions are exposed to grinding for longer times than others. Levenspiel (*Chemical Reaction Engineering*, Wiley, New York, 1962) shows several types of residence-time distribution that can be observed. Data on large mills indicate that a curve like that of Fig. 8-15 is typical (Keienberg et al., *3d European Symposium on Size Reduction*, op. cit., 1972, p. 629). This curve can be accurately expressed as a series of arbitrary functions (Merz and Molerus, *3d European Symposium on Size Reduction*, op. cit., 1972, p. 607). A good fit is more easily obtained

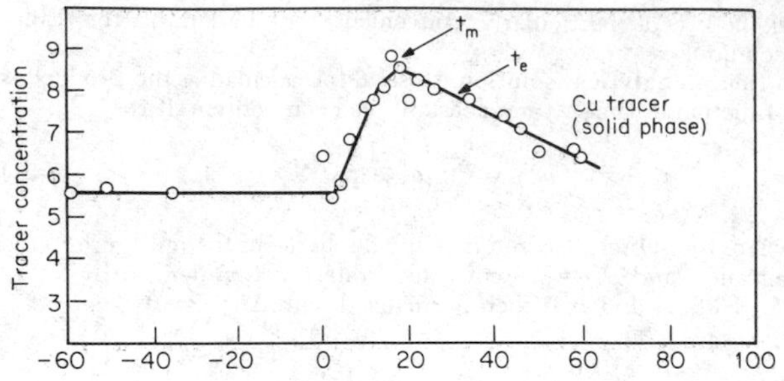

FIG. 8-15 Ore transit through a ball mill. Feed rate is 500 lb/h. *(Courtesy Phelps Dodge Corporation.)*

if we choose a function that has the right shape since then only the first two moments are needed. The log-normal probability curve fits most available mill data, as was demonstrated by Mori [*Chem. Eng. (Japan),* **2**(2), 173 (1964)]. Two examples are shown in Fig. 8-16. The log-normal plot fails only when the mill acts nearly as a perfect mixer.

To measure a residence-time distribution, a pulse of tagged feed is inserted into a continuous mill and the effluent is sampled on a schedule. If it is a dry mill, a soluble tracer such as salt or dye may be used and the samples analyzed conductimetrically or colorimetrically. If it is a wet mill, the tracer must be a solid of similar density to the ore. Materials like copper concentrate, chrome brick, or barites have been used as tracers and analyzed by x-ray fluorescence. To plot results in log-normal coordinates, the concentration data must first be normalized from the form of Fig. 8-15 to the form of cumulative percent discharged, as in Fig. 8-16. For this, one must either know the total amount of pulse fed or determine it by a simple numerical integration by using a time-sharing computer. The data are then plotted as in Fig. 8-16, and the coefficients in the log-normal formula of Mori can be read directly from the graph. Here $t_e = t_{50}$ is the time when 50 percent of the pulse has emerged. The standard deviation σ is the time between t_{16} and t_{50} or between t_{50} and t_{84}. Knowing t_e and σ, one can reconstruct the straight line in log-normal coordinates. One can also calculate the Peclet number, Dt_e/L^2, which is a measure of the sharpness of the pulse (Levenspiel, op. cit.). Here D is the particle diffusivity. A few available data are summarized [Snow, *International Conference on Particle Technology,* IIT Research Institute, Chicago, Ill. 60616, 1973, p. 28) for wet mills. Other experiments are presented for dry mills [Hogg et al., *Trans. Am. Inst. Min. Metall. Pet. Eng.,* **258,** 194 (1975)]. The most important variables affecting the Peclet number are L/diameter of the mill, ball size, mill speed, scale expressed either as diameter or as throughput, degree of ball filling, and degree of material filling.

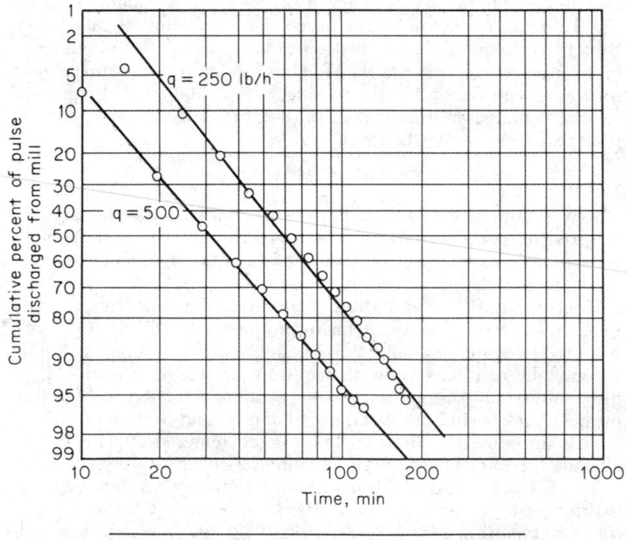

q, lb/h	σ	t_e, min	Dt_e/L^2
250	0.29	39	0.23
500	0.32	18	0.28

FIG. 8-16 Log-normal plot of residence-time distribution in Phelps Dodge mill.

Solution for Continuous Mill In the method of Mori (op. cit.) the residence-time distribution is broken up into a number of segments, and the batch-grinding equation is applied to each of them. The resulting size distribution at the mill discharge is

$$\mathbf{w}(L) = \mathbf{w}(t)\,\Delta\varphi \tag{8-27}$$

where $\mathbf{w}(t)$ is a matrix of solutions of the batch equation for the series of times t, with corresponding segments of the cumulative residence-time curve.

Using the Reid solution, Eq. (8-21), this becomes

$$\mathbf{w}(L) = \mathbf{R}\,\mathbf{Z}\,\Delta\varphi \tag{8-28}$$

where Z is a matrix of exponentials exp $(-\overline{S}t)$ from Eq. (8-21).

Closed-Circuit Milling In closed-circuit milling the tailings from a classifier are mixed with fresh feed and recycled to the mill. Calculations can be based on a material balance and an explicit solution such as Eq. (8-26). Material balances for the normal circuit arrangement (Fig. 8-17) give

$$q = q_F + q_R \tag{8-29}$$

where q = total mill throughput, q_F = rate of feed of new material, and q_R = recycle rate. A material balance on each size gives

$$w_{ok} = \frac{q_F w_{Fk} + \dfrac{q_R}{R}\eta_k w_{Lk}}{q} \tag{8-30}$$

where w_{Ok} = fraction of size k in the mixed feed streams, R = the recycle ratio, and η_k = classifier selectivity for size k. With these conditions a calculation of the transient behavior of the mill can be performed by using any method of solving the milling equation and iterating over intervals of time τ = residence time in the mill. This information is important for evaluating mill-circuit-control stability and strategies. If the throughput q is controlled to be a constant, as is often the case, then τ is constant, and a closed-form matrix solution can be found for the steady state [Callcott, *Trans. Inst. Min. Metall.,* **76**(1), C1–11 (1967)]. The resulting flow rates and composition vectors are given in Fig. 8-17. Equations for the reverse-circuit case, in which the feed is classified before it enters the mill, are given by Calcott (loc. cit). These results can be used to investigate the effects of changes in feed composition on the product. Separate calculations can be made to find the effects of classifier selectivity, mill throughput or recycle, and grindability (rate function) to determine optimum mill-classifier combinations [Lynch, Whiten, and Draper, *Trans. Inst. Min. Metall.,* **76,** C169, 179 (1967)].

Data on Behavior of Grinding Functions Although several breakage functions were early suggested [Gardner and Austin, *1st European Symposium on Size Reduction,* op. cit., 1962, p. 217; Broadbent and Calcott, *J. Inst. Fuel,* **29,** 524 (1956); 528 (1956); 18 (1957); **30,** 21 (1957)], the simple Gates-Gaudin-Schumann equation [Eq. (8-2) and Fig. 8-13] has been most widely used to fit ball-mill data. For example, this form was assumed by Herbst and Fuerstenau [*Trans. Am. Inst. Min. Metall. Pet. Eng.,* **241**(4), 538 (1968)] and Kelsall et al. [*Powder Technol.,* **1**(5), 291 (1968); **2**(3), 162 (1968); 3(3), 170 (1970)]. More recently it has been observed that when the Schumann equation is used, the amount of coarse fragments cannot be made to agree with the mill-product distribution regardless of the choice of rate function. This points to the need for a breakage function that has more coarse fragments, such as the function used by

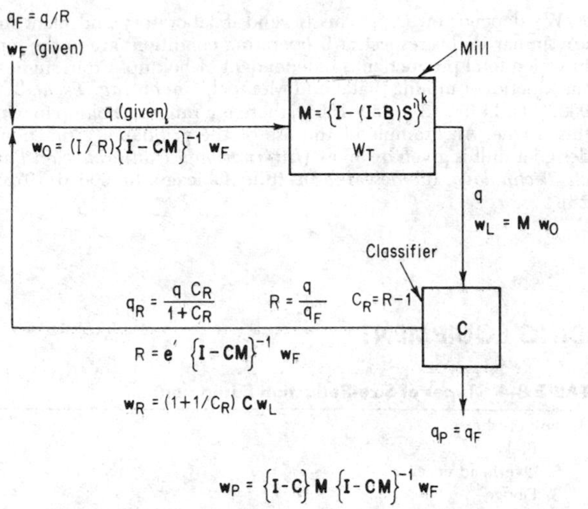

Nomenclature

C_R = circulating load, $R - 1$
$\mathbf{C}$ = classifier selectivity matrix, which has classifier selectivity-function values η on diagonal, zeros elsewhere
I = identity matrix, which has ones on diagonal, zeros elsewhere
$\mathbf{M}$ = mill matrix, which transforms mill-feed-size distribution into mill-product-size distribution
q = flow rate of a material stream
R = recycle ratio q/q_F
$\mathbf{w}$ = vector of differential size distribution of a material stream
W_T = holdup, total mass of material in mill
Subscripts:
0 = inlet to mill
F = feed stream
L = mill-discharge stream
P = product stream
R = recycle stream, classifier tailings

FIG. 8-17 Normal closed-circuit continuous grinding system with stream flows and composition matrices, obtained by solving material-balance equations. [*Callcott*, Trans. Inst. Min. Metall., *76(1), C1–11 (1967).*]

Reid and Stewart (Chemica meeting, 1970) and Stewart and Restarick [*Proc. Australas. Inst. Min. Metall.* (239), 81 (1971)] and shown in Fig. 8-18. This graph can be fitted by a *double Schumann equation*

$$B(X) = A\left(\frac{X}{X_0}\right)^s + (1 - A)\left(\frac{X}{X_0}\right)^r \tag{8-31}$$

where A is a coefficient less than 1.

In the investigations mentioned earlier the breakage function was assumed to be normalizable; i.e., the shape was independent of X_0. Austin and Luckie [*Powder Technol.*, **5**(5), 267 (1972)] allowed the coefficient A to vary with the size of particle breaking when grinding soft feeds.

Grinding-Rate Functions These were determined by tracer experiments in laboratory mills by Kelsall et al. (op. cit.) as shown in Fig. 8-19 and in similar work by Szantho and Fuhrmann [*Aufbereit. Tech.*, **9**(5), 222 (1968)]. These curves can be fitted by the following equation:

$$\frac{S}{S_{max}} = \left(\frac{X}{X_{max}}\right)^\alpha \exp\left(-\frac{X}{X_{max}}\right) \tag{8-32}$$

That a maximum must exist should be apparent from the observation of Coghill and Devaney (U.S. Bur. Mines Tech. Pap., 1937, p. 581) that there is an optimum ball size for each feed size. Figure 8-19

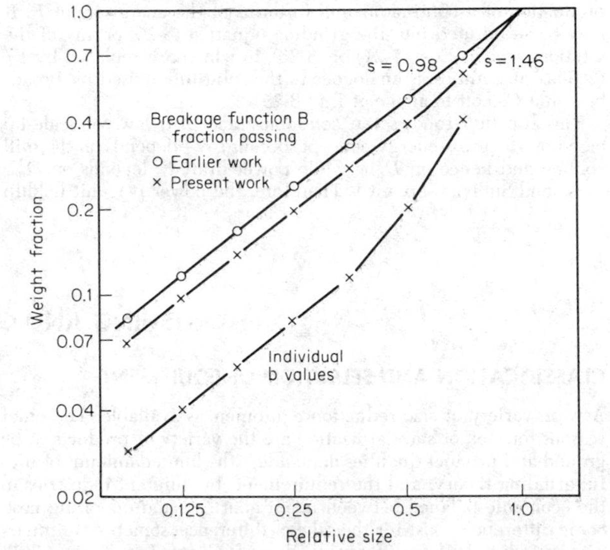

FIG. 8-18 Experimental breakage functions. *(Reid and Stewart, Chemica meeting, 1970.)*

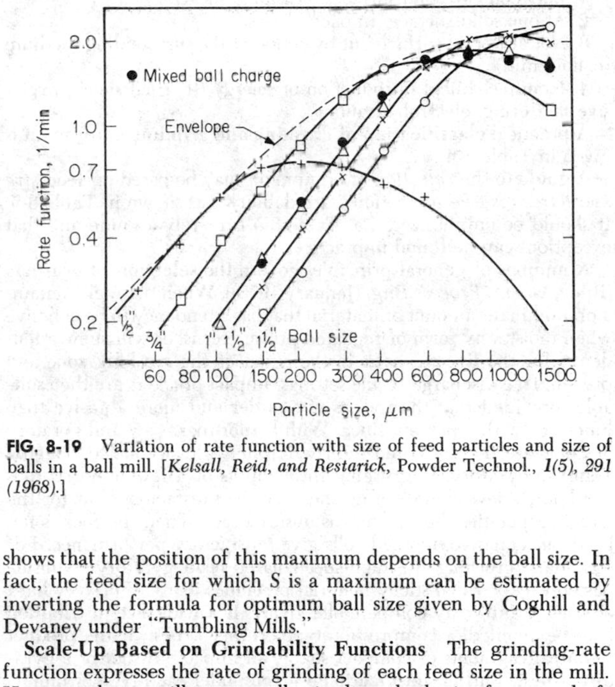

FIG. 8-19 Variation of rate function with size of feed particles and size of balls in a ball mill. [*Kelsall, Reid, and Restarick*, Powder Technol., *1(5), 291 (1968).*]

shows that the position of this maximum depends on the ball size. In fact, the feed size for which S is a maximum can be estimated by inverting the formula for optimum ball size given by Coghill and Devaney under "Tumbling Mills."

Scale-Up Based on Grindability Functions The grinding-rate function expresses the rate of grinding of each feed size in the mill. However, large mills are usually sized on the basis of power draft (see subsection "Energy Laws"). A new function called the **grindability function** was defined by Snow and Meloy (*3d European Symposium on Size Reduction*, op. cit., 1972, p. 535) and also by Herbst and Fuerstenau [*Trans. Am. Inst. Min. Metall. Pet. Eng.*, **254**, 343 (1973)] for a batch mill:

$$S_G(X) = \frac{S(X)W_T}{E} \times \frac{60}{1000} \tag{8-33}$$

where $S_G(x)$ is the grindability function, metric tons/kWh; $S(X)$ is the rate function, fraction/min; W_T is the mill holdup, kg; and E is the mean power draft of the mill, kW. This function expresses the production of a continuous mill in terms of the energy input E. It may be substituted into the grinding equation (8-20) or any of the solutions such as Eqs. (8-21) or (8-26), in which S is replaced by E/Q. This substitution is analogous to the substitution used by Broadbent and Calcott to arrive at Eq. (8-25).

This function follows the commonly accepted laws of scale-up based on the gross-energy concept, because W_T depends on the mill volume and hence on D_{mill}^2, while power draft E depends on $D_{mill}^{2.6}$ (Rose and Sullivan, op. cit.). Therefore, the power per unit holdup E/W_T depends on $D_{mill}^{0.6}$. This is valid if laboratory and plant mills are similar in this respect or if operating conditions are in the range in which total production is independent of holdup. From studies of the kinetics of milling [Patat and Mempel, *Chem. Ing. Tech.*, **37**(9), 933; (11), 1146; (12), 1259 (1965)] there is a range of holdup in which this is true. An example of the use of the grindability function to design a mill is given by Snow (*International Conference on Particle Technology*, IIT Research Institute, Chicago, Ill. 60616, 1973, p. 28).

CRUSHING AND GRINDING EQUIPMENT

CLASSIFICATION AND SELECTION OF EQUIPMENT

A wide variety of size-reduction equipment is available. The chief reasons for lack of standardization are the variety of products to be ground and product qualities demanded, the limited amount of useful grinding theory, and the requirements by different industries in the economic balance between investment cost and operating cost. Some differences exist for the sake of difference; sometimes similarities are advertised as differences [Rumpf, *Chem. Ing. Tech.*, **37**(3), 187–202 (1965)].

Equipment may be classified according to the way in which forces are applied, as follows (Rumpf, loc. cit.):

1. Between two solid surfaces (crushing, shearing)
2. At one solid surface (impact)
3. Not at a solid surface but by action of the surrounding medium (colloid mill)
4. Nonmechanical introduction of energy (thermal shock, explosive shattering, electrohydraulic)

A practical **classification of crushing and grinding equipment** is given in Table 8-4.

A guide to the *selection* of equipment may be based on feed *size* and *hardness* (see subsection "Grindability") as shown in Table 8-5. It should be emphasized that Table 8-5 is merely a guide and that exceptions can be found in practice.

A number of general principles govern the **selection of crushers** [Riley, *Chem. Process Eng.* (January 1953)]. When the rock contains a predominant amount of material that has a tendency to be cohesive when moist, any form of repeated pressure crusher will show a tendency for the fines to pack in the outlet of the crushing zone and prevent free discharge at fine settings. Impact breakers are then suitable, provided that the rock is not harder and more abrasive than limestone with 5 percent silica. With harder rocks, jaw and gyratory crushers are required, and the jaw crusher is less prone to clogging than the gyratory. In crushing throughputs of a few hundred tons per hour, a jaw or impact crusher may be satisfactory, but for the largest capacities the gyratory is unsurpassed. When the rock is not hard but cohesive, toothed rolls give satisfactory performance. For secondary crushing the high-speed flared-head gyratory is unsurpassed except when sticky material precludes its use. For very hard ores a rod mill may compete effectively. If a wide size distribution is to be avoided, a compression-type crusher is best; if the product requires fragments of compact shape, an impact crusher or a gyratory is best. Further information is given under each type of mill.

TABLE 8-4 Types of Size-Reduction Equipment

A. Jaw crushers:
  1. Blake
  2. Overhead eccentric
  3. Dodge
B. Gyratory crushers:
  1. Primary
  2. Secondary
  3. Cone
C. Heavy-duty impact mills:
  1. Rotor breakers
  2. Hammer mills
  3. Cage impactors
D. Roll crushers:
  1. Smooth rolls (double)
  2. Toothed rolls (single and double)
E. Dry pans and chaser mills
F. Shredders:
  1. Toothed shredders
  2. Cage disintegrators
  3. Disk mills
G. Rotary cutters and dicers
H. Media mills:
  1. Ball, pebble, rod, and compartment mills:
    a. Batch
    b. Continuous
  2. Autogenous tumbling mills
  3. Stirred ball and sand mills
  4. Vibratory mills
I. Medium peripheral-speed mills:
  1. Ring-roll and bowl mills
  2. Roll mills, cereal type
  3. Roll mills, paint and rubber types
  4. Buhrstones
J. High-peripheral-speed mills:
  1. Fine-grinding hammer mills
  2. Pin mills
  3. Colloid mills
  4. Wood-pulp beaters
K. Fluid-energy superfine mills:
  1. Centrifugal jet
  2. Opposed jet
  3. Jet with anvil

JAW CRUSHERS

Design and Operation These crushers may be divided into three main groups (Fig. 8-20), the *Blake*, with a movable jaw pivoted at the top, giving greatest movement to the smallest lumps; the *Dodge*, with the movable jaw pivoted at the bottom, giving greatest movement to the largest lumps; and the *overhead eccentric*, which is hinged at the top similarly to the Blake, with the movable jaw suspended on the eccentric shaft.

The **Blake** has a removable crushing plate, usually corrugated, fixed in a vertical position at the front end of a hollow rectangular frame. A similar plate, at a suitable angle, is attached to a swinging lever (movable jaw) suspended from a shaft resting in the sides of the frame. Movement is accomplished through a knuckle action by the rising and falling of a second lever (pitman) carried by an eccentric shaft. The vertical movement is communicated horizontally to the jaw by two plates (toggles).

Crushing angles in standard Allis-Chalmers Blake-type machines generally are near 0.47 rad (27°) (see Fig. 8-21). The reduction ratios

TABLE 8-5 Guide to Selection of Crushing and Grinding Equipment

Size-reduction operation	Hardness of material	Size*: Range of feeds, in.† Max.	Range of feeds, in.† Min.	Range of products, in.† Max.	Range of products, in.† Min.	Reduction ratio‡	Types of equipment
Crushing:							
Primary	Hard	60	12	20	4	3 to 1	*A* to *D*
		20	4	5	1	4 to 1	
Secondary	Hard	5	1	1	0.2	5 to 1	*A* to *F*
		1.5	0.25	0.185 (4)	0.033 (20)	7 to 1	
	Soft	20	4	2	0.4	10 to 1	*C* to *G*
Grinding:							
Pulverizing:							
Coarse	Hard	0.185 (4)	0.033 (20)	0.023 (28)	0.003 (200)	10 to 1	*D* to *I*
Fine	Hard	0.046 (14)	0.0058 (100)	0.003 (200)	0.00039 (1250)	15 to 1	*H* to *K*
Disintegration:							
Coarse	Soft	0.5	0.065	0.023	0.003	20 to 1	*F, I*
Fine	Soft	0.156 (5)	0.0195 (32)	0.003 (200)	0.00039 (1250)	50 to 1	*I* to *K*

*85% by weight smaller than the size given.
†Sieve number in parentheses.
‡Higher reduction ratios for closed-circuit operations.
NOTE: To convert inches to centimeters, multiply by 2.54.

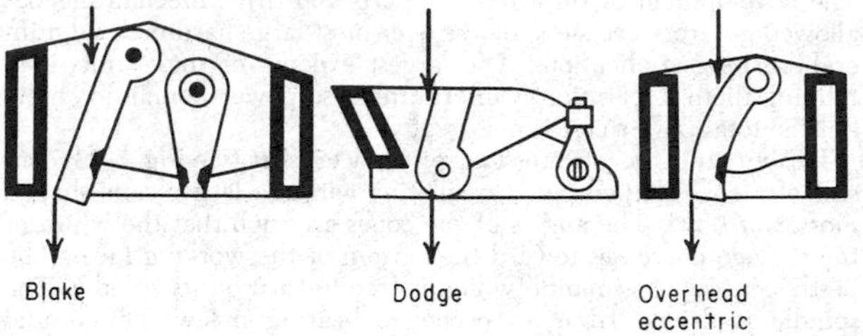

FIG. 8-20 Jaw-crusher designs.

at minimum recommended settings and with straight jaw plates average about 8:1. Curved (or concave) jaw plates are designed to minimize choking.

The **Dodge** type has the advantage of a larger feed opening for the same cost as a Blake, and it is useful for low-production intermittent service to produce a uniform product in sizes having a smaller than 11- by 15-in feed opening. It is seldom used now.

The **overhead eccentric jaw crusher** *(Kennedy Van Saun, Telsmith)* and the similar **fine-reduction machine** *(Allis-Chalmers)* fall into the third type. These are single-toggle machines having the swing jaw mounted directly on the eccentric shaft so that it receives a downward as well as a forward motion. The lower end of the swing jaw is held in position against the toggle by a tension rod. Greater wear caused by this motion and direct transmittal of shocks to the bearing limit its use to readily breakable material. However, a large reduction ratio, useful for simplified low-tonnage circuits with fewer grinding steps is possible.

Jaw crushers are usually rated by the dimensions of their feed area. This depends on the width of the crushing jaws and the gap, which is the maximum distance between the fixed and movable jaws at the feed opening.

The choice among the three types of jaw crushers is generally dictated by the feed characteristics, tonnage, and product requirements (Pryon, *Mineral Processing*, Mining Publications, London, 1960; Wills, *Mineral Processing Technology*, Pergamon, Oxford, 1979). Double-toggle Blake crushers are most commonly used for primary crushing of hard rocks. Double-toggle-type crushers cost about 50 percent more than similar overhead-eccentric-type crushers. However, the jaw wear and operating costs of the overhead eccentric are higher for the crushing of hard rocks. Overhead eccentric crushers are generally preferred for crushing rocks with a hardness equal to or lower than that of limestone. The Dodge crusher is suitable when a close-sized product is desired and throughput is less important. It is more commonly used in the laboratory and is rarely seen in milling flow sheets.

Performance Jaw crushers are applied to the primary crushing of hard materials and are usually followed by other types of crushers. In smaller sizes they are used as single-stage machines.

The setting of a jaw crusher may be stated as being the close or the wide opening between the moving jaws at the outlet end. The

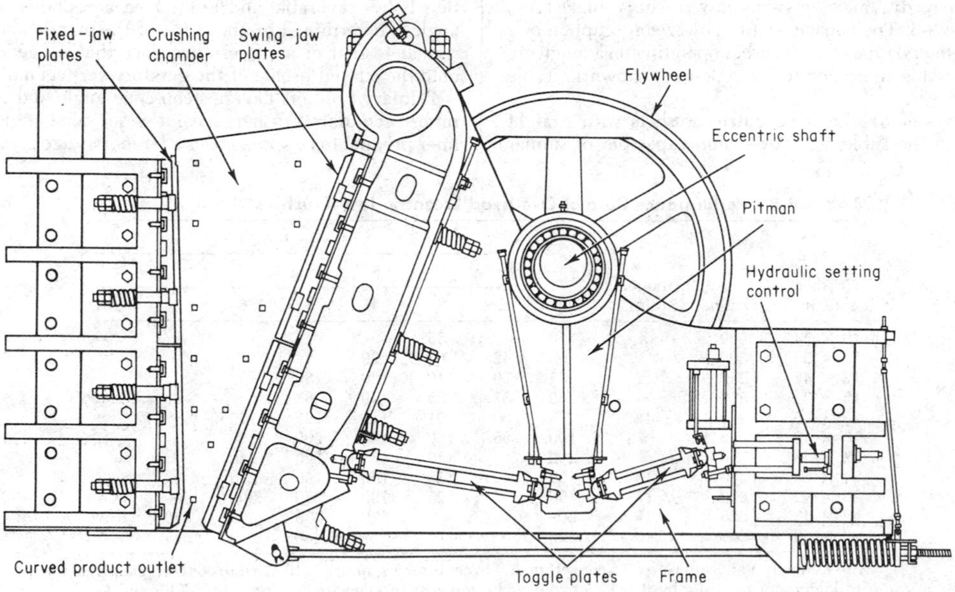

FIG. 8-21 Blake jaw crusher. *(Allis-Chalmers Corporation.)*

TABLE 8-6 Performance Data of Blake Swing Jaw Crusher*

Crusher size receiving opening, in.	Approx. rough wt., lb.	Approx. speed, r.p.m.	Required hp.	Capacity at setting, tons/hr.	Setting, in.
7 × 10	6,000	275	7	7–10	2½
10 × 24	18,500	275	15	25–30	2½
14 × 24	25,000	275	25	45–55	4
18 × 36	51,000	250	40	70–90	5
24 × 36	80,000	200	75	150–175	5
				180–210	7
30 × 42	115,000	200	100	190–230	6
				240–270	8
36 × 48	160,000	200	125–150	230–270	6
				280–320	8
48 × 60	350,000	175	175–200	400–475	6
				525–600	8

*Kennedy Van Saun Corp. To convert inches to centimeters, multiply by 2.54; to convert pounds to kilograms, multiply by 0.4535; to convert horsepower to kilowatts, multiply by 0.746; and to convert tons per hour to kilograms per hour, multiply by 907.

TABLE 8-7 Performance Data of Dodge Jaw Crusher*

Crusher size, in.	R.p.m.	Jaw motion, in.	Hp.	Capacity, tons/hr. Setting, in. ½	¾	1	1½
4 × 6	275	½	3	¼	½	1	
7 × 9	235	13/16	6	. . .	1	2	3
8 × 12	220	19/32	10	. . .	1½	3	4
11 × 15	200	1⅝	15	. . .	2	4	6

*Allis-Chalmers Corporation. To convert inches to centimeters, multiply by 2.54; to convert pounds to kilograms, multiply by 0.4535; to convert horsepower to kilowatts, multiply by 0.746; and to convert tons per hour to kilograms per hour, multiply by 907.

reciprocating motion of the jaws causes the opening to vary between close and wide. Specifications usually are based on close settings. The setting is adjustable.

Capacities of Kennedy Van Saun **swing jaw crushers** (Blake type) are given in Table 8-6. The capacities and power consumption of a **Dodge** *(Allis-Chalmers)* type of jaw crusher operating on a tough ore that offers considerable resistance to reduction are shown in Table 8-7.

Performance data of **overhead eccentric** crushers with straight jaw plates are given in Table 8-8. Power and capacities of similar crushers with curved jaw plates are 50 percent greater. The **Kue-Ken** jaw crusher *(Pennsylvania Crusher Corp.)* is a version of the overhead eccentric in which the eccentric is positioned well above the swinging jaw plate to which it is attached. Consequently the moving jaw meets the rock firmly and squarely. There is no rubbing action or vertical motion, and so abrasion is eliminated and the jaw plates last longer. Kue-Ken jaw crushers are most advantageous for the primary and secondary reduction of hard, abrasive rock to a relatively coarse but uniform product. Depending on machine size, feed may be as large as 1.2 m (48 in), and product top size may be as small as 1.3 cm (½ in). Capacities range from a few hundred pounds per hour up to 634 Mg/h (700 tons/h) when crushing hard dry quartz.

Jaw crushers are preferred when the crusher gap is more important than the throughput. They can accept a larger size feed than gyratory crushers of the same capacity. If the required throughput in tons per hour is less than the square of the gap in inches, a jaw crusher is more economical (Taggart, *Handbook of Mineral Dressing*, Wiley, New York, 1945). Gyratory crushers are preferred for larger throughputs. Jaw crushers can also handle clayey material better than other types of primary crushers.

Crusher Product Sizes Table 8-9 relates product size to the discharge setting of the crusher in terms of the percent smaller than that size in the product. Size-distribution curves differ for various types of materials crushed, and a general set of curves is not valid.

GYRATORY CRUSHERS

The development of improved supports and drive mechanisms has allowed gyratory crushers to take over most large hard-ore and mineral-crushing applications. The largest expense of these units is in relining them. Operation is intermittent; so power demand is high, but the total power cost is not great.

Design and Operation The gyratory crusher (see Fig. 8-22) consists of a cone-shaped pestle oscillating within a larger cone-shaped mortar or bowl. The angles of the cones are such that the width of the passage decreases toward the bottom of the working faces. The pestle consists of a mantle which is free to turn on its spindle. The spindle is driven from an eccentric bearing below. Differential motion causing attrition can occur only when pieces are caught simultaneously at the top and bottom of the passage owing to different radii at these points.

The circular geometry of the crusher gives a favorably small **nip angle** in the horizontal direction. The nip angle in the vertical direction is less favorable and limits feed acceptance. The vertical nip angle is determined by the shape of the mantle and bowl liner; it is similar to that of a jaw crusher since the jaw geometry is mapped onto the curved profile of the gyratory vertical outline.

Primary crushers have a steep cone angle and a small reduction ratio. **Secondary** crushers have a wider cone angle; this allows the finer product to be spread over a larger passage area and also spreads

TABLE 8-8 Performance Data of Overhead Eccentric Jaw Crushers*

Crusher size, in	r/min	Jaw motion, in	hp	Capacity, tons/h Setting, in 2	3	4	5	6	8	10
10 × 24	300	½	50	32	55					
10 × 26	300	1	60	42	67	90				
12 × 48	250	½	75	70	110	140	180			
15 × 24	300	9/16	60	37	55	70	88			
15 × 36	275	1⅛	75	60	91	110	125	150		
18 × 48	250	⅝	100	95	130	150	200	225		
22 × 48	250	⅝	100		140	165	220	240		
24 × 36	260	1¼	125		115	150	190	230		
30 × 42	230	1⅜	150		125	190	250	300	400	500
42 × 48	225	1⅜	200			270	335	405	540	675
48 × 60	200	1½	300				440	570	645	770

*Kennedy Van Saun Corp. To convert inches to centimeters, multiply by 2.54; to convert pounds to kilograms, multiply by 0.4535; to convert horsepower to kilowatts, multiply by 0.746; and to convert tons per hour to kilograms per hour, multiply by 907.

TABLE 8-9 Relation of Product Size to Discharge Setting of Crusher*

Setting measured on:	Kind of feed	% of product passing a square opening equal to discharge setting of crusher			
		Limestone	Granite	Trap rock	Ores
Open side:					
Primary service:					
Jaw crusher†	Quarry-run	85–90	70–75	65–70	85–90
	Prescalped	80–85	65–70	60–65	80–85
Gyratory crusher	Quarry-run	85–90	75–80	65–70	85–90
	Prescalped	80–85	70–75	60–65	80–85
Secondary service:					
Gyratory crusher‡	Screened	85–90	80–85	75–80	85–90
Close side:					
Reduction service:					
Gyratory§	Screened	70–75	65–70	65–70	70–75

*From "Crushing Theory and Practice," Allis-Chalmers Mfg. Co.

†Blake type, or crushers with equivalent speeds and throws: opening measured from tip of corrugations on one jaw plate to bottom of corrugations in opposing plate.

‡For standard, or reduction, types with non-choking concaves. Single-toggle jaw product, on screened feed, will approximate that of gyratory-type secondary crushers with non-choking concaves.

§For high-speed short-throw crushers, such as Newhouse or Type R.

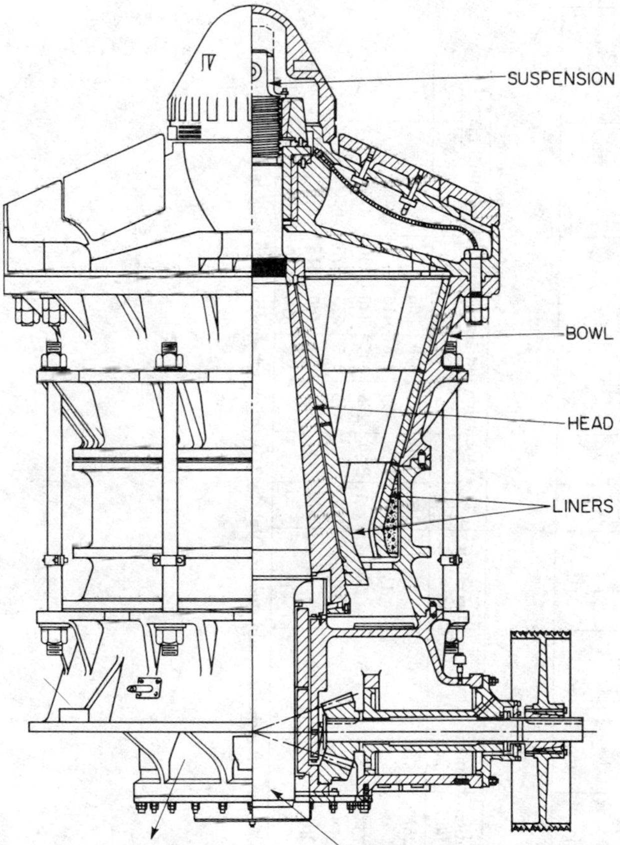

FIG. 8-22 Primary gyratory crusher with spider suspension. *(Nordberg Industrial.)*

the wear over a wider area. Wear occurs to the greatest extent in the lower, fine-crushing zone.

The three general types of gyratory crusher are the **suspended-spindle,** the **supported-spindle,** and the **fixed-spindle** types. Primary gyratories are designated by the size of feed opening and secondary or reduction crushers by the diameter of the head in feet and inches. There is a close opening and a wide opening as the mantle gyrates with respect to the concave ring at the outlet end. The close opening is known as the close setting or the close-side setting and sometimes as the closed-side setting, while the wide opening is known as the wide-side or open-side setting. Specifications usually are based on close settings. The setting is adjustable by raising or lowering the mantle.

The length of the crushing stroke greatly affects the capacity and the screen analysis of the crushed product. A very short stroke will give a very evenly crushed product but will not give the greatest capacity. A very long stroke will give the greatest capacity, but the product will contain some very fine particles, a larger amount of medium-size pieces, and a greater proportion of large fragments.

Performance Crushing occurs through the full cycle in a gyratory crusher, and this produces a higher crushing capacity than a similar-sized jaw crusher, which crushes only in the shutting half of the cycle. Gyratory crushers also tend to be cheaper and are easier to operate. They operate most efficiently when they are fully charged. These crushers are designed to operate with the main shaft fully buried in charge. Power consumption for gyratory crushers is also lower than that of jaw crushers. These are preferred over jaw crushers when capacities of 800 Mg/h (900 tons/h) or higher are required.

The crushing rate of a gyratory generally depends on the hardness of the material being crushed and on the amount of product-size material in the feed. For this reason gyratories are often operated in parallel with a scalping grizzly screen, provided the added cost of the screen is less than the cost of increased crusher capacity.

Primary gyratories will accept feed directly from truck or railcar. Most manufacturers make both mechanical and hydraulically supported types. Figure 8-22 shows a Nordberg primary gyratory crusher with spider suspension. It is available in 1- to 1.5-m (42-, 48-, 54-, and 60-in) feed sizes. Table 8-10 gives capacity data for the Superior gyratory crusher *(Allis-Chalmers)*.

Gearless primary gyratory crushers *(Kennedy Van Saun)* are driven by a multi-V belt from an electric motor. They are available in a larger series of 10 sizes with feed openings from 0.3 to 1.4 m (12 to 54 in). Capacity ranges from 68 to 1360 Mg/h (75 to 1500 tons/h), while power ranges from 37 to 336 kW (50 to 450 hp) on ordinary rock. A finer series of 10 sizes has feed openings from 4.4 to 43 cm (1¾ to 17 in) and capacities from 1 to 363 Mg/h (1 to 400 tons/h). Secondary crushers are driven through a gear from the motor.

The **multistage fine-reduction crusher** *(Traylor)* features upper and lower stages. The upper stage functions as a distributing feeder for the lower or finishing stage. The upper stage accomplishes about half of the crushing and provides feed of proper size for the lower stage.

TABLE 8-10 Performance Data for Superior Gyratory Crushers*

Capacities for crushing limestone, tons/hr.

A. Primary Crushers

Crusher size	Approx. feed opening, in.	Gyrations per min.	Pinion r.p.m.	Max. hp.	Eccentric throw, in.	Open-side setting of discharge opening, in.																			
						$2\frac{1}{2}$	3	$3\frac{1}{2}$	4	$4\frac{1}{2}$	5	$5\frac{1}{2}$	6	$6\frac{1}{2}$	7	$7\frac{1}{2}$	8	$8\frac{1}{2}$	9	$9\frac{1}{2}$	10	$10\frac{1}{2}$	11	$11\frac{1}{2}$	12
30–55	30 × 78	175	585	150	$\frac{5}{8}$	150	205	270	335	390	450	510													
					$1\frac{1}{4}$	...	...	...	605	675	735	800													
36–55	36 × 90	175	585	180	$\frac{3}{4}$	...	...	...	270	310	350	380													
					$1\frac{1}{4}$	...	...	...	...	...	515	585													
42–65	42 × 108	150	497	265	1	...	...	...	...	540	660	790	920	1040	1170										
					$1\frac{1}{2}$	...	...	...	...	...	...	....	1040	1260	1490										
48–74	48 × 120	135	497	300	1	...	...	...	...	...	930	1000	1080	1150	1230	1300	1380								
					$1\frac{5}{8}$	...	...	...	...	...	...	....	....	1920	2060	2180	2340								
54–74	54 × 132	135	497	300	1	...	...	...	...	...	...	960	1040	1100	1160	1240	1330								
					$1\frac{5}{8}$	...	...	...	...	...	...	....	....	....	1950	2070	2210								
60–89	60 × 145	110	435	330	1	...	...	...	...	...	...	....	1130	1170	1240	1310	1400	1480	1610						
					$1\frac{13}{16}$	...	...	...	...	...	...	....	....	....	....	....	2540	2600	2700						
60–109	60 × 150	100	400	1000	$1\frac{1}{2}$	...	...	...	...	...	...	....	....	....	....	....	....	3250	3500	3750	4000	4250	4500	4750	5000

B. Secondary Crushers

Crusher size	Approx. feed opening, in.	Gyrations per min.	Pinion r.p.m.	Max. hp.	Eccentric throw, in.	Open-side setting of discharge opening, in.								
						$1\frac{1}{2}$	2	$2\frac{1}{2}$	3	$3\frac{1}{2}$	4	$4\frac{1}{2}$	5	$5\frac{1}{2}$
16–50	16 × 55	225	764	100	$\frac{3}{4}$	150	190	250	310	360				
					$1\frac{1}{4}$	...	...	350	376	415				
24–60	24 × 66	175	585	175	$\frac{3}{4}$	195	230	268	305	340	380	415		
					$1\frac{1}{4}$	...	...	415	475	525	575	620		
30–70	30 × 84	150	497	200	$\frac{3}{4}$	...	300	350	400	460	515	570	620	680
					$1\frac{1}{2}$	...	...	...	...	810	940	1060	1180	1310

*Allis-Chalmers Corporation. To convert inches to centimeters, multiply by 2.54; to convert horsepower to watts, mulitply by 746.

Gyratory crushers of smaller size feature wide cone angles, making them suitable for throughput of a finer product. They can be used as **secondary crushers** or as primaries when quarrying produces suitable feed sizes. Usually the diameter of or height of the mantle is given as the nominal size rather than the feed opening. These crushers are generally available with either a spring-release mechanism or a hydraulic setting and release mechanism.

Hydrocone *(Allis-Chalmers)* crushers are made in sizes of 0.56 to 2.13 m (22, 30, 36, 45, 51, 60, and 84 in). Feed size varies from ¾ to 11 in, power from 22 to 373 kW (30 to 500 hp), and capacity from 9 to 900 Mg/h (10 to 1000 tons/h). Three shapes of crushing chamber are available.

The **Gyra-Kone** crusher (*Kennedy Van Saun*) is available in 1- and 1.3-m (40- and 52-in) sizes for the secondary crushing of stone and aggregates. Capacities range from 91 to 272 Mg/h (100 to 300 tons/h) depending on feed size. This crusher is available in either geared or gearless types.

The fixed-spindle or pillar-shaft gyratory, known as the **Telsmith breaker** *(Smith Engineering Works, Barber-Greene Co.)* has a rigid shaft that does not rotate or gyrate; the full stroke is exerted on the largest particles as they enter the bowl. Seven standard sizes of Telsmith breaker are made, from 11.2 to 93.2 kW (15 to 125 hp), and receiving opening from 0.17 by 0.89 m to 0.63 by 2.7 m (6¾ by 35 in to 25 by 106 in). Capacities range from 15 to 16 Mg/h (17 to 18 tons/h) for the smallest size to 272 to 317 Mg/h (300 to 350 tons/h) for the largest, discharge opening from 2.54 to 10.2 cm (1 to 4 in).

The **Telsmith Gyrasphere** is a crusher of long-stroke, spring-relief type with a spherical crushing head. Five sizes are available, in standard or fine style, with coarse, medium, or fine bowl. Power ranges from 18.7 to 224 kW (25 to 300 hp), feed opening from 3.2 to 38 cm (1¼ to 15 in), discharge setting from 0.32 to 6.4 cm (⅛ to 2½ in), in capacities from 5.4 to 413 Mg/h (6 to 455 tons/h), depending on type of crusher and setting. The **Telsmith Intercone** crusher is a comparatively low-cost machine of rugged construction. Two standard sizes are made, of 15 to 37 kW (20 to 50 hp), 5.7- to 10.1-cm (2¼- to 4-in) feed opening, 1.3- to 2.9-cm (½- to 1⅛-in) discharge opening, and capacities ranging from 14 to 62 Mg/h (15 to 68 tons/h).

The **Kue Ken** gyratory crusher *(Pennsylvania Crusher Corp.)* has an adjustable bowl with spring relief. A 1.3-m (51-in) unit with 7-cm (2¾-in) feed opening and 1.3-cm (½-in) discharge setting crushes 82 Mg/h (90 ton/h) of medium-hard rock with an optimum amount of minus 0.64-cm (¼-in) product. Three types of bowl are available.

The **Symons Standard cone crusher** *(Nordberg Industrial)* (Fig. 8-23) is a version of gyratory crusher having a still wider cone angle, making it suitable for throughput of a finer product. Shapes of cone heads available vary from fine to extra coarse; shape determines the minimum size as well as the maximum feed opening. Capacities and power consumptions are given in Table 8-11.

These crushers are also available in the **short-head** version. The bowl is less curved in shape; so the sides of the crushing cavity are more parallel, and the feed opening is more restricted. They produce a finer product and may be operated in closed circuit or in parallel

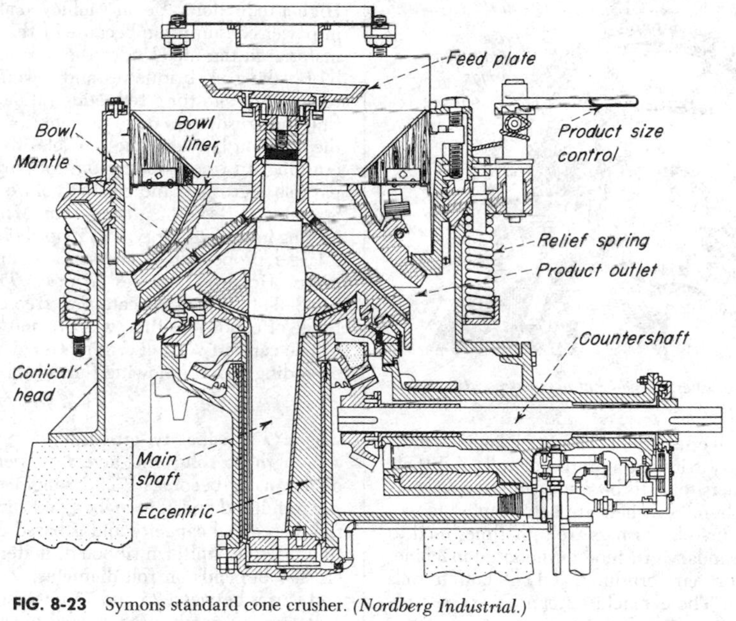

FIG. 8-23 Symons standard cone crusher. *(Nordberg Industrial.)*

TABLE 8-11 Performance of Symons Standard Cone Crushers in Open Circuit*

Size, ft.	Open-side feed opening, in.	Hp.	R.p.m.	Capacities, tons/hr. at indicated discharge setting, in.										
				¼	⅜	½	⅝	¾	⅞	1	1¼	1½	2	2½
2	2¾–4	30	575	15	20	25	30	35	40	45	50	60		
3	4⅛–7½	60	580	...	35	40	55	70	75	80	85	90		
4	5⅝–9¾	100	485	...	60	80	100	120	135	150	170	180	185	
4¼	5⅜–11¼	150	485	...	...	100	125	140	150	175	190	220	250	
5	7½–12¼	200	485	...	...	...	145	175	200	230	250	275	300	
5½	7¾–14½	200	485	...	...	...	160	200	235	275	320	365	430	[illegible]
7	11–18⅛	300	435	...	...	...	...	370	400	500	620	[illegible]	1000	1050

*Nordberg Industrial. To convert feet to meters, multiply by 0.3048; to convert inches to centimeters, multiply by 2.54; to convert horsepower to kilowatts, multiply by 0.746; and to convert tons per hour to megagrams per hour, multiply by 0.907.

with scalping screens for multistage size reduction. The **Gyradisc** crusher *(Nordberg Industrial)* is available in 0.9- to 1.4-m (36- and 54-in) diameters. It features the crushing of a multilayered bed of particles, and this is said to result in reduced wear. The cones are of wide-angle shape, with a relatively acute angle between them. Feed of 1- to 2.5-cm (⅜- to 1-in) size is reduced to 8 to 16 mesh. Power requirement is 75 to 150 kW (100 or 200 hp) for the two sizes of crusher. The Gyradisc is said to do the job of a rod mill for one-third of the cost.

ROLL CRUSHERS

Once popular for coarse crushing, these devices long ago lost favor to gyratory and jaw crushers because of their poorer wear characteristics with hard rocks. Roll crushers are still commonly used for both primary and secondary crushing of coal and other friable rocks such as oil shale and phosphate. The roll surface is smooth, corrugated, or toothed, depending on the application. *Smooth rolls* tend to wear ring-shaped corrugations that interfere with particle nipping, although some designs provide a mechanism to move one roll from side to side to spread the wear. *Corrugated rolls* give a better bite to the feed, but wear is still a problem. *Toothed rolls* are still practical for rocks of not too high silica content, since the teeth can be regularly resurfaced with hard steel by electric-arc welding. Toothed rolls are frequently used for crushing coal and chemicals (Fig. 8-24).

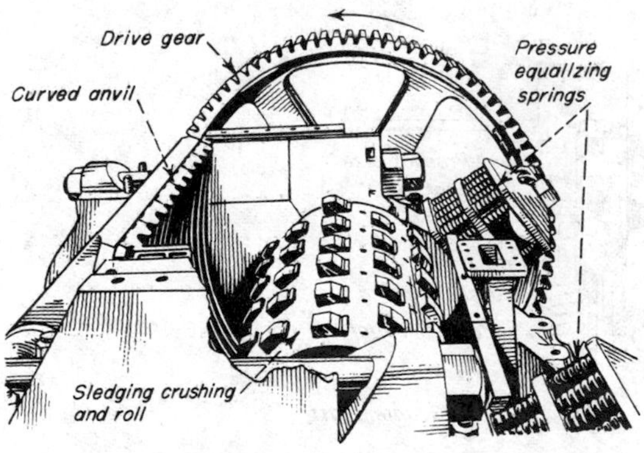

FIG. 8-24 Fairmount single-roll crusher. *(Allis-Chalmers Corporation.)*

Design and Operation Roll crushers can consist of single or multiple rolls. Crushing forces vary with the number of rolls and their construction. Single-roll crushers tend to be more common for primary crushing, whereas double-roll crushers are better suited to secondary crushing. Three- or four-roll crushers are sometimes used to achieve both primary and secondary crushing in the same machine. One or two rolls are used for primary crushing, and two bottom rolls perform secondary reduction. These machines can accept run-of-mine feed and produce material as fine as 1 cm (½ in) in one pass.

The single-roll crusher is one of the oldest and simplest crushers (Fig. 8-24). It consists of a sturdy hopper with an internally mounted, renewable breaker plate opposed to a frame-mounted crushing roll. Feed is crushed between the revolving roll and the breaker plate. Crushing action in a toothed roll crusher is a combination of impact, shear, and compression.

The rolls of the double-roll crusher are rotated toward each other at the same or different speeds. One of the shafts moves in fixed bearings; the other, in movable bearings. Product size in both crushers can be changed by adjusting the clearance. Primary crushers have a single tooth size. Secondary crushers may have long and short teeth to achieve higher reduction ratios. The long teeth act as feeders and split lumps into smaller pieces, whereas the short teeth produce secondary crushing to result in the required product size.

The tension springs exert pressures on the rolls up to 10.5 kN/cm (6000 lb/in) of roll face for light duty to as high as 70 kN/cm (40,000 lb/in) for heavy duty. This is equivalent to crushing strengths of 125 to 838 MPa (18,000 to 120,000 lbf/in^2) based on effective face length equal to one-third of actual length.

The **angle of nip,** the angle formed by the tangents to the roll faces at the point of contact with a particle to be crushed, is determined by $\cos (N/2) = (r + a)/(r + b)$, where r = radius of rolls, a = one-half of the distance between rolls, b = radius of particle, and N = angle of nip. The angle of nip varies for different operations, but seldom exceeds 0.52 rad (30°). The required roll diameter is determined by the maximum size of feed that can be nipped without slippage: $b_{max} = 0.04r + a$; all dimensions usually are in centimeters (inches).

The **peripheral speed** at which rolls normally operate is from 61 to 366 m/min (200 to 1200 ft/min), occasionally as high as 457 m/min (1500 ft/min). A **reduction ratio** of 4 should not be exceeded for hard materials. For large pieces of hard materials, 3 or 2.5 gives better results. The economical range of reduction usually is limited to a No. 12 to No. 16 sieve product. The capacity of rolls can be increased if the particles are dropped from a sufficient height so that they fall into the nip at the roll peripheral speed. Then the rolls can be run at higher speed so that a pilot-size roll has the capacity of large heavy rolls [Adamski, *Chem. Process Eng.*, 343–347 (July 1964)].

Reduction ratio and differential roll speed affect production rate and energy consumed per unit surface produced [Ohe, *Chem. Ing. Tech.*, **39**(5,6), 357 (1967)]. Differential speeds increase power consumption without helping performance for reduction ratios up to 4. Higher reduction ratios are achieved at the expense of great increases in power consumption because of the briquetting of already broken material in the nip. Under these conditions differential speed helps by breaking up briquettes and lowers power usage, but the power required is high for a reduction ratio of 8:1. When the rolls are kept full, the crushing is done not only by the action of the rolls but by the attrition between the particles themselves. This is called choke crushing. In free crushing the rolls are fed at such a rate that each particle is crushed and ejected before the next is nipped (Carey and Stairmand, *Recent Advances in Mineral Dressing*, Institution of Mining and Metallurgy, 1953, pp. 117–136). Free crushing produces a larger proportion of coarser sizes and is generally more advantageous. However, studies [Hiorns, *Trans. Inst. Min. Metall.*, **75,** C343–344 (1966)] indicate that free crushing prevails over a wider range of conditions than was previously supposed.

The **capacity** of roll crushers is calculated from the ribbon theory, according to the following formula:

$$Q = dLs/2.96 \tag{8-34}$$

where Q = capacity, cm^3/min; d = distance between rolls, cm; L = length of rolls, cm; and s = peripheral speed, cm/min. The denominator becomes 1728 in engineering units for Q in cubic feet per minute, d and L in inches, and s in inches per minute. This gives the theoretical capacity and is based on the rolls' discharging a continuous, solid uniform ribbon of material. The actual capacity of the crusher depends on roll diameter, feed irregularities, and hardness and varies between 25 and 75 percent of theoretical capacity.

Power is directly proportional to capacity and ratio of reduction. When crushing 4:1, a typical power requirement is 0.27 kW/(Mg·h) [0.4 hp/(ton·h)] for soft rock and 0.68 kW (1 hp) for hard rock.

Performance By virtue of its positive discharge, the single-roll crusher can handle wet and sticky material, including stone containing a substantial admixture of loam or clay, which is sometimes present in cement rock. In this one respect it outperforms the jaw crusher, which in turn clears more readily than the gyratory crusher. Soft materials such as limestones, dolomites, phosphate rock, cement rock of the Lehigh Valley district, shale, and similar deposits provide suitable feed for this toothed-roll crusher. The compressive strength of the rock should not exceed 104 MPa (15,000 lbf/in^2). Operating data are given in Table 8-12.

McNally single-roll and double-roll crushers can accept run-of-mine feed and produce reduction ratios of 4:1 to 10:1 for coal crushing. They are available as **Gearmatic** or heavy-duty belt-driven units. Individual rolls of the heavy-duty units are provided with sep-

TABLE 8-12 Operating Data for Single-Roll Crushers
Roll speed, 1200 r/min

Crusher size, in	Approximate feed size, in	Approximate capacity, tons/h, for discharge opening, in 2	4	6	8	hp
20 × 18	12	70	100			20
20 × 30	14	95	140	190		25
24 × 24	16	80	125	175		25
24 × 48	16	200	270	330	360	40
36 × 36	18	200	270	330	360	50
36 × 66	20	380	520	660	730	100

NOTE: To convert inches to centimeters, multiply by 2.54; to convert horsepower to kilowatts, multiply by 0.746.

arate motors and overload protection. There are four lines of **Pennsylvania** single-roll crushers; these are capable of accepting 1-m (42-in) cubes of material.

Koppers single-roll crushers are available for both primary and secondary crushing. They are also available in both standard and heavy-duty types for crushing chemicals, limestone, shale, and mine rock. Capacities vary from 135 to 1000 Mg/h (150 to 1130 tons/h) with power ranging from 37 to 187 kW (50 to 250 hp).

IMPACT BREAKERS

Impact breakers include heavy-duty hammer crushers and rotor impact breakers. Fine hammer mills are described in a subsequent subsection.

Hammer Crusher (Fig. 8-25) Pivoted hammers are mounted on a horizontal shaft, and crushing takes place by impact between the hammers and breaker plates.

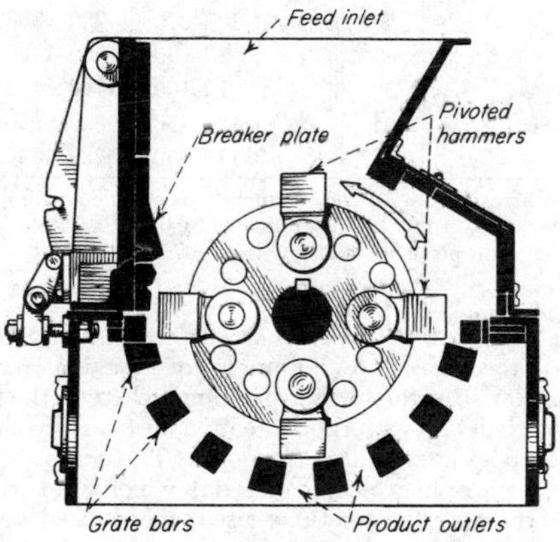

FIG. 8-25 Hammer crusher. (*Jeffrey Mfg. Co.*)

Particles acquire high velocities, and this leads to little control on particle size and a much higher proportion of fines. A cylindrical grating is provided beneath the rotor for product discharge. Some hammer crushers are symmetrically designed so that the direction of rotation can be reversed to distribute wear evenly on the hammer and breaker plates. Each hammer may weigh several hundred kilograms (pounds). Speed varies between 500 to 1800 r/min, depending on the size of the machine. Heavy-duty hammer crushers are frequently used in the quarrying industry, for processing municipal solid waste, and scrap automobiles.

Koppers offers three lines of impact breakers in various sizes. These are suitable for the crushing of limestone, gypsum, cement rocks, coal, and chemicals. Koppers one-direction hammer mills can accept feed as large as 91 cm (36 in). Capacities of one-direction and reversible hammer mills vary between 9 and 770 Mg/h (10 and 850 tons/h). Their impact crushers have a maximum capacity of 1700 Mg/h (1900 tons/h). Performance data of reversible hammer mills are shown in Table 8-13.

The *Pennsylvania nonclog hammer mill* incorporates the continuous movement of the traveling breaker plate, which forces the feed into the crushing path. This feature makes it impossible for wet, sticky material to build up on the breaker plate out of reach of the hammers. A traveling rear element can be provided for material which will not flow freely through the screen openings after it has been crushed.

TABLE 8-13 Performance Data for Reversible Hammer Mills

Model no.	Rotor dimensions, in	Maximum feed size, in	Maximum speed, r/min	hp	Capacity, tons/h
505	30 × 30	2½	1200	100–200	40–60
605	36 × 30	4	1200	200–300	80–100
708	42 × 48	8	900	300–550	140–180
815	48 × 90	10	900	900–1200	330–400
1014	60 × 84	12	720	1100–1500	450–500
1217	72 × 102	14	600	1550–2000	620–685
1221	72 × 126	14	600	1900–2500	760–850

NOTE: To convert inches to centimeters, multiply by 2.54; to convert horsepower to kilowatts, multiply by 0.746; and to convert tons per hour to megagrams per hour, multiply by 0.907.

Kennedy Van Saun heavy-duty hammer mills reduce nonabrasive materials from 30-cm (12-in) feed size down to a uniformly fine cubical product. Internal grid bars permit closed-circuit multi-impact operation in one machine. These mills are available in standardized sizes.

Rotor Impactors The rotor of these machines is a cylinder to which is affixed a tough steel bar (see Fig. 8-26). Breakage can occur against this bar or on rebound from the walls of the device. Free impact breaking is the principle of the rotor breaker, and it does not rely on pinch crushing or attrition grinding between rotor hammers and breaker plates. The result is lower wear and lower power requirement.

Not all rocks shatter well by impact. Impact breaking is best suited for the reduction of relatively nonabrasive and low-silica-content

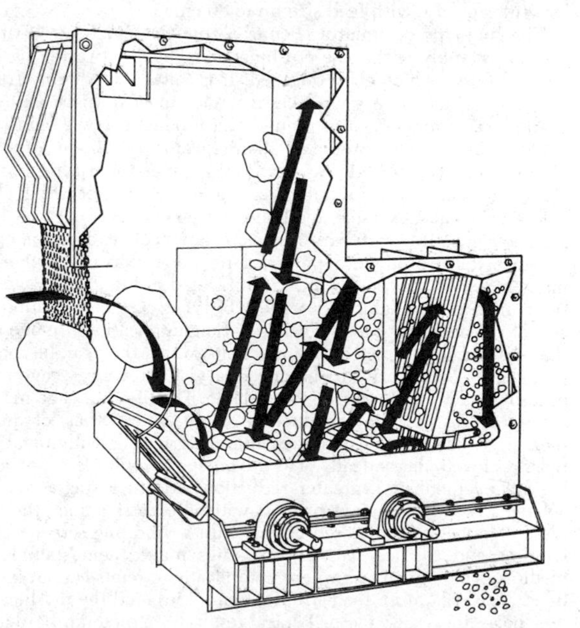

FIG. 8-26 Dual-rotor impact breaker. (*Kennedy Van Saun Corp.*)

materials such as limestone, dolomite, anhydrite, shale, and cement rock, the most popular application being on limestone.

Kennedy Van Saun **dual-rotor impact breakers** are built to handle these types of rock when low production costs are important. Both rotors rotate in the same direction (see Fig. 8-26). The geometry of the crushing chamber promotes three-stage crushing. The feed is struck upward into an expansion chamber, where shattered particles fall apart. Subsequent blows occur between the rotors and internal baffles of manganese steel. The result is a high reduction ratio and elimination of secondary and tertiary crushing stages. The investment cost may be a third of that for a two-stage jaw and gyratory crusher plant producing 180 Mg/h (200 tons/h) and half for a plant crushing 540 Mg/h (600 tons/h) [Godfrey, *Quarry Managers' J.*, 405–416 (October 1964)]. Variation in product grading is accomplished by changes in rotor speed and baffle adjustments. Table 8-14 gives capacities for reducing well-shot quarry-run limestone. Results will vary with the crushing resistance of material processed. With many types of rock, 100 percent of the product of these impactors will pass 2.5 cm (1 in) when operating in a closed circuit. By adding a screen on a portable mounting, a complete, compact mobile crushing plant of high capacity and efficiency for use in any location is provided. A single-rotor impactor is also available.

TABLE 8-14 Performance of Dual-Rotor Impact Breakers*

Model no.	Feed opening, in	Speed, r/min	hp	Hammer weight, lb	Product size, in	Capacity, tons/h
3648	36 × 48	550–990	250–300	300	2	300
4850	48 × 50	550–990	300–400	400	3	500
5462	54 × 62	480–750	400–500	500	4–5	700
6072	60 × 72	300–600	500–600	600	5–6	1200

*Kennedy Van Saun Corp. To convert pounds to kilograms, multiply by 0.4535; to convert inches to centimeters, multiply by 2.54; to convert horsepower to kilowatts, multiply by 0.746; and to convert tons per hour to megagrams per hour, multiply by 0.907.

The Pennsylvania **twin-rotor impactor** was developed specifically as a secondary crusher for wet, sticky materials which would normally plug other crushers and has proved to be an excellent crusher for reducing clays and shale. It is available in sizes to 180 Mg/h (200 tons/h) capacity, with feed size up to 30 cm (12 in).

The **ring-type granulator** *(Pennsylvania Crusher Corp.)* features a rotor assembly with loose crushing rings, held outwardly by centrifugal force, which chop the feed. It is suitable for highly friable materials which may give excessive fines in an impact mill. For example, bituminous coal is ground to a product below 2 cm (¾ in).

Cage Mills The **Stedman disintegrator** *(Stedman Machine Co.)*, commonly referred to as a cage mill, is used for crushing quarry rock, phosphate rock, and fertilizer and for disintegrating clays, colors, press cake, asbestos, and bones. Cages of one, two, three, four, six, and eight rows, with bars of special alloy steel, revolving in opposite directions, produce a powerful impact action that pulverizes many materials. A two-row mill is illustrated in Fig. 8-27. Cages for these mills are available in a variety of types designed to minimize cost of replacement due to wear in various applications. In one type the pins are held in place by single bolts so that they can be rotated periodically and replaced when necessary. A throwaway cage made of a single alloy-steel casting is available. The life of a cage may be a few months and may produce 9000 Mg (10,000 tons) of quarry rock. A gray-iron cage is used for alumina grinding, with metal particles removed magnetically. The advantage of multiple-row cages is the achievement of a greater reduction ratio in a single pass. To achieve desired grading of product without excessive fines, the mills are generally used in closed circuit with a vibrating screen. These features and the low cost of the mills make them suitable for medium-scale operations where complicated circuits cannot be justified. The maximum feed size is 20 cm (8 in), and the product size may be as fine as 325 mesh. Typical results for a medium-hard limestone are given in Table 8-15.

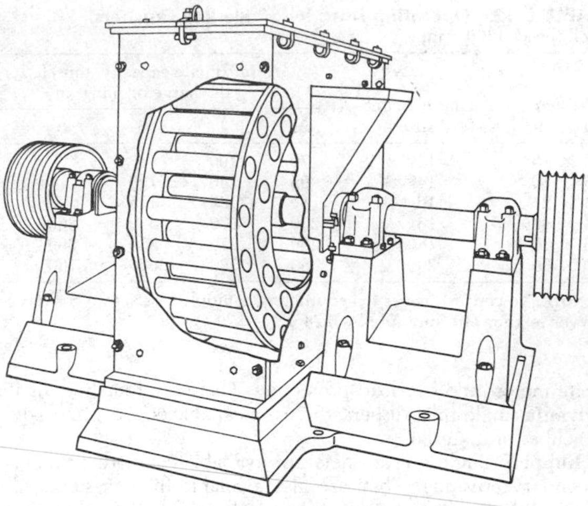

FIG. 8-27 Two-cage disintegrator. *(Stedman Machine Co.)*

TABLE 8-15 Performance of Two- and Four-Row Cage Mills Crushing Medium-Hard Limestone, Open Circuit*

No. of rows	Product size, 95% through mesh	Mill size, in.	Feed size, in.	Total h.p., 2 motors	Speed, r.p.m.	Approx. capacity, tons/hr.
2	−20	30	¾	45–70	1200–1500	10–15
		36	1	90–120	1000–1250	20–30
		44	1¼	125–200	820–1000	30–45
		50	1½	180–270	720– 900	40–60
4	−6	30	½	35– 50	1200–1400	5– 7
		36	½	70–100	1100–1160	10–14
		44	¾	120–250	820– 950	16–22
		50	¾	175–350	720– 840	25–30

*Stedman Machine Co. To convert inches to centimeters, multiply by 2.54; to convert horsepower to kilowatts, multiply by 0.746; and to convert tons per hour to megagrams per hour, multiply by 0.907.

Prebreakers Aside from the normal problems of grinding, there are special procedures and equipment for breaking large masses of feed to smaller sizes for further grinding. There is the breaking or shredding of bales, as with rubber, cotton, or hay, in which the compacted mass does not readily come apart. There also is often caking in bags of plastic or hygroscopic materials which were originally fine. Although crushers are sometimes used, the desired size-reduction ratio often is not obtainable. Furthermore, a lower capital investment may result through choosing a less rugged device which progressively attacks the large mass to remove only small amounts at a time. In structure such a device comprises a toothed rotating shaft in a casing.

The **sawtooth crusher** *(Sprout, Waldron)* has two shafts geared together at differential speeds normally in the ratio of 2¼:1. Each shaft carries sawtooth and spacer assemblies. The size of product can be controlled by the spacing of the saws and the peripheral speeds. Models are available for sheets up to 7.6-cm (3-in) thickness and 152-cm (60-in) width. Feed can be in continuous sheets.

Sawtooth crushers are also used for friable lump stocks up to 6-in ring size. Applications include the processing of press cakes, phenolic plastics, alkali cellulose sheets, sheet glue, naphthalene, resins, sulfur, bark, lump pitch, calcium chloride, and asphalt floor tiling.

Prater Industries, Inc., offers a double roll with heavy-duty teeth as a precrusher feeder. The **Mikro roll crusher** *(Pulverizing Machinery Co.)* serves as grinder or prebreaker and is of similar type. The **Rietz prebreaker** differs somewhat from these.

The **horizontal rotary crusher** *(Sprout Waldron Companies)* has a toothed cone supported on a horizontal shaft for preliminary crushing; final crushing takes place between close-fitting sections at the base of the cone. Clearance is adjusted with a handwheel. The horizontal crusher is used for friable material such as pitch, rosin, mica, coconut shells, and compacted inorganic salts. Its output ranges from 0.9 to 9 Mg/h (1 to 10 tons/h) to a product that may all be able to pass through a No. 5 sieve and contain material finer than a No. 100 sieve, depending on the feed quality. Power is less than 11 kW.

Rotary Cutters These are used with tough or fibrous materials, with which successive shear actions are more effective than pressure or impact. The feedstock should not exceed the cutter-knife length, and the thickness is less than 2.5 cm (1 in). The usual structure involves a rotor with knives spaced uniformly on the periphery so as to cut against stationary knives on the casing. Product is passed through screens; its maximum size is controlled by the screen aperture and by the design and operation of the mill. From the 20-mesh screen, in some cases down to 80 mesh, the collection system is pneumatic.

The data shown in Table 8-16 are for a unit operating at 920 r/min with 25-cm (10-in) rotor and knife length at 46 cm (18 in) with five moving and five fixed knives.

In general, rotary cutters are available in steel or stainless steel and may be specially supplied in other corrosion-resistant metals. Knives for shear cutting are also generally available, leading to reduction in shock loading. Laboratory units, using a few horsepower and having capacities up to a few hundred pounds per hour, are generally available, while production units of varying size use between 4 and 48 kW (5 and 60 hp), are around 30 to 60 cm (1 to 2 ft) in diameter, have knife lengths from 30 to 76 cm (12 to 30 in), and have capacities up to 0.9 to 1.8 Mg/h (1 to 2 tons/h). Specifications by Paul O. Abbe (Table 8-17) and a description of the Sprout, Waldron heavy-duty rotary knife cutter illustrate details.

Sprout, Waldron offers two series of **rotary knife cutters** as follows: 25-cm-diameter, 920-r/min rotor units with 46-, 61-, and 76-cm (18-, 24-, and 30-in) knife lengths mounted in a cast frame (iron, steel, stainless); and 50-cm- (20-in) diameter, 750 r/min rotor units of 25- and 76-cm (10- and 30-in) knife lengths with an all-welded steel frame. A conventional arrangement of the largest of these units (model F-11) is illustrated in Fig. 8-28. These basic units are varied considerably according to application. Entry of feed is by hopper, slot, or compression-feed rolls. Generally five rotor knives are specified, and these are set at a slight angle with the shaft to provide shear cuts with direction reversed on alternate knives to avoid conveying the charge against the end of the cutter. Two to seven stationary knives alternating with screen sections around the cage may be specified, to provide maximum discharge area and to keep fines at a minimum. Variations in construction permit such widely different applications as sheet-plastic granulation, flocking, tobacco-leaf threshing, etc. Models are powered by 7.5- to 45-kW (10- to 60-hp) motors through V belts and employ shear-pin safety hubs.

TABLE 8-16 Performance of Rotary Knife Cutter

Material	Screen opening	Feed rate, lb./hr.	Hp.	Air	Remarks on product
Amosite asbestos pencils.	1½″	1000	11	Yes	Finer fiber bundles average length 2″
Cellophane bags..	1⅓₂″	200	10	Yes	Finer than 5⁄16″
Cork............	3⁄16″	525	16	Yes	90% 4/24* sieve
Chemical cotton..	60 mesh	120	15	Yes	Flock; 35% under No. 100 sieve
Leather scrap....	¾″	600	20	Yes	Precutting before shredding
Fiberglas	3⁄16″	300	18	Yes	1″ (approx.) lengths
Waste paper.....	5⁄16″	338	13	Yes	Through No. 4 sieve and finer
Sheet pulp.......	40 mesh	150	15	Yes	Flock; 85%, 40/100 sieve
Tenite scrap.....	5⁄16″	340	12	No	Granulated for reuse
Vinylite scrap....	7⁄32″	300	15	Yes	35%, 6/10 sieve; granular
⅛″ Geon sheet...	5⁄16″	540	11	No	99%, 4/20 sieve; for molding granules
Cotton rags......	¾″	500	11	Yes	No linting
Buna scrap......	10 mesh	264	12	Yes	Granular
Neoprene scrap..	30 mesh	90	14	Yes	20°F. temperature rise
Soft-wood chips..	⅛″	960	12	Yes	90%, 10/50 sieve
Hard-wood chips.	1⁄16″	290	11	Yes	83%, 20/100 sieve

NOTE: To convert inches to centimeters, multiply by 2.54; to convert horsepower to kilowatts, multiply by 0.746.

°90 percent 4⁄24 sieve, i.e., 90 percent is through No. 4 and on No. 24 sieve.

TABLE 8-17 Rotary-Cutter Specifications

Machine No.	Floor space required, in.	Shipping weight, lb.	Speed, r.p.m.	Hp.	Screen size, in.
0	37 × 17	500	900–1200	2– 5	10 × 17
1	54 × 34	1,500	600– 900	5–15	20 × 24
2	68 × 42	4,000	600– 900	15–40	20 × 28
2½	96 × 39	6,000	500– 800	20–45	35 × 36
3	102 × 43	12,000	500– 750	30–60	51 × 30

NOTE: To convert inches to centimeters, multiply by 2.54; to convert pounds to kilograms, multiply by 0.4535; and to convert horsepower to kilowatts multiply by 0.746.

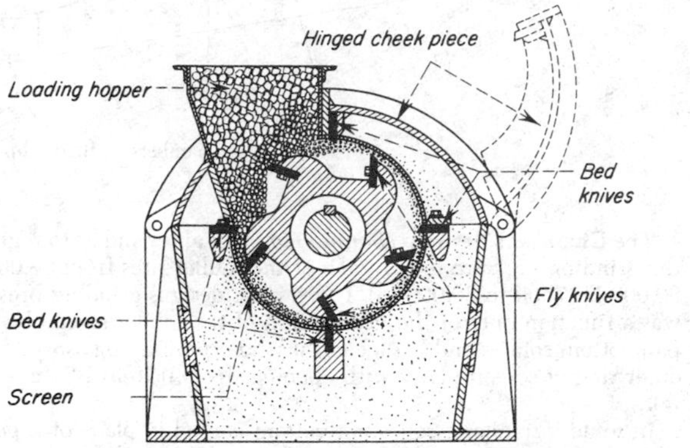

FIG. 8-28 Fabricated-steel rotary knife cutter. *(Sprout, Waldron Companies.)*

Precision Cutters and Slitters Often it becomes desirable to reduce the size of a solid mass to regular smaller sizes. Examples that typify the range would be punching or cutting of metal plate; cutting or dicing of rubber or plastic masses, from an extruder, rolled sheet, or random pieces; and slicing of bread. With metal and other resistant solids, shearing action in which there is positive bearing area on the front edge of each cutting surface is generally employed. A punch and die would serve as an example, as would also shearing rolls. With paper, rubber, plastics, and bread, a sharp-edged device is effective. In some cases, the direct pressure of a knife is employed, but generally some sliding or sawing action is helpful. For bread, direct pressure would be injurious to the structure, and the cutting knives, pinned to a rotating shaft, are curved for substantial sliding motion in the cut.

Precision knife cutters differ from random cutting mills in that a feeder is synchronized with the knives. This ensures the exact size, whether it be slit widths in a sheet, fiber length from a strand, or both width and length from a sheet, as in dicing. In the **Giant dicing cutter,** the sheet stock is first slit lengthwise with opposing sets of circular knives. The slit strands then pass between pressure rolls to a rotary cutter which operates against an adjustable bed knife. Capacities range up to 18 Mg/h (20 tons/h), with sheet stock up to 60 cm (24 in) wide.

PAN CRUSHERS

Design and Operation The pan crusher (Fig. 8-29) consists of one or more grinding wheels or mullers revolving in a pan; the pan may remain stationary and the mullers be driven, or the pan may be driven while the mullers revolve by friction. The mullers are made of tough alloys such as Ni-Hard. Iron scrapers or plows at a proper angle feed the material under the mullers.

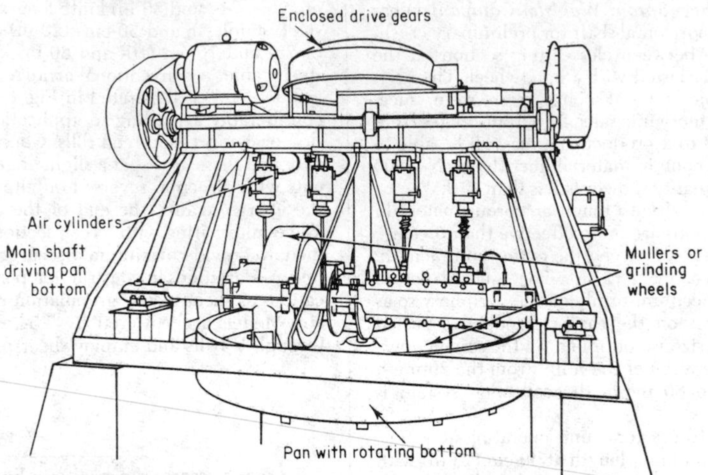

FIG. 8-29 Chambers 10-ft, 100-hp dry pan. *(Bonnot Co.)*

The **Chambers dry pan** *(Bonnot Co.)* uses air cylinders to regulate the grinding pressure under each of the muller tires from 33,000 to 90,000 N (7500 to 20,000 lb). In previous designs grinding pressure was a function only of the deadweight of the muller assembly. The pan bottom rotates and has a central solid crushing ring as well as an outer ring of screen plates with openings from 0.16 to 1.3 cm (1/16 to ½ in).

In some instances a solid pan bottom is used in place of a perforated screen bottom, and the ground material is discharged through a slot in the rim. In such an application the machine is described as a **rim-discharge grinder** rather than a dry pan. It will give a greater throughput with wetter materials but will require greater screening area and a higher recirculating load, as the rim of the grinder must necessarily be set wider than the screen openings in the bottom of the dry pan.

Performance The dry pan is useful for crushing medium-hard and soft materials such as clays, shales, cinders, and soft minerals such as barites. Materials fed should normally be 7.5 cm (3 in) or smaller, and a product able to pass No. 4 to No. 16 sieves can be delivered, depending on the hardness of the material. Finer products can be obtained by operating a pan in closed circuit with a vibrating screen. A circulating load of 75 percent is common.

High reduction ratios with low power and maintenance are features of pan crushers.

The Chambers dry pan is available from 1.8 to 3 m (6 to 10 ft) pan diameter with mullers ranging from 0.71 to 1.6 m (28 to 62 in) in diameter with 13- to 46-cm (5- to 18-in) face. Power ranges from 11 to 75 kW (15 to 100 hp) or from 0.8 to 4 kW/Mg (1 to 5 hp/ton) of product. Production rate varies from 1 to 54 Mg/h (1 to 60 tons/h) according to pan size and hardness of material as well as fineness of feed and product.

The **wet pan** is used for developing plasticity or molding qualities in ceramic feed materials. The abrasive and kneading actions of the mullers blend finer particles with the coarser particles as they are crushed (Greaves-Walker, Am. Refract. Inst. Tech. Bull. 64, 1937).

TUMBLING MILLS

Ball, pebble, rod, tube, and compartment mills have a cylindrical or conical shell, rotating on a horizontal axis, and are charged with a grinding medium such as balls of steel, flint, or porcelain or with steel rods. The **ball mill** differs from the tube mill by being short in length; its length, as a rule, is not far from its diameter (Fig. 8-30). Feed to ball mills can be as large as 2.5 to 4 cm (1 to 1½ in) for very fragile materials, although the top size is generally 1 cm (½ in). Most ball mills operate with a reduction ratio of 20 to 200:1. The largest balls are typically 13 cm (5 in) in diameter.

The tube mill is generally long in comparison with its diameter, uses smaller balls, and produces a finer product. The compartment mill consists of a cylinder divided into two or more sections by perforated partitions; preliminary grinding takes place at one end and finish grinding at the discharge end. These mills have a length-to-diameter ratio in excess of 2 and operate with a reduction ratio of up to 600:1. Rod mills deliver a more uniform granular product than other revolving mills while minimizing the percentage of fines, which are sometimes detrimental.

The **pebble mill** is a tube mill with flint or ceramic pebbles as the grinding medium and may be lined with ceramic or other nonmetallic liners. The **rock-pebble** mill is an autogenous mill in which the medium consists of larger lumps scalped from a preceding step in the grinding flow sheet.

The ball mill and the pebble mill are simple to operate and versatile in use. A steel or stone-lined cylindrical steel shell, containing a charge of steel balls or stone pebbles, is rotated horizontally about its axis so that size reduction or pulverizing is effected by the tumbling of the balls or pebbles on the material between them. The mills may be operated wet or dry, in either batch or open-circuit use or in closed circuit with size classifiers (see subsection "Dry versus Wet Grinding").

Design The conventional type of **batch mill** consists of a cylindrical steel shell with flat steel-flanged heads. Openings through which the grinding medium and the process material can be loaded and discharged are provided. Mill length is equal to or less than the diameter [Coghill, De Vaney, and O'Meara, *Trans. Am. Inst. Min. Metall. Pet. Eng.*, **112**, 79 (1934)]. The discharge opening is often opposite the loading manhole and for wet grinding usually is fitted with a valve. One or more vents are provided to release any pressure developed in the mill, to introduce inert gas, or to supply pressure to assist discharge of the mill. In dry grinding, the material is discharged into a hood through a grate over the manhole while the mill rotates. Jackets can be provided for heating and cooling.

Material is fed and discharged through hollow trunnions at opposite ends of **continuous mills** (Fig. 8-30). A grate or diaphragm just inside the discharge end may be employed to regulate the slurry level in wet grinding and thus control retention time. In the case of **air-swept mills,** provision is made for blowing air in at one end and removing the ground material in air suspension at the same or other end.

Ball mills usually have **liners** which are replaceable when they wear. Liners may have a baffling action because of a wave shape or

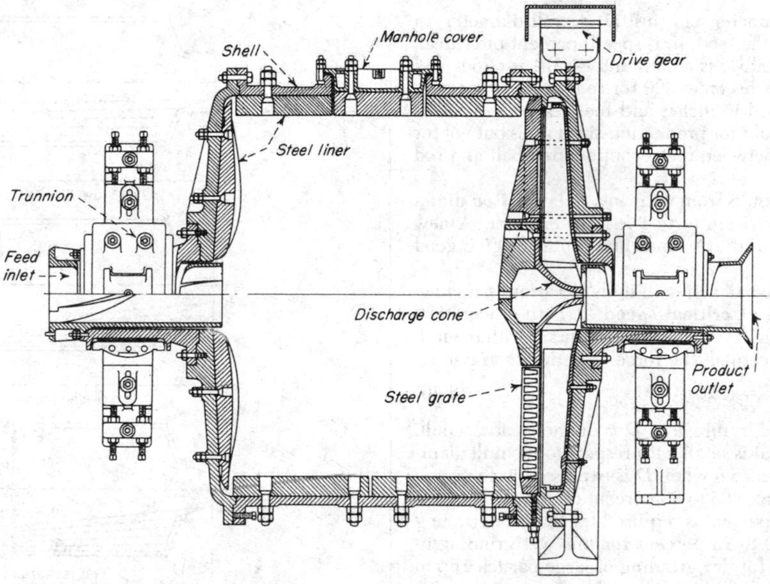

FIG. 8-30 Marcy grate-type continuous ball mill. *(Mine & Smelter.)*

because of **lifter** inserts which key the ball charge to the shell and prevent slippage. Typical liner shapes are illustrated in Fig. 8-31. Special operating problems occur with smooth-lined mills owing to erratic slip of the charge against the wall. At low speeds the charge may **surge** from side to side without actually tumbling; at higher speeds tumbling with **oscillation** occurs. The use of lifters prevents this [Rose, *Proc. Inst. Mech. Eng. (London)*, **170**(23), 773–780 (1956)]. Power consumption in a smooth mill depends in a complex way on operating conditions such as feed viscosity, whereas it is more predictable in a mill with lifters [Kitchener and Clarke, *Br. Chem. Eng.*, **13**(7), 991 (1968)].

Grinding balls can be made of forged steel, cast steel, or cast iron. Heat-treated forged-steel balls generally give optimum wear characteristics. Balls vary considerably in hardness, with soft balls having Brinell hardness in the range of 350 to 450 and hard balls having hardness in excess of 700.

The most reliable test results on **wear** [Norman and Loeb, *Trans. Am. Inst. Min. Metall. Pet. Eng.*, **183**, 330 (1949)] indicate that a matrix of martensite or low-temperature bainite plus retained austenite shows the best wear resistance of steel alloys.

Rubber block linings have become accepted for large ball mills. Wear and production performance are similar to those of steel liners, but replacement labor is less because of easier handling [Snow, *Ind. Eng. Chem.*, **62**(11), 39 (1970)].

Pebble mills are frequently lined with nonmetallic materials when iron contamination would harm a product such as a white pigment

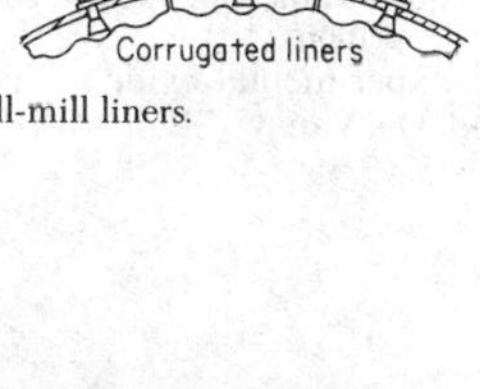

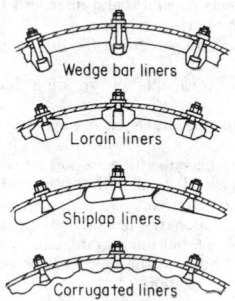

FIG. 8-31 Types of ball-mill liners.

or cement. Belgian silex (silica) or porcelain block are popular linings. Silica linings and ball media have proved to wear better than other nonmetallic materials. The higher density of silica media increases the production capacity and power draft of a given mill.

Capacities of pebble mills are generally 30 to 50 percent of the capacity of the same size of ball mill with steel grinding media and liners; this depends directly on the density of the media.

Smaller mills (up to about 0.19-m^3 (50-gal) capacity) are made in one piece with a cover. U.S. Stoneware Co. makes these in wear-resistant Burundum-fortified ceramic and also makes larger three-piece units, in a metal protective case, up to 0.8-m^3 (210-gal) capacity. A handbook on pebble milling is available from Paul O. Abbe, Inc.

Operation **Cascading** and **cataracting** are the terms applied to the motion of grinding media. The former applies to the rolling of balls or pebbles from top to bottom of the heap, and the latter refers to the throwing of the balls through the air to the toe of the heap. Ball action has been studied and given mathematical consideration [Langemann, *Chem. Ing. Tech.*, **34**, 615–624 (1962), NTC translation 70-14797; Gow, Campbell, and Coghill, *Trans. Am. Inst. Min. Metall. Pet. Eng.*, **87**, 51 (1930)]. These extensive mathematical constructs are based on speculative hypotheses about the shape of the ball mass.

The chief factors determining the **size of grinding balls** are fineness of the material being ground and maintenance cost for the ball charge. A coarse feed requires a larger ball than a fine feed; a relation has been proposed: $D_b^2 = KX_p$, where D_b is ball diameter and X_p is the size of the coarser feed particles, both in centimeters (inches); and K is a grindability constant varying from 140 for chert to 90 for dolomite (or 55 to 35 in) (Coghill and De Vaney, U.S. Bur. Mines Tech. Publ. 581, 1937).

The need for a calculated ball-size feed distribution is open to question; however, methods have been proposed for calculating a rationed ball charge [Bond, *Trans. Am. Inst. Min. Metall. Pet. Eng.*, **153**, 373 (1943)].

The recommended optimum size of makeup rods and balls is [Bond, *Min. Eng.*, **10**, 592–595 (1958)]

$$D_b = \sqrt{\frac{X_p E_i}{K n_r}} \sqrt{\frac{\rho_s}{\sqrt{D}}} \tag{8-35}$$

where D_b = rod or ball diameter, cm (in); D = mill diameter, m (ft); E_i is the work index of the feed; n_r is speed, percent of critical; ρ_s is feed specific gravity; and K is a constant = 214 for rods and 143 for balls. The constant K becomes 300 for rods and 200 for balls when D_b and D are expressed in inches and feet respectively. This formula gives reasonable results for production-sized mills but not for laboratory mills. The ratio between the recommended ball and rod sizes is 1.23.

A graded charge of rods results from wear in a rod mill. **Rod diameter** may range from 10 to 2.5 cm (4 to 1 in), for example. A new rod load usually is patterned after a used one found to give good results.

The criterion by which the ball action in mills of various sizes may be compared is the concept of **critical speed.** It is the theoretical speed at which the centrifugal force on a ball in contact with the mill shell at the height of its path equals the force on it due to gravity:

$$N_c = 42.3/\sqrt{D} \tag{8-36}$$

where N_c is the critical speed, r/min, and D is diameter of the mill, m (ft), for a ball diameter that is small with respect to the mill diameter. The numerator becomes 76.6 when D is expressed in feet.

Actual mill speeds vary from 65 to 80 percent of critical. It might be generalized that 65 to 70 percent is required for fine wet grinding in viscous suspension and 70 to 75 percent for fine wet grinding in low-viscosity suspension and for dry grinding of large particles up to 1-cm (½-in) size. The speeds might be increased by 5 percent of critical for unbaffled mills to compensate for slip.

Tumbling-Mill Circuits Tumbling mills may be operated in **normal closed circuit,** as shown in Fig. 8-12, or in **reverse** arrangement in which the feed passes through the classifier before entering the mill (see secondary mill in Fig. 8-40). These arrangements may also be used with compartment mills, the material being air-classified between stages of grinding in compartments of the same mill.

Material and Ball Charges The load of a grinding medium can be expressed in terms of the percentage of the volume of the mill that it occupies; i.e., a bulk volume of balls half filling a mill is a 50 percent ball charge. The void space in a static bulk volume of balls is approximately 41 percent. Since the medium expands as the mill is rotated, the actual running volume is unknown.

Simple relationships govern the amount of balls and voids in the mill. The weight of balls = $\rho_b \epsilon_b V_m$, where ρ_b = bulk density of balls, g/cm^3 (lb/ft^3); ϵ_b = apparent ball filling fraction; and V_m = volume of mill = $\pi D^2 L/4$. Steel balls have a bulk density of approximately 4.8 g/cm^3 (300 lb/ft^3); stone pebbles, 1.68 g/cm^3 (100 lb/ft^3); and alumina balls, 2.4 g/cm^3 (150 lb/ft^3).

The amount of material in a mill can be expressed conveniently as the ratio of its volume to that of the voids in the ball load. This is known as the **material-to-void ratio.** If the solid material and its suspending medium (water, air, etc.) just fill the ball voids, the M/V ratio is 1, for example. Grinding-media loads vary from 20 to 50 percent in practice, and M/V ratios are usually near 1.

The solids concentration in a pebble-mill slurry should be high enough to give a slurry viscosity of at least 0.2 Pa·s (200 cP) for best grinding efficiency [Creyke and Webb, *Trans. Br. Ceram. Soc.*, **40,** 55 (1941)], but this may have been required to key the charge to the walls of the smooth mill used. The material charge of continuous mills cannot be set directly but is indirectly determined by operating conditions. There is a maximum throughput rate that depends on the shape of the mill, the flow characteristics of the feed, the speed of the mill, and the type of feed and discharge arrangement.

Feed and Discharge Feed and discharge arrangements for ball and rod mills depend on their mode of operation. Various feed and discharge mechanisms are shown in Fig. 8-32.

Mill feeders attached to the feed trunnion of the conical mill and used to pass the feed into the mill without backspill are of several types. A feed chute is generally used for dry grinding, this consisting of an inclined chute which is sealed at the outer edge of the trunnion and down which the material slides to pass through the trunnion and into the mill. A screw feeder, consisting of a short section of screw conveyor which extends partway into the opening in the feed trunnion and conveys the material into the mill, may also be used when

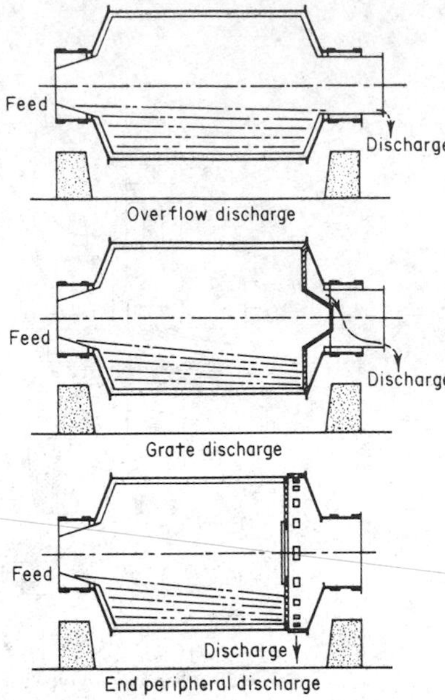

FIG. 8-32 Continuous ball-mill discharge arrangements for wet grinding. *(Koppers Co., Inc.)*

dry grinding. For wet grinding, several different types of feeders are available: the scoop feeder, which is attached to and rotates with the mill trunnion and which dips into a stationary box to pick up the material and pass it into the mill; a drum feeder attached to and rotating with the feed trunnion, having a central opening into which the material is fed and an internal deflector or lifter to pass the material through the trunnion into the mill; or a combination drum and scoop feeder, in which the new feed to the mill is fed through the central opening of the drum while the scoop picks up the oversize being returned from a classifier to a scoop box well below the centerline of the mill. The mill feeder must be able to handle any quantity of material which the mill may be capable of grinding and, in addition, a circulating load which may be as high as 1000 percent of the new feed rate.

Control of pulp level to obtain high circulating load is accomplished by use of **grate-discharge mills.** In one case an 18 percent increase in capacity resulted from conversion of an overflow mill to a grate-discharge mill despite a loss of 10 percent of the mill volume due to the change. The main reason was the removal of fines from the mill because of the increased recycle ratio. The grates allowed passage of sufficient pulp to maintain the circulating load at 400 percent (Duggan, *Min. Technol.*, Tech. Publ. 1456, March 1942).

Mill Efficiencies The controlling factors conceded to govern the ore-grinding efficiency of cylindrical mills are as follows:

1. Speed of mill affects capacity, also liner and ball wear, in direct proportion up to 85 percent of critical speed.
2. Ball charge equal to 50 percent of the mill volume gives the maximum capacity.
3. Minimum-size balls capable of grinding the feed give maximum efficiency.
4. Grooved liners of the wave type have found much favor among operators.
5. Classifier efficiency becomes more important in multiple-stage grinding.
6. Higher circulating loads tend to increase production and decrease the amount of unwanted fine material.
7. Low-level or grate discharge has increased grinding capacity over the center or overflow discharge, but liner, grate, and media wear is higher.
8. Ratio of solids to liquids in the mill must be considered on the basis of ore gravity and volumetric relation.

Mill Parameters Experimental evidence presented in a classic paper by Coghill and De Vaney ("Ball Mill Grinding," U.S. Bur.

Mines Tech. Publ. 581, 1937) causes the authors to draw the following conclusions:

1. In wet-batch ball milling with ore charges from 90 to 160 kg (200 to 350 lb) [about 35 kg (75 lb) of ore was required to fill the interstices of the balls at rest] and speeds from 30 to 80 percent critical, the slow speed gave the same type of grinding as high speed. Heavy ore charges yielded a little more selective grinding of the coarse particles than light charges. Best capacities were obtained with light charges, and slightly better efficiencies were obtained with heavy ore charges.

2. Some of the characteristics of dry-batch milling were unlike those of wet grinding. In the dry work, efficiency as well as capacity was best with the light ore charge. Power decreased with a decrease in the amount of ore in the mill; in wet grinding it increased with a decrease in the amount of ore in the mill. In dry grinding high speed was more efficient than low speed.

3. In comparing wet and dry grinding the tests were paired so that all the set variables were the same except pulp consistency (wet or dry). With an intermediate weight of ore charge, selective grinding was of the same degree; with a heavy ore charge, wet grinding was more selective, and with the light ore charge, dry grinding was more selective.

4. In comparing wet and dry open-circuit ball milling, wet grinding gave 39 percent more capacity and 26 percent more efficiency.

5. A small ball volume was not satisfactory in the overflow type of dry mill because too much ore built up in the mill. When building up of the ore was prevented by simulating the low-pulp-level mill, the small ball volume did good work.

6. With 60 percent solids, pebbles the same size of balls did about the same type of work as balls when dolomite was ground, but they failed in selective grinding of chert. Pebbles gave about 35 percent of the capacity and 81 percent of the efficiency shown by the balls.

7. For hard and medium-hard ores, tetrahedrons were unsatisfactory as media for coarse grinding.

8. Very hard balls (Ni-Hard) were better than ordinary balls; this was particularly so when the ore was very hard.

9. The efficiency of battered reject balls was about 11 percent less than that of new spherical balls.

10. A ball mill as small as 48 by 91 cm (19 by 36 in) duplicated the work of a plant-size mill. The tests led to the belief that if each of a variety of mills, large or small, is run under the same conditions and if each applies a unit of work to a unit of ore, the effect (comminution), as indicated by the products, will be the same; i.e., the same relation between cause and effect will maintain. These results have been graphically presented (Rose, loc. cit.).

Selection of Mill The selection of a ball- or rod-mill grinding unit is based on pilot-mill experiments, scaled up on the basis that production is proportional to energy input. When pilot experiments cannot be undertaken, performance is based on published data for similar types of materials, expressed in terms of either grindability or an energy requirement (see subsections "Grindability" and "Energy Laws"). Newer methods of sizing mills and determining operating conditions for optimum circuit performance are based on computer solutions of the grinding equations with values of rate and breakage functions determined from pilot or full-scale tests (see subsection "Simulation of Milling Circuits").

The ball mill is used for fine or coarse, wet or dry grinding in closed circuit with classifier, screen, or air separator. It is available in diaphragm (grate) or overflow discharge types.

The choice between wet and dry grinding is generally dictated by the end use of product. When the material can be ground either wet or dry, power consumption, liner wear, and capital costs determine the design. Grinding-media and liner-wear consumption per ton of ground product is lower for a dry-grinding system. However, power consumption for dry grinding is about 30 percent larger than for wet grinding and requires the use of dust-collecting equipment.

Capacity and Power Consumption One of the methods of mill sizing is based on the observation that the amount of grinding depends on the amount of energy expended, if one assumes comparable good practice of operation in each case. The energy applied to a ball mill is primarily determined by the size of mill and load of balls. Theoretical considerations show the net power to drive a ball mill to be proportional to $D^{2.5}$, but this exponent may be used without modification in comparing two mills only when operating conditions are identical [Gow, Guggenehim, Campbell, and Coghill, *Trans. Am. Inst. Min. Metall. Pet. Eng.*, **112,** 24 (1934)]. The net power to drive a ball mill was found to be

$$E = [(1.64L - 1)K + 1][(1.64D)^{2.5}E_2] \qquad (8\text{-}37)$$

where L is the inside length of the mill, m (ft); D is the mean inside diameter of the mill, m (ft); E_2 is the net power used by a 0.6- by 0.6-m (2- by 2-ft) laboratory mill under similar operating conditions; and K is 0.9 for mills less than 1.5 m (5 ft) long and 0.85 for mills over 1.5 m long. This formula may be used to scale up pilot-milling experiments in which the diameter and length of the mill are changed but the size of balls and the ball loading as a fraction of mill volume are unchanged. Reliable results have been obtained from this procedure.

Motor and Drive The power consumption of ball and rod mills is essentially constant and depends mainly on the diameter and ball charge. Hence, synchronous motors are best suited for this purpose. Large ball mills are now powered with motors of over 7500 kW (10,000 hp). Such high power requirements make proper selection of drive and gear systems extremely important [Schwedes, *Engineering/Mining Journal Operating Handbook of Mineral Processing,* Thomas (ed.), McGraw-Hill, New York, 1977]. Transmission of larger torques from the pinion to a mill gear becomes either unreliable or cost-prohibitive. Larger mills are powered by a multiple pinion arrangement with direct-axis load-share regulators.

Performance of Proprietary Equipment

Allis Chalmers Corporation Allis-Chalmers **grate-discharge ball mills** produce finely ground product of 28 to 325 mesh from a feed size of about 1 cm (½ in). Diameters range from 2.7 to 4.9 m (9 to 16 ft); lengths, from 2.4 to 7.3 m (8 to 24 ft); and power, from 110 to 2500 kW (150 to 3300 hp). They are generally recommended for the following service: wet or dry grinding (dry in closed circuit with a classifier to prevent overgrinding), performing a relatively coarse grind with top size of product around 48 mesh. A fine crusher product with a top size in the 0.6- to 1-cm (¼- to ⅜-in) range makes an excellent feed for low-level diaphragm mills, and they will handle up to 2.5-cm (1-in) feed if provided with extra-thick shell liners and large-diameter discharge. For intermediate diaphragm mills, top-size feed should be in the 0.3- to 0.6-cm (⅛- to ¼-in) range. Both low-level and intermediate-level mills are available.

The **overflow-type mill** is generally recommended for the following service: wet grinding in closed circuit with classifier to prevent overgrinding, performing a fine grind with top size of product no greater than 65 mesh. A rod-mill product or other feed in the 8-mesh or finer range makes an excellent ball-mill feed. Feed is usually 8 mesh or below. Diameters range from 2.7 to 4.9 m (9 to 16 ft); lengths, from 2.4 to 7.3 m (8 to 24 ft); and power from 22 to 1200 kW (300 to 1600 hp).

Multicompartmented mills feature grinding of coarse feed to finished product in a single operation, wet or dry. The primary grinding compartment carries large grinding balls or rods; one or more secondary compartments carry smaller media for finer grinding. Diameters range from 1.5 to 4.9 m (5 to 16 ft); lengths, up to 16 m (52 ft); and power, up to 3300 kW (4400 hp) with Twinducer drive. **Pebble mills** produce finely divided products which must be free from iron contamination. They are widely used for grinding glass sand, high-grade sands for scouring powders, and applications in the talc and ceramics industries. The grinding charge consists of flint pebbles. Diameters range from 0.9 to 2.7 m (3 to 9 ft); lengths, from 1.8 to 8.5 m (6 to 28 ft). Allis-Chalmers **rod mills** produce a 6- to 35-mesh product with a minimum of fines. Because a rod mill can use a 2.5-cm (1-in) slot-size feed, it has supplanted the last stage of crushing in many plants. The center-peripheral-discharge type is widely used for production of fine-specification aggregates, raw brick mix, and roofing granules. Rod mills are built by Allis-Chalmers in either end or center peripheral types and in overflow types for wet-grinding applications. The length of rod mills must be at least 1.25 times the working diameter. Diameters range from 2.7 to 4.3 m (9 to 14 ft); lengths, from 3.7 to 5.5 m (12 to 18 ft); and power, from 335 to 1040 kW (450 to 1400 hp) with direct drive; 930 to 3300 kW (1250 to 4440 hp) with Twinducer drive.

Kennedy Van Saun Corp. Kennedy Van Saun ball and tube mills are engineered to perform at greater than rated capacity for each grinding application, wet or dry. Mechanical-discharge mills are available in sizes of 1.8 to 4.9-m (6- to 16-ft) diameter and 2.4 to 10 m (8 to 33 ft) in length; wet overflow mills are available in sizes of 1.5- to 4.5-m (5- to 15-ft) diameter and 2.1 to 7.6 m (7 to 25 ft) in length. These mills are powered with 37- to 4500-kW (50- to 6000-hp) motors.

Kennedy **air-swept and air-lift grinding systems** are used extensively for pulverizing power-plant coal, for firing cement kilns and metallurgical furnaces, and for grinding phosphates and other ores. They simultaneously grind and dry materials in a closed circuit with nonmechanical air separators for kiln feed and kiln firing. This system can be adjusted to produce and maintain 38 to 90 percent passing 200 mesh with control of fine and coarse ends. For firing systems the cheapest fuel obtainable may be used, such as high-ash, high-moisture, low-grindability, and low-Btu coal. Wear and maintenance are low, and foreign material cannot damage the system.

Kennedy **rod mills** are built in sizes from 0.9 by 1.8 to 4 by 6.1 m (3 by 6 to 13 by 20 ft) for both wet and dry grinding.

Since rod mills avoid packing, they are particularly useful for the reduction of damp or sticky materials. Ordinarily they are used to produce materials in the 6- to 20-mesh range, although both finer and coarser products are readily obtained. They are used to grind ores, cement clinker, and for numerous other materials.

Marcy Ball Mill (Fig. 8-30) This is traditionally a grate-discharge mill, used to give a high throughput rate for a high circulating load in the wet and dry grinding of ores. The data in Table 8-18 must not be used for design but for orientation only. Mill design must be based on pilot experiments or other techniques previously discussed.

Koppers Co., Inc. The Hardinge **conical mill** is used extensively for both wet and dry grinding in open and closed circuits. The conical mill is similar to the cylindrical mill in that it consists of a drum rotating about its horizontal axis and operating in much the same way, but unlike the cylindrical mill, it has conical ends instead of straight ends.

Hardinge wet-grinding mills are supplied with discharge arrangements illustrated in Fig. 8-32 for high, medium, or low pulp levels, the use of which depends on the particular problem under consideration. For dry grinding, a vertical grate with low-pulp-level discharge lifters is used.

Hardinge mills are available in sizes from 0.9 to 4.3 m (3 to 14 ft) in diameter with a length 1½ to 2½ times the diameter. These are used for the wet grinding of sandstone, quartzite, and granites and the dry grinding of abrasives and coke.

Autogenous Tumbling Mills The principle of the ball mill has been employed in some cases in which coarse lump feed will serve as the grinding medium while it is itself being ground. The **Cascade mill** *(Koppers Co., Inc.)* is a wet or dry autogenous mill. It has been built in diameters up to 11 m (36 ft). For all sizes the length-to-diameter ratio is 1:3. A realtively slow speed is used to promote cascading action and prevent segregation of large lumps into the center of the mill. A grate discharges through a trommel screen in the discharge trunnion. The latter scalps oversize pieces, which are conveyed to the feed end of the mill. The feed can be run-of-mine or primary crusher product, with provision made in the bin and feeder arrangements to assure uniform, constant feed-size distribution. For an application, see Fig. 8-59.

The **Aerofall mill** *(Aerofall Mills Ltd.)* is a dry autogenous mill of a shape similar to the Cascade mill. It is an air-swept mill and thus does not need a discharge grate, but it does require an air-handling system and cyclone. Run-of-mine feed is reduced in closed circuit to final product sizes. Autogenous mills eliminate wear of ball media, although frequently a 5 percent charge of large balls is included. Their small length-to-diameter ratios make them suitable for very high circulating loads in closed-circuit operations.

Allis-Chalmers **Rockcyl mills** are used for either dry or wet autogenous grinding.

Rockcyl mills with a length-to-diameter ratio of approximately 1:3 eliminate all except primary crushing stages, all rod-mill grinding, and part or all of the ball-milling stages in a conventional flow sheet.

Intermediate rock grinding reduces, to a desired size, the minus 2-cm (¾-in) product from a closed-circuit crushing operation. Sized rock from the primary crusher product serves as the medium. Rockcyl mills with length-to-diameter ratios of approximately 1:2 are used in intermediate rock grinding.

Secondary rock grinding is the reduction, to a desired size, of a rod-mill or primary rock-mill product, using sized media from either a crushing stage or a primary mill. This is often referred to as **rock-pebble milling.**

Rockpeb mills with a length-to-diameter ratio of approximately 2:1 are used in secondary rock grinding. Because Rockcyl lifters wear faster than liner plates, lifter bars are designed for easy removal and replacement. A large-diameter spout feeder provides free flow of feed into the mill through the short-length trunnion bearing.

NONROTARY BALL OR BEAD MILLS

These include **stirred** and **vibratory** types. In the first, a central paddle wheel or impeller armature stirs the media at speeds from 100 to 1500 r/min. In the second an eccentric motion is imparted either to an armature or to the shell at frequencies up to 1800/min. The media oscillate in one or more planes and commonly rotate very slowly. Stirred mills use media 0.6 cm (¼ in) in size or smaller, whereas vibratory mills use larger media for the same power input. Vibratory mills may grind dry, but most stirred mills are restricted to wet milling. Solids vary from 25 to 70 percent, depending on the feed size and rheology. Media load varies from 3 to 6 times the wet charge mass. Unlike in rotary ball mills, some sedimentation may occur.

Although applications overlap, vibrational equipment is generally used for hard-grinding operations ($ZrSiO_4$, SiO_2, TiO_2, Al_2O_3, etc.), while stirred grinders are mainly used for dispersion and soft grinding (dyes, clays, $CaCO_3$, etc.). Stirred mills are also called **sand mills** when Ottawa sand is used as media.

Contamination and grinder body wear may be minimized in both types by the use of resilient coatings.

Stirred Mills The **Sweco dispersion mill** *(Sweco, Inc.)* has counterrotating radial armatures to move the grinding medium in a vibrating chamber. The high-amplitude DM-70 has a working vol-

TABLE 8-18 Illustrative Performance of Marcy Ball Mills

Size, ft.	Ball charge, tons	Hp. to run	Mill speed, r.p.m.	Capacity, tons/24 hr. (based on medium-hard ore): No. 8 sieve* 20% −200	No. 20 sieve 35% −200	No. 35 sieve 50% −200	No. 48 sieve 60% −200	No. 65 sieve 70% −200	No. 80 sieve 80% −200	No. 100 sieve 85% −200	No. 150 sieve 93% −200	No. 200 sieve 97% −200
3 × 2	0.85	5– 7	35	19	15	12	10	8	6½	5	4	3
4 × 3	2.73	20– 24	30	80	64	53	45	36	28	22	18	14
5 × 4	5.25	44– 50	27	180	145	120	102	82	63	51	41	32
6 × 4½	8.90	85– 95	24	375	300	250	210	170	135	105	85	66
7 × 5	13.10	135–150	22½	640	510	425	360	290	225	180	145	113
8 × 6	20.2	220–245	21	1100	885	735	625	500	390	310	250	195
9 × 7	30.0	345–380	20	1800	1450	1200	1020	815	635	505	410	315
10 × 10	56.50	700–750	18	3680	2960	2450	2100	1700	1325	1050	850	655
12 × 12	90.5	1260–1345	16.4	7125	5725	4750	4070	3290	2570	2035	1650	1275

*Sieve through which substantially all the material can pass.

NOTE: To convert horsepower to kilowatts, multiply by 0.746; to convert tons to megagrams, multiply by 0.907; and to convert tons per 24 hours to megagrams per day, multiply by 0.907.

ume of 0.65 m^3 (23 ft^3) and has a 30-kW (40-hp) motor. Recirculation is possible with an external pump.

In the **Attritor** *(Union Process, Inc.)* a single fixed-axis armature rotates several long radial arms (see Fig. 8-33). These mills are available in batch, continuous, and circulation types. Grinding action is affected by the continuous but irregular approach and retreat of the media in the vicinity of the arms. However, group movement and wall impact are suppressed. See Table 8-19.

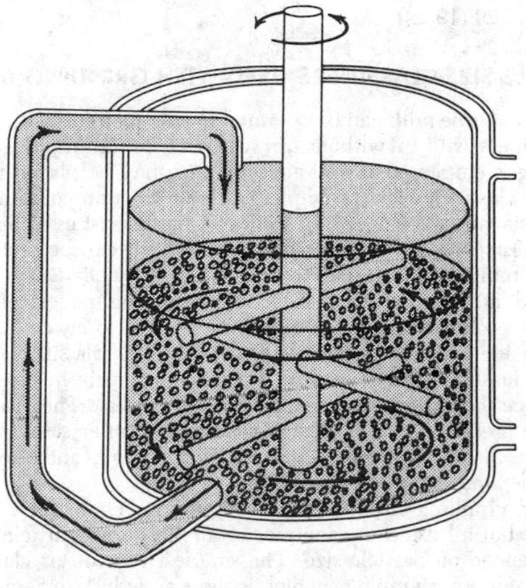

FIG. 8-33 The Attritor. *(Union Process, Inc.)*

Koppers tower mills are available in several sizes for wet-grinding applications. Feed along with steel balls moves down through the height of the mill. Attrition occurs between the feed, the balls, and the screw-flight agitator.

TABLE 8-19 Characteristics of Attritors

Designation	Capacity, gal	Liquid capacity, gal	Motor hp
1-S-WC	1½	¾	2
15-S-WC	20	10–12	15
30-S-WC	43	20–25	20
100-S-WC	113	50–60	50
200-S-WC	245	125–160	100

NOTE: To convert gallons to cubic meters, multiply by 3.785 × 10^{-3}; to convert horsepower to kilowatts, multiply by 0.746.

The **Bureau of Mines mill** (U.S. Patent 3,075,710) consists of a cylindrical, vaned vertical armature in a squirrel-cage configuration rotating in close proximity within a parallel vaned shell. Grinding action occurs dominantly in the vicinity of the vanes, which also impart a vibratory motion to the system as they pass.

Vibratory Mills The **Vibro-Energy** *(Sweco, Inc.)* and **Podmore-Boulton** mills are pedestal-mounted top-loading grinders set in vibration by means of off-center weighting of a base-mounted motor. The grinding chamber is spring-supported to minimize floor vibration (see Fig. 8-34). Grinding is achieved by three-dimensional vibration at a frequency of about 20 Hz of the contained media, usually alumina spheres or cylinders. Other characteristics appear in Table 8-20.

Vibracron mills *(Bepex Corporation)* are available in single- or multiple-tube types for both dry and wet grinding. Feed to the mill can be as large as 5 cm (2 in) in diameter.

Another mill with horizontally induced vibrations is made by Allis-Chalmers. Three parallel cylinders contain the charge (center) and eccentric cams (outer) driven by two separate but interconnected motors at 1200 r/min. The mill is top-loaded through a flexible port. Other information appears in Table 8-21.

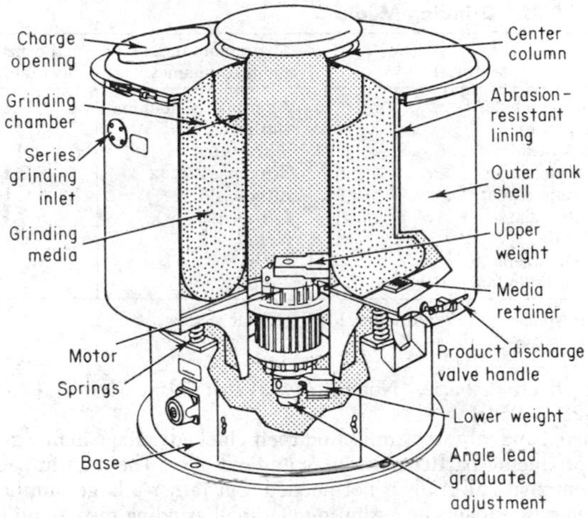

FIG. 8-34 Vibro-Energy mill. *(Sweco, Inc.)*

TABLE 8-20 Characteristics of Sweco Vibratory Mills

Designation	Capacity	Typical sample charge, lb	Motor	Mill diameter, in
M-18	2.6 gal	5–20	¼	18
M-45	20 gal	50–200	5	45
M-60	70 gal	200–1000	10	60
M-80	182 gal	500–2000	40	80
DM-1	0.125 ft^3	3–5	⅛	24
DM-3	0.5 ft^3	20–60	1¼	30
DM-10	3 ft^3	100–400	5	45
DM-20	65 ft^3	200–800	10	60
DM-70	23 ft^3	900–3000	40	95

NOTE: To convert gallons to cubic meters, multiply by 3.785 × 10^{-3}; to convert pounds to kilograms, multiply by 0.4535; to convert horsepower to kilowatts, multiply by 0.746; and to convert inches to centimeters, multiply by 2.54.

Performance The grinding-media diameter should preferably be 10 times that of the feed and should not exceed 100 times the feed diameter. To obtain improved efficiency when reducing size by several orders of magnitude, several stages should be used with different media diameters. As fine grinding proceeds, rheological factors alter the charge ratio, and power requirements may increase.

A variety of grinding media are available, as shown in Table 8-22. Size availability varies, ranging from 1.3 cm (½ in) down to 325 mesh.

Although there are no definitive data on media shape and grinding, spheres and cylinders generate less impurity from attrition than irregular particles. Ball-milling data indicate that spheres are the

TABLE 8-21 Characteristics of Allis-Chalmers Vibratory Mills

Designation	Capacity, gal.	Height, in.	Length, in.	Width, in.	Total motor hp.
1518-D	14	39	70	53	15
3034-D	100	68	116	87	100
3640-D	176	76	125	104	150
4248-D	286	80	159	117	250

NOTE: To convert gallons to cubic meters, multiply by 3.785 × 10^{-3}; to convert pounds to kilograms, multiply by 0.4535; to convert horsepower to kilowatts, multiply by 0.746; and to convert inches to centimeters, multiply by 2.54.

TABLE 8-22 Grinding Media

Material	Common and/or trade names	Forms available*
Aluminum oxide	Alumina, corundum	S,C,I
Silicon carbide	Carborundum	C,I
Silicon dioxide	Silica, sand	I
Zirconium oxide	Zirconia, Zircoa	S,C,I
Zirconium silicate	Zircon	S,C,I
Annealed glass	Ceramedia	S
Polyamide	Nylon	S,C,
Divinylbenzene	DVB	S,C
Polyfluoroethylene	Teflon	C

*S = spheres, C = cylinders, I = irregular shapes.

most effective shape. [Norris, *Trans. Inst. Min. Metall.*, **63**(567), 197–209 (1954)].

Stirred and vibratory mills find their chief advantage in fine grinding, producing particle sizes of 1 μm and finer. The high impact of conventional ball mills is not needed, but rather a large number of low-energy impacts necessitating (1) small grinding media and (2) a high vibration or rotation rate.

A typical improvement in efficiency is shown in Fig. 8-35 for the submicrometer grinding of zircon. Although each machine has its peculiar characteristics and time requirements for various types of grinding, Fig. 8-36 illustrates some typical results obtained under optimum conditions for several materials.

Planetary Ball Milling This is a method of increasing the gravitational force acting on balls in a ball mill. For example, refractory metals and carbides can be ground to 1 to 2.6 μm in 5 to 20 min in

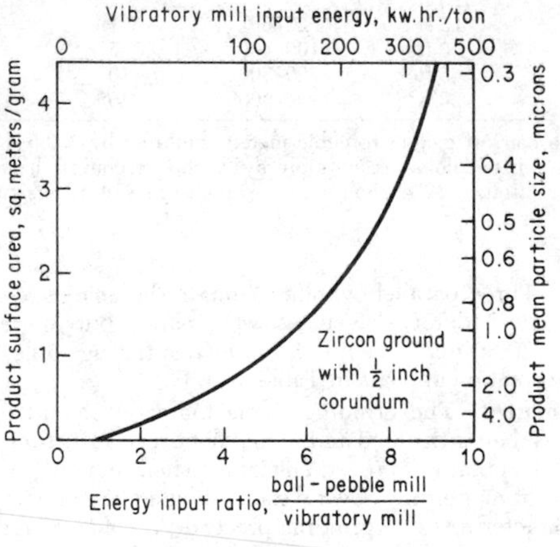

FIG. 8-35 Energy comparison: conventional versus vibratory ball mills versus product fineness. *(Sweco, Inc.)*

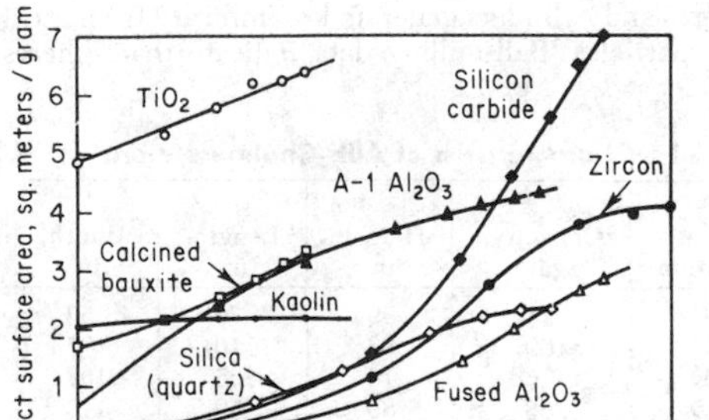

FIG. 8-36 Vibratory-mill typical performance. *(Sweco, Inc.)*

an apparatus capable of applying a centrifugal force of 10 to 50 G. [Dobrovol'skii, *Poroshk. Metall.*, **7**(6), 1–7 (1967)]. **Pulverit** planetary mills are available from Geoscience Inc.

High-speed planetary ball mills can be used to perform rapid tests to simulate ball milling of materials [Vock, *DECHEMA-Monogr.*, **69**, III-8 (1972)]. The size of high-speed mills will be much smaller than the size of same-capacity ball mills [Bradley, *S. Afr. Mech. Eng.*, **22**, 129 (1972)].

PARTICLE-SIZE CLASSIFIERS USED WITH GRINDING MILLS

Ball mills or tube mills can be operated in closed circuit with external air classifiers with or without air sweeping being employed. If air sweeping is employed, a cyclone separator may be placed between mill and classifier. (The principles of size reduction combined with size classification are discussed under "Characteristics of Size Classifiers.") Likewise other types of grinding mill can be operated in closed circuit with external size classifiers (Fig. 8-12), as will be described at appropriate places on succeeding pages. However, many types of grinders are air-swept and are so closely coupled with their classifiers that the latter are termed internal classifiers.

Dry Classifiers Dry **screens** are used primarily in crusher circuits, since they are most effective down to 4 mesh. They can sometimes be used to 35 mesh. Examples are *Hummer* screens *(W. S. Tyler, Inc.)*, *Rotex* screens *(Orvill-Simpson Co.)*, and the *Vibro-Energy separator (Sweco, Inc.)*

Most dry-milling circuits use **air classifiers.** There are a number of types, but all use the principles of air drag and particle inertia, which depend on particle size. The simplest type of air classifier is an elutriator, an example of which is the Kennedy Van Saun expansion-type classifier. A countercurrent multielement elutriator is the Zig-Zag classifier *(Alpine American Corp.)*. The sharpness of separation increases with the number of elements. These devices are effective in the 30- to 80-mesh range.

Another type of classifier directs an air stream across a stream of the particles to be classified. An example is the radial-flow classifier *(Kennedy Van Saun Corp.)*, which features adjustable elements to control the flow and classification. A further development on this principle is the Vari-Mesh classifier *(Kennedy Van Saun Corp.)*, which controls classification by adjustable flow baffles. A change in direction of air flow is the operating principle of the reverse-flow Superfine classifier *(Koppers Co., Inc.)* depicted in Fig. 8-37.

Rotating blades are the main elements of several types of classifiers. The blades set up a centrifugal motion that tends to throw coarser particles outward. An example is the Mikro-Atomizer (Fig. 8-47), in which an external fan forces air inwardly through the blades, carrying with it the fines. Centrifugal motion returns coarse particles to the hammers. The whizzer blades shown on the Raymond high-side mill (Fig. 8-45) have a similar centrifugal effect, throwing coarse particles to the wall of the chamber, where the lower-boundary-layer air velocity allows them to fall back to the grinding zone.

Rotor blades also form an element of several external classifiers that are used in closed-circuit dry milling. These are generally called mechanical air separators or classifiers. Examples are the Whirlwind classifier *(Sturtevant Mill Co.)*, the Gayco centrifugal separator *(Universal Road Machinery Co.* (see Fig. 8-37), and the whizzer separator *(Raymond Division of Combustion Engineering Inc.)*. The feed enters these devices through a chute at the top and is distributed between two rotating feed plates. The coarse particles fall into an interior cone, while the fines must pass inward through rotor blades to move up over the top plate. A fan at the top of the unit circulates air and fines outward and down between the central cone and an outer conical shell until it passes inward through a set of fixed vanes, across the descending coarse material, and upward again, elutriating the coarse material. Thus these classifiers make use of several separation principles.

Some mechanical air classifiers are designed so that the fine product must pass radially inward through rotor blades instead of spirally moving across them as with whizzer blades. Examples are the Mikron separator *(Pulverizing Machinery Co.)* and the Majac classifier shown attached to the Majac jet mill (Fig. 8-57).

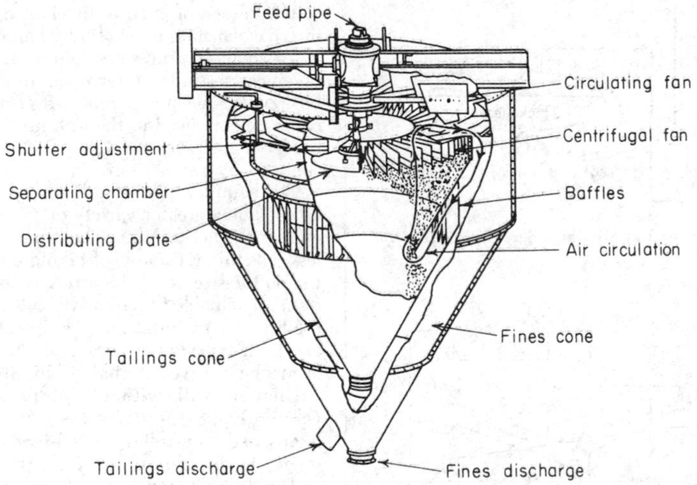

FIG. 8-37 Gayco centrifugal separator. *Universal Road Machinery Co.)*

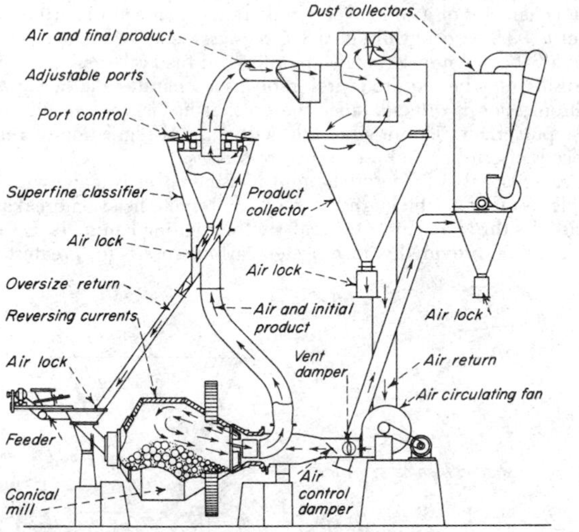

FIG. 8-38 Hardinge conical mill with reversed-current air classifier.

There are several mechanical air classifiers designed to operate in the superfine 10- to 90-μm range. Two of these are the Mikroplex spiral air classifier MPVI *(Alpine American Corp.)* and the classifier which is an intregal part of the Hurricane pulverizer-classifier *(Bauer Bros. Co.)* described under "Hammer Mills." Another is the Donaldson classifier. When mechanical air classifiers form an integral part of a mill, the rotating vanes, air fan, and grinding elements may be mounted all on the same shaft or on different shafts with separate drives. The former arrangement gives mechanical simplicity and often a simpler air-flow path. Separate drives allow independent adjustment of separator and mill speeds, hence more versatile and often more effective classifier operation. Many examples of these arrangements are given in the following paragraphs. In addition, the classifier can be completely separate, connected with the mill in closed circuit by ductwork. An example is shown in Fig. 8-38. The sweep air enters and leaves the Hardinge ball mill at the same end, while the coarse recycle and feed enter at the other. The fine product is carried by the air out of the top of the classifier and into the cyclone, where product and air are separated. A negative pressure is maintained in the air-classifying system to prevent dusting.

Performance Each type of classifier has a range of sizes which it can separate, although the ranges can be extended by design changes that sacrifice capacity. Deflector-type classifiers without rotating elements may give a product 85 percent through a 250-μm sieve, although more typically they give a product 95 percent below 74 μm. Mechanical air classifiers with rotating elements can give a product from 85 percent through 250 μm to as fine as 99.9 percent below 44 μm. The single whizzer is designed for operation where finenesses range to about 95 percent passing 74 μm, whereas the double-whizzer separator is intended for use where higher-fineness products, in the range of 99.9 percent or better passing 44 μm, are required. Sizes of mechanical air classifiers range from 1 to 7 m (3 to 24 ft) in diameter, with power requirements from 2 to 450 kW (3 to 600 hp). Superfine types can give a product 98 percent through 10 μm.

Typical separation efficiency curves of an air classifier versus particle size are given in Fig. 8-14. The amount of top size in the fines may be very low, but there is typically 10 to 30 percent fines in the coarse product. In addition, the separation at the cut size is typically a gradual curve. Data of this sort, which are needed to evaluate closed-circuit mill performance, are seldom available. See subsection on characteristics of size classifiers for a testing method.

Wet Classifiers Closed-circuit wet milling is the rule in large-scale operations because of its greater production and economy. The simplest wet classifier is a **settling basin** arranged so that the fines do not have time to settle but are drawn off while the thickened coarse product is raked to a central discharge. Examples are the Hardinge Hydro-Classifier and the Dorr thickener. For classification near micrometer size a **continuous centrifuge** such as the Sharpless super-centrifuge or the Bird centrifuge is effective. The separation is not sharp in settlers, and the large space requirement is a detraction. **Rake classifiers** and **screw classifiers** are described in Sec. 21. The action is countercurrent, so separation of coarse grains is more effective. Examples are the Hardinge countercurrent (screw) classifier and the Dorr-Oliver rake classifier. Typical circuits used with these classifiers in cement- and ore-processing plants are shown in Figs. 8-39 and 8-40. **Hydrocyclones** have become the most popular wet classifiers in ore operations owing to their compact design and economy of operation. Control is effected by feeding at a constant rate from a sump, in which the liquid level is maintained by varying water addition as the slurry feed rate varies (see Fig. 8-41).

In the 1930s there were attempts to use **screens** for wet closed-circuit milling, but operating cost was prohibitive. Recently screens have been developed that are practical for mill circuits. The first of these was the Dutch State Mines screen, which has the vibrating screen cloth on a curved incline, with the dilute slurry flowing over and through it.

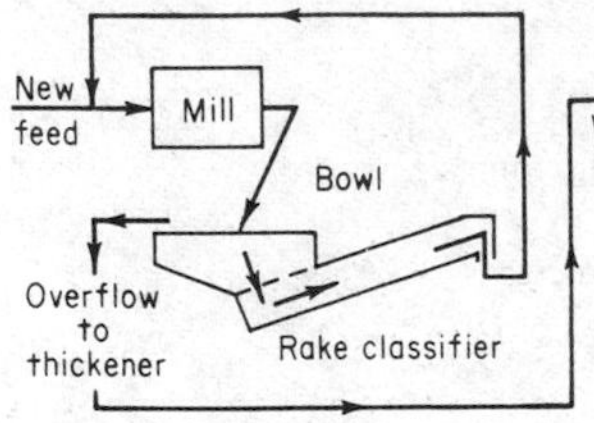

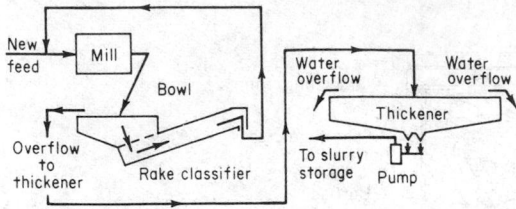

FIG. 8-39 Single-stage closed-circuit wet-grinding system. [*Tonry,* Pit Quarry *(February–March 1959).*]

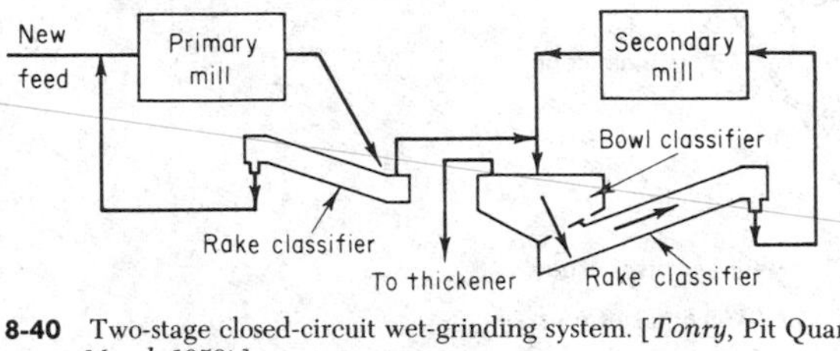

FIG. 8-40 Two-stage closed-circuit wet-grinding system. [*Tonry,* Pit Quarry *(February–March 1959).*]

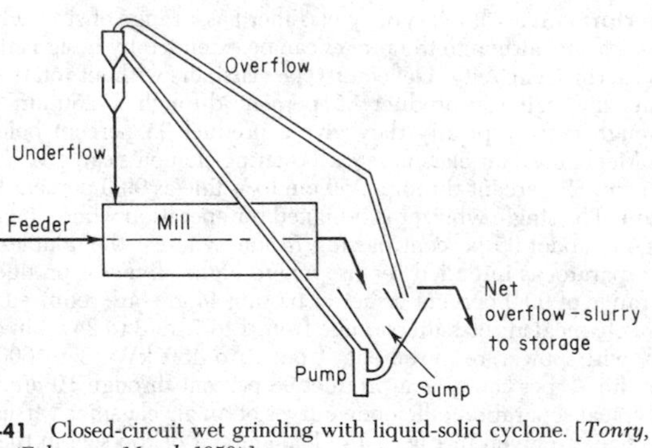

FIG. 8-41 Closed-circuit wet grinding with liquid-solid cyclone. [*Tonry,* Pit Quarry *(February–March 1959).*]

The use of rubber screen cloths solves problems of blinding [Wessel, *Aufbereit. Tech.,* 8(2), 53–62; (5), 167–80; (8), 417–428 (1967); Michel, *Min. Mag. (London),* annual review issue (5), 189–193, 207 (1968)]. An upper layer of rubber is perforated with fine slots for particle sizes from 0.2 to 2.5 mm., and this is supported by a lower layer with coarse holes. The vibration rate is 2500 to 3000 cycles/min. The advantage of screens is that a considerably sharper separation can be effected, and less fines are returned to the mill. Screen separation is considerably less than perfect, although there are few published data.

HAMMER MILLS

Hammer mills for pulverizing and disintegration are operated at high speeds. The rotor shaft may be vertical or horizontal, generally the latter. The shaft carries hammers, sometimes called beaters. The hammers may be T-shaped elements, stirrups, bars, or rings fixed or pivoted to the shaft or to disks fixed to the shaft. The rotor runs in a housing containing grinding plates or liners. The clearance maintained between the liners and rotor is important with respect to the fineness of product. The grinding action results from **impact** and **attrition** between lumps or particles of the material being ground, the housing, and the grinding elements. A cylindrical screen or grating usually encloses all or part of the rotor. The fineness of product can be regulated by changing rotor speed, feed rate, or clearance between hammers and grinding plates, as well as by changing the number and type of hammers used and the size of discharge openings.

The **screen** or **grating discharge** for a hammer mill serves as an internal classifier, but its limited area does not permit effective usage when small apertures are required. To meet critical maximum-size specifications in the intermediate-size range, the hammer mill may be operated in closed circuit with external screens of larger area than could be employed in the mill itself. The mill discharge screen then has large apertures to retain grossly oversize material in the grinding zone.

The hammer mill is made in a great many types and sizes and can be used on a greater variety of soft materials than any other type of machine. The feed must be nonabrasive with a hardness of 1.5 or less. The mill is capable of taking 2-cm (¾-in) feed material, depending on the size of the feed throat, and reducing it to a product substantially all able to pass a No. 200 sieve. For producing materials in the fine-size range, it may be operated in conjunction with external air classifiers. Such an arrangement is shown in Fig. 8-38. A number of machines have internal air classifers.

Hammer Mills without Internal Air Classifiers The **Mikro-Pulverizer** (Fig. 8-42) *(Pulverizing Machinery Co.)* is a close-clearance, high-speed, controlled sealed-feed hammer mill used for a wide range of nonabrasive materials, the major applications being sugar, carbon black, chemicals, pharmaceuticals, plastics, dyestuffs, dry colors, and cosmetics. For performance see Table 8-23. Speeds, types of hammers, feed devices, housing variations, and perforations of screens are all varied to fit applications, with the result that finenesses and character of grind cover a wide range. Some of the grinds are as fine as 99.9 percent through a 325-mesh screen. Feed material should usually be down to 4 cm (1½ in) or finer. If feed is larger, an auxiliary crusher may be required, preferably as a separate unit, because synchronization is difficult since the crusher has larger capacities than the pulverizer. Tie-in is possible with careful regulation of relative speeds of crusher and feed screw or screws.

A replaceable liner for the mill housing cover is made with multiple serrations, which are designed to promote head-on breakage of particles thrown against the wall by the rotating hammers. Hammer tips can be provided with tungsten carbide inserts for greatest wear

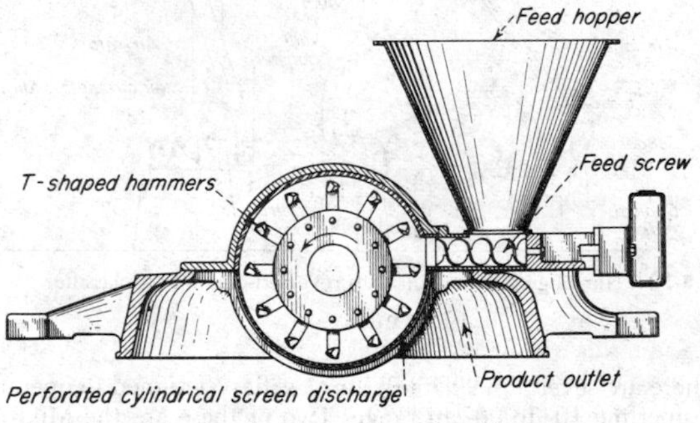

FIG. 8-42 Mikro-Pulverizer hammer mill. *(Mikropul Corporation.)*

TABLE 8-23 Mikro-Pulverizer Performance

Size	Rotor diam., in.	Max. r.p.m.	Hp.	Avg. capacities, lb./hr.: 6X sugar	Clay-graphite water slurry	Pigments and colors (dry)
Bantam	5	16,000	¾–1	75–100	75–100	70–90
1	8	9,600	3–5	350–550	550	300–500
2	12	6,900	7½–15	800–1500	750–1600	800–2000
3	18	4,600	20–40	2000–4500	4800	2500–4500
4	24	3,450	40–100	4000–9000	7000	4500–7000

NOTE: To convert inches to centimeters, multiply by 2.54; to convert horsepower to kilowatts, multiply by 0.746; and to convert pounds per hour to kilograms per hour, multiply by 0.4535.

resistance or with Hastellite tipping. An air-injector feeder can be supplied to project the feed particles directly in front of the hammer tips, to provide a more direct blow and thus increase mill efficiency. Wet feed can be charged with feed screws or pumps for wet grinding.

A cryogenic grinding system is available for the grinding of resilient and heat-sensitive materials. It consists of a precooler and feeder unit in which liquid nitrogen is sprayed on the material to be ground. The material is embrittled at −200°C prior to grinding.

Mikro-Pulverizers are made in five sizes shown in Table 8-23. The smallest size is the Bantam, which is widely used in laboratories for development and pilot work. Results may be extrapolated and translated into what may be expected of full-scale production units.

The **Blue Streak dual-screen pulverizer** *(Prater Industries, Inc.)* is used for the grinding of resins, chemical salts, plastic scrap, food products, and similar materials to a granular uniform powder of No. 30 or No. 40 sieve fineness. Feed enters opposite ends of the rotor and undergoes three stages of size reduction by hammers of decreasing size. Two perforated screens cover more than 70 percent of the area of the final sizing drum through which the product passes.

The **Atrita pulverizer** *(Riley Stoker Corp.)* is available in several single and duplex types. Capacities vary from 3400 to 25,000 kg/h (7500 to 54,000 lb/h). This type of pulverizer utilizes a series of swing hammers pivoted to the rotor hub, around which is a stationary grid, cut away at one section so that foreign material is thrown out. After passing through this first effect, the coal is carried in a current of air into the second effect, which contains alternate rows of moving and stationary pegs, where most of the pulverizing is done. Leaving the second effect, the coal is passed through a rejector, a number of scooplike blades on the main shaft, where the heaviest particles are thrown back into the pulverizing compartment, permitting the passage of the finer particles only, which enter the fan inlet and are carried into the furnace. Hot air can be introduced into the machine for drying the coal. Air at 150°C dries coal with 8 percent moisture down to about 1 percent.

The **Aero pulverizer** *(Foster Wheeler Corp.)* is used for coal, pitch, and coke, blowing the ground material directly into the furnace. The housing is divided into two or three short cylindrical pulverizing chambers. Primary air is admitted at the feed end and between the last chamber and fan. The horizontal shaft carries disks to which are fixed hammers, a set for each chamber. Coal is pulverized by impact and attrition. Annular baffles of increasing diameter between the chambers cause particles to be retained until properly reduced in size for discharge from the final chamber in suspension in the air stream. Hot gases can be introduced to dry the fuel being pulverized. Refractory material such as tramp iron is removed in the first pulverizing chamber and eliminated through a tramp-iron pocket.

Disintegrator The Rietz machine (Fig. 8-43) consists of a rotor running inside a 360° screen enclosure. The rotating shaft is usually vertical. The rotor includes a number of hammers designed to run at fairly close clearance relative to the inside of the cylindrical screen enclosing the disintegration chamber. The hammers are normally fixed rigidly to the shaft by keyways, pins, or welding, but swing hammers are used when indicated.

Rietz disintegrators are supplied in three types. *In-line disintegrators* (RI series) are designed for in-line installation, in which they function without impeding the process flow. Their primary applications are the mixing, delumping, and dissolving of fluids, slurries, and pastes and the grinding and separation of high-fiber solids.

Angle disintegrators (RA and RP series) are used for the fine pulping of many food products and for fine dispersion and homogenizing in the food and chemical industries. *Vertical disintegrators* (RD series) are used for dry pulverizing, wet grinding to produce slurries or pastes, shredding, defibrizing, and the fine pulping of soft fruits and vegetables.

Rietz disintegrators are normally supplied in rotor diameters from 10 to 60 cm (4 to 24 in), with rotational speeds to produce hammer tip speeds in ranges of 300 to 6700 m/min (1000 to 22,000 ft/min) and power ranges from 0.4 to 150 kW (½ to 200 hp). Higher speeds and higher power are available. Models are available in various

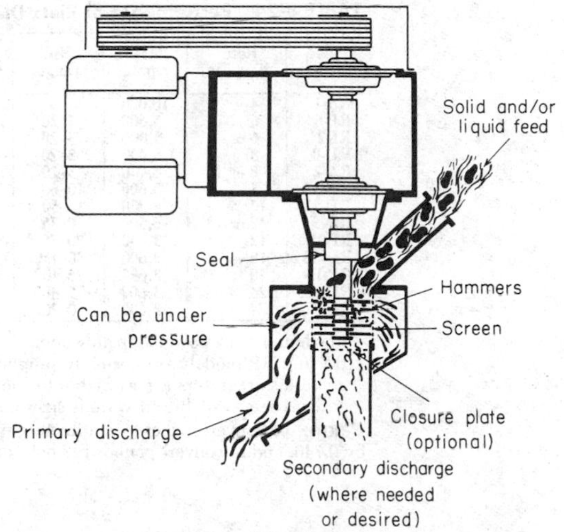

FIG. 8-43 Rietz disintegrator. *(Rietz Mfg. Co.)*

materials of construction and in highly sanitary, easy-cleaning models or heavy-duty industrial construction (Table 8-24).

Fitz mills *(Fitzpatrick Co.)* consist of several series of hammer mills in configurations adapted to a variety of uses in food processing. There are high-speed screen hammer mills with flat hammers for impact, and narrow hammers or sharp hammers for tough plastic or fibrous materials. There are long, small-diameter rotating mills for processing pastes and two-shaft toothed masticators. There are also single-roll toothed choppers and shredders with fixed knives.

Prater Industries, Inc., manufactures narrow swing-hammer screened mills for oilseeds and fibrous materials.

Turbo pulverizers and **turbo mills** *(Pallmann Pulverizer Co.)* combine the action of hammer and attrition mills, finding special application for grinding plastic materials that would be softened under high-energy warm-mill conditions.

Pin Mills In contrast to peripheral hammers of the rigid or swing types, there is a class of high-speed mills having pin breakers in the grinding circuit. These may be on a rotor with stator pins between circular rows of pins on the rotor disk, or they may be on rotors operating in opposite directions, thereby securing an increased differential of speed. See also the Mikro-ACM pulverizer described later.

Kollopex mills *(Alpine American Corp.)* are high-speed impact mills with one stationary and one rotating stud disk. The mills are operated without a sieve and hence can be used with materials that tend to block (see Fig. 8-44). The **Contraplex wide chamber** is a similar mill with both disks rotating. It is suitable for grinding materials that tend to form deposits or for greasy, heat-sensitive products. These mills are used in the grinding of food, pesticides, pigments, and soft minerals, the wet grinding of PVC suspensions, and the crushing of cacao beans, etc. A laboratory model is also available.

Entoleter impact mills *(Entoleter, Inc.)* are a class of vertical-shaft devices in which feed at the shaft is caused to move rotationally and is thrown outward from the rotor to impact on an outer ring. Pin-type structures have been found effective; and in these the pins on the rotor do primary breakage, while the outer ring of pins gives further reduction. A wide range of speeds is employed, the higher ones being used for fine pulverizing. These mills grind a great variety of free-flowing or semi-free-flowing substances to controlled, preset sizes. Among these are plastics, rubber, asbestos to fiber, grain and flour, coal, clay, slag, and salts. In some cases, external classification is required to remove oversize for return to the mill. Plastic materials are embrittled by liquid nitrogen or other suitable refrigerants to reduce their elasticity. For the highest speeds, the stator pins are mounted on a ring which is moving in reverse rotation to the central rotor. The mills are characterized by low power, low heating, and high capacity.

TABLE 8-24 Performance of Rietz Disintegrators

Model	Rotor diam., in.	Max. r.p.m.	Hp. range	Screen perforation, in.	Typical applications	
					Material	Capacity
RA-1	4	16,000	½–5	1/32–¼	General lab use	1–10 lb./min.
RP-6	6	3,600	1–20	3/16	Horseradish	300 lb./hr.
RI-2	6 or 8	5,000	3–20	1/16	Detergent delumping	100 gal./min.
RD-8	8	8,400	3–20	⅛	Color coat	3600 lb./hr.
RA-2	8 or 12	8,400	3–20	1/32	Meat, cooked	3000–5000 lb./hr.
RP-8	8	3,600	10–60	¼	Blood declotting	20 gal./min.
RD-12	12	7,200	15–50	¼–¾	Polystyrene	3000–10,000 lb./hr.
RA-3	12 or 18	6,500	10–75	3/64	Corn, heated	350 lb./min.
RP-12	12	3,600	20–75	⅜	Asbestos-cement slurry	200 gal./min.
RD-18	18	3,600	30–150	⅜	Chemical-fertilizer delumping	15 tons/hr.
RP-18	18	3,600	25–100	¼	Animal fat (90°F.)	15,000 lb./hr.
RD-24	24	3,600	75–400	1	Wood-chip shredding	30 tons/hr.
RI-4	24	3,600	50–200	¼	Bagasse depithing	30 tons/hr. (dry)

Maximum horsepower depends upon maximum speed.
RA and RP models are normally supplied with stainless-steel contact parts.
Some disintegrators are available for operation under pressure.
Screens are available in various sizes and types of perforations down to 0.006 in.

NOTE: To convert inches to centimeters, multiply by 2.54; to convert horsepower to kilowatts, multiply by 0.746; and to convert pounds per hour to kilograms per hour, multiply by 0.4535.

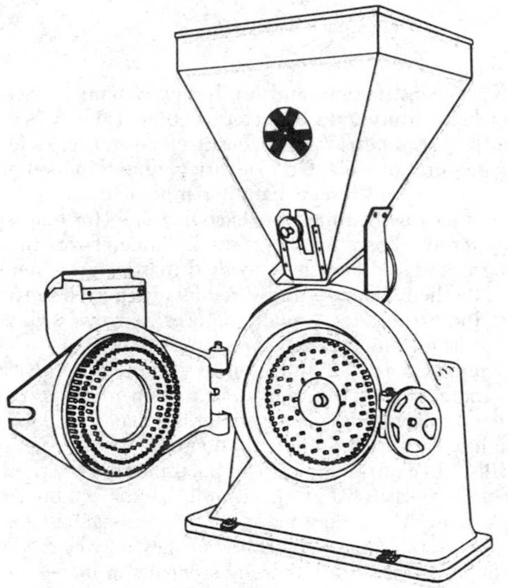

FIG. 8-44 Alpine Kolloplex mill. *(Alpine American Corp.)*

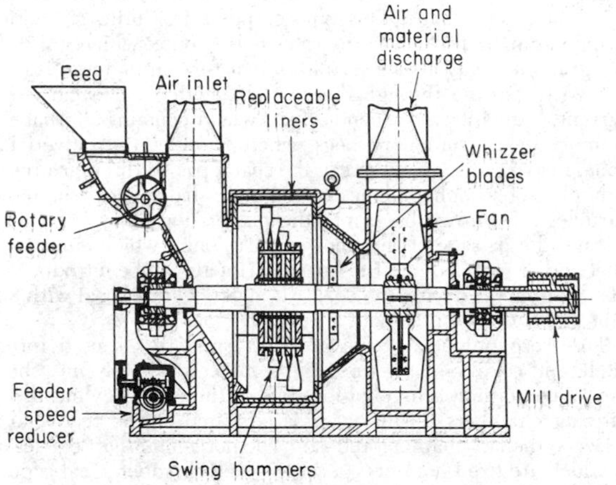

FIG. 8-45 Whizzer air classification applied to the Raymond Imp mill.

Hammer Mills with Internal Air Classifiers The **Imp pulversizer** *(Raymond Division, Combustion Engineering Inc.)* is an air-swept hammer mill, as illustrated in Fig. 8-45. This machine is made in many sizes from the smallest, having one row of hammers using 7.5 kW (10 hp), to the largest size, with six rows of hammers and requiring 150 kW (200 hp) to drive it. The machines are equipped with a hopper below which is the star feeder, actuated by a pawl-and-ratchet mechanism.

A fan is placed on one end of the hammer shaft; between the fan and the hammers is the whizzer, consisting of two or more thin blades with tips tapered to conform to the housing. Distance between blades and housing is regulated by moving the whizzer along the shaft. As the whizzer is moved toward the hammers, a coarser product results. The action of the whizzer is described under "Particle-Size Classifiers with Grinding Mills." The classified product passes through the fan and is blown to a cyclone collector, where it is discharged into bins or containers. The air goes back to the pulverizer, completing the cycle.

It is necessary to vent a small amount of surplus air to a final dust collector. If proper care is used in feed and product handling, the operation can be relatively dust-free.

These Imp units are excellent drying devices and are widely used for simultaneous drying, pulverizing, and classifying.

The **Automatic pulverizer** *(Raymond Division)* is a hammer-type machine equipped with air classifier of the vacuum multivane type (see "Particle-Size Classifiers with Grinding Mills") or the double-whizzer type. It has a horizontal shaft on which may be mounted one or more disks fitted with hammers. A star feeder with a pawl-and-ratchet mechanism receives the raw material from a stock bin and drops it into the pulverizing chamber, on top of which is mounted the air classifier. The air enters the pulverizing chamber at the rear and removes the pulverizer material. Particles of proper fineness are blown to the cyclone, which discharges to bins or containers, while oversize is returned to the pulverizer through the bottom valve of the inner cone. On the door of the pulverizing chamber is mounted an automatic throwout, the function of which is to remove resistant materials contained in the feed, such as sand and gravel from clay. Impurities in the oversize accumulate in the grinding chamber until they are picked up by the rapidly revolving hammers and thrown through the slot on the door into the throwout chamber, where they are finally rejected through the flap valve. The slide damper on top of the throwout may be adjusted to admit air from the atmosphere, which enters the pulverizer through the slot in the door. In its travel through the throwout the air cleans the rejects and blows fine particles back into the pulverizing chamber.

The rotating components of the Raymond **vertical mill** are carried on its vertical shaft. They are the grinding element, double-whizzer classifier, and fan, as shown in Fig. 8-46. This mill has a hammer tip speed of 7600 m/min (25,000 ft/min), so that it is effective for finer grinding than the Imp mill, which has a tip speed of 6400 m/min (21,000 ft/min). The vertical mill also has a more effective classifier.

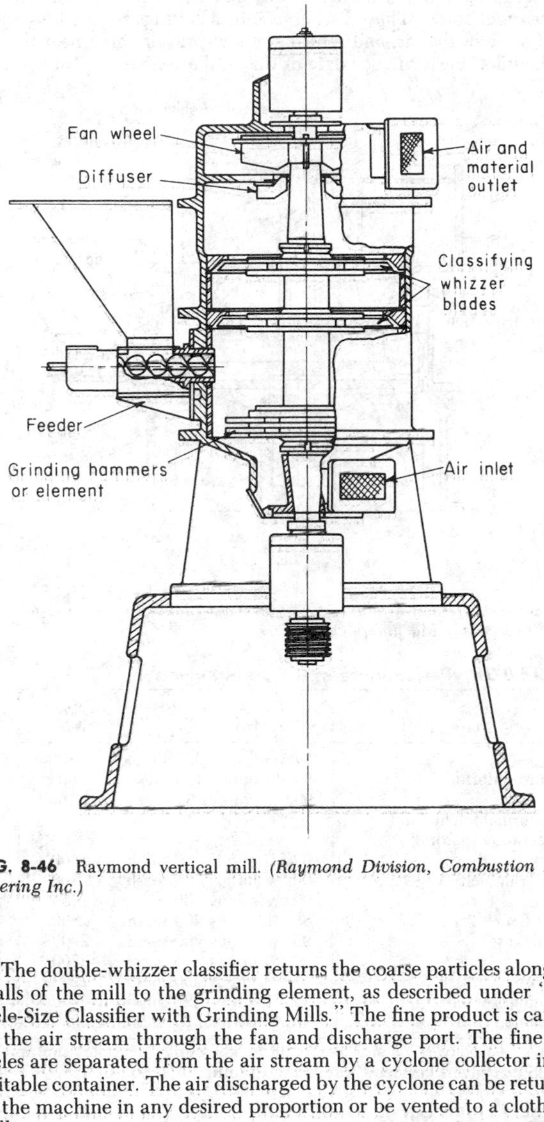

FIG. 8-46 Raymond vertical mill. *(Raymond Division, Combustion Engineering Inc.)*

The double-whizzer classifier returns the coarse particles along the walls of the mill to the grinding element, as described under "Particle-Size Classifier with Grinding Mills." The fine product is carried in the air stream through the fan and discharge port. The fine particles are separated from the air stream by a cyclone collector into a suitable container. The air discharged by the cyclone can be returned to the machine in any desired proportion or be vented to a cloth-bag collector.

Machines are available with rotor diameters of 45.7 and 88.9 cm (18 and 35 in), driven by 15- and 110-kW (20- and 150-hp) motors respectively. The larger mill is directly connected to a vertical motor. Normal rotor speed for the 45.7-cm (18-in) Raymond vertical mill is 6000 r/min and 3600 r/min for the 88.9-cm (35-in) machine.

The field of application of the Raymond vertical mill is the production of materials that range in size from those having 99 percent passing a 44-μm sieve to those having 99 percent smaller than 15 μm, depending on the state of aggregation of the feed. A production rate of 227 kg/h (500 lb/h) is achieved with a chemical in a 45.7-cm (18-in) machine, consuming 13.4 kW (18 hp) when the product is substantially smaller than 15 μm. In a talc operation on a 88.9-cm (35-in) machine requiring 110 kW (150 hp), a production rate of 320 kg/h (700 lb/h) is obtained. At a production rate of 2250 kg/h (5000 lb/h), a sample of the product leaves only a trace of talc on a No. 325 testing sieve.

The Bauer **Hurricane pulverizer-classifier** is a hammer mill with a whizzer classifier on a common shaft. The tip speed is 6700 m/min (20,000 ft/min). The classifier is effective to particle sizes down to 10 μm, and the mill finds use in the asbestos industry and for grinding kaolins. It is available in two sizes requiring from 45 to 93 kW (60 to 125 hp).

The **Mikro-Atomizer** *(Mikropul Corporation)* is a built-in classifier unit as per Fig. 8-47, and has a horizontal rotor shaft carrying hammers, classifier wheels, and fan wheels. Material is fed into the unit through a screw feed and comes in contact with the T-shaped hammers and is divided into two streams with a spiral circular motion to either side of the hammers as grinding takes place between the high-speed hammers and a ridged main liner. Air is drawn through the unit by fan blades. The action of the air classifier is described under "Particle-Size Classifier with Grinding Mills." Fine product is carried out through the classifier and fans, through the product outlets, which usually converge into a single duct, and thence either directly to a dust filter alone or to a cyclone or a combination of a cyclone and a bag filter. Higher rotor classifier and fan speeds, larger vanes on the separator wheel, and smaller fan-wheel diameter all contribute toward obtaining particles of the finest order, and various combinations of these factors are used to obtain variations in results.

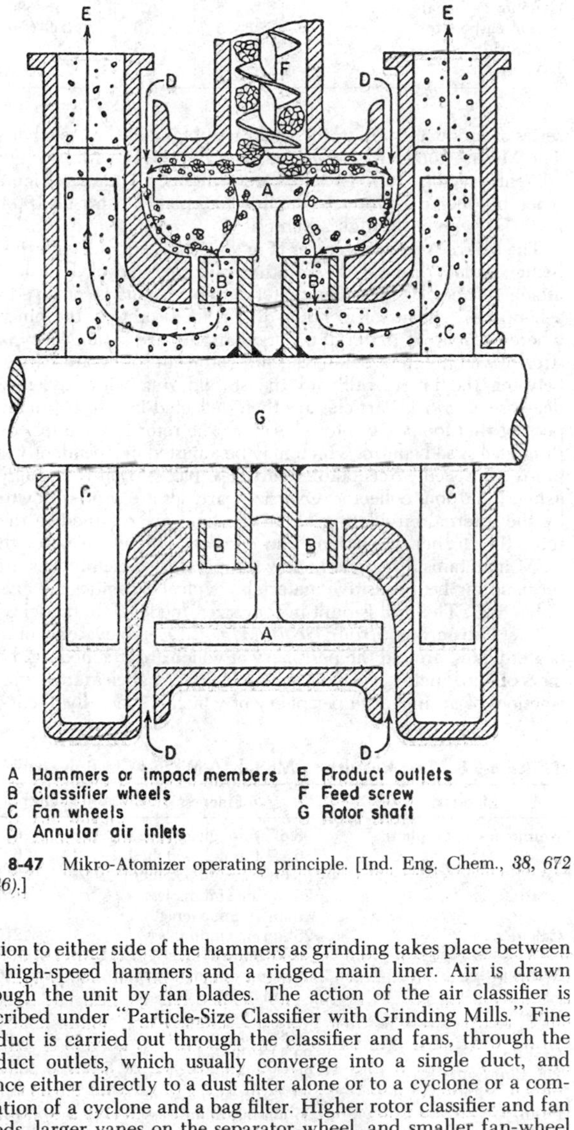

FIG. 8-47 Mikro-Atomizer operating principle. [Ind. Eng. Chem., 38, 672 *(1946)*.]

Mikro-Atomizers are made in three sizes, with characteristics given in Table 8-25. Feed size is limited to 1.9 cm (¾ in) and smaller. Table 8-26 gives performance for the No. 6 Mikro-Atomizer on a

TABLE 8-25 Mikro-Atomizer Operating Characteristics

Machine No.	Rotor diam., in.	Max. rotor r.p.m.	Hp.	Relative capacity
5	8	14,000	5	1
6	12	7,000	20	4
8	24	3,450	75	18

Table 8-26 Performance of the No. 6 Mikro-Atomizer

Material	Particle size, μ		Production rate, lb./hr.
	Avg.	Max.	
Sugar	19	40	500
Polyvinyl chloride	10–12	20–30	125
Calcium carbonate	5	25	600
Nickel carbonate	2.5– 5	10–20	300–650
Lead oxide	2	5	1250
Dry colors	4	15	500

series of products; similar finenesses are obtainable on the other sizes. The Mikro-Atomizer is also used to grind cocoa with cocoa-butter content varying from 12 to 23 percent but requires refrigeration when producing a product which is 99.5 percent through 100 mesh and 97.5 percent through 200 mesh.

The **Mikro-ACM pulverizer** (Fig. 8-48) is a pin mill with the feed being carried through the rotating pins and recycled through an attached vane classifier. The material to be ground is conveyed from a hopper by means of a motor-driven feed screw to the pin rotor, where breakup of material occurs. Particles are entrained by an air stream which enters below the pin rotor, and they are carried up between the inner wall and the shroud ring with baffles which decrease air swirl. Particles are then deflected inward by an air-dispersing ring to a vane-rotor classifier. The rotor is separately driven through a speed control which may be adjusted independently of the pin-rotor speed. Acceptable particles pass upward through the exhaust and to a collector. Oversize particles are carried downward by the internal circulating air stream and are returned to the pin rotor for further reduction. The constant flow of air through the ACM maintains a reasonable low temperature which makes it ideal for handling heat-sensitive materials. Typical capacities are given in Table 8-27. The mill is built in four sizes: model 10 to model 60.

The **Pulvocron** *(Strong Scott Mfg. Co.)* employs one or more beater plates, around the periphery of which are attached rigid hammers of hard metal. It is driven within a casing at clearances of small fractions of an inch, the periphery of which is generally V-cut (Fig. 8-49). The grinding ring has provision for cooling with liquid in direct contact with its periphery. Feed enters around the driving shaft and is first broken by breaker plates on the first disk. It then travels circumferentially with an axial component to a classifying chamber, in which is a separately driven and controlled rotor with vanes. The volume of air carries the fine particles inward to an axial discharge opening, while the coarse particles are kept outward by centrifugal force. They discharge into a tailings-return line, along with some of the air, and return to a low-pressure area near the axis of the inlet. Performance data of this mill are given in Table 8-28.

TABLE 8-27 Test Results on Model 10 Mikro-ACM Pulverizer*

Material	Fineness	Output, lb/h
Ammonium phosphate	98.8% through 200 mesh	665
	86.7% through 350 mesh	665
	63.6% through 350 mesh	862
Apatite	25 μm maximum, less than 10 μm average	67
Calcite	25 μm maximum, less than 5 μm average	210
Dyestuff (black)	9 μm maximum, less than 2 μm average	360
Glue	95.9% through 200 mesh	32
	34.8% through 200 mesh	68.5
Graphite	99.82% through 300 mesh	110
Potassium sulfate	92.8% through 350 mesh	685
	80.6% through 350 mesh	1,070
Resin	25 μm maximum, less than 10 μm average	140
Resin (thermoset)	15 μm maximum, less than 2–4 μm average	52
Resin (ultratough)	97.6% through 300 mesh	29.5
Shellac	99.1% through 200 mesh	77.5

*Power consumption, 6 to 10 hp. To convert pounds per hour to kilograms per hour, multiply by 0.4535.

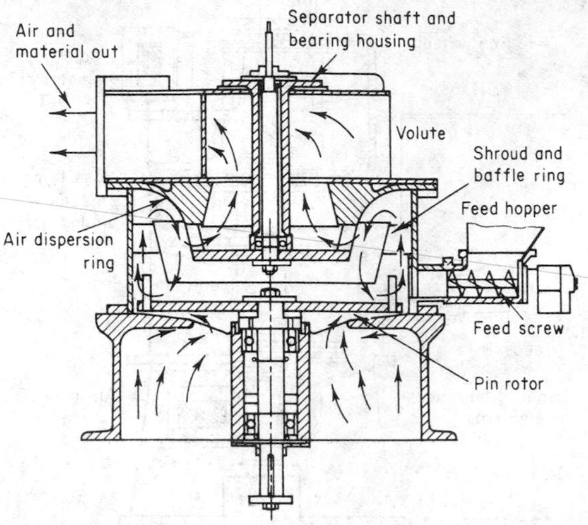

FIG. 8-48 Section of a Mikro-ACM pulverizer, illustrating air and material flow. *(Pulverizing Machinery Co.)*

TABLE 8-28 Performance of the 20-in Pulvocron

Material	Particle analysis, by weight	Capacity, lb./hr.	Hp.
Sucrose	97.5% minus 325 mesh	1800	60
Sodium chloride	99.4% minus 100 mesh	3600	50
	99.95% minus 325 mesh	160	45
Urea-formaldehyde and melamine molding compounds	99.2% minus 80 mesh	1600	45
Paraformaldehyde	99.7% minus 325 mesh	1300	40
Casein	99 % minus 80 mesh	650	50
Corn flour	88 % minus 200 mesh	800	35
Soy flakes	95 % minus 200 mesh	2000	60
Sterols	100 % minus 5 μ	700	60
Lactose	98.5% minus 200 mesh	1200	40
Alumina, hydrated	99 % minus 325 mesh	700	30
Cinnamon	99.7% minus 60 mesh	1000	50

NOTE: To convert pounds per hour to kilograms per hour, multiply by 0.4535; to convert horsepower to kilowatts, multiply by 0.746.

RING-ROLLER MILLS

Ring-roller mills (Fig. 8-50) are equipped with rollers that operate against grinding rings. Grinding takes place between the surfaces of the grinding element, i.e., the ring and rollers. Pressure may be applied with heavy springs or by centrifugal force of the rollers against the ring. Either the ring or the rollers may be stationary. The grinding ring may be in a vertical or a horizontal position. Ring-roller mills also are referred to as ring-roll mills or roller mills or medium-speed mills. The ball-and-ring and bowl mills are types of ring-roller mill.

Ring-roller mills should be distinguished from roller mills. Paint roll mills are described under "Disk Attrition Mills," and flour roll mills under "Cereals and Other Vegetable Products."

Ring-Roller Mills without Internal Classification The **Sturtevant mill** has a concave vertical grinding ring and is used for non-

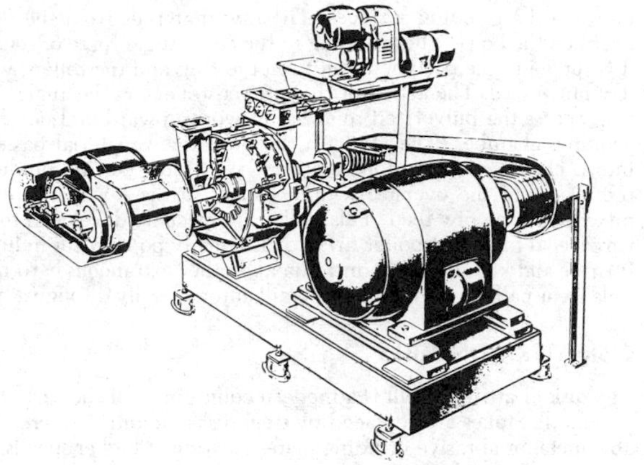

FIG. 8-49 The Pulvocron. *(Strong Scott Mfg. Co.)*

metallics, especially phosphate rock. A No. 1 mill with external air classifier grinds 1.8 to 3.6 Mg/h (2 to 4 tons/h) of limestone or phosphate rock to 90 percent through a No. 80 sieve. The **Kent Maxecon mill** is used for bauxite, coke, limestone, magnesite, and phosphate rock. The rotating grinding ring has a horizontal axis, and the feed falls successively through several nips. The open construction minimizes contamination in changing from one feed material to another. Capacity in closed circuit with external screen or air classifier is 3.6 Mg/h (4 tons/h) of phosphate rock for acidulation or 9.1 Mg/h (10 tons/h) of limestone for agricultural use.

Ring Mills with Internal Screen Classification The grinding action of the Hercules *(Bradley Pulverizer Co.)* is that of three rolls that are revolved around and against a die to create grinding pressures of approximately 100 MPa (15,000 lbf/in^2). It can produce minus-20-mesh agricultural limestone or phosphate rock from minus-5-cm (−2-in) feed. The material is discharged from the grinding chamber through a surrounding screen. Capacity is relatively high, being 23 to 45 Mg/h (25 to 50 tons/h) of average-hardness dry limestone. Other product sizes may be obtained by changing the screen aperture.

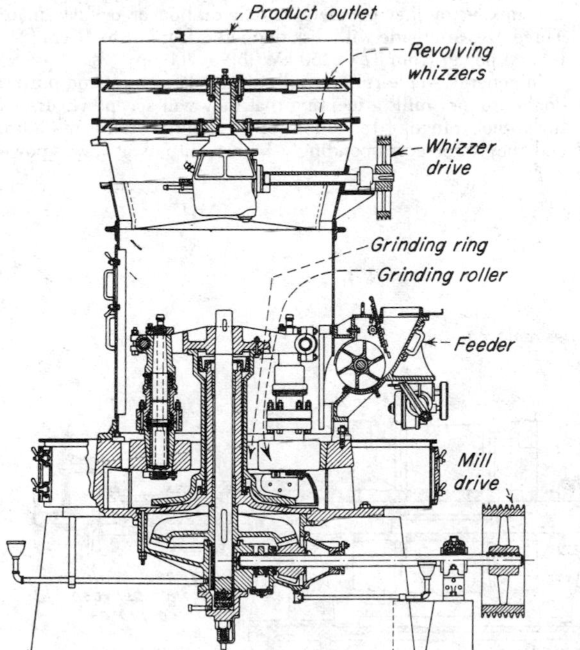

FIG. 8-50 Raymond high-side mill with an internal whizzer classifier.

Ring-Roller Mills with Internal Air Classification The Babcock & Wilcox pulverizers, Type B, 100-series, consist of a single row of balls operating between a stationary bottom ring and a rotating top ring. The Type B, 200- and 300-series, are designed with multiple rows of balls to produce maximum capacity in the space occupied. The 200-series pulverizer consists of two rows of balls, one above the other. The top and bottom rings are stationary with the intermediate ring rotating. Externally adjustable springs load the grinding elements to the pressure required. The 300-series pulverizers include a third row of balls to increase the capacity still further.

The B & W pulverizer, Type E, consists of a single row of balls operating between a rotating bottom ring and a stationary top ring (Fig. 8-51). Externally adjusted springs apply pressure to the top ring to give the required loading for proper pulverizing. In operation wet raw coal is admitted inside the ball row and is fed through the grinding elements by centrifugal force. The Type E pulverizer is particularly suited to the direct firing of rotary kilns and industrial furnaces when close temperature control is required and long periods of continuous operation are essential. It is built in 17 sizes with capacities up to 12.6 Mg/h (14 tons/h) or more.

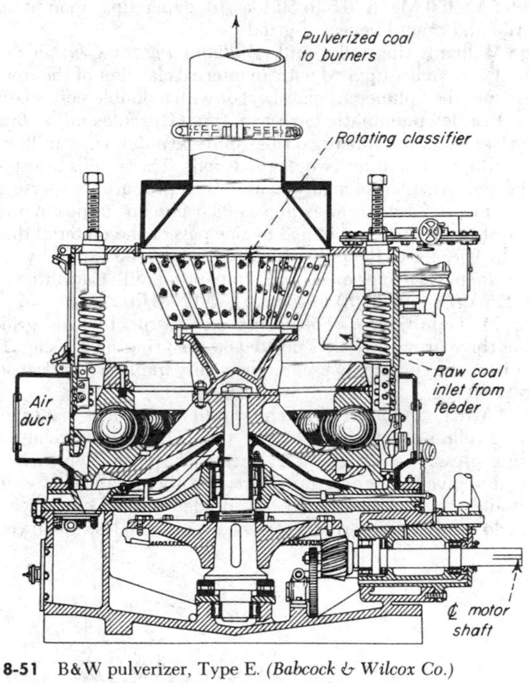

FIG. 8-51 B&W pulverizer, Type E. *(Babcock & Wilcox Co.)*

The Raymond ring-roller mill (Fig. 8-50) is of the internal air-classification type. The base of the mill carries the grinding ring, rigidly fixed in the base and lying in the horizontal plane. Underneath the grinding ring are tangential air ports through which the air enters the grinding chamber. A vertical shaft driven from below carries the roller journals. The rollers on the bottom rotate on their own bearings while traveling around the ring. Centrifugal force urges the pivoted rollers against the ring. The raw material from the feeder drops between the rolls and ring and is crushed. Both centrifugal air motion and plows move the coarse feed to the nips. The air entrains fines and conveys them up from the grinding zone, providing some classification at this point. An air classifier is also mounted above the grinding zone to return oversize.

The method of classification used with Raymond mills depends on the fineness desired. If a medium-fine product is required (up to 85 or 90 percent through a No. 100 sieve), a single-cone air classifier is used. This consists of a housing surrounding the grinding elements with an outlet on top through which the finished product is discharged. This is known as the low-side mill. For a finer product and when frequent changes in fineness are required, the whizzer-type classifier is used. Its mode of operation is described under "Particle-Size Classifier Used with Grinding Mills." This type of mill is known as the *high-side mill* (Fig. 8-50).

The Raymond ring-roll mill with internal air classification is used for the large-capacity fine grinding of most of the softer nonmetallic minerals. Materials with a Mohs-scale hardness up to and including 5 are handled economically on these units. Typical natural materials handled include barites, bauxite, clay, gypsum, magnesite, phosphate rock, iron oxide pigments, sulfur, talc, graphite, and a host of similar materials. Many of the manufactured pigments and a variety of chemicals are pulverized to high fineness on such units. Included are such materials as calcium phosphates, sodium phosphates, organic insecticides, powdered cornstarch, and many similar materials. When properly operated under suction, these mills are entirely dust-free and automatic. They are available in six basic sizes. Connected power ranges from 28 to 500 kW (40 to 700 hp). Capacities range from 0.5 to 450 Mg/h (0.5 to 50 tons/h), depending upon nature of material and exact fineness of grind.

The Williams ring-roller mill *(Williams Patent Crusher & Pulverizer Co.)* can be supplied with an internal classifier of the rotating-blade type (the Spinner air classifier) or with a double-cone classifier.

The **Bradley pneumatic** (air-swept type) **Hercules mills** *(Bradley Pulverizer Co.)* are centrifugal ring-roll-type pulverizing mills which can be fitted with either two or three rolls. These mills are suitable for the pulverization of many materials to produce as coarse as 98 percent minus-20 mesh to as fine as 99.5 percent minus-325 mesh. The finished product is obtained by one pass of the material through the mill. The size of the pulverized product can be varied by adjusting the fineness selector mounted on top of the mill. Capacities range from 225 kg/h (500 lb/h) to 45 Mg/h (50 tons/h).

The **MBF pulverizer** *(Foster Wheeler Corp.)* for coal grinding also has three grinding rollers pivoted off the grinding housing. These pulverizers are commonly used in the utility industry, and capacities of up to 80 Mg/h (90 tons/h) are available.

Bowl Mills In the Raymond bowl mill the journals that carry the grinding rollers are stationary while the grinding ring rotates. The grinding pressure is produced by means of springs, which may be adjusted to give the required pressure, and the distance between the rollers and the ring may be set to a predetermined clearance. The rollers do not touch the ring, there being no metal-to-metal contact between the grinding surfaces. The raw materials from the feeder drops on the bowl, where, owing to the centrifugal force of rotation, it is forced to the periphery between the ring and the rollers, where it is pulverized. The action of the tapered rollers on the angle of the ring causes the pulverized material to work upward and out of the grinding chamber. The air with the pulverized material passes up into a classifier of the double-cone type. Here the fine product is removed and the oversize dropped back to the bowl, where it is mixed with the raw feed. This mill was especially developed to pulverize coal for direct boiler firing. It is equally popular for industrial furnace and kiln firing. Tramp iron and other extraneous hard materials are usually thrown out of the mill automatically through a spout.

DISK ATTRITION MILLS

The disk or attrition mill is a modern counterpart of the early buhrstone mill. Stones are replaced by steel disks mounting interchangeable metal or abrasive grinding plates rotating at higher speeds, thus permitting a much broader range of application. They have a place in the grinding of tough organic materials, such as wood pulp and corn grits. Grinding takes place between the plates, which may operate in a vertical or a horizontal plane. One or both disks may be rotated; if both, then in opposite directions. The assembly, comprising a shaft, disk, and grinding plate, is called a **runner.** Feed material enters a chute near the axis, passes between the grinding plates, and is discharged at the periphery of the disks. The grinding plates are bolted to the disks; the distance between them is adjustable.

The Sprout-Waldron **attrition mill** (Fig. 8-52) is available in single- and double-runner models with 48- to 122-cm- (12- to 48-in-) diameter disks and with power ranging up to 1100 kW (1500 hp). By the use of a variety of plates and shell constructions these units are represented in such applications as coarse granulating, pulverizing, and shredding. A single-runner model, having plates with concentric circular rows of projecting spikes on the rotating plate meshing with those on the stationary plate, acts much like a hammer mill, the spikes being the fixed hammers, and can serve for such applications.

Double-disk mills *(Bauer Bros. Co.)* are used for the grinding of fibrous and nonfibrous substances, fluffing of fibrous materials, intensive mixing of fine powders, and hydration of cellular materials. Three sizes are made with disk diameters from 61 to 91 cm (24 to 36 in) and power from 37 to 150 kW (50 to 200 hp).

In general, single-runner mills are used for the same purposes as double-runner mills, excepting that they will accept a coarser feedstock, their range of reduction for a given material is more limited, and they offer correspondingly higher outputs at lower power. In

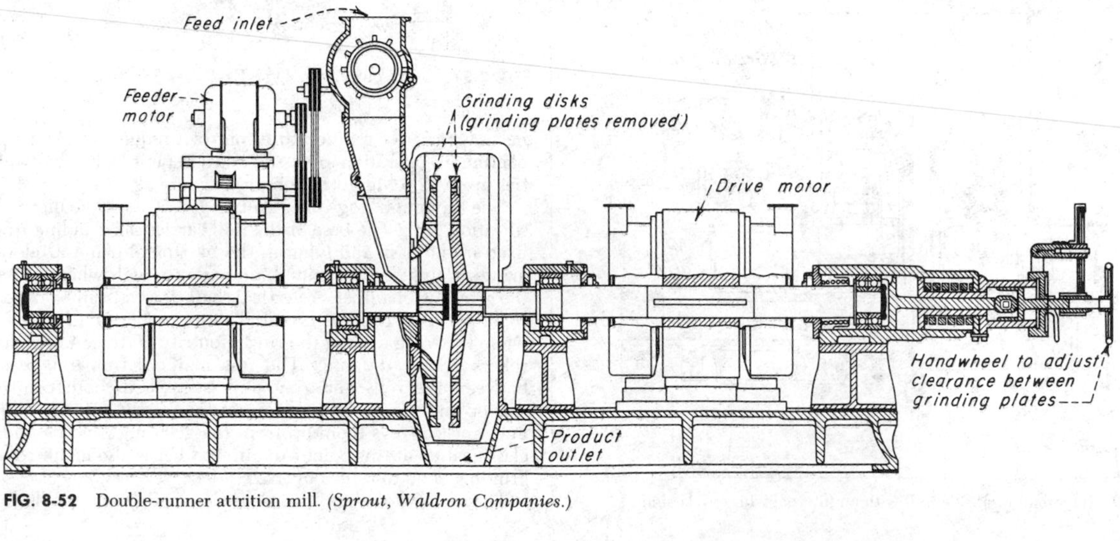

FIG. 8-52 Double-runner attrition mill. *(Sprout, Waldron Companies.)*

addition, there are a number of applications unique to this unit such as the fluffing of sheet pulp from continuous rolls, to which the inlet provisions of a double-runner mill are not suited. The same range of plate types can be used on both single- and double-runner mills. While spike-tooth plates can be used in certain applications to simulate hammer-mill action, they are more generally applied to specialized tasks involving tearing, shredding, or controlled shattering, as in dehulling. The performance data presented in Table 8-29 typify the applications of the attrition mill.

TABLE 8-29 Performance of Disk Attrition Mills

Material	Size-reduction details	Unit*	Capacity lb./hr.	Hp.
Alkali cellulose...	Shredding for xanthation	B	4,860	5
Asbestos.........	Fluffing and shredding	C	1,500	50
Bagasse..........	Shredding	B	1,826	5
Bronze chips......	⅛ in. to No. 100 sieve size	A	50	10
Carnauba wax....	No. 4 sieve to 65% < No. 60 sieve	D	1,800	20
Cast-iron borings.	¼ in. to No. 100 sieve	A	100	10
Cast-iron turnings	¼ in. to No. 100 sieve	E	500	50
Cocoanut shells...	2 × 2 × ¼ in. to 5/100 sieve	B	1,560	17
	5/100 sieve to 43% < No. 200 sieve	D	337	20
Cork.............	2/20† sieve to 20/120 < No. 200 sieve	D	145	15
Corn cobs........	1 in. to No. 10 sieve	F	1,500	150
Cotton seed oil and solvent	Oil release from 10/200 sieve product	B	2,400	30
Mica.............	4 × 4 × ¼ in. to 3/60 sieve	B	2,800	6
	8/60 to 75% < 60/200 sieve	D	510	7.5
Oil-seed cakes (hydraulic)........	1-½ in. to No. 16 sieve	F	15,000	100
Oil-seed residue (screw press)...	1 in. to No. 16 sieve size	F	25,000	100
Oil-seed residue (solvent).......	¼ in. to No. 16 sieve	F	35,000	100
Rags.............	Shredding for paper stock	B	1,440	11
Ramie............	Shredding	B	820	10
Sodium sulfate....	35/200 sieve to 80/325 sieve	B	11,880	10
Sulfite pulp sheet.	Fluffing for acetylation, etc.	C	1,500	50
Wood flour.......	10/50 sieve to 35% < 100 sieve	D	130	15
Wood rosin.......	4 in. max. to 45% < 100 sieve	B	7,200	15

*A—8 in. single-runner mill D—20 in. double-runner mill
B—24 in. single-runner mill E—24 in. double-runner mill
C—36 in. single-runner mill F—36 in. double-runner mill
†2/20, or smaller than No. 2 and larger than No. 20 sieve size.

NOTE: To convert inches to centimeters, multiply by 2.54; to convert pounds per hour to kilograms per hour, multiply by 0.4535; and to convert horsepower to kilowatts, multiply by 0.746.

The **Frigidisc grinder** *(Young Machinery Co.)* is a rugged single-runner disk attrition mill developed for the rubber-reclaiming industry. The mill is suitable for materials that must be ground with a minimum rise in temperature, such as reclaimed tire-tread scrap, synthetic rubber, and other materials of a tough, resilient nature. Both the stationary and the moving grinding disks are cooled with a circulating liquid so that a high pressure can be exerted on them.

Buhrstone mills are attrition mills with hard circular stones serving as grinding media, generally French, American, or Esopus buhrstones; rock emery or a combination of French buhr and Esopus or of pebble grit and emery rock are also used. Buhrstone mills are still employed for grinding special cereals and grains. Feed enters the mill through a center hole in one of the stones. It is distributed between the stone faces and ground while working its way to the periphery.

The buhrstone or stone mill for "paint grinding" has been replaced by the **roller mill** (Fig. 8-53). Paint-grinding roller mills consist of two to five smooth rollers (sometimes called rolls) operating at differential speeds. A paste is fed between the first two, or low-speed, rollers and is discharged from the final, or high-speed, roller by a scraping blade. The paste passes from the surface of one roller to that of the next because of the differential speed, which also applies shear stress to the film of material passing between the rollers. Roller-mill technique and action have been studied by Hummel [*J. Oil Colour Chem. Assoc.*, 270–277 (June 1950)], and the breakup of agglomerates in this mill has been discussed by Krekel [*Chem. Ing. Tech.*, 38(3), 229 (1966)].

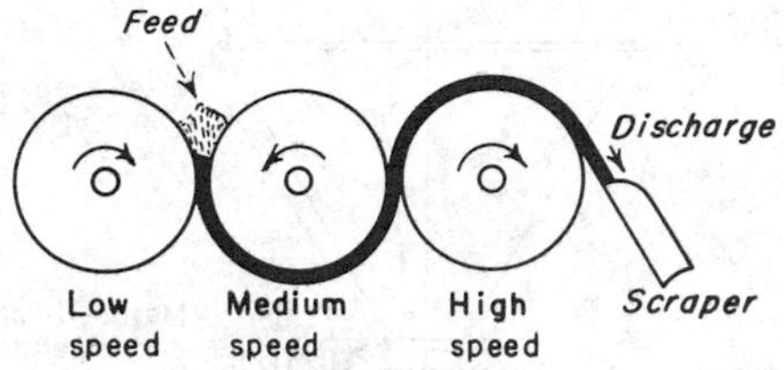

FIG. 8-53 Roller mill for paint grinding.

DISPERSION AND COLLOID MILLS

When there is to be very little breakdown of individual particles and when the problem is to disrupt lightly bonded clusters or agglomerates, a new aspect of fine grinding enters. This may be illustrated by the breakdown of pigments to incorporate them in liquid vehicles in the making of paints. Other comparatively weak structures are amenable to reduction in this way. Purees, food pastes, pulps, and the like are processed by this type of mill. Dispersion is also associated with the formation of emulsions which are basically two-fluid systems. Sirups, sauces, milk, ointments, creams, lotions, and asphalt and water-paint emulsions are in this category. There is a special class of mills employed for dispersion and colloidal operations. They operate on the principle of high-speed **fluid shear.** Although they are classed as grinders, they do not do much actual grinding. Their value lies in ensuring a breakdown of agglomerates or, in the case of emulsions, the shearing of fluid phases to produce dispersed droplets of fine size, around 1 μm.

A mathematical analysis of the action in **Kady** and other colloid mills checks well with experimental performance [Turner and McCarthy, *Am. Inst. Chem. Eng. J.*, 12(4), 784 (1966)]. Various models of the Kady mill have been described, and capacities and costs given by Zimmerman and Lavine [*Cost Eng.*, 12(1), 4–8 (1967)]. Energy requirements differ so much with the materials involved that other devices are often used to obtain the same end. These include high-speed stirrers, turbine mixers, pebble mills, and vibratory mills as well as buhrstone, disk, hammer, and roll mills. In some cases, sonic devices are effective.

Colloid mills which are employed for dispersion or for emulsification fall into four main groups: the hammer or turbine, the smooth-surface disk, the rough-surface type, and valve or orifice devices. The principle of their action is to create a fluid stream of high velocity with very great shear forces existing within the fluid, which serve to disrupt the particles. Chemical aid in the form of dispersing agents very often is valuable.

The concentration of energy in mills of this class is high, and there is a considerable amount of heating. This is materially reduced by use of a cooling-water jacket. In other cases, as when emulsions are made hot, the jacket is employed for heating.

The **Morehouse mill** *(Morehouse Industries, Inc.)* is a high-speed disk type of mill (Fig. 8-54). The undispersed phase is fed at the top and passes between converging disks, being thrown outward at the periphery. As the larger particles are broken and dispersion becomes finer, the stream is subjected to still higher energy between the narrowest zone of disk spacing to complete the disintegration and to ensure essential freedom from coarser particles.

In the **Premier mill** *(Premier Mill Corp.)*, the rotor is shaped like the frustrum of a cone. Surfaces are smooth, and adjustment of the clearance can be made from 25 μm (0.001 in) upward. The mill is jacketed for temperature control. Direct-connected liquid-type mills are available with 15- to 38-cm (6- to 15-in) rotors. These mills operate at 3600 r/min at capacities up to 5.7 m^3/h (1500 gal/h). They are powered up to 75kW (100 hp). Working parts are made of Invar alloy, which does not expand enough to change the grinding gap if heating occurs. The rotor is faced with Stellite or silicon carbide for wear resistance. For pilot-plant operation, the Premier mill is available with 7.5- and 10-cm (3- and 4-in) rotors. These mills are belt-driven and operate at 7200 to 17,000 r/min with capacities of 0.02 to 0.6 m^3/h (5 to 150 gal/h).

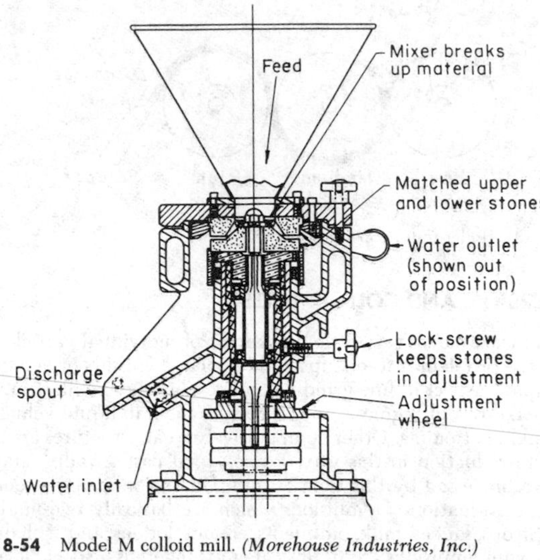

FIG. 8-54 Model M colloid mill. *(Morehouse Industries, Inc.)*

The **Charlotte mill** *(Chemiolloid Corp.)* also employs high speed of rotation with the fluid flowing between a grooved conical rotor and a corresponding grooved conical stator. Clearance between them is regulated by an external calibrated adjustment device.

The whirling currents set up within the grooves subject the product to both hydraulic shear and impact. All models operate at 3600 r/min. The following sizes are available:

Power		Capacity	
kW	hp	L/min	gal/h
2	3	1.3–3.2	20–50
5	7	3.2–6.3	50–100
15	20	6.3–25	100–400
37	50	25–63	400–1000
56	75	63–315	1000–5000
93	125	190–440	3000–7000

Laboratory model W-10 operates at 0.75 kW (1 hp) with a capacity of 4 to 190 L/h (1 to 50 gal/h). The mills are available in several materials, including stainless steel, nickel, Monel, bronze, and cast iron. A special model ND is designed for mayonnaise and salad oils. Sanitary models are available for processing foodstuffs.

The **Tri-Homo disperser-homogenizer** has a stator head and a high-speed rotor in which several designs of grooves are available as well as the smooth and abrasive types.

The **Gaulin colloid mill** has a smooth rotor, shaped like a discus. Entering material is first thrown outward along the disk and then around the edge and inward, thus giving a *two-stage* action. The gap setting between the rotor and housing is adjustable down to 25 μm (0.001 in). The rotor is made of stainless steel and operates at 3600 r/min. The mill is jacketed for temperature control.

The **Manton-Gaulin mill** uses a valve and an impactor. In this device the rough suspension is pumped into a narrow orifice to increase its velocity to levels approaching sonic. This gives high shear forces for reduction, and these are further implemented as this high-velocity stream strikes an impact ring, where its direction is changed. This is accomplished by a high order of turbulence which is converted into work of dispersion.

FLUID-ENERGY OR JET MILLS

A detailed description of mills of this type has been presented by Gossett [*Chem. Process.* (Chicago), **29**(7), 29 (1966)]. **Fluid-energy mills** may be classified in terms of the nature of the mill action. In one class of mills, the fluid energy is admitted in fine high-velocity streams at an angle around a portion or all of the periphery of a grinding and classifying chamber. In this class are the Micronizer, jet pulverizer, Reductionizer, Jet-O-Mizer, and others of somewhat similar structure. In the other class the fluid streams convey the particles at high velocity into a chamber where two streams impact upon each other. The Donaldson and other mills are in this class. Whether the particles are conveyed with the jet or are intercepted with angle jets as they travel around the periphery of the grinding-classifying chamber, there is a high energy release and a high order of turbulence which causes the particles to grind upon themselves and to be ruptured. Not all the particles are fully ground; so it is necessary to carry out a classifying operation and to return the oversize for further grinding. Most of these mills utilize the energy of the flowing fluid stream to effect a centrifugal classification. The Donaldson mill differs, using a mechanical air classifier.

The **Micronizer** *(Sturtevant Mill Co.)* consists of a shallow circular grinding chamber in which the material to be pulverized is acted upon by a number of gaseous fluid jets issuing through orifices spaced around the periphery of the chamber, as shown in Fig. 8-55. The rotating gas must discharge at the center, carrying the fines with it, while the coarse particles are thrown toward the wall, where they are subjected to further reduction by impact from particles in incoming jets. The outlet from the grinding chamber leads directly into a centrifugal product collector. The action in these mills has been studied photographically and mathematically [Rumpf, *Chem. Ing. Tech.*, **32**(3), 129–135; (5), 335–342 (1960); *ATS translations*, 668GJ, 844GJ.]

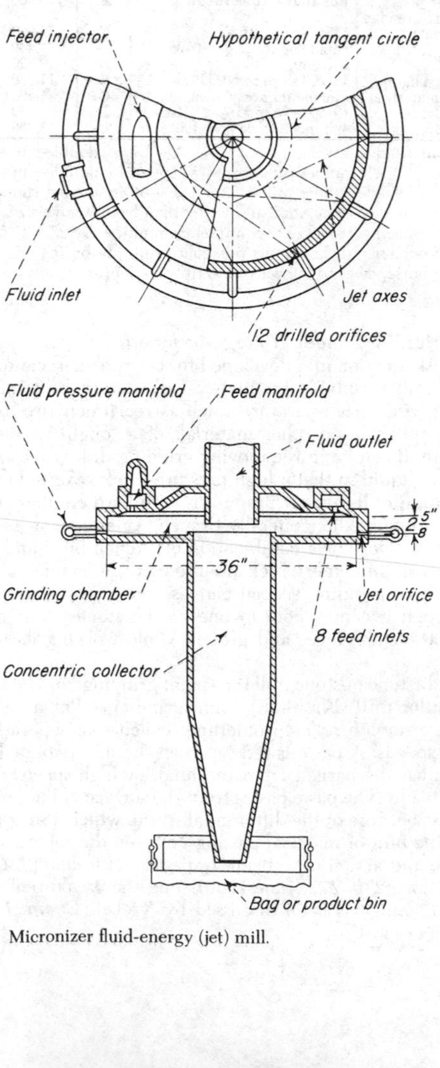

FIG. 8-55 Micronizer fluid-energy (jet) mill.

Micronizer mills are constructed in nine standard sizes from 5 to 107 cm (2 to 42 in) in diameter, with capacities from 250 g/h to 1.8 Mg/h (½ lb/h to 2 tons/h).

The feed size should be smaller than 1 cm (¼ in). Production rate, fluid consumption, and fineness figures are shown in Table 8-30.

The **jet pulverizer** *(Jet Pulverizer Co.)* is another mill of the shallow-pan, angle-jet, and radial-inward-classification type, like the Micronizer.

TABLE 8-30 Micronizer Performance*

Material	Product average size, μm	Feed: Size sieve no.	Feed: Rate, lb/h	Fluid consumption, g fluid/g solid: Air	Fluid consumption, g fluid/g solid: Steam
Ceylon graphite	2	3	200	. . .	8.5
Cryolite	3	60	900	. . .	4.0
Limestone	3.5	80	1000	. . .	4.0
Hard talc	3.5	20	1000	. . .	4.0
Silica gel	5.5	8	500	. . .	3.5
Soft talc	6.5	20	1800	. . .	2.5
Barite	3.5	40	1800	. . .	2.2
Bituminous coal	2	10	1300	. . .	1.2
Copal resin	5	2	600	7.5	
Wolframite ore	5.5	10	800	5.6	
Sulfur	3.5	3	1300	3.5	

Ind. Eng. Chem., 38, 672 (1946). To convert pounds per hour to kilograms per hour, multiply by 0.4535.

The **Jet-O-Mizer** *(Fluid Energy Processing & Equipment Co.)* is one of a group employing a hollow elongated torus which is placed vertically. The operating principle is similar to that of the Micronizer, with the feed entering tangentially to the whirling fluid stream and the fines leaving centrally.

Trost air mills from *Colt Industries* are available in five sizes (Fig. 8-56). The smallest mill (Gem T) is a research unit and can be used for fine-grinding studies. Capacities of 1 to 2300 kg/h (2 to 5000 lb/h) are available. Air-flow rates vary from 0.2 to 28 m^3/min (6.5 to 1000 ft^3/min). Encapsulation of particles is possible by the injection of coating material into the feed.

The **Majac jet pulverizer** *(Donaldson Company, Inc.)* is an opposed-jet type with a mechanical classifier (Fig. 8-57). A screw or other type of feeder discharges the material to be pulverized into the impact zone or into the classifier, depending on the feed material. Fluid and powder from the jets pass into a mechanical classifier above. The oversize material flows downward through an annular space against elutriating air, through two downcomer legs to the nozzles, where they are accelerated by the opposing high-velocity fluid streams and impacted against each other. Fineness is controlled primarily by the classifier speed and the amount of fan air delivered to the classifier, but other effects can be achieved by variation of nozzle pressure, distance between the muzzles of the gun barrels, and position of the classifier disk. These pulverizers are available in 30 sizes, operated on quantities of compressed air ranging from approximately 0.6 to 13.0 m^3/min (20 to 4500 ft^3/min). In most applications, the economics of the use of this type of jet pulverizer becomes attractive in the range of 98 percent through 200 mesh or finer; and as finer products are required, this equipment becomes increasingly attractive.

Materials illustrated in Table 8-31 are shown because of their wide range and type. Alumina represents a very hard and abrasive material, whereas diphenyl phthalate represents a material with a melting point of approximately 60°C. Although the jet pulverizer is normally used for quite fine grinds, feldspar, silica, and coal represent common materials which are ordinarily processed to fairly coarse particle sizes. Mica demonstrates a material that is difficult to handle because of its very flaky structure. Graphite and rare earth illustrate fairly fine pulverization.

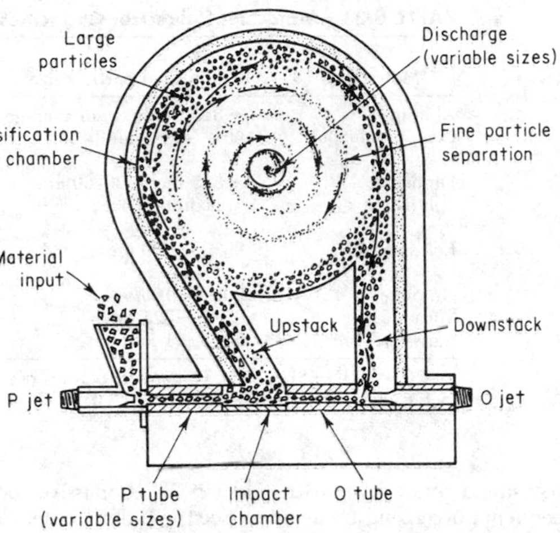

FIG. 8-56 Trost jet mill. *(Colt Industries.)*

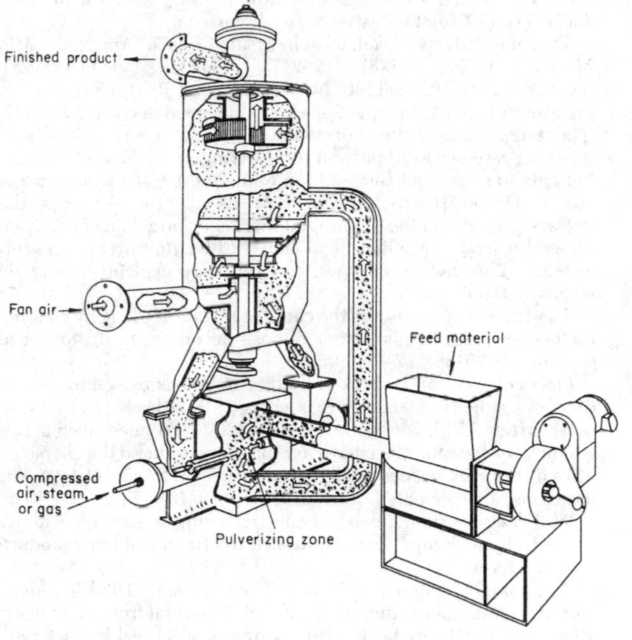

FIG. 8-57 Majac jet pulverizer. *(Donaldson Company, Inc.)*

NOVEL METHODS

Only once in 15 years does a truly novel method of size reduction become successful. However, many more methods are proposed and studied; some of these are described below. The information may be useful to judge other novel methods that may be proposed.

Avoiding Size Reduction Since size reduction is a difficult and inefficient operation, it is sometimes better to avoid it and use another approach. Thus rather than make large crystals and then grind them, one may be able to precipitate or crystallize material in the desired fine size. It may even be possible to control the process to give a more narrow size distribution than would be possible by size reduction.

Some materials that are prepared in the molten state are converted advantageously to flake form by cooling a thin layer continuously on the surface of a rotating drum. Another way is to spray cool from the

TABLE 8-31 Majac Jet Pulverizer Capacities*

Material	Finished particle size	Mill size	Production, lb/h	Grinding fluid used
Alumina	−325 mesh, 3 μm average	15	12,000	6300 lb/h steam at 100 psig, 750°F.
Coal, bituminous	90%, −325 mesh	20	8,000	3000 ft^3/min air at 100 psig, 70°F.
Diphenyl phthalate	−325 mesh, 20–30 μm maximum, 4.2 μm average	2–6	435	300 ft^3/min air at 100 psig, 70°F.
Feldspar, silica	99%, −200 mesh	15	8,500	1350 ft^3/min air at 100 psig, 800°F.
Graphite	90%, −10 μm	S.5–2.5	50	75 ft^3/min air at 100 psig, 70°F.
Mica	95%, −325 mesh	8	1,600	720 ft^3/min air at 100 psig, 800°F.
Rare-earth ore	60%, −1 μm	8–15	400	720 ft^3/min air at 100 psig, 800°F.

*°C = (°F − 32) × 5/9. To convert pounds per hour to kilograms per hour, multiply by 0.4535; to convert pounds per square inch gauge to kilopascals, multiply by 7.0.

melt, using a spray dryer with cold air. Thus, massive cooling and subsequent pulverizing are avoided. See the Index for details of these other methods.

Ultrafine powders can be prepared in high-temperature plasmas. Particles below 1 μm and larger particles with unusual surface structures are formed according to Waldie [*Trans. Inst. Chem. Eng.*, **48**(3), T90 (1970)]. Energy costs are discussed.

Thermal Effects Daellenbach et al. [*Trans. Am. Inst. Min. Metall. Pet. Eng.*, **250**(3), 212–217 (1971)] investigated improvement of taconite grindability by pretreatment. Roasting in a reducing atmosphere has the greatest effect, but quench cooling from the quartz crystal transition temperature is also effective. The work index is decreased to 13 percent for grinding to −355 μm (48 mesh), but only to 60 percent for −38 μm (400 mesh). Calculations carried out by Devore [*Contrib. Geol.*, **8**(1), 21–26 (1969)] suggest that stresses arising from the differential thermal contraction of coexisting mineral crystals are sufficient to cause brittle rupture of rocks subjected to temperature changes, and this may explain some of the results of Daellenbach.

Thermal expansion is also the cause of surface spalling by an exotic method—the use of giant laser pulses—according to Bristow et al. [*Nature*, **222**(5188), 27–29 (1969)].

Pieces of rock can be broken by thermal shock caused by absorption of radio-frequency (RF) energy. Matthaei [*Bergbauwissenschaften*, **15**(9), 338–346; (11) 411–420 (1968)] described a Yagi antenna for focusing the energy on the piece and also the characteristics of the rock needed to absorb the energy. Results of tests with limestone are presented. Pickarski [*Can. Min. J.*, **87**(6), 66–69 (1966)] describes similar experiments fracturing asbestos ore with RF currents. Large lumps are open to such treatment, and fine products are not expected.

Sarapuu (U.S. Patents 3,179,187, 3,169,577, and 3,460,766) demonstrates electrical heating to break rock in several iron-ore mines in Missouri and Minnesota. Heating is due to electrical-contact resistance and internal resistance, and performance depends on the electrical conductivity of the rock. Equipment is commercially available. Multipoint electrodes and electrical drill bits are used. Electric power consumption is 1 to 2 kWh/ton rock.

Electrohydraulic crushing is an exotic method that has not yet been found practical. The method applies a spark under water in the presence of pieces to be broken. Ohme [*Freiberg. Forschungsh.*, **A425**, 31–84 (1968)] reported an extensive investigation to try to make the method practical, but the energy consumption is still 10 times that of conventional methods.

Tensile Comminution Bond [*Min. Eng. (London)*, **60**(1), 63–64 (1968)] reviewed attempts to induce breakage without wastefully applying pressure and concluded that inherent practical limitations have been found for the following methods: spinning particles, resonant vibration, electrohydraulic crushing, induction heating, sudden release of gas pressure, and chisel-effect breakers.

Bergstrom [*Miner. Process.*, **8**(4), 14–18 (1967)] achieved comminution under tension by spinning a disk, as in the failure of a grinding wheel. The energy input is low, but the fragments are large. In fact, the new surface versus energy plots on a straight line extrapolated from published results of compression experiments, indicating that tensile comminution is not more efficient.

Explosive Shattering When permeable materials contain gas or liquid under pressure, the sudden release of that pressure may result in explosive shattering. The rupture of wood has been developed as the Masonite process (U.S. Patent 1,578,609). IIT Research Institute has developed a method for shattering coal by subjecting it to supercritical water and expanding continuously through a nozzle (Massey, Brabets, and Abel, U.S. Patent 4,313,737). The coal particles are ground finer than 4 μm, while the minerals remain relatively coarse.

Special Tumbling Mills Plank [*Zem.-Kalk-Gips*, **60**(3), 95–197 (1971)] described slanting liner plates which are supposed to cause large balls in a ball mill to go to the feed end and smaller ones to the discharge end. If the grist in the mill is not well mixed, then this old idea would put the largest balls where they could act on the largest particles, and this would increase the grinding rate. However, no data have been published to prove that such liners actually cause ball classification.

CRUSHING AND GRINDING PRACTICE

CEREALS AND OTHER VEGETABLE PRODUCTS

Flour and Feed Meal The **roller mill** is the traditional machine for grinding wheat and rye into high-grade flour. A typical mill used for this purpose is fitted with two pairs of rolls, capable of making two separate reductions. After each reduction the product is taken to a bolting machine to separate the fine flour, the coarse product being returned for further reduction. Feed is supplied at the top, where a vibratory shaker spreads it out in a thin stream across the full width of the rolls.

Rolls are made with various types of corrugation. Two standard types are most generally used: the dull and the sharp, the former mainly on wheat and rye, and the latter for corn and feed. Under ordinary conditions, a sharp roll is used against a sharp roll for very tough wheat, a sharp fast roll against a dull slow roll for moderately tough wheat, a dull fast roll against a sharp slow roll for slightly brittle wheat, and a dull roll against a dull roll for very brittle wheat. The speed ratio usually is 2½:1 for corrugated rolls and 1¼:1 for smooth rolls. By examining the marks made on the grain fragments it has been concluded (Scott, *Flour Milling Processes*, Chapman &

Hall, London, 1951) that the differential action of the rolls actually can open up the berry and strip the endosperm from the hulls.

High-speed **hammer** or **pin mills** result in some *selective grinding*. Such mills combined with air classification can produce fractions with controlled protein content. An example of such a combination is a Bauer hurricane hammer mill combined with the Alpine Mikroplex superfine classifier. Flour with different protein content is needed for the baking of breads and cakes, and these types of flour were formerly available only by selection of the type of wheat, which is limited by growing conditions prevailing in particular locations [Wichser, *Milling*, 3(5), 123–125 1958)].

Roll mills and **disk attrition** mills are still used for the grinding of wheat, however. Performance of a single-runner disk attrition mill is given in Table 8-32.

TABLE 8-32 Operating Characteristics of a Single-Runner Robinson Attrition Mill, Grinding Grain

	Size of mill							
	16 in.	18 in.	20 in.	24 in.	26 in.	30 in.	32 in.	36 in.
Speed, r.p.m.[1]	2500	2250	2200	1800	1600	1400	1300	1200
Speed, r.p.m.[2]	1000	950	900	800	750			
Capacity[3]	1200	1600	2000	3300	4000	5000	5300	6300
Capacity[4]	1200	1300	1500	1900	1900	2200		
Capacity[5]	65	80	100	150	200			
Hp[6]	9–12	10–15	12–18	20–30	22–32	25–35	28–38	30–50
Hp[7]	5– 8	6– 9	8–10	9–12	10–15	12–18		

[1]R.p.m. when grinding feed or corn meal.
[2]R.p.m. when cracking corn.
[3]Grinding feed, pounds per hour.
[4]Grinding corn meal, pounds per hour.
[5]Cracking corn, bushels per hour.
[6]Power when grinding feed or corn meal.
[7]Power for cracking corn.

NOTE: To convert inches to centimeters, multiply by 2.54; to convert pounds per hour to kilograms per hour, multiply by 0.4535; and to convert horsepower to kilowatts, multiply by 0.746.

Soybeans, Soybean Cake, and Other Pressed Cakes After granulation on rolls the granules are generally treated in presses or solvent extracted to remove the oil. The product from the presses goes to attrition mills or flour rolls and then to bolters, depending upon whether the finished product is to be a feed meal or a flour. If the whole cake is to be pulverized without removal of fibrous particles, it may be ground in a hammer mill with or without air classification. A 15-kW (20-hp) hammer mill with an air classifier, grinding pressed cake, had a capacity of 136 kg/h (300 lb/h), 90 percent through No. 200 sieve; a 15-kW (20-hp) screen-hammer mill grinding to 0.16-cm (1/16-in) screen produced 453 kg/h (1000 lb/h).

The method used for grinding pressed cakes depends upon the nature of the cake, its purity, residual oil, and moisture content. Many of these materials are treated in hammer mills, especially when no fine reduction is required. In many cases the hammer mill is used merely as a preliminary disintegrator, followed by an attrition mill. Typical performance of the attrition mill is given in Table 8-29. A finer product may be obtained in a hammer mill in closed circuit with an external screen or classifier. Table 8-33 gives the results obtained with hammer mills disintegrating linseed cake, cottonseed cake, and an expeller cake.

TABLE 8-33 Operating Results with Williams Hammer Mills, Disintegrating Various Seed Cakes

	Capacity, tons/h			
Material	Pea meal	Pea and finer	Extra fine	Hp
Cottonseed cake	1	¾–1	½	8–12
Expeller cake	2½–3	2½–3	2	25–30
Linseed cake	6–8	5–6	4–5	50–60

NOTE: To convert tons per hour to megagrams per hour, multiply by 0.907; to convert horsepower to kilowatts, multiply by 0.746.

High-speed hammer mills are extensively used for the grinding of soya flour. For example, the Raymond Imp mill with an air classifier is used, primarily with solvent-extracted soya.

Starch and Other Flours Grinding of starch is not particularly difficult, but precautions must be taken against explosions; starches must not come in contact with hot surfaces, sparks, or flame when suspended in air. See "Properties of Solids: Safety" for safety precautions. When a product of medium fineness is required, a hammer mill of the screen type is employed. Potato flour, tapioca, banana, and similar flours are handled in this manner. For finer products a high-speed impact mill such as the Entoleter pin mill is used in closed circuit with bolting cloth, an internal air classifier, or vibrating screens.

ORES AND MINERALS

Metalliferous Ores The most extensive grinding operations are done in the ore-processing and cement industries. Grinding is one of the major problems in milling practice and one of the main items of expense. Mill manufacturers, operators, and engineers find it necessary to compare grinding practice in one plant with that of another, attempting to evaluate circuits and practices (Arbiter, *Milling in the Americas*, 7th International Mineral Processing Congress, Gordon and Breach, New York, 1964). Direct-shipping ores are high in metal assay, and require only preliminary crushing before being fed to a blast furnace or smelter. As these high-grade ores have been depleted, it has become necessary to concentrate ores of lower mineral value. The native copper ores of Michigan have given way to porphyry copper ores of the southwest. Initially the deposits containing 2 to 4 percent copper were worked, but now ores of 0.40 percent must be processed by grinding and flotation. The effectiveness of closed-circuit milling with wet classifiers reopened the Iron Range of Minnesota by permitting economic beneficiation of taconite iron ores, which contain up to two-thirds hard cherty gangue. By grinding, magnetic separation or froth flotation and pelletizing, a blast-furnace feed is produced that is more uniform and gives a higher iron yield than the direct-shipping ores.

A flow sheet for one iron ore process is shown in Fig. 8-58. For the grinding of softer copper ore the rod mill might be eliminated, both coarse-crushing and ball-milling ranges being extended to fill the gap.

Currently there is great interest in **autogenous milling** of iron and copper ores. When successful, this method results in economies due to elimination of media wear. Probably another reason for efficiency is the use of higher circulating loads and better classification. These improvements resulted from the need to use larger-diameter mills to obtain grinding with rock media that have a lower density than steel balls. The major difficulty is in arranging the crushing circuits and the actual mining so as to assure a steady supply of large ore lumps to serve as grinding media. With rocks that are too friable this cannot be achieved. A flow sheet for a typical wet autogenous circuit is shown in Fig. 8-59.

Nonmetallic Minerals Dry and wet grinding processes are used. **Dry grinding** is less expensive than wet grinding in the coarser sizes because a dry product is obtained without a final drying step. Dry grinding is carried out with ball, pebble, roller, and hammer mills with closed-circuit air classification. The product may be 99.8 percent through a 325-mesh sieve. Jet mills can produce a product in the range 5 to 15 μm, but at greater cost.

Wet processes use continuous ball and pebble mills, stirred and vibratory media mills, and pug mills. Wet processes take advantage of more effective classification in water, using bowl and cone classifiers, hydroseparators, thickeners, continuous centrifuges and cyclones, vacuum filters, and rotary, tray, or tunnel dryers. After drying, the cake generally has to be broken up in some type of disintegrator or pulverizer unless it was spray-dried. The objective of a process may be to obtain many grades of the same material by tying classifiers and screens into the grinding system to remove various-sized products. Choice of equipment generally depends on (1) hardness and (2) contaminations. Capacity of any system decreases rap-

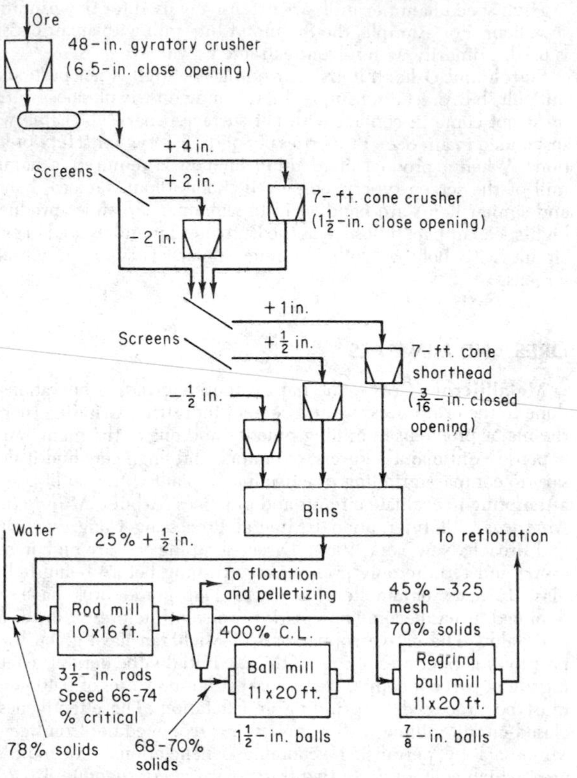

FIG. 8-58 Simplified flow sheet of the Cleveland-Cliffs Iron Co. Republic mine iron-ore concentrator. To convert inches to centimeters, multiply by 2.54; to convert feet to centimeters, multiply by 30.5. (*Johnson and Bjorne,* Milling in the Americas, *Gordon and Breach, New York, 1964.*)

idly with increasing fineness of the material; this applies particularly to nonmetallics, for which extreme fineness is usually required.

Clays and Kaolins There are two processes for the preparation of clays: the dry and the wet. Clay particles occur naturally in fine sizes, but grinding is required to break up lumps and agglomerates. This deagglomeration process is more complete in the wet process.

In the **dry process,** mined clay with 22 percent moisture is broken up into pieces of less than 5 cm (2 in) in a rotary impact mill without screen, and fed to a rotary gas-fired kiln for drying (see Fig. 8-60). The moisture content is then 8 to 10 percent, and this material is fed to a mill, usually a Raymond ring-roll mill with an internal whizzer classifier. Hot gases introduced to the mill complete the drying while the material is being pulverized to the required fineness.

These mills are equipped with automatic throwout devices to eliminate impurities in the oversize form. A high percentage of an impurity such as silica sand can be eliminated. On an average grade of unwashed clay a No. 5057 Raymond ring-roller mill equipped with whizzer classification will grind from 2.7 to 3.2 Mg/h (3 to 3½ tons/h) to a fineness of about 99.95 percent through a No. 325 sieve, while on washed clay the capacity will be from 30 to 40 percent higher. To grind 3.2 Mg/h (3½ tons/h) of a raw clay, power consumption will be about 75 kW (100 hp), and it takes about 31 m^3 (1100 ft^3) of natural gas 3.7 MJ/m^3 (1000 Btu/ft^3) to dry the clay from 10 percent moisture down to about 1 percent. The product is used in paint pigments and rubber fillers.

In the **wet process,** the clay is masticated in a pug mill to break up lumps and then dispersed with a dispersing aid and water to make a 40 percent solids slip of low viscosity. A Cowles dissolver is used for this purpose. Sands are settled out, and then the clay is classified into two size fractions in either a Hydrosettler or a continuous Sharples or Bird centrifuge. The fine fraction, with sizes of less than 1 μm, is used as a pigment and for paper coating, while the coarser fraction is used as a paper filler.

A process for upgrading kaolin by grinding in a stirred bead mill has been reported (Stanczyk and Feld, U.S. Bur. Mines Rep. Invest. 6327 and 6694, 1965). By this means the clay particles are delaminated, and the resulting platelets give a much improved surface on coated paper.

Talc and Soapstone Generally these are easily pulverized, although certain fibrous and foliated talcs may offer greater resistance to reduction to impalpable powder.

Talc milling is largely a grinding operation accompanied by air separation. Most of the industrial talcs are dry-ground. Dryers are

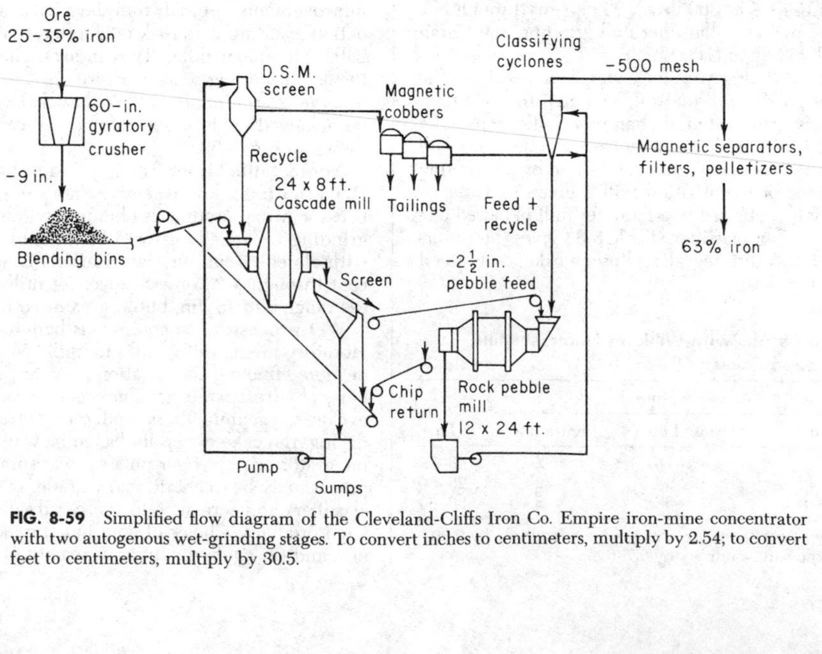

FIG. 8-59 Simplified flow diagram of the Cleveland-Cliffs Iron Co. Empire iron-mine concentrator with two autogenous wet-grinding stages. To convert inches to centimeters, multiply by 2.54; to convert feet to centimeters, multiply by 30.5.

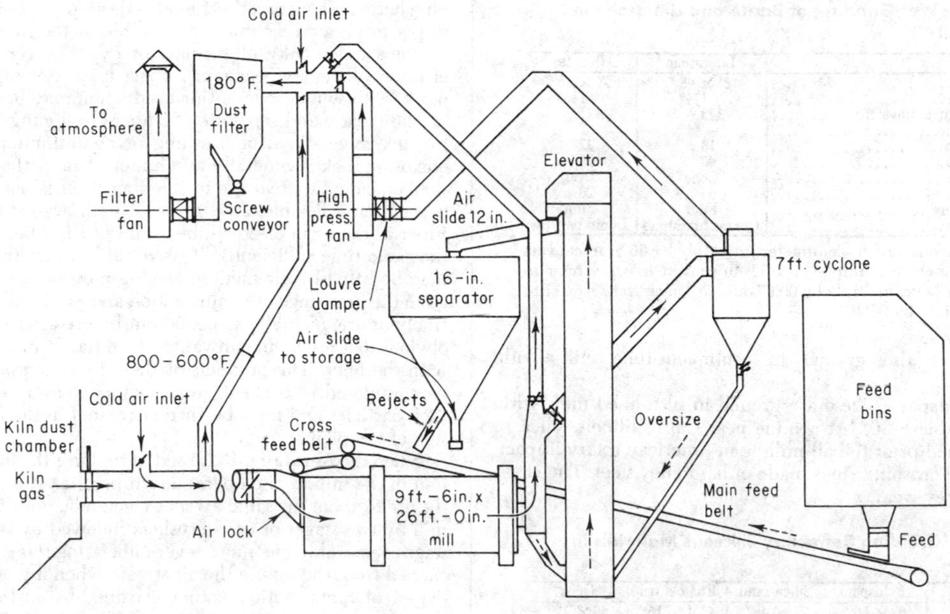

FIG. 8-60 Combined drying-grinding system using ball-mill and hot kiln exhaust gases. To convert inches to centimeters, multiply by 2.54; to convert feet to centimeters, multiply by 30.5; °C = (°F − 32) × 5/9. [*Tonry,* Pit Quarry *(February–March 1959).*]

commonly employed to predry ahead of the milling operation because the wet material reduces mill capacity by as much as 30 percent.

Conventionally, in talc milling, rock taken from the mines is crushed in primary and then in secondary crushers to at least 1.25 cm (½ in) and frequently as fine as 0.16 cm (1/16 in).

Ring-roll mills with internal air separation are widely used for the large-capacity fine grinding of the softer talcs. Pebble mills and mechanical air separators are usually used for the tough fibrous varieties. Fluid-energy mills also have application on these tough talcs. High-speed hammer mills with internal air separation are also an outstanding success on some of the softer high-purity talcs for very fine fineness. For a ring-roller mill receiving 2.54-cm (1-in) feed, production rates range from 1360 to 2720 kg/h (3000 to 6000 lb/h) for 60-kW (80-hp) grinding to 99 to 99.5 percent able to pass a No. 200 sieve.

A mill in the Gouverneur, New York, district, working on a tough fibrous variety, uses primary crushing with a jaw crusher on the bottom level of the mine, secondary crushing with a gyratory crusher at the shaft-head house, and tertiary crushing with Symons short-head and Gyradisc crushers in closed circuit with 16-mesh screens. Coarser-product grades are ground in Hardinge pebble mills with synthetic grinding media. These mills are in closed circuit with Raymond and Sturtevant separators. The finest grades are ground with Wheeler fluid-energy mills.

The mills in the western United States, in which generally are ground softer talcs than those of New York, have simpler flow sheets. Single-stage crushing is employed, and the talc is merely ground in Raymond roller mills in closed circuit with air separators.

Most industrial talcs are ground to one of three very general size groups: 98 percent minus 200-mesh screen, 98.5 percent minus 325-mesh, and 99.5 percent minus 325-mesh. The product may contain a high proportion of fines 15 to 35 μm in size. Extremely fine grinds are obtained by means of fluid-energy mills. The relatively soft grades of crude mined in California, Nevada, and Montana can be ground in fluid-energy mills to an average particle size of 5 μm and a specific surface of 30 m^2/g. These talcs of extreme fineness and high surface area are rapidly attaining industrial importance and are used for various purposes in the paint, paper, plastics, and rubber industries.

Extremely fine grinding necessarily increases the cost per ton of talc to industrial consumers, but if appropriate methods are used, net costs are lower when the fine grinding remains an integrated phase of the initial milling of talc rather than when it is attempted as an independent secondary process instituted by an industrial consumer. Accordingly, the trend is toward finer grinding by the talc producers.

Carbonates and Sulfates Carbonates include limestone, calcite, marble, marls, chalk, dolomite, and magnesite; the most important sulfates are barite, celestite, anhydrite, and gypsum; these are used as fillers in paint, paper, and rubber. (Gypsum and anhydrite are discussed below as part of the cement, lime, and gypsum industries.)

Table 8-34 shows the performance of a Raymond 5057 ring-roller mill pulverizing **limestone.** A complete range of larger mill sizes is available.

TABLE 8-34 Performance of Raymond 5057 Ring-Roller Mill Pulverizing Limestone

Fineness, % through	Fineness, Test sieve	Capacity, lb./hr.	Operating hp.
75	200	14,000	140
85	200	11,500	135
95	200	9,000	130
99	200	7,000	125
99	300	6,000	125
99	325	5,000	125
99.5	325	4,000	120

NOTE: To convert pounds per hour to kilograms per hour, multiply by 0.4535; to convert horsepower to kilowatts, multiply by 0.746.

These results may be applied to practically the entire group. When a material is very soft, such as some high-grade barite, the capacities may be about 25 to 35 percent higher. Of the carbonates, magnesite is generally the hardest to pulverize (see subsection "Refractories").

Table 8-35 gives the results obtained in a Hardinge mill wet-grinding barite and limestone to be used as a paint filler.

A Raymond 5057 ring-roller mill pulverizing fluorspar produces 4500 kg (10,000 lb) of product per hour 95 percent minus 200 mesh with 104 kW (104-hp) operating power, or 26 kWh/Mg (23.4 kWh/

TABLE 8-35 Wet Grinding of Barite and Limestone in Hardinge Mill

	Limestone	Barytes
Size of mill	8′ × 48″	7′ × 36″
Size of feed	1½″	1½″
Fineness of product, sieve No.	325	325
Capacity, tons/hr	¾	2
Mill speed, r.p.m.	18	22
Power for mill, hp	40	25
Classifier system	Cone	Drag
Moisture in mill, %	30	28
Type of mill lining	Flint	Flint
Grinding mediums	Coarse limestone	Lump barytes

NOTE: To convert feet to centimeters, multiply by 30.5; to convert inches to centimeters, multiply by 2.54; to convert tons per hour to megagrams per hour, multiply by 0.907; and to convert horsepower to kilowatts, multiply by 0.746.

ton). Fluorspar is also ground in continuous-tube mills with classification.

Silica and Feldspar These are ground in silex-lined mills with flint balls (see Table 8-36). At a mine near Cairo, Illinois, silica is successfully crushed prior to ball-milling in American rotary impact mills having loose crushing rings made of hard alloy steel. The rings are replaced as they wear.

TABLE 8-36 Grinding Refractory Siliceous Materials in Pebble Mills

	Feldspar	Silica sand	Enamel frit	Grog
Size of mill	8′ × 60″	8′ × 48″	4½′ × 16″	5′ × 22″
Feed size	2″	20 mesh	⅛″	1½″
Size of product	99% through No. 200 sieve	98% through No. 325 sieve	97% through No. 100 sieve	95% through No. 10 sieve
Capacity, tons/hr	1.75	1.25	0.225	5
Power for mill, hp	68	58	8.5	28
Power for auxiliaries, hp	21	20		
Pebble load, lb	10,000	12,000	2000	2800
Speed of mill, r.p.m.	22	18	30	30
Moisture, %	1	1	0	1
Type of classifier	Hardinge	Air	Trommel screen on mill	
Lining and grinding mediums	Flint blocks and flint pebbles			Steel

NOTE: To convert feet to centimeters, multiply by 30.5; to convert inches to centimeters, multiply by 2.54; to convert tons per hour to megagrams per hour, multiply by 0.907; and to convert horsepower to kilowatts, multiply by 0.746.

Feldspar for the ceramic and chemical industries is ground finer than for the glass industry. The following description of a feldspar mill is abstracted from U.S. Bur. Mines Cir. 6488, 1931. The fine-grinding department consists of three silex-lined Hardinge mill units with flint pebbles, 27 to 28 r/min, each fed from a 45-Mg (50-ton) surge bin by James automatic belt feeders. Unit 1 has a 2.4- by 1.2-m (8- by 4-ft) Hardinge mill, discharging by an elevator to a Gayco air classifier or a James vibrating screen, according to the kind of product desired, a fine (No. 120 to 250 sieve) or a coarse (No. 20 sieve). An intermediate product (No. 40 to 100 sieve) is made by removing fines in the Gayco centrifugal classifier and screening the coarse.

As a spar free from fines is desired by the glass industry, crushing is done with rolls by a process of gradual reduction on three sets of Sturtevant rolls close-circuited with Hummer vibratory screens [cf. Carey and Bosanquet, *J. Soc. Glass Technol.*, **17**, 384–410 (1933)]. The material is first crushed in jaw crushers at 1.25-cm (½-in) setting. The rolls ares set at 1 cm (⅜ in), ½ cm (³⁄₁₆ in), and close, with screening at 4, 8, and 20 mesh. Fines are removed from the product by a Gayco centrifugal classifier.

With slight modifications the systems may be used to produce a fine grade of granules from the following materials: quartz, slate, marble, corundum, silicon carbide, tripoli, pumice, and volcanic ash. Practically all abrasive silicates are handled in ball and tube mills followed by air classifiers. Table 8-36 gives the results obtained with Hardinge pebble mills, grinding several siliceous refractory materials.

Asbestos and Mica The choice of crusher for asbestos depends on whether a long or a short fiber is desired. Crushing is done in slow stages to preserve as much as possible of the fiber length. Primary crushers are usually of the jaw or gyratory type, with secondary crushers of the jaw, cone, or impact type. Corrugated rolls are also used. With some grades a third reduction may be required.

The release and separation of fiber from gangue is accomplished by successive stages of crushing or comminution by impact. The enclosing rock is generally much more friable than the fiber so that the coincidental breakage in the fiber itself is not great if the latter is tough and flexible, as is the case with a good-quality chrysolite. Fiber in the form of cross-fiber veins or slip fiber thus released is at the same time sufficiently "teased out" so that the portion so freed may be lifted by air suction, leaving most of the rock as a reject to go to the next stage of comminution and eventually to tailings. Finer fractions are generally screened out before air separation, since it is obvious that much fine unwanted material would otherwise be lifted with the fiber. The products of these first stages of separation may be considered as concentrates. They contain a large percentage of rock particles and must be further treated in the fiber-grading division of the plant.

When the **Aerofall mill** is used, it replaces the first stages of reduction in the mill in one operation. In practice it is adjusted to reduce the ore to about the same extent as would be done in another asbestos mill by two stages of cone crushers followed by two mild fiberizing stages or possibly one more severe fiberizing stage. The ore, which is carried from the mill in the air stream when it reaches the necessary degree of comminution, is then classified by air followed by screening, and the fiber concentrates are lifted by suction similar to the common practice.

In the **grading mill** the fiber is further separated into the approximate length brackets required for specific grades and is subjected to several stages of screenings, using shaking screens, rotating screens, conventional trommels, trommel-like graders and rotary dusters. When well-opened or fluffed-out grades are called for, the fiber is subject to special treatment in one or more of a variety of machines ranging from graders or willows (fixed-shell trommels having a rotating center shaft to which beater arms are attached) to one of several types of high-speed hammer mills and disk grinders. The type of machine or machines used depends on the length and type of fiber to be processed and the degree of opening or fluffing up required. This additional treatment is generally given to the shorter fibers.

Asbestos is often pulverized. This is the case when it is used for molded products. The pulverizing is usually accomplished by a high-speed screen mill with air-transport system. A mill with a 400-μm (¹⁄₆₄-in) screen pulverized 180 kg/h (400 lb/h) with 9-7 kW (13-hp) power consumption.

Asbestos is no longer mined in the United States because of the severe health hazard, but it is still mined and processed in Canada. It is now used only in products where the fibers are encased in other material, as in asbestos-cement pipe and in brake linings.

The **micas,** as a class, are difficult to grind to a fine powder; one exception is disintegrated schist, in which the mica occurs in minute flakes. The material pulverized is generally the waste from production of sheets and scrap from punching and trimming.

For dry grinding, hammer mills equipped with an air-transport system are generally used. Maintenance is often high. The material, after dropping through a perforated screen into the intake of an exhauster, is collected in a cyclone followed by bolting reels.

It has been established that the method of milling has a definite effect on the particle characteristics of the final product. Dry grinding of mica is customary for the coarser sizes down to 100 mesh. Water-ground mica produced by a slow frictional grinding process is claimed to give maximum flakiness, high sheen, and slip. Micronized mica, produced by high-pressure steam jets, is considered to consist of highly delaminated particles which are even finer than water-ground grades.

Refractories Refractory bricks are made from fireclay, alumina, magnesite, chrome, forsterite, and silica ores. These materials are crushed and ground, wetted, pressed into shape, and fired. To obtain the maximum brick density, furnishes of several sizes are prepared and mixed. Thus a magnesia brick may be made from 40 percent coarse, 40 percent middling, and 20 percent fines. Theorems have been proposed for calculating weight ratios of sizes to produce maximum packing density of powder mixtures [Lewis and Goldman, *J. Am. Ceram. Soc.*, **49**(6), 323 (1966)]. Preliminary crushing is done in jaw or gyratories, intermediate crushing in pan mills or ring rolls, and fine grinding in open-circuit ball mills. Since refractory plants must make a variety of products in the same equipment, pan mills and ring rolls are preferred over ball mills because the former are more easily cleaned.

Sixty percent of refractory magnesite is made synthetically from Michigan brines. When calcined, this material is one of the hardest refractories to grind. Gyratory crushers, jaw crushers, pan mills, and ball mills are used.

Crushed Stone and Aggregate Crushed stone for road building must be relatively strong and inert, and must meet specifications regarding **size distribution** and **shape.** Both size and shape are determined by the crushing operation. Table 8-37 lists specifications for a few size ranges. Sometimes a product that does not meet these requirements must be adjusted by adding a specially crushed fraction. No crushing device available will give any arbitrary size distribution, and so crushing with a small reduction ratio and recycle of oversize is practiced when necessary. In general, the size distribution is less broad when the crushing energy is the minimum needed to rupture the particle, but even then there is a distribution of sizes.

The claim of various crushers to produce a **cubical product** is exaggerated. However, there are differences. If an impact mill is designed to apply an excess of energy at each blow, then production of slivers can be avoided but a larger amount of fines is produced. A distinction between gyratory and jaw crushers is that a slab produced in a jaw crusher can pass through the discharge, whereas it cannot pass the curved opening of a gyratory without being broken again.

Primary crushers used are jaw, gyratory, impact, and toothed roll crushers. Impact crushers are limited to limestone and softer stone. With rocks containing more than 5 percent quartz, maintenance of hammers may become prohibitive. Gyratory and cone crushers dominate the field for secondary crushing of hard and tough stone. Impact mills are extensively used. Rod mills have been employed to manufacture stone sand when natural sands are not available.

A survey of research on product shape and size distribution **(grading)** was given by Rösslein [translation by Shergold, *Quarry Managers' J.*, 207–222 (October 1946)]. The main study involved tests on 30 jaw crushers, with shapes of over 1 million particles being measured. Concerning **particle shape,** there is a tendency for hard rocks to produce more numerous flaky chippings than soft rocks. The size of feed to a crusher does not affect the shape of products. With jaw crushers, the largest and finest sizes contain the highest proportion of flaky pieces, but even the intermediate sizes are irregular. An increase in the reduction ratio of jaw crushers increases the flakiness of the product. Smooth jaws produce more numerous flaky pieces than corrugated jaws. Curved jaws produce less fines but more numerous flaky particles. Crusher speed has little effect. The presence of material too small to be crushed has a deleterious effect on the shape of products. Secondary crushers with a small reduction ratio can improve the shape of primary crushed material, but secondary crushers are not inherently different from primary crushers. Slotted screens can remove flaky particles from the product. Impact crushers produce fewer flaky particles than any other type.

Grading of the product (i.e., size distribution) depends on the discharge opening, which is difficult to measure and adjust owing to wear of the plates in a jaw crusher. In addition to the main-size peak, a definite peak of very fine product is generally found, because of abrasion occurring between feed and jaws. Size of feed and wear of plates do not affect grading significantly. Table 8-9 relates sizes of product to crusher settings.

FERTILIZERS AND PHOSPHATES

Many of the materials used in the fertilizer industry are pulverized, such as those serving as sources for calcium, phosphorus, potassium, and nitrogen. The most commonly used for their lime content are limestone, oyster shells, marls, lime, and, to a small extent, gypsum. Limestone is generally ground in hammer mills, ring-roller mills, and ball mills. Fineness required varies greatly from No. 10 sieve to 75 percent through No. 100 sieve.

Oyster Shells and Lime Rock Operating characteristics for hammer mills grinding oyster shells and burned lime for agricultural purposes are given in Table 8-38.

Phosphates Phosphate rock is generally ground for one of two major purposes: for direct application to the soil or for acidulation with mineral acids in the manufacture of fertilizers. Because of larger capacities and fewer operating-personnel requirements, plant installations involving production rates over 900 Mg/h (100 tons/h) have used ball-mill grinding systems. Ring-roll mills are used in smaller applications. Rock for direct use as fertilizer is usually ground to various specifications, ranging from 40 percent minus 200 mesh to 70 percent minus 200 mesh. For manufacture of normal and concentrated superphosphates the fineness of grind ranges from 65 percent minus 200 mesh to 85 percent minus 200 mesh. Typical ball-mill performance in grinding phosphate rock is shown in Table 8-39.

Grindability of phosphate rocks from different areas varies widely; in Table 8-40 typical work-index data are shown.

Grinding-media wear in phosphate ball-mill grinding systems ranges from 5 to 25 g/Mg (0.05 to 0.20 lb/ton) ground; ball-mill liners show an average consumption of 2.5 to 100 g/Mg (0.01 to 0.05 lb/ton) ground.

Capacities from 27 to 45 Mg/h (30 to 50 tons/h) are obtained with 1.8-m (73-in) Raymond ring-roller mill with internal air separation when pulverizing phosphate rock, concentrates, pebble, or any mixture thereof to 60 percent through 200 mesh. In general, phosphate rock from Tunisia, Morocco, and the Pacific Islands is easier to pulverize. Capacities will be 10 to 30 percent greater than for Florida rock.

Some average results obtained in grinding various organic and inorganic raw materials for fertilizers are given in Table 8-41.

Inorganic salts often do not require fine pulverizing, but they frequently become lumpy. In such cases, they are passed through a double-cage mill or some type of hammer mill with the screen or cage bars removed. In other cases, a coarse screen in the hammer mill gives better operation than no screen at all. This is done with ammonium sulfate from by-product ovens and sodium nitrate. When used as an ingredient of fertilizer the latter is generally mixed with other raw materials, and the mixture is later disintegrated. The various potassium salts used in fertilizers are generally shipped ready for use, but if they have become caked in transit, they are broken in a disintegrator.

Basic slag is often used as a source of phosphorus. Its grinding resistance depends largely upon the way in which it has been cooled, slowly cooled slag generally being more easily pulverized. The most common method for grinding basic slag is in a ball mill, followed by a tube mill or a compartment mill. Both systems may be in closed circuit with an air classifier. A 2.1- by 1.5-m (7- by 5-ft) mill, requiring 94 kW (125 hp), operating with a 4.2-m (14-ft) 22.5-kW (30-hp) classifier, gave a capacity of 4.5 Mg/h (5 tons/h) from the classifier, 95 percent through a No. 200 sieve. Mill product was 68 percent through a No. 200 sieve, and circulating load 100 percent.

CEMENT, LIME, AND GYPSUM

Portland-cement manufacture requires grinding on a very large scale and entails a large use of electric power. Raw materials consist of sources of lime, alumina, and silica such as limestone, clay, shale, or cement rock. Crushing is done in gyratories, jaw crushers, impact crushers, or toothed rolls, depending on the nature of the material.

In the grinding of the raw materials, two processes are used: the dry process in which the materials are dried to less than 1 percent moisture and then ground to a fine powder, and the wet process in which the grinding takes place with addition of water to the mills to produce a slurry. The two processes are used about equally in the United States.

Dry-Process Cement The grinding from a size of 5 to 6.3 cm (2 to 2½ in) to a powder of 75 to 90 percent passing a 200-mesh sieve may be done in one or several stages. The first stage, reducing the material size to approximately 20 mesh, may be done in vertical,

TABLE 8-37 Selected Standard Sizes of Coarse Aggregate*

Size number	Nominal size, square openings	Amounts finer than each laboratory sieve (square openings), percent by weight														
		4 in	3½ in	3 in	2½ in	2 in	1½ in	1 in	¾ in	½ in	⅜ in	No. 4 (4760 μm)	No. 8 (2380 μm)	No. 16 (1190 μm)	No. 50 (297 μm)	No. 100 (149 μm)
1	3½ to 1½ in.	100	90 to 100	....	25 to 60	....	0 to 15	....	0 to 5							
2	2½ to 1½ in.	....	....	100	90 to 100	35 to 70	0 to 15	....	0 to 5							
3	2 to 1 in.	....	....	....	100	90 to 100	35 to 70	0 to 15	....	0 to 5						
4	1½ to ¾ in.	....	....	....	....	100	90 to 100	20 to 55	0 to 15	....	0 to 5					
5	1 to ½ in.	....	....	....	....	....	100	90 to 100	20 to 55	0 to 10	0 to 5					
6	¾ to ⅜ in.	....	....	....	....	....	....	100	90 to 100	20 to 55	0 to 15	0 to 5				
7	½ in. to No. 4	....	....	....	....	....	....	....	100	90 to 100	40 to 70	0 to 15	0 to 5			
8	⅜ in. to No. 8	....	....	....	....	....	....	....	....	100	85 to 100	10 to 30	0 to 10	0 to 5		
9	No. 4 to No. 16	...	....	..	....	....	....	....	....	....	100	85 to 100	10 to 40	0 to 10	0 to 5	
10	No. 4 to 0	...	....	..	....	....	....	....	....	....	100	85 to 100	......	....	......	10 to 30

*ASTM *Standards on Mineral Aggregates and Concrete,* March 1956.
NOTE: To convert inches to centimeters, multiply by 2.54.

TABLE 8-38 Operating Data for Grinding Oyster Shells and Burned Lime in Hammer Mills

Type of mill	Material	Size, in.	Capacity, tons/hr.	Hp.
Jeffrey	Oyster shells	15 × 8	0.5– 0.75	8
		20 × 12	1 – 1.5	12
		24 × 18	2 – 3	20
		30 × 24	4 – 5	30
		36 × 24	8 –10	40
Stedman	Burned lime	12 × 9	1.5	8
		20 × 12	4	20
		24 × 20	8	40
		30 × 30	12	60
		36 × 36	20	100

NOTE: To convert inches to centimeters, multiply by 2.54; to convert tons per hour to megagrams per hour, multiply by 0.907; and to convert horsepower to kilowatts, multiply by 0.746.

TABLE 8-39 Typical Ball-Mill Performance Grinding Florida Phosphate Rock*

Analysis, BPL†	Feed top size, in.	Mill size, ft.	Connected hp.	Capacity, tons/hr.	Product size, % minus 200 mesh	Energy,‡ hp.-hr./ton
72	1/4	11½ × 30	2000	110	60	22.5
68	1/2	11½ × 24	1600	100	60	20.0
68	1/4	11½ × 30	2000	120	60	20.8
68	3/4	11½ × 24	1750	99	60	23.5
72	3/8	11½ × 24	1750	66	80	35.3

*Kennedy Van Saun Corp. ball-mill grinding systems, air-lift type.
†Bone phosphate of lime, *i.e.*, calcium phosphate content.
‡Based on total connected horsepower for system, including fan.
NOTE: To convert inches to centimeters, multiply by 2.54; to convert feet to centimeters, multiply by 30.5; to convert horsepower to kilowatts, multiply by 0.746; to convert tons per hour to megagrams per hour, multiply by 0.907; and to convert horsepower-hours per ton to kilowatthours per megagram, multiply by 0.822.

TABLE 8-40 Ball-Mill Grindability of Various Phosphate Rocks

Rock type	Calcium phosphate content, %	Work index, kw.-hr./ton
Central Florida, pebble	68	14.5–16.5
North Florida, pebble	72	18.0–21.0
North Carolina, calcined concentrate	62–67	14.2–18.0
Central Florida, concentrate	72–74	16.0–23.0
Morocco	80–82	10.1–23.6
Idaho	....	19.0–24.7

NOTE: To convert kilowatthours per ton to kilowatthours per megagram, multiply by 1.1.

TABLE 8-41 Results Obtained in Grinding Raw Materials for Fertilizers

Material	Hammer mill	Type or size	Bar opening, in.	Capacity, lb./hr.	Hp.
Acid phosphate	Stedman	36 in.	. .	12,000	40
Steamed bone	Jeffrey	A-30 × 24 in.	1/8	10,000	40
Dry kelp	Williams	Shredder 2	1/4	12,000	35
Guano (Peruvian)	Jeffrey	A-24 × 18 in.	1/8	7,000	40

NOTE: To convert inches to centimeters, multiply by 2.54; to convert pounds per hour to kilograms per hour, multiply by 0.4535; and to convert horsepower to kilowatts, multiply by 0.746.

roller, ball-race, or ball mills. The last-named rotate from 15 to 18 r/min and are charged with grinding balls 5 to 13 cm (2 to 5 in) in diameter. The second stage is done in tube mills charged with grinding balls of 2 to 5 cm (¾ to 2 in).

Frequently ball and tube mills are combined into a single machine consisting of two or three compartments, separated by perforated-steel diaphragms and charged with grinding media of different size. Rod mills are hardly ever used in cement plants.

Dry-process ball mills, compartment mills, and tube mills may all be operated in closed circuit with air separators which separate the mill stream into coarse and fine fractions. Ball mills currently in use in the cement industry range in size from 4 m (13 ft) in diameter and 5.2 m (17 ft) in length, operated with 130-kW (175-hp) motors, to 4.5 m (15 ft) in diameter and 15 m (50 ft) in length, with 4900-kW (6600-hp) motors.

Drying is usually necessary before grinding. Such drying may take place in cylindrical dryers, typically 1.8 to 2.4 m (6 to 8 ft) in diameter and 18 to 46 m (60 to 150 ft) long. Figure 8-60 shows an installation in which kiln exhaust gases are used for drying in a ball mill. Drying may also be done in a raw-mix circuit with part of the hot kiln gases going into the mill and 2 to 3 times as much going into the classifier.

The compartments of a tube mill may be combined in various circuit arrangements with classifiers, as shown in Fig. 8-61. The first compartment usually contains larger balls than the second.

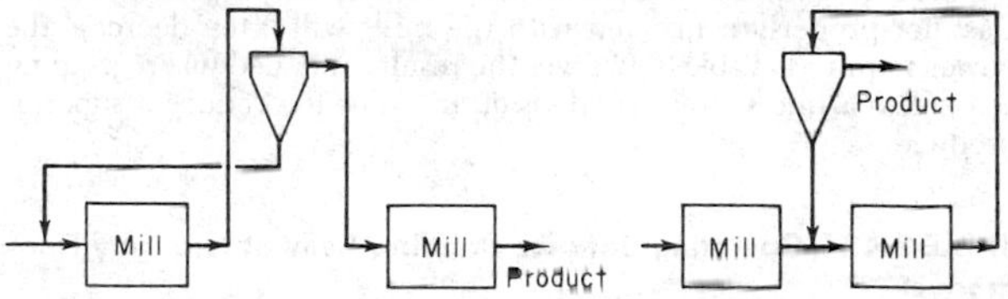

FIG. 8-61 Two cement-milling circuits. [*For others, see Tonry,* Pit Quarry *(February–March 1959).*]

A dry-process plant has been described by Bergstrom [*Rock Prod.*, 59–62 (August 1968)]. Limestone and shale are crushed in a 1.06- by 1.22-m (42- by 48-in) jaw crusher with scalping grizzly feeder. The minus 15-cm (6-in) product is reduced to minus 4 cm (1½ in) in a hammer crusher. Iron ore and sand are added as needed. Raw materials are dried in a 3- by 6-m (10- by 20-ft) double-shell rotary dryer and stored in bins. Weigh feeders proportion materials from these bins into the raw mill. This mill is 4 by 10 m (13 by 33 ft) in size, with a 2250-kW (3000-hp) drive. It is operated in closed circuit with a 5.5-m- (18-ft-) diameter air separator. The product is fed to a surge tank and thence to the kiln. Each unit in the raw-materials plant is provided with a dust collector. Clinker is cooled, stored, and then ground. The finish mill is 4 by 11 m (13 by 36 ft) in size with a 2625-kW (3500-hp) drive. It operates in closed circuit with a 6-m- (20-ft-) diameter air separator. The cyclone dust collector provides an air-sweep circuit for cooling the mill.

Wet-Process Cement Ball, tube, and compartment mills of essentially the same construction as for the dry process are used for grinding. Water or clay slip is added at the feed end of the initial grinder, together with the roughly proportioned amounts of limestone and other components. Vertical mills are rarely suitable for wet grinding. The use of closed circuit involves vibrating screens to remove oversize particles from the ball-mill product for return to the ball mill and rake classifiers, hydroseparators, and thickeners for the tube-mill product. In modern installations wet grinding is sometimes accomplished in ball mills alone, operating with excess water in closed circuit with classifiers and hydroseparators.

Figures 8-39 to 8-41 illustrate single-stage and two-stage closed-circuit wet-grinding systems. The circuits of Fig. 8-61 may also be used as a closed-circuit wet-grinding system incorporating a liquid-solid cyclone as the classifier.

A wet-process plant making cement from shale and limestone has been described by Bergstrom [*Rock Prod.*, 64–71 (June 1967)]. There are separate facilities for grinding each type of stone. Primary hammer crushers with 1.5- by 1.8-m (60- by 72-in) openings are provided with a scalping grizzly. The 5-cm (2-in) limestone product is given a second hammer-mill crushing to minus 1 cm (⅜ in). The product is fed with water to the raw mill. This is a three-compartment ball mill. Ball sizes range from 10 to 2.5 cm (4 to 1 in) in the compartments, and balls occupy 43 percent of the mill volume. The mill discharges through a trommel screen into a circulating sump and operates in closed circuit with a battery of Dutch State Mines screens. Material passing the screens is 85 percent minus 200 mesh.

The minus 2.5-cm (1-in) crushed shale goes to an overflow rod mill, with product 85 percent minus 200 mesh. The product is blended with limestone slurries of appropriate analyses and fed directly to the kiln, where it is converted to cement clinker. The clinker passes from a cooler to two finish mills, which allow simultaneous grinding of two types of cement. Each operates at 300 percent circulating load with a cooler and an air separator. The entire process is extensively instrumented and controlled by computer. Automatic devices sample crushed rock, slurries, and finished product for chemical analysis by x-ray fluorescence. Mill circuit feed rates and water additions are governed by conventional controllers.

Closed-Circuit Grinding of Cement Clinker Clinker is ground by the same equipment in both processes. The time of setting and the strength of cements vary with the fineness to which the cement has been ground. A size distribution between 10 and 50 μm is often required to obtain a cement having the desired properties. An air classifier properly connected with the mills will often decrease the power required. Table 8-42 gives the results obtained when grinding cement to a fine state of subdivision in order to produce a superior product.

TABLE 8-42 Operating Data for Grinding Cement to a Very Fine Product

Type and size of mill, ft	Three-compartment, 7.5 × 43
Diameter of classifier, ft	18
Hourly capacity, bbl	46
Feed to classifier, % through No. 200 sieve	98.7
Tailings from classifier, % through No. 200 sieve	83.2
Finished product, % through No. 200 sieve	99.8
through No. 325 sieve	97.9
—30 μm	92.1
—25 μm	83.8
—20 μm	72.9
Total power required, hp	746
hp/bbl of product	12.1

NOTE: To convert feet to centimeters, multiply by 30.5; to convert barrels to cubic meters, multiply by 0.159; and to convert horsepower per barrel to kilowatts per cubic meter, multiply by 4.7.

When this cement was ground in open circuit to produce a comparable material, the power required was 92.5 kW/m^3 (19.7 hp/bbl) cement. A 3- by 2-m (10- by 6.5-ft) Hardinge mill with air classifier, grinding a 1.25-cm (½-in) clinker to 82 percent through No. 200 sieve, showed a power consumption of 23.2 kW/m^3 (4.95 hp/bbl) with an hourly capacity of 16.2 m^3 (102 bbl). Grinding to 88 percent through No. 200 sieve, the consumption was 26.5 kW/m^3 (5.65 hp/bbl) and the capacity 14 m^3/h (88 bbl/h). Ball load in both cases was 27 Mg (60,000 lb) and mill speed 18 r/min.

The specific surface of cement is determined by air permeability, generally of the Blaine type. A product having a high surface area contains a higher percentage of impalpable powder than the same material ground to a lower surface area. Use of the Blaine device has become standard in the cement industry. Surface area of ordinary cement may run 3200 cm^2/g, whereas surface area of the high early cement may run from 5000 to 6000 cm^2/g.

Surface areas of products are being determined more and more in connection with the pulverizing of many other minerals. A product should, of course, not be ground to a surface much higher than is absolutely necessary, since the power for pulverizing and classifying increases rapidly as surface area increases. This is indicated in Table 8-43. The total cost would be 2 or 3 times the power cost.

Lime for agricultural purposes generally is ground in hammer mills. It includes burnt, hydrated, and raw limestone. When a fine product is desired, as in the building trade and for chemical manufacture, ring-roller mills, ball mills, and certain types of hammer mills are used. Table 8-44 gives operating data for a Hardinge mill grinding quicklime.

Ring-roller mills with air classification produce quicklime for the beet-sugar industry such that 99 to 99.9 percent will pass through a No. 200 sieve. Power requirements vary between 17 and 25 kWh/Mg [20 and 30 (hp·h)/ton], depending on size of mill being used.

TABLE 8-43 Pulverizing Slate to Different Blaine Surface Areas

Surface area, sq. cm./g.	Production rate, tons/hr.	Power, kw.-hr./ton	Power ratio
3050	4.30	21.0	1.0
3330	3.40	25.0	1.2
3850	2.38	33.0	1.6
4100	2.00	38.0	1.8
4500	1.54	49.0	2.3
5040	1.10	68.0	3.2
5250	0.90	84.0	4.0
5580	.70	107.0	5.1
5950	.55	135.0	6.4
6300	.39	192.0	9.2
6600	.29	258.0	12.3
7050	.22	341.0	16.2
7300	.18	417.0	19.9

NOTE: To convert tons per hour to megagrams per hour, multiply by 0.907; to convert kilowatthours per ton to kilowatthours per megagram, multiply by 1.1.

TABLE 8-44 Operating Data for Grinding Quicklime in Ball Mill

Size of mill, ft.	7 × 3
Type of classifier	Hardinge air classifier
Size of feed, % through No. 100 sieve	90
Capacity, tons/hr.	12.5
Power required for mill, hp.	100
Power required for auxiliaries, hp.	35
Ball load, lb.	20,000
Speed of mill, r.p.m.	20

NOTE: To convert feet to centimeters, multiply by 30.5; to convert pounds to kilograms, multiply by 0.4535; to convert horsepower to kilowatts, multiply by 0.746; and to convert tons per hour to megagrams per hour, multiply by 0.907.

Lime coming from the hydrator is often pulverized without separating out the impurities by the use of ring-roller mills or ball mills. As a rule it is not really pulverized but is air-classified to remove impurities such as sand, overburned, underburned, and core. The hydrate is passed through an automatic pulverizer with air classification and throwout. There is a tendency to handle this material with air classifiers in series, the tailings from the final classifier being discarded or sold for agricultural use.

Gypsum is usually calcined in kettles. A very few rotary calciners are used. These, of course, use crushed and screened raw gypsum.

Gypsum is pulverized 85 to 95 percent minus 100 mesh ahead of the calcining kettles. There is a growing trend toward requiring a Blaine surface area of 3000 or better. Ring-roller mills with internal classification are used exclusively for grinding raw gypsum ahead of the kettles. Table 8-45 indicates the performance of a 6058 Raymond mill grinding raw gypsum rock from various sources; a complete range of other mill sizes is available. The majority of the new mills grinding gypsum are set up with heat so as to dry and grind in one operation.

Gypsum calcined in rotary kilns is pulverized after calcining. Tube mills are usually used. These impart plasticity and workability. Occa-

TABLE 8-45 Performance of Raymond 6058 Ring-Roller Mill Grinding Raw Gypsum Rock from Various Sources, 90 Percent Minus 100 Mesh

Source	Tons/hr.	Kw. input	Kw.-hr./ton
Fort Dodge, Iowa	32	200	6.25
Jamaica	23	205	8.9
Shoals, Indiana	16	210	13.1
Nova Scotia (Dingwall)	10	230	23

NOTE: To convert tons per hour to megagrams per hour, multiply by 0.907; to convert kilowatthours per ton to kilowatthours per megagram, multiply by 1.1.

sionally such calcined gypsum is passed through ring-roller mills ahead of the tube mills.

In special cases, Raymond Imp pulverizers are used to grind and calcine in one operation. The result is a rapid-setting stucco. It is ideal for the manufacture of wallboard and similar structural gypsum products. A Raymond No. 50 Imp mill with 56 kW (75 hp) will produce about 3200 kg/h (7000 lb/h) of stucco, 90 percent minus 100 mesh, when starting with an average raw gypsum.

COAL, COKE, AND OTHER CARBON PRODUCTS

Bituminous Coal The grinding characteristics of bituminous coal are affected by impurities contained, such as inherent ash, slate, gravel, sand, and sulfur balls. The grindability of coal is determined by grinding it in a standard laboratory mill and comparing the results with the results obtained under identical conditions on a coal selected as a standard. This standard coal is a low-volatile coal from Jerome Mines, Upper Kittaning bed, Somerset County, Pennsylvania, and is assumed to have a grindability of 100. Thus a coal with a grindability of 125 could be pulverized more easily than the standard, while a coal with a grindability of 70 would be more difficult to grind. (Grindability and grindability methods are discussed under "Properties of Solids.") The capacity obtained with a mill will be a function of grindability.

There are two general methods of burning pulverized coal: the unit system, in which the coal is blown directly into the furnace as it is pulverized; and the central grinding system, in which the coal is stored in a bin before it is used. Direct firing of boilers and rotary kilns is replacing the storage system. Mills of the ball, ring-roller, bowl, and ball-and-ring type are used for the direct firing of large installations.

The Kennedy air-swept tube mill is of relatively short length. No screens are used either inside or out; the fine coal is mixed thoroughly with air and air-floated to the burners. The mill rotates slowly at a speed of 19 to 50 r/min, depending on the size. Hot air is used if drying is desired.

The Raymond bowl mill is used primarily for pulverizing coal and blowing it directly into industrial furnaces, boilers, or rotary kilns. Hot gases may be used for drying the coal while it is being pulverized.

Thermostatic control through a tempering device may be used to maintain a constant temperature in the mill irrespective of the moisture in the coal.

The Babcock and Wilcox type E pulverizer is used for direct-firing industrial furnaces and has application in direct-fired circulating systems.

Anthracite Anthracite is harder to reduce than bituminous coal. It is pulverized for foundry-facing mixtures in ball mills or hammer mills followed by air classifiers. Only to a lesser extent is it used for fuel in powdered form.

A 3- by 1.65-m (10-ft by 66-in) Hardinge mill in closed circuit with an air classifier as shown in Fig. 8-38, grinding 4-mesh anthracite with 3.5 percent moisture, produced 10.8 Mg/h (12 tons/h), 82 percent through No. 200 sieve. The power required for the mill was 278 kW (370 hp); for auxiliaries, 52.5 kW (70 hp); speed of mill, 19 r/min; ball load, 25.7 Mg (28.5 tons). Data for a similar pilot circuit are given in Table 8-46 (Sanner, U.S. Bur. Mines Rep. Invest. 7170, 1968).

TABLE 8-46 Closed-Circuit Continuous Grinding of Anthracite in an Air-Swept Ball Mill

Production rate, lb/h	Mean product size, μm	Circulating load, %	Recirculated material, −37 μm, %	Energy, kWh/ton
19.8	7.3	277	42	330
23.0	6.6	283	59	280
27.0	6.1	757	31	246

NOTE: To convert pounds per hour to kilograms per hour, multiply by 0.454; to convert kilowatthours per ton to kilowatthours per megagram, multiply by 1.1.

A 2.4- by 3.0-m (8- by 10-ft) tube mill, in closed curcuit with a 4.2-m (14-ft) centrifugal air classifier, operating on a feed with 90 percent through No. 40 sieve, pulverized 1.8 Mg/h (2 tons/h), 99 percent through No. 200 sieve, with 75 kW (100 hp) for the mill and 22.5 kW (30 hp) for the classifier.

Anthracite for use in the manufacture of electrodes is calcined, and the degree of calcination determines the grinding characteristics. Calcined anthracite is generally ground in ball and tube mills or ring-roller mills equipped with air classification. A Raymond high-side ring-roller mill grinding calcined anthracite for electrode manufacture has a capacity of 2.1 Mg/h (4600 lb/h) for a product fineness of 76 percent passing a No. 200 sieve and 52.5-kW (70-hp) power requirement.

Coke The grinding characteristics of coke vary widely. Petroleum coke is generally easier to grind than coke derived from bituminous coal. By-product coke is hard and abrasive, while certain foundry and retort coke is extremely hard to grind. For certain purposes it may be necessary to produce a uniform granule with minimum fines. This is best accomplished in rod or ball mills in closed circuit with screens. Hourly capacity of a 1.2- by 3-m (4- by 10-ft) rod mill with screens, operating on by-product-coke breeze, was 8.1 Mg (9 tons), 100 percent through No. 10 sieve, and 73 percent on No. 200 sieve; power requirement, 30 kW (40 hp).

Petroleum coke is generally pulverized for the manufacture of electrodes; ring-roller mills with air classification and tube mills are generally used. A No. 5057 Raymond ring-roller mill gave an hourly output of 3.8 tons, 78.5 percent through No. 200 sieve, with 67 kW (90 hp).

Other Carbon Products **Pitch** may be pulverized as a fuel or for other commercial purposes; in the former case the unit system of burning is generally employed, and the same equipment is used as described for coal. Grinding characteristics vary with the melting point, which may be anywhere from 50 to 175°C.

Natural graphite may be divided into three grades in respect to grinding characteristics: flake, crystalline, and amorphous. Flake is generally the most difficult to reduce to fine powder, and the crystalline variety is the most abrasive. Graphite is ground in ball mills, tube mills, ring-roller mills, and jet mills with or without air classification. For large capacities, ball and tube mills are used, particularly on the flake and crystalline varieties. Beneficiation by flotation is an essential part of most current procedures.

Majac jet-pulverizer performance on natural graphite is given in Table 8-31. Graphite for pencils has 47, 83, 91, and 94 percent by weight smaller than 4, 9, 18, and 31 μm respectively.

Artificial graphite has been ground in ball mills in closed circuit with air classifiers. For lubricants the graphite is ground wet in a paste in which water is eventually replaced by oil. The colloid mill is used for production of graphite paint.

Mineral black, a shale sometimes erroneously called "rotten stone," contains a large amount of carbon and is used as a filler for paints and other chemical operations. It is pulverized and classified with the same equipment as shale, limestone, and barite.

Bone black is sometimes ground very fine for paint, ink, or chemical uses. A tube mill often is used, the mill discharging to a fan which blows the material to a series of cyclone collectors in tandem. The discharge from the first cyclone is usually returned to the mill for further grinding; the discharge from the last goes to an air filter where the finest grades are obtained. The number of cyclones used depends on the grades required.

Decolorizing carbons of vegetable origin should not be ground too fine. Standard fineness varies from 100 percent through No. 30 sieve to 100 percent through No. 50, with 50 to 70 percent on No. 200 sieve as the upper limit. Ball mills, hammer mills, and rolls, followed by screens, are used. When the material is used for filtering, a product of uniform size must be used.

Charcoal usually is ground in hammer mills with screen or air classification. For absorption of gases it is usually crushed and graded to about No. 16 sieve size. Care should be taken to prevent it from igniting during grinding.

Gilsonite sometimes is used in place of asphalt or pitch. It is easily pulverized and is generally reduced on hammer mills with air classification.

Carbon mixtures (green mix) are generally made from flour of petroleum coke, graphite, and lampblack, mixed with a binder such as pitch; solvents such as benzol are incorporated in the mix. After cooling, the mixture is caked and therefore generally reground. It is generally ground in air-swept hammer mills, since the material must be kept cool.

Carbon blacks when manufactured are usually very fine. The gas is passed through baffled chambers or ducts in which the various grades are precipitated; the coarser grades are often pulverized for the carbon-brush industry. Grinding may be done in ball, hammer, ring-roller, or jet mills with or without air classification. When an extremely fine product is required, the same system as described for bone black may be used. A hammer mill equipped with air classification ground about 90 kg/h (200 lb/h) to a fineness of 95 percent No. 200 sieve with a power consumption of 15 kW (20 hp).

CHEMICALS AND SOAPS

Colors and Pigments **Dry colors and dyestuffs** generally are pulverized in hammer mills (see Tables 8-23 and 8-26). The jar mill or a large pebble mill is often used for small lots. There is a special problem with some dyes which are coarsely crystalline. These are ground to the desired fineness with hammer or jet mills using air classification to limit the size.

Easily dispersible colors are not ordinarily ground fine, since they are subsequently processed in a liquid medium in pebble mills, rolls, or colloid mills. There is, however, a tendency to grind them wet with a dispersing agent, then drying and pulverizing, after which they are easily dispersed in the vehicle in which they are used.

White pigments are the basic commodities processed in large quantities. Titanium dioxide is the most important. The problem of cleaning the mill between batches does not exist as with different colors. These pigments are finish-ground to sell as dry pigments using mills with air classification. For the denser, low-oil-absorption grades, roller and pebble mills are employed. For looser, fluffier products, hammer and jet mills are used. Often a combination of the two mill actions is used to set the finished quality. Jet-mill performance for a number of extenders is given in Table 8-30. Figure 8-62 shows the grinding characteristics of a mill with whizzer classification when a titanium pigment is ground to different finenesses.

Milling techniques may be modified by the use of grinding aids and additives. The use of stearates, resinates, and other surfactants has an effect not only upon the efficiency and cost of milling but also upon the shape and nature of the comminuted particles, which are important in their final use as fillers. As an example, it is reported that ordinary $CaCO_3$ fillers delay vulcanization in natural rubber because of basicity and poor rubber contact but that when coated with modified stearic acid $CaCO_3$ becomes a satisfactory inert filler, causing little alteration of mechanical properties.

Table 8-47 gives operating characteristics of a Raymond No. 5057 high-side mill grinding various oxides of iron. Synthetic mineral pigments are usually fine agglomerates. They may be disintegrated with hammer or jet mills without elaborate pregrinding.

Power consumption includes power required for grinding, classifying, and conveying product to bins above the baggers. A 1.5- by 0.4-m (4.5-ft by 16-in) Hardinge conical mill in closed circuit with classifier, grinding 50-mesh iron oxide with 33 percent moisture for the paint trade, showed a capacity of 22.5 Mg/day (25 tons/day), 100 percent through No. 200 sieve. Power consumption was 15 kW (20 hp), mill speed 30 r/min, ball load 1800 kg (4000 lb). The conditions necessary for dispersing pigments in paint by means of steel ball mills have been investigated [Fischer, *Ind. Eng. Chem.*, 33(12), 1465–1472 (1941)].

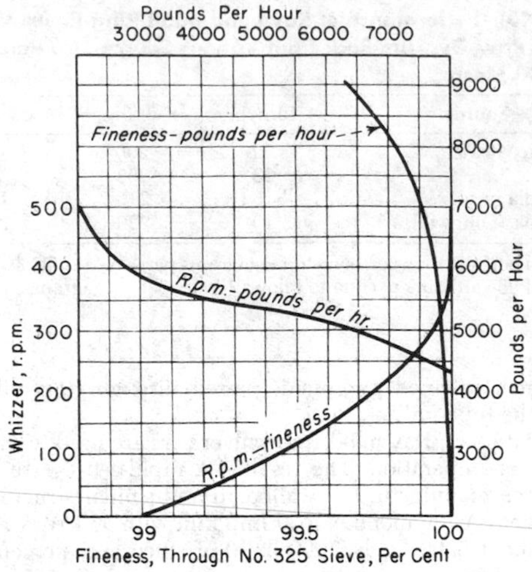

FIG. 8-62 Grinding characteristics of a ring-roller mill with a whizzer classifier when grinding a titanium pigment. To convert pounds per hour to kilograms per hour, multiply by 0.4535.

TABLE 8-47 Grinding Iron Oxides in a Ring-Roller Mill

Material	Fineness	Capacity, lb./hr.	Total hp.-hr./ton
Raw sienna	99% through No. 200 sieve	5950	23.5
Burnt sienna	99.5% through No. 200 sieve	5800	22.1
Raw umber	99% through No. 200 sieve	5200	26.9
Burnt umber	99.5% through No. 200 sieve	5400	25.9
Natural ocher	99.9% through No. 200 sieve	4500	31.0
Iron oxide (ore)	99% through No. 325 sieve	3100	45.0
Iron oxide (precipitated)	99.9% through No. 325 sieve	1800	72.5

NOTE: To convert pounds per hour to kilograms per hour, multiply by 0.4535; to convert horsepower-hours per ton to kilowatthours per megagram, multiply by 0.82.

Lead Oxides Leady litharge containing 25 to 30 percent free lead is required for storage-battery plates. It is processed on Raymond Imp mills. They have the ability to produce litharge that has a desired low density of 1.1 to 1.3 g/cm^3 (18 to 22 g/in^3). A 56-kW (75-hp) unit produces 860 kg/h (1900 lb/h) of material having this density.

The processing of **diatomite** is unique, since its particle-size control is effected by calcination treatments and air classification.

Chemicals The fineness obtainable with a hammer mill on rock salt and chemicals is given in Tables 8-27 and 8-28.

Sometimes it is necessary to produce a granular product of definite size limits, such as a granulated monocalcium phosphate that will all pass through a No. 50 sieve and, within a few percent, remain on a No. 200 sieve.

Soft materials, such as fuller's earth, sodium bicarbonate, and monocalcium phosphate, are generally ground in flour mills to obtain a finely divided product. A ratio of reduction of 25:1 or even 1.5:1 is generally used; the material passes through a series of rolls with bolting reels. Sometimes air classifiers are used in the circuit for removal of fines. Table 8-48 gives the results obtained in granulating soft materials.

Sulfur The ring-roller mill can be used for the fine grinding of sulfur. Inert gases are supplied instead of hot air (see "Properties of Solids: Safety" for use of inert gas). Performance of a Raymond No. 5057 ring-roller mill is given in Table 8-49. If power cost is 2 cents per kilowatthour, the total cost might be 3 to 4 times the power cost and include labor, inert gas, maintenance, and fixed charges.

TABLE 8-48 Granulation of Soft Materials

	Fuller's earth	Sodium bicarbonate
Size of product required, sieve No.	− 80 +150	−120 +200
Type of mill used	Two roller	Two roller
Number of mills in series	12	5
Size of rolls, in.	7 × 16	7 × 20
Capacity, lb./hr.	3500	2500
Recovery from original feed, %	80	55
Horsepower required, total	80	30

NOTE: To convert inches to centimeters, multiply by 2.54; to convert pounds per hour to kilograms per hour, multiply by 0.4535.

TABLE 8-49 Grinding Sulfur

Fineness, % through No. 325 sieve	Capacity, tons/hr.	Power, kw.-hr./ton
90	6.0	13.7
95	5.0	16.4
99	3.5	23.4
99.9	2.5	32.7

NOTE: To convert tons per hour to megagrams per hour, multiply by 0.907; to convert kilowatthours per ton to kilowatthours per megagram, multiply by 1.1.

Grinding of Soaps Soaps in a finely divided form may be classified as soap powder, powdered soap, and chips or flakes. The term "soap powder" is applied to a granular product, No. 12 to No. 16 sieve size with a certain amount of fines, which is produced in hammer mills with perforated or slotted screens.

Pulverizing of the metallic soaps, stearates, palmitates, resinates, laurates, and erucates is not difficult by using modern equipment with provision for keeping the material cool and in rapid motion. Batch grinding is not practicable, as the material tends to cake, particularly if a fine product is needed. Oleates are usually most troublesome, as they tend to become plastic and creamy. Listed in order of their resistance to pulverization some of the metallic soaps are lead, silver, zinc, copper, and nickel stearate; zinc, lead, and copper palmitate; lead, copper, and zinc laurate; silver, mercury, and lead erucate; and silver and lead oleate.

The oleates and erucates are best pulverized by multicage mills; laurates and palmitates, in cage mills and also in hammer mills if particularly fine division is not required; stearates may generally be pulverized in multicage mills, screen mills, and air-classification hammer mills. Table 8-50 gives the operating characteristics of hammer mills when grinding zinc stearate and aluminum stearate to a finely divided powder.

ORGANIC POLYMERS

The grinding characteristics of various resins, gums, waxes, hard rubbers, and molding powders depend greatly upon their softening temperatures. When a finely divided product is required, it is often necessary to use a water-jacketed mill or a pulverizer with an air

TABLE 8-50 Performance of Screen-Type Hammer Mill Grinding Zinc and Aluminum Stearates

	Zinc stearate	Aluminum stearate
Capacity, lb./hr.	500	300
Fineness, % through No. 325 sieve	60	70
In closed circuit with air classifier:		
Capacity, lb./hr.	100	75
Fineness, % through No. 325 sieve	99.5	99.7
Horsepower required	25	20

NOTE: To convert pounds per hour to kilograms per hour, multiply by 0.4535; to convert horsepower to kilowatts, multiply by 0.746.

classifier in which cooled air is introduced into the system. Not all waxes can be ground, inasmuch as some of them are soft at the temperatures obtainable. However, a great many of them can be powdered if precautions are taken to prevent overheating. Hammer and cage mills are used for this purpose. Some low-softening-temperature resins can be ground by mixing with 15 to 50 percent by weight of dry ice before grinding. Refrigerated air sometimes is introduced into the hammer mill to prevent softening and agglomeration [Dorris, *Chem. Metall. Eng.*, **51**, 114 (July 1944)].

Most **gums and resins,** natural or artificial, when used in the paint, varnish, or plastic industries, are not ground very fine, and hammer or cage mills will produce a suitable product. Typical performance of the attrition mill is given in Table 8-29. Roll crushers will often give a sufficiently fine product. Certain resins used in the phenolic-resin industries must be pulverized very fine; pebble mills, cooled with water or brine, in closed circuit with an air classifier, are used. In general, one may say that the grinding of the thermoplastic resins or of the thermosetting resins after they have set is difficult except in the very coarse range. In the case, however, of the thermosetting resins, before setting a very high fineness may be obtained readily, although these materials, particularly when ground on a hammer mill, may cake up very rapidly after leaving the mill. Phenolformaldehyde resins have been ground in the Raymond Imp mill to about 80 percent through No. 325 sieve at a rate of 18 kg/kWh [30 lb/(hp·h)]. This is about 99 percent through No. 100 sieve.

The Raymond ring-roll mill with its internal air separation is widely used to pulverize **phenolformaldehyde resins.** The usual fineness of grind is finer than 99 percent minus 200 mesh. Air at 4°C (40°F) is usually introduced into the mill to limit temperature rise. A typical 3036 Raymond mill using 34 kW (45 hp) will produce better than 900 kg/h (2000 lb/h) at 99 percent minus 200 mesh.

Hard rubber is one of the few combustible materials which is generally ground on heavy steam-heated rollers. The raw material passes to a series of rolls in closed circuit with screens and air classifiers. Farrel-Birmingham rolls are used extensively for this work. There is a differential in the roll diameters, and the particular size best suited for the average hard rubber is one having rolls with 33- and 43-cm (13- and 17-in) diameters and 51-cm (20-in) face. The motor should be separated from the grinder by a fire wall. It is also desirable to run these machines at rather low speed and low differential between the rolls because it is very easy to overheat hard rubber in grinding, making it smolder, which necessitates the shutting down of the grinder until it cools off before clearing out the charred material. The performance of a series of rolls grinding hard rubber, producing a finished product through an air classifier, is given in Table 8-51. A larger production could probably be attained, but operation at the lower rate is advisable to prevent generation of an excessive amount of heat.

Specifications for **molding powders** vary widely, from a No. 8 to a No. 60 sieve product; generally the coarser products are No. 12, 14, or 20 sieve material. Specifications usually prescribe a minimum of fines (below No. 100 and No. 200 sieve). For most purposes the ideal molding powder would consist of particles testing smaller than No. 20 and larger than No. 100 sieve size. Molding powders are produced with hammer mills, either of the screen type or equipped with air classifiers.

TABLE 8-51 Grinding Hard Rubber

Number of roller mills	3
Size of each mill, in.	13 and 17 × 20
Motor on each mill, hp.	50
Size of vacuum air classifier, ft.	4.5
Size of motor on classifier fan, hp.	20
Fineness of feed to classifier, % through No. 100 sieve	32
Fineness of product from classifier, % through No. 100 sieve	95
Production, lb./hr.	250

NOTE: To convert inches to centimeters, multiply by 2.54; to convert feet to centimeters, multiply by 30.5; to convert horsepower to kilowatts, multiply by 0.746; and to convert pounds per hour to kilograms per hour, multiply by 0.4535.

TABLE 8-51a Objectives of Size Enlargement

Reduce dusting losses.
Reduce handling hazards, particularly with irritating or obnoxious powders.
Render powders free-flowing.
Densify materials for more convenient storage or shipment.
Prevent caking and lump formation.
Provide definite quantity units suitable for metering, dispensing, and administering.
Produce useful structural forms.
Create uniform blends of solids which do not segregate.
Improve the appearance of products.
Permit control over properties of finely divided solids (e.g., solubility, porosity, surface-volume ratio, heat-transfer rates).
Separate multicomponent particle mixtures by selective wetting and agglomeration.
Remove particles from liquids.

Curves *A* and *B* of Fig. 8-63 give the screen analysis of a molding powder produced with screen pulverizers fitted with an 8-mesh screen. Curve *C* gives the data obtained with an air-classification pulverizer unit operated to give a minimum of No. 100 sieve material, which amounted to only 12 percent. This material was passed over an 8-mesh screen to remove oversize, and the resulting product passed through an air classifier to remove the No. 100 sieve size particles. Curve *D* gives the screen analysis of the final granular product.

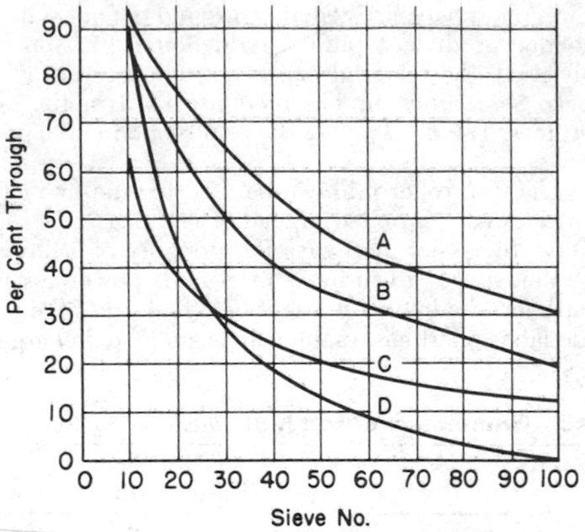

FIG. 8-63 Sieve analyses of molding powders produced by various installations.

The following materials may be ground at ordinary temperatures if only the regular commercial fineness is required: amber, arabac, tragacanth, rosin, olibanum, gum benzoin, myrrh, guaiacum, and montan wax. If a finer product is required, hammer mills or attrition mills in closed circuit, with screens or air classifiers, are used. Attrition-mill performance is given in Table 8-29.

SIZE ENLARGEMENT

GENERAL REFERENCES: Ball et al., *Agglomeration of Iron Ores*, Heinemann, London, 1973. Capes, *Particle Size Enlargement*, Elsevier, New York, 1980. King, "Tablets, Capsules and Pills," in *Remington's Pharmaceutical Sciences*, Mack Pub. Co., Easton Pa., 1970. Knepper (ed.), *Agglomeration*, Interscience, New York, 1962. Mead (ed.), *Encyclopedia of Chemical Process Equipment*, Reinhold, New York, 1964. Pietsch, *Roll Pressing*, Heyden, London, 1976. Sastry (ed.), *Agglomeration 77*, AIME, New York, 1977. Sauchelli (ed.), *Chemistry and Technology of Fertilizers*, Reinhold, New York, 1960. Sherrington and Oliver, *Granulation*, Heyden, London, 1981.

SCOPE AND APPLICATIONS

Size enlargement is any process whereby small particles are gathered into larger, relatively permanent masses in which the original particles can still be identified. Applications include formation of useful shapes (e.g., brick and tile) and irregular pellets or balls for the industrial beneficiation of finely divided materials.

Numerous benefits result from size-enlargement processes, as will be appreciated from Table 8-51a. A wide variety of size-enlargement methods and applications are available; a classification of these is given in Table 8-52.

STRENGTH OF AGGLOMERATES

Bonding Mechanisms Agglomerate bonding mechanisms may be divided into five major groups [Rumpf, in Knepper (ed.), *Agglomeration*, op. cit., p. 379]. More than one mechanism may apply during a given size-enlargement operation.

Solid bridges can form between particles by the sintering of ores, the crystallization of dissolved substances during drying as in the granulation of fertilizers, and the hardening of bonding agents such as glue and resins.

Mobile liquid binding produces cohesion through interfacial forces and capillary suction. Three states can be distinguished in an assembly of particles held together by a mobile liquid (Fig. 8-64). Small amounts of liquid are held as discrete lens-shaped rings at the points of contact of the particles; this is the **pendular state.** As the liquid content increases, the rings coalesce and there is a continuous network of liquid interspersed with air; this is the **funicular state.** When all the pore spaces in the agglomerate are completely filled, the **capillary state** has been reached. When a mobile liquid bridge fails, it constricts and divides without fully exploiting the adhesion and cohesive forces in the bridge.

By contrast, **immobile liquid bridges** formed from highly viscous materials such as asphalt or pitch fail by tearing apart the weakest bond. Then adhesion and/or cohesion forces are fully exploited, and binding ability is much larger.

Intermolecular and electrostatic forces bond very fine particles without the presence of material bridges. Such bonding is responsible for the tendency of particles less than about 1 μm diameter to form agglomerates spontaneously under agitation. With larger particles, however, these short-range forces are insufficient to counterbalance the weight of the particle, and adhesion does not occur.

Mechanical interlocking of particles may occur during the agitation or compression of, for example, fibrous particles, but it is probably only a minor contributor to agglomerate strength in most cases.

TABLE 8-52 Size-Enlargement Methods and Applications*

Method	Equipment	Representative applications
Pressure compaction	Piston or molding press	Plastic preforms, small machine parts from metal powders (cams, gears, gaskets), metal borings and turnings
	Tableting press	Pharmaceuticals, catalysts, industrial chemicals, ceramics, metal powders
	Roll-type press	Clay-type minerals, potassium chloride, sodium chloride, organic compounds, metal powders, ores, charcoal, lime, magnesia, titanium sponge, phosphate rock
	Pellet mill	Pharmaceuticals, plastics, clays, carbon, charcoal, industrial chemicals, fertilizers, rubber products, animal feeds
	Screw extruder	Bauxite, plastics, rare-earth fluorides, clays, catalysts
Tumbling and mixer agglomeration	Inclined pan or disk; rotary-drum agglomerator	Fertilizers, iron ores, nonferrous ores, mineral and clay products, carbon black, various finely divided solid-waste products
	Paddle mixer; horizontal pan	Fertilizers, premixing for balling, conditioning steel-plant fines
	Powder blenders; flow-jet mixing	"Instant" foods, detergent granulation
Thermal processes	Sintering and heat hardening in traveling grate, rotary kiln, grate-kiln, shaft furnace	Ferrous and nonferrous ores, minerals, cement clinker, solid-waste products
	Drying and solidification in drum dryers, flakers, endless-belt systems	Sulfur slates, urea, ammonium nitrate, caustic, various resins, hot-melt adhesives
Spray methods	Spray dryers	Instant foods, washing powders, dyestuffs, press feeds
	Prilling towers	Urea, ammonium nitrates, resins, coal-tar pitch, etc.
	Fluidized and spouted beds	Fertilizers, clays, sulfur, nuclear and other wastes
	Flash dryers	Clays, diatomaceous earths, starch, waste by-products
Liquid systems	Immiscible-liquid wetting in various high-shear and turbine mixers	Coal fines, soot and oil removal from water
	Sol-gel process in spray column	Metal dicarbide spheroids
	Pellet flocculation in drums and stirred vessels	Waste sludge, mud and clay slurries, sewage sludge

*Cf. Browning, *Chem. Eng.*, 74(25), 147 (1967).

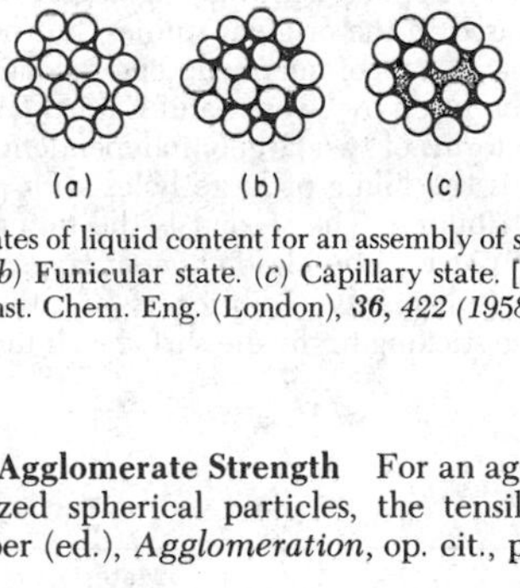

FIG. 8-64 Three states of liquid content for an assembly of spherical particles. (*a*) Pendular state. (*b*) Funicular state. (*c*) Capillary state. [*Newitt and Conway-Jones,* Trans. Inst. Chem. Eng. (London), *36, 422 (1958).*]

Calculation of Agglomerate Strength For an agglomerate composed of equal-sized spherical particles, the tensile strength i is [Rumpf, in Knepper (ed.), *Agglomeration*, op. cit., p. 379]

$$i = \frac{9}{8}\left(\frac{1-\epsilon}{\pi X^2}\right) NF \tag{8-38}$$

where X is the particle diameter; F is the bonding force per point of contact; N is the mean coordination number, i.e., average number of points of contact between one sphere and its neighbors; and ϵ is the volume fraction of voids in the agglomerate. Values of X and ϵ can be obtained from a size-distribution analysis of the powder and the bulk density of the packed particles. As an approximation, the coordination number N is π/ϵ (Rumpf, loc. cit.) or $N = 2 \exp 2.4(1 - \epsilon)$ [Meissner, *Ind. Eng. Chem. Process Des. Dev.*, **3**, 202 (1964)].

For mobile liquid binders in the pendular state

$$i = 2.8\left(\frac{1-\epsilon}{\epsilon}\right)\frac{\sigma}{Xf(\delta)} \tag{8-39}$$

where σ is the surface tension of the binding liquid and $f(\delta)$ is a function of the angle of contact [Newitt and Conway-Jones, *Trans. Inst. Chem. Eng. (London)*, **36**, 422 (1958)].

If wetting is complete, $f(\delta) = 1$. For the capillary state

$$i = 8.0\left(\frac{1-\epsilon}{\epsilon}\right)\frac{\sigma}{Xf(\delta)} \tag{8-40}$$

The tensile strength of an agglomerate in the pendular state is about one-third of that in the capillary state, while the funicular state has intermediate strengths. A decrease in particle size and porosity yields greater strength. To improve agglomerate strength, the importance of correct particle-size distribution in attaining minimum porosity should be recognized [Ridgway and Tarbuck, *Chem. Process Eng.* (February 1968)].

For the other binding mechanisms calculated values of tensile strength shown in Fig. 8-65 indicate the strength to be expected in various size-enlargement processes.

Strength-Testing Methods Concepts of **fracture mechanics** (see subsection "Properties of Solids") are applicable to the methods of testing the strength of agglomerates.

Compression tests, in which agglomerates are crushed between parallel platens, are used for quick production checking. Various means of distributing the applied force uniformly over the agglomerate surface are used, including shaving off opposite sides, fitting them with hardening plastic, or covering the platen surface with compressive board.

A log-log plot of load at failure against pellet diameter for approximately spherically pellets produced under the same conditions often yields a straight line with slope approximately equal to 2. The intercept of such a plot at unit diameter yields a compressive-strength factor.

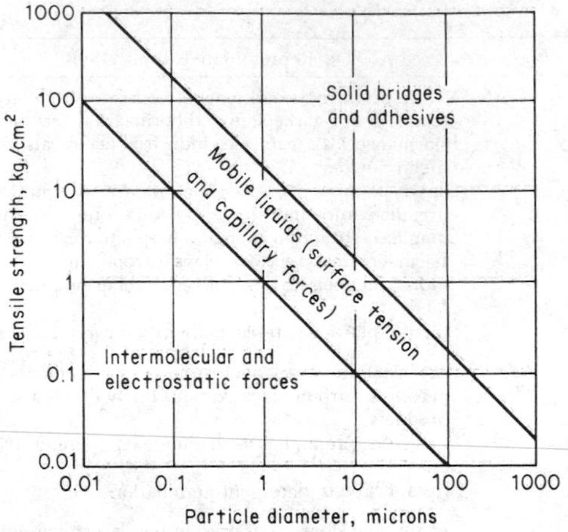

FIG. 8-65 Theoretical tensile strength of agglomerates. [*Adapted from Rumpf, "Strength of Granules and Agglomerates," in Knepper (ed.),* Agglomeration, *Wiley, New York, 1962.*]

Abrasion tests are carried out by tumbling a quantity of agglomerates in a drum with a wire-mesh shell. The proportion of fines produced in a definite time and speed measures the abrasion. Abrasion resistance is very important for agglomerates processed in metering and packaging machines [Pietsch, *Aufbereit. Tech.*, **7**, 177 (1966)]. The **Linder test** [Linder, *J. Iron Steel Inst.*, **189**, 233 (1958)] measures the resistance of agglomerates to breakup under reducing conditions simulating their descent through a blast furnace.

Drop tests are used to measure the resistance to fall of agglomerates, for example, during handling on conveyors. A test might consist of dropping a number of agglomerates from a height of 30.5 cm (12 in) onto a steel plate. The average number of drops to cause fracture is designated the drop number.

Caking tests as used for fertilizer granules consist of two parts (Bookey and Raistrick, in Sauchelli (ed.), *Chemistry and Technology of Fertilizers*, p. 454). A cake of the granules is first formed in a compression chamber under controlled conditions of humidity, temperature, etc. The crushing strength of the cake is then measured to determine the degree of caking.

Tablet-disintegration tests consist of cyclical immersion in a suitable dissolving fluid of pharmaceutical tablets contained in a basket. Acceptable tablets disintegrate completely by the end of the specified test period (*United States Pharmacopeia*, 17th rev., Mack Pub. Co., Easton, Pa., 1965, p. 919).

PRESSURE COMPACTION

Agglomeration may occur when force is applied to a particulate system in a confined space. The success of the operation depends partly on the effective utilization and transmission of the applied external force and partly on the physical properties of the particulate material.

Transmission of Forces As pressure is applied to a powder in a die, various zones in the compact are subjected to differing intensities of pressure and are in different stages of compaction. The pressure and density distributions [Train, *Trans. Inst. Chem. Eng. (London)*, **35**, 258 (1957)] are shown in Fig. 8-66. There is a dense axial core in the lower part of the compact. A low-density region is apparent just below the ram and may be responsible for the type of cracking or delamination known as **capping.**

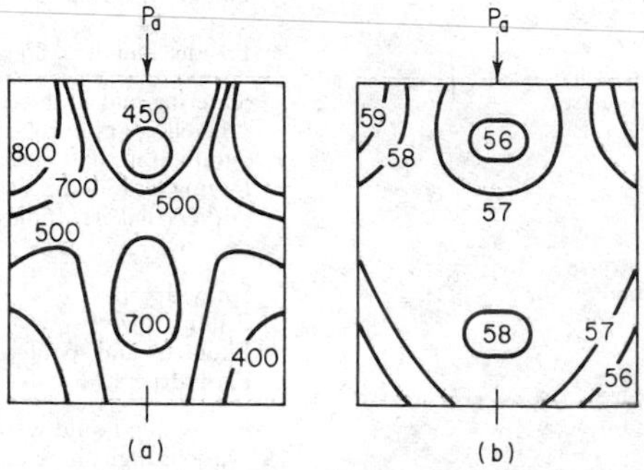

FIG. 8-66 Reactions in compacts of magnesium carbonate when pressed (P_a = 671 kg/cm²). (*a*) Contour levels in kilograms per square centimeter. (*b*) Contours represent percent solid present. [*Train,* Trans. Inst. Chem. Eng. (London), *35, 258 (1957).*]

Table 8-53 summarizes the equations that have been used to represent the compaction of powders to cylindrical die cavities. The equations have been applied with varying degrees of success over the pressure ranges used in the original studies, but no generally applicable equation can be recommended. The mechanisms of compaction have been discussed by Cooper and Eaton [*J. Am. Ceram. Soc.*, **45**, 97 (1962)] in terms of two largely independent probabilistic processes. The first is the filling of large holes with particles from the original size distribution. The second is the filling of holes smaller than the original particles by plastic flow or fragmentation.

Binders and Lubricants Lubricants aid the transmission of forces and reduce sticking to the die surfaces. Internal lubricants are

TABLE 8-53 Equations Used to Represent Compression of Powders

Equation	References	Materials compacted
$\log p = mV_R + b$	Huffine, Ph.D. Thesis, Columbia University, 1953 Stewart, *Engineering* **169**, 175, 203 (1950)	Variety of non-metallic powders Particles of sulfur, sodium chloride, T.N.T.
$\log \dfrac{1}{1-\rho_R} = mp + b$	Spencer, Gilmore, and Wiley, *J. Appl. Phys.*, **21**, 527 (1950) Heckel, *Trans. Met. Soc.* A.I.M.E., **221**, 1001 (1961)	Polystyrene particles Variety of metallic powders
$\dfrac{V_0 - V}{V_0 - V_s} = a_1 \exp - \left(\dfrac{k_1}{p}\right) + a_2 \exp - \left(\dfrac{k_2}{p}\right)$	Cooper and Eaton, *J. Am. Ceram. Soc.*, **45**, 97 (1962)	Four ceramic powders
$\log p = m \log \rho_a + b$	Jones, "Fundamental Principles of Powder Metallurgy," E. Arnold, London, 1960	Industrial metal powders

p = pressure applied to compact
V = volume of compact at p
V_0 = volume of compact at zero pressure
V_S = solid material volume (void-free)
$V_R = V/V_S$
ρ_a = apparent or bulk density of compact
ρ_S = true density of solid material
$\rho_R = \rho_a/\rho_S$
m, b, a_1, a_2, k_1, k_2 are constants.

mixed with the material being agglomerated, and external lubricants are applied to the die surface. Internal lubricants sometimes weaken bonding properties.

Binders improve the strength of compacts. They may be classified as **matrix type, film type,** and **chemical** [Komarek, *Chem. Eng.*, **74**(25), 154 (1967)]. A classification of binders and lubricants used in the tableting of various materials is offered in Table 8-54. Experimental evaluation is necessary for each application.

Compacting Equipment Compacting is carried out in two classes of equipment. These are **confined-pressure devices** (molding, tableting, roll presses), in which internal motion and shear of the particulates are incidental to their consolidation in closed molds or between two surfaces; and **extrusion devices** (pellet mills, screw extruders), in which material undergoes definite shear and mixing as it is consolidated while being pressed through a die.

Piston or molding presses are used to create uniform and sometimes intricate compacts, especially in powder metallurgy and plastics forming. Equipment comprises a mechanically or hydraulically operated press and, attached to the platens of the press, a two-part mold consisting of top (male) and bottom (female) portions. The action of pressure and heat on the particulate charge causes it to flow and take the shape of the cavity of the mold. Compacts of metal powders are then sintered to develop metallic properties, whereas compacts of plastics are essentially finished products on discharge from the molding machine.

In the automotive industry and other metal-working industries, coarse scrap-metal particulates are compressed and recycled to melting operations through piston-type briquetting presses. Feed materials are typically cast-iron and steel borings and turnings which tend to bond under pressure at least partially by mechanical interlocking. This operation should be distinguished from the much finer metal powders molded into parts such as gears by pressure in the powder-metallurgy field.

Tableting presses are employed in applications having strict specifications for weight, thickness, hardness, density, and appearance in the agglomerated product. They produce simpler shapes at higher production rates than do molding presses. A **single-punch press** is one that will take one station of tools consisting of an upper punch, a lower punch, and a die. A **rotary press** employs a rotating round die table with multiple stations of punches and dies. Older rotary

TABLE 8-54 Additives Used in Tableting*

Additive	% dry wt.	Application					
		Catalyst	Ceramic	Chemical	Food	Metal powders	Pharmaceutical
Binders:							
Agar, algin	½–3	. . .	A–B	B–C	B–C	. . .	B
Dextrine	1–4	. . .	A	B	B	. . .	A
Dextrose	5–20	. . .	. . .	A–B	A–B	. . .	A
Gelatine	1–3	. . .	. . .	A–B	A–B	. . .	A
Glucose	1–5	. . .	A	A	B	. . .	A
Glues	1–5	. . .	D	A			
Gums	1–5	. . .	A	A	A	. . .	A
Lactose	5–20	. . .	. . .	B	B	. . .	A
Pitch, asphaltum	2–50	D	. . .	A			
Resins	½–5	A	A	A	. . .	A	
Salts	5–20	. . .	. . .	B	. . .	. . .	A
Silicate of soda	1–4	. . .	C	B			
Starch (paste)	1–3	. . .	A	B	A	. . .	A
Sugar (sucrose)	2–20	. . .	. . .	A–B	A–B	. . .	A
Sulfite liquor	1–5	. . .	A	A			
Waxes	2–5	. . .	C	C	D	A	D
Water	½–25	A	C	. . .	C	. . .	C
Lubricants:							
Benzoate of soda	1–4	. . .	. . .	D	. . .	. . .	D
Boric acid	2–5	. . .	. . .	D	. . .	. . .	D
Graphite	¼–2	A	. . .	A	. . .	B–C	
Oils	¼–1	. . .	A	C	C	C	C
Soaps	½–2	. . .	. . .	C	. . .	. . .	D
Starch	1–5	. . .	. . .	C	C	. . .	B–C
Stearates:							
Aluminum	¼–2	B	B	B	. . .	B	
Magnesium, calcium	¼–2	A	A	A	A	B	A
Sodium	¼–2	. . .	. . .	B	. . .	. . .	B
Lithium, zinc	¼–2	A	A	A	. . .	A	
Stearic acid	¼–2	B	B	A	A	B	A
Sterotex	¼–2	A	A	A	A	A	A
Talc	1–5	C	B	B	. . .	. . .	A
Waxes	1–5	. . .	B–C	C	. . .	B–C	D
Water	0.1–5	A	A	C	C	. . .	C

Effectiveness: A—Excellent
B—Good
C—Fair
D—Poor

*Pennwalt Chemicals Corporation.

machines are **single-sided;** that is, there is one fill station and one compression station to produce one tablet per station at every revolution of the rotary head. Modern high-speed rotary presses are **double-sided;** that is, there are two feed and compression stations to produce two tablets per station at every revolution of the rotary head. Some characteristics of tableting presses are shown in Table 8-55.

TABLE 8-55 Characteristics of Tableting Presses*

	Single-punch	Rotary
Tablets per minute	8–140	72–6,000
Tablet diameter, in.	$\frac{1}{8}$–4	$\frac{5}{8}$–$2\frac{1}{2}$
Pressure, tons	$1\frac{1}{2}$–100	4–100
Horsepower	$\frac{1}{4}$–15	$1\frac{1}{2}$–50

*Browning, *Chem. Eng.*, **74**(25), 147 (1967).

NOTE: To convert inches to centimeters, multiply by 2.54; to convert tons to megagrams, multiply by 0.907; and to convert horsepower to kilowatts, multiply by 0.746.

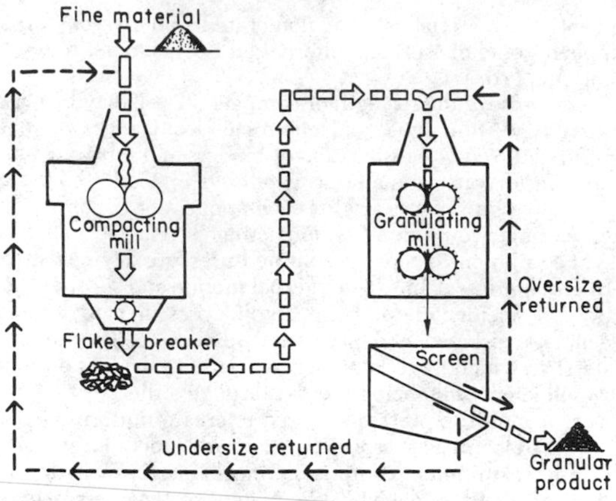

FIG. 8-67 Compacting-process flow diagram. *(Allis-Chalmers Corporation.)*

For successful tableting, a material must have suitable flow properties to allow it to be fed to the tableting machine. Wet or dry granulation is used to improve the flow properties of materials. In **wet granulation,** the blended dry ingredients are mixed with a liquid solution. The wet mass is then extruded through a coarse mesh screen, and the resulting crumb is dried. The dried material is screened to obtain the particle size required for tableting. In **dry granulation,** the blended dry ingredients are first densified in a heavy-duty rotary tableting press which produces "slugs" 1.9 to 2.5 cm ($\frac{3}{4}$ to 1 in) in diameter. These are subsequently crushed into particles of the size required for tableting. Predensification can also be accomplished by using a rotary compactor-granulator system. A third technique, **direct compaction,** uses sophisticated devices to feed the blended dry ingredients to a high-speed rotary press.

An excellent account of tableting in the pharmaceutical industry has been given by Kibbe [*Chem. Eng. Prog.*, **62**(8), 112 (1966)].

Roll presses compact raw material as it is carried into the gap between two rolls rotating at equal speeds. The size and shape of the agglomerates are determined by the geometry of the roll surfaces. Pockets or indentations in the roll surfaces form **briquettes** the shape of eggs, pillows, teardrops, or similar forms from a few grams up to 2 kg (5 lb) or more in weight. Smooth or corrugated rolls produce a solid sheet which can be granulated into the desired particle size on conventional grinding equipment. A typical compacting-granulating circuit is shown in Fig. 8-67.

Roll presses can produce large quantities of materials at low cost, but the product is less uniform than that from molding or tableting presses. The introduction of the proper quantity of material into each of the rapidly rotating pockets in the rolls is the most difficult problem in the briquetting operation. Various types of feeders have helped to overcome much of this difficulty.

The impacting rolls can be either solid or divided into segments. Segmented rolls are preferred for hot briquetting, as the thermal expansion of the equipment can be controlled more easily.

The most important factor that must be determined in a given application is the pressing force required for the production of acceptable compacts. Roll **loadings** (i.e., roll separating force divided by roll width) in commercial installations vary from 4.4 MN/m to more than 440 MN/m (1000 lb/in to more than 100,000 lb/in). Roll sizes up to 91 cm (36 in) in diameter by 61 cm (24 in) wide are in use.

The allowable roll width is inversely related to the required pressing force because of mechanical-design considerations. The throughput of a roll press at constant roll speed decreases as pressing force increases since the allowable roll width is less. Machines with capacities up to 45 Mg/h (50 tons/h) are available. Some average figures for the pressing force and energy necessary to compress a number of materials on roll-type briquette machines are given in Table 8-56. Typical capacities are given in Table 8-57.

A design procedure has been presented by Johanson (*Proc. Inst. Briquet. Agglom. Bien. Conf.*, **9**, 17 (1965)] to predict successfully the performance of and to design roll-type briquetting presses from small-scale laboratory measurements of the compressibility and flow properties of granular materials.

Pellet mills operate on the principle shown in Fig. 8-68. Moist plastic feed is pushed through holes in dies of various shapes. The friction of the material in the die holes supplies the resistance nec-

TABLE 8-56 Pressure and Energy Requirements to Briquette Various Materials*

Pressure range, lb./sq. in.	Approximate energy required, kw.-hr./ton	Type of material being briquetted or compacted: Without binder	With binder	Hot
Low 500–20,000	2–4	Mixed fertilizers, phosphate ores, shales, urea	Coal, charcoal, coke, lignite, animal feed, candy	Phosphate ores, urea
Medium 20,000–50,000	4–8	Acrylic resins, plastics, PVC, ammonium chloride, DMT, copper compounds, lead	Ferroalloys, fluorspar, nickel	Iron, potash, glass-making mixtures
High 50,000–80,000	8–16	Aluminum, copper, zinc, vanadium, calcined dolomite, lime, magnesia, magnesium carbonates, sodium chloride, sodium and potassium compounds	Flue dust, natural and reduced iron ores	Flue dust, iron oxide, natural and reduced iron ores, scrap metals
Very high >80,000	>16	Metal powders, titanium	. . . .	Metal chips

*Bepex Corporation. To convert pounds per square inch to newtons per square meter, multiply by 6895; to convert kilowatthours per ton to kilowatthours per megagram, multiply by 1.1.

TABLE 8-57 Some Typical Capacities (tons/h) for a Range of Roll Presses*

Roll diameter, in	10	16	12	10.3	13	20.5	28	36
Maximum roll-face width, in	3.25	6	4	6	8	13.5	27	10
Roll-separating force, tons	25	50	40	50	75	150	300	360
Carbon								
Coal, coke		2 ·	1		3	6	25	
Charcoal			8			13		
Activated					3	7		
Metal and ores								
Alumina					5	10	28	
Aluminum				2	4	8	20	
Brass, copper	0.5			1.5	3	6	16	
Steel-mill waste					5	10		
Iron				3	6	15	40	
Nickel powder					2.5	5.0		
Nickel ore						20	40	
Stainless steel				2	5	10		
Steel								25
Bauxite		1.5				10	20	
Ferrometals						10		
Chemicals								
Copper sulfate	0.5	1.5		1	3	6	15	
Potassium hydroxide				1	4	8		
Soda ash	0.5				3	6	15	
Urea	0.25					10		
DMT	0.25				2	6		
Minerals								
Potash						20	80	
Salt				2	5	9		
Lime					4	8	15	
Calcium sulphate							13	40
Fluorspar						5	10	28
Magnesium oxide						1.5	5	
Asbestos						1.5	3	
Cement						5		
Glass batch						5	12	

*Courtesy Bepex Corporation. To convert inches to centimeters, multiply by 2.54; to convert tons to megagrams, multiply by 0.907; and to convert tons per hour to megagrams per hour, multiply by 0.907.

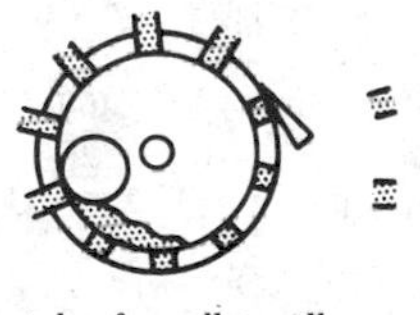

FIG. 8-68 Operating principle of a pellet mill.

essary for compaction. Adjustable knives shear the rodlike extrudates into pellets of the desired length. Although several designs are in use, the most commonly used pellet mills operate by applying power to the die and rotating it around a freely turning roller with fixed horizontal or vertical axis.

Pellet quality and capacity vary with properties of the feed such as moisture, lubricating characteristics, particle size, and abrasiveness, as well as die characteristics and speed. A readily pelleted material will yield about 122 kg/kWh [200 lb/(hp·h)] by using a die with 0.6-cm (¼-in) holes. Some characteristics of pellet mills are given in Table 8-58.

Screw extruders employ a screw to force material in a plastic state continuously through a die. If the die hole is round, a compact in the form of a rod is formed, whereas if the hole is a thin slit, a film or sheet is formed. Many other forms are also possible.

A common use of screw extruders is in the forming and compounding of plastics. Table 8-59 shows typical outputs that can be expected per horsepower for various plastics and the characteristics of several popular extruder sizes.

Deairing pug-mill extruders which combine mixing, densification, and extrusion in one operation are available for agglomerating clays, catalysts, fertilizers, etc. Table 8-60 gives data on screw extruders for the production of catalyst pellets.

TABLE 8-58 Characteristics of Pellet Mills

Horsepower range	10–250
Capacity, lb/(hp·h)	75–300
Die characteristics	
Size	Up to 26 in inside diameter × approximately 8 in wide
Speed range	75–500 r/min
Hole-size range	1/16–1¼ in inside diameter
Rollers	As many as three rolls; up to 10-in diameter

NOTE: To convert horsepower to kilowatts, multiply by 0.746; to convert pounds per horsepower-hour to kilograms per kilowatthour, multiply by 0.6; and to convert inches to centimeters, multiply by 2.54.

TUMBLING AND MIXER AGGLOMERATION

Powders with the correct amount of a liquid binder can be formed into regular agglomerates by **tumbling, vibrating, shaking,** or **paddle mixing.** Rotating drums and disks are the equipment most commonly used industrially. Binding is usually by mobile liquid (but intermolecular forces operate in the spheronizing process for carbon blacks and other very fine materials).

Tumbling agglomeration is known by several names. Fertilizers are **granulated** with solution phase and hardened by drying. Ores are **balled** or **wet-pelletized** to form a green agglomerate that may be sintered. In a process known as **nodulizing** the green pellet is mixed with the fuel and sintered.

Agglomerate Growth Agglomerate growth can take place in two fairly clear-cut ways. **Agglomeration** is the formation of aggregates by the sticking together of feed and/or recycle materials and includes the formation of agglomerate nuclei. **Layering** is the deposition of layers of the raw materials onto a previously formed nucleus (Slack, in *Encyclopedia of Chemical Technology,* vol. 9, Intersci-

TABLE 8-59 Characteristics of Plastics Extruders*

Efficiencies	lb/(hp·h)
Rigid PVC	7–10
Plasticized PVC	10–13
Impact polystyrene	8–12
ABS polymers	5–9
Low-density polyethylene	7–10
High-density polyethylene	4–8
Polypropylene	5–10
Nylon	8–12

Relation of size, power, and output

hp	Diameter, in	Diameter, mm	Output, lb/h, low-density polyethylene
15	2	45	Up to 125
25	2½	60	Up to 250
50	3½	90	Up to 450
100	4½	120	Up to 800

The Encyclopedia of Plastics Equipment, Simonds (ed.), Reinhold, New York, 1964.

NOTE: To convert inches to centimeters, multiply by 2.54; to convert horsepower to kilowatts, multiply by 0.746; to convert pounds per hour to kilograms per hour, multiply by 0.4535; and to convert pounds per horsepower-hour to kilograms per kilowatthour, multiply by 0.6.

TABLE 8-60 Characteristics of Pelletizing Screw Extruders for Catalysts*

Screw diameter, in	Drive hp	Typical capacity, lb/h
2.25		60
4	7.5–15	200–600
6	Up to 60	600–1500
8	75–100	Up to 2000

*Courtesy The Bonnot Co. To convert inches to centimeters, multiply by 2.54; to convert horsepower to kilowatts, multiply by 0.746; and to convert pounds per hour to kilograms per hour, multiply by 0.4535.

NOTE:
1. Typical feeds are high alumina, kaolin carriers, molecular sieves, and gels.
2. Water-cooled worm and barrel, variable-speed drive.
3. Die orifices as small as ¹⁄₁₆ in.
4. Vacuum-deairing option available.

ence, 1966, p. 122). The layering process requires higher recycle ratios and lower moisture contents than the agglomeration process; it can be made to predominate by the control of operating conditions.

The size distribution changes as growth proceeds in ways that are characteristic of the mechanism of growth [Sastry and Fuerstenau, *Agglomeration 77*, op. cit., p. 381]. During ball growth there is a balance between the destructive forces produced within the charge and the cohesive forces holding the pellets together. Pellet strength must be sufficient to withstand these destructive forces. Equation (8-38) indicates that there is a maximum particle size from which pellets can be formed satisfactorily under given conditions. The top size of feed to disk pelletizers is usually 30 to 50 mesh, while at least 25 percent should be minus 200 mesh [Engelleitner, *Ceram. Age*, **22**, 24 (1966)]. For iron-ore pelletization, the particulate feed normally contains 40 to 80 percent minus 325-mesh material. The particle-size specifications noted earlier apply to systems in which water is used as the binding liquid. Other liquids of lower surface tension or liquid-solid systems in which the particle surface is imperfectly wetted require finer particle sizes to make successful balling possible.

Equation (8-41) may be used to calculate, very approximately, the amount of liquid required for agglomeration from a knowledge of the bulk density of the feed material:

$$x = \frac{1}{1 + (1 - \epsilon)\rho_s/\epsilon\rho_l} \qquad (8\text{-}41)$$

where x is the weight fraction of liquid, ϵ is the porosity of close-packed material, ρ_s is the true particle density, and ρ_l is the liquid density [Capes, Germain, and Coleman, *Ind. Eng. Chem. Process Des. Dev.*, **16**, 517 (1977)].

If possible, however, the liquid requirement should be measured in a balling test on the material in question, since unusual packing and wetting effects, particle internal porosity and solubility, air inclusions, etc., may cause error. Approximate moisture requirements for balling several systems are given in Table 8-61.

TABLE 8-61 Moisture Requirements for Balling Various Materials*

Raw material	Approximate size of raw material, less than indicated mesh	Moisture content of balled product, % H_2O
Precipitated calcium carbonate	200	29.5–32.1
Hydrated lime	325	25.7–26.6
Pulverized coal	48	20.8–22.1
Calcined ammonium metavanadate	200	20.9–21.8
Lead-zinc concentrate	20	6.9–7.2
Iron pyrite calcine	100	12.2–12.8
Specular hematite concentrate	150	8.0–10.0
Taconite concentrate	150	8.7–10.1
Magnetic concentrate	325	9.8–10.2
Direct-shipping open-pit iron ores	10	10.3–10.9
Underground iron ore	¼ in.	10.4–10.7
Basic oxygen converter fume	1 μ	9.2–9.6
Raw cement meal	150	13.0–13.9
Fly ash	150	24.9–25.8
Fly ash–sewage sludge composite	150	25.7–27.1
Fly ash–clay slurry composite	150	22.4–24.9
Coal-limestone composite	100	21.3–22.8
Coal–iron ore composite	48	12.8–13.9
Iron ore–limestone composite	100	9.7–10.9
Coal–iron ore–limestone composite	14	13.3–14.8

*Dravo Corp.

Importance of Correct Rolling Action Optimum agglomeration is obtained in drum and disk equipment when the correct tumbling, cascading motion occurs in the charge. This motion is caused by centrifugal force and is related to the critical speed of the agglomerator [see Eq. (8-36)]. If the device is operated at an angle from the horizontal, then the formula for critical speed is multiplied by $\sqrt{\sin \beta}$, where β is the angle of inclination.

Tumbling Agglomerators An **inclined-pan or disk agglomerator** consists basically of an inclined rotating disk equipped with a rim to contain the agglomerating charge (Fig. 8-69). Pan angle from the

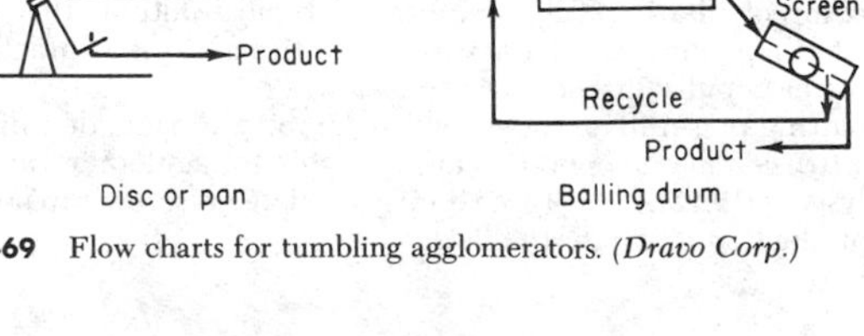

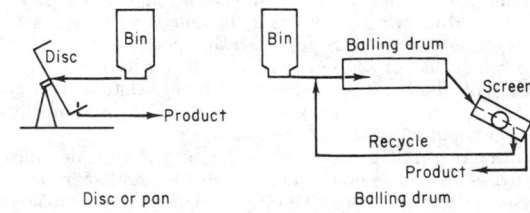

FIG. 8-69 Flow charts for tumbling agglomerators. *(Dravo Corp.)*

horizontal can be adjusted over a wide range to obtain best results, but a range of 0.78 to 0.96 rad (45 to 55°) appears to be normal for most applications. Either constant-speed or variable-speed motors are available at disk drives.

To promote the lifting and cascading of the material in the pan and prevent the whole mass from sliding, the inner surface is lined with expanded metal or an abrasive coating. A uniform cake of the processed material is maintained by adjustable scrapers and plows which may oscillate mechanically. This layer of cake protects the pan from abrasive wear and promotes the correct tumbling action. The locations of the plows, as well as the feed and liquid introduction points, are all important to the operation of the equipment and are usually adjustable to suit specific applications.

The inclined pan has a depth that is proportioned to the diameter of the disk. Typically, $H \approx 0.20D$, where H is the height of the rim and D the disk diameter. Pan depth can normally be adapted to specific applications. Variations of the simple pan shape are used, including:

1. An outer reroll ring which allows pellets to be firmed or coating agents to be applied
2. Multistepped sidewalls
3. A pelletizer in the form of a truncated cone

Balling disks range from laboratory-scale models 30 cm (1 ft) in diameter up to production models 6 m (20 ft) in diameter. There is thus a wide variation in speed, power consumption, capacity, etc. Pietsch [*Aufbereit. Tech.*, 7, 177 (1966)] has surveyed scale-up for inclined-pan agglomerators and notes that the following equations apply for the approximate design of these devices, where D is the pan diameter in meters (feet): power consumption, $P = k_1D^2$ kW (hp); throughput, $Q = k_2D^2$ Mg/h (tons/h).

For materials such as cement raw meal and iron ore, Pietsch refers to work which indicates that both k_1 and k_2 are approximately 0.13 in these equations.

It should be emphasized that these equations are best used in combination with actual experimental data on the system in question in order to indicate, approximately, the effect of scale-up. Some production-scale data collected by a manufacturer of pelletizing disks are given in Table 8-62, which may be used as a guide to capacity and power requirements for various systems.

TABLE 8-62 Characteristics of Disk Agglomerators*

	70 lb./cu. ft. material		125 lb./cu. ft. material			
	Pelletizing		Pelletizing		Mixing	
Disk size, ft.	Approx. capacity, tons/hr.	Horse-power	Approx. capacity, tons/hr.	Horse-power	Approx. capacity, tons/hr.	Horse-power
18	30	40	40	50	250	100
15	18	25	25	30	150	60
12	10	12	15	16	75	30
9	5	6	10	$7\frac{1}{2}$	30	20
6	3	3	5	5	10	10
$3\frac{1}{4}$	$\frac{1}{2}$	1	1	1	1	1

*Dravo Corp. To convert feet to centimeters, multiply by 30.5; to convert tons per hour to megagrams per hour, multiply by 0.907; to convert horsepower to kilowatts, multiply by 0.746; and to convert pounds per cubic foot to kilograms per cubic meter, multiply by 16.

A **drum agglomerator** consists of an inclined rotary cylinder powered by a variable-speed drive (Fig. 8-69). The pitch of the drum, which may be up to about 0.175 rad (10°) from the horizontal, is sufficient to ensure movement of material down the length of the drum. The cylinder may be either open-ended or fitted with annular retaining rings at each end. Feed to the drum may be moistened either before its introduction or in the feed end of the cylinder. In the former case, various mixers are used to form ball nuclei, which then pass to the drum. In the latter case, liquid may be sprayed onto the tumbling load or may be introduced under the bed by means of horizontal multiple-outlet distributor pipes such as are used in the ammoniation of fertilizer blends.

As with the inclined pan, the inside surface of the drum must be rough to ensure proper tumbling action. A uniform buildup of feed material is maintained by scraper or cutter bars which run longitudinally down the inside of the cylinder. Characteristics of some typical drum agglomerators are given in Table 8-63. The length of the drum is an important design consideration, since this influences the retention time of material in the equipment. Difficult systems require longer retention times than those that agglomerate more readily. An average figure is 1- to 2-min retention time.

A variation of the basic cylindrical shape, the multicone drum which contains a series of compartments formed by annular baffles, is in use [Stirling, in Knepper (ed.), *Agglomeration*, op. cit., p. 177].

Relative Merits of Inclined-Pan and Drum Agglomerators No set rules have yet been formulated for determining which type of balling device should be used in a given situation. The final choice rests on a careful consideration of the particular application by individuals experienced in the field.

The principal difference between pan and drum agglomerators is the classifying effect of the pan. Under centrifugal action, the agglomerating material in a pan travels in spiral rings of decreasing diameter until balls of the required size discharge over the lip. Fine material sifts down through the larger balls and remains in the pan. Inclined pans normally give a sufficiently uniform product so that they can be operated without screens, whereas drums, as seen in Fig. 8-69, are usually closed-circuited with screens; undersize and oversize product is recycled.

Other advantages claimed for the inclined-pan agglomerator include low equipment cost, sensitivity to operating controls, and easy observation of the balling action, all of which lend versatility in agglomerating many different materials into agglomerates from 0.16 to 4 cm ($\frac{1}{16}$ to $1\frac{1}{2}$ in) in diameter. Disadvantages are that dusty materials and chemical reactions such as the ammoniation of fertilizer can be handled less readily in the pan agglomerator than in the drum.

Advantages claimed for the drum agglomerator over the pan are greater capacity, longer retention time for materials difficult to agglomerate, and less sensitivity to upsets in the system due to the damping effect of a large recirculating load.

System variables and operating variables affect the process of agglomeration by tumbling. With proper control of these variables it is possible, within limits, to influence agglomerate properties such as **size, shape,** and **porosity.** A summary of these factors has been given by Pietsch [*Aufbereit. Tech.*, 7, 177 (1966)], with particular reference to the operation of pan granulators.

The two most important operating variables are liquid content and retention time in the device. In the normal operating range an increase in liquid leads to an approximately exponential increase in agglomerate size. Thus the process is very sensitive to liquid content,

TABLE 8-63 Characteristics of Typical Drum Agglomerators

Application	Diameter, ft.	Length, ft.	Speed, r.p.m.	Production capacity,* tons/hr.	Installed horsepower
Granulation of fertilizers	5–11	7–25	15–9	15–40	25–100
Balling of iron ore	9–10	25–30	13–12	30–35	50–60

*This figure excludes recycle; actual drum throughput may be two to seven times this value.

NOTE: To convert feet to centimeters, multiply by 30.5; to convert tons per hour to megagrams per hour, multiply by 0.907; and to convert horsepower to kilowatts, mulitply by 0.746.

which, in turn, is a very effective control variable. For materials containing soluble constituents, such as fertilizer formulations, the total solution phase is the controlling factor, and not simply the amount of water used. In addition, solubility is affected by temperature.

Increased retention time leads to larger and denser agglomerates of higher wet strength. Interaction of agglomerate size, retention time, throughput rate, and moisture content is shown qualitatively in Fig. 8-70. Numerical values vary with the particular application. With the inclined-pan agglomerator retention time can also be increased by increasing pan depth and speed or by decreasing pan slope. Retention time in drum agglomerators can be influenced by internal baffling, as in the multicone drum previously referred to.

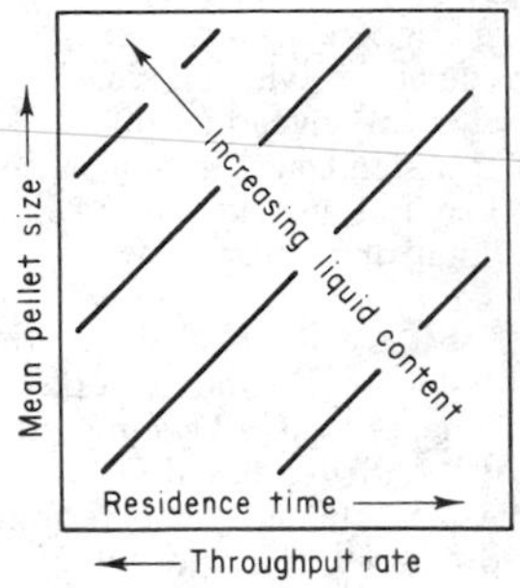

FIG. 8-70 Relationship between mean pellet size and residence time (throughput rate) as a function of liquid content for pan agglomerators. [*Adapted from Pietsch,* Aufbereit. Tech., *7, 177 (1966).*]

A uniform product with constant properties is obtained only when the agglomeration device is operated uniformly, with a constant feed of raw material of uniform properties as well as constant liquid flow rate (or a flow rate in constant proportion to the solids rate).

Mixer Agglomerators Many types of solid-liquid mixers can be used to form agglomerates. Control of the amount of liquid phase and the intensity and duration of mixing determine agglomerate size.

Pug mills and other **intensive mixers** use agitators and other tools to provide a positive rubbing and shearing action. Advantages claimed for this method over tumbling include harder, stronger agglomerates owing to the kneading action, the ability to process plastic sticky materials, and greater tolerance in accommodating variations in operating conditions (Hignett, in Sauchelli (ed.), *Chemistry and Technology of Fertilizers,* p. 269). Because of the greater compaction achieved, less liquid phase is required for agglomeration in a mixer than in a tumbling device [Sherrington, *Chem. Eng. (London),* no. 220, CE201 (1968)]. Disadvantages include generally higher maintenance and power requirements and an irregular product form which may require further shaping, such as in a tumbling dryer.

Pug mills (blungers, pug mixers, and paddle mixers) consist of a horizontal trough or troughs containing a mixing shaft or shafts. Attached to the shaft are mixing blades of bar, rod, paddle, and other designs. The vessel may be of single- or double-trough design, although the latter configuration is more popular. Twin shafts rotate in opposite directions, throwing the materials forward and to the center as the pitched blades on the shaft pass through the charge. Characteristics of a range of pug mixers available for fertilizer granulation are given in Table 8-64.

In applications such as the preparation of tableting feeds, the manufacture of detergent powders, and the preparation of "instant" food products, the aim is to produce small agglomerates (usually 2-mm diameter and less) with improved flow, wetting, dispersing, or dissolution properties. This is accomplished by superficially wetting the feed powder, often with less than 5 percent of bridging liquid in the form of a spray, steam, mist, etc. The wetting is carried out in a relatively dry state in standard or specialized powder mixers in which the mass becomes moist rather than wet or pasty.

TABLE 8-64 Characteristics of Pug Mixers for Fertilizer Granulation*

Model	Material bulk density, lb/ft³	Approximate capacity, tons/h	Size (width × length), ft	Plate thickness, in	Shaft diameter, in	Speed, r/min	Drive, hp
A	25	8	2 × 8	¼	3	56	15
	50	15	2 × 8	¼	3	56	20
	75	22	2 × 8	¼	3	56	25
	100	30	2 × 8	¼	3	56	30
B	25	30	4 × 8	⅜	4	56	30
	50	60	4 × 8	⅜	4	56	50
	75	90	4 × 8	⅜	4	56	75
	100	120	4 × 8	⅜	5	56	100
C	25	30	4 × 12	⅜	5	56	50
	50	60	4 × 12	⅜	5	56	100
	75	90	4 × 12	⅜	6	56	150
	100	120	4 × 12	⅜	6	56	200
	125	180	4 × 12	⅜	7	56	300

*Feeco International, Inc. To convert pounds per cubic foot to kilograms per cubic meter, multiply by 16; to convert tons per hour to megagrams per hour, multiply by 0.907; to convert feet to centimeters, multiply by 30.5; to convert inches to centimeters, multiply by 2.54; and to convert horsepower to kilowatts, multiply by 0.746.

A continuous **flow-mixing system** used to cluster-agglomerate a wide range of food powders is represented in Fig. 8-71. The feed powder is introduced to the moistening zone by means of a pneumatic conveyor and a rotary valve. The dry powder falls in a narrow stream between two jet tubes, which inject the agglomerating fluid in a highly dispersed state. Steam, water, or solvents or a combination of these are used. In addition, air at ambient temperature is introduced through radial wall slots in the moistening chamber to induce a vortex motion. Control of this air flow controls the flow pattern and particle temperature. The reduced temperature serves to condense fluid onto the particles, while the vortexing motion induces particle-particle collisions. The clustered material then drops through an air-heated chamber onto a conditioning conveyor, where it is allowed sufficient time to reach a uniform moisture distribution. The material then passes to an afterdryer, cooler, and sifter, followed by bagging of the selected product.

Other types of mixer agglomerators are described by Capes (*Particle Size Enlargement,* op. cit., p. 83).

THERMAL PROCESSES

Bonding and agglomeration by temperature elevation or reduction are applied either in conjunction with other size-enlargement processes or as a separate process. Agglomeration occurs through one or more of the following mechanisms:

1. Drying of a concentrated slurry or wet mass of fines
2. Fusion
3. High-temperature chemical reaction
4. Solidification and/or crystallization of a melt or concentrated slurry during cooling

Sintering and Heat Hardening In powder metallurgy compacts are sintered with or without the addition of binders. In ore processing the agglomerated mixture is either sintered or indurated. **Sintering** refers to a process in which fuel is mixed with the ore and burned on a grate. The product is a porous cake. **Induration,** or **heat hardening,** is accomplished by combustion of gases passed through the bed. The aim is to harden the pellets without fusing them together, as is done in the sintering process.

Ceramic bond formation and grain growth by diffusion are the two prominent reactions for bonding at the high temperature (1100 to 1370°C, or 2000 to 2500°F, for iron ore) employed.

In addition to agglomeration, other useful processes may occur during sintering and heat hardening. For example, carbonates and sulfates may be decomposed, or sulfur may be eliminated. Although the major application is in ore beneficiation, other applications, such as the preparation of lightweight aggregate from fly ash and the formation of clinker from cement raw meal, are also possible. Nonferrous sinter is produced from oxides and sulfides of manganese, zinc, lead, and nickel. An excellent account of the many possible appli-

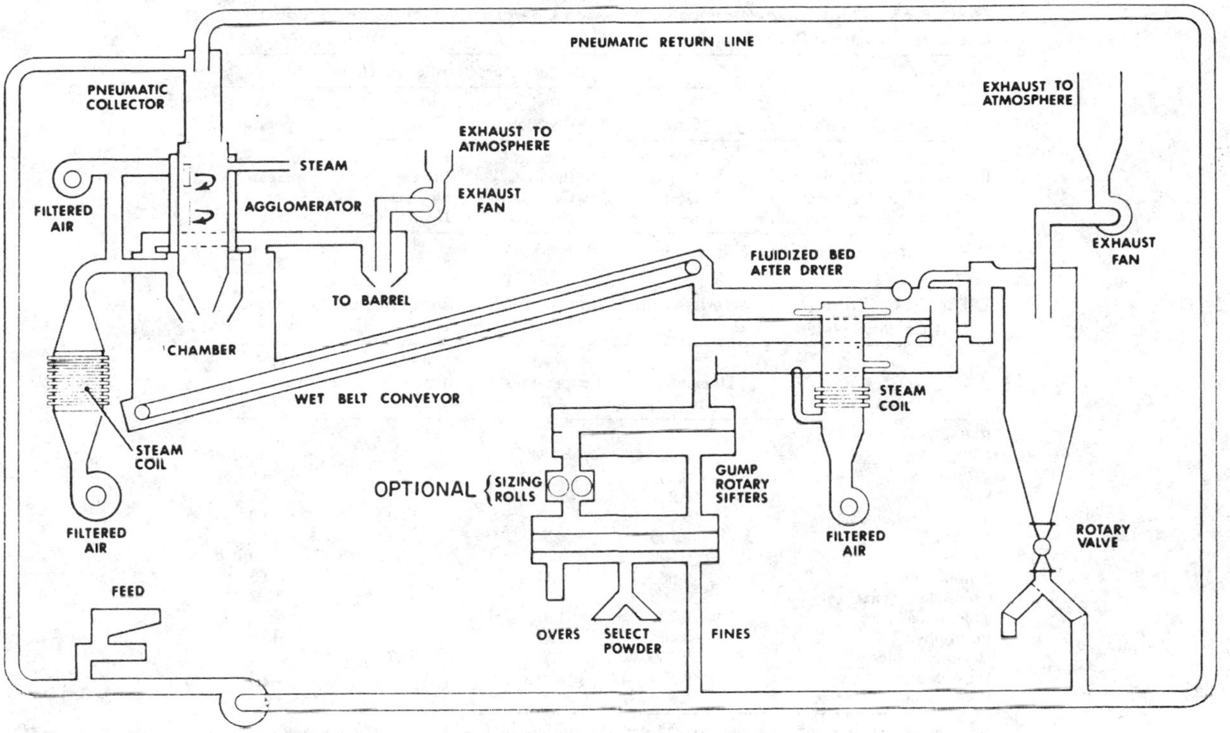

FIG. 8-71 Flow-diagram of Blaw-Knox instantizer-agglomerator. *(Courtesy Blaw-Knox Food & Chemical Equipment Co.)*

cations is given by Ban et al. [Knepper (ed.), *Agglomeration*, op. cit., p. 511]. The highest tonnage use at present is in the beneficiation of iron ore.

The machine most commonly used for sintering iron ores is a **traveling grate** (Fig. 8-72), which is a modification of the Dwight-Lloyd continuous sintering machine formerly used only in the lead and zinc industries. Modern sintering machines may be 4 m (13 ft) wide by 60 m (200 ft) long and have capacities of 7200 Mg/day (8000 tons/day).

The productive capacity of a sintering strand is related directly to the rate at which the burning zone moves downward through the bed. This rate, which is of the order of 2.5 cm/min (1 in/min), is controlled by the air rate through the bed, with the air functioning as the heat-transfer medium.

Heat hardening of green iron-ore pellets is accomplished in a vertical shaft furnace, a traveling-grate machine, or a grate-plus-kiln combination (see Ball et al., op. cit.). Nodulizing of phosphate rock and iron ores to prepare them for subsequent processing is noted in Sec. 20 in the discussion of direct-fired rotary kilns.

Drying and Solidification Granular free-flowing solid products are often an important result of the drying of concentrated slurries and pastes and the cooling of melts. Size enlargement of originally finely divided solids results. Pressure agglomeration including extrusion, pelleting, and briquetting is used to preform wet material into forms suitable for drying in through-circulation and other types of dryers. Details are given in Sec. 20 in the account of solids-drying equipment.

Rotating-drum-type and **belt-type** heat-transfer equipment forms granular products directly from fluid pastes and melts without intermediate preforms. These processes are described in Sec. 11 as examples of indirect heat transfer to and from the solid phase. When solidification results from melt freezing, the operation is known as

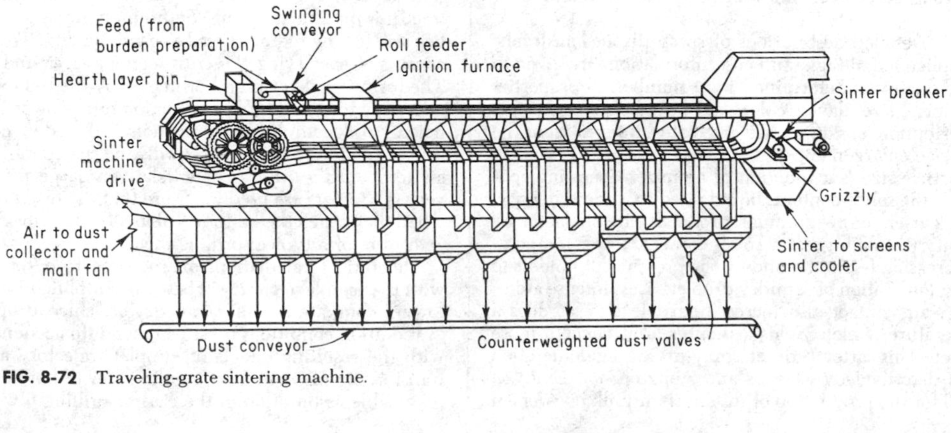

FIG. 8-72 Traveling-grate sintering machine.

TABLE 8-65 Qualitative Relationship between Operating Variables and Product Characteristics for a Drum Flaker*

	Product characteristics affected			
Operating variables	Capacity	Flake thickness	Flake size	Flake temperature
Increased drum speed	Increase	Decrease	Decrease	Increase
Increased drum immersion	Increase	Increase	Increase	Increase
Increased feed temperature	Decrease	Decrease	Decrease	Increase
Increased cooling temperature	Decrease	Decrease	Decrease	Increase

*Blaw-Knox Food & Chemical Equipment Co., Buflovak Division.

TABLE 8-66 Product Characteristics and Capacity Data for Some Materials Treated in Belt Cooling Systems*

Product	Thickness, in	Feed temperature, °F	Discharge temperature, °F	Capacity, $lb/(h \cdot ft^2)$
Resins				
Phenolic	0.062	275	110	46
Phenolic	0.048–0.051	280	92	56.8
Sulfur	0.25	290	150	55
Tetrachlorobenzene	0.06	320	70	90
Asphalt	0.125	425	125	18.5
Urea	0.093	375	140	39
Ammonium nitrate	0.063	400	160	90
Chlorinated wax	0.063	300	100	62
Sodium acetate	0.125	180	100	37.5
Butyl phenol	0.050–0.068	230	97	46.5
Hot-melt adhesive	0.436	330	103	14.4
Wax blend	0.024	270	85	26.4
Epoxy resin	0.040	350	100	40

*Sandvik, Inc. To convert inches to centimeters, multiply by 2.54; to convert pounds per hour per square foot to kilograms per hour per square meter, multiply by 4.88; °C = (°F − 32) × 5/9.

flaking. If evaporation occurs, solidification is by drying. Production capacity and characteristics of the finished product vary between materials and have a complex relationship (see Table 8-65 and Fig. 11-31 in Sec. 11). Some typical products treated in belt cooling systems (Fig. 11-32 in Sec. 11) are listed in Table 8-66, together with product characteristics and capacity data.

SPRAY METHODS

Spray dryers, prilling towers, spouted and fluid bed dryers, and flash dryers can all be used to produce granular solids from fines as either a primary or a secondary function. Feed solids in a fluid state (solution, gel, paste, emulsion, slurry, or melt) are dispersed in a gas and converted to granular solid products by heat and/or mass transfer. These methods are readily automated to continuous large-scale operation. Product diameter is limited to a maximum of about 5 mm and is often much smaller. In addition to the requirement of a pumpable and dispersible feed, it should also be noted that attrition is usually a problem, so that carry-over fines must be recovered and/or recycled.

Spray Drying Detailed descriptions of spray dispersion dryers, together with application, design, and cost information, are given in Sec. 20. Product quality is determined by a number of properties such as particle form, size, flavor, color, and heat stability. Particle size and size distribution, of course, are of greatest interest from the point of view of size enlargement.

In general, particle size is a function of atomizer-operating conditions and also of the solids content, liquid viscosity, liquid density, and feed rate. Coarser, more granular products can be made by increasing viscosity (through greater solids content, lower temperature, etc.), by increasing feed rate, and by the presence of binders to produce more agglomeration of semidry droplets. Less intense atomization and spray-air contact also increase particle size, as does a lower exit temperature, which yields a moister (and hence a more coherent) product. This latter type of spray-drying agglomeration system has been described by Masters and Stoltze [*Food Eng.*, 64 (February 1973)] for the production of instant skim-milk powders in which the completion of drying and cooling takes place in vibrating conveyors (see Sec. 20) downstream of the spray dryer.

Prilling The prilling process is similar to spray drying and consists of spraying droplets of liquid into the top of a tower and allowing these to fall against a countercurrent stream of atmospheric air. During their fall the droplets are solidified into approximately spherical particles or prills which are up to about 3 mm in diameter, or larger than those formed in spray drying. The process also differs from spray drying since the droplets are formed from a melt which solidifies primarily by cooling with little, if any, contribution from drying. Traditionally, ammonium nitrate, urea, and other materials of low viscosity and melting point and high surface tension have been treated in this way. Improvements in the process now allow viscous and high-melting point materials and slurries containing undissolved solids to be treated as well.

The design of a prilling unit first must take into account the properties of the material and its sprayability before the tower design can proceed. By using data on the melting point, viscosity, surface tension, etc., of the material, together with laboratory-scale spraying tests, it is possible to specify optimum temperature, pressure, and orifice size for the required prill size and quality. Tower sizing basically consists of specifying the cross-sectional area and the height of fall. The former is determined primarily by the number of spray nozzles necessary for the desired production rate. Tower height must be sufficient to accomplish solidification and is dependent on the heat-transfer characteristics of the prills and the operating conditions (e.g., air temperature). Because of relatively large prill size, narrow but very tall towers are used to ensure that the prills are sufficiently solid when they reach the bottom. Table 8-67 describes the principal characteristics of a typical prilling tower.

Theoretical calculations are possible to determine tower height with reasonable accuracy. These are simplified in the case of prilling towers compared with spray dryers, since droplet motion under evaporation conditions must be taken into account in the latter case, with the resulting effects on droplet trajectory and heat and mass transfer. Simple parallel streamline flow of both droplets and air is a reasonable assumption in the case of prilling towers compared with

TABLE 8-67 Some Characteristics of a Typical Prilling Operation*

Tower size		
Prill tube height, ft	130	
Rectangular cross section, ft	11 by 21.4	
Cooling air		
Rate, lb/h	360,000	
Inlet temperature	Ambient	
Temperature rise, °F	15	
Melt		
Type	Urea	Ammonium nitrate
Rate, lb/h	35,200 (190 lb H_2O)	43,720 (90 lb H_2O)
Inlet temperature, °F	275	365
Prills		
Outlet temperature, °F	120	225
Size, mm	Approximately 1 to 3	

*HPD Incorporated. To convert feet to centimeters, multiply by 30.5; to convert pounds per hour to kilograms per hour, multiply by 0.4535; °C = (°F − 32) × 5/9.

the more complex rotational flows produced in spray dryers. For velocity of fall, see, for example, Becker [*Can. J. Chem.*, **37,** 85 (1959)]. For heat transfer, see, for example, Kramers [*Physica*, **12,** 61 (1946)]. Specific design procedures for prilling towers are available in the *Proceedings of the Fertilizer Society (England);* see Berg and Hallie, no. 59, 1960; and Carter and Roberts, no. 110, 1969.

Fluid- and Spouted-Bed Granulation The design and uses of fluidized beds are described in detail in Sec. 20. Size enlargement occurs when atomized feed liquid (solution, melt, slurry, etc.) impinges on hot bed material and solids are deposited by drying, together with chemical reaction in some cases.

In a fluidized bed, liquid may be sprayed either onto the surface of the bed or directly into the bed. Depending on operating conditions, two particle-growth mechanisms are possible: (1) layering of solid onto individual particles and (2) agglomeration of several particles into a larger particle. In a spouted bed liquid is injected as a spray into the base together with the hot spouting gas so that thin solids layers deposit onto the recirculating seed particles.

Because multiple layers can be deposited on the seed particles in these systems, coarser granular products can be made compared with those produced in spray dryers. The product is thus less dusty, and the longer residence times possible mean that larger dryer loads with less concentrated feed liquors can be handled. Since the drying particles are less dispersed in fluid beds, smaller equipment is needed [Pictor, *Process Eng.*, 66 (June 1974)].

The ease of remote control and the absence of moving parts with attendant seals, packing, etc., make the fluid-bed calcination process attractive in the disposal of liquid radioactive nuclear-reactor fuel wastes [Lee et al., *Am. Inst. Chem. Eng. J.*, **8,** 53 (1962)]. Table 8-68 contains information on the fluid-bed incineration process for the disposal of waste sludge to yield a granular ash by-product. The fluid-bed granulation process [Scott et al., *J. Pharm. Sci.*, **53,** 314 (1964)] combines granulation, blending, and drying into one operation in the production of tablet granulations in the pharmaceutical industry. Advantages claimed over conventional wet granulation include good temperature control for the processing of heat-sensitive materials, less material handling, and shorter operating cycles.

For further details on fluid-bed spray granulators, see Davies and Goor [*J. Pharm. Sci.*, **60,** 1869 (1971); **61,** 618 (1972)] and Mortensen and Hovmand [in Keairns (ed.), *Fluidization Technology*, vol. II, Hemisphere Publishing, Washington, 1976, p. 519]. For details on spouted-bed systems, see Mathur and Epstein (*Spouted Beds*, Academic, New York, 1974).

Flash Drying Special designs of pneumatic-conveyor dryers, described in Sec. 20, can handle filter and centrifuge cakes and other sticky or pasty feeds to yield granular size-enlarged products. The dry product is recycled and mixed with fresh cohesive feed, followed by disintegration and dispersion of the mixed feed in the drying-air stream.

LIQUID SYSTEMS

Traditional flocculation procedures (using electrolytes, trivalent metal ions, or polymeric flocculants) rely on relatively small interparticle forces to form rather weak, voluminous cluster agglomerates (cf. Sec. 19: "Gravity Sedimentation Operations"). Methods summarized here use stronger interparticle bonding and specialized equipment to form larger and more permanent agglomerates in liquid suspensions.

Wetting with Immiscible Liquid Finely divided solids in liquid suspension can be agglomerated and separated from the suspending liquid by the addition of a small amount of a second liquid which preferentially wets the solid and is immiscible with the first liquid. Under appropriate agitation, such as that in a rotating drum or a propeller mixer, compact spherical agglomerates of the solid particles held together by the second liquid are formed. The agglomerates are readily and completely separated from the suspending liquid and, in addition, have good handling properties.

The simplest application of this process is in the simultaneous separation and pelletization of a single solid component in liquid suspension. Zweiderweg and Van Lookeren Compagne [*Chem. Eng. (London)*, CE223 (July–August 1968)] have described the purification of water containing 1 to 3 percent soot in which a small quantity of oil was used to form soot pellets which were readily separated, yielding water with ~5 ppm of soot. Agitation was provided by a system of two coaxial cylinders of which the inner one could rotate. Operating data were given.

Because preferential wetting of the solid surface by a second liquid is a basic requirement of the process, it is possible to separate multicomponent systems by selective agglomeration. This can be accomplished with the aid of surface-conditioning agents (as in flotation) to allow the desired components to be wetted by the bridging liquid. Possible applications are summarized by Capes et al. [in Sastry (ed.), *Agglomeration 77*, op. cit., p. 910].

Probably the most important application has been the recovery of fine coal in water suspension by agglomeration with oil in the coal-cleaning industry. The coal is simultaneously recovered, dewatered, and purified when ash impurities, which are not wetted by the oil, remain in the water suspension. Examples include the Trent process [Perrott and Kinney, *Chem. Metall., Eng.*, **25,** 182 (1921)], the Con-

TABLE 8-68 Granular Products from Fluidized-Bed Incineration*

Type of sludge	Incinerator size	Bed temperature	Capacity	Granular-product composition
Oil-refinery waste sludge (85–95% water)	40 ft high; 20 ft inside diameter at base, increasing to 28 ft at top	1330°F	31×10^3 lb/h of sludge	Start-up material was silica sand; replaced by nodules of various ash components such as $CaSO_4$, Na, Ca, Mg silicates, and Al_2O_3 after operation of incinerator
Paper-mill* waste liquor (40% solids)	20 ft inside diameter at top	1350°F	31×10^3 lb/h	Sulfur added to produce 90 to 95% Na_2SO_4 and some Na_2CO_3

*Wall et al., *Chem. Eng.*, **82**(8), 77 (1975). To convert feet to centimeters, multiply by 30.5; to convert pounds per hour to kilograms per hour, multiply by 0.4535; °C = (°F − 32) × 5/9.

vertol process [Brisse and McMorris, *Min. Eng.*, **10**, 258 (1958)], and the spherical agglomeration process (Capes et al., op. cit., p. 910), which have all been developed and used in this century.

Oil-coal contacting is critical to this coal-agglomeration process. Standard turbine agitators or special high-shear mixers are typically used to disperse the oil and to contact oil droplets and coal particles so that agglomerates can form between oil-coated particles. The required intensity and duration of mixing are determined by the oil and coal characteristics and by the solids concentration and oil usage. Predispersion of the oil as an emulsion appears to be helpful. The range of operating conditions used in the oil agglomeration of fine bituminous coals can be summarized:

Coal-water feed slurry	
Weight % solids	3 to 50
Particle size	Typically, minus 200 mesh
Ash content, weight % dry basis	10 to 50
Oil usage (light fuel oil), % of solids weight	2 to 30
Turbine agitator	
Tip speed, m/s	~10 to 30
Mixing time	30 s to several min
Power consumption, kW/m^3	~10 to 40
Product agglomerates	
Weight % recovery of solid combustible matter	>90
Ash content, weight % dry basis	5 to 10

Sol-Gel Process Kelly et al. [*Ind. Eng. Chem. Process Des. Dev.*, **4**, 212 (1965)] have described the use of this process in the production of uranium dicarbide particles (approximately spherical in shape with diameters of 100 to 300 μm), which are of interest as fuel material in the nuclear power field. High-temperature techniques and mechanical shaping of the hard dicarbides were thus avoided.

The basic steps involved are preparation of the sol, formation of droplets of sol in a liquid, and setting the sol droplets into spherical gel particles. For example, an aqueous sol may be dispersed into droplets by agitation in carbon tetrachloride. Isopropyl alcohol (15 percent by weight) added to the dispersion then gels the droplets by extracting some of their water into the organic phase. Normally, heat-hardening or high-temperature reaction follows the formation of the gel particles.

The following properties are required of the medium in which the sol droplets are dispersed:

1. Immiscibility with the sol phase
2. High interfacial tension between the two phases
3. Ability to gel or partially set the sol particles
4. Fluid properties that yield a satisfactory settling rate for the particles

On a small scale the dispersion operation can take place in an agitated vessel. For larger operation a spray-column technique is possible.

Pellet Flocculation This technique combines relatively large amounts of polymeric flocculants with gentle rolling mixing to consolidate settled flocs into compact agglomeratelike sludges of low liquid content. Choice of flocculant is of prime importance in these processes. Criteria for choosing a flocculant include the degree of floc formation and effect on water clarification, the amount of water contained in the settled and dewatered flocs, and the dosage required in terms of cost per unit weight of dry solids. Organic polyelectrolytes provide the best results with many materials, and it can be anticipated that a cationic flocculant will be most useful with organic sludges while an anionic or nonionic flocculant will be best for inorganic and mineral sludges (*Flocpress*, Bull. DB845, Infilco Degremont Inc., September 1976).

One type of equipment effective in forming these densified flocculated agglomerates has been developed in Japan [*Dehydrum Continuous Pelletizing Dehydrator*, Ebara-Infilco Co. Ltd., Tokyo]. A slowly revolving (1-m/min-peripheral-speed) horizontal drum is used to treat the preflocculated solids. The drum interior is divided into three sections for successively pelletizing, decanting, and consolidating the sludge. Typical sludge-treating capacities for a 2.4-m-diameter unit are 5.4 to 6.3 Mg/h (6 to 9 tons/h) for gravel-waste sludge, 1.3 to 2.0 Mg/h (1.4 to 2.2 tons/h) for a dredged-mud sludge, and 0.36 Mg/h (0.4 tons/h) for a mixed-waste sludge from an automobile factory.

Other approaches involve rotary drums and related devices to dewater and consolidate initially the flocculated solids, followed by drainage on a filter belt and/or pressing between filter belts.

Section 9

Energy Utilization, Conversion, and Resource Conservation

Richard C. Corey, B.S., *Department Staff (Retired), Energy Systems Engineering Department, The MITRE Corporation, Metrek Division; Member, American Society of Mechanical Engineers and Air Pollution Control Association. (Solid and Gaseous Fuels, Combustion, Fuel and Energy Costs, Coal Conversion, Fired Process Equipment, Heat Transport and Regeneration). (Section Editor)*

Richard Barrett, M.S., *Projects Manager, Battelle Memorial Institute, Columbus Laboratories; Member, American Society of Mechanical Engineers, American Society of Heating, Refrigerating, and Air Conditioning Engineers; Institute of Energy (United Kingdom); Registered Professional Engineer (Ohio). (Steam Systems)*

Robert C. Amero, B.S., *Staff Engineer, Gulf Science and Technology Company (Retired); Member, American Society of Mechanical Engineers, American Institute of Chemical Engineers, International Association for Hydrogen Energy; Registered Professional Engineer (Pennsylvania). (Liquid Fuels, Combustion)*

Harold F. Chambers, Jr., Ph.D., *Supervisory Mechanical Engineer, Pittsburgh Energy Technology Center, U.S. Department of Energy; Registered Professional Engineer (Ohio). (Coal Liquefaction)*

Ezekail L. Clark, B.S., *Consultant; Fellow, American Institute of Chemical Engineers; Member, American Chemical Society, American Association for the Advancement of Science; Registered Professional Engineer (Pennsylvania). (Associate Section Editor; Coal Gasification)*

Neil H. Coates, B.S., *Department Head, Energy Systems Engineering Department, The MITRE Corporation, Metrek Division. (Fluidized-Bed Combustion)*

Willard E. Fraize, Sc.D., *Senior Energy Systems Engineer, Energy and Resources Division, The MITRE Corporation, Metrek Division; Member, American Society of Mechanical Engineers, American Institute of Aeronautics and Astronautics. (Cogeneration)*

Yuan C. Fu, Ph.D., *Project Manager, Pittsburgh Energy Technology Center, U.S. Department of Energy; Member, American Chemical Society, The Chemical Society of Japan, Catalysis Society. (Coal Liquefaction)*

H. A. Grabowski, B.S., *Senior Engineering Consultant, C-E Environmental Systems, Combustion Engineering Inc.; Member, American Society of Mechanical Engineers, American Society for Testing and Materials. (Steam Generators)*

Eugene Mezey, Ph.D., *Senior Chemist, Battelle Memorial Institute, Columbus Laboratories; Member, American Chemical Society, The Society of Sigma Xi, International Microwave Power Institute, American Ceramic Institute; Fellow, American Association for the Advancement of Science. (Electric Heating)*

David E. Stutz, B.S., *Research Scientist, Battelle Memorial Institute, Columbus Laboratories; Member, International Microwave Power Institute. (Electric Heating)*

FUELS

HEAT GENERATION

HEAT TRANSPORT

HEAT REGENERATION

Nomenclature and Units

Symbol	Definition	SI units	U.S. customary units
Ar	Archimedes number	Dimensionless	Dimensionless
D	Diameter	m	ft
D	Rate of power input	J/s	Btu/h
d	Depth	cm	ft
E	Electric field	V/cm	V/in
g	Acceleration of gravity	m/s^2	ft/s^2
h	Isentropic enthalpy	kJ/kg	Btu/lb
L	Height	m	ft
P	Pressure	Pa	lbf/in^2
P	Rate of power input	W/cm^3	W/in^3
p	Resistivity	Ω·cm	Ω·ft
Q	Heating value	J/kg	Btu/lb
Re	Reynolds number	Dimensionless	Dimensionless
R	Power-to-heat ratio	Dimensionless	Dimensionless
U	Velocity	m/s	ft/s
v	Velocity of light	cm/s	in/s
W	Flow	kg/h	lb/h
Z	Supercompressibility factor		
	Greek symbols		
ϵ	Voidage	Dimensionless	Dimensionless
η	Thermal efficiency	Dimensionless	Dimensionless
μ	Viscosity	kg/(m·s)	lb/(ft·s)
μ	Relative magnetic permeability	Dimensionless	Dimensionless
ρ	Density	kg/m^3	lb/ft^3
ϕ	Shape factor		

FUELS

RESOURCES AND RESERVES

The resources and reserves of the principal fossil fuels in the United States—coal, petroleum, and natural gas—follow:

Fuel	Known reserves	Potential economic resources	Submarginal resources
	Quintillion (10^{18}) Btu		
Coal	4.8	3.0	25.0
Petroleum liquids (crude and natural-gas liquids)	0.26	2.7	14.0
Natural gas	0.30	2.1	4.5

Conversion factors		
Bituminous and anthracite coal	30.2 J/kg	(26×10^6 Btu/ton)
Lignite and subbituminous coal	23.2 J/kg	(20×10^6 Btu/ton)
Crude oil	38.4 J/L	(5.8×10^6 Btu/bbl)*
Natural-gas liquids	30.5 J/L	(4.6×10^6 Btu/bbl)*
Natural gas	36×10^6 J/m^3	(1032 Btu/ft^3)

* 1 bbl = 42 gal = 159 L = 0.159 m^3.

SOLID FUELS

Coal

GENERAL REFERENCES: Elliott (ed.), *Chemistry of Coal Utilization*, 2d suppl. vol., Wiley, New York, 1981. Van Krevelen, *Coal*, Elsevier, Amsterdam, 1961.

Origin Coal originated from the arrested decay of the remains of trees, bushes, ferns, mosses, vines, and other forms of plant life which flourished in huge swamps and bogs many millions of years ago during prolonged periods of humid, tropical climate and abundant rainfall. The precursor of coal was peat, which was formed by bacterial and chemical action on the plant debris. Subsequent actions of heat, pressure, and other physical phenomena metamorphosed the peat to the various ranks of coal as we know them today. Because of the various degrees of the metamorphic changes during this process, coal is not a uniform substance; no two coals are ever the same in every respect.

Classification of Coal Coals are classified by rank, i.e., according to the degree of metamorphism in the series from lignite to anthracite. Table 9-1 shows the classification system adopted by the American Society for Testing and Materials, D388-77. The heating value on the moist, mineral-matter-free (mmf) basis and the fixed carbon, on the dry mmf basis, are the bases of this system. The lower-rank coals are classified according to the heating value, J/kg (Btu/lb), on the moist mmf basis. The agglomerating character is used to differentiate between adjacent groups. Coals are considered agglomerating if the coke button remaining from the test for volatile matter will support a weight of 500 g or if the button swells or has a porous cell structure.

The Parr formulas, Eqs (9-1) to (9-3), or the approximation formulas, Eqs. (9-4) and (9-5), are used for classifying coals according to rank. The Parr formulas are employed in litigation cases.

$$F' = \frac{100(F - 0.15S)}{100 - (M + 1.08A + 0.55S)} \tag{9-1}$$

TABLE 9-1 Classification of Coals by Rank[a]

Class	Group	Abbreviation	Fixed carbon limits, % (dry mmf basis): Equal or greater than	Fixed carbon: Less than	Volatile matter limits, % (dry mmf basis): Greater than	Volatile matter: Equal or less than	Calorific value limits, Btu/lb[b] (moist[c] mmf basis): Equal or greater than	Calorific value: Less than	Agglomerating character
Anthracitic	Metaanthracite	ma	98	...	...	2			Nonagglomerating
	Anthracite	an	92	98	2	8			
	Semianthracite[d]	sa	86	92	8	14			
Bituminous	Low-volatile bituminous coal	lvb	78	86	14	22			Commonly agglomerating[f]
	Medium-volatile bituminous coal	mvb	69	78	22	31			
	High-volatile A bituminous coal	hvAb	...	69	31	...	14,000[e]		
	High-volatile B bituminous coal	hvBb	...	...	...	...	13,000[e]	14,000	
	High-volatile C bituminous coal	hbCb	...	...	...	...	11,500	13,000	
							10,500	11,500	Agglomerating
Subbituminous	Subbituminous A coal	SubA	...	...	...	...	10,500	11,500	Nonagglomerating
	Subbituminous B coal	SubB	...	...	...	...	9,500	10,500	
	Subbituminous C coal	SubC	...	...	...	...	8,300	9,500	
Lignitic	Lignite A	ligA	...	...	...	...	6,300	8,300	
	Lignite B	ligB	...	...	...	...	...	6,300	

[a] *Annual Book of ASTM Standards*, part 26, D 388-77, 1977. This classification does not include a few coals, principally nonbanded varieties, which have unusual physical and chemical properties and which come within the limits of fixed carbon or calorific value of the high-volatile bituminous and subbituminous ranks. All these coals either contain less that 48% dry mmf fixed carbon or have more that 15,500 moist mmf Btu/lb.

[b] To convert British thermal units per pound to joules per kilogram, multiply by 2326.

[c] Moist refers to coal containing its natural inherent moisture but not including visible water on the surface of the coal.

[d] If agglomerating, classify in low-volatile group of the bituminous class.

[e] Coals having 69% or more fixed carbon on the dry mmf basis are classified according to fixed carbon regardless of calorific value.

[f] There may be nonagglomerating varieties in these groups of the bituminous class, and there are notable exceptions in the high-volatile C bituminous group.

$$V' = 100 - F' \tag{9-2}$$

$$Q' = \frac{100(Q - 50S)}{100 - (1.08A + 0.55S)} \tag{9-3}$$

$$F' = \frac{100F}{100 - (M + 1.1A + 0.1S)} \tag{9-4}$$

$$Q' = \frac{100Q}{100 - (1.1A + 0.1S)} \tag{9-5}$$

where M, F, A, and S are weight percentages, on a moist basis, respectively of moisture, fixed carbon, ash, and sulfur; F' and V' are weight percentages, on a dry mmf basis, respectively of fixed carbon and volatile matter; Q and Q' are calorific values, Btu/lb (× 2326 = J/kg), respectively on a moist non-mmf basis and a moist mmf basis.

Table 9-2 shows the principal ranks of coal mined in the major coal-producing states of the United States.

Composition and Heating Value of Coal The composition of coal is generally reported in two different ways, the proximate analysis and the ultimate analysis, both expressed in weight percent. The **proximate analysis** is the determination by prescribed methods of moisture, volatile matter, fixed carbon, and ash. Figure 9-1 gives the proximate analyses and heating values, on a moist mmf basis, of the various ranks of coal. It is seen that the fixed carbon and heating values increase with an increase in rank and that the moisture and volatile matter decrease.

The **total moisture** in coal consists of **inherent** moisture and **bed** moisture. Inherent moisture, also referred to as **bed and equilibrium** moisture, exists as a quality of a coal seam in its natural state of deposition. Free moisture, also referred to as **surface** moisture, is the part of the total moisture that is lost when coal is air-dried under standard conditions.

The **volatile matter** is the portion of the coal which, when heated

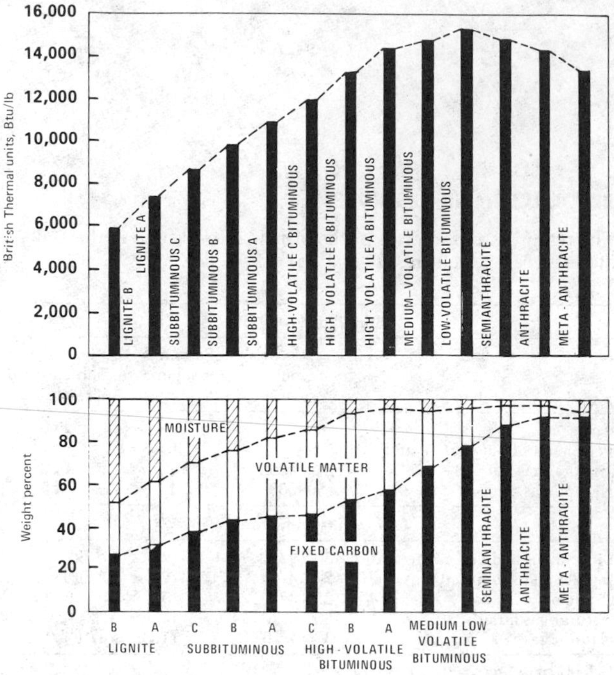

FIG. 9-1 Heating values and proximate analyses of coal on a moist, mineral-matter-free basis. To convert British thermal units per pound to joules per kilogram, multiply by 2326.

in the absence of air under prescribed conditions, is liberated as gases and vapors. Volatile matter does not exist by itself in coal, except for a little absorbed methane, but results from thermal decomposition of the coal substance.

Fixed carbon is the residue left after the volatile matter is driven off and is calculated by subtracting from 100 the percentages of moisture, volatile matter, and ash of the proximate analysis. In addition to carbon, it may contain several tenths of a percent of hydrogen and oxygen, 0.4 to 1.0 percent nitrogen, and about half of the sulfur that was in the coal.

Ash is the inorganic residue that remains after the coal has been burned under specified conditions, and it is composed largely of compounds of silicon, aluminum, iron, and calcium and of minor amounts of compounds of magnesium, sodium, potassium, and titanium. Ash may vary considerably from the original mineral matter, which is largely kaolinite, illite, montmorillonite, quartz, pyrites, and gypsum.

The **ultimate analysis** is the determination by prescribed methods of the ash, carbon, hydrogen, nitrogen, oxygen (by difference), and sulfur. Along with these analyses, the heating value, expressed as joules per kilogram (British thermal units per pound), is also determined. This is the heat produced at constant volume by the complete combustion of a unit quantity of coal in an oxygen-bomb calorimeter under specified conditions. The result includes the latent heat of vaporization of the water in the combustion products and is called the gross heating or **high heating value** (HHV), Q_h. The heating value when the water is not condensed is called the **low heating value** (LHV), Q_ℓ, and is obtained from

$$Q_\ell = Q_h - 1030W \tag{9-6}$$

where W = lb water formed/lb of fuel. The factor 1030 converts the high heating value at constant volume to low heating value at constant pressure. When dealing with gases, a useful equation is

$$Q_\ell = Q_h - (859\ pv/T) \tag{9-6a}$$

TABLE 9-2 Principal Ranks of Coal Mined in Various States*

State	Anthracite	Semianthracite	Low-volatile bituminous	Medium-volatile bituminous	High-volatile *A* bituminous	High-volatile *B* bituminous	High-volatile *C* bituminous	Subbituminous *A*	Subbituminous *B*	Subbituminous *C*	Lignite
Alabama	...	...	...	x	x						
Alaska	...	...	...	...	...	x	x	x	x	x	x
Arkansas	...	x	x	x	x	...	...	...	...	...	x
Colorado	...	...	...	...	x	x	x	x	x	x	x
Illinois	...	...	...	...	x	x	x				
Indiana	...	...	...	...	...	x	x				
Iowa	...	...	...	...	...	...	x				
Kansas	...	...	...	...	x	x					
Kentucky:											
Eastern	...	...	...	...	x	x					
Western	...	...	...	...	x	x	x				
Maryland	...	...	x	x	x						
Missouri	...	...	...	...	...	...	x				
Montana	...	...	...	...	...	x	x	x	x	x	x
New Mexico	...	...	...	...	x	x	x	...	x		
North Dakota	...	...	...	...	...	...	...	...	...	...	x
Ohio	...	...	...	...	x	x	x				
Oklahoma	...	...	x	x	x	x	x				
Pennsylvania	x	x	x	x	x						
South Dakota	...	...	...	...	...	...	...	...	...	...	x
Tennessee	...	...	...	x	x						
Texas	...	...	...	...	...	...	x	x	...	...	x
Utah	...	...	...	...	x	x	x	x	x		
Virginia	...	x	x	x	x						
Washington	...	...	...	...	x	x	x	...	x	...	x
West Virginia	...	...	x	x	x						
Wyoming	...	...	...	...	...	x	x	x	x	x	

*Compiled largely from Typical Analyses of Coal of the United States, *U.S. Bur. Mines Bull.* 446, and Coal Reserves of the United States, *U.S. Geol. Survey Bull.* 1136.

where Q_ℓ and Q_h = Btu/ft^3 of gas

p = pressure, of mercury

v = ft^3 water in combustion products/ft^3 fuel

T = temperature, °R

For 60F and 29.92 inHg, the equation simplifies to

$$Q_\ell = Q_h - 49.4v \qquad (9\text{-}6b)$$

Q_h in Btu/lb (× 2326 = J/kg) can be approximated by a formula developed by the Institute of Gas Technology;

$$Q_h = 146.58C + 568.78H + 29.4S - 6.58A - 51.53\,(O + N) \qquad (9\text{-}7)$$

where C, H, S, A, O, and N are respectively the weight percentages on a dry basis of carbon, hydrogen, sulfur, ash, oxygen, and nitrogen. The standard deviation for 775 coal samples is 127.

Coal analyses are reported on several bases, and it is customary to select the basis best suited to the application. The **as-received** basis represents the weight percentage of each constituent in the sample as received in the laboratory. The sample itself may be coal as fired, as mined, or as prepared for a particular use. The **moisture-free** (dry) basis is generally the most useful basis because performance calculations can be easily corrected for the actual moisture content at the point of use. The **dry, ash-free** basis is frequently used to approximate the rank and source of a coal. For example, the heating value of coals of a given source and rank is remarkably constant when calculated on this basis. Use of these bases is illustrated in Table 9-3.

Laboratory procedures for proximate and ultimate analyses are given in the *Annual Book of ASTM Standards*, part 26, 1977; and in *Methods of Analyzing and Testing Coal and Coke*, U.S. Bureau of Mines Bulletin 638, 1967.

Sulfur in Coal Efforts to abate atmospheric pollution have drawn considerable attention to the sulfur content of coal, since the combustion of coal results in the discharge to the atmosphere of sulfur oxides. Sulfur occurs in coal in three forms: as pyrites (FeS_2); as organic sulfur, which is a part of the coal substance; and as sulfate sulfur. The sulfate sulfur comprises at the most only a few hundredths of a percent of the coal. The organic sulfur may comprise from 20 to 80 percent of the total sulfur. Since organic sulfur is chemically bound to the coal substance in a complex manner, drastic treatment is necessary to break the chemical bonds before the sulfur can be removed. There is no *economic* method known at present that will remove organic sulfur, but progress is being made with so-called chemical methods for cleaning coal. Pyritic sulfur can be partially removed by using standard coal-washing equipment. The degree of pyrite removal depends on the size of the coal and the size and distribution of the pyrite particles.

The sulfur content of United States coals varies widely, ranging from a low of 0.2 percent to as much as 7 percent by weight, on a dry basis. The estimated remaining United States coal reserves of all ranks, by sulfur content, are shown in Fig. 9-2. Extensive data on sulfur in United States coals are given in U.S. Bureau of Mines Information Circular 8312. The sulfur reduction potential of United States coals is described in U.S. Bureau of Mines Information Circular 8118, which gives washability data for 455 raw-coal samples.

Coal-Ash Characteristics and Composition When coal is to be burned or gasified, it is important to determine the ash fusibility, comprising the initial deformation, softening, and fluid temperatures. The difference between the softening and initial-deformation temperatures is called the **softening interval,** and that between the fluid temperature and the softening temperature is called the **fluid interval.** The fusibility of coal ash is determined by ASTM D 1857, *Annual Book of ASTM Standards*, part 26, 1977. The softening temperature is most often used as a rough qualitative guide to the behavior of ash on a grate and on furnace heat-transfer surfaces, with respect to the tendency to form large masses of sintered or fused ash, which impair heat transfer and impede gas flow. Likewise, the fluid temperature and the fluid interval are qualitative guides to the "flowability" of ash in a slag-tap furnace. However, because ash fusibility is not an infallible index of ash behavior in practice, care is needed in using ash-fusibility data for designing and operating purposes. There is an excellent discussion of this subject in *Steam: Its Generation and Use*, Babcock & Wilcox Co., New York, 1978.

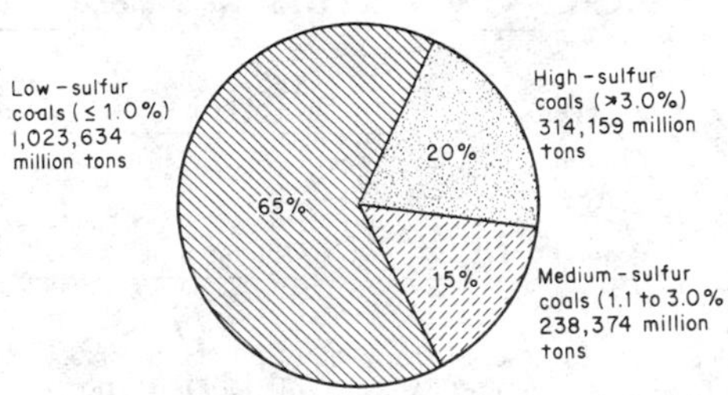

FIG. 9-2 Estimated remaining coal reserves of all ranks, by sulfur content, in the United States, Jan. 1, 1965. *(From U.S. Bur. Mines Inf. Circ. 8312.)*

The **composition of coal ash** varies widely. Calculated as oxides, the composition (percent by weight) varies as follows:

SiO_2	20–60
Al_2O_3	10–35
Fe_2O_3	5–35
CaO	1–20
MgO	0.3–4
TiO_2	0.5–2.5
Na_2O and K_2O	1–4
SO_3	0.1–12

Knowledge of the composition of coal ash is useful for estimating slagging and clinkering in fuel beds, predicting the flow properties of coal-ash slag (U.S. Bur. of Mines Bull. 618) in slag-tap and cyclone furnaces, and predicting, to a limited extent, the fouling and corrosion of heat-exchange surfaces in pulverized-coal-fired furnaces.

Multiple correlations for ash composition and ash fusibility are discussed in the *Coal Conversion Systems Technical Data Book*, part IA, U.S. Department of Energy, 1978.

The slag viscosity-temperature relationship for completely melted slag is

$$\text{Log viscosity} = 10^7\,M/(T - 150)^2 + C \qquad (9\text{-}8)$$

where viscosity is in poises (× 0.1 = Pa·s), M = 0.00835 (SiO_2) + 0.00601 (Al_2O_3) − 0.109, C = 0.0415 (SiO_2) + 0.0192 (Al_2O_3) + 0.0276 (equivalent Fe_2O_3) + 0.0160 (CaO) − 3.92, and T = °C.

TABLE 9-3 Comparison of Bases for Coal Analyses; High-Volatile A Bituminous Coal, Allegheny County, Pa., Pittsburgh Bed

	Proximate weight %				Ultimate weight %*					
Basis	Moisture	Volatile matter	Fixed carbon	Ash	Carbon	Hydrogen	Oxygen	Nitrogen	Sulfur	Heating value, B.t.u./lb.
As-received	2.4	36.6	53.2	7.8	75.8	5.1	8.2	1.5	1.6	13,560
Dry	. . .	37.5	54.5	8.0	77.7	5.0	6.2	1.5	1.6	13,890
Dry, ash-free	. . .	40.8	59.2	. . .	84.4	5.4	6.7	1.7	1.8	15,100

*On the as-received basis, the hydrogen and oxygen include the hydrogen and oxygen of the moisture.

NOTE: To convert British thermal units per pound to joules per kilogram, multiply by 2326.

TABLE 9-4 Specimen Free-Swelling and Hardgrove Grindability Indices

		Free-swelling index (FSI)			Hardgrove grindability index (HGI)		
State	Name of bed	High	Low	Average	High	Low	Average
Alabama	Mary Lee	8	1½	4	70	46	53
Illinois	No. 6	6½	1	4½	66	52	60
Kentucky	Winifrede	4	2	3	47	43	45
Pennsylvania	Upper Kittanning	9	2½	8	111	58	96

The oxides in parentheses are the weight percentages of these oxides when $SiO_2 + Al_2O_3 + Fe_2O_3 + CaO + MgO = 100$.

Physical Properties of Coal The **free-swelling index** (FSI) measures the tendency of a coal to swell when burned or gasified in fixed or fluidized bed. Coals with a high free-swelling index (>4) can usually be expected to cause difficulties in such beds. Details of the test are given in ASTM D 720, *Annual Book of ASTM Standards*, part 26, 1977; and U.S. Bureau of Mines Report of Investigations 3989.

The **Hardgrove grindability index** (HGI) indicates the ease (or difficulty) of grinding coal and is complexly related to physical properties such as hardness, fracture, and tensile strength. The Hardgrove machine is usually employed (see ASTM D 409). It determines the relative grindability or ease of pulverizing coal in comparison with a standard coal, chosen as 100 grindability (see Sec. 8). The FSI and HGI of some United States coals are given in Table 9-4. See Bureau of Mines Information Circular 8025 for FSI and HGI data for 2812 and 2339 samples respectively.

The **bulk density** of broken coal varies according to the specific gravity, size distribution, and moisture content of the coal and the amount of settling when the coal is piled. Following are some useful approximations of the bulk density of various ranks of coal:

	lb/ft^3	kg/m^3
Anthracite	50–58	801–929
Bituminous	42–57	673–913
Lignite	40–54	641–865

Size stability refers to the ability of coal to withstand breakage during handling and shipping. It is determined by twice dropping a 22.7-kg (50-lb) sample of coal from a height of 1.83 m (6 ft) onto a steel plate. From the size distribution before and after the test, the size stability is reported as a percentage factor (see ASTM D 440). The *friability* test measures the tendency of coal to break during repeated handling. It is actually the complement of size stability and is determined by the standard tumbler test (ASTM D 441-45).

Spiers's *Technical Data on Fuels* gives the specific heat of dry, ash-free coal as follows:

	Btu/(lb·°F)	J/(kg·K)
Anthracite	0.22–0.23	921–963
Bituminous	0.24–0.26	1004–1088

The relationship between specific heat and water content and between specific heat and ash content is linear. Given the specific heat on a dry, ash-free basis, it can be corrected to an as-received basis. The specific heat and enthalpy of coal to 1090°C (2000°F) are given in *Coal Conversion Systems Technical Data Book*, part IA, U.S. Department of Energy, 1978.

The **mean specific heat** of coal ash and slag, which is used for calculating heat balances on furnaces, gasifiers, and other coal-consuming systems, follows:

Temperature range		Mean specific heat	
°C	°F	Btu/(lb·°F)	J/(kg·K)
0– 38	32–100	0.212	888
0– 813	32–1500	0.224	938
0–1038	32–1900	0.232	971
0–1093	32–2000	0.235	984
0–1371	32–2500	0.272	1139

Coke Coke is the solid, cellular, infusible material remaining after the carbonization of coal, pitch, petroleum residues, and certain other carbonaceous materials. The varieties of coke, other than those from coal, generally are identified by prefixing a word to indicate the source, e.g., petroleum coke. To indicate the process by which a coke is manufactured, a prefix also is often used, e.g., *oven* coke.

Transformation of coal into coke. The mechanism of the formation of coke when coal is carbonized is a complex of physical and chemical phenomena that are not perfectly understood. Some of the physical changes, which are interrelated when certain ranks of coal or blends are heated, are softening, devolatilization, swelling, and resolidification. Some of the accompanying chemical changes are cracking, depolymerization, polymerization, and condensation. More detailed theoretical information is given in *Coal*, by Van Krevelen (Elsevier, Amsterdam, 1961), and *Chemistry of Coal Utilization*, by Lowry (Wiley, New York, 1945 and 1963).

High-temperature coke (900 to 1150°C). This type is most commonly used in the United States; nearly 20 percent of the total bituminous coal consumed is used to make high-temperature coke for metallurgical applications. About 99 percent of this type of coke is made in slot-type recovery ovens, and the remainder in beehive and other types of ovens. Blast furnaces use about 90 percent of the production, the rest going mainly to foundries and gas plants.

A U.S. Bureau of Mines survey of 12 blast-furnace coke plants, whose capacity is 30 percent of the total production in the United States, provides an excellent picture of the acceptable chemical and physical properties of such coke. The ranges of properties are given in Table 9-5.

The typical by-product yields per ton of dry coal from high-temperature carbonization in ovens, with inner-wall temperatures from 1000 to 1150°C (1832 to 2102°F), are: coke, 653 kg; gas, 154 kg (11,200 ft^3); tar, 44 kg (10 gal); water, 38 kg; light oil, 11 kg (3.3 gal); and ammonia, 2.2 kg.

Foundry coke. This coke has different requirements from blast-furnace coke. The volatile matter should not exceed 2.0 percent, the sulfur should not exceed 0.7 percent, the ash should not exceed 12.0 percent, and the size should exceed 3 in.

Low- and medium-temperature coke (500 to 750°C). Cokes of this type are not now produced in the United States to a significant extent. However, there is now interest in low-temperature carbonization as a source of both hydrocarbon liquids and gases to supplement petroleum and natural-gas resources.

The **Fischer assay** is an arbitrary but precise analytical tool for determining the yield of products from low-temperature carbonization. A known weight of coal is heated at a controlled rate in the absence of air to 500°C, and the products are collected and weighed.

TABLE 9-5 Chemical and Physical Properties of High-Temperature Cokes Used in the United States*

Property	Range
Volatile matter	0.6–1.4 wt. %, as-received
Ash	7.5–10.7 wt. %, as-received
Sulfur	0 6–1.1 wt. %, as-received
Stability factor	39–58 (1-in. tumbler)
Hardness factor	60–68 (¼-in. tumbler)
Apparent specific gravity (water = 1.0)	0.80–0.99

Comparison of Properties of Coke Produced by BM-AGA and Industrial Methods, U.S. Bur. Mines Rep. Invest. 6354. To convert inches to centimeters, multiply by 2.54.

TABLE 9-6 Fischer-Assay Yields from Various Ranks of Coal (As-Received Basis)

ASTM classification by rank: Class	Group	Coke, weight %	Tar, gal/ton	Light oil, gal/ton	Gas, ft^3/ton	Water, weight %
Bituminous	1. Low-volatile bituminous	90	8.6	1.0	1760	3
	2. Medium-volatile bituminous	83	18.9	1.7	1940	4
	3. High-volatile A bituminous	76	30.9	2.3	1970	6
	4. High-volatile B bituminous	70	30.3	2.2	2010	11
	5. High-volatile C bituminous	67	27.0	1.9	1800	16
Subbituminous	1. Subbituminous A	59	20.5	1.7	2660	23
	2. Subbituminous B	58	15.4	1.3	2260	28
Lignite	1. Lignite A	37	15.2	1.2	2100	44

NOTE: To convert gallons per ton to liters per kilogram, multiply by 0.004; to convert cubic feet per ton to cubic meters per kilogram, multiply by 3.1×10^{-5}.

Table 9-6 gives the approximate yields of products for various ranks of coal.

Pitch coke and petroleum coke. Pitch coke is made from coal-tar pitch, and petroleum coke is made from petroleum residues from petroleum refining. Pitch coke has about 1.0 percent volatile matter, 1.0 percent ash, and less than 0.5 percent sulfur on the as-received basis. There are two kinds of petroleum coke: delayed coke and fluid coke. Since they contain the impurities from the original crude oil, the sulfur is usually high, and appreciable vanadium salts may be present. Ranges of composition and properties are as follows:

Composition and properties	Delayed coke	Fluid coke
Volatile matter, wt. %	8–18	3.7–7.0
Ash, wt. %	0.05–1.6	0.1–2.8
Sulfur, wt. %	....	1.5–10.0
Grindability index	40–60	20–30
True density, g./ml.	1.28–1.42	1.5–1.6

Other Solid Fuels

Char Char is the nonagglomerated, nonfusible residue from the thermal treatment of solid carbonaceous material. Coal chars are obtained as a residue or a coproduct from low-termperature carbonization processes and from processes being developed to convert coal to liquid and gaseous fuels and to chemicals. Such chars have a substantial heating value. The net amount of char from a conversion process varies widely; in some instances it may represent between about 30 and 55 percent of the weight of coal feed; in others no net or excess char is produced; i.e., the entire char yield is consumed as in-plant fuel. The analyses of feed coals and resulting chars from two coal-conversion processes are given in Table 9-7. The volatile matter, sulfur, and heating values of the chars are lower, and the ash is higher, than in the original coal.

Wood Higher heating values are $20,004 \times 10^3$ J/oven-dried kg of hardwood species and $20,930 \times 10^3$ J/oven-dried kg of softwood species. These values are accurate enough for most engineering purposes. U.S. Department of Agriculture Handbook 72 (revised 1974) gives the specific gravity of the important softwoods and hardwoods if heating value on a volume basis is needed.

Peat Peat is partially decomposed plant matter that has accumulated underwater or in a water-saturated environment. It was the precursor of coal but is not classified as coal. Peat is sold under the term "peat moss" or "moss peat" and currently is used in the United States mainly for horticultural and agricultural applications. Interest is growing in its use as a fuel in certain local areas. Although analyses of peat vary widely, a typical high-grade peat has 90 percent water, 3 percent fixed carbon, 5 percent volatile matter, 1.5 percent ash, and 0.10 percent sulfur. The moisture-free heating value is approximately $20,930 \times 10^3$ J/kg (9000 Btu/lb).

Charcoal Charcoal is the residue from the destructive distillation of wood. It absorbs moisture readily, often containing as much as 10 to 15 percent water. In addition, it usually contains about 2 to 3 percent ash and 0.5 to 1.0 percent hydrogen. The heating value of charcoal is about $27,912 \times 10^3$ to $30,238 \times 10^3$ J/kg (12,000 to 13,000 Btu/lb).

Tanbark Tanbark is the residue remaining after bark has been used in tanning operations. It usually contains from 60 to 70 percent water and has a heating value of 5815×10^3 to 6978×10^3 J/kg (2500 to 3000 Btu/lb).

Bagasse Bagasse is the solid residue remaining after sugarcane has been crushed by pressure rolls. It usually contains from 40 to 50 percent water. The dry bagasse has a heating value of $18,608 \times 10^3$ to $20,934 \times 10^3$ J/kg (8000 to 9000 Btu/lb).

Solid Wastes and Biomass Large and increasing quantities of solid wastes generated per capita are a significant feature of affluent societies. In the United States the rate exceeds 5 lb per capita per day, or nearly 200 million tons per year, and it is growing rapidly. Table 9-8 shows the composition of various solid wastes. On a moisture-free basis, the composition of miscellaneous refuse is surprisingly uniform, but size and moisture variations cause major difficulties in efficient, economical disposal.

TABLE 9-7 Examples of Analyses of Coal Feeds and Resulting Chars from Various Coal-Conversion Processes*

Process	FMC†				IGT‡	
Coal bed	Pittsburgh-Federal		Illinois No. 6		Pittsburgh	
Composition and properties	Coal, dry basis	Char, dry basis	Coal, dry basis	Char, dry basis	Coal, dry basis	Char, dry basis
Analysis, wt. %:						
Volatile matter	36.8	3.7	38.6	3.5	32.7	1.2
Fixed carbon	57.0	86.8	50.0	76.4	52.3	77.5
Ash	6.2	9.5	11.4	20.1	14.1	21.3
Sulfur	2.9	1.9	3.8	3.1	4.3	1.7
Heating value, B.t.u./lb.	14,470	13,400	12,600	11,870	13,200	12,200

*Nelson et al., *Study of the Identification and Assessment of Potential Markets for Chars from Various Processing Systems*, Battelle Memorial Institute, Columbus. To convert British thermal units per pound to joules per kilogram, multiply by 2326.

†FMC process involves multistage fluidized-bed pyrolysis of coal to produce a liquid, residual char, and some gas.

‡IGT process involves hydrogasification of coal to produce a gas of pipe-line quality (about 1000 Btu/ft^3) and char.

TABLE 9-8 Waste-Fuel Analyses*

Type of waste	Heating value, Btu/lb	Volatiles	Percentage composition by weight: Moisture	Ash	Sulfur	Dry combustible	Density, lb/ft³
Paper	7,572	84.6	10.2	6.0	0.20		
Wood	8,613	84.9	20.0	1.0	0.05		
Rags	7,652	93.6	10.0	2.5	0.13		
Garbage	8,484	53.3	72.0	16.0	0.52		
Coated fabric: rubber	10,996	81.2	1.04	21.2	0.79	78.80	23.9
Coated felt: vinyl	11,054	80.87	1.50	11.39	0.80	88.61	10.7
Coated fabric: vinyl	8,899	81.06	1.48	6.33	0.02	93.67	10.1
Polyethylene film	19,161	99.02	0.15	1.49	0	98.51	5.7
Foam: scrap	12,283	75.73	9.72	25.30	1.41	74.70	9.1
Tape: resin-covered glass	7,907	15.08	0.51	56.73	0.02	43.27	9.5
Fabric: nylon	13,202	100.00	1.72	0.13	0	99.87	6.4
Vinyl scrap	11,428	75.06	0.56	4.56	0.02	95.44	23.4

*From Hescheles, *MECAR Conference on Waste Disposal*, New York, 1968; and *Refuse Collection Practice*, 3d ed., American Public Works Association, Chicago, 1966. To convert British thermal units per pound to joules per kilogram, multiply by 2326; to convert pounds per cubic foot to kilograms per cubic meter, multiply by 16.02.

The fuel value of most solid wastes is usually sufficient to enable self-supporting combustion, leaving only the incombustible residue and reducing the volume of waste eventually consigned to sanitary landfill to only 10 to 15 percent of the original volume. The heat released by the combustion of waste can be recovered and utilized, although the cost of the recovery equipment or the distance to a suitable point of use for the heat may make heat recovery economically infeasible.

Biomass (e.g., crop residues and cattle manure) is an important source of energy currently being examined for engineering feasibility and economics. *Energy from Bioconversion of Waste Materials*, by D. J. DeRenzo (Noyes Data Corp., Park Ridge, N.J., 1977), is a useful reference.

Coal-Oil Mixture (COM) COM is viewed as a near-term method for extending domestic fuel supplies and for reducing foreign oil imports. The most attractive feature of the COM concept is that the technology for producing and burning coal-oil slurries exists; they can be burned in many existing boilers with few modifications.

Coal for the mixture is pulverized wet or dry until about 80 percent passes through a 200-mesh screen. Then it is blended with oil and a surfactant, which helps keep the coal in suspension. Boiler tests indicate that COM can be burned in boilers and some fired process equipment designed for oil and gas without significant derating or expensive retrofit work, with about 50 percent coal by weight. It is generally agreed that COM is not economically attractive for boilers producing less steam than 45,360 kg/h (100,000 lb/h).

COMs are nonnewtonian fluids in the pseudoplastic classification; that is, their viscosities decrease with increasing velocity gradient. Once COMs start flowing, their viscous nature approaches that of a heavy oil.

A useful reference for COMs is "Can Coal-Oil Mixtures Make It as Industrial Fuels?" by J. W. Eberle and R. H. Hickman [*Mech. Eng.*, 24–28 (March 1978)].

LIQUID FUELS

Liquid Petroleum Fuels The principal liquid fuels are made by fractional distillation of crude petroleum (a mixture of hydrocarbons and hydrocarbon derivatives ranging from methane to heavy bitumen). As many as one-quarter to one-half of the molecules in crude may contain sulfur atoms; and some contain nitrogen, oxygen, vanadium, nickel, or arsenic. Desulfurization, hydrogenation, cracking (to lower molecular weight), and other refining processes may be performed on selected fractions before they are blended and mar-

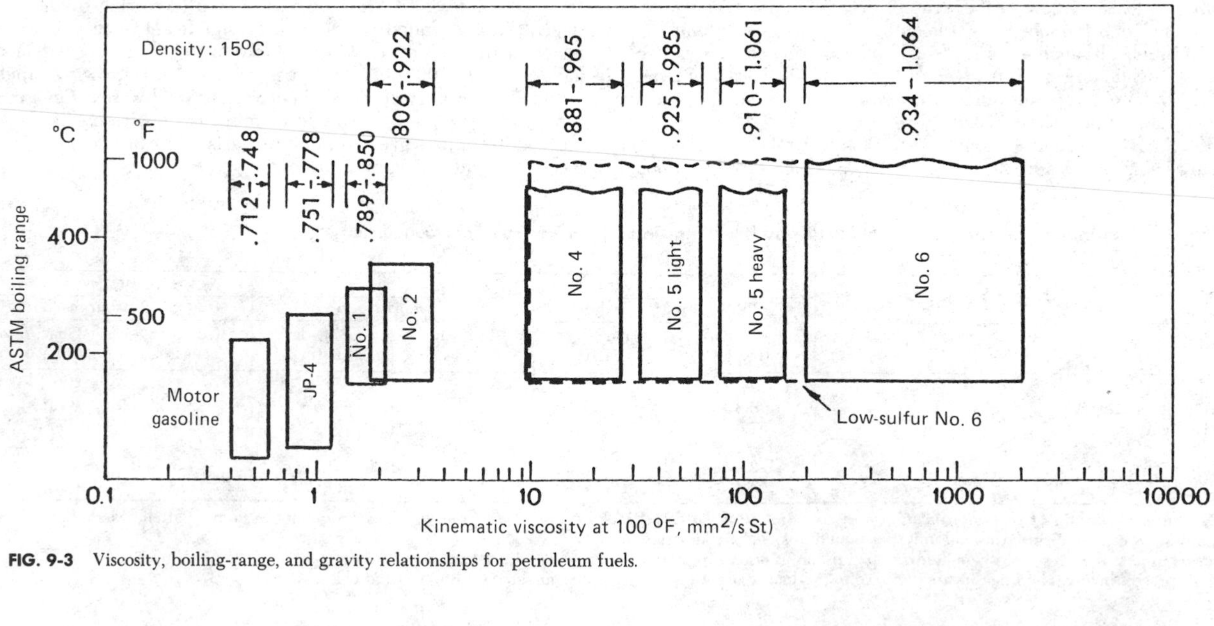

FIG. 9-3 Viscosity, boiling-range, and gravity relationships for petroleum fuels.

keted as fuels. Viscosity–gravity–boiling-range relationships of common fuels are shown in Fig. 9-3.

The highly viscous oil extracted from tar sands is also a grade of crude petroleum which can be hydrogenated and refined into conventional petroleum products.

Specifications Specifications developed by the American Society for Testing and Materials are widely used to classify fuels. Table 9-9 shows ASTM Fuel Oil Specification D 396. Note footnote *F*, covering low-sulfur residual fuels of high wax content, which require the same heated storage and handling as No. 6 but, when warm, are in the No. 4 or No. 5 viscosity range. (D 396 omits kerosine, a low-sulfur, clean-burning No. 1 fuel for lamps and freestanding flueless domestic heaters that is covered by Federal Specification VV-K-211d.)

In drawing contracts and making acceptance tests, refer to ASTM Standards, Parts 23 and 24 (American Society for Testing and Materials, Philadelphia, annually). Part 23 contains test methods and specifications (classifications) for burner fuel, motor and aviation gasoline, diesel fuel, and aviation and gas-turbine fuel. It also contains ASTM D 270, with sampling procedures for bulk oil in tanks, barges, etc. Part 24 contains test procedures for water and sediment and alternative procedures for sulfur determination.

Fuel specifications from different sources may differ in test limits on sulfur, density, etc., but the same general categories are recognized worldwide: kerosine-type vaporizing fuel, distillate (or "gas oil") for atomizing burners, and more viscous blends and residuals for commerce and heavy industry. Typical specifications are as follows:

Specifier	Number	Category
Canadian Government Specification Board, Department of Defence Production, Ottawa, Canada	3-GP-2	Fuel oil, heating
Deutschen Normenauschusses, Berlin 15	DIN 51603	Heating (fuel) oils
British Standards Institution, British Standards House, 2 Park Street, London, W1A 2BS	B.S. 2869	Petroleum fuels for oil engines and burners
Japan	JIS K2203	Kerosine
	JIS K2204	Gas oil
	JIS K2205	Fuel oils
Federal specifications, United States	VV-F-815	Fuel oil, burner

Foreign specifications are generally available from the American National Standards Institute, New York; United States federal specifications, at Naval Publications and Forms, Philadelphia.

Equipment manufacturers and large-volume users often write fuel specifications to suit particular equipment, operating conditions, and economics. Nonstandard test procedures and restrictive test limits should be avoided; they reduce the availability of fuel and increase its cost.

Bunker-fuel specifications for merchant vessels often cover only viscosity, flash point, water and sediment, and ash content. Deep-draft vessels carry residual (e.g., No. 6 fuel oil) or distillate-residual blend for main propulsion, plus distillate for startup, shutdown, maneuvering, deck engines, and diesel generators. Main-propulsion fuel is identified by its viscosity in centistokes at 50°C (e.g., "marine diesel 40"). Obsolete designations include those based on Redwood No. 1 seconds at 100°F (e.g., "MD 1500") and the designations "Bunker A" for No. 5 fuel oil and "Bunker B and C" for No. 6 fuel oil in the lower- and upper-viscosity ranges respectively.

Chemical and Physical Properties Petroleum fuels contain paraffins, isoparaffins, aromatics, and naphthenes, plus organic sulfur, oxygen, and nitrogen compounds that were not removed by refining. Olefins are absent or negligible except when created by severe refining. Vacuum-tower distillates with a final boiling point equivalent to 454 to 566°C (850 to 1050°F) at atmospheric pressure may contain from 0.1 to 0.5 ppm vanadium and nickel, but these metal-bearing compounds do not distill into No. 1 and 2 fuel oils.

Black, viscous residuum directly from the still at 200°C or higher may serve as fuel in nearby furnaces or may be cooled and blended to make commercial fuels. Diluted with 5 to 20 percent distillate, the blend is No. 6 fuel oil. With 20 to 50 percent distillate, it becomes No. 4 and No. 5 fuel oils for commercial use, as in schools and apartment houses. Distillate-residual blends also serve as diesel fuel in large stationary and marine engines. However, distillates with inadequate solvent power will precipitate asphaltenes and other high-molecular-weight colloids from "visbroken" (severely heated) residuals. A blotter test, ASTM D 2781, will detect sludge in pilot blends. Tests employing centrifuges, filtration, or microscopic examination have also been used.

No. 6 fuel oil contains from 10 to 500 ppm vanadium and nickel in complex organic molecules, principally porphyrins. These cannot be removed economically, except incidentally during severe hydrodesulfurization (Amero, Silver, and Yanik, "Hydrodesulfurized Residual Oils as Gas Turbine Fuels," ASME Pap. 75-WA/GT-8). Salt, sand, rust, and dirt may also be present, giving No. 6 a typical ash content of 0.01 to 0.5 percent by weight.

Ultimate analyses of some typical fuels are shown in Table 9-10.

Hydrogen content of petroleum fuels can be calculated from density with the following formula, with an accuracy of about 1 percent for petroleum liquids that contain no sulfur, water, or ash:

$$H = 26 - 15s \tag{9-9}$$

where H = percent hydrogen and s = relative density at 15°C (specific gravity at 60°/60°F). Schmidt (*Fuel Oil Manual*, 3d ed., Industrial Press, New York, 1969) claims improved precision of the formula by replacing 26 with different constants:

Relative fuel density at 15°C	API gravity	Constant
1.0754–1.0065	0– 9	24.50
1.0065–0.9935	10–20	25.00
0.9935–0.8757	21–30	25.20
0.8757–0.8013	31–45	25.45

Gravity is usually determined at ambient temperature with specialized hydrometers and expressed in degrees API (at 60°F), a scale that relates inversely to specific gravity s at 60°/60°F as follows (see also the abscissa scale of Fig. 9-4):

$$\text{Degrees API} = 141.5/s - 131.5 \tag{9-10}$$

Specific gravity is widely used in countries outside the United States, and with the adoption of SI units the American Petroleum Institute favors density at 15°C instead of degrees API.

The hydrogen content, heat of combustion, thermal expansion, specific heat, and thermal-conductivity data herein were abstracted from Bureau of Standards Miscellaneous Publication 97, *Thermal Properties of Petroleum Products*. These data are widely used, although newer correlations have appeared, notably that by Linden and Othmer [*Chem. Eng.*, **54**(4, 5) (April and May 1947)].

Heat of combustion can be estimated within 1 percent from fuel gravity by using Fig. 9-4. Corrections for water and sediment must be applied for residual fuels, but they are insignificant for clean distillates.

Figure 9-5 shows **viscosity-temperature relationships** for typical petroleum fuels. Between the cloud point and the boiling point and at pressures below 6.9 MPa (1000 lbf/in^2), they are practically newtonian liquids. At low temperatures where solids begin to separate, viscosity depends on the rate of shear.

Pour point ranges from −60°C (−80°F) for some kerosine-type jet fuels to 46°C (115°F) for waxy No. 6 fuel oils. **Cloud point** (which is not measured on opaque fuels) is 5 or 10°C higher than pour point unless the pour has been depressed by additives.

The drop in viscosity with increasing temperature is greater for some fuels than for others. Generalized viscosity charts become less reliable at temperatures substantially removed from the specification temperatures of 38 or 50°C (100 or 122°F).

TABLE 9-9 ASTM D 396-Note 79: Detailed Requirement for Fuel Oils*

Grade of fuel oil	Flash point, °C (°F)	Pour point, °C (°F)	Water and sediment, volume %	Carbon residue on 10% bottoms, %	Ash, weight %	Distillation temperatures, °C (°F)		
						10% point	90% point	
	Min.	Max.	Max.	Max.	Max.	Max.	Min.	Max.
No. 1: a distillate oil intended for vaporizing pot-type burners and other burners requiring this grade of fuel	38 (100)	-18^{C} (0)	0.05	0.15	....	215 (420)	....	288 (550)
No. 2: a distillate oil for general-purpose heating for use in burners not requiring No. 1 fuel oil	38 (100)	-6^{C} (20)	0.05	0.35	....	....	282^{C} (540)	338 (460)
No. 4: preheating not usually required for handling or burning	55 (130)	-6^{C} (20)	0.50	....	0.10	....	....	....
No. 5 (light): preheating may be required, depending on climate and equipment	55 (130)	....	1.00	....	0.10	....	....	....
No. 5 (heavy): preheating may be required for burning and, in cold climates, for handling	55 (130)	....	0.00	....	1.10	....	....	....
No. 6: preheating required for burning and handling	60 (140)	...[G]	2.00^{E}	....	....	....	....	....

*Reprinted, with permission, from the *Annual Book of Standards*.
[A] It is the intent of these classifications that failure to meet any requirement of a given grade does not automatically place an oil in the next lower grade unless in fact it meets all requirements of the lower grade.
[B] In countries outside the United States other sulfur limits may apply.
[C] Lower or higher pour points may be specified whenever required by conditions of storage or use. When a pour point less than −18°C (0°F) is specified, the minimum viscosity for grade No. 2 shall be 1.7 cSt (31.5 SUS) and the minimum 90% point shall be waived.
[D] Viscosity values in parentheses are for information only and not necessarily limiting.
[E] The amount of water by distillation plus the sediment by extraction shall not exceed 2.00%. The amount of sediment by extraction shall not exceed 0.50%. A deduction in quantity shall be made for all water and sediment in excess of 1.0%.
[F] Where low-sulfur fuel oil is required, fuel oil falling in the viscosity range of a lower numbered grade down to and including No. 4 may be supplied by agreement between purchaser and supplier. The viscosity range of the initial shipment shall be identified, and advance notice shall be required when changing from one viscosity range to another. This notice shall be in sufficient time to permit the user to make the necessary adjustments.
[G] Where low-sulfur fuel oil is required, Grade 6 fuel oil will be classified as low pour +15°C (60°F) max, or high pour (no max.). Low-pour fuel oil should be used unless all tanks and lines are heated.

Thermal expansion of petroleum fuels can be estimated as volume change per unit volume per degree:

Density		Volume change/unit volume	
kg/dm^3, 15°C	API gravity	Coefficient/°F	Coefficient/°C
>0.9660	Below 14.9	0.00035	0.00063
−0.8499	15.0– 34.9	0.00040	0.00072
−0.7754	35.0– 50.9	0.00050	0.00090
−0.7239	51.0– 63.9	0.00060	0.00108
−0.6724	64.0– 78.9	0.00070	0.00126
−0.6419	79.0– 88.9	0.00080	0.00144
−0.6277	89.0– 93.9	0.00085	0.00153
−0.6112	94.0–100.0	0.00090	0.00162

ASTM-IP Petroleum Measurement Tables (ASTM D 1250 IP 200) are used for volume corrections in commercial transactions.

Specific heat capacity of petroleum liquids between 0 and 205°C (32 and 400°F) having a relative density of 0.75 to 0.96 at 15°C can be calculated within 2 to 4 percent of the experimental values from the following equations:

$$c = (1.685 + 0.0039 \times °C)/s \qquad (9\text{-}11)$$

$$c' = (0.388 + 0.00045 \times °F)/s \qquad (9\text{-}12)$$

where c = heat capacity, kJ/(kg · °C)
c' = heat capacity, Btu/(lb · °F)
s = relative density at 15°C (specific gravity, 60/60°F)

Specific heat varies with temperature, and an arithmetic average of the specific heats at the initial and final temperatures can be used for calculations related to the heating or cooling of oil.

Table 9-10. Typical Ultimate Analyses of Petroleum Fuels

Composition, %	No. 1 fuel oil (41.5°A.P.I.)	No. 2 fuel oil (33°A.P.I.)	No. 4 fuel oil (23.2°A.P.I.)	Low sulfur, No. 6 F.O. (12.6°A.P.I.)	High sulfur, No. 6 (15.5°A.P.I.)
Carbon	86.4	87.3	86.47	87.26	84.67
Hydrogen	13.6	12.6	11.65	10.49	11.02
Oxygen	0.01	0.04	0.27	0.64	0.38
Nitrogen	0.003	0.006	0.24	0.28	0.18
Sulfur	0.09	0.22	1.35	0.84	3.97
Ash	<0.01	<0.01	0.02	0.04	0.02
C/H Ratio	6.35	6.93	7.42	8.31	7.62

NOTE: The C/H ratio is a weight ratio.

Saybolt viscosity, s^D				Kinematic viscosity, cSt^D						Specific gravity 60/60°F (°API)	Copper strip corrosion	Sulfur, %
Universal at 38°C (100°F)		Furol at 50°C (122°F)		At 38°C (100°F)		At 40° C (104°F)		At 50°C (122°F)				
Min.	Max.	Min.	Max.	Min.	Max.	Min.	Max.	Min.	Max.	Max.	Max.	Max.
.....	.....	.....	....	1.4	2.2	1.3	2.1	....	....	0.8499 (35 min)	No. 3	0.5
(32.6)	(37.9)	.....	....	2.0^C	3.6	1.9^C	3.4	....	....	0.8762 (30 min)	No. 3	0.5^B
(45)	125	.....	....	5.8	26.4^F	5.5	24.0^F					
(>125)	(300)	.....	....	>26.4	65^F	>24.0	58^F					
(>300)	(900)	(23)	(40)	>65	194^F	>58	168^F	(42)	(81)			
(>900)	(9000)	(>45)	(300)	.....	....	.....	....	>92	638^F			

The **thermal conductivity** of liquid petroleum products is given in Fig. 9-6. Thermal-conductivity coefficients for asphalt and paraffin wax in their solid states are 0.17 and 0.23 W/(m·K) respectively for temperatures above 0°C [1.2 and 1.6 Btu/(h·ft²)(°F/in)].

Fuel systems for No. 1 and No. 2 fuel oil are not heated. Systems for No. 6 fuel oil are usually designed to preheat the fuel to 32 to 49°C (90 to 120°F) to reduce viscosity for handling and to 74 to 93°C (165 to 200°F) to reduce viscosity further for proper atomization. No. 5 fuel oil may also be heated. (See Table 9-9.) Steam or electric heating is employed as dictated by economics, climatic conditions, length of line, and frequency of use. (See Fig. 9-7.) Pressure-relief arrangements are recommended on sections of heated pipe lines when fuel could be inadvertently closed between valves. Fuel can expand several percent between 15 and 100°C (59° and 212°F). Fuel-pump capacity exceeds burner capacity; excess flow recirculates through pressure-relief valves in the pump (for simple No. 2 fuel systems) or in the fuel header (for multiple-burner heavy-fuel systems).

Commercial Considerations Fuels are sold in gallons and in multiples of the 42-gal barrel in the United States. Table 9-11 shows units used elsewhere. Transactions exceeding 5000 to 10,000 gal usually involve volume corrections back to 15.6°C (60°F) for accounting purposes. Fuel passes through an air eliminator and mechanical meter when loaded into or dispensed from trucks. Larger transfers such as pipe-line, barge, or tanker movements are measured by fuel depth and "strapping tables" (calibration tables) in tanks and vessels, but positive-displacement meters that are "proved" (calibrated) frequently are gaining acceptance. After an appropriate settling period, water in the tank bottom is measured with a plumb bob or stick smeared with water-detecting paste.

Receipts of tank-car quantities or larger are usually checked for gravity, appearance, and flash point to confirm product identification and absence of contamination.

Safety Considerations Design and location of storage tanks, vents, piping, and connections are specified by state fire marshals, Underwriters codes, and local ordinances. In NFPA 30, *Flammable and Combustible Liquids Code, 1977*, published by National Fire Protection Association, Quincy, Massachusetts, liquid petroleum fuels are classified as follows for safety in handling:

Class I (flammable) liquid has a flash point below 37.8°C (100°F) and a vapor pressure not exceeding 0.28 MPa at 37.8°C (40 lbf/in² absolute at 100°F).

Class IA includes those liquids having flash points below 22.8°C (73°F) and boiling points below 37.8°C (100°F).

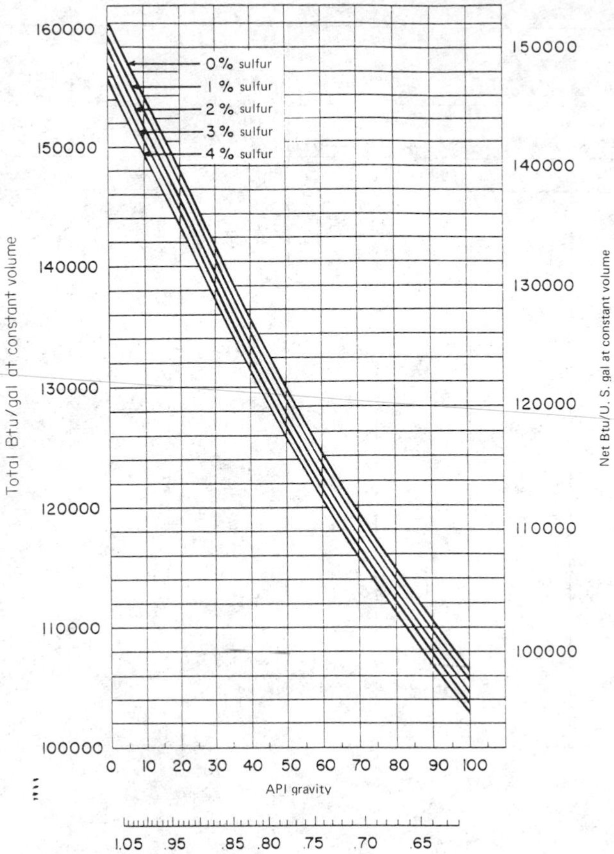

FIG. 9-4 Heat of combustion of petroleum fuels. To convert British thermal units per U.S. gallon to kilojoules per cubic meter, multiply by 2.787 163 E+02.

Class IB liquids have flash points below 22.8°C (73°F); boiling points, at or above 37.8°C (100°F).

Class IC includes those liquids having flash points at or above 22.8°C (73°F) and below 37.8°C (100°F).

Class II combustible liquids have flash points at or above 37.8°C (100°F) and below 60°C (140°F).

Class IIIA combustible liquids have flash points at or above 60°C (140°F) and below 93.4°C (200°F).

Class IIIB liquids flash at or above 93.4°C (200°F).

NFPA 30 details the design features and safe placement of handling equipment for flammable and combustible liquids.

Crude oils with flash points below 37.8°C (100°F) have been used in place of No. 6 fuel oil. Different pumps may be required because of low fuel viscosity. Handling precautions related to flammability are covered in Petroleum Safety Data Sheet PSD 2215, *Crude Oil as Burner Fuel*, American Petroleum Institute, Washington.

Nonpetroleum Liquid Fuels Table 9-12 shows typical data on liquid fuels from tar sands, oil shale, and coal.

Tar Sands Canadian **tar sands** are strip-mined and extracted with hot water to recover heavy oil (bitumen). The oil is processed into naphtha, kerosine, and gasoline fractions, which are hydrotreated, in addition to gas, which is recovered. Tar sands are being developed in Utah also.

Oil Shale Oil shale is nonporous rock containing organic kerogen. Raw shale oil is extracted by pyrolysis in a surface retort, after mining, or in situ after "rubblizing" the rock with explosives. Pyrolysis cracks the kerogen, yielding raw shale oil high in nitrogen, oxygen, and sulfur.

Shale oil has been hydrotreated and refined in demonstration tests into relatively conventional fuels. Refining in petroleum facilities is possible, and blending with petroleum is most likely.

Coal-Derived Fuels Liquid fuels derived from coal range from highly aromatic coal tars to liquids resembling petroleum. Raw liquids from different processes show variations that reflect the degree of hydrogenation achieved. Also, the raw liquids can be further hydrogenated to refined products. Properties and cost depend on the degree of hydrogenation and the boiling range of the fraction selected. A proper balance between fuel upgrading and equipment modification is essential for the most economical use of coal liquids in boilers, industrial furnaces, diesels, and stationary gas turbines.

Coal-tar fuels are high-boiling fractions of crude tar from pyrolysis in coke ovens and coal retorts. Grades range from free-flowing liquids to pulverizable pitch. Low in sulfur and ash, they contain hydrocarbons, phenols, and heterocyclic nitrogen and oxygen compounds. Being more aromatic than petroleum fuels, they burn with a more luminous flame. From 15 to 205°C (60 to 400°F) properties include:

Specific heat capacity	1.47–1.67 kJ/(kg·°C) [0.35–0.40 Btu/(lb·°F)]
Thermal conductivity	0.14–0.15 W(m·K) [0.080–0.085 Btu/ft^2 · (°F/ft)]
Heat of vaporization	349 kJ/kg (150 Btu/lb)
Heat of fusion	Nil

See the monograph edited by W. H. Huxtable, *Coal Tar Fuels* (Association of Tar Distillers, London, 1961). Properties of hydrogenated-coal liquids, largely unreported as yet, will no doubt fall between coal-tar fuel and petroleum.

The **EDS, H-Coal,** and **SRC-II processes** hydrogenate coal to yield hydrocarbon liquids. Distillates between 177 and 510°C (350 and 950°F) are low-sulfur liquids in the viscosity range of No. 2 to No. 5 fuel oil. Ash content ranges from nil to a few ppm. High-ash residuals are minor fractions, often to be consumed in process rather than marketed.

SASOL in South Africa is using coal gasification followed by synthesis to produce gasoline and diesel fuel (Fischer-Tropsch process). There are proposals in the United States to use the same gasification process to provide for the synthesis of **methanol** as a liquid fuel. Fermentation **ethanol** can be considered as a fuel supplement in countries with sufficient land, sunshine, and water to grow food, feed, and fiber at surplus levels. With flash points of 11 and 14°C (51.8 and 57.2°F) respectively, methanol and ethanol must be handled with due regard to flammability.

Coal-derived fuels are discussed in greater detail in a later subsection.

GASEOUS FUELS

Natural Gas Natural gas is a combustible gas that occurs in porous rock of the earth's crust and is found with or near accumulations of crude oil. Being a gas, it may occur alone in separate reservoirs. More commonly, it forms a gas cap entrapped between petroleum and an impervious, capping rock layer in a petroleum reservoir. Under high-pressure conditions, it is mixed with or dissolved in crude oil.

Natural gas consists of hydrocarbons with a very low boiling point. Methane is the main constituent, with a boiling point of −154°C (−245°F). Ethane, with a boiling point of −89°C (−128°F) may be present in amounts up to 10 percent; propane, with a boiling point up to −42°C (−44°F), up to 3 percent. Butane, pentane, hexane, heptane, and octane may also be present. Physical properties of these gases are given in Table 9-13.

Although there is no single composition that may be called "typical" natural gas, Table 9-14 shows analyses of natural gas in 14 large cities in the United States.

Natural gas termed "dry" has less than 0.013 L of gasoline/m^3 (0.1 gal/1000 ft^3). Above this amount, it is termed "wet." The terms "sweet" and "sour" are used to denote the absence or presence of H_2S.

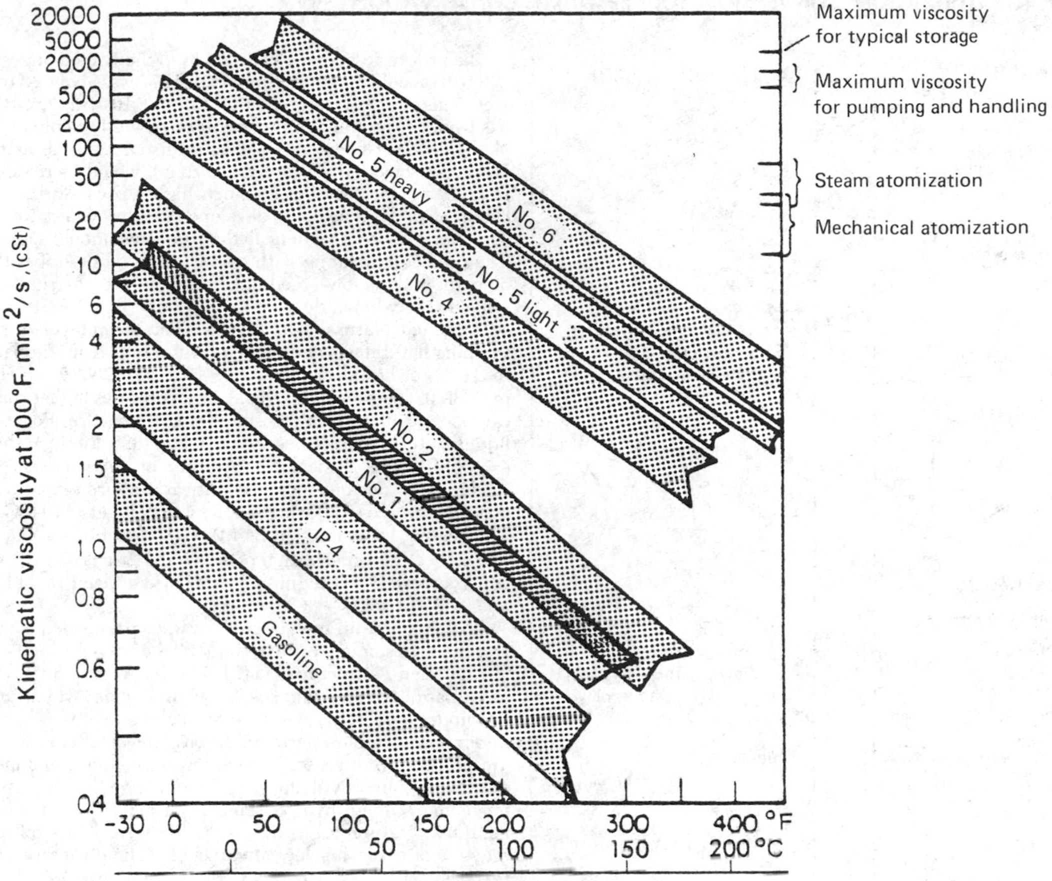

FIG. 9-5 Viscosity-temperature relationships for typical petroleum fuels.

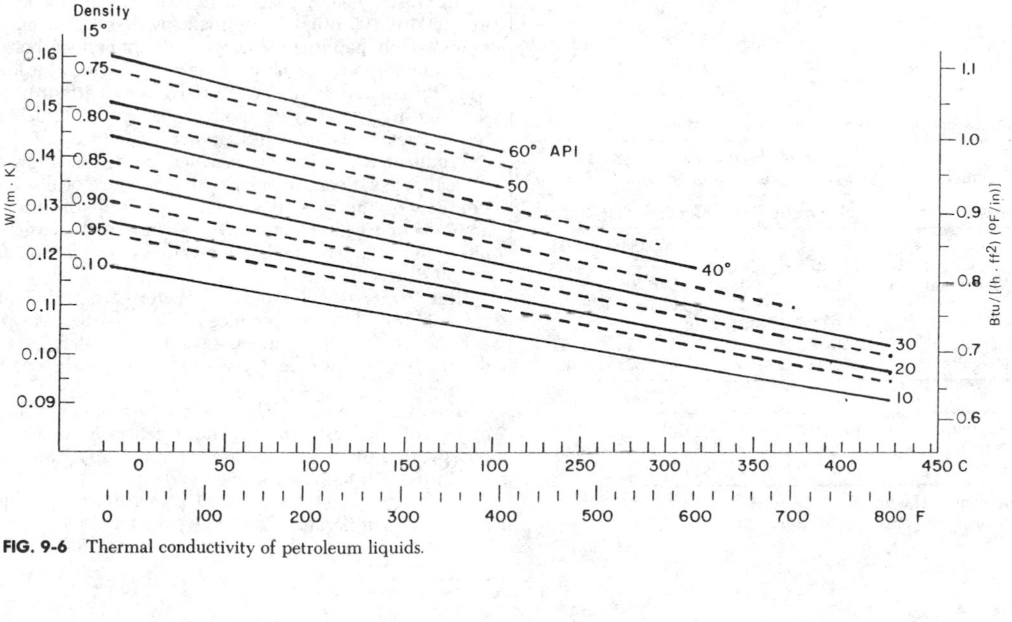

FIG. 9-6 Thermal conductivity of petroleum liquids.

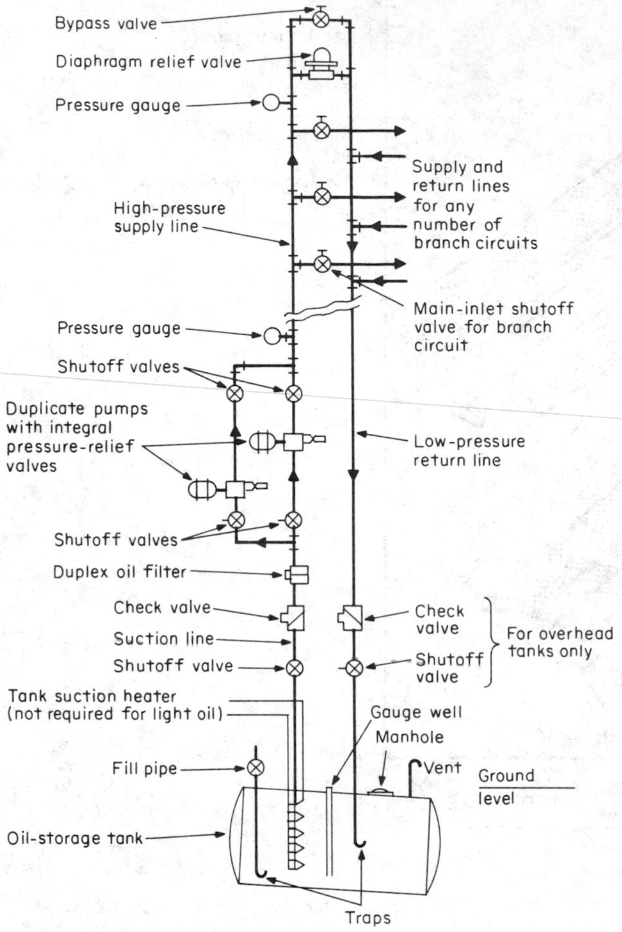

FIG. 9-7 Oil-storage tank and main circulating loop. (Combustion Handbook, *North American Mfg. Co., Cleveland, Ohio 44105.*)

TABLE 9-11 Commercial Units of Fuel Measurement

Customary unit	Multiply by this factor to get API-preferred metric unit	
Volume	Cubic meter, m^3	Cubic decimeter, dm^3
m^3 (264.2 U.S. gal)	1	1.000 000 E + 03
bbl (42 U.S. gal)	1.589 873 E − 01	
Imperial gal (1.201 U.S. gal)	4.546 092 E − 03	4.546 092 E + 00
U.S. gal	3.785 412 E − 03	3.785 412 E + 00
L (0.2642 U.S. gal)	1.000 000 E − 03	1
Imperial qt (0.3002 U.S. gal)	1.136 523 E − 03	1.136 523 E + 00
U.S. qt (0.2500 U.S. gal)	9.463 529 E −04	9.463 529 E − 01
Mass	Kilogram, kg	Tonne, t*
Long (U.K.) ton (2240 lb)	1.016 047 E + 03	1.016 047 E + 00
Short (U.S.) ton (2000 lb)†	9.071 847 E + 02	9.071 847 E − 01
Metric ton (tonne; 2204.616 lb)	1.000 000 E + 03	1

*Allowable (not official SI) unit.
†Not commonly used for petroleum measurement.

The proven reserves of natural gas in the United States total about 210 trillion cubic feet. Production in 1977 was about 20 trillion cubic feet. This reserves-to-production ratio of 10.5 suggests that natural gas from conventional sources will be seriously depleted by the end of the 1980s. Unconventional natural sources, such as tight sands, Devonian shales, and geopressure zones, would increase natural-gas reserves significantly but at much higher costs. Surface coal gasification processes now being developed may provide substitute natural gas that is more economical than unconventional natural sources. The subsection dealing with coal gasification describes the technologies for making low-, medium-, and high-Btu gas (substitute natural gas) currently being developed.

Liquefied Natural Gas (LNG) The advantages of storing and shipping natural gas in liquefied form derive from the fact that 0.03 m^3 (1 ft^3) of liquid methane at −162°C (−260°F) equals about 18 m^3 (630 ft^3) of gaseous methane. Temperatures higher than −162°C can be used if the liquid is stored under pressure. For example, the liquid state is maintained at 22.4 bar (325 lbf/in^2) and −103°C (−155°F). The critical temperature of methane is −82°C (−116°F), and the corresponding critical pressure is 46.4 bar (673 lbf/in^2). One m^3 (264 gal) weighs 412 kg (910 lb) at −164°C (−263°F). The heating value is about 24 MJ/L (86,000 Btu/gal).

The heat of vaporization of LNG at 1 bar is about 300 J/m^3 (10 Btu/standard ft^3). It requires 232 MJ to vaporize 1 m^3 of liquid methane (6575 Btu/ft^3).

LNG is stored in metal double-wall or prestressed concrete tanks, frozen-earth storage, or mined quarries or caverns.

Liquefied Petroleum Gas (LPG) The term "liquefied petroleum gas" is applied to certain specific hydrocarbons which can be liquefied under moderate pressure at normal temperatures but are gaseous under normal atmospheric conditions. The chief constituents of LPG are propane, propylene, butane, butylene, and isobutane, mixed in any proportion or with air. LPG produced in the separation of heavier or more dense hydrocarbons from natural gas is mainly of the paraffinic (saturated) series. LPG derived from oil-refinery gas may contain varying low amounts of olefinic (unsaturated) hydrocarbons.

LPG is widely used for domestic service, supplied either in tanks or by pipe lines. It is also used to augment natural-gas deliveries on peak days and by some industries as a standby fuel.

Re-formed Gas Although applicable to any gas transformed by suitable treatment, the term "re-formed" is ordinarily applied to lower-thermal-value gases obtained by the pyrolysis and steam decomposition of high-thermal-value gases such as natural gas, propane, butane, or refinery oil gas. It is sometimes used to meet peak-load requirements.

Oil Gases These gases, with heating values from 30 to 41 MJ/m^3 (300 to 1100 Btu/ft^3), are made by thermal decomposition of oils ranging from naphtha to heavy-residuum high-carbon oils. Although of minor importance now in many countries, the lower-Btu gases were distributed as manufactured gas by certain utilities, whereas the high-Btu gases were used primarily to supplement peak-load requirements by companies serving natural gas.

Producer Gas This gas is generated by blasting a deep, hot bed of coal or coke continuously with a mixture of air and steam. The products of the process are CO, N_2 (from the use of air), small amounts of H_2, and some CO_2. Because of the large percentage of nitrogen in the gas, the heating value is low [4.4 to 5.3 MJ/m^3 (125 to 150 Btu/ft^3)].

Blue Water Gas, Carbureted Water Gas, and Coal Gas These are combustible gases produced from coal or coke (in some cases enriched with oil, natural gas, or LPG) which have continuously declined in use in recent years to an extremely small share of the market.

Blast-Furnace Gas This gas is a by-product in the manufacture of pig iron in blast furnaces and is generally used for heating purposes within the plant. The heating value, approximately 3.2 MJ/m^3 (90 Btu/ft^3), is too low for sale to others.

Acetylene Acetylene is used primarily in operations requiring high flame temperature, such as welding and metal cutting. In order

TABLE 9-12 Characteristics of Typical Nonpetroleum Fuels

	Conventional coal-tar fuels from retorting[a]		Typical coal-derived fuels with different levels of hydrogenation[b]					Synthetic crude oils, by hydrogenation	
	CTF 50	CTF 400	Minimal		Mild	Mild[c]	Severe	Oil shale	Tar sands[d]
Distillation range, °C			175–280	280–500	160–415	175–400	125–495		
Density, kg/m^3, 15°C	1.018	1.234	.974	1.072	0.964	0.9607	0.914	0.817	0.864
lb/U.S. gal, 60°F	8.5	10.3	8.1	8.9	8.0	8.0	7.6	6.8	7.2
Viscosity, mm^2/s	2–9	9–18	3.1–3.4	50–90	3.6	....	2.18		
	At 38°C	At 121°C	At 38°C	At 38°C	At 38°C	....	At 38°C		
Ultimate analysis, %									
Carbon	87.4	90.1	86.0	89.1	87.8	89.6	89.0	86.1	87.1
Hydrogen	7.9	5.4	9.1	7.5	9.7	10.1	11.1	13.84	12.69
Oxygen	3.6	2.4	3.6–4.3	1.4–1.8	2.4	0.3	0.5	0.12	0.04
Nitrogen	0.9	1.4	0.9–1.1	1.2–1.4	0.6	0.04	0.09	0.01	0.07
Sulfur	0.2	0.7	<0.2	0.4–0.5	0.07	0.004	0.04	0.02	0.10
Ash[e]	Trace	0.15	<0.001	[f]					
C/H ratio, weight	11.0	16.5	9.4	11.9	9.1	8.9	8.0	6.2	6.9
Gross calorific value, MJ/kg	38.4–40.7	36.8–37.9							
Btu/lb	16,500 to 17,500	15,800 to 16,300							

[a]CTF 50, 100, etc., indicate approximate preheat temperature, °F, for atomization of fuel in burners (terminology used in British Standard B.S. 1469).
[b]Properties depend on distillation range, as shown, and to a lesser extent on coal source.
[c]Using recycle-solvent process.
[d]Tar sands, although a form of petroleum, are included in this table for comparison.
[e]Inorganic mineral constituents of coal tar fuel:
5 to 50 ppm: Ca, Fe, Pb, Zn (Na, in tar treated with soda ash)
0.05 to 5 ppm: Al, Bi, Cu, Mg, Mn, K, Si, Na, Sn
Less than 0.05 ppm: As, B, Cr, Ge, Ti, V, Mo
Not detected: Sb, Ba, Be, Cd, Co, Ni, Sr, W, Zr
[f]Inherent ash is "trace" or "<0.1%," although entrainment in distillation has given values as high as 0.03 to 0.1%.

TABLE 9-13 Physical Properties of Light Hydrocarbons*

	Methane	Ethane	Propane	Isobutane	Butane	Pentane
Molecular volume of gas, cu. ft.†	378.7	375.8	372.7	366.7	365.4	
Molecular weight of gas	16.04	30.07	44.09	58.12	58.12	72.15
Gal./lb.-mole at 60°F	6.4‡	9.64	10.41	12.38	11.94	13.71
Weight:						
% carbon	74.88	79.88	81.72	82.66		
% hydrogen	25.12	20.12	18.28	17.34	17.34	
Specific gravity:						
Of liquid (water = 1)	0.248	0.377	0.508	0.563	0.584	0.631
Of liquid, °A.P.I.	340†	247	147	120	111	93
Of gas (air = 1)	0.555	1.048	1.550	2.077	2.084	2.490
Weights and volumes:						
Lb./gal. liquid	2.5†	3.145	4.235	4.694	4.873	5.250
Cu. ft. gas/gal. liquid	59.0†	39.69	36.28	30.65	31.46	27.67
Cu. ft. gas/lb. liquid	24.8	12.50	8.55	6.50		
Ratio, gas volume to liquid volume§	443†	293.4	272.7	229.3	237.8	207.0
Initial boiling point (atmospheric pressure)	−259	−128.2	−43.7	10.9	31.1	97
Heat value (gross):						
B.t.u./cu. ft. gas	1,012	1,786	2,522	3,163	3,261	4,023
B.t.u./lb. liquid	23,885	22,323	21,560	20,732	21,180	21,110
B.t.u./gal. liquid		70,210	91,500	103,750	102,600	110,800
Vapor pressure, lb./sq. in. abs.						
At −44°F.		88	0	−9	−12	−14
At 0°F.		206	38	12	−7	−13
At 33°F.		343	54	17	0	−11
At 70°F.		563	124	45	31	−6
At 90°F.		710	165	62	44	
At 100°F.			189	72	52	4
At 130°F.			275	110	81	11
At 150°F.			346	138	87	21
Latent heat of vaporization at boiling point:						
B.t.u./lb.	221	211	185	158	167	153
B.t.u./gal.	553	664	785	742	808	802
Specific heat:						
Of liquid, at C_p and 60°F., B.t.u./(lb.)(°F.)		0.780	0.588	0.560	0.549	
Of gas, at C_p and 60°F., B.t.u./(lb.)(°F.)	0.526	0.413	0.390	0.406	0.396	0.402
Of gas, at C_v and 60°F., B.t.u./(lb.)(°F.)	0.402	0.347	0.346	0.373	0.363	0.376

*Johnson and Auth (eds.), *Fuels and Combustion Handbook*, McGraw-Hill, New York, 1951. To convert British thermal units per cubic foot to megajoules per cubic meter, multiply by 0.0373; to convert British thermal units per pound to megajoules per kilogram, multiply by 0.00232; to convert British thermal units per gallon to megajoules per cubic meter, multiply by 0.277; and to convert cubic feet to cubic meters, multiply by 0.0283. Gal/(lb·mol)(at 60°F) × 0.008 = m^3/(kg·mol)(at 16°C).

†Ideal gas = 379.5 ft^3.
‡Apparent values for dissolved methane at 60°F.
§Based on "perfect gas."

TABLE 9-14 Analyses of Natural Gas*,†

City	Components of gas, % by volume									Heating value,‡ B.t.u./cu. ft.	Specific gravity
	Methane	Ethane	Propane	Butanes	Pentanes	Hexanes plus	CO_2	N_2	Misc.		
Baltimore, Md.	94.40	3.40	0.60	0.50	0.00	0.00	0.60	0.50	...	1051	0.590
Birmingham, Ala.	93.14	2.50	.67	.32	.12	.05	1.06	2.14	...	1024	.599
Boston, Mass.	93.51	3.82	.93	.28	.07	.06	0.94	0.39	...	1057	.604
Columbus, Ohio	93.54	3.58	0.66	.22	.06	.03	.85	1.11	...	1028	.597
Dallas, Texas	86.30	7.25	2.78	.48	.07	.02	.63	2.47	...	1093	.641
Houston, Texas	92.50	4.80	2.00	.30	...	...	.27	0.13	...	1031	.623
Kansas City, Mo.	72.79	6.42	2.91	.50	.06	Trace	.22	17.10	...	945	.695
Los Angeles, Calif.	86.50	8.00	1.90	.30	.10	.10	.50	2.60	...	1084	.638
Milwaukee, Wis.	89.01	5.19	1.89	.66	.44	.02	.00	2.73	.06 He	1051	.627
New York, N.Y.	94.52	3.29	0.73	.26	.10	.09	.70	0.31	...	1049	.595
Phoenix, Ariz.	87.37	8.11	2.26	.13	.00	.00	.61	1.37	...	1071	.633
Salt Lake City, Utah	91.17	5.29	1.69	.55	.16	.03	.29	0.82	...	1082	.614
San Francisco, Calif.	88.69	7.01	1.93	.28	.03	.00	.62	1.43	.01 He	1086	.624
Washington, D.C.	95.15	2.84	0.63	.24	.05	.05	.62	0.42	...	1042	.586

*Reproduced by permission from "Gas Engineers Handbook," American Gas Association, Industrial Press, New York, 1965.

†Average analyses (1954 data) obtained from the operating utility company(s) supplying the city. The gas supply may vary considerably from these data—especially where more than one pipe line supplies the city. Also, as new supplies may be received from other sources, the analyses may change. Peak shaving (if used) is not accounted for in these data.

NOTE: To convert British thermal units per cubic foot to megajoules per cubic meter, multiply by 0.0373.

to transport acetylene, it is dissolved in acetone under pressure and drawn into small containers which are filled with porous material.

Hydrogen Used primarily in the production of ammonia and chemicals, in the hydrogenation of fats and oils, and as an oven reducing atmosphere, hydrogen is limited as a fuel to certain special industrial purposes such as cutting and welding. It is made industrially by electrolysis of water, by thermal cracking of natural gas and other hydrocarbons, and by the steam–re-forming reaction. Hydrogen as a nonpolluting fuel gas has been getting considerable attention.

Sulfur Impurities Most natural gas is free from sulfur compounds; however, some wells deliver gas containing levels of hydrogen sulfide and sulfur which must be removed before delivery. Pipe-line-company purchase contracts specify maximum limits of impurities; actual H_2S and sulfur (after purification) seldom exceed 0.002 and 0.02 g/m^3 (0.1 and 1.0 gr/100 ft^3) respectively.

Hydrogen sulfide in manufactured gases may range from approximately 2.30 g/m^3 (100 gr/100 ft^3) in blue and carbureted water gas to several hundred grains in coal and coke-oven gases. Another important sulfur impurity is carbon disulfide, which may be present in amounts varying from 0.007 to 0.07 percent by volume. Smaller amounts of carbon oxysulfide, mercaptans, and thiophene may be found. However, most of the impurities are removed during the purification process and either do not exist in the finished product or are present in only trace amounts.

Supercompressibility of Natural Gas All gases deviate from the simple gas laws to a varying extent. This deviation is called "supercompressibility" and must be taken into account in gas measurement, particularly at high line pressure. For example, since natural gas is more compressible under high pressure at ordinary temperatures than is called for by Boyle's law, gas purchased at an elevated pressure gives a greater volume when the pressure is reduced than it would if the gas were ideal.

The supercompressibility factor may be expressed as follows:

$$Z = (RT/PV)^{0.5}$$

where Z = supercompressibility factor
R = universal gas constant
T = gas temperature, absolute
P = gas pressure, absolute
V = molar gas volume

For determining supercompressibility factors of natural-gas mixtures, see A.G.A. *Manual for the Determination of Supercompressibility Factors for Natural Gas*, American Gas Association, New York, 1963; and A.G.A. Gas Measurement Committee Report No. 3, 1969.

FUEL AND ENERGY COSTS

Fuel costs vary widely from one area to another because of the cost of the fuel itself and the cost of transportation. Any meaningful cost comparison between fuels would require current costs based on the amounts used at a particular geographical location, utilization efficiencies or energy-ratio data for the equipment involved, the effects of "form value," etc. Although the costs given in Table 9-15 will not apply to specific locations, they do give fuel-cost trends.

Figures 9-8 and 9-9 are charts for comparing fuel and energy costs.

COAL CONVERSION TO LIQUID AND GASEOUS FUELS AND FEEDSTOCKS FOR THE CHEMICAL-PROCESS INDUSTRIES

Converting coal to environmentally acceptable synthetic gaseous and liquid hydrocarbons would supplement the dwindling reserves of petroleum and natural gas of the United States and help reduce national dependence on imported fuels.

Coal gasification yields a wide variety of useful products for the residential, utility, and industrial markets. A medium-Btu gas (MBG), produced by gasifying coal with oxygen and steam, can be upgraded to substitute natural gas (SNG) for distribution in existing natural-gas pipe lines, or MBG can be used for firing boilers and kilns and as a feedstock for the catalytic synthesis of chemicals. A low-Btu gas (LBG), produced by gasifying coal with air and steam, can be used close to gasifier sites to fire boilers, kilns, and several other industrial processes that now depend on petroleum and natural gas.

Coal liquefaction, effected by hydrogenating coal at high temperature and pressure, yields low-ash, low-sulfur boiler fuels for electric

TABLE 9-15 Time-Price Relationships for Fossil Fuels*

	Bituminous coal and lignite, $/ton	Natural gas at wellhead, cents/1000 standard ft^3	Crude petroleum at wellhead, $/bbl
1955	4.50	10.4	2.77
1960	4.69	14.0	2.88
1965	4.44	15.6	2.86
1970	6.26	17.1	3.18
1975	18.75	52.0	6.85†

SOURCE: *Mineral Facts and Problems*, U.S. Bur. Mines Bull. 667, 1975.
*All average annual prices.
†1974.

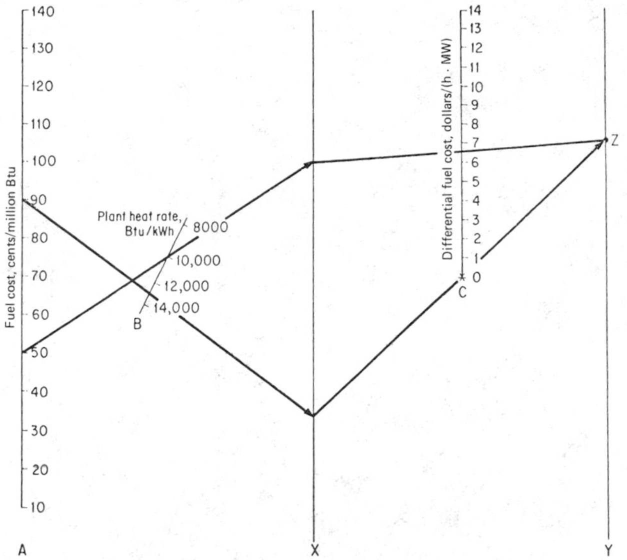

FIG. 9-8 Comparing power-plant fuel costs. The difference in fuel costs between alternative generating plants is found easily with the nomograph. Knowing fuel costs in cents per million British thermal units and plant heat rates in British thermal units per kilowatthour makes it possible to find a fuel-cost differential factor in dollars per hour-megawatt. This number, when multiplied by unit output in megawatts and unit operating hours, gives the differential fuel cost in dollars. (*Courtesy* Power.)

power generation, process heating, and making high-grade fuels such as gasoline and chemical feedstocks. Coal pyrolysis, or thermal decomposition, and the catalytic hydrogenation of carbon monoxide (from coal gasification) are also sources of liquid fuels and chemical feedstocks.

Bodle, Vyas, and Talwalker (*Clean Fuels from Coal Symposium II*, Institute of Gas Technology, Chicago, 1975) presented the chart in Fig. 9-10, which shows very simply the different routes from coal to clean gases and liquids.

Coal gasification and liquefaction are old technologies; current research, development, and demonstration efforts are aimed toward technical and economic improvements in some of the old, or first-generation, technologies and toward seeking new ways to accomplish the same ends: cheap and "clean" coal-conversion processes.

COAL GASIFICATION

General References: *Fuel Gasification*, symposium sponsored by American Chemical Society, 152d meeting, Sept. 12–13, 1966. Von Fredersdorff and Elliott, *Chemistry of Coal Utilization*, suppl. vol., ed. by H. H. Lowry, Wiley, New York, 1963. *Chemistry of Coal Utilization*, 2d suppl. vol., ed. by M.A. Elliott, Wiley, New York, 1981.

Background Converting coal to combustible gas has been practiced commercially since early in the nineteenth century. The first gas-producing companies were chartered in 1812 in England and 1816 in the United States to produce gas for illumination by the heating or pyrolysis of coal. This method of producing gas is still in use as a by-product of the carbonization of coal to produce coke for metallurgical purposes. The advantages of a gaseous fuel as a source of heat and power increased the use of coal for gas generation. The gas producer, in which a downward-moving bed of coal or coke is reacted with air and steam, was extensively used to prepare a gas of relatively low thermal content per unit volume, 3.7 to 6.0 MJ/m^3.

The development of the cyclic water-gas process in 1873 permitted the continuous production of gas of higher thermal content, 11.2 to 13.0 MJ/m^3. By adding oil to the reactor, the thermal content of the gas was increased to from 18.6 to 20.5 MJ/m^3. This type of gas, carbureted water gas, was distributed in urban areas of the United States as a fuel gas for residential and commercial uses until its displacement by lower-cost natural gas began in the 1940s. At approximately that time, development of oxygen-based gasification processes was initiated in both the United States and other countries. An early gasification process developed by Lurgi Kohle u Mineralöltechnik GmbH at that time, which operated at elevated pressure, is still in use, as are many producer-gas systems. Table 9-16 lists the composition of the coal-derived gases produced by these methods.

Theoretical Considerations The chemistry of coal gasification can be depicted by conveniently assuming coal as carbon and by listing the several well-known reactions involved; see Table 9-17. Reaction (9-13) is the combustion of carbon and oxygen, which is highly exothermic. This reaction supplies most of the thermal energy for the gasification process. The oxygen may be used as pure oxygen or be contained in air. Reactions (9-15) and (9-16) are endothermic and represent the conversion of carbon to combustible gases. These are driven by the heat energy supplied by Reaction (9-13).

As hydrogen and carbon monoxide are produced by the gasification reaction, these gases react with each other and with carbon. The reaction of hydrogen with carbon as shown in Reaction (9-14) is exothermic and can contribute heat energy. Similarly, the methanation reaction (9-18) can contribute heat energy to the gasification. These equations are interrelated by the water-gas-shift reaction (9-17), the equilibrium of which controls the extent of Reactions (9-15) and (9-16).

It has been shown by several authors (cf. Gumz, *Gas Producers and Blast Furnaces*, Wiley, New York, 1950; and chapter on gasification by Elliott and von Fredersdorff, *Chemistry of Coal Utilization*, suppl. vol., ed. by Lowry, Wiley, New York, 1965) that all the reactions could be based on three fundamental reactions: the Boudouard reaction (9-16), the heterogeneous water-gas reaction (9-15), and the equation for methane formation by the hydrogasification reaction (9-14). The equilibrium constants for these reactions are sufficient to calculate all the reactions listed.

It is not possible to calculate accurately actual gas composition by using the relationships of Reactions (9-13) to (9-18) in Table 9-17. Since the gasification of coal always takes place at elevated temperatures, thermal decomposition, or pyrolysis, takes place as coal enters the gasification reactor and is exposed to these temperatures. Reaction (9-19) treats coal as a compound of carbon and hydrogen and postulates its thermal disintegration to produce carbon (coke) and methane. Reaction (9-20) assumes the stoichiometry of hydrogasifying part of the carbon to produce methane and carbon. These reactions as well as those involving the other elemental constituents of coal take place during gasification.

It is possible to utilize these reactions and their relationships with each other for predicting the effects of changes in the operating parameters of gasification. At higher temperatures, endothermic reactions are favored at the expense of exothermic reactions. We can expect a decrease in methane production as Reactions (9-14) and (9-18) will proceed at a lower rate, CO production will be favored, and all reaction *rates* will increase in the direction in which heat absorption takes place. An increase in pressure will favor those reactions in which a smaller number of moles (or volumes) is formed. At higher pressures, production of CO_2 will be favored as well as that of methane. The knowledge of stoichiometry, equilibrium conditions, and rates for these gasification reactions provides a sound basis for modeling and extrapolating gasification systems.

A great deal depends on the gasifier system, coal reactivity and particle size, and method of contacting coal with gaseous reactants (steam and air or oxygen). It is generally believed that oxygen reacts completely in very short distances from the point at which it is mixed or comes in contact with coal or char. The heat evolved acts to pyrolyze the coal, and the char formed will then react with carbon dioxide, steam, or other gases formed by the combustion and pyrolysis of coal and char. The assumption made in Table 9-17 that the solid reactant is carbon is probably close to being correct. The conversion of coal to char and the type of char formed affect the kinetics of gas-solid reactions. While the reaction rate does vary with tem-

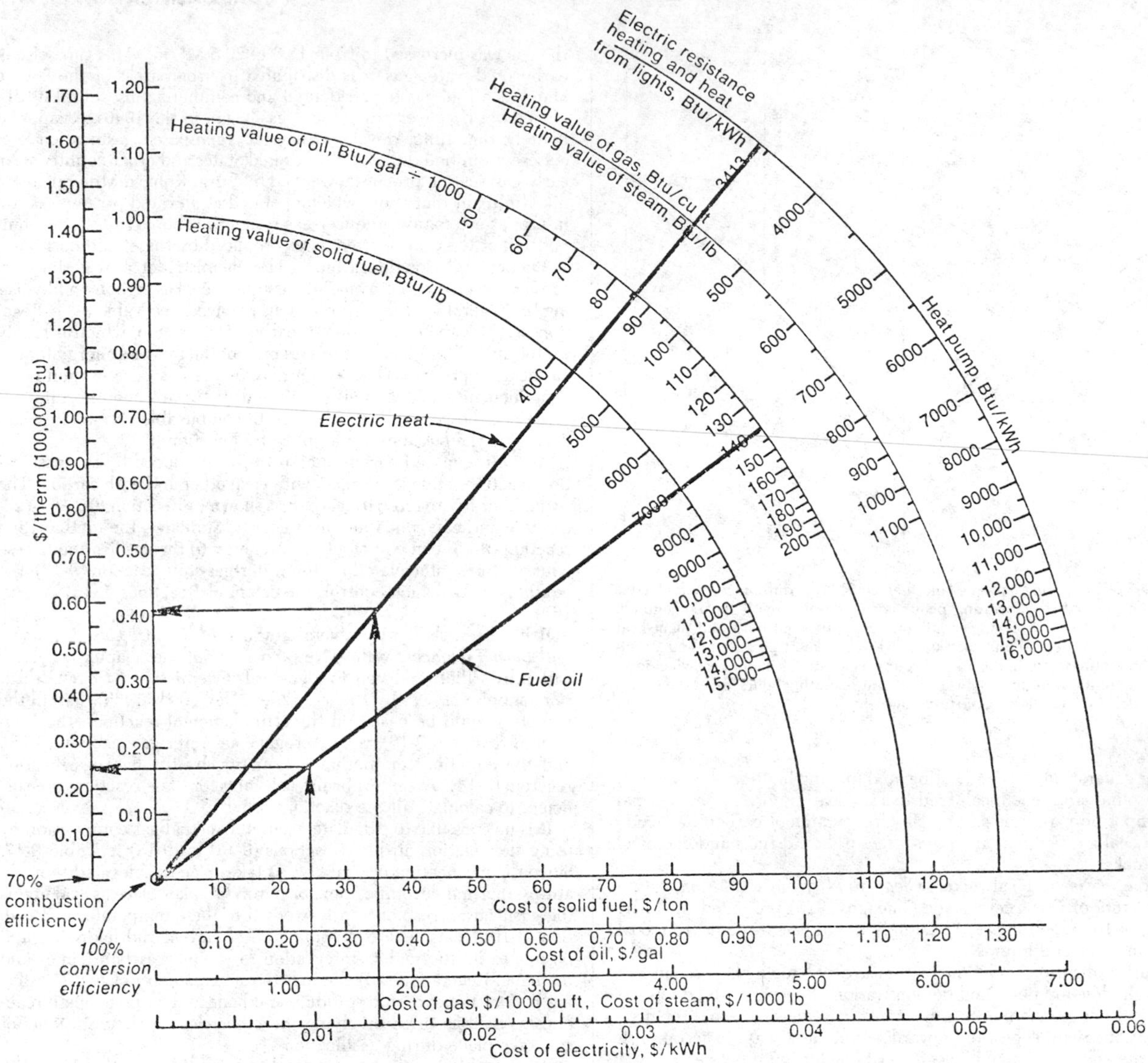

Typical Heating Values°	
Fuel oil, Btu/gal	
No. 1	137,400
No. 2	139,600
No. 4	145,100
No. 5	148,800
No. 6	152,400
Propane, Btu/gal	91,500
Natural gas, Btu/ft³	1,035
Coal, Btu/lb	
Bituminous	11,500–14,000
Subbituminous	8,300–11,500
Lignite	6,300–8,300
Steam, Btu/lb	1,012
Electricity, Btu/kWh	
Resistance heating	3,413
Heat pumps	Up to 13,000

° To convert British thermal units per gallon to kilograms per cubic meter, multiply by 278.7; to convert British thermal units per cubic foot to kilojoules per cubic meter, multiply by 37.3; to convert British thermal units per pound to kilojoules per kilogram, multiply by 2.326; and to convert British thermal units per kilowatthour to kilojoules per kilowatthour, multiply by 1.055.

FIG. 9-9 Comparing process-energy costs. (*Courtesy* Power.)

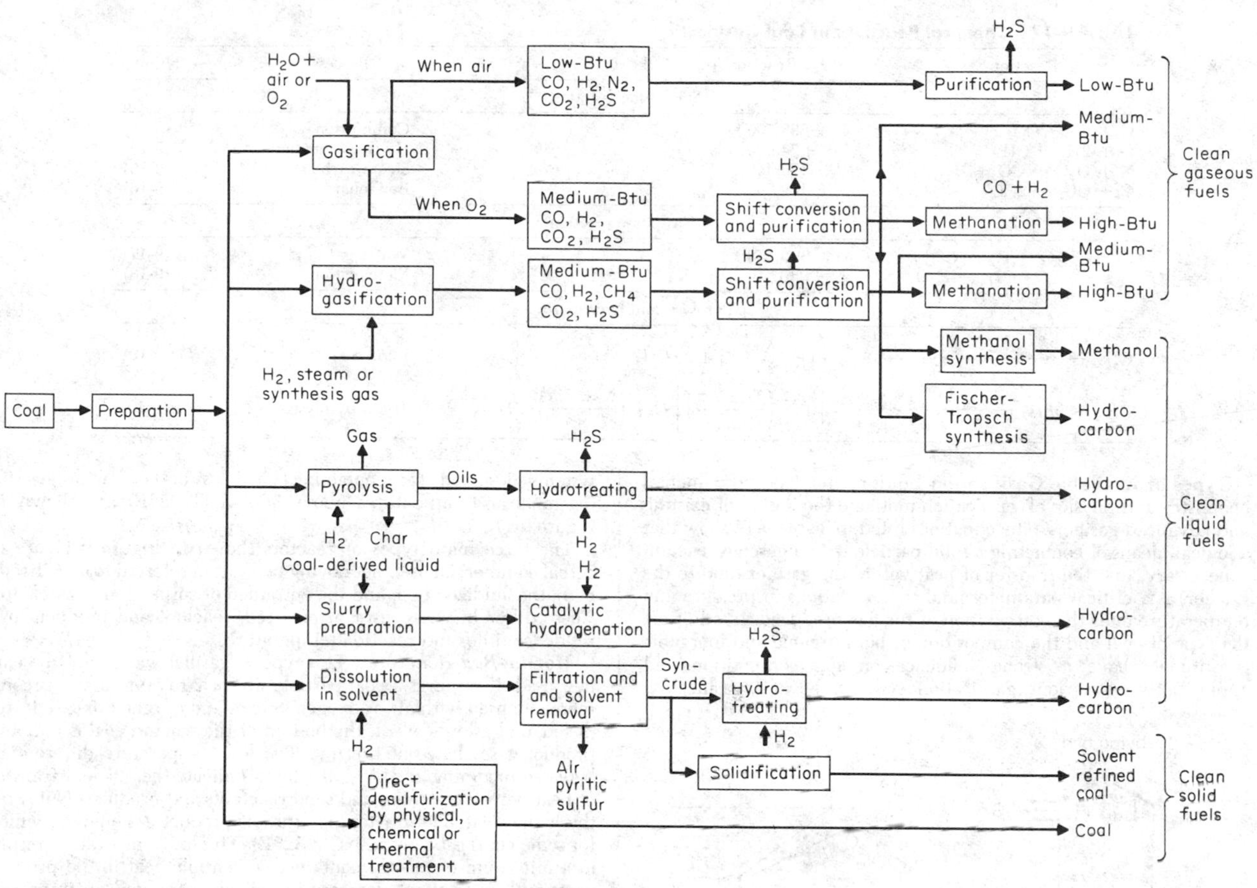

FIG. 9-10 The production of clean fuels from coal. (*Based on W. Bodle, K. Vyas, and A. Talwalker,* Clean Fuels from Coal Symposium II, *Institute of Gas Technology, Chicago, 1975.*)

perature as in all chemical reactions, the overall rate of reaction is controlled probably by the chemical reaction rate below 1000°C (1832°F). Above this, pore diffusion has an overriding effect, and at very high temperatures surface-film diffusion probably controls. Thus, for many gasification processes the reactivity of the char is quite important. This may depend not only on parent-coal characteristics but also on the method of heating, rate of heating, and particle-gas dynamics.

The importance of these concepts can be illustrated by the extent to which the pyrolysis reactions contribute to gas production. In a moving-bed gasifier (e.g., producer-gas gasifier), the particle is heated through several distinct zones of carbonization, char devolatilization, char gasification, and fixed-carbon combustion. As much as 15 to 17 percent of total gas production is liberated in the coal carbonization, almost 23 percent may be produced by char devolatilization, and some 60 percent is generated by actual gasification of the remaining char. This emphasizes the importance of coal quality or reactivity.

TABLE 9-16 Properties of Coal-Derived Gases

	Coke-oven gas	Producer gas	Water gas	Carbureted water gas	Synthetic coal gas
Reactant system	Pyrolysis	Air plus steam	Steam (cyclic-air)	Steam (cyclic-air)	Oxygen plus steam at pressure
Analysis, volume %*					
Carbon monoxide, CO	6.8	27.0	42.8	33.4	15.8
Hydrogen, H_2	47.3	14.0	49.9	34.6	40.6
Methane, CH_4	33.9	3.0	0.5	10.4	10.9
Carbon dioxide, CO_2	2.2	4.5	3.0	3.9	31.3
Nitrogen, N_2	6.0	50.9	3.3	7.9	
Other†	3.8	0.5	0.5	9.8	2.4
Fuel value, MJ/m³ (Btu/fts³)	22.0 (590)	5.6 (150)	11.5 (308)	20.0 (536)	10.8 (290)
	Fuel, chemicals	Fuel	Fuel, chemicals	Fuel	Fuel, chemicals

*Analyses and fuel values vary with the type of coal and operating conditions.
†Other contents include hydrocarbon gases other than methane, hydrogen sulfide, and small amounts of other impurities

TABLE 9-17 Chemical Reactions in Coal Gasification

Reaction	Reaction heat, kJ/(kg·mol)	Process	Number
	Solid—gas reactions		
$C + O_2 \rightarrow CO_2$	+393,790	Combustion	(9-13)
$C + 2H_2 \rightarrow CH_4$	+74,900	Hydrogasification	(9-14)
$C + H_2O \rightarrow CO + H_2$	−175,440	Steam-carbon	(9-15)
$C + CO_2 \rightarrow 2CO$	−172,580	Boudouard	(9-16)
	Gas-phase reaction		
$CO + H_2O \rightarrow H_2 + CO_2$	+2,853	Water-gas shift	(9-17)
$CO + 3H_2 \rightarrow CH_4 + H_2O$	+250,340	Methanation	(9-18)
	Pyrolysis and hydropyrolysis		
$CH_X \longrightarrow$	$\left(1 - \frac{X}{4}\right)C + \left(\frac{X}{4}\right)CH_4$	Pyrolysis	(9-19)
$CH_X + m\,H_2 \longrightarrow$	$\left[1 - \left(\frac{X + 2m}{4}\right)\right]C + \left(\frac{X + 2m}{4}\right)CH_4$	Hydropyrolysis	(9-20)

Types of Available Gasification Equipment The fundamental chemistry and physics of gasification motivate the design of existing and advanced gasifiers. The equations listed in Table 9-17 show that a logical means of contacting a solid particle with a gaseous reactant is necessary, that the transfer of heat within the gasifier (and to the gasifier) is a critical parameter, and that variations in pressure and temperature alter the composition of the gas produced. In addition, the type of coal and the composition of both organic and inorganic constituents have a strong influence on gas composition and applicability of various gasification systems. Several gasification systems designed for contacting particles of coal with gaseous reactants at temperatures above 700°C (1292°F) are shown in Figure 9-11.

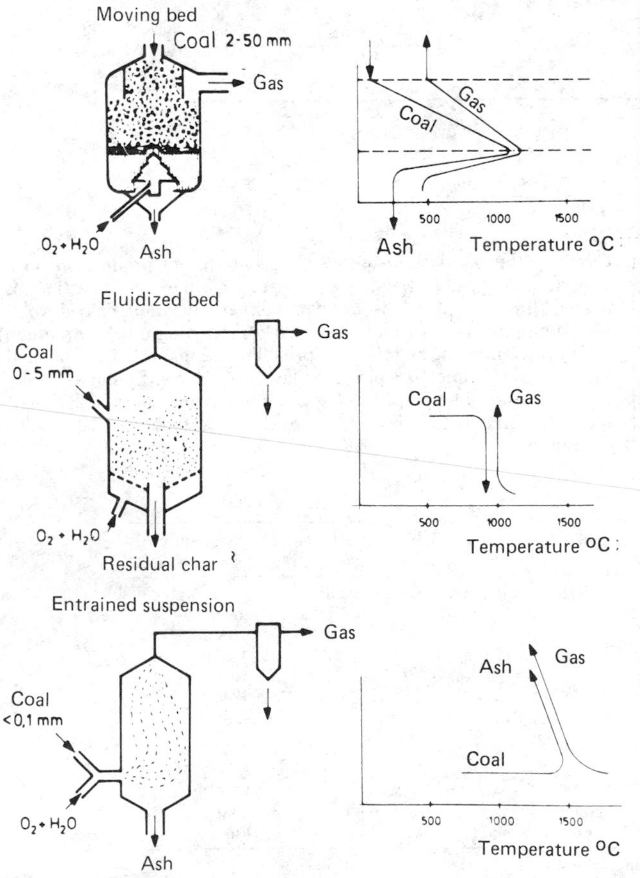

FIG. 9-11 Coal-gasification systems.

The three main types of reactors shown in Figure 9-11 are in actual commercial use: the moving bed (often referred to as a "fixed" bed), the fluidized bed, and the entrained or suspension flow reactor. These differ in size consist of coal fed, reactant and product flow, residence time, and reaction temperatures.

Moving-Bed Reactors This type of gasifier was one of the earliest used. It requires a coal particle size of 2 to 50 mm and is commercially used with air or oxygen. Steam and oxygen (air) are introduced, and ashes leave at the bottom of the reactor; coal is fed, and product gases leave at the top. The low temperature differentials between incoming and outgoing flows indicate the efficient transfer of heat between gaseous and solid reactants and products. Noteworthy is the relatively low temperature of product-gas effluent, which for some coals is below 500°C (932°F). This low temperature results in a minimum of energy contained as sensible heat in the product gases with a resultant decrease in oxygen usage and an improved thermal efficiency.

The countercurrent flow, low exit-gas temperature, and use of a mechanical grate at the bottom of these moving-bed systems cause some disadvantages. To protect the grate, steam in greater quantity than required for reaction is injected to reduce the bottom temperature. This impresses a negative thermal input and reduces overall efficiency. The coal as it enters is subject to pyrolysis as it passes down the gasifier, and tars and oils which must be recovered and/or utilized are produced. The resultant mixture of organic liquids and water condensate must be separated and the water purified prior to discharge. Finally, the need to dispose of the portion of the run-of-mine coal which is too fine to be used for feed can be an economic burden.

Reasonable procedures are available for meeting some of these problems. Some of these currently being utilized in other countries may require modification to meet environmental conditions in the United States. Considerable progress has been made toward this end by the Lurgi Kohle u Mineralöltechnik GmbH for their Lurgi gasifier. While many moving-bed gasifiers are available, the Lurgi is currently the only gasifier that is commercially available for operation at elevated pressure. Units are routinely being operated at pressures of 30 bar (450 psig). An outline sketch of the pressurized Lurgi gasifier is shown in Fig. 9-12. Also shown is the outline of the first step in purifying the gas: a multiple-venturi water scrubber.

The maximum reactor size for the Lurgi gasifier is currently 3.9 m (12.8 ft) in internal diameter. The capacity of this size of gasifier depends on coal quality and has been able to handle approximately 750-GJ coal input/h, equivalent to more than 25,000 kg/h (55,115 lb/h) of coal or 70,000 m³/h (245,700 ft³/h) of gas production. While stirrers have been designed to break up agglomerates so that caking coals can be handled, limited commercial experience is available with such coals.

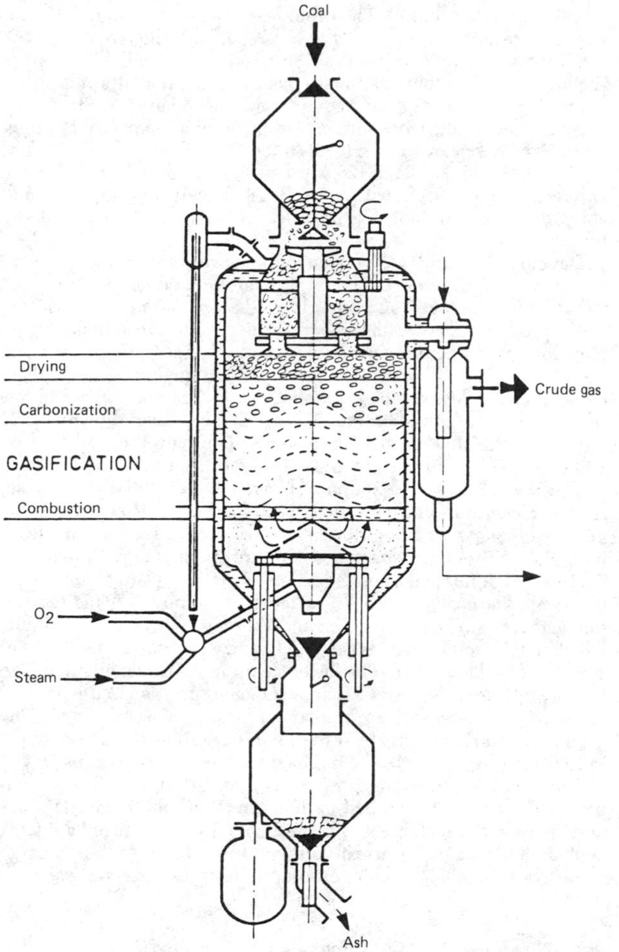

FIG. 9-12 Lurgi gasifier.

Fluidized-Bed Reactors The commercial gasification of coal with oxygen (air) and steam began with the use of a fluidized-bed reactor developed by Winkler (German Patent 437,970, filed Sept. 28, 1922). The coal is usually ground to sizes below 8 mm (5/16 in), which allows the entire run-of-mine coal output to be utilized. In contrast to the moving-bed reactor, the fluidized-bed reactor is essentially a completely mixed, partially cocurrent reactor. In the Winkler generator, coal is injected into the fluidized bed and the reactant gases are injected at two levels in the fluidized bed to maximize carbon conversion. The present commercial units are operated at essentially atmospheric pressure and are used primarily with reactive coal or with chars derived from lignite carbonization. Figure 9-13 shows an outline sketch of a Winkler gasifier.

The fluidized-bed reactor normally operates at higher temperatures than the moving-bed reactor and, being a completely mixed system, results in a higher outlet-gas temperature. This reduces efficiency but does destroy tars and oils and reduces the water-contamination problem. The major problem in any fluidized-bed system is the extent of carbon conversion and the carry-over of fine particles from the reactor. Normally, this carry-over has a high carbon content and must be collected and used for boiler fuel to avoid economic losses. The gasifiers currently in commercial use have internal diameters in excess of 4 m and are 20 m in height. Production of more than 50,000 m^3/h of gas is achieved by the Winkler gasifier (equivalent to 540 GJ/h or 18,000 kg/h of bituminous coal) even though the gasifier is operated at atmospheric pressure.

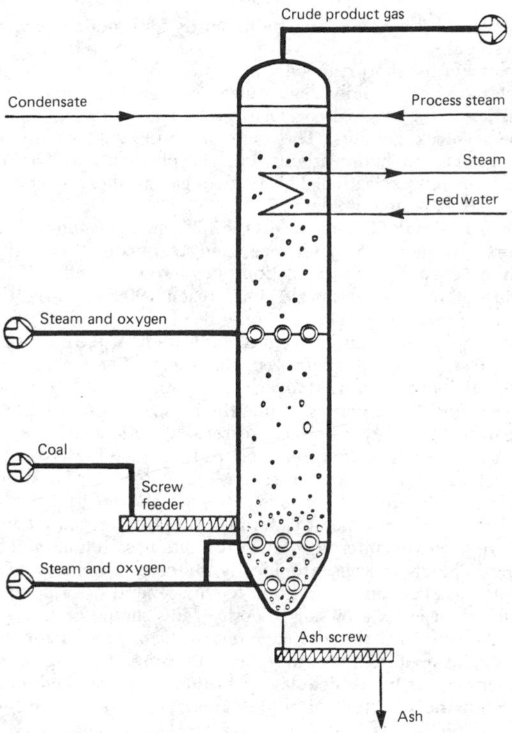

FIG. 9-13 Winkler gasifier.

The Winkler gasifier cannot handle agglomerating coals, as the formation of such agglomerates would disrupt the uniformity of the fluidized bed. Similarly, coals containing ash of low fusion temperature could cause difficulties by forming ash agglomerates. Efforts in solving these problems and increasing operating pressure are being actively pursued.

Entrained or Suspension Gasification Entrained gasification in the Koppers-Totzek gasifier, shown in Fig. 9-14, takes place at temperatures close to 1500°C (2732°F), using fine particles of coal at short residence times. The process is cocurrent with coal particles entrained in the turbulent reactant gases. The Koppers-Totzek process, currently operated at atmospheric pressure, is the most widely utilized commercial example of entrained gasification. More than 50

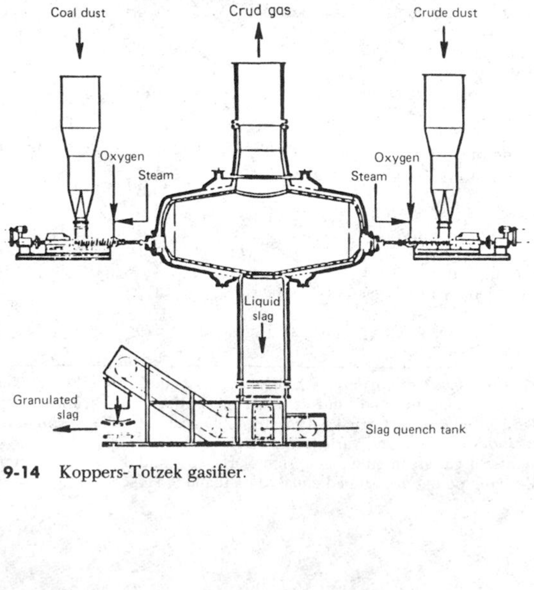

FIG. 9-14 Koppers-Totzek gasifier.

gasifiers of this type have been built, mostly for producing ammonia-synthesis gas.

Because of the short reaction time, usually only a few seconds, coal particles are very rapidly devolatilized and lose any inherent characteristics of the original coal. All types of coal can be handled in Koppers-Totzek gasifiers. The high operating temperature effectively gasifies all hydrocarbons, tars, or phenolics which may be formed during gasification. This reduces gas-purification and water-condensate-handling problems.

Pulverized coal (90 percent less than 200-mesh) is screw-fed from hoppers to burners in which oxygen and steam convey coal at velocities in excess of flame-propagation rates into the gasifier. The temperature at the burner nozzles may reach 1900°C (3452°F). Gas temperatures at the reactor exit may be close to 1500°C (2732°F). At these high temperatures, the mineral content of the coal is in molten form, and part of it is removed from the gasifier as a slag or melt. Part of the ash is carried overhead in particulate form and is removed in water scrubbers. While the Koppers-Totzek type of gasifier can handle all types of coal, careful attention must be paid to ash characteristics. Generally, one can characterize two general classes of slags produced from coal ash: a "short" slag, in which the temperature differential between softening and flow point is small; and "long" slags, in which this temperature differential is large and there are phases of sintering, clinker formation, softening, and finally flowing. The short slags are ideally suited for high-temperature entrained gasification. In addition to this general description, there are important criteria of slag viscosity. This should be less than 25 Pa·s (250 P) for suitable drainage of slag from powdered-fuel-slagging systems operating at atmospheric pressure. The slag chemistry is also critical, as high-silica slags present problems of both high viscosity and the potential of silica vaporization at temperatures of 1850°C (3362°F). [For detailed data on slag behavior and characterizations, see Hoy, Roberts, and Wilkins, *Inst. Gas Eng. J.*, 444–469 (June 1965)].

The entrained gasifier offers many advantages: the ability to handle a wide variety of coals, the elimination of tar and oil formation, and a conveniently disposable water-condensate and solid residue. The latter is generally free of carbonaceous matter and low in leachability. There are problems of handling high temperatures, refractory life, and slag control. The rapid reaction rate requires good control to prevent an excess of oxygen should the coal feed be interrupted. There is no reservoir of carbon, as is available in either the moving- or the fluidized-bed reactors. Operation at atmospheric pressure is a handicap, and considerable development effort is in progress to provide a pressurized system. Present capacity at atmospheric pressure can reach 380 GJ/h (360 × 10^6 Btu/h) per gasifier, equivalent to 13,000 kg/h (28,660 lb/h) of bituminous coal for the two-headed gasifier shown in Fig. 9-14. Recent development and commercial adaptation of a four-headed unit have doubled these throughputs.

Development and Utilization Both government and industry are sponsoring large development programs aimed at alleviating the various disadvantages of presently available commercial gasifiers. These programs involve all types of gasifiers and have as major objectives increased operating pressures, reduced reactant consumption, decreased capital costs, and improved product quality. Many of these developments have reached the pilot-plant stage and appear to justify commercial consideration. Table 9-18 lists several commercial and developmental gasification systems. The data are all based on 100 kg (2205 lb) of moisture-and-ash-free (maf) coal.

The two moving-bed systems shown have essentially the same reactor configuration except that the BG/Lurgi unit eliminates the grate, utilizing a slagging bottom with a large reduction in steam input. This raises the average temperature and increases throughput. The critical difference is the reduction in specific steam consumption (kg steam/kg coal gasified) by a factor of almost 5. This reduces unreacted steam in the gasifier and thereby the total gas volume, which in turn reduces solids carry-over and gas-handling costs downstream from the gasifier. The lower steam input results in a reduced H_2/CO ratio and reduced CH_4 and CO_2 concentration in the product gas. Carbon conversions are similar for both moving-bed units. The problems of tars and oils are also similar, but progress has been made in recycling heavy tars and contained fines back to the gasifier.

The effect of increasing pressure is shown dramatically by comparing the two fluidized-bed gasifiers in Table 9-18. The HYGAS unit operates at much higher pressures and, by using staged fluidized beds, achieves a low gas-exit temperature. The result is a large decrease in specific oxygen consumption (kg oxygen/kg coal gasi-

TABLE 9-18 Commercial and Developmental Gasification Processes, Oxygen-Blown Reactant Consumption, and Gas Production

	Moving bed		Fluidized bed		Entrained flow	
	Lurgi grate; commercial	BG/Lurgi slagging; pilot plant[a]	Winkler; commercial	HYGAS; pilot plant[a]	Koppers-Totzek; commercial	Texaco; pilot plant[a]
Gas exit temperature, °C	580	440	700	340	1290	1290
Pressure, bar	30	20	atm	70	atm	40
Feed capacity, GJ/(h·gasifier)	750	2100	800	6900	380	410
Reactants and products per 1000 kg coal[b]						
Oxygen, nM^3	247	337	490	213	587	657
Oxygen, kg	352	481	700	304	838	953
Steam, kg	1336	275	516	684	300	. . .[c]
Dry raw product gas, nM^3	2050	1744	1900	1642	2018	2138
Gas analysis, volume %						
Carbon dioxide, CO_2	29.7	2.5	20.0	24.7	7.1	10.6
Carbon monoxide, CO_2	18.9	60.6	34.0	24.0	58.7	51.6
Hydrogen, H_2	39.1	27.8	41.0	30.5	32.8	35.1
Methane, CH_4	11.3	7.6	3.0	19.4	. . . .	0.1
Other[d]	1.0	1.5	2.0	1.4	1.4	2.6
Fuel value, MJ/nM^3	12.3	14.2	10.7	14.5	11.6	11.1
H_2/CO, volume ratio	2.1	0.46	1.21	1.27	0.56	0.68
Product utilization						
Equivalent synthesis gas,[e] nM^3/1000 kg coal	1884	1939	1596	1850	1845	1988
Equivalent SNG, (CH_4),[f] nM^3/1000 kg coal	529	517	413	542	462	470

[a]Pilot plants are of various sizes: BG/Lurgi, 12,000 kg/h; HYGAS, 2500 kg/h; Texaco, 6500 kg/h.
[b]Based on moisture-and-ash-free (maf) coal with fuel value approximately 30 GJ/1000 kg.
[c]Texaco feeds coal slurried with water (60% coal); no other steam added.
[d]Includes nitrogen and various impurities (H_2S, COS, NH_3, etc.).
[e]Synthesis gas assumed equal to sum of H_2, CO, and 3 × CH_4 concentration.
[f]Methane potential assumed equal to methane (CH_4) concentration plus ¼ (H_2 + CO) concentration.

fied). The much higher pressure favors methane production and a higher CO_2 level in exit gas. The results shown for the developmental HYGAS processes are tentative; additional operation to confirm them is required. Winkler is operating a single-stage fluidized-bed pilot plant at pressures up to 10 bar.

While elevated pressure will decrease compression costs (oxygen volumes are one-fourth or less of the volume of product gas), carbon conversion is not appreciably improved. The HYGAS process, by utilizing several fluidized beds, does achieve some improvement. Two processes, **Westinghouse** and **U gas,** report success in agglomerating the ash particles by allowing higher temperature at the bottom of the bed. The low-carbon agglomerates are discharged preferentially owing to their larger size. Several processes utilize the high heat-transfer capability and solids mobility to supply external heat to the gasification reaction. The CO_2 acceptor process uses the exothermic absorption of CO_2 by calcined limestone or dolomite to supply heat for the steam-carbon reaction. Pilot-plant tests have demonstrated technical feasibility but not economic viability. The **COGAS** process supplies heat by burning part of the char externally with air and recycling the hot char to the steam-carbon reactor. The use of nuclear heat has been suggested in Germany, where high-cost coal could be saved by this means. All processes operated without the use of oxygen can effect a saving since the oxygen plant is of an order of magnitude of capital cost similar to that of the gasification system. However, these processes cannot be considered as operational in the commercial sense.

Operation at elevated pressure can benefit the performance of an entrained-flow gasifier by reducing compression cost. Krupp-Koppers, the parent firm supplying the Koppers-Totzek gasifier, has joined with Shell Oil to develop a pressurized version. This version should provide a gasifier of higher throughput and gas production at a lower cost primarily through a reduction of compression cost. In Table 9-18 data are provided for a pressurized Texaco unit. Both atmospheric and pressurized entrained-flow gasifiers have an exit-gas temperature above 1250°C (2282°F), and this results in a high oxygen requirement. Efficient heat recovery and utilization of thermal energy in the gasifier exit gases are mandatory for recovery of the energy required for oxygen production and gas compression.

For elevated-pressure operation of entrained-flow gasifiers, a reliable coal-feed system is a critical requirement. No carbon inventory in the gasifier is available to avoid high oxygen concentrations if coal feed is interrupted. The Texaco gasifier uses a pumped coal-water-slurry feed. The water content of such a slurry must be minimized to avoid high thermal penalties. The higher oxygen usage due to evaporating the water of a 65 percent coal-water slurry may be noted in Table 9-18 when the Texaco results are compared with those of Koppers-Totzek.

The use of coal-derived gases as a fuel is a well-known technique, but other important uses of such gases are addressed in Table 9-18. Conversion of raw product gas to a synthesis gas by re-forming its methane content is calculated for each of the processes. Similarly, conversion of product gas to methane is estimated by adjusting the hydrogen-to-carbon-monoxide ratio by the water-gas-shift reaction followed by methanation. Those processes which produce high methane concentration in the raw product gas are best suited for producing synthetic natural gas or methane (see HYGAS data).

For such conversions, the ratio of hydrogen to carbon monoxide is an important factor, and this ratio is listed for each process. The processes using high temperatures and low steam-to-coal ratios produce a gas rich in carbon monoxide. For producing hydrogen or hydrogen-rich synthesis gases, the carbon monoxide may be reacted with steam in a catalytic reactor external to the gasifier. The added steam must be considered in any overall evaluation of a gasification process. For example, if a hydrogen-to-carbon monoxide ratio of 2 is required in using a BG/Lurgi gasifier, an additional 437 kg (961 lb) of steam would be required for converting carbon monoxide to hydrogen. This would increase total steam consumption from 275 to 712 kg/1000 kg coal (0.275 to 0.712 lb/lb coal). While this would still be less steam than required by the Lurgi grate-type gasifier, the advantage of dependence on product-gas requirements is an important factor to be considered.

For this reason, any attempt to calculate the efficiency of gasification should be related to final product requirements and plant design. All the data are available in Table 9-18 to calculate the heating value of products and relate them to the energy input in the 1000 kg of coal used as a basis. However, use of the crude product gas as a final product can be misleading, as shown by the added steam required for producing a 2:1 synthesis gas. In addition, the energy used in the actual process must be considered, and this will depend on the use of waste heat to produce steam, power, or other needed energy forms for process use. Also given in Table 9-18 is an estimate of the possible coal throughput for a single gasifier of each type. The commercial extrapolations for coal throughput for pilot-plant processes are based on published data. The process proponents have shown varying optimism.

The oxygen-blown gasifiers shown in Table 9-18 are probably of greatest industrial interest. In addition to producing a fuel gas undiluted by nitrogen, they offer a potential of supplying synthesis gas for chemical use and synthetic natural gas (SNG) for pipe-line transport, which extends the marketability of oxygen-blown gasification systems.

In some cases, air-blown gasifiers may offer an advantage when fuel gas is desired. The use of air as a reactant does eliminate the capital cost and power required for operating an oxygen plant. This saving is negated to some extent by the need to purify larger volumes of product gas diluted with nitrogen. The lower fuel value of these product gases prevents economical pipe-line transport and mandates use of the fuel gases close to the point of production. Some estimates do show a 10 to 15 percent lower cost per unit of heating value for the products of air-blown gasification versus oxygen-blown product gases. Data are provided for two air-blown systems in Table 9-19. The gases produced are similar in characteristics to the producer gas shown in Table 9-16.

This table can be conveniently compared with Table 9-18, so that downstream gas volumes and purification-unit sizes can be estimated and compared for oxygen- and air-blown gasifiers. Owing to the use of air, larger volumes of lower-fuel-content product gas are obtained. As in the case of oxygen systems, a higher gas exit temperature requires more air usage and results in a lower fuel value. It should be noted that the Combustion Engineering gasifier shown in Table 9-19 is not a commercial system, but the data are based on pilot-plant operation which has been extrapolated to a commercial size. In an attempt to improve the heat recovery of the high-temperature-gasifier product, the Combustion Engineering gasifier injects coal feed

TABLE 9-19 Commercial and Developmental Gasifiers; Air-Blown-Reactant Consumption and Gas Production

	Moving bed Lurgi, grate-type; commercial	Entrained flow, Combustion Engineering; pilot plant*
Gas exit temperature, °C	535	927
Pressure, bar	20	atm
Feed capacity, GJ/h/gasifier	750	2700
Reactants and products per 1000 kg coal†		
Air, nM^3	1920	3561
Air, kg	2560	4880
Steam, kg	1650	None
Dry, raw product gas, nM^3	2695	5322
Gas analysis, volume %		
Carbon dioxide, CO_2	13.5	5.1
Carbon monoxide, CO	16.5	22.3
Hydrogen, H_2	23.8	11.2
Methane, CH_4	4.0	
Nitrogen, N_2	41.3	57.4
Other	0.9	4.0
Fuel value, MJ/nM‡	6.4	4.0

*Pilot plant has capacity of 4500 kg/h of coal.
†Based on moisture-and-ash-free (maf) coal with a fuel value of 30 GJ/1000 kg.
‡Includes some hydrocarbons, moisture, and sulfur compounds.

into these hot gases to enrich the product gas through pyrolysis of the coal.

It is not possible to identify all the gasifiers under development or in use throughout the world in this brief discussion. It should be noted that many brands of small air-blown moving-bed gasifiers are commercially available. Most of these are analogous to the Lurgi shown in Table 9-18, except that they operate at atmospheric pressure. This decrease in pressure would reduce methane formation and decrease gasifier throughput. Other gasifiers under development utilize molten salts, metals, or slags as a medium to improve contact between solid and gaseous reactants. Many of these systems are in too early a stage of development to provide meaningful product-gas and reactant usage information. The use of alkali catalysts to increase the rate of steam-carbon and methane formation is reported as showing promise in bench-scale and process-development units (PDUs). The data indicate that lower operating temperatures may be achieved and the use of oxygen eliminated because of the exothermic methane-forming reactions. Further data on catalyst recovery and downstream processing are needed to establish the technical and engineering feasibility of such processes.

Application of Coal Gasification While Table 9-19 mentions some of the products that can be obtained by refining the raw-gas output of the gasifier, application of coal gasification involves considerable processing downstream of the gasifier. Figure 9-15 is a composite flow plan indicating some of the complexities of such processing. The area denoted as "Gasification" includes the supply of coal and gaseous reactants (air or oxygen and steam). The particulate matter, dust, and tars must be removed and the sulfur content lowered to meet regulations for the combustion of gaseous fuels. At this point, the gasifier output has been converted to a usable fuel: low-Btu fuel gas for air-blown gasifiers and medium-Btu fuel gas for oxygen-blown gasifiers.

The area denoted as "Product Refinement" includes more complete removal of sulfur to avoid deactivation of catalysts; adjustment of gas composition using the water-gas-shift reaction [see Reaction (9-17), Table 9-17]; and, to produce high-Btu or pipe-line gas (also called SNG), complete conversion to methane [see Reaction (9-18), Table 9-17]. As can be noted from Table 9-17, the water-gas-shift reaction is slightly exothermic but requires the addition of steam, which results in an energy drain on the overall process. The methanation reaction is highly exothermic, and the heat released is usually utilized to produce steam, which may be used for heating or for providing power.

The centrality of gasification to all coal-conversion processes is illustrated by the end products and their potential uses. The mixture of hydrogen and carbon monoxide produced as synthesis gas may be used to prepare chemicals or converted to liquid hydrocarbons (indirect liquefaction). By altering the steam-to-gas ratio in the shift reactor, pure hydrogen may be produced for use in ammonia manufacture and petroleum refining. Finally, for direct liquefaction, or hydrogenation of coal, hydrogen must be supplied generally through gasification of coal or unreacted residua of the hydrogenation process.

Gas Purification and Refinement Most of the processing steps shown in Fig. 9-15 upstream and downstream of the actual coal-gasification unit are of conventional design. The handling, storage, size reduction, and feeding of coal are discussed in other portions of this section and in other sections of this handbook (see "Solid Fuels: Coal" and "Solid-Fuels Combustion on Stokers and in Suspension" in this section and also Secs. 7 and 8). The removal of particulates and droplets of oil and tar from raw product gases is also a generally well known processing step. Figure 9-12, depicting a Lurgi gasifier, includes a gas washer of relatively simple but effective design. Additional information on gas scrubbing is available in Secs. 14 and 18, where the engineering of such systems is discussed.

Adjustment of gas composition by the water-gas-shift reaction, using an external catalytic reactor, is shown in Fig. 9-15 and has been previously discussed. The recent development of sulfur-resistant catalysts for this reaction allows some choice as to the location of this processing step. Designers of total plant systems have shown a preference for locating the shift reactor immediately after the particulate-removal units; in a pressurized system, this allows for a high gas exit temperature. This location permits the use of steam in the gasifier product gas in the shift reaction and eliminates the need for reheating gases after complete sulfur removal.

An important function of the gas-purification section of a coal-gasification plant is the removal of CO_2 and H_2S (acid gases). A variety of chemical and physical solvents are offered by various proprietary processes for removing acid gases. Table 9-20 lists one or two typical solvents for each class of absorbents. The alkaline amines and hot

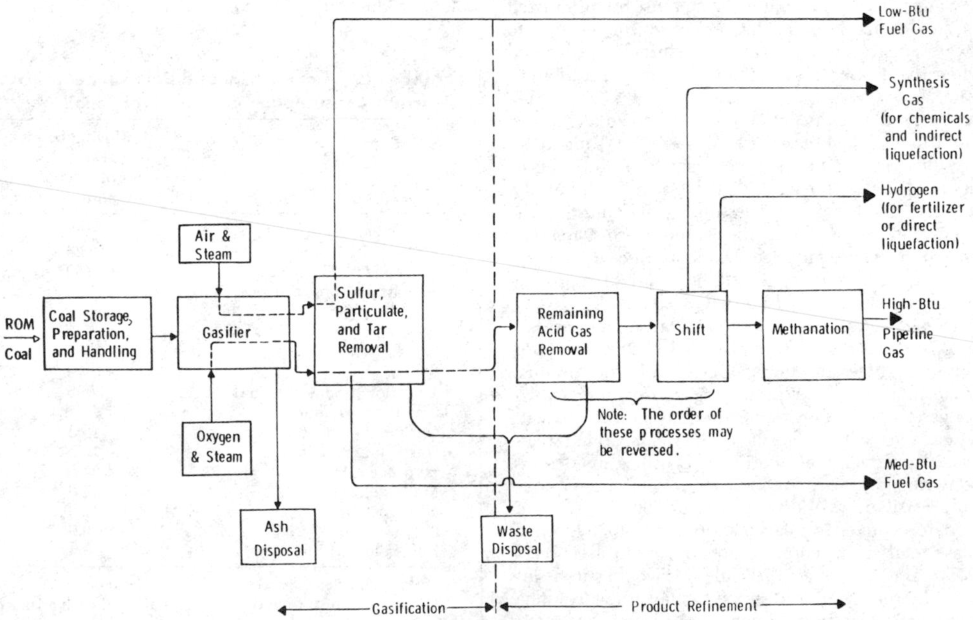

FIG. 9-15 General systems description: surface-coal gasification.

TABLE 9-20 Acid-Gas-Removal Systems for Coal Gasification

Solvents	Operating ranges, CO_2 partial pressure	Selectivity* H_2S/CO_2	Selectivity* CO_2/HC	Utility requirements†
Amines				
MEA	Low	1	8	10
TEA	Low	3	8	7
Hot carbonate				
Benfield	Moderate	6	9	4
Giammarco Vetrocoke	Moderate	9	9	4
Physical solvents				
Methanol (Rectisol)	High	7	2	2
Selexol	High	9	3	1
Mixed solvents				
Methanol plus DGA	Moderate	1	3	5

SOURCE: Fleming and Primack (IGT), presented at 81st National Meeting, AIChE, Kansas City, Mo., April 1976.

*Arbitrary number scale, 1 to 10: 1 = poor and 10 = excellent selectivity.

†Utility consumption, 1 to 10: 1 = low and 10 = high utility use.

‡MEA, methylethanolamine; TEA, triethanolamine.

carbonate solutions react chemically with acidic gases. These solvents require heating to regenerate them and remove the acid gases. As a result, utility consumption is high. Physical solvents can be regenerated with only minor use of heat energy but operate only at elevated pressure.

An important characteristic is solvent selectivity. For cleansing a gas to be used as a fuel, the only acidic components which must be removed are those containing sulfur. For such systems, a physical solvent such as Selexol (trade name of the Allied Chemical Company) would be ideal. The amine solvents, which were originally developed for use with natural gas, have excellent selectivity vis-à-vis hydrocarbons. For facilities which require total removal of sulfur compounds and carbon dioxide for downstream processing over sensitive catalysts, more than one removal step may be required to avoid catalyst damage.

Economics of Coal Gasification Many estimates of the costs of coal gasification have been provided in the literature since the early 1970s. Usually, these estimates were outdated by the time they were published. Double-digit inflation increased the costs of almost all capital equipment, labor, and materials. Even though the prices of petroleum and natural gas increased sharply, fuel gases made from coal have not yet become competitive with these naturally occurring products.

Unfortunately, the accuracy of these published estimates is not dependable. A very large portion (as much as 70 percent) of production cost is related to capital investment for the plant. It is in estimating capital investment that most published estimates go astray. These capital costs are obtained by a system of factors. The costs of major capital-equipment items are estimated and these costs multiplied by various factors to generate costs of installation, piping, instrumentation, electrical needs, etc., associated with these items. These factors have been shown to be not representative of plants handling large volumes of gases and solids but to be more typical of liquid-processing plants such as oil refineries or chemical plants. As a result, errors of as much as 30 percent have been found when a factored estimate was compared with a line-by-line material-takeoff estimate.

Using 1978 dollars and assuming utility financing in which 75 percent of the capital cost is borrowed, it is estimated that synthetic natural gas produced from lignite will cost from \$5 to \$6.50/10^6 Btu (1.05 GJ). Such a plant, producing 125 to 250 million scfd (3.5 to 7.0 m^3), will require a capital investment of \$9000 to \$10,000 per daily 10^6 Btu of production. This is equivalent to approximately \$2.5 billion for a plant producing 250 million scfd of SNG. It should be noted that these costs will vary with inflation and also with the costs of borrowed funds. This results in a doubling of the effects of inflation, since interest rates are driven upward by rising costs and decreasing currency values.

In spite of major research and development efforts, it appears doubtful that major reductions in cost can be achieved. Almost 80 percent of the capital cost of coal-gasification plants consists of conventional equipment, which will not be altered appreciably by new gasification technology. Improvements will be small in size and slow in being used and will depend on actual commercial implementation of state-of-the-art systems. Only through profit-oriented operation will reliable and practical improvements be realized.

DIRECT COAL LIQUEFACTION

GENERAL REFERENCES: Donath, *Chemistry of Coal Utilization*, suppl. vol., ed. by Lowry, Wiley, New York, 1963. Gorin, *Chemistry of Coal Utilization*, 2d suppl. vol., ed. by Elliott, Wiley, New York, 1981. Wu and Storch, *Hydrogenation of Coal and Tar*, U.S. Bur. Mines Bull. 633, 1968.

Background The primary objective of any coal-liquefaction process is to increase the hydrogen-carbon ratio and remove sulfur, nitrogen, oxygen, and ash from the coal. Table 9-21 shows the hydrogen-carbon ratios in proceeding from coal to crude petroleum and gasoline, together with the four techniques for accomplishing these objectives. Direct coal liquefaction refers to any process approach in which coal and hydrogen are reacted at high pressure and temperature either with or without hydrogen donor solvent and catalyst. The first three techniques in Table 9-21, direct hydrogenation, solvent extraction, and pyrolysis, follow this approach. The fourth, catalytic hydrogenation of carbon monoxide, or indirect liquefaction, first converts coal to a synthesis gas, followed by purification and use of a catalyst to form liquid products. An alternative indirect-liquefaction route converts purified synthesis gas to methanol, for fuel use or conversion to gasoline.

The technology of coal liquefaction to synthetic fuels is not new. In 1913 Dr. Friedrich Bergius discovered the technique of adding hydrogen to coal at a pressure of 200 atm (2940 lbf/in^2) and a temperature of about 450°C (842°F). Under these conditions most oxygen was hydrogenated to water, some nitrogen to ammonia, and most sulfur to hydrogen sulfide. Hydrogen was also chemically combined with the coal to produce a liquid similar to petroleum. Later research at I. G. Farben led to the discovery of catalysts which increased the speed of hydrogen addition to coal. Interest by the U.S. Bureau of Mines began in 1924 with A. C. Fieldner visiting several

TABLE 9-21 Basic Approaches of Coal Conversion to Liquid Hydrocarbons

1. Direct hydrogenation at elevated temperature and pressure, with or without catalysts
2. Solvent extraction (hydrogen donor)
3. Pyrolysis
4. Catalytic hydrogenation of carbon monoxide

	Bituminous coal	Lignite	Crude petroleum	Gasoline
H/C	0.8	0.7	1.8	1.9

facilities in Germany, France, and England to become familiar with their coal research. Production of synthetic liquid fuels and chemicals from coal in Germany became very noticeable in the mid-1920s. In 1927 the U.S. Bureau of Mines started a laboratory research program on the thermodynamics and kinetics of chemical reactions for the production of alcohols and hydrocarbons from carbon monoxide and hydrogen. By 1939 gasoline was being produced by coal hydrogenation in Germany at 1 million tons/year and in England at 150,000 tons/year.

Research at the Bureau of Mines was also conducted to verify claims of Fischer and Tropsch in 1926 on the discovery of a catalytic technique to produce liquid hydrocarbons similar to petroleum by using water gas at atmospheric pressure and a temperature of 135 to 250°C. By using a cobalt-copper-manganese catalyst at 275°C, about 1.7 gal of oil/1000 ft^3 of gas which contained 34 weight percent of motor fuel could be produced. German commercial plants operated at from 1 to 10 atm (14.7 to 1470 lbf/in^2) at catalyst temperatures of 180° to 210°C.

In the early 1950s, the Bureau of Mines constructed and operated a demonstration plant at Louisiana, Missouri, using the technology of catalytic hydrogenation of coal and carbon monoxide to produce liquid fuels. Research in the chemistry of coal liquefaction since then has produced several modifications of the four types of liquefaction processes. These include catalytic hydrogenation in fixed and ebullated beds, zinc halide hydrocracking, reaction of carbon monoxide–rich synthesis gas and steam with coal, and a combination of solvent extraction and catalytic hydrogenation in two-stage liquefaction.

Currently, the most advanced direct coal liquefaction processes are solvent-refined coal (SRC), Exxon donor solvent (EDS), and H-Coal. These are compared in a generalized flow diagram, Fig. 9-16, and discussed in greater detail later.

Direct-Liquefaction Kinetics All direct-liquefaction processes may be considered to consist of three basic steps: (1) coal slurrying in a vehicle solvent, (2) coal dissolution under high pressure and temperature, and (3) transfer of hydrogen to the dissolved-coal products.

In the temperature range of 350 to 450°C (662 to 842°F) the coal is dissolved in about 1 min. Solvents are assumed to promote thermal cracking of coal into smaller, more readily dissolved fragments. These may be stabilized through reactions with each other or with hydrogen supplied either by a donor solvent or in a gas phase.

Data on Illinois, River King (hvCb), and Kentucky No. 9 (hvBb) coals were used by Wen and Hahn to obtain a rate equation for coal dissolution under hydrogen pressure. These data included a temperature range of 375 to 500°C and pressures up to 136 atm (0–2000 psig). An empirical rate expression was proposed as

Rate of dissolution = rate constant
× fraction of undissolved solid organics × coal-solvent ratio

$$r_A = k_0 \exp -E/RT \exp (0.000684\, P_{H_2})(C_{so})(1 - x)(C/S) \tag{9-21}$$

where r_A = rate of dissolution, g/(h·cm^3) reactor volume
C_{so} = weight fraction of organics in unreacted coal
k_0 = rate constant, g/(h·cm^3) reactor volume
P_{H2} = h· drogen partial pressure, psia
x = conversion, solid organics/solid organics in original coal
C/S = coal-solvent weight ratio

By assuming an Arrhenius temperature dependency, calculated and experimentally reported conversions agreed well for the following values:

Constant	Illinois, River King (hvCb)	Kentucky No. 9 (hvBb)
k_0	2125 g/(h·cm^3)	15.3 g/(h·cm^3)
E	11 kcal/(g·mol)	4.5 kcal/(g·mol)

The small values of activation energy suggest that the dissolution rate is controlled by counterdiffusion of organic components from the coal surface and dissolved hydrogen from the solvent-hydrogen mixture. Also, the rate of dissolution appears to be exponentially dependent on hydrogen partial pressure.

A free-radical mechanism was also proposed for coal dissolution in hydrogen donor solvents. Initial dissolution was considered a thermal process, with a net rate dependent upon the type of solvent and its effectiveness in stabilizing free radicals. The greater a solvent hydrogen donor capability, the more effective it is in terminating radicals. An overall rate-limiting step in the process appears to be rehydrogenation of the donor solvent.

A mixture of Kentucky No. 9/14 (hvBb) coal and creosote oil was used in the study at a 3:1 solvent-to-coal ratio. Hydrogen solubility was experimentally measured over 100 to 400°C and 500 to 3000 psia, simulating the high temperature and hydrogen partial pressure of coal-liquefaction reactors. The first-order-reaction model obtained from hydrogen-absorption data expressed total hydrogen consumption as

$$H_c = H_{T\alpha}[1 - \exp(-\alpha k_L t)] \tag{9-22}$$

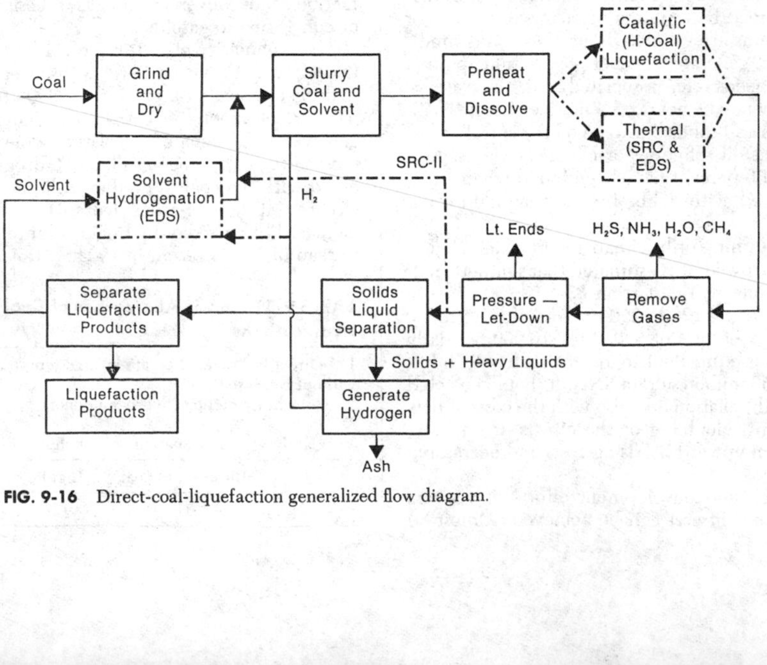

FIG. 9-16 Direct-coal-liquefaction generalized flow diagram.

where α = H_L/H_T
H_c = hydrogen consumed in reaction, g
H_{To} = total hydrogen, $t = o$, g
H_L = hydrogen in liquid phase, g
H_T = total hydrogen at time t, g
t = time, min
k_L = first-order rate constant, min^{-1}

Solubility of hydrogen in the coal-liquefaction solvent was represented by a Henry's-law coefficient:

$$\beta = S/P_{H2}$$

where S = solubility, g H_2/g oil
P_{H2} = hydrogen partial pressure, psia

$\beta \times 10^7$, g H_2/(g oil·psia)	T, °C	α	T, °C
5.95	100	.183	385
6.94	200	.193	400
7.75	300	.196	410
9.65	400	.211	435

An Arrhenius plot of hydrogen-transfer rates gave an activation energy of 21 kcal/mol and a frequency factor of 1.06×10^5 min^{-1}. From these values it appears that mass-transfer effects are small and that the hydrogen consumption rate was kinetically controlled.

Solvents and heat facilitate thermal degradation of coal to form relatively low-molecular-weight free radicals. These may be stabilized by hydrogen transfer from a hydroaromatic solvent. The functions of dissolved gaseous hydrogen and catalyst are to rehydrogenate the solvent. Catalysts are considered to be primarily responsible for nitrogen and sulfur removal and the formation of light liquid products.

A first-order hydrogen absorption in creosote oil was proposed with creosote oil being considered a good representative for steady-state, coal-derived solvent. The progress of coal conversion was measured on the basis of a fraction of organic products soluble in benzene or pyridine. At conversion above 90 percent these results will be very similar.

The rate of consumption of gas-phase hydrogen in the liquefaction reactor was expressed as

$$-rH_2 = \frac{1}{S}\frac{dN_{H2}}{dt} = k_g(P_{H2} - P_{H2i}) \qquad (9\text{-}23)$$

where rH_2 = rate of consumption, (g·mol)/(cm^2·h)
S = surface area for absorption, cm^2
NH_2 = mole of hydrogen, g·mol
k_g = gas-side mass-transfer coefficient, (g·mol)/(cm^2·h·atm)
P_{H2} = hydrogen partial pressure, atm
P_{H2i} = hydrogen partial pressure at gas-liquid interface, atm

Integrating this over the reactor length,

$$-\ln (Y_{H2O}/y_{H2i}) = k_g a R_l P_T V_R / F_G \qquad (9\text{-}24)$$

where $k_g a$ = overall gas-absorption coefficient, (g·mol)/(h·cm^3·atm)
R_l = liquid holdup
P_T = total pressure, atm
V_R = reactor volume, cm^3
F_G = total gas flow rate, (g·mol)/h

The liquid holdup R_l may be estimated by use of the Lockhart-Martinelli correlation for large-diameter reactors or the Nicklin correlation for smaller diameters, less than 7 cm.

Temperature, reactor hydrodynamics, and catalyst affect the overall gas-absorption coefficient ($k_g a$). Values of ($k_g a$) are presented in Fig. 9-17 and show that increasing temperature and turbulence increase $k_g a$ for both catalyzed and noncatalyzed systems. As shown in Fig. 9-18, increasing turbulence, which reduces diffusion barriers, has more effect on catalyzed systems when reaction rates are higher. The pressure effects on the coal-dissolution coefficient were observed to be $P_{H2}^{0.5}$ for $P_{H2} > 50$ atm.

Bituminous coals begin to soften in the 325–350°C temperature range. Rapid mixing will increase both heat and mass transfer in this regime by reducing diffusion resistance to hydrogen at the solid-liquid boundary. The effect of the slurry Reynolds number on the coal-dissolution rate is shown for three coals in Fig. 9-19. Increased hydrogen absorption for catalytic as compared with noncatalytic systems is due to the further hydrogenation of coal liquids to lighter products and to sulfur and nitrogen removal.

The conversion reaction from coal to oil has been modeled as a series of steps:

Coal + solvent → preasphaltene → asphaltene → oil

with some gas formation accompanying each step. In a study using Illinois No. 6 coal, Burning Star Mine (hvCb), and process-derived heavy distillate (232 to 454°C), data were obtained at 13.8 MPa (2000 lbf/in²) and 400 to 475°C. Activation energies for the steps of the reaction series were determined to be:

Activation energy, kcal/(g·mol)	Reaction step
15	Preasphaltene → asphaltene
21	Asphaltene → oil
32	Coal → preasphaltene

At 450°C (843°F) stoichiometries for the reaction steps were represented as

Coal + 3 solvent → preasphaltene
Preasphaltene → 2 asphaltene
asphaltene → 3 oil

which appears consistent with published molecular-weight values of solvent (250), preasphaltene (1000), and coal (2250).

A more complex reaction model was proposed from the results of a kinetic study of thermal liquefaction of Belle Ayr subbituminous coal. Data were obtained over a temperature range of 400 to 470°C (752 to 878°F) at 13.8 MPa (2000 psig) by using two solvents, hydrogenated anthracene oil (HAO) and hydrogenated phenanthrene oil (HPO), at a coal-solvent ratio of 1:15. Results were correlated with the following model;

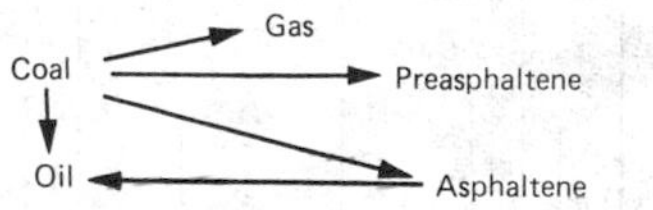

Activation energies and frequency factors for the various steps of this model were determined as follows:

Reactor	Rate constant	Activation energy, kcal/(g·mol)		Frequency factor, min^{-1}	
		Coal-HAO	Coal-HPO	Coal-HAO	Coal-HPO
Coal → oil	K_o	14.1	28.9	3.11×10^3	2.1×10^8
Coal → preasphaltene	k_p	13.8	4.3	2.81×10^3	4.94
Coal → asphaltene	K_a	15.6	8.6	1.12×10^4	9.63×10^1
Coal → gas	K_g	21.5	10.5	8.72×10^5	3.85×10^2
Preasphaltene → asphaltene	K_{pa}	12.8	33.9	9.66×10^2	2.48×10^9
Asphaltene → oil	K_{ao}	16.0	25.6	1.42×10^3	1.53×10^7

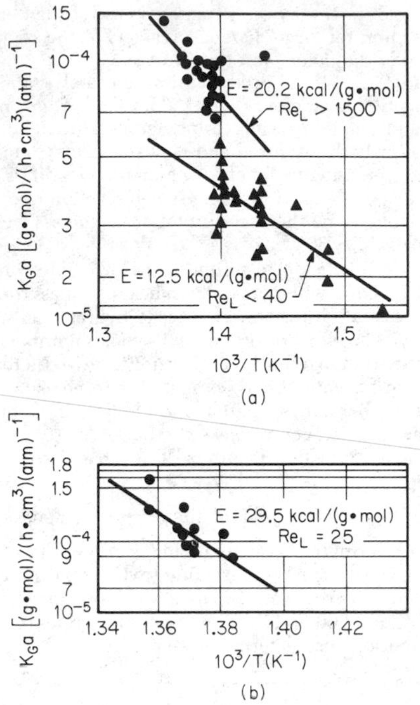

FIG. 9-17 Effect of temperature on the coal-dissolution-rate coefficient including hydrodynamic influence. (*a*) Noncatalytic systems. (*b*) Catalytic systems.

Magnitudes of K_g, K_p, K_a, and K_o indicate the importance of direct reactions with coal, where K_{pa} and K_{ao} are hydrocracking reactions in the conversion process. Data for K_o and K_{ao} from the experiments with HPO show the effect of an effective hydrogen donor solvent on product yield and indicate that oil production from coal will be increased by the use of a good hydrogen donor solvent.

The ability of coal minerals to catalyze the hydrogenation creosote oil and Kentucky coal has been investigated. Catalytic materials investigated are listed in Table 9-22. Hydrogen partial pressure was used as a measurement of hydrogenation activity for these materials as shown in Fig. 9-20. Desulfurization results are contained in Fig. 9-21. Data for both tests of catalytic activity are in general agreement for all except two materials. First, pyrite was determined to be an effective hydrogenation catalyst but poor for desulfurization. Second, contrasting results were obtained in the catalytic-activity experimental results for iron. Thus, differences in the composition of coal-ash minerals will determine their effectiveness for hydrogenation or hydrodesulfurization activity in coal liquefaction.

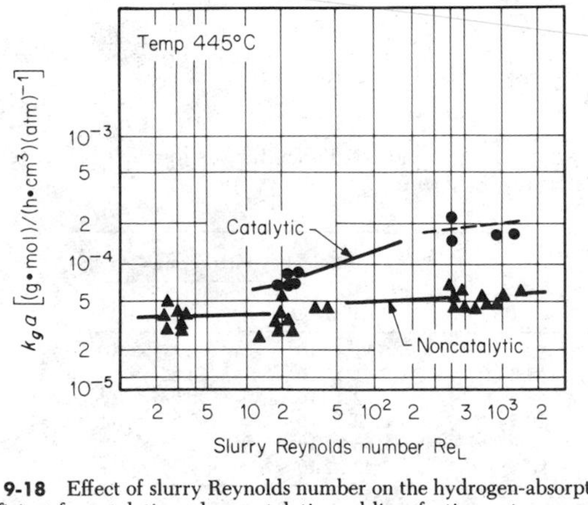

FIG. 9-18 Effect of slurry Reynolds number on the hydrogen-absorption-rate coefficient for catalytic and noncatalytic coal-liquefaction systems.

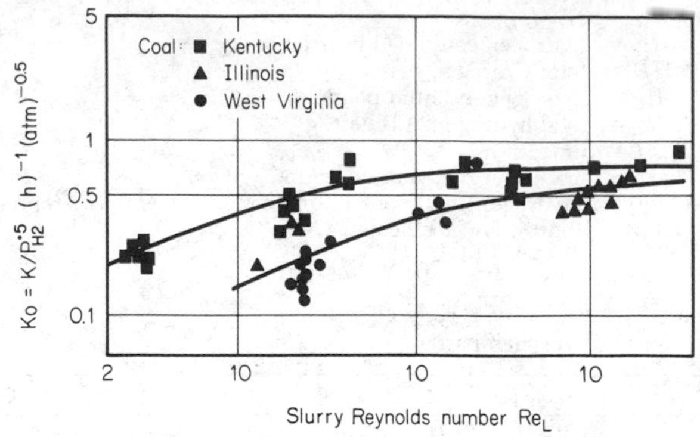

FIG. 9-19 Effect of the slurry Reynolds number on the rate of coal dissolution for Kentucky, Illinois, and West Virginia coals.

Direct-Liquefaction Processes

Solvent-Refined Coal (SRC) This process was initiated by the Pittsburgh & Midway Coal Mining Co. in the early 1960s.

SRC-I process. In the SRC-1 (solid-fuel) operating mode, shown in the flow diagram of Fig. 9-22, pulverized coal is mixed with process-derived solvent. After adding hydrogen, the slurry is pumped through a preheater and into a dissolver which operates at 425 to 470°C (800 to 878°F) and a pressure of 10 to 14 MPa (1450 to 2030 lb/in^2). Coal conversions of 92 to 95 percent are achieved during the residence time of 20 to 60 min. The dissolver effluent is separated into gases and slurry in a separator. The raw gas is scrubbed with a dilute caustic solution to remove hydrogen sulfide and carbon dioxide. A portion of the scrubbed gas is purged to prevent accumulation of noncondensables (C_1–C_4 hydrocarbon gases and CO), but most of the gas is recycled back into the process. The slurry product, after being cooled to 315°C (600°F) and depressurized to 0.8 MPa (116 lbf/in^2), is passed through a precoated rotary drum or leaf filter to separate undissolved coal and mineral matter from the SRC solution. In a commercial plant, these residues would be sent to a gasifier to produce process hydrogen. The filtrate is vacuum-distilled to separate the solvent from the nondistillable product (SRC), which is then solidified by cooling. The SRC product is almost ash-free ($<$0.16 percent) and has a sulfur content of 0.6 to 0.9 percent and a heating value of approximately 37.2 MJ/kg (16,000 Btu/lb). Typical operating conditions and product yields for SRC-I are shown in Table 9-23.

Studies have indicated that bituminous coals of Kentucky No. 9 and No. 14, Illinois No. 6, and Indiana V are attractive for use in the SRC process but that Wyodak subbituminous coal and Utah bituminous coal are not. It appears that certain components of mineral matter in coal, iron compounds in particular, may have a catalytic effect on hydrogen transfer, resulting in an increase in coal conversion. High-vitrinite and low-oxygen contents of the coal are also desirable to achieve high SRC yield at low hydrogen consumption.

SRC-II process. The SRC-II process is an improved version of the SRC process, recycling a portion of the reactor effluent slurry in place of the distillate solvent of the original SRC-I process. Because of increased severity of operating conditions in the SRC-II process, the primary product is a liquid distillate fuel with a 217 to 455°C (423 to 851°F) boiling range. The flow diagram of the SRC-II operating mode is shown in Fig. 9-23. The general process is the same for both modes from the slurry-mix tank through the dissolver. In the SRC-II mode, however, the dissolver slurry is split into two streams; one is recycled to the slurry mix tank, while the other is passed into a vacuum-flash unit. The vacuum-flash condensate is fractionated to

TABLE 9-22 Description of Coal Minerals or Catalytic Agents Studied

Species	Classification	Description
Ankerite (ferriferrous dolomite)°	Carbonate	An isomorphorus mixture of $CaMg(CO_3)_2$ and $CaFe(CO_3)_2$.
Calcite°	Carbonate	A crystalline form (hexagonal scalenohedral class of the hexagonal system) of $CaCO_3$. Often, to a small extent the Ca is replaced by iron, magnesium, and manganese. Clay, sand, bitumen, and other mechanical impurities may be present.
Dolomite°	Carbonate	A double salt with equal molecular quantities of $CaCO_3$ and $MgCO_3$ and not an isomorphous mixture of these two compounds; usually found in a curved rhombohedral form.
Kaolinite°	Kaolin	A common type of clay; often found in minute pseudohexagonal (monoclinic) crystals; chemically, an acid aluminum silicate, $H_4Al_2Si_2O_9$ or $2H_2O\text{-}Al_2O_3\text{-}2SiO_2$ (H_2O = 14%). Iron is often present is small amounts.
Muscovite°	Shale	A lamina-type silica substance, having a monoclinic crystal structure, and chemically classified as an acid potassium aluminum orthosilicate, $H_2KAl_3(SiO_4)_3$ or $2H_2O\text{-}K_2O\text{-}3Al_2O_3\text{-}6SiO_2$($H_2O$ = 4.5%). Often, the potassium is partially replaced by sodium, and some varieties contain an excess of silicon over that indicated.
Pyrite	Sulfide	A cubic structure of FeS_2, having in its crystalline structure a rock-salt type of arrangement of Fe^{2+} and S_2^{2} ions, with iron being octahedrally surrounded by S and each S atom having one S and three Fe atoms as neighbors. Uncommonly, Ni, Co, or sometimes both are found substituted for Fe. Obtained from Matheson, Coleman, and Bell (90–95% pure).
Quartz°	Accessory	A crystalline form of SiO_2; a member of the triagonal trapezohedral class of the hexagonal system.
Siderite°	Carbonate	A crystalline form of $FeCO_3$, with the brown to gray crystals usually being rhombohedral. Calcium, magnesium, and manganese are usually present in small amounts as replacing elements.
Co-Mo-Al	....	Commercial catalyst from Laporte Industries, Inc. (Comox 451, 1.5-mm extrudute); surface area = 300 m^2/g; pore volume = 0.66 mL/g. Chemical analysis: 3.7, 12.8, 0.06, 1.4, and 0.03% CoO, MoO_3, Na_2O, Na_2O + K_2, SiO_2, and SO_3 respectively.
Iron	....	Reagent-grade hydrogen-reduced iron from Mallinckrodt, Inc.
Reduced pyrite	....	Solid residue from hydrogenation-of-creosote oil in presence of 15% by weight of FeS_2 at 425°C, stirrer setting of 2000 r/min, and 3000 psig of initial hydrogen pressure.
SRC residue	....	Obtained from filter cake from Wilsonville SRC pilot plant. Analysis: 55.2% ash content and 13.6% S for −325-mesh material; and, prior to screening, 30% filter aid, 53.6% ash, and 2.9% S.
Coal ash	....	Obtained by burning Kentucky No. 9/14 mixture (7.2% ash) in a muffle furnace at 1000°C; analysis: 13.7% iron.
Kaolin	....	Obtained from W. H. Curtis and Co.

°Minerals obtained from David New, Minerals and Books, Providence, Utah.

NOTE: All agents were ground to −325 mesh prior to use, except muscovite, which was ground only to −80 mesh because of its laminar silica structure, and except as indicated.

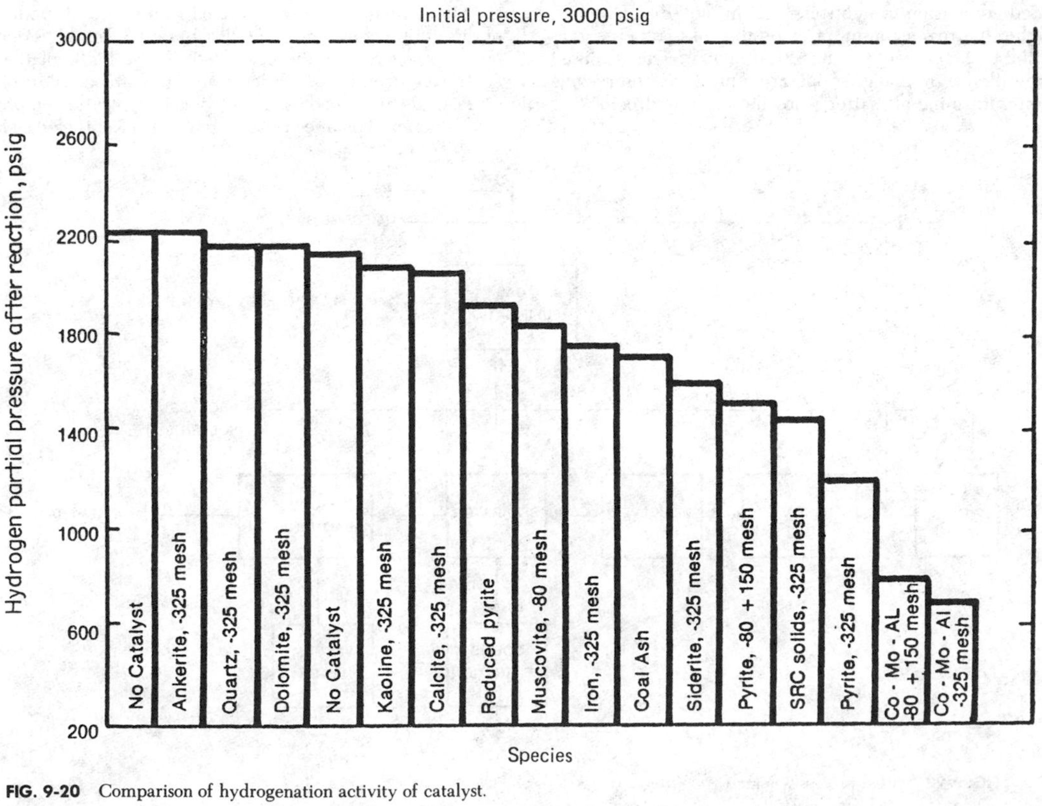

FIG. 9-20 Comparison of hydrogenation activity of catalyst.

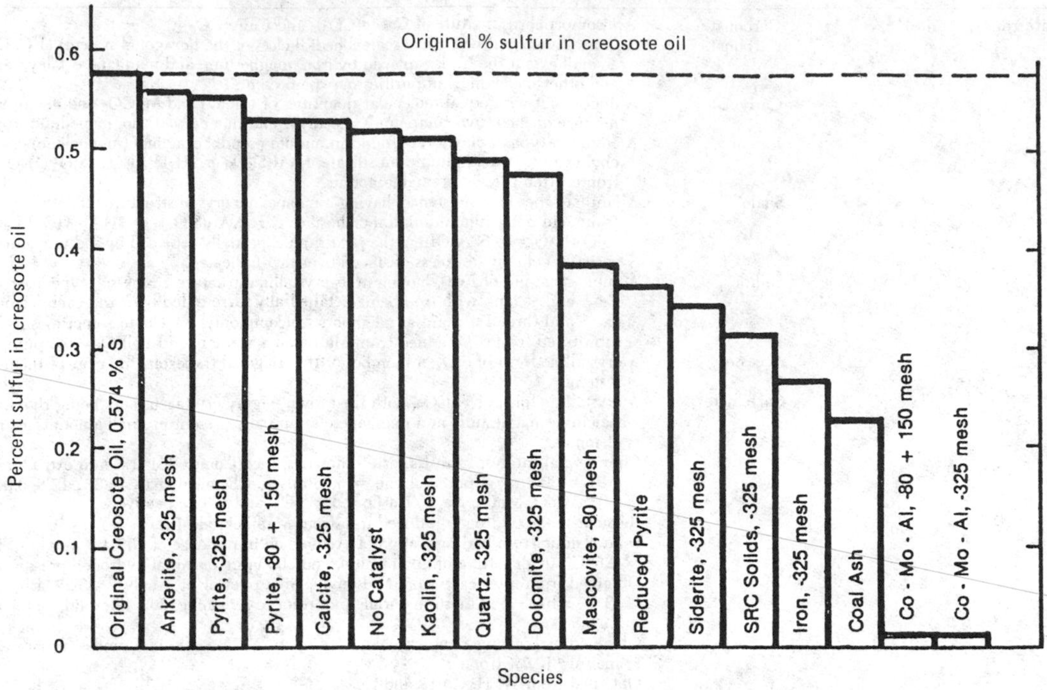

FIG. 9-21 Comparison of desulfurization activity of catalyst.

yield naphtha and middle and heavy distillates. The vacuum residue is heavy oil, unreacted coal, and mineral matter.

The increased conversion to light products in the SRC-II process results in a higher hydrogen consumption of about 4 to 5 percent, as compared with 2 to 2.5 percent for the SRC-I process. This is caused by the longer residence time, higher operating pressure, and recycle of coal solution with mineral matter, allowing more hydrocracking in the dissolver.

Because the mineral residue is separated as vacuum-flash bottoms along with the undissolved coal, the solid-liquid separation step of the SRC-I mode can be eliminated in the SRC-II mode. The degree of dissolution of coal and the distribution of products depend greatly on the reactivity of the coal feed. Three high-volatile bituminous coals have been tested to determine the effects of dissolver temperature, coal-slurry feed rate, and coal concentration on product yields.

The material balance runs with western Kentucky coal, Illinois No.

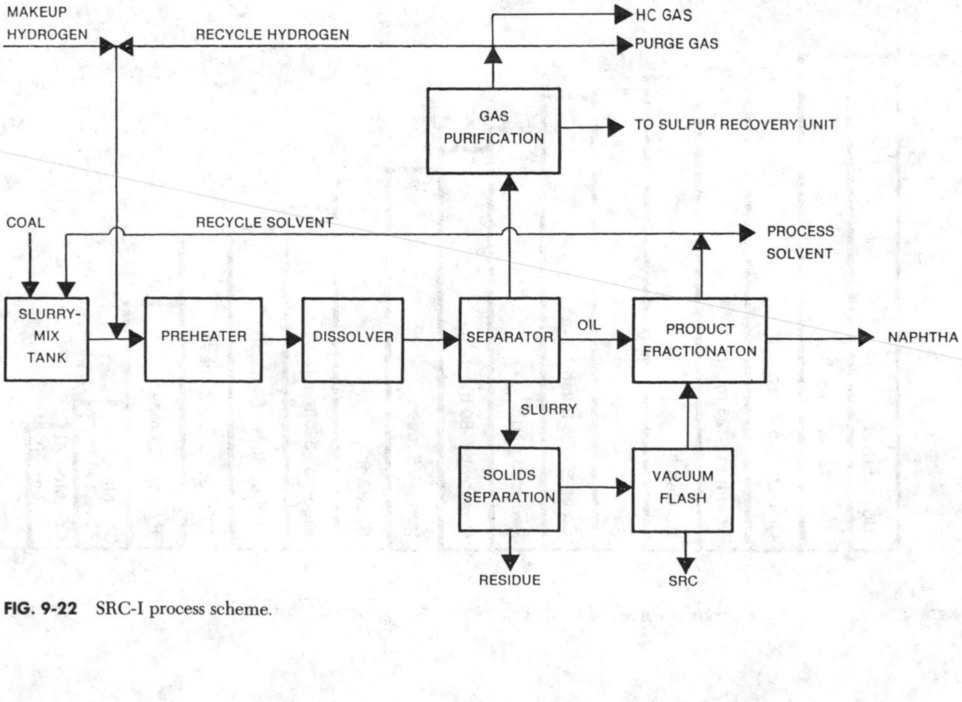

FIG. 9-22 SRC-I process scheme.

TABLE 9-23 SRC-I Product Yields from Pilot Plant

Operating conditions	
Coal	Western Kentucky 9 and 14
Coal (maf) feed rate	1837 kg/h (4050 lb/h)
Solvent feed rate	2860 kg/h (6305 lb/h)
Hydrogen purity in feed gas	94 mole %
Dissolver outlet temperature	724 K (842°F)
Dissolver pressure	10.3 MPa (1500 lbf/in^2)
H_2 consumption	2.4 weight % maf coal
Product yield, weight % maf coal	
C_1–C_4 hydrocarbon gas	3.7
Light oil, 193°C$^-$ (385°F)	5.1
Wash solvent, 193–249°C (385–480°F)	4.0
Process solvent, 249–254°C (480–860°F)	4.4
SRC	63.0
Ash	9.6
Unreacted coal	5.4

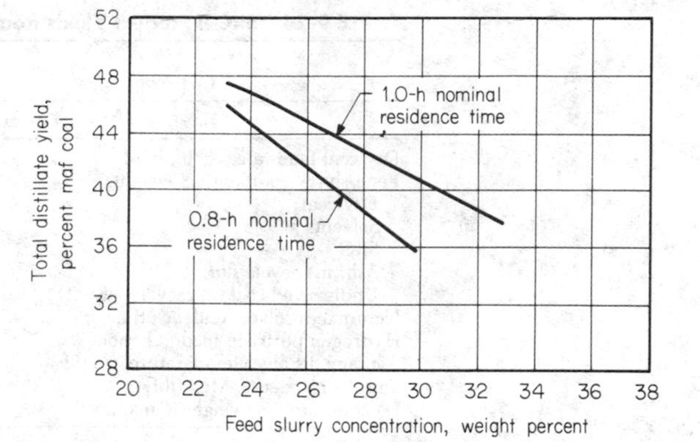

FIG. 9-24 Effect of coal concentration and nominal dissolver residence time on total distillate yield in the SRC-II operating mode (western Kentucky coal, 460°C, 13.1 MPa).

6 coal, and Blacksville No. 2 Pittsburgh Seam coal indicated that a higher distillate yield was obtained at decreased coal concentration and increased dissolver residence time. This is shown in Fig. 9-24 for western Kentucky coal. The test results with the three coals are summarized in Table 9-24. The western Kentucky and Illinois No. 6 coals are more reactive than the Blacksville No. 2 Pittsburgh Seam coal.

Exxon Donor Solvent (EDS) Process The Exxon Research & Engineering Co. has developed the EDS process, which also liquefies coal with a hydrogen donor solvent under hydrogen pressure. The solvent, however, is a catalytically hydrogenated recycle stream fractionated from the middle boiling range, 201 to 455°C (395 to 850°F), of the liquid product. The liquefaction system is an upflow, tubular plug-flow reactor which is divided into four sections illustrated in Fig. 9-25. The hydrogenated recycle solvent is mixed with fresh coal feed and pumped through a tubular preheat furnace into the liquefaction reactor. The reactor operates at 427 to 471°C (800 to 880°F) and 10.3 to 13.8 MPa (1490 to 2000 lbf/in^2), similar to the conditions of SRC-II.

The reactor effluent is separated by distillation into light hydrocarbon gases, naphtha, distillates, and vacuum bottoms. The vacuum bottom, 638°C+ (1180°F+), may be treated with air and steam in Exxon's proprietary Flexicoking process to produce some heavy oil, a low-heating-value flue gas, and a mineral-ash residue. The process hydrogen is produced by the steam re-forming of C_1–C_4 hydrocarbon gases produced in the hydrogenation reactors.

The hydrogenation of the recycle solvent is conducted in a conventional fixed-bed catalytic reactor by using hydrotreating catalysts such as cobalt molybdate or nickel molybdate. Coal conversion and liquid yield strongly depend on the molecular composition, boiling-point range, and other properties of the solvent. Exxon uses its proprietary Solvent Quality Index (SQI) as the main criterion of solvent quality and correlates product yields with SQI. The SQI can be controlled by adjusting the hydrotreater temperature, and coal conversion, product yields, and hydrogen consumption decrease significantly when the SQI is less than a certain value. This critical SQI appears to depend on the coal type and liquefaction conditions, and solvent quality requirements are significantly reduced when molecular hydrogen is present in the reactor.

The solvent has two major roles in the liquefaction process: (1) to stabilize the liquid products by donating hydrogen to the coal and (2) to disperse and transport the coal particles through the liquefac-

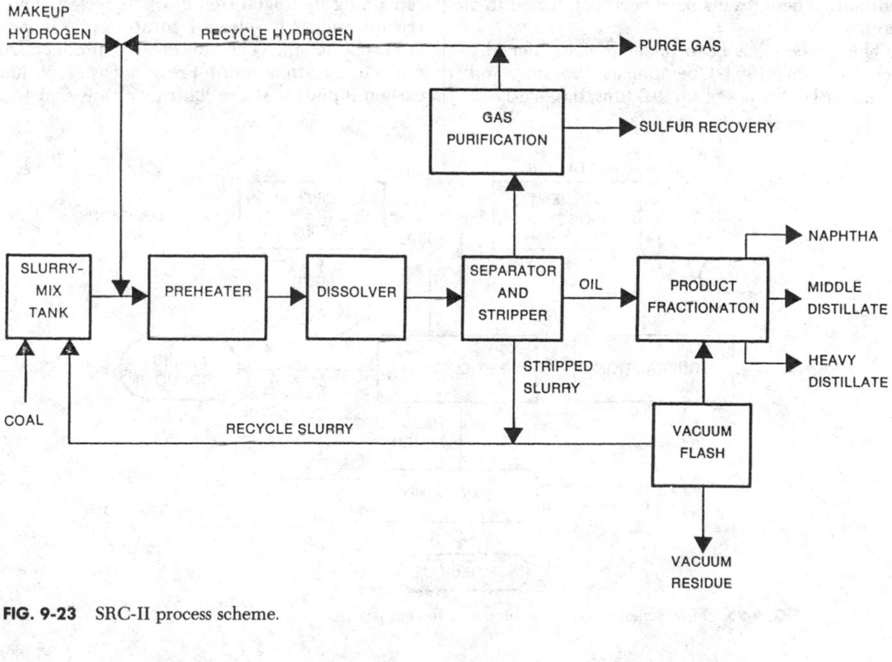

FIG. 9-23 SRC-II process scheme.

TABLE 9-24 SRC-II Product Yields from Pilot Plant

Coal	Western Kentucky Nos. 9 and 14	Illinois No. 6 River King	Blacksville No. 2 Pittsburgh Seam
Operating conditions			
Dry coal feed rate, kg/h (lb/h)	900 (1984)	911 (2008)	906 (1997)
Feed-slurry composition, weight %			
Dry coal	29.5	29.5	30.3
Solvent	33.4	35.6	29.8
SRC	23.9	20.0	24.1
Ash (in recycle slurry)	8.2	11.0	8.3
Undissolved coal (in recycle solvent)	5.0	3.9	7.5
Nominal dissolver residence time, h	0.98	0.97	1.00
Hydrogen purity in feed gas, mole %	89.8	93.7	91.6
Average dissolver temperature, °C (°F)	461 (862)	457 (855)	456 (853)
Dissolver pressure, MPa (lbf/in^2)	13.34 (1930)	13.44	14.09 (2044)
H_2 consumption, weight % maf coal	4.8	4.7	3.5
Product yield, weight % maf coal			
C_1–C_4 hydrocarbon gas	18.4	15.8	13.5
Naphtha, 177°C− (341°F−)	14.2	17.0	11.9
Middle distillate, 177–280°C (341–536°F) + heavy distillate, 288–454°C (550–849°F)	28.2	30.3	22.3
SRC, 454°C+ (849°F+)	26.1	23.0	36.8
Undissolved coal	6.6	5.0	11.9

tion system. The specific feature of the EDS process is that the catalyst used in the process is not in contact with the high-molecular-weight asphaltenes, preasphaltenes, or mineral matter, thereby preventing fast deactivation of the catalyst. The use of vacuum distillation provides a residue, which in addition to Flexicoking, may be used as a fuel or feed to gasification for hydrogen generation.

Several coals including bituminous, subbituminous, and lignite coals have been tested. Different coals gave different liquid-yield response with a change in operating conditions. For Illinois No. 6 bituminous, Wyoming subbituminous, and Texas lignite, the conversion increased only slightly after 40 min. The low-rank coals gave slightly higher conversion, owing to higher yield of water and carbon oxides. The C_4 638°C (811 K) liquid yields, on the other hand, is adversely affected at long residence times owing to the cracking of liquids to gases. The optimum residence time is about 40 min for Illinois No. 6, 60 min for Wyoming, and 25 to 40 min for Texas lignite. Table 9-25 shows the liquefaction product yields at these preferred operating conditions. These results have been confirmed in a 250-ton/day pilot plant.

H-Coal Process This process has been under development by Hydrocarbon Research, Inc., since 1964. Operation at laboratory and process-development-unit (PDU) scales of up to 3 tons/day has been confirmed by a 600-ton/day pilot plant using various coals and demonstrating satisfactory control of catalyst bed expansion.

This process can be operated in either a fuel-oil mode or a syncrude mode, depending on the type of fuel desired. The flow scheme of the H-Coal process is shown in Fig. 9-26. Coal, which has been ground to −60 mesh and dried, is mixed with recycle oil, pumped to 20 MPa (2900 lbf/in^2), and mixed with compressed hydrogen. The mixture is preheated to 371°C (700°F) and fed to the bottom of the ebullated-bed catalytic reactor. The catalyst is kept in a fluidized state by the upward flow of the slurry and gas through the reactor. Catalyst activity is maintained by the addition and withdrawal of small quantities of catalyst to and from the reactor. The reactor also contains an internal tube for recirculating the reaction mixture through the catalyst bed. The reactor temperature is kept at 454°C (849°F). Vapor products leaving the top of the reactor are cooled to separate a liquid condensate. The gas stream is sent to a scrubber to absorb light hydrocarbons, ammonia, and hydrogen sulfide. The remaining hydrogen-rich gas is recycled. The condensate from the reactor vapors is released to atmospheric pressure and fed to an atmospheric still. The liquid-solid product from the reactor is let down to the atmospheric-pressure flash separator. The lighter hydrocarbon liquids that are flashed off are sent to the atmospheric still.

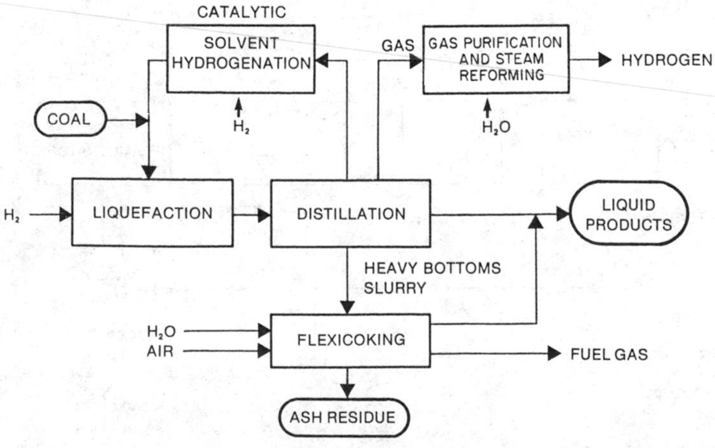

FIG. 9-25 Flow scheme of the Exxon donor solvent process.

TABLE 9-25 EDS Product Yields*

	Illinois No. 6 bituminous	Wyoming subbituminous	Texas lignite
Residence time, min	40	60	25–40
Yields, weight % maf coal			
H_2	−4.3	−4.6	−3.9
$H_2O + CO_x$	12.2	22.3	21.7
$H_2S + NH_3$	4.2	0.9	1.7
C_1–C_3 gas	7.3	9.3	9.1
C_4, 638°C (1180°F) liquid	38.8	33.3	33.3
638°C+ (1180°F+) bottoms	41.8	38.8	38.1

*10.3 MPa, 449°C.

TABLE 9-26 H-Coal Product Yields

	Fuel-oil mode	Syncrude mode
Coal feed rate, Mg/(h·m³)	1.35	0.53
Recycle oil-coal ratio	2.1	2.1
Reactor temperature, °C (°F)	453 (847)	453 (847)
Hydrogen partial pressure, MPa (lbf/in²)	12.1 (1750)	12.6 (1827)
Yields, weight % of dry coal		
C_1–C_3	7.71	9.97
C_4, 204°C (399°F)	16.90	23.66
205–524°C (401–975°F)	18.28	23.21
524°C+ (975°F+)	32.45	19.25
Unconverted coal	6.75	5.68
Ash	10.95	11.67
H_2O	6.67	7.37
NH_3	0.53	0.84
H_2S	2.55	2.65
$Co + CO_2$	1.01	0.95
Total	103.80	105.25
H_2 consumption, weight % of dry coal	3.80	5.25
Sulfur in 205°C+ (399°F+) oil, weight %	0.49	0.26

The bottoms product from the flash separator is further separated with a hydroclone, with a liquid-solid separator, and by vacuum distillation. The atmospheric still yields light and heavy distillates, and the vacuum still yields heavy distillates and residual fuel. A portion of the heavy distillate and hydroclone overflow is recycled as the slurry vehicle.

The severity of operating conditions of the H-Coal process affects the type of fuel produced. In the fuel-oil mode, a relatively high slurry space velocity of 1.25 Mg/(h·m³) of reactor is used. The reactor space velocity can be related to the residence time of the slurry phase in the H-Coal reactor. The hydrogen partial pressure is lowered to about 12 MPa (1740 lbf/in²); hydrogen consumption is usually less than 4 weight percent based on dry coal. In general, at lower hydrogen partial pressures there is less H_2 pickup by the oils, resulting in higher residuum and lower distillate yields. Table 9-26 shows the product yields obtained at PDU by using an Illinois No. 6 bituminous coal. The catalyst was cobalt molybdate on an alumina support. A 0.49 percent sulfur fuel oil [205°C (399°F+)] was produced at 3.8 weight percent hydrogen consumption.

Table 9-26 also shows the results obtained in the syncrude mode operating at a space velocity of 0.53 Mg/(h·m³). In general, with all other variables held constant, the decrease of the space velocity decreases the residuum yield but increases the naphtha yield. The hydrogen consumption is higher in the syncrude mode, at 5.25 weight percent based on dry coal. The total liquid yield was 73 weight percent and 76 weight percent on an maf coal basis for the syncrude and fuel-oil modes respectively.

As with the SRC processes, one of the major problems has been solids separation. The H-Coal process uses hydroclones to remove about two-thirds of the solids from recycled liquid employed as the vehicle. In the syncrude mode, the separation of unreacted coal and ash from the liquid can be accomplished by vacuum distillation. In

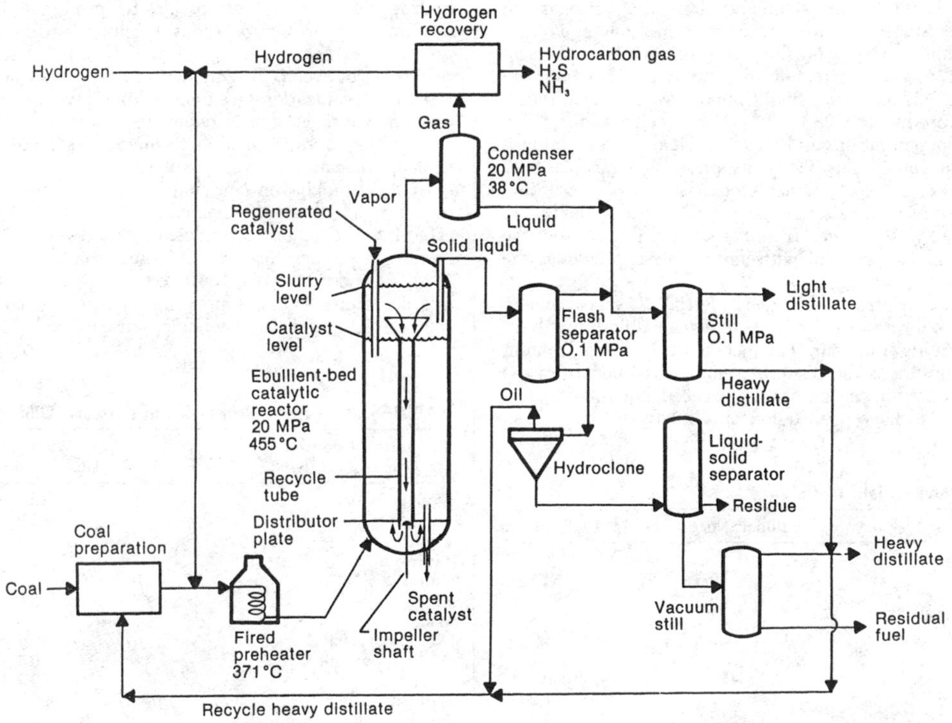

FIG. 9-26 H-Coal process scheme.

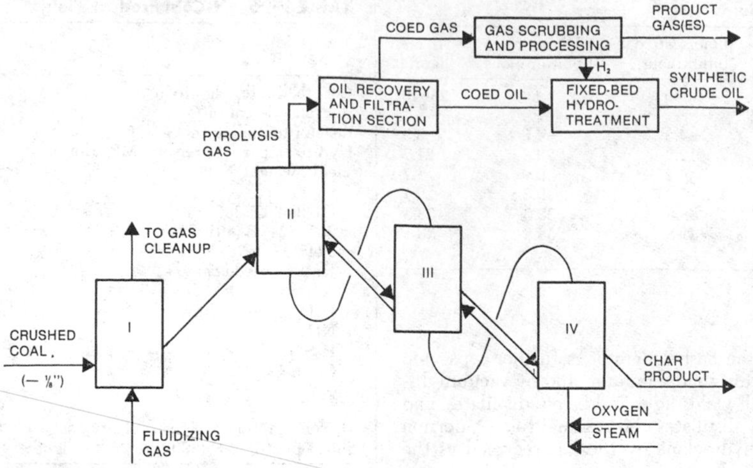

FIG. 9-27 COED coal pyrolysis.

the fuel-oil mode, however, an efficient solid-liquid separation method is needed. An antisolvent-precipitation method has been selected.

Coal-Pyrolysis Processes Another approach to coal liquefaction is pyrolysis, which produces synthetic crude oil, gas, and char as demonstrated in the COED process shown schematically in Fig. 9-27. Crushed coal is dried and heated to successively higher temperatures in a series of fluidized-bed reactors. In each stage, a portion of the coal's volatile matter is released, with each stage maintained slightly below the temperature at which it would agglomerate and defluidize. Typically, four stages of 316, 454, 538, and 816°C (600, 850, 1000, and 1500°F) were used, but operations varied owing to the different coal-agglomerating properties. Process heat was generated by burning char in the last stage and circulating hot char and gases to the other stages. Volatile products were condensed in a recovery system and pyrolysis oil filtered to remove fines. The filtered-oil product was hydrotreated in a fixed-bed reactor with hydrogen at 371 to 427°C (700 to 800°F) to remove sulfur, nitrogen, and oxygen, thus producing a 25 to 30° API synthetic crude.

Typical operating conditions and pyrolysis yields for two bituminous coals, Utah A and Illinois No. 6, are presented in Tables 9-27 and 9-28. These results were obtained from a 36-ton/day coal pilot plant at Princeton, New Jersey. Higher net oil product yield, 1.23 versus 1.10 bbl/ton, was obtained from the Utah coal than from the Illinois coal. The major problem with any pyrolysis process is the high yield of char.

Flash Pyrolysis and Hydropyrolysis Pyrolysis of coal at elevated temperatures for a short residence time, called flash pyrolysis, produces gases, liquids, and char. The increase in hydrogen content in the gases and liquids is the result of removing carbon from the process as a char containing a significantly reduced amount of hydrogen. Several processes have been tested on a relatively small scale. None have demonstrated economic potential, although the technical concepts may be valid.

Flash hydropyrolysis, which is the rapid pyrolysis of coal in the presence of pressurized hydrogen, could improve the oil product quality and liquid yields. It requires large usage of hydrogen.

The **Occidental Research Corporation (ORC) flash pyrolysis** process consists of rapidly pyrolyzing coal particles at a temperature of 621 to 679°C (1149 to 1254°F) in an entrained stream of hot coal char and gas. The process takes advantage of the high heating rates that are made possible by recycling the hot char, and it gives the highest liquid yield of the pyrolysis processes. The process has been tested in a 3-ton/day PDU.

Flash hydropyrolysis, the noncatalytic short-residence-time hydropyrolysis of coal, is a process for producing light aromatic liquids and gaseous hydrocarbons. Product distribution and yields are greatly influenced by temperature and time because of competitive reactions between fragmentation of repolymerization. Several research organizations are undertaking developmental programs by using bench-scale units to determine operating conditions for maximizing the production of high-quality distillate liquids by rapid-heat-up noncatalytic hydropyrolysis.

In the Brookhaven National Laboratory unit pulverized coal is mixed with hydrogen preheated to 775°C (1427°F), and the mixture is fed into the top of a downflow tubular reactor. With North Dakota lignite at temperatures in the range of 725 to 750°C (1337 to 1382°F) and 13.8 MPa, the maximum yield of the heavier liquid hydrocarbons ($>C_8$) is approximately equal to that of the BTX yield.

The Institute of Gas Technology used a coil reactor in its bench-

TABLE 9-27 Pyrolysis Yield Data

	Illinois No. 6 seam	Utah A seam
Net yields, weight % dry coal		
Char	59.5	54.5
Oil	19.3	21.5
Gas	15.1	18.3
Liquor	6.1	5.7
Net process yields		
Char, lb/ton	1190	1090
Oil, bbl/ton	1.10	1.23
Gas, scf/ton	8810	8545
Liquor, gal/ton	14.6	13.7

TABLE 9-28 Typical Properties of Pyrolysis Oils

Properties of derived coal: elements analysis	Utah A seam	Illinois No. 6 seam
Weight %, dry		
Carbon	83.8	79.6
Hydrogen	9.5	7.1
Nitrogen	0.9	1.1
Sulfur	0.4	2.8
Oxygen	5.0	8.5
Ash	0.3	0.9
API gravity, 60°F	−3.5	−4
Moisture, weight %	0.5	0.8
Pour point, °F	100	100
Viscosity, SUS 210°F	390	1333
Solids, weight %, dry basis	3.8	4.0
Gross heating value, Btu/lb	16,100	15,050

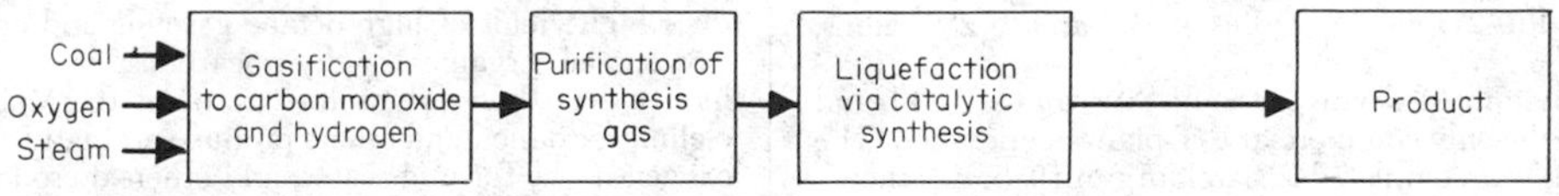

FIG. 9-28 Generalized flow diagram for indirect liquifaction.

scale unit to investigate hydrocarbon yields from North Dakota lignite at operating pressures of 3.5 to 13.8 MPa. At 13.8 MPa, carbon conversions and yields of hydrocarbon liquids were higher. During normal operations, pulverized coal is gravity-fed to the reactor from a feed hopper, and carrier gas is supplied on a once-through basis. The residence time of the reactants is varied by altering the coal and the carrier-gas flow rates. On the basis of the results of bench-scale studies, a 23- to 45-kg/h riser reactor PDU has been designed. Recent bench-scale studies with Illinois No. 6 bituminous coal showed that yields of hydrocarbon liquids were greater than those obtained from North Dakota lignite under similar processing conditions. To explore methods for handling caking coals, silica sand has been mixed with coal as a dry diluent.

Cities Service Co. tested several bituminous coals and lignites in a bench-scale unit to determine operability in a small-diameter, entrained-flow hydropyrolysis reactor and relative reactivity toward carbon conversion. The tests were made at an average temperature of 704°C (1299°F), hydrogen partial pressure of 15.2 MPa, and a residence time of about 3 s. Carbon conversion ranging from 35 to 75 weight percent was obtained with carbon selectivities of about 30 to 35 percent to liquids.

Rocketdyne has been testing its flash-hydropyrolysis process in a 1-ton/h PDU to study residence time and reactor hydrogen requirements. Pulverized coal is fed from a high-pressure feeder into the entrained-flow reactor, where it is mixed rapidly with high-temperature hydrogen by using a rocket-engine injector element. The reactor is generally operated at temperatures in the range of 760 to 1037°C (1400 to 1898°F). The residence time is 20 to 200 ms, and the operating pressure is in the range of 3.4 to 10.3 MPa (493 to 1493 lbf/in^2). A hydrogen-to-coal-feed ratio as low as 0.16 by weight was achieved. Overall conversions of about 60 percent can be obtained, yielding a plant thermal efficiency of about 75 percent if the char is gasified to supply the process hydrogen.

INDIRECT COAL LIQUEFACTION

GENERAL REFERENCES: Lowry (ed.), *Chemistry of Coal Utilization*, vols. I and II, Wiley, New York, 1945. Lowry (ed.), *Chemistry of Coal Utilization*, suppl. vol., Wiley, New York, 1963. Storch, Golumbic, and Anderson, *The Fischer-Tropsch and Related Syntheses*, Wiley, New York, 1951.

Fischer-Tropsch Synthesis A general schematic approach for all indirect-liquefaction processes is shown in Fig. 9-28. All use the basic Fischer-Tropsch (F-T) reactions to produce liquids from hydrogen and carbon monoxide.

The F-T synthesis differs from the direct-liquefaction processes described earlier in that it involves catalytic reactions between mixtures of hydrogen and carbon monoxide, so-called synthesis gas (syngas), which can be made by steam-oxygen gasification of coal or other hydrocarbons. The F-T synthesis produces largely paraffinic and olefinic hydrocarbons instead of the predominantly aromatic hydrocarbons from direct coal liquefaction.

The basic reactions in the F-T synthesis follow:

$$(2n + 1)H_2 + nCO = C_nH_{2n+2} + nH_2O \qquad (9\text{-}25)$$

$$2nH_2 + nCO = C_nH_{2n} + nH_2O \qquad (9\text{-}26)$$

$$(n + 1)H_2 + 2nCO = C_nH_{2n+2} + nCO_2 \qquad (9\text{-}27)$$

$$nH_2 + 2nCO = C_nH_{2n} + nCO_2 \qquad (9\text{-}28)$$

$$2nH_2 + nCO = C_nH_{2n+1}OH + (n - 1)H_2O \qquad (9\text{-}29)$$

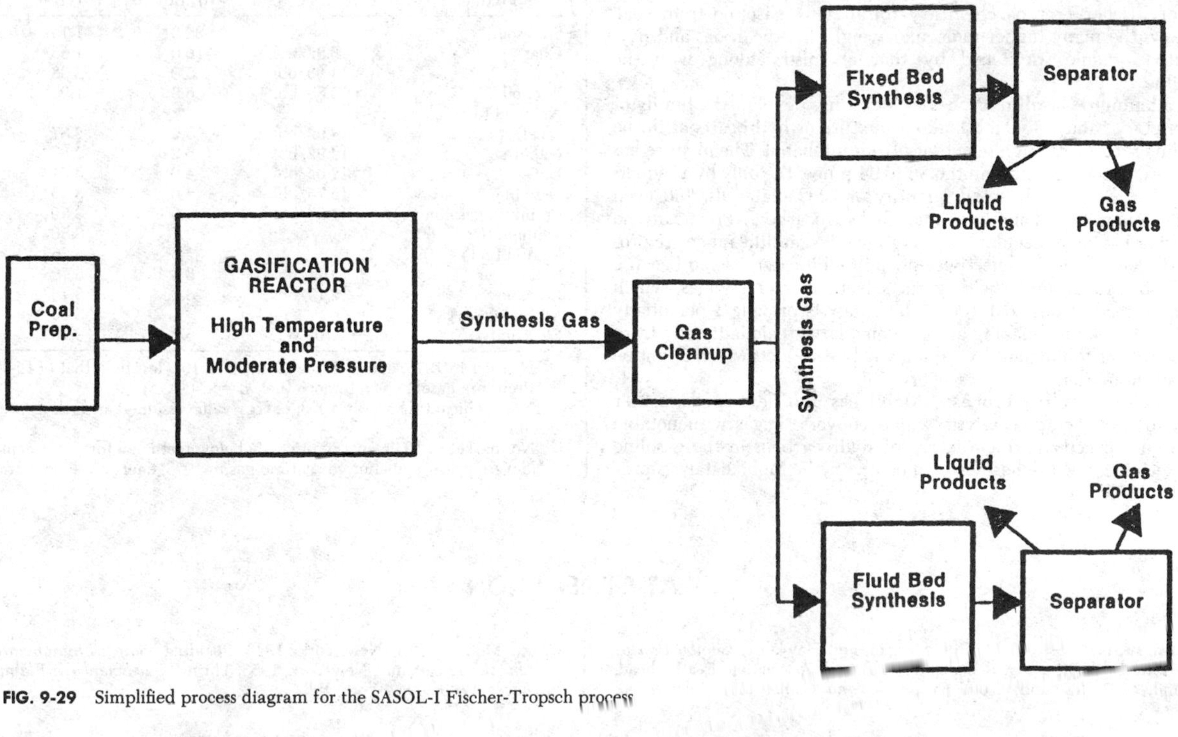

FIG. 9-29 Simplified process diagram for the SASOL-I Fischer-Tropsch process

Reactions (9-25) and (9-27) yield paraffins, (9-26) and (9-28) olefins, and (9-29) alcohols.

SASOL (an acronym in Afrikaans for South African Coal, Oil, and Gas Corporation) is the only commercial F-T plant in operation. The first plant, SASOL-I, was completed in Sasolburg in 1955; it is shown schematically in Fig. 9-29. Two additional plants, SASOL II and III, are now in operation.

SASOL-I includes both fixed- and fluidized-bed Synthol reactors. SASOL-II uses only fluidized-bed reactors since they produce a higher percentage of transportation fuels.

A typical Synthol product slate follows:

Product	Yield, percent/weight
Methane	11.0
Ethane and ethylene	7.5
Propane and propylene	13.0
Butanes and butylenes	11.0
C_5 to 375°F fraction	37.0
375 to 750°F fraction*	11.0
750 to 970°F fraction	3.0
Heavier than 970°F fraction	0.5
Chemicals	6.0
Total	100.0

*Gasoline.

It has been estimated that the installed cost of a SASOL-II type of plant in the United States would range from $2.5 to $3.6 billion (based on 1979 dollars), depending on where it was constructed. This corresponds approximately to $60,000 per daily barrel.

Other approaches have been investigated for selectively making specific materials from syngas. Some have culminated in well-known, widely used processes such as those for making methanol and acetic acid. More particularly, the approaches being pursued by the Mobil Research and Development Corporation are especially attractive because they show great promise of being the basis for making high-octane gasoline on a large scale from coal via syngas and methanol or directly from syngas without simultaneously making by-products for which it may be difficult to find markets. Thus Mobil's approaches appear to avoid one of the major stumbling blocks to the use of Fischer-Tropsch chemistry for making gasoline from coal, namely, the many higher-molecular-weight hydrocarbons and oxygenated organics produced by that chemistry along with the gasoline.

Methanol-to-Gasoline Process Mobil has developed a family of proprietary catalysts which make a mixture of hydrocarbons in the gasoline range with high selectivity from methanol. The mixture has a research octane number in the 93 to 94 range; the only by-products are water, a relatively small quantity of LPG materials, and even smaller quantities of methane and C_2 hydrocarbons. Practically no materials having molecular weights above the gasoline range are produced. Because of the selectivity of the Mobil catalysts and because methanol can be made with great selectivity from syngas, which could be made from coal, the Mobil catalysts provide a potentially promising route for making gasoline on a large scale indirectly from coal without simultaneously making any by-products which might be difficult to market.

Syngas-to-Gasoline Process Mobil has also developed another group of proprietary catalysts which convert syngas with notable selectivity directly into a mixture of hydrocarbons in the gasoline range. Unlike the Fischer-Tropsch catalysts, the Mobil catalysts produce high yields of high-octane gasoline and only trace amounts of oxygenated organics or hydrocarbons having molecular weights higher than those in gasoline. Some of the Mobil catalysts produce olefinic gasolines, and others produce aromatic gasolines. Some reject oxygen in the form of water, and others do so in the form of carbon dioxide. The other by-products of the Mobil catalysts are LPG materials, C_2 hydrocarbons, and methane.

Preliminary aging tests suggest that some of the catalysts can be regenerated, and some also exhibit relatively long service lives. This preliminary work had been done in microscale units having the capacity for about 10 cm^3 of catalyst in a fixed bed. Larger units (100 cm^3 of catalyst) for both fixed- and fluid-bed operation have now been built and are in shakedown testing.

ECONOMICS OF COAL LIQUEFACTION

An example of process economics is included here to show a method for this type of calculation. Results obtained should be regarded as estimates based on conceptual plant designs and on a relative rather than an absolute basis.

A product-value technique was used in determining product cost from a plant with a variety of products having different properties. This technique calculates a price for a reference fuel, in this case premium gasoline, and all others are determined from this reference. The basic assumption here is that particular product prices will remain in a fixed ratio with time. Solid fuels were priced on their heating value relative to fuel oil. Fuel oil is defined as industrial No. 6 or bunker C, and No. 2 is considered the same as diesel fuel. Naphtha is a general distillate product used for the production of gasoline, diesel, or other products. All naphthas were assigned values related to a cost of upgrading to gasoline. Liquid petroleum gas (LPG) consists primarily of propane and butane, and gasoline is low-octane as obtained from a simple Fischer-Tropsch process. Value factors for these products are shown in Table 9-29.

TABLE 9-29 Value Factors for Energy Products (Reference to Premium Gasoline)

Product	1978 price, $	Heating value, 10^6 Btu	Energy price, $/ 10^6 Btu	Value factor, f_i
SRC solid*	. . . .	32.0	1.75	.50
Char	26.32/ton	16.0	1.65	.47
Tar oil	14.88/bbl	5.0	2.98	.85
Fuel oil	12.35/bbl	6.3	1.95	.56
No. 2 oil (diesel fuel)	14.92/bbl	5.2	2.87	.82
Naphtha	14.92/bbl	5.2	2.87	.82
LPG	12.04/bbl	4.0	3.01	.86
Gasoline	15.75/bbl	5.0	3.15	.90
Premium gasoline†	17.50/bbl	5.0	3.50	1.00
Methanol‡	. . . .	2.7	3.50	1.00
Methyl fuel§	. . . .	2.6	3.36	.96
Butane	13.11/bbl	3.5	3.75	1.07
Propane	12.10/bbl	3.2	3.78	1.08
All fuel gases¶	. . . .	. . .	3.50	1.00
Electricity	31.06/10^6 Wh	3.4	9.10	2.60

*No sales in 1978; energy price assumed as 10% less than that of fuel oil.
†Premium gasoline is reference fuel; f_i = 1.00
‡No significant market in 1978; energy value assumed as same as premium gasoline.
§No market in 1978; energy price 4% below methanol for water content.
¶Severe price regulation for natural gas in 1978; energy value set at 1.00.

HEAT GENERATION

GENERAL REFERENCES: Field, Gill, Morgan, and Hawksley, *Combustion of Pulverized Coal*, British Coal Utilization Research Association, Leatherhead, England, 1967. Jost and Croft, *Explosions and Combustion Processes in Gases*, McGraw-Hill, New York, 1946. Spalding, *Some Fundamentals of Combustion*, Academic, New York, 1955. Thring, *The Science of Flames and Furnaces*, Wiley, New York, 1962.

TABLE 9-30 Combustion Constants*,†

No.	Substance	Formula	Molecular weight	Lb./cu. ft.	Cu. ft./lb.	Sp. gr. air 1.000	Heat of combustion				Moles/mole or cu. ft./cu. ft. combustible						Lb./lb. combustible					
							B.t.u./cu. ft.		B.t.u./lb.		Required for combustion			Flue products			Required for combustion			Flue products		
							Gross	Net	Gross	Net	O_2	N_2	Air	CO_2	H_2O	N_2	O_2	N_2	Air	CO_2	H_2O	N_2
1	Carbon‡	C	12.01	......	....	......	....	....	14,093	14,093	1.0	3.76	4.76	1.0	...	3.76	2.66	8.86	11.53	3.66	....	8.86
2	Hydrogen	H_2	2.016	0.0053	187.723	0.0696	325	275	61,100	51,623	0.5	1.88	2.38	....	1.0	1.88	7.94	26.41	34.34	....	8.94	26.41
3	Oxygen	O_2	32.000	0.0846	11.819	1.1053																
4	Nitrogen (atm.)	N_2	28.016	0.0744	13.443	0.9718																
5	Carbon monoxide	CO	28.01	0.0740	13.506	0.9672	322	322	4,347	4,347	0.5	1.88	2.38	1.0	...	1.88	0.57	1.90	2.47	1.57	....	1.90
6	Carbon dioxide	CO_2	44.01	0.1170	8.548	1.5282																
Paraffin series																						
7	Methane	CH_4	16.041	0.0424	23.565	0.5543	1013	913	23,879	21,520	2.0	7.53	9.53	1.0	2.0	7.53	3.99	13.28	17.27	2.74	2.25	13.28
8	Ethane	C_2H_6	30.067	0.0803	12.455	1.0488	1792	1641	22,320	20,432	3.5	13.18	16.68	2.0	3.0	13.18	3.73	12.39	16.12	2.93	1.80	12.39
9	Propane	C_3H_8	44.092	0.1196	8.365	1.5617	2590	2385	21,661	19,944	5.0	18.82	23.82	3.0	4.0	18.82	3.63	12.07	15.70	2.99	1.63	12.07
10	n-Butane	C_4H_{10}	58.118	0.1582	6.321	2.0665	3370	3113	21,308	19,680	6.5	24.47	30.97	4.0	5.0	24.47	3.58	11.91	15.49	3.03	1.55	11.91
11	Isobutane	C_4H_{10}	58.118	0.1582	6.321	2.0665	3363	3105	21,257	19,629	6.5	24.47	30.97	4.0	5.0	24.47	3.58	11.91	15.49	3.03	1.55	11.91
12	n-Pentane	C_5H_{12}	72.144	0.1904	5.252	2.4872	4016	3709	21,091	19,517	8.0	30.11	38.11	5.0	6.0	30.11	3.55	11.81	15.35	3.05	1.50	11.81
13	Isopentane	C_5H_{12}	72.144	0.1904	5.252	2.4872	4008	3716	21,052	19,478	8.0	30.11	38.11	5.0	6.0	30.11	3.55	11.81	15.35	3.05	1.50	11.81
14	Neopentane	C_5H_{12}	72.144	0.1904	5.252	2.4872	3993	3693	20,970	19,396	8.0	30.11	38.11	5.0	6.0	30.11	3.55	11.81	15.35	3.05	1.50	11.81
15	n-Hexane	C_6H_{14}	86.169	0.2274	4.398	2.9704	4762	4412	20,940	19,403	9.5	35.76	45.26	6.0	7.0	35.76	3.53	11.74	15.27	3.06	1.46	11.74
Olefin series																						
16	Ethylene	C_2H_4	28.051	0.0746	13.412	0.9740	1614	1513	21,644	20,295	3.0	11.29	14.29	2.0	2.0	11.29	3.42	11.39	14.81	3.14	1.29	11.39
17	Propylene	C_3H_6	42.077	0.1110	9.007	1.4504	2336	2186	21,041	19,691	4.5	16.94	21.44	3.0	3.0	16.94	3.42	11.39	14.81	3.14	1.29	11.39
18	n-Butene	C_4H_8	56.102	0.1480	6.756	1.9336	3084	2885	20,840	19,496	6.0	22.59	28.59	4.0	4.0	22.59	3.42	11.39	14.81	3.14	1.29	11.39
19	Isobutene	C_4H_8	56.102	0.1480	6.756	1.9336	3068	2869	20,730	19,382	6.0	22.59	28.59	4.0	4.0	22.59	3.42	11.39	14.81	3.14	1.29	11.39
20	n-Pentene	C_5H_{10}	70.128	0.1852	5.400	2.4190	3836	3686	20,712	19,363	7.5	28.23	35.73	5.0	6.0	28.23	3.42	11.39	14.81	3.14	1.29	11.39
Aromatic series																						
21	Benzene	C_6H_6	78.107	0.2060	4.852	2.6920	3751	3601	18,210	17,480	7.5	28.23	35.73	6.0	3.0	28.23	3.07	10.22	13.30	3.38	0.69	10.22
22	Toluene	C_7H_8	92.132	0.2431	4.113	3.1760	4484	4284	18,440	17,620	9.0	33.88	42.88	7.0	4.0	33.88	3.13	10.40	13.53	3.34	0.78	10.40
23	Xylene	C_8H_{10}	106.158	0.2803	3.567	3.6618	5230	4980	18,650	17,760	10.5	39.52	50.02	8.0	5.0	39.52	3.17	10.53	13.70	3.32	0.85	10.53
Miscellaneous gases																						
24	Acetylene	C_2H_2	26.036	0.0697	14.344	0.9107	1499	1448	21,500	20,776	2.5	9.41	11.91	2.0	1.0	9.41	3.07	10.22	13.30	3.38	0.69	10.22
25	Naphthalene	$C_{10}H_8$	128.162	0.3384	2.955	4.4208	5854	5654	17,298	16,708	12.0	45.17	57.17	10.0	4.0	45.17	3.00	9.97	12.96	3.43	0.56	9.97
26	Methyl alcohol	CH_3OH	32.041	0.0846	11.820	1.1052	868	768	10,259	9,078	1.5	5.65	7.15	1.0	2.0	5.65	1.50	4.98	6.48	1.37	1.13	4.98
27	Ethyl alcohol	C_2H_5OH	46.067	0.1216	8.221	1.5890	1600	1451	13,161	11,929	3.0	11.29	14.29	2.0	3.0	11.29	2.08	6.93	9.02	1.92	1.17	6.93
28	Ammonia	NH_3	17.031	0.0456	21.914	0.5961	441	365	9,668	8,001	0.75	2.82	3.57	...	1.5	3.32	1.41	4.69	6.10	...	1.59	5.51
														SO						SO		
29	Sulfur‡	S	32.06	......	....	......	....	....	3,983	3,983	1.0	3.76	4.76	1.0	...	3.76	1.00	3.29	4.29	2.00	....	3.29
30	Hydrogen sulfide	H_2S	34.076	0.0911	10.979	1.1898	647	596	7,100	6,545	1.5	5.65	7.15	1.0	1.0	5.65	1.41	4.69	6.10	1.88	0.53	4.69
31	Sulfur dioxide	SO_2	64.06	0.1733	5.770	2.264																
32	Water vapor	H_2O	18.016	0.0476	21.017	0.6215																
33	Air	....	28.9	0.0766	13.063	1.0000																

*From American Gas Association. To convert pounds per cubic foot to kilograms per cubic meter, multiply by 16.02; to convert cubic feet per pound to cubic meters per kilogram, multiply by 0.062; to convert British thermal units per cubic foot to joules per cubic meter, multiply by 37.3×10^3; and to convert British thermal units per pound to joules per kilogram, multiply by 2329. 60°F and 30 inHg dry = 15.6°C and 76.2 cm.

†All gas volumes corrected to 60°F and 30 inHg dry.

‡Carbon and sulfur are considered as gases for molal calculations only.

COMBUSTION STOICHIOMETRY

Theoretical Oxygen and Air for Combustion The amount of oxygen or air just sufficient to burn the carbon, net hydrogen,° and sulfur in a fuel to carbon dioxide, water vapor, and sulfur dioxide is the theoretical oxygen or air. The general expression for combustion of a fuel is

$$C_mH_n + \left(\frac{4m + n}{4}\right) O_2 = mCO_2 + \left(\frac{n}{2}\right) H_2O \qquad (9\text{-}30)$$

m and n being the number of atoms of carbon and net hydrogen, respectively, in the fuel. For example, this relationship shows that 1 mol of methane (CH_4) requires 2 mol of oxygen for complete combustion to 1 mol of carbon dioxide and 2 mol of water. If air is used, each mole of oxygen is accompanied by 3.76 mol of nitrogen.

The theoretical weight or volume of oxygen or air required to burn a given weight of a fuel is of primary interest in engineering calculations for the design of equipment. The volume of theoretical oxygen needed to burn any fuel can be calculated from the ultimate analysis of the fuel as follows:

$$359\left(\frac{C}{12} + \frac{H_2}{4} - \frac{O_2}{32} + \frac{S}{32}\right) = \text{ft}^3 \text{ oxygen/lb fuel} \qquad (9\text{-}31)$$

where C, H_2, O_2, and S are the decimal weights of these elements in 1 lb of fuel. (To convert cubic feet to cubic meters, multiply by 0.0283; to convert cubic feet per pound to cubic meters per kilogram, multiply by 0.0623.) The coefficient 359 is the volume in cubic feet of 1 mol of oxygen at 0°C (32°F) and 1 atm. The *weight* of oxygen in pounds is obtained by multiplying cubic feet by 0.0891, the density of oxygen at the same conditions. The *volume* of theoretical air is obtained by using a coefficient of 1710 instead of 359 in Eq. (9-31).

°Net hydrogen = total hydrogen − (0.125 × oxygen in fuel).

Table 9-30 gives the theoretical oxygen and air requirements and the gaseous products of combustion (POC) for a wide variety of combustible substances.

Table 9-31 gives the theoretical air requirements on the basis of the higher heating value and the maximum, or theoretical, carbon dioxide content of the POC for several industrial fuels. The close agreement among theoretical air requirements suggests that a good approximation of the air requirements is 7.7 lb/10,000 Btu (HHV) for coal and coke and 7.4 for petroleum oils. If only the lower heating value of the fuel is known, the HHV can be calculated from Eq. (9-6) or (9-7).

TABLE 9-31 Theoretical Air and CO_2 for Combustion of Industrial Fuels*

Fuel	Combustion conditions at zero excess air			
	Atmospheric air required (lb./10,000 B.t.u.)†		Max. CO_2 %	
	Range	Avg.	Range	Avg.
Anthracite:				
New Mexico	....	7.83	....	19.5
Colorado	....	7.85	....	19.3
Pennsylvania	7.81–7.93	6.87	20.0–20.0	20.0
Semianthracite	7.68–7.82	7.74	19.1–19.2	19.1
Bituminous coal:				
Low-volatile	7.62–7.76	7.69	18.5–18.9	18.7
Medium-volatile	....	7.77	....	18.5
High-volatile A	7.51–7.73	7.63	17.7–18.7	18.4
High-volatile B	7.56–7.73	7.66	18.0–18.7	18.4
High-volatile C	7.54–7.67	7.60	18.0–18.5	18.2
Subbituminous coal	7.56–7.57	7.56	19.1–19.2	19.1
Lignite:				
North Dakota	....	7.47	....	19.5
Texas	....	7.52	....	19.2
Coke:				
High-temperature	....	7.96	....	20.7
Low-temperature	....	7.63	....	19.3
Beehive	....	8.05	....	20.5
By-product	....	8.01	....	20.5
Gasworks coke	8.02–8.10	8.06	20.4–20.6	20.6
Petroleum coke	....	7.73	....	19.5
Pitch coke	....	8.13	....	20.7
Wood:				
Softwoods	7.02–7.22	7.11	18.7–20.4	19.8
Hardwoods	7.09–7.28	7.15	19.5–20.5	20.0
Bagasse	6.25–6.99	6.59	19.4–20.5	20.3
Petroleum oils:				
Gasoline (60°A.P.I.)	....	7.46	....	14.9
Kerosene (45°-A.P.I.)	....	7.42	....	15.1
Gas oil (30°A.P.I.)	....	7.45	....	15.5
Fuel oil (15°A.P.I.)	....	7.58	....	15.9
Gaseous fuels:				
Natural gas	7.32–7.41	7.37	6.9–15.2	12.2
Refinery and oil gas	6.52–7.38	7.44	10.7–13.6	12.8
Blast-furnace gas	5.73–6.27	5.82	20.0–26.9	24.7
Coke-oven gas	6.66–7.02	6.80	9.5–12.7	11.1
Propane	....	7.24	....	13.7
Butane	....	7.26	....	14.0
Methane	....	7.20	....	11.7

*From Johnson and Auth (eds.), *Fuels and Combustion Handbook*, McGraw-Hill, New York, 1951, p. 355. (lb/10,000 Btu) × 43 = kg/kJ.

†Higher heating value.

Excess Air for Combustion More than the theoretical amount of air is necessary in practice to achieve complete combustion. This excess air is expressed either as a percentage of the theoretical air or as the total air divided by the theoretical air. The former is most frequently used. The latter is sometimes called the **excess air number.** If, for example, it is 1.25, there is 25 percent excess air. Figure 9-30 shows the relationship of oxygen in the POC to excess air for various industrial fuels.

If it is desired to know the percentage of excess air A_x under the operating conditions of a particular combustion process, it can be calculated from

$$A_x = \left(\frac{O_2}{0.266N_2 - O_2}\right) 100 \qquad (9\text{-}32)$$

where O_2 and N_2 are the percentages by volume of these gases in the dry POC, as determined from an Orsat analysis or other volumetric methods for analyzing POC. Equation (9-32) is applicable only when the nitrogen in the fuel is negligible and there are no combustible gases, such as carbon monoxide or hydrogen, in the POC. If these gases are present, the percentage excess air is

$$A_x = \frac{O_2 - 0.5(C + H_2)}{0.266N_2 - O_2 + 0.5(CO + H_2)} (100) \qquad (9\text{-}33)$$

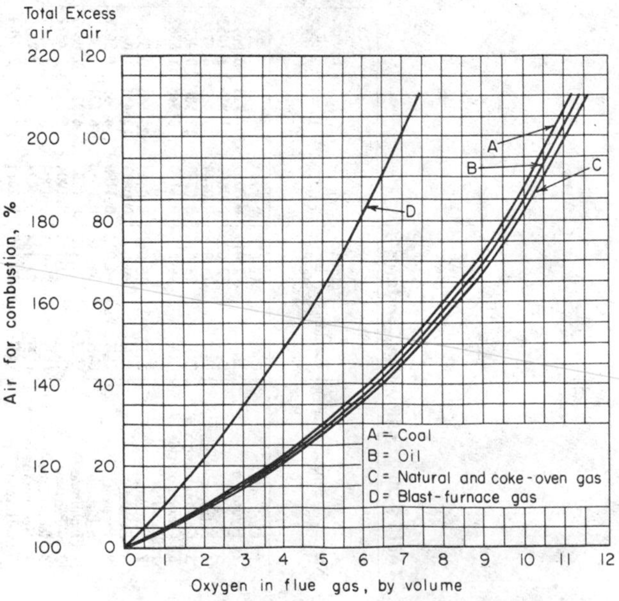

FIG. 9-30 Relationship of oxygen in the flue gas to excess air for various fuels. *(Bailey Meter Co.)*

If only the carbon dioxide content of the POC is known,

$$A_x = \frac{7900[(CO_2)_t - CO_2]}{CO_2[100 - (CO_2)_t]} \tag{9-34}$$

where $(CO_2)_t$ is the maximum volume percentage of carbon dioxide obtainable in the dry POC for a given fuel, and CO_2 is the actual percentage found in the dry POC. Tables 9-30 and 9-31 give the values of $(CO_2)_t$ for several fuels and other combustible substances.

Products of Combustion The products of combustion consist, in the case of solid fuels, of solid residue which may contain unburned combustible and gases previously referred to as POC.

The amount of residue will exceed the ash content of the original solid fuel by an amount equal to the unburned combustible in the residue. It may be estimated by determining the ash content of the residue.

The weight of dry gas in the POC may be calculated from

$$G_d = \frac{C_f - C_r}{100}\left[\frac{4(CO_2) + O_2 + 700}{3(CO_2 + O_2)}\right] \tag{9-35}$$

where G_d = lb dry gas/lb fuel fired, C_f and C_r are the percentages by weight of carbon in the fuel and the residue respectively, and CO_2 and O_2 are the percentages by volume of these gases in the POC. Equation (9-35) assumes hydrogen to be absent or negligible in the dry POC.

The weight of wet gas in the POC is

$$G_w = G_d + 0.09H_2 \tag{9-36}$$

where G_w = lb wet gas/lb fuel fired; and H_2 is the total hydrogen, percent by weight, in the as-fired fuel.

ENTHALPY OF COMBUSTION PRODUCTS

Figure 9-31 gives the enthalpy of combustion products above 15.6°C (60°F). To illustrate its use, assume that a flue gas at 1093°C (2000°F) contains, in percent per volume, 13.7 CO_2, 3.9 O_2, 7.7 H_2O, and 74.7 N_2. The enthalpy of the flue gas at the specified temperature would be (0.137 × 59) + (0.039 × 38) + (0.077 × 48) + (0.747 × 38) = 41.65 Btu/ft³, or 1553 kJ/m³.

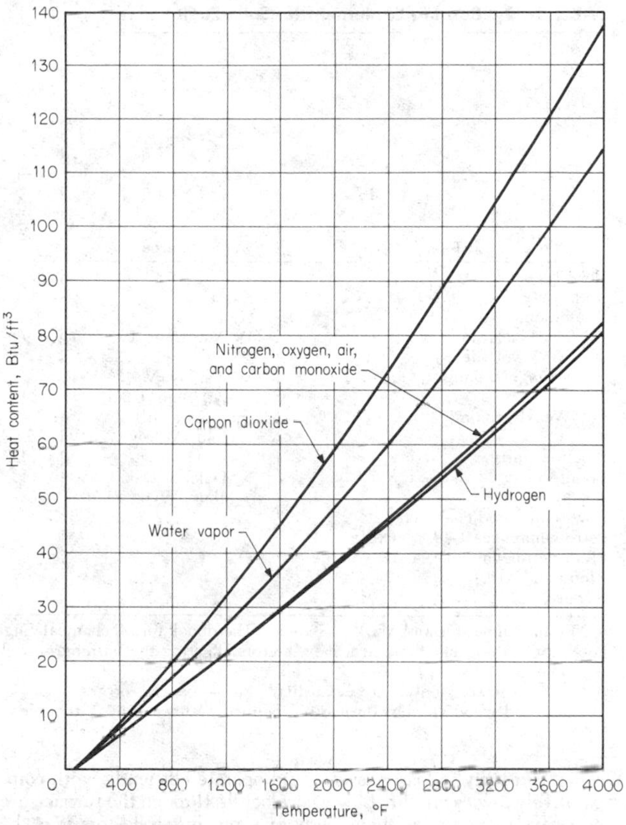

FIG. 9-31 Heat content of various gases above 60°F in British thermal units per cubic foot (to convert to kilojoules per cubic meter, multiply by 37.3). °C = (°F − 32) × 5/9.

SOLID-FUELS COMBUSTION ON STOKERS AND IN SUSPENSION

There are basically two modes for burning solid fuels: in a fuel bed or in suspension. There are several variations of both modes, each suited to a particular system. Table 9-32 lists the kinds of equipment used to burn different types of solid fuels.

Fuel-Bed Firing Fuel-bed firing is accomplished with mechanical stokers, which are designed to achieve continuous or intermittent fuel feed, fuel ignition, proper distribution of the combustion air, free release of the gaseous combustion products, and continuous or intermittent disposal of the unburned residue. These aims are met conventionally with three types of stokers: underfeed, crossfeed, and overfeed, which differ mainly in the relative directions of the flow of fuel and air. The principles of these types of stokers are shown schematically in Fig. 9-32, and capacities are given in Table 9-33.

Both fuel and air have the same relative direction in the **underfeed stoker,** which is built in single-retort and multiple-retort designs. In the **single-retort,** side-dump stoker, a ram pushes coal into the retort toward the end of the stoker and upward toward the tuyere blocks, where air is admitted to the bed. This type of stoker will handle most bituminous coals and anthracite, preferably 1.9 to 5 cm (¾ to 2 in) and no more than 50 percent through a 0.6-cm (¼-in) screen. Overfire air or steam jets are frequently used in the bridgewall at the end of the stoker to promote turbulence.

In the **multiple-retort stoker,** rams feed coal to the top of sloping grates between banks of tuyeres. Auxiliary small sloping rams perform the same function as the pusher rods in the single retort. Air is admitted along the top of the banks of tuyeres, and on the largest units the tuyeres themselves are given a slight reciprocating action to agitate the bed further. This type of stoker operates best with caking coals having a relatively high ash-softening temperature. Coal sizing is up to 5 cm (2 in) with 30 to 50 percent through a 0.6-cm (¼-in) screen.

The fuel flows at right angles to the air flow in the **crossfeed stoker.** The most common of this type is the **traveling-grate stoker,** which is built in two types, the **chain grate** and the **bar grate.** Coal feeds by gravity into a hopper at one end of the moving grate. As the grate passes under the hopper, it carries a bed of fresh coal toward the furnace. Only a small amount of air is fed at the front of the stoker, to keep the fuel mixture rich, but as the coal moves toward the middle of the furnace, the amount of air is increased, and most of the coal is burned by the time it gets halfway down the length of the grate. Fuel-bed depth varies from 10 to 20 cm (4 to 8 in), depending on the fuel, which can be coke breeze, anthracite, or any noncoking bituminous coal.

Figure 9-33 illustrates a variation of the traveling-grate stoker, which is widely used for sintering ore. Combustion air is drawn downward through the bed of ore and coal, and combustion products are removed beneath the grate.

The fuel and air flow in opposite directions in the **overfeed stoker.** Except for certain types of gas producers, in which coal moves downward toward a grate against upward movement of the air (or oxygen and steam) through the grate, there is no combustion system that operates purely in the overfeed mode. However, the **spreader stoker** approximates overfeed action. A portion of the fuel burns in suspension, and the remainder burns on a dump, undulating, or moving grate.

Spreader stokers with oscillating, pulsating, or traveling grates have been widely employed because they will burn all types of coal,

TABLE 9-32 Burning Equipment for Solid Fuels*

Fuel	Source	Stokers: Underfeed	Stokers: Chain grate	Stokers: Chain grate, jet ignition	Stokers: Spreader	Pulverized fuel: U flame, pulverized fuel	Pulverized fuel: Horizontal burners, pulverized fuel	Crushed: Cyclone (where fusion temp. is suitable)	Cell furnaces
Coke breeze	...	...	√						
Anthracite	E. Pa.	...	√c†	...	...	√			
Bituminous coal:									
17–27% volatile	W. Va., Cent. Pa.	√	...	...	√	√	√	√	
27–35% volatile									
Strongly coking	W. Pa., W. Va., Ky., Ohio, Utah	√	√	√	√	√	√	√	
Weakly coking	Ind., Iowa, Ill., Colo., W. Ky.	...	√	√	√	√	√	√	
Pipeline slurry	...	...	...	...	...	√	√	√	
Lignite	N. Dak., S. Dak., Mont., Wyo., Tex.	...	√	√	√	√	√	√	
Low-temp. fluid-coal char	...	...	...	...	...	√	√ Aux†	√	
Petroleum coke, 9–14% volatile	...	...	...	...	...	√	√	√	
Fluid petroleum coke, 4–5% volatile	...	...	...	...	...	√	√ Aux†	√	
Wood and bark‡	...	...	...	...	√	...	......	√ Aux†	√
Bagasse	...	...	...	...	√	...	......	......	√

*From Baumeister and Marks, "Standard Handbook for Mechanical Engineers," 7th ed., McGraw-Hill, 1967. Equipment indicated will usually result in a good application, but there are many factors affecting the burning of fuel, and variations of fuel properties that guide the individual selection of burning equipment.

†c = coarse sizes only. Aux = auxiliary fuel—coal, oil, or gas.

‡Bark and wood are also burned on inclined grates and in Dutch-oven pile furnaces.

respond rapidly to load changes, and operate efficiently, with comparatively low excess air. Low gas velocities through the furnace are necessary to minimize fly-ash erosion, and dust collectors must be used to minimize dust emissions from the stack. (Since spreader firing combines the features of fuel-bed and suspension firing, the combination is discussed later under "Spreader-Stoker Firing.")

Suspension Firing Suspension firing of pulverized coal is used much more often than fuel-bed firing of coarse coal in the United States. This mode of firing coal affords higher steam-generation capacity, is independent of the caking characteristics of the coal, and responds quickly to load changes. However, pulverized-coal firing is

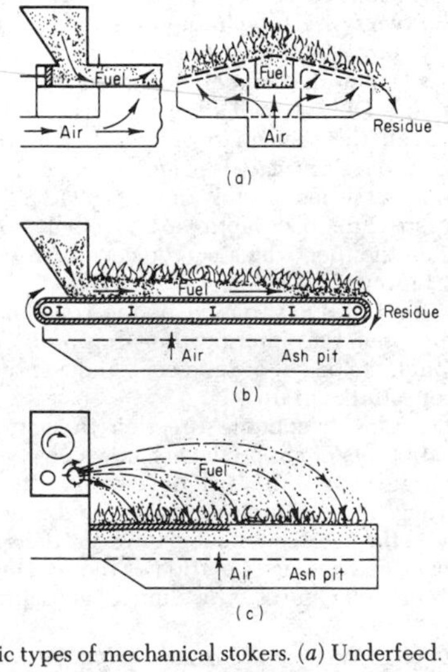

FIG. 9-32 Basic types of mechanical stokers. (*a*) Underfeed. (*b*) Crossfeed. (*c*) Overfeed.

TABLE 9-33 Capacities of Mechanical Stokers

Type	Steam generation, 1000 lb/h*	Maximum grate heat release, 1000 Btu/(h·ft²)†
Single retort (underfeed)	5–50	200
Multiple retort (underfeed)	40–300	300
Traveling (crossfeed)	10–300	300
Spreader (overfeed)	10–300	1000

*To obtain kg/s, multiply by 1.26×10^{-4}.
†To obtain J/(m²·s), multiply by 3.16.

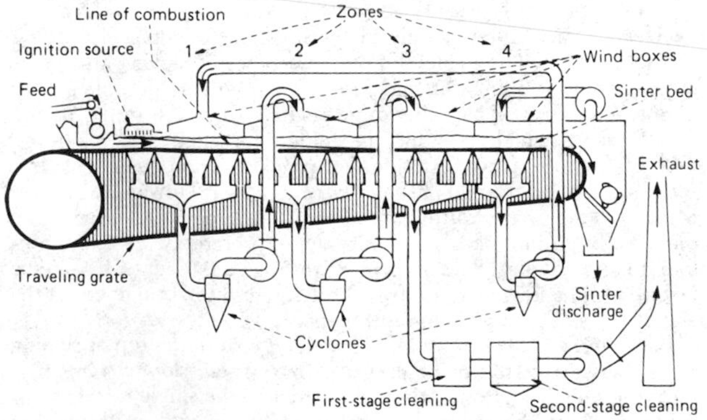

FIG. 9-33 Improved sintering process. (*Courtesy of* Chemical Engineering.)

rarely used on boilers of less than 45,360 kg/h (100,000 lb/h) steam capacity, since stokers are more economical for units of lower capacity.

The various burner and furnace configurations for pulverized-coal firing are shown schematically in Fig. 9-34. The U-shaped flame, designated as **fantail vertical firing** (Fig. 9-34*a*), was developed initially for pulverized coal before the advent of water-cooled furnace walls. Because a large percentage of the total combustion air is withheld from the fuel stream until it projects well down into the furnace, this type of firing is well suited for solid fuels that are difficult to ignite, such as those with less than 15 percent volatile matter. Although this configuration is no longer used in central-station power plants, it may find favor again if low-volatile chars from coal-conversion processes are used for steam generation or process heating.

Modern central stations use the other burner-furnace configurations shown in Fig. 9-34, in which the coal and air are mixed as rapidly as possible in and close to the burner. The **primary air,** used to transport the pulverized coal to the burner, comprises 10 to 20 percent of the total combustion air. The **secondary air,** comprising the remainder of the total air, mixes in the burner with the primary air and coal in a manner to promote rapid mixing. The velocity of the mixture leaving the burner must be high enough to prevent flashback in the primary air-coal piping. In practice, the velocity in the primary air-coal pipe is maintained at about 31 m/s (100 ft/s).

In **tangential firing** (Fig. 9-34*b*) the burners are arranged in vertical banks at each corner of a square (or nearly square) furnace and directed toward an imaginary circle in the center of the furnace. This results in the formation of a large vortex with its axis on the vertical centerline. The burners consist of an arrangement of slots one above the other, admitting, through alternate slots, primary air-fuel mixture and secondary air. It is possible to tilt the burners upward or downward, the maximum inclination to the horizontal

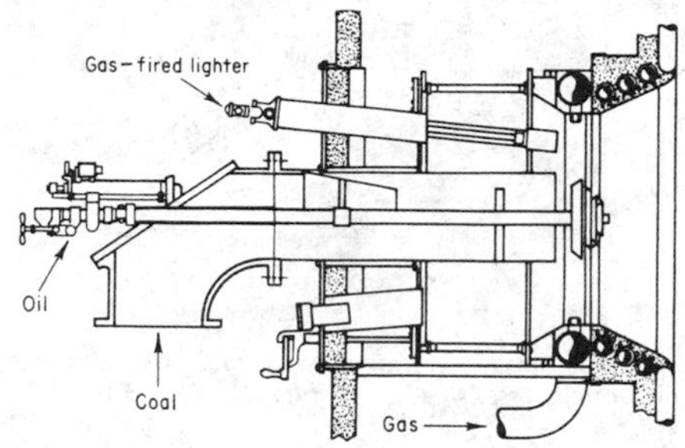

FIG. 9-35 Circular burner for pulverized coal, oil, or gas. (*From Baumeister,* Marks' Standard Handbook for Mechanical Engineers, *7th Ed., McGraw-Hill, New York, 1967.*)

being 30°, enabling the operator to control superheat and to permit selective utilization of furnace heat-absorbing surfaces. Basically the turbulence needed for mixing is generated in the furnace instead of in the burners.

The **circular burner** shown in Fig. 9-35 is widely used in horizontally fired furnaces and is capable of firing coal, oil, or gas in capacities as high as 174 GJ/h (165 million Btu/h).

In **cyclone firing** (Fig. 9-34*d*) the coal is not pulverized but is crushed to 4-mesh size, admitted with the primary air in a tangential manner to a horizontal primary, cylindrical chamber called a cyclone furnace. The finer particles burn in suspension, while the coarser ones are thrown by centrifugal force to the outer wall of the cyclone furnace. The wall surface, having a sticky coating of molten slag, retains most of the coal particles until they burn. The secondary air, which is admitted tangentially along the top of the cyclone fur-

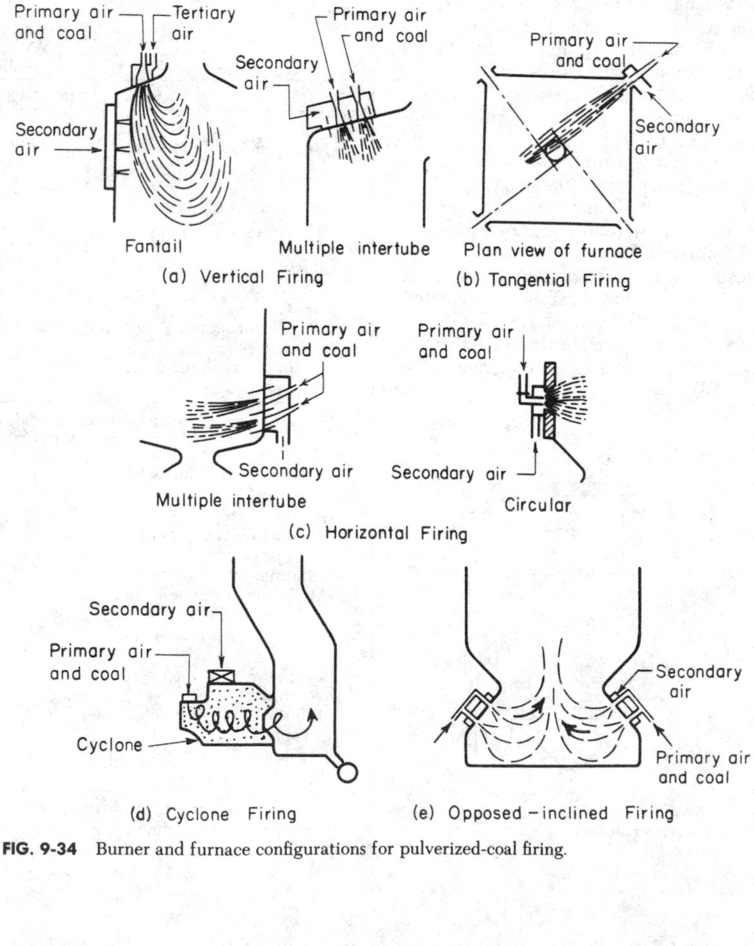

FIG. 9-34 Burner and furnace configurations for pulverized-coal firing.

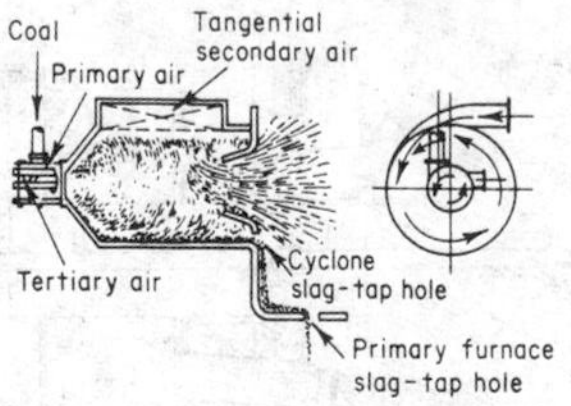

FIG. 9-36 Cyclone furnace. (*From Baumeister,* Marks' Standard Handbook for Mechanical Engineers, *7th ed., McGraw-Hill, New York, 1967.*)

nace, completes the combustion of the coarse particles at a firing rate of about 18.6 GJ/(h·m^3) [500,000 Btu/(h·ft^3)]. Slag drains continuously into the main boiler furnace and then into a quenching tank. Figure 9-36 shows the cyclone furnace schematically.

Spreader-Stoker Firing **Spreader stokers** burn coal by propelling it into the furnace. A portion of the coal burns in suspension (the percentage depending on the coal fineness), while the rest burns on a grate. In most units coal is pushed off a plate under the storage hopper onto revolving paddles (either overthrow or underthrow) which distribute the coal on the grate (Fig. 9-32*c*). The angle and speed of the paddles control coal distribution. The largest coal particles travel the farthest, while the smallest ones become partially consumed during their trajectory and fall on the forward half of the grate.

A few spreaders use air to transport the coal to the furnace and distribute it, while others use mechanical means to transport the coal to a series of pneumatic jets.

The **performance** of spreader stokers is affected by changes in coal sizing, the main problem being the size consist of the fuel fired. The equipment can distribute a wide range of fuel sizes, but it distributes each particle on the basis of the size and weight of the particle. Normal size specifications call for 1.9-cm (¾-in) nut and slack with not more than 30 percent less than 0.64 cm (¼ in).

Approximately 30 to 50 percent of the coal is burned in suspension. If excessive fines are present, more coal will be burned in suspension, more coal particles will be carried out of the furnace, and very little ash will be available to provide a protective cover for the grate surface. If sufficient fines are not present, the capacity of the unit will be reduced because the grate is not designed to burn the entire capacity of the furnace, usually resulting in excessive live coals being dumped to the ash hopper.

The **grates** used with spreader stokers are of several types, as shown in Fig. 9-37. **Stationary grates** are the cheapest to install, but they must be divided into zones for cleaning and ash removal. In the **dumping-grate stoker** the grates themselves can be dumped, thereby eliminating the hoeing of ashes. This provides some reduction in the time necessary for cleaning the grates. **Continuous-cleaning grates** do not need to be zoned for ash removal; the higher grate heat releases possible make such grates economical. Small units can be equipped with grates installed at a slight slope and using a vibrating or oscillating motion to propel the ash to the end of the grate. The stoker is in motion only a small portion of the time, and the fuel bed moves forward as one mass, so there is no serious intermixing of ash and burning coals. These units are restricted by the size of grate, weight of fuel bed, and physical size of the driving apparatus. Large spreader stokers are normally equipped with traveling grates.

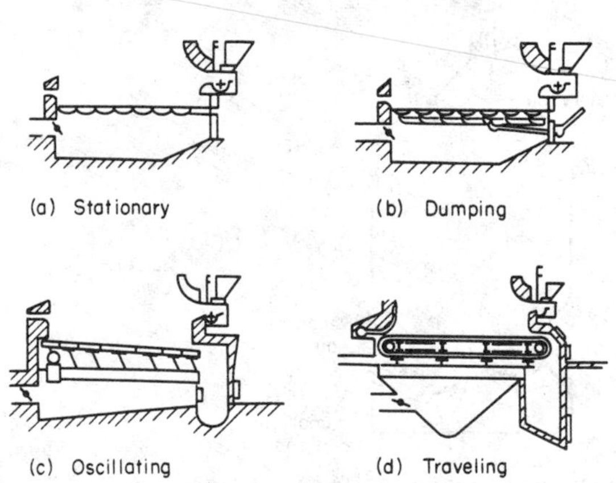

FIG. 9-37 Types of grates used with spreader stokers.

Grate **heat-release** rates used in sizing spreader stokers are listed in Table 9-34. The figures given in the table are for normal designs and will vary for a particular installation, depending upon the coal used and the load factor of the particular unit.

Excess air is usually 30 to 40 percent for stationary and dumping grates, while continuous-cleaning grates are operated with from 22 to 30 percent excess air. Preheat air can be supplied for all types of grates, but the temperature is usually limited to 120 to 150°C (250 to 300°F) to prevent any excessive slagging of the fuel bed.

Overfire air nozzles are located in the front wall underneath the spreaders and in the rear wall from 0.3 to 0.9 m (12 to 36 in) above the grate level. These nozzles use air directly from a fan or inspirate air with steam to provide **turbulence** above the grate for most effective mixing of fuel and air. They supply about 15 percent of the total combustion air.

The **size** range of spreader stokers extends into that for which multiple-fuel firing is generally considered necessary. It is usually easy to increase the height of the furnace and install oil or gas burners in the upper portions of the furnace. During firing of this auxiliary fuel

TABLE 9-34 Grate Heat Releases in Spreader Stokers

Grate type	Grate heat release 1000 B.t.u./(hr.)(sq. ft.) Without dust collector	With dust collector
For bituminous coals		
Stationary:		
Maximum continuous load	350	400
2-hr. peak	400	450
Dumping:		
Maximum continuous load	375	475
2-hr. peak	450	525
Oscillating:		
Maximum continuous load	...	555
2-hr. peak	...	625
Traveling:		
1000 lb./hr. continuous boiler capacity	...	100 200 300
Maximum continuous load	...	650 700 725
2-hr. peak	...	725 775 800
For Iowa coal, subbituminous coal, lignite, wood, and bagasse		
Stationary:		
Maximum continuous load	450	600
2-hr. peak	550	750
Dumping:		
Maximum continuous load	500	650
2-hr. peak	600	750
Oscillating:		
Maximum continuous load	...	650 Refractory with clinker chill
Maximum continuous load	...	700 Water-cooled furnace
2-hr. peak	...	800 All furnaces
Traveling:		
1000 lb./hr. continuous boiler capacity	...	100 200 300
Maximum continuous load	...	750 850 850
2-hr. peak	...	850 900 1000

NOTE: To convert British thermal units per hour–square foot to joules per square meter–second, multiply by 3.16.

TABLE 9-35 Factors for Selecting Mechanical Stokers

Factors	Chain or traveling grate	Spreader	Underfeed
Initial capital investment	Least	Least	Most
Handling fine coal	Poor	Poor	Poor
Handling caking coal	Poor	Good	Poor
Handling ash	Good	Excellent	Fair
Maintenance and operating costs	Intermediate	Lowest	Largest
Fly ash and carbon loss	Lowest	Largest	Lowest
Draft loss through fuel bed	Intermediate	Lowest	Largest
Minimum excess air, percent	30	25	45
Ability to handle load changes	Fair	Excellent	Fair

the grate must be protected from overheating by a very deep bed of ash or a firebrick cover, both with slight air leakages.

The carbon content of the ash passing out of the furnace varies from 30 to 50 percent. Overall efficiency of a spreader stoker can be increased by **reburning** this **fly ash.** It is returned to the stoker grate by a gravity or pneumatic feed system. The reinjected ash must be retained on the fuel bed to keep the fly-ash loading in the boiler passes low.

Table 9-35 shows some of the factors for selecting mechanical stokers.

Pulverizers The heart of any solid-fuel suspension-firing system is the pulverizer. Air is used to dry the coal, transport it through the pulverizer, classify it, and transport the specified fines to the burner, where the transport air provides part of the air for combustion. The pulverizers themselves are classified according to whether they operate under positive or negative pressure and whether they are slow-, medium-, or high-speed.

Pulverization occurs by impact, attrition, or crushing. The capacity of a pulverizer depends on the grindability of the coal and the fineness desired, as shown by Fig. 9-38. Capacity can also be affected by excessive moisture in the coal. Capacity can be restored by increasing the temperature of the primary air; Fig. 9-39 indicates the temperatures needed. For pulverized coal-fired boilers, the coal size usually is 65 to 80 percent through a 200-mesh screen, which is equivalent to 74 μm. The kinds of pulverizers and their characteristics are discussed in Sec. 8.

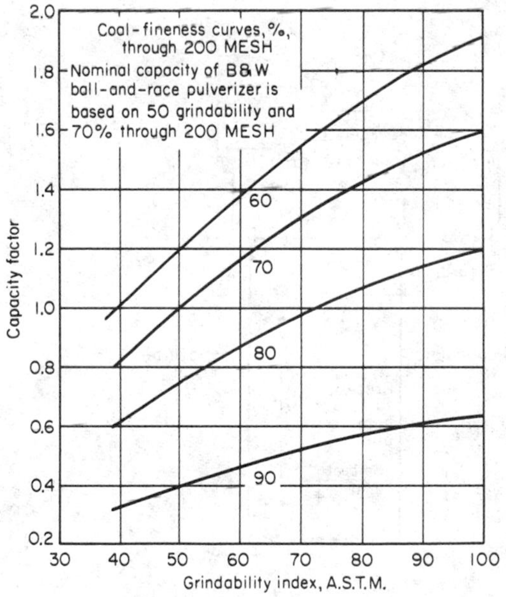

FIG. 9-38 Variation of pulverizer capacity with the grindability of the coal and the fineness to which the coal is ground. *(Babcock & Wilcox Co.)*

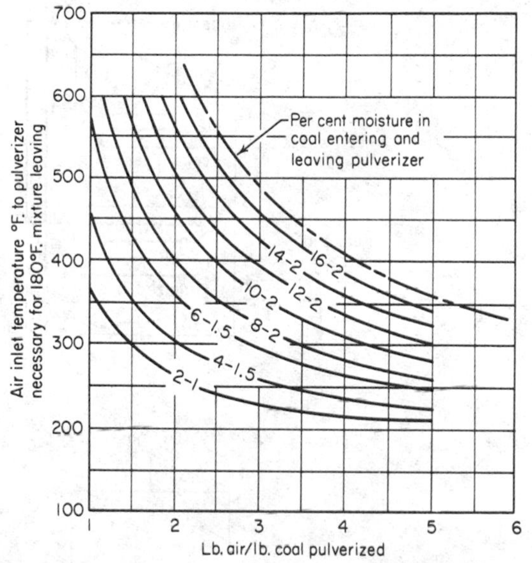

FIG. 9-39 Effect of moisture in coal on pulverizer capacity. Sufficient drying can be accomplished to restore capacity if air temperatures are high enough. (Combustion Engineer, *Combustion Engineering Inc., New York, 1966.)*

SOLID-FUELS COMBUSTION IN FLUIDIZED BEDS

GENERAL REFERENCES: *Proceedings of the Fourth International Conference on Fluidized-Bed Combustion,* Dec. 9–11, 1975, The MITRE Corporation, McLean, Va. 22102. *Proceedings of the Workshop on Utility/Industrial Implementation of Fluidized Bed-Combustion Systems,* Atlanta, Ga., Apr. 27–28, 1976, The MITRE Corporation, McLean, Va. 22102. *Proceedings of the Fluidized-Bed Combustion Technology Exchange Workshop,* Apr. 13–15, 1977, The MITRE Corporation, McLean, Va. 22102. *Proceedings of the Fifth International Conference on Fluidized-Bed Combustion,* December 1977, The MITRE Corporation, McLean, Va. 22102. *An Assessment of the Status of Fluidized-Bed Combustion Based on the Papers of the Fifth International Conference,* December 1979, The MITRE Corporation, McLean, Va. 22102. *Proceedings of the Sixth International Conference on Fluidized-Bed Combustion,* Atlanta, Ga., Apr. 9–11, 1980, U.S. Department of Energy, CONF-800428, vols. 1–3.

Atmospheric and Pressurized Fluidized Beds (AFBC and PFBC) Fluidized-bed combustion (FBC) involves the combustion of fuel in a bed of solid particles, which is fluidized (held in suspension) by the injection of air at the bottom of the bed. When coal is burned in this manner, the bed can consist of inert solids, coal ash, or a sorbent such as limestone or dolomite. Limestone or dolomite in the bed reacts with the sulfur dioxide formed during combustion of the coal and forms a solid sulfate which can be discarded as a dry solid. Variations in the technology include atmospheric-pressure fluidized-bed combustion (AFBC) and pressurized fluidized-bed combustion (PFBC). Figure 9-40 shows various applications of AFBC, and Fig. 9-41 shows PFBC-cycle concepts.

In the early 1960s, Great Britain's National Coal Board, the British Coal Utilization Research Association, and the Central Electricity Generating Board began investigating FBC. After the energy crisis in 1973, AFBC and PFBC of coal developed rapidly, mostly in the United States, Great Britain, and the Federal Republic of Germany, for use in utilities, process industries, and the institutional sectors. FBC has great potential to burn efficiently and cleanly a wide variety of fuels, ranging from high-sulfur, high-ash coals of all ranks and grades to industrial-process wastes and municipal sludges.

Some of the potential advantages of FBC, compared with pulverized-coal firing, are:

1. Less sulfur dioxide emissions, because much of the sulfur is captured by fuel-bed additives such as limestone or dolomite

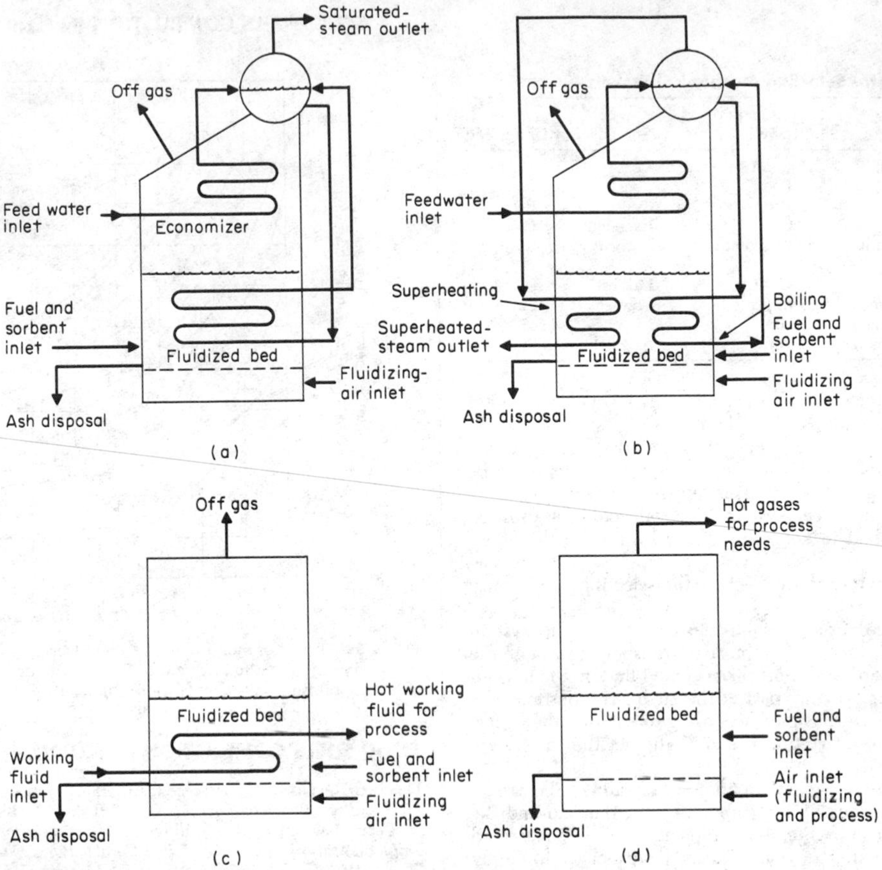

FIG. 9-40 Atmospheric-fluidized-bed-combustion boiler-combustor arrangements. (*a*) Saturated-steam boiler. (*b*) Superheater-steam generator. (*c*) Indirect-process heater. (*d*) Direct-process heater.

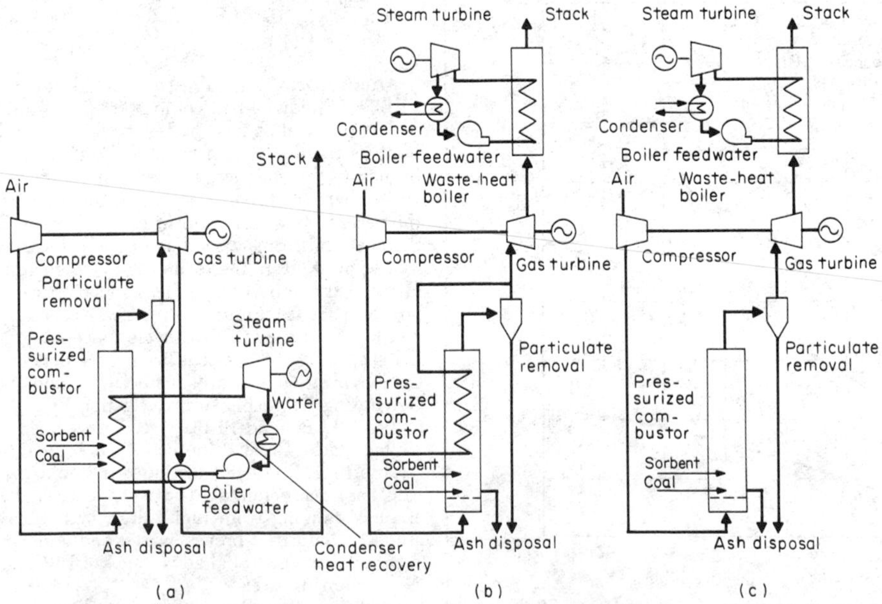

FIG. 9-41 Pressurized-fluidized-bed-combustion concepts. (*a*) Steam-cooled tubes in bed. (*b*) Air-cooled tubes in bed. (*c*) Bed cooled by excess air (300 percent); no in-bed tubes.

2. Low fuel-bed temperatures, up to about 954°C (1750°F), which mean:

a. Less thermal nitrogen oxide emissions

b. Fewer problems caused by ash agglomeration

c. Less volatilization of sodium and potassium in the coal and, consequently, less deposition on and corrosion of downstream components, such as boiler superheaters and reheaters and turbine blades in PFBC gas-turbine cycles

3. High heat-transfer rates to heat-absorption surfaces, as high as 3.154 kW/m^2 [100,000 Btu/(ft^2·h)], and high heat-transfer coefficients between the bed and tubes immersed in the bed of 398 W/(m^2·°C) [70 Btu/(ft^2·h·°F)]

4. High volumetric energy release rates of 5.2 MW/m^3 [500,000 Btu/(h·ft^3)], as compared with 0.2 MW/m^3 [20,000 Btu/(h·ft^3)] in a conventional pulverized coal-fired boiler

An AFBC steam-generating system is shown schematically in Fig. 9-42. Air enters through a distributor plate, and crushed coal, 0.6 to 1.2 cm (¼ to ½ in), and SO_2 sorbent, limestone or dolomite, are injected directly into the bed or by overbed feeders. At any given instant, the bed contains coal ash, a small inventory of coal (½ to 2 percent of bed material), plus sorbent particles calcined and sulfated to varying degrees. The bed material is agitated and set in motion by the air flowing upward through the distributor at a superficial velocity of 0.6 to 4.6 m/s (2 to 15 ft/s). The turbulent and recirculating motion of the bed provides good mixing.

Heat released in the bed is absorbed by cooled tubes in the walls and in the bed, and heat in the products of combustion (POC) is absorbed by convection tubes in the freeboard above the bed. The amount of in-bed heat-transfer surface depends on the fuel heating value; for some low-heating-value fuels, there is no surface in the bed. Water, steam, or air can be used as coolant, depending on the application, but water heating and steam generating or superheating constitute the major applications.

AFBC technology is more highly developed than PFBC, and some manufacturers are offering AFBC units in industrial sizes. Much is known about PFBC with regard to combustion and environmental performance, but integrated operation of a PFBC system, including the gas cleanup and gas turbine, has not been demonstrated. PFBC likely will be better suited for utility and large industrial applications; AFBC has significant potential for use in the industrial sector. Because of its state of development and commercial availability, AFBC is discussed with regard to some basic aspects.

AFBC Plant Design An AFBC plant consists of several subsystems including (1) fuel and sorbent receiving, storage, and preparation; (2) solids feeding; (3) combustor, including spent-bed removal; and (4) particulate removal from flue gas. Except for the combustor, most components for the subsystems are commercially available.

Two important considerations in AFBC are carbon utilization and sorbent performance (SO_2 removal). Both carbon utilization and sorbent performance depend on many interrelated variables, and a clear understanding of AFBC performance is further complicated by the fact that both coal and sorbents are heterogeneous materials. Combustor performance can be expected to vary with changes in operating conditions, different combinations of coal and sorbent, and design-specific combustor features (i.e., distributor, heat-exchanger configuration, and freeboard height). Therefore, the design of an AFBC combustor requires trade-offs among several considerations to achieve desired performance. These design considerations determine the extent to which the combustor will meet its heat-generation and emissions-control requirements. Additional important performance characteristics include heat release, heat-transfer rates, and NO_x emissions. Although fuel and sorbent properties have direct effects on performance, an AFBC boiler is tolerant of a wide range of fuels. Detailed in the following paragraphs are important operating characteristics and the effects of operating or design variables on these characteristics.

Carbon-Combustion Efficiency Achieving a high carbon-combustion efficiency is important for overall plant efficiency and consequently for plant economics. Carbon losses occur through the carry-over of unburned carbon particles and, to a much smaller degree, the incomplete combustion of carbon to carbon monoxide and unburned carbon withdrawn with spent bed material. Operating variables that affect carbon-combustion efficiency include bed temperature, excess air, and fluidizing velocity. Typical data showing the bed temperature and excess-air effects are given in Fig. 9-43. Increased carbon-combustion efficiency is achieved through increased bed temperature and increased excess air. However, limits are placed on these variables by considerations of SO_2 capture, NO_x formation, and ash-fusion characteristics.

The effect of fluidizing velocity on carbon-combustion efficiency is shown in Fig. 9-44. Increasing velocity increases carry-over of fine unburned carbon particles, which decreases carbon-combustion efficiency. Combustion efficiency can be increased by designing for lower velocity, but lower-velocity operation requires a larger bed.

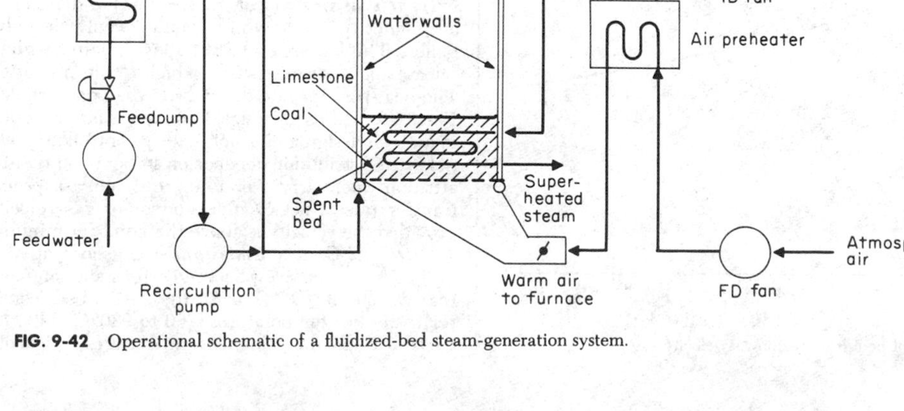

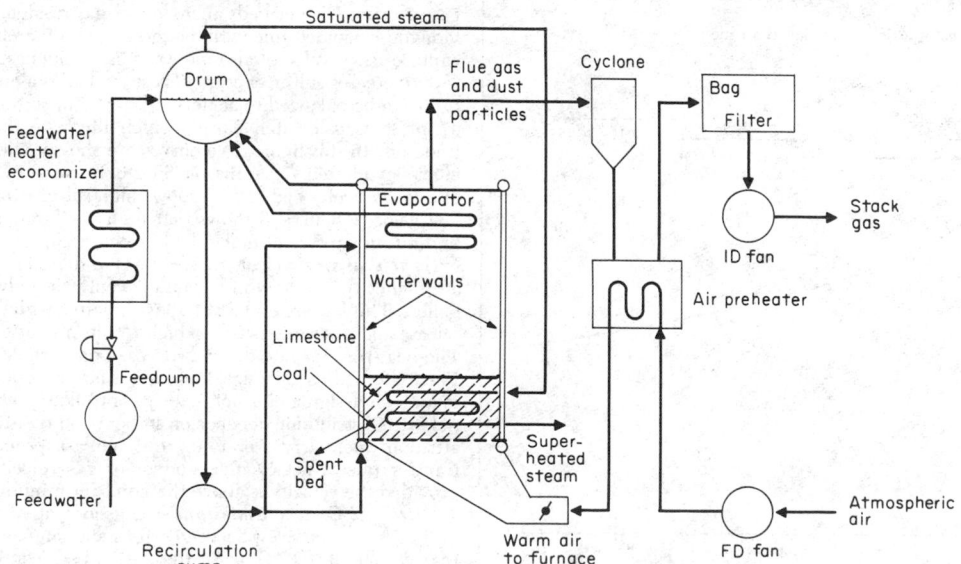

FIG. 9-42 Operational schematic of a fluidized-bed steam-generation system.

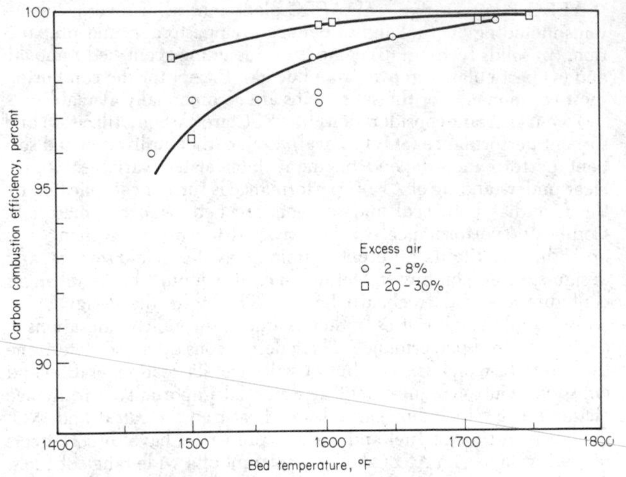

FIG. 9-43 Carbon-combustion efficiency as a function of bed temperature.

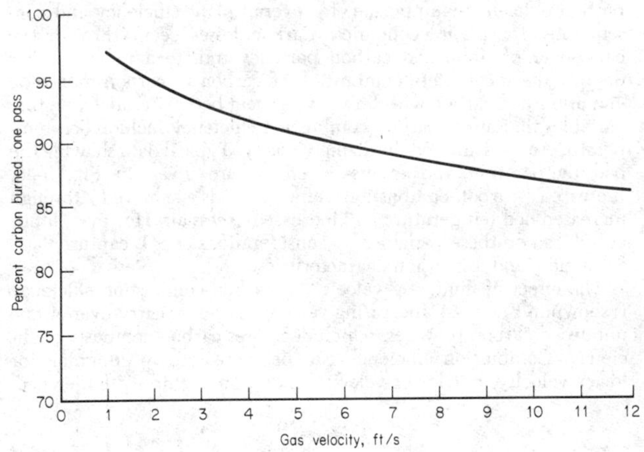

FIG. 9-44 Carbon-combustion efficiency as a function of gas velocity.

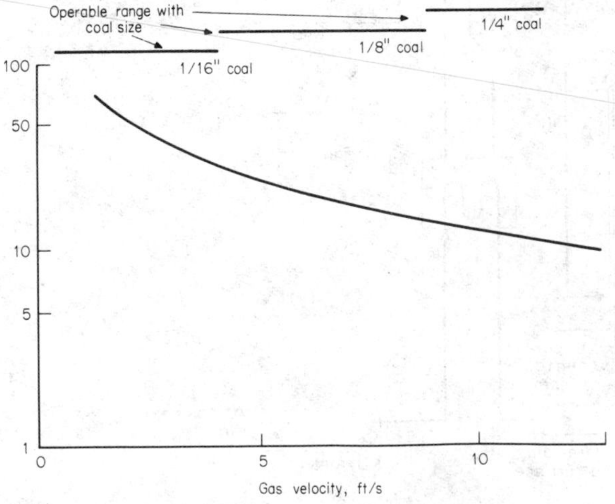

FIG. 9-45 Dependency of bed grid area on gas velocity.

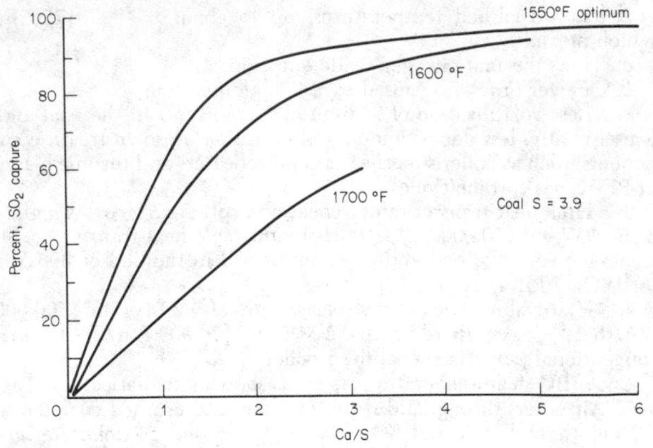

FIG. 9-46 Calcium-sulfur-ratio effect on SO_2 capture.

The relationship of bed area to velocity is illustrated in Fig. 9-45. Decreasing velocity rapidly increases the required square feet of grid or bed area needed per unit of output.

SO_2 Removal SO_2 removal or S retention in situ is a unique feature of the AFBC process. Sulfur contained in the fuel is oxidized to SO_2 during the combustion process. This SO_2 chemically reacts, in the presence of oxygen, with the sorbent to form a stable, solid sulfate in the bed. SO_2 emissions are thereby controlled in the combustion vessel itself, eliminating the need for downstream SO_2 cleanup of the flue gas.

Sulfur capture depends on a number of design and operating variables as well as on the sorbent used. A primary variable, which can be readily controlled, is the calcium-to-sulfur mole ratio. Also of importance is the bed temperature, since the SO_2-sorbent reaction is temperature-dependent. These two effects are illustrated in Fig. 9-46. Typical experimental trends are shown for SO_2 capture plotted against the Ca/S ratio for several bed temperatures. An optimum sulfur-capture temperature of around 843°C (1550°F) is common for many sorbents. Increasing the Ca/S mole ratio increases sulfur capture, but it is desirable to minimize calcium addition because this represents greater sorbent costs, parasitic heat loss through calcination, and increased disposal problems.

Other variables that affect sulfur capture include bed depth and gas velocity. Deeper beds allow longer gas residence time, i.e., provide more contact time with the sorbent, and thereby increase sulfur capture. Low velocity also increases gas residence time and, therefore, increases sulfur capture. These variables along with the Ca/S ratio can be balanced to achieve efficiency of sulfur capture. Fig. 9-47 presents calculated curves which illustrate this sulfur-capture trade-off. In this figure, two curves are shown, with sulfur-removal efficiency plotted versus the Ca/S ratio. In curve A, the performance of a low-velocity bed (4 ft/s) shows high sulfur capture at relatively low velocity. Curve *B* shows that high Ca/S ratios are needed for high sulfur capture at higher bed velocities.

Heat Transfer Heat-transfer rates to tubes immersed in AFBCs are higher than those to tubes in conventional pulverized-coal-fired boilers. This feature of AFBCs allows relatively high volumetric heat release and absorption rates, which result in relatively small boilers. The heat-transfer rate depends on the bed particle size as shown in Fig. 9-48. Bed particle size, however, is largely a dependent variable in a given design and is not readily controlled or changed. Bed-particle-size distribution depends on sorbent and fuel-feed size, sorbent-attrition rate, carry-over rates, and fuel-ash properties. The heat-transfer rate is nearly independent of gas velocity in the AFBC, provided the velocity is above the point for minimum fluidization.

Nitrogen Oxides Emissions Emissions of nitrogen oxides are limited by current New Source Performance Standards (NSPS) to less than 0.7 lb $NO_2/10^6$ Btu. In the AFBC bed, combustion occurs at relatively low temperatures [760 to 930°C (1400 to 1700°F)], compared with the much higher temperatures typical of pulverized-coal-fired boilers. These low temperatures do not allow significant oxidation of atmospheric nitrogen, so the AFBC process is characterized by low NO_x emissions. Nitrogen oxides formation in the AFBC results primarily from oxidation of fuel-bound nitrogen. The reaction of carbon monoxide and NO_x in the gas-burnout zone apparently further reduces NO_x emission from the AFBC. Fuel nitrogen content and percent of excess oxygen are the two main variables affecting NO_x emission. Over the excess-oxygen range of 0 to 10 percent, NO_x emissions are generally below the current NSPS level as well as the proposed level of 0.6 lb $NO_2/10^6$ Btu.

Experimental data on NO_x emissions taken since the late 1960s from major FBC experimental facilities are shown in Fig. 9-49. Superimposed on the graph are three levels of emission control: moderate, intermediate, and stringent. These three levels are those that

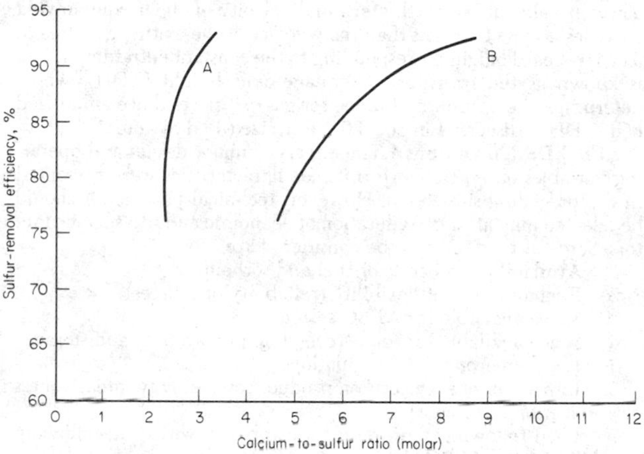

Base Operating Conditions

Sorbent type	AFBC limestone A	AFBC limestone B
Average diameter, μm	500	1000
Pressure, kPa	101	101
Bed temperature, °C	840	840
Excess air, percent	20	20
Velocity, m/s	1.38	3.50
Bed depth, m	1.22	1.22

FIG. 9-47 Sulfur-capture trade-off.

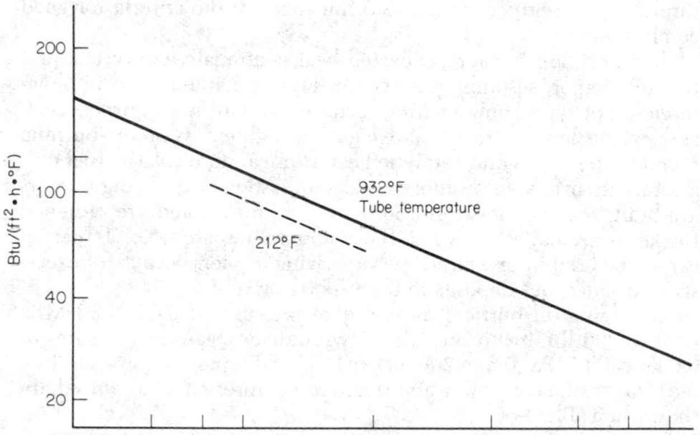

FIG. 9-48 Effect of particle size on bed-to-tube heat transfer.

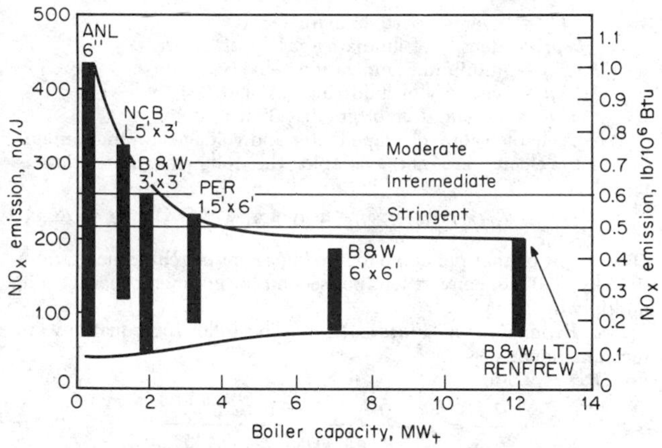

FIG. 9-49 NO_x emissions from AFBC facilities.

can be achieved by (1) normal or routine operation, (2) a well-maintained system operating with some optimization of design conditions, and (3) optimization of all design parameters so that the system is performing at its technological limit.

Figure 9-49 shows that as AFBC systems increase in size, the range of NO_x emissions decreases. When capacities approach those expected for commercial use (i.e., greater than 5 to 10 MW_t), the range of NO_x emissions falls below even the most stringent levels identified for possible emission standards.

Fluidization Design Factors In addition to the AFBC process considerations, the design of an AFBC combustor must follow the design principles associated with fluidization as discussed in Sec. 20; "Fluidized-Bed Systems." Information useful in the design of coal-fired AFBCs is presented in the following paragraphs. The information is based on the *Coal Conversion Systems Data Book*, prepared by the Institute of Gas Technology for the U.S. Department of Energy (HCP/T2286-01).

Shape Factors of Various Materials Shape factors for nonspherical particles are often required in fluidization studies. Shape factor ϕ is defined as the ratio of the area of a sphere, with a volume equivalent to that of the particles, divided by the actual surface area of the particles. Some values follow:

Material	Shape factor ϕ
Coal dust, pulverized	0.73
Coal, anthracite	0.63
Coal, bituminous	0.63
Limestone	0.45

Relation of Minimum Fluidized-Bed Voidage to Average Particle Diameter Minimum fluidized-bed voidage ϵ_{mf} is often required in fluidization studies. Fig. 9-50 gives ϵ_{mf} as a function of particle diameter for various materials.

Minimum Fluidization Velocity The Kunii-Levenspiel correlation is recommended for estimating minimum fluidization velocities.

$$\frac{1.75}{(\varphi \epsilon_{mf}^3)}(\mathrm{Re}_{mf})^2 + 150\left(\frac{1-\epsilon_{mf}}{\varphi^2 \epsilon_{mf}^3}\right)\mathrm{Re}_{mf} - \mathrm{Ar} = 0 \qquad (9\text{-}37)$$

where $\mathrm{Re}_{mf} = \dfrac{D_p \rho_g U_{mf}}{\mu}$ = Reynolds number at minimum fluidization velocity

$\mathrm{Ar} = \dfrac{D_p^3 \rho_g (\rho_s - \rho_g) g}{\mu^2}$ = Archimedes number

φ = shape factor

ϵ_{mf} = minimum fluidized-bed voidage

ρ_s = particle density of fluidizing solids, lb/ft^3

D_p = average particle diameter, ft
ρ_g = density of fluidizing gas, lb/ft^3
U_{mf} = minimum fluidization velocity, ft/s
μ = viscosity of fluidizing gas, lb/(ft·s)
g = acceleration of gravity, 32.2 ft/s^2

When reliable values of shape factor and voidage at minimum fluidization velocity are not available, the following correlation is useful:

$$U_{mf} = (\mu/\rho_g D_p)\,[(25.25)^2 + 0.0651\ \mathrm{Ar}]^{0.5} - 25.25 \tag{9-38}$$

(To convert pounds per cubic foot to kilograms per cubic meter, multiply by 16.02; to convert feet per second to centimeters per second, multiply by 30.48.)

Bed Expansion on Fluidization The following equations are recommended:

For $D_{bed} \leq 2.5$ in,

$$\frac{L_f}{L_{mf}} = 1 + \frac{0.762(U - U_{mf})^{0.570}\rho_g^{0.083}}{\rho_s^{0.166}U_{mf}^{0.063}D_{bed}0.445} \tag{9-39}$$

and for $D_{bed} > 2.5$ in,

$$\frac{L_f}{L_{mf}} = 1 + \frac{10.978(U - U_{mf})^{0.738}D_p^{1.006}\rho_s^{0.376}}{U_{mf}^{0.937}\rho_g^{0.126}} \tag{9-40}$$

where D_{bed} = column diameter, ft
L_{mf} = height of minimum fluidized bed, ft
L_f = height of fluidized bed, ft
U = superficial gas velocity, ft/s
U_{mf} = minimum fluidization velocity, ft/s
ρ_s = particle density of fluidizing solids, lb/ft^3
ρ_g = density of fluidizing gas, lb/ft^3
D_p = average particle diameter, ft = $1/[\Sigma(X_i/D_{pi})]$

The experimental data on which the correlation is based are limited to a maximum bed diameter of 1 ft. Extrapolations to very large bed diameters must be made with caution. Second, the correlations in their present form do not account for the fact that there would be no bed (infinite expansion) at superficial gas velocities equal to or greater than the terminal velocity of the particles. Superficial velocities in the data used are less than 60 percent of the terminal velocity for the average particle diameter for the bed. Projections to higher relative velocities should be made cautiously.

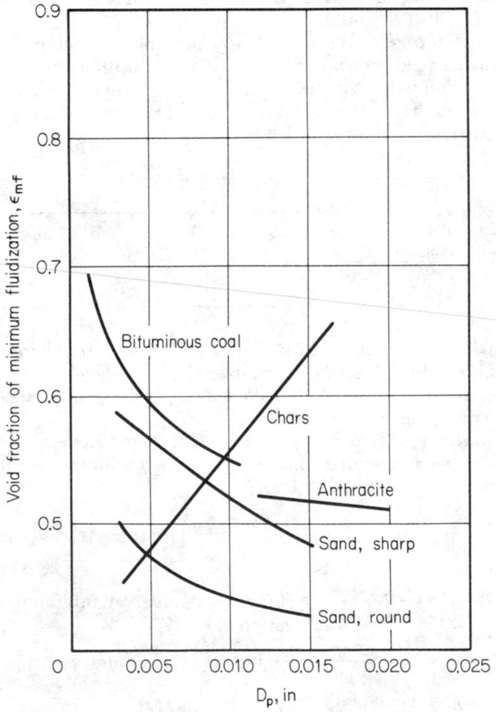

FIG. 9-50 Relation of minimum fluidized-bed voidage to average particle diameter.

Transport Disengagement Height (TDH) Fluidized beds are usually operated with a wide size distribution of solids, containing a substantial portion of fines. More fines are also created by attrition and by virtue of reacting solids, and the exiting gases may exceed the terminal velocity of many of these fine particles, thereby carrying them out of the reactor. In addition to the entrainment of fines, solids are carried into the freeboard by erupting bubbles. However, as the particles ascending the freeboard space lose their kinetic energy, some particles will return to the bed, depending on particle size and particle density. As a result, particle loading in the escaping freeboard gas drops rapidly to a certain point beyond which it attains a constant value at which the terminal velocity of the accompanying particles is close to or less than the velocity of the exiting gas stream. The freeboard height corresponding to the constant entrainment rate is known as the **transport disengagement height** (TDH), which determines the optimum distance for gas exit ports above a fluidized bed. TDH is discussed in Sec. 20: "Fluidized-Bed Systems."

AFBC Design Options The effects of major design and operating variables on AFBC performance characteristics were presented in the preceding subsection. However, technical parameters should be selected only after consideration of economic and site-specific factors. Some of the factors to be considered are:

1. Availability and costs of fuel and sorbent
2. Fuel and sorbent flexibility (reliability of sources)
3. Space available for AFBC system
4. Space available for solids receiving, preparation, and storage
5. Local environmental regulations
6. Impacts of energy costs on product costs (energy intensiveness of application or process)
7. Load-following requirements compared with load-following capability of the AFBC system

COMBUSTION OF LIQUID FUELS

Burners for Liquid Fuels For combustion, liquid fuels are vaporized or atomized in the combustion air. Distillate fuel oil can be burned with a blue flame if it is completely vaporized and homogeneously dispersed in the air before burning. Blue-flame combustion is a two-step mechanism, the first step being hydroxylation of the fuel. Yellow flame indicates glowing carbon from fuel pyrolysis in oxygen-deficient parts of the flame. Droplets of fuel may be partly volatilized, leaving the residue to decompose and burn as coke particles. Either a blue or a yellow flame may be preferred, depending on the need for conductive versus radiant heat transfer. Both types can give complete combustion if the flame is not quenched prematurely. Time, temperature, and turbulence are the criteria for good combustion.

In **vaporizing burners,** reflected heat continually converts liquid fuel into vapor, sustaining the flame. This principle is used in blowtorches, pot-type home-heating furnaces, and all wick burners such as kerosine lamps, stoves, and cigarette lighters. Gasoline-burning mantle-type lamps and catalytic heaters use a more volatile fuel plus a catalytic surface to promote rapid combustion. Vaporizing burners are built in capacities up to 30 to 40 dm^3 fuel/h and are designed for kerosine, naphtha, No. 1 fuel oil, gasoline, etc. (No. 1 fuel oil cannot be used in pressure-type vaporizing burners because of excessive carbonaceous deposits in the vaporizing tube.)

Atomizing oil burners spray fuel at pressures of 0.69 to 2.1 MPa (100 to 300 lbf/in^2) or atomize it with air or steam at pressures of 0.003 to 1.4 MPa (0.5 to 200 lbf/in^2). Several types are shown in Fig. 9-51. Quantities of power and steam or compressed air required are shown in Table 9-36.

Combustion air is blown into the furnace with the fuel spray. Vanes and baffles are built into the air stream to assure proper air-fuel mixing. The air-handling parts, the fuel-spray pattern, and the furnace shape are matched to avoid flame impingement, a cause of

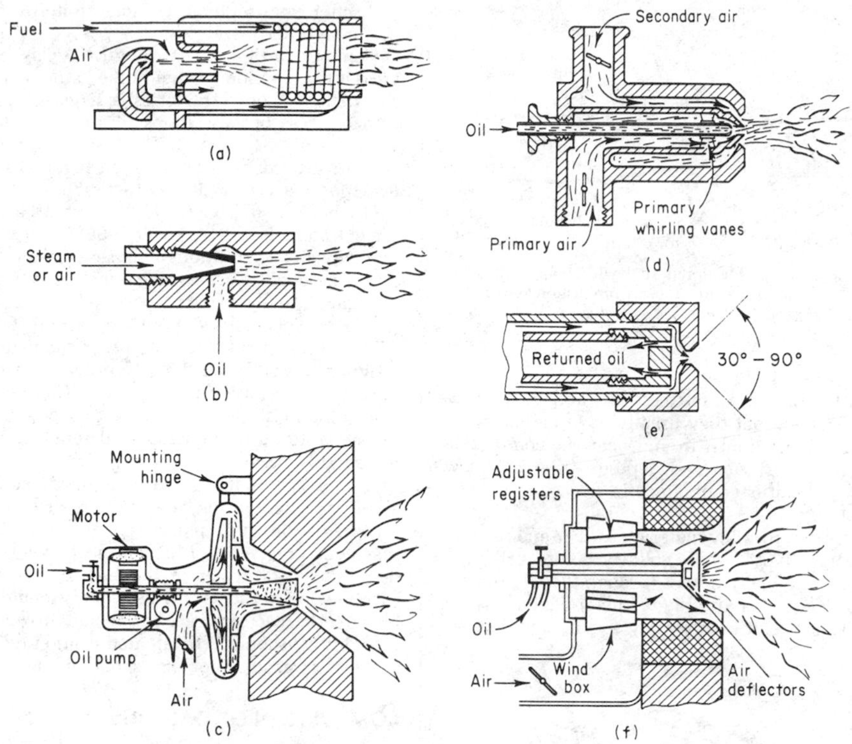

FIG. 9-51 Principal types of oil burners. (*a*) Pressure-type vaporizing burner; oil travels through coil of pipe. Uses kerosine or gasoline, 0.1 to 8 gal/h at 5- to 50-lbf/in² fuel pressure, with a turndown ratio of 3:1. For blowtorches, lamps, and portable torch-type burner equipment. (*b*) High-pressure steam- or air-atomizing burner, injector or venturi type. Uses all grades of oil, No. 1 to No. 6, with heavy oil heated to flow. Oil pressure, low; steam pressure, 40 to 175 lbf/in². Uses 5 to 10 percent of air or 2 to 4 lb of steam for atomizing. On boilers, 2 percent of the steam output is used for atomization. Steam helps heat oil and assists in combustion to reduce soot. (*c*) Horizontal rotary-cup atomizing oil burner. Uses any grade of fuel, No. 1 to No. 6. Heavy oil must be heated to 150 to 330 SSU. Oil pressure, low; air pressure, 0.25 to 3 lbf/in². Turndown ratio of 5:1. Capacities of 1 to 250 gal/h. For use on automatic-fired boilers. (*d*) Low-pressure air-atomizing burner, variable-pressure type. Uses all grades of oil, No. 1 to No. 6, when supplied at viscosity of 80 to 90 SSU. Oil pressure, 5 to 20 lbf/in²; air pressure, 0.5 to 5 lbf/in². Primary air pressure, constant; secondary air pressure varies. Turndown ratio of 4:1. Capacity of 1 to 200 gal/h. (*e*) Mechanical or oil-pressure atomizing burner, return-flow type, showing general operating principle and typical design. Uses all grades of fuel oil, No. 1 to No. 6, with heavy oil heated to 150 SSU. Oil pressure, 300 lbf/in²; air pressure, low or natural draft. Turndown ratio of 10:1. Capacities of 10 to 1000 gal/h. (*f*) Complete mechanical or oil-pressure atomizing burner unit. Air supplied by natural draft of low-pressure blower. Used on boilers and rotary kilns. Domestic burners of this type use oil at 100-lbf/in² pressure. *(Hauck Mfg. Co.)*

soot or hard-carbon deposits and/or refractory spalling, washout, or slagging. Except in small domestic furnaces, oil spray and air usually enter the furnace through an **ignition tile.** The tile configuration, matched with the oil-spray pattern, stabilizes the flame. Furnace volume is established to allow time for complete combustion. Heat-release rates depend on fuel properties, excess-air concentration, air-fuel mixing, and allowable smoke levels.

TABLE 9-36 Compressed-Air, Steam, and Power Requirements for Atomizing Burners

	Requirements per cm³/s (0.951 gal oil/h)	
Atomizing medium	Power, kW	Fluid
Low-pressure air, 6.9 kPa	0.071	4.5–6.7 dm³/s
High-pressure air, 0.52 MPa	0.284	1.1–1.4 dm³/s
Steam	0.851*	0.85–3.5 kg
Mechanical	0.0227	
Rotary-cup burners	0.0355	

*Hydraulic-power equivalent.

Coal- and shale-derived liquids generally contain more nitrogen than petroleum and form more NO_x when burned. Also the high *C*/*H* ratio in coal-derived liquids gives luminous flames and potentially more smoke. Combustion modifications are being developed to control stack emissions without requiring drastic upgrading of the fuel.

Staged combustion involves primary burning with insufficient air (to prevent NO_x formation in the highest temperature zone), followed by the introduction of more air to complete combustion. In multiburner boilers, this may be accomplished by firing some burners rich and others lean. (See KVB, Inc., "Reference Guideline for Industrial Boiler Manufacturers to Control Pollution with Combustion Modification," EPA 600/8-77-003b, November 1977.)

For gas turbines, staged combustion in each burner is a promising prospect. See Fig. 9-52.

High-intensity burners permit clean combustion at low excess-air levels through finer fuel atomization and better air-oil mixing. Figure 9-53 shows nozzles in which oil is sprayed through a gas, steam, or air whistle and atomized by sonic energy. Typical oil pressure might be below 0.4 MPa (60 lbf/in²); atomizing air or steam, at 0.34 to 0.4 MPa (50 to 60 lbf/in²); windbox (combustion air), at 0.025 to 2.5 kPa (0.1 to 10 in water).

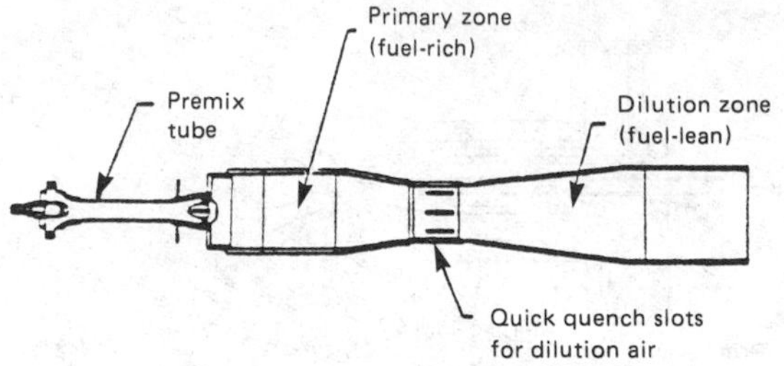

FIG. 9-52 Rich-lean-burner concept for liquid fuel with bound nitrogen. (*Adapted from Pierce, Mosier, and Smith,* Advanced Combustion Systems for Stationary Gas Turbine Engines, *vol. II:* Bench Scale Evaluation, *EPA-600/7-80-017b.*)

Combustion Deposits Residual fuels cost less and yield more heat per gallon than distillates, but they usually have higher sulfur content plus **ash-forming ingredients.** In and near the combustion zone, molten ash can cause corrosion and deposits; in areas below 180°C (350°F), water and sulfur compounds in flue gas condense into corrosive acid solutions.

During combustion, the ash-forming materials oxidize and interact to form a variety of compounds. If they solidify before striking a solid surface, ash particles are likely to pass through the equipment. However, parts of a boiler (or gas turbine or diesel engine) which operate above the ash fusion temperature may accumulate deposits and suffer catastrophic corrosion. Molten vanadium compounds are particularly corrosive.

Combustion air often carries **dust.** A classical example in limestone areas is the diopside formed in the flame by limestone and silica dust and deposited on refractories as a glass. Spalling occurs when the furnace cools because diopside and refractory shrink at different rates.

Ashes are mixtures, and fusion is not sharply defined. Final liquefaction may occur at 100 to 125°C higher than initial sintering. V_2O_5 melts at 673°C (1243°F), but **oil-ash fusion temperatures** range from below 540 to over 1095°C (1000 to 2000°F), depending on relative concentrations of fluxes (principally sodium) and refractory compounds (such as silica, magnesia, and alumina). Vanadium corrosion usually occurs above 675°C (1250°F), and sulfidation (attack of nickel alloys by sulfates) above 900°C (1650°F).

Magnesia, epsom salts, and other compounds are added at Mg/V weight ratios of 3:1 or 3.5:1 to prevent corrosion and deposition by raising the ash fusion temperature. There is disagreement over the value of alumina as a coadditive to overcome the slight tendency of magnesia to form deposits. Calcium compounds are considered undesirable because they form hard, adherent, insoluble deposits. Manganese compounds and sometimes lead and copper compounds are used as combustion catalysts to reduce soot and smoke. There are also many **additives** sold to benefit the fuel-handling system. These may contain solvents or dispersants to combat sludge, emulsifiers or demulsifiers for water in the fuel, corrosion inhibitors, and other functional ingredients. Fuel suppliers should be consulted for possible adverse reactions between the additive and fuel, and claims for the additive should be evaluated cautiously, but their potential usefulness for specific problems should not be overlooked.

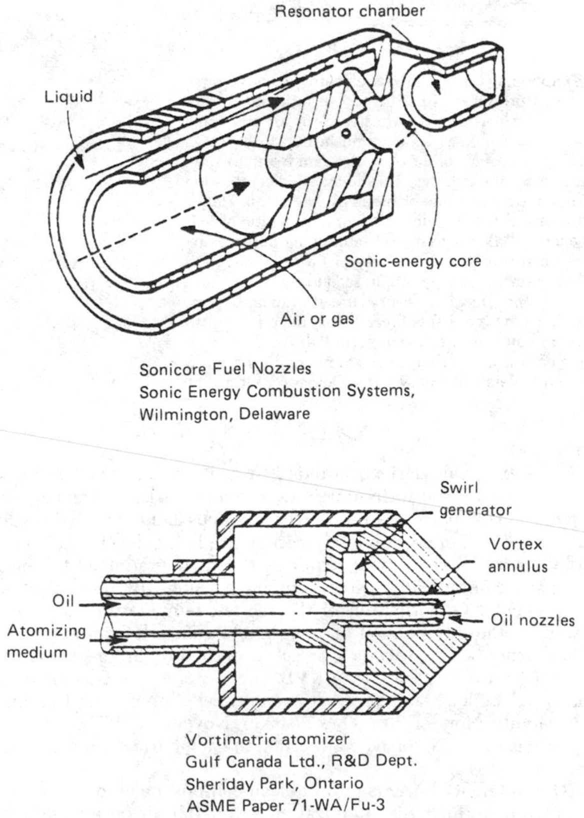

FIG. 9-53 Sonic fuel nozzles. Oil is sheared by pressure waves of sonic frequency in a stream of air, gas, or steam. (*a*) Sonicore fuel nozzle. (*Sonic Energy Combustion Systems, Wilmington, Del.*) (*b*) Vortimetric atomizer. (*R&D Department, Gulf Canada Limited, Sheridan Park, Ont., ASME Pap. 71-WA/Fu-3.*)

COMBUSTION OF GASEOUS FUELS

Combustion of gas takes place in two ways, depending upon when gas and air are mixed. When gas and air are mixed *before* ignition, as in a bunsen burner, burning proceeds by hydroxylation. The hydrocarbons and oxygen form hydroxylated compounds which become aldehydes; the addition of heat and additional oxygen breaks down the aldehydes to H_2, CO, CO_2, and H_2O. As carbon is converted to aldehydes in the initial stages of mixing, no soot can be developed even if the flame is quenched.

"Cracking" occurs when oxygen is added to hydrocarbons *after* they have been heated, decomposing the hydrocarbons into carbon and hydrogen, which, when combined with sufficient oxygen, form CO_2 and H_2O. Soot and carbon black are formed if insufficient oxygen is present or if the combustion process is arrested before completion.

Gas Burners A gas burner is primarily a proportioner and mixing device. In industrial furnaces in which long, "lazy" flames are essential, slow and gradual mixing of air and gas is necessary. When a short, bushy flame is needed, the burner should be designed to achieve rapid and thorough mixing of air and gas. Figure 9-54 illus-

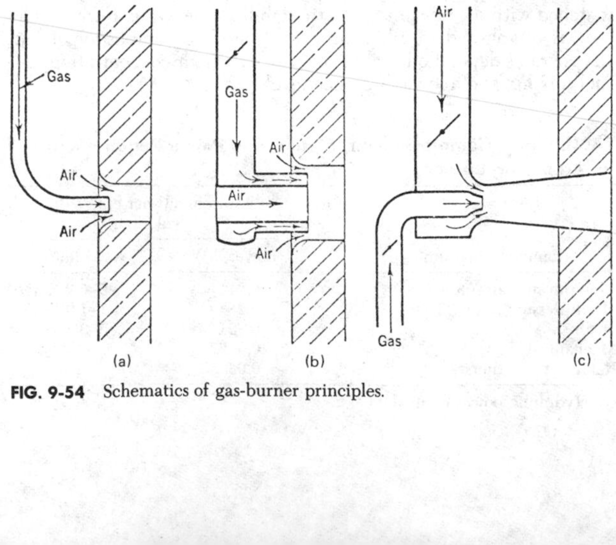

FIG. 9-54 Schematics of gas-burner principles.

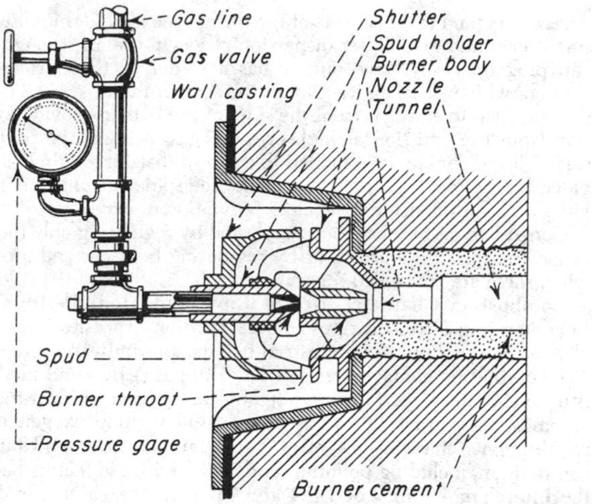

FIG. 9-55 Atmospheric industrial gas-burner installation with individual air control to each burner.

trates simply the principles of gas burners. In burner A, the gas enters the furnace through a burner port and induces a flow of air through the port. Mixing is poor, and a long flame results. The flame can be shortened with a ring burner B, in which gas flows through an annular ring and induces air flow both around and within the annulus of gas. When both air and gas are under pressure, burner C may be used.

In each of these burners, the gas must flow through the port into the furnace at a velocity high enough to prevent the flame from burning back (flashback) into the burner. Burner shapes and arrangements should avoid flame impingement on furnace walls or heat-transfer surfaces. Variations of these principles are found in commercially available burners such as the following.

Premix burners burn by hydroxylation and are used for many natural-draft applications when accurate furnace conditions must be maintained. Figure 9-55 indicates a common natural-draft industrial-type burner with air being aspirated at the spud and burner throat. Figure 9-56 shows another type of premix burner, in which high-pressure air is used to aspirate the gas. The governor diaphragm controls the amount of gas admitted to the aspirator.

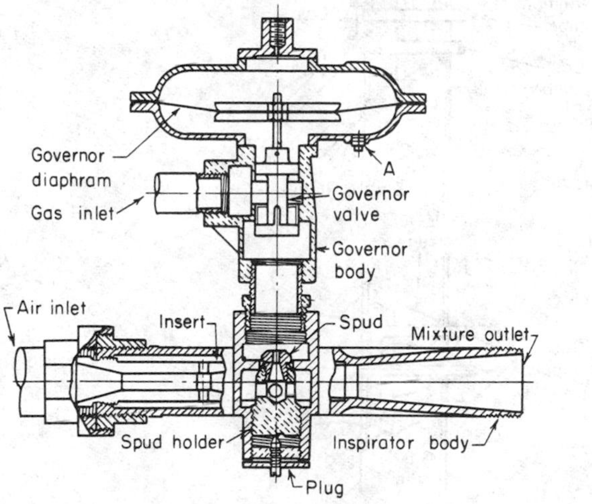

FIG. 9-56 Premix burner in which a proportional mixer uses air velocity to draw in a measured amount of gas. *(Surface Combustion Division, Midland-Ross Corp.)*

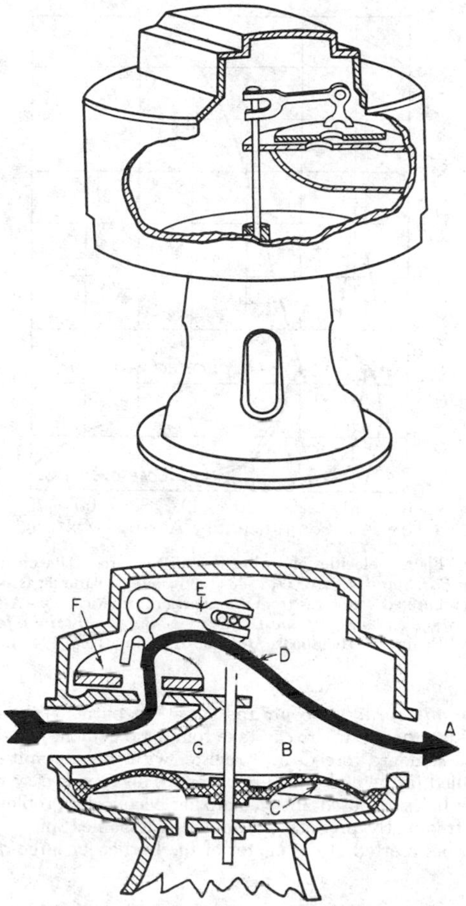

FIG. 9-57 Carburetor for maintaining a preset ratio of gas and air over a wide load range, adjusting to the total volume handled. *(C. M. Kemp Mfg. Co.)*

Burners used for close air control, such as for generating inert gas, must exercise close control over both the fuel and the air admitted to the burners. Figure 9-57 shows a typical carburetor arrangement used for this control. The high-velocity burner shown in Figure 9-58 can be adapted for use with various gaseous fuels. Although it is not strictly a premix burner, its temperatures and mixing produce results similar to those of premix burners.

The **rate of flame propagation** must be exceeded in premix burners to assure that ignition cannot travel back into the burner. Figure 9-59 indicates the rate of flame propagation at various air-gas ratios for several gases.

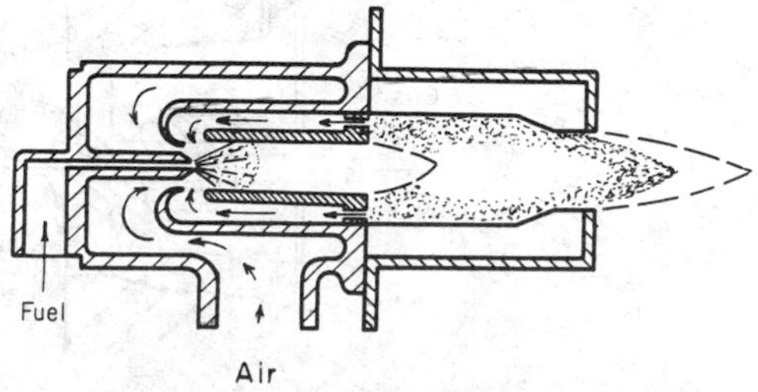

FIG. 9-58 High-velocity burner. High heat release produces combustion results similar to hydroxylation. *(Thermal Research and Engineering Co.)*

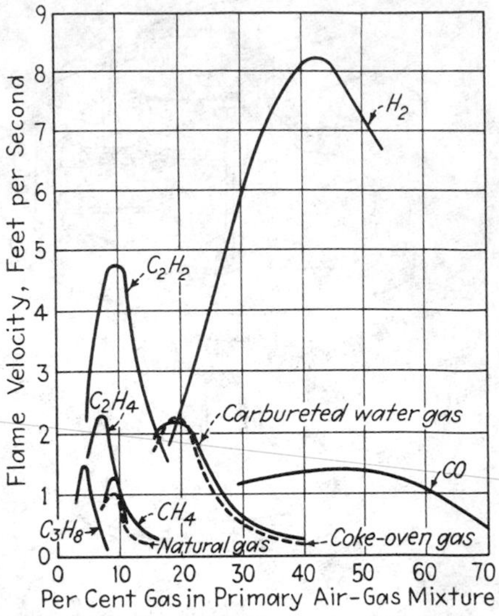

FIG. 9-59 Flame velocities of various gas-air mixtures. Data on individual gases from *Bur. Stand. J. Res.*, **17,** 7–43 (1936); data on natural gas, coke-oven gas, and carbureted water gas from *Combustion*, American Gas Association, 1932. (*As compiled by Elliott and Denues, U.S. Bureau of Mines, for* Marks, Mechanical Engineers' Handbook, *McGraw-Hill, New York, 1941.*)

Nozzle-mix burners mix air and gas at the burner tile. As shown in Fig. 9-60, these burners can take four arrangements. The burner may be a standard forced-draft register with the gas emitted from holes drilled in the end of a supply pipe. This type is easy to build, but large holes are used and gas mixing becomes a problem; these burners frequently produce a luminous gas flame. Small-diameter pipe can be inserted at the center of the burner, or large-diameter rings can extend to the outside of the burner tile. These rings use very small holes and give better dispersion of gas in the air, though they can plug up easily. One burner has a spider in the burner inlet through which gas is emitted in all the several radial arms. The spider is drilled to emit gas from the sides of the bars to provide a reaction from the emission of high-pressure gas, causing the spider to turn. The spider can be attached to a fan so that forced draft is provided by the movement of the spider. The spider arrangement provides high turbulence for the close regulation of excess air.

Control of gas burners is accomplished by regulating only the flow of gas in aspirating burners or by regulating both gas and air flows when these are controlled separately.

Combustion Characteristics of Low- and Medium-Btu Gases Low-energy gas from various coal-gasification processes is a source of clean fuel for utilities, industrial boilers, and industrial processes. The combustion characteristics of low-Btu gas (LBG) and medium-Btu gas (MGB) can best be described by comparison with natural gas. Combustion characteristics for comparison include oxygen or air requirements, flue-gas flow rates, flame temperatures, and flue-gas composition, including pollutant emissions. All these factors bear on the **interchangeability** of LBG and MGB in the context of flame characteristics and stability. Results of experiments by the Institute of Gas Technology* on utility-type burners and industrial burners follow.

Table 9-37 shows adiabatic flame temperatures calculated for natural gas and manufactured gases. MBGs have adiabatic flame temperatures near that of natural gas. Adiabatic flame temperatures for LBGs are lower than that of natural gas, owing to the nitrogen in the air and fuel. The lower flame temperatures will result in lower heat-transfer rates per unit load surface area in radiant furnaces.

Figure 9-61 shows that for a given flue-gas temperature significantly less heat is available from the flame of a low-Btu gas than from the flame of a fuel gas from a medium-Btu gasifier or natural gas.

**Coal Conversion Systems Technical Data Book*, Institute of Gas Technology, Chicago, November 1980.

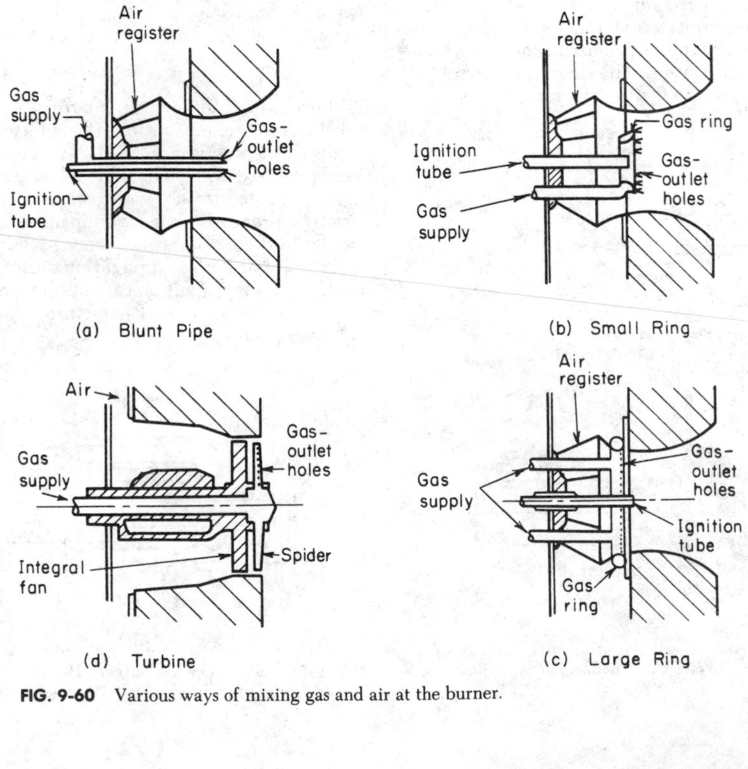

FIG. 9-60 Various ways of mixing gas and air at the burner.

TABLE 9-37 Typical Properties of Cleaned Manufactured Gases

		MBGs				LBGs	
	Natural gas	Koppers-Totzek oxygen	Lurgi oxygen	Winkler oxygen	Wellman-Galusha oxygen	Wellman-Galusha air	Winkler air
Composition, %							
CH_4	96.6	0.5	9.4	3.0	0.9	2.6	0.6
CO	0.0	52.9	18.5	32.9	39.2	26.9	21.4
CO_2	0.52	9.2	29.4	20.0	16.2	7.4	6.9
H_2	0.0	34.5	40.2	41.2	40.4	14.3	12.7
H_2O	0.0	1.9	1.9	1.9	1.9	1.9	1.9
N_2	0.46	1.0	0.6	1.0	1.4	46.9	56.5
C_2H_6	1.82	. . .°	. . .°	. . .°	. . .°	. . .°	. . .°
C_3H_8	0.3	. . .°	. . .°	. . .°	. . .°	. . .°	. . .°
C_4H_{10}	0.70	. . .°	. . .°	. . .°	. . .°	. . .°	. . .°
Higher heating value, Btu/scf	1020	288	285	270	267	159	116
Specific gravity (calculated)	0.62	0.68	0.70	0.67	0.66	0.83	0.85
Adiabatic flame temperature, °F (stoichiometric)	3551	3684	3360	3511	3574	3114	2818
Wobbe no., Btu†	1300	350	335	325	325	175	125

°Small amounts may be present after cleaning processes.
†Higher heating value/(specific gravity)$^{1/2}$.
NOTE: To convert British thermal units per standard cubic foot to kilocalories per cubic meter (15°C, saturated), multiply by 8.98; °C = (°F − 32) × 5/9.

For example, if furnaces fired by MBG from a Koppers-Totzek oxygen-blown generator (KTO) and by LBG from a Wellman-Galusha air-blown generator (WGA) each had an exit flue-gas temperature of 1316°C (2400°F), the KTO-fired furnace would be losing approximately 74 percent.

Conversely, for a given percentage of the input enthalpy to be transferred in the furnace, flue gas from LGB would have to be discharged from the furnace at a significantly lower temperature. For example, to have 30 percent of the input enthalpy available in the furnace, the WGA fuel gases would have to be cooled to 1099°C (2010°F), while the KTO fuel gases could be discharged at 1543°C (2810°F).

Table 9-38 shows the combustion air required and volume of combustion products per 1000 Btu of manufactured gases and natural gas. Manufactured gases require slightly less combustion air than natural gas, and air blowers and air ducts sized for natural-gas operation will be more than adequate for LBG.

The larger gas volumes required for a *given* Btu input for MBG and LBG will make gas piping, pressure regulators, and control valves undersized for most conversions. The combustion-product flows for low-Btu gases are considerably larger than for natural gas. This requires larger flues, stacks, and induced-draft fans.

Table 9-38 also gives the flammability limits calculated by the method given in the *Gas Engineers Handbook* (1977). The values were calculated by using the fuel-gas compositions in Table 9-37. These calculated limits of flammability are approximate values only. Accurate flammability predictions require that the fuel in question be subjected to experimental measurement.

Experimental Evaluations of LBG and MBG Research programs have provided information on utilizing manufactured gases in the following areas:

1. In industrial-process burners designated for natural gas
2. As utility boiler fuel
3. Pollutant emission

The results of these investigations may be summarized as follows:

1. When used in burners designated for natural gas, manufactured MBG generally produce stable flames and give thermal performance, flame temperature, and heat-transfer rates similar to those of natural-gas flames. Flame lengths may be longer or shorter, depending on the exact burner design.

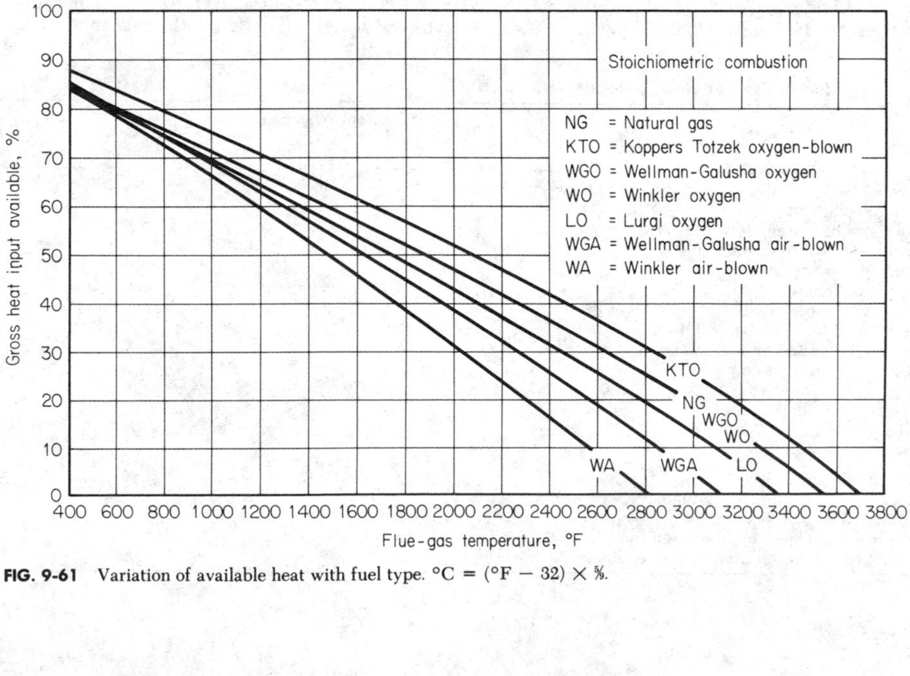

FIG. 9-61 Variation of available heat with fuel type. °C = (°F − 32) × 5/9.

TABLE 9-38 Stoichiometric Combustion Quantities

	scf/1000 Btu			Calculated flammability limits, % fuel in air	
	Fuel	Air	Products	Lower	Higher
Natural gas	1.0	9.44	10.4	5.0	15.0
Koppers-Totzek oxygen	3.5	7.40	9.4	7.5	71.0
Lurgi oxygen	3.5	8.10	10.6	7.5	50.5
Winkler oxygen	3.7	7.60	10.0	7.5	62.5
Wellman-Galusha oxygen	3.7	6.90	9.3	7.5	68.5
Wellman-Galusha air	6.3	7.80	12.8	16.5	64.5
Winkler air	8.6	7.50	14.7	20.5	68.5

2. When used in natural-gas burners, manufactured LBG may need enlarged fuel nozzles, continuous piloting, or downrating to prevent blowoff. They give reduced flame temperatures and heat-transfer rates compared with natural-gas flames in the same furnace. Preheating the fuel and/or combustion air can increase flame temperatures and heat-transfer rates but would then require specially designed burner equipment with refractory linings and very large fuel flow passages.

3. Using hot manufactured gases directly from the gasifier, without cleanup to avoid loss of the sensible heat, can produce NO_x and SO_x emissions greatly above those for cleaned fuels or natural gas.

ELECTRIC HEATING

Industrial electric heating can be divided into three basic categories: resistance heating, both direct and indirect; induction heating; and dielectric heating. The choice of which method to use depends on many factors. Resistance heating is the generation of heat by Joule's law, that is, I^2R losses in relatively conducting materials by passing current through them by direct contact. Induction heating is also I^2R heating, but the current is induced in the conductor by placing it in an alternating magnetic field. Dielectric heating is the generation of heat in a nonconductor by placing it in an alternating electric field, with the mechanisms of heating being primarily dipole rotation losses. A discussion of laser heating is also included in this analysis because it is powered by electric energy. This is not an in-depth analysis but a guide to making choices. For further study see Brown, Hoyler, and Bierworth, *Radio Frequency Heating*, Van Nostrand, New York, 1947; Copson, *Microwave Heating*, AVI Publishing Co., Westport, Conn., 1962; and Davies and Simpson, *Induction Heating Handbook*, McGraw-Hill, London, 1979.

The applications for electric heating are many. In general, direct-resistance heating is most often used for heating metal rods and billets prior to rolling or forging, for melting glass in combination with other heat, in boilers for hot water, and for heating salt baths for the heat treatment of metals. Indirect-resistance heat in ovens and furnaces has many applications, from drying to melting. Induction heating is used to heat metals for melting, forging, brazing, and hardening, as well as in many other less frequent applications. Dielectric and microwave heating is used for drying many materials from wood to foods, for processing plastic materials, and for an increasingly large application of microwaves in residential and commercial food heating. Table 9-39 presents the major advantages and disadvantages of electric heating. It should be emphasized that one of the major advantages often listed for electric heating is high efficiency. This is usually very true for electrical efficiency, that is, the percentage of electrical energy converted to usable heat. There is therefore less wasted heat in the user's plant. However, the generation of electricity from fossil fuel is only about 30 percent efficient; therefore, overall efficiencies of electric heat vary from 15 percent in a dielectric-heating application to 30 percent in a direct-resistance-heating application.

Direct-Resistance Heating Resistance losses (I^2R) in conductors are in general undesirable, but they can be desirable when the conductor is the load to be heated. This method is used in the iron and steel industry to preheat metal rods and billets prior to rolling or forging and for annealing similar shapes. Billets to be heated efficiently should have a length much greater than the diameter; otherwise, relative heat losses to the contacts will be large. For low-resistance metals such as copper, the L/D ratio should be greater than about 10. For relatively high resistance metals such as steel, the ratio can be lower. Cross sections of the rod should also be uniform for uniform heating. Energy consumption for heating to forging temperatures is approximately 300 kWh/ton. Comparable induction heating systems seldom realize better than 350 to 450 kWh/ton. Resistance heating systems also cost about 75 percent of the cost of an induction heating system.

The power supply for direct-resistance heating can be either three-phase or single-phase. It consists essentially of step-down transformers and contactors with power-factor-correction capacitors often included. The whole system must also be strong to withstand the forces between high-current-carrying conductors, as currents of the order of 10^5 A are not uncommon in the load circuit.

Contacts through which the current is transferred give rise to major design problems. Large areas of contact, with consequently low resistance, are good heat sinks and produce cold ends. Small contact areas with little heat withdrawal have high resistance and cause power losses and local overheating. Contacts also often must support the load. Careful design with some compromises is therefore called

TABLE 9-39 Comparison of Electric-Heating Methods

	Resistance heating		Induction heating	Dielectric heating
	Direct	Indirect		
Advantages				
Clean: no pollutants	X	X	X	X
Easily controlled; accurately controlled	X	X	X	X
Efficient: less heat to surroundings	X		X	X
Lower capital investment	X	X	X	X
Lower maintenance	X	X	X	X
Electricity readily available	X	X	X	X
Performing task impossible with fossil fuel	X		X	X
Electricity generated by coal or nuclear reaction	X	X	X	X
Heating electrical nonconductors				X
Heat generated internally	X		X	X
High electrical efficiency possible	X		X	
No direct contact			X	X
Disadvantages				
High cost of energy	X	X	X	X
Uniformity of heat sometimes difficult	X		X	X
Only limited energy sometimes available	X	X	X	X
Low conversion efficiency from fossil fuel	X	X	X	X
High cost of conversion from present equipment	X	X	X	X

for. Multiple-end contacts have been used with both peripheral and end contact.

Direct-resistance heating of liquids such as molten glass or molten salts is also used. Glass above 1100°C is molten and has a resistivity sufficiently low for direct heating; current is passed between electrodes immersed in the melt. The electrodes must withstand both the high temperatures and the movement of the melt. Molybdenum or tin oxide electrodes are normally used. Salt baths used for metal treatment have similar electrode problems. Graphite or corrosion-resistant steel is normally used. Generation of steam by passing current directly through water is also now common practice.

The preceding discussion assumed 60-Hz resistance heating. High-frequency resistance heating is also commonly used. This method takes advantage of both the "skin effect," or shallow penetration of high-frequency current, and the "proximity effect," in which the current follows the path of lowest reactance rather than that of lowest resistance. Such large applications as tube-seam welding and such small applications as selective area-surface hardening are uses of high-frequency resistance heating.

Indirect-Resistance Heating Indirect-resistance heating includes surface resistance heaters, immersion heaters, uninsulated heating elements in furnaces and ovens, and infrared heaters. Factors affecting selection include the temperature coefficient of resistance, creep resistance, atmosphere of use, thermal-shock resistance, and need for high resistivity.

Industrial heating elements, for use in air, fall into four groups: nickel-base and iron-base, for metallic elements; and silicon carbide and molybdenum disilicide for nonmetallic elements.

The most widely used nickel-base heating elements consist of 80 percent nickel and 20 percent chromium and are generally recommended for a maximum element operating temperature of 1200°C. They form a coating of chromium oxide which protects the material from excessive oxidation. However, they are not recommended for use in furnace atmospheres which contain traces of sulfur because of a low-temperature reaction with nickel.

For slightly higher element temperatures, up to 1375°C (2500°F), iron-base alloys are widely used. These consist mainly of iron, chromium, and aluminum. In air, the aluminum oxidizes and forms a tightly adhering film of alumina which protects the material from oxidation. These alloys, as well as the nickel-chromium series, can readily be shaped into coils or sinuated strips.

For element temperatures above 1375°C (2500°F) nonmetallic materials are needed. Silicon carbide elements can be used up to 1600°C (2912°F). These elements are made in the form of a straight rod, sometimes with a spiral cut. One of the characteristics of silicon carbide is that resistance increases with time (this property is called aging), making it necessary to use electrical equipment which provides for numerous voltage settings.

Another type of nonmetallic element, molybdenum disilicide, provides element temperatures up to 1700°C (3092°F) while avoiding the aging problems of silicon carbide. These elements are made of a cermet containing 90 percent $MoSi_2$ and 10 percent ceramic additives. They soften at high temperatures and in general are made in U-shaped sections. Table 9-40 summarizes the data on the types previously discussed and also includes others such as graphite and platinum. Platinum has the disadvantage of high cost but the advantage of low corrosion. Graphite has the advantage of highest temperature use, high blackbody radiation coefficient, high resistance, high thermal-shock resistance, and low thermal expansion but the disadvantage of oxidation at low temperatures in an oxidizing atmosphere.

Resistance heaters are used for many relatively low temperature applications in which direct contact is made with the insulated element. These include applications from plastic-sealing heat sources to melting soft metals. As flexible insulated heaters, these heaters are also used for heating pipes in many industrial areas.

Electric furnaces and ovens are used for a wide range of processes ranging from annealing, hardening, and forging metals to drying many materials in hot air. Ovens are usually defined as operating up to about 450°C (842°F) and furnaces as operating above this temperature. The dominant mode of heat transfer in ovens and

TABLE 9-40 Materials Used for Resistance-Heating Elements

Material	Maximum operating temperature in dry air, °C	Resistivity at 20°C, Ω·m	Mean temperature coefficient of resistivity over operating range	Principal applications
Copper	350	1.72×10^{-8}		Low-power surface heaters
Nickel-based alloys*				
80-20 Ni-Cr	1200	108×10^{-8}	6×10^{-5}	Furnace heating elements; resistance heaters
80-20 Ni-Cr + Al	1250	124×10^{-8}	-2×10^{-5}	Furnace heating elements; lower cost
60-15-25 Ni-Cr-Fe	1100	112×10^{-8}	13×10^{-5}	Furnace heating elements; lower cost
50-18-32 Ni-Cr-Fe	1075	111×10^{-8}	17×10^{-5}	Furnace heating elements; lower cost
37-18-43-2 Ni-Cr-Fe-Si	1050	105×10^{-8}	24×10^{-5}	Furnace heating elements; lower cost
44-56 Ni-Cu	400	49×10^{-8}		Resistance heaters; domestic appliances
Iron-based alloys*				
72-22-4 Fe-Cr-Al	1050	139×10^{-8}	$4{\cdot}7 \times 10^{-5}$	Furnace heating elements
72-22-4 Fe-Cr-Al + Co	1375	145×10^{-8}	$3{\cdot}2 \times 10^{-5}$	Furnace heating elements
78-16-4 Fe-Cr-Al + Yt, C	1300	134×10^{-8}	12×10^{-5}	Furnace heating elements
Refractory metals				
Platinum	1300	11×10^{-8}	$3{\cdot}92 \times 10^{-3}$	Small muffle furnaces
90-10 Pt-Rh	1550	$19{\cdot}2 \times 10^{-8}$	$2{\cdot}0 \times 10^{-3}$	Small muffle furnaces
60-4-0 Pt-Rh	1800	$17{\cdot}4 \times 10^{-8}$	$2{\cdot}0 \times 10^{-3}$	Small muffle furnaces
Molybdenum	1750†	$5{\cdot}7 \times 10^{-8}$	$5{\cdot}5 \times 10^{-3}$	Vacuum furnaces; small muffle furnaces in hydrogen atmosphere
Tantalum	2500	$12{\cdot}5 \times 10^{-8}$	$3{\cdot}2 \times 10^{-3}$	Vacuum furnaces
Tungsten	1800†	$5{\cdot}6 \times 10^{-8}$	$5{\cdot}94 \times 10^{-3}$	Infrared lamps; vacuum furnaces
Nonmetals				
Graphite	3000†	1000×10^{-8}	$-2{\cdot}66 \times 10^{-4}$	Vacuum furnaces; reducing atmospheres
Molybdenum disilicide	1800	40×10^{-8}	$1{\cdot}02 \times 10^{-2}$	Small glass-melting furnaces; forehearths
Silicon carbide	1600	$1{\cdot}1 \times 10^{-3}$	$-2{\cdot}63 \times 10^{-4}$	Furnace heating elements; oxidizing and reducing atmospheres
Lanthanum chromite	1800	2×10^{-5}		

*Approximate compositions only.
†Not in air.

furnaces is radiation, although convection and conduction from air heated by the elements make a large contribution, depending upon the furnace type.

A large number of different construction methods are used for furnaces. The most common include box furnaces with heating elements in the furnace cavity, forced-convection furnaces with heaters external to the cavity, hot-retort furnaces with the load sealed from the outside air and the heating elements, low-thermal-mass furnaces using fiber insulation rather than ceramic refractories, cold-retort furnaces with heat reflectors protecting the heat from the walls, and muffle furnaces with the heater outside a refractory tube. The choice for various applications includes such factors as the atmosphere desired, start-up and shutdown times, and the maintenance scheduling required.

Furnace insulation consists mostly of oxides of aluminum, silicon, magnesium, and zirconium. Furnaces for heat treatment generally use high-grade aluminosilicate bricks for lining. Fiber insulation of aluminates and silicates has a very low thermal conductivity and low heat capacity and is used for low-thermal-inertia designs suitable for batch processing. High-purity alumina and zirconia fibers have recently pushed the upper temperature limits of fibers to about 1700°C (3092°F). However, if temperatures above this level must be used, carbon powder, either in submicrometer size or pelletized, is the only satisfactory insulation. Its thermal conductivity is somewhat higher than that of fibrous insulation, but it is still a satisfactory insulator.

Infrared (IR) heating, or radiant heating, is advantageous because the work need not be contacted by the heating elements or by circulating air. The heaters must have a line of sight to the surface to be heated. Although generation of heat and delivery to the load are more efficient than in other furnace or oven methods, infrared heating can be less efficient overall because the portion of the energy used depends upon the absorptivity of the load. The radiation is controlled by reflectors, which also contribute to the efficiency. The intensity distribution of an IR source varies with the temperature, and more power is radiated at shorter wavelengths as the temperature increases. IR sources of relatively low temperature are normally used for heating nonmetallic materials which absorb at long wavelengths. High-temperature sources with most output at shorter wavelengths are required to heat metals which absorb them better.

The advantages of IR heating are:

1. It can heat up portions of a process or product.
2. Less floor space is needed than for other ovens.
3. Production can be increased easily by adding heating units to current units to heat more at local areas or more overall.
4. High-flow exhaust systems can be replaced with lower-flow systems because air is not heated excessively.
5. Product temperatures up to 650°C (1202°F) are practical.

Early IR systems were simply on-off controlled heaters. Current silicon-controlled-rectifier (SCR) controls make possible closed-loop automatic temperature control.

There are three major types of IR heaters: wide-beam, with about a 100° reflector and low power density; multibeam, with several reflector angles available to establish desired power densities and coverage; and heavy-duty, which have power ratings up to 20 or 30 kW, have a medium beamwidth of about 60°, and are selected for mechanical ruggedness. Three types of heating elements are available:

1. Quartz lamps, which are the most radiation-efficient, and are tungsten elements in a clear evacuated quartz enclosure
2. Quartz tubes, which have a nickel-chrome element in a quartz enclosure and are used for wide-beam heaters and a broad selection of power levels
3. Metal rods, which consist of a nickel-chrome element surrounded by magnesium oxide insulation with an outside metal sheath and are used in heavy-duty heaters

Induction Heating Induction heating provides a means for the precise heating of electrically conducting objects. In some cases, it is the only practical method of supplying heat to the work material. It is clean, fast, and repeatable and lends itself to automatic cycling. No contact is required between the workload and the heat source, and heat may be restricted to localized areas or to a surface zone of the load.

TABLE 9-41 Major Applications of Induction Heating

Application	Approximate frequency range, Hz
Melting	60–10 k
Surface hardening	10 k–10 M
Brazing and soldering	10 k–10 M
Forging	60–3 k
Tube wlding	10–500 k
Annealing	60–10 k
Strip heating	10–500 k
Semiconductor-zone refining	50–500 k

Induction heating occurs when electrically conductive materials, such as metal workpieces, are immersed in an alternating magnetic field. This field is usually produced by an electrical coil energized by a suitable source of alternating-current electric energy. The ac magnetic field induces voltages in the conductive material, and these voltages cause circulating currents (eddy currents). The magnitude of the induced currents is determined by the effective magnitude of the induced voltage and the impedance of the workpiece. The flow of induced current generates I^2R losses and heat in the workpiece. Additional heat is produced in magnetic workpieces as the result of hysteresis losses; this heat is usually small, but in some cases involving strong magnetic fields the heat resulting from hysteresis losses can become significant. Table 9-41 lists major applications of induction heating.

Induction heating is efficient and practical if certain basic relationships dealing with the frequency of the magnetic field and the properties of the workpiece are satisfied. Although the relationships are not sharply critical, they must be satisfied to the extent that a suitable degree of skin effect is produced in the workpiece. Skin effect is the phenomenon by which the currents flowing in the workpiece tend to be most intense at the surface, while currents at the center are near zero. As a consequence of this distribution the currents produce a greater rate of heating near the surface. Skin effect is present in every successful induction-heating application.

Consideration of the operating frequency should be the first step in the design of a successful induction-heating installation. To achieve efficient induction heating, there must be a proper ratio of workpiece diameter (or thickness) to reference depth. For a given workpiece diameter, the thinner the current-carrying layers, the greater the rate of heat generation in the surface, other factors remaining constant. If minor effects are ignored, the effective depth of the current-carrying layer depends on the frequency of the ac magnetic field and on the electrical resistivity and magnetic permeability of the workpiece. Frequency is the only one of these factors that can be readily manipulated.

The current density in a workpiece decreases exponentially from

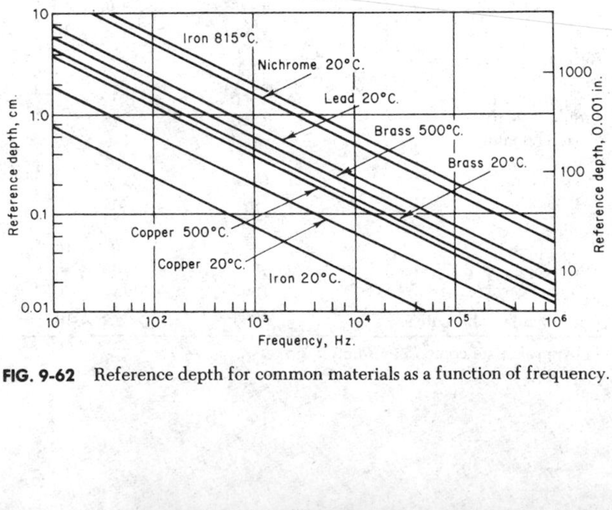

FIG. 9-62 Reference depth for common materials as a function of frequency.

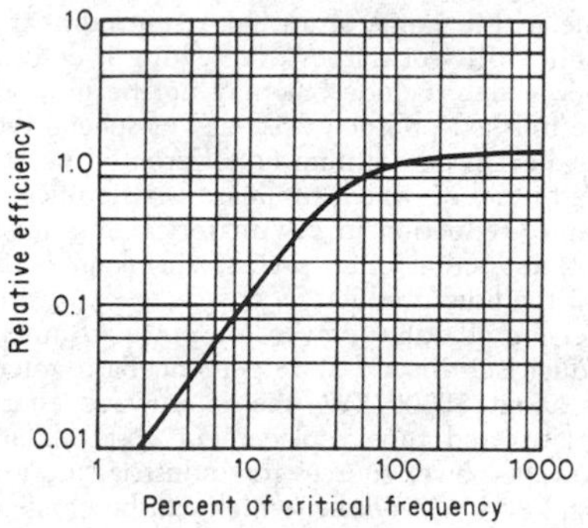

FIG. 9-63 Relationship between relative efficiency and critical frequency.

the surface. The rate of decrease can be compared from one application to another by means of reference depth. Reference depth, which has actual physical significance in certain special cases, is defined by

$$d = 5000\sqrt{p/\mu f} \tag{9-41}$$

where d is the reference depth, cm; p is the resistivity of the work, $\Omega \cdot m$; μ is the relative magnetic permeability of the work, dimensionless; and f is the frequency of the alternating magnetic field of the work coil, Hz. Figure 9-62 shows reference depth versus frequency for various common metals. It can be shown that if the ratio of workpiece diameter to reference depth drops below about 4:1, the efficiency of heating decreases. The critical frequency is defined as the frequency at which the workpiece-to-reference-depth ratio is 4.5:1 for round bars; if heating sheet from both sides, the ratio is 2.25:1. Figure 9-63 shows the efficiency of heating as a function of this critical frequency.

For a shallow depth in a large workpiece, a high frequency is selected. There is no concern about critical frequency, since the diameter will be many times the skin depth. However, for the fastest through-heating, a frequency close to the critical frequency must be chosen, and the calculations of reference depth and efficiency become more important.

The rate of heat generation for induction heating is proportional to the coil ampere-turns squared. The electrical resistivity of the work governs the rate of heat generation in it by I^2R. In a magnetic load, there are also developed hysteresis losses which are negligible compared with the I^2R losses unless exceptionally strong fields are present. In heating a nonmagnetic load that has a low resistivity, large currents must be used in the work coil to achieve high heating rates. Coil losses therefore tend to be high and efficiency low when heating a low-resistivity workpiece. The efficiencies for heating various types of loads with various types of coils are given in Table 9-42.

Induction furnaces operating at frequencies from 60 to 1000 Hz are useful for obtaining temperatures up to 3000°C (5432°F). In one type of furnace, the currents are induced directly in the charge; in others, the currents are induced in a conductive case containing the charge, and the heat is radiated inward to the furnace charge. Since the coil currents may be as high as 15,000 A, the coil conductors are usually hollow to permit water circulation for cooling. Compared with resistance heating, the power factor of induction heating is relatively poor at 60 Hz because of the reactance effects. This tends to lower the overall plant power factor, which may incur power penalties from the utility company. Power-factor-correction capacitors are normally used.

TABLE 9-42 Average Coupling Efficiencies for Induction Heating of Close-Coupled Loads

Type of coil	Magnetic steel below curie	Steel above curie, stainless steel	Brass, titanium, aluminum, bronze	Copper
Helical-around work .	0.90	0.65	0.50	0.30
Helical-internal	0.70	0.40	0.30	0.20
One turn-around work	0.85	0.60	0.45	0.25
One turn-internal . . .	0.65	0.35	0.25	0.15
Hairpin	0.85	0.60	0.45	0.25
Pancake	0.70	0.40	0.30	0.20

Power sources to the induction coil fall into four categories, and the type to be used usually depends upon the frequency range desired. The four types are supply frequency (60 Hz), solid-state-converted systems, motor-alternator systems, and vacuum-tube systems. Table 9-43 summarizes the four systems with their features and frequency limits. Some supply-frequency magnetic multipliers are still used, but they are generally being replaced by solid-state systems. Since about 1965 solid-state systems have also been increasingly replacing motor-alternator systems in the middle-frequency range. The main advantages of solid-state systems are possibly higher efficiency and no warm-up time, and the main advantages of motor-alternator systems are fixed frequency and easier maintenance. The features may be advantages or disadvantages depending upon applications. Vacuum-tube radio-frequency power supplies have long been used for shallow, rapid surface-heating applications. The upper frequency limit of solid-state systems is about 50 kHz, and these systems cannot currently compete with vacuum-tube supplies above that frequency.

Dielectric Heating Dielectric heating is the term applied to the generation of heat in nonconducting materials by their losses when subject to an alternating electric field of high frequency. The **frequencies** necessary range from 1 to 200 MHz. Heating of nonconductors by this method is extremely rapid. This form of heating is applied by placing the nonconducting load between two electrodes, across which the high-frequency voltage is applied. This arrangement in effect constitutes an electric capacitor, with the load acting as the dielectric. Although ideally a capacitor has no losses, practical losses do occur, and sufficient heat is generated at high frequencies to make this a practical form of heat source.

The frequency used in dielectric heating is a function of the power

TABLE 9-43 Induction Power Sources

Source	Frequency range	Power range	Effeciency, %	Features
Line-frequency	60 Hz	100 kW–100 MW	90–95	High efficiency; low cost; no complex equipment; deep current penetration
Motor-alternator	500 Hz–10 kHz	10 kW–1 MW	75–85	Low sensitivity to ambient heat; low sensitivity to line surge; fixed frequency; low-cost maintenance; spares not needed
Solid-state	500 Hz–50 kHz	10 kW–1 mW	75–95	No standby current; high efficiency; no moving parts; protection needed outdoors; no warm-up time; impedance matching the changing loads
Vacuum-tube	50 kHz–10 MHz	1–500 kW	65–75	Shallow heating depth; localized heating; highest cost; impedance matching the changing loads; lowest efficiency

desired and the size of the work material. Practical values of voltages applied to the electrodes are 2000 to 5000 V/in of thickness of the work material. The source of power is exclusively by electronic (vacuum-tube) oscillators which are capable of generating the very high frequencies desirable. Units with up to 500-kW output are in commercial use. Some of the more common uses of dielectric heating are wood drying, curing, and gluing; preheating of plastics; processing of rubber and synthetic materials; food processing; and drying and heat treatment of textiles.

The basic requirement for dielectric heating is the establishment of a high-frequency alternating electric field within the material or load to be heated. Once the electric field has been established, the second requirement involves dielectric-loss properties of the material to be heated. The **dielectric loss** of a given material occurs as a result of electrical polarization effects in the material itself. There are at least four recognized types of such polarization: two which occur as a result of the field itself, or induced polarization; and two which are inherent and are determined by the arrangement of the component particles of the material itself. These loss mechanisms are (1) electronic polarization, (2) atomic polarization, (3) dipole orientation, and (4) space-charge polarization. The latter two are most prominent in dielectric heating.

The volume rate of power input P_v to a dielectric material can be expressed in watts per cubic inch as

$$P_v = 1.41E^2 f \epsilon_r \tan\delta \times 10^{-12} \qquad (9\text{-}42)$$

where E = electric field, V/in; f = frequency of electric field, Hz; ϵ_r = relative dielectric constant of the work; and $\tan\delta$ = dielectric loss tangent of the work, or the complement of the power factor angle, that is, the angle between the voltage and current in the dielectric material. The product of ϵ_r and $\tan\delta$, called the loss factor, is often listed in tables of dielectric properties of materials for various frequencies (see Table 9-44).

Microwave Heating Microwave heating is simply dielectric heating at still higher frequencies, done in a cavity where the wavelength is of the order of the cavity size rather than large compared with the applicator, as in low-frequency heating. The two predominant frequencies used in microwave heating are 915 and 2450 MHz. Since microwave heating is 10 to 100 times higher in frequency than the usual dielectric heating, from Eq. (9-42) it can be seen that a lower voltage is needed if the loss factor is constant. However, the loss factor is generally greater at microwave frequencies. Also at microwave frequencies there is a definite **skin depth** in high-loss-factor materials analogous to the skin depth in induction heating. The skin depth is

$$d = \frac{v}{2\pi f}\left[\frac{2}{\epsilon_r(\sqrt{1+\tan^2\delta}-1)}\right]^{0.5} \qquad (9\text{-}43)$$

where d is the penetration depth, in; v is the velocity of light, in/s; and f, ϵ_r, and $\tan\delta$ are as in Eq. (9-42). Skin depth is important in heating foods, for example, in which it is quite large in frozen foods with a small loss factor but small in room-temperature foods such as unfrozen meat.

Microwave heating is rapidly becoming the key to new techniques and processes. This form of heating has begun to proliferate as its cost has come within range of an increasing number of users. The **range of applicability** of microwave heating is determined by economic factors, which include but may not be limited to price per pound of the finished product, existence of special features such as instantaneous heat programming or differential heating, savings in storage space or tooling which result from a significant reduction in heat-cycle time, reduction in raw-material costs for equal-quality final product, reduction of in-process shrinkage and/or loss, and superiority of the final product.

The conversion of utility-power-line power to microwave power into the product is estimated at 50 percent. Basic microwave-power sources cost about \$2500/kW, the microwave energy applicator about \$500/kW, and tube replacement costs about \$0.06/h (all 1980). Microwave-power sources for industrial use are available in 1-, 5-, 20-, and 30-kW modules, which can be ganged to meet specific process-power requirements.

Two distinct areas for the use of microwave energy in chemical processing are developing. One area utilizes the direct absorption of microwaves to effect rapid and efficient **heating of dielectric materials** without heating the surroundings or associated low-dielectric-loss materials. This form of radiant energy is being studied or used for large-scale food processing (cooking, sterilization, freeze drying, etc.) and for such industrial processes and products as ceramics, chemicals, coatings, electronics, forest products, graphic arts, paper, pharmaceuticals, plastics, rubber, and textiles. Typical functions performed are bonding, curing, deinfestation, drying, foaming, fusing, heat treating, polymerization, and sealing. Utilization in many cases goes beyond just a replacement for or complement to a heating cycle. This is, in part, due to the ability of microwaves to penetrate dielectric materials and to be absorbed throughout the entire exposed volume and thereby to generate heat uniformly. It is also, in part, due to the **selectivity** inherent in microwave heating. For example, in drying moist materials, the microwave power is absorbed in the wettest regions and essentially passes through the dried regions. This selectivity of heating offers a means for self-limiting the energy taken up by heterogeneous materials, and overheating is unlikely. With these combined effects, microwave heating often can do a heating job more quickly, in less space, without undue superficial heating, without thermal lag, and, therefore, with rapid control.

The other area of microwave energy under development involves chemical synthesis by means of microwave-coupled plasmas. As an accurately controlled, monochromatic source of electromagnetic radiation, microwaves supply the alternating electric field necessary for the continuous ionization of a gaseous substance passed through this field. The ionized gas, or plasma, offers several routes to chemical synthesis: the plasma contains ionized species and electrons; it can supply radicals and/or excited atomic and molecular species; and it can be used as a precise ultraviolet-light source for photochemical reactions.

Electric-Arc (Plasma) Heating Electric heating by means of plasma formation has rather significant usage. The plasma is formed from gases or vapors of reactive or diluent constituents of a process stream and electrons in voltage gradient. A plasma is a partially ionized gas containing molecules, atoms, ions, electrons, and free radicals, each moving with a certain velocity. The plasma is in thermal equilibrium when the average energy of each species is the same and the laws of thermodynamics apply. Thermal equilibrium is approached at pressures of 101.325 kPa (1 atm) and at power levels sufficiently high to maintain an electron concentration of $>10^{14}$ cm^{-3}. Plasmas having these properties are known as **arc discharges.** Devices that supply the electric power to form and employ arc discharges are the electric-arc furnace, the induction plasma, and the plasma jet.

The principle of **electric-arc heating** is that when an electric circuit is interrupted (by an air gap), current will continue to flow if the current is high enough to vaporize some of the conductor and thus fill the gap with a conducting vapor path. The electric-arc furnace is capable of operating at temperatures up to the order of 3500°C (6332°F). At such temperatures, the electric furnace has no economic competitors. The main uses are in the purification of ores, changing an undesirable crystalline structure of an ore, and synthesis

TABLE 9-44 Typical Dielectric Loss Factors for Various Materials

Material	Loss factors
Spruce wood, dry	0.11
6% moisture	0.175
10% moisture	0.29
Paper	0.20
Textolite	0.62
Hard rubber	0.015
Porcelain	0.044
Cellulose acetate	0.25
Cellulose nitrate	0.50
Phenol formaldehyde	0.45
Urea formaldehyde	0.16

of compounds not available in the natural state by fusing different raw materials. The arc furnace is therefore used extensively in the metallurgical, abrasive, refractory, and electrochemical industries.

The arc is established by momentarily bringing together two electrodes, or one electrode and the charge (which frequently acts as a second electrode), and then separating them. The vapor produced on separation provides the conducting path. The preferable method of operation is with a short arc which gives rise to high currents. Voltages up to 250 V are normally used in arc furnaces in conjunction with tap-changing transformers to obtain the most suitable voltage. Typical **electrode materials** are graphite and carbon. Graphite is more expensive and purer and has greater strength and higher current-carrying capacity than carbon. Severity of operation and purity requirements dictate the type of electrode to be used. Carbon arcs are the most intense source of heat, with a temperature of about 5500°C (9932°F). Electrodes are consumed in the heating process by oxidation and by contact with the furnace charge. In the direct-arc furnace, the voltage is applied between two or more electrodes located above the charge, which in most cases is a nonconductor of electricity at room temperatures but becomes a conductor at higher temperatures. To fuse the charge, layers of coke are placed on the surface between electrodes. When the electrodes are lowered into the coke bed, a current flows, generating heat, which in turn melts the furnace charge near it. The charge then becomes conductive, and current will flow from electrode to electrode through the molten bath. Heat is generated by radiation from the electric arc between electrodes and bath and through the resistance-heating effect in the bath.

Arc furnaces may be single-phase or three-phase. For balanced electrical loads on the power supplies three-phase is desirable. However, single-phase furnaces properly connected to different single-phase power can also maintain an overall balanced system. Furnaces with capacities of 250 to 10,000 kVA are in commercial use. Sizes in the range of 1000 to 2000 kVA are most common. Since the position of the electrode is critical, automatic regulators are used to control the position of the electrodes to maintain constant current, voltage, or power—whichever is desired.

The **plasma jet** can be operated with direct or alternating current. The water-cooled metal electrodes are nonconsumable, and the gap between the electrodes is made conductive by passing an easily ionized, unreactive gas such as argon or helium between them. The gas velocity attains supersonic levels. For the purpose of heating, reactive gases can be added either to the inert gas stream at low levels before passing through the electrode region or, at high levels, to the heated inert gas outside the electrode region. Pure reactive gases cannot be used because of the reactivity of the electrodes with them. Similar restrictions exist when a particulate solid is a reactant. Because of the high velocities encountered in the plasma jet device, contact time in the maximum heat-flux region is extremely short. Temperatures in excess of 8000°C (14,432°F) exist in gas issuing from such a device.

The limitations set by the electrodes in the electric-arc and plasma-jet devices are eliminated in the **induction plasma** generating device. Power from a radio-frequency source operating up to 20 MHz is transferred to the thermally ionized gas by inductively coupling to the partially conductive gas. The energy transfer occurs by inducing high-frequency current in a very thin annulus on the outside of the load, much as current would be induced in a thin-walled steel tube located in the same position within the multiturn coil.

At a frequency of 4 MHz and power levels up to 10 kW only monatomic gases can be used to initiate the arc. Reactive diatomic gases can be metered into an argon plasma and heated. The plasma is extinguished at a critical ratio fixed by the type of gas added and the reactor geometry. At higher frequencies and output power levels, larger amounts of a diatomic gas can be metered into the argon plasma without the gas being extinguished.

Since no electrodes are exposed to the plasma region, more types of reactions at elevated temperatures can be undertaken than in the plasma jet or the electric (carbon)-arc devices. The induction plasma device is useful as a clean heat source for crystal growth or spheroidization and for high-temperature chemical processes which are endothermic overall. For exothermic reactions, rapid quenching of the heated products is required.

Laser Heating Laser heating is accomplished by the absorption of light energy on a material surface, except in the case of semitransparent materials, which can absorb light energy beneath the surface. The absorption of light energy by a material surface may vary from 1 to 99 percent, depending on the specific material and the condition of mechanical and/or chemical surface preparation. The uniqueness of laser heating results from the high level of energy shaping which may be accomplished optically with the monochromatic, coherent laser light energy. Energy densities delivered to a material surface can easily exceed 10^6 W/cm^2 in a small focal spot or may be reduced to much lower levels for broad-area coverage, depending on the heating requirement.

While the cutting and drilling of metals, plastics, ceramics, and organics are the most widely practiced applications for laser heating, the greatest amount of laser energy is consumed in the transformation hardening (heat treating) of steels and cast irons to improve the wear performance of specific locations on mechanical components. Laser welding of automotive-transmission components is also beginning to consume larger amounts of laser energy.

Although hundreds of different laser types have been successfully operated, continuous-wave CO_2 lasers provide by far the greatest amount of industrial laser energy. These machine tools provide output powers of 50 to 15,000 W in the far-infrared (10.6-μm) wavelength and may be operated continuously or in pulsed fashion. The lasing medium is a mixture of low-pressure CO_2, N_2, He, and sometimes O_2 or CO which is excited by an electric-glow discharge. Laser light energy is extracted from the excited medium by means of an optical "resonator" composed of at least two mirror surfaces, one of which is partially transparent and provides the output beam. Electrical conversion efficiencies for CO_2 lasers range from 5 to 14 percent, and their operating costs per watt of delivered power are the lowest for any industrially significant laser-heating equipment.

FIRED PROCESS EQUIPMENT

Twenty-four major energy-intensive industries depend on direct-fired or indirect-fired equipment for drying, heating, calcining, melting, and chemical processing. Some of these industries used coal and producer gas at first and switched to natural gas or oil when these fuels were less expensive than coal. There is a strong movement now to return to coal and coal-derived gases and to adapt to coal processes that conventionally used other fuels. This subsection will deal with both direct- and indirect-fired equipment, with the greatest emphasis on indirect firing for chemical-process industries.

Steam generation for process heat and electric power production is the largest user of indirect firing and is handled separately under "Steam Generators."

Direct-fired combustion equipment is that in which the flame and/or products of combustion are used to achieve the desired result by direct contact with another material. Common examples are rotary kilns, open-hearth furnaces, and submerged-combustion evaporators.

Indirect-fired combustion equipment is that in which the flame and products of combustion are separated from any contact with the

TABLE 9-45 Industrial Applications of Direct and Indirect Firing

Industry	Direct	Indirect
Food	x	
Lumber		x
Paper		x
Petroleum		x
Rubber and plastics		x
Glass	x	
Cement	x	
Lime	x	
Ore beneficiation	x	
Coke		x
Iron and steel	x	

TABLE 9-46 Average Energy Consumption Rate in Process Industries

Industry	10^6 Btu consumed/ton product*
Glass containers	12.0
Cement	7.0
Lime	6.3
Steel (ore beneficiation, blast stove, blast furnace, open hearth, soaking pits)	24.0
Copper (roasting, smelting, refining, melting)	40.2
Structural clay products	4.9
Plastic products	20.6

SOURCE: M. E. Fejer et al., *Assessment Application for Direct Coal Combustion*, Institute of Gas Technology, Chicago; National Science Foundation Cont. NSF-C924, October 1976.
*To convert 10^6 British thermal units per ton to megajoules per kilogram, multiply by 1.16.

principal material in the process by metallic or refractory walls. Examples are steam boilers, vaporizers, heat exchangers, and melting pots. Table 9-45 shows the 11 industrial applications of both these types of fired process equipment.

Direct-Fired Equipment Table 9-46 gives the average energy consumption in seven major industries that use direct firing. Table 9-47 shows the kinds of equipment used in the direct-fired cement and lime industries and the respective energy consumption per ton of product. Details of the equipment are described in a report entitled "Assessment Application for Direct Coal Combustion," issued in 1977 by the National Science Foundation. Section 20 of this edition describes and illustrates rotary dryers, rotary kilns, and hearth and multiple-hearth furnaces.

Indirect-Fired Equipment The following discussion is based largely on articles by Herbert L. Berman ["Fired Heaters I, II, III, and IV," *Chem. Eng.* (June 19, July 31, Aug. 14, and Sept. 11, 1978)]. Process-industry requirements for fired heaters are divided into a half-dozen general service categories.

Column Reboilers The charge stock taken from a distillation column is a recirculating liquid that is partially vaporized in the fired heater. The mixed vapor-liquid stream reenters the column, where the vapor condenses and releases the heat of vaporization. Depending on the particular application, reboiler heater outlet temperatures generally fall in the range of 204 to 288°C (400 to 550°F).

Fractionating-Column Feed Preheaters The charge stock (usually all liquid, although some feeds may contain a nominal amount of vapor at the inlet) is sent to the fired heater following upstream preheating in unfired equipment.

A typical example of this service is the feed heater for an atmospheric distillation column in the crude-oil unit of a petroleum refinery. Here, crude oil entering the fired heater as a 371°C (700°F) liquid might exit near 232°C (450°F) with about 60 percent of the charge stock vaporized.

Reactor-Feed Preheaters Fired heaters in this application raise the charge-stock temperature to a level necessary for controlling a chemical reaction taking place in an adjoining reactor vessel. The nature of the charge stock and the heater operating temperatures and pressures can vary considerably, depending on the process. The following examples illustrate the diversity of the applications performed by reactor-feed preheaters.

- *Single-phase–single-component heating such as steam superheating in the reaction sections of styrene-manufacturing processes.* In this service, the fluid temperature across the fired heater increases from an inlet temperature of about 371°C (700°F) to an exit temperature of about 816°C (1500°F).
- *Single-phase–multicomponent heating, such as the heating of mixtures of vaporized hydrocarbons and recycle hydrogen gas prior to catalytic re-forming in a refinery.* In this service, the charge stock enters the fired heater at about 427°C (800°F) and exits at approximately 538°C (1000°F). In re-formers, the fluid pressure may range from about 1.723×10^6 Pa (250 psig) to 4.137×10^6 Pa (600 psig). Severe restrictions on fluid pressure drop are normally associated with this service.
- *Mixed-phase–multicomponent heating, such as the heating of mixtures of liquid hydrocarbons and recycle hydrogen gas for reaction in a refinery hydrocracker.* Fluid temperatures typically run from 371°C (700°F) at the inlet to 454°C (850°F) at the outlet. Operating pressures may reach 20.685×10^6 Pa (3000 psig), depending on the process.

Heat Supplied to Heat-Transfer Media Many plants furnish heat to individual users via an intermediate heat-transfer medium. A fired heater is generally employed to elevate the temperature of the recirculating medium, which is typically a heating oil: Dowtherm, Therminol, molten salt, etc. Fluids flowing through the fired heater in these systems almost always remain in the liquid phase from inlet to outlet.

Heat Supplied to Viscous Fluids Often heavy oil must be pumped from one location to another for processing. At low temperatures, at which the oil may have so high a viscosity as to render pumping infeasible, a fired heater is employed to warm the oil to a temperature that will facilitate pumping.

Fired Reactors In this category are heaters in which a chemical reaction occurs within the tube coil. As a class, these units represent the fired-heater industry's most sophisticated technology. The following two applications typify the majority of installations.

- *Steam hydrocarbon–re-former heaters, in which the tubes of the combustion chamber function individually as vertical reaction vessels filled with nickel-bearing catalyst.* In re-formers that yield hydrogen, fluid outlet temperatures range from 788 to 899°C (1450 to 1650°F).
- *Pyrolysis heaters, used to produce olefins from gaseous feedstocks such as ethane and propane and from liquid feedstocks such as naphtha and gas oil.* In cracking heaters, in which chemical reactions occur in the coil, the tubes and burners are arranged so as to assure pinpoint firing control. Fluid outlet temperatures in heaters designed for liquid feedstocks are in the 816 to 899°C (1500 to 1650°F) range.

The principal classification of fired heaters relates to the orientation of the heating coil in the radiant section, i.e., whether the tubes are vertical or horizontal. Vertical arrangements are shown in Fig. 9-64; horizontal arrangements, in Fig. 9-65. Salient features of each follow.

Vertical-Cylindrical; All Radiant Here the tube coil is placed vertically along the walls of the combustion chamber. Firing is also vertical, from the floor of the heater.

Heaters of this type represent a low-cost, low-efficiency design that requires a minimum of plot area. Typical duties are 528×10^6 J/h to 21 GJ/h (0.5 to 20 million Btu/h).

Vertical-Cylindrical; Helical Coil In these units, the coil is arranged helically along the walls of the combustion chamber, and firing is vertical from the floor. Although these heaters are grouped

TABLE 9-47 Energy Consumption of Various Cement- and Lime-Manufacturing Processes

Process	10^6 Btu/ton*
Cement	
Wet	
Long kiln†	5.9
Calcinator and short kiln	4.7
Semiwet: preheater and short kiln	3.6
Dry	
Long kiln	4.7
Suspension preheater and short kiln	3.2
Semidry: grate preheater and short kiln	3.4
Lime	
Rotary long kiln†	7–8
Rotary short kiln†	5–6
Vertical kiln	4.5–5.0
Rotary-hearth kiln	4.5–5.0

*To convert 10^6 British thermal units per ton to megajoules per kilogram, multiply by 1.16.
†Most common in the United States.

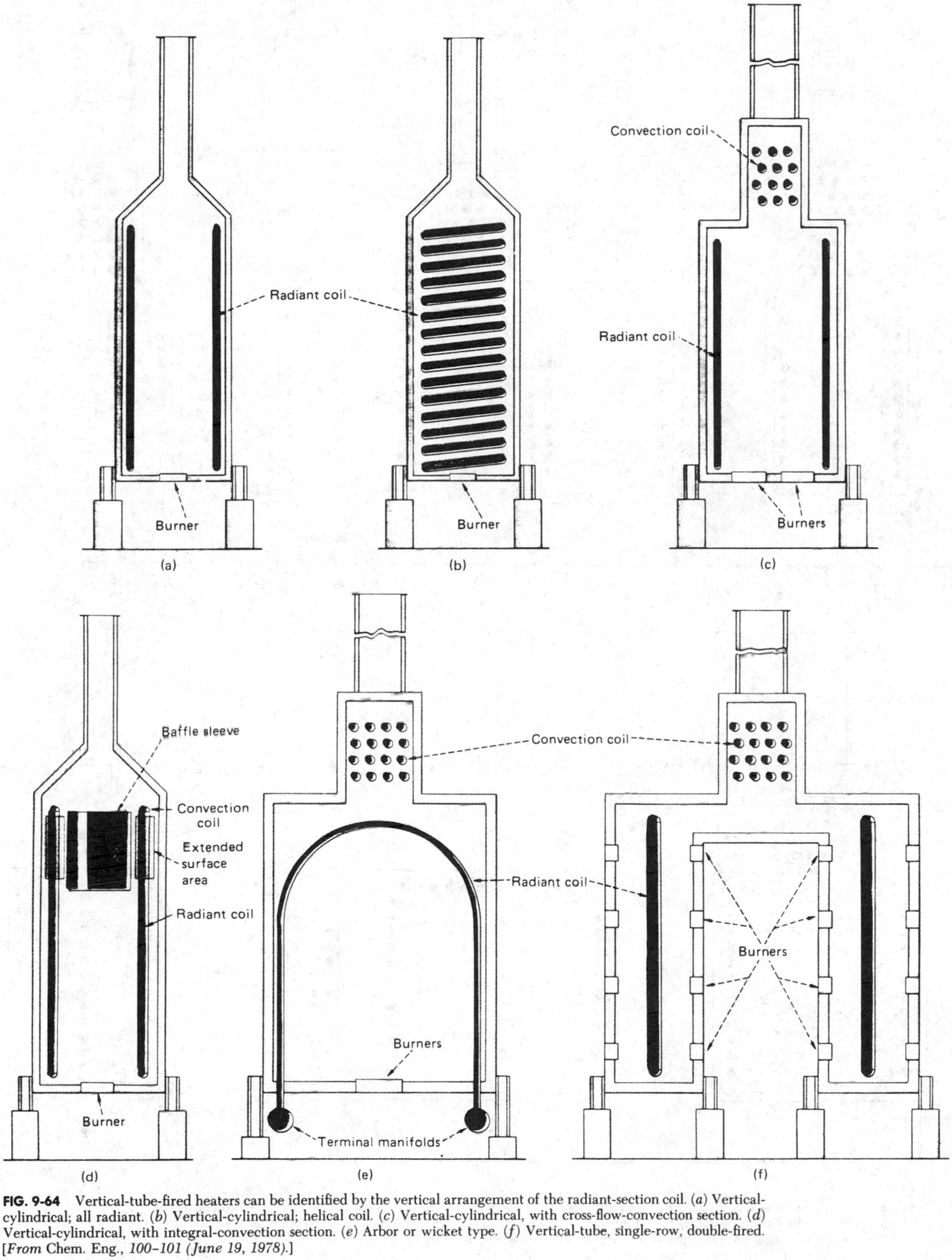

FIG. 9-64 Vertical-tube-fired heaters can be identified by the vertical arrangement of the radiant-section coil. (*a*) Vertical-cylindrical; all radiant. (*b*) Vertical-cylindrical; helical coil. (*c*) Vertical-cylindrical, with cross-flow-convection section. (*d*) Vertical-cylindrical, with integral-convection section. (*e*) Arbor or wicket type. (*f*) Vertical-tube, single-row, double-fired. [*From* Chem. Eng., *100–101 (June 19, 1978).*]

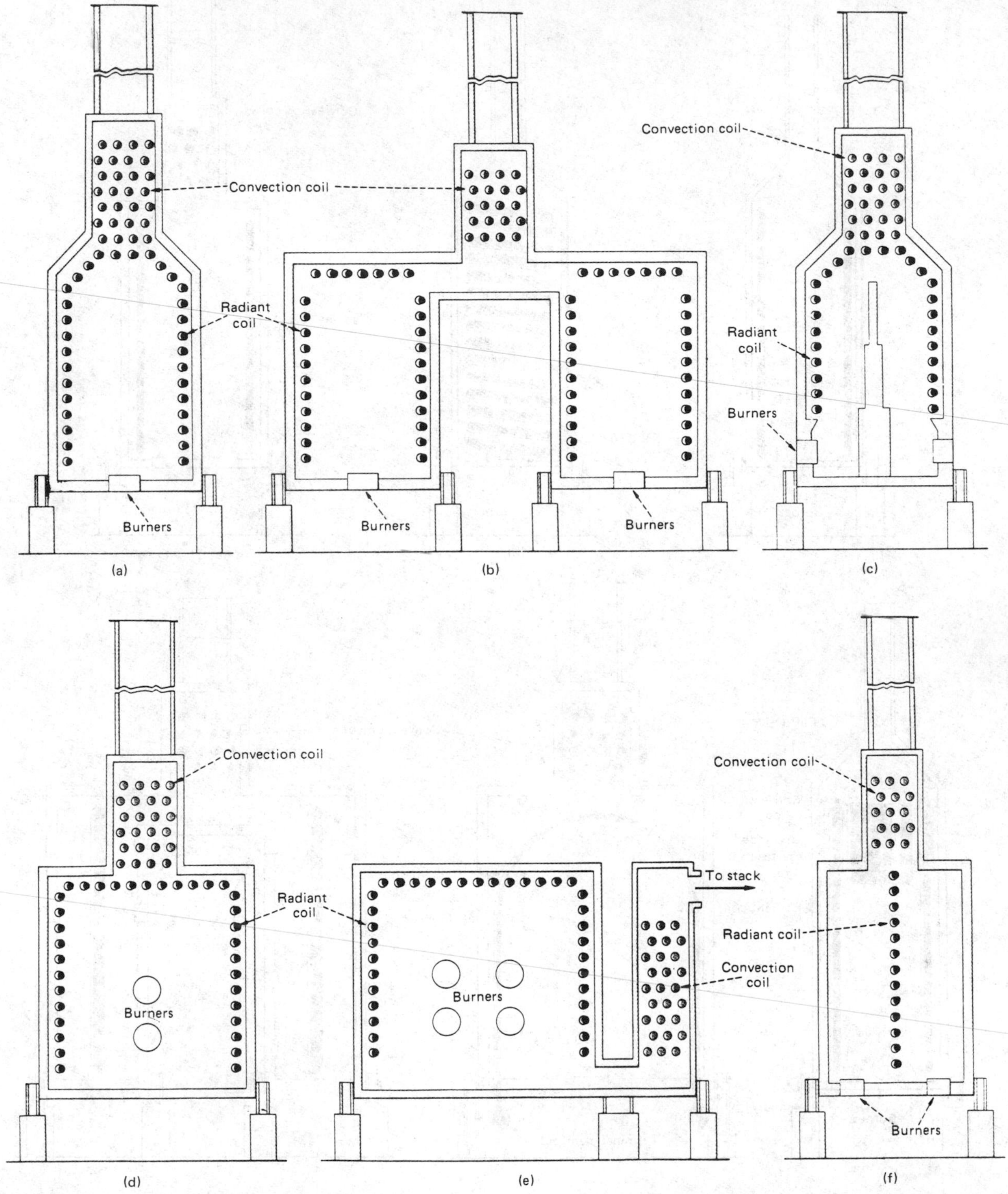

FIG. 9-65 Six basic designs used in horizontal-tube-fired heaters. Radiant-section coil is horizontal. (*a*) Cabin. (*b*) Two-cell box. (*c*) Cabin with dividing bridgewall. (*d*) End-fired box. (*e*) End-fired box, with side-mounted convection section. (*f*) Horizontal-tube, single-row, double-fired. [*From* Chem. Eng., *102–103 (June 19, 1978).*]

with others having vertical-tube designs, their in-tube characteristics resemble those of horizontal-tube heaters.

This design also represents low cost and low efficiency and requires a minimum of plot area. Heating duties are the same as for the preceding heater.

Vertical-Cylindrical, with Cross-Flow Convection These heaters, also fired vertically from the floor, feature both radiant and convection sections. The radiant-section tube coil is disposed in a vertical arrangement along the walls of the combustion chamber. The convection-section tube coil is arranged as a horizontal bank of tubes positioned above the combustion chamber.

This configuration provides an economical, high-efficiency design that requires a minimum of plot area. The majority of new vertical-tube fired-heater installations fall into this category. Typical duty range is 10.6 to 212 GJ/h (10 to 200 million Btu/h).

Vertical-Cylindrical, with Integral Convection Although this design is rarely chosen for new installations, the vast number of existing units of this type warrants its mention in any review of fired heaters.

As with the preceding types, this design is vertically fired from the floor, with its tube coil installed in a vertical arrangement along the walls. The distinguishing feature of this type is the use of added surface area on the upper reaches of each tube to promote convection heating. This surface area extends into the annular space formed between the convection coil and a central baffle sleeve. Medium efficiency can be achieved with a minimum of plot area. Typical duty for this design is 10.6 to 106 GJ/h (10 to 100 million Btu/h).

Arbor or Wicket This is a specialty design in which the radiant heating surface is provided by U tubes connecting the inlet and outlet terminal manifolds. This type is especially suited for heating large flows of gas under conditions of low pressure drop. Typical applications are found in petroleum refining, in which this design is often employed in the catalytic-re-former charge heater, and in various reheat services. Firing modes are usually either vertical from the floor or horizontal between the riser portions of the U tubes.

This design type can be expanded to accommodate several arbor coils within one structure. Each coil can be separated by dividing walls so that individual firing control can be attained. In addition, a cross-flow convection section is normally installed to provide supplementary heating capacity for chores such as steam generation. Typical duties for each arbor coil of this design are about 53 to 106 GJ/h (50 to 100 million Btu/h).

Vertical-Tube, Double-Fired In these units, vertical radiant tubes are arranged in a single row in each combustion cell (there are often two cells) and are fired from both sides of the row. Such an arrangement yields a highly uniform distribution of heat-transfer rates (heat flux) about the tube circumference.

Another variation of these heaters uses multilevel sidewall firing, which gives maximum control of the heat-flux profile along the length of the tubes. Multilevel-sidewall-firing units are often employed in fired-reactor services and in critical reactor-feed heating services. In addition to the twin-cell furnaces already mentioned, single-cell models are available for smaller duties. As a group, these represent the most expensive fired-heater configuration. The typical duty range for each cell runs from about 21 to 133 GJ/h (20 to 125 million Btu/h).

Horizontal-Tube Cabin The radiant-section tube coils of these heaters are arranged horizontally so as to line the sidewalls of the combustion chamber and the sloping roof, or "hip." The convection-section tube coil is positioned as a horizontal bank of tubes above the combustion chamber. Normally the tubes are fired vertically from the floor, but they can also be fired horizontally by sidewall-mounted burners located below the tube coil. This economical, high-efficiency design currently represents the majority of new horizontal-tube fired-heater installations. Duties run from 10.6 to 106 GJ/h (10 to 100 million Btu/h).

Two-Cell Horizontal-Tube Box Here the radiant-section tube coil is deployed in a horizontal arrangement along the sidewalls and roof of the two combustion chambers. Vertically fired from the floor, this is again an economical, high-efficiency design. Typical duties range from 106 to 266 GJ/h (100 to 250 million Btu/h).

Horizontal-Tube Cabin with Dividing Bridgewall Again the radiant-section tube coil is arranged horizontally along the sidewalls of the combustion chamber and along the hip. The convection-section tube coil takes the form of a horizontal bank of tubes positioned above the combustion chamber. A dividing bridgewall between the cells allows for individual firing control over each cell in the combustion chamber. A typical duty range for this design is 21 to 105 GJ/h (20 to 100 million Btu/h).

End-Fired Horizontal-Tube Box The radiant-section tube coil is disposed in a horizontal arrangement along the sidewalls and roof of the combustion chamber. The convection-section tube coil is arranged as a horizontal bank of tubes positioned above the combustion chamber. These furnaces are horizontally fired by burners mounted in the end walls. A typical duty range for this design is 5.3 to 53 GJ/h (5 to 50 million Btu/h).

End-Fired Horizontal-Tube Box, with Side-Mounted Convection Section Here the radiant-section tube coil is disposed in a horizontal arrangement along the sidewalls and roof of the combustion chamber. The convection-section coil is arranged as a horizontal bank of tubes positioned alongside the chamber. The unit is horizontally fired from burners mounted on the end wall.

These furnaces are found in many older installations and occasionally in new facilities that burn particularly poor grades of fuel oil containing a high ash concentration. This relatively expensive design provides duties ranging from 53 to 212 GJ/h (50 to 200 million Btu/h).

Horizontal-Tube, Double-Fired Horizontal radiant tubes are arranged in a single row and are fired from both sides to achieve a uniform distribution of heat-transfer rates around the tube circumference. Such heaters are normally fired vertically from the floor. They are often selected for critical reactor-feed heating services. For increased capacity, the concept can be expanded to provide for a dual combustion chamber. A typical duty range for each cell of this design is about 21 to 53 GJ/h (20 to 50 million Btu/h).

Fuel-Saving Methods for Existing Heaters Because of increased fuel costs, industries that operate process heaters must give serious consideration to improving fuel-burning efficiency. They can now justify increased costs for more efficient major equipment and control instrumentation. In today's environment, industrial-boiler and process-heater efficiency not only is measured in British thermal units but must also reflect the much greater dollar cost of those units. Figure 9-66 illustrates one of the two basic methods of improving design for fuel conservation:

1. *Optimize air for combustion.* Reduction of total air to the left of the zone of maximum combustion efficiency will result in increasing unburned-fuel loss. Heat loss is rapid when all the fuel is not burned. To the right of the zone of maximum combustion efficiency, the losses rise with increasing excess air.

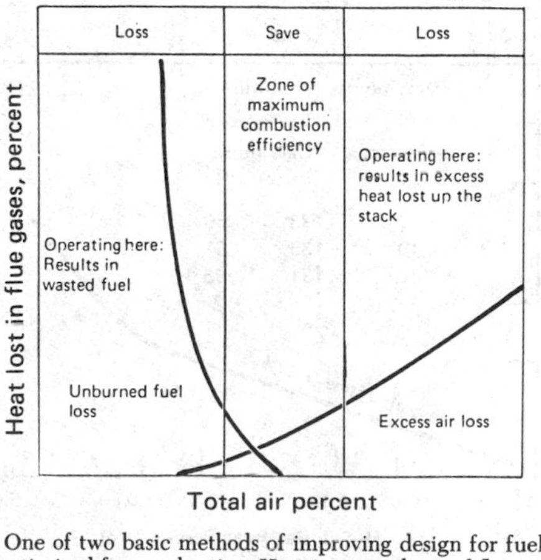

FIG. 9-66 One of two basic methods of improving design for fuel conservation. Air is optimized for combustion. Heat is trapped out of flue gas (with air heaters or economizers). [*From* Combustion, *10–16 (November 1978).*]

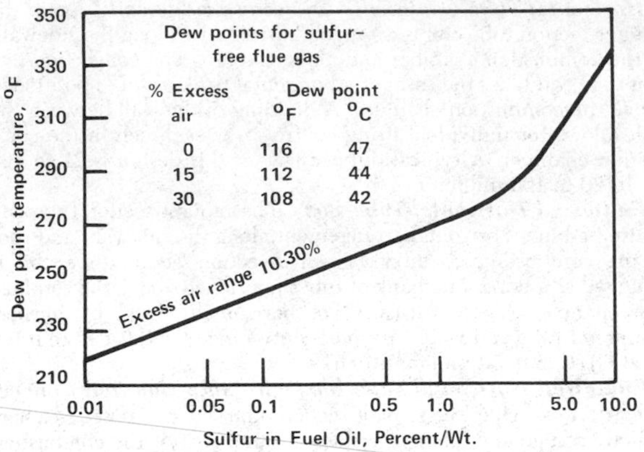

FIG. 9-67 Maximum flue-gas dew point versus percent of sulfur in typical oil fuels.

An additional factor is that when excess air is increased, flue-gas temperature rises, and this compounds with excess-air loss. Excess-air loss rises at a slower rate than does the loss from unburned combustibles by a factor of about 6 (316°C stack temperature); therefore, it is important to keep the flue gas out of the combustibles' range. There is a zone, then, of maximum combustion efficiency where the sum of the losses is at a minimum. The extent of overlap between combustible gas and excess oxygen depends on the particulars of a given installation, the type of burners, and the fuel fired.

2. *Recover heat in flue gas.* If the flue-gas temperature is high, this thermal-energy loss can be reduced by installing heat exchangers that absorb energy from the stack gases. Air heaters and feed preheaters are well-known examples of such equipment.

Corrosion problems in air heaters are associated with high flue-gas dew points because of SO_3 formation from sulfur in the fuel. The curves in Figs. 9-67 and 9-68 can be used to determine safe upper limits for dew points in fired heaters burning liquid or gaseous fuels.

STEAM GENERATORS

Steam generators are designed to produce steam for process requirements, for process needs with electric power generation, and solely for electric power generation. In each case, the prime incentive is to design the most efficient and reliable boiler for the least cost. Many factors influence the design and selection of the type of steam generator, and some of these will be discussed later in connection with industrial and utility boilers.

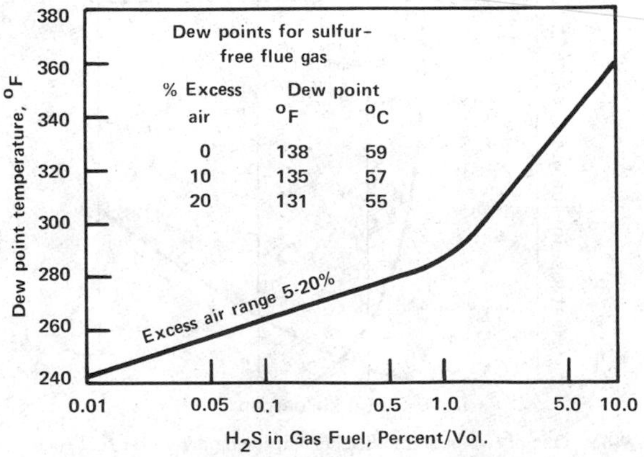

FIG. 9-68 Maximum flue-gas dew point versus percent of H_2S in typical gas fuels.

In the industrial market, boilers have been designed to burn a wide range of fuels and operate at pressures up to 12.4 MPa (1800 lbf/in^2) and steaming rates extending to 455,000 kg/h (1,000,000 lb/h). High-capacity shop-assembled boilers range in capacity from 4500 kg/h (10,000 lb/h) to about 250,000 kg/h (550,000 lb/h). These units are designed for operation at pressures up to 11.1 MPa (1650 psig) and 783 K (950°F). Figure 9-69 shows a gaseous-liquid fuel-fired unit. While most shop-assembled boilers are gas- or oil-fired, designs are available to burn pulverized coal. The significant increase in the cost of fuels and the greater reliance on coal have provided the incentive toward large-capacity field-erected boilers operating at higher pressures with superheat and possibly reheat. A field-erected coal-fired industrial boiler is shown in Fig. 9-70.

Boilers designed for service in electric power utility systems operate at both subcritical-pressure [pressures below 221.1 bar (3206 lbf/in^2)] and supercritical-pressure steam conditions. Subcritical-pressure boilers range in design pressures up to about 18.6 MPa (2700 lbf/in^2) and in steaming capacities up to about 2948 Mg/h (6,500,000 lb/h). A subcritical-pressure boiler is shown in Fig. 9-71. Supercritical-pressure boilers have been designed to operate at pressures up to 344.5 bar (5000 psig). In practice, the 241.2-bar (3500-psig) cycle has been firmly established in the utility industry, and boilers with steaming capacities up to 4219 Mg/h (9,300,000 lb/h) and superheat and reheat temperatures of 814 K (1005°F) are in service.

Some Fundamentals of Boiler Design Boiler design involves the interaction of many variables: water-steam circulation, fuel characteristics, firing systems and heat input, and heat transfer. The furnace enclosure is one of the most critical components of a steam generator and must be conservatively designed to assure high boiler availability. The furnace configuration and its size are determined by combustion requirements, fuel characteristics, emission standards for particulate matter, and the need to provide a uniform gas flow and temperature entering the convection heat-absorbing surfaces to minimize ash deposits and excessive superheater metal temperatures. Discussion of some of these factors follows.

Circulation and Heat Transfer Circulation, as applied to a steam generator, is the movement of water or steam or a mixture of both through the heated tubes. The circulation objective is to absorb heat from the tube metal at a rate that assures sufficient cooling of the furnace-wall tubes during all operating conditions, with an adequate margin of reserve for transient upsets. Adequate circulation prevents excessive metal temperatures or temperature differentials that would cause failures due to overstressing, overheating, or corrosion.

The heat transfer from the tubes to the fluid depends primarily on turbulence and heat flux. Turbulence is a function of mass velocity of the fluid and tube roughness. Turbulence has been achieved by designing for high mass velocities, which ensure that nucleate boiling takes place at the inside surface of the tube. If sufficient turbulence

FIG. 9-69 Shop-assembled boiler.

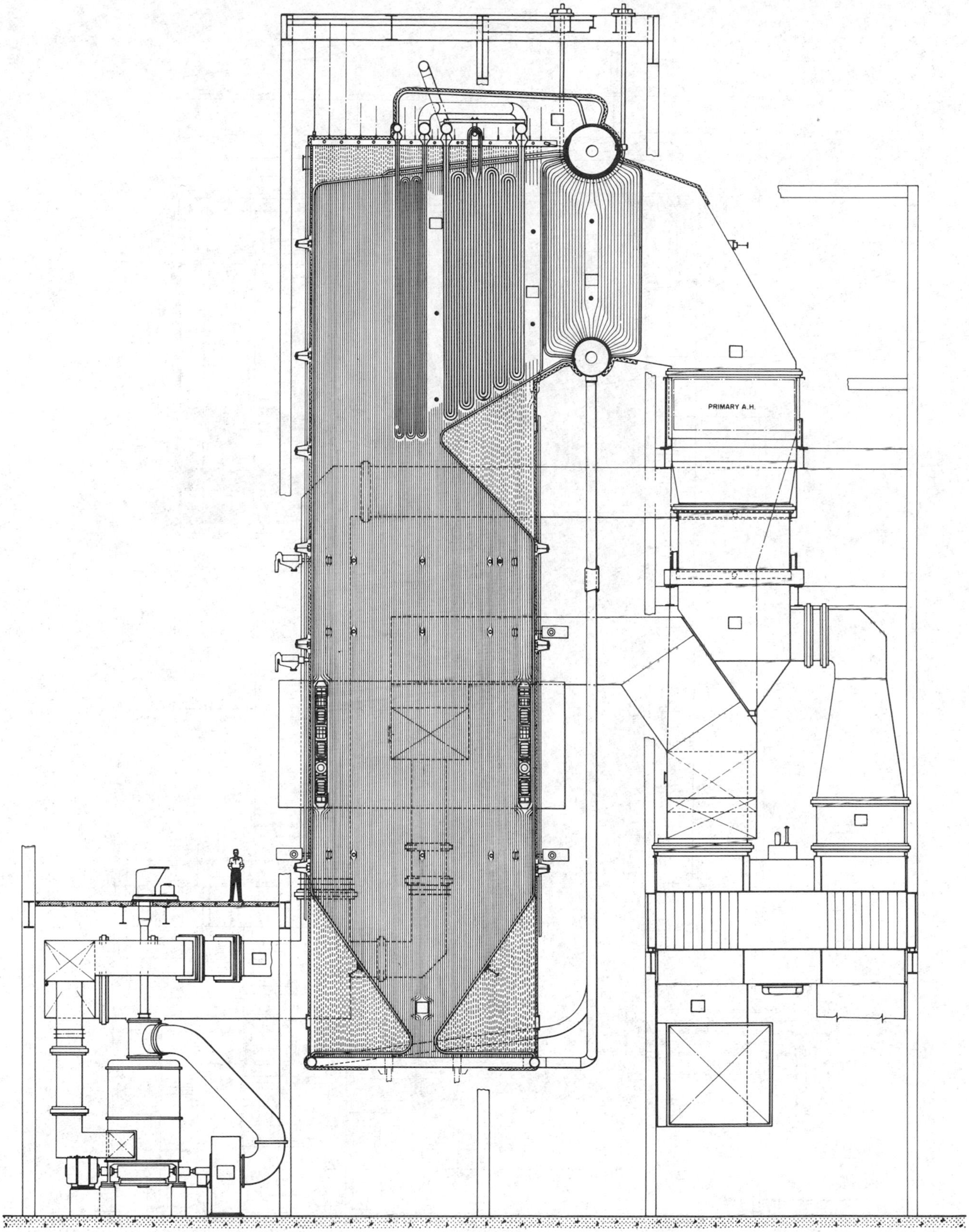

FIG. 9-70 Coal-fired industrial boiler.

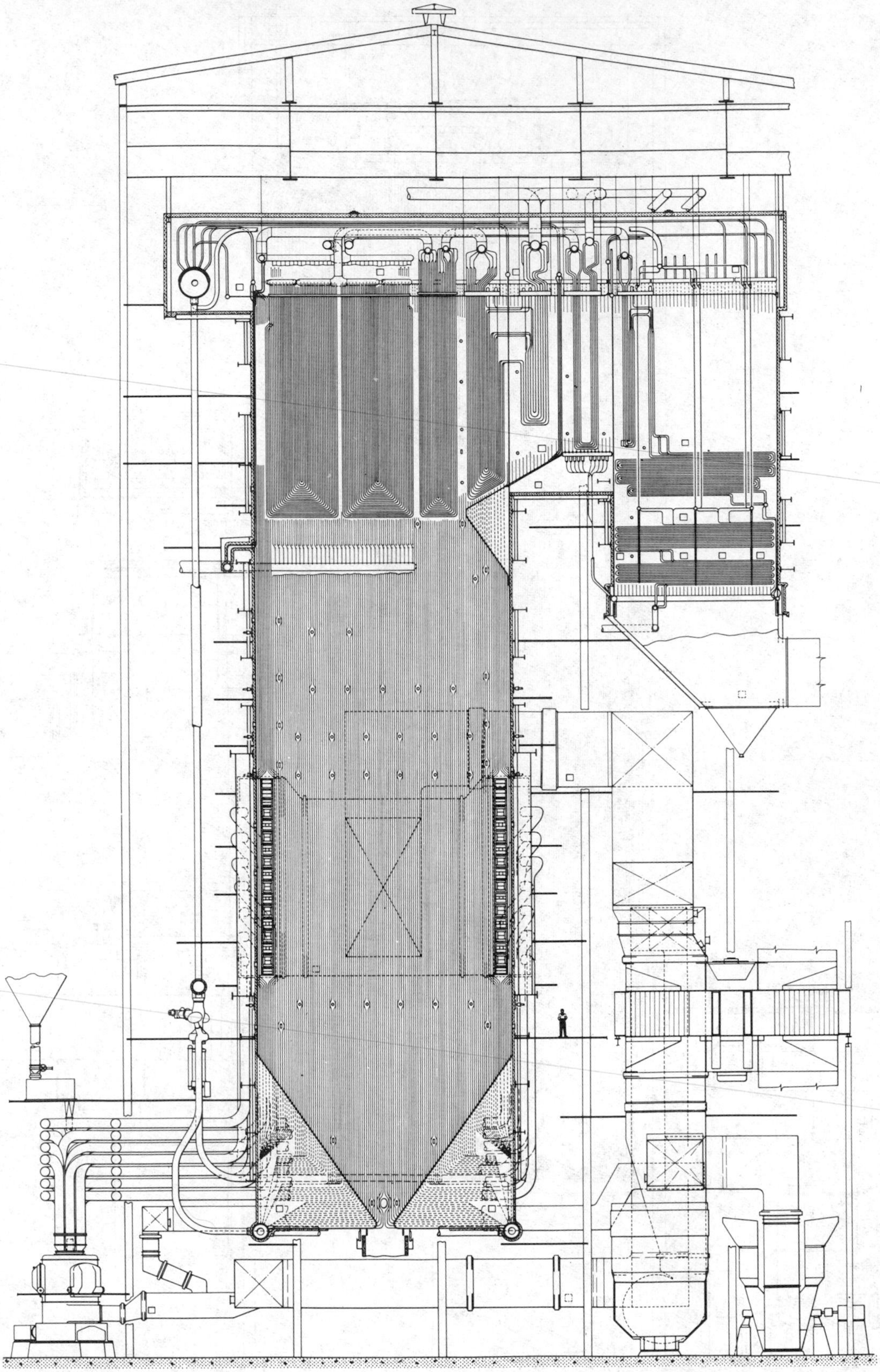

FIG. 9-71 Subcritical-pressure utility boiler.

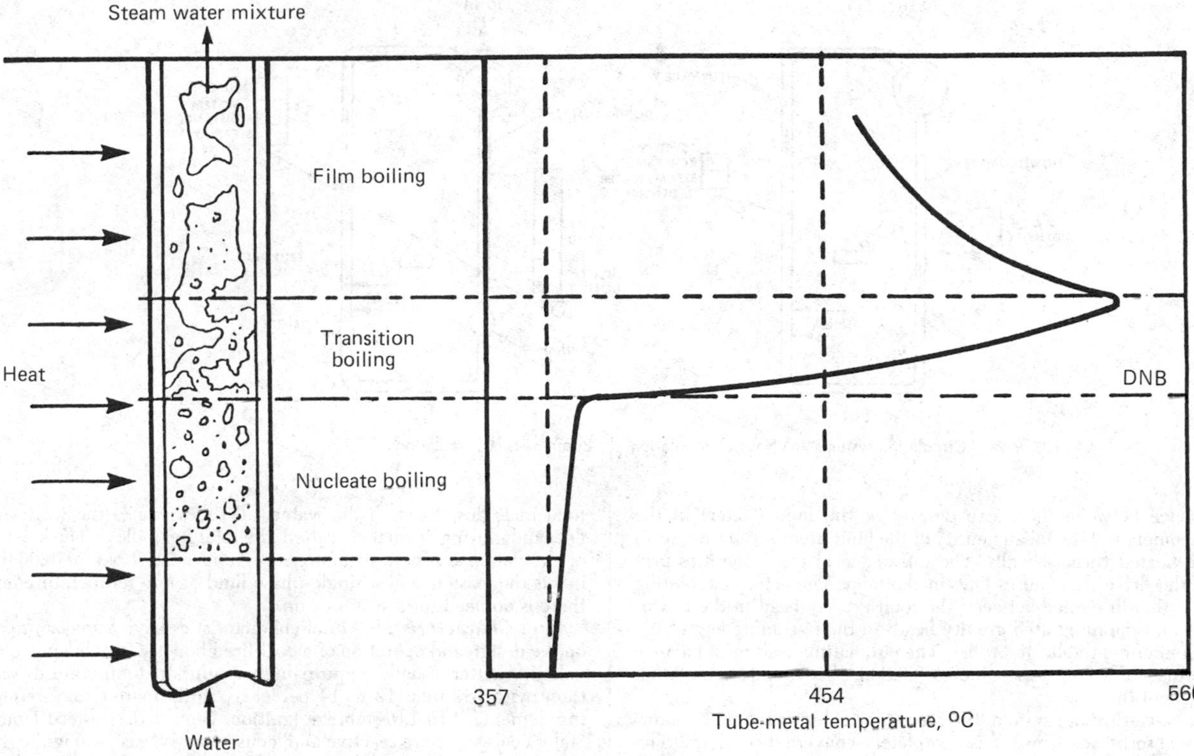

FIG. 9-72 Effect of departure from nucleate boiling (DNB) on tube-metal temperature.

is not provided, departure from nucleate boiling (DNB) occurs. DNB is the production of a film of steam on the tube surface that impedes the flow of heat to cool the tubes. This phenomenon is illustrated in Fig. 9-72.

Satisfactory performance is obtained with tubes having helical ribs on the inside surface, which generate a swirling flow. The resulting centrifugal action forces the water droplets toward the inner tube surface and prevents the formation of a steam film. The internally rifled tube maintains nucleate boiling at much higher steam qualities and with much lower mass velocities than those in smooth tubes. This improvement is shown in Fig. 9-73.

Circulation ratio, defined as the weight of circulating flow divided by the weight of steam generated, is an empirical criterion for evaluating the performance of circulation systems. In modern practice, however, the most important criterion in drum boilers is the prevention of conditions that lead to DNB.

Utility Steam Generators

Steam-Generator Circulation System Circulation systems for utility application are generally classified as natural circulation and forced or pump-assisted circulation in drum-type boilers and as once-through flow in subcritical- and supercritical-pressure boilers. The circulation systems for natural- and pump-assisted circulation boilers are illustrated schematically in Fig. 9-74.

Natural circulation in a boiler circulation loop relies only on the

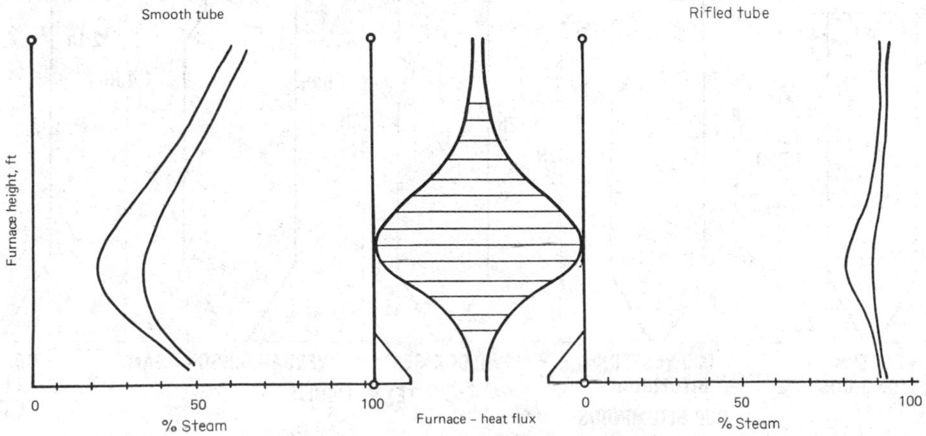

FIG. 9-73 Maximum allowable percent of steam to avoid departure from nucleate boiling (DNB).

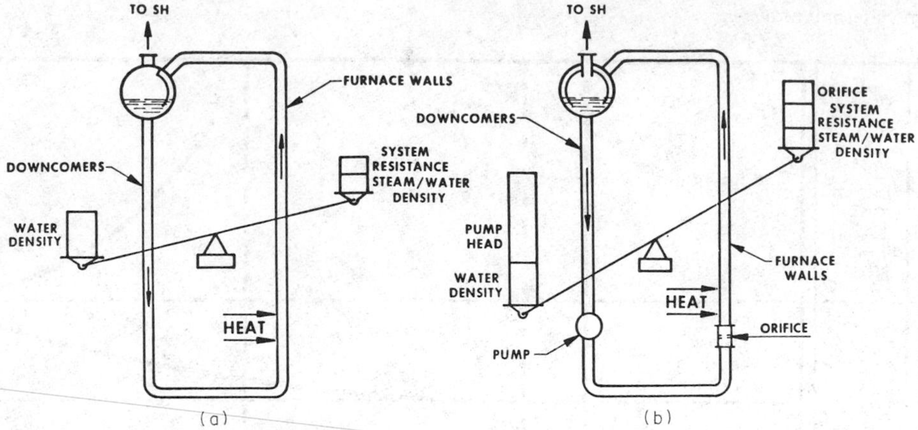

FIG. 9-74 Circulation systems. (*a*) Natural circulation. (*b*) Pump-assisted circulation.

difference between the mean density of the fluid (water) in the downcomers and the mean density of the fluid (steam-water mixture) in the heated furnace walls. The difference in these densities provides the drive that causes flow in the loop. The actual circulating head is the difference between the total gravity head in the downcomer and the integrated gravity heads in the upcoming legs of the loop containing the heated tubes. The circulating head must balance the sum of the flow losses due to friction, shock, and acceleration throughout the loop.

In a **once-through** system, the feedwater entering the unit continuously absorbs heat until it is completely converted to steam. The total mass flow through the waterwall tubes equals the feedwater flow and, during normal operation, the total steam flow. The key fact of circulation at supercritical pressures (above 221.1 bar, or 3206 lbf/in^2) is the existence of a single-phase fluid at any temperature, and there is no need for a steam drum.

Fuel Characteristics Fuel characteristics have a major impact on the design and operation of a coal-fired boiler. Coals having a low volatile matter usually require higher ignition temperatures, and those with less than 12 to 14 percent volatile matter may require supplementary fuel to stabilize ignition. Generally, western United States coals are more reactive and, consequently, easier to ignite, but

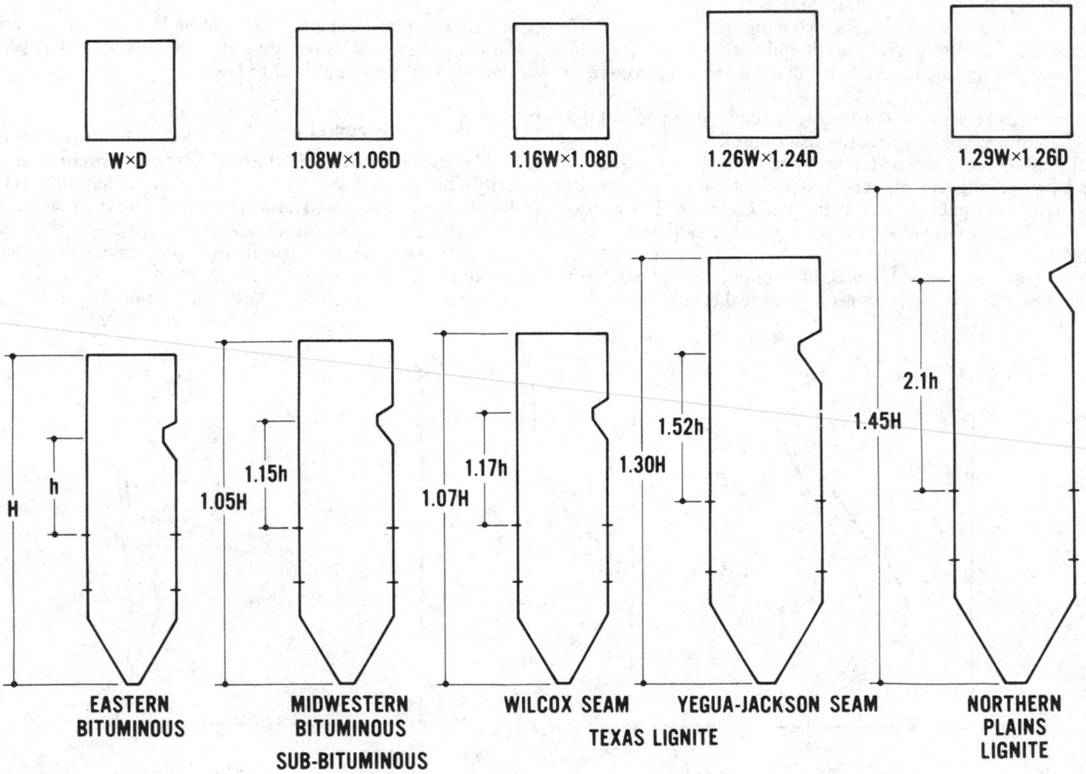

FIG. 9-75 Effect of coal rank on furnace sizing.

because of high moisture content they require higher air temperatures to the mills for drying the coal to attain proper pulverization. Extremely high-ash-content coal could create problems in ignition and stabilization. The ash constituents and the quantity of ash will have a decided influence on sizing the furnace. Accordingly, a thorough review of coal characteristics is needed to establish the effect on the design and operation of a boiler.

In designing and sizing furnaces, particular attention is required for the following fuel-ash properties:

1. Ash fusion temperatures in terms of their absolute values and the spread between initial deformation temperature and fluid temperature
2. Ratio of basic to acidic ash constituents
3. Iron-to-calcium ratio
4. Ash content, kg/1,055,000 kJ (lb/million Btu)
5. Ash friability

These characteristics translate into the furnaces shown in Fig. 9-75 for five United States coals. Note the progressive increase in furnace size in terms of plan area, volume, and burning zone as lower-grade fuels are fired to develop equivalent steam-flow capacity. There are wide variations in coal properties within the coal ranks as well as several subclassifications (e.g., subbituminous A, B, C) of these fuels. These variations require a different size of furnace for each classification.

Coal properties influence pulverizer capacity, the sizing of the air heater, and other heat-recovery sections of a steam generator. Furnace size and heat-release rates are designed to control slagging characteristics. Consequently, heat-release rates, in terms of net heat input to plan area, range from 4,420,000 W/m^2 [1.4 million Btu/(h·ft^2)] for severely slagging coals to 6,620,000 W/m^2 [2,100,000 Btu/(h·ft^2)] for low-slagging fuels.

Firing-System Heat Input Heat input is a critical design criterion in establishing the combustion of fuel with regard to minimizing the formation of the oxides of nitrogen (NO_x) in the combustion process. To accomplish these functions, the designer considers the following design parameters:

- Design and size of burners and their arrangement in the furnace
- Clearances from burners to water-cooled surfaces
- Furnace release rate, (kg·cal)/m^2 (Btu/ft^2) effective projected radiant surface or EPRS
- Furnace-gas residence time for combustion and NO_x emissions
- Furnace exit-gas temperatures, K (°F)
- Combustion rate, kJ/m^3 (Btu/ft^3)
- Burner-zone release rate, W/m^2 [Btu/(h·ft^2)]

Firing systems can be classified into two groups:

- Wall firing, in which individual flames are formed by burners in the front, rear, and/or side walls of the furnace
- Tangential firing, in which fuel and air are introduced into the furnace from delivery nozzles located in the corners of the unit

A major difference between the systems is that in wall firing there is little interaction among the flames from individual burners; in tangential firing, reactants from all coal nozzles interact, forming a cyclonic flame mass and making the furnace a single "burner."

The furnace of large coal-fired steam generators absorbs about one-half of the heat released, so that the gas temperature leaving the furnace is about 1376 K (2000°F).

Superheaters and Reheaters The superheater raises the temperature of the steam generated by the boiler to some point above saturation. The thermodynamic advantage of superheating steam is shown in Fig. 9-76, which compares the available head above a given exhaust condition with the total heat in the steam. Another important point is to minimize moisture in the last stages of a turbine to avoid blade erosion. With continued increase of evaporation temperatures and pressures, a point is reached at which the available superheat temperatures are insufficient to prevent excessive moisture from forming in the low-pressure turbine stages. This condition is resolved by removing the vapor for reheat at constant pressure in the boiler and returning it to the turbine for continued expansion to condenser pressure. The thermodynamic cycle using this modification of the Rankine cycle is called the reheat cycle. Figure 9-77 illustrates the surface arrangements in a high-pressure boiler.

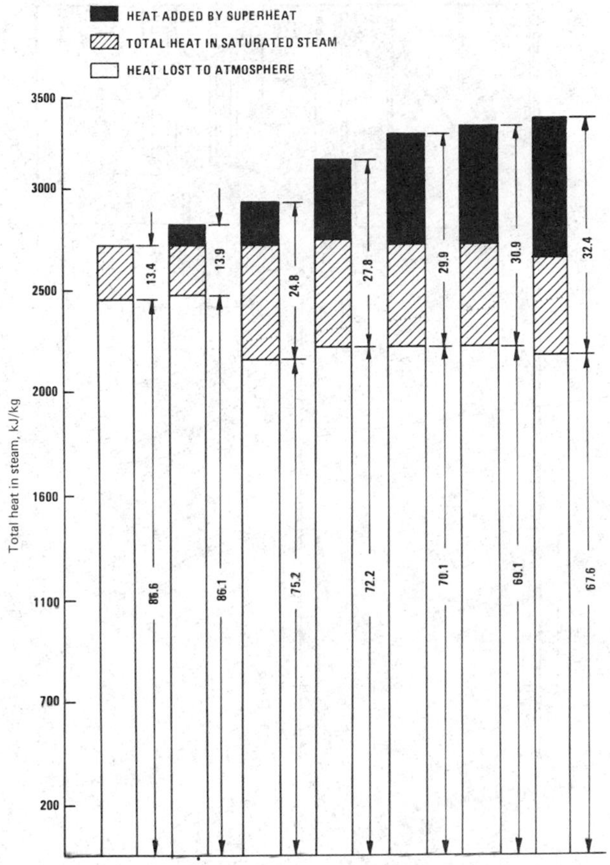

FIG. 9-76 Improved prime-mover performance with superheat.

Economizers Economizers improve boiler efficiency by extracting heat from the discharged flue gases and transferring it to feedwater, which enters the steam generator at a temperature appreciably lower than the saturation-steam temperature.

Industrial Boilers A common definition of an industrial boiler is a stationary water-tube boiler in which some of the steam is generated in a convective tube bank. The percentage of total heat absorption in the boiler bank varies considerably with boiler steam pressure and temperature and feedwater temperature. For a typical coal-fired boiler producing 90,720 kg/h (200,000 lb/h):

%, total steam	Boiler pressure kg/cm^2	Boiler pressure lb/in^2	Steam temperature, °C	Feedwater temperature, °C
45	14	200	187	116
30	42	600	399	116
16	105	1500	510	177
10	126	1800	538	177

The thicker plate for operation at higher pressures increases the cost of boiler drums. As a result, it is normally not economical to use a boiler bank for heat absorption at pressures above 109 kg/cm^2 (1550 psig).

Industrial boilers are used over a wide range of applications, ranging from large power-generating units, for which emphasis is placed on maximum efficiency and sophisticated control systems, to small low-pressure units for space or process heating, for which the principal aims are simplicity and low first cost.

While an industrial boiler's primary function is usually to provide

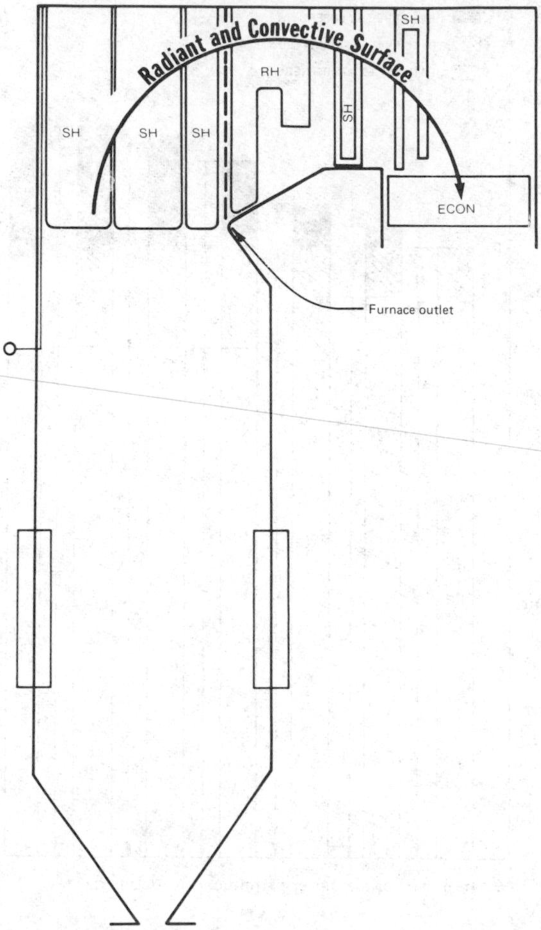

FIG. 9-77 Typical boiler superheater-reheater arrangement.

energy in the form of steam, there are a number of applications in which steam generation is incidental to a chemical process, e.g., a chemical recovery unit in the paper industry, a carbon monoxide boiler in an oil refinery, or a gas-cooling waste-heat boiler in an open-hearth furnace. Within a given industrial plant, it is not unusual for industrial boilers to serve a multiplicity of functions. For example, in a paper-pulp mill the chemical-recovery boiler is used to convert black liquor into useful chemicals and to generate process steam. At the same plant a bark-burning unit recovers heat from otherwise wasted material and also generates power. Industrial boilers burn oil, gas, coal, and a wide range of product and/or waste fuels as shown in Table 9-48.

Current statistics show that pulverized-coal firing is the choice for larger boilers, above 113,398 kg/h (250,000 lb/h). For boilers in the medium-size range, 45,359 to 113,398 kg/h·(100,000 to 250,000 lb/h), stoker-fired boilers dominate, although the percentage of pulverized-coal-fired boilers in this range is increasing. The higher thermal efficiency of a pulverized-coal-fired boiler makes it attractive toward the upper limit of the medium-size range.

A major factor to consider when comparing a stoker-fired boiler with a pulverized-coal-fired boiler is the reduction in efficiency due to carbon loss. A properly designed pulverized-coal-fired boiler can maintain an efficiency loss due to unburned carbon of less than 0.4 percent. A continuous-ash-discharge spreader-stoker-fired unit will typically have a carbon loss of 4 to 8 percent, depending on the amount of reinjection. Dump-grate, underfeed, or overfeed stokers

TABLE 9-48 Solid-Waste Fuels Burned in Industrial Boilers

Waste	HHV, kJ/kg*
Bagasse	8374–11,630
Furfural residue	11,630–13,956
Bark	9304–11,630
General wood wastes	10,467–18,608
Coffee grounds	11,397–15,119
Nut hulls	16,282–18,608
Rich hulls	12,095–15,119
Corncobs	18,608–19,306
Rubber scrap	26,749–45,822
Leather	27,912–45,822
Cork scrap	27,912–30,238
Paraffin	39,077
Cellophane plastics	27,912
Polyvinyl chloride	40,705
Vinyl scrap	40,705
Sludges	4652–27,912
Paper wastes	13,695–18,608

*To convert kilojoules per kilogram to British thermal units per pound, multiply by 4.299×10^{-1}.

produce an even higher carbon loss. The lower carbon loss produced with the pulverized-coal unit results from increased combustion efficiency obtained from the finer coal particles that enter the furnace (normally 70 to 80 percent will pass through a 200-mesh screen). In contrast, the coal particles that enter the furnace with a spreader stoker are much coarser, with a 19-mm top size and not more than 50 percent passing through a 6.35-mm screen.

The coal itself also affects the total fuel-cost difference between these firing methods. For efficient operation of a spreader-stoker-fired boiler, the coal must have the proper mixture of coarse and fine particles. Normally, double-screened coal is purchased to obtain the proper mixture. One cannot depend on run-of-the-mine coal to have the optimum balance of coarse and fine material. The fine coal particles are burned in suspension above the grate, while the larger particles burn on the grate surface. If the amount of fine particles is too great, excessive suspension burning will occur with the following possible results:

- Higher stack particulate-matter loading
- Furnace pulsations
- Possible heat damage to the distributors
- Higher carbon loss and smoke

If the amount of coarse coal is too great, there will be insufficient suspension burning with these possible results:

- Loss of flame stability
- Higher carbon loss and smoke

If it is decided not to use double-screened coal, there will be a loss in efficiency as the coal sizing diverges from the optimum. The question of whether the reduced efficiency obtained with unsized coal is offset by the higher cost for double-screened coal cannot be answered on a general basis. The answer depends on such factors as degree of coal-sizing variation, quality of operators, boiler or stoker size, and coal classification.

A pulverized-coal-fired boiler requires only that the coal entering a mill be 1½-in top size. Coal of any size can be purchased and crushed at the plant site if the proper size is not available at the mine.

In addition to designing for a range of coals, liquid, gaseous, or other solid fuels may be considered. Because pulverized coal is suspension-fired, as are conventional liquid and gaseous fuels, the furnace is properly proportioned for this function. Oil or gas is required for unit warm-up; therefore, the required piping, combustion controls, and fuel-firing equipment are in place. These fuels can be fired to carry load with little or no design or equipment modifications. Firing a liquid fuel derived from coal or a coal-oil slurry would also be possible because the furnace has been designed to cool the gases below the ash-softening temperature before these gases enter the closely spaced generating bank section.

Well-prepared solid wastes may also be burned in suspension. Up to 20 percent of the total heat input may be added. Bark, bagasse, and refuse have all been successfully suspension-fired in a pulverized-coal-fired boiler. These wastes are normally blown into the furnace pneumatically through ports located above the burners.

For a stoker-fired boiler, firing liquid or gaseous fuels is more difficult. The firing equipment must be placed high above the stoker to protect it from the radiant heat produced by the auxiliary burners. The placement of firing equipment high in the furnace reduces the effective furnace volume. For long-term operation, it may be necessary to cover the stoker with refractory. For short-term operation, considerable quantities of air are required to cool the stoker, resulting in large quantities of excess air, which is inefficient. While firing liquid or gaseous fuels in a stoker-fired boiler is more difficult, the stoker is ideally suited for firing other unsized solid fuels. Bark, bagasse, or refuse can normally be fired on a stoker to supplement the coal with a minimum amount of additional equipment. Also, these waste fuels can comprise a higher percentage of the total heat input in a stoker-fired boiler than in a pulverized-coal-fired boiler.

Design Criteria Industrial-boiler designs are tailored to the fuels and firing systems involved. Some of the more important design criteria include:

- Furnace heat-release rates, both W/m³ and W/m² of effective projected radiant surface [Btu/(h·ft³) and Btu/(h·ft²)]
- Heat release on grates
- Flue-gas velocities through tube banks
- Tube spacings

Table 9-49 gives typical values or ranges of these criteria for gas, oil, and coal. The furnace release rates are important and establish maximum local absorption rates within safe limits. They have a bearing on completeness of combustion and therefore on efficiency and particulate emissions. Limiting heat release on grates (in stoker firing) will minimize carbon loss, control smoke, and avoid excessive fly ash.

Limits on flue-gas velocities for gas- or oil-fired industrial boilers are usually determined by the need to limit draft loss. For coal firing, design gas velocities are established to minimize fouling and plugging of tube banks in high-temperature zones and erosion in low-temperature zones.

Convection tube spacing is important when the fuel is residual oil or coal, especially coals with low ash-fusion or high ash-fouling tendencies. The amount of the ash and, even more important, the characteristics of the ash must be specified for design. The recommended minimum tube spacing for industrial solid-fuel-fired boilers is shown in Table 9-50.

Natural-circulation and convection boiler banks are the basic design features on which a line of standard industrial boilers has been developed to accommodate the diverse steam, water, and fuel requirements of the industrial market.

Figure 9-78 shows the amount of energy available for power by using a fire-tube boiler, an industrial boiler, and subcritical- and supercritical-pressure boilers. Condensing losses decrease substantially, and regeneration of air and feedwater becomes increasingly important in the most advanced central-station boilers.

The boiler designer must proporton heat-absorbing and heat-recovery surfaces in a way to make the best use of heat released by the fuel. Waterwalls, superheaters, and reheaters are exposed to convection and radiant heat, whereas convection heat transfer predominates in air heaters and economizers. The relative amounts of these surfaces vary with the size and operating conditions of the boiler.

Factors Influencing Performance Some of the dominant factors influencing performance are fouling, slagging, moisture in the coal, grindability of the coal, and heating value of the coal.

The volatile constituents in coal ash (i.e., Na_2SO_4 or $CaSO_4 \cdot Na_2SO_4$) cause fouling and can be used as a fouling index of a given coal. Two factors that affect fouling are deposit hardness and the rate of deposition. Deposit hardness is affected by chemical composition, temperature, and, to some extent, time. The rate of deposition depends on the volatile constituents of coal and the amount of ash in the coal. In general, fouling is related to the sodium and potassium compounds in the ash, and it becomes severe as the level increases above 4 to 5 percent Na_2O and K_2O for high-calcium, low-iron ash coals. Slag deposits are caused primarily by the physical transport of molten or partially fused ash particles entrained by the gas stream. When the particles strike the wall or tube surfaces, they become

TABLE 9-49 Typical Design Parameters for Industrial Boilers

Furnace			Heat-release rate, W/m²* of EPRS†
Natural gas-fired			630,800
Oil-fired			551,900–630,800
Coal: pulverized coal			220,780–378,480
Spreader stoker			252,320–410,020
Stoker, coal-fired			**Grate heat-release rate, W/m²**
Continuous-discharge spreader			2,050,000–2,207,800
Dump-grade spreader			1,419,300–1,734,700
Overfeed traveling grate			1,261,000 –1,734,700
Flue-gas velocity: type	**Single-pass**	**Baffled**	
Fuel-fired	**Boiler m/s**	**Boiler m/s**	**Economizer m/s**
Gas or distillate oil	30.5	30.5	30.5
Residual oil	30.5	22.9	30.5
Coal (not lignite)			
Low-ash	19.8–21.3	15.2	15.2–18.3
High-ash	15.2	NA‡	12.2–15.2

*To convert watts per square meter to British thermal units per hour-foot, multiply by 0.317.
†Effective projected radiant surface.
‡Not available.

TABLE 9-50 Recommended Minimum Tube Spacing for Industrial Boilers, mm

	Superheater		Boiler		
	Platen	Spaced	Front	Rear	Economizer
Oil and clean gases	305	102	25.4	25.4	76
Coal					
Low ash; high AST*	305	102	25.4	25.4	76
High ash; low AST*	305	152	50.8	38	76
Bark, bagasse	305	152	38	25.4	76

*Ash-softening temperature.

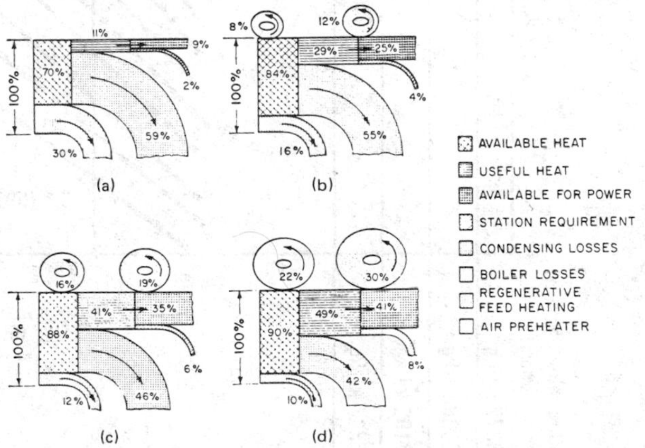

FIG. 9-78 Sankey diagrams for various types of boilers. (*a*) Fire-tube boiler. (*b*) Industrial boiler. (*c*) Subcritical-pressure boiler. (*d*) Supercritical-pressure boiler.

chilled and solidify. The strength of their attachment is influenced by the temperature and physical contour of the surface, direction, force of impact, and melting characteristics of the slag. Coals with low ash-fusion temperatures have a high potential for slagging [below 2200°F (1477 K)]. Normally, slagging is confined to the radiant surfaces, but it also occurs in the convection superheater if the gas temperature exceeds good design practice. Slagging characteristics affect the furnace design, requiring a significant increase in furnace volume to reduce the potential for deposit formation.

High moisture in coal influences furnace performance. Excess surface moisture compounds the difficulties of handling coal at the exit of the bunker into the inlet of the feeder and at the feeder outlet into the pulverizer. Wet coal within the pulverizer causes it to ball up and inhibit grinding. Such problems are typical with coal having a high inherent moisture in the fuel, up to 30 percent. In these cases, providing sufficient high-temperature air will obviate pulverization problems.

The high moisture in the fuel lowers the flame temperature, reducing local furnace absorption and ignition stability, especially during start-up. High moisture also reduces boiler efficiency, owing to heat losses up the stack.

The grindability index is a measure of the relative ease of pulverizing coal. The fineness required for both good ignition and complete combustion of coal depends on a number of factors, including volatile matter, agglomerating properties, and particle reactivity. In terms of rank, low-volatile bituminous coals must be pulverized to a higher degree of fineness than subbituminous coals and lignites to ensure carbon burnout. The grindability of coals varies over a wide range. Pulverizer output related to pulverizer base capacities for various grindability indices is shown in Fig. 9-79. Grindability can be a significant factor in limiting rated boiler output when coal having a lower grindability than that for which the pulverizer was sized is burned.

The heating value of coal has a significant influence on boiler performance. The greatest potential for variations in the pulverizer output rate is the result of fuel heating-value variations. The lower heating value of lignites and subbituminous coals creates a potential limitation in boiler output because it takes more coal for a given heat input. Heating values range from 12,000 Btu/lb (27,912 kJ/kg) for an eastern bituminous coal to 6800 Btu/lb (15,817 kJ/kg) for northern plains lignite. The range of firing rates of various United States coals to yield comparable heat inputs for a given unit is shown in Table 9-51.

Stacks and Chimneys

Theoretical Draft The theoretical draft is that which is produced by the manometric difference in static head of equal columns of atmospheric air and combustion gases. The theoretical draft of a smokestack or a chimney can be calculated readily from the formula

$$\Delta p_t = 0.256\, LP \left(\frac{1}{T} - \frac{1}{T_1}\right) \tag{9-44}$$

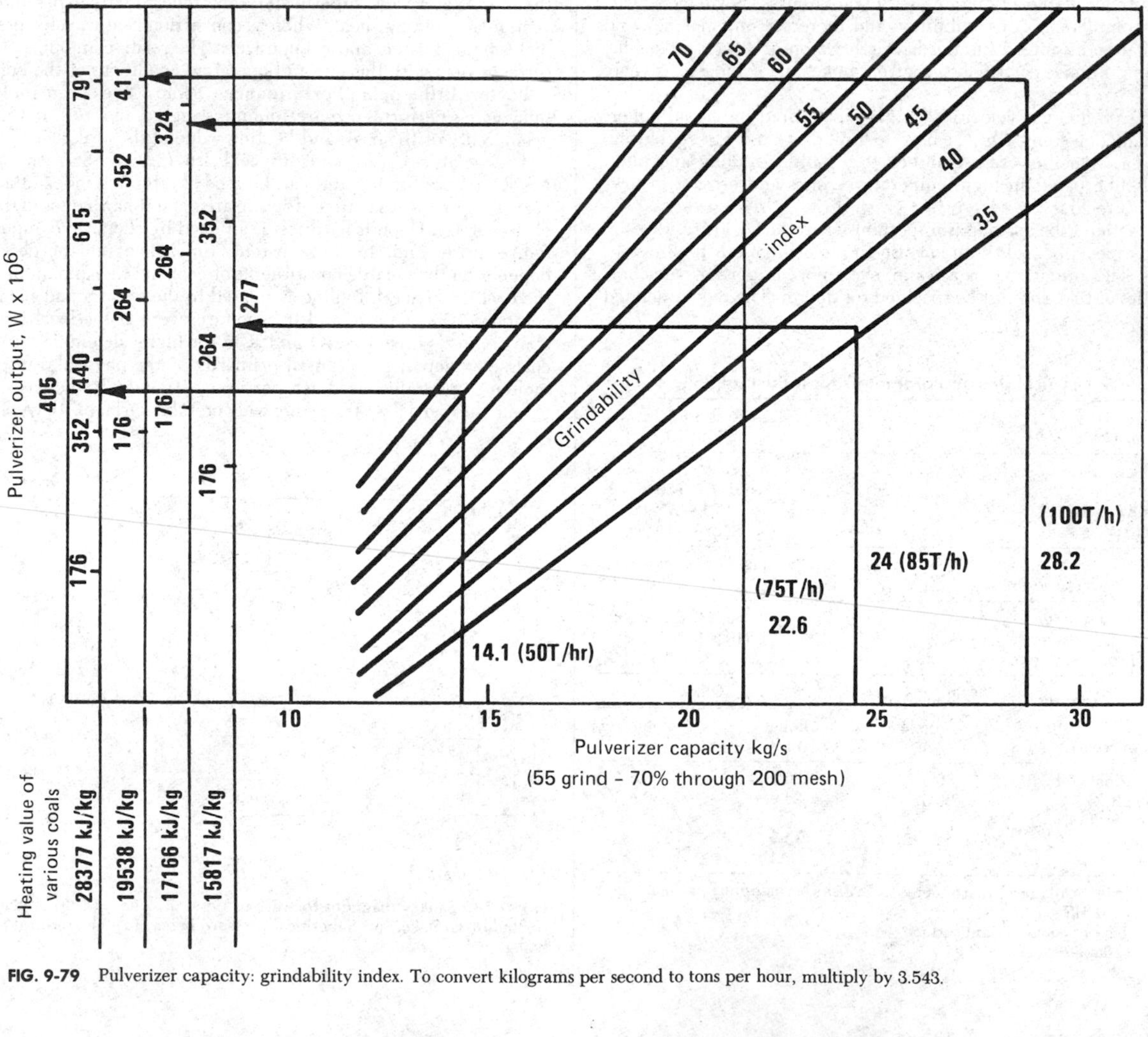

FIG. 9-79 Pulverizer capacity: grindability index. To convert kilograms per second to tons per hour, multiply by 3.543.

TABLE 9-51 Range of United States Coal Properties*

	Eastern bituminous	Midwestern bituminous	Subbituminous C	Texas lignite	Northern lignite
HHV, kJ/kg	27,900	23,300	19,540	16,980	15,820
Moisture, %	6	12	27	32	37
Moisture, kg H_2O/GJ	11	26	71	97	119
Fuel fired, kg/h	204,300	145,160	291,900	336,000	360,500

*Based on Q_F = 12,500 GJ/kg.

where Δp_t = theoretical stack draft, in water; L = stack height above furnace, ft; P = barometric pressure, inHg; T = ambient temperature, °R; and T_1 = average stack temperature, °R. Figure 9-80*a* may be used to estimate the exit-gas temperature required to obtain the average stack temperature for masonry or steel stacks.

Actual Draft To obtain the actual draft for a chimney or stack, the losses that occur with flow must be deducted from the theoretical draft. These losses include both friction and exit velocity. They may be evaluated by the following equation:

$$\Delta p_f = \frac{u^2}{2g}\left(1 + \frac{fL}{D}\right)\left(\frac{\rho_g}{5.2}\right) \tag{9-45}$$

where Δp_f = stack flow loss, in water; u = stack exit velocity, ft/s; g = dimensional constant, 32.17; f = friction factor from Fig. 9-80*b*; L = stack height above furnace, ft; D = stack diameter, ft; and ρ_g = average density of gases in the stack, lb/ft³. To apply Fig. 9-80*b* the Reynolds number may be approximated by the equation

$$N_{Re} = \frac{24{,}000 \times W}{D(T_1 + 715)} \tag{9-46}$$

where T_1 = average stack temperature, °F
W = gas flow, lb/h

The actual draft may be obtained by subtracting the frictional losses from the theoretical draft. This quantity is known as the **nat-**

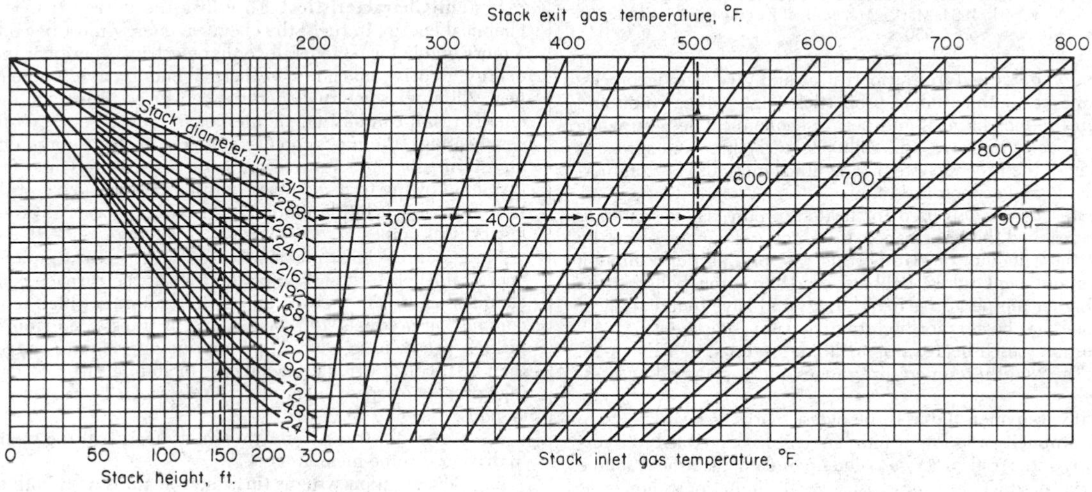

FIG. 9-80*a* Exit-gas temperatures from stacks.

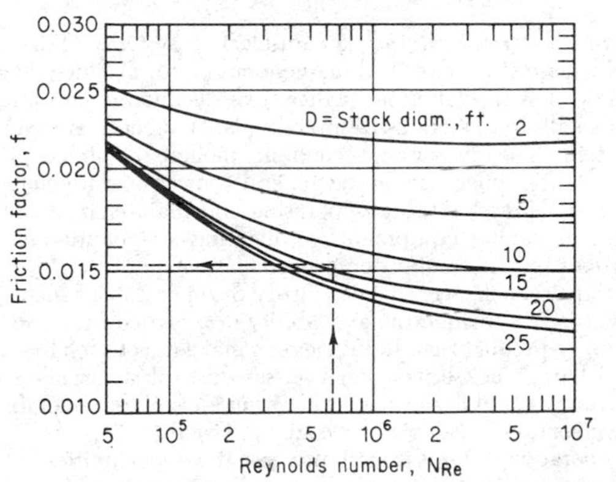

FIG. 9-80*b* Friction factors for stacks. *(Babcock & Wilcox Co.)*

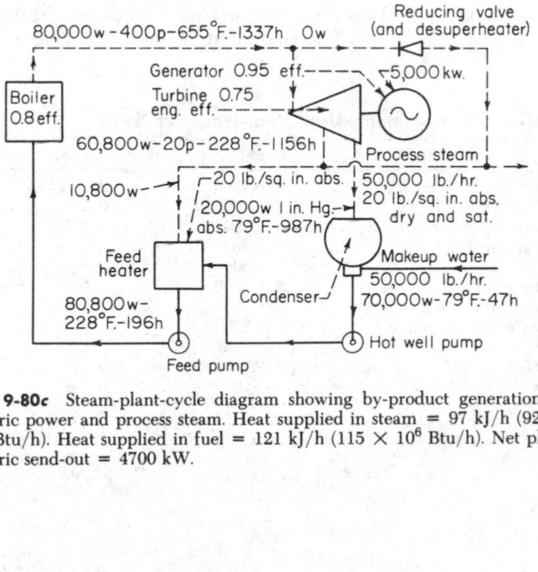

FIG. 9-80*c* Steam-plant-cycle diagram showing by-product generation of electric power and process steam. Heat supplied in steam = 97 kJ/h (92 × 10^6 Btu/h). Heat supplied in fuel = 121 kJ/h (115 × 10^6 Btu/h). Net plant electric send-out = 4700 kW.

ural draft of the chimney, since it is produced without mechanical means. When it is insufficient to overcome the flow resistance of the system, fans must be added. If the fans supply air at pressures above atmospheric to the burners and combustion space, they are called **forced-draft** fans. If the fans handle the products of combustion, they usually operate at pressures below atmospheric and are called **induced-draft** fans.

Special Problems Frequently stacks are required to handle gases of a highly corrosive nature. Even the gases of conventional fuels may contain sufficient amounts of contaminants such as sulfur trioxide to attack steel and concrete rapidly. Efforts should be made to maintain temperatures above the dew point or to reduce partial pressures by dilution when condensable acid gases occur. When **corrosive conditions** warrant the extra expense, chimneys and stacks are frequently lined with glass, acid-resistant brick, or acid-resisting cement. Steel stacks may be purchased lined with bonded glass for erection as a unit, or they may be lined in place with brick after erection. Masonry stacks are erected in place of either brick or reinforced-concrete design. Either steel or masonry stacks may be lined after erection with special coatings applied by hand or by pressure spraying (guniting). It is advisable to inspect stacks for internal corrosion at periodic intervals. Water washing after cooling is advisable to expose the surface for inspection if the stack is heavily coated with soot.

HEAT TRANSPORT

GENERAL REFERENCES: *The Dowtherm Handbook*, Dow Chemical Company, Midland, Mich., 1960. Geiringer, *Handbook of Heat Transfer Media*, Reinhold, New York, 1962. "High Temperature Heating Media," *Chem. Eng. Prog.*, **59**(5), 33–53 (1963). Keenan and Keyes, *Thermodynamic Properties of Steam*, Wiley, New York, 1956. Kent, *Mechanical Engineers' Handbook*, vol. 2: *Power*, 12th ed., Wiley, New York, 1950. Kern, *Process Heat Transfer*, McGraw-Hill, New York, 1950. McAdams, *Heat Transmission*, 3d ed., McGraw-Hill, New York, 1954. Marks, *Mechanical Engineers' Handbook*, 7th ed., McGraw-Hill, New York, 1950.

Liquids and **vapors** used to transport heat and give it up to process require a unique combination of properties. High boiling points and low pressures reduce hazards and permit economical designs, while high heat capacity improves the ability to carry heat and spread it evenly through the process system. The fluids must also be commercially available and economical to use.

Solid materials are also used for heat transport. These materials have never reached the importance of liquids and vapors, primarily because of the difficulty of transporting the solid particles through small pipes and bent tubes. Pebbles, sand, and iron balls can be heated to higher temperatures than liquids and have found application in air and gas heaters for high-temperature equipment.

The commonly used heat-transport fluids are listed in Table 9-52, along with their useful temperature ranges and corresponding pressure ranges.

Historically, the **open flame** is the oldest source of heat known to man. Basic simplicity and low equipment cost have enabled it to survive in many industrial processes. The open flame as a heat source for industrial-product heating has several limitations, however. These include low fuel efficiency, minimum control over uniformity of heat absorption, and correspondingly poor control over local temperatures in the product material. The burners, fuel supply, and controls must be at the operating point, making the units unwieldy and not suitable to locations of advantage to the manufacturing operations. As a result of these limitations, fluids adaptable to much closer control long ago supplanted flame for process work.

TABLE 9-52 Commonly Used Heat-Transport Fluids*

Fluid	Temp., °F.	Pressure, lb./sq. in. gage
Steam	200–1100	0–4500
Water	300–400	90–230
Dowtherm A	450–750	0–145
Dowtherm E	300–500	0–72
Oil	30–600	0
Molten salts	290–1100	0
Silicon compounds	100–700	0
Chlorinated biphenyls	0–600	0
NaK	100–1400	0
Mercury	600–1000	0–180
Flue gas or air	30–2000	0–100

*See Table 9-54 for more detailed data. To convert pounds per square inch gauge to kilopascals, multiply by 0.15; °C = (°F − 32) × 5/9.

STEAM SYSTEMS

System Characteristics To utilize the energy that is stored as chemical energy in fuels, the chemical energy must be converted to a more usable form, generally either electrical energy or mechanical energy, to drive machines, or thermal energy, as a source of process heat. Although some fuel energy is converted directly to mechanical energy (as in combustion engines) or to thermal energy (direct-process heating), the most popular systems involve the use of steam as an intermediate medium. Steam is the most widely used heat-transport fluid owing to its nontoxic nature, stability, low costs, and high heat capacity. It has limitations, too: its high vapor pressure and its low critical point.

Most commercial sytems start with energy in the form of chemical energy in fuel. Combustion of the fuel in the furnace converts the chemical energy to thermal energy in the form of high-temperature combustion products. In a boiler, heat is transferred from the combustion products to water, thus producing steam. The steam may then be used for driving machinery (including turbines) or for heating (process or space). The use of steam can be via three types of systems:

1. Steam-electric systems (in which nearly all the steam is used to drive a turbine-generator)
2. Cogeneration systems (in which steam is used both to drive a turbine-generator and for heating, usually in series)
3. Steam systems (in which steam is used as a heat-transfer medium for heating purposes only)

Steam-electric and steam systems are discussed in the following paragraphs. Cogeneration systems are dealt with in another part of this subsection.

Steam-Electric Systems Steam-electric systems utilize superheated steam to drive turbine-generators to produce electrical energy. Unless special circumstances exist, economics are unfavorable for small- to moderate-size process plants to consider steam-electric systems. The unfavorable economics include the high capital cost of the furnace-boiler system, the generally lower operating efficiency of small boiler systems, high operating and maintenance costs, and the cost of capital equipment for meeting air-pollution-emissions regulations for the combustion process.

Special circumstances that can justify the operation of small steam-electric systems include the availability of a low-cost fuel, such as a waste or by-product fuel. Industries can make use of such by-product fuels as bark, black liquor, bagasse, sawdust, blast-furnace gas and coke-oven gas, and fluid coke, either as main fuels or in combination with conventional fuels such as coal, oil, or gas.

Cogeneration systems greatly increase the opportunities for a process plant to generate electric power economically.

Steam Systems* Steam systems include those systems in which steam is generated and used for process and/or space heating and, less frequently, for the direct driving of machinery. Although industrial steam can be generated at high pressure (up to 12.4 MPa) and temperature, most industrial steam systems are limited to about 1.724-MPa (250-lbf/in^2) steam pressure and 204°C (400°F) steam temperature. In an industrial steam system, the major component is the furnace-boiler unit; another major component is the condenser, if one is used.

Furnace-boiler units are constructed as integrated units; design varies with size and fuel. Small gas- and oil-fired boilers, with capacities up to 31.7-Mg (70,000-lb) steam per hour, are usually fire-tube boilers. Most solid-fuel-fired boilers and larger gas- and oil-fired boilers are water-tube design. Water-tube boilers for industrial steam generation are of drum-type design. Water-tube boilers may be shop-assembled packaged boilers or field-erected. Gas- and oil-fired boilers of capacities to 113.4-Mg (250,000-lb) steam/h can be installed as package units (to 550,000 lb/h if shipment is possible). Coal-fired boilers with capacities above about 11.3-Mg (25,000-lb) steam per hour are partially or totally field-erected. Because of the high cost of field erection, it may be more economical to install several smaller boilers than one large boiler. Also, because the minimum firing rate of boilers is about one-third of the design firing rate, the use of several smaller boilers provides a much higher turndown ratio.

A typical steam-plant cycle for generation of process steam and electric power is shown in Fig. 9-80*c*. This is the cycle of an industrial power plant that requires the delivery of 5000 kW of electric energy and 22.7 Mg/h (50,000 lb/h) of dry saturated process steam at 137.9 kPa absolute (20 psia). Because all power cannot here be generated as a by-product of the process steam flow, a condensing element is added to the turbine and operated at 2.54 cmHg (1 inHg) absolute pressure. The heat balance of Fig. 9-80*c* is for one set of load conditions. Careful analysis of load curves is necessary for the selection of the most economical cycle.

Most industrial steam systems utilize saturated steam; thus, superheaters are rarely used. Also, when low steam temperatures are used, there is no need for an economizer or an air preheater. However, pulverized-coal firing requires air preheat to achieve ignition and combustion stability.

The cost of coal-fired boilers is about 4 times the cost of packaged gas- and oil-fired boilers. Modern industrial boilers operate at efficiencies of about 85 percent.

Including a condenser in the steam system provides opportunities for reusing the clean steam condensate, as opposed to providing clean boiler water continuously. The choice of whether or not to include a condenser in the system will vary with plant design and size, water availability, and water quality. The cost of return-water piping and a condenser becomes less attractive for widely dispersed plant facilities and as steam use decreases. Conversely, high steam (and, therefore, water) use and limited water availability may make a condenser attractive.

Factors that must be considered when selecting an industrial boiler include:

- Fuels, including projected costs and availability
- Steam requirements: temperature, rate of delivery, and pressure
- Load profile; turndown
- Boiler feedwater: source and treatment
- Space requirements, including fuel storage
- Air pollution emissions and regulations
- Energy to drive auxiliaries
- Operating personnel required

Water Constituents Water, as the working fluid of steam systems, is one of the most widely dispersed natural substances but is never found in a pure state, suitable for direct feed to a boiler. Water in its natural state is usually turbid with solid matter in fine suspension. Even when clear, natural water contains solutions of salts and acids that will quickly damage steel or copper-bearing metals in steam systems. **Recycling** steam condensate from process heating is desirable to take advantage of the relatively pure condensate. Because of atmospheric dissipation and contamination from process equipment, some raw makeup is constantly required.

Various **constituents** in waters may be classed in accordance with the troubles that may result from their presence:

1. Corrosive substances
2. Scale-forming substances
3. Foam-producing substances

Corrosive substances are usually in the form of acid solutions or as dissolved gases such as carbon dioxide, oxygen, hydrogen sulfide, or ammonia. Oxygen and carbon dioxide are dissolved in the feedwater by aeration and unavoidable contact with the atmosphere. Since the solubility of oxygen decreases with an increase in water temperature, the most common method of removal is the deaeration of water, in which the water is heated to the boiling point by direct contact with steam and the heated water is allowed to cascade over trays. The trays increase the exposed surface and permit easier dissipation of the oxygen. Deaeration is also effective in removing other dissolved gases, and all modern steam systems use any of several types of deaerators. In addition to deaeration, use is made of chemicals, such as sodium sulfite, which combines with oxygen and is introduced into the boiler with a chemical feed pump. At higher boiler pressures sodium sulfite is less desirable, because of an increase in the dissolved solids produced by the end-product sodium sulfate and a decomposition into sulfur dioxide and hydrogen sulfide which contribute to corrosion. Hydrazine removes dissolved oxygen without increasing dissolved solids at high pressures with the following reaction:

$$N_2H_4 + O_2 \rightarrow 2H_2O + N_2$$

High residuals of hydrazine in the water must be avoided to prevent the decomposition-product ammonia from attacking copper-bearing alloys in the system.

Reused water that is high in acidity must be treated to maintain a proper alkaline environment in which the pH is between 10.5 and 11.0. Bicarbonate alkalinity in the boiler can hydrolyze under the action of heat, and liberate carbon dioxide, which will be carried along with the steam to form a corrosive carbonic acid product with the condensate in process heat exchangers or condensate piping. Present-day practice calls for water treatment to prevent corrosion in the recycling system by means of neutralizing or filming amines. Neutralizing amines combine with CO_2 and neutralize its acidity. Filming amines do not combine chemically but act by forming an impervious, nonwettable film on metal surfaces which acts as a barrier between metal and condensate, preventing both oxygen and carbon dioxide attack.

Steam systems in which the bulk of condensate is unrecoverable are more often subject to difficulties caused by **scale-forming** substances. The makeup water invariably has constituents which will be scale-forming when present in the water in concentrations in excess of their solubility. Some materials exhibit a decrease in solubility with an increase in temperature, and the scales commonly deposited in boilers belong to this class. Chemical treatment in the preboiler system is successful in reducing most scale-forming substances to a soft sludge, which is removed before it can enter the boiler, while sludges formed by internal treatment may be collected in quiescent zones of the boiler and removed through blowdown pipes.

Water recovered from process heating causes **foaming** within the boiler from the presence of organic, inorganic, or insoluble materials, when they are present in sufficiently large quantities. Oil and the products of decomposition of sewage and humic matter are the chief causes of foaming, and these products should be strictly excluded from condensate returns. The foaming effect of these materials is especially true in the presence of high alkalinities, and the alkaline concentration of boiler water should be limited for this reason.

Table 9-53 summarizes recommended limits for impurities in water used in boilers. Continuous or intermittent blowdown of the boiler water to keep concentrations below recommended limits is the most effective way of preventing foaming. When the blowdown

*See "Steam Generators" for further details on equipment.

TABLE 9-53 Recommended Limits for Boiler-Water Constituents

Parts per million

Pressure, lb./sq. in.	Total dis-solved solids	Alka-linity	Hard-ness	Silica	Tur-bidity	Oil	Phos-phate residual
0–300	3500	700	0	100–60	175	7	140
301–450	3000	600	0	60–45	150	7	120
451–600	2500	500	0	45–35	125	7	100
601–750	2000	400	0	35–25	100	7	80
751–900	1500	300	0	25–15	75	7	60
901–1000	1250	250	0	15–12	63	7	50
1001–1500	1000	200	0	12–2	50	7	40

NOTE: To convert parts per million (by weight) to micrograms per liter, multiply by 1/198; to convert pounds per square inch to kilopascals, multiply by 6.895.

results in large quantities of heat being lost to the system, much of this heat may be recovered by passing the blowdown through heat exchangers used to heat feedwater or air, which returns some of the heat to the unit.

THERMAL-LIQUID SYSTEMS

Liquids and Their Properties Thermal liquids used for process heating and cooling may be in the form of liquids, vapors, or a combination of both. In addition to steam, thermal liquids include hot water, mercury, NaK, diphenyldiphenyl oxide (Dowtherm A), *o*-dichlorobenzene (Dowtherm E), molten salt mixtures, mineral oils, arylaryloxysilane (Hydrotherm 750–200), tetraaryl silicate (Hydrotherm 700–160), and chlorinated biphenyls (Therminols). **Physical properties** of these materials are given in Table 9-54.

High-temperature hot water is a favorable system for process temperatures of 149 to 204°C (300 to 400°F) but requires pump pressures greater than the saturation pressure of 17.5 kg/cm^2 (250 lbf/in^2) to maintain the water in liquid form. Hot-water systems are stable, with simple equipment, and easy to control. Corrosion is at a minimum when air is excluded from the system. Deleterious solids in the water do not build up to high concentrations in the absence of evaporation, and scale formation and foaming are at a minimum.

Dowtherm A (Dow Chemical Company) presently dominates the 204 to 399°C (400 to 750°F) field of indirect process heating. This fluid is an organic compound of high heat stability, a eutectic mixture containing 73.5 percent diphenyl oxide and 26.5 percent diphenyl by weight. At its freezing point of 12.2°C (54°F) Dowtherm A contracts slightly, thereby removing the possibility of damage to process equipment when shut down under cold-weather conditions. At room temperatures it is a clear, almost colorless liquid, which darkens rapidly in use without change in physical characteristics, and has a characteristic rose-geranium odor. It does not react chemically with metals commonly used in heat-transport systems and is not toxic to humans, presenting no appreciable hazard to health in heat-transfer use and requiring no special precautions.

Dowtherm A is quite stable at moderately high temperatures. Many vaporizers and accessories have operated for years at Dowtherm temperatures of 343°C (650°F) with no decomposition. At higher temperatures **decomposition** may occur in two ways. Above about 399°C (750°F) two molecules of diphenyl may react to yield one molecule of *p*-diphenylbenzene and one of benzene. The *p*-diphenylbenzene dissolves in Dowtherm A, but the benzene, being a noncondensing vapor in practical Dowtherm heating installations, escapes into vent pipes. There is a similar reaction in the case of diphenyl oxide.

The second form of decomposition is more troublesome. When Dowtherm A is severely overheated, such as by flame impingement on the tubes of a vaporizer or by forcing the heater beyond its rated capacity, complete decomposition into carbon and hydrogen may take place. The formulation of carbon occurs when inadequate circulation caused by the accumulation of materials holds the Dowtherm in a stagnated condition. The lighter fractions then distill off, leaving behind the higher-boiling-point fractions, which carbonize. When this begins, the carbon forms a skin on the heating surface of the vaporizer, and this increases the thermal resistance so that decomposition is greatly accelerated. In this manner, a Dowtherm vaporizer may be filled completely with carbon in a few hours.

To prevent overheating and decomposition, Dowtherm vaporizers are of liquid-tube or fire-tube natural-circulation design or liquid-tube forced-circulation design. Small laboratory vaporizers are frequently heated by electricity. A natural-circulation vaporizer as in Fig. 9-81 is arranged with few bends or restrictions to allow fast recirculation of the Dowtherm liquid, and is designed with ample furnace capacity. Good flame-shape control prevents hot spots from forming along the vaporizer tubes.

Dowtherm E, a specially processed *o*-dichlorobenzene which boils at 177°C (350°F) and has a freezing point below zero, is commonly used between 177 and 260°C (350 and 500°F). No trace of decomposition of Dowtherm E has been noted in tests conducted at temperatures and heat loads considerably higher than those recommended for commercial installations. There is some evidence that aluminum can catalyze the decomposition of Dowtherm E to form hydrochloric acid. This acid is likely to corrode severely the polished surfaces of precision tools, machines, and sheet or formed metals. Aluminum should not be used with Dowtherm E.

Inorganic Salts Molten mixtures of inorganic salts, one of which is a eutectic consisting of 40 percent $NaNO_2$, 7 percent $NaNO_3$, and 53 percent KNO_3, are widely used in salt baths and petroleum refining when high temperatures are maintained and when the system is kept in continuous operation. A melting point of 146°C (288°F) precludes their use in low-temperature systems and requires that the circulating fluid be kept hot and molten throughout its flow path. The salt mixture need not be pressurized higher than required to overcome the pressure drop of the system.

Inorganic-salt mixtures are nontoxic and chemically stable up to 427 to 454°C (800 to 850°F) in the absence of contaminants. Between 454 and 593°C (850 and 1100°F), which is the maximum operating temperature, the salt undergoes a slow thermal decomposition, which is largely a thermal breakdown of the nitrite to nitrate, alkali metal oxide, and nitrogen:

$$5NaNO_2 \rightarrow 3NaNO_3 + Na_2O + N_2$$

This reaction is evidenced by the slow evolution of nitrogen gas, and it is accompanied by a gradual rise in the freezing point of the mixture [Alexander and Hindin, *Ind. Eng. Chem.*, **39**, 1044 (1947)].

The nitrite is also subject to slow oxidation by atmospheric oxygen above 454°C (850°F) with formation of sodium nitrate. This reaction is eliminated by excluding air or blanketing the salt with an atmosphere of inert gas such as nitrogen.

Other reactions will gradually alter the composition of the salt: (1) absorption of carbon dioxide to form carbonates which may settle out in the system; and (2) absorption of water vapor to form alkali metal hydroxides. These reactions do not interfere with process operation but if allowed to continue will ultimately affect utility of the system. They may be eliminated by blanketing the molten salt with nitrogen.

Ordinary carbon steel may be used successfully in molten-salt equipment up to 454°C (850°F). Above this temperature more resistant alloys are recommended. Copper equipment has been used satisfactorily at moderate temperatures, 316°C (600°F), but cast iron is not recommended because of a reaction between the molten salt and the iron which results in embrittlement or fissuring.

Mineral Oils Conventional mineral oils are of low cost and readily available for process use and are valuable in systems operating from −1.1 to 316°C (30 to 600°F). They need not be pressurized in this range. The paraffinic-type cylinder oils are often employed in open systems to about 232°C (450°F), such as are used in tempering metal. At higher temperatures the conventional mineral oils are used in closed systems, up to 316°C (600°F), in which temperature region the oils become susceptible to thermal cracking. This decomposition, similar to the controlled cracking process used to produce gasoline from heavy oils, is not nearly so severe as the cracking that occurs in petroleum refining but will produce volatile materials that reduce

TABLE 9-54 Physical Properties of Thermal Fluids

Property	Water	Dowtherm A*	Dowtherm E*	Fused Salt Hi Tec†	Oil Mobiltherm 600‡	Oil Mobiltherm light‡	Hydrotherm§ 750-200	Hydrotherm§ 700-160	Therminol¶ FR-2	Mercury	NaK
Chemical formula	H_2O	$(C_6H_5)_2O$ $(C_6H_5)_2$	$C_6H_4Cl_2$ ...	$NaNO_2$ $NaNO_3$ KNO_3	...	...	...	...	...	Hg	44 wt.% K
Molecular weight	18	165	147	92	...	...	...	...	...	200	
Specific gravity at 212°F.	0.958	0.997	1.181	1.98(300°F.)	0.90	0.930	1.11	1.08	1.38	13.35	0.84(600°F.)
Melting point, °F.	32	53.6	−6.7	288	20 (pour point)	−20 (pour point)	5 (pour point)	−40 (pour point)	20 (pour point)	−38.2	65
Boiling point, °F. (atm. pressure)	212	495.8	352	...	>600	>400	...	...	644	674.4	1518
Flash point, COC, °F.	...	255	155	...	360	250	475	...	379		
Specific heat of liquid, B.t.u./(lb.)(°F.)	1.005(212°F.)	0.526(496°F.)	0.412(352°F.)	0.373(300°F.)	0.580(500°F.)	0.53(300°F.)	0.56(600°F.)	0.64(500°F.)	0.333(500°F.)	0.033(212°F.)	0.25(600°F.)
Heat of vaporization, B.t.u./lb.	970.2	125.0	119.0	...	...	...	...	...	...	117.0	
Heat of fusion, B.t.u./lb.	143.3	64	38	35	...	...	...	...	...	5.1	
Cubical expansion coefficient	0.0024	0.00043	...	0.00020	0.00035	0.00035	...	...	0.00039	0.000101	
Absolute viscosity of liquid, centipoise	0.284(212°F.)	0.30(600°F.)	0.30(400°F.)	1.7(400°F.)	0.595(500°F.)	0.873(300°F.)	0.572(600°F.)	0.605(500°F.)	0.63(500°F.)	1.23(200°F.)	0.24(600°F.)
Surface tension (contact with air), dynes/cm.	72.8	43	37	...	...	...	37	...	...	487	105
Thermal conductivity liquid, B.t.u./(hr.)(sq. ft.)(°F./ft.)	0.393	0.076	0.064	0.35	0.067	0.0652	0.0590	0.072	0.057	4.85	15.6

*The Dow Chemical Company.

†E. I. du Pont de Nemours & Co., Explosives Department, Wilmington, Del.

‡Mobil Oil Corp.

§American Hydrotherm Corp.

¶Monsanto Co.

NOTE: To convert British thermal units per pound–degree Fahrenheit to joules per kilogram-kelvin, multiply by 4.187×10^3; to convert atmospheres to kilograms per square centimeter, multiply by 1.0333; to convert British thermal units per pound to kilojoules per kilogram, multiply by 2.326; to convert centipoises to grams per centimeter-second, multiply by 0.01; and to convert British thermal units per hour–square feet–degrees Fahrenheit per foot to watts per square meter–kelvin, multiply by 5.678. $°C = (°F - 32) \times 5/9$.

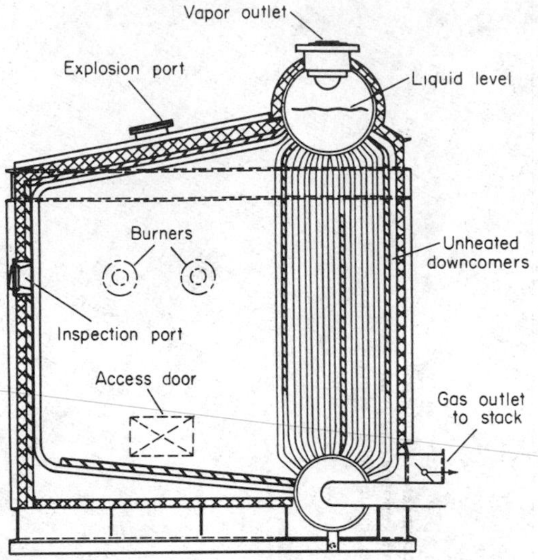

FIG. 9-81 Dowtherm vaporizer. *(Foster Wheeler Corp.)*

the flash point of the oil. At the same time, the decomposition yields heavier products, which after long periods of operation flow less readily, leading to formation of coke deposits.

Mobiltherm 600 and Mobiltherm Light (Mobil Oil Corp.) are **aromatic mineral oils** of lower viscosity than conventional mineral oils and can operate between −1.1 and 316°C (30 and 600°F) for the former and at −25°C (−15°F) but not above (400°F) for the latter. They will not be broken down by temperature when used in the recommended range. Their flash points thus remain unchanged after many hours of service life. When the aromatic oils are subjected to excessive temperatures, thermal cracking will occur in a form similar to the decomposition of conventional mineral oils. Sludge and coke deposits do not readily occur with such aromatic mineral oils, as they have a powerful solvent action, and some installations have operated for years without changing oil or cleaning the system.

Neither aromatic mineral oil is suitable for operation in an open system, in direct contact with air. Oxidation results in deterioration of the oil, a chemical breakdown that is accelerated when the oil is at an elevated temperature. All process systems must include a "cold-oil" expansion tank, in which the temperature of the oil will not exceed 54°C (130°F). The expansion tank prevents the hot oil of the process system from coming into contact with air. The aromatic oils should not be used with copper or copper-bearing-alloy parts, since these metals are powerful catalysts which promote oxidation and sludging. Iron and carbon are preferred for the entire system. Other oils, whether lubricating or mineral oils, should not come into contact with aromatic oils, as this causes the solvent power of the oil to be lowered and may result in harmful sludge being deposited in the system.

Silicon Compounds Three types of silicon compounds are in use as heat-transfer media: silanes, silicones, and silicates. Silanes are substituted hydrides of silicon; silane itself has the form SiH_4. The silicone fluids in use are usually polymers. The basic structure is of the form Si-O-Si, with organic radicals attached directly to the silicon atoms. Silicates are the salts or esters of the silicon acid in which the central atom is silicon. The organic heat-transfer fluids are generally esters. The organic constituents usually comprise alkyl, aryl, or alkaryl groups.

Hydrotherm 750-200 is an arylaryloxy- (mixed) silane used at temperatures from 63 to 371°C (145 to 700°F). It is a transparent amber fluid with a slightly phenolic odor. There appear to be no chronic or other pathological effects from exposure to vapors; skin tests show dermatitic effects similar to those of phenol. The substance is lethal on injection or ingestion. There is negligible corrosion in ferrous metals, copper, and copper alloys. There is a slight attack on aluminum and magnesium metals.

Hydrotherm organosilicate heat-transfer fluids are applicable for heating and cooling operations from −4.56 to 357°C (−50 to 675°F). Since these fluids do not freeze and stay pumpable below −17.8°C (0°F), neither reheating nor steam tracing is required. Hydrotherm silicate heat-transfer fluids are noncorrosive toward mild steel and copper and its alloys, even at high temperatures. However, magnesium, aluminum, zinc, titanium, and their alloys are attacked. Organosilicate liquids do not present a potentially serious fire hazard. **Hydrotherm 700-160,** a tetraaryl silicate, is an amber-colored liquid with a slight phenolic odor that has good thermal and chemical stability. It does react with oxygen, and so contact with the atmosphere should be avoided. In the presence of moisture hydrolysis takes place; free phenols are released and a silicate sludge is deposited. Thermally, a re-forming reaction occurs at elevated temperatures, above 357°C (675°F). Low-boiling products, primarily monohydric or dihydric phenols, are formed and a gradual increase in viscosity occurs.

Therminol FR heat-transfer liquids are a series of chlorinated biphenyls of increasing chlorine content. Many important physical characteristics such as viscosity, density, and pour point vary from lowest- to highest-molecular-weight members of the series. One reason for their success as a heat-transfer medium is that they do not support combustion. Their spontaneous ignition temperature is about 649°C (1200°F). Therefore they can be considered fire-resistant. These compounds are relatively inert liquids, resistant to the action of water, dilute alkalies, dilute acid solutions, air, and oxygen. They are readily soluble in most of the common organic solvents and drying oils. All these fluids are stable below 316°C (600°F). At higher temperatures thermal decomposition takes place, higher polyphenyls are formed, gaseous HCl is liberated, and viscosity increases. The vapors at high temperatures are at a level of toxicity such that adequate provision must be made for ventilation when leakage occurs, and system vents should be remote from normal operating areas. Although the compounds are not severe skin irritants, they should not be in repeated contact with the skin. Because of its low pour point, **Therminol-FR-2** is widely used at temperatures between 10 and 316°C (50 and 600°F).

Mercury is useful in thermal-liquid systems. Its stability as an element makes it suitable for high temperatures, in the range of 316 to 538°C (600 to 1000°F). Experimental work shows that it has no corrosive effect on metals commonly used in practice. Mercury is toxic to humans, and mercury systems must include elaborate precautions to prevent the escape of mercury vapor to the surrounding atmosphere. The low latent heat of vaporization of mercury makes it unattractive to use as a vapor at low temperatures when heat must be given up during condensation. The high cost of mercury, in comparison with other thermal liquids commercially available, precludes it use at temperatures below 316°C (600°F).

Sodium-potassium alloys are used for heat transfer instead of the pure elements because these mixtures have lower melting points. The two most popular alloys are, by mass percent, 56 Na-44 K and 22 Na-78 K. The latter is approximately the eutectic composition. In comparison with the pure elements, NaK alloys have a lower thermal conductivity, whereas vapor pressure, density, specific heat, and viscosity fall between values of the two elements. The chemical properties of NaK alloys are almost identical to those of sodium; however, the alloys are more reactive. Increased chemical activity can be ascribed to the presence of potassium, which has a greater reactivity than sodium. Upon contact with the atmosphere, potassium is oxidized to the superoxide, KO_2; and explosions have been caused by reaction of the superoxide with hydrocarbon gases used as cleaning media. The use of hydrocarbons in NaK systems should be discouraged. Experiments at 204°C (400°F) have revealed that sodium monoxide, rather than potassium superoxide, precipitates from NaK alloys. Other reactions of potassium not duplicated by sodium are an attack on silicon and the formation of an explosive carbonyl on contact with carbon monoxide. Carbon steel and low-alloy steels show good resistance to attack up to 538°C (1000°F), and they can supplant more expensive stainless-steel alloys. Oxygen impurities above

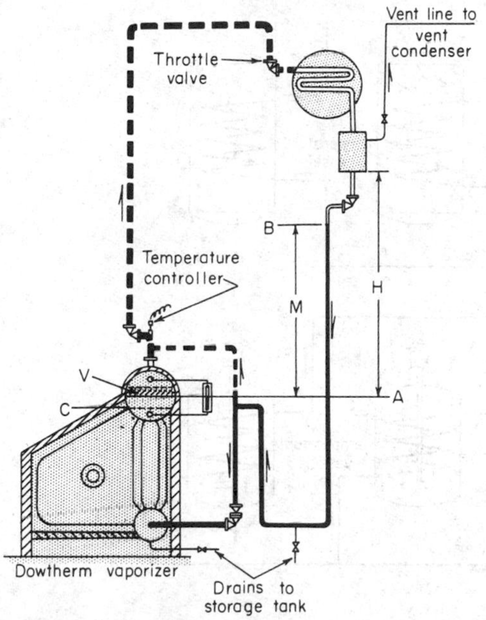

FIG. 9-82 Dowtherm gravity-return system. V, C = maximum and minimum liquid levels in vaporizer drum; M, H = range of liquid heights in condensate leg between A and B to cause flow through system. Dotted lines show the vapor path; solid lines, the liquid path. *(Foster Wheeler Corp.)*

100 ppm markedly accelerate corrosion. Alkali metals react with the skin and eyes, causing thermal and alkali burns. Oxide smoke or hydroxide mist is irritating and corrosive to the throat and lungs.

Toxicity Many instances of potential hazards have been described in this discussion of heat-transfer fluids, and this paragraph is intended to emphasize the problems. Toxicity and ecology are extremely important from both an operating and a process standpoint. There is always a chance that a heat-transfer fluid will leak from a system, e.g., packing glands on valves, pumps, and heat exchangers. If this happens, operators, maintenance personnel, and the environment in general may be overexposed to fluids known to be hazardous. Polychlorinated biphenyls (PCBs), especially, must be used with great caution.

Process Systems

Dowtherm Dowtherm process sytems are either gravity-return or pumped-return. The most desirable system is the gravity system, in which the vapor rises from the vaporizer to the heated vessels, condenses, and flows back to the vaporizer by gravity. No moving parts are required. It is essential that the gravity-return-system piping be suitably proportioned to the limitations of the headroom available between the bottom of the heated vessel and the liquid level in the vaporizer. This involves the frictional loss in the vapor piping, the user, and the condensate return. Figure 9-82 is a **gravity-return system,** and Fig. 9-83 shows the **pumped-return system.** Pumps should generally be of cast-steel construction with a deep water-cooled stuffing box designed for metallic-foil packing.

Transport of heat at **two different temperatures** may be accomplished with Dowtherm in a single vaporizer by supplying the higher-temperature process with vapor and the lower-temperature process with Dowtherm liquid. Close control of the vapor temperature is achieved by maintaining the pressure in the vaporizer, while the liquid temperature is controlled by circulating only part of the process returns through the heating unit. In such an installation, illustrated in Fig. 9-84, the Dowtherm vaporizer provides vapor at the desired temperature for two high-temperature users. A liquid-Dowtherm circulating system heats the low-temperature users. In this system, the hot liquid is withdrawn from the vaporizer and returned to it after passing through the heating elements of the vessels. A three-way valve, which divides the return-liquid flow, provides automatic control of the liquid circuit. Part of this flow returns to the vaporizer to be reheated, while the remainder bypasses the vaporizer and flows directly to the circulating pump section. The amount of heat put into the liquid system is controlled by varying that part of the flow returning to the vaporizer.

As Dowtherms have extremely low surface tension and viscosity at high temperature, more than ordinary care is necessary in the fabrication and erection of equipment to **prevent leakage.** There is ordinarily more or less evidence of leakage at pump stuffing boxes, relief-valve outlets, etc. Welded construction in accordance with ASME specifications is advisable whenever possible. The wide range of temperature requires adequate provision for expansion of piping. The high temperature renders ordinary relief-valve springs unsafe, and only special tungsten-steel-alloy relief-valve springs are suitable. Other relief-valve parts and other fittings should be steel rather than brass or bronze.

The condensate line of a gravity-return system should include the so-called **Hartford loop.** This consists of a line connecting the lower vaporizer drum and the vapor space above the upper drum. The con-

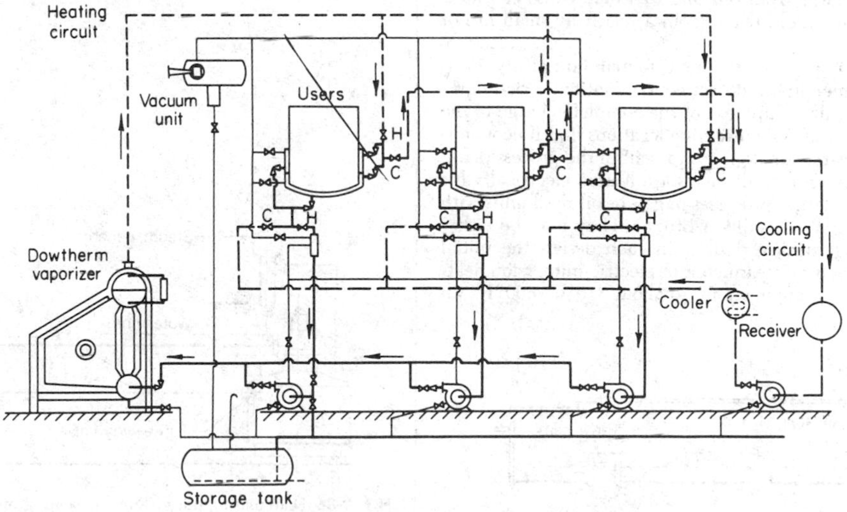

FIG. 9-83 Dowtherm pumped-return system. *(Foster Wheeler Corp.)*

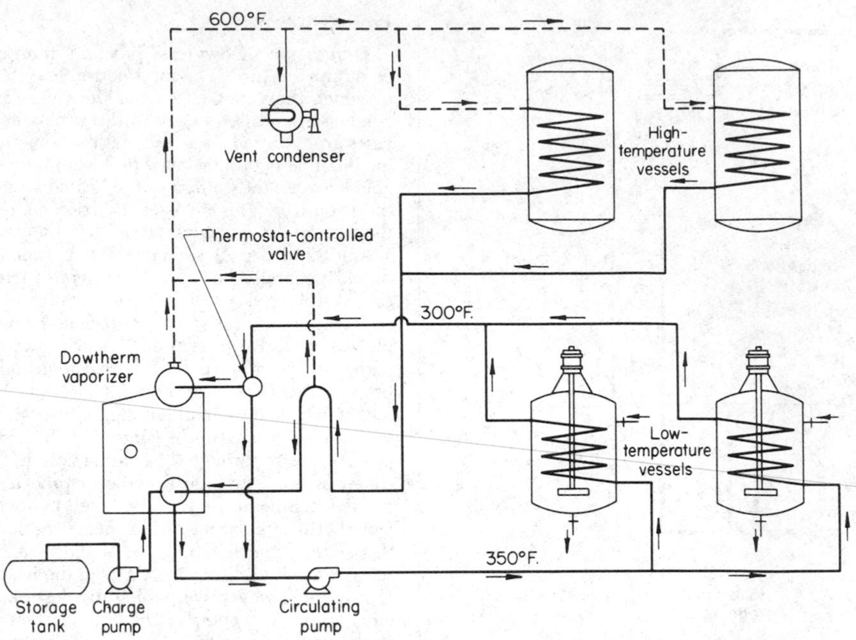

FIG. 9-84 Combination vapor and liquid heating with Dowtherm. *(Foster Wheeler Corp.)*

densate return is brought into this line at a point just above the top of the tubes in the vaporizer. With this loop it is impossible to draw liquid out of the vaporizer up through the condensate line after the liquid level in the vaporizer falls below the point at which the condensate line enters this connection. If the liquid level is drawn lower than this point, vapor will be drawn back into the condensate line, and the resulting water-hammer effect will give warning that the liquid level is too low.

Other recommended **safety features** include installation of a storage tank of sufficient capacity to contain the entire system charge. It should be buried or otherwise located where not exposed to fire. The drain valve from the vaporizer should be accessible so that, in the event of an uncontrollable vaporizer fire from a tube leak, the liquid in the vaporizer can be drawn off promptly. Emergency drainage of the vaporizer requires considerable judgment, however, since damage to dry heating surfaces from burning fuel may cause considerably more damage than would result from a relatively small fire or tube leak.

Recycled Dowtherm needs no treatment to maintain purity. Periodic analysis is recommended to detect contamination or deterioration. Repurification requires shipment of the complete charge to the reprocessing plant. For this reason remote locations find it economical to install continuous-reclamation equipment in the process plant.

Inorganic Salts Inorganic-salt mixes are heated electrically for pilot-plant units and for larger processes in gas- or oil-fired units with capacities up to 4396 kW (15 million Btu/h). They may be either fire-tube or circulating furnace design. In each design the initial charge is melted by means of steam coils or electric immersion heaters. For moderate heating applications, to about 316°C (600°F), an

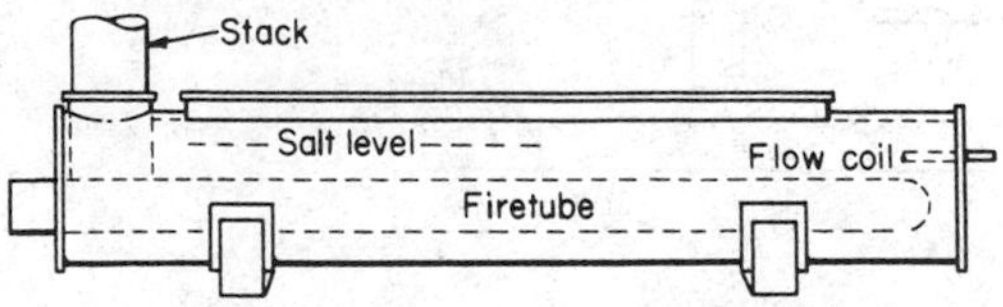

FIG. 9-85 Fire-tube salt-bath heater. *(E. I. du Pont de Nemours & Co.)*

indirect salt-bath heater, including both the molten salt and a flow coil for the product to be heated, is placed in a fire-tube design (Fig. 9-85). Gases passing through the fire tube maintain the salt in a molten state with precise control over the temperature of the product passing through the flow coil. Advantages for this indirect heater are its safety of operation, even heat distribution to the flow coil, high efficiency, and elimination of flow-coil failure because of flame impingement or localized overheating. The maximum size of these heaters of 2344 kW (8 million Btu/h) can be supplemented by installing several heaters in battery fashion.

Heating and cooling are accomplished in one unit, shown in Fig. 9-86, for processes in which reactions taking place at a high temperature level require removal of exothermic heat. Once the salt initiates the exothermic-process reaction, it maintains the reaction temperature by switching the flow of molten salt from the heating unit

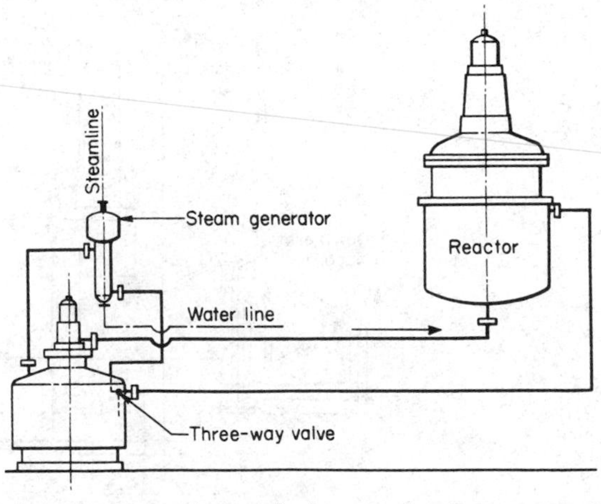

FIG. 9-86 Circulating-salt system for heating and cooling operation. *(Bethlehem Corp.)*

to the steam generator that is mounted integrally with the heating unit. By means of a three-way valve, all or a portion of the salt flow can go to the cooler. Controls on the system maintain temperature levels by forestalling pump operation when the salt temperature falls below the desired level and sounds an alarm if the salt temperature exceeds a safe limit of the process.

Because of the high melting point of the inorganic-salt mix, salt lines and valves must be traced or steam-jacketed to prevent **freezing** by solidified salt, especially for intermittent operation. Submerged centrifugal pumps are used to circulate the salt and are of a type which permits no contact of the salt with the packing gland.

Remelting of salt, or fusing of the initial charge, is done with electric immersion heaters or steam coils that pass below the surface of the bath. Heating a solid bath of salt from the bottom alone can develop sufficient pressure to rupture equipment or expel the molten salt through the solid surface.

Combustible solids, such as wood, coke, paper, plastics, cyanides, chlorates, and ammonium salts, and active metals, such as aluminum and magnesium, are potential **hazards.** Magnesium, except as an alloying agent in low concentration, must not come in contact with the salt mix. The salt itself is not flammable but will support combustion. Water from spray sprinklers or low-velocity fog nozzles is recommended as fire protection.

Mineral Oils Conventional mineral oils are not affected by contact with air at temperatures of about 232°C (450°F), but aromatic mineral oils must always be used in a closed system with a cold-oil expansion tank to prevent the hot oil from contact with air. Figure 9-87 is a layout for indirect heating with hot-oil recirculation.

The **heater** may be direct-fired, with combustion gases passing over the tubes through which the heated oil circulates, or electrically heated, with the oil flowing through narrow channels over the heating source. Other heater designs utilize high-temperature steam passing through heating coils with the circulating oil outside. The heat source must be such that excessive temperatures are avoided. Large heating-surface areas and moderate temperature differences between the oil and the heat source are preferred to minimize thermal breakdown of the aromatic oil. In electric heaters this is achieved with a power input of not more than 1.9 W/cm^2 (12 W/in^2) of heating surface; for oil- or gas-fired radiant heaters the input should not exceed 126 to 158 kW/m^2 [40,000 to 50,000 Btu/(h·ft^2)] of coil surface.

All hot-oil systems use **forced circulation** because of the absence of gravity-head differences between heated and cooled oil. Centrifugal pumps are usually supplied with exterior lubrication of the shaft bearings to prevent lubricating oil from entering the system. The pump should have sufficient capacity to provide a velocity through the heater of not less than 1.2 m/s (4 ft/s) for oil temperaures up to 204°C (400°F) and about 3.05 m/s (10 ft/s) for 204°C (600°F).

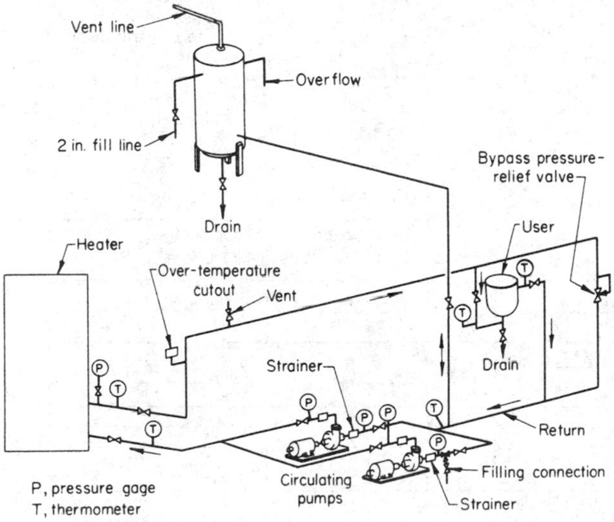

FIG. 9-87 Typical hot-oil circulating system. *(Mobil Oil Corp.)*

A thermostat maintains the desired temperature in the process vessel and acts to cut out burners or electric current in the heaters when oil temperatures rise too high. To prevent constant on-off switching of the heaters a relief valve is supplied to act as a bypass and maintain a constant flow of oil through the pump and heater when the thermostat reduces the flow of hot oil to the processing vessel.

Reuse of aromatic mineral oils is continuous, with losses from the system being replaced by identical oil added to expansion tank or pump suction. Periodic analysis and comparison with initial properties will determine service life.

Therminol Fluids Two basic heater designs for Therminol fluids are available: the liquid-tube types and the fire-tube types. In the former, the Therminol is pumped through the tubes at a definite flow rate as it is heated. In fire-tube types, the fluid flows through the shell. When operating temperatures above 260°C (500°F) are required, a liquid-tube-type heater is preferable unless a specific design is devised to force a steady flow over the heat-source surface. The fluid should be pumped over the heating surface so that no areas of stagnant fluid which might present hot spots occur. The fluid velocities over the heat-source surface should be in in the range of 1.2 to 3.05 m/s (4 to 10 ft/s). When electric heating is used, maximum energy density should not exceed 1.6 W/m^2 (10 W/in^2) at the minimum velocity, although 3.2 W/m^2 (20 W/in^2) can be used with proper design and sufficiently high velocity.

For large flow rates, the fluid circulating pump should generally be the centrifugal type. When gear pumps are used, care should be taken to select a capacity that will give adequate flow through the heater even after wear. If expansion loops are used in the pump suction piping, they should be arranged to prevent trapping air or noncondensables.

Like most organic liquids, Therminols expand in volume by about 4 percent per 37.8°C (100°F) temperature rise. Thus, in heating from room temperature to 316°C (600°F), the fluid in the system expands by about 20 percent. The **expansion tank** must therefore accommodate this increase in total fluid volume; doubling the minimum size is recommended. It should be on the pump suction side, above the highest point in the system, and should be connected to the system by a sufficiently small line to avoid thermal recirculation. Major venting can be done from the expansion tank. The vent should be outdoors, away from working areas. The entry of atmospheric moisture on "breathing" should be prevented by use of an air dryer, such as a calcium chloride pot or other desiccant. An alternative would be to blanket the expansion tank with nitrogen, but this system must be thoroughly dried.

In systems using Therminol above 316°C (600°F), some provision must be made to vent any HCl gas formed in the system and to assure that the system remains dry. It is advisable to provide a pipe to feed a small sidestream of hot Therminol from near the pump discharge to the vented expansion tank, above the highest liquid level. It is advisable to trace and insulate the expansion tank and vent so that any moisture accidentally entering the system will remain as vapor and not condense.

COGENERATION

Definition and General Description Cogeneration is an energy-production process involving the simultaneous generation of thermal (e.g., process-steam) and electric energy by using a single primary heat source. It can be employed whenever there is a need for the two energy forms and whenever on-site electric power generation is justified or when thermal-energy users are in close proximity to an electric-power-generation site.

Industrial use of cogeneration leads to small, dispersed electric-power-generation installations—an alternative to complete reliance on large central power plants. Because of the relatively short distances over which thermal energy can be transported, process-heat generation is characteristically an on-site process, with or without cogeneration.

Fuel saving is the major incentive for the use of cogeneration. Since all heat-engine-based electric power systems reject heat to the environment, that rejected heat can frequently be used to meet all or part of the on-site or local thermal-energy needs. Use of rejected heat usually has no effect on the amount of primary fuel used, yet it leads to a saving in all or part of the fuel that would otherwise be used for the thermal-energy process. Heat engines also require a high-temperature thermal input and in some situations can obtain the input thermal energy as the rejected heat from a higher-temperature thermal-energy process. In the former case, the cogeneration process employs a heat-engine topping cycle; in the latter case, a bottoming cycle is used. The fuel-savings benefits in a topping-cycle configuration are illustrated in Fig. 9-88.

Cogeneration systems can be designed from at least two perspectives: they can be sized to meet the process-heat needs of industrial or institutional users, so that the electric power produced is treated as a by-product which must either be used on site or sold to the local utility; or cogeneration systems can be sized to meet electric power demand, and the rejected heat is then used to supply process-heat needs at or near the site. The latter approach is the likely one if utility ownership is involved.

Furthermore, cogeneration systems will usually not match the varying power and heat demands at all times for most applications. Thus, a cogeneration system's output frequently must be supplemented by the separate on-site generation of heat or the purchase of utility-supplied electric power. Another option is the sale of excess locally generated electric power to the local utility grid—a situation which can occur when the cogeneration system is matched to the heat load and on-site electric power demand is low. These options for on-site power and heat use are also illustrated by Fig. 9-88.

Fuel-use options for cogeneration systems are determined by the primary heat-engine cycle. Closed-cycle power systems which are externally fired, such as the steam turbine, the indirectly fired open-cycle gas turbine, and the closed-cycle gas-turbine power system, can use virtually any fuel which can be burned in a safe and environmentally acceptable manner; fuels such as coal, municipal solid waste, biomass, and industrial waste are burnable with closed power systems. Internal-combustion engines, such as diesel engines and open-cycle (combustion) gas turbines, are restricted to fuels which have combustion characteristics compatible with the engine type and which yield combustion products clean enough to pass through the engine without damaging it; refined liquid and gaseous fuels derived from petroleum, shale, coal, or biomass are included in this category. These heat engines and fuel options are illustrated in Fig. 9-89.

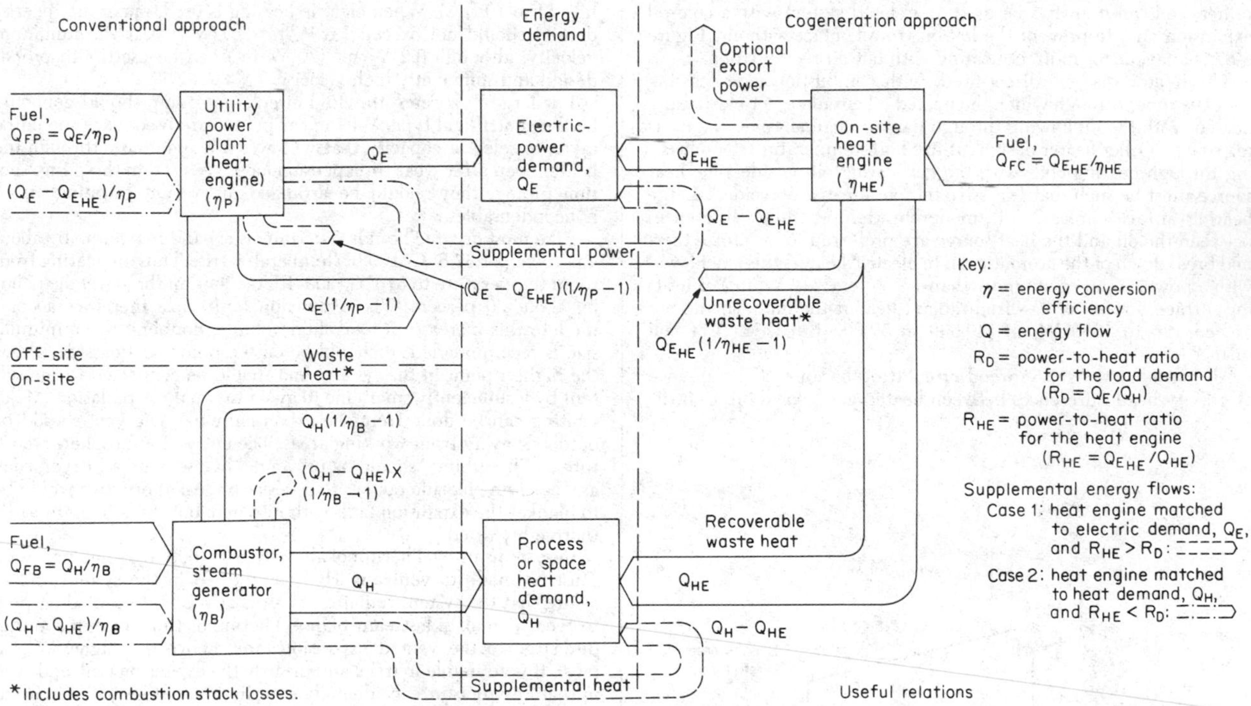

Useful relations

Parameter	Conventional approach	Cogeneration approach (1) $Q_{E_{HE}} = Q_E; R_{HE} > R_D$	Cogeneration approach (2) $Q_{HE} = Q_H; R_{HE} < R_D$
Total fuel used, Q_F	$\frac{Q_E}{\eta_P} + \frac{Q_H}{\eta_B}$	$\frac{Q_E}{\eta_{HE}} + \frac{Q_H - Q_E/R_{HE}}{\eta_B}$	$\frac{Q_H R_{HE}}{\eta_{HE}} + \frac{Q_E - Q_H R_{HE}}{\eta_P}$
Overall efficiency $\eta_{OV} = \left(\frac{Q_E + Q_H}{Q_F}\right)$	$\frac{R_D + 1}{\frac{R_D}{\eta_P} + \frac{1}{\eta_B}}$	$\frac{R_D + 1}{\left(\frac{R_D}{\eta_{HE}} + \frac{1 - R_D/R_{HE}}{\eta_B}\right)}$	$\frac{R_D + 1}{\left(\frac{R_{HE}}{\eta_{HE}} + \frac{R_D - R_{HE}}{\eta_P}\right)}$
Fractional fuel savings, $\Delta Q_F/Q_F$	———	$1 - \frac{\left(\frac{R_D}{\eta_{HE}} + \frac{1 - R_D/R_{HE}}{\eta_B}\right)}{\left(\frac{R_D}{\eta_P} + \frac{1}{\eta_B}\right)}$	$1 - \frac{\left(\frac{R_{HE}}{\eta_{HE}} + \frac{R_D - R_{HE}}{\eta_P}\right)}{\left(\frac{R_D}{\eta_P} + \frac{1}{\eta_B}\right)}$

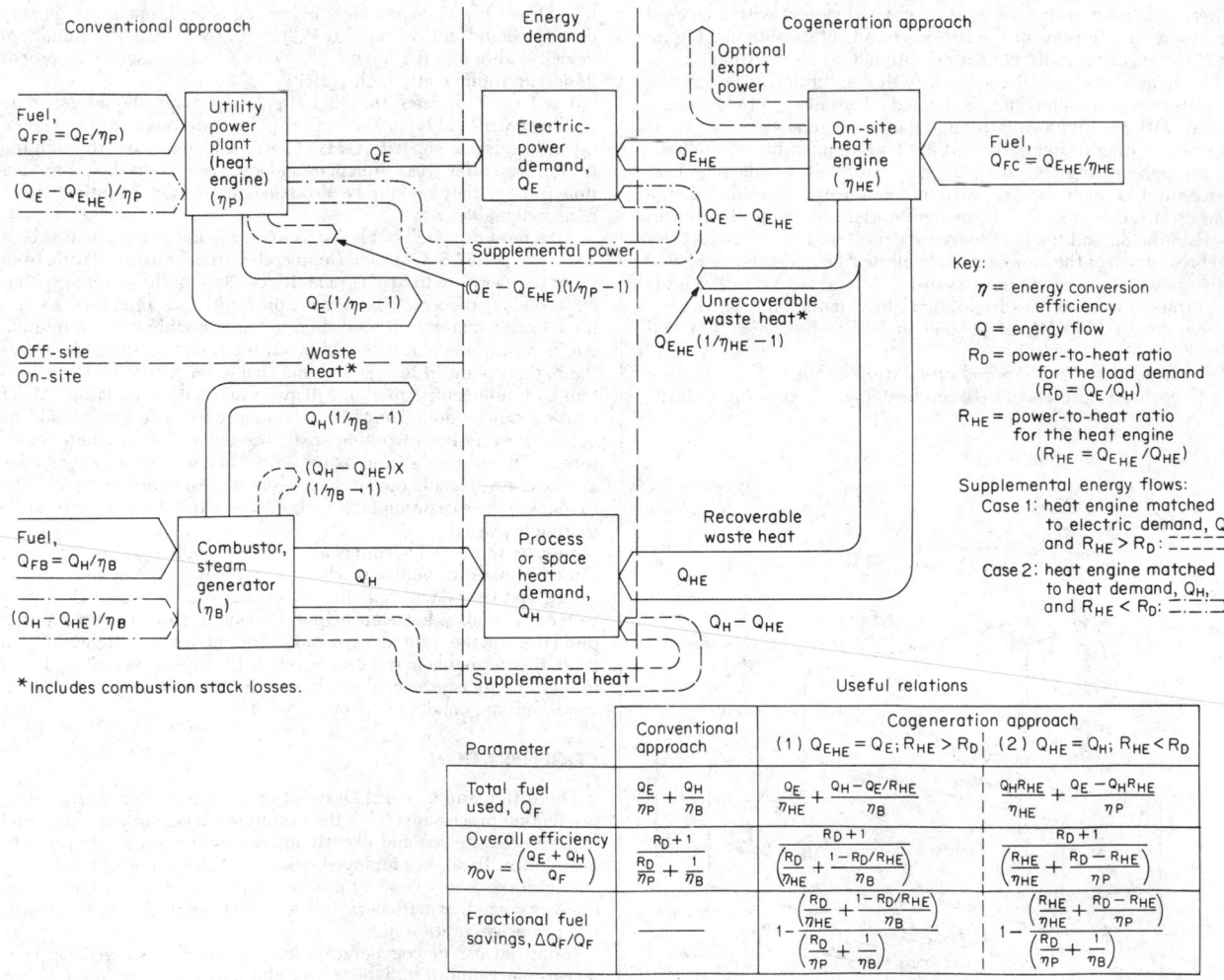

FIG. 9-88 Illustrative fuel savings with cogeneration.

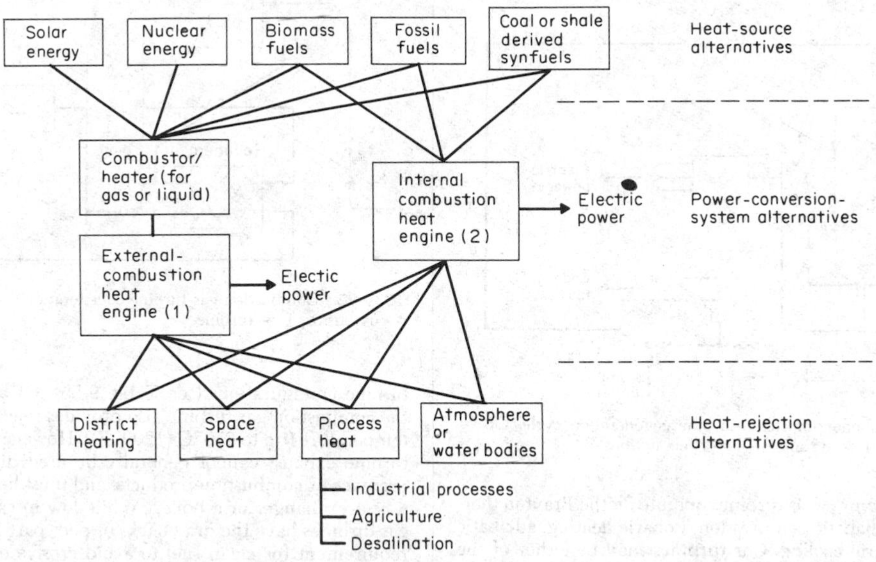

FIG. 9-89 Alternative configurations for cogenerating heat engines.

There are at least three broad classes of application for topping-cycle cogeneration systems:

• Utilities or municipal power systems supplying electric power and low-grade heat (e.g., 149°C, or 300°F) for local-district heating systems

• Large residential, commercial, or institutional complexes requiring space heat, hot water, and electricity

• Large industrial operations with on-site needs for electricity and heat in the form of process steam, direct heat, and/or space heat

Typical Systems All cogeneration systems involve the operation of a heat engine for the production of mechanical work which, in nearly all cases, is used to drive an electric generator. The most common heat-engine types appropriate for topping-cycle cogeneration systems are:

• Steam turbines (back-pressure and extraction configurations)
• Open-cycle (combustion) gas turbines
• Indirectly fired gas turbines: open cycles and closed cycles
• Diesel engines

Each heat-engine type has unique characteristics making it better suited for some cogeneration applications than for others. For example, engine types can be characterized by:

• Power-to-heat ratio at design point
• Efficiency at design point
• Capacity range
• Power-to-heat-ratio variability
• Off-design (part-load) efficiency
• Multifuel capability

The major heat-engine types are described in terms of these characteristics in Table 9-55.

Representative cogeneration systems appropriate for each engine type are described in the following paragraphs.

Steam-Turbine Systems Both back-pressure and extraction-condensing steam-turbine systems can be used in cogeneration applications. Typical configurations are illustrated in Figs. 9-90 and 9-91. The back-pressure configuration is better suited for applications in which the power-to-heat ratio is low and relatively fixed. Extraction-condensing steam turbines offer greater variability of power-to-heat ratio (through changes in the amount of extracted steam) and a higher power-to-heat ratio (when extracted steam is minimized). The extraction-condensing cycle must always condense the steam flow which passes through the full-pressure ratio of the turbine; the extracted steam may or may not need to pass through the condenser, depending on the temperature limits of the heat load and whether or not it condenses the steam flow. For the back-pressure configuration, the need for a condenser, separate from the heat load, depends, first, on whether the spent steam is available for recycling to the steam generator, and second, on whether the heat load itself condenses the steam flow.

TABLE 9-55 Cogeneration Characteristics for Heat Engines

Engine type	Size range, MWe/unit	Efficiency at design point	Part-load efficiency	Multifuel capability	Maximum temperature of recoverable heat, °F(°C)	Recoverable heat, Btu/kWh*	Typical power-to-heat ratio
Steam turbine							
Extraction-condensing type	30–300	0.25–0.30	Fair	Excellent	200 (93)–600 (315)†	11,000–35,000	0.1–0.3
Back-pressure type	20–200	0.20–0.25	Fair	Excellent	200 (93)–600 (315)†	17,000–70,000	0.05–0.2
Combustion gas turbines	10–100	0.25–0.30	Poor	Poor	1000 (538)–1200 (649)	3000–11,000	0.3–0.45
Indirectly fired gas turbines							
Open-cycle turbines	10–85	0.25–0.30	Poor	Good	700 (371)–900 (482)	3500–8500	0.4–1.0
Closed-cycle turbines	5–350	0.25–0.30	Excellent	Good	700 (371)–900 (482)	3500–8500	0.4–1.0
Diesel engines	0.05–25	0.35–0.40	Good	Fair to poor	500 (260)–700 (371)	4000–6000	0.6–0.85

*1 Btu = 1055 J.
†Saturated steam.

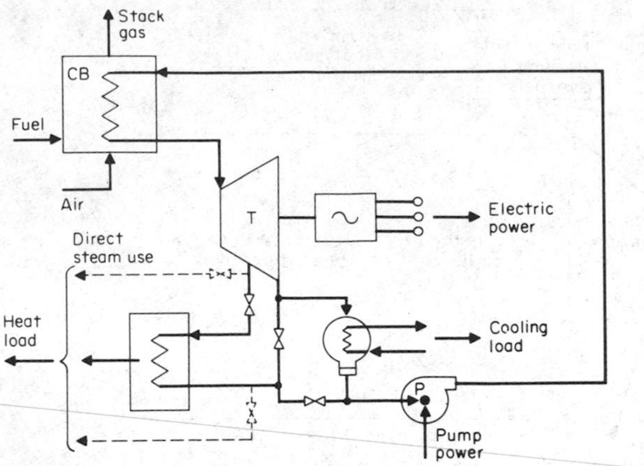

FIG. 9-90 Extraction-condensing-steam-turbine cogeneration cycle. CB = combustor; P = pump; T = turbine; --- = cycle options.

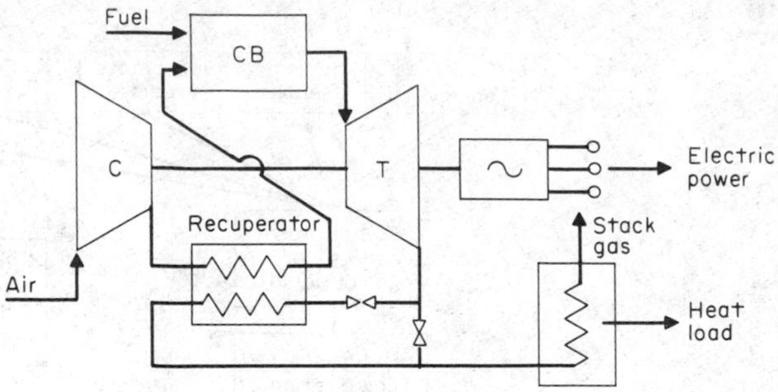

FIG. 9-92 Combustion-gas-turbine cogeneration cycle. C = compressor; CB = combustor; T = turbine.

Gas-Turbine Systems Gas turbines operate on the Brayton thermodynamic cycle: adiabatic compression, isobaric heating, adiabatic expansion, and isobaric cooling. Gas turbines may be either of the combustion or the indirectly fired type. In a combustion turbine a clean fuel* is burned directly in the compressed-air working fluid, and the combustion products are then expanded through the turbine and exhausted to the atmosphere. Indirectly fired turbines receive their thermal input via an external combustor and a heat exchanger, so that none of the combustion products pass through the turbine. The combustion turbine must be an open cycle; that is, one in which the working fluid (in this case, air mixed with combustion products in the turbine) is not recycled and compressor inlet and turbine exhaust are at atmospheric pressure. Indirectly fired gas turbines do not contaminate their working fluid with combustion products and can therefore recycle the working fluid, allowing the cycle to be either open (to the atmospheric intake and exhaust) or closed. Closed gas-turbine cycles offer distinct advantages in terms of choice of working fluid and system pressure level. The two classes of gas-turbine cogeneration systems are illustrated in Figs. 9-92 and 9-93.

Combustion gas turbines. The combustion-gas-turbine system (Fig. 9-92) provides a higher power-to-heat ratio than typical steam-turbine configurations (see Table 9-55). Of all systems considered, the combustion gas turbine also provides rejected heat at the highest temperature [up to 650°C (1200°F)]. However, the waste heat in the turbine exhaust cannot generally be used directly, because of the presence of combustion products, and must be removed by means of a heat exchanger or a boiler. While low in capital cost, combustion gas turbines have the drawbacks of poor part-load efficiency and the requirement for clean fuel to avoid erosive or corrosive damage to the turbine.

Indirectly fired gas turbines. Indirectly fired gas turbines avoid the problems of fuel sensitivity of combustion gas turbines. Otherwise, the major advantage of indirect firing is the possibility of using a closed- (pressurized-) cycle configuration. Closed-cycle gas turbines, as illustrated in Fig. 9-93, can be coupled to a gas-management system which adjusts the working-fluid inventory (hence, mean pressure level) so that load fluctuations can be accommodated with little or no loss in efficiency. This latter feature is a consequence of the fact that load variations are met by changes in density without the need to change temperature, pressure ratios, or relative velocities in the turbomachinery.

Power-to-heat-ratio variations over a relatively wide range (see Table 9-55) are accommodated by varying the recuperator-bypass-flow fraction. With no recuperator bypass, a maximum power-to-heat ratio is obtained. Combustion gas turbines may similarly be recuperated (see Fig. 9-92).

Diesel Engines Diesel engines have long been used in small-scale (e.g., 1- to 10-MWe) cogeneration applications. They are characterized (see Table 9-55) by high engine efficiency, good part-load

*Usually a gas or light liquid with virtually no ash, sulfur, or heavy metals present.

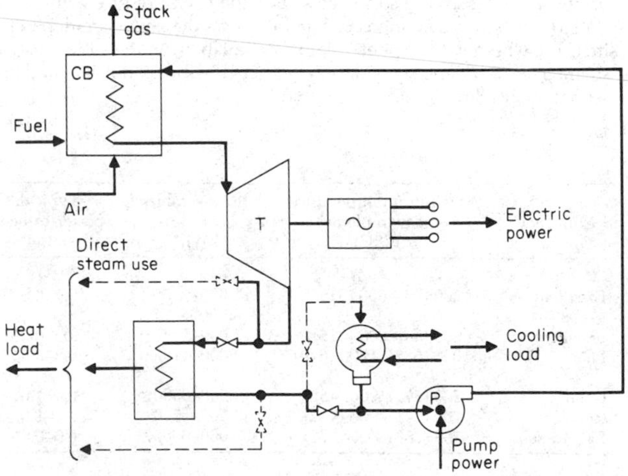

FIG. 9-91 Back-pressure-steam-turbine cogeneration cycle. CB = cumbustor; P = pump; T = turbine; --- = cycle options.

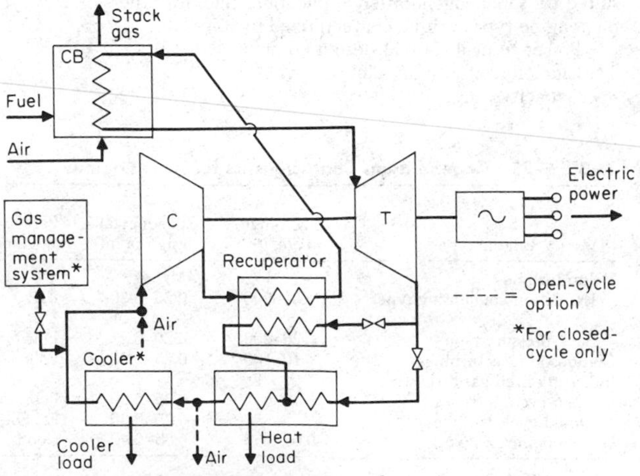

FIG. 9-93 Indirectly-fired-gas-turbine cogeneration cycle (closed-cycle configuration shown). C = compressor; CB = combustor; T = turbine.

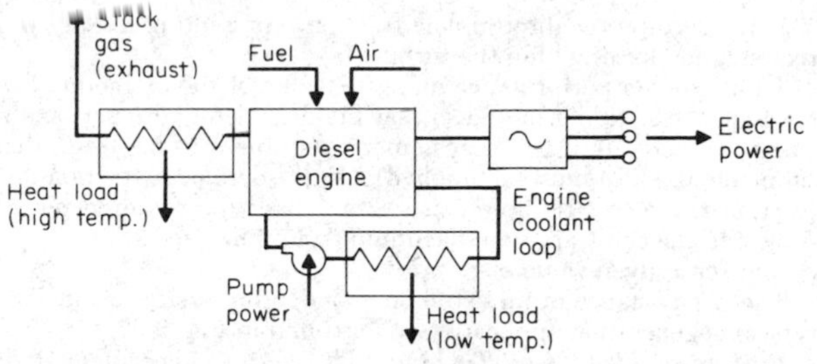

FIG. 9-94 Diesel-engine cogeneration cycle.

characteristics, a relatively low rejected-heat temperature, and a high power-to-heat ratio. As an internal-combustion engine (the working fluid is, in part, made up of the combustion products), the diesel offers only a limited tolerance for fuel variability.

Diesels supply heat in two forms: low-temperature (93 to 121°C, or 200 to 250°F) engine coolant; and higher-temperature (260 to 371°C, or 500 to 700°F) exhaust gas. The diesel engine as a cogeneration system is illustrated in Fig. 9-94.

Energy-Saving Potential The energy-saving potential of an on-site cogeneration system can be analyzed in the manner illustrated by Fig. 9-88. An on-site energy demand for electricity and process heat is conventionally met through the purchase of electric power, generated by an off-site heat engine (usually a steam-turbine cycle) at a central utility, and the burning of fossil fuel in an on-site boiler-steam generator. When cogenerating, an on-site heat engine supplies all or a portion of the electricity demand, while the heat demand can be supplied, at least in part, by the rejected heat from the heat engine.

The fuel-saving potential for a heat engine having an efficiency [from fuel (HHV)* input-to-shaft power output] η_E can best be estimated by the overall efficiency [fuel (HHV) to useful work and power] η_{ov} of the cogeneration cycle, which is, in turn, a function of the power-to-heat ratio R supplied by the heat engine. This relationship is given by

$$\eta_{ov} = \eta_E\left(1 + \frac{1}{R}\right) \tag{9-47}$$

and is plotted as a function of R, for fixed values of η_E, in Fig. 9-95. For large values of R, η_{ov} approaches η_E, as the useful-heat portion of the engine output goes to zero and the engine reverts to a conventional, noncogenerating operating mode. For low values of R, η_{ov} rises to unity as the power-to-heat ratio drops to the minimum value R_{min}, when all the rejected heat from the engine is fully utilized. R_{min} is given by

$$R_{min} = \eta_E/(1 - \eta_E) \tag{9-48}$$

Equation (9-48) applies to an internal-combustion engine, such as a combustion gas turbine or a diesel engine, or to an external-combustion (i.e., indirectly fired) engine in which all rejected thermal energy, relative to the ambient reference, can be removed, including the energy in the stack gases from an external combustor. In most cases it is not practical to extract 100 percent of the available thermal energy, especially from the combustion-product stream of a dirty-fuel combustor. If that combustor has an efficiency [fuel (HHV) to heat delivered to cycle working fluid] η_B, then the maximum overall efficiency is η_B, not unity, and the minimum value for the power-to-heat ratio R_{min} for such an external combustion cycle is given by

$$R'_{min} = \eta_E/(\eta_B - \eta_E) \tag{9-49}$$

For the case of a heat-engine efficiency of 30 percent, including combustor losses as defined by a value of $\eta_B = 0.85$, values for R_{min} and R'_{min} are illustrated in Fig. 9-95.

Figure 9-95 is useful as a means for estimating overall-fuel-use efficiency and its dependence on the delivered power-to-heat ratio for a given heat-engine efficiency. It is not intended as a substitute for a cogeneration-performance prediction for a particular cogeneration system. The limitation of Fig. 9-95, in this regard, is due to the fact that for most of the heat-engine cycles considered here (the diesel engine being the exception) the efficiency of the heat engine η_E is a function of the power-to-heat ratio R. For instance, the engine effi-

*High heating value.

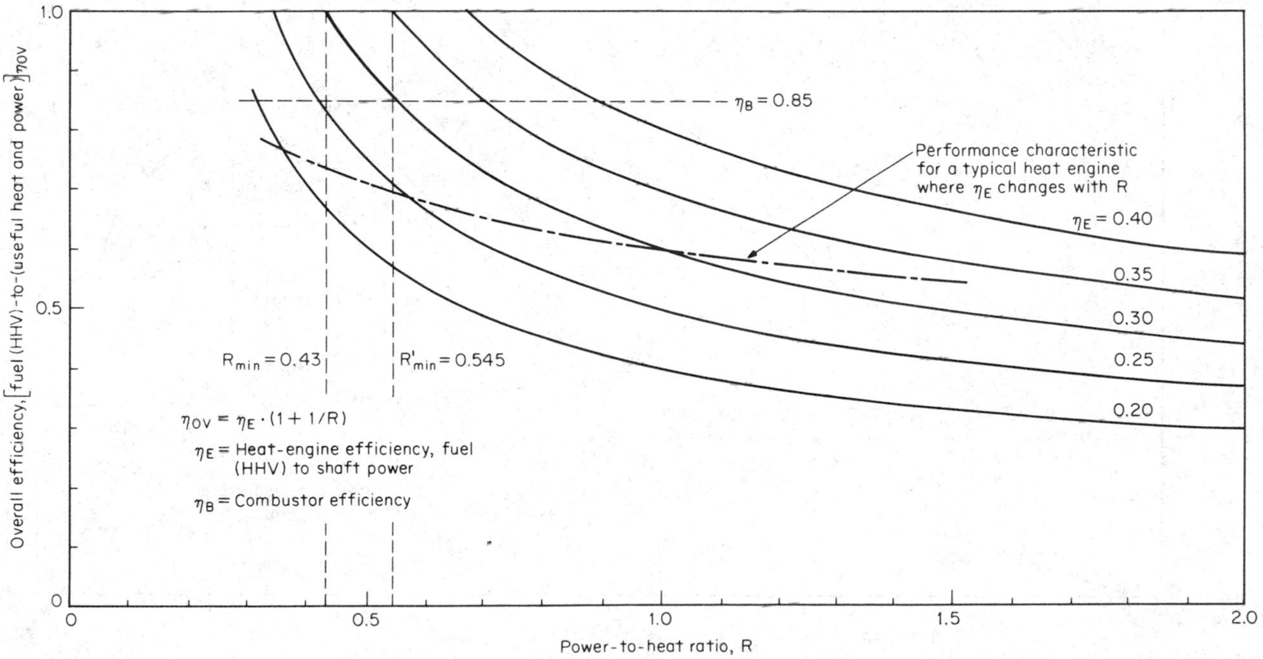

FIG. 9-95 Overall efficiency of a heat engine in cogeneration mode.

ciency of an extraction-condensing steam turbine (Fig. 9-91) is a strong function of how much steam is extracted and at what pressure level. A similar argument applies to gas-turbine cycles in which the recuperator-bypass-flow fraction is varied to accommodate a changing power-to-heat demand. Typically, heat-engine efficiency η_E decreases as the demand for heat rises (R decreases), and heat-engine performance must be compromised to meet it. Such a characteristic is illustrated in Fig. 9-95.

Cogeneration Performance of Representative Systems The engineer designing a cogeneration system for a particular application must analyze the performance capabilities of the most appropriate heat-engine types, selected on the basis of the characteristics listed in Table 9-55. Examples for two-engine types, an extraction-condensing steam turbine and an indirectly-fired-gas-turbine cycle, both chosen because of their compatibility with coal firing, are presented in the following paragraphs.

Extraction-Condensing Steam Turbine The performance of a representative extraction-condensing steam turbine, allowing for variable extraction to permit variation in the delivered power-to-heat ratio, is illustrated in Fig. 9-96. The performance envelope is seen to be specified by:

f_o, the full-throttle flow for no extraction

γ, the ratio of maximum-extraction flow to full-throttle flow, f_o

η_H and η_L, the efficiencies for the high- and low-pressure portions of the turbine

h_1, h_{2s}, h_{3s}, the isentropic enthalpies defined by the steam-supply conditions (state 1), the extraction conditions (state 2), and the condensing conditions (state 3)

β, the zero-power throttle-flow fraction representing leakage and recirculation losses within the turbine

To use such a performance map, the value of the extraction-flow fraction r is first determined so that the heat demand is satisfied (r cannot exceed 1.0; if the heat demand requires r to be >1.0, then supplemental heat must be supplied). For the defined extraction-flow fraction r, the throttle flow f_t is set to deliver the required power, subject to the constraints on maximum and minimum turbine power output for a given value of r.

The performance of an extraction-condensing steam turbine in a typical cogeneration application is illustrated in Fig. 9-97.

Indirectly Fired Gas Turbine The performance of an indirectly fired gas turbine with variable recuperator bypass as a means for matching a power-to-heat-ratio variation is illustrated in Fig. 9-98. The illustrated cycle schematic is for a closed cycle gas turbine. However, the performance envelope is applicable also to open cycles (which would have no cooler or cooling heat load Q_c) with the same cycle parameters (temperatures, pressure ratio, component efficiencies, etc.).

Since the assumed combustor efficiency is 0.85, this value marks the upper limit on overall efficiency (refer to the discussion for Fig. 9-95). Overall efficiency is determined as a function of the power-to-heat ratio R, recuperator-bypass fraction λ, and gas temperature T_7 at the exit of the heater supplying the useful heat load Q_W. In a particular cogeneration application, the power-to-heat demand ratio can be supplied by a range of combinations of recuperator bypass fraction λ and heater exit temperature T_7. However, the desired operating parameters are given by that combination of λ and T_7 which

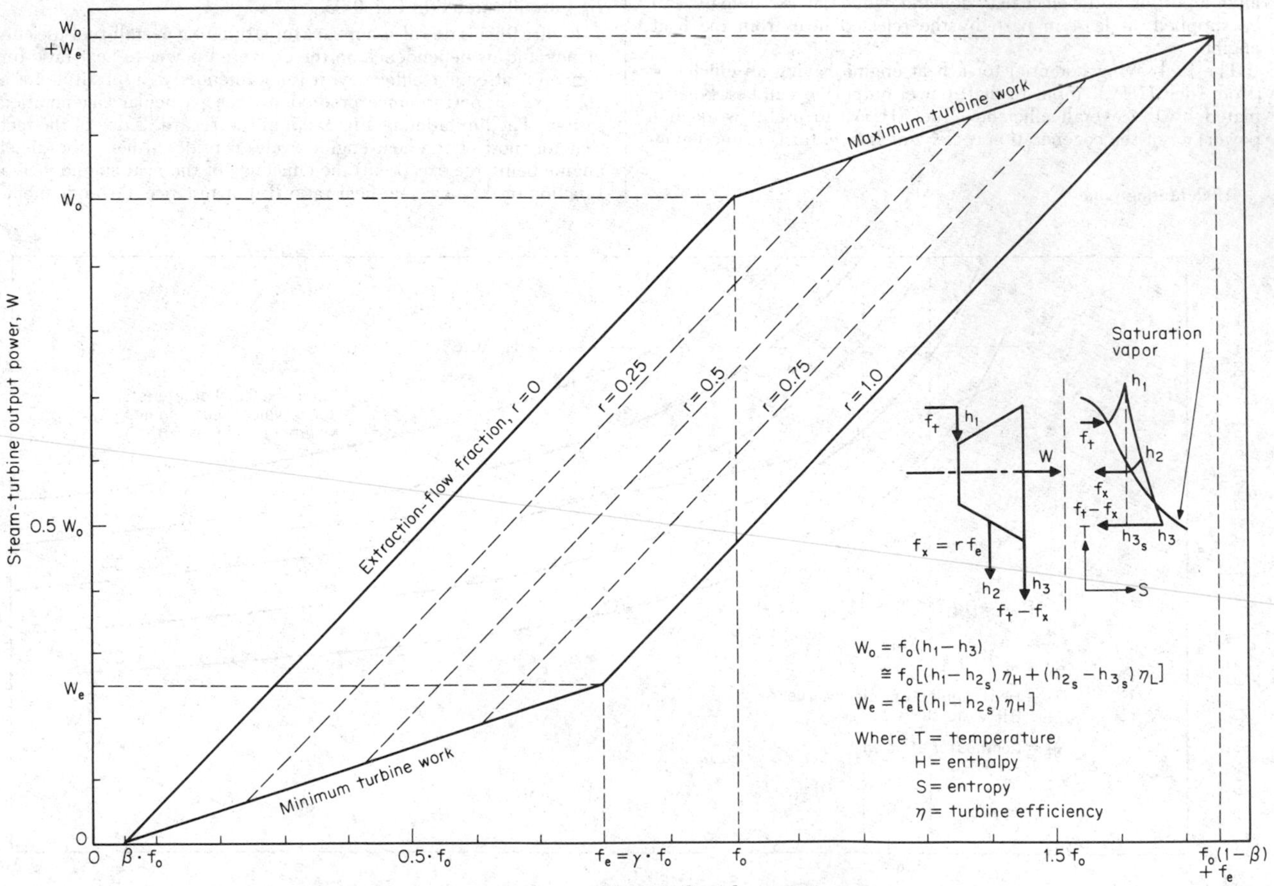

FIG. 9-96 Performance envelope for a typical steam extraction-condensing turbine. (*Derived from* Elliott Multivalve Turbines, *Bull. H-37, Carrier Corporation, 1973.*)

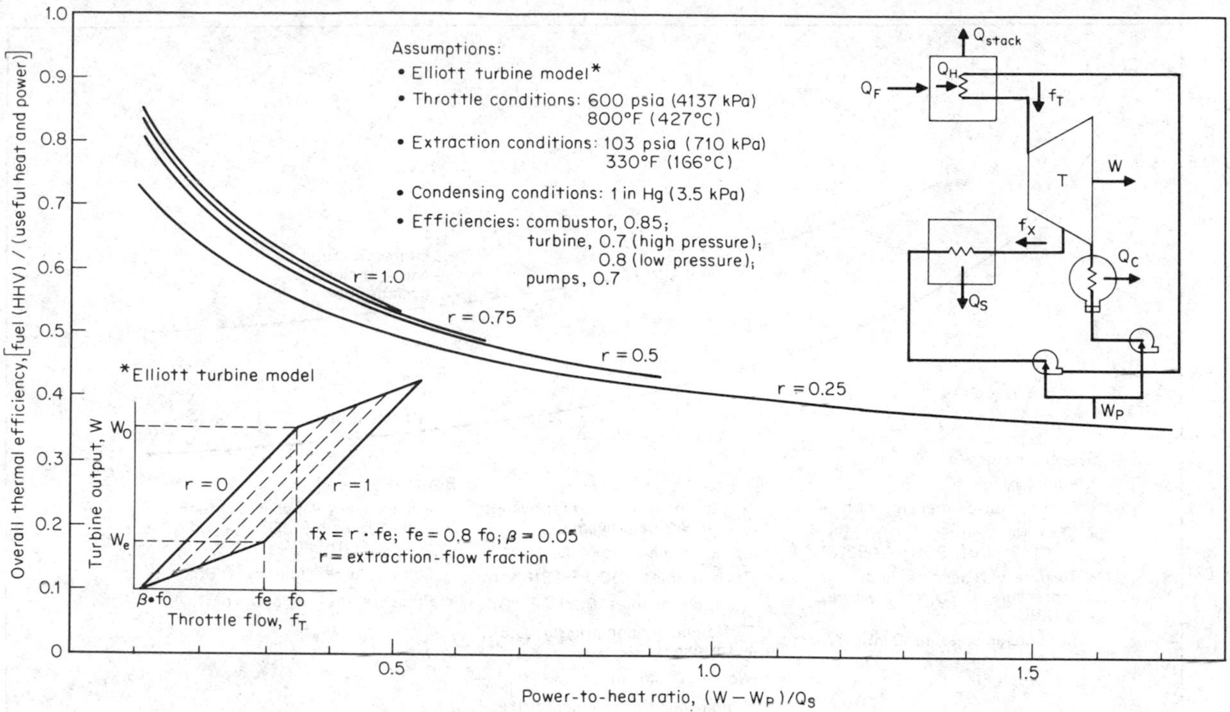

FIG. 9-97 Steam extraction-condensing cogeneration-cycle performance characteristics.

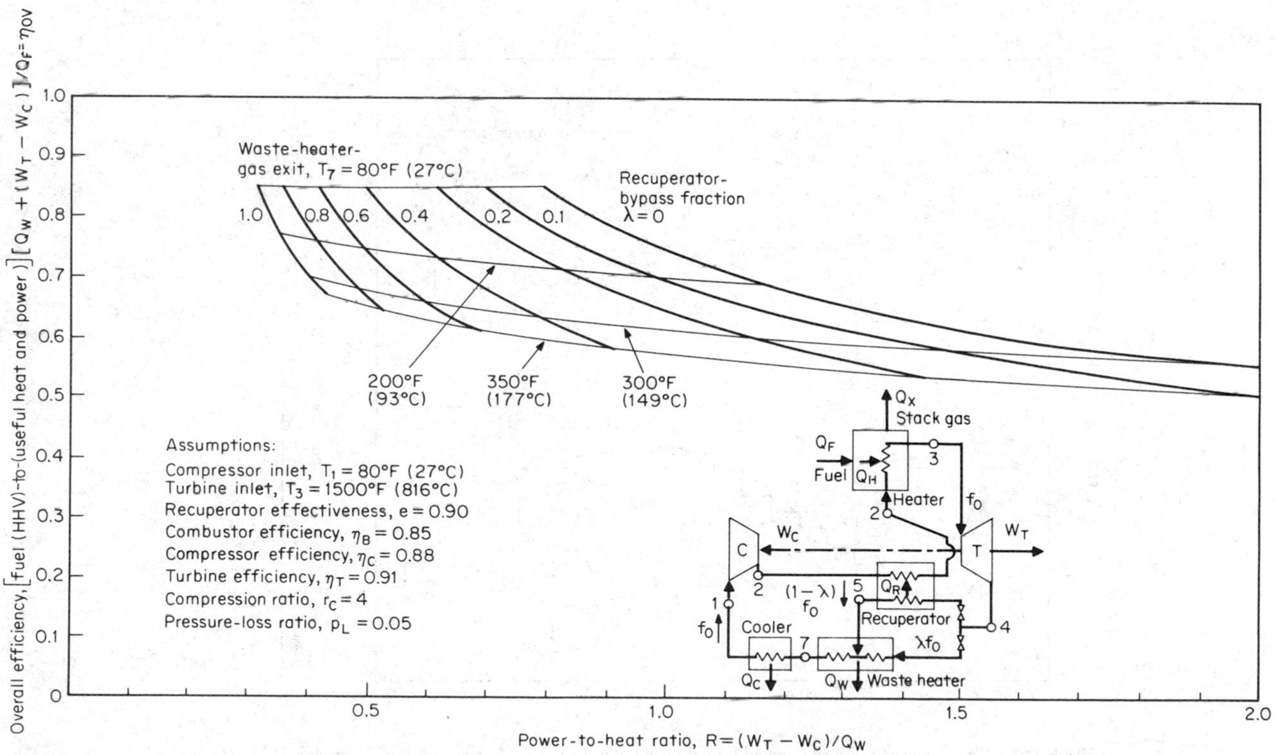

FIG. 9-98 Typical performance envelope for an indirectly fired, open- or closed-cycle gas-turbine cogeneration system: air-working fluid; recuperation with variable bypass.

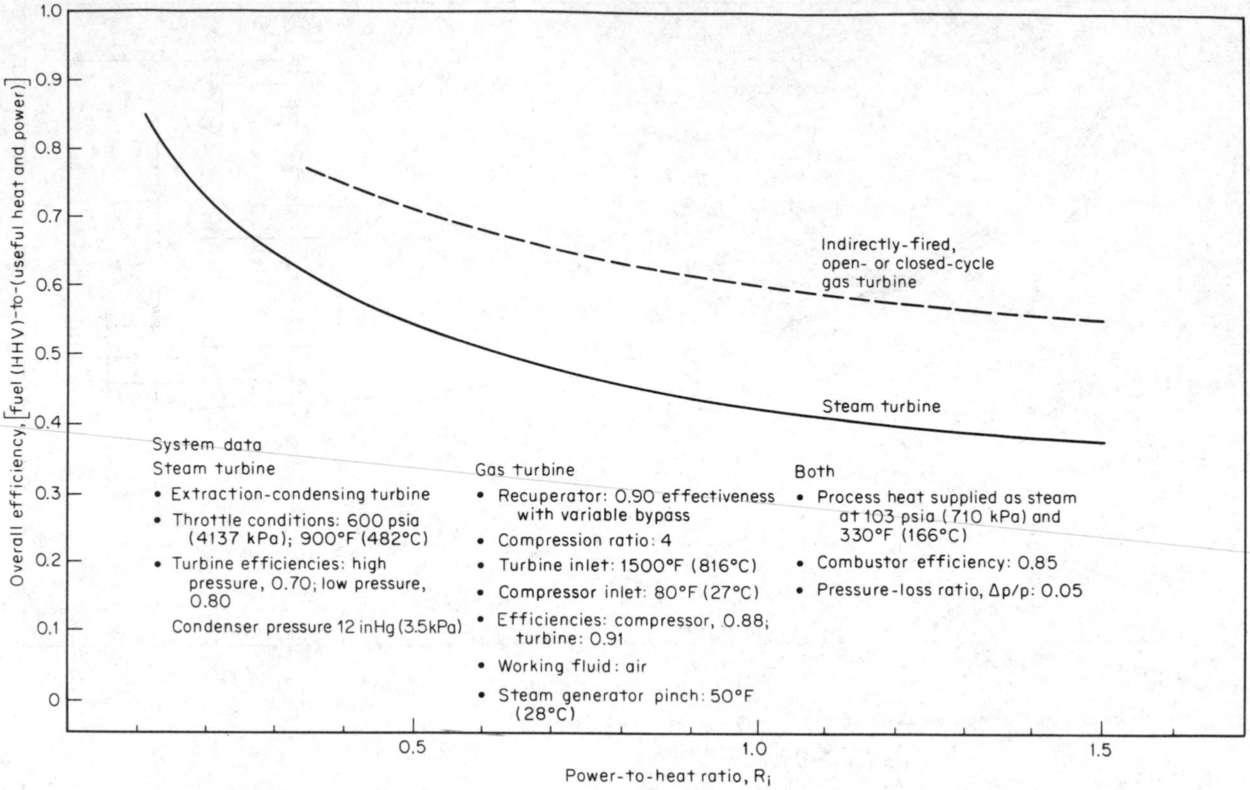

FIG. 9-99 Comparative performance of steam-turbine and gas-turbine (indirectly fired) cogeneration cycles supplying 166°C (330°F) steam and electric power.

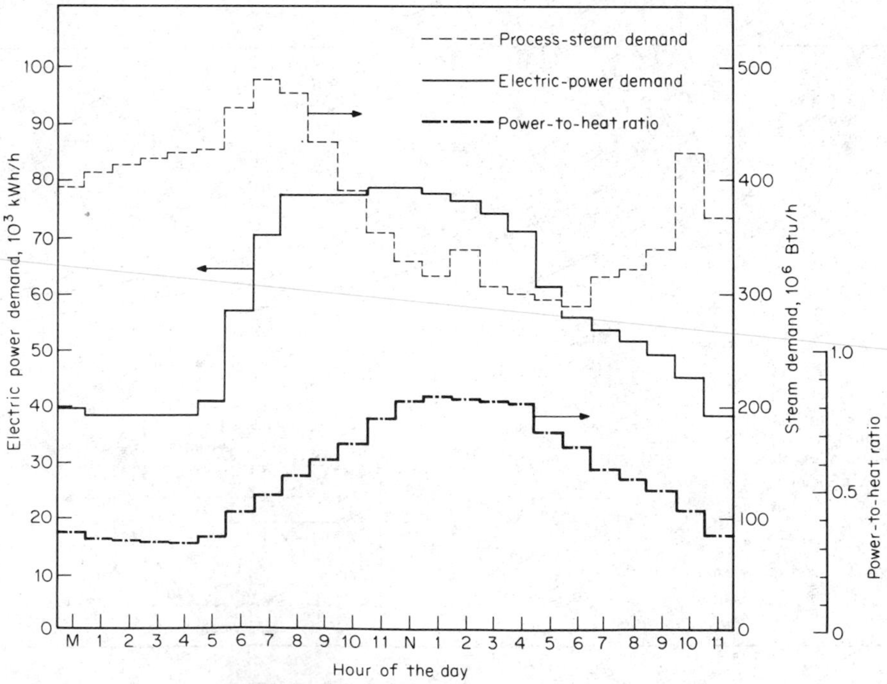

FIG. 9-100 Typical power and heat demand for a typical fall weekday in a large institutional complex. 10^6 Btu/h = 1.055 × 10^9 J/h; 10^3 kWh/h = 3.143 × 10^6 Btu/h thermal equivalent.

just satisfies the requirement that the heat Q_W be supplied with no less than a specified minimum temperature difference (called the "pinch") between the gas-turbine working fluid, which is being cooled, and the heat-supply-system fluid (usually water or steam), which is being heated.

Comparative Performance of Steam and Indirectly Fired Gas-Turbine Systems As noted in Table 9-55, gas-turbine systems can be expected to operate at higher overall efficiency for a given power-to-heat ratio than will a steam-turbine system; conversely, for a given overall efficiency, a gas turbine will yield a higher delivered power-to-heat ratio than will a steam-turbine cogeneration system. This comparison is illustrated in Fig. 9-99, in which the two cycles described by Figs. 9-96, 9-97, and 9-98 have been compared as they supply electric power and 166°C (330°F) steam in varying proportions. For a given power-to-heat ratio, the gas turbine yields an overall efficiency 15 to 18 points higher than that of the assumed steam cycle. Similarly, to achieve an overall efficiency of 0.60 or greater, the steam turbine is restricted to power-to-heat ratios of less than 0.375, whereas for the same efficiency limit the indirectly fired gas turbine can realize power-to-heat ratios as high as 1.0. Thus, the major determinant in whether a gas or a steam turbine is the more appropriate is not the efficiency but rather the power-to-heat ratio which the system must supply.

In addition, if the gas-turbine cycle is a closed (pressurized) cycle, it offers the additional very valuable feature of accommodating large load swings (down to 15 to 20 percent of design power level) with no significant effect on efficiency. As noted earlier, this benefit is achieved by virtue of the ability to change working-fluid inventory (and, hence, pressure level) in the system. Finally, closed-cycle gas turbines also offer the benefit of selection of working fluids (other than air) which may yield enhanced turbine and heat-exchanger performance.

Operational Issues

Varying, Noncoincident Load Demands for Power and Heat For most practical cogeneration applications, the demand for heat and power varies considerably during the course of a typical day. Such a variation for a typical large institutional complex is illustrated in Fig. 9-100. For a cogeneration system to meet the entire power and heat demand for the typical day shown, it would have to have the following performance range capability:

- Power-to-heat ratio R, 0.31 to 0.84
- Power level, 38 MWe (at $R = 0.31$) to 79 MWe (at $R = 0.84$)

If an extraction-condensing steam-turbine system were chosen for this application, it could not efficiently meet the high power-to-heat demand portion to the daily load; either the system would be designed to meet the steam demand, in which case large quantities of electric power would have to be purchased during the middle of the day, or the system would be designed to meet the electrical load, in which case system efficiency would be greatly reduced during the high-power-to-heat-ratio portion of the day.

On the other hand, an indirectly fired gas turbine could efficiently meet (at >60 percent overall efficiency) all the heat and power demand except, possibly, in the early morning hours, when the power-to-heat ratio dips below 0.35. In this case, the system can be designed to meet either the steam demand, while selling excess power to the utility during the period from 1 to 5 A.M. (when the utility is least likely to want it), or the power demand, while supplying the small unsatisfied steam demand with a separate fired boiler.

Cost Considerations Ultimately, cost saving is the objective in most cogeneration installations. The exceptions are those applications which are remote from utility power grids or which have special reliability or security concerns (e.g., military installations, hospitals, etc.). Cogeneration involves additional fixed costs associated with buying and maintaining a larger, generally more complex, on-site physical plant. Against this added fixed cost can be traded the reduction in total fuel costs when purchased electricity is considered a high-grade fuel for on-site use. The reduction in total fuel costs must be greater than the increased annualized fixed cost associated with the cogeneration plant in order to justify the facility economically. While a treatment of capital costs for various cogeneration alternatives is beyond the scope of this subsection, an analysis of fuel-cost savings, as a function of fuel-price ratios and cogeneration-system parameters can be useful. Figure 9-101 presents the fractional reduction in total fuel costs (purchased electricity plus fossil fuel burned on site) as a function of power-to-heat ratio, heat-engine efficiency, and the ratio of electricity to fossil-fuel cost on a thermal equivalent ($/Btu or $/J) basis. By way of illustration, a cogeneration application with a power-to-heat ratio of 0.8, using a heat engine at 30 percent efficiency, with a fuel-cost ratio of 6, would realize a fuel-cost-saving fraction of 0.56, or 56 percent.

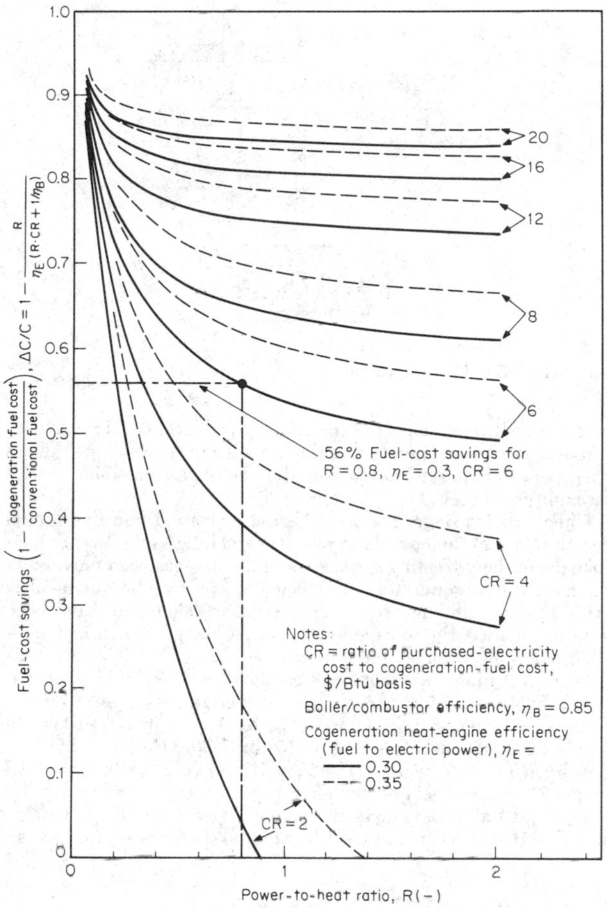

FIG. 9-101 Fuel-cost savings for heat-engine-based cogeneration systems.

HEAT REGENERATION

Storage of heat is a temporary operation since perfect thermal insulators are unknown; thus heat is temporarily absorbed in solids or liquids as sensible or latent heat to be released later at designated times and conditions. Examples of collecting and releasing heat on a

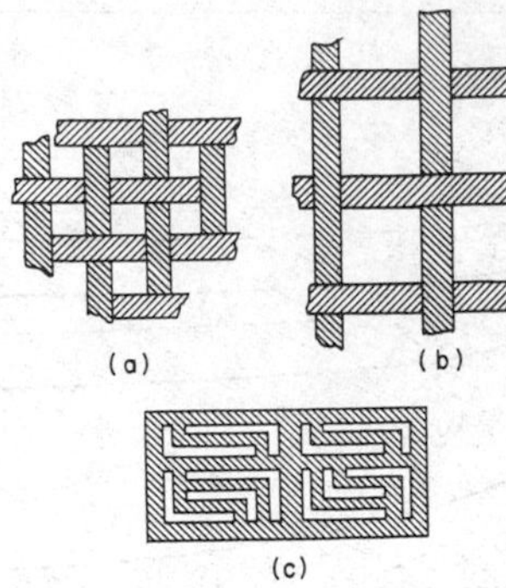

FIG. 9-102 Checkerwork designs.

batch or continuous basis are the checkerwork **regenerator** for blast furnaces as a batch operation or the Ljungstrom type as a continuous operation. **Recuperators** are covered in the discussion on heat-transfer equipment (Sec. 11).

Checkerbrick Regenerators Preheating combustion air in open-hearth furnaces, ingot-soaking pits, glass-melting tanks, by-product-coke ovens, heat-treating furnaces, and the like has been universally carried out in **regenerators** constructed of fireclay, chrome, or silica brick shapes. Although many geometric arrangements have been used in practice, the so-called basketweave design shown in Fig. 9-102*a* and *b* has been adopted in many applications.

In **blast-furnace** stove construction, standard 22.9- by 11.4- by 6.4-cm (9- by 4.5- by 2.5-in) firebrick, assembled in basketweave design, forms square flues 8.3 by 8.3 cm (3.25 by 3.25 in) (Fig. 9-102*a*). In **open-hearth regenerators,** 46- by 15- by 7-cm (18- by 6- by 3-in) tiles form flues 19 by 19 cm (7.5 by 7.5 in) (Fig. 9-102*b*). Special shapes have been devised for more complicated, if frequently less rugged, heat-absorbing elements, e.g., coke-oven tiles (Fig. 9-102*c*). Standard firebricks are cheaper than special shapes, and this fact has tended to confine regenerator design to the readily available and less expensive standard refractories.

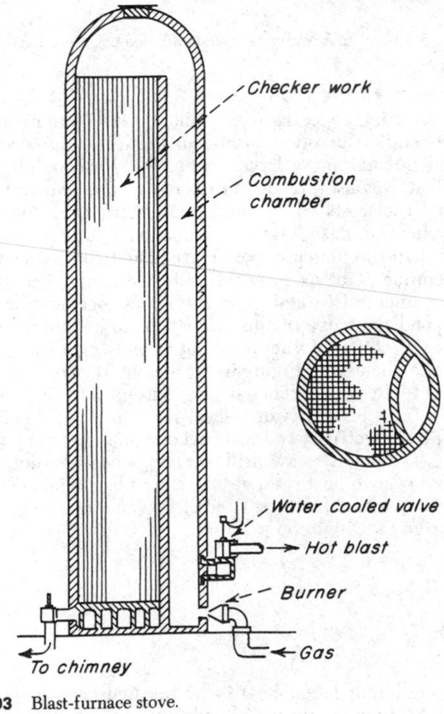

FIG. 9-103 Blast-furnace stove.

Blast-Furnace Stoves A typical blast furnace, producing 1650 tons of pig iron per day, will be blown with 47 m^3/s (100,000 standard ft^3/min) of atmospheric air, preheated to temperatures ranging in normal practice from 482 to 649°C (900 to 1200°F), with 538°C (1000°F) close to an average. To preheat this blast volume, a set of four stoves is usually provided. A vertical and horizontal section of one such stove is shown in Fig. 9-103. Each stove consists of a vertical steel cylinder 7.3 m (24 ft) in diameter, 33m (110 ft) high, topped with a spherical dome. A side combustion chamber is separated by a bridgewall with a lens-shaped horizontal cross section. The remaining volume is filled with heat-absorbing checkerwork.

The **heat-exchanging surface** in each stove is just under 11,500 m^2 (124,000 ft^2). In operation, each stove is carried through a two-step 4-h cycle. In one 3-h "on-gas" step, the checkers are heated by the combustion of blast-furnace gas. In the alternating "on-wind" 1-h step, the checkers are cooled by the passage of cold air through the stove. At any given time, three stoves are simultaneously on gas, while a single stove is on wind.

After 3 h on gas, an on-wind step is initiated. At the start, about one-half of the air, entering at 93°C (200°F), passes through the checkers, the other half being bypassed around the stove through the cold-blast mixer valve. The gas passing through the stove exhausts initially at 1093°C (2000°F). Mixing this with the unheated air produces a blast temperature of 538°C (1000°F). The temperature of the heated air from the stove falls rapidly, minute by minute, throughout the on-wind step. The fraction of total air volume bypassed through the mixer valve is continually decreased by progressive closing of this valve, its operation being automatically regulated under control of a thermocouple and potentiometer. At the end of 60 min of on-wind operation, in usual practice the cold-blast mixer valve is practically closed, the entire blast then passing through the checkers.

Satisfactory approach to uniform blast temperature can readily be realized by this automatic control of the mixer valve, provided that the uniform blast temperature does not greatly exceed one-half of the combustion temperature in the preceding on-gas step. The rapid decrease in temperature exhibited by the air discharging from the checkers is a characteristic feature of classical checkerwork heat transfer. The thickness of the refractory flue walls retards the flow of heat by thermal diffusion into the central portions of the brick. Although the heat removed in an on-wind step is less than 5 percent of the total sensible heat stored in the stove refractories, the introduction and removal of heat from such large-dimensioned refractory elements is sluggish, with the result that the in-and-out movement of heat is largely a skin effect confined closely to the refractory surface.

Open-Hearth and Glass-Tank Regenerators Because of the higher working temperatures, more drastic thermal shock, and dirtier gases encountered in open-hearth and glass-tank regenerators, checkerwork construction in these furnace units, while somewhat similar to that employed in blast-furnace stoves, requires considerable modification. The vertical height of the flues is limited by the elevation of the furnace above plant level. Short flues from 3 to 4.9 m (10 to 16 ft) are common in contrast to the 26- to 29-m (85- to 95-ft) flue lengths in blast-furnace stoves. Larger brick shapes (Fig. 9-102*b*) form flue cross sections 5 times as large as the stove flues, and the percentage of voids in the checkerwork is 51 percent, in contrast with the stove 32 percent voids. In a typical open hearth (Fig. 9-104), a checker volume of 210 m^3 (7500 ft^3) contains 810 flues.

As a result of the larger dimensions of flues and the restricted surface per unit gas passed, regenerators employed with this type of reverberatory furnace exhibit much lower efficiency than would be realized with smaller flue dimensions. In view, however, of the large amount of iron oxide contained in the open-hearth exhaust gas and the alkali fume present in glass-tank stack gases, resort to smaller checker dimensions has appeared impractical.

Coke-Oven Regenerators In the by-product coke oven, waste-heat recovery is effected in the standard Siemens manner, although, as seen in Fig. 9-105, the dimensions of the upstream and down-

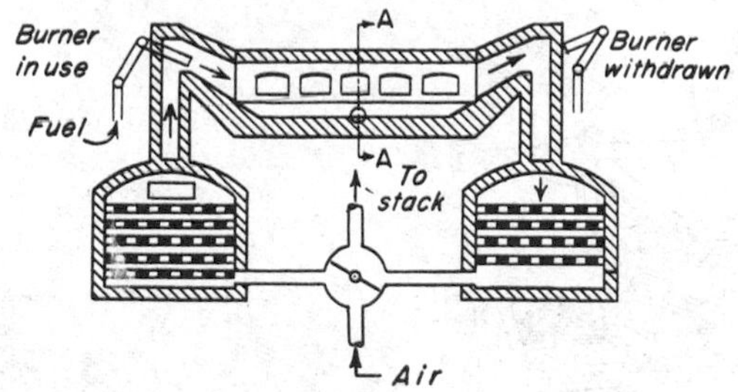

FIG. 9-104 Cross section of an open-hearth steel furnace, including regenerators.

stream regenerators show little outward resemblance either to the blast-furnace stove or to the reverberatory-furnace regenerators. From structural necessity, the coke-oven regenerator is located under the oven itself and must assume the dimensions of an extremely narrow parallelepiped. Fortunately, the design problem is simplified because of the absence of fume and dust in the flue system. Special regenerator blocks are commonly employed, a typical design being shown in Fig. 9-102*c*. An oven 12 by 3.7 by 0.5 m (40 ft by 12 ft by 16 in) carbonizing 24 tons/day coal to produce 17 tons/day coke will be provided with a pair of regenerators having a horizontal cross section of 6.5 m^2 (70 ft^2) containing 210 flues ("slots") and an overall volume of 8.4 m^3 (300 ft^3). Because fuels used in underfiring are either cleaned coke-oven gas, clean blast-furnace gas, or mixtures of the two, difficulty with dirt and fume accumulation in the flues is not encountered, and because of the lower working temperatures in coking, this intricate type of flue has been found satisfactory. Attempts to duplicate coke-oven–regenerator construction in other Siemens units have not been successful.

Pebble Stove Although considerable ingenuity has been applied to the design of the Cowper checkerwork, involving variations in

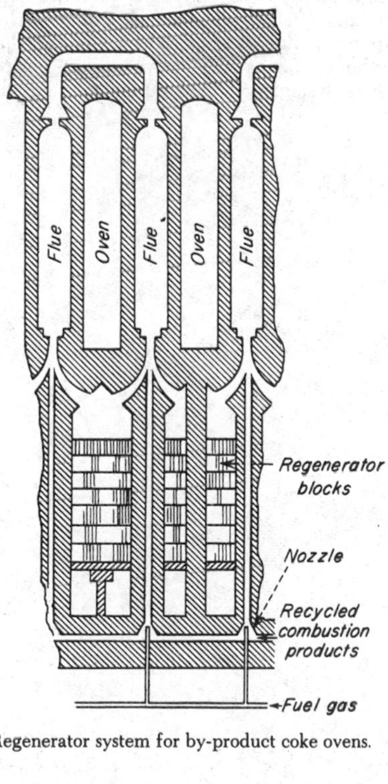

FIG. 9-105 Regenerator system for by-product coke ovens.

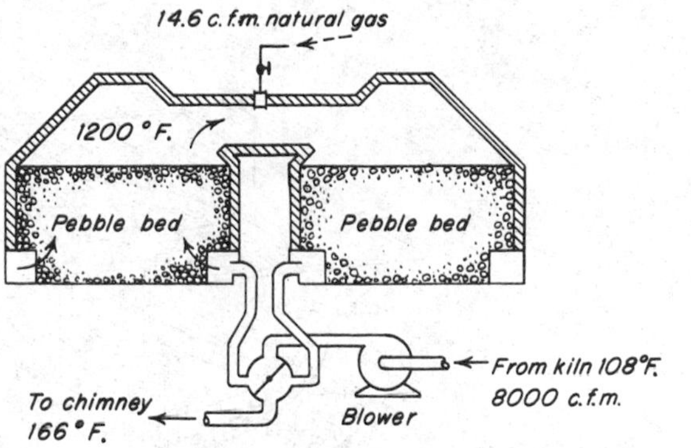

FIG. 9-106 Pebble-bed regenerator.

bricklaying patterns and special ceramic shapes, the checkerbrick regenerators still exhibit five engineering defects: (1) high initial cost of construction; (2) unsatisfactory thermal efficiency with clean heat-exchanging surfaces; (3) tendency to lose efficiency with dust-and fume-coated surfaces inevitable with dirty gas; (4) inaccessibility of surface causing lost time and high labor cost in any cleaning operation; and (5) danger of fusion, slagging, and spalling when subjected to high maximum temperatures and rapid temperature changes. It is possible that one or even all of these difficulties are inherent in the Cowper checkerbrick system, as such. No serious attempt, however, to alter U.S. construction appears to have been made prior to 1929, when the Department of Agriculture undertook to substitute the so-called **pebble-bed** heat-exchanging structure for the classical checkerwork in order to provide higher air preheat in the blast-furnace smelting of phosphate rock.

In this departure from regenerator precedent, a mass of small refractory particles, enclosed in a brick-lined steel shell, was substituted as a functional equivalent of standard checkerwork (see Fig. 9-106). In the operation of an experimental phosphate blast furnace, blast temperatures as high as 1093°C (2000°F) were readily obtained. In later operation of a 25-ton/day blast furnace producing pig iron, ferromanganese, ferrochromium, and ferrosilicon, air preheat temperatures of 1538°C (2800°F) were attained and maintained. In connection with the process of converting air into NO [Daniels and Gilbert, *Ind. Eng. Chem.*, **40,** 1719 (1948)] air was preheated to 1980°C (3600°F) in magnesia refractory pebble stoves.

Aside from the engineering value of the elevated temperatures attained with pebble-stove regenerators, extremely high thermal efficiencies are an inherent characteristic of this type of heat interchanger.

Ljungstrom Heater The continuous-regenerative type of air heater or recuperator is familiarly known as the Ljungstrom heater (Fig. 9-107). The heater assembly consists of a slow-moving rotor packed with closely spaced metal plates or wires. At each end of the rotor is a housing divided by partitions to confine the hot gas to one half of the rotor and the cold gas to the other. Radial and circumferential seals sliding on the rotor limit the leakage between streams. The rotor is divided into sectors and each sector packed with a filling to promote high heat transfer at low pressure drop. The packing may be divided into layers of different materials to suit the temperature and corrosion conditions.

Leakage between streams comes from (1) entrainment in the rotor passages (this can be reduced by providing for a blowout section between the hot and cold zones), (2) leakage around the circumference of the rotor through the annular space between rotor shell and housing, and (3) leakage of radial seals.

These heaters are available with rotors up to 6-m (20-ft) diameter. Gas temperatures up to 816°C (1500°F) can be handled [a higher temperature of 982°C (1800°F) with special alloys], and gas face velocities are usually around 2.5 m/s (500 ft/min). Thickness of rotor depends on service, efficiency, and operating conditions but usually

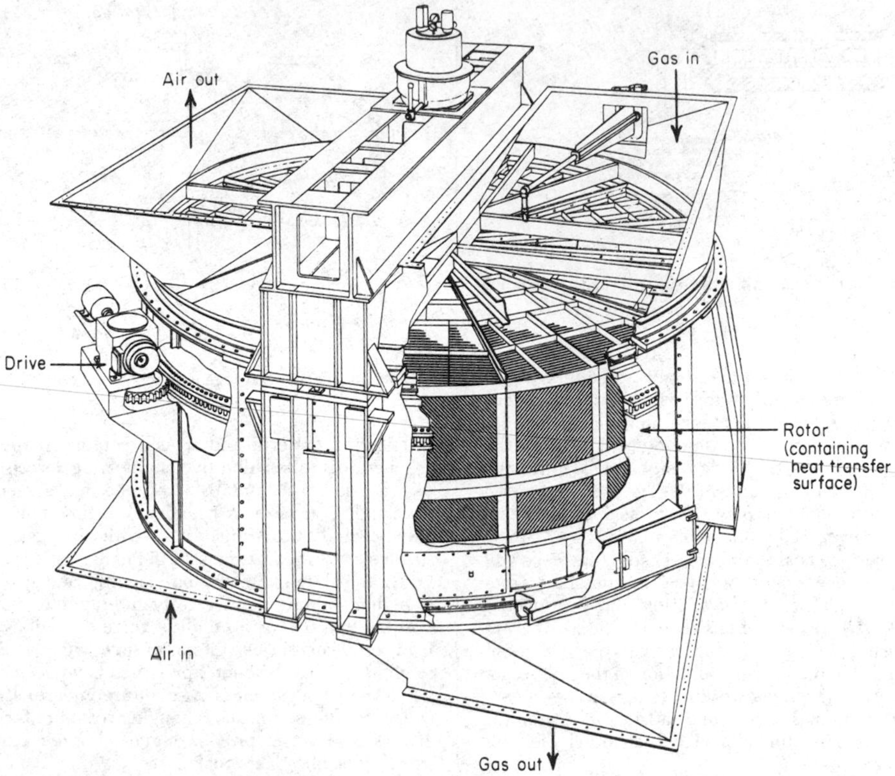

FIG. 9-107 Ljungstrom air heater.

ranges from 20.3 to 91.4 cm (8 to 36 in). Rotors are driven by small motors with rotor speed up to 10 to 20 r/min. The effectiveness of these heaters can be as high as 85 to 90 percent.

Ljungstrom-type heaters are widely used in power-plant boilers. Use in process industries is increasing for air conditioning and building heating by transferring heat between the fresh and exhaust air streams and for process-heat recovery.

Miscellaneous Systems Many systems have been proposed for transferring heat regeneratively, such as described earlier, plus the use of high-temperature liquids and fluidized beds for direct contact with gases, but other problems which limit industrial application are encountered. These systems are covered by methods described in Sec. 11.

Section 10

Heat Transmission

James G. Knudsen, Ph.D., *Professor of Chemical Engineering, Oregon State University; Member, American Institute of Chemical Engineers, American Chemical Society, National Society of Professional Engineers; Registered Professional Engineer (Oregon). (Conduction and Convection; Section Editor)*

Kenneth J. Bell, Ph.D., *Regents Professor of Chemical Engineering, Oklahoma State University; Member, American Institute of Chemical Engineers. (Thermal Design of Heat Exchangers, Condensers, Reboilers)*

Arthur D. Holt, *Formerly Consultant, Processing Equipment Division, The Jeffrey Manufacturing Co.; Member, American Institute of Chemical Engineers, American Society of Mechanical Engineers, National Society of Professional Engineers; Registered Professional Engineer (Ohio). (Thermal Design for Solids Processing)*

Hoyt C. Hottel, S.M., *Professor Emeritus of Chemical Engineering, Massachusetts Institute of Technology; Member, National Academy of Sciences, American Academy of Arts and Sciences, American Institute of Chemical Engineers, American Chemical Society, Combustion Institute. (Radiation)*

Adel F. Sarofim, Sc.D., *Professor of Chemical Engineering and Assistant Director, Fuels Research Laboratory, Massachusetts Institute of Technology; Member, American Institute of Chemical Engineers, American Chemical Society, Combustion Institute. (Radiation)*

F. C. Standiford, M.S., *President, W. L. Badger Associates, Inc.; Member, American Institute of Chemical Engineers, American Chemical Society; Registered Professional Engineer (Michigan). (Thermal Design of Evaporators)*

David Stuhlbarg, Ch.E., *Heat-Transfer Consultant; Member, American Institute of Chemical Engineers. (Thermal Design of Tank Coils)*

Vincent W. Uhl, Ph.D., *Professor of Chemical Engineering, University of Virginia; Member, American Institute of Chemical Engineers, American Chemical Society. (Agitated Vessels)*

INTRODUCTION

HEAT TRANSMISSION BY CONDUCTION

HEAT TRANSFER BY CONVECTION

HEAT TRANSFER WITH CHANGE OF PHASE

THERMAL DESIGN OF HEAT-TRANSFER EQUIPMENT

HEAT TRANSMISSION BY RADIATION

Nomenclature and Units

Specialized nomenclature used for radiative heat transfer is defined in the subsection "Heat Transmission by Radiation."

Symbol	Definition	SI units	U.S. customary units
a	Proportionality coefficient	Dimensionless	Dimensionless
a_x	Cross-sectional area of a fin	m^2	ft^2
a'	Proportionality factor		
A	Area of heat transfer surface; A_i for inside; A_o for outside; A_m for mean; A_{avg} for average A_1, A_2, and A_3 for points 1, 2, and 3 respectively; A_B for bare surface of finned tube; A_f for finned portion of tube; A_{uf} for external area of unfinned portion of finned tube; A_{of} for external area of finned tube before fins are attached, equals A_o; A_{oe} for effective area of finned surface; A_T for total external area of finned tube; A_d for surface area of dirt (scale) deposit	m^2	ft^2
b	Proportionality coefficient		
b'	Proportionality factor		
b_f	Height of fin	m	ft
B	Material constant $= 5D^{-0.5}$		
c_1, c_2, etc.	Constants of integration		
c, Cp	Specific heat at constant pressure; c_s for specific heat of solid; c_g for specific heat of gas	J/(kg·K)	Btu/(lb·°F)
C	Thermal conductance, equals kA/x, hA, or UA; C_1, C_2, C_3, C_n, thermal conductance of sections 1, 2, 3, and n respectively of a composite body	J/(s·K)	Btu/(h·°F)
C_r	Correlating constant; proportionality coefficient	Dimensionless	Dimensionless
d_m	Depth of divided solids bed	m	ft
D	Diameter; D_o for outside; D_i for inside; D_r for root diameter of finned tube	m	ft
D_c	Diameter of a coil or helix	m	ft
D_e	Equivalent diameter of a cross section, usually 4 times free area divided by wetted perimeter; D_w for equivalent diameter of window	m	ft
D_j	Diameter of a jacketed cylindrical vessel	m	ft
D_{otl}	Outside diameter of tube bundle	m	ft
D_p	Diameter of packing in a packed tube	m	ft
D_s	Inside diameter of heat-exchanger shell	m	ft
D_t	Solids-processing vessel diameter	m	ft
D_1, D_2	Diameter at points 1 and 2 respectively; inner and outer diameter of annulus respectively	m	ft
E_H	Eddy conductivity of heat	J/(s·m·K)	Btu/(h·ft·°F)
E_M	Eddy viscosity	Pa·s	lb/(ft·h)
f	Friction factor; f_1 for inner wall and f_2 for outer wall of annulus; f_k for ideal tube bank	Dimensionless	Dimensionless
F	Entrance factors		
F_a	Dry solids feed rate	kg/(s·m^2)	lb/(h·ft^2)
F_g	Gas volumetric flow rate	m^3 (s·m^2 of bed area)	ft^3/(h·ft^2 of bed area)
F_c	Fraction of total tubes in cross-flow: F_{bp} for fraction of cross-flow area available for bypass flow		
F_t	Factor, ratio of temperature difference across tube-side film to overall mean temperature difference	Dimensionless	Dimensionless
F_s	Factor, ratio of temperature difference across shell-side film to overall mean temperature difference	Dimensionless	Dimensionless
F_w	Factor, ratio of temperature difference across retaining wall to overall mean temperature difference between bulk fluids	Dimensionless	Dimensionless
F_D	Factor, ratio of temperature difference across combined dirt or scale films to overall mean temperature difference between bulk fluids	Dimensionless	Dimensionless
F_T	Temperature-difference correction factor		
g, g_L	Acceleration due to gravity	981 m/s^2	(4.18)(10^8) ft/h^2
g_c	Conversion factor	1.0 (kg·m)/(N·s^2)	(4.17)(10^8)(lb·ft)/(lbf·h^2)
G	Mass velocity, equals $V\rho$ or W/S; G_v for vapor mass velocity	kg/(m^2·s)	lb/(h·ft^2)
G_{max}	Mass velocity through minimum free area between rows of tubes normal to the fluid stream	kg/(m^2·s)	lb/(h·ft^2)
G_{mf}	Minimum fluidizing mass velocity	kg/(m^2·s)	lb/(h·ft^2)
h	Local individual coefficient of heat transfer, equals $dq/(dA)(\Delta T)$	J/(m^2·s·K)	Btu/(h·ft^2·°F)

Nomenclature and Units (*continued*)

Symbol	Definition	SI units	U.S. customary units
h_{am}, h_{lm}	Film coefficient based on arithmetic-mean temperature difference and logarithmic-mean temperature difference respectively	$J/(m^2 \cdot s \cdot K)$	$Btu/(h \cdot ft^2 \cdot °F)$
h_b	Film coefficient delivered at base of fin	$J/(m^2 \cdot s \cdot K)$	$Btu/(h \cdot ft^2 \cdot °F)$
h_{cg}	Effective combined coefficient for simultaneous gas-vapor cooling and vapor condensation	$J/(m^2 \cdot s \cdot K)$	$Btu/(h \cdot ft^2 \cdot °F)$
$h_c + h_r$	Combined coefficient for conduction, convection, and radiation between surface and surroundings	$J/(m^2 \cdot s \cdot K)$	$Btu/(h \cdot ft^2 \cdot °F)$
h_{do}, h_{di}	Film coefficient for dirt or scale on outside or inside respectively of a surface	$J/(m^2 \cdot s \cdot K)$	$Btu/(h \cdot ft^2 \cdot °F)$
h_f	Film coefficient for air film of air-cooled finned-tube exchangers based on total external surface	$J/(m^2 \cdot s \cdot K)$	$Btu/(h \cdot ft^2 \cdot °F)$
h_{fi}	Effective outside film coefficient of a finned tube based on inside area	$J/(m^2 \cdot s \cdot K)$	$Btu/(h \cdot ft^2 \cdot °F)$
h_{fo}	Film coefficient for air film of an air-cooled finned-tube exchanger based on external bare surface	$J/(m^2 \cdot s \cdot K)$	$Btu/(h \cdot ft^2 \cdot °F)$
h_F, h_s	Effective film coefficient for dirt or scale on heat-transfer surface	$J/(m^2 \cdot s \cdot K)$	$Btu/(h \cdot ft^2 \cdot °F)$
h_i, h_o	Film coefficient for heat transfer for inside and outside surface respectively	$J/(m^2 \cdot s \cdot K)$	$Btu/(h \cdot ft^2 \cdot °F)$
h_k	Film coefficient for ideal tube bank; h_s for shell side of baffled exchanger; h_{sv} for coefficient at liquid-vapor interface	$J/(m^2 \cdot s \cdot K)$	$Btu/(h \cdot ft^2 \cdot °F)$
h_1	Condensing coefficient on top tube; h_N coefficient for N tubes in a vertical row	$J/(m^2 \cdot s \cdot K)$	$Btu/(h \cdot ft^2 \cdot °F)$
h'	Film coefficient for enclosed spaces	$J/(m^2 \cdot s \cdot K)$	$Btu/(h \cdot ft^2 \cdot °F)$
h_{lm}	Film coefficient based on log-mean temperature difference	$J/(m^2 \cdot s \cdot K)$	$Btu/(h \cdot ft^2 \cdot °F)$
h_r	Heat-transfer coefficient for radiation	$J/(m^2 \cdot s \cdot K)$	$Btu/(h \cdot ft^2 \cdot °F)$
h_T	Coefficient of total heat transfer by conduction. convection, and radiation between the surroundings and the surface of a body subject to unsteady-state heat transfer	$J/(m^2 \cdot s \cdot K)$	$Btu/(h \cdot ft^2 \cdot °F)$
h_w	Equivalent coefficient of retaining wall, equals k/x	$J/(m^2 \cdot s \cdot K)$	$Btu/(h \cdot ft^2 \cdot °F)$
j	Ordinate, Colburn j factor, equals f/2; j_H for heat transfer; j_{H1} for inner wall of annulus; j_{H2} for outer wall of annulus; j_k for heat transfer for ideal tube bank	Dimensionless	Dimensionless
J	Mechanical equivalent of heat	$1.0(N \cdot m)/J$	$778(ft \cdot lbf)/Btu$
J_b, J_c, J_l, J_r	Correction factors for baffle bypassing, baffle configuration, baffle leakage, and adverse temperature gradient respectively		
k	Thermal conductivity; k_1, k_2, k_3, thermal conductivities of bodies 1, 2, and 3	$J/(m \cdot s \cdot K)$	$(Btu \cdot ft)/(h \cdot ft^2 \cdot °F)$
k_v	Thermal conductivity of vapor; k_l for liquid thermal conductivity; k_s for thermal conductivity of solid	$J/(m \cdot s \cdot K)$	$(Btu \cdot ft)/(h \cdot ft^2 \cdot °F)$
k_{avg}, k_m	Mean thermal conductivity	$J/(m \cdot s \cdot K)$	$(Btu \cdot ft)/(h \cdot ft^2 \cdot °F)$
k_f	Thermal conductivity of fluid at film temperature	$J/(m \cdot s \cdot K)$	$(Btu \cdot ft)/(h \cdot ft^2 \cdot °F)$
k_w	Thermal conductivity of retaining-wall material	$J/(m \cdot s \cdot K)$	$(Btu \cdot ft)/(h \cdot ft^2 \cdot °F)$
l_c	Baffle cut; l_s for baffle spacing	m	ft
L	Length of heat-transfer surface	m	ft
L_o	Flow rate	kg/s	lb/h
L_u	Undisturbed length of path of fluid flow	m	ft
L_F	Thickness of dirt or scale deposit	m	ft
L_H	Depth of fluidized bed	m	ft
L_p	Diameter of agitator blade	m	ft
m	Ratio, term, or exponent as defined where used		
M	Molecular weight	kg/mol	lb/mol
M	Weight of fluid	kg	lb
n	Position ratio or number	Dimensionless	Dimensionless
n_t	Number of tubes in parallel in a heat exchanger		
n_r	Number of rows in a vertical plane		
n'	Flow-behavior index for nonnewtonian fluids		
n_b	Number of baffle-type coils		
N_r	Speed of agitator	rad/s	r/h
N	Number of tubes in a vertical row; or number of tubes in a bundle; N_b for number of baffles; N_T for total number of tubes in exchanger; N_c for number of tubes in one cross-flow section; N_{cw} for number of cross-flow rows in each window		

Nomenclature and Units (*continued*)

Symbol	Definition	SI units	U.S. customary units
N_B	Biot number, $h_T \Delta x/k$		
N_d	Proportionality coefficient	Dimensionless	Dimensionless
N_{Gr}	Grashof number, $L^3\rho^2 g\beta \Delta t/\mu^2$		
N_{Nu}	Nusselt number, hD/k or hL/k		
N_{Pe}	Peclet number, DGc/k		
N_{Pr}	Prandtl number, $c\mu/k$		
N_{Re}	Reynolds number, DG/μ		
N_{St}	Stanton number, $N_{Nu}/N_{Re}N_{Pr}$		
N_{ss}	Number of sealing strips		
p	Pressure	kPa	lbf/ft^2 abs
p_f	Perimeter of a fin	m	ft
p'	Center-to-center spacing of tubes in tube bundle (tube pitch); p_n for tube pitch normal to flow; p_p for tube pitch parallel to flow	m	ft
Δp	Pressure of the vapor in a bubble minus saturation pressure of a flat liquid surface	kPa	lbf/ft^2 abs
P	Absolute pressure; P_c for critical pressure	kPa	lbf/ft^2
P'	Spacing between adjacent baffles on shell side of a heat exchanger (baffle pitch)	m	ft
ΔP_{bk}, ΔP_{wk}	Pressure drop for ideal-tube-bank cross-flow and ideal window respectively; ΔP_s for shell side of baffled exchanger	kPa	lbf/ft^2
q	Rate of heat flow, equals Q/θ	W, J/s	Btu/h
q'	Rate of heat generation	$J/(s \cdot m^3)$	$Btu/(h \cdot ft^3)$
$(q/A)_{max}$	Maximum heat flux in nucleate boiling	$J/(s \cdot m^2)$	$Btu/(h \cdot ft^2)$
Q	Quantity of heat; rate of heat transfer	J/s	Btu/h
Q	Quantity of heat; Q_T for total quantity	J	Btu
r	Radius; cylindrical and spherical coordinate; distance from midplane to a point in a body; r_1 for inner wall of annulus; r_2 for outer wall of annulus; r_i for inside radius of tube; r_m for distance from midplane or center of a body to the exterior surface of the body	m	ft
r_j	Inside radius	Dimensionless	Dimensionless
R	Thermal resistance, equals x/kA, $1/UA$, $1/hA$; R_1, R_2, R_3, R_n for thermal resistance of sections 1, 2, 3, and n of a composite body; R_T for sum of individual resistances of several resistances in series or parallel; R_{di} and R_{do} for dirt or scale resistance on inner and outer surface respectively	$(s \cdot K)/J$	$(h \cdot °F)/Btu$
R_f	Ratio of total outside surface of finned tube to area of tube having same root diameter		
S	Cross-sectional area; S_m for minimum cross-sectional area between rows of tubes, flow normal to tubes; S_{tb} for tube-to-baffle leakage area for one baffle; S_{sb} for shell-to-baffle area for one baffle; S_w for area for flow through window; S_{wg} for gross window area; S_{wt} for window area occupied by tubes	m^2	ft^2
S_r	Slope of rotary shell		
s	Specific gravity of fluid referred to liquid water		
t	Bulk temperature; temperature at a given point in a body at time θ	K	°F
t_1, t_2, t_n	Temperature at points 1, 2, and n in a system through which heat is being transferred	K	°F
t'	Temperature of surroundings	K	°F
t'_1, t'_2	Inlet and outlet temperature respectively of hotter fluid	K	°F
t''_1, t''_2	Inlet and outlet temperature respectively of colder fluid	K	°F
t_b	Initial uniform bulk temperature of a body; bulk temperature of a flowing fluid	K	°F
t_H, t_L	High and low temperature respectively on tube side of a heat exchanger	K	°F
t_s	Surface temperature	K	°F
t_{sv}	Saturated-vapor temperature	K	°F
t_w	Wall temperature	K	°F
t_∞	Temperature of undisturbed flowing stream	K	°F
T_H, T_L	High and low temperature respectively on shell side of a heat exchanger	K	°F

Nomenclature and Units ***(continued)***

Symbol	Definition	SI units	U.S. customary units
T	Absolute temperature; T_b for bulk temperature; T_w for wall temperature; T_v for vapor temperature; T_c for coolant temperature; T_e for temperature of emitter; T_r for temperature of receiver	K	°F
ΔT, Δt	Temperature difference; Δt_1, Δt_2, and Δt_3 temperature difference across bodies 1, 2, and 3 or at points 1, 2, and 3; ΔT_o, Δt_o for overall temperature difference; Δt_b for temperature difference between surface and boiling liquid	K	°F
Δt_{am}, Δt_{lm}	Arithmetic- and logarithmic-mean temperature difference respectively	K	°F
Δt_{om}	Mean effective overall temperature difference	K	°F
ΔT_H, Δt_H	Greater terminal temperature difference	K	°F
ΔT_L, Δt_L	Lesser terminal temperature difference	K	°F
ΔT_m, Δt_m	Mean temperature difference	K	°F
u	Velocity in x direction	m/s	ft/h
u^*	Friction velocity	m/s	ft/h
U	Overall coefficient of heat transfer; U_o for outside surface basis; U' for overall coefficient between liquid-vapor interface and coolant	J/(s·m²·K)	Btu/(h·ft²·°F)
U_1, U_2	Overall coefficient of heat transfer at points 1 and 2 respectively	J/(s·m²·K)	Btu/(h·ft²·°F)
U_{co}, U_{cv}, U_{ct}, U_{ra}	Overall coefficients for divided solids processing by conduction, convection, contact, and radiation mechanism respectively	J/(s·m²·K)	Btu/(h·ft²·°F)
U_m	Mean overall coefficient of heat transfer	J/(s·m²·K)	Btu/(h·ft²·°F)
v	Velocity in y direction	m/s	ft/h
V_r	Volume of rotating shell	m³	ft³
V	Velocity	m/s	ft/h
V', V_s	Velocity	m/s	ft/s
V_F	Face velocity of a fluid approaching a bank of finned tubes	m/s	ft/h
V_g, V_l	Specific volume of gas, liquid	m³/kg	ft³/lb
V'_{max}	Maximum velocity through minimum free area between rows of tubes normal to the fluid stream	m/s	ft/h
w	Velocity in z direction	m/s	ft/h
w	Flow rate	kg/s	lb/h
W	Total mass rate of flow; mass rate of vapor generated; W_T for total rate of vapor condensation in one tube	kg/s	lb/h
W_r	Weight rate of flow	kg/(s·tube)	lb/(h·tube)
W_1, W_o	Total mass rate of flow on tube side and shell side respectively of a heat exchanger	kg/s	lb/h
x_q	Vapor quality, x_i for inlet quality, x_o for outlet quality	kg/s	lb/h
x	Coordinate direction; length of conduction path; x_s for thickness of scale; x_1, x_2, and x_3 at positions 1, 2, and 3 in a body through which heat is being transferred	m	ft
X	Factor	Dimensionless	Dimensionless
y	Coordinate direction	m	ft
y^+	Wall distance	Dimensionless	Dimensionless
Y	Factor	Dimensionless	Dimensionless
z	Coordinate direction	m	ft
z_p	Distance (perimeter) traveled by fluid across fin	m	ft
Z_H	Ratio of sensible heat removed from vapor to total heat transferred	Dimensionless	Dimensionless
	Greek symbols		
α	Thermal diffusivity, equals $k/\rho c$; α_e for effective thermal diffusivity of powdered solids	m²/s	ft²/h
β	Volumetric coefficient of thermal expansion	K^{-1}	$°F^{-1}$
β'	Contact angle of a bubble	°	°
γ	Fluid consistency	$kg/(s^{2-n'}\cdot m)$	$lb/(ft\cdot s^{2-n'})$
Γ	Mass rate of flow of a falling film from a tube or surface per unit perimeter, equals $w/\pi D$ for vertical tube, $w/2L$ for horizontal tube	kg/(s·m)	lb/(h·ft)
δ_s	Correction factor, ratio of nonnewtonian to newtonian shear rates		
δ	Cell width		
δ_{sb}	Diametral shell-to-baffle clearance	m	ft

Nomenclature and Units ***(continued)***

Symbol	Definition	SI units	U.S. customary units
ϵ	Eddy diffusivity; ϵ_M for eddy diffusivity of momentum; ϵ_H for eddy diffusivity of heat	m^2/s	ft^2/h
ϵ_v	Fraction of voids in porous bed		
η	Fluidization efficiency		
θ	Time	s	h
θ_b	Baffle cut		
λ	Latent heat (enthalpy) of vaporization (condensation)	J/kg	Btu/lb
λ_m	Radius of maximum velocity	m	ft
μ	Viscosity; μ_w for viscosity at wall temperature; μ_b for viscosity at bulk temperature; μ_f for viscosity at film temperature; μ_G, μ_g, and μ_v for viscosity of gas or vapor: μ_L, μ_l for viscosity of liquid; μ_w for viscosity at wall; μ_l for viscosity of fluid at inner wall of annulus	Pa·s	lb/(h·ft)
ν	Kinematic viscosity	m^2/s	ft^2/h
ρ	Density; ρ_L, ρ_l for density of liquid; ρ_G, ρ_v for density of gas or vapor; ρ_s for density of solid	kg/m^3	lb/ft^3
σ	Surface tension between a liquid and its vapor	N/m	lbf/ft
Σ	Term indicating summation of variables		
τ	Shear stress τ_w for shear stress at the wall	N/m^2	lbf/ft^2
ϕ	Velocity-potential function		
ϕ_p	Particle sphericity		
Φ	Viscous-dissipation function		
ω	Angle of repose of powdered solid	rad	rad
Ω	Fin efficiency	Dimensionless	Dimensionless

INTRODUCTION

General References: Bird, Stewart, and Lightfoot, *Transport Phenomena*, Wiley, New York, 1960. Carslaw and Jaeger, *Conduction of Heat in Solids*, Clarendon Press, Oxford, 1959. Chapman, *Heat Transfer*, 2d ed., Macmillan, New York, 1967. Drew and Hoopes, *Advances in Chemical Engineering*, Academic, New York, vol. 1, 1956; vol. 2, 1958; vol. 5, 1964; vol. 6, 1966; vol. 7, 1968. Dusinberre, *Heat Transfer Calculations by Finite Differences*, International Textbook, Scranton, Pa., 1961. Eckert and Drake, *Heat and Mass Transfer*, 2d ed., McGraw-Hill, New York, 1959. Gebhart, *Heat Transfer*, McGraw-Hill, New York, 1961. Irvine and Hartnett, *Advances in Heat Transfer*, Academic, New York, vol. 1, 1964; vol. 2, 1965; vol. 3, 1966. Jakob, *Heat Transfer*, Wiley, New York, vol. 1, 1949; vol. 2, 1957. Jakob and Hawkins, *Elements of Heat Transfer*, 3d ed., Wiley, New York, 1957. Kakac, Bergles, and Mayinger, *Heat Exchangers: Thermal Hydraulic Fundamentals and Design*, Hemisphere Publishing, Washington, 1981. Kakac and Yener, *Convective Heat Transfer*, Hemisphere Publishing, Washington, 1980. Kay, *An Introduction to Fluid Mechanics and Heat Transfer*, 2d ed., Cambridge University Press, Cambridge, England, 1963. Kays, *Convective Heat and Mass Transfer*, McGraw-Hill, New York, 1966. Kays and London, *Compact Heat Exchangers*, 2d ed., McGraw-Hill, New York, 1964. Kern, *Process Heat Transfer*, McGraw-Hill, New York, 1950. Knudsen and Katz, *Fluid Dynamics and Heat Transfer*, McGraw-Hill, New York, 1958. Kraus, *Analysis and Evaluation of Extended Surface Thermal Systems*, Hemisphere Publishing, Washington, 1982. Kutatladze, *A Concise Encyclopedia of Heat Transfer*, 1st English ed., Pergamon, New York, 1966. Lykov, *Heat and Mass Transfer in Capillary Porous Bodies*, translated from Russian, Pergamon, New York, 1966. McAdams, *Heat Transmission*, 3d ed., McGraw-Hill, New York, 1954. Mickley, Sherwood, and Reed, *Applied Mathematics in Chemical Engineering*, 2d ed., McGraw-Hill, New York, 1957. Rohsenow and Choi, *Heat, Mass, and Momentum Transfer*, Prentice-Hall, Englewood Cliffs, N.J., 1961. Schlünder (ed.), *Heat Exchanger Design Handbook*, Hemisphere Publishing, Washington, 1983. Skelland, *Non-Newtonian Flow and Heat Transfer*, Wiley, New York, 1967. Taborek and Bell, *Process Heat Exchanger Design*, Hemisphere Publishing, Washington, 1984. Taborek, Hewitt, and Afghan, *Heat Exchangers: Theory and Practice*, Hemisphere Publishing, Washington, 1983. TSederberg, *Thermal Conductivity of Liquids and Gases*, M.I.T., Cambridge, Mass., 1965. Welty, Wicks, and Wilson, *Fundamentals of Momentum, Heat and Mass Transfer*, Wiley, New York, 1969. Zenz and Othmer, *Fluidization and Fluid Particle Systems*, Reinhold, New York, 1960.

MODES OF HEAT TRANSFER

There are three fundamental types of heat transfer: conduction, convection, and radiation. All three types may occur at the same time, and it is advisable to consider the heat transfer by each type in any particular case.

Conduction is the transfer of heat from one part of a body to another part of the same body, or from one body to another in physical contact with it, without appreciable displacement of the particles of the body.

Convection is the transfer of heat from one point to another within a fluid, gas, or liquid by the mixing of one portion of the fluid with another. In natural convection, the motion of the fluid is entirely the result of differences in density resulting from temperature differences; in forced convection, the motion is produced by mechanical means. When the forced velocity is relatively low, it should be realized that "free-convection" factors, such as density and temperature difference, may have an important influence.

Radiation is the transfer of heat from one body to another, not in contact with it, by means of wave motion through space.

HEAT TRANSMISSION BY CONDUCTION

Fourier's Law Fourier's law is the fundamental differential equation for heat transfer by conduction:

$$dQ/d\theta = -kA(dt/dx) \tag{10-1}$$

where $dQ/d\theta$ (quantity per unit time) is the rate of flow of heat, A is the area at right angles to the direction in which the heat flows, and $-dt/dx$ is the rate of change of temperature with the distance in the direction of the flow of heat, i.e., the temperature gradient. The factor k is called the thermal conductivity; it is a characteristic property of the material through which the heat is flowing and varies with temperature.

Three-Dimensional Conduction Equation Equation (10-1) is used as a basis for derivation of the unsteady-state three-dimensional energy equation for **solids or static fluids:**

$$c\rho\frac{\partial t}{\partial \theta} = \frac{\partial}{\partial x}\left(k\frac{\partial t}{\partial x}\right) + \frac{\partial}{\partial y}\left(k\frac{\partial t}{\partial y}\right) + \frac{\partial}{\partial z}\left(k\frac{\partial t}{\partial z}\right) + q' \tag{10-2}$$

where x, y, z are distances in the rectangular coordinate system and q' is the rate of heat generation (by chemical reaction, nuclear reaction, or electric current) in the solid per unit of volume. Solution of Eq. (10-2) with appropriate boundary and initial conditions will give the temperature as a function of time and location in the material. Equation (10-2) may be transformed into spherical or cylindrical coordinates to conform more closely to the physical shape of the system.

Thermal Conductivity Thermal conductivity varies with temperature but not always in the same direction. The thermal conductivities for many materials, as a function of temperature, are given in Sec. 3. Additional and more comprehensive information may often be obtained from suppliers of the materials. Impurities, especially in metals, can give rise to variations in thermal conductivity of from 50 to 75 percent. In using thermal conductivities, engineers should remember that conduction is not the sole method of transferring heat and that, particularly with liquids and gases, radiation and convection may be much more important.

The thermal conductivity at a given temperature is a function of the apparent, or bulk, density. Thus, at 0°C (32°F), k for asbestos wool is 0.09 J/(m·s·K) [0.052 Btu/(hr·ft·°F)] when the bulk density is 400 kg/m^3 (24.9 lb/ft^3) and is 0.19 (0.111) for a density of 700 (43.6).

In determining the apparent thermal conductivities of **granular solids,** such as granulated cork or charcoal grains, Griffiths (Spec. Rep. 5, Food Investigation Board, H. M. Stationery Office, 1921) found that air circulates within the mass of granular solid. Under a certain set of conditions, the apparent thermal conductivity of a charcoal was 9 percent greater when the test section was vertical than when it was horizontal. When the apparent conductivity of a mixture of cellular or porous nonhomogeneous solid is determined, the observed temperature coefficient may be much larger than for the homogeneous solid alone, because heat is transferred not only by the mechanism of conduction but also by convection in the gas pockets and by radiation from surface to surface of the individual particles. If internal radiation is an important factor, a plot of the apparent conductivity as ordinates versus temperature should show a curve concave upward, since radiation increases with the fourth power of the absolute temperature. Griffiths noted that cork, slag, wool, charcoal, and wood fibers, when of good quality and dry, have thermal conductivities about 2.2 times that of still air, whereas a highly cellular form of rubber, 112 kg/m^3 (7 lb/ft^3), had a thermal conductivity only 1.6 times that of still air. In measuring the apparent thermal conductivity of diathermanous substances such as quartz (especially when exposed to radiation emitted at high temperatures), it should be remembered that a part of the heat is transmitted by radiation.

Bridgman [*Proc. Am. Acad. Arts Sci.*, **59,** 141 (1923)] showed that the thermal conductivity of **liquids** is increased by only a few percent under a pressure of 100,330 kPa (1000 atm). The thermal conductivity of some liquids varies with temperature through a maximum. It is often necessary for the engineer to estimate thermal conductivities; methods are indicated in Sec. 3.

Equation (10-2) considers the thermal conductivity to be variable. If k is expressed as a function of temperature, Eq. (10-2) is nonlinear and difficult to solve analytically except for certain special cases. Usually in complicated systems numerical solution by means of computer is possible. A complete review of heat conduction has been given by Davis and Akers [*Chem. Eng.*, **67**(4), 187, (5), 151 (1960)] and by Davis [*Chem. Eng.*, **67**(6), 213, (7), 135 (8), 137 (1960)].

STEADY-STATE CONDUCTION

For steady flow of heat, the term $dQ/d\theta$ in Eq. (10-1) is constant and may be replaced by Q/θ or q. Likewise, in Eq. (10-2) the term $\partial t/\partial\theta$ is zero. Hence, for constant thermal conductivity, Eq. (10-2) may be expressed as

$$\nabla^2 t = (q'/k) \tag{10-3}$$

One-Dimensional Conduction Many heat-conduction problems may be formulated into a one-dimensional or pseudo-one-dimensional form in which only one space variable is involved. Forms of the conduction equation for rectangular, cylindrical, and spherical coordinates are, respectively,

$$\frac{\partial^2 t}{\partial x^2} = -\frac{q'}{k} \tag{10-4a}$$

$$\frac{1}{r}\frac{d}{dr}\left(r\frac{dt}{dr}\right) = -\frac{q'}{k} \tag{10-4b}$$

$$\frac{1}{r^2}\frac{d}{dr}\left(r^2\frac{dt}{dr}\right) = -\frac{q'}{k} \tag{10-4c}$$

These are second-order differential equations which upon integration become, respectively,

$$t = -(q'x^2/2k) + c_1 x + c_2 \tag{10-5a}$$

$$t = -(q'r^2/4k) + c_1 \ln r + c_2 \tag{10-5b}$$

$$t = -(q'r^2/6k) - (c_1/r) + c_2 \tag{10-5c}$$

Constants of integration c_1 and c_2 are determined by the boundary conditions, i.e., temperatures and temperature gradients at known locations in the system.

For the case of a solid surface exposed to surroundings at a different temperature and for a finite surface coefficient, the **boundary condition** is expressed as

$$h_T(t_s - t') = -k\,(dt/dx)_{surf} \tag{10-6}$$

Inspection of Eqs. (10-5a), (10-5b), and (10-5c) indicates the form of temperature profile for various conditions and geometries and also reveals the effect of the heat-generation term q' upon the temperature distributions.

In the **absence of heat generation,** one-dimensional steady-state conduction may be expressed by integrating Eq. (10-1):

$$q\int_{x1}^{x2}\frac{dx}{A} = -\int_{t1}^{t2} k\,dt \tag{10-7}$$

Area A must be known as a function of x. If k is constant, Eq. (10-7) is expressed in the integrated form

$$q = kA_{avg}(t_1 - t_2)/(x_2 - x_1) \tag{10-8}$$

where

$$A_{avg} = \frac{1}{x_2 - x_1}\int_{x1}^{x2}\frac{dx}{A} \tag{10-9}$$

Examples of values of A_{avg} for various functions of x are shown in the following table.

Area proportional to	A_{avg}
Constant	$A_1 = A_2$
x	$\dfrac{A_2 - A_1}{\ln (A_2/A_1)}$
x^2	$\sqrt{A_2 A_1}$

Usually, thermal conductivity k is not constant but is a function of temperature. In most cases, over the ranges of values used the relation is linear. Integration of Eq. (10-7), with k linear in t, gives

$$q\int_{x1}^{x2}\frac{dx}{A} = k_{avg}(t_1 - t_2) \tag{10-10}$$

where k_{avg} is the arithmetic-average thermal conductivity between temperatures t_1 and t_2. This average probably gives results which are correct within the precision of the data in the majority of cases, though a special integration can be made whenever k is known to be greatly different from linear in temperature.

Conduction through Several Bodies in Series Figure 10-1 illustrates diagrammatically the temperature gradients accompanying the steady conduction of heat in series through three solids.

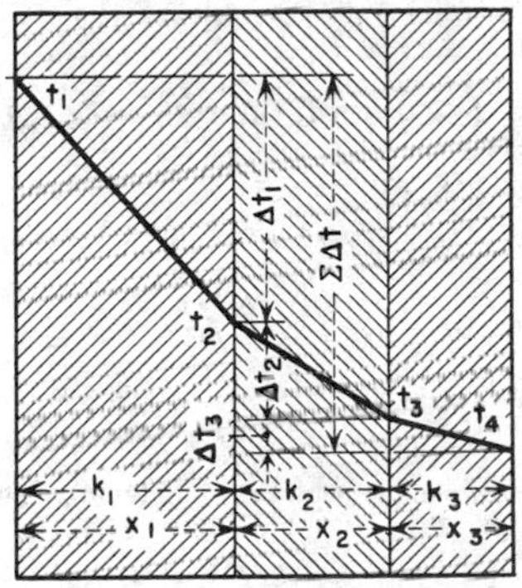

FIG. 10-1 Temperature gradients for steady heat conduction in series through three solids.

Since the heat flow through each of the three walls must be the same,

$$q = (k_1A_1\,\Delta t_1/x_1) = (k_2A_2\,\Delta t_2/x_2) = (k_3A_3\,\Delta t_3/x_3) \tag{10-11}$$

Since, by definition, individual thermal resistance

$$R = x/kA \tag{10-12}$$

then

$$\Delta t_1 = qR_1 \quad \Delta t_2 = qR_2 \quad \Delta t_3 = qR_3 \tag{10-13}$$

Adding the individual temperature drops, noting that q is uniform,

$$q(R_1 + R_2 + R_3) = \Delta t_1 + \Delta t_2 + \Delta t_3 = \Sigma\,\Delta t \tag{10-14}$$

or

$$q = \Sigma\,\Delta t/R_T = (t_1 - t_4)/R_T \tag{10-15}$$

where R_T is the overall resistance and is the sum of the individual resistances in series, then

$$R_T = R_1 + R_2 + \cdots + R_n \tag{10-16}$$

When a wall is constructed of several layers of solids, the joints at adjacent layers may not perfectly exclude air spaces, and these additional resistances should not be overlooked.

Conduction through Several Bodies in Parallel For n resistances in parallel, the rates of heat flow are additive:

$$q = \Delta t/R_1 + \Delta t/R_2 + \cdots + \Delta t/R_n \tag{10-17a}$$

$$q = \left(\frac{1}{R_1} + \frac{1}{R_2} + \cdots + \frac{1}{R_n}\right)\Delta t \qquad (10\text{-}17b)$$

$$q = (C_1 + C_2 + \cdots + C_n)\,\Delta t = \Sigma C\,\Delta t \qquad (10\text{-}17c)$$

where R_1 to R_n are the individual resistances and C_1 to C_n are the individual conductances; $C = kA/x$.

Several Bodies in Series with Heat Generation The simple Fourier type of equation indicated by Eq. (10-15) may *not* be used when heat generation occurs in one of the bodies in the series. In this case, Eq. (10-5*a*), (10-5*b*), or (10-5*c*) must be solved with appropriate boundary conditions.

Example 1 A plate-type nuclear fuel element, consisting of a uranium-zirconium alloy $(3.2)(10^{-3})$ m (0.125 in) thick clad on each side with a $(6.4)(10^{-4})$-m- (0.025-in-) thick layer of zirconium, is cooled by water under pressure at 200°C (400°F), the heat-transfer coefficient being 42,600 J/(m²·s·K) [7500 Btu/(h·ft²·°F)]. If the temperature at the center of the fuel must not exceed 570°C (1050°F), determine the maximum rate of heat generation in the fuel. The zirconium and zirconium alloy have a thermal conductivity of 21 J/(m·s·K) [12 Btu/(h·ft²)(°F/ft)].

Solution. Equation (10-4*a*) may be integrated for each material. The heat generation is zero in the cladding, and its value for the fuel may be determined from the integrated equations. Let $x = 0$ at the midplane of the fuel. Then $x_1 = (1.6)(10^{-3})$ m (0.0625 in) at the cladding-fuel interface and $x_2 = (2.2)(10^{-3})$ m (0.0875 in) at the cladding-water interface. Let the subscripts c, f refer to cladding and fuel respectively.

The boundary conditions are:

For fuel, at $x = 0$, $t = 570$°C (1050°F), $dt/dx = 0$ (this follows if the temperature is finite at the midplane).

For fuel and cladding, at $x = x_1$, $t_f = t_c$,

$$k_f(dt/dx) = k_c(dt/dx)$$

For cladding, at $x = x_2$,

$$t_c - 400 = -(k_c/42{,}600)(dt/dx)$$

For the fuel, the first integration of Eq. (10-4*a*) gives

$$dt_f/dx = -(q'/k_f)x + c_1$$

which gives $c_1 = 0$ when the boundary condition is applied. Thus the second integration gives

$$t_f = -(q'/2k_f)x^2 + c_2$$

from which c_2 is determined to be 570 (1050) upon application of the boundary condition. Thus the temperature profile in the fuel is

$$t_f = -(q'/2k_f)x^2 + 570$$

The temperature profile in the *cladding* is obtained by integrating Eq. (10-4*a*) twice with $q' = 0$. Hence

$$(dt_c/dx) = c_1 \quad \text{and} \quad t_c = c_1x + c_2$$

There are now three unknowns, c_1, c_2, and q', and three boundary conditions by which they can be determined.

At $x = x_1$,

$$q'x_1^2/2k_f + 570 = c_1x_1 + c_2 - k_f\,q'x_1/k_f = k_cc_1$$

At $x = x_2$,

$$c_1x_2 + c_2 - 200 = -(k_c/42{,}600)c_1$$

From which $q' = (2.53)(10^9)$ J/(m³·s) [(2.38)(10⁸)Btu/(h·ft³)]
$c_1 = -(1.92)(10^5)$
$c_2 = 724$

Two-Dimensional Conduction If the temperature of a material is a function of two space variables, the two-dimensional conduction equation is (assuming constant k)

$$\partial^2t/\partial x^2 + \partial^2t/\partial y^2 = -q'/k \qquad (10\text{-}18)$$

When q' is zero, Eq. (10-18) reduces to the familiar Laplace equation. The analytical solution of Eq. (10-18) as well as of Laplace's equation is possible for only a few boundary conditions and geometric shapes. Carslaw and Jaeger (*Conduction of Heat in Solids*, Clarendon Press, Oxford, 1959) have presented a large number of analytical solutions of differential equations applicable to heat-conduction problems. Generally, graphical or numerical **finite-difference methods** are most frequently used. Other numerical and relaxation methods may be found in the general references in the "Introduction." The methods may also be extended to three-dimensional problems.

UNSTEADY-STATE CONDUCTION

When temperatures of materials are a function of both time and space variables, more complicated equations result. Equation (10-2) is the three-dimensional unsteady-state conduction equation. It involves the rate of change of temperature with respect to time $\partial t/\partial\theta$. Solutions to most practical problems must be obtained through the use of digital computers. Numerous articles have been published on a wide variety of transient conduction problems involving various geometrical shapes and boundary conditions.

One-Dimensional Conduction The one-dimensional transient conduction equations are (for constant physical properties)

$$\partial t/\partial\theta = \alpha(\partial^2t/\partial x^2) + q'/c\rho \quad \text{(rectangular coordinates)} \qquad (10\text{-}19a)$$

$$\frac{\partial t}{\partial\theta} = \frac{\alpha}{r}\frac{\partial}{dr}\left(r\frac{\partial t}{\partial r}\right) + \frac{q'}{c\rho} \quad \text{(cylindrical coordinates)} \qquad (10\text{-}19b)$$

$$\frac{\partial t}{\partial\theta} = \frac{\alpha}{r^2}\frac{\partial}{\partial r}\left(r^2\frac{\partial t}{\partial r}\right) + \frac{q'}{c\rho} \quad \text{(spherical coordinates)} \qquad (10\text{-}19c)$$

These equations have been solved analytically for solid slabs, cylinders, and spheres. The solutions are in the form of infinite series, and usually the results are plotted as curves involving four ratios [Gurney and Lurie, *Ind. Eng. Chem.*, **15**, 1170 (1923)] defined as follows with $q' = 0$:

$$Y = (t' - t)/(t' - t_b) \qquad X = k\theta/\rho c r_m^2 \qquad (10\text{-}20a,b)$$

$$m = k/h_Tr_m \qquad n = r/r_m \qquad (10\text{-}20c,d)$$

Since each ratio is dimensionless, any consistent units may be employed in any ratio. The significance of the symbols is as follows: t' = temperature of the surroundings; t_b = initial uniform temperature of the body; t = temperature at a given point in the body at the time θ measured from the start of the heating or cooling operations; k = uniform thermal conductivity of the body; ρ = uniform density of the body; c = specific heat of the body; h_T = coefficient of total heat transfer between the surroundings and the surface of the body expressed as heat transferred per unit time per unit area of the surface per unit difference in temperature between surroundings and surface; r = distance, in the direction of heat conduction, from the midpoint or midplane of the body to the point under consideration; r_m = radius of a sphere or cylinder, one-half of the thickness of a slab heated from both faces, the total thickness of a slab heated from one face and insulated perfectly at the other; and x = distance, in the direction of heat conduction, from the surface of a semi-infinite body (such as the surface of the earth) to the point under consideration. In making the integrations which lead to the curves shown, the following factors were assumed constant: c, h_T, k, r, r_m, t', x, and ρ.

The working curves are shown in Figs. 10-2 to 10-5 for **cylinders of infinite length, spheres, slabs of infinite faces,** and **semi-infinite solids** respectively, with Y plotted as ordinates on a logarithmic scale versus X as abscissas to an arithmetic scale, for various values of the ratios m and n. To facilitate calculations involving instantaneous rates of cooling or heating of the semi-infinite body, Fig. 10-5 shows also a curve of dY/dX versus X. Similar plots to a larger scale are given in McAdams, Brown and Marco, Schack, and Stoever (see "Introduction: General References"). For a solid of infinite thickness (Fig. 10-5) and with $m = 0$,

$$Y = \frac{2}{\sqrt{\pi}}\int_0^z \exp(-z^2)\,dz \qquad (10\text{-}21)$$

where $z = 1/\sqrt{2X}$ and the "error integral" may be evaluated from standard mathematical tables.

Various numerical and graphical methods are used for unsteady-state conduction problems, in particular the Schmidt graphical

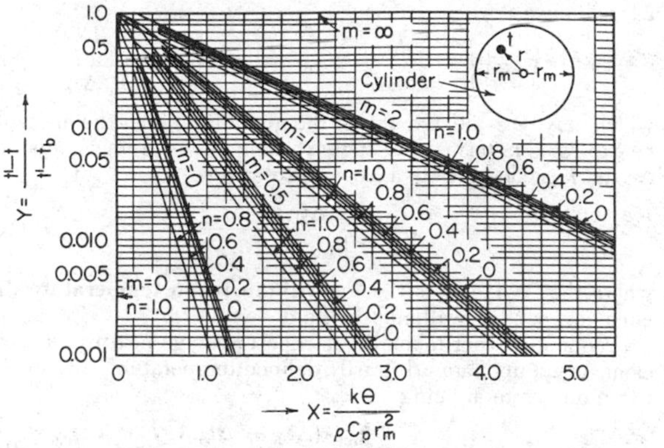

FIG. 10-2 Heating and cooling of a solid cylinder having an infinite ratio of length to diameter.

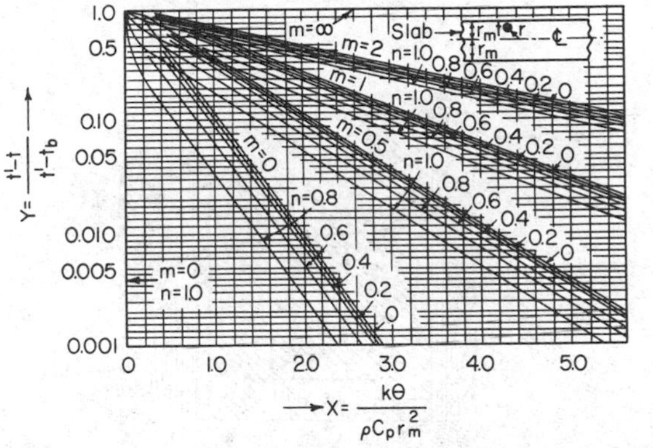

FIG. 10-4 Heating and cooling of a solid slab having a large face area relative to the area of the edges.

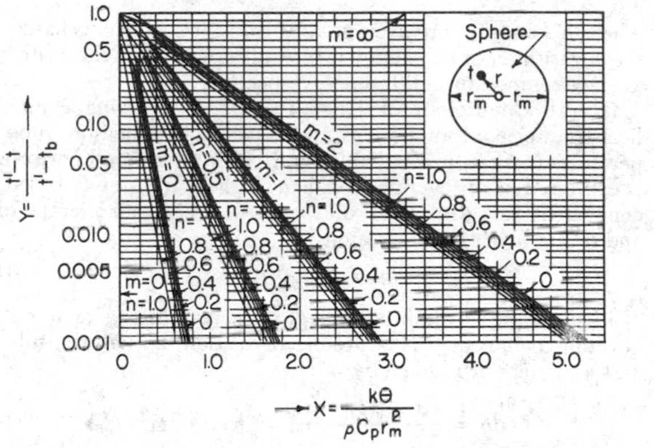

FIG. 10-3 Heating and cooling of a solid sphere.

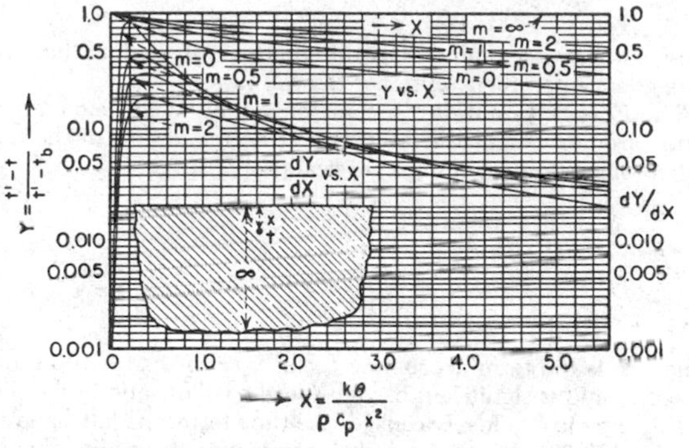

FIG. 10-5 Heating and cooling of a solid of infinite thickness, neglecting edge effects. (This may be used as an approximation in the zone near the surface of a body of finite thickness.)

method (*Foppls Festschrift,* Springer-Verlag, Berlin, 1924). These methods are very useful because any form of initial temperature distribution may be used.

Two-Dimensional Conduction The governing differential equation for two-dimensional transient conduction is

$$\frac{\partial t}{\partial \theta} = \alpha\left(\frac{\partial^2 t}{\partial x^2} + \frac{\partial^2 t}{\partial y^2}\right) + \frac{q'}{c\rho} \tag{10-22}$$

McAdams (*Heat Transmission,* 3d ed., McGraw-Hill, New York, 1954) gives various forms of transient difference equations and methods of solving transient conduction problems. The availability of computers and a wide variety of computer programs permits virtually routine solution of many rather complicated conduction problems.

Conduction with Change of Phase A special type of transient problem (the Stefan problem) involves conduction of heat in a material when freezing or melting occurs. The liquid-solid interface moves with time, and in addition to conduction, latent heat is either generated or absorbed at the interface. Various problems of this type are discussed by Bankoff [in Drew et al. (eds.), *Advances in Chemical Engineering,* vol. 5, Academic, New York, 1964].

HEAT TRANSFER BY CONVECTION

COEFFICIENT OF HEAT TRANSFER

In many cases of heat transfer involving either a liquid or a gas, convection is an important factor. In the majority of heat-transfer cases met in industrial practice, heat is being transferred from one fluid through a solid wall to another fluid. Assume a hot fluid at a temperature t_1 flowing past one side of a metal wall and a cold fluid at t_7 flowing past the other side to which a scale of thickness x_s adheres. In such a case, the conditions obtaining at a given section are illustrated diagrammatically in Fig. 10-6.

For turbulent flow of a fluid past a solid, it has long been known that, in the immediate neighborhood of the surface, there exists a relatively quiet zone of fluid, commonly called the **film.** As one approaches the wall from the body of the flowing fluid, the flow tends to become less turbulent and develops into laminar flow immediately adjacent to the wall. The film consists of that portion of the flow which is essentially in laminar motion (the laminar sublayer) and through which heat is transferred by molecular conduction. The resistance of the laminar layer to heat flow will vary according to its thickness and can range from 95 percent of the total resistance for some fluids to about 1 percent for other fluids (liquid metals). The turbulent core and the buffer layer between the laminar sublayer and turbulent core each offer a **resistance to heat transfer** which is a function of the turbulence and the thermal properties of the flowing

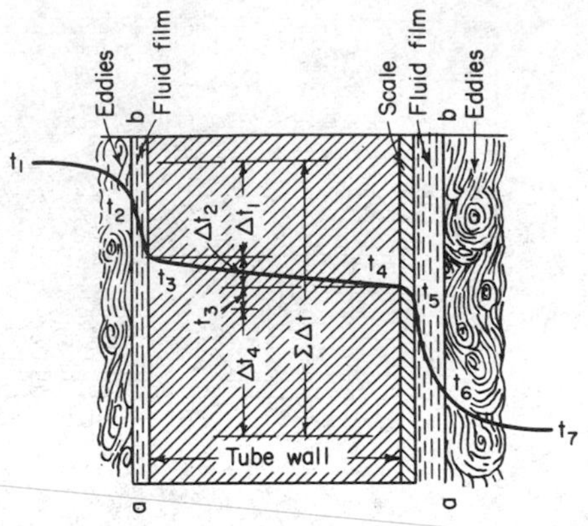

FIG. 10-6 Temperature gradients for a steady flow of heat by conduction and convection from a warmer to a colder fluid separated by a solid wall.

fluid. The relative temperature difference across each of the layers is dependent upon their resistance to heat flow.

The Energy Equation A complete energy balance on a flowing fluid through which heat is being transferred results in the energy equation (assuming constant physical properties):

$$c\rho\left(\frac{\partial t}{\partial \theta} + u\frac{\partial t}{\partial x} + v\frac{\partial t}{\partial y} + w\frac{\partial t}{\partial z}\right) = k\left(\frac{\partial^2 t}{\partial x^2}\frac{\partial^2 t}{\partial y^2} + \frac{\partial^2 t}{\partial z^2}\right) + q' + \Phi \qquad (10\text{-}23)$$

where Φ is the term accounting for energy dissipation due to fluid viscosity and is significant in high-speed gas flow and in the flow of highly viscous liquids. Except for the time term, the left-hand terms of Eq. (10-23) are the so-called **convective terms** involving the energy carried by the fluid by virtue of its velocity. Therefore, the solution of the equation is dependent upon the solution of the momentum equations of flow. Solutions of Eq. (10-23) exist only for several simple flow cases and geometries and mainly for laminar flow. For turbulent flow the difficulties of expressing the fluid velocity as a function of space and time coordinates and of obtaining reliable values of the effective thermal conductivity of the flowing fluid have prevented solution of the equation unless simplifying assumptions and approximations are made.

Individual Coefficient of Heat Transfer Because of the complicated structure of a turbulent flowing stream and the impracticability of measuring thicknesses of the several layers and their temperatures, the **local rate of heat transfer** between fluid and solid is defined by the equations

$$dq = h_i\, dA_i\,(t_1 - t_3) = h_o\, dA_o\,(t_5 - t_7) \qquad (10\text{-}24)$$

where h_i and h_o are the local heat-transfer coefficients inside and outside the wall, respectively, and temperature t is defined by Fig. 10-6.

The definition of the heat-transfer coefficient is arbitrary, depending on whether *bulk-fluid temperature, centerline temperature,* or *some other reference temperature* is used for t_1 or t_7. Equation (10-24) is an expression of Newton's law of cooling and incorporates all the complexities involved in the solution of Eq. (10-23). The **temperature gradients** in both the fluid and the adjacent solid at the fluid-solid interface may also be related to the heat-transfer coefficient:

$$dq = h_i\, dA_i\,(t_1 - t_3) = \left(-k\frac{dt}{dx}\right)_{\text{fluid}} = \left(-k\frac{dt}{dx}\right)_{\text{solid}} \qquad (10\text{-}25)$$

Equation (10-25) holds for the liquid *only* if laminar flow exists immediately adjacent to the solid surface. The integration of Eq. (10-24) will give

$$A_i = \int_{\text{in}}^{\text{out}} \frac{dq}{h_i\,\Delta t_i} \qquad \text{or} \qquad A_o = \int_{\text{in}}^{\text{out}} \frac{dq}{h_o\,\Delta t_o} \qquad (10\text{-}26)$$

which may be evaluated only if the quantities under the integral can be expressed in terms of a single variable. If q is a linear function of Δt and h is constant, then Eq. (10-26) gives

$$q = \frac{hA(\Delta t_{\text{in}} - \Delta t_{\text{out}})}{\ln\,(\Delta t_{\text{in}}/\Delta t_{\text{out}})} \qquad (10\text{-}27)$$

where the Δt factor is the **logarithmic-mean temperature difference** between the wall and the fluid.

Frequently experimental data report average heat-transfer coefficients based upon an arbitrarily defined temperature difference, the two most common being

$$q = \frac{h_{lm}A(\Delta t_{\text{in}} - \Delta t_{\text{out}})}{\ln(\Delta t_{\text{in}}/\Delta t_{\text{out}})} \qquad (10\text{-}28a)$$

$$q = \frac{h_{am}A(\Delta t_{\text{in}} + \Delta t_{\text{out}})}{2} \qquad (10\text{-}28b)$$

where h_{lm} and h_{am} are average heat-transfer coefficients based upon the logarithmic-mean temperature difference and the arithmetic-average temperature difference, respectively.

Overall Coefficient of Heat Transfer In testing commercial heat-transfer equipment, it is not convenient to measure tube temperatures (t_3 or t_4 in Fig. 10-6), and hence the overall performance is expressed as an overall coefficient of heat transfer U based on a convenient area dA, which may be dA_i, dA_o, or an average of dA_i and dA_o; whence, by definition,

$$dq = U\, dA\,(t_1 - t_7) \qquad (10\text{-}29)$$

U is called the "overall coefficient of heat transfer," or merely the "overall coefficient." The rate of conduction through the tube wall and scale deposit is given by

$$dq = \frac{k dA_{\text{avg}}(t_3 - t_4)}{x} = h_d\, dA_d(t_4 - t_5) \qquad (10\text{-}30)$$

Upon eliminating t_3, t_4, t_5 from Eqs. (10-24), (10-29), and (10-30), the complete expression for the **steady rate of heat flow** from one fluid through the wall and scale to a second fluid, as illustrated in Fig. 10-6, is

$$dq = \frac{t_1 - t_7}{\dfrac{1}{h_i\, dA_i} + \dfrac{x}{k\, dA_{\text{avg}}} + \dfrac{1}{h_d\, dA_d} + \dfrac{1}{h_o\, dA_o}} = U\, dA\,(t_1 - t_7) \qquad (10\text{-}31)^*$$

Representation of Heat-Transfer Film Coefficients There are two general methods of expressing film coefficients: (1) dimensionless relations and (2) dimensional equations.

The dimensionless relations are usually indicated in either of two forms, each yielding identical results. The preferred form is that suggested by Colburn [*Trans. Am. Inst. Chem. Eng.*, **29**, 174–210 (1933)]. It relates, primarily, three dimensionless groups: the Stanton number h/cG, the Prandtl number $c\mu/k$, and the Reynolds number DG/μ. For more accurate correlation of data (at Reynolds number $<$10,000), two additional dimensionless groups are used: ratio of length to diameter L/D and ratio of viscosity at wall (or surface) temperature to viscosity at bulk temperature. Colburn showed that the product of the Stanton number and the two-thirds power of the Prandtl number (and, in addition, power functions of L/D and μ_w/μ for Reynolds number $<$10,000) is approximately equal to half of the Fanning friction factor $f/2$. This product is called the **Colburn *j* factor.** Since the Colburn type of equation relates heat transfer and

*Normally, dirt and scale resistance must be considered on both sides of the tube wall. The area dA is any convenient reference area.

fluid friction, it has greater utility than other expressions for the heat-transfer coefficient.

The classical (and perhaps more familiar) form of dimensionless expressions relates, primarily, the Nusselt number hD/k, the Prandtl number $c\mu/k$, and the Reynolds number DG/μ. The L/D and viscosity-ratio modifications (for Reynolds number <10,000) also apply.

The **dimensional equations** are usually expansions of the dimensionless expressions in which the terms are in more convenient units and in which all numerical factors are grouped together into a single numerical constant. In some instances, the combined physical properties are represented as a linear function of temperature, and the dimensional equation resolves into an equation containing only one or two variables.

NATURAL CONVECTION

Natural convection occurs when a solid surface is in contact with a fluid of different temperature from the surface. Density differences provide the body force required to move the fluid. Theoretical analyses of natural convection require the simultaneous solution of the coupled equations of motion and energy. Details of theoretical studies are available in several general references (Brown and Marco, *Introduction to Heat Transfer*, 3d ed., McGraw-Hill, New York, 1958; and Jakob, *Heat Transfer*, Wiley, New York, vol. 1, 1949; vol. 2, 1957) but have generally been applied successfully to the simple case of a vertical plate. Solution of the motion and energy equations gives temperature and velocity fields from which heat-transfer coefficients may be derived. The general type of equation obtained is the so-called **Nusselt equation:**

$$\frac{hL}{k} = a\left(\frac{L^3\rho^2 g\beta\,\Delta t}{\mu^2}\,\frac{c\mu}{k}\right)^m \qquad (10\text{-}32a)$$

$$N_{Nu} = a(N_{Gr}N_{Pr})^m \qquad (10\text{-}32b)$$

Nusselt Equation for Various Geometries Natural-convection coefficients for various bodies may be predicted from Eq. (10-32). The various numerical values of a and m have been determined experimentally and are given in Table 10-1. Fluid properties are evaluated at $t_f = (t_s + t')/2$. For **vertical plates and cylinders** and $1 < N_{Pr} < 40$, Kato, Nishiwaki, and Hirata [*Int. J. Heat Mass Transfer*, **11**, 1117 (1968)] recommend the relations

$$N_{Nu} = 0.138N_{Gr}^{0.36}(N_{Pr}^{0.175} - 0.55) \qquad (10\text{-}33a)$$

for $N_{Gr} > 10^9$, and

$$N_{Nu} = 0.683N_{Gr}^{0.25}N_{Pr}^{0.25}\,[N_{Pr}/(0.861 + N_{Pr})]^{0.25} \qquad (10\text{-}33b)$$

for $N_{Gr} < 10^9$.

Simplified Dimensional Equations Equation (10-32) is a dimensionless equation, and any consistent set of units may be used. Simplified dimensional equations have been derived for air, water, and organic liquids by rearranging Eq. (10-32) into the following form by collecting the fluid properties into a single factor:

$$h = b(\Delta t)^m L^{3m-1} \qquad (10\text{-}34)$$

Values of b in SI and U.S. customary units are given in Table 10-1 for air, water, and organic liquids.

Simultaneous Loss by Radiation The heat transferred by radiation is often of significant magnitude in the loss of heat from surfaces to the surroundings because of the diathermanous nature of atmospheric gases (air). It is convenient to represent radiant-heat transfer, for this case, as a **radiation film coefficient** which is added to the film coefficient for convection, giving the combined coefficient for convection and radiation ($h_c + h_r$). In Fig. 10-7 values of the film coefficient for radiation h_r are plotted against the two surface temperatures for emissivity = 1.0.

Table 10-2 shows values of ($h_c + h_r$) from single horizontal oxidized pipe surfaces.

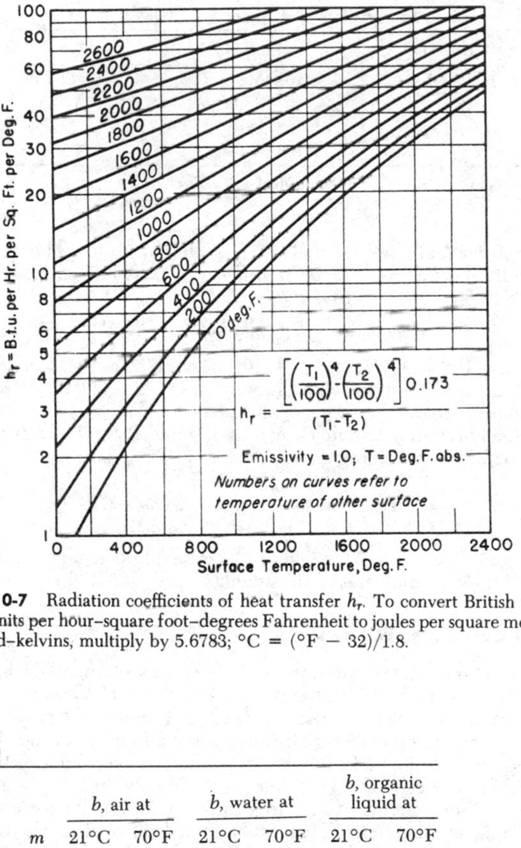

FIG. 10-7 Radiation coefficients of heat transfer h_r. To convert British thermal units per hour–square foot–degrees Fahrenheit to joules per square meter–second–kelvins, multiply by 5.6783; °C = (°F − 32)/1.8.

TABLE 10-1 Values of *a*, *m*, and *b* for Eqs. (10-32) and (10-34)

Configuration	$Y = N_{Gr}N_{Pr}$	a	m	b, air at 21°C	b, air at 70°F	b, water at 21°C	b, water at 70°F	b, organic liquid at 21°C	b, organic liquid at 70°F
Vertical surfaces	$<10^4$	1.36	1/5						
L = vertical dimension <3 ft	$10^4 < Y < 10^9$	0.59	1/4	1.37	0.28	127	26	59	12
	$>10^9$	0.13	1/3	1.24	0.18				
Horizontal cylinder	$<10^{-5}$	0.49	0						
L = diameter <8 in	$10^{-5} < Y < 10^{-3}$	0.71	1/25						
	$10^{-3} < Y < 1$	1.09	1/10						
	$1 < Y < 10^4$	1.09	1/5						
	$10^4 < Y < 10^9$	0.53	1/4	1.32	0.27				
	$>10^9$	0.13	1/3	1.24	0.18				
Horizontal flat surface	$10^5 < Y < 2 \times 10^7$(FU)	0.54	1/4	1.86	0.38				
	$2 \times 10^7 < Y < 3 \times 10^{10}$(FU)	0.14	1/3						
	$3 \times 10^5 < Y < 3 \times 10^{10}$(FD)	0.27	1/4	0.88	0.18				

NOTE: FU = facing upward; FD = facing downward. b in SI units is given in °C column; b in U.S. customary units, in °F column.

TABLE 10-2 Values of $(h_c + h_r)$*

Btu/(h·ft²·°F from pipe to room)
For horizontal bare standard steel pipe of various sizes in a room at 80°F

Nominal pipe diameter, in	Temperature difference, °F														
	30	50	100	150	200	250	300	350	400	450	500	550	600	650	700
1	2.16	2.26	2.50	2.73	3.00	3.29	3.60	3.95	4.34	4.73	5.16	5.60	6.05	6.51	6.98
3	1.97	2.05	2.25	2.47	2.73	3.00	3.31	3.69	4.03	4.43	4.85	5.26	5.71	6.19	6.66
5	....	1.95	2.15	2.36	2.61	2.90	3.20	3.54	3.90						
10	1.80	1.87	2.07	2.29	2.54	2.82	3.12	3.47	3.84						

*Bailey and Lyell [*Engineering*, **147**, 60 (1939)] give values for $(h_c + h_r)$ up to Δt_s of 1000°F. °C = (°F − 32)/1.8; 5.6783 Btu/(h·ft²·°F) = J/(m²·s·/K).

Enclosed Spaces The rate of heat transfer across an enclosed space is calculated from a special coeffient h' based upon the temperature difference between the two surfaces, where $h' = (q/A)/(t_{s1} - t_{s2})$. The value of $h'L/k$ may be predicted from Eq. (10-32) by using the values of a and m given in Table 10-3.

TABLE 10-3 Values of *a* and *m* for Eq. (10-32)

Configuration	$N_{Gr}N_{Pr}(\delta/L)^3$	a	m
Vertical spaces	2×10^4 to 2×10^5	$0.20\,(\delta/L)^{-5/36}$	¼
	2×10^5 to 10^7	$0.071\,(\delta/L)^{1/9}$	⅓
Horizontal spaces	10^4 to 3×10^5	$0.21\,(\delta/L)^{-1/4}$	¼
	3×10^5 to 10^7	0.075	⅓

δ = cell width, L = cell length.

For **vertical enclosed cells** 10 in high and up to 2-in gap width, Landis and Yanowitz (*Proc. Third Int. Heat Transfer Conf.*, Chicago, 1966, vol. II, p. 139) give

$$q\delta/k\,\Delta t = 0.123(\delta/L)^{0.84}(N_{Gr}N_{Pr})^{0.28} \qquad (10\text{-}35)$$

for $2 \times 10^3 < N_{Gr}N_{Pr}(\delta/L)^3 < 10^7$, where q is the uniform heat flux and Δt is the temperature difference at $L/2$. Equation (10-35) is applicable for air, water, and silicone oils.

For **horizontal annuli** Grugal and Hauf (*Proc. Third Int. Heat Transfer Conf.*, Chicago, 1966, vol. II, p 182) report

$$\frac{h\delta}{k} = \left(0.2 + 0.145\frac{\delta}{D_1}N_{Gr}\right)^{0.25} \exp\left(-0.02\frac{\delta}{D_1}\right) \qquad (10\text{-}36)$$

for $0.55 < \delta/D_1 < 2.65$, where N_{Gr} is based upon gap width δ and D_1 is the core diameter of the annulus.

FORCED CONVECTION

Forced-convection heat transfer is the most frequently employed mode of heat transfer in the process industries. Hot and cold fluids, separated by a solid boundary, are pumped through the heat-transfer equipment, the rate of heat transfer being a function of the physical properties of the fluids, the flow rates, and the geometry of the system. Flow is generally turbulent, and the flow duct varies in complexity from circular tubes to baffled and extended-surface heat exchangers. Theoretical analyses of forced-convection heat transfer have been limited to relatively simple geometries and laminar flow. Analyses of turbulent-flow heat transfer have been based upon some mechanistic model and have not generally yielded relationships which were suitable for design purposes. Usually for complicated geometries only empirical relationships are available, and frequently these are based upon limited data and special operating conditions. Heat-transfer coefficients are strongly influenced by the mechanics of flow occurring during forced-convection heat transfer. Intensity of turbulence, entrance conditions, and wall conditions are some of the factors which must be considered in detail as greater accuracy in prediction of coefficients is required.

Analogy between Momentum and Heat Transfer The interrelationship of momentum transfer and heat transfer is obvious from examining the equations of motion and energy. For constant fluid properties, the equations of motion must be solved before the energy equation is solved. If fluid properties are not constant, the equations are coupled, and their solutions must proceed simultaneously. Considerable effort has been directed toward deriving some simple relationship between momentum and heat transfer. The methodology has been to use easily observed velocity profiles to obtain a measure of the diffusivity of momentum in the flowing stream. The analogy between heat and momentum is invoked by assuming that diffusion of heat and diffusion of momentum occur by essentially the same mechanism so that a relatively simple relationship exists between the diffusion coefficients. Thus, the diffusivity of momentum is used to predict temperature profiles and thence by Eq. (10-25) to predict the heat-transfer coefficient.

The analogy has been reasonably successful for simple geometries and for fluids of very low Prandtl number (liquid metals). For high-Prandtl-number fluids the **empirical analogy of Colburn** [*Trans. Am. Inst. Chem. Eng.*, **29**, 174 (1933)] has been very successful. A j factor for momentum transfer is defined as $j = f/2$, where f is the friction factor for the flow. The j factor for heat transfer is assumed to be equal to the j factor for momentum transfer

$$j = h/cG(c\mu/k)^{2/3} \qquad (10\text{-}37)$$

More involved analyses for **circular tubes** reduce the equations of motion and energy to the form

$$\frac{\tau g_c}{\rho} = -\frac{(\nu + \epsilon_M)\,du}{dy} \qquad (10\text{-}38a)$$

$$\frac{q/A}{c\rho} = -\frac{(\alpha + \epsilon_H)\,dt}{dy} \qquad (10\text{-}38b)$$

where ϵ_H is the eddy diffusivity of heat and ϵ_M is the eddy diffusivity of momentum. The units of diffusivity are L^2/θ. The eddy viscosity is $E_M = \rho\epsilon_M$, and the eddy conductivity of heat is $E_H = \epsilon_H c\rho$. Values of ϵ_M are determined via Eq. (10-38*a*) from experimental velocity-distribution data. By assuming ϵ_H/ϵ_M = constant (usually unity), Eq. (10-38*b*) is solved to give the temperature distribution from which the heat-transfer coefficient may be determined. The major difficulties in solving Eq. (10-38*b*) are in accurately defining the thickness of the various flow layers (laminar sublayer and buffer layer) and in obtaining a suitable relationship for prediction of the eddy diffusivities. For assistance in predicting eddy diffusivities, see Reichardt (NACA Tech. Memo 1408, 1957) and Strunk and Chao [*Am. Inst. Chem. Eng. J.*, **10**, 269 (1964)].

Internal and External Flow Two main types of flow are considered in this subsection: internal or conduit flow, in which the fluid completely fills a closed stationary duct, and external or immersed flow, in which the fluid flows past a stationary immersed solid. With **internal flow,** the heat-transfer coefficient is theoretically infinite at the location where heat transfer begins. The local heat-transfer coefficient rapidly decreases and becomes constant, so that after a certain length the average coefficient in the conduit is independent of the length. The local coefficient may follow an irregular pattern, however, if obstructions or turbulence promoters are present in the duct. For **immersed flow,** the local coefficient is again infinite at the point where heating begins, after which it decreases and may show various irregularities depending upon the configuration of the body. Usually in this instance the local coefficient never becomes constant as flow proceeds downstream over the body.

When heat transfer occurs during immersed flow, the rate is dependent upon the configuration of the body, the position of the body, the proximity of other bodies, and the flow rate and turbulence of the stream. The heat-transfer coefficient varies over the immersed body, since both the thermal and the momentum boundary layers vary in thickness. Relatively simple relationships are available for simple configurations immersed in an infinite flowing fluid. For complicated configurations and assemblages of bodies such as are found on the shell side of a heat exchanger, little is known about the local heat-transfer coefficient; empirical relationships giving average coefficients are all that are usually available. Research that has been conducted on local coefficients in complicated geometries has not been extensive enough to extrapolate into useful design relationships.

Laminar Flow Normally, laminar flow occurs in closed ducts when $N_{Re} < 2100$ (based on equivalent diameter $D_e = 4 \times$ free area $\div$ perimeter). Laminar-flow heat transfer has been subjected to extensive theoretical study. The energy equation has been solved for a variety of boundary conditions and geometrical configurations. However, true laminar-flow heat transfer very rarely occurs. Natural-convection effects are almost always present, so that the assumption that molecular conduction alone occurs is not valid. Therefore, empirically derived equations are most reliable.

Data are most frequently correlated by the Nusselt number $(N_{Nu})_{lm}$ or $(N_{NU})_{am}$, the Graetz number $N_{Gz} = (N_{Re}N_{Pr}D/L)$, and the Grashof (natural-convection effects) number N_{Gr}. Some correlations consider only the variation of viscosity with temperature, while others also consider density variation. Theoretical analyses indicate that for very long tubes $(N_{Nu})_{lm}$ approaches a limiting value. Limiting Nusselt numbers for various closed ducts are shown in Table 10-4.

TABLE 10-4 Values of Limiting Nusselt Number in Laminar Flow in Closed Ducts

Configuration	Limiting Nusselt number $N_{Gr} < 4.0$: Constant wall temperature	Constant heat flux
Circular tube	3.66	4.36
Concentric annulus	....	Eq. (10-42)
Equilateral triangle	....	3.00
Rectangles		
Aspect ratio:		
1.0 (square)	2.89	3.63
0.713	....	3.78
0.500	3.39	4.11
0.333	....	4.77
0.25	....	5.35
0 (parallel planes)	7.60	8.24

Circular Tubes For **horizontal tubes** several relationships are applicable, depending upon the value of the Graetz number. For $N_{Gz} < 100$, Hausen's [*Z. Ver. Dtsch. Beih. Verfahrenstech.*, no. 4, 91 (1943)] equation is recommended:

$$(N_{Nu})_{lm} = 3.66 + \frac{0.085N_{Gz}}{1 + 0.047N_{Gz}^{2/3}}\left(\frac{\mu_b}{\mu_w}\right)^{0.14} \tag{10-39}$$

For $N_{Gz} > 100$, the Sieder-Tate relationship [*Ind. Eng. Chem.*, **28**, 1429 (1936)] is satisfactory for small diameters and Δt's:

$$(N_{Nu})_{am} = 1.86N_{Gz}^{1/3}\,(\mu_b/\mu_w)^{0.14} \tag{10-40}$$

A more general expression covering all diameters and Δt's is obtained by including an additional factor $0.87(1 + 0.015N_{Gr}^{1/3})$ on the right side of Eq. (10-40). The diameter should be used in evaluating N_{Gr}. An equation published by Oliver [*Chem. Eng. Sci.*, **17**, 335 (1962)] is also recommended.

For laminar flow in **vertical tubes** a series of charts developed by Pigford [*Chem. Eng. Prog. Symp. Ser.* 17, **51**, 79 (1955)] may be used to predict values of h_{am}.

Annuli Approximate heat-transfer coefficients for laminar flow in annuli may be predicted by the equation of Chen, Hawkins, and Solberg [*Trans. Am. Soc. Mech. Eng.*, **68**, 99 (1946)]:

$$(N_{Nu})_{am} = 1.02N_{Re}^{0.45}N_{Pr}^{0.5}\left(\frac{D_e}{L}\right)^{0.4}\left(\frac{D_2}{D_1}\right)^{0.8}\left(\frac{\mu_b}{\mu_1}\right)^{0.14}N_{Gr}^{0.05} \tag{10-41}$$

Limiting Nusselt numbers for **slug-flow annuli** may be predicted (for constant heat flux) from Trefethen (*General Discussions on Heat Transfer*, London, ASME, New York, 1951, p. 436):

$$(N_{Nu})_{lm} = \frac{8(m-1)(m^2-1)^2}{4m^4 \ln m - 3m^4 + 4m^2 - 1} \tag{10-42}$$

where $m = D_2/D_1$. The Nusselt and Reynolds numbers are based on the equivalent diameter, $D_2 - D_1$.

Limiting Nusselt numbers for laminar flow in annuli have been calculated by Dwyer [*Nucl. Sci. Eng.*, **17**, 336 (1963)]. In addition, theoretical analyses of laminar-flow heat transfer in concentric and eccentric annuli have been published by Reynolds, Lundberg, and McCuen [*Int. J. Heat Mass Transfer*, **6**, 483, 495 (1963)]. Lee [*Int. J. Heat Mass Transfer*, **11**, 509 (1968)] presented an analysis of turbulent heat transfer in entrance regions of concentric annuli. Fully developed local Nusselt numbers were generally attained within a region of 30 equivalent diameters for $0.1 < N_{Pr} < 30$, $10^4 < N_{Re} < 2 \times 10^5$, $1.01 < D_2/D_1 < 5.0$.

Parallel Plates and Rectangular Ducts The limiting Nusselt number for parallel plates and flat rectangular ducts is given in Table 10-4. Norris and Streid [*Trans. Am. Soc. Mech. Eng.*, **62**, 525 (1940)] report for constant wall temperature

$$(N_{Nu})_{lm} = 1.85N_{Gz}^{1/3} \tag{10-43}$$

for $N_{Gz} > 70$. Both Nusselt number and Graetz numbers are based on equivalent diameter. For large temperature differences it is advisable to apply the correction factor $(\mu_b/\mu_w)^{0.14}$ to the right side of Eq. (10-43).

For **rectangular ducts** Kays and Clark (Stanford Univ., Dept. Mech. Eng. Tech. Rep. 14, Aug. 6, 1953) published relationships for heating and cooling of air in rectangular ducts of various aspect ratios. For most **noncircular ducts** Eqs. (10-39) and (10-40) may be used if the equivalent diameter (= 4 × free area/wetted perimeter) is used as the characteristic length. See also Kays and London, *Compact Heat Exchangers*, 2d ed., McGraw-Hill, New York, 1964.

Immersed Bodies When flow occurs over immersed bodies such that the boundary layer is completely laminar over the whole body, laminar flow is said to exist even though the flow in the mainstream is turbulent. The following relationships are applicable to single bodies immersed in an infinite fluid and *are not valid for assemblages of bodies.*

In general, the average heat-transfer coefficient on immersed bodies is predicted by

$$N_{Nu} = C_r(N_{Re})^m(N_{Pr})^{1/3} \tag{10-44}$$

Values of C_r and m for various configurations are listed in Table 10-5. The characteristic length is used in both the Nusselt and the Reynolds numbers, and the properties are evaluated at the film temperature $= (t_w + t_\infty)/2$. The velocity in the Reynolds number is the undisturbed free-stream velocity.

Heat transfer from immersed bodies is discussed in detail by Eckert and Drake, Jakob, and Knudsen and Katz (see "Introduction: General References"), where equations for local coefficients and the effects of unheated starting length are presented. Equation (10-44) may also be expressed as

$$N_{St}N_{Pr}^{2/3} = C_rN_{Re}^{m-1} = f/2 \tag{10-45}$$

where f is the *skin-friction drag coefficient* (not the form drag coefficient).

Falling Films When a liquid is distributed uniformly around the periphery at the top of a vertical tube (either inside or outside) and allowed to fall down the tube wall by the influence of gravity, the fluid does not fill the tube but rather flows as a thin layer. Similarly, when a liquid is applied uniformly to the outside and top of a horizontal tube, it flows in layer form around the periphery and falls

TABLE 10-5 Laminar-Flow Heat Transfer over Immersed Bodies [Eq. (10-44)]

Configuration	Characteristic length	N_{Re}	N_{Pr}	C_r	m
Flat plate parallel to flow	Plate length	10^3 to 3×10^5	>0.6	0.648	0.50
Circular cylinder axes perpendicular to flow	Cylinder diameter	1–4		0.989	0.330
		4–40		0.911	0.385
		40–4000	>0.6	0.683	0.466
		4×10^3–4×10^4		0.193	0.618
		4×10^4–2.5×10^5		0.0266	0.805
Non-circular cylinder, axis	Square, short diameter	5×10^3–10^5		0.104	0.675
Perpendicular to flow, characteristic	Square, long diameter	5×10^3–10^5		0.250	0.588
Length perpendicular to flow	Hexagon, short diameter	5×10^3–10^5	>0.6	0.155	0.638
	Hexagon, long diameter	5×10^3–2×10^4		0.162	0.638
		2×10^4–10^5		0.0391	0.782
Sphere*	Diameter	1–7×10^4	0.6–400	0.6	0.50

*Replace N_{Nu} by $N_{Nu} - 2.0$ in Eq. (10-44).

off the bottom. In both these cases the mechanism is called gravity flow of liquid layers or falling films.

For the turbulent flow of **water** in layer form down the walls of **vertical tubes** the dimensional equation of McAdams, Drew, and Bays [*Trans. Am. Soc. Mech. Eng.*, **62**, 627 (1940)] is recommended:

$$h_{lm} = b\Gamma^{1/3} \qquad (10\text{-}46)$$

where b = 9150 (SI) or 120 (U.S. customary) and is based on values of $\Gamma = W_F/\pi D$ ranging from 0.25 to 6.2 kg/(m·s) [600 to 15,000 lb/(h·ft)] of wetted perimeter. This type of water flow is used in vertical vapor-in-shell ammonia condensers, acid coolers, cycle water coolers, and other process-fluid coolers.

The following dimensional equations may be used for **any liquid** flowing in layer form down **vertical surfaces:**

$$\text{For } \frac{4\Gamma}{\mu} > 2100 \quad h_{lm} = 0.01\left(\frac{k^3\rho^2 g}{\mu^2}\right)^{1/3}\left(\frac{c\mu}{k}\right)^{1/3}\left(\frac{4\Gamma}{\mu}\right)^{1/3} \qquad (10\text{-}47a)$$

$$\text{For } \frac{4\Gamma}{\mu} < 2100 \; h_{am} = 0.50\left(\frac{k^2\rho^{4/3}cg^{2/3}}{L\mu^{1/3}}\right)^{1/3}\left(\frac{\mu}{\mu_w}\right)^{1/4}\left(\frac{4\Gamma}{\mu}\right)^{1/9} \qquad (10\text{-}47b)$$

$$\text{where} \qquad B = (3\mu\Gamma/\rho 2g)^{1/3}$$

Equation (10-47*b*) is based on the work of Bays and McAdams [*Ind. Eng. Chem.*, **29**, 1240 (1937)]. The significance of the term L is not clear. When $L = 0$, the coefficient is definitely not infinite. When L is large and the fluid temperature has not yet closely approached the wall temperature, it does not appear that the coefficient should necessarily decrease. Within the finite limits of 0.12 to 1.8 m (0.4 to 6 ft), this equation should give results of the proper order of magnitude.

For falling films applied to the **outside of horizontal tubes,** the Reynolds number rarely exceeds 2100. Equations may be used for falling films on the outside of the tubes by substituting $\pi D/2$ for L.

For **water** flowing over a **horizontal tube,** data for several sizes of pipe are roughly correlated by the dimensional equation of McAdams, Drew, and Bays [*Trans. Am. Soc. Mech. Eng.*, **62**, 627 (1940)].

$$h_{am} = b\,(\Gamma/D_0)^{1/3} \qquad (10\text{-}48)$$

where b = 3360 (SI) or 65.6 (U.S. customary) and Γ ranges from 0.94 to 4 kg/m·s) [100 to 1000 lb/(h·ft)].

Falling films are also used for evaporation in which the film is both entirely or partially evaporated (juice concentration). This principle is also used in crystallization (freezing).

The advantage of high coefficient in falling-film exchangers is partially offset by the difficulties involved in distribution of the film, maintaining complete wettability of the tube, and pumping costs required to lift the liquid to the top of the exchanger.

Transition Region Turbulent-flow equations for predicting heat transfer coefficients are usually valid only at Reynolds numbers greater than 10,000. The transition region lies in the range $2000 < N_{Re} < 10{,}000$. No simple equation exists for accomplishing a smooth mathematical transition from laminar flow to turbulent flow. Of the relationships proposed, Hausen's equation [Z. *Ver. Dtsch. Ing. Beih. Verfahrenstech.*, No. 4, 91 (1934)] fits both the laminar extreme and the fully turbulent extreme quite well.

$$(N_{Nu})_{am} = 0.116(N_{Re}^{2/3} - 125)N_{Pr}^{1/3}\left[1 + \left(\frac{D}{L}\right)^{2/3}\right]\left(\frac{\mu_b}{\mu_w}\right)^{0.14} \qquad (10\text{-}49)$$

between 2100 and 10,000. It is customary to represent the probable magnitude of coefficients in this region by hand-drawn curves (Fig. 10-8). Equation (10-40) is plotted as a series of curves (j factor versus

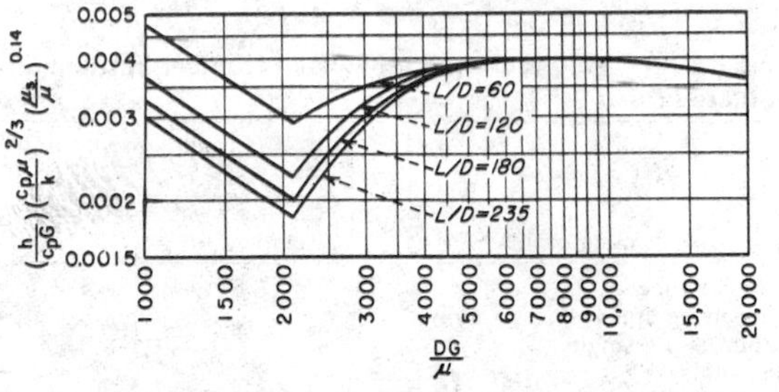

FIG. 10-8 Graphical representation of the Colburn j factor for the heating and cooling of fluids inside tubes. The curves for N_{Re} below 2100 are based on Eq. (10-40). L is the length of each pass in feet. The curves for N_{Re} between 2100 and 10,000 are represented by Eq. (10-49). The curve for N_{Re} above 10,000 is represented by Eq. (10-51).

Reynolds number with L/D as parameters) terminating at Reynolds number = 2100. Continuous curves for various values of L/D are then hand-drawn from these terminal points to coincide tangentially with the curve for forced-convection, fully turbulent flow [Eq. (10-51)].

Turbulent Flow

Circular Tubes Numerous relationships have been proposed for predicting turbulent flow in tubes. For high-Prandtl-number fluids, relationships derived from the equations of motion and energy through the momentum–heat-transfer analogy are more complicated and no more accurate than many of the empirical relationships that have been developed.

For $N_{Re} > 10{,}000$, $0.7 < N_{Pr} < 700$, $L/D > 60$ and properties based on bulk temperature, the **Sieder-Tate equation** is recommended:

$$N_{Nu} = 0.023 N_{Re}^{0.8} N_{Pr}^{1/3}\,(\mu_b/\mu_w)^{0.14} \qquad (10\text{-}50)$$

The **Colburn form** of Eq. (10-50) is

$$j_H = N_{St} N_{Pr}^{2/3} (\mu_w/\mu_b)^{0.14} = 0.023 N_{Re}^{-0.2} \qquad (10\text{-}51)$$

In Eq. (10-51) the viscosity-ratio factor may be neglected if properties are evaluated at the film temperature $(t_b + t_w)/2$.

Approximate predictions for **rough pipes** may be obtained from Eq. (10-51) if the right-hand term is replaced by $f/2$ for the rough pipe. For air, Nunner (Z. *Ver. Dtsch. Ing. Forsch.*, 1956, p. 455) obtains

$$\frac{(N_{Nu})_{rough}}{(N_{Nu})_{smooth}} = \frac{f_{rough}}{f_{smooth}} \tag{10-52}$$

Dippery and Sabersky [*Int. J. Heat Mass Transfer*, **6**, 329 (1963)] present a complete discussion of the influence of roughness on heat transfer in tubes.

Dimensional Equations for Various Conditions For gases at ordinary pressures and temperatures based on $c\mu/k = 0.78$ and $\mu = (1.76)(10^{-5})$ Pa·s [0.0426 lb/(ft·h)]

$$h = bc\rho^{0.8}(V^{0.8}/D^{0.2}) \tag{10-53}$$

where $b = (3.04)(10^{-3})$ (SI) or $(1.44)(10^{-2})$ (U.S. customary). For air at atmospheric pressure

$$h = b(V^{0.8}/D^{0.2}) \tag{10-54}$$

where $b = 3.52$ (SI) or $(4.35)(10^{-4})$ (U.S. customary). For water [based on a temperature range of 5 to 104°C (40 to 220°F)]

$$h = 1057\ (1.352 + 0.02t)\ (V^{0.8}/D^{0.2}) \tag{10-55a}$$

in SI units with t = °C, or

$$h = 0.13(1 + 0.011t)(V^{0.8}/D^{0.2}) \tag{10-55b}$$

in U.S. customary units with t = °F.

For organic liquids, based on c = 2.092 J/kg·K) [0.5 Btu/(lb·°F)], k = 0.14 J/(m·s·K) [0.08 Btu/(h·ft·°F)], $\mu_b = (1)(10^{-3})$ Pa·s (1.0 cP), and ρ = 810 kg/m³ (50 lb/ft³),

$$h = b(V^{0.8}/D^{0.2}) \tag{10-56}$$

where $b = 423$ (SI) or $(5.22)(10^{-2})$ (U.S. customary). Within reasonable limits, coefficients for organic liquids are about one-third of the values obtained for water.

Entrance effects are usually not significant industrially if $L/D > 60$. Below this limit Nusselt recommended the conservative equation for $10 < L/D < 400$ and properties evaluated at bulk temperature

$$N_{Nu} = 0.036N_{Re}^{0.8}N_{Pr}^{1/3}(L/D)^{-0.054} \tag{10-57}$$

It is common to correlate entrance effects by the equation

$$h_m/h = 1 + F(D/L) \tag{10-58}$$

where h is predicted by either Eq. (10-49) or Eq. (10-50) and h_m is the mean coefficient for the pipe in question. Values of F are reported by Boelter, Young, and Iverson [NACA Tech. Note 1451, 1948] and tabulated by Kays and Knudsen and Katz (see "Introduction: General References"). Selected values of F are as follows:

Fully developed velocity profile	1.4
Abrupt contraction entrance	6
90° right-angle bend	7
180° round bend	6

For **large temperature differences** different equations are necessary and usually are specifically applicable to either gases or liquids. Gambill (*Chem. Eng.*, Aug. 28, 1967, p. 147) provides a detailed review of high-flux heat transfers to gases. He recommends

$$N_{Nu} = \frac{0.021N_{Re}^{0.8}N_{Pr}^{0.4}}{(T_w/T_b)^{0.29+0.0019(L/D)}} \tag{10-59}$$

for $10 < L/D < 240$, $110 < T_b < 1560$ K ($200 < T_b < 2800$°R), $1.1 < (T_w/T_b) < 8.0$, and properties evaluated at T_b. For liquids, Eq. (10-50) is generally satisfactory.

Annuli For diameter ratios $D_1/D_2 > 0.2$, Monrad and Pelton's equation [*Trans. Am. Inst. Chem. Eng.*, **38**, 593 (1942)] is recommended for either or both the inner and outer tube:

$$N_{Nu} = 0.020N_{Re}^{0.8}N_{Pr}^{1/3}(D_2/D_1)^{0.53} \tag{10-60}$$

The Colburn form of relationship may be employed for the individual walls of the annulus by using the individual friction factor for each wall [see Knudsen, *Am. Inst. Chem. Eng. J.*, **8**, 566 (1962)]:

$$j_{H1} = (N_{St})_1N_{Pr}^{2/3} = f_{1/2} \tag{10-61a}$$

$$j_{H2} = (N_{St})_2N_{Pr}^{2/3} = f_{2/2} \tag{10-61b}$$

Rothfus, Monrad, Sikchi, and Heideger [*Ind. Eng. Chem.*, **47**, 913 (1955)] report that the friction factor f_2 for the outer wall bears the same relation to the Reynolds number for the outer portion of the annular stream $2(r_2^2 - \lambda^2)V\rho/r_2\mu$ as the friction factor for circular tubes does to the Reynolds number for circular tubes, where r_2 is the radius of the outer tube and λ is the position of maximum velocity in the annulus, estimated from

$$\lambda^2 = \frac{r_2^2 - r_1^2}{\ln (r_2/r_1)^2} \tag{10-62}^*$$

To calculate the friction factor f_1 for the inner tube use the relation

$$f_1 = \frac{f_2r_2(\lambda^2 - r_1^2)}{r_1(r_2^2 - \lambda^2)} \tag{10-63}$$

There have been several analyses of turbulent heat transfer in annuli: for example, Deissler and Taylor (NACA Tech. Note 3451, 1955), Kays and Leung [*Int. J. Heat Mass Transfer*, **6**, 537 (1963)], Lee [*Int. J. Heat Transfer*, **11**, 509 (1968)], Sparrow, Hallman and Siegel [*Appl. Sci. Res.*, **7A**, 37 (1958)], and Johnson and Sparrow [*Am. Soc. Mech. Eng. J. Heat Transfer*, **88**, 502 (1966)]. The reader is referred to these for details of the analyses.

For **annuli containing externally finned tubes** the heat-transfer coefficients are a function of the fin configurations. Knudsen and Katz (*Fluid Dynamics and Heat Transfer*, McGraw-Hill, New York, 1958) present relationships for transverse finned tubes, spined tubes, and longitudinal finned tubes in annuli.

Noncircular Ducts Equation (10-50) may be employed for noncircular ducts by using the equivalent diameter $D_e = 4 \times$ free area per wetted perimeter. For high temperature differences the exponent on the Prandtl number is changed from ⅓ to 0.4. Kays and London (*Compact Heat Exchangers*, 2d ed., McGraw-Hill, New York, 1965) give charts for various noncircular ducts encountered in compact heat exchangers.

Vibrations and pulsations generally tend to increase heat-transfer coefficients.

Example 2 Calculate the heat-transfer j factors for both walls of an annulus for the following conditions: D_1 = 0.0254 m (1.0 in); D_2 = 0.0635 m (2.5 in); water at 15.6°C (60°F); $\mu/\rho = (1.124)(10^{-6})$ m²/s [$(1.21)(10^{-5})$ ft²/s]; velocity = 1.22 m/s (4 ft/s).

$$\lambda^2 = \frac{0.0635^2 - 0.0254^2}{4 \ln (0.0635/0.0254)^2} = (4.621)(10^{-4})\ \text{m}^2\ (0.716\ \text{in}^2)$$

$$\frac{2(r_2^2 - \lambda^2)V\rho}{r_2\mu} = \frac{2[0.0318^2 - (4.621)(10^{-4})(1.22)]}{(0.0318)(1.124)(10^{-6})} = (3.74)(10^4)$$

From the f versus N_{Re} relation for smooth circular tubes given in Sec. 5, f_2 = 0.0055. Hence

$$j_{H2} = (N_{St})_2\ N_{Pr}^{2/3} = 0.00275$$

From Eq. (10-63),

$$f_1 = \frac{(0.0055)(0.0318)[(4.621)(10^{-4}) - 0.0127^2]}{(0.0127)[0.0318^2 - (4.621)(10^{-4})]} = 0.00754$$

from which $jH_1 = (N_{St})_1N_{Pr}^{2/3} = 0.00377$.

These results indicate that for this system the heat-transfer coefficient on the inner tube is about 40 percent greater than on the outer tube.

Coils For flow *inside* **helical coils,** Reynolds number above 10,000, multiply the value of the film coefficient obtained from the applicable equation for straight tubes by the term $(1 + 3.5\ D_i/D_c)$.

For flow inside helical coils, Reynolds number less than 10,000,

*Equation (10-62) predicts the point of maximum velocity for *laminar* flow in annuli and is only an approximate equation for turbulent flow. Brighton and Jones [*Am. Soc. Mech. Eng. Basic Eng*, **86**, 835 (1964)] and Macagno and McDougall [*Am. Inst. Chem. Eng. J.*, **12**, 437 (1966)] give more accurate equations for predicting the point of maximum velocity for turbulent flow.

substitute the term $(D_c/D_i)^{1/2}$ for (L/D_i) where the latter appears in the applicable equation for straight tubes (frequently as part of the Graetz number).

For **flat spiral (pancake) coils,** in which the ratio D_c/D_i varies for each turn, a different value of coefficient will be obtained for each turn; a weighted average based on length per turn is used.

For flow *outside* **helical coils** use the equation for flow normal to a bank of tubes, in-line flow.

Finned Tubes (Extended Surface) When the film coefficient on the outside of a metal tube is much lower than that on the inside, as when steam condensing in a pipe is being used to heat air, externally finned (or extended) heating surfaces are of value in increasing substantially the rate of heat transfer per unit length of tube. The data on extended heating surfaces, for the case of air flowing outside and at right angles to the axes of a bank of finned pipes, can be represented approximately by the dimensional equation derived from

$$h_f = b\frac{V_F^{0.6}}{D_0^{0.4}}\left(\frac{p'}{p' - D_0}\right)^{0.6} \tag{10-64}$$

where b = 5.29 (SI) or (5.39)(10^{-3}) (U.S. customary); h_f is the film coefficient of heat transfer on the air side; V_F is the face velocity of the air; p' is the center-to-center spacing, m, of the tubes in a row; and D_0 is the outside diameter, m, of the bare tube (diameter at the root of the fins).

In atmospheric air-cooled finned tube exchangers, the air-film coefficient from Eq. (10-64) is sometimes converted to a value based on outside bare surface as follows:

$$h_{fo} = h_f\frac{A_f + A_{uf}}{A_{of}} = h_f\frac{A_T}{A_o} \tag{10-65}$$

in which h_{fo} is the air-film coefficient based on external bare surface; h_f is the air-film coefficient based on total external surface; A_T is total external surface, and A_o is external bare surface of the unfinned tube; A_f is the area of the fins; A_{uf} is the external area of the unfinned portion of the tube; and A_{of} is area of tube before fins are attached.

Fin efficiency is defined as the ratio of the mean temperature difference from surface to fluid divided by the temperature difference from fin to fluid at the base or root of the fin. Graphs of fin efficiency for extended surfaces of various types are given by Gardner [*Trans. Am. Soc. Mech. Eng.*, **67**, 621 (1945)]. See Figs. 10-38 and 10-39.

Heat-transfer coefficients for finned tubes of various types are given in a series of papers [*Trans. Am. Soc. Mech. Eng.*, **67**, 601 (1945)]. See also Eq. (10-159).

For flow of air normal to fins in the form of **short strips or pins,** Norris and Spofford [*Trans. Am. Soc. Mech. Eng.*, **64**, 489 (1942)] correlate their results for air by the dimensionless equation of Pohlhausen:

$$\frac{h_m}{c_pG_{max}}\left(\frac{c_p\mu}{k}\right)^{2/3} = 1.0\left(\frac{z_pG_{max}}{\mu}\right)^{-0.5} \tag{10-66}$$

for values of z_pG_{max}/μ ranging from 2700 to 10,000.

For the general case, the treatment suggested by Kern (*Process Heat Transfer*, McGraw-Hill, New York, 1950, p. 512) is recommended. Because of the wide variations in fin-tube construction, it is convenient to convert all film coefficients to values based on the inside bare surface of the tube. Thus to convert the film coefficient based on outside area (finned side) to a value based on inside area Kern gives the following relationship:

$$h_{fi} = (\Omega A_f + A_o)(h_f/A_i) \tag{10-67}$$

in which h_{fi} is the effective outside film coefficient based on the inside area, h_f is the outside film coefficient calculated from the applicable equation for bare tubes, A_f is the surface area of the fins, A_o is the surface area on the outside of the tube which is not finned, A_i is the inside area of the tube, and Ω is the fin efficiency defined as

$$\Omega = (\tanh mb_f)/mb_f \tag{10-68}$$

in which

$$m = (h_fp_f/ka_x)^{1/2}\ \text{m}^{-1}\ (\text{ft}^{-1}) \tag{10-69}$$

and b_f = height of fin. The other symbols are defined as follows: p_f is the perimeter of the fin, a_x is the cross-sectional area of the fin, and k is the thermal conductivity of the material from which the fin is made.

Fin efficiencies and fin dimensions are available from manufacturers. Ratios of finned to inside surface are usually available so that the terms A_f, A_o, and A_i may be obtained from these ratios rather than from the total surface areas of the heat exchangers. See Figs. 10-38 and 10-39.

Banks of Tubes For heating and cooling of fluids flowing normal to a bank of circular tubes at least 10 rows deep the following equations are applicable:

Colburn type:

$$\frac{h}{cG_{max}}\left(\frac{c\mu}{k}\right)^{2/3} = \frac{a}{(DG_{max}/\mu)^{0.4}} = j \tag{10-70}$$

Nusselt type:

$$\frac{hD}{k} = a\left(\frac{DG_{max}}{\mu}\right)^{0.6}\left(\frac{c\mu}{k}\right)^{1/3} \tag{10-71}$$

The dimensionless constant a in these equations varies depending upon conditions.

Conditions, Reynolds number > 3000	Value of a
Flow normal to apex of diamond, staggered arrangement	
No leakage	0.330
Normal leakage in baffled exchanger	0.198
Flow normal to flat side of diamond, not staggered (in-line) arrangement	
No leakage	0.260
Normal leakage in baffled exchanger	0.156

For Reynolds number less than 3000, Eq. (10-70) would give conservative results, but greater accuracy (if desired) may be obtained by using the following equation.

$$\frac{h}{cG_{max}}\left(\frac{c\mu}{k}\right)^{2/3} = \frac{a}{(D_oG_{max}/\mu)^m} = j \tag{10-72}$$

in which the constant a and exponent m are as follows:

Reynolds number	m	Tube pitch	Leakage	a
100–300	0.492	Staggered	None	0.695
			Normal	0.416
		In-line	None	0.548
			Normal	0.329
1–100	0.590	Staggered	None	1.086
			Normal	0.650
		In-line	None	0.855
			Normal	0.513

The following **dimensional equations** (10-73 to 10-77) are based on flow normal to a bank of staggered tubes without leakage. Multiply the values obtained for h by 0.6 for normal leakage and, in addition, by 0.79 for in-line (not staggered) tube arrangement.

$$h = b\frac{c^{1/3}k^{2/3}\rho^{0.6}\,V_{max}^{0.6}}{\mu^{0.267}\,D_0^{0.4}} \tag{10-73}$$

where b = 0.33 (SI) or 0.261 (U.S. customary). For gases at ordinary pressures and temperatures, based on $c\mu/k$ = 0.78; μ = (1.76)(10^{-5}) Pa·s [0.0426 lb/(ft·h)],

$$h = bc\frac{G_{max}^{0.6}}{D_0^{0.4}} \tag{10-74}$$

where b = (4.82)(10^{-3}) (SI) or 0.109 (U.S. customary). For air at atmospheric pressure

$$h = b\frac{V_{max}^{0.6}}{D_0^{0.4}} \tag{10-75}$$

where b = 5.33 (SI) or (5.44)(10^{-3}) (U.S. customary). For water based on a temperature range 7 to 104°C (40 to 220°F)

$$h = 986(1.21 + 0.0121t)\frac{V_{max}^{0.6}}{D_0^{0.4}} \tag{10-76a}$$

in SI units and t in °C.

$$h = 1.01(1 + 0.0067t)\frac{V_{max}^{0.6}}{D_0^{0.4}} \tag{10-76b}$$

in U.S. customary units and t in °F. For organic liquids, based on c = 2.22 J/(kg·K) [0.53 Btu/(lb·°F)], k = 0.14 J/(m·s·K) [0.08 Btu/(h·ft·°F)], μ_b = (1)(10^{-3}) Pa·s (1.0 cP), ρ = 810 kg/m³ (50 lb/ft³),

$$h = b\frac{V_{max}^{0.6}}{D_0^{0.4}} \tag{10-77}$$

where b = 400 (SI) or 0.408 (U.S. customary).

JACKETS AND COILS OF AGITATED VESSELS

Most of the correlations for heat transfer from the agitated liquid contents of vessels to **jacketed walls** have been of the form:

$$\frac{hD_j}{k} = a\left(\frac{L_p^2N_r\rho}{\mu}\right)^b\left(\frac{c\mu}{k}\right)^{1/3}\left(\frac{\mu_b}{\mu_w}\right)^m \tag{10-78}$$

The film coefficient h is for the inner wall; D_j is the inside diameter of the mixing vessel. The term $L_p^2N_r\rho/\mu$ is the Reynolds number for mixing in which L_p is the diameter and N_r the speed of the agitator. Recommended values of the constants a, b, and m are given in Table 10-6.

TABLE 10-6 Values of Constants for Use in Eq. (10-78)

Agitator	a	b	m	Range of Reynolds number
Paddle[a]	0.36	⅔	0.21	300–3 × 10^5
Pitched-blade turbine[b]	0.53	⅔	0.24	80–200
Disk, flat-blade turbine[c]	0.54	⅔	0.14	40–3 × 10^5
Propeller[d]	0.54	⅔	0.14	2 × 10^3 (one point)
Anchor[b]	1.0	½	0.18	10–300
Anchor[b]	0.36	⅔	0.18	300–40,000
Helical ribbon[e]	0.633	½	0.18	8–10^5

[a] Chilton, Drew, and Jebens, *Ind. Eng. Chem.*, **36**, 510 (1944), with constant m modified by Uhl.
[b] Uhl, *Chem. Eng. Progr., Symp. Ser.* 17, **51**, 93 (1955).
[c] Brooks and Su, *Chem. Eng. Progr.*, **55**(10), 54 (1959).
[d] Brown *et al.*, *Trans. Inst. Chem. Engrs. (London)*, **25**, 181 (1947).
[e] Gluz and Pavlushenko, *J. Appl. Chem. U.S.S.R.*, **39**, 2323 (1966).

A wide variety of configurations exists for coils in agitated vessels. Correlations of data for heat transfer to **helical coils** have been of two forms, of which the following are representative:

$$\frac{hD_j}{k} = 0.87\left(\frac{L_p^2N_r\rho}{\mu}\right)^{0.62}\left(\frac{c\mu}{k}\right)^{1/3}\left(\frac{\mu_b}{\mu_w}\right)^{0.14} \tag{10-79}$$

where the agitator is a paddle, the Reynolds number range is 300 to 4 × 10^5 [Chilton, Drew, and Jebens, *Ind. Eng. Chem.*, **36**, 510 (1944)], and

$$\frac{hD_o}{k} = 0.17\left(\frac{L_p^2N_r\rho}{\mu}\right)^{0.67}\left(\frac{c\mu}{k}\right)^{0.37}\left(\frac{L_p}{D_j}\right)^{0.1}\left(\frac{D_o}{D_j}\right)^{0.5} \tag{10-80}$$

where the agitator is a disk flat-blade turbine, and the Reynolds number range is 400 to (2)(10^5) [Oldshue and Gretton, *Chem. Eng. Prog.*, **50**, 615 (1954)]. The term D_O is the outside diameter of the coil tube.

The most comprehensive correlation for heat transfer to **vertical baffle-type coils** is for a disk flat-blade turbine over the Reynolds number range 10^3 to (2)(10^6):

$$\frac{hD_O}{k} = 0.09\left(\frac{L_p^2N_r\rho}{\mu}\right)^{0.65}\left(\frac{c\mu}{k}\right)^{1/3}\left(\frac{L_p}{D_j}\right)^{1/3}\left(\frac{2}{n_b}\right)^{0.2}\left(\frac{\mu}{\mu_f}\right)^{0.4} \tag{10-81}$$

where n_b is the number of baffle-type coils and μ_f is the fluid viscosity at the mean film temperature [Dunlop and Rushton, *Chem. Eng. Prog. Symp. Ser.* 5, **49**, 137 (1953)].

For predicting heat-transfer coefficients for turbulent flow **inside pipe coils**, the multiplier (1 + 3.5 D_i/D_c) can be included on the right-hand side of Eq. (10-50).

Chapman and Holland (*Liquid Mixing and Processing in Stirred Tanks*, Reinhold, New York, 1966) review heat transfer to low-viscosity fluids in agitated vessels. Uhl ["Mechanically Aided Heat Transfer," in Uhl and Gray (eds.), *Mixing: Theory and Practice*, vol. I, Academic, New York, 1966, chap. V] surveys heat transfer to low- and high-viscosity agitated fluid systems. This review includes scraped-wall units and heat transfer on the jacket and coil side for agitated vessels.

NONNEWTONIAN FLUIDS

A wide variety of nonnewtonian fluids are encountered industrially. They may exhibit Bingham-plastic, pseudoplastic, or dilatant behavior and may or may not be thixotropic. For design of equipment to handle or process nonnewtonian fluids, the properties must usually be measured experimentally, since no generalized relationships exist to predict the properties or behavior of the fluids. Details of handling nonnewtonian fluids are described completely by Skelland (*Non-Newtonian Flow and Heat Transfer*, Wiley, New York, 1967). The generalized shear-stress rate-of-strain relationship for nonnewtonian fluids is given as

$$n' = \frac{d\ln(D\,\Delta P/4L)}{d\ln(8V/D)} \tag{10-82}$$

as determined from a plot of shear stress versus velocity gradient.

For **circular tubes,** N_{Gz} > 100, n' > 0.1, and laminar flow

$$(N_{Nu})_{lm} = 1.75\,\delta_s^{1/3}N_{Gz}^{1/3} \tag{10-83}$$

where $\delta_s = (3n' + 1)/4n'$. When natural-convection effects are considered, Metzer and Gluck [*Chem. Eng. Sci.*, **12**, 185 (1960)] obtained the following for **horizontal tubes:**

$$(N_{Nu})_{lm} = 1.75\,\delta_s^{1/3}\left[N_{Gz} + 12.6\left(\frac{N_{Pr}N_{Gr}D}{L}\right)^{0.4}\right]^{1/3}\left(\frac{\gamma_b}{\gamma_w}\right)^{0.14} \tag{10-84}$$

where properties are evaluated at the wall temperature, i.e., $\gamma = g_cK'8^{n'-1}$ and $\tau_w = K'(8V/D)^{n'}$. Nonnewtonian properties are evaluated experimentally.

Metzner and Friend [*Ind. Eng. Chem.*, **51**, 879 (1959)] present relationships for turbulent heat transfer with nonnewtonian fluids. Relationships for heat transfer by natural convection and through laminar boundary layers are available in Skelland's book (op. cit.).

LIQUID METALS

Liquid metals constitute a class of heat-transfer media having Prandtl numbers generally below 0.01. Heat-transfer coefficients for liquid metals cannot be predicted by the usual design equations applicable to gases, water, and more viscous fluids with Prandtl numbers greater than 0.6. Relationships for predicting heat-transfer coefficients for liquid metals have been derived from solution of Eqs. (10-38*a*) and (10-38*b*). By the momentum-transfer–heat-transfer analogy, the eddy conductivity of heat is $kN_{Pr}(E_M/\mu) \approx k$ for small N_{Pr}. Thus in the solution of Eqs. (10-38*a*) and (10-38*b*) the knowledge of the thickness of various layers of flow is not critical. In fact, assumption of slug flow and constant conductivity (= k) across the duct gives reasonable values of heat-transfer coefficients for liquid metals.

For **constant heat flux:**

$$N_{Nu} = 5 + 0.025(N_{Re}N_{Pr})^{0.8} \quad (10\text{-}85)$$

For **constant wall temperature:**

$$N_{Nu} = 7 + 0.025(N_{Re}N_{Pr})^{0.8} \quad (10\text{-}86)$$

For most liquid metals except mercury Lubarsky and Kaufman (NACA Tech. Note 3336, 1966) obtained the empirical equation for **uniform heat flux:**

$$N_{Nu} = 0.625(N_{Re}N_{Pr})^{0.4} \quad (10\text{-}87)$$

For **parallel plates and annuli** with $D_2/D_1 < 1.4$ and uniform heat flux, Seban [*Trans. Am. Soc. Mech. Eng.*, **72**, 789 (1950)] obtained the equation

$$N_{Nu} = 5.8 + 0.020(N_{Re}N_{Pr})^{0.8} \quad (10\text{-}88)$$

For annuli only, application of a factor of $0.70(D_2/D_1)^{0.53}$ is recommended for Eqs. (10-85) and (10-86). For more accurate semiempirical relationships for tubes, annuli, and rod bundles, refer to Dwyer [*Am. Inst. Chem. Eng. J.*, **9**, 261 (1963)].

Hsu [*Int. J. Heat Mass Transfer*, **7**, 431 (1964)] and Kalish and Dwyer [*Int. J. Heat Mass Transfer*, **10**, 1533 (1967)] discuss heat transfer to liquid metals flowing across **banks of tubes.** Hsu recommends the equations

$$N_{Nu} = 0.81 N_{Re}N_{Pr}\,(\phi/D)^{1/2} \quad \text{(for uniform heat flux)} \quad (10\text{-}89)$$

$$N_{Nu} = 0.096 N_{Re}N_{Pr}\,(\phi/D)^{1/2} \quad \text{(for cosine surface temperature)} \quad (10\text{-}90)$$

where the heat-transfer coefficient is based on the average circumferential temperature around the tubes, the Reynolds number is based on the superficial velocity through the tube bank, D is the tube outside diameter, and ϕ is a velocity potential function having the following values:

D/p'	ϕ/D square pitch	ϕ/D equilateral triangular pitch
0	2.00	2.00
0.1	2.02	2.02
0.2	2.07	2.06
0.3	2.16	2.15
0.4	2.30	2.27
0.5	2.52	2.45
0.6	2.84	2.71
0.7	3.34	3.11
0.8	4.23	3.80

Equations (10-89) and (10-90) are useful in calculating tube-surface temperatures.

Further information on liquid-metal heat transfer in tube banks is given by Hsu for spheres and elliptical rod bundles [*Int. J. Heat Mass Transfer*, **8**, 303 (1965)] and by Kalish and Dwyer for oblique flow across tube banks [*Int. J. Heat Mass Transfer*, **10**, 1533 (1967)]. For additional details of heat transfer with liquid metals for various systems see Dwyer (1968 ed., Na and Nak supplement to *Liquid Metals Handbook*) and Stein ("Liquid Metal Heat Transfer," in *Advances in Heat Transfer*, vol. 3, Academic, New York, 1966).

HEAT TRANSFER WITH CHANGE OF PHASE

In any operation in which a material undergoes a change of phase, provision must be made for the addition or removal of heat to provide for the latent heat of the change of phase plus any other sensible heating or cooling that occurs in the process. Heat may be transferred by any one or a combination of the three modes—conduction, convection, and radiation. The process involving change of phase involves mass transfer simultaneous with heat transfer.

CONDENSATION

Condensation Mechanisms Condensation occurs when a saturated vapor comes in contact with a surface whose temperature is below the saturation temperature. Normally a film of condensate is formed on the surface, and the thickness of this film, per unit of breadth, increases with increase in extent of the surface. This is called **film-type condensation.**

Another type of condensation, called **dropwise,** occurs when the wall is not uniformly wetted by the condensate, with the result that the condensate appears in many small droplets at various points on the surface. There is a growth of individual droplets, a coalescence of adjacent droplets, and finally a formation of a rivulet. Adhesional force is overcome by gravitational force, and the rivulet flows quickly to the bottom of the surface, capturing and absorbing all droplets in its path and leaving dry surface in its wake.

Film-type condensation is more common and more dependable. Dropwise condensation normally needs to be promoted by introducing an impurity into the vapor stream. Substantially higher (6 to 18 times) coefficients are obtained for dropwise condensation of steam, but design methods are not available. Therefore, the development of equations for condensation will be for the film type only.

The physical properties of the liquid, rather than those of the vapor, are used for determining the film coefficient for condensation. Nusselt [*Z. Ver. Dtsch. Ing.*, **60**, 541, 569 (1916)] derived theoretical relationships for predicting the film coefficient of heat transfer for condensation of a pure saturated vapor. A number of simplifying assumptions were used in the derivation.

The **Reynolds number** of the condensate film (falling film) is $4\Gamma/\mu$, where Γ is the weight rate of flow (loading rate) of condensate per unit perimeter kg/(s·m) [lb/(h·ft)]. The thickness of the condensate film for Reynolds number less than 2100 is $(3\mu\Gamma/\rho^2 g)^{1/3}$.

Condensation Coefficients

Vertical Tubes For the following cases Reynolds number < 2100 and is calculated by using $\Gamma = W_F/\pi D$. The **Nusselt equation** for the heat-transfer coefficient for condensate films may be written in the following ways (using liquid physical properties and where L is the heated length and Δt is $t_{sv} - t_s$):

Colburn type:

$$\frac{h}{cG}\frac{c\mu}{k} = \frac{5.35}{4\Gamma/\mu} \quad (10\text{-}91)$$

where $G = \dfrac{\Gamma}{(3\mu\Gamma/\rho^2 g)^{1/3}} = \left(\dfrac{W_F^2\rho^2 g}{29.6D^2\mu}\right)^{1/3}$ kg/(s·m²) [lb/(h·ft²)]

Nusselt type:

$$\frac{hL}{k} = 0.943\left(\frac{L^3\rho^2 g\lambda}{k\mu\,\Delta t}\right)^{1/4} = 0.925\left(\frac{L^3\rho^2 g}{\mu\Gamma}\right)^{1/3} \quad (10\text{-}92)$$

Dimensional:

$$h = b(k^3\rho^2 D/\mu_b W_F)^{1/3} \quad (10\text{-}93)$$

where b = 127 (SI) or 756 (U.S. customary). For steam at atmospheric pressure, k = 0.682 J/(m·s·K) [0.394 Btu/(h·ft·°F)], ρ = 960 kg/m³ (60 lb/ft³), μ_b = (0.28)(10⁻³) Pa·s (0.28 cP),

$$h = b(D/W_F)^{1/3} \tag{10-94}$$

where b = 2954 (SI) or 6978 (U.S. customary). For organic vapors at normal boiling point, k = 0.138 J/(m·s·K) [0.08 Btu/(h·ft·°F)], ρ = 720 kg/m³ (45 lb/ft³), μ_b = (0.35)(10⁻³) Pa·s (0.35 cP),

$$h = b(D/W_F)^{1/3} \tag{10-95}$$

where b = 457 (SI) or 1080 (U.S. customary).

Horizontal Tubes For the following cases Reynolds number <2100 and is calculated by using $\Gamma = W_F/2L$.

Colburn type:

$$\frac{h}{cG}\frac{c\mu}{k} = \frac{4.4}{4\Gamma/\mu} \tag{10-96}$$

$$G = \frac{\Gamma}{(3\mu\Gamma/\rho^2 g)^{1/3}} = \left(\frac{W_F^2\rho^2 g}{12L^2\mu}\right)^{1/3} \text{kg/(s·m}^2\text{) [lb/(h·ft}^2\text{)]}$$

Nusselt type:

$$\frac{hD}{k} = 0.73\left(\frac{D^3\rho^2 g\lambda}{k\mu\,\Delta t}\right)^{1/4} = 0.76\left(\frac{D^3\rho^2 g}{\mu\Gamma}\right)^{1/3} \tag{10-97}^*$$

Dimensional:

$$h = b(k^3\rho^2 L/\mu_b W_F)^{1/3} \tag{10-98}$$

where b = 205.4 (SI) or 534 (U.S. customary). For steam at atmospheric pressure

$$h = b(L/W_F)^{1/3} \tag{10-99}$$

where b = 2080 (SI) or 4920 (U.S. customary). For organic vapors at normal boiling point

$$h = b(L/W_F)^{1/3} \tag{10-100}$$

where b = 324 (SI) or 766 (U.S. customary).

Figure 10-9 is a nomograph for determining coefficients of heat transfer for condensation of pure vapors.

Banks of Horizontal Tubes (N_{Re} < 2100) In the idealized case of N tubes in a vertical row where the total condensate flows smoothly from one tube to the one beneath it, without splashing, and still in laminar flow on the tube, the mean condensing coefficient h_N for the entire row of N tubes is related to the condensing coefficient for the top tube h_1 by

$$h_N = h_1 N^{-1/4} \tag{10-101}$$

Dukler Theory The preceding expressions for condensation are based on the classical Nusselt theory. It is generally known and conceded that the film coefficients for steam and organic vapors calculated by the Nusselt theory are conservatively low. Dukler [*Chem. Eng. Prog.*, **55**, 62 (1959)] developed equations for velocity and temperature distribution in thin films on vertical walls based on expressions of Deissler (NACA Tech. Notes 2129, 1950; 2138, 1952; 3145, 1959) for the eddy viscosity and thermal conductivity near the solid boundary. According to the Dukler theory, three fixed factors must be known to establish the value of the average film coefficient: the terminal Reynolds number, the Prandtl number of the condensed phase, and a dimensionless group N_d defined as follows:

$$N_d = (0.250\mu_L^{1.173}\mu_G^{0.16}/(g^{2/3}D^2\rho_L^{0.553}\rho_G^{0.78}) \tag{10-102}$$

Graphical relationships of these variables are available in Document 6058, ADI Auxiliary Publications Project, Library of Congress, Washington. If rigorous values for condensing-film coefficients are desired, especially if the value of N_d in Eq. (10-102) exceeds (1)(10⁻⁵), it is suggested that these graphs be used. For the case in which interfacial shear is zero, Fig. 10-10 may be used. It is interesting to note that, according to the Dukler development, there is no definite transition Reynolds number; deviation from Nusselt theory is less at low Reynolds numbers; and when the Prandtl number of a fluid is less than 0.4 (at Reynolds number above 1000), the predicted values for film coefficient are lower than those predicted by the Nusselt theory.

*If the vapor density is significant, replace ρ^2 with $\rho_l(\rho_l - \rho_v)$.

The Dukler theory is applicable for condensate films on horizontal tubes and also for falling films, in general, i.e., those not associated with condensation or vaporization processes.

Vapor Shear Controlling For **vertical in-tube condensation** with vapor and liquid flowing cocurrently downward, if gravity controls, Figs. 10-9 and 10-10 may be used. If vapor shear controls, the Carpenter-Colburn correlation (*General Discussion on Heat Transfer*, London, 1951, ASME, New York, p. 20) is applicable:

$$h\mu_l/k_l\rho_l^{1/2} = 0.065(N_{Pr})_l^{1/2}F_{vc}^{1/2} \tag{10-103}$$

where

$$F_{vc} = fG_{vm}^2/2\rho_v \tag{10-103a}$$

$$G_{vm} = \left(\frac{G_{vi}^2 + G_{vi}G_{vo} + G_{vo}^2}{3}\right)^{1/2} \tag{10-103b}$$

and f is the Fanning friction factor evaluated at

$$(N_{Re})_{vm} = D_i G_{vm}/\mu_v \tag{10-103c}$$

An alternative formulation, directly in terms of the friction factor, is

$$h = 0.065\,(c\rho kf/2\mu\rho_v)^{1/2}G_{vm} \tag{10-103d}$$

expressed in consistent units.

Another correlation for vapor-shear-controlled condensation is the Boyko-Kruzhilin correlation [*Int. J. Heat Mass Transfer*, **10**, 361 (1967)], which gives the mean condensing coefficient for a stream between inlet quality x_i and outlet quality x_o:

$$\frac{hD_i}{k_l} = 0.024\left(\frac{D_i G_T}{\mu_l}\right)^{0.8}(N_{Pr})_l^{0.43}\,\frac{\sqrt{(\rho/\rho_m)_i} + \sqrt{(\rho/\rho_m)_o}}{2} \tag{10-104}$$

where G_T = total mass velocity in consistent units

$$\left(\frac{\rho}{\rho_m}\right)_i = 1 + \frac{\rho_l - \rho_v}{\rho_v}x_i \tag{10-104a}$$

and

$$\left(\frac{\rho}{\rho_m}\right)_o = 1 + \frac{\rho_l - \rho_v}{\rho_v}x_o \tag{10-104b}$$

For **horizontal in-tube condensation** at low flow rates Kern's modification (*Process Heat Transfer*, McGraw-Hill, New York, 1950) of the Nusselt equation is valid:

$$h_m = 0.761\left[\frac{Lk_l^3\rho_l(\rho_l - \rho_v)g}{W_T\mu_l}\right]^{1/3} = 0.815\left[\frac{k_l^3\rho_l(\rho_l - \rho_v)g\lambda}{\pi\mu_l D_i\,\Delta t}\right]^{1/4} \tag{10-105}$$

where W_T is the total vapor condensed in one tube and Δt is $t_{sv} - t_s$. A more rigorous correlation has been proposed by Chaddock [*Refrig. Eng.*, **65**(4), 36 (1957)]. Use consistent units.

At high condensing loads, with vapor shear dominating, tube orientation has no effect, and Eq. (10-104) may also be used for horizontal tubes.

Condensation of pure vapors under laminar conditions in the presence of noncondensable gases, interfacial resistance, superheating, variable properties, and diffusion has been analyzed by Minkowycz and Sparrow [*Int. J. Heat Mass Transfer*, **9**, 1125 (1966)].

BOILING (VAPORIZATION) OF LIQUIDS

Boiling Mechanisms Vaporization of liquids may result from various mechanisms of heat transfer, singly or combinations thereof. For example, vaporization may occur as a result of heat absorbed, by radiation and convection, at the surface of a pool of liquid; or as a result of heat absorbed by natural convection from a hot wall beneath the disengaging surface, in which case the vaporization takes

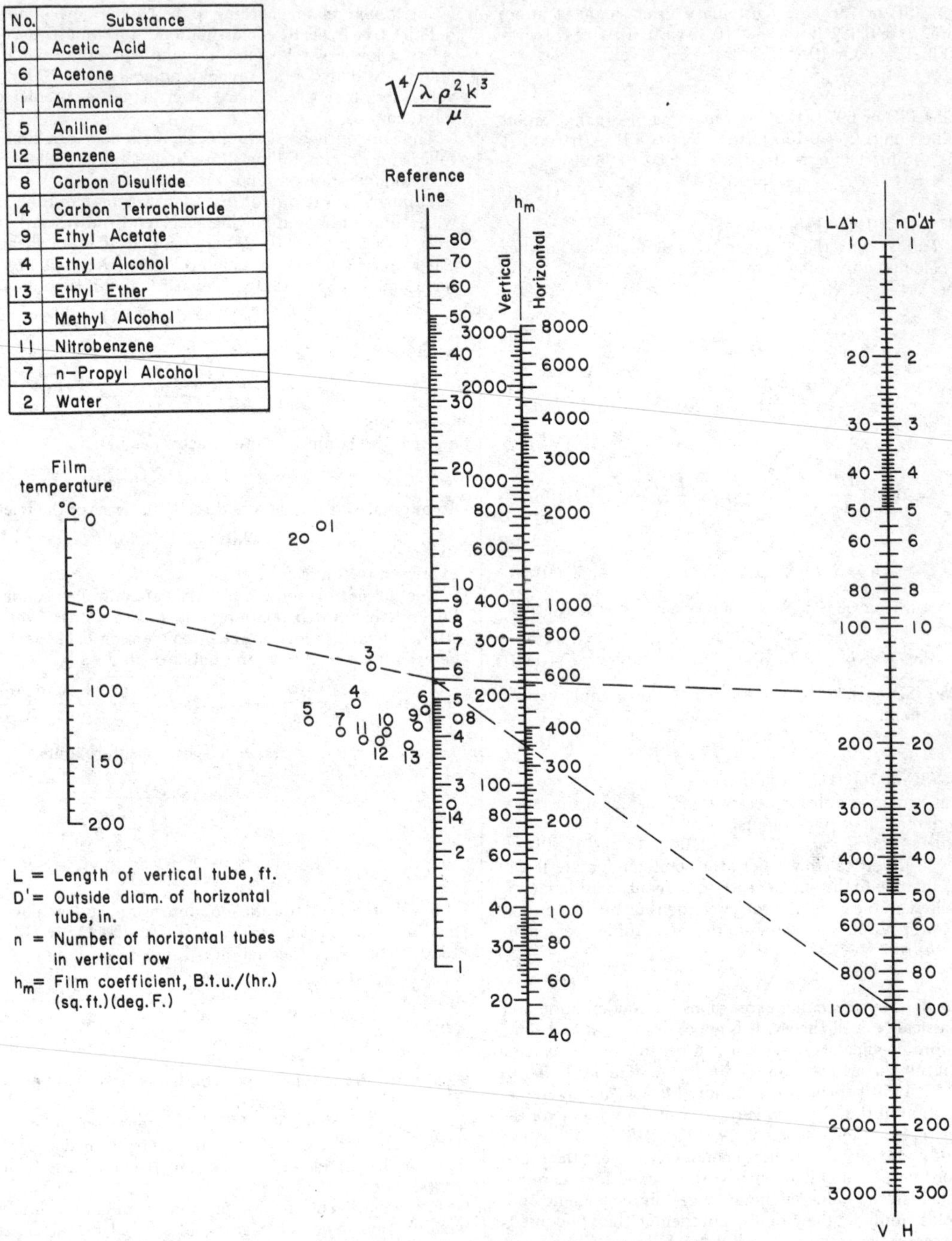

No.	Substance
10	Acetic Acid
6	Acetone
1	Ammonia
5	Aniline
12	Benzene
8	Carbon Disulfide
14	Carbon Tetrachloride
9	Ethyl Acetate
4	Ethyl Alcohol
13	Ethyl Ether
3	Methyl Alcohol
11	Nitrobenzene
7	n-Propyl Alcohol
2	Water

FIG. 10-9 Chart for determining film coefficient h_m for film-type condensation of pure vapor, based on Eqs. (10-92) and (10-97). For vertical tubes multiply h_m by 1.2. If $4\Gamma/\mu_f$ exceeds 2100, use Fig. 10-10. $\sqrt[4]{\lambda\rho^2k^3/\mu}$ is in U.S. customary units; to convert feet to meters, multiply by 0.3048; to convert inches to centimeters, multiply by 2.54; and to convert British thermal units per hour–square foot–degrees Fahrenheit to watts per square meter–kelvins, multiply by 5.6780.

place when the superheated liquid reaches the pool surface. Vaporization also occurs from falling films (the reverse of condensation) or from the flashing of liquids superheated by forced convection under pressure.

Pool boiling refers to the type of boiling experienced when the heating surface is surrounded by a relatively large body of fluid which is not flowing at any appreciable velocity and is agitated only by the motion of the bubbles and by natural-convection currents. Two types of pool boiling are possible: subcooled pool boiling, in which the bulk fluid temperature is below the saturation temperature, resulting in collapse of the bubbles before they reach the surface, and saturated pool boiling, with bulk temperature equal to saturation temperature, resulting in net vapor generation.

The general shape of the curve relating the heat-transfer coeffi-

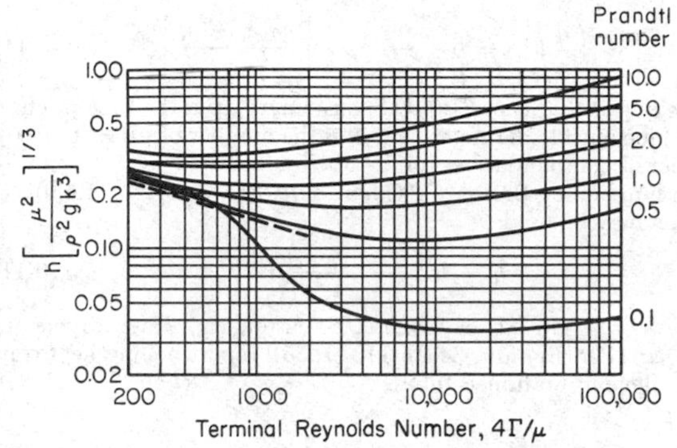

FIG. 10-10 Dukler plot showing average condensing-film coefficient as a function of physical properties of the condensate film and the terminal Reynolds number. (Dotted line indicates Nusselt theory for Reynolds number < 2100.) [*Reproduced by permission from* Chem. Eng. Prog., *55, 64 (1959).*]

cient to Δt_b, the temperature driving force (difference between the wall temperature and the bulk fluid temperature) is one of the few parametric relations that are reasonably well understood. The familiar boiling curve was originally demonstrated experimentally by Nukiyama [*J. Soc. Mech. Eng. (Japan)*, **37**, 367 (1934)]. This curve points out one of the great dilemmas for boiling-equipment designers. They are faced with at least six heat-transfer regimes in pool boiling: natural convection (+), incipient nucleate boiling (+), nucleate boiling (+), transition to film boiling (−), stable film boiling (+), and film boiling with increasing radiation (+). The signs indicate the sign of the derivative $d(q/A)/d\,\Delta t_b$. In the transition to film boiling, heat-transfer rate *decreases* with driving force. The regimes of greatest commercial interest are the *nucleate-boiling* and *stable-film-boiling regimes.*

Heat transfer by **nucleate boiling** is an important mechanism in the vaporization of liquids. It occurs in the vaporization of liquids in kettle-type and natural-circulation reboilers commonly used in the process industries. High rates of heat transfer per unit of area (heat flux) are obtained as a result of bubble formation at the liquid-solid interface rather than from mechanical devices external to the heat exchanger. The mechanism has not yet been clearly established, but as a result of considerable activity in the field of nucleate boiling there are available several expressions from which reasonable values of the film coefficients may be obtained. These expressions do not yield exactly the same numerical results even though the correlations are based upon much of the same data. There is thus neither a prominent nor a unique equation for nucleate boiling. Either convenience or familiarity will govern the user's selection of one of the equations in the next subsection.

The boiling curve, particularly in the nucleate-boiling region, is significantly affected by the temperature driving force, the total system pressure, the nature of the boiling surface, the geometry of the system, and the properties of the boiling material. In the nucleate-boiling regime, heat flux is approximately proportional to the cube of the temperature driving force. Designers in addition must know the minimum Δt (the point at which nucleate boiling begins), the critical Δt (the Δt above which transition boiling begins), and the maximum heat flux (the heat flux corresponding to the critical Δt). For designers who do not have experimental data available, the following equations may be used.

Boiling Coefficients For the **nucleate-boiling coefficient** the Mostinski equation [*Teplenergetika*, **4**, 66 (1963)] may be used:

$$h = bP_c^{0.69}\left(\frac{q}{A}\right)^{0.7}\left[1.8\left(\frac{P}{P_c}\right)^{0.17} + 4\left(\frac{P}{P_c}\right)^{1.2} + 10\left(\frac{P}{P_c}\right)^{10}\right] \qquad (10\text{-}106)$$

where $b = (3.75)(10^{-5})$(SI) or $(2.13)(10^{-4})$ (U.S. customary), P_c is the critical pressure and P the system pressure, q/A is the heat flux, and h is the nucleate-boiling coefficient. The McNelly equation [*J. Imp. Coll. Chem. Eng. Soc.*, 7(18), (1953)] may also be used:

$$h = 0.225\left(\frac{qc_l}{A\lambda}\right)^{0.69}\left(\frac{Pk_l}{\sigma}\right)^{0.31}\left(\frac{\rho_l}{\rho_v} - 1\right)^{0.33} \qquad (10\text{-}107)$$

where c_l is the liquid heat capacity, λ is the latent heat, P is the system pressure, k_l is the thermal conductivity of the liquid, and σ is the surface tension.

An equation of the Nusselt type has been suggested by Rohsenow [*Trans. Am. Soc. Mech. Eng.*, **74**, 969 (1952)].

$$hD/k = C_r(DG/\mu)^{2/3}(c\mu/k)^{-0.7} \qquad (10\text{-}108)$$

in which the variables assume the following form:

$$\frac{h\beta'}{k}\left[\frac{g_c\sigma}{g(\rho_L - \rho_v)}\right]^{1/2} = C_r\left[\frac{\beta'}{\mu}\left(\frac{g_c\sigma}{g(\rho_L - \rho_v)}\right)^{1/2}\frac{W}{A}\right]^{2/3}\left(\frac{c\mu}{k}\right)^{-0.7} \qquad (10\text{-}108a)$$

The constant C_r is not truly constant but varies from 0.006 to 0.015.* It is possible that the nature of the surface is partly responsible for the variation in the constant. The only factor in Eq. (10-108*a*) not readily available is the value of the contact angle β'.

Another Nusselt-type equation has been proposed by Forster and Zuber:†

$$N_{Nu} = 0.0015N_{Re}^{0.62}N_{Pr}^{1/3} \qquad (10\text{-}109)$$

which takes the following form:

$$\frac{c\rho_L\sqrt{\pi\alpha}}{k\rho_v}\frac{W}{A}\left(\frac{2\sigma}{\Delta p}\right)^{1/2}\left(\frac{\rho_L}{\Delta p g_c}\right)^{1/4} = 0.0015\left[\frac{\rho_L}{\mu}\left(\frac{c\rho_L\,\Delta T\,\sqrt{\pi\alpha}}{\lambda\rho_v}\right)^2\right]^{0.62}\left(\frac{c\mu}{k}\right)^{1/2} \qquad (10\text{-}109a)$$

where α = $k/\rho c$ (all liquid properties)
Δp = pressure of the vapor in a bubble minus saturation pressure of a flat liquid surface

Equations (10-108*a*) and (10-109*a*) have been arranged in dimensional form by Westwater.

These equations will give conservative results because they are based on a considerable amount of data from various sources. The numerical constant may be adjusted to suit any particular set of data if one desires to use a certain criterion. However, surface conditions vary so greatly that deviations may be as large as ±25 percent from results obtained.

The **maximum heat flux** may be predicted by the Kutateladse-Zuber [*Trans. Am. Soc. Mech. Eng.*, **80**, 711 (1958)] relationship, using consistent units:

$$\left(\frac{q}{A}\right)_{max} = 0.18g_c^{1/4}\rho_v\lambda\left[\frac{(\rho_l - \rho_v)\sigma g}{\rho_v^2}\right]^{1/4} \qquad (10\text{-}110)$$

Alternatively, Mostinski presented an equation which approximately represents the Cichelli-Bonilla [*Trans. Am. Inst. Chem. Eng.*, **41**, 755 (1945)] correlation:

$$\frac{(q/A)_{max}}{P_c} = b\left(\frac{P}{P_c}\right)^{0.35}\left(1 - \frac{P}{P_c}\right)^{0.9} \qquad (10\text{-}111)$$

where b = 0.368(SI) or 5.58 (U.S. customary); P_c is the critical pres-

*Reported by Westwater in Drew and Hoopes, *Advances in Chemical Engineering*, vol. I, Academic, New York, 1956, p. 15.

†Forster, *J. Appl. Phys.*, **25**, 1067 (1954); Forster and Zuber, *J. Appl. Phys.*, **25**, 474 (1954); Forster and Zuber, Conference on Nuclear Engineering, University of California, Los Angeles, 1955; excellent treatise on boiling of liquids by Westwater in Drew and Hoopes, *Advances in Chemical Engineering*, vol. I, Academic, New York, 1956.

sure, Pa absolute; P is the system pressure; and $(q/A)_{max}$ is the maximum heat flux.

The lower limit of applicability of the nucleate-boiling equations is from 0.1 to 0.2 of the maximum limit and depends upon the magnitude of natural-convection heat transfer for the liquid. The best method of determining the lower limit is to plot two curves: one of h versus Δt for natural convection, the other of h versus Δt for nucleate boiling. The intersection of these two curves may be considered the lower limit of applicability of the equations.

These equations apply to single tubes or to flat surfaces in a large pool. In tube bundles the equations are only approximate, and designers must rely upon experiment. Palen and Small [*Hydrocarbon Process.*, **43**(11), 199 (1964)] have shown the effect of tube-bundle size on maximum heat flux.

$$\left(\frac{q}{A}\right)_{max} = b\frac{p}{D_o\sqrt{N}}\rho_v\lambda\left[\frac{g\sigma(\rho_l - \rho_v)}{\rho_v^2}\right]^{1/4} \qquad (10\text{-}112)$$

where b = 0.43 (SI) or 61.6 (U.S. customary), p is the tube pitch, D_o is the tube outside diameter, and N is the number of tubes (twice the number of complete tubes for U-tube bundles).

For film boiling Bromley's [*Chem. Eng. Prog.*, **46**, 221 (1950)] correlation may be used:

$$h = b\left[\frac{k_v^3(\rho_l - \rho_v)\rho_v g}{\mu_v D_o \Delta t_b}\right]^{1/4} \qquad (10\text{-}113)$$

where b = 4.306 (SI) or 0.620 (U.S. customary). Katz, Myers, and Balekjian [*Pet. Refiner*, **34**(2), 113 (1955)] report boiling heat-transfer coefficients on finned tubes.

THERMAL DESIGN OF HEAT-TRANSFER EQUIPMENT

INTRODUCTION TO THERMAL DESIGN

Design methods for several important classes of process heat-transfer equipment are presented in the following portions of Sec. 10. Mechanical descriptions and specifications of equipment are given in Sec. 11, which should be read in conjunction with the use of this material. It is impossible to present here a comprehensive treatment of heat-exchanger selection, design, and application. The best general references in this field are Kern, *Process Heat Transfer*, McGraw-Hill, New York, 1950; and Schlünder (ed.), *Heat Exchanger Design Handbook*, Hemisphere Publishing, Washington, 1983.

Approach to Heat-Exchanger Design The proper use of basic heat-transfer knowledge in the design of practical heat-transfer equipment is an art. Designers must be constantly aware of the differences between the idealized conditions for and under which the basic knowledge was obtained and the real conditions of the mechanical expression of their design and its environment. The result must satisfy process and operational requirements (such as availability, flexibility, and maintainability) and do so economically. An important part of any design process is to consider and offset the consequences of error in the basic knowledge, in its subsequent incorporation into a design method, in the translation of design into equipment, or in the operation of the equipment and the process. Heat-exchanger design is not a highly accurate art under the best of conditions.

The **design of a process heat exchanger** usually proceeds through the following steps:

1. Process conditions (stream compositions, flow rates, temperatures, pressures) must be specified.
2. Required physical properties over the temperature and pressure ranges of interest must be obtained.
3. The type of heat exchanger to be employed is chosen.
4. A preliminary estimate of the size of the exchanger is made, using a heat-transfer coefficient appropriate to the fluids, the process, and the equipment.
5. A first design is chosen, complete in all details necessary to carry out the design calculations.
6. The design chosen in step 5 is evaluated, or *rated*, as to its ability to meet the process specifications with respect to both heat transfer and pressure drop.
7. On the basis of the result of step 6, a new configuration is chosen if necessary and step 6 is repeated. If the first design was inadequate to meet the required heat load, it is usually necessary to increase the size of the exchanger while still remaining within specified or feasible limits of pressure drop, tube length, shell diameter, etc. This will sometimes mean going to multiple-exchanger configurations. If the first design more than meets heat-load requirements or does not use all the allowable pressure drop, a less expensive exchanger can usually be designed to fulfill process requirements.
8. The final design should meet process requirements (within reasonable expectations of error) at lowest cost. The lowest cost should include operation and maintenance costs and credit for ability to meet long-term process changes, as well as installed (capital) cost.

Exchangers should not be selected entirely on a lowest-first-cost basis, which frequently results in future penalties.

Overall Heat-Transfer Coefficient The basic design equation for a heat exchanger is

$$dA = dQ/U\,\Delta T \qquad (10\text{-}114)$$

where dA is the element of surface area required to transfer an amount of heat dQ at a point in the exchanger where the overall heat-transfer coefficient is U and where the overall bulk temperature difference between the two streams is ΔT. The overall heat-transfer coefficient is related to the individual film heat-transfer coefficients and fouling and wall resistances by Eq. (10-31). Basing U_o on the outside surface area A_o results in

$$U_o = \frac{1}{1/h_o + R_{do} + xA_o/k_w A_{wm} + (1/h_i + R_{di})A_o/A_i} \qquad (10\text{-}115)$$

Equation (10-114) can be formally integrated to give the outside area required to transfer the total heat load Q:

$$A_o = \int_0^Q \frac{dQ}{U_o\,\Delta T} \qquad (10\text{-}116)$$

To integrate Eq. (10-116), U_o and ΔT must be known as functions of Q. For some problems, U_o varies strongly and nonlinearly throughout the exchanger. In these cases, it is necessary to evaluate U_o and ΔT at several intermediate values and numerically or graphically integrate. For many practical cases, it is possible to calculate a constant mean overall coefficient U_{om} from Eq. (10-115) and define a corresponding mean value of ΔT_m, such that

$$A_o = Q_T/U_{om}\,\Delta T_m \qquad (10\text{-}117)$$

Care must be taken that U_o does not vary too strongly, that the proper equations and conditions are chosen for calculating the individual coefficients, and that the mean temperature difference is the correct one for the specified exchanger configuration.

Mean Temperature Difference The temperature difference between the two fluids in the heat exchanger will, in general, vary from point to point. The mean temperature difference (MTD) can be calculated from the terminal temperatures of the two streams if the following assumptions are valid:

1. All elements of a given fluid stream have the same thermal history in passing through the exchanger.*

*This assumption is vital but is usually omitted or less satisfactorily stated as "each stream is well mixed at each point." In a heat exchanger with substantial bypassing of the heat-transfer surface, e.g., a typical baffled shell-and-tube exchanger, this condition is not satisfied. However, the error is in some degree offset if the same MTD formulation used in reducing experimental heat-transfer data to obtain the basic correlation is used in applying the correlation to design a heat exchanger. The compensation is not in general exact, and insight and judgment are required in the use of the MTD formulations. Particularly, in the design of an exchanger with a very close temperature approach, bypassing may result in an exchanger that is inefficient and even thermodynamically inoperable.

2. The exchanger operates at steady state.
3. The specific heat is constant for each stream (or if either stream undergoes an isothermal phase transition).
4. The overall heat-transfer coefficient is constant.
5. Heat losses are negligible.

Countercurrent or Cocurrent Flow If the flow of the streams is either *completely* countercurrent or completely cocurrent or if one or both streams are isothermal (condensing or vaporizing with negligible pressure change), the correct MTD is the logarithmic-mean temperature difference (LMTD), defined as

$$\Delta t_{lm} = \frac{(t_1' - t_2'') - (t_2' - t_1'')}{\ln\left(\dfrac{t_1' - t_2''}{t_2' - t_1''}\right)} \qquad (10\text{-}118a)$$

for *countercurrent flow* (Fig. 10-11*a*) and

$$\Delta t_{lm} = \frac{(t_1' - t_1'') - (t_2' - t_2'')}{\ln\left(\dfrac{t_1' - t_1''}{t_2' - t_2''}\right)} \qquad (10\text{-}118b)$$

for *cocurrent flow* (Fig. 10-11*b*).

If U is not constant but a linear function of ΔT, the correct value of $U_{om}\,\Delta T_m$ to use in Eq. (10-117) is [Colburn, *Ind. Eng. Chem.*, **25**, 873 (1933)]

$$U_{om}\,\Delta T_m = \frac{U_o''(t_1' - t_2'') - U_o'(t_2' - t_1'')}{\ln\left(\dfrac{U_o''(t_1' - t_2'')}{U_o'(t_2' - t_1'')}\right)} \qquad (10\text{-}119a)$$

for *countercurrent flow*, where U_o'' is the overall coefficient evaluated when the stream temperatures are t_1' and t_2'' and U_o' is evaluated at t_2' and t_1''. The corresponding equation for *cocurrent flow* is

$$U_{om}\,\Delta T_m = \frac{U_o''(t_1' - t_1'') - U_o'(t_2' - t_2'')}{\ln\left(\dfrac{U_o''(t_1' - t_1'')}{U_o'(t_2' - t_2'')}\right)} \qquad (10\text{-}119b)$$

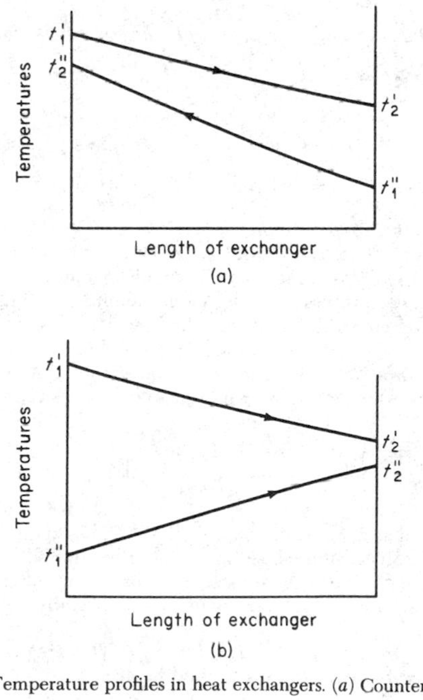

FIG. 10-11 Temperature profiles in heat exchangers. (*a*) Countercurrent. (*b*) Cocurrent.

where U_o' is evaluated at t_2' and t_2'' and U_o'' is evaluated at t_1' and t_1''. To use these equations, it is necessary to calculate two values of U_o.*

The use of Eq. (10-119) will frequently give satisfactory results even if U_o is not strictly linear with temperature difference.

Reversed, Mixed, or Cross-Flow If the flow pattern in the exchanger is not completely countercurrent or cocurrent, it is necessary to apply a **correction factor** F_T by which the LMTD is multiplied to obtain the appropriate MTD. These corrections have been mathematically derived for flow patterns of interest, still by making assumptions 1 to 5 [see Bowman, Mueller, and Nagle, *Trans. Am. Soc. Mech. Eng.*, **62**, 283 (1940) or Kern, op. cit.]. For a common flow pattern, the 1-2 exchanger (Fig. 10-12), the correction factor F_T

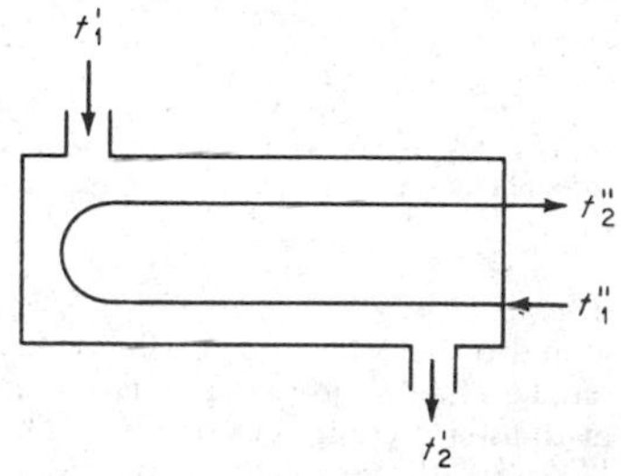

FIG. 10-12 Diagram of a 1-2 exchanger (one well-baffled shell pass and two tube passes with an equal number of tubes in each pass).

is given in Fig. 10-14*a*, which is also valid for finding F_T for a 1-2 exchanger in which the shell-side flow direction is reversed from that shown in Fig. 10-12. Figure 10-14*a* is also applicable with negligible error to exchangers with one shell pass and any number of tube passes. Values of F_T less than 0.8 (0.75 at the very lowest) are generally unacceptable because the exchanger configuration chosen is inefficient; the chart is difficult to read accurately; and even a small violation of the first assumption underlying the MTD will invalidate the mathematical derivation and lead to a thermodynamically inoperable exchanger.

Correction-factor charts are also available for exchangers with more than one shell pass provided by a longitudinal shell-side baffle. However, these exchangers are seldom used in practice because of mechanical complications in their construction. Also thermal and physical leakages across the longitudinal baffle further reduce the mean temperature difference and are not properly incorporated into the correction-factor charts. Such charts are useful, however, when it is necessary to construct a multiple-shell exchanger train such as that shown in Fig. 10-13 and are included here for two, three, four, and six *separate* shells and two or more tube passes per shell in Fig. 10-14*b*, *c*, *d*, and *e*. If only one tube pass per shell is required, the piping can and should be arranged to provide pure countercurrent flow, in which case the LMTD is used with no correction.

Cross-flow exchangers of various kinds are also important and require correction to be applied to the LMTD calculated by assuming countercurrent flow. Several cases are given in Fig. 10-14*f*, *g*, *h*, *i*, and *j*.

Many other MTD correction-factor charts have been prepared for various configurations. The F_T charts are often employed to make approximate corrections for configurations even in cases for which they are not completely valid.

THERMAL DESIGN FOR SINGLE-PHASE HEAT TRANSFER

Double-Pipe Heat Exchangers The design of double-pipe heat exchangers is straightforward. It is generally conservative to neglect

*This task can be avoided if a hydrocarbon stream is the limiting resistance by the use of the caloric temperature charts developed by Colburn [*Ind. Eng. Chem.*, **25**, 873 (1933)] and discussed in Kern (op. cit.).

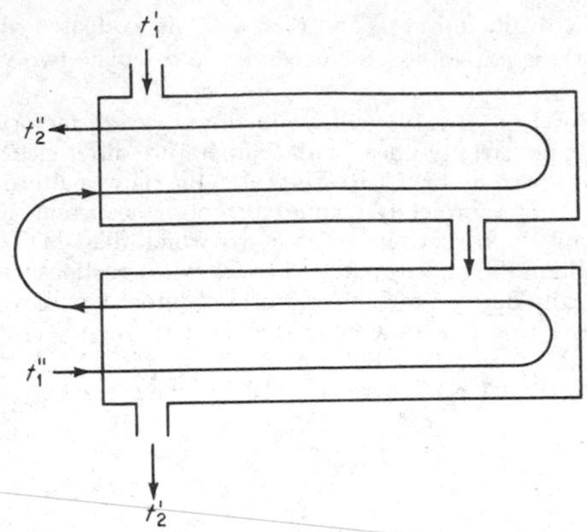

FIG. 10-13 Diagram of a 2-4 exchanger (two separate identical well-baffled shells and four or more tube passes).

natural-convection and turbulent-entrance effects. In laminar flow, the length term in the Graetz number is usually taken to be the length of one heat-transfer section between return bends. Pressure drop is calculated by using the correlations given in Sec. 5.

If the inner tube is longitudinally finned on the outside surface, the equivalent diameter is used as the characteristic length in both the Reynolds-number and the heat-transfer correlations. The fin efficiency must also be known to calculate an *effective* outside area to use in Eq. (10-115).

Fittings contribute strongly to the pressure drop on the annulus side. General methods for predicting this are not reliable, and manufacturer's data should be used when available.

Double-pipe exchangers are often piped in complex series-parallel arrangements on both sides. The MTD to be used has been derived for some of these arrangements and is reported in Kern (*Process Heat Transfer*, McGraw-Hill, New York, 1950). More complex cases may require trial-and-error balancing of the heat loads and rate equations for subsections or even for individual exchangers in the bank.

Baffled Shell-and-Tube Exchangers The method given here is based on the research summarized in Final Report, Cooperative Research Program on Shell and Tube Heat Exchangers, Univ. Del. Eng. Exp. Sta. Bull. 5 (June 1963). The method assumes that the shell-side heat transfer and pressure-drop characteristics are equal to those of the ideal tube bank corresponding to the cross-flow sections of the exchanger, modified for the distortion of flow pattern introduced by the baffles and the presence of leakage and bypass flow through the various clearances required by mechanical construction.

It is assumed that process conditions and physical properties are known and the following are known or specified: tube outside diameter D_o, tube geometrical arrangement (unit cell), shell inside diameter D_s, shell outer tube limit D_{otl}, baffle cut l_c, baffle spacing l_s, and number of sealing strips N_{ss}. The effective tube length between tube sheets L may be either specified or calculated after the heat-transfer coefficient has been determined. If additional specific information (e.g., tube-baffle clearance) is available, the exact values (instead of estimates) of certain parameters may be used in the calculation with some improvement in accuracy. To complete the rating, it is necessary to know also the tube material and wall thickness or inside diameter.

This rating method, though apparently generally the best in the open literature, is not extremely accurate. An exhaustive study by Palen and Taborek [*Chem. Eng. Prog. Symp. Ser.* 92, **65,** 53 (1969)] showed that this method predicted shell-side coefficients from about 50 percent low to 100 percent high, while the pressure-drop range was from about 50 percent low to 200 percent high. The mean error for heat transfer was about 15 percent low (safe) for all Reynolds numbers, while the mean error for pressure drop was from about 5 percent low (unsafe) at Reynolds numbers above 1000 to about 100 percent high at Reynolds numbers below 10.

Calculation of Shell-Side Geometrical Parameters

1. *Total number of tubes in exchanger N_t.* If not known by direct count, find in the tube-count table, Table 11-3.

2. *Tube pitch parallel to flow p_p and normal to flow p_n.* These quantities are needed only for estimating other parameters. If a detailed drawing of the exchanger is available, it is better to obtain these other parameters by direct count or calculation. The pitches are described by Fig. 10-15 and read therefrom for common tube layouts.

3. *Number of tube rows crossed in one cross-flow section N_c.* Count from exchanger drawing or estimate from

$$N_c = \frac{D_s[1 - 2(l_c/D_s)]}{p_p} \tag{10-120}$$

4. *Fraction of total tubes in cross-flow F_c*

$$F_c = \frac{1}{\pi}\left[\pi + 2\frac{D_s - 2l_c}{D_{otl}} \sin\left(\cos^{-1}\frac{D_s - 2l_c}{D_{otl}}\right) - 2\cos^{-1}\frac{D_s - 2l_c}{D_{otl}}\right] \tag{10-121}$$

F_c is plotted in Fig. 10-16. This figure is strictly applicable only to split-ring, floating-head construction but may be used for other situations with minor error.

5. *Number of effective cross-flow rows in each window N_{cw}*

$$N_{cw} = \frac{0.8l_c}{p_p} \tag{10-122}$$

6. *Cross-flow area at or near centerline for one cross-flow section S_m*

a. For rotated and in-line square layouts:

$$S_m = l_s\left[D_s - D_{otl} + \frac{D_{otl} - D_o}{p_n}(p' - D_o)\right] \text{ m}^2 \text{ (ft}^2) \tag{10-123a}$$

b. For triangular layouts:

$$S_m = l_s\left[D_s - D_{otl} + \frac{D_{otl} - D_o}{p'}(p' - D_o)\right] \text{ m}^2 \text{ (ft}^2) \tag{10-123b}$$

7. *Fraction of cross-flow area available for bypass flow F_{bp}*

$$F_{bp} = \frac{(D_s - D_{otl})l_s}{S_m} \tag{10-124}$$

8. *Tube-to-baffle leakage area for one baffle S_{tb}.* Estimate from

$$S_{tb} = bD_oN_T(1 + F_c) \text{ m}^2 \text{ (ft}^2) \tag{10-125}$$

where b = (6.223)(10^{-4}) (SI) or (1.701)(10^{-4}) (U.S. customary). These values are based on Tubular Exchanger Manufacturers Association (TEMA) Class R construction which specifies 1/32-in diametral clearance between tube and baffle. Values should be modified if extra tight or loose construction is specified or if clogging by dirt is anticipated.

9. *Shell-to-baffle leakage area for one baffle S_{sb}.* If diametral shell-baffle clearance δ_{sb} is known, S_{sb} can be calculated from

$$S_{sb} = \frac{D_s\,\delta_{sb}}{2}\left[\pi - \cos^{-1}\left(1 - \frac{2l_c}{D_s}\right)\right] \text{ m}^2 \text{ (ft}^2) \tag{10-126}$$

where the value of the term $\cos^{-1}(1 - 2l_c/D_s)$ is in radians and is between 0 and $\pi/2$. S_{sb} is plotted in Fig. 10-17, based on TEMA Class R standards. Since pipe shells are generally limited to diameters below 24 in, the larger sizes are shown by using the rolled-shell specification. Allowance should be made for especially tight or loose construction.

10. *Area for flow through window S_w.* This area is obtained as the difference between the gross window area S_{wg} and the window

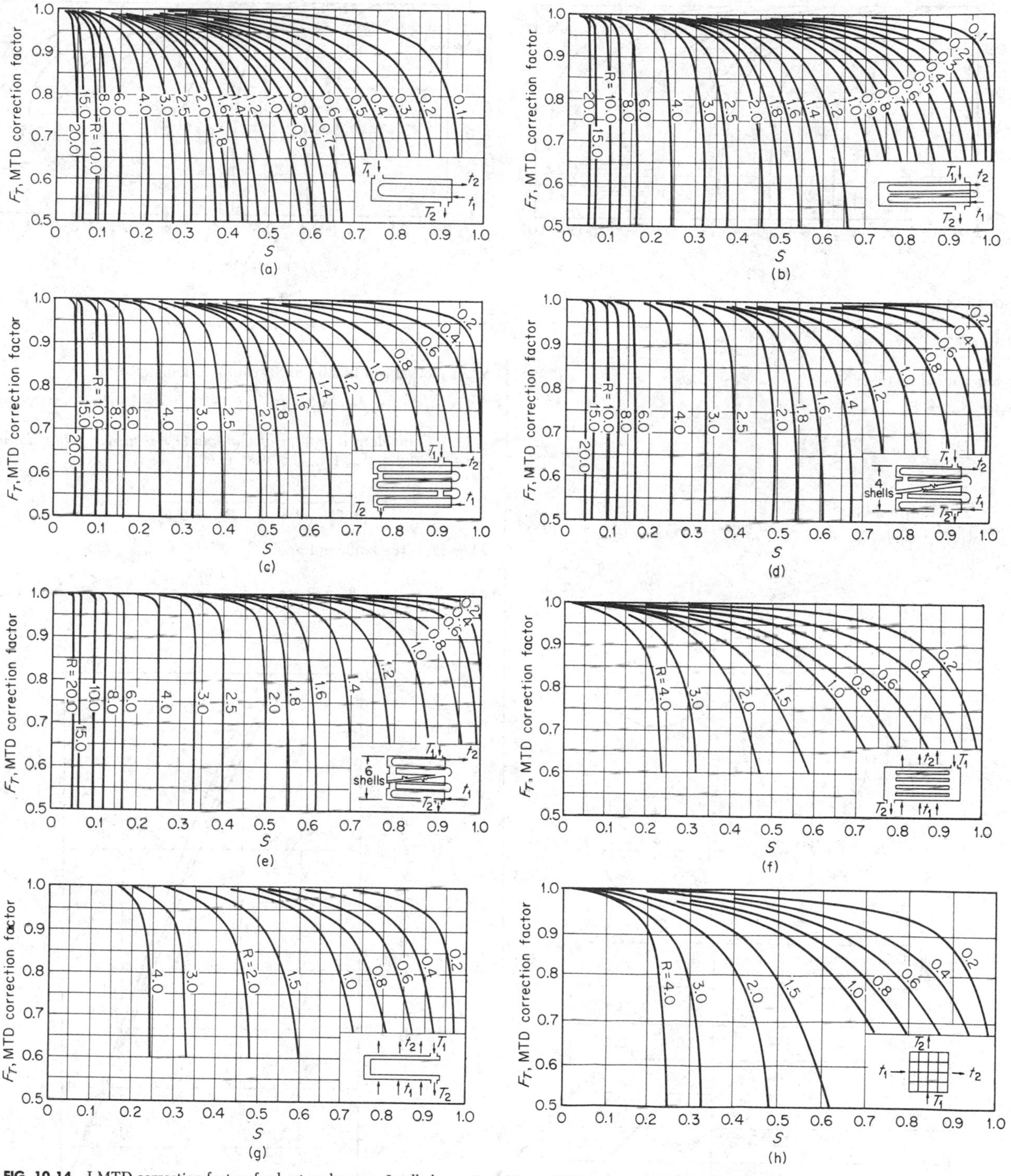

FIG. 10-14 LMTD correction factors for heat exchangers. In all charts, $R = (T_1 - T_2)/(t_2 - t_1)$ and $S = (t_2 - t_1)/(T_1 - t_1)$. (*a*) One shell pass, two or more tube passes. (*b*) Two shell passes, four or more tube passes. (*c*) Three shell passes, six or more tube passes. (*d*) Four shell passes, eight or more tube passes. (*e*) Six shell passes, twelve or more tube passes. (*f*) Cross-flow, one shell pass, one or more parallel rows of tubes. (*g*) Cross-flow, two passes, two rows of tubes; for more than two passes, use $F_T = 1.0$. (*h*) Cross-flow, one shell pass, one tube pass, both fluids unmixed. (*i*) Cross-flow (drip type), two horizontal passes with U-bend connections (trombone type). (*j*) Cross-flow (drip type), helical coils with two turns.

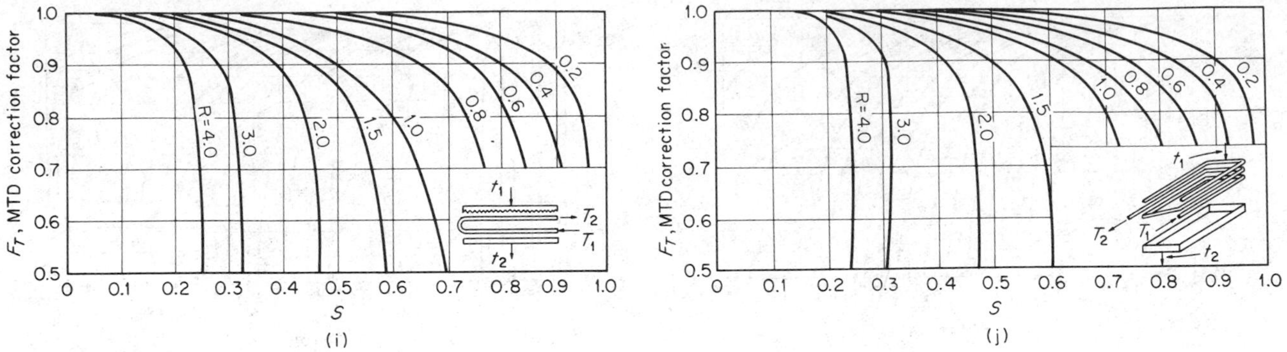

FIG. 10-14 (*continued*).

area occupied by tubes S_{wt}:

$$S_w = S_{wg} - S_{wt}$$

$$S_{wg} = \frac{D_s^2}{4}\left[\cos^{-1}\left(1 - 2\frac{l_c}{D_s}\right)\right. \tag{10-127}$$

$$\left. - \left(1 - 2\frac{l_c}{D_s}\right)\sqrt{1 - \left(1 - 2\frac{l_c}{D_s}\right)^2}\right] \quad \text{m}^2\ (\text{ft}^2) \tag{10-128}$$

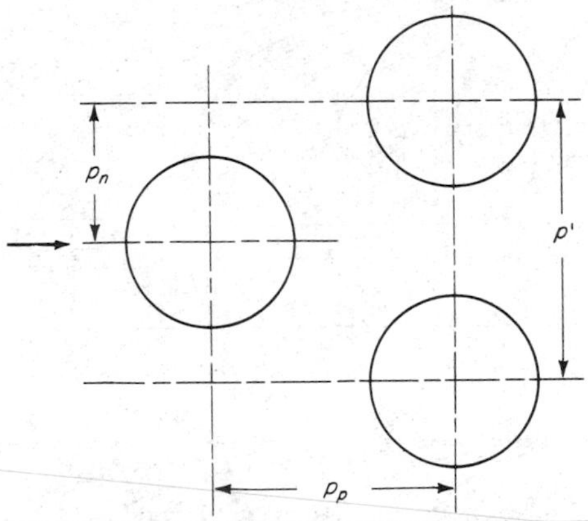

Tube O.D. D_o, in.	Tube pitch p', in.	Layout	p_p, in.	p_n, in.
0.625	0.812	→ △	0.704	0.406
0.750	0.938	→ △	0.814	0.469
0.750	1	→ □	1.000	1.000
0.750	1	→ ◇	0.707	0.707
0.750	1	→ △	0.866	0.500
1.000	1.250	→ □	1.250	1.250
1.000	1.250	→ ◇	0.884	0.884
1.000	1.250	→ △	1.082	0.625

FIG. 10-15 Values of tube pitch for common tube layouts. To convert inches to meters, multiply by 0.0254. Note that D_O, p', p_p, and p_n have units of inches.

S_{wg} is plotted in Fig. 10-18. S_{wt} can be calculated from

$$S_{wt} = (N_T/8)(1 - F_c)\pi D_o^2 \quad \text{m}^2\ (\text{ft}^2) \tag{10-129}$$

11. *Equivalent diameter of window D_w [required only if laminar flow, defined as $(N_{Re})_s \le 100$, exists]*

$$D_w = \frac{4S_w}{(\pi/2)N_T(1 - F_c)D_o + D_s\theta_b} \quad \text{m (ft)} \tag{10-130}$$

where θ_b is the baffle-cut angle given by

$$\theta_b = 2\cos^{-1}\left(1 - \frac{2l_c}{D_s}\right) \quad \text{rad} \tag{10-131}$$

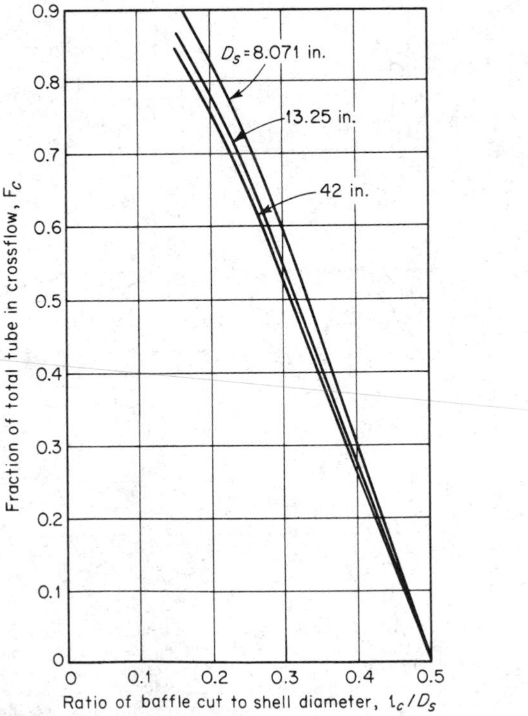

FIG. 10-16 Estimation of fraction of tubes in cross-flow F_c [Eq. (10-121)]. To convert inches to meters, multiply by 0.0254. Note that ℓ_c and D_s have units of inches.

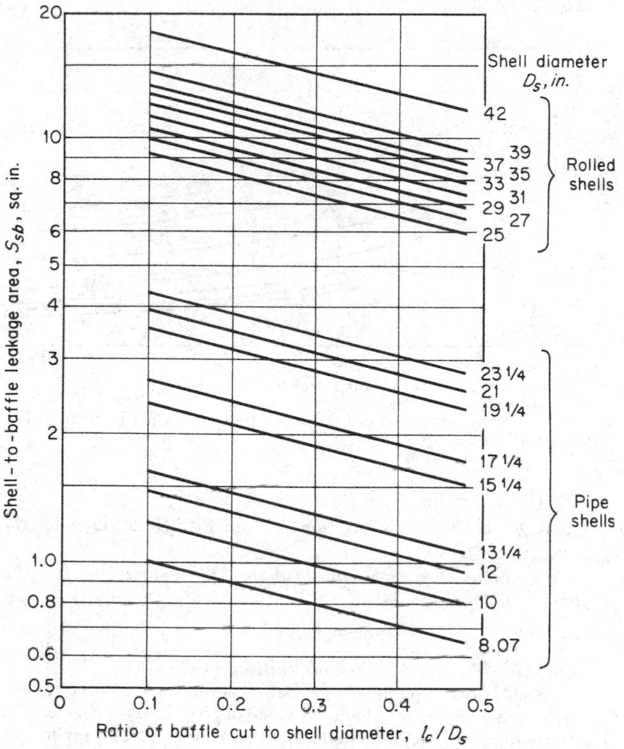

FIG. 10-17 Estimation of shell-to-baffle leakage area [Eq. (10-126)]. To convert inches to meters, multiply by 0.0254; to convert square inches to square meters, multiply by (6.45)(10^{-4}). Note that ℓ_c and D_s have units of inches.

12. *Number of baffles* N_b

$$N_b = (L/l_s) - 1 \tag{10-132}$$

The effective tube length L must be known to calculate N_b, which is needed to calculate shell-side pressure drop. In designing an exchanger, the shell-side coefficient may be calculated and the required exchanger length for heat transfer obtained before N_b is calculated.

Shell-Side Heat-Transfer Coefficient Calculation

1. Calculate the *shell-side Reynolds number* $(N_{Re})_s$.

$$(N_{Re})_s = D_o W/\mu_b S_m \tag{10-133}$$

where W = weight flow rate and μ_b = viscosity at bulk temperature. The arithmetic mean bulk shell-side fluid temperature is usually adequate to evaluate all bulk properties of the shell-side fluid. For large temperature ranges or for viscosity that is very sensitive to temperature change, special care must be taken, such as using Eq. (10-119).

2. Find j_k from the ideal-tube bank curve for a given tube layout at the calculated value of $(N_{Re})_s$, using Fig. 10-19, which is adapted from ideal-tube-bank data obtained at Delaware by Bergelin et al. [*Trans. Am. Soc. Mech. Eng.*, **74,** 953 (1952) and the Grimison correlation [*Trans. Am. Soc. Mech. Eng.*, **59,** 583 (1937)].

3. Calculate the shell-side *heat-transfer coefficient for an ideal tube bank* h_k.

$$h_k = j_k c \frac{W}{S_m}\left(\frac{k}{c\mu}\right)^{2/3}\left(\frac{\mu_b}{\mu_w}\right)^{0.14} \tag{10-134}$$

where c is the specific heat, k is the thermal conductivity, and μ_w is the viscosity evaluated at the mean surface temperature.

4. Find the correction factor for baffle-configuration effects J_c from Fig. 10-20.

5. Find the correction factor for baffle-leakage effects J_l from Fig. 10-21.

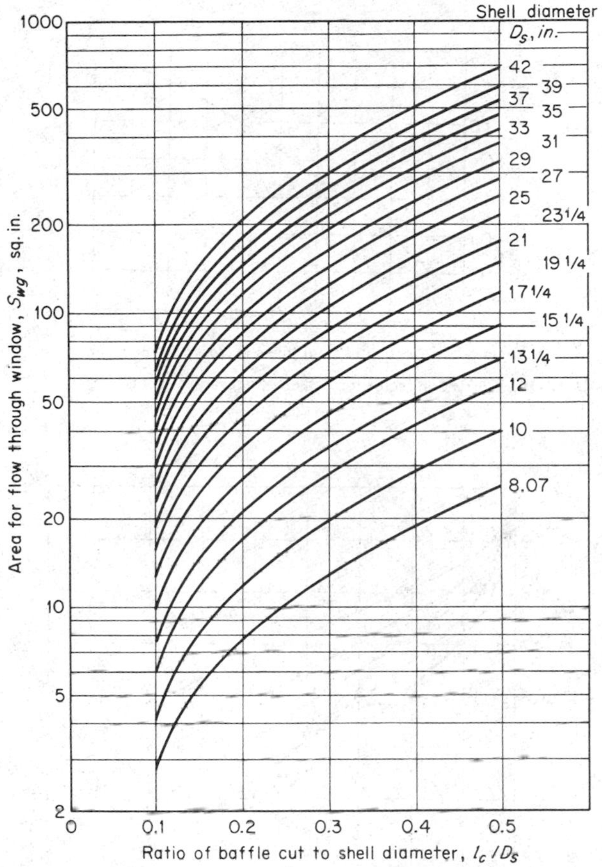

FIG. 10-18 Estimation of window cross-flow area [Eq. (10-128)]. To convert inches to meters, multiply by 0.0254. Note that ℓ_c and D_s have units of inches.

6. Find the correction factor for bundle-bypassing effects J_b from Fig. 10-22.

7. Find the correction factor for adverse temperature-gradient buildup at low Reynolds number J_r:

a. If $(N_{Re})_s < 100$, find J_r° from Fig. 10-23, knowing N_b and $(N_c + N_{cw})$.

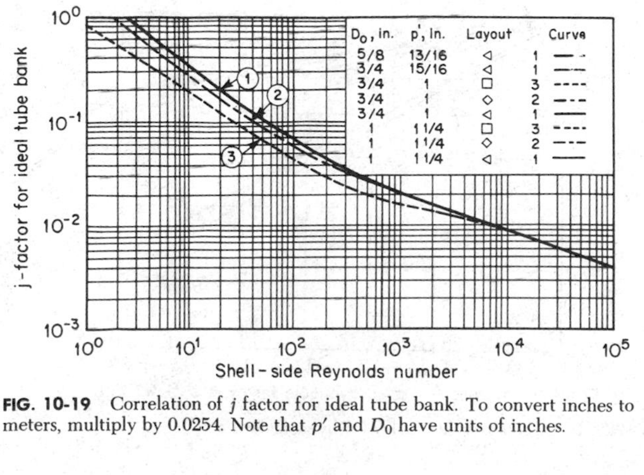

FIG. 10-19 Correlation of j factor for ideal tube bank. To convert inches to meters, multiply by 0.0254. Note that p' and D_0 have units of inches.

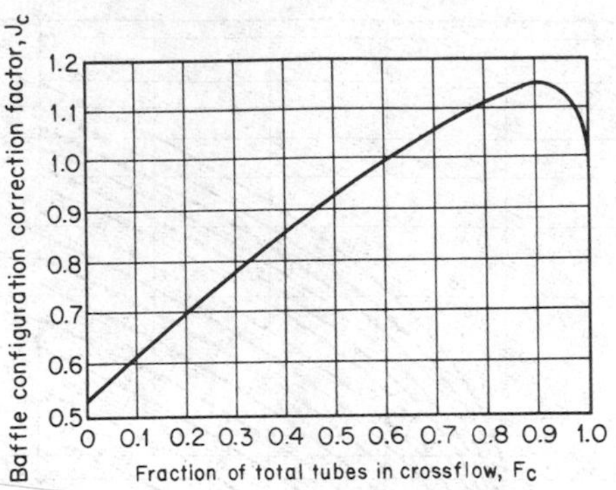

FIG. 10-20 Correction factor for baffle-configuration effects.

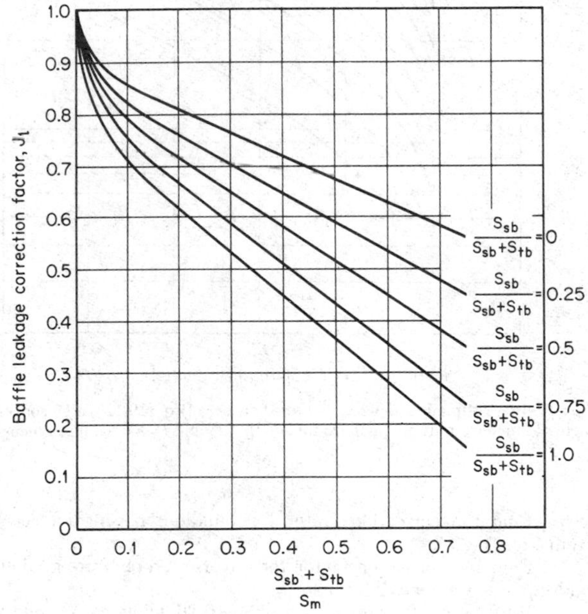

FIG. 10-21 Correction factor for baffle-leakage effects.

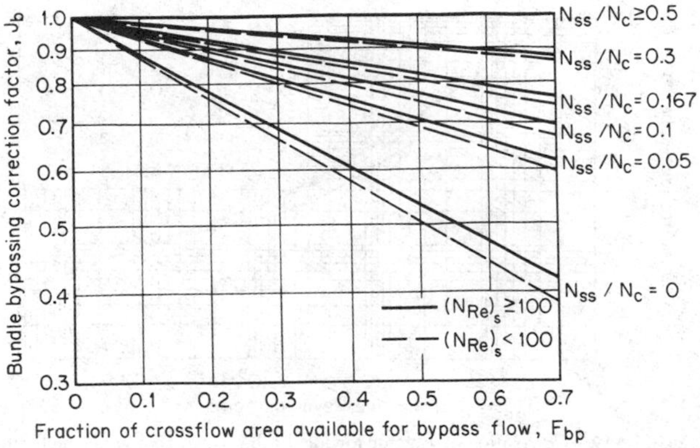

FIG. 10-22 Correction factor for bypass flow.

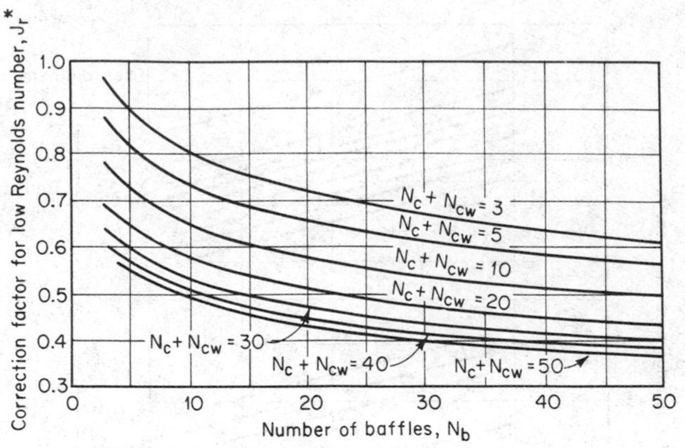

FIG. 10-23 Basic correction factor for adverse temperature gradient at low Reynolds numbers.

b. If $(N_{Re})_s \leq 20$, $J_r = J_r^*$.

c. If $20 < (N_{Re})_s < 100$, find J_r from Fig. 10-24, knowing J_r^* and $(N_{Re})_s$.

8. Calculate the *shell-side heat-transfer coefficient for the exchanger* h_s from

$$h_s = h_k J_c J_l J_b J_r \tag{10-135}$$

Shell-Side Pressure-Drop Calculation

1. Find f_k from the ideal-tube-bank friction-factor curve for the given tube layout at the calculated value of $(N_{Re})_s$, using Fig. 10-25*a* for triangular and rotated square arrays and Fig. 10-25*b* for in-line square arrays. These curves are adapted from Bergelin et al. and Grimison (loc. cit.).

2. Calculate the *pressure drop for an ideal cross-flow section.*

$$\Delta P_{bk} = b\frac{f_k W^2 N_c}{\rho S_m^2}\left(\frac{\mu_w}{\mu_b}\right)^{0.14} \tag{10-136}$$

where $b = (2.0)(10^{-3})$ (SI) or $(9.9)(10^{-5})$ (U.S. customary).

3. Calculate the *pressure drop for an ideal window section.* If $(N_{Re})_s \geq 100$,

$$\Delta P_{wk} = b\frac{W^2(2 + 0.6N_{cw})}{S_m S_{w}\rho} \tag{10-137a}$$

where $b = (5)(10^{-4})$ (SI) or $(2.49)(10^{-5})$ (U.S. customary).

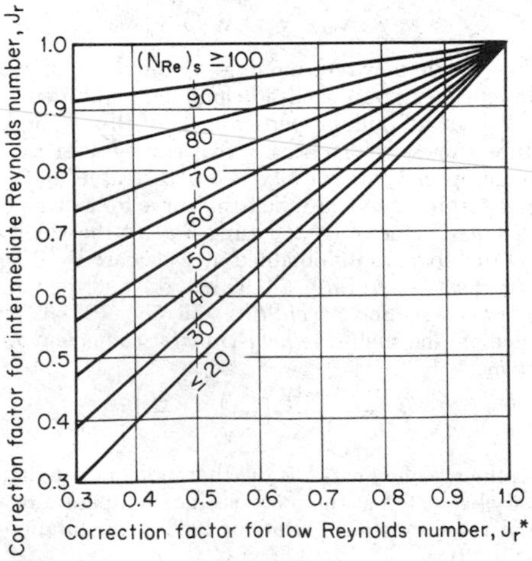

FIG. 10-24 Correction factor for adverse temperature gradient at intermediate Reynolds numbers.

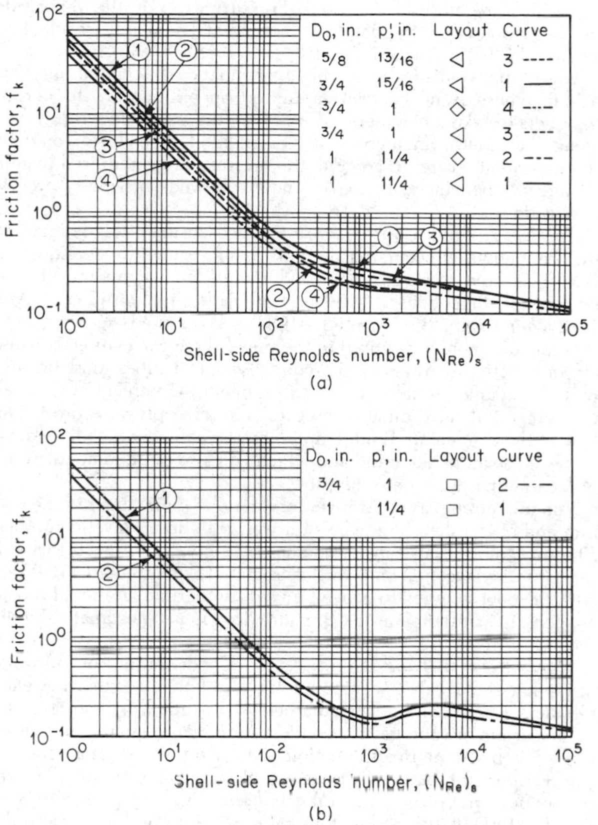

FIG. 10-25 Correction of friction factors for ideal tube banks. (*a*) Triangular and rotated square arrays. (*b*) In-line square arrays.

If $(N_{Re})_s < 100$,

$$\Delta P_{wk} = b_1 \frac{\mu_b W}{S_m S_w \rho}\left(\frac{N_{cw}}{p' - D_o} + \frac{l_s}{D_w^2}\right) + b_2 \frac{W^2}{S_m S_w \rho} \qquad (10\text{-}137b)$$

where $b_1 = (1.681)(10^{-5})$ (SI) or $(1.08)(10^{-4})$ (U.S. customary), and $b_2 = (9.99)(10^{-4})$ (SI) or $(4.97)(10^{-5})$ (U.S. customary).

4. Find the correction factor for the effect of baffle leakage on pressure drop R_l from Fig. 10-26. Curves shown are not to be extrapolated beyond the points shown.

5. Find the correction factor for bundle bypass R_b from Fig. 10-27.

6. Calculate the *pressure drop across the shell side* (excluding nozzles).

$$\Delta P_s = [(N_b - 1)(\Delta P_{bk})R_b + N_b\, \Delta P_{wk}]R_l + 2\,\Delta P_{bk} R_b \left(1 + \frac{N_{cw}}{N_c}\right) \qquad (10\text{-}138)$$

The values of h_s and ΔP_s calculated by this procedure are for clean exchangers and are intended to be as accurate as possible, not conservative (i.e., low). A fouled exchanger will generally give lower heat-transfer rates, as reflected by the dirt resistances incorporated into Eq. (10-115), and higher pressure drops. Some estimate of **fouling effects** on pressure drop may be made by using the methods just given by assuming that the fouling deposit blocks the leakage and possibly the bypass areas. The fouling may also decrease the clearance between tubes and significantly increase the pressure drop in cross-flow.

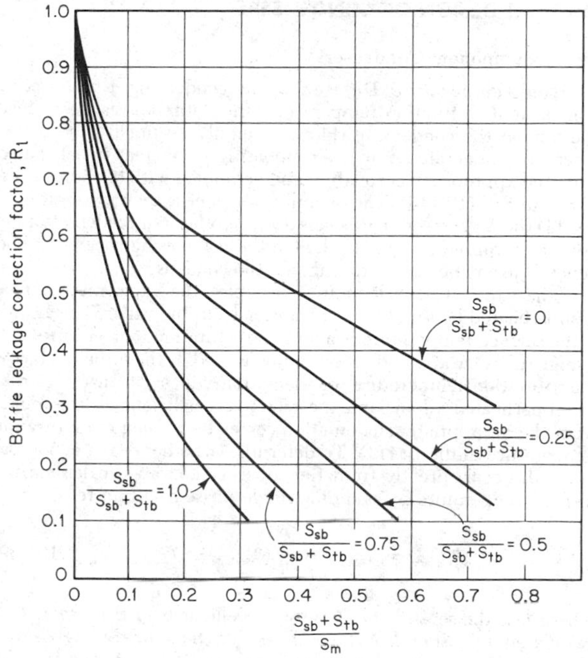

FIG. 10-26 Correction factor for baffle-leakage effect on pressure drop.

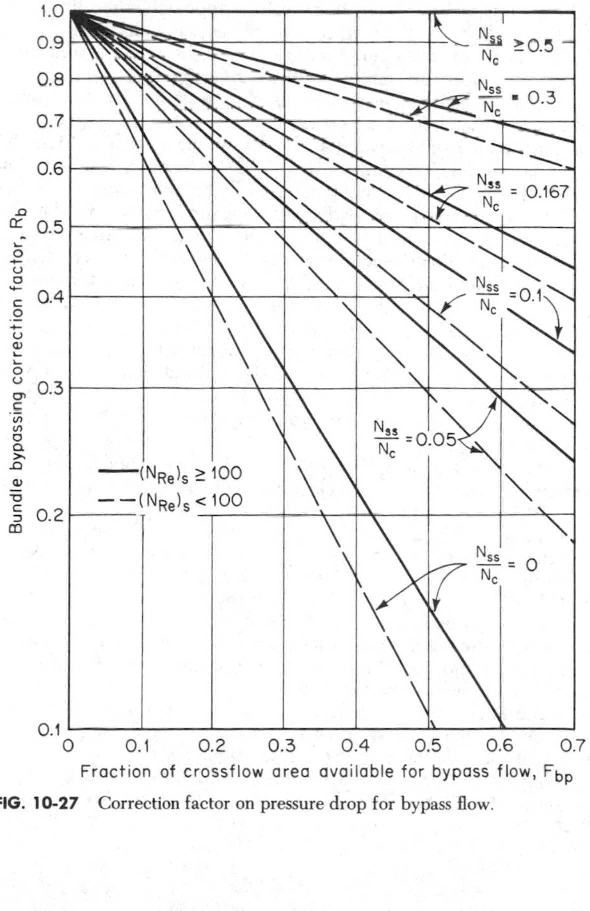

FIG. 10-27 Correction factor on pressure drop for bypass flow.

THERMAL DESIGN OF CONDENSERS

Single-Component Condensers

Mean Temperature Difference In condensing a single component at its saturation temperature, the entire resistance to heat transfer on the condensing side is generally assumed to be in the layer of condensate. A mean condensing coefficient is calculated from the appropriate correlation and combined with the other resistances in Eq. (10-115). The overall coefficient is then used with the LMTD (no F_T correction is necessary for isothermal condensation) to give the required area, even though the condensing coefficient and hence U are not constant throughout the condenser.

If the vapor is **superheated** at the inlet, the vapor may first be desuperheated by sensible heat transfer from the vapor. This occurs if the surface temperature is above the saturation temperature, and a single-phase heat-transfer correlation is used. If the surface is below the saturation temperature, condensation will occur directly from the superheated vapor, and the effective coefficient is determined from the appropriate condensation correlation, using the saturation temperature in the LMTD. To determine whether or not condensation will occur directly from the superheated vapor, calculate the surface temperature by assuming single-phase heat transfer.

$$T_{\text{surface}} = T_{\text{vapor}} - \frac{U}{h}(T_{\text{vapor}} - T_{\text{coolant}}) \qquad (10\text{-}139)$$

where h is the sensible heat-transfer coefficient for the vapor, U is calculated by using h, and both are on the same area basis. If $T_{\text{surface}} > T_{\text{saturation}}$, no condensation occurs at that point and the heat flux is actually higher than if $T_{\text{surface}} \le T_{\text{saturation}}$ and condensation did occur. It is generally conservative to design a pure-component desuperheater-condenser as if the entire heat load were transferred by condensation, using the saturation temperature in the LMTD.

The design of an integral **condensate subcooling section** is more difficult, especially if close temperature approach is required. The condensate layer on the surface is on the average subcooled by one-third to one-half of the temperature drop across the film, and this is often sufficient if the condensate is not reheated by raining through the vapor. If the condensing-subcooling process is carried out inside tubes or in the shell of a vertical condenser, the single-phase subcooling section can be treated separately, giving an area that is added onto that needed for condensation. If the subcooling is achieved on the shell side of a horizontal condenser by flooding some of the bottom tubes with a weir or level controller, the rate and heat-balance equations must be solved for each section to obtain the area required.

Pressure drop on the condensing side reduces the final condensing temperature and the MTD and should always be checked. In designs requiring close approach between inlet coolant and exit condensate (subcooled or not), underestimation of pressure drop on the condensing side can lead to an exchanger that cannot meet specified terminal temperatures. Since pressure-drop calculations in two-phase flows such as condensation are relatively inaccurate, designers must consider carefully the consequences of a larger-than-calculated pressure drop.

Horizontal In-Shell Condensers The mean **condensing coefficient** for the outside of a bank of horizontal tubes is calculated from Eq. (10-97) for a single tube, corrected for the number of tubes in a vertical row. For undisturbed laminar flow over all the tubes, Eq. (10-101) is, for realistic condenser sizes, overly conservative because of rippling, splashing, and turbulent flow (*Process Heat Transfer*, McGraw-Hill, New York, 1950). Kern proposed an exponent of $-\frac{1}{6}$ on the basis of experience, while Freon-11 data of Short and Brown (*General Discussion on Heat Transfer*, Institute of Mechanical Engineers, London, 1951) indicate independence of the number of tube rows. It seems reasonable to use no correction for inviscid liquids and Kern's correction for viscous condensates. For a cylindrical tube bundle, where N varies, it is customary to take N equal to two-thirds of the maximum or centerline value.

Baffles in a horizontal in-shell condenser are oriented with the cuts vertical to facilitate drainage and eliminate the possibility of flooding in the upward cross-flow sections. **Pressure drop** on the vapor side can be estimated by the data and method of Diehl and Unruh [*Pet. Refiner*, **36**(10), 147 (1957); **37**(10), 124 (1958)].

High vapor velocities across the tubes enhance the condensing coefficient. There is no correlation in the open literature to permit designers to take advantage of this. Since the vapor flow rate varies along the length, an incremental calculation procedure would be required in any case. In general, the pressure drops required to gain significant benefit are above those allowed in most process applications.

Vertical In-Shell Condensers Condensers are often designed so that condensation occurs on the outside of vertical tubes. Equation (10-92) is valid as long as the condensate film is laminar. When it becomes turbulent, Fig. 10-10 or Colburn's equation [*Trans. Am. Inst. Chem. Eng.*, **30**, 187 (1933–1934) may be used.

Some judgment is required in the use of these correlations because of construction features of the condenser. The tubes must be supported by baffles, usually with maximum cut (45 percent of the shell diameter) and maximum spacing to minimize pressure drop. The flow of the condensate is interrupted by the baffles, which may draw off or redistribute the liquid and which will also cause some splashing of free-falling drops onto the tubes.

For **subcooling,** a liquid inventory may be maintained in the bottom end of the shell by means of a weir or a liquid-level-controller. The subcooling heat-transfer coefficient is given by the correlations for natural convection on a vertical surface [Eqs. (10-33*a*), (10-33*b*)], with the pool assumed to be well mixed (isothermal) at the subcooled condensate exit temperature. Pressure drop may be estimated by the shell-side procedure.

Horizontal In-Tube Condensers Condensation of a vapor inside horizontal tubes occurs in kettle and horizontal thermosiphon reboilers and in air-cooled condensers. In-tube condensation also offers certain advantages for condensation of multicomponent mixtures, discussed in the subsection "Multicomponent Condensers." The various in-tube correlations are closely connected to the **two-phase flow pattern** in the tube [*Chem. Eng. Prog. Symp. Ser.*, **66**(102), 150 (1970)]. At low flow rates, when gravity dominates the flow pattern, Eq. (10-105) may be used. At high flow rates, the flow and heat transfer are governed by vapor shear on the condensate film, and Eq. (10-104) is valid. A simple and generally conservative procedure is to calculate the coefficient for a given case by both correlations and use the *larger* one.

Pressure drop during condensation inside horizontal tubes can be computed by using the correlations for two-phase flow given in Sec. 5 and neglecting the pressure recovery due to deceleration of the flow.

Vertical In-Tube Condensation Vertical-tube condensers are generally designed so that vapor and liquid flow cocurrently downward; if pressure drop is not a limiting consideration, this configuration can result in higher heat-transfer coefficients than shell-side condensation and has particular advantages for multicomponent condensation. If gravity controls, the mean heat-transfer coefficient for condensation is given by Figs. 10-9 and 10-10. If vapor shear controls, Eq. (10-103) is applicable. It is generally conservative to calculate the coefficients by both methods and choose the *higher* value. The pressure drop can be calculated by using the Lockhart-Martinelli method [*Chem. Eng. Prog.*, **45**, 39 (1945)] for friction loss, neglecting momentum and hydrostatic effects.

Vertical in-tube condensers are often designed for **reflux or knock-back application** in reactors or distillation columns. In this case, vapor flow is upward, countercurrent to the liquid flow on the tube wall; the vapor shear acts to thicken and retard the drainage of the condensate film, reducing the coefficient. Neither the fluid dynamics nor the heat transfer is well understood in this case, but Soliman, Schuster, and Berenson [*J. Heat Transfer*, **90**, 267–276 (1968)] discuss the problem and suggest a computational method. The Diehl-Koppany correlation [*Chem. Eng. Prog. Symp. Ser.* 92, **65** (1969)] may be used to estimate the maximum allowable vapor velocity at the tube inlet. If the vapor velocity is great enough, the liquid film will be carried upward; this design has been employed in a few cases in which only part of the stream is to be condensed. This velocity cannot be accurately computed, and a very conservative (high) outlet velocity must be used if unstable flow and flooding are to be avoided; 3 times the vapor velocity given by the Diehl-Koppany correlation for incipient flooding has been suggested as the design value for completely stable operation.

Multicomponent Condensers

Thermodynamic and Mass-Transfer Considerations *Multicomponent vapor mixture* includes several different cases: all the components may be liquids at the lowest temperature reached in the condensing side, or there may be components which dissolve substantially in the condensate even though their boiling points are below the exit temperature, or one or more components may be both noncondensable and nearly insoluble.

Multicomponent condensation always involves sensible-heat changes in the vapor and liquid along with the latent-heat load. Compositions of both phases in general change through the condenser, and **concentration gradients** exist in both phases. Temperature and concentration profiles and transport rates at a point in the condenser usually cannot be calculated, but the binary cases have been treated: condensation of one component in the presence of a completely insoluble gas [Colburn and Hougen, *Ind. Eng. Chem.*, **26,** 1178–1182 (1934); and Colburn and Edison, *Ind. Eng. Chem.*, **33,** 457–458 (1941)] and condensation of a binary vapor [Colburn and Drew, *Trans. Am. Inst. Chem. Eng.*, 33, 196–215 (1937)]. It is necessary to know or calculate diffusion coefficients for the system, and a reasonable approximate method to avoid this difficulty and the reiterative calculations is desirable. To integrate the point conditions over the total condensation requires the temperature, composition enthalpy, and flow-rate profiles as functions of the heat removed. These are calculated from component thermodynamic data if the vapor and liquid are assumed to be in equilibrium at the local vapor temperature. This assumption is not exactly true, since the condensate and the liquid-vapor interface (where equilibrium does exist) are intermediate in temperature between the coolant and the vapor.

In calculating the condensing curve, it is generally assumed that the vapor and liquid flow collinearly and in intimate contact so that composition equilibrium is maintained between the total streams at all points. If, however, the condensate drops out of the vapor (as can happen in horizontal shell-side condensation) and flows to the exit without further interaction, the remaining vapor becomes excessively enriched in light components with a decrease in condensing temperature and in the temperature difference between vapor and coolant. The result may be not only a small reduction in the amount of heat transferred in the condenser but also an inability to condense totally the light ends even at reduced throughput or with the addition of more surface. To prevent the liquid from segregating, in-tube condensation is preferred in critical cases.

Thermal Design If the controlling resistance for heat and mass transfer in the vapor is sensible-heat removal from the cooling vapor, the following design equation is obtained:

$$A = \int_0^Q \frac{1 + U'Z_H/h_{sv}}{U'(T_v - T_c)} dQ \qquad (10\text{-}140)$$

U' is the overall heat-transfer coefficient between the vapor-liquid interface and the coolant, including condensate film, dirt and wall resistances, and coolant. The condensate film coefficient is calculated from the appropriate equation or correlation for pure vapor condensation for the geometry and flow regime involved, using mean liquid properties. Z_H is the ratio of the sensible heat removed from the vapor-gas stream to the total heat transferred; this quantity is obtained from thermodynamic calculations and may vary substantially from one end of the condenser to the other, especially when removing vapor from a noncondensable gas. The sensible-heat-transfer coefficient for the vapor-gas stream h_{sv} is calculated by using the appropriate correlation or design method for the geometry involved, neglecting the presence of the liquid. As the vapor condenses, this coefficient decreases and must be calculated at several points in the process. T_v and T_c are temperatures of the vapor and of the coolant respectively. This procedure is similar in principle to that of Ward [*Petro/Chem. Eng.*, **32**(11), 42–48 (1960)]. It may be nonconservative for condensing steam and other high-latent-heat substances, in which case it may be necessary to increase the calculated area by 25 to 50 percent.

Pressure drop on the condensing side may be estimated by judicious application of the methods suggested for pure-component condensation, taking into account the generally nonlinear decrease of vapor-gas flow rate with heat removal.

THERMAL DESIGN OF REBOILERS

For a **single-component reboiler design,** attention is focused upon the mechanism of heat and momentum transfer at the hot surface. In *multicomponent systems*, the light components are preferentially vaporized at the surface, and the process becomes limited by their rate of diffusion. The net effect is to decrease the effective temperature difference between the hot surface and the bulk of the boiling liquid. If one attempts to vaporize too high a fraction of the feed liquid to the reboiler, the temperature difference between surface and liquid is reduced to the point that nucleation and vapor generation on the surface are suppressed and heat transfer to the liquid proceeds at the lower rate associated with single-phase natural convection. The only safe procedure in design for wide-boiling-range mixtures is to vaporize such a limited fraction of the feed that the boiling point of the remaining liquid mixture is still at least 5.5°C (10°F) below the surface temperature. Positive flow of the unvaporized liquid through and out of the reboiler should be provided.

Kettle Reboilers It has been generally assumed that kettle reboilers operate in the pool boiling mode, but with a lower peak heat flux because of vapor binding and blanketing of the upper tubes in the bundle. There is some evidence that vapor generation in the bundle causes a high circulation rate through the bundle. The result is that, at the lower heat fluxes, the kettle reboiler acutally gives higher heat-transfer coefficients than a single tube. Present understanding of the recirculation phenomenon is insufficient to take advantage of this in design. Available nucleate pool boiling correlations are only very approximate, failing to account for differences in the nucleation characteristics of different surfaces. The Mostinski correlation [Eq. (10-106)] and the McNelly correlation [Eq. (10-107)] are generally the best for single components or narrow-boiling-range mixtures at low fluxes, though they may give errors of 40 to 50 percent. Experimental heat-transfer coefficients for pool boiling of a given liquid on a given surface should be used if available. The bundle **peak heat flux** is a function of tube-bundle geometry, especially of tube-packing density; in the absence of better information, the Palen-Small modification [Eq. (10-112)] of the Zuber maximum-heat-flux correlation is recommended.

If the boiling range of the fraction vaporized is more than about one-quarter to one-third of the mean temperature difference available in the reboiler, the method described may give generally nonconservative designs. In this case, the method recommended by Kern (*Process Heat Transfer*, McGraw-Hill, New York, 1950, pp. 474–475) should be used.

Kettle reboilers are generally assumed to require negligible pressure drop. It is important to provide good longitudinal liquid flow paths within the shell so that the liquid is uniformly distributed along the entire length of the tubes and excessive local vaporization and vapor binding are avoided.

This method may also be used for the thermal design of **horizontal thermosiphon reboilers.** The recirculation rate and pressure profile of the thermosiphon loop can be calculated by the methods of Fair [*Pet. Refiner*, **39**(2), 105–123 (1960)].

Vertical Thermosiphon Reboilers Vertical thermosiphon reboilers operate by natural circulation of the liquid from the still through the downcomer to the reboiler and of the two-phase mixture from the reboiler through the return piping. The flow is induced by the hydrostatic pressure imbalance between the liquid in the downcomer and the two-phase mixture in the reboiler tubes. Thermosiphons do not require any pump for recirculation and are generally regarded as less likely to foul in service because of the relatively high two-phase velocities obtained in the tubes. Heavy components are not likely to accumulate in the thermosiphon, but they are more difficult to design satisfactorily than kettle reboilers, especially in vacuum operation. Several shortcut methods have been suggested for thermosiphon design, but they must generally be used with caution. The method due to Fair (loc. cit.), based upon two-phase flow correlations, is the most complete in the open literature but requires a computer for practical use. Fair also suggests a shortcut method that is satisfactory for preliminary design and can be reasonably done by hand.

Forced-Recirculation Reboilers In forced-recirculation reboilers, a pump is used to ensure circulation of the liquid past the heat-transfer surface. Force-recirculation reboilers may be designed so that boiling occurs inside vertical tubes, inside horizontal tubes, or on the shell side. For forced boiling inside vertical tubes, Fair's method (loc. cit.) may be employed, making only the minor modification that the recirculation rate is fixed and does not need to be balanced against the pressure available in the downcomer. Excess pressure required to circulate the two-phase fluid through the tubes and back into the column is supplied by the pump, which must develop a positive pressure increase in the liquid.

Fair's method may also be modified to design forced-recirculation reboilers with horizontal tubes. In this case the hydrostatic-head-pressure effect through the tubes is zero but must be considered in the two-phase return lines to the column.

The same procedure may be applied in principle to design of forced-recirculation reboilers with shell-side vapor generation. Little is known about two-phase flow on the shell side, but a reasonable estimate of the friction pressure drop can be made from the data of Diehl and Unruh [*Pet. Refiner*, **36**(10), 147 (1957); **37**(10), 124 (1958)]. No void-fraction data are available to permit accurate estimation of the hydrostatic or acceleration terms. These may be roughly estimated by assuming homogeneous flow.

THERMAL DESIGN OF EVAPORATORS

Sizing of evaporator heating surfaces can in many cases make use of individual film coefficients developed earlier in this section. However, most evaporators deal with concentrated solutions, frequently containing high concentrations of crystallizing solids in suspension, that have properties far different from those of pure liquids. Care must be exercised in defining the temperature differences used in computing the heat-transfer characteristics (see Sec. 11). In the data that follow, the true temperature difference is used when dealing with film coefficients and the apparent temperature difference corrected for boiling-point rise when dealing with overall coefficients. The normal practice for evaporators is to define the heating-surface area as that in contact with the liquid being evaporated, whereas most other heat-exchange equipment is rated on the basis of the outside area of the tubes.

Forced-Circulation Evaporators In evaporators of this type in which hydrostatic head prevents boiling at the heating surface, **heat-transfer coefficients** can be predicted from the usual correlations for condensing steam (Fig. 10-10) and forced-convection sensible heating [Eq. 10-50)]. The liquid film coefficient is improved if boiling is not completely suppressed. When only the film next to the wall is above the boiling point, Boarts, Badger, and Meisenberg [*Ind. Eng. Chem.*, **29**, 912 (1937)] found that results could be correlated by Eq. (10-50) by using a constant of 0.0278 instead of 0.023. In such cases, the course of the liquid temperature can still be calculated from known circulation rate and heat input.

When the bulk of the liquid is boiling in part of the tube length, the film coefficient is even higher. However, the liquid temperature starts dropping as soon as full boiling develops, and it is difficult to estimate the course of the temperature curve. It is certainly safe to estimate heat transfer on the basis that no bulk boiling occurs. Fragen and Badger [*Ind. Eng. Chem.*, **28**, 534 (1936)] obtained an **empirical correlation** of overall heat-transfer coefficients in this type of evaporator, based on the ΔT at the heater inlet:

In U.S. customary units

$$U = 2020D^{0.57}(V_s)^{3.6/L}/\mu^{0.25}\,\Delta T^{0.1} \qquad (10\text{-}141)$$

where D = mean tube diameter, V_s = inlet velocity, L = tube length, and μ = liquid viscosity. This equation is based primarily on experiments with copper tubes of 0.022 m (⅞ in) outside diameter, 0.00165 m (16 gauge), 2.44 m (8 ft) long, but it includes some work with 0.0127-m (½-in) tubes 2.44 m (8 ft) long and 0.0254-m (1-in) tubes 3.66 m (12 ft) long.

Long-Tube Vertical Evaporators In the rising-film version of this type of evaporator, there is usually a nonboiling zone in the bottom section and a boiling zone in the top section. The length of the nonboiling zone depends on heat-transfer characteristics in the two zones and on pressure drop during two-phase flow in the boiling zone. The work of Martinelli and coworkers [Lockhart and Martinelli, *Chem. Eng. Prog.*, **45**, 39–48 (January 1949); and Martinelli and Nelson, *Trans. Am. Soc. Mech. Eng.*, **70**, 695–702 (August 1948)] permits a prediction of pressure drop, and a number of correlations are available for estimating film coefficients of heat transfer in the two zones. In estimating pressure drop, integrated curves similar to those presented by Martinelli and Nelson are the easiest to use. The curves for pure water are shown in Figs. 10-28 and 10-29, based on the assumption that the flow of both vapor and liquid would be turbulent if each were flowing alone in the tube. Similar curves can be prepared if one or both flows are laminar or if the properties of the liquid differ appreciably from the properties of pure water. The **acceleration pressure drop** ΔP_a is calculated from the equation

$$\Delta P_a = br_2G^2/32.2 \qquad (10\text{-}142)$$

where b = (2.6)(10^7)(SI) and 1.0 (U.S. customary) and using r_2 from Fig. (10-28). The frictional pressure drop is derived from Fig. 10-29, which shows the ratio of two-phase pressure drop to that of the entering liquid flowing alone.

Pressure drop due to hydrostatic head can be calculated from liquid holdup R_1. For nonfoaming dilute aqueous solutions, R_1 can be estimated from $R_1 = 1/[1 + 2.5(V/L)(\rho_1/\rho_v)^{1/2}]$. Liquid holdup, which represents the ratio of liquid-only velocity to actual liquid velocity, also appears to be the principal determinant of the convective coefficient in the boiling zone (Dengler, Sc.D. thesis, MIT, 1952). In other words, the convective coefficient is that calculated from Eq. (10-50) by using the liquid-only velocity divided by R_1 in the Reynolds number. Nucleate boiling augments convective heat transfer, primarily when ΔT's are high and the convective coefficient is low [Chen, *Ind. Eng. Chem. Process Des. Dev.*, **5**, 322 (1966)].

Film coefficients for the **boiling of liquids other than water** have been investigated. Coulson and McNelly [*Trans. Inst. Chem. Eng.*,

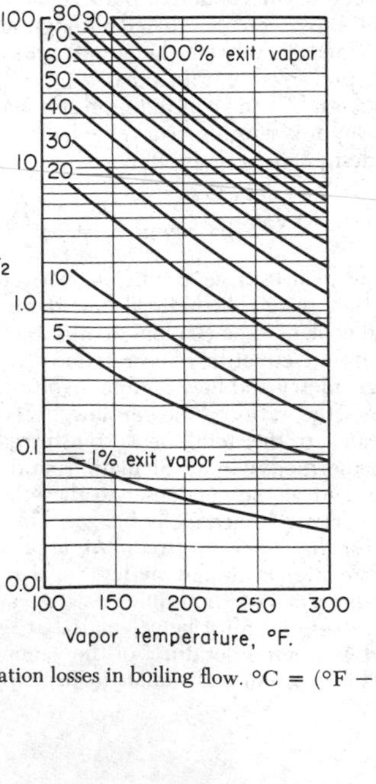

FIG. 10-28 Acceleration losses in boiling flow. °C = (°F − 32)/1.8.

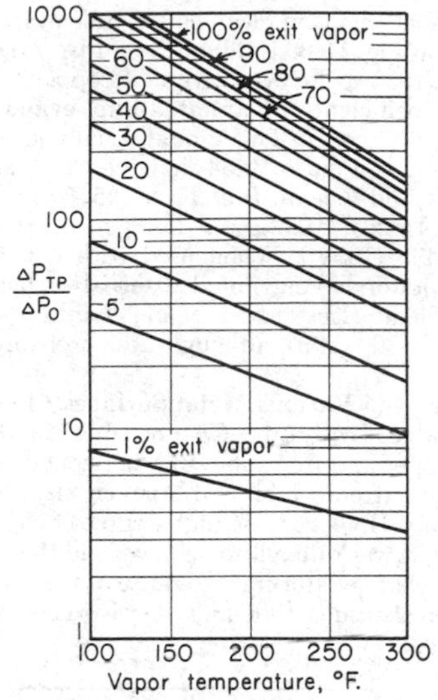

FIG. 10-29 Friction pressure drop in boiling flow. °C = (°F − 32)/1.8.

34, 247 (1956)] derived the following relation, which also correlated the data of Badger and coworkers [*Chem. Metall. Eng.*, **46**, 640 (1939); *Chem. Eng.*, **61**(2), 183 (1954); and *Trans. Am. Inst. Chem. Eng.*, **33**, 392 (1937); **35**, 17 (1939); **36**, 759 (1940)] on water:

$$N_{Nu} = (1.3 + b\,D)(N_{Pr})_l^{0.9}(N_{Re})_l^{0.23}(N_{Re})_g^{0.34}\left(\frac{\rho_l}{\rho_g}\right)^{0.25}\left(\frac{\mu_g}{\mu_l}\right) \qquad (10\text{-}143)$$

where b = 128 (SI) or 39 (U.S. customary), N_{Nu} = Nusselt number based on liquid thermal conductivity, D = tube diameter, and the remaining terms are dimensionless groupings of liquid Prandtl number, liquid Reynolds number, vapor Reynolds number, and ratios of densities and viscosities. The Reynolds numbers are calculated on the basis of each fluid flowing by itself in the tube.

Additional corrections must be applied when the fraction of vapor is so high that the remaining liquid does not wet the tube wall or when the velocity of the mixture at the tube exits approaches sonic velocity. McAdams, Woods, and Bryan (*Trans. Am. Soc. Mech. Eng.*, 1940), Dengler and Addoms (loc. cit.), and Stroebe, Baker, and Badger [*Ind. Eng. Chem.*, **31**, 200 (1939)] encountered dry-wall conditions and reduced coefficients when the weight fraction of vapor exceeded about 80 percent. Schweppe and Foust [*Chem. Eng. Prog.*, **49**, *Symp. Ser.* **5**, 77 (1953)] and Harvey and Foust (ibid., p. 91) found that "sonic choking" occurred at surprisingly low flow rates.

The simplified method of calculation outlined includes no allowance for the **effect of surface tension.** Stroebe, Baker, and Badger (loc. cit.) found that by adding a small amount of surface-active agent the boiling-film coefficient varied inversely as the square of the surface tension. Coulson and Mehta [*Trans. Inst. Chem. Eng.*, **31**, 208 (1953)] found the exponent to be −1.4. The higher coefficients at low surface tension are offset to some extent by a higher pressure drop, probably because the more intimate mixture existing at low surface tension causes the liquid fraction to be accelerated to a velocity closer to that of the vapor. The pressure drop due to acceleration ΔP_a derived from Fig. 10-28 allows for some slippage. In the limiting case, such as might be approached at low surface tension, the acceleration pressure drop in which "fog" flow is assumed (no slippage) can be determined from the equation

$$\Delta P_a' = \frac{y(V_g - V_l)G^2}{g_c} \qquad (10\text{-}144)$$

where y = fraction vapor by weight
V_g, V_l = specific volume gas, liquid
G = mass velocity

While the foregoing methods are valuable for detailed evaporator design or for evaluating the effect of changes in conditions on performance, they are cumbersome to use when making preliminary designs or cost estimates. Figure 10-30 gives the general range of **overall long-tube vertical- (LTV) evaporator heat-transfer coefficients** usually encountered in commercial practice. The higher coefficients are encountered when evaporating dilute solutions and the lower range when evaporating viscous liquids. The dashed curve represents the approximate lower limit, for liquids with viscosities of about 0.1 Pa·s (100 cP). The LTV evaporator does not work well at low temperature differences, as indicated by the results shown in Fig. 10-31 for seawater in 0.051-m (2-in), 0.0028-m (12-gauge) brass tubes 7.32 m (24 ft) long (W. L. Badger Associates, Inc., U.S. Department of the Interior, Office of Saline Water Rep. 26, December 1959, OTS Publ. PB 161290). The feed was at its boiling point at the vapor-head pressure, and feed rates varied from 0.025 to 0.050 kg/(s·tube) [200 to 400 lb/(h·tube)] at the higher temperature to 0.038 to 0.125 kg/(s·tube) [300 to 1000 lb/(h·tube)] at the lowest temperature.

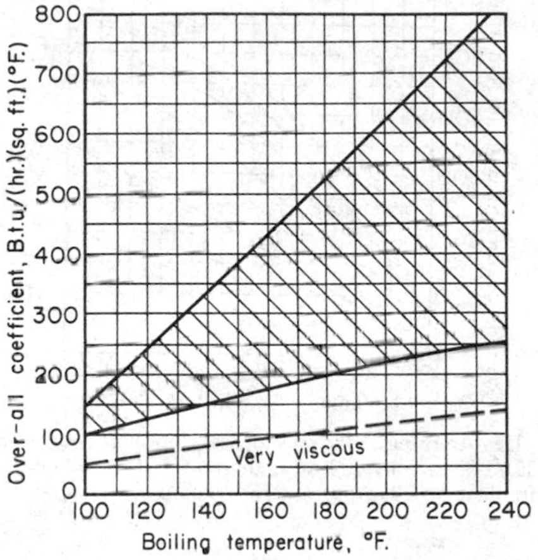

FIG. 10-30 General range of long-tube vertical- (LTV) evaporator coefficients. °C = (°F − 32)/1.8; to convert British thermal units per hour–square foot–degrees Fahrenheit to joules per square meter–second–kelvins, multiply by 5.6783.

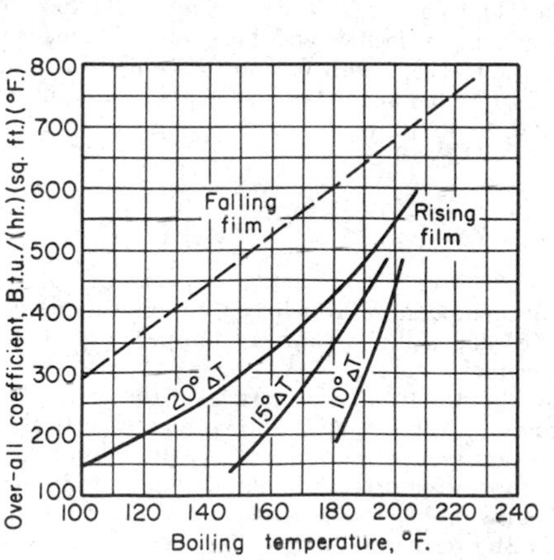

FIG. 10-31 Heat-transfer coefficients in LTV seawater evaporators. °C = (°F − 32)/1.8; to convert British thermal units per hour-square foot-degrees Fahrenheit to joules per square meter-second-kelvins, multiply by 5.6783.

Falling-film evaporators exhibit approximately the same performance as falling-film heat exchangers for nonboiling liquids. The liquid generally flows down the tube as a film and is not mixed with or accelerated by the vapor core. The boiling point in the tubes is higher than in the vapor head because of frictional pressure drop of the vapor and the superheat required to form vapor bubbles in the film (Sinek, Ph.D. thesis, University of Michigan, 1961). Both of these factors can be predicted. They tend to make overall coefficients lower than those for nonboiling conditions. Figure 10-31 shows overall heat-transfer coefficients determined in a falling-film seawater evaporator using the same tubes and flow rates as for the rising-film tests (W. L. Badger Associates, Inc., loc. cit.).

Short-Tube Vertical Evaporators Coefficients can be estimated by the same detailed method described for recirculating LTV evaporators. Performance is primarily a function of temperature level, temperature difference, and viscosity. While liquid level can also have an important influence, this is usually encountered only at levels lower than considered safe in commercial operation. **Overall heat-transfer coefficients** are shown in Fig. 10-32 for a basket-type evaporator (one with an annular downtake) when boiling water with 0.051-m (2-in) outside-diameter 0.0028-m-wall (12-gauge), 1.22-m- (4-ft-) long steel tubes [Badger and Shepard, *Chem. Metall. Eng.*, **23,** 281 (1920)]. Liquid level was maintained at the top tube sheet. Foust, Baker, and Badger [*Ind. Eng. Chem.*, **31,** 206 (1939)] measured recirculating velocities and heat-transfer coefficients in the same evaporator except with 0.064-m (2.5-in) 0.0034-m-wall (10-gauge), 1.22-m- (4-ft-) long tubes and temperature differences from 7 to 26°C (12 to 46°F). In the normal range of liquid levels, their results can be expressed as

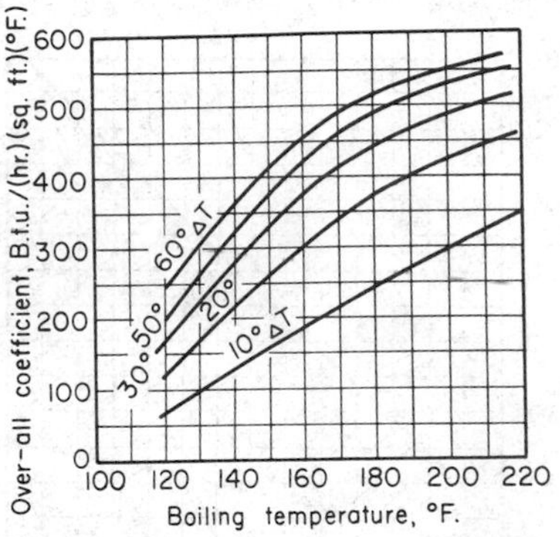

FIG. 10-32 Heat-transfer coefficients for water in short-tube evaporators. °C = (°F − 32)/1.8; to convert British thermal units per hour–square foot–degrees Fahrenheit to joules per square meter–second–kelvins, multiply by 5.6783.

$$U_c = \frac{b(\Delta T_c)^{0.22} N_{\mathrm{Pr}}^{0.4}}{(V_g - V_l)^{0.37}} \tag{10-145}$$

where b = 153 (SI) or 375 (U.S. customary) and the subscript c refers to true liquid temperature, which under these conditions was about 0.56°C (1°F) above the vapor-head temperature. This work was done with water.

No detailed tests have been reported for the performance of propeller calandrias. Not enough is known regarding the performance of the propellers themselves under the cavitating conditions usually encountered to permit predicting circulation rates. In many cases, it appears that the propeller does no good in accelerating heat transfer over the transfer for natural circulation (Fig. 10-32).

Miscellaneous Evaporator Types **Horizontal-tube evaporators** behave in much the same way as short-tube verticals, and heat-transfer coefficients are of the same order of magnitude. Some test results for water were published by Badger [*Trans. Am. Inst. Chem. Eng.*, **13,** 139 (1921)].

Heat-transfer coefficients in clean coiled-tube evaporators for seawater are shown in Fig. 10-33 [Hillier, *Proc. Inst. Mech. Eng. (London)*, **1B**(7), 295 (1953)]. The tubes were of copper.

Heat-transfer coefficients in **agitated-film evaporators** depend primarily on liquid viscosity. This type is usually justifiable only for very viscous materials. Figure 10-34 shows general ranges of overall coefficients [Hauschild, *Chem. Ing. Tech.*, **25,** 573 (1953); Lindsey, *Chem. Eng.*, **60**(4), 227 (1953); and Leniger and Veldstra, *Chem. Ing. Tech.*, **31,** 493 (1959)]. When used with nonviscous fluids, a wiped-film evaporator having fluted external surfaces can exhibit very high coefficients (Lustenader et al., *Trans. Am. Soc. Mech. Eng.*, Paper 59-SA-30, 1959), although at a probably unwarranted first cost.

Heat Transfer from Various Metal Surfaces In an early work, Pridgeon and Badger [*Ind. Eng. Chem.*, **16,** 474 (1924)] published test results on copper and iron tubes in a horizontal-tube evaporator that indicated an extreme **effect of surface cleanliness** on heat-transfer coefficients. However, the high degree of cleanliness needed for high coefficients was difficult to achieve, and the tube layout and liquid level were changed during the course of the tests so as to make direct comparison of results difficult. Other workers have found little

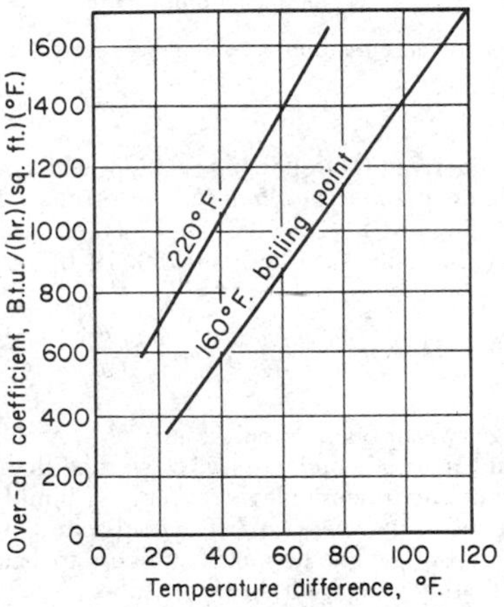

FIG. 10-33 Heat-transfer coefficients for seawater in coil-tube evaporators. °C = (° F − 32)/1.8; to convert British thermal units per hour–square foot–degrees Fahrenheit to joules per square meter–second–kelvins, multiply by 5.6783.

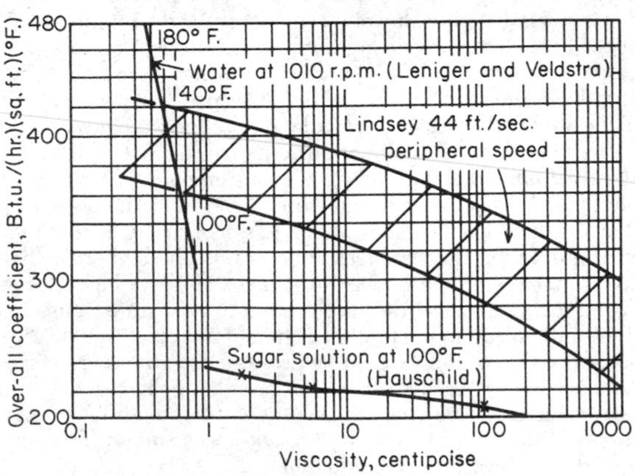

FIG. 10-34 Overall heat-transfer coefficients in agitated-film evaporators. °C = (°F − 32)/1.8; to convert British thermal units per hour–square foot–degrees Fahrenheit to joules per square meter–second–kelvins, multiply by 5.6783; to convert centipoises to pascal-seconds, multiply by 10^{-3}.

or no effect of conditions of surface or tube material on boiling-film coefficients in the range of commercial operating conditions [Averin, *Izv. Akad. Nauk SSSR Otd. Tekh. Nauk*, no. 3, p. 116, 1954; and Coulson and McNelly, *Trans. Inst. Chem. Eng.*, **34,** 247 (1956)].

Work in connection with desalination of seawater has shown that **specially modified surfaces** can have a profound effect on heat-transfer coefficients in evaporators. Figure 10-35 (Alexander and Hoffman, Oak Ridge National Laboratory TM-2203) compares overall coefficients for some of these surfaces when boiling fresh water in 0.051-m (2-in) tubes 2.44 m (8 ft) long at atmospheric pressure in both upflow and downflow. The area basis used was the nominal outside area. Tube 20 was a smooth 0.0016-m- (0.062-in-) wall aluminum brass tube that had accumulated about 6 years of fouling in seawater service and exhibited a fouling resistance of about $(2.6)(10^{-5})$ $(m^2 \cdot s \cdot K)/J$ [0.00015 $(ft^2 \cdot h \cdot °F)/Btu$]. Tube 23 was a clean aluminum tube with 20 spiral corrugations of 0.0032-m (⅛-in) radius on a 0.254-m (10-in) pitch indented into the tube. Tube 48 was a clean copper tube that had 50 longitudinal flutes pressed into the wall (General Electric double-flute profile, Diedrich, U.S. Patent 3,244,601, Apr. 5, 1966). Tubes 47 and 39 had a specially patterned porous sintered-metal deposit on the boiling side to promote nucleate boiling (Minton, U.S. Patent 3,384,154, May 21, 1968). Both of these tubes also had steam-side coatings to promote dropwise condensation—parylene for tube 47 and gold plating for tube 39.

Of these special surfaces, only the **double-fluted tube** has seen extended services. Most of the gain in heat-transfer coefficient is due to the condensing side; the flutes tend to collect the condensate and leave the lands bare [Carnavos, *Proc. First Int. Symp. Water Desalination*, **2,** 205 (1965)]. The condensing-film coefficient (based on the actual outside area, which is 28 percent greater than the nominal area) may be approximated from the equation

$$h = b\left(\frac{k^3\rho^2 g}{\mu^2}\right)^{1/3}\left(\frac{\mu\lambda}{L}\right)^{1/3}\left(\frac{q}{A}\right)^{-0.833} \qquad (10\text{-}146a)$$

where b = 2100 (SI) or 1180 (U.S. customary). The boiling-side coefficient (based on actual inside area) for salt water in downflow may

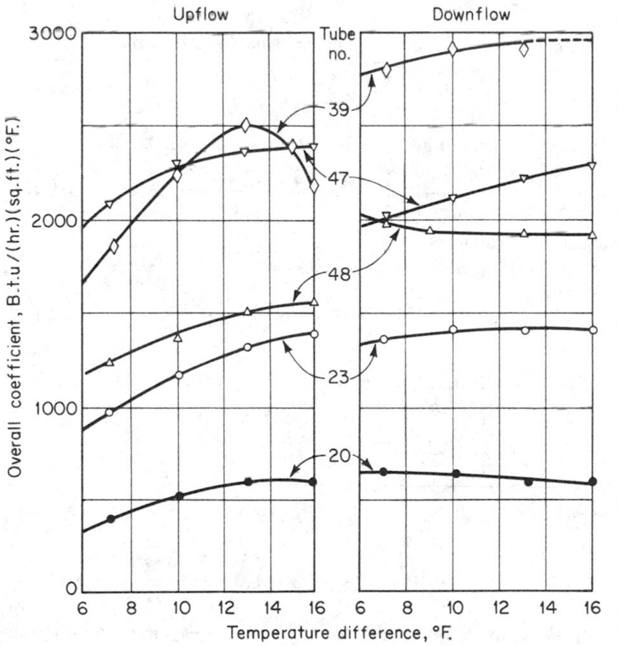

FIG. 10-35 Heat-transfer coefficients for enhanced surfaces. °C = (°F − 32)/1.8; to convert British thermal units per hour–square foot–degrees Fahrenheit to joules per square meter–second–kelvins, multiply by 5.6783. *(By permission from Oak Ridge National Laboratory TM-2203.)*

be approximated from the equation

$$h = 0.035(k^3\rho^2 g/\mu^2)^{1/3}(4\Gamma/\mu)^{1/3} \qquad (10\text{-}146b)$$

The boiling-film coefficient is about 30 percent lower for pure water than it is for salt water or seawater. There is as yet no accepted explanation for the superior performance in salt water. This phenomenon is also seen in evaporation from smooth tubes.

Effect of Fluid Properties on Heat Transfer Most of the heat-transfer data reported in the preceding paragraphs were obtained with water or with dilute solutions having properties close to those of water. Heat transfer with other materials will depend on the type of evaporator used. For forced-circulation evaporators, methods have been presented to calculate the effect of changes in fluid properties. For natural-circulation evaporators, **viscosity** is the most important variable as far as aqueous solutions are concerned. Badger (*Heat Transfer and Evaporation*, Chemical Catalog, New York, 1926, pp. 133–134) found that, as a rough rule, overall heat-transfer coefficients varied in inverse proportion to viscosity if the boiling film was the main resistance to heat transfer. When handling molasses solutions in a forced-circulation evaporator in which boiling was allowed to occur in the tubes, Coates and Badger [*Trans. Am. Inst. Chem. Eng.*, **32,** 49 (1936)] found that from 0.005 to 0.03 Pa·s (5 to 30 cP) the overall heat-transfer coefficient could be represented by $U = b/\mu_f^{1.24}$, where b = 2.55 (SI) or 7043 (U.S. customary). Fragen and Badger [*Ind. Eng. Chem.*, **28,** 534 (1936)] correlated overall coefficients on sugar and sulfite liquor in the same evaporator for viscosities to 0.242 Pa·s (242 cP) and found a relationship that included the viscosity raised only to the 0.25 power.

Little work has been published on the effect of viscosity on heat transfer in the long-tube vertical evaporator. Cessna, Leintz, and Badger [*Trans. Am. Inst. Chem. Eng.*, **36,** 759 (1940)] found that the overall coefficient in the nonboiling zone varied inversely as the 0.7 power of viscosity (with sugar solutions). Coulson and Mehta [*Trans. Inst. Chem. Eng.*, **31,** 208 (1953)] found the exponent to be −0.44, and Stroebe, Baker, and Badger (loc. cit.) arrived at an exponent of −0.3 for the effect of viscosity on the film coefficient in the boiling zone.

Kerr (Louisiana Agr. Exp. Sta. Bull. 149) obtained plant data shown in Fig. 10-36 on various types of full-sized evaporators for cane sugar. These are invariably forward-feed evaporators concentrating to about 50° Brix, corresponding to a viscosity on the order of 0.005 Pa·s (5 cP) in the last effect. In Fig. 10-36 curve *A* is for short-tube verticals with central downtake, *B* is for standard horizontal tube evaporators, *C* is for Lillie evaporators (which were horizontal-tube machines with no liquor level but having recirculating liquor showered over the tubes), and *D* is for long-tube vertical evaporators. These curves show apparent coefficients, but sugar solutions

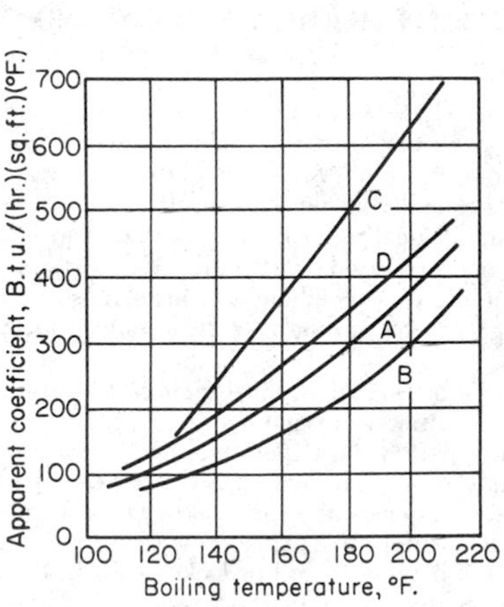

FIG. 10-36 Kerr's tests with full-sized sugar evaporators. °C = (°F − 32)/1.8; to convert British thermal units per hour–square foot–degrees Fahrenheit to joules per square meter–second–kelvins, multiply by 5.6783.

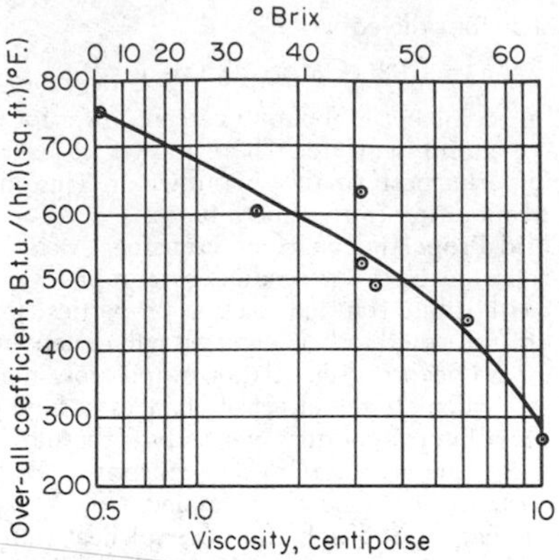

FIG. 10-37 Effect of viscosity on heat transfer in short-tube vertical evaporator. To convert centipoises to pascal-seconds, multiply by 10^{-3}; to convert British thermal units per hour–square foot–degrees Fahrenheit to joules per square meter–second–kelvins, multiply by 5.6783.

have boiling-point rises low enough not to affect the results noticeably. Kerr also obtained the data shown in Fig. 10-37 on a laboratory short-tube vertical evaporator with 0.44- by 0.61-m (1¾- by 24-in) tubes. This work was done with sugar juices boiling at 57°C (135°F) and an 11°C (20°F) temperature difference.

Effect of Noncondensables on Heat Transfer Most of the heat transfer in evaporators does not occur from pure steam but from vapor evolved in a preceding effect. This vapor usually contains inert gases—from air leakage if the preceding effect was under vacuum, from air entrained or dissolved in the feed, or from gases liberated by decomposition reactions. To prevent these inerts from seriously impeding heat transfer, the gases must be channeled past the heating surface and vented from the system while the gas concentration is still quite low. The influence of inert gases on heat transfer is due partially to the effect on ΔT of lowering the partial pressure and hence condensing temperature of the steam. The primary effect, however, results from the formation at the heating surface of an insulating blanket of gas through which the steam must diffuse before it can condense. The latter effect can be treated as an added resistance or fouling factor equal to 6.5×10^{-5} times the local mole percent inert gas (in $J^{-1} \cdot s \cdot m^2 \cdot K$) [Standiford, *Chem. Eng. Prog.*, **75**, 59–62 (July 1979)]. The effect on ΔT is readily calculated from Dalton's law. Inert-gas concentrations may vary by a factor of 100 or more between vapor inlet and vent outlet, so these relationships should be integrated through the tube bundle.

BATCH OPERATIONS: HEATING AND COOLING OF VESSELS

Nomenclature (Use consistent units.) A = heat-transfer surface; C, c = specific heats of hot and cold fluids respectively; L_0 = flow rate of liquid added to tank; M = weight of fluid in tank; T, t = temperature of hot and cold fluids respectively; T_1, t_1 = temperatures at beginning of heating or cooling period or at inlet; T_2, t_2 = temperature at end of period or at outlet; T_0, t_0 = temperature of liquid added to tank; U = coefficient of heat transfer; and W, w = flow rate through external exchanger of hot and cold fluids respectively.

Applications One typical application in heat transfer with batch operations is the heating of a reactor mix, maintaining temperature during a reaction period, and then cooling the products after the reaction is complete. This subsection is concerned with the heating and cooling of such systems in either unknown or specified periods.

The technique for deriving expressions relating time for heating or cooling agitated batches to coil or jacket area, heat-transfer coefficients, and the heat capacity of the vessel contents was developed by Bowman, Mueller, and Nagle [*Trans. Am. Soc. Mech. Eng.*, **62**, 283–294 (1940)] and extended by Fisher [*Ind. Eng. Chem.*, **36**, 939–942 (1944)] and Chaddock and Sanders [*Trans. Am. Inst. Chem. Eng.*, **40**, 203–210 (1944)] to external heat exchangers. Kern (*Process Heat Transfer*, McGraw-Hill, New York, 1950, chap. 18) collected and published the results of these investigators.

The assumptions made were that (1) U is constant for the process and over the entire surface, (2) liquid flow rates are constant, (3) specific heats are constant for the process, (4) the heating or cooling medium has a constant inlet temperature, (5) agitation produces a uniform batch fluid temperature, (6) no partial phase changes occur, and (7) heat losses are negligible. The developed equations are as follows. If any of the assumptions do not apply to a system being designed, new equations should be developed or appropriate corrections made. Heat exchangers are counterflow except for the 1-2 exchangers, which are one-shell-pass, two-tube-pass, parallel-flow counterflow.

Coil-in-Tank or Jacketed Vessel: Isothermal Heating Medium

$$\ln (T_1 - t_1)/(T_1 - t_2) = UA\theta/Mc \qquad (10\text{-}147)$$

Cooling-in-Tank or Jacketed Vessel: Isothermal Cooling Medium

$$\ln (T_1 - t_1)/(T_2 - t_1) = UA\theta/MC \qquad (10\text{-}147a)$$

Coil-in-Tank or Jacketed Vessel: Nonisothermal Heating Medium

$$\ln \frac{T_1 - t_1}{T_1 - t_2} = \frac{WC}{Mc}\left(\frac{K_1 - 1}{K_1}\right)\theta \qquad (10\text{-}147b)$$

where $K_1 = e^{UA/WC}$

Coil-in-Tank: Nonisothermal Cooling Medium

$$\ln \frac{T_1 - t_1}{T_2 - t_1} = \frac{wc}{MC}\left(\frac{K_2 - 1}{K_2}\right)\theta \qquad (10\text{-}147c)$$

where $K_2 = e^{UA/wc}$

External Heat Exchanger: Isothermal Heating Medium

$$\ln \frac{T_1 - t_1}{T_1 - t_2} = \frac{wc}{Mc}\left(\frac{K_2 - 1}{K_2}\right)\theta \qquad (10\text{-}147d)$$

External Exchanger: Isothermal Cooling Medium

$$\ln \frac{T_1 - t_1}{T_2 - t_1} = \frac{WC}{MC}\left(\frac{K_1 - 1}{K_1}\right)\theta \qquad (10\text{-}147e)$$

External Exchanger: Nonisothermal Heating Medium

$$\ln \frac{T_1 - t_1}{T_1 - t_2} = \left(\frac{K_3 - 1}{M}\right)\left(\frac{wWC}{K_3 wc - WC}\right)\theta \qquad (10\text{-}147f)$$

where $K_3 = e^{UA(1/wc - 1/WC)}$

External Exchanger: Nonisothermal Cooling Medium

$$\ln \frac{T_1 - t_1}{T_2 - t_1} = \left(\frac{K_4 - 1}{M}\right)\left(\frac{Wwc}{K_4 WC - wc}\right)\theta \qquad (10\text{-}147g)$$

where $K_4 = e^{UA(1/WC - 1wc)}$

External Exchanger with Liquid Continuously Added to Tank: Isothermal Heating Medium

$$\ln \frac{t_1 - t_0 - \dfrac{w}{L_0}\left(\dfrac{K_2 - 1}{K_2}\right)(T_1 - t_1)}{t_2 - t_0 - \dfrac{w}{L_0}\left(\dfrac{K_2 - 1}{K_2}\right)(T_1 - t_2)} = \left[\frac{w}{L_0}\left(\frac{K_2 - 1}{K_2}\right) + 1\right] \ln \frac{M + L_0\theta}{M} \qquad (10\text{-}147h)$$

If the addition of liquid to the tank causes an average endothermic or exothermic heat of solution, $\pm q_s$ J/kg (Btu/lb) of makeup, it may be included by adding $\pm q_s/c_0$ to both the numerator and the denominator of the left side. The subscript 0 refers to the makeup.

External Exchanger with Liquid Continuously Added to Tank: Isothermal Cooling Medium

$$\ln \frac{T_0 - T_1 - \dfrac{W}{L_0}\left(\dfrac{K_1 - 1}{K_1}\right)(T_1 - t_1)}{T_0 - T_2 - \dfrac{W}{L_0}\left(\dfrac{K_1 - 1}{K_1}\right)(T_2 - t_1)} = \left[1 - \frac{W}{L_0}\left(\frac{K_1 - 1}{K_1}\right)\right] \ln \frac{M + L_0\theta}{M} \qquad (10\text{-}147i)$$

The heat-of-solution effects can be included by adding $\pm q_s/C_0$ to both the numerator and the denominator of the left side.

External Exchanger with Liquid Continuously Added to Tank: Nonisothermal Heating Medium

$$\ln \frac{t_0 - t_1 + \dfrac{wWC(K_5 - 1)(T_1 - t_1)}{L_0(K_5WC - wc)}}{t_0 - t_2 + \dfrac{wWC(K_5 - 1)(T_1 - t_2)}{L_0(K_5WC - wc)}} = \left[\frac{wWC(K_5 - 1)}{L_0(K_5Wc - wc)} + 1\right] \ln \frac{M + L_0\theta}{M} \qquad (10\text{-}147j)$$

where $K_5 = e^{(UA/wc)(1 - wc/WC)}$

The heat-of-solution effects can be included by adding $\pm q_s/C_0$ to both the numerator and the denominator of the left side.

External Exchanger with Liquid Continuously Added to Tank: Nonisothermal Cooling Medium

$$\ln \frac{T_0 - T_1 - \dfrac{Wwc(K_6 - 1)(T_1 - t_1)}{L_0(K_6wc - WC)}}{T_0 - T_2 - \dfrac{Wwc(K_6 - 1)(T_2 - t_1)}{L_0(K_6wc - WC)}} = \left[\frac{Wwc(K_6 - 1)}{L_0(K_6wc - WC)} + 1\right] \ln \frac{M + L_0\theta}{M} \qquad (10\text{-}147k)$$

where $K_6 = e^{(UA/WC)(1 - WC/we)}$

The heat-of-solution effects can be included by adding $\pm q_s/C_0$ to both the numerator and the denominator of the left side.

Heating and Cooling Agitated Batches: 1-2 Parallel Flow-Counterflow

$$\frac{UA}{wc} = \frac{1}{\sqrt{R^2 + 1}} \ln \frac{2 - S(R + 1 - \sqrt{R^2 + 1})}{2 - S(R + 1 + \sqrt{R^2 + 1})} \qquad (10\text{-}147l)$$

$$R = \frac{T_1 - T_2}{t' - t} = \frac{wc}{WC} \quad \text{and} \quad S = \frac{t' - t}{T_1 - t}$$

$$\frac{2 - S(R + 1 - \sqrt{R^2 + 1})}{2 - S(R + 1 + \sqrt{R^2 + 1})} = e^{(UA/wc)\sqrt{R^2+1}} = K_7 \qquad (10\text{-}147m)$$

$$S = \frac{2(K_7 - 1)}{K_7(R + 1 + \sqrt{R^2 + 1}) - (R + 1 - \sqrt{R^2 + 1})}$$

External 1-2 Exchanger: Heating

$$\ln (T_1 - t_1)/(T_1 - t_2) = (Sw/M)\theta \qquad (10\text{-}147n)$$

External 1-2 Exchanger: Cooling

$$\ln (T_1 - t_1)/(T_2 - t_1) = S(wc/MC)\theta \qquad (10\text{-}147o)$$

The cases of multipass exchangers with liquid continuously added to the tank are covered by Kern, as cited earlier. An alternative method for all multipass-exchanger gases, including those presented as well as cases with two or more shells in series, is as follows:

1. Determine UA for using the applicable equations for counterflow heat exchangers.
2. Use the initial batch temperature T_1 or t_1.
3. Calculate the outlet temperature from the exchanger of each fluid. (This will require trial-and-error methods.)
4. Note the F_T correction factor for the corrected mean temperature difference. (See Fig. 10-14.)
5. Repeat steps 2, 3, and 4 by using the final batch temperature T_2 and t_2.
6. Use the average of the two values for F, then increase the required multipass UA as follows:

$$UA(\text{multipass}) = UA(\text{counterflow})/F_T$$

In general, values of F_T below 0.8 are uneconomical and should be avoided. F_T can be raised by increasing the flow rate of either or both of the flow streams. Increasing flow rates to give values well above 0.8 is a matter of economic justification.

If F_T varies widely from one end of the range to the other, F_T should be determined for one or more intermediate points. The average should then be determined for each step which has been established and the average of these taken for use in step 6.

Effect of External Heat Loss or Gain If heat loss or gain through the vessel walls cannot be neglected, equations which include this heat transfer can be developed by using energy balances similar to those used for the derivations of equations given previously. Basically, these equations must be modified by adding a heat-loss or heat-gain term.

A simpler procedure, which is probably acceptable for most practical cases, is to ratio UA or θ either up or down in accordance with the required modification in total heat load over time θ.

Another procedure, which is more accurate for the external-heat-exchanger cases, is to use an equivalent value for MC (for a vessel being heated) derived from the following energy balance:

$$Q = (Mc)_e(t_2 - t_1) = Mc(t_2 - t_1) + U'A'(MTD')\theta \qquad (10\text{-}147p)$$

where Q is the total heat transferred over time θ, $U'A'$ is the heat-transfer coefficient for heat loss times the area for heat loss, and MTD' is the mean temperature difference for the heat loss.

A similar energy balance would apply to a vessel being cooled.

Internal Coil or Jacket Plus External Heat Exchanger This case can be most simply handled by treating it as two separate problems. M is divided into two separate weights, M_1 and $(M - M_1)$, and the appropriate equations given earlier are applied to each part of the system. Time θ, of course, must be the same for both parts.

Equivalent-Area Concept The preceding equations for batch operations, particularly Eq. (10-147), can be applied for the calculation of heat loss from tanks which are allowed to cool over an extended period of time. However, different surfaces of a tank, such as the top (which would not be in contact with the tank contents) and the bottom, may have coefficients of heat transfer which are different from those of the vertical tank walls. The simplest way to resolve this difficulty is to use an equivalent area A_e in the appropriate equations where

$$A_e = A_bU_b/U_s + A_tU_t/U_s + A_s \qquad (10\text{-}147q)$$

and the subscripts b, s, and t refer to the bottom, sides, and top respectively. U is usually taken as U_s. Table 10-7 lists typical values for U_s and expressions for A_e for various tank configurations.

Nonagitated Batches Cases in which vessel contents are vertically stratified, rather than uniform in temperature, have been treated by Kern (op. cit.). These are of little practical importance except for tall, slender vessels heated or cooled with external exchangers. The result is that a smaller exchanger is required than for an equivalent agitated batch system that is uniform.

Storage Tanks The equations for batch operations with agitation may be applied to storage tanks even though the tanks are not agitated. This approach gives conservative results. The important cases (nonsteady state) are:

1. *Tanks cool; contents remain liquid.* This case is relatively simple and can easily be handled by the equations given earlier.
2. *Tanks cool, contents partially freeze, and solids drop to bottom or rise to top.* This case requires a two-step calculation. The first step is handled as in case 1. The second step is calculated by

TABLE 10-7 Typical Values for Use with Eqs. (10-148) to (10-157)*

Application	Fluid	U_s	A_s
Tanks on legs, outdoors, not insulated	Oil	3.7	$0.22\ A_t + A_b + A_s$
	Water at 150°F.	5.1	$0.16\ A_t + A_b + A_s$
Tanks on legs, outdoors, insulated 1 in.	Oil	0.45	$0.7\ A_t + A_b + A_s$
	Water	0.43	$0.67\ A_t + A_b + A_s$
Tanks on legs, indoors, not insulated	Oil	1.5	$0.53\ A_t + A_b + A_s$
	Water	1.8	$0.35\ A_t + A_b + A_s$
Tanks on legs, indoors, insulated 1 in.	Oil	0.36	$0.8\ A_t + A_b + A_s$
	Water	0.37	$0.73\ A_t + A_b + A_s$
Flat-bottom tanks,† outdoors, not insulated	Oil	3.7	$0.22\ A_t + A_s + 0.43\ D_t$
	Water	5.1	$0.16\ A_t + A_s + 0.31\ D_t$
Flat-bottom tanks,† outdoors, insulated 1 in.	Oil	0.36	$0.7\ A_t + A_s + 3.9\ D_t$
	Water	0.37	$0.16\ A_t + A_s + 3.7\ D_t$
Flat-bottom tanks, indoors, not insulated	Oil	1.5	$0.53\ A_t + A_s + 1.1\ D_t$
	Water	1.8	$0.35\ A_t + A_s + 0.9\ D_t$
Flat-bottom tanks, indoors, insulated 1 in.	Oil	0.36	$0.8\ A_t + A_s + 4.4\ D_t$
	Water	0.37	$0.73\ A_t + A_s + 4.5\ D_t$

*Based on typical coefficients.
†The ratio $(t - t_g)(t - t')$ assumed at 0.85 for outdoor tanks. °C = (°F − 32)/1.8; to convert British thermal units per hour-square foot-degrees Fahrenheit to joules per square meter-second-kelvins, multiply by 5.6783.

assuming an isothermal system at the freezing point. It is possible, given time and a sufficiently low ambient temperature, for tank contents to freeze solid.

3. *Tanks cool and partially freeze; solids form a layer of self-insulation.* This complex case, which has been known to occur with heavy hydrocarbons and mixtures of hydrocarbons, has been discussed by Stuhlbarg [*Pet. Refiner*, 38, 143 (Apr. 1, 1959)]. The contents in the center of such tanks have been known to remain warm and liquid even after several years of cooling.

It is very important that a melt-out riser be installed whenever tank contents are expected to freeze on prolonged shutdown. The purpose is to provide a molten chimney through the crust for relief of thermal expansion or cavitation if fluids are to be pumped out or recirculated through an external exchanger. An external heat tracer, properly located, will serve the same purpose but may require more remelt time before pumping can be started.

Batch Operations: Optimum Exchanger Size Five variables are involved in an external exchanger design, such as that covered by Eq. (10-147*g*). These are time θ, recirculation rate W, coolant rate w, heat-transfer coefficient U, and exchanger area A. Since U and A are interdependent, it is convenient to consider the product UA as a single variable, reducing the number of variables to four. Any three of these must be set before the fourth can be determined. A practical procedure is:

1. Set θ arbitrarily or as required to meet process demands.
2. Find w from the equation

$$wc = \frac{MC(T_1 - T_2)}{\theta(t' - t_1)} \qquad (10\text{-}148)$$

where t' is a reasonable average exchanger outlet temperature.

3. Find the minimum flow rate from

$$W_{\min}C = \frac{MC}{\theta}\left(\ln \frac{T_1 - t_1}{T_2 - t_1}\right) \qquad (10\text{-}148a)$$

4. Set W at a series of values greater than $W_{\min}$, then calculate UA from Eq. (10-147*g*) for each value of W. The range of flows chosen should be from about 1.25 $W_{\min}$ to over 2.0 $W_{\min}$.
5. Size and price the exchangers, pump, and piping. The cost of pumping power must be taken into consideration.
6. An economic evaluation is next made. One suggested method is to preset a payout period n and then calculate the function $(P + jC_E)$ for each value of W, where

$$j = \frac{1 - r}{1/n + (1 - r)E - rD}$$

and C_E is the annual pumping cost; P is the total installed cost for the exchanger pump, piping, and any other item of equipment which might be affected; r is the income tax rate; E is actual expense rate for maintenance and insurance; and D is the depreciation rate. Both E and D are expressed as fractions of installed cost.

Similar equations for $W_{\min}$ can be developed for heating and for other external exchanger cases. Some of these and an illustrative example are given by Stuhlbarg [*Hydrocarbon Process.*, 149 (January 1970)].

The optimum occurs at the point where the specified function is a minimum. It is possible to graph $(P - jC_E)$ on the ordinate axis versus W to find the minimum, but such graphs will be discontinuous at points where there is a step change in pipe size, pump size, or pump motor. By carefully selecting values of W which are just within the capacities of each size of pipe or item of equipment, it is possible to determine the optimum without using the graphical approach.

THERMAL DESIGN OF TANK COILS

The thermal design of tank coils involves the determination of the area of heat-transfer surface required to maintain the contents of the tank at a constant temperature or to raise or lower the temperature of the contents by a specified magnitude over a fixed time.

Nomenclature A = area; A_b = area of tank bottom; A_c = area of coil; A_e = equivalent area; A_s = area of sides; A_t = area of top; A_1 = equivalent area receiving heat from external coils; A_2 = equivalent area not covered with external coils; D_t = diameter of tank; F = design (safety) factor; h = film coefficient; h_a = coefficient of ambient air; h_c = coefficient of coil; h_h = coefficient of heating medium; h_l = coefficient of liquid phase of tank contents or tube-side coefficient referred to outside of coil; h_z = coefficient of insulation; k = thermal conductivity; k_g = thermal conductivity of ground below tank; M = weight of tank contents when full; t = temperature; t_a = temperature of ambient air; t_d = temperature of dead-air space; t_f = temperature of contents at end of heating; t_g = temperature of ground below tank; t_h = temperature of heating medium; t_0 = temperature of contents at beginning of heating; U = overall coefficient; U_b = coefficient at tank bottom; U_c = coefficient of coil; U_d = coefficient of dead air to the tank contents; U_i = coefficient through insulation; U_s = coefficient at sides; U_t = coefficient at top; and U_2 = coefficient at area A_2.

Typical coil coefficients are listed in Table 10-8. More exact values can be calculated by using the methods for natural convection or forced convection given elsewhere in this section.

Maintenance of Temperature Tanks are often maintained at temperature with internal coils if the following equations are assumed to be applicable:

$$q = U_s A_e (T - t') \qquad (10\text{-}149)$$

TABLE 10-8 Overall Heat-Transfer Coefficients for Coils Immersed in Liquids
U Expressed as Btu/(h·ft²·°F)

Substance inside coil	Substance outside coil	Coil material	Agitation	U
Steam	Water	Lead	Agitated	70
Steam	Sugar and molasses solutions	Copper	None	50–240
Steam	Boiling aqueous solution	....	....	600
Cold water	Dilute organic dye intermediate	Lead	Turboagitator at 95 r.p.m.	300
Cold water	Warm water	Wrought iron	Air bubbled into water surrounding coil	150–300
Cold water	Hot water	Lead	0.40 r.p.m. paddle stirrer	90–360
Brine	Amino acids	....	30 r.p.m.	100
Cold water	25% oleum at 60°C.	Wrought iron	Agitated	20
Water	Aqueous solution	Lead	500 r.p.m. sleeve propeller	250
Water	8% NaOH	....	22 r.p.m.	155
Steam	Fatty acid	Copper (pancake)	None	96–100
Milk	Water	....	Agitation	300
Cold water	Hot water	Copper	None	105–180
60°F. water	50% aqueous sugar solution	Lead	Mild	50–60
Steam and hydrogen at 1500lb./sq. in.	60°F. water	Steel	....	100–165
Steam 110–146 lb./sq. in. gage	Vegetable oil	Steel	None	23–29
Steam	Vegetable oil	Steel	Various	39–72
Cold water	Vegetable oil	Steel	Various	29–72

NOTES: Chilton, Drew, and Jebens [*Ind. Eng. Chem.*, **36**, 510 (1944)] give film coefficients for heating and cooling agitated fluids using a coil in a jacketed vessel.

Because of the many factors affecting heat transfer, such as viscosity, temperature difference, and coil size, the values in this table should be used primarily for preliminary design estimates and checking calculated coefficients.

°C = (°F − 32)/1.8; to convert British thermal units per hour-square foot-degrees Fahrenheit to joules per square meter-second-kelvins, multiply by 5.6783.

and

$$A_c = q/U_c(MTD) \qquad (10\text{-}149a)$$

These make no allowance for unexpected shutdowns. One method of allowing for shutdown is to add a safety factor to Eq. (10-149*a*). A better method is to establish a design shutdown time and a design reheat or recool time and then to use the appropriate equations from the subsection "Batch Operations: Optimum Exchanger Size." Modification by use of Eqs. (10-147*p*) and (10-147*q*) may also be required.

In the case of a tank maintained at temperature with external coils, the coils are usually designed to cover only a portion of the tank. The temperature t_d of the dead-air space between the coils and the tank is obtained from

$$U_dA_1(t_d - t) = U_2A_2(t - t') \qquad (10\text{-}150)$$

The heat load is

$$q = U_dA_1(t_d - t) + A_1U_i(t_d - t') \qquad (10\text{-}151)$$

The coil area is

$$A_c = \frac{qF}{U_c(t_h - t_d)_m} \qquad (10\text{-}152)$$

where F is a safety factor.

Heating

Heating with Internal Coil from Initial Temperature for Specified Time

$$Q = Wc(t_f - t_o) \qquad (10\text{-}153)$$

$$A_c = \left[\frac{Q}{\theta_h} + U_sA_e\left(\frac{t_f + t_o}{2} - t'\right)\right]\left[\frac{1}{U_c[t_h - (t_f + t_o)/2]}\right](F) \qquad (10\text{-}154)$$

where θ_h is the length of heating period. This equation may also be used when the tank contents have cooled from t_f to t_o and must be reheated to t_f. If the contents cool during a time θ_c, the temperature at the end of this cooling period is obtained from

$$\ln\left(\frac{t_f - t'}{t_o - t'}\right) = \frac{U_sA_e\theta_c}{Wc} \qquad (10\text{-}155)$$

Heating with External Coil from Initial Temperature for Specified Time The temperature of the dead-air space is obtained from

$$U_dA_1[t_d - 0.5(t_f - t_o)] = U_2A_2[0.5(t_f - t_o) - t'] + Q/\theta_h \qquad (10\text{-}156)$$

The heat load is

$$q = U_iA_1(t_d - t') + U_2A_2[0.5(t_f - t_o) - t'] + Q/\theta_h \qquad (10\text{-}157)$$

The coil area is obtained from Eq. (10-152).

The safety factor used in the calculations is a matter of judgment based on confidence in the design. A value of 1.10 is normally not considered excessive. Typical design parameters are shown in Tables 10-7 and 10-8.

EXTENDED OR FINNED SURFACES

Finned-Surface Application Extended or finned surfaces are often used when one film coefficient is substantially lower than the other, the goal being to make $h_o A_{oe} \approx h_i A_i$. A few typical fin configurations are shown in Fig. 10-38. Longitudinal fins are used in double-pipe exchangers. Transverse fins are used in cross-flow and shell-and-tube configurations. High transverse fins are used mainly with low-pressure gases; low fins are used for boiling and condensation of nonaqueous streams as well as for sensible-heat transfer. Finned surfaces should not be used for severely fouling services, especially if mechanical cleaning is required, or for streams carrying large amounts of particulate matter.

The area added by the fin is not as efficient for heat transfer as bare tube surface owing to resistance to conduction through the fin. The effective heat-transfer area is

$$A_{oe} = A_{uf} + A_f \Omega \qquad (10\text{-}158)$$

The fin efficiency is found from mathematically derived relations, in which the film heat-transfer coefficient is assumed to be constant over the entire fin and temperature gradients across the thickness of the fin have been neglected (see Kraus, *Extended Surfaces*, Spartan Books, Baltimore, 1963). The efficiency curves for some common fin configurations are given in Figs. 10-38 and 10-39.

High Fins To calculate heat-transfer coefficients for cross-flow to a transversely finned surface, it is best to use a correlation based on experimental data for that surface. Such data are not often available, and a more general correlation must be used, making allowance for the possible error. Probably the best general correlation for bundles of finned tubes is given by Schmidt [*Kaltetechnik*, **15**, 98–102, 370–378 (1963)]:

$$hD_r/k = K(D_r\rho V'_{max}/\mu)^{0.625} R_f^{-0.375} N_{Pr}^{1/3} \qquad (10\text{-}159)$$

where $K = 0.45$ for staggered tube arrays and 0.30 for in-line tube arrays: D_r is the root or base diameter of the tube; V'_{max} is the maximum velocity through the tube bank, i.e., the velocity through the minimum flow area between adjacent tubes; and R_f is the ratio of the total outside surface area of the tube (including fins) to the surface of a tube having the same root diameter but without fins.

Pressure drop is particularly sensitive to geometrical parameters, and available correlations should be extrapolated to geometries different from those on which the correlation is based only with great caution and conservatism. The best correlation is that of Robinson and Briggs [*Chem. Eng. Prog.*, **62**, *Symp. Ser.* 64, 177–184 (1966)].

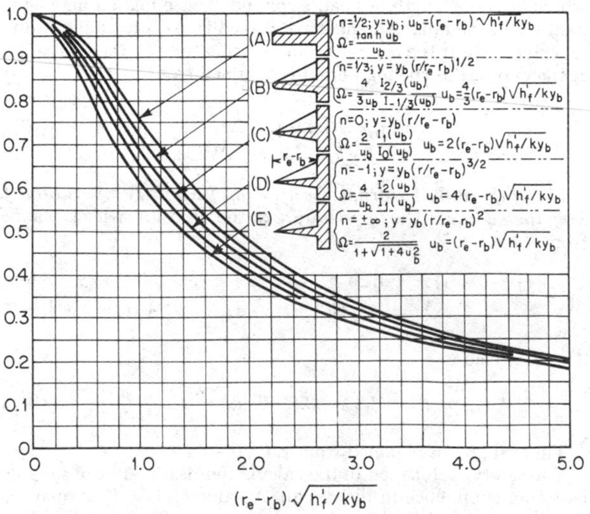

FIG. 10-38 Efficiencies for several longitudinal fin configurations.

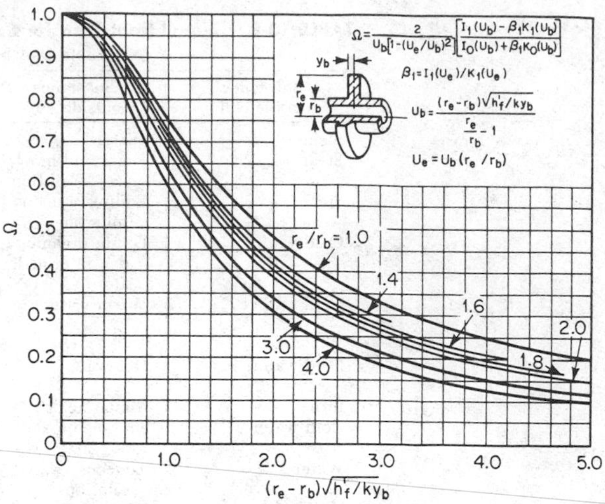

FIG. 10-39 Efficiencies for annular fins of constant thickness.

Low Fins Low-finned tubing is generally used in shell-and-tube configurations. For sensible-heat transfer, only minor modifications are needed to permit the shell-side method given earlier to be used for both heat transfer and pressure [see Briggs, Katz, and Young, *Chem. Eng. Prog.*, **59**(11), 49–59 (1963)]. For condensing on low-finned tubes in horizontal bundles, the Nusselt correlation [Eq. (10-97)] is generally satisfactory for low-surface-tension [$\sigma < (3)(10^{-6})$N/m (30 dyn/cm)] condensates; finned surfaces should not be used for high-surface-tension condensates (notably water), which do not drain easily.

The modified Palen-Small method can be employed for reboiler design using finned tubes, but the maximum flux [Eq. (10-112)] is calculated from A_o, the total outside heat-transfer area including fins. The resulting value of q_{max} refers to A_o.

FOULING AND SCALING

Fouling refers to any layer or deposit of extraneous material on a heat-transfer surface. These materials usually have low thermal conductivity, resulting in a major resistance to heat transfer. A number of different kinds of fouling occur in process heat-transfer equipment. Silting or sedimentation is the deposition of finely divided materials from the process stream. Scaling is frequently caused by crystallization of a material whose solubility at the wall temperature is lower than at the bulk liquid temperature; calcium sulfate scaling is the most common case of this kind. Many process streams polymerize, and the resulting less soluble material is deposited on the surface as a film, often of considerable thickness and toughness. Corrosion products may form a serious resistance to the transfer of heat. Biological growth such as algae is a serious problem with many cooling-water streams and in the fermentation industry.

Control of Fouling Various procedures have proved effective in eliminating or limiting fouling. Inhibitors are frequently added to cooling streams to minimize the tendency for salt deposition, corrosion, and algal growth. Designing to give high velocity of process streams is quite effective in reducing fouling but is limited in application because of the increased pressure drop required and because of erosion damage to the surface. As a general rule, in-tube water velocities below 0.91 m/s (3 ft/s) are never recommended, and there is increasing incentive to go as high as 2.44 to 3.05 m/s (8 to 10 ft/s). Excessive temperature differences should be avoided, particularly if solubility limits are exceeded at the surface. In reboiler design, it is very important to keep temperature differences low enough to avoid the transition and film-boiling regimes. It is important to limit the fraction vaporized to keep solids in solution and to maintain positive circulation in order to avoid solids building up in the reboiler inventory liquid.

Fouling Transients and Operating Periods Two common behaviors are noted in the development of a fouling film over a period of time. One is the so-called **asymptotic fouling** in which the resistance builds up very rapidly at the start but becomes asymptotic to a steady-state value if conditions remain unchanged. This kind of fouling occurs most commonly when the wall temperature is constant and the heat flux falls in response to the increased fouling resistance.

The other transient behavior is that of a more or less **linear increase** in the fouling resistance during the entire operating period. This is often encountered when the heat flux is kept constant by increasing the temperature difference, resulting in a nearly constant temperature at the fluid-fouling deposit interface.

The optimum operating period between cleanouts depends upon the rate of fouling, the effect of the fouling on the overall heat-transfer rate, and the ease with which the heat exchanger may be removed from service without affecting the process. It is often necessary to overdesign a heat exchanger so that it will continue to transfer sufficient heat until plant shutdown. Care must be taken that the overdesign of the heat exchanger does not introduce factors such as excessive shell diameter and resulting low velocity that will accelerate fouling and defeat the purpose of overdesign. In the most extreme cases, it may be necessary to provide 100 percent standby heat-transfer capacity so that one exchanger may be cleaned while the other exchanger is on stream.

Removal of Fouling Deposits Chemical removal of fouling can be achieved in some cases by weak acid, special solvents, etc. Other deposits adhere weakly and can be washed off by periodic operation at very high velocities or by flushing with a high-velocity steam or water jet or a sand-water slurry. These methods can be applied to both the shell side and the tube side without pulling the bundle. Most fouling deposits, however, must be removed by positive mechanical action such as rodding, turbining, or scraping the surface. These techniques can be applied inside tubes without pulling the bundle but can be applied on the shell side only after bundle removal and even then with limited success because of the closeness of the tubes. The provision of cleaning lanes by using rotated square or large-pitch-ratio triangular arrays is helpful.

Fouling Resistance There are no published methods for predicting fouling resistances a priori. The accumulated experience of exchanger designers and users is published in the *Standards of Tubular Exchanger Manufacturers Association* (5th ed., New York, 1968). The table in Kern (*Process Heat Transfer*, McGraw-Hill, New York, 1950), taken from an earlier edition of the *Standards*, is essentially identical. In the absence of more pertinent experience, these values may be used. See Table 10-9.

TABLE 10-9 Heat-Transfer Coefficients for Deposits*

Btu/(h · ft² · °F)

Water

Temperature of heating medium	Up to 240°F.		240°–400°F.	
Temperature of water	125°F. or less		Above 125°F.	
Water velocity, ft./sec.	≤3	>3	≤3	>3
Distilled	2000	2000	2000	2000
Sea water	2000	2000	1000	1000
Treated boiler feed water	1000	2000	1000	1000
Treated makeup for cooling tower	1000	1000	500	500
City, well, Great Lakes	1000	1000	500	500
Brackish, clean river water	500	1000	330	500
River water, muddy, silty	330	500	250	330
Hard (over 15 g./gal.)	330	330	200	200
Chicago Sanitary Canal	125	170	100	125

Chemicals

Inorganic:	
Gases (oil-bearing or dirty)	500
Liquids (heating or vaporization)	500
Refrigerant brines	1000
Organic:	
Gases	
Process	1000
Utility (oil-bearing, refrigerant, etc.)	500
Condensing vapors (condensers)	1000
Liquids	
Process	1000
Vaporizing liquids (reboilers)	500
Heat-transfer media	1000
Refrigerant liquids	1000
Polymer-forming liquids	200
Oils (vegetable and heavy gas oil)	330
Asphalt and residuum	100

*Rearranged and reproduced, by permission, from *Standards of Tubular Exchanger Manufacturers Association*, 5th ed., New York, 1968.

NOTE: °C = (°F − 32)/1.8; to convert feet per second to meters per second, multiply by 0.3048; to convert British thermal units per hour-square foot-degrees Fahrenheit to joules per square meter-second-kelvins, multiply by 5.6783.

TYPICAL HEAT-TRANSFER COEFFICIENTS

Typical overall heat-transfer coefficients are given in Tables 10-10 through 10-15. Values from these tables may be used for preliminary estimating purposes. They should not be used in place of the design methods described elsewhere in this section, although they may serve as a useful check on the results obtained by those design methods.

THERMAL DESIGN FOR SOLIDS PROCESSING

Solids in divided form, such as powders, pellets, and lumps, are heated and/or cooled in chemical processing for a variety of objectives such as solidification or fusing (Sec. 11), drying and water removal (Sec. 20), solvent recovery (Secs. 13 and 20), sublimation (Sec. 17), chemical reactions (Sec. 20), and oxidation. For process and mechanical-design considerations, see the referenced sections.

Thermal design concerns itself with sizing the equipment to effect the heat transfer necessary to carry on the process. The design equation is the familiar one basic to all modes of heat transfer, namely,

$$A = Q/U\,\Delta t \tag{10-160}$$

where A = effective heat-transfer surface, Q = quantity of heat required to be transferred, Δt = temperature difference of the process, and U = overall heat-transfer coefficient. It is helpful to define the modes of heat transfer and the corresponding overall coefficient as U_{co} = overall heat-transfer coefficient for (indirect through-a-wall) *conduction*, U_{cv} = overall heat-transfer coefficient for the little-used *convection* mechanism, U_{ct} = heat-transfer coefficient for the *contactive* mechanism in which the gaseous-phase heat carrier passes directly through the solids bed, and U_{ra} = heat-transfer coefficient for *radiation*.

There are two general methods for determining numerical values for U_{co}, U_{cv}, U_{ct}, and U_{ra}. One is by analysis of actual operating data. Values so obtained are used on geometrically similar systems of a size not too different from the equipment from which the data were obtained. The second method is predictive and is based on the material properties and certain operating parameters. Relative values of the coefficients for the various modes of heat transfer at temperatures up to 980°C (1800°F) are as follows (Holt, Paper 11, Fourth National Heat Transfer Conference, Buffalo, 1960):

Convective	1
Radiant	2
Conductive	20
Contactive	200

Because heat-transfer equipment for solids is generally an adaptation of a primarily material-handling device, the area of heat transfer is often small in relation to the overall size of the equipment. Also peculiar to solids heat transfer is that the Δt varies for the different heat-transfer mechanisms. With a knowledge of these mechanisms, the Δt term generally is readily estimated from temperature limitations imposed by the burden characteristics and/or the construction.

TABLE 10-10 Typical Overall Heat-Transfer Coefficients in Tubular Heat Exchangers

U = Btu/(°F·ft²·h)

Shell side	Tube side	Design U	Includes total dirt
Liquid-liquid media			
Aroclor 1248	Jet fuels	100–150	0.0015
Cutback asphalt	Water	10–20	.01
Demineralized water	Water	300–500	.001
Ethanol amine (MEA or DEA) 10–25% solutions	Water or DEA, or MEA solutions	140–200	.003
Fuel oil	Water	15–25	.007
Fuel oil	Oil	10–15	.008
Gasoline	Water	60–100	.003
Heavy oils	Heavy oils	10–40	.004
Heavy oils	Water	15–50	.005
Hydrogen-rich reformer stream	Hydrogen-rich reformer stream	90–120	.002
Kerosene or gas oil	Water	25–50	.005
Kerosene or gas oil	Oil	20–35	.005
Kerosene or jet fuels	Trichlorethylene	40–50	.0015
Jacket water	Water	230–300	.002
Lube oil (low viscosity)	Water	25–50	.002
Lube oil (high viscosity)	Water	40–80	.003
Lube oil	Oil	11–20	.006
Naphtha	Water	50–70	.005
Naphtha	Oil	25–35	.005
Organic solvents	Water	50–150	.003
Organic solvents	Brine	35–90	.003
Organic solvents	Organic solvents	20–60	.002
Tall oil derivatives, vegetable oil, etc.	Water	20–50	.004
Water	Caustic soda solutions (10–30%)	100–250	.003
Water	Water	200–250	.003
Wax distillate	Water	15–25	.005
Wax distillate	Oil	13–23	.005
Condensing vapor-liquid media			
Alcohol vapor	Water	100–200	.002
Asphalt (450°F.)	Dowtherm vapor	40–60	.006
Dowtherm vapor	Tall oil and derivatives	60–80	.004
Dowtherm vapor	Dowtherm liquid	80–120	.0015
Gas-plant tar	Steam	40–50	.0055
High-boiling hydrocarbons V	Water	20–50	.003
Low-boiling hydrocarbons A	Water	80–200	.003
Hydrocarbon vapors (partial condenser)	Oil	25–40	.004
Organic solvents A	Water	100–200	.003
Organic solvents high NC, A	Water or brine	20–60	.003
Organic solvents low NC, V	Water or brine	50–120	.003
Kerosene	Water	30–65	.004
Kerosene	Oil	20–30	.005
Naphtha	Water	50–75	.005
Naphtha	Oil	20–30	.005
Stabilizer reflux vapors	Water	80–120	.003
Steam	Feed water	400–1000	.0005
Steam	No. 6 fuel oil	15–25	.0055
Steam	No. 2 fuel oil	60–90	.0025
Sulfur dioxide	Water	150–200	.003
Tall-oil derivatives, vegetable oils (vapor)	Water	20–50	.004
Water	Aromatic vapor-stream azeotrope	40–80	.005
Gas-liquid media			
Air, N_2, etc. (compressed)	Water or brine	40–80	.005
Air, N_2, etc., A	Water or brine	10–50	.005
Water or brine	Air, N_2 (compressed)	20–40	.005
Water or brine	Air, N_2, etc., A	5–20	.005
Water	Hydrogen containing natural-gas mixtures	80–125	.003
Vaporizers			
Anhydrous ammonia	Steam condensing	150–300	.0015
Chlorine	Steam condensing	150–300	.0015
Chlorine	Light heat-transfer oil	40–60	.0015
Propane, butane, etc.	Steam condensing	200–300	.0015
Water	Steam condensing	250–400	.0015

NC = noncondensable gas present.
V = vacuum.
A = atmospheric pressure.
Dirt (or fouling factor) units are (h·ft²·°F)/Btu.
To convert British thermal units per hour–square foot–degrees Fahrenheit to joules per square meter–second–kelvins, multiply by 5.6783; to convert hours per square foot–degree Fahrenheit–British thermal units to square meters per second–kelvin–joules, multiply by 0.1761.

TABLE 10-11 Typical Overall Heat-Transfer Coefficients in Refinery Service

Btu/(°F·ft²·h)

Fluid	API gravity	Fouling factor (one stream)	Reboiler, steam-heated	Condenser, water-cooled*	Exchangers, liquid to liquid (tube-side fluid designation appears below)			Reboiler (heating liquid designated below)			Condenser (cooling liquid designated below)			
					C	G	H	C	G†	K	D	F	G	J
A Propane	...	0.001	160	95	85	85	80	110	95	35				
B Butane	...	.001	155	90	80	75	75	105	90	35	80	55	40	30
C 400°F. end-point gasoline	50	.001	120	80	70	65	60	65	50	30				
D Virgin light naphtha	70	.001	140	85	70	55	55	75	60	35	75			
E Virgin heavy naphtha	45	.001	95	75	65	55	50	55	45	30	70	50	35	30
F Kerosene	40	.001	85	60	60	55	50	...	45	25	...	50	35	30
G Light gas oil	30	.002	70	50	60	50	50	...	40	25	70	45	30	30
H Heavy gas oil	22	.003	60	45	55	50	45	50	40	20	70	40	30	20
J Reduced crude	17	.005	...	...	55	45	40							
K Heavy fuel oil (tar)	10	.005	...	...	50	40	35							

Fouling factor, water side 0.0002; heating or cooling streams are shown at top of columns as C, D, F, G, etc.; to convert British thermal units per hour-square foot–degrees Fahrenheit to joules per square meter–second–kelvins, multiply by 5.6783; to convert hours per square foot–degree Fahrenheit–British thermal units to square meters per second–kelvin–joules, multiply by 0.1761.

*Cooler, water-cooled, rates are about 5 percent lower.

†With heavy gas oil (H) as heating medium, rates are about 5 percent lower.

TABLE 10-12 Overall Coefficients for Air-Cooled Exchangers on Bare-Tube Basis

Btu/(°F·ft²·h)

Condensing	Coefficient	Liquid cooling	Coefficient
Ammonia	110	Engine-jacket water	125
Freon-12	70	Fuel oil	25
Gasoline	80	Light gas oil	65
Light hydrocarbons	90	Light hydrocarbons	85
Light naphtha	75	Light naphtha	70
Heavy naphtha	65	Reformer liquid streams	70
Reformer reactor effluent	70	Residuum	15
Low-pressure steam	135	Tar	7
Overhead vapors	65		

Gas cooling	Operating pressure, lb./sq. in. gage	Pressure drop, lb./sq. in.	Coefficient
Air or flue gas	50	0.1 to 0.5	10
	100	2	20
	100	5	30
Hydrocarbon gas	35	1	35
	125	3	55
	1000	5	80
Ammonia reactor stream	....	...	85

Bare-tube external surface is 0.262 ft²/ft.
Fin-tube surface/bare-tube surface ratio is 16.9.
To convert British thermal units per hour-square foot-degrees Fahrenheit to joules per square meter-second-kelvins, multiply by 5.6783; to convert pounds-force per square inch to kilopascals, multiply by 6.895.

Conductive Heat Transfer Heat-transfer equipment in which heat is transferred by conduction is so constructed that the solids load (burden) is separated from the heating medium by a wall.

For a high proportion of applications, Δt is the log-mean temperature difference. Values of U_{co} are reported in Secs. 11, 15, 17, and 19. A *predictive* equation for U_{co} is

$$U_{co} = \left(\frac{h}{h - 2ca/d_m}\right)\left(\frac{2ca}{d_m}\right) \tag{10-161}$$

where h = wall film coefficient, c = volumetric heat capacity, d_m = depth of the burden, and α = thermal diffusivity. Relevant thermal properties of various materials are given in Table 10-16. For details of terminology, equation development, numerical values of terms in typical equipment and use, see Holt [*Chem. Eng.*, **69**, 107 (Jan. 8, 1962)].

Equation (10-161) is applicable to burdens in the solid, liquid, or gaseous phase, either static or in laminar motion; it is applicable to solidification equipment and to divided-solids equipment such as metal belts, moving trays, stationary vertical tubes, and stationary-shell fluidizers.

Fixed (or packed) bed operation occurs when the fluid velocity is low or the particle size is large so that fluidization does not occur. For such operation, Jakob (*Heat Transfer*, vol. 2, Wiley, New York, 1957) gives

$$hD_t/k = b_1 b D_t^{0.17}(D_p G/\mu)^{0.83}(c\mu/k) \tag{10-162}$$

where b_1 = 1.22 (SI) or 1.0 (U.S. customary), $h = U_{co}$ = overall coefficient between the inner container surface and the fluid stream,

TABLE 10-13 Panel Coils Immersed in Liquid: Overall Average Heat-Transfer Coefficients*

U expressed in Btu/(h·ft²·°F)

Hot side	Cold side	Clean-surface coefficients: Natural convection	Clean-surface coefficients: Forced convection	Design coefficients, considering usual fouling in this service: Natural convection	Design coefficients, considering usual fouling in this service: Forced convection
Heating applications:					
Steam	Watery solution	250–500	300–550	100–200	150–275
Steam	Light oils	50–70	110–140	40–45	60–110
Steam	Medium lube oil	40–60	100–130	35–40	50–100
Steam	Bunker C or No. 6 fuel oil	20–40	70–90	15–30	60–80
Steam	Tar or asphalt	15–35	50–70	15–25	40–60
Steam	Molten sulfur	35–45	45–55	20–35	35–45
Steam	Molten paraffin	35–45	45–55	25–35	40–50
Steam	Air or gases	2–4	5–10	1–3	4–8
Steam	Molasses or corn sirup	20–40	70–90	15–30	60–80
High temperature hot water	Watery solutions	115–140	200–250	70–100	110–160
High temperature heat-transfer oil	Tar or asphalt	12–30	45–65	10–20	30–50
Dowtherm or Aroclor	Tar or asphalt	15–30	50–60	12–20	30–50
Cooling applications:					
Water	Watery solution	110–135	195–245	65–95	105–155
Water	Quench oil	10–15	25–45	7–10	15–25
Water	Medium lube oil	8–12	20–30	5–8	10–20
Water	Molasses or corn sirup	7–10	18–26	4–7	8–15
Water	Air or gases	2–4	5–10	1–3	4–8
Freon or ammonia	Watery solution	35–45	60–90	20–35	40–60
Calcium or sodium brine	Watery solution	100–120	175–200	50–75	80–125

*Tranter Manufacturing, Inc.

NOTE: To convert British thermal units per hour–square foot–degrees Fahrenheit to joules per square meter–second–kelvins, multiply by 5.6783.

TABLE 10-14 Jacketed Vessels: Overall Coefficients

Jacket fluid	Fluid in vessel	Wall material	Overall U* Btu/(h·ft²·°F)	J/(m²·s·K)
Steam	Water	Stainless steel	150–300	850–1700
Steam	Aqueous solution	Stainless steel	80–200	450–1140
Steam	Organics	Stainless steel	50–150	285– 850
Steam	Light oil	Stainless steel	60–160	340– 910
Steam	Heavy oil	Stainless steel	10– 50	57– 285
Brine	Water	Stainless steel	40–180	230–1625
Brine	Aqueous solution	Stainless steel	35–150	200– 850
Brine	Organics	Stainless steel	30–120	170– 680
Brine	Light oil	Stainless steel	35–130	200– 740
Brine	Heavy oil	Stainless steel	10– 30	57– 170
Heat-transfer oil	Water	Stainless steel	50–200	285–1140
Heat-transfer oil	Aqueous solution	Stainless steel	40–170	230– 965
Heat-transfer oil	Organics	Stainless steel	30–120	170– 680
Heat-transfer oil	Light oil	Stainless steel	35–130	200– 740
Heat-transfer oil	Heavy oil	Stainless steel	10– 40	57– 230
Steam	Water	Glass-lined CS	70–100	400– 570
Steam	Aqueous solution	Glass-lined CS	50– 85	285– 480
Steam	Organics	Glass-lined CS	30– 70	170– 400
Steam	Light oil	Glass-lined CS	40– 75	230– 425
Steam	Heavy oil	Glass-lined CS	10– 40	57– 230
Brine	Water	Glass-lined CS	30– 80	170– 450
Brine	Aqueous solution	Glass-lined CS	25– 70	140– 400
Brine	Organics	Glass-lined CS	20– 60	115– 340
Brine	Light oil	Glass-lined CS	25– 65	140– 370
Brine	Heavy oil	Glass-lined CS	10– 30	57– 170
Heat-transfer oil	Water	Glass-lined CS	30– 80	170– 450
Heat-transfer oil	Aqueous solution	Glass-lined CS	25– 70	140– 400
Heat-transfer oil	Organics	Glass-lined CS	25– 65	140– 370
Heat-transfer oil	Light oil	Glass-lined CS	20– 70	115– 400
Heat-transfer oil	Heavy oil	Glass-lined CS	10– 35	57– 200

*Values listed are for moderate nonproximity agitation. CS = carbon steel.

D_p = particle diameter, D_t = vessel diameter, b = coefficient dependent on D_p/D_t (Fig. 10-40; note that D_p/D_t has units of inches per foot in the figure), G = superficial mass velocity, k = fluid thermal conductivity, μ = fluid viscosity, and c = fluid specific heat. Other correlations are those of Leva [*Ind. Eng. Chem.*, **42**, 2498 (1950)]:

$$h = 0.813 \frac{k}{D_t} e^{-6D_p/D_t} \left(\frac{D_pG}{\mu}\right)^{0.90} \quad \text{for } \frac{D_p}{D_t} < 0.35 \qquad (10\text{-}163a)$$

$$h = 0.125 \frac{k}{D_t} \left(\frac{D_pG}{\mu}\right)^{0.75} \quad \text{for } 0.35 < \frac{D_p}{D_t} < 0.60 \qquad (10\text{-}163b)$$

and Calderbank and Pogerski [*Trans. Inst. Chem. Eng.* (London), **35**, 195 (1957)]:

$$hD_p/k = 3.6(D_pG/\mu\epsilon_v)^{0.365} \qquad (10\text{-}164)$$

where ϵ_v = fraction voids in the bed.

TABLE 10-15 External Coils; Typical Overall Coefficients*

U expressed in Btu/(h·ft²·°F)

Type of coil	Coil spacing, in.†	Fluid in coil	Fluid in vessel	Temp. range, °F.	U‡ without cement	U with heat-transfer cement
⅜ in. o.d. copper tubing attached with bands at 24-in. spacing	2	5 to 50 lb./sq. in. gage steam	Water under light agitation	158–210	1-5	42-46
	3⅛			158–210	1-5	50-53
	6¼			158–210	1-5	60-64
	12½ or greater			158–210	1-5	69-72
⅜ in. o.d. copper tubing attached with bands at 24-in. spacing	2	50 lb./sq. in. gage steam	No. 6 fuel oil under light agitation	158–258	1-5	20-30
	3⅛			158–258	1-5	25-38
	6¼			158–240	1-5	30-40
	12½ or greater			158–238	1-5	35-46
Panel coils		50 lb./sq. in. gage steam	Boiling water	212	29	48-54
		Water	Water	158–212	8–30	19-48
		Water	No. 6 fuel oil	228–278	6–15	24-56
			Water	130–150	7	15
			No. 6 fuel oil	130–150	4	9-19

*Data courtesy of Thermon Manufacturing Co.
†External surface of tubing or side of panel coil facing tank.
‡For tubing, the coefficients are more dependent upon tightness of the coil against the tank than upon either fluid. The low end of the range is recommended.
NOTE: To convert British thermal units per hour–square foot–degrees Fahrenheit to joules per square meter–second–kelvins, multiply by 5.6783; to convert inches to meters, multiply by 0.0254; and to convert pounds-force per square inch to kilopascals, multiply by 6.895.

TABLE 10-16 Thermal Properties of Various Materials as Affecting Conductive Heat Transfer

Material	Thermal conductivity, B.t.u./(hr.)(sq. ft.)(°F./ft.)	Volume specific heat, B.t.u./(cu. ft.)(°F.)	Thermal diffusivity, sq. ft./hr.
Air	0.0183	0.016	1.143
Water	0.3766	62.5	0.0755
Double steel plate, sand divider	0.207	19.1	0.0108
Sand	0.207	19.1	0.0108
Powdered iron	0.0533	12.1	0.0044
Magnetite iron ore . . .	0.212	63	0.0033
Aerocat catalysts . . .	0.163	20	0.0062
Table salt	0.168	12.6	0.0133
Bone char . . .	0.0877	16.9	0.0051
Pitch coke . . .	0.333	16.2	0.0198
Phenolformaldehyde resin granules	0.0416	10.5	0.0042
Phenolformaldehyde resin powder .	0.070	10	0.0070
Powdered coal	0.070	15	0.0047

To convert British thermal units per hour–square foot–degrees Fahrenheit to joules per meter–second–kelvins, multiply by 1.7307; to convert British thermal units per cubic foot–degrees Fahrenheit to joules per cubic meter–kelvins, multiply by $(6.707)(10^4)$; and to convert square feet per hour to square meters per second, multiply by $(2.581)(10^{-5})$.

A technique for calculating radial temperature gradients in a packed bed is given by Smith (*Chemical Engineering Kinetics*, McGraw-Hill, New York, 1956).

Fluidization occurs when the fluid flow rate is great enough so that the pressure drop across the bed equals the weight of the bed. As stated previously, the solids film thickness adjacent to the wall d_m is difficult to measure and/or predict. Wen and Fau [*Chem. Eng.*, **64**(7), 254 (1957)] give for *external walls:*

$$h = bk(c_s\rho_s)^{0.4}(G\eta/\mu N_f)^{0.36} \tag{10-165a}$$

where $b = 0.29$ (SI) or 11.6 (U.S. customary), c_s = heat capacity of solid, ρ_s = particle density, η = fluidization efficiency (Fig. 10-41), and N_f = bed expansion ratio (Fig. 10-42). *For internal walls,* Wen and Fau give

$$h_i = bhG^{-0.37} \tag{10-165b}$$

where $b = 0.78$ (SI) or 9 (U.S. customary), h_i is the coefficient for internal walls, and h is calculated from Eq. (10-165a). G_{mf}, the minimum fluidizing velocity, is defined by

$$G_{mf} = \frac{b\rho_g^{1.1}(\rho_s - \rho_g)^{0.9}D_p^2}{\mu} \tag{10-165c}$$

where $b = (1.23)(10^{-2})$ (SI) or $(5.23)(10^5)$ (U.S. customary).

Wender and Cooper [*Am. Inst. Chem. Eng. J.*, **4**, 15 (1958)] developed an empirical correlation for *internal walls:*

$$\frac{hD_p/k}{1-\epsilon_v}\left(\frac{k}{c\rho}\right)^{0.43} = bC_R\left(\frac{D_pG}{\mu}\right)^{0.23}\left(\frac{c_s}{c_g}\right)^{0.80}\left(\frac{\rho_s}{\rho_g}\right)^{0.66} \tag{10-166a}$$

where $b = (3.51)(10^{-4})$ (SI) or 0.033 (U.S. customary) and C_R = correction for displacement of the immersed tube from the axis of the vessel (see the reference). For external walls:

$$\frac{hD_p}{k_g(1-\epsilon_v)(c_s\rho_s/c_g\rho_g)} = f(1 + 7.5e^{-x}) \tag{10-166b}$$

where $x = 0.44L_Hc_s/D_tc_g$ and f is given by Fig. 10-43. An important feature of this equation is inclusion of the ratio of bed depth to vessel diameter L_H/D_t.

For **dilute fluidized beds** on the shell side of an unbaffled tubular bundle Genetti and Knudsen [*Inst. Chem. Eng. (London) Symp. Ser.* 3,172 (1968)] obtained the relation:

$$\frac{hD_p}{k} = \frac{5\phi(1-\epsilon_v)}{\left[1 + \frac{580}{N_{Re}}\left(\frac{k_s}{D_p^{1.5}c_s\rho_s g^{0.5}}\right)\left(\frac{\rho_s}{\rho_g}\right)^{1.1}\left(\frac{G_{mf}}{G}\right)^{7/3}\right]^2} \tag{10-167a}$$

where ϕ = particle surface area per area of sphere of same diameter. When particle transport occurred through the bundle, the heat-trans-

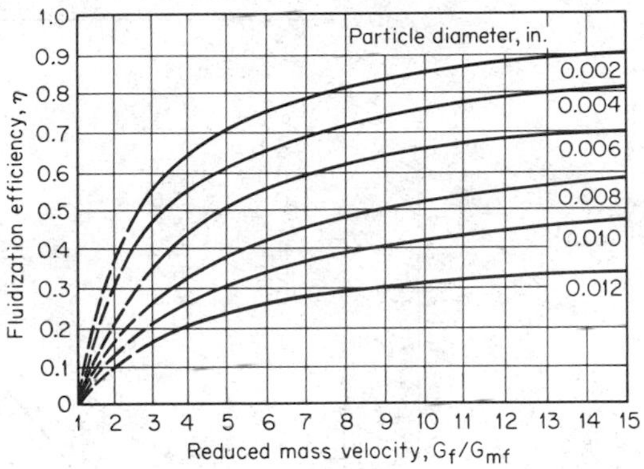

FIG. 10-41 Fluidization efficiency.

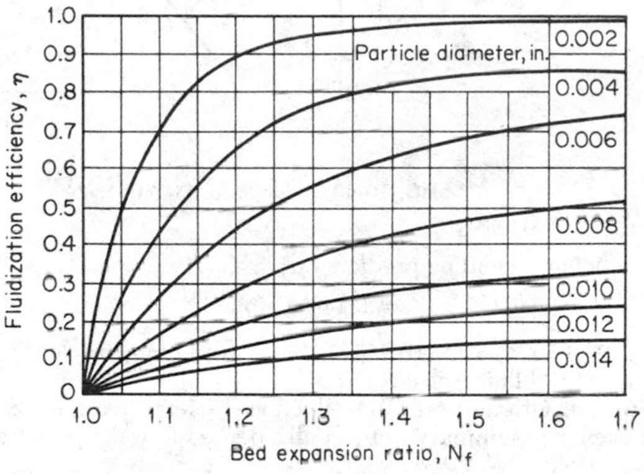

FIG. 10-42 Bed expansion ratio.

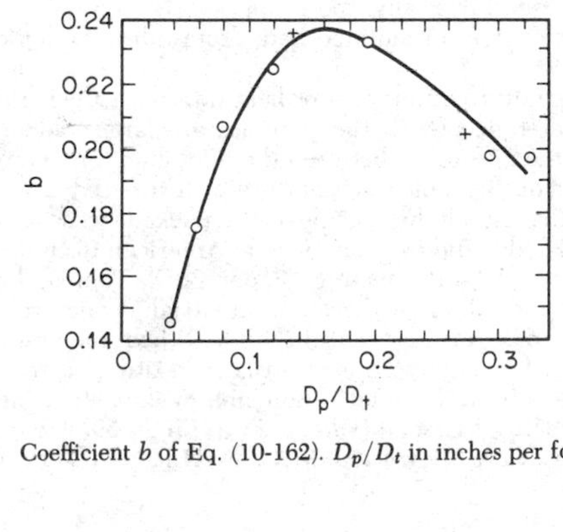

FIG. 10-40 Coefficient b of Eq. (10-162). D_p/D_t in inches per foot.

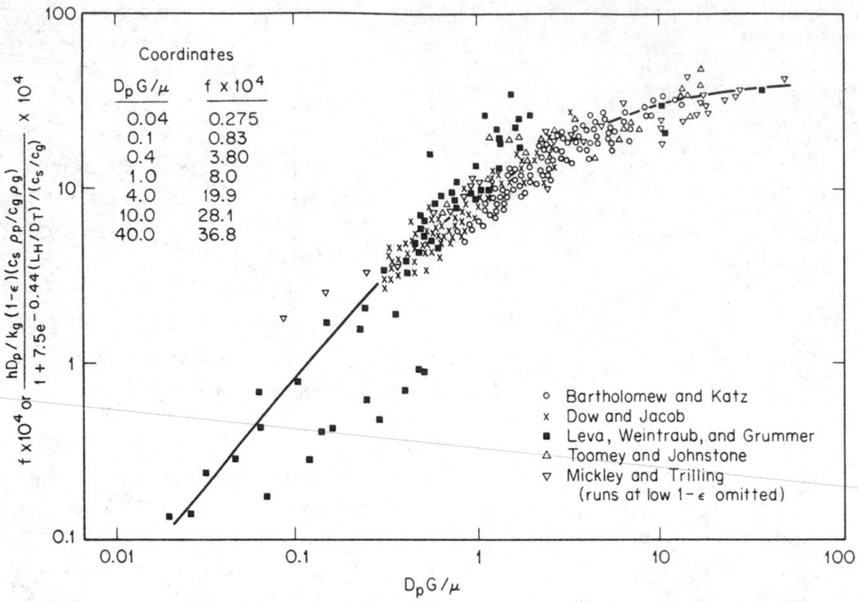

FIG. 10-43 *f* factor for Eq. (10-166*b*).

fer coefficients could be predicted by

$$j_H = 0.14(N_{Re}/\phi)^{-0.68} \qquad (10\text{-}167b)$$

In Eqs. (10-167*a*) and (10-167*b*), N_{Re} is based on particle diameter and superficial fluid velocity.

Zenz and Othmer (see "Introduction: General References") give an excellent summary of fluidized bed-to-wall heat-transfer investigations.

Solidification involves heavy heat loads transferred essentially at a steady temperature difference. It also involves the varying values of liquid- and solid-phase thickness and thermal diffusivity. When these are substantial and/or in the case of a liquid flowing over a changing solid layer interface, Siegel and Savino (ASME Paper 67-WA/Ht-34, November 1967) offer equations and charts for prediction of the layer-growth time. For solidification (or melting) of a slab or a semi-infinite bar, initially at its transition temperature, the position of the interface is given by the one-dimensional Newmann's solution given in Carslaw and Jaeger (*Conduction of Heat in Solids*, Clarendon Press, Oxford, 1959).

Later work by Hashem and Sliepcevich [*Chem. Eng. Prog.*, **63**, *Symp. Ser.* 79, 35, 42 (1967)] offers more accurate second-order finite-difference equations.

The heat-transfer rate is found to be substantially higher under conditions of **agitation.** The heat transfer is usually said to occur by combined conductive and convective modes. A discussion and explanation are given by Holt [*Chem. Eng.*, **69**(1), 110 (1962)]. Prediction of U_{co} by Eq. (10-161) can be accomplished by replacing α by α_e, the effective thermal diffusivity of the bed. To date so little work has been performed in evaluating the effect of mixing parameters that few predictions can be made. However, for agitated liquid-phase devices Eq. (10-79) is applicable. Holt (loc. cit.) shows that this equation can be converted for solids heat transfer to yield

$$U_{co} = a'c_sD_t^{-0.3}N^{0.7}(\cos\omega)^{0.2} \qquad (10\text{-}168)$$

where D_t = agitator or vessel diameter; N = turning speed, r/min; ω = effective angle of repose of the burden; and a' is a proportionality constant. This is applicable for such devices as agitated pans, agitated kettles, spiral conveyors, and rotating shells.

The solids passage time through **rotary devices** is given by Saemann [*Chem. Eng. Prog.*, **47**, 508, (1951)]:

$$\theta = 0.318L\sin\omega/S_rND_t \qquad (10\text{-}169a)$$

and by Marshall and Friedman [*Chem. Eng. Prog.*, **45**, 482–493, 573–588 (1949)]:

$$\theta = (0.23L/S_rN^{0.9}D_t) \pm (0.6BLG/F_a) \qquad (10\text{-}169b)$$

where the second term of Eq. (10-169*b*) is positive for counterflow of air, negative for concurrent flow, and zero for indirect rotary shells. From these equations a predictive equation is developed for rotary-shell devices, which is analogous to Eq. (10-168):

$$U_{co} = \frac{b'c_sD_tN^{0.9}Y}{(\Delta t)L\sin\omega} \qquad (10\text{-}170)$$

where θ = solids-bed passage time through the shell, min; S_r = shell slope; L = shell length; Y = percent fill; and b' is a proportionality constant.

Vibratory devices which constantly agitate the solids bed maintain a relatively constant value for U_{co} such that

$$U_{co} = a'c_s\alpha_e \qquad (10\text{-}171)$$

with U_{co} having a nominal value of 114 J/(m²·s·K) [20 Btu/(h·ft²·°F)].

Contactive (Direct) Heat Transfer Contactive heat-transfer equipment is so constructed that the particulate burden in solid phase is directly exposed to and permeated by the heating or cooling medium (Sec. 20). The carrier may either heat or cool the solids. A large amount of the industrial heat processing of solids is effected by this mechanism. Physically, these can be classified into packed beds and various degrees of agitated beds from dilute to dense fluidized beds.

The temperature difference for heat transfer is the log-mean temperature difference when the particles are large and/or the beds packed, or the difference between the inlet fluid temperature t_3 and average exhausting fluid temperature t_4, expressed Δ_3t_4, for small particles. The use of the log mean for packed beds has been confirmed by Thodos and Wilkins (Second American Institute of Chemical Engineers–IIQPR Meeting, Paper 30D, Tampa, May 1968). When fluid and solid flow directions are axially concurrent and particle size is small, as in a vertical-shell fluid bed, the temperature of the exiting solids t_2 (which is also that of exiting gas t_4) is used as Δ_3t_2, as shown by Levenspiel, Olson, and Walton [*Ind. Eng. Chem.*, **44**, 1478 (1952)], Marshall [*Chem. Eng. Prog.*, **50**, *Monogr. Ser.* 2, 77 (1954)], Leva (*Fluidization*, McGraw-Hill, New York, 1959), and

Holt (Fourth Int. Heat Transfer Conf. Paper 11, American Institute of Chemical Engineers–American Society of Mechanical Engineers, Buffalo, 1960). This temperature difference is also applicable for well-fluidized beds of small particles in cross-flow as in various vibratory carriers.

The **packed-bed-to-fluid heat-transfer coefficient** has been investigated by Baumeister and Bennett [*Am. Inst. Chem. Eng. J.*, **4,** 69 (1958)], who proposed the equation

$$j_H = (h/cG)(c\mu/k)^{2.3} = aN_{\mathrm{Re}}^m \tag{10-172}$$

where N_{Re} is based on particle diameter and superficial fluid velocity. Values of a and m are as follows:

D_t/D_p (dimensionless)	a	m
10.7	1.58	−0.40
16.0	0.95	−0.30
25.7	0.92	−0.28
>30	0.90	−0.28

Glaser and Thodos [*Am. Inst. Chem. Eng. J.*, **4,** 63 (1958)] give a correlation involving individual particle shape and bed porosity. Kunii and Suzuki [*Int. J. Heat Mass Transfer*, **10,** 845 (1967)] discuss heat and mass transfer in packed beds of fine particles.

Particle-to-fluid heat-transfer coefficients in gas **fluidized beds** are predicted by the relation (Zenz and Othmer, op. cit.)

$$\frac{hD_p}{k} = 0.017(D_pG_{mf}/\mu)^{1.21} \tag{10-173a}$$

where G_{mf} is the superficial mass velocity at incipient fluidization.

A more general equation is given by Frantz [*Chem. Eng.*, **69**(20), 89 (1962)]:

$$hD_p/k = 0.015(D_pG/\mu)^{1.6}(c\mu/k)^{0.67} \tag{10-173b}$$

where h is based on true gas temperature.

Bed-to-wall coefficients in dilute-phase transport generally can be predicted by an equation of the form of Eq. (10-50). For example, Bonilla et al. (American Institute of Chemical Engineers Heat Transfer Symp., Altantic City, N.J., December 1951) found for 1- to 2-μm chalk particles in water up to 8 percent by volume that the coefficient on Eq. (10-50) is 0.029 where k, ρ, and c were arithmetic weighted averages and the viscosity was taken equal to the coefficient of rigidity. Farber and Morley [*Ind. Eng. Chem.*, **49,** 1143 (1957)] found the coefficient on Eq. (10-50) to be 0.025 for the upward flow of air transporting silica-alumina catalyst particles at rates less than 2 kg solids/kg air (2 lb solids/lb air). Physical properties used were those of the transporting gas. See Zenz and Othmer (op. cit.) for additional details covering wider porosity ranges.

The thermal performance of **cylindrical rotating shell** units is based upon a volumetric heat-transfer coefficient

$$U_{ct} = \frac{Q}{V_r(\Delta t)} \tag{10-174a}$$

where V_r = volume. This term indirectly includes an area factor so that thermal performance is governed by a cross-sectional area rather than by a heated area. Use of the heated area is possible, however:

$$U_{ct} = \frac{Q}{(\Delta_3 t_2)A} \quad \text{or} \quad \frac{Q}{(\Delta_3 t_4)A} \tag{10-174b}$$

For **heat transfer directly to solids,** predictive equations give directly the volume V or the heat-transfer area A, as determined by heat balance and airflow rate. For devices with gas flow normal to a fluidized-solids bed,

$$A = \frac{Q}{\Delta t_p(c\rho_g)(F_g)} \tag{10-175}$$

where $\Delta t_p = \Delta_3 t_4$ as explained above, $c\rho$ = volumetric specific heat, and F_g = gas flow rate. For air, $c\rho$ at normal temperature and pressure is about 1100 J/(m³·K) [0.0167 Btu/(ft³·°F)]; so

$$A = \frac{bQ}{(\Delta_3 t_4)F_g} \tag{10-176}$$

where b = 0.0009 (SI) or 60 (U.S. customary). Another such equation, for stationary vertical-shell and some horizontal rotary-shell and pneumatic-transport devices in which the gas flow is parallel with and directionally concurrent with the fluidized bed, is the same as Eq. (10-176) with $\Delta_3 t_4$ replaced by $\Delta_3 t_2$. If the operation involves drying or chemical reaction, the heat load Q is much greater than for sensible-heat transfer only. Also, the gas flow rate to provide moisture carry-off and stoichiometric requirements must be considered and simultaneously provided. A good treatise on the latter is given by Pinkey and Plint (*Miner. Process.*, June 1968, p. 17).

Evaporative cooling is a special patented technique that often can be advantageously employed in cooling solids by contactive heat transfer. The drying operation is terminated *before* the desired final moisture content is reached, and solids temperature is at a moderate value. The cooling operation involves contacting the burden (preferably fluidized) with air at normal temperature and pressure. The air adiabatically absorbs and carries off a large part of the moisture and, in doing so, picks up heat from the warm (or hot) solids particles to supply the latent heat demand of evaporation. For entering solids at temperatures of 180°C (350°F) and less with normal heat-capacity values of 0.85 to 1.0 kJ/(kg·K) [0.2 to 0.25 Btu/(lb·°F)], the effect can be calculated by:

1. Using 285 m³ (1000 ft³) of airflow at normal temperature and pressure at 40 percent relative humidity to carry off 0.45 kg (1 lb) of water [latent heat 2326 kJ/kg (1000 Btu/lb)] and to lower temperature by 22 to 28°C (40 to 50°F).
2. Using the lowered solids temperature as t_3 and calculating the remainder of the heat to be removed in the regular manner by Eq. (10-176). The required air quantity for (2) must be equal to or greater than that for (1).

When the solids heat capacity is higher (as is the case for most organic materials), the temperature reduction is inversely proportional to the heat capacity.

A nominal result of this technique is that the required airflow rate and equipment size is about two-thirds of that when evaporative cooling is not used. See Sec. 20 for equipment available.

Convective Heat Transfer Equipment using the true convective mechanism when the heated particles are mixed with (and remain with) the cold particles is used so infrequently that performance and sizing equations are not available. Such a device is the pebble heater as described by Norton (*Chem. Metall. Eng.*, July 1946). For operation data, see Sec. 9.

Convective heat transfer is often used as an adjunct to other modes, particularly to the conductive mode. It is often more convenient to consider the agitative effect a performance-improvement influence on the thermal diffusivity factor α, modifying it to α_e, the effective value.

A *pseudo-convective heat-transfer* operation is one in which the heating gas (generally air) is passed over a bed of solids. Its use is almost exclusively limited to drying operations (see Sec. 20, tray and shelf dryers). The operation, sometimes termed direct, is more akin to the conductive mechanism. For this operation, Tsao and Wheelock [*Chem. Eng.*, **74**(13), 201 (1967)] predict the heat-transfer coefficient when radiative and conductive effects are absent by

$$h = bG^{0.8} \tag{10-177}$$

where b = 14.31 (SI) or 0.0128 (U.S. customary), h = convective heat transfer, and G = gas flow rate.

The **drying rate** is given by

$$K_{cv} = \frac{h(T_d - T_w)}{\lambda} \tag{10-178}$$

where K_{cv} = drying rate, for constant-rate period, kg/(m²·s) [lb/(h·ft²)]; T_d and T_w = respective dry-bulb and wet-bulb temperatures of the air; and λ = latent heat of evaporation at temperature T_w. Note here that the temperature-difference determination of the operation is a simple linear one and of a steady-state nature. Also note that the operation is a function of the airflow rate. Further, the solids are granular with a fairly uniform size, have reasonable capillary voids, are of a firm texture, and have the particle surface wetted.

The coefficient h is also used to predict (in the constant-rate period) the total overall **air-to-solids heat-transfer coefficient** U_{cv} by

$$1/U_{cv} = 1/h + x/k \qquad (10\text{-}179)$$

where k = solids thermal conductivity and x is evaluated from

$$x = \frac{z(X_c - X_o)}{X_c - X_e} \qquad (10\text{-}179a)$$

where z = bed (or slab) thickness and is the total thickness when drying and/or heat transfer is from one side only but is one-half of the thickness when drying and/or heat transfer is simultaneously from both sides; X_o, X_c, and X_e are respectively the initial (or feedstock), critical, and equilibrium (with the drying air) moisture contents of the solids, all in kg H_2O/kg dry solids (lb H_2O/lb dry solids). This coefficient is used to predict the *instantaneous* drying rate

$$-\frac{W}{A}\frac{dX}{d\theta} = \frac{U_{cv}(T_d - T_w)}{\lambda} \qquad (10\text{-}180)$$

By rearrangement, this can be made into a design equation as follows:

$$A = -\frac{W\lambda(dX/d\theta)}{U_{cv}(T_d - T_w)} \qquad (10\text{-}181)$$

where W = weight of dry solids in the equipment, λ = latent heat of evaporation, and θ = drying time. The reader should refer to the full reference article by Tsao and Wheelock (loc. cit.) for other solids conditions qualifying the use of these equations.

Radiative Heat Transfer Heat-transfer equipment using the radiative mechanism for divided solids is constructed as a "table" which is stationary, as with trays, or moving, as with a belt, and/or agitated, as with a vibrated pan, to distribute and expose the burden in a plane parallel to (but not in contact with) the plane of the radiant-heat sources. Presence of air is not necessary (see Sec. 20 for vacuum-shelf dryers and Sec. 17 for resublimation). In fact, if air in the intervening space has a high humidity or CO_2 content, it acts as an energy absorber, thereby depressing the performance.

For the radiative mechanism, the temperature difference is evaluated as

$$\Delta t = T_e^4 - T_r^4 \qquad (10\text{-}182)$$

where T_e = absolute temperature of the radiant-heat source, K (°R); and T_r = absolute temperature of the bed of divided solids, K (°R).

Numerical values for U_{ra} for use in the general design equation may be calculated from experimental data by

$$U_{ra} = \frac{Q}{A(T_e^4 - T_r^4)} \qquad (10\text{-}183)$$

The literature to date offers practically no such values. However, enough proprietary work has been performed to present a reliable evaluation for the comparison of mechanisms (see "Introduction: Modes of Heat Transfer").

For the radiative mechanism of heat transfer to solids, the rate equation for parallel-surface operations is

$$q_{ra} = b(T_e^4 - T_r^4)i_f \qquad (10\text{-}184)$$

where b = (5.67)(10^{-8}) (SI) or (0.172)(10^{-8}) (U.S. customary), q_{ra} = radiative heat flux, and i_f = an interchange factor which is evaluated from

$$1/i_f = 1/e_s - 1/e_r - 1 \qquad (10\text{-}184a)$$

where e_s = coefficient of emissivity of the source and e_r = "emissivity" (or "absorptivity") of the receiver, which is the divided-solids bed. For the emissivity values, particularly of the heat source e_s, an important consideration is the wavelength at which the radiant source emits as well as the flux density of the emission. Data for these values are available from Polentz [*Chem. Eng.*, **65**(7), 137; (8), 151 (1958)] and Adlam (*Radiant Heating*, Industrial Press, New York, p. 40). Both give radiated flux density versus wavelength at varying temperatures. Often, the seemingly cooler but longer wavelength source is the better selection.

Emitting sources are (1) pipes, tubes, and platters carrying steam, 2100 kPa (300 lbf/in^2); (2) electrical-conducting glass plates, 150 to 315°C (300 to 600°F) range; (3) light-bulb type (tungsten-filament resistance heater); (4) modules of refractory brick for gas burning at high temperatures and high fluxes; and (5) modules of quartz tubes, also operable at high temperatures and fluxes. For some emissivity values see Table 10-17.

For *predictive work*, where U_{ra} is desired for sizing, this can be obtained by dividing the flux rate q_{ra} by Δt:

$$U_{ra} = q_{ra}/(T_e^4 - T_r^4) = i_f b \qquad (10\text{-}185)$$

where b = (5.67)(10^{-8}) (SI) or (0.172)(10^{-8}) (U.S. customary). Hence:

$$A = \frac{Q}{U_{ra}(T_e^4 - T_r^4)} \qquad (10\text{-}186)$$

where A = bed area of solids in the equipment.

Important considerations in the application of the foregoing equations are:

1. Since the temperature of the emitter is generally known (preselected or readily determined in an actual operation), the absorptivity value e_r is the unknown. This absorptivity is partly a measure of the ability of radiant heat to penetrate the body of a solid particle (or a moisture film) instantly, as compared with diffusional heat transfer by conduction. Such instant penetration greatly reduces processing time and case-hardening effects. Moisture release and other mass transfer, however, still progress by diffusional means.
2. In one of the major applications of radiative devices (drying), the surface-held moisture is a good heat absorber in the 2- to 7-μm wavelength range. Therefore, the absorptivity, color, and nature of the solids are of little importance.
3. For drying, it is important to provide a small amount of venting air to carry away the water vapor. This is needed for two reasons. First, water vapor is a good absorber of 2- to 7-μm energy. Second, water-vapor accumulation depresses further vapor release by the solids. If the air over the solids is kept fairly dry by venting, very little heat is carried off, because dry air does not absorb radiant heat.
4. For some of the devices, when the overall conversion efficiency has been determined, the application is primarily a matter of computing the required heat load. It should be kept in mind, however, that there are two conversion efficiencies that must be differentiated. One measure of efficiency is that with which the source converts input energy to output radiated energy. The other is the overall efficiency that measures the proportion of input energy that is actually absorbed by the solids. This latter is, of course, the one that really matters.

Other applications of radiant-heat processing of solids are the toasting, puffing, and baking of foods and the low-temperature roasting and preheating of plastic powder or pellets. Since the determination of heat loads for these operations is not well established, bench and pilot tests are generally necessary. Such processes require a fast input of heat and higher heat fluxes than can generally be provided by indirect equipment. Because of this, infrared-equipment size and space requirements are often much lower.

Although direct contactive heat transfer can provide high temperatures and heat concentrations and at the same time be small in size, its use may not always be preferable because of undesired side effects such as drying, contamination, case hardening, shrinkage, off color, and dusting.

When radiating and receiving surfaces are not in parallel, as in rotary-kiln devices, and the solids burden bed may be only intermittently exposed and/or agitated, the calculation and procedures become very complex, with photometric methods of optics requiring

TABLE 10-17 Normal Total Emissivity of Various Surfaces

A. Metals and Their Oxides

Surface	t, °F.*	Emissivity*
Aluminum		
Highly polished plate, 98.3% pure	440–1070	0.039–0.057
Polished plate	73	0.040
Rough plate	78	0.055
Oxidized at 1110°F	390–1110	0.11–0.19
Aluminum-surfaced roofing	100	0.216
Calorized surfaces, heated at 1110°F.		
Copper	390–1110	0.18–0.19
Steel	390–1110	0.52–0.57
Brass		
Highly polished:		
73.2% Cu, 26.7% Zn	476–674	0.028–0.031
62.4% Cu, 36.8% Zn, 0.4% Pb, 0.3% Al	494–710	0.033–0.037
82.9% Cu, 17.0% Zn	530	0.030
Hard rolled, polished:		
But direction of polishing visible	70	0.038
But somewhat attacked	73	0.043
But traces of stearin from polish left on	75	0.053
Polished	100–600	0.096
Rolled plate, natural surface	72	0.06
Rubbed with coarse emery	72	0.20
Dull plate	120–660	0.22
Oxidized by heating at 1110°F	390–1110	0.61–0.59
Chromium; see Nickel Alloys for Ni-Cr steels	100–1000	0.08–0.26
Copper		
Carefully polished electrolytic copper	176	0.018
Commercial, emeried, polished, but pits remaining	66	0.030
Commercial, scraped shiny but not mirror-like	72	0.072
Polished	242	0.023
Plate, heated long time, covered with thick oxide layer	77	0.78
Plate heated at 1110°F	390–1110	0.57
Cuprous oxide	1470–2010	0.66–0.54
Molten copper	1970–2330	0.16–0.13
Gold		
Pure, highly polished	440–1160	0.018–0.035
Iron and steel		
Metallic surfaces (or very thin oxide layer):		
Electrolytic iron, highly polished	350–440	0.052–0.064
Polished iron	800–1880	0.144–0.377
Iron freshly emeried	68	0.242
Cast iron, polished	392	0.21
Wrought iron, highly polished	100–480	0.28
Cast iron, newly turned	72	0.435
Polished steel casting	1420–1900	0.52–0.56
Ground sheet steel	1720–2010	0.55–0.61
Smooth sheet iron	1650–1900	0.55–0.60
Cast iron, turned on lathe	1620–1810	0.60–0.70
Oxidized surfaces:		
Iron plate, pickled, then rusted red	68	0.612
Completely rusted	67	0.685
Rolled sheet steel	70	0.657
Oxidized iron	212	0.736
Cast iron, oxidized at 1100°F	390–1110	0.64–0.78
Steel, oxidized at 1100°F	390–1110	0.79
Smooth oxidized electrolytic iron	260–980	0.78–0.82
Iron oxide	930–2190	0.85–0.89
Rough ingot iron	1700–2040	0.87–0.95
Sheet steel, strong rough oxide layer	75	0.80
Dense shiny oxide layer	75	0.82
Cast plate:		
Smooth	73	0.80
Rough	73	0.82
Cast iron, rough, strongly oxidized	100–480	0.95
Wrought iron, dull oxidized	70–680	0.94
Steel plate, rough	100–700	0.94–0.97
High temperature alloy steels (see Nickel Alloys).		
Molten metal		
Cast iron	2370–2550	0.29
Mild steel	2910–3270	0.28
Lead		
Pure (99.96%), unoxidized	260–440	0.057–0.075
Gray oxidized	75	0.281
Oxidized at 390°F.	390	0.63
Mercury	32–212	0.09–0.12
Molybdenum filament	1340–4700	0.096–0.292
Monel metal, oxidized at 1110°F	390–1110	0.41–0.46
Nickel		
Electroplated on polished iron, then polished	74	0.045
Technically pure (98.9% Ni, + Mn), polished	440–710	0.07–0.087
Electropolated on pickled iron, not polished	68	0.11
Wire	368–1844	0.096–0.186
Plate, oxidized by heating at 1110°F	390–1110	0.37–0.48
Nickel oxide	1200–2290	0.59–0.86
Nickel alloys		
Chromnickel	125–1894	0.64–0.76
Nickelin (18–32 Ni; 55–68 Cu; 20 Zn), gray oxidized	70	0.262
KA-2S alloy steel (8% Ni; 18% Cr), light silvery, rough, brown, after heating	420–914	0.44–0.36
After 42 hr. heating at 980°F.	420–980	0.62–0.73
NCT-3 alloy (20% Ni; 25% Cr.), brown, splotched, oxidized from service	420–980	0.90–0.97
NCT-6 alloy (60% Ni; 12% Cr), smooth, black, firm adhesive oxide coat from service	520–1045	0.89–0.82
Platinum		
Pure, polished plate	440–1160	0.054–0.104
Strip	1700–2960	0.12–0.17
Filament	80–2240	0.036–0.192
Wire	440–2510	0.073–0.182
Silver		
Polished, pure	440–1160	0.0198–0.0324
Polished	100–700	0.0221–0.0312
Steel, see Iron.		
Tantalum filament	2420–5430	0.194–0.31
Tin—bright tinned iron sheet	76	0.043 and 0.064
Tungsten		
Filament, aged	80–6000	0.032–0.35
Filament	6000	0.39
Zinc		
Commercial, 99.1% pure, polished	440–620	0.045–0.053
Oxidized by heating at 750°F.	750	0.11
Galvanized sheet iron, fairly bright	82	0.228
Galvanized sheet iron, gray oxidized	75	0.276

B. Refractories, Building Materials, Paints, and Miscellaneous

Surface	t, °F.*	Emissivity*
Asbestos		
Board	74	0.96
Paper	100–700	0.93–0.945
Brick		
Red, rough, but no gross irregularities	70	0.93
Silica, unglazed, rough	1832	0.80
Silica, glazed, rough	2012	0.85
Grog brick, glazed	2012	0.75
See Refractory Materials below.		
Carbon		
T-carbon (Gebr. Siemens) 0.9% ash (this started with emissivity at 260°F. of 0.72, but on heating changed to values given)	260–1160	0.81–0.79
Carbon filament	1900–2560	0.526
Candle soot	206–520	0.952
Lampblack-waterglass coating	209–362	0.959–0.947

TABLE 10-17 Normal Total Emissivity of Various Surfaces *(Continued)*

Surface	t, °F.*	Emissivity*
Same	260–440	0.957–0.952
Thin layer on iron plate	69	0.927
Thick coat	68	0.967
Lampblack, 0.003 in. or thicker	100–700	0.945
Enamel, white fused, on iron	66	0.897
Glass, smooth	72	0.937
Gypsum, 0.02 in. thick on smooth or blackened plate	70	0.903
Marble, light gray, polished	72	0.931
Oak, planed	70	0.895
Oil layers on polished nickel (lube oil)	68	
Polished surface, alone	.....	0.045
+0.001-in. oil	.....	0.27
+0.002-in. oil	.....	0.46
+0.005-in. oil	.....	0.72
Infinitely thick oil layer	.....	0.82
Oil layers on aluminum foil (linseed oil)		
Al foil	212	0.087†
+1 coat oil	212	0.561
+2 coats oil	212	0.574
Paints, lacquers, varnishes		
Snowhite enamel varnish or rough iron plate	73	0.906
Black shiny lacquer, sprayed on iron	76	0.875
Black shiny shellac on tinned iron sheet	70	0.821
Black matte shellac	170–295	0.91
Black lacquer	100–200	0.80–0.95
Flat black lacquer	100–200	0.96–0.98
White lacquer	100–200	0.80–0.95
Oil paints, sixteen different, all colors	212	0.92–0.96
Aluminum paints and lacquers		
10% Al, 22% lacquer body, on rough or smooth surface	212	0.52
26% Al, 27% lacquer body, on rough or smooth surface	212	0.3
Other Al paints, varying age and Al content	212	0.27–0.67
Al lacquer, varnish binder, on rough plate	70	0.39
Al paint, after heating to 620°F.	300–600	0.35
Paper, thin		
Pasted on tinned iron plate	66	0.924
On rough iron plate	66	0.929
On black lacquered plate	66	0.944
Plaster, rough lime	50–190	0.91
Porcelain, glazed	72	0.924
Quartz, rough, fused	70	0.932
Refractory materials, 40 different	1110–1830	
poor radiators	......	[0.65, 0.70] –0.75
good radiators	......	0.80–0.85; 0.85–0.90
Roofing paper	69	0.91
Rubber		
Hard, glossy plate	74	0.945
Soft, gray, rough (reclaimed)	76	0.859
Serpentine, polished	74	0.900
Water	32–212	0.95–0.963

*When two temperatures and two emissivities are given, they correspond, first to first and second to second, and linear interpolation is permissible. °C = (°F − 32)/1.8.

†Although this value is probably high, it is given for comparison with the data by the same investigator to show the effect of oil layers. See Aluminum, Part A of this table.

consideration. The following equation for heat transfer, which allows for convective effects, is commonly used by designers of **high-temperature furnaces:**

$$q_{ra} = Q/A = b\sigma[(T_g/100)^4 - (T_s/100)^4] \qquad (10\text{-}187)$$

where b = 5.67 (SI) or 0.172 (U.S. customary); Q = total furnace heat transfer; σ = an emissivity factor with recommended values of 0.74 for gas, 0.75 for oil, and 0.81 for coal; A = effective area for absorbing heat (here the solids burden exposed area); T_g = exiting-combustion-gas absolute temperature; and T_s = absorbing surface temperature.

In rotary devices, reradiation from the exposed shell surface to the solids bed is a major design consideration. A treatise on furnaces, including radiative heat-transfer effects, is given by Ellwood and Danatos [*Chem. Eng.*, 73(8), 174 (1966)]. For discussion of radiation heat-transfer computational methods, heat fluxes obtainable, and emissivity values, see Schornshort and Viskanta (ASME Paper 68-H 7-32), Sherman (ASME Paper 56-A-111), and the following subsection.

HEAT TRANSMISSION BY RADIATION

General References: Much of the pertinent literature on radiative heat transfer has been surveyed in the following texts: Goody, *Atmospheric Radiation*, Clarendon Press, Oxford, 1964. Hottel and Sarofim, *Radiative Transfer*, McGraw-Hill, New York, 1967. Kreith, *Radiation Heat Transfer*, International Textbook, Scranton, Pa., 1962. Love, *Radiative Heat Transfer*, Merrill, Columbus, 1968. Siegel and Howell, *Thermal Radiation Heat Transfer*, NASA SP-164, GPO, Washington, 1968. Sparrow and Cess, *Radiation Heat Transfer*, Brooks/Cole Publishing Company, Belmont, Calif., 1966. Wiebelt, *Engineering Radiation Heat Transfer*, Holt, New York, 1966.

Additional sources are the *Journal of Applied Optics* and the *Journal of the Optical Society of America*, particularly for surface properties; the *Journal of Quantitative Spectroscopy and Radiative Transfer* for gas properties; the *Journal of Heat Transfer* and the *International Journal of Heat and Mass Transfer* for broad coverage; and the *Journal of the Institute of Fuel* for applications to industrial furnaces. Reference in this subsection is mainly to Hottel and Sarofim (op. cit.), since the material covered here is to a large extent abstracted from it.

Thermal radiation—electromagnetic energy in transport—is emitted within matter excited by temperature; it is absorbed in other matter at distances from the source which depend on the mean free path of the photons emitted. The ratio of the mean free path involved in an energy-transport process to a characteristic dimension of the system of interest determines the mathematical structure of the formulation. In **molecular conduction** this ratio is minute (unless the system or the density of matter is minute, which is the case of free molecular flow), and a differential equation of energy diffusion is involved. In **gas radiation** the ratio is generally large enough to give rise to an integral equation, with an unknown function inside the integral. Solids generally have small enough photon mean free paths (high enough absorption coefficients) for the radiation escaping through the surface to have originated close to the surface; radiative loss is then identifiable with its surface temperature, but an integral equation is still involved if all the surfaces of an enclosure filled with a diathermanous medium like air are not specified as to temperature or are not black.

Radiation differs from conduction and convection not only in mathematical structure but in its much higher sensitivity to temperature. It is of dominating importance in furnaces because of their temperature, and in cryogenic insulation because of the vacuum existing between particles. The temperature at which it accounts for roughly half of the total heat loss from a surface in air depends on such factors as surface emissivity and the convection coefficient. For pipes in free convection, this is room temperature; for fine wires of low emissivity it is above red heat. Gases at combustion-chamber temperatures lose more than 90 percent of their energy by radiation from the carbon dioxide, water vapor, and particulate matter.

NOMENCLATURE FOR RADIATIVE TRANSFER

Terms which are defined at specific places in the text are excluded.

a = effective energy fraction of blackbody spectrum in which a nongray gas absorbs.
A = area.
c = number concentration of particles in a cloud.
c_1, c_2 = first and second Planck-law constants.
C = axis-to-axis distance of separation of tubes.
C_b = mean specific heat of combustion products from base temperature T_o to leaving-gas temperature T_G.
C_S = cold-surface fraction of a furnace enclosure.
C_{SO} = correction factor for spectral-band overlap.
C_W = correction factor for pressure broadening of radiation from water vapor.
d = particle diameter.
D = tube diameter; characteristic dimension; a part of reduced firing density D'.
D' = reduced firing density.
E = hemispherical emissive power of a blackbody.
f = fraction of blackbody radiation lying below λ.
f_v = volume fraction of space occupied by particles.
F = direct view factor; F_{ij}, fraction of isotropic radiation from A_i intercepted directly by A_j.
$\bar{F}$ = total view factor from black source to black sink, with allowance for refractory surfaces (subscripts identify source and sink).
$\mathcal{F}_{ij}$ = total view factor, radiation from i to j both directly and indirectly, expressed as fraction of blackbody radiation from A_i.
$\overline{gs}$ = direct-exchange area between gas volume and surface.
$\overline{GS}$ = total-exchange area between gas and surface; subscript R indicates allowance for radiatively adiabatic surfaces.
h = coefficient of convective heat transfer.
H_F = enthalpy of fuel plus air entering combustion chamber.
I = intensity, radiant-energy-flux density per unit solid angle of divergence.
$\overline{ij}$ = shorthand for $\overline{s_i s_j}$.
k = absorption or emission coefficient; or thermal conductivity.
K = constant defined in connection with Eq. (10-208).
L = mean beam length; L_0, at vanishingly small optical thickness; L_m, average value.
L' = wall-loss group.
$\dot{m}$ = mass flow rate.
n = refractive index.
P = total pressure or, with subscript, partial pressure, atm.
q = heat-flux density, energy per time-area.
$\dot{Q}$ = heat flux, energy per time.
Q' = reduced furnace efficiency.
r = separating distance; or electrical resistivity; or refractory (radiatively adiabatic) surface.
$\overline{ss}$ ≡ AF, direct-exchange area (subscripts identify surface zones).
$\overline{SS}$ ≡ $A\mathcal{F}$, total-exchange area.
T = absolute temperature.
U = overall coefficient of heat transfer, gas convection to wall to outside air.
W = total leaving-flux density.
α = absorptivity or absorptance; α_{12}, absorptance of surface 1 for radiation from surface 2.
Δ = difference between radiating temperature and leaving-gas temperature.
ϵ = emissivity or emittance.
θ = polar angle.
λ = wavelength.
μm = micrometer (m^{-6}).
ρ = reflectance; ρ_s, specular reflectance.
σ = Stefan-Boltzmann constant.
τ = transmittance; or reduced sink temperature.
Ω = solid angle.
ω = albedo of a surface.

NATURE OF THERMAL RADIATION

Consider a pencil of radiation, defined as all the rays passing through each of two small widely separated areas dA_1 and dA_2. The rays at dA_1 will have a solid angle of divergence $d\Omega_1$ equal to the apparent area of dA_2 viewed from dA_1, divided by the square of the separating distance. Let the normal to dA_1 make the angle θ_1 with the pencil. The flux density q (energy per time-area) normal to the beam and per unit solid angle of its divergence is called the **intensity** I, and the flux $d\dot{Q}_1$ (energy per time) through the area dA_1 (of apparent area $dA_1 \cos\theta_1$ normal to the beam) is therefore given by

$$d\dot{Q}_1 = dA_1(\cos\theta_1)q = I\,dA_1(\cos\theta_1)\,d\Omega_1 \tag{10-188}$$

The intensity I along a pencil, in the absence of absorption or scatter, is constant (unless the beam passes into a medium of different refractive index n; then $I_1/n_1^2 = I_2/n_2^2$).

The **emissive power**° of a surface is the flux density (energy per time–surface area) due to emission from it throughout a hemisphere. If the intensity I of emission from a surface is independent of the angle of emission, Eq. (10-188) may be integrated to show that the surface emissive power is πI, though the emission is throughout 2π sr.

Blackbody Radiation Engineering calculations of thermal radiation from surfaces are best keyed to the radiation characteristics of the **blackbody,** or ideal radiator. The characteristic properties of a blackbody are that it absorbs all the radiation incident on its surface and that the quality and intensity of the radiation it emits are completely determined by its temperature. The total radiative flux throughout a hemisphere from a black surface of area A and absolute temperature T is given by the **Stefan-Boltzmann law:**

$$\dot{Q} = A\sigma T^4 \quad \text{or} \quad q = \sigma T^4 \tag{10-189}$$

The **Stefan-Boltzmann constant** σ has the value $(0.1713)(10^{-8})$ Btu/($ft^2 \cdot h \cdot °R^4$); $(1.00)(10^{-8})$ CHU/($ft^2 \cdot h \cdot K^4$); $(4.88)(10^{-8})$ kcal/($m^2 \cdot h \cdot K^4$); $(1.356)(10^{-12})$ cal ($cm^2 \cdot s \cdot K^4$); $(5.67)(10^{-12})$ W/($cm^2 \cdot K^4$); $(5.67)(10^{-8})$ W/($m^2 \cdot K^4$); or in terms of Planck constants, $c_1(\pi/c_2)^4/15$. From the definition of emissive power, σT^4 is the total emissive power of a blackbody, called E; the intensity I_B of blackbody emission is E/π or a $\sigma T^4/\pi$.

The spectral distribution of energy flux from a black body is expressed by **Planck's law:**

$$E_\lambda\,d\lambda = (2\pi hc^2 n^2 \lambda^{-5})/(e^{hc/k\lambda T} - 1)\,d\lambda \tag{10-190}$$

$$\equiv (n^2 c_1 \lambda^{-5})/(e^{c_1/\lambda T} - 1)\,d\lambda \tag{10-191}$$

where $E_\lambda\,d\lambda$ is the hemispherical flux density lying in the wavelength range λ to $\lambda + d\lambda$; h is Planck's constant, $(6.6256)(10^{-27})$ erg·s; c is the velocity of light in vacuo, $(2.9979)(10^{10})$ cm/s; k is the Boltzmann constant, $(1.3805)(10^{-16})$ erg/K; λ is the wavelength measured in vacuo; and n is the refractive index of the emitter ($\lambda = n\lambda_m$, where λ_m is the wavelength measured in the medium; $E_\lambda\,d\lambda = E_{\lambda m}\,d\lambda_m$, where E_λ and $E_{\lambda m}$ are both measured in the medium; engineers commonly use E_λ). Equation (10-191) may be written

$$\frac{E_\lambda}{n^2 T^5} = \frac{c_1(\lambda T)^{-5}}{e^{c_2/\lambda T} - 1} \tag{10-191a}$$

The first and second **Planck-law constants** c_1 and c_2 are respectively $(3.740)(10^{-16})$ ($J \cdot m^2$)/s and $(1.4388)(10^{-2})$ m·K. The term E_λ/n^2T^5, clearly a function only of the product λT, is given in Fig. 10-44,

°Variously called, in the literature, emittance, total hemispherical intensity, or radiant flux density.

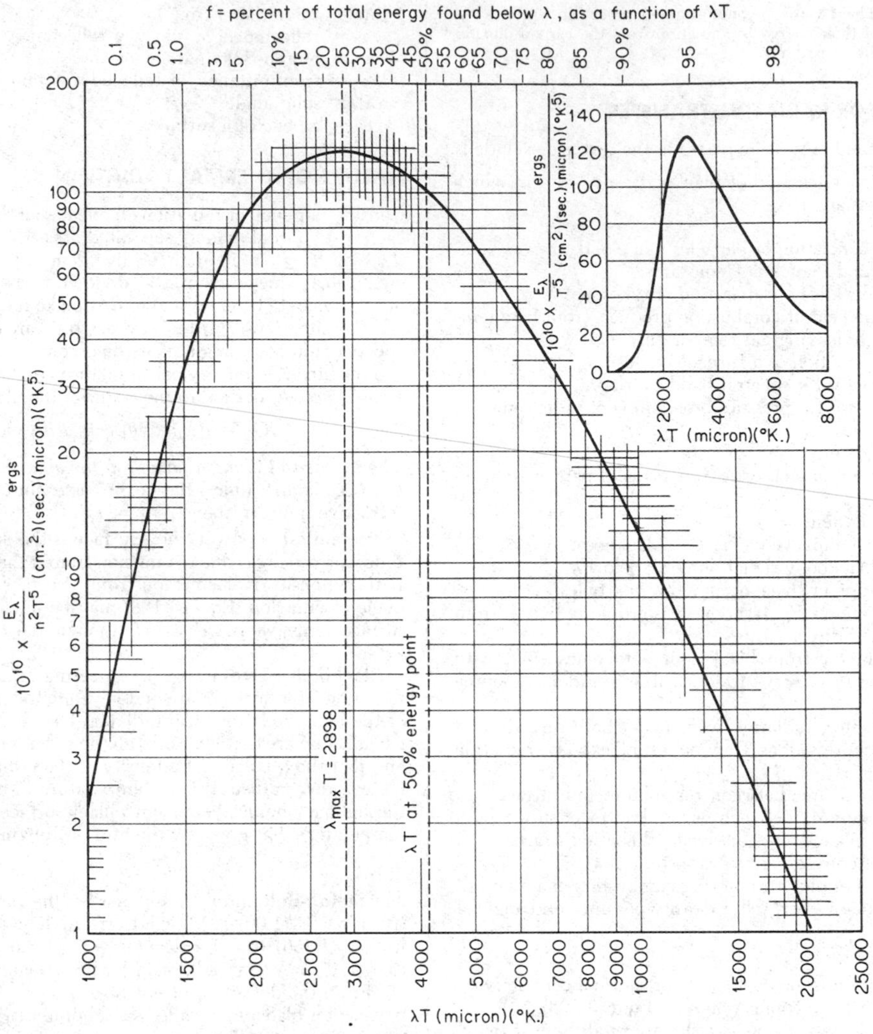

FIG. 10-44 Distribution of energy in the spectrum of a blackbody. To convert microns to micrometers, multiply by unity. To convert ergs per square centimeter–second–micron–K^5 to joules per square meter–second–meter–K^5, multiply by 10^{-3}.

which may be visualized as the monochromatic emissive power versus wavelength measured in vacuo of a black surface at 1 K discharging in vacuo.

The wavelength of maximum intensity is seen to be inversely proportional to the absolute temperature. The relation is known as **Wien's displacement law:** $\lambda_{max}T = (2.898)(10^{-3})$ m·K. This can be misleading, however, since the wavelength of maximum intensity depends on whether intensity is defined in terms of wavelength interval or frequency interval. More useful displacement laws refer to the value of λT corresponding to maximum energy per unit *fractional change* in wavelength or frequency [$(3.67)(10^{-3})$ m·K] or to the value of λT corresponding to half of the energy [$(4.11)(10^{-3})$ m·K]. Figure 10-44 carries, at the top, a scale giving the fraction f of the total energy in the spectrum that lies below λT. A generalization useful for identifying the spectral range of greatest interest in evaluations of radiative transfer is that roughly half of the energy from a black surface lies within the twofold range of λT geometrically centered on 0.367, i.e., from $\lambda T = (367/\sqrt{2})(10^{-3})$ to $(367/\sqrt{2})(10^{-3})$ m·K.

One limiting form of the Planck equation, approached as $\lambda T \to 0$, is the **Wien equation** [Eqs. (10-190) and (10-191)] with the 1 missing in the denominator. The error is less than 1 percent when $\lambda T < (3)(10^{-3})$ m·K or when $T < 4800$ K if an optical pyrometer with red screen ($\lambda = 0.65\mu$m) is used.

RADIATIVE EXCHANGE BETWEEN SURFACES OF SOLIDS

Emittance and Absorptance The ratio of the total radiating power of a real surface to that of a black surface at the same temperature is called the **emittance** of the surface (for a perfectly plane surface, the **emissivity**), designated by ϵ. Subscripts λ, θ, and n may be assigned to differentiate monochromatic-directional and surface-normal values respectively from the total hemispherical value. If radiation is incident on a surface, the fraction absorbed is called the **absorptance (absorptivity),** a term to which two subscripts may be appended, the first to identify the temperature of the surface and the second to identify the quality of the incident radiation.

According to **Kirchhoff's law,** the emissivity and absorptivity of a surface *in surroundings at its own temperature* are the same for both monochromatic and total radiation. When the temperatures of the surface and its surroundings differ, the total emissivity and absorptivity of the surface often are found to be different, but because absorptivity is substantially independent of irradiation density, the monochromatic emissivity and absorptivity of surfaces are for all practical purposes the same. The difference between total emissivity and absorptivity depends on the variation, with wavelength, of ϵ_λ and on the difference between the emitter temperature and the effective source temperature.

Consider radiative exchange between a body of area A_1 and temperature T_1 and black surroundings at T_2. The net interchange is given by

$$\dot{Q}_{1\rightleftharpoons 2} = A_1 \int_0^\infty [\epsilon_\lambda E_\lambda(T_1) - \alpha_\lambda E_\lambda(T_2)]\, d\lambda$$
$$= A_1(\epsilon_1 \sigma T_1^4 - \alpha_{12}\sigma T_2^4) \qquad (10\text{-}192)$$

where $\epsilon_1 = \int_0^1 \epsilon_\lambda\, df_{\lambda T_1}$ and $\alpha_{12} = \int_0^1 \epsilon_\lambda\, df_{\lambda T_2}$ (10-192*a*,*b*)

The value of ϵ_1 (or α_{12}) is the area under a curve of ϵ_λ versus f, the latter read as a function of λT_1 (or λT_2) from the top ordinate of Fig. 10-44. If ϵ_λ does not change with wavelength, the surface is called **gray,** and $\epsilon_1 = \alpha_{12} = \epsilon_\lambda$. A **selective** surface is one whose ϵ_λ changes dramatically with wavelength. If this change is unidirectional, ϵ_1 and α_{12} are, according to Eqs. (10-192) and (10-192*a*,*b*), markedly different when the absolute-temperature ratio is far from 1; e.g., when T_1 = 294 K (530°R; ambient temperature), and T_2 = 6000 K (10,800°R; effective solar temperature), $\epsilon_1 = 0.9$ and $\alpha_{12} = 0.1$ to 0.2 for a white paint, but ϵ_1 can be as low as 0.12 and α_{12} above 0.9 for a thin layer of copper oxide on bright aluminum.

The **effect of radiation-source temperature** on the low-temperature absorptivity of a number of additional materials is presented in Fig. 10-45. It will be noted that polished aluminum (curve 15) and anodized (surface-oxidized) aluminum (curve 13), representative of metals and nonmetals respectively, respond oppositely to a change in the temperature of the radiation source. The absorptance of surfaces for solar radiation may be read from the right of Fig. 10-45, if solar radiation is assumed to consist of blackbody radiation from a source at 6000 K (10,800°R).

Although values of emittance and absorptance depend in very complex ways on the real and imaginary components of the refractive index and on the geometrical structure of the surface layer, the generalizations that follow are possible.

Polished Metals

1. ϵ_λ in the infrared is governed by free-electron contributions, is quite low, and is a function of the resistivity-wavelength quotient r/λ (Fig. 10-46). For $\lambda > 8\mu m$, $\epsilon_{\lambda,n}$ is approximately $0.0365\sqrt{r/\lambda}$, where r is in ohm-meters and λ in micrometers (the Drude or Hagen-Rubens relation). At shorter wavelengths, bound-electron contributions become significant and ϵ_λ increases, sometimes exhibiting maxima; values of 0.4 to 0.8 are common in the visible spectrum (0.4 to 0.7μm). ϵ_λ is approximately proportional to the square root of the absolute temperature ($\epsilon_\lambda \propto \sqrt{T}$, and $r \propto T$) in the far infrared ($\lambda > 8\mu m$), is temperature-insensitive in the near infrared (0.7 to 1.5μm), and decreases slightly as temperature increases in the visible.

2. Total emittance is substantially proportional to absolute temperature; at moderate temperature, $\epsilon_n = 0.058T\sqrt{rT}$, where T is in kelvins.

3. The total absorptance of a metal at T_1 for radiation from a black or gray source at T_2 is equal to the emissivity evaluated at the geometric mean of T_1 and T_2. Figure 10-46 gives values of ϵ_λ, $\epsilon_{\lambda,n}$, and their ratio as a function of r/λ (dashed lines); and total emissivities ϵ, ϵ_n and their ratio as a function of rT (solid lines). Although the figure is based on free-electron contributions to emissivity in the far infrared, the relations for total emissivity are remarkably good even at high temperatures. Unless extraordinary pains are taken to prevent oxidation, however, a metallic surface may exhibit several times the emittance or absorptance of a polished specimen. The emittance of iron and steel, for example, varies widely with degree of oxidation and roughness; clean metallic surfaces have an emittance of from 0.05 to 0.45 at ambient temperatures to 0.4 to 0.7 at high temperatures; oxidized and/or rough surfaces range from 0.6 to 0.95 at low temperatures to 0.9 to 0.95 at high temperatures.

Refractory Materials Grain size and concentration of trace impurities are important.

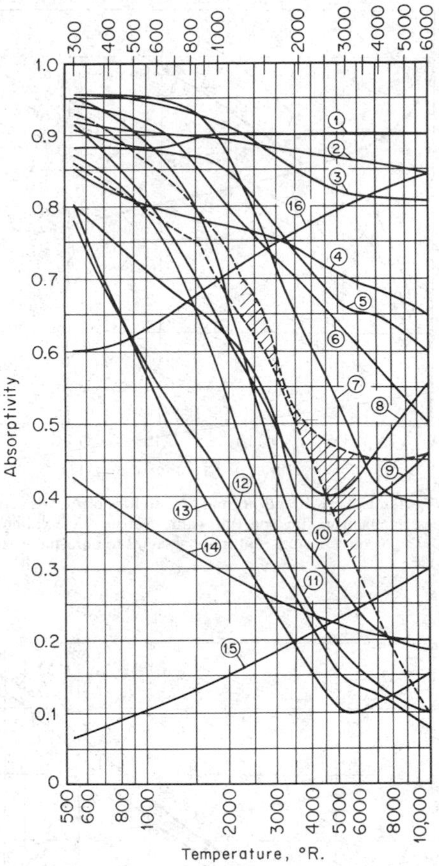

FIG. 10-45 Variation of absorptivity with temperature of radiation source. (1) Slate composition roofing. (2) Linoleum, red brown. (3) Asbestos slate. (4) Soft rubber, gray. (5) Concrete. (6) Porcelain. (7) Vitreous enamel, white. (8) Red brick. (9) Cork. (10) White dutch tile. (11) White chamotte. (12) MgO, evaporated. (13) Anodized aluminum. (14) Aluminum paint. (15) Polished aluminum. (16) Graphite. The two dashed lines bound the limits of data on gray paving brick, asbestos paper, wood, various cloths, plaster of paris, lithopone, and paper. To convert degrees Rankine to kelvins, multiply by $(5.556)(10^{-1})$.

1. Most refractory materials have an ϵ_λ of 0.8 to 1.0 at wavelengths beyond 2 to 4 μm; ϵ_λ decreases rapidly toward shorter wavelengths for materials that are white in the visible but retains its high value for black materials such as FeO and Cr_2O_3. Small concentrations of FeO and Cr_2O_3 or other colored oxides can cause marked increases in the emittance of materials that normally are white. The sensitivity of the emittance of refractory oxides to small additions of absorbing materials is demonstrated by the results of calculations, shown in Fig. 10-47, of the emittance of a semi-infinite absorbing-scattering medium as a function of its albedo: the ratio of the scatter coefficient to the sum of scatter and absorption coefficients. The results, pertinent to the radiative properties of fibrous materials, paints, oxide coatings, and refractories, show that when absorption accounts for only 0.5 percent (10 percent) of the total attenuation within the medium, the emittance is greater than 0.15 (0.5). ϵ_λ for refractory materials varies little with temperature, with the exception of some white oxides which at high temperatures become good emitters in the visible spectrum as a consequence of the induced electronic transitions.

2. Refractory materials generally have a total emittance which is high (0.7 to 1.0) at ambient temperatures and decreases with increase in temperature; a change from 1000 to 1570°C (1850 to 2850°F) may cause a decrease in ϵ of one-fourth to one-third.

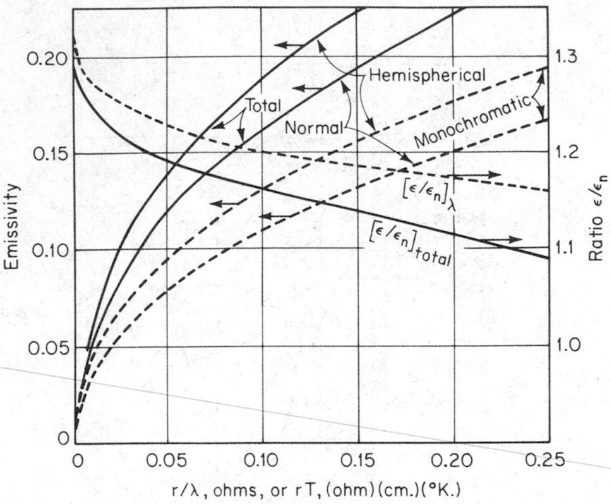

FIG. 10-46 Hemispherical and normal emissivities of metals and their ratio. Dashed lines: monochromatic (spectral) values versus r/λ. Solid lines: total values versus rT. To convert ohm-centimeter-kelvins to ohm-meter-kelvins, multiply by 10^{-2}.

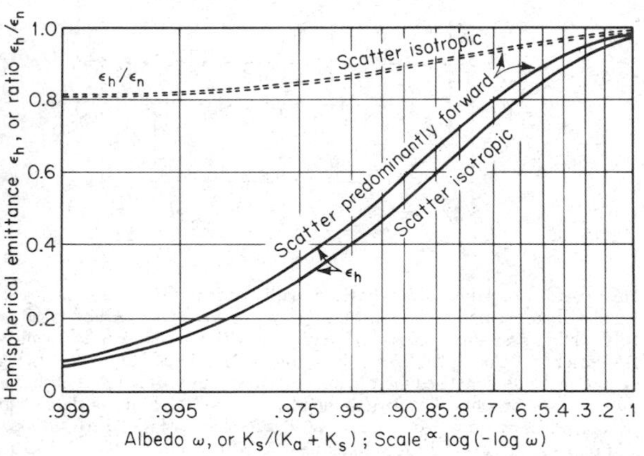

FIG. 10-47 Hemispherical emittance ε_h and the ratio of hemispherical to normal emittance $\varepsilon_h/\varepsilon_n$ for a semi-infinite absorbing-scattering medium.

3. The emittance and absorptance increase with increase in grain size over a grain-size range of 1 to 200 μm.
4. The ratio ϵ/ϵ_n of hemispherical to normal emissivity of polished surfaces varies with refractive index n from 1 at $n = 1.0$ to 0.93 at $n = 1.5$ (common glass) and back to 0.96 at $n = 3$.
5. The ratio ϵ/ϵ_n for a surface composed of particulate matter which scatters isotropically varies with ϵ from 1 when $\epsilon = 1$ to 0.8 when $\epsilon = 0.07$ (see Fig. 10-47).
6. The total absorptance shows a decrease with increase in temperature of the radiation source similar to the decrease in emittance with increase in the specimen temperature.

Figure 10-45 shows a regular variation of α_{12} with T_2. When T_2 is not very different from T_1, α_{12} may be expressed as $\epsilon_1(T_2/T_1)^m$. It may be shown that Eq. (10-192) is then approximated by

$$\dot{Q}_{1,\text{net}} = \sigma A_1 \epsilon_{av}\left(1 + \frac{m}{4}\right)(T_1^4 - T_2^4) \qquad (10\text{-}193)$$

where ϵ_{av} is evaluated at the arithmetic mean of T_1 and T_2. For metals m is about 0.5; for nonmetals it is small and negative.

Table 10-17, based on a critical evaluation of early data, is illustrative of the emittance of **materials encountered in engineering practice;** it shows the wide variation possible in the emissivity of a particular material due to variations in surface roughness and thermal pretreatment. (With few exceptions the values refer to emission normal to the surface; see above for conversion to hemispherical values.) More recent data support the range of emittance values given in Table 10-17 and their dependence on surface conditions. Extensive compilations of data are provided by Schmidt and Furthmann (*Mitt. Kaiser-Wilhelm-Inst. Eisenforsch.*, **109**, 225), covering data to 1928; by Gubareff, Jansen, and Torborg (*Thermal Radiation Properties Survey*, Honeywell Research Center, Minneapolis, 1960), covering data to 1940; and by Goldsmith, Waterman, and Hirschhorn (*Thermophysical Properties of Solid Materials*, WADC, TR58-476, vols. I–V, Wright-Patterson Air Force Base, Ohio, 1960).

For opaque materials, the **reflectance** ρ is the complement of the absorptance. The directional distribution of the reflected radiation depends on the material, its degree of roughness or grain size, and, if a metal, its state of oxidation. Polished surfaces of homogeneous materials reflect **specularly.** In contrast, the intensity of the radiation reflected from a perfectly **diffuse,** or Lambert, surface is independent of direction. The directional distribution of reflectance of many oxidized metals, refractory materials, and natural products approximates that of a perfectly diffuse reflector. A better model, adequate for many calculational purposes, is achieved by assuming that the total reflectance ρ is the sum of diffuse and specular components ρ_D and ρ_S.

Black-Surface Enclosures

View Factor and Direct-Exchange Area When several surfaces are present, the need arises for evaluating a geometrical factor F, called the **direct view factor.** In the following discussion, restriction is to black surfaces, the intensity from which is independent of angle of emission. Define F_{12} as the fraction of the radiation leaving surface A_1 in all directions which is intercepted by surface A_2. Since the net interchange between A_1 and A_2 must be zero when their temperatures are alike, it follows that $A_1F_{12} = A_2F_{21}$. This product, having the dimensions of area, is called the **direct-exchange area** and is designated for brevity by $\overline{12}(\equiv\overline{21})$. It is sometimes designated $\overline{s_1s_2}$. Clearly, $\overline{11} + \overline{12} + \overline{13} + \cdots = A_1$; and when A_1 cannot "see" itself, $\overline{11} = 0$.

From Eq. (10-188) and the definition of F:

$$A_1F_{12} \equiv \overline{s_1s_2} \equiv \frac{Q_{1\to2}}{E_1} = \int_{A1}\int_{A2} \frac{dA_1(\cos\theta_1)\,d\Omega_1}{\pi} = \int_{A1}\int_{A2} \frac{dA_1(\cos\theta_1)\,dA_2\,(\cos\theta_2)}{\pi r^2} \qquad (10\text{-}194)$$

where r is the distance from dA_1 to dA_2. Values of $\overline{s_1s_2}$ (or of F_{12}) may be obtained by integrating either Eq. (10-194) or an equivalent contour integral (see Hottel and Sarofim, *Radiative Transfer*, McGraw-Hill, New York, 1967, chap. 2). Such values are given for opposed parallel disks or rectangles in Fig. 10-48, for perpendicular adjacent rectangles in Fig. 10-49, for finite coaxial coextensive cylinders in Fig. 10-50, and for an infinite plane parallel to a system of rows of parallel tubes as curves 1 and 3 of Fig. 10-51. When the two surfaces are small relative to their center-to-center separating distance r, Eq. (10-194) gives

$$\overline{12} = \frac{A_1(\cos\theta_1)A_2(\cos\theta_2)}{\pi r^2} \qquad (10\text{-}194a)$$

and when, in addition, the normals to A_1 and A_2 are in a common plane,

$$\overline{12} = A_1A_2n_1n_2/\pi r^4 \qquad (10\text{-}194b)$$

where n_1 is the normal-to-A_1 component of the distance to A_2. Equation (10-194b) is, for example, in error only by +7 percent for the case of opposed squares separated by 3 times their side dimension.

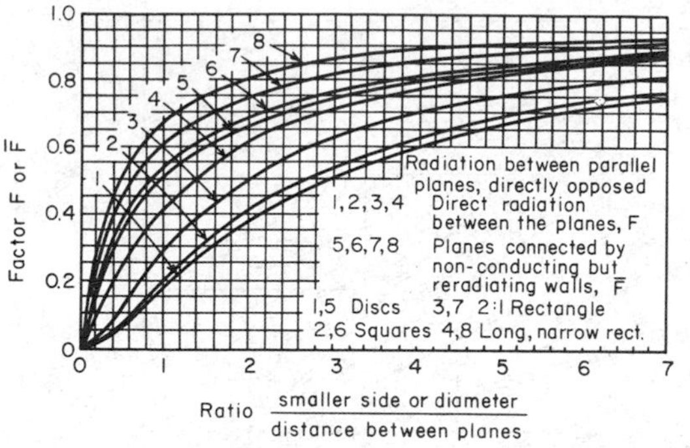

FIG. 10-48 Radiation between parallel planes, directly opposed.

The **exchange area** between any two area elements **of a sphere** is independent of their relative shape and position and is simply the product of the areas divided by the area of the whole sphere; i.e., any spot on a sphere has equal views of all other spots.

For surfaces in **two-dimensional systems** (with third dimension infinite), A_1F_{12} per unit length in the third dimension may be obtained simply by evaluating, in a cross-sectional view, the sum of lengths of **crossed strings** from the ends of A_1 to the ends of A_2 less the sum of **uncrossed strings** from and to the same points, all divided by 2. The strings must be so drawn that all the flux from one surface to the other must cross each of a pair of crossed strings and neither of a pair of uncrossed ones. If one surface can see the other around both sides of an obstruction, two more pairs of strings are involved.

Example 3 Evaluate the view factor between two parallel circular tubes long enough compared with their diameter D or their axis-to-axis separating distance C to make the problem two-dimensional. With reference to Fig. 10-52, the crossed-strings method yields

$$A_1F_{12} = \frac{2(EFGH - HJ)}{2} = D\left\{\sin^{-1}\frac{D}{C} + \left[\left(\frac{C}{D}\right)^2 - 1\right]^{1/2} - \frac{C}{D}\right\}$$

Results for a large number of other cases are given by Hottel and Sarofim (op. cit., chap. 2) and Hamilton and Morgan (NACA-TN2836, December 1952). A comprehensive bibliography is provided by Siegel and Howell (*Thermal Radiation Heat Transfer*, NASA SP-164, GPO, Washington, 1968).

The view factor F may often be evaluated from that for simpler configurations by the application of three principles: that of **reciprocity,** $A_iF_{ij} = A_jF_{ji}$; that of **conservation,** $\Sigma F_{ij} = 1$; and that due to **Yamauti** [*Res. Electrotech. Lab. (Tokyo),* **148,** 1924; **194,** 1927; **250,** 1929], showing that the exchange areas AF between two pairs of surfaces are equal when there is a one-to-one correspondence for all sets of symmetrically placed pairs of elements in the two surface combinations.

Example 4 The exchange area between the two squares 1 and 4 of Fig. 10-53 is to be evaluated. The following exchange areas may be obtained from the values, in Fig. 10-48, of F for common-side rectangles: $\overline{13} = 0.24$, $\overline{24} = 2 \times 0.29 = 0.58$, $\overline{(1+2)(3+4)} = 3 \times 0.32 = 0.96$. Expression of $\overline{(1+2)(3+4)}$ in terms of its components yields $\overline{(1+2)(3+4)} = \overline{13} + \overline{14} + \overline{23} + \overline{24}$. And by the Yamauti principle $\overline{14} = \overline{23}$, since for every pair of elements in 1 and 4 there is a corresponding pair in 2 and 3. Therefore,

$$\overline{14} = \frac{\overline{(1+2)(3+4)} - \overline{13} - \overline{24}}{2} = 0.07$$

Figures 10-48 and 10-50 may be used in the same way.

Non-Black-Surface Enclosures

In the following discussion we are concerned with enclosures containing gray sources and sinks, radiatively adiabatic surfaces, and no absorbing gas. The calculation of interchange between a source and a sink under conditions involving successive multiple reflections from other source-sink surfaces in the enclosure, as well as reradiation from refractory surfaces which are in radiative equilibrium, can become complicated.

Zone Method Let a **zone** of a furnace enclosure be an area small enough to make all elements of itself have substantially equivalent "views" of the rest of the enclosure. (In a furnace containing a symmetry plane, parts of a single zone would lie on either side of the

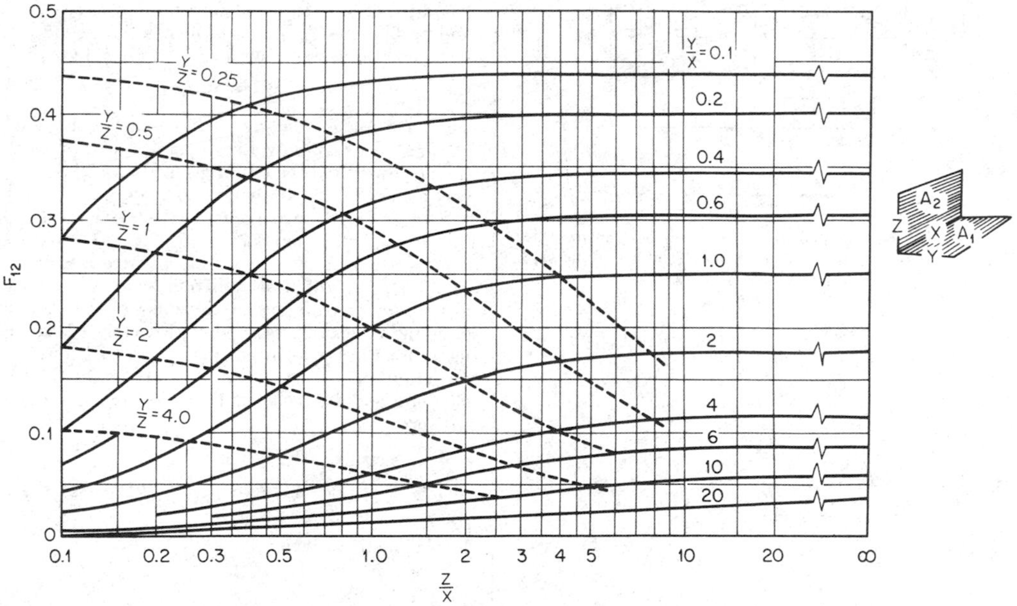

FIG. 10-49 View factor between two rectangles in perpendicular planes and having a common edge. For maximum accuracy define dimensions to make $Y < Z$. For values off plot to the left, use crossed-string method; for values in lower right corner, use $A_1F_{12} \equiv \overline{12} + \dfrac{X^2}{4\pi}\left[3 - \ln\left(\dfrac{X^2}{Y^2} + \dfrac{X^2}{Z^2}\right) + \dfrac{3}{2}\left(\dfrac{X^2}{Y^2} + \dfrac{X^2}{Z^2} - \dfrac{X^2}{Y^2 + Z^2}\right)\right]$

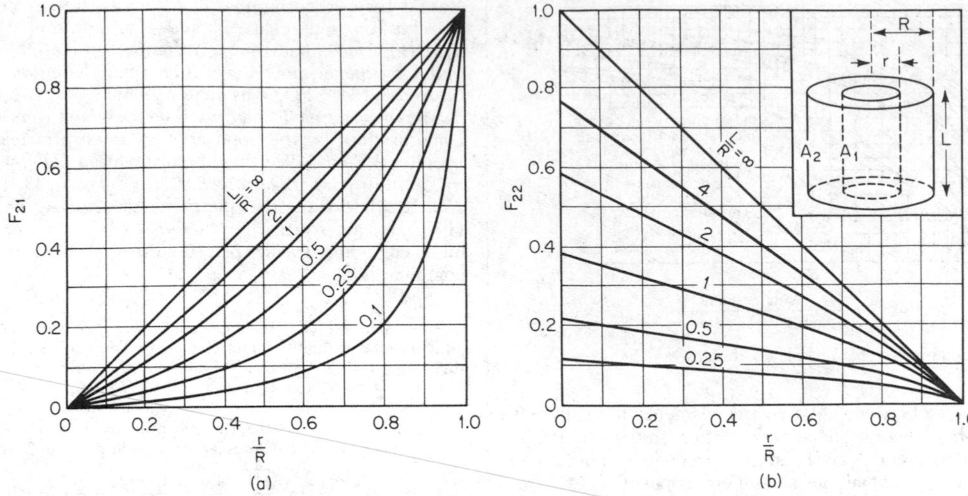

FIG. 10-50 View factors for a system of two concentric coaxial cylinders of equal length. (*a*) Inner surface of outer cylinder to inner cylinder. (*b*) Inner surface of outer cylinder to itself.

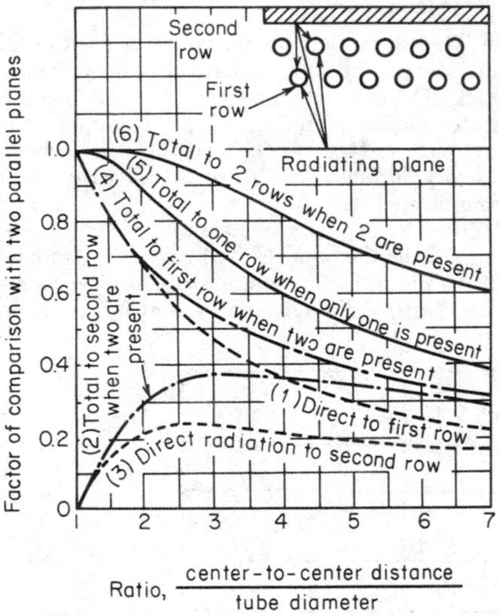

FIG. 10-51 Distribution of radiation to rows of tubes irradiated from one side. Dashed lines: direct view factor F from plane to tubes. Solid lines: total view factor $\bar{F}$ for black tubes backed by a refractory surface.

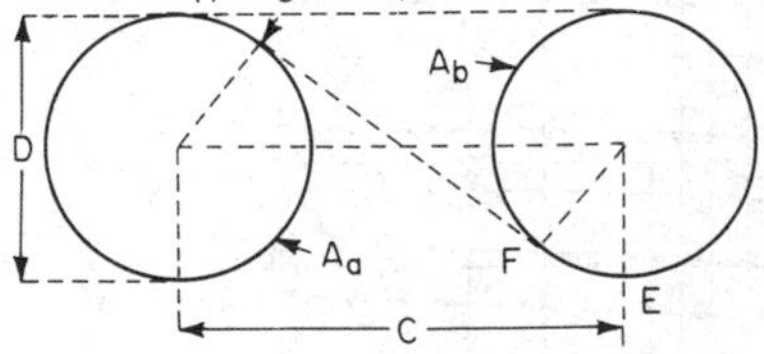

FIG. 10-52 Direct exchange between parallel circular tubes.

plane.) Zones are of two classes: source-sink surfaces, designated by numerical subscripts and having areas A_1, A_2, . . . , and emissivities ϵ_1, ϵ_2, . . . ; and surfaces at which the net radiant-heat flux is zero (fulfilled by the average refractory wall in which difference between internal convection and external loss is minute compared with inci-

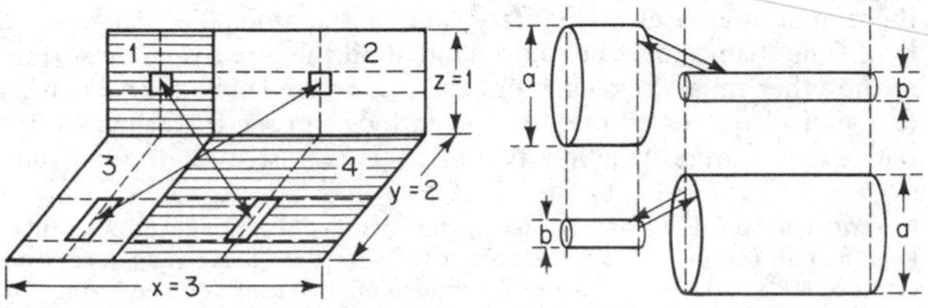

FIG. 10-53 Illustration of the Yamauti principle.

dent radiation), designated by letter subscripts starting with r, and having areas A_r, A_s, . . . It may be shown (see, for example, Hottel and Sarofim, op. cit., chap. 3) that the net radiation interchange between source-sink zones i and j is given by

$$\dot{Q}_{i \rightleftharpoons j} = A_i \mathcal{F}_{ij} \sigma T_i^4 - A_j \mathcal{F}_{ji} \sigma T_j^4 \tag{10-195}$$

The term $A_i\mathcal{F}_{ij}$, sometimes designated $\overline{S_iS_j}$, is called the **total interchange area** shared by areas A_i and A_j and depends on the shape of the enclosure and the emissivity and absorptivity of the source and sink zones. Restriction here is to gray source-sink zones, for which $A_i\mathcal{F}_{ij} = A_j\mathcal{F}_{ji}$; the more general case is treated elsewhere (Hottel and Sarofim, op. cit., chap. 5).

Evaluation of the $A\mathcal{F}$'s that characterize an enclosure involves solution of a system of radiation balances on the surfaces. If the assumption is made that all the zones of the enclosure are gray and emit and reflect diffusely,* then the direct-exchange area $\overline{ij}$, as evaluated for the black-surface pair A_i and A_j, applies to emission *and* reflections between them. If at a surface the total **leaving-flux density,** emitted plus reflected, is denoted by W, radiation balances take the form:

For source-sink j,

$$A_j \epsilon_j E_j + \rho_j \sum_i (\overline{ij}) W_i = A_j W_j \tag{10-196a}$$

For adiabatic surface r,

$$\sum_i (\overline{ir}) W_i = A_r W_r \tag{10-196b}$$

*So-called **Lambert surfaces,** which emit or reflect with an intensity independent of angle; approximately satisfied by most nonmetallic, tarnished, oxidized, or rough surfaces.

where ρ is reflectance and the summation is over all surfaces in the enclosure. In matrix notation, Eq. (10-196) becomes, with source or sink zones represented by 1, 2, 3 . . . and adiabatic zones by r, s, t . . . ,

$$\begin{bmatrix} \overline{11}-\dfrac{A_1}{\rho_1} & \overline{12} & \overline{1r} & \overline{1s} \\ \overline{12} & \overline{22}-\dfrac{A_2}{\rho_2} & \overline{2r} & \overline{2s} \\ \overline{1r} & \overline{2r} & \overline{rr}-A_r & \overline{rs} \\ \overline{1s} & \overline{2s} & \overline{rs} & \overline{ss}-A_s \end{bmatrix} \begin{bmatrix} W_1 \\ W_2 \\ W_r \\ W_s \end{bmatrix} = \begin{bmatrix} -\dfrac{A_1\epsilon_1}{\rho_1}E_1 \\ -\dfrac{A_2\epsilon_2}{\rho_2}E_2 \\ 0 \\ 0 \end{bmatrix} \qquad (10\text{-}196c)$$

This represents a system of simultaneous equations equal in number to the number of rows of the square matrix. Each equation consists, on the left, of the sum of the products of the members of a row of the square matrix and the corresponding members of the W-column matrix and, on the right, of the member of that row in the third matrix. With this set of equations solved for W_i, the net flux at any surface A_i is given by

$$\dot{Q}_{i,\text{net}} = (A_i\epsilon_i/\rho_i)(E_i - W_i) \qquad (10\text{-}197)$$

Refractory temperature is obtained from $W_r = E_r = \sigma T_r^4$.

The more general use of Eq. (10-196c) is to obtain the set of total interchange areas $A\mathcal{F}$ which *constitute a complete description of the effect of shape, size, and emissivity on radiative flux, independent of the presence or absence of other transfer mechanisms.* It may be shown that

$$A_i\mathcal{F}_{ij} \equiv A_j\mathcal{F}_{ji} \equiv \overline{S_iS_j} = \frac{A_i\epsilon_i}{\rho_i}\frac{A_j\epsilon_j}{\rho_j}\left(-\frac{D'_{ij}}{D}\right) \qquad (10\text{-}198)$$

where D is the determinant of the square coefficient matrix in Eq. (10-196c) and D'_{ij} is the cofactor of its ith row and jth column, or $(-1)^{i+j}$ times the minor of D formed by crossing out the ith row and jth column.

As an example, consider radiant interchange between concentric gray spheres of inner and outer radii r_1 and r_2.

$$A_1\mathcal{F}_{12} = \frac{A_1\epsilon_1}{\rho_1}\frac{A_2\epsilon_2}{\rho_2} \frac{\overline{12}}{\begin{vmatrix} \overline{11}-\dfrac{A_1}{\rho_1} & \overline{12} \\ \overline{12} & \overline{22}-\dfrac{A_2}{\rho_2} \end{vmatrix}}$$

$\overline{12} = A_1$; $\overline{11} = 0$; $\overline{22} = A_2 - \overline{21} = A_2 - \overline{12} = A_2 - A_1$. Substitution gives

$$A_1\mathcal{F}_{12} = \frac{A_1\epsilon_1}{\rho_1}\frac{A_2\epsilon_2}{\rho_2} \frac{A_1}{\begin{vmatrix} -A_1/\rho_1 & A_1 \\ A_1 & (A_2 - A_1 - A_2/\rho_2) \end{vmatrix}} = \frac{1}{\dfrac{1}{A_1\epsilon_1} + \dfrac{1}{A_2}\left(\dfrac{1}{\epsilon_2} - 1\right)} \qquad (10\text{-}199)$$

This case includes that of infinite parallel planes [$\mathcal{F} = 1/(1/\epsilon_1 + 1/\epsilon_2 - 1)$] and that of a small body A_1 enclosed in a large one ($A_1\mathcal{F}_{12} = A_1\epsilon_1$).

Many furnace problems are adequately handled by dividing the enclosure into but two source-sink zones A_1 and A_2 and any number of no-flux zones A_r, A_s, ... For this case Eq. (10-198) yields

$$\frac{1}{A_1\mathcal{F}_{12}}\left(\equiv \frac{1}{A_2\mathcal{F}_{21}}\right) = \frac{1}{A_1}\left(\frac{1}{\epsilon_1} - 1\right) + \frac{1}{A_2}\left(\frac{1}{\epsilon_2} - 1\right) + \frac{1}{A_1\overline{F}_{12}} \qquad (10\text{-}200)$$

Here the expression $A_1\overline{F}_{12}(\equiv A_2\overline{F}_{21})$ represents the **total interchange area** for the limiting case of a **black source and black sink** (the refractory emissivity is of no moment). The factor $\overline{F}$ is known exactly for a few geometrically simple cases and may be approximated for others. If A_1 and A_2 are equal parallel disks, squares, or rectangles, connected by nonconducting but reradiating refractory walls, then $\overline{F}$ is given by Fig. 10-48, curves 5 to 8. If A_1 represents an infinite plane and A_2 is one or two rows of infinite parallel tubes in a parallel plane and if the only other surface is a refractory surface behind the tubes, $\overline{F}_{12}$ is given by curve 5 or 6 of Fig. 10-51.

If an enclosure may be divided into several radiant-heat sources or sinks A_1, A_2, etc., and the rest of the enclosure (reradiating refractory surface) may be lumped together as A_r at a uniform temperature T_r, then the total interchange area for zone pairs in the black system is given by

$$A_1\overline{F}_{12}(\equiv A_2\overline{F}_{21}) = \overline{12} + \frac{(\overline{1r})(\overline{r2})}{A_r - \overline{rr}} \qquad (10\text{-}201)$$

For the two-source-sink-zone system to which Eq. (10-200) applies, Eq. (10-201) simplifies to

$$A_1\overline{F}_{12} = \overline{12} + 1/(1/\overline{1r} + 1/\overline{2r}) \qquad (10\text{-}201a)$$

and if A_1 and A_2 each can see none of itself, there is further simplification to

$$A_1\overline{F}_{12} = \overline{12} + \frac{1}{1/(A_1 - \overline{12}) + 1/(A_2 - \overline{12})} = \frac{A_1A_2 - (\overline{12})^2}{A_1 + A_2 - 2(\overline{12})} \qquad (10\text{-}201b)$$

which necessitates the evaluation of but one geometrical factor F.

Equation (10-200) covers many of the problems of radiant-heat interchange between source and sink in a furnace enclosure. The error due to single zoning of source and sink is small even if the views of the enclosure from different parts of each zone are quite different, provided the emissivity is fairly high; the error in $\overline{F}$ is zero if it is obtainable from Fig. 10-48 or 10-51, small if Eq. (10-201) is used and the variation in temperature over the refractory is small. An approach to any desired accuracy can be made by use of Eqs. (10-196) and (10-198) with division of the surfaces into more zones.

From the definitions of F, $\overline{F}$, and $\mathcal{F}$ it is to be noted that

$$F_{11} + F_{12} + F_{13} + \cdots + F_{1r} + F_{1s} + \cdots = 1$$

$$\overline{F}_{11} + \overline{F}_{12} + \overline{F}_{13} + \cdots = 1$$

$$\mathcal{F}_{11} + \mathcal{F}_{12} + \mathcal{F}_{13} + \cdots = \epsilon_1$$

Example 5 A furnace chamber of rectangular parallelepipedal form is heated by the combustion of gas inside vertical radiant tubes lining the sidewalls. The tubes are of 0.127-m (5-in) outside diameter on 0.305-m (12-in) centers. The stock forms a continuous plane on the hearth. Roof and end walls are refractory. Dimensions are shown in Fig. 10-54. The radiant tubes and stock are gray bodies having emissivities of 0.8 and 0.9 respectively. What is the net rate of heat transmission to the stock by radiation when the mean temperature of the tube surface is 816°C (1500°F) and that of the stock is 649°C (1200°F)?

This problem must be broken up into two parts, first considering the walls with their refractory-backed tubes. To imaginary planes A_2 of area 1.83 by 3.05 m (6 by 10 ft) and located parallel to and inside the rows of radiant tubes,

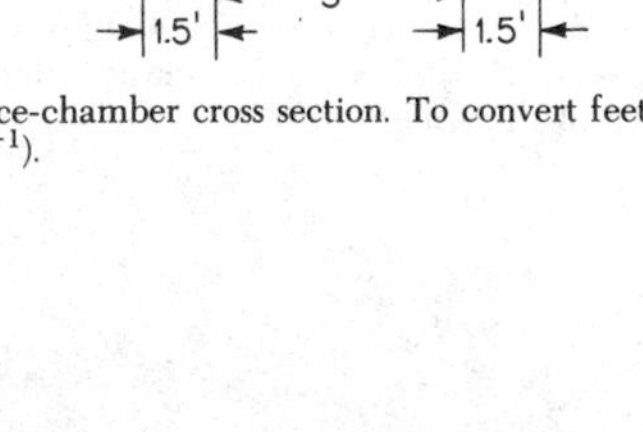

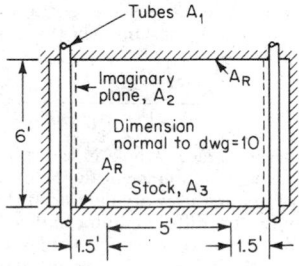

FIG. 10-54 Furnace-chamber cross section. To convert feet to meters, multiply by (3.048)(10^{-1}).

the tubes emit radiation $\sigma T_1^4 A_1 \mathcal{F}_{12}$, which equals $\sigma T_1^4 A_2 \mathcal{F}_{21}$. To find $\mathcal{F}_{21}$ use Fig. 10-51, curve 5, from which $\bar{F}_{21} = 0.81$. Then from Eq. (10-200)

$$\mathcal{F}_{21} = \frac{1}{(1/0.81) + (1/1 - 1) + (12/5\pi)(1/0.8 - 1)} = 0.702$$

This amounts to saying that the system of refractory-backed tubes is equal in radiating power to a continuous plane A_2 replacing the tubes and refractory back of them, having a temperature equal to that of the tubes and an equivalent or effective emissivity of 0.702.

The new simplified furnace now consists of an enclosure formed by two 1.83- by 3.05-m (6- by 10-ft) radiating sidewalls (area A_2, emissivity 0.702), a 1.52- by 3.05-m (5- by 10-ft) receiving plane on the floor A_3, and refractory surfaces A_r to complete the enclosure (ends, roof, and floor side strips). The desired heat transfer is

$$q_{2\rightleftharpoons 3} = \sigma(T_1^4 - T_3^4)A_2\mathcal{F}_{23}$$

To evaluate $\mathcal{F}_{23}$, start with the direct interchange factor F_{23}. $F_{23} = F$ from A_2 to (A_3 + a strip of A_r alongside A_3, which has a common edge with A_2) minus F from A_2 to the strip only. These two F's may be evaluated from Fig. 10-54. For the first F, $Y/X = 1.83/3.05$, $Z/X = 1.98/3.05$, and $F = 0.239$; for the second F, $Y/X = 1.83/3.05$, $Z/X = 0.46/3.05$, and $F = 0.100$. Then $F_{23} = 0.239 - 0.10 = 0.139$. Now $\bar{F}$ may be evaluated. From Eq. (10-201) et seq.,

$$A_2\bar{F}_{23} = \overline{23} + \frac{1}{1/\overline{2r} + 1/\overline{3r}} \quad \bar{F}_{23} = F_{23} + \frac{1}{(1/F_{2r}) + (A_2/A_3)(1/F_{3r})}$$

Since A_2 "sees" A_r, A_3, and some of itself (the plane opposite), $F_{2r} = 1 - F_{22} - F_{23}$. F_{22}, the direct interchange factor between parallel 1.83- by 3.05-m (6- by 10-ft) rectangles separated by 2.44 m (8 ft), may be taken as the geometric mean of the factors for 1.83-m (6-ft) squares separated by 2.44 m (8 ft) and for 3.05-m (10-ft) squares separated by 2.44 m (8 ft). These come from Fig. 10-48, curve 2, according to which $F_{22} = \sqrt{0.13 \times 0.255} = 0.182$. Then $F_{2r} = 1 - 0.182 - 0.139 = 0.679$. The other required direct factor is $F_{3r} = 1 - F_{32} = 1 - F_{23}\ A_2/A_3 = 1 - (0.139)(11.14)/4.65 = 0.666$. Then

$$\bar{F}_{23} = 0.139 + \frac{1}{(1/0.679) + (11.14/4.65)(1/0.666)} = 0.336$$

Having $\bar{F}_{23}$, we may now evaluate the factor $\mathcal{F}_{23}$:

$$\mathcal{F}_{23} = \frac{1}{(1/0.336) + [1/0.702) - 1] + (11.14/4.65)(1/0.9) - 1]} = 0.273$$

$$q_{net} = \sigma(T_1^4 - T_3^4)A_2\mathcal{F}_{23} = 5.67(10.89^4 - 9.22^4)(11.15)(0.273)$$
$$= 118{,}000 \text{ J/s } (402{,}000 \text{ Btu/h})$$

A result of interest is obtained by dividing the term $A_2\mathcal{F}_{23}(11.15 \times 0.273$, or 3.04) by the actual area A_1 of the radiating tubes $(0.127\pi)(18.3)(2) = 14.6$ m^2 (157 ft^2). Thus $3.04/14.6 = 0.208$, which means that the net radiation from a tube to the stock is 20.8 percent as much as if the tube were black and completely surrounded by black stock.

Integral Formulation The zone method has the purpose of dodging the solution of an integral equation. If in Eq. (10-196*a*) the zone on which the radiation balance is formulated is decreased to a differential element, that equation becomes

$$dA_j\epsilon_j E_j + \rho_j \int \frac{dA_i\, dA_j(\cos\theta_i)(\cos\theta_j)W_i}{r^2} = dA_j W_j \qquad (10\text{-}202)$$

which is an integral equation with the unknown function W inside the integral. Integration is over the entire surface area. Exact solutions have been carried out for only a few simple cases. One of these is the evaluation of emittance of an **isothermal spherical cavity,** for which $dA_i\, dA_j(\cos\theta_i)(\cos\theta_j)/r^2$ in the integral of Eq. (10-202) becomes $dA_i\, dA_j/4\pi R^2$, where R is the sphere radius. For this special case W is from Eq. (10-202) constant over the inner surface of the cavity and is given by

$$W = \frac{\epsilon E}{1 - \rho(1 - A_1/4\pi R^2)} \qquad (10\text{-}203)$$

where A_1 is the curved area of a hole in the sphere's surface. The ratio W/E is the effective emittance of the hole as sensed by a narrow-angle receiver viewing the cavity interior. If the material of construction of the cavity is a diffuse emitter and reflector and has an emissivity of 0.5 and the cavity is to appear at least 98 percent black, the curved area A_1 of the hole must be smaller than 2 percent of the total surface area of the sphere.

Enclosures of Surfaces That Are Not Diffuse Reflectors If no restriction that the surfaces be diffuse emitters and reflectors is imposed, Eq. (10-202) becomes much more complex. The W's are replaced by πI's and ϵ_j, I_i, and I_j all become functions of the angle of the leaving beam, and ρ_j goes inside the integral and becomes a function of angles of incidence and reflection. Seldom are such details of reflectance known. When they are and a solution is needed, the Monte Carlo method of tracing the history of a large number of beams emitted from random positions and in random initial directions is probably the best method of obtaining a solution. Another approach is possible, however, because of the tendency of most surfaces to fit a simpler reflection model. The total reflectance $\rho(\equiv 1 - \epsilon)$ can be represented by the sum of a diffuse component ρ_D and a specular component ρ_S. For applications see Hottel and Sarofim (op. cit., chap. 5). The method yields the following relation for exchange between concentric spheres or infinite cylinders:

$$A_1\mathcal{F}_{12} \equiv \overline{S_1S_2}$$
$$= \frac{1}{1/A_1\epsilon_1 + (1/A_2)(1/\epsilon_2 - 1) + [\rho_{s2}/(1 - \rho_{s2})](1/A_1 - 1/A_2)} \qquad (10\text{-}204)$$

When there is no specular reflectance, the third term in the denominator drops out, in agreement with Eq. (10-199). When the reflectance is exclusively specular, the denominator becomes $1/A_1\epsilon_1 + \rho_{s2}/A_1(1 - \rho_{s2})$, easily derivable from first principles.

EMISSIVITIES OF COMBUSTION PRODUCTS

The **radiation from a flame** is due to radiation from burning soot particles of microscopic and submicroscopic dimensions, from suspended larger particles of coal, coke, or ash, and from the water vapor and carbon dioxide in the hot gaseous combustion products. The contribution of radiation emitted by the combustion process itself, so-called chemiluminescence, is relatively negligible. Common to these problems is the effect of the shape of the emitting volume on the radiative flux; this is considered first.

Mean Beam Lengths Evaluation of radiation from a nonisothermal volume is beyond the scope of this section (see Hottel and Sarofim, *Radiative Transfer*, McGraw-Hill, New York, 1967, chap. 11). Consider an isothermal gas confined within the volume bounded by the solid angle $d\Omega$ with vertex at dA and making the angle θ with the normal to dA. The ratio of the emission to dA from the gas to that from a blackbody at the gas temperature and filling the field of view $d\Omega$ is called the **gas emissivity ϵ.** Clearly, ϵ depends on the path length L through the volume to dA. A hemispherical volume radiating to a spot on the center of its base represents the only case in which L is independent of direction. Flux at that spot relative to hemispherical blackbody flux is thus an alternative way to visualize emissivity.

The flux density to an area of interest on the envelope of an emitter volume of any shape can be matched by that at the base of a hemispherical volume of some radius L, which is called the **mean beam length.** It is found that although the ratio of L to a characteristic dimension D of the shape varies with opacity, the variation is small enough for most engineering purposes to permit use of a constant ratio L_M/D, where L_M is the **average mean beam length.** L_M can be defined to apply either to a spot on the envelope or to any finite portion of its area. An important limiting case is that of opacity approaching zero ($PD \to 0$, where P = partial pressure of the emitter constituent). For this case, L (called L_0) equals $4V/A$ (V = gas volume; A = bounding area) when interest is in radiation to the entire envelope. For the range of PD encountered in practice, L (now L_M) is always less. For various shapes, 0.8 to 0.95 times L_0 has been found optimum (see Table 10-18); for shapes not reported in Table 10-18, a factor of 0.88 (or $L_M = 0.88L_0 = 3.5V/A$) is recommended.

Instead of using the average-mean-beam-length concept to approximate $A_1[\epsilon(L_m)]$ (the flux per unit black emissive power from a gas volume to a surface of area A_1), one may calculate the flux rigorously by integration, over the gas volume and over A_1, of the

TABLE 10-18 Mean Beam Lengths for Volume Radiation

Shape	Characteristic dimension, D	L_0/D	L_M/D
Sphere	Diameter	0.67	0.63
Infinite cylinder	Diameter	1.0	0.94
Semi-infinite cylinder, radiating to:			
Center of base	Diameter	1.0	0.90
Entire base	Diameter	0.81	0.65
Right-circle cylinder, ht. = diam., radiating to:			
Center of base	Diameter	0.76	0.71
Whole surface	Diameter	0.67	0.60
Right-circle cylinder, ht. = 0.5 diam., radiating to:			
End	Diameter	0.47	0.43
Side	Diameter	0.52	0.46
Total surface	Diameter	0.50	0.45
Right-circle cylinder, ht. = 2 × diam., radiating to:			
End	Diameter	0.73	0.60
Side	Diameter	0.82	0.76
Total surface	Diameter	0.80	0.73
Infinite cylinder, half-circle cross section, radiating to spot on middle of flat side	Radius	....	1.26
Rectangular parallelepipeds:			
1:1:1 (cube)	Edge	0.67	0.60
1:1:4, radiating to:			
1 × 4 face	Shortest edge	0.90	0.82
1 × 1 face	Shortest edge	0.86	0.71
Whole surface	Shortest edge	0.89	0.81
1:2:6, radiating to:			
2 × 6 face	Shortest edge	1.18	
1 × 6 face	Shortest edge	1.24	
1 × 2 face	Shortest edge	1.18	
Whole surface	Shortest edge	1.2	
Infinite parallel planes	Clearance	2.00	1.76
Space outside infinite bank of tubes, centers on equilateral triangles; tube diam. = clearance	Clearance	3.4	2.8
Same, except tube diam. = 0.5 clearance	Clearance	4.45	3.8
Same, except tube centers on squares, diam. = clearance	Clearance	4.1	3.5

expression $4k\ dv\ \tau(r)\ dA \cos\theta/\pi r^2$. Here k is the emission coefficient of the gas, and $\tau(r)$ is the transmittance through the distance r between dv and dA. The result has the dimensions of area and, by analogy to $\overline{ss}$, is called $\overline{gs}_1$, the direct-exchange area between the gas zone and the surface zone (Hottel and Sarofim, op. cit., chap. 7). The use of $A_1[\epsilon(L_m)]$ instead of $\overline{gs}_1$ is adequate when the problem is such that all the gas can be treated as a single zone having a mean radiating temperature, but the $\overline{gs}$ concept is clearly useful if allowance is to be made for temperature variations within the gas.

Gaseous Combustion Products Radiation from water vapor and carbon dioxide occurs in spectral bands in the infrared. In magnitude it overshadows convection at furnace temperatures.

Carbon Dioxide The emissivity ϵ_G of a CO_2-containing gas volume depends on gas temperature T_G, on the CO_2 partial-pressure–beam-length product P_cL and, to a much lesser extent, on the total pressure. Figure 10-55 gives ϵ_G for carbon dioxide at a total pressure of 101.3 kPa (1 atm). The gas absorptivity α_G equals the emissivity when the absorbing gas and the emitter are at the same temperature. When the emitter surface temperature is T_s, α_G is $(T_G/T_S)^{0.65}$ times the ϵ_G read from Fig. 10-55 at T_S instead of T_G and at P_cLT_S/T_G instead of P_cL. Line broadening due to increases either in total pressure or in partial pressure of CO_2, makes necessary a correction to the emissivity read from Fig. 10-55. However, at a total pressure of 101.3 kPa (1 atm) the correction factor may be ignored, since it decreases with increase in temperature and is never more than 4 percent at temperatures above 1111 K (2000°R). Estimations of the correction in systems up to 1013.3 kPa (10 atm) are given by Hottel and Sarofim (op. cit., p. 228), and by Edwards [*J. Opt. Soc. Am.*, **50,** 617 (1960)], who in addition presents data on CO_2-band emission for use in calculations involving spectrally selective surfaces. The principal emission bands of CO_2 are at about 2.64 to 2.84, 4.13 to 4.5, and 13 to 17 μm.

Water Vapor The gas emissivity depends on T_G and P_wL and on total pressure P_T and on the partial pressure of water vapor P_w. Emissivity due to water vapor is given in Fig. 10-56 as a function of T_G and P_wL, for the special case of $P_w = 0$ and $P_T = 1$. Allowance for departure from these special conditions is made by multiplying ϵ_G from Fig. 10-56 by a factor C_w read from Fig. 10-57 as a function of $(P_w + P_T)$ and P_wL. The absorptivity α_G of water vapor for blackbody radiation is ϵ_G from Fig. 10-56, read at T_S and at $P_wL(T_S/T_G)$ instead of P_wL, then multiplied by $(T_G/T_S)^{0.45}$. The correction factor C_w still applies. Spectral data for water vapor, tabulated for 371 wavelength intervals from 1 to 40 μm, are also available [Ferriso, Ludwig, and Thompson, *J. Quant. Spectros. Radiat. Transfer*, **6,** 241–273 (1966)]. The principal emission is in bands at about 2.55 to 2.84, 5.6 to 7.6, and 12 to 25 μm.

Carbon Dioxide-Water-Vapor Mixtures When these gases are present together, the total radiation due to both is somewhat less than the sum of the separately calculated effects, because each gas is somewhat opaque to radiation from the other in the wavelength regions 2.7 and 15 μm. The **spectral-overlap correction factor** C_{SO} by which to reduce the sum of ϵ_G for CO_2 and ϵ_G for H_2O (each evaluated as if the other were absent) to obtain the ϵ_G due to the two together is read from Fig. 10-58. The same type of correction applies in calculating α_G.

To summarize, the gas emissivity and absorptivity due to carbon dioxide and water vapor are formulated from the following:

$$\epsilon_G = (\epsilon_{CO2,T_G,P_cL} + \epsilon_{H2O,T_G,P_wL}C_W)(1 - C_{SO}) \qquad (10\text{-}205)$$

$$\alpha_G = (\alpha_{CO2} + \alpha_{H2O})(1 - C_{SO}) \qquad (10\text{-}206)$$

$$\alpha_{CO2} = (\epsilon_{CO2,T_S,P_cLT_S/T_G})(T_G/T_S)^{0.65} \qquad (10\text{-}206a)$$

$$\alpha_{H2O} = (\epsilon_{H2O,T_S,P_wLT_S/T_G})(T_G/T_S)^{0.45}C_w \qquad (10\text{-}206b)$$

The long subscripts indicate the values of T and PL at which Figs. 10-55 and 10-56 are read.

Effective use can sometimes be made of the fact that, at furnace temperatures and for gases containing H_2O and CO_2 in a fixed ratio, ϵ_G decreases with rising temperature in such a way that the product ϵ_GT_G depends almost exclusively on $(P_c + P_w)L$. Figure 10-59, constructed on this basis, gives the (total emissivity) (temperature) prod-

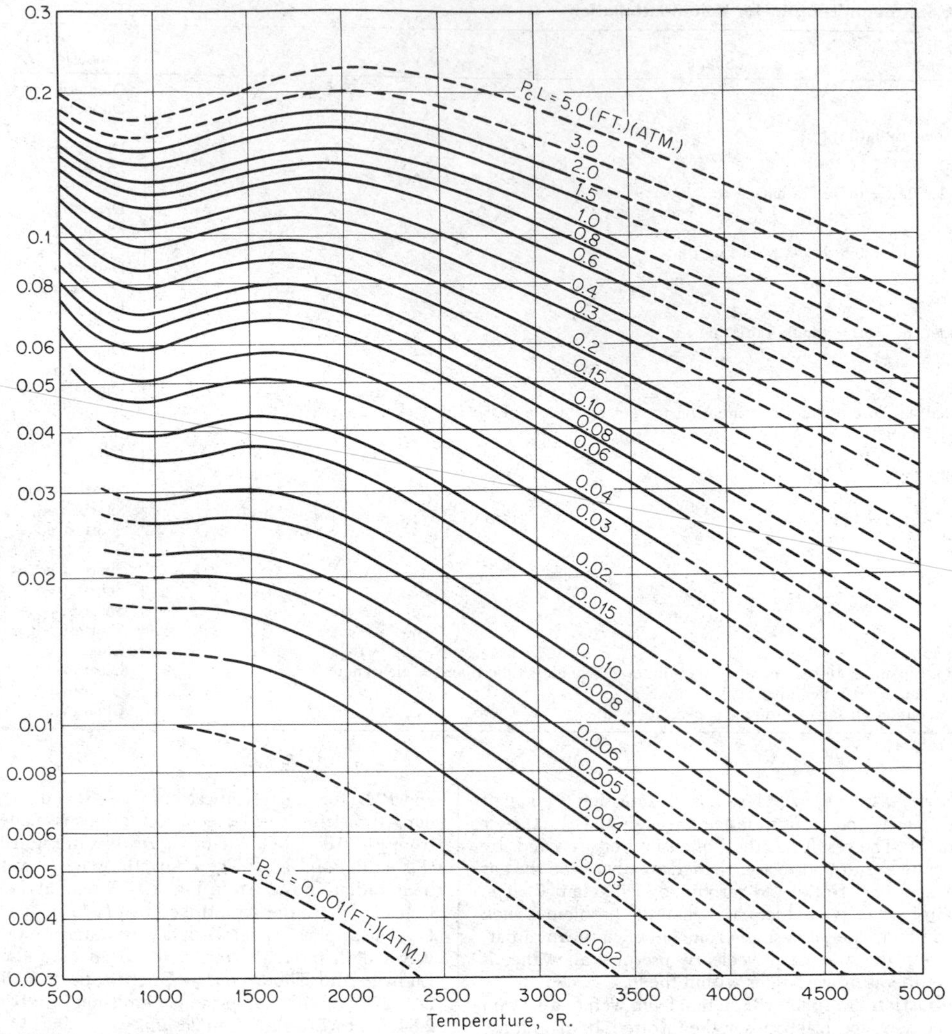

FIG. 10-55 Emissivity of carbon dioxide, 1-atm (101.3-kPa) total pressure. To convert degrees Rankine to kelvins, multiply by $(5.556)(10^{-1})$; to convert atmosphere-feet to kilopascal-meters, multiply by 30.89.

uct versus $(P_c + P_w)L$ for combustion products with $P_w/P_c = 1$ (dashed line) and 2 (solid line); it shows as well the temperature range (shaded zone) within which the error in total emissivity is less than 4 percent. The P_w/P_c range of 1 to 2 almost brackets industrial hydrocarbon fuels. Figure 10-59 has value, however, only for problems in which the gas temperature is so much higher than the surface temperature as to make the value of gas absorptivity unimportant.

Example 6 Flue gas containing 6 percent CO_2 and 11 percent H_2O vapor, wet basis, flows through a bank of tubes of 0.102-m (4-in) outside diameter on equilateral 0.203-m (8-in) triangular centers. In a section in which the gas and tube surface temperatures are 691°C (964 K, 1275°F) and 413°C (686 K, 775°F), what are the emissivity and absorptivity of the gas? From Table 10-18, $L_m = (2.8)(0.102) = 0.286$ m (0.94 ft), $P_cL = 1.73$ kPa·m (0.056 ft·atm), and $P_wL = 3.18$ kPa·m (0.103 ft·atm). From Fig. 10-55, $\epsilon_{CO_2} = 0.065$; $\alpha_{CO_2} = 0.054\,(964/686)^{0.65} = 0.067$. From Fig. 10-56, $\epsilon_w = 0.063$ and $\alpha_w = 0.069\,(964/686)^{0.45} = 0.080$ at $P_w = 0$. From Fig. 10-57, $C_w = 1.08$. From Fig. 10-58, overlap correction = 0.02.

$$\epsilon_{G,\text{total}} = [0.065 + (0.063)(1.08)](1 - 0.02) = 0.130$$
$$\alpha_{G,\text{total}} = [0.067 + (0.080)(1.08)](1 - 0.02) = 0.149$$

If the approximation represented by Fig. 10-59 is accepted, then for $P_w/P_c = 11/6$ and $(P_c + P_w)L = 4.91$, the temperature range for 4 percent accuracy is found to be 1500 to 890 K (2700 to 1600°R); and at $T = 964$ K (1735°R), $\epsilon_{c+w}T = 133.3$, $\epsilon_{c+w} = 133.3/964 = 0.138$ (versus 0.130).

Other Gases Because of their practical importance, the emissivities of CO_2 and H_2O have been studied much more extensively than those of other gases, and the values summarized in the preceding paragraphs are based on extensive measurement of both total and integrated spectral values. A summary of the less adequate information on other gases appears in Table 10-19.

Flames and Particle Clouds

Luminous Flames Luminosity conventionally refers to **soot** radiation; it is important when combustion occurs under such conditions that the hydrocarbons in the flame are subject to heat in the absence of sufficient air well mixed on a molecular scale. Because soot particles are small relative to the wavelength of the radiation of interest [diameters $(2)(10^{-8})$ to $(1.4)(10^{-7})$ m (200 to 1400 Å)], the monochromatic emissivity ϵ_λ depends on the total particle volume

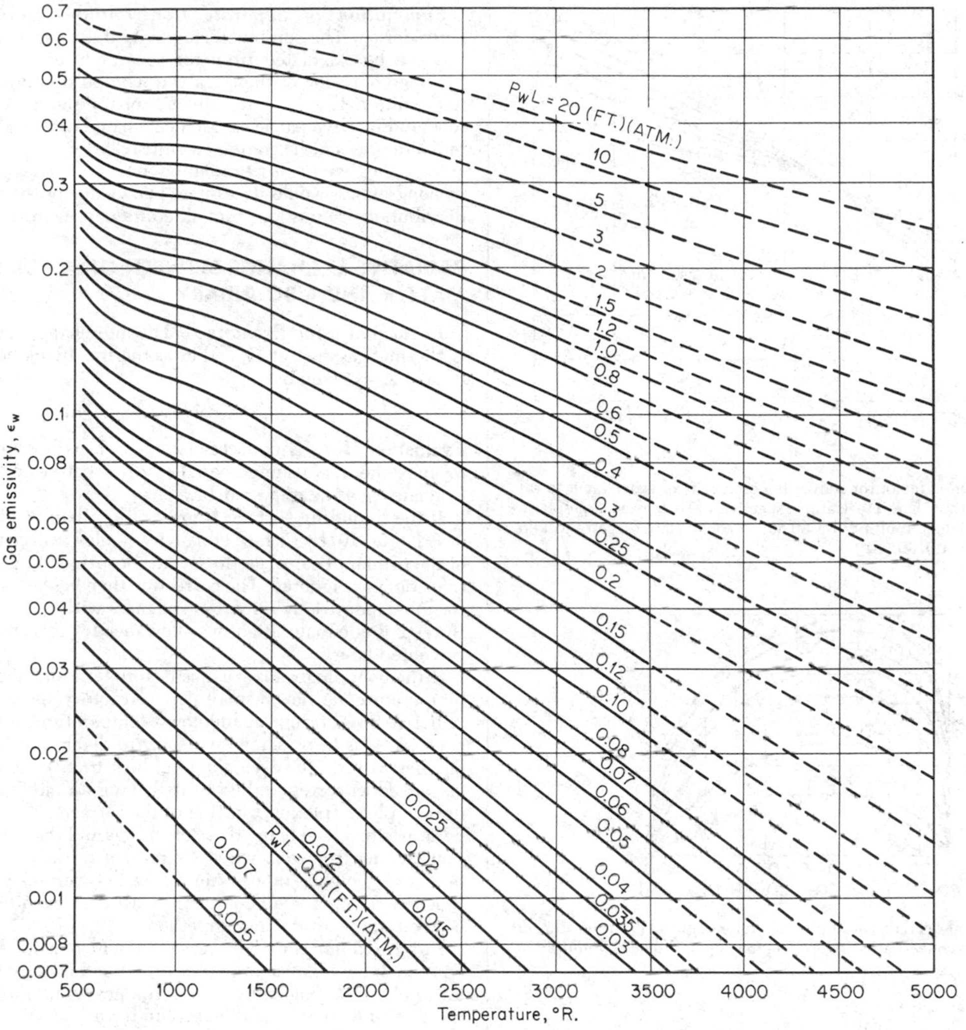

FIG. 10-56 Emissivity of water vapor at zero partial pressure and 1-atm (101.3-kPa) total pressure. To convert degrees Rankine to kelvins, multiply by (5.556)(10^{-1}).

per unit volume of space f_v regardless of particle size. It is given by

$$\epsilon_\lambda = 1 - e^{-Kf_vL/\lambda} \tag{10-207}$$

where L is the path length. Integration of this over the energy spectrum gives the total emissivity ϵ as

$$\epsilon = 1 - \frac{1}{(1 + KTf_vL/c_2)^4} \tag{10-208}$$

where c_2 is the second Planck constant. K/c_2 can be obtained from the complex refractive index of soot, in turn dependent on its hydrogen-carbon ratio H/C. According to a study of coals, K/c_2 varies from 480 $m^{-1} \cdot K^{-1}$ at H/C = 0 to 2.4 at H/C = 0.4; some experimental work at IJmuiden (International Flame Foundation) on two oil-flame types leads to 4.4 and 9.8. A tentative value of 500 $m^{-1} \cdot K^{-1}$ (85 $ft^{-1} \cdot °R^{-1}$) is recommended.

There is at present no method of predicting soot concentration of a luminous flame analytically; reliance must be placed on experimental measurement on flames similar to that of interest. Visual observation is misleading; a flame so bright as to hide the wall behind it may be far from a "black" radiator. The International Flame Foundation has recorded data on many luminous flames from gas, oil, and coal (see *J. Inst. Fuel*, 1956 to present). Addition of 0.1 to nonluminous-gas emissivity to allow for soot luminosity is often sufficient if calculations of total flame emission are to be based on a mean flame temperature; this is true because emission from the flame comes more from its cool envelope than from its hot core, especially as its emissivity goes up.

Clouds of Large Black Particles The emissivity ϵ_c of a cloud of particles with a perimeter large compared with wavelength λ is

$$\epsilon_c = 1 - e^{-(a/v)L} \tag{10-209}$$

where a/v is the projected area of the particles in unit volume of space. If the particles have no negative curvature (a particle can see none of itself) and are randomly oriented, a is $a'/4$, where a' is the actual surface area; and if the particles are uniform, $a/v = cA = cA'/4$ where A and A' are the projected and total areas of each particle and c is the number concentration of particles. For spherical particles this gives

$$\epsilon_c = 1 - e^{-(\pi/4)cd^2L} = 1 - e^{-1.5f_vL/d} \tag{10-209a}$$

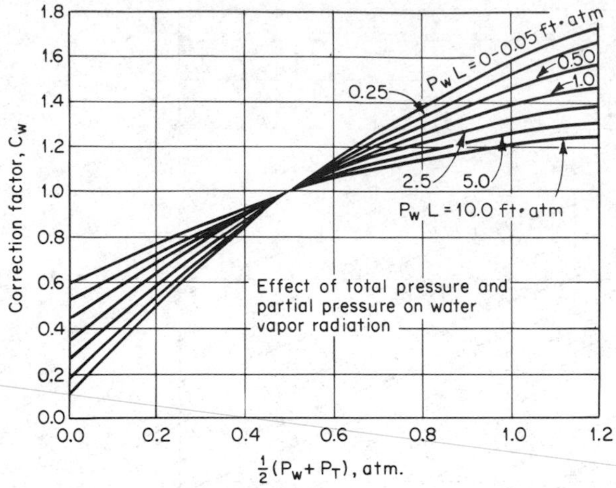

FIG. 10-57 Correction factor for converting emissivity of water vapor to values of P_w and P_T other than 0 to 1 atm respectively. To convert atmosphere-feet to kilopascal-meters, multiply by 30.89; to convert atmospheres to kilopascals, multiply by (1.0133)(10^2).

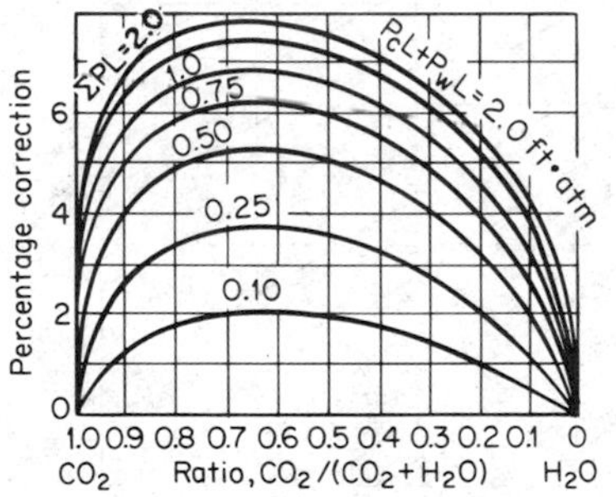

FIG. 10-58 Spectral-overlap correction C_{so} for mixtures of carbon dioxide and water vapor. To convert atmosphere-feet to kilopascal-meters, multiply by 30.89.

As an example, consider heavy fuel oil ($CH_{1.5}$, specific gravity, 0.95) atomized to a surface mean particle diameter of d, burned with 20 percent excess air to produce coke-residue particles having the original drop diameter and suspended in combustion products at 1204°C (2200°F). The flame emissivity due to the particles along a path of L m will be, with d in micrometers,

$$\epsilon = 1 - e^{-24.3L/d} \qquad (10\text{-}209b)$$

With 200-μm particles and an L of 3.05 m (10 ft), the particle contribution to emissivity will be 0.31. Soot luminosity will increase this; particle burnout will decrease it.

Clouds of Nonblack Particles The correction for nonblackness of the particles is complicated by multiple scatter of the radiation reflected by each particle. The emissivity ϵ_c of a cloud of gray particles of individual surface emissivity ϵ_S can be estimated by the use of Eq. (10-209), with its exponent multiplied by ϵ_S, if the optical thickness $(a/v)L$ does not exceed about 2. Modified Eq. (10-209) would predict an approach of ϵ_c to 1 as $L \rightarrow \infty$, an impossibility in a scattering system; the asymptotic value of ϵ_c can be read from Fig. 10-44 as ϵ_h, with albedo ω given by particle-surface reflectance $1 - \epsilon_S$. Particles with a perimeter lying between 0.5 and 5 times the wavelength of interest can be handled with difficulty by use of the Mie equations (see Hottel and Sarofim, op. cit., chaps. 12 and 13).

Summation of Separate Contributions to Gas or Flame Emissivity The combined emissivity due to several kinds of emitters can be calculated from the separately calculated emissivities, provided only one of these (gaseous combustion products) is a selective emitter. If ϵ_G, ϵ_{C1}, ϵ_{C2} are the separate emissivities due to gas, soot, and massive particles, each calculated as though no other emitter were present, the combined emissivity is $1 - (1 - \epsilon_G)(1 - \epsilon_{C1})(1 - \epsilon_{C2})$. This assumes soot radiation to be gray; rigorous allowance for its nongrayness would necessitate knowledge of the detailed spectral distribution of emission from all emitters present.

RADIATIVE EXCHANGE BETWEEN GASES OR SUSPENDED MATTER AND A BOUNDARY

Local Radiative Exchange The interchange rate $\dot{Q}$ between an isothermal gas mass at T_G and its isothermal **black bounding surface** of area A_1 is given by

$$\dot{Q} = A_1\sigma(T_G^4\epsilon_G - T_1^4\alpha_G) \qquad (10\text{-}210)$$

Evaluation of α_G is unnecessary when T_1 is less than one-half T_G; α_G may then be assumed equal to ϵ_G. A better approximation is to evaluate α_G as ϵ_G at T_1 and PL.

If the **bounding surface is gray** rather than black, multiplication of Eq. (10-210) by surface emissivity ϵ_1 allows properly for reduction of the primary beams, gas-to-surface or surface-to-gas, but secondary reflections are ignored. The correction then lies between ϵ_1 and 1, and for most industrially important surfaces with $\epsilon_1 > 0.8$ a value of $(1 + \epsilon_1)/2$ is adequate. Rigorous allowance for this and other factors is presented later.

If the bounding walls are mostly **sink-type surfaces** of area A_1 and temperature T_1, but in small part **refractory surfaces** of area A_r in radiative equilibrium at unknown temperature T_r, an energy balance on A_r is in principle necessary to determine T_r and the effect on energy flux. However, the total heat transfer to the sink may be visualized as corresponding to its having an effective area equal to its own plus a fraction x of that of the refractory, with the only temperatures involved being those of the gas and the heat sink. The fraction x varies from zero when the ratio of refractory to heat-sink surface is very high to unity when the ratio is very low and the value of ϵ_G is low. If A_r is small compared with A_1, a value for x of 0.7 may be used in the approximate method.

Long Exchanger This case, in which axial radiative flux is ignored, includes most radiatively modified heat exchangers of interest to chemical engineers. When the gas temperature transverse to the flow direction is reasonably uniform and the chamber is long compared with its mean hydraulic radius, the opposed upstream and downstream fluxes through the flow cross section will substantially cancel (hot combustion products through tubes or across tube banks, tunnel kilns, billet-reheating furnaces, Example 7). Under these conditions, the radiative contribution to local flux density q may be formulated in terms of local temperatures and beam lengths or exchange areas evaluated for a two-dimensional system infinite in the flow direction. The **local flux density** at the sink A_1 is then

$$q(T_G,T_1) = q_r(T_G,T_1) + h(T_G - T_1) \qquad (10\text{-}211)$$

where h is the local convective heat-transfer coefficient and $q_r(T_G,T_1)$ the radiation contribution calculated from T_G, T_1, ϵ_G, and ϵ_1 by using the approximate treatment in the preceding subsection or the more rigorous treatment in the following subsection. If $\dot{m}C_p$ is the hourly heat capacity of the gas stream, the temperature of which changes by dT_G over the sink-area increment dA_1, then

$$[q(T_G,T_1)]\, dA_1 = -\dot{m}C_p\, dT_G \qquad (10\text{-}212)$$

from which

$$A_1 = \dot{m}\int_{T_{G,\text{outlet}}}^{T_{G,\text{inlet}}} \frac{C_p\, dT_G}{q(T_G,T_1)} \qquad (10\text{-}213)$$

The area under a curve of C_p/q versus T_G or $1/q$ versus the specific enthalpy i may be used to solve for the area A_1 required to obtain a

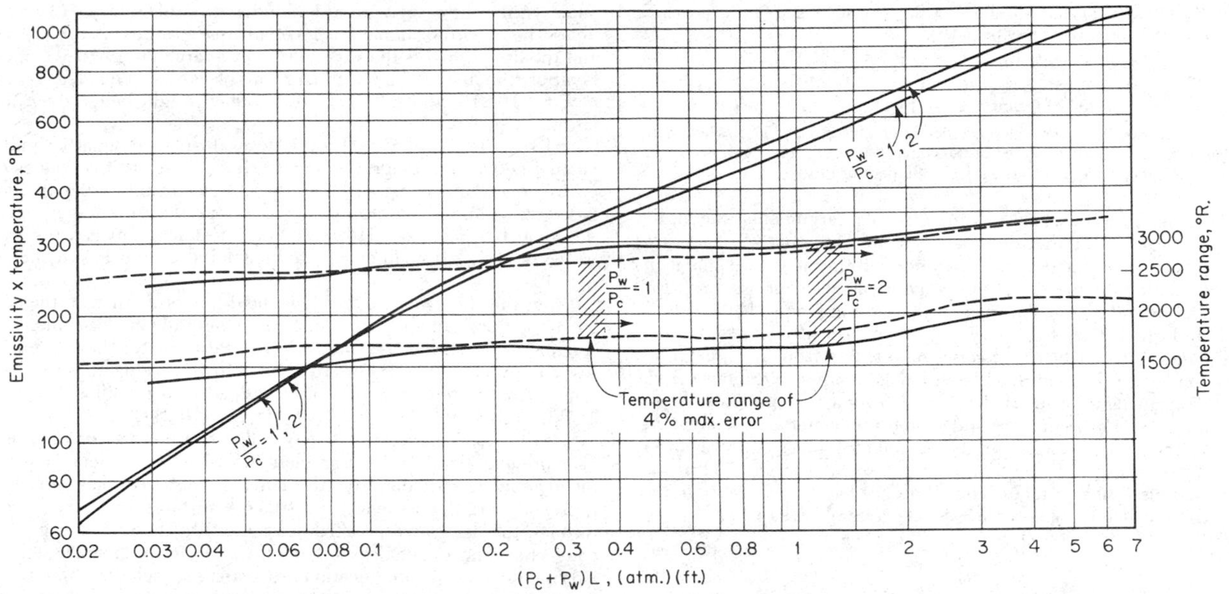

FIG. 10-59 Simplified emissivity chart for CO_2-H_2O mixture with $P_w/P_c = 1$ and 2. Error is less than 4 percent within temperature range given by shaded band. To convert atmosphere-feet to kilopascal-meters, multiply by 30.89; to convert degrees Rankine to kelvins, multiply by $(5.556(10^{-1})$.

given outlet temperature or to obtain the outlet temperature given A_1. Three points generally suffice to determine the area under the curve within 10 percent.

Instead of using graphical integration, which can handle any complexity of variation of flux density q with T_G and T_1 along an interchanger flow path, one may evaluate a mean flux density based on mean gas and sink temperatures, based in turn on terminal temperatures. It has been found empirically that fair results are obtained by the use of a mean surface temperature equal to the arithmetic mean of the terminal surface temperatures and by the use of a mean gas temperature equal to the mean surface temperature plus the logarithmic mean of the temperature difference, gas to surface, at the two ends of the exchanger. When radiation dominates the transfer process, however, graphical integration is safer.

Example 7 Flue gas containing 6 percent carbon dioxide and 11 percent water vapor by volume (wet basis) flows through the convection bank of an oil tube still consisting of rows of 0.102-m (4-in) tubes on 0.203-m (8-in) centers, nine 7.62-m (25-ft) tubes in a row, the rows staggered to put the tubes on equilateral triangular centers. The flue gas enters at 871°C (1144 K, 1600°F) and leaves at 538°C (811 K, 1000°F). The oil flows in a countercurrent direction to the gas and rises from 316 to 427°C (600 to 800°F). Tube surface emissivity is 0.8. What is the average heat-input rate, due to gas radiation alone, per square meter of external tube area?

With each row of tubes there is associated $(0.203)(\sqrt{3}/2) = 0.176$ m (0.577 ft) of wall height, of area $[(0.203)(9)(2) + (7.62)(2)]0.176 - (9)(2)(\pi)(0.0508)^2 = 3.18$ m^2 (34.2 ft^2). One row of tubes has an area of $(\pi)(0.102)(7.62)(9) = 22.0$ m^2 (236 ft^2). If the recommended factor of 0.7 on the refractory area is used, the effective area of the tubes is $[22.0 + (0.7)(3.18)]/22.0 = 1.10$ m^2/m^2 of actual area. The exact evaluation of the outside tube temperature from the known oil temperature would involve a knowledge of the oil-film coefficient, tube-wall resistance, and rate of heat flow into the tube, the evaluation usually involving trial and error. However, for the present purpose the temperature drop through the tube wall and oil film will be assumed to be 41.7°C (75°F), making the tube surface temperatures 357°C (675°F) and 468°C (875°F); the average is 412°C (775°F). The radiating gas temperature is

$$t_g = 412 + \frac{(871 - 468) - (538 - 357)}{2.3 \log [(871 - 468)/(538 - 357)]}$$
$$= 412 + 278 = 690°\text{C } (1274°\text{F})$$

These temperatures, partial pressures, and dimensions were used in Example 6 to determine gas emissivity and absorptivity. $\epsilon_G = 0.130$; $\alpha_G = 0.149$. The approximate effective emissivity of the boundary is $(0.8 + 1)/2 = 0.9$. Then

TABLE 10-19 Total Emissivities of Some Gases

Temperature	1000°R.			1600°R.			2200°R.			2800°R.		
P_xL, (atm.)(ft.)	0.01	0.1	1.0	0.01	0.1	1.0	0.01	0.1	1.0	0.01	0.1	1.0
NH_3[a]	0.047	0.20	0.61	0.020	0.120	0.44	0.0057	0.051	0.25	(0.001)	(0.015)	(0.14)
SO_2[b]	0.020	0.13	0.28	0.013	0.090	0.32	0.0085	0.051	0.27	0.0058	0.043	0.20
CH_4[c]	0.020	0.060	0.15	0.023	0.072	0.194	0.022	0.070	0.185	0.019	0.059	0.17
CO[d]	0.011	0.031	0.061	0.022	0.057	0.10	0.022	0.050	0.080	(0.012)	(0.035)	(0.050)
NO[e]	0.0046	0.018	0.060	0.0046	0.021	0.070	0.0019	0.010	0.040	0.00078	0.004	0.025
HCl[f]	0.00022	0.00079	0.0020	0.00036	0.0013	0.0033	0.00037	0.0014	0.0036	0.00029	0.0010	0.0027

NOTE: Figures in this table are taken from plots in Hottel and Sarofim, *Radiative Transfer*, McGraw-Hill, New York, 1967, chap. 6. Values in parentheses are extrapolated. To convert degrees Rankine to kelvins, multiply by $(5.556)(10^{-1})$. To convert atmosphere-feet to kilopascal-meters, multiply by 30.89.

[a]Total-radiation measurements of Port (Sc.D. thesis in chemical engineering, MIT, 1940) at 1-atm total pressure, L = 1.68 ft, T to 2000°R.

[b]Calculations of Guerrieri (S.M. thesis in chemical engineering, MIT, 1932) from room-temperature absorption measurements of Coblentz (*Investigations of Infrared Spectra*, Carnegie Institution, Washington, 1905) with poor allowance for temperature.

[c]Band measurements of Lee and Happel [*Ind. Eng. Chem. Fundam.*, 3, 167 (1964)] at T up to 2050°R plus calculations to extrapolate temperature to 3800°R.

[d]Total-radiation measurements of Ullrich (Sc.D. thesis in chemical engineering, MIT, 1953) at 1-atm total pressure, L = 1.68 ft, T to 2200°R.

[e]Calculations of Malkmus and Thompson [*J. Quant. Spectros. Radiat. Transfer*, **2**, 16 (1962)], to T = 5400°R and PL = 30 atm·ft.

[f]Calculations of Malkmus and Thompson [*J. Quant. Spectros. Radiat. Transfer*, **2**, 16 (1962)], to T = 5400°R and PL = 300 atm·ft.

from Eq. (10-210), modified to allow for sink emissivity and for the presence of a small amount of refractory boundary,

$$\dot{Q}/A_1 = q = (0.9)(1.10)\sigma(T_G^4\epsilon_G - T_1^4\alpha_G)$$
$$= (0.9)(1.10)(5.67)[(9.63)^4(0.13) - (6.85)^4(0.149)]$$
$$= 4440 \text{ J/(m}^2 \text{ tube area}\cdot\text{s) [1408 Btu/(ft}^2 \text{ tube area}\cdot\text{h)]}$$

This is equivalent to a convection coefficient of 4440/278, or 11.8, which is of the order of magnitude expected of the convection coefficient itself. Radiation rapidly becomes dominant as the system temperature rises.

Total-Exchange Areas $\overline{SS}$ and $\overline{GS}$ The arguments leading to the development of the interchange factor $A_i\mathcal{F}_{ij}$ ($\equiv\overline{S_iS_j}$) between surfaces apply to the case of absorption within the gas volume if in the evaluation of the direct-exchange areas allowance is made for attenuation in the gas. This necessitates nothing more than redefinition, in Eqs. (10-196) to (10-198), of every term $\overline{ij}(\equiv\overline{S_iS_j}\equiv A_iF_{ij})$ to represent, per unit black emissive power, flux from A_i through an absorbing gas to A_j. This may be visualized as multiplication of A_iF_{ij} by the mean gas transmittance $\tau_{ij}(=1-\epsilon$ for a gray gas). In a system containing an isothermal gas and source-sink boundaries of areas $A_1 \ldots A_n$, the total emission from A_1 per unit of its black emissive power is $\epsilon_1 A_1$, of which $\overline{S_1S_1} + \overline{S_1S_2} + \cdots + \overline{S_1S_n}$ is absorbed in the various source-sink surfaces. The difference has been absorbed in the gas and is called the **gas-surface total-exchange area** $\overline{GS_1}$

$$\overline{GS_1} = A_1\epsilon_1 - \sum_i \overline{S_1S_i} \qquad (10\text{-}214)$$

If the gas volume is not isothermal and is zoned, an additional magnitude, the gas-to-gas total-exchange area $\overline{G_iG_j}$, arises (see Hottel and Sarofim, *Radiative Transfer*, McGraw-Hill, New York, 1967, chap. 11). Space does not permit derivations of special cases; only the single-gas-zone system is treated here.

If an enclosure consists of an **isothermal gray gas** of such shape and size that its emissivity is ϵ_G, bounded by two surface zones, a sink of area A_1 and a radiatively adiabatic zone of area A_r, $\overline{GS_1}$ is given by

$$\frac{1}{(\overline{GS_1})_R} = \frac{1}{A_1}\left(\frac{1}{\epsilon_1} - 1\right) + \epsilon_G\left[A_1 + \frac{A_r}{1 + [\epsilon_G/(1-\epsilon_G)F_{r1}]}\right] \qquad (10\text{-}215)$$

where F_{r1} represents the view factor from A_r to A_1 and the subscript R on $\overline{GS_1}$ reminds the reader that allowance has been made for the contribution of refractory surface, through reflection or absorption and reemission, to the net flux between gas and A_1. If the enclosure is assumed to be speckled, with its sink surface A_1 and its refractory surface A_r intimately mixed so that $F_{r1} = A_1/(A_1 + A_r)$, the relation simplifies to

$$(\overline{GS_1})_R = \frac{A_T}{1/C_S\epsilon_1 + 1/\epsilon_G - 1} \qquad (10\text{-}216)$$

where C_S is the "cold" surface fraction A_1/A_T and A_T is the total surface area $A_1 + A_r$. The model whose performance is given by Eq. (10-216) may be referred to as the **gray-gas speckled-furnace model.** It is to be noted that the refractory emissivity does not enter.

Rigorous allowance for gas nongrayness is made elsewhere (Hottel and Sarofim, op. cit., chap. 8). If a simple model of real gas—the **gray-plus-clear-gas model**—is accepted and if the refractory surface is assumed to reflect diffusely and completely, Eq. (10-216) becomes

$$(\overline{GS_1})_R = \frac{aA_T}{1/C_S\epsilon_1 + a/\epsilon_G - 1} \qquad (10\text{-}217)$$

where a is the effective energy fraction of the blackbody spectrum in which the gas absorbs and is calculated from ϵ_G evaluated at the mean beam length L_m and at twice that value,

$$a = \frac{[\epsilon_G(L_m)]^2}{2\epsilon_G(L_m) - \epsilon_G(2L_m)} \qquad (10\text{-}217a)$$

The model whose performance is given by Eq. (10-217) may be referred to as the **real-gas white-refractory speckled-furnace model.** Of more interest than Eq. (10-216) or Eq. (10-217) would be one covering the case of a **real gas** and a **gray refractory.** However, unless the radiating characteristics of a furnace refractory are known and the design merits the expenditure of a large effort, $(\overline{GS_1})_R$ may be approximated as the weighted mean of values calculated by use of Eqs. (10-216) and (10-217), with double weight being given the latter.

Strictly, the use of Eq. (10-217) to evaluate net radiative interchange between a nongray gas and sink A_1 involves two values of $(\overline{GS_1})_R$, one multiplying σT_G^4 to represent gas-to-sink radiation, and based on ϵ_G, and one multiplying σT_1^4 to represent sink-to-gas radiation, and based on α_{G1}. However, the difference between the two values of $(\overline{GS_1})_R$ is small enough to be ignored if T_1 is less than one-half T_G.

From this discussion of $(\overline{GS_1})_R$ it should be clear that, in the calculation of performance of a "long" exchanger in which a significant fraction of the surface is radiatively adiabatic and/or the sink emissivity is low, Eq. (10-216) or (10-217), with $\overline{GS}$ evaluated for a two-dimensional system, would be used instead of a modified Eq. (10-210) in the formulation of $q_r(T_G, T_1)$ in Eq. (10-211).

Combustion Chambers The so-called **radiant section** of a furnace presents a heat-transfer problem in which there enters the combined action of radiation from the flame to the stock or heat sink and radiation from the flame to the refractory surfaces and thence back through the flame (with partial absorption) to the sink, along with convection and external losses. Solutions of the problem based on varying degrees of simplification are available, including allowance for temperature variation in both gas and refractory walls (Hottel and Sarofim, *Radiative Transfer*, McGraw-Hill, New York, 1967, chap. 14). A less rigorous treatment suffices, however, for handling many problems. There are two limiting cases: the long chamber with gas temperature varying only in the direction of gas flow (already treated) and the compact chamber containing a gas or a flame to which can be assigned an effective or average radiating temperature. Let us consider the latter.

Stirred-Chamber Model; Refractory Wall Loss Negligible What furnace engineers most need is a closed-form solution of the problem, theoretically sound in structure and therefore containing a minimum number of parameters and no empirical constants and, preferably, physically visualizable. They can then (1) correlate data on existing furnaces, (2) develop a performance equation for standard design, or (3) estimate performance of a new furnace type on which no data are available.

Let the furnace-chamber walls be divided into just two classes of surfaces, the heat sink of area A_1 and specified temperature T_1 and the radiatively adiabatic surface of area A_r and unknown temperature T_r. Refractory surfaces are substantially radiatively adiabatic when the convective flux to their inside face differs from the external loss through the walls by an amount small compared with incident radiation (this is true of most refractory furnace walls but not, for example, of the roof of an open-hearth furnace). When the chamber heat sink is a continuous surface, no ambiguity exists about A_1; when it consists of tubes mounted near and parallel to a refractory wall and when convection due to gas flow behind and between the tubes can be ignored relative to radiation, the tube-back-wall system can be replaced by an equivalent gray plane of area A_1 equal to that of the tube-covered wall (not just the projected tube area), of temperature T_1 equal to that of the tube surface, and of effective emissivity and absorptivity ϵ_1 equal to the fraction of the radiation, incident on the tube-refractory system from the furnace-chamber side, which is absorbed both directly and by aid of the refractory backing. This is obtained from Eq. (10-200) as illustrated in Example 5. The gas-surface total-exchange-area concept of the preceding subsection may be used to formulate the net flux from the gas. From it

$$\dot{Q}_{G,\text{net}} = (\overline{GS_1})_R\sigma(T_G^4 - T_1^4) + h_1A_1(T_G - T_1) \qquad (10\text{-}218)$$

Convection is usually a small part of the total transfer rate. Its linearization in T^4 allows replacement of $h_1(T_G - T_1)$ by $\sigma(T_G^4 - T_1^4)h_1/4\sigma T_{G1}^3$, where T_{G1} may be approximated as the arithmetic mean of T_G and T_1 (guessed for the first time in calculating). Equation (10-218) then becomes

$$\dot{Q}_{G,\text{net}} = \left[(\overline{GS_1})_R + \frac{h_1 A_1}{4\sigma T_{G1}^3} \right] \sigma(T_G^4 - T_1^4) \equiv (\overline{GS_1})_{R,c}\sigma(T_G^4 - T_1^4) \tag{10-219}$$

The last part of the equation defines a pseudo total-exchange area which includes allowance for gas convection. With this equation of flux, there must be combined an energy balance. If T_0 is the base temperature, H_F the entering hourly enthalpy of the feed stream (air and fuel) above that base, Δ the amount by which the leaving-gas temperature is less than the radiating temperature T_G (the apparent clairvoyance needed to choose Δ is considered later), and $\dot{m}C_p$ the hourly heat capacity of the combustion products over the temperature interval $(T_G - \Delta)$ to T_0, the energy balance is

$$H_F - \dot{Q}_G = (T_G - \Delta - T_0)\dot{m}C_p$$

If the same mean specific heat is used to define T_{AF} (a kind of adiabatic flame temperature given by $T_0 + H_F/mC_p$ and much higher than the true value), these relations may be combined with Eq. (10-219) to eliminate the unknown T_G and, by suitable manipulation, to give the dimensionless relation

$$\dot{Q}'D' + \tau^4 = (1 + \Delta' - \dot{Q}')^4 \tag{10-220}$$

where $\dot{Q}'$ = reduced furnace efficiency, the actual efficiency $\dot{Q}_G/H_F$ times the temperature ratio $(T_{AF} - T_0)/T_{AF}$; D' = reduced firing density,

$$\frac{H_F}{\sigma A_T T_{AF}^3 (T_{AF} - T_0)} \frac{A_T}{(\overline{GS_1})_{R,c}} \quad \text{or} \quad D\frac{A_T}{(\overline{GS_1})_{R,c}}$$

τ = reduced sink temperature, T_1/T_{AF}; $\Delta' = \Delta/T_{AF}$, the ratio of the temperature correction, from mean radiating point to chamber outlet, to the adiabatic flame temperature

Equation (10-220) says that the efficiency of the chamber is a function of the firing density, the heat-sink temperature, and the gas-temperature change Δ (at this point the one arbitrary quantity); and the firing-density term makes due allowance for any operating variables such as fuel type, excess air, or air preheat which affect flame temperature or gas emissivity, for fractional occupancy of the walls by sink surfaces, and for sink emissivity. If the furnace gas is well stirred, Δ' approaches zero.

The relation of reduced values of furnace efficiency $\dot{Q}'$ to firing density D' and sink temperature τ, with $\Delta = 0$, is shown in Fig. 10-60, based on Eq. (10-220). The shaded areas indicate the operating regimes of a wide range of furnace types. Note the significant properties of the function presented: (1) As firing rate D' goes down, the efficiency rises and approaches $1 - \tau$ in the limit. (This conclusion is modified if wall losses are significant.) (2) Changes in sink temperature have little effect if $\tau < 0.3$. (3) As the furnace walls approach complete coverage by a black sink [$C_S\epsilon_1 \rightarrow 1$ in Eqs. (10-216) and (10-217)] and as convection becomes unimportant, the effect of flame emissivity on D' becomes one of inverse proportionality; thus at very high firing rates at which $\dot{Q}'$ approaches inverse proportionality to D', the efficiency of heat transfer varies directly as ϵ_G (gas-turbine chambers), but at low firing rates ϵ_G has relatively little effect. (4) When $C_S\epsilon_1 \ll 1$ because of a nonblack sink or much refractory surface, the effect of changing flame emissivity is to produce a much less than proportional effect on heat flux.

The factor Δ, the allowance for imperfect stirring, must be estimated. Values in the range of 93 to 149°C (200 to 300°F) have been found to produce data correlation for a series of tests on marine boilers. Δ increases with increase in $\dot{Q}'$ or decrease in D'; in the absence of data on the type of furnace of interest, a tentative recommendation is that $\Delta' = \dot{Q}'/4$.

Equation (10-220) and Fig. 10-60 serve as a framework for correlating the performance of furnaces with flow patterns—plug flow, parabolic profile, and recirculatory flow—which differ from the well-stirred model [Hottel and Sarofim, *Int. J. Mass Heat Transfer*, 8, 1153 (1965)]. As expected, plug-flow furnaces show somewhat higher efficiency, mild-recirculation types somewhat lower efficiency, and strong-recirculation furnaces a performance closely similar to that of the well-stirred model. Equation (10-220) has also been used to correlate data on the radiant section of a tube still in which τ varied from 0.34 to 0.47. The assumption of a mean τ of 0.4 and a value of Δ' of $Q'/4$ leads, from Eq. (10-220), to

$$\frac{\dot{Q}'D'A_T}{(\overline{GS_1})_{R,c}} = \left(1 - \frac{3}{4}\dot{Q}'\right)^4 - (0.4)^4$$

Though $(\overline{GS_1})_{R,c}$ varied with gas temperature and excess air, its mean value was used. The average deviation of 10 measured performance points from the above relation was only 3.8 percent even though no constants of the equation were determined empirically from furnace data [Hottel, *J. Inst. Fuel*, 34, 220 (1961)].

Allowance for Wall Losses Steam-boiler furnaces, tube stills, and reforming furnaces are typical of systems in which, in the interest of simplicity of the governing equation, wall losses are ignored, as they have been in Eq. (10-220). In many high-temperature furnaces, however, particularly those containing large refractory-wall areas, wall losses cannot be ignored. If the wall loss through refractory areas A_r is approximated by $\dot{Q}_{\text{loss}} = UA_r(T_G - T_0)$, with U given by $1/(1/h_{c,\text{inside}} + \Sigma W_i/k_i + 1/h_{c+r,\text{outside}})$ where W_i is that thickness of wall having thermal conductivity k_i, it will be found that

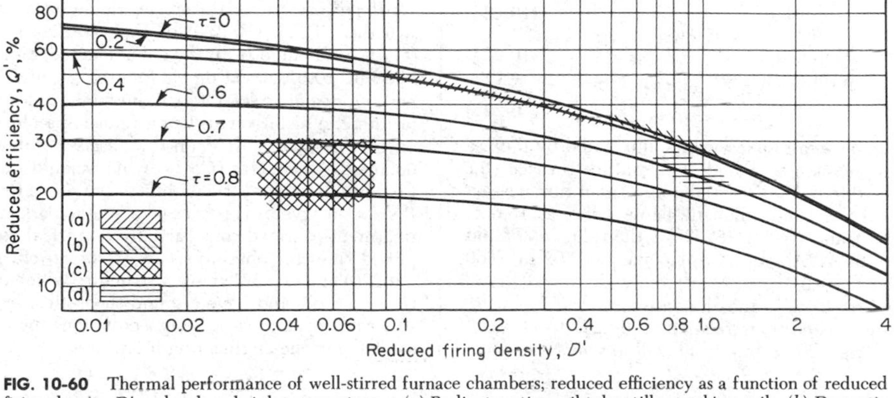

FIG. 10-60 Thermal performance of well-stirred furnace chambers; reduced efficiency as a function of reduced firing density D' and reduced sink temperature τ. (*a*) Radiant section, oil tube stills, cracking coils. (*b*) Domestic boiler combustion chambers. (*c*) Open-hearth furnaces. (*d*) Soaking pits.

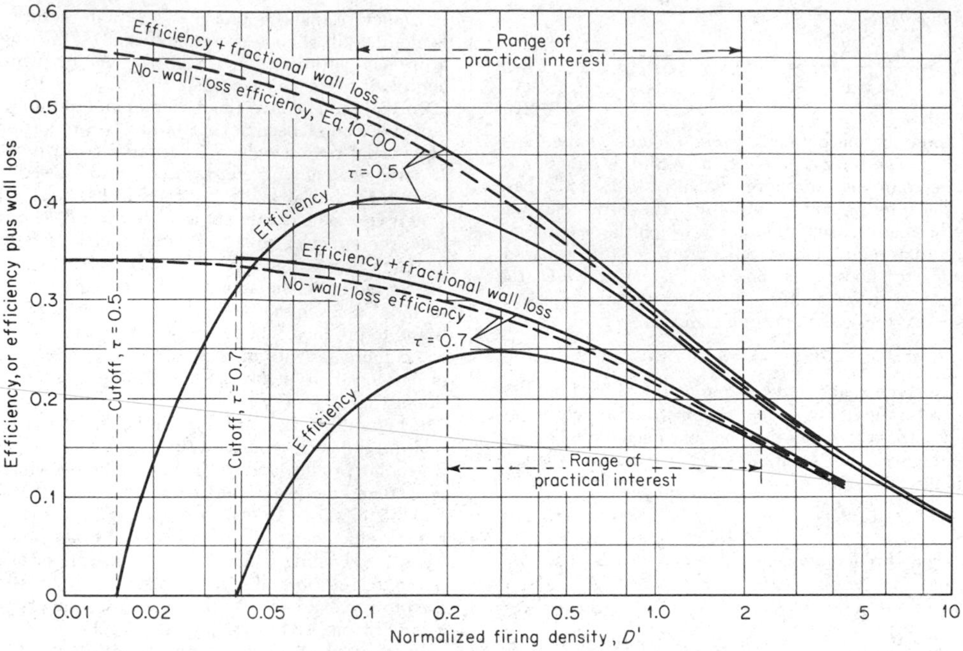

FIG. 10-61 Effect of wall-loss factor L on combustion-chamber performance; $L' = 0.02$, and $\tau = 0.5$ and 0.7.

the only change in Eq. (10-220) is to add to its right-hand side the term $L'(1 + \Delta' - \dot{Q}' - T_0')$ to give

$$\dot{Q}'D' + \tau^4 = (1 + \Delta' - \dot{Q}')^4 + L'(1 + \Delta' - \dot{Q}' - T_0') \tag{10-221}$$

L', called the wall-loss group, and T_0', the normalized base temperature, are two new dimensionless parameters:

$$L' = \frac{UA_r}{T_{AF}^3\sigma(\overline{GS_1})_{R,c}} \tag{10-221a}$$

$$T_0' = T_0/T_{AF} \tag{10-221b}$$

In the modified analysis summarized by Eq. (10-221), $\dot{Q}'$ refers to loss of energy from the gases, as before, but not to useful gain by the stock; rather, it is now the stock gain plus the wall loss. These quantities, expressed as ratios to the enthalpy in the fuel and air entering the furnace, are given by

$$\text{Furnace efficiency} + \text{fractional loss through refractory walls} = \frac{\dot{Q}'}{1 - T_0'} \tag{10-222}$$

$$\text{Fractional loss through refractory walls} = \frac{L'}{D'}\left(1 + \frac{\Delta' - \dot{Q}'}{1 - T_0'}\right) \tag{10-223}$$

$$\text{Furnace efficiency} = \text{difference} \tag{10-224}$$

$$\text{Normalized leaving-gas temperature} \equiv T_G/T_{AF} = 1 + \Delta' - \dot{Q}' \tag{10-225}$$

The **effect of refractory-wall losses** will be illustrated by the use of Eq. (10-221) in application to a high-temperature furnace (for which Δ will be assumed to be 0). With temperatures of furnace surroundings, stock in furnace, and adiabatic flame taken as 15.6°C (60°F), 882°C (1620°F), and 2038°C (3700°F), the values of T_0' and τ become ⅛ and ½. With A_1, A_r, and $(\overline{GS_1})_{R,c}$ taken as 37.2 m² (400 ft²), 92.9 m² (1000 ft²), and 20.9 m² (225 ft²) and with U = 3.151 J/(m²·s·K) [0.555 Btu/(ft²·h·°F)], the dimensionless loss factor L' becomes 0.02. For these conditions relations (10-221) to (10-224) yield the values for efficiency and fractional wall loss versus normalized firing density given in Fig. 10-61, top set of three curves. At very high firing rates the efficiency is low, and the wall loss is minute compared with the transfer rate to the stock. As the firing rate is decreased, the total fractional transfer from the gas (furnace efficiency + fractional loss through hot walls) increases continuously, reaching the value $(1 + \Delta' - \tau)$ in the limit. During that decrease in firing rate the efficiency climbs, passes through a maximum, and goes to zero at the firing rate at which the gases and refractory surfaces are just enough hotter and colder respectively than the stock to decrease the net useful heat transfer to zero. Figure 10-61 includes a second set of curves (the lower three) for a furnace with the same adiabatic flame temperature and loss factor L' as before but with a higher stock temperature, T_{stock} = 1344°C (2452°F), making τ = 0.7. The efficiency is of course less, and the cutoff point of zero efficiency occurs at a higher firing rate.

This model of heat transfer in combustion chambers may be used in various ways: (1) If the performance of a new type of furnace is to be estimated, the value for $(\overline{GS_1})_{R,c}$ to be used in Eq. (10-220) or Eq. (10-221) must be estimated from first principles or, with some simplifying assumptions, from equations like (10-215) and (10-216). (2) If performance data are available on a particular furnace, the structure of Eq. (10-220) or Eq. (10-221) may be used, but with $(\overline{GS_1})_{R,c}$ evaluated from the data. The technique is to prepare a plot, like Fig. 10-61, based on values of L', τ, and T_0' specific to the furnace. A second plot of experimental values of furnace efficiency (arithmetic scale) versus firing rate (log scale) is then superimposed on the first and shifted on the abscissa scale until a best match is obtained. A value for $(\overline{GS_1})_{R,c}$ which should vary somewhat with air-fuel ratio but very little with firing rate can thereby be obtained. (3) If data on several furnaces of a single class are available, a similar treatment can lead to a partially empirical equation based on simplified rules for obtaining $(\overline{GS_1})_{R,c}$ or an effective Δ. Because Eqs. (10-220) and (10-221) have a structure which covers a wide range of furnace types and have a sound theoretical basis, they suggest safer structures of empirical design equations than many such equations available in the engineering literature.

Heat-Transfer Equipment

Frank L. Rubin, B.A., B.Ch.E., *Engineering Consultant, Practical Heat Transfer Consultants, Houston; Member, American Institute of Chemical Engineers, American Society of Mechanical Engineers; Registered Professional Engineer. (Shell-and-Tube Heat Exchangers, Other Heat Exchangers for Liquids and Gases; Section Editor)*

Herbert A. Moak, B.S., *Project Engineer, E. I. du Pont de Nemours & Co.; Member, American Society for Testing and Materials; Registered Professional Engineer. (Thermal Insulation)*

Arthur D. Holt, *Formerly Consultant, Processing Equipment Division, The Jeffrey Manufacturing Co.; Registered Professional Engineer; Retired. (Heat Exchangers for Solids)*

F. C. Standiford, M. S., *President, W. L. Badger Associates, Inc.; Member, American Institute of Chemical Engineers, American Chemical Society; Registered Professional Engineer. (Evaporators)*

David Stuhlbarg, Ch.E., *Heat-Transfer Consultant; Member, American Institute of Chemical Engineers; Retired. (Tank Coils)*

SHELL-AND-TUBE HEAT EXCHANGERS

OTHER HEAT EXCHANGERS FOR LIQUIDS AND GASES

EVAPORATORS

HEAT EXCHANGERS FOR SOLIDS

THERMAL INSULATION

SHELL-AND-TUBE HEAT EXCHANGERS

TYPES AND DEFINITIONS

Shell-and-tube-type exchangers constitute the bulk of the unfired heat-transfer equipment in chemical-process plants, although increasing emphasis has been developing in other designs. The principal types of exchangers are illustrated in Fig. 11-1, and their features are summarized in Table 11-1.

Size Numbering and Type Designation Recommended practice for the designation of conventional shell-and-tube heat exchangers by numbers and letters has been established by the Tubular Exchanger Manufacturers Association (TEMA). This information from the sixth edition of the TEMA Standards is reproduced in the following paragraphs.

It is recommended that heat-exchanger size and type be designated by numbers and letters.

1. *Size.* Sizes of shells (and tube bundles) shall be designated by numbers describing shell (and tube-bundle) diameters and tube lengths as follows:
2. *Diameter.* The nominal diameter shall be the inside diameter of the shell in inches, rounded off to the nearest integer. For kettle reboilers the nominal diameter shall be the port diameter followed by the shell diameter, each rounded off to the nearest integer.
3. *Length.* The nominal length shall be the tube length in inches. Tube length for straight tubes shall be taken as the actual overall length. For U tubes the length shall be taken as the straight length from end of tube to bend tangent.
4. *Type.* Type designation shall be by letters describing stationary head, shell (omitted for bundles only), and rear head, in that order, as indicated in Fig. 11-1.

Typical Examples (*A*) Split-ring floating-heat exchanger with removable channel and cover, single-pass shell, 591-mm (23¼-in) inside diameter with tubes 4.9 m (16 ft) long. SIZE 23-192 TYPE AES.

(*B*) U-tube exchanger with bonnet-type stationary head, split-flow shell, 483-mm (19-in) inside diameter with tubes 21-m (7-ft) straight length. SIZE 19-84 TYPE GBU.

(*C*) Pull-through floating-heat-kettle-type reboiler having stationary head integral with tube sheet, 584-mm (23-in) port diameter and 940-mm (37-in) inside shell diameter with tubes 4.9 m (16 ft) long. SIZE 23/37-192 TYPE CKT.

(*D*) Fixed-tube sheet exchanger with removable channel and cover, bonnet-type rear head, two-pass shell, 841-mm (33⅛-in) diameter with tubes 2.4 m (8 ft) long. SIZE 33-96 TYPE AFM.

(*E*) Fixed-tube sheet exchanger having stationary and rear heads integral with tube sheets, single-pass shell, 432-mm (17-in) inside diameter with tubes 4.9 m (16 ft) long. SIZE 17-192 TYPE CEN.

Functional Definitions Heat-transfer equipment can be designated by type (e.g., fixed tube sheet, outside packed head, etc.) or by function (chiller, condenser, cooler, etc.). Almost any type of unit can be used to perform any or all of the listed functions. Many of these terms have been defined by Donahue [*Pet. Process.*, 103 (March 1956)].

Equipment	Function
Chiller	Cools a fluid to a temperature below that obtainable if water only were used as a coolant. It uses a refrigerant such as ammonia or Freon.
Condenser	Condenses a vapor or mixture of vapors, either alone or in the presence of a noncondensable gas.
Partial condenser	Condenses vapors at a point high enough to provide a temperature difference sufficient to preheat a cold stream of process fluid. This saves heat and eliminates the need for providing a separate preheater (using flame or steam).
Final condenser	Condenses the vapors to a final storage temperature of approximately 37.8°C (100°F). It uses water cooling, which means that the transferred heat is lost to the process.
Cooler	Cools liquids or gases by means of water.
Exchanger	Performs a double function: (1) heats a cold fluid by (2) using a hot fluid which it cools. None of the transferred heat is lost.
Heater	Imparts sensible heat to a liquid or a gas by means of condensing steam or Dowtherm.
Reboiler	Connected to the bottom of a fractionating tower, it provides the reboil heat necessary for distillation. The heating medium may be either steam or a hot-process fluid.
Thermosiphon reboiler	Natural circulation of the boiling medium is obtained by maintaining sufficient liquid head to provide for circulation.
Forced-circulation reboiler	A pump is used to force liquid through the reboiler.
Steam generator	Generates steam for use elsewhere in the plant by using the available high-level heat in tar or a heavy oil.
Superheater	Heats a vapor above the saturation temperature.
Vaporizer	A heater which vaporizes part of the liquid.
Waste-heat boiler	Produces steam; similar to steam generator, except that the heating medium is a hot gas or liquid produced in a chemical reaction.

GENERAL DESIGN CONSIDERATIONS

Selection of Flow Path In selecting the flow path for two fluids through an exchanger, several general approaches are used. The tube-side fluid is more corrosive or dirtier or at a higher pressure. The shell-side fluid is a liquid of high viscosity or a gas.

When alloy construction for one of the two fluids is required, a carbon steel shell combined with alloy tube-side parts is less expensive than alloy in contact with the shell-side fluid combined with carbon steel headers.

Cleaning of the inside of tubes is more readily done than cleaning of exterior surfaces.

For gauge pressures in excess of 2068 kPa (300 lbf/in^2) for one of the fluids, the less expensive construction has the high-pressure fluid in the tubes.

For a given pressure drop, higher heat-transfer coefficients are obtained on the shell side than on the tube side.

Heat-exchanger shutdowns are most often caused by fouling, corrosion, and erosion.

Construction Codes "Rules for Construction of Pressure Vessels, Division 1," which is part of Section VIII of the ASME Boiler and Pressure Vessel Code (American Society of Mechanical Engineers), serves as a construction code by providing minimum standards. New editions of the code are usually issued every 3 years. Interim revisions are made semiannually in the form of addenda. Compliance with ASME Code requirements is mandatory in much of the United States and Canada. Originally these rules were not prepared for heat exchangers. However, the welded joint between tube sheet and shell of the fixed-tube-sheet heat exchanger is now included. A nonmandatory appendix on tube-to-tube-sheet joints is also included. Additional rules for heat exchangers are being developed.

Standards of Tubular Exchanger Manufacturers Association 6th ed., 1978 (commonly referred to as the TEMA Standards), serve to

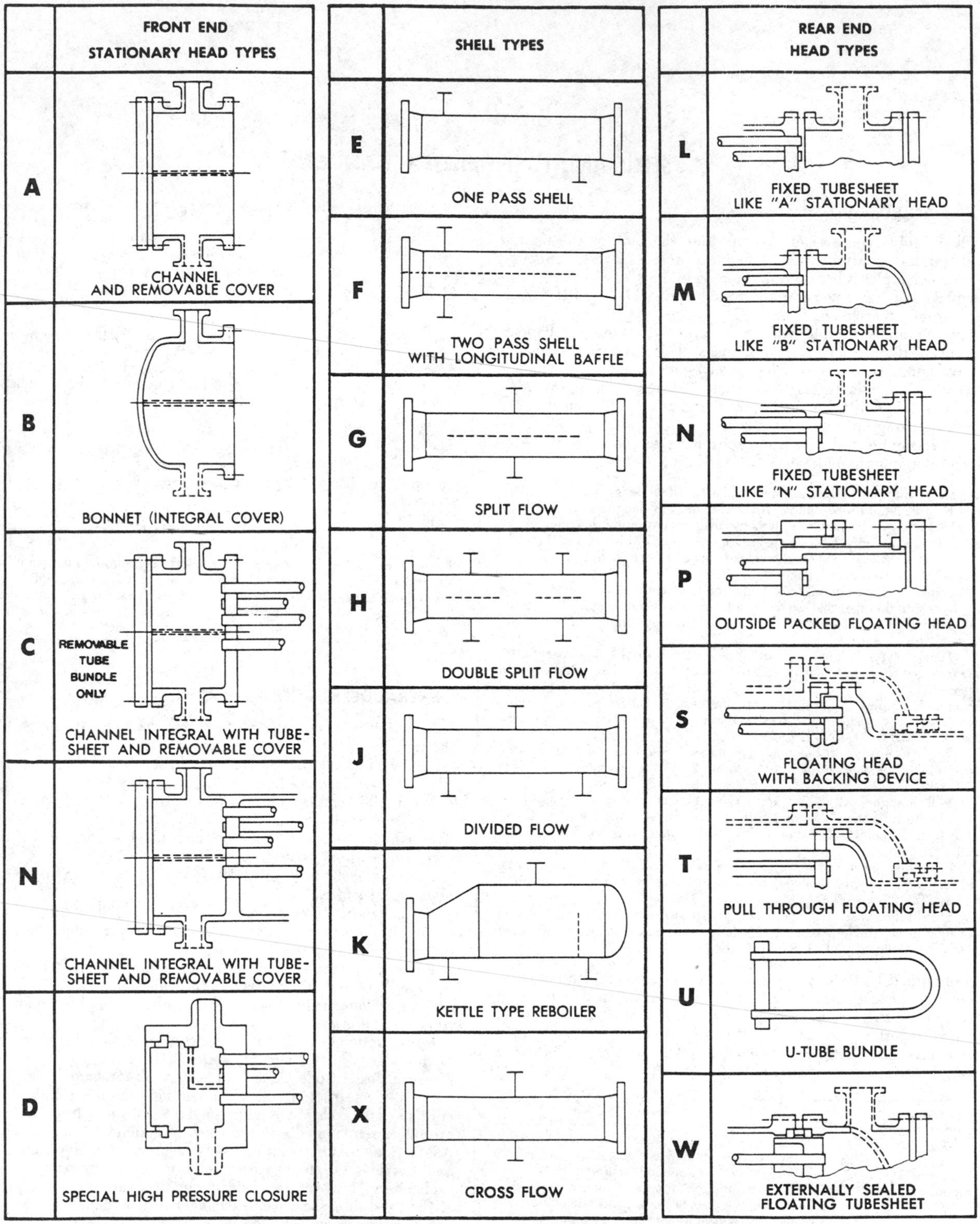

FIG. 11-1 TEMA-type designations for shell-and-tube heat exchangers. (Standards of Tubular Exchanger Manufacturers Association, *6th ed., 1978.*)

TABLE 11-1 Features of Principal Shell-and-Tube-Type Exchangers*

Type of design	Fixed tube sheet	U-tube	Packed lantern-ring floating head	Internal floating head (split backing ring)	Outside-packed floating head	Pull-through floating head
T.E.M.A. rear-head type	L or M or N	U	W	S	P	T
Relative cost increases from A (least expensive) through E (most expensive)	B	A	C	E	D	E
Provision for differential expansion	Expansion joint in shell	Individual tubes free to expand	Floating head	Floating head	Floating head	Floating head
Removable bundle	No	Yes	Yes	Yes	Yes	Yes
Replacement bundle possible	No	Yes	Yes	Yes	Yes	Yes
Individual tubes replaceable	Yes	Only those in outside row†	Yes	Yes	Yes	Yes
Tube cleaning by chemicals inside and outside	Yes	Yes	Yes	Yes	Yes	Yes
Interior tube cleaning mechanically	Yes	Special tools required	Yes	Yes	Yes	Yes
Exterior tube cleaning mechanically:						
Triangular pitch	No	No‡	No‡	No‡	No‡	No‡
Square pitch	No	Yes	Yes	Yes	Yes	Yes
Hydraulic-jet cleaning:						
Tube interior	Yes	Special tools required	Yes	Yes	Yes	Yes
Tube exterior	No	Yes	Yes	Yes	Yes	Yes
Double tube sheet feasible	Yes	Yes	No	No	Yes	No
Number of tube passes	No practical limitations	Any even number possible	Limited to one or two passes	No practical limitations§	No practical limitations	No practical limitations§
Internal gaskets eliminated	Yes	Yes	Yes	No	Yes	No

NOTE: Relative costs A and B are not significantly different and interchange for long lengths of tubing.

* Modified from page a-8 of the Patterson-Kelley Co. Manual No. 700A, Heat Exchangers.

† U-tube bundles have been built with tube supports which permit the U-bends to be spread apart and tubes inside of the bundle replaced.

‡ Normal triangular pitch does not permit mechanical cleaning. With a wide triangular pitch, which is equal to 2 (tube diameter plus cleaning lane)/ $\sqrt{3}$, mechanical cleaning is possible on removable bundles. This wide spacing is infrequently used.

§ For odd number of tube side passes, floating head requires packed joint or expansion joint.

supplement and define the ASME Code for all shell-and-tube-type heat-exchanger applications (other than double-pipe construction). TEMA Class R design is "for the generally severe requirements of petroleum and related processing applications. Equipment fabricated in accordance with these standards is designed for safety and durability under the rigorous service and maintenance conditions in such applications." TEMA Class C design is "for the generally moderate requirements of commercial and general process applications," while TEMA Class B is "for chemical process service."

The mechanical-design requirements are identical for all three classes of construction. The differences between the TEMA classes are minor and were listed by Rubin [*Hydrocarbon Process.*, **59**, 92 (June 1980)].

Among the topics of the TEMA Standards are nomenclature, fabrication tolerances, inspection, guarantees, tubes, shells, baffles and support plates, floating heads, gaskets, tube sheets, channels, nozzles, end flanges and bolting, material specifications, and fouling resistances.

Shell and Tube Heat Exchangers for Geneal Refinery Services, API Standard 660, 4th ed., 1982, is published by the American Petroleum Institute to supplement both the TEMA Standards and the ASME Code. Many companies in the chemical and petroleum processing fields have their own standards to supplement these various requirements. *The Interrelationships between Codes, Standards, and Customer Specifications for Process Heat Transfer Equipment* is a symposium volume which was edited by F. L. Rubin and published by ASME in December 1979. (See discussion of pressure-vessel codes in Sec. 6.)

Design pressures and temperatures for exchangers usually are specified with a margin of safety beyond the conditions expected in service. Design pressure is generally about 172 kPa (25 lbf/in^2) greater than the maximum expected during operation or at pump shutoff. Design temperature is commonly 14°C (25°F) greater than the maximum temperature in service.

Tube-Bundle Vibration Damage from tube vibration has become an increasing phenomenon as heat-exchanger sizes and quantities of flow have increased. The shell-side flow, baffle configuration, and unsupported-tube span are of prime consideration. Mechanisms of tube vibration are as follows:

Vortex Shedding The vortex-shedding frequency of the fluid in cross-flow over the tubes may coincide with a natural frequency of the tubes and excite large resonant vibration amplitudes.

Fluid-Elastic Coupling Fluid flowing over tubes causes them to vibrate with a whirling motion. The mechanism of fluid-elastic coupling occurs when a "critical" velocity is exceeded and the vibration then becomes self-excited and grows in amplitude. This mechanism frequently occurs in process heat exchangers which suffer vibration damage.

Pressure Fluctuation Turbulent pressure fluctuations which develop in the wake of a cylinder or are carried to the cylinder from upstream may provide a potential mechanism for tube vibration. The tubes respond to the portion of the energy spectrum that is close to their natural frequency.

Acoustic Coupling When the shell-side fluid is a low-density gas, acoustic resonance or coupling develops when the standing waves in the shell are in phase with vortex shedding from the tubes. The standing waves are perpendicular to the axis of the tubes and to the direction of cross-flow. Damage to the tubes is rare. However, the noise can be extremely painful.

Testing Upon completion of shop fabrication and also during maintenance operations it is desirable hydrostatically to test the shell side of tubular exchangers so that visual examination of tube ends can be made. Leaking tubes can be readily located and serviced. When leaks are determined without access to the tube ends, it is necessary to reroll or reweld all the tube-to-tube-sheet joints with possible damage to the satisfactory joints.

Testing for leaks in heat exchangers was discussed by Rubin [*Chem. Eng.*, **68**, 160–166 (July 24, 1961)].

Performance testing of heat exchangers is described in the American Institute of Chemical Engineers' *Standard Testing Procedure for Heat Exchangers*, sec. 1. "Sensible Heat Transfer in Shell-and-Tube-Type Equipment."

PRINCIPAL TYPES OF CONSTRUCTION

Figure 11-2 shows details of the construction of the principal types of shell-and-tube heat exchangers. These and other types are discussed in the following paragraphs.

Fixed-Tube-Sheet Heat Exchangers Fixed-tube-sheet exchangers (Fig. 11-2*b*) are used more often than any other type, and the frequency of use has been increasing in recent years. The tube sheets are welded to the shell. Usually these extend beyond the shell and serve as flanges to which the tube-side headers are bolted. This construction requires that the shell and tube-sheet materials be weldable to each other.

When such welding is not possible, a "blind"-gasket type of construction is utilized. The blind gasket is not accessible for maintenance or replacement once the unit has been constructed. This construction is used for steam surface condensers, which operate under vacuum.

The tube-side header (or channel) may be welded to the tube sheet, as shown in Fig. 11-1 for type C and N heads. This type of construction is less costly than types B and M or A and L and still offers the advantage that tubes may be examined and replaced without disturbing the tube-side piping connections.

There is no limitation on the number of tube-side passes. Shell-side passes can be one or more, although shells with more than two shell-side passes are rarely used.

Tubes can completely fill the heat-exchanger shell. Clearance between the outermost tubes and the shell is only the minimum necessary for fabrication. Between the inside of the shell and the baffles some clearance must be provided so that baffles can slide into the shell. Fabrication tolerances then require some additional clearance between the outside of the baffles and the outermost tubes. The edge distance between the outer tube limit (OTL) and the baffle diameter must be sufficient to prevent vibration of the tubes from breaking through the baffle holes. The outermost tube must be contained within the OTL. Clearances between the inside shell diameter and OTL are 13 mm ($\frac{1}{2}$ in) for 635-mm-(25-in-) inside-diameter shells and up, 11 mm ($\frac{7}{16}$ in) for 254- through 610-mm (10- through 24-in) pipe shells, and slightly less for smaller-diameter pipe shells.

Tubes can be replaced. Tube-side headers, channel covers, gaskets, etc., are accessible for maintenance and replacement. Neither the shell-side baffle structure nor the blind gasket is accessible. During tube removal, a tube may break within the shell. When this occurs, it is most difficult to remove or to replace the tube. The usual procedure is to plug the appropriate holes in the tube sheets.

Differential expansion between the shell and the tubes can develop because of differences in length caused by thermal expansion. Various types of **expansion joints** are used to eliminate excessive stresses

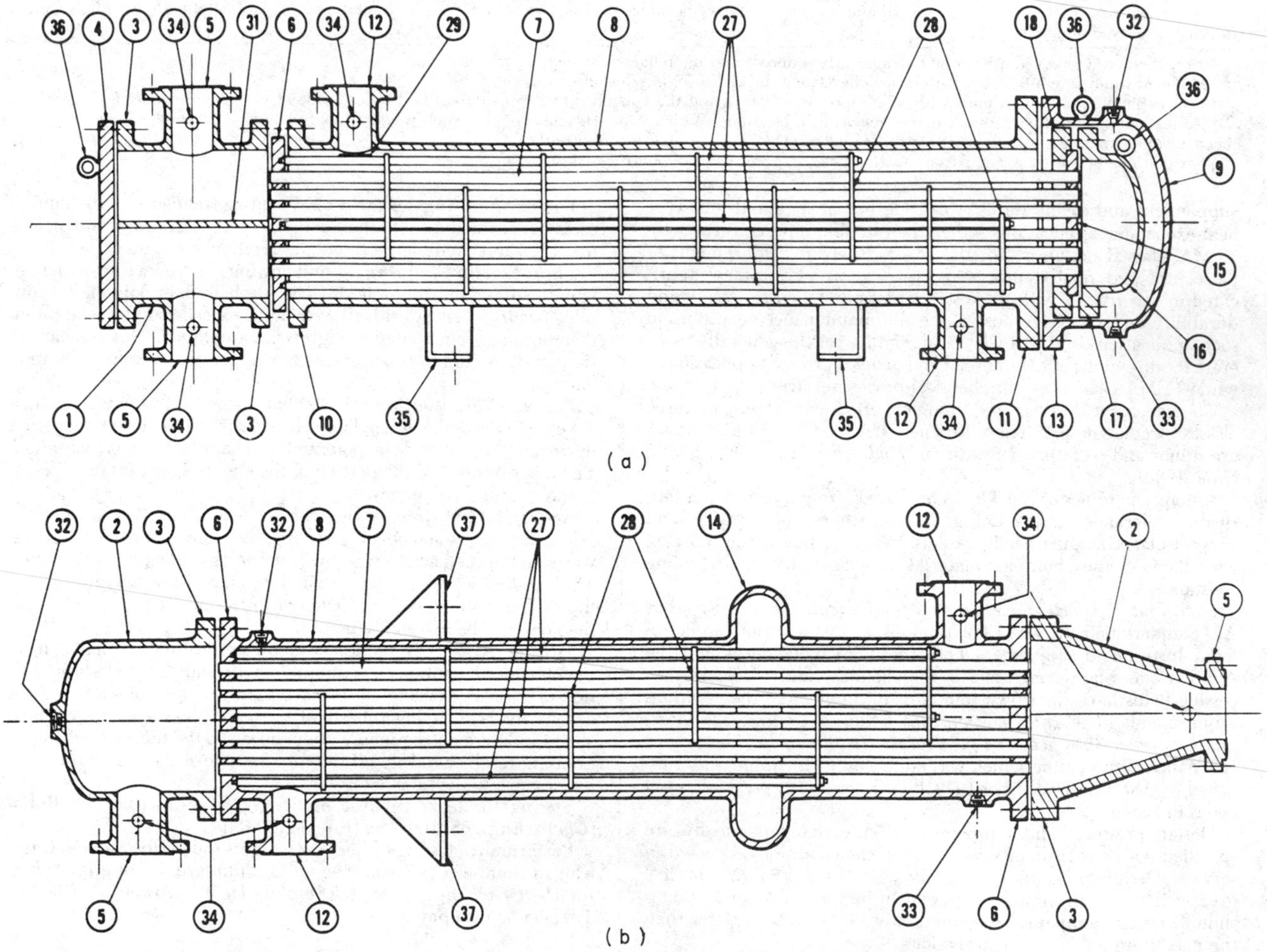

FIG. 11-2 Heat-exchanger-component nomenclature. (*a*) Internal-floating-head exchanger (with floating-head backing device). Type AES. (*b*) Fixed-tube-sheet exchanger. Type BEM. (*c*) Outside-packed floating-head exchanger. Type AEP. (*d*) U-tube heat exchanger. Type CFU. (*e*) Kettle-type floating-head reboiler. Type AKT. (*f*) Exchanger with packed floating tube sheet and lantern ring. Type AJW. (Standards of Tubular Exchanger Manufacturers Association, *6th ed., 1978*.)

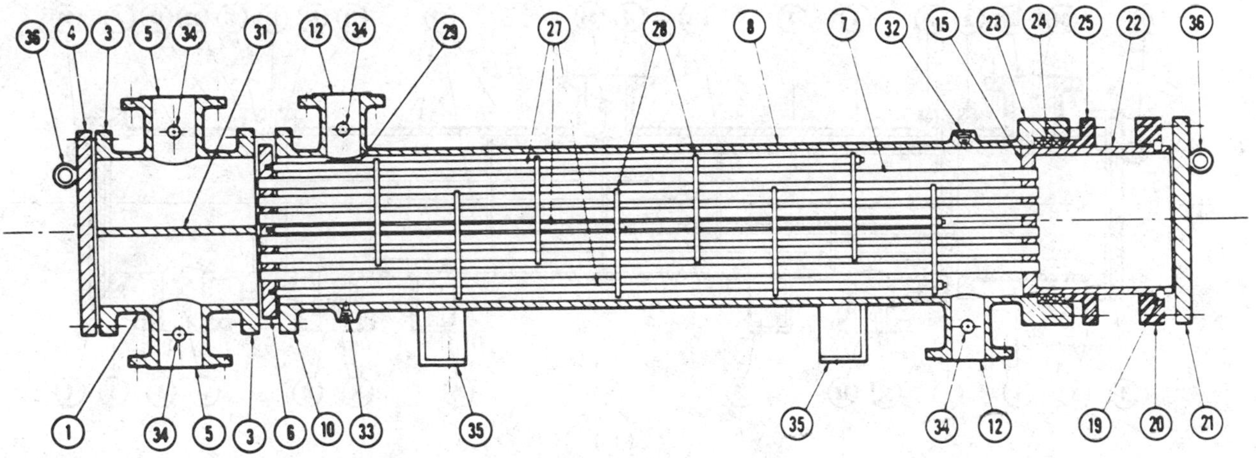

(c)

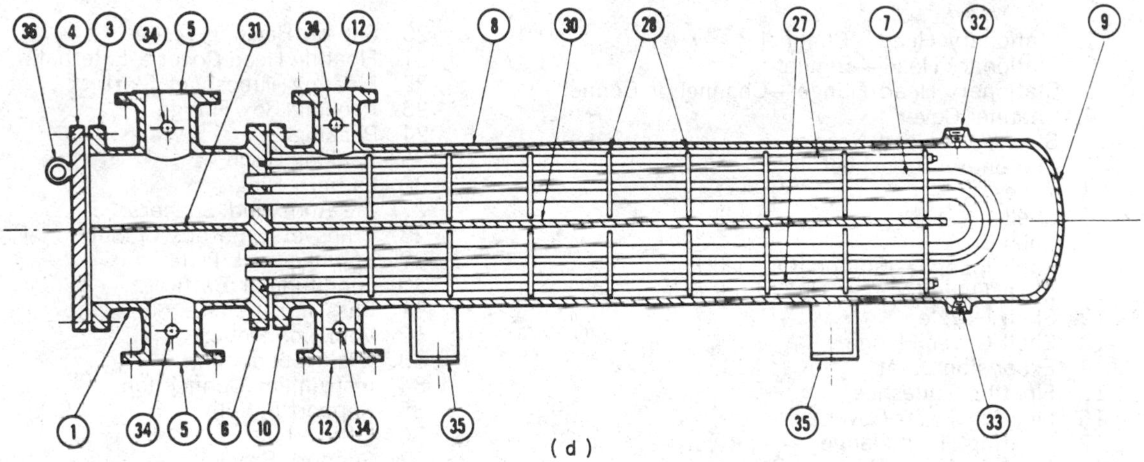

(d)

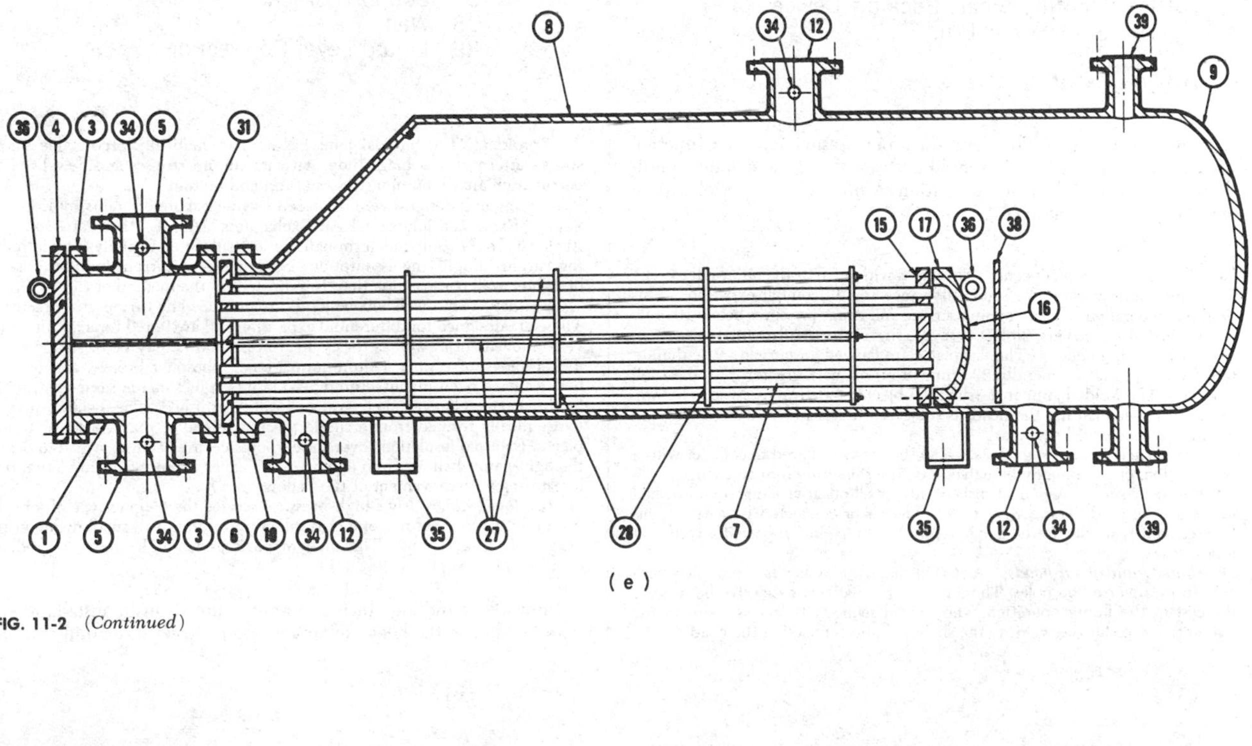

(e)

FIG. 11-2 *(Continued)*

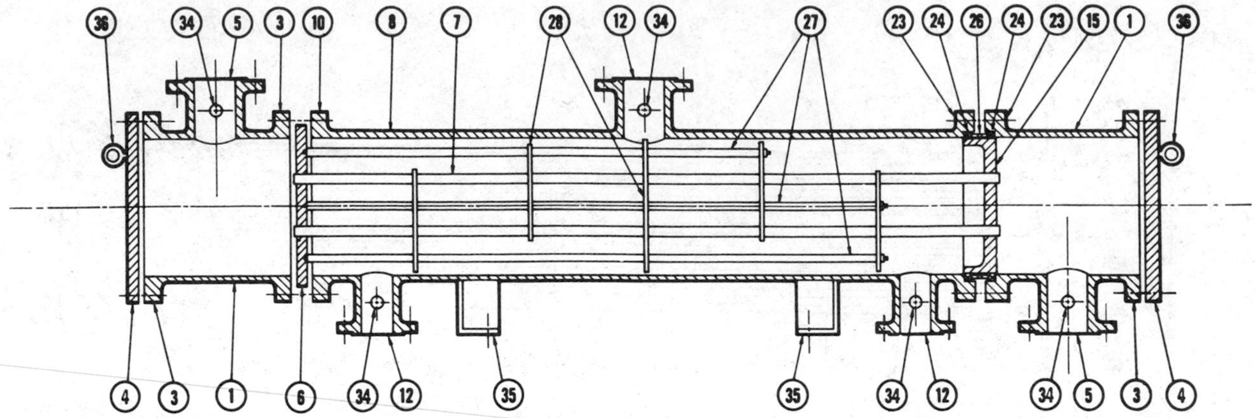

(f)

1. Stationary Head—Channel
2. Stationary Head—Bonnet
3. Stationary Head Flange—Channel or Bonnet
4. Channel Cover
5. Stationary Head Nozzle
6. Stationary Tubesheet
7. Tubes
8. Shell
9. Shell Cover
10. Shell Flange—Stationary Head End
11. Shell Flange—Rear Head End
12. Shell Nozzle
13. Shell Cover Flange
14. Expansion Joint
15. Floating Tubesheet
16. Floating Head Cover
17. Floating Head Flange
18. Floating Head Backing Device
19. Split Shear Ring
20. Slip-on Backing Flange
21. Floating Head Cover—External
22. Floating Tubesheet Skirt
23. Packing Box Flange
24. Packing
25. Packing Gland
26. Lantern Ring
27. Tie Rods and Spacers
28. Transverse Baffles or Support Plates
29. Impingement Plate
30. Longitudinal Baffle
31. Pass Partition
32. Vent Connection
33. Drain Connection
34. Instrument Connection
35. Support Saddle
36. Lifting Lug
37. Support Bracket
38. Weir
39. Liquid Level Connection

FIG. 11-2 *(Continued)*

caused by expansion. The need for an expansion joint is a function of both the amount of differential expansion and the cycling conditions to be expected during operation. A number of types of expansion joints are available (Fig. 11-3).

a. Flat plates. Two concentric flat plates with a bar at the outer edges. The flat plates can flex to make some allowance for differential expansion. This design is generally used for vacuum service and gauge pressures below 103 kPa (15 lbf/in^2). All welds are subject to severe stress during differential expansion.

b. Flanged-only heads. The flat plates are flanged (or curved). The diameter of these heads is generally 203 mm (8 in) or more greater than the shell diameter. The welded joint at the shell is subject to the stress referred to before, but the joint connecting the heads is subjected to less stress during expansion because of the curved shape.

c. Flared shell or pipe segments. The shell may be flared to connect with a pipe section, or a pipe may be halved and quartered to produce a ring.

d. Formed heads. A pair of dished-only or elliptical or flanged and dished heads can be used. These are welded together or connected by a ring. This type of joint is similar to the flanged-only-head type but apparently is subject to less stress.

e. Flanged and flued heads. A pair of flanged-only heads is provided with concentric reverse flue holes. These heads are relatively expensive because of the cost of the fluing operation. The curved shape of the heads reduces the amount of stress at the welds to the shell and also connecting the heads.

f. Toroidal. The toroidal joint has a mathematically predictable smooth stress pattern of low magnitude, with maximum stresses at sidewalls of the corrugation and minimum stresses at top and bottom.

The foregoing designs were discussed as ring expansion joints by Kopp and Sayre, "Expansion Joints for Heat Exchangers" (ASME Misc. Pap., vol. 6, no. 211). All are statically indeterminate but are subjected to analysis by introducing various simplifying assumptions. Some joints in current industrial use are of lighter wall construction than is indicated by the method of this paper.

g. Bellows. Thin-wall bellows joints are produced by various manufacturers. These are designed for differential expansion and are tested for axial and transverse movement as well as for cyclical life. Bellows may be of stainless steel, nickel alloys, or copper. (Aluminum, Monel, phosphor bronze, and titanium bellows have been manufactured.) Welding nipples of the same composition as the heat-exchanger shell are generally furnished. The bellows may be hydraulically formed from a single piece of metal or may consist of welded pieces. External insulation covers of carbon steel are often provided to protect the light-gauge bellows from damage. The cover also prevents insulation from interfering with movement of the bellows (see *h*).

i. Toroidal bellows. For high-pressure service the bellows type of joint has been modified so that movement is taken up by thin-wall small-diameter bellows of a toroidal shape. Thickness of parts under high pressure is reduced considerably (see *f*).

Improper handling during manufacture, transit, installation, or maintenance of the heat exchanger equipped with the thin-wall-bel-

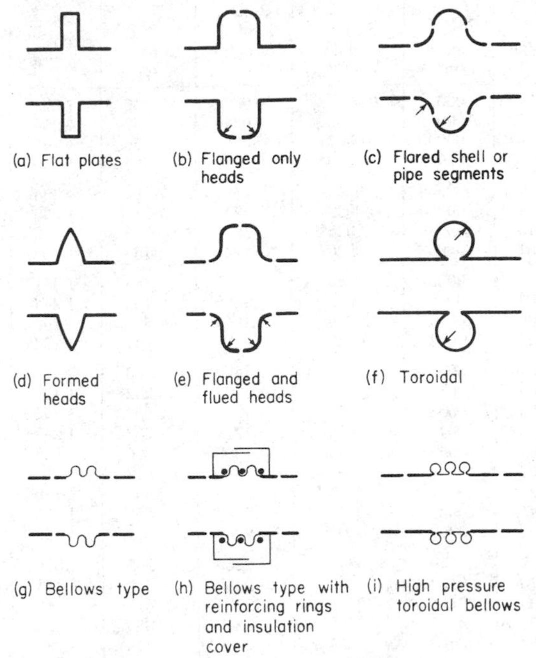

FIG. 11-3 Expansion joints.

lows type or toroidal type of expansion joint can damage the joint. In larger units these light-wall joints are particularly susceptible to damage, and some designers prefer the use of the heavier walls of formed heads.

Chemical-plant exchangers requiring expansion joints most commonly have used the flanged-and-flued-head type. There is a trend toward more common use of the light-wall-bellows type.

U-Tube Heat Exchanger (Fig. 11-2*d*) The tube bundle consists of a stationary tube sheet, U tubes (or hairpin tubes), baffles or support plates, and appropriate tie rods and spacers. The tube bundle can be removed from the heat-exchanger shell. A tube-side header (stationary head) and a shell with integral shell cover, which is welded to the shell, are provided. Each tube is free to expand or contract without any limitation being placed upon it by the other tubes.

The U-tube bundle has the advantage of providing minimum clearance between the outer tube limit and the inside of the shell for any of the removable-tube-bundle constructions. Clearances are of the same magnitude as for fixed-tube-sheet heat exchangers.

The number of tube holes in a given shell is less than that for a fixed-tube-sheet exchanger because of limitations on bending tubes of a very short radius.

The U-tube design offers the advantage of reducing the number of joints. In high-pressure construction this feature becomes of considerable importance in reducing both initial and maintenance costs. The use of U-tube construction has increased significantly with the development of hydraulic tube cleaners, which can remove fouling residues from both the straight and the U-bend portions of the tubes.

Mechanical **cleaning** of the inside of the tubes was described by John [*Chem. Eng.*, **66**, 187–192 (Dec. 14, 1959)]. Rods and conventional mechanical tube cleaners cannot pass from one end of the U tube to the other. Power-driven tube cleaners, which can clean both the straight legs of the tubes and the bends, are available.

Hydraulic jetting with water forced through spray nozzles at high pressure for cleaning tube interiors and exteriors of removal bundles is reported by Canaday ("Hydraulic Jetting Tools for Cleaning Heat Exchangers," ASME Pap. 58-A-217, unpublished).

The **tank suction heater,** as illustrated in Fig. 11-4, contains a U-tube bundle. This design is often used with outdoor storage tanks for heavy fuel oils, tar, molasses, and similar fluids whose viscosity must be lowered to permit easy pumping. Uusally the tube-side heating medium is steam. One end of the heater shell is open, and the liquid being heated passes across the outside of the tubes. Pumping costs can be reduced without heating the entire contents of the tank. Bare tube and integral low-fin tubes are provided with baffles. Longitudinal fin-tube heaters are not baffled.

Kettle-type reboilers, evaporators, etc., are often U-tube exchangers with enlarged shell sections for vapor-liquid separation. The U-tube bundle replaces the floating-heat bundle of Fig. 11-2*e*.

The U-tube exchanger with copper tubes, cast-iron header, and other parts of carbon steel is used for **water and steam services** in office buildings, schools, hospitals, hotels, etc. Nonferrous tube sheets and admiralty or 90-10 copper-nickel tubes are the most frequently used substitute materials. These standard exchangers are available from a number of manufacturers at costs far below those of custom-built process-industry equipment.

Packed-Lantern-Ring Exchanger (Fig. 11-2*f*) This construction is the least costly of the straight-tube removal-bundle types. The shell- and tube-side fluids are each contained by separate rings of packing separated by a lantern ring and are installed at the floating tube sheet. The lantern ring is provided with weep holes. Any leakage passing the packing goes through the weep holes and then drops to the floor. Leakage at the packing will not result in mixing within the exchanger of the two fluids.

The width of the floating tube sheet must be great enough to allow for the packings, the lantern ring, and differential expansion. Sometimes a small skirt is attached to a thin tube sheet to provide the required bearing surface for packings and lantern ring.

The clearance between the outer tube limit and the inside of the shell is slightly larger than that for fixed-tube-sheet and U-tube exchangers. The use of a floating-tube-sheet skirt increases this clearance. Without the skirt the clearance must make allowance for tube-hole distortion during tube rolling near the outside edge of the tube sheet or for tube-end welding at the floating tube sheet.

The packed-lantern-ring construction is generally limited to design temperatures below 191°C (375°F) and to the mild services of water, steam, air, lubricating oil, etc. Design gauge pressure does not exceed 2068 kPa (300 lbf/in^2) for pipe shell exchangers and is limited to 1034 kPa (150 lbf/in^2) for 610- to 1067-mm- (24- to 42-in-) diameter shells.

Outside-Packed Floating-Head Exchanger (Fig. 11-2*c*) The shell-side fluid is contained by rings of packing, which are compressed within a stuffing box by a packing follower ring. This construction was frequently used in the chemical industry, but in recent years usage has decreased. The removable-bundle construction accommodates differential expansion between shell and tubes and is used for shell-side service up to 4137 kPa gauge pressure (600 lbf/in^2) at 316°C (600°F). There are no limitations upon the number of tube-side passes or upon the tube-side design pressure and temperature. The outside-packed floating-head exchanger was the most commonly used type of removable-bundle construction in chemical-plant service.

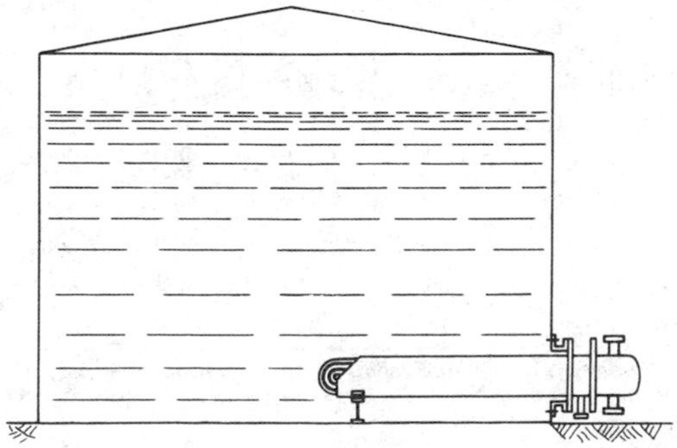

FIG. 11-4 Tank suction heater.

TABLE 11-2 Tubing Characteristics*

Tubing OD, in	BWG gauge	Wall thickness, in	Inside cross-sectional area, in^2	Ft^2 external surface per ft length	Ft^2 internal surface per ft length	Weight per ft length, steel, lb†	Tubing ID, in	Moment of inertia, in^4	Section modulus, in^3	Radius of gyration, in	Constant C‡	OD/ID	Transverse metal area, in^2
¼	22	0.028	0.0295	0.0655	0.0508	0.066	0.194	0.00012	0.00098	0.0792	46	1.289	0.0195
¼	24	.022	.0333	.0655	.0539	.054	.206	.00011	.00083	.0810	52	1.214	.0159
¼	26	.018	.0360	.0655	.0560	.045	.214	.00009	.00071	.0824	56	1.168	.0131
¼	27	.016	.0373	.0655	.0570	.040	.218	.00008	.00064	.0829	58	1.146	.0117
⅜	18	.049	.0603	.0982	.0725	.171	.277	.00068	.0036	.1164	94	1.354	.0502
⅜	20	.035	.0731	.0982	.0798	.127	.305	.00055	.0029	.1213	114	1.233	.0374
⅜	22	.028	.0799	.0982	.0835	.104	.319	.00046	.0025	.1227	125	1.176	.0305
⅜	24	.022	.0860	.0982	.0867	.083	.331	.00038	.0020	.1248	134	1.133	.0244
½	16	.065	.1075	.1309	.0969	.302	.370	.0022	.0086	.1556	168	1.351	.0888
½	18	.049	.1269	.1309	.1052	.236	.402	.0018	.0072	.1606	198	1.244	.0694
½	20	.035	.1452	.1309	.1126	.174	.430	.0014	.0056	.1649	227	1.163	.0511
½	22	.028	.1548	.1309	.1162	.141	.444	.0012	.0046	.1671	241	1.126	.0415
⅝	12	.109	.1301	.1636	.1066	.602	.407	.0061	.0197	.1864	203	1.536	.177
⅝	13	.095	.1486	.1636	.1139	.537	.435	.0057	.0183	.1903	232	1.437	.158
⅝	14	.083	.1655	.1636	.1202	.479	.459	.0053	.0170	.1938	258	1.362	.141
⅝	15	.072	.1817	.1636	.1259	.425	.481	.0049	.0156	.1971	283	1.299	.125
⅝	16	.065	.1924	.1636	.1296	.388	.495	.0045	.0145	.1993	300	1.263	.114
⅝	17	.058	.2035	.1636	.1333	.350	.509	.0042	.0134	.2016	317	1.228	.103
⅝	18	.049	.2181	.1636	.1380	.303	.527	.0037	.0118	.2043	340	1.186	.089
⅝	19	.042	.2298	.1636	.1416	.262	.541	.0033	.0105	.2068	358	1.155	.077
⅝	20	.035	.2419	.1636	.1453	.221	.555	.0028	.0091	.2089	377	1.126	.065
¾	10	.134	.1825	.1963	.1262	.884	.482	.0129	.0344	.2229	285	1.556	.260
¾	11	.120	.2043	.1963	.1335	.809	.510	.0122	.0326	.2267	319	1.471	.238
¾	12	.109	.2223	.1963	.1393	.748	.532	.0116	.0309	.2299	347	1.410	.220
¾	13	.095	.2463	.1963	.1466	.666	.560	.0107	.0285	.2340	384	1.339	.196
¾	14	.083	.2679	.1963	.1529	.592	.584	.0098	.0262	.2376	418	1.284	.174
¾	15	.072	.2884	.1963	.1587	.520	.606	.0089	.0238	.2410	450	1.238	.153
¾	16	.065	.3019	.1963	.1623	.476	.620	.0083	.0221	.2433	471	1.210	.140
¾	17	.058	.3157	.1963	.1660	.428	.634	.0076	.0203	.2455	492	1.183	.126
¾	18	.049	.3339	.1963	.1707	.367	.652	.0067	.0178	.2484	521	1.150	.108
¾	20	.035	.3632	.1963	.1780	.269	.680	.0050	.0134	.2532	567	1.103	.079
⅞	10	.134	.2892	.2291	.1589	1.061	.607	.0221	.0505	.2662	451	1.441	.312
⅞	12	.109	.3390	.2291	.1720	.891	.657	.0196	.0449	.2736	529	1.332	.262
⅞	13	.095	.3685	.2291	.1793	.792	.685	.0180	.0411	.2778	575	1.277	.233
⅞	14	.083	.3948	.2291	.1856	.704	.709	.0164	.0374	.2815	616	1.234	.207
⅞	16	.065	.4359	.2291	.1950	.561	.745	.0137	.0312	.2873	680	1.174	.165
⅞	18	.049	.4742	.2291	.2034	.432	.777	.0109	.0249	.2925	740	1.126	.127
⅞	20	.035	.5090	.2291	.2107	.313	.805	.0082	.0187	.2972	794	1.087	.092
1	8	.165	.3526	.2618	.1754	1.462	.670	.0392	.0784	.3009	550	1.493	.430
1	10	.134	.4208	.2618	.1916	1.237	.732	.0350	.0700	.3098	656	1.366	.364
1	11	.120	.4536	.2618	.1990	1.129	.760	.0327	.0654	.3140	708	1.316	.332
1	12	.109	.4803	.2618	.2047	1.037	.782	.0307	.0615	.3174	749	1.279	.305
1	13	.095	.5153	.2618	.2121	.918	.810	.0280	.0559	.3217	804	1.235	.270
1	14	.083	.5463	.2618	.2183	.813	.834	.0253	.0507	.3255	852	1.199	.239
1	15	.072	.5755	.2818	.2241	.714	.856	.0227	.0455	.3291	898	1.167	.210
1	16	.065	.5945	.2618	.2278	.649	.870	.0210	.0419	.3314	927	1.149	.191
1	18	.049	.6390	.2618	.2361	.496	.902	.0166	.0332	.3366	997	1.109	.146
1	20	.035	.6793	.2618	.2435	.360	.930	.0124	.0247	.3414	1060	1.075	.106
1¼	7	.180	.6221	.3272	.2330	2.057	.890	.0890	.1425	.3836	970	1.404	.605
1¼	8	.165	.6648	.3272	.2409	1.921	.920	.0847	.1355	.3880	1037	1.359	.565
1¼	10	.134	.7574	.3272	.2571	1.598	.982	.0741	.1186	.3974	1182	1.273	.470
1¼	11	.120	.8012	.3272	.2644	1.448	1.010	.0688	.1100	.4018	1250	1.238	.426
1¼	12	.109	.8365	.3272	.2702	1.329	1.032	.0642	.1027	.4052	1305	1.211	.391
1¼	13	.095	.8825	.3272	.2775	1.173	1.060	.0579	.0926	.4097	1377	1.179	.345
1¼	14	.083	.9229	.3272	.2838	1.033	1.084	.0521	.0833	.4136	1440	1.153	.304
1¼	16	.065	.9852	.3272	.2932	.823	1.120	.0426	.0682	.4196	1537	1.116	.242
1¼	18	.049	1.042	.3272	.3016	.629	1.152	.0334	.0534	.4250	1626	1.085	.185
1¼	20	.035	1.094	.3272	.3089	.456	1.180	.0247	.0395	.4297	1707	1.059	.134
1½	10	.134	1.192	.3927	.3225	1.955	1.232	.1354	.1806	.4853	1860	1.218	.575
1½	12	.109	1.291	.3927	.3356	1.618	1.282	.1159	.1546	.4933	2014	1.170	.476
1½	14	.083	1.398	.3927	.3492	1.258	1.334	.0931	.1241	.5018	2181	1.124	.370
1½	16	.065	1.474	.3927	.3587	.996	1.370	.0756	.1008	.5079	2299	1.095	.293
2	11	.120	2.433	.5236	.4608	2.410	1.760	.3144	.3144	.6660	3795	1.136	.709
2	14	.038	2.642	.5236	.4801	1.699	1.834	.2300	.2300	.6784	4121	1.090	.500
2½	9	.148	3.815	.6540	.5770	3.719	2.204	.7592	.6074	.8332	5951	1.134	1.094

**Standards of Tubular Exchanger Manufacturers Association*, 5th ed., 1968, and 6th ed., 1978.

†Weights are based on low-carbon steel with a density of 0.2833 lb/in^3. For other metals multiply by the following factors:

Metal	Factor
Aluminum	0.35
Titanium	0.58
AISI 400 series stainless steels	0.99
AISI 300 series stainless steels	1.02
Aluminum bronze	1.04
Aluminum brass	1.06
Nickel-chrome-iron	1.07
Admiralty	1.09
Nickel and nickel-copper	1.13
Copper and cupronickels	1.14

‡Liquid velocity $= \dfrac{\text{lb per tube per h}}{C \times \text{sp gr of liquid}}$ in ft/s [sp gr of water at 16°C (60°F) = 1.0]

Conversion Factors for Table 11-2 Tubing Characteristics

Multiply tabulated value by constant shown below to obtain metric units.

	Constant	New unit
Tubing ID, in Wall thickness, in Tubing ID, in Radius of gyration, in	25.4	mm
External surface, ft^2/ft Internal surface, ft^2/ft	0.3048	m^2/m
Weight, lb/ft	1.488	kg/m
Moment of inertia, in^4	416230	mm^4
Section modulus, in^3	16387	mm^3
Metal area, in^2	645	mm^2

The floating-tube-sheet skirt, where in contact with the rings of packing, has fine machine finish. A split shear ring is inserted into a groove in the floating-tube-sheet skirt. A slip-on backing flange, which in service is held in place by the shear ring, bolts to the external floating-head cover.

The floating-head cover is usually a circular disk. With an odd number of tube-side passes, an axial nozzle can be installed in such a floating-head cover. If a side nozzle is required, the circular disk is replaced by either a dished head or a channel barrel (similar to Fig. 11-2*f*) bolted between floating-head cover and floating-tube-sheet skirt.

The outer tube limit approaches the inside of the skirt but is farther removed from the inside of the shell than for any of the previously discussed constructions. Clearances between shell diameter and bundle OTL are 22 mm (⅞ in) for small-diameter pipe shells, 44 mm (1¾ in) for large-diameter pipe shells, and 58 mm (2$\frac{5}{16}$ in) for moderate-diameter plate shells.

Internal Floating-Head Exchanger (Fig. 11-2*a*) The internal floating-head design is used extensively in petroleum-refinery service, but in recent years there has been a decline in usage.

The tube bundle is removable, and the floating tube sheet moves (or floats) to accommodate differential expansion between shell and tubes. The outer tube limit approaches the inside diameter of the gasket at the floating tube sheet. Clearances (between shell and OTL) are 29 mm (1⅛ in) for pipe shells and 37 mm (1$\frac{7}{16}$ in) for moderate-diameter plate shells.

A split backing ring and bolting usually hold the floating-head cover at the floating tube sheet. These are located beyond the end of the shell and within the larger-diameter shell cover. Shell cover, split backing ring, and floating-head cover must be removed before the tube bundle can pass through the exchanger shell.

With an even number of tube-side passes the floating-head cover serves as return cover for the tube-side fluid. With an odd number of passes a nozzle pipe must extend from the floating-head cover through the shell cover. Provision for both differential expansion and tube-bundle removal must be made.

Pull-Through Floating-Head Exchanger (Fig. 11-2*e*) Construction is similar to that of the internal-floating-head split-backing-ring exchanger except that the floating-head cover bolts directly to the floating tube sheet. The tube bundle can be withdrawn from the shell without removing either shell cover or floating-head cover. This feature reduces maintenance time during inspection and repair.

The large clearance between the tubes and the shell must provide for both the gasket and the bolting at the floating-head cover. This clearance is about 2 to 2½ times that required by the split-ring design. Sealing strips or dummy tubes are often installed to reduce bypassing of the tube bundle.

TUBE-SIDE CONSTRUCTION

Tube-Side Header The tube-side header (or stationary head) contains one or more flow nozzles.

The **bonnet** (Fig. 11-1*B*) bolts to the shell. It is necessary to remove the bonnet in order to examine the tube ends. The fixed-tube-sheet exchanger of Fig. 11-2*b* has bonnets at both ends of the shell.

The **channel** (Fig. 11-1*A*) has a removable channel cover. The tube ends can be examined by removing this cover without disturbing the piping connections to the channel nozzles. The channel can bolt to the shell as shown in Fig. 11-2*a* and *c*. The Type C and Type N channels of Fig. 11-1 are welded to the tube sheet. This design is comparable in cost with the bonnet but has the advantages of permitting access to the tubes without disturbing the piping connections and of eliminating a gasketed joint.

Special High-Pressure Closures (Fig. 11-1*D*) The channel barrel and the tube sheet are generally forged. The removable channel cover is seated in place by hydrostatic pressure, while a shear ring subjected to shearing stress absorbs the end force. For pressures above 6205 kPa (900 lbf/in^2) these designs are generally more economical than bolted constructions, which require larger flanges and bolting as pressure increases in order to contain the end force with bolts in tension. Relatively light-gauge internal pass partitions are provided to direct the flow of tube-side fluids but are designed only for the differential pressure across the tube bundle.

Tube-Side Passes Most exchangers have an even number of tube-side passes. The fixed-tube-sheet exchanger (which has no shell cover) usually has a return cover without any flow nozzles as shown in Fig. 11-1*M*; Types *L* and *N* are also used. All removable-bundle designs (except for the U tube) have a floating-head cover directing the flow of tube-side fluid at the floating tube sheet.

Tubes Standard heat-exchanger tubing is ¼, ⅜, ½, ⅝, ¾, 1, 1¼, and 1½ in in outside diameter (in × 25.4 = mm). Wall thickness is measured in Birmingham wire gauge (BWG) units. (Tubing characteristics appear in Table 11-2.) The most commonly used tubes in chemical plants and petroleum refineries are 19- and 25-mm (¾- and 1-in) outside diameter. Standard tube lengths are 8, 10, 12, 16, and 20 ft, with 20 ft now the most common (ft × 0.3048 = m).

Manufacturing tolerances for steel, stainless-steel, and nickel-alloy tubes are such that the tubing is produced to either average or minimum wall thickness. Seamless carbon steel tube of minimum wall thickness may vary from 0 to 20 percent above the nominal wall thickness. Average-wall seamless tubing has an allowable variation of plus or minus 10 percent. Welded carbon steel tube is produced to closer tolerances (0 to plus 18 percent on minimum wall; plus or minus 9 percent on average wall). Tubing of aluminum, copper, and their alloys can be drawn easily and usually is made to minimum wall specifications.

Common practice is to specify **exchanger surface** in terms of total external square feet of tubing. The effective outside heat-transfer surface is based on the length of tubes measured between the inner faces of tube sheets. In most heat exchangers there is little difference between the total and the effective surface. Significant differences are usually found in high-pressure and double-tube-sheet designs.

Integrally finned tube, which is available in a variety of alloys and sizes, is being used in shell-and-tube heat exchangers. The fins are radially extruded from thick-walled tube to a height of 1.6 mm ($\frac{1}{16}$ in) spaced at 1.33 mm (19 fins per inch) or to a height of 3.2 mm (⅛ in) spaced at 2.3 mm (11 fins per inch). External surface is approximately 2½ times the outside surface of a bare tube with the same outside diameter. Also available are 0.93-mm- (0.037-in-) high fins spaced 0.91 mm (28 fins per inch) with an external surface about 3.5 times the surface of the bare tube. Bare ends of nominal tube diameter are provided, while the fin height is slightly less than this diameter. The tube can be inserted into a conventional tube bundle and rolled or welded to the tube sheet by the same means used for bare tubes. An integrally finned tube rolled into a tube sheet with double serrations and flared at the inlet is shown in Fig. 11-5. Internally finned tubes have been manufactured but have limited application.

Longitudinal fins are commonly used in double-pipe exchangers upon the outside of the inner tube. U-tube and conventional removable tube bundles are also made from such tubing. The ratio of external to internal surface generally is about 10 or 15:1.

Transverse fins upon tubes are used in low-pressure gas services. The primary application is in air-cooled heat exchangers (as discussed under that heading), but shell-and-tube exchangers with these tubes are in service.

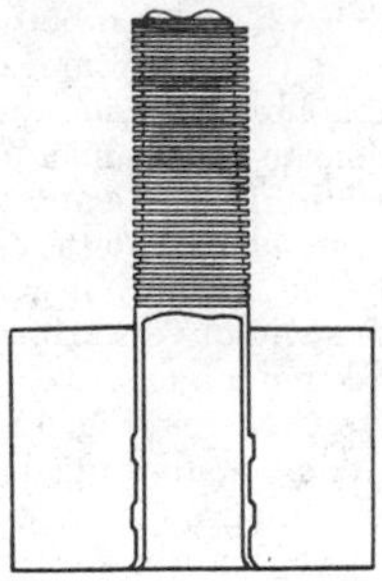

FIG. 11-5 Integrally finned tube rolled into tube sheet with double serrations and flared inlet. *(Wolverine Division, UOP, Inc.)*

Rolled Tube Joints Expanded tube-to-tube-sheet joints are standard. Properly rolled joints have uniform tightness to minimize tube fractures, stress corrosion, tube-sheet ligament pushover and enlargement, and dishing of the tube sheet. Tubes are expanded into the tube sheet for a length of two tube diameters, or 50 mm (2 in), or tube-sheet thickness minus 3 mm (⅛ in). Generally tubes are rolled for the last of these alternatives. The expanded portion should never extend beyond the shell-side face of the tube sheet, since removing such a tube is extremely difficult. Methods and tools for tube removal and tube rolling were discussed by John [*Chem. Eng.*, **66**, 77–80 (Dec. 28, 1959)], and rolling techniques by Bach [*Pet. Refiner*, **39**, 8, 104 (1960)].

Tube ends may be projecting, flush, flared, or beaded (listed in order of usage). The flare or bell-mouth tube end is usually restricted to water service in condensers and serves to reduce erosion near the tube inlet.

For moderate general process requirements at gauge pressures less than 2058 kPa (300 lbf/in^2) and less than 177°C (350°F), tube-sheet holes without grooves are standard. For all other services with expanded tubes at least two grooves in each tube hole are common. The number of grooves is sometimes changed to one or three in proportion to tube-sheet thickness.

Expanding the tube into the **grooved tube holes** provides a stronger joint but results in greater difficulties during tube removal.

Welded Tube Joints When suitable materials of construction are used, the tube ends may be welded to the tube sheets. Welded joints may be seal-welded "for additional tightness beyond that of tube rolling" or may be strength-welded. Strength-welded joints have been found satisfactory in very severe services. Welded joints may or may not be rolled before or after welding.

The variables in tube-end welding were discussed in two unpublished papers (Emhardt, "Heat Exchanger Tube-to-Tubesheet Joints," ASME Pap. 69-WA/HT-47; and Reynolds, "Tube Welding for Conventional and Nuclear Power Plant Heat Exchangers," ASME Pap. 69-WA/HT-24), which were presented at the November 1969 meeting of the American Society of Mechanical Engineers.

Tube-end rolling before welding may leave lubricant from the tube expander in the tube hole. Fouling during normal operation followed by maintenance operations will leave various impurities in and near the tube ends. Satisfactory welds are rarely possible under such conditions, since tube-end welding requires extreme cleanliness in the area to be welded.

Tube **expansion after welding** has been found useful for low and moderate pressures. In high-pressure service tube rolling has not been able to prevent leakage after weld failure.

Double-Tube-Sheet Joints This design prevents the passage of either fluid into the other because of leakage at the tube-to-tube-sheet joints, which are generally the weakest points in heat exchangers. Any leakage at these joints admits the fluid to the gap between the tube sheets. Mechanical design, fabrication, and maintenance of double-tube-sheet designs require special consideration.

SHELL-SIDE CONSTRUCTION

Shell Sizes Heat-exchanger shells are generally made from standard-wall steel pipe in sizes up to 305-mm (12-in) diameter; from 9.5-mm- (⅜-in-) wall pipe in sizes from 356 to 610 mm (14 to 24 in); and from steel plate rolled at discrete intervals in larger sizes. Clearances between the outer tube limit and the shell are discussed elsewhere in connection with the different types of construction.

Tube-count data for heat exchangers appear in Table 11-3. Design criteria used to determine **tube counts** are as follows:

1. Tubes have been eliminated to provide entrance area for a nozzle equal to 0.2 × shell diameter.
2. Tube layouts are symmetrical about both horizontal and vertical axes.
3. Distance from tube OD to centerline of pass partition is 7.9 mm (5/16 in) for shell diameter less than 559 mm (22 in) and 9.5 mm (⅜ in) for larger shells.

Note: "These tables do not necessarily reflect the maximum count nor do they necessarily reflect the actual counts in use. There are too many parameters which influence the final count in a heat exchanger—baffle cut, tie-rod location, cleaning lanes, seal strips, etc. . . ." (Wasielewski, personal communication.)

Tube counts for high-pressure heat-exchanger shells are often less than those tabulated. High-pressure pipe shells maintain outside diameter, while increased shell thickness decreases outer tube limit and tube count. High-pressure bolting for Type S or a thick skirt for Type P may increase required clearance between shell ID and OTL, with a consequent reduction in tube count.

Shell-Side Arrangements The **one-pass shell** (Fig. 11-1*E*) is the most commonly used arrangement. Condensers for single-component vapors often have the nozzles moved to the center of the shell for vacuum and steam services.

A solid longitudinal baffle is provided to form a **two-pass shell** (Fig. 11-1*F*). It may be insulated to improve thermal efficiency. (See further discussion on baffles.) A two-pass shell can improve thermal effectiveness at a cost lower than for two shells in series.

For **split flow** (Fig. 11-1*G*), the longitudinal baffle may be solid or perforated. The latter feature is used with condensing vapors.

A **double-split-flow** design is shown in Fig. 11-1*H*. The longitudinal baffles may be solid or perforated.

In the **divided-flow design** (Fig. 11-1*J*), mechanically this is like the one-pass shell except for the addition of a nozzle. Divided flow is used to meet low-pressure-drop requirements.

The **kettle-type reboiler** is shown in Fig. 11-1*K*. When vaporization occurs on the shell side, this common design provides adequate dome space for separation of vapor and liquid above the tube bundle and surge capacity beyond the weir near the shell cover.

BAFFLES AND TUBE BUNDLES

The **tube bundle** is the most important part of a tubular heat exchanger. The tubes generally constitute the most expensive component of the exchanger and are the one most likely to corrode. Tube sheets, baffles, or support plates, tie rods, and usually spacers complete the bundle.

Minimum **baffle spacing** is generally one-fifth of the shell diameter and not less than 50.8 mm (2 in). Maximum baffle spacing is limited by the requirement to provide adequate support for the tubes. The maximum unsupported tube span in inches equals 74 $d^{0.75}$ (where d is the outside tube diameter in inches). The unsupported tube span is reduced by about 12 percent for aluminum, copper, and their alloys.

Baffles are provided for heat-transfer purposes. When shell-side baffles are not required for heat-transfer purposes, as may be the case in condensers or reboilers, tube supports are installed.

Segmental Baffles Segmental or cross-flow baffles are standard. Single, double, and triple segmental baffles are used. Baffle cuts are illustrated in Fig. 11-6. The double segmental baffle reduces cross-flow velocity for a given baffle spacing. The triple segmental baffle

TABLE 11-3 Tube-Sheet Tube Hole Count*

A. ⅝-in OD tubes on 13/16-in square pitch

Shell ID		TEMA P or S				TEMA U		
		Number of passes				Number of passes		
mm	in	1	2	4	6	2	4	6
203	8	55	48	34	24	52	40	32
254	10	88	78	62	56	90	80	74
305	12	140	138	112	100	140	128	108
337	13¼	178	172	146	136	180	164	148
387	15¼	245	232	208	192	246	232	216
438	17¼	320	308	274	260	330	312	292
489	19¼	405	392	352	336	420	388	368
540	21¼	502	484	442	424	510	488	460
591	23¼	610	584	536	508	626	596	562
635	25	700	676	618	600	728	692	644
686	27	843	812	742	716	856	816	780
737	29	970	942	868	840	998	956	920
787	31	1127	1096	1014	984	1148	1108	1060
838	33	1288	1250	1172	1148	1318	1268	1222
889	35	1479	1438	1330	1308	1492	1436	1388
940	37	1647	1604	1520	1480	1684	1620	1568
991	39	1840	1794	1700	1664	1882	1816	1754
1067	42	2157	2112	2004	1968	2196	2136	2068
1143	45	2511	2458	2326	2288	2530	2464	2402
1219	48	2865	2808	2686	2656	2908	2832	2764
1372	54	3656	3600	3462	3404	3712	3624	3556
1524	60	4538	4472	4310	4256	4608	4508	4426

B. ¾-in OD tubes on 15/16-in triangular pitch

Shell ID		TEMA L or M				TEMA P or S				TEMA U		
		Number of passes				Number of passes				Number of passes		
mm	in	1	2	4	6	1	2	4	6	2	4	6
203	8	64	48	34	24	34	32	16	18	32	24	24
254	10	85	72	52	50	60	62	52	44	64	52	52
305	12	122	114	94	96	109	98	78	68	98	88	78
337	13¼	151	142	124	112	126	120	106	100	126	116	108
387	15¼	204	192	166	168	183	168	146	136	180	160	148
438	17¼	264	254	228	220	237	228	202	192	238	224	204
489	19¼	332	326	290	280	297	286	258	248	298	280	262
540	21¼	417	396	364	348	372	356	324	316	370	352	334
591	23¼	495	478	430	420	450	430	392	376	456	428	408
635	25	579	554	512	488	518	498	456	444	534	500	474
686	27	676	648	602	584	618	602	548	532	628	600	570
737	29	785	762	704	688	729	708	650	624	736	696	668
787	31	909	878	814	792	843	812	744	732	846	812	780
838	33	1035	1002	944	920	962	934	868	840	978	928	904
889	35	1164	1132	1062	1036	1090	1064	990	972	1100	1060	1008
940	37	1304	1270	1200	1168	1233	1196	1132	1100	1238	1200	1152
991	39	1460	1422	1338	1320	1365	1346	1266	1244	1390	1336	1290
1067	42	1703	1664	1578	1552	1611	1580	1498	1464	1632	1568	1524
1143	45	1960	1918	1830	1800	1875	1834	1736	1708	1882	1820	1770
1219	48	2242	2196	2106	2060	2132	2100	1998	1964	2152	2092	2044
1372	54	2861	2804	2682	2660	2730	2684	2574	2536	2748	2680	2628
1524	60	3527	3476	3360	3300	3395	3346	3228	3196	3420	3340	3286
1676	66	4292	4228	4088	4044							
1829	72	5116	5044	4902	4868							
1981	78	6034	5964	5786	5740							
2134	84	7005	6934	6766	6680							
2286	90	8093	7998	7832	7708							
2438	96	9203	9114	8896	8844							
2743	108	11696	11618	11336	11268							
3048	120	14459	14378	14080	13984							

*Courtesy of Nooter Corp.

TABLE 11-3 Tube-Sheet Tube Hole Count* *(Continued)*

C. ¾-in OD tubes on 1-in square pitch

Shell ID		TEMA P or S				TEMA U		
		Number of passes				Number of passes		
mm	in	1	2	4	6	2	4	6
203	8	28	26	16	12	28	24	12
254	10	52	48	44	24	52	44	32
305	12	80	76	66	56	78	72	70
337	13¼	104	90	70	80	96	92	90
387	15¼	136	128	128	114	136	132	120
438	17¼	181	174	154	160	176	176	160
489	19¼	222	220	204	198	224	224	224
540	21¼	289	272	262	260	284	280	274
591	23¼	345	332	310	308	348	336	328
635	25	398	386	366	344	408	392	378
686	27	477	456	432	424	480	468	460
737	29	554	532	510	496	562	548	530
787	31	637	624	588	576	648	636	620
838	33	730	712	682	668	748	728	718
889	35	828	812	780	760	848	820	816
940	37	937	918	882	872	952	932	918
991	39	1048	1028	996	972	1056	1044	1020
1067	42	1224	1200	1170	1140	1244	1224	1212
1143	45	1421	1394	1350	1336	1436	1408	1398
1219	48	1628	1598	1548	1536	1640	1628	1602
1372	54	2096	2048	2010	1992	2108	2084	2068
1524	60	2585	2552	2512	2476	2614	2584	2558

D. ¾-in OD tubes on 1-in triangular pitch

Shell ID		TEMA L or M				TEMA P or S				TEMA U		
		Number of passes				Number of passes				Number of passes		
mm	in	1	2	4	6	1	2	4	6	2	4	6
203	8	42	40	26	24	31	26	16	12	32	24	24
254	10	73	66	52	44	56	48	42	40	52	48	40
305	12	109	102	88	80	88	78	62	68	84	76	74
337	13¼	136	128	112	102	121	106	94	88	110	100	98
387	15¼	183	172	146	148	159	148	132	132	152	140	136
438	17¼	237	228	208	192	208	198	182	180	206	188	182
489	19¼	295	282	258	248	258	250	228	220	226	248	234
540	21¼	361	346	318	320	320	314	290	276	330	316	296
591	23¼	438	416	382	372	400	384	352	336	400	384	356
635	25	507	486	448	440	450	442	400	392	472	440	424
686	27	592	574	536	516	543	530	488	468	554	528	502
737	29	692	668	632	604	645	618	574	556	648	616	588
787	31	796	774	732	708	741	716	666	648	744	716	688
838	33	909	886	836	812	843	826	760	740	852	816	788
889	35	1023	1002	942	920	950	930	878	856	974	932	908
940	37	1155	1124	1058	1032	1070	1052	992	968	1092	1056	1008
991	39	1277	1254	1194	1164	1209	1184	1122	1096	1224	1180	1146
1067	42	1503	1466	1404	1372	1409	1378	1314	1296	1434	1388	1350
1143	45	1726	1690	1622	1588	1635	1608	1536	1504	1652	1604	1560
1219	48	1964	1936	1870	1828	1887	1842	1768	1740	1894	1844	1794
1372	54	2519	2466	2380	2352	2399	2366	2270	2244	2426	2368	2326
1524	60	3095	3058	2954	2928	2981	2940	2932	2800	3006	2944	2884
1676	66	3769	3722	3618	3576							
1829	72	4502	4448	4324	4280							
1981	78	5309	5252	5126	5068							
2134	84	6162	6108	5964	5900							
2286	90	7103	7040	6898	6800							
2438	96	8093	8026	7848	7796							
2743	108	10260	10206	9992	9940							
3048	120	12731	12648	12450	12336							

TABLE 11-3 Tube-Sheet Tube Hole Count* *(Continued)*

E. 1-in OD tubes on 1¼-in square pitch

Shell ID		TEMA P or S				TEMA U		
		Number of passes				Number of passes		
mm	in	1	2	4	6	2	4	6
203	8	17	12	8	12	14	8	6
254	10	30	30	16	18	30	24	12
305	12	52	48	42	24	44	40	32
337	13¼	61	56	52	50	60	48	44
387	15¼	85	78	62	64	80	72	74
438	17¼	108	108	104	96	104	100	100
489	19¼	144	136	130	114	132	132	120
540	21¼	173	166	154	156	172	168	148
591	23¼	217	208	194	192	212	204	198
635	25	252	240	230	212	244	240	230
686	27	296	280	270	260	290	284	274
737	29	345	336	310	314	340	336	328
787	31	402	390	366	368	400	384	372
838	33	461	452	432	420	456	444	440
889	35	520	514	494	484	518	504	502
940	37	588	572	562	548	584	576	566
991	39	661	640	624	620	664	644	640
1067	42	776	756	738	724	764	748	750
1143	45	900	882	862	844	902	880	862
1219	48	1029	1016	984	972	1028	1008	1004
1372	54	1310	1296	1268	1256	1320	1296	1284
1524	60	1641	1624	1598	1576	1634	1616	1614

F. 1-in OD tubes on 1¼-in triangular pitch

Shell ID		TEMA L or M				TEMA P or S				TEMA U		
		Number of passes				Number of passes				Number of passes		
mm	in	1	2	4	6	1	2	4	6	2	4	6
203	8	27	26	8	12	18	14	8	12	14	12	6
254	10	42	40	34	24	33	28	16	18	28	24	24
305	12	64	66	52	44	51	48	42	44	52	40	40
337	13¼	81	74	62	56	73	68	52	44	64	56	52
387	15¼	106	106	88	92	93	90	78	76	90	80	78
438	17¼	147	134	124	114	126	122	112	102	122	112	102
489	19¼	183	176	150	152	159	152	132	136	152	140	136
540	21¼	226	220	204	186	202	192	182	172	196	180	176
591	23¼	268	262	236	228	249	238	216	212	242	224	216
635	25	316	302	274	272	291	278	250	240	286	264	246
686	27	375	360	336	324	345	330	298	288	340	320	300
737	29	430	416	390	380	400	388	356	348	400	380	352
787	31	495	482	452	448	459	450	414	400	456	436	414
838	33	579	554	520	504	526	514	484	464	526	504	486
889	35	645	622	586	576	596	584	548	536	596	572	548
940	37	729	712	662	648	672	668	626	608	668	636	614
991	39	808	792	744	732	756	736	704	692	748	728	700
1067	42	947	918	874	868	890	878	834	808	890	856	830
1143	45	1095	1068	1022	1000	1035	1008	966	948	1028	992	972
1219	48	1241	1220	1176	1148	1181	1162	1118	1092	1180	1136	1100
1372	54	1577	1572	1510	1480	1520	1492	1436	1416	1508	1468	1442
1524	60	1964	1940	1882	1832	1884	1858	1800	1764	1886	1840	1794
1676	66	2390	2362	2282	2260							
1829	72	2861	2828	2746	2708							
1981	78	3368	3324	3236	3216							
2134	84	3920	3882	3784	3736							
2286	90	4499	4456	4370	4328							
2438	96	5144	5104	4986	4936							
2743	108	6546	6494	6360	6300							
3048	120	8117	8038	7870	7812							

TABLE 11-3 Tube-Sheet Tube Hole Count* *(Continued)*

G. 1¼-in OD tubes on 1 9/16-in square pitch

Shell ID		TEMA P or S				TEMA U		
		Number of passes				Number of passes		
mm	in	1	2	4	6	2	4	6
203	8	12	12	4	0	4	4	6
254	10	21	12	8	12	12	8	12
305	12	29	28	16	18	26	20	12
337	13¼	38	34	34	24	36	28	15
387	15¼	52	48	44	48	44	44	32
438	17¼	70	66	56	50	60	60	56
489	19¼	85	84	70	80	82	76	79
540	21¼	108	108	100	96	100	100	100
591	23¼	136	128	128	114	128	120	120
635	25	154	154	142	136	154	148	130
686	27	184	180	158	172	176	172	160
737	29	217	212	204	198	212	204	198
787	31	252	248	234	236	242	240	234
838	33	289	276	270	264	280	280	274
889	35	329	316	310	304	324	312	308
940	37	372	368	354	340	358	352	350
991	39	420	402	402	392	408	400	392
1067	42	485	476	468	464	480	476	464
1143	45	565	554	546	544	558	548	550
1219	48	653	636	628	620	644	628	632
1372	54	837	820	812	804	824	808	808
1524	60	1036	1028	1012	1008	1028	1016	1008

H. 1¼-in OD tubes on 1 9/16-in triangular pitch

Shell ID		TEMA L or M				TEMA P or S				TEMA U		
		Number of passes				Number of passes				Number of passes		
mm	in	1	2	4	6	1	2	4	6	2	4	6
203	8	15	10	8	12	13	10	4	0	6	4	6
254	10	27	22	16	12	18	20	8	12	14	12	12
305	12	38	36	26	24	33	26	26	18	28	20	18
337	13¼	55	44	42	40	38	44	34	24	34	28	30
387	15¼	66	64	52	50	57	58	48	44	52	48	40
438	17¼	88	82	78	68	81	72	62	68	72	68	64
489	19¼	117	106	98	96	100	94	86	80	90	84	78
540	21¼	136	134	124	108	126	120	116	102	118	112	102
591	23¼	170	164	146	148	159	146	132	132	148	132	120
635	25	198	188	166	168	183	172	150	148	172	160	152
686	27	237	228	208	192	208	206	190	180	200	188	180
737	29	268	266	242	236	249	238	224	220	242	228	216
787	31	312	304	284	276	291	282	262	256	282	264	250
838	33	357	346	322	324	333	326	298	296	326	308	292
889	35	417	396	372	364	372	368	344	336	362	344	336
940	37	446	446	422	408	425	412	394	384	416	396	384
991	39	506	490	472	464	478	468	442	432	472	444	428
1067	42	592	584	552	544	558	546	520	512	554	524	510
1143	45	680	676	646	632	646	634	606	596	636	624	592
1219	48	788	774	736	732	748	732	704	696	736	708	692
1372	54	1003	980	952	928	962	952	912	892	946	916	890
1524	60	1237	1228	1188	1152	1194	1182	1144	1116	1176	1148	1116
1676	66	1520	1496	1448	1424							
1829	72	1814	1786	1736	1724							
1981	78	2141	2116	2068	2044							
2134	84	2507	2470	2392	2372							
2286	90	2861	2840	2764	2744							
2438	96	3275	3246	3158	3156							
2743	108	4172	4136	4046	4020							
3048	120	5164	5128	5038	5000							

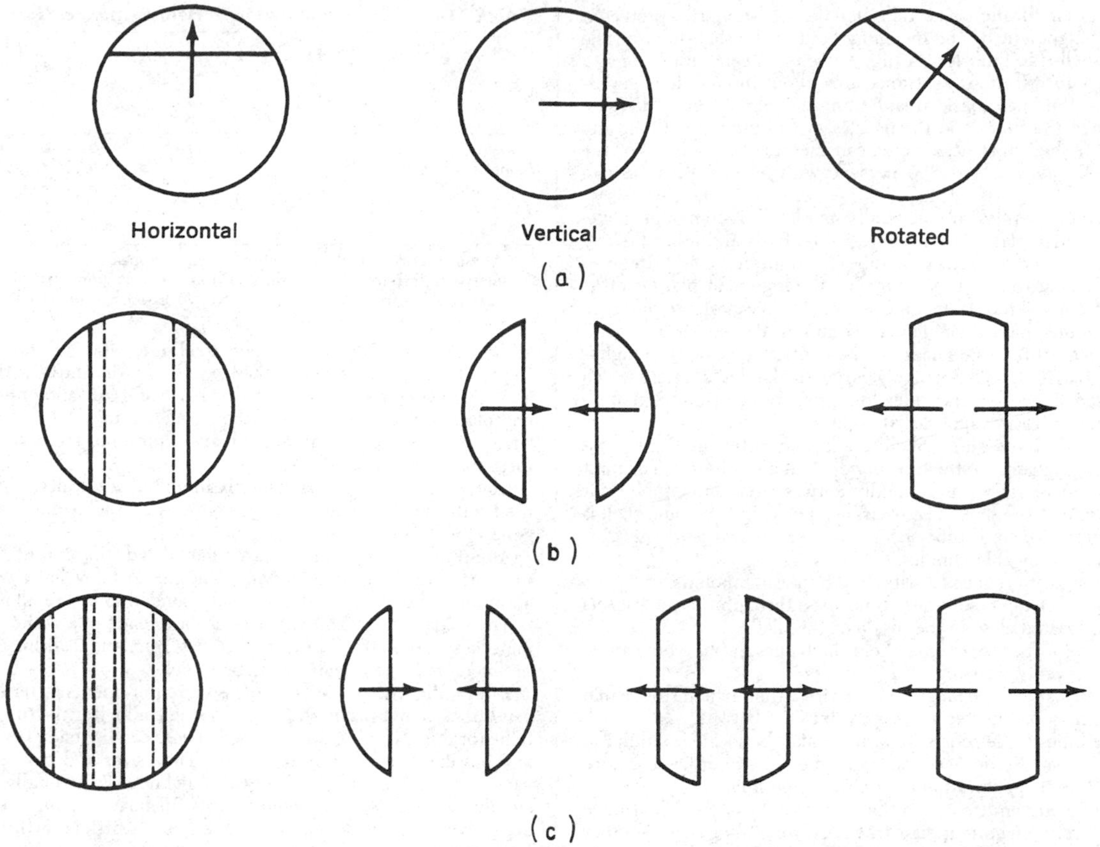

FIG. 11-6 Baffle cuts. (*a*) Baffle cuts for single segmental baffles. (*b*) Baffle cuts for double segmental baffles. (*c*) Baffle cuts for triple segmental baffles.

reduces both cross-flow and long-flow velocities and has been identified as the "window-cut" baffle.

Baffle cuts are expressed as the ratio of segment opening height to shell inside diameter. Cross-flow baffles with horizontal cut are shown in Fig. 11-2*a*, *c*, and *f*. This arrangement is not satisfactory for horizontal condensers, since the condensate can be trapped between baffles, or for dirty fluids in which the dirt might settle out. Vertical-cut baffles are used for side-to-side flow in horizontal exchangers with condensing fluids or dirty fluids. Baffles are notched to assure complete drainage when the units are taken out of service. (These notches permit some bypassing of the tube bundle during normal operation.)

Tubes are most commonly arranged on an equilateral triangular pitch. Tubes are arranged on a square pitch primarily for mechanical cleaning purposes in removable-bundle exchangers.

Maximum baffle cut is limited to about 45 percent for single segmental baffles so that every pair of baffles will support each tube. Tube bundles are generally provided with baffles cut so that at least one row of tubes passes through all the baffles or support plates. These tubes hold the entire bundle together. In pipe-shell exchangers with a horizontal baffle cut and a horizontal pass rib for directing tube-side flow in the channel, the maximum baffle cut, which permits a minimum of one row of tubes to pass through all baffles, is approximately 33 percent in small shells and 40 percent in larger pipe shells.

Maximum shell-side heat-transfer rates in forced convection are apparently obtained by cross-flow of the fluid at right angles to the tubes. In order to maximize this type of flow some heat exchangers are built with segmental-cut baffles and with "no tubes in the window" (or the baffle cutout). Maximum baffle spacing may thus equal maximum unsupported-tube span, while conventional baffle spacing is limited to one-half of this span.

The maximum baffle spacing for no tubes in the window of single segmental baffles is unlimited when intermediate supports are provided. These are cut on both sides of the baffle and therefore do not affect the flow of the shell-side fluid. Each support engages all the tubes; the supports are spaced to provide adequate support for the tubes.

Rod Baffles Rod or bar baffles have either rods or bars extending through the lanes between rows of tubes. A baffle set can consist of a baffle with rods in all the vertical lanes and another baffle with rods in all the horizontal lanes between the tubes. The shell-side flow is uniform and parallel to the tubes. Stagnant areas do not exist.

A recently patented device uses four baffles in a baffle set. Only half of either the vertical or the horizontal tube lanes in a baffle have rods. The new design apparently provides a maximum shell-side heat-transfer coefficient for a given pressure drop.

Tie Rods and Spacers Tie rods are used to hold the baffles in place with spacers, which are pieces of tubing or pipe placed on the rods to locate the baffles. Occasionally baffles are welded to the tie rods, and spacers are eliminated. Properly located tie rods and spacers serve both to hold the bundle together and to reduce bypassing of the tubes.

In very large fixed-tube-sheet units, in which concentricity of shells decreases, baffles are occasionally welded to the shell to eliminate bypassing between the baffle and the shell.

Metal baffles are standard. Occasionally plastic baffles are used either to reduce corrosion or in vibratory service, in which metal baffles may cut the tubes.

Impingement Baffle The tube bundle is customarily protected against impingement by the incoming fluid at the shell inlet nozzle when the shell-side fluid is at a high velocity, is condensing, or is a two-phase fluid. Minimum entrance area about the nozzle is generally equal to the inlet nozzle area. Exit nozzles also require adequate area between the tubes and the nozzles. A full bundle without any provision for shell inlet nozzle area can increase the velocity of the inlet fluid by as much as 300 percent with a consequent loss in pressure.

Impingement baffles are generally made of rectangular plate, although circular plates are more desirable. Rods and other devices are sometimes used to protect the tubes from impingement. In order to maintain a maximum tube count the impingement plate is often placed in a conical nozzle opening or in a dome cap above the shell.

Impingement baffles or flow-distribution devices are recommended for axial tube-side nozzles when entrance velocity is high.

Vapor Distribution Relatively large shell inlet nozzles, which may be used in condensers under low pressure or vacuum, require provision for uniform vapor distribution.

Tube-Bundle Bypassing Shell-side heat-transfer rates are maximized when bypassing of the tube bundle is at a minimum. The most significant bypass stream is generally between the outer tube limit and the inside of the shell. The clearance between tubes and shell is at a minimum for fixed-tube-sheet construction and is greatest for straight-tube removable bundles.

Arrangements to reduce tube-bundle bypassing include:

Dummy tubes. These tubes do not pass through the tube sheets and can be located close to the inside of the shell.

Tie rods with spacers. These hold the baffles in place but can be located to prevent bypassing.

Sealing strips. These longitudinal strips either extend from baffle to baffle or may be inserted in slots cut into the baffles.

Dummy tubes or tie rods with spacers may be located within the pass partition lanes (and between the baffle cuts) in order to ensure maximum bundle penetration by the shell-side fluid.

When tubes are omitted from the tube layout to provide entrance area about an impingement plate, the need for sealing strips or other devices to cause proper bundle penetration by the shell-side fluid is increased.

Longitudinal Baffles In fixed-tube-sheet construction with multipass shells, the baffle is usually welded to the shell and positive assurance against bypassing results. Removable tube bundles have a sealing device between the shell and the longitudinal baffle. Flexible light-gauge sealing strips and various packing devices have been used. Removable U-tube bundles with four tube-side passes and two shell-side passes can be installed in shells with the longitudinal baffle welded in place.

In split-flow shells the longitudinal baffle may be installed without a positive seal at the edges if design conditions are not seriously affected by a limited amount of bypassing.

Fouling in petroleum-refinery service has necessitated rough treatment of tube bundles during cleaning operations. Many refineries avoid the use of longitudinal baffles, since the sealing devices are subject to damage during cleaning and maintenance operations.

CORROSION IN HEAT EXCHANGERS

Some of the special considerations in regard to heat-exchanger corrosion are discussed in this subsection. A more extended presentation in Sec. 23 covers corrosion and its various forms as well as materials of construction.

Materials of Construction The most common material of construction for heat exchangers is carbon steel. Stainless-steel construction throughout is sometimes used in chemical-plant service and on rare occasions in petroleum refining. Many exchangers are constructed from dissimilar metals. Such combinations are functioning satisfactorily in certain services. Extreme care in their selection is required since electrolytic attack can develop.

Carbon steel and alloy combinations appear in Table 11-4. "Alloys" in chemical- and petrochemical-plant service in approximate order of use are stainless-steel series 300, nickel, Monel, copper alloy, aluminum, Inconel, stainless-steel series 400, and other alloys. In petroleum-refinery service the frequency order shifts, with copper alloy (for water-cooled units) in first place and low-alloy steel in second place. In some segments of the petroleum industry copper alloy, stainless series 400, low-alloy steel, and aluminum are becoming the most commonly used alloys.

TABLE 11-4 Dissimilar Materials in Heat-Exchanger Construction

Part	Relative use	1	2	3	4	5	6
	Relative cost	A	B	C	D	C	E
Tubes		•	•	•	•	•	•
Tube sheets			•	•	•	•	•
Tube-side headers				•	•		
Baffles					•	•	•
Shell							•

Carbon steel replaced by an alloy when • appears.
Relative use: from 1 (most popular) through 6 (least popular) combinations.
Relative cost: from A (least expensive) to E (most expensive).

Copper-alloy tubing, particularly inhibited admiralty, is generally used with cooling water. Copper-alloy tube sheets and baffles are generally of naval brass.

Aluminum alloy (and in particular alclad aluminum) tubing is sometimes used in water service. The alclad alloy has a sacrificial aluminum-alloy layer metallurgically bonded to a core alloy.

Tube-side headers for water service are made in a wide variety of materials: carbon steel, copper alloy, cast iron, and lead-lined or plastic-lined or specially painted carbon steel.

Bimetallic Tubes When corrosive requirements or temperature conditions do not permit the use of a single alloy for the tubes, bimetallic (or duplex) tubes may be used. These can be made from almost any possible combination of metals. Tube sizes and gauges can be varied. For thin gauges the wall thickness is generally divided equally between the two components. In heavier gauges the more expensive component may comprise from a fifth to a third of the total thickness.

The component materials comply with applicable ASTM specifications, but after manufacture the outer component may increase in hardness beyond specification limits, and special care is required during the tube-rolling operation. When the harder material is on the outside, precautions must be exercised to expand the tube properly. When the inner material is considerably softer, rolling may not be practical unless ferrules of the soft material are used.

In order to eliminate galvanic action the outer tube material may be stripped from the tube ends and replaced with ferrules of the inner tube material. When the end of a tube with a ferrule is expanded or welded to a tube sheet, the tube-side fluid can contact only the inner tube material, while the outer material is exposed to the shell-side fluid.

Bimetallic tubes are available from a small number of tube mills and are manufactured only on special order and in large quantities.

Clad Tube Sheets Usually tube sheets and other exchanger parts are of a solid metal. Clad or bimetallic tube sheets are used to reduce costs or because no single metal is satisfactory for the corrosive conditions. The alloy material (e.g., stainless steel, Monel) is generally bonded or clad to a carbon steel backing material. In fixed-tube-sheet construction a copper-alloy-clad tube sheet can be welded to a steel shell, while most copper-alloy tube sheets cannot be welded to steel in a manner acceptable to ASME Code authorities.

Clad tube sheets in service with carbon steel backer material include stainless-steel types 304, 304L, 316, 316L, and 317, Monel, Inconel, nickel, naval rolled brass, copper, admiralty, silicon bronze, and titanium. Naval rolled brass and Monel clad on stainless steel are also in service.

Ferrous-alloy-clad tube sheets are generally prepared by a weld overlay process in which the alloy material is deposited by welding upon the face of the tube sheet. Precautions are required to produce a weld deposit free of defects, since these may permit the process fluid to attack the base metal below the alloy. Copper-alloy-clad tube sheets are prepared by brazing the alloy to the carbon steel backing material.

Clad materials can be prepared by bonding techniques, which involve rolling, heat treatment, explosive bonding, etc. When properly manufactured, the two metals do not separate because of thermal-expansion differences encountered in service. Applied tubesheet facings prepared by tack welding at the outer edges of alloy and base metal or by bolting together the two metals are in limited use.

Nonmetallic Construction Shell-and-tube exchangers with glass tubes 14 mm (0.551 in) in diameter and 1 mm (0.039 in) thick with tube lengths from 2.015 m (79.3 in) to 4.015 m (158 in) are available. Steel shell exchangers have a maximum design pressure of 517 kPa (75 lbf/in^2). Glass shell exchangers have a maximum design gauge pressure of 103 kPa (15 lbf/in^2). Shell diameters are 229 mm (9 in), 305 mm (12 in), and 457 mm (18 in). Heat-transfer surface ranges from 3.16 to 51 m^2 (34 to 550 ft^2). Each tube is free to expand, since a Teflon sealer sheet is used at the tube-to-tube-sheet joint.

Impervious graphite heat-exchanger equipment is made in a variety of forms, including outside-packed-head shell-and-tube exchangers. They are fabricated with impervious graphite tubes and tubeside headers and metallic shells. Single units containing up to 1300 m^2 (14,000 ft^2) of heat-transfer surface are available.

Teflon heat exchangers of special construction are described later in this section.

Fabrication Expanding the tube into the tube sheet reduces the tube wall thickness and work-hardens the metal. The induced stresses can lead to **stress corrosion.** Differential expansion between tubes and shell in fixed-tube-sheet exchangers can develop stresses, which lead to stress corrosion.

When austenitic stainless-steel tubes are used for corrosion resistance, a close fit between the tube and the tube hole is recommended in order to minimize work hardening and the resulting loss of corrosion resistance.

In order to facilitate removal and replacement of tubes it is customary to roller-expand the tubes to within 3 mm (⅛ in) of the shell-side face of the tube sheet. A 3-mm- (⅛-in-) long gap is thus created between the tube and the tube hole at this tube-sheet face. In some services this gap has been found to be a focal point for corrosion.

It is standard practice to provide a chamfer at the inside edges of tube holes in tube sheets to prevent cutting of the tubes and to remove burrs produced by drilling or reaming the tube sheet. In the lower tube sheet of vertical units this chamfer serves as a pocket to collect material, dirt, etc., and to serve as a corrosion center.

Adequate **venting** of exchangers is required both for proper operation and to reduce corrosion. Improper venting of the water side of exchangers can cause alternate wetting and drying and accompanying chloride concentration, which is particularly destructive to the series 300 stainless steels.

Certain corrosive conditions require that special consideration be given to complete **drainage** when the unit is taken out of service. Particular consideration is required for the upper surfaces of tube sheets in vertical heat exchangers, for sagging tubes, and for shell-side baffles in horizontal units.

SHELL-AND-TUBE EXCHANGER COSTS

Basic costs of shell-and-tube heat exchangers made in the United States of carbon steel construction in 1958 are shown in Fig. 11-7. Cost data for shell-and-tube exchangers from 15 sources were correlated and found to be consistent when scaled by the Marshall and Swift index [Woods et al., *Can. J. Chem. Eng.*, **54**, 469–489 (December 1976)].

Costs of shell-and-tube heat exchangers can be estimated from Fig. 11-7 and Tables 11-5 and 11-6. These 1960 costs should be updated by use of the Marshall and Swift Index, which appears in each issue of *Chemical Engineering*. Note that during periods of high and low demand for heat exchangers the prices in the marketplace may vary significantly from those determined by this method.

Small heat exchangers and exchangers bought in small quantities are likely to be more costly than indicated.

Standard heat exchangers (which are in some instances off-the-shelf items) are available in sizes ranging from 1.9 to 37 m^2 (20 to 400 ft^2) at costs lower than for custom-built units. Steel costs are approximately one-half, admiralty tube-side costs are two-thirds, and stainless costs are three-fourths of those for equivalent custom-built exchangers.

Kettle-type-reboiler costs are 15 to 25 percent greater than for equivalent internal-floating-head or U-tube exchangers. The higher extra is applicable with relatively large kettle-to-port-diameter ratios and with increased internals (e.g., vapor-liquid separators, foam breakers, sight glasses).

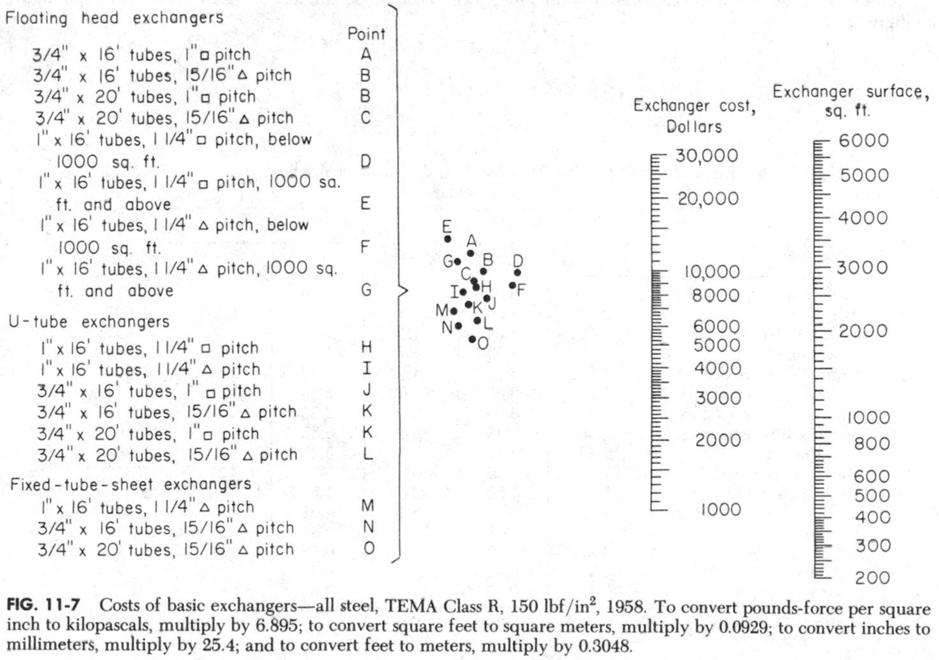

FIG. 11-7 Costs of basic exchangers—all steel, TEMA Class R, 150 lbf/in^2, 1958. To convert pounds-force per square inch to kilopascals, multiply by 6.895; to convert square feet to square meters, multiply by 0.0929; to convert inches to millimeters, multiply by 25.4; and to convert feet to meters, multiply by 0.3048.

TABLE 11-5 Extras for Pressure and Alloy Construction and Surface and Weights*

Percent of steel base price, 1500-lbf/in^2 working pressure

		Shell diameters, in												
		12	14	16	18	20	22	24	27	30	33	36	39	42
Pressure†	300 lbf/in^2	7	7	8	8	9	9	10	11	11	12	13	14	15
	450 lbf/in^2	18	19	20	21	22	23	24	27	29	31	32	33	35
	600 lbf/in^2	28	29	31	33	35	37	39	40	41	32	44	45	50
Alloy	All-steel heat exchanger	100	100	100	100	100	100	100	100	100	100	100	100	100
	Tube sheets and baffles													
	Naval rolled brass	14	17	19	21	22	22	22	22	22	23	24	24	25
	Monel	24	31	35	37	39	39	40	40	41	41	41	41	42
	1¼ Cr, ½ Mo	6	7	7	7	8	8	8	8	9	10	10	10	11
	4–6 Cr, ½ Mo	19	22	24	25	26	26	26	25	25	25	26	26	26
	11–13 Cr (stainless 410)	21	24	26	27	27	27	27	27	27	27	27	27	28
	Stainless 304	22	27	29	30	31	31	31	31	30	30	30	31	31
	Shell and shell cover													
	Monel	45	48	51	52	53	52	52	51	49	47	45	44	44
	1¼ Cr, ½ Mo	20	22	24	25	25	25	24	22	20	19	18	17	17
	4–6 Cr ½ Mo	28	31	33	35	35	35	34	32	30	28	27	26	26
	11–13 Cr (stainless 410)	29	33	35	36	36	36	35	34	32	30	29	27	27
	Stainless 304	32	34	36	37	38	37	37	35	33	31	30	29	28
	Channel and floating-head cover													
	Monel	40	42	42	43	42	41	40	37	34	32	31	40	30
	1¼ Cr, ½ Mo	23	24	24	25	24	24	23	22	21	21	21	20	20
	4–6 Cr, ½ Mo	36	37	38	38	37	36	34	31	29	27	26	25	24
	11–13 Cr (stainless 410)	37	38	39	39	38	37	35	32	30	28	27	26	25
	Stainless 304	37	39	39	39	38	37	36	33	31	29	28	26	26
Surface	Surface, ft^2, internal floating head, ¾-in OD by 1-in square pitch, 16 ft 0 in, tube‡	251	302	438	565	726	890	1040	1470	1820	2270	2740	3220	3700
	1-in OD by 1¼-in square pitch, 16-ft 0-in tube§	218	252	352	470	620	755	876	1260	1560	1860	2360	2770	3200
	Weight, lb, internal floating head, 1-in OD, 14 BWG tube	2750	3150	4200	5300	6600	7800	9400	11,500	14,300	17,600	20,500	24,000	29,000

*Modified from E. N. Sieder and G. H. Elliot, *Pet. Refiner*, **39,** 5, 223 (1960).

†Total extra is 0.7 × pressure extra on shell side plus 0.3 × pressure extra on tube side.

‡Fixed-tube-sheet construction with ¾-in OD tube on 15/16-in triangular pitch provides 36 percent more surface.

§Fixed-tube-sheet construction with 1-in OD tube on 1¼-in triangular pitch provides 18 percent more surface.

For an all-steel heat exchanger with mixed design pressures the total extra for pressure is 0.7 × pressure extra on shell side plus 0.3 × pressure extra tube side.

For an exchanger with alloy parts and a design pressure of 150 lbf/in^2, the alloy extras are added. For shell and shell cover the combined alloy-pressure extra is the alloy extra times the shell-side pressure extra/100. For channel and floating-head cover the combined alloy-pressure extra is the alloy extra times the tube-side pressure extra/100. For tube sheets and baffles the combined alloy-pressure extra is the alloy extra times the higher-pressure extra times 0.9/100. (The 0.9 factor is included since baffle thickness does not increase because of pressure.)

NOTE: To convert pounds-force per square inch to kilopascals, multiply by 6.895; to convert square feet to square meters, multiply by 0.0929; and to convert inches to millimeters, multiply by 25.4.

TABLE 11-6 Base Quantity Extra Cost for Tube Gauge and Alloy

Dollars per square foot

	¾-in OD tubes			1-in OD tubes		
	16 BWG	14 BWG	12 BWG	16 BWG	14 BWG	12 BWG
Carbon steel	0	0.02	0.06	0	0.01	0.07
Admiralty	0.78	1.20	1.81	0.94	1.39	2.03
(T-11) 1¼ Cr, ½ Mo	1.01	1.04	1.11	0.79	0.82	0.95
(T-5) 4–6 CR	1.61	1.65	1.74	1.28	1.32	1.48
Stainless 410 welded	2.62	3.16	4.12	2.40	2.89	3.96
Stainless 410 seamless	3.10	3.58	4.63	2.84	3.31	4.47
Stainless 304 welded	2.50	3.05	3.99	2.32	2.83	3.88
Stainless 304 seamless	3.86	4.43	5.69	3.53	4.08	5.46
Stainless 316 welded	3.40	4.17	5.41	3.25	3.99	5.36
Stainless 316 seamless	7.02	7.95	10.01	6.37	7.27	9.53
90-10 cupronickel	1.33	1.89	2.67	1.50	2.09	2.90
Monel	4.25	5.22	6.68	4.01	4.97	6.47
Low fin						
Carbon steel	0.22	0.23	....	0.18	0.19	
Admiralty	0.58	0.75	....	0.70	0.87	
90-10 cupronickel	0.72	0.96	....	0.86	1.06	

NOTE: To convert inches to millimeters, multiply by 25.4.

To estimate exchanger costs for varying construction details and alloys, first determine the base cost of a similar heat exchanger of basic construction (carbon steel, Class R, 150 lbf/in^2) from Fig. 11-7. From Table 11-5, select appropriate extras for higher pressure rating and for alloy construction of tube sheets and baffles, shell and shell cover, and channel and floating-head cover. Compute these extras in accordance with the notes below the table. For tubes other than welded carbon steel, compute the extra by multiplying the exchanger surface by the appropriate cost per square foot from Table 11-6.

When points for 20-ft-long tubes do not appear in Fig. 11-7, use 0.95 times the cost of the equivalent 16-ft-long exchanger. Length variation of steel heat exchangers affects costs by approximately $1 per square foot. Shell diameters for a given surface are approximately equal for U-tube and floating-head construction.

Low-fin tubes ($\frac{1}{16}$-in-high fins) provide 2.5 times the surface per lineal foot. Surface required should be divided by 2.5; then use Fig. 11-7 to determine basic cost of the heat exchanger. Actual surface times extra costs (from Table 11-6) should then be added to determine cost of fin-tube exchanger.

OTHER HEAT EXCHANGERS FOR LIQUIDS AND GASES

DOUBLE-PIPE AND MULTITUBE SECTIONS

Double-pipe heat exchangers have been used for many years primarily for low flow rates and high temperature ranges. Multitube sections are of similar construction but have seven or more tubes within an outer shell. Double-pipe and multitube sections permit true countercurrent flow, which is of particular advantage for long temperature ranges of fluid temperature and when close temperature approaches are required.

Commercially available **double-pipe sections** range from 51- to 152-mm- (2- to 6-in-) pipe shells with inner tubes varying from 19-mm ($\frac{3}{4}$-in) tubing through 102-mm (4-in) pipe. About 12 percent of the double-pipe sections are bare tube or pipe, and the rest have longitudinal fins upon the inner tube. **Multitube sections** have shells made of 4-, 6-, 8-, and 12-in pipe (in × 25.4 = mm). The tubes are fabricated from straight portions of pipe or tubing. Longitudinal fins are generally welded to the tubes. Bare-tube or bare-pipe U bends are welded to two straight lengths of tube to produce the hairpin tube used in both double-pipe and multitube sections.

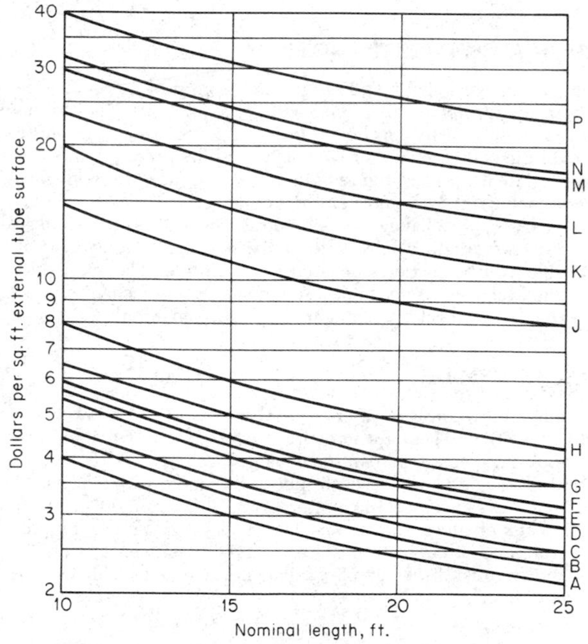

FIG. 11-8 Costs of double-pipe sections (refer to Table 11-7 for descriptive data). To convert feet to meters, multiply by 0.3048; to convert dollars per square foot to dollars per square meter, multiply by 10.764. *(Brown Fintube Co.)*

Several manufacturers offer standard double-pipe and multitube sections. These have removable tubes and provision for differential expansion between shell and tube. The closures to seal the shell-side fluid are of varying types. The simplest type permits interstream leakage upon failure of the seal. The best design uses two sets of through-bolted stud bolts, and interstream leakage is not possible. Multitube sections have the ends of the hairpin tubes rolled into the tube sheets.

Costs of double-pipe sections (as illustrated in Fig. 11-9) in 1971 with both bare tubes and finned tubes are shown in Fig. 11-8 with the curves identified by the specifications shown in Table 11-7.

PLATE-TYPE EXCHANGERS

Plate-type exchangers are available in several distinctively different forms: spiral, plate (and frame), brazed-plate fin, and plate fin-and-tube types.

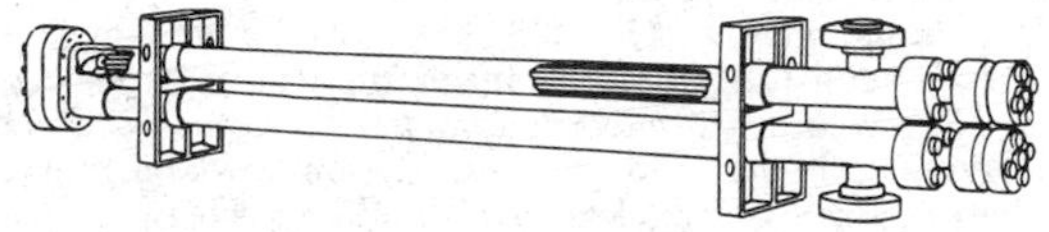

FIG. 11-9 Double-pipe-exchanger section with longitudinal fins. *(Brown Fintube Co.)*

TABLE 11-7 Double-Pipe Hairpin-Section Data (for Costs Shown in Fig. 11-8)

Curve	Outer pipe, in.	Inner pipe o.d., in.	Inner pipe material	Inner pipe thickness, in.	No. of fins on inner tube	Surface, sq. ft. Nominal tube length, ft. 10	20	25
P	4	1.900	Admiralty	0.109	None	10.9	20.9	25.9
N	3	1.900	Admiralty	0.109	None	10.9	20.9	25.9
M	4	1.900	Steel	0.145	None	10.9	20.9	25.9
L	3	1.900	Steel	0.145	None	10.9	20.9	25.9
K	4	1.000	Admiralty	0.065	None	37.6	74.3	92.6
J	4	1.000	Steel	0.134	None	37.6	74.3	92.6
H	3	1.900	Admiralty	0.109	24	50.9	100.9	125.9
G	4	1.900	Admiralty	0.109	24	90.9	180.9	225.9
F	4	0.875	Admiralty	0.065	24	150.7	300.3	375.2
E	3	1.900	Steel	0.145	24	50.9	100.9	125.9
D	4	2.875	Steel	0.203	36	76	151.1	188.7
C	4	0.875	Steel	0.083	20	131.1	261.1	326.2
B	4	2.875	Steel	0.203	48	96	191.1	238.6
A	4	1.900	Steel	0.145	24	90.9	180.9	225.9

Outer pipe is schedule 40 carbon steel, iron pipe size.

Admiralty inner tube has longitudinal yellow-brass fins. Steel tube has longitudinal steel fins.

NOTE: To convert inches to millimeters, multiply by 25.4; to convert feet to meters, multiply by 0.3048.

Spiral-Plate Exchangers The spiral-plate exchanger (Alfa-Laval, Thermal Division) is made from a pair of plates rolled to provide two relatively long rectangular passages for fluids in countercurrent flow. The continuous path eliminates flow reversals (and accompanying pressure drop), bypassing, and differential-expansion problems. Solids can be maintained in suspension. Turbulence occurs at a lower Reynolds number than in straight tubes.

Spiral design is **compact** in that 167 m^2 (1800 ft^2) of heat-transfer surface can be provided in a 1422-mm- (56-in-) diameter unit. Spiral height is generally 1524 mm (60 in).

Each path of the exchanger is accessible by removing the respective cover. Spirals can be designed for gauge pressure as high as 1034 kPa (150 lbf/in^2). Materials of construction include carbon steel, stainless steel types 304, 316, and 430F, alloy 20, Inconel, Monel, nickel, Hastelloy B and C, Everdur, and titanium.

Plate-and-Frame Exchangers Plate-and-frame exchangers consist of standard plates, which serve as heat-transfer surfaces, and a frame to support them and were described by Lawry [*Chem. Eng.*, **66**, 89–94 (June 29, 1959)]. The design principle is much like that of the plate-and-frame filter press. Pressure drop is low and interleakage of fluids is impossible.

Standard heat-transfer plates (normally of stainless-steel types 304 and 316, but titanium, nickel, Monel, Incoloy 825, Hastelloy C, phosphor bronze, and cupronickel are also available) pressed in a single piece from 1.3- to 6.4-mm (0.05- to 0.125-in) material are provided with grooves for elastomer gaskets. Corrugated-plate design imparts rigidity to the plate, induces turbulence in the fluids, and assures complete flow distribution. The frame and supporting members are available in clad stainless steel or enamel-coated mild steel. Plates are easily cleaned and replaced. The area can be readily adjusted by adding or subtracting plates.

Plate heat exchangers can be used for **multiple duty;** several different fluids can flow through different parts of the exchanger. Viscous fluids (up to 30,000 cP) also yield relatively high transfer rates since many plate-type heat exchangers secure turbulent flow at a Reynolds number as low as 150.

Design limitations include 2068-kPa (300-lbf/in^2) maximum gauge pressure and 149°C (300°F) maximum temperature; condensation of large volumes of vapor is impractical and not satisfactory for true gases; and with solid-liquid suspensions the largest particle should be 0.5 mm (0.02 in) less than the distance between plates. Asbestos fiber, which is not an elastomer, is used as a gasketing material at temperatures up to 250°C (482°F). Stronger frames are required for gasket compression.

Figure 11-10 presents a multiplate exchanger. The hot fluid enters at the top on the right side of the fixed end cover and flows downward through alternate channels between the plates. The hot fluid leaves the exchanger through a connection at the bottom of the fixed end cover. The cold fluid enters at the bottom on the left side of the fixed end cover. The cold fluid flows in countercurrent flow through alternate channels between the plates and leaves the exchanger at the top on the left side of the fixed end cover. At the stationary end of the unit a "frame," or fixed end cover plate, is provided. At the other end a "movable end cover," or "pressure" plate, serves to compress the "plate pack" (which contains plates and gaskets) as the nuts on the compression bolts are tightened.

When multiple-service or stainless-steel tube-side construction is specified, the plate type competes with the tubular design. If all-stainless-steel construction is required, the plate type will be less expensive than the tubular units.

The upper limit of a standard plate exchanger is reported as 650 m^2 (7000 ft^2) of heat-transfer surface, with the unit measuring 1.1 m (42 in) wide by 4.2 m (165 in) long by 2.8 m (111 in) high with 400 plates.

Brazed-Plate-Fin Heat Exchanger Brazed-aluminum-plate-fin heat exchangers were first manufactured for aircraft applications during World War II. In 1950 the first tonnage air-separation plant with these compact, lightweight reversing heat exchangers began producing oxygen for a steel mill. Aluminum-plate-fin exchangers are used in the process industries, particularly for services below −45.6°C (−50°F), and in gas-separation processes operating between 204 and −268°C (400 and −450°F).

Plate-fin heat-transfer surface is made up of a stack of layers, with each layer consisting of a corrugated fin between flat metal sheets sealed off on two sides by channels or bars to form one passage for the flow of fluid. An exploded view of a typical layer is shown in Fig. 11-11. The simple straight corrugation shown here may be replaced by a wavy or herringbone pattern or a serrated or strip fin when higher performance is required.

Maximum design condition has been 5171 kPa (750 lbf/in^2) gauge pressure. Typical design is for lower pressures at subzero temperatures.

The most widely used plate-fin heat-exchanger core is the 914- by 914- by 6096-mm (36- by 36- by 240-in) reversing unit with approximately 6968 m^2 (75,000 ft^2) of total heat-transfer surface. A range of core sizes and densities are available with up to 46.5 m^2 (500 ft^2) of total heat-transfer surface per cubic foot of heat-exchanger volume. The surface may be arranged for countercurrent or parallel flow or both with several different process streams. Exchangers handling as many as seven different streams have been built.

These heat exchangers are used with gases, liquids, and liquid-vapor mixtures for sensible heat-transfer, evaporation, and condensation.

The **cost** of aluminum-plate-fin process heat exchangers varies considerably with core size, quantity, and design complexity. The 1977 range was from $0.65 to $3 per square foot of total heat-transfer surface, with two-stream low-pressure cores in large quantities priced at the low side and multiple-stream high-pressure cores in small quantities at the higher figure. These exchangers are economical when surface requirements exceed 372 m^2 (4000 ft^2) and are marginal for services with less than 186 m^2 (2000 ft^2).

Plate Fin-and-Tube Surface Rectangular fins are pierced, formed, belled, and stacked before tubes are inserted into the fin collars and expanded to produce the plate fin-and-tube surface. No solder or brazing metal is used. Tube diameters range from 9.5- to 38-mm (⅜- to 1½-in) outside diameter; fin spacing varies from 38 to 2 mm (8 to 156 fins per foot); and the external to internal surface ratio extends up to 40:1.

GRAPHITE-BLOCK EXCHANGERS

Cubic exchangers of impervious graphite consist of solid cubes perforated by rows of parallel holes which are at right angles to those above and below. Headers bolted to the opposite sides of the vertical faces of the cube provide for flow of process fluid through the block. Appropriate headers on the remaining vertical faces direct the heating or cooling medium through the exchangers.

A block-type exchanger, which consists of a series of cylindrical impervious-graphite blocks, with radial and axial passages, is also available (Carbone Lorraine Industries Corp.). The cubic-block graphite exchanger is not subject to damage from mechanical shock, as is the shell-and-tube exchanger of the same material.

CASCADE COOLERS

Cascade coolers consist of a series of tubes mounted horizontally, one above the other. These are sometimes referred to as trombone coolers, trickle coolers, or serpentine coolers. Cooling water from a distributing trough drips over each tube and then flows to a drain. The hot fluid flows generally in countercurrent flow from bottom to top of the bank of tubes. Cascade coolers of glass, impervious graphite, cast iron, and other materials are available. Glass coolers use 38-, 51-, and 76-mm (1½-, 2-, and 3-in) glass pipe elements with 3-m- (10-ft-) long tube.

ATMOSPHERIC COOLERS

Atmospheric sections consist of bare tubes arranged in rectangular tube bundles. These are installed above the water basin at the bottom

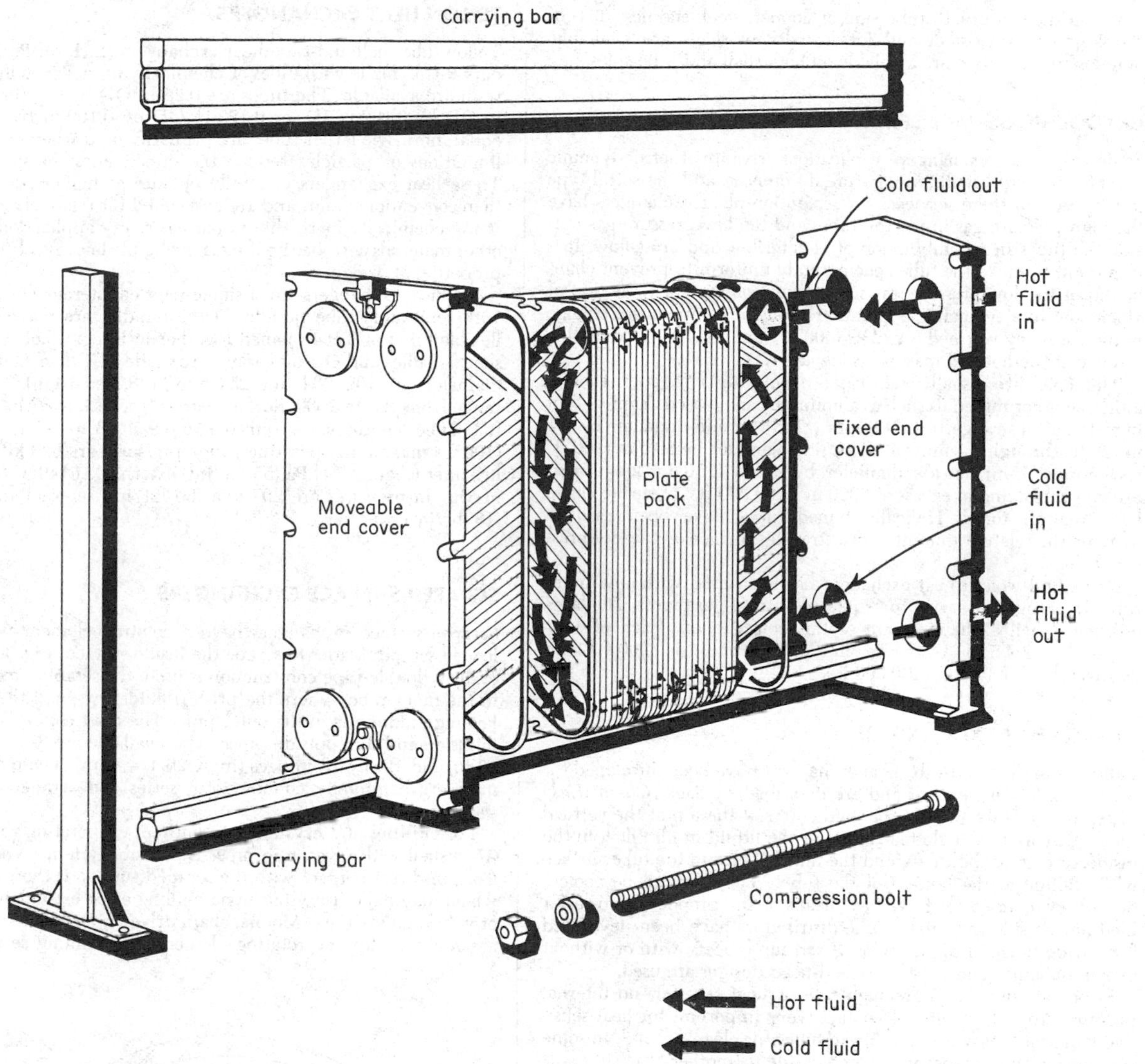

FIG. 11-10 Plate-and-frame heat exchanger. Hot fluid flows down between alternate plates, and cold fluid flows up between alternate plates. *(Thermal Division, Alfa-Laval, Inc.)*

of a cooling tower. A process fluid or the primary cooling water flows inside the tubes.

BAYONET-TUBE EXCHANGERS

This type of exchanger is useful when there is an extreme temperature difference between shell- and tube-side fluids, since all parts subject to differential expansion are free to move independently of each other. This unique construction does not suffer failure due to freezing of steam condensate, since the steam in the inner tube melts any ice which may form during periods of intermittent operation. Costs are relatively high, since only the outer bundle tubes transfer heat to the shell-side fluid. The inner tubes are unsupported. The outer tubes are supported by conventional baffles or support plates.

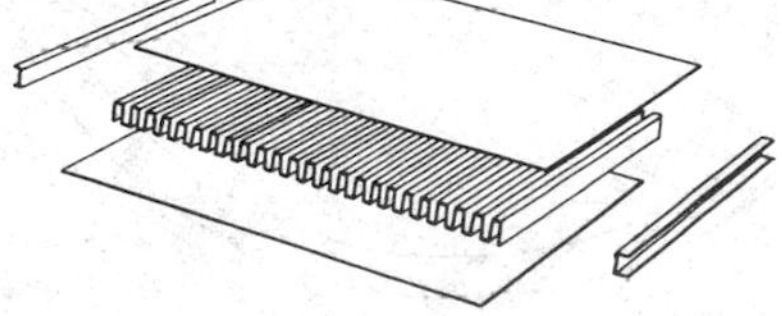

FIG. 11-11 Exploded view of a typical plate-fin arrangement. *(Trane Co.)*

SPIRAL-TUBE EXCHANGERS

Spiral-tube exchangers consist of a group of concentric spirally wound coils which are generally connected by manifolds. Features include countercurrent flow, elimination of differential-expansion difficulties, constant velocity, and compactness.

For general-process exchanger applications, sizes available range from 0.2 to 33 m^2 (2 to 356 ft^2). The largest of these with bare tubes requires a space of 0.94-m (37-in) diameter by 1.07 m (42 in) deep. Tubing sizes used are ¼-, ⅜-, ½-, ⅝-, and ¾-in diameter (in × 25.4 = mm). Integrally finned tubes have been used to increase the heat-transfer surface. Standard constructions of Graham Mfg. Co. include cast-iron shell side with tube side of copper, steel, stainless 304, or stainless 316. Cast-steel and fabricated-steel shells are available. Stainless 304 and 316 are available on both shell and tube sides.

CRYOGENIC-SERVICE SPIRAL-TUBE EXCHANGERS

Some cryogenic-exchanger applications require thermodynamic reversibility with small temperature differences, and the spiral-type unit is used in these services. Air-separation-plant exchangers have the high-pressure gas inside the tubes and the low-pressure gas outside the tubes in a combination of counterflow and cross-flow. It is important to make the tube spacing quite uniform to prevent channeling of the low-pressure gas stream and consequent reduction of efficiency. In a typical air-separation plant the temperature change in the gas being warmed is 222°C (400°F), and the temperature difference at the hot end may be as low as 1.7°C (3°F).

The coils are usually fabricated from drawn copper tubing, although other materials such as aluminum can be used. Typical tubing dimensions range from 4.8-mm (3/16-in) outside diameter by 6.1 m (20 ft) through 6.4-mm (1/4-in) outside diameter by 30.5 m (100 ft) to 9.5-mm (3/8-in) outside diameter by 46 m (150 ft) long. Heat exchangers having in excess of 9300 m^2 (100,000 ft^2) of surface have been manufactured. Helically finned tubes have been used to increase the relative amount of heat-transfer surface on the low-pressure side.

The spiral-wound heat exchanger has been used with aluminum tubes in cryogenic service for liquefied-natural-gas plants. The largest single-shell exchangers are of this construction. Each of four exchangers is 4.5 m (14 ft 8 in) in diameter by 61 m (200 ft) long and weighs about 180 t (400,000 lb).

FALLING-FILM EXCHANGERS

Falling-film shell-and-tube heat exchangers have been developed for a wide variety of services and are described by Sack [*Chem. Eng. Prog.*, **63,** 55 (July 1967)]. The fluid enters at the top of the vertical tubes. Distributors or slotted tubes put the liquid in film flow in the inside surface of the tubes, and the film adheres to the tube surface while falling to the bottom of the tubes. The film can be cooled, heated, evaporated, or frozen by means of the proper heat-transfer medium outside the tubes. Tube distributors have been developed for a wide range of applications. Fixed tube sheets, with or without expansion joints, and outside-packed-head designs are used.

Principal advantages are high rate of heat transfer, no internal pressure drop, short time of contact (very important for heat-sensitive materials), easy accessibility to tubes for cleaning, and, in some cases, prevention of leakage from one side to another.

These falling-film exchangers are used in various services as described in the following paragraphs.

Liquid Coolers and Condensers Dirty water can be used as the cooling medium. The top of the cooler is open to the atmosphere for access to tubes. These can be cleaned without shutting down the cooler by removing the distributors one at a time and scrubbing the tubes.

Evaporators These are used extensively for the concentration of ammonium nitrate, urea, and other chemicals sensitive to heat when minimum contact time is desirable. Air is sometimes introduced in the tubes to lower the partial pressure of liquids whose boiling points are high. These evaporators are built for pressure or vacuum and with top or bottom vapor removal.

Absorbers These have a two-phase flow system. The absorbing medium is put in film flow during its fall downward on the tubes as it is cooled by a cooling medium outside the tubes. The film absorbs the gas which is introduced into the tubes. This operation can be cocurrent or countercurrent.

Freezers By cooling the falling film to its freezing point, these exchangers convert a variety of chemicals to the solid phase. The most common application is the production of sized ice and paradichlorobenzene. Selective freezing is used for isolating isomers. By melting the solid material and refreezing in several stages, a higher degree of purity of product can be obtained.

TEFLON HEAT EXCHANGERS

Teflon tube shell-and-tube heat exchangers (E. I. du Pont de Nemours & Co.) made with tubes of chemically inert Teflon fluorocarbon resin are available. The tubes are 0.25-in OD by 0.20-in ID, 0.175-in OD by 0.160-in ID, or 0.125-in OD by 0.100-in ID (in × 25.4 equal mm). The larger tubes are primarily used when pressure-drop limitations or particles reduce the effectiveness of smaller tubes. These heat exchangers generally operate at higher pressure drops than conventional units and are best suited for relatively clean fluids. Being chemically inert, the tubing has many applications in which other materials corrode. Fouling is negligible because of the antistick properties of Teflon.

The heat exchangers are of single-pass, countercurrent-flow design with removable tube bundles. Tube bundles are made of straight flexible tubes of Teflon joined together in integral honeycomb tube sheets. Baffles and O-ring gaskets are made of Teflon. Standard shell diameters are 102, 204, and 254 mm (4, 8, and 10 in). Tube counts range from 105 to 2000. Surface varies from 1.9 to 87 m^2 (20 to 940 ft^2). Tube lengths vary from 0.9 to 4.9 m (3 to 16 ft). At 37.8°C (100°F) maximum operating gauge pressures are 690 kPa (100 lbf/in^2) internal and 379 kPa (55 lbf/in^2) external. At 149°C (300°F) the maximum pressures are 207 kPa (30 lbf/in^2) internal and 124 kPa (18 lbf/in^2) external.

SCRAPED-SURFACE EXCHANGERS

Scraped-surface exchangers have a rotating element with spring-loaded scraper blades to scrape the inside surface (Fig. 11-12). Generally a double-pipe construction is used; the scraping mechanism is in the inner pipe, where the process fluid flows; and the cooling or heating medium is in the outer pipe. The most common size has 6-in inside and 8-in outside pipes. Also available are 3- by 4-in, 8- by 10-in, and 12- by 14-in sizes (in × 25.4 = mm). These double-pipe units are commonly connected in series and arranged in double stands.

For **chilling** and **crystallizing** with an evaporating refrigerant, a 27-in shell with seven 6-in pipes is available (Henry Vogt Machine Co.). In direct contact with the scraped surface is the process fluid, which may deposit crystals upon chilling or be extremely fouling or of very high viscosity. Motors, chain drives, appropriate guards, etc., are required for the rotating element. For chilling service with a

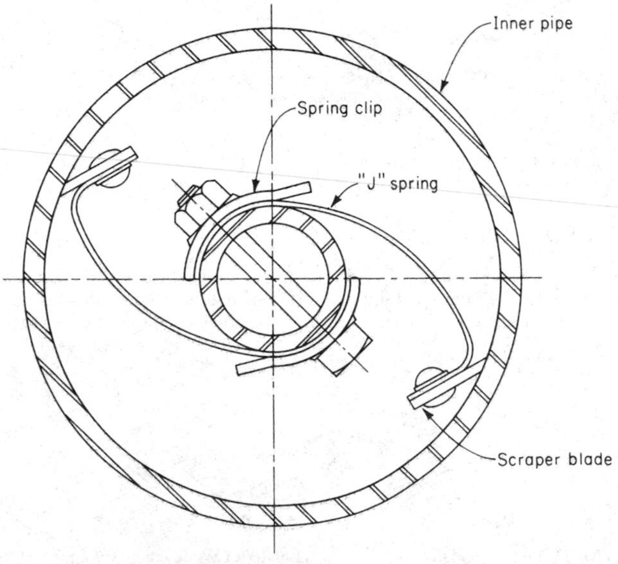

FIG. 11-12 Scraper blade of scraped-surface exchanger. *(Henry Vogt Machine Co., Inc.)*

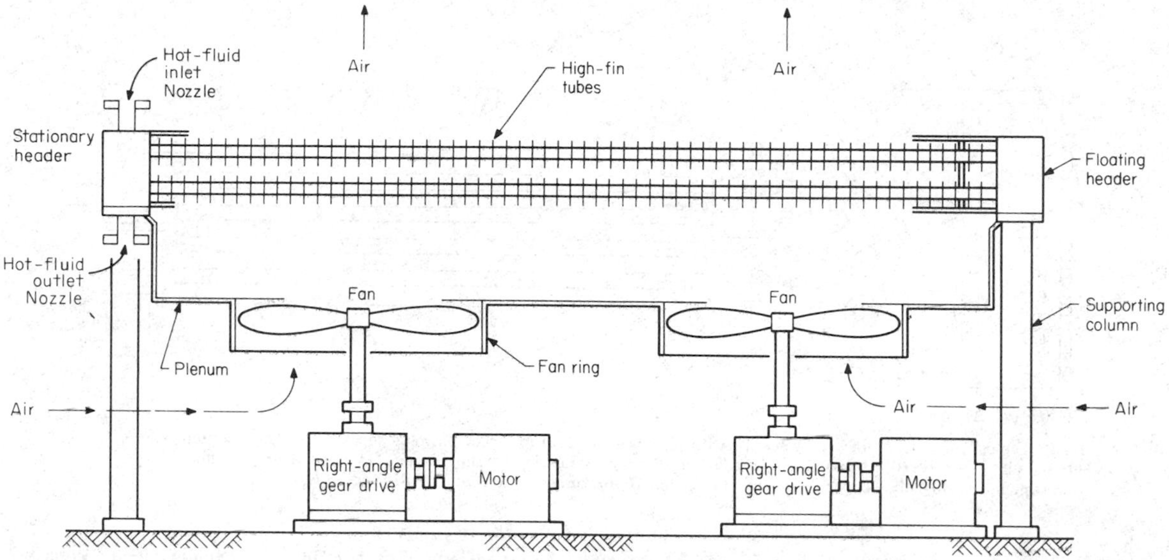

FIG. 11-13 Forced-draft air-cooled heat exchanger. [Chem. Eng., *114 (Mar. 27, 1978).*]

refrigerant in the outer shell, an accumulator drum is mounted on top of the unit.

Scraped-surface exchangers are particularly suitable for heat transfer with crystallization, heat transfer with severe fouling of surfaces, heat transfer with solvent extraction, and heat transfer of high-viscosity fluids. They are extensively used in paraffin-wax plants and in petrochemical plants for crystallization.

AIR-COOLED HEAT EXCHANGERS

Atmospheric air has been used for many years to cool and condense fluids in areas of water scarcity. During the 1960s the use of air-cooled heat exchangers grew rapidly in the United States and elsewhere. In Europe, where seasonal variations in ambient temperatures are relatively small, air-cooled exchangers are used for the greater part of process cooling. In some new plants all cooling is done with air. Increased use of air-cooled heat exchangers has resulted from lack of available water, significant increases in water costs, and concern for water pollution.

Air-cooled heat exchangers include a tube bundle, which generally has spiral-wound fins upon the tubes, and a fan, which moves air across the tubes and is provided with a driver. Electric motors are the most commonly used drivers; typical drive arrangements require a V belt or a direct right-angle gear. A plenum and structural supports are basic components. Louvers are often used.

A bay generally has two tube bundles installed in parallel. These may be in the same or different services. Each bay is usually served by two (or more) fans and is furnished with a structure, a plenum, and other attendant equipment.

The location of air-cooled heat exchangers must consider the large space requirements and the possible recirculation of heated air because of the effect of prevailing winds upon buildings, fired heaters, towers, various items of equipment, and other air-cooled exchangers. Inlet air temperature at the exchanger can be significantly higher than the ambient air temperature at a nearby weather station. See *Air-Cooled Heat Exchangers for General Refinery Services*, API Standard 661, 2d ed., January 1978, for information on refinery-process air-cooled heat exchangers.

Forced and Induced Draft The forced-draft unit, which is illustrated in Fig. 11-13, pushes air across the finned tube surface. The fans are located below the tube bundles. The induced-draft design has the fan above the bundle, and the air is pulled across the finned tube surface. In theory, a primary advantage of the forced-draft unit is that less power is required. This is true when the air-temperature rise exceeds 30°C (54°F).

Air-cooled heat exchangers are generally arranged in banks with several exchangers installed side by side. The height of the bundle aboveground must be one-half of the tube length to produce an inlet velocity equal to the face velocity. This requirement applies both to ground-mounted exchangers and to those pipe-rack-installed exchangers which have a fire deck above the pipe rack.

The forced-draft design offers better accessibility to the fan for on-stream maintenance and fan-blade adjustment. The design also provides a fan and V-belt assembly, which are not exposed to the hot-air stream that exits from the unit. Structural costs are less, and mechanical life is longer.

Induced-draft design provides more even distribution of air across the bundle, since air velocity approaching the bundle is relatively low. This design is better suited for exchangers designed for a close approach of product outlet temperature to ambient-air temperature.

Induced-draft units are less likely to recirculate the hot exhaust air, since the exit air velocity is several times that of the forced-draft unit. Induced-draft design more readily permits the installation of the air-cooled equipment above other mechanical equipment such as pipe racks or shell-and-tube exchangers.

In a service in which sudden temperature change would cause upset and loss of product, the induced-draft unit gives more protection in that only a fraction of the surface (as compared with the forced-draft unit) is exposed to rainfall, sleet, or snow.

Tube Bundle The principal parts of the tube bundle are the finned tubes and the header. Most commonly used is the plug header, which is a welded box that is illustrated in Fig. 11-14. The finned tubes are described in a subsequent paragraph. The components of a tube bundle are identified in the figure.

The second most commonly used header is a cover-plate header. The cover plate is bolted to the top, bottom, and end plates of the header. Removing the cover plate provides direct access to the tubes without the necessity of removing individual threaded plugs.

Other types of headers include the bonnet-type header, which is constructed similarly to the bonnet construction of shell-and-tube heat exchangers; manifold-type headers, which are made from pipe and have tubes welded into the manifold; and billet-type headers, made from a solid piece of material with machined channels for distributing the fluid. Serpentine-type tube bundles are sometimes used for very viscous fluids. A single continuous flow path through pipe is provided.

Tube bundles are designed to be rigid and self-contained and are mounted so that they expand independently of the supporting structure.

The face area of the tube bundle is its length times width. The net

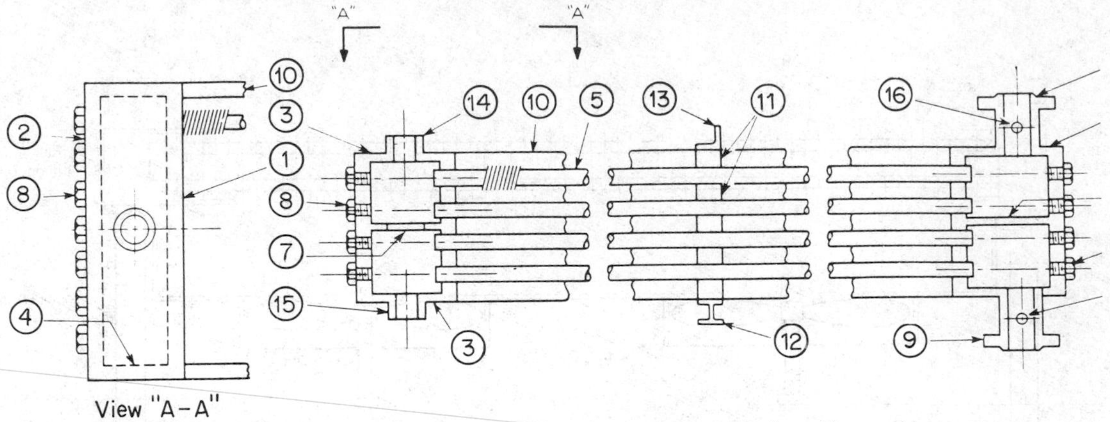

FIG. 11-14 Typical construction of a tube bundle with plug headers: (1) tube sheet; (2) plug sheet; (3) top and bottom plates; (4) end plate; (5) tube; (6) pass partition; (7) stiffener; (8) plug; (9) nozzle; (10) side frame; (11) tube spacer; (12) tube-support cross member; (13) tube keeper; (14) vent; (15) drain; (16) instrument connection. *(API Standard 661.)*

free area for air flow through the bundle is about 50 percent of the face area of the bundle.

The standard air **face velocity** (FV) is the velocity of standard air passing through the tube bundle and generally ranges from 1.5 to 3.6 m/s (300 to 700 ft/min).

Tubing The 25.4-mm (1-in-) outside-diameter tube is most commonly used. Fin heights vary from 12.7 to 15.9 mm (0.5 to 0.625 in), fin spacing from 3.6 to 2.3 mm (7 to 11 per linear inch), and tube triangular pitch from 50.8 to 63.5 mm (2.0 to 2.5 in). Ratio of extended surface to bare-tube outside surface varies from about 7 to 20. The 38-mm (1½-in) tube has been used for flue-gas and viscous-oil service. Tube size, fin heights, and fin spacing can be further varied.

Tube lengths vary and may be as great as 18.3 m (60 ft). When tube length exceeds 12.2 m (40 ft), three fans are generally installed in each bay. Frequently used tube lengths vary from 6.1 to 12.2 m (20 to 40 ft).

Finned-Tube Construction The following are descriptions of commonly used finned-tube constructions (Fig. 11-15).

1. *Embedded.* Rectangular-cross-section aluminum fin which is wrapped under tension and mechanically embedded in a groove 0.25 ± 0.05 mm (0.010 ± 0.002 in) deep, spirally cut into the outside surface of a tube.
2. *Integral.* An aluminum outer tube from which fins have been formed by extrusion, mechanically bonded to an inner tube or liner.
3. *Overlapped footed.* L-shaped aluminum fin wrapped under tension over the outside surface of a tube, with the tube fully covered by the overlapped feet under and between the fins.
4. *Footed.* L-shaped aluminum fin wrapped under tension over the outside surface of a tube with the tube fully covered by the feet between the fins.
5. *Bonded.* Tubes on which fins are bonded to the outside surface by hot-dip galvanizing, brazing, or welding.

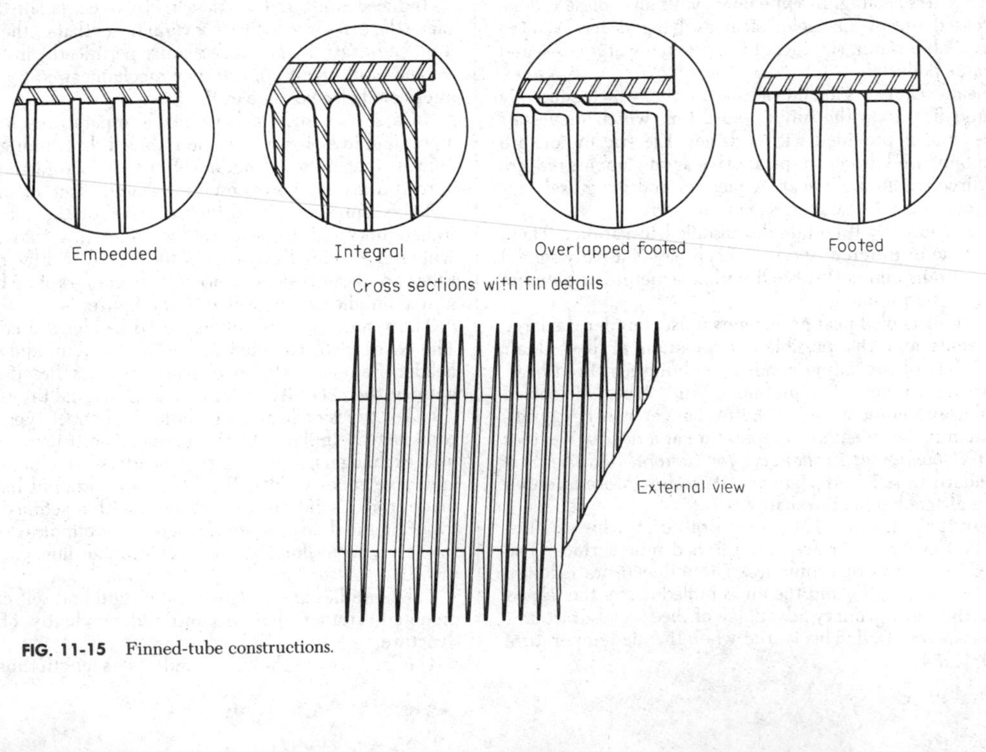

FIG. 11-15 Finned-tube constructions.

Typical metal design temperatures for these finned-tube constructions are 399°C (750°F) embedded, 288°C (550°F) integral, 232°C (450°F) overlapped footed, and 177°C (350°F) footed.

Tube ends are left bare to permit insertion of the tubes into appropriate holes in the headers or tube sheets. Tube ends are usually roller-expanded into these tube holes.

Fans Axial-flow fans are large-volume, low-pressure devices. Fan diameters are selected to give velocity pressures of approximately 2.5 mm (0.1 in) of water. Total fan efficiency (fan, driver, and transmission device) is about 75 percent, and fan drives usually have a minimum of 95 percent mechanical efficiency.

Usually fans are provided with four or six blades. Larger fans may have more blades. Fan diameter is generally slightly less than the width of the bay.

At the fan-tip speeds required for economical performance, a large amount of noise is produced. The predominant source of noise is vortex shedding at the trailing edge of the fan blade. Noise control of air-cooled exchangers is required by the Occupational Safety and Health Act (OSHA). API Standard 661 (*Air-Cooled Heat Exchangers for General Refinery Services*, 2d ed., January 1978) has the purchaser specifying sound-pressure-level (SPL) values per fan at a location designated by the purchaser and also specifying sound-power-level (PWL) values per fan. These are designated at the following octave-band-center frequencies: 63, 125, 250, 1000, 2000, 4000, 8000, and also the dBa value (the dBa is a weighted single-value sound-pressure level).

Reducing the fan-tip speed results in a straight-line reduction in air flow while the noise level decreases. The API Standard limits fan-tip speed to 61 m/s (12,000 ft/min) for typical constructions. Fan-design changes which reduce noise include increasing the number of fan blades, increasing the width of the fan blades, and reducing the clearance between fan tip and fan ring.

Both the quantity of air and the developed static pressure of fans in air-cooled heat exchangers are lower than indicated by fan manufacturers' test data, which are applicable to testing-facility tolerances and not to heat-exchanger constructions.

The axial-flow fan is inherently a device for moving a consistent volume of air when blade setting and speed of rotation are constant. Variation in the amount of air flow can be obtained by adjusting the blade angle of the fan and the speed of rotation. The blade angle can be either (1) permanently fixed, (2) hand-adjustable, or (3) automatically adjusted. Air delivery and power are a direct function of blade pitch angle.

Fan mounting should provide a minimum of one-half to three-fourths diameter between fan and ground on a forced-draft heat exchanger and one-half diameter between tubes and fan on an induced-draft cooler.

Fan blades can be made of aluminum, molded plastic, laminated plastic, carbon steel, stainless steel, and Monel.

Fan Drivers Electric motors or steam turbines are most commonly used. These connect with gears or V belts. (Gas engines connected through gears and hydraulic motors either direct-connected or connected through gears are in use. Fans may be driven by a prime mover such as a compressor with a V-belt takeoff from the flywheel to a jack shaft and then through a gear or V belt to the fan. Direct motor drive is generally limited to small-diameter fans.

V-belt drive assemblies are generally used with fans 3 m (10 ft) and less in diameter and motors of 22.4 kW (30 hp) and less.

Right-angle gear drive is preferred for fans over 3 m (10 ft) in diameter, for electric motors over 22.4 kW (30 hp), and with steam-turbine drives.

Fan Ring and Plenum Chambers The air must be distributed from the circular fan to the rectangular face of the tube bundle. The air velocity at the fan is between 3.8 and 10.2 m/s (750 and 2000 ft/min). The plenum-chamber depth (fron fan to tube bundle) is dependent upon the fan dispersion angle (Fig. 11-16), which should have a maximum value of 45°.

The fan ring is made to commercial tolerances for the relatively large diameter fan. These tolerances are greater than those upon closely machined fan rings used for small-diameter laboratory-performance testing. Fan performance is directly affected by this

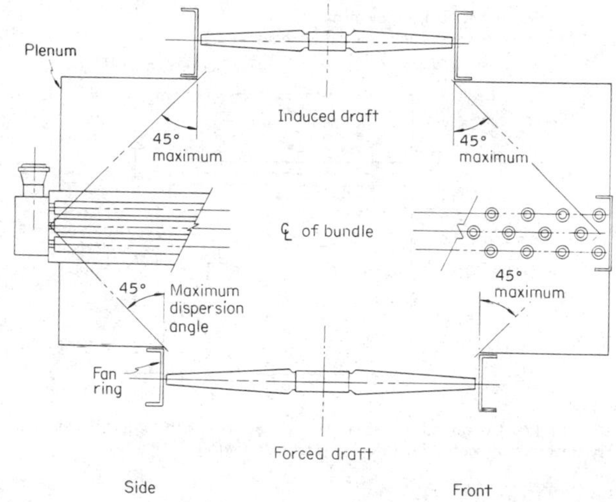

FIG. 11-16 Fan dispersion angle. *(API Standard 661.)*

increased clearance between the blade tip and the ring, and adequate provision in design must be made for the reduction in air flow. API Standard 661 requires that fan-tip clearance be a maximum of 0.5 percent of the fan diameter for diameters between 1.9 and 3.8 m (6.25 and 12.5 ft). Maximum clearance is 9.5 mm (⅜ in) for smaller fans and 19 mm (¾ in) for larger fans.

The depth of the fan ring is critical. Worsham (ASME Pap. 59-PET-27, Petroleum Mechanical Engineering Conference, Houston, 1959) reports an increase in flow varying from 5 to 15 percent with the same power consumption when the depth of a fan ring was doubled. The percentage increase was proportional to the volume of air and static pressure against which the fan was operating.

When making a selection, the stall-out condition, which develops when the fan cannot produce any more air regardless of power input, should be considered.

Air-Flow Control Process operating requirements and weather conditions are considered in determining the method of controlling air flow. The most common methods include simple on-off control, on-off step control (in the case of multiple-driver units), two-speed-motor control, variable-speed drivers, controllable fan pitch, manually or automatically adjustable louvers, and air recirculation.

Winterization is the provision of design features, procedures, or systems for air-cooled heat exchangers to avoid process-fluid operating problems resulting from low-temperature inlet air. These include fluid freezing, pour point, wax formation, hydrate formation, laminar flow, and condensation at the dew point (which may initiate corrosion). Freezing points for some commonly encountered fluids in refinery service include: benzene, 5.6°C (42°F); *p*-xylene 15.5°C (55.9°F); cyclohexane, 6.6°C (43.8°F); phenol, 40.9°C (105.6°F); monoethanolamine, 10.3°C (50.5°F); and diethanolamine, 25.1°C (77.2°F). Water solutions of these organic compounds are likely to freeze in air-cooled exchangers during winter service. Paraffinic and olefinic gases (C_1 through C_4) saturated with water vapor form hydrates when cooled. These hydrates are solid crystals which can collect and plug exchanger tubes.

Air-flow control in some services can prevent these problems. Cocurrent flow of air and process fluid during winter may be adequate to prevent problems. (Normal design has countercurrent flow of air and process fluid.) In some services when the hottest process fluid is in the bottom tubes, which are exposed to the lowest-temperature air, winterization problems may be eliminated.

Following are references which deal with problems in low-temperature environments: Brown and Benkley, "Heat Exchangers in Cold Service—A Contractor's View," *Chem. Eng. Prog.*, **70**, 59–62 (July 1974); Franklin and Munn, "Problems with Heat Exchangers in Low Temperature Environments," *Chem. Eng. Prog.*, **70**, 63–67

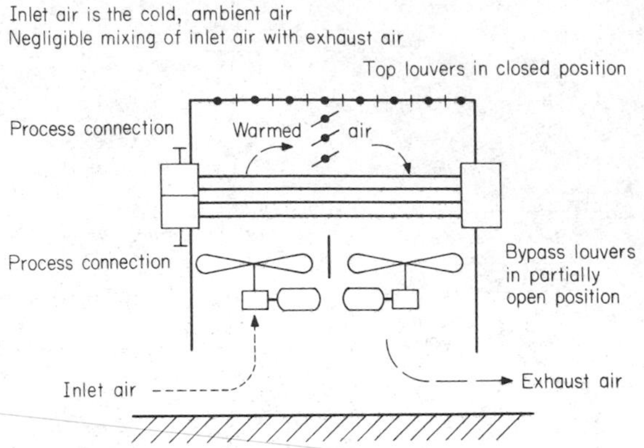

FIG. 11-17 Contained internal recirculation (with internal louvers). [Hydrocarbon Process., *59, 148–149 (October 1980).*]

(July 1974); Newell, "Air-Cooled Heat Exchangers in Low Temperature Environments: A Critique," *Chem. Eng. Prog.*, **70**, 86–91 (October 1974); Rubin, "Winterizing Air Cooled Heat Exchangers," *Hydrocarbon Process.*, **59**, 147–149 (October 1980); Shipes, "Air-Cooled Heat Exchangers in Cold Climates," *Chem. Eng. Prog.*, **70**, 53–58 (July 1974).

Air Recirculation Recirculation of air which has been heated as it crosses the tube bundle provides the best means of preventing operating problems due to low-temperature inlet air. Internal recirculation is the movement of air within a bay so that the heated air which has crossed the bundle is directed by a fan with reverse flow across another part of the bundle. Wind skirts and louvers are generally provided to minimize the entry of low-temperature air from the surroundings. Contained internal recirculation uses louvers within the bay to control the flow of warm air in the bay as illustrated in Fig. 11-17. Note that low-temperature inlet air has access to the tube bundle.

External recirculation is the movement of the heated air within the bay to an external duct, where this air mixes with inlet air, and the mixture serves as the cooling fluid within the bay. Inlet air does not have direct access to the tube bundle; an adequate mixing chamber is essential. Recirculation over the end of the exchanger is illustrated in Fig. 11-18. Over-the-side recirculation also is used. External recirculation systems maintain the desired low temperature of the air crossing the tube bundle.

Trim Coolers Conventional air-cooled heat exchangers can cool the process fluid to within 8.3°C (15°F) of the design dry-bulb temperature. When a lower process outlet temperature is required, a trim cooler is installed in series with the air-cooled heat exchanger. The water-cooled trim cooler can be designed for a 5.6 to 11.1°C (10 to 20°F) approach to the wet-bulb temperature (which in the United States is about 8.3°C (15°F) less than the dry-bulb temperature). In arid areas the difference between dry- and wet-bulb temperatures is much greater.

Humidification Chambers The air-cooled heat exchanger is provided with humidification chambers in which the air is cooled to a close approach to the wet-bulb temperature before entering the finned-tube bundle of the heat exchanger.

Evaporative Cooling The process fluid can be cooled by using evaporative cooling with the sink temperature approaching the wet-bulb temperature.

Steam Condensers Air-cooled steam condensers have been fabricated with a single tube-side pass and several rows of tubes. The bottom row has a higher temperature difference than the top row, since the air has been heated as it crosses the rows of tubes. The bottom row condenses all the entering steam before the steam has traversed the length of the tube. The top row, with a lower temperature driving force, does not condense all the entering steam. At the exit header, uncondensed steam flows from the top row into the bottom row. Since noncondensable gases are always present in steam, these accumulate within the bottom row because steam is entering from both ends of the tube. Performance suffers.

Various solutions have been used. These include orifices to regulate the flow into each tube, a "blow-through steam" technique with a vent condenser, complete separation of each row of tubes, and inclined tubes.

Air-Cooled Overhead Condensers Air-cooled overhead condensers (AOC) have been designed and installed above distillation columns as integral parts of distillation systems. The condensers generally have inclined tubes, with air flow over the finned surfaces induced by a fan. Prevailing wind affects both structural design and performance.

AOC provide the additional advantages of reducing ground-space requirements and piping and pumping requirements and of providing smoother column operation.

The **downflow condenser** is used mainly for nonisothermal condensation. Vapors enter through a header at the top and flow downward. The **reflux condenser** is used for isothermal and small-temperature-change conditions. Vapors enter at the bottom of the tubes.

AOC usage first developed in Europe but became more prevalent in the United States during the 1960s. A state-of-the-art article was published by Dehne [*Chem. Eng. Prog.*, **64**, 51 (July 1969)].

Air-Cooled Heat-Exchanger Costs The cost data that appear in Table 11-8 are unchanged from those published in the 1963 edition of this *Handbook*. In 1969 Guthrie [*Chem. Eng.*, **75**, 114 (Mar. 24, 1969)] presented cost data for field-erected air-cooled exchangers. These costs are only 25 percent greater than those of Table 11-8 and include the costs of steel stairways, indirect subcontractor charges, and field-erection charges. Since minimal field costs would be this high (i.e., 25 percent of purchase price), the basic costs appear to be unchanged. (Guthrie indicated a cost band of plus or minus 25 percent.) Preliminary design and cost estimating of air-cooled heat exchangers have been discussed by J. E. Lerner ["Simplified Air Cooler Estimating," *Hydrocarbon Process.*, **52**, 93–100 (February 1972)].

Design Considerations

1. *Design dry-bulb temperature.* The typically selected value is the temperature which is equaled or exceeded 2½ percent of the time during the warmest consecutive 4 months. Since air tempera-

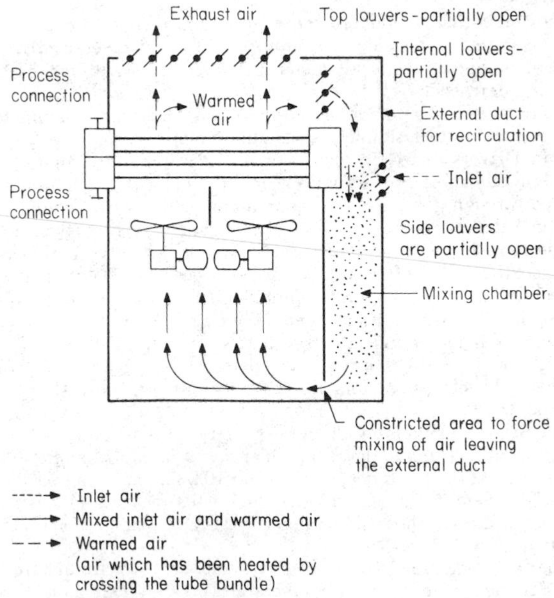

FIG. 11-18 External recirculation with adequate mixing chamber. [Hydrocarbon Process., *59, 148–149 (October 1980).*]

TABLE 11-8 Air-Cooled Heat-Exchanger Costs (1970)

Surface (bare tube), sq. ft.	500	1000	2000	3000	5000
Cost for 12-row-deep bundle, dollars/square foot	9.0	7.6	6.8	5.7	5.3
Factor for bundle depth:					
6 rows	1.07	1.07	1.07	1.12	1.12
4 rows	1.2	1.2	1.2	1.3	1.3
3 rows	1.25	1.25	1.25	1.5	1.5

Base: Bare-tube external surface 1 in. o.d. by 12 B.W.G. by 24 ft. 0 in. steel tube with 8 aluminum fins per inch ⅝-in. high. Steel headers. 150 lb./sq. in. design pressure. V-belt drive and explosion-proof motor. Bare-tube surface 0.262 sq. ft./ft. Fin-tube surface/bare-tube surface ratio is 16.9.

Factors: 20 ft. tube length 1.05
30 ft. tube length 0.95
18 B.W.G. admiralty tube 1.04
16 B.W.G. admiralty tube 1.12

NOTE: To convert feet to meters, multiply by 0.3048; to convert square feet to square meters, multiply by 0.0929; and to convert inches to millimeters, multiply by 25.4.

tures at industrial sites are frequently higher than those used for these weather-data reports, it is good practice to add 1 to 3°C (2 to 6°F) to the tabulated value.

2. *Air recirculation.* Prevailing winds and the locations and elevations of buildings, equipment, fired heaters, etc., require consideration. All air-cooled heat exchangers in a bank are of one type, i.e., all forced-draft or all induced-draft. Banks of air-cooled exchangers must be placed far enough apart to minimize air recirculation.

3. *Wintertime operations.* In addition to the previously discussed problems of winterization, provision must be made for heavy rain, strong winds, freezing of moisture upon the fins, etc.

4. *Noise.* Two identical fans have a noise level 3 dBa higher than one fan, while eight identical fans have a noise level 9 dBa higher than a single fan. Noise level at the plant site is affected by the exchanger position, the reflective surfaces near the fan, the hardness of these surfaces, and noise from adjacent equipment. The extensive use of air-cooled heat exchangers contributes significantly to plant noise level.

5. *Ground area and space requirements.* Comparisons of the overall space requirements for plants using air cooling versus water cooling are not consistent. Some air-cooled units are installed above other equipment—pipe racks, shell-and-tube exchangers, etc. Some plants avoid such installations because of safety considerations, as discussed later.

6. *Safety.* Leaks in air-cooled units are directly to the atmosphere and can cause fire hazards or toxic-fume hazards. However, the large air flow through an air-cooled exchanger greatly reduces any concentration of toxic fluids. Segal [*Pet. Refiner*, 38, 106 (April 1959)] reports that air-fin coolers "are not located over pumps, compressors, electrical switchgear, control houses and, in general, the amount of equipment such as drums and shell-and-tube exchangers located beneath them are minimized."

Pipe-rack-mounted air-cooled heat exchangers with flammable fluids generally have concrete fire decks which isolate the exchangers from the piping.

7. *Atmospheric corrosion.* Air-cooled heat exchangers should not be located where corrosive vapors and fumes from vent stacks will pass through them.

8. *Air-side fouling.* Air-side fouling is generally negligible.

9. *Process-side cleaning.* Either chemical or mechanical cleaning on the inside of the tubes can readily be accomplished.

10. *Process-side design pressure.* The high-pressure process fluid is always in the tubes. Tube-side headers are relatively small as compared with water-cooled units when the high pressure is generally on the shell side. High-pressure design of rectangular headers is complicated. The plug-type header is normally used for design gauge pressures to 13,790 kPa (2000 lbf/in^2) and has been used to 62,000 kPa (9000 lbf/in^2). The use of threaded plugs at these pressures creates problems. Removable cover plate headers are generally limited to gauge pressures of 2068 kPa (300 lbf/in^2). The expensive billet-type header is used for high-pressure service.

11. *Bond resistance.* Vibration and thermal cycling affect the bond resistance of the various types of tubes in different manners and thus affect the amount of heat transfer through the fin tube.

12. *Approach temperature.* The approach temperature, which is the difference between the process-fluid outlet temperature and the design dry-bulb air temperature, has a practical minimum of 8 to 14°C (15 to 25°F). When a lower process-fluid outlet temperature is required, an air-humidification chamber can be provided to reduce the inlet air temperature toward the wet-bulb temperature. A 5.6°C (10°F) approach is feasible. Since typical summer wet-bulb design temperatures in the United States are 8.3°C (15°F) lower than dry-bulb temperatures, the outlet process-fluid temperature can be 3°C (5°F) below the dry-bulb temperature.

13. *Mean-temperature-difference (MTD) correction factor.* When the outlet temperatures of both fluids are identical, the MTD correction factor for a 1:2 shell-and-tube exchanger (one pass shell side, two or more passes tube side) is approximately 0.8. For a single-pass air-cooled heat exchanger the factor is 0.91. A two-pass exchanger has a factor of 0.96, while a three-pass exchanger has a factor of 0.99 when passes are arranged for counterflow.

14. *Maintenance cost.* Maintenance for air-cooled equipment as compared with shell-and-tube coolers (complete with cooling-tower costs) indicates that air-cooling maintenance costs are approximately 0.3 to 0.5 those for water-cooled equipment.

15. *Operating costs.* Power requirements for air-cooled heat exchangers can be lower than at the summer design condition provided that an adequate means of air-flow control is used. The annual power requirement for an exchanger is a function of the means of air-flow control, the exchanger service, the air-temperature rise, and the approach temperature.

When the mean annual temperature is 16.7°C (30°F) lower than the design dry-bulb temperature and when both fans in a bay have automatically controllable pitch of fan blades, annual power required has been found to be 22, 36, and 54 percent respectively of that needed at the design condition for three process services [Frank L. Rubin, "Power Requirements Are Lower for Air-Cooled Heat Exchangers with AV Fans," *Oil Gas J.*, 165–167 (Oct. 11, 1982)]. Alternatively, when fans have two-speed motors, these deliver one-half of the design flow of air at half speed and use only one-eighth of the power of the full-speed condition.

HEATING AND COOLING OF TANKS

Tank Coils **Pipe tank coils** are made in a wide variety of configurations, depending upon the application and shape of the vessel. **Helical** and **spiral** coils are most commonly shop-fabricated, while the **hairpin** pattern is generally field-fabricated. The helical coils are used principally in process tanks and pressure vessels when large areas for rapid heating or cooling are required. In general, heating coils are placed low in the tank, and cooling coils are placed high or distributed uniformly through the vertical height.

Stocks which tend to solidify on cooling require uniform coverage of the bottom or agitation. A maximum spacing of 0.6 m (2 ft) between turns of 50.8-mm (2-in) and larger pipe and a close approach to the tank wall are recommended. For smaller pipe or for low-temperature heating media, closer spacing should be used. In the case of the common hairpin coils in vertical cylindrical tanks, this means adding an encircling ring within 152 mm (6 in) of the tank wall (see Fig. 11-19 for this and other typical coil layouts). The coils should be set directly on the bottom or raised not more than 50.8 to 152 mm (2 to 6 in), depending upon the difficulty of remelting the solids, in order to permit free movement of product within the vessel. The coil inlet should be above the liquid level (or an internal melt-out riser installed) to provide a molten path for liquid expansion or venting of vapors.

Coils may be sloped to facilitate drainage. When it is impossible to do so and remain close enough to the bottom to get proper remelting, the coils should be blown out after usage in cold weather to avoid damage by freezing.

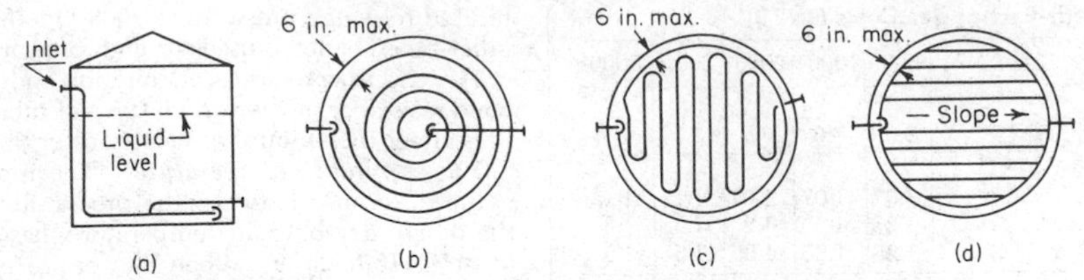

FIG. 11-19 Typical coil designs for good bottom coverage. (*a*) Elevated inlet on spiral coil. (*b*) Spiral with recircling ring. (*c*) Hairpin with encircling ring. (*d*) Ring header type.

Most coils are firmly clamped (but not welded) to supports. **Supports** should allow expansion but be rigid enough to prevent uncontrolled motion (see Fig. 11-20). Nuts and bolts should be securely fastened. Reinforcement of the inlet and outlet connections through the tank wall is recommended, since bending stresses due to thermal expansion are usually high at such points.

In general, 50.8- and 63.4-mm (2- and 2½-in) coils are the most economical for shop fabrication and 38.1- and 50.8-mm (1½- and 2-in) for field fabrication. The tube-side heat-transfer coefficient, high-pressure, or layout problems may lead to the use of smaller-size pipe.

The wall thickness selected varies with the service and material. Carbon steel coils are often made from schedule 80 or heavier pipe to allow for corrosion. When stainless-steel or other high-alloy coils are not subject to corrosion or excessive pressure, they may be of schedule 5 or 10 pipe to keep costs at a minimum, although high-quality welding is required for these thin walls to assure trouble-free service.

Methods for calculating heat loss from tanks and the sizing of tank coils have been published by Stuhlbarg [*Pet. Refiner*, **38**, 143 (April 1959)].

Fin-tube coils are used for fluids which have poor heat-transfer characteristics to provide more surface for the same configuration at reduced cost. Fin tubing is not generally used when bottom coverage is important. **Fin-tube tank heaters** are compact prefabricated bundles which can be brought into tanks through manholes. These are normally installed vertically with longitudinal fins to produce good convection currents. To keep the heaters low in the tank, they can be installed horizontally with helical fins or with perforated longitudinal fins to prevent entrapment. Fin tubing is often used for heat-sensitive material because of the lower surface temperature for the same heating medium.

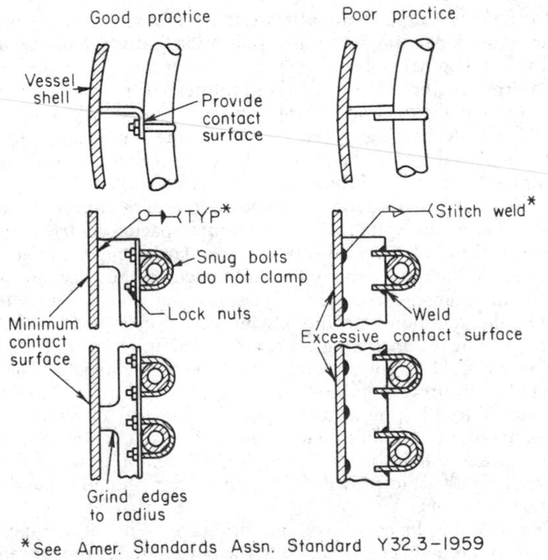

FIG. 11-20 Right and wrong ways to support coils. [Chem. Eng., *172 (May 16, 1960).*]

Plate or panel coils made from two metal sheets with one or both embossed to form passages for a heating or cooling medium can be used in lieu of pipe coils. Panel coils are relatively light in weight, easy to install, and easily removed for cleaning. They are available in a range of standard sizes and in both flat and curved patterns. Process tanks have been built by using panel coils for the sides or bottom. A serpentine construction is generally utilized when liquid flows through the unit. Header-type construction is used with steam or other condensing media.

Standard **glass coils** with 0.18 to 11.1 m^2 (2 to 120 ft^2) of heat-transfer surface are available. Also available are plate-type units made of **impervious graphite.**

Teflon Immersion Coils Immersion coils made of Teflon fluorocarbon resin are available with 2.5-mm (0.10-in) ID tubes to increase overall heat-transfer efficiency. The flexible bundles are available with 100, 160, 280, 500, and 650 tubes with standard lengths varying in 0.6-m (2-ft) increments between 1.2 and 4.8 m (4 and 16 ft). These coils are most commonly used in metal-finishing baths and are adaptable to service in reaction vessels, crystallizers, and tanks where corrosive fluids are used.

Bayonet Heaters A bayonet-tube element consists of an outer and an inner tube. These elements are inserted into tanks and process vessels for heating and cooling purposes. Often the outer tube is of expensive alloy or nonmetallic (e.g., glass, impervious graphite), while the inner tube is of carbon steel. In glass construction, elements with 50.8- or 76.2-mm (2- or 3-in) glass pipe [with lengths to 2.7 m (9 ft)] are in contact with the external fluid, with an inner tube of metal.

External Coils and Tracers Tanks, vessels, and pipe lines can be equipped for heating or cooling purposes with external coils. These are generally 9.8 to 19 mm (⅜ to ¾ in) so as to provide good distribution over the surface and are often of soft copper or aluminum, which can be bent by hand to the contour of the tank or line. When necessary to avoid "hot spots," the tracer is so mounted that it does not touch the tank.

External coils spaced away from the tank wall exhibit a coefficient of around 5.7 W/($m^2 \cdot °C$) [1 Btu/($h \cdot ft^2$ of coil surface $\cdot °F$)]. Direct contact with the tank wall produces higher coefficients, but these are difficult to predict since they are strongly dependent upon the degree of contact. The use of **heat-transfer cements** does improve performance. These puttylike materials of high thermal conductivity are troweled or caulked into the space between the coil and the tank or pipe surface.

Costs of the cements (in 1960) varied from 37 to 63 cents per pound, with requirements running from about 0.27 lb/ft of ⅜-in-outside-diameter tubing to 1.48 lb/ft of 1-in pipe. Panel coils require ½ to 1 lb/ft^2. A rule of thumb for preliminary estimating is that the per foot installed cost of tracer with cement is about double that of the tracer alone.

Jacketed Vessels Jacketing is often used for vessels needing frequent cleaning and for glass-lined vessels which are difficult to equip with internal coils. The jacket eliminates the need for the coil yet gives a better overall coefficient than external coils. However, only a limited heat-transfer area is available. The conventional jacket is of simple construction and is frequently used. It is most effective with a condensing vapor. A liquid heat-transfer fluid does not maintain uniform flow characteristics in such a jacket. Nozzles, which set up a swirling motion in the jacket, are effective in improving heat transfer. Wall thicknesses are often high unless reinforcement rings are installed.

Spiral baffles, which are sometimes installed for liquid services to improve heat transfer and prevent channeling, can be designed to serve as reinforcements. A spiral-wound channel welded to the vessel wall is an alternative to the spiral baffle which is more predictable in performance, since cross-baffle leakage is eliminated, and is reportedly lower in cost [Feichtinger, *Chem. Eng.*, **67**, 197 (Sept. 5, 1960)].

The half-pipe jacket is used when high jacket pressures are required. The flow pattern of a liquid heat-transfer fluid can be controlled and designed for effective heat transfer. The dimple jacket offers structural advantages and is the most economical for high jacket pressures. The low volumetric capacity produces a fast response to temperature changes.

EVAPORATORS

GENERAL REFERENCES: Badger and Banchero, *Introduction to Chemical Engineering*, McGraw-Hill, New York, 1955. Standiford, *Chem. Eng.*, **70**, 158–176 (Dec. 9, 1963). *Testing Procedure for Evaporators*, American Institute of Chemical Engineers, 1979. *Upgrading Evaporators to Reduce Energy Consumption*, ERDA Technical Information Center, Oak Ridge, Tenn., 1977.

PRIMARY DESIGN PROBLEMS

Heat transfer is the most important single factor in evaporator design, since the heating surface represents the largest part of evaporator cost. Other things being equal, the type of evaporator selected is the one having the highest heat-transfer cost coefficient under desired operating conditions in terms of kilowatts per degree kelvin (British thermal units per hour per degree Fahrenheit) per dollar of installed cost. When power is required to induce circulation past the heating surface, the coefficient must be even higher to offset the cost of power for circulation.

Vapor-liquid separation may be important for a number of reasons. The most important is usually prevention of entrainment because of value of product lost, pollution, contamination of the condensed vapor, or fouling or corrosion of the surfaces on which the vapor is condensed. Vapor-liquid separation in the vapor head may also be important when spray forms deposits on the walls, when vortices increase head requirements of circulating pumps, and when short circuiting allows vapor or unflashed liquid to be carried back to the circulating pump and heating element.

Evaporator performance is rated on the basis of **steam economy**—kilograms of solvent evaporated per kilogram of steam used. Heat is required (1) to raise the feed from its initial temperature to the boiling temperature, (2) to provide the minimum thermodynamic energy to separate liquid solvent from the feed, and (3) to vaporize the solvent. The first of these can be changed appreciably by reducing the boiling temperature or by heat interchange between the feed and the residual product and/or condensate. The greatest increase in steam economy is achieved by reusing the vaporized solvent. This is done in a **multiple-effect evaporator** by using the vapor from one effect as the heating medium for another effect in which boiling takes place at a lower temperature and pressure. Another method of increasing the utilization of energy is to employ a **thermocompression** evaporator, in which the vapor is compressed so that it will condense at a temperature high enough to permit its use as the heating medium in the same evaporator.

Selection Problems Aside from heat-transfer considerations, the selection of type of evaporator best suited for a particular service is governed by the characteristics of the feed and product. Points that must be considered are crystallization, salting and scaling, product quality, corrosion, and foaming. In the case of a **crystallizing evaporator,** the desirability of producing crystals of a definite uniform size usually limits the choice to evaporators having a positive means of circulation. **Salting,** which is the growth on body and heating-surface walls of a material having a solubility that increases with increase in temperature, is frequently encountered in crystallizing evaporators. It can be reduced or eliminated by keeping the evaporating liquid in close or frequent contact with a large surface area of crystallized solid. **Scaling** is the deposition and growth on body walls, and especially on heating surfaces, of a material undergoing an irreversible chemical reaction in the evaporator or having a solubility that decreases with an increase in temperature. Scaling can be reduced or eliminated in the same general manner as salting. Both salting and scaling liquids are usually best handled in evaporators that do not depend on boiling to induce circulation. **Fouling** is the formation of deposits other than salt or scale and may be due to corrosion, solid matter entering with the feed, or deposits formed by the condensing vapor.

Product-quality considerations may require low holdup time and low-temperature operation to avoid thermal degradation. The low holdup time eliminates some types of evaporators, and some types are also eliminated because of poor heat-transfer characteristics at low temperature. Product quality may also dictate special materials of construction to avoid metallic contamination or a catalytic effect on decomposition of the product. **Corrosion** may also influence evaporator selection, since the advantages of evaporators having high heat-transfer coefficients are more apparent when expensive materials of construction are indicated. Corrosion and erosion are frequently more severe in evaporators than in other types of equipment because of the high liquid and vapor velocities used, the frequent presence of solids in suspension, and the necessary concentration differences.

EVAPORATOR TYPES AND APPLICATIONS

Evaporators may be classified as follows:

1. Heating medium separated from evaporating liquid by tubular heating surfaces
2. Heating medium confined by coils, jackets, double walls, flat plates, etc.
3. Heating medium brought into direct contact with evaporating liquid
4. Heating by solar radiation

By far the largest number of industrial evaporators employ tubular heating surfaces. Circulation of liquid past the heating surface may be induced by boiling or by mechanical means. In the latter case, boiling may or may not occur at the heating surface.

Forced-Circulation Evaporators (Fig. 11-21*a*, *b*, *c*) Although it may not be the most economical for many uses, the forced-circulation (FC) evaporator is suitable for the widest variety of evaporator applications. The use of a pump to ensure circulation past the heating surface makes possible separating the functions of heat transfer, vapor-liquid separation, and crystallization. The pump withdraws liquor from the flash chamber and forces it through the heating element back to the flash chamber. Circulation is maintained regardless of the evaporation rate; so this type of evaporator is well suited to **crystallizing operation,** in which solids must be maintained in suspension at all times. The liquid velocity past the heating surface is limited only by the pumping power needed or available and by accelerated corrosion and erosion at the higher velocities. **Tube velocities** normally range from a minimum of about 1.2 m/s (4 ft/s) in salt evaporators with copper or brass tubes and liquid containing 5 percent or more solids up to about 3 m/s (10 ft/s) in caustic evaporators having nickel tubes and liquid containing only a small amount of solids. Even higher velocities can be used when corrosion is not accelerated by erosion.

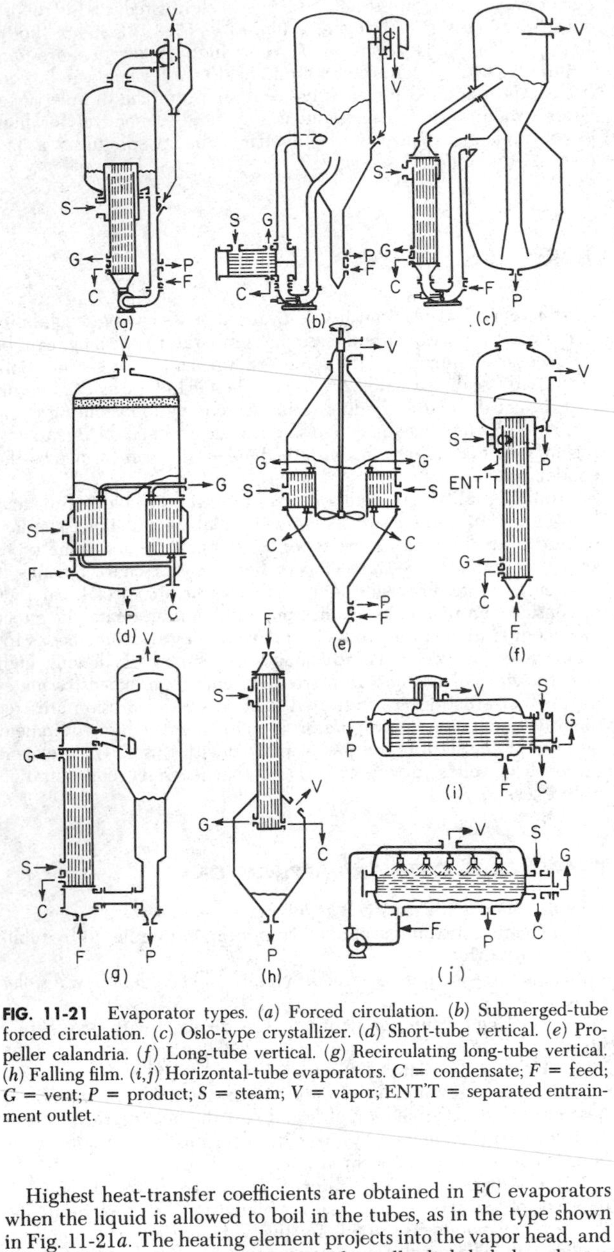

FIG. 11-21 Evaporator types. (*a*) Forced circulation. (*b*) Submerged-tube forced circulation. (*c*) Oslo-type crystallizer. (*d*) Short-tube vertical. (*e*) Propeller calandria. (*f*) Long-tube vertical. (*g*) Recirculating long-tube vertical. (*h*) Falling film. (*i,j*) Horizontal-tube evaporators. C = condensate; F = feed; G = vent; P = product; S = steam; V = vapor; ENT'T = separated entrainment outlet.

Highest heat-transfer coefficients are obtained in FC evaporators when the liquid is allowed to boil in the tubes, as in the type shown in Fig. 11-21*a*. The heating element projects into the vapor head, and the liquid level is maintained near and usually slightly below the top tube sheet. This type of FC evaporator is not well suited to salting solutions because boiling in the tubes increases the chances of salt deposit on the walls and the sudden flashing at the tube exits promotes excessive nucleation and production of fine crystals. Consequently, this type of evaporator is seldom used except when there are headroom limitations or when the liquid forms neither salt nor scale.

By far the largest number of forced-circulation evaporators are of the submerged-tube type, as shown in Fig. 11-21*b*. The heating element is placed far enough below the liquid level or return line to the flash chamber to prevent boiling in the tubes. Preferably, the hydrostatic head should be sufficient to prevent boiling even in a tube that is plugged (and hence at steam temperature), since this prevents salting of the entire tube. Evaporators of this type sometimes have horizontal heating elements (usually two-pass), but the vertical single-pass heating element is used whenever sufficient headroom is available. The vertical element usually has a lower friction loss and is easier to clean or retube than a horizontal heater. The submerged-tube forced-circulation evaporator is relatively immune to salting in the tubes, since no supersaturation is generated by evaporation in the tubes. The tendency toward scale formation is also reduced, since supersaturation in the heating element is generated only by a controlled amount of heating and not by both heating and evaporation.

The type of **vapor head** used with the FC evaporator is chosen to suit the product characteristics and may range from a simple centrifugal separator to the crystallizing chambers shown in Fig. 11-21*b* and *c*. Figure 11-21*b* shows a type frequently used for common salt. It is designed to circulate a slurry of crystals throughout the system. Figure 11-21*c* shows a submerged-tube FC evaporator in which heating, flashing, and crystallization are completely separated. The crystallizing solids are maintained as a fluidized bed in the chamber below the vapor head and little or no solids circulate through the heater and flash chamber. This type is well adapted to growing coarse crystals, but the crystals usually approach a spherical shape, and careful design is required to avoid production of fines in the flash chamber.

In a submerged-tube FC evaporator, all heat is imparted as sensible heat, resulting in a temperature rise of the circulating liquor that reduces the overall temperature difference available for heat transfer. Temperature rise, tube proportions, tube velocity, and head requirements on the circulating pump all influence the selection of circulation rate. Head requirements are frequently difficult to estimate since they consist not only of the usual friction, entrance and contraction, and elevation losses when the return to the flash chamber is above the liquid level but also of increased friction losses due to flashing in the return line and vortex losses in the flash chamber. Circulation is sometimes limited by vapor in the pump suction line. This may be drawn in as a result of inadequate vapor-liquid separation or may come from vortices near the pump suction connection to the body or may be formed in the line itself by short circuiting from heater outlet to pump inlet of liquor that has not flashed completely to equilibrium at the pressure in the vapor head.

Advantages of forced-circulation evaporators:
1. High heat-transfer coefficients
2. Positive circulation
3. Relative freedom from salting, scaling, and fouling

Disadvantages of forced-circulation evaporators:
1. High cost
2. Power required for circulating pump
3. Relatively high holdup or residence time

Best applications of forced-circulation evaporators:
1. Crystalline product
2. Corrosive solutions
3. Viscous solutions

Frequent difficulties with forced-circulation evaporators:
1. Plugging of tube inlets by salt deposits detached from walls of equipment
2. Poor circulation due to higher than expected head losses
3. Salting due to boiling in tubes
4. Corrosion-erosion

Short-Tube Vertical Evaporators (Fig. 11-21*d*) This is one of the earliest types still in widespread commercial use. Its principal use at present is in the evaporation of cane-sugar juice. Circulation past the heating surface is induced by boiling in the tubes, which are usually 50.8 to 76.2 mm (2 to 3 in) in diameter by 1.2 to 1.8 m (4 to 6 ft) long. The body is a vertical cylinder, usually of cast iron, and the tubes are expanded into horizontal tube sheets that span the body diameter. The circulation rate through the tubes is many times the feed rate; so there must be a return passage from above the top tube sheet to below the bottom tube sheet. Most commonly used is a central well or **downtake** as shown in Fig. 11-21*d*. So that friction losses through the downtake do not appreciably impede circulation up through the tubes, the area of the downtake should be of the same order of magnitude as the combined cross-sectional area of the tubes. This results in a downtake almost half of the diameter of the tube sheet.

Circulation and heat transfer in this type of evaporator are strongly affected by the liquid "level." Highest heat-transfer coefficients are achieved when the level, as indicated by an external gauge glass, is only about halfway up the tubes. Slight reductions in level below the optimum result in incomplete wetting of the tube walls with a consequent increased tendency to foul and a rapid reduction in capacity. When this type of evaporator is used with a liquid that can deposit salt or scale, it is customary to operate with the liquid level appreciably higher than the optimum and usually appreciably above the top tube sheet.

Circulation in the standard short-tube vertical evaporator is dependent entirely on boiling, and when boiling stops, any solids present settle out of suspension. Consequently, this type is seldom used as a crystallizing evaporator. By installing a propeller in the downtake, this objection can be overcome. Such an evaporator, usually called a **propeller calandria,** is illustrated in Fig. 11-21*e*. The propeller is usually placed as low as possible to reduce cavitation and is shrouded by an extension of the downtake well. The use of the propeller can sometimes double the capacity of a short-tube vertical evaporator. The evaporator shown in Fig. 11-21*e* includes an elutriation leg for salt manufacture similar to that used on the FC evaporator of Fig. 11-21*b*. The shape of the bottom will, of course, depend on the particular application and on whether the propeller is driven from above or below. To avoid salting when the evaporator is used for crystallizing solutions, the liquid level must be kept appreciably above the top tube sheet.

Advantages of short-tube vertical evaporators:

1. High heat-transfer coefficients at high temperature differences
2. Low headroom
3. Easy mechanical descaling
4. Relatively inexpensive

Disadvantages of short-tube vertical evaporators:

1. Poor heat transfer at low temperature differences and low temperature
2. High floor space and weight
3. Relatively high holdup
4. Poor heat transfer with viscous liquids

Best applications of short-tube vertical evaporators:

1. Clear liquids
2. Crystalline product if propeller is used
3. Relatively noncorrosive liquids, since body is large and expensive if built of materials other than mild steel or cast iron
4. Mild scaling solutions requiring mechanical cleaning, since tubes are short and large in diameter

---

Long-Tube Vertical Evaporators (Fig. 11-21*f*, *g*, *h*) More total evaporation is accomplished in this type than in all others combined because it is normally the **cheapest per unit of capacity.** The long-tube vertical (LTV) evaporator consists of a simple one-pass vertical shell-and-tube heat exchanger discharging into a relatively small vapor head. Normally, no liquid level is maintained in the vapor head, and the residence time of liquor is only a few seconds. The tubes are usually about 50.8 mm (2 in) in diameter but may be smaller than 25.4 mm (1 in). Tube length may vary from less than 6 to 10.7 m (20 to 35 ft). The evaporator is usually operated single-pass, concentrating from the feed to discharge density in just the time that it takes the liquid and evolved vapor to pass through a tube. An extreme case is the caustic high concentrator, producing a substantially anhydrous product at 370°C (700°F) from an inlet feed of 50 percent NaOH at 149°C (300°F) in one pass up 22-mm- (⅞-in-) outside-diameter nickel tubes 6 m (20 ft) long. The largest use of LTV evaporators is for concentrating black liquor in the pulp and paper industry. Because of the long tubes and relatively high heat-transfer coefficients, it is possible to achieve higher single-unit capacities in this type of evaporator than in any other.

The LTV evaporator shown in Fig. 11-21*f* is typical of those commonly used, especially for black liquor. Feed enters at the bottom of the tube and starts to boil partway up the tube, and the mixture of liquid and vapor leaving at the top at high velocity impinges against a deflector placed above the tube sheet. This deflector is effective both as a primary separator and as a foam breaker.

In many cases, as when the ratio of feed to evaporation or the ratio of feed to heating surface is low, it is desirable to provide for **recirculation of product** through the evaporator. This can be done in the type shown in Fig. 11-21*f* by adding a pipe connection between the product line and the feed line. Higher recirculation rates can be achieved in the type shown in Fig. 11-21*g*, which is used widely for condensed milk. By extending the enlarged portion of the vapor head still lower to provide storage space for liquor, this type can be used as a batch evaporator.

Liquid temperatures in the tubes of an LTV evaporator are far from uniform and are difficult to predict. At the lower end, the liquid is usually not boiling, and the liquor picks up heat as sensible heat. Since entering liquid velocities are usually very low, true heat-transfer coefficients are low in this nonboiling zone. At some point up the tube, the liquid starts to boil, and from that point on the liquid temperature decreases because of the reduction in static, friction, and acceleration heads until the vapor-liquid mixture reaches the top of the tubes at substantially vapor-head temperature. Thus the true temperature difference in the boiling zone is always less than the total temperature difference as measured from steam and vapor-head temperatures.

Although the true heat-transfer coefficients in the boiling zone are quite high, they are partially offset by the reduced temperature difference. The point in the tubes at which boiling starts and at which the maximum temperature is reached is sensitive to operating conditions, such as feed properties, feed temperature, feed rate, and heat flux. Figure 11-22 shows typical variations in liquid temperature in tubes of an LTV evaporator operating at a constant terminal temperature difference. Curve 1 shows the normal case in which the feed is not boiling at the tube inlet. Curve 2 gives an indication of the temperature difference lost when the feed enters at the boiling point. Curve 3 is for exactly the same conditions as curve 2 except that the feed contained 0.01 percent Teepol to reduce surface tension [Coulson and Mehta, *Trans. Inst. Chem. Eng.*, 31, 208 (1953)]. The surface-active agent yields a more intimate mixture of vapor and liquid, with the result that liquid is accelerated to a velocity more nearly approaching the vapor velocity, thereby increasing the pressure drop in the tube. Although the surface-active agent caused an increase of more than 100 percent in the true heat-transfer coefficient, this was more than offset by the reduced temperature difference so that the net result was a reduction in evaporator capacity. This sensitivity of the LTV evaporator to changes in operating conditions is less pronounced at high than at low temperature differences and temperature levels.

The **falling-film** version of the LTV evaporator (Fig. 11-21*h*) eliminates these problems of hydrostatic head. Liquid is fed to the tops of the tubes and flows down the walls as a film. Vapor-liquid separation usually takes place at the bottom, although some evaporators of this type are arranged for vapor to rise through the tube countercurrently to the liquid. The pressure drop through the tubes is usually very small, and the boiling-liquid temperature is substantially the same as the vapor-head temperature. The falling-film evaporator is widely used for concentrating **heat-sensitive materials,** such as fruit juices, because the holdup time is very small, the liquid

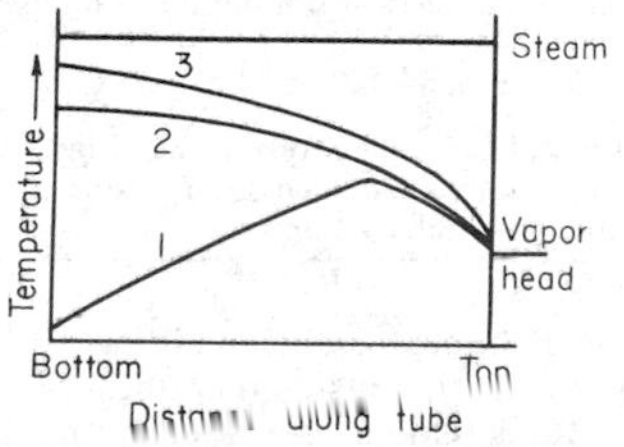

FIG. 11-22 Temperature variations in a long-tube vertical evaporator.

is not overheated during passage through the evaporator, and heat-transfer coefficients are high even at low boiling temperatures.

The principal problem with the falling-film LTV evaporator is that of **feed distribution** to the tubes. It is essential that all tube surfaces be wetted continually. This usually requires recirculation of the liquid unless the ratio of feed to evaporation is quite high. An alternative to the simple recirculation system of Fig. 11-21*h* is sometimes used when the feed undergoes an appreciable concentration change and the product is viscous. The feed chamber and vapor head are divided into a number of liquor compartments, and separate pumps are used to pass the liquor through the various banks of tubes in series, all in parallel as to steam and vapor pressures. The actual distribution of feed to the individual tubes of a falling-film evaporator may be accomplished by orifices at the inlet to each tube, by a number of perforated plates above the tube sheet, or by one or more spray nozzles.

Both rising-film and falling-film LTV evaporators are generally unsuited to salting or severely scaling liquids. However, the rising-film evaporator is widely used for black liquor, which presents a mild scaling problem, and also is used to carry solutions beyond saturation with respect to a crystallizing salt. In the latter case, deposits can usually be removed quickly by increasing the feed rate or reducing the steam rate in order to make the product unsaturated for a short time. The falling-film evaporator is not generally suited to liquids containing solids because of difficulty in plugging the feed distributors. However, it has been applied to the evaporation of saline waters saturated with $CaSO_4$ and containing 5 to 10 percent $CaSO_4$ seeds in suspension for scale prevention (Anderson, ASME Pap. 76-WA/Pwr-5, 1976).

Because of their simplicity of construction, compactness, and generally high heat-transfer coefficients, LTV evaporators are well suited to service with corrosive liquids. An example is the reconcentration of rayon spin-bath liquor, which is highly acid. These evaporators employ impervious graphite tubes, lead, rubber-covered or impervious graphite tube sheets, and rubber-lined vapor heads. Polished stainless-steel LTV evaporators are widely used for food products. The latter evaporators are usually similar to that shown in Fig. 11-21*g*, in which the heating element is at one side of the vapor head to permit easy access to the tubes for cleaning.

Advantages of long-tube vertical evaporators:
1. Low cost
2. Large heating surface in one body
3. Low holdup
4. Small floor space
5. Good heat-transfer coefficients at reasonable temperature differences (rising film)
6. Good heat-transfer coefficients at all temperature differences (falling film)

Disadvantages of long-tube vertical evaporators:
1. High headroom
2. Generally unsuitable for salting and severely scaling liquids
3. Poor heat-transfer coefficients of rising-film version at low temperature differences
4. Recirculation usually required for falling-film version

Best applications of long-tube vertical evaporators:
1. Clear liquids
2. Foaming liquids
3. Corrosive solutions
4. Large evaporation loads
5. High temperature differences—rising film, low temperature differences—falling film
6. Low-temperature operation—falling film

Frequent difficulties with long-tube vertical evaporators:
1. Sensitivity of rising-film units to changes in operating conditions
2. Poor feed distribution to falling-film units

Horizontal-Tube Evaporators (Fig. 11-21*i*) In these types the steam is inside and the liquor outside the tubes. The submerged-tube version of Fig. 11-21*i* is seldom used except for the preparation of boiler feedwater. Low entrainment loss is the primary aim: the horizontal cylindrical shell yields a large disengagement area per unit of vessel volume. Special versions use deformed tubes between restrained tube sheets that crack off much of a scale deposit when sprayed with cold water. By showering liquor over the tubes in the version of Fig. 11-21*j*, hydrostatic head losses are eliminated and heat-transfer performance is improved to that of the falling-film tubular type of Fig. 11-21*h*. Originally called the Lillie, this evaporator is now also called the spray-film or simply the horizontal-tube evaporator. Liquid distribution over the tubes is accomplished by sprays or perforated plates above the topmost tubes. Maintaining this distribution through the bundle to avoid overconcentrating the liquor is a problem unique to this type of evaporator. It is now used primarily for seawater evaporation.

Advantages of horizontal-tube evaporators:
1. Very low headroom
2. Large vapor-liquid disengaging area—submerged-tube type
3. Relatively low cost in small-capacity straight-tube type
4. Good heat-transfer coefficients
5. Easy semiautomatic descaling—bent-tube type

Disadvantages of horizontal-tube evaporators:
1. Unsuitable for salting liquids
2. Unsuitable for scaling liquids—straight-tube type
3. High cost—bent-tube type
4. Maintaining liquid distribution—film type

Best applications of horizontal-tube evaporators:
1. Limited headroom
2. Small capacity
3. Nonscaling nonsalting liquids—straight-tube type
4. Severely scaling liquids—bent-tube type

Miscellaneous Forms of Heating Surface Special evaporator designs are sometimes indicated when heat loads are small, special product characteristics are desired, or the product is especially difficult to handle. **Jacketed kettles,** frequently with agitators, are used when the product is very viscous, batches are small, intimate mixing is required, and/or ease of cleaning is an important factor. Evaporators with steam in **coiled tubes** may be used for small capacities with scaling liquids in designs that permit "cold shocking," or complete withdrawal of the coil from the shell for manual scale removal. Other designs for scaling liquids employ flat-plate heat exchangers, since in general a scale deposit can be removed more easily from a flat plate than from a curved surface. One such design, the **channel-switching evaporator,** alternates the duty of either side of the heating surface periodically from boiling liquid to condensing vapor so that scale formed when the surface is in contact with boiling liquid is dissolved when the surface is next in contact with condensing vapor.

Agitated thin-film evaporators employ a heating surface consisting of one large-diameter tube that may be either straight or tapered, horizontal or vertical. Liquid is spread on the tube wall by a rotating assembly of blades that either maintain a close clearance from the wall or actually ride on the film of liquid on the wall. The expensive construction limits application to the most difficult materials. High agitation [on the order of 12 m/s (40 ft/s) rotor-tip speed] and power intensities of 2 to 20 kW/m^2 (0.25 to 2.5 hp/ft^2) permit handling extremely viscous materials. Residence times of only a few seconds permit concentration of heat-sensitive materials at temperatures and temperature differences higher than in other types [Mutzenberg, Parker, and Fischer, *Chem. Eng.*, **72,** 175–190 (Sept. 13, 1965)]. High feed-to-product ratios can be handled without recirculation.

Economic and process considerations usually dictate that agitated thin-film evaporators be operated in single-effect mode. Very high temperature differences can then be used; many are heated with Dowtherm or other high-temperature media. This permits achieving reasonable capacities in spite of the relatively low heat-transfer coefficients and the small surface that can be provided in a single tube [to about 20 m^2 (200 ft^2)]. The structural need for wall thicknesses of 6 to 13 mm (¼ to ½ in) is a major reason for the relatively low heat-transfer coefficients when evaporating water-like materials.

Evaporators without Heating Surfaces The **submerged-combustion** evaporator makes use of combustion gases bubbling through the liquid as the means of heat transfer. It consists simply of a tank to hold the liquid, a burner and gas distributor that can be lowered into the liquid, and a combustion-control system. Since there are no heating surfaces on which scale can deposit, this evaporator is well suited to use with severely scaling liquids. The ease of constructing the tank and burner of special alloys or nonmetallic materials makes practical the handling of highly corrosive solutions. However, since the vapor is mixed with large quantities of noncondensable gases, it is impossible to reuse the heat in this vapor, and installations are usually limited to areas of low fuel cost. One difficulty frequently encountered in the use of submerged-combustion evaporators is a high entrainment loss. Also, these evaporators cannot be used when control of crystal size is important.

Disk or cascade evaporators are used in the pulp and paper industry to recover heat and entrained chemicals from boiler stack gases and to effect a final concentration of the black liquor before it is burned in the boiler. These evaporators consist of a horizontal shaft on which are mounted disks perpendicular to the shaft or bars parallel to the shaft. The assembly is partially immersed in the thick black liquor so that films of liquor are carried into the hot-gas stream as the shaft rotates.

Some forms of **flash evaporators** require no heating surface. An example is a recrystallizing process for separating salts having normal solubility curves from salts having inverse solubility curves, as in separating sodium chloride from calcium sulfate [Richards, *Chem. Eng.*, **59**(3), 140 (1952)]. A suspension of raw solid feed in a recirculating brine stream is heated by direct steam injection. The increased temperature and dilution by the steam dissolve the salt having the normal solubility curve. The other salt remains undissolved and is separated from the hot solution before it is flashed to a lower temperature. The cooling and loss of water on flashing cause recrystallization of the salt having the normal solubility curve, which is separated from the brine before the brine is mixed with more solid feed for recycling to the heater. This system can be operated as a multiple effect by flashing down to the lower temperature in stages and using flash vapor from all but the last stage to heat the recycle brine by direct injection. In this process no net evaporation occurs from the total system, and the process cannot be used to concentrate solutions unless heating surfaces are added.

HEAT TRANSFER IN EVAPORATORS

While the rate of heat transfer in evaporators is expressed most conveniently in terms of the usual equation, $q = UA\,\Delta T$, the *heat-transfer coefficient U* in most types of evaporators is a strong function of *temperature difference* ΔT. Unless otherwise specified, the *area A* used in reporting evaporator sizes or heat-transfer coefficients is the surface through which heat transfer takes place, measured on the liquid side of the surface.

The **temperature difference** used in computing heat transfer in evaporators is frequently an arbitrary figure, since it is quite difficult to determine the temperature of the liquid at all parts of the heating surface of most types of evaporators. The condensing temperature of steam, the most common heating medium, can usually be determined simply and accurately from a measurement of pressure in the steam side of the heating element, together with use of the *Steam Tables*. No allowances are made for superheat in the steam or subcooling of the condensate when calculating steam temperature. In a similar manner, a pressure measurement in the vapor space above the boiling liquid will give the saturated-vapor temperature, which, if a negligible boiling-point rise is assumed, would be substantially the same as the boiling-liquid temperature. Temperature differences calculated on the basis of this assumption are called **apparent temperature differences,** and heat-transfer coefficients are called **apparent coefficients.**

Boiling-point rise (BPR) is the difference between the boiling point of a solution and the boiling point of water at the same pressure. Data on BPR of a number of commonly encountered materials can be estimated from Fig. 11-23. When the boiling-point rise is deducted from the apparent temperature difference, the terms **temperature difference corrected for boiling-point rise** and **heat-transfer coefficient corrected for boiling-point rise** are used. This is the most common basis of reporting evaporator heat-transfer data and is the basis understood in the absence of any qualifying statement. It is also the best basis for comparing performances of various evaporator types—at the difference between the temperature at which heat is accepted from steam or a preceding effect and that at which vapor is discharged to the next effect or a condenser.

In most types of evaporators, the liquid at the heating surface is at a temperature higher than its saturation temperature at the pressure in the vapor space. In the submerged-tube, natural-circulation evaporator, **hydrostatic head** raises the temperature at the heating surface. In the forced-circulation evaporator, heat is absorbed as sensible heat, which results in a temperature rise through the heater and represents a loss in available temperature difference. In this type, liquid temperatures into and out of the heater can be measured and a true temperature difference determined. Alternatively, the heater inlet temperature may be taken as that in equilibrium with the vapor space and the outlet temperature calculated from known heat input and circulation rate. However, this type of evaporator is sometimes prone to another loss, termed **short circuiting,** which is failure of the liquid to flash completely to equilibrium; hence it is returned superheated to the heater inlet. Circulation rate is usually chosen, for economic reasons, to give a temperature rise through the heater of 2.8°C (5°F) or more, and in some cases short-circuiting losses have been of the same magnitude. These two losses may then represent a loss of almost half of the temperature difference available. Even in the falling-film evaporator, there is some loss in temperature difference due to pressure drop from friction and acceleration of the vapor generated in the tubes. Care must thus be taken in reporting or using heat-transfer data to make sure of the temperature-difference basis used.

For estimation of **heat-transfer coefficients** for evaporators, see Sec. 10.

UTILIZATION OF TEMPERATURE DIFFERENCE

Temperature difference is the driving force for evaporator operation and usually is limited, as by compression ratio in vapor-compression evaporators and by available steam-pressure and heat-sink temperature in single- and multiple-effect evaporators. A fundamental objective of evaporator design is to make as much of this total temperature difference available for heat transfer as is economically justifiable. Some losses in temperature difference, such as those due to BPR, are unavoidable. However, even these can be minimized, as by passing the liquor through effects or through different sections of a single effect in series so that only a portion of the heating surface is in contact with the strongest liquor. The principal reducible loss in ΔT is that due to friction and to entrance and exit losses in vapor piping and entrainment separators. Pressure-drop losses here correspond to a reduction in condensing temperature of the vapor and hence a loss in available ΔT. These losses become most critical at the low-temperature end of the evaporator, both because of the increasing specific volume of the vapor and because of the reduced slope of the vapor-pressure curve. Sizing of vapor lines is part of the economic optimization of the evaporator, extra costs of larger vapor lines being balanced against savings in ΔT, which correspond to savings in heating-surface requirements. It should be noted that entrance and exit losses in vapor lines usually exceed by severalfold the straight-pipe friction losses, so they cannot be ignored.

VAPOR-LIQUID SEPARATION

Product losses in evaporator vapor may result from foaming, splashing, or entrainment. Primary separation of liquid from vapor is accomplished in the vapor head by making the horizontal plan area large enough so that most of the entrained droplets can settle out against the rising flow of vapor. Allowable velocities are governed by the Souders-Brown equation: $V = k\sqrt{(\rho_l - \rho_v)/\rho_v}$, in which k depends on the size distribution of droplets and the decontamination factor F desired. For most evaporators and for F between 100 and

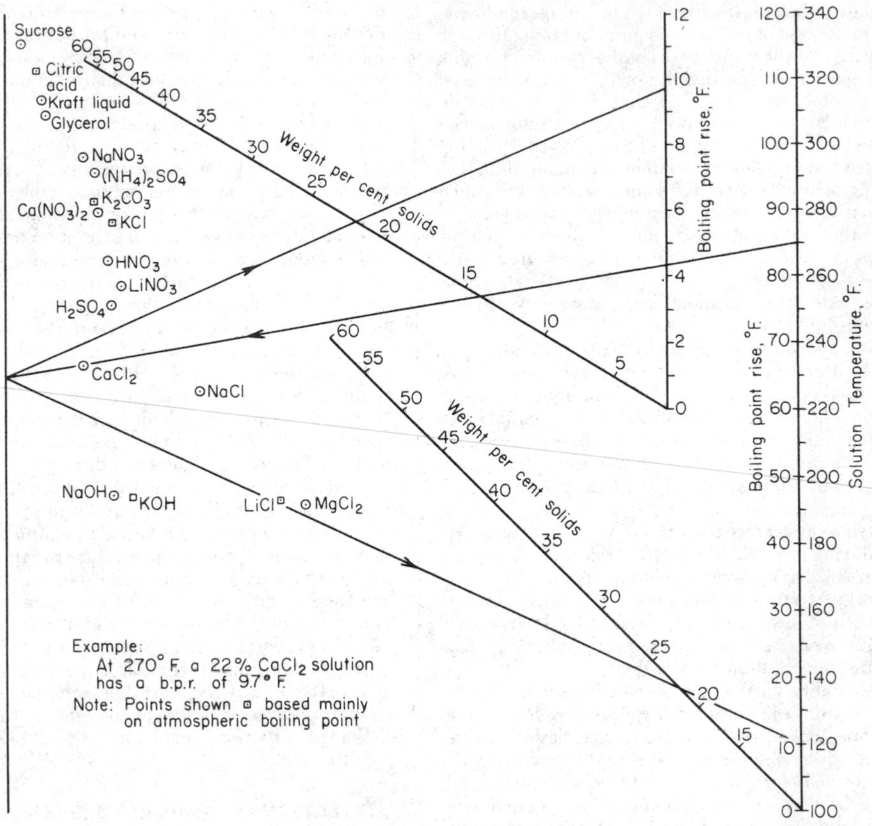

FIG. 11-23 Boiling-point rise of aqueous solutions. °C = 5/9 (°F − 32).

10,000, $k \cong 0.245/(F - 50)^{0.4}$ (Standiford, *Chemical Engineers' Handbook*, 4th ed., McGraw-Hill, New York, 1963, p. 11–35). Higher values of k (to about 0.15) can be tolerated in the falling-film evaporator, where most of the entrainment separation occurs in the tubes, the vapor is scrubbed by liquor leaving the tubes, and the vapor must reverse direction to reach the outlet.

Foaming losses usually result from the presence in the evaporating liquid of colloids or of surface-tension depressants and finely divided solids. Antifoam agents are often effective. Other means of combating foam include the use of steam jets impinging on the foam surface, the removal of product at the surface layer, where the foaming agents seem to concentrate, and operation at a very low liquid level so that hot surfaces can break the foam. Impingement at high velocity against a baffle tends to break the foam mechanically, and this is the reason that the long-tube vertical, forced-circulation, and agitated-film evaporators are particularly effective with foaming liquids.

Splashing losses are usually insignificant if a reasonable height has been provided between the liquid level and the top of the vapor head. The height required depends on the violence of boiling. Heights of 2.4 to 3.6 m (8 to 12 ft) or more are provided in short-tube vertical evaporators, in which the liquid and vapor leaving the tubes are projected upward. Less height is required in forced-circulation evaporators, in which the liquid is given a centrifugal motion or is projected downward as by a baffle. The same is true of long-tube vertical evaporators, in which the rising vapor-liquid mixture is projected against a baffle.

Entrainment losses by flashing are frequently encountered in an evaporator. If the feed is above the boiling point and is introduced above or only a short distance below the liquid level, entrainment losses may be excessive. This can occur in a short-tube-type evaporator if the feed is introduced at only one point below the lower tube sheet [Kerr, Louisiana Agric. Expt. Stn. Bull. 149, 1915]. The same difficulty may be encountered in forced-circulation evaporators having too high a temperature rise through the heating element and thus too wide a flashing range as the circulating liquid enters the body. Poor vacuum control, especially during startup, can cause the generation of far more vapor than the evaporator was designed to handle, with a consequent increase in entrainment.

Entrainment separators are frequently used to reduce product losses. There are a number of specialized designs available, practically all of which rely on a change in direction of the vapor flow when the vapor is traveling at high velocity. Typical separators are shown in Fig. 11-21, although not necessarily with the type of evaporator with which they may be used. The most common separator is the cyclone, which may have either a top or a bottom outlet as shown in Fig. 11-21*a* and *b* or may even be wrapped around the heating element of the next effect as shown in Fig. 11-21*f*. The separation efficiency of a cyclone increases with an increase in inlet velocity, although at the cost of some pressure drop, which means a loss in available temperature difference. Pressure drop in a cyclone is from 10 to 16 velocity heads [Lawrence, *Chem. Eng. Prog.*, **48**, 241 (1952)], based on the velocity in the inlet pipe. Such cyclones can be sized in the same manner as a cyclone dust collector (using velocities of about 30 m/s (100 ft/s) at atmospheric pressure) although sizes may be increased somewhat in order to reduce losses in available temperature difference.

Knitted wire mesh serves as an effective entrainment separator when it cannot easily be fouled by solids in the liquor. The mesh is available in woven metal wire of most alloys and is installed as a blanket across the top of the evaporator (Fig. 11-21*d*) or in a monitor of reduced diameter atop the vapor head. These separators have low pressure drops, usually on the order of 13 mm (½ in) of water, and collection efficiency is above 99.8 percent in the range of vapor velocities from 2.5 to 6 m/s (8 to 20 ft/s) [Carpenter and Othmer, *Am. Inst. Chem. Eng. J.*, **1**, 549 (1955)].

EVAPORATOR ARRANGEMENT

Single-Effect Evaporators Single-effect evaporators are used when the required capacity is small, steam is cheap, the material is so corrosive that very expensive materials of construction are required, or the vapor is so contaminated that it cannot be reused. Single-effect evaporators may be operated in batch, semibatch, or continuous-batch modes or continuously. Strictly speaking, **batch evaporators** are ones in which filling, evaporating, and emptying are consecutive steps. This method of operation is rarely used since it requires that the body be large enough to hold the entire charge of feed and the heating element be placed low enough so as not to be uncovered when the volume is reduced to that of the product. The more usual method of operation is **semibatch,** in which feed is continually added to maintain a constant level until the entire charge reaches final density. **Continuous-batch evaporators** usually have a continuous feed and, over at least part of the cycle, a continuous discharge. One method of operation is to circulate from a storage tank to the evaporator and back until the entire tank is up to desired concentration and then finish in batches. **Continuous evaporators** have essentially continuous feed and discharge, and concentrations of both feed and product remain substantially constant.

Thermocompression The simplest means of reducing the energy requirements of evaporation is to compress the vapor from a single-effect evaporator so that the vapor can be used as the heating medium in the same evaporator. The compression may be accomplished by mechanical means or by a steam jet. In order to keep the compressor cost and power requirements within reason, the evaporator must work with a fairly narrow temperature difference, usually from about 5.5 to 11°C (10° to 20°F). This means that a large evaporator heating surface is needed, which usually makes the vapor-compression evaporator more expensive in first cost than a multiple-effect evaporator. However, total installation costs may be reduced when purchased power is the energy source, since the need for boiler and heat sink is eliminated. Substantial savings in operating cost are realized when electrical or mechanical power is available at a low cost relative to low-pressure steam, when only high-pressure steam is available to operate the evaporator, or when the cost of providing cooling water or other heat sink for a multiple-effect evaporator is high.

Mechanical thermocompression may employ reciprocating, rotary positive-displacement, centrifugal, or axial-flow compressors. Positive-displacement compressors are impractical for all but the smallest capacities, such as portable seawater evaporators. Axial-flow compressors can be built for capacities of more than 472 m^3/s (1 × 10^6 ft^3/min). Centrifugal compressors are usually cheapest for the intermediate-capacity ranges that are normally encountered. In all cases, great care must be taken to keep entrainment at a minimum, since the vapor becomes superheated on compression and any liquid present will evaporate, leaving the dissolved solids behind. In some cases a vapor-scrubbing tower may be installed to protect the compressor. A mechanical recompression evaporator usually requires more heat than is available from the compressed vapor. Some of this extra heat can be obtained by preheating the feed with the condensate and, if possible, with the product. Rather extensive heat-exchange systems with close approach temperatures are usually justified, especially if the evaporator is operated at high temperature to reduce the volume of vapor to be compressed. When the product is a solid, an elutriation leg such as that shown in Fig. 11-21*b* is advantageous, since it cools the product almost to feed temperature. The remaining heat needed to maintain the evaporator in operation must be obtained from outside sources.

While theoretical compressor power requirements are reduced slightly by going to lower evaporating temperatures, the volume of vapor to be compressed and hence compressor size and cost increase so rapidly that low-temperature operation is more expensive than high-temperature operation. The requirement of low temperature for fruit-juice concentration has led to the development of an evaporator employing a **secondary fluid,** usually Freon or ammonia. In this evaporator, the vapor is condensed in an exchanger cooled by boiling Freon. The Freon, at a much higher vapor density than the water vapor, is then compressed to serve as the heating medium for the evaporator. This system requires that the latent heat be transferred through two surfaces instead of one, but the savings in compressor size and cost are enough to justify the extra cost of heating surface or the cost of compressing through a wider temperature range.

Steam-jet thermocompression is advantageous when steam is available at a pressure appreciably higher than can be used in the evaporator. The steam jet then serves as a reducing valve while doing some useful work. The efficiency of a steam jet is quite low and falls off rapidly when the jet is not used at the vapor-flow rate and terminal pressure conditions for which it was designed. Consequently multiple jets are used when wide variations in evaporation rate are expected. Because of the low first cost and the ability to handle large volumes of vapor, steam-jet thermocompressors are used to increase the economy of evaporators that must operate at low temperatures and hence cannot be operated in multiple effect. The steam-jet thermocompression evaporator has a heat input larger than that needed to balance the system, and some heat must be rejected. This is usually done by venting some of the vapor at the suction of the compressor.

Multiple-Effect Evaporation Multiple-effect evaporation is the principal means in use for economizing on energy consumption. Most such evaporators operate on a continuous basis, although for a few difficult materials a continuous-batch cycle may be employed. In a multiple-effect evaporator, steam from an outside source is condensed in the heating element of the first effect. If the feed to the effect is at a temperature near the boiling point in the first effect, 1 kg of steam will evaporate almost 1 kg of water. The first effect operates at (but is not controlled at) a boiling temperature high enough so that the evaporated water can serve as the heating medium of the second effect. Here almost another kilogram of water is evaporated, and this may go to a condenser if the evaporator is a double-effect or may be used as the heating medium of the third effect. This method may be repeated for any number of effects. Large evaporators having six and seven effects are common in the pulp and paper industry, and evaporators having as many as 17 effects have been built. As a first approximation, the **steam economy** of a multiple-effect evaporator will increase in proportion to the number of effects and usually will be somewhat less numerically than the number of effects.

The increased steam economy of a multiple-effect evaporator is gained at the expense of evaporator first cost. The total heat-transfer surface will increase substantially in proportion to the number of effects in the evaporator. This is only an approximation since going from one to two effects means that about half of the heat transfer is at a higher temperature level, where heat-transfer coefficients are generally higher. On the other hand, operating at lower temperature differences reduces the heat-transfer coefficient for many types of evaporator. If the material has an appreciable boiling-point elevation, this will also lower the available temperature difference. The only accurate means of predicting the changes in steam economy and surface requirements with changes in the number of effects is by detailed heat and material balances together with an analysis of the effect of changes in operating conditions on heat-transfer performance.

The approximate **temperature distribution** in a multiple-effect evaporator is under the control of the designer, but once built, the evaporator establishes its own equilibrium. Basically, the effects are a number of series resistances to heat transfer, each resistance being approximately proportional to $1/U_nA_n$. The total available temperature drop is divided between the effects in proportion to their resistances. If one effect starts to scale, its temperature drop will increase at the expense of the temperature drops across the other effects. This provides a convenient means of detecting a drop in heat-transfer coefficient in an effect of an operating evaporator. If the steam pressure and final vacuum do not change, the temperature in the effect that is scaling will decrease and the temperature in the preceding effect will increase.

The feed to a multiple-effect evaporator is usually transferred from one effect to another in series so that the ultimate product concentration is reached only in one effect of the evaporator. In **backward-feed** operation, the raw feed enters the last (coldest) effect, the discharge from this effect becomes the feed to the next-to-the-last effect, and so on until product is discharged from the first effect. This method of operation is advantageous when the feed is cold, since much less liquid must be heated to the higher temperature existing in the early effects. It is also used when the product is so viscous that high temperatures are needed to keep the viscosity low enough to give reasonable heat-transfer coefficients. When product viscosity is high but a hot product is not needed, the liquid from the first effect is sometimes flashed to a lower temperature in one or more stages and the flash vapor added to the vapor from one or more later effects of the evaporator.

In **forward-feed** operation, raw feed is introduced in the first effect and passed from effect to effect parallel to the steam flow. Product is withdrawn from the last effect. This method of operation is advantageous when the feed is hot or when the concentrated product would be damaged or would deposit scale at high temperature. Forward feed simplifies operation when liquor can be transferred by pressure difference alone, thus eliminating all intermediate liquor pumps. When the feed is cold, forward feed gives a low steam economy since an appreciable part of the prime steam is needed to heat the feed to the boiling point and thus accomplishes no evaporation. If forward feed is necessary and feed is cold, steam economy can be improved markedly by preheating the feed in stages with vapor bled from intermediate effects of the evaporator. This usually represents little increase in total heating surface or cost since the feed must be heated in any event and shell-and-tube heat exchangers are generally less expensive per unit of surface area than evaporator heating surface.

Mixed-feed operation is used only for special applications, as when liquor at an intermediate concentration and a certain temperature is desired for additional processing.

Parallel feed involves the introduction of raw feed and the withdrawal of product at each effect of the evaporator. It is used primarily when the feed is substantially saturated and the product is a solid. An example is the evaporation of brine to make common salt. Evaporators of the types shown in Fig. 11-21*b* or *e* are used, and the product is withdrawn as a slurry. In this case, parallel feed is desirable because the feed washes impurities from the salt leaving the body.

Heat-recovery systems are frequently incorporated in an evaporator to increase the steam economy. Ideally, product and evaporator condensate should leave the system at a temperature as low as possible. Also, heat should be recovered from these streams by exchange with feed or evaporating liquid at the highest possible temperature. This would normally require separate liquid-liquid heat exchangers, which add greatly to the complexity of the evaporator and are justifiable only in large plants. Normally, the loss in thermodynamic availability due to flashing is tolerated since the flash vapor can then be used directly in the evaporator effects. The most commonly used is a **condensate flash** system in which the condensate from each effect but the first (which normally must be returned to the boiler) is flashed in successive stages to the pressure in the heating element of each succeeding effect of the evaporator. Product flash tanks may also be used in a backward- or mixed-feed evaporator. In a forward-feed evaporator, the principal means of heat recovery may be by use of **feed preheaters** heated by vapor bled from each effect of the evaporator. In this case, condensate may be either flashed as before or used in a separate set of exchangers to accomplish some of the feed preheating. A feed preheated by last-effect vapor may also materially reduce condenser water requirements.

Seawater Evaporators The production of potable water from saline waters represents a large and growing field of application for evaporators. Extensive work done in this field to 1972 was summarized in the annual *Saline Water Conversion Reports* of the Office of Saline Water, U.S. Department of the Interior. **Steam economies** on the order of 10 kg evaporation/kg steam are usually justified because (1) unit production capacities are high, (2) fixed charges are low on capital used for public works (i.e., they use long amortization periods and have low interest rates, with no other return on investment considered), (3) heat-transfer performance is comparable with that of pure water, and (4) properly treated seawater causes little deterioration due to scaling or fouling.

Figure 11-24*a* shows a **multiple-effect** (falling-film) flow sheet as used for seawater. Twelve effects are needed for a steam economy of 10. Seawater is used to condense last-effect vapor, and a portion is then treated to prevent scaling and corrosion. Treatment usually consists of acidification to break down bicarbonates, followed by deaeration, which also removes the carbon dioxide generated. The treated seawater is then heated to successively higher temperatures by a portion of the vapor from each effect and finally is fed to the evaporating surface of the first effect. The vapor generated therein and the partially concentrated liquid are passed to the second effect, and so on until the last effect. The feed rate is adjusted relative to the steam rate so that the residual liquid from the last effect can carry away all the salts in solution, in a volume about one-third of that of the feed. Condensate formed in each effect but the first is flashed down to the following effects in sequence and constitutes the product of the evaporator.

As the feed-to-steam ratio is increased in the flow sheet of Fig. 11-24*a*, a point is reached where all the vapor is needed to preheat the feed and none is available for the evaporator tubes. This limiting case is the **multistage flash evaporator,** shown in its simplest form in Fig. 11-24*b*. Seawater is treated as before and then pumped through a number of feed heaters in series. It is given a final boost in temperature with prime steam in a **brine heater** before it is flashed down in series to provide the vapor needed by the feed heaters. The amount of steam required depends on the approach-temperature difference in the feed heaters and the flash range per stage. Condensate from the feed heaters is flashed down in the same manner as the brine.

Since the flow being heated is identical to the total flow being flashed, the temperature rise in each heater is equal to the flash range in each flasher. This temperature difference represents a loss from the temperature difference available for heat transfer. There are thus two ways of increasing the steam economy of such plants: increasing the heating surface and increasing the number of stages. Whereas the number of effects in a multiple-effect plant will be about 20 percent greater than the steam economy, the number of stages in a flash plant will be 3 to 4 times the steam economy. However, a large number of stages can be provided in a single vessel by means of internal bulkheads. The heat-exchanger tubing is placed in the same vessel, and the tubes usually are continuous through a number of stages. This requires ferrules or special close tube-hole clearances where the tubes pass through the internal bulkheads. In a plant for a steam economy of 10, the ratio of flow rate to heating surface is usually such that the seawater must pass through about 152 m of 19-mm (500 ft of ¾-in) tubing before it reaches the brine heater. This places a limitation on the physical arrangement of the vessels.

Inasmuch as it requires a flash range of about 61°C (110°F) to produce 1 kg of flash vapor for every 10 kg of seawater, the multistage flash evaporator requires handling a large volume of seawater relative to the product. In the flow sheet of Fig. 11-24*b*, all this seawater must be deaerated and treated for scale prevention. In addition, the last-stage vacuum varies with the ambient seawater temperature, and ejector equipment must be sized for the worst condition. These difficulties can be eliminated by using the **recirculating multistage flash** flow sheet of Fig. 11-24*c*. The last few stages, called the **reject stages,** are cooled by a flow of seawater that can be varied to maintain a reasonable last-stage vacuum. A small portion of the last-stage brine is blown down to carry away the dissolved salts, and the balance is recirculated to the **heat-recovery stages.** This arrangement requires a much smaller makeup of fresh seawater and hence a lower treatment cost.

The multistage flash evaporator is similar to a multiple-effect forced-circulation evaporator, but with all the forced-circulation

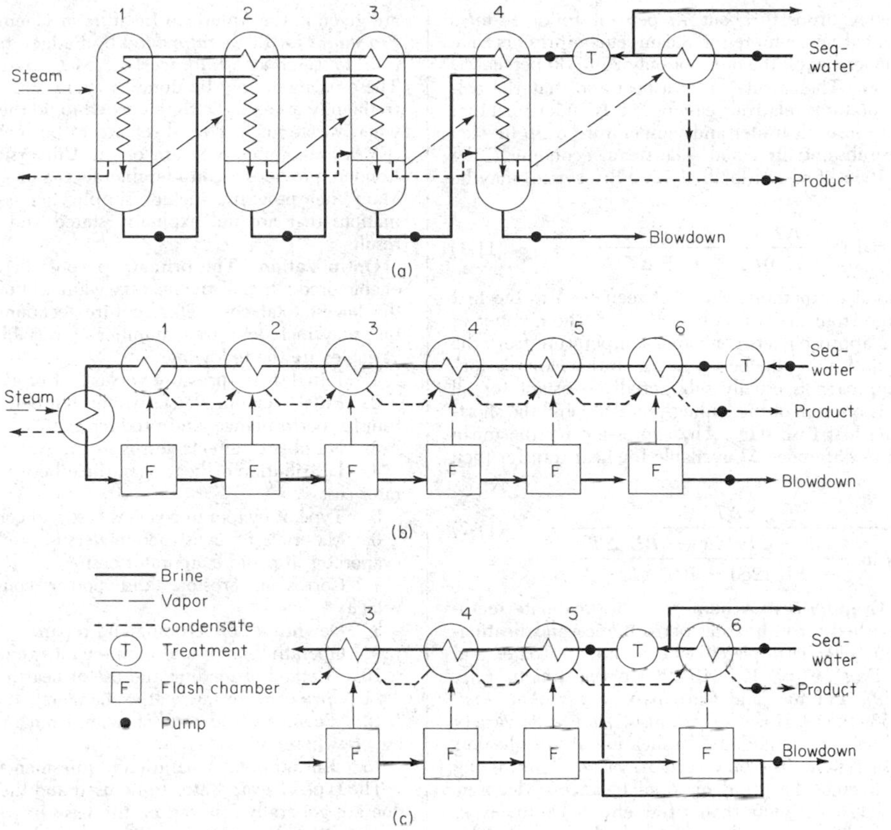

FIG. 11-24 Flow sheets for seawater evaporators. (*a*) Multiple effect (falling film). (*b*) Multistage flash (once-through). (*c*) Multistage flash (recirculating).

heaters in series. This has the advantage of requiring only one large-volume forced-circulation pump, but the sensible heating and short-circuiting losses in available temperature differences remain. A disadvantage of the flash evaporator is that the liquid throughout the system is at almost the discharge concentration. This has limited its industrial use to solutions in which no great concentration differences are required between feed and product and to where the liquid can be heated through wide temperature ranges without scaling. A partial remedy is to arrange several multistage flash evaporators in series, the heat-rejection section of one being the brine heater of the next. This permits independent control of concentration but eliminates the principal advantage of the flash evaporator, which is the small number of pumps and vessels required. An unusual feature of the flash evaporator is that fouling of the heating surfaces reduces primarily the steam economy rather than the capacity of the evaporator. Capacity is not affected until the heat-rejection stages can no longer handle the increased flashing resulting from the increased heat input.

EVAPORATOR CALCULATIONS

Single-Effect Evaporators The **heat requirements** of a single-effect continuous evaporator can be calculated by the usual methods of stoichiometry. If enthalpy data or specific heat and heat-of-solution data are not available, the heat requirement can be estimated as the sum of the heat needed to raise the feed from feed to product temperature and the heat required to evaporate the water. The latent heat of water is taken at the vapor-head pressure instead of at the product temperature in order to compensate partially for any heat of solution. If sufficient vapor-pressure data are available for the solution, methods are available to calculate the true latent heat from the slope of the Dühring line [Othmer, *Ind. Eng. Chem.*, **32,** 841 (1940)].

The heat requirements in batch evaporation are the same as those in continuous evaporation except that the temperature (and sometimes pressure) of the vapor changes during the course of the cycle. Since the enthalpy of water vapor changes but little relative to temperature, the difference between continuous and batch heat requirements is almost always negligible. More important usually is the effect of variation of fluid properties, such as viscosity and boiling-point rise, on heat transfer. These can only be estimated by a step-by-step calculation.

In selecting the **boiling temperature,** consideration must be given to the effect of temperature on heat-transfer characteristics of the type of evaporator to be used. Some evaporators show a marked drop in coefficient at low temperature—more than enough to offset any gain in available temperature difference. The condenser **cooling-water** temperature and cost must also be considered.

Thermocompression Evaporators Thermocompression-evaporator calculations [Pridgeon, *Chem. Metall. Eng.*, **28,** 1109 (1923); Peter, *Chimia (Switzerland)*, **3,** 114 (1949); Petzold, *Chem. Ing. Tech.*, **22,** 147 (1950); and Weimer, Dolf, and Austin, *Chem. Eng. Prog.*, **76**(11), 78 (1980)] are much the same as single-effect calculations with the added complication that the heat supplied to the evaporator from compressed vapor and other sources must exactly balance the heat requirements. Some knowledge of compressor efficiency is also required. Large axial-flow machines on the order of 236-m^3/s (500,000-ft^3/min) capacity may have efficiencies of 80 to

85 percent. Efficiency drops to about 75 percent for a 14-m^3/s (30,000-ft^3/min) centrifugal compressor. Steam-jet compressors have thermodynamic efficiencies on the order of only 25 to 30 percent.

Flash Evaporators The calculation of a heat and material balance on a flash evaporator is relatively easy once it is understood that the temperature rise in each heater and temperature drop in each flasher must all be substantially equal. The steam economy E, kg evaporation/kg of 1055-kJ steam (lb/lb of 1000-Btu steam) may be approximated from

$$E = \left(1 - \frac{\Delta T}{1250}\right)\frac{\Delta T}{Y + R + \Delta T/N} \tag{11-1}$$

where ΔT is the total temperature drop between feed to the first flasher and discharge from the last flasher, °C; N is the number of flash stages; Y is the approach between vapor temperature from the first flasher and liquid leaving the heater in which this vapor is condensed, °C (the approach is usually substantially constant for all stages); and R, °C, is the sum of the boiling-point rise and the short-circuiting loss in the first flash stage. The expression for the mean effective temperature difference Δt available for heat transfer then becomes

$$\Delta t = \frac{\Delta T}{N \ln \dfrac{1 - \Delta T/1250 - RE/\Delta T}{1 - \Delta T/1250 - RE/\Delta T - E/N}} \tag{11-2}$$

Multiple-Effect Evaporators A number of approximate methods have been published for estimating performance and heating-surface requirements of a multiple-effect evaporator [Coates and Pressburg, *Chem. Eng.*, **67**(6), 157 (1960); Coates, *Chem. Eng. Prog.*, **45**, 25 (1949); and Ray and Carnahan, *Trans. Am. Inst. Chem. Eng.*, **41**, 253 (1945)]. However, because of the wide variety of methods of feeding and the added complication of feed heaters and condensate flash systems, the only certain way of determining performance is by detailed heat and material balances. Algebraic solutions may be used, but if more than a few effects are involved, trial-and-error methods are usually quicker. These frequently involve trial-and-error within trial-and-error solutions. Usually, if condensate flash systems or feed heaters are involved, it is best to start at the first effect. The basic steps in the calculation are then as follows:

1. Estimate temperature distribution in the evaporator, taking into account boiling-point elevations. If all heating surfaces are to be equal, the temperature drop across each effect will be approximately inversely proportional to the heat-transfer coefficient in that effect.
2. Determine total evaporation required, and estimate steam consumption for the number of effects chosen.
3. From assumed feed temperature (forward feed) or feed flow (backward feed) to the first effect and assumed steam flow, calculate evaporation in the first effect. Repeat for each succeeding effect, checking intermediate assumptions as the calculation proceeds. Heat input from condensate flash can be incorporated easily since the condensate flow from the preceding effects will have already been determined.
4. The result of the calculation will be a feed to or a product discharge from the last effect that may not agree with actual requirements. The calculation must then be repeated with a new assumption of steam flow to the first effect.
5. These calculations should yield liquor concentrations in each effect that make possible a revised estimate of boiling-point rises. They also give the quantity of heat that must be transferred in each effect. From the heat loads, assumed temperature differences, and heat-transfer coefficients, heating-surface requirements can be determined. If the distribution of heating surface is not as desired, the entire calculation may need to be repeated with revised estimates of the temperature in each effect.
6. If sufficient data are available, heat-transfer coefficients under the proposed operating conditions can be calculated in greater detail and surface requirements readjusted.

Such calculations require considerable judgment to avoid repetitive trials but are usually well worth the effort. Sample calculations are given in the American Institute of Chemical Engineers *Testing Procedure for Evaporators* and by Badger and Banchero, *Introduction to Chemical Engineering*, McGraw-Hill, New York, 1955. These balances may be done by computer but programming time frequently exceeds the time needed to do them manually, especially when variations in flow sheet are to be investigated. The PAPSYS (EVAP) computer program of the University of Western Ontario, London, provides a considerable degree of flexibility in this regard. Many such programs include simplifying assumptions and approximations that are not explicitly stated and can lead to erroneous results.

Optimization The primary purpose of evaporator design is to enable production of the necessary amount of satisfactory product at the lowest total cost. This requires economic-balance calculations that may include a great number of variables. Among the possible variables are the following:

1. Initial steam pressure versus cost or availability.
2. Final vacuum versus water temperature, water cost, heat-transfer performance, and product quality
3. Number of effects versus steam, water, and pump power cost
4. Distribution of heating surface between effects versus evaporator cost
5. Type of evaporator versus cost and continuity of operation
6. Materials of construction versus product quality, tube life, evaporator life, and evaporator cost
7. Corrosion, erosion, and power consumption versus tube velocity
8. Downtime for retubing and repairs
9. Operating-labor and maintenance requirements
10. Method of feeding and use of heat-recovery systems
11. Size of recovery heat exchangers
12. Possible withdrawal of steam from an intermediate effect for use elsewhere
13. Entrainment separation requirements

The type of evaporator to be used and the materials of construction are generally selected on the basis of past experience with the material to be concentrated. The method of feeding can usually be decided on the basis of known feed temperature and the properties of feed and product. However, few of the listed variables are completely independent. For instance, if a large number of effects is to be used, with a consequent low temperature drop per effect, it is impractical to use a natural-circulation evaporator. If expensive materials of construction are desirable, it may be found that the forced-circulation evaporator is the cheapest and that only a few effects are justifiable.

The variable having the greatest influence on total cost is the number of effects in the evaporator. An economic balance can establish the optimum number where the number is not limited by such factors as viscosity, corrosiveness, freezing point, boiling-point rise, or thermal sensitivity. Under present United States conditions, savings in steam and water costs justify the extra capital, maintenance, and power costs of about seven effects in large commercial installations when the properties of the fluid are favorable, as in black-liquor evaporation. Under governmental financing conditions, as for plants to supply fresh water from seawater, evaporators containing over 12 effects can be justified.

As a general rule, the optimum number of effects increases with an increase in steam cost or plant size. Larger plants favor more effects, partly because they make it easier to install heat-recovery systems that increase the steam economy attainable with a given number of effects. Such recovery systems usually do not increase the total surface needed but do require that the heating surface be distributed between a greater number of pieces of equipment.

The most common evaporator design is based on the use of the same heating surface in each effect. This is by no means essential since few evaporators are "standard" or involve the use of the same patterns. In fact, there is no reason why all effects in an evaporator must be of the same type. For instance, the cheapest salt evaporator might use propeller calandrias for the early effects and forced-circulation effects at the low-temperature end, where their higher cost per unit area is more than offset by higher heat-transfer coefficients.

Bonilla [*Trans. Am. Inst. Chem. Eng.*, **41,** 529 (1945)] developed a simplified method for distributing the heating surface in a multiple-effect evaporator to achieve minimum cost. If the cost of the evaporator per unit area of heating surface is constant throughout, then minimum cost and area will be achieved if the ratio of area to temperature difference $A/\Delta T$ is the same for all effects. If the cost per unit area z varies, as when different tube materials or evaporator types are used, then $zA/\Delta T$ should be the same for all effects.

EVAPORATOR ACCESSORIES

Condensers The vapor from the last effect of an evaporator is usually removed by a condenser. **Surface condensers** are employed when mixing of condensate with condenser cooling water is not desired. They are for the most part shell-and-tube condensers with vapor on the shell side and a multipass flow of cooling water on the tube side. Heat loads, temperature differences, sizes, and costs are usually of the same order of magnitude as for another effect of the evaporator. Surface condensers use more cooling water and are so much more expensive that they are never used when a direct-contact condenser is suitable.

The most common type of direct-contact condenser is the countercurrent **barometric condenser,** in which vapor is condensed by rising against a rain of cooling water. The condenser is set high enough so that water can discharge by gravity from the vacuum in the condenser. Such condensers are inexpensive and are economical on water consumption. They can usually be relied on to maintain a vacuum corresponding to a saturated-vapor temperature within 2.8°C (5°F) of the water temperature leaving the condenser [How, *Chem. Eng.*, **63**(2), 174 (1956)]. The ratio of water consumption to vapor condensed can be determined from the following equation:

$$\frac{\text{Water flow}}{\text{Vapor flow}} = \frac{H_v - h_2}{h_2 - h_1} \tag{11-3}$$

where H_v = vapor enthalpy and h_1 and h_2 = water enthalpies entering and leaving the condenser. Another type of direct-contact condenser is the **jet** or **wet condenser,** which makes use of high-velocity jets of water both to condense the vapor and to force noncondensable gases out the tailpipe. This type of condenser is frequently placed below barometric height and requires a pump to remove the mixture of water and gases. Jet condensers usually require more water than the more common barometric-type condensers and cannot be throttled easily to conserve water when operating at low evaporation rates.

Vent Systems Noncondensable gases may be present in the evaporator vapor as a result of leakage, air dissolved in the feed, or decomposition reactions in the feed. When the vapor is condensed in the succeeding effect, the noncondensables increase in concentration and impede heat transfer. This occurs partially because of the reduced partial pressure of vapor in the mixture but mainly because the vapor flow toward the heating surface creates a film of poorly conducting gas at the interface. (See Sec. 10 for means of estimating the effect of noncondensable gases on the steam-film coefficient.) The most important means of reducing the influence of noncondensables on heat transfer is by properly channeling them past the heating surface. A positive vapor-flow path from inlet to vent outlet should be provided, and the path should preferably be tapered to avoid pockets of low velocity where noncondensables can be trapped. Excessive clearances and low-resistance channels that could bypass vapor directly from the inlet to the vent should be avoided [Standiford, *Chem. Eng. Prog.*, **75,** 59–62 (July 1979)].

In any event, noncondensable gases should be vented well before their concentration reaches 10 percent. Since gas concentrations are difficult to measure, the usual practice is to overvent. This means that an appreciable amount of vapor can be lost.

To help conserve steam economy, venting is usually done from the steam chest of one effect to the steam chest of the next. In this way, excess vapor in the vents does useful evaporation at a steam economy only about one less than the overall steam economy. Only when there are large amounts of noncondensable gases present, as in beet-sugar evaporation, is it desirable to pass the vents directly to the condenser to avoid serious losses in heat-transfer rates. In such cases, it can be worthwhile to recover heat from the vents in separate heat exchangers, which preheat the entering feed.

The noncondensable gases eventually reach the condenser (unless vented from an effect above atmospheric pressure to the atmosphere or to auxiliary vent condensers). These gases will be supplemented by air dissolved in the condenser water and by carbon dioxide given off on decomposition of bicarbonates in the water if a barometric condenser is used. These gases may be removed by the use of a water-jet-type condenser but are usually removed by a separate vacuum pump.

The vacuum pump is usually of the steam-jet type if high-pressure steam is available. If high-pressure steam is not available, more expensive mechanical pumps may be used. These may be either a water-ring (Hytor) type or a reciprocating pump.

The primary source of noncondensable gases usually is air dissolved in the condenser water. Figure 11-25 shows the dissolved-gas content of fresh water and seawater, calculated as equivalent air. The lower curve for seawater includes only dissolved oxygen and nitrogen. The upper curve includes carbon dioxide that can be evolved by complete breakdown of bicarbonate in seawater. Breakdown of bicarbonates is usually not appreciable in a condenser but may go almost to completion in a seawater evaporator. The large increase in gas volume as a result of possible bicarbonate breakdown is illustrative of the uncertainties involved in sizing vacuum systems.

By far the largest load on the vacuum pump is water vapor carried with the noncondensable gases. Standard power-plant practice assumes that the mixture leaving a surface condenser will have been cooled 4.2°C (7.5°F) below the saturation temperature of the vapor. This usually corresponds to about 2.5 kg of water vapor/kg of air. One advantage of the countercurrent barometric condenser is that it can cool the gases almost to the temperature of the incoming water and thus reduce the amount of water vapor carried with the air.

Salt Removal When an evaporator is used to make a crystalline product, a number of means are available for concentrating and removing the salt from the system. The simplest is to provide settling space in the evaporator itself. This is done in the types shown in Fig. 11-21*b*, *c*, and *e* by providing a relatively quiescent zone in which the salt can settle. Sufficiently high slurry densities can usually be achieved in this manner to reach the limit of pumpability. The evaporators are usually placed above barometric height so that the slurry can be discharged intermittently on a short time cycle. This permits the use of high velocities in large lines that have little tendency to plug.

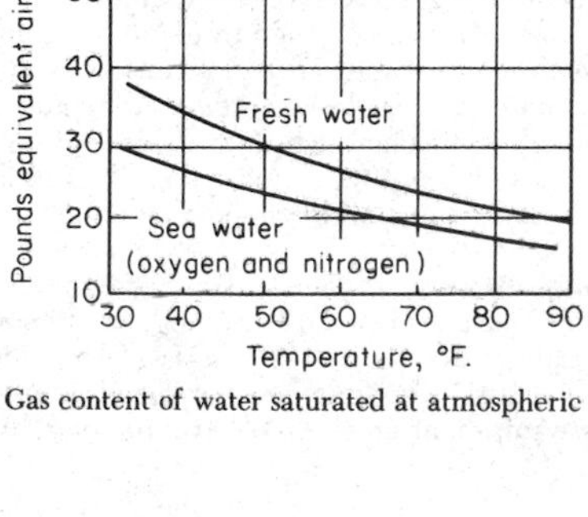

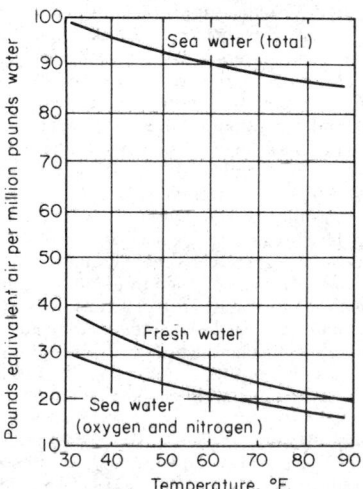

FIG. 11-25 Gas content of water saturated at atmospheric pressure. $°C = \frac{5}{9}(°F - 32)$.

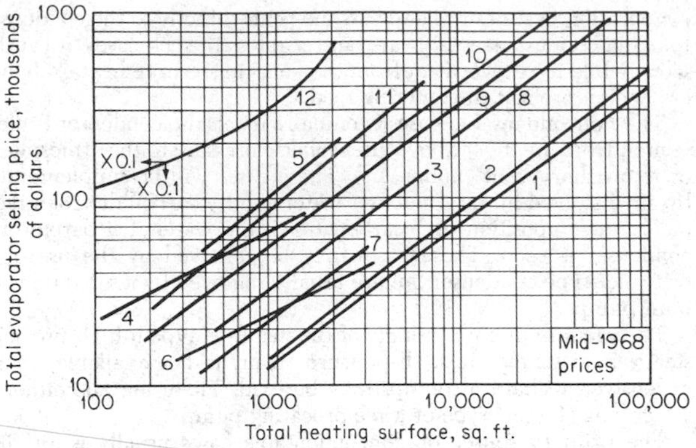

FIG. 11-26 Approximate evaporator selling prices. (1) Long-tube vertical, steel body, steel tubes (12-gauge). (2) Long-tube vertical, steel body, aluminum-brass tubes (16-gauge). (3) Long-tube vertical, cast-iron body, copper tubes. (4) Long-tube vertical, rubber-lined body, impervious graphite tubes. (5) Long-tube vertical, all nickel (700°F, 300 lbf/in^2). (6) Horizontal or short-tube vertical, cast-iron body, copper tubes. (7) Horizontal-tube, steel shell, admiralty tubes (100-lbf/in^2 shell). (8) Propeller calandria, cast-iron body, copper tubes. (9) Forced-circulation, cast-iron body, copper tubes. (10) Forced-circulation, all-Monel body, 90-10 Cu-Ni tubes. (11) Forced-circulation, nickel–cast-iron body, nickel tubes. (12) Agitated film, 316 stainless steel. To convert square feet to square meters, multiply by 0.0929; to convert pounds-force per square inch to kilopascals, multiply by 6.895; °C = 5/9 (°F − 32).

If the amount of salts crystallized is on the order of a ton an hour or less, a salt trap may be used. This is simply a receiver that is connected to the bottom of the evaporator and is closed off from the evaporator periodically for emptying. Such traps are useful when insufficient headroom is available for gravity removal of the solids. However, traps require a great deal of labor, give frequent trouble with the shutoff valves, and also can upset evaporator operation completely if a trap is reconnected to the evaporator without first displacing all air with feed liquor.

EVAPORATOR COSTS AND OPERATION

Capital Cost Approximate 1968 **selling prices** of various types of evaporators are shown in Fig. 11-26 [Zimmerman and Lavine, *Chemical Engineering Costs*, Industrial Research Service, Dover, N.H., 1950; Gushin, *Chem. Eng.*, **65**(12), 187 (1958); and private communication]. These prices include all auxiliary equipment that a manufacturer would normally supply, such as vapor piping, barometric condenser, steam jets, condensate flash tanks, and, in some cases, liquor piping and pumps. Costs are correlated against total heating surface in all effects of the evaporator. This permits inclusion of such equipment as a vacuum system. Appreciable errors can result if the evaporator under consideration does not contain roughly the same number of effects as the ones on which the curves of Fig. 11-26 are based. These range from single effects for curves 4, 5, and 12 through six and seven effects for curves 1 and 2.

Estimated 1968 installed costs of a number of types of evaporators are shown in Fig. 11-27 [Chilton, *Chem. Eng.*, **56**(6), 97 (1949); and private communication]. These costs include foundation, steelwork, evaporator assembly, pumps, piping, insulation, painting, and a moderate degree of instrumentation. When costs of evaporators constructed of other materials are desired, the incremental cost will be approximately equal to the change in tube cost if only the tubes are changed. It is usually difficult to estimate the effect of a change in body material—in some cases, welded alloy bodies are cheaper than cast-iron bodies.

Seawater-evaporator costs shown in Fig. 11-28 are derived from quotations, public bids, and desalting-cost updates prepared by Oak Ridge National Laboratory (ORNL/TM-6912, 1979). The smallest-capacity units, to about 50 m^3/day, are completely shop-assembled, and costs are for equipment only. An electric-motor prime mover is

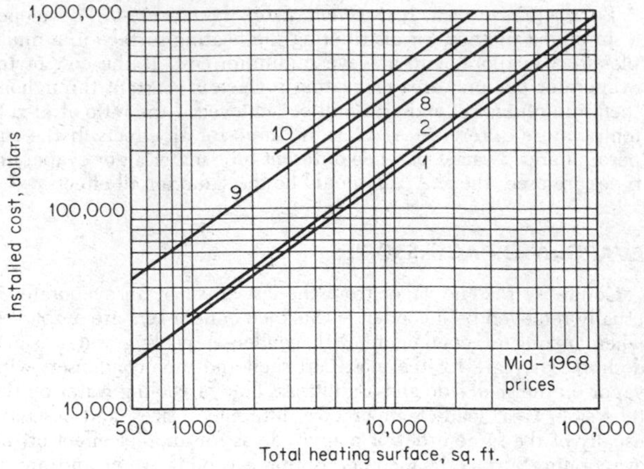

FIG. 11-27 Approximate installed cost of evaporators. (For code, see legend for Fig. 11-26.) To convert square feet to square meters, multiply by 0.0929.

included for thermocompression units. The heat-using units are generally run on waste heat such as diesel-jacket-cooling water. There is but little variation in cost with energy efficiency, apparently because most of the cost is in the auxiliaries rather than in the heating surface.

The larger units are generally one-of-a-kind, field-erected designs, and Fig. 11-28 costs cover the entire evaporator with its pumps, piping, and controls, delivered and erected. Costs of the large thermocompression units are based on the use of doubly fluted tubes, with twice the heat-transfer coefficients of smooth tubes, and include the cost of the gas-turbine driver and waste-heat boiler. The economy of these units is about 8 kg water per MJ gross heating value of gas fuel (19 lb/1000 Btu) (Off. Saline Water Res. Dev. Rep. 377, 1969). Costs of desalination plants in general have increased more rapidly than inflationary trends, primarily because it has been found advantageous to use more corrosion-resistant construction materials. Carbon steel stage dividers and water boxes have given particular difficulty in MSF plants, as have pumps and the less expensive grades of copper-alloy tubing (Kutbi, *Proc. NWSIA*, 1982). Much more extensive

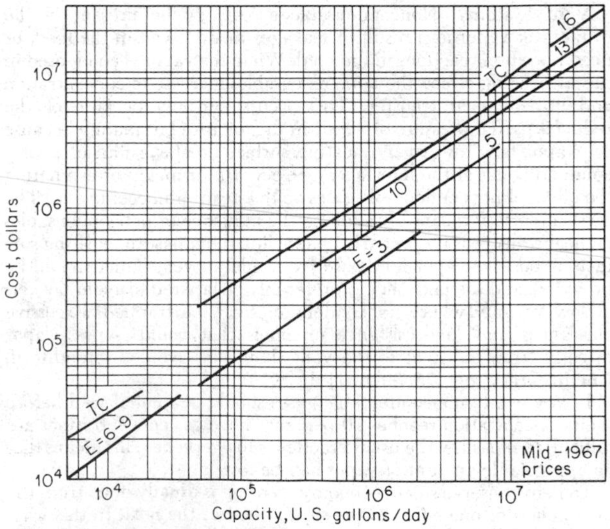

FIG. 11-28 Seawater-evaporator costs. *TC* = thermocompression units; *E* = multiple-effect or multistage flash units with the indicated steam economy. To convert gallons per day to cubic meters per day, multiply by 3.79×10^{-3}

use is now being made of cupronickel, stainless steel, and titanium. However, relatively inexpensive aluminum has been found well suited in some applications and promises an appreciable reduction in desalination-plant costs.

Operating Cost **Operating-labor** requirements depend mainly on the proximity of the evaporator to other process units where occasional assistance and maintenance help can be obtained. Normally, one person can easily control an evaporator of any number of effects—several such evaporators if a moderate number of control instruments are used [Bergstrom and Lientz, *Pap. Trade J.*, **124**(1), 6 (1947)]. Occasional maintenance labor is required for the repacking of pumps and valves and the repair of piping. The main maintenance requirement is for mechanical cleaning and for tube replacement. **Total maintenance cost,** including retubing cost, is about 1 percent of installed equipment cost per year for evaporators in noncorrosive service [Leonard, *Chem. Eng.*, **58**(9), 149 (1951)]. Cast-iron and copper evaporators in the salt industry, where corrosion is moderately severe, show maintenance costs on the order of 3 percent per year.

Evaporator Operation Control of an evaporator requires more than proper instrumentation. Operator logs should reflect changes in basic characteristics, as by use of **pseudo heat-transfer coefficients,** which can detect obstructions to heat flow, hence to capacity. These are merely the ratio of any convenient measure of heat flow to the temperature drop across each effect. **Dilution** by wash and seal water should be monitored since it absorbs evaporative capacity. Detailed tests, routine measurements, and operating problems are covered more fully in *Testing Procedure for Evaporators* (loc. cit.) and by Standiford [*Chem. Eng. Prog.*, **58**(11), 80 (1962)].

By far the best application of computers to evaporators is for working up operators' data into the basic performance parameters such as heat-transfer coefficients, steam economy, and dilution.

HEAT EXCHANGERS FOR SOLIDS

This section describes equipment for heat transfer to or from solids by the **indirect mode.** Such equipment is so constructed that the solids load (burden) is separated from the heat-carrier medium by a wall; the two phases are never in direct contact. Heat transfer is by conduction based on diffusion laws. Equipment in which the phases are in direct contact is covered in other sections of this *Handbook*, principally in Sec. 20.

Some of the devices covered here handle the solids burden in a **static or laminar-flowing bed.** Other devices can be considered as **continuously agitated kettles** in their heat-transfer aspect. For the latter, unit-area performance rates are higher.

Computational and graphical **methods for predicting performance** are given for both major heat-transfer aspects in Sec. 10. In solids heat processing with indirect equipment, the engineer should remember that the heat-transfer capability of the wall is many times that of the solids burden. Hence the solids properties and bed geometry govern the rate of heat transfer. This is more fully explained in Sec. 10. Only limited resultant (not predictive) and "experience" data are given here.

EQUIPMENT FOR SOLIDIFICATION

A frequent operation in the chemical field is the removal of heat from a material in a molten state to effect its conversion to the solid state. When the operation is carried on batchwise, it is termed **casting,** but when done continuously, it is termed **flaking.** Because of rapid heat transfer and temperature variations, jacketed types are limited to an initial melt temperature of 232°C (450°F). Higher temperatures [to 316°C (600°F)] require extreme care in jacket design and cooling-liquid flow pattern. Best performance and greatest capacity are obtained by (1) holding precooling to the minimum and (2) optimizing the cake thickness. The latter cannot always be done from the heat-transfer standpoint, as size specifications for the end product may dictate thickness.

Table Type This is a simple flat metal sheet with slightly upturned edges and jacketed on the underside for coolant flow. For many years this was the mainstay of food processors. Table types are still widely used when production is in small batches, when considerable batch-to-batch variation occurs, for pilot investigation, and when the cost of continuous devices is unjustifiable. Slab thicknesses are usually in the range of 13 to 25 mm (½ to 1 in). These units are homemade, with no standards available. Initial cost is low, but operating labor is high.

Agitated-Pan Type A natural evolution from the table type is a circular flat surface with jacketing on the underside for coolant flow and the added feature of a stirring means to sweep over the heat-transfer surface. This device is the agitated-pan type (Fig. 11-29). It is a batch-operation device. Because of its age and versatility it still serves a variety of heat-transfer operations for the chemical-process industries. While the most prevalent designation is **agitated-pan dryer** (in this mode, the burden is heated rather than cooled), considerable use is made of it for solidification applications. In this field, it is particularly suitable for processing burdens that change phase (1) slowly, by "thickening," (2) over a wide temperature range, (3) to an amorphous solid form, or (4) to a soft semigummy form (versus the usual hard crystalline structure).

The stirring produces the end product in the desired divided-solids form. Hence, it is frequently termed a "granulator" or a "crystallizer." A variety of factory-made sizes in various materials of construction are available. Initial cost is modest, while operating cost is rather high (as is true of all batch devices), but the ability to process "gummy" burdens and/or simultaneously effect two unit operations often yields an economical application.

Vibratory Type This construction (Fig. 11-30) takes advantage of the burden's special needs and the characteristic of **vibratory actuation.** A flammable burden requires the use of an inert atmosphere over it and a suitable nonhazardous fluid in the jacket. The vibratory action permits construction of rigid self-cleaning chambers with simple flexible connections. When solidification has been completed and vibrators started, the intense vibratory motion of the whole deck structure (as a rigid unit) breaks free the friable cake [up to 76 mm (3 in) thick], shatters it into lumps, and conveys it up over

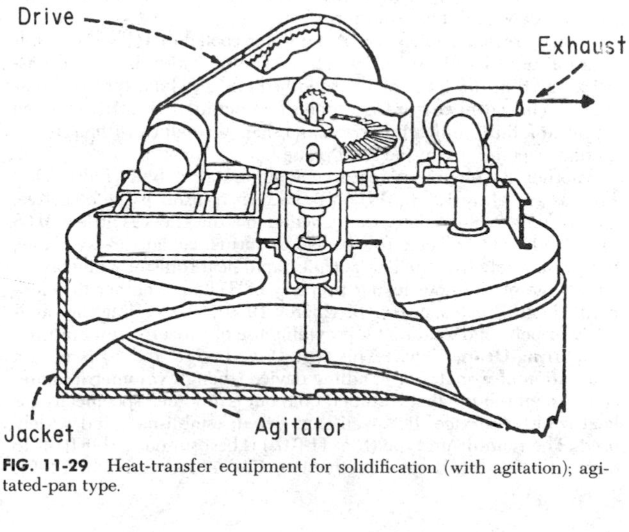

FIG. 11-29 Heat-transfer equipment for solidification (with agitation); agitated-pan type.

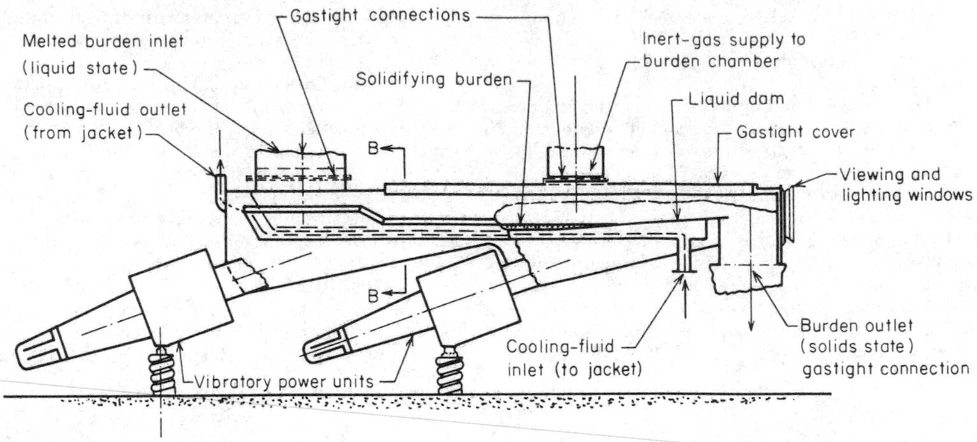

FIG. 11-30 Heat-transfer equipment for batch solidification; vibrating-conveyor type. *(Courtesy of Jeffrey Mfg. Co.)*

the dam to discharge. Heat-transfer performance is good, with overall coefficient U of about 68 W/(m^2·°C) [12 Btu/(h·ft^2·°F)] and values of heat flux q in the order of 11,670 W/m^2 [3700 Btu/(h·ft^2)]. Application of timing-cycle controls and a surge hopper for the discharge solids facilitates automatic operation of the caster and continuous operation of subsequent equipment.

Belt Types The patented **metal-belt** type (Fig. 11-31*a*), termed the "water-bed" conveyor, features a thin wall, a well-agitated fluid side for a thin water film (there are no rigid welded jackets to fail), a stainless-steel or Swedish-iron conveyor belt "floated" on the water with the aid of guides, no removal knife, and cleanability. It is mostly used for cake thicknesses of 3.2 to 15.9 mm (⅛ to ⅝ in) at speeds up to 15 m/min (50 ft/min), with 45.7-m (150-ft) pulley centers common. For 25- to 32-mm (1- to 1¼-in) cake, another belt on top to give two-sided cooling is frequently used. Applications are in food operations for cooling to harden candies, cheeses, gelatins, margarines, gums, etc.; and in chemical operations for solidification of sulfur, greases, resins, soaps, waxes, chloride salts, and some insecticides. Heat transfer is good, with sulfur solidification showing values of q = 5800 W/m^2 [1850 Btu/(h·ft^2)] and U = 96 W/(m^2·°C) [17 Btu/(h·ft^2·°F)] for a 7.9-mm (5⁄16-in) cake.

The **submerged metal belt** (Fig. 11-31*b*) is a special version of the metal belt to meet the peculiar handling properties of pitch in its solidification process. Although adhesive to a dry metal wall, pitch will not stick to the submerged wetted belt or rubber edge strips. Submergence helps to offset the very poor thermal conductivity through two-sided heat transfer.

A fairly recent application of the water-cooled metal belt to solidification duty is shown in Fig. 11-32. The operation is termed **pastillizing** from the form of the solidified end product, termed "pastilles." The novel feature is a one-step operation from the molten liquid to a fairly uniformly sized and shaped product without intermediate operations on the solid phase.

Another development features a **nonmetallic belt** [*Plast. Des. Process.*, 13 (July 1968)]. When rapid heat transfer is the objective, a glass-fiber, Teflon-coated construction in a thickness as little as 0.08 mm (0.003 in) is selected for use. No performance data are available, but presumably the thin belt permits rapid heat transfer while taking advantage of the nonsticking property of Teflon. Another development [*Food Process. Mark.*, 69 (March 1969)] is extending the capability of belt solidification by providing use of subzero temperatures.

Rotating-Drum Type This type (Fig. 11-31*c* and *d*) is not an adaptation of a material-handling device (though volumetric material throughput is a first consideration) but is designed specifically for heat-transfer service. It is well engineered, established, and widely used. The twin-drum type (Fig. 11-31*d*) is best suited to thin [0.4- to 6-mm (1⁄64- to ¼-in)] cake production. For temperatures to 149°C (300°F) the coolant water is piped in and siphoned out. Spray application of coolant water to the inside is employed for high-temperature work, permitting feed temperatures to at least 538°C (1000°F), or double those for jacketed equipment. Vaporizing refrigerants are readily applicable for very low temperature work.

The burden must have a definite **solidification temperature** to assure proper pickup from the feed pan. This limitation can be overcome by side feeding through an auxiliary rotating spreader roll. Application limits are further extended by special feed devices for burdens having oxidation-sensitive and/or supercooling characteristics. The standard double-drum model turns downward, with adjustable roll spacing to control sheet thickness. The newer twin-drum model (Fig. 11-31*d*) turns upward and, though subject to variable cake thickness, handles viscous and indefinite solidification-temperature-point burden materials well.

Drums have been successfully applied to a wide range of chemical products, both inorganic and organic, pharmaceutical compounds, waxes, soaps, insecticides, food products to a limited extent (including lard cooling), and even flake-ice production. A novel application is that of using a water-cooled roll to pick up from a molten-lead bath and turn out a 1.2-m- (4-ft-) wide continuous sheet, weighing 4.9 kg/m^2 (1 lb/ft^2), which is ideal for a sound barrier. This technique is more economical than other sheeting methods [*Mech. Eng.*, 631 (March 1968)].

Heat-transfer performance of drums, in terms of reported heat flux is: for an 80°C (176°F) melting-point wax, 7880 W/m^2 [2500 Btu/(h·ft^2)]; for a 130°C (266°F) melting-point organic chemical, 20,000 W/m^2 [6500 Btu/(h·ft^2)]; and for high- [318°C (604°F)] melting-point caustic soda (water-sprayed in drum), 95,000 to 125,000 W/m^2 [30,000 to 40,000 Btu/(h·ft^2)], with overall coefficients of 340 to 450 W/(m^2·°C) [60 to 80 Btu/(h·ft^2·°F)]. An innovation that is claimed often to increase these performance values by as much as 300 percent is the addition of hoods to apply impinging streams of heated air to the solidifying and drying solids surface as the drums carry it upward [*Chem. Eng.*, **74**, 152 (June 19, 1967)]. Similar rotating-drum indirect heat-transfer equipment is also extensively used for drying duty on liquids and thick slurries of solids (see Sec. 20).

Rotating-Shelf Type The patented **Roto-shelf** type (Fig. 11-31*e*) features (1) a large heat-transfer surface provided over a small floor space and in a small building volume, (2) easy floor cleaning, (3) nonhazardous machinery, (4) stainless-steel surfaces, (5) good control range, and (6) substantial capacity by providing as needed 1 to 10 shelves operated in parallel. It is best suited for thick-cake production and burden materials having an indefinite solidification temperature. Solidification of liquid sulfur into 13- to 19-mm- (½- to ¾-in-) thick lumps is a successful application. Heat transfer, by liquid-coolant cir-

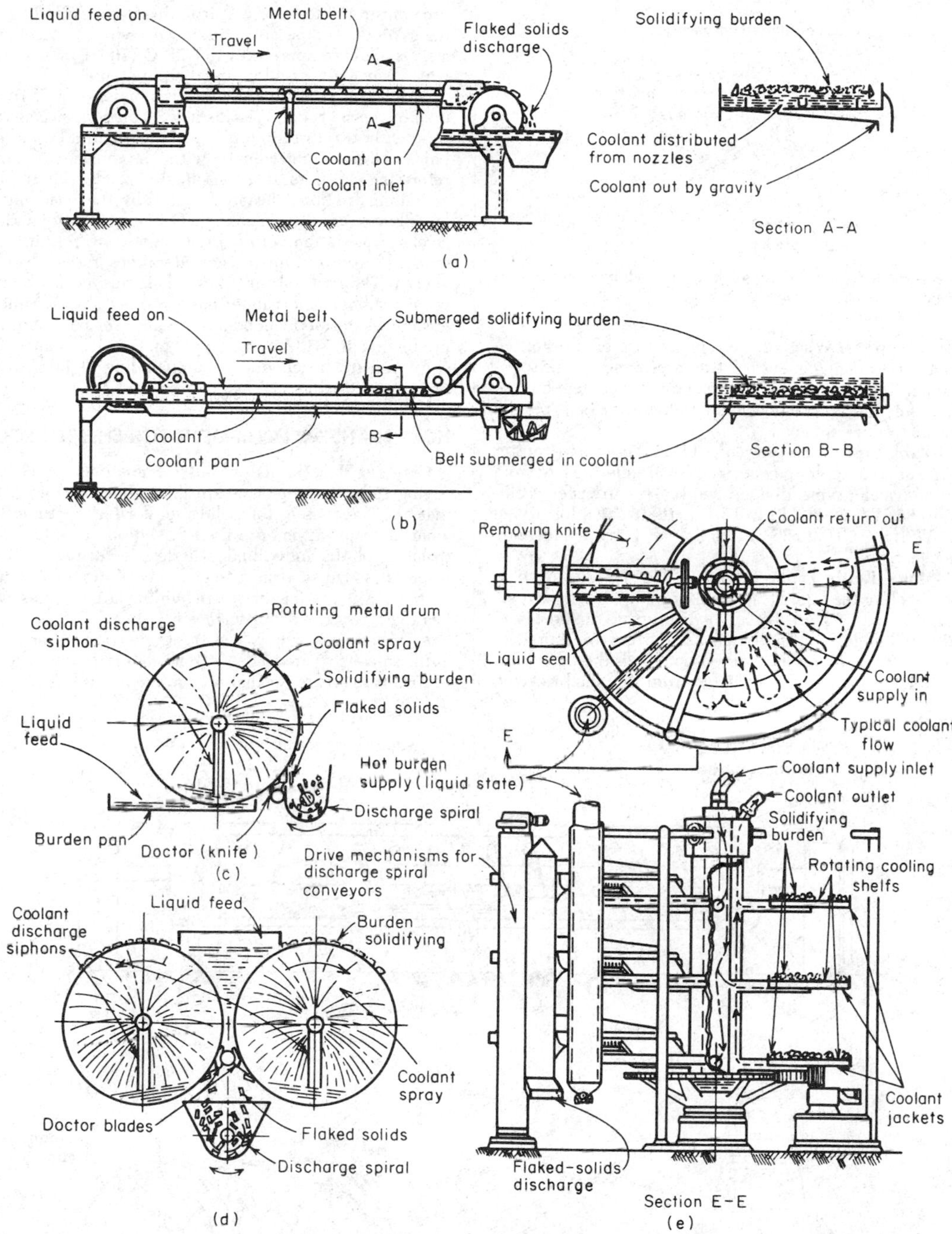

FIG. 11-31 Heat-transfer equipment for continuous solidification. (*a*) Cooled metal belt. *(Courtesy of Sandvik, Inc.)* (*b*) Submerged metal belt. *(Courtesy of Sandvik, Inc.)* (*c*) Single drum. (*d*) Twin drum. (*e*) Roto-shelf. (c, d, e, *courtesy of Buflovak Division, Blaw-Knox Food & Chemical Equipment, Inc.)*

culation through jackets, limits feed temperatures to 204°C (400°F). Heat-transfer rate, controlled by the thick cake rather than by equipment construction, should be equivalent to the belt type. Thermal performance is aided by applying water sprayed directly to the burden top to obtain two-sided cooling.

EQUIPMENT FOR FUSION OF SOLIDS

The thermal duty here is the opposite of solidification operations. The indirect heat-transfer equipment suitable for one operation is not suitable for the other because of the material-handling rather

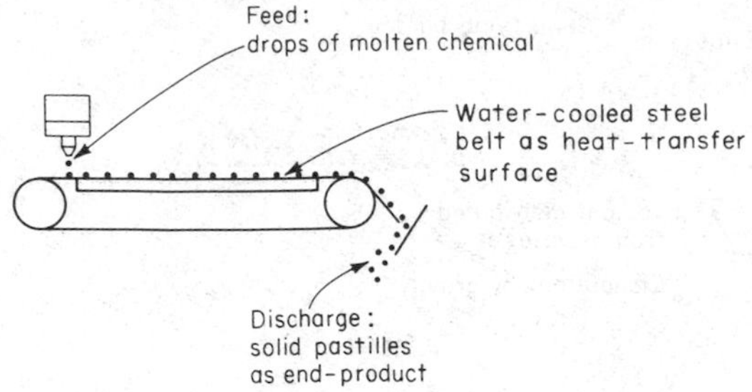

FIG. 11-32 Heat-transfer equipment for solidification; belt type for the operation of pastillization. *(Courtesy of Sandvik, Inc.)*

than the thermal aspects. Whether the temperature of transformation is a definite or a ranging one is of little importance in the selection of equipment for fusion. The burden is much agitated, but the beds are deep. Only fair overall coefficient values may be expected, although heat-flux values are good.

Horizontal-Tank Type This type (Fig. 11-33*a*) is used to transfer heat for melting or cooking dry powdered solids, rendering lard from meat-scrap solids, and drying divided solids. Heat-transfer coefficients are 17 to 85 $W/(m^2 \cdot °C)$ [3 to 15 $Btu/(h \cdot ft^2 \cdot °F)$] for drying and 28 to 140 $W/(m^2 \cdot °C)$ [5 to 25 $Btu/(h \cdot ft^2 \cdot °F)$] for vacuum and/or solvent recovery.

Vertical Agitated-Kettle Type Shown in Fig. 11-33*b*, this type is used to cook, melt to the liquid state, and provide or remove reaction heat for solids that vary greatly in "body" during the process so that material handling is a real problem. The virtues are simplicity and 100 percent cleanability. These often outweigh the poor heat-transfer aspect. These devices are available from the small jacketed type illustrated to huge cast-iron direct-underfired bowls for calcining gypsum. Temperature limits vary with construction; the simpler jackets allow temperatures to 371°C (700°F) (as with Dowtherm), which is not true of all jacketed equipment.

Mill Type Figure 11-33*c* shows one model of roll construction used. Note the ruggedness, as it is a *power device* as well as one for *indirect* heat transfer, employed to knead and heat a mixture of dry powdered-solid ingredients with the objective of reacting and reforming via fusion to a consolidated product. In this compounding operation, frictional heat generated by the kneading may require heat-flow reversal (by cooling). Heat-flow control and temperature-level considerations often predominate over heat-transfer performance. Power and mixing considerations, rather than heat transfer, govern. The two-roll mill shown is employed in compounding raw plastic, rubber, and rubberlike elastomer stocks. Multiple-roll mills less knives (termed calenders) are used for continuous sheet or film production in widths up to 2.3 m (7.7 ft). Similar equipment is employed in the chemical compounding of inks, dyes, paint pigments, and the like.

HEAT-TRANSFER EQUIPMENT FOR SHEETED SOLIDS

Cylinder Heat-Transfer Units Sometimes called "can" dryers or drying rolls, these devices are differentiated from drum dryers in that they are used for solids in flexible continuous-sheet form, whereas drum dryers are used for liquid or paste forms. The construction of the individual cylinders, or drums, is similar in most respects to that of drum dryers. Special designs are used to obtain uniform distribution of steam within large drums when uniform heating across the drum surface is critical.

A cylinder dryer may consist of one large cylindrical drum, such as the so-called Yankee dryer, but more often it comprises a **number of drums** arranged so that a continuous sheet of material may pass

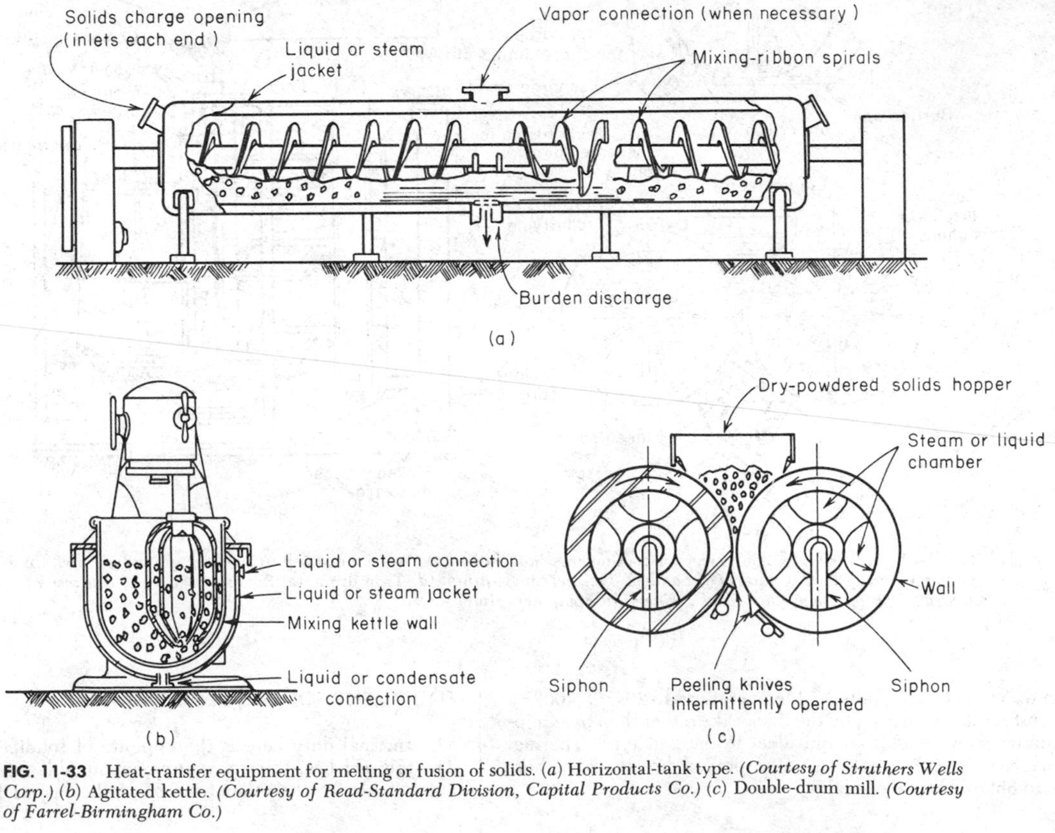

FIG. 11-33 Heat-transfer equipment for melting or fusion of solids. (*a*) Horizontal-tank type. *(Courtesy of Struthers Wells Corp.)* (*b*) Agitated kettle. *(Courtesy of Read-Standard Division, Capital Products Co.)* (*c*) Double-drum mill. *(Courtesy of Farrel-Birmingham Co.)*

over them in series. Typical of this arrangement are Fourdrinier-paper-machine dryers, cellophane dryers, slashers for textile piece goods and fibers, etc. The multiple cylinders are arranged in various ways. Generally they are staggered in two horizontal rows. In any one row, the cylinders are placed close together. The sheet material contacts the undersurface of the lower rolls and passes over the upper rolls, contacting 60 to 70 percent of the cylinder surface. The cylinders may also be arranged in a single horizontal row, in more than two horizontal rows, or in one or more vertical rows. When it is desired to contact only one side of the sheet with the cylinder surface, unheated guide rolls are used to conduct the sheeting from one cylinder to the next. For sheet materials that shrink on processing, it is frequently necessary to drive the cylinders at progressively slower speeds through the dryer. This requires elaborate individual electric drives on each cylinder.

Cylinder dryers usually operate at atmospheric pressure. However, the Minton paper dryer is designed for operation under vacuum. The drying cylinders are usually heated by steam, but occasionally single cylinders may be gas-heated, as in the case of the Pease blueprinting machine. Upon contacting the cylinder surface, wet sheet material is first heated to an equilibrium temperature somewhere between the wet-bulb temperature of the surrounding air and the boiling point of the liquid under the prevailing total pressure. The heat-transfer resistance of the vapor layer between the sheet and the cylinder surface may be significant.

These cylinder units are applicable to almost any form of sheet material that is not injuriously affected by contact with steam-heated metal surfaces. They are used chiefly when the sheet possesses certain properties such as a tendency to shrink or lacks the mechanical strength necessary for most types of continuous-sheeting air dryers. Applications are to dry films of various sorts, paper pulp in sheet form, paper sheets, paperboard, textile piece goods and fibers, etc. In some cases, imparting a special finish to the surface of the sheet may be an objective.

The **heat-transfer performance capacity** of cylinder dryers is not easy to estimate without a knowledge of the sheet temperature, which, in turn, is difficult to predict. According to published data, steam temperature is the largest single factor affecting capacity. Overall evaporation rates based on the total surface area of the dryers cover a range of 3.4 to 23 kg water/(h·m^2) [0.7 to 4.8 lb water/(h·ft^2)].

The value of the **coefficient of heat transfer** from steam to sheet is determined by the conditions prevailing on the inside and on the surface of the dryers. Low coefficients may be caused by (1) poor removal of air or other noncondensables from the steam in the cylinders, (2) poor removal of condensate, (3) accumulation of oil or rust on the interior of the drums, and (4) accumulation of a fiber lint on the outer surface of the drums. In a test reported by Lewis et al. [*Pulp Pap. Mag. Can.*, 22 (February 1927)] on a sulfite-paper dryer, in which the actual sheet temperatures were measured, a value of 187 W/(m^2·°C) [33 Btu/(h·ft^2·°F)] was obtained for the coefficient of heat flow between the steam and the paper sheet.

Operating-cost data for these units are meager. Power costs may be estimated by assuming 1 hp per cylinder for diameters of 1.2 to 1.8 m (4 to 6 ft). Data on labor and maintenance costs are also lacking.

The size of commercial cylinder dryers covers a wide range. The individual rolls may vary in diameter from 0.6 to 1.8 m (2 to 6 ft) and up to 8.5 m (28 ft) in width. In some cases, the width of rolls decreases throughout the dryer in order to conform to the shrinkage of the sheet. A single-cylinder dryer, such as the Yankee dryer, generally has a diameter between 2.7 and 4.6 m (9 and 15 ft).

HEAT-TRANSFER EQUIPMENT FOR DIVIDED SOLIDS

Most equipment for this service is some adaptation of a *material-handling* device whether or not the transport ability is desired. The old vertical tube and the relatively new vertical shell (fluidizer) are exceptions. Material-handling problems, plant transport needs, power, and maintenance are prime considerations in equipment selection and frequently overshadow heat-transfer and capital-cost considerations. Material handling is generally the most important aspect. Material-handling characteristics of the divided solids may vary during heat processing. The body changes are usually important in drying, occasionally significant for heating, and only on occasion important for cooling. The ability to minimize effects of changes is a major consideration in equipment selection. Dehydration operations are better performed on contactive apparatus (see Sec. 20) that provides air to carry off released water vapor before a semiliquid form develops.

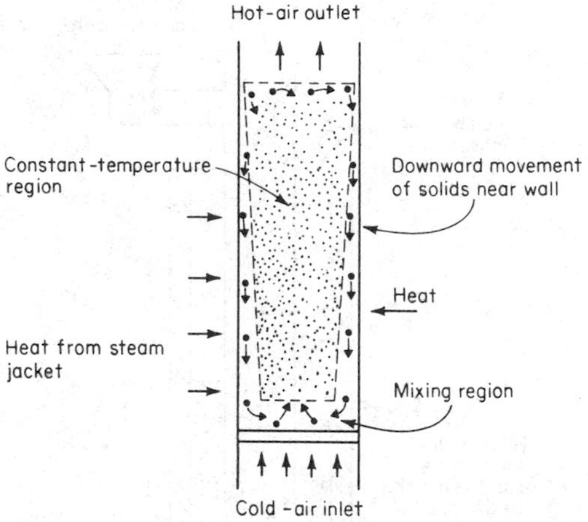

FIG. 11-34 Heat-transfer equipment for divided solids; stationary vertical-shell type. The indirect fluidizer.

Some types of equipment are convertible from heat removal to heat supply by simply changing the temperature level of the fluid or air. Other types require an auxiliary change. Others require constructional changes. Temperature limits for the equipment generally vary with the thermal operation. The kind of thermal operation has a major effect on heat-transfer values. For drying, overall coefficients are substantially higher in the presence of substantial moisture for the constant-rate period than in finishing. However, a stiff "body" occurrence due to moisture can prevent a normal "mixing" with an adverse effect on the coefficient.

Fluidized-Bed Type Known as the cylindrical fluidizer, this operates with a bed of **fluidized solids** (Fig. 11-34). It is an indirect heat-transfer version of the contactive type in Sec. 20. An application disadvantage is the need for batch operation unless some short circuiting can be tolerated. Solids-cooling applications are few, as they can be more effectively accomplished by the fluidizing gas via the contactive mechanism that is referred to in Sec. 10. Heating applications are many and varied. These are subject to one shortcoming, which is the dissipation of the heat input by carry-off in the fluidizing gas. Heat-transfer performance for the indirect mode to solids has been outstanding, with overall coefficients in the range of 570 to 850 W/(m^2·°C) [100 to 150 Btu/(h·ft^2·°F)]. This device with its thin film does for solids what the falling-film and other thin-film techniques do for fluids, as shown by Holt (Pap. 11, 4th National Heat-Transfer Conference, August 1960). In a design innovation with high heat-transfer capability, heat is supplied indirectly to the fluidized solids through the walls of in-bed, horizontally placed, finned tubes [Petrie, Freeby, and Buckham, *Chem. Eng. Prog.*, **64**(7), 45 (1968)].

Moving-Bed Type This concept uses a single-pass tube bundle in a vertical shell with the divided solids flowing by gravity in the tubes. It is little used for solids. A major difficulty in divided-solids applications is the problem of charging and discharging with uniformity. A second is poor heat-transfer rates. Because of these limitations, this tube-bundle type is not the workhorse for solids that it is for liquid and gas-phase heat exchange.

However, there are applications in which the nature of a specific chemical reactor system requires indirect heating or cooling of a moving bed of divided solids. One of these is the segregation process

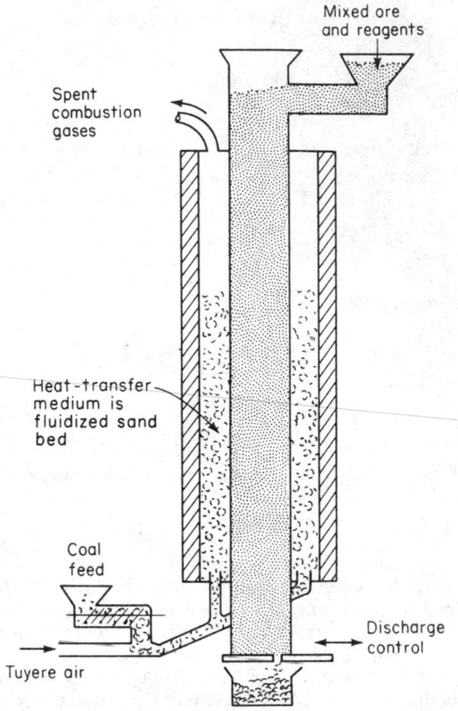

FIG. 11-35 Stationary vertical-tube type of indirect heat-transfer equipment with divided solids inside tubes; laminar solids flow and steady-state heat conditions.

which through a gaseous reaction frees chemically combined copper in an ore to a free copper form which permits easy, efficient subsequent recovery [Pinkey and Plint, *Miner. Process.*, 17–30 (June 1968)]. The apparatus construction and principle of operation are shown in Fig. 11-35. The functioning is abetted by a novel heat-exchange provision of a fluidized sand bed in the jacket. This provides a much higher unit heat-input rate (coefficient value) than would the usual low-density hot-combustion-gas flow.

Agitated-Pan Type This device (Fig. 11-29) is not an adaptation of a material-handling device but was developed many years ago primarily for heat-transfer purposes. As such, it has found wide application. In spite of its batch operation with high attendant labor costs, it is still used for processing divided solids when no phase change is occurring. Simplicity and easy cleanout make the unit a wise selection for handling small, experimental, and even some production runs when quite a variety of burden materials are heat-processed. Both heating and cooling are feasible with it, but greatest use has been for drying [see Sec. 20 and Uhl and Root, *Chem. Eng. Prog.*, **63**(7), 8 (1967)]. This device, because it can be readily covered (as shown in the illustration) and a vacuum drawn or special atmosphere provided, features versatility to widen its use. For drying granular solids, the heat-transfer rate ranges from 28 to 227 W/(m^2·°C) [5 to 40 Btu/(h·ft^2·°F)]. For atmospheric applications, thermal efficiency ranges from 65 to 75 percent. For vacuum applications, it is about 70 to 80 percent. These devices are available from several sources, fabricated of various metals used in chemical processes.

Kneading Devices These are closely related to the agitated pan but differ as being primarily mixing devices with heat transfer a secondary consideration. Heat transfer is provided by jacketed construction of the main body and is effected by a coolant, hot water, or steam. These devices are applicable for the compounding of divided solids by mechanical rather than chemical action. Application is largely in the pharmaceutical and food-processing industries. For a more complete description, illustrations, performance, and power requirements, refer to Sec. 19.

Shelf Devices Equipment having heated and/or cooled shelves is available but is little used for divided-solids heat processing. Most extensive use of stationary shelves is freezing of packaged solids for food industries and for **freeze drying by sublimation** (see Sec. 17).

Rotating-Shell Devices These (see Fig. 11-36) are installed horizontally, whereas stationary-shell installations are vertical. Material-handling aspects are of greater importance than thermal performance. Thermal results are customarily given in terms of overall coefficient on the basis of the total area provided, which varies greatly with the design. The effective use, chiefly percent fill factor, varies widely, affecting the reliability of stated coefficient values. For performance calculations see Sec. 10 on heat-processing theory for solids. These devices are variously used for cooling, heating, and drying and are the workhorses for heat-processing divided solids in the large-capacity range. Different modifications are used for each of the three operations.

The **plain** type (Fig. 11-36*a*) features simplicity and yet versatility through various end-construction modifications enabling wide and varied applications. Thermal performance is strongly affected by the "body" characteristics of the burden because of its dependency for material handling on frictional contact. Hence, performance ranges from well-agitated beds with good thin-film heat-transfer rates to poorly agitated beds with poor thick-film heat-transfer rates. Temperature limits in application are (1) low-range cooling with shell dipped in water, 400°C (750°F) and less; (2) intermediate cooling with forced circulation of tank water, to 760°C (1400°F); (3) primary cooling, above 760°C (1400°F), water copiously sprayed and loading kept light; (4) low-range heating, below steam temperature, hot-water dip; and (5) high-range heating by tempered combustion gases or ribbon radiant-gas burners.

The **flighted** type (Fig. 11-36*b*) is a first-step modification of the plain type. The simple flight addition improves heat-transfer performance. This type is most effective on semifluid burdens which slide readily. Flighted models are restricted from applications in which soft-cake sticking occurs, breakage must be minimized, and abrasion is severe. A special flighting is one having the cross section compartmented into four lesser areas with ducts between. Hot gases are drawn through the ducts en route from the outer oven to the stack to provide about 75 percent more heating surface, improving efficiency and capacity with a modest cost increase. Another similar unit has the flights made in a triangular-duct cross section with hot gases drawn through.

The **tubed-shell** type (Fig. 11-36*c*) is basically the same device more commonly known as a "steam-tube rotary dryer" (see Sec. 20). The rotation, combined with slight inclination from the horizontal, moves the shell-side solids through it continuously. This type features good mixing with the objective of increased heat-transfer performance. Tube-side fluid may be water, steam, or combustion gas. Bottom discharge slots in the shell are used so that heat-transfer-medium supply and removal can be made through the ends; these restrict wide-range loading and make the tubed type inapplicable for floody materials. These units are seldom applicable for sticky, soft-caking, scaling, or heat-sensitive burdens. They are not recommended for abrasive materials. This type has high thermal efficiency because heat loss is minimized. **Heat-transfer coefficient** values are: water, 34 W/(m^2·°C) [6 Btu/(h·ft^2·°F)]; steam, same, with heat flux reliably constant at 3800 W/m^2 [1200 Btu/(h·ft^2)]; and gas, 17 W/(m^2·°C) [3 Btu/(h·ft^2·°F)], with a high temperature difference. Although from the preceding discussion the device may seem rather limited, it is nevertheless widely used for drying, with condensing steam predominating as the heat-carrying fluid. But with water or refrigerants flowing in the tubes, it is also effective for cooling operations. The units are custom-built by several manufacturers in a wide range of sizes and materials. A few fabricators that specialize in this type of equipment have accumulated a vast store of data for determining application sizing.

The patented **deep-finned** type in Fig. 11-36*d* is named the "Roto-fin cooler." It features loading with a small layer thickness, excellent mixing to give a good effective diffusivity value, and a thin fluid-side film. Unlike other rotating-shell types, it is installed horizontally, and the burden is moved positively by the fins acting as an Archimedes

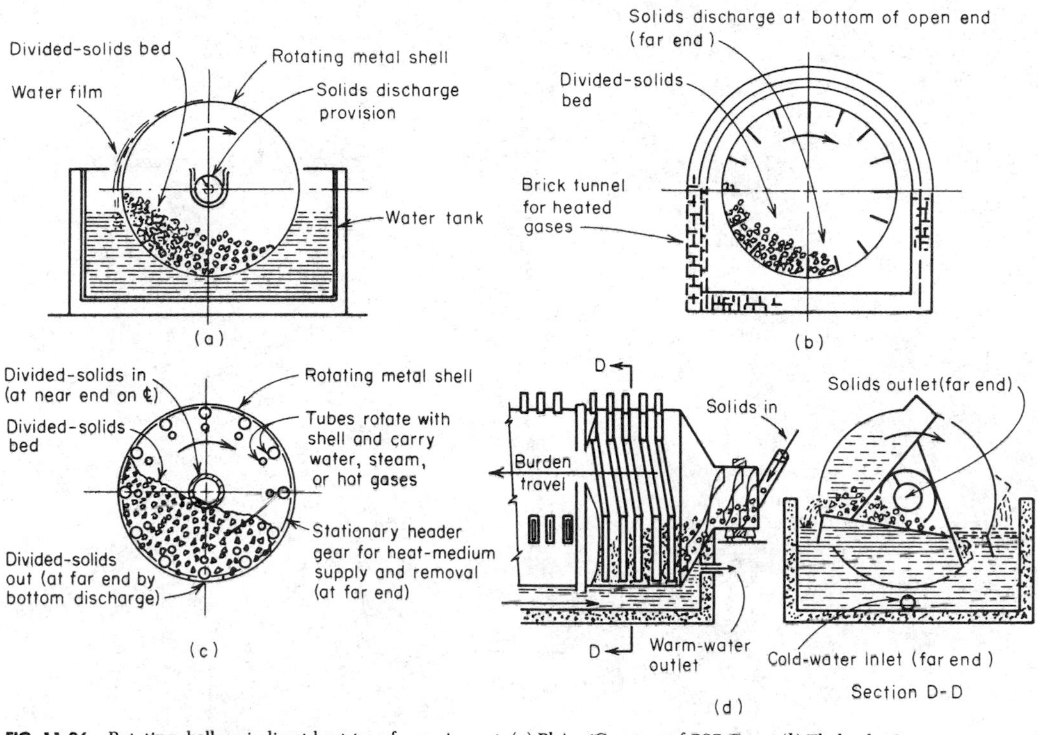

FIG. 11-36 Rotating shells as indirect heat-transfer equipment. (*a*) Plain. *(Courtesy of BSP Corp.)* (*b*) Flighted. *(Courtesy of BSP Corp.)* (*c*) Tubed. *(Courtesy of General American Transportation Corp.)* (*d*) Deep-finned type. *(Courtesy of Link-Belt Co.)*

spiral. Rotational speed and spiral pitch determine travel time. For cooling, this type is applicable to both secondary and intermediate cooling duties. Applications include solids in small lumps [9 mm ($\frac{3}{8}$ in)] and granular size [6 mm and less ($\frac{1}{4}$ to 0 in)] with no larger pieces to plug the fins, solids that have a free-flowing body characteristic with no sticking or caking tendencies, and drying of solids that have a low moisture and powder content unless special modifications are made for substantial vapor and dust handling. Thermal performance is very good, with overall coefficients to 110 W/(m^2·°C) [20 Btu/(h·ft^2·°F)], with one-half of these coefficients nominal for cooling based on the total area provided (nearly double those reported for other indirect rotaries).

Conveyor-Belt Devices The metal-belt type (Fig. 11-32) is the only device in this classification of material-handling equipment that has had serious effort expended on it to adapt it to indirect heat-transfer service with divided solids. It features a lightweight construction of a large area with a thin metal wall. Indirect-cooling applications have been made with poor thermal performance, as could be expected with a static layer. Auxiliary plowlike mixing devices, which are considered an absolute necessity to secure any worthwhile results for this service, restrict applications.

Spiral-Conveyor Devices Figure 11-37 illustrates the major adaptations of this widely used class of material-handling equipment to indirect heat-transfer purposes. These conveyors can be considered for heat-transfer purposes as continuously agitated kettles. The adaptation of Fig. 11-37*d* offers a batch-operated version for evaporation duty. For this service, all are package-priced and package-shipped items requiring few, if any, auxiliaries.

The **jacketed solid-flight** type (Fig. 11-37*a*) is the standard low-cost (parts-basis-priced) material-handling device, with a simple jacket added and employed for secondary-range heat transfer of an incidental nature. Heat-transfer coefficients are as low as 11 to 34 W/(m^2·°C) [2 to 6 Btu/(h·ft^2·°F)] on sensible heat transfer and 11 to 68 W/(m^2·°C) [2 to 12 Btu/(h·ft^2·°F)] on drying because of substantial static solids-side film.

The **small-spiral–large-shaft** type (Fig. 11-37*b*) is inserted in a solids-product line as pipe banks are in a fluid line, solely as a heat-transfer device. It features a thin burden ring carried at a high rotative speed and subjected to two-sided conductance to yield an estimated heat-transfer coefficient of 285 W/(m^2·°C) [50 Btu/(h·ft^2·°F)], thereby ranking thermally next to the shell-fluidizer type. This device for powdered solids is comparable with the Votator of the fluid field.

Figure 11-37*c* shows a fairly new spiral device with a **medium-heavy annular solids bed** and having the combination of a jacketed, stationary outer shell with moving paddles that carry the heat-transfer fluid. A unique feature of this device to increase volumetric throughput, by providing an overall greater temperature drop, is that the heat medium is supplied to and withdrawn from the rotor paddles by a parallel piping arrangement in the rotor shaft. This is a unique flow arrangement compared with the usual series flow. In addition, the rotor carries burden-agitating spikes which give it the trade name of Porcupine Heat-Processor (*Chem. Equip. News*, April 1966; and Uhl and Root, AIChE Prepr. 21, 11th National Heat-Transfer Conference, August 1967).

The **large-spiral hollow-flight** type (Fig. 11-37*d*) is an adaptation, with external bearings, full fill, and salient construction points as shown, that is highly versatile in application. Heat-transfer coefficients are 34 to 57 W/(m^2·°C) [6 to 10 Btu/(h·ft^2·°F)] for poor, 45 to 85 W/(m^2·°C) [8 to 15 Btu/(h·ft^2·°F)] for fair, and 57 to 114 W/(m^2·°C) [10 to 20 Btu/(h·ft^2·°F)] for wet conductors. A popular version of this employs two such spirals in one material-handling chamber for a pugmill agitation of the deep solids bed. The spirals are seldom heated. The shaft and shell are heated.

Another deep-bed spiral-activated solids-transport device is shown by Fig. 11-37*e*. The flights carry a heat-transfer medium as well as the jacket. A unique feature of this device which is purported to increase heat-transfer capability in a given equipment space and cost is the dense-phase fluidization of the deep bed that promotes agitation and moisture removal on drying operations.

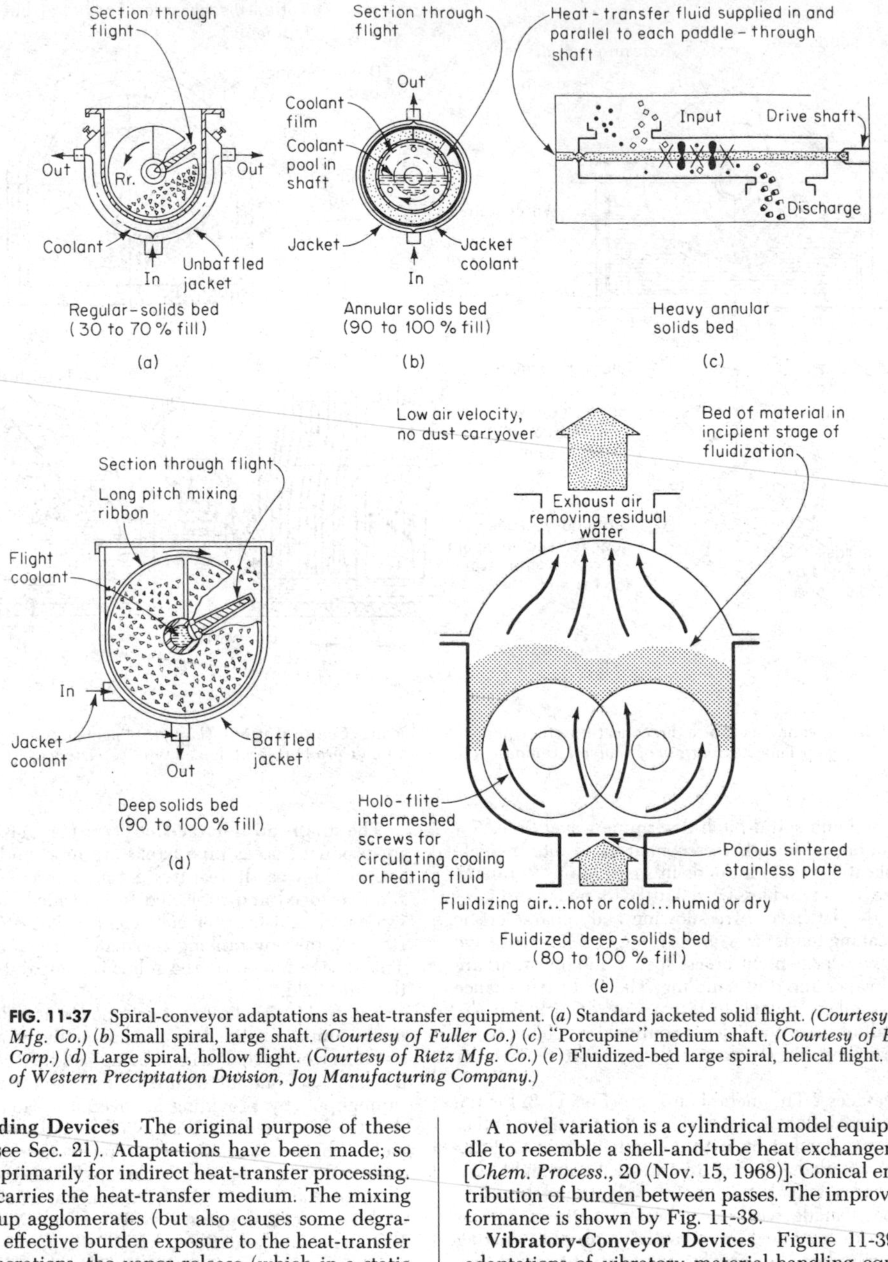

FIG. 11-37 Spiral-conveyor adaptations as heat-transfer equipment. (*a*) Standard jacketed solid flight. *(Courtesy of Jeffrey Mfg. Co.)* (*b*) Small spiral, large shaft. *(Courtesy of Fuller Co.)* (*c*) "Porcupine" medium shaft. *(Courtesy of Bethlehem Corp.)* (*d*) Large spiral, hollow flight. *(Courtesy of Rietz Mfg. Co.)* (*e*) Fluidized-bed large spiral, helical flight. *(Courtesy of Western Precipitation Division, Joy Manufacturing Company.)*

Double-Cone Blending Devices The original purpose of these devices was mixing (see Sec. 21). Adaptations have been made; so many models now are primarily for indirect heat-transfer processing. A jacket on the shell carries the heat-transfer medium. The mixing action, which breaks up agglomerates (but also causes some degradation), provides very effective burden exposure to the heat-transfer surface. On drying operations, the vapor release (which in a static bed is a slow diffusional process) takes place relatively quickly. To provide vapor removal from the burden chamber, a hollow shaft is used. Many of these devices carry the hollow-shaft feature a step further by adding a rotating seal and drawing a vacuum. This increases thermal performance notably and makes the device a natural for solvent-recovery operations.

These devices are replacing the older tank and spiral-conveyor devices. Better provisions for speed and ease of fill and discharge (without powered rotation) minimize downtime to make this batch-operated device attractive. Heat-transfer coefficients ranging from 28 to 200 W/(m$^2\cdot$°C) [5 to 35 Btu/(h$\cdot$ft$^2\cdot$°F)] are obtained. However, if caking on the heat-transfer walls is serious, then values may drop to 5.5 or 11 W/(m$^2\cdot$°C) [1 or 2 Btu/(h$\cdot$ft$^2\cdot$°F)], constituting a misapplication. The double cone is available in a fairly wide range of sizes and construction materials. The users are the fine-chemical, pharmaceutical, and biological-preparation industries.

A novel variation is a cylindrical model equipped with a tube bundle to resemble a shell-and-tube heat exchanger with a bloated shell [*Chem. Process.*, 20 (Nov. 15, 1968)]. Conical ends provide for redistribution of burden between passes. The improved heat-transfer performance is shown by Fig. 11-38.

Vibratory-Conveyor Devices Figure 11-39 shows the various adaptations of vibratory material-handling equipment for indirect heat-transfer service on divided solids. The basic vibratory-equipment data are given in Sec. 7. These indirect heat-transfer adaptations feature simplicity, nonhazardous construction, nondegradation, nondusting, no wear, ready conveying-rate variation [1.5 to 4.5 m/min (5 to 15 ft/min)], and good heat-transfer coefficient—115 W/(m$^2\cdot$°C) [20 Btu/(h$\cdot$ft$^2\cdot$°F)] for sand. They usually require feed-rate and distribution auxiliaries. They are suited for heating and cooling of divided solids in powdered, granular, or moist forms but no sticky, liquefying, or floody ones. Terminal-temperature differences less than 11°C (20°F) on cooling and 17°C (30°F) on heating or drying operations are seldom practical. These devices are for medium and light capacities.

The **heavy-duty jacketed** type (Fig. 11-39*a*) is a special custom-built adaptation of a heavy-duty vibratory conveyor shown in Fig. 11-37. Its application is continuously to cool the crushed material [from about 177°C (350°F)] produced by the vibratory-type "caster"

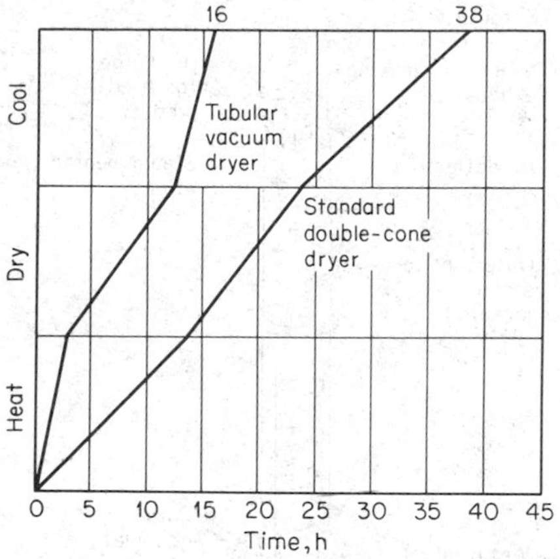

FIG. 11-38 Performance of tubed blender heat-transfer device.

of Fig. 11-30. It does not have the liquid dam and is made in longer lengths that employ L, switchback, and S arrangements on one floor. The capacity rate is 27,200 to 31,700 kg/h (30 to 35 tons/h) with heat-transfer coefficients in the order of 142 to 170 W/(m^2·°C) [25 to 30 Btu/(h·ft^2·°F)]. For heating or drying applications, it employs steam to 414 kPa (60 lbf/in^2).

The **jacketed or coolant-spraying** type (Fig. 11-39*b*) is designed to assure a very thin, highly agitated liquid-side film and the same initial coolant temperature over the entire length. It is frequently employed for transporting substantial quantities of hot solids, with cooling as an incidental consideration. For heating or drying applications, hot water or steam at a gauge pressure of 7 kPa (1 lbf/in^2) may be employed. This type is widely used because of its versatility, simplicity, cleanability, and good thermal performance.

The **light-duty jacketed** type (Fig. 11-39*c*) is designed for use of air as a heat carrier. The flow through the jacket is highly turbulent and is usually counterflow. On long installations, the air flow is parallel to every two sections for more heat-carrying capacity and a fairly uniform surface temperature. The outstanding feature is that a wide range of temperature control is obtained by merely changing the heat-carrier temperature level from as low as atmospheric moisture condensation will allow to 204°C (400°F). On heating operations, a very good thermal efficiency can be obtained by insulating the machine and recycling the air. While heat-transfer rating is good, the heat-removal capacity is limited. Cooler units are often used in series with like units operated as dryers or when clean water is unavailable. Drying applications are for heat-sensitive [49 to 132°C (120° to 270°F)] products; when temperatures higher than steam at a gauge pressure of 7 kPa (1 lbf/in^2) can provide are wanted but heavy-duty equipment is too costly; when the jacket-corrosion hazard of steam is unwanted; when headroom space is at a premium; and for highly abrasive burden materials such as fritted or crushed glasses and porcelains.

The **tiered arrangement** (Fig. 11-39*d*) employs the units of Fig. 11-39 with either air or steam at a gauge pressure of 7 kPa (1 lbf/in^2) as a heat medium. These are custom-designed and built to provide a large amount of heat-transfer surface in a small space with the minimum of transport and to provide a complete processing system. These receive a damp material, resize while in process by granulators or rolls, finish dry, cool, and deliver to packaging or tableting. The applications are primarily in the fine-chemical, food, and pharmaceutical manufacturing fields.

The **Mix-R-Step** type in Fig. 11-39*e* is an adaptation of a vibratory conveyor. It features better heat-transfer rates, practically doubling the coefficient values of the standard flat surface and trebling heat-flux values, as the layer depth can be increased from the normal 13

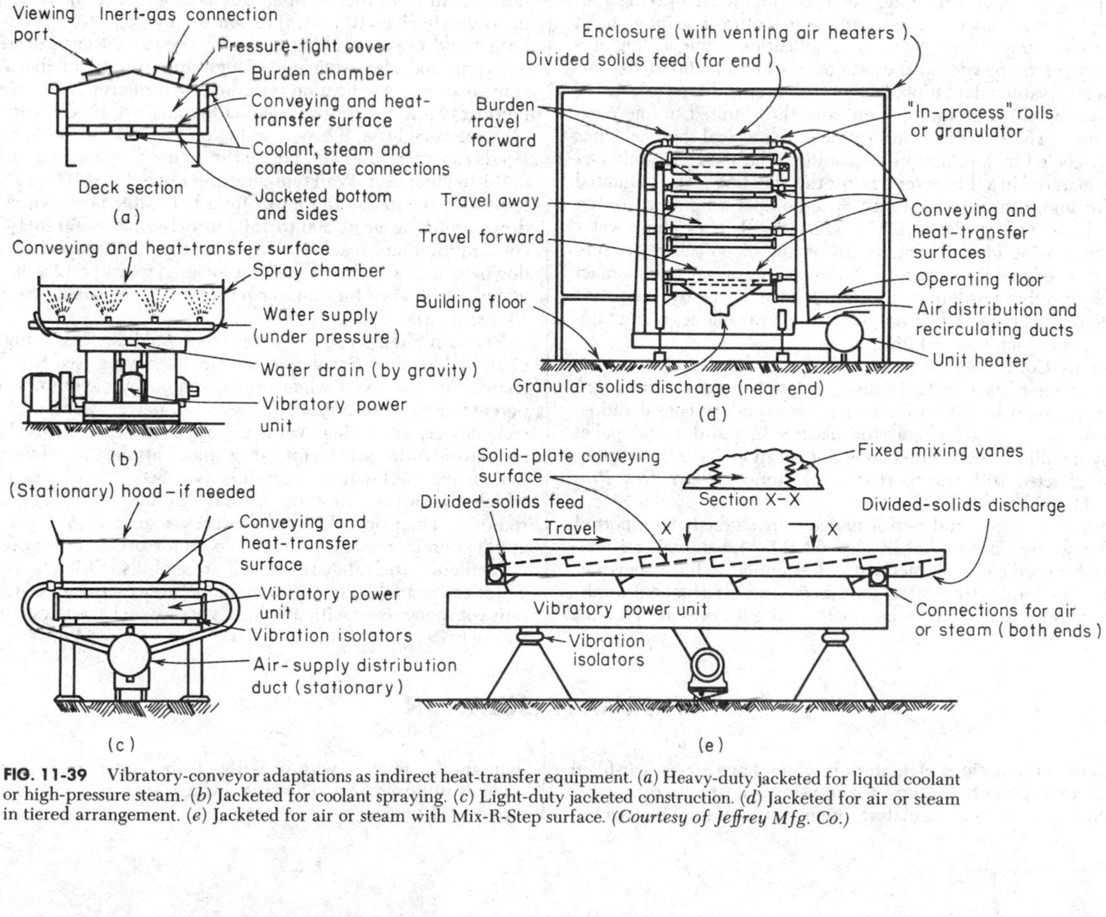

FIG. 11-39 Vibratory-conveyor adaptations as indirect heat-transfer equipment. (*a*) Heavy-duty jacketed for liquid coolant or high-pressure steam. (*b*) Jacketed for coolant spraying. (*c*) Light-duty jacketed construction. (*d*) Jacketed for air or steam in tiered arrangement. (*e*) Jacketed for air or steam with Mix-R-Step surface. *(Courtesy of Jeffrey Mfg. Co.)*

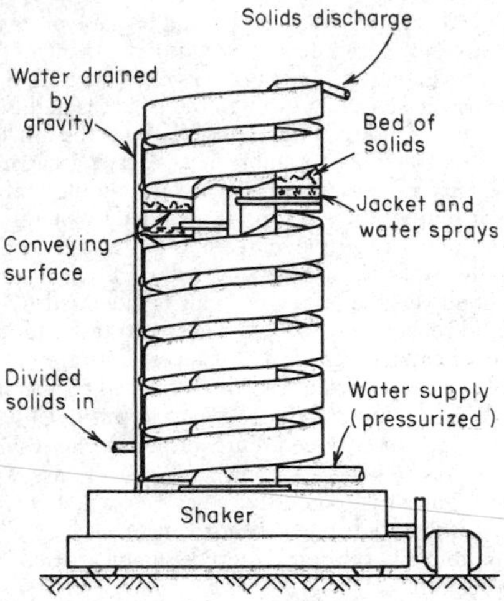

FIG. 11-40 Elevator type as heat-transfer equipment. *(Courtesy of Carrier Conveyor Corp.)*

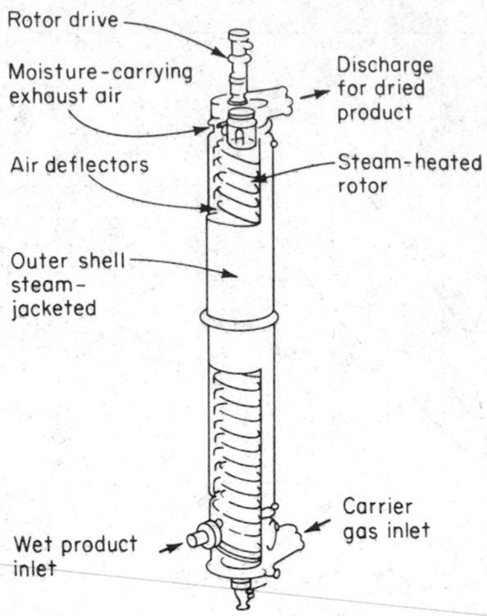

FIG. 11-41 A pneumatic-transport adaptation for heat-transfer duty. *(Courtesy of Werner & Pfleiderer Corp.)*

to 25 and 32 mm (½ to 1 and 1¼ in). It may be provided on decks jacketed for air, steam, or water spray. It is also often applicable when an infrared heat source is mounted overhead to supplement the indirect or as the sole heat source.

Elevator Devices The **vibratory elevating-spiral** type (Fig. 11-40) adapts divided-solids-elevating material-handling equipment to heat-transfer service. It features a large heat-transfer area over a small floor space and employs a reciprocating shaker motion to effect transport. Applications, layer depth, and capacities are restricted, as burdens must be of such "body" character as to convey uphill by the microhopping-transport principle. The type lacks self-emptying ability. Complete washdown and cleaning is a feature not inherent in any other elevating device. A typical application is the cooling of a low-density plastic powder at the rate of 544 kg/h (1200 lb/h).

Another elevator adaptation is that for a **spiral-type elevating** device developed for ground cement and thus limited to fine powdery burdens. The spiral operates inside a cylindrical shell, which is externally cooled by a falling film of water. The spiral not only elevates the material in a thin layer against the wall but keeps it agitated to achieve high heat-transfer rates. Specific operating data are not available [*Chem. Eng. Prog.*, **68**(7), 113 (1968)]. The falling-water film, besides being ideal thermally, by virtue of no jacket pressure very greatly reduces the hazard that the cooling water may contact the water-sensitive burden in process. Surfaces wet by water are accessible for cleaning. A fair range of sizes is available, with material-handling capacities to 60 tons/h.

Pneumatic-Conveying Devices See Sec. 7 for descriptions, ratings, and design factors on these devices. Use is primarily for transport purposes, and heat transfer is a very secondary consideration. Applications have largely been for plastics in powder and pellet forms. By modifications, needed cooling operations have been simultaneously effected with transport to stock storage [*Plast. Des. Process.*, 28 (December 1968)].

Heat-transfer aspects and performance were studied and reported on by Depew and Farbar (ASME Pap. 62-HT-14, September 1962). Heat-transfer coefficient characteristics are similar to those shown in Sec. 10 for the indirectly heated fluid bed. Another frequent application on plastics is a small, rather incidental but necessary amount of drying required for plastic pellets and powders on receipt when shipped in bulk to the users. Pneumatic conveyors modified for heat transfer can handle this readily.

A pneumatic-transport device designed primarily for heat-sensitive products is shown in Fig. 11-41. This was introduced into the United States after 5 years' use in Europe [*Chem. Eng.*, **76,** 54 (June 16, 1969)].

Both the shell and the rotor carry steam as a heating medium to effect indirect transfer as the burden briefly contacts those surfaces rather than from the transport air, as is normally the case. The rotor turns slowly (1 to 10 r/min) to control, by deflectors, product distribution and prevent caking on walls. The carrier gas can be inert, as nitrogen, and also recycled through appropriate auxiliaries for solvent recovery. Application is limited to burdens that (1) are fine and uniformly grained for the pneumatic transport, (2) dry very fast, and (3) have very little, if any, sticking or decomposition characteristics. Feeds can carry 5 to 100 percent moisture (dry basis) and discharge at 0.1 to 2 percent. Wall temperatures range from 100 to 170°C (212 to 340°F) for steam and lower for a hot-water-heat source. Pressure drops are in order of 500 to 1500 mmH_2O (20 to 60 inH_2O). Steam consumption approaches that of a contractive-mechanism dryer down to a low value of 2.9 kg steam/kg water (2.9 lb steam/lb water). Available burden capacities are 91 to 5900 kg/h (200 to 13,000 lb/h).

Vacuum-Shelf Types These are very old devices, being a version of the table type. Early-day use was for drying (see Sec. 20). Heat transfer is slow even when supplemented by vacuum, which is 90 percent or more of present-day use. The newer vacuum blender and cone devices are taking over many applications. The slow heat-transfer rate is quite satisfactory in a major application, freeze drying, which is a sublimation operation (see Sec. 17 for description) in which the water must be retained in the solid state during its removal. Then slow diffusional processes govern. Another extensive application is freezing packaged foods for preservation purposes.

Available sizes range from shelf areas of 0.4 to 67 m^2 (4 to 726 ft^2). These are available in several manufacturers' standards, either as system components or with auxiliary gear as packaged systems.

THERMAL INSULATION

Materials or combinations of materials which have air- or gas-filled pockets or void spaces that retard the transfer of heat with reasonable effectiveness are thermal insulators. Such materials may be particulate and/or fibrous, with or without binders, or may be assembled, such as multiple heat-reflecting surfaces that incorporate air- or gas-filled void spaces.

TABLE 11-9 Thicknesses of Piping Insulation

in / mm	Outer diameter		Insulation, nominal thickness: 1 / 25		1½ / 38		2 / 51		2½ / 64		3 / 76		3½ / 89		4 / 102	
	Approximate wall thickness															
Nominal iron-pipe size, in	in	mm	in	mm	in	mm	in	mm	in	mm	in	mm	in	mm	in	mm
½	0.84	21	1.01	26	1.57	40	2.07	53	2.88	73	3.38	86	3.88	99	4.38	111
¾	1.05	27	0.90	23	1.46	37	1.96	50	2.78	71	3.28	83	3.78	96	4.28	109
1	1.32	33	1.08	27	1.58	40	2.12	54	2.64	67	3.14	80	3.64	92	4.14	105
1¼	1.66	42	0.91	23	1.66	42	1.94	49	2.47	63	2.97	75	3.47	88	3.97	101
1½	1.90	48	1.04	26	1.54	39	2.35	60	2.85	72	3.35	85	3.85	98	4.42	112
2	2.38	60	1.04	26	1.58	40	2.10	53	2.60	66	3.10	79	3.60	91	4.17	106
2½	2.88	73	1.04	26	1.86	47	2.36	60	2.86	73	3.36	85	3.92	100	4.42	112
3	3.50	89	1.02	26	1.54	39	2.04	52	2.54	65	3.04	77	3.61	92	4.11	104
3½	4.00	102	1.30	33	1.80	46	2.30	58	2.80	71	3.36	85	3.86	98	4.36	111
4	4.50	114	1.04	26	1.54	39	2.04	52	2.54	65	3.11	79	3.61	92	4.11	104
4½	5.00	127	1.30	33	1.80	46	2.30	58	2.86	73	3.36	85	3.86	98	4.48	114
5	5.56	141	0.99	25	1.49	38	1.99	51	2.56	65	3.06	78	3.56	90	4.18	106
6	6.62	168	0.96	24	1.46	37	2.02	51	2.52	64	3.02	77	3.65	93	4.15	105
7	7.62	194	. . .	. .	1.52	39	2.02	51	2.52	64	3.15	80	3.65	93	4.15	105
8	8.62	219	. . .	. .	1.52	39	2.02	51	2.65	67	3.15	80	3.65	93	4.15	105
9	9.62	244	. . .	. .	1.52	39	2.15	55	2.65	67	3.15	80	3.65	93	4.15	105
10	10.75	273	. . .	. .	1.58	40	2.08	53	2.58	66	3.08	78	3.58	91	4.08	104
11	11.75	298	. . .	. .	1.58	40	2.08	53	2.58	66	3.08	78	3.58	91	4.08	104
12	12.75	324	. . .	. .	1.58	40	2.08	53	2.58	66	3.08	78	3.58	91	4.08	104
14	14.00	356	. . .	. .	1.46	37	1.96	50	2.46	62	2.96	75	3.46	88	3.96	101
Over 14, up to and including 36			. . .	. .	1.46	37	1.96	50	2.46	62	2.96	75	3.46	88	3.96	101

The ability of a material to retard the flow of heat is expressed by its thermal **conductivity** (for unit thickness) or **conductance** (for a specific thickness). Low values for thermal conductivity or conductance (or high thermal resistivity or resistance value) are characteristics of thermal insulation.

Heat is transferred by radiation, conduction, and convection. Radiation is the primary mode and can occur even in a vacuum. The amount of heat transferred for a given area is relative to the temperature differential and emissivity from the radiating to the absorbing surface. Conduction is due to molecular motion and occurs within gases, liquids, and solids. The tighter the molecular structure, the higher the rate of transfer. As an example, steel conducts heat at a rate approximately 600 times that of typical thermal-insulation materials. Convection is due to mass motion and occurs only in fluids. The prime purpose of a thermal-insulation system is to minimize the amount of heat transferred.

INSULATION MATERIALS

Materials Thermal insulations are produced from many materials or combinations of materials in various forms, sizes, shapes, and thickness. The most commonly available materials fall within the following categories:

Fibrous or cellular—mineral. Alumina, asbestos, glass, perlite, rock, silica, slag, or vermiculite.

Fibrous or cellular—organic. Cane, cotton, wood, and wood bark (cork).

Cellular organic plastics. Elastomer, polystyrene, polyisocyanate, polyisocyanurate, and polyvinyl acetate.

Cements. Insulating and/or finishing.

Heat-reflecting metals (reflective). Aluminum, nickel, stainless steel.

Available forms. Blanket (felt and batt), block, cements, loose fill, foil and sheet, formed or foamed in place, flexible, rigid, and semirigid.

The actual thicknesses of piping insulation differ from the nominal values. Dimensional data of ASTM Standard C585 appear in Table 11-9.

Thermal Conductivity (*K* Factor) Depending on the type of insulation, the thermal conductivity (*K* factor) can vary with age, manufacturer, moisture content, and temperature. Typical published values are shown in Fig. 11-42. Mean temperature is equal to the arithmetic average of the temperatures on both sides of the insulating material.

Actual system heat loss (or gain) will normally exceed calculated values because of projections, axial and longitudinal seams, expansion-contraction openings, moisture, workers' skill, and physical abuse.

Finishes Thermal insulations require an external covering (finish) to provide protection against entry of water or process fluids, mechanical damage, and ultraviolet degradation of foamed materials. In some cases the finish can reduce the flame-spread rating and/or provide fire protection.

The finish may be a coating (paint, asphaltic, resinous, or polymeric), a membrane (coated felt or paper, metal foil, or laminate of plastic, paper, foil or coatings), or sheet material (fabric, metal, or plastic).

Finishes for systems operating below 2°C (35°F) must be sealed and retard vapor transmission. Those from 2°C (35°F) through 27°C (80°F) should retard vapor transmission (to prevent surface condensation), and those above 27°C (80°F) should prevent water entry and allow moisture to escape.

Metal finishes are more durable, require less maintenance, reduce heat loss, and, if uncoated, increase the surface temperature on hot systems.

SYSTEM SELECTION

A combination of insulation and finish produces the thermal-insulation system. Selection of these components depends on the purpose for which the system is to be used. No single system performs satisfactorily from the cryogenic through the elevated-temperature range. Systems operating below freezing have a low vapor pressure, and atmospheric moisture is pushed into the insulation system, while the reverse is true for hot systems. Some general guidelines for system selection follow.

Cryogenic [−273 to −101°C (−459 to −150°F)] High Vacuum This technique is based on the Dewar flask, which is a double-walled vessel with reflective surfaces on the evacuated side to

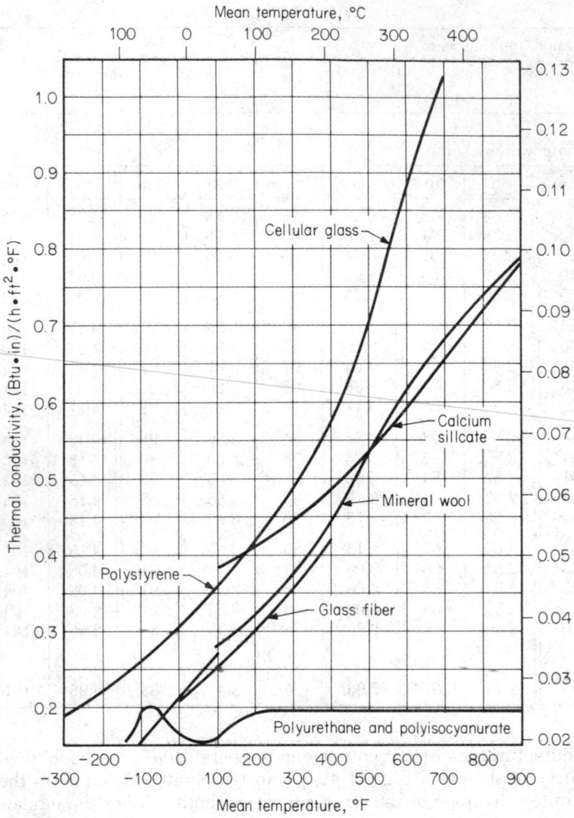

FIG. 11-42 Thermal conductivity of insulating materials.

reduce radiation losses. Figure 11-43 shows a typical laboratory-size Dewar. Figure 11-44 shows a semiportable type. Radiation losses can be further reduced by filling the cavity with powders such as perlite or silica prior to pulling the vacuum.

Multilayer Multilayer systems consist of series of radiation-reflective shields of low emittance separated by fillers or spacers of very low conductance and exposed to a high vacuum.

Foamed or Cellular Cellular plastics such as polyurethane and polystyrene do not hold up or perform well in the cryogenic temperature range because of permeation of the cell structure by water vapor, which in turn increases the heat-transfer rate. Cellular glass holds up better and is less permeable.

Low Temperature [−101 to −1°C (−150 to +30°F)] Cellular glass, glass fiber, polyurethane foam, and polystyrene foam are frequently used for this service range. A vapor-retarder finish with a perm rating less than 0.02 is required. In addition, it is good practice to coat all contact surfaces of the insulation with a vapor-retardant mastic to prevent moisture migration when the finish is damaged or is not properly maintained. Closed-cell insulation should not be relied on as the vapor retarder. Hairline cracks can develop, cells can break down, glass-fiber binders are absorbent, and moisture can enter at joints between all materials.

Moderate and High Temperature [over 2°C (36°F)] Cellular or fibrous materials are normally used. See Fig. 11-45 for nominal temperature range. Nonwicking insulation is desirable for systems operating below 100°C (212°F).

Other Considerations **Autoignition** can occur if combustible fluids are absorbed by wicking-type insulations. **Chloride stress corrosion** of austenitic stainless steel can occur when chlorides are concentrated on metal surfaces at or above approximately 60°C (140°F). The chlorides can come from sources other than the insulation. Some calcium silicates are formulated to exceed the requirements of the MIL-I-24244A specification. **Fire resistance** of insulations varies widely. Calcium silicate, cellular glass, glass fiber, and mineral wool are fire-resistant but do not perform equally under actual fire conditions. A steel jacket provides protection, but aluminum does not.

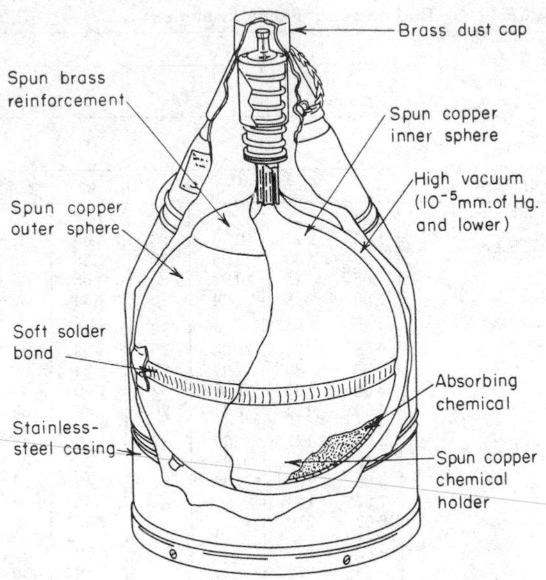

FIG. 11-43 Dewar flask.

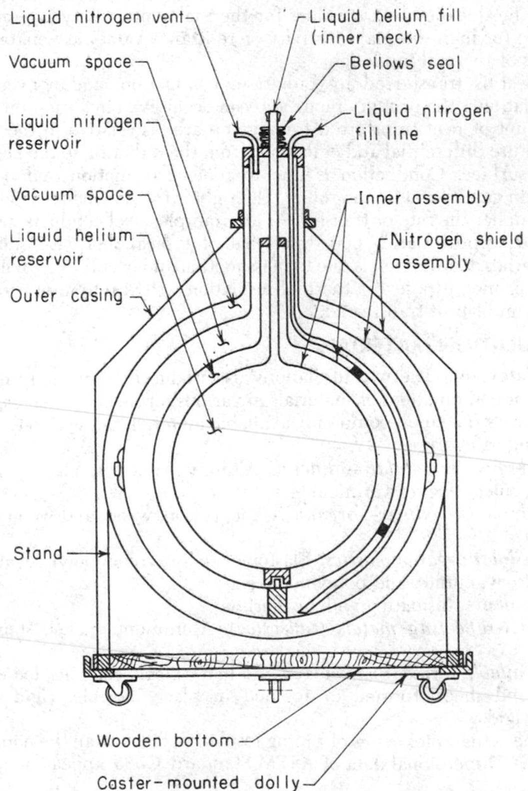

FIG. 11-44 Hydrogen bottle.

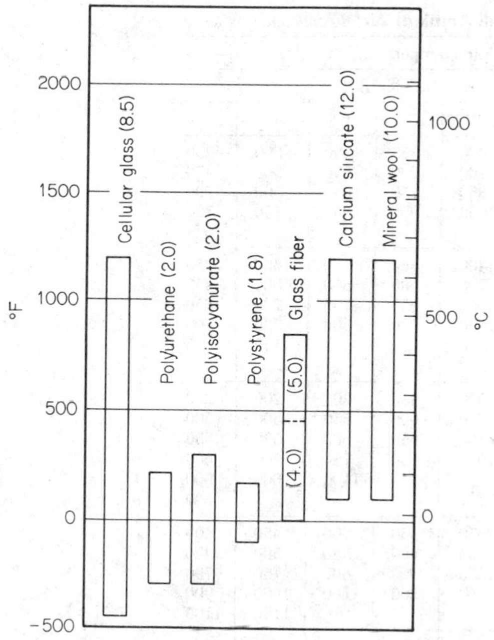

FIG. 11-45 Insulating materials and applicable temperature ranges.

Traced pipe performs better with a nonwicking insulation which has low thermal conductivity. **Underground systems** are very difficult to keep dry permanently. Methods of insulation include factory-preinsulated pouring types and conventionally applied types. **Corrosion** can occur under wet insulation. A protective coating, applied directly to the metal surface, may be required.

ECONOMIC THICKNESS OF INSULATION

Optimal economic insulation thickness may be determined by various methods. Two of these are the minimum-total-cost method and the incremental-cost method (or marginal-cost method). The minimum-total-cost method involves the actual calculations of lost energy and insulation costs for each insulation thickness. The thickness producing the lowest total cost is the optimal economic solution. The optimum thickness is determined to be the point where the last dollar invested in insulation results in exactly $1 in energy-cost savings ("ETI—Economic Thickness for Industrial Insulation," Conservation Pap. 46, Federal Energy Administration, August 1976). The incremental-cost method provides a simplified and direct solution for the least-cost thickness.

The total-cost method does *not* in general provide a satisfactory means for making most insulation investment decisions, since an economic return on investment is required by investors and the method does not properly consider this factor. Return on investment is considered by Rubin ("Piping Insulation—Economics and Profits," in *Practical Considerations in Piping Analysis*, ASME Symposium, vol. 69, 1982, pp. 27–46). The incremental method used in this reference requires that each incremental ½ in of insulation provide the predetermined return on investment. The minimum thickness of installed insulation is used as a base for calculations. The incremental installed capital cost for each additional ½ in of insulation is determined. The energy saved for each increment is then determined. The value of this energy varies directly with the temperature level [e.g., steam at 538°C (1000°F) has a greater value than condensate at 100°C (212°F)]. The final increment selected for use is required either to provide a satisfactory return on investment or to have a suitable payback period.

Recommended Thickness of Insulation Indoor insulation thickness appears in Table 11-10, and outdoor thickness appears in Table 11-11. These selections were based upon calcium silicate insulation with a suitable aluminum jacket. However, the variation in thickness for fiberglass, cellular glass, and rock wool is minimal. Fiberglass is available for maximum temperatures of 260, 343, and 454°C (500, 650, and 850°F). Rock wool, cellular glass, and calcium silicate are used up to 649°C (1200°F).

The tables were based upon the cost of energy at the end of the first year, a 10 percent inflation rate on energy costs, a 15 percent interest cost, and a present-worth pretax profit of 40 percent per annum on the last increment of insulation thickness. Dual-layer insulation was used for 3½-in and greater thicknesses. The tables and a full explanation of their derivation appear in a paper by F. L. Rubin (op. cit.). Alternatively, the selected thicknesses have a payback period on the last nominal ½-in increment of 1.44 years as presented in a later paper by Rubin ["Can You Justify More Piping Insulation?" *Hydrocarbon Process.*, 152–155 (July 1982)].

Example 1 For 24-in pipe at 371°C (700°F) with an energy cost of $4/million Btu, select 2-in thickness for indoor and 2½-in for outdoor locations. [A 2½-in thickness would be chosen at 399°C (750°F) indoors and 3½-in outdoors.]

Example 2 For 16-in pipe at 343°C (650°F) with energy valued at $5/million Btu, select 2½-in insulation indoors [use 3-in thickness at 371°C

TABLE 11-10 Indoor Insulation Thickness, 80°F Still Ambient Air*

		Minimum pipe temperature, °F							
		Energy cost, $/million Btu							
Pipe size, in	Insulation thickness, in	1	2	3	4	5	6	7	8
¾	1½	950	600	550	400	350	300	250	250
	2	....	....	....	1100	1000	900	800	750
	2½	....	....	....	1750	1050	950	850	800
	3	....	....	....	....	....	....	....	1200
1	1½	1200	800	600	500	450	400	350	300
	2	....	....	1200	1000	900	800	700	700
	2½	....	....	....	....	1200	1050	1000	900
	3	....	....	....	....	....	1100	1150	950
1½	1½	1100	750	550	450	400	400	350	300
	2	....	....	1000	850	700	650	600	500
	2½	....	....	....	1050	900	800	750	650
	3	....	....	....	....	....	1150	1100	1000
2	1½	1050	700	500	450	400	350	300	300
	2	....	....	1050	850	750	700	600	600
	2½	....	....	1100	950	1000	750	700	650
	3	....	....	....	1200	1050	950	850	800

TABLE 11-10 Indoor Insulation Thickness, 80°F Still Ambient Air* *(Continued)*

		Minimum pipe temperature, °F							
		Energy cost, $/million Btu							
Pipe size, in	Insulation thickness, in	1	2	3	4	5	6	7	8
3	1½	950	650	500	400	350	300	300	250
	2	....	1100	900	700	600	550	500	450
	2½	....	....	1050	850	750	650	500	500
	3	....	....	....	1050	950	800	750	700
		....	....	....	....	....	....	....	....
4	1½	950	600	500	400	350	300	300	250
	2	....	1100	850	700	600	550	500	450
	2½	....	....	1200	1000	850	750	700	650
	3	....	....	....	1050	900	800	750	700
	3½	....	....	....	....	....	....	1150	1050
		....	....	....	....	....	....	....	
6	1½	600	350	300	250	250	200	200	200
	2	....	1100	850	700	600	550	500	500
	2½	....	....	900	800	650	600	550	550
	3	....	....	1150	1000	850	750	700	600
	3½	....	....	....	....	....	1100	1000	900
	4	....	....	....	....	....	....	....	1200
8	2	....	1000	800	650	550	500	450	400
	2½	....	1050	850	700	600	550	500	450
	3	....	....	....	1050	900	800	750	700
	3½	....	....	....	....	1200	1100	1000	900
	4	....	....	....	....	....	....	1150	1100
10	2	....	1100	850	700	650	550	500	450
	2½	....	1200	900	750	700	600	550	500
	3	....	....	1050	900	750	700	600	550
	3½	....	....	....	....	1200	1050	950	900
	4	....	....	....	....	....	....	....	1200
		....	....	....	....	....	....	....	....
12	2	1150	750	600	500	400	400	350	300
	2½	....	1000	800	650	550	500	450	400
	3	....	....	1200	1000	900	800	700	650
	3½	....	....	....	....	1200	1100	1000	900
	4	....	....	....	....	....	1150	1050	950
	4½	....	....	....	....	....	1200	1100	1000
14	2	1050	650	550	450	400	350	300	300
	2½	....	1000	800	650	550	500	450	400
	3	....	....	1100	950	800	700	650	600
	3½	....	....	....	....	1150	1000	950	850
	4	....	....	....	....	1200	1050	1000	900
	4½	....	....	....	....	....	1200	1100	1000
16	2	950	650	500	400	350	300	300	300
	2½	....	1000	800	700	600	550	500	450
	3	....	1200	950	800	700	600	550	500
	3½	....	....	....	....	1150	1050	950	850
	4	....	....	....	....	1200	1100	1000	900
	4½	....	....	....	....	....	1150	1050	950
18	2	1000	650	500	400	350	350	300	300
	2½	....	950	750	600	550	500	450	400
	3	....	1150	900	750	650	550	500	500
	3½	....	....	....	....	1200	1100	1000	900
	4	....	....	....	....	....	1150	1050	950
	4½	....	....	....	....	....	1200	1100	1000
20	2	1050	700	550	450	400	350	350	300
	2½	....	1000	800	600	550	500	450	400
	3	....	1150	900	750	650	550	500	500
	3½	....	....	....	....	....	1100	1000	950
	4	....	....	....	....	....	1150	1050	1000
	4½	....	....	....	....	....	....	1200	1100
24	2	950	600	500	400	350	300	300	250
	2½	....	1150	900	750	650	550	500	450
	3	....	....	1050	900	750	700	600	550
	3½	....	....	....	....	1100	1000	900	800
	4	....	....	....	....	1150	1050	950	850
	4½	....	....	....	....	....	1150	1050	950

*Aluminum-jacketed calcium silicate insulation with an emissivity factor of 0.05. To convert inches to millimeters, multiply by 25.4; to convert dollars per 1 million British thermal units to dollars per 1 million kilojoules, multiply by 0.948; °C = 5⁄9 (°F − 32).

TABLE 11-11 Outdoor Insulation Thickness, 7.5-mi/h Wind, 60°F Air*

		Minimum pipe temperature, °F							
		Energy cost, $/million Btu							
Pipe size, in	Thickness, in	1	2	3	4	5	6	7	8
¾	1	450	300	250	250	200	200	150	150
	1½	800	500	400	300	250	250	200	200
	2	....	....	1150	950	850	750	700	650
	2½	....	....	1100	1000	900	800	750	700
1	1	400	300	250	200	200	150	150	150
	1½	1000	650	500	400	350	300	300	250
	2	....	....	1100	900	800	700	600	600
	2½	....	....	....	1200	1050	950	850	800
	3	....	....	....	....	1100	1000	900	850
1½	1	350	250	200	200	150	150	150	150
	1½	900	600	450	350	300	300	250	250
	2	....	100	850	700	600	550	500	450
	2½	....	....	1150	950	800	750	700	600
	3	....	....	....	....	1200	1050	1000	900
2	1	350	250	200	150	150	150	150	150
	1½	900	550	450	400	300	300	250	250
	2	....	1150	900	750	650	600	550	500
	2½	....	....	1000	850	750	650	600	550
	3	....	....	....	1050	950	850	750	700
3	1	300	200	150	150	150	150	150	150
	1½	750	500	400	300	250	250	250	200
	2	....	950	750	600	500	450	400	350
	2½	....	1150	950	750	650	600	500	500
	3	....	....	1150	1000	850	750	650	600
	3½	....	....	....	....	....	....	....	1150
4	1	250	200	150	150	150	150	150	150
	1½	750	500	350	300	250	250	200	200
	2	....	950	750	600	500	450	400	350
	2½	....	....	1050	900	700	650	600	550
	3	....	....	1100	950	750	700	650	600
	3½	....	....	....	....	....	1200	1100	1000
6	1	250	150	150	150	150	150	150	150
	1½	450	300	200	200	150	150	150	150
	2	....	900	700	600	500	450	400	350
	2½	....	1050	800	650	600	500	450	400
	3	....	....	1050	900	750	700	600	550
	3½	....	....	....	....	1150	1050	950	850
	4	....	....	....	....	....	....	1200	1150
	4½	....	....	....	....	....	....	....	1200
8	1	250	200	150	150	150	150	150	150
	2	....	850	650	550	450	400	350	350
	2½	....	900	700	600	500	450	400	400
	3	....	....	1100	950	800	750	700	600
	3½	....	....	....	....	1150	1000	950	850
	4	....	....	....	....	....	....	1050	1000
10	2	200	150	150	150	150	150	150	150
	2½	....	1000	800	650	550	500	450	400
	3	....	1200	950	800	700	600	550	500
	3½	....	....	....	....	1100	1000	900	800
	4	....	....	....	....	....	....	1150	1050
	4½	....	....	....	....	....	....	1200	1100
12	1½	250	150	150	150	150	150	150	150
	2	950	600	500	400	350	300	250	250
	2½	....	900	700	550	500	400	400	350
	3	....	....	1100	900	800	700	650	550
	3½	....	....	....	....	1100	1000	900	850
	4	....	....	....	....	1150	1050	950	900
	4½	....	....	....	....	1200	1100	1000	950

TABLE 11-11 Outdoor Insulation Thickness, 7.5-mi/h Wind, 60°F Air* *(Continued)*

		Minimum pipe temperature, °F							
		Energy cost, $/million Btu							
Pipe size, in	Thickness, in	1	2	3	4	5	6	7	8
14	1½	250	150	150	150	150	150	150	150
	2	850	550	400	350	300	250	250	250
	2½	....	850	650	550	500	400	400	400
	3	....	....	1000	850	700	650	550	500
	3½	....	....	....	1200	1000	950	850	800
	4	....	....	....	....	1050	1000	900	850
	4½	....	....	....	....	....	1100	1000	950
		....	....	....	....	....	....	....	....
16	1½	250	150	150	150	150	150	150	150
	2	800	500	350	300	300	250	250	200
	2½	....	900	700	550	500	450	400	350
	3	....	1000	850	700	600	500	450	400
	3½	....	....	....	1200	1000	950	850	800
	4	....	....	....	....	1100	1000	900	850
	4½	....	....	....	....	1150	1000	950	900
		....	....	....	....	....	....	....	....
18	1½	250	150	150	150	150	150	150	150
	2	850	550	400	350	300	250	250	200
	2½	....	800	650	500	450	400	350	350
	3	....	1000	800	650	550	500	450	400
	3½	....	....	....	....	1100	1000	900	850
20	1½	150	150	150	150	150	150	150	150
	2	900	550	450	350	300	300	250	250
	2½	....	850	650	550	450	400	350	350
	3	....	1000	800	650	550	500	450	400
	3½	....	....	....	....	1150	1050	950	900
	4	....	....	....	....	1200	1100	1000	950
	4½	....	....	....	....	....	1200	1100	1050
		....	....	....	....	....	....	....	....
24	1½	150	150	150	150	150	150	150	150
	2	800	500	400	300	250	250	200	200
	2½	....	950	750	650	550	500	450	400
	3	....	1150	950	750	650	600	550	500
	3½	....	....	....	1150	1000	900	800	750
	4	....	....	....	1200	1050	950	850	800
	4½	....	....	....	....	....	1050	950	850
	5	....	....	....	....	....	....	....	....

*Aluminum-jacketed calcium silicate insulation with an emissivity factor of 0.05. To convert inches to millimeters, multiply by 25.4; to convert miles per hour to kilometers per hour, multiply by 1.609; and to convert dollars per 1 million British thermal units to dollars per 1 million kilojoules, multiply by 0.948; °C = $\frac{5}{9}$ (°F − 32).

(700°F)]. Outdoors choose 3-in insulation [use 3½-in dual-layer insulation at 538°C (1000°F)].

Example 3 For 12-in pipe at 593°C (1100°F) with an energy cost of $6/million Btu, select 3½-in thickness for an indoor installation and 4½-in thickness for an outdoor installation.

INSTALLATION PRACTICE

Pipe Depending on diameter, pipe is insulated with cylindrical half, third, or quarter sections or with flat segmental insulation. Fittings and valves are insulated with preformed insulation covers or with individual pieces cut from sectional straight pipe insulation.

Method of Securing Insulation with factory-applied jacketing may be secured with adhesive on the overlap, staples, tape, or wire, depending on the type of jacket and the outside diameter. Insulation which has a separate jacket is wired or banded in place before the jacket (finish) is applied.

Double Layer Pipe expansion is a significant factor at temperatures above 600°F (316°C). Above this temperature, insulation should be applied in a double layer with all joints staggered to prevent excessive heat loss and high surface temperature at joints opened by pipe expansion. This procedure also minimizes thermal stresses in the insulation.

Finish Covering for cylindrical surfaces ranges from asphalt-saturated or saturated and coated organic and asbestos paper, through laminates of such papers and plastic films or aluminum foil, to medium-gauge aluminum, galvanized steel, or stainless steel. Fittings and irregular surfaces may be covered with fabric-reinforced mastics or preformed metal or plastic covers. Finish selection depends on function and location. Vapor-barrier finishes may be in sheet form or a mastic, which may or may not require reinforcing, depending on the methd of application, and additional protection may be required to prevent mechanical abuse and/or provide fire resistance. Criteria for selecting other finishes should include protection of insulation against water entry, mechanical abuse, or chemical attack. Appearance, life-cycle cost, and fire resistance may also be determining factors. Finish may be secured with tape, adhesive, bands, or screws. Fasteners which will penetrate vapor-retarder finishes should not be used.

Tanks, Vessels, and Equipment Flat, curved, and irregular surfaces such as tanks, vessels, boilers, and breechings are normally insulated with flat blocks, beveled lags, curved segments, blankets, or spray-applied insulation. Since no general procedure can apply to all materials and conditions, it is important that manufacturers' specifications and instructions be followed for specific insulation applications.

Method of Securing On small-diameter cylindrical vessels, the insulation may be secured by banding around the circumference. On larger cylindrical vessels, banding may be supplemented with angle-iron ledges to support the insulation and prevent slipping. On large flat and cylindrical surfaces, banding or wiring may be supplemented with various types of welded studs or pins. Breather springs may be required with bands to accommodate expansion and contraction.

Finish The materials are the same as for pipe and should satisfy the same criteria. Breather springs may be required with bands.

ADDITIONAL REFERENCES: *ASHRAE Handbook and Product Directory: Fundamentals*, American Society of Heating, Refrigerating and Air Conditioning Engineers, Atlanta, 1981. Turner and Malloy, *Handbook of Thermal Insulation Design Economics for Pipes and Equipment*, Krieger, New York, 1980. Turner and Malloy, *Thermal Insulation Handbook*, McGraw-Hill, New York, 1981.

Section 12

Psychrometry, Evaporative Cooling, Refrigeration, and Cryogenic Processes

Eno Bagnoli, M.S., *Senior Research Associate, E. I. du Pont de Nemours & Co.; Member, American Institute of Chemical Engineers. (Psychrometry)*

Robert W. Norris, B.S., P.E., *President, Robert W. Norris and Associates, Inc; Fellow, American Society of Heating, Refrigerating and Air-Conditioning Engineers. (Evaporative Cooling and Refrigeration)*

Thomas M. Flynn, Ph.D., P.E., *Consultant, Cryogenic Engineering, Formerly National Bureau of Standards, Boulder, Colorado, Laboratories; Member, American Institute of Chemical Engineers. (Cryogenic Processes)*

Klaus D. Timmerhaus, Ph.D., P.E., *Associate Dean of Engineering and Director of Engineering Research Center, University of Colorado, Boulder, Colorado; Member, American Institute of Chemical Engineers, American Society for Engineering Education, American Association for the Advancement of Science, American Astronautical Society, National Academy of Engineering, Austrian Academy of Science, International Institute of Refrigeration, American Society of Heating, Refrigerating and Air Conditioning Engineers, Sigma Xi, The Research Society. (Cryogenic Processes)*

PSYCHROMETRY

EVAPORATIVE COOLING

REFRIGERATION

CRYOGENIC PROCESSES

PSYCHROMETRY

GENERAL REFERENCES: Brown et al., *Unit Operations*, Wiley, New York, 1950. Coulson and Richardson, *Chemical Engineering*, vol. 1, McGraw-Hill, New York, 1955. *Handbook of Fundamentals*, American Society of Heating, Refrigerating and Air-Conditioning Engineers, New York, 1967. Penman, *Humidity*, Institute of Physics Monographs, London, 1955. Quinn, "Humidity: The Neglected Parameter," *Test Eng.* (July 1968). Sherwood and Pigford, *Absorption and Extraction*, 2d ed., McGraw-Hill, New York, 1952. Treybal, *Mass-Transfer Operations*, McGraw-Hill, New York, 1955. Walker, Lewis, McAdams, and Gilliland, *Principles of Chemical Engineering*, 3d ed., McGraw-Hill, New York, 1937. Wexler, *Humidity and Moisture*, vol. I, Reinhold, New York, 1965. Wexler and Bombacher, NBS Circ. 512, September 1951. Zimmerman and Lavine, Psychrometric Charts and Tables, Industrial Research Service, 2d ed., Dover, N.H., 1964.

Psychrometry is concerned with determination of the properties of gas-vapor mixtures. The air-water vapor system is by far the system most commonly encountered.

Principles involved in determining the properties of other systems are the same as with air-water vapor, with one major exception. Whereas the psychrometric ratio (ratio of heat-transfer coefficient to product of mass-transfer coefficient and humid heat, terms defined in the following subsection) for the air-water system can be taken as 1, the ratio for other systems in general does not equal 1. This has the effect of making the adiabatic-saturation temperature different from the wet-bulb temperature. Thus, for systems other than air-water vapor, calculation of psychrometric and drying problems is complicated by the necessity for point-to-point calculation of the temperature of the evaporating surface. For example, for the air-water system the temperature of the evaporating surface will be constant during the constant rate drying period even though temperature and humidity of the gas stream change. For other systems, the temperature of the evaporating surface would change.

TERMINOLOGY

Terminology and relationships pertinent to psychrometry are:

Absolute humidity H equals the pounds of water vapor carried by 1 lb of dry air. If ideal-gas behavior is assumed, $H = M_w p/[M_a(P - p)]$, where M_w = molecular weight of water; M_a = molecular weight of air; p = partial pressure of water vapor, atm; and P = total pressure, atm.

When the partial pressure p of water vapor in the air at a given temperature equals the vapor pressure of water p_s at the same temperature, the air is saturated and the absolute humidity is designated the **saturation humidity** H_s.

Percentage absolute humidity (percentage saturation) is defined as the ratio of absolute humidity to saturation humidity and is given by $100\ H/H_s = 100p(P - p_s)/[p_s(P - p)]$.

Percentage relative humidity is defined as the partial pressure of water vapor in air divided by the vapor pressure of water at the given temperature. Thus RH = $100p/p_s$.

Dew point, or **saturation temperature,** is the temperature at which a given mixture of water vapor and air is saturated, i.e., the temperature at which water exerts a vapor pressure equal to the partial pressure of water vapor in the given mixture.

Humid heat c_s is the heat capacity of 1 lb of dry air and the moisture it contains. For most engineering calculations, $c_s = 0.24 + 0.45H$, where 0.24 and 0.45 are the heat capacities of dry air and water vapor, respectively, and both are assumed constant.

Humid volume is the volume in cubic feet of 1 lb of dry air and the water vapor it contains.

Saturated volume is the humid volume when the air is saturated.

Wet-bulb temperature is the dynamic equilibrium temperature attained by a water surface when the rate of heat transfer to the surface by convection equals the rate of mass transfer away from the surface. At equilibrium, if negligible change in the dry-bulb temperature is assumed, a heat balance on the surface is

$$k_g\lambda(p_s - p) = h_c(t - t_w) \qquad (12\text{-}1a)$$

where k_g = mass-transfer coefficient, lb/(h·ft²·atm); λ = latent heat of vaporization, Btu/lb; p_s = vapor pressure of water at wet-bulb temperature, atm; p = partial pressure of water vapor in the environment, atm; h_c = heat-transfer coefficient, Btu/(h·ft²·°F); t = temperature of air-water vapor mixture (dry-bulb temperature), °F; and t_w = wet-bulb temperature, °F. Under ordinary conditions the partial pressure and vapor pressure are small relative to the total pressure, and the wet-bulb equation can be written in terms of humidity differences as

$$H_s - H = (h_c/\lambda k')(t - t_w) \qquad (12\text{-}1b)$$

where k' = lb/(h·ft²) (unit humidity difference) = $(M_a/M_w)k_g = 1.6k_g$.

Adiabatic-Saturation Temperature, or Constant-Enthalpy Lines If a stream of air is intimately mixed with a quantity of water at a temperature t_s in an adiabatic system, the temperature of the air will drop and its humidity will increase. If t_s is such that the air leaving the system is in equilibrium with the water, t_s will be the adiabatic-saturation temperature, and the line relating the temperature and humidity of the air is the adiabatic-saturation line. The equation for the adiabatic-saturation line is

$$H_s - H = (c_s/\lambda)(t - t_s) \qquad (12\text{-}2)$$

RELATION BETWEEN WET-BULB AND ADIABATIC-SATURATION TEMPERATURES

Experimentally it has been shown that for air-water systems the value of $h_c/k'c_s$, the **psychrometric ratio,** is approximately equal to 1. Under these conditions the wet-bulb temperatures and adiabatic-saturation temperatures are substantially equal and can be used interchangeably. The difference between adiabatic-saturation temperature and wet-bulb temperature increases with increasing humidity, but this effect is unimportant for most engineering calculations.

For systems other than air-water vapor, the value of $h_c/k'c_s$ may differ appreciably from unity, and the wet-bulb and adiabatic-saturation temperatures are no longer equal. For these systems the psychrometric ratio may be obtained by determining h_c/k' from heat- and mass-transfer analogies such as the Chilton-Colburn analogy [*Ind. Eng. Chem.*, **26,** 1183 (1934)]. For low humidities this analogy gives

$$\frac{h_c}{k'} = c_s\left(\frac{\mu/\rho D_v}{c_s\mu/k}\right)^{2/3} \qquad (12\text{-}3)$$

where c_s = humid heat, Btu/(lb·°F); μ = viscosity, lb/(ft·h); ρ = density, lb/ft³; D_v = diffusivity, ft²/h; and k = thermal conductivity, Btu/(h·ft²)(°F/ft). All properties should be evaluated for the gas mixture.

For the case of **flow past cylinders,** such as a wet-bulb thermometer, Bedingfield and Drew [*Ind. Eng. Chem.*, **42,** 1164 (1950)] obtained a correlation for their data on sublimation of cylinders into

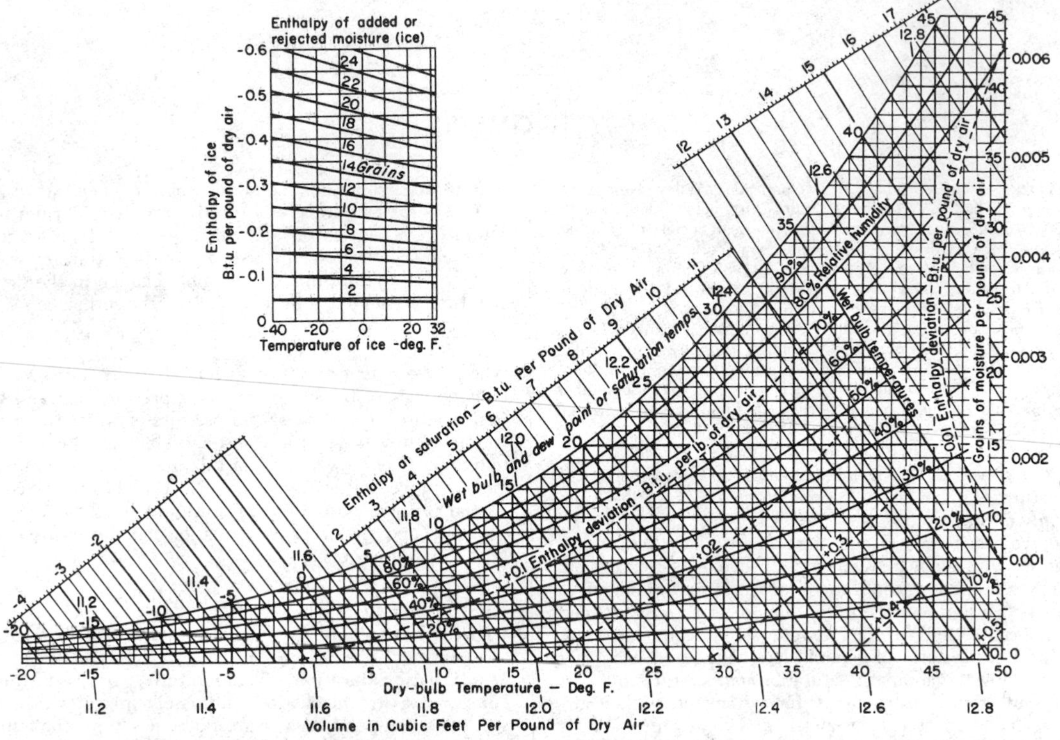

FIG. 12-1 Psychrometric chart—low temperatures. Barometric pressure, 29.92 inHg.

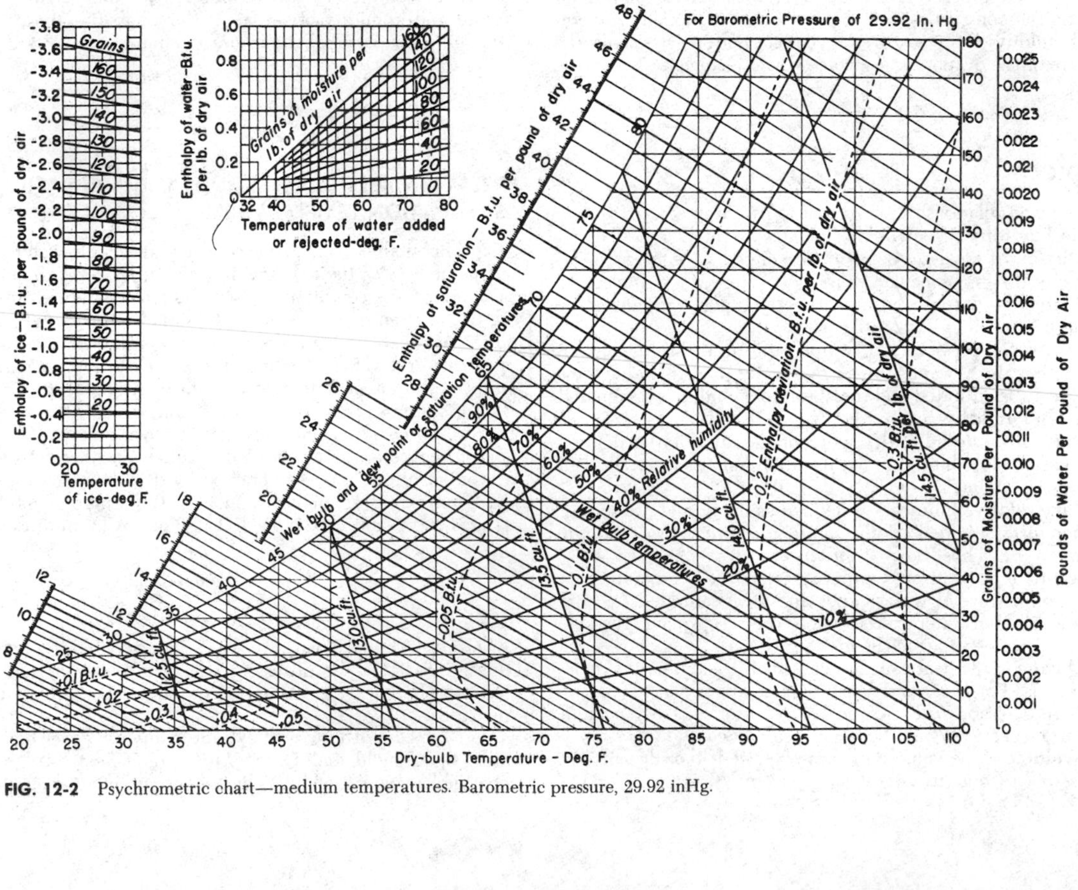

FIG. 12-2 Psychrometric chart—medium temperatures. Barometric pressure, 29.92 inHg.

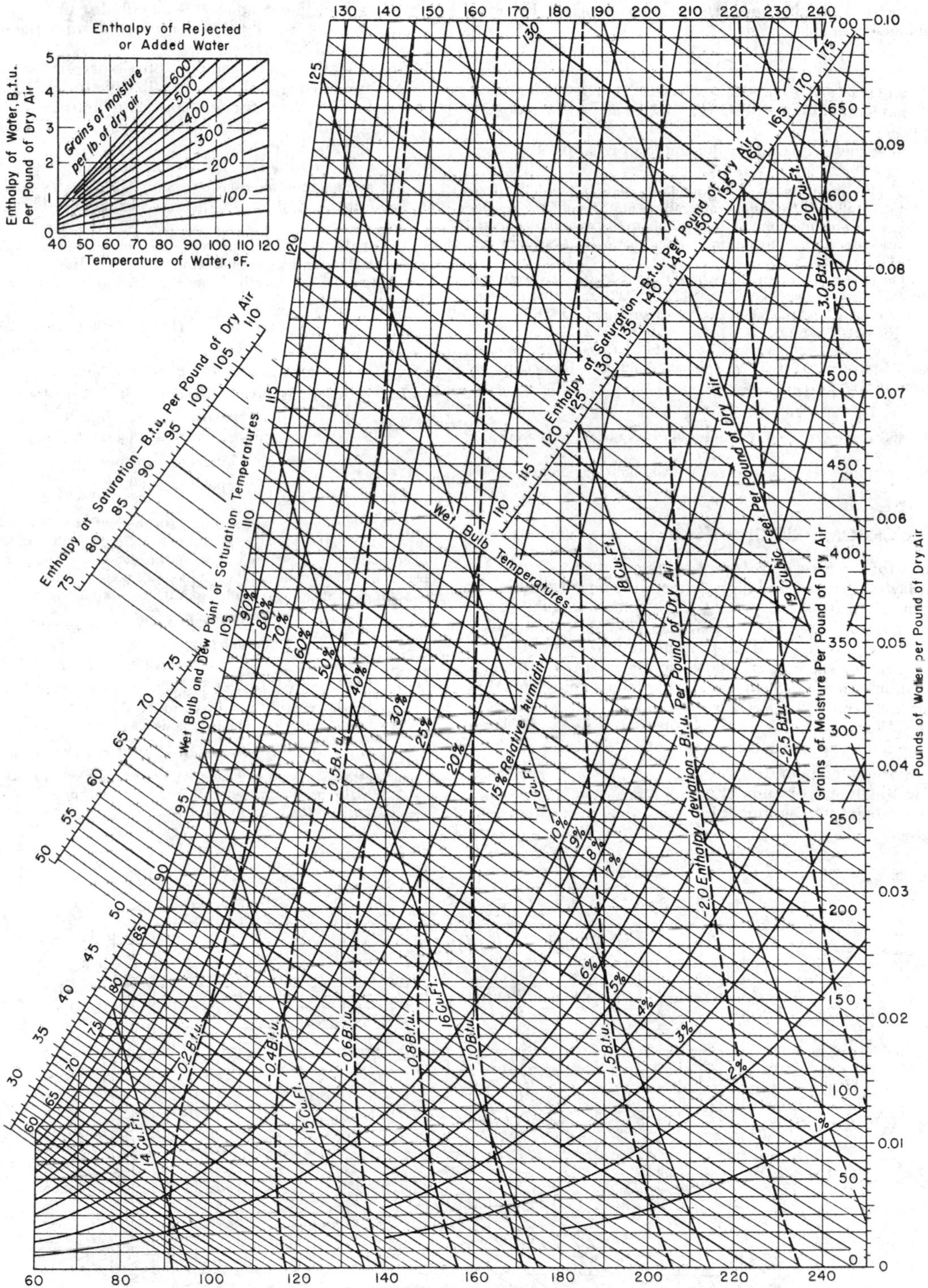

FIG. 12-3 Psychrometric chart—high temperatures. Barometric pressure, 29.92 inHg.

air and for the data of others on wet-bulb thermometers. For wet-bulb thermometers in air they give

$$h_c/k' = 0.294(\mu/\rho D_v)^{0.56} \qquad (12\text{-}4)$$

where the nomenclature is identical to that in Eq. (12-3). For evaporation into gases other than air, Eq. (12-3) with an exponent of 0.56 would apply.

Application of these equations is illustrated in Example 1.

Example 1 For the air-water system at atmospheric pressure, the measured values of dry-bulb and wet-bulb temperatures are 85 and 72°F respectively. Determine the absolute humidity and compare the wet-bulb temperature and adiabatic-saturation temperature. Assume that h_c/k' is given by Eq. (12-4).

Solution. For relatively dry air the Schmidt number $\mu/\rho D_v$ is 0.60, and from Eq. (12-4) $h_c/k' = 0.294(0.60)^{0.56} = 0.221$. At 72°F the vapor pressure of water is 20.07 mmHg, and the latent heat of vaporization is 1051.6 Btu/lb. From Eq. (12-1*b*), $[20.07/(760 - 20.07)](18/29) - H = (0.221/1051.6)(85 - 72)$, or $H = 0.0140$ lb water/lb dry air. The humid heat is calculated as $c_s = 0.24 + 0.45(0.0140) = 0.246$. The adiabatic-saturation temperature is obtained from Eq. (12-2) as

$$H_s - 0.0140 = (0.246/1051.6)(85 - t_s)$$

Values of H_s and t_s are given by the saturation curve of the psychrometric chart, such as Fig. 12-2. By trial and error, $t_s = 72.1$°F, or the adiabatic-saturation temperature is 0.1°F higher than the wet-bulb temperature.

USE OF PSYCHROMETRIC CHARTS

Three **charts for the air-water vapor system** are given as Figs. 12-1 to 12-3 for low-, medium-, and high-temperature ranges. Figure 12-4 shows a modified Grosvenor chart, which is more familiar to the chemical engineer. These charts are for an absolute pressure of 1 atm. The corrections required at pressures different from atmospheric are given in Table 12-2. Figure 12-5 shows a psychrometric chart for combustion products in air. The thermodynamic properties of moist air are given in Table 12-1.

Examples Illustrating Use of Psychrometric Charts In these examples the following nomenclature is used:

t = dry-bulb temperatures, °F
t_w = wet-bulb temperature, °F
t_d = dew-point temperature, °F
H = moisture content, lb water/lb dry air
ΔH = moisture added to or rejected from the air stream, lb water/lb dry air
h' = enthalpy at saturation, Btu/lb dry air
D = enthalpy deviation, Btu/lb dry air
h = $h' + D$ = true enthalpy, Btu/lb dry air
h_w = enthalpy of water added to or rejected from the system, Btu/lb dry air
q_a = heat added to the system, Btu/lb dry air
q_r = heat removed from system, Btu/lb dry air

Subscripts 1, 2, 3, etc., indicate entering and subsequent states.

Example 2 Find the properties of moist air when the dry-bulb temperature is 80°F and the wet-bulb temperature is 67°F.

Solution. Read directly from Fig. 12-2 (Fig. 12-6 shows the solution diagrammatically).

Moisture content H = 78 gr/lb dry air
= 0.011 lb water/lb dry air
Enthalpy at saturation h' = 31.6 Btu/lb dry air
Enthalpy deviation D = −0.1 Btu/lb dry air
True enthalpy h = 31.5 Btu/lb dry air
Specific volume v = 13.8 ft³/lb dry air
Relative humidity = 51 percent
Dew point t_d = 60.3°F

Example 3 Air is heated by a steam coil from 30°F dry-bulb temperature and 80 percent relative humidity to 75°F dry-bulb temperature. Find the relative humidity, wet-bulb temperature, and dew point of the heated air. Determine the quantity of heat added per pound of dry air.

Solution. Reading directly from the psychrometric chart (Fig. 12-2),

Relative humidity = 15 percent
Wet-bulb temperature = 51.5°F
Dew point = 25.2°F

The enthalpy of the inlet air is obtained from Fig. 12-2 as $h_1 = h_1' + D_1 = 10.1 + 0.06 = 10.16$ Btu/lb dry air; at the exit, $h_2 = h_2' + D_2 = 21.1 - 0.1 = 21$ Btu/lb dry air. The heat added equals the enthalpy difference, or

$$q_a = \Delta h = h_2 - h_1 = 21 - 10.16 = 10.84 \text{ Btu/lb dry air}$$

If the enthalpy deviation is ignored, the heat added q_a is $\Delta h = 21.1 - 10.1 = 11$ Btu/lb dry air, or the result is 1.5 percent high. Figure 12-7 shows the heating path on the psychrometric chart.

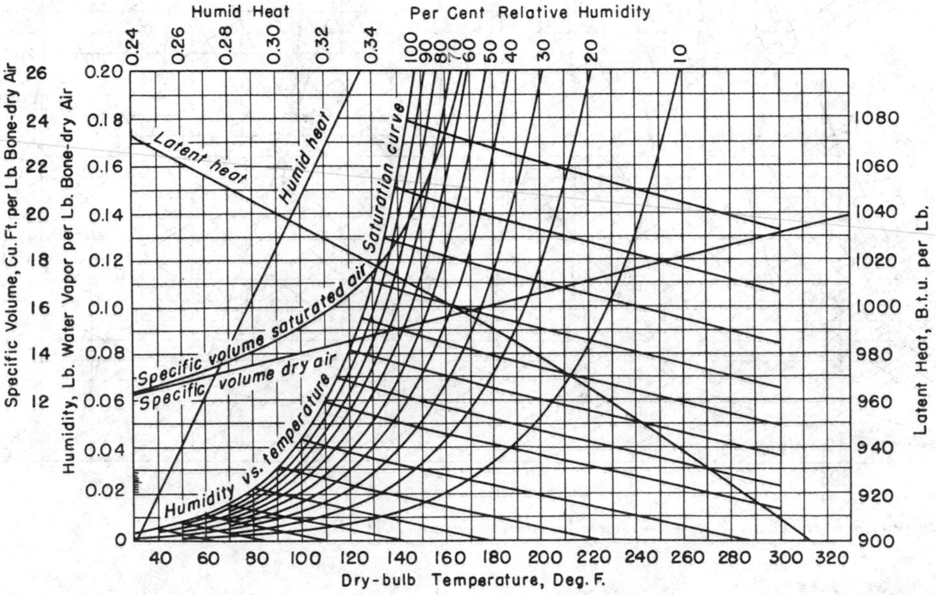

FIG. 12-4 Humidity chart for air-water vapor mixtures.

TABLE 12-1 Thermodynamic Properties of Moist Air (Standard Atmospheric Pressure, 29.921 inHg)

Temp. t, °F.	Saturation humidity $H_s \times 10^8$	Volume, cu. ft./lb. dry air			Enthalpy, B.t.u./lb. dry air			Entropy, B.t.u./(°F.)(lb. dry air)			Condensed water			Temp. t, °F.
		v_a	v_{as}	v_s	h_a	h_{as}	h_s	s_a	s_{as}	s_s	Enthalpy, B.t.u./lb. h_w	Entropy, B.t.u./(lb.)(°F.) s_w	Vapor press., in. Hg $p_s \times 10^6$	
−160	0.2120	7.520	0.000	7.520	−38.504	0.000	−38.504	−0.10300	0.00000	−0.10300	−222.00	−0.4907	0.1009	−160
−155	.3869	7.647	.000	7.647	−37.296	.000	−37.296	−0.09901	.00000	−0.09901	−220.40	−0.4853	.1842	−155
−150	.6932	7.775	.000	7.775	−36.088	.000	−36.088	−0.09508	.00000	−0.09508	−218.77	−0.4800	.3301	−150
−145	1.219	7.902	.000	7.902	−34.881	.000	−34.881	−0.09121	.00000	−0.09121	−217.12	−0.4747	.5807	−145
−140	2.109	8.029	.000	8.029	−33.674	.000	−33.674	−0.08740	.00000	−0.08740	−215.44	−0.4695	1.004	−140
−135	3.586	8.156	.000	8.156	−32.468	.000	−32.468	−0.08365	.00000	−0.08365	−213.75	−0.4642	1.707	−135
−130	6.000	8.283	.000	8.283	−31.262	.000	−31.262	−0.07997	.00000	−0.07997	−212.03	−0.4590	2.858	−130
	$H_s \times 10^7$												$p_s \times 10^5$	
−125	0.9887	8.411	.000	8.411	−30.057	.000	−30.057	−0.07634	.00000	−0.07634	−210.28	−0.4538	0.4710	−125
−120	1.606	8.537	.000	8.537	−28.852	.000	−28.852	−0.07277	.00000	−0.07277	−208.52	−0.4485	.7653	−120
−115	2.571	8.664	.000	8.664	−27.648	.000	−27.648	−0.06924	.00000	−0.06924	−206.73	−0.4433	1.226	−115
−110	4.063	8.792	.000	8.792	−26.444	.000	−26.444	−0.06577	.00000	−0.06577	−204.92	−0.4381	1.939	−110
−105	6.340	8.919	.000	8.919	−25.240	.001	−25.239	−0.06234	.00000	−0.06234	−203.09	−0.4329	3.026	−105
−100	9.772	9.046	.000	9.046	−24.037	.001	−24.036	−0.05897	.00000	−0.05897	−201.23	−0.4277	4.666	−100
	$H_s \times 10^6$												$p_s \times 10^4$	
−95	1.489	9.173	.000	9.173	−22.835	.002	−22.833	−0.05565	.00000	−0.05565	−199.35	−0.4225	0.7111	−95
−90	2.242	9.300	.000	9.300	−21.631	.002	−21.629	−0.05237	.00001	−0.05236	−197.44	−0.4173	1.071	−90
−85	3.342	9.426	.000	9.426	−20.428	.003	−20.425	−0.04913	.00001	−0.04912	−195.51	−0.4121	1.597	−85
−80	4.930	9.553	.000	9.553	−19.225	.005	−19.220	−0.04595	.00001	−0.04594	−193.55	−0.4069	2.356	−80
−75	7.196	9.680	.000	9.680	−18.022	.007	−18.015	−0.04280	.00002	−0.04278	−191.57	−0.4017	3.441	−75
−70	10.40	9.806	.000	9.806	−16.820	.011	−16.809	−0.03969	.00003	−0.03966	−189.56	−0.3965	4.976	−70
−65	14.91	9.932	.000	9.932	−15.617	.015	−15.602	−0.03663	.00005	−0.03658	−187.53	−0.3913	7.130	−65
	$H_s \times 10^5$												$p_s \times 10^3$	
−60	2.118	10.059	.000	10.059	−14.416	.022	−14.394	−0.03360	.00006	−0.03354	−185.47	−0.3861	1.0127	−60
−55	2.982	10.186	.000	10.186	−13.214	.031	−13.183	−0.03061	.00009	−0.03052	−183.39	−0.3810	1.4258	−55
−50	4.163	10.313	.001	10.314	−12.012	.043	−11.969	−0.02766	.00012	−0.02754	−181.29	−0.3758	1.9910	−50
−45	5.766	10.440	.001	10.441	−10.811	.060	−10.751	−0.02474	.00015	−0.02459	−179.16	−0.3707	2.7578	−45
−40	7.925	10.566	.001	10.567	−9.609	.083	−9.526	−0.02186	.00021	−0.02165	−177.01	−0.3655	3.7906	−40
−35	10.81	10.693	.002	10.695	−8.408	.113	−8.295	−0.01902	.00028	−0.01874	−174.84	−0.3604	5.1713	−35
	$H_s \times 10^4$												$p_s \times 10^2$	
−30	1.464	10.820	.002	10.822	−7.207	.154	−7.053	−0.01621	.00038	−0.01583	−172.64	−0.3552	0.70046	−30
−25	1.969	10.946	.004	10.950	−6.005	.207	−5.798	−0.01342	.00051	−0.01291	−170.42	−0.3500	.94212	−25
−20	2.630	11.073	.005	11.078	−4.804	.277	−4.527	−0.01067	.00068	−0.00999	−168.17	−0.3449	1 2587	−20
−15	3.491	11.200	.006	11.206	−3.603	.368	−3.235	−0.00796	.00089	−0.00707	−165.90	−0.3398	1.6706	−15
−10	4.606	11.326	.008	11.334	−2.402	.487	−1.915	−0.00529	.00115	−0.00414	−163.60	−0.3346	2.2035	−10
−5	6.040	11.452	.011	11.463	−1.201	.639	−0.562	−0.00263	.00149	−0.00114	−161.28	−0.3295	2.8886	−5
	$H_s \times 10^3$													
0	0.7872	11.578	.015	11.593	0.000	.835	0.835	0.00000	.00192	0.00192	−158.93	−0.3244	3.7645	0
5	1.020	11.705	.019	11.724	1.201	1.085	2.286	.00260	.00246	.00506	−156.57	−0.3193	4.8779	5
10	1.315	11.831	.025	11.856	2.402	1.401	3.803	.00518	.00314	.00832	−154.17	−0.3141	6.2858	10
15	1.687	11.958	.032	11.990	3.603	1.800	5.403	.00772	.00399	.01171	−151.76	−0.3090	8.0565	15
20	2.152	12.084	.042	12.126	4.804	2.302	7.106	.01023	.00504	.01527	−149.31	−0.3039	10.272	20
25	2.733	12.211	.054	12.265	6.005	2.929	8.934	.01273	.00635	.01908	−146.85	−0.2988	13.032	25
30	3.454	12.338	.068	12.406	7.206	3.709	10.915	.01519	.00796	.02315	−144.36	−0.2936	16.452	30
32	3.788	12.388	.075	12.463	7.686	4.072	11.758	.01617	.00870	.02487	−143.36	−0.2916	18.035	32
32*	3.788	12.388	.075	12.463	7.686	4.072	11.758	.01617	.00870	.02487	0.04	0.0000	18.037	32*
34	4.107	12.438	.082	12.520	8.167	4.418	12.585	.01715	.00940	.02655	2.06	.0041	19.546	34
													p_s	
36	4.450	12.489	.089	12.578	8.647	4.791	13.438	.01812	.01016	.02828	4.07	.0081	0.21166	36
38	4.818	12.540	.097	12.637	9.128	5.191	14.319	.01909	.01097	.03006	6.08	.0122	.22904	38
40	5.213	12.590	.105	12.695	9.608	5.622	15.230	.02005	.01183	.03188	8.09	.0162	.24767	40
42	5.638	12.641	.114	12.755	10.088	6.084	16.172	.02101	.01275	.03376	10.09	.0202	.26763	42
44	6.091	12.691	.124	12.815	10.569	6.580	17.149	.02197	.01373	.03570	12.10	.0242	.28899	44
46	6.578	12.742	.134	12.876	11.049	7.112	18.161	.02293	.01478	.03771	14.10	.0282	.31185	46
48	7.100	12.792	.146	12.938	11.530	7.681	19.211	.02387	.01591	.03978	16.11	.0321	.33629	48
50	7.658	12.843	.158	13.001	12.010	8.291	20.301	.02481	.01711	.04192	18.11	.0361	.36240	50
52	8.256	12.894	.170	13.064	12.491	8.945	21.436	.02575	.01839	.04414	20.11	.0400	.39028	52
54	8.894	12.944	.185	13.129	12.971	9.644	22.615	.02669	.01976	.04645	22.12	.0439	.42004	54
56	9.575	12.995	.200	13.195	13.452	10.39	23.84	.02762	.02121	.04883	24.12	.0478	.45176	56
58	10.30	13.045	.216	13.261	13.932	11.19	25.12	.02855	.02276	.05131	26.12	.0517	.48558	58
60	11.08	13.096	.233	13.329	14.413	12.05	26.46	.02948	.02441	.05389	28.12	.0555	.52159	60
62	11.91	13.147	.251	13.398	14.893	12.96	27.85	.03040	.02616	.05656	30.12	.0594	.55994	62
64	12.80	13.197	.271	13 468	15.374	13.94	29.31	.03132	.02803	.05935	32.12	.0632	.60073	64
66	13.74	13.247	.292	13.539	15.855	14.98	30.83	.03223	.03002	.06225	34.11	.0670	.64411	66
68	14.75	13.298	.315	13.613	16.335	16.09	32.42	.03314	.03213	.06527	36.11	.0708	.69019	68
-	$H_s \times 10^2$													
70	1.582	13.348	.339	13.687	16.816	17.27	34.09	.03405	.03437	.06842	38.11	.0746	.73915	70
72	1.697	13.398	.364	13.762	17.297	18.53	35.83	.03495	.03675	.07170	40.11	.0784	.79112	72
74	1.819	13.449	.392	13.841	17.778	19.88	37.66	.03585	.03928	.07513	42.10	.0821	.84624	74
76	1.948	13.499	.422	13.921	18.259	21.31	39.57	.03675	.04197	.07872	44.10	.0859	.90470	76
78	2.086	13.550	.453	14.003	18.740	22.84	41.58	.03765	.04482	.08247	46.10	.0896	.96665	78

NOTE: Compiled by John A. Goff and S. Gratch. See also Keenan and Kaye. *Thermodynamic Properties of Air*, Wiley, New York, 1945. Enthalpy of dry air taken as zero at 0°F. Enthalpy of liquid water taken as zero at 32°F.

*Entrapolated to represent metastable equilibrium with undercooled liquid.

TABLE 12-1 Thermodynamic Properties of Moist Air (Standard Atmospheric Pressure, 29.921 inHg) *(Continued)*

Temp. t, °F.	Saturation humidity $H_s \times 10^2$	Volume, cu. ft./lb. dry air v_a	v_{as}	v_s	Enthalpy, B.t.u./lb. dry air h_a	h_{as}	h_s	Entropy, B.t.u./(°F.)(lb. dry air) s_a	s_{as}	s_s	Condensed water: Enthalpy B.t.u./lb. h_w	Entropy, B.t.u./(lb.)(°F.) s_w	Vapor press., in. Hg p_s	Temp. t, °F.
80	2.233	13.601	0.486	14.087	19.221	24.47	43.69	0.03854	0.04784	0.08638	48.10	0.0933	1.0323	80
82	2.389	13.651	.523	14.174	19.702	26.20	45.90	.03943	.05105	.09048	50.09	.0970	1.1017	82
84	2.555	13.702	.560	14.262	20.183	28.04	48.22	.04031	.05446	.09477	52.09	.1007	1.1752	84
86	2.731	13.752	.602	14.354	20.663	30.00	50.66	.04119	.05807	.09926	54.08	.1043	1.2529	86
88	2.919	13.803	.645	14.448	21.144	32.09	53.23	.04207	.06189	.10396	56.08	.1080	1.3351	88
90	3.118	13.853	.692	14.545	21.625	34.31	55.93	.04295	.06596	.10890	58.08	.1116	1.4219	90
92	3.330	13.904	.741	14.645	22.106	36.67	58.78	.04382	.07025	.11407	60.07	.1153	1.5135	92
94	3.556	13.954	.795	14.749	22.587	39.18	61.77	.04469	.07480	.11949	62.07	.1188	1.6102	94
96	3.795	14.005	.851	14.856	23.068	41.85	64.92	.04556	.07963	.12519	64.06	.1224	1.7123	96
98	4.049	14.056	.911	14.967	23.548	44.68	68.23	.04643	.08474	.13117	66.06	.1260	1.8199	98
100	4.319	14.106	.975	15.081	24.029	47.70	71.73	.04729	.09016	.13745	68.06	.1296	1.9333	100
102	4.606	14.157	1.043	15.200	24.510	50.91	75.42	.04815	.09591	.14406	70.05	.1332	2.0528	102
104	4.911	14.207	1.117	15.324	24.991	54.32	79.31	.04900	.1020	.1510	72.05	.1367	2.1786	104
	$H_s \times 10$													
106	0.5234	14.258	1.194	15.452	25.472	57.95	83.42	.04985	.1085	.1584	74.04	.1403	2.3109	106
108	.5578	14.308	1.278	15.586	25.953	61.80	87.76	.05070	.1153	.1660	76.04	.1438	2.4502	108
110	.5944	14.359	1.365	15.724	26.434	65.91	92.34	.05155	.1226	.1742	78.03	.1472	2.5966	110
112	.6333	14.409	1.460	15.869	26.915	70.27	97.18	.05239	.1302	.1826	80.03	.1508	2.7505	112
114	.6746	14.460	1.560	16.020	27.397	74.91	102.31	.05323	.1384	.1916	82.03	.1543	2.9123	114
116	.7185	14.510	1.668	16.178	27.878	79.85	107.73	.05407	.1470	.2011	84.02	.1577	3.0821	116
118	.7652	14.561	1.782	16.343	28.359	85.10	113.46	.05490	.1562	.2111	86.02	.1612	3.2603	118
120	.8149	14.611	1.905	16.516	28.841	90.70	119.54	.05573	.1659	.2216	88.01	.1646	3.4474	120
122	.8678	14.662	2.034	16.696	29.322	96.66	125.98	.05656	.1763	.2329	90.01	.1681	3.6436	122
124	.9242	14.712	2.174	16.886	29.804	103.0	132.8	.05739	.1872	.2446	92.01	.1715	3.8493	124
126	.9841	14.763	2.323	17.086	30.285	109.8	140.1	.05821	.1989	.2571	94.01	.1749	4.0649	126
128	1.048	14.813	2.482	17.295	30.766	117.0	147.8	.05903	.2113	.2703	96.00	.1783	4.2907	128
130	1.116	14.864	2.652	17.516	31.248	124.7	155.9	.05985	.2245	.2844	98.00	.1817	4.5272	130
132	1.189	14.915	2.834	17.749	31.729	133.0	164.7	.06067	.2386	.2993	100.00	.1851	4.7747	132
134	1.267	14.965	3.029	17.994	32.211	141.8	174.0	.06148	.2536	.3151	102.00	.1885	5.0337	134
136	1.350	15.016	3.237	18.253	32.692	151.2	183.9	.06229	.2695	.3318	104.00	.1918	5.3046	136
138	1.439	15.066	3.462	18.528	33.174	161.2	194.4	.06310	.2865	.3496	106.00	.1952	5.5878	138
	H_s													
140	0.1534	15.117	3.702	18.819	33.655	172.0	205.7	.06390	.3047	.3686	107.99	.1985	5.8838	140
142	.1636	15.167	3.961	19.128	34.136	183.6	217.7	.06470	.3241	.3888	109.99	.2018	6.1930	142
144	.1745	15.218	4.239	19.457	34.618	196.0	230.6	.06549	.3449	.4104	111.99	.2051	6.5160	144
146	.1862	15.268	4.539	19.807	35.099	209.3	244.4	.06629	.3672	.4335	113 99	.2084	6.8532	146
148	.1989	15.319	4.862	20.181	35.581	223.7	259.3	.06708	.3912	.4583	115.99	.2117	7.2051	148
150	.2125	15.369	5.211	20.580	36.063	239.2	275.3	.06787	.4169	.4848	117.99	.2150	7.5722	150
152	.2271	15.420	5.587	21.007	36.545	255.9	292.4	.06866	.4445	.5132	119.99	.2183	7.9550	152
154	.2430	15.470	5.996	21.466	37.026	273.9	310.9	.06945	.4743	.5438	121.99	.2216	8.3541	154
156	.2602	15.521	6.439	21.960	37.508	293.5	331.0	.07023	.5066	.5768	123.99	.2248	8.7701	156
158	.2788	15.571	6.922	22.493	37.990	314.7	352.7	.07101	.5415	.6125	125.99	.2281	9.2036	158
160	.2990	15.622	7.446	23.068	38.472	337.8	376.3	.07179	.5793	.6511	128.00	.2313	9.6556	160
162	.3211	15.672	8.020	23.692	38.954	363.0	402.0	.07257	.6204	.6930	130.00	.2345	10.125	162
164	.3452	15.723	8.648	24.371	39.436	390.5	429.9	.07334	.6652	.7385	132.00	.2377	10.614	164
166	.3716	15.773	9.339	25.112	39.918	420.8	460.7	.07411	.7142	.7883	134.00	.2409	11.123	166
168	.4007	15.824	10.098	25.922	40.400	454.0	494.4	.07488	.7680	.8429	136.01	.2441	11.652	168
170	.4327	15.874	10.938	26.812	40.882	490.6	531.5	.07565	.8273	.9030	138.01	.2473	12.203	170
172	.4682	15.925	11.870	27.795	41.364	531.3	572.7	.07641	.8927	.9691	140.01	.2505	12.775	172
174	.5078	15.975	12.911	28.886	41.846	576.5	618.3	.07718	.9654	1.0426	142.02	.2537	13.369	174
176	.5519	16.026	14.074	30.100	42.328	627.1	669.4	.07794	1.047	1.125	144.02	.2568	13.987	176
178	.6016	16.076	15.386	31.462	42.810	684.1	726.9	.07870	1.137	1.216	146.03	.2600	14.628	178
180	.6578	16.127	16.870	32.997	43.292	748.5	791.8	.07946	1.240	1.319	148.03	.2631	15.294	180
182	.7218	16.177	18.565	34.742	43.775	821.9	865.7	.08021	1.357	1.437	150.04	.2662	15.985	182
184	.7953	16.228	20.513	36.741	44.257	906 2	950.5	.08096	1.490	1.571	152.04	.2693	16.702	184
186	.8805	16.278	22.775	39.053	44.740	1004	1049	.08171	1.645	1.727	154.05	.2724	17.446	186
188	.9802	16.329	25.427	41.756	45.222	1119	1164	.08245	1.825	1.907	156.06	.2755	18.217	188
190	1.099	16.379	28.580	44.959	45.704	1255	1301	.08320	2.039	2.122	158.07	.2786	19.017	190
192	1.241	16.430	32.375	48.805	46.187	1418	1464	.08394	2.296	2.380	160.07	.2817	19.845	192
194	1.416	16.480	37.036	53.516	46.670	1619	1666	.08468	2.609	2.694	162.08	.2848	20.704	194
196	1.635	16.531	42.885	59.416	47.153	1871	1918	.08542	3.002	3.087	164.09	.2879	21.594	196
198	1.917	16.581	50.426	67.007	47.636	2195	2243	.08616	3.507	3.593	166.10	.2910	22.514	198
200	2.295	16.632	60.510	77.142	48.119	2629	2677	.08689	4.179	4.266	168.11	.2940	23.468	200

TABLE 12-2 Additive Corrections for *H*, *h*, and *v* When Barometric Pressure Differs from Standard Barometer

Wet-bulb temp. t_w	Sat. vapor press., in. Hg	Approximate altitude in feet: −900		900		1800		2700		3700		4800		5900	
		$\Delta p = +1$		$\Delta p = -1$		$\Delta p = -2$		$\Delta p = -3$		$\Delta p = -4$		$\Delta p = -5$		$\Delta p = -6$	
		ΔH_s	Δh	ΔH_s	Δh	ΔH_s	Δh	ΔH_s	Δh	ΔH_s	Δh	ΔH_s	Δh	ΔH_s	Δh
−10	0.022	−0.10	−0.02	0.11	0.02	0.23	0.03	0.36	0.05	0.50	0.07	0.64	0.10	0.81	0.12
−8	.025	−0.12	−0.02	.12	.02	.26	.04	.40	.06	.55	.08	.72	.11	.90	.13
−6	.027	−0.13	−0.02	.14	.02	.29	.04	.44	.07	.62	.09	.80	.12	1.00	.15
−4	.030	−0.14	0.02	.15	.02	.32	.05	.50	.07	.69	.10	.89	.13	1.12	.17
−2	.034	−0.16	−0.02	.17	.02	.35	.05	.55	.08	.76	.11	.99	.15	1.24	.19
0	.038	−0.18	−0.03	.19	.03	.39	.06	.61	.09	.85	.13	1.10	.17	1.38	.21
2	.042	−0.20	−0.03	.21	.03	.44	.07	.68	.10	.94	.14	1.22	.19	1.53	.23
4	.046	−0.22	−0.03	.23	.03	.48	.07	.75	.11	1.05	.16	1.36	.21	1.70	.26
6	.051	−0.24	−0.04	.26	.04	.54	.08	.83	.13	1.16	.18	1.51	.23	1.89	.29
8	.057	−0.27	−0.04	.29	.04	.59	.09	.93	.14	1.28	.19	1.67	.25	2.09	.32
10	.063	−0.30	−0.04	.32	.05	.66	.10	1.03	.16	1.42	.22	1.85	.28	2.31	.35
12	.069	−0.33	−0.05	.35	.05	.73	.11	1.13	.17	1.57	.24	2.04	.31	2.56	.39
14	.077	−0.36	−0.05	.39	.06	.81	.12	1.25	.19	1.74	.26	2.26	.34	2.82	.43
16	.085	−0.40	−0.06	.43	.06	.89	.14	1.38	.21	1.92	.29	2.49	.38	3.12	.48
18	.093	−0.44	−0.07	.47	.07	.98	.15	1.53	.23	2.12	.32	2.75	.42	3.44	.53
20	.103	−0.49	−0.08	.52	.08	1.08	.17	1.68	.26	2.33	.36	3.03	.46	3.79	.58
22	.113	−0.5	−0.08	.6	.09	1.2	.18	1.9	.29	2.6	.40	3.4	.52	4.2	.64
24	.124	−0.6	−0.09	.6	.10	1.3	.20	2.1	.32	2.8	.43	3.7	.57	4.6	.71
26	.137	−0.7	−0.10	.7	.11	1.4	.22	2.3	.35	3.1	.48	4.1	.63	5.1	.78
28	.150	−0.7	−0.11	.8	.12	1.6	.24	2.5	.38	3.4	.52	4.5	.69	5.6	.86
30	.165	−0.8	−0.12	.8	.13	1.7	.27	2.7	.42	3.8	.58	4.9	.75	6.1	.92
32	.180	−0.9	−0.13	.9	.14	1.9	.29	3.0	.45	4.1	.63	5.3	.82	6.6	1.01
34	.197	−0.9	−0.14	1.0	.15	2.1	.32	3.2	.49	4.4	.68	5.7	.88	7.2	1.11
36	.212	−1.0	−0.15	1.1	.17	2.2	.35	3.5	.53	4.8	.74	6.2	.96	7.8	1.20
38	.229	−1.1	−0.17	1.2	.18	2.4	.37	3.8	.58	5.2	.80	6.8	1.05	8.4	1.30
40	.248	−1.2	−0.18	1.3	.20	2.6	.41	4.1	.63	5 7	.88	7.4	1.14	9.2	1.42
42	.268	−1.3	−0.20	1.4	.21	2.8	.44	4.4	.69	6.1	.94	8.0	1.23	10.0	1.54
44	.289	−1.4	−0.22	1.5	.23	3.1	.47	4.8	.74	6.7	1.04	8.7	1.34	10.8	1.67
46	.312	−1.5	−0.23	1.6	.25	3.3	.51	5.2	.80	7.2	1.11	9.4	1.45	11.7	1.81
48	.336	−1.6	−0.25	1.8	.27	3.6	.56	5.6	.87	7.8	1.21	10.2	1.58	12.6	1.95
50	.3624	−1.7	−0.27	1.9	.29	3.9	.60	6.1	.94	8.4	1.30	10.9	1.69	13.6	2.11
52	.3903	−1.9	−0.29	2.0	.32	4.2	.65	6.5	1.01	9.0	1.40	11.8	1.83	14.7	2.28
54	.4200	−2.0	−0.31	2.2	.34	4.5	.70	7.0	1.09	9.7	1.50	12.7	1.97	15.8	2.45
56	.4518	−2.2	−0.34	2.4	.37	4.9	.76	7.6	1.18	10.5	1.63	13.7	2.13	17.1	2.66
58	.4856	−2.3	−0.37	2.5	.39	5.3	.82	8.2	1.27	11.3	1.76	14.7	2.28	18.4	2.86
60	.522	−2.5	−0.40	2.7	.42	5.7	.88	8.8	1.37	12.2	1.90	15.9	2.47	19.9	3.09
62	.560	−2.7	−0.43	2.9	.46	6.1	.95	9.5	1.48	13.2	2.05	17.1	2.66	21.4	3.33
64	.601	−2.9	−0.46	3.2	.49	6.5	1.02	10.2	1.59	14.2	2.21	18.4	2.87	23.1	3.60
66	.644	−3.2	−0.50	3.4	.53	7.1	1.10	11.0	1.72	15.3	2.38	19.8	3.09	24.8	3.87
68	.690	−3.4	−0.53	3.7	.57	7.6	1.18	11.8	1.84	16.4	2.56	21.3	3.32	26.7	4.16
70	.739	−3.7	−0.57	3.9	.61	8.1	1.27	12.7	1.98	17.6	2.75	22.9	3.58	28.7	4.48
72	.791	−3.9	−0.61	4.2	.66	8.7	1.36	13.6	2.13	18.8	2.94	24.6	3.84	30.9	4.82
74	.846	−4.2	−0.66	4.6	.71	9.4	1.46	14.6	2.28	20.2	3.16	26.4	4.14	33.1	5.18
76	.905	−4.5	−0.71	4.9	.77	10.0	1.57	15.7	2.46	21.7	3.39	28.3	4.42	35.5	5.56
78	.967	−4.9	−0.76	5.2	.82	10.8	1.69	16.9	2.65	23.3	3.65	30.5	4.77	38.2	5.98
80	1.032	−5.2	−0.82	5.6	.88	11.6	1.82	18.1	2.84	25.1	3.93	32.7	5.13	41.0	6.43
82	1.102	−5.6	−0.88	6.0	.94	12.5	1.96	19.5	3.06	27.0	4.24	35.1	5.51	44.0	6.90
84	1.175	−6.0	−0.94	6.4	1.00	13.3	2.10	20.9	3.28	28.9	4.54	37.7	5.92	47.2	7.41
86	1.253	−6.4	−1.00	6.9	1.08	14.3	2.24	22.3	3.50	30.9	4.85	40.4	6.34	50.6	7.94
88	1.335	−6.9	−1.08	7.4	1.16	15.3	2.40	23.9	3.75	33.1	5.20	43.2	6.79	54.2	8.51
90	1.422	−7.4	−1.16	7.9	1.24	16.5	2.59	25.7	4.04	35.6	5.60	46.4	7.29	58.2	9.15
92	1.514	−7.9	−1.24	8.5	1.34	17.6	2.77	27.5	4.33	38.2	6.01	49.8	7.83	62.5	9.83
94	1.610	−8.5	−1.34	9.1	1.43	18.9	2.98	29.5	4.64	41.0	6.46	53.4	8.41	67.0	10.55
96	1.712	−9.1	−1.43	9.8	1.54	20.2	3.18	31.5	4.96	43.8	6.90	57.2	9.01	71.7	11.30
98	1.820	−9.7	−1.53	10.4	1.64	21.7	3.42	33.8	5.33	47.0	7.41	61.3	9.67	76.8	12.11
100	1.933	−10.4	−1.64	11.2	1.77	23.2	3.66	36.3	5.73	50.4	7.95	65.7	10.37	82.5	13.02
102	2.053	−11.1	−1.75	12.0	1.90	24.8	3.92	38.9	6.14	54.1	8.54	70.5	11.13	88.5	13.98
104	2.179	−11.9	−1.88	12.8	2.02	26.6	4.20	41.6	6.58	57.9	9.15	75.5	11.93	94.8	14.98
106	2.311	−12.8	−2.02	13.7	2.17	28.6	4.52	44.6	7.06	62.1	9.82	81.1	12.83	101.7	16.09
108	2.450	−13.7	−2.17	14.7	2.33	30.6	4.84	47.7	7.55	66.5	10.53	87.0	13.77	109.1	17.27
110	2.597	−14.7	−2.33	15.8	2.50	32.8	5.20	51.3	8.13	71.3	11.30	93.1	14.75	117.0	18.54
112	2.751	−15.7	−2.49	16.9	2.68	35.2	5.58	55.0	8.72	76.4	12.11	99.9	15.84	125.9	19.96
114	2.913	−16.9	−2.68	18.1	2.87	37.7	5.98	58.9	9.50	82.0	13.01	107.3	17.03	135.0	21.42
116	3.082	−18.0	−2.86	19.4	3.08	40.4	6.42	63.2	10.03	88.0	13.97	115.1	18.28	144.7	22.98
118	3.260	−19.3	−3.07	20.8	3.31	43.3	6.88	67.8	10.77	94.4	15.00	123.5	19.63	155.4	24.73
120	3.448	−20.7	−3.29	22.4	3.56	46.6	7.41	72.8	11.58	101.4	16.13	132.7	21.10	167.1	26.58
122	3.644	−22.2	−3.53	24.0	3.82	50.0	7.96	78.2	12.45	109.0	17.35	142.6	22.70	179.6	28.58
124	3.850	−23.8	−3.79	25.8	4.11	53.7	8.55	84.0	13.38	117.1	18.65	153.3	24.42	193.2	30.77
126	4.065	−25.6	−4.08	27.6	4.40	57.7	9.20	90.3	14.39	125.9	20.07	165.0	26.30	208.0	33.15
128	4.291	−27.5	−4.39	29.7	4.74	62.0	9.89	97.1	15.49	135.5	21.61	177.6	28.33	224.0	35.73
130	4.527	−29.5	−4.71	32.0	5.11	66.7	10.64	104.5	16.68	145.9	23.29	191.4	30.55	241.5	38.55
132	4.775	−31.8	−5.08	34.4	5.50	71.8	11.47	112.6	17.99	157.2	25.11	206.3	32.96	260.6	41.63
134	5.034	−34.2	−5.47	37.1	5.93	77.4	12.37	121.4	19.41	169.6	27.12	222.7	35.60	281.4	44.99
136	5.305	−36.8	−5.89	40.0	6.40	83.4	13.34	130.9	20.94	183.1	29.30	240.5	38.48	304.2	48.67
138	5.588	−39.7	−6.36	43.2	6.92	90.0	14.41	141.4	22.64	197.8	31.67	260.1	41.65	329.3	52.73
140	5.884	−42.8	−6.86	46.5	8.45	97.3	15.59	152.8	24.48	214.0	34.29	281.6	45.12	356.8	57.17

TABLE 12-2 Additive Corrections for *H*, *h*, and *v* When Barometric Pressure Differs from Standard Barometer *(Continued)*

t = dry-bulb temperature, °F
t_w = wet-bulb temperature, °F
p = barometric pressure, inHg
Δp = pressure difference from standard barometer (inHg)
H = moisture content of air, gr/lb dry air
H_s = moisture content of air saturated at wet-bulb temperature (t_w), gr/lb dry air
ΔH = moisture-content correction of air when barometric pressure differs from standard barometer, gr/lb dry air
ΔH_s = moisture-content correction of air saturated at wet-bulb temperature when barometric pressure differs from standard barometer, gr/lb dry air

NOTE: To obtain ΔH reduce value of ΔH_s by 1 percent where $t - t_w = 24°F$ and correct proportionally when $t - t_w$ is not 24°F.

h = enthalpy of moist air, Btu/lb dry air
Δh = enthalpy correction when barometric pressure differs from standard barometer, for saturated or unsaturated air, Btu/lb dry air
v = volume of moist air, ft³/lb dry air

$$= \frac{0.754(t + 459.8)}{p}\left(1 + \frac{H}{4360}\right)$$

Example At a barometric pressure of 25.92 with 220°F dry-bulb and 100°F wet-bulb temperatures, determine H, h, and v. $\Delta p = -4$, and from table $\Delta H_s = 50.4$. From note,

$$\Delta H = \Delta H_s - \left(\frac{120}{24} \times 0.01 \times 50.4\right) = 50.4 - 2.5 = 47.9$$

Therefore H = 102 (from chart) + 47.9 = 149.9 gr/lb. dry air. From table $\Delta h = 7.95$. Therefore, h = saturation enthalpy from chart + deviation + 7.95 = 71.7 − 2.0 + 7.95 = 77.65 Btu/lb dry air. From previous equation

$$v = \frac{0.754(220 + 459.7)}{25.92}\left(1 + \frac{149.9}{4360}\right) = 20.43 \text{ ft}^3\text{/lb dry air}$$

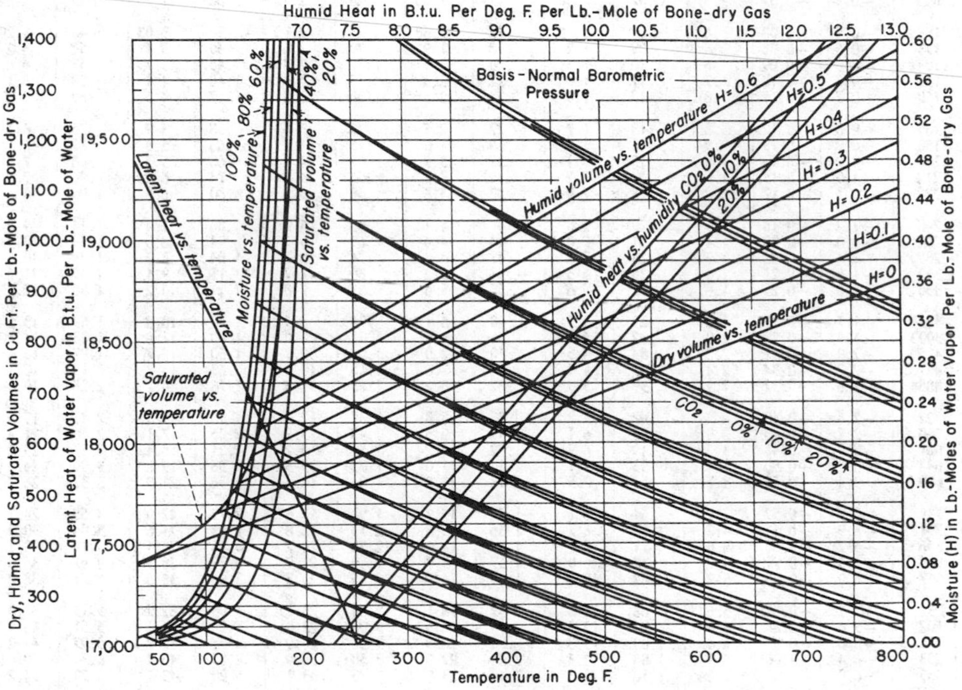

FIG. 12-5 Revised form of high-temperature psychrometric chart for air and combustion products, based on pound-moles of water vapor and dry gases. [*Hatta*, Chem. Metall. Eng., *37*, *64 (1930)*.]

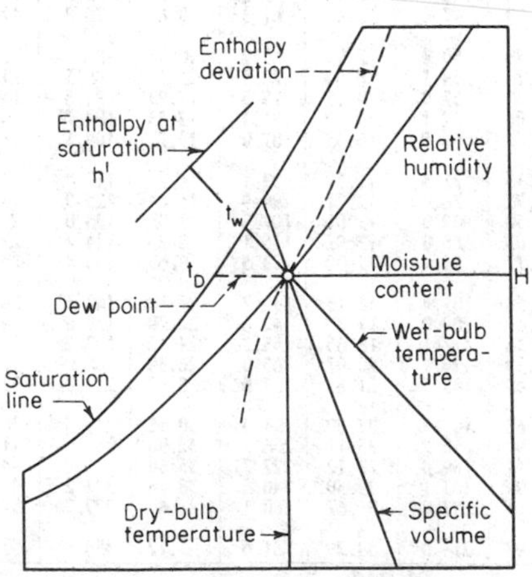

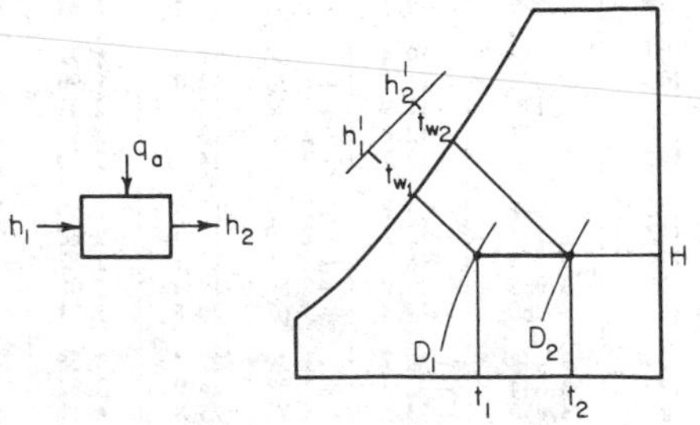

FIG. 12-7 Heating process.

FIG. 12-6 Diagram of psychrometric chart showing the properties of moist air.

Example 4 Air at 95°F dry-bulb temperature and 70°F wet-bulb temperature contacts a water spray, where its relative humidity is increased to 90 percent. The spray water is recirculated; makeup water enters at 70°F. Determine exit dry-bulb temperature, wet-bulb temperature, change in enthalpy of the air, and quantity of moisture added per pound of dry air.

Solution Figure 12-8 shows the path on a psychrometric chart. The leaving dry-bulb temperature is obtained directly from Fig. 12-2 as 72.2°F. Since

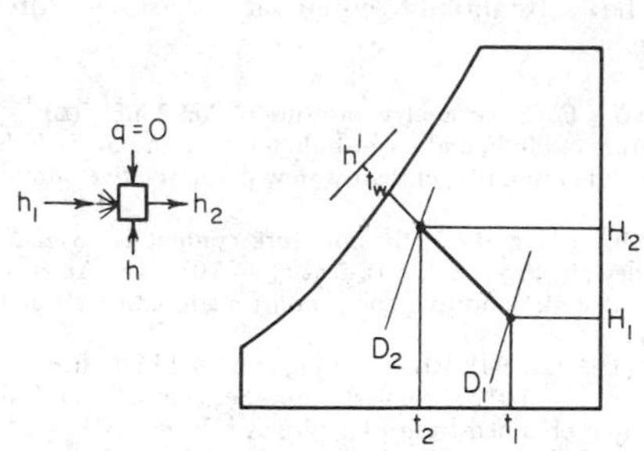

FIG. 12-8 Spray or evaporative cooling.

the spray water enters at the wet-bulb temperature of 70°F and there is no heat added to or removed from it, this is by definition an adiabatic process and there will be no change in wet-bulb temperature. The only change in enthalpy is that from the heat content of the makeup water. This can be demonstrated as follows:

$$\text{Inlet moisture } H_1 = 70 \text{ gr/lb dry air}$$
$$\text{Exit moisture } H_2 = 107 \text{ gr/lb dry air}$$
$$\Delta H = 37 \text{ gr/lb dry air}$$
$$\text{Inlet enthalpy } h_1 = h_1' + D_1 = 34.1 - 0.22$$
$$= 33.88 \text{ Btu/lb dry air}$$
$$\text{Exit enthalpy } h_2 = h_2' + D_2 = 34.1 - 0.02$$
$$= 34.08 \text{ Btu/lb dry air}$$
$$\text{Enthalpy of added water } h_w = 0.2 \text{ Btu/lb dry air (from small diagram, 37 gr at 70°F)}$$

Then

$$q_a = h_2 - h_1 + h_w$$
$$= 34.08 - 33.88 + 0.2 = 0$$

Example 5 Find the cooling load per pound of dry air resulting from infiltration of room air at 80°F dry-bulb temperature and 67°F wet-bulb temperature into a cooler maintained at 30°F dry-bulb and 28°F wet-bulb temperature, where moisture freezes on the coil, which is maintained at 20°F.

Solution. The path followed on a psychrometric chart is shown in Fig. 12-9.

$$\text{Inlet enthalpy } h_1 = h_1' + D_1 = 31.62 - 0.1$$
$$= 31.52 \text{ Btu/lb dry air}$$
$$\text{Exit enthalpy } h_2 = h_2' + D_2 = 10.1 + 0.06$$
$$= 10.16 \text{ Btu/lb dry air}$$
$$\text{Inlet moisture } H_1 = 78 \text{ gr/lb dry air}$$
$$\text{Exit moisture } H_2 = 19 \text{ gr/lb dry air}$$

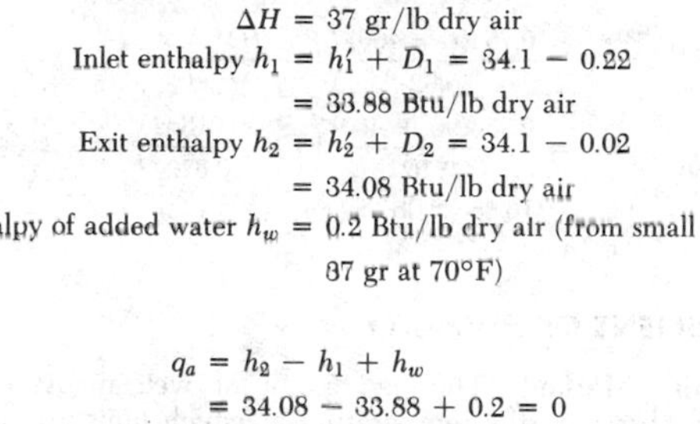

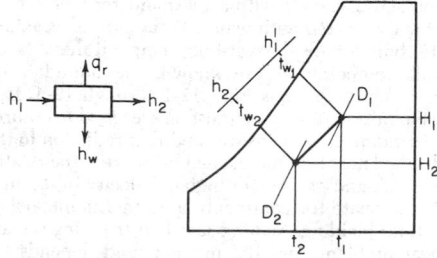

FIG. 12-9 Cooling and dehumidifying process.

$$\text{Moisture rejected } \Delta H = 59 \text{ gr/lb dry air}$$
$$\text{Enthalpy of rejected moisture} = -1.26 \text{ Btu/lb dry air (from small diagram of Fig. 12-2)}$$
$$\text{Cooling load } q_r = 31.52 - 10.16 + 1.26$$
$$= 22.62 \text{ Btu/lb dry air}$$

Note that if the enthalpy deviations were ignored, the calculated cooling load would be about 5 percent low.

Example 6 Determine water consumption and amount of heat dissipated per 1000 ft³/min of entering air at 90°F dry-bulb temperature and 70°F wet-bulb temperature when the air leaves saturated at 110°F and the makeup water is at 75°F.

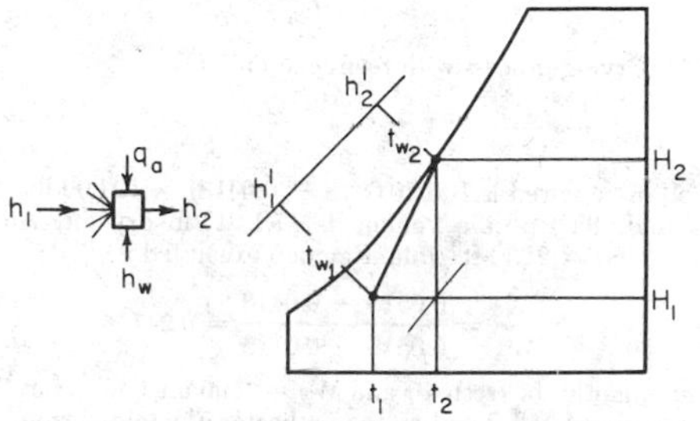

FIG. 12-10 Cooling tower.

Solution. The path followed is shown in Fig. 12-10.

$$\text{Exit moisture } H_2 = 416 \text{ gr/lb dry air}$$
$$\text{Inlet moisture } H_1 = 78 \text{ gr/lb dry air}$$
$$\text{Moisture added } \Delta H = 338 \text{ gr/lb dry air}$$
$$\text{Enthalpy of added moisture } h_w = 2.1 \text{ Btu/lb dry air (from small diagram of Fig. 12-3)}$$

If greater precision is desired, h_w can be calculated as

$$h_w = (338/7000)(1)(75 - 32)$$
$$= 2.08 \text{ Btu/lb dry air}$$
$$\text{Enthalpy of inlet air } h_1 = h_1' + D_1 = 34.1 - 0.18$$
$$= 33.92 \text{ Btu/lb dry air}$$
$$\text{Enthalpy of exit air } h_2 = h_2' + D_2 = 92.34 + 0$$
$$= 92.34 \text{ Btu/lb dry air}$$
$$\text{Heat dissipated} = h_2 - h_1 - h_w$$
$$= 92.34 - 33.92 - 2.08$$
$$= 56.34 \text{ Btu/lb dry air}$$
$$\text{Specific volume of inlet air} = 14.1 \text{ ft}^3\text{/lb dry air}$$
$$\text{Total heat dissipated} = \frac{(1000)(56.34)}{14.1} = 3990 \text{ Btu/min}$$

Example 7 A dryer is removing 100 lb water/h from the material being dried. The air entering the dryer has a dry-bulb temperature of 180°F and a wet-bulb temperature of 110°F. The air leaves the dryer at 140°F. A portion of the air is recirculated after mixing with room air having a dry-bulb temperature of 75°F and a relative humidity of 60 percent. Determine quantity of air required, recirculation rate, and load on the preheater if it is assumed that the system is adiabatic. Neglect heatup of the feed and of the conveying equipment.

Solution. The path followed is shown in Fig. 12-11.

$$\text{Humidity of room air } H_1 = 0.0113 \text{ lb/lb dry air}$$
$$\text{Humidity of air entering dryer } H_3 = 0.0418 \text{ lb/lb dry air}$$
$$\text{Humidity of air leaving dryer } H_4 = 0.0518 \text{ lb/lb dry air}$$
$$\text{Enthalpy of room air } h_1 = 30.2 - 0.3$$
$$= 29.9 \text{ Btu/lb dry air}$$
$$\text{Enthalpy of entering air } h_3 = 92.5 - 1.3$$
$$= 91.2 \text{ Btu/lb dry air}$$
$$\text{Enthalpy of leaving air } h_4 = 92.5 - 0.55$$
$$= 91.95 \text{ Btu/lb dry air}$$

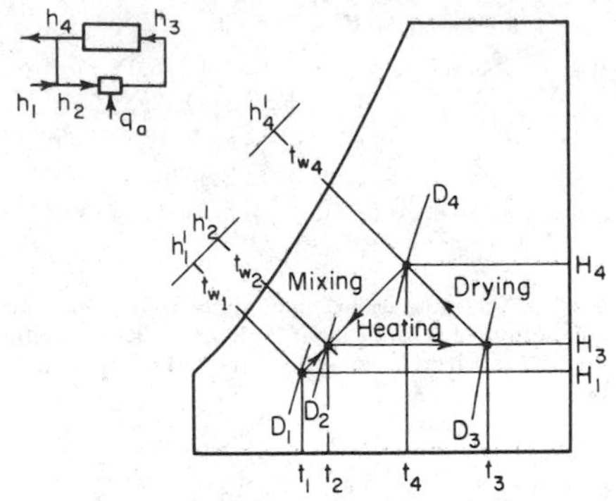

FIG. 12-11 Drying process with recirculation.

Quantity of air required is 100/(0.0518 − 0.0418) = 10,000 lb dry air/h. At the dryer inlet the specific volume is 17.1 ft^3/lb dry air. Air volume is (10,000)(17.1)/60 = 2850 ft^3/min. Fraction exhausted is

$$\frac{X}{W_a} = \frac{0.0518 - 0.0418}{0.0518 - 0.0113} = 0.247$$

where X = quantity of fresh air and W_a = total air flow. Thus 75.3 percent of the air is recirculated. Load on the preheater is obtained from an enthalpy balance

$$\begin{aligned} q_a &= 10{,}000(91.2) - 2470(29.9) - 7530(91.95) \\ &= 146{,}000 \text{ Btu/h} \end{aligned}$$

Use of Psychrometric Charts at Pressures Other Than Atmospheric The psychrometric charts shown as Figs. 12-1 through 12-4 and the data of Table 12-1 are based on a system pressure of 1 atm (29.92 inHg). For other system pressures, these data must be corrected for the effect of pressure. Additive corrections to be applied to the atmospheric values of absolute humidity and enthalpy are given in Table 12-2.

The **specific volume** of moist air in cubic feet per pound of dry air can be determined for other pressures, if ideal-gas behavior is assumed, by the following equation:

$$v = \frac{0.754(t + 460)}{P}\left(1 + \frac{HM_a}{M_w}\right) \tag{12-5}$$

where v = specific volume, ft^3/lb dry air; t = dry-bulb temperature, °F; P = pressure, inHg; H = absolute humidity, lb water/lb dry air; M_a = molecular weight of air, lb/(lb·mol); and M_w = molecular weight of water vapor, lb/(lb·mol).

Relative humidity and **dew point** can be determined for other than atmospheric pressure from the partial pressure of water in the mixture and from the vapor pressure of water vapor. The partial pressure of water is calculated, if ideal-gas behavior is assumed, as

$$p = \frac{HP}{(M_w/M_a) + H} \tag{12-6}$$

where p = partial pressure of water vapor, inHg; P = total pressure, inHg; H = absolute humidity, lb water/lb dry air, corrected to the actual pressure; M_a = molecular weight of air, lb/(lb·mol); and M_w = molecular weight of water vapor, lb/(lb·mol). The dew point of the mixture is then read directly from a table of vapor pressures as the temperature corresponding to the calculated partial pressure.

The relative humidity is obtained by dividing the calculated partial pressure by the vapor pressure of water at the dry-bulb temperature. Thus:

$$\text{Relative humidity} = 100p/p_s \tag{12-7}$$

where p = calculated partial pressure, inHg; and p_s = vapor pressure at dry-bulb temperature, inHg.

The preceding equations, which have assumed that both the air and the water vapor behave as ideal gases, are sufficiently accurate for most engineering calculations. If it is desired to remove the restriction that water vapor behave as an ideal gas, the actual density ratio should be used in place of the molecular-weight ratio in Eqs. (12-5) and (12-6).

Since the Schmidt number, Prandtl number, latent heat of vaporization, and humid heat are all essentially independent of pressure, the adiabatic-saturation-temperature and wet-bulb-temperature lines will be substantially equal at pressures different from atmospheric.

Example 8 For a barometric pressure of 25.92 inHg ($\Delta p = -4$), a dry-bulb temperature of 90°F, and a wet-bulb temperature of 70°F determine the following: absolute humidity, enthalpy, dew point, relative humidity, and specific volume.

Solution. From Fig. 12-2, the moisture content is 78 gr/lb dry air = 0.0114 lb/lb dry air. From Table 12-2 at t_w = 70°F and $\Delta p = -4$ read ΔH_s = 17.6 gr/lb dry air (additive correction for air saturated at the wet-bulb temperature).

$\Delta H = 17.6[1 - (20/24)(0.01)] = 17.4$, or actual humidity is 78 + 17.4 = 95.4 gr/lb dry air, or 0.01362 lb/lb dry air. (See footnotes for Table 12-2).

The enthalpy is obtained from Fig. 12-2 as $h = h' + D = 34.1 - 0.18 = 33.92$. To this must be added the correction of 2.75 read from Table 12-2 for $\Delta p = -4$ and t_w = 70°F, giving the true enthalpy as 33.92 + 2.75 = 36.67 Btu/lb dry air.

The partial pressure of water vapor is calculated from Eq. (12-6) as

$$p = \frac{HP}{(M_w/M_a) + H} = \frac{0.01362 \times 25.92}{0.622 + 0.01362} = 0.556 \text{ inHg}$$

From a table of vapor pressure, this corresponds to a dew point of 61.8°F.

Relative humidity is obtained from Eq. (12-7) as $100\,p/p_s = (100 \times 0.556)/1.422 = 39.1$ percent.

The specific volume in cubic feet per pound of dry air is obtained from Eq. (12-5):

$$\begin{aligned} v &= \frac{0.754(t + 460)}{25.92}\left(1 + \frac{HM_a}{M_w}\right) \\ &= \frac{0.754(90 + 460)}{25.92}\left(1 + \frac{0.01362}{0.622}\right) \\ &= 16.35 \text{ ft}^3\text{/lb dry air} \end{aligned}$$

MEASUREMENT OF HUMIDITY

Dew-Point Method The dew point of wet air is measured directly by observing the temperature at which moisture begins to form on an artificially cooled polished surface. The polished surface is usually cooled by evaporation of a low-boiling solvent such as ether, by vaporization of a condensed permanent gas such as carbon dioxide or liquid air, or by a temperature-regulated stream of water.

Although the dew-point method may be considered a fundamental technique for determining humidity, several uncertainties occur in its use. It is not always possible to measure precisely the temperature of the polished surface or to eliminate gradients across the surface. It is also difficult to detect the appearance or disappearance of fog; the usual practice is to take the dew point as the average of the temperatures when fog first appears on cooling and disappears on heating.

Wet-Bulb Method Probably the most commonly used method for determining the humidity of a gas stream is the measurement of wet- and dry-bulb temperatures. The wet-bulb temperature is measured by contacting the air with a thermometer whose bulb is covered by a wick saturated with water. If the process is adiabatic, the thermometer bulb attains the wet-bulb temperature. When the wet- and dry-bulb temperatures are known, the humidity is readily obtained from charts such as Figs. 12-1 through 12-4. In order to obtain reliable information, care must be exercised to ensure that the wet-bulb thermometer remains wet and that radiation to the bulb is minimized. The latter is accomplished by making the relative velocity between wick and gas stream high [a velocity of 4.6 m/s (15 ft/s) is usually adequate for commonly used thermometers] or by the use of radiation shielding. Making sure that the wick remains wet is a mechanical problem, and the method used depends to a large extent on the particular arrangement. Again, as with the dew-point method, errors associated with the measurement of temperature can cause difficulty.

For measurement of atmospheric humidities the **sling psychrometer** is widely used. This is composed of a wet- and dry-bulb thermometer mounted in a sling which is whirled manually to give the desired gas velocity across the bulb. In the **Assmann psychrometer** the air is drawn past the bulbs by a motor-driven fan.

In addition to the mercury-in-glass thermometer, other temperature-sensing elements may be used for psychrometers. These include resistance thermometers, thermocouples, bimetal thermometers, and thermistors.

Mechanical Hygrometers Materials such as human hair, wood fiber, and plastics have been used to measure humidity. These methods rely on a change in dimension with humidity.

Electric hygrometers measure the electrical resistance of a film of moisture-absorbing materials exposed to the gas. A wide variety of sensing elements have been used.

The **gravimetric method** is accepted as the most accurate humidity-measuring technique. In this method a known quantity of gas is passed over a moisture-absorbing chemical such as phosphorus pentoxide, and the increase in weight is determined.

EVAPORATIVE COOLING

GENERAL REFERENCES: *ASHRAE Handbook and Product Directory: Equipment*, American Society of Heating, Refrigerating and Air-Conditioning Engineers, Atlanta, 1983. *Counterflow Cooling Tower Performance*, Pritchard Corporation, Kansas City, Mo., 1957. Hensley, "Cooling Tower Energy," *Heat. Piping Air Cond.* (October 1981). Kelley and Swenson, *Chem. Eng. Prog.*, **52**, 263 (1956). Lewis, *Trans. Am. Soc. Mech. Eng.*, **44**, 329 (1922). Lichtenstein, *Trans. Am. Soc. Mech. Eng.*, **65**, 779 (1943). London, Mason, and Boelter, *Trans. Am. Soc. Mech. Eng.*, **62**, 41 (1940). McAdams, *Heat Transmission*, 3d ed., McGraw-Hill, New York, 1954, pp. 356–365. Merkel, Z. *Ver. Dtsch. Ing. Forsch.*, no. 275 (1925). *The Parallel Path Wet-Dry Cooling Tower*, Marley Co., Mission Woods, Kan., 1972. *Performance Curves*, Cooling Tower Institute, Houston, 1967. *Plume Abatement and Water Conservation with Wet-Dry Cooling Tower*, Marley Co., Mission Woods, Kan., 1973. Tech. Bull. R-54-P-5, R-58-P-5, Marley Co., Mission Woods, Kan., 1957. Wood and Betts, *Engineer*, **189** (4912), 377, (4913), 349 (1950). Zivi and Brand, *Refrig. Eng.*, **64**, 8, 31–34, 90 (1956).

PRINCIPLES

The processes of cooling water are among the oldest known. Usually water is cooled by exposing its surface to air. Some of the processes are slow, such as the cooling of water on the surface of a pond; others are comparatively fast, such as the spraying of water into air. These processes all involve the exposure of water surface to air in varying degrees.

The heat-transfer process involves (1) latent heat transfer owing to vaporization of a small portion of the water and (2) sensible heat transfer owing to the difference in temperature of water and air. Approximately 80 percent of this heat transfer is due to latent heat and 20 percent to sensible heat.

Theoretical possible heat removal per pound of air circulated in a cooling tower depends on the temperature and moisture content of air. An indication of the moisture content of the air is its wet-bulb temperature. Ideally, then, the wet-bulb temperature is the lowest theoretical temperature to which the water can be cooled. Practically, the cold-water temperature approaches but does not equal the air wet-bulb temperature in a cooling tower; this is so because it is impossible to contact all the water with fresh air as the water drops through the wetted fill surface to the basin. The magnitude of approach to the wet-bulb temperature is dependent on tower design. Important factors are air-to-water contact time, amount of fill surface, and breakup of water into droplets. In actual practice, cooling towers are seldom designed for approaches closer than 2.8°C (5°F).

COOLING-TOWER THEORY

The most generally accepted theory of the cooling-tower heat-transfer process is that developed by Merkel (op. cit.). This analysis is based upon **enthalpy potential difference** as the driving force.

Each particle of water is assumed to be surrounded by a film of air, and the enthalpy difference between the film and surrounding air provides the driving force for the cooling process. In the integrated form the Merkel equation is

$$\frac{KaV}{L} = \int_{T_2}^{T_1} \frac{dT}{h' - h} \tag{12-8}$$

where K = mass-transfer coefficient, lb water/(h·ft^2); a = contact area, ft^2/ft^3 tower volume; V = active cooling volume, ft^3/ft^2 of plan area; L = water rate, lb/(h·ft^2); h' = enthalpy of saturated air at water temperature, Btu/lb; h = enthalpy of air stream, Btu/lb; and T_1 and T_2 = entering and leaving water temperatures, °F. The right-hand side of Eq. (12-8) is entirely in terms of air and water properties and is independent of tower dimensions.

Figure 12-12 illustrates water and air relationships and the driving potential which exist in a counterflow tower, where air flows parallel but opposite in direction to water flow. An understanding of this diagram is important in visualizing the cooling-tower process.

The water operating line is shown by line AB and is fixed by the inlet and outlet tower water temperatures. The air operating line begins at C, vertically below B and at a point having an enthalpy corresponding to that of the entering wet-bulb temperature. Line BC represents the initial driving force ($h' - h$). In cooling water 1°F, the enthalpy per pound of air is increased 1 Btu multiplied by the ratio of pounds of water per pound of air. The liquid-gas ratio L/G

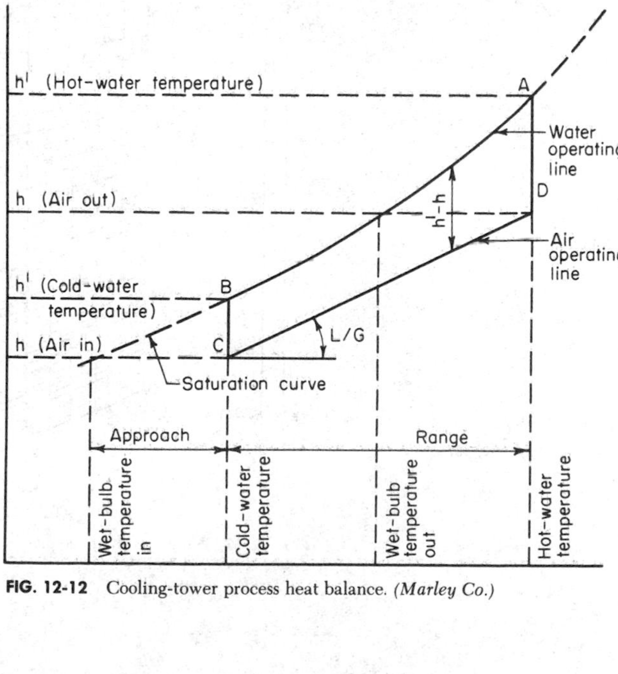

FIG. 12-12 Cooling-tower process heat balance. *(Marley Co.)*

is the slope of the operating line. The air leaving the tower is represented by point D. The cooling range is the projected length of line CD on the temperature scale. The cooling-tower approach is shown on the diagram as the difference between the cold-water temperature leaving the tower and the ambient wet-bulb temperature.

The coordinates refer directly to the temperature and enthalpy of any point on the water operating line but refer directly only to the enthalpy of a point on the air operating line. The corresponding wet-bulb temperature of any point on CD is found by projecting the point horizontally to the saturation curve, then vertically to the temperature coordinate. The integral [Eq. (12-8)] is represented by the area $ABCD$ in the diagram. This value is known as the **tower characteristic,** varying with the L/G ratio.

For example, an increase in entering wet-bulb temperature moves the origin C upward, and the line CD shifts to the right to maintain a constant KaV/L. If the cooling range increases, line CD lengthens. At a constant wet-bulb temperature, equilibrium is established by moving the line to the right to maintain a constant KaV/L. On the other hand, a change in L/G ratio changes the slope of CD, and the tower comes to equilibrium with a new KaV/L.

In order to predict tower performance it is necessary to know the required tower characteristics for fixed ambient and water conditions.

The tower characteristic KaV/L can be determined by integration. Normally used is the Chebyshev method for numerically evaluating the integral, whereby

$$\frac{KaV}{L} = \int_{T2}^{T1} \frac{dT}{h_w - h_a} \cong \frac{T_1 - T_2}{4}\left(\frac{1}{\Delta h_1} + \frac{1}{\Delta h_2} + \frac{1}{\Delta h_3} + \frac{1}{\Delta h_4}\right)$$

where h_w = enthalpy of air-water vapor mixture at bulk water temperature, Btu/lb dry air

h_a = enthalpy of air-water vapor mixture at wet-bulb temperature, Btu/lb dry air

Δh_1 = value of $(h_w - h_a)$ at $T_2 + 0.1\,(T_1 - T_2)$

Δh_2 = value of $(h_w - h_a)$ at $T_2 + 0.4\,(T_1 - T_2)$

Δh_3 = value of $(h_w - h_a)$ at $T_1 - 0.4\,(T_1 - T_2)$

Δh_4 = value of $(h_w - h_a)$ at $T_1 - 0.1\,(T_1 - T_2)$

Example 9 Determine the theoretically required KaV/L value for a cooling duty from 105°F inlet water, 85°F outlet water, 78°F ambient wet-bulb temperature, and an L/G ratio of 0.97.

From air-water vapor-mixture tables, the enthalpy h_1 of the ambient air at 78°F wet-bulb temperature is 41.58 Btu/lb.

$$h_2 \text{ (leaving air)} = 41.58 + 0.97(105 - 85) = 60.98 \text{ Btu/lb}$$

T, °F.	h_{water}	h_{air}	$h_w - h_a$	$1/\Delta h$
$T_2 = 85$	49.43	$h_1 = 41.58$		
$T_2 + 0.1(20) = 87$	51.93	$h_1 + 0.1L/G(20) = 43.52$	$\Delta h_1 = 8.41$	0.119
$T_2 + 0.4(20) = 93$	60.25	$h_1 + 0.4L/G(20) = 49.34$	$\Delta h_2 = 10.91$	0.092
$T_1 - 0.4(20) = 97$	66.55	$h_2 - 0.4L/G(20) = 53.22$	$\Delta h_3 = 13.33$	0.075
$T_1 - 0.1(20) = 103$	77.34	$h_2 - 0.1L/G(20) = 59.04$	$\Delta h_4 = 18.30$	0.055
$T_1 = 105$	81.34	$h_2 = 60.98$		0.341

$$\frac{KaV}{L} = \frac{105 - 85}{4}(0.341) = 1.71$$

A quicker but less accurate method is by the use of a nomograph (Fig. 12-13) prepared by Wood and Betts (op. cit.).

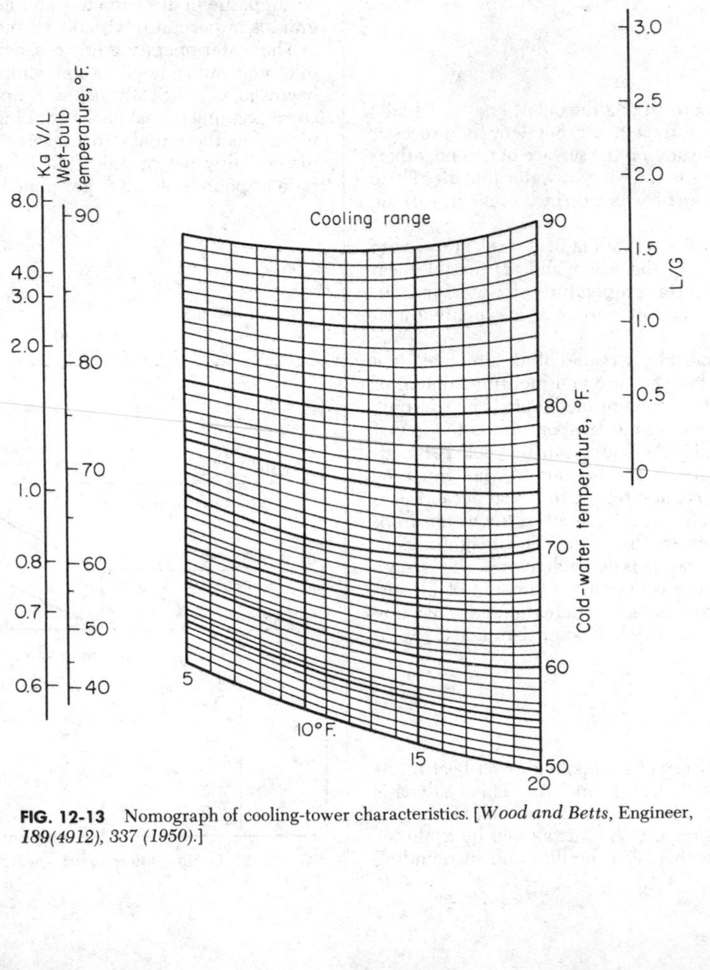

FIG. 12-13 Nomograph of cooling-tower characteristics. [*Wood and Betts,* Engineer, *189(4912), 337 (1950).*]

Mechanical-draft cooling towers normally are designed for L/G ratios ranging from 0.75 to 1.50; accordingly, the values of KaV/L vary from 0.50 to 2.50. With these ranges in mind, an example of the use of the nomograph will readily explain the effect of changing variables.

Example 10 If a given tower is operating with 20°F range, a cold-water temperature of 80°F, and a wet-bulb temperature of 70°F, a straight line may be drawn on the nomograph. If the L/G ratio is calculated to be 1.0, then KaV/L may be established by a line drawn through $L/G = 1.0$ and parallel to the original line. The tower characteristic KaV/L is thus established at 1.42. If the wet-bulb temperature were to drop to 50°F, KaV/K and L/G ratios may be assumed to remain constant. A new line parallel to the original will then show that for the same range the cold-water temperature will be 70°F.

The nomograph provides an approximate solution; degree of accuracy will vary with changes in cooling as well as from tower to tower. Once the theoretical cooling-tower characteristic has been determined by numerical integration or from the nomograph for a given cooling duty, it is necessary to design the cooling-tower fill and air distribution to meet the theoretical tower characteristic. The Pritchard Corporation (op. cit.) has developed performance data on various tower-fill designs. These data are too extensive to include here, and those interested should consult this reference. See also Baker and Mart (Marley Co., Tech. Bull. R-52-P-10, Mission Woods, Kan.) and Zivi and Brand (loc. cit.).

MECHANICAL-DRAFT TOWERS

Two types of mechanical-draft towers are in use today: the forced-draft and the induced-draft. In the **forced-draft tower** the fan is mounted at the base, and air is forced in at the bottom and discharged at low velocity through the top. This arrangement has the advantage of locating the fan and drive outside the tower, where it is convenient for inspection, maintenance, and repairs. Since the equipment is out of the hot, humid top area of the tower, the fan is not subjected to corrosive conditions. However, because of the low exit-air velocity, the forced-draft tower is subjected to excessive recirculation of the humid exhaust vapors back into the air intakes. Since the wet-bulb temperature of the exhaust air is considerably above the wet-bulb temperature of the ambient air, there is a decrease in performance evidenced by an increase in cold (leaving)-water temperature.

The **induced-draft tower** is the most common type used in the United States. It is further classified into counterflow and cross-flow design, depending on the relative flow directions of water and air. Thermodynamically, the **counterflow arrangement** is more efficient, since the coldest water contacts the coldest air, thus obtaining maximum enthalpy potential. The greater the cooling ranges and the more difficult the approaches, the more distinct are the advantages of the counterflow type. For example, with an L/G ratio of 1, an ambient wet-bulb temperature of 25.5°C (78°F), and an inlet water temperature of 35°C (95°F), the counterflow tower requires a KaV/L characteristic of 1.75 for a 2.8°C (5°F) approach, while a cross-flow tower requires a characteristic of 2.25 for the same approach. However, if the approach is increased to 3.9°C (7°F), both types of tower have approximately the same required KaV/L (within 1 percent).

The **cross-flow-tower** manufacturer may effectively reduce the tower characteristic at very low approaches by increasing the air quantity to give a lower L/G ratio. The increase in air flow is not necessarily achieved by increasing the air velocity but primarily by lengthening the tower to increase the air-flow cross-sectional area. It appears then that the cross-flow fill can be made progressively longer in the direction perpendicular to the air flow and shorter in the direction of the air flow until it almost loses its inherent potential-difference disadvantage. However, as this is done, fan power consumption increases.

Ultimately, the economic choice between counterflow and cross-flow is determined by the effectiveness of the fill, design conditions, and the costs of tower manufacture.

Performance of a given type of cooling tower is governed by the ratio of the weights of air to water and the time of contact between water and air. In commercial practice, the variation in the ratio of air to water is first obtained by keeping the air velocity constant at about 350 ft/(min·ft^2 of active tower area) and varying the water concentration, gal/(min·ft^2 of tower area). As a secondary operation, air velocity is varied to make the tower accommodate the cooling requirement.

Time of contact between water and air is governed largely by the time required for the water to discharge from the nozzles and fall through the tower to the basin. The time of contact is therefore obtained in a given type of unit by varying the height of the tower. Should the time of contact be insufficient, no amount of increase in the ratio of air to water will produce the desired cooling. It is therefore necessary to maintain a certain minimum height of cooling tower. When a wide approach of 8 to 11°C (15 to 20°F) to the wet-bulb temperature and a 13.9 to 19.4°C (25 to 35°F) cooling range are required, a relatively low cooling tower will suffice. A tower in which the water travels 4.6 to 6.1 m (15 to 20 ft) from the distributing system to the basin is sufficient. When a moderate approach and a cooling range of 13.9 to 19.4°C (25 to 35°F) are required, a tower in which the water travels 7.6 to 9.1 m (25 to 30 ft) is adequate. Where a close approach of 4.4°C (8°F) with a 13.9 to 19.4°C (25 to 35°F) cooling range is required, a tower in which the water travels from 10.7 to 12.2 m (35 to 40 ft) is required. It is usually not economical to design a cooling tower with an approach of less than 2.8°C (5°F), but it can be accomplished satisfactorily with a tower in which the water travels 10.7 to 12.2 m (35 to 40 ft).

Figure 12-14 shows the relationship of the hot-water, cold-water, and wet-bulb temperatures to the water concentration.* From this, the **minimum area** required for a given performance of a well-designed counterflow induced-draft cooling tower can be obtained. Figure 12-15 gives the horsepower per square foot of tower area required for a given performance. These curves do not apply to parallel or cross-flow cooling, since these processes are not so efficient as the counterflow process. Also, they do not apply when the approach to the cold-water temperature is less than 2.8°C (5°F). These charts should be considered approximate and for preliminary estimates only. Since many factors not shown in the graphs must be included in the computation, the manufacturer should be consulted for final design recommendations.

The cooling performance of any tower containing a given depth of filling varies with the **water concentration.** It has been found that maximum contact and performance are obtained with a tower having a water concentration of 2 to 5 gal/(min·ft^2 of ground area). Thus the problem of calculating the size of a cooling tower becomes one of determining the proper concentration of water required to

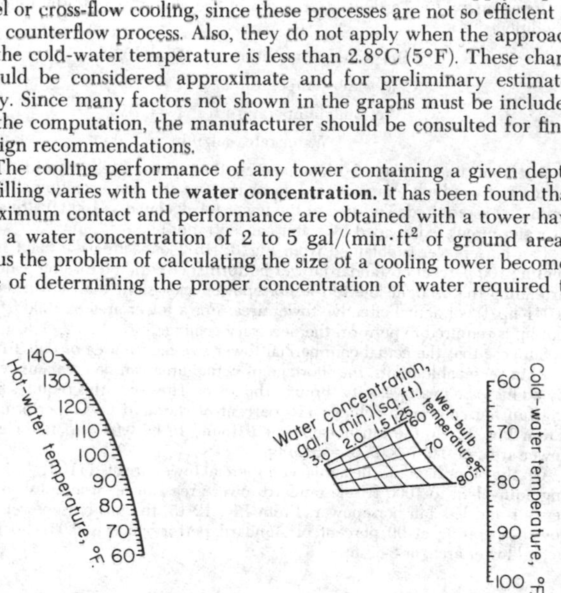

FIG. 12-14 Sizing chart for a counterflow induced-draft cooling tower, for induced-draft towers with (1) an upspray distributing system with 24 ft of fill or (2) a flume-type distributing system and 32 ft of fill. The chart will give approximations for towers of any height. [*Fluor Corp. (now Ecodyne Corp.)*]

*See also London, Mason, and Boelter, loc. cit.; Lichtenstein, loc. cit.; Simpson and Sherwood, *J. Am. Soc. Refrig. Eng.*, **52**, 535, 574 (1946); Simons, *Chem. Metall. Eng.*, **49**(5), 138; (6), 83 (1942); **46**, 208 (1939); and Hutchinson and Spivey, *Trans. Inst. Chem. Eng.*, **20**, 14 (1942).

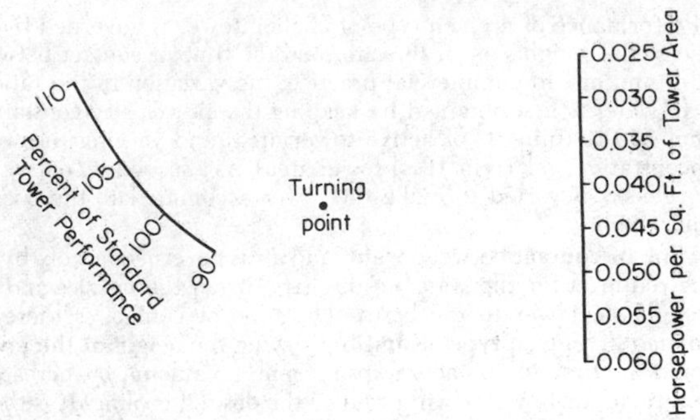

FIG. 12-15 Horsepower chart for a counterflow induced-draft cooling tower. [*Fluor Corp. (now Ecodyne Corp.)*]

obtain the desired results. Once the necessary water concentration has been established, tower area can be calculated by dividing the gallons per minute circulated by the water concentration in gallons per minute–square foot. The required tower size then is a function of the following:

1. Cooling range (hot-water temperature minus cold-water temperature)
2. Approach to wet-bulb temperature (cold-water temperature minus wet-bulb temperature)
3. Quantity of water to be cooled
4. Wet-bulb temperature
5. Air velocity through the cell
6. Tower height

Example 11 To illustrate the use of the charts, assume the following conditions:

Hot-water temperature T_1, °F = 102

Cold-water temperature T_2, °F = 78

Wet-bulb temperature t_w, °F = 70

Water rate, gal/min = 2000

A straight line on Fig. 12-14, connecting the points representing the design water and wet-bulb temperatures, shows that a water concentration of 2 gal/(min·ft^2) is required. The area of the tower is calculated as 1000 ft^2 (quantity of water circulated divided by water concentration).

Fan horsepower is obtained from Fig. 12-15. Connecting the point representing 100 percent of standard tower performance with the turning point and extending this straight line to the horsepower scale show that it will require 0.041 hp/ft^2 of actual effective tower area. For a tower area of 1000 ft^2 41.0 fan hp is required to perform the necessary cooling.

Suppose that the actual commercial tower size has an area of only 910 ft^2. Within reasonable limits, the shortage of actual area can be compensated for by an increase in air velocity through the tower. However, this requires boosting fan horsepower to achieve 110 percent of standard tower performance. From Fig. 12-15, the fan horsepower is found to be 0.057 hp/ft^2 of actual tower area, or 0.057 × 910 = 51.9 hp.

On the other hand, if the actual commercial tower area is 1110 ft^2, the cooling equivalent to 1000 ft^2 of standard tower area can be accomplished with less air and less fan horsepower. From Fig. 12-15, the fan horsepower for a tower operating at 90 percent of standard performance is 0.031 hp/ft^2 of actual tower area, or 34.5 hp.

This example illustrates the sensitivity of fan horsepower to small changes in tower area. The importance of designing a tower that is slightly oversize in ground area and of providing plenty of fan capacity becomes immediately apparent.

Example 12 Assume the same cooling range and approach as used in Example 11 except that the wet-bulb temperature is lower. Design conditions would then be

Water rate, gal/min = 2000

Temperature range ($T_1 - T_2$), °F = 24

Temperature approach ($T_2 - t_w$), °F = 8

Hot-water temperature T_1, °F = 92

Cold-water temperature T_2, °F = 68

Wet-bulb temperature t_w, °F = 60

From Fig. 12-14, the water concentration required to perform the cooling is 1.75 gal/(min·ft^2), giving a tower area of 1145 ft^2 versus 1000 ft^2 for a 70°F wet-bulb temperature. This shows that the lower the wet-bulb temperature for the same cooling range and approach, the larger is the area of the tower required and therefore the more difficult is the cooling job.

Figure 12-16 illustrates the type of **performance curve** furnished by the cooling-tower manufacturer. This shows the variation in performance with changes in wet-bulb and hot-water temperatures while the water quantity is maintained constant.

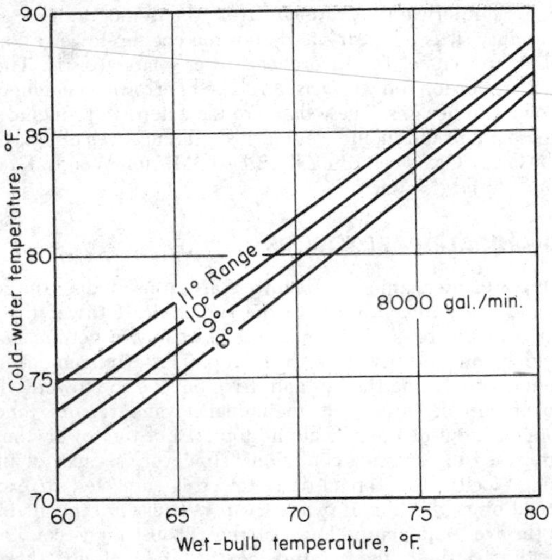

FIG. 12-16 Typical cooling-tower performance curve.

COOLING-TOWER OPERATION

Water Makeup Makeup requirements for a cooling tower consist of the summation of evaporation loss, drift loss, and blowdown. Therefore,

$$W_m = W_e + W_d + W_b \tag{12-9}$$

where W_m = makeup water, W_d = drift loss, and W_b = blowdown [consistent units, m^3/(h·gal·min)].

Evaporation loss can be estimated by the following equation:

$$W_e = 0.00085\ W_c(T_1 - T_2) \tag{12-10}$$

where W_c = circulating-water flow, gal/min at tower inlet

$T_1 - T_2$ = inlet-water temperature minus outlet-water temperature, °F

Drift is entrained water in the tower discharge vapors. Drift loss is a function of the drift-eliminator design, which typically varies between 0.1 and 0.2 percent of the water supplied to the tower. New developments in eliminator design make it possible to reduce drift loss well below 0.1 percent.

Blowdown discards a portion of the concentrated circulating water due to the evaporation process in order to lower the system solids concentration. The amount of blowdown can be calculated according to the number of cycles of concentration required to limit scale formation. Cycles of concentration are the ratio of dissolved solids in the recirculating water to dissolved solids in the makeup water. Since chlorides remain soluble on concentration, cycles of concentration are best expressed as the ratio of the chloride content of the circulating and makeup waters. Thus, the blowdown quantities required are determined from

$$\text{Cycles of concentration} = (W_e + W_b)/W_b \qquad (12\text{-}11)$$

or

$$W_b = W_e/(\text{cycles} - 1) \qquad (12\text{-}12)$$

Cycles of concentration involved with cooling-tower operation normally range from three to five cycles. Below three cycles of concentration, excessive blowdown quantities are required and the addition of acid to limit scale formation should be considered.

Example 13 Determine the amount of makeup required for a cooling tower with following conditions:

Inlet water flow, m^3/h (gal/min)	2270	(10,000)
Inlet water temperature, °C (°F)	37.77	(100)
Outlet water temperature, °C (°F)	29.44	(85)
Drift loss, percent	0.2	
Concentration, cycles	5	

Evaporation loss

$$W_e,\ m^3/h = 0.00085 \times 2270 \times (37.77 - 29.44) \times 1.8 = 28.9$$

$$W_e,\ \text{gal/min} = 127.5$$

Drift loss

$$W_d,\ m^3/h = 2270 \times 0.002 = 4.54$$

$$W_d,\ \text{gal/min} = 20$$

Blowdown

$$W_b,\ m^3/h = \frac{W_e}{(S-1)} = \frac{28.9}{4} = 7.24$$

$$W_b,\ \text{gal/min} = 31.9$$

Makeup

$$W_m,\ m^3/h = 28.9 + 4.54 + 7.24 = 40.7$$

$$W_m,\ \text{gal/min} = 179.4$$

Fan Horsepower In evaluating cooling-tower owning and operating costs, fan-horsepower requirements can be a significant factor. Large air quantities are circulated through cooling towers at exit velocities of about 10.2 m/s (2000 ft/min) maximum for induced-draft towers. Fan air-flow quantities depend upon tower-design factors, including such items as type of fill, tower configuration, and thermal-performance conditions.

The effective output of the fan is static air horsepower (SAHP), which is obtained by the following equation:

$$\text{SAHP} = \frac{Q(h_s)(d)}{33{,}000(12)} \qquad (12\text{-}13)$$

where Q = air volume, ft^3/min; h_s = static head, in of water; and d = density of water at ambient temperature, lb/ft^3.

Cooling-tower fan horsepower can be reduced substantially as the ambient wet-bulb temperature decreases if two-speed fan motors are used. Theoretically, operating at half speed will reduce air flow by 50 percent while decreasing horsepower to one-eighth of full-speed operation. However, actual half-speed operation will require about 17 percent of the horsepower at full speed as a result of the inherent motor losses at lighter loads.

Figure 12-17 shows a typical plot of outlet-water temperatures when a cooling tower is operated (1) in the fan-off position, (2) with the fan at half speed, and (3) with the fan at full speed. Note that at decreasing wet-bulb temperatures the water leaving the tower during half-speed operation could meet design water-temperature requirements of, say, 85°F. For example, for a 60°F wet-bulb, 20°F range, a leaving-water temperature slightly below 85°F is obtained with design water flow over the tower. If the fan had a 100-hp motor, 83 hp would be saved when operating it at half speed. In calculating savings, one should not overlook the advantage of having colder tower water available for the overall water-circulating system.

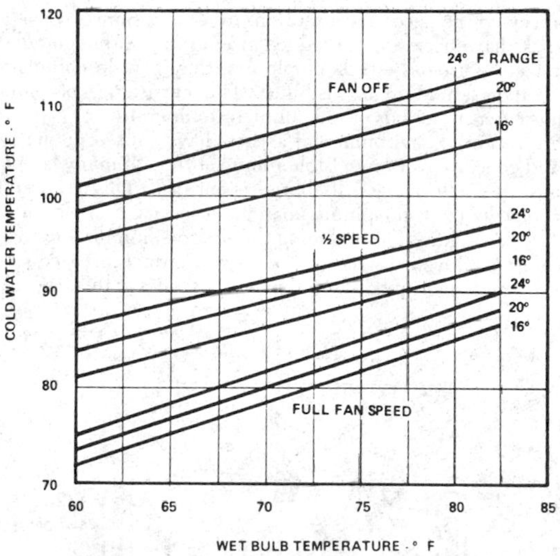

FIG. 12-17 Typical plot of cooling-tower performance at varying fan speeds.

Recent developments in cooling-tower fan energy management also include automatic variable-pitch propeller-type fans and inverter-type devices to permit variable fan speeds. These schemes involve tracking the load at a *constant* leaving-water temperature.

The variable-pitch arrangement at constant motor speed changes the pitch of the blades through a pneumatic signal from the leaving-water temperature. As the thermal load and/or the ambient wet-bulb temperature decreases, the blade pitch reduces air flow and less fan energy is required.

Inverters make it possible to control a variable-speed fan by changing the frequency modulation. Standard alternating-current fan motors may be speed-regulated between 0 and 60 Hz. In using inverters for this application, it is important to avoid frequencies that would result in fan critical speeds.

Even though tower-fan energy savings can result from these arrangements, they may not constitute the best system approach. Power-plant steam condensers and refrigeration units, for example, can take advantage of colder tower water to reduce power consumption. Invariably, these system savings are much larger than cooling-tower fan savings with constant leaving-water temperatures. A refrigeration-unit condenser can utilize inlet-water temperatures down to 12.8°C (55°F) to reduce compressor energy consumption by 25 to 30 percent.

Pumping Horsepower Another important factor in analyzing cooling-tower selections, especially in medium to large sizes, is the portion of pump horsepower directly attributed to the cooling tower. A counterflow type of tower with spray nozzles will have a pumping head equal to static lift plus nozzle pressure loss. A cross-flow type of tower with gravity flow enables a pumping head to equal static lift. A reduction in tower height therefore reduces static lift, thus reducing pump horsepower:

$$\text{Pump bhp} = \frac{\text{gal/min}(h_t)}{3960\ (\text{pump efficiency})} \qquad (12\text{-}14)$$

where h_t = total head, ft.

Fogging and Plume Abatement A phenomenon that occurs in cooling-tower operation is fogging, which produces a highly visible plume and possible icing hazards. Fogging results from mixing warm, highly saturated tower discharge air with cooler ambient air that lacks the capacity to absorb all the moisture as vapor. While in the past visible plumes have not been considered undesirable, properly locating towers to minimize possible sources of complaints has now received the necessary attention. In some instances, guyed high fan stacks have been used to reduce ground fog. Although tall stacks minimize the ground effects of plumes, they can do nothing about water-vapor saturation or visibility. The persistence of plumes is much greater in periods of low ambient temperatures.

More recently, environmental aspects have caused public awareness and concern over any visible plume, although many lay persons misconstrue cooling-tower discharge as harmful. This has resulted in a new development for plume abatement known as a wet-dry cooling-tower configuration. Reducing the relative humidity or moisture content of the tower discharge stream will reduce the frequency of plume formation. Figure 12-18 shows a "parallel path" arrangement

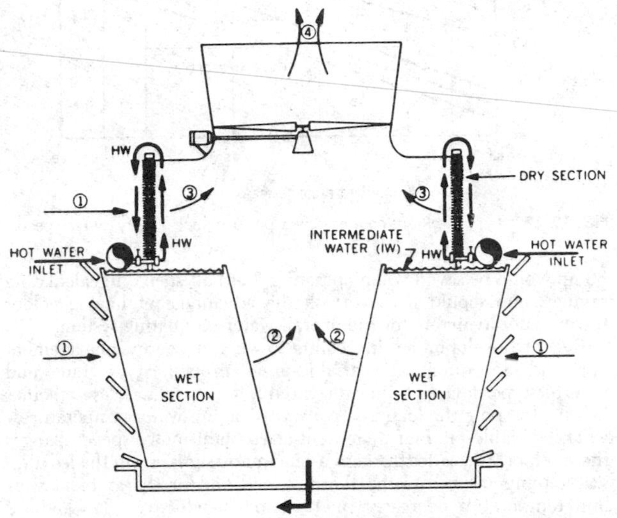

FIG. 12-18 Parallel-path cooling-tower arrangement for plume abatement. *(Marley Co.)*

that has been demonstrated to be technically sound but at substantially increased tower investment. Ambient air travels in parallel streams through the top dry-surface section and the evaporative section. Both sections benefit thermally by receiving cooler ambient air with the wet and dry air streams mixing after leaving their respective sections. Water flow is arranged in series, first flowing to the dry-coil section and then to the evaporation-fill section. A "series path" airflow arrangement, in which dry coil sections can be located before or after the air traverses the evaporative section, also can be used. However, series-path air flow has the disadvantage of water impingement, which could result in coil scaling and restricted air flow.

Wet-dry cooling towers incorporating these designs are being used for large-tower industrial applications. At present they are not available for commercial applications.

Energy Management With today's emphasis on energy management, cooling towers have not been overlooked. During periods below 50°F ambient wet-bulb temperatures, cooling towers have the temperature capability to furnish chilled water directly to air-conditioning systems. For existing refrigeration–cooling-tower systems, piping can be installed to bypass the chiller to allow tower effluent to flow directly to cooling coils. After heat has been removed from the air stream, water returns directly to the cooling tower. Water temperature leaving the cooling tower is controlled between 8.9 and 12.2°C (48 and 54°F), usually by cycling cooling-tower fans. Depending upon the cleanliness of the cooling-tower water, it may be necessary to install a side-stream or full-flow filter to minimize contamination of the normally closed chilled-water circuit. Figure 12-19 shows the general arrangement of this system. Substantial savings can be realized during colder months by eliminating refrigeration-compressor energy.

Several other methods involving cooling towers have been used to reduce refrigeration energy consumption. These systems, as applied to centrifugal and absorption refrigeration machines, are known as thermocycle or free cooling systems. When water leaving the cooling tower is available below 10°C (50°F), the thermocycle system permits shutting down the compressor prime mover or reducing steam flow to an absorption unit. Figure 12-20 shows the arrangement for a centrifugal refrigeration unit.

The thermocycle system can be operated only when condensing water is available at a temperature lower than the required chilled-water-supply temperature. Modifications for a centrifugal refrigeration unit include the installation of a small liquid-refrigerant pump, cooler spray header nozzles, and a vapor bypass line between the cooler and the condenser. Without the compressor operating, a thermocycle capacity up to 35 percent of the refrigeration-unit rating can be produced.

The cooling-tower fan is operated at full speed to produce the coldest water temperature possible for a given ambient wet-bulb temperature. The large vapor bypass between the cooler and the condenser is opened along with the compressor suction damper or prerotational vanes. The heat removed from the chilled-water stream boils off refrigerant vapor from the cooler. This vapor flows mainly through the bypass line to the condenser, where it is condensed to a liquid. (Units having hot-gas bypass lines cannot be used for this purpose, as the pipe size is too small.) The liquid then returns to the cooler as in the normal refrigeration cycle. If the refrigeration unit contains internal float valves, they are bypassed manually or held open by an adjusting stem.

Thermocycle capacity is a function of the temperature difference between the chilled-water outlet temperature leaving the cooler and the inlet condenser water. The cycle finally stops when these two temperatures approach each other and there is not sufficient vapor pressure difference to permit flow between the heat exchangers.

Precise control of the outlet chilled-water temperature does not occur with thermocycle operation. This temperature is dependent on ambient wet-bulb-temperature conditions. Normally, during cold winter days little change occurs in wet-bulb temperatures, so that only slight water-temperature variations may occur. This would not be true of many spring and fall days, when relatively large climatic temperature swings can and do occur.

Refrigeration units modified for free cooling do not include the liquid-refrigerant pump and cooler spray header nozzles. Without the cooler refrigerant agitation for improved heat transfer, this arrangement allows up to about 20 percent of rated capacity. Expected capacities for both thermocycle and free cooling are indicated in Fig. 12-21.

In operating a cooling tower in the thermocycle or free-cooling mode, some precautions are necessary to minimize icing problems. These include fan reversals to circulate air down through the tower inlet louvers, proper water distribution, constant water flow over the tower, heat tracing of lines such as makeup lines as required, and maximum loading per tower cell.

NATURAL-DRAFT TOWERS

Natural-draft, or hyperbolic-type, towers have been in use since about 1916 in Europe and have become standard equipment for the water-cooling requirements of British power stations. They are primarily suited to very large cooling-water quantities, and the reinforced-concrete structures used are as large as diameters of 80.7 m (265 ft) and heights of 103.6 m (340 ft).

The design convenience obtained from the steady air flow of mechanical-draft towers is not realized in natural-draft-tower design. Air flow through a natural-draft tower is due largely to the difference in density between the cool inlet air and the warm exit air. The air leaving the stack is lighter than the ambient air, and a draft is created by chimney effect, thus eliminating the need for mechanical fans. McKelvey and Brooke (*The Industrial Cooling Tower*, Elsevier, New York, 1959, p. 108) note that natural-draft towers commonly operate at air-pressure differences in the region of 0.2-in water gauge when under full load. The mean velocity of the air above the tower packing is generally about 1.2 to 1.8 m/s (4 to 6 ft/s).

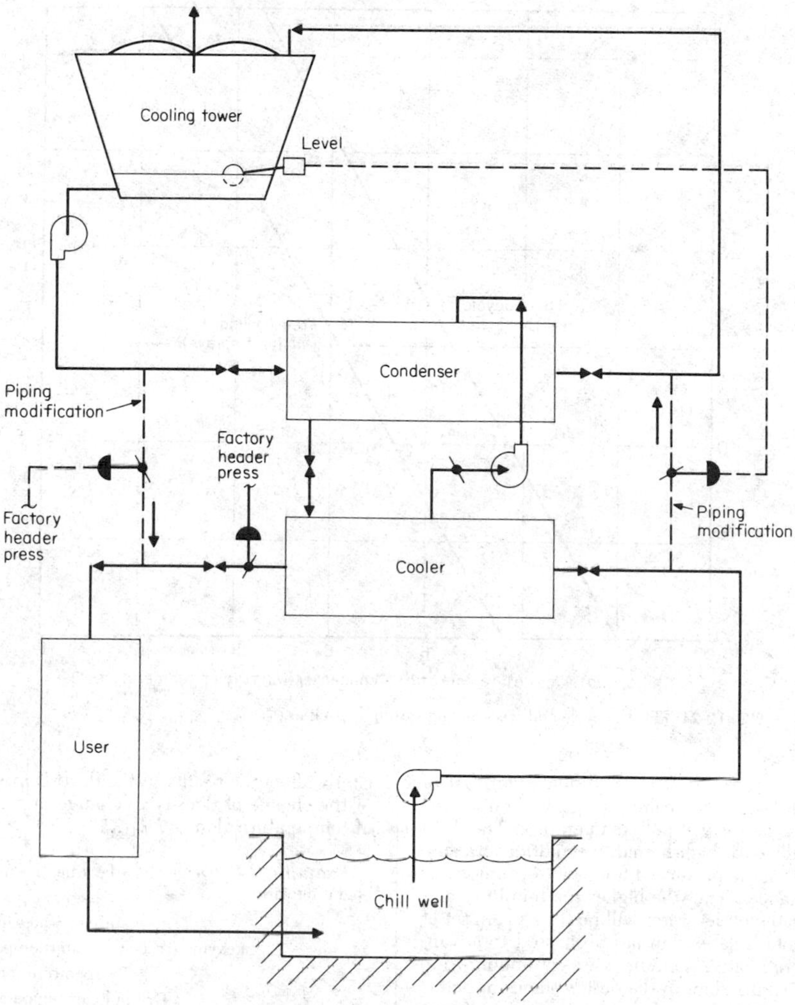

FIG. 12-19 Cooling-tower water for direct cooling during winter months.

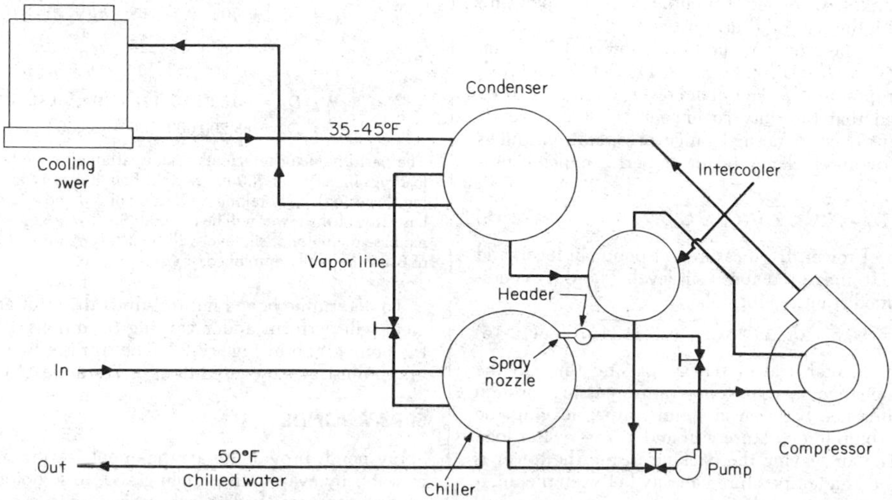

FIG. 12-20 Cooling-tower use on a thermocycle system during winter months.

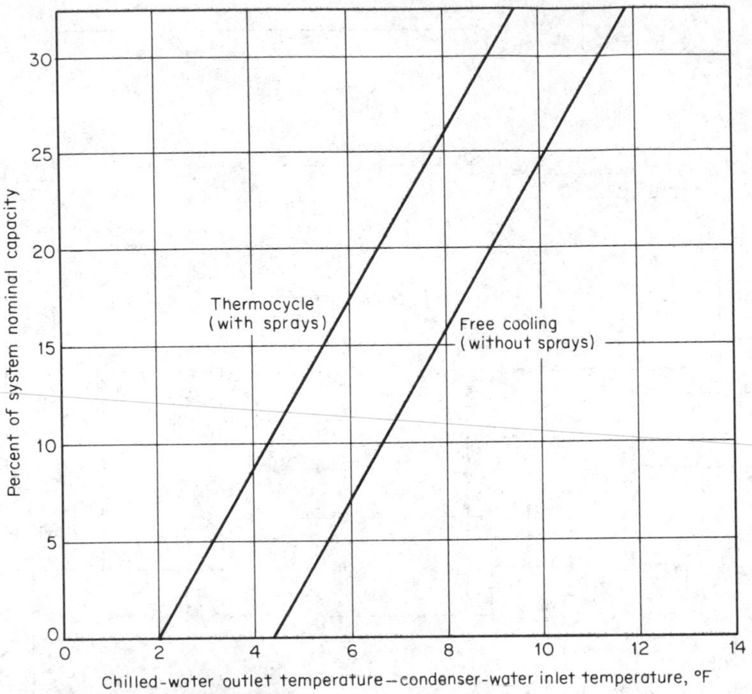

FIG. 12-21 Thermocycle and free-cooling-system capacities.

The performance of the natural-draft tower differs from that of the mechanical-draft tower in that the cooling is dependent upon the relative humidity as well as on the wet-bulb temperature. The draft will increase through the tower at high-humidity conditions because of the increase in available static pressure difference to promote air flow against internal resistances. Thus the higher the humidity at a given wet bulb, the colder the outlet water will be for a given set of conditions. This fundamental relationship has been used to advantage in Great Britain, where relative humidities are commonly 75 to 80 percent. Therefore, it is important in the design stages to determine correctly and specify the density of the entering and effluent air in addition to the usual tower-design conditions of range, approach, and water quantity. The performance relationship to humidity conditions makes exact control of outlet-water temperature difficult to achieve with the natural-draft tower.

Data for determining the size of natural-draft towers have been presented by Chilton [*Proc. Inst. Elec. Eng.*, **99**, 440 (1952)] and Rish and Steel (ASCE Symposium on Thermal Power Plants, October 1958). Chilton showed that the duty coefficient D_t of a tower is approximately constant over its normal range of operation and is related to tower size by an efficiency factor or performance coefficient C_t as follows:

$$D_t = (A\sqrt{Z_t})/(C_t\sqrt{C_t}) \qquad (12\text{-}15)$$

where A = base area of tower, ft², measured at pond sill level; and Z_t = height of tower, ft, measured above sill level. The duty coefficient may be determined from the formula

$$(W_L/D_t) = 90.59(\Delta h/\Delta T)\sqrt{\Delta t + 0.3124\Delta h} \qquad (12\text{-}16)$$

where Δh = change in total heat of the air passing through the tower, Btu/lb; ΔT = change of water temperature passing through tower, °F; Δt = difference between air temperature leaving the packing and inlet dry-bulb temperature, °F; and W_L = water load in the tower, lb/h. The air leaving the packing inside the tower is assumed to be saturated at a temperature halfway between the inlet- and outlet-water temperatures. A divergence between theory and practice of a few degrees in this latter assumption does not significantly affect the results, as the draft component depends on the ratio of the change of density to change of total heat and not on change of temperature alone.

Example 14 Determine the duty coefficient for a hyperbolic tower operating with

Temperature of water to tower, °F = 82
Leaving (recooled) water temperature, °F = 70
Temperature range ΔT, °F = 12
Dry-bulb air temperature t_2, °F = 57
Aspirated (ambient) wet-bulb air temperature t_{w2}, °F = 51.7
Water loading to tower W_L, lb/h = 38,2000,000

t_1 = (82° + 70°)/2 = 76° h_1 = 39.8 (from Fig. 12-2)
t_2 = 57° h_2 = 21.3
Δt = 19° Δh = 18.5

$$W_L/D_t = 90.59(18.5/12)\sqrt{19 + 0.3124(18.5)} = 696$$
$$D_t = 38{,}200{,}000/696 = 55{,}000$$

The performance coefficients usually attained have been about 5.2 for water loadings in excess of 750 lb/(h·ft²), though new types of packing are improving (lowering) it. By taking a C_t value of 5.0 and a tower height of 320 ft, the base area of the tower will be $(55{,}000)(5\sqrt{5})/\sqrt{320}$ = 34,600 ft², or the internal base diameter at sill level will be 210 ft. A ratio of height to base diameter of 3:2 is normally employed.

To determine how a natural-draft tower of any given duty coefficient will perform under varying conditions, Rish and Steel plotted the nomograph in Fig. 12-22. The straight line shown on the nomograph illustrates the conditions of Example 14.

SPRAY PONDS

Spray ponds provide an arrangement for lowering the temperature of water by evaporative cooling and, in so doing, greatly reduce the cooling area required in comparison with a cooling pond. A spray pond uses a number of nozzles which spray water into contact with

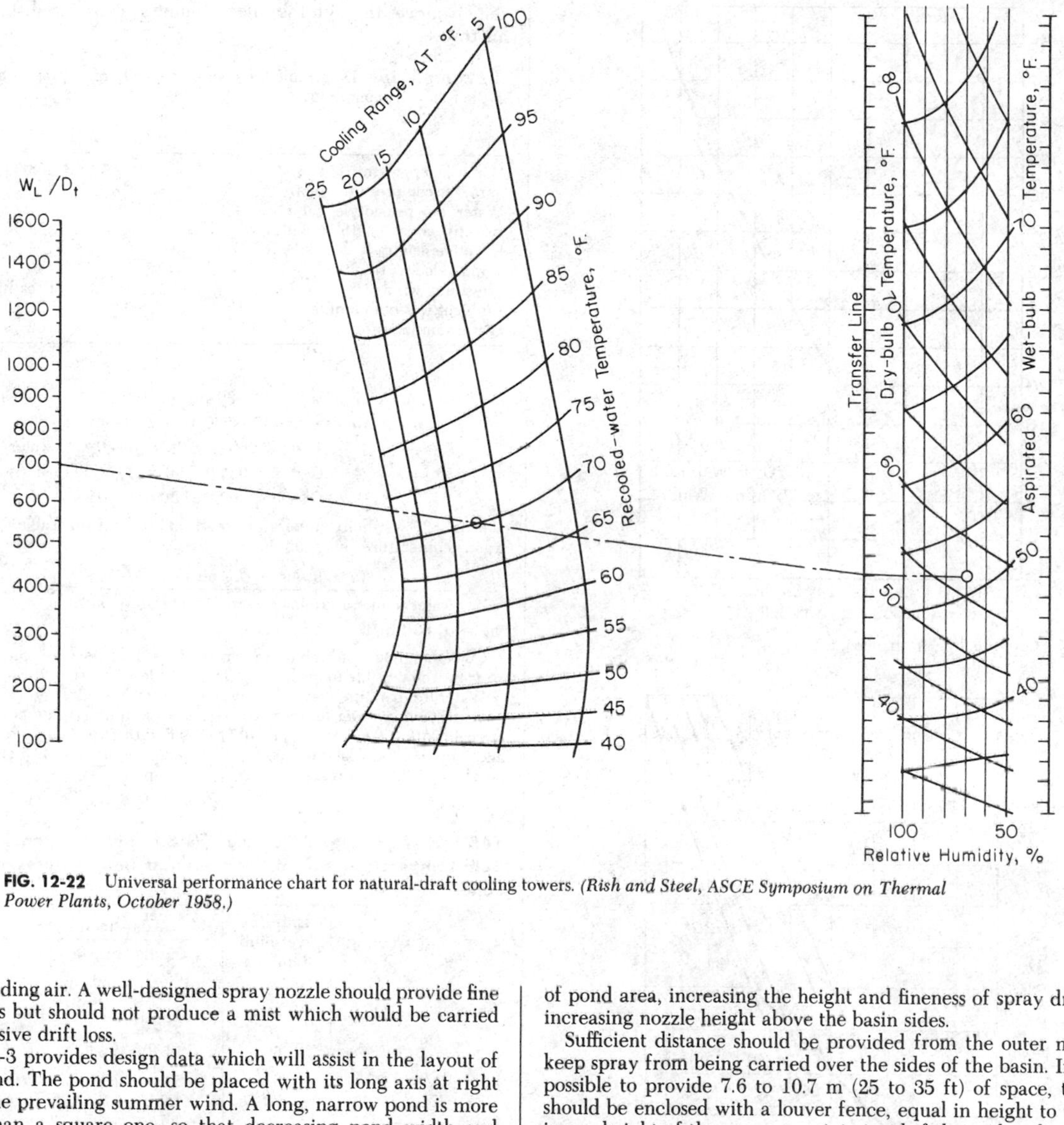

FIG. 12-22 Universal performance chart for natural-draft cooling towers. *(Rish and Steel, ASCE Symposium on Thermal Power Plants, October 1958.)*

the surrounding air. A well-designed spray nozzle should provide fine water drops but should not produce a mist which would be carried off as excessive drift loss.

Table 12-3 provides design data which will assist in the layout of a spray pond. The pond should be placed with its long axis at right angles to the prevailing summer wind. A long, narrow pond is more effective than a square one, so that decreasing pond width and increasing pond length will improve performance. Performance can also be increased by decreasing the amount of water sprayed per unit of pond area, increasing the height and fineness of spray drops, and increasing nozzle height above the basin sides.

TABLE 12-3 Spray-Pond Engineering Data and Design*

Recommendations	Usual	Minimum	Maximum
Nozzle capacity, gal./min. each	35–50	10	60
Nozzles per 12-ft. length of pipe	5–6	4	8
Height of nozzles above sides of basin, ft.	7–8	2	10
Nozzle pressure, lb./sq. in.	5–7	4	10
Size of nozzles and nozzle arms, in.	2	1¼	2½
Distance between spray lateral piping, ft.	25	13	38
Distance of nozzles from side of pond, unfenced, ft.	25–35	20	50
Distance of nozzles from side of pond, fenced, ft.	12–18	10	25
Height of louver fence, ft.	12	6	18
Depth of pond basin, ft.	4–5	2	7
Friction loss per 100 ft. pipe, in. of water	1–3	....	6
Design wind velocity, m.p.h.	5	3	10

*From Spray Pond Bull. SP-51, Marley Co., Mission Woods, Kan., p. 3.

Sufficient distance should be provided from the outer nozzles to keep spray from being carried over the sides of the basin. If it is not possible to provide 7.6 to 10.7 m (25 to 35 ft) of space, the pond should be enclosed with a louver fence, equal in height to the maximum height of the spray, to minimize drift loss. Also, during cold-weather periods, fogging can occur from the spray pond, so that consideration should be given to possible hazards to roadways or buildings in the immediate vicinity.

The physical designs and operating conditions of spray-pond installations vary greatly, and it is difficult to develop exact rating data that can be used for determining cooling performance in all cases. However, Fig. 12-23 shows the performance that can be obtained with a well-designed spray pond, based on a 21.1°C (70°F) wet-bulb temperature and a 2.2-m/s (5-mi/h) wind. This curve shows that a 3.3°C (6°F) approach to the wet bulb is possible at a 2.2°C (4°F) range, but at higher ranges the obtainable approach increases. If it is necessary to cool water through a large temperature range to a reasonably close approach, the spray pond could be staged. With this method, the water is initially sprayed, collected, and then resprayed in another part of a sectionalized pond basin.

Figure 12-24 shows performance curves for a spray pond used in steam-condensing service at varying wet-bulb and range conditions. Spray-pond performance can be calculated within reasonable accuracy on the basis of the leaving wet-bulb temperature of the air passing through the spray-filled volume. The air temperature leaving

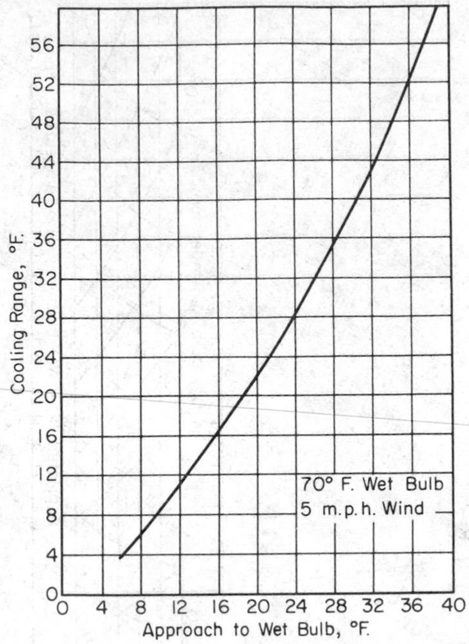

FIG. 12-23 Spray-pond performance curve.

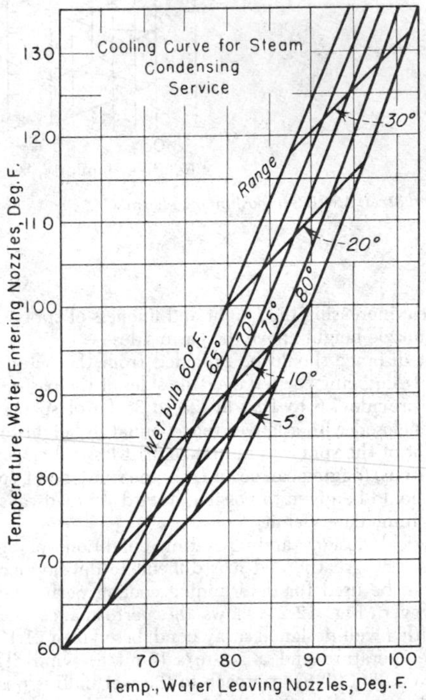

FIG. 12-24 Spray ponds: cooling curves for steam-condensing service.

cannot exceed the warm water to the pond, and the closeness of approach will depend on the pond layout. In calculating the cooling obtained, the spray-filled volume is figured from a height equal to the elevation of the nozzles above the pond surface plus 0.3 m (1 ft) for each 7-kPa (lbf/in^2) nozzle pressure and a plan area extending 3 m (10 ft) beyond the outer nozzles. The air area involved is the projected area of a vertical plane through the filled volume and broadside to the direction of air movement. The horizontal distance that the air moves through the filled volume is considered the length of air travel.

Example 15 Determine the cooling capacity of a spray pond operating at the following conditions:

Water flow, gal/min	46,000
Spray-nozzle pressure, lb/in^2	7
Water flow per nozzle, gal/min	42.5
Effective area, length × width, ft^2	434 × 100
Effective height, ft	7 + 7 spray height
Wind velocity, ft/min	440
Prevailing wind	Broadside to pond
Ambient wet-bulb temperature, °F	78
Water temperature in, °F	102

$$\text{Effective air area} = 434 \times 14 = 6080 \text{ ft}^2$$
$$\text{Air flow} = 440 \times 6080 = 2{,}680{,}000 \text{ ft}^3/\text{min}$$
$$L = 46{,}000 \times 8.33 = 384{,}000 \text{ lb water/min}$$
$$G = 2{,}680{,}000/14.3 = 187{,}500 \text{ lb air/min}$$
$$L/G = 384{,}000/187{,}500 = 2.05$$

h' at 78°F wet-bulb temperature = 41.58 Btu/lb (from Table 12-1). Assume water temperature out = 92°F.

$$\frac{L}{G} = \frac{(h_2') \text{ air out} - (h_1') \text{ air in}}{\text{water temperature in} - \text{water temperature out}} = 2.05 = \frac{h_2' - 41.58}{10}$$

$h_2' = 61.63$ Btu/lb

Corresponding wet-bulb temperature = 94°F air leaving pond.
Approach possible to air leaving (from Table 12-4) = −2°F.
Water temperature leaving spray pond = 94°F − 2°F = 92°F.
Since leaving-water temperature checks assumption, spray pond is capable of cooling 46,000 gal/min from 102 to 92°F with 78°F wet-bulb temperature and 5-mi/h wind. Total of 1080 spray nozzles required at 42.5 gal/min each, nozzles at 7-lbf/in^2 pressure.

TABLE 12-4 Degree Adjustment to Be Applied to Leaving-Air Wet-Bulb Temperature to Find Cooled-Water Temperatures of Spray Ponds*

Cooling range, °F.	Entering wet-bulb temp., † °F.	Adjustment, °F. Length of air travel, ft.‡ 100	50	25
10	80	−3	+2	+4
	70	−2	+3	+5
	60	−1.5	+3.5	+5.5
15	80	−5.0	+1	+5
	70	−4.0	+2	+5.5
	60	−3.5	+2.5	+6
20	80	−7	0	+6
	70	−6	+1	+7
	60	−5.5	+1.5	+7.5

Cooled-water temperature = wet-bulb temperature of leaving air plus values shown.

*From "Heating, Ventilating, Air Conditioning Guide," 38th ed., p. 598 American Society of Heating, Refrigerating and Air Conditioning Engineers 1960.

†Wet-bulb temperature of air entering spray-filled volume.

‡Length of air travel through spray-filled volume.

COOLING PONDS

When large ground areas are available, cooling ponds offer a satisfactory method of removing heat from water. A pond may be constructed at a relatively small investment by pushing up an earth dike 1.8 to 3.1 m (6 to 10 ft) high. For a successful pond installation, the soil must be reasonably impervious, and location in a flat area is desirable. Four principal heat-transfer processes are involved in obtaining cooling from an open pond. Heat is lost through evaporation, convection, and radiation and is gained through solar radiation.

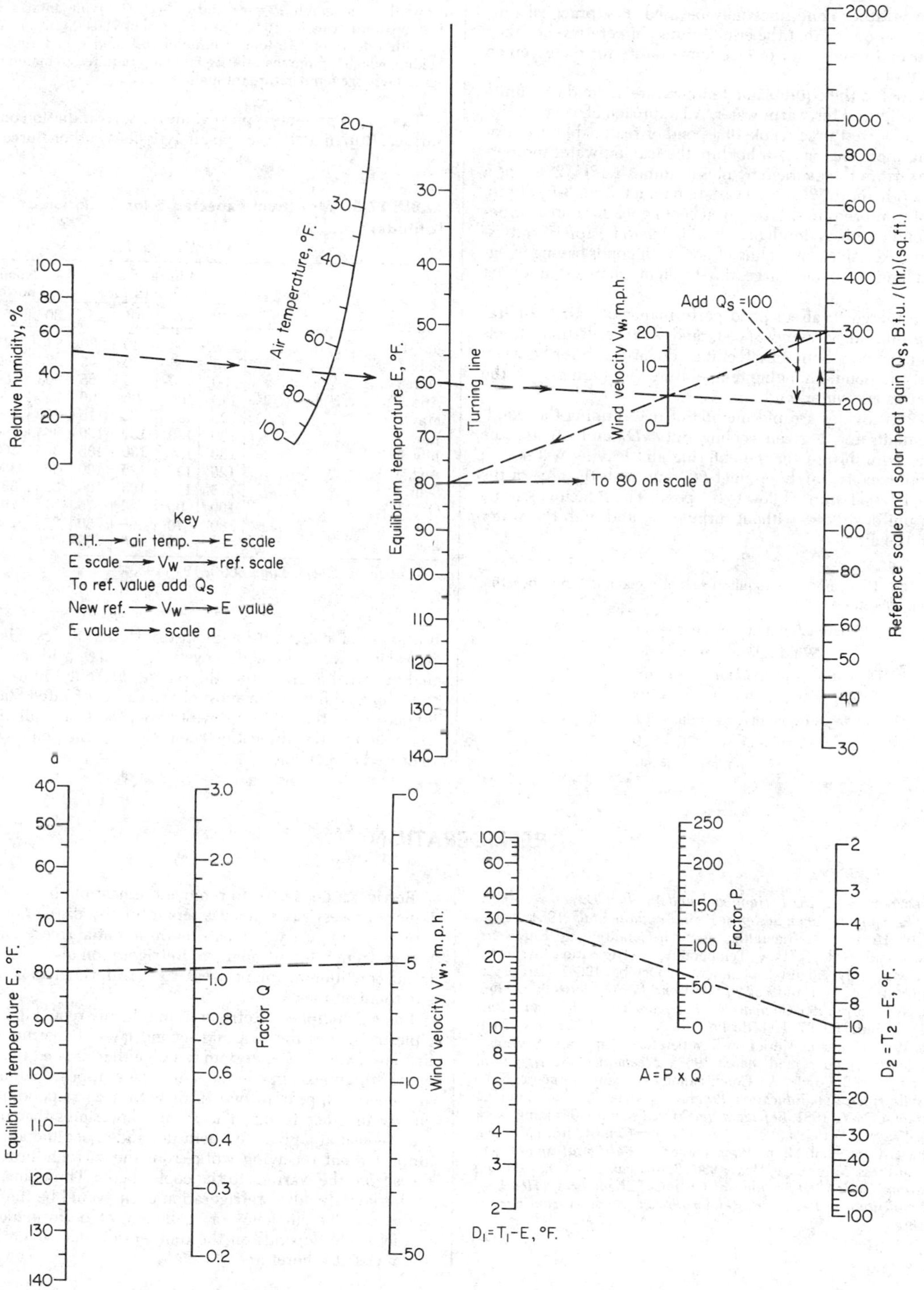

FIG. 12-25 Nomograph for determining cooling-pond performance and size. [*Langhaar*, Chem. Eng., *60(8), 194 (1953).*]

The required pond area depends on the number of degrees of cooling required and the net heat loss from each square foot of pond surface.

Langhaar [*Chem. Eng.*, **60**(8), 194 (1953)] states that under given atmospheric conditions a body of water would eventually come to a temperature at which heat loss would equal heat gain. This temperature is referred to as the equilibrium temperature and is designated as *E* in Fig. 12-25, a nomograph of cooling-pond performance. The equilibrium temperature is greatly affected by the amount of solar radiation, which is usually not known very accurately and which varies throughout the day. If a pond has at least a 24-h holdup, then daily average weather conditions may be used. For practical purposes, it is recommended that the equilibrium temperature be taken as equal to normal river-water or lake temperature for the specified weather conditions.

In order to cool to the equilibrium temperature, a pond of infinite size would be required for warm water. An approach of 1.7 to 2.2°C (3 to 4°F) is the lowest practicable in a pond of reasonable size. For a pond having more than a 24-h holdup, the leaving-water temperature will vary from the average by plus or minus 1.1°C (2°F) for a 0.9-m (5-ft) depth and 1.7°C (3°F) variation for a 0.9-m (3-ft) depth.

The area of pond required for a given cooling load is almost independent of pond depth. A depth of at least 0.9 m (3 ft) appears advisable to prevent excessive channeling of flow with ponds having irregular bottoms and to avoid large day-to-night changes in outlet temperature.

Factors considered to affect pond performance are air temperature, relative humidity, wind speed, and solar radiation. Items appearing to have only a minor effect include heat transfer between the earth and the pond, changing temperature and humidity of the air as it traverses the water, and rain.

Figure 12-25 provides a rapid method of determining the pond-area requirements for a given cooling duty. D_1 and D_2 are the approaches to equilibrium for the entering and leaving water, °F; V_w is the wind velocity, mi/h; product PQ represents the area of the pond surface, ft^2/(gal·min) of flow to the pond. The P factor assumes a pond with uniform flow, without turbulence, and with the water warmer than the air.

Example 16 Determine the required size of a cooling pond operating at the following conditions:

Relative humidity, percent = 50
Wind velocity, mi/h = 5
Dry-bulb air temperature, °F = 68
Solar heat gain, Btu/(h·ft^2) = 100
Water quantity, gal/min = 10,000
Water inlet, °F = 110
Water outlet, °F = 90

From the nomograph, P = 68 and Q = 1.07, giving an area required of 73 ft^2/(gal·min). Area for 10,000 gal/min is thus 730,000 ft^2, or 17 acres.

With a depth of 5 ft, total volume of the pond would amount to a 45.5-h holdup, which is more than the 24-h holdup required to maintain a fairly constant discharge temperature throughout the day.

Table 12-5 presents typical values of solar radiation on a horizontal surface, Btu/(h·ft^2), based on analysis of Weather Bureau records for a number of stations throughout the United States. These are clear-day values, rarely exceeded even in the high arid regions. The normal or actual average monthly values are only 50 to 60 percent of the tabulated figures for most of the eastern United States and 80 to 90 percent in the arid southwest. Also, the solar radiation should be multiplied by the absorption coefficient for the pond, which appears to exceed 95 percent.

TABLE 12-5 Maximum Expected Solar Radiation at Various North Latitudes*

B.t.u./(hr.)(sq. ft.)

	24-hr. avg. at north latitude				Noon value at north latitude			
	30°	35°	40°	45°	30°	35°	40°	45°
Jan. 1	65	50	40	30	240	205	170	135
Feb. 1	75	65	55	45	270	240	210	175
Mar. 1	90	80	75	65	305	285	255	230
Apr. 1	110	105	95	90	340	320	300	280
May 1	120	120	120	115	360	350	335	320
June 1	130	130	130	130	365	360	345	335
July 1	130	130	130	130	365	360	350	340
Aug. 1	125	125	125	120	360	350	340	325
Sept. 1	115	110	105	100	350	335	315	300
Oct. 1	100	90	80	75	315	295	270	245
Nov. 1	80	70	60	50	270	245	215	185
Dec. 1	65	55	45	35	240	210	175	140

*Langhaar, *Chem. Eng.*, **60**(8), 194 (1953).

REFRIGERATION

GENERAL REFERENCES: *ASHRAE Handbook and Product Directory: Applications*, 1978; *Equipment*, 1983; *Systems*, 1980; *Fundamentals*, 1981; American Society of Heating, Refrigerating and Air-Conditioning Engineers, Atlanta. Brown and Briley, "Low Temperature Refrigeration Systems," *Refrig. Serv. Contr.* (July, August, September, and October 1968). Carrier Air Conditioning Company, *A Handbook of Air Conditioning System Design*, McGraw-Hill, New York, 1980. *Comparative Refrigeration Performance Ratings*, Freon Tech. Bull. RT-27, E. I. du Pont de Nemours & Co., Inc., Wilmington, Del., 1966. Holman, "Ultra Low-Temperature Compound Systems," *Refrig. Serv. Contr.* (January and March 1968). *Mechanical Refrigeration Principles*, parts 1 and 2, Carrier Air Conditioning Company, Syracuse, N.Y., 1963. *Process Refrigeration Education Program*, Carrier Air Conditioning Company, Syracuse, N.Y., 1981. *Refrigeration*, York Technical Training, York Division, Borg-Warner, York, Pa., 1972. Trane Co., *Air Conditioning Manual*, McGill Graphic Arts, St. Paul, Minn., 1980. Trane Co., *Refrigerating Manual*, North American Press, Milwaukee, Wis., 1980. *Periodicals. Air-Conditioning and Refrigeration News; ASHRAE Journal; Chemical Engineering; Heating/Piping/Air Conditioning; Petroleum Refiner; Power; Refrigeration Service and Contracting.*

INTRODUCTION

Mechanical refrigeration is the process of lowering the temperature of a substance below that of its surroundings. The chemical-process industry is a major user of refrigeration facilities. Typical large refrigeration users in this field include the manufacturing of synthetic rubber and textiles, refrigerants, chlorine, plastics, hydrogen fluoride, napthalene intermediates, dyes, dimethyl terephthalate, acrylonitrile, and caprolactam.

Refrigeration is used to remove the heat of chemical reactions, to liquefy process gases, for gas separation by distillation and condensation, and to purify products by preferential freeze-out of one component from a liquid mixture. Refrigeration also is used extensively in air-conditioning plant areas for comfort, process, and thermal-environment uses.

Basic Principles Refrigeration is firmly rooted in two basic principles known as the first and second laws of thermodynamics. The first law states that energy may be neither created nor destroyed. If energy disappears in one form, it must reappear in another. Nor can any energy appear in one form with a corresponding decrease of energy in other forms. The second law states that no system can receive heat at a given temperature and reject it at a higher temperature without receiving work from the surroundings. Heat always flows from the warmer to the cooler body. Through a consideration of this law, the ideal refrigeration cycle would be the reversed Carnot cycle. The efficiency or coefficient of performance (COP) of a Carnot cycle depends on the temperatures at which heat is added and rejected. Therefore,

$$\text{COP} = T_1/(T_2 - T_1) \tag{12-17}$$

where T_1 = evaporator temperature, absolute
T_2 = condensing temperature, absolute

In an actual refrigeration cycle reversibility does not exist and therefore there will be losses, causing the COP to be less than that for the ideal cycle.

Section 4 includes a thorough and comprehensive approach to the theory of thermodynamics.

Definitions A **ton of refrigeration** is the refrigeration produced by melting 1 ton of ice at a temperature of 0°C (32°F) in 24 h. It is a rate of removing heat equivalent to the removal of 11,376 kJ/h (12,000 Btu/h).

A **British thermal unit** (Btu) is the heat required to produce a temperature rise of 1°F in 1 lb of water.

Liquids with low boiling points are used as refrigerants in mechanical refrigeration. Liquids that change from a liquid to a gas after absorbing heat are known as **primary refrigerants.** Brine, air, and water, which act only as heat carriers, are known as **secondary refrigerants.** The chemical constitution of various refrigerants with their numerical designations are listed in Table 12-6.

TABLE 12-6 Numerical Designations and Chemical Composition of Various Refrigerants

R-11	Trichlorofluoromethane
R-12	Dichlorodifluoromethane
R-13	Chlorotrifluoromethane
R-13B1	Bromotrifluoromethane
R-14	Carbon tetrafluoride
R-21	Dichlorofluoromethane
R-22	Chlorodifluoromethane
R-23	Trifluoromethane
R-40	Methyl chloride
R-50	Methane
R-113	Trichlorotrifluoroethane
R-114	Dichlorotetrafluoroethane
R-115	Chloropentafluoroethane
R-142b	Chlorodifluoroethane
R-152a	Difluoroethane
R-170	Ethane
R-216	1,3-Dichloro-1,1,2,2,3,3-hexafluoropropane
R-290	Propane
R-C 318	Octafluorocyclobutane
R-500	Azeotrope of R-12 and R-152a
R-502	Azeotrope of R-12 and R-115
R-503	Azeotrope of R-23 and R-13
R-504	Azeotrope of R-32 and R-115
R-600	*n*-Butane
R-600a	Isobutane
R-717	Ammonia
R-744	Carbon dioxide
R-1150	Ethylene
R-1270	Propylene

REFRIGERANT PROPERTIES

A number of refrigerants which permit optimum selection for a specific application have been developed. Factors that are important include (1) chemical, thermodynamic, and physical properties, (2) system capacity required, (3) compressor type, (4) desired temperature level, and (5) safety considerations.

The halogenated hydrocarbons are predominantly used for both air-conditioning and low-temperature service. Primary advantages are their nonflammable, nonexplosive, and nontoxic properties. Consequently, these refrigerants have replaced to a great extent those formerly used, such as methyl chloride, carbon dioxide, sulfur dioxide, propane, propylene, and ethylene. Ammonia (R-717) finds application in low-temperature work with reciprocating-type compressors and when high toxicity is not a critical factor.

Important properties for a refrigerant include the following:

Boiling Temperature and Pressure It is desirable to maintain a pressure above atmospheric to avoid air and moisture leakage into the system. Therefore, the refrigerant boiling point should be lower than the desired system temperature level.

Freezing Temperature The refrigerant selected must have a freezing temperature well below the minimum temperature at which the system will be operated.

Critical Temperature and Pressure The operating system pressure and temperature must be well below the critical values. The critical temperature is the temperature above which no amount of pressure will liquefy a specific gas. Above the critical condition liquid and gaseous phases have identical properties.

Condenser and Evaporator Pressure The condenser pressure should be low enough to permit the use of relatively lightweight equipment. The higher the system operating pressure, the greater the expense for equipment and piping. The evaporator pressure should not be excessively low so that the compression ratio becomes abnormally high.

Specific Volume This property relates directly to the size of the compressor when multiplied by the weight flow. It is desirable to have low suction volumes for reciprocating compressors and high suction volumes for centrifugal compressors. Reciprocating compressors normally use R-12, R-22, R-500, R-502, R-13, and R-717. Centrifugal compressors are adaptable to R-11, R-12, R-114, R-113, and, in very large tonnages, R-22.

Latent Heat A high latent heat of vaporization is important since it affects the size of the refrigerating effect, the amount of refrigerant circulated, and the size and cost of auxiliary piping and equipment. However, it must not be considered alone but along with other properties such as specific volume of the vapor and specific heat of the liquid.

Specific Heat of Liquid A low value is desired. Otherwise, too much cooling is required for the hot liquid entering the evaporator.

Molecular Weight This property is directly related to vapor specific volume: the higher the molecular weight, the higher the specific volume. For centrifugal-compressor applications, requiring large gas quantities, a refrigerant should have a high molecular weight.

Theoretical Horsepower per Ton At air-conditioning levels, this value is approximately the same for most refrigerants. It becomes more important at lower temperatures.

Discharge Temperature Refrigerants that have relatively high compressor discharge temperatures are likely to cause oil breakdown and formation of sludges. R-502, for instance, enables a much lower discharge temperature for reciprocating compressors than R-22.

Miscibility Miscibility aids in the return of oil from the evaporator to the compressor in reciprocating applications, thus minimizing this type of problem. R-12 and R-500 are highly miscible, R-22 and R-502 are less miscible, and R-717 is not miscible with oil.

Safety Aspects Refrigerants are grouped according to toxicity and flammability. Halogenated hydrocarbons such as R-12, R-22, R-502, and R-13 are classified by ASA Standard B9.1 as Group 1. Group 1 is the least hazardous relative to flammability and explosiveness and has negligible toxicity. As the group number increases, the hazard increases. R-717, methyl chloride, and sulfur dioxide are included among the refrigerants in Group 2, which are toxic or flammable or both. Group 3 refrigerants are highly flammable and explosive and include propane, propylene, ethylene, ethane, methane, butane, and isobutane.

Other Desirable Properties In addition to having the properties listed in the preceding paragraphs, a refrigerant should be stable and noncorrosive and have high thermal conductivity and low viscosity. A low cost per pound is desirable, but this consideration seldom plays an important part in the final evaluation of commercially available refrigerants.

There has been increased interest in azeotrope refrigerant mixtures. An azeotrope is a mixture, usually of two compounds, which behaves physically as though it were a single pure compound. R-500, an azeotrope of R-12 and R-152a, with a composition ratio of 73.8 to 26.2 weight percent respectively, has been used in reciprocating compressors to provide capacity between R-12 and R-22. A newer azeotrope, R-502, has been developed for reciprocating compressors primarily for application at lower temperatures between −17.8 and −28.9°C (0 and −20°F). This is a mixture of R-22 and R-115 in the ratio of 48.5 to 51.2 weight percent. An existing R-22 reciprocating unit, operating within this range, can be increased in capacity by 4 to 15 percent by charging with R-502.

Figure 12-26 is a vapor pressure-temperature chart for various refrigerants. Table 12-7 presents the comparative performance of specific refrigerants at various operating temperatures. Section 3 contains complete thermodynamic properties for many refrigerants.

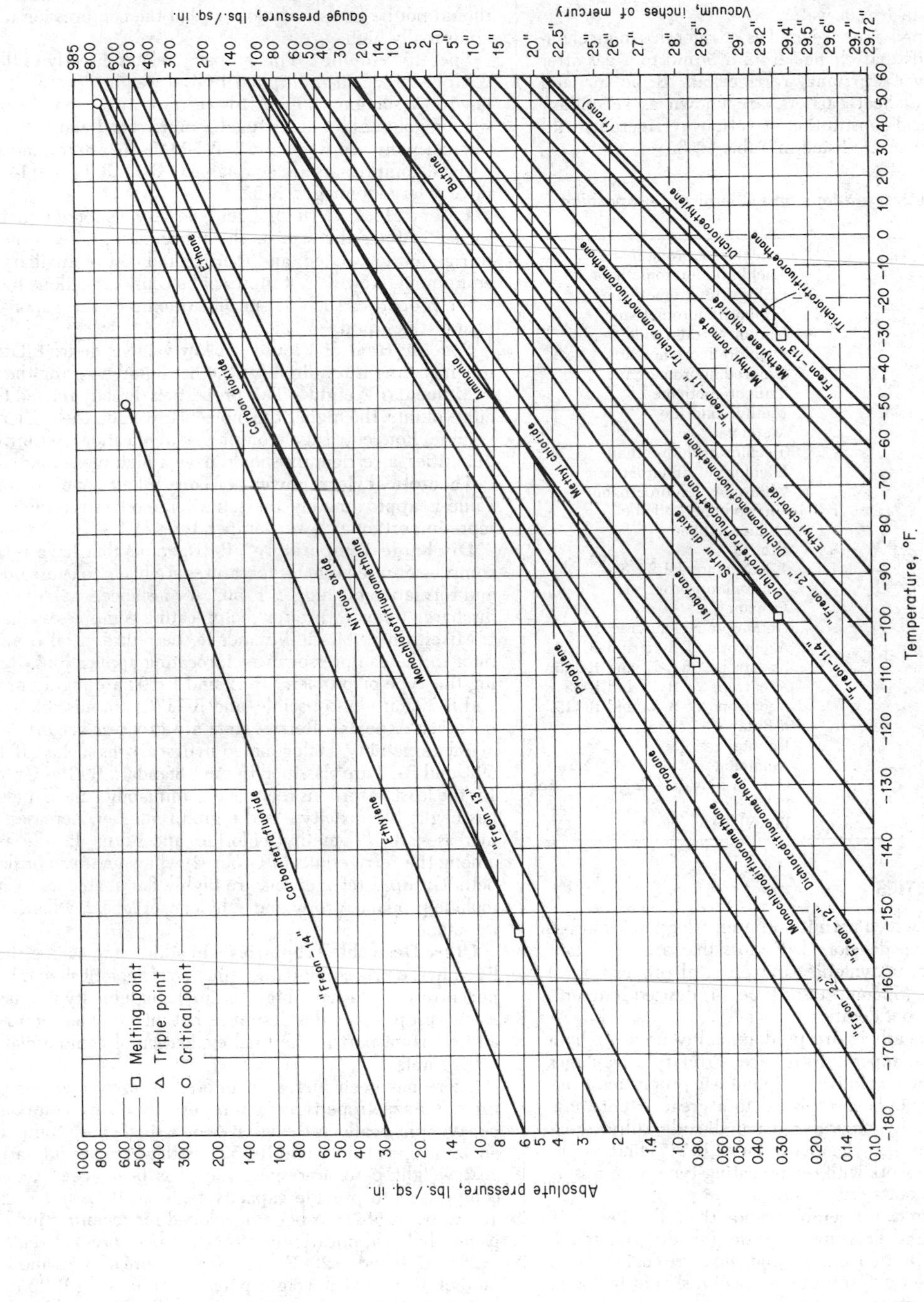

FIG. 12-26 Pressure-temperature relationship of refrigerants. (*E. I. du Pont de Nemours & Co., Inc.*)

TABLE 12-7 Theoretical Performance of Halocarbon Refrigerants

Based on one ton of refrigeration

+40°F. Evaporating, 100°F. Condensing

Property	R-11	R-12	R-13B1	R-21	R-22	R-113	R-114	R-115	R-C318	R-502
Superheat temperature, °F.	65	65	65	65	65	65	65	65	65	65
Evaporator pressure, p.s.i.g.	15.6*	37.0	123.9	4.8*	68.5	24.5*	0.4	57.9	7.4	80.2
Condenser pressure, p.s.i.g.	8.8	117.2	300.6	25.3	195.9	8.6*	31.2	166.2	52.7	214.4
Compression ratio	3.34	2.55	2.27	3.25	2.53	3.95	3.04	2.49	3.04	2.41
Net refrigerating effect,† B.t.u./lb.	71.48	54.27	30.56	93.44	73.27	59.40	62.60	34.50	37.35	49.61
Refrigerant circulated, lb./min.	2.798	3.685	6.545	21.40	2.729	3.367	3.195	5.797	5.354	4.032
Specific volume of vapor, cu. ft./lb.	5.724	0.828	0.232	4.749	0.706	11.240	2.109	0.456	1.202	0.480
Compressor displacement, cu. ft./min.	16.014	3.050	1.516	10.164	1.928	37.841	6.738	2.645	6.435	1.936
Horsepower	0.63	0.66	0.76	0.69	0.68	0.63	0.50	0.75	0.69	0.72
C.O.P.	7.530	7.100	6.203	6.863	6.901	7.499	9.397	6.267	6.840	6.578
Discharge temperature, °F.	139.2	133.0	137.5	167.0	152.3	118.2	110.8	109.6	103.7	133.9

−30°F. Evaporating, 100°F. Condensing

Property	R-11	R-12	R-13B1	R-21	R-22	R-113	R-114	R-115	R-C318	R-502
Superheat temperature, °F.	65	65	65	65	65	65	65	65	65	65
Evaporator pressure, p.s.i.g.	27.8*	5.5*	24.9	26.1*	4.9	29.3*	24.6*	3.0	22.2*	9.4
Condenser pressure, p.s.i.g.	8.8	117.2	300.6	25.3	195.9	8.6*	31.2	166.3	52.7	214.4
Compression ratio	22.58	10.99	7.97	21.21	10.76	35.09	17.68	10.21	17.74	9.51
Net refrigerating effect,† B.t.u./lb.	71.92	56.22	32.96	93.84	76.47	59.50	48.71	36.89	38.25	52.06
Refrigerant circulated, lb./min.	2.781	3.558	6.068	2.131	2.615	3.361	4.106	5.421	5.229	3.842
Specific volume of vapor, cu. ft./lb.	39.436	3.811	0.918	28.962	3.255	100.480	12.693	2.002	7.344	2.081
Compressor displacement, cu. ft./min.	109.66	13.56	5.57	61.73	8.514	337.745	52.117	10.854	38.402	7.997
Horsepower	1.81	1.91	2.18	1.90	2.05	1.76	1.82	2.03	1.93	2.08
C.O.P.	2.600	2.468	2.159	2.483	2.302	2.673	2.595	2.320	2.449	2.271
Discharge temperature, °F.	262.5	231.0	227.6	318.0	283.0	206.7	179.1	171.4	155.9	221.5

−100°F. Evaporating, −30°F. Condensing

Property	Saturated conditions				Superheated vapor			
	R-13B1	R-13	R-22	R-23	R-13B1	R-13	R-22	R-23
Superheat temperature, °F.	...	...	...	...	−50	−50	−50	−50
Evaporator pressure, p.s.i.g.	16.6*	7.5	25.0*	9.0	16.6*	7.5	25.0*	9.0
Condenser pressure, p.s.i.g.	24.9	90.9	4.9	111.3	24.9	90.9	4.9	111.3
Compression ratio	6.05	4.75	8.16	5.31	6.05	4.75	8.16	5.31
Net refrigerating effect,† B.t.u./lb.	41.86‡	46.39‡	90.82‡	78.05‡	46.86†	55.69†	97.39†	86.16†
Refrigerant circulated, lb./min.	4.778	4.311	2.202	2.562	4.268	3.591	2.054	2.321
Specific volume of vapor, cu. ft./lb.	3.889	1.564	18.433	2.191	4.456	1.950	21.241	2.559
Compressor displacement, cu. ft./min.	18.58	6.74	40.59	5.61	19.02	7.002	43.619	5.941
Horsepower	1.08	1.13	1.07	1.13	1.10	1.15	1.13	1.19
C.O.P.	4.378	4.190	4.408	4.167	4.306	4.101	4.175	3.974
Discharge temperature, °F.	10.9	−0.6	53.8	36.7	61.8	74.5	112.1	93.0

*Inches of mercury below 1 atm.
†Superheated vapor and saturated liquid.
‡Saturated vapor and liquid.

EQUIPMENT SELECTION

There is no universal refrigerant that can be used for all applications. The same is true in selecting the type of refrigeration equipment to meet a given cooling duty.

A number of variables must be studied. These include (1) refrigeration load, (2) temperature level to which process fluid must be cooled, (3) energy source for driving the refrigeration unit, (4) quantity available and temperature of condensing media, and (5) space. On occasions more than one type may be technically suited. Therefore, it becomes necessary to make a selection from practical considerations of the unit size or capacity available, investment and operating costs, operating flexibility, maintenance expense, and reliability.

Reciprocating equipment finds its widest use up to 528-kW (150-tons) capacity at air-conditioning levels. However, reciprocating compressors have higher maintenance costs and require more space per ton, placing them at a disadvantage in the larger sizes. Centrifugal machines are usually high-capacity units between 528 and 29,920 kW (150 and 8500 tons). Maintenance costs are lower than for reciprocating machines, and reliability is good.

Absorption units find chilled-water application when low-pressure steam is available at low cost. Capacities range from 352 to 4224 kW (100 to 1200 tons) in a single unit when chilling water to 7.2 to 10°C (45 to 50°F). The main disadvantage is the difficulty in maintaining a tight system with the highly corrosive lithium bromide and an operating vacuum of 0.2 inHg absolute in the evaporator and absorber. Lithium bromide absorption units cannot be applied to brine cooling duty.

Steam-jet units also are used on chilled-water applications, normally in the range of 176 to 5280 kW (50 to 1500 tons) per machine. These machines offer low investment and maintenance costs, especially if a barometric-type condenser is used. This is particularly suitable for relatively high chilled-water outlet-temperature applications when excess steam is available at about 689 kPa (100 psig) and relatively cool condensing water is also available. The main disadvantages are the physical size of the equipment and high steam and water requirements.

VAPOR-COMPRESSION CYCLES

Single-Stage Refrigeration Cycle Figure 12-27 shows the basic refrigeration cycle used for single-stage vapor compression. The four basic components in the system are the compressor, condenser, expansion, valve, and evaporator. The cycle involves two pressures, high and low, to enable a continuous process to produce a cooling effect.

As liquid refrigerant flows through the evaporator, heat is absorbed from a fluid being cooled and the refrigerant boils. The low-pressure vapor is compressed. The pressure and temperature levels are increased to a point at which the superheated vapor can be condensed by the cooling media available. In compressing the gas, heat of compression is added to vapor as the pressure is raised. The vapor then flows to the condenser, where the gas is liquefied. The liquid refrigerant flows from the condenser to an expansion valve, where its pressure and temperature are reduced to those in the evaporator. The cycle is thus completed. Figure 12-28 is a schematic diagram of a typical centrifugal refrigeration cycle.

Refrigeration cycles can best be analyzed by means of a Mollier diagram or pressure-enthalpy chart. Figure 12-29 is a typical *P-h* diagram for R-12 with a typical refrigeration cycle indicated. When work is done or heat transferred, the refrigerant undergoes a change in enthalpy. The curve on the left of the diagram is the saturated-liquid line, and the curve on the right is the saturated-vapor line. In

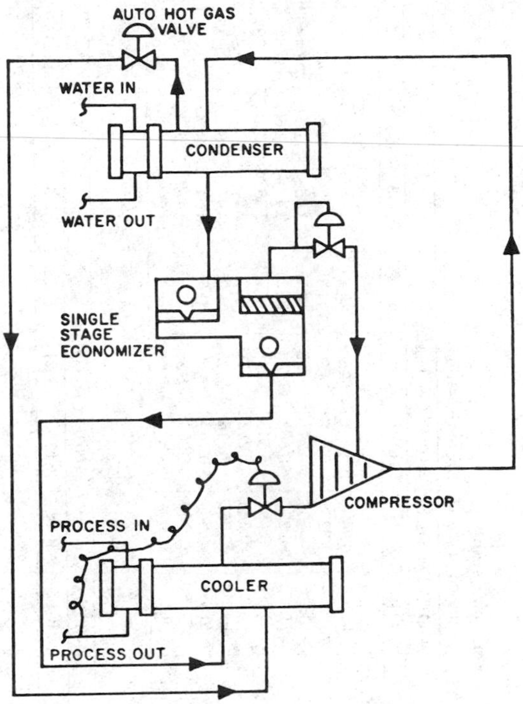

FIG. 12-28 Centrifugal refrigeration system. *(Carrier Air Conditioning Company.)*

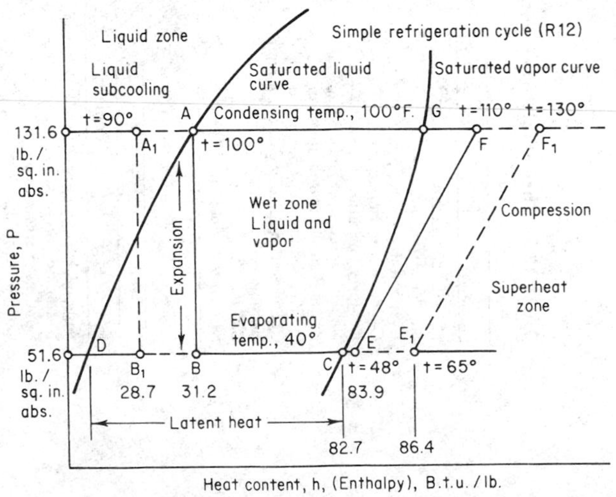

FIG. 12-29 Typical *P-h* diagram for R-12. [*Holman,* Refrig. Serv. Contr. *(January 1968).*]

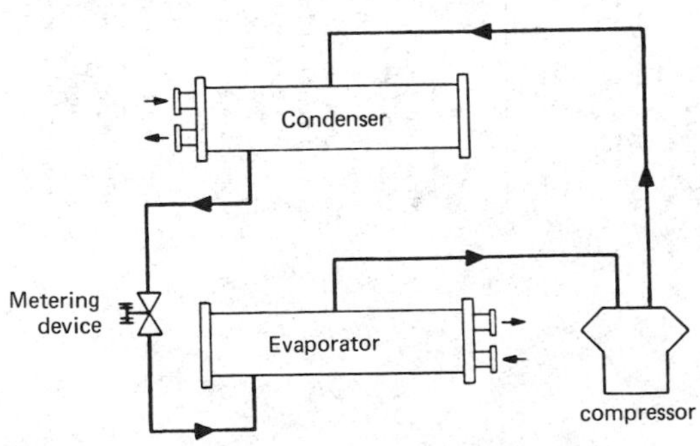

FIG. 12-27 Basic refrigeration cycle. *(York, Division of Borg-Warner.)*

the area between these saturation curves, the refrigerant exists as a mixture of liquid and vapor. All points to the right of the saturated-vapor line represent superheated-vapor conditions. Those to the left of the liquid-saturation line represent refrigerant in a liquid phase at a temperature lower than that of the saturation temperature for the existing pressure. This area is known as the subcooled region. Evaporation and condensation are considered constant-pressure processes, so they are represented by horizontal lines on the *P-h* diagram. Ideally, compressing a gas is an isentropic process (constant entropy), which assumes no loss of heat and the absence of friction. Constant-entropy lines are shown on the diagram in the superheat zone, designated "compression."

The theoretical cycle starts at point *A*, where saturated liquid enters the expansion valve from the condenser. Flow of refrigerant through an expansion valve is a throttling process without change in enthalpy, so a vertical line from *A* to *B* represents this process. As the liquid refrigerant expands to evaporator pressure, some of it flashes and cools the remainder of the liquid to evaporator temperature. After absorbing heat from the fluid being cooled in the evaporator, refrigerant vapor leaves the coil at point *E* and enters the compressor.

The gas is then compressed from a low pressure to a high pressure isentropically from *E* to *F*. In the condenser, the superheated gas is initially desuperheated from *F* to *G* until the saturated-vapor curve is again reached, at which time condensation begins. Condensation occurs along line *GA*. At point *A* the vapor has been completely condensed into a liquid. This completes the theoretical single-stage cycle.

By means of a *P-h* diagram, calculations of cycle conditions can be readily understood. The **net refrigerating effect** accomplished in the evaporator is

$$\text{RE} = h_g - h_f \qquad (12\text{-}18)$$

where RE = refrigerating effect, Btu/lb
h_g = enthalpy of vapor leaving the evaporator, Btu/lb
h_f = enthalpy of liquid leaving the condenser, Btu/lb

Subscripts *g* and *f* are standard nomenclature respectively for gas and fluid streams and do not refer to Fig. 12-29, where points *E* and *A* respectively are analogous.

The **weight of the refrigerant circulated per ton** of capacity is determined by

$$\text{Weight flow, lb/(min}\cdot\text{ton)} = \frac{200\ \text{Btu/(min}\cdot\text{ton)}}{\text{RE, Btu/lb}} \qquad (12\text{-}19)$$

The **theoretical volume of vapor** to be handled per ton is

$$\text{ft}^3/(\text{min}\cdot\text{ton}) = \text{weight flow} \times V_g \qquad (12\text{-}20)$$

where V_g = specific volume of the suction vapor entering the compressor, ft^3/lb. The **heat of compression** is the difference in enthalpy between the discharge h_d and inlet compressor gas conditions h_a, i.e., points *F* and *E*, respectively, in Fig. 12-29.

$$\text{Heat of compression} = h_d - h_g \qquad \text{Btu/lb} \qquad (12\text{-}21)$$

The **work of compression** per ton then is determined by multiplying the heat of compression by the weight flow or

$$\text{Work of compression, Btu/(min}\cdot\text{ton)} = (h_d - h_g) \times \text{weight flow} \qquad (12\text{-}22)$$

The work required normally is expressed as horsepower per ton. Since 1 hp equals 42.4 Btu/min,

$$\text{hp/ton} = \frac{\text{work of compression, Btu/min}}{42.4\ \text{Btu/min}} \qquad (12\text{-}23)$$

Condenser heat load is determined by subtracting the enthalpy of the saturated liquid leaving the condenser from the enthalpy of the superheated vapor entering the condenser, or

$$\text{Condenser heat load, Btu/lb} = h_d - h_f$$

Note also that condenser heat load is equal to refrigerating effect plus heat of compression.

The cycle **coefficient of performance** is defined as the ratio of the refrigeration produced to the work supplied, each expressed in the same thermal units.

$$\text{COP} = \frac{\text{net refrigerating effect, Btu/lb}}{\text{heat of compression, Btu/lb}} \qquad (12\text{-}24)$$

On the *P-h* diagram (Fig. 12-29) the basic single-stage cycle is represented by *ABCEFGA*. However, the basic equipment is often supplemented by adding a liquid-to-suction heat exchanger to the system. This heat exchanger is located usually in the suction line between the evaporator and the compressor. This results in subcooling the warm liquid leaving the condenser and superheating the vapor returning from the evaporator. The cycle then becomes $A_1B_1CE_1F_1GA_1$.

The effect of subcooling the liquid is to increase the net refrigerating effect and the capacity of the system. When the liquid is cooled outside the evaporator, less of the latent heat of evaporation is used up in cooling the liquid to the evaporation temperature. By referring to the conditions of Fig. 12-29, the increase in net refrigeration effect for subcooling alone would be

$$\text{RE} = h_E - h_{B1} = 83.9 - 28.7 = 55.2\ \text{Btu/lb}$$

rather than 83.9 − 31.2 or 52.7 Btu/lb with the basic cycle. The percentage increase in capacity thus becomes

$$(2.5 \times 100)/51.5 = 4.1\ \text{percent}$$

The effect of superheating the vapor is to reduce slightly the capacity of the compressor and the capacity of the system. The specific volume of the vapor will increase from 0.80 ft^3/lb at *E* to 0.828 ft^3/lb at E_1. The weight of refrigerant circulated without subcooling would be 200/52.7 = 3.79 lb/min for a theoretical compressor displacement of 3.79 × 0.80 × 3.03 ft^3/(min·ton). With suction vapor superheating from 48 to 65°F the compressor displacement would be 3.03 × 0.828/0.80 × 3.09 ft^3/(min·ton).

To calculate the decrease in capacity assume that the compressor is fixed and will handle 3.03 ft^3/(min·ton). With vapor entering at 65°F this would be 3.03/0.828 × 3.66 lb/min weight flow.

If the net refrigeration effect is assumed to be 52.7 Btu/lb, the capacity of the system would then be

$$52.7 \times 3.66 \times 192.9\ \text{Btu/min}$$

The loss in capacity is 200 − 192.9 = 7.1 Btu/min, or 7.1 × 100/200 = 3.5 percent. The overall effect, then, of liquid subcooling plus vapor superheating on the cycle efficiency in this case is 4.1 − 3.5 = 0.6 percent net increase.

Superheating the vapor returning to the compressor with a heat exchanger is an important feature from an operating standpoint regardless of the overall cycle efficiency. Liquid carry-over of refrigerant from the evaporator is vaporized to minimize "liquid slugging," which can cause serious compressor damage. Liquid carry-over can occur as a result of malfunctioning of the expansion valve or for other off-standard operating reasons.

Multistage Cycles To avoid high compression ratios it is necessary for low-temperature applications to have several stages of compression. A refrigeration system that consists of more than one stage of compression is defined as a multistage system. Multistage systems are of two basic types, **compound** and **cascade.** Some applications between −78.3 and −101.1°C (−100 and −150°F) combine the advantages of both types. Equipment manufacturers stipulate that compression ratios not exceed 10:1 because of resulting high discharge temperatures, oil problems, and bearing considerations.

Economics, as well as mechanical-equipment considerations, dictates the use of multistage systems below about −28.9 to −34.4°C (−20 to −30°F) suction temperatures. In this temperature range, two-stage systems can enable power savings because of increased compressor volumeteric efficiency with lower compression ratios. Investment savings also are possible with smaller compressor and motor sizes.

With reciprocating or rotary-type compressors, two-stage plants are practical from about −28.9 to −57.7°C (−20 to −70°F).

Three-stage systems are used at temperatures down to about −87.2°C (−125°F). A temperature of about −101.1°C (−150°F) is considered a minimum for normal refrigeration equipment service. Centrifugal compressors combine several impeller wheels in series so that it is possible to have one multistage centrifugal compressor operating down to about −42.8°C (−45°F).

Refrigerants commonly used for multistage applications include R-12, R-22, R-502, R-717, R-13, R-14, and occasionally ethane, ethylene, propylene, or propane.

Compound Cycles The general arrangement of a compound cycle using reciprocating compressors is shown in Fig. 12-30. A booster or first-stage compressor and a gas-liquid interstage cooler have been added to the basic refrigeration cycle. Only one refrigerant is used in a compound cycle. The low-stage compressor discharges gas directly in series to the suction of the high-stage compressor after desuperheating in an interstage cooler. The interstage cooler also contains a submerged cooling coil for liquid subcooling of the refrigerant to the low-stage evaporator. Some of the high-pressure liquid is flashed at the intermediate pressure to enable this cooling.

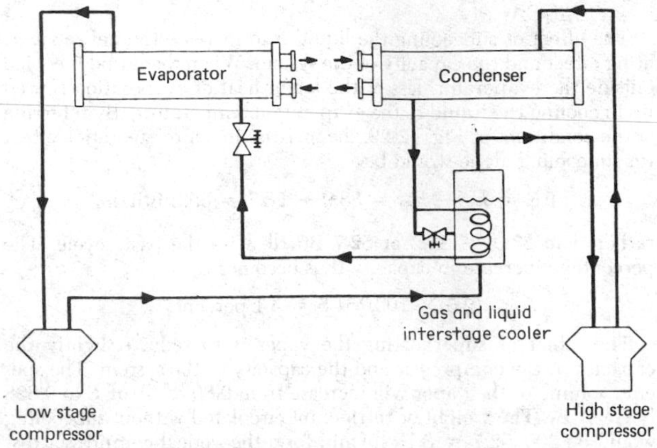

FIG. 12-30 Compound compression. *(York, Division of Borg-Warner.)*

The work of compression should be divided equally between the two stages, and under these conditions the intermediate pressure is determined by the following equations:

$$P_2 = \sqrt{P_1 \times P_3} \tag{12-25}$$

where P_2 = absolute intermediate pressure, kPa psia
P_1 = absolute suction pressure, low-stage compressor, kPa psia
P_3 = absolute discharge pressure, high-stage compressor, kPa psia

The *P-h* diagram is useful in determining the capacity required of the second-stage compressor. The capacity of the low-stage compressor is fixed by the cooling duty required. The high-stage compressor must be adequate to handle the evaporator load plus the flash gas developed in the intercooler. As a general rule, the intercooling duty will require that the high-stage compressor have a capacity about 15 to 35 percent higher than that of the low-stage compressor. This does not mean a direct proportional increase in compressor displacement by these percentages since gas density is much greater at intermediate conditions than at evaporator conditions.

A compound centrifugal system (Fig. 12-31) also uses series flow of a single refrigerant. The main difference is in the use of a flash-type desuperheater instead of the vertical gas-liquid interstage cooler as before. Interstage intercoolers with internal float valves also are incorporated into this system to flash gas off to intermediate centrifugal-compressor stages. These intercoolers (economizers) may be external vessels or be located in the side of the cooler shell.

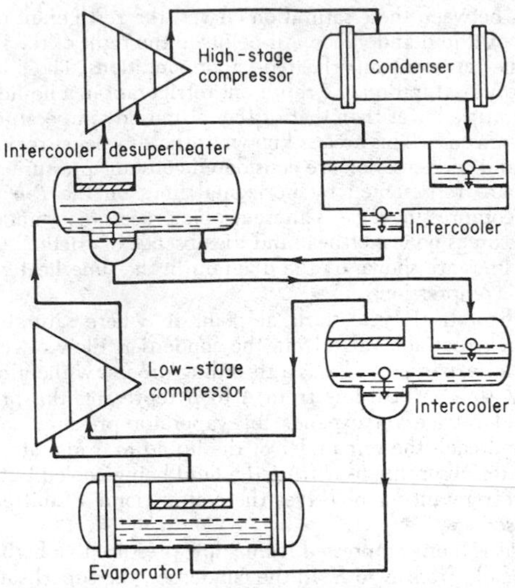

FIG. 12-31 Multistage centrifugal system. [*Brown and Briley,* Refrig. Serv. Contr. *(September 1968).*]

Cascade Cycles This multistage application involves two separate refrigeration systems which are interconnected in such a manner that one provides a means of heat rejection for the other. Cascade systems permit the use of different refrigerants in each cycle of the cascade to produce low temperatures. This has an advantage over compound systems in that refrigerants with low cubic feet per minute per ton and low compression ratios can be used in the low stage to produce very low temperatures. For example, using R-13 on the low stage of a cascade cycle to obtain −76.1°C (−105°F) will require 6.9 ft^3/(min·ton) and a 5.6 compression ratio. R-22, on the other hand, would require 48.8 ft^3/(min·ton) for the low stage and a 10.0 compression ratio. These figures are based on a high-stage condensing temperature of 26.7°C (80°F). Therefore, in this instance the low-stage compressor could be selected more economically by using a suitable refrigerant for a given duty. Cascade cycles are normally considered for applications between −45.6 and −101.1°C (−50 and −150°F).

Figure 12-32 presents the general arrangement for a two-stage cascade system. A cascade condenser is shown between stages. This heat exchanger serves as the condenser of the low stage and the evaporator for the high stage. Another advantage of the cascade cycle is that low temperatures are theoretically unlimited. Any number of cascaded stages are possible by the use of an additional cascade condenser and an additional stage of compression.

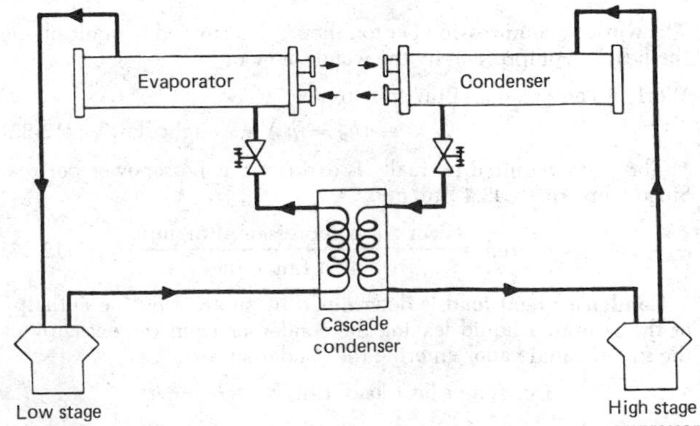

FIG. 12-32 Basic two-stage cascade system. *(York, Division of Borg-Warner.)*

The cascade condenser is usually sized so that the evaporating temperature for the high stage is 5.5°C (10°F) below the condensing temperature of the low stage. This requirement of temperature overlap for the cascade condenser reduces overall cycle efficiency in comparison with a compound system. This is the main disadvantage of the cascade cycle. The cascade condenser also requires a higher initial investment than the intercoolers used on the compound system. However, the possibility of using more economical refrigerants with the cascade arrangement may offset this additional heat exchanger cost.

The heat load on the cascade condenser, or the high-stage evaporator load, is equal to the low-stage tonnage plus the power of compression for the low-stage compressor.

A *P-h* diagram for a two-stage centrifugal cascade cycle is shown in Fig. 12-33. This consists simply of overlapping two basic single-stage vapor cycles upon each other.

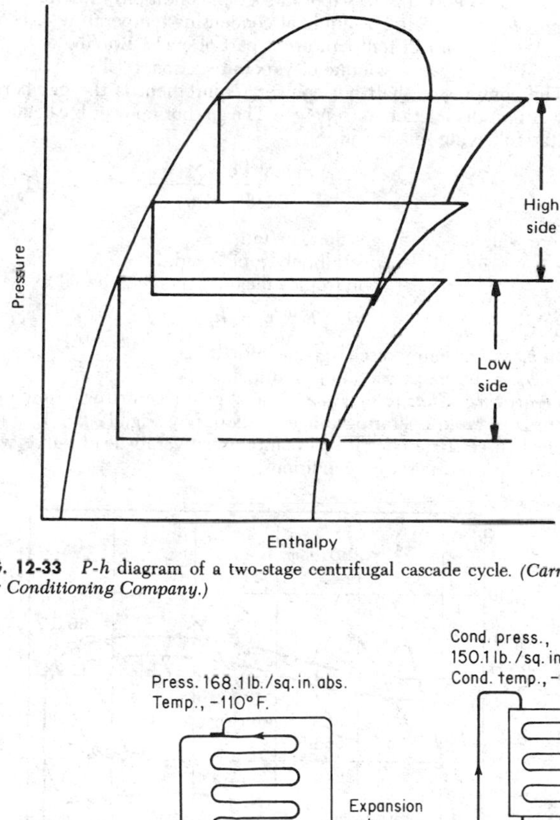

FIG. 12-33 *P-h* diagram of a two-stage centrifugal cascade cycle. *(Carrier Air Conditioning Company.)*

Figure 12-34 presents a common cascade arrangement used to produce −78.9°C (−110°F) refrigerant for a high-altitude test chamber. This system uses R-13 on the low stage and R-22 on the high stage. R-13 requires the use of an expansion tank because of high ambient pressures. For example, with R-13, a pressure of about 3445 kPa (500 psig) exists at 23.9°C (75°F) ambient temperature. It is desirable to be able to charge the system with a "fade-out" charge such that, on shutdown, all liquid in the system will evaporate into gas and the gas can warm up to ambient temperature without resulting in excessive system pressure. The usual method of handling this situation is to provide an expansion tank which is connected to the evaporating side of the refrigerating system. The expansion tank must be sized so that it, in combination with the volume of the system, can contain all the refrigerant in vapor form within the design tank pressure based on expected ambient temperature.

Combination Compound-and-Cascade Cycles It is sometimes desirable to combine the advantages of both a compound and a cascade cycle. Such an arrangement is shown on Fig. 12-35, using centrifugal compressors. It also can be adapted to reciprocating compressors. R-13 is used in the low stage, condensing the R-13 with a two-stage compound system using R-12.

VAPOR-COMPRESSION EQUIPMENT

Introduction Three main types of compressors are used for basic refrigeration vapor cycles utilizing common refrigerants. These comprise the dynamic-type centrifugal compressor, the displacement-type reciprocating compressor, and the wet rotary screw compressor. Steam jets also have been used to compress low-pressure water vapor to produce water for air-conditioning applications.

In recent years, there has been considerable debate over the relative merits of centrifugal versus reciprocating compressors. Actually, the two compressor types complement each other, and each has its own field of application. Large-capacity systems are most economically handled by centrifugal machines. Reciprocating compressors can be properly applied for systems of 150 tons or less on air-conditioning requirements and for specialized low-temperature work when inlet-gas volumes are not too great.

Centrifugal Compressors The main component in this type of compressor is the impeller wheel, which imparts energy to the gas being compressed. The impeller blades supply energy to the gas owing to centrifugal force and high velocity, which is subsequently transformed into pressure head.

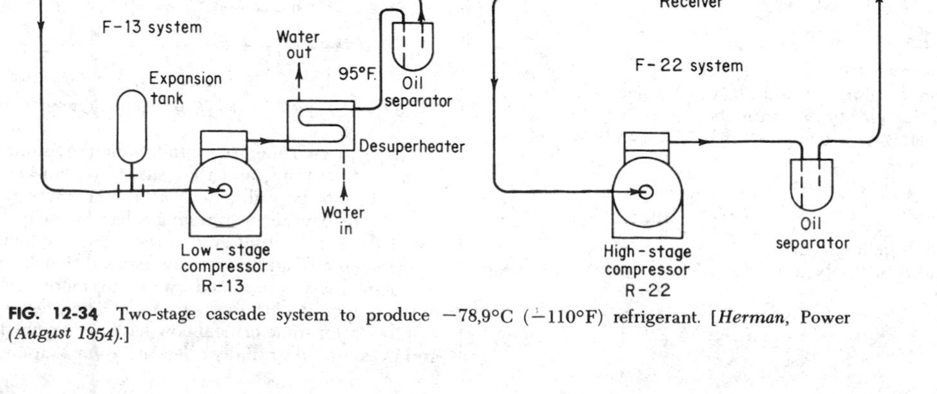

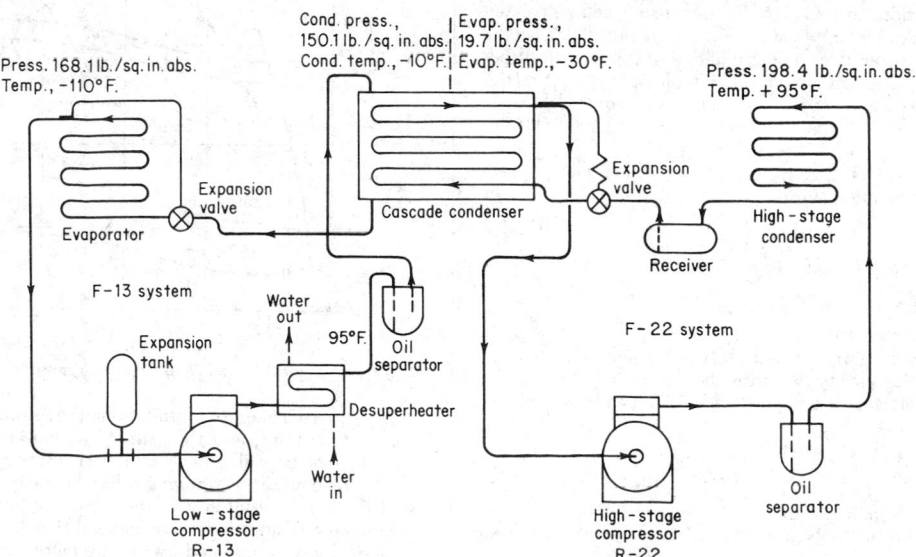

FIG. 12-34 Two-stage cascade system to produce −78,9°C (−110°F) refrigerant. [*Herman*, Power *(August 1954)*.]

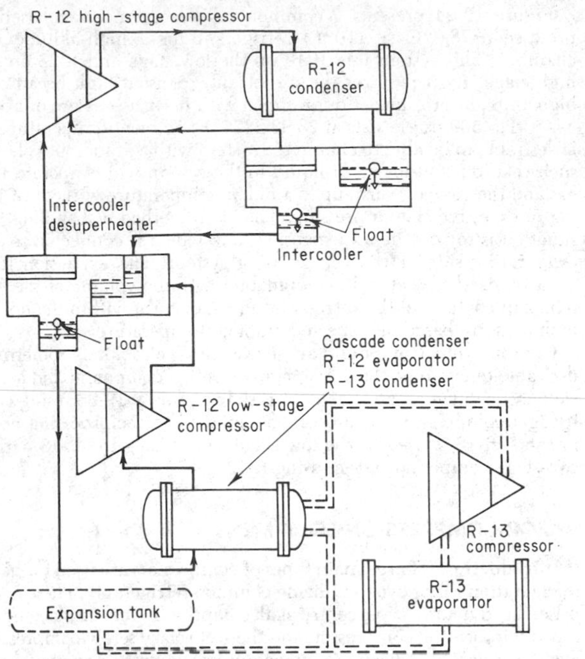

FIG. 12-35 Combination compound-and-cascade refrigeration system. [*Brown and Briley,* Refrig. Serv. Contr. *(September 1968).*]

The pressure lift through which a fluid may be raised is a function of the square of the velocity and the density. The overall head or lift against which the compressor must operate is determined by the suction and condensing properties of the refrigerant selected. The number of impeller wheels required is a direct function of the overall head requirements and the ability of a given size of impeller to produce a certain head.

The physical compressor size and thus the allocation of plant space by the engineer for the equipment are dependent upon (1) the number of centrifugal impeller wheels (or stages) and (2) the wheel diameter. R-113 will require a polytropic head of 8000 ft when operating at 0.6°C (33°F) suction, 46.1°C (115°F) condensing, and 75 percent compressor efficiency. This application could be handled by a single-stage impeller, as it is possible to obtain about 10,000 ft of head per stage. On the other hand, ammonia, a low-molecular-weight refrigerant, would require eight stages to develop the required head to operate between −6.7°C (20°F) suction and 43.3°C (110°F) condensing.

The polytropic head per stage is a direct function of the tip speed, according to the following formula:

$$W = (\mu_p \times U^2)/g \tag{12-26}$$

where W = polytropic head, ft
U = tip speed, ft/s
g = 32.2 ft/s²
μ_p = pressure coefficient

The pressure coefficient μ_p varies slightly according to relative gas volumes and acoustic velocity and must be determined by test. An average value for refrigerant compressors is 0.50, and tip speeds are normally held to 800 ft/s.

The impeller diameter is determined by

$$D = (U \times 12 \times 60)\ (\pi \times \text{r/min}) \tag{12-27}$$

where D = impeller diameter, in.

The overall compressor polytropic head W_o is calculated by the following formula:

$$W_0 = 144 \log_e \frac{P_2}{P_1} \left[\frac{P_2V_2 - P_1V_1}{\log_e(P_2V_2/P_1V_1)} \right] \tag{12-28}$$

where W_0 = overall polytropic head, ft
P_2 = absolute discharge pressure, kPa (psia)
P_1 = absolute suction pressure, kPa (psia)
V_2 = specific volume discharge gas, m³/kg (ft²/lb)
V_1 = specific volume suction gas, m³/kg (ft²/lb)

In determining the compressor overall head requirements, polytropic analysis is preferred to isentropic analysis, which does not consider machine losses due to turbulence, impact, and fluid friction.

In selecting a compressor-impeller size to meet design requirements, it is necessary also to determine the flow volume Q_0 as well as the head. This value is calculated by the following equation:

$$Q_0 = \left(\frac{200 \times q}{h_g - h_f} \right) V_{gs} \tag{12-29}$$

where Q_0 = quantity of refrigerant at compressor intake, ft³/min
q = heat load, tons
h_g = enthalpy of vapor at suction conditions, Btu/lb
h_f = enthalpy of liquid at condensing temperature or economizer temperature if part of cycle, Btu/lb
V_{gs} = specific volume of vapor at suction, ft³/lb

The compressor shaft-horsepower requirement is the gas horsepower plus the friction horsepower. The gas horsepower is calculated by the following relationship:

$$\text{Gas hp} = \frac{\text{weight flow} \times W_0}{33{,}000 \times \eta_p} \tag{12-30}$$

where weight flow = gas flow, lb/min
W_0 = overall polytropic head, ft
η_p = polytropic efficiency, percent, equal to

$$W_0/[778(h_d - h_g)]$$

with h_d = enthalpy discharge gas, Btu/lb
h_g = enthalpy suction gas, Btu/lb

Operating Characteristics For a given centrifugal impeller, there is a stable operating range as shown by Fig. 12-36. This is a typical three-stage centrifugal-compressor overall head curve with R-12 at low-temperature conditions.

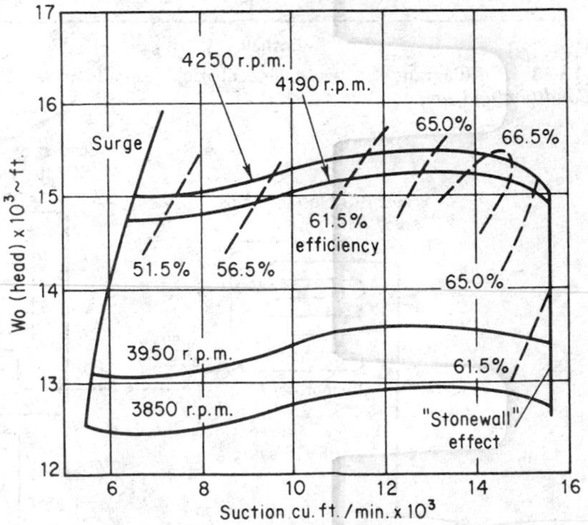

FIG. 12-36 Typical overall centrifugal-compressor curve for R-12.

There are two operating limitations: the "stone-wall" region at the high-gas-flow limit, and the "surge" region at low-gas-flow conditions. The stone-wall effect will occur when the design of the impeller reaches its maximum gas-handling capabilities regardless of speed thus preventing any further capacity increase. At the stone-wall region any additional flow beyond that for which the impeller is designed is accompanied by rapidly increasing losses. This results in the reduction of delivery pressure. The pressure drop is gradual at first, but after some critical flow has been reached, the system chokes and pressure falls rapidly. (See Sec. 6 for a discussion of surging.)

As can be noted from Fig. 12-36, centrifugal-compressor curves are plotted with polytropic head versus suction gas flow for parameters of compressor speed. Suction gas flow is directly dependent on capacity required, refrigerating effect, and specific volume of a particular refrigerant. Impellers must have a minimum gas flow of approximately 708 to 944 L/s (1500 to 2000 ft^3/min) to minimize friction losses and to eliminate higher manufacturing costs.

Also, the efficiency of a centrifugal compressor is at maximum when operating at conditions somewhat less than the stone-wall line and at selected head conditions. As conditions approach the surge line, efficiency declines. For this reason, it always is best to keep centrifugal machines loaded up to design capacity if possible. Centrifugal-compressor efficiencies will vary according to suction temperatures and flows, head required, tip speed, and the acoustical velocity of the refrigerant used.

Compressor-Capacity Control There are five basic methods of controlling the capacity of centrifugal machines. These include speed control, inlet guide vanes, and suction dampers, which are discussed in Sec. 6.

In addition, capacity can be controlled by:

Constant head. This supplements the five basic methods by maintaining a constant condenser pressure on the compressor by throttling condenser water flow. This enables water conservation during winter operation but at some expense in compressor brake horsepower per ton. It also prevents the discharge pressure from dropping too low, which could cause reduced capacity of system float valves to control liquid-refrigerant flow.

Care must be exercised in selecting the control points, as too high a pressure set point will result in compressor surging if the prerotation vanes are fairly well closed during light loads.

Hot-gas bypass. Hot gas is recirculated from the condenser to the cooler or compressor suction to enable additional gas flow at very light load conditions. This will prevent compressor surging, and it will also permit operation even down to no load if the hot-gas bypass is properly sized.

Reciprocating Compressors Present-day design of this type of compressor is primarily of the vertical, single-acting arrangement. In the refrigeration field, double-acting horizontal reciprocating compressors are considered obsolete. Reciprocating compressors are further classified into two basic categories:

1. *Open type.* The compressor is driven externally through a crankshaft extending outside the crankcase. This arrangement makes it possible to use electric-motor drives, internal-combustion engines, or reciprocating steam engines. Open-type machines may be either direct-driven or arranged for belt drive.

2. *Hermetic type.* The compressor and motor are connected by a common shaft, and both are located internally within the compressor housing. This eliminates the need for a mechanical shaft seal, thus removing a major potential source for leaks. Small hermetic compressors are normally of welded construction and are not serviceable in the field, as are the larger hermetic units.

Reciprocating compressors also can be categorized according to speed, whether slow or high speed; number of cylinders; arrangement of cylinders with respect to relative angles such as V type or W type; stages of compression, whether single- or two-stage; and also cycle location such as a booster compressor. The majority of the reciprocating compressors produced today are of the high-speed [up to 188.4 rad/s (1800 r/min)] multicylinder design primarily because of initial cost and space considerations.

The size of a reciprocating compressor is usually expressed according to the number of cylinders, bore diameter, length of stroke, operating speed, and refrigerant designation. A large multicylinder reciprocating compressor, therefore, would be identified, for example, as a 16-cylinder, 3¼- by 4-in compressor operating at 183.2 rad/s (1750 r/min) with R-12.

Compression Fundamentals All reciprocating designs are essentially alike, consisting of a reciprocating piston with the necessary intake and discharge valves. Fundamentally, a reciprocating compressor cycle can be shown on a pressure-volume diagram (Fig. 12-37). The compression stroke is represented by the curved line *AB*.

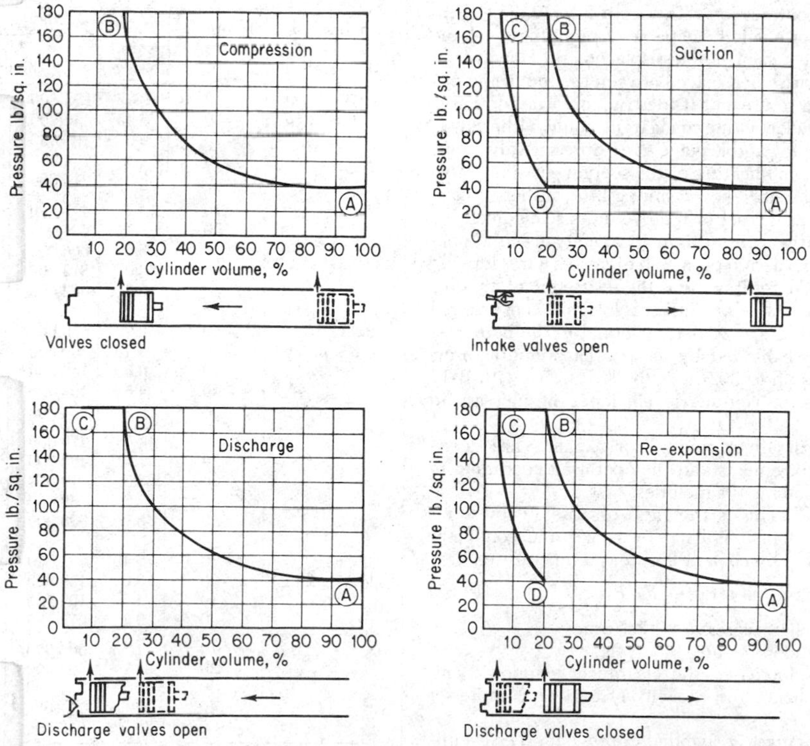

FIG. 12-37 Basic reciprocating-compressor cycle. *(Carrier Air Conditioning Company.)*

As the piston moves upward, the suction valve closes and the gas is compressed to a higher pressure and a reduced volume. At point *B* the discharge valve opens, and gas is expelled from the cylinder until the top of the stroke is reached at point *C*. At this point, the suction valve is closed and the discharge valve is open. As the piston moves downward from point *C*, the discharge valve closes and reexpansion of the high-pressure gas remaining in the clearance space occurs from *C* to *D*, at which point the suction valve opens. From point *D* vapor is drawn into the cylinder until the piston reaches the bottom of its stroke at point *A*. The cycle is thus completed.

Factors Affecting Compressor Capacity The chemical engineer should understand that the capacity of a given compressor is affected by mechanical-design factors as well as application factors. These factors are important in evaluating competitive bids and analyzing refrigeration plant operating problems.

Mechanical-design factors are those inherent to the compressor design and include:

1. *Piston displacement.* See Sec. 6.

2. *Total volumetric efficiency.* The term that expresses compressor efficiency due to various causes is called total volumetric efficiency. Total volumetric efficiency is defined as the ratio of the volume of vapor actually pumped to the compressor displacement. Total volumetric efficiency can be determined by formula from published compressor ratings as follows:

$$\text{VE total} = \frac{\text{rated capacity}}{\text{theoretical capacity}}$$

Theoretical capacity is calculated by the following formula:

$$\text{Theoretical capacity, Btu/h} = \frac{\text{compressor displacement} \times \text{refrigerating effect} \times 60}{V_g} \qquad (12\text{-}31)$$

where compressor displacement is in ft^3/min; refrigerating effect $= h_g - h_g$, Btu/lb; and V_g = specific volume at suction conditions, ft^3/lb.

Typical reciprocating-compressor volumetric-efficiency curves are shown in Fig. 12-38 for compression ratios from 2 to 10. The theoretically perfect compressor would be capable of pumping a volume of refrigerant equal to the design piston displacement. However, it is necessary to provide a clearance space between the top of the piston and the machine head to prevent its destruction. The top curve in the figure shows the efficiency due to clearance only. Other losses occur, including wire drawing and leakage of compressor valves, piston leakage, and cylinder-wall heating of the suction gas. The curve that expresses the practical efficiency resulting after all losses have been considered is shown as the net or total volumetric efficiency.

3. *Suction and discharge valve design.* Capacity also is a function of the compressor valve design by which excessive wire-drawing losses must be prevented. Wire drawing is the restriction of area for a flowing fluid, causing a loss in pressure by internal and external friction. For design purposes, gas velocity is equal to the bore area times average piston speed divided by valve area. Manufacturers have found that velocities up to 30.5 m/s (6000 ft/min) with R-12 result in no material loss in volumetric efficiency or increase in horsepower.

Application factors that affect compressor capacity are determined by the cooling requirements and the operating conditions to be encountered. Application factors include:

1. *Compression ratio.* This is the ratio of absolute pressures before and after compression. It is fixed by cooler and condenser heat-transfer rates as well as by coolant flow and temperatures.

$$\text{Compression ratio} = P_d/P_s \qquad (12\text{-}32)$$

where P_d = absolute discharge pressure, kPa (psia)
P_s = absolute suction pressure, kPa (psia)

As the compression ratio increases, the clearance volumetric efficiency will decrease and hence the capacity of a given compressor will decrease.

High compression ratios must be avoided to prevent excessive discharge temperatures along with lubrication problems. (See Sec. 6 for a calculation of discharge temperature.) Normally, a maximum compression ratio is limited to about 10 for refrigeration compressors. At higher-compression-ratio requirements, compound or cascade system arrangements are used.

The compressor discharge gas temperature normally should not exceed about 135°C (275°F).

2. *Suction pressure.* Compressor capacity changes rapidly with suction pressure, which depends on evaporator temperature and pressure selected for a given cooling duty. The specific volume of the vapor increases rapidly as the suction pressure decreases. For a given speed the reciprocating compressor has a constant displacement. Therefore, at reduced suction pressure and temperatures, the compressor will handle a reduced weight of refrigerant owing to increased specific volume. The volumetric efficiency also decreases at lower suction pressures because of more expansion of high-pressure gas in the clearance volume.

Table 12-8 is a typical rating chart for a single-acting multicylinder compressure using R-12. Note that at 4.4°C (40°F) suction the capacity of a given compressor operating at 183.2 rad/s (1750 r/min) and 37.8°C (100°F) condensing is 914,000 Btu/h or 76.2 tons. At −6.7°C (20°F) suction the capacity has decreased to 49 tons. Compressor horsepower also is affected by suction temperatures, with the brake horsepower per ton ratio increasing rapidly at lower conditions.

3. *Discharge pressure.* The discharge pressure is established by the size of the condenser selected and the inlet temperature available for the condensing medium. An increase in discharge pressure decreases compressor capacity but at a much lesser effect than decreasing suction pressure. From Table 12-8 a compressor operating at 4.4°C (40°F) suction, 183.2 rad/s (1750 r/min), R-12 will have a capacity of 81.7 tons at 32.2°C (90°F) condensing. At 40.6°C

TABLE 12-8 Performance Data for a Single-Acting Multicylinder Reciprocating Compressor Using R-12

Saturated discharge temperature	Saturated suction: Temperature, °F.	Saturated suction: Pressure, p.s.i.g.	Operating at 1450 r.p.m.: B.t.u./hr.	Operating at 1450 r.p.m.: Brake hp.	Operating at 1750 r.p.m.: B.t.u./hr.	Operating at 1750 r.p.m.: Brake hp.
90°F. (99.6 p.s.i.g.)	−40	10.92	72,800	18.6	87,600	23.3
	−30	5.45	115,200	24.8	139,000	31.1
	−20	0.6	168,000	30.6	202,400	38.3
	−10	4.5	234,000	36.2	282,000	45.3
	0	9.2	316,000	41.4	381,600	51.9
	10	14.7	416,000	46.0	502,000	57.7
	20	21.1	434,000	49.8	644,000	62.3
	30	28.5	666,000	52.3	804,000	65.5
	40	37.0	812,000	53.8	980,000	67.3
	50	46.7	962,000	53.8	1,160,000	67.3
100°F. (116.9 p.s.i.g.)	−40	10.92	559,600	18.0	72,000	22.5
	−30	5.45	99,600	24.7	120,000	30.9
	−20	0.6	148,000	31.0	178,000	38.9
	−10	4.5	208,400	37.3	252,000	46.7
	0	9.2	284.200	43.0	343,000	53.9
	10	14.7	378,000	48.3	456,000	60.5
	20	21.1	488,000	52.5	588,000	65.9
	30	28.5	616,000	56.1	744,000	70.3
	40	37.0	758,000	59.2	914,000	74.1
	50	46.7	908,000	61.2	1,092,000	76.7
105°F. (126.2 p.s.i.g.)	−30	5.45	91,500	23.8	110,400	29.8
	−20	0.6	139,800	30.6	168,600	38.4
	−10	4.5	198,700	37.3	240,000	46.7
	0	9.2	272,100	43.7	328,800	54.7
	10	14.7	363,000	49.4	438,000	61.8
	20	21.1	469,000	54.1	566,000	67.8
	30	28.5	594,000	56.2	717,000	72.9
	40	37.0	731,000	61.6	882,000	77.1
	50	46.7	875,000	64.0	1,054,000	80.1

(105°F) condensing the capacity will have decreased to 73.5 tons. Also, as can be noted from the rating table, compressor horsepower requirements increase with an increase in discharge pressure.

4. *Compressor speed.* For constant evaporator and condensing temperatures, compressor capacity is directly proportional to a decrease in speed. This effect of lower capacity at lower speeds also can be noted from Table 12-8 when comparing a given compressor operating at 151.8 rad/s (1450 r/min) versus 183.2 rad/s (1750 r/min). Compressor horsepower also decreases in almost direct proportion to speed.

Slow-speed two-cylinder vertical single-acting compressors operate at speeds from 31.4 to 41.9 rad/s (300 to 400 r/min). Direct-drive multicylinder high-speed compressors are selected usually at 183.2 rad/s (1750 r/min). Some small-size compressors operate up to 376.9 rad/s (3600 r/min).

5. *Refrigerant selection.* The capacity of a given size of compressor also changes with the refrigerant selected. The commonly used refrigerants for reciprocating compressors are R-12, R-22, R-500, R-502, and R-717. The various refrigerants have different physical properties which vary the refrigerating capacities per pound of refrigerant. This makes it possible to have greater flexibility in selecting a compressor that will more nearly meet the requirements of the cooling duty at reduced investment.

All components of the refrigeration system have to be selected for a given refrigerant.

A typical example of the use of the preceding equations and data for a reciprocating-refrigerator-unit problem will be useful.

Example 17 Given a plant process that requires cooling of 120 gal/min of water from 15.6 to 10°C (60 to 50°F), assume that the cooler heat-transfer surface area will enable a 5.5°C (10°F) differential between the chilled water leaving the cooler and the R-12 evaporating temperature. Also assume that the condenser heat-transfer surface area will enable a 5.5°C (10°F) differential between the condenser water out and the R-12 condensing temperature. Water is available for the condensing medium at 29.4°C (85°F) inlet and 35°C (95°F) outlet. Assume no liquid subcooling or suction gas superheating.

Find:

a. Tons of refrigeration
b. Evaporator pressure, psig
c. Saturated condensing temperature, °F
d. Refrigerating effect, Btu/lb
e. R-12 weight flow circulated, lb/min
f. Theoretical compressor displacement, 100 percent efficiency
g. Compression ratio
h. Actual compressor displacement required, ft³/min
i. Actual compressor discharge temperature, °F with exponent $n = 1.19$
j. Compressor size at 1450-r/min speed
k. Horsepower required and hp/ton
l. Condenser water quantity, gal/min

Solution

a.
$$\text{Tons} = \frac{120 \times 8.33 \times (60°\text{F} - 50°\text{F}) \times 1.0}{200} = 50.0$$

b. From R-12 tables,* evaporator pressure corresponding to 4.4°C (40°F) evaporator temperature is 36.9 psig.

c. The saturated condensing temperature will be 35°C (95°F) water outlet plus a 5.6°C (10°F) leaving temperature difference or 40.6°C (105°F). From R-12 tables, this corresponds to a condenser pressure of 126.5 psig.

d. Refrigerating effect $= h_g - h_f$:

$$h_g = 81.4 \text{ Btu/lb} \quad \text{(R-12 tables)}$$
$$h_f = 32.3 \text{ Btu/lb} \quad \text{(R-12 tables)}$$
$$\text{RE} = 81.4 - 32.3 = 49.1 \text{ Btu/lb}$$

e. Weight flow:

$$\text{lb/min} = \frac{50 \text{ tons} \times 200 \text{ Btu/min}}{49.1} = 208$$

f. Theoretical compressor displacement:

$$\text{ft}^3\text{/min} = \text{weight flow} \times V_g$$
$$V_g = 0.773 \text{ ft}^3\text{/lb (R-12 tables)}$$
$$\text{ft}^3\text{/min} = 208 \times 0.773 \times 161$$

*See Fig. 12-27 and tables in Sec. 3.

g. Compression ratio:

$$\text{CR} = \frac{P_d}{P_s} = \frac{126.5 + 14.7}{36.9 + 14.7} = 2.73$$

h. From Fig. 12-38, the total volumetric efficiency at 2.73 compression ratio is 73.5 percent. Therefore, the actual compressor displacement equals 161/0.735, or 219 ft³/min required.

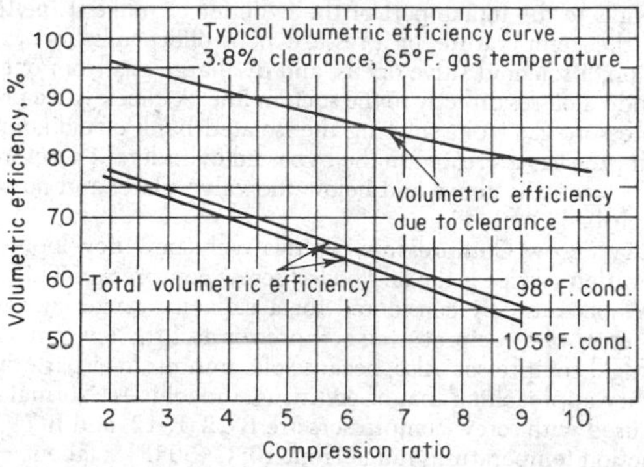

FIG. 12-38 Volumetric efficiency versus compression ratio for reciprocating compressors. *(Carrier Air Conditioning Company.)*

i. Compressor discharge temperature:

$$T_d = 40°\text{F} + 460(2.73)^{\frac{1.19 - 1.0}{1.19}}$$
$$T_d = 500(2.73)^{0.1595} = 586°\text{F absolute}$$
$$= 126°\text{F actual}$$

j. Manufacturer has standard single-acting compressor available with 3¼-in bore × 2¾-in stroke.

$$\text{No. of cylinders required} = \frac{219 \text{ ft}^3\text{/min} \times 4 \times 1728}{3.1416 \times (3.25)^2 \times 2.75 \times 1450} = 11.4$$

Use 12-cylinder compressor.

k. Power of compression:

$$\text{Btu/min} = (h_d - h_g) \times \text{weight flow}$$
$$h_d = 91.5 \text{ Btu/lb} \quad \text{(R-12 tables, superheated-vapor)}$$
$$h_g = 81.4 \text{ Btu/lb} \quad \text{(R-12 tables)}$$
$$\text{Btu/min} = (91.5 - 81.4) \times 208 = 2100$$
$$\text{hp} = 2100/42.4 = 49.6$$
$$\text{hp/ton} = 49.6/50.0 = 0.99$$

l. Condenser water quantity:

$$\text{gal/min} = \frac{(50 + 49.6 \times 2545/12{,}000) \times 200}{10 \times 8.33}$$
$$\text{gal/min} = (60.5 \times 200)/83.3 = 145$$

Reciprocating-Compressor Capacity Control There are four basic methods of controlling the capacity of reciprocating compressors. In addition to cylinder unloaders, discussed at length in Sec. 6, they are:

1. *Hot-gas bypass.* This arrangement artificially loads the compressor and reduces system capacity. Discharge gas flows to the low side of the system through a constant-pressure valve. As the evaporator pressure is reduced, a constant-pressure valve opens to maintain suction pressure constant. Hot gas can be introduced between the thermal-expansion valve and the evaporator coil inlet; at the evaporator exit ahead of the thermal-expansion bulb; or directly into the suction return ahead of the compressor, but then requiring liquid quenching for desuperheating the discharge gas.

Hot-gas bypass can be combined with cylinder unloading to give capacity control from 0 to 100 percent of full load. With hot-gas bypass capacity control the compressor brake horsepower remains fairly constant, whereas with cylinder unloading power decreases almost proportionally to capacity reduction. For this reason hot-gas bypass is used mainly for control between 0 to 25 percent of rating.

2. *Variable-speed motors.* This method of control varies the compressor speed by using electric-motor drives having two or more speeds or by using an internal-combustion engine. Multispeed motors are more expensive than single-speed motors and consequently are not commonly used for capacity control.

3. *Cylinder-head bypass.* This method uses a bypass of discharge gas to the intake port of the cylinder to make it ineffective. When the temperature or pressure controller calls for capacity reduction, a solenoid valve opens, and discharge gas from one block of cylinders passes directly to the suction line. A check valve prevents high-pressure gas from entering the isolated bank of cylinders, and no high pressure is created in the bypassed cylinders. Therefore, suction pressure exits above and below the valve plate, and no work is accomplished.

Rotary Screw Compressors A relatively new development for refrigeration service is the oil-flooded screw compressor. This type of compressor is usually considered for a capacity range between the upper limits of a reciprocating compressor and the lower limits of a centrifugal compressor. Also, because of a minimum of parts, it offers reliability approaching that of centrifugal machines. Normal refrigerants used with screw compressors are R-22, R-12, and R-717. Normal suction temperatures range from 10°C (50°F) to about −50°C (−58°F).

The screw compressor is a positive-displacement compressor, consisting of two rotors (male and female) with asymmetric profiles. A slide valve, hydraulically actuated, is used to modulate capacity between 10 percent loading and full load. Internal parts also include four main sleeve bearings, two thrust bearings of the angular-contact ball type, balance pistons, and a mechanical shaft seal. Screw compressor components are shown in Fig. 12-39.

In the screw compressor, the refrigerant gas is compressed by the meshing action of the male lobes and the female flutes. Direct metal contact is prevented by submerging the rotor assembly with oil lubrication. As the rotor turns and meshes, suction gas is trapped in the interlobe space. This results in reduced internal volume and increased pressure. When the enclosed vapor reaches the outlet port of the compressor, the refrigerant is transported by the rotating rotor into the discharge outlet.

Owing to the relatively large quantities of oil involved with the wet-screw design, an efficient oil separator is required to recover oil from the discharge gas. Both horizontal and vertical separator designs are used. Because multistage oil separators are not 100 percent efficient, some oil is carried over to the evaporator. Oil return with direct-expansion evaporators is normally accomplished by entraining oil in the suction gas. A flooded evaporator will require additional considerations for oil return.

Screw-Compressor Performance The screw compressor can operate satisfactorily over a wide range of condensing temperatures, similarly to the reciprocating compressor. The screw machine is capable of achieving compression ratios approaching 20, although it is seldom designed at that maximum condition. Consequently, the screw compressor is particularly suitable for heat-recovery schemes and air-cooled-condenser applications.

The screw compressor requires more brake horsepower per ton than the centrifugal or the reciprocating compressor. As the suction temperature decreases, the horsepower penalty of the screw compressor rapidly increases in comparison with other types.

In order to improve screw-compressor energy consumption on low-temperature refrigeration applications, manufacturers have developed an economizer cycle. Figure 12-40 shows a typical arrangement using an intermediate heat exchanger to allow vapor flow to an economizer port on the compressor. This is similar to the economizer cycle used on multistage centrifugal machines. By using the economizer cycle, liquid to the evaporator is subcooled to a point about 5.5°C (10°F) above intermediate-saturation temperature. Thus, overall cycle efficiency is improved, resulting in increased capacity and lower brake horsepower per ton. Table 12-9 presents

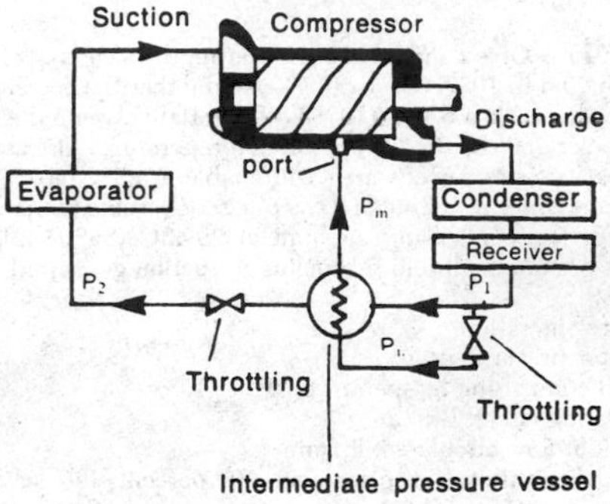

FIG. 12-40 Screw-compressor intermediate port with economizer cycle. *(Vilter Mfg. Corp.)*

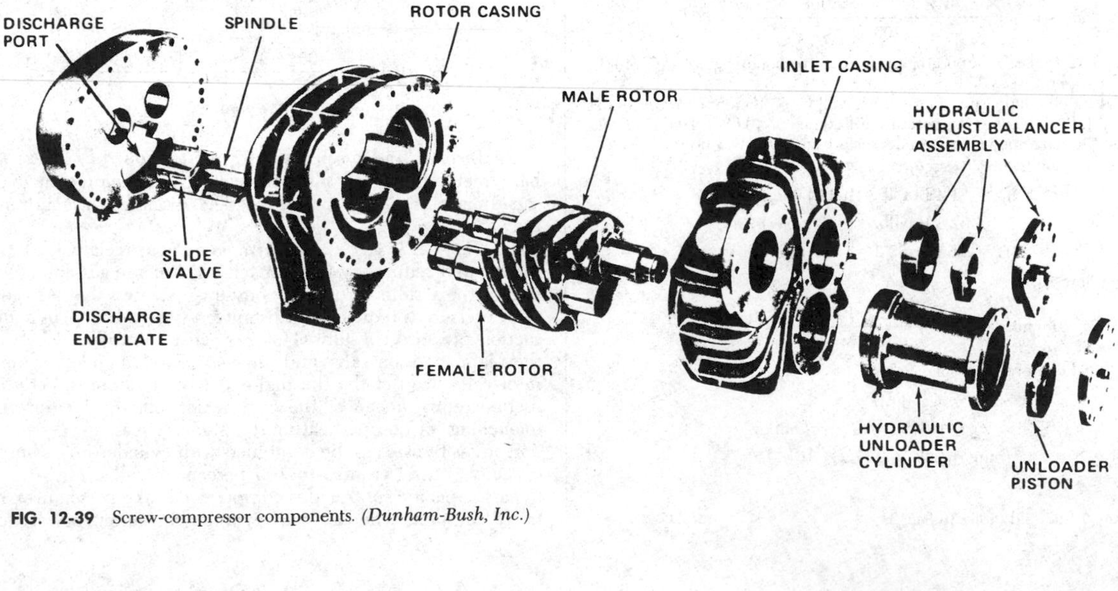

FIG. 12-39 Screw-compressor components. *(Dunham-Bush, Inc.)*

TABLE 12-9 Brake Horsepower per Ton Comparison at Varying Suction Temperatures, 40.5°C (105°F) Condensing; Halocarbon Refrigerant

	Type of compressor		
Saturation suction temperature, °C	Screw, bhp/ton°	Reciprocating, bhp/ton	Centrifugal, bhp/ton
4.4	1.1	1.05	0.9
−17.8	2.1	2.0	1.8
−40.0	4.3	....	3.0

°Screw compressor equipped with intermediate port and economizer cycle.

energy data on three types of compressors based upon a constant condensing temperature and varying suction temperatures from 4.4 to −40°C (40 to −40°F).

The economizer cycle is only in effect between 75 percent load and full load. Once the slide valve moves to its 25 percent open (75 percent travel) position, the intermediate pressure port will be at suction pressure. At this point, the compressor capacity of a unit equipped with an economizer cycle becomes identical with that of a conventional unit without an intermediate port. The part-load energy consumption of the screw compressor also exceeds that of reciprocating and centrifugal compressors.

Manufacturers also can optimize screw-compressor selection to give the lowest horsepower for design requirements. Three different internal-volume ratios can be furnished for a given screw-compressor selection. The volume ratio is the ratio of the gas volume when the gas is trapped in the rotors (at the beginning of compression) to the volume when the rotors open to the discharge port. By locating the discharge-port position in the slide valve, to open earlier or later, the volume ratio can be varied.

Screw-Compressor Capacity Control As previously mentioned, a slide valve is used to regulate capacity. This is achieved by means of a moving slide valve located in the compressor casing below the rotors and operated by a piston in a hydraulic cylinder. The piston is actuated by high-pressure oil flowing to either side of the piston, thus moving the slide valve and altering the point in the rotor length at which compression begins. This allows internal gas recirculation without any compression having occurred.

When the compressor is started, the slide valve is fully open and the compressor is unloaded. The compressor capacity mechanism is operated by means of two solenoid valves. If both solenoid valves are deenergized, the compressor will be in the unloaded mode. The slide valve is spring-loaded and opposed by a conventional hydraulic system. To increase capacity, a solenoid valve is energized to permit oil to one side of the piston while opening another valve to the oil return. This hydraulically causes the piston to move toward a position for more capacity.

Steam Jets A steam jet may be substituted for the compressor in refrigeration systems. The refrigerant is then water vapor, and the steam jet removes it from the flash tank (evaporator), compresses it, and delivers it to the condenser.

The operating principle for a steam-jet refrigeration system is relatively simple. If water at a temperature higher than, say, 10°C (50°F) is delivered to a tank in which the pressure is maintained at 0.178 psia, the water will immediately cool to 10°C (50°F). In so doing, it surrenders latent heat, which vaporizes a portion of the water delivered to the tank. The cooling of water in this way depends upon maintaining the low pressure that corresponds to the desired water temperature.

A typical arrangement for the steam-jet refrigeration cycle is shown in Fig. 12-41. It consists of the following components:

1. *Primary steam ejector.* This is essentially a kinetic device that utilizes the momentum of a high-velocity jet to entrain and accelerate a slower-moving medium into which it is directed. High-pressure steam is delivered to the nozzle of the ejector. The steam expands while flowing through the nozzle where the velocity increases rapidly. The velocity leaving the nozzle is around 1219 m/s (4000 ft/s).

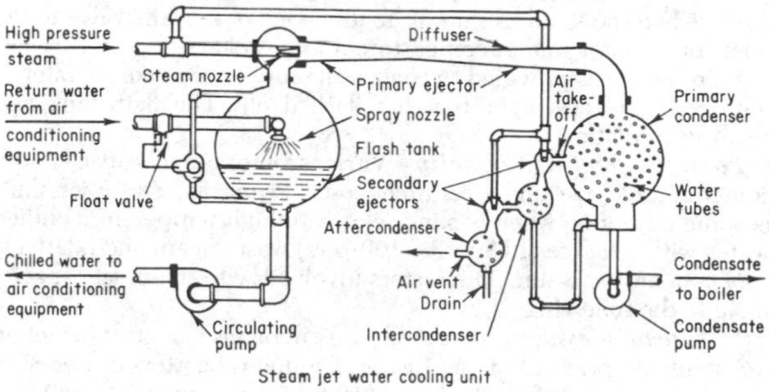

FIG. 12-41 Steam-jet refrigeration cycle. (Air Conditioning Manual, *Trane Co., 1966.*)

Because of this high velocity, flash vapor from the tank is continually aspirated into the moving steam. The mixture of steam and flash vapor then enters the diffuser section, where the velocity is gradually reduced because of increasing cross-sectional area. The energy of the high-velocity steam compresses the vapor during its passage through the diffuser. The steam pressure will have been increased in this example from 0.178 psia [corresponding to 10°C (50°F)] at the diffuser entrance to 1.1 psia at the condenser [corresponding to a condensing temperature of 40.6°C (105°F)].

2. *Condenser.* As in any compression-type refrigeration system, the purpose of the condenser is to liquefy the vaporized refrigerant. In the steam jet, a mixture of high-pressure steam and flash vapor is liquefied. The condenser heat rejection Q is equal to

$$Q_{cond} = (W_s + W_{wv})h_{fg}$$

where Q = heat rejection, Btu/h
W_s = primary booster steam rate, lb/h
W_{wv} = flash vapor, lb/h
h_{fg} = latent heat of steam, Btu/lb

The latent heat of steam at low pressure is approximately 1060 Btu/lb. Therefore, 11.3 lb/h of water must be vaporized in the flash tank per ton of capacity.

The condenser design surface area and water quantity should be based on the highest water temperature likely to be encountered. If the inlet-water temperature becomes hotter than design, the primary booster (ejector) may cease functioning because of the increase in condenser pressure.

Two types of condensers are used: the surface condenser and the barometric or jet condenser. The surface condenser is of shell-and-tube design with water flowing through the tubes and steam condensed on the outside surface. In the jet condenser, condenser water and the steam being condensed are mixed directly, and no tubes are provided. The jet condenser can be barometric or a low-level type. The barometric condenser requires a height of 10.4 m (34 ft) above the level of water in the hot well. A tailpipe of this length is needed so that condenser water and condensate can drain by gravity. In the low-level jet type, the tailpipe is eliminated, and it becomes necessary to remove the condenser water and condensate by pumping from the condenser to the hot well. The main advantages of the jet condenser are low maintenance with the absence of tubes and the fact that condenser water of varying degrees of cleanliness may be used.

3. *Flash tank.* This is the evaporator of a steam-jet system in which the evaporation of water takes place and the cooling effect is obtained. Warm water returning from process is sprayed into the flash chamber through nozzles, and the cooled effluent is pumped from the bottom of the flash tank.

A compartmental flash tank is frequently used. When the steam supply to one ejector of a group is closed, some means must be provided for preventing the pressure in the condenser and flash tank from equalizing through that ejector. With this arrangement, partitions are provided so that each booster is operating on its own flash tank. When the steam is shut off to any one booster, the valve to the inlet spray water to that compartment also is closed.

A float valve is provided to control the supply of makeup water to replace the water vapor that has flashed off. The flash tank also requires insulation.

Factors Affecting Capacity Various factors affect capacity and, consequently, size and operating costs. Primarily, steam-jet units become attractive when cooling relatively high-temperature chilled water with a source of 689-kPa (100-psig) waste steam and relatively cool condensing water. The factors involved with steam-jet capacity include the following:

1. *Steam pressure.* The steam rate depends to a great extent on the available pressure of the steam. The main boosters can operate on steam pressures from as low as 13.8 kPa (2 psig) up to 689 kPa (100 psig). However, the quantity of steam required increases rapidly as the steam pressure drops. Best steam rates are obtained with about 689 kPa (100-psig) steam. With pressures above 689 kPa (100 psig), the decrease in the quantity of steam required is practically negligible. Ejectors must be designed for the highest available steam pressure, to take advantage of the lower steam consumption with a higher pressure. Figure 12-42 shows steam consumption for various steam-inlet pressures.

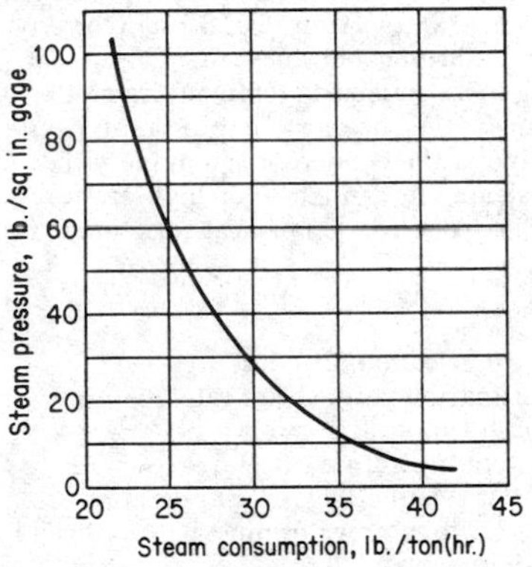

FIG. 12-42 Steam-jet consumption for varying steam pressures. [*Spencer,* Heat. Piping Air Cond. *(August 1954).*]

The secondary ejector systems, used for removing air, require steam pressures of 241 kPa (35 psig) or greater. When the available steam pressure is lower than this, an electrically driven vacuum pump is used for either the final secondary ejector or for the entire secondary group. The secondary ejectors normally require 1.5 to 2.5 lb/h of steam per ton of refrigeration capacity.

2. *Condenser water temperature.* In comparison with other vapor-compression systems, steam-jet machines require relatively large water quantities for condensing. The higher the inlet-water temperature, the higher the water requirements. Also, condensing-water temperature has an important effect on steam rate per ton, rapidly decreasing with colder water. Figure 12-43 presents data on steam rate versus condenser water inlet for a given chilled-water outlet temperature and steam pressure.

3. *Chilled-water temperature.* The lower the chilled-water outlet temperature, the more difficult the application becomes for steam-jet units. Justification is improved with outlet chilled-water temperatures of 12.8°C (55°F) or higher rather than lower values down to a minimum of 1.7°C (35°F).

As the chilled-water outlet temperature decreases, more pounds per hour per ton of steam are required, accompanied by increased condensing-water requirements. Figure 12-44 shows steam consumption for varying chilled-water outlet temperatures.

Unlike centrifugal and reciprocating refrigeration machines, chilled-water flow rate is of no particular importance in steam-jet-system design. Steam vacuum refrigeration is an evaporating process and does not depend on high evaporator-tube velocities, as do other

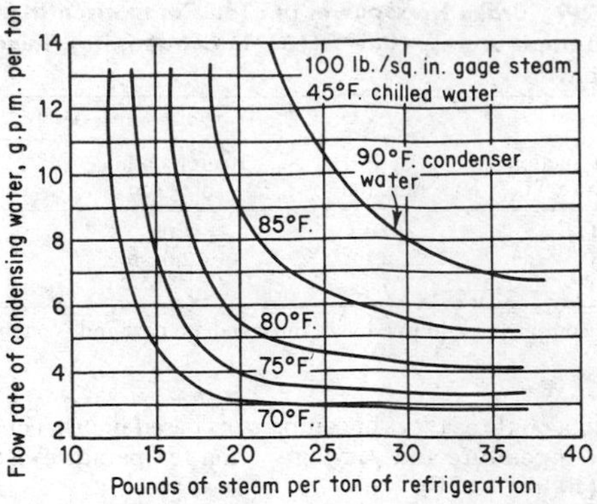

FIG. 12-43 Steam-jet consumption versus condenser-water flow rate. [*Spencer,* Heat. Piping Air Cond. *(August 1954).*]

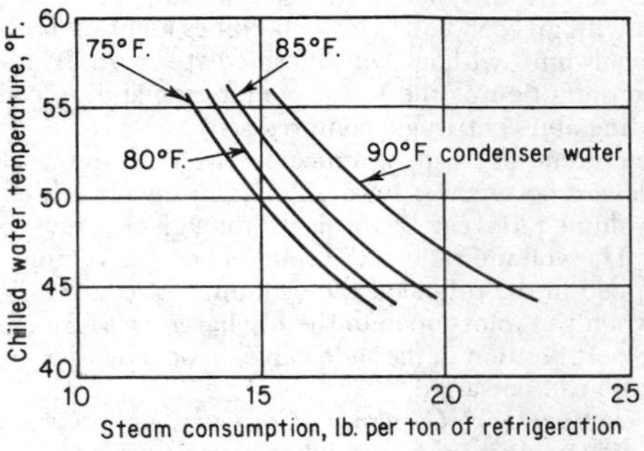

FIG. 12-44 Steam-jet consumption for varying chilled-water outlet temperatures. [*Spencer,* Heat. Piping Air Cond. *(August 1954).*].

systems for good heat-transfer rates. Also, widely varying return chilled-water temperatures have little effect on steam-jet equipment.

Capacity Control The simplest way to regulate the capacity of most steam vacuum refrigeration systems is to furnish several primary boosters in parallel and operate only those required to handle the heat load. It is not uncommon to have as many as four main boosters on larger units for capacity variation. A simple automatic on-off type of control may be used for this purpose. By sensing the chilled-water temperature leaving the flash tank, a controller can turn steam on and off to each ejector as required.

Additionally, two other control systems which will regulate steam flow or condenser water flow to the machine are available. As the condenser-water temperature decreases during various periods of the year, the absolute condenser pressure will decrease also. This will permit the ejectors to operate on less steam because of the reduced discharge pressure. Either the steam flow or the condenser water quantities can be reduced to lower operating costs at other than design periods. The arrangement selected depends on cost considerations between the two flow quantities. Some systems have been arranged for a combination of the two, automatically reducing steam flow down to a point, followed by reduction in condenser-water flow. For maximum operating efficiency, automatic control systems are usually justifiable in keeping operating costs to a minimum without excessive operator attention. In general, steam savings of about 10 percent of rated booster flow are realized for each 2.8°C (5°F) reduction in condensing-water temperature below the design point.

In some cases with relatively cold inlet condenser water, it has been possible to adjust automatically the steam inlet pressure in response to chilled-water outlet temperatures. In general, however, this type of control is not possible because of differences in temperature between the flash tank and the condenser. Under usual conditions of warm condenser-water temperatures, the main ejectors must compress water vapor over a relatively high ratio, requiring an ejector with an entirely different operating characteristic. In most cases, when the ejector steam pressure is throttled, the capacity of the jet remains almost constant until the steam pressure is reduced to a point at which there is a sharp capacity decrease. At this point, the ejectors are unstable, and the capacity is severely curtailed. With a sufficient increase in steam pressure, the ejectors will once again become stable and operate at their design capacity. In effect, steam jets have a vapor-handling capacity fixed by the pressure at the suction inlet. In order for the ejector to operate along its characteristic pumping curve, it requires a certain minimum amount of steam which is fixed for any particular pressure in the condenser. (For further information on the design of ejectors, see Sec. 6.)

ABSORPTION SYSTEMS

Introduction Present-day refrigeration systems of the absorption type mainly use water as the refrigerant and lithium bromide as the absorbent. Consequently, this type of machine is limited to chilled-water duty. For low-temperature service, an ammonia-water absorption unit is applicable. Renewed interest in ammonia-water absorption equipment has occurred in the past several years in petrochemical and chemical plants. These plants generate large quantities of process waste heat which, in some instances, can be economically recovered to produce absorption refrigeration. Ammonia-water absorption units for this industrial use can be quite large and in themselves become small chemical plants. At the other end of the spectrum, small ammonia-water packaged absorption units in the 3- to 10-ton sizes are available for commercial chilled-water applications.

Description Water–lithium bromide absorption units utilize two basic factors in producing a refrigeration effect: (1) Water will boil and flash-cool itself at low temperatures when it is maintained at a high vacuum. (2) Certain substances, such as a salt, will absorb water vapor. Lithium bromide solution is a hygroscopic salt solution which has been found to have the best solubility–vapor-pressure relationship to enable high cycle efficiency. In the absorption unit water flashes off to a vapor, and the temperature of the remaining water is lowered. The affinity of water to salt is measured by the depression of the water-vapor pressure, being more pronounced as the salt concentration increases.

The complete cycle is shown in Fig. 12-45. An absorption machine consists of five main components:

1. *Evaporator.* Tube section in which returning chilled water is cooled indirectly by water sprayed over the tubes. Since this shell is maintained at a low absolute pressure, water flashes and cools the remaining water to a temperature closely corresponding to shell pressure.

2. *Absorber.* A strong salt solution is used in this part of the system to absorb the water vapor flashed in the evaporator. A solution pump sprays the lithium bromide solution over the absorber tube

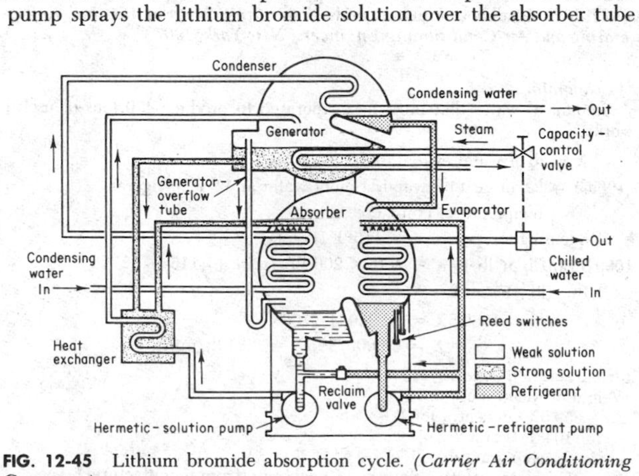

FIG. 12-45 Lithium bromide absorption cycle. *(Carrier Air Conditioning Company.)*

section through which relatively cool water flows. The absorber work or total heat load—consisting of refrigeration load, heat of dilution, cooling of condensed water, and solution sensible cooling—is thus transferred to cooling water normally circulated from a cooling tower.

3. *Solution heat exchanger.* This component is used to improve cycle efficiency by exchanging heat between the weak solution leaving the absorber and the strong, hot solution returning from the generator. Steam and condenser water is reduced by the use of this exchanger.

4. *Generator.* A steam-heated tube section used to restore the solution concentration by boiling off the water vapor absorbed.

5. *Condenser.* The water vapor boiled off in the generator is condensed in this tube section and returned to the cooler.

A discussion of the steps in the cycle will aid in understanding the absorption-unit operation. The absorption cycle is a two-pressure cycle, normally maintaining, with 7.2 to 8.3°C (45 to 47°F) outlet chilled water, 0.27 inHg absolute pressure in the evaporator-generator section and 3.0 inHg absolute pressure in the generator-condenser section. Three circuits are involved: (1) Water as a refrigerant is pumped to the evaporator. (2) Lithium bromide as the absorbent is circulated over the absorber tubes, through the heat exchanger, and to the generator. (3) Cooling water flows in series initially through the absorber tubes and partially through the condenser tubes at about two-thirds of design flow rate.

Water to be cooled enters the evaporator (cooler) tube bundle, where it is cooled indirectly by spray water. The water vaporized is absorbed by a strong solution of lithium bromide at a low pressure. The lithium bromide that has absorbed the water vapor is then pumped through the solution heat exchanger to the generator so as to reconstitute the weak solution. Low-pressure steam [55 to 96 kPa (8 to 14 psig)] is used in the generator to boil off the water vapor, thus concentrating the salt solution before reentering the absorber. The solution flow from the generator to the absorber is the result of gravity and pressure difference and not of pumping. The water boiled off in the generator then is condensed to a liquid in the condenser section, and the condensate is returned to the evaporator.

The heat dissipated in the absorber and condenser is removed by the cooling water, which is returned normally to a cooling tower. It is necessary for the inlet cooling-water temperature to be controlled so that the proper cycle concentrations will result.

Equilibrium Diagram The absorption machine operation is analyzed by the use of a lithium bromide–water equilibrium diagram, as shown in Fig. 12-46. The equilibrium diagram is useful in determining the solution concentration of the unit so that maximum operating efficiency can be maintained.

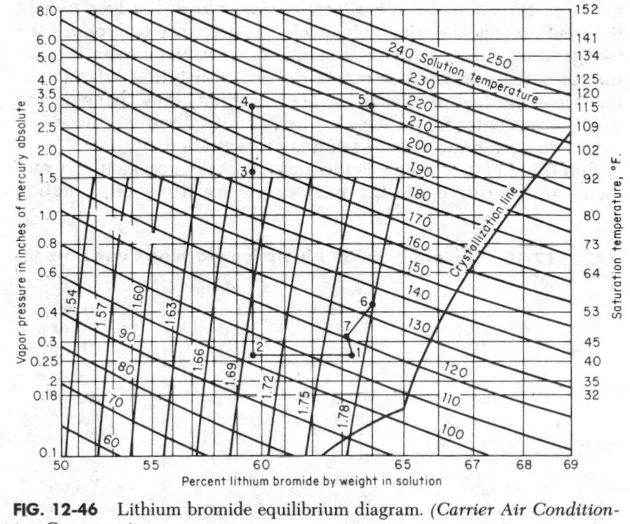

FIG. 12-46 Lithium bromide equilibrium diagram. *(Carrier Air Conditioning Company.)*

Vapor pressure in inches of mercury absolute is plotted against the percent of lithium bromide by weight in solution. The corresponding saturation temperature for a given vapor pressure is shown on the right-hand side of the diagram. The curved lines running from left to right are solution-temperature lines. The curved lines extending upward from the bottom of the diagram are specific-gravity lines for determining solution concentration. By measuring the specific gravity and temperature of the salt solution, the concentration can be determined by plotting these two points on the diagram.

The curved line in the lower right-hand corner is called the crystallization line. This line indicates the point at which the solution will begin to change from a liquid to a solid. Consequently, the crystallization line sets the limits of the cycle. If the solution becomes overconcentrated or if excessive noncondensables are present, the absorption cycle will be interrupted owing to solidification, and capacity will not be restored until the unit is desolidified. This normally requires the addition of steam heat to the outside of the solution heat exchanger and the solution pump. An automatic dilution or reclaim valve is installed in the piping hookup to prevent minor off-standard operation.

A typical machine absorption cycle is plotted on the equilibrium diagram. Points 1 through 7 represent a complete absorption cycle. Specific point values of temperatures, pressure, and solution concentration are shown in Table 12-10. An explanation of each point and the lines drawn between is as follows:

Point 1. The condition of the strong solution in the absorber as it starts to absorb refrigerant.

Point 2. The condition of the weak solution as it leaves the absorber and enters the heat exchanger. Line 1–2 represents the decrease in solution concentration due to absorption of water vapor from the evaporator.

Point 3. The condition of the weak solution after it has passed through the solution heat exchanger. Line 2–3 represents the amount of heat gained by the solution in the heat exchanger.

Point 4. The condition of the weak solution as it enters the generator and is being heated. Line 3–4 represents the amount of heat required to start the weak solution to boil.

Point 5. The condition of the strong solution as it leaves the generator. Line 4–5 represents the amount of heat required to boil off the water vapor contained in the solution.

Point 6. The condition of the strong solution after it leaves the heat exchanger. Line 5–6 represents the flow of solution from the generator to the heat exchanger without a change in concentration.

Point 7. The condition of the strong solution entering the spray nozzles of the absorber.

Capacity Control Capacity control of the water–lithium bromide absorption machine for part-load conditions can be arranged in one of three ways: (1) throttling of condenser-water flow to hold the solution concentration only as high as necessary to produce the refrigeration capacity at the design chilled-water temperature; (2) throttling the solution flow with a three-way diverting valve in the line from the heat exchanger to the generator, thus changing the solution concentration in the absorber; and (3) throttling the steam flow to the generator to vary solution concentration.

Throttling of steam flow is the most generally accepted arrangement, although it is not the most economical from an operating standpoint. Throttling solution flow gives lower steam consumption, but the cost of the special solution control valve limits the desirability of this method. Condenser-water throttling for capacity control has the disadvantage of high water temperatures resulting at very light loads, which accelerates the buildup of deposits in the condenser tubes.

TABLE 12-10 Lithium Bromide Absorption Machine Cycle Data*

Point	Solution temperature, °F.	Vapor pressure, in. Hg abs.	Lithium bromide solution, %	Saturated temperature, °F.
1	115	0.27	63.3	42
2	104	0.27	59.5	42
3	167	1.65	59.5	95
4	192	3.0	59.5	115
5	215	3.0	64.0	115
6	135	0.45	64.0	55
7	120	0.32	63.3	46

*See Fig. 12-46.

Unit Efficiency The absorption machine efficiency can be conveniently expressed in terms of performance ratio:

$$R = \frac{\text{useful refrigerating effect, Btu/h}}{\text{heat input, Btu/h}}$$

A performance ratio of 0.60 to 0.70 is normal. The usual steam consumption is about 19 to 20 lb/(h·ton). Cooling-water flow requirements are normally 3.0 to 3.5 gal/(min·ton).

Absorption Machine Calculations By using the data shown in Table 12-10 for a typical operating cycle, calculations can be made to determine theoretical performance. Data on specific gravity and specific heat for aqueous solutions of lithium bromide are shown in Figs. 12-47 and 12-48. Other data required for a representative problem will assume the following:

Refrigeration capacity required, tons	450
Entering chilled-water temperature, °F	55
Leaving chilled-water temperature, °F	45
Entering condenser-water temperature, °F	85
Cooling-water flow, gal/min	1575
Steam pressure, psig	12

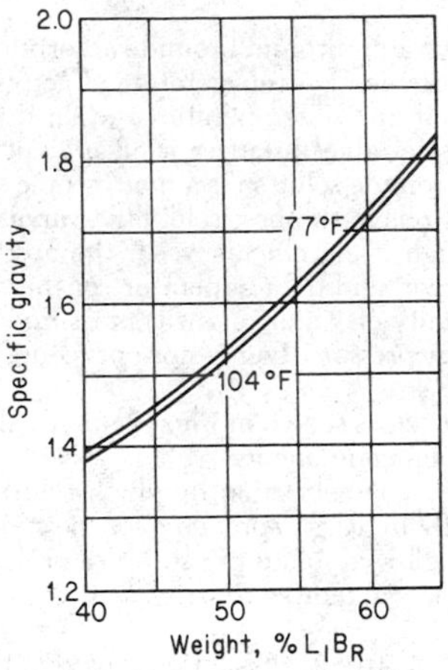

FIG. 12-47 Specific gravity of aqueous solutions of lithium bromide. (ASHRAE Handbook: Fundamentals, *American Society of Heating, Refrigerating and Air-Conditioning Engineers, New York, 1967.*)

Evaporator

Amount of water that must be evaporated to produce 450 tons of useful work:

X = lb of water evaporated

1069.5 = latent heat of evaporation of water at 42°F refrigerant temperature (Point 1)

Condensing temperature = 115°F (Point 4)

$1069.5 \text{ Btu/lb} \times \text{lb/min} = 450 \times 200 \text{ Btu/min} + (115 - 42) \times 1 \text{ Btu/(lb}\cdot\text{°F)} \times X$

$$1069.5X = 90{,}000 + 73X$$

$$X = 90{,}000/996.5 = 90.4 \text{ lb/min}$$

Strong solution concentration

Weight of weak solution:

59.5% concentration

104°F (Point 2)

Specific gravity = 1.71

Generator solution flow = 110 gal/min (from manufacturer)

$$110 \text{ gal/min} \times 8.33 \text{ lb/gal} \times 1.71 = 1570 \text{ lb}$$

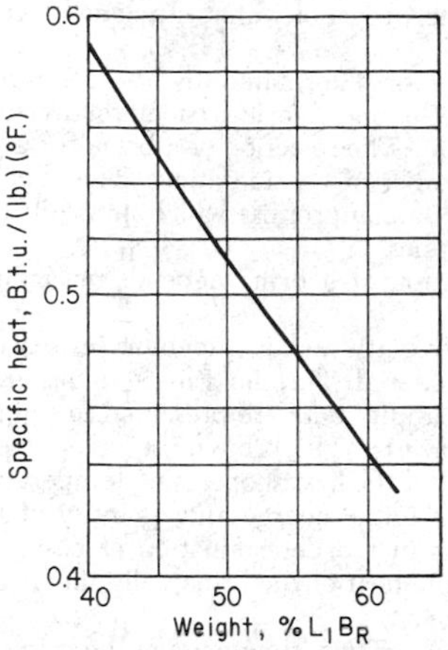

FIG. 12-48 Specific heat of aqueous solutions of lithium bromide. (ASHRAE Handbook: Fundamentals, *American Society of Heating, Refrigerating and Air-Conditioning Engineers, New York, 1967.*)

Weight of Li-Br

$$59.5\% \times 1570 = 935 \text{ lb}$$

Weight of strong solution
Must remove 90.4 lb of water from weak solution:

$$1570 - 90.4 = 1479.6$$

Concentration of strong solution:

$$935/1479.6 = 63.3 \text{ (64.0\% from table)}$$

Absorber heat rejection
1. Heat that must be removed in cooling recirculated solution:

$$t = 120 - 104 = 16°F$$

Average temperature = 112°F
Average concentration = 61.4%
Specific gravity = 1.73 (Fig. 12-47)
Specific heat = 0.44 Btu/(lb·°F) (Fig. 12-48)
Absorber solution flow = 145 gal/min (from manufacturer)
245 × 8.33 × 16 × 0.44 × 1.73 = 24,800 Btu/min

2. Exothermic heat of dilution that must be removed:
Average cycle concentration = 61.8%
Heat of dilution = 202 Btu/lb water
lb/min of water absorbed = 90.4

$$90.4 \times 202 = 18{,}150 \text{ Btu/min}$$

3. Heat of condensation of 90.4 lb/min of water vapor at 42°F that must be removed:

Latent heat at 42°F = 1069.5 Btu/lb

$$90.4 \times 1069.5 = 96{,}500 \text{ Btu/min}$$

4. Heat that must be added to warm 90.4 lb/min of water from 42 to 103°F:

t = 61°F
Specific heat = 0.44 Btu/(lb·°F)

$$90 \times 0.44 \times 61 = 2420 \text{ Btu/min}$$

Total absorber heat rejection = 1 + 2 + 3 − 4 = 24,800 + 18,150 + 96,500 − 2420 = 137,030 Btu/min.

Absorber tonnage = 137,030/200 = 685 tons

$$\text{Absorber ratio} = \frac{\text{absorber work}}{\text{evaporator work}} = \frac{685}{450} = 1.52$$

$$\text{Cooling-water rise} = \frac{137{,}030 \text{ Btu/min}}{1 \text{ Btu/(lb}\cdot°\text{F)} \times 1575 \text{ gal/min} \times 8.33 \text{ lb/gal}} = 10.4°F$$

Generator calculations
1. Heat absorbed in preheating weak solution:
Weight of weak solution = 1570 lb

$$t = 192 - 167 = 25°F$$

Concentration = 59.5%
Specific heat = 0.442

$$1570 \times 0.442 \times 25 = 17{,}300 \text{ Btu/min}$$

2. Heat absorbed in change of boiling point from 192 to 215°F:
Average concentration = 61.8%
Specific heat = 0.44

$$\text{Average flow} = \frac{1570 + 1479.6}{2} = 1525 \text{ lb/min}$$

t = 23°F

$$1525 \times 0.44 \times 23 = 15{,}500 \text{ Btu/min}$$

3. Endothermic heat of solution that must be added:

$$90.4 \text{ lb/min} \times 202 \text{ Btu/lb} = 18{,}150 \text{ Btu/min}$$

4. Heat absorbed in evaporating 90.4 lb/min of water at condensing temperature of 115°F:
Latent heat at 115°F = 1028 Btu/lb

$$90.4 \times 1028 = 92{,}600 \text{ Btu/min}$$

Total heat absorbed in generator = 1 + 2 + 3 + 4
= 143,500 Btu/min
Generator tonnage = 143,550/200 = 717.7 tons
Generator ratio = 717.7/450 = 1.59

Steam consumption
Latent heat (12-psig, steam) = 949.7 Btu/lb

Steam requirement = 143,550 Btu/min × 60 min/h
= 8,600,000 Btu/h

Useful work = 450 tons

$$\text{Steam rate} = \frac{8{,}600{,}000}{949.7 \times 450} = 20.2 \text{ lb/(h}\cdot\text{ton)}$$

Machine efficiency or coefficient of performance

$$\text{Machine efficiency} = \frac{\text{useful work output}}{\text{steam input}} \times 100\%$$

$$= \frac{\text{evaporator work}}{\text{generator work}} \times 100\%$$

$$= \frac{450}{717.7} \times 100\% \times 62.8\%$$

Condenser calculations
1. Heat rejected in condensing 90.4 lb/min of water at condensing temperature of 115°F:
Latent heat at 115°F = 1028 Btu/lb

$$90 \times 1028 = 92{,}600 \text{ Btu/min}$$

2. Heat rejected in cooling 90.4 lb/min of superheated vapor from 215 to 115°F
t = 100°F
Specific heat = 0.425

$$90.4 \times 100 \times 0.425 = 3840 \text{ Btu/min}$$

Total condenser heat rejection = 1 + 2 = 96,440 Btu/min
Condenser tonnage = 96,440/200 = 482 tons
Condenser ratio = 482/450 = 1.07

Condenser water rise:
Cooling water flow = 1575 gal/min × ⅔ = 1055 gal/min

$$\frac{96{,}440}{1 \text{ Btu/(lb}\cdot°\text{F)} \times 8.33 \times 1055} = 10.9°F$$

Overall heat balance

Heat input = evaporator work + generator work
= 450 + 717.7 = 1167.7 tons
Heat output = absorber work + condenser work
= 685 + 482 = 1167 tons

Ammonia-Water Absorption This type of absorption equipment was fairly common until about 1950. Thereafter the development of lithium bromide absorption machines drastically reduced the production of ammonia-water units. However, with the great emphasis now on process-energy conservation, ammonia-water units are applicable if large quantities of waste heat are available. Installations in petrochemical and chemical plants can have capacities in the range of 1000 to 5000 tons on low-temperature brine or direct cooling. These sizes are too large for packaged equipment, so considerable field piping is required at the site, thus substantially increasing the investment over other types of refrigeration.

Figure 12-49 shows a simplified ammonia-water absorption cycle. The refrigerant is ammonia, and the absorbent is a dilute aqueous solution of ammonia. A strong aqua solution, resulting from mixing ammonia vapors from the evaporator and weak aqua from the generator, is produced in the absorber. Strong aqua then is separated by a distillation column, producing (1) ammonia overhead product for recycling to the evaporator and (2) water-rich bottoms. Waste heat to the generator vaporizes ammonia in the aqua, and a weak aqua stream returns to the absorber. Ammonia refrigerant in the evaporator absorbs heat from the process stream or indirect brine, thus cooling it for refrigeration use. Cooling-tower or river water flows through the condenser to condense vapors and then in series to the absorber. Water flowing through the absorber tubes removes the heat generated by combining weak aqua and anhydrous ammonia.

Table 12-11 presents utility requirements for ammonia-water absorption refrigeration. Evaporator temperatures are shown for applications between 10 and −45.6°C (50 and −50°F). When producing evaporator temperatures below the ammonia atmospheric boiling point of −33°C (−28°F), air in-leakage is possible under vacuum operating conditions. This can result in the formation of corrosive ammonium carbonate. Ammonia-water units can be arranged for single-stage or cascaded two-stage operation. The advantage of two-staging is a decrease in the temperature required for the generator. However, two-staging increases investment and results in a lower coefficient of performance owing to increased quantities of generator heat required. This is true because it is necessary to separate ammonia-water vapor from strong aqua solutions at two different pressure levels.

BRINES

Introduction Increased requirements for low temperature in industrial-process cooling have led to the development of a number of brines for this service. In refrigeration terminology, a brine is any liquid cooled by a refrigerant and circulated as a heat-transfer fluid. The design or plant engineer now has a wide choice in the selection of a secondary coolant in operating temperature ranges from 0° to −67.7°C (+32 to −90°F). Brines may be:

1. An aqueous solution of inorganic salts, such as sodium chloride and calcium chloride.
2. An aqueous solution of organic compounds, such as alcohols or glycols. These are water solutions of various concentrations, including methanol water, ethanol water, ethylene glycol, and propylene glycol.
3. Chlorinated or fluorinated hydrocarbons and halocarbons. These include methylene chloride, trichloroethylene, and R-11.

Brine Selection The selection of a brine for a given application depends on a number of considerations. The choice of a particular brine is invariably a compromise which best suits the specific application and economics.

The final selection of a brine depends on a number of factors, including:

1. *Safety.* Toxicity and flammability are two very important factors to be considered. Threshold limit values are shown in Table 12-12, and flammability data are presented in Table 12-13.
2. *Freezing point.* The brine must have a freezing point sufficiently below that of the lowest operating temperature of the system.
3. *Cost.* The initial charge and quantity of makeup required are considerations in the determination of costs. Also, the specific gravity and specific heat of the brine will influence pumping power costs.
4. *Application.* The ultimate use of the brine is important as to whether the process equipment is installed indoors or outdoors and whether the system will be of open or closed design. Highly toxic brines should not be used with air-conditioning coils when year-round operation occurs and when leaks into the air system are possible.
5. *Thermal performance.* The heat-transfer properties of the brine circulated through the refrigeration unit evaporator play an important part in determining the surface area required and the resulting evaporating temperature. Viscosity, specific heat, specific gravity, and thermal conductivity are values affecting heat transfer. Figures 12-50 through 12-57 show these data for two common "brines" used for refrigeration systems today: calcium chloride and ethylene glycol.
6. *Corrosiveness.* Piping and system equipment material of construction require a stable and relatively corrosion-free brine.

Salt Brines

Sodium Chloride Water solutions of this salt are commonly used for refrigeration service. Sodium chloride offers the lowest cost per gallon of solution of any brine available. Also, it can be used in applications involving contact with foods and in open systems because of its low toxicity. The properties of this brine also enable high brine film coefficients, which are desirable to reduce the amount of heat-exchanger surface required.

However, it has two main drawbacks: (1) It has a relatively high freezing point, which limits its application to operating levels from about −9.4°C (15°F) up. (2) It is highly corrosive, requiring inhib-

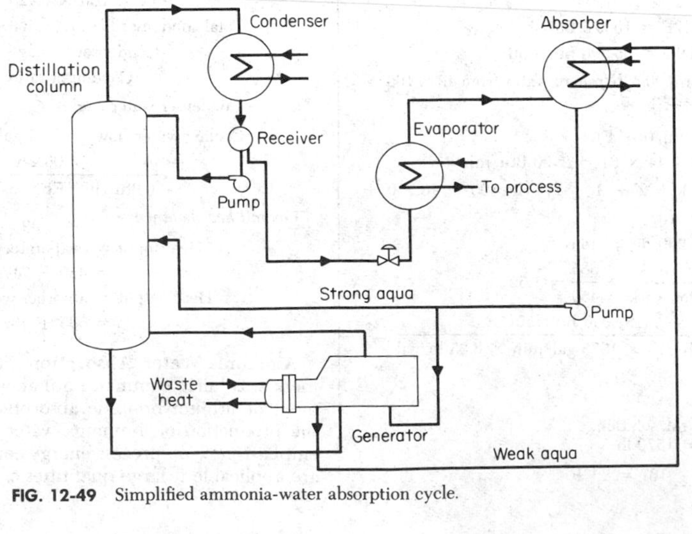

FIG. 12-49 Simplified ammonia-water absorption cycle.

TABLE 12-11 Utilities Required for Ammonia-Water Absorption at Varying Suction Temperatures*

Evaporation temperature, °F	50	40	30	20	10	0	−10	−20	−30	−40	−50
Single-stage											
Steam pressure, psia	14.1	19.8	24.9	30.9	41.8	53.2	67.0	83.1	103.1	134.6	173.3
Steam-saturation temperature, °F (or waste-heat exit temperature)	210	225	240	255	270	285	300	315	330	350	370
Generator heat required, Btu/(min·tons registered)	300	325	347	373	400	430	466	511	571	645	754
Steam rate, lb/(h·tons registered)	18.9	20.2	21.8	23.6	25.8	28.1	30.9	34.1	39.1	44.6	53.2
Water required through condenser and absorber, gal/(min·tons registered), 80°F on, 105°F off	3.6	3.7	3.8	4.0	4.3	4.5	5.0	5.5	6.1	7.1	8.8
Two-stage											
Generator steam-saturation temperature, °F, exit temperature	175	180	190	195	205	210	220	230	240	250	265
Steam pressure, psia	6.7	7.5	9.3	10.4	12.3	14.1	17.1	20.7	24.9	29.8	38.1
Generator heat required, Btu/(min·tons registered)	550	577	605	637	670	711	753	799	850	905	970
Steam rate, lb/(h·tons registered)	33.2	34.9	37.0	39.1	41.3	43.9	46.7	49.9	53.6	57.5	62.3
Water required through condenser and absorber, gal/(min·tons registered), 87°F on, 105°F off	4.0	4.2	4.3	4.5	4.9	5.3	5.8	6.4	7.2	8.3	10.2

*Lewis Refrigeration Company.

NOTE: All data approximated. Detailed selections can vary by ±10 percent, depending upon heat source, water available, evaporator constructions, etc. A detailed analysis is required for each system.

TABLE 12-12 Brine-Toxicity Data

Type of brine	Threshold limit values, p.p.m. (A.C.G.I.H.,* 1966)	Remarks
Sodium chloride	. . .	Non-toxic. Suitable for direct contact with food
Propylene glycol	. . .	Non-toxic. Suitable for direct contact with food
Calcium chloride	. . .	Essentially non-toxic but not suitable for direct contact with food
Freon-11®	1000	Non-toxic
Ethanol water	. . .	Non-toxic
Methylene chloride	500	Very few toxic effects
Ethylene glycol	. . .	As toxic as methanol water but considered less dangerous because it is not absorbed as readily
Methanol water	200	Toxic. Outdoor use normally recommended
Trichloroethylene	100	Toxic. Outdoor use normally recommended

*American Conference of Governmental Industrial Hygienists.

itors that must be checked on a regular schedule and replenished to prevent an acid condition from occurring in the system.

Calcium Chloride Aqueous solutions of calcium chloride find wide use as a circulating brine. This is the second-lowest-cost brine solution, and the lower freezing point of calcium chloride solution makes its use more convenient than that of sodium chloride. Calcium chloride brine has been used at temperature levels as low as −37.2°C (−35°F).

The main disadvantages are that (1) it is highly corrosive, (2) it has rapidly reduced heat-transfer coefficients below −20.6°C (−5°F), and (3) it cannot be used in direct contact with foods.

Organic Compounds

Ethylene Glycol Commonly used as an antifreeze, ethylene glycol also has found use for refrigeration service. Common applications include process cooling at lower temperatures.

TABLE 12-13 Brine-Flammability Data

Type of brine	Ratings*
1. Sodium chloride	Non-flammable
2. Calcium chloride	Non-flammable
3. Freon-11®	Non-flammable
4. Trichloroethylene	Non-flammable at ordinary temperatures
5. Methylene chloride	Practically non-flammable at ordinary temperatures.
6. Propylene glycol	Flammable, moderate fire hazard. Flash point 210 to 225°F. undiluted
7. Ethylene glycol	Flammable—moderate fire hazard. Flash point 232 to 240°F. undiluted
8. Methanol water	Highly flammable—fire hazard. Flash point 54 to 60°F. undiluted. Flash point 75°F., 30% solution
9. Ethanol water	Highly flammable—fire hazard. Flash point 55°F. undiluted

*The higher the flash point, the more safely the liquid can be handled. Liquids with flash points under 70°F should be regarded as highly flammable.

REFERENCES: *ASHRAE Guide and Data Book: Handbook of Fundamentals*, American Society of Heating, Refrigerating and Air-Conditioning Engineers, New York, 1967 (items 1–5). Sax (ed.), *Dangerous Properties of Industrial Materials*, 5th ed., Van Nostrand Reinhold, New York, 1979 (items 6–9).

Ethylene glycol is colorless and practically odorless and is completely miscible with water. When properly inhibited, it has a relatively low corrosivity, which is a major advantage when compared with salt brines. It also has the advantages of lowering the freezing point of water and having low volatility. Ethylene glycol solutions can be used as a refrigeration-system brine as low as −34.4°C (−30°F). However, it is considered a relatively high-temperature brine, finding best application at 9.4°C (15°F) and above.

Its main disadvantage is its poor heat-transfer coefficients at decreasing temperatures resulting from high viscosities. Ethylene glycol is somewhat toxic but less harmful than methanol water solutions. It should not be allowed to stand in open containers, and it is not suitable for contact with foods.

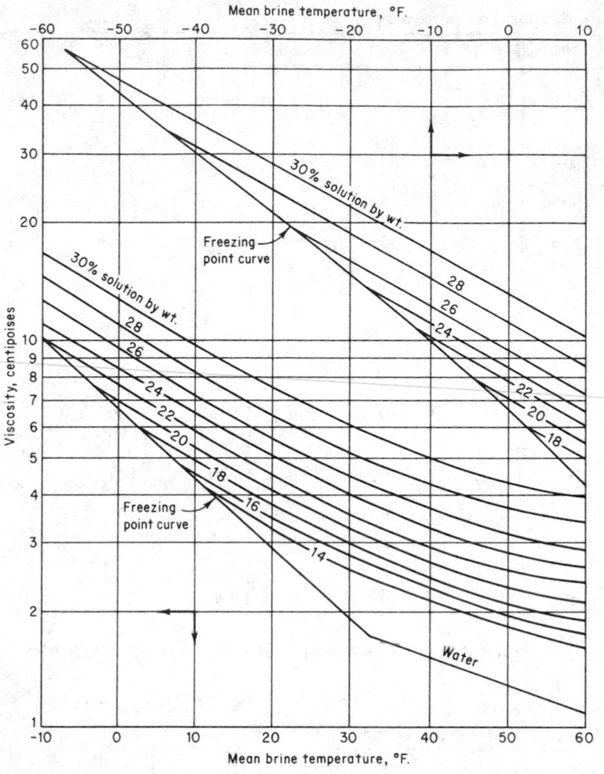

FIG. 12-50 Viscosity of calcium chloride. *(Carrier Air Conditioning Company.)*

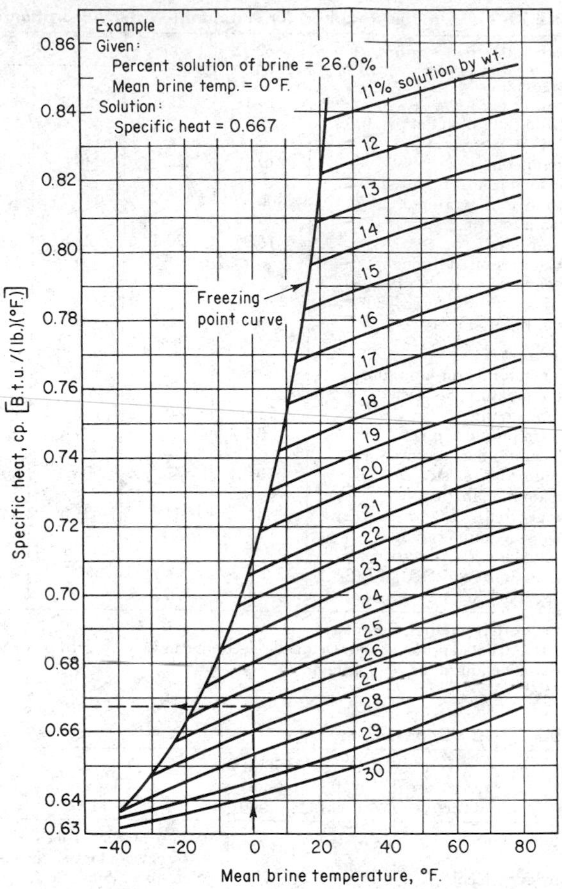

FIG. 12-51 Specific heat of calcium chloride. *(Carrier Air Conditioning Company.)*

The quality of the water used to prepare an ethylene glycol solution is important. Preferably waters that are classified as soft and are low in chloride and sulfate ions should be used. A monitoring schedule should be maintained to assure that inhibitor depletion is avoided. Ethylene glycol normally has a pH of 8.8 to 9.2 and should not be used below about 7.5. Addition of more inhibitor cannot restore the solution to its original condition. Once the inhibitor has been depleted, it is recommended that the old glycol be removed from the system and a new charge be installed.

Propylene Glycol In its inhibited form propylene glycol has the same advantages of low corrosivity and low volatility shown by ethylene glycol. It is not considered toxic, and propylene glycol has been used in direct contact with foods. Other than lack of toxicity, it has no advantages over ethylene glycol, being higher in cost at temperature ranges normally considered and more viscous, and it has much poorer heat-transfer brine film coefficients. Propylene glycol normally would be used for brine service of −9.4°C (15°F) and above.

Methanol Water This is an alcohol-base compound which has been used widely as an antifreeze solution. It is lower in cost than other organic compounds, finding use as refrigeration-service brines, and offers important heat-transfer properties over other organic compounds. In its inhibited form it is not considered corrosive. Methanol water has a wide application range from −1.1° to −34.4°C (30 to −30°F). It is particularly suitable for brine service from −20.6 to −34.4°C (−5 to −30°F) owing to its relatively high rate of heat transfer in this temperature range, in which halocarbon fluids are usually too expensive for consideration.

Its main disadvantages as a brine are its toxicological considerations. It is considered more harmful than ethylene glycol and consequently has found use only for process applications located outdoors. Also, methanol is a flammable liquid and, as such, introduces a potential fire hazard where it is stored, handled, or used.

Ethanol Water This is an aqueous solution of denatured grain alcohol. Its main advantage is that it is nontoxic. Corrosion inhibitors are used with ethanol to make it noncorrosive for brine service. Ethanol water has found application in breweries, chemical plants, and food-freezing plants.

Its heat-transfer properties are not as good as those of methanol water because of higher viscosities, and it is more expensive than methanol water but less expensive than ethylene glycol. As an alcohol derivative, it is a flammable liquid requiring certain fire precautions for storage and use.

Hydrocarbons and Halocarbons When circulating brine systems are required for −37.2°C (−35°F) or lower-temperature service, salt-base and organic brines become unusable because of freezing-point limitations or very high viscosities of the water solution. Consequently, many of the common refrigerants can be employed as secondary coolants. They have favorable properties for heat-transfer fluids, including low freezing point, low viscosities, nonflammability, and good stability. For low-temperature work, the following have been commonly used:

Methylene chloride (CH_2Cl_2). This chlorohydrocarbon has found extensive use as a brine from −37.2 to −84.4°C (−35 to −120°F) operating range. Methylene chloride gives excellent heat-transfer rates. It presents no problems of fire hazard under normal conditions of use and is one of the least toxic of the chlorinated hydrocarbons. Although methylene chloride is nonflammable, contact with open flames or smoldering fire may cause decomposition and the formation of toxic or corrosive substances. There should be adequate ventilation, and smoking should be prohibited in working areas. Also, methylene chloride has a major advantage in being low in initial cost as compared with halocarbon refrigerants.

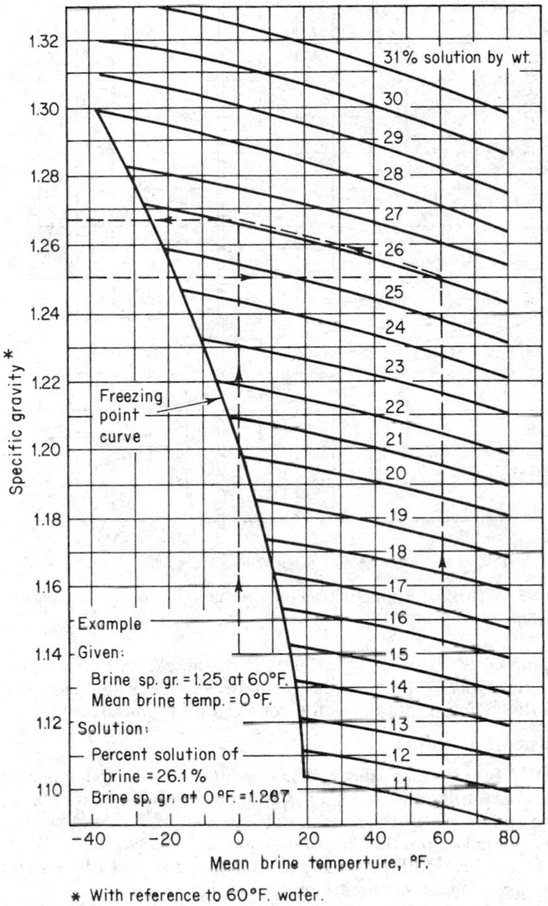

FIG. 12-52 Specific gravity of calcium chloride. *(Carrier Air Conditioning Company.)*

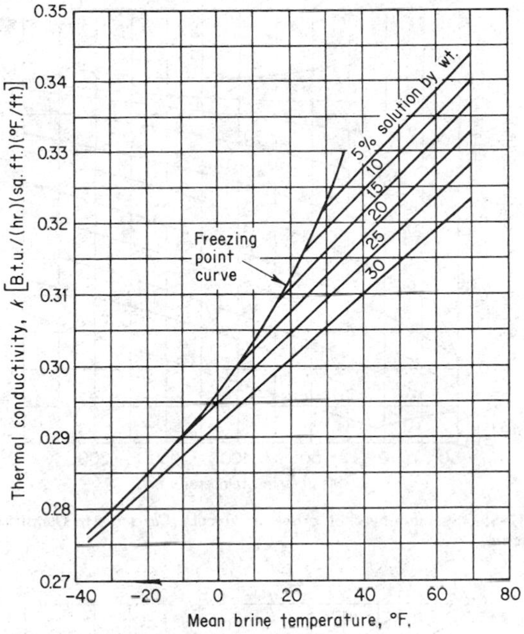

FIG. 12-53 Thermal conductivity of calcium chloride. *(Carrier Air Conditioning Company.)*

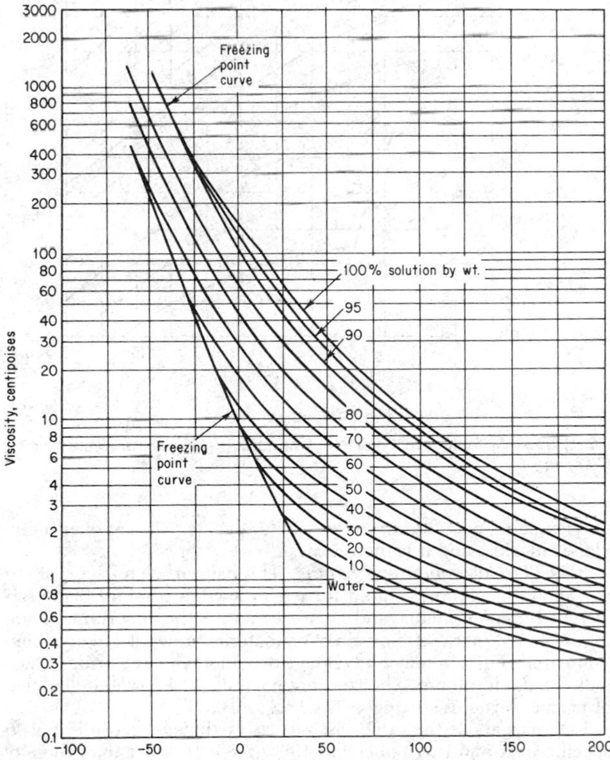

FIG. 12-54 Viscosity of ethylene glycol. *(Carrier Air Conditioning Company.)*

Small leaks of methylene chloride are difficult to detect. A good practice is to dissolve an azo-oil red dye in the fluid in concentrations of about 0.3 to 0.5 lb/ton of fluid. The colored dye remains at the point of leakage after evaporation of the fluid.

Methylene chloride brine systems require the use of adequate dryers to remove moisture. Methylene chloride is only slightly soluble in water, especially at low temperatures. Entrance of moisture into a system without dryers will invariably cause icing at the brine cooler and loss of heat transfer. Methylene chloride should not be used in direct contact with aluminum. The possibility of leakage of methylene chloride into the refrigerant side should not be overlooked and may dictate the use of R-11 or a compressor equipped with other than aluminum parts.

Trichloroethylene ($CHCl{:}CCl_2$). This heat-transfer fluid has not been as widely used as methylene chloride, which exhibits better properties for a low-temperature heat-transfer fluid. While it has a more favorable vapor pressure and evaporation rate than methylene chloride, these are comparatively insignificant for continuous low-temperature operations. Trichloroethylene is the preferred convection fluid when the upper temperature range exceeds 60°C (140°F). Superior stability makes it particularly suitable for use at temperatures up to 121°C (250°F). The lower limit of trichloroethylene is about −73.3°C (−100°F).

Comparison of data with methylene chloride shows that trichloroethylene has less favorable factors in respect to density, viscosity, specific heat, thermal conductivity, and water solubility. Trichloro-

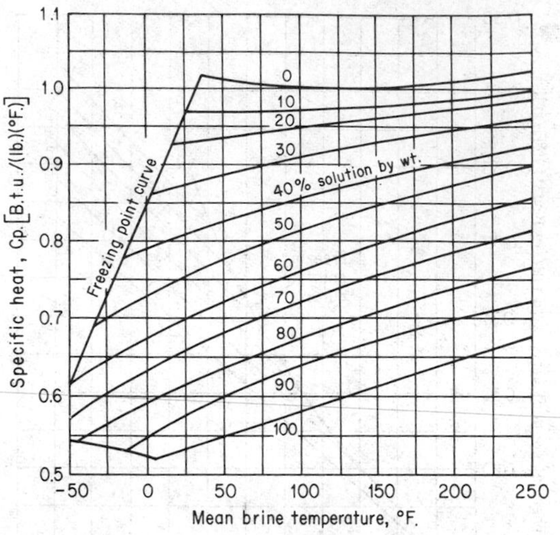

FIG. 12-55 Specific heat of ethylene glycol. *(Carrier Air Conditioning Company.)*

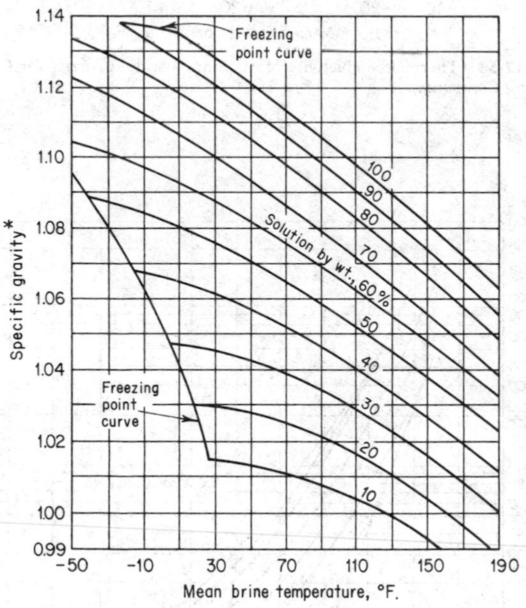

*With reference to 60°F. water.

FIG. 12-56 Specific gravity of ethylene glycol. *(Carrier Air Conditioning Company.)*

ethylene does not offer any appreciable advantage in cost over methylene chloride, and it is more toxic.

R-11 (Trichlorofluoromethane). This halocarbon refrigerant is a very good heat-transfer fluid for use in low-temperature brine service. It is nonflammable and nontoxic and has the important advantage of being inert, with no effect on materials unless it becomes contaminated. This is a major advantage over methylene chloride, which will attack aluminum. The freezing point of R-11 is well below that of normal brine use, being −111.1°C (−168°F).

R-11 has about the same viscosity as methylene chloride, but its specific heat and thermal conductivity are both lower than those of methylene chloride. The specific gravity of R-11 is also 10 to 15 percent higher than that of methylene chloride, which would increase

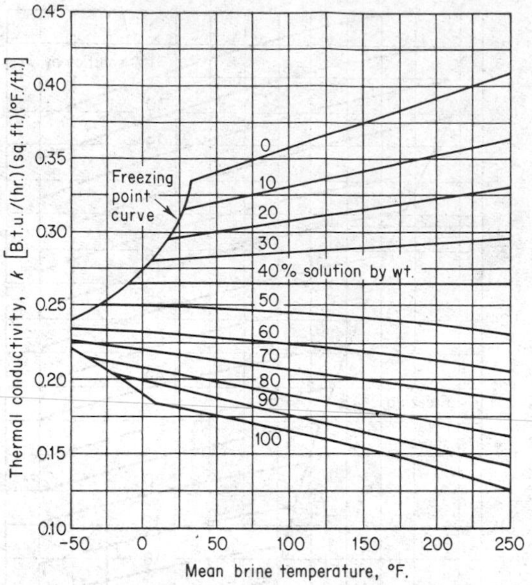

FIG. 12-57 Thermal conductivity of ethylene glycol. *(Carrier Air Conditioning Company.)*

pumping costs. R-11 has a further disadvantage: an initial product cost about twice as high as that of methylene chloride. This would be relatively important only on extensive piping installations.

Corrosion Prevention

Salt Brines Salt brines can cause high corrosion damage and must be carefully checked and maintained on a regular schedule. The corrosion process requires a continuous supply of oxygen, which normally comes from the atmosphere and dissolves in the brine solution. Dilute brines dissolve oxygen more readily and are generally more corrosive than concentrated brines.

An open brine system is one in which oxygen enters the brine circuit. However, it is believed that a closed brine system will still not prevent the infiltration of oxygen.

To retard corrosion most effectively, it is desirable to maintain salt brines in a slightly alkaline condition (pH, 7.5 to 8.0). Alkaline brines are generally less corrosive than neutral or acid brines, although with high alkalinity the activity may increase.

Sodium dichromate is generally considered to be the most effective and economical method of retarding brine corrosion. The dichromate has a bright orange color, is granular in form, and can be readily dissolved in warm water. It dissolves very slowly in cold water and should never be put directly into a brine tank in crystal form. The best plan is to dissolve the dichromate in warm water and then pour this into the brine.

The quantities recommended by ASHRAE for long-term protection are as follows:

1. 125 lb of sodium dichromate per 1000 ft^3 of calcium chloride brine.
2. 200 lb of sodium dichromate per 1000 ft^3 of sodium chloride brine

In addition, caustic soda should be added in sufficient quantity to convert the dichromate to a neutral chromate. Sodium dichromate when dissolved in water or brine makes the solution acid. Before a brine is treated with dichromate, the pH should be determined. If the brine is acid or neutral (pH, 7.0 or lower), sufficient caustic soda should be added to increase the pH to 8.0. If the untreated brine has the proper pH value, the acidifying effect of the dichromate may be neutralized by adding 27 lb of 76 percent commercial flake caustic soda for each 100 lb of sodium dichromate used. Caustic soda must be thoroughly dissolved in warm water before it is added to the brine.

Inhibited Glycols and Alcohols In using glycols or alcohols as brines for refrigeration service, it is necessary to purchase them with inhibitors included in the solution. In general, properly inhibited and properly maintained solutions should provide advantages in corrosion protection as compared with salt-brine solutions. An indefinite service life, however, should not be expected, and the inhibitor should be checked yearly. Indiscriminate mixing of inhibited formulations should be avoided unless these are known to be compatible. In this connection, replacement of salt-brine systems with inhibited glycol or alcohol should be approached with caution because of the potential incompatibility of the brine components with these formulations. Chromate treatment should *never* be used with ethylene glycol because of the rapid formation of sludge. Also, it is suggested that system piping be thoroughly cleaned and flushed by means of a heated trisodium phosphate solution before the introduction of the water-glycol mixture. Ethylene glycol will react with pipe dope, cutting oils, solder flux, and dirt.

Hydrocarbons and Halogens Methylene chloride, trichloroethylene, and halocarbon refrigerants do not show general corrosive tendencies and require no further corrosion-preventive measures. It is important that dryers be installed to prevent buildup of moisture in the system.

Materials of Construction Steel, iron, copper, or red brass can be used with brine circulating systems. Calcium chloride systems are generally equipped with all-iron-and-steel pumps and valves to prevent electrolysis in event of acidity. However, calcium chloride evaporators have successfully used copper and red brass for tubing. Sodium chloride systems are usually equipped with all-iron or all-bronze pumps. For ethylene glycol, copper tubing is often used for pipe sizes under 3 in, while valves, pumps, cooler tubes, or coils are made of iron, steel, brass, copper, or aluminum. Ethylene glycol should not be used with galvanized piping because of the reaction of the inhibitor with the zinc in the galvanizing. Methanol water solutions are compatible with most materials but in sufficient concentration will badly corrode aluminum.

Methylene chloride and trichloroethylene must not be used with aluminum or zinc. They also attack most rubber compounds, plastics, and electric-motor varnishes.

Low-temperature service requires special considerations with regard to materials of construction, depending on operating temperatures.

Toxicity and Flammability Data Toxicity and flammability criteria are important considerations in the selection of a brine.

Threshold limit values (TLV) are considered the best means of evaluating the relative toxicity of various air contaminants. They should be used as guides in the control of health hazards and should not be regarded as fine lines between safe and dangerous concentrations. TLV values are reviewed and published annually by the American Conference of Governmental Industrial Hygienists (ACGIH).

The higher the threshold limit value, the less toxic the substance is considered. For example, halocarbon refrigerants are considered nontoxic with a TLV of 1000. Methanol is considered toxic with a TLV of 200. TLV data have not been developed for all brines normally used in refrigeration service as indicated in Table 12-12. However, remarks are included for guiding purposes.

Flammability data are based normally on flash-cup tests. The higher the flash point, the more safely the material can be handled. A nonflammable material would be one in which no flash point was demonstrated. Flammability data, primarily for refrigerants, are given in the *ASHRAE Guide and Data Book: Handbook of Fundamentals* and more extensively in Sax's *Dangerous Properties of Industrial Materials* (5th ed., Van Nostrand Reinhold, New York, 1979).

CYROGENIC PROCESSES

INTRODUCTION

Cryogenics, the production and utilization of low temperatures, has grown spectacularly since World War II. It is now a major business in the United States, with an annual value in excess of $10 billion. This estimate is based on a generally accepted definition of cryogenics which ascribes to it the unusual and unexpected property variations that appear at low temperatures and make extrapolations from ambient to low temperatures unreliable. A more limited definition ascribes cryogenics to a temperature range below 125 K.

Cryogenics is a very diverse supporting technology, a means to an end and not an end in itself. For example, gases such as oxygen and nitrogen, obtained by the cryogenic separation of air, are very important industrial gases. Some 50 percent of the oxygen obtained in this manner is used in the production of steel, and 20 percent is used in the chemical-processing industry. Since the early 1950s liquid-hydrogen production has risen from laboratory quantities to a level of over 2.1 kg/s, first spurred by nuclear-weapons development and later by the United States space program. Similarly, the space age has increased the need for liquid helium by more than a factor of 10, requiring the construction of large plants to separate helium from natural gas by cryogenic means. The demands for energy have likewise accelerated the construction of large-base-load liquefied-natural-gas (LNG) plants around the world and have been responsible for the associated domestic LNG industry with its use of peak-shaving plants and increasing imports of overseas LNG.

Freezing as a means of preserving food dates back to 1840. Today the food industry uses large quantities of liquid nitrogen for this purpose and as a refrigerant in frozen-food transport systems. In biological applications liquid-nitrogen-cooled containers are routinely used to preserve whole blood, tissue, bone marrow, and animal semen for extended periods of time. Cryogenic surgery has become accepted in curing such involuntary disorders as Parkinson's disease. Finally, one must recognize the role that cryogenics plays in the chemical-processing industry with the recovery of valuable feedstock from natural-gas streams, the upgrading of the heat content of fuel gas, the recovery of useful components from air, the purification of various process and waste streams, the production of ethylene, etc.

PROPERTIES OF ENGINEERING MATERIALS AT LOW TEMPERATURES

A knowledge of the properties and behavior of materials used in any cryogenic system is essential for proper design considerations. Often the choice of materials for the construction of cryogenic equipment will be dictated by considerations besides mechanical properties, such as, for example, thermal conductivity (heat transfer along a structural member), thermal expansivity (expansion and contraction during cycling between ambient and low temperatures), and density (mass of system). Since properties at low temperatures are often significantly different from those at ambient temperature, there is no substitute for test data on a truly representative sample specimen when designing for the limit of effectiveness of a cryogenic material or structure. For example, some metals including elements, intermetallic compounds, and alloys exhibit the phenomenon of superconductivity at very low temperatures. The properties that are affected when a material becomes superconducting include specific heat, thermal conductivity, electrical resistance, magnetic permeability, and thermoelectric effect. As a result, the use of superconducting metals in the construction of equipment for temperatures lower than 10 K needs to be evaluated carefully.

Strength, Ductility, and Elastic Modulus It is most convenient to classify metals by their lattice symmetry for low-temperature mechanical-properties considerations. The face-centered-cubic (fcc) metals and their alloys are most often used in the construction of cryogenic equipment. Al, Cu, Ni, their alloys, and the austenitic stainless steels of the 18-8 type are fcc and do not exhibit an impact ductile-to-brittle transition at low temperatures. As a general rule, the mechanical properties of these metals improve as the temperature is reduced. The yield strength at 20 K is quite a bit larger than at ambient temperature; Young's modulus is 5 to 20 percent larger at the lower temperatures, while fatigue properties, with the exception of 2024-T4 aluminum, are also improved at the lower temperatures. Since the annealing of these metals and alloys can affect both ultimate and yield strengths, care must be exercised under these conditions.

The body-centered-cubic (bcc) metals and alloys are normally classified as undesirable for low-temperature construction. This class includes Fe, the martensitic steels (low-carbon steels and the 400 series of stainless steels), Mo, and Nb. If not brittle at room temperature, these materials exhibit a ductile-to-brittle transition at low temperatures. Cold working of some steels, in particular, can induce the austenite-to-martensite transition.

The hexagonal-close-packed (hcp) metals exhibit mechanical properties intermediate between those of fcc and bcc metals. For example, Zn suffers a ductile-to-brittle transition, whereas Zr and pure Ti do not. The latter and its alloys, having an hcp structure, remain reasonably ductile at low temperatures and have been used for many applications for which weight reduction and reduced heat leakage through the material have been important. However, small impurities of O, N, H, and C can have a detrimental effect on the low-temperature ductility properties of Ti and its alloys.

Plastics increase in strength as the temperature is decreased, but this increase is accompanied by a rapid decrease in elongation in a tensile test and a decrease in impact resistance. Teflon and glass-reinforced plastics retain appreciable impact resistance as the temperature is lowered. The glass-reinforced plastics also have high-strength-to-weight and strength-to-thermal-conductivity ratios. All elastomers, on the other hand, become brittle at low temperatures. Nevertheless, many of these materials including rubber, Mylar, and nylon can be used for static-seal gaskets provided they are highly compressed at room temperature prior to cooling.

The strength of glass under constant loading also increases with a decrease in temperature. Since failure occurs at a lower stress when the glass surface contains surface defects, strength can be improved by tempering the surface.

Specific Heat This physical property can be predicted fairly accurately by mathematical models through statistical mechanics and quantum theory. For solids, the Debye model gives a satisfactory representation of the specific heat with temperature. Procedures for calculating values of Θ_D, the Debye characteristic temperature, using either elastic constants, compressibility, the melting point, or the temperature dependence of the expansion coefficient, are outlined by Dillard et al. [*Chem. Eng. Prog. Symp. Ser.*, **64**(87), 1 (1968)].

Thermal Conductivity Adequate predictions of this property for pure metals can be made by means of the Wiedemann-Franz law, which states that the ratio of thermal conductivity to the product of electrical conductivity and the absolute temperature is a constant. This ratio for high-conductivity metals extrapolates to approximately the Sommerfeld value of 2.449×10^{-8} $(W \cdot \Omega)/K^2$ at 0 K but falls considerably below it at higher temperatures. It should be noted that high-purity aluminum and copper exhibit peaks in thermal conductivity between 20 to 50 K, but these peaks are rapidly suppressed with increased impurity levels and cold work of the metal. In fact, aluminum alloys show a steady decrease in thermal conductivity with a decrease in temperature. This behavior is also demonstrated by other metals. Inconel, Monel, and stainless steel, structural alloys exhibiting these properties, are thus useful in cryogenic service, which requires low thermal conductivity over the entire temperature range.

All cryogenic liquids except hydrogen and helium have thermal conductivities that increase as the temperature is decreased. For these two exceptions, thermal conductivity decreases with a decrease in temperature. The kinetic theory of gases correctly predicts the decrease in thermal conductivity of all gases as the temperature is lowered.

Thermal Expansivity The expansion coefficient of a solid can be estimated with the aid of an approximate thermodynamic equation of state for solids which equates the thermal-expansion coefficient β with the quantity $\gamma C_v \rho/B$, where γ is the Grüneisen dimensionless ratio, C_v is the specific heat of the solid, ρ is the density of the material, and B is the bulk modulus. For fcc metals the average value of the Grüneisen constant is near 2.3. However, there is a tendency for this constant to increase with atomic number.

Electrical Resistivity The electrical resistivity of most pure metallic elements at ambient and moderately low temperatures is approximately proportional to the absolute temperature. At very low temperatures, however, the resistivity (with the exception of superconductors) approaches a residual value almost independent of temperature. Alloys, on the other hand, have resistivities much higher than those of their constituent elements and resistance-temperature coefficients that are quite low. Electrical resistivity as a consequence is largely independent of temperature and may often be of the same magnitude as the room-temperature value.

The insulating quality of solid electrical conductors usually improves as the temperature is lowered. In fact, all the common cryogenic fluids are good electrical insulators. In the temperature region from 1 to 5 K, the electrical resistivity of many semiconductors increases quite rapidly with a small decrease in temperature and forms the basis for the development of numerous sensitive semiconductor resistance thermometers for very low temperature measurements.

PROPERTIES OF CRYOGENIC FLUIDS

There are available numerous collections of thermodynamic-properties data for fluids commonly associated with low-temperature processing. These have conveniently been reviewed and referenced in the literature (Steward, *Advances in Cryogenic Engineering*, vol. 17, Plenum, New York, 1972, p. 8).

Helium-4 Liquid helium-4 can exist in two different liquid phases, liquid helium I, the normal liquid, and liquid helium II, the superfluid, since under certain conditions the fluid acts as if it had no viscosity. The phase transition between the two liquid phases is identified as the lambda line, and where this line intersects the vapor-pressure curve is labeled the lambda point. Thus, there is no triple point for this fluid as for other fluids. In fact, solid helium can exist only under a pressure of 2.5 MPa or more. The most recent tabulation of thermodynamic data is given by McCarty (NBS Tech. Note 631, 1972).

Hydrogen A unique property of hydrogen is that it can exist in two different molecular forms, orthohydrogen and parahydrogen. (This is also true of deuterium, an isotope of hydrogen with an atomic mass of 2.) The thermodynamic equilibrium composition of the ortho and para varieties is temperature-dependent. The equilibrium mixture of 75 percent orthohydrogen and 25 percent parahydrogen at ambient temperatures is recognized as normal hydrogen. A tabulation of thermodynamic data for parahydrogen to 300 K and 100 MPa is presented by Weber (NBS IR 74-374/NASA SP-3088, 1974).

Nitrogen The most recent compilation of thermodynamic properties for nitrogen is given by Jacobson et al. (NBS Tech. Note 648, 1973). The tabulations in this revised study extend from the fusion line to 1950 K and pressures to 1.03 GPa.

Air Since air is a mixture of predominantly oxygen, nitrogen, and a host of lesser impurities, there has been less interest in developing precise thermodynamic properties. The only recent correlation of thermodynamic properties is that published by Vasserman et al. (Barouch, Israel Program for Scientific Translations, Jerusalem, 1970); it is based on the principle of corresponding states because of the scarcity of experimental data.

Oxygen In contrast to other cryogenic fluids, liquid oxygen is slightly magnetic. It is also chemically reactive, particularly with hydrocarbon materials. Oxygen thus presents a safety problem and requires extra precautions in handling.

Thermodynamic properties based on an extensive experimental program have been published by McCarty et al. (NBS Tech. Note

384, 1971). The properties in this tabulation extend from the freezing line to 333 K and pressures to 34.5 MPa.

Methane The rapid growth of the LNG industry has created a flurry of activity seeking more reliable thermodynamic data for methane. Numerous correlations and studies have been published in the literature. The most useful publication for process calculations is that by Goodwin (NBS Tech. Note 653, 1974). Thermodynamic tabulations extend from 90 to 500 K and pressures to 70 MPa. The data, equations, and computer programs provided in this report have been used by Mollerup in calculations of the phase equilibria, orthobaric densities, and thermodynamic properties of LNG mixtures (*Advances in Cryogenic Engineering*, vol. 20, Plenum, New York, 1975, p. 172).

REFRIGERATION AND LIQUEFACTION PRINCIPLES

A process for producing refrigeration at cryogenic temperatures usually involves equipment at ambient temperature in which the process fluid is compressed and heat is rejected to a coolant. During the ambient-temperature compression process, the enthalpy and entropy of the process fluid are decreased. At the cryogenic temperature at which heat is absorbed, the enthalpy and entropy are increased. The reduction in temperature of the process fluid is usually accomplished by heat exchange between the cooling and the warming fluid followed by an expansion. This expansion may take place through either a throttling device (isenthalpic expansion), by which there is a reduction in temperature only, or in a work-producing device (isentropic expansion), by which both temperature and enthalpy are decreased.

In a continuous refrigeration process, there is no accumulation of refrigerant in any part of the system. This contrasts with a gas-liquefying system, in which liquid accumulates and is withdrawn. Thus, a liquefying system experiences an unbalanced flow in the heat exchangers, while a refrigeration system usually operates with a balanced flow in the heat exchangers, except when a portion of the flow is diverted through a work-producing expander.

Performance Criteria The performance of a real refrigerator is measured by the coefficient of performance (COP) and is defined as the ratio of the refrigeration effect to the work input. Thus,

$$\text{COP} = \frac{Q}{W} = \frac{\text{heat absorbed from low-temperature source}}{\text{net work input}} \tag{12-33}$$

Another means of comparing the performance of a practical refrigerator is by the use of the figure of merit (FOM), defined as

$$\text{FOM} = \text{COP}/\text{COP}_i \tag{12-34}$$

where COP is the coefficient of performance of the actual refrigerator system and COP_i is the coefficient of performance of the thermodynamically ideal system. This figure of merit for a liquefier is generally rewritten as

$$\text{FOM} = (W_i/\dot{m})/(W/\dot{m}_f) \tag{12-35}$$

where W_i is the work of compression for the ideal cycle, W is the work of compression for the actual cycle, $\dot{m}$ is the mass flow rate through the compressor (and is also the mass rate liquefied in the ideal cycle), and $\dot{m}_f$ is the mass rate liquefied in the actual cycle.

REFRIGERATION AND LIQUEFACTION METHODS

Only three methods of refrigeration and/or liquefaction have come into practical use: (1) vaporization of a liquid, (2) the Joule-Thomson effect in a gas, and (3) expansion of a gas in a work-producing engine. Normal commercial refrigeration generally is accomplished in a vapor-compression process. Temperatures to about 200 K can be obtained by cascading vapor-compression processes in which refrigeration is accomplished by liquid evaporation. Below this temperature, isenthalpic or isentropic expansions are generally used either singly or in combination. With few exceptions, refrigerators using these methods also absorb heat by liquid evaporation.

If refrigeration is to be accomplished at a temperature at which no suitable liquid exists to absorb heat by evaporation, then a cold gas must be produced to absorb the heat. This is generally accomplished by using a work-producing expansion engine.

Isenthalpic Expansion A thermodynamic process utilizing isenthalpic expansion to obtain cyrogenic temperatures, commonly referred to as the simple Linde or J-T cycle, is shown schematically with its corresponding temperature-entropy diagram in Fig. 12-58.

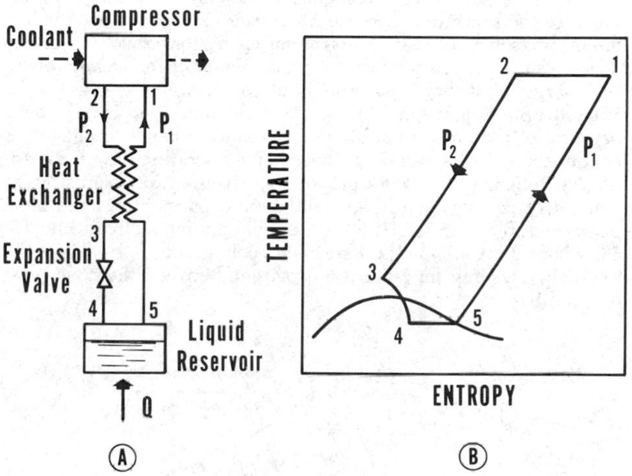

FIG. 12-58 Refrigerator using the simple Linde cycle.

The gaseous refrigerant is compressed at ambient temperature while essentially rejecting heat isothermally to a coolant. The compressed refrigerant is cooled in a heat exchanger by the stream returning to the compressor intake until it reaches the throttling valve. Joule-Thomson cooling upon expansion further reduces the temperature until, in the steady state, a portion of the refrigerant is liquefied. For a refrigerator, the unliquefied fraction and the vapor formed by liquid evaporation from the absorbed heat Q are warmed in the heat exchanger as they are returned to the compressor intake. If it is assumed that there are no heat in-leaks as well as negligible kinetic and potential energy changes in the fluid, the refrigeration duty is equivalent to $\dot{m}\ (h_1 - h_2)$, where the subscripts refer to the points on Fig. 12-58. Thus, the coefficient of performance for the ideal simple Linde refrigerator is given by

$$\text{COP} = \frac{h_1 - h_2}{T_1\,(s_1 - s_2) - (h_1 - h_2)} \tag{12-36}$$

For a simple Linde liquefier, the liquefied portion is continuously withdrawn from the reservoir, and only the unliquefied portion of the fluid is warmed in the countercurrent heat exchanger and returned to the compressor. The fraction y that is liquefied is obtained by applying the first law to the heat exchanger, J-T valve, and liquid reservoir. This results in

$$y = (h_1 - h_2)/(h_1 - h_f) \tag{12-37}$$

where h_f is the specific enthalpy of the liquid being withdrawn. Note that maximum liquefaction occurs when h_1 and h_2 refer to the same temperature. To account for heat in-leak q_L, the relation needs to be modified to

$$y = (h_1 - h_2 - q_L)/(h_1 - h_f) \tag{12-38}$$

with a resultant decrease in the fraction liquefied.

Refrigerants used in this process have a critical temperature well below ambient; consequently liquefaction by direct compression is not possible. In addition, the inversion temperature of the refrigerant must be above ambient temperature to provide cooling as the process is started. Auxiliary refrigeration is required if the simple Linde cycle is to be used to liquefy neon, hydrogen, and helium, whose inversion temperatures are below ambient. Liquid nitrogen is the optimum refrigerant for hydrogen and neon liquefaction systems, while liquid hydrogen is the normal refrigerant for helium liquefaction systems.

To reduce the work of compression in this cycle a two-stage or dual-pressure process by which the pressure is reduced by two successive isenthalpic expansions may be used. Since the work of compression is approximately proportional to the logarithm of the pressure ratio and since Joule-Thomson cooling is roughly proportional to the pressure difference, there is a much greater reduction in compressor work than in refrigerating performance. Hence the dual-pressure process produces a given amount of refrigeration with less energy input than the simple Linde process.

Isentropic Expansion In a work-producing expansion, the temperature of the process fluid is always reduced; hence, cooling does not depend on being below the inversion temperature prior to expansion. Additionally, the work-producing expansion results in a larger amount of cooling than in an isenthalpic expansion over the same pressure difference. This is illustrated diagrammatically in Fig. 12-59, where $T_A - T_B$ is the isentropic cooling and $T_A - T_C$ is the isenthalpic cooling for adiabatic expansion between the same pressure limits.

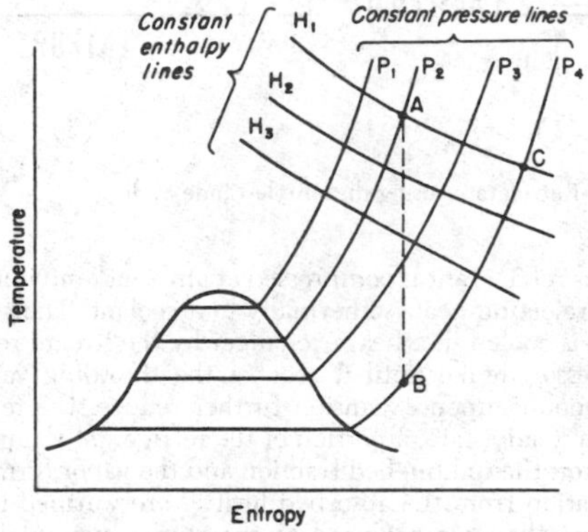

FIG. 12-59 Comparison of the cooling effects in isentropic and isenthalpic expansions.

In large systems utilizing expanders, the work produced during expansion is conserved. In small refrigerators, the energy from the expansion is usually expended in a gas or hydraulic pump or other suitable device. A schematic of a simple refrigerator using this expansion principle and the corresponding temperature-entropy diagram are shown in Fig. 12-60. Gas compressed isothermally at ambient temperature is cooled in a heat exchanger by gas being warmed on its way to the compressor intake. Further cooling takes place during the engine expansion. In practice, this expansion is never truly isentropic and is reflected by path 3–4 on the temperature-entropy diagram. This specific refrigerator produces a cold gas which absorbs heat from 4–5 and provides a method of refrigeration that can be used to obtain temperatures between those of the boiling points of the lower-boiling cryogens.

Combined Isenthalpic and Isentropic Expansion It is not uncommon to combine the isentropic and isenthalpic expansions to allow the formation of liquid in the refrigerator. This is done because of technical difficulties associated with forming liquid in the expander. The Claude cycle is an example of a combination of these methods; it is shown in Fig. 12-61 along with the corresponding temperature-entropy diagram.

Mixed-Refrigerant Cycle With the advent of large natural-gas liquefaction plants, particular emphasis must be devoted to still another cycle which is used principally in LNG production, namely, the mixed-refrigerant cycle. This cycle resembles the classic cascade

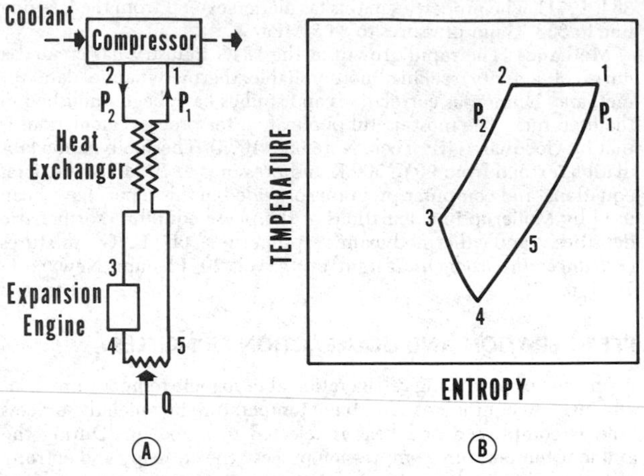

FIG. 12-60 Isentropic-expansion refrigerator.

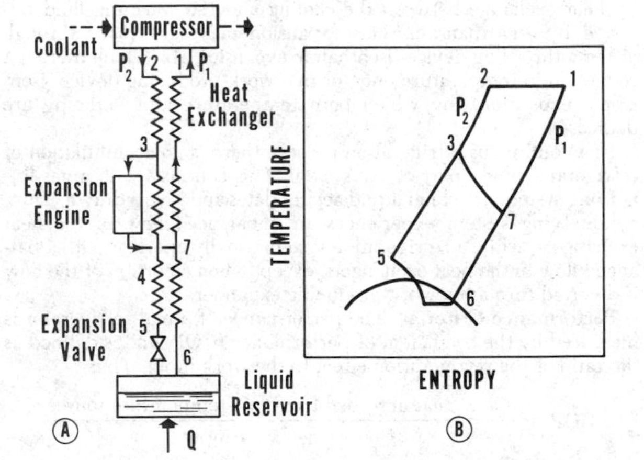

FIG. 12-61 Claude-cycle refrigerator: combined isentropic and isenthalpic expansion.

cycle in principle and may best be understood by referring to that cycle.

A simplified flow sheet of the classical cascade process is shown in Fig. 12-62. After purification, the natural-gas stream is cooled successively by vaporization of propane, ethylene, and methane. These gases have each been liquefied in a conventional refrigeration loop. Each refrigerant may be vaporized at two or three pressure levels to increase natural-gas-cooling efficiency, but at a cost of considerably increased process complexity.

Cooling curves for natural-gas liquefaction by the cascade process are shown in Fig. 12-63. It is evident that cascade-cycle efficiency can be considerably improved by increasing the number of refrigerants employed. The actual work required for the nine-level cascade cycle depicted is approximately 80 percent of that required by the three-level cascade cycle for the same throughput. This increase in efficiency is achieved by minimizing the temperature difference throughout each increment of the cooling curve.

The cascade system can be adapted to any cooling curve; i.e., the quantity of refrigeration supplied at the various temperature levels can be chosen so that the temperature differences in the evaporators and heat exchangers approach a practical minimum (small temperature differences mean low irreversibility and therefore lower power consumption).

The mixed-refrigerant cycle is a variation of the cascade cycle and

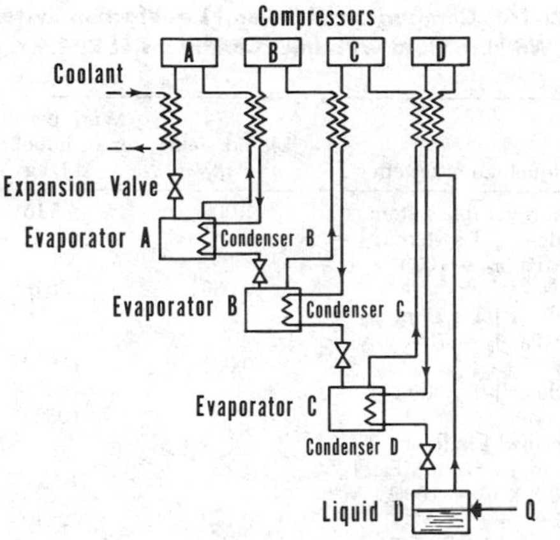

FIG. 12-62 Cascade compressed-vapor refrigerator.

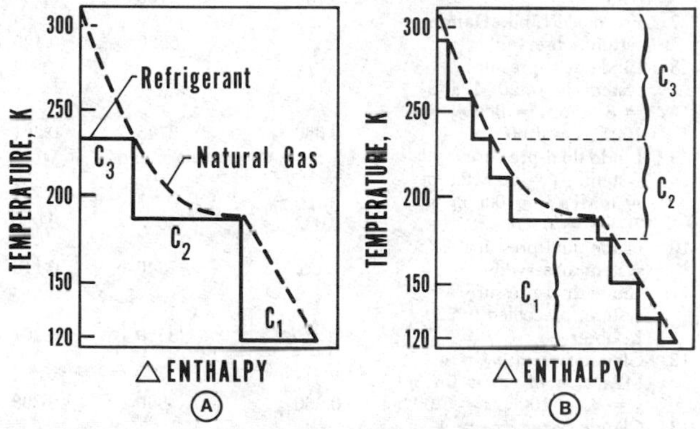

FIG. 12-63 (*a*) Three-level cascade-cycle cooling curve for natural gas. (*b*) Nine-level cascade-cycle cooling curve for natural gas.

involves the circulation of a single multicomponent refrigerant stream. The simplification of the compression and heat-exchange services in such a cycle may, under certain circumstances, offer potential for reduced capital expenditure over the conventional cascade cycle.

Figure 12-64 shows the basic concepts for a mixed-refrigerant cycle. Variations of the cycle are proprietary with those cryogenic engineering firms that have developed the technology. However, all the mixed-refrigerant processes use a carefully prepared refrigerant mix which is repeatedly condensed, vaporized, separated, and expanded. Thus, these processes require more sophisticated design methods and more thorough knowledge of the thermodynamic properties of gaseous mixtures than those required in the expander or classical cascade cycles. This is particularly evident when cooling curves similar to the one shown in Fig. 12-65 are desired. An inspection of the mixed-refrigerant cycle also shows that these processes must routinely handle two-phase flows in the heat exchangers.

Numerous excellent discussions of LNG cascade and mixed-refrigerant cycles are available in the literature (Hale, *LNG Economics and Technology,* 2d ed., Energy Communications, Inc., Dallas, 1974; and Wenzel, *Advances in Cryogenic Engineering,* vol. 20, Plenum, New York, 1975, p. 103).

THERMODYNAMIC ANALYSES OF REFRIGERATION AND LIQUEFACTION SYSTEMS

The thermodynamic-quality measure of either a low-temperature piece of equipment or an entire process is its reversibility. The second law or, more precisely, the entropy increase is an effective guide

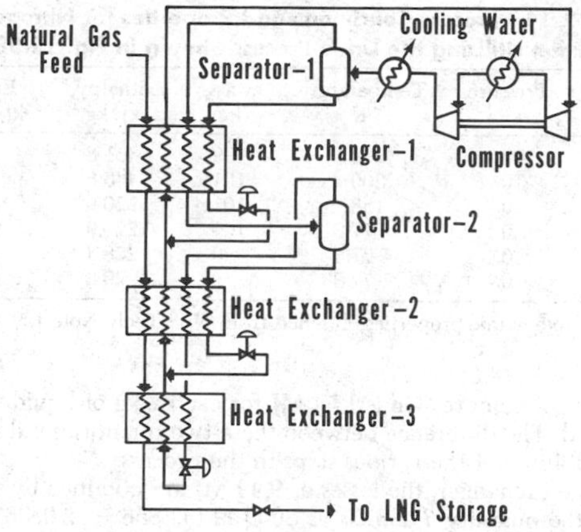

FIG. 12-64 Mixed-refrigerant cycle.

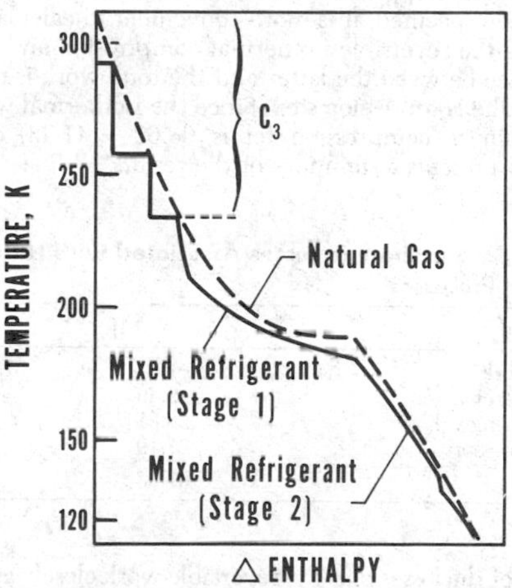

FIG. 12-65 Propane precooled mixed-refrigerant-cycle cooling curve for natural gas.

to the degree of irreversibility associated with such an item or process. However, to obtain a clearer picture of what these entropy increases mean, it has become convenient to relate such an analysis to the additional work that is required to overcome these irreversibilities. The fundamental equation for such an analysis is

$$W = W_{\text{rev}} + T_o \Sigma\, \dot{m}\Delta s \qquad (12\text{-}39)$$

where the total work is the sum of the reversible work plus a summation of the losses in availability for various steps in the analysis.

Application of this method to an actual process can best be demonstrated by a numerical example. Consider a simple nitrogen-liquefaction process using the same concept as the refrigerator shown in Fig. 12-58. Conditions for this process are indicated in Table 12-14. If it is assumed, for simplicity, that there is no heat leak to the process, the fraction of nitrogen liquefied is 0.0198 as calculated from Eq. (12-37). The total work involved in the compressor, if a three-stage adiabatic compression, 75 percent compression efficiency, and an initial temperature of 300 K are assumed, is 68.62 MJ for each 2 kg of liquid nitrogen produced.

The reversible work for this process is obtained from the expression

$$-W/\dot{m}_f = T_o\,(s_1 - s_f) - (h_1 - h_f) \qquad (12\text{-}40)$$

where the subscript 1 refers to the initial conditions and subscript f to the saturated fluid. For an initial temperature of 300 K, the revers-

TABLE 12-14 Process Conditions and Properties for Nitrogen Liquefaction Utilizing the Linde Process Shown in Fig. 12-58

Location	Pressure, kPa	Temperature, K	Mass, kg	Enthalpy,* kJ/kg	Entropy,* kJ/(kg · K)
1	0.1	290	99	451.8	4.3856
2	10	300	101	443.6	2.9849
3	10	158	101	220.9	1.9438
4	0.1	77.3	101	220.9	2.8945
5	0.1	77.3	99	228.4	2.9936
Liquid	0.1	77.3	2	29.3	0.4180

*Thermodynamic properties obtained from NBS Tech. Note 129.

ible work of compression is 1.54 MJ for each 2 kg of liquid nitrogen produced. The difference between these two quantities is due to the irreversibilities of the various steps in the process.

For the exchanger, the losses of 9.49 MJ are obtained by substitution in the quantity $T_0 \Sigma \dot{m} \Delta s$ as 300 [99 (4.3856 − 2.9936) − 101 (2.9849 − 1.9438)]. For the valve, the losses of 28.81 MJ are evaluated by substitution in $T_0 \Sigma \dot{m} \Delta s$ as 300 [101 (2.8945 − 1.9438)]. Since temperature approaches in the intercoolers for the compressor have not been specified, it is more convenient to calculate the work involved for the reversible isothermal compression and assume that the difference between the latter and the total work is the loss associated with the compression step. Since the isothermal work is 41.43 MJ, the loss in the compression step is (68.62 − 41.43), or 27.19 MJ. Table 12-15 presents a summary of the results.

TABLE 12-15 Summary of Losses Associated with Nitrogen-Liquefaction Processes

Location of losses	Losses, MJ	%
Reversible work	1.54	2.3
Loss in heat exchanger	9.49	14.2
Loss in throttling valve	28.81	42.9
Loss in compressor	27.19	40.6
	67.03	100.0

The sum of the losses plus the reversible work closely approximates the total compression work evaluated earlier and shows at a glance where the greatest gains in efficiency can be made. Since all low-temperature processes encounter heat leaks, this can be included in the calculations by evaluating the fraction liquefied from Eq. (12-38) and recalculating the work per mass of liquid produced. This calculation will show that even small heat leaks can contribute significantly to the losses and thus greatly increase the work of compression.

Numerous analyses and comparisons of refrigeration and liquefaction cycles are available in the literature. Great care must be exercised in accepting these comparisons since it is quite difficult to put all processes on a strictly comparable basis. Many assumptions need to be made in the course of the calculations, and these can have considerable effect on the conclusions. Chief factors upon which assumptions generally have to be made include heat leak, temperature differences in the exchangers, efficiencies of compressors and expanders, number of stages of compression, fraction of expander work recovered, state of expander exhaust, purity and condition of inlet gases, pressure drop due to fluid flow, etc. In view of this fact, differences in power requirements of 10 to 20 percent can readily be due to differences in assumed variables and can negate the advantage of one cycle over another.

A comparison that demonstrates this point rather well is the one made by Barron (*Advances in Cryogenic Engineering*, vol. 17, Plenum Press, New York, 1972, p. 20) in his analysis of some of the more common liquefaction systems described earlier using air as the working fluid. Table 12-16 lists some of the results of this analysis based on an inlet-gas temperature of 294.4 K and pressure of 100 kPa.

TABLE 12-16 Comparison of Several Liquefaction Systems Using Air as a Working Fluid with Inlet Conditions of 294.4 K and 100 kPa

Air liquefaction system	Liquid yield, $y = \dot{m}_f/\dot{m}$	Work per unit mass liquefied, kJ/kg	Figure of merit
1. Ideal reversible system	1.000	715	1.000
2. Linde-or J-T system, p_2 = 20 MPa, η_c = 100%, ϵ = 1.0	0.086	5240	0.137
3. Linde-or J-T system, p_2 = 20 MPa, η_c = 70%, ϵ = 0.95	0.061	10620	0.068
4. Linde-or J-T system observed	....	10320	0.070
5. Precooled Linde-or J-T system, p_2 = 20 MPa, T_3 = 228 K η_c = 100% ϵ = 1.00	0.179	2240	0.320
6. Precooled Linde-J-T system, p_2 = 20 MPa, T_3 = 228 K, η_c = 70%, ϵ = 0.95	0.158	3700	0.194
7. Precooled Linde-Hampson system, observed	....	5580	0.129
8. Linde dual-pressure system, p_3 = 20 MPa, p_2 = 6 MPa, i = 0.8, η_c = 100%, ϵ = 1.00	0.060	2745	0.261
9. Linde dual-pressure system, p_3 = 20 MPa, p_2 = 6 MPa, i = 0.8, η_c = 70%, ϵ = 0.95	0.032	8000	0.090
10. Linde dual-pressure system, observed	....	6340	0.113
11. Linde dual-pressure system, precooled to 228 K, observed	....	3580	0.201
12. Claude system, p_2 = 4 MPa, $x = \dot{m}_e/\dot{m}$ = 0.7, $\eta_c = \eta_e$ = 100%, ϵ = 1.00	0.260	890	0.808
13. Claude system, p_2 = 4 MPa, $x = \dot{m}_e/\dot{m}$ = 0.7, η_c = 70%, $\eta_{e,ad}$ = 80%, $\eta_{e,m}$ = 90%, ϵ = 0.95	0.189	2020	0.356
14. Claude system, observed	—	3580	0.201
15. Cascade system, observed	—	3255	0.221

NOTE: η_c = compressor overall efficiency; η_e = expander overall efficiency; $\eta_{e,ad}$ = expander adiabatic efficiency; $\eta_{e,m}$ = expander mechanical efficiency; and ϵ = heat-exchanger effectiveness.

Another comparison of low-temperature refrigeration besides the ones discussed here involves examining the ratio of W/Q for a Carnot refrigerator with the W/Q quantity for the actual refrigerator. This ratio indicates the extent to which an actual refrigerator approaches ideal performance. (The same ratio can be formed for liquefiers as well.) In Fig. 12-66 this approach to the ideal performance in percent as given by this ratio is plotted by Strobridge (NBS Tech. Note 655, 1974) as a function of refrigeration capacity in watts for actual refrigerators and liquefiers, either already installed or under development. The capacity of the liquefiers included has been converted to equivalent refrigeration capacity by determining the percent of Carnot performance that these units achieved as liquefiers and then calculating the refrigeration output of a refrigerator operating at the same efficiency with the same input power. In some instances, equivalent refrigeration capacity was given by the manufacturers and was used directly. In all instances, the input power was taken as the installed drive power, not the power measured at the input to the drive motor.

Historically, the contention has been that higher-temperature refrigerators (or liquefiers) are more efficient. These data by Strobridge for refrigeration temperatures between 10 and 30 K (and 30 to 90 K) appear to refute that notion. However, care must be exercised in such an analysis when nonisothermal refrigerators are compared.

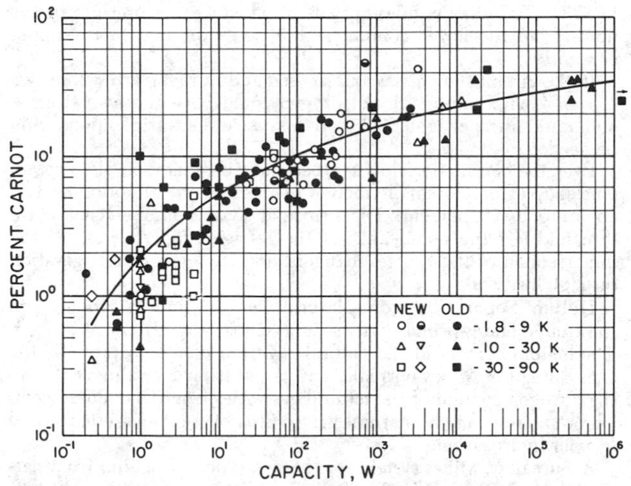

FIG. 12-66 Efficiency of low-temperature refrigerators and liquefiers as a function of refrigeration capacity.

The data presented for these existing low-temperature refrigerators cover a wide range of capacity and temperatures. It appears that the performance (input power) can be fairly well predicted. It does not appear that any significant increase in efficiency has been achieved for the entire class of devices since the early 1970s, although in certain instances good performance has been realized. The more efficient facilities are in the large sizes and use complex thermodynamic cycles. The same performance potential may exist in the smaller units, but costs are rather prohibitive, and the savings in electrical power would not justify the greater capital outlay.

SEPARATION AND PURIFICATION OF GASES

The energy expended to separate gas mixtures reversibly is the energy required to compress each component at constant temperature from the partial pressure of the gas in the mixture to the final pressure of the mixture. This reversible isothermal work is given by the familiar relation

$$-W_i/\dot{m} = T(s_1 - s_2) - (h_1 - h_2) \qquad (12\text{-}41)$$

where s_1 and h_1 refer to conditions before the separation and s_2 and h_2 refer to conditions after the separation. For a binary system, and by assuming a perfect gas for both components, Eq. (12-41) simplifies to

$$-W_i/\dot{m} = RT\left(n_A \ln \frac{\pi}{p_A} + n_B \ln \frac{\pi}{p_B}\right) \qquad (12\text{-}42)$$

where n_A and n_B are the moles of A and B in the mixture and p_A and p_B are the partial pressure of these two components in the mixture.

The figure of merit for a separation system is defined similarly to that for a liquefaction system, namely,

$$\text{FOM} = (W_i/\dot{m})/(W/\dot{m})_{\text{act}} \qquad (12\text{-}43)$$

If the mixture to be separated is essentially a binary, both the McCabe-Thiele and the Ponchon-Savarit graphical methods can be used to obtain the ideal number of stages required. It should be noted, however, that it is not satisfactory in the separation of air to treat it as a binary mixture of oxygen and nitrogen if high-purity (99 percent or better) oxygen is desired. The separation of oxygen from argon is a more difficult separation than that of oxygen from nitrogen and would require correspondingly many more plates. In fact, if the argon is not extracted from air, only 95 percent oxygen would be produced. The other rare-gas constituents of air—helium, neon, krypton, and xenon—are present in such small quantities and have boiling points so far removed from those of oxygen and nitrogen that they introduce no important complications.

Linde Single-Column System for Separating Air Of available separation schemes the simplest is known as the Linde single-column system, shown in Fig. 12-67 and first introduced in 1902. In it, purified compressed air passes through a precooling heat exchanger (if oxygen gas is the desired product, a three-channel exchanger for air,

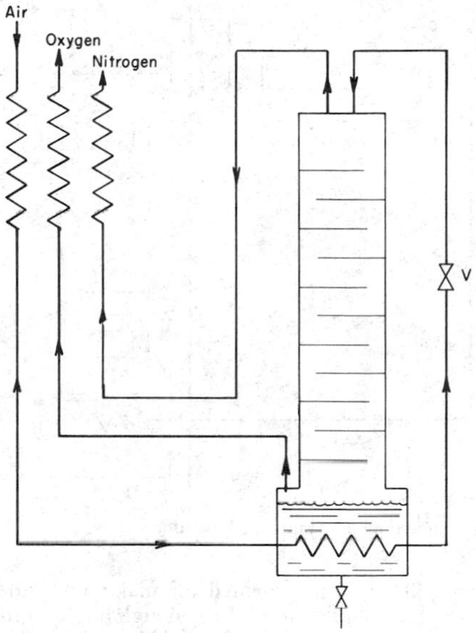

FIG. 12-67 Simple Linde air-separation column.

waste nitrogen, and oxygen gas is used; if liquid oxygen is to be recovered from the bottom of the column, a two-channel exchanger for air and waste nitrogen is employed), then through a coil in the boiler of the rectifying column, where it is further cooled (acting at the same time as the boiler-heat source); following this, it expands essentially to atmospheric pressure through a J-T valve, becoming mostly liquid, and reenters the column at the top. Rectification then takes place and either gaseous or liquid oxygen is the final product. If oxygen gas is to be the product, the air must be compressed and delivered at 3 to 6 MPa; if it is to be liquid oxygen, 20 MPa is necessary.

Note that the Linde single-column separation system is just the simple Linde liquefaction system with a rectification column introduced instead of the liquid reservoir. Any of the other liquefaction systems could be used to furnish liquid for the column, however.

Unfortunately, in the simple single-column process, although the oxygen purity is high, the nitrogen effluent stream is impure. The equilibrium vapor concentration for an initial liquid mixture of 21 percent oxygen–79 percent nitrogen at 100 kPa is about 6 to 7 percent oxygen. In other words, the nitrogen waste-gas stream will have an oxygen impurity of this magnitude, making it unusable for any pure-nitrogen-gas application.

Linde Double-Column System for Separating Air The impurity problem noted in the preceding subsection was solved by the introduction of the Linde double-column system shown in Fig. 12-68. Two rectification columns are placed one on top of the other (hence the name "double-column system").

In this system the liquid air is introduced at an intermediate point B along the lower column, and a condenser-evaporator at the top of the lower column makes the arrangement a complete reflux distillation column which delivers almost pure nitrogen at E. In order for the column simultaneously to deliver pure oxygen, the oxygen-rich liquid (about 45 percent O_2) from the bottom boiler is introduced at an intermediate level C in the upper column. The reflux and rectification in the upper column produce pure oxygen at the bottom and

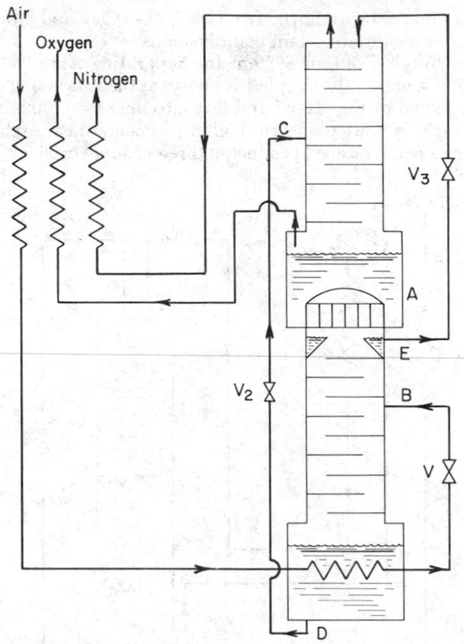

FIG. 12-68 Linde double-column air separator.

pure nitrogen at the top, provided all major impurities are first removed from the column. More than enough liquid nitrogen is produced in the upper column as needed reflux. Since the condenser must condense nitrogen vapor by evaporating liquid oxygen, it is necessary to operate the lower column at a higher pressure, about 500 kPa, while the upper column is operated at approximately 100 kPa. This requires regulating valves to reduce the pressure of the fluids from the lower column as they are admitted to the upper column.

In the circuit shown, gaseous oxygen and nitrogen are withdrawn at room temperature. Liquid oxygen could be withdrawn from *A* and liquid nitrogen from *E*, but in this case more refrigeration would be needed.

Even the best modern low-temperature air-separation plant has an efficiency of only a small fraction of the theoretical optimum, i.e., about 15 to 20 percent. The principal sources of inefficiency are threefold: (1) the nonideality of the refrigerating process, (2) the imperfection of the heat exchangers, and (3) losses of refrigeration through heat leak.

Helium Separation from Natural Gas Helium is produced in the United States primarily by separation of helium-rich natural gas. The helium content of the natural gas from plants operated by the U.S. Bureau of Mines normally has varied from 1 to 2 percent, and the nitrogen content of the natural gas varies from 12 to 80 percent. The remainder of the natural gas consists of methane, ethane, and heavier hydrocarbons.

A Bureau of Mines system for the separation of helium from natural gas is shown in Fig. 12-69. Since the major constituents of natural gas have boiling points very much different from that of helium, a distillation column is not necessary and the separation can be accomplished with condenser-evaporators. Natural gas compressed to 4.1 MPa and treated to remove carbon dioxide, hydrogen sulfide, and water vapor enters heat exchanger *A*, where it is almost totally condensed by the cold outgoing gas. The pressure is then reduced to 1.7 MPa, and the gas is admitted to heat exchanger–separator *B*, where it is further cooled with nitrogen vapor. In the separator about 98 percent of the gas is liquefied, and the part remaining in the vapor phase consists of about 60 percent helium and 40 percent nitrogen with a very small amount of methane. The cold nitrogen vapor from a separate refrigeration cycle passes down cooling tubes of exchanger *B* and, in addition to its cooling function, causes some rectification of the gas phase which increases the helium content. Both liquid and gas are continuously withdrawn from the separator. The crude-

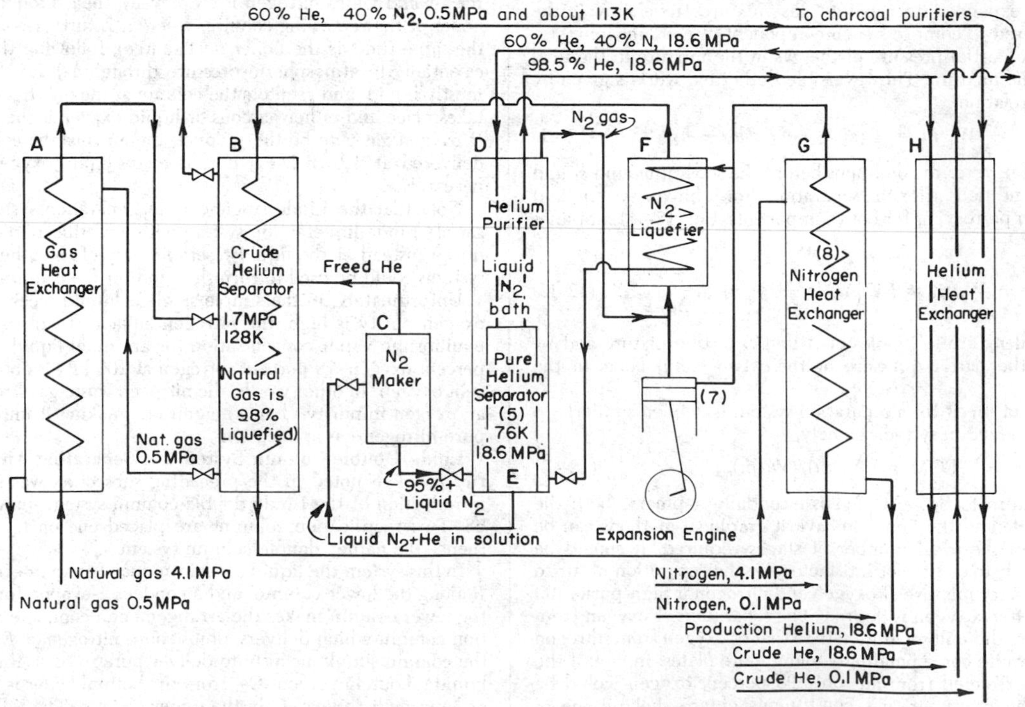

FIG. 12-69 Typical helium-separation plant as operated by the U.S. Bureau of Mines.

helium-gas phase is warmed to near room temperature in heat exchanger *H* and sent to temporary storage pending further purification. The liquid phase, which has been depleted of helium, passes through heat exchanger *A* and furnishes the refrigeration to cool and condense the incoming gas. Finally, the processed gas is recompressed and returned to the natural-gas pipe line. The process improves the value of the natural gas as a fuel since noncumbustible constituents have been removed.

The purification of the crude helium is accomplished by compressing the gas to 18.6 MPa and cooling it first by passage through heat exchanger *H* and then in the separator pot *E*, which is immersed in a bath of liquid nitrogen. In the separator pot nearly all the nitrogen in the crude helium is condensed and removed as liquid. This liquid contains some dissolved helium, which is largely removed and returned to the process gas by reducing the pressure to 1.7 MPa and separating the resultant liquid and vapor phases in nitrogen maker *C*. Helium from separator *E* has a purity of about 98.5 percent. Final purification is accomplished by passing the cold helium through charcoal adsorption purifiers to remove the remaining nitrogen.

Natural-Gas Processing The need to obtain greater recoveries of the C_2, C_3, and C_4's in natural gas has resulted in the expanded use of low-temperature processing of these streams. Most natural-gas processing at low temperatures to recover light hydrocarbons is now accomplished by using the turboexpander cycle. Feed gas is normally available from 1 to 10 MPa. The gas is first dehydrated to dew points of 200 K and lower. After dehydration the feed is cooled with cold residue gas. Liquid produced at this point is separated before entering the expander and sent to the condensate stabilizer. The gas from the separator flows to the expander. The expander exhaust stream can contain as much as 23 weight percent liquid. This two-phase mixture is sent to the top section of the stabilizer, which separates the two phases. The liquid is used as reflux in this unit while the cold gas exchanges heat with fresh feed and is recompressed by the expander-driven compressor. Many variations of this cycle are possible and have been used in actual plants.

Purification The nature and concentration of impurities to be removed depend entirely on the type of process involved. For example, in the production of large quantities of oxygen, various impurities must be removed to avoid plugging the cold-process lines or to avoid a buildup of hazardous contaminants. The impurities in air that would contribute most to plugging would be water and carbon dioxide. Helium, hydrogen, and neon, on the other hand, will accumulate on the condensing side of the oxygen reboiler and will reduce the rate of heat transfer unless removed by intermittent purging. The buildup of acetylene, however, can prove to be dangerous even if the feed concentration in the air is no greater than 0.04 ppm.

Refrigeration purification is a relatively simple method for removing water, carbon dioxide, and certain other contaminants from a process stream by condensation or freezing. (Either regenerators or reversing heat exchangers may be used for this purpose since a flow reversal is periodically necessary to reevaporate and remove the solid deposits.) The effectiveness of this method depends upon the vapor pressure of the impurities relative to that of the major components of the process stream at the refrigeration temperature. Thus, if ideal-gas behavior is assumed, the maximum impurity content in a gas stream after refrigeration would be inversely proportional to its vapor pressure. However, owing to departure from ideality at higher pressures, the impurity content will be considerably higher than predicted for the ideal situation. For example, the actual water-vapor content in air will be over 4 times that predicted by ideal-gas behavior at a temperature of 228 K and a pressure of 20 MPa.

Purification by a solid adsorbent is one of the most common low-temperature methods for removing impurities. Materials such as silica gel, carbon, and synthetic zeolites (molecular sieves) are widely used as adsorbents because of their extremely large effective surface areas. Most of the gels and carbon have pores of varying sizes in a given sample, but the synthetic zeolites are manufactured with closely controlled pore-size openings ranging from about 0.4 to 1.3 nm. This makes them even more selective than other adsorbents since it permits separation of gases on the basis of molecular size.

Information needed in the design of low-temperature adsorbers includes the equilibrium between the solid and gas and the rate of adsorption. Equilibrium data for the common systems generally are available from suppliers of such material. The rate of adsorption is usually very rapid, and adsorption is essentially complete in a relatively narrow zone of the adsorber. If the concentration of the adsorbed gas is considerably more than a trace, then heat of adsorption may also be a factor of importance in the design. (The heat of adsorption is usually of the same order or larger than the normal heat of phase change.) Under such situations it is generally advisable to design the purification in two steps, i.e., first removing a significant portion of the impurity either by condensation or by chemical reaction and then completing the purification with a low-temperature adsorption system. A scheme combining condensation and adsorption is shown in Fig. 12-70.

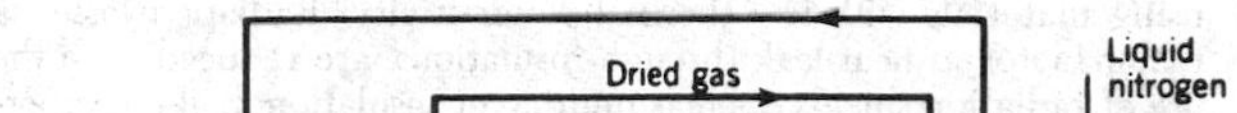

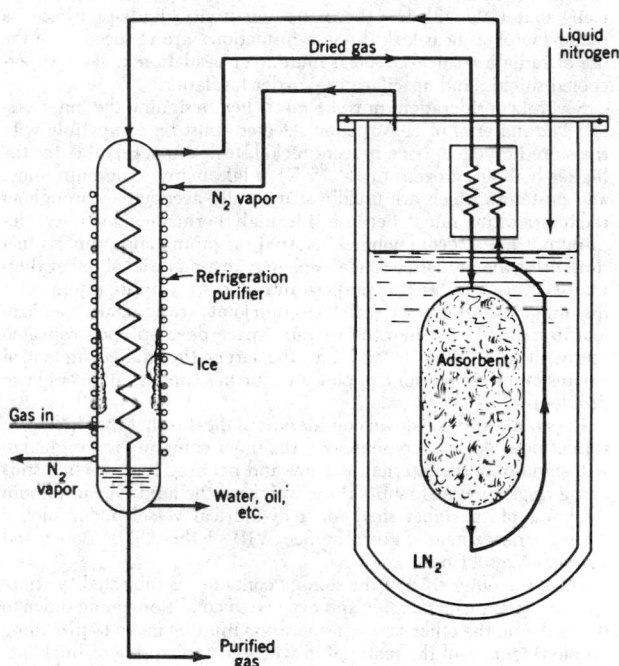

FIG. 12-70 Purifier using refrigeration and adsorption schemes in series.

In usual plant operation at least two adsorption purifiers are employed: one is in service, while the other is being desorbed of its impurities. In some cases there is an advantage in using three purifiers: one adsorbing, one desorbing, and one being cooled, with the latter two units in series. The cooling of the purifier must be effected with some of the purified gas to avoid adsorption during this period.

Chemical purification in air-separation plants involves a rather timeworn method for removing carbon dioxide from the process-air stream by passing the stream through a caustic solution of sodium hydroxide. In the reaction sodium carbonate is formed, and with the passage of time the original solution is weakened or diluted to a condition in which it must be renewed. "Caustic scrubbing" is the term usually given to this process, and although it is still used in many places, the method is being replaced in newer plants whenever possible by the more desirable refrigerative or adsorptive techniques just described.

STORAGE SYSTEMS

Storage vessels range in type from low-performance containers, insulated by rigid foam or fibrous insulation, in which the liquid in the container boils away in a few hours, up to high-performance containers, insulated with multilayer insulations, in which less than 0.1 percent of the fluid contents is evaporated per day. In the more effective units, the storage container consists of an inner vessel, which encloses the cryogenic fluid to be stored, and an outer vessel, or vacuum jacket. The latter contains the vacuum necessary to make the insulation effective and at the same time serves as a vapor barrier to the migration of water vapor and other condensables to the cold surface of the inner vessel. Improvements have been made in the insulation used in these containers, but the vacuum-insulated double-walled Dewar is still the basic idea for high-performance cryogenic-fluid container design.

In general, heat leak into a storage system for cryogens is by (1) radiation and conduction through the insulation and (2) conduction through the inner-shell supports, piping, instrumentation leads, and access ports. Conduction losses are reduced by introducing long heat-leak paths, by making the cross sections for heat flow small, and by using materials with low thermal conductivity. Radiation losses, a major factor in heat leak through insulations, are reduced with the use of radiation shields, such as multilayer insulation, boil-off vapor-cooled shields, and opacifiers in powder insulation.

Several considerations must be met when designing the inner vessel. The material of construction selected must be compatible with the stored cryogen. Nine percent nickel steels are acceptable for the higher-boiling cyrogens ($T > 75$ K), while many aluminum alloys and austenitic steels are usually structurally acceptable throughout the temperature range. Because of its high thermal conductivity, aluminum is not a recommended material for piping and supports that must cross the insulation space. A change to a material of lower thermal conductivity for this purpose introduces a transition joint of a dissimilar material. Since such transition joints are generally mechanical in nature, leaks into the vacuum space develop upon repeated temperature cycling. In addition, the larger thermal coefficient of expansion of aluminum can pose still further support and cooldown problems.

Economic and cooldown considerations dictate that the shell be as thin as possible. As a consequence, the inner container is designed to withstand only the internal pressure and bending forces, while stiffening rings are used to support the weight of the fluid. The minimum thickness of the inner shell for a cylindrical vessel under such a design arrangement is given by Sec. VIII of the ASME Boiler and Pressure Vessel Code.

Since the outer shell of the storage container is subjected to atmospheric pressure on one side and evacuated conditions going down to 0.13 mPa on the other side, consideration must be given to providing ample thickness of the material to withstand collapsing or buckling. Failure by elastic instability is covered by the ASME Code, in which design charts are available for the design of cylinders and spheres subjected to external pressure. Stiffening rings are also used on the outer shell to support the weight of the inner container and its contents and to hold the outer shell circular.

The outer shell is normally constructed of carbon steel for economic reasons, unless aluminum is required to reduce the weight. Stainless-steel standoffs must be provided on the carbon steel outer shell for all piping penetrations to avoid direct contact with these penetrations when they are cold.

There are a variety of methods for supporting the inner shell within the outer shell. Materials that have a high strength-to-thermal-conductivity ratio are selected for these supports. Design of these supports must allow for shipping loads which may be several orders higher than in-service loads. Compression supports such as legs or pads may be used, but tension supports are more common. These may take the form of cables, welded straps, threaded bars, or a combination of these to provide restraints of the inner shell in several directions.

Most storage containers for cryogens are designed for a 10 percent ullage volume. The latter permits reasonable vaporization of the contents owing to heat leak without incurring too rapid a buildup of the pressure in the container. This, in turn, permits closure of the container for short periods of time either to avoid partial loss of the contents or to transport flammable or hazardous cryogens safely from one location to another.

Insulation As noted earlier, the effectiveness of a liquefier or refrigerator is highly dependent upon the heat leak entering such a system. Since heat removal becomes more costly with reduction in temperature, as demonstrated by the Carnot limitation, most cryogenic systems employ some form of insulation to minimize the effect.

Heat can flow through an insulation by the simultaneous action of several different mechanisms including (1) solid conduction, (2) gas conduction, and (3) radiation. Because these heat-transfer mechanisms operate simultaneously and interact with each other, it is common practice to use an apparent thermal conductivity to characterize the insulation. This is measured experimentally during steady-state heat transfer and evaluated from the basic one-dimensional Fourier equation. Typical k_a values for a variety of insulations used in cryogenic service are shown in Table 12-17.

TABLE 12-17 Characteristics of Insulation

Type of insulation	Apparent thermal conductivity, k_a, J/(s·m²·K) (between 77 and 300 K)	Bulk density, kg/m³
Pure gas at 0.1 MPa, 180 K		
H_2	34.07×10^{-2}	0.080
N_2	5.67×10^{-2}	1.121
Pure vacuum, 0.13 mPa	1.70×10^{-2}	Nil
Straight insulation		
Polystyrene foam	8.52×10^{-2}	32–48
Polyurethane foam	10.79×10^{-2}	80–128
Glass foam	11.36×10^{-2}	144
Evacuated powder		
Perlite (13.3 mPa)	$0.34–0.68 \times 10^{-2}$	144–64
Silica (13.3 mPa)	$0.57–0.68 \times 10^{-2}$	64–96
Combination insulation		
Al foil and fiberglass		
12–28 layers/64-112 cm, 1.33 mPa	$1.14–2.27 \times 10^{-4}$	64–112
30–60 layers/120 cm, 1.33 mPa	0.57×10^{-4}	120
Al foil and nylon net, 132 layers/89 cm, 1.33 mPa)	5.68×10^{-4}	89

Cyrogenic insulations have generally been divided into five general categories: high-vacuum, multilayer, powder, foam, and special insulations. Each is discussed in turn in the following subsections.

Vacuum Insulation Heat transport across an evacuated space (0.13 mPa or better) is by radiation and by conduction through the residual gas. The heat transfer by radiation generally is predominant and can be approximated by

$$Q_r/A_1 = \sigma(T_2^4 - T_1^4)\left[\frac{1}{e_1} + \frac{A_1}{A_2}\left(\frac{1}{e_2} - 1\right)\right]^{-1} \tag{12-44}$$

where Q_r/A_1 is the heat transfer by radiation per unit area, σ is the Stefan-Boltzmann constant, and e is the emissivity of the surfaces. Subscript 2 refers to the hot surface, and subscript 1 refers to the cold surface. The bracketed term on the right-hand side of this relation is generally known as the emissivity factor.

For normal gaseous conduction with constant thermal conductivity, a linear temperature gradient exists within the medium between the warm and cold surfaces. However, when the mean free path of the gas molecules becomes large relative to the distance between the two surfaces as the pressure is reduced in an evacuated space, free molecular conduction is encountered. The gaseous heat conduction under free molecular conduction for concentric spheres, coaxial cylinders, and parallel plates is given by

$$Q_{gc}/A_1 = \frac{\gamma + 1}{\gamma - 1}\left[\frac{R}{8\pi MT}\right]^{1/2} \alpha p\,(T_2 - T_1) \tag{12-45}$$

where α, the overall accommodation coefficient, is defined by

$$\alpha = \frac{\alpha_1\alpha_2}{\alpha_2 + \alpha_1\,(1 - \alpha_2)\,(A_1/A_2)} \tag{12-46}$$

and γ is the ratio of the heat capacities, R is the molar gas constant, M is the molecular weight of the gas, and T is the temperature of the gas at the point where the pressure p is measured. The subscripted A_1 and A_2, T_1 and T_2, and α_1, and α_2 are the areas, temperatures, and accommodation coefficients of the cold and warm surfaces respectively. The accommodation coefficient depends upon the specific gas-surface combination and the surface temperature. Because of the great variations in accommodation coefficients, it is difficult to make accurate estimates of conductive heat transfer through a residual gas in a vacuum-insulated system. Fortunately, this is usually not a serious problem because, as a rule, the objective is to obtain a vacuum of such quality that heat transfer by residual gas does not contribute significantly to overall heat transfer.

In order for free molecular conduction to occur, the mean free path of the gas molecules must be large compared with the distance between the two surfaces. To check this condition, the mean free path λ may be determined from

$$\lambda = \frac{3\mu}{p}\left(\frac{\pi RT}{8M}\right)^{1/2} \tag{12-47}$$

where μ is the gas viscosity at temperature T.

Heat transport by radiation, on the other hand, can be effectively reduced by the insertion of floating shields having surfaces of low emissivity within the evacuated space. The effect of the shields is to greatly reduce the emissivity factor. For example, for N shields or $(N + 2)$ surfaces, an emissivity of the outer and inner surface of e_o, and an emissivity of the shields of e_s, the emissivity factor reduces to

$$\left[2\left(\frac{1}{e_o} + \frac{1}{e_s} - 1\right) + \frac{(N-1)(2-e_s)}{e_s}\right]^{-1} \tag{12-48}$$

In essence, one properly located low-emissivity shield can reduce radiant-heat transfer to around one-half of the rate without the shield, two shields can reduce this to around one-fourth of the rate without the shield, etc.

Multilayer Insulation Multilayer insulation consists of alternating layers of highly reflecting material, such as aluminum foil or aluminized Mylar, and a low-conductivity spacer material or insulator, such as fiberglass mat or paper, glass fabric, or nylon net, all under high vacuum. When properly applied at the optimum density, this type of insulation can have an apparent thermal conductivity of as low as 10 to 50 μW/(m·K) between 20 and 300 K.

The very low thermal conductivity of multilayer insulations may be attributed to the fact that all modes of heat transfer—conductive, convective, and radiative—are reduced to a bare minimum. Since radiant-heat transfer is inversely proportional to the number of intermediate reflecting shields and directly proportional to the emissivity of the shields, radiation is minimized by using many shields or layers of a low-emissivity material. Convection is eliminating by lowering the pressure so that the mean free path of the gas molecules is much larger than the spacing between the insulation layers. Heat transfer through the spacer material is proportional to the thermal conductivity of the material used and inversely proportional to the resistance of the heat flow at the points of contact between the spacer and the shield. The low conductivity, size, geometry, and discontinuous nature of the materials generally selected for spacers contribute in reducing solid conduction to a minimum.

For a highly evacuated multilayer insulation (on the order of 0.13 mPa), heat is transferred primarily by radiation and solid conduction through the spacer material. The apparent thermal conductivity of the insulation under these conditions may be determined from

$$k_a = \frac{1}{N/\Delta x}\left\{h_s + \frac{\sigma e T_2^3}{2-e}\left[1 + \left(\frac{T_1}{T_2}\right)^2\right]\left(1 + \frac{T_1}{T_2}\right)\right\} \tag{12-49}$$

where $N/\Delta x$ is the number of complete layers (reflecting shield plus spacer) of insulation per unit thickness, h_s is the solid conductance of the spacer material, σ is the Stefan-Boltzmann constant, e is the effective emissivity of the reflecting shield, and T_2 and T_1 are the temperatures of the warm and col l sides of the insulation respectively. It is evident that apparent thermal conductivity can be reduced by increasing the layer density up to a certain point. It is not obvious from Eq. (12-49) that a compressive load affects apparent thermal conductivity and thus the performance of a multilayer insulation. However, under a compressive load the solid conductance increases much more rapidly than $N/\Delta x$, resulting in an overall increase in k_a. Plots of heat flux versus compressive load on a logarithmic scale result in straight lines with slopes between 0.5 and 0.67 (Glaser et al., *Thermal Insulation Systems*, NASA SP-5027, 1967).

The effective thermal-conductivity values generally obtained in practice are greater by at least a factor of 2 than the one-dimensional thermal-conductivity values measured in the laboratory with carefully controlled techniques. This degradation in insulation thermal performance is caused by the combined presence of edge exposure to isothermal boundaries, gaps, joints, or penetrations in the insulation blanket required for structural supports, fill and vent lines, and the high lateral thermal conductivity of these insulation systems.

Powder Insulation A method of realizing some of the benefits of multiple floating shields without incurring the difficulties of awkward structural complexities is to use evacuated-powder insulation. The penalty incurred in the use of this type of insulation, however, is a tenfold reduction in the overall thermal effectiveness of the insulation system over that obtained for multilayer insulation. In applications for which this is not a serious factor, such as LNG storage facilities, and investment cost is of major concern, even unevacuated-powder insulation systems have found useful applications.

A powder insulation system consists of a finely divided particulate material such as perlite, expanded SiO_2, calcium silicate, diatomaceous earth, or carbon black packed between the surfaces to be insulated. When used at 0.1-MPa gas pressure (generally with an inert), the powder reduces both convection and radiation and, if the particle size is sufficiently small, can also reduce the mean free path of the gas molecules. When the powders are evacuated to pressures of 1.33 to 0.133 Pa, gas conduction becomes very small and heat transfer is chiefly by radiation and solid conduction. The variation in apparent mean thermal conductivity of several powders as a function of interstitial gas pressure is shown in the familiar S-shaped curves of Fig. 12-71.

To calculate heat transfer for porous powder insulations, the following relationship can be used for concentric spheres

$$Q = k_a(T_2 - T_1)\sqrt{A_1A_2}/t \tag{12-50}$$

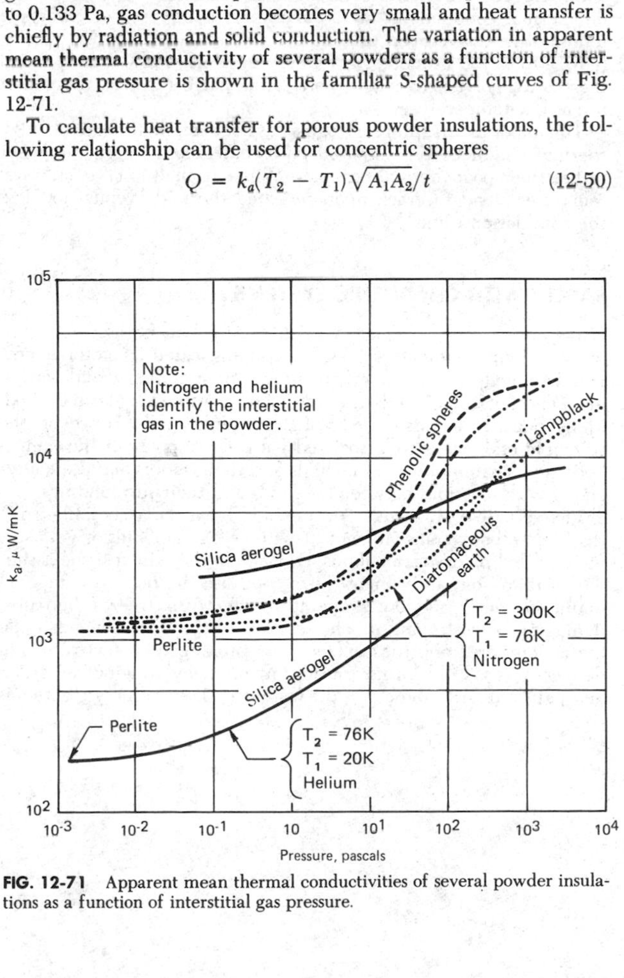

FIG. 12-71 Apparent mean thermal conductivities of several powder insulations as a function of interstitial gas pressure.

where k_a is the apparent thermal conductivity of the evacuated-powder insulation and t is the thickness of the insulation. The apparent thermal conductivity of the powder insulation at cryogenic temperature is generally obtained from

$$k_a = k_g/[1 - v(1 - k_g/k_s)] \quad (12\text{-}51)$$

where k_g is the thermal conductivity of the gas within the insulation, k_s is the thermal conductivity of the powder, and v is the ratio of solid volume to total volume. Most suitable powder insulations will have a k_a value of 2.5×10^3 μW/(m·K) or less if sufficiently evacuated.

The amount of heat transport due to radiation through the powders can be reduced by the addition of metallic powders. A mixture containing approximately 40 to 50 weight percent metal powder gives optimum performance. From a safety point of view, metallic copper as an opacifier is preferable to aluminum because the latter has a large heat of combustion in combination with oxygen. It is also prudent to choose a noncombustible powder for temperatures below the condensation temperature of oxygen.

Foam Insulation Since foams are not homogeneous materials, their apparent thermal conductivity is dependent upon the bulk density of the insulation, the gas used to foam the insulation, and the mean temperature of the insulation. Heat conduction through a foam is determined by convection and radiation within the cells and by conduction in the solid structure. Evacuation of a foam is effective in reducing its thermal conductivity, indicating a partially open cellular structure, but the resulting values are still considerably higher than either multilayer or evacuated-powder insulations. The opposite effect, diffusion of atmospheric gases into the cells, can cause an increase in apparent thermal conductivity. This is even more significant with the diffusion of hydrogen and helium into the cells. Data on thermal conductivity for a variety of foams used at cryogenic temperatures have been presented by Kropschot [in R. W. Vance, (ed.), *Cryogenic Technology*, Wiley, New York, 1963, p. 239]. Of all the foams, polyurethane and polystyrene have received the widest use at low temperatures.

The major disadvantage of foams has not been their relatively high thermal conductivity compared with that of other insulations but rather their poor thermal behavior. When applied to cryogenic systems, they tend to crack upon repeated thermal cycling and lose their insulation value.

SAFETY WITH CRYOGENIC SYSTEMS

Experience has shown that cryogenic fluids can be used safely in industrial environments as well as in sophisticated laboratories provided all facilities are properly designed and maintained and personnel handling these fluids are adequately trained and supervised. Many hazards are associated with cryogenic fluids. However, the principal ones are those associated with the response of the human body and surroundings to the fluids and their vapors and those associated with reactions between the fluids and their surroundings.

Physiological Hazards Severe cold "burns" may be inflicted if the human body comes in contact with cryogenic fluids or with surfaces cooled by cryogenic fluids. Damage to the skin or tissue is similar to that from an ordinary burn. Because the body is composed mainly of water, the low temperature effectively freezes the tissue, damaging or destroying it. The severity of the burn depends upon the contact area and the contact time; prolonged contact results in deeper burns. Cold burns are accompanied by stinging sensations and pain similar to those of ordinary burns. The ordinary reaction is to withdraw the portion of the body that is in contact with the cold surface. Severe burns are seldom sustained if withdrawal is possible. Cold gases may not be damaging if the turbulence in the gas is low, particularly since the body can normally take care of a heat loss of 95 J/(m^2·s) for an area of limited exposure. If the heat loss becomes much greater than this, the skin temperature drops and freezing of the affected area may ensue. Freezing of facial tissue will occur in about 100 s if the heat loss is 2300 J/(m^2·s).

Materials and Construction Hazards Construction materials for noncryogenic service usually are chosen on the basis of tensile strength, fatigue life, weight, cost, ease of fabrication, corrosion resistance, etc. When working with low temperatures the designer must consider the ductility of the material, since low temperatures, as noted earlier, have the effect of making some construction materials brittle or less ductile. Some materials become brittle at low temperatures but still can absorb considerable impact, while others become brittle and lose their impact strength.

Flammability and Explosion Hazards In order to have a fire or an explosion there must exist in combination an oxidant, a fuel, and an ignition source. Generally the oxidizer will be oxygen. The latter may be available from a variety of sources including leakage or spillage, condensation of air on cryogenically cooled surfaces below 90 K, and buildup as a solid impurity in liquid hydrogen. The fuel may be almost any noncompatible material or flammable gas; compatible materials can also act as fuels in the presence of extreme heat (strong ignition sources). The ignition source may be a mechanical or an electrostatic spark, flame, impact, heat by kinetic effects, friction, chemical reaction, etc. Certain combinations of oxygen, fuel, and ignition sources will always result in fire or explosion. The order of magnitude of flammability and detonability limits for fuel-oxidant gaseous mixtures of two widely used cryogens is shown in Table 12-18.

TABLE 12-18 Flammability and Detonability Limits of Hydrogen and Methane Gas*

Mixture	Flammability limits, mol %	Detonability limits, mol %
H_2-air	4–75	20–65
H_2-O_2	4–95	15–90
CH_4-air	5–15	6–14
CH_4-O_2	5–61	10–50

*Zabetakis, *Safety with Cryogenic Fluids*, Plenum, New York, 1967.

High-Pressure Gas Hazards Potential hazards also exist in highly compressed gases because of the stored energy. In cryogenic systems such high pressures are obtained by gas compression during liquefaction or refrigeration, by the pumping of liquids to high pressure followed by evaporation, and by the confinement of cryogenic liquids with subsequent evaporation. If this confined gas is suddenly released through a rupture or break in a line, a significant thrust may be experienced. For example, the force generated on a 13.9-MPa gas cylinder caused by breaking off a 1-in-nominal-diameter valve would be over 6672 N.

Summary It is obvious that the best-designed facility is no better than the attention that is paid to safety. Safety is not considered once and forgotten. Rather it is an ongoing activity that requires constant attention to every conceivable hazard that might be encountered. Because of its importance, safety, particularly at low temperatures, has received a large focus in the literature with its own safety manual (British Cryogenics Council, *Cryogenics Safety Manual*, Institution of Chemical Engineers, London, 1970).

Section 13

Distillation

J. D. Seader, Ph.D., *Professor of Chemical Engineering, University of Utah, Salt Lake City, Utah; Member, American Institute of Chemical Engineers, American Society for Engineering Education, CACHE Committee. (Section Editor*)*

Zdzislaw M. Kurtyka, D. Sc., *(deceased), formerly Department of Chemical Engineering, The University of the West Indies, St. Augustine, Trinidad. (Azeotropy)*

CONTINUOUS-DISTILLATION OPERATIONS

THERMODYNAMIC DATA

DEGREES OF FREEDOM AND DESIGN VARIABLES

SINGLE-STAGE EQUILIBRIUM-FLASH CALCULATIONS

GRAPHICAL METHODS FOR BINARY DISTILLATION

*Certain portions of this section draw heavily on the work of Buford D. Smith, editor of this section in the fifth edition.

APPROXIMATE MULTICOMPONENT DISTILLATION METHODS

RIGOROUS METHODS FOR MULTICOMPONENT DISTILLATION-TYPE SEPARATIONS

Nomenclature and Units

Symbol	Definition	SI units	U.S. customary units
A	Absorption factor		
A	Area	m^2	ft^2
C	Number of chemical species		
D	Disillate flow rate	(kg·mol)/s	(lb·mol)/h
E	Deviation from set point		
E	Residual of phase equilibrium expression	(kg·mol)/s	(lb·mol)/h
F	Feed flow rate	(kg·mol)/s	(lb·mol)/h
F	Vector of stage functions		
G	Volume holdup of liquid	m^3	ft^3
H	Residual of energy balance	kW	Btu/h
H	Height of a transfer unit	m	ft
H	Enthalpy	J/(kg·mol)	Btu/(lb·mol)
K	Vapor-liquid equilibrium ratio (K value)		
K_C	Controller gain		
K_D	Chemical equilibrium constant for dimerization		
K_d	Liquid-liquid distribution ratio		
L	Liquid flow rate	(kg·mol)/s	(lb·mol)/h
M	Residual of component material balance	(kg·mol)/s	(lb·mol)/h
M	Liquid holdup	kg·mol	lb·mol
N	Number of transfer units		
N	Number of equilibrium stages		
N_c	Number of relationships		
N_i	Number of design variables		
N_m	Minimum number of equilibrium stages		
N_p	Number of phases		
N_r	Number of repetition variables		
N_v	Number of variables		
P	Pressure	Pa	psia
P^{sat}	Vapor pressure	Pa	psia
Q	Heat-transfer rate	kW	Btu/h
Q_c	Condenser duty	kW	Btu/h
Q_r	Reboiler duty	kW	Btu/h
R	External-reflux ratio		
R_m	Minimum-reflux ratio		
S	Sidestream flow rate	(kg·mol)/s	(lb·mol)/h
S	Stripping factor		
S	Vapor-sidestream ratio		
T	Temperature	K	°R
U	Liquid-sidestream rate	(kg·mol)/s	(lb·mol)/h
V	Vapor flow rate	(kg·mol)/s	(lb·mol)/h
W	Vapor-sidestream rate	(kg·mol)/s	(lb·mol)/h
X	Vector of stage variables		
Z	Azeotropic range	K	°F
Z_{ij}	Half value of azeotropic range	K	°F
Z_p	Packed-column height	m	ft
a	Activity		
b	Component flow rate in bottoms	(kg·mol)/s	(lb·mol)/h
d	Component flow rate in distillate	(kg·mol)/s	(lb·mol)/h
f	Fraction of feed leaving in bottoms		
f	Fugacity	Pa	psia
g	Residual of energy balance	kW	Btu/h
h	Factor in SB method		
h	Height	m	ft
ℓ	Component flow rate in liquid	(kg·mol)/s	(lb·mol)/h
p	Pressure	kPa	psia
q	Measure of thermal condition of feed		
q_c	Condenser duty	kW	Btu/h
q_r	Reboiler duty	kW	Btu/h
s	Liquid-sidestream ratio		
t	Time	s	h
v	Component flow rate in vapor	(kg·mol)/s	(lb·mol)/h
w	Weight fraction		
x	Mole fraction in liquid		
y	Mole fraction in vapor		
z	Mole fraction in feed		
	Greek symbols		
α	Relative volatility		
γ	Activity coefficient		
δ	Azeotropic temperature deviation	K	°F
Δ	Absolute boiling-point difference	K	°F
ΔS	Molar vaporization entropy	J/(kg·mol·K)	Btu/(lb·mol·°F)
ϵ	Convergence criterion		
ξ	Scale factor		
η	Murphree-stage efficiency		
θ	Time for distillation	s	h
Θ	Parameter in Underwood equations		
Θ	Holland theta factor		
λ	Eigenvalue		
τ	Sum of squares of residuals		

Nomenclature and Units *(Continued)*

Symbol	Definition	SI units	U.S. customary units
τ	Feedback-reset time	s	h
Φ	Fugacity coefficient of pure component		
$\hat{\Phi}$	Fucagity coefficient in mixture		
Φ_A	Fraction of a component in feed vapor that is not absorbed		
Φ_s	Fraction of a component in entering liquid that is not stripped		
Ψ	Factor in Gilliland correlation		

General References: Billet, *Distillation Engineering,* Chemical Publishing, New York, 1979. Fair and Bolles, "Modern Design of Distillation Columns," *Chem. Eng.,* **75**(9), 156 (Apr. 22, 1968). Fredenslund, Gmehling, and Rasmussen, *Vapor-Liquid Equilibria Using UNIFAC, a Group Contribution Method,* Elsevier, Amsterdam, 1977. Friday and Smith, "An Analysis of the Equilibrium Stage Separation Problem—Formulation and Convergence," *Am. Inst. Chem. Eng. J.,* **10,** 698 (1964). Hengstebeck, *Distillation—Principles and Design Procedures,* Reinhold, New York, 1961. Henley and Seader, *Equilibrium-Stage Separation Operations in Chemical Engineering,* Wiley, New York, 1981. Hoffman, *Azeotropic and Extractive Distillation,* Wiley, New York, 1964. Holland, *Fundamentals and Modeling of Separation Processes,* Prentice-Hall, Englewood Cliffs, N.J., 1975. Holland, *Fundamentals of Multicomponent Distillation,* McGraw-Hill, New York, 1981. King, *Separation Processes,* 2d ed., McGraw-Hill, New York, 1980. Robinson and Gilliland, *Elements of Fractional Distillation,* 4th ed., McGraw-Hill, New York, 1950. Smith, *Design of Equilibrium Stage Processes,* McGraw-Hill, New York, 1963. Treybal, *Mass Transfer Operations,* 3d ed., McGraw-Hill, New York, 1980. Van Winkle, *Distillation,* McGraw-Hill, New York, 1967.

CONTINUOUS-DISTILLATION OPERATIONS

General Principles Separation operations achieve their objective by the creation of two or more coexisting zones which differ in temperature, pressure, composition, and/or phase state. Each molecular species in the mixture to be separated reacts in a unique way to differing environments offered by these zones. Consequently, as the system moves toward equilibrium, each species establishes a different concentration in each zone, and this results in a separation between the species.

The separation operation called *distillation* utilizes vapor and liquid phases at essentially the same temperature and pressure for the coexisting zones. Various kinds of devices such as *dumped* or *ordered packings* and *plates* or *trays* are used to bring the two phases into intimate contact. Trays are stacked one above the other and enclosed in a cylindrical shell to form a *column.* Packings are also generally contained in a cylindrical shell between hold-down and support plates. A typical tray-type distillation column plus major external accessories is shown schematically in Fig. 13-1.

The *feed* material, which is to be separated into fractions, is introduced at one or more points along the column shell. Because of the difference in gravity between vapor and liquid phases, liquid runs down the column, cascading from tray to tray, while vapor flows up the column, contacting liquid at each tray.

Liquid reaching the bottom of the column is partially vaporized in a heated *reboiler* to provide *boil-up,* which is sent back up the column. The remainder of the bottom liquid is withdrawn as *bottoms,* or bottom product. Vapor reaching the top of the column is cooled and condensed to liquid in the *overhead condenser.* Part of this liquid is returned to the column as *reflux* to provide liquid overflow. The remainder of the overhead stream is withdrawn as *distillate,* or overhead product.

This overall flow pattern in a distillation column provides countercurrent contacting of vapor and liquid streams on all the trays through the column. Vapor and liquid phases on a given tray approach thermal, pressure, and composition equilibriums to an extent dependent upon the efficiency of the contacting tray.

The *lighter* (lower-boiling) components tend to concentrate in the vapor phase, while the *heavier* (higher-boiling) components tend toward the liquid phase. The result is a vapor phase that becomes richer in light components as it passes up the column and a liquid phase that becomes richer in heavy components as it cascades downward. The overall separation achieved between the distillate and the bottoms depends primarily on the *relative volatilities* of the components, the number of contacting trays, and the ratio of the liquid-phase flow rate to the vapor-phase flow rate.

If the feed is introduced at one point along the column shell, the column is divided into an upper section, which is often called the *rectifying* section, and a lower section, which is often referred to as the *stripping* section. These terms become rather indefinite in *multiple-feed columns* and in columns from which a product *sidestream* is withdrawn somewhere along the column length in addition to the two end-product streams.

Equilibrium-Stage Concept Energy and mass-transfer processes in an actual distillation column are much too complicated to be readily modeled in any direct way. This difficulty is circumvented by the *equilibrium-stage model,* in which vapor and liquid streams leaving an equilibrium stage are in complete equilibrium with each other

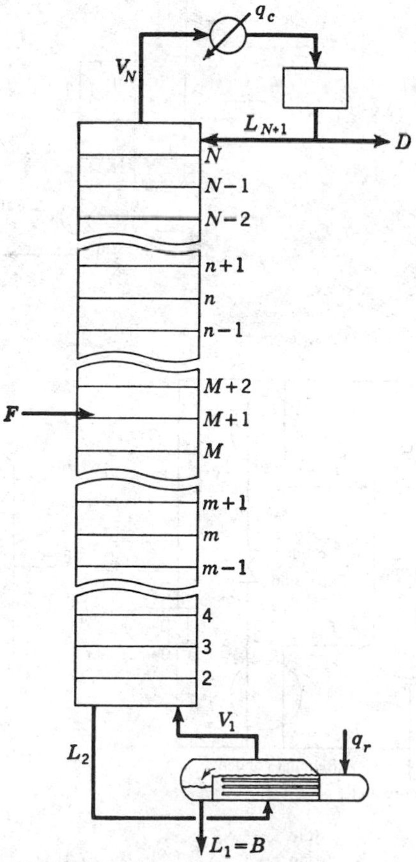

FIG. 13-1 Schematic diagram and nomenclature for a simple distillation column with one feed, a total overhead condenser, and a partial reboiler.

and thermodynamic relations can be used to determine the temperature of and relate the concentrations in the equilibrium streams at a given pressure. A hypothetical column composed of equilibrium stages (instead of actual contact trays) is designed to accomplish the separation specified for the actual column. The number of hypothetical equilibrium stages required is then converted to a number of actual trays by means of *tray efficiencies*, which describe the extent to which the performance of an actual contact tray duplicates the performance of an equilibrium stage.

Use of the equilibrium-stage concept separates the design of a distillation column into three major steps: (1) Thermodynamic data and methods needed to predict equilibrium-phase compositions are assembled. (2) The number of equilibrium stages required to accomplish a specified separation, or the separation that will be accomplished in a given number of equilibrium stages, is calculated. (3) The number of equilibrium stages is converted to an equivalent number of actual contact trays or height of packing, and the column diameter is determined. This section deals primarily with the second step. Section 18 covers the third step. Sections 3 and 4 cover the first step, but a summary of methods and some useful data are included in this section.

Complex Distillation Operations All separation operations require energy input in the form of heat or work. In the conventional distillation operation, as typified in Fig. 13-1, energy required to separate the species is added in the form of heat to the reboiler at the bottom of the column, where the temperature is highest. Also, heat is removed from a condenser at the top of the column, where the temperature is lowest. This frequently results in a large energy-input requirement and low overall thermodynamic efficiency, which was of little concern (except for cryogenic and high-temperature processes) when energy costs were low. With recent dramatic increases in energy costs, complex distillation operations that offer higher ther-

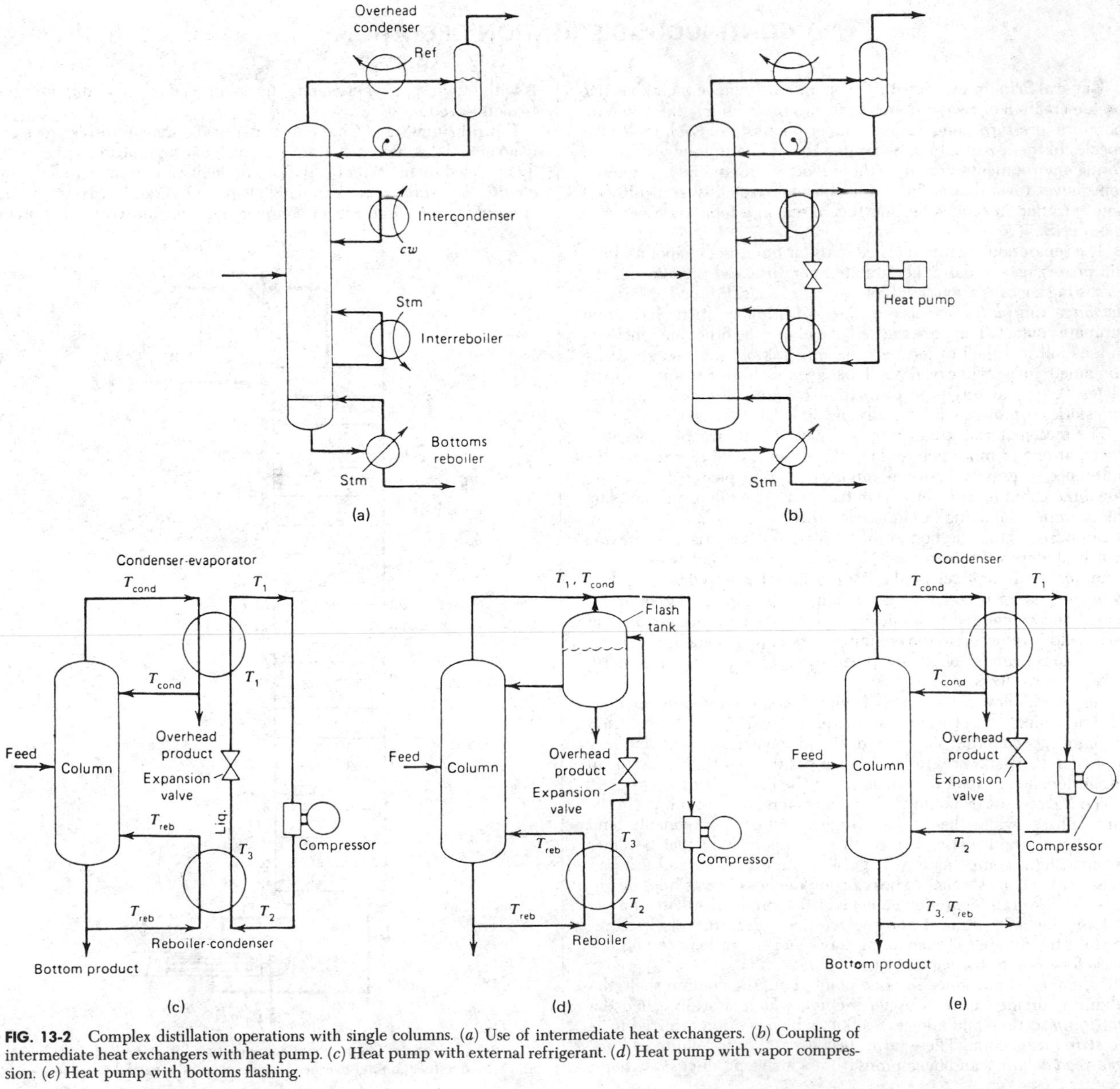

FIG. 13-2 Complex distillation operations with single columns. (*a*) Use of intermediate heat exchangers. (*b*) Coupling of intermediate heat exchangers with heat pump. (*c*) Heat pump with external refrigerant. (*d*) Heat pump with vapor compression. (*e*) Heat pump with bottoms flashing.

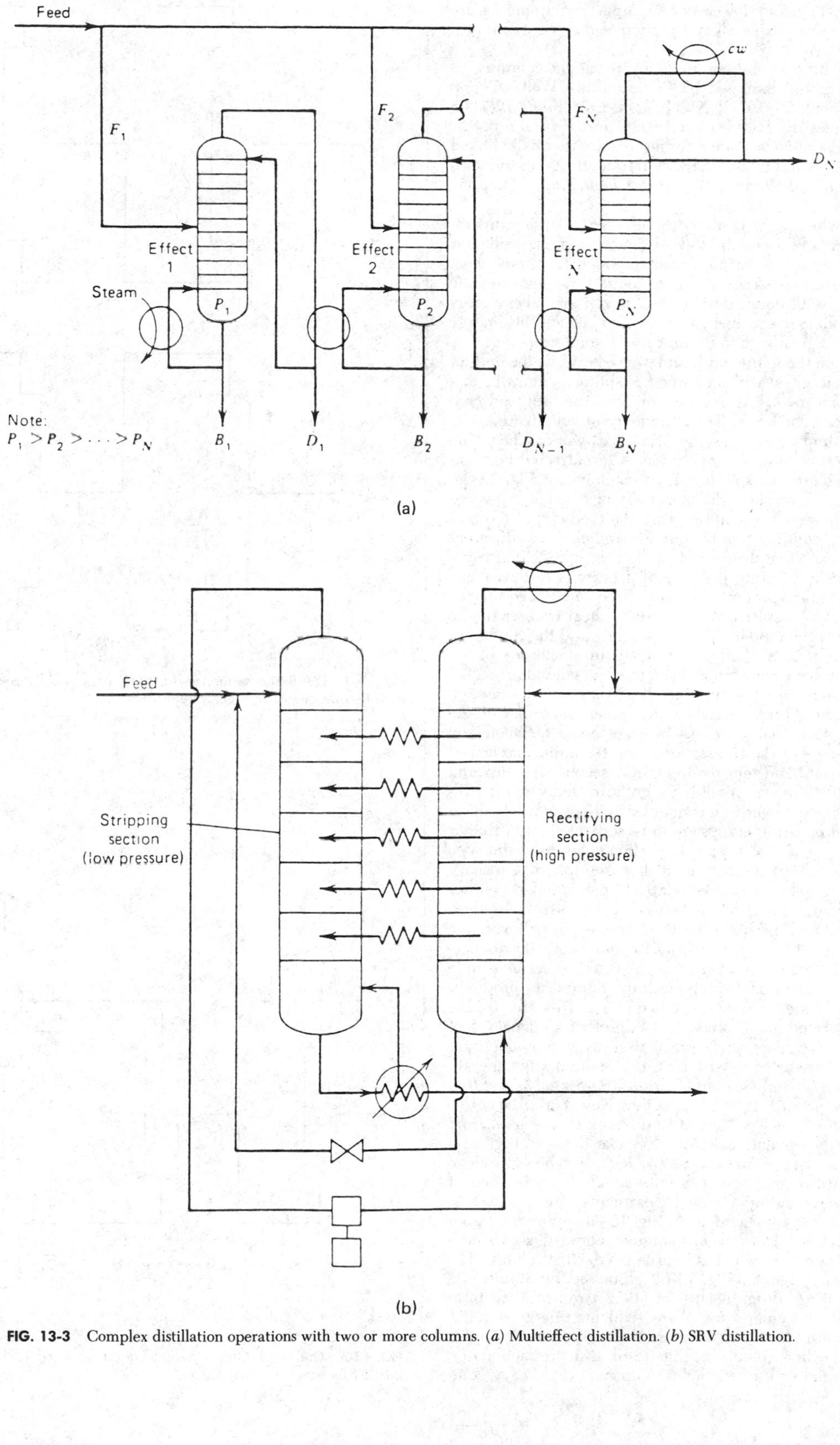

FIG. 13-3 Complex distillation operations with two or more columns. (*a*) Multieffect distillation. (*b*) SRV distillation.

modynamic efficiency and lower energy-input requirements are being explored. In some cases, all or a portion of the energy input is as work.

Complex distillation operations may utilize single columns, as shown in Fig. 13-2 and discussed by Petterson and Wells [*Chem. Eng.*, **84**(20), 78 (Sept. 26, 1977)], Null [*Chem. Eng. Prog.*, **72**(7), 58 (1976)], and Brannon and Marple [*Am. Inst. Chem. Eng. Symp. Ser. 76*, **192**, 10 (1980)], or two or more columns that are thermally linked as shown in Fig. 13-3 and discussed by Petterson and Wells (op. cit.) and Mah, Nicholas, and Wodnik [*Am. Inst. Chem. Eng. J.*, **23**, 651 (1977)].

In Fig. 13-2*a*, which is particularly useful when a large temperature difference exists between the ends of the column, interreboilers add heat at lower temperatures and/or intercondensers remove heat at higher temperatures. As shown in Fig. 13-2*b*, these intermediate heat exchangers may be coupled with a heat pump that takes energy from the intercondenser and uses shaft work to elevate this energy to a temperature high enough to transfer it to the interreboiler.

Particularly when the temperature difference between the ends of the column is not large, any of the three heat-pump systems in Fig. 13-2*c*, *d*, and *e* that involve thermal coupling of the overhead condenser and bottoms reboiler might be considered to eliminate external heat transfer almost entirely, substituting shaft work as the prime energy input for achieving the separation. Alternatively, the well-known multiple-column or split-tower arrangement of Fig. 13-3*a*, which corresponds somewhat to the energy-saving concept employed in multieffect evaporation, might be used. The feed is split more or less equally among columns that operate in parallel, but at different pressures, in a cascade that decreases from left to right. With proper selection of column-operating pressure, this permits the overhead vapor from the higher-pressure column to be condensed in the reboiler of the lower-pressure column. External heat-transfer media are needed only for the reboiler of the first effect and the condenser of the last effect. Thus, for N effects, utility requirements are of the order $1/N$ of those for a conventional single-effect column.

In another alternative, shown in Fig. 13-3*b*, the rectifying section may be operated at a pressure sufficiently higher than that of the stripping section such that heat can be transferred between any desired pairs of stages of the two sections. This technique, described by Mah et al. (op. cit.) and referred to as SRV (secondary reflux and vaporization) distillation, can result in a significant reduction in utility requirements for the overhead condenser and bottoms reboiler.

When multicomponent mixtures are to be separated into three or more products, sequences of simple distillation columns of the type shown in Fig. 13-1 are commonly used. For example, if a ternary mixture is to be separated into three relatively pure products, either of the two sequences in Fig. 13-4 can be used. In the direct sequence, shown in Fig. 13-4*a*, all products but the heaviest are removed one by one as distillates. The reverse is true for the indirect sequence, shown in Fig. 13-4*b*. The number of possible sequences of simple distillation columns increases rapidly with the number of products. Thus, although only the 2 sequences shown in Fig. 13-4 are possible for a mixture separated into 3 products, 14 different sequences, 1 of which is shown in Fig. 13-5, can be synthesized when 5 products are to be obtained. Methods of determining the optimal sequence are discussed by Henley and Seader (*Equilibrium-Stage Separation Operations in Chemical Engineering*, Wiley, New York, 1981).

As shown in a study by Tedder and Rudd [*Am. Inst. Chem. Eng. J.*, **24**, 303 (1978)], conventional sequences like those of Fig. 13-4 may not always be the optimal choice, particularly when species of intermediate volatility are present in large amounts in the feed or need not be recovered at high purity. Of particular interest are thermally coupled systems. For example, in Fig. 13-6*a*, an impure-vapor sidestream is withdrawn from the first column and purified in a sidecut rectifier, the bottoms of which is returned to the first column. The thermally coupled system in Fig. 13-6*b*, discussed by Stupin and Lockhart [*Chem. Eng. Prog.*, **68**(10), 71 (1972)] and referred to as Petlyuk towers, is particularly useful for reducing energy requirements when the initial feed contains close-boiling species. Shown for a ternary feed, the first column in Fig. 13-6*b* is a prefractionator, which sends essentially all of the light component and heavy com-

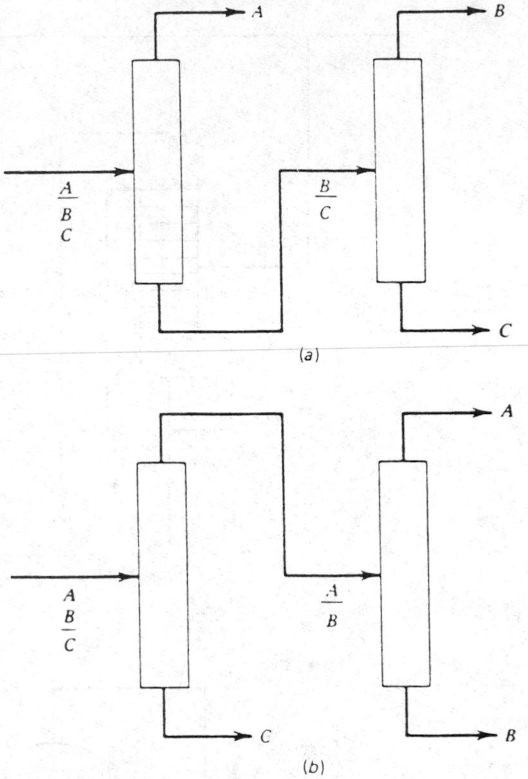

FIG. 13-4 Distillation sequences for the separation of three components. (*a*) Direct sequence. (*b*) Indirect sequence.

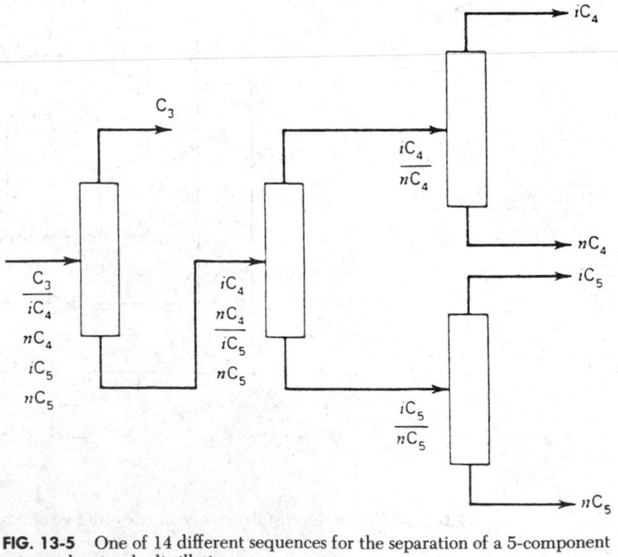

FIG. 13-5 One of 14 different sequences for the separation of a 5-component mixture by simple distillation.

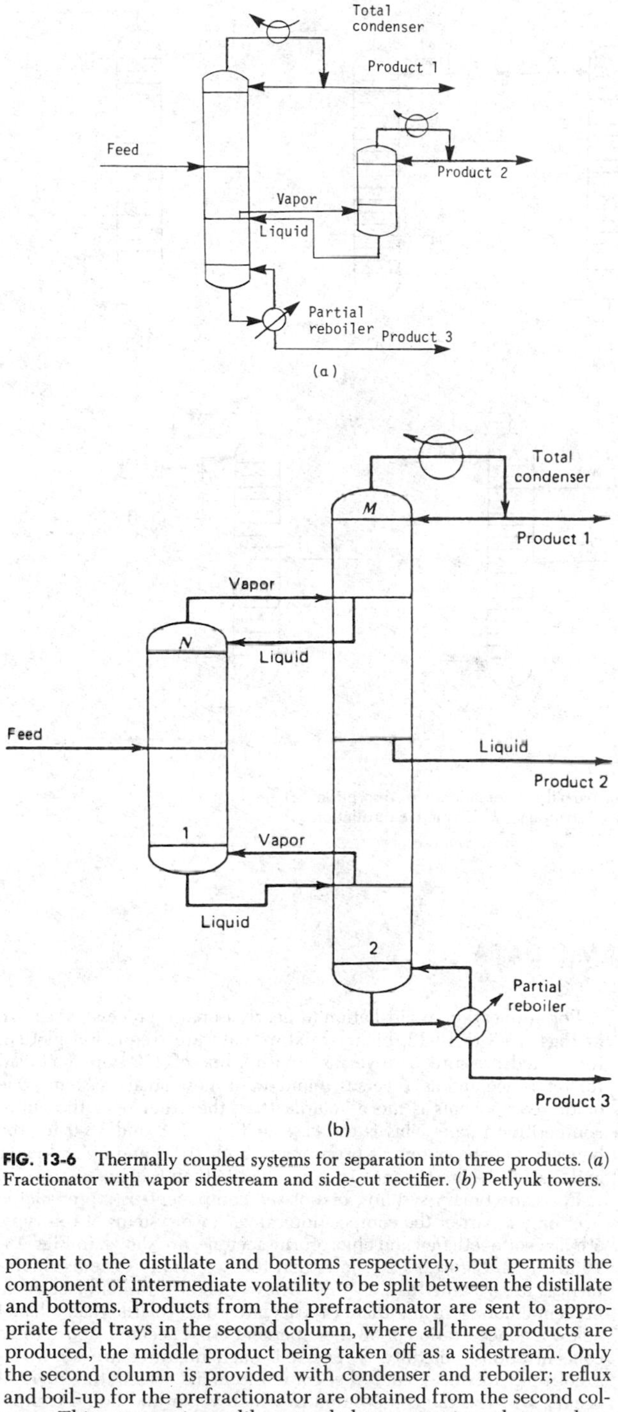

FIG. 13-6 Thermally coupled systems for separation into three products. (*a*) Fractionator with vapor sidestream and side-cut rectifier. (*b*) Petlyuk towers.

ponent to the distillate and bottoms respectively, but permits the component of intermediate volatility to be split between the distillate and bottoms. Products from the prefractionator are sent to appropriate feed trays in the second column, where all three products are produced, the middle product being taken off as a sidestream. Only the second column is provided with condenser and reboiler; reflux and boil-up for the prefractionator are obtained from the second column. This concept is readily extended to separations that produce more than three products.

Related Separation Operations The simple and complex distillation operations just described all have two things in common: (1) both rectifying and stripping sections are provided so that a separation can be achieved between two components that are adjacent in volatility; and (2) the separation is effected only by the addition and removal of energy and not by the addition of any mass separating agent (MSA) such as in liquid-liquid extraction. Sometimes, alternative single- or multiple-stage vapor-liquid separation operations, of the types shown in Fig. 13-7, may be more suitable than distillation for the specified task.

A single-stage flash, as shown in Fig. 13-7*a*, may be appropriate if (1) the relative volatility between the two components to be separated is very large; (2) the recovery of only one component, without regard to the separation of the other components, in one of the two product streams is to be achieved; or (3) only a partial separation is to be made. A common example is the separation of light gases such as hydrogen and methane from aromatics. The desired temperature and pressure of a flash may be established by the use of heat exchangers, a valve, a compressor, and/or a pump upstream of the vessel used to separate the product vapor and liquid phases. Depending on the original condition of the feed, it may be partially condensed or partially vaporized in a so-called flash operation.

If the recovery of only one component is required rather than a sharp separation between two components of adjacent volatility, absorption or stripping in a single section of stages may be sufficient. If the feed is vapor at separation conditions, absorption is used either with a liquid MSA absorbent of relatively low volatility as in Fig. 13-7*b* or with reflux produced by an overhead partial condenser as in Fig. 13-7*c*. The choice usually depends on the ease of partially condensing the overhead vapor or of recovering and recycling the absorbent. If the feed is liquid at separation conditions, stripping is used, either with an externally supplied vapor stripping agent of relatively high volatility as shown in Fig. 13-7*d* or with boil-up produced by a partial reboiler as in Fig. 13-7*e*. The choice depends on the ease of partially reboiling the bottoms or of recovering and recycling the stripping agent.

If a relatively sharp separation is required between two components of adjacent volatility, but either an undesirably low temperature is required to produce reflux at the column-operating pressure or an undesirably high temperature is required to produce boil-up, then refluxed stripping as shown in Fig. 13-7*g* or reboiled absorption as shown in Fig. 13-7*f* may be used. In either case, the choice of MSA follows the same consideration given for simple absorption and stripping.

When the volatility difference between the two components to be separated is so small that a very large number of stages would be required, then extractive distillation, as shown in Fig. 13-7*h*, should be considered. Here, an MSA is selected that increases the volatility difference sufficiently to reduce the stage requirement to a reasonable number. Usually, the MSA is a polar compound of low volatility that leaves in the bottoms, from which it is recovered and recycled. It is introduced in an appreciable amount near the top stage of the column so as to affect the volatility difference over most of the stages. Some reflux to the top stage is utilized to minimize the MSA content in the distillate. An alternative to extractive distillation is azeotropic distillation, which is shown in Fig. 13-7*i* in just one of its many modes. In a common mode, an MSA that forms a heterogeneous minimum-boiling azeotrope with one or more components of the feed is utilized. The azeotrope is taken overhead, and the MSA-rich phase is decanted and returned to the top of the column as reflux.

Numerous other multistaged configurations are possible. One important variation of a stripper, shown in Fig. 13-7*d*, is a refluxed stripper, in which an overhead condenser is added. Such a configuration is sometimes used to steam-strip sour water containing NH_3, H_2O, phenol, and HCN.

All the separation operations shown in Fig. 13-7, as well as the simple and complex distillation operations described earlier, are referred to here as distillation-type separations because they have much in common with respect to calculations of (1) thermodynamic properties, (2) vapor-liquid equilibrium stages, and (3) column sizing. In fact, as will be evident from the remaining treatment of this section, the trend is toward single generalized digital-computer-program packages that will compute many or all distillation-type separation operations.

This section does not include a treatment of distillation-type separations from a mass-transfer or a transfer-unit point of view. However, Sec. 14 does present details of that subject as applied to absorption and stripping.

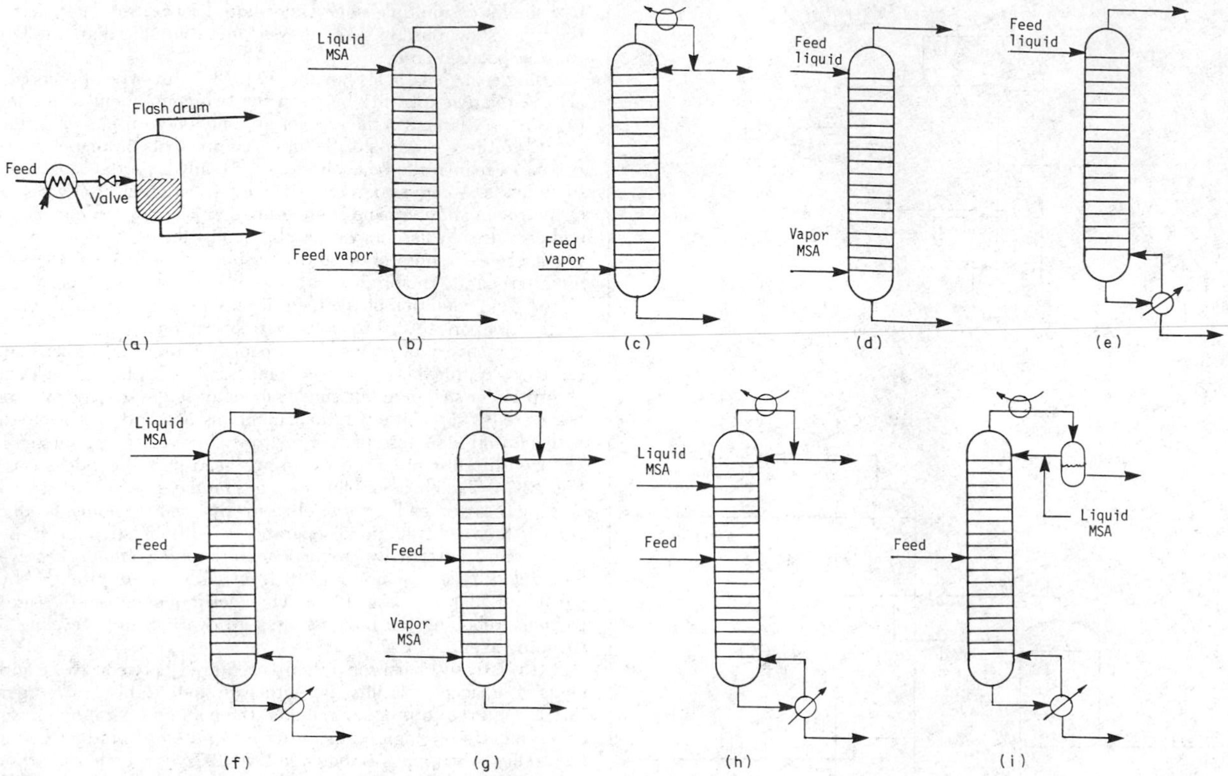

FIG. 13-7 Separation operations related to distillation. (*a*) Flash vaporization or partial condensation. (*b*) Absorption. (*c*) Rectifier. (*d*) Stripping. (*e*) Reboiled stripping. (*f*) Reboiled absorption. (*g*) Refluxed stripping. (*h*) Extractive distillation. (*i*) Azeotropic distillation.

THERMODYNAMIC DATA

Introduction Reliable thermodynamic data are essential for the accurate design or analysis of distillation columns. Failure of equipment to perform at specified levels is often attributable, at least in part, to the lack of such data.

This subsection summarizes and presents examples of phase equilibrium data currently available to the designer. The thermodynamic concepts utilized are presented in the subsection "Thermodynamics" of Sec. 4.

Phase Equilibrium Data For a binary mixture, pressure and temperature fix the equilibrium vapor and liquid compositions. Thus, experimental data are frequently presented in the form of tables of vapor mole fraction y and liquid mole fraction x for one constituent over a range of temperature T for a fixed pressure P or over a range of pressure for a fixed temperature. A compilation of such data, mainly at a pressure of 101.3 kPa (1 atm, 1.013 bar), for binary systems (mainly nonideal) is given in Table 13-1. More extensive presentations and bibliographies of such data may be found in Hala, Wichterle, Polak, and Boublik (*Vapour-Liquid Equilibrium Data at Normal Pressures*, Pergamon, Oxford, 1968); Hirata, Ohe, and Nagahama (*Computer Aided Data Book of Vapor-Liquid Equilibria*, Elsevier, Amsterdam, 1975); Wichterle, Linek, and Hala (*Vapor-Liquid Equilibrium Data Bibliography*, Elsevier, Amsterdam, 1973, Supplement I, 1976, Supplement II, 1979); and, particularly, Gmehling and Onken (*Vapor-Liquid Equilibrium Data Collection*, DECHEMA Chemistry Data ser., vol. 1 (parts 1–10), Frankfort, 1977).

For application to distillation (a nearly isobaric process), as shown in Figs. 13-8 to 13-13, binary-mixture data are frequently plotted, for a fixed pressure, as y versus x with a line of 45° slope included for reference and as T versus y and x. In most binary systems, one of the components is more volatile than the other over the entire composition range. This is the case in Figs. 13-8 and 13-9 for the benzene-toluene system at pressures of both 101.3 and 202.6 kPa (1 and 2 atm), where benzene is more volatile than toluene.

For some binary systems, one of the components is more volatile over only a part of the composition range. Two systems of this type, ethyl acetate–ethanol and chloroform-acetone, are shown in Figs. 13-10 to 13-12. Figure 13-10 shows that for two binary systems chloroform is less volatile than acetone below a concentration of 66 mole percent chloroform and that ethyl acetate is less volatile than ethanol below a concentration of 53 mole percent ethyl acetate. Above these concentrations, volatility is reversed. Such mixtures are known as azeotropic mixtures, and the composition in which the reversal occurs, which is the composition in which vapor and liquid compositions are equal, is the azeotropic composition, or azeotrope. The azeotropic liquid may be homogeneous or heterogeneous (two immiscible liquid phases). Many of the binary mixtures of Table 13-1 form homogeneous azeotropes. Non-azeotrope-forming mixtures such as benzene and toluene in Figs. 13-8 and 13-9 can be separated by simple distillation into two essentially pure products. By contrast, simple distillation of azeotropic mixtures will at best yield the azeotrope and one essentially pure species. The distillate and bottoms

TABLE 13-1 Constant-Pressure Liquid-Vapor Equilibrium Data for Selected Binary Systems

Component		Temperature, °C	Mole fraction *A* in		Total pressure, kPa	Reference
A	*B*		Liquid	Vapor		
Acetone	Chloroform	62.50	0.0817	0.0500	101.3	1
		62.82	0.1390	0.1000		
		63.83	0.2338	0.2000		
		64.30	0.3162	0.3000		
		64.37	0.3535	0.3500		
		64.35	0.3888	0.4000		
		64.02	0.4582	0.5000		
		63.33	0.5299	0.6000		
		62.23	0.6106	0.7000		
		60.72	0.7078	0.8000		
		58.71	0.8302	0.9000		
		57.48	0.9075	0.9500		
Acetone	Methanol	64.65	0.0	0.0	101.3	2
		61.78	0.091	0.177		
		59.60	0.190	0.312		
		58.14	0.288	0.412		
		56.96	0.401	0.505		
		56.22	0.501	0.578		
		55.78	0.579	0.631		
		55.41	0.687	0.707		
		55.29	0.756	0.760		
		55.37	0.840	0.829		
		55.54	0.895	0.880		
		55.92	0.954	0.946		
		56.21	1.000	1.000		
Acetone	Water	74.80	0.0500	0.6381	101.3	3
		68.53	0.1000	0.7301		
		65.26	0.1500	0.7716		
		63.59	0.2000	0.7916		
		61.87	0.3000	0.8124		
		60.75	0.4000	0.8269		
		59.95	0.5000	0.8387		
		59.12	0.6000	0.8532		
		58.29	0.7000	0.8712		
		57.49	0.8000	0.8950		
		56.68	0.9000	0.9335		
		56.30	0.9500	0.9627		
Carbon tetrachloride	Benzene	80.0	0.0	0.0	101.3	4
		79.3	0.1364	0.1582		
		78.8	0.2157	0.2415		
		78.6	0.2573	0.2880		
		78.5	0.2944	0.3215		
		78.2	0.3634	0.3915		
		78.0	0.4057	0.4350		
		77.6	0.5269	0.5480		
		77.4	0.6202	0.6380		
		77.1	0.7223	0.7330		
Chloroform	Methanol	63.0	0.040	0.102	101.3	5
		60.9	0.095	0.215		
		59.3	0.146	0.304		
		57.8	0.196	0.378		
		55.9	0.287	0.472		
		54.7	0.383	0.540		
		54.0	0.459	0.580		
		53.7	0.557	0.619		
		53.5	0.636	0.646		
		53.5	0.667	0.655		
		53.7	0.753	0.684		
		54.4	0.855	0.730		
		55.2	0.904	0.768		
		56.3	0.937	0.812		
		57.9	0.970	0.875		
Ethanol	Benzene	76.1	0.027	0.137	101.3	6
		72.7	0.063	0.248		
		70.8	0.100	0.307		
		69.2	0.167	0.360		
		68.4	0.245	0.390		
		68.0	0.341	0.422		
		67.9	0.450	0.447		
		68.0	0.578	0.478		
		68.7	0.680	0.528		

TABLE 13-1 Constant-Pressure Liquid-Vapor Equilibrium Data for Selected Binary Systems *(Continued)*

Component A	Component B	Temperature, °C	Mole fraction A in Liquid	Mole fraction A in Vapor	Total pressure, kPa	Reference
		69.5	0.766	0.566		
		70.4	0.820	0.615		
		72.7	0.905	0.725		
		76.9	0.984	0.937		
Ethanol	Water	95.5	0.0190	0.1700	101.3	7
		89.0	0.0721	0.3891		
		86.7	0.0966	0.4375		
		85.3	0.1238	0.4704		
		84.1	0.1661	0.5089		
		82.7	0.2337	0.5445		
		82.3	0.2608	0.5580		
		81.5	0.3273	0.5826		
		80.7	0.3965	0.6122		
		79.8	0.5079	0.6564		
		79.7	0.5198	0.6599		
		79.3	0.5732	0.6841		
		78.74	0.6763	0.7385		
		78.41	0.7472	0.7815		
		78.15	0.8943	0.8943		
Ethyl acetate	Ethanol	78.3	0.0	0.0	101.3	8
		76.6	0.050	0.102		
		75.5	0.100	0.187		
		73.9	0.200	0.305		
		72.8	0.300	0.389		
		72.1	0.400	0.457		
		71.8	0.500	0.516		
		71.8	0.540	0.540		
		71.9	0.600	0.576		
		72.2	0.700	0.644		
		73.0	0.800	0.726		
		74.7	0.900	0.837		
		76.0	0.950	0.914		
		77.1	1.000	1.000		
Ethylene glycol	Water	69.5	0.0	0.0	30.4	9
		76.1	0.23	0.002		
		78.9	0.31	0.003		
		83.1	0.40	0.010		
		89.6	0.54	0.020		
		103.1	0.73	0.06		
		118.4	0.85	0.13		
		128.0	0.90	0.22		
		134.7	0.93	0.30		
		145.0	0.97	0.47		
		160.7	1.00	1.00		
n-Hexane	Ethanol	78.30	0.0	0.0	101.3	10
		76.00	0.0100	0.0950		
		73.20	0.0200	0.1930		
		67.40	0.0600	0.3650		
		65.90	0.0800	0.4200		
		61.80	0.1520	0.5320		
		59.40	0.2450	0.6050		
		58.70	0.3330	0.6300		
		58.35	0.4520	0.6400		
		58.10	0.5880	0.6500		
		58.00	0.6700	0.6600		
		58.25	0.7250	0.6700		
		58.45	0.7650	0.6750		
		59.15	0.8980	0.7100		
		60.20	0.9550	0.7450		
		63.50	0.9900	0.8400		
		66.70	0.9940	0.9350		
		68.70	1.0000	1.0000		
Methanol	Benzene	70.67	0.026	0.267	101.3	11
		66.44	0.050	0.371		
		62.87	0.088	0.457		
		60.20	0.164	0.526		
		58.64	0.333	0.559		
		58.02	0.549	0.595		
		58.10	0.699	0.633		
		58.47	0.782	0.665		
		59.90	0.898	0.760		
		62.71	0.973	0.907		

TABLE 13-1 Constant-Pressure Liquid-Vapor Equilibrium Data for Selected Binary Systems *(Continued)*

Component A	Component B	Temperature, °C	Mole fraction A in Liquid	Mole fraction A in Vapor	Total pressure, kPa	Reference
Methanol	Ethyl acetate	76.10	0.0125	0.0475	101.3	12
		74.15	0.0320	0.1330		
		71.24	0.0800	0.2475		
		67.75	0.1550	0.3650		
		65.60	0.2510	0.4550		
		64.10	0.3465	0.5205		
		64.00	0.4020	0.5560		
		63.25	0.4975	0.5970		
		62.97	0.5610	0.6380		
		62.50	0.5890	0.6560		
		62.65	0.6220	0.6670		
		62.50	0.6960	0.7000		
		62.35	0.7650	0.7420		
		62.60	0.8250	0.7890		
		62.80	0.8550	0.8070		
		63.21	0.9160	0.8600		
		63.90	0.9550	0.9290		
Methanol	Water	100.0	0.0	0.0	101.3	13
		96.4	0.020	0.134		
		93.5	0.040	0.230		
		91.2	0.060	0.304		
		89.3	0.080	0.365		
		87.7	0.100	0.418		
		84.4	0.150	0.517		
		81.7	0.200	0.579		
		78.0	0.300	0.665		
		75.3	0.400	0.729		
		73.1	0.500	0.779		
		71.2	0.600	0.825		
		69.3	0.700	0.870		
		67.5	0.800	0.915		
		66.0	0.900	0.958		
		65.0	0.950	0.979		
		64.5	1.000	1.000		
Methyl acetate	Methanol	57.80	0.173	0.342	101.3	14
		55.50	0.321	0.477		
		55.04	0.380	0.516		
		53.88	0.595	0.629		
		53.82	0.648	0.657		
		53.90	0.710	0.691		
		54.50	0.849	0.788		
		56.86	1.000	1.000		
1-Propanol	Water	100.00	0.0	0.0	101.3	15
		98.59	0.0030	0.0544		
		95.09	0.0123	0.1790		
		91.05	0.0322	0.3040		
		88.96	0.0697	0.3650		
		88.26	0.1390	0.3840		
		87.96	0.2310	0.3970		
		87.79	0.3110	0.4060		
		87.66	0.4120	0.4280		
		87.83	0.5450	0.4650		
		89.34	0.7300	0.5670		
		92.30	0.8780	0.7210		
		97.18	1.0000	1.0000		
2-Propanol	Water	100.00	0.0	0.0	101.3	16
		97.57	0.0045	0.0815		
		96.20	0.0069	0.1405		
		93.66	0.0127	0.2185		
		87.84	0.0357	0.3692		
		84.28	0.0678	0.4647		
		82.84	0.1330	0.5036		
		82.52	0.1651	0.5153		
		81.52	0.3204	0.5456		
		81.45	0.3336	0.5489		
		81.19	0.3752	0.5615		
		80.77	0.4720	0.5860		
		80.73	0.4756	0.5886		
		80.58	0.5197	0.6033		
		80.52	0.5945	0.6330		
		80.46	0.7880	0.7546		
		80.55	0.8020	0.7680		

TABLE 13-1 Constant-Pressure Liquid-Vapor Equilibrium Data for Selected Binary Systems *(Continued)*

Component A	Component B	Temperature, °C	Mole fraction A in Liquid	Mole fraction A in Vapor	Total pressure, kPa	Reference
		81.32	0.9303	0.9010		
		81.85	0.9660	0.9525		
		82.39	1.0000	1.0000		
Tetrahydrofuran	Water	73.00	0.0200	0.6523	101.3	17
		66.50	0.0400	0.7381		
		65.58	0.0600	0.7516		
		64.94	0.1000	0.7587		
		64.32	0.2000	0.7625		
		64.27	0.3000	0.7635		
		64.23	0.4000	0.7643		
		64.16	0.5000	0.7658		
		63.94	0.6000	0.7720		
		63.70	0.7000	0.7831		
		63.54	0.8000	0.8085		
		63.53	0.8200	0.8180		
		63.57	0.8400	0.8260		
		63.64	0.8600	0.8368		
		63.87	0.9000	0.8660		
		64.29	0.9400	0.9070		
		65.07	0.9800	0.9625		
		65.39	0.9900	0.9805		
Water	Acetic acid	118.3	0.0	0.0	101.3	18
		110.6	0.1881	0.3063		
		107.8	0.3084	0.4467		
		105.2	0.4498	0.5973		
		104.3	0.5195	0.6580		
		103.5	0.5824	0.7112		
		102.8	0.6750	0.7797		
		102.1	0.7261	0.8239		
		101.5	0.7951	0.8671		
		100.8	0.8556	0.9042		
		100.8	0.8787	0.9186		
		100.5	0.9134	0.9409		
		100.2	0.9578	0.9708		
		100.0	1.0000	1.0000		
Water	1-Butanol	117.6	0.0	0.0	101.3	19
		111.4	0.049	0.245		
		106.7	0.100	0.397		
		102.0	0.161	0.520		
		101.0	0.173	0.534		
		98.5	0.232	0.605		
		96.7	0.288	0.654		
		95.2	0.358	0.693		
		93.6	0.487	0.739		
		93.1	0.551	0.751		
		93.0	0.580	0.752		
		92.9	0.628	0.758		
		92.9	0.927	0.758		
		93.2	0.986	0.760		
		95.2	0.993	0.832		
		96.8	0.996	0.883		
		100.0	1.000	1.000		
Water	Formic acid	102.30	0.0405	0.0245	101.3	20
		104.60	0.1550	0.1020		
		105.90	0.2180	0.1620		
		107.10	0.3210	0.2790		
		107.60	0.4090	0.4020		
		107.60	0.4110	0.4050		
		107.60	0.4640	0.4820		
		107.10	0.5220	0.5670		
		106.00	0.6320	0.7180		
		104.20	0.7400	0.8360		
		102.90	0.8290	0.9070		
		101.80	0.9000	0.9510		
		100.00	1.0000	1.0000		
Water	Glycerol	278.8	0.0275	0.9315	101.3	21
		247.0	0.0467	0.9473		
		224.0	0.0690	0.9563		
		219.2	0.0767	0.9743		
		210.0	0.0901	0.9783		
		202.5	0.1031	0.9724		

TABLE 13-1 Constant-Pressure Liquid-Vapor Equilibrium Data for Selected Binary Systems *(Continued)*

Component		Temperature, °C	Mole fraction *A* in		Total pressure, kPa	Reference
A	*B*		Liquid	Vapor		
		196.5	0.1159	0.9839		
		175.2	0.1756	0.9899		
		149.3	0.3004	0.9964		
		137.2	0.3847	0.9976		
		136.8	0.3895	0.9878		
		131.8	0.4358	0.9976		
		121.5	0.5633	0.9984		
		112.8	0.7068	0.9993		
		111.3	0.7386	0.9994		
		106.3	0.8442	0.9996		
		100.0	1.0000	1.0000		

NOTE: To convert degrees Celsius to degrees Fahrenheit, °C = (°F − 32)/1.8. To convert kilopascals to pounds-force per square inch, multiply by 0.145.

[1]Kojima, Kato, Sunaga, and Hashimoto, *Kagaku Kogaku*, **32**, 337 (1968).
[2]Marinichev and Susarev, *Zh. Prikl. Khim.*, **38**, 378 (1965).
[3]Kojima, Tochigi, Seki, and Watase, *Kagaku Kogaku*, **32**, 149 (1968).
[4]*International Critical Tables*, McGraw-Hill, New York, 1928.
[5]Nagata, *J. Chem. Eng. Data*, **7**, 367 (1962).
[6]Ellis and Clark, *Chem. Age India*, **12**, 377 (1961).
[7]Carey and Lewis, *Ind. Eng. Chem.*, **24**, 882 (1932).
[8]Chu, Getty, Brennecke, and Paul, *Distillation Equilibrium Data*, New York, 1950.
[9]Trimble and Potts, *Ind. Eng. Chem.*, **27**, 66 (1935).
[10]Sinor and Weber, *J. Chem. Eng. Data*, **5**, 243 (1960).
[11]Hudson and Van Winkle, *J. Chem. Eng. Data*, **14**, 310 (1969).
[12]Murti and Van Winkle, *Chem, Eng. Data Ser.*, **3**, 72 (1958).
[13]Dunlop, M.S. thesis, Brooklyn Polytechnic Institute, 1948.
[14]Dobroserdov and Bagrov, *Zh. Prikl. Khim. (Leningrad)*, **40**, 875 (1967).
[15]Smirnova, *Vestn. Leningr. Univ. Fiz. Khim.*, **81** (1959).
[16]Kojima, Ochi, and Nakazawa, *Int. Chem. Eng.*, **9**, 342 (1964).
[17]Shnitko and Kogan, *J. Appl. Chem.*, **41**, 1236 (1968).
[18]Brusset, Kaiser, and Hocquel, *Chim. Ind., Genie Chim.* **99**, 207 (1968).
[19]Boublik, *Collect. Czech. Chem. Commun.*, **25**, 285 (1960).
[20]Ito and Yoshida, *J. Chem. Eng. Data*, **8**, 315 (1963).
[21]Chen and Thompson, *J. Chem. Eng. Data*, **15**, 471 (1970).

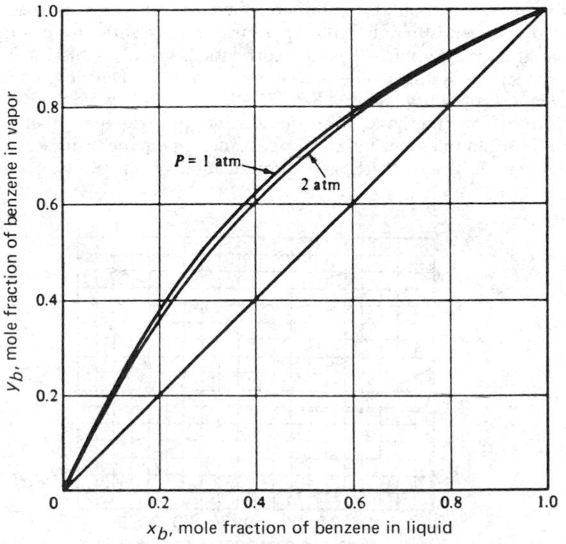

FIG. 13-8 Isobaric *y-x* curves for benzene-toluene. (*Brian,* Staged Cascades in Chemical Processing, *Prentice-Hall, Englewood Cliffs, N.J., 1972.*)

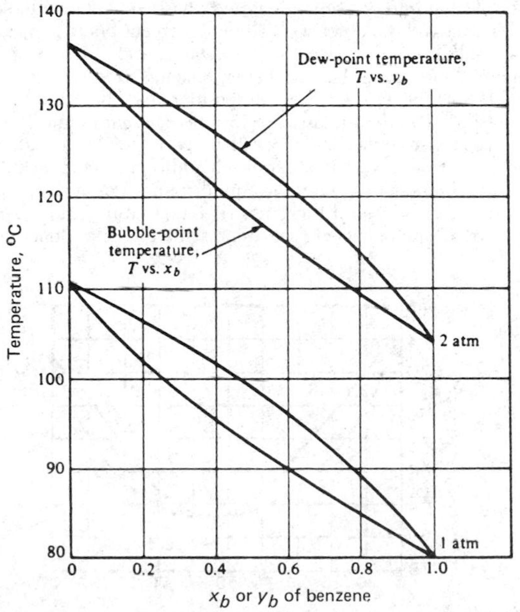

FIG. 13-9 Isobaric vapor-liquid equilibrium data for benzene-toluene. (*Brian,* Staged Cascades in Chemical Processing, *Prentice-Hall, Englewood Cliffs, N.J.,* 1972.)

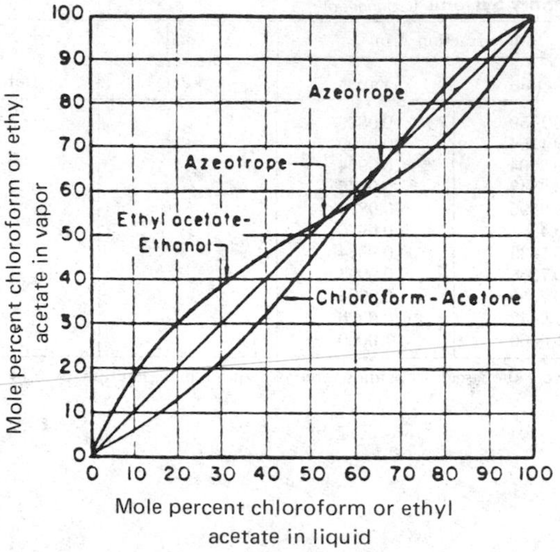

FIG. 13-10 Vapor-liquid equilibriums for the ethyl acetate–ethanol and chloroform-acetone systems at 101.3 kPa (1 atm).

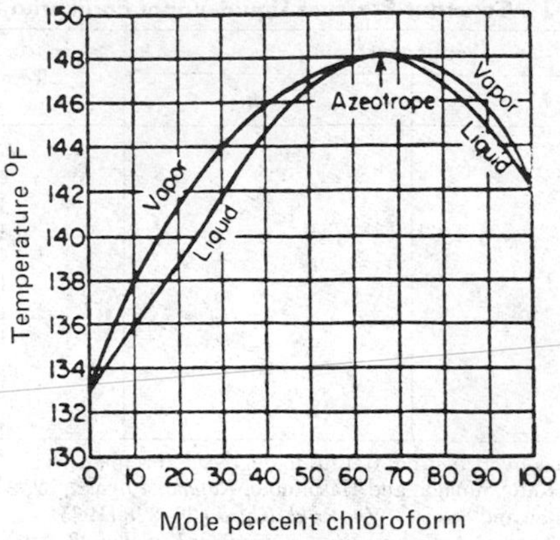

FIG. 13-12 Liquid boiling points and vapor condensation temperatures for maximum-boiling azeotrope mixtures of chloroform and acetone at 101.3-kPa (1-atm) total pressure.

products obtained depend upon the feed composition and whether a minimum-boiling azeotrope is formed as with the ethyl acetate–ethanol mixture in Fig. 13-11 or a maximum-boiling azeotrope is formed as with the chloroform-acetone mixture in Fig. 13-12. For example, if a mixture of 30 mole percent chloroform and 70 mole percent acetone is fed to a simple distillation column, such as that shown in Fig. 13-1, operating at 101.3 kPa (1 atm), the distillate could approach pure acetone and the bottoms could approach the azeotrope.

An example of heterogeneous-azeotrope formation is shown in Fig. 13-13 for the water–normal-butanol system at 101.3 kPa. At liquid compositions between 0 and 3 mole percent butanol and between 40 and 100 mole percent butanol, the liquid phase is homogeneous. Phase splitting into two separate liquid phases (one with 3 mole percent butanol and the other with 40 mole percent butanol) occurs for any overall liquid composition between 3 and 40 mole percent butanol. A minimum-boiling heterogeneous azeotrope occurs at 92°C (198°F) when the vapor composition and the overall composition of the two liquid phases are 75 mole percent butanol.

For mixtures containing more than two species, an additional degree of freedom is available for each additional component. Thus, for a four-component system, the equilibrium vapor and liquid compositions are only fixed if the pressure, temperature, and mole fractions of two components are set. Representation of multicomponent vapor-liquid equilibrium data in tabular or graphical form of the type shown earlier for binary systems is either difficult or impossible. Instead, such data, as well as binary-system data, are commonly represented in terms of K values (vapor-liquid equilibrium ratios), which are defined by

$$K_i = y_i/x_i \tag{13-1}$$

and are correlated empirically or theoretically in terms of temperature, pressure, and phase compositions in the form of tables, graphs, and equations. K values are widely used in multicomponent-distillation calculations, and the ratio of the K values of two species, called the relative volatility,

$$\alpha_{ij} = K_i/K_j \tag{13-2}$$

is a convenient index of the relative ease or difficulty of separating components i and j by distillation. Rarely is distillation used on a large scale if the relative volatility is less than 1.05, with i more volatile than j.

Graphical *K*-Value Correlations As discussed in Sec. 4, the K value of a species is a complex function of temperature, pressure, and equilibrium vapor- and liquid-phase compositions. However, for mixtures of compounds of similar molecular structure and size, the K value depends mainly on temperature and pressure. For example, several major graphical K-value correlations are available for light-hydrocarbon systems. The easiest to use are the DePriester charts [*Chem. Eng. Prog. Symp. Ser.* 7, **49**, 1 (1953)], which cover 12 hydrocarbons (methane, ethylene, ethane, propylene, propane, isobutane, isobutylene, *n*-butane, isopentane, *n*-pentane, *n*-hexane, and *n*-heptane). These charts are simplification of the Kellogg charts

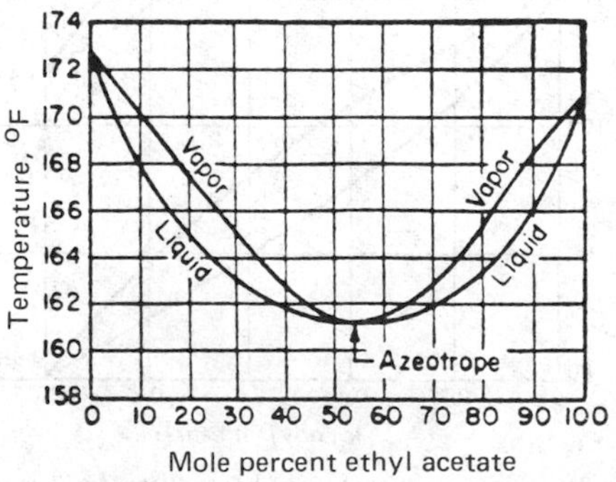

FIG. 13-11 Liquid boiling points and vapor condensation temperatures for minimum-boiling azeotrope mixtures of ethyl acetate and ethanol at 101.3-kPa (1-atm) total pressure.

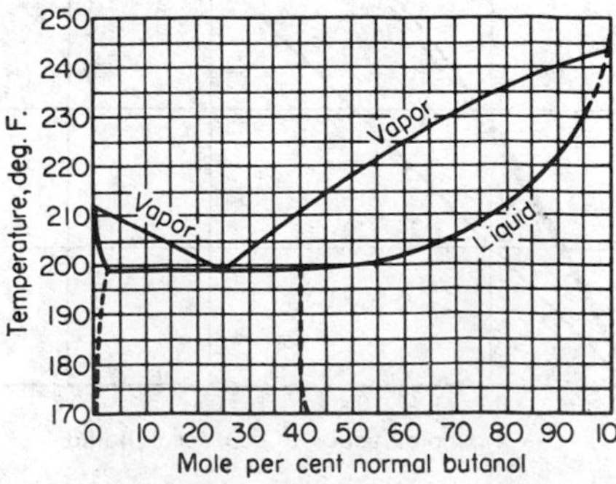

FIG. 13-13 Vapor-liquid equilibrium data for an n-butanol–water system at 101.3 kPa (1 atm); phase splitting and heterogeneous-azeotrope formation.

(*Liquid-Vapor Equilibria in Mixtures of Light Hydrocarbons, MWK Equilibrium Constants, Polyco Data*, 1950) and include additional experimental data. The Kellogg charts, and hence the DePriester charts, are based primarily on the Benedict-Webb-Rubin equation of state [*Chem. Eng. Prog.*, **47,** 419 (1951); **47,** 449 (1951)], which can represent both the liquid and the vapor phases and can predict K values quite accurately when the equation constants are available for the components in question. Edmister and Ruby [*Chem. Eng. Prog.*, **51,** 95-F (1955)] presented a generalized extension of the Kellogg charts that can be used for paraffin and olefin hydrocarbons other than the 12 covered by the Kellogg and DePriester charts.

A trial-and-error procedure is required with any K-value correlation that takes into account the effect of composition. One cannot calculate K values until phase compositions are known, and those cannot be known until the K values are available to calculate them. For K as a function of T and P only, DePriester provided nomographs, which furnish good starting values for the iteration. These nomographs are shown in Fig. 13-14*a* and *b*. SI versions of these charts have been developed by Dadyburjor [*Chem. Eng. Prog.*, **74** (4), 85 (1978)].

The Kellogg, DePriester, and Edmister-Ruby charts and their subsequent extensions and generalizations use the molar average boiling points of the liquid and vapor phases to represent the composition effect. An alternative measure of composition is the convergence pressure of the system, which is defined as that pressure at which the K values for all the components in an isothermal mixture converge to unity. It is analogous to the critical point for a pure component in the sense that the two phases become indistinguishable. The behavior of a complex mixture of hydrocarbons for a convergence pressure of 34.5 MPa (5000 psia) is illustrated in Fig. 13-15.

Two major graphical correlations based on convergence pressure as the third parameter (besides temperature and pressure) are the charts published by the Gas Processors Association (GPA, *Engineering Data Book*, 9th ed., Tulsa, 1981) and the charts of the American Petroleum Institute (API, *Technical Data Book—Petroleum Refining*, New York, 1966) based on the procedures from Hadden and Grayson [*Hydrocarbon Process., Pet. Refiner*, **40**(9), 207 (1961)]. The former uses the method proposed by Hadden [*Chem. Eng. Prog. Symp. Ser. 7*, **49,** 53 (1953)] for the prediction of convergence pressure as a function of composition. The basis for Hadden's method is illustrated in Fig. 13-16, where it is shown that the critical loci for various mixtures of methane-propane-pentane fall within the area circumscribed by the three binary loci. (This behavior is not always typical of more nonideal systems.) The critical loci for the ternary mixtures vary linearly, at constant temperature, with weight percent propane on a methane-free basis. The essential point is that critical loci for mixtures are independent of the concentration of the lightest component in a mixture. This permits representation of a multicomponent mixture as a pseudo binary. The light component in this

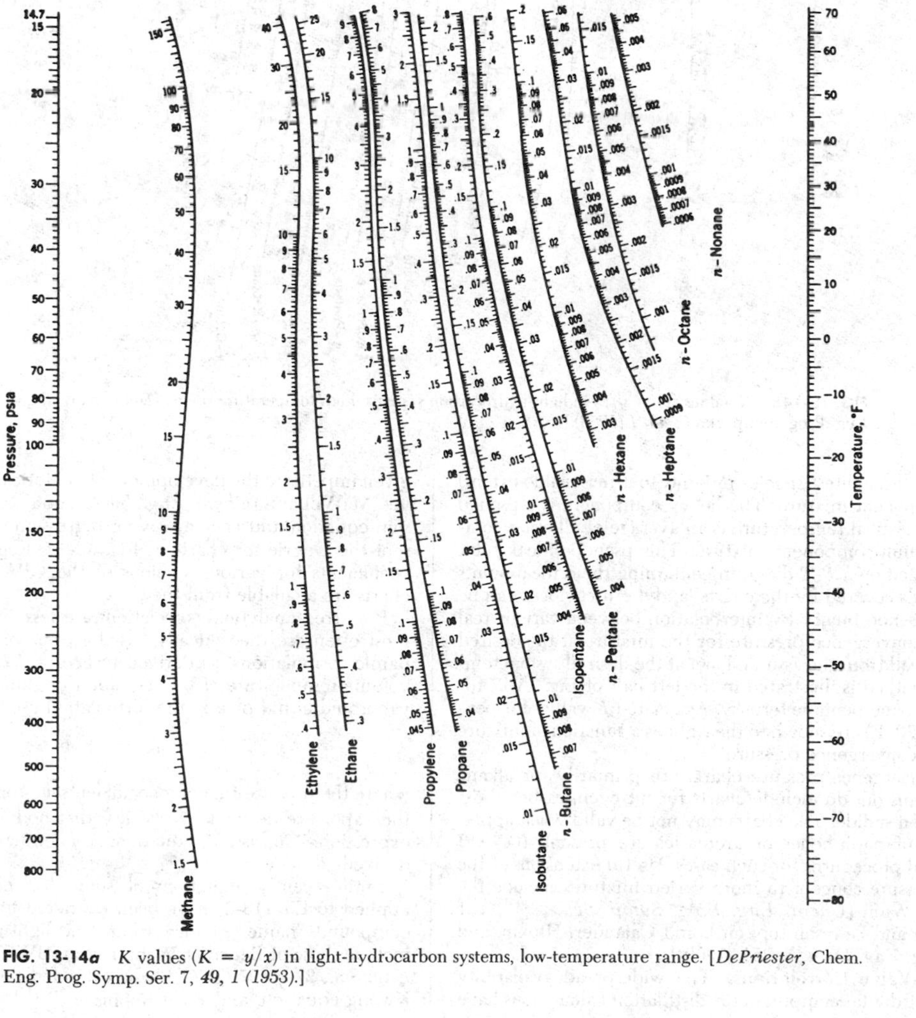

FIG. 13-14*a* K values ($K = y/x$) in light-hydrocarbon systems, low-temperature range. [*DePriester*, Chem. Eng. Prog. Symp. Ser. 7, *49, 1 (1953).*]

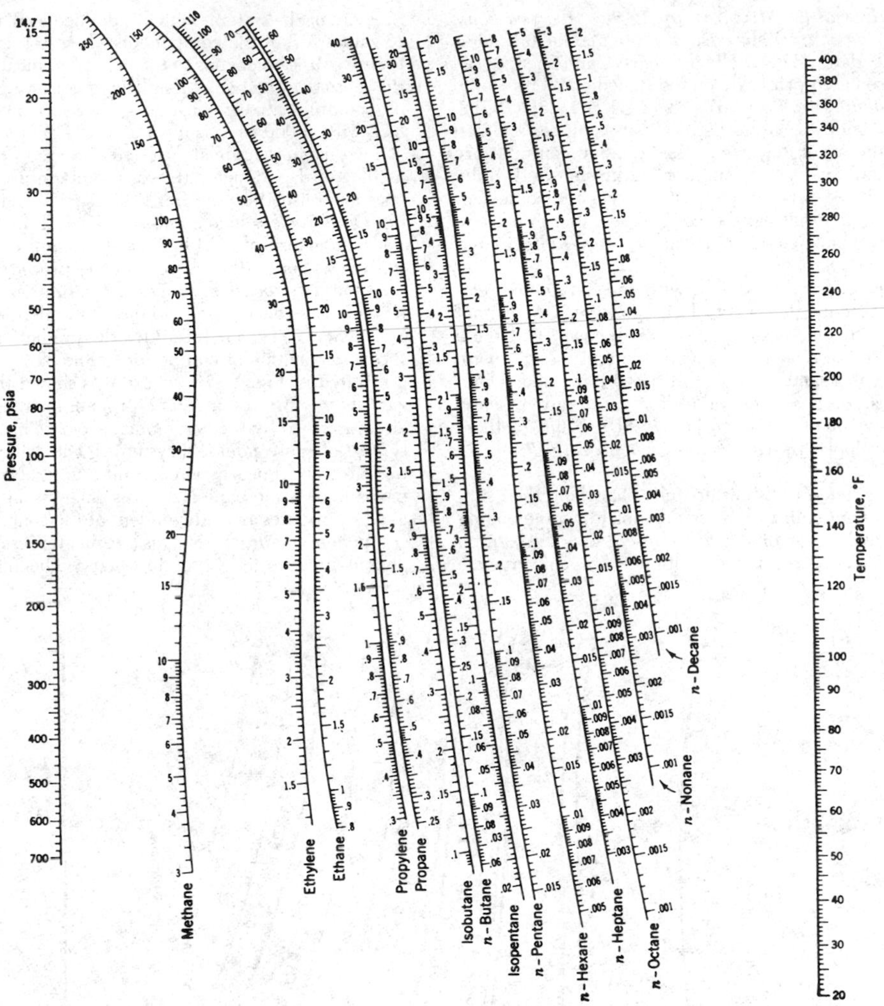

FIG. 13-14b *K* values ($K = y/x$) in light-hydrocarbon systems, high-temperature range. [*DePriester,* Chem. Eng. Prog. Symp. Ser. 7, *49, 1 (1953).*]

pseudo binary is the lightest species present (to a reasonable extent) in the multicomponent mixture. The heavy component is a pseudo substance whose critical temperature is an average of all other components in the multicomponent mixture. This pseudocritical point can then be located on a *P-T* diagram containing the critical points for all compounds covered by the charts, and a critical locus can be drawn for the pseudo binary by interpolation between various real binary curves. Convergence pressure for the mixture at the desired temperature is read from the assumed loci at the desired system temperature. This method is illustrated in the left half of Fig. 13-17 for the methane-propane-pentane ternary. Associated *K* values for pentane at 104°C (220°F) are shown to the right as a function of mixture composition (or convergence pressure).

The GPA convergence-pressure charts are primarily for alkane and alkene systems but do include charts for nitrogen, carbon dioxide, and hydrogen sulfide. The charts may not be valid when appreciable amounts of naphthenes or aromatics are present; the API charts use special procedures for such cases. Useful extensions of the convergence-pressure concept to more varied mixtures include the nomographs of Winn [*Chem. Eng. Prog. Symp. Ser.* 2, **48**, 121 (1952)], Hadden and Grayson (op. cit.), and Cajander, Hipkin, and Lenoir [*J. Chem. Eng. Data,* **5**, 251 (1960)].

Analytical *K*-Value Correlations The widespread availability and utilization of digital computers for distillation calculations have given impetus to the development of analytical expressions for *K* values. McWilliams [*Chem. Eng.,* **80**(25), 138 (1973)] presents a regression equation and accompanying regression coefficients that represent the DePriester charts of Fig. 13-14. Regression equations and coefficients for various versions of the GPA convergence-pressure charts are available from the GPA.

Preferred analytical correlations are less empirical in nature and most often are theoretically based on one of two exact thermodynamic formulations, as derived in Sec. 4. When a single pressure-volume-temperature *(PVT)* equation of state is applicable to both vapor and liquid phases, the formulation used is

$$K_i = \hat{\Phi}_i^L / \hat{\Phi}_i^V \qquad (13\text{-}3)$$

where the mixture fugacity coefficients $\hat{\Phi}_i^L$ for the liquid and $\hat{\Phi}_i^V$ for the vapor are derived by classical thermodynamics from the *PVT* expression. Consistent equations for enthalpy can similarly be derived.

Until recently, equations of state that have been successfully applied to Eq. (13-3) have been restricted to mixtures of nonpolar compounds, namely, hydrocarbons and light gases. These equations include those of Benedict-Webb-Rubin (BWR), Soave (SRK) [*Chem. Eng. Sci.,* **27**, 1197 (1972)], who extended the remarkable Redlich-Kwong equation, and Peng-Robinson (PR) [*Ind. Eng. Chem. Fun-*

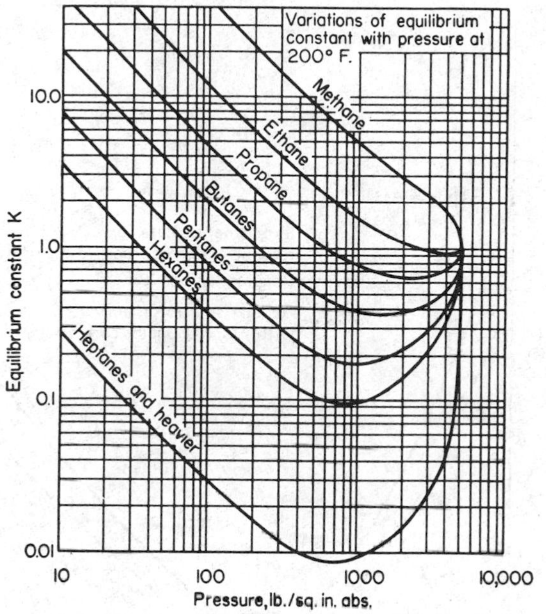

Fig. 13-15 Typical variation of K values with total pressure at constant temperature for a complex mixture. Light hydrocarbons in admixture with crude oil. [*Katz and Hachmuth*, Ind. Eng. Chem., **29**, *1072 (1937)*.]

dam., **15**, 59 (1976)]. The Starling extension of the BWR equation (*Fluid Thermodynamic Properties for Light Petroleum Systems*, Gulf, Houston, 1973) predicts K values and enthalpies of the normal paraffins up through n-octane, as well as isobutane, isopentane, ethylene, propylene, nitrogen, carbon dioxide, and hydrogen sulfide, including the cryogenic region. Computer programs for K values derived from the SRK and PR equation of state are available from the GPA. The ability of the SRK correlation to predict K values even when the pressure approaches the convergence pressure is shown for a multicomponent system in Fig. 13-18. Similar results are achieved with the PR correlation.

Gmehling, Liu, and Prausnitz [*Chem. Eng. Sci.*, **34**, 951 (1979)] developed an equation of state, based on perturbed-hard-chain the-

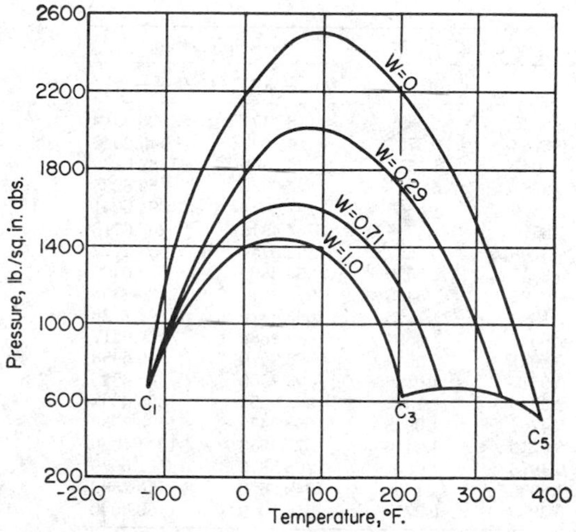

FIG. 13-16 Critical loci for a methane-propane-pentane system according to Hadden. [Chem. Eng. Prog. Symp. Ser. 7, *49, 58 (1953)*.] Parameter W is weight fraction propane on a methane-free basis.

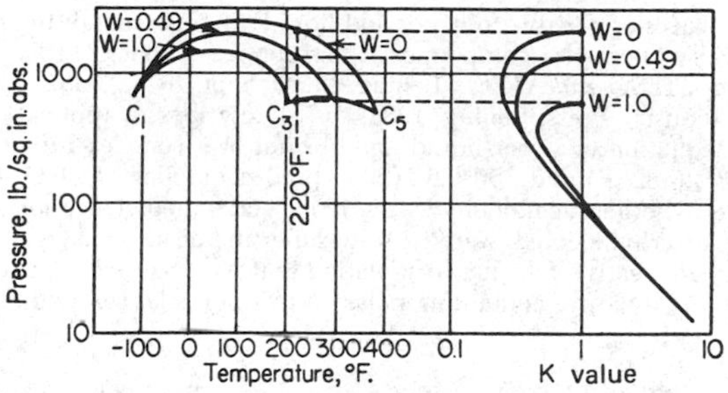

FIG. 13-17 Effect of mixture composition upon K value for n-pentane at 104°C (220°F). K values are shown for various values of W, weight fraction propane on a methane-free basis for the methane-propane-pentane system. [*Hadden*, Chem. Eng. Prog. Symp. Ser. 7, *49, 58 (1953)*.]

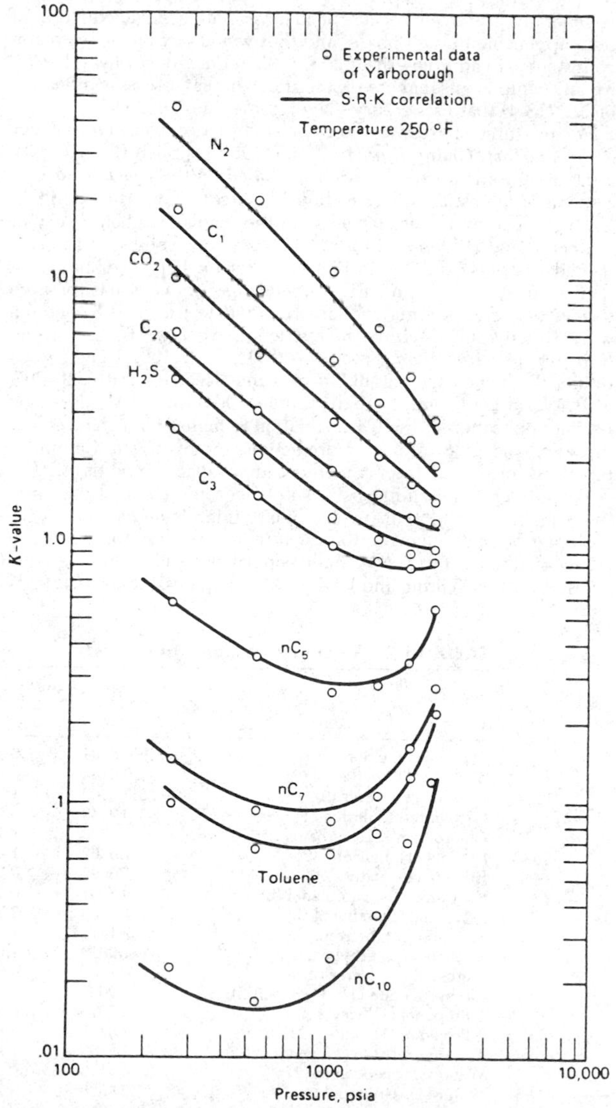

FIG. 13-18 Comparison of experimental K-value data and SRK correlation. [*Henley and Seader*, Equilibrium-Stage Separation Operations in Chemical Engineering, *Wiley, New York, 1981; data of Yarborough*, J. Chem. Eng. Data, *17, 129 (1972)*.]

ory, that is applicable to vapor and liquid mixtures containing polar as well as nonpolar compounds at high pressures. Buck (*Proc. Joint Meet. CIESC and AIChE*, Beijing, China, Sept. 19–22, 1982, Chemical Industry Press, Beijing, 1982, p. 75) describes an application of that equation to vapor-liquid equilibrium of a complex mixture at 250°C (482°F) and 150 bars (2200 psia) containing H_2, CO, CH_4, dimethyl ether, acetaldehyde, methyl acetate, methyl alcohol, ethyl acetate, ethyl alcohol, water, 1,4-dioxane, and acetic acid.

An alternative K-value formulation that has received wide application to mixtures containing polar and/or nonpolar compounds is

$$K_i = \gamma_i^L \Phi_i^L / \hat{\Phi}_i^V \tag{13-4}$$

where different equations of state may be used to predict the pure-component liquid fugacity coefficient Φ_i^L and the vapor-mixture fugacity coefficient, and any one of a number of mixture free-energy models may be used to obtain the liquid activity coefficient γ_i^L. At low to moderate pressures, accurate prediction of the latter is crucial to the application of Eq. (13-4).

When either Eq. (13-3) or Eq. (13-4) can be applied, the former is generally preferred because it involves only a single equation of state applicable to both phases and thus would seem to offer greater consistency. In addition, the quantity Φ_i^L in Eq. (13-4) is hypothetical for any components that are supercritical. In that case, a modification of Eq. (13-4) that uses Henry's law is sometimes applied.

For mixtures of hydrocarbons and light gases, Chao and Seader (CS) [*Am. Inst. Chem. Eng. J.*, **7**, 598 (1961)] applied Eq. (13-4) by using an empirical expression for Φ_i^L based on the generalized corresponding-states *PVT* correlation of Pitzer et al., the Redlich-Kwong equation of state for $\hat{\Phi}_i^V$, and the regular solution theory of Scatchard and Hildebrand for γ_i^L. The predictive ability of the last-named theory is exhibited in Fig. 13-19 for the heptane-toluene system at 101.3 kPa (1 atm). Five pure-component constants for each species (T_c, P_c, ω, δ, and v^L) are required to use the CS method, which when applied within the restrictions discussed by Lenoir and Koppany [*Hydrocarbon Process.*, **46**(11), 249 (1967)] gives good results. Revised coefficients of Grayson and Streed (GS) (Pap. 20-P07, Sixth World Pet. Conf., Frankfurt, June, 1963) for the Φ_i^L expression permit application of the CS correlation to higher temperatures and pressures and give improved predictions for hydrogen. Computer programs for the CS and GS methods are available from the GPA.

For mixtures containing polar substances, more complex predictive equations for γ_i^L that involve binary-interaction parameters for each pair of components in the mixture are required for use in Eq. (13-4), as discussed in Sec. 4. Four popular expressions are the Margules, van Laar, Wilson, and UNIQUAC equations. Extensive listings

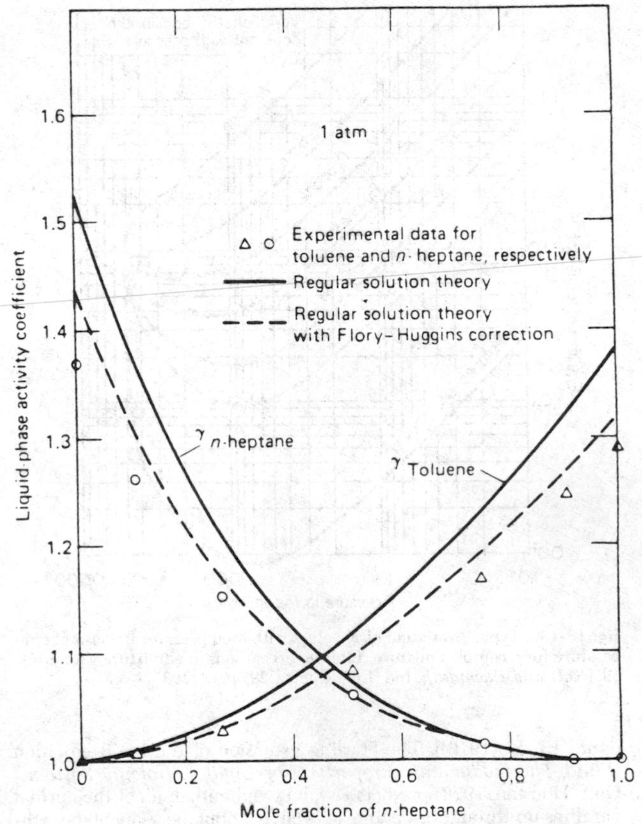

Fig. 13-19 Liquid-phase activity coefficients for an n-heptane–toluene system at 101.3 kPa (1 atm). [*Henley and Seader*, Equilibrium-Stage Separation Operations in Chemical Engineering, *Wiley, New York*, 1981; *data of Yerazunis et al.*, Am. Inst. Chem. Eng. J., *10, 660 (1964).*]

TABLE 13-2 Binary-Interaction Parameters*

System	Margules		van Laar		Wilson (cal/mol)	
	$\bar{A}_{12}$	$\bar{A}_{21}$	A_{12}	A_{21}	$(\lambda_{12} - \lambda_{11})$	$(\lambda_{21} - \lambda_{22})$
Acetone (1), chloroform (2)	−0.8404	−0.5610	−0.8643	−0.5899	116.1171	−506.8519
Acetone (1), methanol (2)	0.6184	0.5788	0.6184	0.5797	−114.4047	545.2942
Acetone (1), water (2)	2.0400	1.5461	2.1041	1.5555	344.3346	1482.2133
Carbon tetrachloride (1), benzene (2)	0.0948	0.0922	0.0951	0.0911	7.0459	59.6233
Chloroform (1), methanol (2)	0.8320	1.7365	0.9356	1.8860	−361.7944	1694.0241
Ethanol (1), benzene (2)	1.8362	1.4717	1.8570	1.4785	1264.4318	266.6118
Ethanol (1), water (2)	1.6022	0.7947	1.6798	0.9227	325.0757	953.2792
Ethyl acetate (1), ethanol (2)	0.8557	0.7476	0.8552	0.7526	58.8869	570.0439
n-Hexane (1), ethanol (2)	1.9398	2.7054	1.9195	2.8463	320.3611	2189.2896
Methanol (1), benzene (2)	2.1411	1.7905	2.1623	1.7925	1666.4410	227.2126
Methanol (1), ethyl acetate (2)	1.0016	1.0517	1.0017	1.0524	982.2689	−172.9317
Methanol (1), water (2)	0.7923	0.5434	0.8041	0.5619	82.9876	520.6458
Methyl acetate (1), methanol (2)	0.9605	1.0120	0.9614	1.0126	−93.8900	847.4348
1-Propanol (1), water (2)	2.7070	0.7172	2.9095	1.1572	906.5256	1396.6398
2-Propanol (1), water (2)	2.3319	0.8976	2.4702	1.0938	659.5473	1230.2080
Tetrahydrofuran (1), water (2)	2.8258	1.9450	3.0216	1.9436	1475.2583	1844.7926
Water (1), acetic acid (2)	0.4178	0.9533	0.4973	1.0623	705.5876	111.6579
Water (1), 1-butanol (2)	0.8608	3.2051	1.0996	4.1760	1549.6600	2050.2569
Water (1), formic acid (2)	−0.2966	−0.2715	−0.2935	−0.2757	−310.1060	1180.8040

*Abstracted from Gmehling and Onken, *Vapor-Liquid Equilibrium Data Collection*, DECHEMA Chemistry Data ser., vol. 1 (parts 1–10), Frankfurt, 1977.

TABLE 13-3 Activity-Coefficient Equations in Binary Form for Use with Parameters and Constants in Tables 13-2 and 13-4

Type of equation	Adjustable parameters	Equations in binary form
Margules	$\overline{A}_{12}$	$\ln \gamma_1 = [\overline{A}_{12} + 2(\overline{A}_{21} - \overline{A}_{12})x_1]x_2^2$
	$\overline{A}_{21}$	$\ln \gamma_2 = [\overline{A}_{21} + 2(\overline{A}_{12} - \overline{A}_{21})x_2]x_1^2$
van Laar	A_{12}	$\ln \gamma_1 = A_{12}\left(\dfrac{A_{21}x_2}{A_{12}x_1 + A_{21}x_2}\right)^2$
	A_{21}	$\ln \gamma_2 = A_{21}\left(\dfrac{A_{12}x_1}{A_{12}x_1 + A_{21}x_2}\right)^2$
Wilson	$\lambda_{12} - \lambda_{11}$	$\ln \gamma_1 = -\ln(x_1 + \Lambda_{12}x_2) + x_2\left(\dfrac{\Lambda_{12}}{x_1 + \Lambda_{12}x_2} - \dfrac{\Lambda_{21}}{\Lambda_{21}x_1 + x_2}\right)$
	$\lambda_{21} - \lambda_{22}$	$\ln \gamma_2 = -\ln(x_2 + \Lambda_{21}x_1) - x_1\left(\dfrac{\Lambda_{12}}{x_1 + \Lambda_{12}x_2} - \dfrac{\Lambda_{21}}{\Lambda_{21}x_1 + x_2}\right)$

where

$$\Lambda_{12} = \frac{v_2^L}{v_1^L}\exp\left(-\frac{\lambda_{12} - \lambda_{11}}{RT}\right) \qquad \Lambda_{21} = \frac{v_1^L}{v_2^L}\exp\left(-\frac{\lambda_{21} - \lambda_{22}}{RT}\right)$$

v_i^L = molar volume of pure-liquid component i

λ_{ij} = interaction energy between components i and j, $\lambda_{ij} = \lambda_{ji}$

of binary-interaction parameters for use in these equations and in the NRTL equation of Renon and Prausnitz [*Ind. Eng. Chem. Process Des. Dev.*, **8**, 413 (1969)] are given by Gmehling and Onken (op. cit.). They obtained the parameters for binary systems at 101.3 kPa (1 atm) from best fits of the experimental T-y-x equilibrium data by setting Φ_i^V and Φ_i^L to their ideal-gas, ideal-solution limits of 1.0 and P^{sat}/P respectively, with the vapor pressure P^{sat} given by a three-constant Antoine equation, whose values they tabulate. Table 13-2 lists their parameters for some of the binary systems included in Table 13-1, based on the binary-system activity-coefficient-equation forms given in Table 13-3. Consistent Antoine vapor-pressure constants and liquid molar volumes are listed in Table 13-4. The Wilson equation is particularly useful for systems that are highly nonideal but do not undergo phase splitting, as exemplified by the ethanol-hexane system, whose activity coefficients are shown in Fig. 13-20. For systems such as this, in which activity coefficients in dilute regions may exceed values of approximately 7.5, the van Laar equation erroneously predicts phase splitting.

Tables 13-1, 13-2, and 13-4 include data on formic acid and acetic acid, two substances that tend to dimerize in the vapor phase according to the chemical-equilibrium expression

$$K_D = P_D/P_M^2 = 10^{A+B/T} \tag{13-5}$$

where K_D is the chemical-equilibrium constant for dimerization, P_D and P_M are partial pressures of dimer and monomer respectively in torr, and T is in K. Values of A and B for the first four normal aliphatic acids are:

	A	B
Formic acid	−10.743	3083
Acetic acid	−10.421	3166
n-Propionic acid	−10.843	3316
n-Butyric acid	−10.100	3040

As shown by Marek and Standart [*Collect. Czech. Chem. Commun.*, **19**, 1074 (1954)], it is preferable to correlate and utilize liquid-phase

TABLE 13-4 Antoine Vapor-Pressure Constants and Liquid Molar Volume*

Species	Antoine constants† A	B	C	Applicable temperature region, °C	v^L, liquid molar volume, cm^3/g·mol
Acetic acid	8.02100	1936.010	258.451	18–118	57.54
Acetone	7.11714	1210.595	229.664	(−13)–55	74.05
Benzene	6.87987	1196.760	219.161	8–80	89.41
1-Butanol	7.36366	1305.198	173.427	89–126	91.97
Carbon tetrachloride	6.84083	1177.910	220.576	(−20)–77	97.09
Chloroform	6.95465	1170.966	226.232	(−10)–60	80.67
Ethanol	7.58670	1281.590	193.768	78–203	58.68
Ethanol	8.11220	1592.864	226.184	20–93	58.68
Ethyl acetate	7.10179	1244.951	217.881	16–76	98.49
Formic acid	6.94459	1295.260	218.000	36–108	37.91
n-Hexane	6.91058	1189.640	226.280	(−30)–170	131.61
Methanol	8.08097	1582.271	239.726	15–84	40.73
Methyl acetate	7.06524	1157.630	219.726	2–56	79.84
1-Propanol	8.37895	1788.020	227.438	(−15)–98	75.14
2-Propanol	8.87829	2010.320	252.636	(−26)–83	76.92
Tetrahydrofuran	6.99515	1202.290	226.254	23–100	81.55
Water	8.07131	1730.630	233.426	1–100	18.07

*Abstracted from Gmehling and Onken, *Vapor-Liquid Equilibrium Data Collection*, DECHEMA Chemistry Data ser., vol. 1 (parts 1–10), Frankfurt, 1977.

†Antoine equation is $\log P^{sat} = A - B/(T + C)$ with P^{sat} in torr and T in °C.

NOTE: To convert degrees Celsius to degrees Fahrenheit, °F = 1.8°C + 32. To convert cubic centimeters per gram-mole to cubic feet per pound-mole, multiply by 0.016.

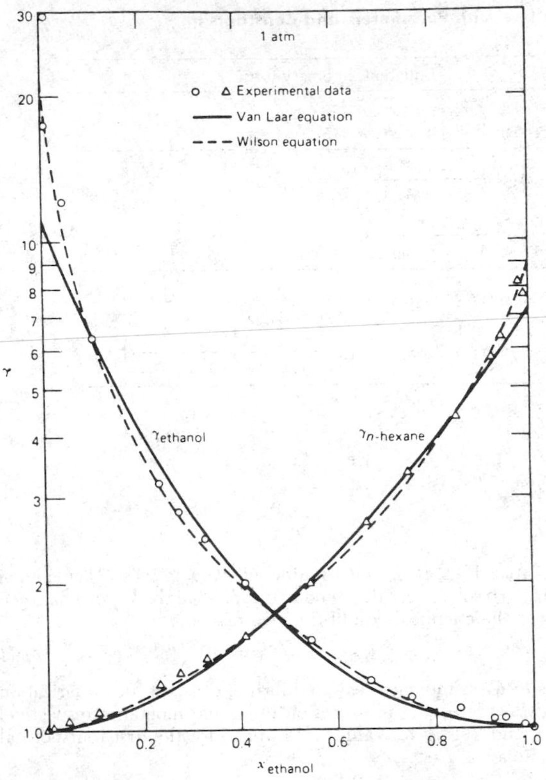

FIG. 13-20 Liquid-phase activity coefficients for an ethanol–n-hexane system. [*Henley and Seader,* Equilibrium-Stage Separation Operations in Chemical Engineering, *Wiley, New York, 1981; data of Sinor and Weber,* J. Chem. Eng. Data, *5, 243–247 (1960).*]

activity coefficients for the dimerizing component by considering separately the partial pressures of the monomer and dimer. For example, for a binary system of components 1 and 2, when only compound 1 dimerizes in the vapor phase, the following equations apply if an ideal gas is assumed:

$$P_1 = P_D + P_M \tag{13-6}$$

$$y_1 = (P_M + 2P_D)/P \tag{13-7}$$

These equations when combined with Eq. (13-5) lead to the following equations for liquid-phase activity coefficients in terms of measurable quantities:

$$\gamma_1 = \frac{Py_1}{P_1^{sat}x_1}\left\{\frac{1 + (1 + 4K_DP_1^{sat})^{0.5}}{1 + [1 + 4K_DPy_1(2 - y_1)]^{0.5}}\right\} \tag{13-8}$$

$$\gamma_2 = \frac{Py_1}{P_2^{sat}x_2}\left(\frac{2\{1 - y_1 + [1 + 4K_DPy_1(2 - y_1)]^{0.5}\}}{(2 - y_1)\{1 + [1 + 4K_DPy_1(2 - y_1)]^{0.5}\}}\right) \tag{13-9}$$

Detailed procedures, including computer programs for evaluating binary-interaction parameters from experimental data and then utilizing these parameters to predict K values and phase equilibria, are given in terms of the UNIQUAC equation by Prausnitz et al. (*Computer Calculations for Multicomponent Vapor-Liquid and Liquid-Liquid Equilibria*, Prentice-Hall, Englewood Cliffs, N.J., 1980) and in terms of the UNIFAC group contribution method by Fredenslund, Gmehling, and Rasmussen (*Vapor-Liquid Equilibria Using UNIFAC*, Elsevier, Amsterdam, 1980). Both use the method of Hayden and O'Connell [*Ind. Eng. Chem. Process Des. Dev.*, **14**, 209 (1975)] to compute $\hat{\Phi}_i^V$ in Eq. (13-4). When the system temperature is greater than the critical temperature of one or more components in the mixture, Prausnitz et al. utilize a Henry's-law constant $H_{i,M}$ in place of the product $\gamma_i^L\Phi_i^L$ in Eq. (13-4). Otherwise Φ_i^L is evaluated from vapor-pressure data with a Poynting saturated-vapor fugacity correction. When the total pressure is less than about 202.6 kPa (2 atm) and all components in the mixture have a critical temperature that is greater than the system temperature, then $\Phi_i^L = P_i^{sat}/P$ and $\hat{\Phi}_i^V = 1.0$. Equation (13-4) then reduces to

$$K_i = \gamma_i^L P_i^{sat}/P \tag{13-10}$$

which is referred to as a modified Raoult's-law K value. If, furthermore, the liquid phase is ideal, then $\gamma_i^L = 1.0$ and

$$K_i = P_i^{sat}/P \tag{13-11}$$

which is referred to as a Raoult's-law K value that is dependent solely on the vapor pressure P_i^{sat} of the components in the mixture.

DEGREES OF FREEDOM AND DESIGN VARIABLES

Definitions For separation processes, a design solution is possible if the number of independent equations equals the number of unknowns.

$$N_i = N_v - N_c$$

where N_v is the total number of variables (unknowns) involved in the process under consideration, N_c is the number of restricting relationships among the unknowns (independent equations), and N_i is termed the number of design variables. In the analogous phase-rule analysis, N_i is usually referred to as the degrees of freedom or variance. It is the number of variables that the designer must specify to define one unique operation (solution) of the process.

The variables N_i with which the designer of a separation process must be concerned are:

1. Stream concentrations (e.g., mole fractions)
2. Temperatures
3. Pressures
4. Stream flow rates
5. Repetition variables N_r

The first three are intensive variables. The fourth is an extensive variable that is not considered in the usual phase-rule analysis. The fifth is neither an intensive nor an extensive variable but is a single degree of freedom that the designer utilizes in specifying how often a particular element is repeated in a unit. For example, a distillation-column section is composed of a series of equilibrium stages, and when the designer specifies the number of stages that the section contains, he or she utilizes the single degree of freedom represented by the repetition variable ($N_r = 1.0$). If the distillation column contains more than one section (such as above and below a feed stage), the number of stages in each section must be specified and as many repetition variables exist as there are sections, that is, $N_r = 2$.

The various restricting relationships N_c can be classified as:

1. Inherent
2. Mass-balance
3. Energy-balance
4. Phase-distribution
5. Chemical-equilibrium

The inherent restrictions are usually the result of definitions and take the form of identities. For example, the concept of the equilibrium stage involves the inherent restrictions that $T^V = T^L$ and $P^V = P^L$ where the superscripts V and L refer to the equilibrium exit streams.

The mass-balance restrictions are the C balances written for the C components present in the system. (Since we will only deal with non-reactive mixtures, each chemical compound present is a phase-rule component.) An alternative is to write (C-1) component balances and one overall mass balance.

The phase-distribution restrictions reflect the requirement that $f_i^V = f_i^L$ at equilibrium where f is the fugacity. This may be expressed by Eq. (13-1). In vapor-liquid systems, it should always be recognized that all components appear in both phases to some extent and there will be such a restriction for each component in the system. In vapor-liquid-liquid systems, each component will have three such restrictions, but only two are independent. In general, when all components exist in all phases, the number of restricting relationships due to the distribution phenomenon will be $C(N_p - 1)$, where N_p is the number of phases present.

For the analysis here, the forms in which the restricting relationships are expressed are unimportant. Only the number of such restrictions is important.

Analysis of Elements An *element* is defined as part of a more complex *unit*. The unit may be all or only part of an operation or the entire *process*. Our strategy will be to analyze all elements that appear in a separation process and determine the number of design variables associated with each. The appropriate elements can then be quickly combined to form the desired units and the various units combined to form the entire process. Allowance must of course be made for the connecting streams *(interstreams)* whose variables are counted twice when elements or units are joined.

The simplest element is a *single homogeneous stream*. The variables necessary to define it are:

	N_v^e
Concentrations	$C - 1$
Temperature	1
Pressure	1
Flow rate	1
	$C + 2$

There are no restricting relationships when the stream is considered only at a point. Henley and Seader (*Equilibrium-Stage Separation Operations in Chemical Engineering*, Wiley, New York, 1981) count all C concentrations as variables, but then have to include

$$\sum_i x_i = 1.0 \quad \text{or} \quad \sum_i y_i = 1.0$$

as a restriction.

A stream divider simply splits a stream into two or more streams of the same composition. Consider Fig. 13-21, which pictures the division of the condensed overhead liquid L_c into distillate D and reflux L_{N+1}. The divider is permitted to operate nonadiabatically if desired. Three mass streams and one possible "energy stream" are involved; so

$$N_v^e = 3(C + 2) + 1 = 3C + 7$$

Each mass stream contributes $C + 2$ variables, but an energy stream has only its rate q as a variable. The independent restrictions are as follows:

	N_c^e
Inherent	
T and P identities between L_{N+1} and D	2
Concentration identities between L_{N+1} and D	$C - 1$
Mass balances	C
Energy balance	1
	$2C + 2$

The number of design variables for the element is given by

$$N_i^e = N_v^e - N_c^e = (3C + 7) - (2C + 2) = C + 5$$

Specification of the feed stream L_c(C + 2 variables), the ratio L_{N+1}/D, the "heat leak" q, and the pressure of either stream leaving the divider utilizes these design variables and defines one unique operation of the divider.

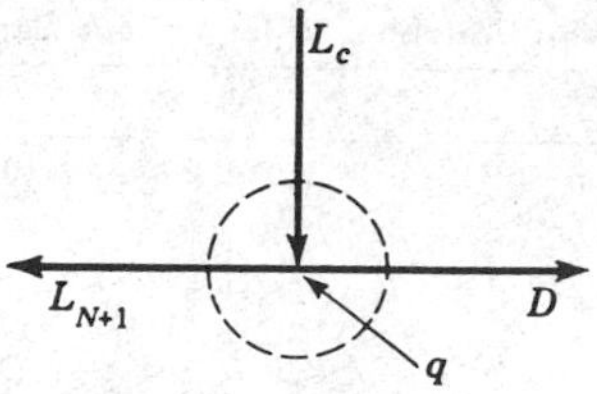

FIG. 13-21 Stream divider.

A simple equilibrium stage (no feed or sidestreams) is depicted in Fig. 13-22. Four mass streams and a heat-leak (or heat-addition) stream provide the following number of variables:

$$N_v^e = 4(C + 2) + 1 = 4C + 9$$

Vapor and liquid streams V_n and L_n respectively are in equilibrium with each other by definition and therefore are at the same T and P. These two inherent identities when added to C-component balances, one energy balance, and the C phase-distribution relationships give

$$N_c^e = 2C + 3$$

Then

$$N_i^e = N_v^e - N_c^e$$
$$= (4C + 9) - (2C + 3) = 2C + 6$$

These design variables can be utilized as follows:

Specifications	N_i^e
Specification of L_{n+1} stream	$C + 2$
Specification of V_{n-1} stream	$C + 2$
Pressure of either leaving stream	1
Heat leak q	1
	$2C + 6$

The results of the analyses for all the various elements commonly encountered in distillation processes are summarized in Table 13-5. Details of the analyses are given by Smith (*Design of Equilibrium Stage Processes*, McGraw-Hill, New York, 1967) and in a somewhat different form by Henley and Seader (op. cit).

Analysis of Units A "unit" is defined as a combination of elements and may or may not constitute the entire process. By definition

$$N_v^u = N_r + \sum_i N_i^e$$

and

$$N_i^u = N_v^u - N_c^u$$

where N_c^u refers to *new* restricting relationships (identities) that may arise when elements are combined. N_c^u does not include any of the restrictions considered in calculating the N_i^e's for the various elements. It includes only the stream identities that exist in each interstream between two elements. The interstream variables (C + 2) were counted in each of the two elements when their respective N_i^e's were calculated. Therefore, (C + 2) new restricting relationships must be counted for each interstream in the combination of elements to prevent redundancy.

The simple absorber column shown in Fig. 13-23 will be analyzed

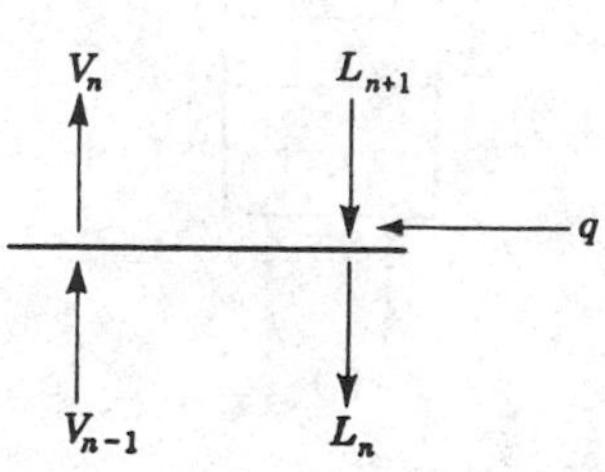

FIG. 13-22 Simple equilibrium stage.

TABLE 13-5 Design Variables N_i^e for Various Elements

Element	N_v^e	N_c^e	N_i^e
Homogeneous stream	$C + 2$	0	$C + 2$
Stream divider	$3C + 7$	$2C + 2$	$C + 5$
Stream mixer	$3C + 7$	$C + 1$	$2C + 6$
Pump	$2C + 5$	$C + 1$	$C + 4$
Heater	$2C + 5$	$C + 1$	$C + 4$
Cooler	$2C + 5$	$C + 1$	$C + 4$
Total condenser	$2C + 5$	$C + 1$	$C + 4$
Total reboiler	$2C + 5$	$C + 1$	$C + 4$
Partial condenser	$3C + 7$	$2C + 3$	$C + 4$
Partial reboiler	$3C + 7$	$2C + 3$	$C + 4$
Simple equilibrium state	$4C + 9$	$2C + 3$	$2C + 6$
Feed stage	$5C + 11$	$2C + 3$	$3C + 8$
Sidestream stage	$5C + 11$	$3C + 4$	$2C + 7$
Adiabatic equilibrium flash	$3C + 6$	$2C + 3$	$C + 3$
Nonadiabatic equilibrium flash	$3C + 7$	$2C + 3$	$C + 4$

here to illustrate the procedure. This unit consists of a series of simple equilibrium stages of the type in Fig. 13-22. Specification of the number of stages N utilizes the single repetition variable and

$$N_v^u = N_r + \sum_i N_i^e = 1 + N(2C + 6)$$

since in Table 13-5 $N_i^e = 2C + 6$ for a simple equilibrium stage. There are $2(N - 1)$ interstreams, and therefore $2(N - 1)(C + 2)$ new identities (not previously counted) come into existence when elements are combined. Subtraction of these restrictions from N_v^u gives N_i^u, the design variables that must be specified.

$$N_i^u = N_v^u - N_c^u = N_r + \sum_i N_i^e - N_c^u$$

$$= [1 + N(2C + 6)] - [2(N - 1)(C + 2)]$$

$$= 2C + 2N + 5$$

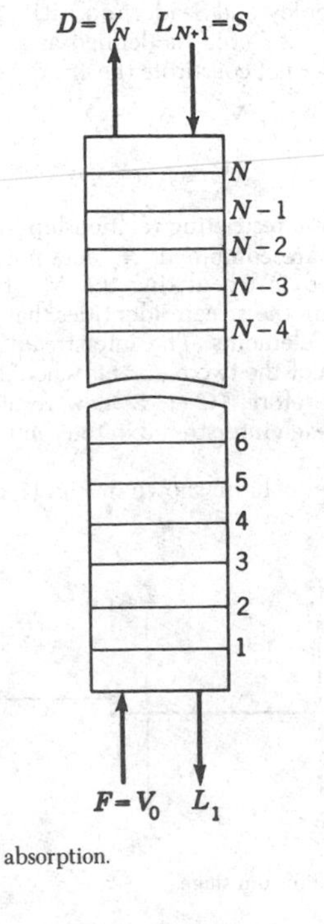

FIG. 13-23 Simple absorption.

These might be used as follows:

Specifications	N_i^u	
Two feed streams	$2C$	$+ 4$
Number of stages N		1
Pressure of either stream leaving each stage	N	
Heat leak for each stage	N	
	$2C + 2N$	$+ 5$

A more complex unit is shown in Fig. 13-24, which is a schematic diagram of a *distillation column* with one feed, a total condenser, and a partial reboiler. Dotted lines encircle the six connected elements (or units) that constitute the distillation operation. The variables N_v^u that must be considered in the analysis of the entire process are just the sum of the N_i^e's for these six elements since here $N_r = 0$. Using Table 13-5,

Element (or unit)	$N_v^u = \sum_i N_i^e$	
Total condenser	C	$+ 4$
Reflux divider	C	$+ 5$
$N - (M + 1)$ equilibrium stages	$2C + 2(N - M - 1)$	$+ 5$
Feed stage	$3C$	$+ 8$
$(M - 1)$ equilibrium stages	$2C + 2(M - 1)$	$+ 5$
Partial reboiler	C	$+ 4$
	$10C + 2N$	$+ 27$

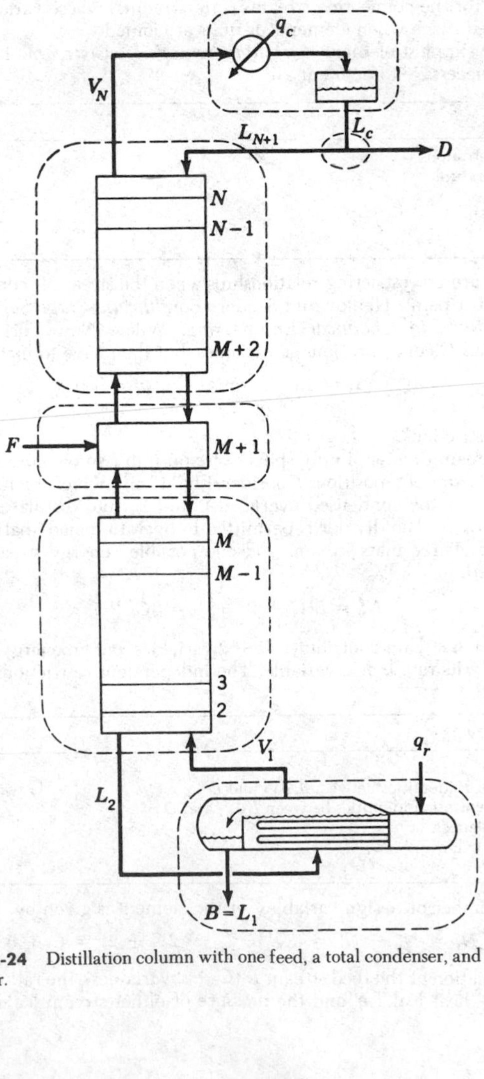

FIG. 13-24 Distillation column with one feed, a total condenser, and a partial reboiler.

Here, the two units of $N - (M + 1)$ and $(M - 1)$ stages are treated just like elements. Nine interstreams are created by the combination of elements; so

$$N_c^u = 9(C + 2) = 9C + 18$$

The number of design variables is

$$N_i^u = C + 2N + 9\ N_v^u - N_c^u = (10C + 2N + 27) - (9C + 18)$$

One set of specifications that is particularly convenient for computer solutions is:

Specifications	N_i^u
Pressure of either stream leaving each stage (including reboiler)	N
Pressure of stream leaving condenser	1
Pressure of either stream leaving reflux divider	1
Heat leak for each stage (excluding reboiler)	$N - 1$
Heat leak for reflux divider	1
Feed stream	$C + 2$
Reflux temperature	1
Total number of stages N	1
Number of stages below feed M	1
Distillate rate D/F	1
Reflux rate (L_{N+1}/D)	1
	$C + 2N + 9$

Other specifications often used in place of one or more of the last four listed are the fractional recovery of one component in either D or B and/or the concentration of one component in either D or B.

TABLE 13-6 Design Variables N_i^u for Separation Units

Unit	N_i^{u*}
Distillation (partial reboiler–total condenser)	$C + 2N + 9$
Distillation (partial reboiler–partial condenser)	$C + 2N + 6$
Absorption	$2C + 2N + 5$
Rectification (partial condenser)	$C + 2N + 3$
Stripping	$2C + 2N + 5$
Reboiled stripping (partial reboiler)	$C + 2N + 3$
Reboiled absorption (partial reboiler)	$2C + 2N + 6$
Refluxed stripping (total condenser)	$2C + 2N + 9$
Extractive distillation (partial reboiler–total condenser)	$2C + 2N + 12$

*N includes reboiler, but not condenser.

Other Units and Complex Processes In Table 13-6, the number of design variables is summarized for several distillation-type separation operations, most of which are shown in Fig. 13-7. For columns not shown in Figs. 13-1 or 13-7 that involve additional feeds and/or sidestreams, add $(C + 3)$ degrees of freedom for each additional feed ($C + 2$ to define the feed and 1 to designate the feed stage) and 2 degress of freedom for each sidestream (1 for the sidestream flow rate and 1 to designate the sidestream-stage location). Any number of elements or units can be combined to form complex processes. No new rules beyond those developed earlier are necessary for their analysis. When applied to the thermally coupled distillation process of Fig. 13-6*b*, the result is $N_i^u = 2(N + M) + C + 18$. Further examples are given in Henley and Seader (op. cit.).

SINGLE-STAGE EQUILIBRIUM-FLASH CALCULATIONS

Introduction The simplest continuous-distillation process is the adiabatic single-stage equilibrium-flash process pictured in Fig. 13-25. Feed temperature and the pressure drop across the valve are adjusted to vaporize the feed to the desired extent, while the drum provides disengaging space to allow the vapor to separate from the liquid. The expansion across the valve is at constant enthalpy, and this fact can be used to calculate T_2 (or T_1 to give a desired T_2).

From Table 13-5 it can be seen that the variables subject to the designer's control are $C + 3$ in number. The most common way to utilize these is to specify the feed rate, composition, and pressure ($C + 1$ variables) plus the drum temperature T_2 and pressure P_2. This operation will give one point on the *equilibrium-flash curve* shown in Fig. 13-26. This curve shows the relation at constant pressure between the fraction V/F of the feed flashed and the drum temperature. The temperature at $V/F = 0.0$ when the first bubble of vapor is about to form (saturated liquid) is the *bubble-point* temperature of the feed mixture, and the value at $V/F = 1.0$ when the first droplet of liquid is about to form (saturated liquid) is the *dew-point* temperature.

Bubble Point and Dew Point For a given drum pressure and feed composition, the bubble- and dew-point temperatures bracket the temperature range of the equilibrium flash. At the bubble-point temperature, the total vapor pressure exerted by the mixture becomes equal to the confining drum pressure, and it follows that $\Sigma y_i = 1.0$ in the bubble formed. Since $y_i = K_i x_i$ and since the x_i's still equal the feed concentrations (denoted by z_i's), calculation of the bubble-point temperature involves a trial-and-error search for the temperature which, at the specified pressure, makes $\Sigma K_i z_i = 1.0$. If instead the temperature is specified, one can find the bubble-point pressure that satisfies this relationship.

At the dew-point temperature y_i still equals z_i, and the relationship $\Sigma x_i = \Sigma z_i / K_i = 1.0$ must be satisfied. As in the case of the bubble point, a trial-and-error search for the dew-point temperature at a specified pressure is involved. Or, if the temperature is specified, the dew-point pressure can be calculated.

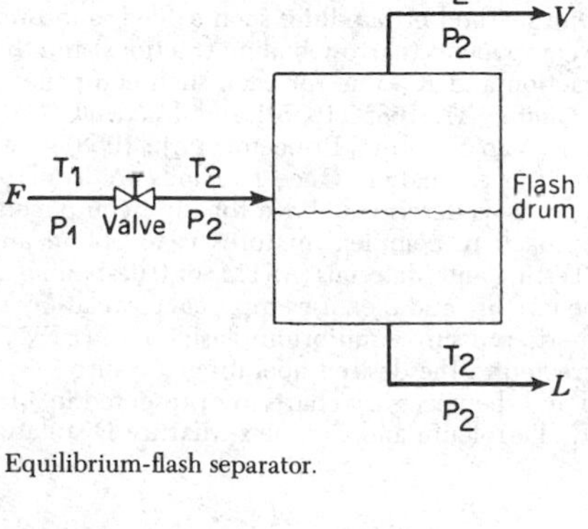

FIG. 13-25 Equilibrium-flash separator.

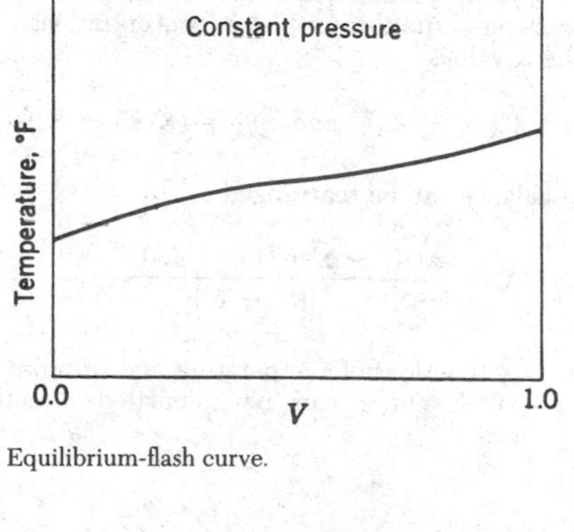

FIG. 13-26 Equilibrium-flash curve.

Isothermal Flash The calculation for a point on the flash curve that is intermediate between the bubble point and the dew point is referred to as an isothermal-flash calculation because T_2 is specified. Except for an ideal binary mixture, procedures for calculating an isothermal flash are iterative. A popular method is the following due to Rachford and Rice [*J. Pet. Technol.*, 4(10), sec. 1, p. 19, and sec. 2, p. 3 (October 1952)]. Taking a basis of F = 1.0 mol, the component mole balance ($Fz_i = Vy_i + Lx_i$), phase-distribution relation ($K_i = y_i/x_i$), and total mole balance ($F = V + L$) can be combined to give

$$x_i = \frac{z_i}{1 + V(K_i - 1)} \qquad (13\text{-}12)$$

$$y_i = \frac{K_i z_i}{1 + V(K_i - 1)} \qquad (13\text{-}13)$$

Since $\Sigma x_i - \Sigma y_i = 0$,

$$f\{V\} = \sum_i \frac{z_i(1 - K_i)}{1 + V(K_i - 1)} = 0 \qquad (13\text{-}14)$$

Equation (13-14) is solved iteratively for V, followed by the calculation of values of x_i and y_i from Eqs. (13-12) and (13-13) and L from the total mole balance. Any one of a number of numerical root-finding procedures such as the Newton-Raphson, secant, false-position, or bisection method can be used to solve Eq. (13-14). Values of K_i are constants if they are independent of liquid and vapor compositions. Then the resulting calculations are straightforward. Otherwise, the K_i values must be periodically updated for composition effects, perhaps after each iteration, using prorated values of x_i and y_i from Eqs. (13-12) and (13-13). Generally, the iterations are continued until the calculated value of V equals to within ± 0.0005 the value of V that was used to initiate that iteration. When converged, Σx_i and Σy_i will each be very close to a value of 1, and, if desired, T_1 can be computed from an energy balance around the valve if no heat exchanger is used. Alternatively, if T_1 is fixed as mentioned earlier, a heat exchanger must be added before, after, or in place of the valve with the required heat duty being calculated from an energy balance. The limits of applicability of Eqs. (13-12) to (13-14) are the bubble point, at which $V = 0$ and $x_i = z_i$, and the dew point, at which $L = 0$ and $y_i = z_i$, at which Eq. (13-2) reduces to the bubble-point equation

$$\sum_i K_i x_i = 1 \qquad (13\text{-}15)$$

and the dew-point equation

$$\sum_i \frac{y_i}{K_i} = 1 \qquad (13\text{-}16)$$

For a *binary feed*, specification of the flash-drum temperature and pressure fixes the equilibrium-phase concentrations, which are related to the K values by

$$x_1 = (1 - K_2)/(K_1 - K_2) \quad \text{and} \quad y_1 = (K_1K_2 - K_1)/(K_2 - K_1)$$

The mole balance can be rearranged to

$$V = \frac{z_1(K_1 - K_2)/(1.0 - K_2) - 1.0}{K_1 - 1.0}$$

If K_1 and K_2 are functions of temperature and pressure only (ideal solutions), the flash curve can be calculated directly without iteration.

Adiabatic Flash In Fig. 13-25, if P_2 and the feed-stream conditions (i.e., F, z_i, T_1, P_1) are known, then the calculation of T_2, V, L, y_i, and x_i is referred to as an adiabatic flash. In addition to Eqs. (13-12) to (13-14) and the total mole balance, the following energy balance around both the valve and the flash drum combined must be included:

$$H^F F = H^V V + H^L L \qquad (13\text{-}17)$$

Taking a basis of F = 1.0 mol and eliminating L with the total mole balance, Eq. (13-17) becomes

$$f_2\{V, T_2\} = H^F - V(H^V - H^L) - H^L = 0 \qquad (13\text{-}18)$$

With T_2 now unknown, Eq. (13-17) becomes

$$f_1\{V, T_2\} = \sum_i \frac{z_i(1 - K_i)}{1 + V(K_i - 1)} = 0 \qquad (13\text{-}19)$$

A number of iterative procedures have been developed for solving Eqs. (13-18) and (13-19) simultaneously for V and T_2. Frequently, and especially if the feed contains components of a narrow range of volatility, convergence is rapid for a tearing method in which a value of T_2 is assumed, Eq. (13-19) is solved iteratively by the isothermal-flash procedure, and, using that value of V, Eq. (13-18) is solved iteratively for a new approximation of T_2, which is then used to initiate the next cycle until T_2 and V converge. However, if the feed contains components of a wide range of volatility, it may be best to invert the sequence and assume a value for V, solve Eq. (13-19) for T_2, solve Eq. (13-18) for V, and then repeat the cycle. If K values and/or enthalpies are sensitive to the unknown phase compositions, it may be necessary simultaneously to solve Eqs. (13-18) and (13-19) by a Newton or other suitable iterative technique. Alternatively, the two-tier method of Boston and Britt [*Comput. Chem. Eng.*, **2**, 109 (1978)], which is also suitable for difficult isothermal-flash calculations, may be applied.

Other Flash Specifications Flash-drum specifications in addition to (P_2, T_2) and (P_2, adiabatic) are also possible but must be applied with care to avoid impossible situations. For example, the K and H, Mod II computer program of the GPA allows the user to specify also (P_2, V/F) and (T_2, V/F).

Three-Phase Flash Single-stage equilibrium-flash calculations become considerably more complex when an additional liquid phase can form, as from mixtures of water with hydrocarbons. Procedures for computing such situations are referred to as three-phase flash methods, which are given for the general case by Henley and Rosen (*Material and Energy Balance Computations*, Wiley, New York, 1968, chap. 8). When the two liquid phases are almost mutually insoluble, they can be considered separately and relatively simple procedures apply as discussed by Smith (*Design of Equilibrium Stage Processes*, McGraw-Hill, New York, 1963). Condensation of such mixtures may result in one liquid phase being formed before the other.

Complex Mixtures Feed analyses in terms of component concentrations are usually not available for complex hydrocarbon mixtures with a final normal boiling point above about 38°C (100°F) (*n*-pentane). One method of handling such a feed is to break it down into pseudo components (narrow-boiling fractions) and then estimate the mole fraction and K value for each such component. Edmister [*Ind. Eng. Chem.*, **47**, 1685 (1955)] and Maxwell (*Data Book on Hydrocarbons*, Van Nostrand, Princeton, N.J., 1958) give charts that are useful for this estimation. Once K values are available, the calculation proceeds as described above for multicomponent mixtures. Another approach to complex mixtures is to obtain an American Society for Testing and Materials (ASTM) or true-boiling point (TBP) curve for the mixture and then use empirical correlations to construct the atmospheric-pressure equilibrium-flash curve (EFV), which can then be corrected to the desired operating pressure. A discussion of this method and the necessary charts are presented in a later subsection entitled "Petroleum and Complex-Mixture Distillation."

GRAPHICAL METHODS FOR BINARY DISTILLATION

INTRODUCTION

Multistage distillation under continuous, steady-state operating conditions is widely used in practice to separate a variety of mixtures. Table 13-7, taken from the study of Mix, Dweck, Weinberg, and Armstrong [*Am. Inst. Chem. Eng. J. Symp. Ser. 76*, **192,** 10 (1980)] lists key components for 27 industrial distillation processes. The design of multistage columns can be accomplished by graphical techniques when the feed mixture contains only two components. The *x-y* diagram [McCabe and Thiele, *Ind. Eng. Chem.*, **17,** 605 (1925)] utilizes only equilibrium and mole-balance relationships but approaches rigorousness only for those systems in which energy effects on vapor and liquid rates leaving the stages are negligible. The enthalpy-concentration diagram [Ponchon, *Tech. Mod.*, **13,** 20, 55 (1921); and Savarit, *Arts Metiers*, 65, 142, 178, 241, 266, 307 (1922)] utilizes the energy balance also and is rigorous when enough calorimetric data are available to construct the diagram without assumptions.

The availability of computers has decreased our reliance on graphical methods. Nevertheless, diagrams are useful for quick approximations and for demonstrating the effect of various design variables. The *x-y* diagram is the most convenient for these purposes, and its use is developed in detail here. The enthalpy-concentration diagram is given by Smith (*Design of Equilibrium Stage Processes*, McGraw-Hill, New York, 1963) and Henley and Seader (*Equilibrium-Stage Separation Operations in Chemical Engineering*, Wiley, New York, 1981).

PHASE EQUILIBRIUM DATA

Three types of binary equilibrium curves are shown in Fig. 13-27. The *y-x* diagram is almost always plotted for the component that is the more volatile (denoted by the subscript 1) in the region where distillation is to take place. Curve *A* shows the most usual case, in which component 1 remains more volatile over the entire composition range. Curve *B* is typical of many systems (ethanol-water, for example) in which the component that is more volatile at low values of x_1 becomes less volatile than the other component at high values of x_1. The vapor and liquid compositions are identical at the azeotrope where curve *B* crosses the 45° diagonal. A heterogeneous azeotrope is formed with two liquid phases by curve *C*.

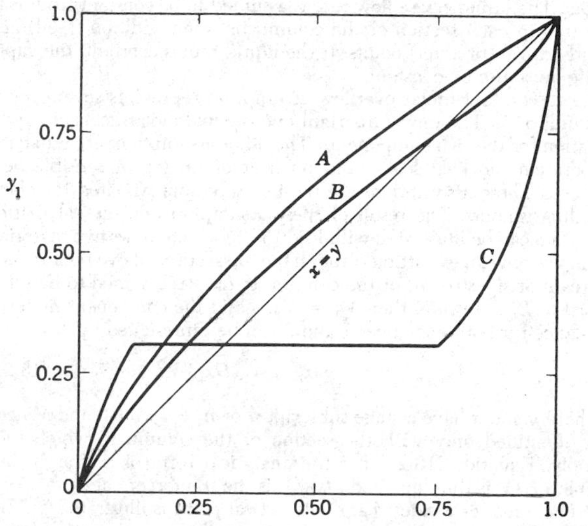

FIG. 13-27 Typical binary equilibrium curves. Curve *A*, system with normal volatility. Curve *B*, system with homogeneous azeotrope (one liquid phase). Curve *C*, system with heterogeneous azeotrope (two liquid phases in equilibrium with one vapor phase).

TABLE 13-7 Key Components for Distillation Processes of Industrial Importance

Key components	Typical number of trays
Hydrocarbon systems	
Ethylene-ethane	73
Propylene-propane	138
Propyne–1–3-butadiene	40
1–3 Butadiene–vinyl acetylene	130
Benzene-toluene	34, 53
Benzene–ethyl benzene	20
Benzene–diethyl benzene	50
Toluene–ethyl benzene	28
Toluene-xylenes	45
Ethyl benzene–styrene	34
o-Xylene–*m*-xylene	130
Organic systems	
Methanol-formaldehyde	23
Dichloroethane-trichloroethane	30
Acetic acid–acetic anhydride	50
Acetic anhydride–ethylene diacetate	32
Vinyl acetate–ethyl acetate	90
Ethylene glycol–diethylene glycol	16
Cumene-phenol	38
Phenol-acetophenone	39, 54
Aqueous systems	
HCN-water	15
Acetic acid–water	40
Methanol-water	60
Ethanol-water	60
Isopropanol-water	12
Vinyl acetate–water	35
Ethylene oxide–water	50
Ethylene glycol–water	16

An azeotrope limits the separation that can be obtained between components by simple distillation. For the system described by curve *B*, the maximum overhead-product concentration that could be obtained from a feed with $x_1 = 0.25$ is the azeotropic composition. Similarly, a feed with $x_1 = 0.9$ could produce a bottom-product composition no lower than the azeotrope.

The phase rule permits only two variables to be specified arbitrarily in a binary two-phase system at equilibrium. Consequently, the curves in Fig. 13-27 can be plotted at either constant temperature or constant pressure but not both. The latter is more common, and data in Table 13-1 are for that case. The *y-x* diagram can be plotted in either mole, weight, or volume fractions. The units used later for the phase flow rates must, of course, agree with those used for the equilibrium data. Mole fractions, which are almost always used, are applied here.

It is sometimes permissible to assume constant *relative volatility* in order to approximate the equilibrium curve quickly. Then by applying Eq. (13-2) to components 1 and 2,

$$\alpha = K_1/K_2 = y_1x_2/x_1y_2$$

which, since $x_2 = 1 - x_1$ and $y_2 = 1 - y_1$, can be rewritten as

$$y_1 = \frac{x_1\alpha}{1 + (\alpha - 1)x_1} \tag{13-20}$$

for use in calculating points for the equilibrium curve.

McCABE-THIELE METHOD

Operating Lines The McCabe-Thiele method is based upon representation of the material-balance equations as operating lines on the y-x diagram. The lines are made straight (and the need for the energy balance obviated) by the assumption of *constant molar overflow*. The liquid-phase flow rate is assumed to be constant from tray to tray in each section of the column between addition (feed) and withdrawal (product) points. If the liquid rate is constant, the vapor rate must also be constant.

The constant-molar-overflow assumption represents several prior assumptions. The most important one is equal molar heats of vaporization for the two components. The other assumptions are adiabatic operation (no heat leaks) and no heat of mixing or sensible heat effects. These assumptions are most closely approximated for close-boiling isomers. The result of these assumptions on the calculation method can be illustrated with Fig. 13-28, which shows two material-balance envelopes cutting through the top section (above the top feed stream or sidestream) of the column. If L_{n+1} is assumed to be identical to L_{n-1} in rate, then $V_n = V_{n-2}$ and the component material balance for both envelopes 1 and 2 can be represented by

$$y_n = (L/V)x_{n+1} + (Dx_D/V) \tag{13-21}$$

where y and x have a stage subscript n or $n + 1$, but L and V need be identified only with the section of the column to which they apply. Equation (13-21) has the analytical form of a straight line where L/V is the slope and Dx_D/V is the y intercept at $x_1 = 0$.

The effect of a sidestream withdrawal point is illustrated by Fig. 13-29. The material-balance equation for the column section below the sidestream is

$$y_n = \frac{L'}{V'}x_{n+1} + \frac{Dx_D + Sx_S}{V'} \tag{13-22}$$

where the primes designate the L and V below the sidestream. Since the sidestream must be a saturated phase, $V = V'$ if a liquid side stream is withdrawn and $L = L'$ if it is a vapor.

If the sidestream in Fig. 13-29 had been a feed, the balance for the section below the feed would be

$$y_n = \frac{L'}{V'}x_{n+1} + \frac{Dx_D - Fx_F}{V'} \tag{13-23}$$

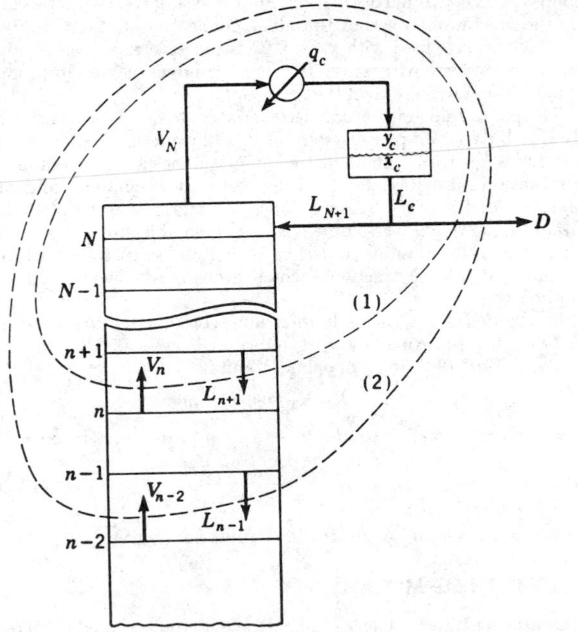

FIG. 13-28 Two material-balance envelopes in the top section of a distillation column.

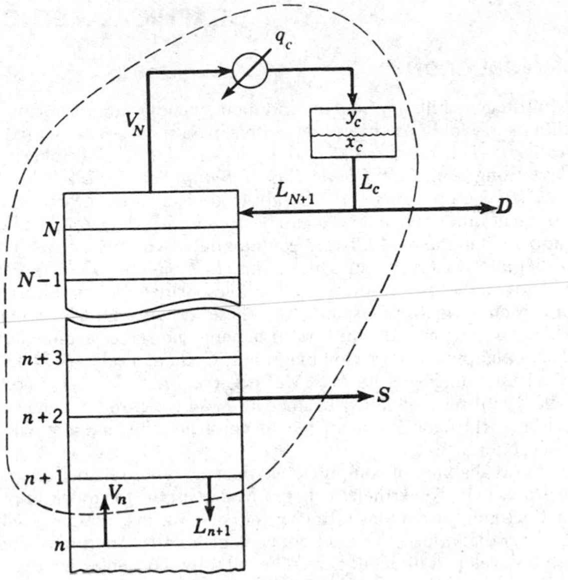

FIG. 13-29 Material-balance envelope which contains two external streams D and S, where S represents a sidestream product withdrawn above the feed plate.

Similar equations can be written for the bottom section of the column. For the envelope shown in Fig. 13-30,

$$y_m = (L''/V'')x_{m+1} - (Bx_B/V) \tag{13-24}$$

where the subscript m is used to identify the stage number in the bottom section.

Equations such as (13-21) through (13-24) when plotted on the y-x diagram furnish a set of *operating lines*. A point on an operating line represents two *passing streams*, and the operating line itself is

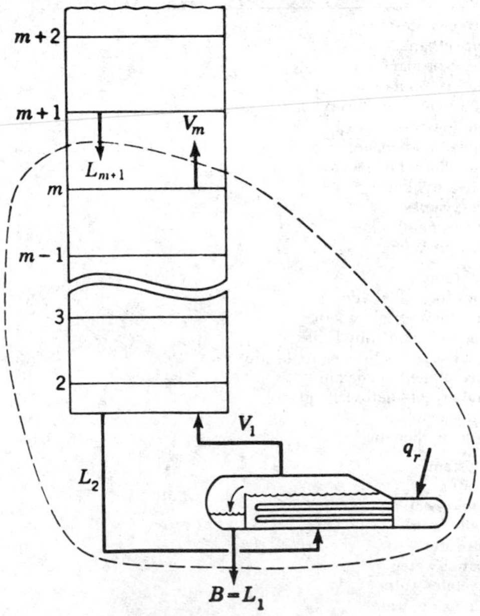

FIG. 13-30 Material-balance envelope around the bottom end of the column. The partial reboiler is equilibrium stage 1.

the locus of all possible pairs of passing streams within the column section to which the line applies.

An operating line can be located on the y-x diagram if (1) two points on the line are known or (2) one point and the slope are known. The known points on an operating line are usually its intersection with the y-x diagonal and/or its intersection with another operating line.

The slope L/V of the operating line is termed the *internal-reflux ratio*. This ratio in the operating-line equation for the top section of the column [see Eq. (13-21)] is related to the *external-reflux ratio* $R = L_{N+1}/D$ by

$$\frac{L}{V} = \frac{L_{N+1}}{V_N} = \frac{RD}{(1+R)D} = \frac{R}{1+R} \qquad (13\text{-}25)$$

when the reflux stream L_{N+1} is a saturated liquid.

Thermal Condition of the Feed The slope of the operating line changes whenever a feed stream or a sidestream is passed. To calculate this change, it is convenient to introduce a quantity q which is defined by the following equations for a feed stream F:

$$L' = L + qF \qquad (13\text{-}26)$$

$$V = V' + (1 - q)F \qquad (13\text{-}27)$$

The primes denote the streams below the stage to which the feed is introduced. The q is a measure of the thermal condition of the feed and represents the moles of saturated liquid formed in the feed stage per mole of feed. It takes on the following values for various possible feed thermal conditions.

Subcooled-liquid feed: $q > 1$
Saturated-liquid feed: $q = 1$
Partially flashed feed: $1 > q > 0$
Saturated-vapor feed: $q = 0$
Superheated-vapor feed: $q < 0$

The q value for a particular feed can be estimated from

$$q = \frac{\text{energy to convert 1 mol of feed to saturated vapor}}{\text{molar heat of vaporization}}$$

Equations analogous to (13-26) and (13-27) can be written for a sidestream, but the q will be either 1 or 0 depending upon whether the sidestream is taken from the liquid or the vapor stream.

The q can be used to derive the "q-line equation" for a feed stream or a sidestream. The q line is the locus of all points of intersection of the two operating lines, which meet at the feed-stream or sidestream stage. This intersection must occur along that section of the q line between the equilibrium curve and the $y = x$ diagonal. At the point of intersection, the same y, x point must satisfy both the operating-line equation above the feed-stream (or sidestream) stage and the one below the feed-stream (or sidestream) stage. Subtracting one equation from the other gives for a feed stage

$$(V - V')y = (L - L')x + Fx_F$$

which when combined with Eqs. (13-26) and (13-27) gives the q-line equation

$$y = \frac{q}{q-1}x - \frac{x_F}{q-1} \qquad (13\text{-}28)$$

A q-line construction for a partially flashed feed is given in Fig. 13-31. It is easily shown that the q line must intersect the diagonal at x_F. The slope of the q line is $q/(q-1)$. All five q-line cases are shown in Fig. 13-32.

The derivation of Eq. (13-28) assumes a single-feed column and no sidestream. However, the same result is obtained for other column configurations. Typical q-line constructions for sidestream stages are shown in Fig. 13-33. Note that the q line for a sidestream must always intersect the diagonal at the composition (y_s or x_s) of the sidestream.

Figure 13-33 also shows the intersections of the operating lines with the diagonal construction line. The top operating line must always intersect the diagonal at the overhead-product composition x_D. This can be shown by substituting $y = x$ in Eq. (13-21) and using $V - L = D$ to reduce the resulting equation to $x = x_D$. Similarly

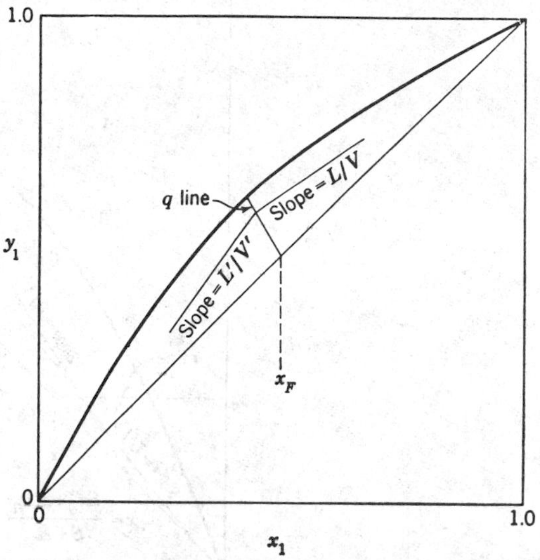

FIG. 13-31 Typical intersection of the two operating lines at the q line for a feed stage. The q line shown is for a partially flashed feed.

(except for columns in which open steam is introduced at the bottom), the bottom operating line must always intersect the diagonal at the bottom-product composition x_B.

Equilibrium-Stage Construction The alternate use of the equilibrium curve and the operating line to "step off" equilibrium stages is illustrated in Fig. 13-34. The plotted portions of the equilibrium curve (curved) and the operating line (straight) cover the composition range existing in the column section shown in the lower right-hand corner. If y_n and x_n represent the compositions (in terms of the more volatile component) of the equilibrium vapor and liquid leaving stage n, then point (y_n, x_n) on the equilibrium curve must represent the equilibrium stage n. The operating line is the locus for compositions of all possible pairs of passing streams within the section, and therefore a horizontal line (dotted) at y_n must pass through the point (y_n, x_{n+1}) on the operating line since y_n and x_{n+1} represent passing streams. Likewise, a vertical line (dashed) at x_n must intersect the operating line at point (y_{n-1}, x_n). The equilibrium stages above and below stage n can be located by a vertical line through (y_n, x_{n+1}) to find (y_{n+1},x_{n+1}) and a horizontal line through (y_{n-1}, x_n) to find (y_{n-1}, x_{n-1}). It can be seen that one can work upward or down-

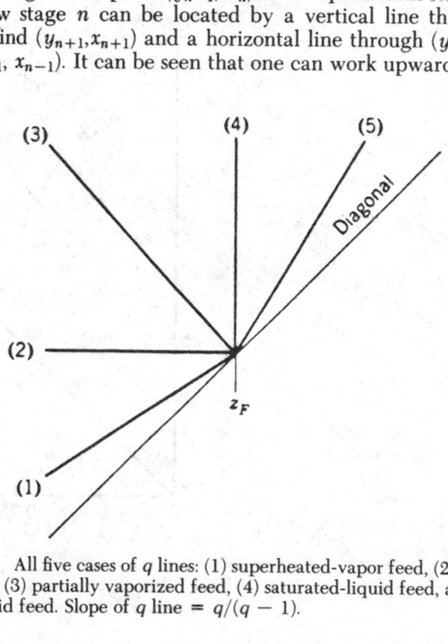

FIG. 13-32 All five cases of q lines: (1) superheated-vapor feed, (2) saturated-vapor feed, (3) partially vaporized feed, (4) saturated-liquid feed, and (5) subcooled-liquid feed. Slope of q line $= q/(q-1)$.

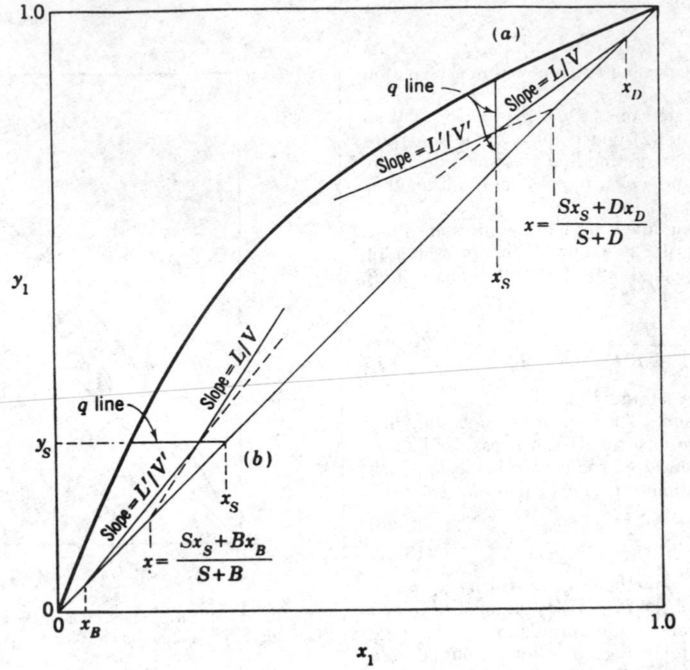

FIG. 13-33 Typical construction for a sidestream showing the intersection of the two operating lines with the q line and with the $x = y$ diagonal. (*a*) Liquid sidestream near the top of the column. (*b*) Vapor sidestream near the bottom of the column.

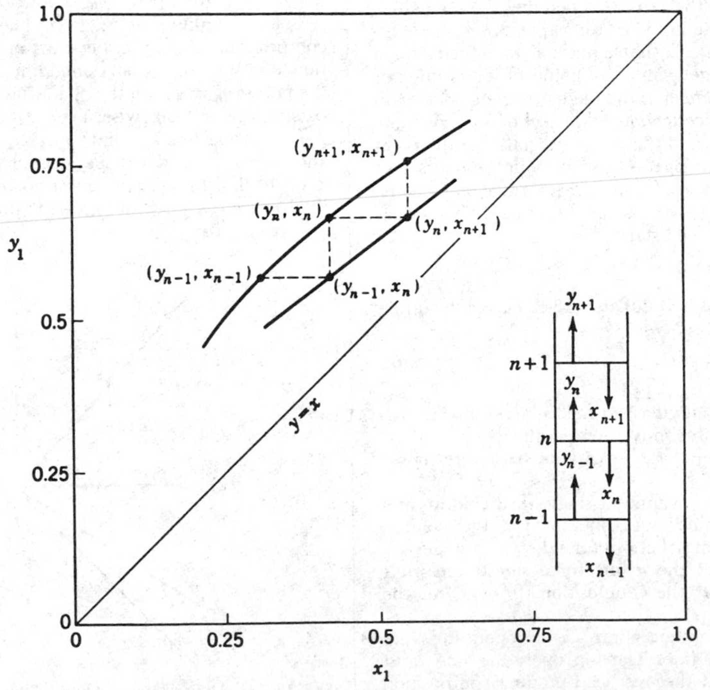

FIG. 13-34 Illustration of how equilibrium stages can be located on the x-y diagram through the alternating use of the equilibrium curve and the operating line.

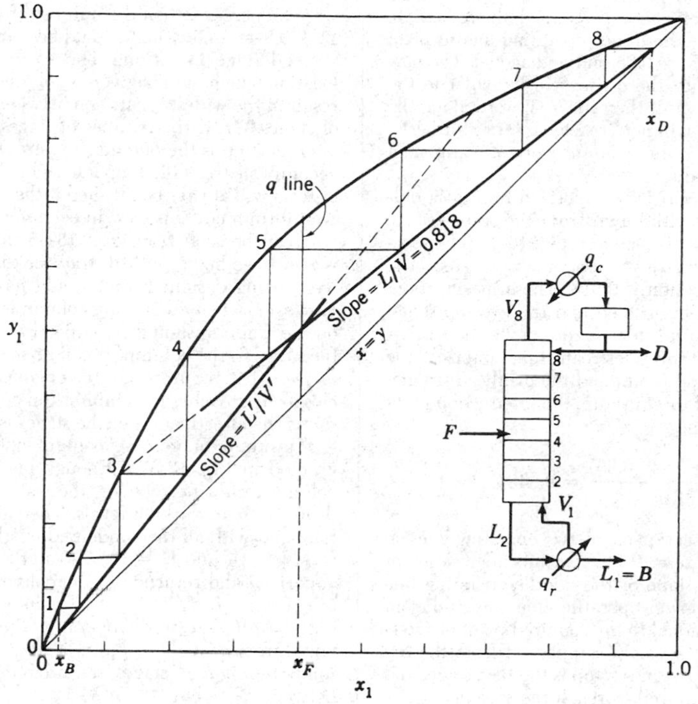

FIG. 13-35 Construction for a column with a bubble-point feed, a total condenser, and a partial reboiler.

ward through the column by alternating the use of equilibrium and operating lines.

Total-Column Construction The graphical construction for an entire column is shown in Fig. 13-35. The process, pictured in the lower right-hand corner of the diagram, is an existing column with a number of actual trays equivalent to eight equilibrium stages. A partial reboiler (equivalent to an equilibrium stage) and a total condenser are used. This column configuration was analyzed earlier (see Fig. 13-24) and shown to have $C + 2N + 9$ design variables (degrees of freedom) which must be specified to define one unique operation. These may be used as follows as the basis for a graphical solution:

Specifications		N_i^u
Stage pressures (including reboiler)		N
Condenser pressure		1
Stage heat leaks (except reboiler)		$N - 1$
Pressure and heat leak in reflux divider		2
Feed stream	C	$+ 2$
Feed-stage location		1
Total number of stages N		1
One overhead purity		1
Reflux temperature		1
External-reflux ratio		1
		$C + 2N + 9$

Pressures can be specified at any level below the safe working pressure of the column. The condenser pressure will be set at 275.8 kPa (40 psia), and all pressure drops within the column will be neglected. The equilibrium curve in Fig. 13-35 represents data at that pressure. All heat leaks will be assumed to be zero. The feed composition is 40 mole percent of the more volatile component 1, and the feed rate is 0.126 (kg·mol)/s [1000 (lb·mol)/h] of saturated liquid ($q = 1$). The feed-stage location is fixed at stage 4 and the total number of stages at eight.

The overhead purity is specified as $x_D = 0.95$. The reflux temperature is the bubble-point temperature (saturated reflux), and the external-reflux ratio is set at $R = 4.5$.

Answers are desired to the following two questions. First, what bottom-product composition x_B will the column produce under these specifications? Second, what will be the top vapor rate V_N in this operation, and will it exceed the maximum vapor-rate capacity for this column, which is assumed to be 0.252 (kg·mol)/s [2000 (lb·mol)/h] at the top-tray conditions?

The solution is started by using Eq. (13-25) to convert the external-reflux ratio of 4.5 to an internal-reflux ratio of $L/V = 0.818$. The $x_D = 0.95$ value is then located on the diagonal, and the upper operating line is drawn as shown in Fig. 13-35.

If the x_B value were known, the bottom operating line could be immediately drawn from the x_B value on the diagonal up to its required intersection point with the upper operating line on the feed q line. In this problem, since the number of stages is fixed, the x_B which gives a lower operating line that will require exactly eight stages must be found by trial and error. An x_B value is assumed, and the resulting lower operating line is drawn. The stages can be stepped off by starting from either x_B or x_D; x_B was used in this case.

Note that the lower operating line is used until the fourth stage is passed, at which time the construction switches to the upper operating line. This is necessary because the vapor and liquid streams passing each other between the fourth and fifth stages must fall on the upper line.

The x_B that requires exactly eight equilibrium stages is $x_1 = 0.026$. An overall component balance gives $D = 0.051$ (kg·mol)/s [405 (lb·mol)/h]. Then,

$$V_N = V_8 = L_{N+1} + D = D(R + 1) = 0.051(4.5 + 1.0)$$
$$= 0.280 \text{ (kg·mol)/s [2230 (lb·mol)/h]}$$

which exceeds the column capacity of 0.252 (kg·mol)/s [2007 (lb·mol)/h]. This means that the column cannot provide an overhead-product yield of 40.5 percent at 95 percent purity. Either the purity specification must be reduced, or we must be satisfied with a lower yield. If the $x_D = 0.95$ specification is retained, the reflux rate must be reduced. This will cause the upper operating line to pivot upward around its fixed point of $x = 0.95$ on the diagonal. The new intersection of the upper line with the q line will lie closer to the equilibrium curve. The x_B value must then move upward along the diagonal because the eight stages will not "reach" as far as formerly. The higher x_B concentration will reduce the recovery of component 1 in the 95 percent overhead product.

Another entire column with a partially vaporized feed, a liquid-sidestream rate equal to D and withdrawn from the second stage from the top, and a total condenser is shown in Fig. 13-36. The specified concentrations are $x_F = 0.40$, $x_B = 0.05$, and $x_D = 0.95$. The specified L/V ratio in the top section is 0.818. These specifications permit the top operating line to be located and the two top stages stepped off to determine the liquid-sidestream composition $x_S = 0.746$. The operating line below the sidestream must intersect the diagonal at the "blend" of the sidestream and the overhead stream. Since S was specified to be equal to D in rate, the intersection point is

$$x = \frac{(1.0)(0.746) + (1.0)(0.95)}{1.0 + 1.0} = 0.848$$

This point plus the point of intersection of the two operating lines on the sidestream q line (vertical at $x_S = 0.746$) permits the location of the middle operating line. (The slope of the middle operating line could also have been used.) The lower operating line must run from the specified x_B value on the diagonal to the required point of intersection on the feed q line. The stages are stepped off from the top down in this case. The sixth stage from the top is the feed stage, and a total of about 11.4 stages is required to reach the specified $x_B = 0.05$.

Fractional equilibrium stages have meaning. The 11.4 will be divided by a tray efficiency, and the rounding to an integral number of actual trays should be done after that division. For example, if the average tray efficiency for the process being modeled in Fig. 13-36 were 80 percent, then the number of actual trays required would be $11.4/0.8 = 14.3$, which would be rounded to 15.

Feed-Stage Location The *optimum* feed-stage location is that location which, with a given set of other operating specifications, will result in the widest separation between x_D and x_B for a given number of stages. Or, if the number of stages is not specified, the optimum feed location is the one that requires the lowest number of stages to accomplish a specified separation between x_D and x_B. Either of these criteria will always be satisfied if the operating line farthest from the equilibrium curve is used in each step as in Fig. 13-35.

It can be seen from Fig. 13-35 that the optimum feed location would have been the fifth tray for that operation. If a new column were being designed, that would have been the designer's choice. However, when an existing column is being modeled, the feed stage on the diagram should correspond as closely as possible to the actual feed tray in the column. It can be seen that a badly mislocated feed (a feed that requires one to remain with an operating line until it closely approaches the equilibrium curve) can be very wasteful insofar as the effectiveness of the stages is concerned.

Minimum Stages A column operating at total reflux is diagramed in Fig. 13-37*a*. Enough material has been charged to the column to fill the reboiler, the trays, and the overhead condensate drum to their working levels. The column is then operated with no feed and with all the condensed overhead stream returned as reflux ($L_{N+1} = V_N$ and $D = 0$). Also all the liquid reaching the reboiler is vaporized and returned to the column as vapor. Since F, D, and B are all zero, $L_{n+1} = V_n$ at all points in the column. With a slope of unity ($L/V = 1.0$), the operating line must coincide with the diagonal throughout the column. Total-reflux operation gives the minimum number of stages required to effect a specified separation between x_B and x_D.

Minimum Reflux The minimum-reflux ratio is defined as that ratio which if decreased by an infinitesimal amount would require an infinite number of stages to accomplish a specified separation between two components. The concept has meaning only if a separation between two components is specified and the number of stages

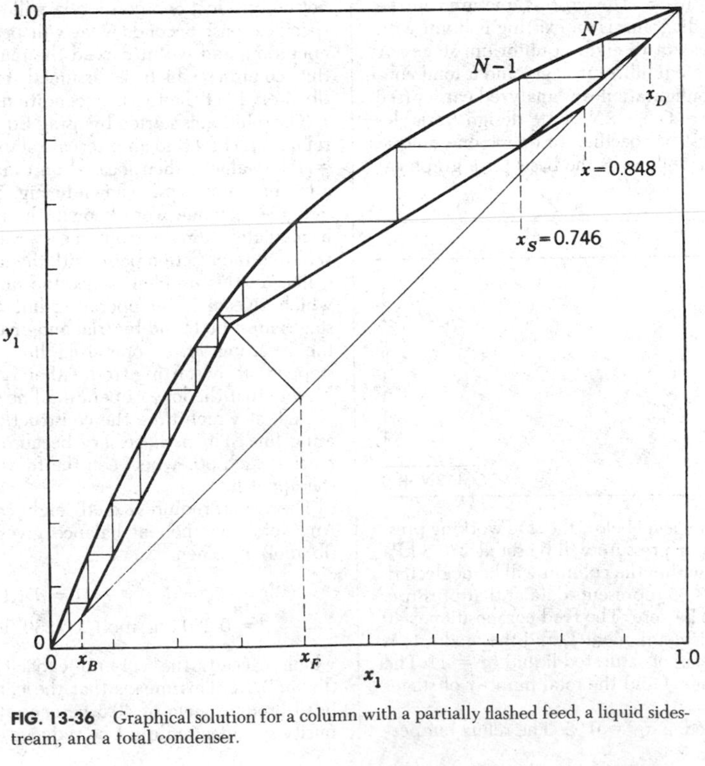

FIG. 13-36 Graphical solution for a column with a partially flashed feed, a liquid sidestream, and a total condenser.

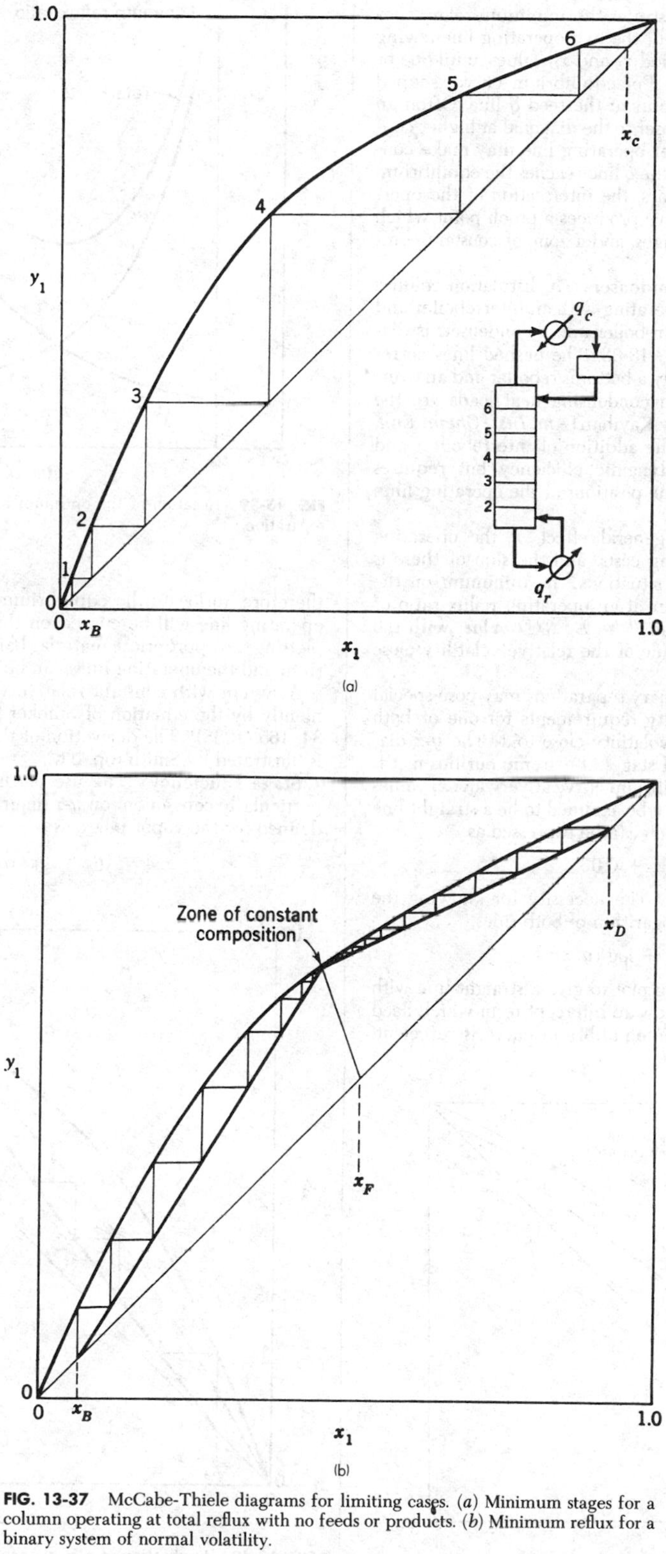

FIG. 13-37 McCabe-Thiele diagrams for limiting cases. (*a*) Minimum stages for a column operating at total reflux with no feeds or products. (*b*) Minimum reflux for a binary system of normal volatility.

is not specified. Figure 13-37*b* illustrates the minimum-reflux condition. As the reflux ratio is reduced, the two operating lines swing upward, pivoting around the specified x_B and x_D values, until one or both touch the equilibrium curve. For equilibrium curves shaped like the one shown, the contact occurs at the feed q line. Often an equilibrium curve will dip down closer to the diagonal at higher concentrations. In such cases, the upper operating line may make contact before its intersection point on the q line reaches the equilibrium curve. Wherever the contact appears, the intersection of the operating line with the equilibrium curve produces a pinch point which contains a very large number of stages, and a zone of constant composition is formed.

Intermediate Reboilers and Condensers A distillation column of the type shown in Fig. 13-2*a*, operating with an interreboiler and an intercondenser in addition to a reboiler and a condenser, is diagramed with the solid lines in Fig. 13-38. The dashed lines correspond to simple distillation with only a bottoms reboiler and an overhead condenser. Total boiling and condensing heat loads are the same for both columns. As shown by Kayihan [*Am. Inst. Chem. Eng. J. Symp. Ser. 76,* **192,** 1 (1980)], the addition of interreboilers and intercondensers increases thermodynamic efficiency but requires additional stages, as is clear from the positions of the operating lines in Fig. 13-38.

Optimum Reflux Ratio The general effect of the operating reflux ratio on fixed costs, operating costs, and the sum of these is shown in Fig. 13-39. In ordinary situations, the minimum on the total-cost curve will generally occur at an operating reflux ratio of from 1.1 to 1.5 times the minimum $R = L_{N+1}/D$ value, with the lower value corresponding to a value of the relative volatility close to 1.

Difficult Separations Some binary separations may pose special problems because of extreme purity requirements for one or both products or because of a relative volatility close to 1. The y-x diagram is convenient for stepping off stages at extreme purities if it is plotted on log-log paper. The equilibrium curve at very low x_1 values on ordinary graph paper can usually be assumed to be a straight line with an intercept term of zero which can be expressed as

$$y = (y/x)x + 0.0$$

where the slope y/x is a constant. The necessity for knowing the slope is eliminated by taking the logarithm of both sides

$$\log y = \log x + \log (y/x)$$

and plotting y versus x on a log-log plot to give a straight line with a slope of unity. The slope y/x is now an intercept term which need not be known. One point from the equilibrium curve is sufficient, therefore, to locate the equilibrium curve on the log-log plot. The operating line will be curved on the log-log plot and is located by plotting the appropriate material-balance equation. Both the equilibrium and the operating lines can be extended to any purity desired.

A system with constant relative volatility can be handled conveniently by the equation of Smoker [*Trans. Am. Inst. Chem. Eng.*, 34, 165 (1938)]. The derivation of the equation is shown, and its use is illustrated by Smith (op. cit.).

Stage Efficiency The use of the *Murphree plate efficiency* is particularly convenient on y-x diagrams. The Murphree efficiency is defined for the vapor phase as

$$\eta = (y_n - y_{n-1})/(y_n^\circ - y_{n-1}) \qquad (13\text{-}29)$$

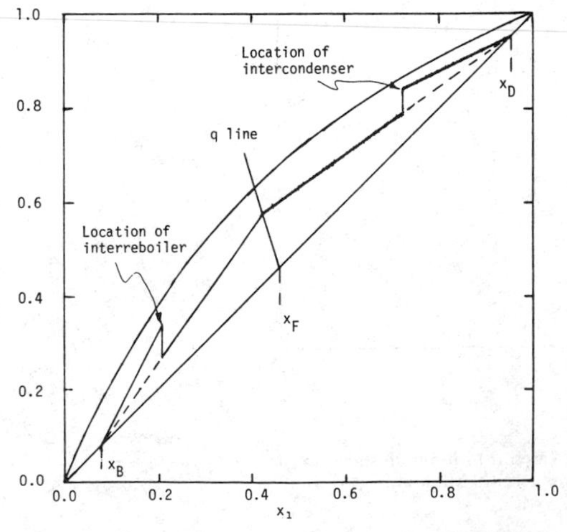

FIG. 13-38 McCabe-Thiele diagram for columns with and without an interreboiler and an intercondenser.

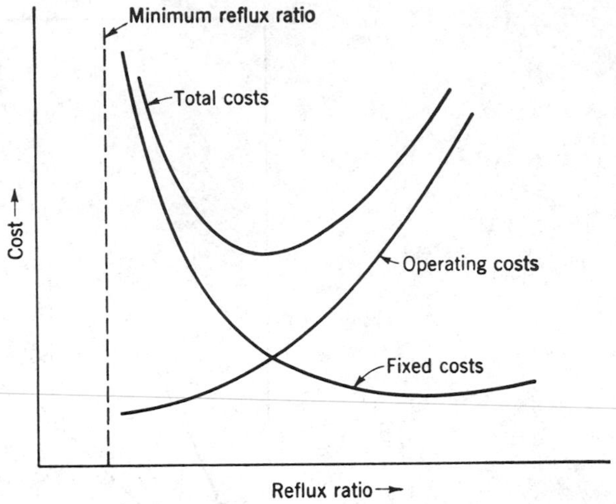

FIG. 13-39 Location of the optimum reflux for a given feed and specified separation.

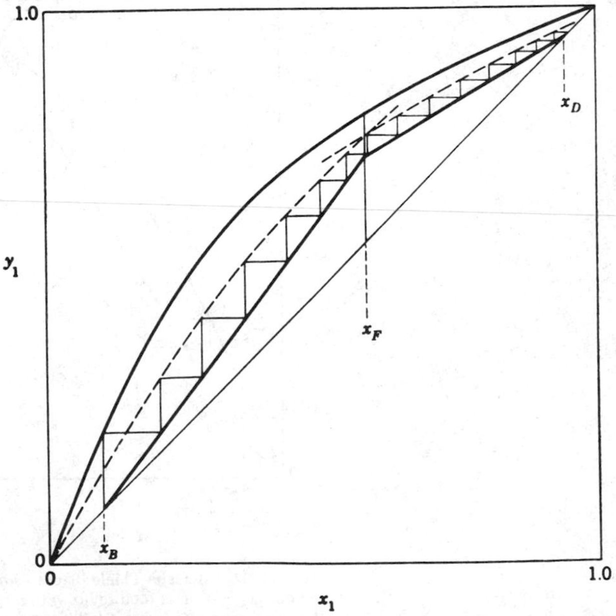

FIG. 13-40 Application of a 50 percent Murphree vapor-phase efficiency to each stage (excluding the reboiler) in the column. Each step in the diagram corresponds to an actual stage.

where y_n° is the composition of the vapor that would be in equilibrium with the liquid leaving stage n and is the value read from the equilibrium curve. The y_{n-1} and y_n are the actual (nonequilibrium) values for vapor streams leaving the $n - 1$ and n stages respectively. Note that the y_{n-1} and y_n values assume that vapor streams are completely mixed and uniform in composition. An analogous efficiency can be defined for the liquid phase.

The application of a 50 percent Murphree vapor-phase efficiency on a y-x diagram is illustrated in Fig. 13-40. A "pseudo-equilibrium" curve is drawn halfway (on a vertical line) between the operating lines and the true-equilibrium curve. The true-equilibrium curve is used for the first stage (the partial reboiler is assumed to be an equilibrium stage), but for all other stages the vapor leaving each stage is assumed to approach the equilibrium value y_n° only 50 percent of the way. Consequently, the steps in Fig. 13-40 represent actual trays.

Application of a constant efficiency to each stage as in Fig. 13-40 will not give, in general, the same answer as obtained when the number of equilibrium stages (obtained by using the true-equilibrium curve) is divided by the same efficiency factor.

The prediction and use of stage efficiencies are described in detail in Sec. 18.

Miscellaneous Operations The y-x diagrams for several other column configurations have not been presented here. The omitted items are *partial condensers*, *rectifying columns* (feed introduced to the bottom stage), *stripping columns* (feed introduced to the top stage), total reflux in the top section but not in the bottom section, multiple feeds, and introduction of *open steam* to the bottom stage to eliminate the reboiler. These configurations are discussed in Smith (op. cit.) and Henley and Seader (op. cit.), who also describe the more rigorous Ponchon-Savarit method, which is not included here.

APPROXIMATE MULTICOMPONENT DISTILLATION METHODS

INTRODUCTION

Approximate calculation methods for the solution of multicomponent, multistage separation problems continue to serve useful purposes even though computers are available to provide more rigorous solutions. Often, the available phase equilibrium and enthalpy data are not accurate enough to justify the longer rigorous methods. Or in extensive design and optimization studies, a large number of cases can be worked quickly and cheaply by an approximate method to define roughly the optimum specifications, which can then be investigated more exactly with a rigorous method. This procedure saves expensive computer time, because the time required for a rigorous method usually exceeds that for an approximate method by a factor of 10 or more. If a computer is not available, the designer may have to rely on approximate methods alone, because the time required for a manual solution by a rigorous method is usually prohibitive.

Two approximate multicomponent shortcut methods for simple distillation are presented here. The Smith-Brinkley (SB) method is based on an analytical solution of the finite-difference equations that can be written for staged separation processes when stages and interstage flow rates are known or assumed. The Fenske-Underwood-Gilliland (FUG) method combines Fenske's total-reflux equation and Underwood's minimum-reflux equation with a graphical correlation by Gilliland that relates actual column performance to total- and minimum-reflux conditions for a specified separation between two key components. Thus the SB and FUG methods are rating and design methods respectively. Both methods work best when mixtures are nearly ideal. Both approaches require about the same time for solution, and it cannot always be predicted which one will give results closer to the rigorous solution. If extensive exploratory work is to be done, the user should first determine which method more closely models the separation operation of interest and then use that method for all subsequent studies on that system.

Extensions of the SB method to nonideal mixtures and complex configurations are developed by Eckert and Hlavacek [*Chem. Eng. Sci.*, **33**, 77 (1978)] and Eckert [*Chem. Eng. Sci.*, **37**, 425 (1982)] respectively but are not discussed here. However, the approximate and very useful method of Kremser [*Nat. Pet. News*, **22**(21), 43 (May 21, 1930)] for application to absorbers and strippers is discussed at the end of this subsection.

SMITH-BRINKLEY GROUP METHOD

Extending the development of Kremser (op. cit.), Smith and Brinkley [*Am. Inst. Chem. Eng. J.*, **6**, 446 (1960)] presented a general equation that can be applied to absorption and extraction processes as well as distillation. The equation form, which applies to the distillation process pictured in Fig. 13-1, is, for each component i,

$$f = \frac{(1 + S_n^{N-M}) + R(1 - S_n)}{(1 - S_n^{N-M}) + R(1 - S_n) + hS_n^{N-M}(1 - S_m^{M+1})} \qquad (13\text{-}30)$$

where R is the external-reflux ratio L_{N+1}/D and $f = (Bx_B/Fx_F)_i$ is the fraction of i leaving in the bottoms product. The S quantity is the stripping factor (for component i) and is defined for each group of stages in the column by $S_{n,i} = K_iV/L$ and $S_{m,i} = K_i'V'/L'$. The K_i, V, and L values are the effective ones for the top column section; K_i', V', and L' are the effective values for the section below the feed stage. The quantity h depends upon whether K_i or K_i' is used for the feed stage. If the feed is mostly liquid, the feed stage is grouped with the lower stages and

$$h_i = \frac{K_i'}{K_i}\frac{L}{L'}\left(\frac{1 - S_n}{1 - S_m}\right)_i \qquad (13\text{-}31)$$

If the feed is mostly vapor,

$$h_i = \frac{L}{L'}\left(\frac{1 - S_n}{1 - S_m}\right)_i \qquad (13\text{-}32)$$

Equation (13-30) is applicable to a distillation column with either a partial reboiler or heat input to the bottom stage. The lowest equilibrium stage must always be numbered 1 regardless of its form. The equation is not strictly applicable to a column with a partial condenser because it ignores any possible difference between the overhead and reflux compositions. The effect of a partial condenser can be approximated by increasing N by 1.0.

In any computer solution of distillation problems, particularly when an existing column is being studied, it is most convenient to include the following four items in the specifications:

1. N, the total number of equilibrium stages
2. M, the number of stages below the feed stage
3. The reflux rate or the maximum vapor rate at some point in the column
4. The total distillate rate D

Equation (13-30) is particularly convenient when these specifications are made. Specification of D and L (or V), along with specification of the feed-stream variables, fixes the assumed phase rates in both sections of the column. Determination of separation factors $S_{n,i}$ and $S_{m,i}$ then depends upon estimation of individual K_i values. If ideal solutions are assumed, K_i values are functions only of temperature and specified column pressure. Estimation of K values and, in turn, S_n and S_m values for each component reduces, in ideal-solution systems, to estimation of the effective temperature in each column section or group of stages.

The effective section temperature is assumed to be some kind of average of the feed-stage and the end-stage temperatures. The simplest average is the arithmetic one,

$$T_n = (T_N + T_{M+1})/2 \qquad (13\text{-}33)$$

$$T_m = (T_{M+1} + T_1)/2 \qquad (13\text{-}34)$$

The final T_N and T_1 values must of course be respectively the dew point of the calculated overhead product and the bubble point of the calculated bottoms product. This leaves only T_{M+1}, which can be varied independently to force the column to meet the specified distillate-product rate and/or the reflux rate. This iteration technique is analogous to that of a column operator who adjusts the control-tray temperature by trial and error until the column produces the desired end products.

Example 1: Calculation of SB Method The use of Eq. (13-30) is illustrated with the following problem. A large butane-pentane splitter is to be shut down for repairs. Some of its feed will be diverted temporarily to an available smaller column, which has only 11 trays plus a partial reboiler. The feed enters on the middle tray. Past experience on similar feeds indicates that the 11 trays plus the reboiler are roughly equivalent to 10 equilibrium stages and that the column has a maximum top vapor capacity of 1.75 times the feed rate on a mole basis. The column will operate at a condenser pressure of 827.4 kPa (120 psia). The feed will be at its bubble point ($q = 1.0$) at the feed-tray conditions and has the following composition on the basis of 0.0126 (kg·mol)/s [100 (lb·mol)/h]:

Component	Fx_F
C_3	5
i-C_4	15
n-C_4	25
i-C_5	20
n-C_5	35
	100

The original column normally has less than 7 mol percent i-C_5 in the overhead and less than 3 mole percent n-C_4 in the bottoms product when operating at a distillate rate of $D/F = 0.489$. Can these product purities be produced on the smaller column at $D/F = 0.489$?

Pressure drops in the column will be neglected, and the K values will be read at 827 kPa (120 psia) in both column sections from the DePriester nomograph in Fig. 13-14*b*. When constant molar overflow is assumed in each section, the rates in pound-moles per hour in the upper and lower sections are as follows:

Top section	Bottom section
$D = (0.489)(100) = 48.9$	$B = 100 - 48.9 = 51.1$
$V = (1.75)(100) = 175$	$V' = V = 175$
$L = 175 - 48.9 = 126.1$	$L' = L + F = 226.1$
$V/L = 1.388$	$V'/L' = 0.7739$

$$L/L' = 126.1/226.1 = 0.5577$$

$$R = 126.1/48.9 = 2.579$$

NOTE: To convert pound-moles per hour to kilogram-moles per second, multiply by 1.26×10^{-4}.

Since the feed enters at the middle of the column, $M = 5$ and $M + 1 = 6$.

Initial top- and bottom-stage temperatures can be estimated by assuming a split on the feed and then making dew- and bubble-point calculations with the resulting overhead and bottom compositions. The i-C_5 and n-C_4 concentrations obtained on the original column plus the specified D of 48.9 are satisfied by the following assumed splits:

Component	Fx_F	Dx_D	Bx_B	x_D	x_B
C_3	5	5.0	0.0	0.102	0.0
i-C_4	15	14.5	0.5	0.296	0.010
n-C_4	25	23.5	1.5	0.481	0.029
i-C_5	20	3.4	16.6	0.070	0.325
n-C_5	35	2.5	32.5	0.051	0.636
	100	48.9	51.1	1.000	1.000

Dew and bubble points of assumed products give top- and bottom-stage temperatures of 347 K (165°F) and 386 K (236°F) respectively. The bubble point of the feed is 358 K (185°F), and this will be used as the initial assumption for the feed-tray temperature. (An arithmetic average of the two assumed end temperatures provides an initial T_{M+1} assumption and in this example would have been closer to the correct value.) The resulting T_n and T_m values obtained from Eqs. (13-33) and (13-34) are 352 K (175°F) and 372 K (210°F).

Evaluation of Eq. (13-30) for all components is best done in tabular form. Table 13-8 shows how initial T_n and T_m values are used to make the first estimate of D and B rates and associated product compositions. The h values for each component are obtained from Eq. (13-31).

The D calculated in Table 13-8 is 43.83, which is considerably lower than the specified value of 48.9. Also,

$$V_N = (1 + R)D = (3.579)(43.8) = 156.8$$

which is below the column capacity of 175. The temperature profile in the column must be raised.

The calculated product compositions in Table 13-8 should not be used to estimate new top and bottom temperatures. The amount taken overhead is so far below specification that calculated compositions cannot be realistic. Calculated overhead composition is too "light" and calculated bottoms composition too "heavy." It is better, therefore, not to make new estimates of end temperatures until the calculated D is closer to the specified value. The temperature profile can be raised by increasing the assumed feed-tray temperature and retaining the old end temperatures.

If $T_{m+1} = 372$ K (210°F) is assumed for the second trial, $T_n = 359.4$ K (187.5°F) and $T_m = 379$ K (223°F). These new temperatures give the following results:

Component	f	Bx_B	Dx_D	x_B	x_D
C_3	0.000789	0.004	4.996	0.000	0.101
i-C_4	0.0232	0.348	14.65	0.007	0.296
n-C_4	0.0900	2.25	22.75	0.045	0.459
i-C_5	0.8125	16.25	3.75	0.322	0.076
n-C_5	0.9031	31.61	3.39	0.627	0.067
		50.46	49.54	1.000	1.000

The new $D = 49.54$ is close enough to 48.9 to permit the use of calculated compositions to estimate new end temperatures. Dew- and bubble-point calculations provide temperatures of 346.1 K (163.5°F) and 385 K (234°F) for the top stage and the reboiler respectively. Averaging these with the previously assumed $T_{M+1} = 372$ K (210°F) gives $T_n = 358.9$ K (186.7°F) and $T_m = 379$ K (222°F). A third trial with these temperatures provides the final results

TABLE 13-8 First Trial in Application of Smith-Brinkley Method

Component	K_{175}	S_n	K_{210}	S_m	$\frac{K'}{K}$	$\frac{1 - S_n}{1 - S_m}$	h	S_n^5	$1 - S_n^5$	$R(1 - S_n)$	Numerator
C_3	3.03	4.21	3.83	2.96	1.26	1.63	1.15	1290	−1289	−8.27	−1297
i-C_4	1.53	2.12	2.07	1.60	1.35	1.87	1.41	43	−42	−2.90	−44.9
n-C_4	1.17	1.62	1.62	1.25	1.38	2.46	1.90	11.6	−10.6	−1.61	−12.2
i-C_5	0.575	0.798	0.825	0.638	1.43	0.558	0.447	0.323	0.677	0.521	1.20
n-C_5	0.485	0.673	0.720	0.557	1.48	0.738	0.611	0.140	0.860	0.843	1.70

Component	S_m^6	$hS_n^5(1 - S_m^6)$	Denominator	f	Bx_B	Dx_D	x_B	x_D
C_3	700	−1,037,000	−1,038,000	0.00125	0.006	4.994	0.000	0.114
i-C_4	17.3	−987.7	−1,032	0.0435	0.625	14.375	0.011	0.328
n-C_4	3.81	−61.7	−73.9	0.165	4.13	20.87	0.074	0.476
i-C_5	0.069	0.1344	1.33	0.899	18.0	2.0	0.320	0.046
n-C_5	0.036	0.0826	1.79	0.953	33.4	1.6	0.595	0.036
					56.16	43.83	1.000	1.000

shown below. For comparison, numbers in parentheses are the results of a rigorous computer solution, which is described in the next subsection.

Component	f	Bx_B	Dx_D
C_3	0.000825	0.004 (0.002)	4.996 (4.998)
i-C_4	0.0253	0.379 (0.330)	14.6 (14.6)
n-C_4	0.0964	2.41 (1.90)	22.6 (23.1)
i-C_5	0.8216	16.4 (16.4)	3.57 (3.56)
n-C_5	0.9098	31.8 (32.4)	3.16 (2.61)
		51.0	48.9

For the SB method, the concentration of n-C_4 in the bottoms product is 4.7 percent, which exceeds the 3 percent obtained on the larger column. Also, the i-C_5 concentration in D is 7.3 percent, which exceeds the 7 percent previously obtained. The smaller column will not, therefore, be able to meet both specifications at any overhead rate. Increasing D above 48.9 would decrease the concentration of n-C_4 in the bottoms product but would also increase the concentration of i-C_5 in the overhead stream. Likewise, decreasing D would remove i-C_5 from the overhead but would also increase the n-C_4 left in the bottoms stream. The rigorous method predicts a bottoms concentration for n-C_4 of 3.7 percent and a distillate concentration of 7.3 percent for i=C_5.

FENSKE-UNDERWOOD-GILLILAND (FUG) SHORTCUT METHOD

In this approach, Fenske's equation [*Ind. Eng. Chem.*, **24**, 482 (1932)] is used to calculate N_m, which is the number of plates required to make a specified separation at total reflux, i.e., the minimum value of N. Underwood's equations [*J. Inst. Pet.*, **31**, 111 (1945); **32**, 598 (1946); **32**, 614 (1946); and *Chem. Eng. Prog.*, **44**, 603 (1948)] are used to estimate the minimum-reflux ratio R_m. The empirical correlation of Gilliland [*Ind. Eng. Chem.*, **32**, 1220 (1940)] shown in Fig. 13-41 then uses these values to give N for any specified R or R for any specified N. Limitations of the Gilliland correlation are discussed by Henley and Seader (*Equilibrium-Stage Separation Operations in Chemical Engineering*, Wiley, New York, 1981). The following equation, developed by Molokanov et al. [*Int. Chem. Eng.*, **12**(2), 209 (1972)] satisfies the end points and fits the Gilliland curve reasonably well:

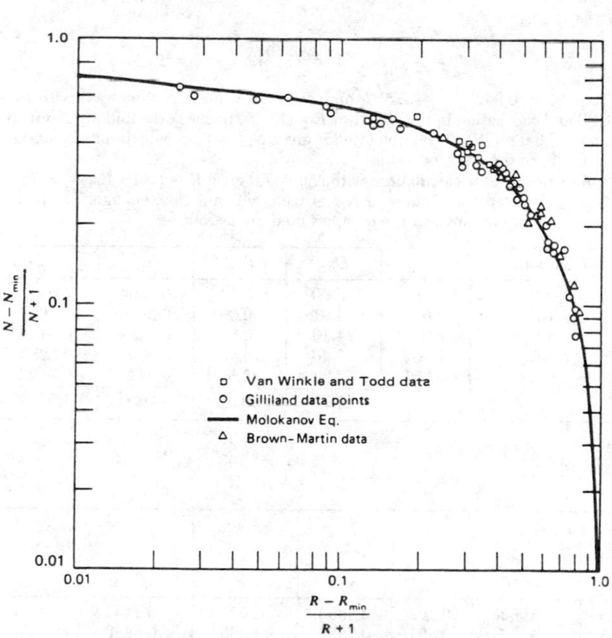

FIG. 13-41 Comparison of rigorous calculations with Gilliland correlation. [*Henley and Seader,* Equilibrium-Stage Separation Operations in Chemical Engineering, *Wiley, New York, 1981; data of Van Winkle and Todd,* Chem. Eng., *78 (21), 136 (Sept. 20, 1971); data of Gilliland,* Elements of Fractional Distillation, *4th ed., McGraw-Hill, New York, 1950; data of Brown and Martin,* Trans. Am. Inst. Chem. Eng., *35, 679 (1939).*]

$$\frac{N - N_m}{N + 1} = 1 - \exp\left[\left(\frac{1 + 54.4\Psi}{11 + 117.2\Psi}\right)\left(\frac{\Psi - 1}{\Psi^{0.5}}\right)\right] \qquad (13\text{-}35)$$

where $\Psi = (R - R_m)/(R + 1)$.

The *Fenske total-reflux equation* can be written as

$$\left(\frac{x_i}{x_r}\right) = (\alpha_i)^{N_m}\left(\frac{x_1}{x_r}\right)_B \qquad (13\text{-}36)$$

or as

$$N_m = \frac{\log\left[\left(\frac{Dx_D}{Bx_B}\right)_i\left(\frac{Bx_B}{Dx_D}\right)_r\right]}{\log \alpha_i} \qquad (13\text{-}37)$$

where i is any component and r is an arbitrarily selected reference component in the definition of relative volatilities,

$$\alpha_i = K_i/K_r = y_i x_r/y_r x_i \qquad (13\text{-}38)$$

The particular value of α_i is the effective value used in Eqs. (13-36) and (13-37) defined in terms of values for each stage in the column by

$$\alpha^N = \alpha_N \alpha_{N-1} \cdots \alpha_2 \alpha_1 \qquad (13\text{-}39)$$

Equations (13-36) and (13-37) are rigorous relationships between the splits obtained for components i and r in a column at total reflux. However, the correct value of α_i must always be estimated, and this is where the approximation enters. It is usually estimated from

$$\alpha = (\alpha_{top}\alpha_{bottom})^{1/2} \qquad (13\text{-}40)$$

or

$$\alpha = (\alpha_{top}\alpha_{middle}\alpha_{bottom})^{1/3} \qquad (13\text{-}41)$$

A reasonably good estimate of the separation that will be accomplished in a plant column often can be obtained by specifying the split of one component (designated as the reference component r), setting N_m equal to from 40 to 60 percent of the number of equilibrium stages (not actual trays), and then using Eq. (13-37) to estimate the splits of all the other components. This is an iterative calculation because the component splits must first be arbitrarily assumed to give end compositions that can be used to give initial end-temperature estimates. The α_{top} and α_{bottom} values corresponding to these end temperatures are used in Eq. (13-40) to give α_i values for each component. The iteration is continued until the α_i values do not change from trial to trial.

The *Underwood minimum-reflux equations* of main interest are those that apply when some of the components do not appear in either the distillate or the bottoms products at minimum reflux. These equations are

$$\sum_i \frac{\alpha_i (x_{i,D})_m}{\alpha_i - \Theta} = R_m + 1 \qquad (13\text{-}42)$$

and

$$\sum_i \frac{\alpha_i x_{i,F}}{\alpha_i - \Theta} = 1 - q \qquad)13\text{-}43)$$

The relative volatilities α_i are defined by Eq. (13-38), R_m is the minimum-reflux ratio $(L_{N+1}/D)_{min}$, and q describes the thermal condition of the feed (e.g., 1.0 for a bubble-point feed and 0.0 for a saturated-vapor feed). The $x_{i,F}$ values are available from the given feed composition. The Θ is the common root for the top-section equations and the bottom-section equations developed by Underwood for a column at minimum reflux with separate zones of constant composition in each section. The common root value must fall between α_{hk} and α_{lk}, where hk and lk stand for *heavy key* and *light key* respectively. The *key components* are the ones that the designer wants to separate. In the butane-pentane splitter problem used in Example 1, the light key is n-C_4 and the heavy key is i-C_5.

The α_i values in Eqs. (13-42) and (13-43) are effective values obtained from Eq. (13-40) or Eq. (13-41). Once these values are available, Θ can be calculated in a straightforward iteration from Eq.

(13-43). Since the $(\alpha - \Theta)$ difference can be small, Θ should be determined to four decimal places to avoid numerical difficulties.

The $(x_{i,D})_m$ values in Eq. (13-42) are minimum-reflux values, i.e., the overhead concentration that would be produced by the column operating at the minimum reflux with an infinite number of stages. When the light key and the heavy key are adjacent in relative volatility and the specified split between them is sharp or the relative volatilities of the other components are not close to those of the two keys, only the two keys will distribute at minimum reflux and the $(x_{i,D})_m$ values are easily determined. This is often the case and is the only one considered here. Other cases in which some or all of the nonkey components distribute between distillate and bottom products are discussed in detail by Henley and Seader (op. cit.).

The FUG method is not as convenient for the study of existing columns as the SB method but is more convenient for new-column design. Convenient specifications for the FUG method may include the following:

1. R/R_m, the ratio of reflux to minimum reflux
2. Split on the reference component (usually chosen as the heavy key)
3. Split on one other component (usually the light key)

However, the total number of equilibrium stages N, N/N_m, or the external-reflux ratio can be substituted for one of these three specifications. It should be noted that the feed location is automatically specified as the optimum one; this is assumed in the Underwood equations. The assumption of saturated reflux is also inherent in the Fenske and Underwood equations. An important limitation on the Underwood equations is the assumption of constant molar overflow. As discussed by Henley and Seader (op. cit.), this assumption can lead to a prediction of the minimum reflux that is considerably lower than the actual value. No such assumption is inherent in the Fenske equation. An exact calculational technique for minimum reflux is given by Tavana and Hansen [*Ind. Eng. Chem. Process Des. Dev.*, **18**, 154 (1979)]. A computer program for the FUG method is given by Chang [*Hydrocarbon Process.*, **60**(8), 79 (1980)].

Example 2: Calculation of FUG Method Application of the FUG method is demonstrated on the butane-pentane-splitter example problem used for the SB method. Specifications necessary to model the existing column are:

1. $N = 10$, total number of equilibrium stages.
2. Optimum feed location (which may or may not reflect the actual location).
3. Maximum V/F at the top tray of 1.75.
4. Split on one component given in the following paragraphs.

As always, the solution starts with an assumed arbitrary split of all the components to give estimates of top and bottom compositions that can be used to get initial end temperatures. The α_i's evaluated at these temperatures are averaged with an assumed feed-stage temperature (assumed to be the bubble point of the feed) by using Eq. (13-41). The initial assumption for the split on i-C_5 will be $Dx_D/Bx_B = 3.15/16.85$. As mentioned earlier, N_m usually ranges from $0.4N$ to $0.6N$, and the initial N_m value assumed here will be $(0.6)(10) = 6.0$. Equation (13-37) can be rewritten as

$$\left(\frac{Dx_D}{Fx_F - Dx_D}\right)_i = \alpha_i^{6.0}\left(\frac{3.15}{16.85}\right) = \alpha_i^{6.0}(0.1869)$$

or

$$Dx_{i,D} = \frac{0.1869\alpha_i^{6.0}}{1 + 0.1869\alpha_i^{6.0}} Fx_{i,F}$$

The evaluation of this equation for each component is as follows:

Component	α_i	$\alpha_i^{6.0}$	$0.1869\alpha_i^{6.0}$	Fx_F	Dx_D	Bx_B
C_3	5.00			5	5.0	0.0
i-C_4	2.63	330	61.7	15	14.8	0.2
n-C_4	2.01	66	12.3	25	25.1	1.9
i-C_5	1.00	1.00	0.187	20	3.15	16.85
n-C_5	0.843	0.36	0.0672	35	2.20	32.80
				100	48.25	51.75

The end temperatures corresponding to these product compositions are 344 K (159°F) and 386 K (236°F). These temperatures plus the feed bubble-point temperature of 358 K (185°F) provide a new set of α_i's which vary only slightly from those used earlier. Consequently, the $D = 48.25$ value is not expected to vary greatly and will be used to estimate a new i-C_5 split. The desired overhead concentration for i-C_5 is 7 percent; so it will be assumed that $Dx_D = (0.07)(48.25) = 3.4$ for i-C_5 and that the split on that component will be 3.4/16.6. The results obtained with the new α_i's and the new i-C_5 split are as follows:

Component	$\alpha_i^{6.0}$	$0.2048\alpha_i^{6.0}$	Fx_F	Dx_D	Bx_B	x_D	x_B
C_3			5	5.0	0.0	0.102	0.000
i-C_4	322	65.9	15	14.8	0.2	0.301	0.004
n-C_4	68	13.9	25	23.3	1.7	0.473	0.033
i-C_5	1.00	0.205	20	3.4	16.6	0.069	0.327
n-C_5	0.415	0.085	35	2.7	32.3	0.055	0.636
			100	49.2	50.8	1.000	1.000

The calculated i-C_5 concentration in the overhead stream is 6.9 percent, which is close enough to the 7.0 figure for now.

Table 13-9 shows subsequent calculations using the Underwood minimum-reflux equations. The α and x_D values in Table 13-9 are those from the Fenske total-reflux calculation. As noted earlier, the x_D values should be those at minimum reflux. This inconsistency may reduce the accuracy of the Underwood method, but to be useful a shortcut method must be fast, and it has not been shown that a more rigorous estimation of x_D values results in an overall improvement in accuracy. The calculated R_m is 0.9426. The actual reflux assumed is obtained from the specified maximum top vapor rate of 0.022 (kg·mol)/s [175 (lb·mol)/h] and the calculated D of 49.2 (from the Fenske equation).

$$L_{N+1} = V_N - D$$

$$R = V_N/D - 1 = 175/49.2 - 1 = 2.557$$

The $R_m = 0.9426$, $R = 2.557$, and $N = 10$ values are now used with the Gilliland correlation in Fig. 13-41 or Eq. (13-35) to check the initially assumed value of 6.0 for N_m. Equation (13-35) gives $N_m = 6.95$, which differs considerably from the assumed value.

Repetition of the calculations with $N_m = 7.0$ gives $R = 2.519$, $R_m = 0.9782$, and a calculated check value of $N_m = 6.85$, which is close enough. The final-product compositions and the α values used are as follows:

Component	α_i	Dx_D	Bx_B	x_D	x_B
C_3	4.98	5.00	0	0.1004	0.0
i-C_4	2.61	14.91	0.09	0.2996	0.0017
n-C_4	2.02	24.16	0.84	0.4852	0.0168
i-C_5	1.00	3.48	16.52	0.0700	0.3283
n-C_5	0.851	2.23	32.87	0.0448	0.6532
		49.78	50.32	1.0000	1.0000

TABLE 13-9 Application of Underwood Equations

				$\theta = 1.36$		$\theta = 1.365$					
Component	x_F	α	αx_F	$\alpha - \theta$	$\frac{\alpha x_F}{\alpha - \theta}$	$\alpha - \theta$	$\frac{\alpha x_F}{\alpha - \theta}$	x_D	αx_D	$\alpha - \theta$	$\frac{\alpha x_D}{\alpha - \theta}$
C_3	0.05	4.99	0.2495	3.63	0.0687	3.625	0.0688	0.102	0.5090	3.6253	0.1404
i-C_4	0.15	2.62	0.3930	1.26	0.3119	1.255	0.3131	0.301	0.7886	1.2553	0.6282
n-C_4	0.25	2.02	0.5050	0.66	0.7651	0.655	0.7710	0.473	0.9555	0.6553	1.4581
i-C_5	0.20	1.00	0.2000	−0.36	−0.5556	−0.365	−0.5479	0.069	0.0690	−0.3647	−0.1892
n-C_5	0.35	0.864	0.3024	−0.496	−0.6097	−0.501	−0.6036	0.055	0.0475	−0.5007	−0.0949
	1.00				−0.0196		+0.0014	1.000			$1.9426 = R_m + 1$

Interpolation gives $\theta = 1.3647$.

These results indicate that the 7 percent i-C_5 in D and the 3 percent n-C_4 in B concentrations obtained in the original column can easily be obtained on the smaller column. This disagrees with the answers obtained from the SB method and from a rigorous computer solution as shown in the following comparison:

	x_D			x_B		
Component	Rigorous	SB	FUG	Rigorous	SB	FUG
C_3	0.102	0.102	0.100	0.0	0.0	0.0
i-C_4	0.299	0.299	0.300	0.006	0.007	0.002
n-C_4	0.473	0.463	0.485	0.037	0.047	0.017
i-C_5	0.073	0.073	0.070	0.322	0.322	0.328
n-C_5	0.053	0.063	0.045	0.635	0.624	0.653
	1.000	1.000	1.000	1.000	1.000	1.000

Comparison of SB and FUG Methods Because of the necessary specifications, the Smith-Brinkley (SB) method is more convenient for analysis of existing columns than for design of new columns when a specific product purity or recovery is specified. The Fenske-Underwood-Gilliland (FUG) method permits the direct specification of component splits and therefore is often more convenient for new-column design than for analysis of an existing column.

The SB method can utilize the actual feed-tray location, whereas the FUG method always assumes the optimum location. This is a serious handicap when analyzing an existing column whose feed tray is badly located for the intended new service.

The SB method uses different effective K values for the two column sections, and this may be an advantage for systems in which relative volatilities vary widely from one end of the column to the other. The FUG method uses one effective value for the entire column. Also, as mentioned earlier, the assumption of constant molar overflow can be a serious problem for the Underwood equation in the FUG method.

The SB method has the advantage of being applicable to other separation processes such as absorption, stripping, extraction, and washing. Another version of the method is given by Edmister [*Am. Inst. Chem. Eng. J.*, 3, 165 (1957)].

KREMSER GROUP METHOD

Starting with the classical method of Kremser (op. cit.), approximate group methods of increasing complexity have been developed to calculate groups of equilibrium stages for a countercurrent cascade, such as is used in simple absorbers and strippers of the type depicted in Fig. 13-7*b* and *d*. However, none of these group methods can adequately account for stage temperatures that are considerably higher or lower than the two entering-stream temperatures for absorption and stripping respectively when appreciable composition changes occur. Therefore, only the simplest form of the Kremser method is presented here. Fortunately, rigorous computer methods described later can be applied when accurate results are required. The Kremser method is most useful for making preliminary estimates of absorbent and stripping-agent flow rates or equilibrium-stage requirements. The method can also be used to extrapolate quickly results of a rigorous solution to a different number of equilibrium stages.

Consider the general adiabatic countercurrent cascade of Fig. 13-42 where v and ℓ are molar component flow rates. Regardless of whether the cascade is an absorber or a stripper, components in the entering vapor will tend to be absorbed and components in the entering liquid will tend to be stripped. If more moles are stripped than absorbed, the cascade is a stripper; otherwise, the cascade is an absorber. The Kremser method is general and applies to either case. Application of component material-balance and phase equilibrium equations successively to stages 1 through $N-1$, 1 through $N-2$, etc., as shown by Henley and Seader (op. cit.), leads to the following equations originally derived by Kremser. For each component i,

$$(v_i)_N = (v_i)_0(\Phi_i)_A + (\ell_i)_{N+1}[1 - (\Phi_i)_S] \tag{13-44}$$

where

$$(\Phi_i)_A = \frac{(A_i)_e - 1}{(A_i)_e^{N+1} - 1} \tag{13-45}$$

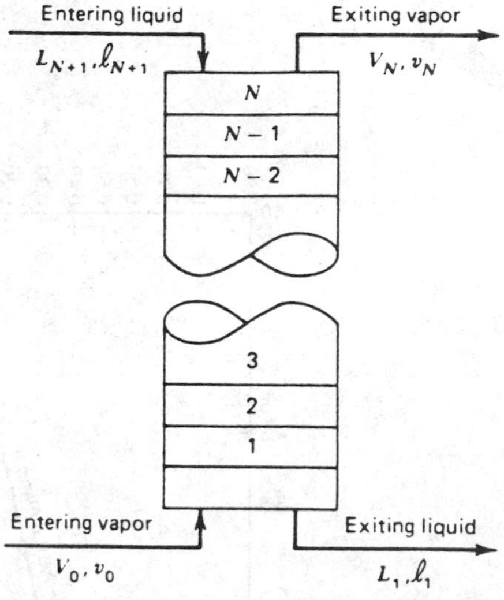

FIG. 13-42 General adiabatic countercurrent cascade for simple absorption or stripping.

is the fraction of component i in the entering vapor that is not absorbed,

$$(\Phi_i)_S = \frac{(S_i)_e - 1}{(S_i)_e^{N+1} - 1} \tag{13-46}$$

is the fraction of component i in the entering liquid that is not stripped,

$$(A_i)_e = (L/K_iV)_e \tag{13-47}$$

is the effective or average absorption factor for component i, and

$$(S_i)_e = 1/(A_i)_e \tag{13-48}$$

is the effective or average stripping factor for component i. When the entering streams are at the same temperature and pressure and negligible absorption and stripping occur, effective component absorption and stripping factors are determined simply by entering-stream conditions. Thus, if K values are composition-independent, then

$$(A_i)_e = 1/(S_i)_e = (L_{N+1}/K_i\{T_{N+1}, P_{N+1}\}V_0) \tag{13-49}$$

When entering-stream temperatures differ and/or moderate to appreciable absorption and/or stripping occurs, values of A_i and S_i should be based on effective average values of L, V, and K_i in the cascade. However, even then Eq. (13-49) with T_{N+1} replaced by $(T_{N+1} + T_0)/2$ may be able to give a first-order approximation of $(A_i)_e$. In the case of an absorber, $L_{N+1} < L_e$ and $V_0 > V_e$ will be compensated to some extent by $K_i\{(T_{N+1} + T_0)/2, P\} < K_i\{T_e, P\}$. A similar compensation, but in opposite directions, will occur in the case of a stripper.

Equations (13-45) and (13-46) are plotted in Fig. 13-43. Components having large values of A_e or S_e absorb or strip respectively to a large extent. Corresponding values of Φ_A and Φ_S approach a value of 1 and are almost independent of the number of equilibrium stages.

An estimate of the minimum absorbent flow rate for a specified amount of absorption from the entering gas of some key component K for a cascade with an infinite number of equilibrium stages is obtained from Eq. (13-45) as

$$(L_{N+1})_{\min} = K_K V_0[1 - (\Phi_K)_A] \tag{13-50}$$

The corresponding estimate of minimum stripping-agent flow rate for a stripper is obtained as

$$(V_0)_{\min} = L_{N+1}[1 - (\Phi_K)_S]/K_K \tag{13-51}$$

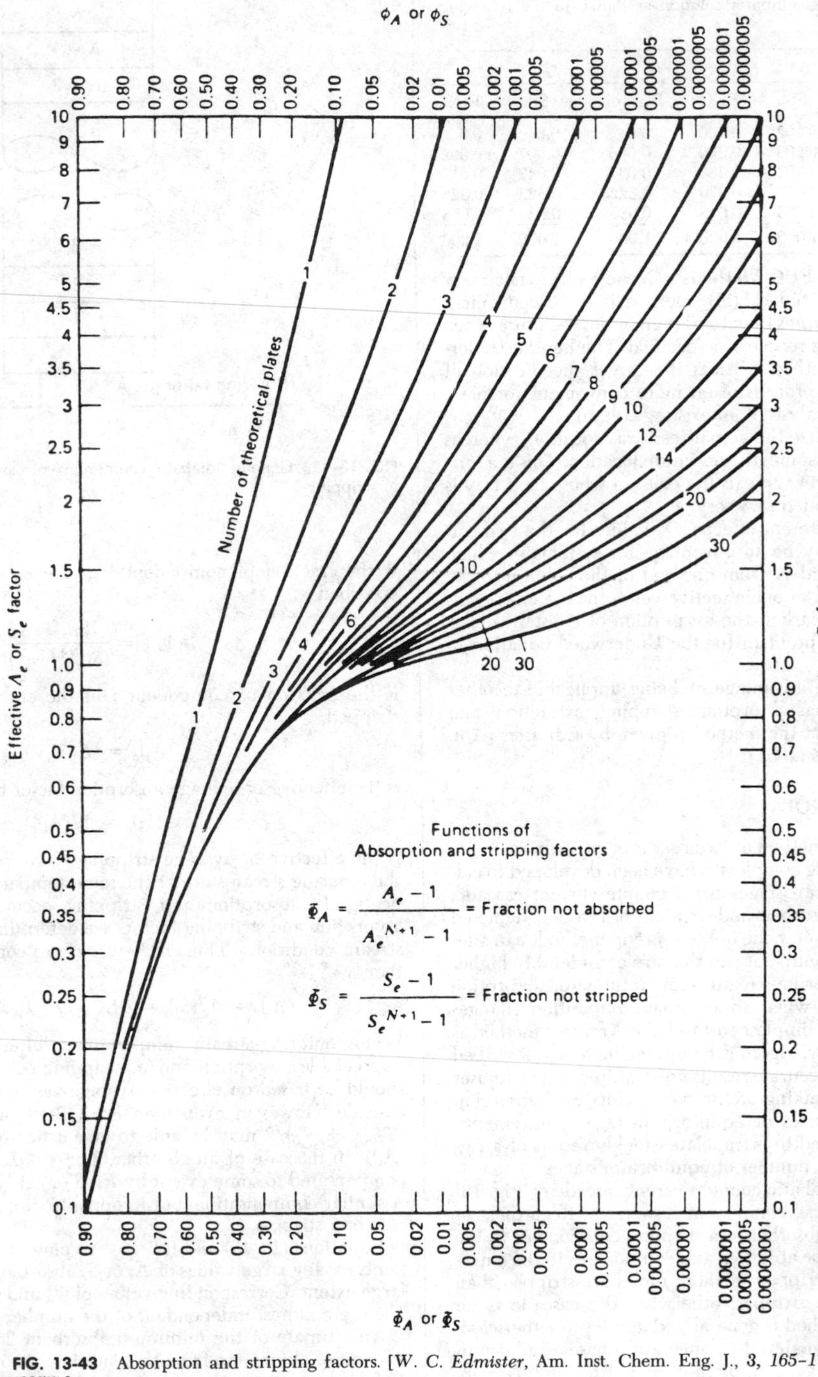

FIG. 13-43 Absorption and stripping factors. [*W. C. Edmister,* Am. Inst. Chem. Eng. J., *3, 165–171 (1957).*]

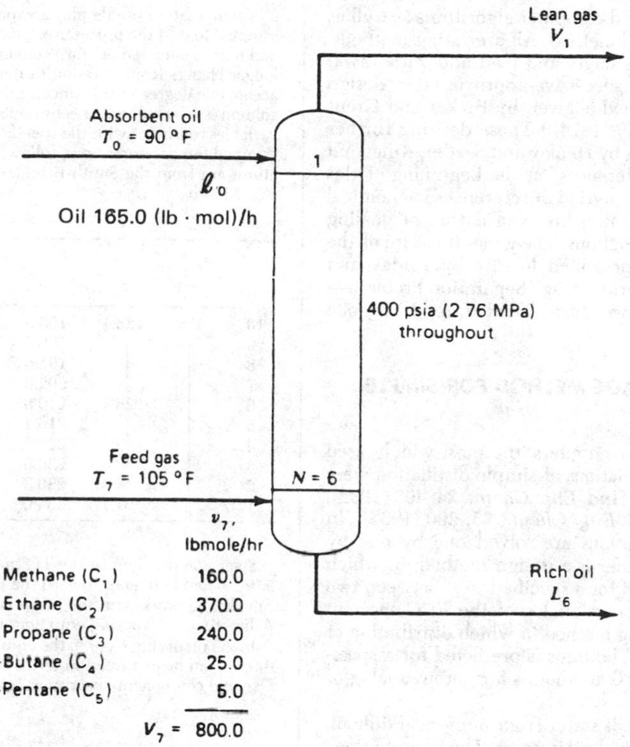

FIG. 13-44 Specifications for the absorber example.

Example 3: Calculation of Kremser Method For the simple absorber specified in Fig. 13-44, a rigorous calculation procedure as described below gives results in Table 13-10. Values of Φ were computed from component-product flow rates, and corresponding effective absorption and stripping factors were obtained by iterative calculations in using Eqs. (13-45) and (13-46) with $N = 6$. Use the Kremser method to estimate component-product rates if N is doubled to a value of 12.

Assume that values of A_e and S_e will not change with a change in N. Application of Eqs. (13-45), (13-46), and (13-44) gives the results in the last four columns of Table 13-10. Because of its small value of A_e, the extent of absorption of C_1 is unchanged. For the other components, somewhat increased amounts of absorption occur. The degree of stripping of the absorber oil is essentially unchanged. Overall, only an additional 0.5 percent of absorption occurs. The greatest increase in absorption occurs for n-C_4, to the extent of about 4 percent.

TABLE 13-10 Results of Calculations for Simple Absorber of Fig. 13-44

	N = 6 (rigorous method)				N = 12 (Kremser method)					
	(lb·mol)/h						(lb·mol)/h			
Component	$(v_i)_6$	$(\ell_i)_1$	$(\Phi_i)_A$	$(\Phi_i)_S$	$(A_i)_e$	$(S_i)_e$	$(v_i)_{12}$	$(\ell_i)_1$	$(\Phi_i)_A$	$(\Phi_i)_S$
C_1	147.64	12.36	0.9228	......	0.0772	.....	147.64	12.36	0.9228	
C_2	276.03	94.97	0.7460	......	0.2541	.....	275.98	94.02	0.7459	
C_3	105.42	134.58	0.4393	......	0.5692	.....	103.46	136.54	0.4311	
nC_4	1.15	23.85	0.0460	......	1.3693	.....	0.16	24.84	0.0063	
nC_5	0.0015	4.9985	0.0003	......	3.6	.....	0	5.0	0.0	
Absorber oil	0.05	164.95	......	0.9997	......	0.0003	0.05	164.95	.....	0.9997
Totals	530.29	435.71					527.29	437.71		

NOTE: To convert pound-moles per hour to kilogram-moles per hour, multiply by 0.454.

RIGOROUS METHODS FOR MULTICOMPONENT DISTILLATION-TYPE SEPARATIONS

INTRODUCTION

Availability of large digital computers has made possible rigorous solutions of equilibrium-stage models for multicomponent, multistage distillation-type columns to an exactness limited only by the accuracy of the phase equilibrium and enthalpy data utilized. Time and cost requirements for obtaining such solutions are very low compared with the cost of manual solutions. Methods are available that can accurately solve almost any type of distillation-type problem quickly and efficiently. The material presented here covers, in some detail, some of the more widely used computer algorithms as well as the classical Thiele-Geddes manual method. All are rating methods, in that the number of equilibrium stages and feed and withdrawal stages are specified. However, a successive-approximation design method that utilizes a rating method is given by Ricker and Grens [*Am. Inst. Chem. Eng. J.*, **20,** 238 (1974).] Those desiring further details are referred to the textbooks by Henley and Seader, King, and Holland cited under "General References" at the beginning of this section. These books, in turn, cite a myriad of references in chemical engineering journals. The mathematics involved is that of dealing with sets of nonlinear algebraic equations. The general nature of the main mathematical problems is presented lucidly by Friday and Smith ["An Analysis of the Equilibrium Stage Separation Problem—Formulation and Convergence," *Am. Inst. Chem. Eng. J.*, **10,** 698 (1964).

THIELE-GEDDES STAGE-BY-STAGE METHOD FOR SIMPLE DISTILLATION

Prior to the availability of digital computers, the most widely used manual methods for rigorous calculations of simple distillation were those of Lewis and Matheson (LM) [*Ind. Eng. Chem.*, **24,** 496 (1932)] and Thiele and Geddes (TG) [*Ind. Eng. Chem.*, **25,** 290 (1933)], in which the equilibrium-stage equations are solved one by one by using tearing techniques. The former is a design method, in which the number of stages is determined for a specified split between two key components. Thus, it is a rigorous analog of the FUG shortcut method. The TG method is a rating method in which distribution of components between distillate and bottoms is predicted for a specified number of stages. Thus, the TG method is a rigorous analog of the SB method.

Both the LM and the TG methods suffer from numerical difficulties that can prevent convergence in certain cases. The stage-to-stage calculation used in the LM method proceeds from the top down and from the bottom up and is subject to large truncation-error buildup if the components differ widely in volatility. The TG method avoids that difficulty, but numerical instabilities arise as soon as the stage-to-stage calculation crosses a feed stage. Then, a difference term appears in the equations, and sometimes this results in a serious loss of significant digits, making the TG method basically unsuited for multiple-feed columns.

All stage-to-stage methods that work from both ends of the column toward the middle suffer from two other disadvantages. First, the top-down and the bottom-up calculations must "mesh" somewhere in the column. Usually the mesh is made at a feed stage, and if more than one feed stage exists, a choice of mesh point must be made for each component. When the components vary widely in volatility, the same mesh point cannot be used for all components if serious numerical difficulties are to be avoided. Second, arbitrary procedures must be set up to handle *nondistributed* components. (A nondistributed component is one whose concentration in one of the end-product streams is smaller than the smallest number carried by the computer.) In the LM and TG equations, the concentrations for these components do not naturally take on nonzero values at the proper point as the calculations proceed through the column.

Because of all these numerical difficulties, neither the LM nor the TG stage-by-stage method is commonly implemented in modern computer algorithms. Nevertheless, the TG method is very instructive and is developed in the following example. For a single narrow-boiling feed, the TG manual method is quite efficient.

Example 4: Calculation of TG Method The TG method will be demonstrated by using the same example problem that was used above for the approximate methods. The example column was analyzed previously and found to have $C + 2N + 9$ design variables. The specifications to be used in this example were also listed at that time and included the total number of stages ($N = 10$), the feed-plate location ($M = 5$), the reflux temperature (corresponding to saturated liquid), the distillate rate ($D = 48.9$), and the top vapor rate ($V = 175$). As before, the pressure is uniform at 827 kPa (120 psia), but a pressure gradient could be easily handled if desired.

A temperature profile plus a vapor-rate profile through the column must be assumed to start the procedure. These variables are referred to as tear variables and must be iterated on until convergence is achieved in which their values no longer change from iteration to iteration and all equations are satisfied to an acceptable degree of tolerance. Each iteration down and then up through the column is referred to as a column iteration. A set of assumed values of the tear variables consistent with the specifications, plus the component K values at the assumed temperatures, is as follows, where assumed end and middle temperatures are from the Smith-Brinkley shortcut solution and K values are from Fig. 13-14*b*:

				K				
Stage	V	L	T	C_3	$i-C_4$	n-C_4	i-C_5	n-C_5
10	175	126.1	163.5	2.77	1.38	1.04	0.500	0.420
9	│	│	178.5	3.10	1.60	1.22	0.590	0.495
8	│	│	191.3	3.40	1.78	1.37	0.685	0.585
7	│	↓	202.0	3.63	1.94	1.49	0.770	0.660
6	│	226.1	210.0	3.84	2.06	1.60	0.825	0.702
5	│	│	216.4	4.00	2.21	1.73	0.895	0.765
4	│	│	221.7	4.15	2.28	1.80	0.925	0.800
3	│	│	226.3	4.28	2.36	1.88	0.965	0.835
2	│	↓	230.3	4.36	2.43	1.94	1.000	0.870
1	↓	51.1	234.0	4.42	2.50	1.99	1.030	0.890

Stage compositions in the TG method are obtained by stage-to-stage calculations from both ends toward the feed stage. With reference to Fig. 13-1, the calculations work with the ratios v_n/d, ℓ_n/d, v_m/b, and ℓ_m/b instead of v or ℓ directly. The working equations are derived as follows:

In the rectifying section, the equilibrium relationship for component i at any stage n can be expressed in terms of component flow rate in the distillate $d = Dx_D$ and component absorption factor $A_n = L_n/K_nV_n$.

$$\begin{aligned} x_n &= y_n/K_n \\ L_nx_n &= (L_n/K_nV_n)V_ny_n \\ \ell_n &= A_nv_n \\ \ell_n/d &= (v_n/d)A_n \end{aligned} \qquad (13\text{-}52)$$

The general component-i balance around a section of stages from stage n to the top of the column is

$$v_n = \ell_{n+1} + d$$

or

$$v_n/d = (\ell_{n+1}/d) + 1 \qquad (13\text{-}53)$$

Increasing the subscripts in Eq. (13-52) by 1 and substituting for ℓ_{n+1}/d in Eq. (13-53) gives the following combined equilibrium and material-balance relationship for component i:

$$v_n/d = (v_{n+1}/d)A_{n+1} + 1 \qquad (13\text{-}54)$$

Or, if v_n/d is eliminated in Eq. (13-53),

$$\frac{\ell_n}{d} = A_n\left(\frac{\ell_{n+1}}{d} + 1\right) \qquad (13\text{-}55)$$

Equation (13-55) is used to calculate, from the previous stage, the (ℓ/d) ratio on each stage in the rectifying section. The assumed temperature and phase-rate-profile assumptions conveniently fix all the A_n values for ideal solutions. The calculations are started by writing the equation for stage N:

$$\frac{\ell_N}{d} = A_N\left(\frac{\ell_{N+1}}{d} + 1\right) \qquad (13\text{-}56)$$

For a total condenser, $x_D = x_{N+1}$ and

$$\ell_{N+1}/d = L_{N+1}/D = R \qquad (13\text{-}57)$$

A knowledge of the reflux ratio (obtained from the specified distillate and top vapor rates) permits the calculation of $(\ell_N/d)_i$ from which $(\ell_{m-1}/d)_i$ is obtained, etc. Equation (13-55) is applied to each stage in succession until the ratio ℓ_{M+2}/d in the overflow from the stage above the feed stage is obtained. The calculations are then switched to the stripping section.

The equilibrium relationship for component i in the stripping section can be expressed in terms of component flow rate in the bottoms, $b = Bx_B$, and $S_m = K_mV_m/L_m$ as

$$\begin{aligned} y_m &= K_mx_m \\ V_my_m &= (K_mV_m/L_m)L_mx_m \\ v_m &= S_m\ell_m \\ v_m/b &= (\ell_m/b)S_m \end{aligned} \qquad (13\text{-}58)$$

Combination with the material balance

$$(\ell_{m+1}/b) = v_m/b + 1 \tag{13-59}$$

gives

$$(\ell_{m+1}/b) = (\ell_m/b)S_m + 1 \tag{13-60}$$

The bottom-up calculations are started by writing Eq. (13-60) for stage 1 as

$$\ell_2/b = V_1K_1/B + 1 = S_1 + 1 \tag{13-61}$$

The S_m values all are fixed by assumed temperature and phase-rate profiles. Equation (13-60) is applied to each of the stripping stages in sequence until the ratio ℓ_{M+2}/b in the liquid entering the feed stage is obtained.

The manner in which rectifying and stripping-section calculations are meshed at the feed stage depends upon the thermal condition of the feed. Figure 13-45 shows three possible ways in which fresh feed can affect the L and V rates between the feed stage and stage $M + 2$. The superscript bar denotes the stream rate when the stream enters a stage, while the lack of a bar denotes the rate when the stream leaves a stage.

Top-down calculations for the example problem are shown in Table 13-11 and bottom-up calculations in Table 13-12. Top-down and bottom-up calculations have provided values of ℓ_{M+2}/d and ℓ_{M+2}/b respectively. For a bubble-point feed,

$$v_{M+1} = \bar{v}_{M+1}$$

and a combination of Eqs. (13-53) and (13-59) provides for each component i

$$\frac{b}{d} = \frac{v_{M+1}/d}{v_{M+1}/b} = \frac{\ell_{M+2}/d + 1}{\ell_{M+2}/b - 1} \tag{13-62}$$

The b/d ratios obtained from this equation can then be used to calculate the individual b and d values as follows. Since

$$d + b = Fx_F \tag{13-63}$$

$$d = \frac{Fx_F}{1 + (b/d)}$$

and

$$b = (b/d)d \tag{13-64}$$

Calculated values of d from the first column iteration in the example problem are as follows:

Component	$\frac{\ell_7}{d} + 1$	$\frac{\ell_7}{b} - 1$	$\frac{b}{d}$	Fx_F	d
C_3	1.26	5450	0.000231	5	5.00
i-C_4	1.71	175	000977	15	14.85
n-C_4	2.26	47.7	0.0474	25	23.88
i-C_5	10.3	2.46	0.19	20	3.85
n-C_5	17.5	1.54	11.4	35	2.82
					50.4

The calculated D is 50.4 instead of 48.9. Before these incorrect d (and b) values are used to calculate the stage concentrations, followed by a new set of values of T, V, and L, convergence of the iteration is aided as follows by using the Θ method developed by Holland (*Fundamentals of Multicomponent Distillation*, McGraw-Hill, New York, 1981) and coworkers. A quantity Θ is defined by

$$d' = \frac{Fx_F}{1 + (b/d)\Theta} \tag{13-65}$$

where values of d' are the ones that satisfy

$$\sum_i d' = D_{\text{specified}}$$

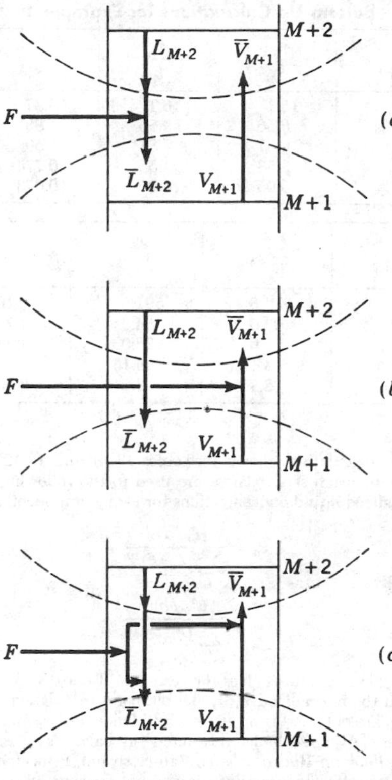

FIG. 13-45 Effect of feed on stream rates just above feed stage $M + 1$. (*a*) Subcooled or bubble-point feed. (*b*) Superheated or dew-point feed. (*c*) Partially flashed feed.

Comparison of Eqs. (13-63) and (13-65) shows that

$$b' = \Theta(b/d)d' \tag{13-66}$$

The value of Θ is found by solving the following nonlinear equation, where D is the specified distillate rate:

$$D - \sum_i \frac{Fx_F}{1 + (b/d)\Theta} = 0 \tag{13-67}$$

For the first column iteration, $\Theta = 1.25$ satisfies this equation. The b/d values and the Θ value are used in Eqs. (13-65) and (13-66) to give the following corrected end concentrations:

Component	$\frac{b}{d}$	d'	b'	x_D	x_1
C_3	0.000231	5.00	0.00144	0.102	0
i-C_4	0.00977	14.82	0.181	0.303	0.004
n-C_4	0.0474	23.60	1.40	0.482	0.027
i-C_5	4.19	3.21	16.79	0.066	0.329
n-C_5	11.4	2.30	32.7	0.047	0.640
		48.9	51.1	1.000	1.000

TABLE 13-11 Top-Down Calculations for Example 4

Component	$R + 1$	A_{10}	$\frac{\ell_{10}}{d}$	$\frac{\ell_{10}}{d} + 1$	A_9	$\frac{\ell_9}{d}$	$\frac{\ell_9}{d} + 1$	A_8	$\frac{\ell_8}{d}$	$\frac{\ell_8}{d} + 1$	A_7	$\frac{\ell_7}{d}$
C_3	3.58	0.260	0.931	1.931	0.232	0.448	1.448	0.212	0.307	1.307	0.198	0.259
i-C_4	3.58	0.522	1.87	2.87	0.450	1.29	2.29	0.405	0.927	1.927	0.371	0.715
n-C_4	3.58	0.693	2.48	3.48	0.590	2.05	3.05	0.526	1.60	2.60	0.484	1.26
i-C_5	3.58	1.44	5.16	6.16	1.22	7.52	8.52	1.05	8.95	9.95	0.936	9.31
n-C_5	3.58	1.72	6.16	7.16	1.46	10.5	11.5	1.23	14.1	15.1	1.09	16.5

TABLE 13-12 Bottom-Up Calculations for Example 4

Component	S_1	$\frac{\ell_2}{b}$	S_2	$\frac{\ell_2}{b}S_2$	$\frac{\ell_3}{b}$	S_3	$\frac{\ell_3}{b}S_3$	$\frac{\ell_4}{b}$	S_4
C_3	15.1	16.1	3.37	54.3	55.3	3.31	183.0	184.0	3.21
i-C_4	8.56	9.56	1.88	18.0	19.0	1.83	34.8	35.8	1.76
n-C_4	6.81	7.81	1.50	11.7	12.7	1.45	18.4	19.4	1.39
i-C_5	3.53	4.53	0.774	3.51	4.51	0.747	3.37	4.37	0.716
n-C_5	3.05	4.05	0.673	2.73	3.73	0.646	2.41	3.41	0.619

Component	$\frac{\ell_4}{b}S_4$	$\frac{\ell_5}{b}$	S_5	$\frac{\ell_5}{b}S_5$	$\frac{\ell_6}{b}$	S_6	$\frac{\ell_6}{b}S_6$	$\frac{\ell_7}{b}$
C_3	590.6	591.6	3.10	1834	1835	2.97	5450	5451
i-C_4	63.0	64.0	1.71	109.4	110.4	1.59	175	176
n-C_4	27.0	28.0	1.34	37.5	38.5	1.24	47.7	48.7
i-C_5	3.13	4.13	0.693	2.86	3.86	0.638	2.46	3.46
n-C_5	2.11	3.11	0.592	1.84	2.84	0.543	1.54	2.54

Stage-to-stage calculations shown in Tables 13-11 and 13-12 provide ℓ/d and ℓ/b values for each stage. These are used in the following equations to calculate normalized liquid concentrations for each component at each stage:

$$x_n = \frac{(\ell_n/d)d'}{\sum_i (\ell_n/d)d'} \tag{13-68}$$

$$x_m = \frac{(\ell_m/b)b'}{\sum_i (\ell_m/b)b'} \tag{13-69}$$

Application of these equations gives the results in Table 13-13. A set of T_n is calculated from the normalized x_n by bubble-point calculations. Corresponding values of y_n are obtained from $y_n = K_n x_n$. Once new x_n and T_n are available, new values of V_n are calculated from energy balances by using data from Maxwell (*Data Book on Hydrocarbons*, Van Nostrand, Princeton, N.J., 1950). First, an estimate of condenser duty is computed from an energy balance around the condenser,

$$Q_c = V_N(H_N^V - H_{N+1}^L) \tag{13-70}$$
$$= 175(18{,}900 - 10{,}750) = 1{,}426{,}000 \text{ Btu/h (417.9 kW)}$$

The reboiler duty Q_r is obtained from an overall energy balance,

$$Q_r = DH_{N+1}^L + BH_1^L + Q_c - FH_F$$
$$= (48.9)(10{,}750) + (51.1)(17{,}080) + 1{,}426{,}000 - 100(13{,}540) \tag{13-71}$$
$$= 1{,}465{,}000 \text{ Btu/h (429.3 kW)}$$

A new set of values of V_m is obtained from energy balances around the bottom section of the column,

$$V_m = \frac{Q_r + B(H_{M+1}^L - H_B^V)}{H_m^V - H_{m+1}^L} \tag{13-72}$$

Similar balances around the top section yield a new set of values of V_n. Corresponding values of L_n and L_m are obtained by material balances around the top and bottom sections respectively. The new V, L, and T profiles are listed in Table 13-14. In this example, they do not differ much from the initial guesses in Table 13-10.

It should be noted in Table 13-14 that it is not necessary to list two values of V, L, and T for the feed stage (stage 6) because the TG procedure gives a perfect match at the feed stage in each trial. This completes the first column iteration.

The new temperature and flow-rate profiles (which would be used as the assumptions to begin the second column iteration) are compared in Fig. 13-46 with the final solution. Both profiles are moving toward the final result.

Figure 13-47 shows the concentration profiles from the final solution. Note the discontinuities at the feed stage and the fact that feed-stage composition differs considerably from feed-stream composition. It can be seen in Fig. 13-47 from the n-C_4 and i-C_5 profiles that the separation between the keys improves rapidly with stage number; additional stages would be worthwhile.

Convergence to the final solution is rapid with the TG method for narrow-boiling feeds but may be slow for wide-boiling feeds. Generally, at least five column iterations are required. Convergence is obtained when successive sets of tear variables are identical to approximately four significant digits. This is accompanied by $\Theta = 1.0$, x = normalized x, and nearly identical successive values of Q_c as well as Q_r.

EQUATION-TEARING PROCEDURES USING THE TRIDIAGONAL-MATRIX ALGORITHM

As seen earlier, the manual Thiele-Geddes method involves solving the equilibrium-stage equations one at a time. More powerful, flexible, and reliable computer programs are based on the application of sparse matrix methods for solving simultaneously all or at least some of the equations. For cases in which combined column feeds represent mixtures that boil within either a narrow range (typical of many

TABLE 13-13 Stage Compositions from First Trial of Example 4

Component	x_1	$\frac{\ell_2}{b}b'$	x_2	$\frac{\ell_3}{b}b'$	x_3	$\frac{\ell_4}{b}b'$	x_4	$\frac{\ell_5}{b}b'$	x_5	$\frac{\ell_6}{b}b'$
C_3	0.000	0.0232	0.000	0.0796	0.000	0.279	0.001	0.852	0.004	2.65
i-C_4	0.004	1.73	0.008	3.44	0.016	6.48	0.030	11.6	0.052	20.0
n-C_4	0.027	10.9	0.049	17.8	0.081	27.2	0.124	39.2	0.176	53.9
i-C_5	0.329	76.1	0.344	75.7	0.346	73.4	0.335	69.3	0.311	64.8
n-C_5	0.640	132.4	0.599	122.0	0.557	111.5	0.510	101.7	0.447	92.9
	1.000	221.1	1.000	219.0	1.000	218.9	1.000	226.6	1.000	234.2

Component	x_6	$\frac{\ell_7}{d}d'$	x_7	$\frac{\ell_8}{d}d'$	x_8	$\frac{\ell_9}{d}d'$	x_9	$\frac{\ell_{10}}{d}d'$	x_{10}
C_3	0.011	1.295	0.012	1.535	0.013	2.240	0.019	4.66	0.038
i-C_4	0.085	10.6	0.097	13.7	0.120	19.1	0.162	27.7	0.228
n-C_4	0.230	29.7	0.271	37.8	0.331	48.4	0.410	58.5	0.481
i-C_5	0.277	29.9	0.273	28.7	0.252	24.1	0.204	16.6	0.136
n-C_5	0.397	37.9	0.347	32.4	0.284	24.1	0.204	14.2	0.117
	1.000	109.4	1.000	114.1	1.000	118.0	1.000	121.7	1.000

TABLE 13-14 New Temperature and Rate Profiles from the First Trial of Example 4

n	New T	$H_D^L - H_{n+1}^L$	$D(H_D^L - H_{n+1}^L) + Q_c$	$H_N^V - H_{n+1}^L$	V	L
10	160.0	0	1,426,000	8150	175.0	124.0
9	175.0	−1220	1,367,000	7900	172.9	119.6
8	186.0	−2190	1,319,000	7830	168.5	117.6
7	194.0	−3010	1,279,000	7680	166.5	114.2
6	200.0	−3490	1,256,000	7700	163.1	214.3
m	New T	$H_{m+1}^L - H_B^L$	$B(H_{m+1}^L - H_B^L) + Q_r$	$H_m^V - H_{m+1}^L$	V	L
5	211.0	−2480	1,338,000	8200	163.2	214.7
4	220.0	−1780	1,374,000	8400	163.6	215.5
3	228.0	−1130	1,407,000	8560	164.4	221.1
2	233.5	− 560	1,438,000	8460	170.0	224.0
1	237.5	− 210	1,454,000	8410	172.9	51.1

distillation operations) or a wide range (typical of absorbers and strippers) and in which great flexibility of problem specifications is not required, equation-tearing procedures that involve solving simultaneously certain subsets of the equations can be applied. Two such equation-tearing procedures are the bubble-point (BP) method for narrow-boiling mixtures suggested by Friday and Smith (op. cit.) and developed in detail by Wang and Henke [*Hydrocarbon Process.*, **45** (8), 155 (1966)], and the sum-rates (SR) method for wide-boiling mixtures proposed by Sujuta [*Hydrocarbon Process.*, **40**(12), **137** (1961)] and further developed by Burningham and Otto [*Hydrocarbon Process.*, **46**(10), 163 (1967)]. Both methods start with the same primitive equations for the theoretical model of an equilibrium stage as presented next.

Consider a general, continuous-flow, steady-state, multicomponent, multistage separation operation. Assume that phase equilibrium between an exiting vapor phase and a single exiting liquid phase is achieved at each stage, that no chemical reactions occur, and that neither of the exiting phases entrains the other phase. A general schematic representation of such a stage j is shown in Fig. 13-48. Entering stage j is a single- or two-phase feed at molal flow rate F_j, temperature T_{Fj}, and pressure P_{Fj} and with overall composition in mole fractions $z_{i,j}$. Also entering stage j is interstage liquid from adjacent stage $j - 1$ above at molal flow rate L_{j-1}, temperature T_{j-1}, pressure P_{j-1}, and mole fractions $x_{i,j-1}$. Similarly, interstage vapor from adjacent stage $j + 1$ below enters at molal flow rate V_{j+1}, T_{j+1}, P_{j+1} and mole fractions $y_{i,j+1}$. Heat is transferred from (+) or to (−) stage j at rate Q_j to simulate a condenser, reboiler, intercooler, interheater, etc. Equilibrium vapor and liquid phases leave stage j at T_j and P_j and with mole fractions $y_{i,j}$ and $x_{i,j}$ respectively. The vapor may be partially withdrawn from the column as a sidestream at a molal flow rate W_j, with the remainder V_j sent to adjacent stage $j - 1$ above. Similarly, exiting liquid may be split into a sidestream at a molal flow rate of U_j, with the remainder L_j sent to adjacent stage j below.

For each stage j, the following $2C + 3$ component material-balance (M), phase-equilibrium (E), mole-fraction-summation (S), and

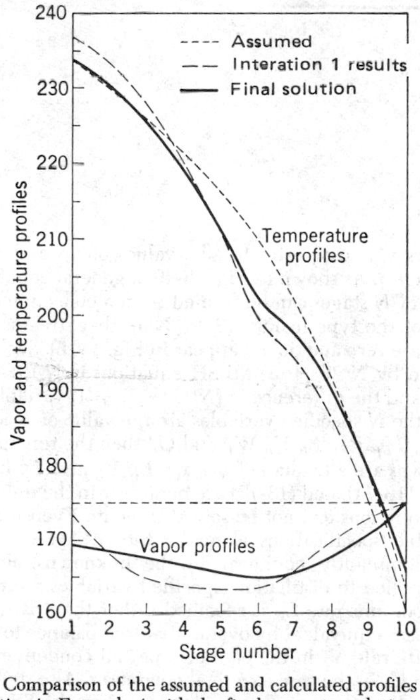

FIG. 13-46 Comparison of the assumed and calculated profiles from the first column iteration in Example 4 with the final computer solution.

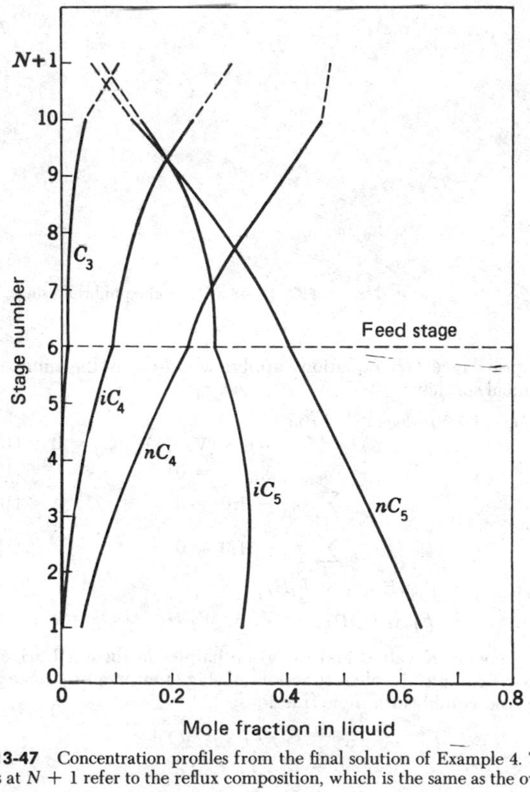

FIG. 13-47 Concentration profiles from the final solution of Example 4. The points at $N + 1$ refer to the reflux composition, which is the same as the overhead vapor.

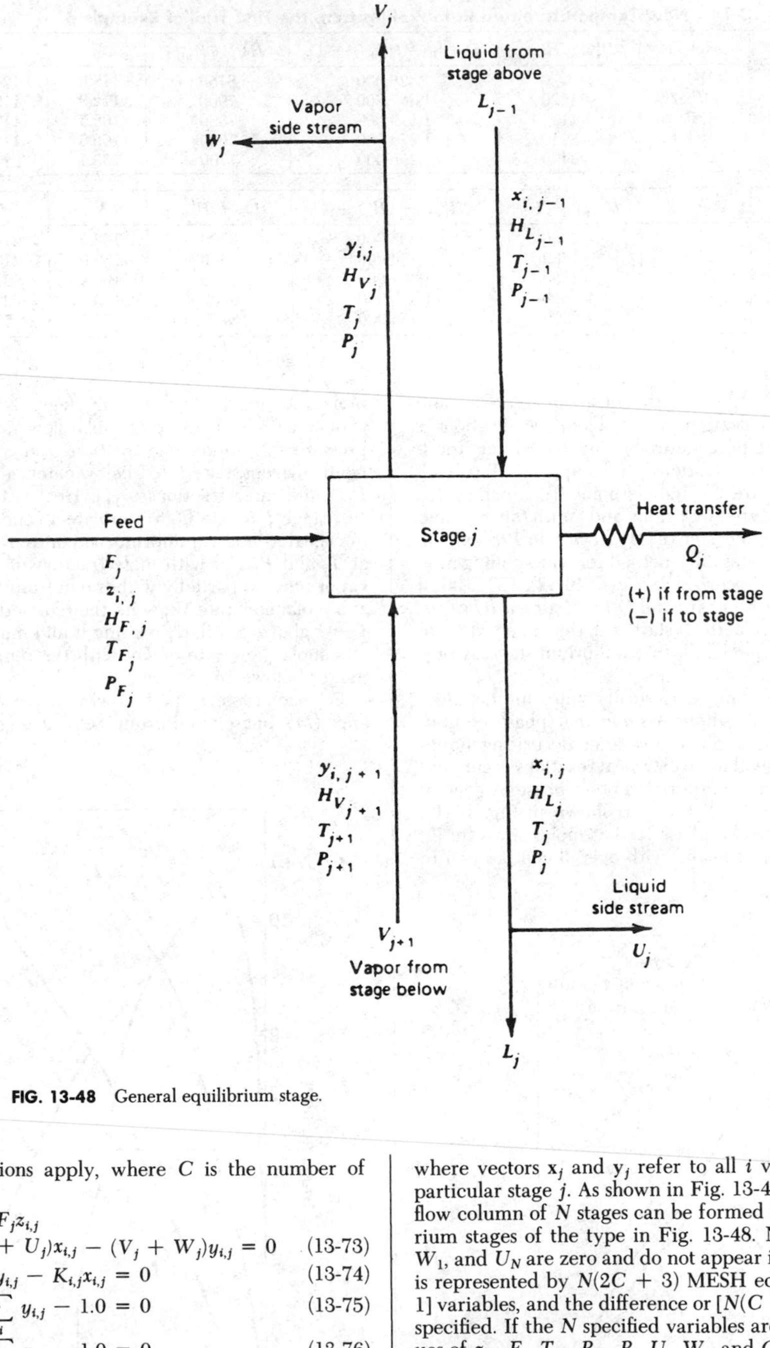

FIG. 13-48 General equilibrium stage.

energy-balance (H) equations apply, where C is the number of chemical species:

$$L_{j-1}x_{i,j-1} + V_{j+1}y_{i,j+1} + F_j z_{i,j} - (L_j + U_j)x_{i,j} - (V_j + W_j)y_{i,j} = 0 \quad (13\text{-}73)$$

$$y_{i,j} - K_{i,j}x_{i,j} = 0 \quad (13\text{-}74)$$

$$\sum_i y_{i,j} - 1.0 = 0 \quad (13\text{-}75)$$

$$\sum_i x_{i,j} - 1.0 = 0 \quad (13\text{-}76)$$

$$L_{j-1}H_{L_{j-1}} + V_{j+1}H_{V_{j+1}} + F_jH_{F_j} - (L_j + U_j)H_{L_j} - (V_j + W_j)H_{V_j} - Q_j = 0 \quad (13\text{-}77)$$

In general, K values and molal enthalpies in these MESH equations are complex implicit functions of stage temperature, stage pressure, and equilibrium mole fractions:

$$K_{i,j} = K_{i,j}\{T_j, P_j, \mathbf{x}_j, \mathbf{y}_j\}$$
$$H_{V_j} = H_{V_j}\{T_j, P_j, \mathbf{y}_j\}$$
$$H_{L_j} = H_{L_j}\{T_j, P_j, \mathbf{x}_j\}$$

where vectors $\mathbf{x}_j$ and $\mathbf{y}_j$ refer to all i values of $x_{i,j}$ and $y_{i,j}$ for the particular stage j. As shown in Fig. 13-49, a general countercurrent-flow column of N stages can be formed from a collection of equilibrium stages of the type in Fig. 13-48. Note that streams L_0, V_{N+1}, W_1, and U_N are zero and do not appear in Fig. 13-49. Such a column is represented by $N(2C + 3)$ MESH equations in $[N(3C + 10) + 1]$ variables, and the difference or $[N(C + 7) + 1]$ variables must be specified. If the N specified variables are the value of N and all values of $z_{i,j}$, F_j, T_{Fj}, P_{Fj}, P_j, U_j, W_j, and Q_j, then the remaining $N(2C + 3)$ unknowns are all values of $y_{i,j}$, $x_{i,j}$, L_j, V_j, and T_j. In this case, Eqs. (13-73), (13-74), and (13-77) are nonlinear in the unknowns and the MESH equations can not be solved directly. Even if a different set of variable specifications is made, the MESH equations still remain predominantly nonlinear in the unknowns. For the BP method as applied to distillation, specified variables are those listed except that bottoms rate L_N is specified rather than partial reboiler duty Q_N. This is equivalent by overall material balance to specifying vapor-distillate rate V_1 in the case of a partial condenser or liquid-distillate rate U_1 in the case of a total condenser. Also, reflux rate L_1 is specified rather than condenser duty Q_1. For the SR method as

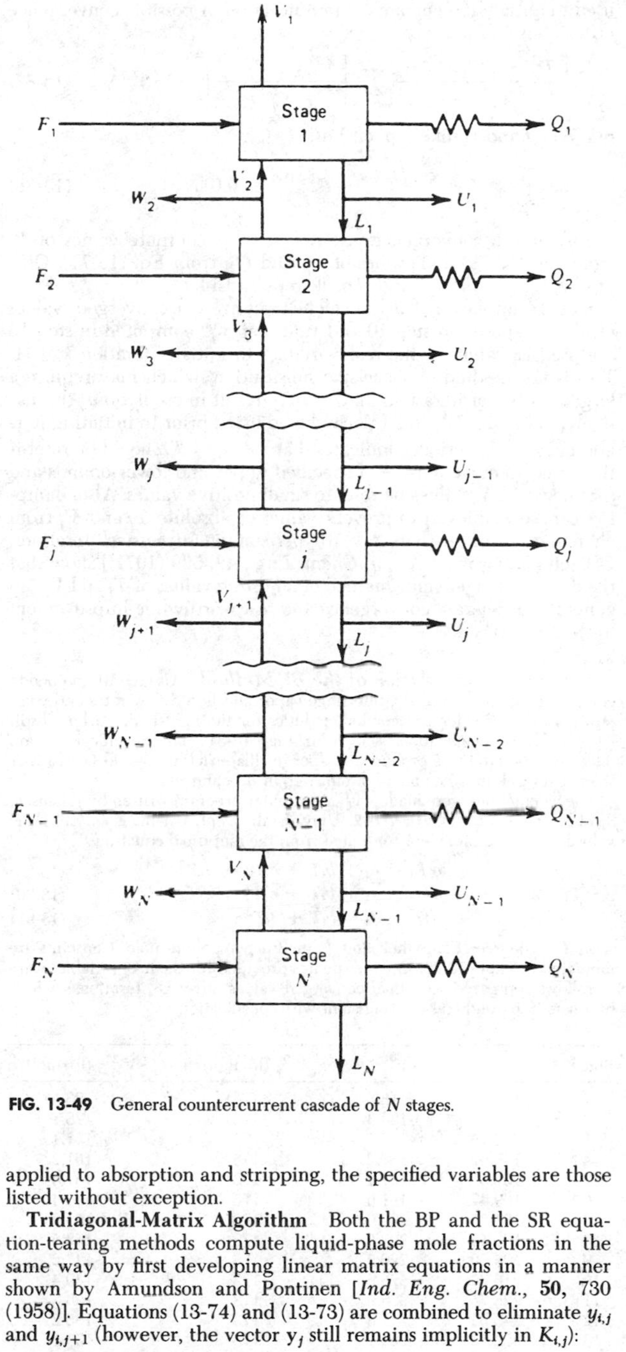

FIG. 13-49 General countercurrent cascade of N stages.

applied to absorption and stripping, the specified variables are those listed without exception.

Tridiagonal-Matrix Algorithm Both the BP and the SR equation-tearing methods compute liquid-phase mole fractions in the same way by first developing linear matrix equations in a manner shown by Amundson and Pontinen [*Ind. Eng. Chem.*, **50**, 730 (1958)]. Equations (13-74) and (13-73) are combined to eliminate $y_{i,j}$ and $y_{i,j+1}$ (however, the vector $\mathbf{y}_j$ still remains implicitly in $K_{i,j}$):

$$L_{j-1}x_{i,j-1} + V_{j+1}K_{i,j+1}x_{i,j+1} + F_j z_{i,j} - (L_j + U_j)x_{i,j} - (V_j + W_j)K_{i,j}x_{i,j} = 0 \qquad (13\text{-}78)$$

Next, Eq. (13-73) is summed over the C components and over stages 1 through j and combined with Eqs. (13-75), (13-76), and $\sum_i z_{i,j} - 1.0 = 0$ to give a total material balance over stages 1 through j:

$$L_j = V_{j+1} + \sum_{m=1}^{j} (F_m - U_m - W_m) - V_1 \qquad (13\text{-}79)$$

By combining Eq. (13-78) with Eq. (13-79), L_j is eliminated to give the following working equations for component material balances:

$$A_j x_{i,j-1} + B_{i,j}x_{i,j} + C_{i,j}x_{i,j+1} = D_{i,j} \qquad (13\text{-}80)$$

where
$$A_j = V_j + \sum_{m=1}^{j-1} (F_m - W_m - U_m) - V_i \qquad (13\text{-}81)$$
$$2 \le j \le N$$

$$B_{i,j} = -\left[V_{j+1} + \sum_{m=1}^{j} (F_m - W_m - U_m) - V_1 + U_j + (V_j + W_j)K_{i,j} \right] \qquad (13\text{-}82)$$
$$1 \ll j \ll N$$

$$C_{i,j} = V_{j+1}K_{i,j+1} \qquad 1 \ll j \ll N - 1 \qquad (13\text{-}83)$$

$$D_{i,j} = -F_j z_{i,j} \qquad 1 \ll j \ll N \qquad (13\text{-}84)$$

The NC equations (13-80) are linearized in terms of the NC unknowns $x_{i,j}$ by selecting unknowns V_j and T_j as tear variables and using values of vectors $\mathbf{x}_j$ and $\mathbf{y}_j$ from the previous iteration to compute values of $K_{i,j}$ for the current iteration. In this manner all values of A_j, $B_{i,j}$, and $C_{i,j}$ can be estimated. Values of $D_{i,j}$ are fixed by feed specifications. Furthermore, the NC equations (13-80) can be partitioned into C sets, one for each component, and solved separately for values of $\mathbf{x}_i$, which pertains to all j values of $x_{i,j}$ for the particular species i. Each set of N equations is a special type of sparse matrix equation called a tridiagonal-matrix equation, which has the form shown in Fig. 13-50a for a five-stage example in which, for convenience, the subscript i has been dropped from the coefficients B, C, and D. For this type of sparse matrix equation, we can apply a highly efficient version of the gaussian elimination procedure called the Thomas algorithm, which avoids matrix inversion, eliminates the need to store the zero coefficients in the matrix, almost always avoids buildup of truncation errors, and rarely produces negative values of $x_{i,j}$.

The Thomas algorithm begins by a forward elimination, row by row starting down from the top row ($j = 1$, the condenser stage), to give the following replacements shown in Fig. 13-50b. For row 1:

$$p_1 = C_1/B_1,\ q_1 = D_1/B_1,\ B_1 \rightarrow 1,\ C_1 \rightarrow p_1,\ D_1 \rightarrow q_1$$

where $\rightarrow$ means "is replaced by."

For all subsequent rows:

$$p_j = C_j/(B_j - A_j p_{j-1}),\ q_j = (D_j - A_j q_{j-1})/(B_j - A_j p_{j-1}),$$
$$A_j \rightarrow 0,\ B_j \rightarrow 1,\ C_j \rightarrow p_j,\ D_j \rightarrow q_j$$

$$\begin{bmatrix} B_1 & C_1 & 0 & 0 & 0 \\ A_2 & B_2 & C_2 & 0 & 0 \\ 0 & A_3 & B_3 & C_3 & 0 \\ 0 & 0 & A_4 & B_4 & C_4 \\ 0 & 0 & 0 & A_5 & B_5 \end{bmatrix} \begin{bmatrix} x_1 \\ x_2 \\ x_3 \\ x_4 \\ x_5 \end{bmatrix} = \begin{bmatrix} D_1 \\ D_2 \\ D_3 \\ D_4 \\ D_5 \end{bmatrix}$$

(a)

$$\begin{bmatrix} 1 & p_1 & 0 & 0 & 0 \\ 0 & 1 & p_2 & 0 & 0 \\ 0 & 0 & 1 & p_3 & 0 \\ 0 & 0 & 0 & 1 & p_4 \\ 0 & 0 & 0 & 0 & 1 \end{bmatrix} \begin{bmatrix} x_1 \\ x_2 \\ x_3 \\ x_4 \\ x_5 \end{bmatrix} = \begin{bmatrix} q_1 \\ q_2 \\ q_3 \\ q_4 \\ q_5 \end{bmatrix}$$

(b)

FIG. 13-50 Tridiagonal-matrix equation for a column with five theoretical stages. (a) Original equation. (b) After forward elimination.

At the bottom row for component i, $x_N = q_N$. The remaining values of x_j for species i are computed recursively by backward substitution:

$$x_{j-1} = q_{j-1} - p_{j-1}x_j$$

BP Method for Distillation The bubble-point method for distillation, particularly when the components involved cover a relatively narrow range of volatility, proceeds iteratively by the following steps, where k is the iteration index for the entire distillation column.

1. Specify N and all values of $z_{i,j}$, F_j, T_{Fj}, P_{Fj}, P_j, U_j, W_j, and Q_j, except Q_1 and Q_N.
2. Specify type of condenser. If total ($U_1 \neq 0$), compute L_N from overall material balance; if partial ($U_1 = 0$), specify V_1 and compute L_N from overall material balance.
3. Specify reflux rate L_1, assuming no subcooling.
4. Compute $V_2 = V_1 + L_1 + U_1 - F_1$.
5. Provide initial guesses ($k = 0$) or values of all tear variables T_j and V_j ($j > 2$). Temperature guesses are readily obtained by linear interpolation between estimates of top- and bottom-stage temperatures. The bottom-stage temperature is estimated by making a bubble-point-temperature calculation by using an estimate of bottoms composition at the specified bottom-stage pressure. A similar calculation is made at the top stage by using an estimate of distillate composition; otherwise, for a partial condenser, a dew-point temperature calculation is made. An estimate of the vapor-rate profile is readily obtained by assuming constant molal overflow down the column.
6. Set index $k = 1$ to initiate the first column iteration.
7. Using specified stage pressures, current estimates of stage temperatures, and current estimates of stage vapor- and liquid-phase compositions, estimate all $K_{i,j}$ values (for $k = 1$, initial estimates of stage phase compositions may be necessary if $K_{i,j}$ values are sensitive to phase compositions).
8. Compute values of $x_{i,j}$ by solving Eqs. (13-80) through (13-84) by the tridiagonal-matrix algorithm once for each component. Unless all mesh equations are converged, $\sum_i x_{i,j} \neq 1$ for each stage j.
9. To force $\sum_i x_{i,j} = 1$ at each stage j, normalize values by the replacement $x_{i,j} = x_{i,j}/\sum_i x_{i,j}$.
10. Compute a new set of values of $T_j^{(k)}$ tear variables by computing, one at a time, the bubble-point temperature at each stage based on the specified stage pressure and corresponding normalized $x_{i,j}$ values. The equation used is obtained by combining Eqs. (13-74) and (13-75) to eliminate $y_{i,j}$ to give

$$\sum_i K_{i,j}\{T_j, P_j, \mathrm{x}_j, \mathrm{y}_j\} = 1.0 \qquad (13\text{-}85)$$

which is a nonlinear equation in $T_j^{(k)}$ and must be solved iteratively by some appropriate root-finding method, such as the Newton-Raphson or the Muller method.

11. Compute values of $y_{i,j}$ one at a time from Eq. (13-74).
12. Compute a new set of values of the V_j tear variables one at a time, starting with V_3, from an energy-balance equation that is obtained by combining Eqs. (13-77) and (13-79), eliminating L_{j-1} and L_j to give

$$V_j = (\tilde{C}_{j-1} - \tilde{A}_{j-1}V_{j-1})/\tilde{B}_{j-1} \qquad (13\text{-}86)$$

where

$$\tilde{A}_{j-1} = H_{Lj-2} - H_{Vj-1}$$
$$\tilde{B}_{j-1} = H_{Vj} - H_{Lj-1}$$
$$\tilde{C}_{j-1} = [\sum_{m=1}^{j-2} (F_m - W_m - U_m) - V_1])(H_{Lj-1} - H_{Lj-2}) + F_{j-1}(H_{Lj-1} - H_{Fj-1}) + W_{j-1}(H_{Vj-1} - H_{Lj-1}) + Q_{j-1}$$

13. Check to determine if the new sets of tear variables $T_j^{(k)}$ and $V_j^{(k)}$ are within some prescribed tolerance of sets $T_j^{(k-1)}$ and $V_j^{(k-1)}$ used to initiate the current column iteration. A possible convergence criterion is

$$\sum_{j=1}^{N} \left[\frac{T_j^{(k)} - T_j^{(k-1)}}{T_j^{(k)}}\right]^2 + \sum_{j=3}^{N} \left[\frac{V_j^{(k)} - V_j^{(k-1)}}{V_j^{(k)}}\right]^2 \leq 10^{-7}N \qquad (13\text{-}87)$$

but Wang and Henke (op. cit.) use

$$\sum_{j=1}^{N} [T_j^{(k)} - T_j^{(k-1)}]^2 \leq 0.01N \qquad (13\text{-}88)$$

14. If the convergence criterion is met, compute values of L_j from Eq. (13-79) and values of Q_1 and Q_N from Eq. (13-77). Otherwise, set $k = k + 1$ and repeat steps 7 to 14.

Step 14 implies that if the calculations are not converged, values of $T_j^{(k)}$ computed in step 10 and values of $V_j^{(k)}$ computed in step 12 are used as values of the tear variables to initiate iteration $k + 1$. This is the method of successive substitution, which may require a large number of iterations and/or may result in oscillation. Alternatively, values of $T_j^{(k)}$ and $V_j^{(k)}$ can be adjusted prior to initiating iteration $k + 1$. Experience indicates that values of T_j should be reset if they tend to move outside of specified upper and lower bounds and that negative V_j values be reset to small positive values. Also, damping can be employed to prevent values of absolute T_j and V_j from changing by more than, say, 10 percent on successive iterations. Orbach and Crowe [*Can. J. Chem. Eng.*, **49**, 509 (1971)] show that the dominant eigenvalue method of adjusting values of T_j and V_j can generally accelerate convergence and is a worthwhile improvement to the BP method.

Example 5: Calculation of the BP Method Use the BP method to compute stage temperatures, interstage vapor and liquid flow rates and compositions, and reboiler and condenser duties for the light-hydrocarbon distillation-column specifications shown in Fig. 13-51. The specifications are selected to obtain three products: a vapor distillate rich in C_2 and C_3, a vapor sidestream rich in nC_4, and a bottoms rich in nC_5 and nC_6.

The calculations were made with a computer program written by Johansen, Seader, and Blauth (WHENDI-2, Univ. Utah, Dept. Chem. Eng., 1981). K values and enthalpies were computed from the empirical equations

$$\ln K_i = a_i + b_iT + c_iT^2 + d_iT^3 \qquad (13\text{-}89)$$
$$H_{V_i} = \bar{a}_i + \bar{b}_iT + \bar{c}_iT^2 \qquad (13\text{-}90)$$
$$H_{L_i} = \tilde{a}_i + \tilde{b}_iT + \tilde{c}_iT^2 \qquad (13\text{-}91)$$

with T in degrees Fahrenheit and H in Btu per pound-mole. Constants are tabulated in Table 13-15. Initial estimates provided for the tear variables were as follows compared with final converged values (after 23 iterations), where numbers in parentheses are consistent with specifications:

Stage	$T^{(0)}$, °F	$T_j^{(23)}$, °F	$V^{(0)}$, (lb·mol)/h	$V^{(23)}$, (lb·mol)/h
1	110.00	118.6	(23)	23
2	121.87	134.3	(173)	173
3	133.75	150.7	173	167.1
4	145.62	167.2	173	161.2
5	157.50	181.1	173	157.8
6	169.37	191.6	173	156.4
7	181.25	200.0	173	155.2
8	193.12	208.6	173	152.5
9	205.00	220.2	173	147.2
10	216.87	228.9	173	149.4
11	228.75	235.6	173	153.3
12	240.62	241.8	173	155.2
13	252.50	249.1	173	155.1
14	264.37	257.7	210	189.7
15	276.25	268.2	210	185.9
16	288.12	280.8	210	182.0
17	300.00	295.3	210	178.7

NOTE: To convert degrees Fahrenheit to degrees Celsius, °C = (°F − 32)/1.8. To convert pound-moles per hour to kilogram-moles per second, multiply by 1.26×10^{-4}.

By employing successive substitution of the tear variables and the criterion of Eq. (13-88), convergence was achieved slowly, but without oscillation, in 23 iterations. Computed products are:

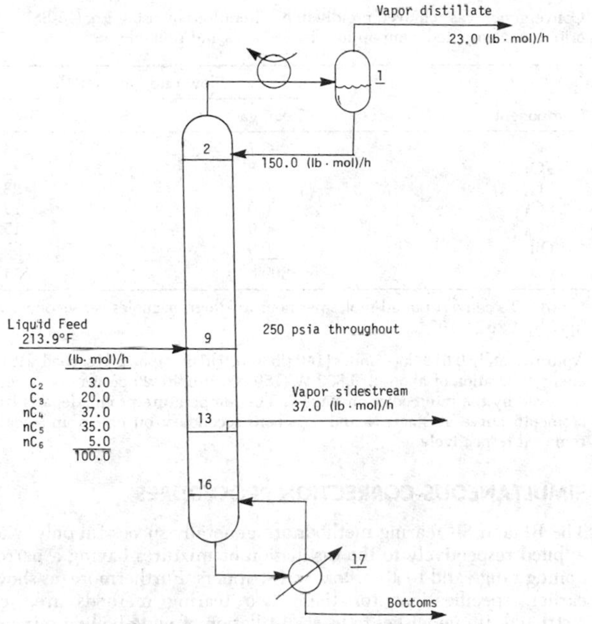

FIG. 13-51 Specifications for the calculation of distillation by the BP method.

Component	Distillate	Sidestream	Bottoms
	Flow rate, (lb·mol)/h		
C_2	3.0	0.0	0.0
C_3	18.8	1.2	0.0
nC_4	1.2	27.2	8.6
nC_5	0.0	8.2	26.8
nC_6	0.0	0.4	4.6
	23.0	37.0	40.0

NOTE: To convert pound-moles per hour to kilogram-moles per second, multiply by 1.26×10^{-4}.

Examination of interstage-composition results showed that maximum nC_4 composition was achieved in the vapor leaving stage 12 rather than stage 13. Therefore, if the sidestream location were moved up one stage, a somewhat higher purity of nC_4 could probably be achieved in that stream. Further improvement in purity of the sidestream as well as the other two products could be achieved by increasing the reflux rate and/or number of stages. Computed condenser and reboiler duties were 271,700 and 425,200 W (927,000 and 1,451,000 Btu/h) respectively.

SR Method for Absorption and Stripping As shown by Friday and Smith (op. cit.), when an attempt is made to apply the BP method to absorption and stripping in which the volatility range of the chemical components in the column is very wide, calculations of stage temperatures from Eq. (13-85) become very sensitive to liquid compositions. This generally causes very oscillatory excursions in temperature from iteration to iteration, making it impossible to obtain convergence. A very successful modification of the BP method for such cases is the sum-rates method, in which new stage temperatures are computed instead from the energy-balance equation. Interstage vapor rates are computed by material balance from new interstage liquid rates that are obtained by multiplying the previous interstage liquid rates by corresponding unnormalized liquid mole-fraction summations computed from the tridiagonal-matrix algorithm. The SR method proceeds by the following steps:

1. Specify N and all values of $z_{i,j}$, F_j, T_{Fj}, P_{Fj}, P_j, U_j, W_j, and Q_j. For an adiabatic operation, all Q_j are zero.
2. Provide initial guesses ($k = 0$) for values of all tear variables T_j and V_j. Temperature guesses are readily obtained by linear interpolation between estimates of the top- and bottom-stage temperatures, taking the top as that of the liquid feed to the top stage and the bottom as that of the vapor feed to the bottom stage. An estimate of the vapor-rate profile is readily obtained by assuming constant molal overflow working up from the bottom in using the specified vapor feed or feeds. Compute corresponding initial values of L_j from Eq. (13-79).
3. Same as step 6 of the BP method.
4. Same as step 7 of the BP method.
5. Same as step 8 of the BP method.
6. Compute a new set of values of L_j from the sum-rates equation:

$$L_j^{(k+1)} = L_j^{(k)} \sum_i x_{i,j} \tag{13-92}$$

7. Compute a corresponding new set of V_j tear variables from the following total material balance, which is obtained by combining Eq. (13-79) with an overall material balance around the column:

$$V_j = L_{j-1} - L_N + \sum_{m=j}^{N} (F_m - W_m - U_m) \tag{13-93}$$

8. Same as step 9 of the BP method.
9. Same as step 11 of the BP method.
10. Normalize values of $y_{i,j}$.
11. Compute a new set of values of the T_j tear variables by solving simultaneously the set of N energy-balance equations (13-77), which are nonlinear in the temperatures that determine the enthalpy values. When linearized by a Newton iterative procedure, a tridiagonal-matrix equation that is solved by the Thomas algorithm is obtained. If we set g_j equal to Eq. (13-77), i.e., its residual, the linearized equations to be solved simultaneously are

$$\left(\frac{\partial g_1}{\partial T_1}\right)^{(r)} \Delta T_1 + \left(\frac{\partial g_1}{\partial T_2}\right)^{(r)} \Delta T_2 = -g_1^{(r)} \tag{13-94}$$

$$\left(\frac{\partial g_j}{\partial T_{j-1}}\right)^{(r)} \Delta T_{j-1} + \left(\frac{\partial g_j}{\partial T_j}\right)^{(r)} \Delta T_j + \left(\frac{\partial g_j}{\partial T_{j+1}}\right)^{(r)} \Delta T_{j+1} = -g_j^{(r)} \qquad 2 \leq j \leq N-1 \tag{13-95}$$

$$\left(\frac{\partial g_N}{\partial T_{N-1}}\right)^{(r)} \Delta T_{N-1} + \left(\frac{\partial g_j}{\partial T_N}\right)^{(r)} \Delta T_N = -g_N^{(r)} \tag{13-96}$$

where $\Delta T_j = T_j^{(r+1)} - T_j^{(r)}$, and thus $T_j^{(r+1)} = T_j^{(r)} + \Delta T_j$, and r is the iteration index. The partial derivatives depend upon the enthalpy correlations utilized and may be obtained analytically or numerically. Simultaneous Eqs. (13-94) to (13-96) are solved iteratively until

TABLE 13-15 Physical-Property Constants at 250 psia for Distillation Example of Fig. 13-51

	K_i				H_{V_i}			H_{L_i}		
Component, i	a_i	$b_i \times 10$	$c_i \times 10^4$	$d_i \times 10^7$	$\bar{a}_i \times 10^{-5}$	$\bar{b}_i \times 10^{-2}$	$\bar{c}_i \times 10^2$	$\tilde{a}_i \times 10^{-4}$	$\tilde{b}_i \times 10^{-2}$	$\tilde{c}_i \times 10$
C_2	−0.13277	0.13069	−0.34699	0.34445	0.08314	0.13189	0.60533	0.3740	0.1960	0.120
C_3	−1.3320	0.13635	−0.21765	0.11496	0.1195	0.12351	2.2093	0.4740	0.26267	0.18667
nC_4	−2.8437	0.20239	−0.41598	0.37815	0.1655	0.09010	3.7800	0.5870	0.31467	0.22667
nC_5	−4.2534	0.25040	−0.51232	0.46928	0.1920	0.2950	0.720	0.6870	0.3830	0.260
nC_6	−5.7430	0.31130	−0.65543	0.60689	0.2200	0.370	0	0.7820	0.44133	0.29333

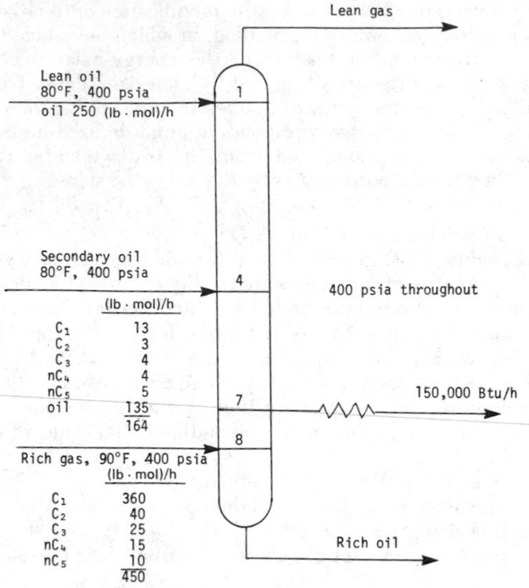

FIG. 13-52 Specifications for the calculation of an absorber by the SR method.

until corrections ΔT_j and, therefore, residual values of g_j approach zero.

12. Same as step 13 of the BP method.
13. If the convergence criterion is not met, set $k = k + 1$ and repeat steps 4 to 13.

With the SR method, convergence is often rapid even when successive substitution of T_j and V_j is used from one iteration to the next.

Example 6: Calculation of the SR Method Use the SR method to compute stage temperatures and interstage vapor and liquid flow rates and compositions for absorber-column specifications shown in Fig. 13-52. Note that a secondary absorber oil is used in addition to the main absorber oil and that heat is withdrawn from the seventh theoretical stage.

Calculations were made with a computer program written by Shinohara, Johansen, Seader, and Blauth (STAB-2, Univ. of Utah, Dept. Chem. Eng., 1981). K values and enthalpies were computed from empirical Eqs. (13-89) to (13-91) by using constants tabulated in Table 13-16. Initial estimates provided for the tear variables were as follows compared with final converged values obtained after five iterations:

Stage	$T^{(0)}$, °F	$T^{(5)}$, °F	$V^{(0)}$, (lb·mol)/h	$V^{(5)}$, (lb·mol)/h
1	80.00	84.9	450	309.8
2	81.43	86.0	450	363.3
3	82.86	86.7	450	366.8
4	84.29	85.8	450	370.1
5	85.71	86.5	450	389.3
6	87.14	87.1	450	393.3
7	88.57	85.0	450	398.1
8	90.00	91.3	450	410.3

NOTE: To convert degrees Fahrenheit to degrees Celsius, °C = (°F − 32)/1.8. To convert pound-moles per hour to kilogram-moles per second, multiply by 1.26×10^{-4}.

Convergence was achieved rapidly in five iterations by using Eq. (13-88) as the criterion. Computed compositions for lean gas and rich oil are:

Component	Flow rate, (lb·mol)/h	
	Lean gas	Rich oil
C_1	298.7	74.3
C_2	10.8	32.2
C_3	0.3	28.7
nC_4	0.0	19.0
nC_5	0.0	15.0
Oil	0.0	385.0
	309.8	554.2

NOTE: To convert pound-moles per hour to kilogram-moles per second, multiply by 1.26×10^{-4}.

Approximately 0.018 (kg·mol)/s [140 (lb·mol)/h] of vapor is absorbed with an energy liberation of about 219,800 W (750,000 Btu/h), 20 percent of which is removed by the intercooler on stage 7. The temperature profile departs from a smooth curve at stages 4 and 7, where secondary oil enters and heat is removed respectively.

SIMULTANEOUS-CORRECTION PROCEDURES

The BP and SR tearing methods are generally successful only when applied respectively to the distillation of mixtures having a narrow boiling range and to absorbers and strippers. Furthermore, as shown earlier, specifications for these two tearing methods are very restricted. If one wishes to treat distillation of wide-boiling mixtures and other operations shown in Fig. 13-7 such as rectification, reboiled stripping, reboiled absorption, and refluxed stripping, it is usually necessary to utilize other procedures. One class of such procedures involves the solution of most or all of the MESH equations or their equivalent simultaneously by some iterative technique such as a Newton or a quasi-Newton method. Such simultaneous-correction (SC) methods are also useful for separations involving very nonideal liquid mixtures including extractive and azeotropic distillation or for cases in which considerable flexibility in specifications is desired.

The development of an SC procedure involves a number of important decisions: (1) What variables should be used? (2) What equations should be used? (3) How should variables be ordered? (4) How should equations be ordered? (5) How should flexibility in specifications be provided? (6) Which derivatives of physical properties should be retained? (7) How should equations be linearized? (8) If Newton or quasi-Newton linearization techniques are employed, how should the Jacobian be updated? (9) Should corrections to unknowns that are computed at each iteration be modified to dampen or accelerate the solution or be kept within certain bounds? (10) What convergence criterion should be applied?

Perhaps because of these many decisions, a large number of SC procedures have been published. Two quite different procedures that have achieved a significant degree of utilization in solving practical problems include the methods of Naphtali and Sandholm [*Am. Inst. Chem. Eng. J.*, **17**, 148 (1971)] and Goldstein and Stanfield [*Ind. Eng. Chem. Process Des. Dev.*, **9**, 78 (1970)]. The former procedure is of particular interest because, in principle, it can be applied to all cases. However, for situations involving large numbers of components, relatively small numbers of stages, and liquid solutions that are not too highly nonideal, the latter procedure is more efficient computationally.

TABLE 13-16 Physical-Property Constants at 450 psia for Absorber Example of Fig. 13-52

	K_i				H_{V_i}			H_{L_i}		
Component, i	a_i	$b_i \times 10$	$c_i \times 10^4$	$d_i \times 10^7$	$\bar{a}_i \times 10^{-4}$	$\bar{b}_i \times 10^{-2}$	$\bar{c}_i \times 10^2$	$\tilde{a}_i$	$\tilde{b}_i \times 10^{-2}$	$\tilde{c}_i \times 10^2$
C_1	1.4792	0.053851	−0.13302	0.17394	0.1604	0.09357	0.1782	0	0.1417	−0.1782
C_2	−0.42648	0.12067	−0.31274	0.45328	0.4661	0.1554	0.3341	0	0.1654	0.3341
C_3	−1.9161	0.17875	−0.43021	0.48004	0.5070	0.2645	0	0	0.2278	0.4899
nC_4	−3.4014	0.23455	−0.62864	0.85027	0.5231	0.3390	0.5812	0	0.3197	0.5812
nC_5	−4.476	0.22883	−0.46519	0.49204	0.5411	0.4209	0.8017	0	0.3968	0.8017
Oil	−11.13	0.24146	−0.57345	0.78973	0.800	0.7467	0.3556	0	0.6933	0.3556

Naphtali-Sandholm SC Method This method employs the equilibrium-stage model of Figs. 13-48 and 13-49 but reduces the number of variables by $2N$ so that only $N(2C + 1)$ equations in a like number of unknowns must be solved. In place of V_j, L_j, $x_{i,j}$, and $y_{i,j}$, component flow rates are used according to their definitions:

$$v_{i,j} = y_{i,j} V_j \tag{13-97}$$

$$\ell_{i,j} = x_{i,j} L_j \tag{13-98}$$

In addition, sidestream flow rates are replaced with sidestream flow ratios by

$$s_j = U_j / L_j \tag{13-99}$$

$$S_j = W_j / V_j \tag{13-100}$$

The MESH equations (13-73) to (13-77) then become the *MEH* functions:

$$M_{i,j} = \ell_{i,j}(1 + s_j) + v_{i,j}(1 + S_j) - \ell_{i,j-1} - v_{i,j+1} - f_{i,j} = 0 \tag{13-101}$$

where $\quad f_{i,j} = F_j z_{i,j}$

$$E_{i,j} = K_{i,j}\ell_{i,j}\left(\sum_{\kappa} v_{\kappa,j} \Big/ \sum_{\kappa} \ell_{\kappa,j}\right) - v_{i,j} = 0 \tag{13-102}$$

$$\begin{aligned} H_j = {} & H_{L_j}(1 + s_j) \sum_i \ell_{i,j} + H_{V_j}(1 + S_j) \sum_i v_{i,j} \\ & - H_{L_{j-1}} \sum_i \ell_{i,j} - H_{V_{j+1}} \sum_i v_{i,j+1} \\ & - H_{F_j} \sum_i f_{i,j} - Q_j \\ = {} & 0 \end{aligned} \tag{13-103}$$

where physical properties are not simplified:

$$K_{i,j} = K_{i,j}\{T_j, P_j, \ell_j, v_j\}$$

$$H_{V_j} = H_{V_j}\{T_j, P_j, v_j\}$$

$$H_{L_j} = H_{L_j}\{T_j, P_j, \ell_j\}$$

Let the order of corrections to the unknowns be according to stage number, which in terms of the corresponding unknowns is

$$\mathbf{X} = [\bar{\mathbf{X}}_1, \bar{\mathbf{X}}_2, \ldots \bar{\mathbf{X}}_j, \ldots, \bar{\mathbf{X}}_N]^T \tag{13-104}$$

where $\quad \bar{\mathbf{X}}_j = [v_{1,j}, v_{2,j}, \ldots v_{C,j}, T_j, \ell_{1,j}, \ell_{2,j}, \ldots \ell_{C,j}]^T \qquad$ (13-105)

Let the order of the linearized *MEH* functions also be according to stage number, which in terms of the corresponding nonlinear functions is

$$\mathbf{F} = [\bar{\mathbf{F}}_1, \bar{\mathbf{F}}_2, \ldots \bar{\mathbf{F}}_j, \ldots \bar{\mathbf{F}}_N]^T \tag{13-106}$$

where $\quad \bar{\mathbf{F}}_j = [H_j, M_{1,j}, M_{2,j}, \ldots M_{C,j}, E_{1,j}, \ldots E_{C,j}]^T \qquad$ (13-107)

Corrections to unknowns for the kth iteration are obtained from

$$\Delta\bar{\mathbf{X}}^{(k)} = -\left[\left(\frac{\partial \bar{\mathbf{F}}}{\partial \mathbf{X}}\right)^{-1}\right]^{(k)} \bar{\mathbf{F}}^{(k)} \tag{13-108}$$

The next approximations to the unknowns are obtained from

$$\bar{\mathbf{X}}^{(k+1)} = \bar{\mathbf{X}}^{(k)} + t\,\Delta\bar{\mathbf{X}}^{(k)} \tag{13-109}$$

where t is a damping ($0 < t < 1$) or acceleration ($t > 1$) factor. By ordering the corrections to the unknowns and the linearized functions in this manner, the resulting Jacobian of partial derivatives of all functions with respect to all unknowns is of a very convenient sparse matrix form of block tridiagonal structure.

$$\left(\frac{\bar{\bar{\mathbf{dF}}}}{\bar{\bar{\mathbf{dX}}}}\right) = \begin{bmatrix} \bar{\bar{\mathbf{B}}}_1 & \bar{\bar{\mathbf{C}}}_1 & 0 & 0 & \cdots & & 0 \\ \bar{\bar{\mathbf{A}}}_2 & \bar{\bar{\mathbf{B}}}_2 & \bar{\bar{\mathbf{C}}}_2 & 0 & \cdots & & 0 \\ 0 & \bar{\bar{\mathbf{A}}}_3 & \bar{\bar{\mathbf{B}}}_3 & \bar{\bar{\mathbf{C}}}_3 & & & 0 \\ \vdots & & \ddots & & \ddots & & \vdots \\ 0 & \cdots & & & & & 0 \\ 0 & \cdots & & 0 & \bar{\bar{\mathbf{A}}}_{N-1} & \bar{\bar{\mathbf{B}}}_{N-1} & \bar{\bar{\mathbf{C}}}_{N-1} \\ 0 & \cdots & & 0 & 0 & \bar{\bar{\mathbf{A}}}_N & \bar{\bar{\mathbf{B}}}_N \end{bmatrix} \tag{13-110}$$

Blocks $\bar{\bar{\mathbf{A}}}_j$, $\bar{\bar{\mathbf{B}}}_j$, and $\bar{\bar{\mathbf{C}}}_j$ are $(2C + 1)$ by $(2C + 1)$ submatrices of partial derivatives of the functions on stage j with respect to unknowns on stage $j - 1$, j, and $j + 1$ respectively. The solution to Eq. (13-108) is readily obtained by a matrix-algebra equivalent of the Thomas algorithm for a tridiagonal-matrix equation. Computer storage requirements are minimized by making the following replacements. Starting at top stage 1, using forward-block elimination,

$$\bar{\bar{\mathbf{C}}}_1 \rightarrow (\bar{\bar{\mathbf{B}}}_1)^{-1}\bar{\bar{\mathbf{C}}}_1, \; \bar{\mathbf{F}}_1 \rightarrow (\bar{\bar{\mathbf{B}}}_1)^{-1}\,\bar{\mathbf{F}}_1$$

$$\text{and} \qquad \bar{\bar{\mathbf{B}}}_1 \rightarrow I \text{ (the identity submatrix)}$$

For stages j from 2 to $(N - 1)$,

$$\bar{\bar{\mathbf{C}}}_j \rightarrow (\bar{\bar{\mathbf{B}}}_j - \bar{\bar{\mathbf{A}}}_j\bar{\bar{\mathbf{C}}}_{j-1})^{-1}\bar{\bar{\mathbf{C}}}_j, \; \bar{\mathbf{F}}_j$$

$$\rightarrow (\bar{\bar{\mathbf{B}}}_j - \bar{\bar{\mathbf{A}}}_j\bar{\bar{\mathbf{C}}}_{j-1})^{-1}(\bar{\mathbf{F}} - \bar{\bar{\mathbf{A}}}_j\bar{\mathbf{F}}_{j-1}), \quad \bar{\bar{\mathbf{A}}}_j \rightarrow 0, \; \bar{\bar{\mathbf{B}}}_j \rightarrow I$$

For final stage N,

$$\bar{\mathbf{F}}_N \rightarrow (\bar{\bar{\mathbf{B}}}_N - \bar{\bar{\mathbf{A}}}_N\bar{\bar{\mathbf{C}}}_{N-1})^{-1}(\bar{\mathbf{F}}_N - \bar{\bar{\mathbf{A}}}_N\bar{\mathbf{F}}_{N-1}), \quad \bar{\bar{\mathbf{A}}}_N \rightarrow 0, \quad \bar{\bar{\mathbf{B}}}_N \rightarrow I$$

This completes the forward steps to give $\Delta\bar{\mathbf{X}}_N = -\bar{\mathbf{F}}_N$. Remaining values of corrections $\Delta\bar{\mathbf{X}}_j$ are obtained by successive backward substitution from $\Delta\bar{\mathbf{X}}_j = -\bar{\mathbf{F}}_j \rightarrow -(\bar{\mathbf{F}}_j - \bar{\bar{\mathbf{C}}}_j\bar{\mathbf{F}}_{j+1})$. Matrix inversions are best done by *LU* decomposition. Efficiency is best for a small number of components C.

The Newton iteration is initiated by providing reasonable guesses for all unknowns. However, these can be generated from guesses of just T, T_N, and one interstage value of F_j or L_j. Remaining values of T_j are obtained by linear interpolation. By assuming constant molal overflow, calculations are readily made of remaining values of V_j and L_j, from which initial values of $v_{i,j}$ and $\ell_{i,j}$ are obtained from Eqs. (13-97) and (13-98) after obtaining approximations of $x_{i,j}$ and $y_{i,j}$ from steps 4, 5, 8, 9, and 10 of the SR method. Alternatively, a much cruder but often sufficient estimate of $x_{i,j}$ and $y_{i,j}$ is obtained by flashing the combined column feeds at average column pressure and a vapor-to-liquid ratio that approximates the ratio of overhead plus vapor-sidestream flows to bottoms plus liquid-sidestream flows. Resulting compositions are used as the initial estimate for every stage.

At the conclusion of each iteration, convergence is checked by employing an approximate criterion such as

$$\tau = \sum_j \left\{\left(\frac{H_j}{\xi}\right)^2 + \sum_i [(M_{i,j})^2 + (E_{i,j})^2]\right\} \le \epsilon \tag{13-111}$$

where ξ is a scale factor that is of the order of the average molal heat of vaporization. If we take

$$\epsilon = N(2C + 1)\left(\sum_j F_j^2\right) 10^{-10} \tag{13-112}$$

converged values of the unknowns will generally be accurate, on the average, to from four or more significant digits.

During early iterations, particularly when initial estimates of the unknowns are poor, τ and corrections to the unknowns will be very large. It is then preferred to utilize a small value of t in Eq. (13-109) so as to dampen changes to unknowns and prevent wild oscillations. However, the use of values of t much less than 0.25 may slow or prevent convergence.

It is also best to reset to zero or small values any negative values of component flow rates before initiating the next iteration. When the neighborhood of the solution is reached, τ will often decrease by one or more orders of magnitude at each iteration, and it is best to set $t = 1$. Because the Newton method is quadratically convergent in the neighborhood of the solution, usually only three or four additional iterations will be required to reach the convergence criterion. Prior to that, it is not uncommon for τ to increase somewhat from one iteration to the next. If the Jacobian tends toward a singular condition, it may be necessary to restart the procedure with different initial guesses or adjust the Jacobian in some manner.

Standard specifications for the Naphtali-Sandholm method are Q_j (including zero values) at each stage at which heat transfer occurs and sidestream flow ratio s_j or S_j (including zero values) at each stage at which a sidestream is withdrawn. However, the desirable block tridiagonal structure of the jacobian matrix can still be preserved when substitute specifications are made if they are associated with the same stage or an adjacent stage. For example, suppose that for a reboiled absorber, as in Fig. 13-7*f*, it is desired to specify a boil-up ratio rather than reboiler duty. Equation (13-103) for function H_N is removed from the $N(2C + 1)$ set of equations and is replaced by the equation

$$\tilde{H}_N = \sum_i v_{i,N} - (V_N/L_N) \sum_i \ell_{i,N} = 0$$

where the value of (V_N/L_N) is specified. Following convergence of the calculations, Q_N is computed from the removed equation.

A number of well-tested computer programs are available for solving any of the distillation-type operations shown in Fig. 13-7. The programs are based on simultaneous-correction procedures of the type discussed earlier or on other suitable techniques and include COLUMN (*Simulation Sciences*, Fullerton, Calif.), DISTIL (Chem Share, Houston, Tex.), and RADFRAC (ASPEN Simulator, National Energy Software Center, Argonne, Ill., and ASPEN Technology, Cambridge, Mass.). The last-named program is based on the novel method of Boston and Sullivan [*Can. J. Chem. Eng.*, **52**, 52 (1974)], in which successive approximation variables are stripping factors and energy and volatility parameters that vary, from iteration to iteration, much less than the primitive variables. A Naphtali-Sandholm type of program, particularly suited for applications to distillation, extractive distillation, and azeotropic distillation, has been published by Fredenslund, Gmehling, and Rasmussen (*Vapor-Liquid Equilibria Using UNIFAC, a Group Contribution Method,* Elsevier, Amsterdam, 1977). Christiansen, Michelsen, and Fredenslund [*Comput. Chem. Eng.*, 3, 535 (1979)] apply a modified Naphtali-Sandholm type of method to the distillation of natural-gas liquids, even near the critical region, using thermodynamic properties computed from the Soave-Redlich-Kwong equation of state. Block and Hegner [*Am. Inst. Chem. Eng. J.*, **22**, 582 (1976)] have extended the Naphtali-Sandholm method to staged separators involving two liquid phases (liquid-liquid extraction) and three coexisting phases (three-phase distillation).

Example 7: Calculation of Naphtali-Sandholm SC Method Use the Naphtali-Sandholm SC method to compute stage temperatures and interstage vapor and liquid flow rates and compositions for the reboiled-stripper specifications shown in Fig. 13-53. The specified bottoms rate is equivalent to removing most of the nC_5 and nC_6 and some of the nC_7 in the bottoms.

Calculations were made with a computer program written by Naphtali and modified by Seader (NAPH, Univ. Utah, Dept. Chem. Eng., 1981). K values were computed by the Chao-Seader method [*Am. Inst. Chem. Eng. J.*, **7**, 598 (1961)], and enthalpies were computed by the method of Edmister, Persyn, and Erbar (paper presented at 42d Annual Convention of NGPA, Houston, Tex., Mar. 20–22, 1963). Initial estimates for stage temperatures and flow rates were as follows, where numbers in parentheses are consistent with specifications:

		(lb·mol)/h	
Stage	T, °F	V	L
1	130	(452.26)	550
8	250	450	(99.33)

NOTE: To convert degrees Fahrenheit to degrees Celsius, °C = (°F − 32)/1.8. To convert pound-moles per hour to kilogram-moles per second, multiply by 1.26×10^{-4}.

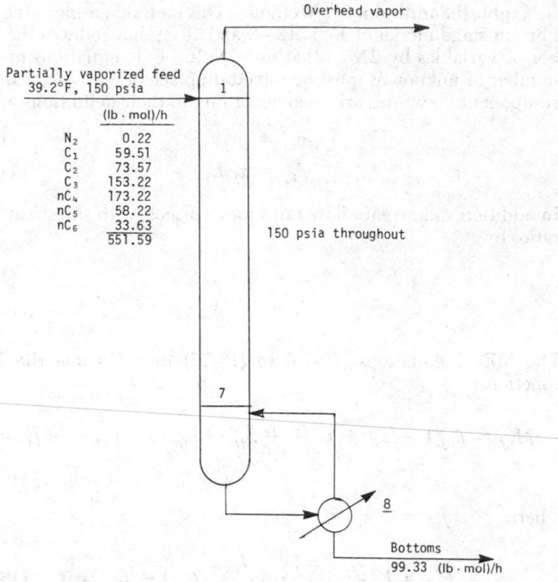

FIG. 13-53 Specifications for the calculation of a reboiled stripper by the Naphtali-Sandholm method.

For specified feed temperature and pressure, an isothermal flash of the feed gave 13.35 percent vaporization.

A rather slow but steady rate of convergence was achieved in 11 iterations by using the following values of t in Eq. (13-109) and obtaining the corresponding values of τ from Eq. (13-111) with $\xi = 1000$:

Iteration	t	τ
1	0.5	543,560
2	1.0	20,016
3	0.6	2,933
4	0.75	388
5	0.62	73.1
6	0.62	13.56
7	0.62	2.462
8	0.62	0.4621
9	0.62	0.0848
10	0.62	0.01579
11	0.62	0.002905

Converged values of temperatures and component flow rates are tabulated in Table 13-17. Computed reboiler duty is 1,254,900 W (4,282,000 Btu/h). Computed temperature, total vapor flow, and nC_4 flow profiles, shown in Fig. 13-54, are not of the shapes that might be expected. Vapor and liquid flow rates for nC_4 change dramatically from stage to stage.

Stage Efficiency The mathematical models presented earlier for rigorous calculations of multistage, multicomponent distillation-type separations assume that equilibrium with respect to both heat and mass transfer is attained at each stage. Unless temperature changes significantly from stage to stage, the assumption that vapor and liquid phases exiting from a stage are at the same temperature is generally valid. However, in most cases, equilibrium with respect to mass transfer is not a valid assumption. If all feed components have the same mass-transfer efficiency, the number of actual stages or trays is simply related to the number of equilibrium stages used in the modeling calculations by an overall stage efficiency. For distillation, as discussed in Sec. 18, this efficiency for well-designed trays typically varies from 40 to 120 percent; the higher value is achieved in some large-diameter towers because of a cross-flow effect. Efficiencies for absorption and extractive distillation can be considerably lower than 40 percent.

When it is desired to compute, with rigorous methods, actual rather than equilibrium stages, Eqs. (13-74) and (13-102) can be modified to include the Murphree vapor-phase efficiency $\eta_{i,j}$, defined

TABLE 13-17 Converged Results for Reboiled Stripper of Fig. 13-53

Variable	Stage							
	1	2	3	4	5	6	7	8
Temperature, °F	132.5	172.3	186.9	195.2	202.85	213.3	229.3	251.9
Component vapor flows, (lb·mol/h								
N_2	0.22	0.01	0.00	0.00	0.00	0.00	0.00	0.00
C_1	59.51	4.31	0.32	0.02	0.00	0.00	0.00	0.00
C_2	73.57	19.83	4.54	0.89	0.16	0.03	0.01	0.00
C_3	153.17	116.08	68.75	34.03	15.49	6.56	2.53	0.84
nC_4	149.03	277.98	372.90	406.99	402.05	363.66	292.94	198.94
nC_5	13.87	25.58	38.09	52.03	72.99	105.32	147.66	185.04
nC_6	2.89	5.27	7.21	8.70	10.97	16.45	31.08	66.23
Total	452.26	449.06	491.81	502.66	501.66	492.02	474.22	451.05
Component liquid flows, (lb·mol)/h								
N_2	0.01	0.00	0.00	0.00	0.00	0.00	0.00	0.00
C_1	4.31	0.32	0.02	0.00	0.00	0.00	0.00	0.00
C_2	19.83	4.54	0.89	0.16	0.03	0.00	0.00	0.00
C_3	116.13	68.80	34.08	15.53	6.60	2.57	0.89	0.05
nC_4	302.17	397.09	431.18	426.25	387.86	317.14	223.13	24.19
nC_5	69.93	82.45	96.38	117.34	149.67	192.01	229.39	44.35
nC_6	36.01	37.95	39.43	41.71	47.19	61.82	96.97	30.74
Total	548.38	591.15	601.98	600.99	591.35	573.54	550.38	99.33

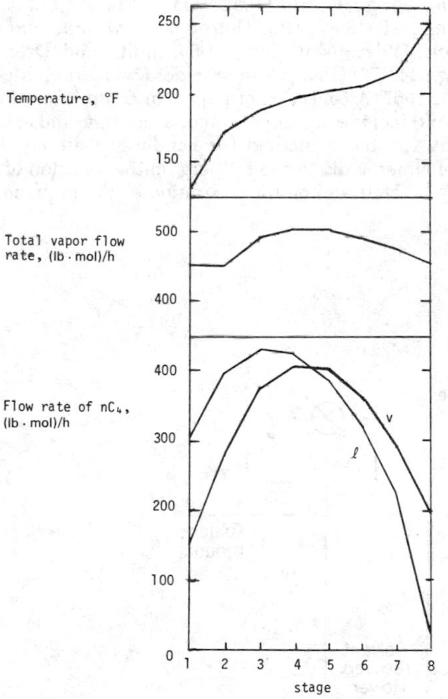

FIG. 13-54 Converged profiles for the reboiled stripper of Fig. 13-53.

by Eq. (13-29). This is particularly desirable for multistage operations involving feeds containing components of a wide range of volatility and/or concentration, in which only a rectification (absorption) or stripping action is provided and all components are not sharply separated. In those cases, the use of a different Murphree efficiency for each component may be necessary to compute recovery accurately. For example, for steam stripping of sour water containing NH_3, H_2S, phenol, and HCN, Won ("Component-Dependent Tray Effiiencies for Sour Water Stripper Calculation," paper presented at AIChE-Anaheim national meeting, June 6–10, 1982) successfully predicted component recoveries for an eight-tray stripper by using the following Murphree efficiencies:

Component	η
NH_3	0.67
H_2S	0.35
phenol	0.65
HCN	0.20

On the other hand, plant data for a large extractive distillation column using furfural to separate butanes from butenes [*Chem. Eng.*, **62**(8), 116 (1955)] showed that Murphree plate efficiencies for isobutane differed from those for *n*-butane by no more than 2 percent. However, efficiencies varied from 19 to 40 percent over the 100 trays in the column.

Departures from the equilibrium-stage model may also occur when entrainment of liquid droplets in the rising vapor or occlusion of vapor in the liquid flow in the downcomer is significant. The former condition may occur at high vapor loading when flooding is approached. The latter condition is possible at high operating pressures when vapor and liquid densities are not drastically different. Entrainment and occlusion effects are not strictly due to mass-transfer inefficiency and are best taken into account by including entrainment terms in the modeling equations, as shown by Loud and Waggoner [*Ind. Eng. Chem. Process Des. Dev.*, **17**, 149 (1978)].

EXTRACTIVE DISTILLATION

Introduction Extractive distillation refers to those processes in which a high-boiling solvent is added to a tray in a column to alter relative volatilities of components in the main feed to the column. The alteration of volatilities is desired because of (1) similarities in the vapor pressures of the feed components or (2) the presence of an azeotrope. The solvent usually boils at a temperature so far above the feed components that formation of new azeotropes is impossible. Also, any troublesome azeotropes present in the untreated feed disappear in the presence of solvent. The absence of azeotropes plus the fact that the solvent can be recovered by simple distillation makes extractive distillation a less complex and more widely useful process than azeotropic distillation.

Extractive-Distillation Processes The typical configuration for an extractive-distillation process is shown in Fig. 13-55, in which methyl cyclohexane and toluene are to be separated. These two components do not form an azeotrope, but their relative volatility is less than 1.01 at low concentrations of toluene. The volatility of methyl cyclohexane relative to toluene is enhanced by the addition of a solvent. This permits separation of these two components in fewer stages than would be required in simple distillation.

The solvent chosen is less volatile than either of the two components and, in order to maintain a high concentration of solvent throughout most of the column, must be introduced to the extractive distillation column above the fresh-feed stage. It is usually introduced a few stages below the top stage, the actual number being determined by the necessity to reduce solvent concentration in the ascending vapor to a negligible amount before overhead product is withdrawn.

The overflow rate of liquid solvent from stage to stage is relatively constant because of its low volatility. The actual concentration of solvent will change abruptly at the fresh-feed stage if a liquid fresh feed is used. A vapor feed is sometimes used to avoid dilution of descending solvent.

Reflux at the top of the extractive-distillation column also tends to dilute the solvent by increasing the amount of nonsolvent material in the liquid overflow. The inherent advantage of higher reflux rates must, in the case of extractive distillation, be balanced against the effect on solvent concentration and changes in relative volatilities that occur.

High solvent concentrations on the trays are usually desirable to maximize the difference in volatilities between the components being separated. However, solubility relationships of the system must be known and care taken to ensure that solvent concentration is maintained in the miscible region. Another limitation on the amount of solvent used is the need to keep the sensible-heat requirement at a reasonable level in the solvent cycle. The solvent-concentration profile in the column is controlled by manipulation of the rates and enthalpies of solvent, fresh feed, and reflux streams.

The choice of solvent determines which of the two components in the fresh feed is removed predominantly in the distillation. For example, assume in Fig. 13-55 that fresh feed to the extractive-distillation column is a mixture of 83 mole percent ethanol in water. If as discussed by Black [*Chem. Eng. Prog.*, **76**(9), 78 (1980)], ethylene glycol is the solvent, the volatility of ethanol is increased more than that of water. Therefore, ethanol is removed as the distillate in the extractive-distillation tower and water is separated from ethylene glycol in the solvent-recovery tower. If a high-boiling hydrocarbon such as octane is used as solvent, the volatility of water is enhanced so that it becomes the distillate in the extractive-distillation tower.

As shown in Fig. 13-55, recovery of solvent is relatively simple for extractive distillation as compared with azeotropic distillation. The solvent chosen does not form an azeotrope with nonsolvent material in the bottoms product from the extractive-distillation tower, and solvent recovery can be accomplished by simple distillation.

The characteristics, design, and operation of extractive-distillation columns have been widely discussed in the literature. Some useful references are as follows: Atkins and Boyer, *Chem. Eng. Prog.*, **45**, 553 (1949); Benedict and Rubin, *Trans. Am. Inst. Chem. Eng.*, **41**, 353 (1945); Black, op. cit.; Buell and Boatwright, *Ind. Eng. Chem.*, **39**, 695 (1947); Carlson et al., *Ind. Eng. Chem.*, **46**, 350 (1954); Chambers, *Chem. Eng. Prog.*, **47**, 555 (1951); Coates, *Chem. Eng.*, **67**(10), 121 (1960); Colburn, *Can. Chem. Proc. Ind.*, **34**, 286 (1950); Drickamer and Hummel, *Trans. Am. Inst. Chem. Eng.*, **41**, 607 (1945); Dunn et al., *Trans. Am. Inst. Chem. Eng.*, **41**, 631 (1945); Dunn and Liedholm, *Pet. Refiner*, **31**, 104 (1952); Hachmuth, *Chem. Eng. Prog.*, **48**, 617 (1952); Happel et al., *Trans. Am. Inst. Chem. Eng.*, **41**, 189 (1946); Hoffman, *Azeotropic and Extractive Distillation*, Wiley, New York, 1964; Smith and Dresser, *Chem. Eng. Prog.*, **44**, 789 (1948); Van Winkle, *Distillation*, McGraw-Hill, New York, 1967. A collection of papers in *Chem. Eng. Prog.*, **65**(9), 43–68 (1969) includes a comparison of azeotropic and extractive distillation by Gerster, a method for handling phase equilibrium by Null and Palmer, a discussion by Berg on the selection of solvents, a case study by Hafslund on the separation of the propane-propylene

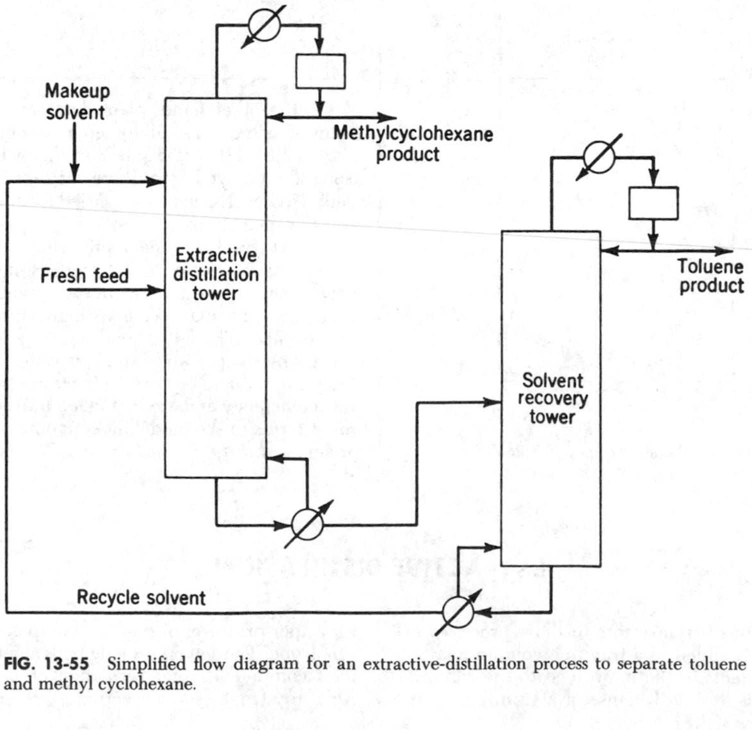

FIG. 13-55 Simplified flow diagram for an extractive-distillation process to separate toluene and methyl cyclohexane.

binary by extractive distillation, and a description by Bannister and Buck of a consistent model for thermal and equilibrium data involved in the recovery of butadiene by extractive distillation.

Selection of Solvent The number of possible solvents available for separation by extractive distillation is usually much larger than for an azeotropic separation because of less severe volatility restrictions, which only include that (1) the solvent boil sufficiently higher than feed components to prevent formation of an azeotrope and (2) the solvent boiling point not be so high that sensible-heat requirements of the solvent cycle become unreasonably large.

A general approach to selection of a solvent is to chose a compound

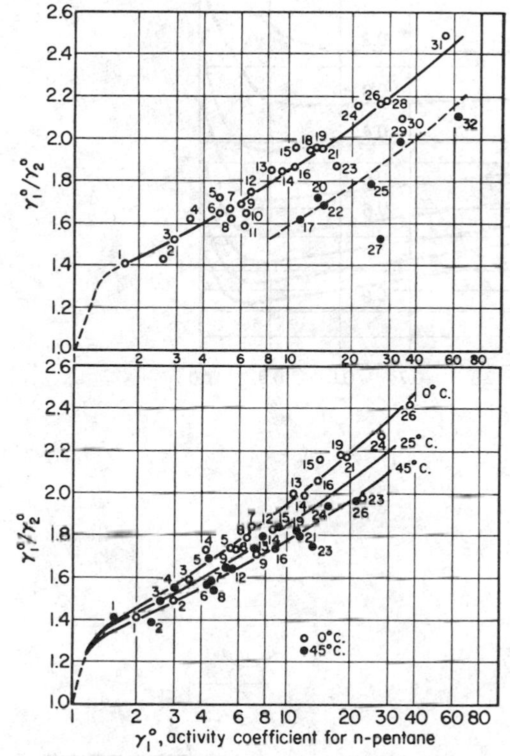

1. Tetrahydrofuran
2. Diethyl ketone
3. Diethyl carbonate
4. Methyl ethyl ketone
5. Pentanedione
6. Cyclopentanone
7. Acetone
8. Butyronitrile
9. Acetyl piperidine
10. Acetophenone
11. Pyridine
12. Diethyl oxalate
13. Propionitrile
14. Dimethyl acetamide
15. n-Methyl pyrrolidone
16. Acetonyl acetone
17. Tetrahydrofurfuryl alcohol
18. Dimethyl sulfolane
19. Dimethyl cyanamide
20. Methyl Carbitol
21. Dimethyl formamide
22. Methyl cellosolve
23. Furfural
24. Acetronitrile
25. Ethylene chlorhydrin
26. γ-Butyrolactone
27. Methanol
28. β-Chloropropionitrile
29. Pyrrolidone
30. Propylene carbonate
31. Nitromethane
32. Ethylene diamine

FIG. 13-56 Effect of an activity coefficient for n-pentane at infinite dilution in extractive agent (γ_1^0) upon selectivity of an agent for n-pentane–1-pentene separation by extractive distillation (γ_1^0/γ_2^0). The upper graph shows results at 25°C (77°F) for various agents listed by number above; solid points and dashed lines are for hydrogen-bonding agents; open points and solid line, for non-hydrogen-bonding agents. The lower graph shows results for 0 and 45°C (32 and 113°F) for non-hydrogen-bonding agents; the solid line from the upper graph, representing results at 25°C (77°F), is reproduced without points in the lower graph. [*Gerster, Gorton, and Eklund,* J. Chem. Eng. Data, *5, 423 (1960).*]

that is more similar to the higher-boiling key and then go up the homologous series for that compound until a homolog is found that boils high enough to make a solvent-nonsolvent azeotrope impossible. This procedure is described by Scheibel [*Chem. Eng. Prog.*, **44**, 927 (1948)].

An actual search for a suitable solvent usually means that many compounds of differing structures must be screened. Dunn et al. (op. cit.) studied the effect of several solvents on volatilities of toluene and nonaromatic compounds boiling between 372 K (210°F) and 386 K (235°F) and chose phenol. Hess, Naragon, and Coghlan [*Chem. Eng. Prog. Symp. Ser. 2*, **48**, 72 (1952)] investigated solvents of many chemical types for the n-butane–2-butene separation.

Some of the results presented by Gerster, Gorto, and Eklund [*J. Chem. Eng. Data*, **5**, 423 (1960)] in their evaluation of solvents for the pentane-pentene separation are shown in Fig. 13-56. The γ_i^0 is the activity coefficient for i at infinite dilution in the solvent. The ratio γ_1^0/γ_2^0 is a convenient way of illustrating the effect of solvent on relative volatilities of the two keys, components 1 and 2. This ratio is plotted versus γ_1^0 in Fig. 13-56 and shows that as the nonideality of the pentane-solvent binary increases, relative volatility of pentane to pentene also increases. If enhancement of relative volatility were the only criterion, one would chose a solvent whose point with respect to the coordinates of Fig. 13-56 lies at the right end of the desired temperature curve.

The use of a vapor-liquid equilibrium still to screen large numbers of possible solvents is expensive. The use of gas-chromatography equipment to determine infinite-dilution activity coefficients of the keys in the solvent can be a faster experimental method and is in common use. An approximate theoretical screening can be accomplished by applying the UNIFAC method to predict liquid-phase activity coefficients.

The effect of a solvent on the volatility of a binary mixture is shown clearly in Figs. 13-57 and 13-58, which are based on vapor-liquid equilibrium data of Drickamer, Brown, and White [*Trans. Am. Inst. Chem. Eng.*, **41**, 555 (1945)] for the ternary system methyl cyclohexane (M)–toluene (T)–phenol (P) at 101.3 kPa (1 atm), where the solvent is phenol. Figure 13-57 shows that the relative volatility α_{M-T} in the absence of phenol varies from only 1.08 for almost

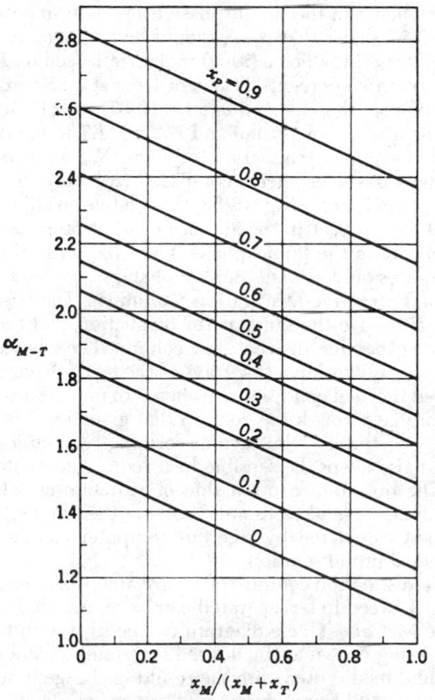

FIG. 13-57 Relative volatility of methyl cyclohexane to toluene.

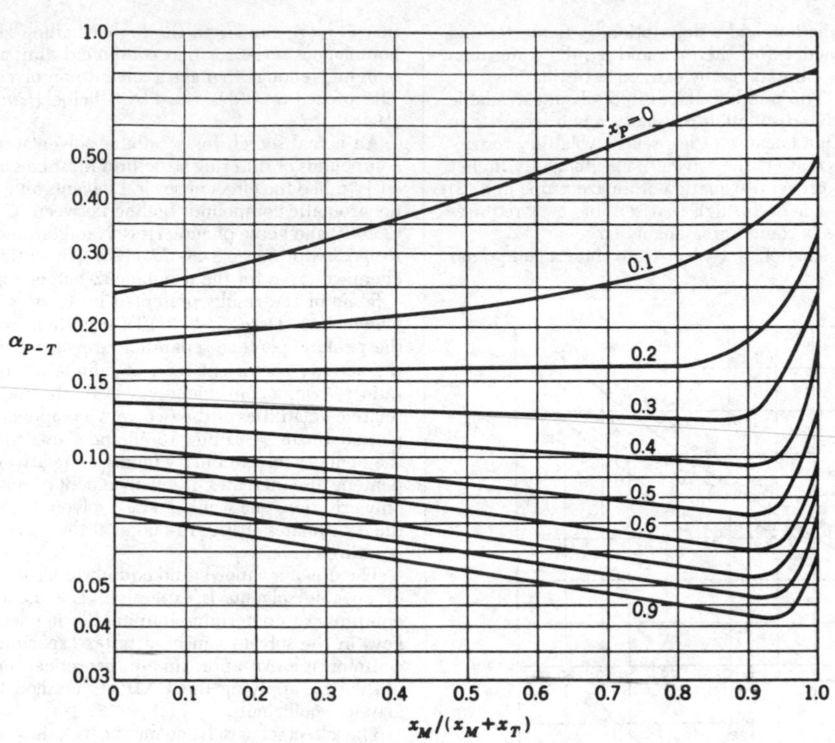

FIG. 13-58 Relative volatility of phenol to toluene.

pure M to 1.54 for almost pure T. Accordingly, if distillation were used to separate M and T into relatively pure products, a large number of stages would be required, particularly in the rectifying section. If extractive distillation were used with approximately 50 mole percent solvent phenol in the liquid phase throughout most of the column, Fig. 13-57 shows that α_{M-T} would be increased to from 1.69 to 2.14. For a case in which a 50-50 molal fresh feed of T and M is distilled to obtain 99 percent T (solvent-free) at a 95 percent recovery, application of the Fenske equations (13-37) and (13-40) gives $N_m = 28.2$ in the absence of phenol and $N_m = 11.7$ in the presence of phenol. If it is assumed that actual trays $= 3N_m$, 85 trays would be needed in the absence of phenol compared with only 35 trays in the presence of phenol. From Fig. 13-58, the relative volatility of phenol with respect to T is 0.11 in the absence of M at 50 percent concentration of phenol in the liquid phase. Thus, the separation between P and T requires only a small number of stages.

Design of Extractive-Distillation Columns The basic calculational procedures are the same as for distillation, and the design is usually simpler than for an azeotropic column. The solvent-introduction point is always within a few trays of the top. Solvent rate, reflux rate, and feed thermal condition are chosen to provide a solvent-concentration profile through the column that gives a satisfactory relative volatility of the keys but avoids forming two liquid phases on any tray and also keeps the sensible-heat requirements at a reasonable level. The optimum combination of variables usually must be found by solving several cases and studying the economics of each. Cases are best calculated by rigorous computer methods that can handle nonideal liquid solutions.

Extractive-distillation columns often are amenable to shortcut calculations. If the feed to be separated can be represented by a binary mixture, the McCabe-Thiele diagram can be used with the equilibrium curve being drawn at the desired (constant) solvent concentration. If a liquid feed is used with the resulting change in solvent concentration, two equilibrium curves will be needed. Stages stepped off

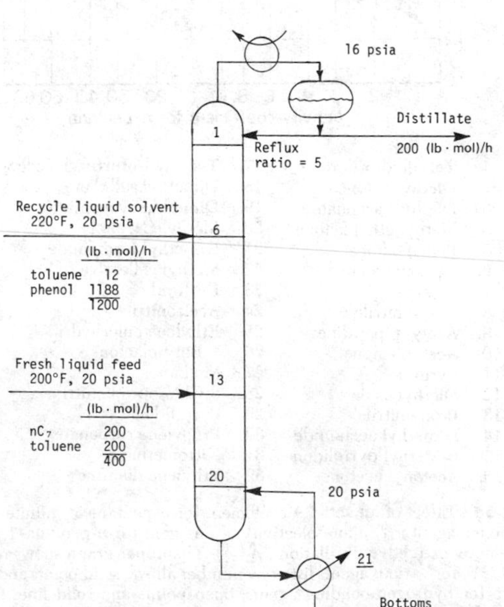

FIG. 13-59 Specifications for the calculation of an extractive-distillation column.

TABLE 13-18 Converged Results for Extractive Distillation Column of Fig. 13-59

Stage	Temperature, °F	Traffic, (lb·mol)/h		Vapor mole fractions			Liquid mole fractions		
		Vapor	Liquid	*H*	*T*	*P*	*H*	*T*	*P*
Distillate	214.5	....	200	.....	.....	.....	0.9858	0.0074	0.0068
Reflux	214.5	....	1000	.....	.....	.....	0.9858	0.0074	0.0068
1	212.6	1200	985	0.9858	0.0074	0.0068	0.9820	0.0083	0.0097
2	213.6	1185	985	0.9826	0.0081	0.0093	0.9770	0.0092	0.0138
3	214.5	1185	983	0.9785	0.0089	0.0126	0.9698	0.0102	0.0200
4	215.6	1183	976	0.9725	0.0098	0.0177	0.9571	0.0115	0.0314
5	217.0	1176	944	0.9619	0.0108	0.0273	0.9236	0.0134	0.0630
6	224.1	1144	2279	0.9345	0.0123	0.0532	0.4364	0.0128	0.5508
7	225.0	1279	2282	0.9317	0.0146	0.0537	0.4342	0.0151	0.5507
8	226.1	1282	2285	0.9270	0.0186	0.0544	0.4305	0.0192	0.5503
9	227.3	1285	2286	0.9190	0.0259	0.0551	0.4230	0.0266	0.5504
10	228.8	1286	2283	0.9049	0.0391	0.0560	0.4086	0.0397	0.5517
11	230.7	1283	2277	0.8803	0.0624	0.0573	0.3839	0.0620	0.5541
12	233.3	1277	2278	0.8387	0.1023	0.0590	0.3488	0.0987	0.5525
13	234.7	1278	2738	0.7759	0.1677	0.0564	0.3667	0.1718	0.4615
14	236.7	1338	2732	0.7482	0.1941	0.0577	0.3417	0.1948	0.4635
15	239.7	1332	2719	0.6987	0.2414	0.0599	0.2991	0.2336	0.4673
16	244.4	1319	2702	0.6153	0.3208	0.0639	0.2371	0.2898	0.4730
17	251.1	1302	2692	0.4897	0.4395	0.0708	0.1617	0.3587	0.4796
18	259.4	1292	2696	0.3348	0.5841	0.0811	0.0927	0.4215	0.4858
19	267.8	1296	2693	0.1906	0.7140	0.0954	0.0439	0.4513	0.5048
20	280.8	1293	2624	0.0895	0.7767	0.1338	0.0144	0.3784	0.6072
21	318.7	1224	1400	0.0286	0.6392	0.3322	0.0018	0.1504	0.8478

on the McCabe-Thiele diagram run from the reboiler to the solvent-introduction point, above which solvent concentration changes rapidly. The rapid change in solvent concentration that usually occurs in the first few stages above the reboiler can be ignored, or those stages can be handled by a stage-to-stage calculation. The use of McCabe-Thiele diagrams is described by Atkins and Boyer (op. cit.), Smith and Dresser (op. cit.), and Chambers (op. cit.).

Example 8: Calculation of Extractive Distillation Use a rigorous method to compute stage temperatures and interstage vapor and liquid flow rates and compositions for the extractive-distillation specifications shown in Fig. 13-59, where phenol (*P*) is used to enhance the separability of *n*-heptane (*H*) from toluene (*T*). The specified distillate rate is equivalent to the quantity of *n*-heptane in the feed. The recycle solvent rate is equivalent to providing mole fractions of phenol of approximately 0.54 and 0.46 in the liquid streams between stages 6 and 13 and stages 13 and 20 respectively. Five stages are provided at the top of the column to prevent appreciable amounts of phenol from appearing in the distillate, and the reflux rate to the top stage is of the same magnitude as the solvent rate.

Calculations were made with a rigorous simultaneous-correction method using the Wilson equation to compute liquid-phase activity coefficients. Tabulated stage-by-stage results for temperature, total phase flows, and compositions are shown in Table 13-18. Liquid-phase mole fractions are plotted in Fig. 13-60, in which it is seen that the phenol mole fraction decreases rapidly in moving up from stage 6, the *n*-heptane mole fraction decreases rapidly in moving down from stage 13, and mole fractions of both phenol and toluene change drastically at the bottom two stages.

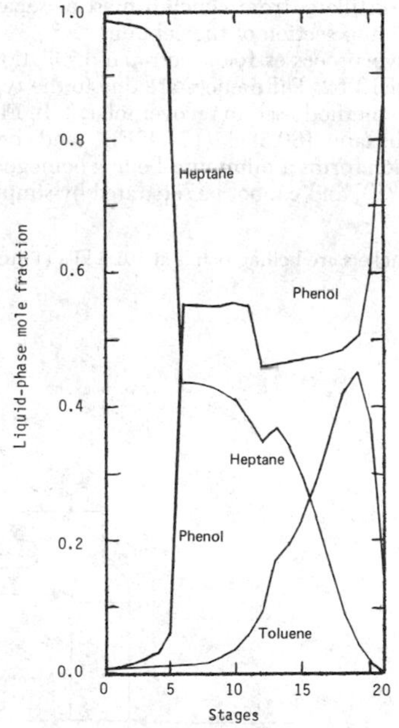

FIG. 13-60 Converged composition profiles for the extractive distillation column of Fig. 13-59.

AZEOTROPIC DISTILLATION

INTRODUCTION

An azeotrope is a liquid mixture that exhibits a maximum or minimum boiling point relative to the boiling points of surrounding mixture compositions. Boiling points of the pure components in the mixture must be sufficiently close to permit formation of an azeotrope. A mixture of close-boiling components may form an azeotrope when only small deviations from ideal liquid solutions occur. A mixture of wide-boiling components may not exhibit an azeotrope even though they form a very nonideal liquid mixture. Azeotropes occur infrequently for mixtures composed of components whose boiling points differ by more than 30°C (54°F).

An azeotrope is homogeneous if only one liquid phase is present. A maximum-boiling-point homogeneous azeotrope may occur if deviations from Raoult's law, as given by Eq. (13-10), are negative ($\gamma_i^L < 1.0$). For a minimum-boiling-point homogeneous azeotrope, deviations from Raoult's law are positive ($\gamma_i^L > 1.0$). If the positive deviations are large enough ($\gamma_i^L \gg 1.0$), phase splitting can occur and a minimum-boiling-point heterogeneous azeotrope may be formed with one vapor phase in equilibrium with two liquid phases. All three types of azeotropes are important. In some literature, mixtures that do not form azeotropes are called zeotropes.

An understanding of the occurrence of azeotropes is important for two reasons. First, azeotropes can make a given separation impossible by simple distillation in a particular pressure range. However, second, azeotropes may be utilized to separate mixtures not ordinarily separable by simple distillation or to increase recovery yield of some components from certain mixtures.

AZEOTROPIC-DISTILLATION PROCESSES

Azeotropic distillation refers to those processes in which a component (called the solvent or entrainer) is added above the main feed tray to form (or nearly form) with one or more of the feed components an azeotrope, which is removed as either the distillate or the bottoms, but usually the former. Azeotropic distillation can also refer to a process in which a solvent is added to break an azeotrope that otherwise would be formed by components in the feed. In this case, the process is distinguished from extractive distillation because the solvent appears in the distillate, from which it must be separated and recycled back to the top section of the column.

Representative processes for azeotropic distillation are shown in Figs. 13-61 and 13-62. Differences are due to the type of azeotrope formed and the method used to recover solvent. In Fig. 13-61, a mixture of cyclohexane [80.8°C (177.4°F)]* and benzene [80.2°C (176.4°F)], which forms a minimum-boiling homogeneous azeotrope [77.4°C (171.3°F)] and cannot be separated by simple distillation, is fed to an azeotropic-distillation column together with acetone as an entrainer, which forms a minimum-boiling binary homogeneous azeotrope [53.1°C (127.6°F)] with acetone [56.4°C (133.5°F)]. Thus near-pure benzene is removed as bottoms. The acetone-cyclohexane near azeotrope is removed as distillate and is treated with water in a liquid-liquid extraction column, where near-pure cyclohexane leaves as overhead. An acetone-water mixture leaves the bottom of the extractor and is separated by simple distillation into separate solvent streams that are recycled. Thus, this process requires azeotropic distillation and two additional multistage separation operations.

Near-pure ethyl alcohol [78.3°C (172.9°F)] cannot be obtained from dilute mixtures with water [100°C (212°F)] by simple distillation at 101.3 kPa (1 atm) because a minimum-boiling homogeneous

*Figures in brackets are boiling points at 101.3 kPa (1 atm).

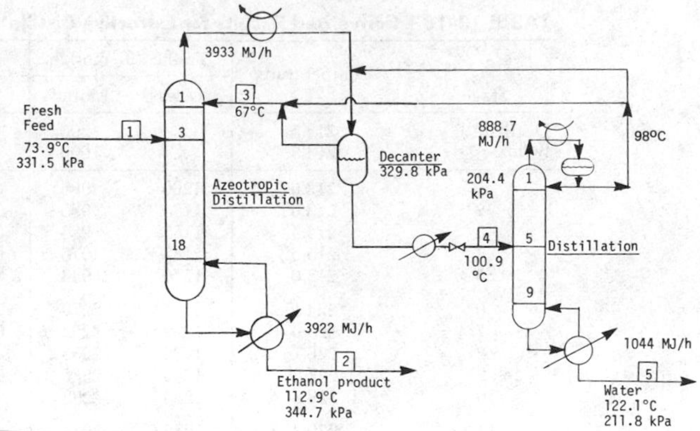

FIG. 13-62 Azeotropic-distillation process for separating ethanol and water by using pentane as the entrainer.

Material Balance, (kg·mol)/h

Stream	1	2	3	4	5
Ethanol	32.555	32.555	13.138	2.630	0.000
Water	6.763	0.001	13.876	7.889	6.762
n-Pentane	0.000	0.000	97.522	0.000	0.000

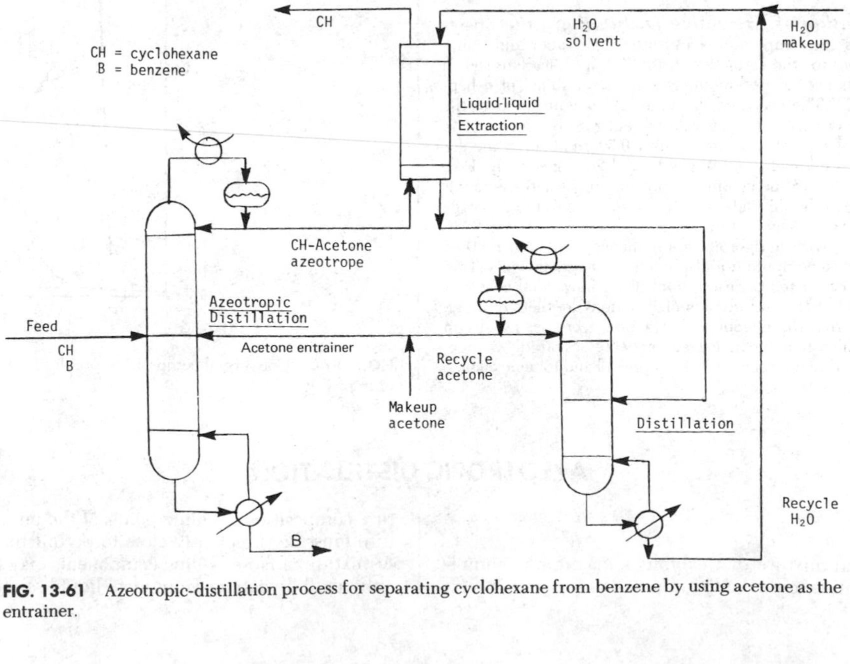

FIG. 13-61 Azeotropic-distillation process for separating cyclohexane from benzene by using acetone as the entrainer.

azeotrope [78.15°C (172.7°F)] is formed. As discussed by Black [*Chem. Eng. Prog.*, **76**(9), 78 (1980)], the separation can be made by low-pressure [<11.5 kPa (1.67 lbf/in^2)] simple distillation, extraction distillation with gasoline or ethylene glycol, or azeotropic distillation with *n*-pentane, benzene, or diethyl ether. His azeotropic-distillation process and accompanying material balance when *n*-pentane is the entrainer are shown in Fig. 13-62, in which stages are theoretical and near-perfect separation of ethanol and water and recovery of *n*-pentane are achieved. In the azeotropic column, nearly pure ethanol is taken as bottoms and a heterogeneous minimum-boiling azeotrope is taken overhead, condensed, and decanted into a pentane-rich reflux stream and an aqueous ethanol stream that is separated by distillation to produce a bottoms product of nearly pure water and a distillate that is combined with the reflux from the decanter.

AZEOTROPY

General Classification of Azeotropes Although the first observations of the appearance of minimum vapor pressure in binary mixtures were made in 1802 by Dalton [*Mem. Manchester Philos. Soc.*, **5**, 585 (1802); *Ann. Philos.*, **9**, 186 (1817)], the term "azeotrope" was

TABLE 13-19 Minimum-Boiling-Point Azeotropic Binary Mixtures*

Pressure 760 mmHg

System A	System B	Mole % A	Temp., °C.
Water	Ethanol	10.57	78.15
	Allyl alcohol	54.50	88.20
	Propionic acid	94.70	99.98
	Propyl alcohol (n)	56.83	87.72
	Isopropyl alcohol	31.46	80.37
	Methyl ethyl ketone	33.00	73.45
	Isobutyric acid	94.50	99.30
	Ethyl acetate (2 phase)	24.00	70.40
	Ethyl ether (2 phase)	5.00	34.15
	Butyl alcohol (n) (2 phase)	75.0	92.25
	Isobutyl alcohol	67.14	89.92
	Butyl alcohol (sec)	66.00	88.50
	Butyl alcohol (tert)	35.41	79.91
	Isoamyl alcohol (2 phase)	82.79	95.15
	Amyl alcohol (tert) (2 phase)	65.00	87.00
	Benzene (2 phase)	29.60	69.25
	Toluene (2 phase)	55.6	84.10
Carbon tetrachloride	Methanol	44.5	55.70
	Ethanol	61.3	64.95
	Allyl alcohol	73.0	72.32
	Propyl alcohol (n)	75.0	72.80
	Ethyl acetate	43.0	74.75
Carbon disulfide	Methanol	72.0	37.65
	Ethanol	86.0	42.40
	Acetone	61.0	39.25
	Methyl acetate	69.5	40.15
Chloroform	Methanol	65	53.5
	Ethanol	84	59.3
	Isopropyl alcohol	92	60.8
Butyl alcohol (n)	Cyclohexane	11	79.8
	Toluene	37	105.5
Isobutyl alcohol	Isoamyl bromide	60.0	103.80
	Benzene	10.0	79.84
	Toluene	50.0	101.15
	Pinene (α)	96.5	107.90
Amyl alcohol (n)	Amyl acetate (iso-)	96.4	131.3
	Butyl propionate (iso-)	85	130.5
Isoamyl alcohol	Chlorobenzene	42	124.3
	Xylene (o)	64	128
	Xylene (m)	58	127
	Xylene (p)	56	126.8
Nitrobenzene	Benzyl alcohol	39	204.3
	Borneol	60	207.75
	Menthol	60	207.9
Phenol	Bromotoluene (p)	58	176.2
	Carvene	49.5	169.0
	Pinene (α)	25	152.75
Aniline	Carvene	48	171.35
Benzyl alcohol	Gualacol	38	204.4
	Naphthalene	64	204.3
Acetic acid	Chlorobenzene	72.5	114.65
	Benzene	2.5	80.05
	Toluene	62.7	105.4
	Xylene (m)	40	115.38
Ethyl alcohol	Methyl ethyl ketone	45	74.8
	Ethyl acetate	46	71.8
	Methyl propionate	67.5	73.2
	Propyl formate (n)	72	73.5
	Benzene	44.8	68.24
	Cyclohexane	44.5	64.9
Ethyl alcohol	Hexane (n)	33.2	58.68
	Toluene	81	76.65
	Heptane (n)	67	72
Allyl alcohol	Benzene	22.2	76.75
	Cyclohexane	26.6	74
	Hexane (n)	6.5	65.5
	Toluene	61.5	92.4
Acetone	Methyl acetate	61	56.1
	Isobutyl chloride	81	55.8
	Diethylamine	43.5	51.5
Propyl alcohol (n)	Ethyl propionate	64	93.4
	Benzene	20.9	77.12
	Hexane (n)	6	65.65
	Toluene	60	92.6
Isopropyl alcohol	Ethyl acetate	30.5	74.8
	Benzene	39.3	71.92
	Hexane (n)	29	61
	Toluene	77	80.6
Tetrachloroethylene	Ethanol	6	77.95
	Allyl alcohol	27	94.0
	Propionic acid	81	118.95
	Propyl alcohol (n)	24	94
	Isopropyl alcohol	8	81.7
	Butyl alcohol (n)	47	110
	Isobutyl alcohol	40	103.05
Trichloroethylene	Allyl alcohol	70	80.95
	Propyl alcohol (n)	69	81.75
	Isopropyl alcohol	54	74
	Isobutyl alcohol	86	85.4
	Butyl alcohol (tert)	74	75
	Amyl alcohol (tert)	83	84
Dichloroethylene	Allyl alcohol	76	79.6
		77	80
Chloral hydrate	Cyclohexane	13	76
Ethylene bromide	Acetic acid	20.7	114.35
	Propionic acid	65	127.75
	Isobutyl alcohol	22	106.2
	Isoamyl alcohol	52	123.2
	Ethyl benzene	83.5	131.1
Methanol	Trichloroethylene	70	60.2
	Acetonitrile	84.5	63.45
	Ethylene dichloride	62	59.5
	1, 1-Dichlorethane	28.5	49.05
	Ethyl bromide	14	34.95
	Chloromethyl methyl ether	57.5	56
	Ethyl iodide	52.5	54.7
	Acetone	20	55.7
	Ethyl formate	30.5	50.95
	Methyl acetate	35	54.0
	Propyl bromide (n)	49	54.1
	Propyl iodide (n)	88	63.5
	Methylal	34.5	41.82
	Trimethyl borate	87	59
	Ethyl acetate	91.7	62.3
	Pentane (n)	13	31
	Pentane (iso-)	9	24.5
	Benzene	61.4	53.84
	Cyclohexene	63.0	55.9
	Cyclohexane	61.0	54.2
	Hexane (n)	51	50.6
	Heptane (n)	83	60.5
	Pinene (d)	98.5	64.5

*Abstracted from *International Critical Tables*, McGraw-Hill.

NOTE: To convert degrees Celsius to degrees Fahrenheit, °F = 1.8°C + 32.

not introduced until 1911 by Wade and Merriman [*J. Chem. Soc. Trans.*, **99**, 997 (1911)]. Azeotrope, from the Greek "to boil unchanged," means literally that the vapor boiling from a liquid has the same composition as the liquid.

For a binary azeotrope, with $x_1 = y_1$ and $x_2 = y_2$, expressions for the activity coefficients γ_1^L and γ_2^L in the case of an ideal-vapor phase take the simple form, namely,

$$\gamma_1^L = P/P_1^{sat} \tag{13-113}$$

and

$$\gamma_2^L = P/P_2^{sat} \tag{13-114}$$

and the relative volatility α_{12} is equal to unity. For a binary system to exhibit azeotropy, it is thus necessary for it to depart sufficiently from ideality that the ratio of the activity coefficients γ_1^L/γ_2^L can for certain values of P and x_2 equal the value of P_2^{sat}/P_1^{sat}.

Conditions for an azeotrope may for some systems be satisfied for all values of temperature (up to the critical state), while for others only in a certain temperature range. Following Lecat [*Ann. Soc. Sci. Bruxelles*, **49B**, 274 (1929)], we may call the first case absolute azeotropy and the second limited azeotropy. Azeotropic systems may be classified broadly in relation to the character of the azeotrope (maximum- or minimum-boiling), number of components in the system, and whether one or more liquid phases are formed.

In line with deviations from Raoult's law, Lecat [*Comptes Rendus*, **183**, 880 (1926)] proposed to divide azeotropes into (1) positive azeotropes, characterized by a minimum-boiling temperature at constant pressure, i.e., a maximum in the total vapor pressure at constant temperature; and (2) negative azeotropes, having a maximum boiling temperature and a minimum total vapor pressure. For positive azeotropes where $P > P_i^{sat}$, the activity coefficients of the components are greater than 1.0, while for negative azeotropes ($P < P_i^{sat}$) they are less than 1.0.

The term azeotrope gives no indication of whether one, two, or more liquid phases are formed. To make this perfectly clear, the terms "homoazeotrope" and "heteroazeotrope" were introduced. In order to indicate that a two-phase (hetero-) azeotrope is formed, it has been proposed by Swietoslawski (*Azeotropy and Polyazeotropy*, Pergamon, Oxford, London, 1963) to add the sign -·-. For example, the symbols (B, E, W -·-) designate a heteroazeotrope of components B, E, and W (benzene-ethanol-water). For a system that forms three liquid phases at the boiling temperature of the azeotrope, a logical proposition seems to be the further addition of ·-. This case is exemplified by the series of ternary heteroazeotropes of nitromethane, N, with water, W, and n-paraffins, H, ranging from heptane to dodecane [Malesinski, *Bull. Acad. Pol. Sci. Ser. Sci. Chim.*, **11**, 475 (1963)]. Such azeotropes are designated by the symbols (N, W, H_i-·-·-).

TABLE 13-20 Maximum-Boiling-Point Azeotropic Binary Mixtures*

A	B	Mole % A	Temp., °C.	Pressure, mm.
Water	Hydrofluoric acid	65.4	120	760
	Hydrochloric acid	88.9	110	
	Perchloric acid	32.0	203	
	Hydrobromic acid	83.1	126	
	Hydriodic acid	84.3	127	
	Nitric acid	62.2	120.5	735
	Formic acid	43.3	107.1	760
Chloroform	Acetone	65.5	64.5	760
Formic acid	Diethyl ketone	48	105.4	
	Methyl propyl ketone	47	105.3	
Phenol	Cyclohexanol	90	182.45	
	Benzaldehyde	54	185.6	
	Benzyl alcohol	8	206.0	
Cresol (o)	Acetophenone	24	203.7	
	Phenyl acetate	42.5	198.6	
	Methyl hexyl ketone	97	191.5	
	Isoamyl butyrate	80	192.0	
Cresol (m)	Acetophenone	54	209.0	
	Isoamyl lactate	60	207.6	
Cresol (p)	Benzyl alcohol	38	207.0	
	Acetophenone	52	208.45	
	Camphor	38	213.15	

*From *International Critical Tables*, McGraw-Hill.
NOTE: To convert degrees Celsius to degrees Fahrenheit, °F = 1.8°C + 32.

Typical vapor-liquid equilibrium data for binary azeotropic systems were discussed earlier and represented in Figs. 13-10 to 13-13. Lists of well-known minimum-boiling-point and maximum-boiling-point binary mixtures, mostly at a pressure of 101.3 kPa (760 torr), are given in Tables 13-19 and 13-20. When heterogeneous azeotropes are formed, compositions given are for the vapor phase and for the overall composition of the two equilibrium liquid phases.

Azeotropic temperature and pressure can be shifted by changing system pressure. Data for the ethanol-water system [Kleinert, *Angew. Chem.*, **46**, 18 (1933) and Wade and Merriman, op. cit.] and for the ethanol-benzene system [Zawiska, *Bull. Acad. Pol. Sci. Ser. Sci. Chim.*, **9**, 141 (1961)] are plotted in Fig. 13-63. For the former system, the azeotropic composition is confined to a narrow range, and an azeotrope does not form below a pressure of approximately 9.2 kPa (70 torr). For the latter system, azeotropes are formed over a wide range of compositions. Multicomponent azeotropes can be considerably more complex. Data for a number of ternary azeotropic mixtures are given in Table 13-21. Total-pressure surface and temperature surfaces for liquid and vapor phases in multicomponent systems often exhibit maximum or minimum points analogous to those found in binary systems. At the high and low points, vapor and liquid temperature surfaces are tangent, and at the point of tangency the two phases will be identical in composition. Multicomponent azeotropes offer the same obstacle to complete separation as do binary azeotropes.

Contours of multicomponent temperature and pressure surfaces can be explored numerically if correlations for activity coefficients as functions of temperature and composition are available. Graphical representation is possible for ternary systems such as the one shown in Fig. 13-64, for the methyl ethyl ketone (MEK)–n-heptane–toluene system, in which the dark lines represent isotherms on the liquid-temperature surface. Two vapor isotherms (dashed lines) are shown, and "tie lines" connecting equilibrium vapor and liquid compositions are shown as the arrows. These arrows point "downhill" and show the direction that a *distillation path* would take at that point. Each arrow represents the separation that would be achieved in an equilibrium stage.

The lowest temperature on the 101.3-kPa (1-atm) liquid surface shown in Fig. 13-64 is at the binary MEK-heptane azeotrope. No maxima or minima occur within the triangle; i.e., no ternary azeotropes exist. As shown in Fig. 13-65, the total-reflux distillation path resulting from batch distillation of a feed whose composition falls within the diagram will produce that binary azeotrope as the overhead product. The dashed line connecting the azeotrope and points 1 and 2 is the material-balance line. Point 1 is the original still-pot charge, and as overhead product is removed, still-pot composition moves toward point 2. The chains of arrows represent total-reflux distillation paths for two different pot compositions. The arrows appear to be connected because passing vapor and liquid streams at total reflux are identical in composition. Actually each arrow lies in a different isothermal plane. Continuous-distillation columns exhibit similar paths. For example, point 1 might be the fresh-feed composition and point 2 the bottom-product composition. Those two points must fall on a straight material-balance line through the overhead product. The arrows in distillation path B would then represent equilibrium stages in the column.

When three or more components are present, additional types of azeotropes are possible. In spite of the fact that a ternary saddle azeotrope was predicted by Ostwald at the end of the nineteenth century, the first azeotrope of this type was found in 1945 by Ewell and Welch [*Ind. Eng. Chem.*, **37**, 1224 (1945)] with the system acetone-chloroform-methanol. Saddle azeotropic systems, also called positive-negative systems, exhibit a hyperbolic point that is neither a minimum nor a maximum in either boiling temperature or total vapor pressure, and are characterized by the presence of a "top-ridge" line. They also exhibit some peculiar properties called distillation anomalies.

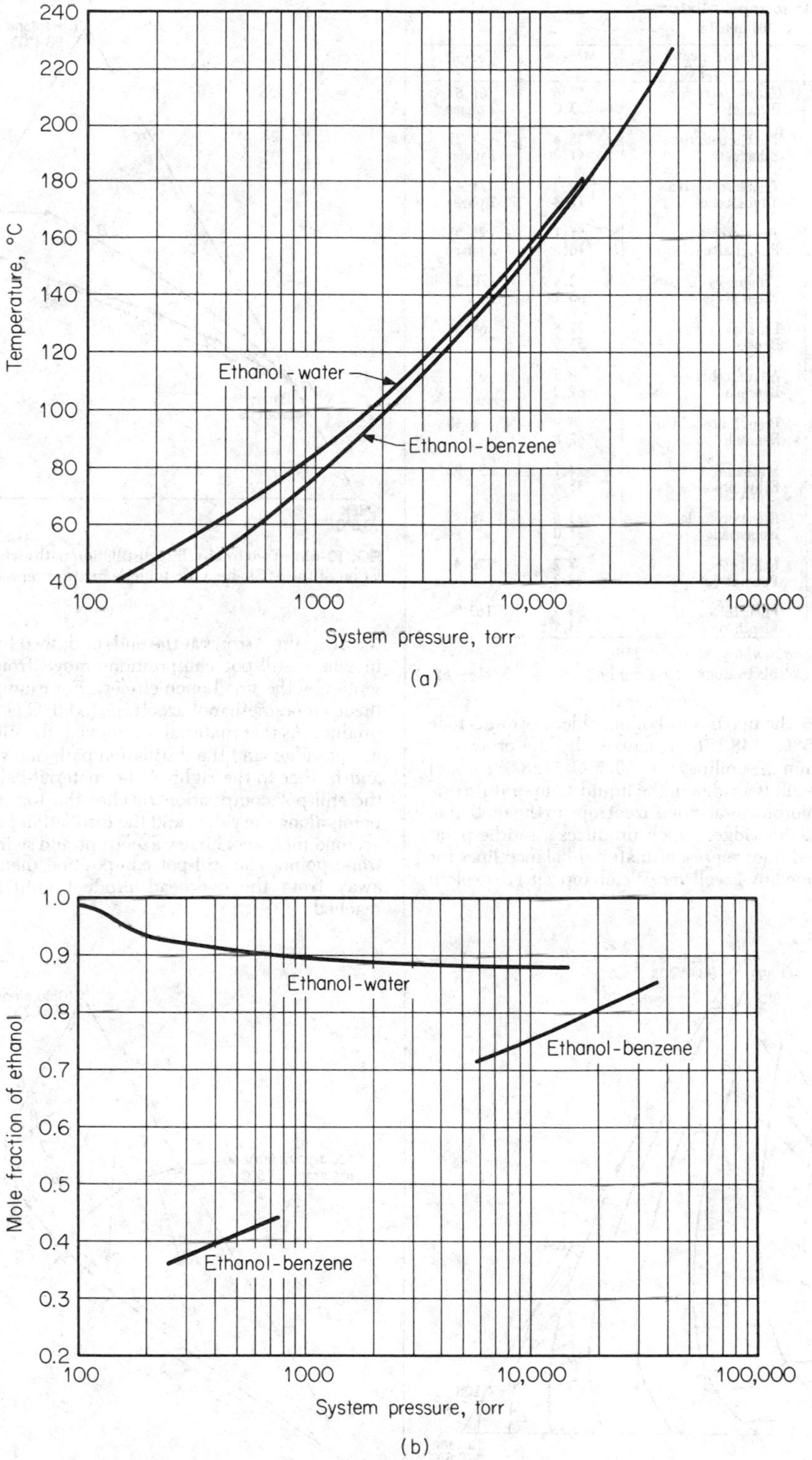

FIG. 13-63 Effect of pressure on azeotrope formation for ethanol-water and ethanol-benzene systems. (*a*) Effect on azeotropic temperature. (*b*) Effect on azeotrope composition.

TABLE 13-21 Ternary Azeotropic Mixtures*

760 mmHg

Component A Mole % A = 100 − (B + C)	Components B and C	Mole % B and C	Temp., °C.
Water	Carbon tetrachloride Ethanol	57.6 23.0	61.8 2 phase
	Trichloroethylene Ethanol	38.4 41.2	67.25 2 phase
	Trichloroethylene Allyl alcohol	49.2 17.3	71.4 2 phase
	Trichloroethylene Propyl alcohol (n)	51.1 16.6	71.55 2 phase
	Ethanol Ethyl acetate	12.4 60.1	70.3
	Ethanol Benzene	22.8 53.9	64.86
	Allyl alcohol Benzene	9.5 62.2	68.3
	Propyl alcohol (n) Benzene	8.9 62.8	68.48
Carbon disulfide	Methanol Ethyl bromide	24.1 35.4	33.92
Methyl formate	Ethyl bromide Isopentane	23.8 31.0	16.95
	Ethyl ether Pentane (n)	7.2 48.2	20.4
Propyl lactate (n)	Phenetol Menthene	35.2 34.1	163.0

*From *International Critical Tables*, McGraw-Hill.

NOTE: To convert degrees Celsius to degrees Fahrenheit, °F = 1.8°C + 32.

As shown in Fig. 13-66, the maximum-boiling chloroform-acetone binary azeotrope at 64.5°C (148.1°F) is flanked by the other two binary azeotropes (minimum-boiling) at 53.5°C (128.3°F) and 54.6°C (130.3°F). The result is a ridge in the liquid-temperature surface running from the chloroform-acetone azeotrope to the methanol corner. There is a sag in this ridge, which produces a saddle-point ternary azeotrope. Dashed lines represent material-balance lines for batch distillations performed by Ewell and Welch (op. cit.) to explore the diagram. Arrows at the ends of dashed lines indicate the direction in which still-pot compositions move from original feeds (represented by the small open circles). For example, feed 2 first produces the acetone-methanol azeotrope [54.6°C (130.3°F)] as the overhead product. As that material is removed, the still-pot composition moves up the ridge, and the distillation path (not shown) bulges out farther and farther to the right of the material-balance line. Finally, when the still-pot composition reaches the top of the ridge, "downhill" points along the ridge, and the distillation path breaks away from the acetone-methanol binary azeotrope and swings into the ternary-azeotrope point. The still-pot composition then moves along the ridge, away from the overhead product until the methanol corner is reached.

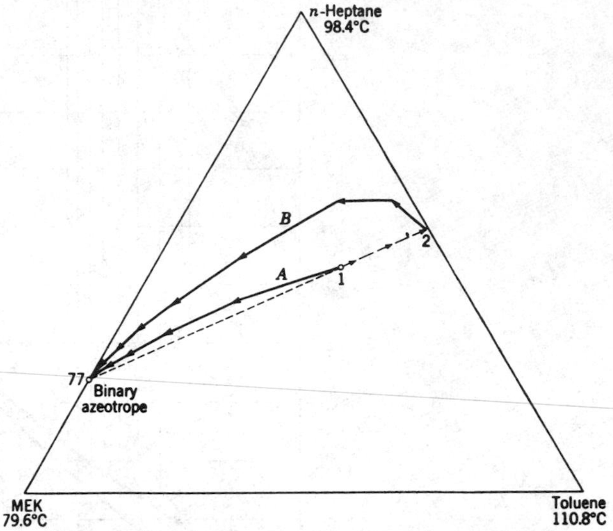

FIG. 13-65 Two total-reflux distillation paths which occur in the batch distillation of the MEK-heptane-toluene mixture represented by point 1.

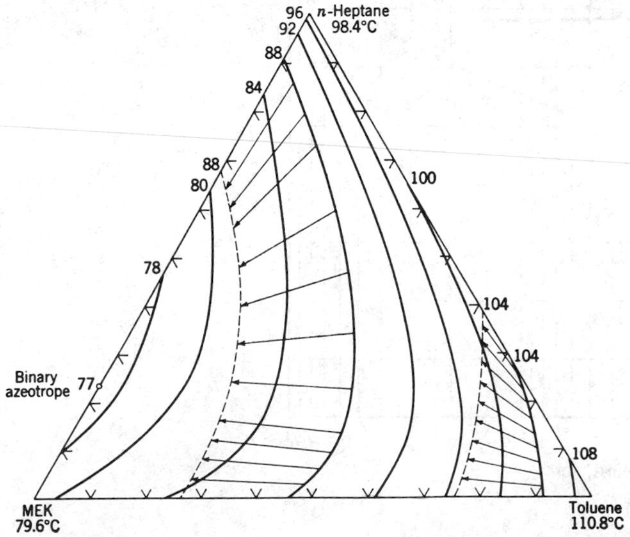

FIG. 13-64 Liquid-temperature contour line (dark) in methyl ethyl ketone(1)–*n*-heptane(2)–toluene(3) system. Vapor-temperature contour lines (dashed) shown at 88 and 104°C (190 and 219°F). Arrows connect equilibrium-phase compositions. [*Data of Steinhauser and White,* Ind. Eng. Chem., *41, 2912 (1949).*]

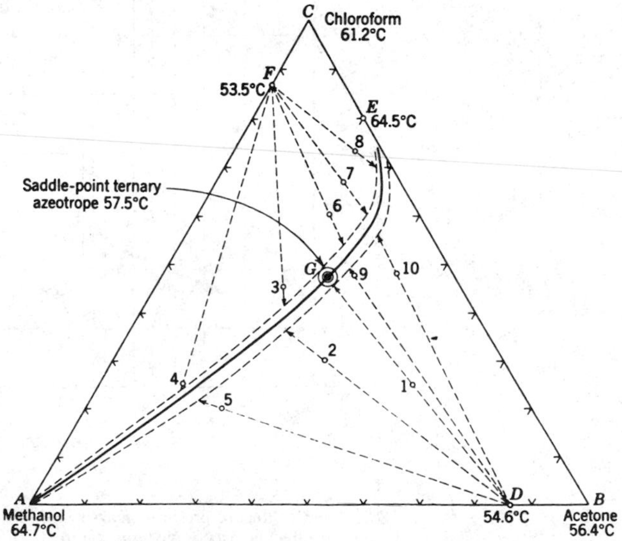

FIG. 13-66 Binary azeotropes and a ternary saddle-point azeotrope in the acetone-chloroform-methanol system. Material balance lines are shown for 10 different feeds.

A ridge in the liquid-temperature surface prevents a distillation path from connecting two points on opposite sides of the ridge because the temperature profile in a distillation column cannot have a maximum or a minimum point between the two end stages.

Triangular diagrams that show the locus of liquid-phase composition as it varies with time are referred to as ternary-residue curve maps by Doherty and Perkins [*Chem. Eng. Sci.*, **34**, 1401 (1979)]. A large number of different types of such maps are shown qualitatively by Doherty and Caldarola ("The Sequencing of Towers for Homogeneous and Extractive Distillations," paper presented at AIChE-Anaheim annual meeting, Nov. 14–19, 1982), who also refer to procedures for quantitatively determining maps for specific cases.

In general, ternary saddle azeotropes are classified according to the number of binary negative systems forming such an azeotrope and are divided into bipositive-negative and binegative-positive. All possible types of ternary bipositive-negative azeotropes have been found as discussed by Kurtyka [*Bull. Acad. Pol. Sci. Ser. Sci. Chim.*, **9**, 741 (1961)].

Bipositive-negative ternary azeotropes may be designated by the symbols $[(-)\ A,\ B\ (+)\ H]$, where A, B, and H are the components forming these azeotropes. For the ternary binegative-positive azeotropes with B, E, and C as the components, the symbols are $[(+)\ B,\ E\ (-)\ C]$; two binary negative azeotropes $[(-)\ C,\ B]$ and $[(-)\ C,\ E]$ occur in the system. This type of ternary saddle azeotrope is rather a rare phenomenon.

Commonness of the Phenomenon of Azeotropy Azeotropes occur in organic and inorganic systems, although most of the known azeotropes are formed by organic compounds because of the very large number of such compounds that boil without decomposition in the easily accessible ranges of temperature and pressure. In the last two decades of the nineteenth century, the appearance of an extreme total vapor pressure and boiling temperature was believed to be a rare phenomenon. This view has survived in certain circles until today despite convincing evidence that the phenomenon of azeotropy is definitely a common one.

The first investigator to provide evidence of the commonness of azeotropy was Ryland [*Am. Chem. J.*, **22**, 384 (1899); *Chem. News*, **81**, 15, 42, 50 (1900).] His work dealt with 80 systems, 45 of which were azeotropic. An additional 80 azeotropes were described in Lecat's doctoral dissertation of 1908–1909. Ten years later Lecat (*L'azéotropisme*, Lamartin, Brussels, 1918) was able to list 1000 azeotropes, among which were numerous binary and ternary heteroazeotropes.

In 1949, Lecat (*Tables azéotropiques*, vol. I, author, Brussels, 1949) published data for 13,290 binary systems, 6287, or 47 percent, of which formed azeotropes.

In 1973, Horsley (*Azeotropic Data—III*, American Chemical Society, Washington, 1973) reported 7945 binary, 371 ternary, 9 quaternary, and 1 quinary azeotropes. He found that 119 binary azeotropes, including 32 negative, occur in systems composed either of two inorganic compounds or an inorganic-organic compound system; 665 binary azeotropic systems contain water as one component. Ternary saddle (positive-negative) azeotropes occur in 40 systems, and 267 ternary azeotropic systems contain water as one component. There are also 4 ternary negative azeotropes. Quaternary systems form 8 positive azeotropes and 1 positive-negative azeotrope.

Since the early 1970s experimental studies of vapor-liquid equilibriums have increased rapidly, mostly at conditions of constant temperature, with the aim of correlating and predicting equilibrium parameters. Unfortunately, a large number of investigators have not reported whether an azeotrope is formed in a particular system. From existing experimental work, one might conclude that the frequency of azeotropy diminishes as the number of components in the system increases and that multicomponent azeotropes should be expected to be rare. Such a point of view, however, may not be justified. Although conditions for the formation of multicomponent azeotropes are complex, as discussed by Malesinski (*Azeotropy and Other Theoretical Problems of Vapour-Liquid Equilibrium*, Wiley-Interscience, New York, 1965), they are not difficult to fulfil in practice.

Azeotropic Range The idea of relating a certain "azeotropic ability" to the chemical character of a substance is evident in the first works of Lecat (op. cit., 1918), in which he arranged experimental data on azeotropes according to the chemical character of the components. The most useful characteristic, as defined by Swietoslawski [*Bull. Acad. Pol. Sci., Ser. A*, **19**, 19 (1950); *Przem. Chem.*, **7**, 363 (1951)], is the azeotropic range. There is also another closely related term known as the "relative azeotropic effect" [Skolnik, *Ind. Eng. Chem.*, **40**, 442 (1948)], but its use is limited.

The azeotropic range Z for a component 1 that forms a series of azeotropes with components $2i$ belonging to a homologous series is determined by estimating the two tangent azeotropes that occur at $x_1 = 0$ and $x_1 = 1.0$ as shown in Fig. 13-67. Here, data at 101.3-kPa (760-torr) pressure for the system ethanol in normal and branched paraffins are plotted as normal boiling temperature of the paraffin versus mole fraction of ethanol in the azeotrope formed with the paraffin. In this case, positive azeotropes are formed that boil at temperatures less than the boiling temperatures of the two constituents of the binary system. The boiling point of ethanol T_1 is shown as 78.3°C (172.9°F). The estimated boiling point T_{2u} of the highest-boiling member of the homologous paraffin series that forms an azeotrope with ethanol is approximately 140°C (284°F). This tangent azeotrope would boil itself at T_1 with a composition $x_1 = 1.0$. The estimated boiling point $T_{2\ell}$ of the lowest-boiling member of the series that forms an azeotrope is approximately 20°C (68°F). This tangent azeotrope would boil at 20°C (68°F) with a composition $x_1 = 0.0$.

The azeotropic range is defined as $Z = T_{2u} - T_{2\ell}$. It consists of two parts: (1) the upper part, $Z_u = T_{2u} - T_1$, and (2) the lower part, $Z_\ell = T_1 - T_{2\ell}$, such that $Z = Z_u + Z_\ell$. In Fig. 13-67, $Z_u = 61.7$°C (143.1°F), $Z_\ell = 58.3$°C (136.9°F), and $Z = 120$°C (248°F).

With regard to the Z_u and Z_ℓ values, three cases have been observed, namely,

$Z_u = Z_\ell$, symmetrical azeotropic range

$Z_u > Z_\ell$, upper asymmetry of azeotropic range

$Z_u < Z_\ell$, lower asymmetry of azeotropic range

The lower asymmetry of the azeotropic range is not as common and is exemplified by the azeotropic range of naphthalene with respect to aromatic hydrocarbons present in petroleum [$Z_\ell \sim 20$°C (68°F)

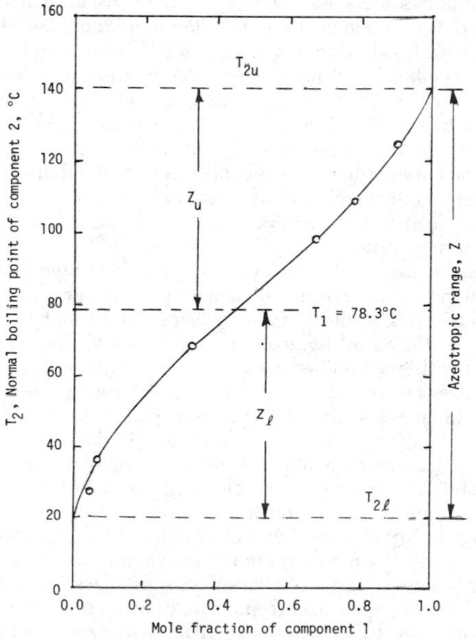

FIG. 13-67 Azeotropic ranges for ethanol in paraffins at 101.3 kPa (760 torr).

and $Z_u \sim 12°C$ (53.6°F)]. The upper asymmetry of the azeotropic range is more common and is often associated with the increase of solubility of the common substance in the higher-boiling homologs [example: acetic acid with *n*-paraffins; $Z_\ell \sim 57°C$ (135°F) and $Z_u \sim 77°C$ (171°F)]. Small azeotropic ranges are mostly symmetrical or show small deviations from that behavior. From the definition of Z_ℓ, the larger its value, the wider the differences in boiling temperatures of azeotropes belonging to a given homologous series. Consequently, for the separation of azeotropes a large Z_ℓ value is a favorable factor.

Malesinski [*Bull. Acad. Pol. Sci. Cl.* III, 3, 601 (1956)] developed equations for predicting the azeotropic range by assuming a simplified regular binary solution that is equivalent to applying the van Laar equation (Table 13-3) with $A_{12} = A_{21}$ = regular solution constant. Then the upper and the lower parts of the azeotropic range, are given by the relations

$$Z_u = (\delta_2/x_1^2) = (A_{12}/\Delta S_2^0) \quad (13\text{-}115)$$

$$Z_\ell = (\delta_1/x_2^2) = (A_{12}/\Delta S_1^0) \quad (13\text{-}116)$$

where x_1 and x_2 are mole fractions of components 1 and 2 respectively in the azeotrope, ΔS_1^0 and ΔS_2^0 are molar vaporization entropies, A_{12} is a regular solution constant, and δ_1 and δ_2 are azeotropic temperature deviations. For positive azeotropes $\delta_1 = T_1 - T^{Az}$ and $\delta_2 = T_2 - T^{Az} - T_2$. For negative azeotropes, $\delta_1 = T^{Az} - T_1$ and $\delta_2 = T^{Az} - T_2$, where T_1 and T_2 are boiling temperatures of the pure components and T^{Az} is the boiling temperature of the azeotrope.

Large differences in vaporization entropies of components usually lead to a considerable asymmetry of the azeotropic range. In the case of equal vaporization entropies, the azeotropic range is symmetrical ($Z_u = Z_\ell = Z_{12}$), which can be computed from

$$(\delta_1^{0.5} + \delta_2^{0.5}) = Z_{12} = 0.5Z \quad (13\text{-}117)$$

where Z_{12} is the half value of the symmetrical azeotropic range Z.

With Eq. (13-117), which is also valid at high pressures (Zawiska, op. cit.), Z may be estimated for a binary azeotropic system from the azeotropic temperature of just one member of the homologous series and the boiling points of the two components. By convention, Z values are positive for positive azeotropic systems and negative for negative ones.

The azeotropic range of agent *A* with respect to the homologous series *H*, their isomers, and closely related substances may also be determined by ebulliometric and distillation methods (Swietoslawski, op. cit., 1963). In addition, the range may be evaluated from values of activity coefficients at infinite dilution for the agent and the representatives of a given series [Kurtyka and Trabczynski, *Rocz. Chem.*, 32, 623 (1958)]. The value of the azeotropic range evaluated by the latter method depends on the accuracy of graphical extrapolation of the curve of ln γ versus concentration to infinity. When this curve is steep in the vicinity of that limiting concentration, its extrapolation may lead to large errors, and this will be reflected in values of the azeotropic range.

It might be expected that values of the azeotropic range of an agent forming binary positive or negative azeotropes with different homologous series would vary to a large extent and that isomers belonging to the same homologous series would not form binary azeotropes with one another. However, this point of view is not justified because azeotropes have been found in such systems as 2,2,3-trimethyl butane–2,4-dimethyl propane [Callingaert and Wojciechowski, *J. Am. Chem. Soc.*, 42, 5310 (1950)]. This azeotrope occurs at a reduced pressure, while the binary system water–deuterium oxide exhibits an azeotrope at an elevated pressure. In the system 3-picoline–2,6-lutidine, an azeotrope occurs at atmospheric pressure.

Some agents are characterized by small and some by large values of azeotropic ranges. Alcohols are typical agents forming large numbers of azeotropes with a large variety of substances. With increasing molecular weight, their azeotropic range decreases. Glycol belongs to a series of powerful azeotropic agents. Phenols and low-molecular-weight organic acids have similar properties, and their azeotropic ranges also decrease with an increase in the number of methyl groups in the molecule. Phenols form azeotropes even with some aromatic hydrocarbons. Their azeotropic ranges, however, decrease rapidly with an increase of methyl groups both in phenols and in aromatic hydrocarbons.

Aromatic amines, and to some extent pyridine and quinoline bases, show relatively small azeotropic ranges, which decrease with an increase of methyl groups substituting for hydrogen atoms in the molecule.

Azeotropic ranges of paraffins in relation to olefins present an interesting case. They are very small (on the order of a few degrees Celsius) and undergo a sharp decrease with an increase in molecular weight of the hydrocarbon. However, present knowledge in this field is insufficient to say whether mixtures of high-molecular-weight paraffins and olefins can form azeotropes.

Without the azeotropic range, some general conditions for the formation of azeotropes could not be formulated. Also, examination of processes taking place in the course of fractional distillation of coal-tar oils and other organic raw materials would not be possible (Swietoslawski, op. cit., 1963; *Physikalische Chemie des Steinkohlenteers*, N. J. Hoffman-Verlag, Cologne, 1959). In addition, as discussed later, azeotropic ranges appear in equations for estimating boiling temperature and composition of a ternary homoazeotrope. These equations have been found useful in predicting the existence of a large number of ternary azeotropes (Z. M. Kurtyka, unpublished data).

Prediction of Azeotropic Data Methods for estimating the composition of a binary azeotrope may be divided into three types: (1) empirical correlations applicable both to homoazeotropes and heteroazeotropes, (2) relations based on regular solution theory, applicable to homoazeotropes only, and (3) relations for heteroazeotropes only.

Lecat (op. cit., 1918) first observed that the composition of a binary azeotrope is related to the difference between the boiling temperatures of its components. He used the following power series to relate these quantities for systems formed by a common substance with members of a homologous series:

$$w_1 = A_0 + A_1\Delta + A_2\Delta^2 + \cdots \quad (13\text{-}118)$$

where w_1 is the weight fraction of the common component in the azeotrope, Δ is the absolute difference between the boiling temperatures of the components, and A_0, A_1, A_2, etc., are constants for a given common substance with a given homologous series. Lecat (*Azéotropisme et distillation: Traité de chimie organique*, ed. by Grignard, vol. I, Masson et Cie, Paris, 1935) later proposed a relation between Δ and the azeotropic deviations δ, between the boiling points of the lower-boiling component and the azeotrope in terms of another set of constants, C_0, C_1, C_2, etc.:

$$\delta = C_0 + C_1\Delta + C_2\Delta^2 + \cdots \quad (13\text{-}119)$$

The existence of an approximate relation of this kind was expected for a series of binary regular solutions (Prigogine and Defay, *Chemical Thermodynamics*, translated by Everett, Longmans, London, 1954). Horsley (op. cit.) presents graphical correlations based on Eqs. (13-118) and (13-119) for many systems.

Other empirical correlating equations and graphical methods have been developed by Mair et al. [*J. Res. Nat. Bur. Stand.*, **27**, 39 (1941)], Meissner and Greenfield [*Ind. Eng. Chem.*, **40**, 438 (1948)], Skolnik (op. cit.), and Johnson and Madonis [*Can. J. Chem. Eng.*, **37**, 71 (1959)]. More recently, Seymour et al. [*Ind. Eng. Chem. Fundam.*, **16**, 200 (1977)] developed the following empirical equation for relating the azeotropic composition of a binary mixture of *A* and *B* at atmospheric pressure to the boiling-point difference:

$$\log_{10}[x_A/(1 - x_A)] = -0.0317\,(T_{refB}/T_{refA})(T_A - T_B) - 0.094 \quad (13\text{-}120)$$

where *A* is the more polar component and T_{ref} is the ratio of the absolute boiling point of the component to a hypothetical normal paraffin of the same molecular weight. For example, for the mixture isopropanol–*n*-hexane, the alcohol is component A, $T_A = 82.45°C$ (180.4°F), $T_B = 68.85°C$ (155.9°F), $T_{refB} = 1.0$, and $T_{refA} = 1.28$.

From Eq. (13-120), a value of $x_A = 0.270$ is obtained, which compares with the value of 0.290 in Table 13-19. Equation (13-120) correlates experimental data at atmospheric pressure for 1108 azeotropes with an average deviation for x_A of 0.046.

Three methods apply the theory of regular solutions for estimating the composition of a binary (positive or negative) azeotrope from certain properties of the pure components and of the azeotrope. Two of these methods involve easily accessible data, namely, boiling temperatures of the pure components T_1 and T_2 and that of the azeotrope T^{Az}, and were developed by Prigogine and Defay (op. cit.) and by Malesinski (op. cit., 1965). The third method is based on activity coefficients of the components at the azeotropic point and is due to Kireev [*Acta Physicochim. URSS*, **14**, 371 (1941)].

The Prigogine equation for equal molar vaporization entropies of the components $\Delta S_1^0 = \Delta S_2^0$ may be written in the form

$$x_2 = \alpha(1 + \alpha)^{-1} \tag{13-121}$$

where α is the square root of the ratio of azeotropic temperature deviations, i.e., $(\delta_1/\delta_2)^{0.5}$, and x_2 is the mole fraction of component 2 in the azeotrope.

When $\Delta S_1^0 = \Delta S_2^0$, Eq. (13-121) takes the form

$$x_2 = \alpha'(1 + \alpha')^{-1} \tag{13-122}$$

where $\alpha' = c\alpha$ and $c = (\Delta S_1^0/\Delta S_2^0)^{0.5}$.

The Malesinski equation for components of equal molar vaporization entropies is given by

$$x_2 = 0.5 + (T_1 - T_2)/2Z_{12} \tag{13-123}$$

where Z_{12} is the half value of the symmetrical azeotropic range. For systems with unequal molar vaporization entropies of the components, Malesinski's equation requires evaluation of Z_u, the quantity that is not easily available.

The simplest and most suitable form of Kireev's equation for the case in which $\Delta S_1^0 = \Delta S_2^0$ is the expression

$$x_2 = (1 + b)^{-1} \tag{13-124}$$

where $b = (\ln \gamma_2/\ln \gamma_1)^{0.5}$. At the azeotropic point (in the case of an ideal vapor phase) activity coefficients γ_1 and γ_2 are given by Eqs. (13-113) and (13-114).

For $\Delta S_1^0 = \Delta S_2^0$, Eq. (13-124) becomes

$$x_2 = (1 + b')^{-1} \tag{13-125}$$

where $b' = cb$ and $c = (\Delta S_2^0/\Delta S_1^0)^{0.5}$.

Equations (13-121), (13-123), and (13-124) were evaluated by Kurtyka and Kurtyka [*Ind. Eng. Chem. Fundam.*, **19**, 225 (1980); and *Ind. Eng. Chem. Fundam.*, **20**, 177 (1981)]. For 30 systems, absolute average deviations between computed and observed azeotropic compositions were 4.5 mole percent for Eqs. (13-123) and (13-124) and 4.9 mole percent for (Eq. 13-121). A distinction to specific systems, i.e., polar-nonpolar and nonpolar-nonpolar, showed no preference of method. A choice among the methods can be made only on the availability of data required by the equations. From this point of view, Eqs. (13-121) and (13-123) are preferable to Eq. (13-124). The possible error in the composition of the binary azeotrope under isobaric conditions is due not only to deviations from the ideal-gas law and deviations of the system from regular solution theory but also to the change of the regular solution constant A_{12} with temperature and to differences in vaporization entropies of the components. Effects of nonideality of the vapor phase and differences in vaporization entropies of the components on azeotropic composition were studied for systems of acetic acid with n-paraffins (Kurtyka and Kurtyka, op. cit., 1981). As discussed in a previous subsection, acetic acid is largely associated (dimerized) in the vapor phase and its vaporization entropy is 14.85 cal/(g·mol) [26.71 Btu/(lb·mol)] compared with ~20 cal/(g·mol) [36 Btu/(lb·mol)], or close to Trouton's rule, for hydrocarbons. Not surprisingly, it was found that when corrections for the real behavior of the vapor phase of acetic acid are taken into account, predicted results for the azeotropic composition improved considerably. The effect of differences in vaporization entropies of the components was found to be small.

A relatively large number of systems exhibit limited solubility and may form positive heteroazeotropes. In binary systems, data on critical solution temperatures, e.g., Francis (*Critical Solution Temperatures*, American Chemical Society, Washington, 1961) enables one to predict whether an azeotrope that occurs at a certain temperature is a homoazeotrope or a heteroazeotrope.

To describe vapor-liquid equilibrium in a heterogeneous system, it is useful to introduce certain simplifications. The simplest case is based on the assumption that the components are completely immiscible in each other and that the vapor phase is ideal. Then the composition of a binary heteroazeotrope is given simply by

$$x_1 = y_1 = \frac{P_1^{sat}\{T^{Az}\}}{P_1^{sat}\{T^{AZ}\} + P_2^{sat}\{T^{Az}\}} \tag{13-126}$$

The boiling temperature of the azeotrope T^{Az} is given implicitly by the relation

$$P = P_1^{sat}\{T^{Az}\} + P_2^{sat}\{T^{Az}\} \tag{13-127}$$

where P is the total pressure at equilibrium.

When the condition of complete immiscibility is not satisfied, the activity a must be introduced to give

$$y_1 = \frac{P_1^{sat}\{T^{Az}\}a_1}{P_1^{sat}\{T^{Az}\}a_1 + P_2^{sat}\{T^{Az}\}a_2} \tag{13-128}$$

and

$$P = a_1 P_1^{sat}\{T^{Az}\} + P_2^{sat}\{T^{Az}\}a_2 \tag{13-129}$$

noting that a_1 is identical in both liquid phases. For systems of nitromethane with n-paraffins, the assumption that $a_1 = a_2$ appears to be valid.

In Table 13-22 are listed calculated azeotropic data (y_1, T^{Az}) using Eqs. (13-126) and (13-127) and corresponding deviations from experiment Δy_1 and ΔT^{Az} for 12 binary systems. Agreement is quite good for systems of aromatic and n-paraffin hydrocarbons with water, but agreement is not as satisfactory for systems in which considerable miscibility of the components occurs, such as n-butanol–water, aniline-water, and acetonitrile–n-paraffins. These relations for binary systems are easily extended to ternary and multicomponent heteroazeotropes.

To separate nonideal ternary and multicomponent liquid mixtures by simple distillation, it is important to establish whether azeotropes are formed. Because experimental data are relatively meager and measurements of azeotropic composition are rather difficult and time-consuming, two computational approaches have been developed to predict the composition of ternary and multicomponent azeotropes. The first, developed by Malesinski [*Bull. Acad. Pol. Sci. Cl. III*, **4**, 701, 709 (1956); **5**, 177, 183 (1957)] for ternary homoazeotropes, is based on the theory of regular solutions. The second is a more general approach based on the use of various equations for predicting liquid-phase activity coefficients.

In the first approach, the composition of a ternary homoazeotrope of components 1, 2, and 3 under isobaric conditions, when the vaporization entropies of the components are equal, is related to any two pairs of binary azeotropes by the following equations.

For pairs (1,2) and (2,3):

$$x_1 = \frac{x_1^{(1,2)} + \hat{a}x_3^{(2,3)}}{1 - \hat{a}\hat{b}} \tag{13-130}$$

$$x_3 = \frac{x_3^{(2,3)} + \hat{b}x_1^{(1,2)}}{1 - \hat{a}\hat{b}} \tag{13-131}$$

where $\hat{a} = (Z_{13} - Z_{23} - Z_{12})/2Z_{12}$, $\hat{b} = (Z_{13} - Z_{23} - Z_{12})/2Z_{23}$, $x_1^{(1,2)}$ and $x_3^{(2,3)}$ are the mole fractions of components 1 and 3 in the binary azeotropes (1,2) and (2,3) respectively, x_1 and x_3 are mole fractions in the ternary azeotrope, and Z_{12}, Z_{13}, and Z_{23} are half values of the symmetrical azeotropic ranges discussed earlier.

When vaporization entropies of components are not equal, regular solution constants A_{12}, A_{13}, and A_{23} take the place of Z_{12}, Z_{13}, and Z_{23} respectively. The effect of differences in vaporization entropies of components on azeotropic composition of a ternary system is small.

TABLE 13-22 Predicted Azeotropic Data (y_1, T^{Az}) Using Eqs. (13-126) and (13-127) and Deviations Δy_1 and ΔT^{Az} for 12 Binary Systems at 760 torr

		Mole percent			
No.	System	y_1	Δy_1	T^{Az}, °C	ΔT^{Az}, °C
1	Water (1)-benzene	29.98	0.41	69.43	0.18
2	Water–*m*-xylene	76.2	−0.47	92.56	0.56
3	Water–ethyl benzene	74.3	−0.07	91.91	−0.09
4	Water–*n*-hexane	21.0	−0.45	61.42	−0.08
5	Water–*n*-heptane	45.6	0.43	79.09	−0.11
6	Water–*n*-octane	67.8	−0.65	89.02	−0.58
7	Water-nitromethane	48.92	−2.31	80.9	−2.85
8	Water-aniline	87.1	−8.50	95.54	−3.06*
9	Water–*n*-butanol	66.75	−8.5	89.15	−3.55
10	Acetonitrile–*n*-hexane	39.0	−6.4	53.94	−2.86
11	Acetonitrile–*n*-heptane–*n*-hexane	63.2	−5.5	67.62	−1.88
12	Acetonitrile–*n*-octane–*n*-hexane	80.8	−3.9	74.98	−1.72

NOTE: $\Delta y_1 = y_1$ (predicted) − y_1 (experiment); $\Delta T^{Az} = T^{Az}$ (predicted) − T^{Az} (experiment).
To convert degrees Celsius to degrees Fahrenheit, °F = 1.8°C + 32. For ΔT, °F = 1.8°C.
*At 98.9 kPa (742 torr).

For pairs (1,3) and (2,3),

$$x_1 = (x_1^{(1,3)} + \hat{c}x_2^{(2,3)})/(1 - \hat{c}\hat{d}) \qquad (13\text{-}132)$$

$$x_2 = (x_2^{(2,3)} + \hat{d}x_1^{(1,3)})/(1 - \hat{c}\hat{d}) \qquad (13\text{-}133)$$

where $\hat{c} = (Z_{12} - Z_{23} - Z_{13})/2Z_{13}$ and $\hat{d} = (Z_{12} - Z_{23} - Z_{13})/2Z_{23}$.

For pairs (1,2) and (1,3),

$$x_2 = (x_2^{(1,2)} + \hat{e}x_3^{(1,3)})/(1 - \hat{e}\hat{f}) \qquad (13\text{-}134)$$

$$x_3 = (x_3^{(1,3)} + \hat{f}x_2^{(1,2)})/(1 - \hat{e}\hat{f}) \qquad (13\text{-}135)$$

where $\hat{e} = (Z_{23} - Z_{13} - Z_{12})/2Z_{12}$ and $\hat{f} = (Z_{23} - Z_{13} - Z_{12})/2Z_{13}$. Values of Z_{12}, Z_{13}, and Z_{23} should be computed from Eq. (13-117).

A ternary system is not azeotropic if, for example, the composition of one component is zero or takes a negative value. If one binary system constituting the ternary system is not azeotropic, the corresponding Z_{ij} value may be estimated from that of any close member of a homologous series, its isomers, or closely related substances.

Sources of error in the computed composition of a ternary azeotrope under isobaric conditions by the Malesinski method are deviations of the system from regularity, differences in vaporization entropies of the components, and variation of A_{ij} with temperature. The last two effects are small compared with the first.

The boiling temperature of a ternary homoazeotrope $T_{1,2,3}^{Az}$ with component 2 as the reference component can be computed by Malesinski's method with the equation

$$T_{1,2,3}^{Az} = T_2 - \frac{\delta_2^{(1,2)} + \left(\dfrac{\delta_2^{(1,2)}}{Z_{12}} \cdot \dfrac{\delta_2^{(2,3)}}{Z_{23}}\right)^{1/2}(Z_{13} - Z_{23} - Z_{12}) + \delta_2^{(2,3)}}{1 - \dfrac{(Z_{13} - Z_{23} - Z_{12})^2}{4Z_{12}Z_{23}}} \qquad (13\text{-}136)$$

where $\delta_2^{(1,2)}$ and $\delta_2^{(2,3)}$ are the azeotropic depressions or elevations in azeotropes (1,2) and (2,3) with respect to component 2 (e.g., $\delta_2^{(1,2)} = T_2 - T^{(1,2)}$) and T_2 and $T^{(1,2)}$ are the boiling temperatures of pure component 2 and the azeotrope (1,2) respectively. By convention, $\delta_1^{(i,j)}$ and Z_{ij} are positive for a positive azeotrope and negative for a negative one. By interchanging components, the boiling temperature of the ternary homoazeotrope may be computed also from the two remaining sets of pairs of the components. For mixtures that exactly fulfill the requirements for regular solutions, these predictions are independent of the choice of the reference component.

For positive azeotropes, Eqs. (13-130) to (13-136) usually give good predictions. In the case of positive-negative (saddle) azeotropes, agreement between predicted and observed azeotropic data is less satisfactory because such azeotropes are formed by complex mixtures containing polar and associated components, e.g., alcohols and low-molecular-weight fatty acids.

Table 13-23 lists predicted and experimental azeotropic data for some ternary saddle systems. For systems 1 and 4, results of calculations based on each of the three possible pairs of components are given. For the series of ternary saddle systems acetic acid–pyridine (2-picoline)–*n*-paraffins, predicted azeotropic compositions are improved when corrections for association (dimerization) of acetic acid in the vapor phase are taken into account.

Equations (13-130) to (13-136) predict the existence of a large number of ternary saddle (positive-negative) azeotropes that may appear in the course of fractional distillation of certain fractions of coal tar. For example, ternary saddle azeotropes are formed in the series of ternary systems phenol–aniline–*n*-paraffins (ranging from nonane to tetradecane). Only one ternary saddle azeotrope of this type has been found experimentally, namely, for the system phenol–aniline–*n*-tridecane by Stadnicki [*Bull. Acad. Pol. Sci. Ser. Sci. Chim.*, **10**, 357 (1962)].

The second approach is a general method for predicting vapor-liquid equilibrium data in ternary and multicomponent systems that can be restricted to the azeotropic point by setting the relative volatility α_{ij} to a value of 1. When applied to an ideal gas,

$$\alpha_{ij} = \frac{y_i x_j}{x_i y_j} = \frac{P_i^{sat}\gamma_i^L}{P_j^{sat}\gamma_j^L} = 1 \qquad (13\text{-}137)$$

The method involves a search for the ternary composition that minimizes a function $\bar{f}$, which for a ternary azeotrope is

$$\bar{f} = |\alpha_{13} - 1| + |\alpha_{23} - 1| \qquad (13\text{-}138)$$

The value $\bar{f}$, sufficiently close to zero, corresponds to the azeotropic composition. Any suitable correlation for predicting the liquid-phase activity coefficient in a multicomponent system may be used to obtain α_{ij} from Eq. (13-137). This method was used by Aristovich and Stepanova [*Zh. Prikl. Khim.*, 43, 2192 (1970)] with the Wilson equation for γ_i^L to predict azeotropic compositions in 19 ternary systems and 1 quaternary system. Their results for nine ternary systems are summarized in Table 13-24. For the quaternary positive-negative system ethanol–chloroform–acetone–*n*-hexane, which is unreported by Horsley, they predicted an azeotrope at 55°C (131°F) with x_i and (Δx_i) of ethanol, 22.8 (2.8); chloroform, 35.1 (1.1), acetone, 1.0 (−1.0), and *n*-hexane, 41.1 mole percent and (−2.9) mole percent respectively.

For ternary and multicomponent heteroazeotropes, this procedure remains essentially unchanged but requires the use of an equation for γ_i^L that is applicable to partially miscible systems, e.g., the nonrandom two-liquid (NRTL) equation or the UNIQUAC equation.

On the basis of previous arguments that are applicable to heteroazeotropic systems of any number of components, expressions for composition and boiling temperature of a ternary heteroazeotrope can be obtained. For the case in which the components are essentially

TABLE 13-23 Predicted Azeotropic Data (x_i, T^{Az}) and Deviations Δx_i and ΔT^{Az} (in Parentheses) for Some Ternary Positive-Negative (Saddle) Systems

No.	System	Mole percent x_1 (Δx_1)	x_2 (Δx_2)	x_3 (Δx_3)	°C T^{Az} (ΔT^{Az})	Pairs from which azeotropic data are computed
1	Methanol(1)-acetone(2)-chloroform(3)	45.6 (0.6)	29.6 (−2.4)	24.8 (1.8)	56.98 (0.52)	(1,2) and (2,3)
		38.4 (−6.6)	44.3 (12.3)	17.3 (−5.7)	57.04 (0.46)	(1,2) and (1,3)
		38.1 (−6.9)	30.9 (−1.1)	31.0 (8.0)	60.73 (−3.23)	(1,3) and (2,3)
2	Ethanol-acetone-chloroform	22.4 (4.1)	30.7 (−2.7)	46.9 (−1.4)	60.3 (−0.87)	(1,3) and (2,3)
3	2,4,6-Collidine–*m*-cresol–naphthalene	15.0 (−4.5)	61.0 (−3.5)	24.0 (8.0)	203.27 (−2.55)	(1.2) and (2,3)
4	2,4,6-Collidine–*o*-cresol–glycol	3.8 (0.8)	49.9 (−0.2)	46.3 (−0.7)	189.9 (0.25)	(1,2) and (2,3)
		nonazeotropic			189.22 (−0.43)	(1,2) and (1,3)
		3.4 (0.4)	51.2 (1.2)	45.4 (−1.6)	190.23 (0.58)	(1,3) and (2,3)
5	Acetic acid–pyridine–*n*-decane	42.5 (−0.5)	33.2 (−6.4)	24.3 (6.9)	131.0 (3.1)	(1,2) and (1,3)
6	Acetic acid–2,6-lutidine–*n*-decane	23.8 (2.7)	64.5 (−5.2)	11.7 (2.5)	145.93 (−1.07)	(1,2) and (1,3)
7	Phenol–3-picoline–glycol (−7.5)	55.5 (1.2)	15.8 (6.3)	28.7 (−2.3)	184.11	(1,2) and (1,3)
8	Acetic acid–2-picoline–*n*-decane	25.0 (−6.0)	50.4 (3.4)	24.6 (2.6)	136.84 (−4.46)	(1,2) and (1,3)
9	Propionic acid–2-picoline–*n*-octane	11.6 (5.6)	11.5 (−0.7)	76.9 (−4.9)	123.05 (−0.65)	(1,2) and (1,3)
10	Propionic acid–2-picoline–*n*-decane	42.6 (3.3)	29.2 (−4.7)	28.2 (1.4)	147.93 (−1.37)	(1,2) and (1,3)

NOTE: $\Delta x_i = x_i$ (predicted) − x_i (experiment); $\Delta T^{Az} = T^{Az}$ (predicted) − T^{Az} (experiment). To convert degrees Celsius to degrees Fahrenheit, °F = 1.8°C + 32.

TABLE 13-24 Azeotropic Compositions Predicted from Eq. (13-138) and Deviation Δx_i (in Parentheses) for Some Ternary Positive and Saddle Systems

No.	System	Mole percent x_1 (Δx_1)	x_2 (Δx_2)	x_3 (Δx_3)	Azeotrope type
1	Acetone(1)-chloroform(2)-methanol(3)	33.3 (1.7)	24.5 (0.5)	42.4 (−2.2)	*s*
2	Acetone-chloroform-ethanol	34.0 (−1.0)	48.5 (2.5)	17.5 (−2.2)	*s*
3	Chloroform-acetone-*n*-hexane	63.1 (2.8)	5.2 (−2.5)	31.7 (−0.3)	*s*
4	Benzene-ethanol-cyclohexane	11.0 (0.0)	43.5 (0.5)	45.5 (−0.5)	*p*
5	Methyl acetate–chloroform–methanol	15.5 (−9.5)	53.4 (24.2)	31.1 (−14.7)	*s*
6	Methyl acetate–methanol–*n*-hexane	32.0 (0.2)	33.2 (2.7)	34.8 (−2.9)	*p*
7	*n*-Butanol–benzene–cyclohexane	2.0 (−2.4)	49.0 (−0.6)	49.0 (3.0)	*p*
8	Cyclohexane–ethanol–ethyl acetate	50.1 (−0.1)	42.1 (0.5)	7.8 (−0.4)	*p*
9	MEK-cyclohexane-isopropanol	20.6 (3.3)	44.1 (−9.4)	35.3 (6.1)	*p*

NOTE: $x_i = x_i$ (predicted) − x_i (experiment); *p* = positive; *s* = saddle.

immiscible in each other, expressions for y_1, y_2, and $T^{Az}_{1,2,3}$ of a ternary heteroazeotrope are given in terms of Dalton's law:

$$y_1 = \frac{P_1^{sat}}{P_1^{sat} + P_2^{sat} + P_3^{sat}} \tag{13-139}$$

$$y_2 = \frac{P_2^{sat}}{P_1^{sat} + P_2^{sat} + P_3^{sat}} \tag{13-140}$$

where the azeotropic temperature satisfies

$$P = P_1^{sat} + P_2^{sat} + P_3^{sat} \tag{13-141}$$

When the condition of complete immiscibility is not fulfilled, activities of the components a_1, a_2, and a_3 may be introduced as shown earlier.

As shown by Malesinski [*Bull. Acad. Pol. Sci. Ser. Sci. Chim.*, **11**, 475 (1963)], a series of ternary heteroazeotropes with three liquid phases occurs in the systems nitromethane–water–*n*-paraffins. For this series, satisfactory agreement was obtained between predicted and observed azeotropic compositions, but predicted boiling temperatures of the azeotropes by Eq. (13-141) were found to be much lower than those observed. This is due to the relatively high miscibility of nitromethane in water at the respective boiling temperatures of the azeotropes. The difference is lower for systems with high-boiling hydrocarbons and tends to the difference between the predicted and observed boiling temperatures for the nitromethane-water system.

In some cases the behavior of a series of ternary heteroazeotropes, in which two components remain unchanged for the series, may be described by the activities a_1, a_2, and a_3, which are common for each component within the entire series. As shown by Trabczynski [*Bull. Acad. Pol. Sci. Ser. Sci. Chim. Geol. Geogr.*, **6**, 269 (1958)], this case is exemplified by the series of ternary heteroazeotropes of water–pyridine–*n*-paraffins. All heteroazeotropes found experimentally have been positive; the existence of negative heteroazeotropes is rather doubtful. Among nine quaternary azeotropes, eight are heteroazeotropes and one a positive-negative homoazeotrope. The only quinary azeotropic system is also a heteroazeotrope.

A tangent azeotrope is defined by Swietoslawski (*Ebulliometric Measurements*, Reinhold, New York, 1945) as one that has the composition and the boiling temperature of one of the pure components. In practice, azeotropes and nonazeotropes that are close to this limiting case exist, and their existence as shown by Anderson [*Ind. Eng. Chem.*, **37**, 541, 1052 (1945)] creates new problems to the purification of liquid mixtures by distillation. Tangent azeotropes commonly occur in the distillation of complex mixtures such as coal tar, petroleum, and products of coal hydrogenation. If the substance to be purified forms tangent or almost tangent azeotropes with contaminants, its purification becomes impossible even with a very efficient fractionating column. The explanation of this is given by Malesinski (op. cit., 1965) and lies in the difference between the composition of the vapor and the liquid $y_2 - x_2$, which in the region of the tangent azeotrope is proportional to $(x_2 - x_2^{Az})^2$. In the neighborhood of an ordinary nontangent azeotrope the differences are larger and proportional to $(x_2 - x_2^{Az})$. For a binary regular mixture, separation of an azeotrope by rectification from a mixture of a composition close to azeotropic is the easier, the closer the azeotropic composition is to equimolar and the larger is the absolute value of the constant A_{12}.

Effect of Pressure on Azeotropic Composition In designing an azeotropic-distillation process, it is important to consider general rules for the variation in composition of an azeotrope with pressure, because pressure is an operating parameter that is readily changed and its change can have a positive effect on the separation of a given mixture.

The first generalization concerning the effect of pressure was made by Roozeboom (*Die heterogenen Gleichgewichte*, vol. II, Leipzig, 1904), who noted that the composition of a binary-positive azeotrope shifts toward the component with the more steeply increasing vapor pressure. By applying the Clausius equation to this observation, with increasing pressure a negative azeotrope becomes enriched or a negative azeotrope becomes impoverished in that component for which the molal heat of vaporization is larger.

A rapid graphical method for indicating the effect of pressure on composition and boiling temperature of a binary azeotrope utilizes the Cox vapor-pressure chart. Total-vapor-pressure curves of azeotropes are essentially straight lines when plotted on a Cox chart of the form log P^{sat} versus $1/(T°C + 230)$, which permits the determination of the complete vapor-pressure curve for the azeotrope from data at two pressures. An azeotrope by definition has either a higher vapor pressure (positive) or a lower vapor pressure (negative) than that of any of its components. Therefore, the azeotropic vapor-pressure curve will always lie above or below the curves of the pure components. At the point of intersection of the azeotropic vapor-pressure curve with the curve of either component, the system becomes nonazeotropic. On the other hand, if the azeotropic curve is nearly parallel to the curves of the components, the system will remain azeotropic up to the critical pressure. Caution should be used in extrapolating vapor-pressure curves to very low pressures where curvature may occur on a Cox chart.

With respect to the effect of pressure, four cases may be distinguished:

1. The system becomes azeotropic at an elevated pressure, e.g., MEK-methanol at 400 kPa (3000 torr).
2. The system becomes azeotropic at a reduced pressure, e.g., aniline–*n*-octane at 53.3 kPa (400 torr).
3. In a rare case the system becomes nonazeotropic at both low and high pressures, e.g., methanol-acetone at 26.7 and 1999.8 kPa (200 and 15,000 torr) respectively.
4. The system remains azeotropic up to the critical pressure, e.g., pyridine–acetic acid.

SELECTION OF AN ENTRAINER

In azeotropic distillation, an entrainer or solvent is added deliberately to the mixture to be separated so as nearly to form an azeotrope or azeotropes. In that way, the vapor-liquid equilibrium is shifted in a favorable direction. Thus, the purpose of the entrainer is either to separate one component of a closely boiling pair or to separate one component of an azeotrope.

General methods involving the selection of entrainers may be summarized as follows. To separate a negative azeotrope or a closely boiling pair of components, select an entrainer that (1) forms a binary positive azeotrope with only one component, or (2) forms binary positive azeotropes with both components of the system when one azeotrope has a sufficiently lower-boiling temperature than the other, or (3) forms a ternary positive azeotrope with a boiling temperature sufficiently below that of any binary azeotrope. The ratio of the original feed components in the ternary azeotrope must be different from that ratio in the feed, and the ternary azeotrope (preferably heterogeneous) must be separable in some way.

To separate a binary positive azeotrope, select an entrainer that forms with one component a binary positive azeotrope that has a boiling temperature sufficiently below that of the original azeotrope or that forms a ternary positive azeotrope with a sufficiently low-boiling temperature and with a different ratio of the original components from that ratio in the original azeotrope.

Entrainers or solvents may be divided into two groups. Those that form azeotropes with some type of compounds but fail to form azeotropes with another type are called selective entrainers or solvents. Those that do not exhibit these properties are nonselective entrainers or solvents. For example, methanol is a nonselective entrainer for all types of hydrocarbons (paraffin, naphthene, olefin, aromatic) boiling in the range from 100 to 110°C (212 to 230°F). Methyl ethyl ketone, on the other hand, is a selective entrainer for hydrocarbons boiling in the same range of temperatures. It forms positive azeotropes with all the hydrocarbons except aromatics. Acetic acid is a nonselective entrainer for *n*-paraffins (ranging from hexane to undecane), aromatics (benzene, toluene, xylenes, ethyl benzene), and pyridine bases (pyridine, picolines, lutidines).

An alternative approach is to use the azeotropic-range values discussed earlier. In the case of a binary system, for instance, it is advisable to compile substances that form azeotropes with either component 1 or component 2, or both. The final choice should consider whether the entrainer has a low molal heat of vaporization, is noncorrosive to the equipment, is nonreactive with the feed components, and is thermally stable, nontoxic, and inexpensive.

In a commercial process, it is necessary to recover the entrainer from the azeotrope and recycle it to the azeotropic-distillation column. This is an economic aspect and requires that the entrainer be separated easily and cheaply. The easiest separation occurs when entrainer and separated substance are insoluble at room temperature. When they are miscible, it is desirable that the entrainer be water-soluble so that water washing can be accomplished. The water-entrainer mixture should be easily separated. When the entrainer is water-insoluble, a different washing component that is inexpensive and easily removed from the entrainer must be used.

The selection of an entrainer to effect a given separation will be illustrated for three binary azeotropes: acetone-methanol, benzene-cyclohexane, and acetone-chloroform. In addition, industrial dehydration of acetic acid and ethanol are described, as well as a process leading to increased recovery of naphthalene from high-temperature coal-tar fractions. Figures in brackets refer to normal boiling points.

At atmospheric pressure, methanol [64.7°C (148.5°F)] and acetone [56.4°C (133.5°F)] form a positive azeotrope boiling at 55.7°C (132.3°F) and containing 80 mole percent acetone. A suitable entrainer is methylene chloride, which forms only with methanol an azeotrope that boils at 37.8°C (100°F) and contains 82.7 mole percent methylene chloride. If methylene chloride is added in an appropriate proportion to the methanol-acetone azeotrope and the resulting mixture subjected to distillation, the distillate will be close to the azeotrope methanol–methylene chloride.

Benzene [80.2°C (176.4°F)] and cyclohexane (80.8°C (177.4°F)] fall into a category of close-boiling substances. At 101.3 kPa (1 atm), the two components form a positive azeotrope boiling at 77.4°C (171.3°F) and containing 54 mole percent benzene. A good entrainer for the separation of this azeotrope is acetone [56.4°C (133.5°F)], which only with cyclohexane forms an azeotrope that boils at 53.1°C (127.6°F) and contains 74.6 mole percent acetone. Benzene lies outside the azeotropic range of acetone. Determination of the required entrainer-to-feed mole ratio is conveniently illustrated by the triangular composition diagram in Fig. 13-68, in which compositions are in mole fractions. The hydrocarbon feed mixture H contains cyclohexane and benzene at a composition that lies between pure benzene B and the binary azeotrope at the right side of the diagram. Acetone

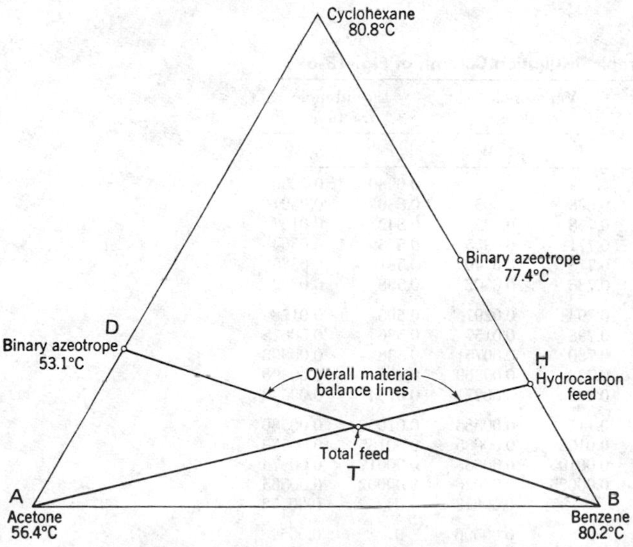

FIG. 13-68 Pure-component and azeotrope boiling points for the acetone-cyclohexane-benzene system at 101.3 kPa (1 atm). Material-balance lines show the solvent treat required to produce pure benzene as the bottoms product.

A represents the entrainer feed to the column. The distillate is the acetone-cyclohexane binary azeotrope D that lies at the left side of the diagram. The bottoms is assumed to be pure benzene. The composition of the two combined column feeds or total feed T is the intersection of straight lines DB and AH, which represent material-balance lines for the column products and column feeds respectively. The molar ratio of acetone feed to hydrocarbon feed is given by $\overline{HT}/\overline{AT}$.

Chloroform and acetone form a negative azeotrope boiling at 64.5°C (148.1°F) and containing 65.5 mole percent chloroform. A possible entrainer is carbon disulfide, which forms a positive azeotrope with acetone. The azeotrope boils at 39.3°C (102.7°F) and contains 76.1 mole percent carbon disulfide.

Acetic acid [118.3°C (244.9°F)] is one of the low-molecular-weight aliphatic acids that do not form an azeotrope with water. However, from data in Table 13-1 it is seen that separation by simple distillation is difficult. Ethylene dichloride [83.5°C (182.3°F)] was the first entrainer used on a large scale for the dehydration of acetic acid. The positive binary azeotrope boils at 71.6°C (160.9°F) and contains 8 weight percent water. Later, in 1932, n-propyl acetate was substituted for ethylene dichloride, and then n-butyl acetate was proposed. Acetic acid produced by hydrocarbon oxidation can be purified by azeotropic distillation with benzene. In a patent by Null (U.S. Patent 3,335,179, Aug. 8, 1967), the separation involves removal of formic acid and water impurities from acetic acid and is based on the formation of binary azeotropes [formic acid–benzene, 71.1°C (160.0°F), 31 weight percent formic acid; water-benzene, 69.25°C (156.7°F), 8.8 weight percent benzene; and acetic acid–benzene, 80.1°C (176.2°F) and 2 weight percent acetic acid].

Young [*J. Chem. Soc. Trans.*, **81**, 707 (1902)] early described a process by which ethanol containing water is dehydrated by the addition of benzene as an entrainer. Benzene forms with ethanol and water a ternary positive heterogeneous azeotrope that has a higher ratio of water to ethanol than the ethanol-water azeotrope. The ternary azeotrope boils at 64.86°C (148.7°F) and contains, on a weight basis, 74.1 percent benzene, 18.5 percent ethanol, and 7.4 percent water. Compositions and boiling temperatures of the two binary azeotropes are 32.4 weight percent ethanol-benzene at 68.24°C (154.8°F) and ethanol with 4 weight percent water at 78.15°C (172.7°F). The azeotrope is the distillate, and pure ethanol is the bottoms product.

In spite of the fact that a simple and efficient method for ethanol dehydration exists, some technological difficulties occur in carrying out the process on a large scale, notably separation of the two liquid phases because of small differences in their densities and unfavorable surface-tension phenomena. To improve this process, Guinot [*Comptes Rendus*, **176**, 1623 (1923)] proposed the addition to the benzene of a gasoline fraction boiling between 101 and 102°C (214 and 216°F). In fact, there are no hydrocarbons boiling within that temperature range. Even if there were, they would not be proper agents for ethanol dehydration. The boiling temperature range of an appropriate gasoline fraction is 93 to 99°C (199 to 210°F). Then ethanol dehydration is an efficient operation owing to formation of a series of quaternary positive heteroazeotropes containing benzene, ethanol, water, and hydrocarbons (paraffinic and naphthenic) boiling between 93 and 99°C (199 and 210°F). In this case, a polycomponent azeotropic agent was used for the first time for a separation purpose. Pressure has a marked effect on the water-ethanol ratio in the ternary azeotrope benzene-ethanol-water. For instance, Karpinski and Swietoslawski [*Comptes Rendus*, **198**, 2166 (1934)] report water-ethanol molal ratios at 101.3, 1013, and 2026 kPa (1, 10, and 20 atm) of 0.4, 0.51, and 0.64 respectively.

Ethanol may also be dehydrated by using trichloroethylene as an entrainer. However, the ethanol bottoms product contains some of the entrainer, which must be removed by an additional distillation. A n-pentane process is shown in Fig. 13-62.

In many cases the opinion that the presence of numerous azeotropes in distillation is undesirable is not quite justified. It has been proved that azeotropy may often be exploited for achieving a considerably higher yield of some high-temperature coal-tar constituents. For example, naphthalene yield may be increased from 35 to 90–93 percent as reported by Swietoslawski (op. cit., 1959).

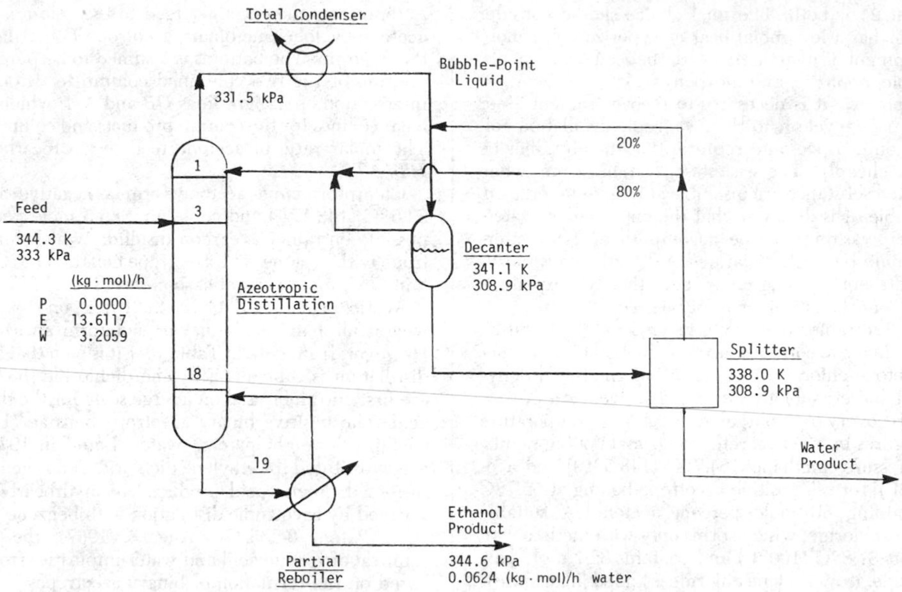

FIG. 13-69 Specifications for an azeotropic-distillation process to separate ethanol from water by using a pentane entrainer.

DESIGN OF AZEOTROPIC-DISTILLATION COLUMNS

Basic computational procedures for azeotropic distillation are the same rigorous methods used for simple distillation as described earlier. However, the design is somewhat more complex because of the additional variables resulting from the additional feed. Both the entrainer-to-feed ratio and the location of the entrainer entry point must be specified. Although in most cases it seems best to introduce the entrainer in the top section of the column, such a generalization is rather dangerous.

If a computer program is used to make the calculations, it must be able to predict and calculate phase splitting if a heterogeneous azeotrope is formed. A simultaneous-correction method or other technique that specifically accounts for strong influences of composition and temperature on liquid-phase activity coefficients is usually essential to obtain convergence of the calculations. Even then, steep concentration and temperature profiles that slow convergence may be encountered. In general, azeotropic distillation represents a severe test of rigorous distillation computer algorithms.

At present, owing to the availability of reasonably reliable computer codes, a thorough investigation of feed and entrainer entry points is possible. Because nonideal liquid solutions are being dealt with, accurate activity-coefficient correlations are essential. In addition, for most cases energy balances must take into account heats of mixing.

Example 9: Calculation of Azeotropic Distillation Use a rigorous method to compute stage temperature and interstage flow rates for the

TABLE 13-25 Converged Results for Azeotropic Distillation Column of Fig. 13-69

Stage	Temperature, K	Traffic, (kg · mol)/h Vapor	Traffic, (kg · mol)/h Liquid	Vapor mole fractions *P*	Vapor mole fractions *W*	Liquid mole fractions *P*	Liquid mole fractions *W*
Reflux	326.6	....	32.2697	....	....	0.6990	0.2350
1	333.9	35.4008	33.0	0.638	0.303	0.830	0.0521
2	341.3	36.1	34.6	0.758	0.135	0.842	0.0167
3	343.0	37.8	51.8	0.771	0.0985	0.573	0.0564
4	344.4	38.1	52.3	0.778	0.0749	0.581	0.0383
5	345.7	38.6	52.8	0.786	0.0502	0.588	0.0232
6	346.7	39.1	53.1	0.794	0.0297	0.593	0.0128
7	347.4	39.4	53.0	0.798	0.0157	0.586	0.00675
8	348.4	39.3	51.4	0.790	0.00751	0.538	0.00392
9	352.2	37.7	46.0	0.732	0.00369	0.338	0.00396
10	367.1	32.3	42.8	0.482	0.00371	0.0754	0.00542
11	382.0	29.1	44.2	0.111	0.00583	0.0106	0.00580
12	385.1	30.5	44.6	0.0153	0.00635	0.00135	0.00580
13	385.6	30.9	44.6	0.00195	0.00635	0.00017	0.00574
14	385.7	31.0	44.6	0.00025	0.00626	0.00002	0.00565
15	385.8	31.0	44.7	0	0.00613	0	0.00553
16	385.9	31.0	44.7	0	0.00596	0	0.00538
17	385.9	31.0	44.7	0	0.00574	0	0.00518
18	386.0	31.0	44.7	0	0.00545	0	0.00491
19	386.1	31.0	13.6736	0	0.00506	0	0.00456

azeotropic-distillation process shown in Fig. 13-69, in which pentane (*P*) is used as the entrainer to form with ethanol (*E*) and water (*W*) a ternary positive heterogeneous azeotrope that has a higher ratio of water to ethanol than the near-azeotropic ethanol-water mixture fed to the azeotropic-distillation tower. Specifications shown in Fig. 13-69 include the water content of the ethanol product, the fraction of the overhead from the splitter that is sent directly as reflux to the azeotropic-distillation column whose theoretical tray requirements are shown, and the temperature of the decanter. Process pressures [just above 303.9 kPa (3 atm)] are set so as to enable the overhead vapor from the azeotropic-distillation column to be totally condensed. The splitter is not modeled rigorously but is assumed to be a near-perfect separator that provides a liquid distillate at 64°C (147°F).

The calculations were made with the program PROCESS^{sm} of Simulation Sciences, Inc., Fullerton, California, by using their (1) rigorous distillation routine to model the 18-plate column plus the partial reboiler; (2) heat-exchanger routine to model the total overhead condenser with outlet condensate at the bubble point; (3) three-phase flash routine to model the decanter, which separates the near-heterogeneous azeotrope into two equilibrium liquid phases; (4) component separator routine to model a specified material balance for the splitter; and (5) mixers and a splitter to model stream mixing and dividing. Liquid-phase activity coefficients were computed from a modified van Laar equation, as discussed by Black et al. (*Extractive and Azeotropic Distillation*, Advances in Chemistry ser. 115, American Chemical Society, Washington, 1972, p. 64). To initiate the calculations, it is necessary to provide initial guesses for component flow rates in at least one stream for each of the two recycle loops. The decanter feed is common to both loops and is, therefore, a convenient choice. Initial guesses in kilogram-moles per hour were (*P*) 25.0, (*E*) 3.0, and (*W*) 7.5 [in pound-moles per hour, (*P*) 55.0, (*E*), 6.6, and (*W*) 16.5].

Tabulated converged results for just the azeotropic-distillation tower and partial reboiler are given by theoretical stage in Table 13-25. As seen, stage temperatures and compositions change only slightly from the reboiler up to stage 13. In this region, pentane concentration builds up from a negligible value to a mole fraction in the liquid of less than 0.02 percent. From stage 13 to stage 8, pentane builds up rapidly to appreciable amounts in both equilibrium liquid and vapor phases, and temperature decreases more rapidly. From stage 8 to the feed entry at stage 3, temperature decreases only a few kelvins, while the water mole fraction in the liquid increases rapidly from 0.00392 to 0.0564. Rather drastic changes occur above the feed entry. Final converged results for the decanter operating at 341.1 K (614°R) and 308.9 kPa (44.8 lbf/in^2) are:

	(kg·mol)/h		
Component	Decanter feed	Organic-rich phase	Water-rich phase
Pentane	22.5568	22.5487	0.0081
Ethanol	2.3962	1.0636	1.3326
Water	12.5821	0.1005	12.4816
	37.5351	23.7128	13.8223

NOTE: To convert kilogram-moles per hour to pound-moles per hour, multiply by 2.20.

Total condenser and partial reboiler duties are 1116.5 and 1135.0 MJ/h (587,900 and 597,000 Btu/h) respectively.

PETROLEUM AND COMPLEX-MIXTURE DISTILLATION

Introduction Although the principles of multicomponent distillation apply to petroleum, synthetic crude oil, and other complex mixtures, this subject warrants special consideration for the following reasons:

1. Such feedstocks are of exceedingly complex composition, consisting, in the case of petroleum, of many different types of hydrocarbons and perhaps of inorganic and other organic compounds. The number of carbon atoms in the components may range from 1 to more than 50, so that the compounds may exhibit atmospheric-pressure boiling points from −162°C (−259°F) to more than 538°C (1000°F). In a given boiling range, the number of different compounds that exhibit only small differences in volatility multiplies rapidly with increasing boiling point. For example, 16 of the 18 octane isomers boil within a range of only 12°C (22°F).

2. Products from the distillation of complex mixtures are in themselves complex mixtures. The character and yields of these products vary widely, depending upon the source of the feedstock. Even crude oils from the same locality may exhibit marked variations.

3. The scale of petroleum-distillation operations is generally large, and, as discussed in detail by Nelson (*Petroleum Refinery Engineering*, 4th ed., McGraw-Hill, New York, 1958) and Watkins (*Petroleum Refinery Distillation*, 2d ed., Gulf, Houston, 1979), such operations are common in several petroleum-refinery processes including atmospheric distillation of crude oil, vacuum distillation of bottoms residuum obtained from atmospheric distillation, main fractionation of gaseous effluent from catalytic cracking of various petroleum fractions, and main fractionation of effluent from thermal coking of various petroleum fractions. These distillation operations are conducted in large pieces of equipment that can consume large quantities of energy. Therefore, optimization of design and operation is very important and frequently leads to a relatively complex equipment configuration.

Characterization of Petroleum and Petroleum Fractions Although much progress has been made in identifying the chemical species present in petroleum, it is generally sufficient for purposes of design and analysis of plant operation of distillation to characterize petroleum and petroleum fractions by gravity, laboratory-distillation curves, component analysis of light ends, and hydrocarbon-type analysis of middle and heavy ends. From such data, as discussed in the *Technical Data Book—Petroleum Refining* [American Petroleum Institute (API), Washington], five different average boiling points and an index of paraffinicity can be determined; these are then used to predict the physical properties of complex mixtures by a number of well-accepted correlations, whose use will be explained in detail and illustrated with examples. Many other characterizing properties or attributes such as sulfur content, pour point, water and sediment content, salt content, metals content, Reid vapor pressure, Saybolt Universal viscosity, aniline point, octane number, freezing point, cloud point, smoke point, diesel index, refractive index, cetane index, neutralization number, wax content, carbon content, and penetration are generally measured for a crude oil or certain of its fractions according to well-specified ASTM tests. But these attributes are of much less interest here even though feedstocks and products may be required to meet certain specified values of the attributes.

Gravity of a crude-oil or petroleum fraction is generally measured by the ASTM D 287 test or the equivalent ASTM D 1298 test and may be reported as specific gravity (SG) 60/60°F [measured at 60°F (15.6°C) and referred to water at 60°F (15.6°C)] or, more commonly, as API gravity, which is defined as

$$\text{API gravity} = 141.5/(\text{SG } 60/60°\text{F}) - 131.5 \qquad (13\text{-}142)$$

Water, thus, has an API gravity of 10.0, and most crude oils and petroleum fractions have values of API gravity in the range of 10 to 80. Light hydrocarbons (*n*-pentane and lighter) have values of API gravity ranging upward from 92.8.

The volatility of crude-oil and petroleum fractions is characterized in terms of one or more laboratory distillation tests that are summarized in Table 13-26. The ASTM D 86 and D 1160 tests are reasonably rapid batch laboratory distillations involving the equivalent of approximately one equilibrium stage and no reflux except for that caused by heat losses. Apparatus typical of the D 86 test is shown in Fig. 13-70 and consists of a heated 100-mL or 125-mL Engler flask containing a calibrated thermometer of suitable range to measure the temperature of the vapor at the inlet to the condensing tube, an inclined brass condenser in a cooling bath using a suitable coolant, and a graduated cylinder for collecting the distillate. A stem correction is not applied to the temperature reading. Related tests using

TABLE 13-26 Laboratory Distillation Tests

Test name	Reference	Main applicability
ASTM (atmospheric)	ASTM D 86	Petroleum fractions or products, including gasolines, turbine fuels, naphthas, kerosines, gas oils, distillate fuel oils, and solvents that do not tend to decompose when vaporized at 760 mmHg
ASTM [vacuum, often 10 torr (1.3 kPa)]	ASTM D 1160	Heavy petroleum fractions or products that tend to decompose in the ASTM D 86 test but can be partially or completely vaporized at a maximum liquid temperature of 750°F (400°C) at pressures down to 1 torr (0.13 kPa)
TBP [atmospheric or 10 torr (1.3 kPa)]	Nelson,* ASTM D 2892	Crude oil and petroleum fractions
Simulated TBP (gas chromatography)	ASTM D 2887	Crude oil and petroleum fractions
EFV (atmospheric, superatmospheric, or subatmospheric)	Nelson†	Crude oil and petroleum fractions

*Nelson, *Petroleum Refinery Engineering*, 4th ed., McGraw-Hill, New York, 1958, pp. 95–99.
†Ibid., pp. 104–105.

similar apparatuses are the D 216 test for natural gasoline and the Engler distillation.

In the widely used ASTM D 86 test, 100 mL of sample is charged to the flask and heated at a sufficient rate to produce the first drop of distillate from the lower end of the condenser tube in from 5 to 15 min, depending on the nature of the sample. The temperature of the vapor at that instant is recorded as the initial boiling point (IBP). Heating is continued at a rate such that the time from the IBP to 5 volume percent recovered of the sample in the cylinder is 60 to 75 s. Again, vapor temperature is recorded. Then, successive vapor temperatures are recorded for from 10 to 90 percent recovered in intervals of 10, and at 95 percent recovered, with the heating rate adjusted so that 4 to 5 mL are collected per minute. At 95 percent recovered, the burner flame is increased if necessary to achieve a maximum vapor temperature referred to as the end point (EP) in from 3 to 5 additional min. The percent recovery is reported as the maximum percent recovered in the cylinder. Any residue remaining in the flask is reported as percent residue, and percent loss is reported as the difference between 100 mL and the sum of the percent recovery and percent residue. If the atmosphere test pressure P is other than 101.3 kPa (760 torr), temperature readings may be adjusted to that pressure by the Sidney Young equation, which for degrees Fahrenheit is

$$T_{760} = T_P + 0.00012(760 - P)(460 + T_P) \qquad (13\text{-}143)$$

Another pressure correction for percent loss can also be applied, as described in the ASTM test method.

Results of a typical ASTM distillation test for an automotive gasoline are given in Table 13-27, in which temperatures have already been corrected to a pressure of 101.3 kPa (760 torr). It is generally assumed that percent loss corresponds to volatile noncondensables that are distilled off at the beginning of the test. In that case, the percent recovered values in Table 13-27 do not correspond to percent evaporated values, which are of greater scientific value. Therefore, it is common to adjust the reported temperatures according to a linear interpolation procedure given in the ASTM test method to obtain corrected temperatures in terms of percent evaporated at the standard intervals as included in Table 13-27. In the example, the corrections are not large because the loss is only 1.5 volume percent.

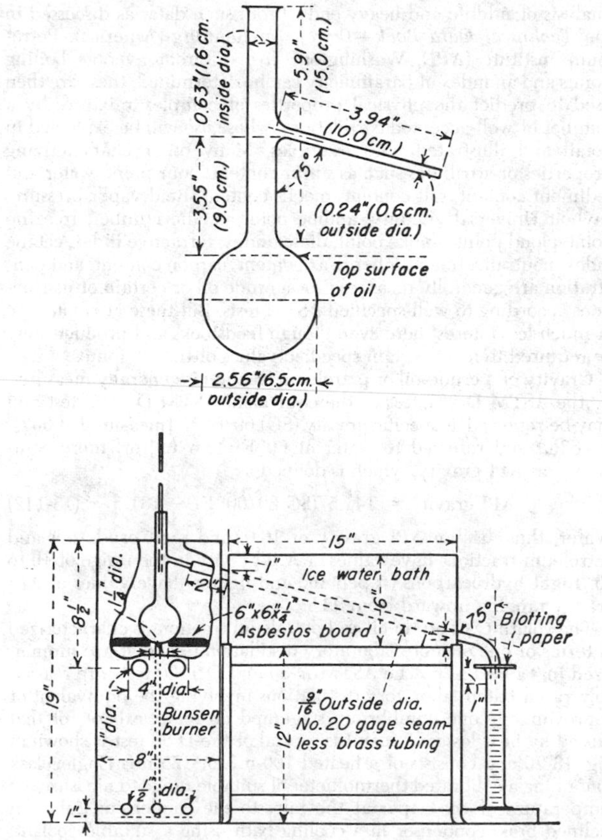

FIG. 13-70 ASTM distillation apparatus; detail of distilling flask is shown in the upper figure.

TABLE 13-27 Typical ASTM D 86 Test Results for Automobile Gasoline Pressure, 760 torr (101.3 kPa)

Percent recovered basis (as measured)			Percent evaporated basis (as corrected)		
Percent recovered	T, °F	Percent evaporated	Percent evaporated	T, °F	Percent recovered
0(IBP)	98	1.5	1.5	98	(IBP)
5	114	6.5	5	109	3.5
10	120	11.5	10	118	8.5
20	150	21.5	20	146	18.5
30	171	31.5	30	168	28.5
40	193	41.5	40	190	38.5
50	215	51.5	50	212	48.5
60	243	61.5	60	239	58.5
70	268	71.5	70	264	68.5
80	300	81.5	80	295	78.5
90	340	91.5	90	334	88.5
95	368	96.5	95	360	93.5
EP	408	. . .	. .	408	(EP)

NOTE: Percent recovery = 97.5; percent residue = 1.0; percent loss = 1.5. To convert degrees Fahrenheit to degrees Celsius, °C = (°F − 32)/1.8.

Although most crude petroleum can be heated to 600°F (316°C) without noticeable cracking, when ASTM temperatures exceed 475°F (246°C), fumes may be evolved, indicating decomposition, which may cause thermometer readings to be low. In that case, the following correction attributed to S. T. Hadden may be applied:

$$\Delta T_{corr} = 10^{-1.587+0.004735T}$$

where T = measured temperature, °F

ΔT_{corr} = correction to be added to T, °F

At 500 and 600°F (260 and 316°C), the corrections are 6 and 18°F (3.3 and 10°C) respectively.

As discussed by Nelson (op. cit.), virtually no fractionation occurs in an ASTM distillation. Thus, components in the mixture do distill one by one in the order of their boiling points but as mixtures of successively higher boiling points. The IBP, EP, and intermediate points have little theoretical significance, and, in fact, components boiling below the IBP and above the EP are present in the sample. Nevertheless, because ASTM distillations are quickly conducted, have been successfully automated, require only a small sample, and are quite reproducible, they are widely used for comparison and as a basis for specifications on a large number of petroleum intermediates and products, including many solvents and fuels. Typical ASTM curves for several such products are shown in Fig. 13-71.

Data from a true-boiling-point (TBP) distillation test provides a much better theoretical basis for characterization. If the sample contains compounds that have moderate differences in boiling points such as in a light gasoline containing light hydrocarbons (e.g., isobutane, n-butane, isopentane, etc.), a plot of overhead-vapor-distillate temperature versus percent distilled in a TBP test would appear in the form of steps as in Fig. 13-72. However, if the sample has a higher average boiling range when the number of close-boiling isomers increases, the steps become indistinct and a TBP curve such as

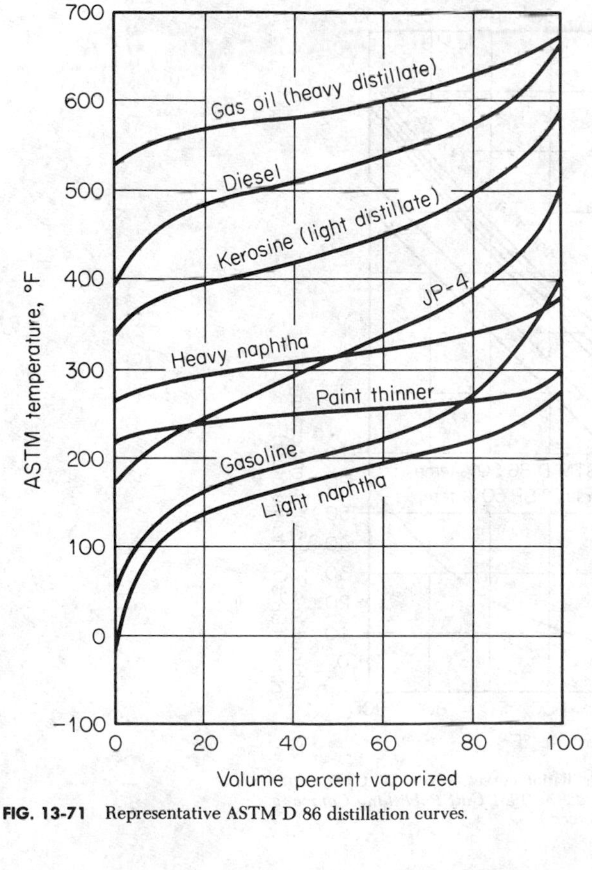

FIG. 13-71 Representative ASTM D 86 distillation curves.

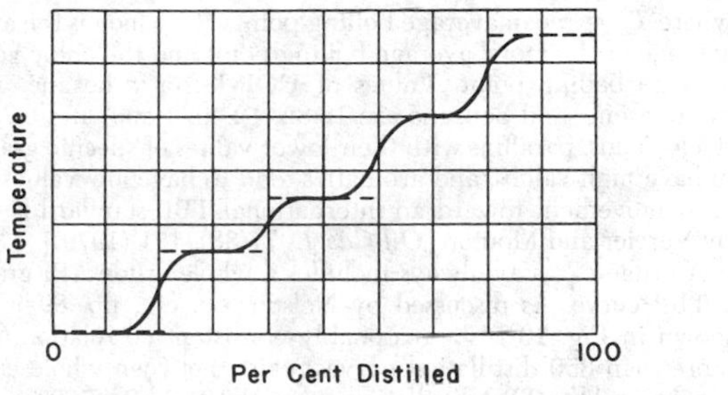

FIG. 13-72 Variation of boiling temperature with percent distilled in true-boiling-point distillation of light hydrocarbons.

that in Fig. 13-73 results. Because the degree of separation for a TBP distillation test is much higher than for an ASTM distillation test, the IBP is lower and the EP is higher for the TBP method as compared with the ASTM method, as shown in Fig. 13-73.

A standard TBP laboratory-distillation-test method has not been well accepted. Instead, as discussed by Nelson (op. cit., pp. 95–99), batch distillation equipment that can achieve a good degree of fractionation is usually considered suitable. In general, TBP distillations are conducted in columns with 15 to 100 theoretical stages at reflux ratios of 5 or greater. Thus, the new ASTM D 2892 test method, which involves a column with from 14 to 17 theoretical stages and a reflux ratio of 5, essentially meets the minimum requirements. Distillate may be collected at a constant or a variable rate. Operation may be at 101.3-kPa (760-torr) pressure or at a vacuum at the top of the column as low as 0.067 kPa (0.5 torr) for high-boiling fractions, with 1.3 kPa (10 torr) being common. Results from vacuum operation are extrapolated to 101.3 kPa (760 torr) by the vapor-pressure correlation of Maxwell and Bonner [*Ind. Eng. Chem.*, **49**, 1187 (1957)], which is given in great detail in the API *Technical Data Book—Petroleum Refining* (op. cit.) and in the ASTM D 2892 test method. It includes a correction for the nature of the sample (paraffin, olefin, napthene, and aromatic content) in terms of the UOP characterization factor, UOP-K, as given by

$$\text{UOP-K} = (T_B)^{1/3}/\text{SG} \tag{13-144}$$

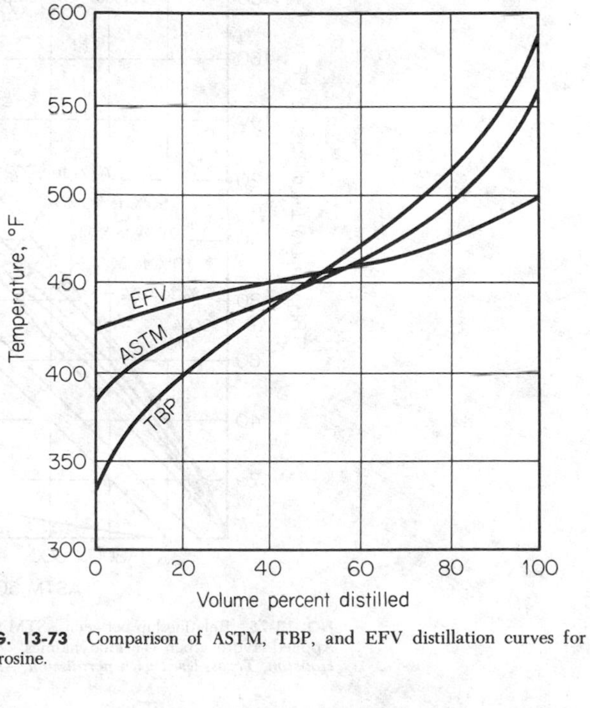

FIG. 13-73 Comparison of ASTM, TBP, and EFV distillation curves for kerosine.

where T_B = mean average boiling point, °R, which is the arithmetic average of the molal average boiling point and the cubic volumetric average boiling point. Values of UOP-K for n-hexane, 1-hexene, cyclohexene, and benzene are 12.82, 12.49, 10.99, and 9.73 respectively. Thus, paraffins with their lower values of specific gravity tend to have high values, and aromatics tend to have low values of UOP-K. A movement toward an international TBP standard is discussed by Vercier and Mouton [*Oil Gas J.*, **77**(38), 121 (1979)].

A crude-oil assay always includes a whole crude API gravity and a TBP curve. As discussed by Nelson (op. cit., pp. 89–90) and as shown in Fig. 13-74, a reasonably consistent correlation (based on more than 350 distillation curves) exists between whole crude API gravity and the TBP distillation curve at 101.3 kPa (760 torr). Exceptions not correlated by Fig. 13-74 are highly paraffinic or naphthenic crude oils.

An alternative to TBP distillation that is receiving widespread use is simulated distillation by gas chromatography. As described by Green, Schmauch, and Worman [*Anal. Chem.*, **36**, 1512 (1965)] and Worman and Green [*Anal. Chem.*, **37**, 1620 (1965)], the method is equivalent to a 100-theoretical-plate TBP distillation, is very rapid, reproducible, and easily automated, requires only a small microliter sample, and can better define initial and final boiling points. The ASTM D 2887 standard test method is based on such a simulated distillation and is applicable to samples having a boiling range greater than 55°C (100°F) for temperature determinations as high as 538°C (1000°F). Typically, the test is conducted with a gas chromatograph having a thermal-conductivity detector, a programmed temperature capability, helium or hydrogen carrier gas, and column packing of silicone gum rubber on a crushed-fire-brick or diatomaceous-earth support.

It is important to note that simulated distillation does not always

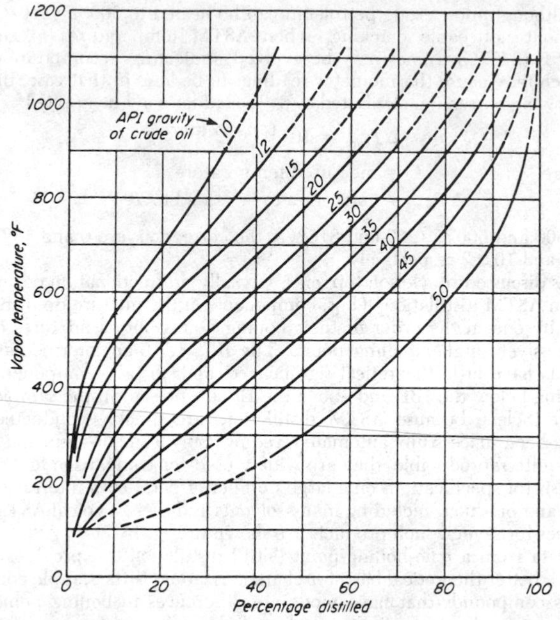

FIG. 13-74 Average true-boiling-point distillation curves of crude oils. (*From W. C. Edmister,* Applied Hydrocarbon Thermodynamics, *vol. 1, 1st ed., © 1961 Gulf Publishing Company, Houston, Texas. Used with permission. All rights reserved.*)

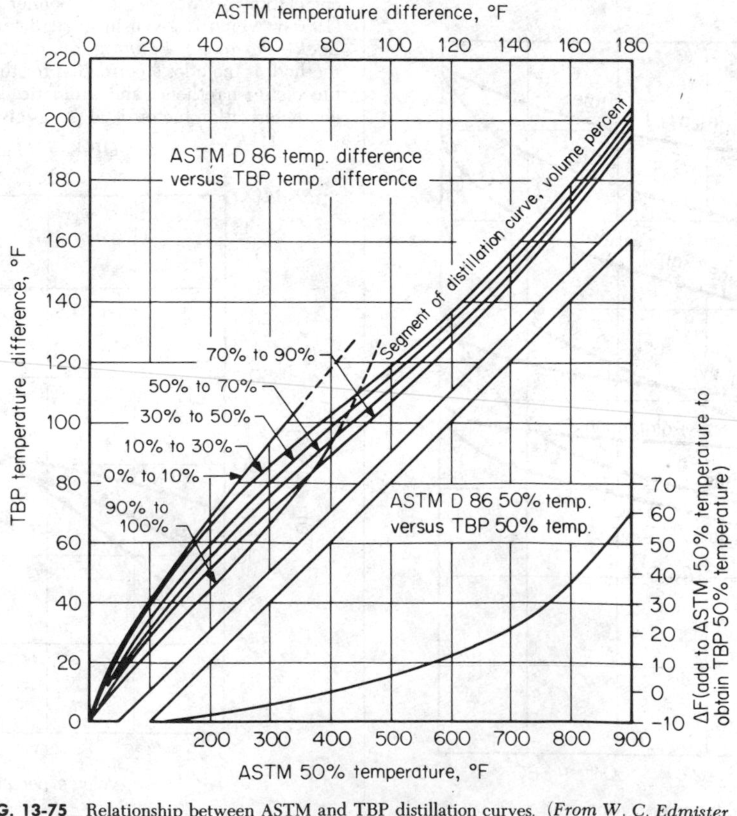

FIG. 13-75 Relationship between ASTM and TBP distillation curves. (*From W. C. Edmister,* Applied Hydrocarbon Thermodynamics, *vol. 1, 1st ed., © 1961 Gulf Publishing Company, Houston, Texas. Used with permission. All rights reserved.*)

separate hydrocarbons in the order of their boiling point. For example, high-boiling multiple-ring-type compounds may be eluted earlier than normal paraffins (used as the calibration standard) of the same boiling point. Gas chromatography is also used in the ASTM D 2427 test method to determine quantitatively ethane through pentane hydrocarbons.

A third fundamental type of laboratory distillation, which is the most tedious to perform of the three types of laboratory distillations, is equilibrium-flash distillation (EFV), for which no standard test exists. The sample is heated in such a manner that the total vapor produced remains in contact with the total remaining liquid until the desired temperature is reached at a set pressure. The volume percent vaporized at these conditions is recorded. To determine the complete flash curve, a series of runs at a fixed pressure is conducted over a range of temperature sufficient to cover the range of vaporization from 0 to 100 percent. As seen in Fig. 13-73, the component separation achieved by an EFV distillation is much less than by the ASTM or TBP distillation tests. The initial and final EFV points are the bubble point and the dew point respectively of the sample. If desired, EFV curves can be established at a series of pressures.

Because of the time and expense involved in conducting laboratory distillation tests of all three basic types, it has become increasingly common to use empirical correlations to estimate the other two distillation curves when either the ASTM, TBP, or EFV curve is available. Preferred correlations given in the API *Technical Data Book—Petroleum Refining* (op. cit.) are based on the work of Edmister and Pollock [*Chem. Eng. Prog.*, **44**, 905 (1948)], Edmister and Okamoto [*Pet. Refiner*, **38**(8), 117 (1959); **38**(9), 271 (1959)], Maxwell (*Data Book on Hydrocarbons*, Van Nostrand, Princeton, N.J., 1950), and Chu and Staffel [*J. Inst. Pet.*, **41**, 92 (1955)]. Because of the lack of sufficiently precise and consistent data on which to develop the correlations, they are, at best, first approximations and should be used with caution. Also, they do not apply to mixtures containing only a few components of widely different boiling points. Perhaps the most useful correlation of the group is Fig. 13-75 for converting between ASTM D 86 and TBP distillations of petroleum fractions at 101.3 kPa (760 torr). The ASTM D 2889 test method, which presents a standard method for calculating EFV curves from the results of an ASTM D 86 test for a petroleum fraction having a 10 to 90 volume percent boiling range of less than 55°C (100°F), is also quite useful.

Applications of Petroleum Distillation Typical equipment configurations for the distillation of crude oil and other complex hydrocarbon mixtures in a crude unit, a catalytic-cracking unit, and a delayed-coking unit of a petroleum refinery are shown in Figs. 13-76, 13-77, and 13-78. The initial separation of crude oil into fractions is conducted in two main columns, shown in Fig. 13-76. In the first column, called the atmospheric tower or topping still, partially vaporized crude oil, from which water, sediment, and salt have been removed, is mainly rectified, at a feed-tray pressure of no more than about 276 kPa (40 psia), to yield a noncondensable light-hydrocarbon gas, a light naphtha, a heavy naphtha, a light distillate (kerosine), a

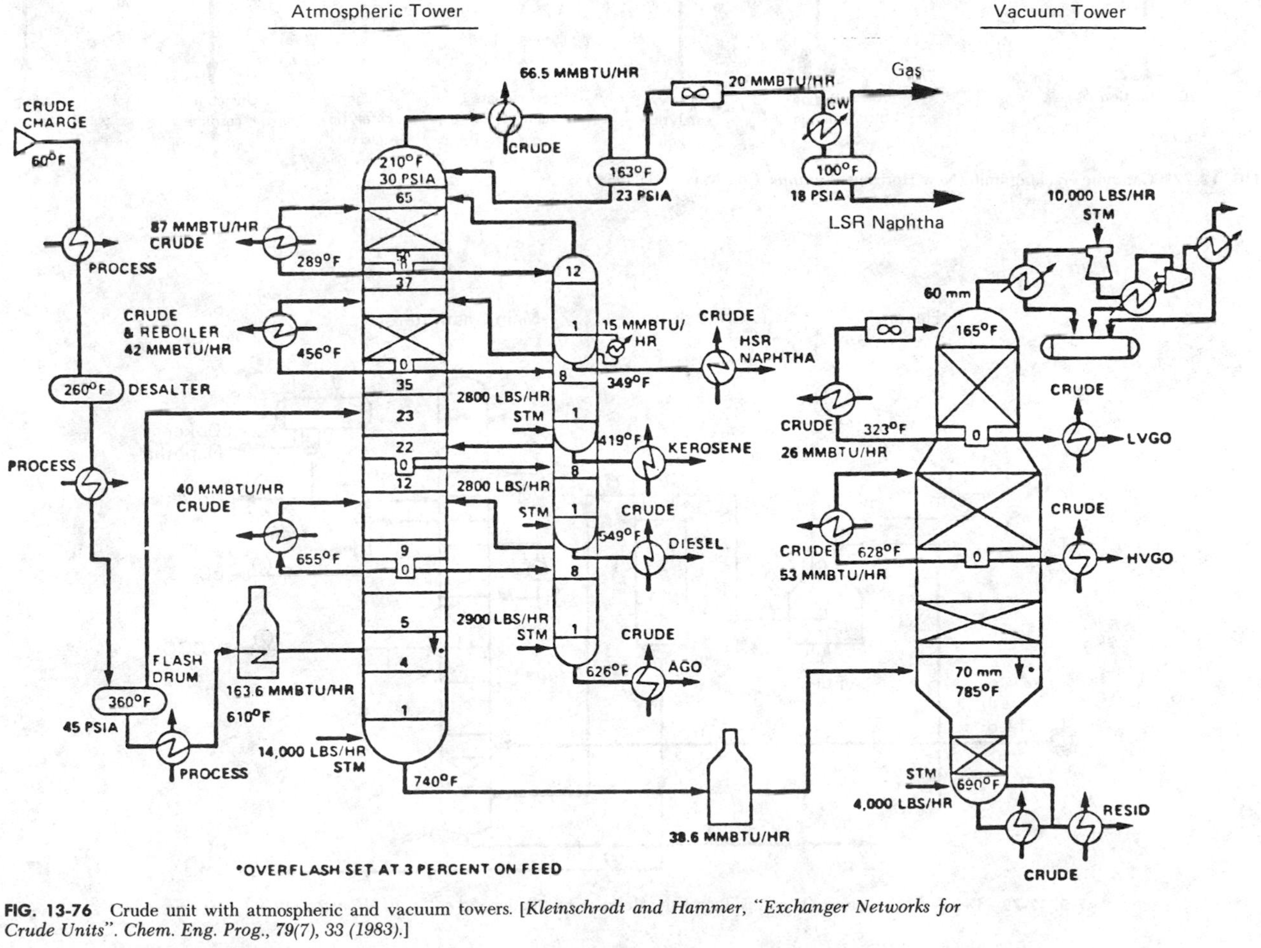

FIG. 13-76 Crude unit with atmospheric and vacuum towers. [*Kleinschrodt and Hammer, "Exchanger Networks for Crude Units". Chem. Eng. Prog., 79(7), 33 (1983).*]

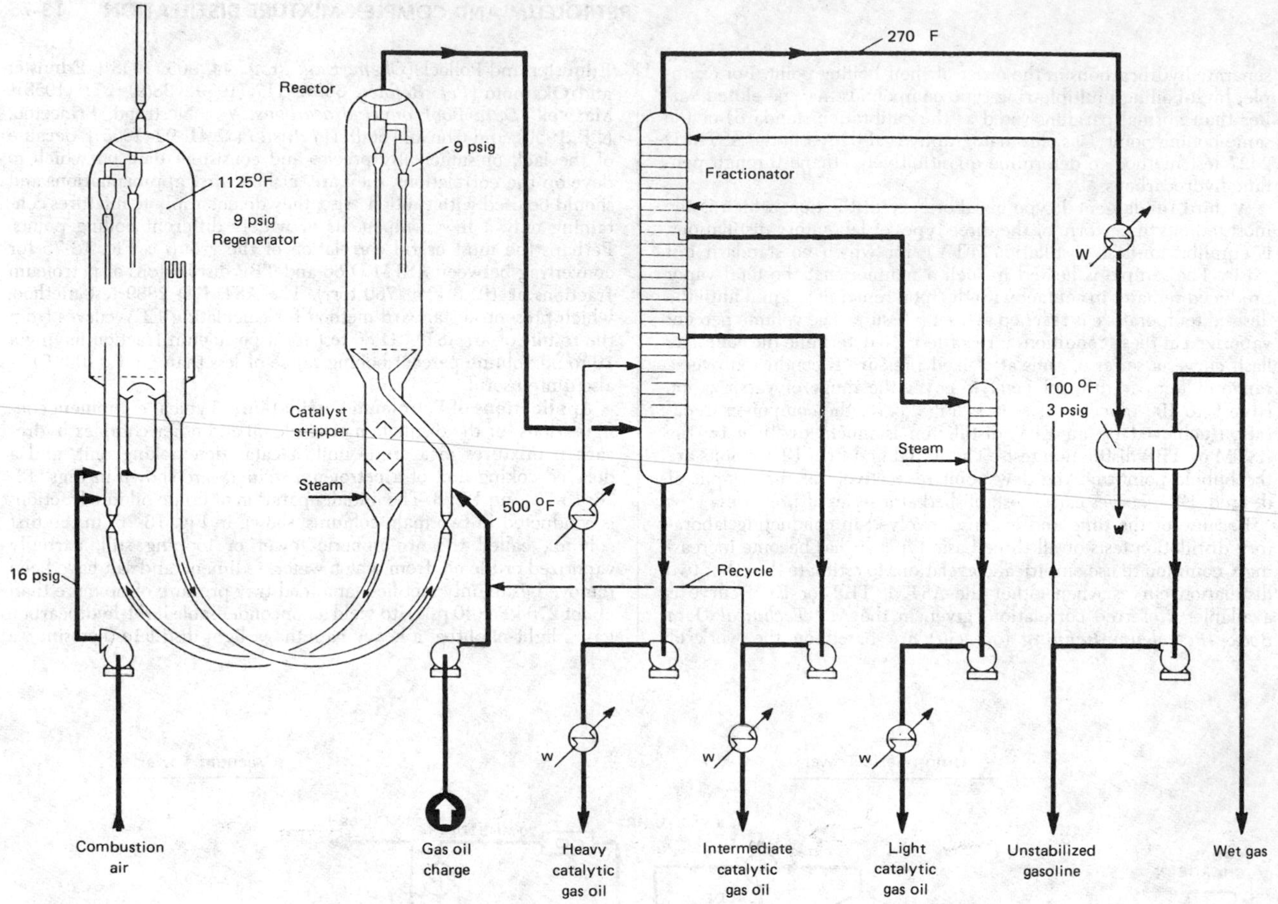

FIG. 13-77 Catalytic cracking unit. (New Horizons, *Lummus Co., New York, 1954.*)

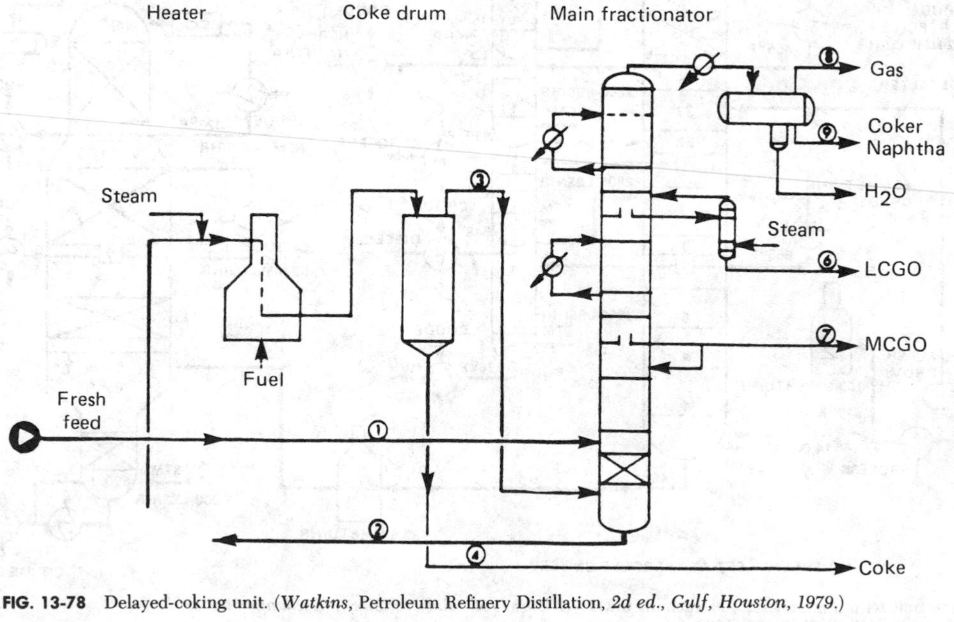

FIG. 13-78 Delayed-coking unit. (*Watkins,* Petroleum Refinery Distillation, *2d ed., Gulf, Houston, 1979.*)

heavy distillate (diesel oil), and a bottoms residual of components whose TBP exceeds approximately 427°C (800°F). Alternatively, other fractions, shown in Fig. 13-71, may be withdrawn. To control the IBP of the ASTM D 86 curves, each of the sidestreams of the atmospheric tower and the vacuum and main fractionators of Figs. 13-76, 13-77, and 13-78 may be sent to side-cut strippers, which use a partial reboiler or steam stripping. Additional stripping by steam is commonly used in the bottom of the atmospheric tower as well as in the vacuum tower and other main fractionators.

Additional distillate in the TBP range of approximately 427 to 593°C (800 to 1100°F) is recovered from bottoms residuum of the atmospheric tower by rectification in a vacuum tower, also shown in Fig. 13-76, at the minimum practical overhead condenser pressure, which is typically 1.3 kPa (10 torr). Use of special low-pressure-drop trays or column packing permits feed-tray pressure to be approximately 5.3 to 6.7 kPa (40 to 50 torr) to obtain the maximum degree of vaporization. Vacuum towers may be designed or operated to produce several different products including heavy distillates, gas-oil feedstocks for catalytic cracking, lubricating oils, bunker fuel, and bottoms residua of asphalt (5 to 8° API gravity) or pitch (0 to 5° API gravity). The catalytic-cracking process of Fig. 13-77 produces a superheated vapor at approximately 538°C (1000°F) and 172 to 207 kPa (25 to 30 psia) of a TBP range that covers hydrogen to compounds with normal boiling points above 482°C (900°F). This gas is sent directly to a main fractionator for rectification to obtain products that are typically gas and naphtha [204°C (400°F) ASTM EP approximately], which are often fractionated further to produce relatively pure light hydrocarbons and gasoline; a light cycle oil [typically 204 to 371°C (400 to 700°F) ASTM D 86 range], which may be used for heating oil, hydrocracked, or recycled to the catalytic cracker; an intermediate cycle oil [typically 371 to 482°C (700 to 900°F) ASTM D 86 range], which is generally recycled to the catalytic cracker to extinction; and a heavy gas oil or bottom slurry oil.

Vacuum-column bottoms, bottoms residuum from the main fractionation of a catalytic cracker, and other residua can be further processed at approximately 510°C (950°F) and 448 kPa (65 psia) in a delayed coker unit, as shown in Fig. 13-78, to produce petroleum coke and gas of TBP range that covers methane (with perhaps a small amount of hydrogen) to compounds with normal boiling points that may exceed 649°C (1200°F). The gas is sent directly to a main fractionator that is similar to the type used in conjunction with a catalytic cracker, except that in the delayed-coking operation the liquid to be coked first enters into and passes down through the bottom trays of the main fractionator to be preheated by and to scrub coker vapor of entrained coke particles and condensables for recycling to the delayed coker. Products produced from the main fractionator are similar, except for more unsaturated cyclic compounds, to those produced in a catalytic-cracking unit and include gas and coker naphtha, which are further processed to separate out light hydrocarbons and a coker naphtha that generally needs hydrotreating; and light and heavy coker gas oils, both of which may require hydrocracking to become suitable blending stocks.

Design Procedures Two general procedures are available for designing fractionators that process petroleum, synthetic crude oils, and complex mixtures. The first, which was originally developed for crude units by Packie [*Trans. Am. Inst. Chem. Eng. J.*, **37**, 51 (1941)], extended to main fractionators by Houghland, Lemieux, and Schreiner [*Proc. API*, sec. III, *Refining*, 385 (1954)], and further elaborated and described in great detail by Watkins (op. cit.), utilizes material and energy balances, with empirical correlations to establish tray requirements, and is essentially a hand-calculation procedure that is a valuable learning experience and is suitable for preliminary designs. Also, when backed by sufficient experience from previous designs, this procedure is adequate for final design.

In the second procedure, which is best applied with a digital computer, the complex mixture being distilled is represented by actual components at the light end and by perhaps 30 pseudo components (e.g., petroleum fractions) over the remaining portion of the TBP distillation curve for the column feed. Each of the pseudo components is characterized by a TBP range, an average normal boiling point, an average API gravity, and an average molecular weight. Rigorous material-balance, energy-balance, and phase equilibrium calculations are then made by an appropriate equation-tearing method as shown by Cecchetti et al. [*Hydrocarbon Process.*, **42**(9), 159 (1963)] or a simultaneous-correction procedure as shown, e.g., by Goldstein and Stanfield [*Ind. Eng. Chem. Process Des. Dev.*, **9**, 78 (1970) and Hess et al. [*Hydrocarbon Process.*, **56**(5), 241 (1977)]. Highly developed procedures of the latter type, suitable for preliminary or final design, include the REFINE^{sm} computer program of the ChemShare Corporation, Houston, Texas, and the PROCESS^{sm} program of Simulation Sciences, Inc., Alhambra, California.

Regardless of the procedure used, certain initial steps must be taken for the determination or specification of certain product properties and yields based on the TBP distillation curve of the column feed, method of providing column reflux, column-operating pressure, type of condenser, and type of side-cut strippers and stripping requirements. These steps are developed and illustrated with several detailed examples by Watkins (op. cit.). Only one example, modified from one given by Watkins, is considered briefly here to indicate the approach taken during the initial steps.

For the atmospheric tower shown in Fig. 13-79, suppose distillation specifications are as follows:

Feed: 50,000 bbl (at 42 U.S. gal each) per stream day (BPSD) of 31.6° API crude oil.

Measured light-ends analysis of feed:

Component	Volume percent of crude oil
Ethane	0.04
Propane	0.37
Isobutane	0.27
n-Butane	0.89
Isopentane	0.77
n-Pentane	1.13
	3.47

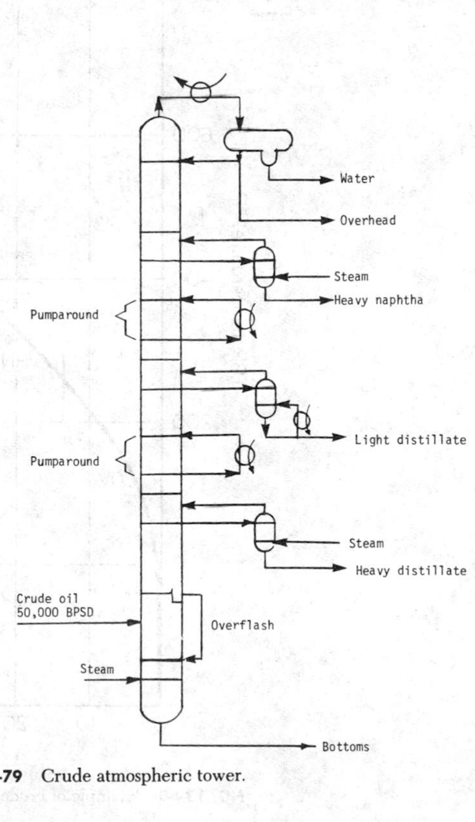

FIG. 13-79 Crude atmospheric tower.

Measured TBP and API gravity of feed, computed atmospheric pressure EFV (from API *Technical Data Book*), and molecular weight of feed:

Volume percent vaporized	TBP, °F	EFV, °F	°API	Molecular weight
0	−130	179		
5	148	275	75.0	91
10	213	317	61.3	106
20	327	394	50.0	137
30	430	468	41.8	177
40	534	544	36.9	223
50	639	619	30.7	273
60	747	696	26.3	327
70	867	777	22.7	392
80	1013	866	19.1	480

Product specifications:

	ASTM D 86, °F		
Desired cut	5%	50%	95%
Overhead (*OV*)	. . .	. . .	253
Heavy naphtha (*HN*)	278	314	363
Light distillate (*LD*)	398	453	536
Heavy distillate (*HD*)	546	589	
Bottoms (*B*)			

NOTE: To convert degrees Fahrenheit to degrees Celsius, °C = (°F − 32)/1.8.

TBP cut point between the heavy distillate and the bottoms = 650°F.
Percent overflash = 2 volume percent of feed.
Furnace outlet temperature = 343°C (650°F) maximum.
Overhead temperature in reflux drum = 49°C (120°F) minimum.

From the product specifications, distillate yields are computed as follows: From Fig. 13-75 and the ASTM D 86 50 percent temperatures, TBP 50 percent temperatures of the three intermediate cuts are obtained as 155, 236, and 316°C (311, 456, and 600°F) for the *HN*, *LD*, and *HD* respectively. The TBP cut points, corresponding volume fractions of crude oil, and flow rates of the four distillates are readily obtained by starting from the specified 343°C (650°F) cut point as follows, where *CP* is the cut point and *T* is the TBP temperature (°F):

$$CP_{HD,B} = 650°F$$

$$(CP_{HD,B} - T_{HD50}) = 650 - 600 = 50°F$$

$$CP_{LD,HD} = T_{HD50} - 50 = 600 - 50 = 550°F$$

$$(CP_{LD,HD} - T_{LD50}) = 550 - 456 = 94°F$$

$$CP_{HN,LD} = T_{LD50} - 94 = 456 - 94 = 362°F$$

$$(CP_{HN,LD} - T_{HN50}) = 362 - 311 = 51°F$$

$$CP_{OV,HN} = T_{HN50} - 51 = 311 - 51 = 260°F$$

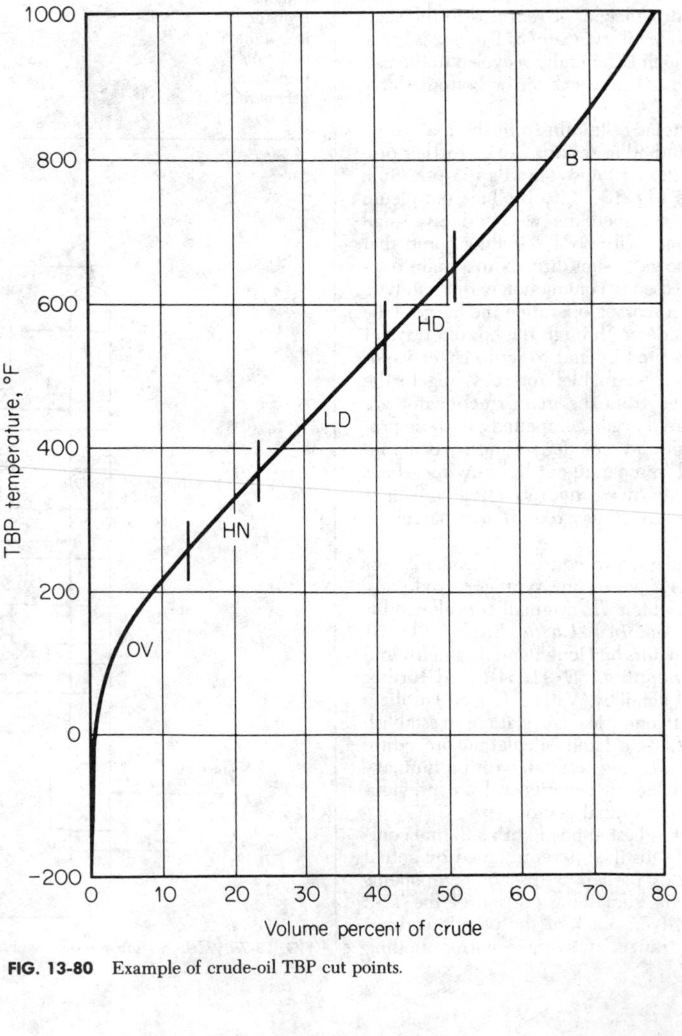

FIG. 13-80 Example of crude-oil TBP cut points.

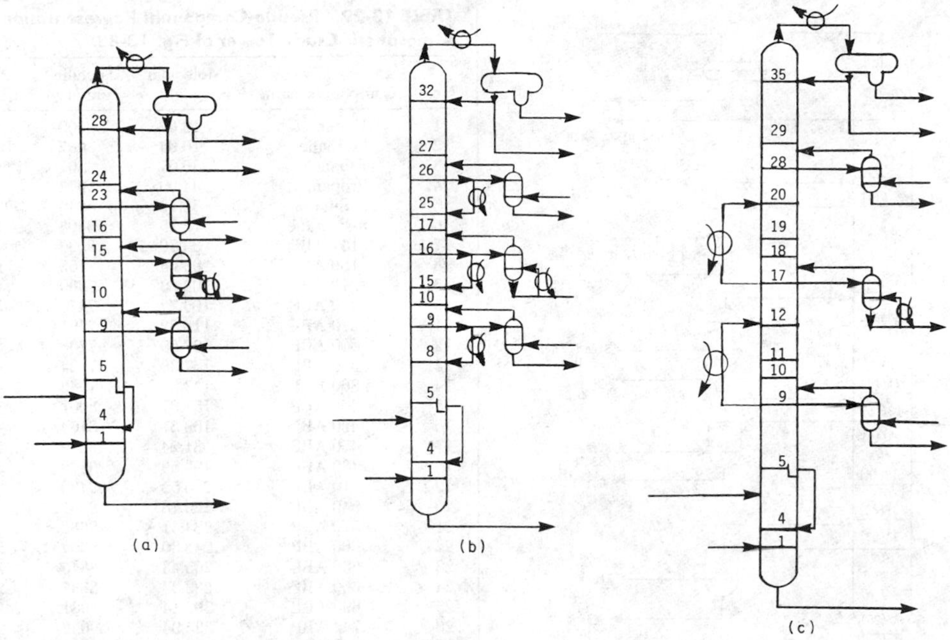

FIG. 13-81 Methods of providing reflux to crude units. (*a*) Top reflux. (*b*) Pump-back reflux. (*c*) Pump-around reflux.

These cut points are shown as vertical lines on the crude-oil TBP plot of Fig. 13-80, from which the following volume fractions and flow rates of product cuts are readily obtained:

Desired cut	Volume percent of crude oil	BPSD
Overhead (*OV*)	13.4	6,700
Heavy naphtha (*HN*)	10.3	5,150
Light distillate (*LD*)	17.4	8,700
Heavy distillate (*HD*)	10.0	5,000
Bottoms (*B*)	48.9	24,450
	100.0	50,000

As shown in Fig. 13-81, methods of providing column reflux include (*a*) conventional top-tray reflux, (*b*) pump-back reflux from side-cut strippers, and (*c*) pump-around reflux. The latter two methods essentially function as intercondenser schemes that reduce the top-tray-reflux requirement. As shown in Fig. 13-82 for the example being considered, the internal-reflux flow rate decreases rapidly from the top tray to the feed-flash zone for case *a*. The other two cases, particularly case *c*, result in better balancing of the column-reflux traffic. Because of this and the opportunity provided to recover energy at a moderate- to high-temperature level, pump-around reflux is the most commonly used technique. However, not indicated in Fig. 13-82 is the fact that in cases *b* and *c* the smaller quantity of

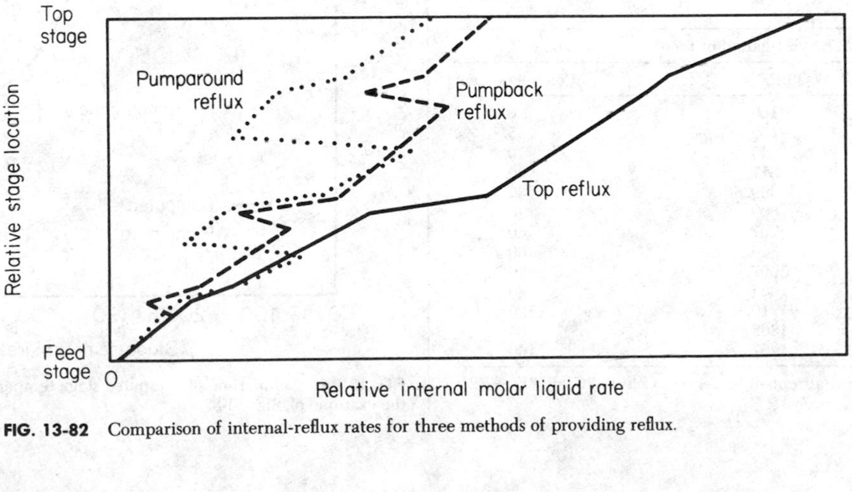

FIG. 13-82 Comparison of internal-reflux rates for three methods of providing reflux.

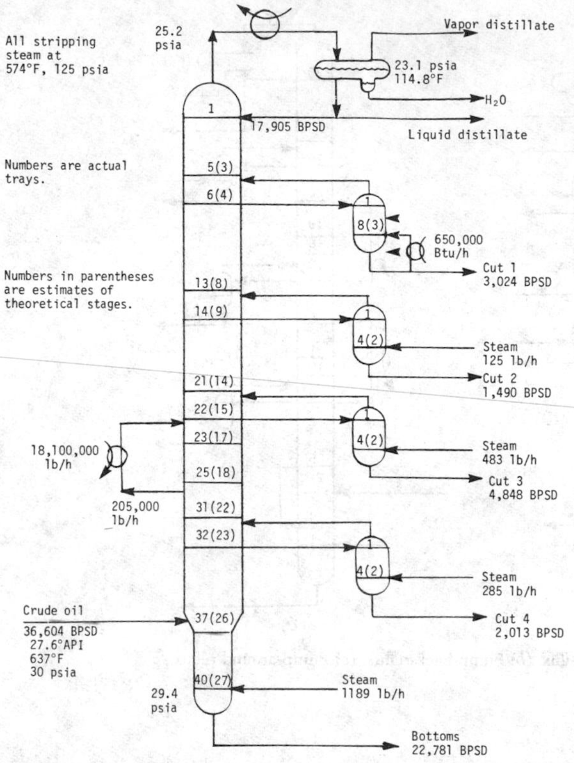

FIG. 13-83 Configuration and conditions for the simulation of the atmospheric tower of a crude unit.

TABLE 13-28 Light-Component Analysis and TBP Distillation of Feed for the Atmospheric Crude Tower of Fig. 13-83

Light-component analysis	
Component	Volume percent
Methane	0.073
Ethane	0.388
Propane	0.618
n-Butane	0.817
n-Pentane	2.05

TBP distillation of feed		
API gravity	TBP, °F	Volume percent
80	−160.	0.1
70	155.	5.
57.5	242.	10.
45.	377.	20.
36.	499.	30.
29.	609.	40.
26.5	707.	50.
23.	805.	60.
20.5	907.	70.
17.	1054.	80.
10.	1210.	90.
−4.	1303.	95.
−22.	1467	100.

NOTE: To convert degrees Fahrenheit to degrees Celsius, °C = (°F − 32)/1.8.

TABLE 13-29 Pseudo-Component Representation of Feed for the Atmospheric Crude Tower of Fig. 13-83

No.	Component name	Molecular weight	Specific gravity	API gravity	(lb·mol)/h
1	Water	18.02	1.0000	10.0	.00
2	Methane	16.04	.3005	339.5	7.30
3	Ethane	30.07	.3561	265.8	24.54
4	Propane	44.09	.5072	147.5	37.97
5	*n*-Butane	58.12	.5840	110.8	43.84
6	*n*-Pentane	72.15	.6308	92.8	95.72
7	131 ABP	83.70	.6906	73.4	74.31
8	180 ABP	95.03	.7152	66.3	66.99
9	210 ABP	102.23	.7309	62.1	65.83
10	240 ABP	109.78	.7479	57.7	70.59
11	270 ABP	118.52	.7591	54.9	76.02
12	300 ABP	127.69	.7706	52.1	71.62
13	330 ABP	137.30	.7824	49.4	67.63
14	360 ABP	147.33	.7946	46.6	64.01
15	390 ABP	157.97	.8061	44.0	66.58
16	420 ABP	169.37	.8164	41.8	63.30
17	450 ABP	181.24	.8269	39.6	59.92
18	480 ABP	193.59	.8378	37.4	56.84
19	510 ABP	206.52	.8483	35.3	59.05
20	540 ABP	220.18	.8581	33.4	56.77
21	570 ABP	234.31	.8682	31.5	53.97
22	600 ABP	248.30	.8804	29.2	52.91
23	630 ABP	265.43	.8846	28.5	54.49
24	660 ABP	283.37	.8888	27.7	51.28
25	690 ABP	302.14	.8931	26.9	48.33
26	742 ABP	335.94	.9028	25.2	109.84
27	817 ABP	387.54	.9177	22.7	94.26
28	892 ABP	446.02	.9288	20.8	74.10
29	967 ABP	509.43	.9398	19.1	50.27
30	1055 ABP	588.46	.9531	17.0	57.12
31	1155 ABP	665.13	.9829	12.5	50.59
32	1255 ABP	668.15	1.0658	1.3	45.85
33	1355 ABP	643.79	1.1618	−9.7	29.39
34	1436 ABP	597.05	1.2533	−18.6	21.19
		246.90	.8887	27.7	1922.43

NOTE: To convert (lb·mol)/h to (kg·mol)/h, multiply by 0.454.

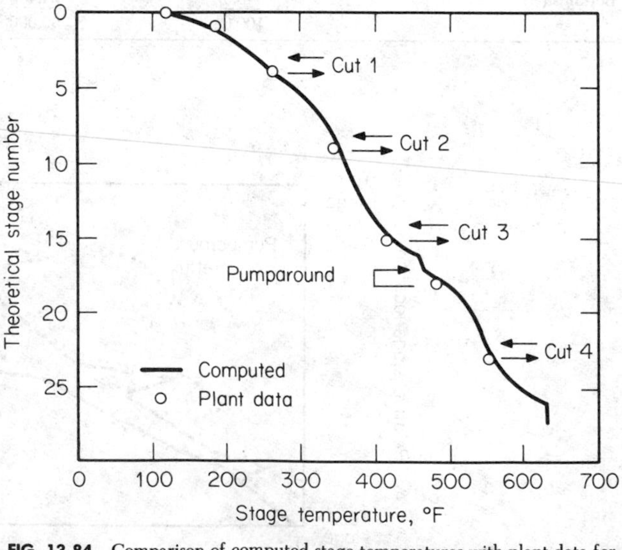

FIG. 13-84 Comparison of computed stage temperatures with plant data for the example of Fig. 13-83.

reflux present in the upper portion of the column increases the tray requirements. Furthermore, the pump-around circuits, which extend over three trays each, are believed to be equivalent for mass-transfer purposes to only one tray each. Representative tray requirements for the three cases are included in Fig. 13-81. In case *c* heat-transfer rates associated with the two pump-around circuits account for approximately 40 percent of the total heat removed in the overhead condenser and from the two pump-around circuits combined.

Bottoms and three side-cut strippers remove light ends from products and may utilize steam or reboilers. In Fig. 13-81 a reboiled stripper is utilized on the light distillate, which is the largest side cut withdrawn. Steam-stripping rates in side-cut strippers and at the bottom of the atmospheric column may vary from 0.45 to 4.5 kg (1 to 10 lb) of steam per barrel of stripped liquid, depending on the fraction of stripper feed liquid that is vaporized.

Column pressure at the reflux drum is established so as to condense totally the overhead vapor or some fraction thereof. Flash-zone pressure is approximately 69 kPa (10 psia) higher. Crude-oil feed temperature at flash-zone pressure must be sufficient to vaporize the total distillates plus the overflash, which is necessary to provide reflux between the lowest sidestream-product draw-off tray and the flash zone. Calculations are made by using the crude-oil EFV curve corrected for pressure. For the example being considered, percent vaporized at the flash zone must be 53.1 percent of the feed.

Tray requirements depend on internal-reflux ratios and ASTM 5-95 gaps or overlaps, and may be estimated by the correlation of Packie (op. cit.) for crude units and the correlation of Houghland, Lemieux, and Schreiner (op. cit.) for main fractionators.

Example 10: Simulation Calculation of an Atmospheric Tower The ability of a rigorous calculation procedure to simulate operation of an atmospheric tower may be illustrated by comparing commercial-test data from an actual operation with results computed with the REFINE program of ChemShare (op. cit.). The tower configuration and plant-operating conditions are shown in Fig. 13-83. Light-component analysis and the TBP and API gravity for the feed are given in Table 13-28. Representation of this feed by pseudo components is given in Table 13-29 based on 16.7°C (30°F) cuts from 82 to 366°C (180°F to 690°F), followed by 41.7°C (75°F) and then 55.6°C (100°F) cuts. Actual tray numbers are shown in Fig. 13-83. Corresponding theoretical-stage numbers, which were determined by trial and error to obtain a reasonable match of computed- and measured-product TBP distillation curves, are shown in parentheses. Overall tray efficiency appears to be approximately 70 percent for the tower and 25 to 50 percent for the side-cut strippers.

Results of rigorous calculations and comparison to plant data, when possible, are shown in Figs. 13-84, 13-85, and 13-86. Plant temperatures are in good agreement with computed values in Fig. 13-84. Computed sidestream-product TBP distillation curves are in reasonably good agreement with values converted from plant ASTM distillations as shown in Fig. 13-85. Exceptions are the initial points of all four cuts and the higher-boiling end of the heavy-distillate curve. This would seem to indicate that more theoretical stripping stages should be added and that either the percent vaporization of the tower feed in the simulation is too high or the internal-reflux rate at the lower draw-off tray is too low. The liquid-rate profile in the tower is shown in Fig. 13-86. The use of two or three pump-around circuits instead of one would result in a better traffic pattern than that shown.

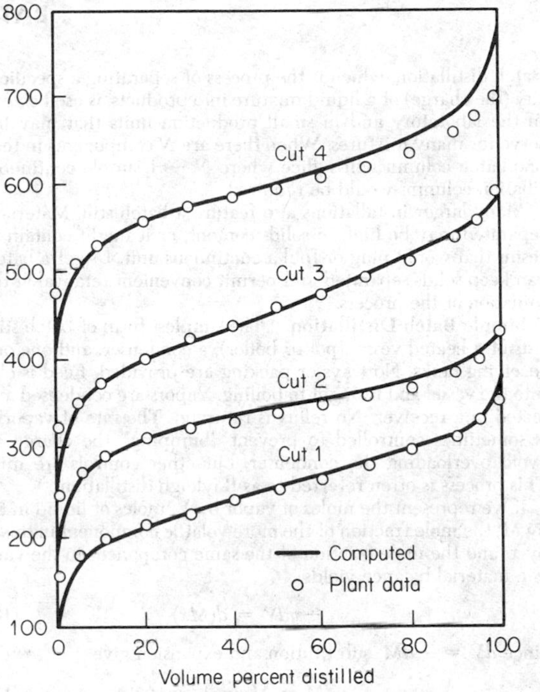

FIG. 13-85 Comparison of computed TBP curves with plant data for the example of Fig. 13-83.

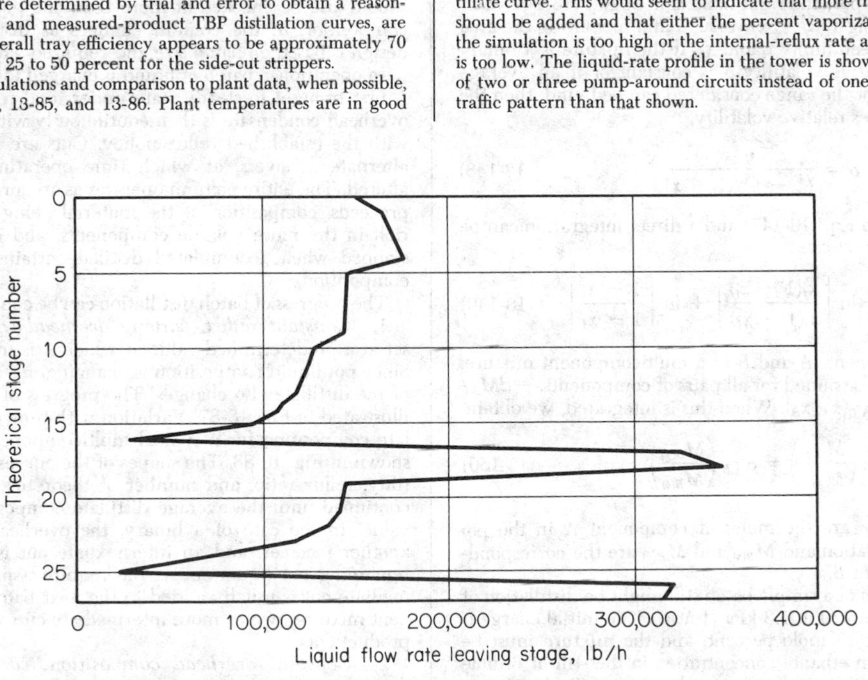

FIG. 13-86 Liquid-rate profile for the example of Fig. 18-83.

BATCH DISTILLATION

Batch distillation, which is the process of separating a specific quantity (the charge) of a liquid mixture into products, is used extensively in the laboratory and in small production units that may have to serve for many mixtures. When there are N components in the feed, one batch column will suffice where $N - 1$ simple continuous-distillation columns would be required.

Many larger installations also feature a batch still. Material to be separated may be high in solids content, or it might contain tars or resins that would plug or foul a continuous unit. Use of a batch unit can keep solids separated and permit convenient removal at the termination of the process.

Simple Batch Distillation. The simplest form of batch still consists of a heated vessel (pot or boiler), a condenser, and one or more receiving tanks. No trays or packing are provided. Feed is charged into the vessel and brought to boiling. Vapors are condensed and collected in a receiver. No reflux is returned. The rate of vaporization is sometimes controlled to prevent "bumping" the charge and to avoid overloading the condenser, but other controls are minimal. This process is often referred to as Rayleigh distillation.

If we represent the moles of vapor by V, moles of liquid in the pot by M, the mole fraction of the more volatile component in this liquid by x, and the mole fraction of the same component in the vapor by y, a material balance yields

$$-y dV = d(Mx) \tag{13-145}$$

Since $dV = -dM$, substitution and expansion give

$$y dM = M\,dx + x dM \tag{13-146}$$

Rearranging and integrating give

$$\ln \frac{M_i}{M_f} = \int_{x_f}^{x_i} \frac{dx}{y - x} \tag{13-147}$$

where subscript i represents the initial condition and f the final condition of the liquid in the still pot. Integration limits have been reversed to obtain a positive integral. If equilibrium is assumed between liquid and vapor, the right-hand side of Eq. (13-147) may be evaluated by plotting $1/(y - x)$ versus x and measuring the area under the curve between limits x_i and x_f. If the mixture is a binary system for which relative volatility α is constant or if an average value that will serve for the range considered can be found, then the relationship that defines relative volatility,

$$\alpha = \frac{y/x}{(1 - y)/(1 - x)} \tag{13-148}$$

can be substituted into Eq. (13-147) and a direct integration can be made:

$$\ln \left(\frac{M_f}{M_i}\right) = \frac{1}{\alpha - 1} \ln \left[\frac{x_f(1 - x_i)}{x_i(1 - x_f)}\right] + \ln \left[\frac{1 - x_i}{1 - x_f}\right] \tag{13-149}$$

For any two components A and B of a multicomponent mixture, if constant α values are assumed for all pairs of components, $-dM_A/-dM_B = y_A/y_B = \alpha_{A,B}(x_A/x_B)$. When this is integrated, we obtain

$$\ln \left(\frac{M_{A(f)}}{M_{A(i)}}\right) = \alpha_{A,B} \left(\frac{M_{B(f)}}{M_{B(i)}}\right) \tag{13-150}$$

where $M_{A(i)}$ and $M_{A(f)}$ are the moles of component A in the pot before and after distillation and $M_{B(i)}$ and $M_{B(f)}$ are the corresponding moles of component B.

A typical application of a simple batch still might be distillation of an ethanol-water mixture at 101.3 kPa (1 atm). The initial charge is 100 mol of ethanol at 18 mole percent, and the mixture must be reduced to a maximum ethanol concentration in the still of 6 mole percent. By using equilibrium data interpolated from Table 13-1,

x	y	$y - x$	$1/(y - x)$
0.18	0.517	0.337	2.97
.16	.502	.342	2.91
.14	.485	.345	2.90
.12	.464	.344	2.90
.10	.438	.338	2.97
.08	.405	.325	3.08
.06	.353	.293	3.41

Plotting $1/(y - x)$ versus x and integrating graphically between the limits of 0.06 and 0.18 for x, the area under the curve is found to be 0.358. Then, $\ln (M_i/M_f) = 0.358$, from which $M_f = 100/1.43 = 70.0$ mol. The liquid remaining consists of (70.0)(0.06) = 4.2 mol of ethanol and 65.8 mol of water. By material balance, the total distillate must contain (18.0 − 4.2) = 13.8 mol of alcohol and (82 − 65.8) = 16.2 mol of water. Total distillate is 30 mol, and distillate composition is 13.8/30 = 0.46 mole fraction ethanol.

The simple batch still provides only one theoretical plate of separation. Its use is usually restricted to preliminary work in which products will be held for additional separation at a later time, when most of the volatile component must be removed from the batch before it is processed further, or for similar noncritical separations.

Batch Distillation with Rectification To obtain products with a narrow composition range, a rectifying batch still is used that consists of a pot (or reboiler), a rectifying column, a condenser, some means of splitting off a portion of the condensed vapor (distillate) as reflux, and one or more receivers. Temperature of the distillate is controlled in order to return the reflux at or near the column temperature to permit a true indication of reflux quantity and to improve column operation. A subcooling heat exchanger is then used for the remainder of the distillate, which is sent to an accumulator or receiver. The column may also operate at elevated pressure or vacuum, in which case appropriate devices must be included to obtain the desired pressure. Equipment-design methods for batch-still components, except for the pot, follow the same principles as those presented for continuous units, but the design should be checked for each mixture if several mixtures are to be processed. It should also be checked at more than one point of a mixture, since composition in the column changes as distillation proceeds. Pot design is based on batch size and required vaporization rate.

In operation, a batch of liquid is charged to the pot and the system is first brought to steady state under total reflux. A portion of the overhead condensate is then continuously withdrawn in accordance with the established reflux policy. Cuts are made by switching to alternate receivers, at which time operating conditions may be altered. The entire column operates as an enriching section. As time proceeds, composition of the material being distilled becomes less rich in the more volatile components, and distillation of a cut is stopped when accumulated distillate attains the desired average composition.

The progress of batch distillation can be controlled in several ways:

1. *Constant reflux, varying overhead composition.* Reflux is set at a predetermined value at which it is maintained for the run. Since pot liquid composition is changing, instantaneous composition of the distillate also changes. The progress of a binary separation is illustrated in Fig. 13-87. Variation with time of instantaneous distillate composition for a typical multicomponent batch distillation is shown in Fig. 13-88. The shapes of the curves are functions of volatility, reflux ratio, and number of theoretical plates. Distillation is continued until the average distillate composition is at the desired value. In the case of a binary, the overhead is then diverted to another receiver, and an intermediate cut is withdrawn until the remaining pot liquor meets the required specification. The intermediate cut is usually added to the next batch. For a multicomponent mixture, two or more intermediate cuts may be taken between product cuts.

2. *Constant overhead composition, varying reflux.* If it is desired to maintain a constant overhead composition in the case of a

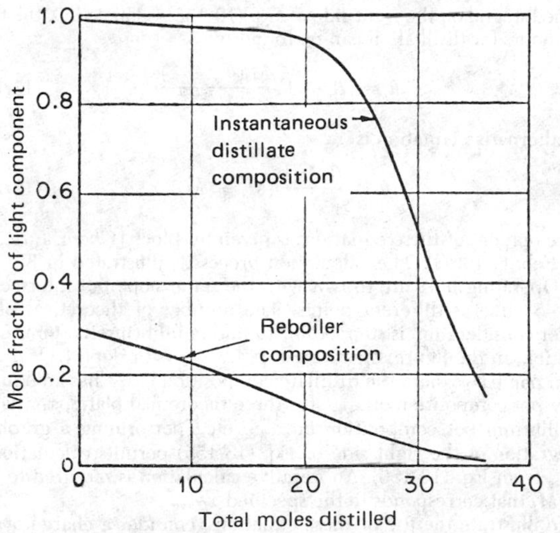

FIG. 13-87 Typical variation in distillate and reboiler compositions with amount distilled in binary batch distillation at a constant-reflux ratio.

binary, the amount of reflux returned to the column must be constantly increased throughout the run. As time proceeds, the pot is gradually depleted of the lighter component. Finally, a point is reached at which the reflux ratio has attained a very high value. The receivers are then changed, the reflux is reduced, and an intermediate cut is taken as before. This technique can also be extended to a multicomponent mixture.

3. *Other control methods*. A cycling procedure can be used to set the pattern for column operation. The unit operates at total reflux until equilibrium is established. Distillate is then taken as total drawoff for a short period of time, after which the column is again returned to total-reflux operation. This cycle is repeated through the course of distillation. Another possibility is to optimize the reflux ratio in order to achieve the desired separation in a minimum of time. Complex operations may involve withdrawal of sidestreams, provision for intercondensers, addition of feeds to trays, and periodic charge addition to the pot.

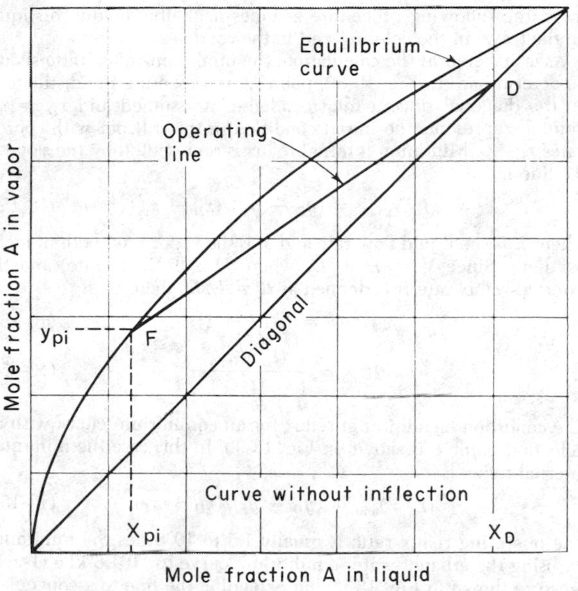

FIG. 13-89 Determination of minimum reflux for normal equilibrium curve.

Approximate Calculation Procedures for Binary Mixtures Because compositions within the column are constantly changing, rigorous calculation methods are extremely complex. An appropriate but useful method for a binary mixture is to use an analysis based on the McCabe-Thiele graphical method. In addition to the usual assumptions of adiabatic column and equimolal overflow on the

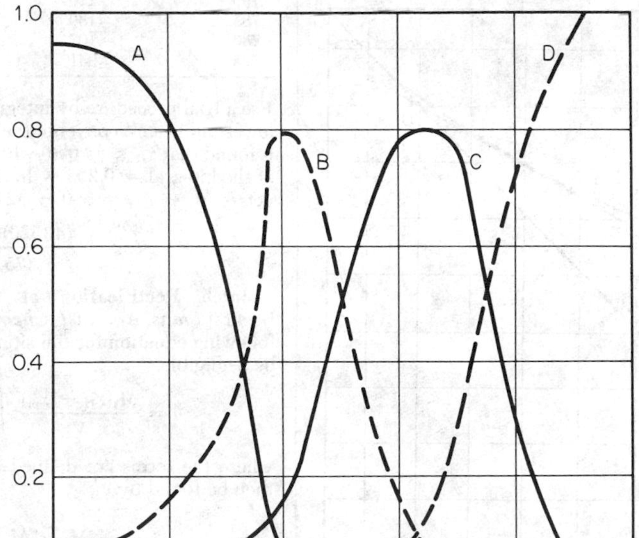

FIG. 13-88 Distillate composition profile for a batch distillation of a four-component mixture.

trays, the following procedure assumes negligible holdup of liquid on the trays, in the column, and in the condenser.

As a first step in the calculation, the minimum-reflux ratio should be determined. In Fig. 13-89, point D, representing the distillate, is on the diagonal since a total condenser is assumed and $x_D = y_D$. Point F represents the initial condition in the still pot with coordinates x_{pi}, y_{pi}. Minimum internal reflux is represented by the slope of the line DF,

$$(L/V)_{min} = (y_D - y_{pi})/(x_D - x_{pi}) \tag{13-151}$$

where L is the liquid flow rate and V is the vapor rate, both in moles per hour. Since $V = L + D$ (where D is distillate rate) and the external-reflux rate R is defined as $R = L/D$, then

$$L/V = R/(R + 1) \tag{13-152}$$

or

$$R_{min} = \frac{(L/V)_{min}}{1 - (L/V)_{min}} \tag{13-153}$$

The condition of minimum reflux for an equilibrium curve with an inflection point P is shown in Fig. 13-90. In this case the minimum internal reflux is

$$(L/V)_{min} = (y_D - y_P)/(x_D - x_P) \tag{13-154}$$

The operating reflux ratio is usually 1.5 to 10 times the minimum. By using the ethanol-water equilibrium curve for 101.3-kPa (1-atm) pressure shown in Fig. 13-90 but extending the line to a convenient point for readability, $(L/V)_{min} = (0.800 - 0.695)/(0.800 - 0.600) = 0.52$ and $R_{min} = 1.083$.

Batch Rectification at Constant Reflux Using an analysis similar to the simple batch still, Smoker and Rose [*Trans. Am. Inst. Chem. Eng.*, **36**, 285 (1940)] developed the following equation:

$$\ln \frac{M_i}{M_f} = \int_{x_{pf}}^{x_{pi}} \frac{dx_p}{x_D - x_p} \tag{13-155}$$

An overall component balance gives the average or accumulated distillate composition $x_{D,avg}$

$$x_{D,avg} = \frac{M_i x_{pi} - M_f x_{pf}}{M_i - M_f} \tag{13-156}$$

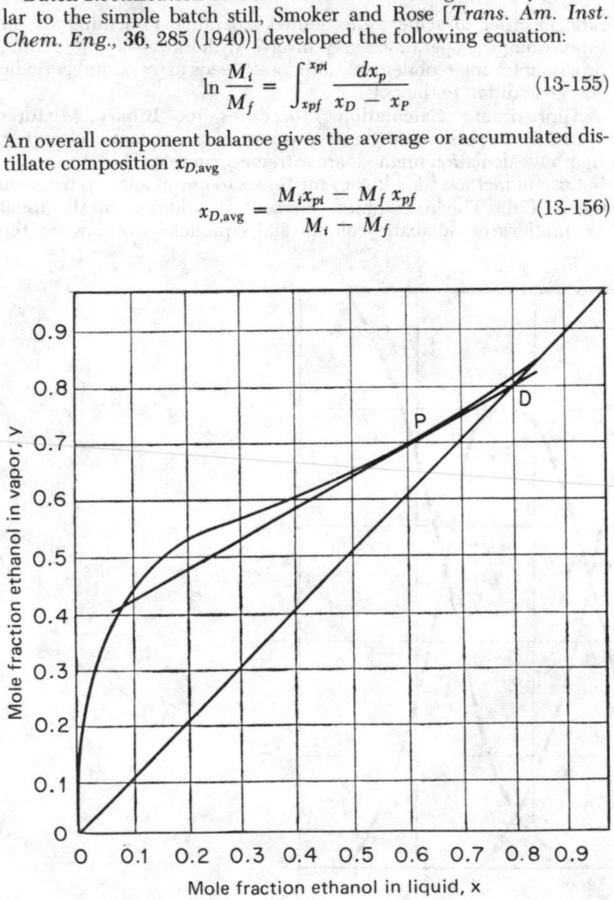

FIG. 13-90 Determination of minimum reflux for equilibrium curve with inflection.

If the integral on the right side of Eq. (13-155) is labeled Q, the time θ in hours for distillation can be found by

$$\theta = (R + 1)\frac{M_i(e^Q - 1)}{Ve^Q} \tag{13-157}$$

An alternative equation is

$$\theta = \frac{R + 1}{V}(M_i - M_f) \tag{13-158}$$

Development of these equations is given by Block [*Chem. Eng.*, **68**, 88 (Feb. 6, 1961)]. The calculation process is illustrated in Fig. 13-91. Operating lines are drawn with the same slope but intersecting the 45° line at different points. The number of theoretical plates under consideration is stepped off to find equilibrium bottoms composition. In the figure, operating line $L - 1$ with slope L/V drawn from point D_1 where the distillate composition is x_{D1} has an equilibrium pot composition of $x_{p1\text{-}3}$ for three theoretical plates, x_{D2} has an equilibrium pot composition of $x_{p2\text{-}3}$, etc.; performing a graphical integration of the right side of Eq. (13-155) permits calculation of $x_{D,avg}$ from Eq. (13-156). An iterative calculation is required to find the M_f that corresponds to the specified $x_{D,avg}$.

To illustrate the use of these equations, consider a charge of 520 mol of an ethanol-water mixture containing 18 mole percent ethanol to be distilled at 101.3 kPa (1 atm). Vaporization rate is 75 mol/h, and the product specification is 80 mole percent ethanol. Let $L/V = 0.75$, corresponding to a reflux ratio $R = 3.0$. If the system has seven theoretical plates, with the pot considered as one of these plates, find how many moles of product will be obtained, what the composition of the residue will be, and the time that the distillation will take.

Using the vapor-liquid equilibrium data, plot a y-x diagram. Draw a number of operating lines at a slope of 0.75. Note the composition at the 45° intersection, and step off seven plates on each to find the equilibrium value of the bottoms. Some of the results are tabulated in the following table:

x_D	x_p	$x_D - x_p$	$1/(x_D - x_p)$
0.800	0.323	0.477	2.097
.795	.245	.550	1.820
.790	.210	.580	1.725
.785	.180	.605	1.654
.780	.107	.673	1.487
.775	.041	.734	1.362

Use a trial procedure by integrating between x_{pi} of 0.18 and various lower limits, and converge the procedure by graphing the results. It is found that $x_{D,avg} = 0.80$ when $x_{pf} = 0.04$, at which time the value of the integral $= 0.205 = \ln (M_i/M_f)$, so that $M_f = 424$ mol. Product $= M_i - M_f = 520 - 424 = 96$ mol. From Eq. (13-157),

$$\theta = \frac{(4)(520)(e^{0.205} - 1)}{(75)(e^{0.205})} = 5.2 \text{ h}$$

Batch Rectification at Constant Overhead Composition Bogart [*Trans. Am. Inst. Chem. Eng.*, **33**, 139 (1937)] developed the following equation for this situation with column holdup assumed to be negligible:

$$\theta = \frac{M_i(x_D - x_{pi})}{V}\int_{x_{pf}}^{x_{pi}} \frac{dx_p}{(1 - L/V)(x_D - x_p)^2} \tag{13-159}$$

where the terms are defined as before. The quantity distilled can then be found by

$$M_i - M_f = \frac{M_i(x_{pi} - x_{pf})}{x_D - x_{pf}} \tag{13-160}$$

The progress of a varying-reflux distillation is shown in Fig. 13-92. Distillate composition is held constant by increasing the reflux as pot composition becomes more dilute. Operating lines with varying slopes ($= L/V$) are drawn from the distillate composition, and the

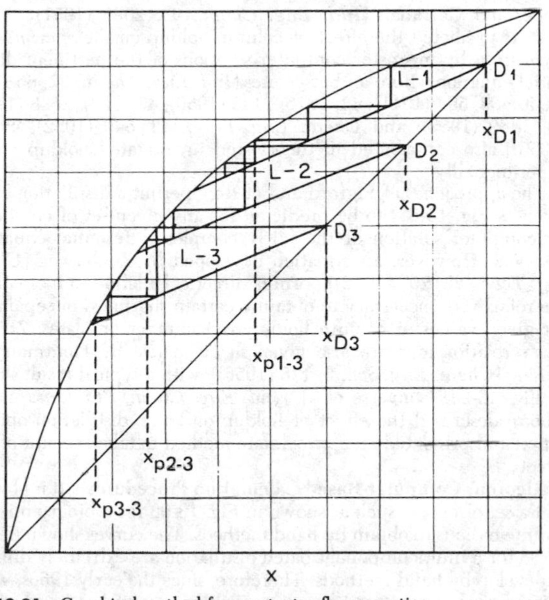

FIG. 13-91 Graphical method for constant-reflux operation.

appropriate number of plates is stepped off to find the corresponding bottoms composition.

As an example, consider distilling at constant composition the same mixture that was used to illustrate constant reflux. The following table is compiled:

L/V	x_p	$x_D - x_p$	$1/1 - L/V)(x_D - x_p)^2$
0.600	0.654	0.147	115.7
.700	.453	.348	27.5
.750	.318	.483	17.2
.800	.143	.658	11.5
.850	.054	.747	11.9
.900	.021	.780	16.4

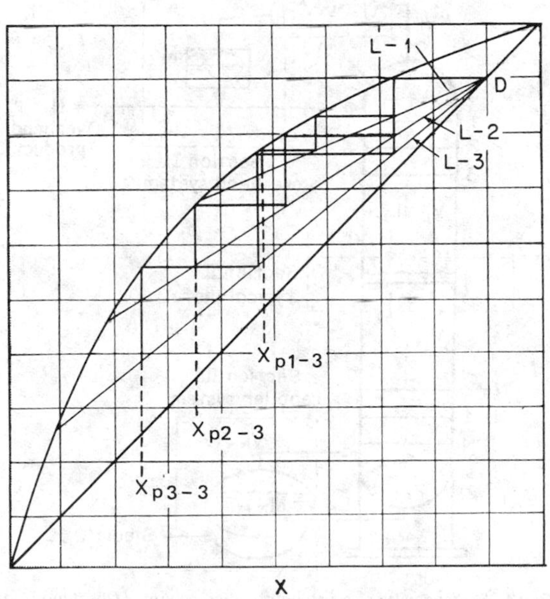

FIG. 13-92 Graphical method for constant-composition operation.

If the right-hand side of Eq. (13-159) is integrated graphically by using a limit for x_{pf} of 0.04, the value of the integral is 1.615 and the time is

$$\theta = \frac{(520)(0.800 - 0.180)(1.615)}{75} = 7.0 \text{ h}$$

The quantity distilled can be found by Eq. (13-160):

$$M_i - M_f = \frac{(520)(0.180 - 0.040)}{0.800 - 0.040} = 96 \text{ mol}$$

Other Operating Methods A useful control method for difficult industrial or laboratory distillations is *cycling operation*. The most common form of cycling control is operating the column at total reflux until equilibrium is established, taking off the complete distillate for a short period of time, and then returning to total reflux. An alternative scheme is to interrupt vapor flow to the column periodically by the use of a solenoid-operated butterfly valve in the vapor line from the pot. In both cases, equations necessary to describe the system are very complex, as shown by Schrodt et al. [*Chem. Eng. Sci.*, **22**, 759 (1967)]. The most reliable method for establishing the cycle relationships is by experimental trial on an operating column. Several investigators have also proposed that batch distillation can be programmed to attain *time optimization* by proper variation of the reflux ratio. A comprehensive discussion is presented by Coward [*Chem. Eng. Sci.*, **22**, 503 (1967)].

The *choice of operating mode* depends upon characteristics of the specific system, the product specifications, and the engineer's preference in setting up a control sequence. Probably the most direct and most common method is *constant reflux*. Operation can be regulated by a timed reflux splitter, a ratio controller, or simply a pair of rotameters. Since composition is changing with time, some way must be found to estimate the average accumulated-distillate composition in order to define the end point. This is no problem when the specification is not critical or the change in distillate composition is sharply defined. When the composition of the distillate changes slowly with time, the cut point is more difficult to determine. Operating with *constant composition* (varying reflux), the specification is automatically achieved if control can be linked to concentration or some concentration-sensitive physical variable. The relative advantage, ratewise, of the two systems depends upon the materials being separated and upon the number of theoretical plates in the column. Results of a comparison of distillation rates by using the same initial and final pot composition for the system benzene-toluene are given in Fig. 13-93. Typical control instrumentation is presented in an article by Block [*Chem. Eng.*, **74**, 147 (Jan. 16, 1967)]. Control procedures for *reflux and vapor-cycling* operation and for the *time-optimal* process are largely a matter of empirical trial.

Effect of Column Holdup When the holdup of liquid on the trays and in the condenser is not negligible compared with the holdup in the pot, the distillate composition at constant-reflux ratio changes with time at a different rate than when the column holdup is negligible because of two separate effects. First, with an appreciable column holdup, composition of the charge to the pot will be higher in the light component than the pot composition at the start of the distillation; the reason for this is that before product takeoff begins, column holdup must be supplied, and its average composition is higher than that of the charge liquid from which it is supplied. Thus, when overhead takeoff begins, the pot composition is lower than it would be if there were no column holdup and separation is more difficult. The second effect of column holdup is to slow the rate of exchange of the components; the holdup exerts an inertia effect, which prevents compositions from changing as rapidly as they would otherwise, and the degree of separation is usually improved. As both these effects occur at the same time and change in importance during the course of distillation, it is difficult, without rigorous calculations, to predict whether the overall effect of holdup will be favorable or detrimental; it is equally difficult to estimate the magnitude of the holdup effect.

A more detailed discussion of holdup effects is given by Pigford,

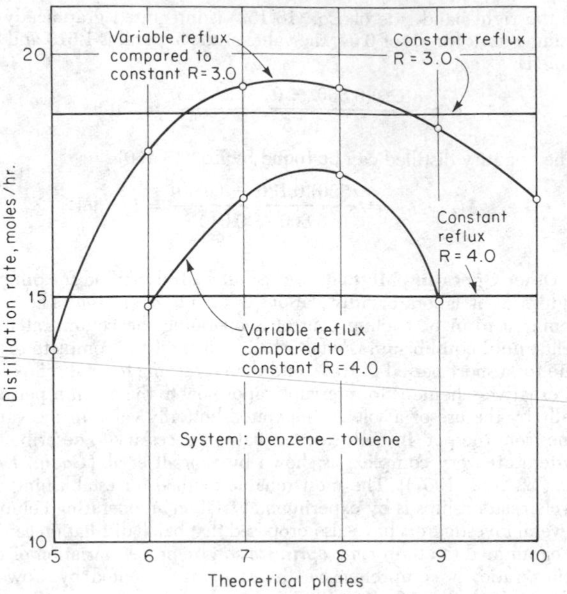

FIG. 13-93 Comparison of operating modes for a batch column.

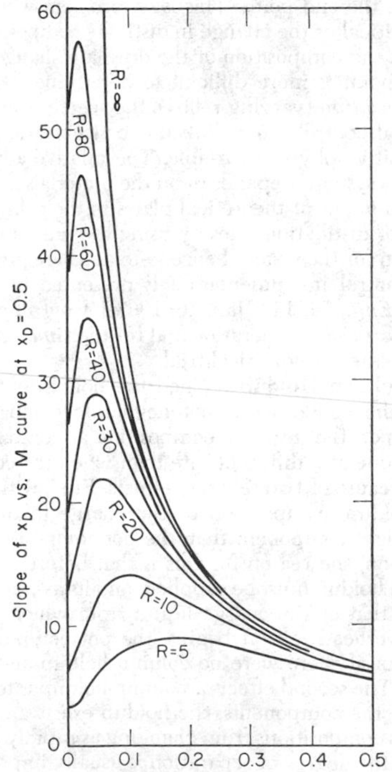

FIG. 13-94 Sharpness of batch separation as a function of operating holdup and reflux ratio (L/D) for cyclohexane–n-heptane equimolar mixture in 25 theoretical-plate columns. [*Houtman and Husain*, Chem. Eng. Sci., *5, 180 (1956).*]

Tepe, and Garrahan [*Ind. Eng. Chem.*, **43**, 2592 (1951)]. These authors predicted the effect of column holdup on the *sharpness of separation* by means of computer solutions of the pertinent differential equations. Rose and associates [*Ind. Eng. Chem.*, **32**, 668, 673 (1940); **33**, 594 (1941); **42**, 1876, 2145 (1950); **43**, 2459, 2608 (1951); **44**, 1480 (1952); and *Chem. Eng. Prog.*, **48**, 549 (1952); **49**, 15 (1953)] also contributed to theory and investigated holdup effects experimentally.

The approaches of Pigford and of Rose permit a distillation curve (such as Fig. 13-87) to be predicted for any given set of conditions by computer solution of the fairly complex differential equations involved. However, an equation developed by Zuiderweg [*Chem. Ing. Tech.*, **25**, 297 (1953)] permits direct calculation to be made of the reflux ratio necessary to obtain a certain sharpness of separation for given values of column holdup and number of plates; Zuiderweg's relationships are also given in an article by Houtman and Husain [*Chem. Eng. Sci.*, **5**, 178 (1956)], with a typical result shown in Fig. 13-94. Converse et al. [*Ind. Eng. Chem. Fundam.*, **4**, 475 (1965)] described the effect of holdup on batch-distillation optimization and presented an approximate method to take account of this effect.

Rigorous Computer-Based Calculation Procedures It is obvious that a set of curves such as shown in Fig. 13-93 for a binary mixture is quite tedious to obtain by hand methods. The curves shown in Fig. 13-88 for a multicomponent batch distillation are extremely difficult to develop by hand methods. Therefore, since the early 1960s, when large digital computers became available, interest has been generated in developing rigorous calculation procedures for binary and multicomponent batch distillation. For binary mixtures of constant relative volatility, Huckaba and Danly [*Am. Inst. Chem. Eng. J.*, **6**, 335 (1960)] developed a computer program that assumed constant-mass tray holdups, adiabatic tray operation, and linear enthalpy relationships but did include energy balances around each tray and permitted use of nonequilibrium trays by means of specified tray efficiencies. Experimental data were provided to validate the simulation. Meadows [*Chem. Eng. Prog. Symp. Ser. 46*, **59**, 48 (1963)] presented a multicomponent-batch-distillation model that included equations for energy, material, and volume balances around theoretical trays. The only assumptions made were perfect

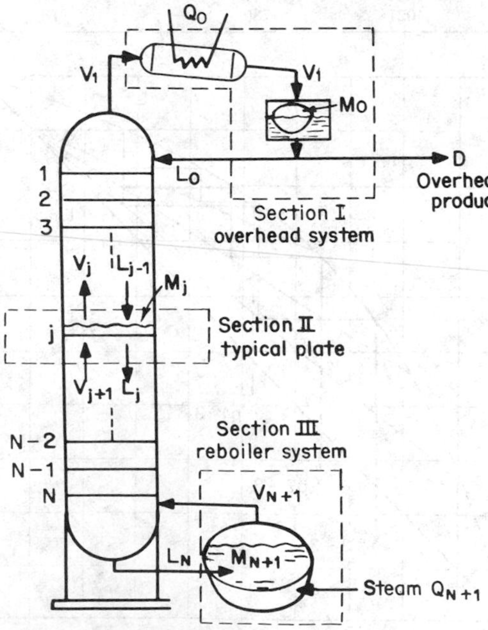

FIG. 13-95 Schematic of a batch-distillation column. [*Distefano*, Am. Inst. Chem. Eng. J., *14, 190 (1968).*]

mixing on each tray, negligible vapor holdup, adiabatic operation, and constant-volume tray holdup. Distefano [*Am. Inst. Chem. Eng. J.*, **14**, 190 (1968)] extended the model and developed a computer-based-solution procedure that was used to simulate successfully several commercial batch-distillation columns. More recently, Boston et al. [*Foundations of Computer-Aided Chemical Process Design*, vol. II, ed. by Mah and Seider, American Institute of Chemical Engineers, New York, 1981, p. 203) further extended the model, provided a variety of practical sets of specifications, and utilized modern numerical procedures and equation formulations to handle efficiently the nonlinear and often stiff nature of the multicomponent-batch-distillation problem. The simpler model of Distefano is used here to illustrate this nonlinear and stiff nature.

Consider the simple batch- or multicomponent-distillation operation in Fig. 13-95. The still consists of a pot or reboiler, a column with N theoretical trays or equivalent packing, and a condenser with an accompanying reflux drum. The mixture to be distilled is charged to the reboiler, to which heat is then supplied. Vapor leaving the top tray is totally condensed and drained into the reflux drum. Initially, no distillate is withdrawn from the system, but instead a total-reflux condition is established at a fixed overhead vapor rate. Then, starting at time $t = 0$, distillate is removed at a constant molal rate and sent to a receiver that is not shown in Fig. 13-95. Simultaneously, a fixed reflux ratio is established such that the overhead vapor rate is not changed from that at total reflux. Alternatively, heat input to the reboiler can be maintained constant and distillate rate allowed to vary accordingly. The equations of Distefano for a batch distillation operated in this manner are as follows (after minor rearrangement), where i,j refers to the ith of C components in the mixture and the jth of N theoretical plates.

Component mole balances for total-condenser-reflux drum, trays, and reboiler, respectively:

$$\frac{dx_{i,0}}{dt} = -\left[\frac{L_0 + D + \dfrac{dM_0}{dt}}{M_0}\right]x_{i,0} + \left[\frac{V_1K_{i,1}}{M_0}\right]x_{i,1} \qquad i = 1 \text{ to } C \qquad (13\text{-}161)$$

$$\frac{dx_{i,j}}{dt} = \left(\frac{L_{j-1}}{M_j}\right)x_{i,j-1} - \left[\frac{L_j + K_{i,j}V_j + \dfrac{dM_j}{dt}}{M_j}\right]x_{i,j} + \left[\frac{K_{i,j+1}V_{j+1}}{M_j}\right]x_{i,j+1} \qquad i = 1 \text{ to } C \quad j = 1 \text{ to } N \qquad (13\text{-}162)$$

$$\frac{dx_{i,N+1}}{dt} = \left(\frac{L_N}{M_{N+1}}\right)x_{i,N} - \left[\frac{V_{N+1}K_{i,N+1} + \dfrac{dM_{N+1}}{dt}}{M_{N+1}}\right]x_{i,N+1} \qquad i = 1 \text{ to } C \qquad (13\text{-}163)$$

where $L_0 = RD$.

Total mole balance for total-condenser-reflux drum and trays respectively:

$$V_1 = D(R + 1) + \frac{dM_0}{dt} \qquad (13\text{-}164)$$

$$L_j = V_{j+1} + L_{j-1} - V_j - \frac{dM_j}{dt} \qquad j = 1 \text{ to } N \qquad (13\text{-}165)$$

Energy balance around jth tray:

$$V_{j+1} = \frac{1}{(H_{Vj+1} - H_{Lj})}\left[V_j(H_{Vj} - H_{Lj}) - L_{j-1}(H_{Lj-1} - H_{Lj}) + M_j\frac{dH_{Lj}}{dt}\right] \qquad j = 2 \text{ to } N + 1 \qquad (13\text{-}166)$$

where H_V and H_L are molar vapor and liquid enthalpies respectively.

Phase equilibriums:

$$y_{i,j} = K_{i,j}x_{i,j} \qquad i = 1 \text{ to } C \qquad j = 1 \text{ to } N + 1 \qquad (13\text{-}167)$$

Mole-fraction sum:

$$\sum_i y_{i,j} = \sum_i K_{i,j}x_{i,j} = 1.0 \qquad j = 0 \text{ to } N + 1 \qquad (13\text{-}168)$$

Molar holdups in condenser-reflux drum, on trays, and in reboiler:

$$M_0 = G_0\rho_0 \qquad (13\text{-}169)$$

$$M_j = G_j\rho_j \qquad j = 1 \text{ to } N$$

$$M_{N+1} = M^0_{N+1} - \sum_{j=0}^{N} M_j - \int_0^t D\,dt \qquad (13\text{-}170)$$

where G is the constant-volume holdup, M^0_{N+1} is the initial molar charge to reboiler, and ρ is the liquid molar density.

Energy balances around condenser and reboiler respectively:

$$Q_0 = V_1(H_{V1} - H_{L0}) - M_0\frac{dH_{L0}}{dt} \qquad (13\text{-}171)$$

$$Q_{N+1} = V_{N+1}(H_{VN+1} - H_{LN+1}) - L_N(H_{LN} - H_{LN+1}) + M_{N+1}\left(\frac{dH_{LN+1}}{dt}\right) \qquad (13\text{-}172)$$

Equation (13-172) is replaced by the following overall energy-balance equation if Q_{N+1} is to be specified rather than D:

$$D = \frac{Q_{N+1} - H_{V1}\left(\dfrac{dM_0}{dt}\right) - \displaystyle\sum_{j=1}^{N+1}\left(\frac{d(M_jH_{Lj})}{dt}\right)}{(R + 1)H_{V1} - RH_{L0}} \qquad (13\text{-}173)$$

With D and R specified, Eqs. (13-161) to (13-173) represent a coupled set of $(2CN + 3C + 4N + 7)$ equations constituting an initial-value problem in an equal number of time-dependent unknown variables, namely, $(CN + 2C)x_{i,j}$, $(CN + C)y_{i,j}$, $(N)L_j$, $(N + 1)V_j$, $(N + 2)T_j$, $(N + 2)M_j$, Q_0, and Q_{N+1}, where initial conditions at $t = 0$ for all unknown variables are obtained by determining the total-reflux steady-state condition for specifications on the number of theoretical stages, amount and composition of initial charge, volume holdups, and molar vapor rate leaving the top stage and entering the condenser.

Various procedures for solving Eqs. (13-161) to (13-173), ranging from a complete tearing method to solve the equations one at a time, as shown by Distefano, to a complete simultaneous method, have been studied. Regardless of the method used, the following considerations generally apply:

1. Derivatives or rates of change of tray and condenser-reflux drum liquid holdup with respect to time are sufficiently small compared with total flow rates that these derivatives can be approximated by incremental changes over the previous time step. Derivatives of liquid enthalpy with respect to time everywhere can be approximated in the same way. The derivative of the liquid holdup in the reboiler can likewise be approximated in the same way except when reflux ratios are low.

2. Ordinary differential equations (13-161) to (13-163) for rates of change of liquid-phase mole fractions are nonlinear because the coefficients of $x_{i,j}$ change with time. Therefore, numerical methods of integration with respect to time must be employed. Furthermore, the equations may be difficult to integrate rapidly and accurately because they may constitute a so-called stiff system as considered by Gear (*Numerical Initial Value Problems in Ordinary Differential Equations*, Prentice-Hall, Englewood Cliffs, N.J., 1971). The choice of time step for simple explicit numerical procedures (such as the Euler and Runge-Kutta methods) of integrating sets of ordinary differential equations in initial-value problems may be governed by either stability or truncation-error considerations. Truncation errors in the dependent variables may be scarcely noticeable and generally accumulate gradually with time. Instability generally causes sudden

and severe errors that are very noticeable. When the equations are stiff, stability controls and extremely small time steps may be necessary to prevent instability. A common measure of the severity of stiffness is the stiffness ratio $|\lambda|_{max}/|\lambda|_{min}$, where λ is an eigenvalue for the jacobian matrix of the set of ordinary differential equations. For Eqs. (13-161) to (13-163), the jacobian matrix is tridiagonal if the equations and variables are arranged by stage (top down) for each component in order. For example, for two components, a condenser, one equilibrium tray, and a reboiler, the matrix is as follows:

For a general jacobian matrix pertaining to C components and N theoretical trays, as shown by Distefano [*Am. Inst. Chem. Eng. J.*, **14**, 946 (1968)], Gerschgorin's circle theorem (Varga, *Matrix Iterative Analysis*, Prentice-Hall, Englewood Cliffs, N.J., 1962) may be employed to obtain bounds on the maximum and minimum absolute eigenvalues. Accordingly,

$$|\lambda|_{max} \leq \max_{j=1,N} \left[\left(\frac{L_{j-1}}{M_j}\right) + \left(\frac{L_j + K_{i,j}V_j + \dfrac{dM_j}{dt}}{M_j}\right) + \left(\frac{K_{i,j+1}V_{j+1}}{M_j}\right) \right]$$

The maximum absolute eigenvalue corresponds to the component with the largest K value ($K_{L,j}$) and the tray with the smallest holdup. Therefore, if the derivative term and any variation in L_j, V_j, and $K_{i,j}$ are neglected,

$$|\lambda|_{max} \simeq 2\left[\frac{L_j + K_{i,j}V_j}{M_j}\right] \tag{13-174}$$

In a similar development, the minimum upper limit on the eigenvalue corresponds to the component with the largest K value and to the largest holdup, which occurs in the reboiler. Thus

$$|\lambda|_{min} = \left[\frac{L_N + K_{L,N}V_N}{M_{N+1}}\right] \tag{13-175}$$

Therefore, the lower bound on the stiffness ratio at the beginning of batch distillation is given approximately by

$$\frac{|\lambda|_{max}}{|\lambda|_{min}} = 2\left(\frac{M_{N+1}^0}{M_N}\right)$$

where M_{N+1} and M_N are the molar holdups in the reboiler initially and on the bottom tray respectively. In the sample problem presented by Distefano (ibid.) for the smallest charge, the approximate initial-stiffness ratio is of the order of 250, which is not considered to be a particularly large value. Using an explicit integration method, almost 600 time increments, which were controlled by stability criteria, were required to distill 98 percent of the charge.

At the other extreme of Distefano's sample problems, for the largest initial charge, the maximum-stiffness ratio is of the order of 1500, which is considered to be a relatively large value. In this case, more than 10,000 time steps are required to distill 90 percent of the initial change, and the problem is better handled by a stiff integrator.

In Distefano's method, Eqs. (13-161) to (13-173) are solved with an initial condition of total reflux at L_0 equal to $D(R + 1)$ from the specifications. At $t = 0$, L_0 is reduced so as to begin distillate withdrawal. The computational procedure is then as follows:

1. Replace L_j^0 by $L_j^0 - D$, but retain V_j^0 and all other initial values from the total-reflux calculation.
2. Replace the holdup derivatives in Eqs. (13-161) to (13-163) by total-stage material-balance equations (e.g., $dM_j/dt = V_{j+1} + L_{j-1} - V_j - L_j$) and solve the resulting equations one at a time by the predictor step of an explicit integration method for a time increment that is determined by stability and truncation considerations. If the mole fractions for a particular stage do not sum to 1, normalize them.
3. Compute a new set of stage temperatures from Eq. (13-168). Calculate a corresponding set of vapor-phase mole fractions from Eq. (13-167).
4. Calculate liquid densities, molar tray and condenser-reflux drum holdups, and liquor and vapor enthalpies. Determine holdup and enthalpy derivatives with respect to time by forward difference approximations.
5. From Eqs. (13-164) to (13-166) compute a new set of values of liquid and vapor molar flow rates.
6. Compute the reboiler molar holdup from Eq. (13-170).
7. Repeat steps 2 through 6 with a corrector step for the same time increment. Repeat again for any further predictor and/or predictor-corrector steps that may be advisable. Distefano (ibid.) discusses and compares a number of suitable explicit methods.
8. Compute condenser and reboiler heat-transfer rates from Eqs. 13-171) and (13-172).
9. Repeat steps 2 through 8 for subsequent time increments until the desired amount of distillate has been withdrawn.

More flexible and efficient methods that can cope with stiffness in batch-distillation calculations utilize stable implicit integration procedures such as the method of Gear (op. cit.). Boston et al. (op. cit.) discuss such a method that also utilizes a two-tier equation-solving technique, referred to as the "inside-outside" algorithm, that can handle both wide-boiling and narrow-boiling charges even when very nonideal mixtures are formed. In addition to the features of the Distefano model, the Boston et al. model permits multiple feeds, sidestream withdrawals, tray heat transfer, and vapor distillate and divides the batch-distillation process into a sequence of operation steps. At the beginning of each step, the reboiler may receive an additional charge, distillate or sidestream receivers may be dumped, and a feed reservoir may be refilled. Specifications for an operation step include feed, sidestream withdrawal, and tray heat-transfer rates. In addition, any two of the following five variables must be specified: reflux ratio, distillate rate, boil-up rate, condenser duty, and reboiler duty. An operation step is terminated when a specified criterion, selected from the following list, is reached: a time duration; a component purity in the reboiler, distillate, or distillate accumulator; an amount of material in the reboiler or distillate accumulator; or a reboiler or condenser temperature. The purity is specified to be met as the purity is increasing or decreasing. Finally, column configuration and operating conditions (number of stages, holdups, tray pressures, and feed, sidestream, and tray heat-transfer rates) can be changed at the beginning of each operation step. In addition, physical properties may be computed from a wide variety of correlations, including equation-of-state and activity-coefficient models.

Example 11: Calculation of Multicomponent Batch Distillation A charge of 45.4 kg·mol (100 lb·mol) of 25 mole percent benzene, 50 mole percent monochlorobenzene (MCB), and 25 mole percent orthodichlorobenzene (DCB) is to be distilled in a batch still consisting of a reboiler, a column containing 10 theoretical stages, a total condenser, a reflux drum, and a distillate accumulator. Condenser-reflux drum and tray holdups are 0.0056 and 0.00056 m^3 (0.2 and 0.02 ft^3) respectively. Pressures are 101.3, 107.6, 117.2, and 120.7 kPa (14.696, 15.6, 17, and 17.5 psia) at the condenser outlet, top stage, bottom stage, and reboiler respectively. Initially, the still is to be brought to total-reflux conditions at a boil-up rate of 45.4 (kg·mol)/h [100 (lb·mol)/h] leaving the reboiler. Then, at $t = 0$, the boil-up rate is to be increased to 90.8 (kg·mol)/h [200 (lb·mol)/h], and the reflux ratio is to be set at 3. The batch is then distilled in three steps. The first step, which is designed to obtain a benzene-rich product, is terminated when the mole fraction of benzene in the distillate being sent to the accumulator has dropped to 0.100 or when 2 h have elapsed. The purpose of the second operation step is to recover an MCB-rich product until the mole fraction of MCB in the distillate drops to 0.400 or 2 h have elapsed since the start of this step. The third step is to be terminated when the mole fraction of DCB in the reboiler reaches 0.98 or 2 h have elapsed since the start of this step. Ideal solutions and an ideal gas are assumed such that Raoult's law can be used to obtain K values. Calculations are made by the method of Boston et al. (op. cit.).

First, the total-reflux condition is computed by making several sets of stage-to-stage calculations from the reboiler to the condenser. For the first set, the reboiler composition is assumed to be that of the initial charge. This composition is adjusted by material balance to initiate each subsequent set of calculations until convergence is achieved. Results are shown in Table 13-30. From these data, the initial-stiffness ratio is approximately [2(99.74)/0.01218] = 16,400. Thus the equations are quite stiff, and an implicit integration method is preferred. Detailed calculated conditions at the end of the first operation step are given in Table 13-31. A short summary of conditions for each of the three operation steps is computed to be as follows:

	Operation step		
	1	2	3
Time of operation step, h	0.5963	0.7944	0.04828
Number of time increments	201	154	39
Accumulated distillate			
Total lb·mol	33.38	41.99	2.361
Mole fractions			
Benzene	0.7360	0.0103	0.74×10^{-10}
MCB	0.2640	0.9537	0.2872
DCB	0.45×10^{-6}	0.036	0.7128
Reboiler holdup			
Total lb·mol	66.40	24.43	22.08
Mole fractions			
Benzene	0.0063	0.63×10^{-11}	0.20×10^{-12}
MCB	0.6172	0.0448	0.0200
DCB	0.3765	0.9552	0.9800
Temperatures, °F			
Condenser outlet	251.58	308.60	330.17
Reboiler outlet	301.67	362.63	366.39
Heat duties, million Btu/h			
Condenser	3.313	3.472	3.456
Reboiler	3.295	3.469	3.433

NOTE: To convert degrees Fahrenheit to degrees Celsius, °C = (°F − 32)/1.8; to convert pound-moles to kilogram-moles, multiply by 0.454; and to convert British thermal units per hour to kilojoules per hour, multiply by 1.055.

TABLE 13-30 Total-Reflux Conditions

Stage	T, °F	L, (lb·mol)/h	M, lb·mol	x Benzene	x MCB	x DCB
Condenser	175.94	116.4	0.1307	1.000	0.525×10^{-7}	0.635×10^{-15}
1	179.46	117.4	0.01304	1.000	0.266×10^{-6}	0.164×10^{-13}
2	180.05	117.5	0.01303	1.000	0.135×10^{-5}	0.424×10^{-12}
3	180.64	117.5	0.01303	1.000	0.683×10^{-5}	0.109×10^{-10}
4	181.22	117.6	0.01302	1.000	0.344×10^{-4}	0.279×10^{-9}
5	181.80	117.7	0.01302	1.000	0.173×10^{-3}	0.711×10^{-8}
6	182.41	117.7	0.01301	0.999	0.871×10^{-3}	0.180×10^{-6}
7	183.14	117.5	0.01300	0.996	0.435×10^{-2}	0.454×10^{-5}
8	184.54	116.4	0.01295	0.979	0.209×10^{-1}	0.112×10^{-3}
9	189.08	111.8	0.01278	0.901	0.965×10^{-1}	0.250×10^{-2}
10	205.91	100.0	0.01218	0.642	0.319	0.389×10^{-1}
Reboiler	250.96	0.0	99.74	0.248	0.501	0.251

NOTE: Reboiler duty = 1,549,000 Btu/h. To convert degrees Fahrenheit to degrees Celsius, °C = (°F − 32)/1.8; to convert pound-moles per hour to kilogram-moles per hour, multiply by 0.454; and to convert pound-moles to kilogram-moles, multiply by 0.454.

TABLE 13-31 Conditions at the End of the First Operation Step

Time = 0.5963 h

Stage	T, °F	V, (lb·mol)/h	L, (lb·mol)/h	M, lb·mol	y Benzene	y MCB	y DCB	x Benzene	x MCB	x DCB
Condenser	251.58	0	154.6	0.1113	.....	....	....	0.100	0.900	0.292×10^{-5}
1	267.69	206.1	157.5	0.01992	0.0994	0.901	0.293×10^{-5}	0.276×10^{-1}	0.972	0.124×10^{-4}
2	271.21	209.0	158.0	0.01088	0.0449	0.955	0.101×10^{-4}	0.121×10^{-1}	0.988	0.405×10^{-4}
3	272.48	209.5	158.1	0.01087	0.0331	0.967	0.312×10^{-4}	0.884×10^{-2}	0.991	0.124×10^{-3}
4	273.28	209.6	158.2	0.01086	0.0306	0.969	0.943×10^{-4}	0.818×10^{-2}	0.991	0.373×10^{-3}
5	273.99	209.7	158.2	0.01085	0.0301	0.970	0.282×10^{-3}	0.806×10^{-2}	0.991	0.111×10^{-2}
6	274.75	209.7	158.1	0.01084	0.0300	0.969	0.839×10^{-3}	0.803×10^{-2}	0.989	0.329×10^{-2}
7	275.72	209.6	157.7	0.01083	0.0300	0.967	0.249×10^{-2}	0.801×10^{-2}	0.982	0.969×10^{-2}
8	277.29	209.2	156.6	0.01080	0.0300	0.963	0.731×10^{-2}	0.794×10^{-2}	0.964	0.280×10^{-1}
9	280.50	208.1	154.0	0.01073	0.0301	0.949	0.211×10^{-1}	0.772×10^{-2}	0.915	0.770×10^{-1}
10	287.47	205.4	148.5	0.01057	0.0302	0.912	0.577×10^{-1}	0.720×10^{-2}	0.803	0.189
Reboiler	301.67	200.0	0	66.40	0.0304	0.829	0.141	0.634×10^{-2}	0.617	0.376

NOTE: To convert pound-moles per hour to kilogram-moles per hour, multiply by 0.454; to convert pound-moles to kilogram-moles, multiply by 0.454.

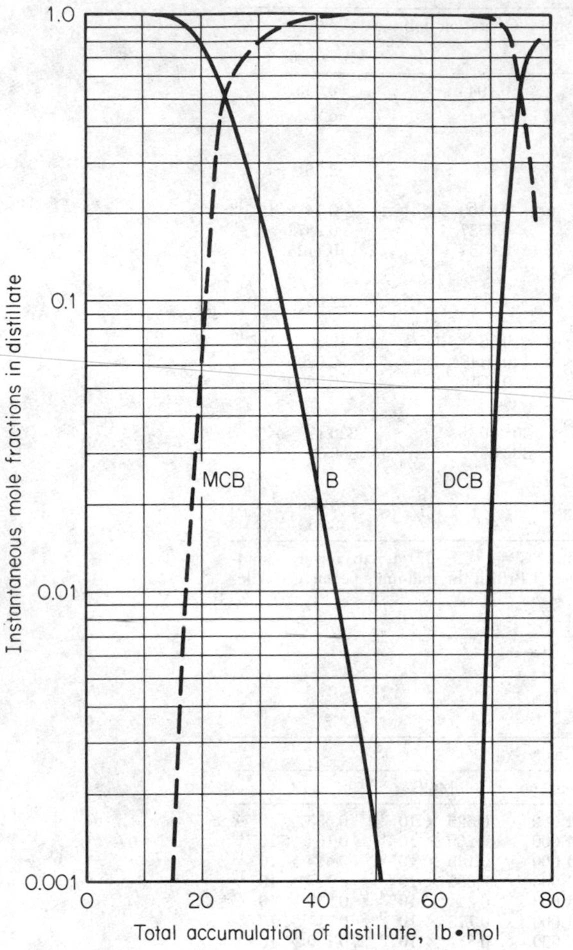

FIG. 13-96 Distillate-composition profile for the multicomponent-batch-distillation example.

From this table, a total of 394 time increments were necessary to distill all but 22.08 lb·mol of the initial charge of 99.74 lb·mol following the establishment of total-reflux conditions. If this problem had to be solved by an explicit integrator, approximately 25,000 time increments would have been necessary.

Instantaneous distillate (or reflux) composition as a function of total accumulated distillate for all three operation steps is plotted in Fig. 13-96. From these results, an alternative schedule of operation steps can be derived to obtain three relatively rich cuts and two intermediate cuts for recycle to the next batch. One example is as follows:

		Composition, mole fractions		
Cut	Amount, lb·mol	Benzene	MCB	DCB
Benzene-rich	17.750	0.99553	0.00447	0.00000
Recycle 1	15.630	0.44120	0.55880	0.00000
MCB-rich	37.195	0.01164	0.98821	0.00015
Recycle 2	7.155	0.00000	0.55430	0.44570
DCB-rich	22.078	0.00000	0.02000	0.98000
Residual holdup	0.192	0.00000	0.11980	0.88020
Total	100.00	0.25000	0.50000	0.25000

NOTE: To convert pound-moles to kilogram-moles, multiply by 0.454.

From these results, 22.98 lb·mol, or almost 23 percent of the charge, would be recycled for redistillation. All three products are at least 98 mole percent pure.

Azeotropic Batch Distillation Batch columns are used extensively for azeotropic separations. Instead of using multiple columns, cuts are segregated in storage for later reuse or distillation. In the separation of *n*-butanol–water, for example, the process starts by charging the original mixture into the still pot of Fig. 13-97. In the first distillation step, the water phase is accumulated in a storage tank for later processing. The second cut is an intermediate mixture rich in butanol with the remaining traces of water. It is also stored to be added to the next fresh batch of feed. The remaining liquors are essentially pure *n*-butanol, but they may also be distilled to remove impurities. If so, the final heel of material is added to successive batches until sufficient impurity builds up to warrant purging. The accumulated water phase from several batches is distilled. The butanol-containing distillate is added to incoming material, and the remaining water in the pot is purged.

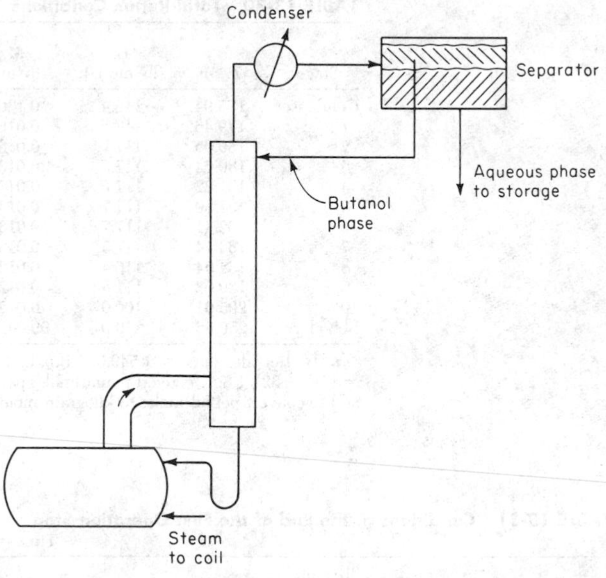

FIG. 13-97 Azeotropic distillation of water from *n*-butanol in a batch column.

DYNAMIC DISTILLATION

As discussed in detail by Archer and Rothfus [*Chem. Eng. Prog. 36*, **57**, 2 (1961)], dynamic or transient behavior of a continuous-distillation operation is important in determining (1) startup and shutdown procedures, (2) transition path between steady states, (3) effect of upsets and fluctuations on controllability, (4) residence times and mass-transfer rates, and (5) operating strategies that may involve deliberate imposition of controlled cyclic fluctuations or oscillations, as summarized by Schrodt [*Ind. Eng. Chem.*, **59**(6), 58 (1967)]. Dynamic behavior may be studied with no controllers in the system to obtain a so-called open-loop response. Alternatively, controllers may be added for certain variables that are to be controlled by manipulating other variables to obtain a so-called closed-loop response. For this latter case, controllers of various levels of complexity [e.g., on-off, proportional (P), proportional with integral action (PI), and proportional with integral and derivative action (PID)] can be considered for various values of tuning parameters, and specific valves of known characteristics may be incorporated if desired. First, an example of open-loop response of an equilibrium-flash-vaporization drum is illustrated. This is followed by the closed-loop response of a simple ideal binary distillation column. Finally, open-loop response of a complex case of azeotropic distillation that involves instability is shown.

Equilibrium-Flash Vaporization Consider an example given by Holland (*Unsteady State Processes with Applications in Multicomponent Distillation*, Prentice-Hall, Englewood Cliffs, N.J., 1966] of a single-stage isothermal equilibrium-flash vaporization at 200°F (93°C) and 300 psia (2068 kPa) of a mixture at a set of initial steady-state conditions as shown in Fig. 13-98*a*, where M is the total pound-moles of liquid holdup in the flash drum and the corresponding residence time in the drum is 45.67 min. At time $t = 0^+$, a substantial step change in feed composition (but not in total molar feed rate) occurs, including the addition of two more components, leading to a second and final set of steady-state conditions for the same temperature and pressure, as shown in Fig. 13-98*b*, where the residence time of the liquid in the drum becomes only 0.59 min for the same total molar liquid holdup M because of the substantial increase in exiting molar liquid flow rate. It is of interest to determine, in the absence of any controllers, the transition time required to approach the fixed steady-state condition and the manner in which the exiting-component flow rates change during the transition.

If the molar liquid holdup in the drum is somehow maintained constant (so-called open-loop condition) and assumed to be perfectly mixed, and the vapor holdup in the drum is negligible, the governing dynamic material-balance and phase equilibrium equations are

$$M(dx_i/dt) = Fz_i - Vy_i - Lx_i \tag{13-176a}$$

$$F = V + L \tag{13-176b}$$

$$K_i = y_i/x_i \tag{13-176c}$$

$$\Sigma x_i - \Sigma y_i = 0 \tag{13-176d}$$

The quantities V and y_i in Eqs.(13-176*a*) and (13-176*d*) can be eliminated by combination with Eqs. (13-176*b*) and (13-176*c*) to obtain

$$dx_i/dt = a_i + b_i x_i + c_i L x_i \tag{13-177}$$

and

$$\Sigma(1 - K_i)x_i = 0 \tag{13-178}$$

where $a_i = Fz_i/M$, $b_i = -FK_i/M$, and $c_i = (K_i - 1)/M$, which constitute a set of $C + 1$ differential algebraic equations. These can

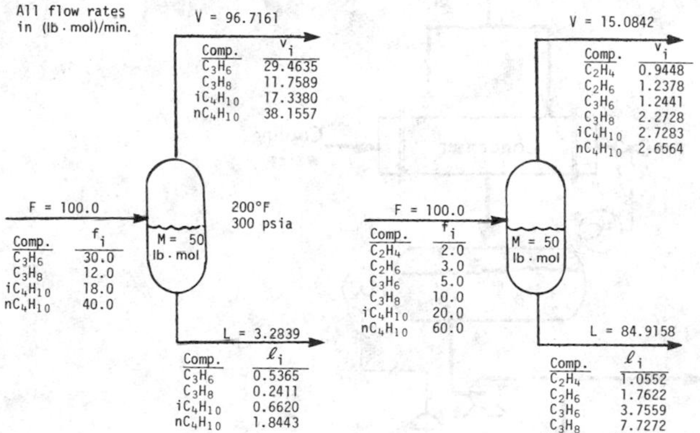

FIG. 13-98 Steady-state equilibrium-flash-vaporization conditions. (*a*) Case 1 initial steady state. (*b*) Case 2 final steady state.

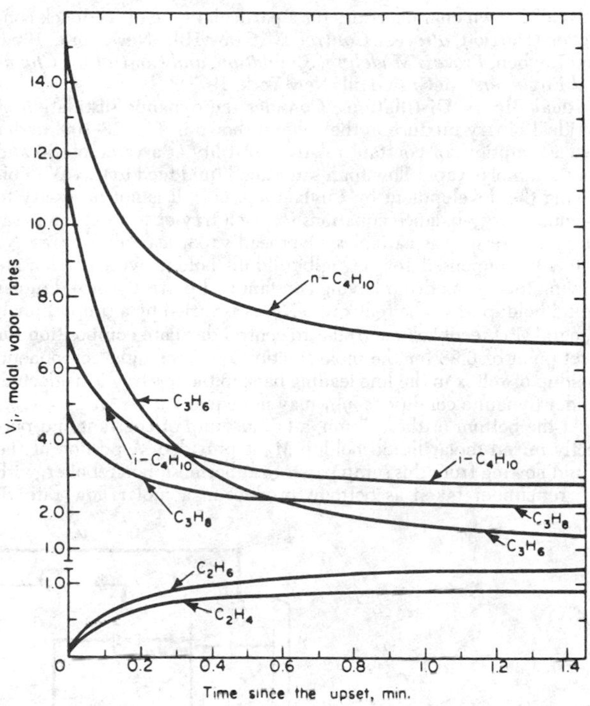

FIG. 13-99 Component vapor flow rates during dynamic conditions between steady-state conditions of Fig. 13-98. (*Holland,* Unsteady State Processes with Applications in Multicomponent Distillation, *Prentice-Hall, Englewood Cliffs, N.J., 1966.*)

be reduced to a set of C differential equations in the following manner. First, note that

$$\Sigma(1 - K_i)\frac{dx_i}{dt} = 0 \tag{13-179}$$

follows from Eq. (13-178) if K values are composition-independent. Now multiply Eq. (13-177) through by $(1 - K_i)$, sum the result over the i components, combine that result with Eq. (13-179), and solve for L to give

$$L = -\frac{\Sigma[(1 - K_i)(a_i + b_i x_i)]}{\Sigma[(1 - K_i)c_i x_i)]} \tag{13-180}$$

Equation (13-180) is then substituted into Eq. (13-177) to give

$$\frac{dx_i}{dt} = a_i + b_i x_i - c_i x_i \frac{\Sigma[(1 - K_i)(a_i + b_i x_i)]}{\Sigma[(1 - K_i)c_i x_i)]} \qquad i = 1 \text{ to } C \tag{13-181}$$

which is the desired set of C differential equations in terms of the C values of x_i. In this particular example, Eqs. (13-181) constitute a stiff system. When solved with a stable-implicit Gear-type integrator, the results shown in Fig. 13-99 are obtained for $v_i = y_i V$. It is important to note that at $t = 0^+$, when values of z_i instantaneously change as shown in Fig. 13-98, values of L and V, as computed from Eqs. (13-180) and (13-176*b*), also instantaneously undergo drastic changes from 3.2839 and 96.7161 respectively to 62.27 and 37.73 respectively. Approximately 98 percent of the change in the total equilibrium vapor flow rate takes place in 0.75 min. This isothermal-flash example is readily extended to adiabatic conditions by adding a transient energy-balance equation. Also, controllability of the holdup can be studied by not specifying a constant value for it, but instead by adding a controller equation for the transient holdup, as determined by liquid level, in terms of the flow of exiting liquid through a control valve of known characteristics for a particular type of feedback controller (Harriott, *Process Control*, McGraw-Hill, New York, 1964; and Luyben, *Process Modeling, Simulation, and Control for Chemical Engineers*, McGraw-Hill, New York, 1973).

Ideal Binary Distillation Consider the dynamic distillation of an ideal binary mixture in the column shown in Fig. 13-100, under two assumptions of constant relative volatility at a value of 2.0 and constant molar vapor flow for a saturated liquid feed to tray N_s. Following the development by Luyben (op. cit.), it is not necessary to include energy-balance equations for each tray or to treat temperature and pressure as variables. Overhead vapor leaving top tray N_T is totally condensed for negligible liquid holdup with condensate flowing to a reflux drum having constant and perfectly mixed molar liquid holdup M_D. The reflux rate L_{NT+1} is varied by a proportional-integral (PI) feedback controller to control distillate composition for a set point of 0.98 for the mole fraction x_D of the light component. Holdup of reflux in the line leading back to the top tray is neglected. Under dynamic conditions, y_{NT} may not equal x_D.

At the bottom of the column, a liquid sump of constant and perfectly mixed molar liquid holdup M_B is provided. A portion of the liquid flowing from this sump passes to a thermosiphon reboiler, with the remainder taken as bottoms product at a molar flow rate B. Vapor boil-up generated in the reboiler is varied by a PI feedback controller to control bottoms composition with a set point of 0.02 for the mole fraction x_B of the light component. Liquid holdups in the reboiler and lines leading from the sump are assumed to be negligible. The composition of the boil-up y_B is assumed to be in equilibrium with x_B.

The liquid holdup M_n on each of the N_T equilibrium trays is assumed to be perfectly mixed but will vary as liquid rates leaving the trays vary. Vapor holdup is assumed to be negligible everywhere. Tray molar vapor rates V vary with time but at any instant in time are everywhere equal.

The dynamic material-balance and phase equilibrium equations corresponding to this description are as follows:

All trays, n:

$$dM_n/dt = F_n + L_{n+1} - L_n \tag{13-182}$$

$$\frac{d}{dt}(M_n x_n) = F_n x_{Fn} + L_{n+1}x_{n+1} + Vy_{n-1} - L_n x_n - Vy_n \tag{13-183}$$

$$L_n = \overline{L}_n + (M_n - \overline{M}_n)/\beta \tag{13-184}$$

$$y_n = \frac{\alpha x_n}{1 + (\alpha - 1)x_n} \tag{13-185}$$

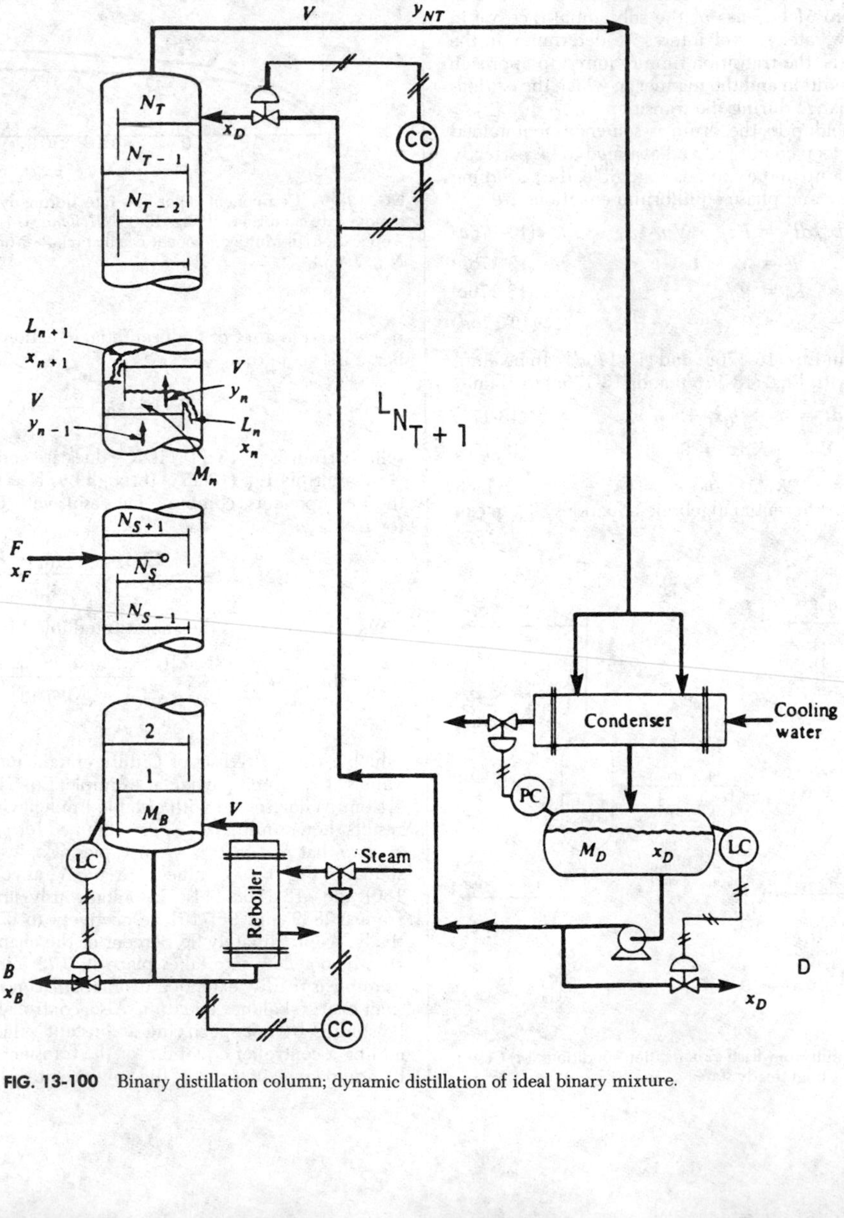

FIG. 13-100 Binary distillation column; dynamic distillation of ideal binary mixture.

Where F_n is nonzero only for tray N_S, y and x refer to the light component only such that the corresponding mole fractions for the heavy component are $(1 - y)$ and $(1 - x)$, $\overline{L}_n$ and $\overline{M}_n$ are the initial steady-state values, and β is a constant that depends on tray hydraulics.

For the condenser–reflux-drum combination:

$$D = V - L_{NT+1} \tag{13-186}$$

$$M_D(dx_D/dt) = Vy_{NT} - Vx_D \tag{13-187}$$

For the reboiler:

$$B = L_1 - V \tag{13-188}$$

$$M_B(dx_B/dt) = L_1x_1 - Vy_B - Bx_B \tag{13-189}$$

$$y_B = \frac{\alpha x_B}{1 + (\alpha - 1)x_B} \tag{13-190}$$

The two PI-controller equations are

$$V = \overline{V} - K_{CB}\left(E_B + \frac{1}{\tau_B}\int E_B\,dt\right) \tag{13-191}$$

$$L_{NT+1} = \overline{L}_{NT+1} + K_{CD}\left(E_D + \frac{1}{\tau_D}\int E_D\,dt\right) \tag{13-192}$$

where $\overline{V}$ and $\overline{L}_{NT+1}$ are initial values, K_C and τ are respectively feedback-controller gain and feedback-reset time for integral action, and E is the error or deviation from the set point as given by

$$E_B = x_B^{set} - x_B \tag{13-193}$$

$$E_D = x_D^{set} - x_D \tag{13-194}$$

These differential equations are readily solved, as shown by Luyben (op. cit.), by simple Euler numerical integration, starting from an initial steady state, as determined. e.g., by the McCabe-Thiele method, followed by some prescribed disturbance such as a step change in feed composition. Typical results for the initial steady-state conditions, fixed conditions, controller and hydraulic parameters, and disturbance given in Table 13-32 are listed in Table 13-33.

Multicomponent Distillation Open-loop behavior of multicomponent distillation may be studied by solving modifications of the multicomponent equations of Distefano [*Am. Inst. Chem. Eng. J.*, **14**, 190 (1968)] as presented in the subsection "Batch Distillation." One frequent modification is to include an equation, such as the Francis weir formula, to relate liquid holdup on a tray to liquid flow rate leaving the tray. Applications to azeotropic-distillation towers are particularly interesting because, as discussed by and illustrated in the following example from Prokopakis and Seider [*Am. Inst. Chem. Eng. J.*, **29**, 1017(1983)], the steep concentration and temperature fronts can be extremely sensitive to small changes in reflux ratio, boil-up rate, product recovery and purity, and feed rate and composition.

Consider azeotropic distillation to dehydrate ethanol with benzene. Initial steady-state conditions are as shown in Fig. 13-101. The

TABLE 13-32 Initial and Fixed Conditions, Controller and Hydraulic Parameters, and Disturbance for Ideal Binary Dynamic-Distillation Example

Other initial conditions		Initial liquid-phase compositions: Tray	x_n
$\overline{F}$	= 100 (lb·mol)/min	Bottoms	0.02
x_F	= 0.50	1	0.035
$\overline{L}_{NT+1}$	= 128.01 (lb·mol)/min	2	0.05719
$\overline{V}$	= 178.01 (lb·mol)/min	3	0.08885
$\overline{D}$	= 50 (lb·mol)/min	4	0.1318
x_D^{set}	= 0.98	5	0.18622
$\overline{B}$	= 50 (lb·mol)/min	6	0.24951
x_B^{set}	= 0.02	7	0.31618
$M_{n,n=1 \text{ to } NT}$	= 10 lb·mol	8	0.37948
Fixed conditions		9	0.43391
N_T	= 20	10	0.47688
N_S	= 10	11	0.51526
M_D	= 100 lb·mol	12	0.56295
M_B	= 100 lb·mol	13	0.61896
Controller and hydraulic parameters		14	0.68052
$K_{CB} = K_{CD}$	= 1000 (lb·mol)/min	15	0.74345
τ_B	= 1.25 min	16	0.80319
τ_D	= 5.0 min	17	0.85603
β	= 0.1 min	18	0.89995
Disturbance at $t = 0^+$		19	0.93458
x_F	= 0.55	20	0.96079
		Distillate	0.98

NOTE: To convert pound-moles per minute to kilogram-moles per minute, multiply by 0.454; to convert pound-moles to kilogram-moles, multiply by 0.454.

TABLE 13-33 Results for Ideal Binary Dynamic-Distillation Example of Table 13-32*

Time, min	Mole fraction of light component in liquid: Bottoms	Stage 5	Stage 10	Stage 15	Distillate	Flow rate, (lb·mol)/min: Reflux	Boli-up
0.00	0.02000	0.18622	0.47688	0.74345	0.98000	128.01	178.01
.50	0.02014	0.19670	0.51310	0.74940	0.98000	128.01	178.16
1.00	0.02107	0.21174	0.52426	0.76049	0.98010	127.91	179.31
1.50	0.02217	0.22038	0.53026	0.76847	0.98034	127.64	181.06
2.00	0.02275	0.22209	0.53229	0.77217	0.98061	127.33	182.65
2.50	0.02268	0.21881	0.53141	0.77222	0.98076	127.11	183.69
3.00	0.02212	0.21287	0.52879	0.76993	0.98077	127.02	184.10
3.50	0.02132	0.20639	0.52560	0.76672	0.98065	127.07	183.99
4.00	0.02051	0.20104	0.52282	0.76381	0.98047	127.19	183.55
4.50	0.01987	0.19777	0.52109	0.76196	0.98030	127.32	182.98
5.00	0.01950	0.19679	0.52057	0.76142	0.98018	127.42	182.47
5.50	0.01939	0.19766	0.52106	0.76198	0.98014	127.45	182.14
6.00	0.01950	0.19956	0.52209	0.76315	0.98016	127.41	182.02
6.50	0.01972	0.20162	0.52320	0.76438	0.98022	127.33	182.08
7.00	0.01995	0.20314	0.52400	0.76525	0.98029	127.24	182.25
7.50	0.02012	0.20380	0.52434	0.76557	0.98034	127.15	182.43
8.00	0.02019	0.20362	0.52422	0.76537	0.98036	127.09	182.56
8.50	0.02016	0.20289	0.52381	0.76484	0.98035	127.07	182.61

*From Luyben, *Process Modeling, Simulation, and Control for Chemical Engineers*, McGraw-Hill, New York, 1973.

NOTE: To convert pound-moles per minute to kilogram-moles per minute, multiply by 0.454.

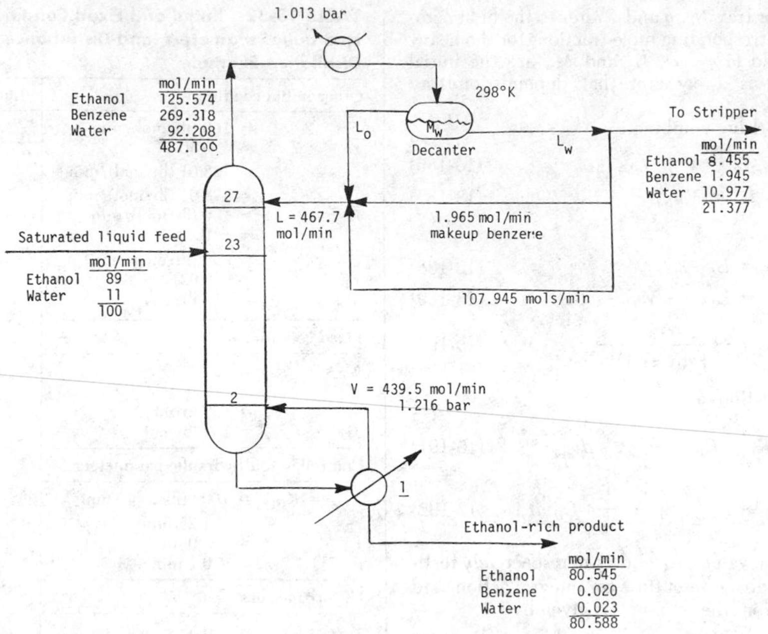

FIG. 13-101 Initial steady state for dynamic azeotropic distillation of ethanol-water with benzene.

overhead vapor is condensed and cooled to 298 K to form two liquid phases that are separated in the decanter. The organic-rich phase is returned to the top tray as reflux together with a portion of the water-rich phase and makeup benzene. The other portion of the water-rich phase is sent to a stripper to recover organic compounds. Ordinarily, vapor from that stripper is condensed and recycled to the decanter, but that coupling is ignored here.

Equations for the decanter are as follows if it is assumed that (1) there are constant holdups in the decanter of both phases in the same ratio as the ratio of the flow rates leaving the decanter, (2) there is a constant decanter temperature, and (3) the two liquid phases in the decanter are in physical equilibrium and each is perfectly mixed.

$$\frac{d}{dt}[M_d(x_i)_d] = V_N y_{i,N} - L_O(x_i)_O - L_w(x_i)_w \qquad (13\text{-}195)$$

$$dM_d/dt = V_N - L_O - L_w \qquad (13\text{-}196)$$

$$K_{d_i} = (x_i)_O/(x_i)_w = (\gamma_i)_w/(\gamma_i)_O \qquad (13\text{-}197)$$

where M_d is the total molar holdup of both phases in the decanter and the total composition in the decanter is

$$(x_i)_d = \frac{(x_i)_O L_O + (x_i)_w L_w}{L_O + L_w} \qquad (13\text{-}198)$$

Combination of Eqs. (13-195), (13-196), and (13-198) gives

$$\frac{d(x_i)_d}{dt} = \frac{V_N}{M_d}[y_{i,N} - (x_i)_d] \qquad (13\text{-}199)$$

These equations, together with those for the tower, constitute a so-called stiff system. They were solved by Prokopakis and Seider (op. cit.), following a prescribed disturbance, using an adaptive semi-implicit Runge-Kutta integration technique by which V_N and the $y_{i,N}$ were obtained by integration of the equations for the tower. Then Eq. (13-199) was integrated to give $(x_i)_d$, which was used with Eqs. (13-197) and (13-198) to obtain the equilibrium compositions $(x_i)_O$ and $(x_i)_w$ leaving the decanter. The UNIQUAC equation was used with data from Gmehling and Onken (*Vapor-Liquid Equilibrium Data Collection*, DECHEMA Chemistry Data ser., vol. 1 (parts 1–10), Frankfurt, 1977) to obtain the activity coefficients needed in Eq. (13-197). Reboiler and decanter volumetric holdups were assumed constant at 1.0 m^3 (35.3 ft^3), and volumetric tray holdups were computed from

$$M_n = (\rho_L)_n A_n (h_{wj} + h_{cj}) \qquad (13\text{-}200)$$

where $(\rho_L)_n$ is the liquid density; A_n, the cross-sectional area of the active portion of the tray = 0.23 m^2 (2.48 ft^2); h_{wj}, the weir height = 0.0254 m (0.0833 ft); and h_{cj}, the weir crest, assumed to be constant at 0.00508 m (0.0167 ft). Accordingly, volumetric tray holdup was constant at 0.007 m^3 (0.247 ft^3).

Assume that at $t = 0^+$ the feed rate to tray 23 is disturbed by increasing it by 30 percent to 130 mol/min without a change in composition. The resulting ethanol liquid mole fraction on several trays is tracked in Fig. 13-102*a*. Above tray 16, ethanol concentration remains very small. Below tray 9, ethanol concentration initially decreases fairly rapidly but then increases slowly and steadies out at significantly higher values than at the initial steady state. Tray 10 is one of the last trays to reach the new steady-state condition, which takes somewhat more than 200 min. This may be compared with initial residence times in the decanter and reboiler of approximately 50 and 250 min respectively. The movement through the column of concentration fronts for all three components is shown in Fig. 13-102*b*. For the first 5 to 10 min, below tray 16, benzene and ethanol fronts shift downward. Then a reversal occurs, and the fronts shift upward until the new steady state is attained. The upward shift is expected because the increased feed rate increases the water-benzene entrainer ratio. The duration of the initial, temporary downward shift is highly dependent on tray holdup and is due to "wash-out" with the feed liquid. This phenomenon is also observed in the dynamic studies of Peiser and Grover [*Chem. Eng. Prog.*, **58**(9), 65 (1962)].

If the feed rate is decreased, the trends of curves in Fig. 13-102

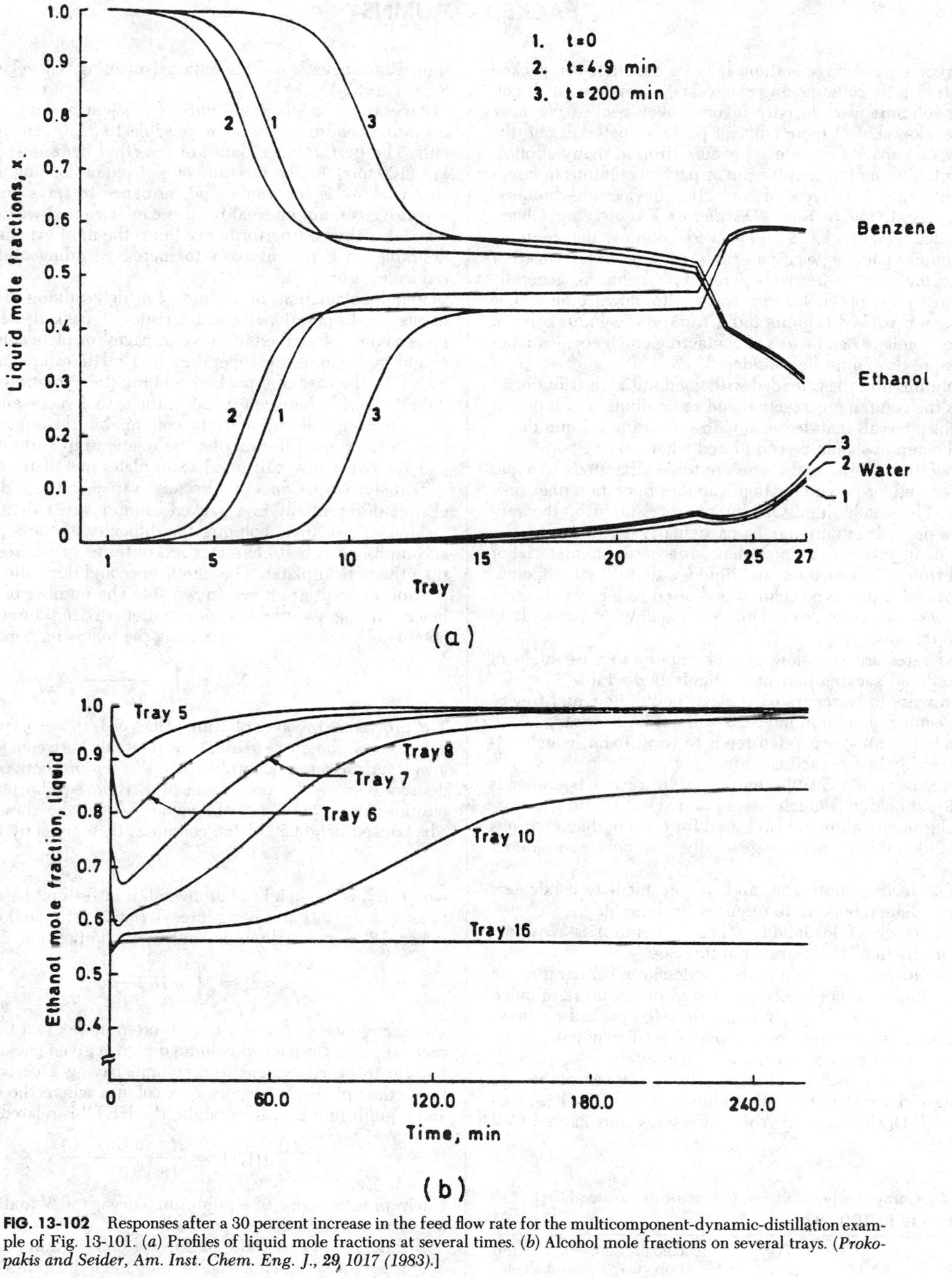

FIG. 13-102 Responses after a 30 percent increase in the feed flow rate for the multicomponent-dynamic-distillation example of Fig. 13-101. (*a*) Profiles of liquid mole fractions at several times. (*b*) Alcohol mole fractions on several trays. (*Prokopakis and Seider, Am. Inst. Chem. Eng. J., 29, 1017 (1983).*]

are reversed. The disturbance of other variables such as feed composition, boil-up ratio, and recycle of water-rich effluent from the decanter produces similar shifts in the steep concentration fronts, indicating that azeotropic towers are among the most sensitive separation operations, for which dynamic studies are essential if reliable process control is to be developed. Such studies indicate the importance of adjusting aqueous-phase recycle and reboiler duty to diminish the movement of steep concentration fronts and the possibility of multiple regimes of operation including unstable regimes, as shown by Magnussen et al. [*Inst. Chem. Eng. Symp. Ser. 56* (1979)].

PACKED COLUMNS

Distillation-type separation operations may be conducted in packed rather than tray-type columns. In prior years, except for small columns, plate columns were heavily favored over packed columns. However, development of more efficient packing materials and the need to increase capacity or reduce pressure drop in many applications has resulted in more extensive use of packed columns in larger sizes in recent years. Both types of contacting devices are discussed extensively in Sec. 18 and by Billet (*Distillation Engineering*, Chemical Publishing, New York, 1979). Packed columns may employ dumped (random) packing, e.g., pall rings, or ordered (arranged or stacked) packing, e.g., wire web. Tray-type columns generally employ valve, sieve, or bubble-cap trays with downcomers. The choice between a packed column and a tray-type column is based mainly on economics when factors of contacting efficiency, loadability, and pressure drop must be considered.

Packed columns must be provided with good initial distribution of liquid across the column cross section and redistribution of liquid at various height intervals that decrease with increasing column diameter. Packed columns should be considered when:

1. Temperature-sensitive mixtures are to be separated. To avoid decomposition and/or polymerization, vacuum operation may then be necessary. The smaller liquid holdup and pressure drop theoretical stage of a packed column may be particularly desirable.
2. Ceramic or plastic (e.g., propylene) is a desirable material of construction from a noncorrosion and liquid-wettability standpoint.
3. Refitting of a tray-type column is desired to increase loading and/or decrease pressure drop. Ordered packing is particularly applicable in this case.
4. Liquid rates are very low and/or vapor rates are high, in which case ordered packing may be particularly desirable.
5. The mixture to be separated is clear, nonfouling, and free of solids, and cleaning of column internals will not be necessary.
6. The mixture to be separated tends to form foam, which collapses more readily in a packed column.
7. High recovery of a volatile component by a batch operation is required. Liquid holdup is much lower in a packed column.

Packed columns are almost always used for column diameters less than 762 mm (30 in) but otherwise generally need not be considered when:

1. Multiple feeds, sidestreams, and/or intermediate condensers and/or intermediate reboilers are required or desirable.
2. A wide range of loadability (turndown ratio) is required. Valve trays are particularly desirable in this case.
3. Design data for separation of the particular or similar mixture in a packed column are not available. Design procedures are much better established for tray-type columns than for packed columns. This is particularly so with respect to separation efficiency since tray efficiency can be estimated much more accurately than packed-height equivalent to a theoretical stage (HETP). A comprehensive survey of data and correlations for tray efficiency and HETP is given by Vital et al. ["Distillation and Absorption Tray and Packed Column Efficiencies," AIChE national meeting, Cleveland (Aug. 29–Sept. 1, 1982)].

TABLE 13-34 Comparative Column Performance Data: Ethyl Benzene–Styrene at 100 torr

Column internal	HETP, m	Tray spacing, m	Specific pressure drop, torr/theoretical stage	Column diameter, m
Wire-web packing	0.27	. .	14.0	1.44
80-mm pall-ring packing	0.68	. .	26.1	1.50
Valve trays	0.70	0.5	53.5	1.63
Sieve trays	0.71	0.5	57.6	1.66
Bubble-cap trays	0.75	0.5	69.4	1.95

NOTE: To convert meters to feet, multiply by 3.28.

An example of the desirability of efficient packing under vacuum operating conditions is shown in Table 13-34, taken from Billet (op. cit.). The performance data are for ethyl benzene–styrene at 13.3 kPa (100 torr), total reflux, and 80 percent of flooding. Dumped pall-ring packing is superior in performance to trays and is the least expensive column internal for this case. Ordered wire-web packing, which has the best performance but is the most expensive, would be desirable on a retrofit basis to increase loading and significantly reduce pressure drop.

Initial calculation procedures for determining design, performance, and operational characteristics of packed columns are the same as those described in previous pages for plate columns; i.e., the equilibrium-curve and operating-line relationships are first established. In the case of a packed column, the operating line describes the relationship between y and x, the actual vapor and liquid compositions at any elevation in the column, but the equations are identical to those used to describe the relationship between values of y and x between any pair of adjacent plates in a plate column.

Transfer-Unit Concept Because vapor and liquid compositions change differentially in a packed column rather than in a stepwise fashion as in a plate column, the difficulty of the separation to be accomplished is best characterized in terms of transfer units rather than theoretical plates. The significance and derivation of the transfer-unit concept are given in Sec. 14. The number of transfer units based on the gas-phase concentration driving force N_{og} can be obtained by graphically integrating the following equation:

$$N_{og} = \int_{y_o}^{y_p} \frac{dy}{y^\circ - y} \tag{13-201}$$

The procedure involves plotting values of $1/(y^\circ - y)$ versus assumed values of y chosen at reasonable intervals between y_o and y_p, the lower and upper concentrations in the column between which it is desired to know N_{og}. Each value of y° is the equilibrium vapor composition corresponding to the value of x for the chosen value of y. The packed height Z_p of the column is then given by

$$Z_p = H_{og} N_{og} \tag{13-202}$$

where H_{og} is the height of an overall transfer unit based on the gas-phase concentration driving force if H_{og}, as estimated by procedures in Sec. 18, is essentially independent of composition. Otherwise,

$$Z_p = \int_0^{y_p} H_{og} \frac{dy}{y^\circ - y} \tag{13-203}$$

Although use of transfer units is often more exact than use of theoretical plates for packed-column design, a given packed column can be said to be equivalent to a column having a certain number of theoretical plates. In regions of a column where the operating line and equilibrium line are straight, the HETP is related to H_{og} by

$$\text{HETP} = \frac{H_{og} \ln (mV/L)}{(mV/L) - 1} \tag{13-204}$$

where m is the slope of equilibrium curve, and V and L are gas and liquid molar flow rates respectively.

If the term mV/L (ratio of slope of equilibrium curve to slope of operating line) is unity, a transfer unit and a theoretical plate become identical (see Sec. 14); if mV/L lies between 0.9 and 1.1, the difference between a transfer unit and a theoretical plate is insignificant. Thus for tall packed columns in which separations are difficult and operating lines are nearly parallel to the equilibrium curve, theoretical plates may be used to characterize the separation obtained.

In the distillation of ideal binary mixtures of low relative volatility at total reflux, as shown by Chilton and Colburn [*Ind. Eng. Chem.*, **27**, 255 (1935)], HETP and H_{og} will be practically identical. For other conditions, they may be substantially different. For binary mixtures, a useful procedure for determining packed height, as described and illustrated in an example by Sherwood, Pigford, and Wilke (*Mass Transfer*, McGraw-Hill, New York, 1975), is as follows:

1. Construct a McCabe-Thiele diagram and step off the theoretical plates in the usual manner.
2. From the vapor-phase composition changes, compute for each step H_g, H_ℓ, m, L, and V.
3. Calculate corresponding values of H_{og} and HTEP (Eq. 13-204). Although H_{og} may not vary appreciably, HETP generally will. Each value of HETP is the increment of packed height required for the corresponding theoretical stage. The sum of the HETP values is the total packed height, which can be separated into rectifying and stripping section zones.

Values of H_g and H_ℓ are given in Sec. 18. Because experimental data often have to be extended, it is helpful to list effects of the controlling variables. Increasing liquid rate causes values of H_g and H_ℓ to decrease, but the effect of the gas rate is minor. Increased temperature causes values to decrease, but the effect of pressure is small or negligible except for its effect upon boiling temperature. The smaller the particle size of packing, the greater the surface area; thus one would expect lower values. In general, such is the case, although the effect is not direct. The throughput capacity for very small packings is too low to make them very useful for large-scale operations.

There is considerable uncertainty regarding the effect of column diameter. It is generally believed that, owing to poorer liquid distribution, values of H_g and H_ℓ become less favorable for a given packing the larger the diameter of the column. In general, values are found to become slightly less favorable for greater heights of packing, possibly because of progressive maldistribution effects. It is usually considered good practice to install liquid redistributors every 3 m (10 ft) of packed height. These considerations are more serious for random packings than for ordered packings.

Mass Transfer and Gas Absorption*

William M. Edwards, Ph.D., P.E., *Senior Staff Research Engineer, Shell Development Company, Westhollow Research Center, Houston, Texas; Member, American Institute of Chemical Engineers.*

*The monumental contributions of Professors Robert L. Pigford and Alan P. Colburn to the development of earlier versions of this section must be acknowledged. Also, the contributions of Drs. R. E. Emmert and R. N. Maddox to the content of the fourth and fifth editions, respectively, should be recognized.

MULTICOMPONENT SYSTEMS

ECONOMIC DESIGN OF ABSORPTION SYSTEMS

ABSORPTION WITH CHEMICAL REACTION

OTHER TYPES OF GAS ABSORBERS

Nomenclature and Units

Symbol	Definition	SI units	U.S. customary units
a	Effective interfacial mass-transfer area per unit volume of tower or apparatus	m^2/m^3	ft^2/ft^3
a_p	External area of packing surfaces per unit volume of tower packing	m^2/m^3	ft^2/ft^3
A	Absorption factor (L_M/mG_M)	Dimensionless	Dimensionless
A^0	Absorption factor for dilute multicomponent systems (L_M^S/mG_M^0)	Dimensionless	Dimensionless
A_e, A'	Edmister's effective absorption factors, Eq. (14-103)	Dimensionless	Dimensionless
B	Function defined in Eq. (14-127)	Dimensionless	Dimensionless
B^0	Bulk-liquid reactant concentration	$kmol/m^3$	$(lb\cdot mol)/ft^3$
c	Solute concentration in bulk liquid	$kmol/m^3$	$(lb\cdot mol)/ft^3$
c_i	Solute concentration in liquid at gas-liquid interface	$kmol/m^3$	$(lb\cdot mol)/ft^3$
c_p	Specific heat	$kJ/(kg\cdot K)$	$Btu/(lb\cdot °F)$
C_i	Cost factors in economic design calculations, Eqs. (14-127) through (14-136)		
d	Characteristic length appropriate to the geometry of the system under consideration	m	ft
d_b	Bubble diameter	m	ft
d_s	Sauter mean diameter	m	ft
d_T	Tower diameter	m	ft
D_A	Diffusion coefficient of reactant A in liquid phase	m^2/s	ft^2/h
D_{AB}	Gas-phase diffusion coefficient of solute A in inert gas B	m^2/s	ft^2/h
D_B	Diffusion coefficient of liquid reactant B	m^2/s	ft^2/h
D_L	Diffusion coefficient of inert solute in the liquid phase	m^2/s	ft^2/h
DF	Design safety factor to correct for assumptions inherent in the use of simplified design equations	Dimensionless	Dimensionless
E	Overall plate efficiency in a plate tower	Dimensionless	Dimensionless
f	Friction factor for fluid flow	Dimensionless	Dimensionless
f_H	Fractional liquid holdup in tower	m^3/m^3	ft^3/ft^3
$F(x)$	Equilbrium function of composition x		
G	Gas-phase mass velocity	$kg/(s\cdot m^2)$	$lb/(h\cdot ft^2)$
G_{opt}	Economic optimum gas mass velocity	$kg/(s\cdot m^2)$	$lb/(h\cdot ft^2)$
G_M	Molar gas-phase mass velocity	$kmol/(s\cdot m^2)$	$(lb\cdot mol)/(h\cdot ft^2)$
G_M'	Molar mass velocity of inert gas	$kmol/(s\cdot m^2)$	$(lb\cdot mol)/(h\cdot ft^2)$
G_M^0	Moles of solute-rich feed gas to be treated per unit time	kmol/s	$(lb\cdot mol)/h$
h	Height coordinate for packed towers	m	ft
h'	Heat-transfer coefficient	$W/(m^2\cdot K) = J/(s\cdot m^2\cdot K)$	$Btu/(h\cdot ft^2\cdot °F)$
h_T	Total height of tower packing required	m	ft
H	Henry's-law constant	kPa/(mole-fraction solute in liquid phase)	(lbf/in^2)/(mole-fraction solute in liquid phase)
H'	Henry's-law constant	kPa/[kmol/(m^3 solute in liquid phase)]	(lbf/in^2)/[(lb·mol)/(ft^3 solute in liquid phase)] or atm/[(lb·mol)/(ft^3 solute in liquid phase)]
H_G	Height of one transfer unit based on gas-phase resistance	m	ft
H_{OG}	Height of one overall gas-phase mass-transfer unit	m	ft
H_L	Height of one transfer unit based on liquid-phase resistance	m	ft
H_{OL}	Height of one overall liquid-phase mass-transfer unit	m	ft
HTU	Height of one transfer unit (general)	m	ft
HETP	Height equivalent to one theoretical plate	m	ft
j_H	Chilton-Colburn j factor for heat transfer	Dimensionless	Dimensionless
j_M	Chilton-Colburn j factor for mass transfer	Dimensionless	Dimensionless
k	Thermal conductivity	$(J\cdot m)/(s\cdot m^2\cdot K) = W/(m\cdot K)$	$Btu/(h\cdot ft\cdot °F)$

Nomenclature and Units (*Continued*)

Symbol	Definition	SI units	U.S. customary units
k_1	First-order-reaction-rate coefficient	s^{-1}	h^{-1}
k_2	Second-order-reaction-rate coefficient	$m^3/(s \cdot kmol)$	$ft^3/(h \cdot lb \cdot mol)$
k_G	Gas-phase mass-transfer coefficient for dilute systems	kmol/[(s·m²)(mole-fraction solute)]	(lb·mol)/[(h·ft²)(mole-fraction solute)]
k_G'	Gas-phase mass-transfer coefficient for dilute systems	kmol/[(s·m²)(kPa solute partial pressure)]	(lb·mol)/[(h·ft²)(lbf/in² solute partial pressure)]
$\hat{k}_G$	Gas-phase mass-transfer coefficient for concentrated systems	kmol/(s·m²)	(lb·mol)/(h·ft²)
$k_G a$	Volumetric gas-phase mass-transfer coefficient for dilute systems	kmol/[(s·m³)(mole fraction)]	(lb·mol)/[(h·ft³)(mole fraction)]
K_G	Overall gas-phase mass-transfer coefficient for dilute systems	kmol/[(s·m²)(mole fraction)]	(lb·mol)/[(h·ft²)(mole fraction)]
$\hat{K}_G$	Overall gas-phase mass-transfer coefficient for concentrated systems	kmol/(s·m²)	(lb·mol)/(h·ft²)
$K_G a$	Overall volumetric gas-phase mass-transfer coefficient for dilute systems	kmol/[(s·m³)(mole-fraction solute in gas)]	(lb·mol)/[(h·ft³)(mole-fraction solute in gas)]
$K_G' a$	Overall volumetric gas-phase mass-transfer coefficient for dilute systems	kmol/[(s·m³)(kPa solute partial pressure)]	(lb·mol)/[(h·ft³)(lbf/in² solute partial pressure)]
$\hat{k}_G a$	Overall volumetric gas-phase mass-transfer coefficient for concentrated systems	kmol/(s·m³)	(lb·mol)/(h·ft³)
k_L	Liquid-phase mass-transfer coefficient for dilute systems	kmol/[(s·m²)(mole-fraction solute in liquid)]	(lb·mol)/[(h·ft²)(mole-fraction solute in liquid)]
k_L'	Liquid-phase mass-transfer coefficient for dilute systems	kmol/[(s·m²)(kmol/m³)] or m/s	(lb·mol)/[(h·ft²)(lb·mol/ft³)] or ft/h
$\hat{k}_L$	Liquid-phase mass-transfer coefficient for concentrated systems	kmol/(s·m²)	(lb·mol)/(h·ft²)
$\hat{k}_L^0$	Liquid-phase mass-transfer coefficient for pure physical absorption, used in the design of chemically reacting systems	kmol/(s·m²)	(lb·mol)/(h·ft²)
$k_L a$	Volumetric liquid-phase mass-transfer coefficient for dilute systems	kmol/[(s·m³)(mole fraction)]	(lb·mol)/[°h·ft³)(mole fraction)]
K_L	Overall liquid-phase mass-transfer coefficient for dilute systems	kmol/[(s·m²) (mole fraction)]	(lb·mol)/[(h·ft²) (mole fraction)]
$\hat{K}_L$	Overall liquid-phase mass-transfer coefficient for concentrated systems	kmol/(s·m²)	(lb·mol)/(h·ft²)
$K_L a$	Overall volumetric liquid-phase mass-transfer coefficient for dilute systems	kmol/[(s·m³)(mole-fraction solute in liquid)]	(lb·mol)/[(h·ft³)(mole-fraction solute in liquid)]
$\hat{K}_L a$	Overall volumetric liquid-phase mass-transfer coefficient for concentrated systems	kmol/(s·m³)	(lb·mol)/(h·ft³)
K	Vapor-liquid equilibrium K value $= y^\circ/x$	Dimensionless	Dimensionless
L	Liquid-phase mass velocity	kg/(s·m²)	lb/(h·ft²)
L_M	Molar liquid-phase mass velocity	kmol/(s·m²)	(lb·mol)/(h·ft²)
L_M'	Molar mass velocity of inert-liquid solvent	kmol/(s·m²)	(lb·mol)/(h·ft²)
L_M^s	Moles of solute-free solvent per unit time	kmol/s	(lb·mol)/h
m	Slope of equilibrium curve $= dy^\circ/dx$ (mole-fraction solute in gas)/(mole-fraction solute in liquid)	Dimensionless	Dimensionless
M	Molecular weight	kg/kmol	lb/(lb·mol)
MWR	Minimum-wetting rate for tower packings	m²/s	ft²/h
n_A	Rate of solute transfer	kmol/s	(lb·mol)/h
N	Number of theoretical stages in a plate-type absorber or stripper	Dimensionless	Dimensionless
N_A	Interphase mass-transfer rate of solute A per unit interfacial area	kmol/(s·m²)	(lb·mol)/(h·ft²)
N_G	Number of gas-phase mass-transfer units	Dimensionless	Dimensionless
N_{Ha}	Hatta number $(\sqrt{k_1 D_A}/k_L^0)$	Dimensionless	Dimensionless
N_{OG}	Number of overall gas-phase mass-transfer units	Dimensionless	Dimensionless
N_L	Number of liquid-phase mass-transfer units	Dimensionless	Dimensionless
N_{OL}	Number of overall liquid-phase mass-transfer units	Dimensionless	Dimensionless
NTU	Number of transfer units	Dimensionless	Dimensionless
N_{Pr}	Prandtl number $(c_p\mu/k)$	Dimensionless	Dimensionless
N_{Re}	Reynolds number (Gd/μ_G)	Dimensionless	Dimensionless

Nomenclature and Units (*Continued*)

Symbol	Definition	SI units	U.S. customary units
N_{Sc}	Schmidt number ($\mu_G/\rho_G D_{AB}$) or ($\mu_L/\rho_L D_L$)	Dimensionless	Dimensionless
N_{Sh}	Sherwood number ($\hat{k}_G RTd/D_{AB}p_T$)	Dimensionless	Dimensionless
N_{St}	Stanton number ($\hat{k}_G/G_M$) or ($\hat{k}_L/L_M$)	Dimensionless	Dimensionless
p	Solute partial pressure in bulk gas	kPa	lbf/in^2
p_A^0	Pure-component vapor pressure of solute A	kPa	lbf/in^2
p_i	Solute partial pressure at gas-liquid interface	kPa	lbf/in^2
p_T	Total system pressure	kPa	lbf/in^2
Δp	Pressure drop	kPa	lbf/in^2
r_A	Rate of absorption per unit interfacial area	kmol/(s·m^2)	(lb·mol)/(h·ft^2)
R	Gas constant	8314 J/(kmol·K) = 8.314 m^{3}kPa/(kmol·K)	(10.73 ft^3·psia)/(lb·mol·°R)
R_A	Volumetric reaction rate	kmol/(s·m^3)	(lb·mol)/(h·ft^3)
s	Fractional surface-renewal rate	s^{-1}	h^{-1}
S	Tower cross-sectional area = $\pi d^2/4$	m^2	ft^2
$\mathcal{S}$	Stripping factor (mG_M/L_M)	Dimensionless	Dimensionless
$\mathcal{S}_e$	Edmister's "effective" stripping factor	Dimensionless	Dimensionless
$\mathcal{S}^0$	Stripping factor (mG_M^0/L_M^s)	Dimensionless	Dimensionless
t	Contact time	s	h
T	Temperature	K	°R
u_L	Superficial liquid velocity in vertical direction	m/s	ft/h
V	Packed volume in tower	m^3	ft^3
V_L	Volumetric liquid velocity	m^3/[s·(m^2 tower cross-sectional area)]	ft^3/[h·(ft^2 tower cross-sectional area)]
x	Mole-fraction solute in bulk-liquid phase	(kmol solute)/(kmol liquid)	(lb·mol solute)/(lb·mol liquid)
x^*	Mole-fraction solute in bulk liquid in equilibrium with bulk-gas solute concentration y	(kmol solute)/(kmol liquid)	(lb·mol solute)/(lb·mol liquid)
x_{BM}	Logarithmic-mean inert-solvent concentration between bulk liquid and interface values	(kmol solvent)/(kmol liquid)	(lb·mol solvent)/(lb·mol liquid)
x_{BM}^*	Logarithmic-mean inert-solvent concentration between bulk-liquid value and value in equilibrium with bulk gas	(kmol solvent)/(kmol liquid)	(lb·mol solvent)/(lb·mol liquid)
x_f	Mole-fraction solute in incoming solvent liquid stream	(kmol solute)/(kmol liquid)	(lb·mol solute)/(lb·mol liquid)
x_i	Mole-fraction solute in liquid at gas-liquid interface	(kmol solute)/(kmol liquid)	(lb·mol solute)/(lb·mol liquid)
X	Moles of solute in liquid per mole of solute-free solvent	(kmol solute)/(kmol solvent)	(lb·mol solute)/(lb·mol solvent)
y	Mole-fraction solute in bulk-gas phase	(kmol solute)/(kmol gas)	(lb·mol solute)/lb·mol gas)
y^*	Mole-fraction solute in bulk gas in equilibrium with bulk-liquid solute concentration x	(kmol solute)/(kmol gas)	(lb·mol solute)/(lb·mol gas)
y_{BM}	Logarithmic-mean inert-gas concentration between bulk gas and interface values	(kmol inert gas)/(kmol gas)	(lb·mol inert gas)/(lb·mol gas)
y_{BM}^*	Logarithmic-mean inert-gas concentration between bulk-gas value and value in equilibrium with bulk liquid	(kmol inert gas)/(kmol gas)	(lb·mol inert gas)/(lb·mol gas)
y_f	Mole-fraction solute in feed gas	(kmol solute)/(kmol feed gas)	(lb·mol solute)/(lb·mol feed gas)
y_i	Mole-fraction solute in gas at gas-liquid interface	(kmol solute)/(kmol gas)	(lb·mol solute)/(lb·mol gas)
y_i^*	Mole-fraction solute in gas at interface in equilibrium with the liquid-phase interfacial solute concentration x_i	(kmol solute)/(kmol gas)	(lb·mol solute)/(lb·mol gas)
Y	Moles of solute in gas per mole of solute-rich feed gas to be treated	kmol/kmol	(lb·mol)/(lb·mol)
	Greek symbols		
α	Constant in Eq. (14-130)		
β	Constant in Eq. (14-130)		
γ	Liquid-phase-activity coefficient	Dimensionless	Dimensionless
δ	Effective thickness of stagnant-film layer	m	ft
ϵ	Void fraction available for gas flow or fractional gas holdup	m^3/m^3	ft^3/ft^3
θ	Process-operating time per year	s/year	s/year
θ'	Process-operating time per year	h/year	h/year
ρ_G	Gas-phase density	kg/m^3	lb/ft^3
ρ_L	Average molar density of liquid phase	kmol/m^3	(lb·mol)/ft^3
μ_G	Gas-phase viscosity	kg/(s·m)	lb/(h·ft)

Nomenclature and Units (*Continued*)

Symbol	Definition	SI units	U.S. customary units
μ_L	Liquid-phase viscosity	kg/(s·m)	lb/(h·ft)
υ	Number of moles of *B* reacting with 1 mol of *A*, a stoichiometric coefficient	Dimensionless	Dimensionless
ϕ	Ratio k_L/k_L^0, the dimensionless reaction or enhancement factor	Dimensionless	Dimensionless
ϕ_∞	Ratio k_L/k_L^0 when $N_{Ha} = \infty$	Dimensionless	Dimensionless
		Subscripts	
1	Tower bottom (either absorber or stripper)		
2	Tower top (either absorber or stripper)		
A	Solute component in liquid or gas phase		
B	Inert-gas or inert-solvent component		
a	Acetone solute in example problems		
G	Gas phase		
L	Liquid phase		

GENERAL REFERENCES: Astarita, G., *Mass Transfer with Chemical Reaction*, Elsevier, New York, 1967. Coulson, J. M., and J. F. Richardson, *Chemical Engineering*, vol. 2: *Unit Operations*, 3d ed., Pergamon, Oxford, 1978, pp. 529–584. Danckwerts, P. V., *Gas-Liquid Reactions*, McGraw-Hill, New York, 1970. Geankoplis, C. J., *Mass Transport Phenomena*, Holt, New York, 1972. Hobler, T., *Mass Transfer and Absorbers*, Pergamon, Oxford, 1966. Kohl, A. L., and F. C. Riesenfeld, *Gas Purification*, 3d ed., Gulf, Houston, 1979. Leva, M., *Tower Packings and Packed Tower Design*, U.S. Stoneware Co., Akron, Ohio, 1953. Norman, W. S., *Absorption, Distillation and Cooling Towers*, Wiley, New York, 1961. Shah, Y. T., *Gas-Liquid-Solid Reactor Design*, McGraw-Hill, New York, 1979. Sherwood, T. K., and R. L. Pigford, *Absorption and Extraction*, McGraw-Hill, New York, 1952. Sherwood, T. K., R. L. Pigford, and C. R. Wilke, *Mass Transfer*, McGraw-Hill, New York, 1975. Treybal, R. E., *Mass Transfer Operations*, 3d ed., McGraw-Hill, New York, 1980.

INTRODUCTION

Definitions Gas absorption is a unit operation in which soluble components of a gas mixture are dissolved in a liquid. The inverse operation, called stripping or desorption, is employed when it is desired to transfer volatile components from a liquid mixture into a gas. This section is concerned principally with the design of commercial equipment for carrying out either of these operations continuously.

Equipment The apparatus used for contacting a liquid and a gas stream continuously may be a packed tower filled with regular or irregular solid packing material, a plate-type unit containing a number of bubble-cap or sieve plates, an empty tower or chamber into which the liquid is sprayed, a wetted-wall column, or a stirred or sparged vessel. Ordinarily, the gas and liquid streams are made to flow countercurrently past each other through the equipment so that the greatest rate of absorption may be obtained.

Design Procedures The procedures to be followed in specifying the principal dimensions of gas-absorption equipment are described in this section by means of illustrative calculations for problems typical of commercial practice. The experimental data required for making such calculations are discussed and are keyed to appropriate references or to other sections of this *Handbook*.

Three main steps are involved in the design of an absorption or stripping tower:

1. Data on the vapor-liquid equilibrium relations for the system are used to determine (*a*) the quantity of liquid needed to absorb the required amount of the soluble components from the gas or (*b*) the quantity of gas needed to strip the required amount of volatile components from the liquid. See **Sec. 3** for detailed data and the following directory for additional data sources.

2. Data on the liquid- and vapor-handling capacity of equipment of the type being considered are used to determine the required cross-sectional area and diameter of the equipment through which the liquid and gas streams flow. Consideration of the economic factors involved may show that it is desirable to set the fluid velocities well below the maximum values that can be employed. See **Sec. 18** for detailed data.

3. Equilibrium data and material balances are used to determine the number of equilibrium stages (theoretical plates or transfer units) required for the separation desired. The difficulty of the separation depends on the degree of recovery that is economically most desirable. The required time of contact between the flowing streams or the required height of the tower can be calculated if data are available for the specific rate of transfer of material between the gas and liquid phases, expressed in terms of the plate efficiency or the height of one transfer unit.

DIRECTORY TO KEY GAS-ABSORPTION DATA

Table 14-1 is a quick-reference directory to data and information contained in other sections of this *Handbook*. Some additional references are discussed later.

Equilibrium Data Finding reliable gas-solubility data probably is the most time-consuming and tedious task, and yet it is the most important task involved in developing reliable designs for gas-absorption systems. An excellent new source of critically evaluated gas-solubility data is the set of volumes edited by Kertes et al., *Solubility Data Series*, published by Pergamon Press (1979 ff.). In the introduction to each volume there is an excellent discussion and definition of the various methods by which gas-solubility data have been reported in the literature, such as the Bunsen coefficient, the Kuenen

TABLE 14-1 Directory to Key Gas-Absorption Data

Type of data	*Handbook* references Sec.	Pages
Equilibrium data		
Gas solubilities	3	3-101–3-103
Pure-component vapor pressures	3	3-45–3-63
Equilibrium *K* values	13	13-16–13-22
Thermal data		
Heats of solution	3	3-157–3-159
Specific heats	3	3-128–3-146
Latent heats	3	3-119–3-128
Transport data		
Diffusion coefficients		
Liquids	3	3-258–3-259
Gases	3	3-256–3-257
Viscosities		
Liquids	3	3-251–3-252
Gases	3	3-247–3-250
Densities		
Liquids	3	3-75–3-96
Gases	3	3-78
Packed-tower data		
Pressure drop and flooding	18	18-22–18-25
Mass-transfer coefficients	18	18-32–18-41
HTU: physical absorption	18	18-32–18-41
HTU: with chemical reaction	14	14-34
HETP	18	18-41
Costs of towers and packings	18	18-45–18-48
Plate-tower data		
Pressure drop and flooding	18	18-5–18-12
Tray efficiencies	18	18-13–18-19
Costs of towers and trays	18	18-45–18-48

coefficient, the Ostwald coefficient, the absorption coefficient, and the Henry's-law constant, with formulas for interrelating these parameters. The new third edition of *The Properties of Gases and Liquids* by Reid, Prausnitz, and Sherwood (McGraw-Hill, New York, 1977) has been completely rewritten and is an excellent source of actual data and information on the proper use of thermodynamic-equilibrium data. Section 13 of this *Handbook* presents a good discussion of equilibrium *K* values.

Prausnitz and Shair [*Am. Inst. Chem. Eng. J.*, **7**, 682 (1961)] have discussed the thermodynamics of low-pressure gas solubility. More recently, Schulze and Prausnitz [*Ind. Eng. Chem. Fundam.*, **20**, 175 (1981)] have presented data and a correlation of gas solubilities in water at high temperatures (0 to 300°C) in terms of Henry's-law constants. They show that Henry's constant is not a monotonic function of temperature and, therefore, that simple extrapolation procedures may lead to large errors.

For systems involving liquid-phase chemical-reaction equilibria, Rivas and Prausnitz [*Am. Inst. Chem. Eng. J.*, **25**, 975 (1979)] present an excellent discussion of the proper definition and use of thermodynamic-equilibrium relations, Henry's-law constants, and chemical equlibria and discuss how to interpolate and extrapolate such data with respect to temperature and pressure.

The following references contain compilations of pure-component vapor pressures and of gas-solubility data:

- Battino, R., et al.: *Chem. Rev.*, **66**, 395 (1966). Documents primary gas-solubility-data references.
- Kertes, A. S., et al.: *Solubility Data Series*, Pergamon, Oxford, 1979 ff.
- Linke, W. F., et al.: *Solubilities of Inorganic and Metal-Organic Compounds*, 4th ed., vol. I, Van Nostrand, New York, 1958.
- ———: *Solubilities of Inorganic and Metal-Organic Compounds*, American Chemical Society, Washington, 1965.
- Stephen, H., et al.: *Solubilities of Inorganic and Organic Compounds*, Pergamon, Oxford, 1963.
- Vargaftic, N. B.: *Tables on the Thermophysical Properties of Liquids and Gases*, Wiley, New York, 1975.
- Wilhelm, E., et al.: *Chem. Rev.*, **73**, 1 (1973). Solubility of gases in nonaqueous solvents.
- ———: *Chem. Rev.*, **77**, 219 (1977). Solubility of gases in water.
- Yaws, C. L.: "Physical Properties: A Guide to the Physical, Thermodynamic and Transport Property Data of Industrially Important Chemical Compounds," in *Chemical Engineering*, McGraw-Hill, New York, 1977.

Thermal Data The Reid, Prausnitz, and Sherwood book (op. cit.) is a good alternative source of information and data on latent heats of vaporization, of heat capacities of gases and liquids, and of heats of formation. Useful discussions of heats of solution are presented by Prausnitz and Shair (op. cit.) and by Wilhelm, Battino, and Wilcock [*Chem. Rev.*, **77**, 219 (1977)].

Transport Data In addition to the Reid, Prausnitz, and Sherwood reference (op. cit.), the book by S. Bretsznajder, *Prediction of Transport and Other Physical Properties of Fluids* (Pergamon Press, Oxford, 1971), is highly recommended. Both of these references give actual data as well as methods for estimating viscosities, diffusion coefficients, densities, thermal conductivities, and surface tensions.

Packed- and Plate-Tower Data Most of the important literature on this subject is referenced in Sec. 18 of this *Handbook* for systems involving purely physical absorption. References to the literature for systems involving gas absorption with chemical reaction are presented later in this section.

NOMENCLATURE AND UNITS CONVERSION

The nomenclature and units employed in this section conform to the American Institute of Chemical Engineers–approved SI system as described by E. Buck (*Chem. Eng. Prog.*, October 1978, p. 73) or the U.S. customary system insofar as possible. For convenience in absorber design work, some of the more frequently used compound conversions from U.S. customary to SI units are listed in Table 14-2. A table of the nomenclature employed in this section is presented at the beginning of the section and is recommended as a guide for potential authors in the field of gas absorption.

INTERPHASE MASS TRANSFER

Transfer of material between phases is important in most separation processes in which gases and liquids are involved. When a pure liquid is being evaporated into a gas, only the gas-phase mass transfer need be calculated; i.e., mass transfer within the pure-liquid phase is not involved. Conversely, when a pure gas is being absorbed into a liquid, only the liquid-phase mass transfer need be considered. Occasionally, mass transfer in one of the two phases may be neglected even though pure components are not involved. This will be the case when the resistance to mass transfer is much larger in one phase than in the other. Understanding the nature and magnitudes of these resis-

TABLE 14-2 Frequently Used Compound Conversion Factors*

Multiply	By	To obtain SI units
lb/h	1.260×10^{-4}	kg/s
(lb·mol)/h	1.260×10^{-4}	kmol/s
lb/(h·ft²)	1.356×10^{-3}	kg/(s·m²)
(lb·mol)/(h·ft²)	1.356×10^{-3}	kmol/(s·m²)
(lb·mol)/(h·ft²·atm)	1.339×10^{-5}	kmol/(s·m²·kPa)
(lb·mol)/(h·ft³)	4.450×10^{-3}	kmol/(s·m³)
(lb·mol)/(h·ft³·atm)	4.391×10^{-5}	kmol/(s·m³·kPa)
lb/ft³	16.02	kg/m³
(lb·mol)/ft³	16.02	kmol/m³
ft²/h	2.581×10^{-5}	m²/s
cm²/s	1.000×10^{-4}	m²/s
atm	101.3	kPa
ft	0.3048	m
ft²	9.290×10^{-2}	m²
ft³	2.832×10^{-2}	m³
ft²/ft³	3.281	m²/m³
cm²/cm³	100	m²/m³
Btu/(lb·°R)	4.187	kJ/(kg·K)
Btu/(lb·mol)	2.326	kJ/kmol
cal/(g·mol)	4.187	kJ/kmol
Btu/(h·ft²·°R)	5.678×10^{-3}	kJ/(s·m²·K)
(Btu·ft)/(h·ft²·°R)	1.731×10^{-3}	(kJ·m)/(s·m²·K)

*Basis: 1 kJ/s = 1 kW; 1 Btu = 1.05506 kJ (ISO/TC 12).

tances is one of the keys to performing reliable gas-absorption calculations. For specific data on gas-phase and liquid-phase mass-transfer coefficients in nonreacting systems see Sec. 18. Chemically reacting systems are discussed later in this section.

Mass-Transfer Principles: Dilute Systems When material is transferred from one phase to another across an interface that separates the two, the resistance to mass transfer in each phase causes a concentration gradient in each, as shown in Fig. 14-1. The concentrations of the diffusing material in the two phases immediately adjacent to the interface generally are unequal, even if expressed in the same units, but usually are assumed to be related to each other by the laws of thermodynamic equilibrium. Thus, it is assumed that the thermodynamic equilibrium is reached at the gas-liquid interface almost immediately when a gas and a liquid are brought into contact.

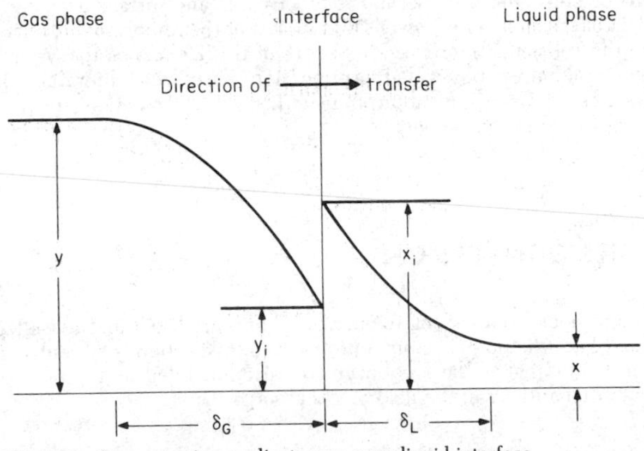

FIG. 14-1 Concentration gradients near a gas-liquid interface.

For systems in which the solute concentrations in the gas and liquid phases are dilute, the rate of transfer may be expressed by equations which predict that the rate of mass transfer is proportional to the difference between the bulk concentration and the concentration at the gas-liquid interface. Thus

$$N_A = k'_G(p - p_i) = k'_L(c_i - c) \tag{14-1}$$

where N_A = mass-transfer rate, k'_G = gas-phase mass-transfer coefficient, k'_L = liquid-phase mass-transfer coefficient, p = solute partial pressure in bulk gas, p_i = solute partial pressure at interface, c = solute concentration in bulk liquid, and c_i = solute concentration in liquid at interface.

The mass-transfer coefficients k'_G and k'_L by definition are equal to the ratios of the molal mass flux N_A to the concentration driving forces $(p - p_i)$ and $(c_i - c)$ respectively. An alternative expression for the rate of transfer in dilute systems is given by

$$N_A = k_G(y - y_i) = k_L(x_i - x) \tag{14-2}$$

where N_A = mass-transfer rate, k_G = gas-phase mass-transfer coefficient, k_L = liquid-phase mass-transfer coefficient, y = mole-fraction solute in bulk-gas phase, y_i = mole-fraction solute in gas at interface, x = mole-fraction solute in bulk-liquid phase, and x_i = mole-fraction solute in liquid at interface.

The mass-transfer coefficients defined by Eqs. (14-1) and (14-2) are related to each other as follows:

$$k_G = k'_G p_T \tag{14-3}$$

$$k_L = k'_L \bar{\rho}_L \tag{14-4}$$

where p_T = total system pressure employed *during the experimental determinations* of k'_G values and $\bar{\rho}_L$ = average molar density of the liquid phase. The coefficient k_G is relatively independent of the total system pressure and therefore is more convenient to use than k'_G, which is inversely proportional to the total system pressure.

The above equations may be used for finding the interfacial concentrations corresponding to any set of values of x and y provided the ratio of the individual coefficients is known. Thus

$$(y - y_i)/(x_i - x) = k_L/k_G = k'_L\bar{\rho}_L/k'_G p_T = L_M H_G/G_M H_L \tag{14-5}$$

where L_M = molar liquid mass velocity, G_M = molar gas mass velocity, H_L = height of one transfer unit based on liquid-phase resistance, and H_G = height of one transfer unit based on gas-phase resistance. The last term in Eq. (14-5) is derived from Eqs. (14-26) and (14-28).

Equation (14-5) may be solved graphically if a plot is made of the equilibrium vapor and liquid compositions and a point representing the bulk concentrations x and y is located on this diagram. A construction of this type is shown in Fig. 14-2, which represents a gas-absorption situation.

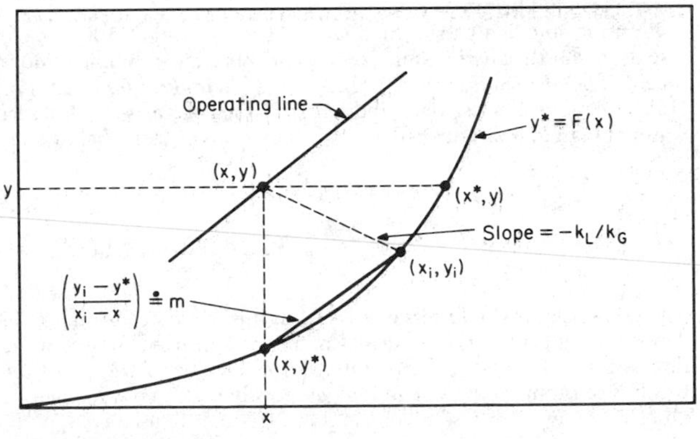

FIG. 14-2 Identification of concentrations at a point in a countercurrent absorption tower.

In the design of equipment, it is necessary to estimate the rate of mass transfer from known or predicted values of the transfer coefficients and the bulk concentrations. This may be done by solving Eq. (14-5) simultaneously with the equilibrium relation $y_i^\circ = F(x_i)$ to obtain y_i and x_i. The rate of transfer may then be calculated from Eq. (14-2).

If the equilibrium relation $y_i^\circ = F(x_i)$ is sufficiently simple, e.g., if a plot of y_i° versus x_i is a straight line, not necessarily through the origin, the rate of transfer is proportional to the difference between the bulk concentration in one phase and the concentration (in that same phase) which would be in equilibrium with the bulk concentration in the second phase. One such difference is $y - y^\circ$, and another is $x^\circ - x$. In this case, there is no need to solve for the interfacial compositions, as may be seen from the following derivation.

The rate of mass transfer may be defined by the equation

$$N_A = K_G(y - y^\circ) = k_G(y - y_i) = k_L(x_i - x) = K_L(x^\circ - x) \tag{14-6}$$

where K_G = overall gas-phase mass-transfer coefficient, K_L = overall liquid-phase mass-transfer coefficient, y° = vapor composition in equilibrium with x, and x° = liquid composition in equilibrium with vapor of composition y. This equation can be rearranged to the formula

$$\frac{1}{K_G} = \frac{1}{k_G}\left(\frac{y - y^\circ}{y - y_i}\right) = \frac{1}{k_G} + \frac{1}{k_G}\left(\frac{y_i - y^\circ}{y - y_i}\right) = \frac{1}{k_G} + \frac{1}{k_L}\left(\frac{y_i - y^\circ}{x_i - x}\right) \tag{14-7}$$

in view of Eq. (14-5). Comparison of the last term in parentheses with the diagram of Fig. 14-2 shows that it is equal to the slope of the chord connecting the points (x,y°) and (x_i,y_i). If the equilibrium curve is a straight line, then it is the slope m. Thus

$$1/K_G = (1/k_G + m/k_L) \tag{14-8}$$

For dilute concentrations of many gases and over a fairly wide range for some gases, the equilibrium relationship is given by **Henry's law,** which relates the partial pressure developed by a dissolved solute A in a liquid solvent B by one of the following two equations:

$$p_A = Hx_A \tag{14-9}$$

or

$$p_A = H'c_A \tag{14-10}$$

where H is the Henry's-law coefficient expressed in units (SI) of kilopascals per mole-fraction solute in liquid and H' is the Henry's-law coefficient expressed in units (SI) of kilopascals per kilomole per cubic meter.

When Henry's law is valid, the slope m can be computed according to the relationship

$$m = H/p_T = H'\bar{\rho}_L/p_T \tag{14-11}$$

where m is defined in terms of mole-fraction driving forces compatible with Eqs. (14-2) through (14-8), i.e., with the definitions of k_L, k_G, and K_G.

If it is desired to calculate the rate of transfer from the overall concentration difference based on bulk-liquid compositions $(x^\circ - x)$, the appropriate overall coefficient K_L is related to the individual coefficients by the equation

$$1/K_L = (1/k_L + 1/mk_G) \tag{14-12}$$

Conversion of these equations to a k'_G, k'_L basis can be accomplished readily by direct substitution of Eqs. (14-3) and (14-4).

Occasionally one will find k'_L or K'_L values reported in units (SI) of meters per second. The correct units for these values are kmol/[(s·m^2)(kmol/m^3)], and Eq. (14-4) is the correct equation for converting them to a mole-fraction basis.

When k'_G and K'_G values are reported in units (SI) of kmol/[(s·m^2)(kPa)], one must be careful in converting them to a mole-fraction basis to multiply by the total pressure actually employed in the original experiments and *not* by the total pressure of the system to be designed. This conversion is valid for systems in which Dalton's law of partial pressures $(p = yp_T)$ is valid.

Comparison of Eqs. (14-8) and (14-12) shows that for systems in which the equilibrium line is straight, the overall mass-transfer coefficients are related to each other by the equation

$$K_L = mK_G \tag{14-13}$$

When the **equilibrium curve is not straight,** there is no logical basis for the use of an overall transfer coefficient, since the value of m will be a function of position in the apparatus, as can be seen from Fig. 14-2. In such cases the rate of transfer must be calculated by solving for the interfacial compositions as described above.

Experimentally observed rates of mass transfer in gas-absorption equipment often are expressed in terms of overall transfer coefficients even when the equilibrium lines are curved. This procedure is purely empirical, since the theory indicates that in such cases the rates of transfer may not vary in direct proportion to the overall bulk concentration differences $(y - y^\circ)$ and $(x^\circ - x)$ at all concentration levels even though the rates may be proportional to the concentration difference in each phase taken separately, i.e., $(x_i - x)$ and $(y - y_i)$.

In most types of gas-absorption equipment such as packed or spray towers, the **interfacial area** that is effective for mass transfer cannot be accurately determined. For this reason it is customary to report experimentally observed rates of transfer in terms of transfer coefficients based on a unit volume of the apparatus rather than on a unit of interfacial area. Such volumetric coefficients are designated as K_Ga, k_La, etc., where a represents the interfacial area per unit volume of the apparatus. Experimentally observed variations in the values of these volumetric coefficients with variations in flow rates, type of packing, etc., may be due as much to changes in the effective value of a as to changes in k. Calculation of the overall coefficients from the individual volumetric coefficients is made by means of the equations

$$1/K_Ga = (1/k_Ga + m/k_La) \tag{14-14}$$

$$1/K_La = (1/k_La + 1/mk_Ga) \tag{14-15}$$

Because of the wide variation in the solubilities of gases in liquids, the variation in the values of m from one system to another can have an important effect on the selection of the type of equipment to use. For example, if it is desired to dissolve a sparingly soluble gas such as oxygen in water, the large value of m for this system would cause the liquid-phase part of the overall resistance to be extremely large in a spray tower where k_L might be relatively small. This kind of reasoning must be applied with caution, however, since gases with different solubilities ordinarily are absorbed under differing conditions of operation. Thus, the effect of solubility changes on the overall resistance to mass transfer may partly be counterbalanced by changes in the individual specific resistances as the flow rates are changed.

Mass-Transfer Principles: Concentrated Systems When solute concentrations in the gas and/or liquid phases are large, the equations derived above for dilute systems no longer are applicable. The correct equations to use for concentrated systems are as follows:

$$\begin{aligned} N_A &= \hat{k}_G(y - y_i)/y_{BM} = \hat{k}_L(x_i - x)/x_{BM} \\ &= \hat{K}_G(y - y^\circ)/y^\circ_{BM} = \hat{K}_L(x^\circ - x)/x^\circ_{BM} \end{aligned} \tag{14-16}$$

where

$$y_{BM} = \frac{(1 - y) - (1 - y_i)}{\ln[(1 - y)/(1 - y_i)]} \tag{14-17}$$

$$y^\circ_{BM} = \frac{(1 - y) - (1 - y^\circ)}{\ln[(1 - y)/(1 - y^\circ)]} \tag{14-18}$$

$$x_{BM} = \frac{(1 - x) - (1 - x_i)}{\ln[(1 - x)/(1 - x_i)]} \tag{14-19}$$

$$x^\circ_{BM} = \frac{(1 - x) - (1 - x^\circ)}{\ln[(1 - x)/(1 - x^\circ)]} \tag{14-20}$$

and where $\hat{k}_G$ and $\hat{k}_L$ are the gas-phase and liquid-phase mass-transfer coefficients for concentrated systems and $\hat{K}_G$ and $\hat{K}_L$ are the overall gas-phase and liquid-phase mass-transfer coefficients for concentrated systems. These coefficients are defined later in the subsection "Definitions of Mass-Transfer Coefficients $\hat{k}_G$ and $\hat{k}_L$."

The factors y_{BM} and x_{BM} arise from the fact that, in the diffusion of a solute through a second stationary layer of insoluble fluid, the resistance to diffusion varies in proportion to the concentration of the insoluble stationary fluid, approaching zero as the concentration of the insoluble fluid approaches zero.

The factors y°_{BM} and x°_{BM} cannot be justified on the basis of the kinetic theory of fluids since they are based on overall resistances.

These factors therefore are included in the equations purely by analogy with the corresponding film equations.

In dilute systems the logarithmic-mean insoluble-gas and nonvolatile-liquid concentrations approach unity, and Eq. (14-16) reduces to the dilute-system formula. In binary distillation, in which both components diffuse simultaneously, these log-mean factors should be omitted (this situation would exist in a reboiled absorber or stripper, for example).

Substitution of Eqs. (14-17) through (14-20) into Eq. (14-16) results in the following simplified formula:

$$\begin{aligned} N_A &= \hat{k}_G \ln [(1 - y_i)/(1 - y)] \\ &= \hat{K}_G \ln [(1 - y^\circ)/(1 - y)] \\ &= \hat{k}_L \ln [(1 - x)/(1 - x_i)] \\ &= \hat{K}_L \ln [(1 - x)/(1 - x^\circ)] \end{aligned} \tag{14-21}$$

Note that the units of $\hat{k}_G$, $\hat{K}_G$, $\hat{k}_L$, and $\hat{K}_L$ are all identical to each other, i.e., kmol/(s·m^2) in SI units.

The equation for computing the interfacial gas and liquid compositions in concentrated systems is

$$\begin{aligned} (y - y_i)/(x_i - x) &= \hat{k}_L y_{BM}/\hat{k}_G x_{BM} \\ &= L_M H_G y_{BM}/G_M H_L x_{BM} = k_L/k_G \end{aligned} \tag{14-22}$$

This equation is identical to the one for dilute systems since $\hat{k}_G = k_G y_{BM}$ and $\hat{k}_L = k_L x_{BM}$. Note, however, that when $\hat{k}_G$ and $\hat{k}_L$ are given, the equation must be solved by trial and error, since x_{BM} contains x_i and y_{BM} contains y_i.

The overall gas-phase and liquid-phase mass-transfer coefficients for concentrated systems are computed according to the following equations:

$$\frac{1}{\hat{K}_G} = \frac{y_{BM}}{y^\circ_{BM}} \frac{1}{\hat{k}_G} + \frac{x_{BM}}{y^\circ_{BM}} \frac{1}{\hat{k}_L} \left(\frac{y_i - y^\circ}{x_i - x} \right) \tag{14-23}$$

$$\frac{1}{\hat{K}_L} = \frac{x_{BM}}{x^\circ_{BM}} \frac{1}{\hat{k}_L} + \frac{y_{BM}}{x^\circ_{BM}} \frac{1}{\hat{k}_G} \left(\frac{x^\circ - x_i}{y - y_i} \right) \tag{14-24}$$

When the equilibrium curve is a straight line, the terms in parentheses can be replaced by the slope m as before. In this case the overall mass-transfer coefficients for concentrated systems are related to each other by the equation

$$\hat{K}_L = m\hat{K}_G(x^\circ_{BM}/y^\circ_{BM}) \tag{14-25}$$

All these equations reduce to their dilute-system equivalents as the inert concentrations approach unity in terms of mole fractions of inert concentrations in the fluids.

HTU (Height Equivalent to One Transfer Unit) Frequently the values of the individual coefficients of mass transfer are so strongly dependent on flow rates that the quantity obtained by dividing each coefficient by the flow rate of the phase to which it applies is more nearly constant than the coefficient itself. The quantity obtained by this procedure is called the height equivalent to one transfer unit, since it expresses in terms of a single length dimension the height of apparatus required to accomplish a separation of standard difficulty.

The following relations between the transfer coefficients and the values of HTU apply:

$$H_G = G_M/k_G a y_{BM} = G_M/\hat{k}_G a \tag{14-26}$$

$$H_{OG} = G_M/K_G a y^\circ_{BM} = G_M/\hat{K}_G a \tag{14-27}$$

$$H_L = L_M/k_L a x_{BM} = L_M/\hat{k}_L a \tag{14-28}$$

$$H_{OL} = L_M/K_L a x^\circ_{BM} = L_M/\hat{K}_L a \tag{14-29}$$

The equations that express the addition of individual resistances in terms of HTUs, applicable to either dilute or concentrated systems, are

$$H_{OG} = \frac{y_{BM}}{y^\circ_{BM}} H_G + \frac{mG_M}{L_M} \frac{x_{BM}}{y^\circ_{BM}} H_L \tag{14-30}$$

$$H_{OL} = \frac{x_{BM}}{x^\circ_{BM}} H_L + \frac{L_M}{mG_M} \frac{y_{BM}}{x^\circ_{BM}} H_G \tag{14-31}$$

These equations are strictly valid only when m, the slope of the equilibrium curve, is constant, as noted previously.

NTU (Number of Transfer Units) The NTU required for a given separation is closely related to the number of theoretical stages or plates required to carry out the same separation in a stagewise or plate-type apparatus. For equimolal counterdiffusion, such as in a binary distillation, the number of overall gas-phase transfer units N_{OG} required for changing the composition of the vapor stream from y_1 to y_2 is

$$N_{OG} = \int_{y_2}^{y_1} \frac{dy}{y - y^\circ} \tag{14-32}$$

When diffusion is in one direction only, as in the absorption of a soluble component from an insoluble gas,

$$N_{OG} = \int_{y_2}^{y_1} \frac{y^\circ_{BM}\, dy}{(1 - y)(y - y^\circ)} \tag{14-33}$$

The total height of packing required is then

$$h_T = H_{OG} N_{OG} \tag{14-34}$$

When it is known that H_{OG} varies appreciably within the tower, this term must be placed inside the integral in Eqs. (14-32) and (14-33) for accurate calculations of h_T. For example, the packed-tower design equation in terms of the overall gas-phase mass-transfer coefficient would then be expressed as follows:

$$h_T = \int_{y_2}^{y_1} \left[\frac{G_M}{K_G a y^\circ_{BM}} \right] \frac{y^\circ_{BM}\, dy}{(1 - y)(y - y^\circ)} \tag{14-35}$$

where the first term under the integral can be recognized as the HTU term. Convenient solutions of these equations for special cases are discussed later.

HETP (Height Equivalent to One Theoretical Plate) HETP is another quantity that is used occasionally to express the efficiency of a packing material for carrying out a separation. Experimental data should be reported as HTUs rather than as HETPs, since the former quantity is theoretically correct for equipment such as packed columns in which mass is transferred by a differential rather than a stagewise action. If the equilibrium and operating lines for a dilute system are parallel, i.e., $mG_M/L_M = 1$, HETPs and HTUs are equal. If the equilibrium and operating lines are straight but not parallel,

$$H_{OG}/\text{HETP} = [(mG_M/L_M) - 1]/[\ln (mG_M/L_M)] \tag{14-36}$$

A further discussion of the use of HETP is presented under "Packed-Tower Design."

Definitions of Mass-Transfer Coefficients $\hat{k}_G$ and $\hat{k}_L$ The mass-transfer coefficient is defined as the ratio of the molal mass flux N_A to the concentration driving force. This leads to many different ways of defining these coefficients. For example, gas-phase mass-transfer rates may be defined as

$$N_A = k_G(y - y_i) = k'_G(p - p_i) = \hat{k}_G(y - y_i)/y_{BM} \tag{14-37}$$

where the units (SI) of k_G are kmol/[(s·m^2)(mole fraction)], the units of k'_G are kmol/[(s·m^2)(kPa)], and the units of $\hat{k}_G$ are kmol/(s·m^2). These coefficients are related to each other as follows:

$$\hat{k}_G = k_G y_{BM} = k'_G p_T y_{BM} \tag{14-38}$$

where p_T is the total system pressure (it is assumed here that Dalton's law of partial pressures is valid).

In a similar way, liquid-phase mass-transfer rates may be defined by the relations

$$N_A = k_L(x_i - x) = k'_L(c_i - c) = \hat{k}_L(x_i - x)/x_{BM} \tag{14-39}$$

where the units (SI) of k_L are kmol/[(s·m^2)(mole fraction)], the units of k'_L are kmol/[(s·m^2)(kmol/m^3)] or meters per second, and the units of $\hat{k}_L$ are kmol/(s·m^2). These coefficients are related as follows:

$$\hat{k}_L = k_L x_{BM} = k'_L \bar{\rho}_L x_{BM} \tag{14-40}$$

where $\bar{\rho}_L$ is the molal density of the liquid phase in units (SI) of kilomoles per cubic meter. Note that, for dilute solutions where x_{BM}

$\doteq 1$, k_L and $\hat{k}_L$ will have identical numerical values. Similarly, for dilute gases $\hat{k}_G \doteq k_G$.

For the special case of steady-state unidirectional diffusion of a component through an inert-gas film in an ideal-gas system, the rate of mass transfer is derived as

$$N_A = \frac{D_{AB}p_T}{RT\,\delta_G}\frac{(y - y_i)}{y_{BM}} = \frac{D_{AB}p_T}{RT\,\delta_G}\ln\frac{1 - y_i}{1 - y} \qquad (14\text{-}41)$$

where D_{AB} = the diffusion coefficient or "diffusivity," δ_G = the "effective" thickness of a stagnant-gas layer which would offer a resistance to molecular diffusion equal to the experimentally observed resistance, and R = the gas constant.

The film thickness δ_G depends primarily on the hydrodynamics of the system and hence on the Reynolds number and the Schmidt number. Thus, various correlations have been developed for different geometries in terms of the following dimensionless variables:

$$N_{Sh} = \hat{k}_G RTd/D_{AB}p_T = f(N_{Re}, N_{Sc}) \qquad (14\text{-}42)$$

where N_{Sh} is the Sherwood number, N_{Re} ($= Gd/\mu_G$) is the Reynolds number based on the characteristic length d appropriate to the geometry of the particular system; and N_{Sc} ($= \mu_G/\rho_G D_{AB}$) is the Schmidt number.

According to this analysis one can see that for gas-absorption problems the most appropriate driving-force expression is of the form $(y - y_i)/y_{BM}$, and the most appropriate mass-transfer coefficient is therefore $\hat{k}_G$. This concept is to be found in all the key equations for the design of mass-transfer equipment.

The Sherwood-number relation for **gas-phase mass-transfer coefficients** as represented in Eq. (14-42) can be rearranged as follows:

$$N_{Sh} = (\hat{k}_G/G_M)N_{Re}N_{Sc} = N_{St}N_{Re}N_{Sc} = f(N_{Re}, N_{Sc}) \qquad (14\text{-}43)$$

where $N_{St} = \hat{k}_G/G_M = k'_G p_{BM}/G_M$ is known as the Stanton number. This equation can now be stated in the alternative functional form

$$N_{St} = \hat{k}_G/G_M = g(N_{Re}, N_{Sc}) \qquad (14\text{-}44)$$

A specific example of this function is the correlation developed by Shulman et al. [*Am. Inst. Chem. Eng. J.*, **1**, 253 (1955)] for gas-phase mass-transfer in Raschig-ring and Berl-saddle packings:

$$j_M = N_{St} \cdot N_{Sc}^{2/3} = 1.195 N_{Re}^{-0.36} \qquad (14\text{-}45)$$

where j_M is the Chilton-Colburn "j factor" for mass transfer (discussed later), N_{St} and N_{Sc} are as defined previously, N_{Re} is defined for these packings as $N_{Re} = Gd/\mu_G(1 - \varepsilon)$, d is the diameter of a sphere possessing the same surface area as a piece of packing, and ε is the void fraction available for gas flow.

The important point to note here is that the gas-phase mass-transfer coefficient $\hat{k}_G$ depends principally upon the transport properties of the fluid (N_{Sc}) and the hydrodynamics of the particular system involved (N_{Re}). It also is important to recognize that specific mass-transfer correlations such as the one given in Eq. (14-45) can be derived only in conjunction with the investigator's particular assumptions concerning the numerical values of the effective interfacial area a of the packing. Shulman's charts for effective interfacial area are discussed in Sec. 18, where similar charts for other packings and geometries also are to be found. Additional considerations concerning the interfacial-area factor are discussed later in this section.

For the **liquid-phase mass-transfer coefficient** Shulman et al. (op. cit.) reported the following function of the Schmidt number and the Reynolds number for the particular geometry involving Raschig rings and Berl saddles in conjunction with the assumptions discussed concerning the effective interfacial area a:

$$k'_L d/D_L = 25.1 N_{Re}^{0.45} N_{Sc}^{0.5} \qquad (14\text{-}46)$$

where $N_{Sc} = \mu_L/\rho_L D_L$; D_L is the diffusion coefficient of the solute in the liquid phase; and N_{Re} is defined as Ld/μ_L, where L is the superficial liquid rate. Equation (14-46) is of interest here primarily because it identifies the key design variables.

The left-hand side of Eq. (14-46) can be rearranged in terms of k_L and μ_L, where $k_L = k'_L \rho_L$, to obtain

$$k'_L d/D_L = (k_L d\overline{M}_L/\mu_L)N_{Sc} = (k_L/L_M)N_{Re}N_{Sc} \qquad (14\text{-}47)$$

where $\overline{M}_L$ = average molecular weight of the liquid phase. For the dilute solutions involved it is probable that $k_L = \hat{k}_L$, since $x_{BM} \doteq 1$. Thus, one can conclude that the appropriate definition of the Stanton number for liquid-phase mass transfer is

$$N_{St} = \hat{k}_L/L_M = k_L x_{BM}/L_M \qquad (14\text{-}48)$$

These relationships will be employed in developing an understanding of the relative effects of temperature and pressure on the mass-transfer coefficients and in describing the Chilton-Colburn analogy. A comprehensive discussion of various empirical correlations for mass-transfer coefficients can be found in Sec. 18.

Effects of Total Pressure on $\hat{k}_G$ and $\hat{k}_L$ The influence of total system pressure on the rate of mass transfer from a gas to a liquid or to a solid has been shown to be the same as would be predicted from stagnant-film theory as defined in Eq. (14-41), where

$$\hat{k}_G = D_{AB}p_T/RT\,\delta_G \qquad (14\text{-}49)$$

Since the quantity $D_{AB}p_T$ is known to be relatively independent of the pressure, it follows that the rate coefficients $\hat{k}_G$, $k_G y_{BM}$, and $k'_G p_T y_{BM}$ ($= k'_G p_{BM}$) do not depend on the total pressure of the system, subject to the limitations discussed later.

Investigators of tower packings normally report $k'_G a$ values measured at very low inlet-gas concentrations, so that $y_{BM} \doteq 1$, and at total pressures close to 100 kPa (1 atm). Thus, the correct rate coefficient for use in packed-tower designs involving the use of the driving force $(y - y_i)/y_{BM}$ is obtained by multiplying the reported $k'_G a$ values by the value of p_T employed in the actual test unit (e.g., 100 kPa) and *not* the total pressure of the system to be designed.

From another point of view one can correct the reported values of $k'_G a$ in kmol/[(s·m³)(kPa)], valid for a pressure of 101.3 kPa (1 atm), to some other pressure by dividing the quoted values of $k'_G a$ by the design pressure and multiplying by 101.3 kPa, i.e., ($k'_G a$ at design pressure p_T) = ($k'_G a$ at 1 atm) × $101.3/p_T$.

One way to avoid a lot of confusion on this point is to convert the experimentally measured $k'_G a$ values to values of $\hat{k}_G a$ straightaway, before beginning the design calculations. A design based on the rate coefficient $\hat{k}_G a$ and the driving force $(y - y_i)/y_{BM}$ will be independent of the total system pressure with the following limitations: caution should be employed in assuming that $\hat{k}_G a$ is independent of total pressure for systems having significant vapor-phase nonidealities, for systems that operate in the vicinity of the critical point where the reduced density (ρ/ρ_c) becomes larger than unity, or for total pressures higher than about 3040 to 4050 kPa (30 to 40 atm).

Experimental confirmations of the relative independence of $\hat{k}_G$ with respect to total pressure have been reported as follows: Gilliland and Sherwood [*Ind. Eng. Chem.*, **26**, 516 (1934)] studied the vaporization of water, aniline, and ethyl acetate over a range of total pressures from 15 to 311 kPa (0.15 to 3.1 atm) and found that for each of these three systems $\hat{k}_G$ was independent of the pressure. A University of California master's thesis (R. J. Fallat, "Effect of Pressure on Mass Transfer in the Gas Phase," March 1959) presents data for the vaporization of naphthalene into either air or helium gas at pressures ranging from 132 to 3830 kPa (1.3 to 37.8 atm), with essentially no effect upon $\hat{k}_G$. Bretsznajder (*Prediction of Transport and Other Physical Properties of Fluids*, Pergamon Press, Oxford, 1971, p. 343) discusses the effects of pressure on the $D_{AB}p_T$ product and presents experimental data on the self-diffusion of CO_2 which show that the D-p product begins to decrease at a pressure of approximately 8100 kPa (80 atm). For reduced temperatures higher than about 1.5, the deviations are relatively modest for pressures up to the critical pressure. However, deviations are large near the critical point.

For the liquid-phase mass-transfer coefficient $\hat{k}_L$, the effects of total system pressure can be ignored for all practical purposes. Thus, when using $\hat{k}_G$ and $\hat{k}_L$ for the design of gas absorbers or strippers, the primary pressure effects to consider will be those which affect the equilibrium curves and the values of m.

Effects of Temperature on $\hat{k}_G$ and $\hat{k}_L$ The Stanton-number relationship for gas-phase mass transfer,

$$N_{St} = \hat{k}_G/G_M = g(N_{Re}, N_{Sc}) \qquad (14\text{-}50)$$

indicates that for a given system geometry the rate coefficient $\hat{k}_G$ depends only on the Reynolds number and the Schmidt number. Since the Schmidt number for a gas is independent of temperature, the principal effect of temperature upon $\hat{k}_G$ arises from changes in the gas viscosity with changes in temperature. For normally encountered temperature ranges, these effects will be small owing to the fractional powers involved in Reynolds-number terms [e.g., Eq. (14-45)]. The effect of pressure on the gas-phase viscosity also is negligible for pressures below about 5060 kPa (50 atm). It thus can be concluded that for all practical purposes $\hat{k}_G$ is independent of temperature and pressure in the normal ranges of these variables.

For modest changes in temperature the influence of temperature upon the interfacial area a may be neglected. For example, in experiments on the absorption of SO_2 in water, Whitney and Vivian [*Chem. Eng. Prog.*, **45,** 323 (1949)] found no appreciable effect of temperature upon $\hat{k}_G a$ over the range from 10 to 50°C.

With regard to the liquid-phase mass-transfer coefficient, Whitney and Vivian found that the effect of temperature upon $k_L a$ could be explained entirely by variations in the liquid-phase viscosity and diffusion coefficient with temperature. Similarly, the oxygen-desorption data of Sherwood and Holloway [*Trans. Am. Inst. Chem. Eng.*, **36,** 39 (1940)] show that the influence of temperature upon H_L can be explained by the effects of temperature upon the liquid-phase viscosity and diffusion coefficients.

It is important to recognize that the effects of temperature on the liquid-phase diffusion coefficients and viscosities can be very large and therefore must be carefully accounted for when using $\hat{k}_L$ or H_L data. For liquids the mass-transfer coefficient $\hat{k}_L$ is correlated in terms of design variables by relations of the form

$$N_{St} = \hat{k}_L/L_M = f(N_{Re}, N_{Sc}) \tag{14-51}$$

of which one particular example involving Shulman's data for Raschig rings and Berl saddles [*Am. Inst. Chem. Eng. J.*, **1,** 253 (1955)] is given by

$$N_{St} = 25.1 N_{Re}^{-0.55} N_{Sc}^{-0.5} \tag{14-52}$$

where definitions of the Reynolds and Schmidt numbers and of the effective interfacial areas are the same as for Eq. (14-46). Relations of this kind are convenient to use in developing temperature-correction factors for $\hat{k}_L$ data as illustrated in the following example for correcting H_L data.

A general relation for H_L which may be used as the basis for applying temperature corrections is as follows:

$$H_L = b N_{Re}^{q} N_{Sc}^{1/2} \tag{14-53}$$

where b is a proportionality constant and the exponent q may range from about 0.2 to 0.5 for different packings and systems. The liquid-phase diffusion coefficients may be corrected from a base temperature T_1 to another temperature T_2 by using the Einstein relation as recommended by Wilke [*Chem. Eng. Prog.*, **45,** 218 (1949)]:

$$D_2 = D_1(T_2/T_1)(\mu_1/\mu_2) \tag{14-54}$$

The Einstein relation can be rearranged to the following equation for relating Schmidt numbers at two temperatures:

$$N_{Sc2} = N_{Sc1}(T_1/T_2)(\rho_1/\rho_2)(\mu_2/\mu_1)^2 \tag{14-55}$$

Substitution of this relation into Eq. (14-53) shows that for a given geometry the effect of temperature on H_L can be estimated as

$$H_{L2} = H_{L1}(T_1/T_2)^{1/2}(\rho_1/\rho_2)^{1/2}(\mu_2/\mu_1)^{1-q} \tag{14-56}$$

In using these relations it should be noted that for equal liquid flow rates

$$H_{L2}/H_{L1} = (\hat{k}_L a)_1/(\hat{k}_L a)_2 \tag{14-57}$$

Effects of System Physical Properties on $\hat{k}_G$ and $\hat{k}_L$ When designing packed towers for nonreacting gas-absorption systems for which no experimental data are available, it is necessary to make corrections for differences in composition between the existing test data and the system in question. For example, the test data of Fellinger for ammonia-water absorption on various packings are reported in Sec. 18. In these tests it is estimated that $H_G = 0.9 H_{OG}$, so that one may wish to use these data as the basis for estimating H_G or $\hat{k}_G a$ values for other systems. This may be done by taking H_G proportional to $N_{Sc}^{0.5}$ and $\hat{k}_G a$ proportional to $N_{Sc}^{-0.5}$, based on a value of N_{Sc} for NH_3-air of 0.66 at 25°C. The coefficient k_G varies as the diffusivity D_{AB} to the 0.5 power. It should be noted, however, that there is conflicting evidence concerning this exponent (⅔ versus ½) as discussed by Yadav and Sharma [*Chem. Eng. Sci.*, **34,** 1423 (1979)].

The existing data indicate that $\hat{k}_L a$ is proportional to the square root of the solute-diffusion coefficient, and since the interfacial area a does not depend on D_L, it follows that $\hat{k}_L$ is proportional to $D_L^{0.5}$. An analysis of the design variables involved indicates that $\hat{k}_L$ should be proportional to $N_{Sc}^{-0.5}$ when the Reynolds number is held constant.

According to the correlation reported by Onda et al. [*J. Chem. Eng. Japan*, **1,** 57 (1968)], the rate coefficient $\hat{k}_L$ is proportional to $\mu_L^{-1/6}$. It should be noted that the influence of substituting solvents of widely differing viscosities upon the interfacial area a can be very large, as discussed later under "The Effective Interfacial Mass-Transfer Area." One therefore should be cautious about extrapolating $\hat{k}_L a$ data to account for viscosity effects between different solvent systems.

Effects of High Solute Concentrations on $\hat{k}_G$ and $\hat{k}_L$ As discussed previously, the stagnant-film model indicates that $\hat{k}_G$ should be independent of y_{BM} and k_G should be inversely proportional to y_{BM}. Figure 14-3 shows the data of Vivian and Behrman [*Am. Inst. Chem. Eng. J.*, **11,** 656 (1965)] for the absorption of ammonia from an inert gas, which strongly suggest that the film model's predicted trend is correct. This is another indication that the most appropriate rate coefficient to use is $\hat{k}_G$ and the proper driving-force term is of the form $(y - y_i)/y_{BM}$.

The use of the rate coefficient $\hat{k}_L$ and the driving force $(x_i - x)/x_{BM}$ is believed to be appropriate, although it has not as yet been

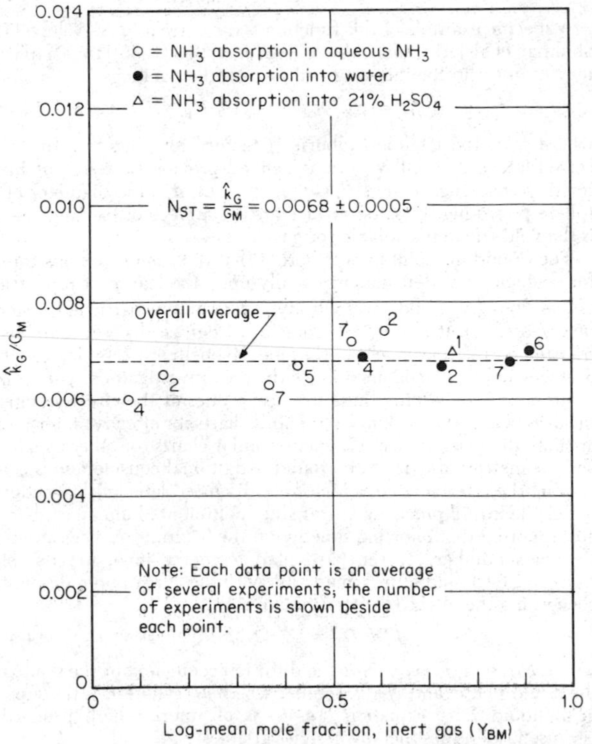

FIG. 14-3 Effect of inert-gas concentration on the rate of gas absorption in a short wetted-wall column. [*Data of Vivian and Behrman from MIT Sc.D. thesis, 1960; also Vivian and Behrman,* Am. Inst. Chem. Eng. J., *11, 656 (1965).*]

demonstrated experimentally. For many practical situations the liquid-phase solute concentrations are low, thus making this assumption unimportant.

Influence of Chemical Reactions on $\hat{k}_G$ and $\hat{k}_L$ When a gas is absorbed by a solvent or by a solution containing a substance with which the dissolving gas reacts, the rate of absorption may be influenced by the chemical reaction in the liquid as well as by the purely physical processes of diffusion and convection within the two phases. One therefore must consider the impacts of chemical equilibrium and reaction kinetics on the absorption rate in addition to accounting for the effects of gas solubility, diffusivity, and system hydrodynamics.

There is no sharp dividing line between pure physical absorption and absorption controlled by the rate of a chemical reaction. Most cases fall in an intermediate range in which the rate of absorption is limited both by the resistance to diffusion and by the finite velocity of the reaction. Even in these intermediate cases the equilibria between the various diffusing species involved in the reaction may affect the rate of absorption.

The gas-phase rate coefficient $\hat{k}_G$ is not affected by the fact that a chemical reaction is taking place in the liquid phase. If the liquid-phase chemical reaction is extremely fast and irreversible, the rate of absorption may be governed completely by the resistance to diffusion in the gas phase. In this case the absorption rate may be estimated by knowing only the gas-phase rate coefficient $\hat{k}_G$ or else the height of one gas-phase transfer unit $H_G = G_M/(\hat{k}_G a)$.

It should be noted that the highest possible absorption rates will occur under conditions in which the liquid-phase resistance is negligible and the equilibrium back pressure of the gas over the solvent is zero. Such situations would exist, for instance, for NH_3 absorption into an acid solution, for SO_2 absorption into an alkali solution, for vaporization of water into air, and for H_2S absorption from a dilute-gas stream into a strong alkali solution, provided there is a large excess of reagent in solution to consume all the dissolved gas. This is known as the **gas-phase mass-transfer limited** condition, when both the liquid-phase resistance and the back pressure of the gas equal zero. Even when the reaction is sufficiently reversible to allow a small back pressure, the absorption may be gas-phase-controlled, and the values of $\hat{k}_G$ and H_G that would apply to a physical-absorption process will govern the rate.

The liquid-phase rate coefficient $\hat{k}_L$ is strongly affected by fast chemical reactions and generally increases with increasing reaction rate. Indeed, the condition for zero liquid-phase resistance $(m/\hat{k}_L)$ implies that either the equlibrium back pressure is negligible, or that $\hat{k}_L$ is very large, or both. Frequently, even though reaction consumes the solute as it is dissolving, thereby enhancing both the mass-transfer coefficient and the driving force for absorption, the reaction rate is slow enough that the liquid-phase resistance must be taken into account. This may be due either to an insufficient supply of a second reagent or to an inherently slow chemical reaction.

In any event the value of $\hat{k}_L$ in the presence of a chemical reaction normally is larger than the value found when only physical absorption occurs, $\hat{k}_L^0$. This has led to the presentation of data on the effects of chemical reaction in terms of the "reaction factor" or "enhancement factor" defined as

$$\phi = \hat{k}_L/\hat{k}_L^0 \geq 1 \tag{14-58}$$

where $\hat{k}_L$ = mass-transfer coefficient with reaction and $\hat{k}_L^0$ = mass-transfer coefficient for pure physical absorption.

It is important to understand that when chemical reactions are involved, this definition of $\hat{k}_L$ is based on the driving force defined as the difference between the concentration of *unreacted* solute gas at the interface and in the bulk of the liquid. A coefficient based on the total of both unreacted and reacted gas could have values *smaller* than the physical-absorption mass-transfer coefficient $\hat{k}_L^0$.

When liquid-phase resistance is important, particular care should be taken in employing any given set of experimental data to ensure that the equilibrium data used conform with those employed by the original author in calculating values of $\hat{k}_L$ or H_L. Extrapolation to widely different concentration ranges or operating conditions should be made with caution, since the mass-transfer coefficient $\hat{k}_L$ may vary in an unexpected fashion, owing to changes in the apparent chemical-reaction mechanism.

Generalized prediction methods for $\hat{k}_L$ and H_L do not apply when chemical reaction occurs in the liquid phase, and therefore one must use actual operating data for the particular system in question. A discussion of the various factors to consider in designing gas absorbers and strippers when chemical reactions are involved is presented later in the subsection "Absorption with Chemical Reaction," wherein one will find that in some of the most important industrial systems (e.g., absorption of CO_2 in alkaline solutions) the liquid-phase resistance is controlling.

Effective Interfacial Mass-Transfer Area *a* In a packed tower of constant cross-sectional area S the differential change in solute flow per unit time is given by

$$-d(G_M S y) = N_A a\, dV = N_A a S\, dh \tag{14-59}$$

where a = interfacial area effective for mass transfer per unit of packed volume and V = packed volume. Owing to incomplete wetting of the packing surfaces and to the formation of areas of stagnation in the liquid film, the effective area normally is significantly less than the total external area of the packing pieces.

The effective interfacial area depends on a number of factors, as discussed in a review by Charpentier [*Chem. Eng. J.*, **11**, 161 (1976)]. Among these factors are (1) the shape and size of packing, (2) the packing material (for example, plastic generally gives smaller interfacial areas than either metal or ceramic), (3) the liquid mass velocity, and (4), for small-diameter towers, the column diameter.

Whereas the interfacial area generally increases with increasing liquid rate, it apparently is relatively independent of the superficial gas mass velocity below the flooding point. According to Charpentier's review, it appears valid to assume that the interfacial area is independent of the column height when specified in terms of unit packed volume (i.e., as a). Also, the existing data for chemically reacting gas-liquid systems (mostly aqueous electrolyte solutions) indicate that the interfacial area is independent of the chemical system. However, this situation may not hold true for systems involving large heats of reaction.

In a recent study Rizzuti et al. [*Chem. Eng. Sci.*, **36**, 973 (1981)] examined the influence of solvent viscosity upon the effective interfacial area in packed columns and concluded that for the systems studied the effective interfacial area a was proportional to the kinematic viscosity raised to the 0.7 power. Thus, the hydrodynamic behavior of a packed absorber is strongly affected by viscosity effects. Surface-tension effects also are important, as expressed in the work of Onda (see Sec. 18).

Detailed data on the effective interfacial area for mass transfer in packed towers are presented in Sec. 18. In using these data it must be understood that, in developing correlations for the mass-transfer coefficients $\hat{k}_G$ and $\hat{k}_L$, the various authors have assumed different but internally compatible correlations for the effective interfacial area a. It therefore would be inappropriate to mix the correlations of different authors unless it has been demonstrated that there is a valid area of overlap between them.

Volumetric Mass-Transfer Coefficients $\hat{K}_G a$ and $\hat{K}_L a$ Experimental determinations of the individual mass-transfer coefficients $\hat{k}_G$ and $\hat{k}_L$ and of the effective interfacial area a involve the use of extremely difficult techniques, and therefore such data are not plentiful. More often, experimental data are reported in terms of overall volumetric coefficients, which normally are defined as follows:

$$K'_G a = n_A/(h_T S p_T\, \Delta y^\circ_{1m}) \tag{14-60}$$

and

$$K_L a = n_A/(h_T S\, \Delta x^\circ_{1m}) \tag{14-61}$$

where $K'_G a$ = overall volumetric gas-phase mass-transfer coefficient, $K_L a$ = overall volumetric liquid-phase mass-transfer coefficient, n_A = overall rate of transfer of solute A, h_T = total packed depth in tower, S = tower cross-sectional area, p_T = total system pressure employed during the experiment, and Δx°_{1m} and Δy°_{1m} are defined as

$$\Delta y^\circ_{1\text{m}} = \frac{(y - y^\circ)_1 - (y - y^\circ)_2}{\ln [(y - y^\circ)_1/(y - y^\circ)_2]} \tag{14-62}$$

and

$$\Delta x^\circ_{1\text{m}} = \frac{(x^\circ - x)_2 - (x^\circ - x)_1}{\ln [(x^\circ - x)_2/(x^\circ - x)_1]} \tag{14-63}$$

where subscripts 1 and 2 refer to the bottom and top of the tower respectively.

Experimental $K'_G a$ and $K_L a$ data are available for most absorption and stripping operations of commercial interest. The solute concentrations employed in these experiments normally are very low, so that $K_L a \doteq \hat{K}_L a$ and $K'_G a p_T \doteq \hat{K}_G a$, where p_T is the total pressure employed in the actual experimental-test system. Unlike the individual gas-film coefficient $\hat{k}_G a$, the overall coefficient $\hat{K}_G a$ will vary with the total system pressure except when the liquid-phase resistance is negligible (i.e., when either $m = 0$, or $\hat{k}_L a$ is very large, or both).

Extrapolation of $K_G a$ data to conditions other than those for which the original measurements were made can be extremely risky, especially in systems involving chemical reactions in the liquid phase. One therefore would be wise to restrict the use of overall volumetric mass-transfer-coefficient data to conditions not too far removed from those employed in the actual tests. The most reliable data for this purpose would be those obtained from an operating commercial unit of similar design.

Chilton-Colburn Analogy When a fluid moves over either a liquid or a solid surface, the eddy motion that causes mass transfer also causes heat transfer and fluid friction owing to the transfer of thermal energy and momentum respectively. This close similarity among the mechanisms for the transfer of mass, heat, and momentum was brought out in the **Reynolds analogy,** which stated that the following dimensionless ratios are equal:

$$\hat{k}_G/G_M = h'/c_pG = f/2 \tag{14-64}$$

where h' = heat-transfer coefficient, c_p = specific heat, G = mass velocity, and f = friction factor.

Experimental data for mass transfer into gas streams agree approximately with Eq. (14-64) when the Schmidt number is close to unity and in smooth, straight tubes or along flat plates when the pressure drop is due entirely to skin friction against the surface. It does not, however, agree for cases involving "form" drag as well as skin friction. Also, it does not account for the mass-transfer resistance of the region of fluid near the liquid or solid boundary in which mass transfer occurs principally by molecular (as opposed to turbulent) motion.

Colburn [*Trans. Am. Inst. Chem. Eng.*, **29,** 174 (1933)] and Chilton and Colburn [*Ind. Eng. Chem.*, **26,** 1183 (1934)] showed empirically that the resistance of the laminar sublayer can be expressed by the following modification of the Reynolds analogy:

$$(\hat{k}_G/G_M)N_{\text{Sc}}^{2/3} = j_M = (h'/c_pG)N_{\text{Pr}}^{2/3} = j_H = f/2 \tag{14-65}$$

for turbulent flow through straight tubes and across plane surfaces, and

$$j_M = j_H \leq f/2 \tag{14-66}$$

for turbulent flow around cylinders, where j_M = mass-transfer factor, j_H = heat-transfer factor, $N_{\text{Pr}} = c_p\mu/k$ = Prandtl number, and k = thermal conductivity; other symbols are as defined earlier.

On occasion one will find that heat-transfer-rate data are available for a system in which mass-transfer-rate data are not readily available. The Chilton-Colburn analogy provides a procedure for developing estimates of the mass-transfer rates based on heat-transfer data. Extrapolation of experimental j_M or j_H data obtained with gases to predict liquid systems (and vice versa) should be approached with caution, however. When pressure-drop or friction-factor data are available, one may be able to place an upper bound on the rates of heat and mass transfer, according to Eq. (14-66).

Penetration Theory The stagnant-film model discussed earlier in this section assumes a steady state in which the local flux across each element of area is constant; i.e., there is no accumulation of the diffusing species within the film. In 1935 Higbie [*Trans. Am. Inst. Chem. Eng.*, **31,** 365 (1935)] pointed out that industrial contactors often operate with repeated brief contacts between phases in which the contact times are too short for the steady state to be achieved. For example, Higbie advanced the theory that in a packed tower the liquid flows across each packing piece in laminar flow and is remixed at the points of discontinuity between the packing elements. Thus, a fresh liquid surface is formed at the top of each piece, and as it moves downward, it absorbs gas at a decreasing rate until it is mixed at the next discontinuity.

If the velocity of the flowing stream is uniform over a very deep region of liquid (total thickness, $\delta_T \gg \sqrt{Dt}$), the time-averaged mass-transfer coefficient according to penetration theory is given by

$$k'_L = 2\sqrt{D_L/\pi t} \tag{14-67}$$

where k'_L = liquid-phase mass-transfer coefficient, D_L = liquid-phase diffusion coefficient, and t = contact time.

In practice, the contact time t is not known except in special cases in which the hydrodynamics are clearly defined. This is somewhat similar to the case of the stagnant-film theory in which the unknown quantity is the thickness of the stagnant layer δ (in film theory, the liquid-phase mass-transfer coefficient is given by $k'_L = D_L/\delta$).

The penetration theory predicts that k'_L should vary by the square root of the molecular diffusivity, as compared with film theory, which predicts a first-power dependency on D. Various investigators have reported experimental powers of D ranging from 0.5 to 0.75, and the Chilton-Colburn analogy suggests a ⅔ power.

Penetration theory often is used in analyzing absorption with chemical reaction because it makes no assumption about the depths of penetration of the various reacting species, and it gives a more accurate result when the diffusion coefficients of the reacting species are not equal. When the reaction process is very complex, however, penetration theory is more difficult to use than film theory, and the latter method normally is preferred.

Surface-Renewal Theory Danckwerts [*Ind. Eng. Chem.*, **42,** 1460 (1951)] proposed an extension of the penetration theory which allows for the eddy motion in the liquid to bring masses of fresh liquid continually from the interior to the surface, where they are exposed to the gas for finite lengths of time before being replaced. In his development, Danckwerts assumed that every element of fluid has an equal chance of being replaced regardless of its age. The Danckwerts model gives

$$k'_L = \sqrt{Ds} \tag{14-68}$$

where s = fractional rate of surface renewal.

Note that both the penetration and the surface-renewal theories predict a square-root dependency on D. Also, it should be recognized that values of the surface-renewal rate s generally are not available, which presents the same problems as do δ and t in the film and penetration models.

The predictions of correlations based on the film model often are nearly identical to predictions based on the penetration and surface-renewal models. Thus, in view of its relative simplicity, the film model normally is preferred for purposes of discussion or calculation. It should be noted that none of these theoretical models has proved adequate for making a priori predictions of absorption rates in packed towers, and therefore empirical correlations such as those discussed in Sec. 18 must be employed.

DESIGN OF GAS-ABSORPTION SYSTEMS

Outline of General Design Procedure The designer ordinarily is required to determine (1) the best solvent; (2) the best gas velocity through the absorber, i.e., the vessel diameter; (3) the height of the vessel and its internal members, e.g., the depth and type of packing or the number of trays; (4) the optimum rate of solvent circulation through the absorber and stripper; (5) the temperatures of streams entering and leaving the absorber and the quantity of heat to be removed to account for heat of solution and other heat effects; (6) the pressures at which the absorber and stripper will operate; and (7) the mechanical design of the absorption and stripping towers, including flow distributors, packing supports, etc. This section is concerned with all these choices except the last, which is discussed in Sec. 18.

The problem presented to the designer of a gas-absorption unit usually specifies the following quantities: (1) gas flow rate; (2) gas composition, at least with respect to the component to be absorbed; (3) operating pressure and allowable pressure drop across the absorber; (4) minimum degree of recovery of one or more solutes; and, possibly, (5) the solvent to be employed. Items 3, 4, and 5 may be subject to economic considerations and therefore are sometimes left up to the designer. For determining the number of variables that must be specified in order to fix a unique solution for the design of an absorber one can use the same phase-rule approach described in Sec. 13 for distillation systems.

Recovery of the solvent, sometimes by chemical means but more often by distillation, is almost always required, and the recovery system ordinarily is considered an integral part of the absorption-system process design. A more efficient solvent-stripping operation normally will result in a less costly absorber because of a smaller concentration of residual dissolved solute in the regenerated solvent; however, this may increase the overall cost of solvent recovery. A more detailed discussion of these and other economic considerations is presented later in this section.

Selection of Solvent When choice is possible, preference is given to liquids with high solubilities for the solute; a high solubility reduces the amount of solvent to be circulated. The solvent should be relatively nonvolatile, inexpensive, noncorrosive, stable, nonviscous, nonfoaming, and preferably nonflammable. Since the exit gas normally leaves saturated with solvent, solvent loss can be costly. Thus, low-cost solvents may be chosen over more expensive ones of higher solubility or lower volatility.

Water generally is used for gases fairly soluble in water, oils for light hydrocarbons, and special chemical solvents for acid gases such as CO_2, SO_2, and H_2S. Sometimes a reversible chemical reaction will result in a very high solubility and a minimum solvent rate. Data on actual systems are desirable when chemical reactions are involved, and those available are referenced later under "Absorption with Chemical Reaction."

Selection of Vapor-Liquid Equilibrium or Solubility Data Solubility values determine the liquid rate necessary for complete or economic solute recovery and so are essential to design. Equilibrium data generally will be found in one of three forms: (1) solubility data expressed either as solubility in weight or mole percent or as Henry's-law constants, (2) pure-component vapor pressures, or (3) equilibrium distribution coefficients (K values). Data for specific systems may be found in Sec. 3; additional references to sources of data are presented in this section.

In order to define completely the solubility of a gas in a liquid, it generally is necessary to state the temperature, the equilibrium partial pressure of the solute gas in the gas phase, and the concentration of the solute gas in the liquid phase. Strictly speaking, the total pressure on the system also should be stated, but for low total pressures, less than about 507 kPa (5 atm), the solubility for a particular partial pressure of solute gas normally will be relatively independent of the total pressure of the system.

Although quite useful when it can be applied, **Henry's law** as expressed in Eqs. (14-9) and (14-10) should be checked experimentally to determine the accuracy with which it can be used. If Henry's law holds, the solubility is defined by stating the value of the constant H (or H') along with the temperature and the solute partial pressure for which it is to be employed.

For quite a number of gases, Henry's law holds very well when the partial pressure of the solute is less than about 100 kPa (1 atm). For partial pressures of the solute gas greater than 100 kPa, H seldom is independent of the partial pressure of the solute gas, and a given value of H can be used over only a narrow range of partial pressures. There is a strongly nonlinear variation of Henry's-law constants with temperature as discussed by Schulze and Prausnitz [*Ind. Eng. Chem. Fundam.*, **20**, 175 (1981)]. Consultation of this reference is recommended before considering temperature extrapolations of Henry's-law data.

Additional data and information on the applicability of Henry's-law constants can be found in the references cited earlier in the subsection "Directory to Key Gas-Absorption Data." The use of Henry's-law constants is illustrated by the following examples.

Example 1 It is desired to find out how much hydrogen can be dissolved in 100 weights of water from a gas mixture when the total pressure is 101.3 kPa (760 torr; 1 atm), the partial pressure of the H_2 is 26.7 kPa (200 torr), and the temperature is 20°C. For partial pressures up to about 100 kPa the value of H is given in Sec. 3 as 6.92×10^6 kPa (6.83×10^4 atm) at 20°C. According to Henry's law,

$$x_{H_2} = p_{H_2}/H_{H_2} = 26.7/6.92 \times 10^6 = 3.86 \times 10^{-6}$$

The mole fraction x is the ratio of the number of moles of H_2 in solution to the total moles of all constituents contained. To calculate the weights of H_2 per 100 weights of H_2O, one can use the following formula, where the subscripts A and w correspond to the solute (hydrogen) and solvent (water):

$$\left(\frac{x_A}{1 - x_A}\right)\frac{M_A}{M_w}100 = \left(\frac{3.86 \times 10^{-6}}{1 - 3.86 \times 10^{-6}}\right)\frac{2.02}{18.02}100$$
$$= 4.33 \times 10^{-5} \text{ weights } H_2/100 \text{ weights } H_2O$$
$$= 0.43 \text{ parts per million weight}$$

Example 2 Oxygen is dissolved in water to the extent of 0.03 weight of O_2 per 100 weights of H_2O. What is the equilibrium partial pressure of O_2 over this solution at 25.9°C?

Solution. We assume a basis of 100 weights of H_2O. Then

$$x_A = (0.03/32)/(0.03/32 + 100/18) = 1.687 \times 10^{-4}$$
$$p_A = H_A x_A \text{ (Henry's law)}$$

A trial-and-error solution is indicated since the value of p_A must be known before the proper value of H_A can be selected. As a first approximation, let us select the value at 25.9°C and 106.7 kPa (800 torr) as shown in Sec. 3:

$$H_A = 4.85 \times 10^6 \text{ kPa } (4.79 \times 10^4 \text{ atm})$$
$$p_A = 4.85 \times 10^6 \times 1.687 \times 10^{-4} = 818.2 \text{ kPa (6137 torr)}$$

For the second approximation with 818.2 kPa assumed, the (interpolated) value of H_A is 5.056×10^6 kPa, and the corresponding value of p_A is 852.9 kPa (6398 torr). Using this new partial-pressure value for the next estimate yields an interpolated value of $H_A = 5.074 \times 10^6$ kPa. Thus, the result of the third approximation is

$$p_A = 5.074 \times 10^6 \times 1.687 \times 10^{-4} = 856.0 \text{ kPa (6421 torr)}$$

This third estimate appears to be close enough; thus, 0.03 weight of O_2 dissolved in 100 weights of H_2O at 25.9°C will exert a partial pressure of approximately 6420 torr (mmHg).

Pure-component vapor pressures can be used for predicting solubilities for systems in which **Raoult's law** is valid. For such systems $p_A = p_A^\circ x_A$, where p_A° is the pure-component vapor pressure of the solute and p_A is its partial pressure. Extreme care should be exercised when attempting to use pure-component vapor pressures to predict gas-absorption behavior. Both liquid-phase and vapor-phase nonidealities can cause significant deviations from the behavior predicted from pure-component vapor pressures in combination with Raoult's law. Vapor-pressure data are available in Sec. 3 for a variety of materials.

Whenever data are available for a given system under similar conditions of temperature, pressure, and composition, **equilibrium distribution coefficients** ($K = y/x$) provide a much more reliable tool for predicting vapor-liquid distributions. A detailed discussion of equilibrium K values is presented in Sec. 13.

Calculation of Liquid-to-Gas Ratio The minimum possible liquid rate is readily calculated from the composition of the entering gas and the solubility of the solute in the exit liquor, saturation being assumed. It may be necessary to estimate the temperature of the exit liquid based on the heat of solution of the solute gas. Values of latent and specific heats and values of heats of solution (at infinite dilution) are given in Sec. 3.

The actual liquid-to-gas ratio (solvent-circulation rate) normally will be greater than the minimum by as much as 25 to 100 percent and may be arrived at by economic considerations as well as by judgment and experience. For example, in some packed-tower applications involving very soluble gases or vacuum operation, the minimum quantity of solvent needed to dissolve the solute may be insufficient to keep the packing surface thoroughly wet, leading to poor distribution of the liquid stream.

Although there is no single flow rate at which a packing material suddenly becomes thoroughly wetted and below which the flow distribution is poor, Morris and Jackson (*Absorption Towers*, Butterworth, London, 1953) recommended a **minimum-wetting rate** (MWR) criterion,

$$MWR = V_L/a_p \tag{14-69}$$

where V_L = volumetric liquid velocity and a_p = external packing surface per unit volume (see Sec. 18 for data). When the net flow of solvent to the packed column is smaller than the MWR, it may be desirable to increase the recirculation of liquid over the packing even at the expense of a reduced mean driving force. An MWR of $2.2 \times 10^{-5} m^2/s$ (0.85 ft^2/h) was recommended for ring packings larger than 0.076 m (3 in) and for grids of pitch greater than 0.051 m (2 in). For other packings an MWR of $3.35 \times 10^{-5} m^2/s$ (1.3 ft^2/h) was recommended. Volumetric flow rates in the range of 0.0034 to 0.0068 $m^3/(s \cdot m^2)$ or 5 to 10 gal/(min·ft^2) are indicated by this criterion. Superficial liquid velocities are in the 0.5-cm/s (1-ft/min) range under these conditions.

For many problems an approximate value for the best liquid-to-gas ratio is satisfactory. For example, if the solute is so dilute that the temperature rise of the scrubbing liquid can be neglected, the value of the dimensionless ratio mG_M/L_M can be taken at about 0.7 for absorption in the dilute end of the tower or 1.4 for stripping. The gas rate is then set at about 50 percent of the estimated flooding velocity at the point of maximum flow in the tower.

In practice, the useful range of values of mG_M/L_M is quite narrow. Figure 14-4 shows that when one has specified the inlet- and outlet-gas concentrations y_1 and y_2, the inlet-solute composition x_2, and the slope m, the values of N_{OG} corresponding to a 95 percent approach to equilibrium between the entering liquid x_2 and the leaving gas y_2 in an absorber increase rapidly for values of mG_M/L_M larger than 0.9. Also, values lower than 0.5 may be impractical.

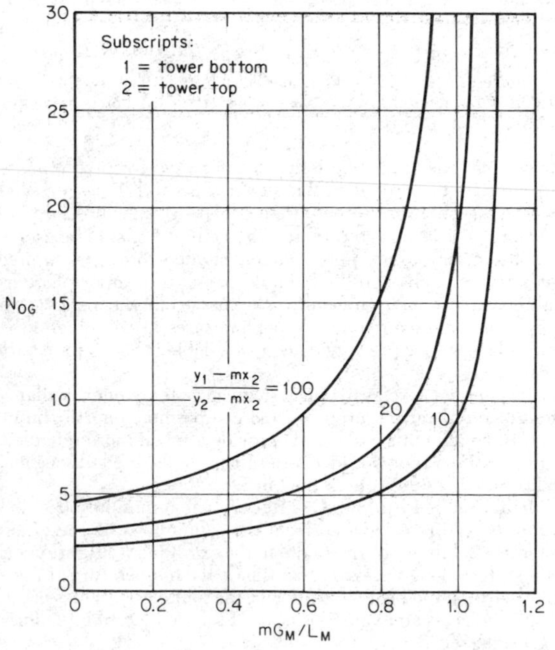

FIG. 14-4 Effect of mG_M/L_M on N_{OG} for recovery efficiencies of 90, 95, and 99 percent.

When the solute concentration in the inlet gas is low and when nearly all the solute is being absorbed (this is the usual case), the approximation

$$y_1 G_M \doteq x_1 L_M \doteq (y_1^\circ/m) L_M \tag{14-70}$$

leads to the conclusion that the ratio mG_M/L_M represents the fractional approach of the exit liquid to saturation with the inlet gas, i.e.,

$$mG_M/L_M \doteq y_1^\circ/y_1 \tag{14-71}$$

Optimization of the liquid-to-gas ratio in terms of total annual costs often suggests that the molar liquid-to-gas ratio L_M/G_M should be about 1.2 to 1.5 times the theoretical minimum corresponding to equilibrium at the rich end of the tower (infinite height), provided flooding is not a problem. This would be an alternative to assuming that $L_M/G_M \doteq m/0.7$, for example.

When the exit-liquor temperature rises owing to the heat of absorption of the solute, the value of m changes through the tower, and the liquid-to-gas ratio must be chosen to give reasonable values of m_1G_M/L_M and m_2G_M/L_M, where the subscripts 1 and 2 refer to the bottom and top of the absorption tower respectively. For this case the value of m_2G_M/L_M will be taken to be somewhat less than 0.7, so that the value of m_1G_M/L_M will not approach unity too closely. This rule-of-thumb approach is useful only when low solute concentrations and mild heat effects are involved.

When the solute has a large heat of solution or when the feed gas contains high percentages of the solute, one should consider the use of internal cooling coils or intermediate external heat exchangers in a plate-type tower to remove the heat of absorption. In a packed tower, one could consider the use of multiple packed sections with intermediate liquid-withdrawal points so that the liquid could be cooled by external heat exchange.

Selection of Equipment Packed columns usually are chosen for very corrosive materials, for liquids that foam badly, for either small- or large-diameter towers involving very low allowable pressure drops, and for small-scale operations requiring diameters of less than 0.6 m (2 ft). The type of packing is selected on the basis of resistance to corrosion, mechanical strength, capacity for handling the required flows, mass-transfer efficiency, and cost. A detailed discussion of the properties and costs of various tower packings is presented in Sec. 18. Economic factors are discussed later in this section.

Plate columns may be economically preferable for large-scale operations and are needed when liquid rates are so low that packing would be inadequately wetted, when the gas velocity is so low (owing to a very high L/G) that axial dispersion or "pumping" of the gas back down the (packed) column can occur, or when intermediate cooling is desired. Also, plate towers have a better turndown ratio and are less subject to fouling by solids than are packed towers. Section 18 should be consulted for additional details on the operating characteristics of plate towers.

Column Diameter and Pressure Drop For packed towers, flooding determines the minimum possible diameter, and the usual design is for 50 to 70 percent of the flooding velocity. Pressure drop at flooding for commonly used packings is around 0.167 m H_2O/m packing (2 in H_2O/ft packing). For operation at about 50 percent of flooding the pressure drop is roughly 0.042 m H_2O/m (0.5 in H_2O/ft). At 70 percent of flooding the pressure drop will be about 0.083 m H_2O/m (1.0 in H_2O/ft). Packed towers should not be designed to operate at pressure drops larger than 1.0 in H_2O/ft since relatively small increases above the design gas feed rate could precipitate flooding conditions within the tower. For systems having a significant foaming tendency, the maximum allowable design pressure drops should be lower than those discussed earlier. These values also are convenient to keep in mind for operating control.

The safe range of operating velocities normally will be close to the velocity one would derive from economic considerations, as discussed later. Detailed procedures for estimating flooding velocities and pressure drops are presented in Sec. 18.

For plate towers the minimum possible diameter may be dictated either by flooding or by excessive liquid entrainment. For very high liquid-to-gas ratios the capacity of the trays and downcomers to handle the liquid flow may set the column diameter. When the gas velocity is too high, entrainment can impair efficiency and/or cause flooding. Vapor velocities in plate towers sometimes are established so as to limit the entrainment to less than 10 percent of the liquid flow rate, for example. Detailed procedures for calculating pressure drops in plate towers are given in Sec. 18.

Computation of Tower Height The required height of a gas-absorption or stripping tower depends on (1) the thermodynamic equilibria involved, (2) the specified degree of removal of the solute from the gas, and (3) the mass-transfer efficiency of the apparatus. These same considerations apply both to plate towers and to packed towers. Items 1 and 2 dictate the required number of theoretical stages (plate tower) or transfer units (packed tower). Item 3 is derived from the tray efficiency and spacing (plate tower) or from the height of one transfer unit (packed tower). Solute-removal specifications normally are derived from economic considerations.

For **plate towers,** the approximate design methods described below may be used in estimating the number of theoretical stages, and the tray efficiencies and spacings for the tower can be specified on the basis of the information given in Sec. 18. Considerations involved in the rigorous design of plate towers are treated in Sec. 13.

For **packed towers,** the continuous differential nature of the contact between gas and liquid leads to a design procedure involving the solution of differential equations, as described in the next subsection.

It should be noted that the design procedures discussed in this section are not applicable to **reboiled absorbers,** which should be designed according to the methods described in Sec. 13.

Caution is advised in distinguishing between systems involving pure physical absorption and those in which a **chemical reaction** can significantly affect design procedures.

Selection of Stripper-Operating Conditions Stripping involves the removal of one or more volatile components from a liquid by contacting it with an inert gas such as steam or nitrogen. The operating conditions chosen for stripping normally result in a low solubility of the solute (i.e., a high value of m), so that the ratio mG_M/L_M will be larger than unity. A value of 1.4 may be used for rule-of thumb calculations involving pure physical desorption. For plate-tower calculations the stripping factor $S = KG_M/L_M$, where $K = y^\circ/x$, usually is specified for each tray.

When the solvent from an absorption operation must be regenerated for recycling back to the absorber, one may employ either a "pressure-swing concept," a "temperature-swing concept," or a combination of both in specifying stripping conditions. In pressure-swing operation the temperature of the stripper is about the same as that of the absorber, but the stripping pressure is much lower. In temperature-swing operation the pressures are about equal, but the stripping temperature is much higher than the absorption temperature.

In pressure-swing operation a portion of the dissolved gas may be "sprung" from the liquid by the use of a flash drum upstream of the stripping-tower feed point. This type of operation is discussed by Burrows and Preece [*Trans. Inst. Chem. Eng.*, **32**, 99 (1954)] and by Langley and Haselden [*Inst. Chem. Eng. Symp. Ser. (London)*, no. 28 (1968)]. If the flashing of the feed liquid takes place inside the stripping tower, this effect must be accounted for in the design of the upper section in order to avoid overloading and flooding near the top of the tower.

More often than not the rate at which residual absorbed gas can be driven from the liquid in a stripping tower is limited by the rate of a chemical reaction, in which case the liquid-phase residence time (and hence, the tower liquid holdup) becomes the most important design factor. Thus, many stripper-regenerators are designed on the basis of liquid holdup rather than on the basis of either K_La or HETP.

Approximate design equations applicable only to the case of pure physical desorption are developed later in this section for both packed and plate stripping towers. A more rigorous approach using distillation concepts may be found in Sec. 13. A brief discussion of desorption with chemical reaction is given in the subsection "Absorption with Chemical Reaction."

Design of Absorber-Stripper Systems The solute-rich liquor leaving a gas absorber normally is distilled or stripped to regenerate the solvent for recirculation back to the absorber, as depicted in Fig. 14-5. It is apparent that the conditions selected for the absorption step (e.g., temperature, pressure, L_M/G_M) will affect the design of the stripping tower, and, conversely, a selection of stripping condi-

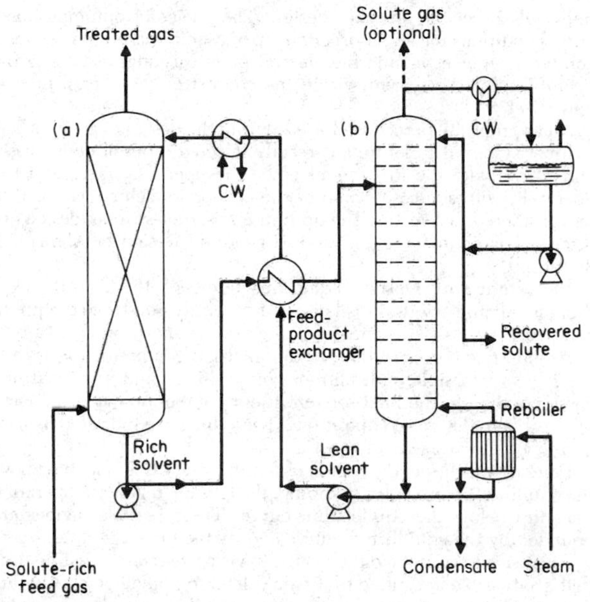

FIG. 14-5 Gas absorber using a solvent regenerated by stripping. (*a*) Absorber. (*b*) Stripper.

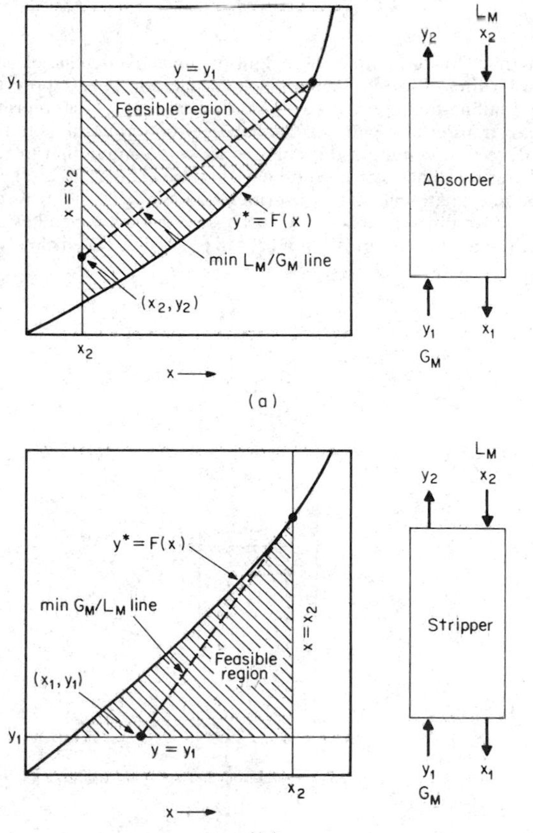

FIG. 14-6 Design diagrams for (*a*) absorption and (*b*) stripping.

tions will affect the absorber design. The choice of optimum operating conditions for an absorber-stripper system therefore involves a combination of economic factors and practical judgments as to the operability of the system within the context of the overall process flow sheet.

An appropriate procedure for executing the design of an absorber-stripper system is to set up a carefully selected series of design cases and to evaluate the investment costs, the operating costs, and the operability of each case. Some of the economic factors that need to be considered in selecting the optimum absorber-stripper design are discussed later in the subsection "Economic Design of Absorption Systems."

Importance of Design Diagrams One of the first things a designer should try to do is lay out a carefully constructed equilibrium curve, $y^\circ = F(x)$, on an xy diagram, as shown in Fig. 14-6. A horizontal line corresponding to the inlet-gas composition y_1 is then the locus of feasible outlet-liquor compositions, and a vertical line corresponding to the inlet-solvent-liquor composition x_2 is the locus of feasible outlet-gas compositions. These lines are indicated as $y = y_1$ and $x = x_2$ respectively on Fig. 14-6.

For gas absorption, the region of feasible operating lines lies above the equilibrium curve; for stripping, the feasible region for operating lines lies below the equilibrium curve. These feasible regions are bounded by the equilibrium curve and by the lines $x = x_2$ and $y = y_1$. By inspection, one should be able to visualize those operating lines that are feasible and those that would lead to "pinch points" within the tower. Also, it is possible to determine if a particular proposed design for solute recovery falls within the feasible envelope.

Once the design recovery for an absorber has been established, the operating curve can be constructed by first locating the point x_2, y_2 on the diagram. The intersection of the horizontal line corresponding to the inlet gas composition y_1 with the equilibrium curve $y^\circ = F(x)$ defines the theoretical minimum liquid-to-gas ratio for systems in which there are no intermediate pinch points. The operating line which connects this point with the point x_2, y_2 corresponds to the minimum value of L_M/G_M. The actual design value of L_M/G_M normally should be around 1.2 to 1.5 times this minimum. Thus, the actual design operating line for a gas absorber will pass through the point x_2, y_2 and will intersect the line $y = y_1$ to the left of the equilibrium curve.

For stripping one begins by using the design specification to locate the point x_1, y_1. Then the intersection of the vertical line $x = x_2$ with the equilibrium curve $y^\circ = F(x)$ defines the theoretical minimum gas-to-liquid ratio. The actual value of G_M/L_M is chosen to be about 20 to 50 percent higher than this minimum, so the actual design operating line will intersect the line $x = x_2$ at a point somewhat below the equilibrium curve.

Design diagrams minimize the possibility of making careless mistakes and allow one to assess easily the effects of operating variable changes on the operability of the system relative to pinch points, etc. Whenever analytical calculations or computer programs are being used for the design of gas-absorption systems, the construction of design diagrams based either on calculation results or on computer printouts may reveal problem areas or even errors in the design concept. It is strongly recommended that design diagrams be employed whenever possible.

PACKED-TOWER DESIGN

Methods for estimating the height of the active section of **counterflow differential contactors** such as packed towers, spray towers, and falling-film absorbers are based on rate expressions representing mass transfer at a point on the gas-liquid interface and on material balances representing the changes in bulk composition in the two phases that flow past each other. The rate expressions are based on the interphase mass-transfer principles described earlier in this section. Combination of such expressions leads to an integral expression for the number of transfer units or to equations related closely to the number of theoretical plates. The paragraphs which follow set forth convenient methods for using such equations, first in a general case and then for cases in which simplifying assumptions are valid.

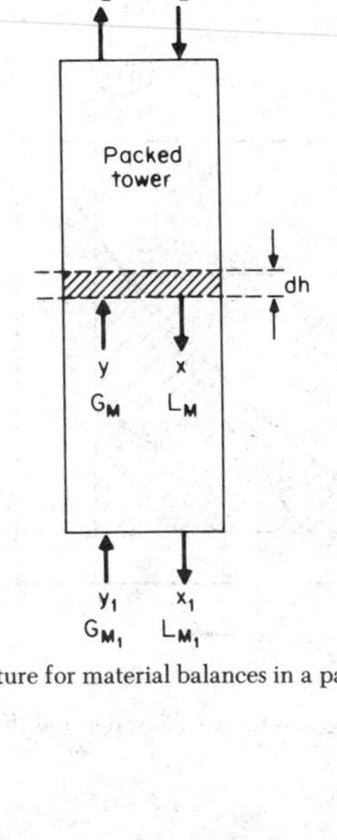

FIG. 14-7 Nomenclature for material balances in a packed-tower absorber or stripper.

Use of Mass-Transfer-Rate Expression Figure 14-7 shows a section of a packed absorption tower together with the nomenclature that will be used in developing the equations which follow. In a differential section dh, we can equate the rate at which solute is lost from the gas phase to the rate at which it is transferred through the gas phase to the interface as follows:

$$-d(G_M y) = -G_M\,dy - y\,dG_M = N_A a\,dh \tag{14-72}$$

When only one component is transferred,

$$dG_M = -N_A a\,dh \tag{14-73}$$

Substitution of this relation into Eq. (14-72) and rearranging yields

$$dh = -\frac{G_M\,dy}{N_A a(1-y)} \tag{14-74}$$

For this derivation we use the gas-phase rate expression $N_A = k_G(y - y_i)$ and integrate over the tower to obtain

$$h_T = \int_{y2}^{y1} \frac{G_M\,dy}{k_G a(1-y)(y-y_i)} \tag{14-75}$$

Multiplying and dividing by y_{BM} place Eq. (14-75) into the $H_G N_G$ format

$$h_T = \int_{y2}^{y1} \left[\frac{G_M}{k_G a y_{BM}}\right] \frac{y_{BM}\,dy}{(1-y)(y-y_i)} \tag{14-76}$$

$$= H_{G,\mathrm{av}} \int_{y2}^{y1} \frac{y_{BM}\,dy}{(1-y)(y-y_i)} = H_{G,\mathrm{av}} N_G$$

The general expression given by Eq. (14-75) is more complex than normally is required, but it must be used when the mass-transfer coefficient varies from point to point, as may be the case when the gas is not dilute or when the gas velocity varies as the gas dissolves. The values of y_i to be used in Eq. (14-75) depend on the local liquid composition x_i and on the temperature. This dependency is best represented by using the operating and equilibrium lines as discussed later.

Example 3 illustrates the use of Eq. (14-75) for scrubbing chlorine from air with aqueous caustic solution. For this case one can make the simplifying assumption that y_i, the interfacial partial pressure of chlorine over the aqueous caustic solution, is zero owing to the rapid and complete reaction of the chlorine after it dissolves. We note that the feed gas is not dilute.

Example 3 Let us compute the height of packing needed to reduce the chlorine concentration of 0.537 kg/(s·m²), or 396 lb/(h·ft²), of a chlorine-air mixture containing 0.503 mole-fraction chlorine to 0.0403 mole fraction. On the basis of test data described by Sherwood and Pigford (*Absorption and Extraction*, McGraw-Hill, 1952, p. 121) the value of $k_G a y_{BM}$ at a gas velocity equal to that at the bottom of the packing is equal to 0.1175 kmol/(s·m³), or 26.4 lb·mol/(h·ft³). The equilibrium back pressure y_i can be assumed to be negligible.

Solution. By assuming that the mass-transfer coefficient varies as the 0.8 power of the local gas mass velocity, we can derive the following relation:

$$\hat{K}_G a = k_G a y_{BM} = 0.1175\left[\frac{71y + 29(1-y)}{71y_1 + 29(1-y_1)}\left(\frac{1-y_1}{1-y}\right)\right]^{0.8}$$

where 71 and 29 are the molecular weights of chlorine and air respectively. Noting that the inert-gas (air) flow rate is given by $G_M' = G_M(1-y) = 5.34 \times 10^{-3}$ kmol/(s·m²), or 3.94 lb·mol/(h·ft²), and introducing these expressions into the integral gives

$$h_T = 1.82\int_{0.0403}^{0.503}\left[\frac{1-y}{29+42y}\right]^{0.8}\frac{dy}{(1-y)^2\ln[1/(1-y)]}$$

This definite integral can be evaluated numerically by the use of Simpson's rule to obtain $h_T = 0.305$ m (1 ft).

Use of Operating Curve Frequently, it is not possible to assume that $y_i = 0$ as in Example 3, owing to diffusional resistance in the liquid phase or to the accumulation of solute in the liquid stream. When the back pressure cannot be neglected, it is necessary to supplement the equations with a material balance representing the operating line or curve. In view of the countercurrent flows into and from the differential section of packing shown in Fig. 14-7, a steady-state material balance leads to the following equivalent relations:

$$d(G_M y) = d(L_M x) \tag{14-77}$$

$$G_M'\frac{dy}{(1-y)^2} = L_M'\frac{dx}{(1-x)^2} \tag{14-78}$$

where L_M' = molar mass velocity of the inert-liquid component and G_M' = molar mass velocity of the inert gas. L_M, L_M', G_M, and G_M' are superficial velocities based on the total tower cross section.

Equation (14-78) is the differential equation of the operating curve, and its integral around the upper portion of the packing is the equation for the operating curve

$$G_M'\left[\frac{y}{1-y} - \frac{y_2}{1-y_2}\right] = L_M'\left[\frac{x}{1-x} - \frac{x_2}{1-x_2}\right] \tag{14-79}$$

For dilute solutions in which the mole fractions of x and y are small, the total molar flows G_M and L_M will be very nearly constant, and the operating-curve equation is

$$G_M(y - y_2) = L_M(x - x_2) \tag{14-80}$$

This equation gives the relation between the bulk compositions of the gas and liquid streams at each level in the tower for conditions in which the operating curve can be approximated by a straight line.

Figure 14-8 shows the relationship between the operating curve and the equilibrium curve $y_i = F(x_i)$ for a typical example involving solvent recovery, where y_i and x_i are the interfacial compositions (assumed to be in equilibrium). Once y is known as a function of x along the operating curve, y_i can be found at corresponding points on the equilibrium curve by Eq. (14-5). Thence the integral in Eq. (14-75) can be evaluated.

Calculation of Transfer Units In the general case the equations described above must be employed in calculating the height of packing required for a given separation. However, if the local mass-transfer coefficient $k_G a y_{BM}$ is approximately proportional to the first power of the local gas velocity G_M, then the height of one gas-phase transfer unit, defined as $H_G = G_M/k_G a y_{BM}$, will be constant in Eq. (14-76). Similar considerations lead to an assumption that the height of one overall gas-phase transfer unit H_{OG} may be taken as constant. The height of packing required is then calculated according to the relation

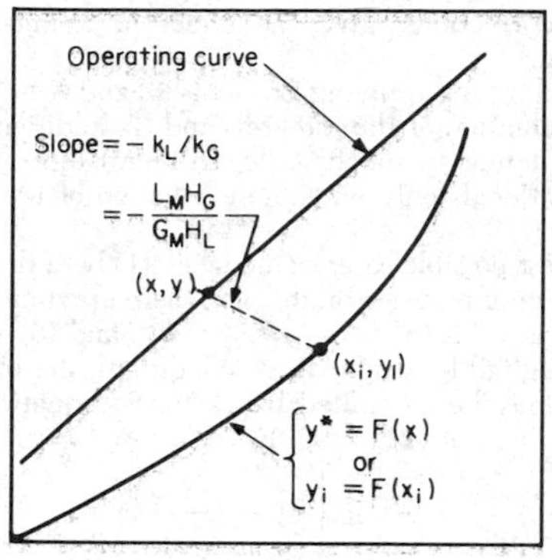

FIG. 14-8 Relationship between equilibrium curve and operating curve in a packed absorber; computation of interfacial compositions.

$$h_T = H_G N_G = H_{OG} N_{OG} \tag{14-81}$$

where N_G = number of gas-phase transfer units and N_{OG} = number of overall gas-phase transfer units. When H_G and H_{OG} are not constant, it may be valid to employ averaged values between the top and bottom of the tower and the relation

$$h_T = H_{G,av} N_G = H_{OG,av} N_{OG} \tag{14-82}$$

In these equations, the terms N_G and N_{OG} are defined by

$$N_G = \int_{y_2}^{y_1}\frac{y_{BM}\,dy}{(1-y)(y-y_i)} \tag{14-83}$$

and by

$$N_{OG} = \int_{y_2}^{y_1}\frac{y_{BM}^\circ\,dy}{(1-y)(y-y^\circ)} \tag{14-84}$$

respectively.

Equation (14-84) is the more useful one in practice: it requires either actual experimental H_{OG} data or values estimated by combining individual measurements of H_G and H_L by Eq. (14-30). Correlations for predicting H_G, H_L, and H_{OG} in nonreacting systems are presented in Sec. 18.

On occasion the changes in gas flow and in the mole fraction of inert gas are so small that the inclusion of terms such as $(1-y)$ and y_{BM}° can be neglected or at least can be included in an approximate way. This leads to some of the simplified procedures described later.

One such simplification was suggested by Wiegand [*Trans. Am. Inst. Chem. Eng.*, **36**, 679 (1940)], who pointed out that the logarithmic-mean mole fraction of inert gas y_{BM}° (or y_{BM}) is often very nearly equal to the arithmetic mean. Thus, substitution of the relation

$$\frac{y_{BM}^\circ}{(1-y)} \doteq \frac{(1-y^\circ)+(1-y)}{2(1-y)} = \frac{y-y^\circ}{2(1-y)} + 1 \tag{14-85}$$

into the equations presented earlier leads to the simplified forms

$$N_G = \frac{1}{2}\ln\frac{1-y_2}{1-y_1} + \int_{y_2}^{y_1}\frac{dy}{y-y_i} \tag{14-86}$$

$$N_{OG} = \frac{1}{2}\ln\frac{1-y_2}{1-y_1} + \int_{y_2}^{y_1}\frac{dy}{y-y^\circ} \tag{14-87}$$

The second (integral) terms represent the numbers of transfer units for an infinitely dilute gas. The first terms, frequently amounting to only small corrections, give the effect of a finite level of gas concentration.

The procedure for applying Eqs. (14-86) and (14-87) involves **two steps:** (1) evaluation of the integrals and (2) addition of the correction corresponding to the first (logarithmic) term. The discussion which follows deals only with the evaluation of the integral terms **(first step).**

The simplest possible case occurs when (1) both the operating and the equilibrium lines are straight (i.e., there are dilute solutions), (2) Henry's law is valid ($y^\circ/x = y_i/x_i = m$), and (3) absorption heat effects are negligible. Under these conditions, the **integral term** in Eq. (14-87) may be computed by Colburn's equation [*Trans. Am. Inst. Chem. Eng.*, **35**, 211 (1939)]:

$$N_{OG} = \frac{1}{1 - (mG_M/L_M)} \ln \left[\left(1 - \frac{mG_M}{L_M}\right) \left(\frac{y_1 - mx_2}{y_2 - mx_2}\right) + \frac{mG_M}{L_M} \right] \tag{14-88}$$

Figure 14-9 is a plot of Eq. (14-88) from which the value of N_{OG} can be read directly as a function of mG_M/L_M and the ratio of concentrations. This plot and Eq. (14-88) are equivalent to the use of a logarithmic mean of terminal driving forces, but they are more convenient because one does not need to compute the exit-liquor concentration x_1.

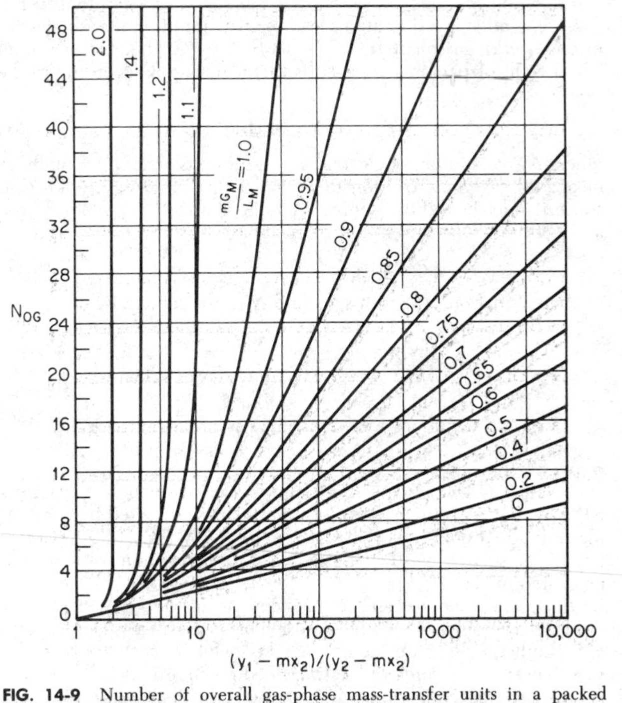

FIG. 14-9 Number of overall gas-phase mass-transfer units in a packed absorption tower for constant mG_M/L_M; solution of Eq. (14-88). (*From Sherwood and Pigford,* Absorption and Extraction, *McGraw-Hill, New York, 1952.*)

In many practical situations involving nearly complete cleanup of the gas, an approximate result can be obtained from the equations just presented even when solutions are concentrated or when absorption heat effects are present. In such cases the driving forces in the upper part of the tower are very much smaller than those at the bottom, and the value of mG_M/L_M used in the equations should be the ratio of the slopes of the equilibrium line m and the operating line L_M/G_M in the low-concentration range near the top of the tower.

Another approach is to divide the tower arbitrarily into a lean section (near the top), where approximate methods are valid, and to deal with the rich section separately. If the heat effects in the rich section are appreciable, consideration could be given to installing cooling units near the bottom of the tower. In any event a design diagram showing the operating and equilibrium curves should be prepared to check on the applicability of any simplified procedure. Figure 14-11, presented in Example 7, is one such diagram for an adiabatic absorption tower.

Stripping Equations Stripping, or desorption, involves the removal of a volatile component from the liquid stream by contact with an inert gas such as nitrogen or steam. In this case the change in concentration of the liquid stream is of prime importance, and it is more convenient to formulate the rate equation analogous to Eq. (14-75) in terms of the liquid composition x. This leads to the following equations defining numbers of transfer units and heights of transfer units based on liquid-phase resistance:

$$h_T = H_L \int_{x2}^{x1} \frac{x_{BM}\, dx}{(1 - x)(x_i - x)} = H_L N_L \tag{14-89}$$

$$h_T = H_{OL} \int_{x2}^{x1} \frac{x^\circ_{BM}\, dx}{(1 - x)(x^\circ - x)} = H_{OL} N_{OL} \tag{14-90}$$

where, as before, subscripts 1 and 2 refer to the bottom and top of the tower respectively (see Fig. 14-7).

In situations in which one cannot assume that H_L and H_{OL} are constant, these terms must be incorporated inside the integrals in Eqs. (14-89) and (14-90), and the integrals must be evaluated graphically or numerically (by using Simpson's rule, for example). In the normal case involving stripping without chemical reactions, the liquid-phase resistance will dominate, making it preferable to use Eq. (14-90) in conjunction with the relation $H_L \doteq H_{OL}$.

The Wiegand approximations of the above integrals in which arithmetic means are substituted for the logarithmic means x_{BM} and x°_{BM} are

$$N_L = \frac{1}{2} \ln \frac{1 - x_1}{1 - x_2} + \int_{x1}^{x2} \frac{dx}{x - x_i} \tag{14-91}$$

$$N_{OL} = \frac{1}{2} \ln \frac{1 - x_1}{1 - x_2} + \int_{x1}^{x2} \frac{dx}{x - x^\circ} \tag{14-92}$$

In these equations, the first term is a correction for finite liquid-phase concentrations, and the integral term represents the numbers of transfer units required for dilute solutions. It would be very unusual in practice to find an example in which the first (logarithmic) term is of any significance in a stripper design.

For dilute solutions in which both the operating and the equilibrium lines are straight and in which heat effects can be neglected, the integral term in Eq. (14-92) is

$$N_{OL} = \frac{1}{(1 - L_M/mG_M)} \ln \left[\left(1 - \frac{L_M}{mG_M}\right) \left(\frac{x_2 - y_1/m}{x_1 - y_1/m}\right) + \frac{L_M}{mG_M} \right] \tag{14-93}$$

This equation is identical in form to Eq. (14-88). Thus, Fig. 14-9 is applicable if the concentration ratio $(x_2 - y_1/m)/(x_1 - y_1/m)$ is substituted for the abscissa and if the parameter on the curves is identified as L_M/mG_M.

Use of HTU and K_Ga Data In estimating the size of a commercial gas absorber it is most desirable to have data on the overall mass-transfer coefficients for the system of interest at the desired conditions of temperature, pressure, solute concentration, and fluid velocities. Such data should be obtained in an apparatus of pilot-plant or semiworks size to avoid abnormal wall effects, the tower diameter being at least 8 to 10 times the packing size.

Hundreds of experimental investigations on the use of packed towers for gas absorption and desorption, both with and without a chemical reaction, have been reported in the literature. Most of these investigations were conducted at ambient temperatures and at pressures near 100 kPa (1 atm) and employed packed towers of very small diameter. One should be extremely cautious in attempting to use such data for the design of large-scale commercial equipment.

When no overall K_Ga or HTU data are available for systems involving pure physical absorption, values of the individual resistances H_G and H_L may be derived from the correlations given in Sec. 18, and these values may then be combined by the use of Eqs. (14-30) and (14-31) to obtain values for H_{OG} and H_{OL}. This procedure is not valid, however, when the rate of absorption is limited by a chemical reaction.

When absorption or desorption occurs simultaneously with chemical reaction, the use of experimental-test data from a prototype unit is particularly important, since prediction methods are not as well developed for these cases. In recent years, P. V. Danckwerts, M. M. Sharma, J. C. Charpentier, and others have stimulated interesting developments along the lines of employing data from small-scale apparatus to predict the performance of industrial absorbers when chemical reactions are involved. This work is discussed later in the subsection "Absorption with Chemical Reaction."

Overall Effects of Gas and Liquid Rates Section 18 presents detailed data on the effects of gas and liquid rates on the interfacial area a, the mass-transfer coefficients $\hat{k}_G$ and $\hat{k}_L$, and the values of H_G and H_L for some of the more common tower packings. The following discussion is a qualitative rule-of-thumb overview of the relative effects of L and G upon packed-tower performance for systems involving purely physical absorption. Specific systems may deviate significantly from these rule-of-thumb guidelines.

1. *Interfacial area.* Below the loading point of the packing the interfacial area a tends to increase with increasing liquid rate L, but it does not appear to be affected significantly by as much as a fivefold variation in the gas rate G. A tenfold increase in the liquid rate (below the loading point) roughly doubles the area available for mass transfer.

2. *Coefficient $\hat{k}_G$.* The gas-phase mass-transfer coefficient does not appear to depend appreciably upon the liquid flow rate L below the loading point of the packing, but it is influenced strongly by the gas rate, increasing roughly as the 0.7 power of G. Thus, a tenfold increase in G will increase $\hat{k}_G$ by a factor of about 5.

3. *Coefficient $\hat{k}_L$.* The liquid-phase mass-transfer coefficient appears to be independent of the gas rate G and increases roughly as the 0.5 power of L. Thus, a tenfold increase in L will increase $\hat{k}_L$ by a factor of about 3.

4. *Coefficient $\hat{k}_G a$.* By combining items 1 and 2, one arrives at the following approximate relation for conditions below the loading point of a tower packing:

$$\hat{k}_G a \propto G^{0.7} L^{0.3}$$

This relation indicates that $\hat{k}_G a$ tends to increase with both gas and liquid rates, with gas velocity having the stronger influence.

5. *Coefficient $\hat{k}_L a$.* Combination of items 1 and 3 indicates that $\hat{k}_L a$ is roughly proportional to $L^{0.8}$ and is relatively independent of the gas velocity G.

6. *Transfer unit height H_G.* Combining item 4 with the definition of the gas-phase transfer unit height leads to the approximate relation $H_G \propto (G/L)^{0.3}$, indicating that H_G increases with increasing gas rate and decreases with increasing liquid rate.

7. *Transfer unit height H_L.* Combination of item 5 with the definition of the liquid-phase transfer unit height indicates that H_L is roughly proportional to $L^{0.2}$ and does not depend appreciably on the gas rate G below the loading point of the packing.

8. *Overall transfer unit heights H_{OL} and H_{OG}.* Combination of items 6 and 7 with the definition of H_{OL} leads to the approximate relation $H_{OL} \doteq b_1 L^{0.2} + b_2(L/G)^{0.7}$, indicating that H_{OL} increases with increasing liquid rate and decreases with increasing gas rate. A similar substitution into the definition of H_{OG} leads to the relation $H_{OG} \doteq b_3(G/L)^{0.3} + b_4(G/L^{0.8})$, indicating that H_{OG} increases with increasing gas rate and decreases with increasing liquid rate. The constants b_1 through b_4 in these equations are proportionality coefficients which would differ for different systems and packings. The relative magnitudes of b_1 versus b_2 and b_3 versus b_4 are dictated by the relative importance of the gas-phase and liquid-phase resistances for each individual system.

Some Typical Examples of HTU Correlations Whitney and Vivian [*Chem. Eng. Prog.*, **45**, 323 (1949)] studied the absorption of SO_2 into water at 21°C using 25-mm (1-in) ceramic Raschig rings. An analysis of these data is presented in the book by Sherwood and Pigford, *Absorption and Extraction* (McGraw-Hill, New York, 1959, p. 294), which leads to the equation

$$H_{OL} = 0.37L^{0.18} + 0.062L^{0.75}/G^{0.7} \tag{14-94}$$

where the units (U.S. customary) of L and G are lb/(h·ft²) and H_{OL} is expressed in feet. The equation for computing H_{OL}, namely,

$$H_{OL} = H_L + (L_M/mG_M)H_G \tag{14-95}$$

indicates that the first term in Eq. (14-94) is equal to H_L and the second term is equal to $(L_M/mG_M)H_G$. By using the given value of $m = dy^\circ/dx = 32.4$ and the molecular weights of air and water (29 and 18 respectively), the following relations were derived for the SO_2-water experiments:

$$H_L = 0.37L^{0.18} \quad \text{ft}$$
$$H_G = 1.24G^{0.3}/L^{0.25} \quad \text{ft}$$

These equations are remarkably similar to the generalized relations for the effects of L and G just described.

Sherwood and Holloway [*Trans. Am. Inst. Chem. Eng.*, **36**, 39 (1940)] studied oxygen desorption on 25-mm ceramic Raschig rings at 25°C. The liquid-phase resistance for these experiments was expressed by the equation (U.S. customary)

$$H_L = 0.13L^{0.25} \quad \text{ft}$$

An equation representing data for the adiabatic vaporization of water into air, the absorption of SO_2 into caustic soda, and the absorption of Cl_2 into caustic soda was reported by Sherwood and Pigford (op. cit., p. 285) as follows:

$$H_G = 1.01G^{0.31}/L^{0.33} \quad \text{ft}$$

These relations also show a remarkable similarity to the generalized relations discussed above.

Example 4 A packed-tower absorber containing 25-mm Raschig rings is to be used for recovering acetone vapors from an air stream using water at 25°C as the solvent. The value of m at 25°C is 2.09, and the design gas and liquid rates are 0.935 kg/(s·m²) and 2.14 kg/(s·m²) respectively. The value of H_{OG} for this system is to be estimated by the use of the SO_2-air-water data of Whitney and Vivian.

Data for the diffusion coefficients for acetone and SO_2 in air and in water are summarized in the following table:

System	t, °C	D_{AB}, cm²/s
Acetone-air	0	0.109
SO_2-air	0	0.122
Acetone-H_2O	20	$(0.9)(10^{-5})$
SO_2-H_2O	25	$(1.7)(10^{-5})$

No adjustment for temperature will need to be made on the gas side because the effects of temperature will be the same for SO_2-air and acetone-air, but the liquid-phase diffusion coefficient for acetone must be corrected to 25°C. By using the Einstein relation given in Eq. (14-54) and taking the viscosities of water at 20°C and 25°C as 1.0050 cP and 0.8937 cP respectively, one obtains

$$D_{25°C} = (0.9)(10^{-5})(298/293)(1.0050/0.8937)$$
$$= (1.03)(10^{-5}) \text{ cm}^2/\text{s}$$

The values of G and L in U.S. customary units are 690 lb/(h·ft²) and 1578 lb/(h·ft²) respectively (see Table 14-2). Thus, at 21°C the values of H_G and H_L for the SO_2-air-water system are given by

$$H_G = 1.24(690)^{0.3}/(1578)^{0.25} = 1.40 \text{ ft}$$
$$H_L = 0.37(1578)^{0.18} = 1.39 \text{ ft}$$

The H_G value will be the same at 25°C as at 21°C since both $\hat{k}_G$ and $\hat{k}_G a$ are independent of the temperature. Correction of H_G for the differences in system diffusion coefficients is derived from the proportionality between H_G and the square root of the Schmidt number:

$$H_G = 1.40(0.122/0.109)^{1/2} = 1.48 \text{ ft}$$

The temperature correction for H_L is estimated by the use of the relation given in Eq. (14-56) and the values $q = 0.18$ and $\mu_{21°C} = 0.9810$ cP:

$$H_L = 1.39(294/298)^{1/2}(0.998/0.997)^{1/2}(0.8937/0.9810)^{1-0.18} = 1.28 \text{ ft}$$

This value of H_L must be adjusted once more for the differences in liquid-phase diffusion coefficients between SO_2-water and acetone-water at 25°C:

$$H_L = 1.28(1.7 \times 10^{-5}/1.03 \times 10^{-5})^{1/2} = 1.64 \text{ ft}$$

Finally, the value of H_{OG} for the acetone-air-water system at 25°C can be computed by Eq. (14-30):

$$H_{OG} = 1.48 + 1.64(2.09)(0.935/2.14)(18/29) = 2.41 \text{ ft} = 0.73 \text{ m}$$

Importance of Good Liquid Distribution The design methods discussed earlier for computing the height of packed-tower sections invariably assume that the liquid flows vertically downward through the packing at a uniform rate throughout the tower cross section. In practice, when the ratio of the tower diameter to the packing size is less than about 10:1, as may be true in small-scale experimental-test units, there is a marked tendency for liquid to migrate to the tower walls. Maldistribution of the liquid also may develop in very tall towers in which the height of packing is more than 3 times the tower diameter. For tall towers, therefore, packing may be divided into several sections with liquid redistributors between the sections. Some of the newer tower packings have been designed to counteract this tendency to maldistribute the liquid within the packing. Section 18 discusses procedures for ensuring adequate liquid distribution in packed towers.

Effects of liquid maldistribution are most severe when there is a close approach to equilibrium, since (owing to saturation) regions in which there is a deficiency of liquid flow will not absorb the gas efficiently. The development of a nonuniform liquid distribution due to "channeling," or formation of rivulets, can lead to bypassing of the gas or to localized regions of very high and very low liquid-to-gas ratios. The net effect will be a reduction in absorption efficiency.

Maldistribution effects will be less severe if there is a large excess of liquid, since it is less likely that the liquid in regions of deficient flow will become completely saturated. Thus, although there may be some effect of liquid distribution on the average mass-transfer coefficient, the most important effect is caused by changes in the average driving force due to premature saturation of the liquid in some parts of the tower.

Effects of Axial Dispersion Packed towers normally are designed by assuming plug flow of the gas and liquid phases. This assumption potentially can lead to an unsafe design, since axial dispersion or "back mixing" of the gas and/or liquid phases within the tower will reduce the average driving force for mass transfer. For example, a very low G and a high L can lead to "pumping" of the gas from top to bottom, thereby canceling the benefits of countercurrent operation. On the other hand, at very high G/L ratios, one should be concerned about the possibility of liquid-phase back mixing created by local upward entrainment of the liquid by the gas.

Axial-dispersion effects are much more serious when a very high degree of solute removal is required, a situation in which high L and low G are likely to arise. If it is suspected that this kind of back mixing could be a problem, consideration should be given to the use of a plate tower instead of a packed tower.

The state of the art on back-mixing effects in gas-liquid reactors, including packed towers, was reviewed by Shah, Stiegel, and Sharma [*Am. Inst. Chem. Eng. J.*, **24**, 369 (1978)]. A design problem illustrating the potential detrimental effects of axial dispersion is presented by Sherwood, Pigford, and Wilke in *Mass Transfer* (McGraw-Hill, New York, 1975, p. 615).

Use of HETP Data for Absorber Design Distillation design methods (see Sec. 13) normally involve determination of the number of theoretical equilibrium stages or plates N. Thus, when packed towers are employed in distillation applications, it is common practice to rate the efficiency of tower packings in terms of the height of packing equivalent to one theoretical plate (HETP).

The HETP of a packed-tower section, valid for either distillation or dilute-gas absorption and stripping systems in which constant molal overflow can be assumed and in which no chemical reactions occur, is related to the height of one overall gas-phase mass-transfer unit H_{OG} by the equation

$$\text{HETP} = H_{OG} \frac{\ln (mG_M/L_M)}{(mG_M/L_M - 1)} \qquad (14\text{-}96)$$

For gas-absorption systems in which the inlet gas is concentrated, the correct equation is

$$\text{HETP} = \left(\frac{y^{\circ}_{BM}}{1 - y}\right)_{\text{av}} H_{OG} \frac{\ln (mG_M/L_M)}{mG_M/L_M - 1} \qquad (14\text{-}97)$$

where the correction term $y^{\circ}_{BM}/(1 - y)$ is averaged over each individual theoretical plate. The equilibrium compositions corresponding to each theoretical plate may be estimated by the methods described in the subsection "Plate-Tower Design." These compositions are used in conjunction with the local values of the gas and liquid flow rates and the equilibrium slope m to obtain values for H_G, H_L, and H_{OG} corresponding to the conditions on each theoretical stage, and the local values of the HETP are then computed by Eq. (14-97). The total height of packing required for the separation is the summation of the individual HETPs computed for each theoretical stage.

For gas-absorption or gas-stripping tower designs in which pure physical absorption or desorption is involved (i.e., no chemical reactions take place), the following HETP values may be used as a rough guide for preliminary estimates:

Nominal packing size, mm (in)	HETP range, m
25 (1)	0.4–0.5
38 (1½)	0.5–0.7
50 (2)	0.7–0.9
75 (3)	0.9–1.0

A detailed discussion of HETP data is presented in Sec. 18.

PLATE-TOWER DESIGN

The design of a plate tower for gas-absorption or gas-stripping operations involves many of the same principles employed in distillation calculations, such as the determination of the number of theoretical plates needed to achieve a specified composition change (see Sec. 13). Distillation differs from gas absorption in that it involves the separation of components based on the distribution of the various substances between a gas phase and a liquid phase when all the components are present in both phases. In distillation, the new phase is generated from the original feed mixture by vaporization or condensation of the volatile components, and the separation is achieved by introducing reflux to the top of the tower.

In gas absorption, the new phase consists of an inert nonvolatile solvent (absorption) or an inert nonsoluble gas (stripping), and normally no reflux is involved. The following paragraphs discuss some of the considerations peculiar to gas-absorption calculations for plate towers and some of the approximate design methods that can be employed when simplifying assumptions are valid.

Graphical Design Procedure Construction of design diagrams (xy diagrams showing the equilibrium and operating curves) should be an integral part of any design involving the distribution of a single solute between an inert solvent and an inert gas. The number of theoretical plates can be stepped off rigorously provided the curvatures of the operating and equilibrium lines are correctly accounted for in the diagram. This procedure is valid even though an insoluble inert gas is present in the gas phase and an inert nonvolatile solvent is present in the liquid phase.

Figure 14-10 illustrates the graphical method for a three-theoretical-plate system. Note that in gas absorption the operating line is above the equilibrium curve, whereas in distillation this does not happen. In gas stripping, the operating line will be below the equilibrium curve.

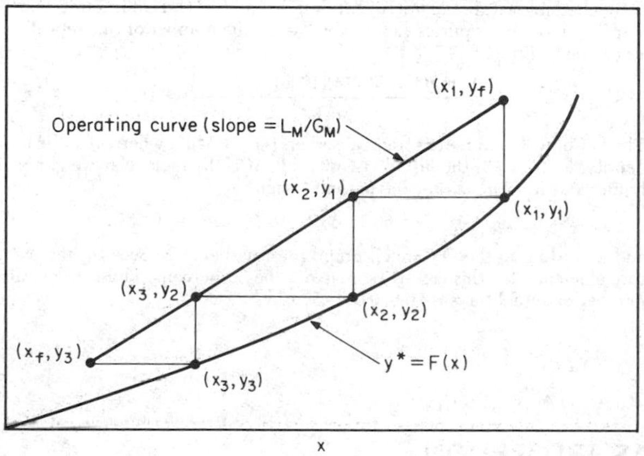

FIG. 14-10 Graphical method for a three-theoretical-plate gas-absorption tower with inlet-liquor composition x_f and inlet-gas composition y_f.

On Fig. 14-10 note that the stepping-off procedure begins on the operating line. The starting point x_f, y_3 represents the compositions of the entering lean wash liquor and of the gas exiting from the top of the tower, as determined by the design specifications. After three steps one reaches the point x_1, y_f, representing the compositions of the solute-rich feed gas y_f and of the solute-rich liquor leaving the bottom of the tower x_1.

Algebraic Method for Dilute Gases By assuming that the operating and equilibrium curves are straight lines and that heat effects are negligible, Souders and Brown [*Ind. Eng. Chem.*, **24,** 519 (1932)] developed the following equation:

$$(y_1 - y_2)/(y_1 - y_2^\circ) = (A^{N+1} - A)/(A^{N+1} - 1) \qquad (14\text{-}98)$$

where N = number of theoretical plates, y_1 = mole-fraction solute in the entering gas, y_2 = mole-fraction solute in the leaving gas, $y_2^\circ = mx_2$ = mole-fraction solute in equilibrium with the incoming solvent liquor (zero for a pure solvent), and A = absorption factor = L_M/mG_M.

When $A = 1$, Eq. (14-98) is indeterminate, and for this case the solution is given by

$$(y_1 - y_2)/(y_1 - y_2^\circ) = N/(N + 1) \qquad (14\text{-}99)$$

Although Eq. (14-98) is convenient for computing the composition of the exit gas as a function of the number of theoretical stages, an alternative equation derived by Colburn [*Trans. Am. Inst. Chem. Eng.*, **35,** 211 (1939)] is more useful when the number of theoretical plates is the unknown:

$$N = \frac{\ln [(1 - A^{-1})(y_1 - y_2^\circ)/(y_2 - y_2^\circ) + A^{-1}]}{\ln (A)} \qquad (14\text{-}100)$$

The numerical results obtained by using either Eq. (14-98) or Eq. (14-100) are identical. Thus, the two equations may be used interchangeably as the need arises.

Comparison of Eqs. (14-100) and (14-88) shows that

$$N_{OG}/N = \ln (A)/(1 - A^{-1}) \qquad (14\text{-}101)$$

thus revealing the close relationship between theoretical stages in a plate tower and mass-transfer units in a packed tower. Equations (14-88) and (14-100) are related to each other by virtue of the relation

$$h_T = H_{OG}N_{OG} = (\text{HETP})N \qquad (14\text{-}102)$$

Albegraic Method for Concentrated Gases When the feed gas is concentrated, the absorption factor, which is defined in general as $A = L_M/KG_M$ where $K = y^\circ/x$, can vary throughout the tower owing to changes in the equilibrium K values due to temperature increases. An approximate solution to this problem can be obtained by substitution of the "effective" absorption factors A_e and A' derived by Edmister [*Ind. Eng. Chem.*, **35,** 837 (1943)] into the equation

$$\frac{y_1 - y_2}{y_1} = \left[1 - \frac{1}{A'}\frac{(L_M x)_2}{(G_M y)_1}\right]\frac{A_e^{N+1} - A_e}{A_e^{N+1} - 1} \qquad (14\text{-}103)$$

where subscripts 1 and 2 refer to the bottom and top of the tower respectively and the absorption factors are defined by the equations

$$A_e = \sqrt{A_1(A_2 + 1) + 0.25} - 0.5 \qquad (14\text{-}104)$$

$$A' = A_1(A_2 + 1)/(A_1 + 1) \qquad (14\text{-}105)$$

This procedure has been applied to the absorption of C_5 and lighter hydrocarbon vapors into a lean oil, for example.

Stripping Equations When the liquid feed is dilute and the operating and equilibrium curves are straight lines, the stripping equations analogous to Eqs. (14-98) and (14-100) are

$$(x_2 - x_1)/(x_2 - x_1^\circ) = (S^{N+1} - S)/(S^{N+1} - 1) \qquad (14\text{-}106)$$

where $x_1^\circ = y_1/m$; $S = mG_M/L_M = A^{-1}$; and

$$N = \frac{\ln [(1 - A)(x_2 - x_1^\circ)/(x_1 - x_1^\circ) + A]}{\ln (S)} \qquad (14\text{-}107)$$

For systems in which the concentrations are large and the stripping factor S may vary along the tower, the following Edmister equations [*Ind. Eng. Chem.*, **35,** 837 (1943)] are applicable:

$$\frac{x_2 - x_1}{x_2} = \left[1 - \frac{1}{S'}\frac{(G_M y)_1}{(L_M x)_2}\right]\frac{S_e^{N+1} - S_e}{S_e^{N+1} - 1} \qquad (14\text{-}108)$$

where

$$S_e = \sqrt{S_2(S_1 + 1) + 0.25} - 0.5 \qquad (14\text{-}109)$$

$$S' = S_2(S_1 + 1)/(S_2 + 1) \qquad (14\text{-}110)$$

and the subscripts 1 and 2 refer to the bottom and top of the tower respectively.

Equations (14-104) and (14-109) represent two different ways of obtaining an effective factor, and a value of A_e obtained by taking the reciprocal of S_e from Eq. (14-109) will not check exactly with a value of A_e derived by substituting $A_1 = 1/S_1$ and $A_2 = 1/S_2$ into Eq. (14-104). Regardless of this fact, the equations generally give reasonable results for approximate design calculations.

It should be noted that throughout this section the subscripts 1 and 2 refer to the bottom and to the top of the apparatus respectively regardless of whether it is an absorber or a stripper. This has been done to maintain internal consistency among all the equations and to prevent the confusion created in some derivations in which the numbering system for an absorber is different from the numbering system for a stripper.

Tray Efficiencies in Plate Absorbers and Strippers Computations of the number of theoretical plates N assume that the liquid on each plate is completely mixed and that the vapor leaving the plate is in equilibrium with the liquid. In actual practice a condition of complete equilibrium cannot exist since interphase mass transfer requires a finite driving-force difference. This leads to the definition of an overall plate efficiency

$$E = N_{\text{theo}}/N_{\text{actual}} \qquad (14\text{-}111)$$

which can be correlated to system design variables.

Mass-transfer theory indicates that for trays of a given design the factors most likely to influence E in absorption and stripping towers are the physical properties of the fluids and the dimensionless ratio mG_M/L_M. Systems in which the mass transfer is gas-film-controlled may be expected to have plate efficiencies as high as 50 to 100 percent, whereas plate efficiencies as low as 1 percent have been reported for the absorption of gases of low solubility (large m) into solvents of relatively high viscosity.

The fluid properties are represented by the Schmidt numbers of the gas and liquid phases. For gases, the Schmidt numbers normally are close to unity and are independent of temperature and pressure. Thus, the gas-phase mass-transfer coefficients are relatively independent of the system.

By contrast, the liquid-phase Schmidt numbers range from about 10^2 to 10^4 and depend strongly on the temperature. The effect of temperature on the liquid-phase mass-transfer coefficient is related primarily to changes in the liquid viscosity with temperature, as indicated by Eq. (14-56), and this derives primarily from the strong dependency of the liquid-phase Schmidt number upon viscosity.

Consideration of the preceding discussion in connection with Eq. (14-8) indicates that variations in the overall resistance to mass transfer in absorbers and strippers are related primarily to variations in the liquid-phase viscosity μ and to variations in the slope m. A correlation of the efficiency of plate absorbers in terms of the viscosity of the liquid solvent and the solubility of the solute gas was developed by O'Connell [*Trans. Am. Inst. Chem. Eng.*, **42,** 741 (1946)]. The O'Connell correlation for plate absorbers is presented in Sec. 18.

The best procedure for making plate-efficiency corrections (which obviously can be quite large) is to use experimental-test data from a prototype system that is large enough to be representative of an actual commercial tower.

Example 5 The number of actual plates required for steam-stripping an acetone-rich liquor containing 0.573 mole percent acetone in water is to be estimated. The design overhead recovery of acetone is 99.9 percent, leaving 18.5 ppm weight of acetone in the stripper bottoms. The design operating temperature and pressure are 101.3 kPa and 94°C respectively, the average liquid-phase viscosity is 0.30 cP, and the average value of $K = y°/x$ for these conditions is 33.

By choosing a value of $mG_M/L_M = S = A^{-1} = 1.4$ and noting that the stripping medium is pure steam (i.e., $x_1^\circ = 0$), the number of theoretical trays according to Eq. (14-107) is

$$N = \frac{\ln[(1 - 0.714)(1000) + 0.714]}{\ln(1.4)} = 16.8$$

The O'Connell parameter for gas absorbers is $\rho_L/KM\mu_L$, where ρ_L is the liquid density, lb/ft^3; μ_L is the liquid viscosity, cP; M is the molecular weight of the liquid; and $K = y°/x$. For the present design

$$\rho_L/KM\mu_L = 60.1/(33 \times 18 \times 0.30) = 0.337$$

and according to the O'Connell graph for absorbers (see Sec. 18) the overall tray efficiency for this case is estimated to be 30 percent. Thus, the required number of actual trays is 16.8/0.3 = 56 trays.

HEAT EFFECTS IN GAS ABSORPTION

Overview One of the most important considerations involved in designing gas-absorption towers is to determine whether or not temperatures will vary along the length of the tower because of heat effects, since the solubility of the solute gas normally depends strongly upon the temperature. When heat effects can be neglected, computation of the tower dimensions and required flows is relatively straightforward, as indicated by the simplified design methods discussed earlier for both packed and plate absorbers and strippers. When heat effects cannot be neglected, the computational problem becomes much more difficult.

Heat effects that may cause temperatures to vary from point to point in a gas absorber are (1) the heat of solution of the solute (including heat of condensation, heat of mixing, and heat of reaction), which can lead to a rise in the liquid temperature; (2) the heat of vaporization or condensation of the solvent; (3) the exchange of sensible heat between the gas and liquid phases; and (4) the loss of sensible heat from the fluids to internal or external cooling coils or to the atmosphere via the tower walls.

Y. T. Shah (*Gas-Liquid-Solid Reactor Design*, McGraw-Hill, New York, 1979, p. 51) has reviewed the literature concerning heat effects in systems involving gas-liquid reactions and concludes that in the majority of the systems involving chemical reactions temperature effects are not very important. For some systems in which large amounts of heat may be liberated, there are compensating effects which decrease the effect on the rate of absorption. For example, increasing temperatures tend simultaneously to increase the rate of chemical reaction and to decrease the solubility of the reactant at the solvent interface. Systems in which compensating effects can occur include absorption of CO_2 in amine solutions and absorption of CO_2 in NaOH solutions.

There are, however, a number of well-known systems in which heat effects definitely cannot be ignored. Examples include absorption of ammonia in water, dehumidification of air with concentrated H_2SO_4, absorption of HCl in water, and absorption of SO_3 in H_2SO_4. Another interesting example is the absorption of acetone in water, in which the heat effects are mild but not negligible.

Some very thorough and knowledgeable discussions of the problems involved in gas absorption with large heat effects have been presented by Coggan and Bourne [*Trans. Inst. Chem. Eng.*, **47,** T96, T160 (1969)], by Bourne, von Stockar, and Coggan [*Ind. Eng. Chem. Process Des. Dev.*, **13,** 115, 124 (1974)], and also by von Stockar and Wilke [*Ind. Eng. Chem. Fundam.*, **16,** 89 (1977)]. The first two of these references discuss plate-tower absorbers and include interesting experimental studies of the absorption of ammonia in water. The third reference discusses the design of packed-tower gas absorbers and includes a shortcut design method based on a semitheoretical correlation of rigorous calculation results. All these authors clearly demonstrate both theoretically and experimentally that when the solvent is volatile, the temperature inside an absorber can go through a maximum. They note that the least expensive of the solvents, water, is capable of exhibiting this unusual "hot-spot" behavior.

From a designer's point of view there are a number of different approaches to be considered in dealing with heat effects, depending on the requirements of the job at hand. For example, one can (1) add internal or external heat-transfer surface to remove heat from the absorber; (2) treat the process as if it were isothermal by arbitrarily assuming that the temperature of the liquid phase is everywhere the same and then add a design safety factor; (3) employ the classical adiabatic model, which assumes that the heat of solution manifests itself only as sensible heat in the liquid phase and that solvent vaporization is negligible; (4) use semitheoretical shortcut methods derived from rigorous calculations; and (5) employ rigorous design procedures requiring the use of a large-scale digital computer.

For preliminary-screening work the simpler methods may be adequate, but for final designs one should seriously consider using a more rigorous approach.

Effects of Operating Variables Conditions that can give rise to significant heat effects are (1) an appreciable heat of solution and (2) absorption of large amounts of solute in the liquid phase. The second of these conditions can arise when the solute concentration in the inlet gas is very large, when the liquid flow rate is relatively low (small L_M/G_M), when the solubility of the solute in the liquid phase is high, and/or when the operating pressure is high.

When the solute is absorbed very rapidly, the rate of heat liberation often is largest near the bottom of the tower. This has the effect of causing the equilibrium line to curve upward near the solute-rich end, although it may remain relatively straight near the lean end, corresponding to the temperature of the lean solvent.

If the solute-rich gas entering the bottom of an absorption tower is cold, the liquid phase may be cooled somewhat by transfer of sensible heat to the gas. A much stronger cooling effect can occur when the solvent is volatile and the entering rich gas is not saturated with respect to the solvent. It is possible to experience a condition in which solvent is being evaporated near the bottom of the tower and condensed near the top. Under these conditions there may develop a pinch point in which the operating and equilibrium curves approach each other at a point inside the tower.

In the references cited previously, the authors discuss the influence of operating variables upon the performance of plate towers when large heat effects are involved. Some general observations are as follows:

Operating Pressure Raising the pressure may increase the separation efficiency considerably. Calculations involving the absorption of methanol from water-saturated air showed that doubling the pressure doubled the concentration of methanol which could be tolerated in the feed gas while still achieving a preset concentration specification in the off gas.

Temperature of Fresh Solvent The temperature of the entering solvent has surprisingly little influence upon the degree of absorption or upon the internal-temperature profiles in an absorber when the heat effects are due primarily to heat of solution or to solvent vaporization. In these cases the temperature profile in the liquid phase apparently is dictated solely by the internal-heat effects.

Temperature and Humidity of Rich Gas Cooling and consequent dehumidification of the feed gas to an absorption tower can be very beneficial. A high humidity (or relative saturation with solvent) limits the capacity of the gas phase to take up latent heat and therefore is unfavorable to absorption. Thus, dehumidification of the inlet gas prior to introducing it into the tower is worth considering in the design of gas absorbers with large heat effects.

Liquid-to-Gas Ratio The L/G ratio can have a significant influence on the development of temperature profiles in gas absorbers. High L/G ratios tend to result in less strongly developed temperature profiles owing to the high heat capacity of the liquid phase. As the L/G is increased, the operating line moves away from the equilibrium line and there is a tendency for more solute to be absorbed per stage. However, there is a compensating effect in that as more heat is liberated at each stage, the plate temperatures will tend to rise, causing an upward shifting of the equilibrium line.

As the L/G is decreased, the concentration of solute tends to build up in the upper parts of the absorber, and the point of highest temperature tends to move upward in the tower until finally the maximum temperature develops only on the topmost plate. Of course, the capacity of the liquid phase to absorb solute falls progressively as the L/G is reduced.

Number of Stages When the heat effects combine to produce an extended zone within the tower where little absorption is taking place (i.e., a pinch zone), the addition of plates to the tower will have no useful effect on separation efficiency. Solutions to these difficulties must be sought by increasing the solvent flow, introducing strategically placed coolers, cooling and dehumidifying the inlet gas, and/or raising the tower operating pressure.

Equipment Considerations When the solute has a large heat of solution and the feed gas contains a large percentage of solute, as in the absorption of HCl in water, the effects of heat release during absorption may be so pronounced that the installation of heat-transfer surface to remove the heat of absorption may be as important as providing sufficient interfacial area for the mass-transfer process itself. The added heat-transfer surface may consist of internal cooling coils on the plates, or else the solvent may be withdrawn from a point intermediate in the tower and passed through an external heat exchanger (intercooler) for cooling.

In many cases the rate of heat liberation is largest near the bottom of the tower, where solute absorption is more rapid, so that cooling surfaces or intercoolers are required only on the first few trays. Coggan and Bourne [*Trans. Inst. Chem. Eng.*, **47,** T96, T160 (1969)] found, however, that the optimal position for a single interstage cooler does not necessarily coincide with the position of the maximum temperature or with the center of a pinch. They found that in a 12-plate tower, two strategically placed interstage coolers tripled the allowable ammonia feed concentration for a given off-gas specification. For a case involving methanol absorption, it was found that more separation was possible in a 12-stage column with two intercoolers than in a simple column with 100 stages and no intercoolers.

In the case of HCl absorption, a shell-and-tube heat exchanger often is employed as a cooled wetted-wall vertical-column absorber so that the exothermic heat of reaction can be removed continuously as it is released into the liquid film.

Installation of heat-exchange equipment to precool and dehumidify the feed gas to an absorber also deserves consideration in order to take advantage of the cooling effects created by vaporization of solvent in the lower sections of the tower.

Classical Isothermal Design Method When the feed gas is sufficiently dilute, the exact design solution may be approximated by the isothermal one over a broad range of L/G ratios, since heat effects generally are less important when washing dilute-gas mixtures. The problem, however, is one of defining the term "sufficiently dilute" for each specific case. For a new absorption duty, the assumption of isothermal operation must be subjected to verification by the use of a rigorous design procedure.

When heat-exchange surface is being provided in the design of an absorber, the isothermal design procedure can be rendered valid by virtue of the exchanger design specifications. With ample surface area and a close approach, isothermal operation can be guaranteed.

For preliminary screening and feasibility studies or for rough cost estimates, one may wish to employ a version of the isothermal method which assumes that the liquid temperatures in the tower are everywhere equal to the inlet-liquid temperature. In their analysis of packed-tower designs, von Stockar and Wilke [*Ind. Eng. Chem. Fundam.* **16**, 89 (1977)] showed that the isothermal method tended to underestimate the required depth of packing by a factor of as much as 1.5 to 2. Thus, for rough estimates one may wish to employ the assumption that the temperature is equal to the inlet-liquid temperature and then apply a design factor to the result.

Another instance in which the constant-temperature method is used involves the direct application of experimental K_Ga values obtained at the desired conditions of inlet temperatures, operating pressure, flow rates, and feed-stream compositions. The assumption here is that, regardless of any temperature profiles that may exist within the actual tower, the procedure of "working the problem in reverse" will yield a correct result. One should be cautious about extrapolating such data very far from the original basis and be careful to use compatible equilibrium data.

Classical Adiabatic Design Method The classical adiabatic method assumes that the heat of solution serves only to heat up the liquid stream and that there is no vaporization of solvent. This assumption makes it feasible to relate increases in the liquid-phase temperature to the solute concentration x by a simple enthalpy balance. The equilibrium curve can then be adjusted to account for the corresponding temperature rise on an xy diagram. The adjusted equilibrium curve will become more concave upward as the concentration increases, tending to decrease the driving forces near the bottom of the tower, as illustrated in Fig. 14-11 in Example 7.

Colburn [*Trans. Am. Inst. Chem. Eng.*, **35,** 211 (1939)] has shown that when the equilibrium line is straight near the origin but curved slightly at its upper end, N_{OG} can be computed approximately by assuming that the equilibrium curve is a parabolic arc of slope m_2 near the origin and passing through the point x_1, K_1x_1 at the upper end. The Colburn equation for this case is

$$N_{OG} = \frac{1}{1 - m_2G_M/L_M} \times \ln\left[\frac{(1 - m_2G_M/L_M)^2}{1 - K_1G_M/L_M}\left(\frac{y_1 - m_2x_2}{y_2 - m_2x_2}\right) + \frac{m_2G_M}{L_M}\right] \tag{14-112}$$

Comparisons by von Stockar and Wilke [*Ind. Eng. Chem. Fundam.*, **16**, 89 (1977)] between the rigorous and the classical adiabatic design methods for packed towers indicated that the simple adiabatic method underestimates the packing depths by as much as a factor of

1.25 to 1.5. Thus, when using the classical adiabatic method, one should consider the possible need to apply a design safety factor.

A slight variation of the above method accounts for increases in the solvent content of the gas stream between the inlet and the outlet of the tower and assumes that the evaporation of solvent tends to cool the liquid. This procedure offsets a part of the temperature rise that would have been predicted with no solvent evaporation and leads to the prediction of a shorter tower.

Rigorous Design Methods A detailed discussion of rigorous methods for the design of packed and plate absorbers when large heat effects are involved is beyond the scope of this section. The references cited previously are believed to contain the most up-to-date discussions of the state of the art available at this time and are recommended reading for anyone needing more information on this topic. The paper of von Stockar and Wilke [*Ind. Eng. Chem. Fundam.*, **16**, 89 (1977)] presents an interesting shortcut method for the design of packed towers which closely approximates rigorous results. This method may be suitable for use on a programmable calculator.

Direct Comparison of Design Methods The following problem, originally solved by Sherwood, Pigford, and Wilke (*Mass Transfer*, McGraw-Hill, New York, 1975, p. 616), was employed by von Stockar and Wilke [*Ind. Eng. Chem. Fundam.*, **16**, 94 (1977)] as the basis for a direct comparison between the isothermal, adiabatic, semitheoretical shortcut, and rigorous design methods for estimating the height of packed towers.

Example 6 Inlet gas to an absorber consists of a mixture of 6 mole percent acetone in air saturated with water vapor at 15°C and 101.3 kPa (1 atm). The scrubbing liquor is pure water at 15°C, and the inlet gas and liquid rates are given as 0.080 and 0.190 kmol/s respectively. The liquid rate corresponds to 20 percent over the theoretical minimum as calculated by assuming a value of x_1 corresponding to complete equilibrium between the exit liquor and the incoming gas. H_G and H_L are given as 0.42 and 0.30 m respectively, and the acetone equilibrium data at 15°C are p_A^0 = 19.7 kPa (147.4 torr), $\gamma_A = 6.46$, and $m_A = 6.46 \times 19.7/101.3 = 1.26$. The heat of solution of acetone is 7656 cal/gmol (32.05 kJ/gmol), and the heat of vaporization of solvent (water) is 10,755 cal/gmol (45.03 kJ/gmol). The problem calls for determining the height of packing required to achieve a 90 percent recovery of the acetone.

The following table compares the results obtained by von Stockar and Wilke (op. cit.) for the various design methods:

Design method used	N_{OG}	Packed height, m	Design safety factor
Rigorous	5.56	3.63	1.00
Shortcut rigorous	5.56	3.73	0.97
Classical adiabatic	4.01	2.38	1.53
Classical isothermal	3.30	1.96	1.85

It should be clear from this example that there is considerable room for error when approximate design methods are employed in situations involving large heat effects, even for a case in which the solute concentration in the inlet gas was only 6 mole percent.

Example 7 Let us consider the absorption of acetone from air at atmospheric pressure into a stream of pure water fed to the top of a packed absorber at 25°C. The inlet gas at 35°C contains 2 percent by volume of acetone and is 70 percent saturated with water vapor (4 percent H_2O by volume). The mole-fraction acetone in the exit gas is to be reduced to 1/400 of the inlet value, or 50 ppmv. For 100 kmol of feed-gas mixture, how many kilomoles of fresh water should be fed to provide a positive driving force throughout the packing? How many transfer units will be needed according to the classical adiabatic method? What is the estimated height of packing required if H_{OG} = 0.70 m?

The latent heats at 25°C are 7656 kcal/kmol for acetone and 10,490 kcal/kmol for water, and the differential heat of solution of acetone vapor in pure water is given as 2500 kcal/kmol. The specific heat of air is 7.0 kcal/(kmol·K).

Acetone solubilities are defined by the equation

$$K = y^\circ/x = \gamma_a p_a^0/p_T \tag{14-113}$$

where the vapor pressure of pure acetone in mmHg (torr) is given by (Sherwood et al., *Mass Transfer*, McGraw-Hill, New York, 1975, p. 537):

$$p_a^0 = \exp(18.1594 - 3794.06/T) \tag{14-114}$$

and the liquid-phase-activity coefficient may be approximated for low concentrations ($x \le 0.01$) by the equation

$$\gamma_a = 6.5 \exp(2.0803 - 601.2/T) \tag{14-115}$$

Typical values of acetone solubility as a function of temperature at a total pressure of 760 mmHg are shown in the following table:

t, °C	25	30	35	40
γ_a	6.92	7.16	7.40	7.63
p_a^0, mmHg	229	283	346	422
$K = \gamma_a p_a^0/760$	2.09	2.66	3.37	4.23

For dry gas and liquid water at 25°C, the following enthalpies are computed for the inlet- and exit-gas streams (basis, 100 kmol of gas entering):

Entering gas:

Acetone	2(2500 + 7656) =	20,312 kcal
Water vapor	4(10,490) =	41,960
Sensible heat	(100)(7.0)(35 − 25) =	7,000
		69,272 kcal

Exit gas (assumed saturated with water at 25°C):

Acetone	(2/400)(94/100)(2500) =	12 kcal
Water vapor	$94\left(\dfrac{23.7}{760 - 23.7}\right)(10{,}490)$ =	31,600
		31,612 kcal

Enthalpy change of liquid = 69,272 − 31,612 = 37,660 kcal/100 kmol gas

Thus, $\Delta t = t_1 - t_2 = 37{,}660/18L_M$, and the relation between L_M/G_M and the liquid-phase temperature rise is

$$L_M/G_M = (37{,}660)/(18)(100)\,\Delta t = 20.92/\Delta t$$

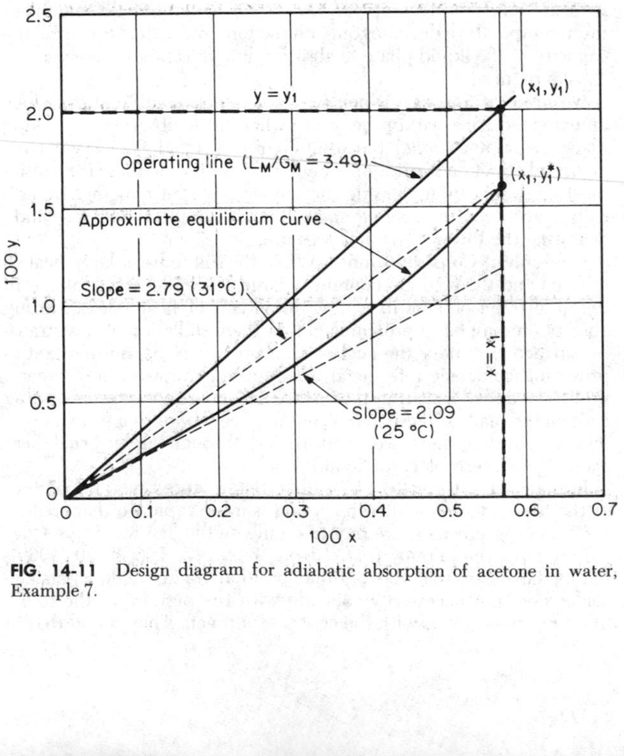

FIG. 14-11 Design diagram for adiabatic absorption of acetone in water, Example 7.

The following table summarizes the critical values for various assumed temperature rises:

Δt, °C	L_M/G_M	K_1	K_1G_M/L_M	m_2G_M/L_M
0	.....	2.09	0.	0.
2	10.46	2.31	0.221	0.200
3	6.97	2.42	0.347	0.300
4	5.23	2.54	0.486	0.400
5	4.18	2.66	0.636	0.500
6	3.49	2.79	0.799	0.599
7	2.99	2.93	0.980	0.699

Evidently a temperature rise of 7°C would not be a safe design because the equilibrium line nearly touches the operating line near the bottom of the tower, creating a pinch. A temperature rise of 6°C appears to give an operable design, and for this case L_M = 349 kmol per 100 kmol of feed gas.

The design diagram for this case is shown in Fig. 14-11, in which the equilibrium curve is drawn with a french curve so that the slope at the origin m_2 is equal to 2.09 and passes through the point $x_1 = 0.02/3.49 = 0.00573$ at $y_1^\circ = 0.00573 \times 2.79 = 0.0160$.

The number of transfer units can be calculated from the adiabatic design equation, Eq. (14-112):

$$N_{OG} = \frac{1}{1 - 0.599} \ln \left[\frac{(1 - 0.599)^2}{(1 - 0.799)} (400) + 0.599 \right] = 14.4$$

The estimated height of tower packing by assuming H_{OG} = 0.70 m and a design safety factor of 1.5 is

$$h_T = (14.4)(0.7)(1.5) = 15.1 \text{ m } (49.6 \text{ ft})$$

For this tower, one should consider the use of two or more shorter packed sections instead of one long section.

MULTICOMPONENT SYSTEMS

When no chemical reactions are involved in the absorption of more than one soluble component from an insoluble gas, the design conditions (pressure, temperature, and liquid-to-gas ratio) normally are determined by the volatility or the physical solubility of the least soluble component for which complete recovery is economical. Components of lower volatility (higher solubility) also will be recovered completely.

The more volatile (i.e., less soluble) components will be only partially absorbed even though the effluent liquid becomes completely saturated with respect to these lighter substances. When a condition of saturation exists, the value of y_1/y_2 will remain finite even for an infinite number of plates or transfer units. This can be seen in Fig. 14-9, in which the asymptotes become vertical for values of mG_M/L_M greater than unity. If the amount of volatile component in the incoming fresh solvent is negligible, then the limiting value of y_1/y_2 for each of the highly volatile components is

$$y_1/y_2 = S/(S - 1) \qquad (14\text{-}116)$$

where $S = mG_M/L_M$ and the subscripts 1 and 2 refer to the bottom and top of the tower respectively.

When the gas stream is dilute, absorption of each constituent can be considered separately as if the other components were absent. The following example illustrates the use of this principle.

Example 8 Air entering a tower contains 1 percent acetaldehyde and 2 percent acetone. The liquid-to-gas ratio for optimum acetone recovery is L_M/G_M = 3.1 mol/mol when the fresh-solvent temperature is 31.5°C. The value of y°/x for acetaldehyde has been measured as 50 at the boiling point of a dilute solution, 93.5°C. What will the percentage recovery of acetaldehyde be under conditions of optimal acetone recovery?

Solution If the heat of solution is neglected, y°/x at 31.5°C is equal to 50(1200/7300) = 8.2, where the factor in parentheses is the ratio of pure-acetaldehyde vapor pressures at 31.5 and 93.5°C respectively. Since L_M/G_M is equal to 3.1, the value of S for the aldehyde is $S = mG_M/L_M = 8.2/3.1 = 2.64$, and $y_1/y_2 = S/(S - 1) = 2.64/1.64 = 1.61$. The acetaldehyde recovery is therefore equal to 100 × 0.61/1.61 = 38 percent recovery.

In **concentrated systems** the change in gas and liquid flow rates within the tower and the heat effects accompanying the absorption of all the components must be considered. A trial-and-error calculation from one theoretical stage to the next usually is required if accurate results are to be obtained, and in such cases calculation procedures similar to those described in Sec. 13 normally are employed. A computer procedure for multicomponent adiabatic absorber design has been described by Feintuch and Treybal [*Ind. Eng. Chem. Process Des. Dev.*, **17**, 505 (1978)]. Also see Holland, *Fundamentals and Modeling of Separation Processes*, Prentice-Hall, Englewood Cliffs, N.J., 1975.

When two or more gases are absorbed in systems involving chemical reactions, the situation is much more complex. This topic is dicussed later in the subsection "Absorption with Chemical Reaction."

Graphical Design Method for Dilute Systems The following notation for multicomponent absorption calculations has been adapted from Sherwood, Pigford, and Wilke (*Mass Transfer*, McGraw-Hill, New York 1975, p. 415):

L_M^s = moles of solvent per unit time

G_M^0 = moles of rich feed gas to be treated per unit time

X = moles of one solute per mole of solute-free solvent fed to the top of the tower

Y = moles of one solute in the gas phase per mole of rich feed gas to be treated

Subscripts 1 and 2 refer to the bottom and top of the tower respectively, and the material balance for any one component may be written as

$$L_M^s(X - X_2) = G_M^0(Y - Y_2) \qquad (14\text{-}117)$$

or else as

$$L_M^s(X_1 - X) = G_M^0(Y_1 - Y) \qquad (14\text{-}118)$$

For the special case of absorption from lean gases with relatively large amounts of solvent, the equilibrium lines are defined for each component by the relation

$$Y^\circ = K'X \qquad (14\text{-}119)$$

Thus, the equilibrium line for each component passes through the origin with slope K', where

$$K' = K(G_M/G_M^0)/(L_M/L_M^s) \qquad (14\text{-}120)$$

and $K = y^\circ/x$. When the system is sufficiently dilute, $K' \doteq K$.

The liquid-to-gas ratio L_M^s/G_M^0 is chosen on the basis of the solubility of the least soluble substance in the feed gas that must be absorbed completely. Each individual component will then have its own operating line with slope equal to L_M^s/G_M^0 (i.e., the operating lines for all the various components will be parallel to each other).

A typical diagram for the complete absorption of pentane and heavier components from a lean gas mixture is shown in Fig. 14-12. The oil used as solvent for this case was assumed to be solute-free (i.e., $X_2 = 0$), and the "key component," butane, was identified as that component absorbed in appreciable amounts whose equilibrium line is most nearly parallel to the operating lines (i.e., the K value for butane is approximately equal to L_M^s/G_M^0).

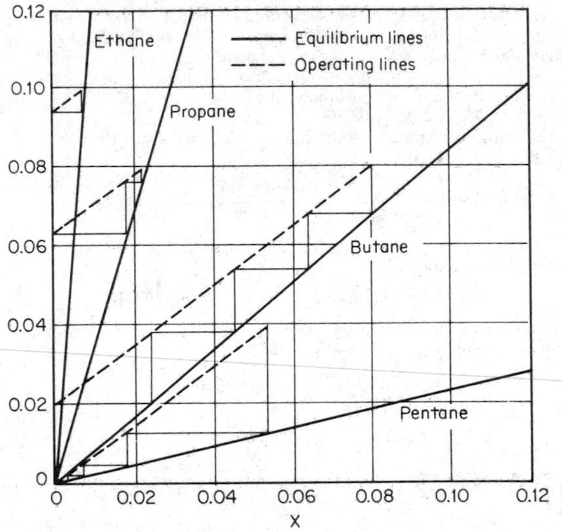

FIG. 14-12 Graphical design method for multicomponent systems; absorption of butane and heavier components in a solute-free lean oil.

In Fig. 14-12 the composition of the gas with respect to components more volatile than butane will approach equilibrium with the liquid phase at the bottom of the tower. The gas composition with respect to components less volatile (heavier) than butane will approach equilibrium with the oil entering the tower, and since $X_2 = 0$, the components heavier than butane will be completely absorbed.

Four theoretical plates have been stepped off for the key component (butane) on Fig. 14-12 and are sufficient to give a 75 percent recovery of butane. The operating lines for the other components were drawn in with the same slope and were placed so as to give the same number of theoretical plates insofar as possible.

The diagram of Fig. 14-12 shows that for the light components equilibrium is achieved easily in fewer than four theoretical plates and that for the heavier components nearly complete recovery is obtained in four theoretical plates. The diagram also shows that absorption of the light components takes place in the upper part of the tower and absorption of the heavier components takes place in the lower section of the tower.

Algebraic Design Method for Dilute Systems The design method described above can be performed algebraically by employing the following modified version of the Kremser formula:

$$\frac{Y_1 - Y_2}{Y_1 - mX_2} = \frac{(A^0)^{N+1} - A^0}{(A^0)^{N+1} - 1} \qquad (14\text{-}121)$$

where for dilute gas absorption $A^0 = L_M^s/mG_M^0$ and $m \doteq K = y^\circ/x$.

The left-hand side of Eq. (14-121) represents the efficiency of absorption of any one component of the feed-gas mixture. If the solvent oil is denuded of solute so that $X_2 = 0$, the left-hand side is equal to the fractional absorption of the component from the rich feed gas. When the number of theoretical plates N and the liquid and gas rates L_M^s and G_M^0 have been fixed, the fractional absorption of each component may be computed directly and the operating lines need not be placed by trial and error as in the graphical approach described earlier.

According to Eq. (14-121), when A^0 is less than unity and N is large,

$$(Y_1 - Y_2)/(Y_1 - mX_2) \doteq A^0 \qquad (14\text{-}122)$$

This equation can be employed for estimating the fractional absorption of the more volatile components whenever the value of A^0 for the component is smaller than the value of A^0 for the key component by a factor of 3 or more.

When A^0 is very much larger than unity and when N is large, the right-hand side of Eq. (14-121) becomes equal to unity. This signifies that the gas will leave the top of the tower in equilibrium with the incoming oil, and when $X_2 = 0$, it corresponds to complete absorption of the component in question. Thus, the least volatile components may be assumed to be at equilibrium with the lean oil at the top of the tower.

When $A^0 = 1$, the right-hand side of Eq. (14-121) becomes indeterminate. The solution for this case is

$$(Y_1 - Y_2)/(Y_1 - mX_2) = N/(N + 1) \qquad (14\text{-}123)$$

For systems in which the absorption factor A^0 for each component is not constant throughout the tower, an effective absorption factor for use in the equations just presented can be estimated by the Edmister formula

$$A_e^0 = \sqrt{A_1^0(A_2^0 + 1) + 0.25} - 0.5 \qquad (14\text{-}124)$$

This procedure is a reasonable approximation only when no pinch points exist within the tower and when the absorption factors vary in a regular manner between the bottom and the top of the tower.

Example 9 A hydrocarbon feed gas is to be treated in an existing four-theoretical-tray absorber to remove butane and heavier components. The recovery specification for the key component, butane, is 75 percent. The composition of the exit gas from the absorber and the required liquid-to-gas ratio are to be estimated. The feed-gas composition and the equilibrium K values for each component at the temperature of the (solute-free) lean oil are presented in the following table:

Component	Mole %	K value
Methane	68.0	74.137
Ethane	10.0	12.000
Propane	8.0	3.429
Butane	8.0	0.833
Pentane	4.0	0.233
C_6 plus	2.0	0.065

For $N = 4$ and $Y_2/Y_1 = 0.25$, the value of A^0 for butane is found to be equal to 0.89 from Eq. (14-121) by using a trial-and-error method. The values of A^0 for the other components are then proportional to the ratios of their K values to that of butane. For example, $A^0 = 0.89(0.833/12.0) = 0.062$ for ethane. The values of A^0 for each of the other components and the exit-gas composition as computed from Eq. (14-121) are shown in the following table:

Component	A^0	Y_2, mol/mol feed	Exit gas, mole %
Methane	0.010	67.3	79.1
Ethane	0.062	9.4	11.1
Propane	0.216	6.3	7.4
Butane	0.890	2.0	2.4
Pentane	3.182	0.027	0.03
C_6 plus	11.406	0.0012	0.0014

The molar liquid-to-gas ratio required for this separation is computed as $L_M^s/G_M^0 = A^0 \times K = 0.89 \times 0.833 = 0.74$.

We note that this example is the analytical solution to the graphical design problem shown in Fig. 14-12, which therefore is the design diagram for this system.

Stripping Equations for Dilute Multicomponent Systems The modified Kremser formula for stripping in a dilute multicomponent system is

$$\frac{X_2 - X_1}{X_2 - X_1^\circ} = \frac{(S^0)^{N+1} - (S^0)}{(S^0)^{N+1} - 1} \qquad (14\text{-}125)$$

where $S^0 = mG_M^0/L_M^s$ and $X_1^\circ = Y_1/m$ and the subscripts 1 and 2 refer to the bottom and top of the tower respectively. The corresponding Edmister formula is

$$S_e^0 = \sqrt{S_2^0(S_1^0 + 1) + 0.25} - 0.5 \qquad (14\text{-}126)$$

ECONOMIC DESIGN OF ABSORPTION SYSTEMS

Rational selection of the design basis and specification of the operating variables in an absorption system involves a number of important economic balances, as pointed out by Colburn (*Collected Papers on the Teaching of Chemical Engineering*, American Institute of Chemical Engineers, New York, 1940, p. 269); also see Sherwood and Pigford, *Absorption and Extraction*, McGraw-Hill, New York, 1952, p. 451.

The design variables to be specified include the type of tower to be used, the tower internals (e.g., packing type and size), the operating pressure, the liquid-to-gas ratio, the tower diameter (i.e., gas velocity), and the tower height (i.e., the exit-gas strength). Once these variables have been selected, the experimental data for transfer coefficients or heights can be established and the appropriate equations can be employed in performing the actual design.

Equipment Selection Detailed cost data for tray towers and packed towers and a discussion of the relative advantages and disadvantages of plate towers versus packed towers from an equipment-performance and operational point of view are to be found in Sec. 18. Often one will find that practical considerations, engineering judgments, or previous commercial experience may outweigh the purely economic factors associated with selecting the type of equipment.

For example, in highly corrosive systems, in systems involving heat-sensitive liquids (in which liquid holdup must be minimized), or in systems involving liquids that have a high foaming tendency, packed towers normally are preferred. On the other hand, when fouling is a problem owing to the formation of solid deposits, plate towers often incur lower maintenance costs than do packed towers.

In systems involving extremely low liquid rates, plate towers are preferable owing to their better gas-liquid-contacting efficiency. In systems in which the liquid-to-gas ratio is extremely large, plate towers are preferred owing to the need to minimize axial-dispersion or back-mixing effects. Plate towers also are preferred when either intercoolers or internal cooling coils are required.

In vacuum operations, packed towers often exhibit a lower pressure drop than plate towers, implying lower capital and operating costs for the vacuum system. Also, other things being equal, packed towers usually will be less expensive than plate towers when the tower diameter is less than about 0.75 m (2½ ft). For larger-diameter towers, the choice must be justified on the basis of detailed cost comparisons.

An interesting point in the selection of packed towers is that the cost of the shell varies approximately in proportion to the tower diameter and the cost of packing pieces per unit volume (dollars per cubic meter) increases with decreasing size. For a given liquid-to-gas ratio and for equal pressure-drop operation, smaller packings require a larger-diameter tower, and for equal values of the pressure drop, the K_Ga values normally will increase only slightly as the packing size decreases. Thus, economic considerations tend to favor the selection of larger packing pieces as long as the tower diameter is more than about 10 times the packing size.

Operating Pressure As a general rule, an absorber should be operated at the highest possible pressure consistent with the requirements of other steps in the process. Thus, the optimum operating pressure normally is determined by considering alternative designs of the complete process. In some applications the gas is compressed before it is fed to the absorber, thus increasing its solubility and increasing the allowable gas mass velocity through the tower. These gains can be obtained only at the expense of the cost of compression, so the optimum operating pressure must be determined by an economic balance in this case.

Liquid-to-Gas Ratio The design parameter mG_M/L_M is of primary importance in determining the height of a transfer unit and the number of transfer units. The liquid-to-gas ratio (L/G) also affects the choice of column diameter. When the solute gas is dilute, the parameter mG_M/L_M is apt to be nearly constant throughout the tower. When the solute gas is not dilute, the heat of solution may cause a temperature rise, resulting in a larger value of m at the bottom of the tower than at the top, and when mG_M/L_M is not constant, the choice of an appropriate value for L/G is difficult. In the case of nearly complete absorption, most of the transfer units will be required in the dilute end near the top of the tower. Thus, conditions at the dilute end may be employed in establishing a preliminary design, using the temperature of the inlet liquid as the basis.

Choice of the magnitude of mG_M/L_M often is based on economic considerations. Larger values of mG_M/L_M lead to reduced solvent-circulation rates and a more concentrated solute-rich liquid leaving the tower. This in turn results in reduced pumping costs and a lower cost of stripping the solute from the solvent. On the other hand, the higher values of mG_M/L_M lead to taller and more expensive absorption towers and to higher losses of solute in the exit gases. A typical calculation of the economic optimum liquid-to-gas ratio in the absorption of acetone from air, using water as the solvent, resulted in a value of $mG_M/L_M = 0.6$. For less valuable solutes, a larger value of this ratio would be employed. In the design of stripping towers, the optimum value of the ratio L_M/mG_M normally will be in the range of 0.5 to 0.8.

Equation (14-127) represents a balance of the costs of absorption and of subsequent stripping to recover the solute in a system such as that shown in Fig. 14-5. In deriving this equation it was assumed that the absorbed solute is stripped from the solute-rich solvent by distillation and that the overhead product from this distillation (stripping) tower is essentially pure solute. Moreover, the liquid-solvent stream leaving the bottom of the stripping tower was assumed to be completely denuded of solute prior to being recirculated back to the top of the absorber.

$$\left(\frac{L_M}{m_2G_M} - 1\right)^2 = \frac{BC_3H_{OG}(K_D - 1)}{C_5\theta rG_Mm_2} \times \text{DF} \qquad (14\text{-}127)$$

where
$$B = \left[1 + n\left(\frac{L_M}{m_2G_M} - 1\right)\right] \times 2.3\log_{10}\left[\left(\frac{y_1}{y_2}\right)\frac{(1 - m_2G_M/L_M)^2}{1 - K_1G_M/L_M}\right] - \frac{(1 - m_2G_M/L_M) - 2(K_1/m_2 - 1)}{1 - K_1G_M/L_M}$$

K_D = $y°/x$ of the feed to the stripping tower at its boiling point

m_2 = slope of the equilibrium curve $y°/x$ at the temperature of the inlet liquid to the absorption tower

K_1 = $y°/x$ at the temperature of the rich absorption liquor as it leaves the bottom of the absorption tower

H_{OG} = height of one overall gas-phase transfer unit, m

C_3 = annual cost of apparatus and power for the absorption tower, $/(year·m³) = $C_1[(G_{opt}/G) + 0.5(G/G_{opt})^2]$ (The terms C_1, G, and G_{opt} are defined in the next subsection.)

C_5 = total cost of stripping operation, expressed as $/kmol of vapor supplied to the stripper (includes fixed charges, cost of cooling water, and cost of steam)

θ = s/year of operation (8000 h = 2.88×10^7 s)

r = (actual reflux ratio in distillation tower) ÷ (minimum reflux ratio, defined as ratio of reflux to product)

G_M = molal gas velocity through absorber, kmol/(s·m²)

n = exponent in the relation $H_{OG} \sim (G/L)^n$

y_1/y_2 = optimum ratio of solute mole fractions in gas stream flowing through the absorber (See later discussion of tower height versus exit-gas strength.)

DF = design safety factor as described earlier under "Heat Effects in Gas Absorption"

In these definitions, SI units have been indicated. However, comparable U.S. customary units also apply. (See Table 14-2.)

Tower Diameter (Gas Velocity) The design gas velocity is selected by considering, first, the safe operating velocity with respect to flooding and, second, the optimum velocity as calculated by an economic balance between column costs and power costs. Data on flooding velocities are given in Sec. 18. The design velocity normally is specified at about 60 percent of the flooding value; this allows for temporary fluctuations in flow rates and leaves a margin of safety to avoid shutdowns of the tower due to flooding.

After checking on the safe velocity with respect to flooding, the economic optimum velocity can be estimated by the equation

$$G_{opt} = (C_1 \rho_G^2 / 2C_2 \theta' b)^{1/3} \tag{14-128}$$

where b is the constant coefficient in the pressure-drop equation

$$\Delta p / h_T = bG^2 / \rho_G \tag{14-129}$$

The coefficient b is obtained by fitting data from the tower-packing pressure-drop performance chart near the vicinity of the expected operating point. For example, Leva [*Chem. Eng. Prog. Symp. Ser.*, **50**(10), 51 (1954)] employed an equation of the following form in fitting packing pressure-drop data:

$$\Delta P / h_T = \alpha 10^{\beta L} G^2 / \rho_G \tag{14-130}$$

The terms in Eqs. (14-128) through (14-130) are defined as follows:

G = gas mass velocity, kg/(s·m²); G_{opt} = optimum
L = liquid mass velocity, kg/(s·m²)
ρ_G = gas density, kg/m³
C_1 = annual cost of tower packing and shell, \$/(year·m³)
C_2 = cost of delivered energy, \$/kWh; 1 kW = 1 kN·m/s
θ' = operating time, h/year
Δp = tower pressure drop, kPa (= kN/m²)
h_T = tower height, m
b = constant in Eq. (14-129), m⁻¹

The coefficients α and β are constants determined by fitting the curves from the pressure-drop performance charts. A convenient factor for these pressure-drop calculations is kPa/m = (in H_2O/ft) × 0.817. SI units have been indicated. However, compatible U.S. customary units could also be used.

The annual cost of the operation in dollars per year per cubic meter for any gas velocity, including the costs of the tower packing and shell plus the cost of power, can be estimated by the equation

$$C_3 = C_1[(G_{opt}/G) + 0.5(G/G_{opt})^2] \tag{14-131}$$

from which it can be seen that when $G = G_{opt}$, the annual cost of tower and energy at the economic optimum is given by

$$C_3 = 1.5C_1 \tag{14-132}$$

Tower Height (Exit-Gas Strength) The optimum solute concentration in the exit gas from an absorber may be determined by an economic balance between the cost of lost solute and the cost of additional tower height. For packed towers operating under isothermal conditions,

$$(y_2 - mx_2)_{opt} = \frac{C_3 H_{OG}(\text{DF})}{C_4 \theta G(1 - mG_M/L_M)} \tag{14-133}$$

where C_4 = value of the solute at its concentration in the exit liquor, \$/kg of solute; and all the other terms are as defined previously.

A similar relation for plate absorbers is as follows:

$$(y_2 - mx_2)_{opt} = \frac{C_6(\text{DF})}{C_4 \theta GE \ln (L_M/mG_M)} \tag{14-134}$$

where C_6 = annual unit cost of tower and energy for pressure drop, \$/(year·plate)(m² cross-section) and E = overall plate efficiency (fractional).

For stripping towers the optimum exit-liquor strength depends upon a balance between the cost of lost solute in the exit liquor and the cost of additional tower height required for improved stripping. For packed stripping towers,

$$(x_2 - y_2/m)_{opt} = \frac{C_3 H_{OL}(\text{DF})}{C_4 L \theta(1 - L_M/mG_M)} \tag{14-135}$$

and for plate stripping towers,

$$(x_2 - y_2/m)_{opt} = \frac{C_6(\text{DF})}{C_4 L \theta E \ln (mG_M/L_M)} \tag{14-136}$$

Temperature of Lean Solvent Continuous operation of an absorber-stripper combination such as that shown in Fig. 14-5 calls for the use of a heat exchanger to recover heat from the stripper effluent before it is returned to the top of the absorber. This stream frequently is cooled further in a second exchanger before entering the absorber or else is cooled by intercoolers located at intermediate points in the absorber. Cooling the liquid stream increases the solubility of the solute (decreases m) and thereby decreases the required liquid circulation rate.

The economic balance is between the savings in stripping costs at the lower liquid flow rates and the cost of additional heat-exchange equipment. Optimum conditions are best found by considering alternative complete designs. Normally there will be little justification for cooling the solvent below the solute-rich inlet-gas temperature. However, cooling of the solute-rich gas and solvent to as low as about −18°C (0°F) often can be justified for large high-pressure natural-gas absorption systems.

Optimization of Multicomponent Systems The simplified equations presented previously can be only rough guides to optimum design conditions, especially when several solutes are being recovered or when the fractional recovery of an important component of the gas is relatively low. In such cases, detailed computations must be made for alternative designs and alternative operating parameters.

In many cases the optimum gas velocity will be near that given by Eq. (14-128) and the optimum value of L_M/G_M will be such that the operating line is nearly parallel to the equilibrium line for the most volatile component that it is desired to recover nearly completely.

Example 10 An economic optimum packed-tower design is desired for reducing the concentration of acetone in an air stream from 2.0 mole percent to 50 ppmv. The following conditions and data are given:

a. Absorption tower

Scrubbing agent = water at 25°C
Packing material = 25-mm Raschig rings
H_{OG} = 0.70 m (2.3 ft)
DF = 1.5 (see "Heat Effects in Gas Absorption")
θ = (2.88)(10⁷) s/year
θ' = 8000 h/year
y_1 = 0.02
y_2 = 0.00005 (design)
p_T = 101.3 kPa (1 atm)
ρ_G = 1.204 kg/m³ (0.075 lb/ft³)
m_2 = 2.09 at 25°C
K_1 = 2.79 at 31°C (see "Heat Effects")
n = 0.5
$b = (\Delta p/h_T)(\rho_G/G^2) = 0.5 \times 10^{0.05L}$ (correlated from performance data on 25-mm Raschig rings)

b. Stripping tower

K_D = 33 at 94°C and 101.3 kPa (1 atm)
r = 1.25

c. Cost data

C_1 = \$160/(year·m³), assuming 10-year amortization
C_2 = \$0.028/(kWh)
$C_3 = 1.5C_1$ = \$240/(year·m³)
C_4 = \$0.264/kg (12 cents/lb)
C_5 = \$0.050/kmol (\$1.26/1000 lb steam)

Finding the best balance between tower height, tower diameter, and liquid-to-gas ratio requires a trial-and-error procedure. For example, since $A = L_M/m_2G_M$, the calculation for liquid mass flow rate is given by

$$L = (18/29)m_2AG = 1.297AG \tag{i}$$

and the Leva equation given previously for the pressure-drop coefficient b can be expressed as

$$b = 0.5 \times 10^{0.05L} = 0.5 \times 10^{0.0649AG} \tag{ii}$$

Substitution of Eq. (ii) into Eq. (14-128) results in the following trial-and-error equation for G_{opt}:

$$G_{opt} = \left(\frac{(160)(1.204)^2}{2(0.028)(8000)b}\right)^{1/3} = (1.035 \times 10^{-0.0649AG})^{1/3} \qquad \text{(iii)}$$

An initial value of the absorption factor A can be obtained by solving the (isothermal) equation for optimum exit-gas concentration with the starting assumption that $G = 1.0$ kg/(s·m^2). The result is as follows:

$$A = \frac{(0.264)(2.88 \times 10^7)(1.0)(0.00005)}{(240)(0.70)(1.5)} = 1.509$$

When this preliminary estimate of A is substituted into Eq. (iii), one obtains by trial and error a value of $G_{opt} = 0.942$ kg/(s·m^2).

The next step is to employ Eq. (14-127) for computing the optimum value of the absorption factor A:

$$(A - 1)^2 = \frac{B(240)(0.70)(33 - 1)(1.5)(29)}{(0.05)(2.88 \times 10^7)(1.25)(2.09)(0.942)} = 0.06599B \qquad \text{(iv)}$$

where $B = [1 + 0.5(A - 1)] \ln [400(1 - A^{-1})^2/(1 - 1.335A^{-1})] - [(0.33 - A^{-1})/(1 - 1.335A^{-1})]$

and $K_1/m_2 = (2.79)/(2.09) = 1.335$. A trial-and-error solution of Eq. (iv) starting with the initial guess of $A = 1.509$ converges to a final value of $A = 1.766$. A recheck of the solution of Eq. (iii) using this new value of A results in $G_{opt} = 0.932$, in good agreement with the assumed value of 0.942 kg/(s·m^2). A summary of the calculation procedure is given in the following table:

Initial guesses		Calculated values		Equation used
G	A	G	A	
1.0	.....	.....	1.509	Exit-gas
1.0	1.509	0.942	.....	(iii)
0.942	1.509	.....	1.766	(iv)
0.942	1.766	0.932	.....	(iii)

The final design values for the tower are $m_2G_M/L_M = 0.566$ and $G_{opt} = 0.935$ kg/(s·m^2).

This solution to the design problem did not assume isothermal operation of the tower but instead employed a value of K_1 corresponding to a temperature rise of 6°C as derived in Example 7. When isothermal operation at 25°C is assumed, $K_1 = m_2 = 2.09$ and Eq. (iv) reduces to

$$(A - 1)^2 = 0.06599B \qquad \text{(v)}$$

where $B = [1 + 0.5(A - 1)] \ln [400(1 - A^{-1})] - 1$

The trial-and-error solution of Eq. (v) is $A = 1.604$, and a recheck of Eq. (iii) results in $G_{opt} = 0.938$ kg/(s·m^2). Thus, for the isothermal case $m_2G_M/L_M = A^{-1} = 0.623$.

As a final check on the validity of the design, the pressure drop is computed by the Leva equation:

$$(\Delta p/h_T) = (0.5)(10^{0.0649AG})G^2/\rho_G \qquad \text{(vi)}$$

The calculated pressure drop for both the nonisothermal and the isothermal cases is 0.46 kPa/m, or 0.56 inH_2O/ft. Flooding may be considered to begin at a pressure drop of approximately 2 inH_2O/ft, and the design percent of flooding can be calculated as

$$\text{Percent flood} = 100(0.56/2)^{1/2} = 53 \text{ percent}$$

Thus the economic optimum design appears to be an acceptable design from a practical standpoint, in that it allows sufficient room for flow-rate fluctuations during operation.

The tower diameter corresponding to the design value selected for G can be calculated by the formula

$$d_T = (4n_G/G)^{1/2} \qquad \text{(14-137)}$$

where d_T = tower diameter, m; and n_G = inlet-gas feed rate, kg/s. In our example, for an air feed rate of 50 kmol/h or 0.403 kg/s (39,500 scfh at 0°C) the tower diameter is calculated as

$$d_T = (4 \times 0.403/3.14 \times 0.935)^{1/2} = 0.74 \text{ m (2 ft 5 in)}$$

For diameters smaller than this, a packed tower is the recommended choice.

ABSORPTION WITH CHEMICAL REACTION

Introduction The majority of present-day commercial gas-absorption processes involve systems in which chemical reactions take place in the liquid phase. Chemical reactions generally enhance the rate of absorption and increase the capacity of the liquid solution to dissolve the solute when compared with physical-absorption systems.

A necessary prerequisite to understanding the subject of absorption accompanied by chemical reaction is the development of a thorough understanding of the principles involved in physical gas absorption. An introductory discussion of physical absorption and of the effects of chemical reaction on the mass-transfer coefficients $\hat{k}_G$ and $\hat{k}_L$ was presented earlier under the heading "Interphase Mass Transfer." A review of that discussion is recommended.

There are a number of excellent classic papers on the subject of absorption with chemical reaction. Many of these are referenced below. A particularly noteworthy classic is the review paper "The Absorption of Carbon Dioxide into Solutions of Alkalis and Amines" by Danckwerts and Sharma [*Chem. Eng. (London)*, **CE 244–280**, October 1966]. This paper is recommended reading for anyone wishing to master both the practical and the theoretical aspects of absorption with chemical reaction.

Recommended Overall Design Strategy When considering the design of a gas-absorption system involving chemical reactions, the following procedure is recommended:

1. Consider the possibility that the physical design methods described earlier in this section may be applicable.

2. Determine whether commercial design overall $\hat{K}_Ga$ values are available for use in conjunction with the traditional design method, being careful to note whether or not the conditions under which the $\hat{K}_Ga$ data were obtained are essentially the same as for the new design. Contact the various tower-packing vendors for information as to whether $\hat{K}_Ga$ data are available for your system and conditions.

3. Consider the possibility of scaling up the design of a new system from experimental data obtained in a laboratory-bench scale or a small pilot-plant unit.

4. Consider the possibility of developing for the new system a rigorous, theoretically based design procedure which will be valid over a wide range of design conditions.

These topics are discussed in the subsections that follow.

Applicability of Physical Design Methods Physical design methods such as the classical isothermal design method or the classical adiabatic design method may be applicable for systems in which chemical reactions are either extremely fast or extremely slow or when chemical equilibrium is achieved between the gas and liquid phases.

If the liquid-phase reaction is **extremely fast** and irreversible, the rate of absorption may in some cases be completely governed by the gas-phase resistance. For practical design purposes one may assume (for example) that this **gas-phase mass-transfer limited** condition will exist when the ratio y_i/y is less than 0.05 everywhere in the apparatus.

From Eq. (14-2) one can readily show that this condition on y_i/y requires that the ratio x/x_i be negligibly small (i.e., a fast reaction) and that the ratio $mk_G/k_L = mk_G/k_L^0\phi$ be less than 0.05 everywhere in the apparatus. The ratio $mk_G/k_L^0\phi$ will be small if the equilibrium back pressure of the solute over the liquid solution is small (i.e., small m; high reactant solubility), or the reaction-enhancement factor $\phi = k_L/k_L^0$ is very large, or both.

As discussed later, the reaction-enhancement factor ϕ will be large for all extremely fast pseudo-first-order reactions and will be large for extremely fast second-order irreversible reaction systems in which there is a sufficiently large excess of liquid-phase reagent. When the rate of an extremely fast second-order irreversible reaction system $A + \nu B \rightarrow$ products is limited by the availability of the liquid-phase reagent B, then the reaction-enhancement factor may be estimated by the formula $\phi = 1 + B^0/\nu c_i$. In systems for which this formula is applicable, it can be shown that the interface concentration y_i will be equal to zero whenever the ratio $k_g y\nu/k_L^0 B^0$ is less than or equal to unity.

Figure 14-13 illustrates the gas-film and liquid-film concentration profiles one might find in an extremely fast (gas-phase mass-transfer limited) second-order irreversible reaction system. The solid curve for reagent B represents the case in which there is a large excess of bulk-liquid reagent B^0. The dashed curve in Fig. 14-13 represents the case in which the bulk concentration B^0 is not sufficiently large to prevent the depletion of B near the liquid interface and for which the equation $\phi = 1 + B^0/\nu c_i$ is applicable.

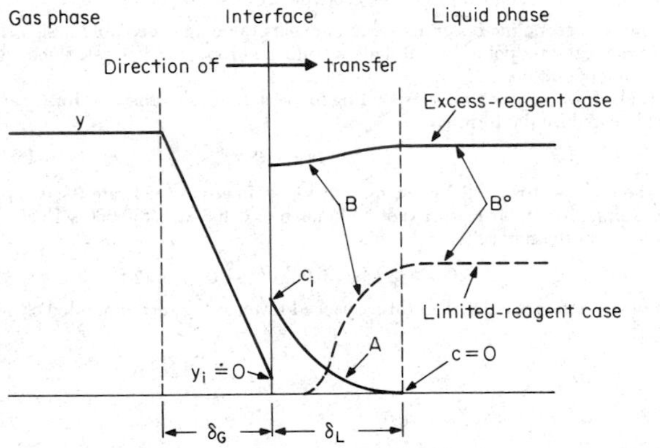

FIG. 14-13 Gas-phase and liquid-phase solute-concentration profiles for an extremely fast (gas-phase mass-transfer limited) irreversible reaction system $A + \nu B \rightarrow$ products.

Whenever these conditions on the ratio y_i/y apply, the design can be based upon the physical rate coefficient k_G or upon the height of one gas-phase mass-transfer unit H_G. The gas-phase mass-transfer limited condition is approximately valid, for instance, in the following systems: absorption of NH_3 into water or acidic solutions, vaporization of water into air, absorption of H_2O into concentrated sulfuric acid solutions, absorption of SO_2 into alkali solutions, absorption of H_2S from a dilute-gas stream into a strong alkali solution, absorption of HCl into water or alkaline solutions, or absorption of Cl_2 into strong alkali.

Example 3 illustrates the design of a gas-phase mass-transfer limited system involving a rapid irreversible chemical reaction.

When liquid-phase chemical reactions are **extremely slow,** the gas-phase resistance can be neglected and one can assume that the rate of reaction has a predominant effect upon the rate of absorption. In this case the differential rate of transfer is given by the equation

$$dn_A = R_A f_H S\, dh = (k_L^0 a/\rho_L)(c_i - c) S\, dh \qquad (14\text{-}138)$$

where n_A = rate of solute transfer, R_A = volumetric reaction rate, a function of c and T, f_H = fractional liquid volume holdup in tower or apparatus, S = tower cross-sectional area, h = vertical distance, k_L^0 = liquid-phase mass-transfer coefficient for pure physical absorption, a = effective interfacial mass-transfer area per unit volume of tower or apparatus, ρ_L = average molar density of liquid phase, c_i = solute concentration in liquid at gas-liquid interface, and c = solute concentration in bulk liquid.

Although the right-hand side of Eq. (14-138) remains valid even when chemical reactions are extremely slow, the mass-transfer driving force may become increasingly small, until finally $c \doteq c_i$. For extremely slow first-order irreversible reactions, the following rate expression can be derived from Eq. (14-138):

$$R_A = k_1 c = k_1 c_i/(1 + k_1 \rho_L f_H/k_L^0 a) \qquad (14\text{-}139)$$

where k_1 = first-order reaction rate coefficient.

For **dilute systems** in countercurrent absorption towers in which the equilibrium curve is a straight line (i.e., $y_i = mx_i$) the differential relation of Eq. (14-138) is formulated as

$$dn_A = -G_M S\, dy = k_1 c f_H S\, dh \qquad (14\text{-}140)$$

where G_M = molar gas-phase mass velocity and y = gas-phase solute mole fraction.

Substitution of Eq. (14-139) into Eq. (14-140) and integration lead to the following relation for an **extremely slow first-order reaction** in an absorption tower:

$$y_2 = y_1 \exp(-\gamma) \qquad (14\text{-}141)$$

where

$$\gamma = \frac{k_1 \rho_L f_H h_T/m G_M}{(1 + k_1 \rho_L f_H/k_L^0 a)} \qquad (14\text{-}142)$$

In Eq. (14-141) the subscripts 1 and 2 refer to the bottom and the top of the tower respectively.

The Hatta number N_{Ha} usually is employed as the criterion for determining whether or not a reaction can be considered extremely slow. For extremely slow reactions a reasonable criterion is

$$N_{Ha} = \sqrt{k_1 D_A}/k_L^0 \leq 0.3 \qquad (14\text{-}143)$$

where D_A = liquid-phase diffusion coefficient of the solute in the solvent. Figure 14-14 illustrates the concentration profiles in the gas and liquid films for the case of an extremely slow chemical reaction.

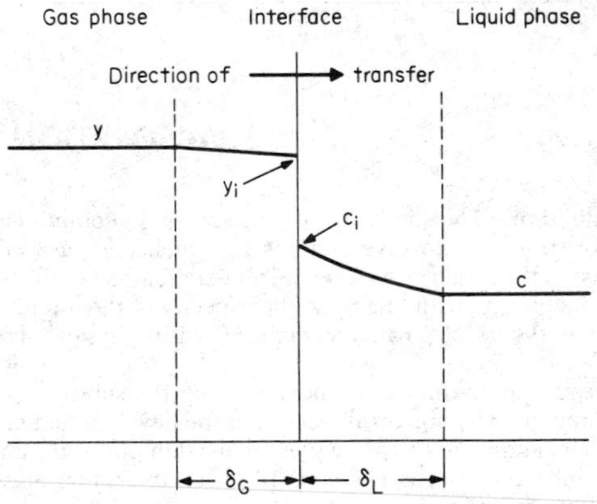

FIG. 14-14 Gas-phase and liquid-phase solute-concentration profiles for an extremely slow (kinetically limited) reaction system for which N_{Ha} is less than 0.3.

We note that when the second term in the denominator of Eq. (14-142) is small, the liquid holdup in the tower can have a significant influence upon the rate of absorption if an extremely slow chemical reaction is involved.

When **chemical equilibrium** is achieved quickly throughout the liquid phase (or can be assumed to exist), the problem becomes one of properly defining the physical and chemical equilibria for the system. It sometimes is possible to design a plate-type absorber by assuming chemical-equilibrium relationships in conjunction with a stage efficiency factor as is done in distillation calculations. Rivas and Prausnitz [*Am. Inst. Chem. Eng. J.*, **25,** 975 (1979)] have presented an excellent discussion and example of the correct procedures to be followed for systems involving chemical equilibria.

Traditional Design Method The traditionally employed conventional procedure for designing packed-tower gas-absorption systems involving chemical reactions makes use of overall volumetric mass-transfer coefficients as defined by the equation

$$K'_G a = n_A/(h_T S p_T \Delta y^\circ_{lm}) \tag{14-60}$$

where $K'_G a$ = overall volumetric mass-transfer coefficient, n_A = rate of solute transfer from the gas to the liquid phase, h_T = total height of tower packing, S = tower cross-sectional area, p_T = total system pressure, and Δy°_{lm} is defined by the equation

$$\Delta y^\circ_{lm} = \frac{(y - y^\circ)_1 - (y - y^\circ)_2}{\ln[(y - y^\circ)_1/(y - y^\circ)_2]} \tag{14-62}$$

in which subscripts 1 and 2 refer to the bottom and top of the absorption tower respectively, y = mole-fraction solute in the gas phase, and y° = gas-phase solute mole fraction in equilibrium with bulk-liquid-phase solute concentration x. When the equilibrium line is straight, $y^\circ = mx$.

The traditional design method normally makes use of overall $K'_G a$ values even when resistance to transfer lies predominantly in the liquid phase. For example, the CO_2-NaOH system most commonly used for comparing the $K'_G a$ values of various tower packings is a liquid-phase-controlled system. When the liquid phase is controlling, extrapolation to different concentration ranges or operating conditions is not recommended since changes in the reaction mechanism can cause k_L to vary unexpectedly and the overall $K'_G a$ values do not explicitly show such effects.

Overall $K'_G a$ data may be obtained from tower-packing vendors for many of the established commercial gas-absorption processes. Such data often are based either upon tests in large-diameter test units or upon actual commercial operating data. Since extrapolation to untried operating conditions is not recommended, the preferred procedure for applying the traditional design method is equivalent to duplicating a previously successful commercial installation. When this is not possible, then a commercial demonstration at the new operating conditions may be required, or else one could consider using some of the more rigorous methods described later.

Aside from the lack of an explicitly defined liquid-phase-resistance term, the limitations on the use of Eq. (14-60) are related to the fact that its derivation implicitly assumes that the system is dilute ($y_{BM} \doteq 1$) and that the operating and equilibrium curves are straight lines over the range of tower operation. Also, Eq. (14-60) is strictly valid only for the temperature and pressure at which the original test was run even though the total pressure p_T appears in the denominator.

The ambiguity of the total pressure effect can be seen by a comparison of the gas-phase- and liquid-phase-controlled cases: when the gas phase controls, the liquid-phase resistance is negligible and $K_G a = K'_G a p_T$ is independent of the total pressure. For this case the coefficient $K'_G a$ is inversely proportional to the total system pressure as shown in Eq. (14-60). On the other hand, when the liquid phase controls, the correct equation is

$$K'_G a = K_G a/p_T = k_L a/H \tag{14-144}$$

where H is the Henry's-law constant defined as $H = p_i/x_i$. This equation indicates that $K'_G a$ will be independent of the total system pressure as long as the Henry's-law constant H does not depend on the total pressure (this will be true only for relatively low pressures). On the basis of this comparison it should be clear that the effects of total system pressure upon $K'_G a$ are not properly defined by Eq. (14-60), especially in cases in which the liquid-phase resistance cannot be neglected.

In using Eq. (14-60), therefore, it should be understood that the numerical values of $K'_G a$ will be a complex function of the pressure, the temperature, the type and size of tower packing employed, the liquid and gas mass flow rates, and the system composition (for example, the degree of conversion of the liquid-phase reactant).

Figure 14-15 illustrates the influence of system composition and **degree of reactant conversion** upon the numerical values of $K'_G a$ for the absorption of CO_2 into sodium hydroxide solutions at constant conditions of temperature, pressure, and type of packing. An excellent experimental study of the influence of operating variables upon overall $K'_G a$ values is that of Field et al. (*Pilot-Plant Studies of the Hot Carbonate Process for Removing Carbon Dioxide and Hydrogen Sulfide,* U.S. Bureau of Mines Bulletin 597, 1962).

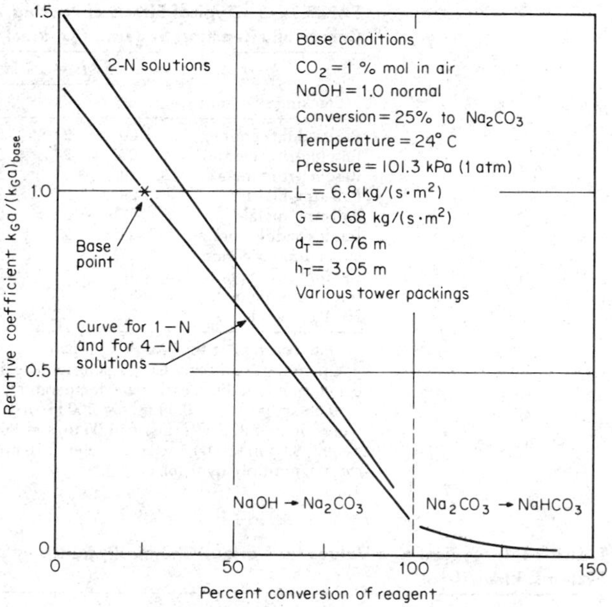

FIG. 14-15 Effects of reagent-concentration and reagent-conversion level upon the relative values of $K_G a$ in the CO_2-NaOH-H_2O system. [*Adapted from Eckert et al.,* Ind. Eng. Chem., *59(2), 41 (1967).*]

Table 14-3 illustrates the observed variations in $\hat{K}_G a$ values for different packing types and sizes for the CO_2-NaOH system at a 25 percent reactant-conversion level for two different liquid flow rates. The lower rate of 2.7 kg/(s·m²) or 2000 lb/(h·ft²) is equivalent to 4 (U.S. gal/min)/ft² and is typical of the liquid rates employed in fume scrubbers. The higher rate of 13.6 kg/(s·m²) or 10,000 lb/(h·ft²) is equivalent to 20 (U.S. gal/min)/ft² and is more typical of absorption towers such as are used in CO_2 removal systems, for example. We note also that two different gas velocities are represented in the table, corresponding to superficial velocities of 0.59 and 1.05 m/s (1.94 and 3.44 ft/s).

Table 14-4 presents a typical range of $\hat{K}_G a$ values for chemically reacting systems. The first two entries in the table represent systems that can be designed by the use of purely physical design methods, for they are completely gas-phase mass-transfer limited. To ensure a negligible liquid-phase resistance in these two tests, the HCl was absorbed into a solution maintained at less than 8 percent weight HCl and the NH_3 was absorbed into a water solution maintained below pH 7 by the addition of acid. The last two entries in Table 14-4 represent liquid-phase mass-transfer limited systems.

The effects of system pressure on these $\hat{K}_G a$ values can be estimated as in Eq. (14-144) by noting that $\hat{K}_G a = K_G a y^\circ_{BM} = K'_G a y^\circ_{BM} p_T$ and recalling that (1) in gas-phase mass-transfer limited systems $\hat{K}_G a = \hat{k}_G a$ and is independent of system pressure, and (2) for liquid-phase mass-transfer limited systems in which H is constant the $\hat{K}_G a$ values can be corrected to other pressures by the relation $\hat{K}_G a$ at p_2 = ($\hat{K}_G a$ at p_1) × p_2/p_1. When both resistances are significant, it is advisable to employ experimentally derived corrections. In any case it is inadvisable to make large pressure corrections by these procedures without experimental verification.

Scaling Up from Laboratory or Pilot-Plant Data For many years it has been thought by practitioners of the art of gas absorption that it would be impossible to carry out an absorption process in a laboratory apparatus or small-scale pilot plant in such a way that the data could be of use in the design of a commercial absorption unit.

TABLE 14-3 Typical Effects of Packing Type, Size, and Liquid Rate on $\hat{K}_G a$ in a Chemically Reacting System, $\hat{K}_G a$, kmol/(h·m³)

	L = 2.7 kg/(s·m²)				L = 13.6 kg/(s·m²)			
Packing size, mm	25	38	50	75–90	25	38	50	75–90
Berl-saddle ceramic	30	24	21	..	45	38	32	..
Raschig-ring ceramic	27	24	21	..	42	34	30	..
Raschig-ring metal	29	24	19	..	45	35	27	..
Pall-ring plastic	29	27	26*	16	45	42	38*	24
Pall-ring metal	37	32	27	21*	56	51	43	27*
Intalox-saddle ceramic	34	27	22	16*	56	43	34	26*
Super-Intalox ceramic	37*	..	26*	..	59*	..	40*	..
Intalox-saddle plastic	40*	..	24*	16*	56*	..	37*	26*
Intalox-saddle metal	43*	35*	30*	24*	66*	58*	48*	37*
Hy-Pak metal	35	32*	27*	18*	54	50*	42*	27*

Data courtesy of the Norton Company.

Operating conditions: CO_2, 1 percent mole in air; NaOH, 4 percent weight (1 normal); 25 percent conversion to sodium carbonate; temperature, 24°C (75°F); pressure, 98.6 kPa (0.97 atm); gas rate = 0.68 kg/(s·m²) = 0.59 m/s = 500 lb/(h·ft²) = 1.92 ft/s except for values with asterisks, which were run at 1.22 kg/(s·m²) = 1.05 m/s = 900 lb/(h·ft²) = 3.46 ft/s superficial velocity; packed height, 3.05 m (10 ft); tower diameter, 0.76 m (2.5 ft). To convert table values to units of (lb·mol)/(h·ft³), multiply by 0.0624.

TABLE 14-4 Typical $\hat{K}_G a$ Values for Various Chemically Reacting Systems, kmol/(h·m³)

Gas-phase reactant	Liquid-phase reactant	$\hat{K}_G a$	Special conditions
HCl	H_2O	353	Gas-phase limited
NH_3	H_2O	337	Gas-phase limited
Cl_2	NaOH	272	8% weight solution
SO_2	Na_2CO_3	224	11% weight solution
HF	H_2O	152	
Br_2	NaOH	131	5% weight solution
HCN	H_2O	114	
HCHO	H_2O	114	Physical absorption
HBr	H_2O	98	
H_2S	NaOH	96	4% weight solution
SO_2	H_2O	59	
CO_2	NaOH	38	4% weight solution
Cl_2	H_20	8	Liquid-phase limited

Data courtesy of the Norton Company.

Operating conditions (see text): 38-mm ceramic Intalox saddles; solute gases, 0.5–1.0 percent mole; reagent conversions = 33 percent; pressure, 101 kPa (1 atm); temperature, 16–24°C; gas rate = 1.3 kg/(s·m²) = 1.1 m/s; liquid rates = 3.4 to 6.8 kg/(s·m²); packed height, 3.05 m; tower diameter, 0.76 m. Multiply table values by 0.0624 to convert to (lb·mol)/(h·ft³).

Indeed, even today most commercial gas-absorption units are designed primarily on the basis of prior commerical experience by using the traditional design methods described previously. Although duplication of a previous commercial design is by far the preferred method, this approach is of little value in developing a completely new process or in attempting to extrapolate an existing design to widely different operating conditions.

Since the early 1960s there have been developed some excellent laboratory experimental techniques, which unfortunately have largely been ignored by the industry. A noteworthy exception was described by Ouwerkerk (*Hydrocarbon Process.*, April 1978, pp. 89–94), in which it was revealed that both laboratory and small-scale pilot-plant data were employed as the basis for the design of an 8.5-m- (28-ft-) diameter commercial Shell Claus off-gas treating (SCOT) plate-type absorber. It is claimed that the cost of developing comprehensive design procedures can be kept to a minimum, especially in the development of a new process, by the use of these modern techniques.

In 1966, in a paper that now is considered a classic, Danckwerts and Gillham [*Trans. Inst. Chem. Eng.*, **44,** T42 (1966)] showed that data taken in a small stirred-cell laboratory apparatus could be used in the design of a packed-tower absorber when chemical reactions are involved. They showed that if the packed-tower mass-transfer coefficient in the absence of reaction (k_L^0) can be reproduced in the laboratory unit, then the rate of absorption in the laboratory apparatus will respond to chemical reactions in the same way as in the packed column even though the means of agitating the liquid in the two systems might be quite different.

According to this method, it is not necessary to investigate the kinetics of the chemical reactions in detail, nor is it necessary to determine the solubilities or the diffusivities of the various reactants in their unreacted forms. To use the method for scaling up, it is necessary independently to obtain data on the values of the interfacial area per unit volume a and the physical mass-transfer coefficient k_L^0 for the commercial packed tower. Once these data have been measured and tabulated, they can be used directly for scaling up the experimental laboratory data for any new chemically reacting system.

Danckwerts and Gillham did not investigate the influence of the gas-phase resistance in their study (for some processes gas-phase resistance may be neglected). However, in 1975 Danckwerts and Alper [*Trans. Inst. Chem. Eng.*, **53,** 34 (1975)] showed that by placing a stirrer in the gas space of the stirred-cell laboratory absorber, the gas-phase mass-transfer coefficient $\hat{k}_G$ in the laboratory unit could be made identical to that in a packed-tower absorber. When this was done, laboratory data obtained for chemically reacting systems having a significant gas-side resistance could successfully be scaled up to predict the performance of a commercial packed-tower absorber.

If it is assumed that the values of $\hat{k}_G$, k_L^0, and a have been measured for the commercial tower packing to be employed, the procedure for using the laboratory stirred-cell reactor is as follows:

1. The gas-phase and liquid-phase stirring rates are adjusted so as to produce the same values of $\hat{k}_G$ and k_L^0 as will exist in the commercial tower.
2. For the reaction system under consideration, experiments are made at a series of bulk-liquid and bulk-gas compositions representing the compositions to be expected at different levels in the commercial absorber (on the basis of a material balance).
3. The rates of absorption $r_A(c_i, B^0)$ are measured at each pair of gas and liquid compositions.

For dilute-gas systems one form of the equation to be solved in conjunction with these experimental data is

$$h_T = \frac{G_M}{a} \int_{y_2}^{y_1} \frac{dy}{r_A} \qquad (14\text{-}145)$$

where h_T = height of commercial tower packing, G_M = molar gas-phase mass velocity, a = effective interfacial area for mass transfer per unit volume in the commercial tower, y = mole-fraction solute in the gas phase, and r_A = experimentally determined rate of absorption per unit of exposed interfacial area.

By using the series of experimentally measured rates of absorption, Eq. (14-145) can be integrated numerically to determine the height of packing required in the commercial tower.

There are a number of different types of experimental laboratory units that could be used to develop design data for chemically reacting systems. Charpentier [*ACS Symp. Ser.*, **72**, 223–261 (1978)] has summarized the state of the art with respect to methods of scaling up laboratory data and tabulated typical values of the mass-transfer coefficients, interfacial areas, and contact times to be found in various commercial gas absorbers as well as in currently available laboratory units.

The laboratory units that have been employed to date for these experiments were designed to operate at a total system pressure of about 100 kPa (1 atm) and at near-ambient temperatures. In practical situations, it may become necessary to design a laboratory absorption unit that can be operated either under vacuum or at elevated pressures and over a reasonable range of temperatures in order to apply the Danckwerts method.

It would be desirable to reinterpret existing data for commercial tower packings to extract the individual values of the interfacial area a and the mass-transfer coefficients k_G and k_L^0 in order to facilitate a more general usage of methods for scaling up from laboratory experiments. Some progress in this direction has already been made, as discussed in Sec. 18. In the absence of such data, it is necessary to operate a pilot plant or a commercial absorber to obtain k_G, k_L^0, and a as described by Ouwerkerk (op. cit.).

Principles of Rigorous Absorber Design Danckwerts and Alper [*Trans. Inst. Chem. Eng.*, **53**, 34 (1975)] have shown that when adequate data are available for the kinetic-reaction-rate coefficients, the mass-transfer coefficients k_G and k_L^0, the effective interfacial area per unit volume a, the physical solubility or Henry's-law constants, and the effective diffusivities of the various reactants, then the design of a packed tower can be calculated from first principles with considerable precision.

For example, the packed-tower design equation for a dilute system in which gas-phase reactant A is being absorbed and reacted with liquid-phase reagent B is

$$r_A a\, dh = \frac{L_M}{\nu \rho_L}\, dB_h^0 = -G_M\, dy_h \qquad (14\text{-}146)$$

where r_A = specific rate of absorption per unit interfacial area, a = interfacial area per unit volume of packing, h = height of packing, L_M = molar liquid mass velocity, ν = number of moles of B reacting with 1 mol of A; ρ_L = average molar density of liquid phase, B_h^0 = bulk-liquid-phase reagent concentration (a function of h), G_M = molar gas-phase mass velocity, and y_h = mole fraction A in gas phase (a function of h).

For dilute systems it can be assumed that G_M, L_M, and ρ_L are constant, and it normally is assumed that the interfacial area a of the packing is constant and is equal to the value that would exist without reaction. This last assumption needs careful consideration, since different methods for measuring a may give different results. Sharma and Danckwerts [*Br. Chem. Eng.*, **15**(4), 522 (1970)] have reviewed various techniques for measuring interfacial areas.

Under the above assumptions for dilute systems Eq. (14-146) can be integrated as follows:

$$h_T = \frac{L_M}{\nu \rho_L a}\int_{B_1^0}^{B_2^0} \frac{dB_h^0}{r_A} = \frac{G_M}{a}\int_{y_2}^{y_1} \frac{dy_h}{r_A} \qquad (14\text{-}147)$$

where h_T = total height of packing and the subscripts 1 and 2 refer to the bottom and the top of the tower packing respectively.

The specific absorption rate $r_A = r_A(c_i, B^0)$ is a function of h and may be computed by combining the rate equation

$$r_A = k_L(x_i - x) = (k_L/\rho_L)(c_i - c) \qquad (14\text{-}148)$$

with the material-balance, or operating-curve, equation

$$G_M(y - y_2) = (L_M/\nu\rho_L)(B_2^0 - B^0) \qquad (14\text{-}149)$$

and with the appropriate relation for computing the interfacial concentration x_i of reactant A. In Eq. (14-148) the mass-transfer coefficient k_L is the coefficient with chemical reaction; i.e., $k_L = \phi k_L^0$.

The interfacial concentration x_i is computed by combining the equilibrium relation $y_i = mx_i$ with the equation $k_G(y - y_i) = k_L(x_i - x)$ to obtain

$$x_i = \frac{y/m + (k_L/mk_G)x}{(1 + k_L/mk_G)} \qquad (14\text{-}150)$$

According to Eq. (14-150), when k_L is very large and the ratio k_L/mk_G is much larger than unity, $x_i - x = yk_G/k_L$ and the specific absorption rate is defined by the equation

$$r_A = k_L(x_i - x) = k_G y \qquad (14\text{-}151)$$

This is the **gas-phase mass-transfer limited condition,** which can be substituted into Eq. (14-147) to obtain the following equation for calculating the height of packing for a dilute system:

$$h_T = (G_M/k_G a) \ln (y_1/y_2) = H_G \ln (y_1/y_2) \qquad (14\text{-}152)$$

At the other extreme, when the ratio k_L/mk_G is much smaller than unity, the interfacial concentration of reactant A may be approximated by the equilibrium relation $x_i = y/m$, and the specific absorption rate expression is

$$r_A = k_L(x_i - x) = k_L(y/m - x) \qquad (14\text{-}153)$$

For **fast chemical reactions** the reactant A is *by definition* completely consumed in the thin film near the liquid interface. Thus, x = 0, and

$$r_A = k_L y/m = (k_L/\rho_L)c_i \qquad (14\text{-}154)$$

This is known as the **liquid-phase mass-transfer limited condition,** as illustrated in Fig. 14-16.

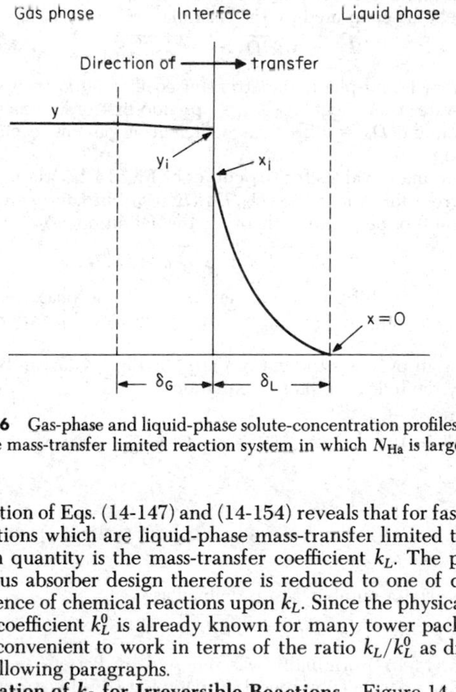

FIG. 14-16 Gas-phase and liquid-phase solute-concentration profiles for a liquid-phase mass-transfer limited reaction system in which N_{Ha} is larger than 3.

Inspection of Eqs. (14-147) and (14-154) reveals that for fast chemical reactions which are liquid-phase mass-transfer limited the only unknown quantity is the mass-transfer coefficient k_L. The problem of rigorous absorber design therefore is reduced to one of defining the influence of chemical reactions upon k_L. Since the physical mass-transfer coefficient k_L^0 is already known for many tower packings, it often is convenient to work in terms of the ratio k_L/k_L^0 as discussed in the following paragraphs.

Estimation of k_L for Irreversible Reactions Figure 14-17 illustrates the influence of either first- or second-order irreversible chemical reactions on the mass-transfer coefficient k_L as developed by Van Krevelen and Hoftyzer [*Rec. Trav. Chim.*, **67**, 563 (1948)] and as later refined by Perry and Pigford and by Brian et al. [*Am. Inst. Chem. Eng. J.*, **7**, 226 (1961)].

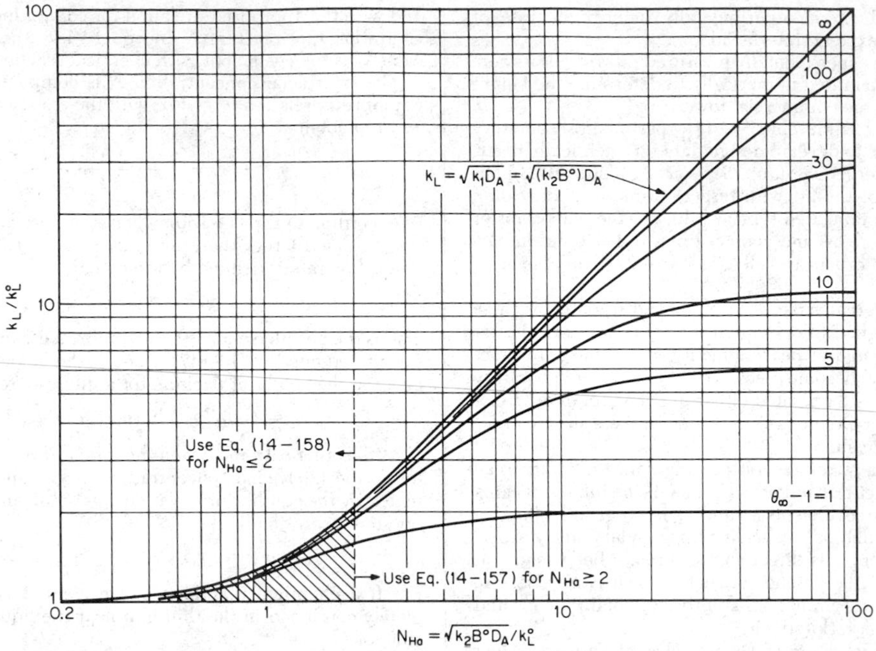

FIG. 14-17 Influence of irreversible chemical reactions on the liquid-phase mass-transfer coefficient k_L. [*Adapted from Van Krevelen and Hoftyzer,* Rec. Trav. Chim., *67*, *563 (1948).*]

First-order and pseudo-first-order reactions are represented by the upper curve in Fig. 14-17. We note that for first-order reactions when the Hatta number N_{Ha} is larger than about 3, the rate coefficient k_L can be computed by the formula

$$k_L = \sqrt{k_1 D_A} = \sqrt{(k_2 B^0) D_A} \tag{14-155}$$

where k_L = liquid-phase mass-transfer coefficient, k_1 = first-order-reaction-rate coefficient, k_2B^0 = pseudo-first-order-reaction-rate coefficient, and D_A = diffusion coefficient of gaseous reactant A in the liquid phase.

The parameter values for the curves of Fig. 14-17 originally were defined from film theory as $(D_B/D_A)(B^0/\nu c_i)$ but later were refined by the results of penetration theory to the definition ($\phi_\infty - 1$), where

$$\phi_\infty = \sqrt{D_A/D_B} + \sqrt{D_B/D_A}(B^0/\nu c_i) \tag{14-156}$$

in which D_B = diffusion coefficient of the liquid-phase reactant B and ϕ_∞ = value of k_L/k_L^0 for large values of N_{Ha} approaching infinity.

For design purposes the entire graph of Fig. 14-17 can be represented by the following pair of equations:

For $N_{Ha} \geq 2$:

$$k_L/k_L^0 = 1 + (\phi_\infty - 1)\{1 - \exp[-(N_{Ha} - 1)/(\phi_\infty - 1)]\} \tag{14-157}$$

For $N_{Ha} \leq 2$:

$$k_L/k_L^0 = 1 + (\phi_\infty - 1)\{1 - \exp[-(\phi_\infty - 1)^{-1}]\} \exp[1 - 2/N_{Ha}] \tag{14-158}$$

where the Hatta number N_{Ha} is defined as

$$N_{Ha} = \sqrt{k_2 B^0 D_A}/k_L^0 \tag{14-159}$$

Equation (14-157) originally was reported by Porter [*Trans. Inst. Chem. Eng.*, **44**(1), T25 (1966)]. Equation (14-158) was derived by the author.

The Van Krevelen-Hoftyzer relationship was tested experimentally for the second-order system in which CO_2 reacts with either NaOH or KOH solutions by Nijsing et al. [*Chem. Eng. Sci.*, **10**, 88 (1959)]. Nijsing's results for the NaOH system are shown in Fig. 14-18 and are in excellent agreement with the second-order-reaction theory. Indeed, these experimental results can be described very well by Eqs. (14-156) and (14-157) when values of ν = 2 and D_A/D_B = 0.64 are employed in the equations.

For fast irreversible chemical reactions, therefore, the principles of rigorous absorber design can be applied by first establishing the effects of the chemical reaction on k_L and then employing the appropriate material-balance and rate equations in Eq. (14-147) to perform the integration to compute the required height of packing.

For an isothermal absorber involving a dilute system in which a **liquid-phase mass-transfer limited** first-order irreversible chemical reaction is occurring, the packed-tower design equation is derived as

$$h_T = (mG_M/\sqrt{k_1 D_A}a) \ln (y_1/y_2) \tag{14-160}$$

For a dilute system in which the **liquid-phase mass-transfer limited** condition is valid, in which a very fast second-order reaction is involved, and for which N_{Ha} is very large, the equation

$$k_L/k_L^0 = \phi_\infty = \sqrt{D_A/D_B} + \sqrt{D_B/D_A}(B^0/\nu c_i) \tag{14-161}$$

is valid and results in the following equation for computing the height of packing in a packed tower:

$$h_T = \frac{mG_M}{k_L^0 a} \sqrt{\frac{D_B}{D_A}} \int_{y_2}^{y_1} \frac{dy}{\dfrac{mB^0 D_B}{\nu \rho_L D_A} + y} \tag{14-162}$$

Evaluation of the integral in Eq. (14-162) requires a knowledge of the liquid-phase bulk concentration of B as a function of y. This relationship is obtained by means of a material balance around the tower, as shown in Eq. (14-149). Numerical integration by a quadrature method such as Simpson's rule normally will be required for this calculation.

Estimation of k_L for Reversible Reactions When the reaction is of the form $A \rightleftharpoons B$, where B is a nonvolatile product and the equilibrium constant is defined by $c_B = K_{eq}c_A$, the expressions for computing k_L become extremely complex. A good discussion of this situation is given in *Mass Transfer* by Sherwood, Pigford, and Wilke

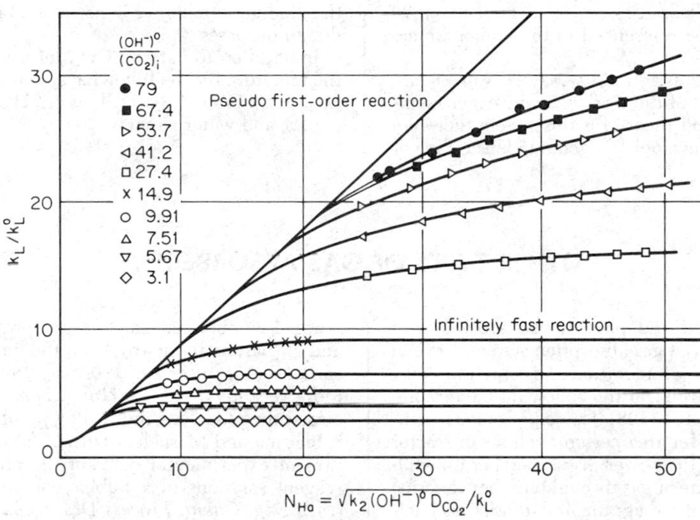

FIG. 14-18 Experimental values of k_L/k_L^0 for absorption of CO_2 into NaOH solutions at 20°C. [*Data of Nijsing et al.,* Chem. Eng. Sci., *10, 88 (1959).*]

(McGraw-Hill, New York, 1975, p. 317). Three limiting cases are listed below:

1. For very slow reactions,

$$\lim_{k_1 \to 0} k_L = k_L^0 \tag{14-163}$$

2. For extremely fast reactions where K_{eq} is very large,

$$\lim_{\substack{k_1 \to \infty \\ K_{eq} = \infty}} k_L = \sqrt{k_1 D_A} \tag{14-164}$$

3. For extremely fast reactions where K_{eq} is finite,

$$\lim_{\substack{k_1 \to \infty \\ K_{eq} = \text{finite}}} k_L = (1 + K_{eq})k_L^0 \tag{14-165}$$

When one of these three conditions is applicable, the appropriate design equation can be obtained by substitution into Eq. (14-147), followed by integration of the resulting relationship.

Some more complex situations involving reversible reactions are discussed in *Mass Transfer* (ibid., pp. 336–343).

Simultaneous Absorption of Two Reacting Gases In multicomponent physical absorption the presence of one gas often does not affect the rates of absorption of the other gases. When chemical reactions in which two or more gases are competing for the same liquid-phase reagent are involved, selectivity of absorption can be affected by the choice of design conditions, and the situation may become extremely complex from a designer's point of view.

The classic work on this subject is that of Ramachandran and Sharma [*Trans. Inst. Chem. Eng.*, **49**, 253 (1971)] and is recommended to those needing further details. The following references also are offered as a sampling of the literature on the subject:

- *CO_2 and H_2S.* Danckwerts and Sharma, *Chem. Eng. (London)*, **CE244–280** (October 1966); Onda, et al., *J. Chem. Eng. Japan*, **5**, 27 (1972); Rivas and Prausnitz, *Am. Inst. Chem. Eng. J.*, **25**, 975 (1979).
- *CO_2 and SO_2.* Goettler and Pigford, *Inst. Chem. Eng. Symp. Ser.*, **28**, 1 (1968); Teramoto et al., *Int. Chem. Eng.*, **18**, 250 (1978).
- *SO_2 and NO_2.* Takeuchi and Yamanaka, *Ind. Eng. Chem. Process Des. Dev.*, **17**, 389 (1978).

Desorption with Chemical Reaction When chemical reactions are involved in a stripping operation, the design problem can become extremely complex. In fact, much less is known about this very important process than is known about absorption. A classic work on this subject is that of Shah and Sharma [*Trans. Inst. Chem. Eng.*, **54**, 1 (1976)], which is recommended to those in need of more details.

In the subsection "Design of Gas-Absorption Systems" it was stated that more often than not the liquid-phase residence time and, hence, the liquid holdup are considered to be the most important design parameters for stripping towers. If Eq. (14-138) is redefined to represent the stripping process for an extremely slow liquid-phase reaction for which $R_A = k_1 c$, then one finds that the liquid holdup will be a factor only when the ratio $k_1 \rho_L f_H / k_L^0 a$ is less than unity. Thus, one can ensure that the liquid-phase reaction rate is not limiting by increasing the temperature and the liquid holdup until this ratio is equal to or greater than unity. The preferred method at present is to base the design on prior commercial experience.

Use of Literature for Specific Systems A large body of experimental data obtained in bench-scale laboratory units and in small-diameter packed towers has been published since the early 1940s. One might wish to consider using such data for a particular chemically reacting system as the basis for scaling up to a commercial design. Extreme caution is recommended in interpreting such data for the purpose of developing commercial designs, as extrapolations of this kind of information can lead to serious errors. Extrapolation to temperatures, pressures, or liquid-phase reagent conversions different from those that were employed by the original investigator definitely should be regarded with caution.

Bibliographies presented in the general references listed at the beginning of this section are an excellent source of information on specific chemically reacting systems. *Mass Transfer* by Sherwood, Pigford, and Wilke (McGraw-Hill, New York, 1975) contains a good bibliography of key publications of experimental data. *Gas-Liquid Reactions* by P. V. Danckwerts (McGraw-Hill, New York, 1970) contains a tabulation of references to specific chemically reacting systems in the introduction (pp. 1–5). Also, *Gas-Liquid-Solid Reactor Design* by Y. T. Shah (McGraw-Hill, New York, 1979) contains a good summary of references to experimental data for chemically reacting systems in Table 2-3 (p. 34). Finally, *Gas Purification* by Kohl and Riesenfeld (Gulf Publishing, Houston, 1979) presents data and references for many chemically reacting systems of current commercial interest.

In searching for data on a particular system, a computerized search of *Chemical Abstracts*, *Engineering Index*, and National Technical Information Service (NTIS) data bases should seriously be considered. Although the NTIS computer contains only information published after 1970, one normally can assume that most pre-1970 publications of merit likely will be referenced in the bibliographies of current articles on the subject.

The experimental data for the system CO_2-NaOH-Na_2CO_3 are unusually well known as the result of the work of many experimenters. A serious study of the data and theory for this system therefore is recommended as the basis for developing a good understanding of the kind and quality of experimental information that is needed for design purposes.

In addition to data on CO_2, information can readily be found in the literature for the following systems: O_2, Cl_2, NH_3, NO_2, NO, SO_2, SO_3, H_2S, COS, CS_2, HCl, HBr, HCN, H_2, $COCl_2$, PCl_3, olefins, dienes, and water vapor.

OTHER TYPES OF GAS ABSORBERS

Introduction Although packed and plate towers are the most commonly used devices in industrial gas-absorption systems, several other types of equipment deserve consideration and sometimes may have distinct advantages, as discussed in the following paragraphs. Nagel et al. [*Int. Chem. Eng.*, **21**, 161 (1981)] have presented a comparison of the effective mass-transfer area per unit volume of reactor for all reactor types in relation to the power dissipation per unit volume of reactor and per unit volume of gas throughput; they describe a procedure for developing working diagrams for translating pilot-plant data into commercial designs. **Section 18** discusses some of the key physical characteristics and mechanical-design parameters associated with these devices.

Stirred Tanks Mechanically stirred vessels may be employed in systems involving chemical reactions in which a gas is first absorbed and then reacts with a component in solution. Stirred tanks are especially advantageous when very slow liquid-phase reactions which require a much longer liquid residence time than conveniently can be provided in a tower-type absorber are involved. Stirred tanks also are advantageous when there are large heat effects, owing to the relatively high rates of heat transfer that can be achieved by using immersed surfaces under agitation. Stirred reactors normally are designed to approach complete back mixing of both the gas and the liquid phases, so that the advantages of countercurrency are lost, and staging may be needed.

Owing to the relatively large $k_L^0 a$ values that can be attained, agitated absorbers can have a slight advantage in size and power consumption over packed towers for some applications, but they are limited to very low gas throughputs under 0.05 m/s (10 ft/min), which is approximately the rate of rise of gas bubbles in the liquid. The key design parameter is the volume-fraction gas holdup ε, since the bubble diameter d_b and the coefficient k_L^0 do not vary widely and $k_L^0 a$ depends primarily on a, which for spherical bubbles is defined as $a = 6\varepsilon/d_b$. The gas holdup rarely exceeds a value of 0.3 to 0.4. An excellent review of this subject was given by Sideman et al. [*Ind. Eng. Chem.*, **58**(7), 32 (1966)]. Uchida et al. [*Can. J. Chem. Eng.*, **56**, 690 (1978)] discuss the theory of gas absorption in stirred-tank reactors and present experimental data on the absorption of SO_2 into a limestone slurry. In this case the resistance to mass transfer between the liquid and the solid particles must be considered in addition to the normally encountered gas-phase and liquid-phase resistances.

Sparged Towers A sparged-tower reactor is similar to a stirred-tank reactor except that the heat- and mass-transfer efficiency will be somewhat lower. Sparged reactors normally are more economical than stirred reactors for high-pressure service owing to the high cost of mechanical seals for use in high-pressure stirred-tank agitators.

In sparged reactors the mixing in the liquid phase normally will be complete even when the height-to-diameter ratio is as large as 10:1, while the gas normally can be assumed to proceed upward in plug flow as a first approximation. Gas contact time is governed by the velocity of the rising bubbles, and gas throughput velocities therefore are relatively low when compared with packed- and plate-tower gas absorbers. As in stirred tanks, k_L^0 varies little with operating conditions, and $k_L^0 a$ depends primarily on a, which varies widely.

Juvekar and Sharma [*Trans. Inst. Chem. Eng.*, **55**, 77 (1977)] discuss and compare the theoretical design relations for bubble-column contactors with those for packed- and plate-tower absorbers. An excellent discussion of the considerations involved in designing gas-sparged commercial reactors was presented by H. W. Prengle, Jr., and N. Barona in an article on the liquid-phase oxidation of hydrocarbons (*Hydrocarbon Process.*, November 1970, pp. 159–175). More recently Deckwer [*Int. Chem. Eng.*, **19**, 21 (1979)] has discussed the modeling and scaling up of bubble-column reactors.

Sharma and Mashelkar [*Inst. Chem. Eng. Symp. Ser.*, **28** (1968)] present experimental data on the absorption of CO_2 into various reagent solutions in a bubble-column reactor. Akita and Yoshida [*Ind. Eng. Chem. Process Des. Dev.*, **12**, 76 (1973)] have reported extensive studies of gas absorption in bubble columns of 0.15 to 0.6 m in diameter with aerated-liquid heights up to 3 m by using a single 0.5-cm orifice sparger.

Spray Chambers or Towers Spray absorbers currently are being applied on a large-scale commercial basis in systems for removing SO_2 from boiler flue gases that are exhausted from large coal-fired power-generating stations. Spray absorbers are particularly advantageous when low pressure drop is essential and when particulate matter may be contained in the incoming gas stream. Normally there is no packing in a spray absorber.

Liquid-phase residence times in spray absorbers are very low, on the order of 1 to 10 s, as are the gas contact times. Thus, spray absorbers are limited to relatively easy absorption duties and are especially applicable to systems in which the rate of transfer is gas-phase mass-transfer limited. This condition will exist whenever the liquid-phase resistance can be neglected and the back pressure of the solute over the liquid is small (there is high solubility).

The surface area for mass transfer between the gas and liquid phases is related to the total volume of liquid holdup per unit volume of spray chamber and to the Sauter mean diameter d_s of the spray droplets. The Sauter mean diameter is defined in terms of the ratio of the total liquid surface area generated inside the absorber divided by the total volume of liquid sprayed, i.e.,

$$\frac{\text{(Liquid surface generated/time)}}{\text{(Liquid volume sprayed/time)}} = \frac{6}{d_s} \tag{14-166}$$

When the effect due to liquid holdup is included, the liquid surface-to-volume ratio inside either a vertical countercurrent or a horizontal cross-flow spray chamber is defined by the equation

$$a = (Q_L/A_L)(6/u_L d_s) \tag{14-167}$$

where a = surface-to-volume ratio inside chamber, Q_L = liquid injection rate, A_L = total cross-sectional area of chamber perpendicular to the direction of liquid flow, d_s = Sauter mean droplet diameter, and u_L = average velocity of individual liquid droplets in the vertical direction.

Although spray nozzles can create very high liquid areas near the nozzles, the surface area normally is rapidly reduced owing to the mechanism of drop coalescence, in which the smaller droplets join together to form larger ones. The correct value of the Sauter mean diameter to use in Eq. (14-167) therefore is an average over the entire chamber volume and not the local value one might measure in the vicinity of the nozzles.

In **vertical spray towers** the gas stream flows vertically upward, and the liquid is sprayed downward at various stations within the tower as well as from the sides. In these towers the height-to-diameter ratio inside the spray section normally is low (on the order of

2:1), and the gas velocity normally cannot exceed about 2.3 m/s (7.5 ft/s) owing to excessive entrainment of fine liquid droplets at higher velocities. The gas tends to become completely back-mixed in a vertical tower owing to entrainment of gas by the action of the sprays. The correct design equation for a gas-phase mass-transfer limited system in a vertical spray tower therefore is

$$y_2/y_1 = 1/(1 + N_G) \quad (14\text{-}168)$$

where y_1 and y_2 are the gas-phase solute mole fractions at the bottom (inlet) and top of the tower respectively and N_G is the number of gas-phase mass-transfer units as defined by the equation

$$N_G = k_G aV/n_G \quad (14\text{-}169)$$

where $\hat{k}_G$ = gas-phase mass-transfer coefficient, a = surface-to-volume ratio inside tower, V = total volume of spray section, and n_G = gas rate.

Mehta and Sharma [*Br. Chem. Eng.*, **15**, 1440 and 1556 (1970)] report measurements in a small vertical spray chamber, as did Pigford and Pyle [*Ind. Eng. Chem.*, **43**(7), 1649 (1951)]. H. N. Head [*National Technical Information Service (NTIS)*, **PB-274,544** (U.S. Dept. of Commerce, September 1977)] has reported performance tests on a pilot-scale vertical spray chamber for SO_2 removal. A brief summary of these tests is given in the Environmental Protection Agency's report EPA 625/2-76-010 (1976).

In **horizontal cross-flow spray chambers** the gas passes horizontally through a duct of either rectangular or circular cross section. The liquid is sprayed vertically downward, perpendicularly to the direction of gas flow, and occasionally also from the sides of the chamber. The allowable gas velocities in horizontal cross-flow chambers are higher than in vertical flow; normally they are around 7 m/s (23 ft/s). The gas normally is well mixed radially by the action of the sprays in a cross-flow chamber, but not axially (in the direction of the gas flow). The design equation for a gas-phase mass-transfer limited system in a cross-flow spray chamber therefore is

$$y_2/y_1 = \exp(-N_G) \quad (14\text{-}170)$$

where all symbols are identical to those defined in Eqs. (14-168) and (14-169).

Comparison of Eqs. (14-168) and (14-170) indicates that for a given solute-removal specification (say, 90 percent removal) the horizontal cross-flow absorber theoretically requires fewer mass-transfer units than does the vertical absorber. As an example, the horizontal unit theoretically requires 2.3 units for 90 percent removal of the solute, while for the vertical unit a value of $N_G = 9$ is needed. One therefore should observe that, for equal removals of solute and for equal values of the volumetric mass-transfer coefficient $\hat{k}_G a$, the total volume of a horizontal cross-flow spray chamber will be significantly lower than that of a vertical spray tower.

Edwards and Huang [*Chem. Eng. Prog.*, **73**(8), 64 (1977)] discuss some of the principles involved in the design of horizontal cross-flow absorbers. Additional data and discussion can be found in the original paper presented by Edwards and Huang [(American Institute of Chemical Engineers, Houston, March 1977, **MS No. 9127,** Lib. Ref. 83N] and in U.S. Patent **4,269,812** (May 26, 1981), in which the basic concepts of horizontal cross-flow scrubber design are discussed.

Venturi Scrubbers Venturi scrubbers normally are preferred for removing particulate matter from a gas stream as opposed to absorbing soluble vapors, although efficient atomization of the liquid and good contacting between the gas and the liquid are important features of these devices. The improved contacting efficiency in a venturi scrubber is obtained at the expense of a relatively large gas-side pressure drop and a consequent larger power consumption when compared with a spray absorber.

In gas-absorption applications venturi scrubbers are limited somewhat by the cocurrent nature of the gas-liquid flow. Figure 14-19 is a typical design diagram for a cocurrent absorber in which gas and liquid are fed to the top of the unit. The fractional solute removal for a dilute system is defined as $(1 - y_1/y_2)$ and is given by the equation

$$\text{Removal} = E(1 - mx_2/y_2)/(1 + mG_M/L_M) \quad (14\text{-}171)$$

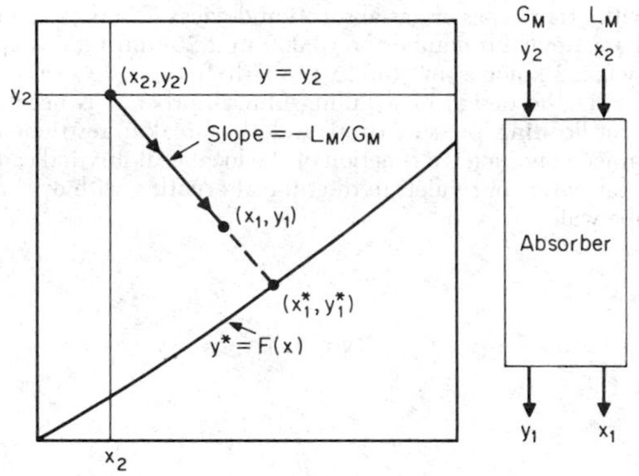

FIG. 14-19 Design diagram for a cocurrent absorber.

where the mass-transfer efficiency E is defined as

$$E = (y_2 - y_1)/(y_2 - y_1^\circ) = 1 - e^{-N_G} \quad (14\text{-}172)$$

and

$$N_G = (K_G aRT/p_T)t_c \quad (14\text{-}173)$$

where N_G = number of overall gas-phase mass-transfer units, $K_G a$ = overall volumetric gas-phase mass-transfer coefficient for dilute systems, R = gas constant, T = gas temperature, p_T = total pressure, and t_c = effective gas-liquid contact time.

According to Fig. 14-19, the maximum solute removal in a venturi scrubber will be equivalent to one equilibrium stage. The mass-transfer efficiency ratio E determines how far the process will move down the operating line toward equilibrium and is related directly to the number of gas-phase mass-transfer units through Eq. (14-172). The number of transfer units that can be achieved in a venturi scrubber typically will be in the range from $N_G = 1$ to $N_G = 2$.

H. N. Head [*National Technical Information Service (NTIS)*, **PB-274,544** (U.S. Dept. of Commerce, September 1977)] has reported performance tests on a pilot-scale venturi scrubber for SO_2 removal. A definitive work on the subject of venturi-scrubber design is that of Yung, Calvert, and Barbarika [*NTIS*, **PB-271,515** (U.S. Dept. of Commerce, August 1977)].

Falling-Film Absorbers These devices are important in commercial applications in which the heat released during absorption is high, dictating the use of a heat-transfer surface adjacent to the liquid. The usual arrangement is to flow the liquid absorbent downward in film flow inside the tubes of a vertically oriented heat-exchanger bundle, with coolant flow on the shell side. The feed gas may be introduced either at the top (cocurrent) or the bottom (countercurrent) of the exchanger. A difficult design problem is the one of distributing the liquid feed uniformly to the top perimeters of a multiplicity of tubes at the top of the exchanger, since bypassing the gas is a serious matter if high solute removals are required.

In designing a falling-film unit it is worthwhile to begin by checking the maximum possible temperature rise in the liquid film by the use of the following equation:

$$\Delta T_{\max} = (-\Delta H)(k_G/h)(y - y_i)_{\max} \quad (14\text{-}174)$$

where $\Delta T_{\max}$ = maximum temperature rise in the liquid film, k_G = gas-phase mass-transfer coefficient, h = heat-transfer coefficient, $-\Delta H$ = heat of absorption of solute, and $(y - y_i)_{\max}$ = maximum gas-phase driving force in the apparatus.

If the temperature rise computed by Eq. (14-174) is small, one can employ either the classical isothermal or the classical adiabatic design methods for packed towers, as appropriate, considering the temperature rise that will exist on the coolant (shell side) of the exchanger.

The surface-to-volume ratio for a falling-film absorber is defined as $a = 4/d_t$. For a 25-mm- (1-in-) inside-diameter tube the surface-to-volume ratio is 160 m^{-1} (49 ft^{-1}), which is quite large in comparison with other types of gas-absorption devices. The $\hat{k}_G a$ value for a gas-phase Reynolds number of 10,000 in a 25-mm tube is approximately 0.223 kmol/(s·m^3)[(50 lb·mol)/(h·ft^3)].

In fixing the design of a falling-film absorber it is necessary to check for flooding, pressure drop, and film breakup restrictions. The last-named condition is a function of the local heat flux and can result in the formation of rivulets in the tube alternating with dry areas on the tube walls.

Hikita et al. [*J. Chem. Eng. Japan*, **11**, 96 (1978)] and [*Can. J. Chem. Eng.*, **57**, 578 (1979)] have published two excellent theoretical and experimental studies of cocurrent and countercurrent wetted-wall columns. The book *Mass Transfer* by Sherwood, Pigford, and Wilke (McGraw-Hill, New York, 1975) contains a good discussion (pp. 203–214). Guerreri and King [*Hydrocarbon Process.*, **53**(1), 131 (1974)] describe an interesting commercial application of a falling-film absorber and a procedure for its design.

Section 15

Liquid-Liquid Extraction

Lanny A. Robbins, Ph.D., *Research Scientist, Dow Chemical Company; Member, American Institute of Chemical Engineers, Sigma Xi*

GENERAL REFERENCES: Foust, Wenzel, Clump, Maus, and Anderson, *Principles of Unit Operations,* 2d ed., Wiley, New York, 1980. Hanson, *Recent Advances in Liquid-Liquid Extraction,* Pergamon, New York, 1971. Schweitzer, *Handbook of Separation Techniques for Chemical Engineers,* McGraw-Hill, New York, 1979. Sorenson and Arlt, *Liquid-Liquid Equilibrium Data Collection,* DECHEMA, Frankfurt, Germany, 1979. Treybal, *Liquid Extraction,* 2d ed., McGraw-Hill, New York, 1963. Treybal, *Mass-Transfer Operations,* 3d ed., McGraw-Hill, New York, 1980. Wisniak and Tamir, *Liquid-Liquid Equilibrium and Extraction: A Literature Source Book,* part A, Elsevier, Amsterdam, 1981.

Nomenclature

Symbol	Definition
a	Mass-transfer area per unit volume of tower
A°	Activity of solute
A_t	Cross-sectional area of tower
C	Heat capacity
E	Weight (or mass flow rate) of extract
E'	Weight (or mass flow rate) of extraction solvent alone in extract
$\mathcal{E}$	Extraction factor (slope of equilibrium line/slope of operating line)
F	Weight (or mass flow rate) of feed
F'	Weight (or mass flow rate) of feed solvent alone in feed
HETS	Height equivalent to a theoretical stage
H_e	Height of a transfer unit attributed to driving force in extract phase
H_{or}	Height of a transfer unit based on overall driving force in raffinate concentrations
H_r	Height of a transfer unit attributed to driving force in raffinate phase
k	Mass-transfer coefficient
K	Distribution coefficient in weight fractions
K°	Distribution coefficient in mole fractions
K'	Distribution coefficient in Bancroft (weight-ratio) coordinates
m	Slope of equilibrium line in Bancroft coordinates
N	Number of theoretical (equilibrium) stages
N_{oh}	Number of heat transfer units based on hot phase
N_{or}	Number of mass transfer units based on overall driving force in raffinate concentration
R	Weight (or mass flow rate) of raffinate
R'	Weight (or mass flow rate) of feed solvent alone in raffinate
S	Weight (or mass flow rate) of extraction-solvent stream
S'	Weight (or mass flow rate) of extraction-solvent alone

Symbol	Definition
t	Temperature in cold (extract) phase
T	Temperature in hot (raffinate) phase
U_o	Overall heat-transfer coefficient
W	Weight (or mass flow rate) of wash phase or stream
W'	Weight (or mass flow rate) of wash solvent alone
x	Weight-fraction solute in feed (raffinate) phase
x°	Mole-fraction solute in feed (raffinate) phase
X	Weight solute/weight feed solvent in feed (raffinate) phase
y	Weight-fraction solute in extract phase
y°	Mole-fraction solute in extract phase
Y	Weight solute/weight extraction solvent in extract
z	Weight-fraction solute
	Greek symbols
α	Relative separation factor (selectivity)
γ	Activity coefficient of solute
	Subscripts
1, 2, etc.	Stream leaving stage 1, 2, etc.
e	Extract phase or stream
f	Feed phase or stream
r	Raffinate phase or stream
s	Extraction solvent phase or stream
m	Mixture
Δ	Delta (or difference) mixture
	Superscripts
A	Case *A*
B	Case *B*
C	Case *C*

INTRODUCTION

Liquid-liquid extraction is a process for separating components in solution by their distribution between two immiscible liquid phases. Such a process can also be simply referred to as **liquid extraction** or **solvent extraction;** however, the latter term may be confusing because it also applies to the leaching of a soluble substance from a solid.

Since liquid-liquid extraction involves the transfer of mass from one liquid phase into a second immiscible liquid phase, the process can be carried out in many different ways. The simplest example involves the transfer of one component from a binary mixture into a second immiscible liquid phase. One example is liquid-liquid extraction of an impurity from wastewater into an organic solvent. This is analogous to stripping or absorption in which mass is transferred from one phase to another. Transfer of the dissolved component (solute) may be enhanced by the addition of "salting out" agents to the feed mixture or by adding "complexing" agents to the extraction solvent. Or in some cases a chemical reaction can be used to enhance the transfer, an example being the use of an aqueous caustic solution to remove phenolics from a hydrocarbon stream. A more sophisticated concept of liquid-liquid fractionation can be used in a process to separate two solutes completely. A primary extraction solvent is used to extract one of the solutes from a mixture (similarly to stripping in distillation), and a wash solvent is used to scrub the extract free from the second solute (similarly to rectification in distillation).

USES FOR LIQUID-LIQUID EXTRACTION

Liquid-liquid extraction is used primarily when distillation is impractical or too costly to use. It may be more practical than distillation when the relative volatility for two components falls between 1.0 and 1.2. Likewise, liquid-liquid extraction may be more economical than distillation or steam-stripping a dissolved impurity from wastewater when the relative volatility of the solute to water is less than 4. In one case discussed by Robbins [*Chem. Eng. Prog.*, **76** (10), 58 (1980)], liquid-liquid extraction was economically more attractive than carbon-bed or resin-bed adsorption as a pretreatment process for wastewater detoxification before biotreatment.

In other cases the components to be separated may be heat-sensitive, like antibiotics, or relatively nonvolatile, like mineral salts, and liquid-liquid extraction may provide the most cost-effective separation process. However, the potential use of distillation should generally be evaluated carefully before considering liquid-liquid extraction. An extraction process usually requires (1) liquid-liquid extraction, (2) solvent recovery, and (3) raffinate desolventizing.

Several examples of cost-effective liquid-liquid extraction processes include the recovery of acetic acid from water (Fig. 15-1), using ethyl ether or ethyl acetate as described by Brown [*Chem. Eng. Prog.*, **59**(10), 65 (1963)], or the recovery of phenolics from water as described by Lauer, Littlewood, and Butler [*Iron Steel Eng.*, **46**(5), 99 (1969)] with butyl acetate, or with isopropyl ether as described by Wurm [*Glückauf*, **12**, 517 (1968)], or with methyl isobutyl ketone as described by Scheibel ["Liquid-Liquid Extraction," in Perry & Weissburg (eds.), *Separation and Purification*, 3d ed., Wiley, New York, 1978, chap. 3]. The solvent is recovered by distillation, and the raffinate is desolventized by steam stripping. In some cases the extraction solvent may have a higher boiling point than the solute to achieve reduced energy consumption, but a buildup of heavies in the recycle solvent can create another problem.

The Udex process (Fig. 15-2) is a cost-effective liquid-liquid fractionation process for the separation of aromatics from aliphatics as described by Grote [*Chem. Eng. Prog.*, **54**(8), 43 (1958)]. In this process the extraction solvent, diethylene or triethylene glycol, is recovered by steam distillation, and the raffinate and extract streams are desolventized by water extraction. Subsequent process modifications described by Symoniak, Ganju, and Vidueira [*Hydrocarbon Process.*, 139 (September 1981)] use tetraethylene glycol as the extraction solvent and a mixture of light aliphatics and benzene as the wash solvent to the main extractor. Water condensate from the steam distillation is used to extract residual extraction solvent from the raffinate and extract streams, so distillation for drying the extraction solvent has been eliminated. Solids are removed from recycle extraction solvent by filtration, while acids and heavies are removed by a solid adsorbent bed. Other processes similar to this use sulfolane (tetrahydrothiophene-1,1-dioxide) or NMP (*N*-methyl-pyrrolidone) as the extraction solvent.

Another example of a cost-effective liquid-liquid extraction process is the one used for recovery of uranium from ore leach liquors (Fig. 15-3). In this case the solvents, alkyl phosphates in kerosine, are recovered by liquid-liquid extraction using a strip solution, and the raffinate requires practically no desolventizing because the solubility of the solvents in water is extremely low. Most of the solvent loss occurs because of the entrainment of small droplets in the water. The economic utility of a liquid-liquid extraction process depends strongly on the solvent selected and on the procedures used for solvent recovery and raffinate desolventizing. After these matters have been considered, the selection and design of an extraction device or assembly can be considered in proper perspective.

DEFINITIONS

The **feed** to a liquid-liquid extraction process is the solution that contains the components to be separated. The major liquid component in the feed can be referred to as the **feed solvent.** Minor components in solution are often referred to as **solutes.** The **extraction solvent,** or just plain **solvent,** is the immiscible liquid added to a process for the purpose of extracting a solute or solutes from the feed. The extraction-solvent phase leaving a liquid-liquid contactor is called the **extract.** The **raffinate** is the liquid phase left from the feed after

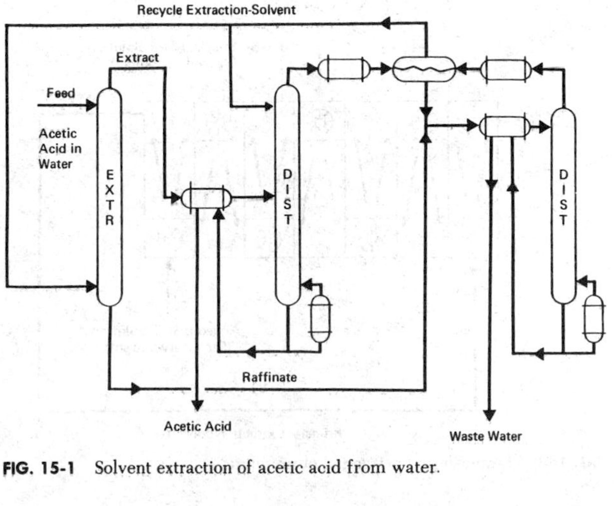

FIG. 15-1 Solvent extraction of acetic acid from water.

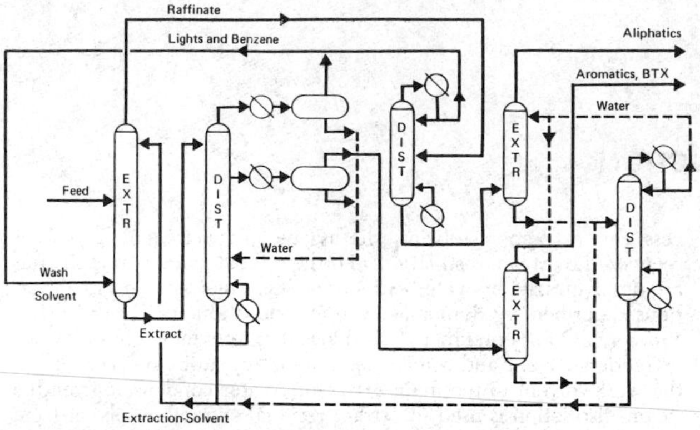

FIG. 15-2 Udex process.

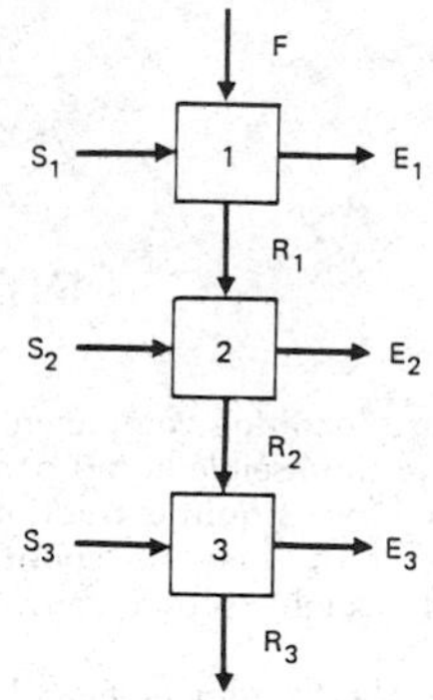

FIG. 15-4 Crosscurrent extraction.

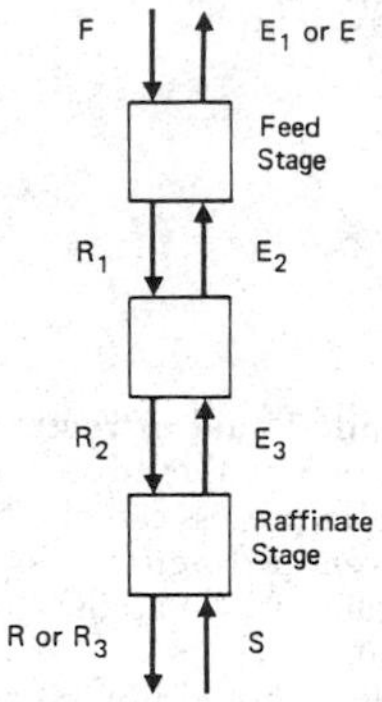

FIG. 15-5 Countercurrent extraction.

being contacted by the second phase. A **wash solvent** is a liquid added to a liquid-liquid fractionation process to wash or enrich the purity of a solute in the extract phase.

A **theoretical** or **equilibrium stage** is a device or combination of devices that accomplishes the effect of intimately mixing two immiscible liquids until equilibrium concentrations are reached, then physically separating the two phases into clear layers. **Crosscurrent extraction** (Fig. 15-4) is a cascade, or series of stages, in which the raffinate *R* from one extraction stage is contacted with additional fresh solvent *S* in a subsequent stage.

Countercurrent extraction (Fig. 15-5) is an extraction scheme in which the extraction solvent enters the stage or end of the extraction farthest from where the feed *F* enters and the two phases pass countercurrently to each other. The objective is to transfer one or more components from the feed solution *F* into the extract *E*. When a **staged contactor** is used, the two phases are mixed with droplets of one phase suspended in the other, but the phases are separated before leaving each stage. When a **differential contactor** is used, one of the phases can remain dispersed as droplets throughout the contactor as the phases pass countercurrently to each other. The dispersed phase is then allowed to coalesce at the end of the device before being discharged.

Liquid-liquid fractionation, or **fractional extraction** (Fig. 15-6), is a sophisticated scheme for nearly complete separation of one solute from a second solute by liquid-liquid extraction. Two immiscible liquids travel countercurrently through a contactor, with the solutes being fed near the center of the contactor. The ratio of immiscible-liquid flow rates is operated so that one of the phases preferentially moves the first solute to one end of the contactor and the other phase moves the second solute to the opposite end of the contactor. Another way to describe the operation is that a primary solvent *S* preferentially extracts, or strips, the first solute from the feed *F* and a wash solvent *W* scrubs the extract free from the unwanted second solute. The second solute leaves the contactor in the raffinate stream.

Dissociation extraction is the process of using chemical reaction to force a solute to transfer from one liquid phase to another. One example is the use of a sodium hydroxide solution to extract phenolics, acids, or mercaptans from a hydrocarbon stream. The opposite transfer can be forced by adding an acid to a sodium phenate stream to **spring** the phenolic back to a **free phenol** that can be extracted into an organic solvent. Similarly, primary, secondary, and tertiary amines can be protonated with a strong acid to transfer the amine into a water solution, for example, as an amine hydrochloride salt. Conversely, a strong base can be added to convert the amine salt back to **free base,** which can be extracted into a solvent. This procedure is quite common in pharmaceutical production.

Fractionation dissociation extraction involves both the chemical reaction and the fractionation scheme for the separation of components by their difference in dissociation constants as described by Colby [in Hanson (ed.), *Recent Advances in Liquid-Liquid Extraction*, Pergamon, New York, 1971, chap. 4].

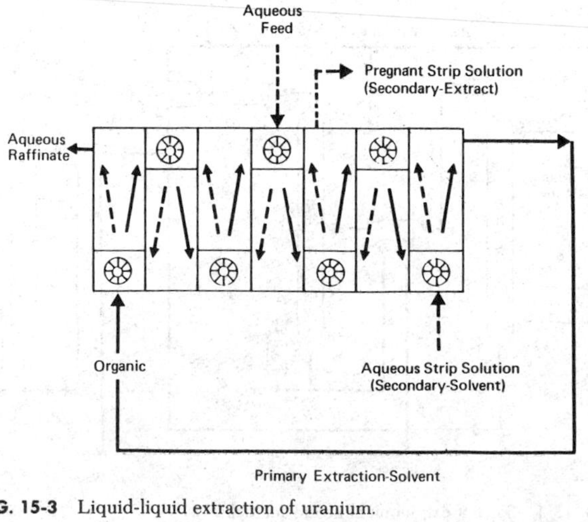

FIG. 15-3 Liquid-liquid extraction of uranium.

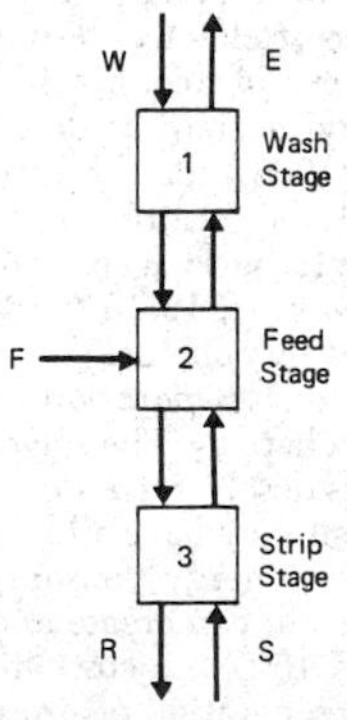

FIG. 15-6 Liquid-liquid fractionation.

PHASE EQUILIBRIUMS

The separation of components by liquid-liquid extraction depends primarily on the thermodynamic equilibrium distribution of those components between the two liquid phases. Knowledge of these distribution relationships is essential for selecting the ratio of extraction solvent to feed that enters an extraction process and for evaluating the mass-transfer rates or theoretical stage efficiencies achieved in process equipment. Since two liquid phases that are immiscible are used, the thermodynamic equilibrium involves considerable evaluation of nonideal solutions. In the simplest case a feed solvent F contains a solute that is to be transferred into an extraction solvent S.

DISTRIBUTION COEFFICIENTS

The weight fraction of solute in the extract phase y divided by the weight fraction of solute in the raffinate phase x at equilibrium is called the **distribution coefficient,** or **partition coefficient,** K [Eq. (15-1)].

$$K = y/x \tag{15-1}$$

Thermodynamically the distribution coefficient $K°$ is derived in mole fractions $y°$ and $x°$ [Eq. (15-2)].

$$K° = y°/x° \tag{15-2}$$

For shortcut calculations the distribution coefficient K' in Bancroft [*Phys. Rev.*, **3**, 120 (1895)] coordinates using the weight ratio of solute to extraction solvent in the extract phase Y and the weight ratio of solute to feed solvent in the raffinate phase X is preferred [Eq. (15-3)].

$$K' = Y/X \tag{15-3}$$

In shortcut calculations the **slope of the equilibrium line** in Bancroft (weight-ratio) coordinates m is also used [Eq. (15-4)].

$$m = dY/dX \tag{15-4}$$

For low concentrations in which the equilibrium line is linear the value of K' is equal to m.

The value of K' is one of the main parameters used to establish the minimum ratio of extraction solvent to feed solvent that can be employed in an extraction process. For example, if the distribution coefficient K' is 4, then a countercurrent extractor would require 0.25 kg or more of extraction-solvent flow to remove all the solute from 1 kg of feed-solvent flow.

The **relative separation,** or **selectivity,** α between two components, b and c, can be described by the ratio of the two distribution coefficients [Eq. (15-5)].

$$\alpha\ (b/c) = K_b°/K_c° = K_b/K_c = K_b'/K_c' \tag{15-5}$$

This is analogous to relative volatility in distillation.

PHASE DIAGRAMS

Ternary-phase equilibrium data can be tabulated as in Tables 15-1 and 15-2 or presented on **equilateral-triangular** diagrams as shown in Fig. 15-7 a and b. The water–acetic acid–methyl isobutyl ketone

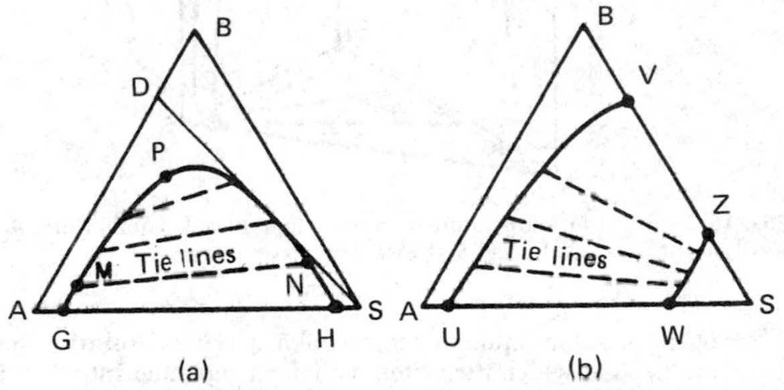

FIG. 15-7 Equilateral-triangular phase diagrams. (a) Type I. (b) Type II. A = feed solvent, B = solute, S = extraction solvent, and P = plait point.

(MIBK) ternary is a **Type I** system in which only one of the binary pairs is immiscible. The ethyl benzene styrene–ethylene glycol ternary is a **Type II** system in which two of the binary pairs are immiscible. The tie lines connect the points that are in equilibrium.

TABLE 15-1 Water–Acetic Acid–Methyl Isobutyl Ketone, 25°C*

Weight % in raffinate				Weight % in extract			
Water	Acetic acid	MIBK	X	Water	Acetic acid	MIBK	Y
98.45	0	1.55	0	2.12	0	97.88	0
95.46	2.85	1.7	0.0299	2.80	1.87	95.33	0.0196
85.8	11.7	2.5	0.1364	5.4	8.9	85.7	0.1039
75.7	20.5	3.8	0.2708	9.2	17.3	73..5	0.2354
67.8	26.2	6.0	0.3864	14.5	24.6	60.9	0.4039
55.0	32.8	12.2	0.5964	22.0	30.8	47.2	0.6525
42.9	34.6	22.5	0.8065	31.0	33.6	35.4	0.9492

*From Sherwood, Evans, and Longcor [*Ind. Eng. Chem.*, **31**, 599 (1939)].

TABLE 15-2 Ethyl Benzene–Styrene–Ethylene Glycol, 25°C*

Weight % in raffinate				Weight % in extract			
Ethyl benzene	Styrene	Ethylene glycol	X	Ethyl benzene	Styrene	Ethylene glycol	Y
90.56	8.63	0.81	0.0953	9.85	1.64	88.51	0.0185
80.40	18.67	0.93	0.2322	9.31	3.49	87.20	0.0400
70.49	28.51	1.00	0.4045	8.72	5.48	85.80	0.0639
60.93	37.98	1.09	0.6233	8.07	7.45	84.48	0.0882
53.55	45.25	1.20	0.8450	7.35	9.25	83.40	0.1109
52.96	45.84	1.20	0.8656	7.31	9.49	83.20	0.1141
43.29	55.32	1.39	1.2779	6.30	12.00	81.70	0.1469
41.51	57.09	1.40	1.3753	6.06	12.54	81.40	0.1541

*From Boobar et al. [*Ind. Eng. Chem.*, **43**, 2922 (1951)].

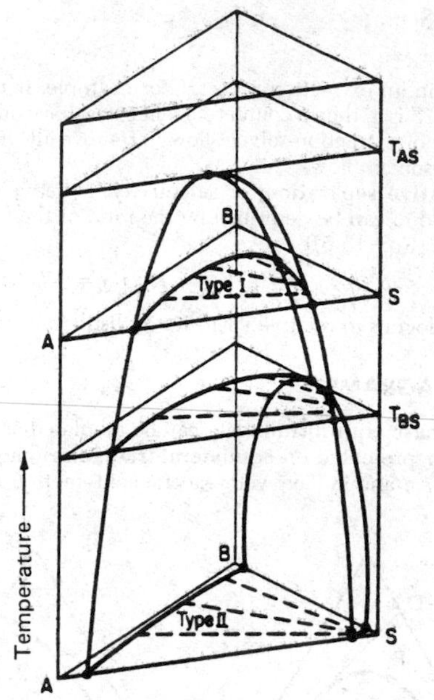

FIG. 15-8 Effect of temperature on ternary liquid-liquid equilibrium. A = feed solvent, B = solute, and S = extraction solvent.

Many immiscible-liquid systems exhibit a **critical solution temperature** beyond which the system no longer separates into two liquid phases. This is shown in Fig. 15-8, in which an increase in temperature can change a Type II system to a Type I system above the critical temperature of the solute and extraction-solvent binary system T_{BS}. The system becomes totally miscible above the critical temperature of the feed solvent and extraction-solvent binary T_{AS}. Occasionally a system can also have a lower critical solution temperature

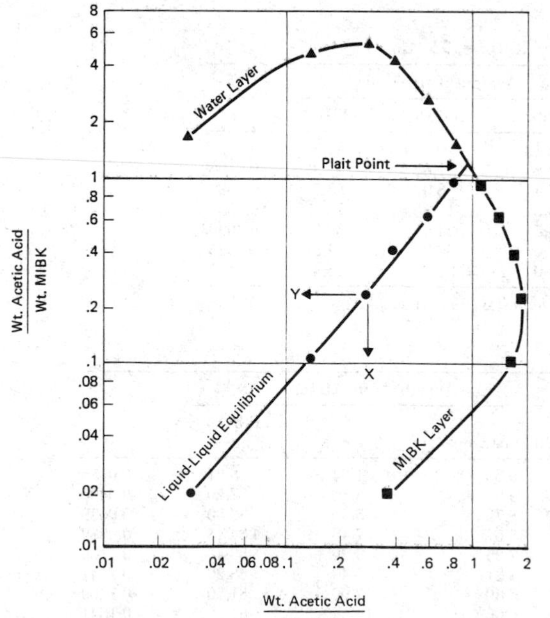

FIG. 15-9 Hand-type ternary diagram for water–acetic acid–methyl isobutyl ketone.

TABLE 15-3 Correlation of Liquid-Liquid Equilibrium Data for Water-Acetic Acid-MIBK System

X	Y (observed)	Y (calculated)
0.0299	0.0196	0.0196
0.1364	0.1039	0.1039
0.2708	0.2354	0.2355
0.3864	0.4039	0.3725
0.5964	0.6525	0.6520
0.8065	0.9492	0.9623

$$\text{Exponent} = \frac{\log(0.6525/0.2354)}{\log(0.5964/0.2708)} = 1.29$$

$$\text{Constant} = 0.6525/(0.5964)^{1.29} = 1.27$$

$$Y(\text{calculated}) = 1.27\,(X)^{1.29},\ \text{for } X \text{ above } 0.25$$

$$\text{Exponent} = \frac{\log(0.1039/0.0196)}{\log(0.1364/0.0299)} = 1.10$$

$$\text{Constant} = 0.1039/(0.1364)^{1.10} = 0.930$$

$$Y(\text{calculated}) = 0.930(X)^{1.10},\ \text{for } X \text{ between } 0.03 \text{ and } 0.25$$

$$Y(\text{calculated}) = K'X = (0.0196/0.0299)X = 0.656X,\ \text{for } X \text{ below } 0.03$$

below which the system will be totally miscible. The methyl ethyl ketone–water binary system provides one example. Changes in pressure ordinarily have a negligible effect on liquid-liquid equilibrium.

For graphical calculation of the number of theoretical stages in a ternary system the **right-triangular diagram** is more convenient to use than an equilateral triangle. The ternary equilibrium data are simply plotted on ordinary rectangular-coordinate graph paper with the weight fraction of the solute on the horizontal axis and the weight fraction of the extraction solvent on the vertical axis. For low-solute concentrations the horizontal scale can be expanded.

For the McCabe-Thiele type of graphical calculations and shortcut methods, the Bancroft (weight-ratio) concentrations can be used on ordinary rectangular-coordinate graph paper. The entire ternary system can be plotted in Bancroft (weight-ratio) concentrations on log-log graph paper as shown by Hand [*J. Phys. Chem.*, **34**, 1961 (1930)], and the equilibrium line can often be correlated by three straight-line segments (Fig. 15-9 and Table 15-3). The plait-point composition for a Type I system can easily be found by using this Hand plot as shown by Treybal, Weber, and Daley [*Ind. Eng. Chem.*, **38**, 817 (1946)]. This type of plot is also helpful for extrapolation and interpolation when data are scarce.

Multicomponent systems containing four or more components become difficult to display graphically. However, process-design calculations can often be made for the extraction of the component with the lowest distribution coefficient K' and treated as a ternary system. The components with higher K' values may be extracted more thoroughly from the raffinate than the solute chosen for design. Or computer calculations can be used to reduce the tedium of multicomponent, multistage calculations.

THERMODYNAMIC BASIS OF LIQUID-LIQUID EQUILIBRIUMS

In a ternary liquid-liquid system, such as the acetic acid–water–MIBK system, all three components are present in both liquid phases. At equilibrium the activity $A°$ of any component is the same in both phases by definition [Eq. (15-6)].

$$A_r^\circ = \gamma_r x^\circ = A_e^\circ = \gamma_e y^\circ \tag{15-6}$$

where $A°$ = activity of solute
γ = activity coefficient of solute
r = raffinate phase
e = extract phase

Consequently, the distribution coefficient in mole-fraction units $K°$ is a result of the ratio of activity coefficients in the two layers [Eq. (15-7)].

$$K^\circ = y^\circ/x^\circ = \gamma_r/\gamma_e \tag{15-7}$$

The **activity coefficient** γ can be defined as the escaping tendency of a component relative to Raoult's law in vapor-liquid equilibrium (see Sec. 4 in this *Handbook* or Null, *Phase Equilibrium in Process Design*, Wiley-Interscience, 1970).

Gmehling and Onken (*Vapor-Liquid Equilibrium Data Collection*, DECHEMA, Frankfurt, Germany, 1979) have reported a large collection of vapor-liquid equilibrium data along with correlations of the resulting activity coefficients. This can be used to predict liquid-liquid equilibrium distribution coefficients as shown in Example 1.

Example 1 Let us estimate the distribution coefficient in weight fractions K for extracting low concentrations of acetone from water into chloroform. The solute is acetone, the feed solvent is water, and the extraction solvent is chloroform in this case.

Gmehling and Onken (op. cit.) give the activity coefficient of acetone in water at infinite dilution γ^∞ as 6.74 at 25°C, depending on which set of vapor-liquid equilibrium data is correlated. From Eqs. (15-1) and (15-7) the distribution coefficient at infinite dilution of solute can be calculated as follows:

$$K = \frac{\gamma_r}{\gamma_e}\,\frac{\text{molecular weight of feed solvent}}{\text{molecular weight of extraction solvent}} = \frac{6.74}{0.80}\,\frac{(18)}{(119.4)} = 3.4$$

Sorenson and Arlt (*Liquid-Liquid Equilibrium Data Collection*, DECHEMA, Frankfurt, Germany, 1979) report several sets of liquid-liquid equilibrium data for the system acetone-water-chloroform, but the lowest solute concentrations reported at 25°C were 3 weight percent acetone in the water layer in equilibrium with 9 weight percent acetone in the chloroform layer. This gives a distribution coefficient K of 3.0.

This example clearly shows good distribution because of a negative deviation from Raoult's law in the extract layer. The activity coefficient of acetone is less than 1.0 in the chloroform layer. However, there is another problem because acetone and chloroform reach a maximum-boiling-point azeotrope composition and cannot be separated completely by distillation at atmospheric pressure.

A higher-boiling solvent, e.g., 1,1,2-trichloroethane, can be used which still gives acetone a negative deviation from Raoult's law (γ_e = 0.732 at 2 mole percent acetone) but does not form a maximum-boiling-point azeotrope according to Treybal, Weber, and Daley [*Ind. Eng. Chem.*, **38**, 817 (1946)].

An activity coefficient greater than 1.0 for a solute in solution is generally considered to be a **positive deviation** from Raoult's law; i.e., the escaping tendency is higher than predicted by Raoult's law. Likewise, an activity coefficient less than 1.0 is considered to be a **negative deviation** from Raoult's law; i.e., the escaping tendency is lower than predicted by ideal-solution behavior. "Positive" and "negative" thus refer to the sign of the logarithm of the activity coefficient.

HYDROGEN-BONDING INTERACTIONS

Deviations from Raoult's law in solution behavior have been attributed to many characteristics such as molecular size and shape, but the strongest deviations appear to be due to hydrogen bonding and electron donor-acceptor interactions. Robbins [*Chem. Eng. Prog.*, **76** (10), 58 (1980)] presented a table of these interactions, Table 15-4, that provides a qualitative guide to solvent selection for liquid-liquid extraction, extractive distillation, azeotropic distillation, or even solvent crystallization. The activity coefficient in the liquid phase is common to all these separation processes.

In Example 1 the solute, acetone, contains a ketone carbonyl group which is a hydrogen acceptor, i.e., solute class 5 according to Table 15-4. This solute is to be extracted from water with chloroform solvent which contains a hydrogen donor group, i.e., solvent class 4. The solute class 5 and solvent class 4 interaction in Table 15-4 is shown to give a negative deviation from Raoult's law.

A negative deviation reduces the activity of the solute in the solvent, which enhances the liquid-liquid distribution coefficient but also leads to maximum-boiling-point azeotropes. Among other classes of solvents shown in Table 15-4 that suppress the escaping tendency of a ketone are classes 1 and 2, i.e., phenolics and acids.

Other ketones, i.e., solvent class 5, are shown to be compatible with acetone, i.e., solute class 5, and tend to give activity coefficients near 1.0, i.e., nearly zero deviation from Raoult's law, and tend to be non-azeotropic. The solvent classes 6 through 12 tend to provide a hostile environment for acetone which increases the escaping tendency, i.e., give activity coefficients greater than 1.0 and tend to form minimum-boiling-point azeotropes. Whenever positive deviations give activity coefficients greater than 7.4, then phase separation, i.e., two liquid phases, can result, as shown by Martin [*Hydrocarbon Process.*, 241 (November 1975)].

Most of the classes in Table 15-4 are self-explanatory, but some can use additional definition. Class 4 includes halogenated solvents that have highly active hydrogens as described by Ewell, Harrison, and Berg [*Ind. Eng. Chem.*, **36**, 871 (1944)]. These are molecules that have two or three halogen atoms on the same carbon as a hydrogen atom, such as methylene chloride, chloroform, 1,1-dichloroethane and 1,1,2,2-tetrachloroethane. Class 4 also includes molecules that have one halogen on the same carbon atom as a hydrogen atom and one or more halogen atoms on an adjacent carbon atom, such as 1,2-dichloroethane and 1,1,2-trichloroethane. Apparently the halogens interact intramolecularly to leave the hydrogen atom highly active.

TABLE 15-4 Organic-Group Interactions Based on 900 Binary Systems*

Solute class		Solvent class 1	2	3	4	5	6	7	8	9	10	11	12
		H donor groups											
1	Phenol	0	0	−	0	−	−	−	−	−	−	+	+
2	Acid, thiol	0	0	−	0	−	−	0	0	0	0	+	+
3	Alcohol, water	−	−	0	+	+	0	−	−	+	+	+	+
4	Active H on multihalogen paraffin	0	0	+	0	−	−	−	−	−	−	0	+
		H acceptor groups											
5	Ketone, amide with no H on N, sulfone, phosphine oxide	−	−	+	−	0	+	+	+	+	+	+	+
6	Tertiary amine	−	−	0	−	+	0	+	+	0	+	0	0
7	Secondary amine	−	0	−	−	+	+	0	0	0	0	0	+
8	Primary amine, ammonia, amide with 2H on N	−	0	−	−	+	+	0	0	+	+	+	+
9	Ether, oxide, sulfoxide	−	0	+	−	+	0	0	+	0	+	0	+
10	Ester, aldehyde, carbonate, phosphate, nitrate, nitrite, nitrile, intramolecular bonding, e.g., *o*-nitrophenol	−	0	+	−	+	+	0	+	+	0	+	+
11	Aromatic, olefin, halogen aromatic, multihalogen paraffin without active H, monohalogen paraffin	+	+	+	0	+	0	0	+	0	+	0	0
		Non-H-bonding groups											
12	Paraffin, carbon disulfide	+	+	+	+	+	0	+	+	+	+	0	0

*From Robbins, *Chem. Eng. Prog.*, **76**(10), 58–61 (1980), by permission.

Monohalogen paraffins like methyl chloride and ethyl chloride are in class 11 along with multihalogen paraffins and olefins without active hydrogen such as carbon tetrachloride and perchloroethylene. Chlorinated benzenes are also in class 11 because they do not have halogens on the same carbon as a hydrogen atom.

Intramolecular bonding on aromatics is another fascinating interaction which gives a net result that behaves much like an ester group, class 10. Examples of this include *ortho*-nitrophenol and *ortho*-hydroxybenzaldehyde (salicylaldehyde). The intramolecular hydrogen bonding is so strong between the hydrogen donor group (phenol) and the hydrogen acceptor group (nitrate or aldehyde) that the molecule ends up by acting as an ester. One result is its low solubility in hot water. By contrast, the *para* derivative is highly soluble in hot water.

Table 15-4 gives a qualitative indication of interactions between classes of molecules but does not give quantitative differences within each class. Taft et al. [*J. Am. Chem. Soc.*, **91**, 4801 (1969)] have quantified the strength of hydrogen acceptors. The quantitative prediction of activity coefficients for solutions is reviewed by Reid, Prausnitz, and Sherwood (*The Properties of Gases and Liquids*, 3d ed., McGraw-Hill, New York, 1977) for the UNIFAC method, the Perotti, Deal, and Derr [*Ind. Eng. Chem.*, **51**, 95 (1959)] method, and the analytical-solution-of-groups (ASOG) method. Leo, Hansch, and Elkins [*Chem. Rev.*, **71**(6), 525 (1971)] also provide methods for predicting distribution coefficients for solutes between water and many solvents. Magnussen, Rasmussen, and Fredenslund [*Ind. Eng. Chem. Process Des. Dev.*, **20**(2), 331 (1981)] have presented a UNIFAC parameter table specifically for predicting liquid-liquid equilibrium.

EXPERIMENTAL EQUILIBRIUM DATA

Several large collections of experimental equilibrium data are now available for liquid-liquid systems. Sorenson and Arlt (*Liquid-Liquid Equilibrium Data Collection*, DECHEMA, Frankfurt, Germany, 1979) have reported several volumes of data that have been correlated with activity-coefficient equations.

Wisniak and Tamir (*Liquid-Liquid Equilibrium and Extraction: A Literature Source Book*, Elsevier, Amsterdam, 1980) have listed many references. Leo, Hansch, and Elkins [*Chem. Rev.*, **71**(6), 525 (1971)] have tabulated distribution coefficients for a large number of solutes between water and solvents. Table 15-5 gives a selected list of distribution coefficients.

DESIRABLE SOLVENT PROPERTIES

The following properties of a potential solvent should be considered before use in a liquid-liquid extraction process.

1. *Selectivity*. The relative separation, or selectivity, α of a solvent is the ratio of two components in the extraction-solvent phase divided by the ratio of the same components in the feed-solvent phase. The separation power of a liquid-liquid system is governed by the deviation of α from unity, analogous to relative volatility in distillation. A relative separation α of 1.0 gives no separation of the components between the two liquid phases. Dilute solute concentrations generally give the highest relative separation factors.

2. *Recoverability*. The extraction solvent must usually be recovered from the extract stream and also from the raffinate stream in an extraction process. Since distillation is often used, the relative volatility of the extraction-solvent to nonsolvent components should be significantly greater or less than unity. A low latent heat of vaporization is desirable for a volatile solvent.

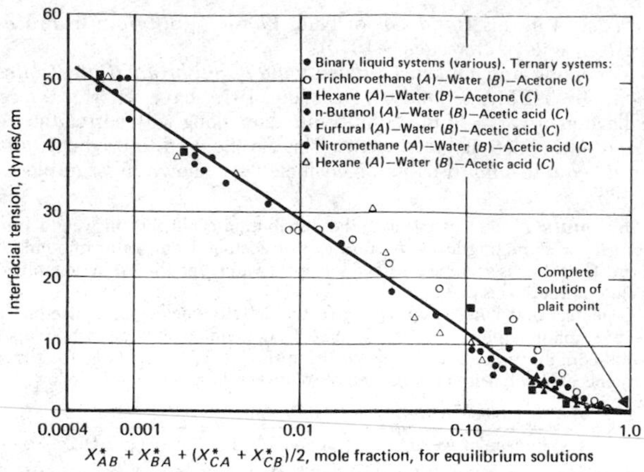

FIG. 15-10 Correlation of interfacial tension with mutual solubility for binary and ternary liquid mixtures. *(From Treybal,* Liquid Extraction, *2d ed., McGraw-Hill, New York, 1963.)*

3. *Distribution coefficient*. The distribution coefficient for a solute should preferably be large so that a low ratio of extraction solvent to feed can be used.

4. *Capacity*. This property refers to the loading of solute per weight of extraction solvent that can be achieved in an extract layer at the plait point in a Type I system or at the solubility limit in a Type II system.

5. *Solvent solubility*. A low solubility of extraction solvent in the raffinate generally leads to a high relative volatility in a raffinate stripper or a low solvent loss if the raffinate is not desolventized. A low solubility of feed solvent in the extract leads to a high relative separation and, generally, to low solute-recovery costs.

6. *Density*. The difference in density between the two liquid phases in equilibrium affects the countercurrent flow rates that can be achieved in extraction equipment as well as the coalescence rates. The density difference decreases to zero at a plait point, but in some systems it can become zero at an intermediate solute concentration (isopycnic, or twin-density, tie line) and can invert the phases at higher concentrations. Differential types of extractors cannot cross such a solute concentration, but mixer-settlers can.

7. *Interfacial tension*. A high interfacial tension promotes rapid coalescence and generally requires high mechanical agitation to produce small droplets. A low interfacial tension allows drop breakup with low agitation intensity but also leads to slow coalescence rates. Interfacial tension usually decreases as solubility and solute concentration increase and falls to zero at the plait point (Fig. 15-10).

8. *Toxicity*. Low toxicity from solvent-vapor inhalation or skin contact is preferred because of potential exposure during repair of equipment or while connections are being broken after a solvent transfer. Also, low toxicity to fish and bioorganisms is preferred when extraction is used as a pretreatment for wastewater before it enters a biotreatment plant and with final effluent discharge to a stream or lake. Often solvent toxicity is low if water solubility is high.

CALCULATION METHODS

SINGLE STAGE

An equilibrium, or theoretical, stage in liquid-liquid extraction as defined earlier is routinely utilized in laboratory procedures. A feed solution is contacted with an immiscible solvent to remove one or more of the solutes from the feed. This can be carried out in a separating funnel, or, preferably, in an agitated vessel that can produce droplets of about 1 mm in diameter. After agitation has stopped and the phases separate, the two clear liquid layers are isolated by decantation.

The equilibrium distribution coefficient can be calculated by material balance, using the weight of the feed F, raffinate R, and

TABLE 15-5 Selected List of Ternary Systems

Component A = feed solvent, component B = solute, and component S = extraction solvent. K_1 is the distribution coefficient in weight-fraction solute y/x for the tie line of lowest solute concentration reported. Ordinarily, K will approach unity as the solute concentration is increased.

Component B	Component S	Temp., °C.	K_1	Ref.
A = cetane				
Benzene	Aniline	25	1.290	47
n-Heptane	Aniline	25	0.0784	47
A = cottonseed oil				
Oleic acid	Propane	85	0.150	46
		98.5	0.1272	46
A = cyclohexane				
Benzene	Furfural	25	0.680	44
Benzene	Nitromethane	25	0.397	127
A = docosane				
1,6-Diphenylhexane	Furfural	45	0.980	11
		80	1.100	11
		115	1.062	11
A = dodecane				
Methylnaphthalene	β,β′-Iminodipropionitrile	*ca.* 25	0.625	92
Methylnaphthalene	β,β′-Oxydipropionitrile	*ca.* 25	0.377	92
A = ethylbenzene				
Styrene	Ethylene glycol	25	0.190	10
A = ethylene glycol				
Acetone	Amyl acetate	31	1.838	86
Acetone	*n*-Butyl acetate	31	1.940	86
Acetone	Cyclohexane	27	0.508	86
Acetone	Ethyl acetate	31	1.850	86
Acetone	Ethyl butyrate	31	1.903	86
Acetone	Ethyl propionate	31	2.32	86
A = furfural				
Trilinolein	*n*-Heptane	30	47.5	15
		50	21.4	15
		70	19.5	15
Triolein	*n*-Heptane	30	95	15
		50	108	15
		70	41.5	15
A = glycerol				
Ethanol	Benzene	25	0.159	62
Ethanol	Carbon tetrachloride	25	0.0667	63
A = *n*-heptane				
Benzene	Ethylene glycol	25	0.300	50
		125	0.316	50
Benzene	β,β′-thiodipropionitrile	25	0.350	92
Benzene	Triethylene glycol	25	0.351	89
Cyclohexane	Aniline	25	0.0815	47
Cyclohexane	Benzyl alcohol	0	0.107	29
		15	0.267	29
Cyclohexane	Dimethylformamide	20	0.1320	28
Cyclohexane	Furfural	30	0.0635	78
Ethylbenzene	Dipropylene glycol	25	0.329	90
Ethylbenzene	β,β′-Oxydipropionitrile	25	0.180	101
Ethylbenzene	β,β′-Thiodipropionitrile	25	0.100	101
Ethylbenzene	Triethylene glycol	25	0.140	89
Methylcyclohexane	Aniline	25	0.087	116
Toluene	Aniline	0	0.577	27
		13	0.477	27
		20	0.457	27
		40	0.425	27
Toluene	Benzyl alcohol	0	0.694	29
Toluene	Dimethylformamide	0	0.667	28
		20	0.514	28
Toluene	Dipropylene glycol	25	0.331	90
Toluene	Ethylene glycol	25	0.150	101
Toluene	Propylene carbonate	20	0.732	39
Toluene	β,β′-Thiodipropionitrile	25	0.150	101
Toluene	Triethylene glycol	25	0.289	89
m-Xylene	β,β′-Thiodipropionitrile	25	0.050	101
o-Xylene	β,β′-Thiodipropionitrile	25	0.150	101
p-Xylene	β,β′-Thiodipropionitrile	25	0.080	101
A = *n*-hexane				
Benzene	Ethylenediamine	20	4.14	23
A = *neo*-hexane				
Cyclopentane	Aniline	15	0.1259	96
		25	0.311	96
A = methylcyclohexane				
Toluene	Methylperfluorooctanoate	10	0.1297	58
		25	0.200	58
A = *iso*-octane				
Benzene	Furfural	25	0.833	44
Cyclohexane	Furfural	25	0.1076	44
n-Hexane	Furfural	30	0.083	78

TABLE 15-5 Selected List of Ternary Systems *(Continued)*

Component *B*	Component S	Temp., °C.	K_1	Ref.
A = perfluoroheptane				
Perfluorocyclic oxide	Carbon tetrachloride	30	0.1370	58
Perfluorocyclic oxide	*n*-Heptane	30	0.329	58
A = perfluoro-*n*-hexane				
n-Hexane	Benzene	30	6.22	80
n-Hexane	Carbon disulfide	25	6.50	80
A = perfluorotri-*n*-butylamine				
Iso-octane	Nitroethane	25	3.59	119
		31.5	2.36	119
		33.7	4.56	119
A = toluene				
Acetone	Ethylene glycol	0	0.286	100
		24	0.326	100
A = triethylene glycol				
α-Picoline	Methylcyclohexane	20	3.87	14
α-Picoline	Diisobutylene	20	0.445	14
α-Picoline	Mixed heptanes	20	0.317	14
A = triolein				
Oleic acid	Propane	85	0.138	46
A = water				
Acetaldehyde	*n*-Amyl alcohol	18	1.43	74
Acetaldehyde	Benzene	18	1.119	74
Acetaldehyde	Furfural	16	0.967	74
Acetaldehyde	Toluene	17	0.478	74
Acetaldehyde	Vinyl acetate	20	0.560	81
Acetic acid	Benzene	25	0.0328	43
		30	0.0984	38
		40	0.1022	38
		50	0.0558	38
		60	0.0637	38
Acetic acid	1-Butanol	26.7	1.613	102
Acetic acid	Butyl acetate	30	0.705	45
			0.391	67
Acetic acid	Caproic acid	25	0.349	73
Acetic acid	Carbon tetrachloride	27	0.1920	91
		27.5	0.0549	54
Acetic acid	Chloroform	*ca.* 25	0.178	70
		25	0.0865	72
		56.8	0.1573	17
Acetic acid	Creosote oil	34	0.706	91
Acetic acid	Cyclohexanol	26.7	1.325	102
Acetic acid	Diisobutyl ketone	25–26	0.284	75
Acetic acid	Di-*n*-butyl ketone	25–26	0.379	75
Acetic acid	Diisopropyl carbinol	25–26	0.800	75
Acetic acid	Ethyl acetate	30	0.907	30
Acetic acid	2-Ethylbutyric acid	25	0.323	73
Acetic acid	2-Ethylhexoic acid	25	0.286	73
Acetic acid	Ethylidene diacetate	25	0.85	104
Acetic acid	Ethyl propionate	28	0.510	87
Acetic acid	Fenchone	25–26	0.310	75
Acetic acid	Furfural	26.7	0.787	102
Acetic acid	Heptadecanol	25	0.312	114
		50	0.1623	114
Acetic acid	3-Heptanol	25	0.828	76
Acetic acid	Hexalin acetate	25–26	0.520	75
Acetic acid	Hexane	31	0.0167	85
Acetic acid	Isoamyl acetate	25–26	0.343	75
Acetic acid	Isophorone	25–26	0.858	75
Acetic acid	Isopropyl ether	20	0.248	31
		25–26	0.429	75
Acetic acid	Methyl acetate	......	1.273	67
Acetic acid	Methyl butyrate	30	0.690	66
Acetic acid	Methyl cyclohexanone	25–26	0.930	75
Acetic acid	Methylisobutyl carbinol	30	1.058	83
Acetic acid	Methylisobutyl ketone	25	0.657	97
		25–26	0.755	75
Acetic acid	Monochlorobenzene	25	0.0435	77
Acetic acid	Octyl acetate	25–26	0.1805	75
Acetic acid	*n*-Propyl acetate	......	0.638	67
Acetic acid	Toluene	25	0.0644	131
Acetic acid	Trichloroethylene	27	0.140	91
		30	0.0549	54
Acetic acid	Vinyl acetate	28	0.294	103
Acetone	Amyl acetate	30	1.228	117
Acetone	Benzene	15	0.940	11
		30	0.862	11
		45	0.725	11
Acetone	*n*-Butyl acetate	......	1.127	67
Acetone	Carbon tetrachloride	30	0.238	12

TABLE 15-5 Selected List of Ternary Systems *(Continued)*

Component *B*	Component *S*	Temp., °C.	K_1	Ref.
Acetone	Chloroform	25	1.830	43
		25	1.720	3
Acetone	Dibutyl ether	25-26	1.941	75
Acetone	Diethyl ether	30	1.00	54
Acetone	Ethyl acetate	30	1.500	117
Acetone	Ethyl butyrate	30	1.278	117
Acetone	Ethyl propionate	30	1.385	117
Acetone	*n*-Heptane	25	0.274	112
Acetone	*n*-Hexane	25	0.343	114
Acetone	Methyl acetate	30	1.153	117
Acetone	Methylisobutyl ketone	25–26	1.910	75
Acetone	Monochlorobenzene	25–26	1.000	75
Acetone	Propyl acetate	30	0.243	117
Acetone	Tetrachloroethane	25–26	2.37	57
Acetone	Tetrachloroethylene	30	0.237	88
Acetone	1,1,2-Trichloroethane	25	1.467	113
Acetone	Toluene	25–26	0.835	75
Acetone	Vinyl acetate	20	1.237	81
		25	3.63	104
Acetone	Xylene	25–26	0.659	75
Allyl alcohol	Diallyl ether	22	0.572	32
Aniline	Benzene	25	14.40	40
		50	15.50	40
Aniline	*n*-Heptane	25	1.425	40
		50	2.20	40
Aniline	Methylcyclohexane	25	2.05	40
		50	3.41	40
Aniline	Nitrobenzene	25	18.89	108
Aniline	Toluene	25	12.91	107
Aniline hydrochloride	Aniline	25	0.0540	98
Benzoic acid	Methylisobutyl ketone	26.7	76.9*	49
iso-Butanol	Benzene	25	0.989	1
iso-Butanol	1,1,2,2-Tetrachloroethane	25	1.80	36
iso-Butanol	Tetrachloroethylene	25	0.0460	7
n-Butanol	Benzene	25	1.263	126
		35	2.12	126
n-Butanol	Toluene	30	1.176	37
tert-Butanol	Benzene	25	0.401	99
tert-Butanol	*tert*-Butyl hypochlorite	0	0.1393	130
		20	0.1487	130
		40	0.200	129
		60	0.539	129
tert-Butanol	Ethyl acetate	20	1.74	5
2-Butoxyethanol	Methylethyl ketone	25	3.05	68
2,3-Butylene glycol	*n*-Butanol	26	0.597	71
		50	0.893	71
2,3-Butylene glycol	Butyl acetate	26	0.0222	71
		50	0.0326	71
2,3-Butylene glycol	Butylene glycol diacetate	26	0.1328	71
		75	0.565	71
2,3-Butylene glycol	Methylvinyl carbinol acetate	26	0.237	71
		50	0.351	71
		75	0.247	71
n-Butylamine	Monochlorobenzene	25	1.391	77
1-Butyraldehyde	Ethyl acetate	37.8	41.3	52
Butyric acid	Methyl butyrate	30	6.75	66
Butyric acid	Methylisobutyl carbinol	30	12.12	83
Cobaltous chloride	Dioxane	25	0.0052	93
Cupric sulfate	*n*-Butanol	30	0.000501	9
Cupric sulfate	*sec*-Butanol	30	0.00702	9
Cupric sulfate	Mixed pentanols	30	0.000225	9
p-Cresol	Methylnaphthalene	35	9.89	82
Diacetone alcohol	Ethylbenzene	25	0.335	22
Diacetone alcohol	Styrene	25	0.445	22
Dichloroacetic acid	Monochlorobenzene	25	0.0690	77
1,4-Dioxane	Benzene	25	1.020	8
Ethanol	*n*-Amyl alcohol	25–26	0.598	75
Ethanol	Benzene	25	0.1191	13
		25	0.0536	115
Ethanol	*n*-Butanol	20	3.00	26
Ethanol	Cyclohexane	25	0.0157	118
Ethanol	Cyclohexene	25	0.0244	124
Ethanol	Dibutyl ether	25–26	0.1458	75
Ethanol	Di-*n*-propyl ketone	25–26	0.592	75
Ethanol	Ethyl acetate	0	0.0263	5
		20	0.500	5
		70	0.455	41
Ethanol	Ethyl isovalerate	25	0.392	13
Ethanol	Heptadecanol	25	0.270	114

TABLE 15-5 Selected List of Ternary Systems *(Continued)*

Component *B*	Component S	Temp., °C.	K_1	Ref.
Ethanol	*n*-Heptane	30	0.274	94
Ethanol	3-Heptanol	25	0.783	76
Ethanol	*n*-Hexane	25	0.00212	111
Ethanol	*n*-Hexanol	28	1.00	56
Ethanol	*sec*-Octanol	28	0.825	56
Ethanol	Toluene	25	0.01816	122
Ethanol	Trichloroethylene	25	0.0682	16
Ethylene glycol	*n*-Amyl alcohol	20	0.1159	59
Ethylene glycol	*n*-Butanol	27	0.412	85
Ethylene glycol	Furfural	25	0.315	18
Ethylene glycol	*n*-Hexanol	20	0.275	59
Ethylene glycol	Methylethyl ketone	30	0.0527	85
Formic acid	Chloroform	25	0.00445	72
		56.9	0.0192	17
Formic acid	Methylisobutyl carbinol	30	1.218	83
Furfural	*n*-Butane	51.5	0.712	42
		79.5	0.930	42
Furfural	Methylisobutyl ketone	25	7.10	19
Furfural	Toluene	25	5.64	53
Hydrogen chloride	*iso*-Amyl alcohol	25	0.170	21
Hydrogen chloride	2,6-Dimethyl-4-heptanol	25	0.266	21
Hydrogen chloride	2-Ethyl-1-butanol	25	0.534	21
Hydrogen chloride	Ethylbutyl ketone	25	0.01515	79
Hydrogen chloride	3-Heptanol	25	0.0250	21
Hydrogen chloride	1-Hexanol	25	0.345	21
Hydrogen chloride	2-Methyl-1-butanol	25	0.470	21
Hydrogen chloride	Methylisobutyl ketone	25	0.0273	79
Hydrogen chloride	2-Methyl-1-pentanol	25	0.502	21
Hydrogen chloride	2-Methyl-2-pentanol	25	0.411	21
Hydrogen chloride	Methylisopropyl ketone	25	0.0814	79
Hydrogen chloride	1-Octanol	25	0.424	21
Hydrogen chloride	2-Octanol	25	0.380	21
Hydrogen chloride	1-Pentanol	25	0.257	21
Hydrogen chloride	Pentanols (mixed)	25	0.271	21
Hydrogen fluoride	Methylisobutyl ketone	25	0.370	79
Lactic acid	*iso*-Amyl alcohol	25	0.352	128
Methanol	Benzene	25	0.01022	4
Methanol	*n*-Butanol	0	0.600	65
		15	0.479	65
		30	0.510	65
		45	1.260	65
		60	0.682	65
Methanol	*p*-Cresol	35	0.313	82
Methanol	Cyclohexane	25	0.0156	125
Methanol	Cyclohexene	25	0.01043	124
Methanol	Ethyl acetate	0	0.0589	5
		20	0.238	5
Methanol	*n*-Hexanol	28	0.565	55
Methanol	Methylnaphthalene	25	0.025	82
		35	0.0223	82
Methanol	*sec*-Octanol	28	0.584	55
Methanol	Phenol	25	1.333	82
Methanol	Toluene	25	0.0099	60
Methanol	Trichloroethylene	27.5	0.0167	54
Methyl-*n*-butyl ketone	*n*-Butanol	37.8	53.4	52
Methylethyl ketone	Cyclohexane	25	1.775	48
		30	3.60	85
Methylethyl ketone	Gasoline	25	1.686	64
Methylethyl ketone	*n*-Heptane	25	1.548	112
Methylethyl ketone	*n*-Hexane	25	1.775	112
		37.8	2.22	52
Methylethyl ketone	2-Methyl furan	25	84.0	109
Methylethyl ketone	Monochlorobenzene	25	2.36	68
Methylethyl ketone	Naphtha	26.7	0.885†	6
Methylethyl ketone	1,1,2-Trichloroethane	25	3.44	68
Methylethyl ketone	Trichloroethylene	25	3.27	68
Methylethyl ketone	2,2,4-Trimethylpentane	25	1.572	64
Nickelous chloride	Dioxane	25	0.0017	93
Nicotine	Carbon tetrachloride	25	9.50	34
Phenol	Methylnaphthalene	25	7.06	82
α-Picoline	Benzene	20	8.75	14
α-Picoline	Diisobutylene	20	1.360	14
α-Picoline	Heptanes (mixed)	20	1.378	14
α-Picoline	Methylcyclohexane	20	1.00	14
iso-Propanol	Benzene	25	0.276	69
iso-Propanol	Carbon tetrachloride	20	1.405	25
iso-Propanol	Cyclohexane	25	0.0282	123
iso-Propanol	Cyclohexene	15	0.0583	124
		25	0.0682	124
		35	0.1875	124

TABLE 15-5 Selected List of Ternary Systems *(Continued)*

Component B	Component S	Temp., °C.	K_1	Ref.
iso-Propanol	Diisopropyl ether	25	0.406	35
iso-Propanol	Ethyl acetate	0	0.200	5
		20	1.205	5
iso-Propanol	Tetrachloroethylene	25	0.388	7
iso-Propanol	Toluene	25	0.1296	121
n-Propanol	*iso*-Amyl alcohol	25	3.34	20
n-Propanol	Benzene	37.8	0.650	61
n-Propanol	*n*-Butanol	37.8	3.61	61
n-Propanol	Cyclohexane	25	0.1553	123
		35	0.1775	123
n-Propanol	Ethyl acetate	0	1.419	5
		20	1.542	5
n-Propanol	*n*-Heptane	37.8	0.540	61
n-Propanol	*n*-Hexane	37.8	0.326	61
n-Propanol	*n*-Propyl acetate	20	1.55	106
		35	2.14	106
n-Propanol	Toluene	25	0.299	2
Propionic acid	Benzene	30	0.598	57
Propionic acid	Cyclohexane	31	0.1955	84
Propionic acid	Cyclohexene	31	0.303	84
Propionic acid	Ethyl acetate	30	2.77	87
Propionic acid	Ethyl butyrate	26	1.470	87
Propionic acid	Ethyl propionate	28	0.510	87
Propionic acid	Hexanes (mixed)	31	0.186	84
Propionic acid	Methyl butyrate	30	2.15	66
Propionic acid	Methylisobutyl carbinol	30	3.52	83
Propionic acid	Methylisobutyl ketone	26.7	1.949*	49
Propionic acid	Monochlorobenzene	30	0.513	57
Propionic acid	Tetrachloroethylene	31	0.167	84
Propionic acid	Toluene	31	0.515	84
Propionic acid	Trichloroethylene	30	0.496	57
Pyridine	Benzene	15	2.19	110
		25	3.00	105
		25	2.73	120
		45	2.49	110
		60	2.10	110
Pyridine	Monochlorobenzene	25	2.10	77
Pyridine	Toluene	25	1.900	120
Pyridine	Xylene	25	1.260	120
Sodium chloride	*iso*-Butanol	25	0.0182	36
Sodium chloride	*n*-Ethyl-*sec*-butyl amine	32	0.0563	24
Sodium chloride	*n*-Ethyl-*tert*-butyl amine	40	0.1792	24
Sodium chloride	2-Ethylhexyl amine	30	0.187	24
Sodium chloride	1-Methyldiethyl amine	39.1	0.0597	24
Sodium chloride	1-Methyldodecyl amine	30	0.693	24
Sodium chloride	*n*-Methyl-1,3-dimethylbutyl amine	30	0.0537	24
Sodium chloride	1-Methyloctyl amine	30	0.589	24
Sodium chloride	*tert*-Nonyl amine	30	0.0318	24
Sodium chloride	1,1,3,3-Tetramethyl butyl amine	30	0.072	24
Sodium hydroxide	*iso*-Butanol	25	0.00857	36
Sodium nitrate	Dioxane	25	0.0246	95
Succinic acid	Ethyl ether	15	0.220	33
		20	0.198	33
		25	0.1805	33
Trimethyl amine	Benzene	25	0.857	51
		70	2.36	51

* Concentrations in lb.-moles/cu. ft.
† Concentrations in volume fraction.

References:

1. Alberty and Washburn, *J. Phys. Chem.*, **49**, 4 (1945).
2. Baker, *J. Phys. Chem.*, **59**, 1182 (1955).
3. Bancroft and Hubard, *J. Am. Chem. Soc.*, **64**, 347 (1942).
4. Barbaudy, *Compt. rend.*, **182**, 1279 (1926).
5. Beech and Glasstone, *J. Chem. Soc.*, **1938**, 67.
6. Berg, Manders, and Switzer, *Chem. Eng. Progr.*, **47**, 11 (1951).
7. Bergelin, Lockhart, and Brown, *Trans. Am. Inst. Chem. Engrs.*, **39**, 173 (1943).
8. Berndt and Lynch, *J. Am. Chem. Soc.*, **66**, 282 (1944).
9. Blumberg, Cejtlin, and Fuchs, *J. Appl. Chem.*, **10**, 407 (1960).
10. Boobar *et al.*, *Ind. Eng. Chem.*, **43**, 2922 (1951).
11. Briggs and Comings, *Ind. Eng. Chem.*, **35**, 411 (1943).
12. Buchanan, *Ind. Eng. Chem.*, **44**, 2449 (1952).
13. Chang and Moulton, *Ind. Eng. Chem.*, **45**, 2350 (1953).
14. Charles and Morton, *J. Appl. Chem.*, **7**, 39 (1957).
15. Church and Briggs, *J. Chem. Eng. Data*, **9**, 207 (1964).
16. Colburn and Phillips, *Trans. Am. Inst. Chem. Engrs.*, **40**, 333 (1944).
17. Conti, Othmer, and Gilmont, *J. Chem. Eng. Data*, **5**, 301 (1960).
18. Conway and Norton, *Ind. Eng. Chem.*, **43**, 1433 (1951).
19. Conway and Phillips, *Ind. Eng. Chem.*, **46**, 1474 (1954).
20. Coull and Hope, *J. Phys. Chem.*, **39**, 967 (1935).
21. Crittenden and Hixson, *Ind. Eng. Chem.*, **46**, 265 (1954).
22. Crook and Van Winkle, *Ind. Eng. Chem.*, **46**, 1474 (1954).
23. Cumming and Morton, *J. Appl. Chem.*, **3**, 358 (1953).
24. Davison, Smith, and Hood, *J. Chem. Eng. Data*, **11**, 304 (1966).
25. Denzler, *J. Phys. Chem.*, **49**, 358 (1945).
26. Drouillon, *J. chim. phys.*, **22**, 149 (1925).
27. Durandet and Gladel, *Rev. Inst. Franc. Pétrole*, **9**, 296 (1954).
28. Durandet and Gladel, *Rev. Inst. Franc. Pétrole*, **11**, 811 (1956).
29. Durandet, Gladel, and Graziani, *Rev. Inst. Franc. Pétrole*, **10**, 585 (1955).
30. Eaglesfield, Kelly, and Short, *Ind. Chemist*, **29**, 147, 243 (1953).
31. Elgin and Browning, *Trans. Am. Inst. Chem. Engrs.*, **31**, 639 (1935).
32. Fairburn, Cheney, and Chernovsky, *Chem. Eng. Progr.*, **43**, 280 (1947).
33. Forbes and Coolidge, *J. Am. Chem. Soc.*, **41**, 150 (1919).
34. Fowler and Noble, *J. Appl. Chem.*, **4**, 546 (1954).

TABLE 15-5 Selected List of Ternary Systems *(Continued)*

35. Frere, *Ind. Eng. Chem.*, **41**, 2365 (1949).
36. Fritzsche and Stockton, *Ind. Eng. Chem.*, **38**, 737 (1946).
37. Fuoss, *J. Am. Chem. Soc.*, **62**, 3183 (1940).
38. Garner, Ellis, and Roy, *Chem. Eng. Sci.*, **2**, 14 (1953).
39. Gladel and Lablaude, *Rev. Inst. Franc. Pétrole*, **12**, 1236 (1957).
40. Griswold, Chew, and Klecka, *Ind. Eng. Chem.*, **42**, 1246 (1950).
41. Griswold, Chu, and Winsauer, *Ind. Eng. Chem.*, **41**, 2352 (1949).
42. Griswold, Klecka, and West, *Chem. Eng. Progr.*, **44**, 839 (1948).
43. Hand, *J. Phys. Chem.*, **34**, 1961 (1930).
44. Henty, McManamey, and Price, *J. Appl. Chem.*, **14**, 148 (1964).
45. Hirata and Hirose, *Kagaku Kogaku*, **27**, 407 (1963).
46. Hixon and Bockelmann, *Trans. Am. Inst. Chem. Engrs.*, **38**, 891 (1942).
47. Hunter and Brown, *Ind. Eng. Chem.*, **39**, 1343 (1947).
48. Jeffreys, *J. Chem. Eng. Data*, **8**, 320 (1963).
49. Johnson and Bliss, *Trans. Am. Inst. Chem. Engrs.*, **42**, 331 (1946).
50. Johnson and Francis, *Ind. Eng. Chem.*, **46**, 1662 (1954).
51. Jones and Grigsby, *Ind. Eng. Chem.*, **44**, 378 (1952).
52. Jones and McCants, *Ind. Eng. Chem.*, **46**, 1956 (1954).
53. Knight, *Trans. Am. Inst. Chem. Engrs.*, **39**, 439 (1943).
54. Krishnamurty, Murti, and Rao, *J. Sci. Ind. Res.*, **12B**, 583 (1953).
55. Krishnamurty and Rao, *J. Sci. Ind. Res.*, **14B**, 614 (1955).
56. Krishnamurty and Rao, *Trans. Indian Inst. Chem. Engrs.*, **6**, 153 (1954).
57. Krishnamurty, Rao, and Rao, *Trans. Indian Inst. Chem. Engrs.*, **6**, 161 (1954).
58. Kyle and Reed, *J. Chem. Eng. Data*, **5**, 266 (1960).
59. Laddha and Smith, *Ind. Eng. Chem.*, **40**, 494 (1948).
60. Mason and Washburn, *J. Am. Chem. Soc.*, **59**, 2076 (1937).
61. McCants, Jones, and Hopson, *Ind. Eng. Chem.*, **45**, 454 (1953).
62. McDonald, *J. Am. Chem. Soc.*, **62**, 3183 (1940).
63. McDonald, Kluender, and Lane, *J. Phys. Chem.*, **46**, 946 (1942).
64. Moulton and Walkey, *Trans. Am. Inst. Chem. Engrs.*, **40**, 695 (1944).
65. Mueller, Pugsley, and Ferguson, *J. Phys. Chem.*, **35**, 1314 (1931).
66. Murty, Murty, and Subrahmanyam, *J. Chem. Eng. Data*, **11**, 335 (1966).
67. Murti, Venkataratnam, and Rao, *J. Sci. Ind. Res.*, **13B**, 392 (1954).
68. Newman, Hayworth, and Treybal, *Ind. Eng. Chem.*, **41**, 2039 (1949).
69. Olsen and Washburn, *J. Am. Chem. Soc.*, **57**, 303 (1935).
70. Othmer, *Chem. Met. Eng.*, **43**, 325 (1936).
71. Othmer, Bergen, Schlechter, and Bruins, *Ind. Eng. Chem.*, **37**, 890 (1945).
72. Othmer and Ku, *J. Chem. Eng. Data*, **4**, 42 (1959).
73. Othmer and Serrano, *Ind. Eng. Chem.*, **41**, 1030 (1949).
74. Othmer and Tobias, *Ind. Eng. Chem.*, **34**, 690 (1942).
75. Othmer, White, and Treuger, *Ind. Eng. Chem.*, **33**, 1240 (1941).
76. Oualline and Van Winkle, *Ind. Eng. Chem.*, **44**, 1668 (1952).
77. Peake and Thompson, *Ind. Eng. Chem.*, **44**, 2439 (1952).
78. Pennington and Marwill, *Ind. Eng. Chem.*, **45**, 1371 (1953).
79. Pilloton, *A.S.T.M. Spec. Tech. Publ.*, **238**, 5 (1958).
80. Pliskin and Treybal, *J. Chem. Eng. Data*, **11**, 49 (1966).
81. Pratt and Glover, *Trans. Inst. Chem. Engrs. (London)*, **24**, 52 (1946).
82. Prutton, Walsh, and Desar, *Ind. Eng. Chem.*, **42**, 1210 (1950).
83. Rao, Ramamurty, and Rao, *Chem. Eng. Sci.*, **8**, 265 (1958).
84. Rao and Rao, *J. Appl. Chem.*, **6**, 270 (1956).
85. Rao and Rao, *J. Appl. Chem.*, **7**, 659 (1957).
86. Rao and Rao, *J. Sci. Ind. Res.*, **14B**, 204 (1955).
87. Rao and Rao, *J. Sci. Ind. Res.*, **14B**, 444 (1955).
88. Rao and Rao, *Trans. Indian Inst. Chem. Engrs.*, **7**, 78 (1954–1955).
89. Rifai, *Riv. Combust.*, **11**, 811 (1957).
90. Rifai, *Riv. Combust.*, **11**, 829 (1957).
91. Saletore, Mene, and Warhadpande, *Trans. Indian. Inst. Chem. Engrs.*, **2**, 16 (1950).
92. Saunders, *Ind. Eng. Chem.*, **43**, 121 (1951).
93. Schott and Lynch, *J. Chem. Eng. Data*, **11**, 215 (1966).
94. Schweppe and Lorah, *Ind. Eng. Chem.*, **46**, 2391 (1954).
95. Selikson and Ricci, *J. Am. Chem. Soc.*, **64**, 2474 (1942).
96. Serjian, Spurr, and Gibbons, *J. Am. Chem. Soc.*, **68**, 1763 (1946).
97. Sherwood, Evans, and Longcor, *Ind. Eng. Chem.*, **31**, 1144 (1939).
98. Sidgwick, Pickford, and Wilsdon, *J. Chem. Soc.*, **99**, 1122 (1911).
99. Simonsen and Washburn, *J. Am. Chem. Soc.*, **68**, 235 (1946).
100. Sims and Bolme, *J. Chem. Eng. Data*, **10**, 111 (1965).
101. Skinner, *Ind. Eng. Chem.*, **47**, 222 (1955).
102. Skrzec and Murphy, *Ind. Eng. Chem.*, **46**, 2245 (1954).
103. Smith, *J. Phys. Chem.*, **45**, 1301 (1941).
104. Smith, *J. Phys. Chem.*, **46**, 229 (1942).
105. Smith, *J. Phys. Chem.*, **46**, 376 (1942).
106. Smith and Bonner, *Ind. Eng. Chem.*, **42**, 896 (1950).
107. Smith and Drexel, *Ind. Eng. Chem.*, **37**, 601 (1945).
108. Smith, Foecking, and Barber, *Ind. Eng. Chem.*, **41**, 2289 (1949).
109. Smith and La Bonte, *Ind. Eng. Chem.*, **44**, 2740 (1952).
110. Smith, Stibolt, and Day, *Ind. Eng. Chem.*, **43**, 190 (1951).
111. Tareşenkov and Paul'sen, *J. Gen. Chem. (U.S.S.R.)*, **7**, 2143 (1937).
112. Treybal and Vondrak, *Ind. Eng. Chem.*, **41**, 1761 (1949).
113. Treybal, Weber, and Daley, *Ind. Eng. Chem.*, **38**, 817 (1946).
114. Upchurch and Van Winkle, *Ind. Eng. Chem.*, **44**, 618 (1952).
115. Varteressian and Fenske, *Ind. Eng. Chem.*, **28**, 928 (1936).
116. Varteressian and Fenske, *Ind. Eng. Chem.*, **29**, 270 (1937).
117. Venkataratnam, Rao, and Rao, *Chem. Eng. Sci.*, **7**, 102 (1957).
118. Vold and Washburn, *J. Am. Chem. Soc.*, **54**, 4217 (1932).
119. Vreeland and Dunlap, *J. Phys. Chem.*, **61**, 329 (1957).
120. Vriens and Medcalf, *Ind. Eng. Chem.*, **45**, 1098 (1953).
121. Washburn and Beguin, *J. Am. Chem. Soc.*, **62**, 579 (1940).
122. Washburn, Beguin, and Beckord, *J. Am. Chem. Soc.*, **61**, 1694 (1939).
123. Washburn, Brockway, Graham, and Deming, *J. Am. Chem. Soc.*, **64**, 1886 (1942).
124. Washburn, Graham, Arnold, and Transue, *J. Am. Chem. Soc.*, **62**, 1454 (1940).
125. Washburn and Spencer, *J. Am. Chem. Soc.*, **56**, 361 (1934).
126. Washburn and Strandskov, *J. Phys. Chem.*, **48**, 241 (1944).
127. Weck and Hunt, *Ind. Eng. Chem.*, **46**, 2521 (1954).
128. Weiser and Geankoplis, *Ind. Eng. Chem.*, **47**, 858 (1955).
129. Westwater, *Ind. Eng. Chem.*, **47**, 451 (1955).
130. Westwater and Audrieth, *Ind. Eng. Chem.*, **46**, 1281 (1954).
131. Woodman, *J. Phys. Chem.*, **30**, 1283 (1926).

extract E, plus the weight-fraction solute in the feed x_f and raffinate x_r, when the weight-fraction solute in the extraction solvent y_s is zero [Eq. (15-8)].

$$K = \frac{y_e}{x_r} = \frac{R}{E}\left[\frac{F}{R}\frac{x_f}{x_r} - 1\right] \tag{15-8}$$

However, an actual analysis of the weight-fraction solute in the extract y_e and raffinate x_r is preferred.

CROSSCURRENT THEORETICAL STAGES

After a single-stage liquid-liquid contact the phase remaining from the feed solution (raffinate) can be contacted with another quantity of fresh extraction solvent. This **crosscurrent extraction scheme** (Fig. 15-4) is an excellent laboratory procedure because the extract and raffinate phases can be analyzed after each stage to generate equilibrium data. Also, the feasibility of solute removal to low levels can be demonstrated.

The number of crosscurrent stages N that are required to reach a specified raffinate composition, in Bancroft coordinates X_n, can be calculated directly if K' is constant, the ratio of extraction solvent to feed solvent S'/F' is kept constant, and fresh extraction solvent $Y_s = 0$ (presaturated with feed solvent) is used in each stage [Eq. (15-9)].

$$N = \frac{\log (X_f/X_n)}{\log (K'S'/F' + 1)} \tag{15-9}$$

The crosscurrent scheme is not generally economically attractive for large commercial processes because solvent usage is high and solute concentration in the combined extract is low.

COUNTERCURRENT THEORETICAL STAGES

The main objective for calculating the number of theoretical stages (or mass-transfer units) in the design of a liquid-liquid extraction process is to evaluate the compromise between the size of the equipment, or number of contactors required, and the ratio of extraction solvent to feed flow rates required to achieve the desired transfer of mass from one phase to the other. In any mass-transfer process there can be an infinite number of combinations of flow rates, number of stages, and degrees of solute transfer. The optimum is governed by economic considerations.

The number of stages that are required can be kept to a minimum

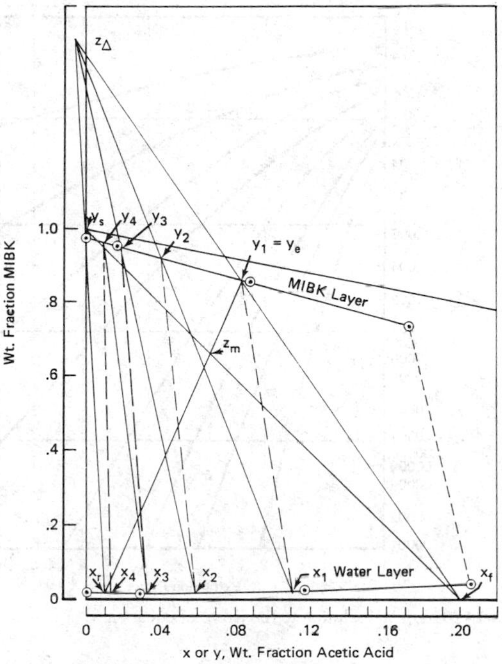

FIG. 15-11 Right-triangular graphical stages.

by selecting a solvent with a high distribution coefficient or by operating with a high ratio of extraction solvent to feed. However, a high solvent flow rate usually requires a high operating cost because of the cost of recovering the solvent. A high solvent flow rate should be carefully compared with an increase in capital cost for taller or more equipment to achieve more theoretical stages (or mass-transfer units) and reduce the required flow of solvent. The operating cost of an extractor is generally quite low in comparison with the operating cost of the solvent-recovery distillation column.

The other common objective for calculating the number of countercurrent theoretical stages (or mass-transfer units) is to evaluate the performance of liquid-liquid extraction test equipment in a pilot plant or to evaluate production equipment in an industrial plant. Most liquid-liquid extraction equipment in common use can be designed to achieve the equivalent of 1 to 8 theoretical countercurrent stages, with some designed to achieve 10 to 12 stages.

Right-Triangular Method This method is a rigorous Ponchon-Savarit type of graphical technique for determining the number of countercurrent theoretical stages of a ternary system (Fig. 15-11). The horizontal axis is the concentration of solute in weight fractions x or y. The vertical axis is the weight fraction of extraction solvent. The weight fraction of feed solvent is simply the amount remaining so that all three weight fractions add up to 1.0.

For the system water–acetic acid–MIBK in Fig. 15-11 the raffinate (water) layer is the solubility curve with low concentrations of MIBK, and the extract (MIBK) layer is the solubility curve with high concentrations of MIBK. The dashed lines are **tie lines** which connect the two layers in equilibrium as given in Table 15-1. Example 2 describes the right-triangular method of calculating the number of theoretical stages required.

Example 2 A 100-kg/h feed stream containing 20 weight percent acetic acid in water is to be extracted with 200 kg/h of recycle MIBK that contains 0.1 percent acetic acid and 0.01 percent water. The aqueous raffinate is to be extracted down to 1 percent acetic acid. How many theoretical stages will be required and what will the extract composition be?

The solute concentration in the feed, $x_f = 0.20$, in the raffinate, $x_r = 0.01$, and in the extraction solvent, $y_s = 0.001$, can be located on the diagram. Then the mix point z_m can be calculated from the feed, $F = 100$ kg/h, and the solvent, $S = 200$ kg/h, entering the extractor [Eq. (15-10)].

$$z_m = (Fx_f + Sy_s)/(F + S) \tag{15-10}$$

The mix point, $z_m = 0.0673$, falls on a straight line connecting x_f and y_s. The extract composition is then determined by drawing a straight line from x_r through z_m until the line intersects the extract line at the final extract composition, $y_e = 0.084$. The delta point z_Δ is then found at the intersection of two lines. One line connects the feed and extract compositions x_f and y_e. The other line connects the raffinate and solvent compositions x_r and y_s.

The graphical stepping off of theoretical stages starts at the extract composition y_e, and a tie line is drawn (parallel to the nearest one) to the raffinate composition leaving stage 1, $x_1 = y_e/K = 0.084\ (0.117/0.089) = 0.1104$. The size of the extract stream can be calculated by the material balance $E = (F + S)(z_m - x_r)/y_e - x_r)$. A straight line is drawn between x_1 and z_Δ to find the extract composition leaving stage 2, $y_2 = 0.0415$. Another tie line is drawn to find the raffinate composition leaving stage 2, x_2, and the stepwise procedure continues until the final raffinate composition, $x_r = 0.01$, is achieved. This requires four theoretical stages plus a fraction. Additional details on the derivation of this procedure are provided by Foust, Wenzel, Clump, Maus, and Anderson (*Principles of Unit Operations*, 2d ed, Wiley, New York, 1980) and Treybal (*Mass-Transfer Operations*, 3d ed., McGraw-Hill, New York, 1980).

Shortcut Methods These methods are often preferred for repetitive calculations of pilot-plant data and numerous design conditions. In distillation calculations the assumption of constant molar vapor and liquid flow rates gave rise to the McCabe-Thiele stepwise calculation method with straight operating lines and a curved equilibrium line. A similar concept can be achieved in liquid-liquid extraction by assuming a constant flow rate of feed solvent F' and a constant flow rate of extraction solvent S' through the extractor. The solute concentrations are then given as the weight ratio of solute to feed solvent X and the weight ratio of solute to extraction solvent Y, i.e., Bancroft coordinates. These concentrations and coordinates will essentially give a straight operating line on an XY diagram for stages 2 through $r - 1$ in Fig. 15-12. Equilibrium data using these weight ratios have already been shown to follow straight-line segments on a log-log plot (see Fig. 15-9). The main problem, then, is to evaluate the primary ratio of extraction solvent to feed solvent passing through the extractor in stages 2 through $r - 1$.

Robbins ("Liquid-Liquid Extraction," in Schweitzer, *Handbook of Separation Techniques for Chemical Engineers*, McGraw-Hill, New York, 1979, sec. 1.9) reported that most liquid-liquid extraction systems can be treated as having either (A) immiscible solvents, (B) partially miscible solvents with a low solute concentration in the extract, or (C) partially miscible solvents with a high solute concentration in the extract.

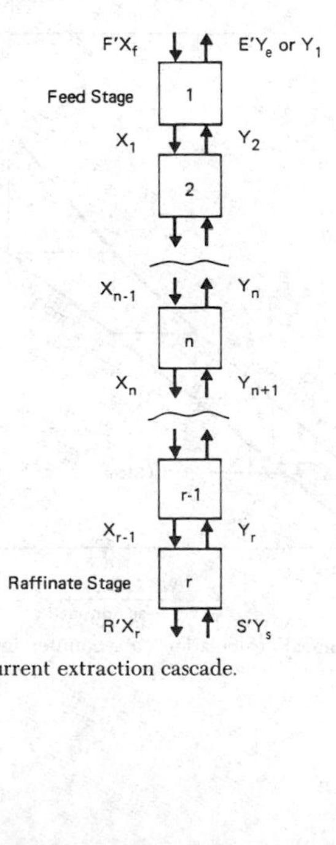

FIG. 15-12 Countercurrent extraction cascade.

In case A the solvents are immiscible, so the rate of feed solvent alone in the feed stream F' is the same as the rate of feed solvent alone in the raffinate stream R'. In like manner, the rate of extraction solvent alone is the same in the stream entering S' as in the extract stream leaving E' (Fig. 15-12). The ratio of extraction-solvent to feed-solvent flow rates is therefore $S'/F' = E'/R'$. A material balance can be written around the feed end of the extractor down to any stage n (see Fig. 15-12) and then rearranged to a McCabe-Thiele type of operating line with a slope of F'/S' [Eq. (15-11)].

$$Y_{n+1} = \frac{F'}{S'} X_n + \frac{E'Y_e - F'X_f}{S'} \qquad (15\text{-}11)$$

Similarly, the same operating line can be derived from a material balance around the raffinate end of the extractor up to stage n [Eq. (15-12)].

$$Y_n = \frac{F'}{S'} X_{n-1} + \frac{S'Y_s - R'X_r}{S'} \qquad (15\text{-}12)$$

The overall extractor material balance is given by Eq. (15-13).

$$Y_e = \frac{F'X_f + S'Y_s - R'X_r}{E'} \qquad (15\text{-}13)$$

The end points of the operating line on an XY plot (Fig. 15-13) are X_r, Y_s and X_f, Y_e, and the number of theoretical stages can be stepped off graphically. The equilibrium curve is taken from the Hand type of correlation shown earlier (Fig. 15-9). When the equilibrium line is straight, its intercept is zero, and the operating line is straight, the number of theoretical stages can be calculated with one of the Kremser equations [Eqs. (15-14) and (15-15)]. When the intercept of the equilibrium line is not zero, the value of Y_s/K'_s should be used instead of Y_s/m, where K'_s is the distribution coefficient in Bancroft coordinates at Y_s.

When $\mathcal{E} \neq 1.0$,

$$N = \frac{\ln\left[\left(\dfrac{X_f - Y_s/m}{X_r - Y_s/m}\right)\left(1 - \dfrac{1}{\mathcal{E}}\right) + \dfrac{1}{\mathcal{E}}\right]}{\ln \mathcal{E}} \qquad (15\text{-}14)$$

When $\mathcal{E} = 1.0$,

$$N = \frac{X_f - Y_s/m}{X_r - Y_s/m} - 1 \qquad (15\text{-}15)$$

The value of m is the slope of the equilibrium line dY/dX [Eq. (15-4)]. This is equal to K' [Eq. (15-3)] at low concentrations where the equilibrium line is straight. The value of $\mathcal{E}$, the **extraction factor,** is calculated by dividing the slope of the equilibrium line m by the slope of the operating line F'/S' [Eq. (15-16)].

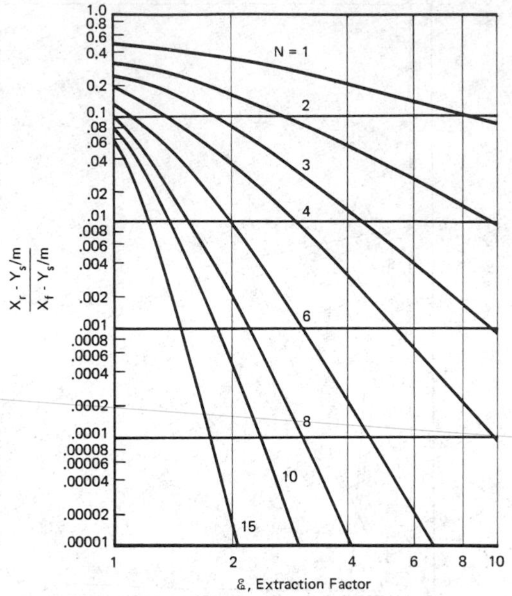

FIG. 15-14 Graphical solution to the Kremser equation.

$$\mathcal{E} = mS'/F' \qquad (15\text{-}16)$$

The solution to the Kremser equation is shown graphically in Fig. 15-14. When a system responds with a constant number of theoretical stages N, the solute concentration in the raffinate X_r can readily be evaluated as the result of changing the ratio of solvent to feed [Eqs. (15-17) and (15-18)].

When $\mathcal{E} \neq 1.0$,

$$\frac{X_r - Y_s/m}{X_f - Y_s/m} = \frac{\mathcal{E} - 1}{\mathcal{E}^{N+1} - 1} \qquad (15\text{-}17)$$

When $\mathcal{E} = 1.0$,

$$\frac{X_r - Y_s/m}{X_f - Y_s/m} = \frac{1}{N + 1} \qquad (15\text{-}18)$$

When the equilibrium line is not straight, Treybal (*Liquid Extraction*, 2d ed., McGraw-Hill, New York, 1963) recommends that the geometric mean value of m be used. The geometric mean of the slope of the equilibrium line at the concentration leaving the feed state m_1 and at the raffinate concentration leaving the raffinate stage m_r is $\sqrt{m_1 m_r}$.

Example 3 Let us solve the problem in Example 2 by using the shortcut calculation method assuming immiscible solvents, case A.

From the problem,

$F' = 100\ (1 - 0.2) = 80$ kg water/h

$X_f = 0.2/0.8 = 0.25$ kg acetic acid/kg water

$X_r = 0.01/0.99 = 0.01$ kg acetic acid/kg water

$S' = 200\ (1 - 0.001) = 199.8$ kg MIBK/h

$Y_s = 0.2/199.8 = 0.001$ kg acetic acid/kg MIBK

If we assume $R' = F'$ and $E' = S'$, calculate Y_e from Eq. (15-13):

$$Y_e = \frac{80\ (0.25) + (199.8)\ (0.001) - 80\ (0.01)}{199.8} = 0.097\ \frac{\text{kg acetic acid}}{\text{kg MIBK}}$$

From the correlation of equilibrium data (Table 15-3), $Y = 0.930(X)^{1.10}$, for X between 0.03 and 0.25.

Calculate $X_1 = (0.097/0.930)^{1/1.10} = 0.128$:

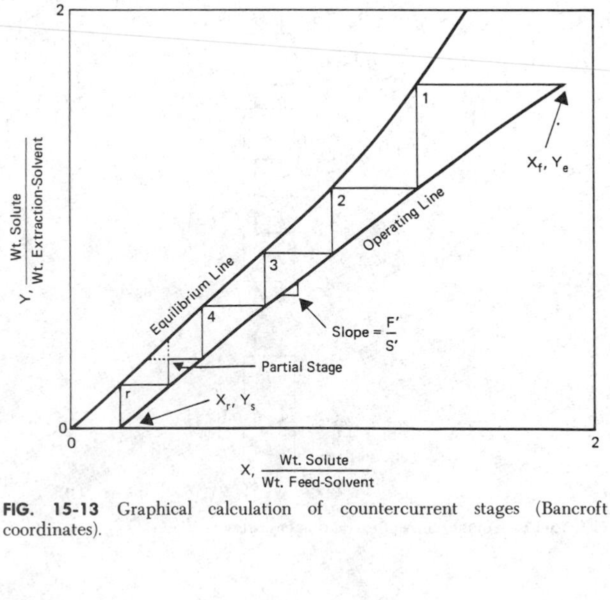

FIG. 15-13 Graphical calculation of countercurrent stages (Bancroft coordinates).

$$m = dY/dX = (0.930)(1.10)(X)^{0.1}, \text{ for } X \text{ between 0.03 and 0.25}$$
$$m_1 = 0.833 \text{ at } X = 0.128$$
$$m_r = dY/dX = K' = 0.656, \text{ for } X \text{ below 0.03}$$
$$K'_s = 0.656 \text{ at } Y_s = 0.001$$
$$\mathcal{E} = \sqrt{m_1 m_r}\, S'/F' = (0.739)(199.8)/80 = 1.85$$

N is determined from Fig. 15-14, Eq. (15-14), or the McCabe-Thiele type of plot (Fig. 15-13):

$$N = \ln\left[\frac{\left(\dfrac{0.25 - 0.001/0.656}{0.01 - 0.001/0.656}\right)\left(1 - \dfrac{1}{1.85}\right) + \dfrac{1}{1.85}}{\ln 1.85}\right] = 4.3$$

From solubility data at $Y = 0.1039$ (Table 15-1) the extract layer contains 5.4/85.7 = 0.0630 kg water/kg MIBK and $y_e = (0.097)/(1 + 0.097 + 0.063) = 0.084$ weight-fraction acetic acid in the extract.

For cases B and C, Robbins ("Liquid-Liquid Extraction," in Schweitzer, *Handbook of Separation Techniques for Chemical Engineers*, McGraw-Hill, New York, 1979, sec. 1.9) developed the concept of pseudo solute concentrations for the feed and solvent streams entering the extractor that will allow the Kremser equations to be used.

In case B the solvents are partially miscible, and the miscibility is nearly constant through the extractor. This frequently occurs when all solute concentrations are relatively low. The feed stream is assumed to dissolve extraction solvent only in the feed stage and to retain the same amount throughout the extractor. Likewise, the extraction solvent is assumed to dissolve feed solvent only in the raffinate stage. With these assumptions the primary extraction-solvent rate moving through the extractor is assumed to be S', and the primary feed-solvent rate is assumed to be F'. The extract rate E' is less than S', and the raffinate rate R' is less than F' because of solvent solubilities.

The slope of the operating line is F'/S', just as in Eqs. (15-11) and (15-12), but only stages 2 through $r - 1$ will fall directly on the operating line. And one knows that X_1 will be on the equilibrium line in equilibrium with Y_e by definition (see Fig. 15-12). One can calculate a pseudo feed concentration X_f that will fall on the operating line [Eq. (15-11)] at $Y_{n+1} = Y_e$ [Eq. (15-19)].

$$X_f^B = X_f + \frac{S' - E'}{F'} Y_e \qquad (15\text{-}19)$$

Likewise, one knows that Y_r will be on the equilibrium line with X_r (see Fig. 15-12). One can therefore calculate a pseudo concentration of solute in the inlet extraction solvent Y_s^B that will fall on the operating line [Eq. (15-12)] where $X_{n-1} = X_r$ [Eq. (15-20)].

$$Y_s^B = Y_s + \frac{F' - R'}{S'} X_r \qquad (15\text{-}20)$$

For case B, the two pseudo inlet concentrations X_f^B and Y_s^B can be used in the Kremser equation with the actual value of X_r and $\mathcal{E} = mS'/F'$ to calculate rapidly the number of theoretical stages required. The graphical stepwise solution shown in Fig. 15-13 can also be used. The operating line will go through points X_r, Y_s^B and X_f^B, Y_e with a slope of F'/S'. In one example studied by Robbins [*Chem. Eng. Prog.*, **76**(10), 58 (1980)], the actual feed and extract compositions gave a point to the left of the equilibrium line on an XY graph like Fig. 15-13 because the solubility of the solvent was so high. But the use of the pseudo feed composition still gave an accurate calculation of the number of theoretical stages as confirmed by a right-triangular graphical calculation.

Example 4 Let us solve the problem in Example 2 by assuming case B. The solute (acetic acid) concentration is low enough in the extract so that we may assume that the mutual solubilities of the solvents remain nearly constant. The material balance can be calculated by an iterative method.

From equilibrium data (Table 15-1) the extraction-solvent (MIBK) loss in the raffinate will be about 0.016/0.984 = 0.0163 kg MIBK/kg water, and the feed-solvent (water) loss in the extract will be about 5.4/85.7 = 0.0630 kg water/kg MIBK.

First iteration: assume $R' = F' = 80$ kg water/h. Then, extraction solvent in raffinate = (0.0163)(80) = 1.30 kg MIBK/h. Estimate $E' = 199.8 - 1.3 = 198.5$ kg MIBK/h. Then feed solvent in extract = (0.063)(198.5) = 12.5 kg water/h.

Second iteration: calculate $R' = 80 - 12.5 = 67.5$ kg water/h. $E' = 199.8 - (0.0163)(67.5) = 198.7$ kg MIBK/h.

Third iteration: converge $R' = 80 - (0.063)(198.7) = 67.5$ kg water/h. Y_e is calculated from the overall extractor material balance [Eq. (15-13)]:

$$Y_e = \frac{(80)(0.25) + (199.8)(0.001) - (67.5)(0.01)}{198.7} = 0.0983 \frac{\text{kg acetic acid}}{\text{kg MIBK}}$$

$$y_e = \frac{0.0983}{1 + 0.0983 + 0.0630} = 0.0846 \text{ weight fraction acetic acid in extract}$$

From the correlation of equilibrium data (Table 15-3),

$$Y_e = 0.930\,(X)^{1.10}, \text{ for } X \text{ between 0.03 and 0.25}$$

The raffinate composition leaving the feed (first stage) is calculated:

$$X_1 = (0.0983/0.930)^{1/1.10} = 0.130$$
$$m_1 = dY/dX = (0.930)(1.10)(X)^{0.1}$$
$$m_r = dY/dX = K' = 0.656$$
$$m_1 = 0.834 \text{ at } X_1 = 0.13$$
$$m_r = 0.656 \text{ at } X_r = 0.01$$
$$K'_s = 0.656 \text{ at } Y_s = 0.001$$
$$\mathcal{E} = \sqrt{m_1 m_r}\, S'/F' = (0.740)(199.8)/80 = 1.85$$

X_f^B is calculated from Eq. (15-19):

$$X_f^B = 0.25 + \frac{(199.8 - 198.7)(0.0983)\ (0.0983)}{80} = 0.251$$

Y_s^B is calculated from Eq. (15-20):

$$Y_s^B = 0.001 + \frac{(80 - 67.5)(0.01)}{199.8} = 0.0016$$

N is determined from Fig. 15-13, Eq. (15-14), or a McCabe-Thiele type of plot (Fig. 15-13) for case B.

$$N = \frac{\ln\left[\left(\dfrac{0.251 - 0.0016/0.656}{0.01 - 0.0016/0.656}\right)\left(1 - \dfrac{1}{1.85}\right) + \dfrac{1}{1.85}\right]}{\ln 1.85}$$

$$= 4.5 \text{ theoretical stages}$$

A less frequent situation, case C, can occur when the solute concentration in the extract is so high that a large amount of feed solvent is dissolved in the extract stream in the "feed stage" but a relatively small amount of feed solvent (say one-tenth as much) is dissolved by the extract stream in the "raffinate stage." The feed stream is assumed to dissolve the extraction solvent only in the feed stage just as in case B. But the extract stream is assumed to dissolve a large amount of feed solvent leaving the feed stage and a negligible amount leaving the raffinate stage. With these assumptions the primary feed-solvent rate is assumed to be R', so the slope of the operating line for case C is R'/S'. Again the extract rate E' is less than S', and the raffinate rate R' is less than F'.

The pseudo feed concentration for case C, X_f^C, can be calculated from Eq. (15-21).

$$X_f^C = \frac{F'}{R'} X_f + \frac{S' - E'}{R'} Y_e \qquad (15\text{-}21)$$

And the value of Y_s will fall on the operating line for case C. The extraction factor for case C is calculated from Eq. (15-22).

$$\mathcal{E}^C = mS'/R' \qquad (15\text{-}22)$$

On an XY diagram for case C the operating line will go through points X_r, Y_s and X_f^C, Y_e with a slope of R'/S' similar to Fig. 15-13. When using the Kremser equation for case C, one uses the pseudo feed concentration X_f^C from Eq. (15-21) and the stripping factor $\mathcal{E}^C$ from Eq. (15-22). One uses the raffinate concentration X_r and inlet solvent concentration Y_s without modification.

For the first time through a liquid-liquid extraction problem, the right-triangular graphical method may be preferred because it is completely rigorous for a ternary system and reasonably easy to understand. However, the shortcut methods with the Bancroft coordinates and the Kremser equations become valuable time-savers for repetitive calculations and for data reduction from experimental runs. The calculation of pseudo inlet compositions and the use of the McCabe-Thiele type of stage calculations lend themselves readily to programmable calculator or computer routines with a simple correlation of equilibrium data.

COUNTERCURRENT MASS-TRANSFER-UNIT CALCULATIONS

The concept of a mass-transfer unit was developed many years ago to represent more rigorously what happens in a differential contactor rather than a stagewise contactor. For a straight operating line and a straight equilibrium line with an intercept of zero, the equation for calculating the number of mass-transfer units based on the overall raffinate phase N_{or} is identical to the Kremser equation except for the denominator when the extraction factor is not equal to 1.0 [Eq. (15-23)].

When $\mathcal{E} \neq 1.0$,

$$N_{or} = \frac{\ln\left[\left(\frac{X_f - Y_s/m}{X_r - Y_s/m}\right)\left(1 - \frac{1}{\mathcal{E}}\right) + \frac{1}{\mathcal{E}}\right]}{1 - 1/\mathcal{E}} \qquad (15\text{-}23)$$

The number of mass-transfer units N_{or} is identical to the number of theoretical stages when the extraction factor $\mathcal{E}$ is 1.0 [Eq. (15-24].

When $\mathcal{E} = 1.0$,

$$N_{or} = [(X_f - Y_s/m)/(X_r - Y_s/m)] - 1 \qquad (15\text{-}24)$$

The differences become pronounced when values of the extraction factor are high [Eq. (15-25)].

$$N_{or} = N \ln \mathcal{E}/(1 - 1/\mathcal{E}) \qquad (15\text{-}25)$$

Even staged equipment may be modeled best by the number of mass-transfer units when the extraction factor is much higher than 1.5, especially if the stage efficiencies are low.

The response of solute concentration in the raffinate X_r to the solvent-to-feed ratio S'/F' can be calculated by Eqs. (15-26) and (15-27) for a constant number of transfer units based on the overall raffinate phase N_{or}.

When $\mathcal{E} \neq 1.0$,

$$\frac{X_r - Y_s/m}{X_f - Y_s/m} = \frac{1 - 1/\mathcal{E}}{e^{N_{or}(1 - 1/\mathcal{E})} - 1/\mathcal{E}} \qquad (15\text{-}26)$$

When $\mathcal{E} = 1.0$,

$$\frac{X_r - Y_s/m}{X_f - Y_s/m} = \frac{1}{N_{or} + 1} \qquad (15\text{-}27)$$

The solution to these equations is shown graphically in Fig. 15-15. Note that the raffinate composition is not reduced appreciably when the extraction factor $\mathcal{E}$ is increased from 5 to infinity. This is true because mass transfer from the raffinate phase limits the performance. This is typical of the performance of many devices including actual staged equipment. However, if there is sufficient residence time in each stage of a staged device so that high stage efficiencies can be achieved, then the raffinate can be reduced substantially by increasing the extraction factor above 5 (see Fig. 15-14). However, the solute concentration in the extract stream would be quite dilute.

Example 5 Let us calculate the number of transfer units required to achieve the separation in Example 3. The solution to the problem is the same as in Example 3 except that the denominator is changed in the final equation [Eq. (15-25)]:

$$N_{or} = 4.5 \frac{\ln 1.85}{1 - 1/1.85} = 6.0 \text{ transfer units}$$

STAGE EFFICIENCY AND HEIGHT OF A THEORETICAL STAGE OR TRANSFER UNIT

The overall stage efficiency of a staged extraction system is simply the number of theoretical stages divided by the number of actual stages times 100 [Eq. (15-28)].

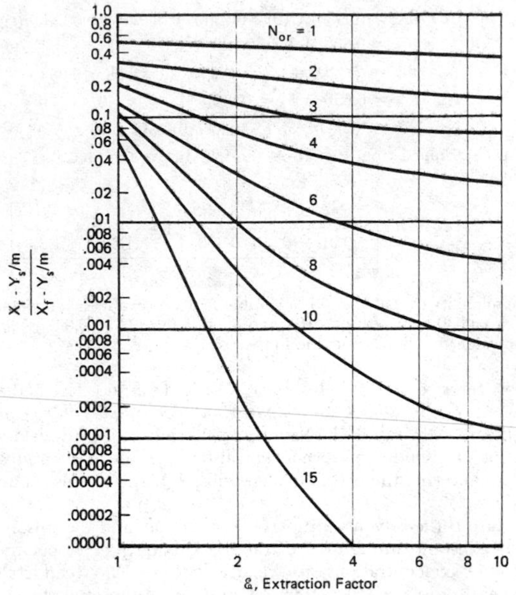

FIG. 15-15 Graphical solution to the mass-transfer-unit equations.

Percent stage efficiency

$$= 100N/\text{number of actual stages} \qquad (15\text{-}28)$$

A similar term of number of transfer units per actual stage could also be envisioned.

The height equivalent to a theoretical stage (HETS) in an extraction tower is simply the height of the tower Z_t divided by the number of theoretical stages achieved [Eq. (15-29)].

$$\text{HETS} = Z_t/N \qquad (15\text{-}29)$$

Likewise, the height of a transfer unit based on raffinate-phase compositions H_{or} is the height of tower divided by the number of transfer units [Eq. (15-30)].

$$H_{or} = Z_t/N_{or} \qquad (15\text{-}30)$$

The contribution to the height of a transfer unit overall based on the raffinate-phase compositions is the sum of the contribution from the resistance to mass transfer in the raffinate phase H_r plus the contribution from the resistance to mass transfer in the extract phase H_e, divided by the extraction factor $\mathcal{E}$ [Eq. (15-31)].

$$H_{or} = H_r + H_e/\mathcal{E} \qquad (15\text{-}31)$$

At high extraction factors the height of a transfer unit is mostly dependent on the resistance to the transfer of solute from the raffinate phase.

Prediction methods attempt to quantify the resistances to mass transfer in terms of the raffinate rate R and the extract rate E, per tower cross-sectional area A_t, and the mass-transfer coefficient in the raffinate phase k_r and the extract phase k_e, times the interfacial (droplet) mass-transfer area per volume of tower a [Eqs. (15-32) and (15-33)].

$$H_r = R/A_t k_r a \qquad (15\text{-}32)$$

$$H_e = E/A_t k_e a \qquad (15\text{-}33)$$

The mass-transfer coefficients depend on complex functions of diffusivity, viscosity, density, interfacial tension, and turbulence. Similarly, the mass-transfer area of the droplets depends on complex functions of viscosity, interfacial tension, density difference, extractor geometry, agitation intensity, agitator design, flow rates, and interfacial rag deposits. Only limited success has been achieved in correlating extractor performance with these basic principles. The lumped parameter H_{or} deals directly with the ultimate design criterion, which is the height of an extraction tower.

FRACTIONATION STAGES

One of the most sophisticated separations achievable by liquid-liquid extraction is fractionation. Two solutes can be separated almost completely by isolating one solute b into the extraction solvent S' and another solute c into a wash solvent W' [Fig. 15-16]. The bottom section of a fractionation extraction is about the same as the countercurrent extractions described earlier, with the extraction solvent S' entering the bottom and extracting, i.e., stripping, one of the solutes b almost completely from the raffinate R'. As the extract stream moves above the feed stage, it is contacted countercurrently with a wash solvent W' that scrubs the unwanted solute c out of the extract stream. This in effect purifies the solute b that is being extracted. The stripping section and the washing (enriching) section will each have its own operating line on a McCabe-Thiele type of XY diagram (Fig. 15-17). The overall material balance must be met at the feed stage.

For the case in which the extraction solvent can be assumed to be totally immiscible with the wash solvent and there is no solvent in the feed, the extraction factor $\mathcal{E}$ must be greater than 1.0 for component b and less than 1.0 for component c [Eq. (15-34)].

$$\mathcal{E} = mS'/W' \tag{15-34}$$

For a symmetrical separation of component b from c, Brian (*Staged Cascades in Chemical Processing*, Prentice-Hall, Englewood Cliffs, N.J., 1972) reported that the ratio of wash solvent to extraction solvent W'/S' should be set equal to the geometric mean of the two slopes of the equilibrium lines [Eq. (15-35)].

$$W'/S' = \sqrt{m_b m_c} \tag{15-35}$$

The ratio of wash solvent to extraction solvent is the same in the enriching section as in the stripping section if no solvent is added in the feed. The degree of separation to be achieved can be chosen for the process design, such as 99 percent of component b into the extract stream and 99 percent of component c into the raffinate stream. Then the feed rate can be chosen so that the solute loadings in the extract stream and the raffinate stream are reasonable. This becomes especially critical near the feed stage, where the solute loadings are highest.

An overall material balance can be calculated around the extractor, and then an XY plot can be constructed for each solute (Figs. 15-17 and 15-18). The solute concentrations at the raffinate end of the extractor, X_{br} and Y_{bs}, can be plotted for component b, and the operating line can be drawn with a slope of W'/S' with no solvent in the

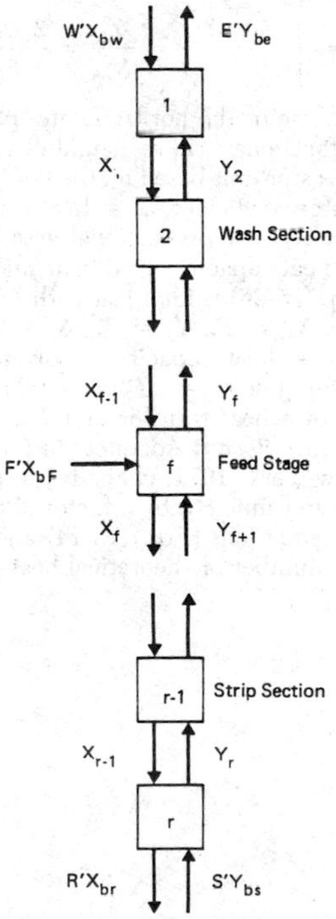

FIG. 15-16 Liquid-liquid fractionation cascade.

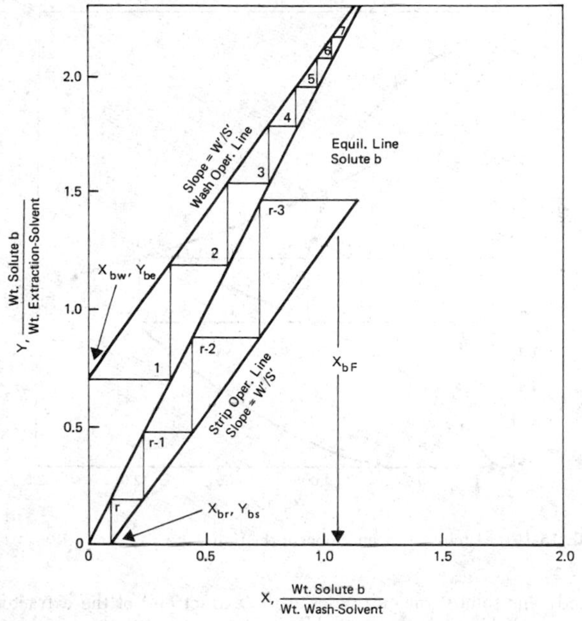

FIG. 15-17 Graphical calculation of fractionation stages for solute b.

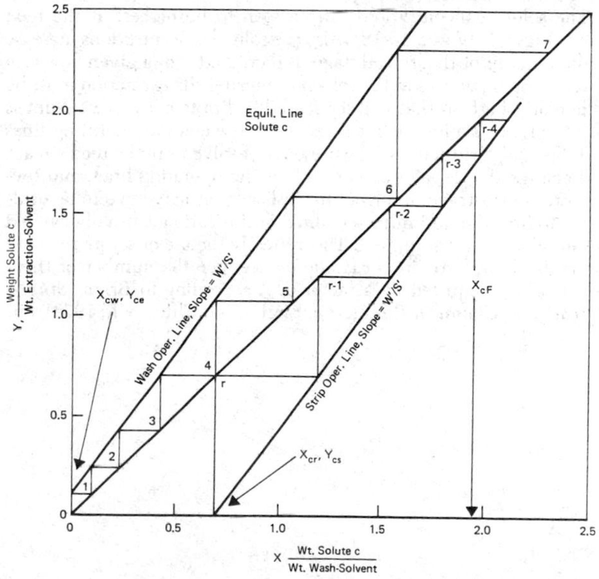

FIG. 15-18 Graphical calculation of fractionation stages for solute c.

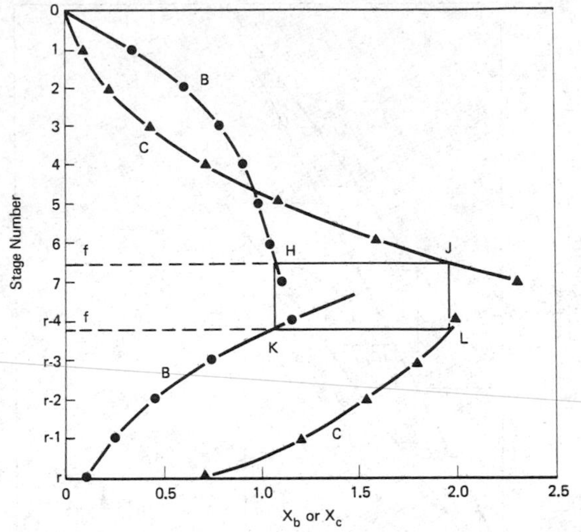

FIG. 15-19 Matching concentrations at the feed stage.

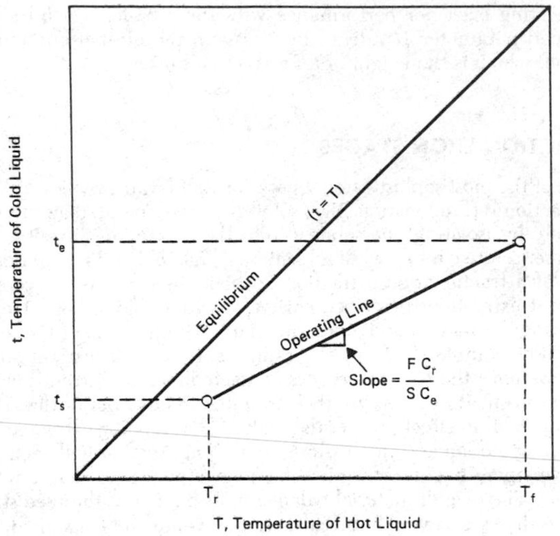

FIG. 15-20 Countercurrent heat transfer.

feed. The solute concentrations at the extract end of the extractor, X_{bw} and Y_{be}, can also be plotted for component *b* with the enriching-section operating line also having the slope W'/S' if no solvents were added with the feed. The theoretical stages can be stepped off for each section of the extractor by starting at the extract end, stage 1, and stepping toward the feed stage *f*, then restarting at the raffinate end, stage *r*, and stepping toward the feed stage *f* (Fig. 15-17). A similar procedure is repeated for component *c* (Fig. 15-18).

The feed-stage number is found by matching the concentrations and stage number. This occurs at the point where the feed should be introduced (see Treybal, *Mass-Transfer Operations*, 3d ed., McGraw-Hill, New York, 1980). The procedure for matching concentrations is carried out by plotting the stage number on the vertical axis and the raffinate concentration *X* for each component (Fig. 15-19). The concentrations are matched when the rectangle *HJLK* can be drawn as shown. The number of stages in the wash section including the feed stage is determined from the position of line *HJ*. The total number of stages *r* is determined from the position of *LK*, which is also at the feed stage.

The solute concentrations can be seen to be highest at the feed stage (Figs. 15-17 and 15-18). Also the solute concentrations increase as the number of theoretical stages is increased. For a given flow rate of feed, the flow rates of the solvents entering the extraction must be sufficiently high so that neither solubility limits nor a plait point is exceeded, nor a pinch point is reached between the operating lines and the equilibrium lines. The presence of solvents in the feed stream will change the slope of one or both of the operating lines, and several ratios of extraction solvent to wash solvent may have to be evaluated to find the optimum. The final optimization is usually carried out in pilot-plant equipment. Theoretically the use of solute reflux to the ends of the extraction cascade can reduce the number of theoretical stages required by a factor of 2 according to Brian (*Staged Cascades in Chemical Processing*, Englewood Cliffs, N.J., 1972), but again the amount of solvent flow rates may have to be increased to avoid a pinch point or plait point near the feed stage.

COUNTERCURRENT LIQUID-LIQUID HEAT TRANSFER

Heat may be transferred between two insoluble liquids in countercurrent flow through an extractor, and the performance can be evaluated in the same general manner as in mass transfer (Fig. 15-20). For a differential contactor the number of overall heat-transfer units based on the hot phase N_{oh} can be derived from the same equations used for the number of mass-transfer units based on the feed (raffinate) phase [Eq. (15-36)].

$$N_{oh} = \int_{T_r}^{T_f} \frac{dT}{T - t} = \frac{Z_t}{N_{oh}} = \frac{Z_t U_o a_r A_t}{F\,C_r} \tag{15-36}$$

where T = temperature of the hot (raffinate) phase, t = temperature of the cold (extract) phase, Z_t = height of tower, H_{oh} = height of an overall heat-transfer unit based on the hot (raffinate) phase, U_o = overall heat-transfer coefficient, a_r = heat-transfer surface of the droplets per volume, A_t = cross-sectional area of tower, F = hot feed rate, and C_r = heat capacity of raffinate and feed. The solution to the integral in Eq. (15-36) is identical with Eqs. (15-23) and (15-24), where $X_f = T_f$, $X_r = T_r$, $Y_s = T_s$, $\mathcal{E} = SC_e/FC_r$, S = cold solvent rate, and C_e = heat capacity of solvent and extract. The slope of the equilibrium line $m = dT/dt = 1.0$ since $t = T$ at equilibrium. The height of a heat transfer unit H_{oh} is reported by Von Berg [in C. Hanson (ed.), *Recent Advances in Liquid-Liquid Extraction*, Pergamon, New York, 1971, chap. 11] to be shorter than the height of a mass-transfer unit H_{or} by a factor of 3 to 20. As an alternative, the Kremser equations [Eqs. (15-14) and (15-15)] could be used to calculate the number of theoretical heat-transfer stages.

Section 16

Adsorption and Ion Exchange

Theodore Vermeulen, Ph.D., *Professor of Chemical Engineering, Faculty Scientist, Lawrence Berkeley Laboratory, and Director of the Water Technology Center, University of California, Berkeley; Fellow, American Institute of Chemical Engineers; Member, American Chemical Society, Combustion Institute, Catalysis Society. (Section Editor)*

M. Douglas LeVan, Ph.D., *Associate Professor of Chemical Engineering, University of Virginia, Charlottesville; Member, American Institute of Chemical Engineers, American Chemical Society. (Section Editor)*

Nevin K. Hiester, Ph.D., *Associate Director, Business Intelligence Program, SRI International, Menlo Park, California; Member, American Institute of Chemical Engineers*

Gerhard Klein, M.S., *Research Engineer, Water Technology Center, University of California, Berkeley; Member, American Institute of Chemical Engineers, American Chemical Society*

SORBENT MATERIALS AND SORPTION-PROCESS ANALYSIS

FLUID-SORBENT EQUILIBRIUM

EQUILIBRIUM-LIMITED TRANSITIONS

RATE-LIMITED CONSTANT-PATTERN TRANSITIONS

LINEAR-EQUILIBRIUM AND OTHER RATE-LIMITED TRANSITIONS

CHANGED-FEED OPERATIONS; REGENERATION; CHROMATOGRAPHY

MULTIVARIANT SYSTEMS: MULTIPLE TRANSITIONS

BATCH AND CONTINUOUS OPERATIONS

Nomenclature and Units

Symbol	Definition	SI units	U.S. customary units
A	Surface area	km^2/kg	ft^2/lb
A_1	Weighting factor for reference adsorption-system concentrations in constant-separation-factor treatment		
a	Slope of isotherm, in bilinear treatment (Table 16-5); also exponent in Eq. (16-142); also term in Eq. (16-179)		
a_p	External surface area of sorbent particles per unit packed volume	mm^{-1}	in^{-1}
a_1, a_2	Coefficients in trinomial fit to isotherm		
B	Ratio of reference q and c for every adsorbed species in constant-separation-factor treatment	L/kg	ft^3/lb
$\mathbf{B}$	Exponent on ratio of NTU values that will give ratio of midheight slopes in breakthrough on ratios of $\mathbf{T} - 1$ for a given $\mathbf{X}$		
b	Slope of isotherm, in bilinear treatment; also a parametric variable		
$\mathbf{b}_f$, $\mathbf{b}_p$	Correction terms for nonlinear addition of mass-transfer resistances, Eq. (16-92)		
b_1, b_2	Coefficients in trinomial fit to isotherm		
C_0	Total concentration in fluid phase in ion exchange	(g·ion equivalent)/L	(lb·ion equivalent)/ft^3
C_f, C_p	Heat capacity: per unit solution volume for fluid phase; per unit packed volume for particle phase	J/(K·L)	Btu/(°F·ft^3)
$C_1 \ldots C_6$	Locally determined constants		
c	Sorbent concentration in fluid phase: for ion exchange, for adsorption	(g·ion equivalent)/L (g·mol)/L	(lb·ion equivalent)/ft^3 (lb·mol)/ft^3
D	Diffusivity	mm^2/s	ft^2/h
$\mathbf{D}$	Partition ratio × void fraction, $\Lambda\epsilon$ (used in 4th ed.)		
d_c	Crystallite diameter	mm	in
d_p	Particle diameter	mm	in
E_0	Superficial axial-dispersion coefficient	mm^2/s	ft^2/h
$\mathbf{E}$	Regenerant efficiency, Eq. (16-157)		
F	Volumetric flow rate of fluid phase		
$\mathbf{F}$	Integral leakage, Eq. (16-134)		
f	Ratio of total gas-phase flow to flow at bed inlet		
$(-\Delta H)$	Heat release due to adsorption at specified sorbent loading	kJ/(g·mol)	Btu/(lb·mol)
h	Length within column, measured from inlet end	m	ft
$\mathbf{h}$	Concentration-path parameter, Eq. (16-197)		
i, j	Cumulative integers in numerical calculations		

Symbol	Definition	SI units	U.S. customary units
$\mathbf{J}$	Bessel-function integral spanning 0 and 1, Eq. (16-124)		
K_{AB}, $\mathbf{K}_{AB}$	Mass-action equilibrium constants, Eq. (16-13a)		
K_F	Similar to K_f, but dimensionless		
K_f	Freudlich adsorption equilibrium constant	$(g \cdot mol^{1-\beta})$ $(kg^{\beta} \cdot L^{-1})$	$(lb \cdot mol^{1-\beta})$ $(lb^{\beta} \cdot ft^{-3})$
K_L	Langmuir adsorption equilibrium constant	bar^{-1}; L/(g·mol)	atm^{-1}; ft^3/(lb·mol)
k	Mass-transfer coefficient	mm/s	ft/s
L	Axial mixing length	mm	ft
L, M, M', N	Constants in batch-sorption equations		
M	Molecular weight	daltons	daltons
m	Exponent in mass-action equilibrium, Eq. (16-13a)		
$\mathbf{N}$	Number of transfer units or reaction units		
$\mathbf{P}_d$	Péclet number for axial dispersion, Eq. (16-105)		
p	Partial pressure	bar	atm
Q	Solid-phase maximum concentration in adsorption or total ion-equivalent concentration in ion exchange	g equivalent/kg or (g·mol)/kg	(lb·mol)/lb
q	Solid-phase concentration	g equivalent/kg or (g·mol)/kg	(lb·mol)/lb
R	Radial distance in bed, for radial-flow case		
R'	Empirical replacement for $\mathbf{R}$ in nonequilibrium conditions		
R_F	Ratio of chromatographic peak velocity to fluid velocity		
R_g	Gas constant	kJ or (m^3·Pa)/[(g·mol)·(K)]	Btu/[(lb·mol)·°R]
$\mathbf{R}$	Separation factor for a single transition		
r	Radial distance within a particle		
$\bar{r}$	Average pore radius within a particle		
$\mathbf{r}$	Separation factor for a sorption system		
S	Superficial cross-sectional area		
s	First variable in argument of $\mathbf{J}$, Eq. (16-124)		
T	Throughput parameter based on total concentrations		
$\overline{T}$	Average value of T		
$\mathbf{T}$	Throughput parameter based on concentration changes for transition of interest		
T_K	Absolute temperature	K	°R
t	Time, with t_a the apparent residence time; also second variable in argument of $\mathbf{J}$		
$\hat{t}$	Time after arrival of the fluid front, t-$(hS\epsilon/F)$	s	s
$\bar{t}$	Mean residence time in semibatch stage	s	s

Nomenclature and Units (*Continued*)

Symbol	Definition	SI units	U.S. customary units
U	Dimensionless concentration velocity; $1/T$		
u_0	Transition velocity, Eq. (16-53)	m/s	ft/min
V	Fluid volume fed to column		
$\hat{V}$	Fluid volume past point of interest, $V - v\epsilon$		
v	Volume of sorbent, from inlet end to point of interest		
W	Weight charge of solid	kg	lb
X	Dimensionless fluid-phase concentration at a point within a pore of a particle; also extent of approach to equilibrium; also mole ratio of solute to carrier gas		
X'	Intermediate function, related to $\mathbf{X}$, for calculation of axial-dispersion case		
$\mathbf{X}$	Dimensionless fluid-phase concentration, between 0 downstream and 1 upstream, within a transition		
$\bar{\mathbf{X}}$	Average leakage, Eq. (16-135)		
x	Dimensionless fluid-phase concentration for sorption or exchange system		
Y	Dimensionless solid-phase concentration at a point within a particle		
$\mathbf{Y}$	Dimensionless solid-phase concentration, within a transition		
$\bar{\mathbf{Y}}$	Resin utilization		
y	Dimensionless solid-phase concentration, for sorption or exchange system		
z	Cross-linkage fractions in polymeric resins		
$\mathbf{Z}$	Throughput parameter $\mathbf{T}$ used in 4th ed.		
z	A number		
	Greek symbols		
α	Separation factor, for a given solute relative to a second solute		
β	Exponent in Freundlich-isotherm relation		
Γ	Ratio of relative concentrations of elutant in sorbent phase and in fluid phase, for a countercurrent column		
$\Gamma(i)$	Gradual transition and corresponding concentration path; $i = 1$ for farthest downstream (first to emerge) transition		
γ	Ratio of relative concentrations of a designated species in stationary phase and in fluid phase, for batch contact		
Δ	Total possible decrease of fluid-phase relative concentration in batch contact, from input level to eventual equilibrium; also increment of $\mathbf{N}$ or $\mathbf{NT}$ in numerical calculations		
δ	Constant of integration in constant-pattern breakthrough equations		
ϵ	Void fraction in a packed column		
ζ	Mechanism parameter, ratio of sorbent-phase NTU to fluid phase NTU		
η	Mechanism parameter, ratio of number of mass-transfer units or reactive units to number of axial dispersion units $\mathbf{N}_d$		
Θ	Temperature change	K	°F
κ	Generalized rate coefficient	s^{-1}	s^{-1}
Λ	System partition ratio for a component or for total solutes, Eq. (16-6)		
$\boldsymbol{\Lambda}$	Transition partition ratio in a binary system, Eq (16-21)		
$\boldsymbol{\Lambda}'$	Local value of $\boldsymbol{\Lambda}$		
μ	Viscosity		
μ_i	Chemical potential of solute i	kJ/(g·mol)	Btu/(lb·mol)
ν	Valence		
ξ	"Traveling" variable integrand, Eq. (16-124)		
π	Spreading pressure	MPa/m	atm/ft
ρ	Density		
$\Sigma(i)$	Abrupt transition		
σ	Surface area per unit weight of dry solid		
$\bar{\tau}$	Tortuosity		
χ	Interior porosity of sorbent particles		
ψ	Correction factor for use of linear driving force inside particles, dependent upon $\mathbf{R}$ value for $\mathbf{R} < 1$		

GENERAL REFERENCES: Aris and Amundson, *Mathematical Methods in Chemical Engineering*, vol. 2: *First-Order Partial Differential Equations with Applications*, Prentice-Hall, Englewood Cliffs, N.J., 1973. Barrer, *Zeolites and Clay Minerals as Sorbents and Molecular Sieves*, Academic, New York, 1978. Clazie, Klein, and Vermeulen, "Multicomponent Diffusion: Generalized Theory with Ion Exchange Applications," U.S. Off. Saline Water Res. Dev. Prog. Rep. 326, 1968. Gregg and Sing, *Adsorption, Surface Area, and Porosity*, Academic, New York, 1982. Hall, Eagleton, Acrivos, and Vermeulen, *Ind. Eng. Chem. Fundam.*, **5**, 212 (1966). Helfferich, *Ion Exchange*, McGraw-Hill, New York, 1962. Helfferich and Klein, *Multicomponent Chromatography*, Maurice Dekker, New York, 1970. Hiester and Vermeulen, *Chem. Eng. Prog.*, **48**, 505 (1952); *Ind. Eng. Chem.*, **44**, 636 (1952); *J. Chem. Phys.*, **22**, 96 (1954); *Am. Inst. Chem. Eng. J.*, **2**, 404 (1956). Holland and Liapis, *Computer Methods for Solving Dynamic Separation Problems*, McGraw-Hill, New York, 1982. Klein, Sinkovic, and Vermeulen, "Weak-Electrolyte Ion Exchange in Waste-Water Reuse," U.S. Interior Dept. Rep. OWRT 82/7, 1982. Kohl and Riesenfeld, *Gas Purification*, 3d ed., Gulf, Houston, 1980. Liberti and Helfferich (eds.), *Mass Transfer and Kinetics in Ion Exchange*, Martinus Nijhoff, Boston, 1982. Rodrigues and Tondeur, *Percolation Processes: Theory and Applications*, Suthoff and Noordhoff, Rockville, Md., 1981. Sherwood, Pigford, and Wilke, *Mass Transfer*, McGraw-Hill, New York, 1975. Tondeur and Klein, *Ind. Eng. Chem. Fundam.*, **6**, 339, 347 (1967). Weber, *Physicochemical Processes for Water Quality Control*, Wiley, New York, 1972.

SORBENT MATERIALS AND SORPTION-PROCESS ANALYSIS

INTRODUCTION

Sorption is the selective transfer of one or more solutes from a fluid phase to a batch of rigid particles. The usual selectivity of a sorbent between solute and carrier fluid or between different solutes makes it possible to separate certain solutes from the carrier or from one another. Similarly a reverse operation, desorption, will often bring about separation of species initially in the solid. This section deals with development and design of separation processes based on the phenomena of sorption and desorption.

Adsorption involves, in general, the accumulation of solute molecules at an interface (including gas-liquid, as in foam fractionation separations; and liquid-liquid). Here we consider only gas-solid and liquid-solid interfaces, with solute from the fluid attaching selectively to the solid. The accumulation per unit area is small; thus, highly porous solids with very large internal area per unit volume are preferred. The surfaces are usually irregular, and the bonding energies (primarily from Van der Waals forces, as in vapor condensation) vary widely from one site to another. However, with "molecular sieves," the adsorptive surfaces are provided by channels or cavities within a macrocrystal structure; the sieves exhibit high uniformity of adsorbent surface with a practically constant binding energy.

A simple example will show the extent of internal area in a typical granular adsorbent. Suppose that the interparticle void fraction is 0.40, the intraparticle porosity is also 0.40, and the effective pore diameter is 10 nm. The surface-to-volume ratio is then 0.1 km^2/L (or 1 mi^2/ft^3) of packed bed. If the solute has formed a complete monomolecular layer, say 0.5 nm deep, it then occupies about 20 percent of the internal pore volume or about 5 percent of the total packed-bed volume.

Ion exchange is usually a three-dimensional effect all through a polymeric solid, the solid being of "gel" type in the sense that it imbibes (or dissolves) some of the fluid-phase solvent. In ion exchange, moleculelike species of positive charge in some cases (cations) and negative charge in others (anions) from the fluid—usually an aqueous solution—replace dissimilar ions of the same charge type initially in the solid. The ion exchanger contains permanently bound "co-ion" groups of opposite charge type. Although many ion exchangers are organic polymers, others are of molecular-sieve type.

In adsorption from the gas phase, usually part of the solid surface is vacant; as the fluid-phase concentration of solute increases, the vacant area diminishes and the sorption level of the solid increases. In the case of ion exchange or of adsorption from a liquid phase, the sorbent has a fixed total capacity and merely exchanges one solute for another.

Sorption could be considered to include gas absorption and liquid extraction. In terms of the solidlike sorbents treated here, such cases still arise. Porous-solid particles in which the pores are impregnated with a durable liquid may absorb from a gas, or extract from a second liquid, to which the particle is exposed. Certain homogeneous "gel" polymers act in similar fashion. So "partition absorption" and "partition extraction" are unit operations to which this section applies also.

Figure 16-1 shows the means of uptake and storage of solutes by a sorbent particle—in its pores and on its pore surfaces, or within a permeable homogeneous gel-like structure, or both. Table 16-1 classifies sorption operations by type of fluid and structure of sorbent.

Adsorption Adsorbents are natural or synthetic materials of amorphous or microcrystalline structure; those used on a large scale include activated carbon, activated alumina, silica gel, fuller's earth, other clays, and molecular sieves as described previously.

At ordinary temperatures, adsorption is usually caused by intermolecular forces rather than by formation of new chemical bonds; it is then called **physical adsorption,** or physisorption. At higher temperatures (above about 200°C, or 400°F), the activation energy is available to make or break chemical bonds, and if such a mechanism prevails, the adsorption is called **chemisorption** or activated adsorption.

Major uses of **liquid-phase adsorption** include:

1. Decolorizing, drying, or degumming of fuel and lubricants, organic solvents, vegetable oils, and animal oils
2. Recovery of biological chemicals (antibiotics, vitamins, flavorings) from fermentation broths or plant extracts

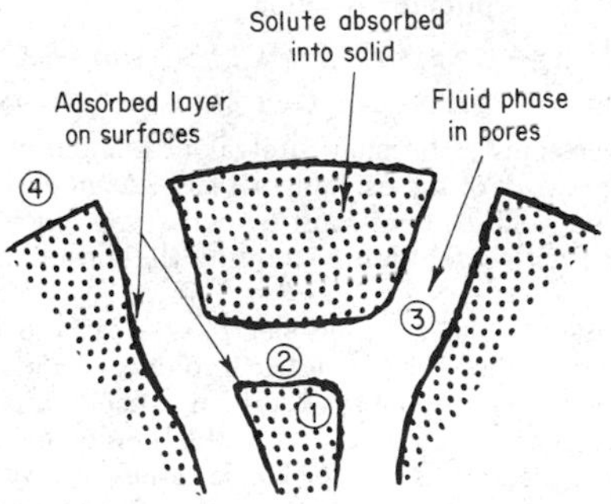

FIG. 16-1 Modes of uptake within adsorbent or ion-exchanger particle. Numerals refer to the subsection "Mass-Transfer Mechanisms."

TABLE 16-1 Classification of Sorption Operations

	Attachment or penetration of solute to sorbent (stationary phase)		
Substrate (fluid phase)	Into homogeneous solid	Between or upon crystallite surfaces	Into liquid permanently held in particle pores
Liquid	Ion exchange Ion retardation Permeation and leaching Ion exclusion Gel dialysis	Exchange adsorption Molecular-sieve adsorption	Partition extraction
Gas	Absorption	Physical adsorption Molecular-sieve adsorption	Partition absorption

3. Clarification of food and drug products
4. Decolorizing of crude sugar syrups
5. Purification of process effluents for pollution control
6. Water-supply treatment for odor, taste, or color improvement
7. Separation of isomeric aromatic or aliphatic hydrocarbons

Major uses of **gas-phase adsorption** include:

1. Drying of gases (in-package desiccation is a special case)
2. Purification of intake, circulating, or exhaust air to remove toxic gases, odors, aerosols, etc.
3. Solvent recovery from air leaving an evaporation chamber (spray painting, textile dry-cleaning, polymer processing)
4. Fractionation of gases: low-molecular-weight hydrocarbons, rare gases, and industrial gases

Major instances of **desorption** are the drying and leaching of particulate-solid raw materials.

Fixed-bed contacting with periodic regeneration is usually employed; countercurrent, batch, and pressure-swing systems find use in individual cases. The apparatus for adsorption and ion-exchange systems is discussed in Sec. 19.

Ion Exchange A solid phase is used that contains bound groups carrying an ionic charge (either + or −) accompanied by displaceable ions of opposite charge ("counterions"). Most ion exchangers in large-scale use are based on synthetic resins, either preformed and then chemically reacted, as for polystyrene, or formed from active monomers (olefinic acids, amines, or phenols). Natural zeolites (inorganic silicate polymers) were the first ion exchangers, and both natural and synthetic zeolites are in use today. The exchangers are permeable only at molecular dimensions unless the particles have a built-in network of "macropore" channels. Ion diffusivities in the resin phase are similar numerically to those in moderately viscous solvents, diminishing sharply with increasing valence or size.

Cation-exchange resins generally contain bound sulfonic acid groups; less commonly, these groups are carboxylic, phosphonic, phosphinic, etc. Anionic resins involve quaternary ammonium groups (strongly basic) or other amino groups (weakly basic).

Ion exchange may be written as a **reversible reaction** involving chemically equivalent quantities; an example for cation exchange is the familiar water-softening reaction

$$Ca^{++}(aq) + 2Na^{+}(resin) \rightleftharpoons Ca^{++}(resin)_2 + 2Na^{+}(aq)$$

or

$$Ca^{++} + 2NaR \rightleftharpoons CaR_2 + 2Na^{+}$$

where R represents a stationary univalent anionic site in the polyelectrolyte network of the exchanger phase. As in nearly all sorption operations, the solid sorbent must be conserved following its use by **regeneration** with a solution containing the ion initially present in the solid.

Ion exchange between strong electrolytes can usually be carried out until most of the **stoichiometric capacity** of the exchanger has been used; consequently the total sorbent capacity is practically constant regardless of the composition of the solution being treated. An apparent exception arises if a weak acid or base is involved, either in the resin or in solution (or in both), when the apparent capacity of the resin may be much less than its stoichiometric value.

Chelating ion-exchange resins display unusually high selectivity for certain cations. One commercial resin (Dowex A-1) contains iminodiacetate groups bound to a polystyrene matrix and shows strong affinity for copper, nickel, cobalt, and iron.

For **ion retardation,** a cationic monomer is polymerized within the structure of a previously formed anion-exchange resin, or vice versa; the resulting structure is called a "snake-in-cage" polyelectrolyte. In its regenerated form, this matrix will have its cationic groups in hydrogen form and anionic groups in hydroxyl. Introduction of a strong electrolyte displaces the H^+ and OH^- and causes their neutralization; the resin thus becomes saturated with ionic species. Regeneration with water causes the resin groups to hydrolyze, with liberation and recovery of the ions.

Other Sorbent Systems **Electron-transfer** (redox) resins perform oxidations and reductions, changing the valence of certain metal ions or eliminating dissolved oxygen. Some such resins incorporate hydroquinone groups which undergo reversible oxidation to quinone; others contain mercaptan groups.

Gel permeation provides a separation based upon molecular size, carried out with a liquid substrate. It involves granules of a permeable and lightly solvated material such as is used in membrane dialysis, e.g., cellulose. The granule admits "crystalloid" solutes, either ionic or molecular, up to around 10 to 15 Å in diameter, but completely rejects "colloidal" material larger than this limiting size. Regeneration with solute-free liquid enables the crystalloid to quit the particle and be recovered in the effluent, leaving the solid ready for reuse.

Absorption- and extraction-type separations conducted with a sorbent solid are brought about by differences in solubility rather than in molecular size. The solute undergoes a true phase change, on the molecular scale; in its "dissolved" state in the solid, each solute molecule is surrounded at close range by molecules of the sorbent material. The usual definitions of absorption and extraction are retained here: in absorption, the solute-carrying phase or substrate is gaseous; in extraction, liquid. Permeation of amorphous polymeric materials is an example of this effect, although usually carried out with membranes rather than with granules; examples are fixed gases or light organic molecules (CH_4, CH_3Br) through polyvinyl chloride, ethyl cellulose, or silicone rubber.

A specific instance of sorbent extraction is **ion exclusion** (or electrolyte exclusion), which can be carried out with ordinary ion-exchange resins. The resin is presaturated with the same mobile ions (cations or anions, depending on resin type) as are in the solution. It will then repel the ionic components of the solution, while extracting neutral nonaqueous materials such as alcohols, carboxylic acids, and ketones of relatively low molecular weight.

Another kind of phase-change sorption, used widely for chemical analysis but with potential industrial applications, is **partition chromatography.** The sorbent is a true liquid, insoluble or nonvolatile with respect to the substrate, contained in the pores of a granular solid supporting material that is usually relatively inert. Since such sorbents can be used for saturation-type operations as well, separations based upon them can be termed **partition absorption** and **partition extraction.**

DESIGN STRATEGY FOR FIXED-BED SEPARATIONS

Fixed-bed operation with the sorbent in granule, bead, or pellet form is the predominant way of conducting sorption separations; downward flow is more often used, but at times upward flow is needed. Although the fixed-bed mode is highly useful, its analysis is unexpectedly complex. Therefore fixed beds are given primary attention in this section with respect to both **interpretation** and **prediction.** Alternative modes, such as batch, slurry-flow, fluidized, and steady-state countercurrent, are treated only briefly.

In a fixed bed or "column" penetrated by a solute-bearing gas or liquid phase in plug flow, coexisting compositions of fluid and solid depend both on elapsed time and on axial distance within the column. Such behavior is shown in Fig. 16-2, where time is plotted as the abscissa and distance from the inlet as the ordinate (measured downward). The bed is shown at the start (t_0) and at two successive times t_1 and t_2. Solvent is identified by sloping-line shading, and transferable solute by horizontal-line shading.

At t_0 neither solvent nor solute has yet entered the bed. At t_1 the solvent has advanced almost to the end of the bed, and the solute (because it is taken up by sorbent) has moved a much shorter distance. At t_2 the void space in the bed has been completely filled by solvent, and the solute "front" (as shown by a lower density of shading) has continued to advance. A plot of this information at any given instant is known as a "concentration profile."

This frontal region, also termed a "transition," "sorption wave," or "sorption zone," is our primary concern. The volume well upstream of it shows the solid equilibrated with the feed, and the volume far downstream shows the emerging fluid equilibrated with the preexisting solid. The transition's center, determined by stoichiometry, is marked by the upper slanting line whose slope corresponds to the mean velocity of the front.

The effluent-concentration "history," or "breakthrough curve,"

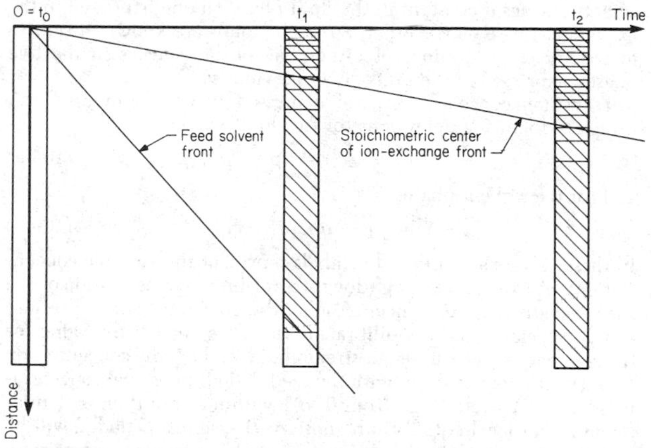

FIG. 16-2 Location of feed-fluid front and sorption zone in a fixed-bed column as a function of elapsed time. Diagonal shading shows the carrier fluid; horizontal shading, the solute.

stays at or near zero for a time. However, as the front approaches the column outlet, the effluent reaches a concentration limit beyond which it becomes unacceptable. This is the "breakthrough point" (in terms of time or fluid volume), where the "saturation," "exhaustion," or "loading" step must stop and "regeneration," "elution," "desorption," or "unloading" must start.

Guiding Principles Design should be guided by full comprehension, with all major aspects of equilibrium and rate behavior considered, rather than merely by massive calculation. From the mathematical standpoint, virtually all sorption operations display the same ranges of characteristic behavior. Therefore, a single unified theoretical treatment which can be adapted to almost any sorption-separation system is outlined here.

Input data should be in the form of complete breakthrough curves, not just breakthrough points. For simplicity and generality, calculational variables are combined into nondimensional groups. This unified approach is implemented by the following recommended procedure:

1. Determine the VARIANCE (DEGREES OF FREEDOM), almost always equal to the number of transitions (i.e., breakthrough steps), by which composition changes between the presaturation condition and the feed condition. (Small capitals identify the relevant text heading within this section.) An n-component ion-exchange or exchange-adsorption system has a variance of $n - 1$. Isothermal n-component adsorption on vacant sites has a variance of n. A bed with all sites initially vacant, with a feed fluid containing only one kind of adsorbable solute, undergoes a single transition. Such "monovariant" cases will be given major attention here; analogous equilibrium and rate considerations can be applied to each transition of most multiple-transition operations.

2. Determine MATERIAL-BALANCE RELATIONS (stoichiometry); that is, for a given volume of bed, the cumulative volume of fluid passing through the bed to the stoichiometric center of each successively emerging transition. The bed is usually a cylinder, fed at one end; occasionally, it is a cylindrical space fed radially from the axis or a spherical space fed radially from the center, as in underground flows. In most calculations, the bed volume is assumed to be monotonic with a single length dimension.

For a constant flow rate, the volume of fluid passed until the transition center emerges is proportional and informationally equivalent to a time-variable value; similarly, the reciprocal of the fluid volume is equivalent to a velocity at which the transition center advances through the bed. Each successive concentration within the transition can thus be identified by a time, a velocity, or a nondimensional ratio of either measure to its respective value at the transition center. Especially note the THROUGHPUT RATIO **T**, which is a time ratio or fluid-volume ratio where a particular composition occurs; its reciprocal, **U**, is a velocity ratio. (Boldface type in this section denotes nondimensional transition parameters.)

To predict the location of a transition center in monovariant systems, one must predetermine the solid capacity from equilibrium (for adsorption), from titration or comparable material balancing (for ion exchange), or from manufacturer's data, and the solution concentration from assumption or measurement. If an effluent-concentration history has been measured that is complete from $\mathbf{X} = 0.02$ to $\mathbf{X} = 0.98$, the area between the curve, the ($\mathbf{X} = 1$) level, and the **X** axis (placed where the fluid front emerges) gives directly the value of fluid volume or time at the stoichiometric center; (see Fig. 16-8). For MULTIVARIANT SYSTEMS, the transition centers can often be located through calculations based on equilibrium data.

The throughput ratio is defined as

$$\mathbf{T} = \frac{V - \epsilon v}{V_{\text{stoic}}} = \frac{t - (\epsilon v/F)}{t_{\text{stoic}}} \tag{16-1}$$

Here v is bulk volume of bed up to the point of interest (often the end of the bed), ϵ the interstitial void fraction, and ϵv the bed void volume. F is the volumetric flow rate; V is the fluid volume that has entered the bed since the start; and $V - \epsilon v$ ($= \hat{V}$) is the volume that has reached the point of interest, corresponding to time $t - (\epsilon v/F)$ ($= \hat{t}$). V_{stoic} is the fluid volume holding enough solute to saturate packed volume v completely if all the solute were transferred to it; t_{stoic} is the corresponding time. V_{stoic} and v are related by the partition ratio:

$$V_{\text{stoic}} = \Lambda v \tag{16-2}$$

[See the text following Eq. (16-6) for further explanation.]

3. For each transition, find whether the FLUID-SORBENT EQUILIBRIUM is favorable (solid concentration a convex-upward function of fluid concentration) or unfavorable (a concave-upward function) relative to the direction of change occurring between feed and bed. For unfavorable equilibrium, LOCAL-EQUILIBRIUM theory is likely to give a good fit or estimate for actual breakthrough behavior. For favorable or near-linear equilibrium, the respective RATE-LIMITED-TRANSITION treatment must be applied.

A reasonably simple "isotherm" function usually suffices for expressing the equilibrium. The **constant-separation-factor** form [constant **R**; see Eq. (16-17)] is used here as a principal model, equivalent to the Langmuir isotherm for gas adsorption and analogous to constant relative volatility in distillation.

4. Consider thoroughly the CYCLIC OPERATION of a process, with review of the process arrangement alternatives available that will lead to optimal combination of exhaustion and regeneration. Has the proper choice of sorbent been made? Also, is forward flow or reverse flow more economical for regeneration?

In multivariant systems, different portions of the outflow may be collected separately, and the exhaustion may be stopped before saturation is complete (borrowing from methods of preparative bulk chromatography). Staging of columns as in a merry-go-round arrangement (see Sec. 19) is often of advantage. If no one sorbent is entirely suitable technically and economically, consider the use of a layered bed or a mixed bed of **two complementary sorbents;** the best-known application, a special case, is the mixture of cation and anion exchangers for deionization of water (see Sec. 19). Optimize the cycle times, noting that shorter cycles should entail lower investment costs (down to the point of a rate-limited "shallow bed") but also higher operating costs.

5. In analyzing RATE-LIMITED TRANSITIONS for the effects of variations in operating and design conditions, begin with the most applicable "speed" method (constant-pattern, linear-equilibrium, or proportionate-pattern). The simplest methods may be based on assumptions of linear driving force, a single controlling resistance, constant diffusivity, plug flow, and fluid-solid boundaries parallel to the flow. When possible, apply an experimentally fitted mass-transfer coefficient: particle-side ("solid" or pore) for most adsorption and for concentrated-solution ion exchange; fluid-side for most ion exchange. If more than one resistance (or other spreading effect) is significant, their combined action may bring about a systematic variation in the

single apparent rate coefficient, since changes in flow velocity have different effects for the different mechanisms (see Fig. 16-12). For "deep bed" cases in which either CONSTANT-PATTERN or PROPORTIONATE-PATTERN treatment is inadequate because the equilibrium is nearly linear (**R** between 0.5 and 3), a NONLINEAR-EQUILIBRIUM REACTION-KINETIC SOLUTION may be most satisfactory.

Once the calculation sequences have been established, the apparent rate coefficients can be compared with predicted values for the assumed mechanism or mechanisms, to try to arrive at a clear-cut interpretation. The refinements potentially attainable through computer calculations will be beneficial only at a late stage in interpretation or design; until then, each hour of semitheoretical analysis will be worth several hours of computer programming (or run time).

Calculation of compositions at the bed outlet (or elsewhere) is based on integration of two differential equations [Eq. (16-51) and a suitable rate equation] that account for the earlier behavior of both fluid- and solid-phase behavior all through the bed upstream. Theoretically the solutions extend to positive infinity in length and time; with or without plug flow, downstream conditions have virtually no effect on upstream behavior. Beyond the column outlet, the concentrations in each element of fluid are assumed not to undergo any further change.

Rate calculations are carried out by using the number of transfer units (NTU), represented by **N**, as a nondimensional bed length. The height of one transfer unit (HTU) is the ratio of the fluid superficial velocity to the appropriate rate coefficient expressed in units of reciprocal time. The overall HTU is larger than any single-resistance HTU, but individual values cannot be added directly (except for linear equilibrium). Only a fluid-side form of NTU is used here, whether fluid-side or particle-side resistance controls, because it is solely fluid flow that defines an apparent residence time (or "space time"). Thus (for linear equilibrum) the reciprocals of $\mathbf{N}_f$ and $\mathbf{N}_p$ are directly additive without being weighted by a "capacity ratio."

Conceptually, the transfer unit is a distance within which every solute molecule has "unit opportunity" to transfer to the solid phase. The fluid-front "first droplet" behavior, usually controlled by fluid-side mass transfer, illustrates this definition; its relative concentration on a scale of 1 for points well upstream and 0 for points well downstream, is given by

$$\mathbf{X} = \exp(-\mathbf{N}_f) \qquad (16\text{-}3)$$

where subscript f denotes fluid-side NTU.

To complement the use of **N** as an effective distance, the calculations utilize either **T** or the product **NT** as the nondimensional fluid volume passed or run time elapsed.

Mathematical Framework For monovariant adsorption, the concentration variables of interest are the solute concentration c or partial pressure p in the fluid phase and the local solute concentration q in the sorbent, expressed, for example, as gram-moles or gram equivalents per liter of fluid and per kilogram of sorbent respectively.

For monovariant ion exchange, there are two exchangeable ion species of interest. With the conditions of negligible solvent transfer between fluid and exchanger phase, absence of phase change or chemical reaction, negligible Donnan uptake of electrolytes by the exchanger, and constant exchange capacity, the total concentration of both species is constant in the fluid phase on one hand and in the exchanger phase on the other. So, if the totals are known, it suffices to specify concentrations of either one of the species in the two phases, expressed in the units given previously.

Transition concentrations are expressed in dimensionless form, using the following definitions for the fluid phase,

$$\mathbf{X} = (c - c')/(c'' - c') = (p - p')/(p'' - p') \qquad (16\text{-}4a)$$

and for the sorbent phase,

$$\mathbf{Y} = (q - q')/(q'' - q') \qquad (16\text{-}4b)$$

In these relations, unprimed symbols represent the variable concentrations; q' is the preexisting (downstream-limit) concentration of the species chosen for reference; c' or p', the concentration or pressure of that species in fluid equilibrated with q'; c'' or p'', the value for the reference species in the upstream fluid; and q'', the concentration for sorbent in equilibrium with the feed. Nondimensional concentrations **X** and **Y** each range from 0 to 1 within a transition. For multicomponent local-equilibrium constant-**R** systems, **X** and **Y** will be exactly the same for all components; in equilibrium systems with other behavior, they will be nearly the same. Thus a single **X** or **Y** will often serve to describe the extent of completion of the transition for all the solutes involved.

The pattern of concentration change is described by relations of differential form, such as

$$\text{fn}(\mathbf{X}) = \text{fn}(\mathbf{Y}, \mathbf{T}, v) \qquad \text{fn}(\mathbf{Y}) = \text{fn}(\mathbf{X}, \mathbf{T}, v) \qquad (16\text{-}5a)$$

Analytic or numeric solution of these equations for given rate and equilibrium behavior (indicated functionally by **R**, which may not be constant) leads in many cases to analytic (sometimes implicit) or tabular solutions:

$$\mathbf{X} = \text{fn}(\mathbf{R}, \mathbf{T}, v) \qquad \mathbf{Y} = \text{fn}(\mathbf{R}, \mathbf{T}, v) \qquad (16\text{-}5b)$$

A partition ratio can be defined in terms of concentrations of any one species:

$$\Lambda = \frac{V_{\text{stoic}}}{v} = \frac{(q'' - q')\rho_b}{(c'' - c')} \qquad (16\text{-}6)$$

For monovariant sorption or for multivariant sorption with constant **R** or α applicable to all binary pairs, the value of Λ from Eq. (16-6) will be the same for all sorbable solutes. When not all the **R**'s are constant, the values will still be close, and an average or effective value can be selected.

Interpretation and Design The underlying principle is that the effluent concentrations c or **X** in a given transition are known numerically from laboratory or plant data (or algebraically from theory), at a constant v (or **N**), as a function of V (or **T**). The physical constants which interconnect the nondimensional and dimensional variables are particle diameter d_p, partition ratio Λ, void fraction ϵ, cross-sectional area S, volumetric flow rate F, and diffusivity D_f (with viscosity μ), D_{pore}, and/or D_p, or the corresponding mass-transfer coefficients.

For the interpretation, experimental points or slopes are matched to theoretically derived graphs or equations in order to determine the applicable **N** and in some cases the **T** scale. For design, the calculation sequence depends on what is known or projected and what

TABLE 16-2 Sequences for Design Calculations

Given: flow rate, breakthrough limit, equilibrium parameter, particle diameter, diffusivities

Specified variable	Throughput ratio **T**	NTU parameter **N**	Apparent contact time hS/F	Bed volume v or hS	Bed content $hSQ\rho_b$	Fluid content $\mathbf{T}hSQ\rho_b$ or $(V - hS\epsilon)C_0$	Fluid volume $V - hS\epsilon$	Running time $\frac{V}{F}$
A. Resin utilization	1	2	3	4	5	6	7	8
B. Run time	4*(9)	8	7	6	5	3	2	1
C. Column volume	4	3	2	1	5	6	7	8

*Assumed.

is left to be determined. Sorbent utilization (**T** at breakthrough point **X**) has economic impact.

Table 16-2 shows calculation sequences for three typical design situations. In most design, cyclic operation is projected, and regeneration (along with exhaustion) must be analyzed fully. For each new trial evaluation, **N** can often be estimated by proportioning with help from Fig. 16-12. Details of these procedures will be made evident in the following paragraphs.

PHYSICAL PROPERTIES OF SORBENT MATERIALS

Data on commercially available materials for the two major types of sorption are given in Tables 16-3 and 16-4, with subsidiary sections. Table 16-3 lists adsorbents, and Table 16-4 covers cationic and anionic exchangers and other closely related materials. The purpose of these tables is twofold: to assist the engineer in identifying materials suitable for a needed application; and to supply typical physical-property values.

For adsorbents, density values are usually reported as **bulk density** ρ_b, or weight of dry material per unit bulk volume as packed in a column. The dry **particle density** ρ_p is related to ρ_b and to the fraction of external voids ϵ in a packed bed as follows:

$$\rho_p(1 - \epsilon) = \rho_b \tag{16-7a}$$

The **crystalline density** of the solid ρ_c, as usually given in property tables for pure chemical compounds, is related to ρ_p and to the internal porosity χ of the particles as

$$\rho_c(1 - \chi) = \rho_p \tag{16-7b}$$

Similarly, the **wet density** of an individual particle ρ_w is related to these factors and to the liquid density ρ_f by

$$\rho_w = \rho_p + \rho_f\chi \tag{16-7c}$$

A less exact relation can be drawn between the pore surface area per unit weight of dry solid σ and the mean pore radius $\bar{r}$, such as

$$\sigma = \frac{(\text{const})\chi}{\rho_p \bar{r}} \tag{16-8}$$

TABLE 16-3 Physical Properties of Adsorbents

Material and uses	Shape* of particles	Size range, U.S. standard mesh†	Internal porosity χ, %	Bulk dry density, kg/L	Average pore diameter, nm	Surface area, km²/kg	Sorptive capacity, kg/kg (dry)
Aluminas							
Low-porosity (fluoride sorbent)	G, S	8–14, etc.	40	0.70	~7	0.32	0.20
High-porosity (drying, separations)	G	Various	57	0.85	4–14	0.25–0.36	0.25–0.33
Desiccant, $CaCl_2$-coated	G	3–8, etc.	30	0.91	4.5	0.2	0.22
Activated bauxite	G	8–20, etc.	35	0.85	5	......	0.1–0.2
Chromatographic alumina	G, P, S	80–200, etc.	30	0.93	......	......	~0.14
Silicates and aluminosilicates							
Molecular sieves	S, C, P	Various					
Type 3A (dehydration)			~30	0.62–0.68	0.3	~0.7	0.21–0.23
Type 4A (dehydration)			~32	0.61–0.67	0.4	~0.7	0.22–0.26
Type 5A (separations)			~34	0.60–0.66	0.5	~0.7	0.23–0.28
Type 13X (purification)			~38	0.58–0.64	1.0	~0.6	0.25–0.36
Mordenite (acid drying)			......	0.88	0.3–0.8	......	0.12
Chabazite (acid drying)			......	0.72	0.4–0.5	......	0.20
Silica gel (drying, separations)	G, P	Various	38–48	0.70–0.82	2–5	0.6–0.8	0.35–0.50
Magnesium silicate (decolorizing)	G, P	Various	~33	~0.50	......	0.18–0.30	
Calcium silicate (fatty-acid removal)	P	......	75–80	~0.20	......	~0.1	
Clay, acid-treated (refining of petroleum, food products)	G	4–8	......	0.85	......		
Fuller's earth (same)	G, P	<200	......	0.80			
Diatomaceous earth	G	Various	......	0.44–0.50	......	~0.002	
Carbons							
Shell-based	G	Various	60	0.45–0.55	2	0.8–1.6	0.40
Wood-based	G	Various	~80	0.25–0.30	......	0.8–1.8	~0.70
Petroleum-based	G, C	Various	~80	0.45–0.55	2	0.9–1.3	0.3–0.4
Peat-based	G, C, P	Various	~55	0.30–0.50	1–4	0.8–1.6	0.5
Lignite-based	G, P	Various	70–85	0.40–0.70	3	0.4–0.7	0.3
Bituminous-coal-based	G, P	8–30, 12–40	60–80	0.40–0.60	2–4	0.9–1.2	0.4
Organic polymers							
Polystyrene (removal of organics, e.g., phenol; antibiotics recovery)	S	20–60	40–50	0.64	4–9	0.3–0.7	
Polyacrylic ester (purification of pulping wastewaters; antibiotics recovery)	G, S	20–60	50–55	0.65–0.70	10–25	0.15–0.4	
Phenolic (also phenolic amine) resin (decolorizing and deodorizing of solutions)	G	16–50	45	0.42	......	0.08–0.12	0.45–0.55

*Shapes: C, cylindrical pellets; F, fibrous flakes; G, granules; P, powder; S, spheres.

†U.S. Standard sieve sizes (given in parentheses) correspond to the following diameters in millimeters: (3) 6.73, (4) 4.76, (8) 2.98, (12) 1.68, (14) 1.41, (16) 1.19, (20) 0.841, (30) 0.595, (40) 0.420, (50) 0.297, (60) 0.250, (80) 0.177, (200) 0.074.

NOTE: To convert kilograms per liter to pounds per cubic foot, multiply by 6.238×10^1; to convert square kilometers per kilogram to square feet per pound, multiply by 4.886×10^6; and to convert kilograms per kilogram to pounds per pound, multiply by 1.0.

TABLE 16-4 Physical Properties of Ion-Exchange Materials

Material	Shape* of particles	Bulk wet density (drained), kg/L	Moisture content (drained), % by weight	Swelling due to exchange, %	Maximum operating temperature,† °C	Operating pH range	Exchange capacity: Dry, equivalent/kg	Exchange capacity: Wet, equivalent/L
Cation exchangers: strongly acidic								
Polystyrene sulfonate								
Homogeneous (gel) resin	S				120–150	0–14		
4% cross-linked		0.75–0.85	64–70	10–12			5.0–5.5	1.2–1.6
6% cross-linked		0.76–0.86	58–65	8–10			4.8–5.4	1.3–1.8
8–10% cross-linked		0.77–0.87	48–60	6–8			4.6–5.2	1.4–1.9
12% cross-linked		0.78–0.88	44–48	5			4.4–4.9	1.5–2.0
16% cross-linked		0.79–0.89	42–46	4			4.2–4.6	1.7–2.1
20% cross-linked		0.80–0.90	40–45	3			3.9–4.2	1.8–2.0
Macroporous structure								
10–12% cross-linked	S	0.81	50–55	4–6	120–150	0–14	4.5–5.0	1.5–1.9
Sulfonated phenolic resin	G	0.74–0.85	50–60	7	50–90	0–14	2.0–2.5	0.7–0.9
Sulfonated coal	G							
Cation exchangers: weakly acidic								
Acrylic (pK 5) or methacrylic (pK 6)								
Homogeneous (gel) resin	S	0.70–0.75	45–50	20–80	120	4–14	8.3–10	3.3–4.0
Macroporous	S	0.67–0.74	50–55	10–100	120	…	~8.0	2.5–3.5
Phenolic resin	G	0.70–0.80	~50	10–25	45–65	0–14	2.5	1.0–1.4
Polystyrene phosphonate	G, S	0.74	50–70	<40	120	3–14	6.6	3.0
Polystyrene aminodiacetate	S	0.75	68–75	<100	75	3–14	2.9	0.7
Polystyrene amidoxime	S	~0.75	58	10	50	1–11	2.8	0.8–0.9
Polystyrene thiol	S	~0.75	45–50	…	60	1–13	~5	2.0
Cellulose								
Phosphonate	F	…	…	…	…	…	~7.0	
Methylene carboxylate	F, P, G	…	…	…	…	…	~0.7	
Greensand (Fe silicate)	G	1.3	1–5	0	60	6–8	0.14	0.18
Zeolite (Al silicate)	G	0.85–0.95	40–45	0	60	6–8	1.4	0.75
Zirconium tungstate	G	1.15–1.25	~5	0	>150	2–10	1.2	1.0
Anion exchangers: strongly basic								
Polystyrene-based								
Trimethyl benzyl ammonium (type I)								
Homogeneous, 8% CL	S	0.70	46–50	~20	60–80	0–14	3.4–3.8	1.3–1.5
Macroporous, 11% CL	S	0.67	57–60	15–20	60–80	0–14	3.4	1.0
Dimethyl hydroxyethyl ammonium (type II)								
Homogeneous, 8% CL	S	0.71	~42	15–20	40–80	0–14	3.8–4.0	1.2
Macroporous, 10% CL	S	0.67	~55	12–15	40–80	0–14	3.8	1.1
Acrylic-based								
Homogeneous (gel)	S	0.72	~70	~15	40–80	0–14	~5.0	1.0–1.2
Macroporous	S	0.67	~60	~12	40–80	0–14	3.0–3.3	0.8–0.9
Cellulose-based								
Ethyl trimethyl ammonium	F	…	…	…	100	4–10	0.62	
Triethyl hydroxypropyl ammonium		…	…	…	100	4–10	0.57	
Anion exchangers: intermediately basic (pK 11)								
Polystyrene-based	S	0.75	~50	15–25	65	0–10	4.8	1.8
Epoxy-polyamine	S	0.72	~64	8–10	75	0–7	6.5	1.7
Anion exchangers: weakly basic (pK 9)								
Aminopolystyrene								
Homogeneous (gel)	S	0.67	~45	8–12	100	0–7	5.5	1.8
Macroporous	S	0.61	55–60	~25	100	0–9	4.9	1.2
Acrylic-based amine								
Homogeneous (gel)	S	0.72	~63	8–10	80	0–7	6.5	1.7
Macroporous	S	0.72	~68	12–15	60	0–9	5.0	1.1
Cellulose-based								
Aminoethyl	P	…	…	…	…	…	1.0	
Diethyl aminoethyl	P	…	…	…	…	…	~0.9	

*Shapes: C, cylindrical pellets; G, granules; P, powder; S, spheres.
†When two temperatures are shown, the first applies to H form for cation, or OH form for anion, exchanger; the second, to salt ion.
NOTE: To convert kilograms per liter to pounds per cubic foot, multiply by 6.238×10^1; °F = 9/5 °C + 32.

The "constant" in this equation may vary with different types of porous sorbents but is often of the order of 3.0.

For ion exchangers, the bulk water-wet density is the usual value given. Ion-exchange materials are not considered porous unless they are inorganic and/or "mineral" (e.g., sulfonated coal), or have had the usual resin structure interlaced with true internal pores (i.e., continuous channels having an average diameter of at least 30 Å) by a separate step in the synthesis. For resinous exchangers in general, one can identify an **internal porosity** χ which measures the uptake of water or other liquid and will vary somewhat with the ion held; but the particle phase approximates a true molecular-scale solution, and no surface area or pore diameter can be defined.

FLUID-SORBENT EQUILIBRIUM

The phase equilibrium for one or several transferable components ("solutes") between fluid and sorbent will now be considered. Often the stoichiometric capacity of a sorbent will depend on fluid concentration and on temperature. The way in which equilibrium varies with rising fluid-phase concentration of solute determines whether equilibrium as such or stoichiometry will dominate the system's performance. In the latter case, equilibrium still has strong influence on the mass-transfer driving force.

In the simplest (monovariant) systems, a single curve can be drawn to show solute concentration in the solid in relation to solid concentration or partial pressure in the fluid. Any such curve may apply at only one temperature and is thus known as an **isotherm.** Typical isotherms are discussed, and equations for several that can be used easily for prediction and design are included. High precision in representing the equilibrium is not usually needed, but serious errors occur when the general nature of the equilibrium is omitted or misstated.

VARIANCE (DEGREES OF FREEDOM)

Variance is a unifying concept that moderates the differences between adsorption and ion exchange and clarifies the behavior of multicomponent sorption systems. It is defined as the number of independent concentration variables in a sorption system at equilibrium—that is, variables that can be changed separately and thereby control the values of all others. Thus variance also equals the difference between the total number of concentration variables and the number of independent relations connecting them. In fixed-bed operation, as mentioned previously, variance generally corresponds to the number of transitions occurring before a bed initially of uniform composition becomes fully saturated by a constant-composition feed stream.

For adsorption systems, the conceptual presence of a dummy component makes the calculations quite comparable to those for ion exchange. Monovariance is typified by adsorption of only one solute or by ion exchange involving only two ion species. By extension, the variance of a system of n adsorbable solutes or of $n + 1$ exchangeable ion species is n.

Numerous cases arise in which ion exchange is accompanied by chemical reaction (neutralization or precipitation, in particular) or in which adsorption is accompanied by evolution of sensible heat. The concept of variance helps greatly to assure correct interpretations and predictions.

Example 1: Calculation of Variance In mixed-bed deionization of a solution of a single salt, there are 8 concentration variables, 2 each for cation, anion, hydrogen, and hydroxide; and also 6 connecting relations, 2 for ion exchange and 1 for neutralization equilibrium and 2 ion-exchanger and 1 solution electroneutrality relations. The variance is therefore $8 - 6 = 2$.

SYSTEM VARIABLES AND TRANSITION VARIABLES

"System" as used here describes a given set of solutes, solvent, and sorbent, referred to the starting conditions for the procedure being conducted. "Transition" refers to only one region of composition change within the procedure that the system undergoes; the change is bounded, before and beyond, by specific equilibrium (or near-equilibrium) compositions.

Concentrations The quantity of each solute can be stated as mass or moles (for adsorption from a gas phase), volume (optionally, for adsorption from liquid), or ion equivalents (for ion exchange). The fluid-phase concentration c is referred to unit volume of fluid. For gases with high solute loading, the reference volume may be that of the solute-free carrier gas, at an arbitrary temperature and pressure; for a constant-pressure gas, partial pressure may replace concentration. For the stationary (solid) phase, concentration q is referred to unit mass in a reference condition (wet or dry, etc.). The convention is to have "solid-phase concentration" measure all solute within the outermost boundaries of the individual granules regardless of the solute's chemical or physical form. The product of q with bulk density of the solid gives the quantity of solute per unit of overall packed volume, a convenient figure. Notably, q times ρ_b is the quantity of solute per unit **mixed volume**, whereas c is quantity per unit **fluid volume.** If more than one solute is present, c (or p) and q are subscripted for each solute.

For brevity and uniformity, the isotherms shown here are given in terms of nondimensional concentrations, ratios of the actual c or q to a reference value. For each solute species, the "system" concentrations are expressed as

$$\mathrm{x} = c/c_{\mathrm{ref}} \qquad \mathrm{y} = q/q_{\mathrm{ref}} \tag{16-9}$$

This choice of symbols stems from the general practice of plotting fluid content as abscissa and solid content as ordinate and from the similarity to separation diagrams (for instance, the McCabe-Thiele plot) for countercurrent systems. For single-solute adsorption, c_{ref} is usually the highest fluid-phase concentration encountered, and q_{ref} is the equilibrium solid-phase concentration coexisting with c_{ref}. For ion exchange, c_{ref} is the total ion-equivalent concentration C_0, and q_{ref} is the total solid-phase exchange capacity Q; in this case, $\Sigma \mathrm{x}_i = 1$ and $\Sigma \mathrm{y}_i = 1$, where successive letters or numerals for i denote the individual ion species being exchanged.

The "transition variables" or fractional extents of change defined by Eqs. (16-4*a*) and (16-4*b*) become

$$\mathbf{X} = (\mathrm{x} - \mathrm{x}')/(\mathrm{x}'' - \mathrm{x}') \qquad \mathbf{Y} = (\mathrm{y} - \mathrm{y}')/(\mathrm{y}'' - \mathrm{y}') \tag{16-10}$$

From the solid-phase (eulerian) viewpoint, a single prime shows the earlier, and a double prime the later, end condition; from the fluid-phase (lagrangian) viewpoint, the timings reverse.

Other Parameters Other nondimensional variables that will differ between a system and a transition are:

Separation factors (**r**, **R**)

Throughput ratio (T, based on c_{ref} and q_{ref}; **T**, based on incremental c and q)

Partition ratios (Λ, $\mathbf{\Lambda}$) based as for T and **T**

For full-range sorption or desorption between (0, 0) and (c_{ref}, q_{ref}), the transition and system parameters become identical.

In Fig. 16-3 plots of the equilibrium y (solid) against x (fluid) are given for a typical system. Superimposed are an upward transition (loading) and a downward transition (unloading), as shown by the respective positions of the (x′, y′) and (x″, y″) coordinates.

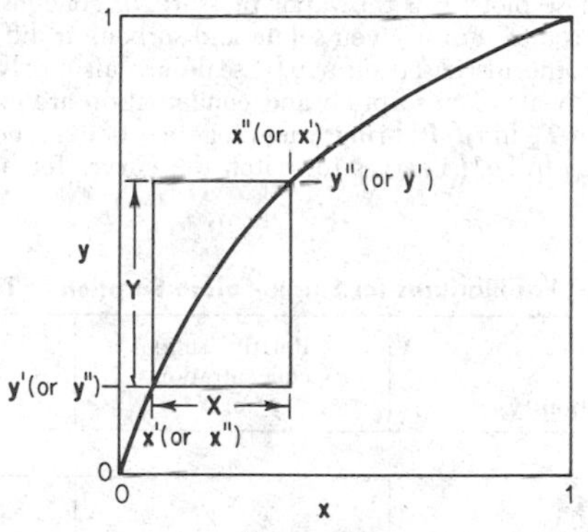

FIG. 16-3 Construction of the isotherm for a partial change in equilibrium concentrations between (x′,y′) and (x″,y″).

MONOVARIANT ISOTHERMS

Equilibrium data may be expressed in the form of (1) a graphical or tabular record based on measurements, (2) empirical algebraic expressions fitted to the data and usually selected for their generality and simplicity of calculational use, or (3) equations reflecting the molecular statistics of the underlying process.

The isotherms for two-component systems are single curves of q° versus c or p or of y° versus x, expressed for the species being taken up by the solid. (The asterisk indicates equilibrium with the coexisting phase. It can be applied instead to c, p, or x or be omitted entirely when no nonequilibrium terms are involved.) A separate plot is usually needed for each different temperature.

Representative isotherms are shown in Fig. 16-4, as classified by Brunauer and coworkers. Curves that are convex upward throughout (type I) are designated as "favorable" to uptake of solute; those that are concave upward throughout (type III), as "unfavorable" to uptake of solute; and those that follow a rising diagonal (not shown), as "linear." (The terms favorable and unfavorable actually refer to the **uptake** step for the component to which the isotherm applies. The opposite designation holds for the discharge step; thus, elution is "unfavorable" for a component having a "favorable" isotherm.) For noninflected isotherms, the part near the origin may slant (as in Fig.

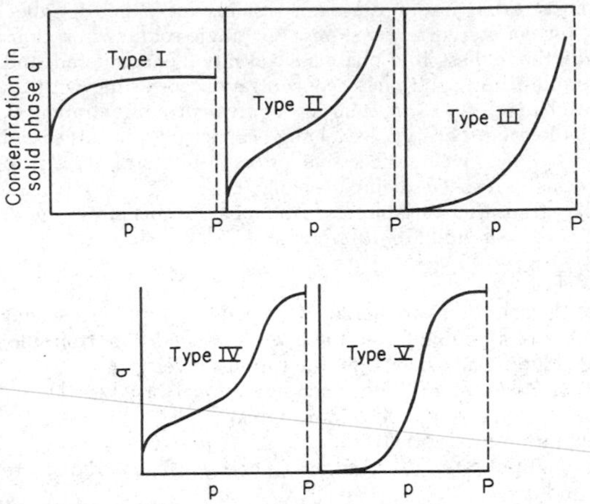

FIG. 16-4 Representative experimental isotherm types for physical adsorption. For type IV equations, see source paper; all others can be fitted by isotherm equations in Table 16-5. [*S. Brunauer,* J. Am. Chem. Soc., *62, 1723 (1940); reprinted by permission.*]

16-3, for Langmuir-type, described later) or may tend to become vertical (for Freundlich-type) or horizontal.

Other isotherms (types II, IV, and V) have one or more inflection points. In these plots, P is the vapor pressure, or condensation pressure, of the solute. For a given solute and sorbent at different temperatures, isotherms based on a p/P scale are all nearly the same because the heats of adsorption and condensation are comparable. The function $T_K \ln (p/P)$ brings them together better, and the function $(T_K/V_m) \ln (p/P)$ serves to match the curves for homologous compounds [Grant and Manes, *Ind. Eng. Chem. Fundam.*, **5**, 490 (1966)]; here, T_K is absolute temperature, and V_m is the molar volume of solute as liquid. This indicates that Freundlich-equation exponents for homologs may be estimated by the relation $\beta_2 = \beta_1 T_{K2} V_{m1}/T_{K1} V_{m2}$.

Equations The simplest relations to describe shapes of isotherms are given in Table 16-5. One of these, the trinomial, can be used to fit all except type IV of the curves shown in Fig. 16-4. The constants of the equations in Table 16-5 must lie within defined ranges in order to fit a specific isotherm type.

The **bilinear isotherm** approximates an equilibrium curve by two intersecting straight lines. For a transition in which equilibrium can be described in this way, sometimes the left-hand line (rising from $\mathbf{X} = 0$) is vertical, and sometimes the right-hand line (tending toward $\mathbf{Y} = 1$) is horizontal. Two other major isotherms described later may each be represented by an empirical bilinear fit. For the constant-separation-factor case, the left-hand line should have the slope $a \sim \mathbf{r}^{-0.75}$, and the right-hand line the slope $b \sim \mathbf{r}^{0.75}$. For the Freundlich case, $a \sim \beta^{-1.50}$ and $b \sim \beta^{0.75}$.

The **Langmuir isotherm** applies to adsorption on completely homogeneous surfaces, with negligible interaction between adsorbed molecules. For a single solute,

$$q^\circ = QK_L c/(1 + K_L c) \qquad (16\text{-}11a)$$

where Q is the asymptotic maximum solid-phase concentration, and K_L is the equilibrium constant. The Langmuir relation for $\mathbf{y}^\circ(\mathbf{x})$ in Table 16-5 is obtained by writing Eq. (16-11*a*) for both q_{ref} in terms of c_{ref} and q° in terms of c, then converting q°/q_{ref} to $\mathbf{y}^\circ$ and c/c_{ref} to $\mathbf{x}$. Plotting $1/q^\circ$ against $1/c$ or $1/p$ (or $1/\mathbf{y}^\circ$ against $1/\mathbf{x}$) gives a straight line which serves to evaluate Q and K_L (or $\mathbf{r}$, defined subsequently and in Table 16-5).

The Brunauer-Emmett-Teller (BET) isotherm, or multilayer Langmuir relation, is useful for those gas-solid systems in which condensation is approached [*J. Am. Chem. Soc.*, **60**, 309 (1938)], fitting

TABLE 16-5 Equilibriums for Single-Solute Sorption or Two-Component Exchange

Equation type	Relative solid concentration $\mathbf{y}^\circ$ or $\mathbf{Y}^\circ$	Relative fluid concentration $\mathbf{x}^\circ$ or $\mathbf{X}^\circ$	Values of constants giving isotherm type: Type I favorable	Type II composite[a]	Type III unfavorable	Type V composite[b]
		Full-range isotherms $\mathbf{y}$ and $\mathbf{x}$				
Linear	$\mathbf{x}$	$\mathbf{y}$	None	None	None	None
Constant separation factor	$\dfrac{\mathbf{x}}{\mathbf{r} + \mathbf{x} - \mathbf{rx}}$	$\dfrac{\mathbf{ry}}{1 + \mathbf{ry} - \mathbf{y}}$	$\mathbf{r} < 1$	None	$\mathbf{r} > 1$	None
Langmuir[c]	$\dfrac{(1 + K_L c_{\text{ref}})\mathbf{x}}{1 + K_L c_{\text{ref}}\mathbf{x}}$	$\dfrac{\mathbf{y}}{1 + K_L c_{\text{ref}}(1 - \mathbf{y})}$	$K_L > 0$	None	$-1 < K_L c_{\text{ref}} < 0$	None
Mass action	Eqs. (16-8), (16-26)	....	$\mathbf{K}_{AB} > 1$	None	$\mathbf{K}_{AB} < 1$	None
Freundlich	$\mathbf{x}^\beta$	$\mathbf{y}^{1/\beta}$	$\beta < 1$	None	$\beta > 1$	None
Trinomial $\mathbf{y}$	$\mathbf{x}(1 - a_1 - a_2)$	....	$a_1 < 0$	$a_1 < 0$	$a_1 > 0$	$a_1 > 0$
$[-2 < a_2 < (1 - a_1)]$	$+ a_1\mathbf{x}^2 + a_2\mathbf{x}^3$	....	$3a_2 < -a_1$	$3a_2 > -a_1$	$3a_2 > -a_1$	$3a_2 < -a_1$
Trinomial $\mathbf{x}$	....	$\mathbf{y}(1 - b_1 - b_2)$	$b_1 > 0$	$b_1 > 0$	$b_1 < 0$	$b_1 < 0$
$[-2 < b_2 < (1 - b_1)]$		$+ b_1\mathbf{y}^2 + b_2\mathbf{y}^3$	$3b_2 > -b_1$	$3b_2 < -b_1$	$3b_2 < -b_1$	$3b_2 > -b_1$
		Limited-range isotherms $\mathbf{Y}$ and $\mathbf{X}$				
Constant separation factor[d]	$\dfrac{\mathbf{X}}{\mathbf{R} + \mathbf{X} - \mathbf{RX}}$	$\dfrac{\mathbf{RY}}{1 + \mathbf{RY} - \mathbf{Y}}$	$\mathbf{R} < 1$	None	$\mathbf{R} > 1$	None
Bilinear	$a\mathbf{X}$, if $\mathbf{Y} < a(1 - b)/(a - b)$;	$\mathbf{Y}/a$, if $\mathbf{X} < (1 - b)/(a - b)$;	$1 < a < \infty$	None	$1 > a > 0$	None
	$(1 - b) + b\mathbf{X}$, if $\mathbf{Y} > a(1 - b)/(a - b)$	$(b - 1 + \mathbf{Y})/b$, if $\mathbf{X} > (1 - b)/(a - b)$	$1 > b > 0$	None	$1 < b < \infty$	None

REFERENCES: Langmuir, *J. Am. Chem. Soc.*, **38**, 221 (1916). Freundlich, "Colloid and Capillary Chemistry," Dutton, New York, 1926. Trinomial, *Advan. Chem. Eng.*, **2**, 157 (1958). Limited-range with **R**, *ibid.*, 187.

[a] Favorable at low, and unfavorable at high, concentrations.

[b] Unfavorable at low, and favorable at high, concentrations.

[c] Same as constant-separation-factor case, if $K_L c_{\text{ref}} = (1/\mathbf{r}) - 1$.

[d] $\mathbf{R} = (\mathbf{r} + \mathbf{x}' - \mathbf{rx}')/(\mathbf{r} + \mathbf{x}'' - \mathbf{rx}'')$.

type II behavior. With pressure p replacing fluid concentration c and with vapor pressure P entering, the BET relation is

$$q^\circ = \frac{QK_Lp}{[1 + K_Lp + (p/P)][1 - (p/P)]} \tag{16-11b}$$

The **Freundlich isotherm** (Table 16-5) corresponds to an exponential distribution of heats of adsorption. Intermediate conditions between the Langmuir and the Freundlich assumptions are more realistic; equations to fit such cases (omitted here because of the complexity of calculating with them) have been given by Sips [*J. Chem. Phys.*, **18,** 1024 (1950)] and by Koble and Corrigan [*Ind. Eng. Chem.*, **44,** 383 (1952)]. With dimensional terms, the Freundlich equation has the form

$$q^\circ = K_f c^\beta \quad \text{or} \quad q^\circ = K_F p^\beta \tag{16-12}$$

Mass-action equilibrium provides a useful model for ion-exchange behavior. The exchange reaction can be written in general terms:

$$mA + R_mB \rightleftharpoons mRA + B$$

where m is an integer or fraction given by the valence ratio for B and A. The associated model equilibrium relation is often of the form

$$K_{AB} = \left(\frac{q_A}{c_A}\right)^m \frac{c_B}{q_B} = \left(\frac{y_A}{x_A}\right)^m \frac{x_B}{y_B}\left(\frac{Q}{C_0}\right)^{m-1} \tag{16-13a}$$

where K_{AB} is the equilibrium constant, or selectivity coefficient, and $K_{AB}(C_0/Q)^{m'-1}$ is dimensionless and can be written as $\mathbf{K}_{AB}$. For A, the ion of smaller charge, with $m > 1$, $\mathbf{K}_{AB}$ increases with C_0. This reflects the fact that ion exchangers exhibit an increasing affinity for ions of lower valence as the solution-phase concentration increases.

Non-ideal-solution behavior will justify the use, in many cases, of activities in place of solution-phase concentrations and also, at times, the use of empirical exponents on q's or y's to describe solid-phase equilibrium behavior. (Solid-phase activity coefficients dependent on q or y may modify the expected mass-action exponents.) Experimental data plot as linear constant-y_A or -y_B contours on activity coordinates of log a_A versus log a_B, or as separate C_0 contours in the conventional y versus x plane.

Mass-action constants involving solution activities have been collected by Marcus and Howery (*Ion Exchange Equilibrium Constants*, IUPAC Rep., Butterworth, London, 1975) for several cation and anion exchangers (many at only one cross-linking) for a variety of counterions. The values given for polystyrene sulfonate cation exchangers over a range of cross-linking Z from 0.01 to 0.25 satisfy the empirical relation (for ion A replacing ion B on the resin):

$$\log K_{AB} = C_1 + C_2Z^{0.5} \tag{16-13b}$$

Table 16-6 gives coefficients derived from Marcus and Howery's report (Vermeulen, private communication), based on H as ion B, with m the reciprocal of ion charge or valence for ion A and with ion quantities measured in ion equivalents. The equilibrium between any two ions A and D is calculated as

$$\log K_{AD} = \log K_{AH} - \log K_{DH} \tag{16-13c}$$

TABLE 16-6 Polystyrene Sulfonate Equilibrium Constants
Coefficients for Eq. (16-13b) at 298 K (77°F)

Ion replacing H	C_1	C_2
Na	−0.05	+0.7
K	+0.05	+1.2
Li	−0.03	−0.2
Cs	+0.09	+1.1
Ag	−0.05	+2.9
NH_4	+0.05	+0.7
0.5 Mg	+0.37	+0.2
0.5 Ca	+0.45	+0.6
0.5 Zn	+0.25	+0.7
0.5 Cu	+0.38	+0.4

Between ions of the same valence at concentrations of 0.1 N or less or between any ions at 0.01 N or less, the solution-phase activity coefficients prorated to unit valence will be similar enough that they can be omitted.

Adsorption from the liquid phase can be identified as **exchange adsorption.** For this case, volume concentrations or volume fractions should be used instead of molar concentrations or mole fractions because the pores lying within the outside surface of the sorbent particles contain a constant total volume of liquid mixture (which of course has an equilibrium composition different from that of the bulk liquid mixture outside). Thus the equilibrium and rate equations for *ion exchange*, with volume fractions replacing ion-equivalent fractions, apply directly for *liquid-phase adsorption*.

Separation Factor By analogy with the mass-action case, a separation factor **r** can be defined:

$$\mathbf{r} = \frac{x_A y_B}{x_B y_A} = \frac{x_A(1 - y_A)}{y_A(1 - x_A)} \tag{16-14}$$

This term is analogous to relative volatility α or its reciprocal in distillations. If $m = 1$ in Eq. (16-13a), $\mathbf{r} = 1/\mathbf{K}_{AB}$, hence is constant. As indicated in Table 16-5, Langmuir adsorption corresponds to constant separation factor (CSF), with

$$\mathbf{r} = 1/(1 + K_L c_{\text{ref}}) \tag{16-15}$$

Hence constant **r** frequently is a useful approximation, throughout the field of sorption operations. Evolution in the shapes of isotherms, for a range of values of **r**, is shown in Fig. 16-5, which can be used to fit **r** values to experimental data by empirical curve matching. (An effective average **r** can thus be found even for curves which deviate from constant-**r** shape.)

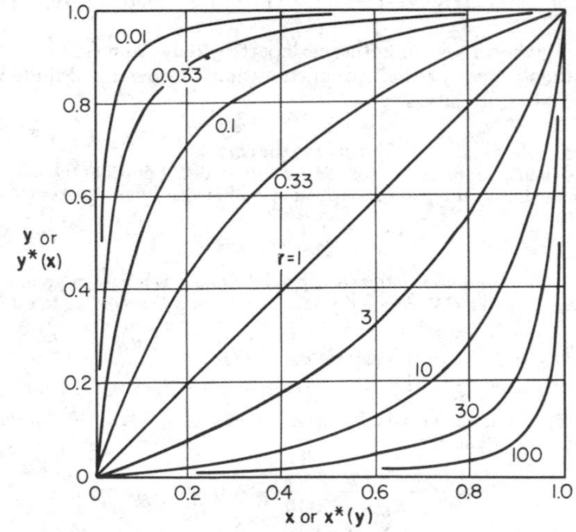

FIG. 16-5 Dimensionless-concentration isotherms as functions of separation factor on linear coordinates.

It is seen that **r** values less than unity give favorable isotherms for uptake, while curves with $\mathbf{r} > 1$ are unfavorable. The linear isotherm, or Raoult-law case, corresponds to $\mathbf{r} = 1$. If **r** is defined for sorption of species A, or for ion A displacing ion B, the **r** for the reverse process, identified by the dagger †, is

$$\mathbf{r}^\dagger = 1/\mathbf{r} \tag{16-16}$$

In this reverse process, A desorbs, or ion B displaces ion A. In Fig. 16-5, the curves for given **r** and 1/**r** are symmetrical about the 45° line.

The separation factor **r** identifies the equilibrium increase in y from 0 to 1 which accompanies an increase in x from 0 to 1. For a concentration change over only a part of the isotherm, a new separation factor $\mathbf{R}$ can be defined which identifies the equilibrium between $\mathbf{Y}$ and $\mathbf{X}$ [see Eqs. (16-4a) and (16-4b)], each ranging from 0 to 1. Figure 16-5 gives $\mathbf{Y}^\circ(\mathbf{X}, \mathbf{R})$ as well as $\mathbf{y}^\circ(\mathbf{x}, \mathbf{r})$. As shown by Vermeulen and Hiester [*J. Chem. Phys.*, **22**, 96 (1954)], the separation factor for a partial transition is

$$\mathbf{R} = \frac{\mathbf{X}(1-\mathbf{Y})}{\mathbf{Y}(1-\mathbf{X})} = \frac{\mathbf{r}+\mathbf{x}'-\mathbf{r}\mathbf{x}'}{\mathbf{r}+\mathbf{x}''-\mathbf{r}\mathbf{x}''} = \frac{\mathbf{y}''\mathbf{x}'}{\mathbf{x}''\mathbf{y}'} \tag{16-17}$$

where $\mathbf{x}'$ is the starting, and $\mathbf{x}''$ the final, value of $\mathbf{x}$. Figure 16-3 shows the geometrical relation of the partial and total isotherms. The partial isotherm is more nearly linear; consequently $\mathbf{R}$ for ascending changes, or $1/\mathbf{R}$ for descending changes, is closer to unity than is $\mathbf{r}$. As $\mathbf{x}'$ may be either larger or smaller than $\mathbf{x}''$, the partial change can occur either upward or downward. From Eq. (16-17), $\mathbf{R}$ for a downward change is the reciprocal of $\mathbf{R}$ for an upward change.

In *trace systems*, there is one gross component. The one or more other components are present in such low concentrations that $\mathbf{R}$, as defined by Eq. (16-17), approaches unity and the partial isotherm becomes linear. A useful definition is that $\Sigma\mathbf{x} < 0.1$ and $\Sigma\mathbf{y} < 0.1$ for all solutes other than the gross component and also that $\mathbf{R}$ for the trace components must differ from unity by less than 0.1. For such cases, the $\mathbf{R}$ for each component is obtained from the respective binary isotherm:

$$\mathbf{R} = \frac{1}{1 + (\mathbf{x}_A)_0\,[(1/\mathbf{r}) - 1]} \tag{16-18}$$

The equilibrium and rate behavior of each trace component are effectively independent of the amounts and properties of other trace components.

The CSF and Langmuir isotherms are *hyperbolic;* both fit the equation $(Z_1 + C_1)(C_2 - Z_2) = C_1C_2$. The respective left-side terms are, for CSF, $\mathbf{x}$, $\mathbf{r}/(1-\mathbf{r})$, $1/(1-\mathbf{r})$, $\mathbf{y}$; for Langmuir, c, $1/K_L$, 1, q/Q.

Even when a system isotherm departs widely from CSF behavior, it is often found that individual transitions are described quite well by "local" CSF curves.

Example 2: Application of Isotherms

a. Thomas [*Ann. N.Y. Acad. Sci.*, **49**, 161 (1948)] provides the following Langmuir isotherm for the adsorption of anthracene from cyclohexane onto alumina:

$$q^\circ = 22c/(1 + 375c)$$

with q° in gram-moles anthracene per kilogram when c is in gram-moles anthracene per liter. What are the values of K_L and Q according to Eq. (16-11a)?

$$K_L = 375 \text{ L/(g·mol anthracene)} \qquad \text{(ans)}$$

$$Q = 22/K_L = 5.87 \times 10^{-2} \text{ (g·mol anthracene)/kg} \qquad \text{(ans)}$$

b. For this isotherm and a feed concentration $c_0 = 8.11 \times 10^{-4}$ (g·mol)/L, what is the value of $\mathbf{r}$?

$$\mathbf{r} = \frac{1}{1 + K_L(8.11 \times 10^{-4})} = 0.766 \qquad \text{(ans)}$$

c. For the isotherm, what feed concentration c_0 would give an $\mathbf{r}$ value greater than 0.9 according to Eq. (16-15)?

$$\mathbf{r} = 1/(1 + 375c_0) > 0.9$$

$$c_0 < 0.111/375 = 2.96 \times 10^{-4} \text{ (g·mol anthracene)/L} \qquad \text{(ans)}$$

d. The equilibrium constant K for the exchange of A^+ with B resin is 3.16. According to Eq. (16-18) what mole fraction should A^+ be of the feed cations for $\mathbf{R}$ to be greater than 0.9?

$$\mathbf{r} = 1/K_{AB} = 0.316$$

$$\mathbf{R} = \frac{1}{1 + (\mathbf{x}_A)_0\,[(1/0.316) - 1]} > 0.9$$

$$(\mathbf{x}_A)_0 < 0.111/2.16 = 0.0515 \qquad \text{(ans)}$$

Partition Ratio For each process arrangement, there is another useful equilibrium property—the dimensionless partition ratio Λ. For the total concentrations in ion exchange or adsorption,

$$\Lambda = Q\rho_b/C_0 = q_{\text{ref}}\rho_b/c_{\text{ref}} \tag{16-19}$$

Under trace conditions, also,

$$\Lambda_i = [\Lambda(\mathbf{y}_i)_0]/[(\mathbf{x}_i)_0] \tag{16-20}$$

The trace condition corresponds to $\mathbf{y}_B \approx 1$, $\mathbf{x}_B \approx 1$, where subscript B designates the bulk component. Hence, for constant separation factor, $\Lambda_i \approx \Lambda K_{AB}$ or $\Lambda/\mathbf{r}$. For mass action generally, $\Lambda_i \approx \Lambda(K_{AB})^{1/m}$. The partition ratio Λ_i is often termed K_D (or $K_D\epsilon$, where ϵ is the void fraction) in the literature on chromatographic separations.

For concentration shifts in binary systems corresponding to partially presaturated sorbent with mixed feed, the applicable partition ratio becomes

$$\Lambda = \frac{q'' - q'}{c'' - c'}\,\rho_b = \frac{\mathbf{y}'' - \mathbf{y}'}{\mathbf{x}'' - \mathbf{x}'}\,\Lambda \tag{16-21}$$

EQUILIBRIUM IN MULTIVARIANT SYSTEMS

As has been noted, a system of any given number of solutes will have one less unit of variance if its sorption behavior is stoichiometric (as in ion exchange, generally) than if it is not (as in gas adsorption, generally). Because the resulting theory for adsorption equilibrium is more complicated, it is useful to review ion exchange ahead of adsorption.

Ion Exchange

Stoichiometry In most applications, except for some weak-electrolyte and some concentrated-solution cases, the following summations apply (where n is the number of components):

$$\sum_{j=1}^{n} c_j = C_0 = \text{const} \qquad \sum_{j=1}^{n} q_j = Q = \text{const} \tag{16-22}$$

In equivalent-fraction terms, by use of Eq. (16-9), the sums become

$$\sum_{j=1}^{n} \mathbf{x}_j = 1 \qquad \sum_{j=1}^{n} \mathbf{y}_j = 1 \tag{16-23}$$

Mass Action Here the equilibrium relations, consistent with Eq. (16-13a), are

$$\left(\frac{\mathbf{y}_i}{\mathbf{x}_i}\right)^{m_{ij}} \frac{\mathbf{x}_j}{\mathbf{y}_j} = \mathbf{K}_{ij} \tag{16-24}$$

With n x's (or y's) known, the n y's (or x's) can be found by simultaneous solution of n-1 independent i,j combinations for Eq. (16-24) using Eqs. (16-22). If one $\mathbf{y}_j/\mathbf{x}_j$ is assumed, the other y/x values can each be calculated by Eq. (16-24). The sum of the trial y's (or x's) is then found. The assumed $\mathbf{y}_j/\mathbf{x}_j$ is reduced for the next trial if $\Sigma\mathbf{x}$ is too large (or $\Sigma\mathbf{x}$ too small) or is increased if $\Sigma\mathbf{y}$ is too small (or $\Sigma\mathbf{x}$ too large). This is based on the fact that all ratios $(\mathbf{y}_i/\mathbf{x}_i)$ will increase or decrease together.

Because an n-component system has n-1 independent concentrations, a three-component equilibrium can be plotted in a plane and a four-component equilibrium in three-dimensional space. Figure 16-6 shows a triangular plot of x contours in equilibrium with the corresponding y coordinates.

Constant-separation-factor Approximation If the valences of all the species are equal, the separation factor α_{ij} applies

$$\frac{\mathbf{y}_i\mathbf{x}_j}{\mathbf{x}_i\mathbf{y}_j} (= \mathbf{K}_{ij}) = \alpha_{ij} \tag{16-25}$$

For a binary system, $\mathbf{r} = \alpha_{BA} = 1/\alpha_{AB}$. The symbol $\mathbf{r}$ applies primarily to the process, while α is oriented toward interactions between pairs of solute species. For each binary pair, $\mathbf{r}_{ij} = \alpha_{ji} = 1/\alpha_{ij}$.

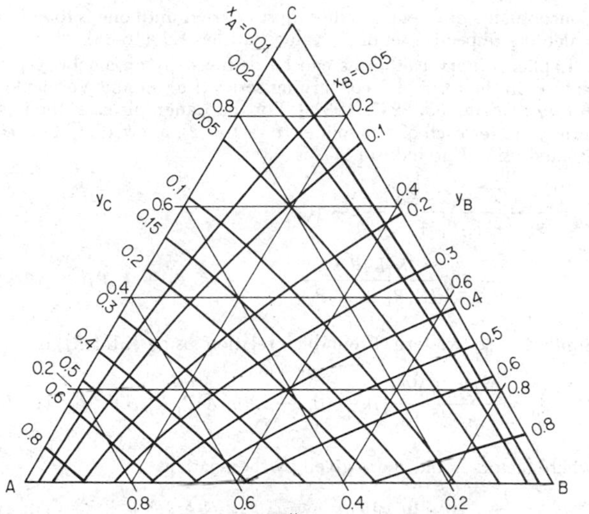

FIG. 16-6 Ideal mass-action equilibrium for three-component ion exchange with unequal valences. $(y_A/x_A)(x_C/y_C)^2 = 8.06$; $(y_B/x_B)(x_C/y_C)^2 = 3.87$. Duolite C-20 polystyrene sulfonate resin, with Ca as A, Mg as B, and Na as C. [*Klein et al.*, Ind. Eng. Chem. Fundam., *6, 339 (1967); American Chemical Society, reprinted by permission.*]

Equilibrium then is given explicitly by

$$y_i = \frac{x_i}{\sum_{j=1}^{n} \alpha_{ji} x_j} = \frac{\alpha_{in} x_i}{\sum_{j=1}^{n} \alpha_{jn} x_j} \tag{16-26}$$

and

$$x_i = \frac{y_i}{\sum_{j=1}^{n} \alpha_{ij} y_j} = \frac{\alpha_{ni} y_i}{\sum_{j=1}^{n} \alpha_{nj} y_j} \tag{16-27}$$

For the CSF case, the x contours in a plot like Fig. 16-6 are linear.

Donnan Uptake At high solution concentrations, above 0.5 g equiv/L with strongly ionized exchangers (or at lower concentrations with those more weakly ionized), the resin's nominal constant exchange capacity is measurably exceeded by the uptake of neutral ion pairs of salt from the solution. With only one co-ion species *S* (matching the charge sign of fixed groups in the resin, as a simplification), its uptake q_S equals the total excess uptake of counterions. With q_S the solid concentration of a counterion *M* having a valence *m*, the Donnan equilibrium of salt pair *SM* is given by a mass-action expression:

$$q_S q_M^{1/m} = (K_{D,M}\gamma_M)^{1/m} \gamma_S c_S c_M^{1/m} \tag{16-28}$$

where the γ's are solution-phase activity coefficients and K_D is the respective Donnan-equilibrium constant. If Eq. (16-28) is written for each of two counterions N' and N'', the ratio of the $(K_D\gamma)$'s is seen to be related directly to the equilibrium constant for exchange between N' and N'' [Eq. (16-13*a*)].

For a bed saturated with a given counterion at one solution concentration, a new feed of the same ion at another concentration will exhibit a transition from the earlier to the later total concentration; thus such a system, in principle, is monovariant. This transition is counted only when it is measurable and significant. It follows that simple binary exchange at high concentrations is likely to be observably bivariant. This area has been explored theoretically at the University of California by F. M. Golden (Ph.D. dissertation, 1973) and D. S. Arnold (M.S. thesis, 1978).

Adsorption

Langmuir-Type Relations For multicomponent systems composed of solutes which individually follow Langmuir isotherms, equilibrium is predicted by the relation

$$q_i^\circ = \frac{K_i Q_i c_i}{1 + \sum_j^{n-1} K_j c_j} \tag{16-29}$$

This equation gives reasonable estimates when the single-solute *Q*'s agree closely. Unequal *Q*'s indicate that the solutes occupy different amounts of surface area; the derivation for this case assumes that vacant area, rather than the number of vacant sites, governs the ease of uptake.

In multicomponent gas adsorption, the ratio of the *KQ*'s for two adsorbable components forms a selectivity factor α or separation factor **r**. An α or **r** so determined for uptake from the gas phase will lie much farther from unity than the respective factor determined for uptake from a liquid phase. The adsorbent's selectivity is exerted mainly in the monolayer at the pore walls, and the fluid in the rest of the pore volume has a composition nearly the same as that of the bulk fluid outside the particles.

Constant-Separation-Factor Treatment The multivariant Langmuir case can be converted to CSF form by the introduction of a "dummy" component which indirectly represents the vacant sites in the sorbent and also the nonsorbing carrier component in the gas; weighting factors, such as c_{ref} and q_{ref} values differing for each component, are also needed. (See Helfferich and Klein, in "General References.") The transformation is similar in its effect to that for a single solute [Eq. (16-15)]. There and here, there is one degree of free choice for a reference value, and the calculational results are not affected by that choice. For variance $n - 1$, there are $(n - 1)$ values of *A* (inversely proportional to q_{ref}) and one *B*, equaling $q_{\text{ref}}/c_{\text{ref}}$ for every solute.

The simplest procedure is one in which *B* is assigned and the *A*'s are fixed and calculable. *B* represents an assumption about the properties of the *n*th, or dummy, component; numerically it is equal to the product $K_n Q_n$, for which separate values of *K* and *Q* do not exist. It is convenient to base *B* on the most weakly adsorbed species in its pure state. Preferably *B* should not yield negative concentration ratios for any species, even the dummy species *n*, but if this happens, no harm is done to the subsequent calculations.

Once *B* is set, the *A*'s can be found from the relation

$$A_i = K_i/(K_i Q_i - B) \tag{16-30}$$

The weighted mole fractions are determined as follows:

$$x_i = c_i/c_{i,\text{ref}} \quad \text{with } c_{i,\text{ref}} = 1/BA_i \tag{16-31}$$

$$y_i = q_i^\circ/q_{i,\text{ref}} \quad \text{with } q_{i,\text{ref}} = 1/A_i \tag{16-32}$$

The summation conditions, representing x's and y's respectively, are

$$\sum_{j=1}^{n} BA_j c_j = 1 \qquad \text{and} \qquad \sum_{j=1}^{n} A_j q_j^\circ = 1 \tag{16-33}$$

However, the only way to determine x_n and y_n for the dummy is to determine them as the residues in these sums after all other x's and y's have been found.

The values of α for the dummy, relative to each regular species, are

$$\alpha_{in} = \frac{A_i}{A_i - (1/Q_i)} = \frac{K_i Q_i}{B} \tag{16-34}$$

Each other α, as noted earlier, is given by the ratio of the respective *KQ* products.

The main reason for exploring a multivariant problem with the CSF assumption is that this is almost the only equilibrium behavior amenable to mathematical solution. Even an inaccurate fit can yield a great deal of useful qualitative information.

"Ideal Adsorbed Solution" Model Theory indicates that if the monolayer capacity Q varies for different molecules and if the monolayer behaves as an ideal solution (which is uncertain), Eq. (16-29) is inconsistent with Gibbs' adsorption equation. For this case, Myers and Prausnitz [*Am. Inst. Chem. Eng. J.*, **11**, 121 (1965)] report a way to predict mixture behavior from single-component data, given either in numeric form or in algebraic form as with a Langmuir or Freundlich relation.

The Gibbs adsorption isotherm for two adsorbable solutes is

$$A\,d\pi = q_1\,d\mu_1 + q_2\,d\mu_2 \tag{16-35}$$

where π is the spreading pressure, A is the surface area per unit mass of sorbent, q is moles of solute i per mass of sorbent, and μ_i is the chemical potential ($d\mu_i = R_g T_K\, d \ln P_i$, with P_i the partial pressure). For each solute, the integral of Eq. (16-35) is

$$\frac{\pi A}{R_g T_K} = \int_0^{P_i} \frac{q_i}{P_i}\,dP_i \tag{16-36}$$

If Raoult's law is followed on the surface, a reference partial pressure P_i° exists for each component in an extrapolated pure state, P_i° being deduced from the actual solute mole fraction y_i ($= q_i/\Sigma q_j$):

$$P_i^\circ = P_i/y_i \tag{16-37}$$

The P_i° for each separate reference state has the same spreading pressure as the actual surface mixture. Also $\Sigma y_j = 1$, and this gives

$$\frac{P_1}{P_1^\circ} + \frac{P_2}{P_2^\circ} + \cdots = 1 \tag{16-38}$$

Conceptually, different π values must be tried until one is found for which the respective set of P_i° values satisfies Eq. (16-38).

Explicit **binary isotherms** can be deduced by an algebraic procedure, in the form of a converging series (LeVan and Vermeulen, *J. Phys. Chem.*, **85**, 3247 (1981)]. For the Langmuir case, the two-term form for each of two solutes ($i = 1$ or 2), alternate to Eq. (16-29) and using K to indicate K_L, is

$$q_i = \frac{(Q_1 + Q_2)K_iP_i}{2(1 + K_1P_1 + K_2P_2)} + (Q_i - Q_j)\frac{K_1K_2P_1P_2}{(K_1P_1 + K_2P_2)^2} \ln (1 + K_1P_1 + K_2P_2) \tag{16-39}$$

Similarly, the two-term Freundlich relation, using K for K_F, is

$$q_i = \frac{P_i(K_i/\beta_i)^{1/\beta_i}}{A^{2-(B/A)}}\,[B + (\beta_i - \beta_j)(\ln A)P_j(K_j/\beta_j)^{1/\beta_j}] \tag{16-40}$$

where factors A and B are given by the relations

$$A = P_1K_1^{1/\beta_1}\beta_1^{-1/\beta_1} + P_2K_2^{1/\beta_2}\beta_2^{-1/\beta_2} \tag{16-41}$$

$$B = P_1K_1^{1/\beta_1}\beta_1^{1-(1/\beta_1)} + P_2K_2^{1/\beta_2}\beta_2^{1-(1/\beta_2)} \tag{16-42}$$

Values from these equations match the exact calculations to within 1 percent in most cases. The resulting effective α's are larger (hence less conservative) than α's obtained directly from ratios of the K's.

EQUILIBRIUM-LIMITED TRANSITIONS

EFFECT OF UNFAVORABLE EQUILIBRIUM

Breakthrough curves in fixed-bed operations divide empirically into nonsharpening, sharpening, intermediate, and compound types. In general, the curve shape is influenced by both equilibrium and transport factors. When curve spreading from transport effects is minimized experimentally or neglected theoretically, curves of gradual, abrupt, and compound type emerge. The gradual, nonsharpening type is identified with **unfavorable equilibrium,** as evidenced by a concave-upward plot of q versus c, or **y** versus **x**, and a separation factor **R** greater than unity. Equilibrium prevents the sorbent from taking up as much solute as its capacity would allow. The spreading from mass-transfer effects is often small compared with that from equilibrum, and thus the breakthrough pattern expands in proportion to the volume of bed through which it passes.

(In contrast to this nonsharpening case, the abrupt, or sharpening, breakthrough has an equilibrium plot that is convex upward and favorable, to be discussed in the next subsection.)

Because many separate effects may influence the calculations, the algebra and the numerical operations are simplified by assembling the various factors into nondimensional ratios or groups. The groups needed in this subsection have already been introduced: **T** by Eq. (16-1); **Λ** by Eqs. (16-6) and (16-21); **x**, **y**, **X**, and **Y** by Eqs. (16-9) and (16-10); and **R** by Eq. (16-17).

A typical distance-time plot for a bed exhibiting equilibrium-limited breakthrough is given in Fig. 16-7; vertical projection to **X** versus time scales shows the proportionate pattern as it emerges through two successive cross sections at h_1 and h_2. Such curves cannot be sharpened by physical effects; if sharper breakthrough is wanted, the equilibrium must be changed by lowering the temperature or by adopting a different sorbent.

In each transition, or partial transition, controlled by **local equilibrium,** the starting composition in the bed is uniformly **Y** = 0, with the coexisting fluid in the bed at **X** = 0. The feed is supplied uniformly at **X** = 1, and if not interrupted it eventually brings any bed of finite length to a uniform saturation level of **Y** = 1. The full breakthrough curve therefore starts at concentration **X** = 0 and ends at **X** = 1. (In practice, the "breakthrough point" is the time of the first nonzero or first nonacceptable value for **X**.)

In order to employ a single "universal" set of algebraic solutions for the breakthrough curves, this dimensionless concentration range is fitted to the particular transition being considered, which may either involve or not involve the initial ("presaturation") state of the column or the entering feed. For concentrated feed entering a completely regenerated column, **x** = 1 for the feed corresponds to **X** = 1, and **x** = 0 for the bed to **X** = 0.

For **regeneration** of a fully saturated column by a solute-free fluid (in ion exchange, for instance, a solution not containing the feed ion now on the column), **x** = 0 corresponds to **X** = 1, and **x** = 1 to **X** = 0. For mixed feed and a partially presaturated column, one may, for example, have **x** = 0.8 at **X** = 1 and **x** = 0.2 at **X** = 0. Also, one may encounter the reverse combination, with a reciprocal separation factor: $\mathbf{R}^\dagger = 1/\mathbf{R}$, the analog of Eq. (16-16). (As earlier, † designates complete reversal of the step first considered.)

MATERIAL-BALANCE RELATIONS: THROUGHPUT RATIO

Unlike **Λ**, which is invariant in any given transition, the time-distance or volume-distance measure **T** can be viewed as the primary independent variable, to which the dependent concentration variables are referred. The breakthrough curve can be expressed nondimensionally either as **X** = fn (**T**) or **x** = fn (**T**) or, equivalently, as **T** = fn (**X**) or **T** = fn (**x**). The **T** for any given transition is measured as the ratio of total quantity of solute brought into the column in the amount of feed stream that has passed the point of interest, to the sorbent's theoretical capacity (upstream of that point) for containing solute. This theoretical capacity defines the **stoichiometric center** of every transition—the time or volume when the degree of incompletion of the transition within the fluid is equal to the degree of incom-

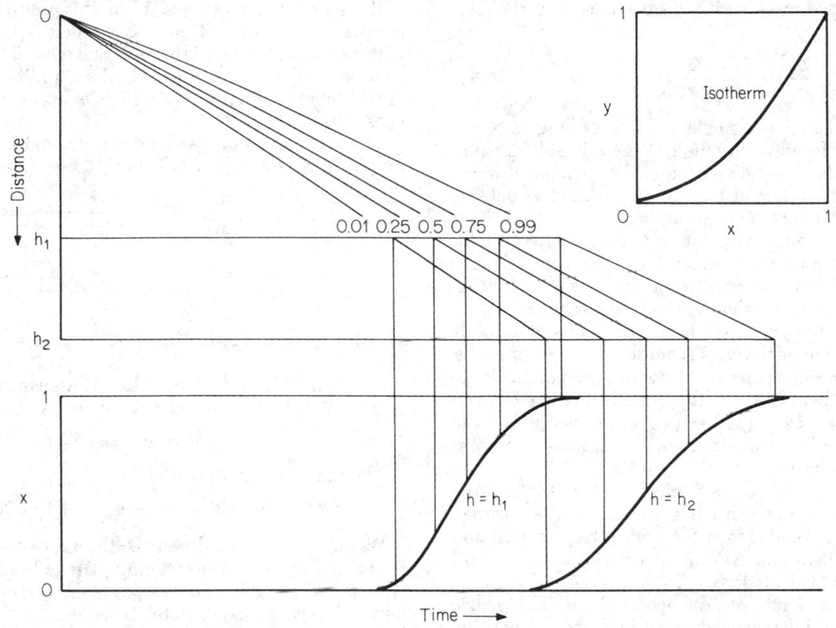

FIG. 16-7 Distance-time plot showing solution-concentration contours for unfavorable equilibrium. Radial extension of the contours represents proportionate-pattern behavior.

pletion within the sorbent, corresponding to the effective V_{stoic} for the transition, as given by the integral

$$\int_{\hat{V}=0}^{\mathbf{X}=1} (1 - \mathbf{X})d\hat{V} = \int_{\mathbf{X}=0}^{\mathbf{X}=1} \hat{V}\, d\mathbf{X} = (V_{stoic})_{effective} \qquad (16\text{-}43)$$

where $\hat{V} = V - \epsilon v$, the feed volume less the bed-filling volume to the point of interest.

For a full-range transition in which c runs from 0 to c_{ref} and q runs from 0 to q_{ref}, the fluid volume given by integrating Eq. (16-43) is the overall V_{stoic} for the system. Thus, $(V_{stoic})_{overall}$ is the fluid volume for which the amount of solute that has "leaked" past the reference point in the bed exactly equals the residual unfilled capacity of the sorbent contained before that point.

The system T can then be defined by any of three equivalent relations:

$$T = \frac{c_{ref}(V - v\epsilon)}{q_{ref}\rho_b v} = \frac{V - v\epsilon}{\Lambda v} = \frac{\hat{V}}{(V_{stoic})_{overall}} \qquad (16\text{-}44)$$

In like manner, the **T** for each transition is given by:

$$\mathbf{T} = \frac{\mathbf{V} - v\epsilon}{\mathbf{\Lambda} v} = \frac{\hat{V}}{(V_{stoic})_{effective}} = \frac{\mathbf{x}'' - \mathbf{x}'}{\mathbf{y}'' - \mathbf{y}'} T \qquad (16\text{-}45)$$

For constant separation factor **r**, Eq. (16-45) becomes

$$\mathbf{T} = [\mathbf{r} + (1 - \mathbf{r})\mathbf{x}''][\mathbf{r} + (1 - \mathbf{r})\mathbf{x}']\frac{T}{\mathbf{r}} \qquad (16\text{-}46)$$

T is the simplest generalized measure of volume, as it becomes 1.0 at the run time or solution volume corresponding to the total sorption capacity of the bed for the particular transition of interest. Equation (16-43) transforms to

$$\int_{\mathbf{T}=0}^{\mathbf{X}=1} (1 - \mathbf{X})\, d\mathbf{T} = \int_{\mathbf{X}=0}^{\mathbf{X}=1} \mathbf{T}\, d\mathbf{X} = 1 \qquad (16\text{-}47)$$

Figure 16-8 shows the positioning of the stoichiometric center, so that the area below the **X** curve to the left of **T** = 1 equals the area above the **X** curve to the right of **T** = 1. If T replaces **T** in Eq. 16-47, the integral is $\overline{T}$. (For later subsections, if **NT** replaces **T**, the integral becomes **N**.) Equations (16-43) and (16-47) are useful for determining $(V_{stoic})_{effective}$ from experiments or for determining the constant of integration (that is, the exact position of the breakthrough curve) in solutions of the differential rate equations.

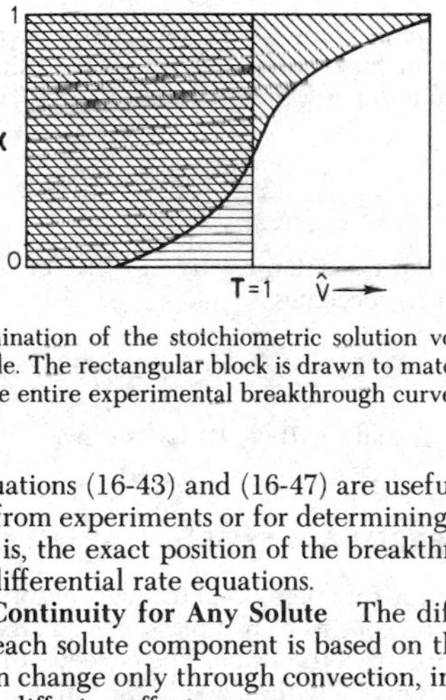

FIG. 16-8 Determination of the stoichiometric solution volume needed to normalize the **T** scale. The rectangular block is drawn to match the area above (or to the left of) the entire experimental breakthrough curve.

Equation of Continuity for Any Solute The differential material balance for each solute component is based on the concept that concentration can change only through convection, interphase transfer, or dispersion-diffusion effects:

$$\frac{\partial c}{\partial v} + \rho_b \frac{\partial q}{\partial V} + \epsilon \frac{\partial c}{\partial V} - \frac{D_{eff} S^2 \epsilon}{F} \frac{\partial^2 c}{\partial v^2} = 0 \qquad (16\text{-}48)$$

Here distance has been replaced by variable sorbent volume v and time by solution volume V. With the dispersion term neglected, and V replaced by $\hat{V}$ as defined following Eq. (16-43), the partial differential equation has the form

$$-\left(\frac{\partial c}{\partial v}\right)_{(V - v\epsilon)} = \rho_b \left[\frac{\partial q}{\partial(V - v\epsilon)}\right]_v \qquad (16\text{-}49)$$

With nondimensional terms (except for v), the equation is

$$-\left(\frac{\partial \mathbf{x}}{\partial v}\right)_{Tv} = \left(\frac{\partial \mathbf{y}}{\partial Tv}\right)_v \qquad (16\text{-}50)$$

For mass-transfer-controlled cases, with **N** proportional to v, the relation becomes

$$-\left(\frac{\partial \mathbf{X}}{\partial \mathbf{N}}\right)_{\mathbf{NT}} = \left(\frac{\partial \mathbf{Y}}{\partial \mathbf{NT}}\right)_{\mathbf{N}} \quad (16\text{-}51)$$

The term Tv or **NT** is treated as a single variable as long as **N** is a variable (in solving the differential equation). When the solved equation is evaluated at fixed **N**, **T** usually becomes the variable of interest. The two nondimensional rates of Eq. (16-51) are each equal to a nondimensional driving force for mass transport.

Composition Velocity If a steady state is reached between solution and sorbent compositions, the coexisting compositions advance through the bed together at a uniform rate, such as is shown by Fig. 16-7. A condition of constant composition velocity is reached as a limiting case with moderately unfavorable equilibrium and as a different limiting case with moderately favorable equilibrium. This condition is known as **coherence** and corresponds mathematically to making Eq. (16-49) a separate total differential equation in each case. For local equilibrium, $U(= 1/T)$ is used as a penetration ratio or relative-velocity coordinate to provide a nondimensional representation of the bed's concentration profile.

The external, visible behavior of a bed, characterized by T, is the object of both measurement and prediction. However, the internal, causative behavior must be understood for both interpretation and design; especially for multivariant systems, a plot of y versus U will often facilitate the needed calculations.

Transition Velocity The rate of advance of the stoichiometric center of a transition may be expressed as an average composition velocity:

$$\overline{U} = \frac{1}{\overline{T}} = \frac{q_{ref}\rho_b v}{c_{ref}(V - v\epsilon)} \quad (16\text{-}52)$$

where v is the variable volume of bed where the transition center lies and V is the volume of solution as before. If the (dimensional) velocity of the transition dv/dt is designated as u_0 and dV/dt is identified with F, then

$$u_0 = \frac{\overline{U}}{\Lambda + \overline{U}\epsilon}\frac{F}{S} \approx \frac{\overline{U}F}{\Lambda S} \quad (16\text{-}53)$$

The velocity of the transition relative to that of the moving fluid, often designated R_F, becomes

$$R_F = \frac{u_0\epsilon}{F} = \frac{1}{1 + \Lambda/\overline{U}\epsilon} \quad (16\text{-}54)$$

Hence, for large Λ and for $\overline{U} = 1$, $R_F \approx \epsilon/\Lambda$.

LOCAL-EQUILIBRIUM ANALYSIS

For all isotherms, a complete breakthrough history for a two-component column run consists of two plateaus and an intermediate single-part or multiple-part transition. The concentration path followed in the transition is given by the slope of the equilibrium curve in the unfavorable case or by the slope of line connecting two equilibrium points (often 0,0 and 1,1) in the favorable case.

Gradual-Transition Behavior The equilibrium theory provides a classical explanation of chromatography, and was first applied successfully by DeVault [*J. Am. Chem. Soc.*, **65,** 532 (1943)]. By a transformation of partial derivatives, Eq. (16-50) leads to

$$\left(\frac{\partial y}{\partial x}\right)_v = \left(\frac{\partial Tv}{\partial v}\right)_x = T_x + v\left(\frac{\partial T}{\partial v}\right)_x \quad (16\text{-}55)$$

The local-equilibrium condition now converts $\partial y/\partial x$ to a total derivative; that is, to a function of x or y alone. The right-hand term also must be a function of x or y only and not of v. The result is

$$\frac{dy^\circ}{dx} = T_x \qquad \frac{d\mathbf{Y}^\circ}{d\mathbf{X}} = \mathbf{T_X} \quad (16\text{-}56)$$

If the second derivative $d^2\mathbf{Y}^\circ/d\mathbf{X}^2$ is positive (conforming to *unfavorable equilibrium*), **T** increases when **X** increases, in accordance with experimental breakthrough behavior. Therefore, with an unfavorable equilibrium (**R** > 1), the equilibrium theory predicts a "real" breakthrough curve which is approached in actual column runs.

For the case of **constant separation factor r**, first treated by Walter [*J. Chem. Phys.*, **13,** 229 (1945)], Eqs. (16-14) and (16-56) lead to

$$\frac{dy^\circ}{dx} = \frac{r}{[(1 - r)x + r]^2} = T \quad (16\text{-}57)$$

or

$$x = \frac{r - \sqrt{r/T}}{r - 1} \quad (16\text{-}58)$$

The limits of validity of Eq. (16-58) are x = 0 at $T = 1/r$; x = 1 at $T = r$.

For an *unfavorable* **Freundlich isotherm** (Table 16-5) where $y = x^\beta$ with $\beta > 1$, the proportionate-pattern result is

$$T = dy^\circ/dx = \beta x^{\beta-1} \quad (16\text{-}59)$$

The **trinomial-isotherm** yields

$$T = dy^\circ/dx = (1 - a_1 - a_2) + 2a_1x + 3a_2x^2 \quad (16\text{-}60)$$

When $a < b$ in a **bilinear isotherm,** two abrupt steps result at $T = a$ and $T = b$; the intervening plateau has the concentration $X = (1 - b)/(a - b)$. In this composite curve, the separation of the two steps exhibits "proportionate" behavior.

Estimation of Isotherms from a Breakthrough Curve Mass-transfer effects will usually cause a spreading of the breakthrough curve beyond that given by local-equilibrium theory. However, for markedly unfavorable equilibrium in "deep" columns, the proportionate-pattern behavior may dominate the breakthrough; then the entire isotherm can be deduced by integrating the breakthrough curve [Glueckauf, *J. Chem. Soc.*, **1949,** 3280]:

$$y^\circ = \int T\,dx \quad (16\text{-}61)$$

Discontinuity (Shock) for Favorable Isotherms If the second derivative is negative (conforming to favorable equilibrium), the algebraic result corresponds to a curve of **X** decreasing as **T** increases. Because **X** = 0 and **X** = 1 are also allowable (are even preferred) values, this curve of reverse slope represents an unattainable or "fictitious" result and gives rise to the theoretical **abrupt transition.** For this step, mathematically

$$\frac{\Delta y}{\Delta x} \equiv \frac{(y'')^\circ - (y')^\circ}{x'' - x'} = T \quad (16\text{-}62)$$

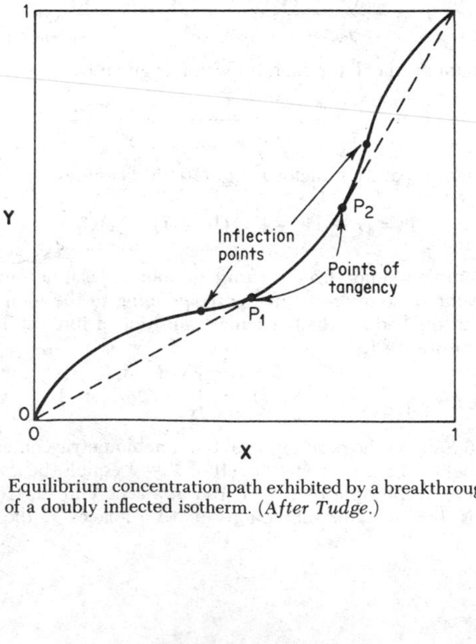

FIG. 16-9 Equilibrium concentration path exhibited by a breakthrough curve in the case of a doubly inflected isotherm. (*After Tudge.*)

In simple cases x and y (or **X** and **Y**) each jump from 0 to 1 at the T (or **T**) value of *unity* as required by material balance.

Composite Isotherms and Transitions If the isotherm for a given transition contains one or more inflection points, $\mathbf{Y}^\circ > \mathbf{X}$ throughout will give a wholly abrupt transition; but if $\mathbf{Y}^\circ < \mathbf{X}$ in any region whatever, the transition will exhibit a proportional pattern which may include abrupt segments. (F. Barnés, private communication.) For an inflected **Y** versus **X** isotherm, partly favorable and partly unfavorable, the concentration path always lies *either on or below* the isotherm. Where a shift occurs between an unfavorable and a favorable region, the region of increasing $d\mathbf{Y}^\circ/d\mathbf{X}$ almost never reaches its inflection point but instead terminates in a tangent line so drawn that it always increases as **X** rises. [Roberts, Ph.D. thesis in chemical engineering, Georgia Institute of Technology, 1951; Tudge, *Can. J. Phys.*, **39,** 1611 (1961); and Golden, Ph.D. thesis in chemical engineering, University of California, 1973.]

Figure 16-9 illustrates the concentration path for a doubly inflected isotherm; the breakthrough curve matching this path will be initially abrupt at a **T** value less than 1, up to the **X** value of point P_1; then gradual, increasing along the isotherm to point P_2; then again abrupt to $\mathbf{X} = 1$, at a **T** value more than 1. Inside the column, the fluid drops its solute level down to the P_2 level, then encounters low sorption capacity due to unfavorable equilibrium; upon reaching P_1, it can drop its remaining solute abruptly. If feed and presaturant were interchanged, for the same isotherm the start and end of the breakthrough curve would be gradual, with an abrupt transition near $\mathbf{T} = 1$.

RATE-LIMITED CONSTANT-PATTERN TRANSITIONS

BREAKTHROUGH BEHAVIOR FOR FAVORABLE EQUILIBRIUM

Local-equilibrium theory would predict, in contrast to reality, that each transition governed by a favorable (convex-upward) isotherm should be vertical at the stoichiometric point ($\mathbf{T} = 1$). In fact, each such transition is "self-sharpening" to a certain extent. If a broader, smaller-slope zone is introduced, it will be sharpened to a constant shape in the time-distance plane; but if an initially abrupt zone front is introduced, it will broaden until the same constant shape is reached. The longer the column, the smaller the fraction of unused bed capacity is for any given breakthrough level; the breakthrough is truly sharpening in this respect.

Equilibrium is an important aspect of favorable-equilibrium behavior. For a given set of mass-transfer coefficients, a more favorable equilibrium (e.g., smaller **R**) gives steeper breakthrough, and the limiting slope is reached in a shorter height of bed. Figure 16-10 shows a typical evolution of fluid-phase composition velocities from the start of a run. When the composition contours have become parallel, **Y** and **X** values have become equal at each point (thus establishing *coherence*), and the projected breakthrough curves appear identical in form.

MASS-TRANSFER MECHANISMS: RATES AND TRANSFER UNITS

Mass transfer and related phenomena become the predominant influence on the shape of composition profiles and effluent concentration histories in all cases except those governed by local equilibrium. It is regrettable that no one complete model exists (except a quite approximate one, based on an analogy to chemical-reaction kinetics); thus, care must be exercised to select the most appropriate calculation method from among those available. A calculation which predicts the **X** value at a given **T** to within 0.01 (i.e., to within experimental accuracy) is generally considered adequate. Figure 16-11 applies this criterion to show the minimum **N** values (number of transfer units) at which proportionate pattern or constant pattern is valid at each **R**

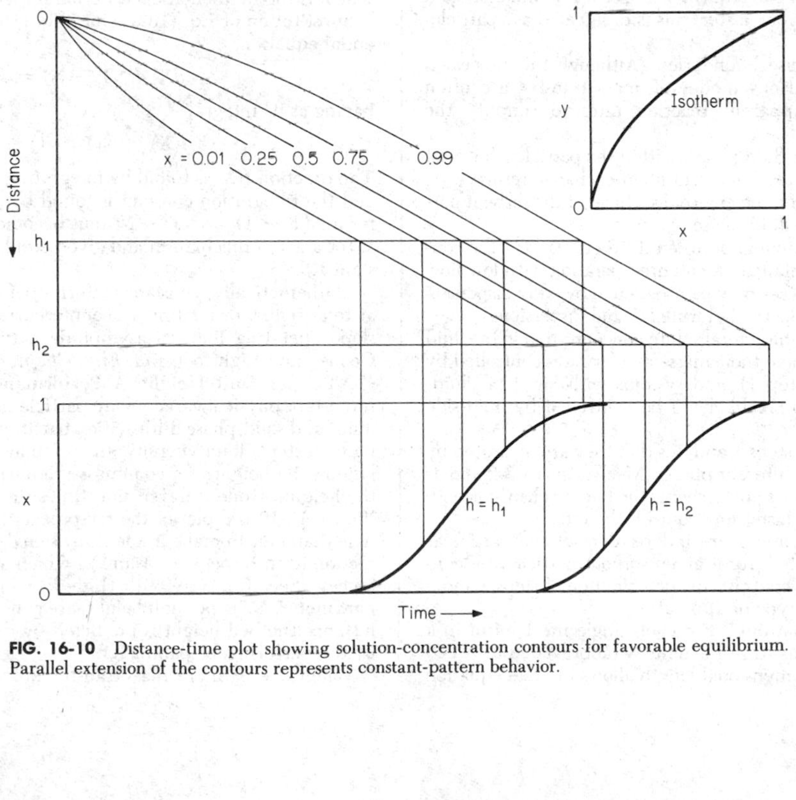

FIG. 16-10 Distance-time plot showing solution-concentration contours for favorable equilibrium. Parallel extension of the contours represents constant-pattern behavior.

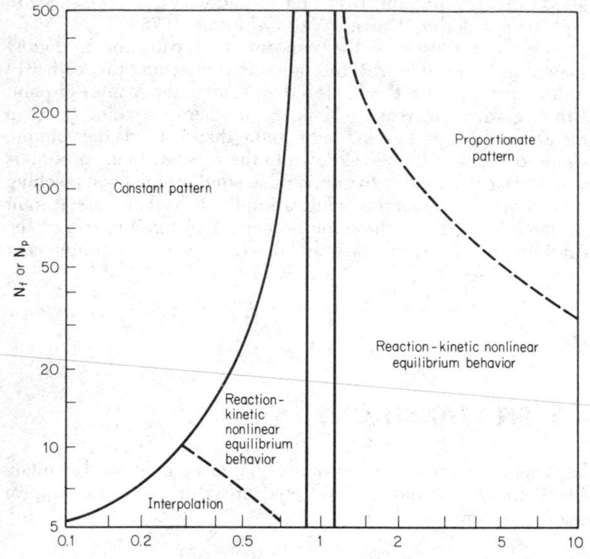

FIG. 16-11 Regions of **R** and $\mathbf{N}_f$ or $\mathbf{N}_p$ in which constant- or proportionate-pattern treatment will apply (mostly within 1 percent of full scale), with alternative treatments shown for other regions.

and thus to show where linear equilibrium, reaction kinetics, or pre-constant-pattern models should instead be used.

Mechanisms Figure 16-1 is of use to indicate the physical nature and location of individual steps in the transport mechanism; steps 1 to 4 are identified by numerals in the figure. Each step will involve a different driving force and will give rise to a somewhat different form of mathematical result; the mathematics of the controlling step is generally used, with adjustment for any other contributory resistances. The individual steps, listed in an order that applies for a desorbing solute, are:

1. Diffusion in the sorbed state (in a uniform liquidlike or solid phase or a pore-surface layer). Later, this is designated as "particle-phase diffusion."
2. Reaction at the phase boundaries. (Although this process is usually very fast, the various modes of mass transfer are often expressed in terms of apparent reaction rates to simplify the mathematics.)
3. Pore diffusion in the fluid phase, within the particles (for most adsorbents, inorganic zeolites, and certain ion-exchange resins).
4. Mass transfer between the external surfaces of the sorbent particles and the surrounding fluid phase.
5. Mixing, or lack of mixing, between different parts of the contacting equipment. For instance, in column operation with low flow rates, the breakthrough curves may be broadened by eddy dispersion or molecular diffusion, collectively termed "axial dispersion."

In general, systems with high total solute concentration in the fluid phase are more likely to have their mass-transfer rates controlled by particle-phase diffusion (step 1), and systems with low total fluid-phase concentration are more likely to be controlled by fluid-side effects (step 3, 4, or 5).

The difference between steps 1 and 3 is that they are separated by step 2 and hence occur in different phases. Mechanisms 4-3-2 and 4-2-1 each occur in series. For some sorbents the two mechanisms both occur, in parallel, and the faster one controls the rate.

The rates of steps 1, 2, and 3 are independent of the particular process arrangement selected, for a given sorbent particle in a constant driving potential. By contrast, the contribution of steps 4 and 5 depends greatly upon the type of apparatus.

Form of the Rate Equations For each single mechanism in a fixed bed, a number of mass-transfer units N (abbreviated as NTU) can be defined. This nondimensional length allows the rate equation to be evaluated by use of a "universal" solution. The counterpart variable **NT**, with reference to Eq. (16-51), is a nondimensional time. Rates for steps 3 and 4, discussed earlier, involve a fluid resistance; for step 1, a solid resistance. The partition ratio $\mathbf{\Lambda}$ (= V_{stoic}/v) enters differently for fluid and for solid resistances. When the rate of change of particle concentration is being measured, a particle-phase rate applies directly, but a fluid-phase rate must be divided by $\mathbf{\Lambda}$.

The rate, in terms of a general coefficient κ_p or κ_f, is

$$d\mathbf{Y}/dt = \kappa_p \,\text{fn}\,(\mathbf{Y}, \mathbf{X}) \quad \text{or} \quad (\kappa_f/\mathbf{\Lambda})\,\text{fn}\,(\mathbf{Y}, \mathbf{X}) \tag{16-63}$$

Here fn (**Y**, **X**) is an applicable nondimensional driving force, often a linear function, which for the solid phase is approximate and subject to empirical correction.

Nondimensional time **NT** is defined equal to the product $\kappa\hat{t}$, where κ is either κ_p or $(\kappa_f/\mathbf{\Lambda})$. At $\mathbf{T} = 1$, $\hat{t} = V_{stoic}/F$, and $\mathbf{N} = \mathbf{NT}$. Therefore,

$$\mathbf{N} = \kappa V_{stoic}/F = \kappa\mathbf{\Lambda}v/F \tag{16-64}$$

Since in all cases **N** and v are related by constants, the material balance of Eq. (16-51) takes the entirely nondimensional form

$$-\left(\frac{\partial\mathbf{X}}{\partial\mathbf{N}}\right)_{\mathbf{NT}} = \left(\frac{\partial\mathbf{Y}}{\partial\mathbf{NT}}\right)_{\mathbf{N}} = \text{fn}\,(\mathbf{Y}, \mathbf{X}) \tag{16-65}$$

where **N** should carry the appropriate subscript designation (p, R, pore, f, or d for steps 1 to 5 respectively).

Constant Pattern For favorable equilibrium, $\mathbf{Y}^\circ$ exceeds **X**, but in reality **Y** can never reach $\mathbf{Y}^\circ$. Material balance makes it impossible for the fluid to supply more solute to the solid than the fluid carries; that is, **Y** cannot exceed **X**. But one can prove that **X** and **Y** do approach and effectively reach each other.

Many breakthrough curves are observed to retain a constant shape as they advance through a column. This means that every part of the curve is moving through the column at the same velocity; for this case, $d\mathbf{T}v/dv$ or $d\mathbf{NT}/d\mathbf{N}$ is constant for all **X**. Then, by Eq. (16-50) or Eq. (16-65), $d\mathbf{Y} = d\mathbf{X}$. Combining this with the limiting behavior that $\mathbf{Y} = 0$ when $\mathbf{X} = 0$ and $\mathbf{Y} = 1$ when $\mathbf{X} = 1$, it follows that

$$\mathbf{Y} = \mathbf{X} \tag{16-66}$$

which is the mathematical criterion for a constant pattern.

Substitution of Eq. (16-66) into Eq. (16-65) gives one total differential equation,

$$d\mathbf{X}/d(\mathbf{NT} - \mathbf{N}) = \text{fn}\,(\mathbf{X}) \tag{16-67}$$

having as its integral

$$f(\mathbf{X}) = \mathbf{N}(\mathbf{T} - 1) + \text{const} \tag{16-68}$$

The function $f(\mathbf{X})$ is found by integrating $d\mathbf{X}/\text{fn}\,(\mathbf{X})$ between limits, and the integration constant is found by applying Eq. (16-47). The term $\mathbf{N}(\mathbf{T} - 1)$, or $\mathbf{NT} - \mathbf{N}$, thus becomes an invariant function of **X** for a given mechanism and given equilibrium behavior (e.g., constant **R**).

Mathematically, constant-pattern operation resembles distillation at total reflux, or continuous countercurrent absorption with a 45°-slope operating line. Its asymptotic nature has been discussed by Cooney and Lightfoot, *Ind. Eng. Chem. Fundam.*, **4**, 233 (1965).

NTU per Unit Height: A Performance Parameter The most important physical variables are particle size and structure, flow rate, fluid- and solid-phase diffusivities, partition ratio, and (to almost negligible extent) fluid viscosity. In customary separation-process calculations, the height of a column is often calculated as the product of the height of one transfer unit, times the number of transfer units. Figure 16-12 is a plot of the reciprocal of HTU, multiplied by particle diameter to make it nondimensional and divided also by a correction term **b** (between 1 and 2) which applies to addition of resistances [See Eq. (16-92)]. Higher values of this **performance parameter,** NTU per unit height, are generally preferred; and in this instance the bed height is predicted by dividing the requisite NTU by the performance parameter.

When one or more mass-transfer processes control the rate, the

NTU per unit height can be expressed as a function of the Péclet number for flow (Reynolds times Schmidt) $d_p(F/S)/D_f$, and a ratio of the controlling diffusivity to the fluid-phase diffusivity. If fluid-side spreading effects control, the effective **N** is given by the uppermost curve (dotted for gas, solid for liquid). If particle-side diffusivities control, the effective **N** is given by a point below the upper envelope on the appropriate diffusional contour (through the ψ's, the contour value depends indirectly on the separation factor **R**). If pore and particle-phase diffusion occur in parallel, **N** is the sum of the separate values for the two mechanisms. Near the intersections of the diffusional contours with the envelope, **N**/**b** is the reciprocal of the sum of reciprocals for the fluid-side and particle-side **N**'s.

Assumptions inherent in these relations include homogeneous particle size and structure (uniform distribution and chemical nature of sites), uniform initial sorbent concentration in the particle, constant feed concentration, uniform flow rate, and temperature constant enough so as to maintain both the equilibrium and the rate coefficient for sorption constant.

Driving Forces, Rates, and Breakthrough Curves Results for several individual mechanisms (or for the corresponding overall mechanisms) will now be considered. In the interpretation of experimental data, each relation will yield a value of **N**. For the prediction of breakthrough curves, each relation requires an input value of **N**, obtained for the proposed bed geometry and flow rate by utilizing either Fig. 16-12 or an estimate based on more direct experimental information.

The equations that follow refer to **local conditions** within the contracting equipment, such as may apply to the average concentrations in the neighborhood of a single sorbent particle; in certain cases they are positioned in still greater detail. They apply specifically to monovariant sorption and to spherical or near-spherical particles of essentially uniform size. (Irregularly shaped particles are treated as "spheres.") To reduce the algebra, they are expressed in dimensionless variables **Y**, **X**, and **R**.

The pathway to solve constant-pattern cases is as follows:

1. Identify the driving force: fluid-side, particle-side, or combined.
2. Use the available equilibrium data to convert the driving force into an equation in a single variable or a set of numbers for use in numerical integration.
3. Complete the integration to obtain the equation or the numerical data for the breakthrough curve.
4. When necessary, apply the integration of Fig. 16-8 so as to center the curve at **T** = 1.

Table 16-7 embodies these steps as carried out for three types of favorable isotherm (constant-separation factor, Freundlich, and bilinear; see Table 16-5) with four types of driving force (fluid-side mass transfer, particle diffusion, pore diffusion, reaction kinetics; the last for CSF only). The first three steps correspond to Eqs. (16-64), (16-66), and (16-67), as indicated in the table. Step 4 has not been carried out for the bilinear case.

For graphical-numerical calculations on a **Y** versus **X** plot, with a linear driving force (LDF), the latter is a horizontal line for fluid-side resistance, a line of slope N_f/N_p for combined resistance, or a vertical for particle-side resistance. The operating line **Y** = **X** is the diagonal drawn between points (0,0) and (1,1).

In interpreting experimental data, the question arises as to which mass-transfer mechanism should be applied. Fortunately, the **N** values calculated by the different forms of Eq. (16-68) are all comparatively close, generally within a factor of 2. Whatever form is chosen, the **N** calculated from **X** versus **T** behavior can be positioned on Fig. 16-12, with the use of approximate values of the physical variables in consistent units: bed height (mm), particle diameter (mm), superficial velocity (mm/s), and fluid diffusivity [near 10^{-3} mm^2/s (4 × 10^{-5} ft^2/h) for liquids, 1 mm^2/s (4 × $10^{-2}/h$) for gases]. Where the point lies in the diagram will suggest which mechanism applies. If the point lies in the internal-diffusion region but does not show the expected effect of changed particle size, then crystallite diffusion, reaction kinetics, or channeling may be the explanation. While predictions of rate have important theoretical uses, it is essential to use experimentally observed rate information in the design of large-scale systems.

The calculation methods given here are based on invariant values for the governing diffusion coefficient and equilibrium constant. When averaged values do not provide a desired degree of precision, it is possible to piece together several partial-range solutions, each calculated with the best constants for its range.

FLUID-SIDE MASS TRANSFER

Transport to or from the Particle Interface Movement of solute between the bulk of the fluid phase and the outer surfaces of the sorbent granules is evidently governed by the molecular or ionic diffusivity and also, in turbulent flow, by the eddy diffusivity which controls the effective thickness of the boundary layer.

Here the rate of material transfer for the solute of interest is given by the classical relation involving the driving potential between the bulk fluid and the granule surface:

$$\frac{d\mathbf{Y}}{dt} = \frac{k_f a_p}{\Lambda}(\mathbf{X} - \mathbf{X}^\circ) \qquad (16\text{-}69)$$

where k_f is the fluid-phase mass-transfer coefficient, $\mathbf{X}^\circ$ is the dimensionless fluid-phase concentration in equilibrium with the outer surface of the solid, and the other variables are as defined previously.

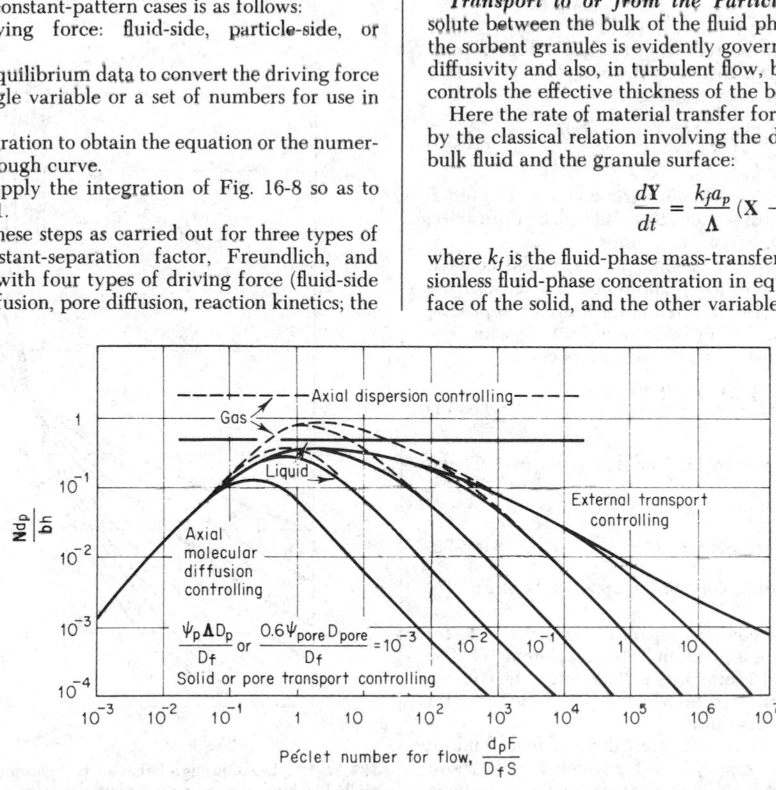

FIG. 16-12 Effects of Péclet group, partition ratio, and diffusivity ratio on performance parameter (particle diameter divided by height of a transfer unit).

TABLE 16-7 Rate Equations for Single Mechanisms, with Constant-Pattern Integrals

	Resistance	1. Fluid side	2. Particle	3. Pore	4. Reaction kinetics
	A. κ (Eq. 16-64)	$3.6\ k_f/d_p\Lambda = 2.62(D_pF/S)^{0.5}/d_p{}^{1.5}$	$\psi_p 60 D_p/d_p^2$	$\psi_{pore} 60 D_{pore}/d_p^2$	$2\kappa/(1 + \mathbf{R})$
Eq. (16-67)	*B.* Linear driving force (LDF)	$\mathbf{X} - \mathbf{X}°$	$\mathbf{Y}° - \mathbf{Y}$	$\mathbf{Y}° - \mathbf{Y}$	
	C. Same, for constant **R**	$\dfrac{\mathbf{X}(1 - \mathbf{Y}) - \mathbf{RY}(1 - \mathbf{X})}{1 + (\mathbf{R} - 1)\mathbf{Y}}$	$\dfrac{\mathbf{X}(1 - \mathbf{Y}) - \mathbf{RY}(1 - \mathbf{X})}{\mathbf{R} + (1 - \mathbf{R})\mathbf{X}}$	Same as for "Particle"	$\mathbf{X}(1 - \mathbf{Y}) - \mathbf{RY}(1 - \mathbf{X})$
	D. Correction factor for LDF, ψ	....	$0.894/[1 - 0.106\mathbf{R}^{0.5}]$	$0.775/[1 - 0.225\mathbf{R}^{0.5}]$	
	E. Alternate DF	....	$\dfrac{\mathbf{Y}°^2 - \mathbf{Y}^2}{2\mathbf{Y}}$	$\dfrac{\mathbf{Y}° - \mathbf{Y}}{[1 + (\mathbf{R} - 1)\mathbf{Y}]^{0.5}}$	
	F. Correction factor for alternate DF, ψ	....	$0.590/[1 - 0.410\mathbf{R}^{0.5}]$	$0.548/[1 - 0.452\mathbf{R}^{0.5}]$	
Eq. (16-68)	*G.* Constant pattern (**NT** − **N**) for LDF, with constant **R**	$\dfrac{\ln \mathbf{X} - \mathbf{R} \ln (1 - \mathbf{X}) + 1}{1 - \mathbf{R}}$	$\dfrac{\mathbf{R} \ln \mathbf{X} - \ln (1 - \mathbf{X}) - 1}{1 - \mathbf{R}}$	Same as for "Particle"	$\dfrac{\ln \mathbf{X} - \ln (1 - \mathbf{X})}{1 - \mathbf{R}}$
	H. Same, Freundlich ($m = 1.65 - 0.15\beta$)	$[\beta/(1 - \beta)] \ln [\mathbf{x}^{1/\beta}/(\mathbf{x} - \mathbf{x}^{1/\beta})] + 1 - \beta[(1 - \beta)^{-m} - 1]$	$[1/(1 - \beta)] \ln [1/(1 - \mathbf{x}^{1-\beta})] - (1 - \beta)^{-m}$	Same as for "Particle"	
	I. Same, bilinear below junction **X**; junction $\mathbf{X} = (1 - b)(a - b)$	$[\ln \mathbf{X}]a/(a - 1) + \text{const}$	$[\ln \mathbf{X}]/(a - 1) + \text{const}$	Same as for "Particle"	
	J. Above junction **X**	$[-\ln (1 - \mathbf{X})]b/(1 - b) + \text{const}$	$[-\ln (1 - \mathbf{X})]/(1 - b) + \text{const}$	Same as for "Particle"	

REFERENCES:
1*A*. Beaton and Furnas, *Ind. Eng. Chem.*, **33**, 1500 (1941); Hiester et al., *Am. Inst. Chem. Eng. J.*, **2**, 404 (1956).
1*BCG*. Michaels, *Ind. Eng. Chem.*, **44**, 1922 (1952).
1*IJ*, 2*IJ*. Tien and Thodos, *Am. Inst. Chem. Eng. J.*, **6**, 364 (1960).
2*ABCG*, 3*AB*. Glueckauf and Coates, *J. Chem. Soc.*, **1947**, 1315; *Trans. Faraday Soc.*, **51**, 1540 (1955).
2*D*, 3*D*. Hall et al., *Ind. Eng. Chem. Fundam.*, **5**, 212 (1966).
2*E*. Vermeulen, *Ind. Eng. Chem.*, **45**, 1664 (1953).
2*H*, 3*H*. Present authors.
3*EF*. Vermeulen and Quilici, *Ind. Eng. Chem. Fundam.*, **9**, 179 (1970).
4*A*. Hiester and Vermeulen, *Chem. Eng. Prog.*, **48**, 505 (1952).
4*CG*. Sillén and Ekedahl, *Arkiv Kemi Mineral. Geol.*, **A22**(15, 16), (1646); Bohart and Adams, *J. Am. Chem. Soc.*, **42**, 523 (1920).

For **packed-bed** operations the following relation, adapted from Wilke and Hougen [*Trans. Am. Inst. Chem. Eng.*, **41**, 445 (1945)], can be used for evaluating the mass-transfer coefficient:

$$k_f a_p = \frac{10.9F(1 - \epsilon)}{d_p S} \left[\frac{D_f}{d_p(F/S)}\right]^{0.51} \left(\frac{D_f \rho_f}{\mu}\right)^{0.16} \quad (16\text{-}70)$$

where F/S is superficial velocity or volumetric flow rate of fluid F per unit area S of contactor cross section; D_f is fluid-phase diffusivity; ρ is density; and μ is viscosity, in consistent units.

Equation (16-70), developed for *laminar-flow* gas-solid contact, appears to give slightly low results for liquid-solid contact. For the latter case, the numerical values are fitted empirically by replacing the right-hand (Schmidt group) term by a numerical constant. For aqueous solutions, with $\epsilon = 0.40$, Eq. (16-70) then becomes

$$k_f a_p = \frac{2.62(D_f F/S)^{0.5}}{d_p^{1.5}} \quad (16\text{-}71)$$

This result is similar to one given by Hiester et al. [*Am. Inst. Chem. Eng. J.*, **2**, 404 (1956)].

With $\mathbf{N}_f = k_f a_p hS/F$, it follows that

$$(\mathbf{N}_f d_p/h) = 2.62(d_p F/D_f S)^{-0.5} \quad (16\text{-}72)$$

which corresponds to the limiting slope in the upper right of Fig. 16-12.

Integrated Results The constant-separation-factor result, for constant-pattern breakthrough, is given in Table 16-7, item 1*G*. The equation is plotted in Fig. 16-13 for constant-**R** contours; the concentration coordinate is a probability scale, which expands the scales for concentrations near zero and near unity.

Pre-Constant-Pattern Behavior The first drop of feed fluid (at $\mathbf{N}_f\mathbf{T} = 0$) passing through the empty bed (for which $\mathbf{Y} = 0$) shows a first-order decline: $\mathbf{X} = \exp(-\mathbf{N}_f)$. The entrance layer of sorbent,

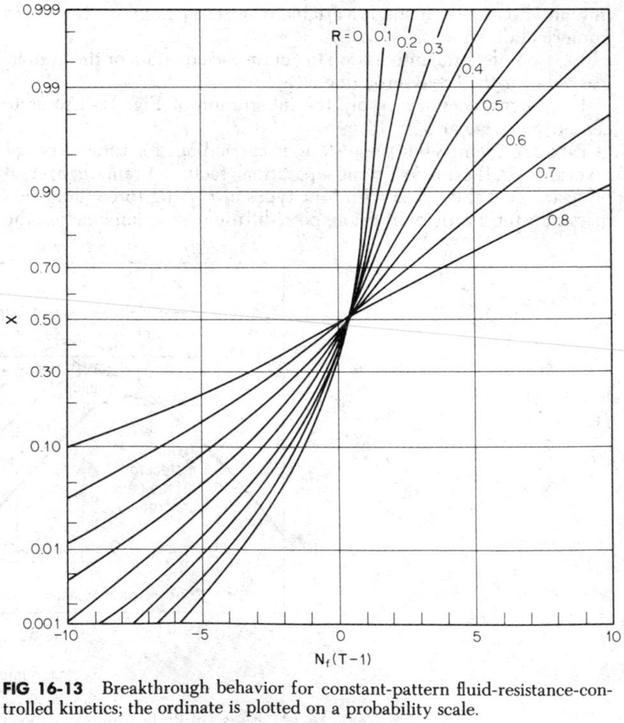

FIG 16-13 Breakthrough behavior for constant-pattern fluid-resistance-controlled kinetics; the ordinate is plotted on a probability scale.

continually exposed to fresh feed ($\mathbf{X} = 1$), has a concentration given by

$$\mathbf{N}_f\mathbf{T} = -\mathbf{R} \ln (1 - \mathbf{Y}) + (1 - \mathbf{R})\mathbf{Y} \qquad (16\text{-}73)$$

At $\mathbf{R} = 0$ and $\mathbf{N}_f = 0$, $\mathbf{N}_f\mathbf{T} = \mathbf{Y}$ between 0 and 1, and the $\mathbf{Y} = 1$ point then moves forward at its constant velocity. However, constant pattern is not considered fully developed until the entire **X** and **Y** curves are equal and constant, to within 0.01. This occurs when $\mathbf{N}_f$ reaches 4.6, and the **X** at $\mathbf{N}_f\mathbf{T} = 0$ has declined to 0.01.

For higher values of **R**, a greater length of bed is needed to attain constant pattern; the limiting NTU ($= \mathbf{N}_{\mathrm{lim}}$) required to reach that condition for any of the mechanisms can be taken as $4.6/(1 - \mathbf{R})^{2.2}$. The pre-constant-pattern curve for **X** may be estimated to have the same shape as is calculated by Eq. (16-68) for the given **R**, but with the $\mathbf{N}(\mathbf{T} - 1)$ scale multiplied by a "scale factor" of $(\mathbf{N}/\mathbf{N}_{\mathrm{lim}})^{-0.32}$. The results of numerical calculations for the approach to constant pattern at $\mathbf{R} = 0.2$, plotted in the **N** versus **NT** plane, are shown in Fig. 16-14.

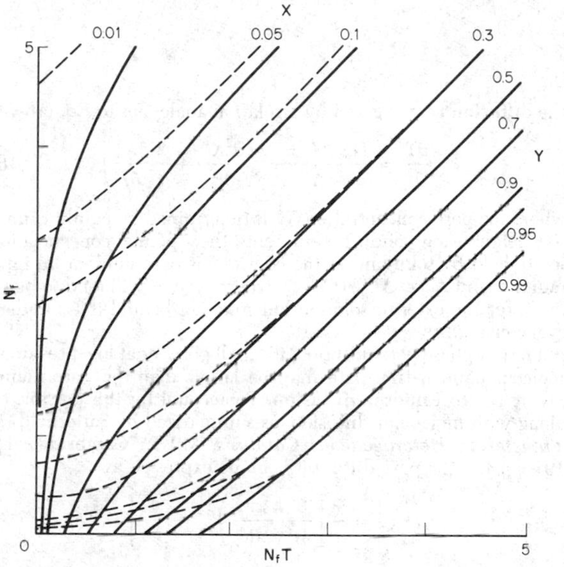

FIG. 16-14 Evolution of **X** (dashed curves) and **Y** (solid curves) from boundary conditions toward constant pattern (linear and parallel, with $\mathbf{X} = \mathbf{Y}$); fluid resistance controlling, $\mathbf{R} = 0.2$.

PARTICLE DIFFUSION

Comparison of Particle and Pore Diffusion Two different modes of intraparticle transport must be distinguished owing to their substantially different magnitudes, mechanisms, and consequent breakthrough shapes. One mode, "pore" diffusion, occurs within the fluid phase inside the particle, in cases of a heterogeneous particle structure enclosing pore space that the external fluid can enter; the pore space then exhibits a solid-fluid partition ratio that is nearly the same as that between the entire particle and the external fluid. In these cases it appears as if pore diffusion occurs before phase change and particle diffusion afterward. In fact, a solute molecule transported mainly by pore diffusion may attach to the sorbent and detach many times along its path; the point is that pore diffusion occurs when the molecule is detached.

Five instances of "particle" diffusion will illustrate its nature:

1. Movement of mobile adsorbed solute molecules along pore surfaces, without detaching
2. Transport in a homogeneously dissolved state, as for a neutral molecule inside a sorbent gel or in a pore-filling fluid that is immiscible with the external fluid
3. Similarly, ion transport in resinous ion-exchanger gel
4. Advance of a solute molecule from one "cage" to another within a zeolitic aluminosilicate
5. In the leaching of plant materials, transport inside cells and through cell walls but not transport through the space between cells

Diffusivity in the particle phase is usually less than in the pores; however, the particle concentration is usually much larger, so it is not at all certain that particle transport will control. If the two modes occur in parallel, the faster one prevails; in series, the slower. More complicated intermediate cases may give rise to breakthrough curves with earlier midheights and longer tails than are normal for a given midheight slope. Particle diffusion can be inferred to control if the apparent diffusivity is either much larger or much smaller than the estimated rate for pore diffusion.

Molecular-sieve zeolites are often a complicated case. The granules, of millimeter to centimeter size, are usually composed of micrometer-size crystallites held together by a bonding agent laced with macropores. The crystallite diffusivity for sorbed gas can range from 10^{-5} to 10^{-10} mm²/s (4×10^{-7} to 4×10^{-12} ft²/h); with microcrystals 1 μm in diameter, the time constant (reciprocal of κ) for crystallite transport (Table 16-7, Item 2*A*) will range from 0.0015 to 150 s. For pore diffusion, with Λ on the order of 50, the time constant may be 2.5 s. The rate of crystallite transport is affected by changes in crystallite size but not in overall particle size.

Diffusivity Numerical data are sparse and disperse, so that locally conducted experiments and interpretation must be used to a great extent. Variability of the diffusivity with composition in a given system has already been mentioned.

In ion-exchange resins, as a qualitative rule, ionic mobility compared with that in water is 1:10 for monovalent ions, 1:100 for divalent ions, and 1:1000 for trivalent ions, showing that resins act much like a liquid phase. However, the rule is oversimplified; great variation is found between homovalent species and at different cross-linkages. Table 16-8 shows typical particle-phase diffusivities in cation and anion resins; for mixtures of unlike ions (the usual case) the diffusivities will be intermediate.

Integration of Rate Equations With spherical symmetry, the transport equation for particle diffusion has the form

$$\frac{\partial Y}{\partial t} = D_p \left(\frac{\partial^2 Y}{\partial r^2} + \frac{2}{r} \frac{\partial Y}{\partial r} \right) \qquad (16\text{-}74)$$

Numerical calculations have been carried out by Hall et al. (see "General References") to evaluate Eq. (16-74) in the constant-pattern case for a family of constant-**R** contours. The integration was carried out between $\mathbf{X} = 0.01$ and $\mathbf{X} = 0.99$ and extrapolated graphically to 0.001 and 0.999 as shown in Fig. 16-15. The calculations, made for concentric shells, were begun by setting **X** (at 0.01) and **Y** in equilibrium with it at the surface. An initial distribution in the particle was developed to make the volume-average **Y** equal to the exterior **X**. Equation (16-74) in finite-difference form was then applied for a time interval estimated to increase **Y** by about 0.005, and the actual rise for this time was determined. For each following step, **X** was set at the new **Y** value augmented by half of the increase in **Y** during the preceding step, and the outer-surface **Y** was correspondingly reset. The curves, when complete, were each centered by use of Eq. (16-47). Thus, only Eq. (16-74) and material balances were needed.

The linear-driving-force result (Table 16-7, Item 2*G*) only deviates significantly in the vicinity of $\mathbf{R} = 0$. The difference can be viewed graphically by inverting Fig. 16-13 and comparing it with Fig. 16-15; items 1*G* and 2*G* have the same algebraic form, $(\mathbf{N}_p\mathbf{T} - \mathbf{N}_p)$ being of opposite sign and $(1 - \mathbf{X})$ replacing **X**. With linear driving force, the boundary conditions are: at $\mathbf{N}_p = 0$, $\mathbf{Y} = 1 - \exp(-\mathbf{N}_p\mathbf{T})$; and along the fluid front ($\mathbf{N}_p\mathbf{T} = 0$):

$$\mathbf{N}_p = -\mathbf{R} \ln \mathbf{X} + (1 - \mathbf{R})(1 - \mathbf{X}) \qquad (16\text{-}75)$$

TABLE 16-8 Self-Diffusion Rates in Polystyrene-Based Ion Exchangers*

Units of 10^{-5} mm^2/s

Temperature	0.3°C					25.0°C				
Cross-linking, %	4	6	8	12	16	4	6	8	12	16°
Cation exchanger (sulfonate): Dowex 50										
Na^+	6.7		3.4	1.15	0.66	14.1		9.44		2.40
Cs^+			6.6		1.11			13.7		3.10
Ag^+			2.62		1.00			6.42		2.75
Zn^{2+}			0.21		0.03			0.63	0.29	0.14
La^{3+}	0.30		0.03		0.002	0.69		0.092		0.005
Anion exchanger (dimethyl hydroxyethyl benzyl ammonium): Dowex 2										
Cl^-		1.25					3.54			
Br^-	(1.8)	1.50	0.63		0.06	(4.3)	3.87	2.04		0.26
I^-		0.35					1.33			
BrO_3^-		1.76					4.55			
WO_4^{2-}		0.60					1.80			
PO_4^{3-}		0.16					0.57			

*Boyd and Soldano, *J. Am. Chem. Soc.*, **75**, 6091, 6099 (1953). To convert square millimeters per second to square feet per hour, multiply by 2.875×10^{-2}; °F = $\frac{9}{5}$ °C + 32.

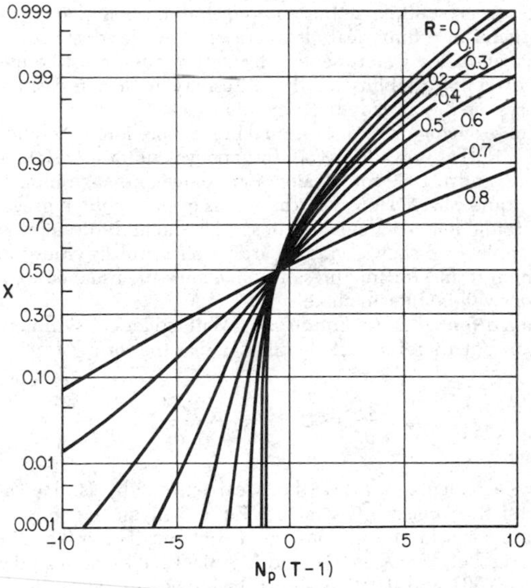

FIG. 16-15 Breakthrough behavior for constant-pattern particle-diffusion kinetics; ordinate on probability scale.

A more exact analytic result can be obtained by using the quadratic driving force (Table 16-7, item 2*E*). For irreversible equilibrium, with a numerical constant of 0.61 instead of 1.0, it is nearly equivalent to the exact solution given by Wicke [*Kolloid* Z., **167,** 289 (1939)], based on Eq. (16-74):

$$\mathbf{X} = 1 - \frac{6}{\pi^2}\sum_{n=1}^{\infty}\frac{1}{n^2}\exp\{-n^2[\frac{\pi^2}{15}\mathbf{N}_p(\mathbf{T}-1)+0.64]\} \qquad (16\text{-}76)$$

This equation was initially derived for linear-equilibrium batch uptake from an infinite reservoir.

PORE DIFFUSION

When fluid-phase transport through a network of fluid-filled pores inside the particles provides access for solute to the adsorption sites, the **diffusion rate** is given by a relation analogous to Eq. (16-74):

$$\frac{\partial Y}{\partial t} = \frac{D_{pore}(1-\epsilon)}{\mathbf{\Lambda}'}\left(\frac{\partial^2 X^\circ}{\partial r^2} + \frac{2}{r}\frac{\partial X^\circ}{\partial r}\right) \qquad (16\text{-}77)$$

where the pore concentration X° is determined by equilibrium with the neighboring solid; Y represents the "point" concentration of solute held by solid and in the contiguous pore fluid, at an internal radius r; and $\mathbf{\Lambda}' = \mathbf{\Lambda}C_0/(C_0)_{pore}$, where $(C_0)_{pore}$ will tend to be equal to C_0 for the exterior solution but may lag behind if C_0 undergoes cyclic changes.

For gas-phase pore diffusion, in small pores or at low pressures, the molecular mean free path may be larger than the pore diameter, giving rise to **Knudsen diffusion.** To account for this possible effect along with molecular diffusion, as summarized by Satterfield (*Mass Transfer in Heterogeneous Catalysis,* M.I.T., Cambridge, Mass., 1970, p.43), the pore diffusivity can be expressed as

$$D_{pore} = \frac{\chi}{\tau}\left[\frac{3}{4\bar{r}}\left(\frac{\pi M}{2R\theta_A}\right)^{1/2} + \frac{1}{D_f}\right]^{-1} \qquad (16\text{-}78)$$

if the pore fluid is a gas. For a liquid, $D_{pore} = D_f\chi/\bar{\tau}$. Here χ is the internal porosity of the particles; $\bar{r}$ is the average pore radius; τ is the tortuosity (usually between 2 and 6). The gas constant R_g, absolute temperature T_K, and molar weight M must be expressed in consistent units.

The cases solved to date, numerically or analytically, are for large $\mathbf{\Lambda}$; that is, the amount of solute held in the pores is assumed to be small compared with the amount adsorbed by solid, even though solute is transported only by pore fluid. As an approximation, pore-diffusion problems may be solved by particle-diffusion methods (see Table 16-7, items 3*GHIJ*). A better analytic approximation to numerical results is given by item 3*EF*.

A numerical solution for constant **R**, analogous to one given previously for particle diffusion, is reported by Hall et al. (see "General References") and shown in Fig. 16-16. The pore-diffusion curves, combining fluid and solid influences, are more symmetric. Relatively, pore diffusion is slower at low **X** values (especially at low **R**) because the solid impounds the solute and impedes its movement; and faster at high **X**, because the fluid retains access to sites still unfilled.

Irreversible Equilibrium For **R** = 0, a shrinking-core model applies, in a method due to Acrivos; the solute inventory is

$$\mathbf{Y} = \mathbf{X} = (r_p^3 - r_i^3)/r_p^3 \qquad (16\text{-}79)$$

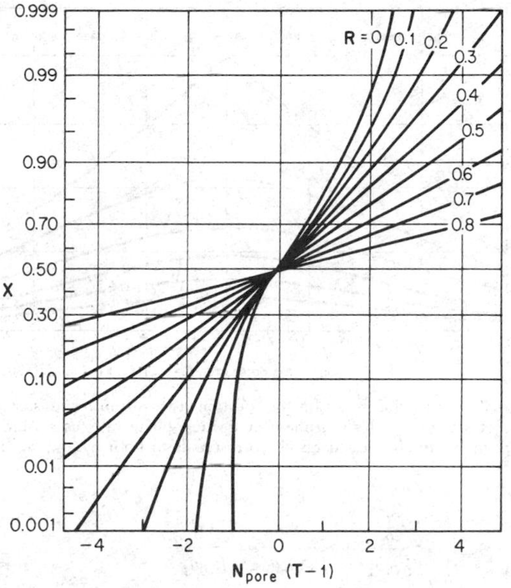

FIG. 16-16 Breakthrough behavior for constant-pattern pore-diffusion kinetics; ordinate on probability scale.

Here, r_i is the radius of the interior spherical shell to which the incoming solute has advanced. The exact equation is

$$\mathbf{N}_{\text{pore}}(\mathbf{T} - 1) + \text{const} = (1/\sqrt{3})[\arctan\{(2r_i + r_p)/r_p\sqrt{3}\}] - 0.5 \ln\{1 + (r_i/r_p) + (r_i/r_p)^2\} \quad (16\text{-}80)$$

Converted to an explicit result for **X**, this becomes:

$$\mathbf{X} = 0.5398\{\mathbf{N}_{\text{pore}}(\mathbf{T} - 1) + 1.235\} - 0.07285\{\mathbf{N}_{\text{pore}}(\mathbf{T} - 1) + 1.235\}^2 \quad (16\text{-}81)$$

The **X** curve leaves zero at $\mathbf{T} = -1.235/\mathbf{N}_{\text{pore}}$ and reaches unity tangentially at $\mathbf{T} = +2.470/\mathbf{N}_{\text{pore}}$.

Example 3 A bed of completely regenerated adsorbent, with $S = 0.093$ m^2 (1.0 ft^2) and $h = 6.1$ m (20 ft), is used to remove acetone vapors initially present at 1.0 mole percent in an air stream at 333 K (60°C) and 1.01 bar (1.00 atm) flowing at 56.0 L/s (120 ft^3/min). Based on this feed level, $\mathbf{r}$ (= $\mathbf{R}$) = 0.5, and Λ (= $\mathbf{\Lambda}$) = 6000. The maximum allowable discharge concentration is 0.030 mole percent. The adsorbent particles are nearly spherical, with an effective d_p = 3.6 mm (0.14 in). Diffusivities are 1.0 mm^2/s for the fluid phase and 0.22 mm^2/s for the particle-pore structure.

1. Estimate $\mathbf{N}_p$. The abscissa in Fig. 16-12, all terms being evaluated in millimeter units, is

$$\frac{d_pF}{D_fS} = \frac{3.6 \times 56.0 \times 10^6}{1.0 \times 9.3 \times 10^4} = 2170$$

For $\mathbf{R} = 0.5$, $\psi_{\text{pore}} = 0.922$, from item 3*D*, Table 16-7. The $\psi_{\text{pore}}D_{\text{pore}}/D_f$ contour is then 0.20, and $\mathbf{N}d_p/\mathbf{b}h$ in Fig. 16-12 is 0.0075. Because ζ is small, $\mathbf{b}_p$ is near unity. (See Fig. 16-17.) Hence,

$$\mathbf{N}_{Op} = 0.0075h/d_p = 0.0075 \times 20 \times 12/0.14 = 12.8$$

This value applies for the ($\mathbf{Y}^\circ - \mathbf{Y}$) driving force. For the particle-integration method, $\mathbf{N}_{Op} = 12.8/0.922 = 13.9$.

2. Solve for time and volume at $\mathbf{T} = 1$. $V_{\text{stoic}} = \Lambda v = 120{,}000$ ft^3; $t \approx \hat{t} = 120{,}000/120 = 1000$ min or 16.7 h.

3. Solve for breakthrough **T**, *t*, and *V* by item 3*G*, Table 16-7. Constant-pattern calculations will be illustrated here, although they are not strictly valid below $N_p = 25$ at $\mathbf{R} = 0.5$. If constant pattern has not been reached, the **T** predicted for breakthrough will be higher than the true value.

$$12.8(\mathbf{T} - 1) = \ln 0.03 - 2 \ln 0.97 - 2.00 = -3.51 + 0.06 - 2.00 = -5.45$$

$$\mathbf{T}_{0.03} = 0.574 \qquad t = 574 \text{ min} \qquad V = 68{,}900 \text{ ft}^3$$

4. Solve for breakthrough **T**, *t* and *V* by Fig. 16-16.

$$13.9(\mathbf{T} - 1) = 5.07 \qquad \mathbf{T}_{0.03} = 0.635 \qquad t = 635 \text{ min} \qquad V = 76{,}000 \text{ ft}^3$$

(In principle, these values are more exact. In practice, the graphs cannot be read to the number of digits shown, and the required value was obtained from a table in the Hall paper.)

REACTION-KINETIC MODEL; COMBINED RESISTANCES

Reaction-Kinetic Treatment The actual adsorption or exchange reaction occurring at pore surfaces or other solid-fluid interfaces usually is extremely rapid. However, in the case of chemisorption in porous catalysts, the reaction step becomes rate-determining [cf. Masamune and Smith, *Am. Inst. Chem. Eng. J.*, **10,** 246 (1964); **11,** 41 (1965)]. In other instances, when diffusion steps control the rate, it may still be convenient to describe a system by its *apparent* second-order kinetic behavior, since this usually provides a good approximation to a more complex exact form. (Compare items 1*C*, 2*C*, and 4*C* in Table 16-7.)

For adsorption of a single component obeying a Langmuir isotherm, the rate (with κ' the rate constant for the adsorption reaction) is

$$\frac{dq}{dt} = \kappa' \left[c(Q - q) - \frac{1}{K_L} q \right] \quad (16\text{-}82)$$

From Eq. (16-15) and the Langmuir equilibrium (Table 16-5), $Q = q_{\text{ref}}/(1 - \mathbf{r})$. This, with $K_Lc_{\text{ref}} = (1 - \mathbf{r})/\mathbf{r}$, and $\hat{\kappa} = \kappa'c_{\text{ref}}/(1 - \mathbf{r}) = \kappa'/(K_L\mathbf{r})$, gives

$$dy/dt = \hat{\kappa}[\mathbf{x}(1 - \mathbf{y}) - \mathbf{r}\mathbf{y}(1 - \mathbf{x})] \quad (16\text{-}83)$$

For **ion exchange** between two ionic species of the same valence or for **exchange adsorption,** the reaction-kinetic relation is

$$dq/dt = \kappa''[c(Q - q) - \mathbf{r}q(C_0 - c)] \quad (16\text{-}84)$$

With $\hat{\kappa} = \kappa''C_0$, this relation also reduces to Eq. (16-83).

For partial-range transitions, Eq. (16-83) can be rewritten with **x**, **y**, and **r** replaced by **X**, **Y**, and **R**, as shown in Table 16-7, Item 4*C*. The integral constant-pattern form is given as item 4*G*. The rate form is best proved by assuming the result and working backward. The coefficient is then

$$\kappa_{\text{m.t.}} = \frac{\hat{\kappa}\mathbf{r}}{\mathbf{r} + (1 - \mathbf{r})\mathbf{x}'} \quad (16\text{-}85)$$

as shown by Vermeulen and Hiester [*J. Chem. Phys.*, **22,** 96 (1954)].

When this relation is developed from linear-driving-force mass-transfer expressions as mentioned previously, κ for particle-side diffusion is

$$\kappa_{\text{m.t.}} = \frac{\psi_pk_pa_p}{\mathbf{R} + (1 - \mathbf{R})\mathbf{X}} \quad (16\text{-}86)$$

and for fluid-side diffusion is

$$\kappa_{\text{m.t.}} = \frac{k_fa_p}{\mathbf{\Lambda}[1 + (\mathbf{R} - 1)\mathbf{Y}]} \quad (16\text{-}87)$$

In each case an average value is obtained for the denominator by determining **X** or **Y** at an intermediate condition (e.g., 0.5, where both denominators become $(1 + \mathbf{R})/2$.)

Combined Resistances

Particle side. When particle-phase and pore diffusion operate in parallel, the effective *rate* is the sum of these two rates. With the combined particle-side coefficient k_{Cp} replacing k_p in the linear-driving-force relation (Table 16-7, item 2*B*) k_{Cp} is given by

$$k_{Cp} = k_p + \frac{\psi_{\text{pore}}}{\psi_p\mathbf{\Lambda}} k_{\text{pore}} \quad (16\text{-}88)$$

If particle-phase diffusion predominates and greater accuracy appears possible through use of the quadratic driving force, k_{Cp} as just given can also replace k_p in item 2*E* of Table 16-7. If pore diffusion predominates, indicating use of item 3*B* or item 3*E*, with an adjusted coefficient $k_{C,\text{pore}}$, the latter is given by

$$k_{C,\text{pore}} = k_{\text{pore}} + \frac{\psi_p \Lambda}{\psi_{\text{pore}}} k_p \tag{16-89}$$

At $\mathbf{R} = 1$, the ratio of ψ's becomes unity.

In a bimodal particle, with uniformly sized crystallites dispersed in a permeable matrix, the rate may be estimated from the series combination of pore diffusion and intracrystalline diffusion. Approximately, then, the effective *resistance* is the sum of these two resistances:

$$\frac{\mathbf{b}_p}{k_{Cp}} = \frac{\psi_p \Lambda}{\psi_{\text{pore}} k_{\text{pore}}} + \frac{1}{(k_p)_{\text{crystallite}}} \tag{16-90}$$

where $\mathbf{b}_p$ is defined subsequently.

Fluid side. External mass transfer and axial dispersion operate in series. However, axial dispersion has a mathematical effect somewhat analogous to that of particle diffusion; it does not add linearly to any other resistance, except at $\mathbf{R} = 1$. Solutions for this case are discussed under "Axial Dispersion and Channeling." When a specific value of combined fluid-side resistance is needed, the midheight slope can be determined from the treatment available there and used to find the equivalent boundary-layer resistance.

Combined NTUs can be calculated ($\mathbf{N}_{Cp}$, $\mathbf{N}_{C,\text{pore}}$, $\mathbf{N}_{Cf}$), proportional to the combined k's. In what follows, pore diffusion will be included with particle diffusion.

Overall Resistances With a near-linear isotherm ($0.7 < \mathbf{R} < 1.5$) the solutions for all mechanisms are nearly identical. Direct addition of the two combined resistance k_{Cf}^{-1} and $(\Lambda k_{Cp})^{-1}$ (together with the reaction resistance a_p/κ if it is appreciable) will give the reciprocal of the overall mass-transfer coefficient or reaction-rate coefficient. With a highly favorable isotherm ($\mathbf{R} \rightarrow 0$), the rate at each point is controlled by the resistance that locally is greater, and there is no addition whatever. For intermediate favorable equilibrium, in the constant-pattern range, several methods are accessible:

1. Graphical construction on the $\mathbf{Y}$ versus $\mathbf{X}$ diagram, as for gas absorption, to follow the change in relative influence of the particle-side and fluid-side resistances.
2. Numerical solutions, much as described for a particle-phase single resistance. (Work of Fleck et al. is cited later as an example.)
3. Additive solutions with a correction factor $\mathbf{b}$ between 1 and 2 that divides the summed resistance or multiplies the k or NTU; an approximate but rapid method. This should be used with the overall-rate equation that corresponds to the predominant resistance, or with the reaction-kinetic methods.

The unfavorable-equilibrium region has not been charted in this respect; however, there are both theoretical and practical reasons to adopt the evaluation given by the reverse case, that is, the correction term found for the reciprocal (i.e., favorable) $\mathbf{R}$.

The correction terms ($\mathbf{b}$'s) have been derived from analysis of the interfacial concentrations $\mathbf{X}_I$ and $\mathbf{Y}_I$ (which are taken to be in mutual equilibrium) and are plotted in Fig. 16-17 as functions of a mechanism parameter ζ:

$$\zeta = \frac{\mathbf{X} - \mathbf{X}_I}{\mathbf{Y}_I - \mathbf{Y}} = \frac{\psi_p \Lambda k_{Cp}}{k_{Cf}} = \frac{\mathbf{N}_{Cp}}{\mathbf{N}_{Cf}} \tag{16-91}$$

where subscript I indicates interfacial concentration and normally implies equilibrium between $\mathbf{Y}_I$ and $\mathbf{X}_I$. As $\mathbf{X}_I \rightarrow \mathbf{X}$, $\zeta \rightarrow 0$, and the particle-side resistance controls; as $\mathbf{X}_I \rightarrow \mathbf{X}^\circ$, $\zeta \rightarrow \infty$, and the fluid-side resistance governs.

The $\mathbf{b}$'s are defined by the following relations, with subscript O indicating overall coefficients:

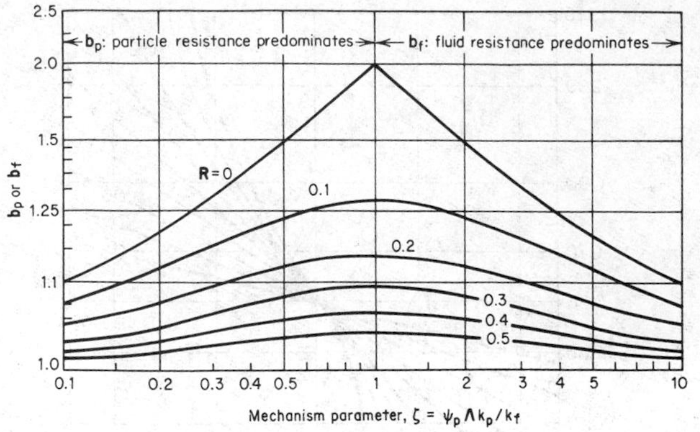

FIG. 16-17 Correction factor for the addition of mass-transfer resistances, relative to effective overall fluid-phase or particle-phase rate, as a function of mechanism parameter. Each curve corresponds to both $\mathbf{b}_f$ and $\mathbf{b}_p$ over its entire range.

$$\frac{\Lambda}{k_{Cf} a_p} + \frac{1}{\psi_p k_{Cp} a_p} = \frac{\Lambda \mathbf{b}_f}{k_{Of} a_p} = \frac{\mathbf{b}_p}{\psi_p k_{Op} a_p} = \frac{1}{\kappa_{\text{m.t.}}} \tag{16-92}$$

It is seen that $\mathbf{b}$ is a correction to the driving force; the addition of resistances shown here is based on linear driving forces. In two-resistance regions, the smoothed $\mathbf{N}$'s of Fig. 16-12 do not include an allowance for $\mathbf{b}$ and thus should be replaced by calculated addition of resistances if $\mathbf{R}$ is less than 0.6 or more than 1.7. In Eq. (16-92) and in Table 16-7 (Item 4*A*), $\kappa_{\text{m.t.}}$ is the contribution of mass-transfer effects to the overall reaction-kinetic coefficient κ_O; if κ_R (the true chemical-kinetic coefficient) is not infinite, the overall kinetic coefficient is

$$\kappa_O = [\{2\kappa_{\text{m.t.}}/(1 + \mathbf{R})\}^{-1} + \kappa_R^{-1}]^{-1} \tag{16-93}$$

For ζ between 0.5 and 2.0 at all $\mathbf{R}$, the reaction-kinetic model is probably closest to an exact constant-pattern solution. For $\mathbf{R}$ between 0.6 and 1.7, even for a single mass-transfer mechanism, the general nonlinear-equilibrium treatment derived for the kinetic model is likely to be more accurate than either the constant-pattern solution for the exact mechanism or the proportionate-pattern solution. Assuming equilibrium at the interface:

$$F\left(\frac{\partial \mathbf{X}}{\partial \mathbf{V}}\right)_v = \psi_p \Lambda k_{Cp} a_p (\mathbf{Y}_I - \mathbf{Y}) = k_{Cf} a_p (\mathbf{X} - \mathbf{X}_I) \tag{16-94}$$

Then, with $\mathbf{Y} = \mathbf{X}$, and with $\zeta = \mathbf{N}_{Cp}/\mathbf{N}_{Cf}$,

$$\zeta(\mathbf{Y}_I - \mathbf{X}) = (\mathbf{X} - \mathbf{X}_I) \tag{16-95}$$

For **irreversible equilibrium** ($\mathbf{R} = 0$ or $\beta = 0$), for external transport with particle diffusion, the rate is wholly dominated by the smaller of the two terms $k_{Cf}(\mathbf{X} - \mathbf{X}^\circ)$ or $\Lambda k_{Cp}(\mathbf{Y}^\circ - \mathbf{Y})$. Fluid-side behavior dominates at the start of a sorption process; but it increases in rate, so that particle-side behavior with a declining rate controls toward the end of the process. The shift from one mechanism to the other occurs at the point where the two rates are equal [Vermeulen, *Ind. Eng. Chem.*, **45**, 1664 (1953)]. The junction value of T is $1 + [(N_p - N_f)/N_p N_f]$.

For irreversible equilibrium, with external transport combined with pore diffusion, the result is a combination of Table 16-7, item 1*G*, and (Eq. 16-81) as obtained by Eagleton (Hall et al., loc. cit.):

$$\mathbf{N}_{\text{pore}}(\mathbf{T} - 1) = \frac{\mathbf{N}_{\text{pore}}}{\mathbf{N}_f}(\ln \mathbf{X} + 1) + 2.470 - 3.705(1 - \mathbf{X})^{1/2} \tag{16-96}$$

For **favorable equilibrium** ($0 < \mathbf{R} < 0.8$), numerical solutions based on exact intraparticle profiles, for particle or pore diffusion combined with external transport, have been reported by Fleck, Kirwin, and Hall, *Ind. Eng. Chem. Fundam.*, **12**, 95 (1973). For a *bilinear isotherm*, the junction occurs at $\mathbf{X}_I = (1 - b)/(a - b)$; $\mathbf{Y}_I = a(1 - b)/(a - b)$. Below the junction, $\mathbf{Y}_I = a\mathbf{X}_I$. From this and from Eq. (16-95), $\mathbf{X}_I = \mathbf{X}(1 + \zeta)/(1 + \zeta a)$; hence

$$\mathbf{X}_{\text{junc}} = \frac{1 + \zeta a}{1 + \zeta}\left(\frac{1 - b}{a - b}\right) \tag{16-97}$$

The solution below the junction has the form of item 1*I*, Table 16-7, but with $(1 + \zeta a)/\zeta$ replacing a. Above the junction, the solution has the form of item 1*J*, but with $(1 + \zeta b)/\zeta$ replacing b.

An overall NTU (either $\mathbf{N}_{Of}$ or $\mathbf{N}_{Op}$, depending on the controlling rate) or an overall NRU ($\mathbf{N}_R$, for combined resistances or at $\mathbf{R}$ values above 0.5) is defined. By analogy to the individual NTUs, one obtains

$$\mathbf{N}_{Of} = k_{Of}a_p \frac{v}{F} = k_{Of}a_p \frac{hS}{F} \tag{16-98}$$

$$\mathbf{N}_{Op} = \psi_p \Lambda k_{Op}a_p \frac{v}{F} \tag{16-99}$$

$$\mathbf{N}_R = \kappa_O \Lambda \frac{v}{F} \tag{16-100}$$

The N's defined are related by the b's:

$$\mathbf{N}_R = \frac{\mathbf{N}_{Op}}{\mathbf{b}_p} = \frac{\mathbf{N}_{Of}}{\mathbf{b}_f} \tag{16-101}$$

Hence the subscript-free quotient N/b will be used to designate all three of these ratios. The general addition of mass-transfer resistances, using b's, can be examined with respect to any one of them; say, $\mathbf{b}_p$. From Eq. (16-92),

$$\frac{\mathbf{b}_p}{k_{Op}a_p} = \frac{1}{k_{Cp}a_p}(1 + \zeta) \tag{16-102}$$

In the special case in which breakthrough behavior is controlled solely by particle-phase diffusion and external transport, Eqs. (16-102) and (16-92) give

$$\frac{\mathbf{b}_p}{k_{Op}a_p} = \frac{d_p^2}{60D_p}\left[1 + 22.9\psi_p\Lambda \frac{D_p}{D_f}\left(\frac{D_f}{d_pF/S}\right)^{1/2}\right] \tag{16-103}$$

The experimental data supporting this correlation have been plotted by Hiester, Radding, et al., *Am. Inst. Chem. Eng. J.*, **2**,404 (1956).

AXIAL DISPERSION AND CHANNELING

"Axial dispersion" as used here refers mainly to velocity distribution and void-space mixing on the scale of a single particle or of a single void space between particles, and it also includes axial molecular diffusion through the voids.

These effects, largely predictable, are shown at the top and left of Fig. 16-12. Axial diffusion, at the top, is characterized by a mixing length of one-half particle diameter for gases and two particle diameters for most liquids, as explained later. It is described empirically by an apparent conductance k_d:

$$k_d a_p = F/SL \tag{16-104}$$

where L is the axial "mixing length," given in turn by

$$L = \frac{d_p}{\mathbf{P}_d} + \frac{D_f}{2(F/S)} \tag{16-105}$$

where the packing Péclet number for axial dispersion $\mathbf{P}_d$ is $d_pF/(SE_0)$, with E_0 the superficial axial-dispersion coefficient. $\mathbf{P}_d$ is approximately 0.5 for liquids with laminar-flow behavior (when $d_p\rho_f F/S\mu$, the Reynolds number based on superficial velocity and particle diameter, is less than 20), 2.0 for liquids in turbulent flow (Reynolds number greater than 200), and 2.0 for gases in either flow regime [McHenry and Wilhelm, *Am. Inst. Chem. Eng. J.*, **3**, 83 (1957); Ebach and White, ibid., **4**, 161 (1958); Carberry and Bretton, ibid., **4**, 367 (1958); Miller and King, ibid., **12**, 767 (1966); and Hennico, Jacques, and Vermeulen, U.S. AEC Rep. UCRL-10696, 1963].

Because axial-diffusionlike effects provide a supplemental term in the differential material balance, constant-pattern axial dispersion departs from the condition of $\mathbf{Y} = \mathbf{X}$ which characterizes all other constant-pattern cases. The material balance given by Acrivos [*Chem. Eng. Sci.*, **15**, 1 (1960)] and by Quilici and Vermeulen (U.S. Off. Saline Water Res. Dev. Prog. Rep. 476, 1969) can be put in the form

$$\frac{\partial^2 \mathbf{X}}{\partial \mathbf{N}_d^2} = \frac{\partial \mathbf{X}}{\partial \mathbf{N}_d} + \frac{\partial \mathbf{Y}}{\partial \mathbf{N}_d\mathbf{T}} \tag{16-106}$$

where $\mathbf{N}_d = hF/(SE_0)$ and E_0 is the axial-dispersion coefficient. The constant-pattern condition is again that $(\partial \mathbf{N}_d\mathbf{T}/\partial \mathbf{N}_d)_\mathbf{X}$ is constant. Hence Eq. (16-106) becomes

$$\frac{\partial^2 \mathbf{X}}{\partial(\mathbf{N}_d\mathbf{T})^2} = \frac{\partial \mathbf{Y}}{\partial \mathbf{N}_d\mathbf{T}} - \frac{\partial \mathbf{X}}{\partial \mathbf{N}_d\mathbf{T}} \tag{16-107}$$

and

$$\frac{\partial \mathbf{X}}{\partial \mathbf{N}_d\mathbf{T}} = \mathbf{Y} - \mathbf{X} \tag{16-108}$$

No Other Resistance When equilibrium is maintained between solid and bulk fluid, Eq. (16-108) is seen to give the same constant-pattern X-curve behavior inside the bed as for particle diffusion controlling without axial dispersion (Table 16-7, item 2*G*), except that their precise location may differ. Thus, with $\mathbf{N}_d$ replacing $\mathbf{N}_p$, the axial-dispersion local-equilibrium constant-pattern solutions for X (but not for Y) are nearly the same as those for particle-diffusion controlling regardless of the type of favorable isotherm.

An end effect at the bed outlet, identified by Coppola and LeVan [*Chem. Eng. Sci.*, **36**, 967 (1981)] through numerical calculations for nearly irreversible cases, makes it necessary to alter the suggested calculation method. The effluent X curve is sharper than the X curve within the bed because the Y curve is always sharper than the X curve. In the irreversible case the Y curve, a vertical front, reaches the outlet at a time barely less than $\mathbf{T} = 1.00$; how much less depends on the partition ratio. At this instant the X curve, which had been moving coherently with the Y curve, suddenly detaches itself and leaves at the fluid velocity. For reversible equilibrium the effect persists, but to a lesser extent. Empirically it is proposed that

$$\text{fn}(\mathbf{X}) = \frac{\Lambda + \epsilon}{\Lambda \mathbf{R}^{0.5} + \epsilon}(\mathbf{NT} - \mathbf{N}) \tag{16-109}$$

where fn (X) has a form such as item 2*G*, *H*, or 1*J* of Table 16-7. For non-CSF isotherms, $\mathbf{R}$ is evaluated at the point where Y and $(1 - \mathbf{X})$ are equal, or (in the Freundlich case) as $\beta^{0.8}$. For mass-transfer resistances accompanying axial dispersion, the exponent on $\mathbf{R}$ becomes $0.5\eta/(1 + \eta)$, where $\eta = \mathbf{N}_{Of}/\mathbf{N}_d$ or $\mathbf{N}_{Op}/\mathbf{N}_d$, as in the following subsection. Above $\mathbf{R} = 0.3$, the correction factor may not be significant.

Irreversible Equilibrium ($\mathbf{R} = 0$ or $\beta = 0$) For **particle diffusion** accompanied by axial dispersion, a mechanism parameter η_p is defined

$$\eta_p = \mathbf{N}_p/\mathbf{N}_d \tag{16-110a}$$

Then

$$\partial \mathbf{Y}/\partial \mathbf{N}_d\mathbf{T} = \eta_p(\mathbf{Y}^\circ - \mathbf{Y}) \tag{16-110b}$$

From Eqs. (16-108) and (16-110*b*) it follows that

$$d\mathbf{Y}/d\mathbf{X} = \eta_p(\mathbf{Y}^\circ - \mathbf{Y})/(\mathbf{Y} - \mathbf{X}) \tag{16-111}$$

The solution for Y behavior with $\mathbf{Y}^\circ = 1$ is nearly the same as for pure particle diffusion at $\mathbf{R} = 0$ (Table 16-7, case 2*G*). X is related to Y and hence to T by the integral of Eq. (16-111) as obtained by Quilici (loc. cit.).

$$\mathbf{X} = \frac{\eta_p[1 - (1 - \mathbf{Y})^{1/\eta_p}] - \mathbf{Y}}{\eta_p - 1} \tag{16-112}$$

For irreversible equilibrium, with **external transport** accompanied by axial dispersion, an analytic solution has been developed by Acrivos. The mechanism parameter is $\eta_f = N_f/N_d$. Then

$$\partial Y/\partial N_d T = \eta_f(X - X^\circ) \tag{16-113}$$

$$dY/dX = \eta_f(X - X^\circ)/(Y - X) \tag{16-114}$$

For $R = 0$, X° is zero until Y has reached unity; this occurs when X reaches $2/[1 + (4\eta_f + 1)^{1/2}]$. Up to this junction point

$$\ln X = \lambda[N_d(T - 1) + C_5] \tag{16-115}$$

and beyond

$$\ln \frac{1}{1 - X} = \lambda[N_d(T - 1) + C_6] \tag{16-116}$$

with $\lambda = 0.5[(4\eta_f + 1)^{1/2} - 1]$. C_6 must be related to C_5 at the junction, and C_5 then determined by use of Eq. (16-43).

Favorable Equilibrium $(0 < R < 0.7)$ With axial dispersion accompanying some one actual or effective transport mechanism, Quilici has developed numerical tables of Y, X, and $N_d(T - 1)$ which permit interpolation for any value of R. Empirical relations for the Y versus X behavior, given in the following paragraphs, can be used to solve Eq. (16-108) by performing a rapid numeric or graphic integration of the reciprocal of the driving force $(Y - X)$ over increments of X.

In the case of **external transport,** X is calculated as a function of Y, through an intermediate function X':

$$X' = \frac{R^{1.40}Y}{1 - (1 - R^{1.40})Y} \tag{16-117}$$

Then

$$X = \frac{1}{\lambda + 1}(Y + \lambda X') \tag{16-118}$$

where $\lambda = 0.5[(4\eta_f + 1)^{1/2} - 1]$ as previously.

With other mechanisms controlling, either X or Y is calculated at assumed values of the other, using a nonequilibrium factor R'; for instance,

$$Y = \frac{X}{R' + (1 - R')X} \tag{16-119}$$

For **particle-phase diffusion,** $R' = R^{\eta'/(1+\eta')}$, where $\eta' = \eta_p/\psi_p$. For **pore diffusion,** $R' = (1 + 4R^{0.75}\eta_{\text{pore}}^{1.25})/(1 + 4\eta_{\text{pore}}^{1.25})$. For the **reaction-kinetic model,** $R' = (1 + 0.9R\eta_R)/(1 + 0.9\eta_R)$, where $\eta_R = N_d/N_R$ and N_R is calculated without including axial dispersion.

Channeling In contrast to axial dispersion, channeling is caused by longer-range variations in local fluid velocity. It is observed especially with nonspherical particles and in beds with broad particle-size distributions and density and orientation of the bed packing. Fingering, caused by density inversion resulting from higher temperature or lower concentration below, is discussed by Helfferich (see "General References"; p. 487). For a case of shallow-bed adsorption using activated carbon, LeVan and Vermeulen (*Am. Inst. Chem. Eng. Symp. Ser.*, 1984) have shown that a radial-dispersion analysis developed by G. I. Taylor [*Proc. R. Soc.*, **A225**, 473 (1954)] applies—that is, that the bed's corrective action reduces and controls the effective (or apparent) axial-dispersion mixing length.

Rinsing (fluid-front spreading) is a single-phase operation controlled by effective axial dispersion and follows an equation akin to those for linear-equilibrium breakthrough.

LINEAR-EQUILIBRIUM AND OTHER RATE-LIMITED TRANSITIONS

SQUARE-ROOT-BROADENING BREAKTHROUGH

The simplest equilibrium is $Y^\circ = X$ (with $X^\circ = Y$), given by $R = 1$ or $\alpha = 1$. It represents the boundary condition between favorable and unfavorable equilibrium; and the governing partial differential equations never converge, even at a very large NTU, to a total differential equation. The breakthrough curve that this equilibrium produces is more complicated algebraically than the preceding types but is also a "landmark" result.

For linear equilibrium, either reaction kinetics or any one of the mass-transfer driving-force models given previously, combined with Eq. (16-51), leads to two mutually independent differential equations which can be written together:

$$-\left(\frac{\partial X}{\partial N}\right)_{NT} = \left(\frac{\partial Y}{\partial NT}\right)_N = X - Y \tag{16-120}$$

The N is used without designation, since at $R = 1$ the results are the same for N_R, N_p, N_{pore}, N_f, N_{Op}, or N_{Of}. For the mass-transfer correlation plot (Fig. 16-12), regardless of mechanism, $b = 1$.

Integrated Equations Anzelius [*Z. Angew. Math. Mech.*, **6**, 291 (1926)] and Schumann [*J. Franklin Inst.*, **208**, 405 (1929)] integrated Eq. (16-120) for the analogous heat-transfer case with boundary conditions $X = 1$ at $N = 0$ and $Y = 0$ at $T = 0$, to obtain

$$X = J(N, NT) \tag{16-121}$$

$$Y = 1 - J(NT, N) \tag{16-122}$$

The driving force $(X - Y)$ is

$$e^{N+NT}(X - Y) = I_0(2N\sqrt{T}) = \sum_{i=0}^{\infty} \frac{(N^2T)^i}{(i!)^2} \tag{16-123}$$

where I_0 is the modified Bessel function of the first kind, and order zero. (The infinite series converges too slowly for practical use above $N^2T = 10$.)

At higher values an alternative series can be used, in which the first term is $[\exp(2NT^{0.5})]/[2(\pi N)^{0.5}T^{0.25}]$. The function J of two variables s and t, as first defined by Hiester and Vermeulen [*Chem. Eng. Prog.*, **48**, 505 (1952)], is given by

$$J(s, t) = 1 - \int_0^s e^{-t-\xi} I_0(2\sqrt{t\xi})\, d\xi \tag{16-124}$$

An approximation to Eq. (16-124), due to Onsager, was reported by Thomas [*Ann. N.Y. Acad. Sci*, **49**, 161 (1948)]:

$$J(s, t) = \frac{1}{2}\left[1 - \text{erf}(\sqrt{s} - \sqrt{t}) + \frac{e^{-(\sqrt{s}-\sqrt{t})^2}}{\sqrt{\pi}(\sqrt{t} + \sqrt[4]{st})}\right] \tag{16-125}$$

Additional terms in the original reference give higher accuracy. For any number z,

$$\text{erf}(\pm|z|) = \pm\frac{2}{\sqrt{\pi}}\int_0^z e^{-\zeta^2}\, d\zeta \tag{16-126}$$

See, for instance, Abramowitz and Stegun, *Handbook of Mathematical Functions*, GPO, Washington, 1970. The error function can be calculated to within 0.0005 by a formula modified from Hastings (*Approximations for Digital Computers*, Princeton, 1955):

$$\text{erf}(\pm|z|) = \pm[1 - (1 + 0.2784|z| + 0.2314|z|^2 + 0.0781|z|^4)^{-4}] \tag{16-127}$$

A study by Klinkenberg [*Ind. Eng. Chem.*, **40**, 1970 (1948); ibid., **46**, 285 (1954); *Chem. Eng. Sci.*, **11**, 260 (1960)] leads to the following approximation, accurate for $N > 2$:

$$X = J(N, NT) = \frac{1}{2}\left[1 - \text{erf}\left(\sqrt{N} - \sqrt{NT} + \frac{1}{4\sqrt[4]{N\,NT}}\right)\right] \tag{16-128}$$

The **Y** term contains the same right-hand term with opposite sign; thus, (**X** − **Y**) at **T** = 1 is given by $\pi^{-0.5}$ times the right-hand term.

A "working value" of **J** is given by the first two terms in the argument of the error function of Eq. (16-125) or (16-128). For *s* or **N** > 50, the working value of **X** or **J** is correct to within 0.01 for **X** < 0.3 and **X** > 0.7; the value at **X** = 0.5 is about 0.02 low at **N** = 50. An increase of *s* or **N** to 150 is needed before the error at **X** = 0.5 drops to 0.01.

The mathematical behavior of **J** is discussed by Goldstein [*Proc. R. Soc. (London),* **A219,** 151, 171 (1953).] A pocket-calculator program for evaluating **J** has been reported by Tan [*Chem. Eng.,* 158 (Oct. 24, 1977)].

A derivation for **particle-phase diffusion,** accompanied by fluid-side mass transfer, has been carried out by Rosen [*J. Chem. Phys.,* **18,** 1587 (1950); ibid., **20,** 387 (1952); *Ind. Eng. Chem.,* **46,** 1590 (1954)], with a limiting form at **N** > 50 essentially equal to Eqs. (16-125) and (16-128).

$$\mathbf{X} = \frac{1}{2}\left\{1 - \operatorname{erf}\left[\frac{1}{2}\sqrt{\mathbf{N}_d}(1 - \mathbf{T})\right]\right\} \qquad (16\text{-}129)$$

Axial dispersion at **R** = 1 gives results very similar in form to Eq. (16-129). [This again reflects mathematical similarity in the two mechanisms, as does Eq. (16-130).]

A more exact relation is given by Lapidus and Amundson, *J. Phys. Chem.,* **56,** 984 (1952). Unidirectional random walk, an alternative model for axial dispersion, gives the closely related result

$$\mathbf{X} = 1 - \mathbf{J}(\mathbf{N}_d\mathbf{T}, \mathbf{N}_d) + \text{const} \qquad (16\text{-}130)$$

(Einstein, Ph.D. dissertation, Eidgenössische Technische Hochschule, Zurich, 1937; and Hennico, Jacques, and Vermeulen, U.S. AEC Rep. UCRL-10696, 1963). At $\mathbf{N}_d\sqrt{\mathbf{T}} \geq 60$, this result again simplifies to Eq. (16-125).

Plot of the J Function Figure 16-18, on logarithmic-probability coordinates, shows the behavior of the **J** function. The concentration histories plotted against time on *linear* scales normally are S-shaped. The probability scale (proportional to the argument of the error function giving the coordinate value) minimizes the curvature of such plots and also makes it possible to plot accurately those values which are either very small or very near to unity. The logarithmic scale for **NT** makes it possible to compare experimental **X** versus time plots directly with the theoretical curves; this curve-fitting technique was utilized in analogous **heat-transfer** calculations by Furnas [*Trans. Am. Inst. Chem. Eng.,* **24,** 1942 (1930)] and in ion-exchange work by Beaton and Furnas [*Ind. Eng. Chem.,* **33,** 1500 (1941)].

The **J** function is also given by Hougen and Marshall [*Chem. Eng. Prog.,* **43,** 197 (1947)] on logarithmic coordinates, by Furnas (loc. cit.) on linear coordinates, and by Klinkenberg (loc. cit.) in nomographic form.

NONLINEAR-EQUILIBRIUM REACTION-KINETIC SOLUTION

The most general relation that has been developed for breakthrough behavior is that of Thomas [*J. Am. Chem. Soc.,* **66,** 1664 (1944)], which includes the separation factor **R** as an independent variable along with the number of reaction units $\mathbf{N}_R$ and the throughput ratio **T.** Equations (16-65) and (16-83) take the dimensionless form

$$-\left(\frac{\partial \mathbf{X}}{\partial \mathbf{N}_R}\right)_{\mathbf{N}_R\mathbf{T}} = \left(\frac{\partial \mathbf{Y}}{\partial \mathbf{N}_R\mathbf{T}}\right)_{\mathbf{N}_R} = \mathbf{X}(1 - \mathbf{Y}) - \mathbf{R}\mathbf{Y}(1 - \mathbf{X}) \qquad (16\text{-}131)$$

These relations have been integrated for the same boundaries as Eq. (16-120) to give

$$\mathbf{X} = \frac{\mathbf{J}(\mathbf{R}\mathbf{N}_R, \mathbf{N}_R\mathbf{T})}{\mathbf{J}(\mathbf{R}\mathbf{N}_R, \mathbf{N}_R\mathbf{T}) + [1 - \mathbf{J}(\mathbf{N}_R, \mathbf{R}\mathbf{N}_R\mathbf{T})]\exp[(\mathbf{R} - 1)\mathbf{N}_R(\mathbf{T} - 1)]} \qquad (16\text{-}132)$$

and

$$\mathbf{Y} = \frac{1 - \mathbf{J}(\mathbf{N}_R\mathbf{T}, \mathbf{R}\mathbf{N}_R)}{\mathbf{J}(\mathbf{R}\mathbf{N}_R, \mathbf{N}_R\mathbf{T}) + [1 - \mathbf{J}(\mathbf{N}_R, \mathbf{R}\mathbf{N}_R\mathbf{T})]\exp[(\mathbf{R} - 1)\mathbf{N}_R(\mathbf{T} - 1)]} \qquad (16\text{-}133)$$

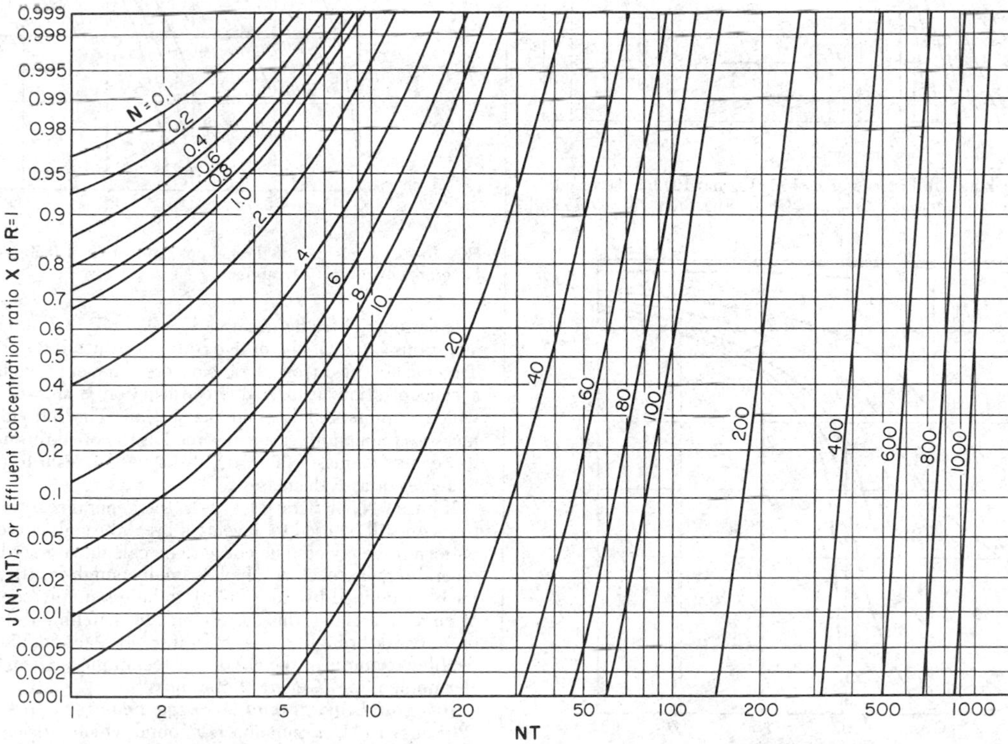

FIG. 16-18 Dependence of **J** function on first argument (**N**, nondimensional bed length, as contour variable) and second argument (**NT**, nondimensional time or fluid volume, as abscissa).

It is apparent that Eq. (16-132) contains Eq. (16-131) as a limiting case, at $\mathbf{R} = 1$. This equation has also been shown to reduce into the constant-pattern result with $\mathbf{R} \ll 1$ and into the proportionate-pattern result with $\mathbf{R} \gg 1$ [Hiester and Vermeulen (see "General References"); and Gilliland and Baddour, *Ind. Eng. Chem.* **45,** 330 (1953)]. Numerical values of Eq. (16-132) have been computed and tabulated by Opler and Hiester (loc. cit.).

Plotted Solutions of the Constant-R Breakthrough Equations Curves evaluated from Eq. (16-132), of concentration **X** versus throughput ratio **T** for contours of constant **N,** are plotted in Fig. 16-19. Cross plots covering the range of **R** from 0 to 2.5 are given in Figs. 16-20 to 16-24 to facilitate use of the theory in design

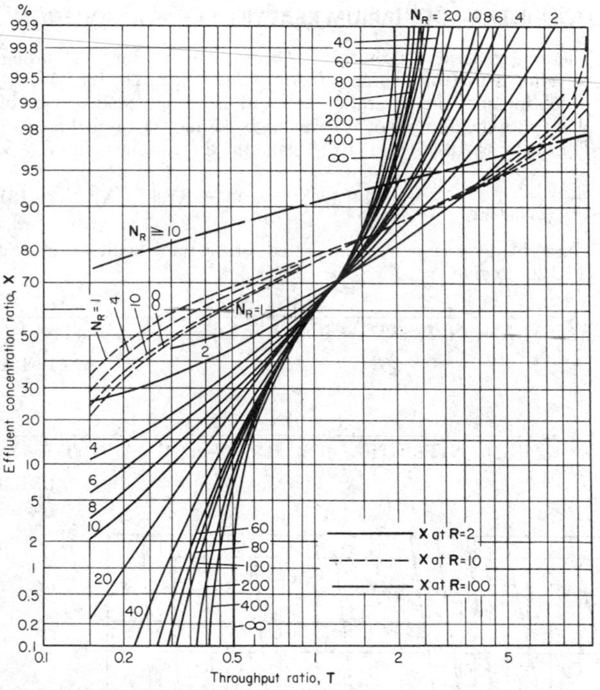

FIG. 16-19 Breakthrough histories at $\mathbf{R} = 2$, 10, and 100 for the reaction-kinetic treatment. [*Hiester and Vermeulen,* Chem. Eng. Prog., *45, 509 (1952).*]

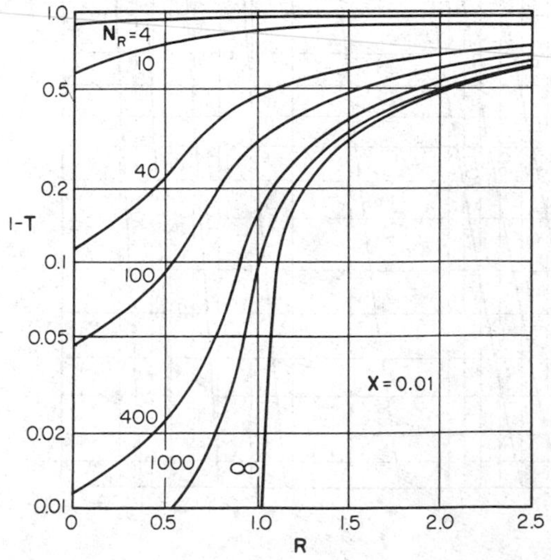

FIG. 16-20 Cross plot of throughput ratio, at $\mathbf{X} = 0.01$, against separation factor; reaction-kinetic treatment.

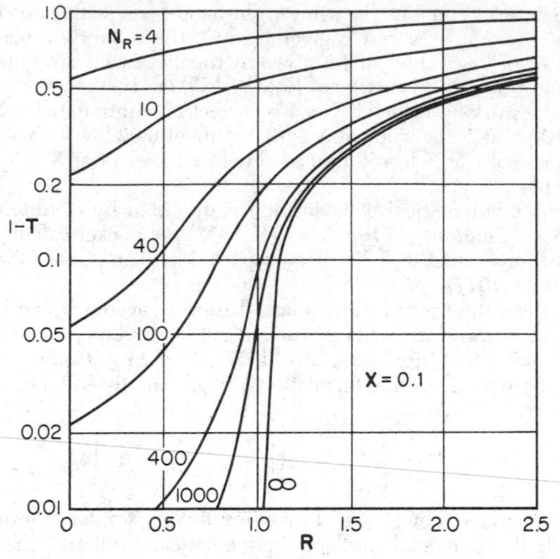

FIG. 16-21 Cross plot of throughput ratio, at $\mathbf{X} = 0.10$, against separation factor; reaction-kinetic treatment.

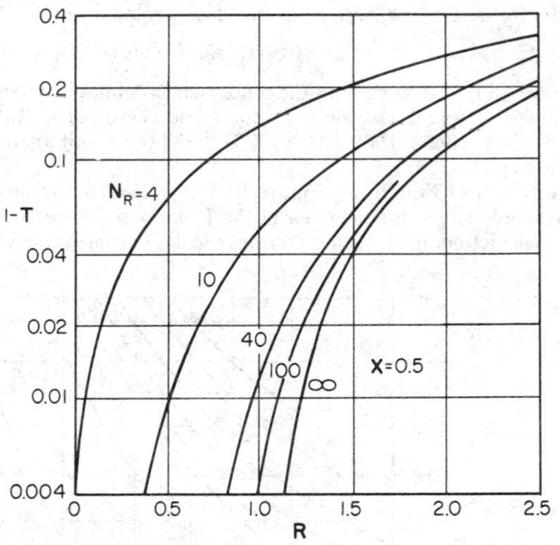

FIG. 16-22 Cross plot of throughput ratio, at $\mathbf{X} = 0.50$, against separation factor; reaction-kinetic treatment.

calculations. These figures give $1 - \mathbf{T}$ (or $\mathbf{T} - 1$) as a function of **R,** at various values of $\mathbf{N_R}$ for five constant values of **X.** The five **X** versus **T** points obtained from these plots are sufficient to define completely a concentration history; this construction is shown in Example 5, which is presented later in the section. The individual cross plots measure the nondimensional spread of the breakthrough curve about its "center of mass" $(\mathbf{T} - 1)$; this spread is seen to diminish as $\mathbf{N_R}$ increases or as **R** decreases.

If either $\mathbf{N_R}$ or **R** varies with **X,** the general **reaction-kinetic solution** can still be used to develop a breakthrough curve for any given pattern of behavior. The curve to be calculated may be divided into several regions, each of which is small enough for the correction factor **b,** as defined by Eq. (16-72), to be essentially constant. In terms of curve matching, this corresponds to matching different segments of a breakthrough curve to different reaction-kinetic curves. A detailed example of the use of this technique is given by Hiester and Vermeulen (see "General References").

Integral Leakage and Average Leakage In solving Eq. (16-120) or (16-131), a function **F** is found which satisfies the relation

$$\mathbf{F} = \frac{1}{\mathbf{N_R}} \int_0^{\mathbf{N_R T}} \mathbf{X}\, d(\mathbf{N_R T}) \qquad (16\text{-}134)$$

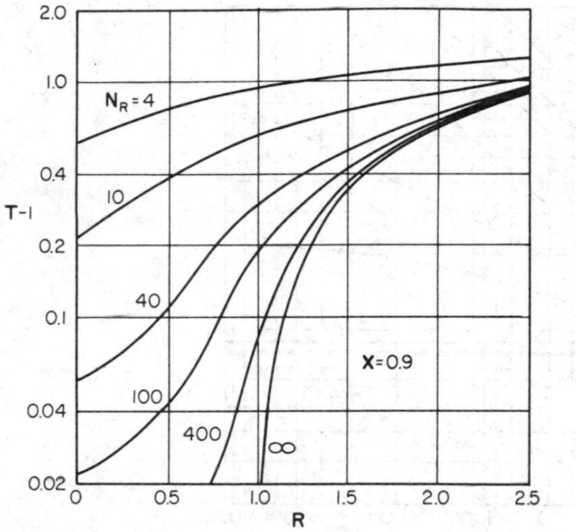

FIG. 16-23 Cross plot of throughput ratio, at $\mathbf{X} = 0.90$, against separation factor; reaction-kinetic treatment.

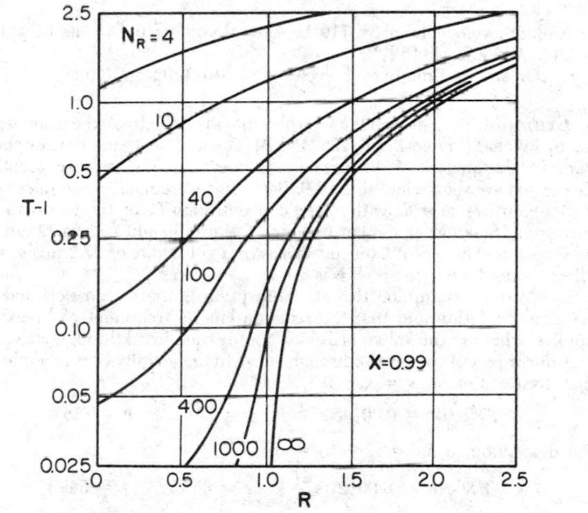

FIG. 16-24 Cross plot of throughput ratio, at $\mathbf{X} = 0.99$, against separation factor; reaction-kinetic treatment.

$\mathbf{F}$ is proportional to the total amount of solute which has emerged from the length of column corresponding to N_R at a time corresponding to $\mathbf{T}$ and therefore is designated as the integral leakage. The average leakage $\overline{\mathbf{X}}$, often a major design criterion, is given by

$$\overline{\mathbf{X}} = \mathbf{F}/\mathbf{T} \tag{16-135}$$

[At high $\mathbf{T}$ values, $\mathbf{F}$ approaches $\mathbf{T} - 1$.]

Interpretation of Experimental Data The use of **experimental breakthrough data** to obtain rate values (or, alternatively, N_R values) is straightforward if the equilibrium constant K or K_L is known and hence the separation factor $\mathbf{R}$ can be calculated. In this case the experimental data can be plotted on a probability ordinate (for $\mathbf{X}$) versus a logarithmic abscissa (for time), and the graph can be shifted horizontally until it matches a specific curve on a theoretical plot of $\mathbf{X}$ versus $\mathbf{T}$ for various N_R's at the proper value of $\mathbf{R}$. The matched curve will give the number of reaction units N_R; and the $\mathbf{T}$ value at any one $\mathbf{X}$ (for instance, $\mathbf{X} = 0.5$) can be used with the matching time (measured at the column exit) to calculate Λ as $Ft/\mathbf{T}v$.

In the event that $\mathbf{R}$ is not known, the complementary nature of saturation and elution can be used in the following procedure.

A saturation run is carried to completion ($\mathbf{X} \approx 1$); the column is then regenerated by a solution of the ion initially present on the resin, at the same concentration level C_0 and at the same flow rate as used in the saturating run. If the saturation data are fitted against one of the standard curves for a given $\mathbf{R}$ value and the elution data are fitted similarly on the related $\mathbf{R}^{\dagger}$chart, apparent values of N_R and $N_R^{\dagger}$ will be obtained. This procedure can be repeated for other $\mathbf{R}$ values. Then, if $N_R^{\dagger}/N_R$ is plotted against the assumed $\mathbf{R}$, intersection of the curve with a line $N^{\dagger}_R/N_R = \mathbf{R}$ will indicate the correct value.

NRUs or Rates Based on Midheight Slopes Determination of rate parameters from the slope of the breakthrough curve at $\mathbf{X} = 0.5$ has been suggested by Thomas (loc. cit.) and Gilliland and Baddour (loc. cit.). In the proportionate-pattern region, for the reaction-kinetic model, differentiation of Eq. (16-58) written for $\mathbf{X}(\mathbf{R}, \mathbf{T})$ shows that

$$\left(\frac{\partial \mathbf{X}_{0.5}}{\partial \mathbf{T}}\right)_{N_R} = \frac{(\mathbf{R}+1)^3}{16\mathbf{R}(\mathbf{R}-1)} \tag{16-136}$$

This result, being independent of N_R, does not give any rate data but does give the $\mathbf{R}$ value. However, in the constant-pattern region, Table 16-7, item 4G, yields

$$\left(\frac{\partial \mathbf{X}_{0.5}}{\partial \mathbf{T}}\right)_{N_R} = \frac{(1-\mathbf{R})N_R}{4} \tag{16-137}$$

These and other values of the **midheight slope** are shown in Fig. 16-25. For $\mathbf{R}$ values less than unity, the slope is seen to depend upon the particular mechanism that is rate-determining.

The dimensionless midheight slope $\partial \mathbf{X}_{0.5}/\partial \mathbf{T}$ is related to the experimental midheight slope, measured graphically on a linear plot of the concentration history:

$$\frac{d\mathbf{X}_{0.5}}{d\mathbf{T}} = \Lambda v \frac{d\mathbf{X}_{0.5}}{d(V-v)} = \frac{\Lambda v}{F}\frac{d\mathbf{X}_{0.5}}{dt} \tag{16-138}$$

The corresponding mass-transfer slope is $N_f(1 - \mathbf{R})/[(1 + \mathbf{R})/2]$. Also, for $\mathbf{R} = 1$, because $\operatorname{erf}(z) \approx 2z/\sqrt{\pi}$ at small z,

$$\left(\frac{\partial \mathbf{X}_{0.5}}{\partial \mathbf{T}}\right)_{N_R} = \frac{\sqrt{N_R}}{2\sqrt{\pi}} \tag{16-139}$$

Radial Flow

Cylinders or Spheres Radial geometry, generally with center feeding, is useful for certain laboratory procedures employing pure or impregnated filter papers; for low-pressure-drop systems in which a large area of shallow bed is needed, as for the annular adsorption canisters used in air conditioning; or for underground seepage of a waste stream from a point or line source. Four separate cases can be identified. When the column profile is needed, N must be assumed locally proportional to v.

1. If local-equilibrium theory applies (with unfavorable equilibrium, giving a proportionate-pattern breakthrough), the behavior is dependent only upon the total volume traversed. For axisymmetric feed $dv = 2\pi hR\, dR$, and for point symmetry $dv = 4\pi R^2\, dR$, where R is the radial distance within the bed and h is the transverse distance.

If the flow is nonuniform (as with a noncentral source), so that different flow paths emerge from the body of sorbent after different residence times ($= S\epsilon_m/F$), the effluent composition at a given time is expressed by the integral

$$\mathbf{X}_m(t) = \left(\frac{1}{F}\right)_{\text{overall}} \int \left(\frac{F}{S\epsilon_m}\right)_{\text{path}} \mathbf{X}_{\text{path}}(t)\, d(S\epsilon_m) \tag{16-140}$$

where $\mathbf{X}_{\text{path}}(t) = \mathbf{X}(\mathbf{T})$

with $\mathbf{T} = \dfrac{t}{(t_{\text{stoic}})_{\text{path}}}$

2. If particle-side resistance controls, the breakthrough curve depends only on the apparent residence time, which is proportional to the volume traversed, as in case 1.

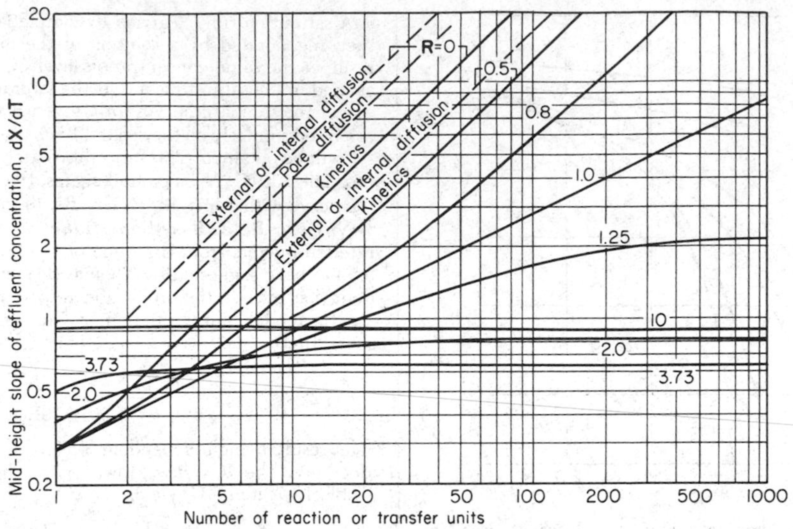

FIG. 16-25 Nondimensional midheight slope as a function of separation factor and NTU (or NRU). [Chem. Eng. Prog., Symp. Ser., *55(24), 65 (1959).*]

3. If linear equilibrium applies and if the fluid-side resistance controls, the breakthrough shape is determined by the total effective **N** the fluid has traversed:

$$\mathbf{N} = \int \frac{k_{Of} a_p}{F}\, dv \qquad (16\text{-}141)$$

4. If favorable equilibrium applies and the fluid-side resistance controls, so that a constant pattern would apply if k_{Of} were constant, the breakthrough curve is calculated as if a value near the exit k_{Of} holds throughout the bed. A working approximation is to use the k at $v = 0.93v_{max}$ for spherical geometry and the k at $v = 0.86v_{max}$ for cylindrical geometry. A more exact solution based on the reaction-kinetic model with nonlinear equilibrium has been given by Lapidus and Amundson [*J. Phys. Chem.*, **54**, 82 (1950); **56**, 373 (1952)].

Example 4 **Decolorization** is desired of 0.47 L/s (1.0 ft^3/min) of sugar solution to 5 percent or less of its initial color level, which is 1.0 g color bodies/kg solution (0.0010 lb/lb). An available column, 6.1 m (20 ft) high and 61 cm (2 ft) in diameter, is charged with activated carbon. Adsorption onto the carbon is described by a Langmuir isotherm with K_L = 100 kg solution/g color bodies (1.0 × 10^5 lb/lb) and by $Q_{\rho b}$ = 16 g color bodies/L (1.0 lb/ft^3). Other properties of the carbon are ϵ = 0.40, d_p = 0.82 mm average (0.032 in), and D_{pore} = 4.56 × 10^{-4} mm^2/s (1.77 × 10^{-5} ft^2/h). Pore diffusion provides the rate-limiting resistance. Solution density is 1.12 kg/L (70 lb/ft^3).

What is the on-stream time between successive regenerations?

This problem corresponds to case C of Table 16-2, and computation steps will be numbered in the same order. The equilibrium parameter **R** is 0.0099; this is essentially equivalent to **R** = 0.

1. Column volume $v = 6.1(0.61)^2\pi/4 \times 10^3 = 6.1 \times 292 = 1780$ L (62.8 ft^3).

2. Superficial linear velocity through the bed $F/S = 470/2920 = 0.161$ cm/s (19.1 ft/h). Hence apparent residence time $hS/F = 610/0.161 = 3790$ s (1.05 h).

3. NTU from item 3A, Table 16-7:

$$N_{pore} = \frac{0.548(60)4.56 \times 10^{-4}(3790)(0.60)}{(0.82)^2} = 50.7$$

4. Throughput ratio **T**, at **X** = 0.05, is obtained from Eq. (16-81):

$$0.07285z^2 - 0.5398z + 0.050 = 0$$

where $z = 50.7(\mathbf{T} - 1) + 1.235$.

Solving, $z = 0.00911$, and **T** = 0.976.

5. Solute capacity of bed = 16 × 0.99 × 1780 = 28,200 g color bodies.

6. Solute content of treated fluid = 0.976 × 28,200 = 27,520 g.

7. Fluid volume through = 27,520/0.893 = 30,820 L (1090 ft^3) = $V - v\epsilon$. Column void volume = 710 L, so total volume fed to the breakthrough point is 31,530 L (1115 ft^3).

8. On-stream time in each cycle = 31,530/1692 = 18.6 h.

Example 5 Adsorption of carbon dioxide on activated carbon was studied by Wicke [*Kolloid Z.*, **86**, 295 (1939)]. The authors thank F. Campbell Williams for this analysis of Wicke's data. Concentration histories for sorption and desorption are shown in Table 16-9. Both runs were made at an inlet flow rate F of 1.85 mL/s, at 0°C, with a feed concentration C_0 of 100 torr with N_2 as a carrier gas making up another 660 torr. Column height h_2 was 72 cm, with a cross-sectional area S of 0.39 cm^2 and a void volume $v\epsilon$ of 11.2 mL so that the effective residence time $v\epsilon/F$ was 6.1 s.

1. On the assumption that the adsorption isotherm can be fitted by the Langmuir relation and that the reaction-kinetic treatment of breakthrough applies, what are the values of **R** and $\mathbf{N_R}$ for each breakthrough curve?

A linear plot of each breakthrough curve in the vicinity of midheight reveals that, for sorption, at **X** = x = 0.5,

$$d\mathbf{X}/dt = 0.00745 \text{ s}^{-1} \qquad t = 734 \text{ s} \qquad \hat{t} = 728 \text{ s}$$

For desorption, at $\mathbf{X}^\dagger$ = (1 − x) = 0.5,

$$d\mathbf{X}^\dagger/dt = 0.00127 \text{ s}^{-1} \qquad t = 655 \text{ s} \qquad \hat{t} = 649 \text{ s}$$

TABLE 16-9 Experimental Breakthrough Histories for Carbon Dioxide on Carbon

Sorption		Desorption	
x	t, s	x†	t† s
0.010	645	0.970	270
0.030	660	0.936	330
0.140	690	0.873	390
0.290	710	0.744	480
0.377	720	0.653	540
0.464	730	0.610	570
0.540	740	0.570	600
0.610	750	0.530	630
0.726	770	0.460	690
0.833	800	0.430	720
0.864	810	0.400	750
0.940	870	0.273	900
0.974	960	0.177	1080
		0.103	1320
		0.070	1500
		0.023	1920

To initiate the calculation, it is assumed that $\mathbf{T} = 1$ when $\mathbf{X} = 0.5$. Then, from Eq. (16-89),

$$\Lambda v/F = \hat{t}/\mathbf{T} = 728 \text{ s}$$

Hence $d\mathbf{X}/dt = 728(0.00745) = 5.4$, and $d\mathbf{X}^\dagger/d\mathbf{T} = 728(0.00127) = 0.92$. From these slopes it is clear that $\mathbf{R} < \mathbf{R}^\dagger$, hence that $\mathbf{R} < 1$ and $\mathbf{R}^\dagger > 1$.

The range of possible $\mathbf{R}$ and $\mathbf{N}$ values is narrowed by applying Fig. 16-22 to the desorption step. At $\mathbf{X}^\dagger = 0.5$, $\mathbf{T} = 646/725 = 0.892$, and $1 - \mathbf{T} = 0.108$. At $\mathbf{R}^\dagger = 1$, $\mathbf{N}_R^\dagger$ would be 5; even for every high $\mathbf{N}_R^\dagger$, $\mathbf{R}^\dagger$ is below 1.95. A set of allowable combinations of $\mathbf{R}^\dagger$ and $\mathbf{N}_R^\dagger$ (and hence also $\mathbf{R}$ and $\mathbf{N}_R$) is obtained from this figure. [The $d\mathbf{X}^\dagger/d\mathbf{T}$ value given earlier if used in Fig. 16-25 suggests that $\mathbf{R}^\dagger$ lies between 1.5 and 2. Equation (16-136) shows that a proportionate pattern for this slope would yield $\mathbf{R}^\dagger = 1.87$].

Corresponding to $\mathbf{R}^\dagger$ between 1.5 and 2, $\mathbf{R}$ will lie between 0.67 and 0.50. Figure 16-25 with the experimental midheight $d\mathbf{X}/d\mathbf{T}$ provides a second set of allowable $\mathbf{R}$ and $\mathbf{N}_R$ values. When the two sets are plotted, their intersection is found to lie at

$$\mathbf{R} = 0.58 \qquad \mathbf{N}_R = 50 \qquad \mathbf{R}^\dagger = 1.72 \qquad \mathbf{N}_R^\dagger = 29$$

These results can be checked further since, in the constant-pattern region that this case approaches (compare Table 16-8), Eq. (16-175) shows that

$$\frac{d\mathbf{X}_{0.5}}{d\mathbf{T}} = \frac{(1 - R)N_R}{4} = \frac{(0.42)(50)}{4} = 5.25$$

which is close to the measured value of 5.4. [The ratio $\mathbf{N}_R^\dagger/\mathbf{N}_R$, evaluated as indicated previously in "Interpretation of Experimental Data," yields $\mathbf{N}_R^\dagger = 30$, in satisfactory agreement with the value of 29.

2. Determine the theoretical breakthrough curves and compare them with the experimental data.

Figures 16-20 and 16-24 can be used to determine the $\mathbf{T}$ values for certain $\mathbf{X}$'s for each of the ($\mathbf{R}$, $\mathbf{N}_R$) combinations. The results are given in Table 16-10.

TABLE 16-10 Calculated Breakthrough Histories for Carbon Dioxide on Carbon

	Sorption $\mathbf{R} = 0.58$, $\mathbf{N}_R = 50$			Desorption $\mathbf{R}^\dagger = 1.72$, $\mathbf{N}_R^\dagger = 29$		
$\mathbf{X}$ or $\mathbf{X}^\dagger$	$\mathbf{T}$	$t - (v\epsilon/F)$, s	t, s	$\mathbf{T}^\dagger$	$t - (v\epsilon/F)$, s	t, s
0.01	0.775	561	567	0.28	205	211
0.1	0.892	645	651	0.46	334	340
0.5	1	723(728)	729	0.904	654(658)	660
0.9	1.11	803	809	1.66	1198	1204
0.99	1.23	885	894	2.33	1683	1689

Note that if $\Lambda v/F$ is taken as 728 s, $\hat{t}$ at $\mathbf{X} = 0.5$ is 728 s and $\mathbf{T}$ at $\mathbf{X}^\dagger = 0.5$ is 658 s, as shown in parentheses. The latter value is 9 s longer than the experimental value. If this difference is split between the adsorption and desorption placement so that the desorption $\hat{t}$ is 654 s, then from the $\mathbf{T}_{0.5}^\dagger/\mathbf{T}_{0.5}$ ratio of 0.904, the sorption $\hat{t}$ is 723 s. This last number is therefore the best average value of $\Lambda v/F$ and is used in calculating the values of t - ($v\epsilon/F$) shown in the table. The experimental points (open circles and triangles) and calculated points (solid circles and triangles) are both plotted in Fig. 16-26 and are found to have good correspondence.

3. What are the values of K_L, Λ, and $Q\rho_b$? From the ideal-gas law,

$$c_0 = p_0/R_g T_K = 100/(760 \times 22.4)$$
$$= 0.00587 \text{ (g·mol)/L } (3.66 \times 10^{-4} \text{ (lb·mol)/ft}^3$$

The Langmuir coefficient, $K_L p_0$ or $K_L' c_0$, is $\mathbf{R}^\dagger - 1 = 0.72$. In pressure units, K_L becomes $0.72/100 = 0.0072$ torr^{-1}, or $0.72/0.132 = 5.47$ bar^{-1} (5.54 atm^{-1}). In concentration units, K_L' becomes $0.72/0.00587 = 123$L/(g·mol) [= 1970 ft^3/(lb·mol)].

The partition ratio $\Lambda = t_{\text{stoic}}(F/v) = 723/15.17 = 47.6$.

Equilibrium content of the column $q_0^\circ \rho_b = \Lambda c_0 = 47.6 \times 0.00587 = 0.279$ (g·mol)/L [0.0174 (lb·mol)/ft^3]. The ultimate column capacity $Q\rho_b = q_0^\circ \rho_b/(1 - R)$, or 0.665 (g·mol)/L [0.0416 (lb·mol)/ft^3].

EMPIRICAL SCALE-UP

The common purpose of all calculation methods listed here is to predict a curve of $\mathbf{X}$ against volumes of fluid per volume of solid ($\hat{V}/v$). For a specified system, the stoichiometry, equilibrium, and type of mechanism usually remain constant during scale-up. Also, the relative shape of the breakthrough curve remains almost constant, with its slope at each $\mathbf{X}$ value remaining proportional to its slope at a central $\mathbf{X}$ point.

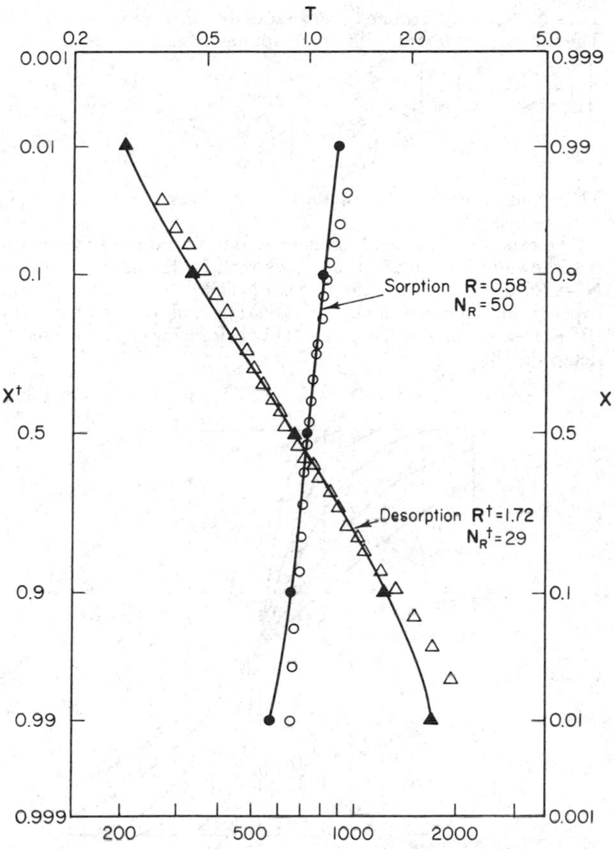

FIG. 16-26 Matching of theoretical breakthrough curves to experimental data of Wicke for the adsorption of carbon dioxide on activated carbon.

Scale-up must be based on a partial or complete experimental curve of $\mathbf{X}$ (or c) versus $\hat{V}/v$, on which the **stoichiometric-equivalence point** (or "center of mass" of the transition $\mathbf{T} = 1$, to be defined later) must be identified. To a first approximation, $\mathbf{X}$ remains constant at this stoichiometric point. If the slope ratio between the experimental curve and the scaled-up curve is known, then all distances (to given $\mathbf{X}$ values, from the stoichiometric point) measured on the $\hat{V}/v$ scale can be expanded or contracted in the proportions given by the slope ratio.

Determining the slope ratio to be applied divides into two parts: first, how $\mathbf{N}$ changes with the changed conditions; second, how the slope ratio changes with $\mathbf{N}$.

Definitions of the NTUs for different mass-transfer mechanisms, with reference to Fig. 16-12, provide the relation:

$$\mathbf{N}d_p/h \propto (DS/d_pF)^{1-a} \tag{16-142}$$

Here $a = 0$ for particle ($D = D_p$) or pore ($D = D_{\text{pore}}$) diffusion; and 0.5 for external mass transfer, 1.0 for axial dispersion, or 2.0 for molecular diffusion ($D = D_f$ for all three). If the mechanism is known, the ratio of any $\mathbf{N}$ to a reference value $\mathbf{N}_{\text{ref}}$ is calculated for ratios of changed v, S, F, d_p, D, or Λ (included in D_p) compared with the respective reference condition. Equation (16-142) is rewritten for both the new and the reference condition as

$$\mathbf{N} \propto vS^{-a}F^{a-1}d^{a-2}D^{1-a} \tag{16-143}$$

Thus N/N_{ref} is determined from ratios for each physical variable. The following proportionality then applies:

$$\left[\left(\frac{\hat{V}}{v}\right)_X - \left(\frac{\hat{V}}{v}\right)_{stoic}\right]_{pred} = \left(\frac{N_{exptl}}{N_{pred}}\right)^B \left[\left(\frac{\hat{V}}{v}\right)_X - \left(\frac{\hat{V}}{v}\right)_{stoic}\right]_{exptl} \quad (16\text{-}144)$$

The terms in brackets can optionally be replaced by respective $[T_X - 1]$ terms.

The exponent **B** depends on both **R** and **N**, and ranges from 1 to 0 as shown in Fig. 16-27; it should be read for the geometric mean, $\bar{N} = (NN_{ref})^{0.5}$. If NRU (rather than NTU) values are found from experiment, the mean value of **N** should be multiplied by the factor $(R + 1)/2$ to convert it into an NTU before Fig. 16-27 is used to determine **B**.

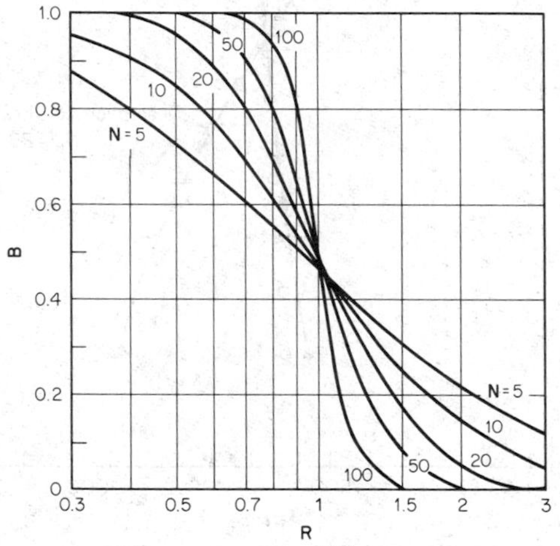

FIG. 16-27 Plot of scale-factor exponent **B** as a function of separation parameter and NTU.

Unfavorable-Equilibrium Pre-Proportionate-Pattern Region The total spreading for a given N_R, with $R > 1$, is somewhat less than the combined effect of the proportionate-pattern spreading [Eq. (16-57)] and the kinetic spreading $(T - 1)$ for the given N_R at $R = 1$. As an approximation, $(T - 1)R^{-1/3}$ may be added to the proportionate-pattern **T** at the given **R** for each particular **X**.

NUMERICAL METHODS

Explicit Procedure: Method of Characteristics For complex rate or equilibrium relations, irregular boundary conditions, or both, numerical integration of Eqs. (16-65) may be carried out. The method of characteristics, described by Acrivos [*Ind. Eng. Chem.*, **48**, 703 (1956), which includes nonisothermal applications] is illustrated here for the simple case involving isothermal conditions and nondimensional rates. (The method does not apply for axial dispersion.)

Lines of constant **NT** or constant **N**, termed "characteristics," are given by the products:

$$NT = i\,\Delta \quad (16\text{-}145)$$

$$N = j\,\Delta \quad (16\text{-}146)$$

where i and j are integers and Δ is a suitable constant usually less than 1. Equations (16-65) in finite-difference form become

$$X_{ij} = X_{i,j-1} - \Delta \cdot \text{fn}\,(X_{i,j-1}, Y_{i,j-1}) \quad (16\text{-}147)$$

$$Y_{ij} = Y_{i-1,j} + \Delta \cdot \text{fn}\,(X_{i-1,j}, Y_{i-1,j}) \quad (16\text{-}148)$$

Here X_{ij} and Y_{ij} are the values sought, the others being known from boundary conditions or prior calculations; the form of the driving-force function depends upon the controlling mass-transfer step.

In each increment, for the first approximation, the upstream **X** and **Y** are used to calculate the driving force; for a second or later approximation, the downstream **X** and **Y** are used similarly, and the average of the upstream and downstream forces is used to calculate better values of downstream **X** and **Y**.

Implicit Procedure: Orthogonal Collocation The leakage F_i (for each solute i), as defined in Eq. (16-134), is a generating function for x_i and y_i, which are a pair of exact differentials. The unknown exact solution can be approximated as a sum (called the trial function) of a series of orthogonal polynomials chosen to satisfy the boundary conditions. The trial function is substituted into the differential equation or equations; the result, called the residual, is required to be zero at the collocation points, namely, the root values for the highest-order orthogonal polynomial in the trial function. Matrix inversion is used to determine the unknown coëfficients in the trial function; the function then becomes an approximate solution in continuous form. As the order of this function is increased, it becomes exact at more (**N**, **NT**) points and thus converges toward an exact solution. Usually this method requires much less computer time than the method of characteristics. For fixed-bed systems that are more than trivariant, both methods are tedious, and more incisive procedures are needed. Key references are Villadsen and Stewart, *Chem. Eng. Sci.*, **22**, 1483 (1967); Liapis and Rippin, *Am. Inst. Chem. Eng. J.*, **25**, 455 (1979); and Holland and Liapis (see "General References"). (The authors thank A. I. Liapis for informative comments.)

CHANGED-FEED OPERATIONS; REGENERATION; CHROMATOGRAPHY

STEP CHANGES IN FEED CONCENTRATION

Single Solute, with Linear Equilibrium It has gradually become evident that the downstream behavior of a solute is described by the **superposition** of upstream effects. Superposition is exact only for a linear isotherm, at all **N** values, and approximate for near-linear isotherms at large **N**; so it is described here in terms of linear behavior ($r = 1$). [A useful reference is Rosen and Winsche, *J. Chem. Phys.*, **18**, 1587 (1950).]

For a bed fed consecutively by feeds of different composition, the fluid front for each new input moving through the bed changes steadily from its input concentration to one determined mainly by the solid phase through which it has passed. The outflow concentrations of succeeding layers passing through any given length of bed evolve toward their initial feed condition.

If the flow rate is constant, **N** is also, and T will be proportional to both t and $\hat{V}$. The calculation normally starts with a uniformly presaturated column, with solid (and fluid) concentrations of y' (and x'), which may be zero. Feed of concentration x'' enters; the volume while it is entering is measured as V'', yielding T'', and the total incremental volume until the feed is changed is V_2, yielding T_2. The next feed has concentration x''', with T''' as its variable throughput-parameter value and T_3 as its total. For the initial feed (that is, after the "first" feed change) the breakthrough curve is equivalent to Eq. (16-121):

$$x = x' + (x'' - x')J(N, NT'') \quad (16\text{-}149)$$

After two feed changes, the breakthrough is given by

$$x = x' + (x'' - x')J[N, N(T_2 + T''')] + (x''' - x'')J(N, NT''') \quad (16\text{-}150)$$

More generally, for constant or nonconstant flow rate, over n feed changes,

$$x = x' + \sum_{i=1}^{n} (x^{(i+1)} - x^{i}) J \left[\overline{N}_{i+1}, \left(N_{n+1} \overline{T}^{(n+1)} + \sum_{j=i+1}^{n} N_j T_j \right) \right] \quad (16\text{-}151)$$

with T proportional to $\hat{V}$ in each step. For constant flow, $N_j = N$ and $\overline{N}_{i+1} = N$. When a different constant flow rate is used in each interval j, the NTU value for that interval is N_j, and the applicable $\overline{N}$ value for the transition at the end of that interval is

$$\overline{N}_{i+1} = \frac{N_{n+1}\overline{T}^{(n+1)} + \sum_{j=i+1}^{n} N_j T_j}{\overline{T}^{(n+1)} + \sum_{j=i+1}^{n} T_j} \quad (16\text{-}152)$$

where $\overline{T}^{(n+1)}$ is the mean value of T in the nth transition. At large times the earlier terms will reach constant values and "erase" the previous memory of the column. Thus, in Eq. (16-150) at large T, the first J term becomes unity, and the equation simplifies to

$$x = x'' + (x''' - x'')J(N, NT''') \quad (16\text{-}153)$$

Memory behavior under nonlinear equilibrium is discussed by Hiester (Ph.D. dissertation, University of California, 1949).

CYCLIC OPERATIONS

Process Efficiency Operating capacities for specific sorption systems under specified flow conditions, as a function of regenerant dosage or time, are frequently quoted by suppliers of the sorbents; compare Kunin, *Ion Exchange Resins*, Wiley, New York, 1956. For the case of binary exchange, with one saturating and one regenerating (or eluting) species, it is desirable to define dimensionless variables that apply to an entire period or cycle of operation rather than to an instantaneous condition of a column [Hiester and Phillips, *Chem. Eng.*, **61**(10), 161 (1954); and private communications from Farnham, Cornaz, De La Rue, Droin, and LeMaguer]. These variables are:

1. Resin utilization $\overline{Y}$—the ratio of operating capacity (or amount of feed ion actually exchanged) to the ultimate capacity for that ion represented by the total amount of sorbent in the column. $\overline{Y}$ is the total amount of solute on the sorbent at the end of a saturating step, less the residual amount already present at the start of that saturating step, divided by the ultimate capacity.
2. Regenerant efficiency E—the quantity of ions removed from the sorbent divided by the total quantity of regenerant ions fed to the column.
3. Average leakage $\overline{X}$—the amount of feed ion which escapes from the column during saturation periods, divided by the total amount of that ion fed.
4. Feed potential T_{sat}—the total amount of feed ion charged, divided by the ultimate capacity of the column.
5. Regenerant potential T_{reg}—the quantity of regenerant ions fed to the column, divided by the ultimate capacity of the column.

Mathematically,

$$\overline{Y} = \frac{1}{N}\int_0^{NR} Y\,dN \qquad \overline{X} = \frac{1}{NT_{sat}}\int_0^{NRT_{sat}} X\,d(NT) \quad (16\text{-}154)$$

In terms of integral leakage [Eq. (16-134)] and with the assumption that the superposition method does not require correction, the resin utilization computed over three half cycles starting and ending with a complete saturation step is

$$\overline{Y} = (T_3' - F_3') - (T_3'' - F_3'') + (T_3''' - F_3''') = 2T_{sat} - F_3' + F_3'' - F_3''' \approx T_{sat} - F_3''' \quad (16\text{-}155)$$

Likewise, the average leakage is

$$\overline{X} = \frac{(F_3' - F_2') - (F_3'' - F_2'') + F_3'''}{T_{sat}} \approx \frac{1 - (F_3'' - F_2'') + F_3'''}{T_{sat}} \quad (16\text{-}156)$$

In these relations, the superscripts on F indicate the half cycle in which the respective leakage integral started, and the subscripts indicate the half cycle at whose conclusion F is being measured.

The regenerant efficiency becomes

$$E = \overline{Y}/T_{reg} = (1 - \overline{X})(T_{sat}/T_{reg}) \quad (16\text{-}157)$$

Undirectional Flows Alternating cycles of saturation (exhaustion) and regeneration (elution) are standard practice for most industrial-scale sorption. The regenerating stream tends to produce concentration levels x', y' throughout the bed, and the saturating stream tends to produce levels x'', y''. The saturating step, as in Example 5, is measured by relative-concentration levels X and Y between 0 and 1 (that is, between x' or y' and x'' or y''). The regenerating step is measured by levels $X^\dagger$ ($= 1 - X$) and $Y^\dagger$ ($= 1 - Y$). The separation factors are designated by R_{sat} and R_{reg} (if they are reciprocals, then designations of $R_{sat} = R$ and $R_{reg} = R^\dagger$ apply). Volume (or throughput ratio) during the saturation period is V (or T), and over the entire period V_{sat} (or T_{sat}); for the regeneration period, respective values of V' (or T') and V_{reg} (or T_{reg}) apply; for the complete cycle, V_{cy} (or T_{cy}).

In cyclic operation with forward flow throughout, the regeneration breakthrough generally overlaps into the saturation period, and the saturation breakthrough always overlaps into the regeneration period. More rarely, the residual effect of one saturation or regeneration will influence the corresponding step one cycle later. Difficult systems sometimes require four or five cycles (each consisting of a saturation followed by a regeneration) before *perfectly periodic behavior* (or "cyclic steady state") is reached. For R values near unity, Eqs. (16-149) to (16-153) can be used, exactly or approximately; farther from linear equilibrium, the problem must be treated empirically or by stepwise numerical simulation, or by the reaction-kinetic model discussed under "Chromatographic Separations."

Linear Equilibrium In a regeneration step affected only by the preceding saturation, the outflow concentration level is

$$X = J(N_{cy}, N_{sat}T_{sat} + N_{reg}T') - J(N_{reg}, N_{reg}T') \quad (16\text{-}158)$$

Similarly, in a saturation step affected only by the preceding regeneration,

$$X = 1 - J(N_{cy}, N_{reg}T_{reg} + N_{sat}T) + J(N_{sat}, N_{sat}T) \quad (16\text{-}159)$$

In these relations, $N_{cy} = (N_{sat}T_{sat} + N_{reg}T_{reg})/T_{cy}$, with $T_{cy} = T_{sat} + T_{reg}$.

Superposition Approximation for Nonlinear Equilibrium A practical test for the validity of superposing breakthrough curves is this: The earlier curve must approach $X = 1$ (or 0) before the later curve approaches $X = 0$ (or 1). Otherwise subtraction of the $X^\dagger$ curve from the X curve, on the given T or V scale, will yield unreal X values. (Pancharatnam, Klein, and Vermeulen, *Design-Optimization Procedure for Cyclic Ion Exchange and Adsorption*, U.S. *Off.* Saline Water Res. Dev. Prog. Rep. 477, 1969.)

Backflow Regeneration For this mode of operation in two-component systems, only one transition (exchange zone or adsorption wave) is involved. This transition is pushed to the product end of the column during saturation, and back to the feed end during regeneration. Generally the transition zone tends to broaden over each successive cycle; this is compensated for by purging a part of the zone by outflow at the feed end during the regeneration cycle. If the equilibrium can be controlled so that both R_{sat} and R_{reg} are less than 1 (that is, both favorable), the transition may retain a constant shape and the purge becomes unnecessary.

Compared with forward-flow regeneration, the reverse-flow system may give the same or lower resin utilization. Its advantage lies in higher regenerant efficiency; if both equilibriums are favorable, 100 percent efficiency is possible.

Prediction of the zone shape will often require stepwise numerical calculation. The effective height of column is the distance within which the stoichiometric center of the transition is cycled. For favorable equilibrium in either direction, constant-pattern calculations based on this distance may suffice for that step. For linear or unfavorable equilibrium in either direction, the midheight breakthrough slope at the end of the preceding half cycle corresponds to a fictitious length outside the column which the new step would appear (mathematically) to have already traversed. If this length or its NTU equivalent is added to the NTU actually traversed, the breakthrough curve can be calculated as if it had developed in a uniformly presaturated bed.

With linear (or unfavorable) equilibrium in both directions, the curve shapes cannot be stabilized except by partial purging of the breakthrough curve in the regeneration step, corresponding to some wastage of regenerant (but less than in forward-flow regeneration). Numerical solution of such cases appears to be needed to describe them accurately.

Layered and Mixed Beds Reverse-flow regeneration can be improved upon in some cases by a technique developed by S. A. Bresler and also by G. Gideon that is based on the reversal of isotherm curvature between the saturation and elution steps for a given sorbent-solute system. This technique utilizes a **split bed,** or two consecutive beds, containing two sorbents in series that have different equilibrium behavior. In the production step, the process stream first meets a sorbent with unfavorable equilibrium, then one with favorable equilibrium. In regeneration, with feed entering at the opposite end of the bed, the sequence is similar; in each case, the sharpening behavior of the farther-downstream part of the bed tends to compensate for the inefficiency of the upstream part.

Layered beds have been analyzed by use of local-equilibrium theory by Klein and Vermeulen [*Am. Inst. Chem. Eng. Symp. Ser.*, **71**(152), 69 (1975)], who have shown that regeneration efficiencies of 100 percent are approached, although sorbent utilization falls; see Fig. 16-28. When two ion exchangers of opposing affinities are available, the performance of a single linear-equilibrium resin is approached when the resin with greater $|\log \mathbf{R}|$ (i.e., the resin with $\mathbf{R}$ farther from 1, measured in logarithmic units) is present in greater quantity. These authors have also found that mixed-resin beds with similar proportions of the resins give similar results (AIChE meeting, New Orleans, November 1981).

TEMPERATURE- AND PRESSURE-SWING DESORPTION

Local-Equilibrium Treatment, Neglecting Heat of Desorption A column to be regenerated in forward flow will require very nearly the same amount of regenerant when partially and when completely saturated. Up to this point only the mathematics to treat a completely presaturated column has been set forth, and the regeneration will be considered in these terms. At the saturation temperature θ_{sat}, the feed partial pressure $(p_A)_0$ produced the equilibrium saturation level $(q_A)_0^\circ$ or $(y_A)_{max} q_{ref}$ in the solid, usually under conditions of a favorable separation factor $\mathbf{R}$. At the regeneration temperature θ_{reg}, the solid saturation level produces a much higher partial pressure $(p_A)_{reg}$, rising toward the total pressure P_{reg}. This discussion neglects the desorption that will occur while the bed is being brought to θ_{reg}.

Relative to the total pressure during regeneration, this solute content is designated $\mathrm{x}_{max}[= (p_A)_{reg}/P_{reg}]$. The stoichiometric volume of effluent required is then

$$V_{stoic} = \frac{(q_A)_0^\circ \rho_b v}{(P_{reg}/R_g T_{K,reg})\mathrm{x}_{max}} \tag{16-160}$$

where R_g is the gas constant, pressure-volume product per mole per degree; and $T_{K,reg}$ is the appropriate absolute temperature. Unfavorable equilibrium for elution is likely to persist at the regeneration temperature, in the way indicated in Fig. 16-29. The total effluent volume to complete removal of solute, by local-equilibrium theory, is then

$$(V_{reg})_{total} = V_{stoic}\mathbf{R}_{reg} \tag{16-161}$$

The carrier-gas volume is less than this, the difference being the vapor-state volume of the solute that the effluent contains. Thus

$$(V_c)_{total} = (V_{reg})_{total} - V_{stoic}\mathrm{x}_{max} \tag{16-162}$$

For the regeneration breakthrough, $\mathbf{X} = 0$ at $\mathrm{x} = \mathrm{x}_{max}$ and $\mathbf{X} = 1$ at $\mathrm{x} = 0$. $\mathbf{X}$ as a function of $\mathbf{R}$ and $\mathbf{T}$ is given by Eq. (16-58); $\mathrm{x}(= p_A/P_{reg})$ is obtained as $\mathrm{x}_{max}(1 - \mathbf{X})$, and the resulting curve for $\mathbf{X}(\mathbf{T})$ is plotted against V_c in Fig. 16-30. The ratio X of moles of solute per mole of carrier gas is given by $X = \mathrm{x}/(1 - \mathrm{x})$; and the effluent-to-feed flow-rate ratio F_{out}/F_{in} is $1/(1 - \mathrm{x})$. Now,

$$V_c = V_{stoic} \int (1 - \mathrm{x})d\mathbf{T} \tag{16-163}$$

integrated from zero to any point x, $\mathbf{T}$

and
$$X = \frac{(\mathbf{R} - 1)/\mathrm{x}_{max}}{[(\mathbf{R} - 1)/\mathrm{x}_{max}] + 1 - \sqrt{\mathbf{R}/\mathbf{T}}} - 1 \tag{16-164}$$

with $\mathbf{T} = 1$ when $\hat{V} = V_{stoic}$, as usual. Figure 16-30 shows $\mathbf{T}$, $\mathbf{X}$, and F_{out}/F_{in} as functions of V_c; if the heated carrier gas is supplied at a constant rate, the abscissa is proportional to the elapsed time.

The same principle applies to desorption by depressuring. However, for complete descriptions, enthalpy and rate effects must also be considered.

If the system pressure in regeneration is **less** than the equilibrium pressure of solute for the q at saturation level, pure solute vapor will emerge until the equilibrium pressure has dropped to the system pressure. Then, because y has dropped and $\mathbf{Y}$ has risen while $\mathbf{X}$ is still zero, regeneration will continue in accordance with Fig. 16-9, show-

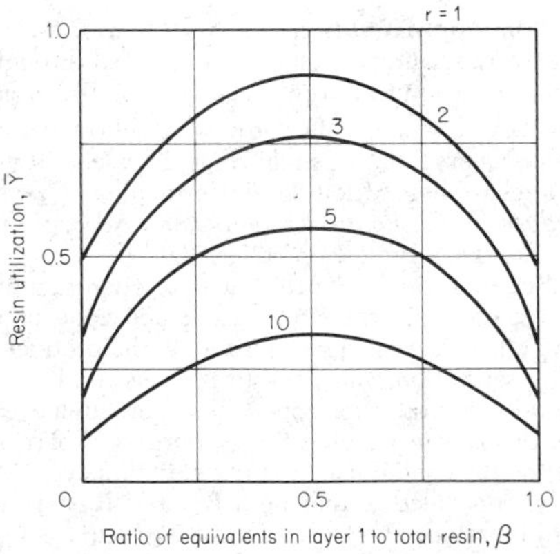

FIG. 16-28 Resin utilization in a layered (or mixed) bed calculated by local-equilibrium theory with 100 percent regeneration efficiency, as a function of $\mathbf{R}_1$, with $\mathbf{R}_1 = 1/\mathbf{R}_1^\dagger = \mathbf{R}_2^\dagger = 1/\mathbf{R}_2$.

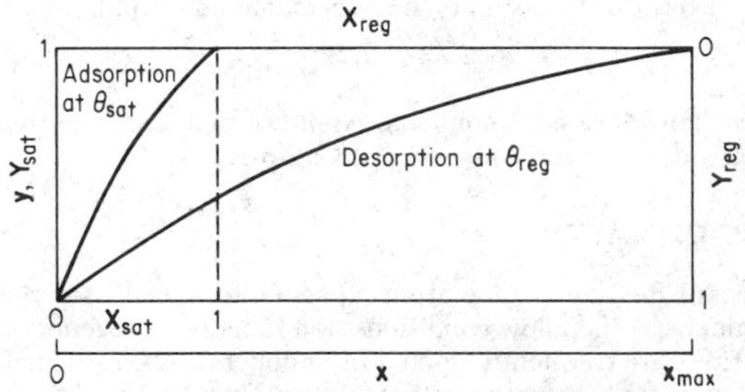

FIG. 16-29 Equilibrium isotherms for adsorption, with desorption at a higher temperature or lower pressure.

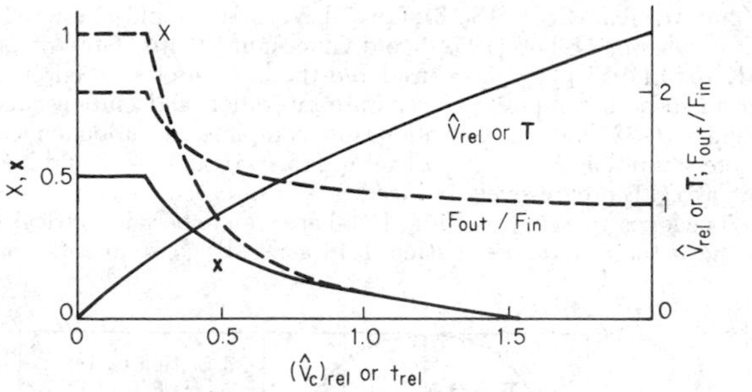

FIG. 16-30 Local-equilibrium isothermal breakthrough behavior for elution from a saturated adsorbent when the solute partial pressure is half the system pressure and $\mathbf{R}_{reg} = 2$. Solute mole fraction x, ratio of effluent to feed F_{out}/F_{in}, and relative integrated effluent volume V are shown as functions of relative carrier volume or relative time. Mole ratio X of solute to carrier is given by the function xF_{out}/F_{in}.

ing an abrupt step to fairly high X and usually finishing with a gradual change.

CHROMATOGRAPHIC SEPARATIONS

Bulk and Trace Modes Chromatography is a discontinuous operation in which a portion of solute mixture ("feed") is introduced at the column inlet. A carrier fluid, which initially is free of these solutes, is then fed continually over an extended period and causes the solutes to travel through the column as bands or zones at slightly different velocities. If the column is long enough, the zones for each individual solute will draw apart from one another sufficiently to be recovered in the effluent as separate solutions. The term "chromatography" arose from the early use of this procedure in separations of plant pigments and the laboratory identification of other separated components by colorimetric tests. Chemical engineers have two interests in chromatography: first, to understand and improve its use as an analytical method; second, to use it as a production-line separation procedure when applicable.

Chromatography can occur in either a bulk mode or a trace mode; usually one or the other of these extremes will be preferred to intermediate conditions. In the bulk mode, a mixture of only a few components is fed in relatively large amount; examples are liquid-phase separations of hydrocarbons on adsorbents and of rare-earth ions on ion exchangers (Fig. 16-31). A number of transitions evolve, and in certain cases the calculation problem reduces to separate calculations for the respective transitions (Helfferich and Klein; see "General References"). The general problem is similar to the consideration of multicomponent systems given later in this section but is still more complex. Bulk-mode chromatography is also called "displacement development."

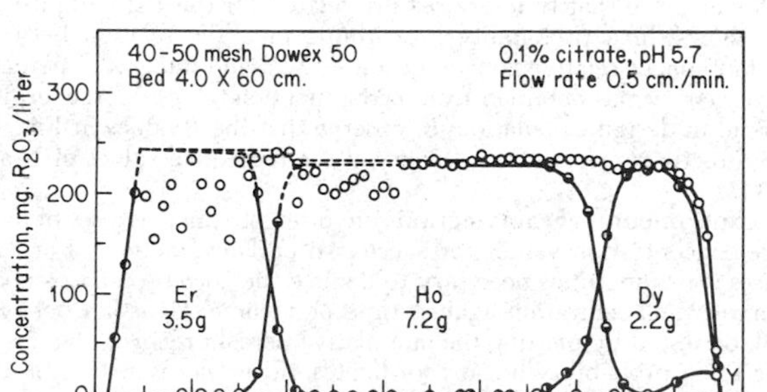

FIG. 16-31 Bulk chromatographic separation of rare-earth ions, showing saturation of resin phase. [*Spedding and Powell,* J. Am. Chem. Soc., *76, 2550 (1954); reprinted by permission.*]

In the **trace** case, only a small amount of solute mixture is admitted to the column. Especially for analytical purposes or in the processing of flavor or perfume extracts, the mixture may contain many different components. On the basis of different affinities for the sorbent, the different solutes tend to travel through the column at different speeds and hence are recovered in the effluent as a series of separate bell-shaped concentration bands, or zones.

With both $\Sigma c_A \ll C_0$ (or c_0°) and $\Sigma q_A \ll Q$ (or q_0°), the interactions between pairs of trace solutes present can be neglected. For trace conditions, the interaction between each trace solute and the carrier or dummy component is the same as if no other trace solute were present. The applicable equilibrium parameter **R** can then be calculated from the **r** involving the individual solute and the carrier component, by Eq. (16-18).

If $\mathbf{R} = 1$, for a trace solute, the chromatographic zone for that component tends to become symmetric about its peak concentration as the zone travels through a relatively long column. Asymmetric chromatograms are usually caused by nontrace condition and resultant nonlinear equilibrium. For a column of given length, the smaller the separation factor for saturation (**R**) or for elution ($\mathbf{R}^\dagger$), the steeper the saturation or elution curve. However, if **R** is small, $\mathbf{R}^\dagger$ is large [by Eq. (16-17)] and vice versa. If $\mathbf{R} < 1$ for saturation, the leading edge of the zone—the first to emerge in the effluent—will be steeper than the trailing edge; if $\mathbf{R} > 1$ for saturation, the leading edge will be more gradual than the trailing edge. Representative zone shapes for $\mathbf{R} < 1$, $\mathbf{R} = 1$, and $\mathbf{R} > 1$ are given as curves of **X** versus **T** in Fig. 16-32.

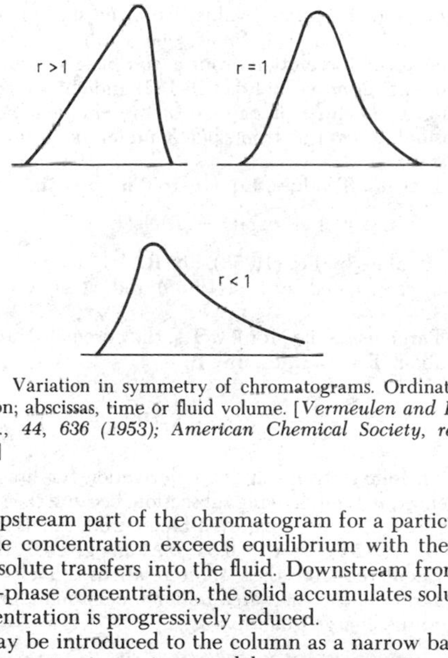

FIG. 16-32 Variation in symmetry of chromatograms. Ordinates represent concentration; abscissas, time or fluid volume. [*Vermeulen and Hiester,* Ind. Eng. Chem., *44, 636 (1953); American Chemical Society, reprinted by permission.*]

In the upstream part of the chromatogram for a particular solute, the particle concentration exceeds equilibrium with the oncoming fluid, and solute transfers into the fluid. Downstream from the peak of particle-phase concentration, the solid accumulates solute and the fluid concentration is progressively reduced.

Feed may be introduced to the column as a narrow band of high nontrace concentration, as in many laboratory separations. In elution, the resulting individual zones will first assume a flattened-top shape, then a peaked asymmetric shape, and ultimately (because of the reduction of the peak concentration level) a peaked symmetric shape consistent with trace conditions. The quantity of material fed in this manner that can be separated at maximum efficiency by a given column, and the eventual chromatogram formed, are nearly identical with those for the same material fed under trace conditions [compare Eq. (16-18)] as an initially broader band of smaller amplitude. The nontrace case has been derived only for a single solute, in simple adsorption; for a mixed feed of a solute and the eluting component, in exchange adsorption; and for multiple solutes in local equilibrium theory with constant **r**'s.

Reaction-Kinetic CSF Model In the simplest mathematical model for the chromatographic process, the flow rate *F* and total concentration level C_0 for elution are taken as the basis for the saturating period as well. Fluid containing the saturating component at C_0 is considered to have passed through the column during a time interval t_{sat} corresponding to volume V_{sat} and throughput ratio $\mathbf{T}_{sat}$. When time t_{sat} is reached, the eluting fluid begins to enter the bed. This elutant stream differs from the previous fluid through the omission of saturating component (or in replacement of saturant by additional eluting component to keep C_0 constant).

The reaction-kinetic model for nonlinear equilibrium applies to single chromatograms (two-component systems) under nontrace conditions. The resulting equations are

$$\mathbf{X}^{-1} = 1 + \frac{1}{\mathbf{J}(\mathbf{R}N_R, N_R\mathbf{T}) - \mathbf{J}(\mathbf{R}N_R, N_R\mathbf{T}')} \{\exp[(\mathbf{R}-1)N_R(\mathbf{T}-1)] \times [1 - \mathbf{J}(N_R, \mathbf{R}N_R\mathbf{T})] + \exp[(\mathbf{R}-1)N_R(\mathbf{T}'-1)] \times [\mathbf{J}(N_R, \mathbf{R}N_R\mathbf{T}')]\} \quad (16\text{-}165)$$

$$\mathbf{Y}^{-1} = 1 + \frac{1}{\mathbf{J}(N_R\mathbf{T}', \mathbf{R}N_R) - \mathbf{J}(N_R\mathbf{T}, \mathbf{R}N_R)} \{\exp[(\mathbf{R}-1)N_R(\mathbf{T}-1)] \times [\mathbf{J}(\mathbf{R}N_R\mathbf{T}, N_R)] + \exp[(\mathbf{R}-1)N_R(\mathbf{T}'-1)] \times [1 - \mathbf{J}(\mathbf{R}N_R\mathbf{T}', N_R)]\} \quad (16\text{-}166)$$

These results were obtained by Hiester and Vermeulen, *J. Chem. Phys.*, **16,** 1087 (1948), and by Goldstein, *Proc. R. Soc. (London),* **A219,** 151, 171 (1953). A related equation applying to axial-dispersion-limited chromatography has been given by Houghton, *J. Phys. Chem.*, **67,** 84 (1963).

The unprimed values of **T** are measured from the start of the saturation period, and the primed values from the start of the elution period. It can be shown that, for $\mathbf{T}_{sat}$ large, these equations will reduce to the form for elution from a completely saturated bed. These solutions are annexed to Eqs. (16-132) and (16-133) at $\mathbf{T}' = 0$, corresponding to the time of passage of the elutant fluid front. **Y** remains constant across this front, but **X** undergoes a usually small discontinuity.

For $\mathbf{R} < 1$, at low **T** values, Eq. (16-165) is approximated by

$$\mathbf{X}^{-1} = 1 + \exp[(1-\mathbf{R})N_R(1-\mathbf{T})] \quad (16\text{-}167)$$

and at high **T** values by Eq. (16-58). For $\mathbf{R} > 1$, at low **T** values, Eq. (16-165) is approximated by Eq. (16-58) and, at high **T** values, by Eq. (16-167).

For low charge levels, i.e., for low $\mathbf{T}_{sat}$, the chromatogram becomes symmetric about $\mathbf{T} = 1/\mathbf{R}$, for any **R**:

$$\mathbf{X} = \frac{\mathbf{R}\mathbf{T}_{sat}\sqrt{N_R}}{2\sqrt{\pi}} \exp[-N_R(\mathbf{R}\mathbf{T}_m - 1)^2] \quad (16\text{-}168)$$

This equation follows from an exact derivation for linear equilibrium, referenced in the following subsection, because the short-range behavior of any constant-separation-factor isotherm is also linear. Here $\mathbf{T}_m = \mathbf{T}' + 0.5\mathbf{T}_{sat} = \mathbf{T} - 0.5\mathbf{T}_{sat}$. At any given **T**, the smallest value of **X** calculated from these three relations is likely to be most correct. In addition, the total area under the chromatogram should correspond to the charge quantity:

$$\int \mathbf{X}\, d\mathbf{T} = \mathbf{T}_{sat} \quad (16\text{-}169)$$

Linear Equilibrium and Gaussian Relation Equation (16-149) gives directly the linear-isotherm chromatographic curve. If $\mathbf{X} = 0$ for $x = x'$, and $\mathbf{X} = 1$ for $x = x''$, with $\mathbf{T} = T_2 + T'''$, $\mathbf{T}_{sat}$ replacing T_2, and $\mathbf{T}'$ replacing T''', the result is

$$\mathbf{X} = \mathbf{J}(\mathbf{N}, \mathbf{N}\mathbf{T}) - \mathbf{J}(\mathbf{N}, \mathbf{N}\mathbf{T}') \quad (16\text{-}170)$$

Equation (16-170) is also obtained from Eq. (16-165) if $\mathbf{R} = 1$. Thus the sequence of saturation and elution can be described by the difference between the concentration curve for saturation starting at **T** = 0 and continuing through the elution period and a second curve representing a nonexistent saturant which starts at $\mathbf{T}' = 0$. Boyd and coworkers [*J. Am. Chem. Soc.*, **69,** 2836, 2849 (1947)], Stene [*Arkiv Kemi Mineral. Geol.*, **18**(18), (1945)], Vermeulen and Hiester [*Ind. Eng. Chem.*, **44,** 636 (1952)], and Glueckauf [*Trans. Faraday Soc.*, **51,** 1540 (1955)] have all considered the linear-isotherm chromatogram to be a composite of separate saturation and elution curves. Figure 16-33 shows the transition from complete saturation to incomplete saturation that is given by Eq. (16-165), at $\mathbf{R} = 1$ and $N_R = 80$, as $\mathbf{T}_{sat}$ is progressively reduced.

The lower two curves of Fig. 16-33 are seen to be symmetrical and to have their **X** values at each **T** in a nearly constant ratio. Such

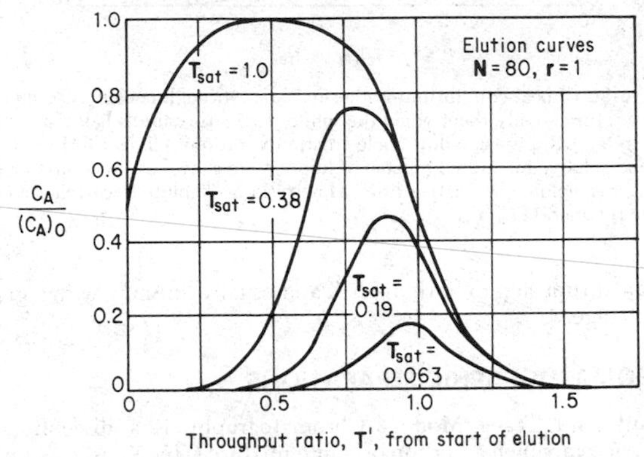

FIG. 16-33 Elution curves for different charge periods.

curves can be fitted by a gaussian distribution, when $N_R > 50$ and also $\mathbf{T}_{sat} < 0.5\sqrt{\pi/N_R}$. Using the error-function approximation to Eq. (16-125), Vermeulen and Hiester (loc. cit.) derived the gaussian relation [Eq. (16-168)] for the case of $\mathbf{R} = 1$. Corresponding to that equation, the peak occurs at $\mathbf{T}_m = 1$, and the concentration relative to the peak value is given by

$$\ln \frac{\mathbf{X}}{\mathbf{X}_{max}} = -\frac{1}{4}\mathbf{N}(\mathbf{T}_m - 1)^2 = -\frac{1}{4}\mathbf{N}\left(\frac{V - V_{peak}}{V_{peak}}\right)^2 \quad (16\text{-}171)$$

Mayer and Tompkins [*J. Am. Chem. Soc.*, **69,** 2868 (1947)] and Matheson [private communication, reported by Tompkins, *J. Chem. Educ.*, **26,** 92 (1949)] have given the equilibrium-stage result

$$\ln \frac{\mathbf{X}}{\mathbf{X}_{max}} = -\frac{1}{2}N_c\frac{\Lambda + \epsilon}{\Lambda}\left(\frac{V - V_{peak}}{V_{peak}}\right)^2 \quad (16\text{-}172)$$

At large Λ, the effective number of equilibrium contacts N_c is half of the number of reaction units or transfer units **N**. With this adaptation, the results given by the two approaches are identical and conform to operational experience.

The instantaneous concentration of a component in the column effluent and the total recovery of that component in all portions of effluent up to that time are readily related for this case. Figure 16-34 shows, in a probability logarithmic plot, the relation between instantaneous relative concentration $\mathbf{X}/\mathbf{X}_{max}$ and integrated percentage recovery, as obtained from error-function tables. Moreover, it is useful in design calculations to observe that the **J** values in Fig. 16-18 directly represent the recovery at the respective values of **N** and $\mathbf{NT}_m$.

Experimental chromatograms are often obtained as a set of average-concentration values for successive outflow fractions. For such cases it is sometimes necessary to deduce the corresponding smooth curve of concentration against time or volume. Gaussian behavior can be tested by plotting the cumulative percent recovery against V (or **T**) on probability-linear coordinates; if the plot is a straight line, gaussian behavior is confirmed and points on the plot can be used with corresponding points on Fig. 16-34 to construct the complete **X** versus V (or **T**) curve. (If the probability-linear plot is curved, it will still aid in interpolating between the measured points. The same data should be plotted as moles versus volume on linear coordinates, and

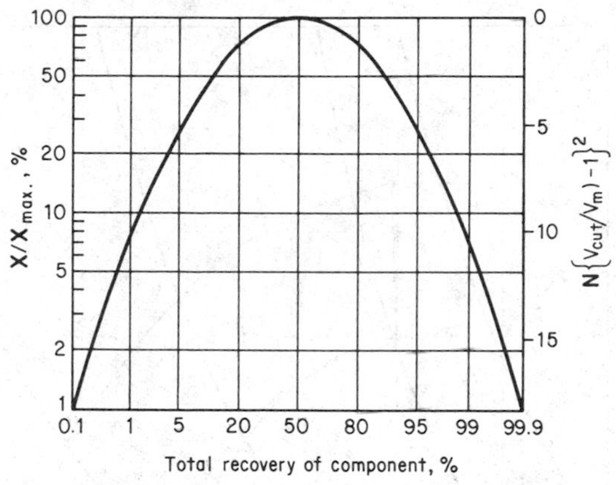

FIG. 16-34 Relation between local solute concentration and integral recovery for a solute band conforming to gaussian distribution.

successive slopes along this curve will provide the instantaneous concentrations.)

Method of Moments A method of mathematical analysis adapted from probability theory, moment analysis, has been applied to chromatographic curves and also to breakthrough curves generated by single-step functions (which can be viewed as the integrals of chromatograms). Foundations for this method were laid by Schneider and Smith [*Am. Inst. Chem. Eng. J.*, **14,** 762 (1968)] and Suzuki and Smith [*Chem. Eng. Sci.*, **26,** 221 (1971)] and were developed further by Carbonell, McCoy, and Razavi [*Chem. Eng. J.*, **9,** 115 (1975); **16,** 211 (1978)].

Design of Trace-Solute Separations To separate two or more solutes on a production-line basis, operating conditions should normally be established in the following order:

1. The adsorbent composition that gives the greatest difference in partition ratio Λ for the different components is selected, V_{peak} being proportional to Λ for each component. This selection should be accompanied by a choice of temperature and of total concentration or pressure level.

2. The average feed-mixture charging rate, molar or volumetric, is fixed by a raw-material supply or by the demand for finished product.

3. The range of **N** values needed to give a reasonable extent of separation is determined from Eq. (16-171). The cut-point volume V_{cut} which represents equal recovery of the two solutes with peaks V_1 and V_2 is given by $2V_1V_2/(V_1 + V_2)$. Thus,

$$\frac{V_{cut} - V_1}{V_1} = \frac{V_2 - V_{cut}}{V_2} = \frac{V_2 - V_1}{V_1 + V_2} \tag{16-173}$$

If the selectivity coefficient α_{21} is equated to V_2/V_1, Eq. (16-171) shows that ln $(\mathbf{X}/\mathbf{X}_{max})$ is proportional to the product $\mathbf{N}[(\alpha_{21} - 1)/(\alpha_{21} + 1)]^2$.

4. To reduce the **N** required while increasing the total feed to the column, two cut points may be set between the two peaks with the intervening low-purity fraction used as a **recycle.**

5. To determine the corresponding length of column, as a ratio to particle diameter, Fig. 16-12 should be used. The optimum Péclet number for flow is around 100 for liquids and around 30 for gases if fluid-side resistance controls or somewhat below these values if particle-side resistance is rate-determining. The allowable pressure drop influences the choice of particle size and helps to determine the column length.

6. When only a few solutes are being separated, they may occupy only a small part of the total column volume at any given instant. In such cases the resin utilization can be improved and the bed area and volume decreased by **multiple cycling**—that is, by adsorbing one (or more) additional feed charge before elution of the first charge has been completed. In this way, the effluent can become an almost continuous succession of trace zones.

7. If the calculated column length is excessive, the question arises whether more than one column should be used. Columns in series, with minimum fluid holdup between, function like segments of a single column. The general rule is that it is wasteful to remix partially separated effluents. Multiple-column operation is desirable in two separate situations—first, when extraneous solutes have been removed in the initial step and closer fractionation remains necessary to remove residual impurities; second, when the desired solute is a very minor fraction of the total feed charge and a change in the scale of operation is needed to complete its recovery and purification (an example of this case is given by Vermeulen and Hiester, loc. cit.).

MULTIVARIANT SYSTEMS: MULTIPLE TRANSITIONS

NONISOTHERMAL (ADIABATIC) ADSORPTION

In the adsorption of a nontrace gaseous component, the heat released may produce a marked departure from isothermal conditions so that a partial breakthrough occurs much ahead of the normal near-stoichiometric time. The heat evolved creates an intermediate plateau zone between the preexisting column condition and the eventual transition to unchanged feed, with higher temperature (a thermal "chromatogram") and reduced equilibrium uptake. This case has been discussed by Grayson, *Ind. Eng. Chem.*, **47,** 41 (1955); Lightfoot et al., in Schoen (ed.), *New Chemical Engineering Separation Techniques*, Interscience, New York, 1962; Leavitt, *Chem. Eng. Prog.*, **58**(8), 54 (1962); Amundson, Aris, and Swanson, *Proc. R. Soc. (London),* **A286,** 129 (1965); Meyer and Weber, *Am. Inst. Chem. Eng. J.*, **13,** 457 (1967); Lee and Cummings, *Chem. Eng. Prog. Symp. Ser.*, **63**(74), 42 (1967); Chase, Gidaspow, and Peck, ibid., **65**(92), 91 (1969); **65**(96), 34 (1969); Rhee et al., *Chem. Eng. J.*, **1,** 241, 279 (1970); **3,** 22, 121 (1972); Pan and Basmadjian, *Chem. Eng. Sci.*, **26,** 45 (1971); and Basmadjian, Ho, and Pan, *Ind. Eng. Chem. Process Des. Dev.*, **14,** 328 (1975). Other topics of interest include adiabatic constant-pattern behavior [Pan and Basmadjian, *Chem. Eng. Sci.*, **22,** 285 (1967); and Ruthven, Garg, and Crawford, *ibid.*, **30,** 803 (1975)]; nonisothermal adsorption with various rate mechanisms [Cooney, *Ind. Eng. Chem. Process Des. Dev.*, **13,** 368 (1974); Basmadjian, ibid., **19,** 137 (1980); and Harwell et al., *Chem. Eng. Sci.*, **35,** 2287 (1980)]; and solute condensation in adsorption beds during thermal regeneration [Friday and LeVan, *Am. Inst. Chem. Eng. J.*, **28,** 86 (1982)].

By analogy to Eq. (16-48) without the dispersion term, the differential enthalpy balance can be written:

$$\frac{\partial(fh_f)}{\partial v} + \frac{\partial h_p}{\partial V} + \epsilon\frac{\partial h_f}{\partial V} = 0 \tag{16-174}$$

with

$$h_p = (C_p + C_p'\rho_b q)\Theta - (-\Delta H)\rho_b q \tag{16-175}$$

$$h_f = C_f\Theta \tag{16-176}$$

Here C_p is the heat capacity of sorbent only, per unit packed volume, and C_p' is the molar heat capacity of solute. C_f is heat capacity of the fluid and contained solute per unit fluid volume. Θ is the temperature relative to a reference value T_{Ki} (e.g., column input in the exhaustion step). $(-\Delta H)$ is the molar integral heat of adsorption at concentration q and temperature $\Theta = 0$. Finally, f is the ratio of the local flow rate to the flow rate at the column entrance.

The method used to solve the coupled material and enthalpy balances takes advantage of the distinct similarity between them. In fact, the enthalpies are analogous to concentrations; and the problem is bivariant, analogous to isothermal adsorption of two components. Application of the method of characteristics (see subsection "Numerical Methods") to Eqs. (16-48) and (16-174) gives the differential relation for a gradual transition:

$$T\Lambda = \rho_b \frac{dq}{d(fc)} = \frac{dh_p}{d(fh_f)} \tag{16-177}$$

A similar equation with differences replacing derivatives applies for an abrupt transition.

To obtain concentration and temperature profiles, the two transitions are assumed to be gradual, and Eq. (16-177) is solved along each path beginning at the respective end points: initial condition of the bed, for the downstream transition, and feed condition, for the upstream transition. If either of the solutions fails to evolve continuously in the expected direction, the difference form is used for such path or paths. The two paths should intersect to give the composition and temperature of the central plateau.

An approximate model can be developed by using constant mean heat capacities C_{pm} and C_{fm}, a mean heat of adsorption, and a constant volumetric flow rate. Equation (16-177) then becomes

$$T\Lambda = \rho_b\left(\frac{\partial q}{\partial c} + \frac{\partial q}{\partial \Theta}\frac{d\Theta}{dc}\right)$$
$$= \frac{C_{pm}}{C_{fm}} - \frac{(-\Delta H)_m \rho_b}{C_{fm}}\left(\frac{\partial q}{\partial \Theta} + \frac{\partial q}{\partial c}\frac{dc}{d\Theta}\right) \tag{16-178}$$

The derivatives $\partial q/\partial c$ and $\partial q/\partial \Theta$ depend only on the adsorption equilibrium conditions. Solving for $dc/d\Theta$ by using the quadratic formula gives

$$dc/d\Theta = [-b \pm (b^2 - 4ad)^{0.5}]/2a \tag{16-179}$$

with $a = [(-\Delta H)_m \rho_b / C_{fm}]\partial q/\partial c \qquad d = \rho_b \partial q/\partial \Theta$

$b = [\rho_b \partial q/\partial c] - [C_{pm}/C_{fm}] + [(-\Delta H)_m \rho_b/C_{fm}]\partial q/\partial \Theta$

The plus sign is taken for the downstream transition, and the minus sign for the upstream one. Thus Eq. (16-178) can be integrated over a gradual transition by using small steps of either Θ or c. For the approximate model with an abrupt transition, Eq. (16-177) with appropriate modifications is rewritten in difference form. Two solution paths pass through each composition point, and care must be taken to ensure that the correct path is obtained.

Example 6: Adiabatic Vapor Adsorption in and Thermal Regeneration of a Fixed Bed of Activated Carbon These processes are examined here, with benzene as solute and nitrogen as carrier. The total system pressure is 10 atm. (Extensive analysis has been made of this system, including some study of coadsorption of cyclohexane; see Rhee et al., Harwell et al., and Friday and LeVan, all cited earlier). The isotherm and physical constants are

$$\rho_b q = \rho_b QKc/(1 + Kc)$$
$$\rho_b Q = 2750 \text{ mol/m}^3$$
$$K = 3.88 \times 10^{-8}(T_{Ki} + \Theta)^{0.5} \exp[5250/(T_{Ki} + \Theta)] \text{ m}^3/(\text{mol}\cdot\text{K}^{1/2})$$
$$C_{pm} = 850 \text{ kJ/(m}^3\cdot\text{K)}$$
$$C_{fm} = 11.3 \text{ kJ/(m}^3\cdot\text{K)}$$
$$(-\Delta H)_m = 43.5 \text{ kJ/mol}$$
$$T_{Ki} = 320 \text{ K}$$
$$c_{ref} = 6 \text{ mol/m}^3$$

From the isotherm, $\rho_b q_{ref} = 2701$ mol/m^3, giving $\Lambda = 450$. In the figures, $\Gamma(k)$ and $\Sigma(k)$ symbolize gradual and abrupt transitions respectively, with $k = 1$ the downstream and $k = 2$ the upstream transition.

Adiabatic adsorption. An initially clean bed at $\Theta = 0$ is fed with an input stream at $\Theta = 0$ and $c = c_{ref}$. Construction of the equilibrium paths is shown in Figs. 16-35 and 16-36. The (1) transition begins at point *A*, the initial condition of the bed. Since $\partial q/\partial \Theta = 0$ here, $dc/d\Theta = 0$ and Eq. (16-178) simplifies to $T = C_{pm}/(\Lambda C_{fm}) = 0.167$, which indicates a pure thermal wave along the $c = 0$ axis of Fig. 16-35, the c versus Θ, or "isostere," plot. The second transition begins at point *B*, the feed condition; a $\Gamma(2)$ transition is attempted but fails, so a $\Sigma(2)$ curve is calculated and plotted in the two figures. The $\Gamma(1)$ and $\Sigma(2)$ curves intersect at point *C*, where $\Theta = 28°$. Breakthrough curves of temperature and concentration are shown in Fig. 16-37.

Thermal regeneration. A bed initially at the reference condition is regenerated with benzene-free nitrogen at $\Theta = 80°$. The feed state, point *D*, provides a successful $\Gamma(2)$ transition; the initial condition, point *B*, yields a $\Gamma(1)$ path which violates the slope criterion, so a $\Sigma(1)$ abrupt transition which differs imperceptibly from $\Gamma(1)$ is determined. The $\Sigma(1)$ and $\Gamma(2)$ paths meet at point *E*, where $\Theta = 0.8°$, $c = 8.93$ mol/m^3, and $\rho_b q = 2716$ mol/m^3; the solid-

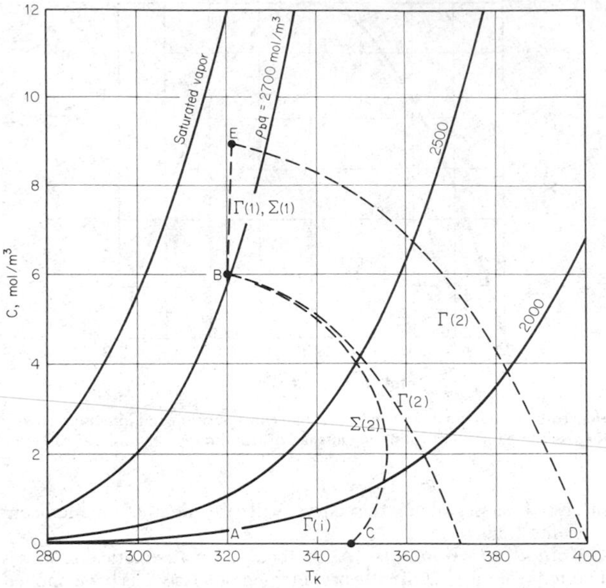

FIG. 16-35 Local-equilibrium transition paths in c versus T_K plane for adiabatic adsorption and thermal regeneration. To convert kilogram-moles per cubic meter to pound-moles per cubic foot, multiply by 6.243×10^{-2}; °R = ⁹⁄₅ K.

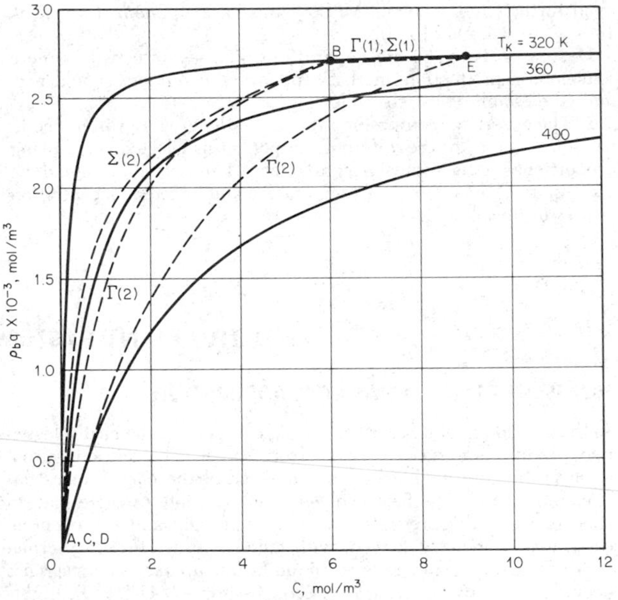

FIG. 16-36 Transition paths in isotherm plane for adiabatic adsorption and thermal regeneration. To convert kilogram-moles per cubic meter to pound-moles per cubic foot, multiply by 6.243×10^{-2}; °R = ⁹⁄₅ K.

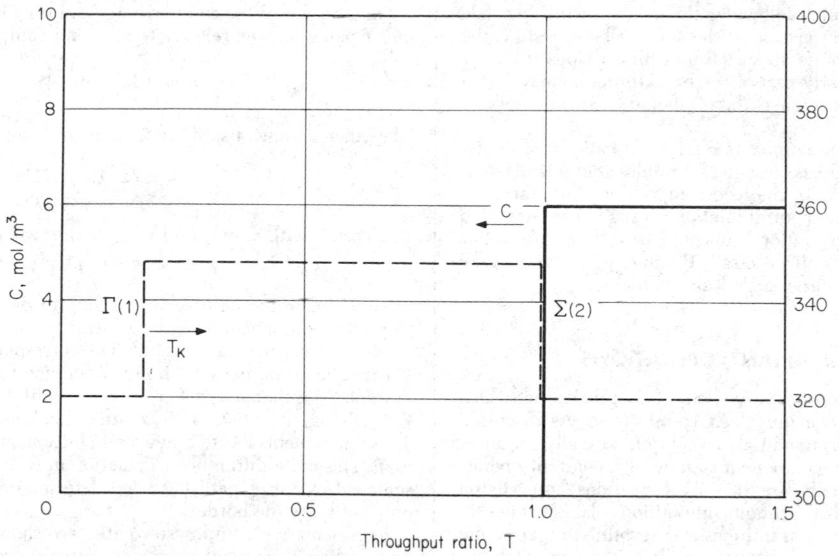

FIG. 16-37 Breakthrough curves for adiabatic adsorption. To convert kilogram-moles per cubic meter to pound-moles per cubic foot, multiply 6.243 × 10^{-2}; °R = 9/5 K.

phase loading here is *greater* than the initial loading. Breakthrough curves for temperature and concentration are shown in Fig. 16-38, with the Σ(1) transition leaving the bed at T = 0.011. Transition Γ(2) begins to emerge at T = 0.169, but regeneration is not complete until [from Eq. (16-178) at Θ = 80°], T = 2.38.

It is possible (with lower initial bed temperature, higher initial loading, or higher regeneration temperature) for the transition paths to contact the saturated-vapor curve in Fig. 16-35 rather than to intersect beneath it. For this case, liquid benzene condenses in the bed and the effluent vapor is saturated during part of the regeneration.

MIXED-BED DEIONIZATION

Stoichiometric Proportions of Cationic and Anionic Resins Two resins of opposite type, one acidic and one in alkaline form, can be used together, so that salts are irreversibly sorbed with displaced H^+ and OH^- ions undergoing mutual neutralization; this method is widely applied to produce "ultrapure" water. Matched amounts of resin yield a single breakthrough, with neutralization overriding the separate equilibriums of the exchange steps, so that **R** = 0 for the combined step. (Exceptions arise only with combinations of weak-acid and weak-base resins, or at concentrations approaching 10^{-7}M.)

Rate behavior will often be determined by fluid-side resistance because of the normally low feed concentrations; here the effective mass-transfer rates will be comparable with those for a single resin at the same superficial velocity of fluid, and Table 16-7, item 1*G* will usually apply. [Data and calculations for this case are given by Frisch and Kunin, *Am. Inst. Chem. Eng. J.*, **6**, 640 (1960).] Nonporous

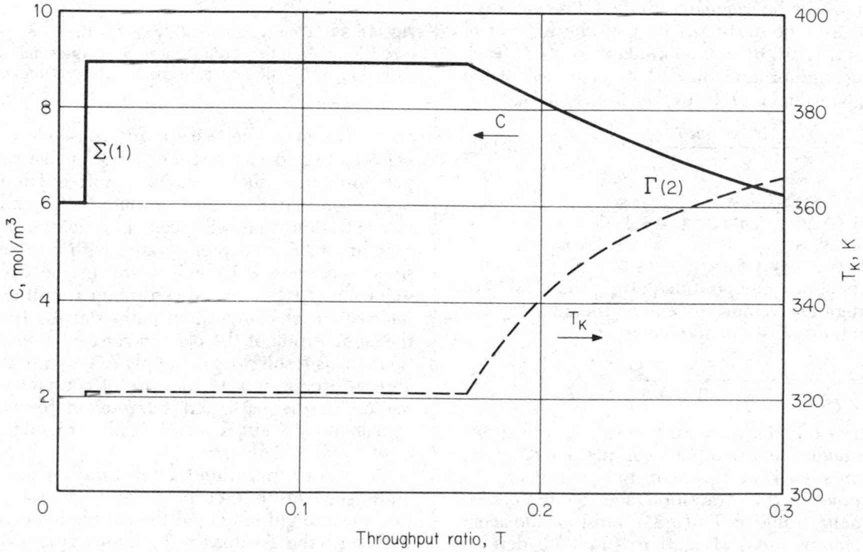

FIG. 16-38 Breakthrough curves for adiabatic regeneration. To convert kilogram-moles per cubic meter to pound-moles per cubic foot, multiply by 6.243 × 10^{-2}; °R = 9/5 K.

weak-acid or weak-base resins will widen the range in which particle diffusion controls. In the latter case, if the acid (or base) resin is the slower, a temporary pH pulse about 0.5 unit higher (lower) than 7 should occur during the early part of the breakthrough curve and a pulse about 0.5 unit lower (higher) than 7 should occur as complete saturation is approached.

Nonstoichiometric Proportions of Resins If cationic (anionic) resin is present in excess, breakthrough of a mildly acid (alkali) solution will precede the arrival of the feed composition. The transition from deionized water to nonneutral solution is usually characterized by $\mathbf{R} = 0$; the upstream transition is uncoupled, and hence exhibits the separate acid-to-salt (or alkali-to-salt) $\mathbf{R}$ value, which may make the equilibrium transition either abrupt or gradual.

THREE-COMPONENT (BIVARIANT) OPERATIONS

The treatment to be given here is restricted to a column with *uniform presaturation* and *constant feed composition*. As discussed earlier, two-component sorption in such a column normally produces a single transition. In an n-component system with constant operating conditions, there usually are $n - 1$ transitions, intervening between n traveling zones of constant composition (plateau zones) in the column. The ensemble of transitions is determined by applying stoichiometry and equilibrium criteria, usually in the form of local-equilibrium theory which considers the stationary and moving phases to be at equilibrium at each separate point in the column.

A transformation of the concentration variables, reported in Helfferich and Klein (see "General References") leads to quantitative prediction of the equilibrium behavior of **constant-separation-factor** and Langmuir-type systems of any number of components, for any feed composition history and column presaturation profile. A related, extensive study of such systems has been given by Rhee, Aris, and Amundson, *Philos. Trans. Soc.*, **267A**, 419–455 (1970).

Mass-Action or Other Equilibrium For arbitrary equilibrium properties not equivalent to constant separation factors, calculations available now are mainly limited to three-component systems. They have been developed furthest for ion-exchange systems with isotherms for each pair of solutes based on ideal mass-action relations in a general form equivalent to Eq. (16-13*a*):

$$(y_i/x_i)^{\nu_j}\,(x_j/y_j)^{\nu_i} = K_{ij} \qquad (16\text{-}180)$$

where i and j are any two exchangeable ion species of the system and ν_i and ν_j are their valences. For such systems, the composition paths in gradual transitions can be calculated in the following way.

For the most general case, the starting point is in the **interior** of the composition diagram (Fig. 16-6) with coordinates x_1, x_2, x_3 (each $\neq 0$). The corresponding equilibrium values y_1, y_2, y_3 are calculated first. One then obtains two values of T from the expression

$$T = \frac{b \pm (b^2 - 4ac)^{0.5}}{2a} \qquad (16\text{-}181)$$

where $a \equiv (\nu_1 y_1 + \nu_2 y_2 + \nu_3 y_3) x_1 x_2 x_3$

$b \equiv (\nu_2 x_2 + \nu_3 x_3) x_1 y_2 y_3 + (\nu_1 x_1 + \nu_3 x_3) x_2 y_1 y_3 + (\nu_1 x_1 + \nu_2 x_2) x_3 y_1 y_2$

$c \equiv (\nu_1 x_1 + \nu_2 x_2 + \nu_3 x_3) y_1 y_2 y_3$

The infinitesimal element of a composition path in the vicinity of the point under consideration is now given by the ratio of the changes of the concentrations of two of the species, as

$$\frac{dx_i}{dx_k} = \frac{dy_i}{dy_k} = \frac{\nu_i\, x_i y_i}{\nu_k\, x_k y_k}\,\frac{Tx_k - y_k}{Tx_i - y_i} \qquad (16\text{-}182)$$

Since there are two values of T, this ratio has two values, so that two composition paths go through every point of a composition diagram. With the transitions numbered from upstream to downstream, the larger value of T corresponds to a 1,2 transition (adjacent to the feed composition) and the smaller value of T to a 2,3 transition (adjacent to the preexisting solid composition). [Equation (16-181) is derived from Eq. (16-182) by way of Eq. (16-192), given later.]

On a **border** of the composition diagram with $x_i = 0$ and $y_i = 0$, one T value is given relative to a second component j:

$$T = \left[K_{ij}\left(\frac{y_j}{x_j}\right)^{\nu_i}\right]^{1/\nu_j} \qquad (16\text{-}183)$$

The other T value, based on components j and k only, is

$$T = \frac{y_j y_k}{x_j x_k}\,\frac{\nu_j x_j + \nu_k x_k}{\nu_j y_j + \nu_k y_k} \qquad (16\text{-}184)$$

In a **corner** with $x_i = 1$ and $y_i = 1$, the two T values are

$$T = K_{ji}^{1/\nu_i} \qquad T = K_{ki}^{1/\nu_i} \qquad (16\text{-}185)$$

To calculate the path corresponding to the appropriate choice of transition and hence of T, an arbitrarily small finite difference δx_k (positive or negative) is selected. The corresponding difference δy_k is $T\,\delta x_k$. The accompanying finite difference for a second solute (for example, i) is then $\delta x_i = (dx_i/dx_k)\,\delta x_k$, with the derivative given by Eq. (16-182). Also, $\delta y_i = T\,\delta x_i$; and $\Sigma(\delta x)$ and $\Sigma(\delta y)$ are both zero. These increments locate a new composition, and with it a new value of T. This finite-difference calculation must thus be repeated until a sufficient extent of path has been determined. (From a corner, the only paths are the borders.)

Representative composition paths are shown in Fig. 16-39. The arrows show the directions of decreasing values of T; a composition change in the direction of an arrow will correspond to a gradual tran-

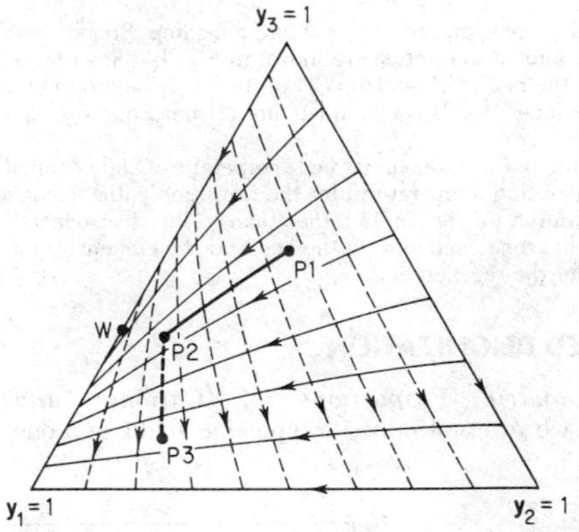

FIG. 16-39 Composition-path grid for three-component ion exchange governed by mass-action equilibrium. Solid lines indicate 1,2 (upstream) paths; dashed lines, 2,3 (downstream) paths. (*After Tondeur.*)

sition. (In principle, except for constant α, a selectivity reversal between two species might occur on such a path, at which α_{ij} would pass through 1; such behavior is infrequent and for simplicity will not be considered here.) The route $P_1P_2P_3$ indicates the sequence of concentrations that will occur in a column receiving a feed corresponding to P_1 and presaturated with composition P_3. The composition sequence P_1P_2 will occur in the upstream transition; the sequence P_2P_3, in the downstream transition. P_2, obtained as the intersection of composition paths starting from P_1 and from P_3, is the composition of the plateau zone intervening between the transitions. [This result does not apply exactly if either of the transitions is abrupt, except only in the case of constant separation factor. Estimation of the path and intersection for an abrupt transition is explained by Klein, Tondeur, and Vermeulen, *Advan. Chem. Eng.*, 2, 147–208 (1958).

Constant-Separation-Factor Equilibrium In the case of **constant separation factors** ($\nu_i = \nu_j = \nu_k$), the composition paths become straight lines, and the calculation is correspondingly simplified. With the T value ($= T_1$) known from Eq. (16-181) for the point at which the path leaves the feed plateau, Eq. (16-182) with feed-plateau conditions gives the slope C_1 ($= dx_i/dx_k$) for the 1,2 transition. Similarly the T value where the 2,3 path leaves the presaturation plateau leads by Eq. (16-182) to the slope C_3 ($= dx_i/dx_k$) for the 2,3 transition. The paths intersect in the plateau-2 composition:

$$x_{k2} = \frac{C_1 x_{k1} - C_3 x_{k3} + x_{i3} - x_i}{C_1 - C_3} \tag{16-186}$$

with

$$x_{i2} = x_{i1} + C_1(x_{k2} - x_{k1}) \tag{16-187}$$

The plateau y's are the equilibrium values [cf. Eqs. (16-24) to (16-26)]. Each transition is then tested by the rules given in a following subsection to determine whether it is abrupt or gradual. If abrupt, Eq. (16-62) again applies. If gradual, the path is calculated from Eq. (16-182) in the form

$$T = \left(\frac{Cx_k - x_i}{Cy_k - y_i}\right)\left(\frac{y_i y_k}{x_i x_k}\right) \tag{16-188}$$

where C is C_1 for path 1,2, or C_3 for path 2,3. After testing whether the transitions are each abrupt or gradual and introducing the equilibrium relations, the concentration profiles and breakthrough behavior (plateaus plus transitions) become completely known.

Graphical calculations for a constant-separation-factor case can be performed easily on an x-value or y-value plot like Fig. 16-6 or 16-39, since the composition-path contours are all linear. The watershed point, on the 13 edge, is located by the relation

$$(x_1)_W = \frac{\alpha_{12} - 1}{\alpha_{13} - 1} \quad \text{or} \quad (y_1)_W = \frac{1 - \alpha_{21}}{1 - \alpha_{31}} \tag{16-189}$$

Each of the various 1,2 composition paths has an intersection with the 13 edge at the same fractional distance between W and the component-1 corner as its other intersection between the 3 corner and the 2 corner along the 32 edge; so the proper 1,2 path through the feed composition P_1 can be found quickly. Likewise, each of the 2,3 paths has a 12-edge intersection at the same fractional distance between W and the component-3 corner as its other intersection along the edge between the 1 corner and the 2 corner. Thus a path through the presaturant composition P_3 can be constructed. The intersection of the respective paths through P_1 and P_3 provides the intermediate-plateau composition P_2.

Parameters for Individual Transitions If a transition is gradual and if its equilibrium is described by **constant separation factors,** the local-equilibrium **X** and **Y** values are identical for all components (calculated in relation to the two adjacent plateau compositions) and are given by relations analogous to Eq. (16-58):

$$\mathbf{X} = \frac{\mathbf{R} - \sqrt{\mathbf{R}\bar{T}/T}}{\mathbf{R} - 1} \tag{16-190}$$

and

$$\mathbf{Y} = \frac{\sqrt{\mathbf{R}T/\bar{T}} - 1}{\mathbf{R} - 1} \tag{16-191}$$

where $x_i = x_i' + \mathbf{X}(x_i'' - x_i')$ and similarly for y_i; $\mathbf{R} = (T''/T')^{1/2} = T''/\bar{T} = \bar{T}/T'$; and T'' (or T') is the appropriate root of the T equation, based on Eq. (16-182), applied to the plateau immediately upstream (or downstream):

$$\sum_{i=1}^{n} \nu_i \frac{x_i y_i}{x_i T - y_i} = 0 \tag{16-192}$$

For constant separation factors, the ν_i are all equal and hence are eliminated. As in two-component behavior, $T'' \leq T'$ or $\mathbf{R} \leq 1$ is the criterion for an abrupt transition.

These parameters are useful in classifying the type of each multicomponent transition, as well as in locating its **R** and $\bar{T}$. Kinetic analysis can then be applied, adopting a single **N** (or D) to describe all the species present in any one transition. Often **N** will be the same for every transition in a given run. Quantitative consideration of rate phenomena in the literature is best applied to individual transitions [Dranoff and Lapidus, *Ind. Eng. Chem.*, **50**, 1648 (1958); **53**, 71 (1961); Cooney and Lightfoot, *Ind. Eng. Chem. Process Des. Dev.*, **5**, 25 (1966); and Clazie et al. (see "General References")]. When two transitions are seen to overlap, their combined behavior may be estimated qualitatively or computed by either an explicit or an implicit numerical method.

MULTICOMPONENT OPERATIONS

In general, the various species are indicated by letters A, B, C . . . or numerals 1, 2, 3 . . . in the order of decreasing affinity for the sorbent. When relative affinities are not identified, the separate species are shown by i, j, k. . . .

Alphabet Rule and Slope Rule For any number of components, if the calculated T values along the path of a transition increase as the concentrations leave those of the earlier (downstream) plateau and approach those of the later (upstream) plateau, the calculated transition is real and hence gradual. If the T values increase in the reverse direction, the calculated transition is fictitious and hence abrupt [compare Eqs. (16-56) and (16-62)]. In many cases the type of transition can be identified before calculations are made by applying the following two predictive rules. In this discussion, component A is the most tightly held, B next, etc.

The **alphabet rule** states that the concentration of A can go to zero only in the 1,2 transition (first from the upstream end; thus, the last to emerge in the effluent); component B in the 1,2 or 2,3 transitions; component C in the 2,3 or 3,4; etc.

The **slope rule** pertains to the slope of a component concentration in the downstream direction, or inside the column; that is, to dx_i/dU or dy_i/dU, where $U = 1/T$. According to the slope rule, the gradual slopes of the concentration profiles of component A in the 1,2 (farthest upstream) transition, of components A and B in the 2,3 transition, etc., are positive. All other gradual slopes are negative. This means that for any transition the direction of concentration change of all components will either match or not match the profile slopes just indicated for the respective components. If the directions match, the transition is gradual; otherwise, it is abrupt. [If selectivity reversal occurs, as mentioned under "Three-component (Bivariant) Operations," caution must be exercised in applying these rules; see Helfferich and Klein (in "General References"), and Tondeur, *Chem. Eng. J.*, **1**, 337 (1970).]

Application of these two rules is illustrated in Fig. 16-40 for a column presaturated with components A, B, C and receiving a feed solution containing components D and E. From the alphabet rule, the concentrations of components A and E become zero respectively in the 1,2 and 4,5 transitions; from the slope rule, these transitions are gradual. The alphabet rule permits component B to vanish in the 1,2 or 2,3 transition. If B were to vanish in the 1,2 transition, the slope rule indicates that this transition would be abrupt, which it is not. Thus B must vanish in the 2,3 transition, which the slope rule now indicates is gradual. Similar reasoning shows that C and D vanish in the 3,4 transition, which also is gradual.

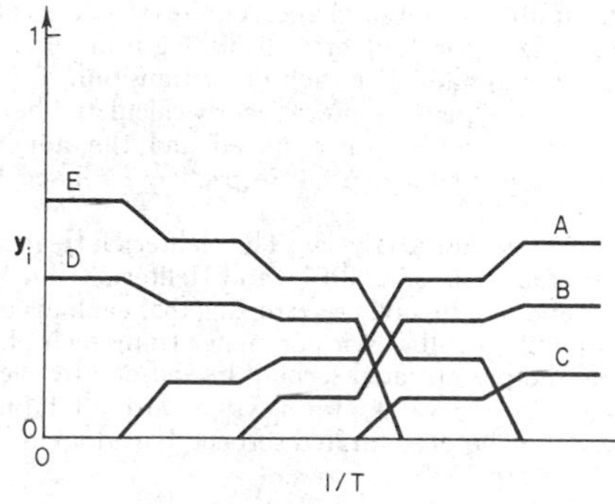

FIG. 16-40 Typical local-equlibrium concentration profiles for column presaturated with A, B, and C and fed with D and E.

Frontal Sorption and Desorption If a feed containing several components is introduced continually into a column presaturated with a single different component, plateau zones are formed which reflect the fact that the more tightly a species is held, the more slowly it advances through the column. In "normal" operation, the presaturant component (say, component H) has the lowest affinity. Then the new plateau zone farthest downstream (first to emerge after the presaturation zone) will contain only G; the next, F and G; etc. Likewise, counting from the upstream end, only A will disappear in the 1,2 transition; B in the 2,3 transition; etc. For this case, all the transitions will be abrupt. Regardless of the number of components, the locations of successive transitions and the compositions of the successive plateau zones can be calculated by starting from the upstream end, as will be illustrated.

The integral material balances for the successive transitions, in the normal situation, reduce, for example, to

$$\overline{T}_{12} = \mathrm{y}_{A1}/\mathrm{x}_{A1} \qquad \overline{T}_{23} = \mathrm{y}_{B2}/\mathrm{x}_{B2} \qquad \ldots \overline{T}_{78} = y_{G7}/x_{G7} \tag{16-193}$$

where the subscripts of $\overline{T}$ denote the plateau zones adjacent to the transition and the second subscripts of x and y denote the known adjacent plateau zone. The composition in each successive downstream plateau zone is calculated from equilibrium relations in conjunction with the integral material balances for the remaining components,

$$(\mathrm{y}_{i,j-1} - \mathrm{y}_{ij})/(\mathrm{x}_{i,j-1} - \mathrm{x}_{ij}) = \overline{T}_{j-1,j} \tag{16-194}$$

where x_{ij} and y_{ij} are the unknown concentrations of component i in plateau zone j. For convenience, the alternative designation $i = 1, 2, \ldots$ is adopted for $i = \mathrm{A}, \mathrm{B}, \ldots$.

In systems with mass-action equilibrium, the composition of a plateau zone adjacent to an upstream plateau zone of known composition satisfies the relation for each component

$$\mathrm{x}_{ij} = \frac{\overline{T}_{j-1,j}\mathrm{x}_{i,j-1} - \mathrm{y}_{i,j-1}}{\overline{T}_{j-1,j} - \mathrm{K}_{i,n-1}^{1/\nu n-1}\,(\mathrm{y}_{n-1,j}/\mathrm{x}_{n-1,j})^{\nu i/\nu n-1}} \tag{16-195}$$

with

$$\sum_{i=j}^{n-1} \mathrm{x}_{ij} = 1 \tag{16-196}$$

By iterative solution, a value can be found for the ratio $\mathrm{y}_{n-1,j}/\mathrm{x}_{n-1,j}$ which will satisfy the summation. The equilibrium y's can then be calculated; see Eq. (16-180). If constant separation factors apply, $(K_{i,n-1})^{1/\nu n-1}$ is replaced by $\alpha_{i,n-1}$, and the ratios ν_i/ν_{n-1} each become unity.

The complementary situation, a column fully presaturated with a mixed feed and then fed continually with a different pure component of greatest affinity, will again develop only abrupt transitions and can be calculated similarly by starting from the downstream end.

Determination of **R** for each of these transitions, for use in calculating the rate-limited transition shape, requires separate calculation. For G displacing H in sorption, or A displacing B in desorption, the respective binary **r** applies. For each other transition, Eq. (16-183) can be applied for the plateau most recently calculated between the component currently added or removed and the adjoining one already (or still) present to give the respective T'. Then **R** can be found as the ratio $\overline{T}/T'$.

CSF Based h-Function Analysis The Helfferich treatment [*Ind. Eng. Chem. Fundam.*, **6**, 362 (1967), and Helfferich and Klein (see "General References")] involves recognizing that each concentration path can be identified fully by a parameter from each of two sets, one defined by the presaturated sorbent $\mathbf{h}_P$ and one by the feed $\mathbf{h}_F$. Each set contains $n - 1$ values, for a system with a total number n of components. For the presaturated sorbent, $\mathbf{h}_P$ values are obtained as the roots of the equation

$$\sum_{i=1}^{n} \frac{\mathrm{y}_{in}}{(1/\mathbf{h}_P) - \alpha_{i1}} = 0 \tag{16-197}$$

where y_{in} is the concentration of component i in the presaturated sorbent. For the feed, $\mathbf{h}_F$ values are roots of the equation

$$\sum_{i=1}^{n} \frac{\mathrm{x}_{i1}}{\mathbf{h}_F - \alpha_{1i}} = 0 \tag{16-198}$$

where x_{i1} is the concentration of component i in the feed.

The roots of Eqs. (16-197) and (16-198) are indexed in accordance with the condition

$$\alpha_{1i} \le \mathbf{h}_i \le \alpha_{1,i+1} \qquad (i = 1, \ldots, n-1) \tag{16-199}$$

which corresponds to the convention that the selectivity of the sorbent for the various components decreases in the order $i = 1(=$ species A), $2(=$ B), $\ldots n$. No specific connection exists between $\mathbf{h}_i$ and the concentrations of component i.

If some of the n components of the system are absent from the presaturated sorbent or the feed, Eq. (16-197) or (16-198) will give fewer than $n - 1$ roots. If k is an absent component, then either $\mathbf{h}_{k-1}$ or $\mathbf{h}_k$ equals α_{1k}. In this manner, $n - 1$ $\mathbf{h}_i$ values can be indexed without ambiguity.

The sets of **h** values are also directly obtainable from roots of T for the corresponding end plateaus. Thus, for the feed conditions,

$$\mathbf{h}_{iF} = \frac{\mathrm{y}_{1F}}{\mathrm{x}_{1F}}\frac{1}{T_{iF}} \tag{16-200}$$

and for the presaturation condition,

$$\mathbf{h}_{iP} = \frac{\mathrm{y}_{1P}}{\mathrm{x}_{1P}}\frac{1}{T_{iP}} \tag{16-201}$$

with the T's calculated by use of Eq. (16-192).

The $\mathbf{h}_i$ have the useful property of varying from the influent value $\mathbf{h}_{iF}$ to the presaturation value $\mathbf{h}_{iP}$ only in transition $i, i+1$ but being otherwise constant. This is illustrated in Fig. 16-41: in plateau zone

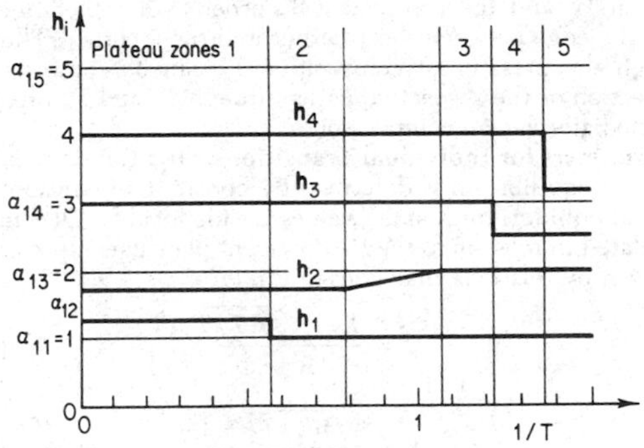

FIG. 16-41 Calculated $\mathbf{h}_i$ profiles corresponding to the example case.

1, the set of $\mathbf{h}_i$ is $\mathbf{h}_{1F}$, $\mathbf{h}_{2F}$, $\mathbf{h}_{3F}$, $\mathbf{h}_{4F}$; in plateau zone 2, $\mathbf{h}_{1P}$, $\mathbf{h}_{2F}$, $\mathbf{h}_{3F}$, $\mathbf{h}_{4F}$; etc. The composition of each plateau zone is now calculated by the relations

$$\mathrm{x}_{jk} = \frac{\prod_i (\mathbf{h}_{ik} - \alpha_{1j})}{\prod_{i \ne j} (\alpha_{1i} - \alpha_{1j})} \qquad \mathrm{y}_{jk} = \frac{\prod_i [(1/\mathbf{h}_{ik}) - \alpha_{j1}]}{\prod_{i \ne j} (\alpha_{i1} - \alpha_{j1})} \tag{16-202}$$

where the $\mathbf{h}_{ik}$ are the values applicable to plateau zone k.

An equivalent **R** value for transition $k, k+1$ is obtained from the relation

$$\mathbf{R}_{k,k+1} = \mathbf{h}_{kP}/\mathbf{h}_{kF} \tag{16-203}$$

Thus the transition is abrupt if $\mathbf{R} < 1$ and gradual if $\mathbf{R} > 1$. The stoichiometric center of transition $k, k+1$ is given by

$$\overline{T}_{k,k+1} = (\mathbf{h}_{kP}\mathbf{h}_{kF}P_k)^{-1} \tag{16-204}$$

where
$$P_k = \prod_{\substack{i=1\\(i\neq k)}}^{n-1} \mathbf{h}_{ik} \prod_{i=1}^{n} \alpha_{i1} \tag{16-205}$$

For an abrupt transition, the $\overline{T}$ value thus found represents the actual position of the abrupt change. For a gradual transition, the local-equilibrium concentration profiles are given by Eqs. (16-190) and (16-191) using the **R** value determined by Eq. (16-203).

Example 7: h-Function Analysis A five-component system is governed by a constant-separation-factor equilibrium with $\alpha_{12} = 1.5$, $\alpha_{13} = 2.0$, $\alpha_{14} = 3.0$, and $\alpha_{15} = 5.0$. What is the set of the $\mathbf{h}_i$ corresponding to the influent and presaturation compositions $\mathbf{x}_{11} = 0.267$, $\mathbf{x}_{21} = 0.149$, $\mathbf{x}_{31} = 0.290$, $\mathbf{x}_{41} = 0.000$, $\mathbf{x}_{51} = 0.294$; $\mathbf{y}_{15} = 0.000$, $\mathbf{y}_{25} = 0.586$, $\mathbf{y}_{35} = 0.000$, $\mathbf{y}_{45} = 0.020$, $\mathbf{y}_{55} = 0.394$?

Because the influent contains more than three components, a numerical method for solving Eq. (16-198) is indicated. A first guess for $\mathbf{h}_1$ is given by the average value of the separation factors α_{11} (= 1) and α_{12} [cf. Eq. (16-199)]; $\mathbf{h}_{1F(1)} = (\alpha_{11} + \alpha_{12})/2 = 1.25$. Here, the first subscript of **h** is the index number, the second denotes the plateau zone, and the third the number of the approximation. With this value of $\mathbf{h}_{1F}$, the sum in Eq. (16-198) now becomes

$$\Sigma_1 = \frac{0.267}{1.25 - 1.00} + \frac{0.149}{1.25 - 1.50} + \frac{0.290}{1.25 - 2.00} + \frac{0.294}{1.25 - 5.00} = 0.0069$$

Newton's method is used for the subsequent iterations. Differentiation of the left side of Eq. (16-198) gives

$$\frac{d\Sigma}{d\mathbf{h}} = \frac{-\sum_{i=1}^{n} \mathbf{x}_i}{(h - \alpha_{1i})^2}$$

With $\mathbf{h}_{1F(1)} = 1.25$, the derivative has a value of -7.19. The second approximation becomes

$$\mathbf{h}_{1F(2)} = \mathbf{h}_{1F(1)} - \frac{\Sigma_1}{d\Sigma/d\mathbf{h}} = 1.25 + 0.0069/7.19 = 1.251$$

This value appears sufficiently accurate. The other two roots of Eq. (16-198), obtained similarly and identified by Eq. (16-199), are $\mathbf{h}_{2F} = 1.710$ and $\mathbf{h}_{4F} = 4.000$. Component 4 is absent, and $\mathbf{h}_{3F}$ is missing. Thus, since either $\mathbf{h}_{3F}$ or $\mathbf{h}_{4F}$ must equal α_{14}, $\mathbf{h}_{3F} = \alpha_{14} = 3.000$.

As the presaturation composition comprises only three components, two **h**-value roots are given by Eq. (16-197) as 2.480 and 3.120. The remaining two roots are given by α_{11} and α_{12}, and the complete set is $\mathbf{h}_{1P} = 1.000$, $\mathbf{h}_{2P} = 2.000$, $\mathbf{h}_{3P} = 2.480$, $\mathbf{h}_{4P} = 3.120$.

The complete concentration history for the effluent is shown in Fig. 16-42. Transition 2,3 can be taken as an illustration of the further calculations. From Eq. (16-205)

$$P_2 = \mathbf{h}_{12}\mathbf{h}_{32}\mathbf{h}_{42}\alpha_{21}\alpha_{31}\alpha_{41}\alpha_{51}$$
$$= (1.0)(3.0)(4.0)(0.677)(0.50)(0.333)(0.20) = 0.2667$$

Then, from Eq. (16-204)

$$\overline{T} = (1.710 \times 2.000 \times 0.2667)^{-1} = 1.094$$

and, from Eq. (16-203)

$$\mathbf{R}_{23} = 2.000/1.710 = 1.170$$

Because the transition is gradual ($\mathbf{R} > 1$), it is found that, from Eq. (16-190), the transition starts ($\mathbf{X} = 0$) at $T = 1.094/1.170 = 0.936$, and ends at $T = 1.170 \times 1.094 = 1.280$.

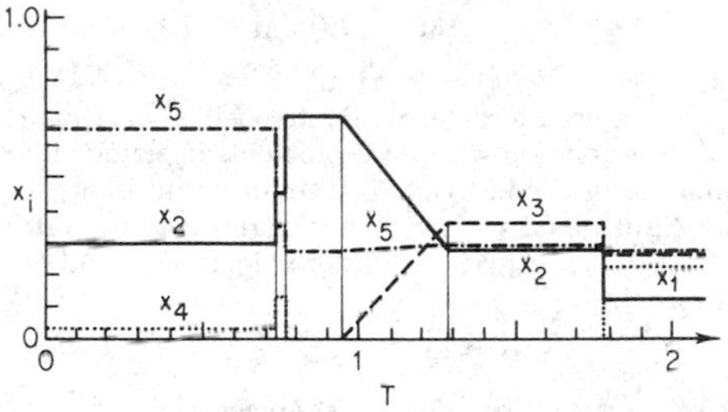

FIG. 16-42 Effluent-concentration history corresponding to the example case.

BATCH AND CONTINUOUS OPERATIONS

SINGLE BATCH STAGE

In a batch run for either adsorption or ion exchange, it is considered that moist sorbent is added to a solution, stirred, and subsequently filtered out. In a continuous-flow mixing stage at steady state, sorbent and solution are fed continuously; mixture is withdrawn and segregated into sorbent and solution. Such stages may be part of an actual countercurrent or fixed-bed operation or of one simulated calculationally by stagewise operation.

Batch treatment is adopted when the capacity and equilibrium of the sorbent are large enough to give nearly complete sorption in a single step, as in decoloration of laboratory preparations. Batch runs are useful in the measurement of equilibrium isotherms and also of particle-side rate behavior, because the simple material balance allows the particle concentration to be calculated from a measured fluid-phase concentration.

A simple material balance describes the operation in either a batch unit or any one mixing stage of a continuous-flow system:

$$W_b(q - q_{in}) = -V_b(c - c_{in}) \tag{16-206}$$

or

$$W_b q_{ref}(\mathbf{y} - \mathbf{y}_{in}) = -V_b c_{ref}(\mathbf{x} - \mathbf{x}_{in}) \tag{16-207}$$

where W_b = weight charge of solid and V_b = volumetric charge of fluid, in the batch system. As reference values, c_{ref} may be set at the inlet concentration of solute or at another larger value; q_{ref} is the solid concentration in equilibrium with c_{ref}. A stoichiometric ratio can be defined:

$$T_b = \frac{V_b c_{ref}}{W_b q_{ref}} = \frac{\mathbf{y} - \mathbf{y}_{in}}{\mathbf{x}_{in} - \mathbf{x}} \tag{16-208}$$

The decrement in **x** from input condition to equilibrium can be designated as Δ, and the extent of approach to equilibrium as X. From the material balance,

$$X = \frac{\mathbf{x}_{in} - \mathbf{x}}{\Delta} = \frac{\mathbf{y} - \mathbf{y}_{in}}{T_b\Delta} \tag{16-209}$$

The dimensionless concentrations are then

$$\mathbf{x} = \mathbf{x}_{in} - X\,\Delta \quad \text{and} \quad \mathbf{y} = \mathbf{y}_{in} + XT_b\,\Delta \tag{16-210}$$

Binary Equilibrium Limit at Constant R Equation (16-208) applies, with Eq. (16-14). Solution of the resulting quadratic relation in Δ yields

$$\Delta = (M - L)/2T_b \tag{16-211}$$

where $M = \mathbf{x}_{in}T_b - \mathbf{y}_{in} + [(1 + T_b\mathbf{r})/(1 - \mathbf{r})]$
$N = [(\mathbf{x}_{in} - \mathbf{y}_{in}\mathbf{r})/(1 - \mathbf{r})] - \mathbf{x}_{in}\mathbf{y}_{in}$
$L = (M^2 - 4NT_b)^{1/2}$

When **r** is not constant, Eq. (16-209) is replaced by graphically or numerically locating the intersection of the equilibrium curve with the material balance given by Eq. (16-208). For mass-action equilibriums, an alternative method is described later under "Multicomponent Equilibriums."

Rate Controlled by Reaction-Kinetic Driving Force The generality of this form of rate expression has already been discussed. Equations (16-83) and (16-210) can be combined to give

$$\frac{dy}{dt} = T_b \Delta \frac{dX}{dt} = \hat{\kappa}(1 - r)(T_b X^2 \Delta^2 - MX\Delta + N) \qquad (16\text{-}212)$$

Integration for batch operation provides the result

$$t = \frac{T_b}{\hat{\kappa}(1 - r)L} \ln \frac{1 - [X(M - L)/(M + L)]}{1 - X} \qquad (16\text{-}213)$$

Solution for perfectly mixed continuous-flow operation gives the mean residence time (when the same value—ratio of holdup volume to volumetric flow rate—holds for both phases):

$$\bar{t} = \frac{XT_b\Delta}{\hat{\kappa}(1 - r)(T_b X^2 \Delta^2 - MX\Delta + N)} \qquad (16\text{-}214)$$

Phase segregation does not complicate the algebra because reaction is first-order within each particle; the extent of approach to equilibrium is

$$X = (M' - L')/(M - L) \qquad (16\text{-}215)$$

where $M' = M + T_b/[(1 - r)\hat{\kappa}\bar{t}]$, and $L' = (M'^2 - 4NT_b)^{1/2}$. These relations can be used to determine rate coefficients from experimental data or to apply known rate coefficients in sizing either a single-stage contactor or the individual stages of a multistage contactor.

Linear Equilibrium If $r = 1$, the dimensionless driving force reduces to $(x - y)$. Equation (16-209) yields the equilibrium condition (with $x = y$):

$$\Delta = (x_{in} - y_{in}/T_b + 1) \qquad (16\text{-}216)$$

Also, for batch runs, Eq. (16-213) is replaced by

$$t = \frac{T_b}{\hat{\kappa}(T_b + 1)} \ln \frac{1}{1 - X} \qquad (16\text{-}217)$$

For perfectly mixed continuous flow, the relation between time and extent is

$$\bar{t} = \frac{T_b X}{\hat{\kappa}(1 + T_b)(1 - X)} \qquad (16\text{-}218)$$

or

$$X^{-1} = 1 + \frac{T_b}{\hat{\kappa}\bar{t}(1 + T_b)} \qquad (16\text{-}219)$$

Multicomponent Equilibriums The single-stage material balance for each component [Eq. (16-208)] can also be expressed as

$$(y_{in})_i + T_b(x_{in})_i = Y_i \qquad (16\text{-}220)$$

where subscript i designates the ith component and Y is the hypothetical solid-phase mole fraction (or ion-equivalent fraction) that includes all the component in both phases. Then, for any extent of approach to equilibrium, $y_i + T_b x_i = Y_i$, and $x_i = (Y_i - y_i)/T_b$. Thus the ratio of mole fractions of i, designated here as Γ_i, takes the form

$$\Gamma_i = y_i/x_i = T_b y_i/(Y_i - y_i) \qquad (16\text{-}221)$$

With certain types of equilibrium relation, the ratios for different species are easily related to one another or to the ratio for a reference component n. Specifically, for systems following constant-separation-factor or mass-action behavior respectively, Γ_i is $(\alpha_{i,n}\Gamma_n)$ or $[K_{i,n}(\Gamma_n)^{mni}]$.

Given the set of Y_i values corresponding to the known material balances and the set of known equilibrium relations, successive trial values of Γ_n can be assumed to find the one that leads to unity as the total for either Σx_i or Σy_i (and hence for both). Thus since $y_i = Y_i\Gamma_i/(1 + \Gamma_i)$, the material-balance criterion is

$$\sum_{i=1}^{n} \frac{Y_i\Gamma_i}{1 + \Gamma_i} = 1 \qquad (16\text{-}222)$$

This stepwise trial-and-error calculation parallels the Lewis and Cope method for multicomponent distillation [*Ind. Eng. Chem.*, **24**, 498 (1932)]. An example of its use in calculating the separation of two minor components by ion exchange has been given by Hiester and coworkers [*Chem. Eng. Prog. Symp. Ser.*, **50**(14), 51 (1954)].

SEMIBATCH STAGE

For this type of operation, it is considered that a batch of sorbent retained in a contacting vessel or zone undergoes **perfect mixing** with coexisting fluid phase, while fluid enters and leaves at a uniform volumetric rate F as in a well-agitated fluidized fixed bed. Solute concentration of the incoming fluid is constant ($\mathbf{X}_{in} = 1$); the levels for sorbent **Y** and for outgoing fluid **X** each increase from 0 toward 1 as the run progresses.

Material balance over a differential time interval gives

$$F(1 - \mathbf{X}) = v\left(\epsilon \frac{d\mathbf{X}}{dt} + \Lambda \frac{d\mathbf{Y}}{dt}\right) \qquad (16\text{-}223)$$

where v = total volume of well-mixed contacting zone, ϵ = liquid-phase volume fraction, and t = elapsed time. The ratio v/F is the mean residence time $\bar{t}$.

For a system **not at equilibrium,** $d\mathbf{Y}/dt$ can be obtained as a function of **X** by using the appropriate rate relation and equilibrium relation from the preceding subsection. The resulting equation for dt as a function of $d\mathbf{X}$ and **X** can be integrated algebraically or numerically. An example of this approach is given by Marchello and Davis, *Ind. Eng. Chem. Fundam.*, **2**, 27 (1963).

For a system at equilibrium, $d\mathbf{Y}/dt$ in Eq. (16-223) is replaced by the product $(d\mathbf{X}/dt)(d\mathbf{Y}^*/d\mathbf{X})$. In the special case of **linear equilibrium, Y* = X,** and the result is

$$\frac{t}{\bar{t}} = \left(1 + \frac{\Lambda}{\epsilon}\right) \ln \frac{1}{1 - \mathbf{X}} \qquad (16\text{-}224)$$

It is also possible to have an internal recirculation of solid that is slow compared with the fluid movement but rapid in relation to the total run time. This situation approaches the idealized condition of plug flow of fluid encountering a perfectly mixed solid. If the fluid closely approaches equilibrium with the solid by the time it leaves the bed, Eq. (16-224) again applies.

COUNTERCURRENT MULTISTAGE OPERATION

A continuous-flow system with countercurrent contact of fluid and solid phases resembles in many respects a constant-molal-overflow distillation column. Its most striking difference, which affects its design and operation, is that the two phases are not cross-convertible, as gas and liquid are in the condenser and reboiler of a distillation unit. Instead the adsorbent must be regenerated with an independent elutant in a separate operation before being returned to the system for reuse. Thus any complete unit must involve both sorption and desorption, with recycle of the solid phase, as shown in Fig. 16-43. For either the sorption section or desorption section of Fig. 16-43, between the upper and lower ends of the section, the material balance for any one component (for instance, **A**) is

$$G_p[(q_A)_{out} - (q_A)_{in}] = -\left(\frac{F}{S}\right)[(c_A)_{out} - (c_A)_{in}] \qquad (16\text{-}225)$$

Between an intermediate cross section and the upper end of the column, also,

$$G_p q_{ref}[\mathbf{y}_A - (\mathbf{y}_A)_{in}] = -\left(\frac{F}{S}\right) c_{ref}[(\mathbf{x}_A)_{out} - \mathbf{x}_A] \qquad (16\text{-}226)$$

Here G_p is the mass velocity of solid phase per unit cross section, and F/S is the superficial velocity of fluid phase.

The ratio of molal flow capacity for fluid to that for solid (the throughput ratio) is given by

$$T = -(c_{ref}F/S)/(q_{ref}G_p) \qquad (16\text{-}227)$$

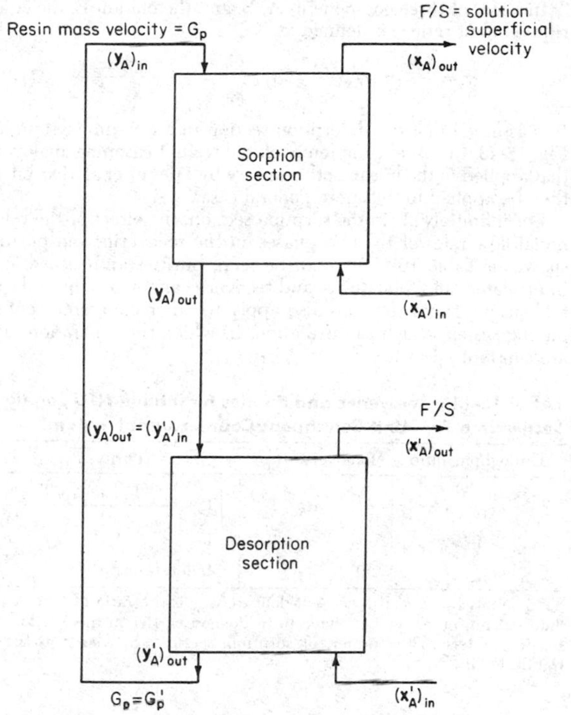

FIG. 16-43 Flow diagram of a two-section continuous countercurrent moving-bed system; primes denote elution variables.

For a particular separation or exchange governed by the entering fluid ($X_{in} = 1$ when $x = x''$) and solid ($Y_{in} = 0$ when $y = y'$), the throughput ratio for the separation is

$$\mathbf{T} = \frac{(F/S)c_{ref}(x'' - x')}{G_p q_{ref}(y'' - y')} \tag{16-228}$$

Linear Equilibrium Where Eq. (16-216) applies for one stage, a sequence of equilibrium stages (N_c in number) involving countercurrent flow can be treated by the Kremser equation [*Nat. Pet. News*, **22**(21), 42 (1930)]. The degree of approach to equilibrium transfer in the leaving fluid is

$$1 - X_{out} = (\mathbf{T}^{N_c} - 1)/(\mathbf{T}^{N_c+1} - 1) \tag{16-229}$$

and the degree of approach in the leaving solid is

$$Y_{out} = \frac{\mathbf{T}^{N_c+1} - \mathbf{T}}{\mathbf{T}^{N_c+1} - 1} = \mathbf{T}(1 - X_{out}) \tag{16-230}$$

From these relations, the number of equilibrium stages is

$$N_c = \frac{1}{\ln \mathbf{T}} \ln \frac{\mathbf{T} - Y_{out}}{\mathbf{T}(1 - Y_{out})} \tag{16-231}$$

If $\mathbf{T} = 1$, these relations simplify to

$$1 - X_{out} = Y_{out} = \frac{N_c}{N_c + 1} \qquad X_{out} = \frac{1}{N_c + 1} \tag{16-232}$$

Use of these relations in the design of a two-section equilibrium-stage contactor, for the separation of several ions having linear isotherms with the elutant, is given by Hiester and coworkers [*Ind. Eng. Chem.*, **45**, 2402 (1953)]. The corresponding flow diagram is given in Fig. 16-44.

Iterative computation can be made for systems of up to 20 components, with any simple nonlinear equilibrium, by adapting a Fortran program developed by Hanson and Somerville, *Adv. Chem. Eng.*, **4**, 279 (1963).

COUNTERCURRENT CONTINUOUS-DIFFERENTIAL OPERATION

Moving-Bed Unit The material balance, given by differentiation of Eq. (16-226) is

$$T = dy/dx \qquad \mathbf{T} = dY/dX \tag{16-233}$$

Two rate relations will be considered in this discussion. The first, the reaction-kinetic form, is presented because of its general applicability. The second, the external-transport form, illustrates the familiar graphical technique for determining the number of transfer units in this type of system (the technique can also be applied to a linear or quadratic driving-force treatment of particle-phase diffusion or to the reaction-kinetic driving force).

In applying the rate expressions, the time scale must be viewed as the residence time for the solid phase ($t_p = h\rho_b/G$). This is related to apparent residence time for the fluid phase ($t_{fa} = hS/F$):

$$\frac{t_{fa}}{t_p} = \frac{hS/F}{h\rho_b/G_p} = \frac{1}{T\Lambda} = \frac{1}{\mathbf{T}\Lambda} \tag{16-234}$$

Hence the derivative expressing the rate is

$$\frac{dY}{dt_p} = \frac{1}{\Lambda}\frac{dX}{dt_{fa}} \tag{16-235}$$

Reaction-Kinetic Form Equation (16-235) can be solved in combination with the reaction-kinetic driving force (see Table 16-7, item 4C) to determine the column height as a function of the proportion X_{out} of unsorbed feed component. With $N_R = \kappa\Lambda hS/F$, the result is

$$N_R = \frac{1}{\mathbf{T}(1 - \mathbf{R})l} \ln\left(\frac{2X_{out} - m + l}{2X_{out} - m - l}\right)\left(\frac{2 - m - l}{2 - m + l}\right) \tag{16-236}$$

with $m = X_{out} + \{(1 - \mathbf{RT})/[\mathbf{T}(1 - \mathbf{R})]\}$; $n = \mathbf{R}X_{out}/(1 - \mathbf{R})$; $l = (m^2 + 4n)^{0.5}$. If l is imaginary, the alternate integral applies:

$$N_R = \frac{2}{\mathbf{T}(1 - \mathbf{R})l'}\left(\tan^{-1}\frac{2 - m}{l'} - \tan^{-1}\frac{2X_{out} - m}{l'}\right) \tag{16-237}$$

with $l' = (-m^2 - 4n)^{0.5}$.

For **linear equilibrium** ($\mathbf{R} = 1$),

$$N = \frac{1}{1 - \mathbf{T}} \ln \frac{1 - \mathbf{T} + \mathbf{T}X_{out}}{X_{out}}$$

$$= \frac{1}{1 - \mathbf{T}} \ln \frac{1 - Y_{out}}{1 - (Y_{out}/\mathbf{T})} \tag{16-238}$$

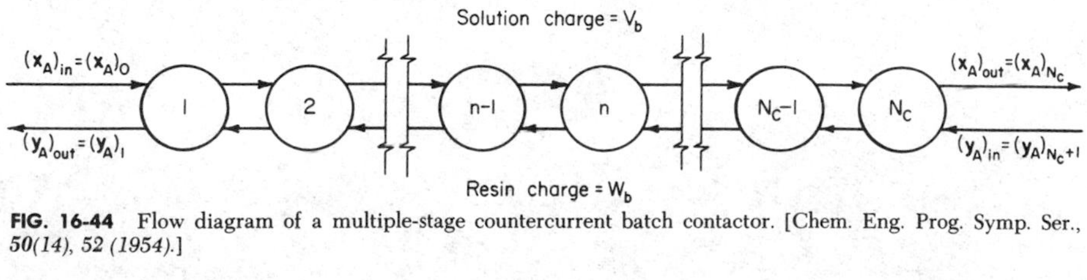

FIG. 16-44 Flow diagram of a multiple-stage countercurrent batch contactor. [Chem. Eng. Prog. Symp. Ser., *50(14), 52 (1954).*]

This relation can be solved explicitly for Y_{out} in terms of **N** and **T**, if desired. This result, the **Colburn equation** [*Ind. Eng. Chem.*, **27**, 255 (1935)], applies also in the linear-equilibrium limit of the external-transport case.

External-Transport Form In the event that the overall kinetics of the sorption step are controlled by **fluid-phase transport** external to the particle, combination of that rate (Table 16-7, item 1*B*) with the material-balance relation [Eq. (16-233)] and the boundary conditions leads to the result:

$$\int_1^{X_{out}} \frac{dX}{X - X^\circ} = N_{Of} = k_f a_p \frac{hS}{F} \tag{16-239}$$

Such calculations have been performed, for both linear and nonlinear isotherms, by Hiester and coworkers [*Chem. Eng. Prog.*, **50**, 139; *Symp. Ser.* 14, 63 (1954)].

Other Cases Solutions for particle-phase-resistance controlling and for heat-transfer controlling have been given by Calderbank and Macrackis, *Trans. Inst. Chem. Eng.*, **34**, 320 (1956), in a study of C_2 and C_3 hydrocarbon uptake by activated carbon in a hypersorption unit.

Ultimate Sorption Efficiency for a Regenerative Moving-Bed System For each set of column operating conditions and each sorption system, there is a limit to the extent of removal or separation possible. Such a limit is achieved in a column having an infinite number of transfer units and is dependent on the equilibrium isotherms of the solutes involved and on the molal flow ratio of the two phases. In a two-step system as shown in Fig. 16-43, with constant separation factors (α's) specified for each pair of components present, the theoretical limits can be deduced analytically.

In terms of three components A, B, and the elutant E, the equilibrium elutant ratio γ is defined as

$$\gamma = \frac{y_E^\circ}{x_E} = \alpha_{EA}\frac{y_A^\circ}{x_A} = \alpha_{EB}\frac{y_B^\circ}{x_B} = \frac{1 - y_A^\circ - y_B^\circ}{1 - x_A - x_B} \tag{16-240}$$

It is assumed that the desorption section in a two-unit system, as in Fig. 16-43, furnishes completely eluted resin. Reasoning analogous to that applied in the linear isotherm case by Hiester et al. (loc. cit.) can then be applied to this more general case.

For infinite NTU in the sorption section, the effect of the relative molal flow rates of the two phases on the recoveries and purities is shown in Table 16-11. As can be seen, purities and recoveries are interrelated, and total purity and recovery cannot be achieved simultaneously. (These relations also apply to other countercurrent contacting systems, such as distillation, in which the separation factors are constant.)

TABLE 16-11 Recoveries and Purities for Infinite-NTU Sorption Section in a Two-Unit Continuous Countercurrent System*

Throughput ratio	Recovery, $\bar{R}_{BU}$	Purity, $\bar{P}_{BU}$
$T > \gamma\alpha_{AE}$	$1 - \gamma\alpha_{BE}/T$	$\left[1 + \frac{1 - (\gamma\alpha_{AE}/T)}{(\phi_{BA})_{in}(1 - \gamma\alpha_{BE}/T)}\right]^{-1}$
$\gamma\alpha_{AE} > T > \gamma\alpha_{BE}$	$1 - \gamma\alpha_{BE}/T$	1
$Y \le \gamma\alpha_{BE}$	0	Indeterminate

*Symbols $\bar{R}_{BU}$ and $\bar{P}_{BU}$ are self-defined; ϕ_{BA} is the ratio of B to A in the fluid stream; other symbols have been defined earlier in this section. Subscript *U* refers to the upper, or sorption, section. Species A is the most tightly held.

Novel Separation Processes

Joseph D. Henry, Jr., Ph.D., P.E., *Professor and Chairman of Chemical Engineering, West Virginia University; Member, American Chemical Society, American Institute of Chemical Engineers. (Section Editor, Crystallization from the Melt, Separation Processes Based Primarily on Action in a Field, and Novel Solid-Liquid Separation Processes)*

William Corder, M.S., P.E., *Assistant Vice President, Consolidation Coal Company; Member, American Institute of Chemical Engineers, American Society of Mechanical Engineers. (Sublimation)*

W. S. Winston Ho, Ph.D., *Engineering Associate, Exxon Research and Engineering Company; Member, American Institute of Chemical Engineers, American Chemical Society. (Membrane Processes)*

Richard L. Hoglund, M.S., *Chemical Engineer, Union Carbide Corporation—Nuclear Division; Member, American Institute of Chemical Engineers. (Diffusional Separation Processes)*

Robert Lemlich, Ph.D., P.E., *Professor of Chemical Engineering, University of Cincinnati; Fellow, American Institute of Chemical Engineers; Member, American Chemical Society, American Society for Engineering Education; Fellow, American Association for the Advancement of Science. (Adsorptive-Bubble Separation Methods)*

Norman N. Li, Sc.D., *Director, Separations Research, UOP, Inc.; Member, American Chemical Society, American Institute of Chemical Engineers, The New York Academy of Science. (Membrane Processes)*

Charles G. Moyers, Jr., Ph.D., P.E., *Principal Engineer, Union Carbide Corporation—Engineering and Technical Services Division; Member, American Institute of Chemical Engineers. (Crystallization from the Melt)*

John Newman, Ph.D., *Professor of Chemical Engineering, University of California, Berkeley; Principal Investigator, Inorganic Materials Research Division, Lawrence Berkeley Laboratory. (Separation Processes Based Primarily on Action in a Field: Theory of Electrical Separations)*

Herbert A. Pohl, Ph.D., *Professor of Physics, Oklahoma State University; Member, American Chemical Society, American Physical Society, American Association for the Advancement of Science, American Association of Physics Teachers. (Separation Processes Based Primarily on Action in a Field: Dielectrophoresis)*

Kent Pollock, Ph.D., *Professor of Physics, Oklahoma State University. (Separation Processes Based Primarily on Action in a Field: Dielectrophoresis)*

Michael E. Prudich, Ph.D., *Research Engineer, Gulf Research and Development Company; Member, American Institute of Chemical Engineers, American Chemical Society. (Novel Solid-Liquid Separation Processes)*

K. S. Spiegler, Ph.D., *Professor of Mechanical Engineering Emeritus, University of California, Berkeley; Charter Member, Association of Energy Engineers, National Water Supply Improvement Association; Member, American Chemical Society. (Separation Processes Based Primarily on Action in a Field: Electrodialysis)*

Edward Von Halle, Ph.D., *Chemical Engineer, Union Carbide Corporation—Nuclear Division; Member, American Institute of Chemical Engineers. (Diffusional Separation Processes)*

CRYSTALLIZATION FROM THE MELT

SUBLIMATION

MEMBRANE PROCESSES

SEPARATION PROCESSES BASED PRIMARILY ON ACTION IN A FIELD

NOVEL SOLID-LIQUID SEPARATION PROCESSES

ADSORPTIVE-BUBBLE SEPARATION METHODS

DIFFUSIONAL SEPARATION PROCESSES

CRYSTALLIZATION FROM THE MELT

GENERAL REFERENCES: Atwood, *Recent Advances in Separation Science*, vol. 1, CRC Press, Cleveland, 1972, pp. 1–35. Mullin, *Crystallization*, 2d ed., Butterworth, London, 1972. Pfann, *Zone Melting*, 2d ed., Wiley, New York, 1966. Schildknecht, *Zone Melting*, Academic, New York, 1966. Zief and Wilcox, *Fractional Solidification*, Marcel Dekker, New York, 1967.

Purification of a chemical species by solidification from a liquid mixture can be termed either **solution** crystallization or crystallization from the **melt.** The distinction between these two operations is somewhat subtle. The term **melt crystallization** has been defined as the separation of components of a binary mixture without addition of solvent, but this definition is somewhat restrictive. In **solution crystallization** a diluent solvent is added to the mixture; the solution is then directly or indirectly cooled, and/or solvent is evaporated to effect crystallization. The solid phase is normally formed and maintained somewhat below its pure-component freezing-point temperature. In melt crystallization no diluent solvent is added to the reaction mixture, and the solid phase is formed by direct or indirect cooling of the melt. Product is frequently maintained near or above its pure-component freezing point in the refining section of the apparatus.

A large number of techniques are available for carrying out crystallization from the melt. An abbreviated list includes partial freezing and solids recovery in cooling crystallizer-centrifuge systems, partial melting (e.g., sweating), staircase freezing, normal freezing, zone melting, and column crystallization. A description of all these methods is not within the scope of this discussion. Zief and Wilcox (op. cit.) have compiled a comprehensive book which describes many of these processes. Current trends in the practice of melt crystallization and ultrahigh purification of organic chemicals are presented by Atwood (op. cit.). Three of the more common methods—normal freezing, zone melting, and column crystallization—are discussed here to illustrate the techniques used for practicing crystallization from the melt.

High or ultrahigh product purity is obtained with many of the melt-purification processes. Table 17-1 compares the product quality and product form that are produced from several of these operations. Zone refining can produce very pure material when operated in a batch mode; however, other techniques also provide high purity and become attractive if continuous high-capacity processing is desired.

TABLE 17-1 Comparison of Processes Involving Crystallization from the Melt

Processes	Approximate upper melting point, °C	Materials tested	Minimum purity level obtained, ppm, weight	Product form
Normal freezing	1500	All types	1	Ingot
Zone melting				
Batch	3500	All types	0.01	Ingot
Continuous	500	SiI_4	100	Melt
Column crystallization				
Continuous end-fed	300	Organic	10	Melt
Continuous center-fed	400	Organic	1	Melt

Abbreviated from Zief and Wilcox, *Fractional Solidification*, Marcel Dekker, New York, 1967, p. 7.

PHASE EQUILIBRIUMS

A brief discussion of solid-liquid phase equilibriums is presented prior to discussing specific crystallization methods. Figures 17-1 and 17-2 illustrate the phase diagrams for binary solid-solution and eutectic systems respectively. In the case of binary solid-solution systems, illustrated in Fig. 17-1, the liquid and solid phases contain equilibrium quantities of both components in a manner similar to vapor-liquid phase behavior. This type of behavior causes separation difficulties since multiple stages are required. In principle, however, high purity and yields of both components can be achieved since no eutectic is present.

If the impurity or minor component is completely or partially soluble in the solid phase of the component being purified, it is convenient to define a distribution coefficient k, defined by Eq. (17-1):

$$k = C_S/C_\ell \tag{17-1}$$

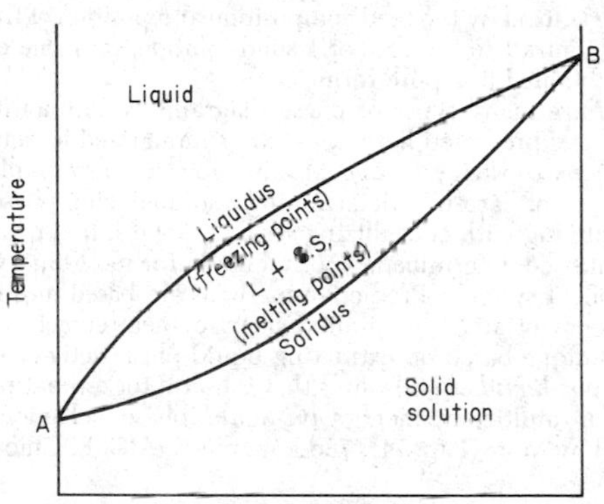

FIG. 17-1 Phase diagram for components exhibiting complete solid solution. (*Zief and Wilcox,* Fractional Solidification, *vol. 1., Marcel Dekker, New York, 1967, p. 31.*)

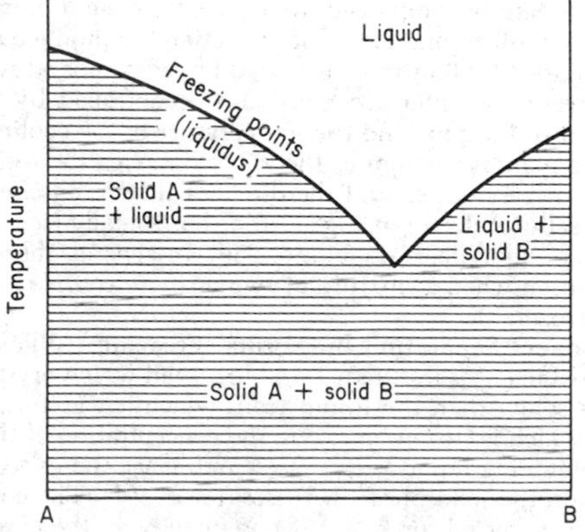

FIG. 17-2 Simple eutectic-phase diagram at constant pressure. (*Zief and Wilcox,* Fractional Solidification, *vol. 1, Marcel Dekker, New York, 1967, p. 24.*)

C_S is the concentration of impurity or minor component in the solid phase, and C_ℓ is the impurity concentration in the liquid phase. The distribution coefficient generally varies with composition. The value of k is greater than 1 when the solute raises the melting point and less than 1 when the melting point is depressed. In the regions near pure A or B the liquidus and solidus lines become linear; i.e., the distribution coefficient becomes constant. This is the basis for the common assumption of constant k in many mathematical treatments of fractional solidification in which ultrapure materials are obtained.

In the case of a simple eutectic system a pure solid phase is obtained by cooling if the composition of the feed mixture is not at the eutectic composition. If liquid composition is eutectic, then separate crystals of both species will form. In practice it is difficult to attain perfect separation of one component by crystallization of a eutectic mixture. The solid phase will always contain trace amounts of impurity because of incomplete solid-liquid separation, slight solubility of the impurity in the solid phase, or volumetric inclusions. It is difficult to generalize on which of these mechanisms is the major cause of contamination because of analytical difficulties in the ultra-high-purity range.

The distribution-coefficient concept is commonly applied to fractional solidification of eutectic systems in the ultrapure portion of the phase diagram. If the quantity of impurity entrapped in the solid phase for whatever reason is proportional to that contained in the melt, then assumption of a constant k is valid. It should be noted that the theoretical yield of a component exhibiting binary eutectic behavior is fixed by the feed composition and position of the eutectic. Also, in contrast to the case of a solid solution, only one component can be obtained in a pure form.

There are many types of phase diagrams in addition to the two simple cases presented here; these are summarized in detail by Zief and Wilcox (op. cit., p. 21). Mullin (op. cit.) also emphasizes the importance of carefully determining and analyzing phase behavior when working with crystallizing systems. Solid-liquid phase equilibriums must be determined experimentally for most binary and multicomponent systems. Predictive methods are based mostly on ideal phase behavior and have limited accuracy near eutectics. A predictive technique based on extracting liquid-phase activity coefficients from *vapor-liquid* equilibriums that is useful for estimating nonideal binary or multicomponent *solid-liquid* phase behavior has been reported by Muir (Pap. 71f, 73d ann. meet., AIChE, Chicago, 1980).

NORMAL FREEZING

Normal freezing, or progressive freezing, is the slow, directional solidification of a melt. Basically, this involves slow solidification at the bottom or sides of a vessel by indirect cooling. The impurity is rejected into the liquid phase by the advancing solid interface. This technique can be employed to concentrate an impurity or, by repeated solidifications and liquid rejections, to produce a very pure ingot. Figure 17-3 illustrates a normal-freezing apparatus. The solidification rate and interface position are controlled by the rate of movement of the tube and the temperature of the cooling medium. There are many variations of the apparatus; e.g., the residual-liquid portion can be agitated and the directional freezing can be carried out vertically as shown in Fig. 17-3 or horizontally (see Richman et al., in Zief and Wilcox, op. cit., p. 259). In general, there is a solute redistribution when a mixture of two or more components is directionally frozen.

Component Separation by Normal Freezing When the distribution coefficient is less than 1, the first solid which crystallizes contains less solute than the liquid from which it was formed. As the fraction which is frozen increases, the concentration of the impurity in the remaining liquid is increased and hence the concentration of impurity in the solid phase increases (for $k < 1$). The concentration gradient is reversed for $k > 1$. Consequently, in the absence of diffusion in the solid phase a concentration gradient is established in the frozen ingot.

One extreme of normal freezing is equilibrium freezing. In this case the freezing rate must be slow enough to permit diffusion in the

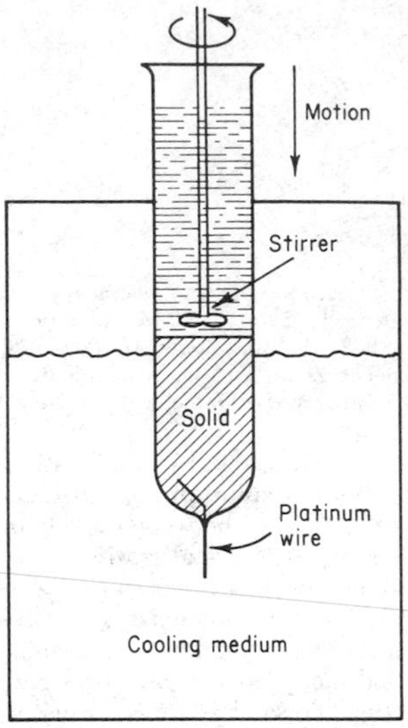

FIG. 17-3 Normal-freezing apparatus.

solid phase to eliminate the concentration gradient. When this occurs, there is no separation if the entire tube is solidified. Separation can be achieved, however, by terminating the freezing before all the liquid has been solidified. Equilibrium freezing is rarely achieved in practice because the diffusion rates in the solid phase are usually negligible (Pfann, op. cit., p. 10).

If the bulk-liquid phase is well mixed and no diffusion occurs in the solid phase, a simple expression relating the solid-phase composition to the fraction frozen can be obtained for the case in which the distribution coefficient is independent of composition and fraction frozen [Pfann, *Trans. Am. Inst. Mech. Eng.*, **194**, 747 (1952)].

$$C_s = kC_0(1 - X)^{k-1} \qquad (17\text{-}2)$$

C_0 is the solution concentration of the initial charge, and X is the fraction frozen. Figure 17-4 illustrates the solute redistribution predicted by Eq. (17-2) for various values of the distribution coefficient.

There have been many modifications of this idealized model to account for variables such as the freezing rate and the degree of mixing in the liquid phase. For example, Burton et al. [*J. Chem. Phys.*, **21**, 1987 (1953)] reasoned that the solid rejects solute faster than it can diffuse into the bulk liquid. They proposed that the effect of the freezing rate and stirring could be explained by the diffusion of solute through a stagnant film next to the solid interface. Their theory resulted in an expression for an effective distribution coefficient k_{eff} which could be used in Eq. (17-2) instead of k.

$$k_{\text{eff}} = \frac{1}{1 + (1/k - 1)e^{-f\delta/D}} \qquad (17\text{-}3)$$

where f_0 = crystal growth rate, cm/s; δ = stagnant film thickness, cm; and D = diffusivity, cm^2/s. No further attempt is made here to summarize the various refinements of Eq. (17-2). Zief and Wilcox (op. cit., p. 69) have summarized several of these models.

Pertinent Variables in Normal Freezing The dominant variables which affect solute redistribution are the degree of mixing in the liquid phase and the rate of solidification. It is important to attain sufficient mixing to facilitate diffusion of the solute away from the solid-liquid interface to the bulk liquid. The film thickness δ decreases as the level of agitation increases. Cases have been reported in which essentially no separation occurred when the liquid was not stirred (Schildknecht, *Zone Melting*, op. cit., p. 30). The freezing rate which is controlled largely by the lowering rate of the tube (see Fig. 17-3) has a pronounced effect on the separation achieved. The separation is diminished as the freezing rate is increased. Also fluc-

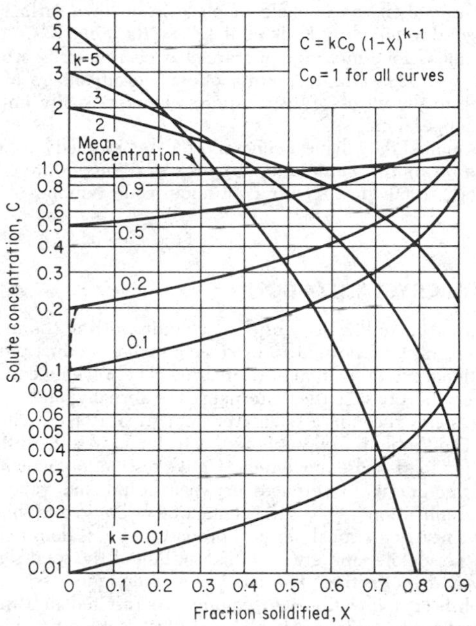

FIG. 17-4 Curves for normal freezing, showing solute concentration C in the solid versus fraction-solidified X. (*Pfann, Zone Melting, 2d ed., Wiley, New York, 1966, p. 12.*)

tuations in the freezing rate caused by mechanical vibrations and variations in the temperature of the cooling medium can decrease the separation.

Applications Normal freezing has been applied to both solid solution and eutectic systems. As Fig. 17-4 illustrates, large separation factors can be attained when the distribution coefficient is favorable. Relatively pure materials can be obtained by removing the desired portion of the ingot. Also in some cases normal freezing provides a convenient method of concentrating the impurities; e.g., in the case of $k < 1$ the last portion of the liquid that is frozen is enriched in the distributing solute.

Normal freezing has been applied on the commercial scale. For example, aluminum has been purified by continuous normal freezing [Dewey, *J. Metals*, **17**, 940 (1965)]. The Proabd refiner described by Molinari (Zief and Wilcox, op. cit., p. 393) is also a commercial example of normal freezing. In this apparatus the mixture is directionally solidified on cooling tubes. Purification is achieved because the impure fraction melts first; this process is called sweating. This technique has been applied to the purification of naphthalene and *p*-dichlorobenzene.

Another commercial process which employs sweating in addition to directional solidification has been described by Saxer and Papp [*Chem. Eng. Prog.*, **76**, 64 (1980)]. Operation is sequential. Steps include partial freezing of a falling film of melt inside 12-m tubes, followed by sweating, and then melting and recovery of refined product. Separation capacity depends on the number of stages, the reflux ratio, and the distribution coefficient. Normal-freezing techniques are applicable to the purification of a wide variety of organic products.

ZONE MELTING

Zone melting also relies on the distribution of solute between the liquid and solid phases to effect a separation. In this case, however, one or more liquid zones are passed through the ingot. This extremely versatile technique, which was invented by W. G. Pfann, has been used to purify hundreds of materials. Zone melting in its simplest form is illustrated in Fig. 17-5. A molten zone can be passed through

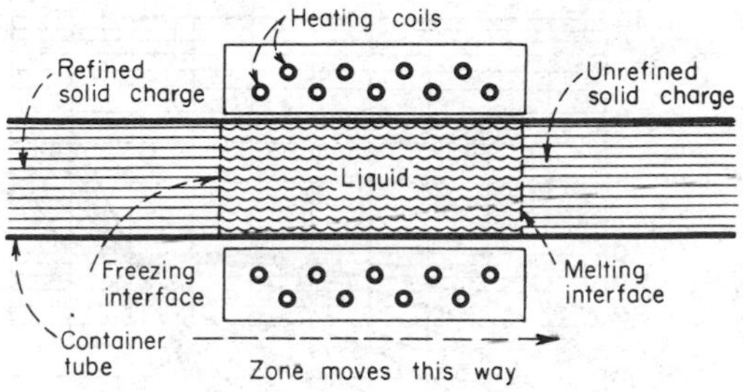

FIG. 17-5 Diagram of zone refining.

an ingot from one end to the other by either a moving heater or by slowly drawing the material to be purified through a stationary heating zone.

Normal freezing can be viewed as a special case of zone melting. If the zone length were equal to the ingot length and if only one pass were used, the operation would become normal freezing. In general, however, when the zone length is only a fraction of the ingot length, zone melting possesses the advantage that a portion of the ingot does not have to be discarded after each solidification. The last portion of the ingot which is frozen in normal freezing must be discarded before a second freezing.

Component Separation by Zone Melting The degree of solute redistribution achieved by zone melting is determined by the zone length l, ingot length L, number of passes n, the degree of mixing in the liquid zone, and the distribution coefficient of the materials being purified. The distribution of solute after one pass can be obtained by material-balance considerations. This is a two-domain problem; i.e., in the major portion of the ingot of length $L - l$ zone melting occurs in the conventional sense. The trailing end of the ingot of length l undergoes normal freezing. For the case of constant-distribution coefficient, perfect mixing in the liquid phase, and negligible diffusion in the solid phase, the solute distribution for a single pass is given by Eq. (17-4) [Pfann, *Trans. Am. Inst. Mech. Eng.*, **194**, 747 (1952)].

$$C_s = C_0[1 - (1 - k)e^{-kx/\ell}] \tag{17-4}$$

The position of the zone x is measured from the leading edge of the ingot. The distribution for multiple passes can also be calculated from a material balance, but in this case the leading edge of the zone encounters solid corresponding to the composition at the point in question for the previous pass. The multiple-pass distribution has been numerically calculated (Pfann, *Zone Melting*, 2d ed., Wiley, New York, 1966, p. 285) for many combinations of k, L/l, and n. Typical solute-composition profiles are shown in Fig. 17-6 for various numbers of passes.

The ultimate distribution after an infinite number of passes is also shown in Fig. 17-6 and can be calculated for $x < (L - l)$ from the following equation (Pfann, op. cit., p. 42):

$$C_s = Ae^{BX} \tag{17-5}$$

where A and B can be determined from the following relations:

$$k = B\ell/(e^{B\ell} - 1) \tag{17-6}$$

$$A = C_0BL/(e^{BL} - 1) \tag{17-7}$$

The ultimate distribution represents the maximum separation that can be attained without cropping the ingot. Equation (17-5) is approximate because it does not include the effect of normal freezing in the last zone length.

As in normal freezing, many refinements of these models have been developed. Corrections for partial liquid mixing and a variable distribution coefficient have been summarized in detail (Zief and Wilcox, op. cit., p. 47).

Pertinent Variables in Zone Melting The dominant variables in zone melting are the number of passes, ingot-length–zone-length ratio, freezing rate, and degree of mixing in the liquid phase. Figure 17-6 illustrates the increased solute redistribution that occurs as the

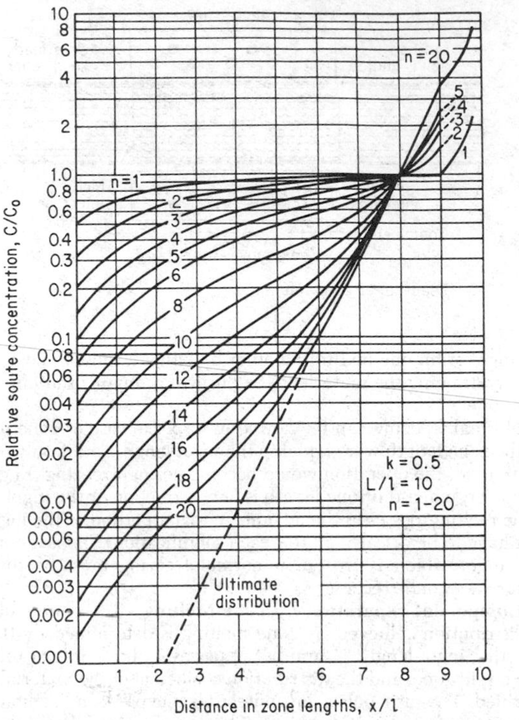

FIG. 17-6 Relative solute concentration C/C_0 (logarithmic scale) versus distance in zone lengths x/ℓ from beginning of charge, for various numbers of passes n. L denotes charge length. (*Pfann,* Zone Melting, *2d ed., Wiley, New York, 1966, p. 290.*)

number of passes increases. Ingot-length–zone-length ratios of 4 to 10 are commonly used (Zief and Wilcox, op. cit., p. 624). An exception is encountered when one pass is used. In this case the zone length should be equal to the ingot length; i.e., normal freezing provides the maximum separation when only one pass is used.

The freezing rate and degree of mixing have effects in solute redistribution similar to those discussed for normal freezing. Zone travel rates of 1 cm/h for organic systems, 2.5 cm/h for metals, and 20 cm/h for semiconductors are common. In addition to the zone-travel rate the heating conditions affect the freezing rate. A detailed summary of heating and cooling methods for zone melting has been outlined by Zief and Wilcox (op. cit., p. 192). Direct mixing of the liquid region is more difficult for zone melting than normal freezing. Mechanical stirring complicates the apparatus and increases the probability of contamination from an outside source. Some mixing occurs because of natural convection. Methods have been developed to stir the zone magnetically by utilizing the interaction of a current and magnetic field (Pfann, op. cit., p. 104) for cases in which the charge material is a reasonably good conductor.

Applications Zone melting has been used to purify hundreds of inorganic and organic materials. Many classes of inorganic compounds including semiconductors, intermetallic compounds, ionic salts, and oxides have been purified by zone melting. Organic materials of many types have been zone-melted. Schildknecht and coworkers (op. cit., p. 710) have explored zone melting of many organic materials. They have also zone-melted aqueous solutions of steam-volatilizable substances, enzymes, bacteria, and plankton. Zief and Wilcox (op. cit., p. 624) have compiled tables which give operating conditions and references for both inorganic and organic materials with melting points ranging from −115°C to over 3000°C.

Some materials are so reactive that they cannot be zone-melted to a high degree of purity in a container. Floating-zone techniques in which the molten zone is held in place by its own surface tension have been developed by Keck et al. [*Phys. Rev.*, **89**, 1297 (1953)].

Continuous-zone-melting apparatus has been described by Pfann (op. cit., p. 171). This technique offers the advantage of a close approach to the ultimate distribution, which is usually impractical for batch operation.

Performance data have been reported by Kennedy et al. (Zief, *Purification of Inorganic and Organic Materials,* Marcel Dekker, New York, 1969, p. 261) for continuous-zone refining of benzoic acid.

COLUMN CRYSTALLIZATION

Conducting crystallization inside a column with a countercurrent flow of crystals and liquid can produce a higher product purity than conventional crystallization or distillation. The working concept is to form a crystal phase, either internally or externally, and then transport the solids through a countercurrent stream of enriched reflux liquid. The problem in practicing this technology is the difficulty of controlling solid-phase movement. Unlike distillation, which exploits the specific-gravity differences between liquid and vapor phases, melt crystallization involves the contacting of liquid and solid phases that have nearly identical physical properties. Phase densities are frequently very close, and gravitational settling of the solid phase may be slow and ineffective. The challenge of designing equipment to accomplish crystallization in a column has resulted in a myriad of configurations to achieve reliable solid-phase movement, high product yield and purity, and efficient heat addition and removal. Relatively few of the equipment embodiments have been commercialized.

Column crystallizers have been systematized into either **end-fed** or **center-fed** devices, depending on whether the feed location is upstream or downstream of the crystal forming section. Figure 17-7 depicts the features of an end-fed commercial column described by McKay et al. [*Chem. Eng. Prog. Symp. Ser.*, no. 25, 55, 163 (1969)] for the separation of xylenes. Crystals are formed by indirect cooling of the melt in scraped-surface heat exchangers, and the resultant slurry is introduced into the column at the top. This type of column has no mechanical internals to transport solids and instead relies upon an imposed hydraulic gradient to force the solids through the column into the melting zone. Residue liquid is removed through a filter directly above the melter. A pulse piston in the product discharge improves washing efficiency and column reliability.

Figure 17-8 shows the features of a horizontal center-fed column [Brodie, *Aust. Mech. Chem. Eng. Trans.*, 37 (May 1979)] which has been commercialized for continuous purification of naphthalene and *para*-dichlorobenzene. Liquid feed enters the column *between* the hot purifying section and the cold freezing or recovery zone. Crystals

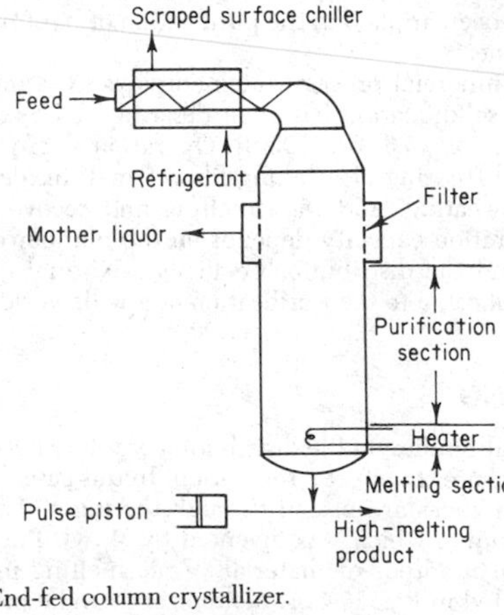

FIG. 17-7 End-fed column crystallizer.

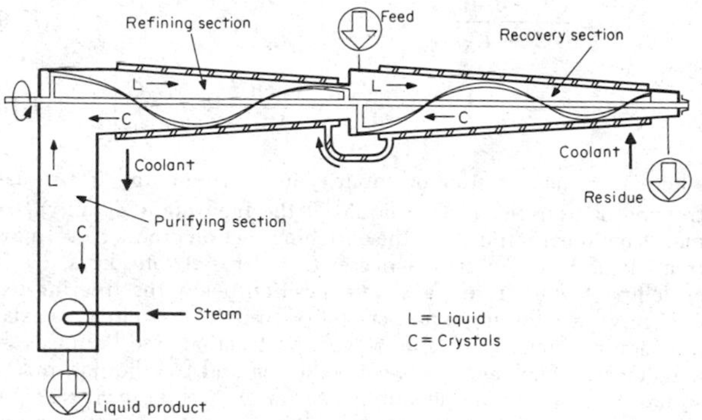

FIG. 17-8 Horizontal center-fed column crystallizer. *(C. W. Nofsinger Co.)*

are formed internally by indirect cooling of the melt through the walls of the refining and recovery zones. Residue liquid that has been depleted of product exits from the coldest section of the column. A spiral conveyor controls the transport of solids through the unit.

Another center-fed design that has been used commercially on a preparative scale is the vertical spiral conveyor column reported by Schildknecht [*Angew. Chem.*, **73**, 612 (1961)]. In this device, a version of which is shown on Fig. 17-9, the dispersed-crystal phase is formed in the freezing section and conveyed downward in a controlled manner by a rotating spiral with or without a vertical oscillation.

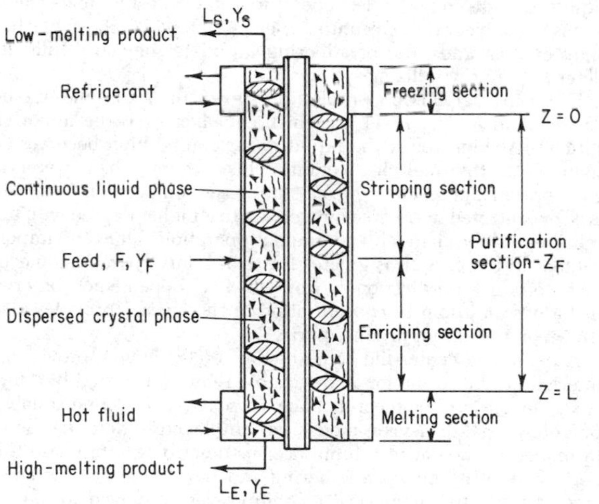

FIG. 17-9 Center-fed column crystallizer with a spiral-type conveyor.

Differences have been observed in the performance of end- and center-fed column configurations. Consequently, discussions of center- and end-fed column crystallizers are presented separately. The design and operation of both columns are reviewed by Powers (Zief and Wilcox, op. cit., p. 343) and Atwood (op. cit.). A comparison of these devices is shown on Table 17-2.

Center-Fed Column Crystallizer Two types of center-fed column crystallizers are illustrated on Figs. 17-8 and 17-9. As in a simple distillation column, these devices are composed of three distinct sections: a freezing or recovery section, where solute is frozen from the impure liquor; the purification zone, where countercurrent contacting of solids and liquid occurs; and the crystal-melting and -refluxing section. Feed position separates the refining and recovery portions of

TABLE 17-2 Comparison of Melt-Crystallizer Performance

Center-fed column	End-fed column
Solid phase is formed internally; thus, only liquid streams enter and exit the column.	Solid phase is formed in external equipment and fed as slurry into the purifier.
Internal reflux can be controlled without affecting product yield.	The maximum internal liquid reflux is fixed by the thermodynamic state of the feed relative to the product stream. Excessive reflux will diminish product yield.
Operation can be continuous or batchwise at total reflux.	Total reflux operation is not feasible.
Center-fed columns can be adapted for both eutectic and solid-solution systems.	End-fed columns are inefficient for separation of solid-solution systems.
Either low- or high-porosity-solids-phase concentrations can be formed in the purification and melting zones.	End-fed units are characterized by low-porosity-solids packing in the purification and melting zones.
Scale-up depends on the mechanical complexity of the crystal-transport system and techniques for removing heat. Vertical oscillating spiral columns are likely limited to about 0.2 m in diameter, whereas horizontal columns of several meters are possible.	Scale-up is limited by design of melter and/or crystal-washing section. Vertical or horizontal columns of several meters in diameter are possible.

the purification zone. The section between feed location and melter is referred to as the refining or enrichment section, whereas the section between feed addition and freezing is called the recovery section. The refining section may have provisions for sidewall cooling. The published literature on column crystallizers connotes stripping and refining in a reverse sense to distillation terminology, since refined product from a melt crystallizer exits at the hot section of the column rather than at the cold end as in a distillation column.

Rate processes that describe the purification mechanisms in a column crystallizer are highly complex since phase transition and heat- and mass-transfer processes occur simultaneously. Nucleation and growth of a crystalline solid phase along with crystal washing and crystal melting are occurring in various zones of the apparatus. Column hydrodynamics is also difficult to describe. Liquid- and solid-phase mixing patterns are influenced by factors such as solids-transport mechanism, column orientation, and, particularly for dilute slurries, the settling characteristics of the solids.

Most investigators have focused their attention on a differential segment of the zone between the feed injection and the crystal melter. Analysis of crystal formation in the recovery section has received scant attention. Table 17-3 summarizes the scope of the literature treatment for center-fed columns for both solid-solution and eutectic forming systems.

Solid-Solution Separation The dominant mechanism of purification for column crystallization of solid-solution systems is recrystallization. The rate of mass transfer resulting from recrystallization is related to the concentrations of the solid phase and free liquid which are in contact. A model based on height-of-transfer-unit (HTU) concepts representing the composition profile in the purification section for the high-melting component of a binary solid-solution system has been reported by Powers et al. (in Zief and Wilcox, op. cit., p. 363) for total-reflux operation. Typical data for the purification of a solid-solution system, azobenzene-stilbene, are shown in Fig. 17-10. The column crystallizer was operated at total reflux. The solid line through the data was computed by Powers et al. (op. cit., p. 364) by using an experimental HTU value of 3.3 cm.

Eutectic-System Separation Most of the analytical treatments of center-fed columns describe the purification mechanism in an adiabatic oscillating spiral column (Fig. 17-9). However, the analyses by Moyers (op. cit.) and Griffin (op. cit.) are for a nonadiabatic

TABLE 17-3 Column-Crystallizer Investigations

	Treatments	
	Theoretical	Experimental
Solid solutions		
Total reflux—steady state	1, 2, 4, 6	1, 4, 6
Total reflux—dynamic	2	
Continuous—steady state	1, 4	4, 8, 9
Continuous—dynamic		
Eutectic systems		
Total reflux—steady state	1, 3, 4, 7	1, 3, 6
Total reflux—dynamic		
Continuous—steady state	1, 5, 10, 11, 12	5, 8, 9, 10, 11, 13
Continuous—dynamic		

1. Powers, *Symposium on Zone Melting and Column Crystallization*, Karlsruhe, 1963.
2. Anikin, *Dokl. Akad. Nauk SSSR*, **151**, 1139 (1969).
3. Albertins et al., *Am. Inst. Chem. Eng. J.*, **15**, 554 (1969).
4. Gates et al., *Am. Inst. Chem. Eng. J.*, **16**, 648 (1970).
5. Henry et al., *Am. Inst. Chem. Eng. J.*, **16**, 1055 (1970).
6. Schildknecht et al., *Angew. Chem.*, **73**, 612 (1961).
7. Arkenbout et al., *Sep. Sci.*, **3**, 501 (1968).
8. Betts et al., *Appl. Chem.*, **17**, 180 (1968).
9. McKay et al., *Chem. Eng. Prog. Symp. Ser.*, no. 25, 55, 163 (1959).
10. Bolsaitis, *Chem. Eng. Sci.*, **24**, 1813 (1969).
11. Moyers et al., *Am. Inst. Chem. Eng. J.*, **20**, 1119 (1974).
12. Giffin, M.S. thesis in chemical engineering, University of Delaware, 1975.
13. Brodie, *Aust. Mech. Chem. Eng. Trans.*, 37 (1971).

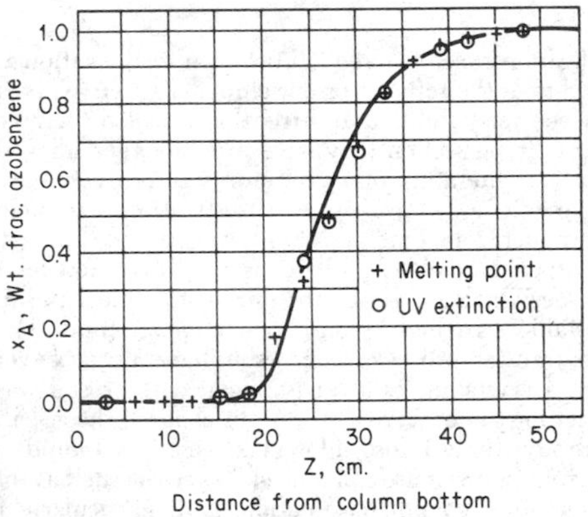

FIG. 17-10 Steady-state separation of azobenzene and stilbene in a center-fed column crystallizer with total-reflux operation. To convert centimeters to inches, multiply by 0.3937. (*Zief and Wilcox,* Fractional Solidification, *vol. 1, Marcel Dekker, New York, 1967, p. 356.*)

dense-bed column. Differential treatment of the horizontal-purifier (Fig. 17-8) performance has not been reported; however, overall material and enthalpy balances have been described by Brodie (op. cit.) and apply equally well to other designs.

Henry (op. cit.) developed a steady-state model for the purification section based on differential countercurrent contacting and the assumptions that internal flow rates are constant and that crystal composition is constant. Constant internal flow rate applies when refreezing of reflux liquid does not occur, as is the case with relatively pure feedstocks. The model includes the effect of axial diffusion of impurity induced by the rotating and oscillating conveyor and mass transfer between an adhering liquid phase and a free enriched wash liquid. The following equation describes impurity distribution in the enriching section.

Enriching section ($z > z_F$):

$$\frac{Y - Y_P}{Y_\phi - Y_P} = e^{-(z-z_F)/\Psi_E} \tag{17-8}$$

$$Y_P \equiv \frac{C\epsilon - L_E Y_E}{C - L_E} \tag{17-9}$$

$$\Psi_E \equiv \frac{1}{C - L_E}\left[D\rho A\eta + \frac{\alpha(\alpha+1)C^2}{KaA\rho} - \frac{\alpha L_E C}{KaA\rho}\right] \tag{17-10}$$

where Y = mass fraction of impurity in the free liquid; Y_ϕ = mass fraction of impurity of free liquid at the feed point Z_F, m; Y_E = mass fraction of impurity in the enriching-section product; ϵ = mass-fraction impurity in crystal phase; C = crystal rate, kg/s; L_E = enriching-product rate, kg/s; z = position below the freezing section, m; α = adhering liquid-crystal rate ratio; D = effective axial-diffusion coefficient, m^2/s; η = volume fraction free liquid; K = mass-transfer coefficient between adhering and free liquids, m/s; a = interfacial area per unit volume, per m; A = column cross section normal to flow direction (window area defined by conveyor and annulus, m^2); and ρ = free-liquid density, kg/m^3.

A similar expression was written for the stripping section, and these equations, along with a constraining material balance on the terminal streams can be solved simultaneously by an iterative technique (ibid.) to give column-composition profiles. To utilize these equations for design the mass-transfer factors Ψ for the refining and stripping sections along with the crystal-phase composition must be available. This implies that measurements or estimates of parameters such as α, D, η, K, and a must be supplied for the system investigated.

Experimental evidence strongly indicates that axial dispersion of mass in the free-liquid phase controls the purification ability of an oscillating-spiral-conveyor center-fed crystallizer column. Mass-transfer terms were shown to be of little importance in the purification process, which implies the absence of an adhering-liquid phase, since the composition difference between the adhering and the free liquid is indistinguishable. The importance of the mass-transfer terms is a subject for conjecture, since parameters such as interfacial transfer area and ratio of adhering liquid to solid are difficult to determine in crystallizing systems.

Experimental column profiles are shown for a cyclohexane-benzene system in Fig. 17-11, in which cyclohexane is the minor constituent. At high reflux the liquid-phase composition becomes constant near the melting section. This occurs when free-liquid composition approaches the crystal-phase composition. Impurities are concentrated in the freezing section to a higher degree with total-reflux operation than with continuous operation. Thus, the impurity content in the crystals is greater for total-reflux operation, and ultimate product purification is limited. This dependence of crystal inclusions on impurity concentration in the freezer provides a constraint on maximum product purity.

A dense-bed center-fed column (Fig. 17-12) having provision for internal crystal formation and variable reflux was tested by Moyers et al. (op. cit.). In the theoretical development (ibid.) a nonadiabatic, plug-flow axial-dispersion model was employed to describe the performance of the entire column. Terms describing interphase transport of impurity between adhering and free liquid are not considered. An iterative numerical technique was employed to solve the resultant second-order differential equation. Figure 17-13 illustrates the typical variation of liquid and solid fluxes computed for a dense-bed column. A jump change in the liquid flux occurs at the feed location. A moderate amount of melting occurs in the purification section of the column. Computed column-temperature profiles shown in Fig. 17-14 indicate the sensitivity of the column-temperature profile to the magnitude of liquid axial mixing. A cusp is predicted at the feed point for low axial dispersion. An abrupt temperature rise was observed experimentally below the feed point, indicating that near-plug-flow conditions are present in the liquid phase.

A comparison of the axial-dispersion coefficients obtained in oscillating-spiral and dense-bed crystallizers is given in Table 17-4. The dense-bed column approaches axial-dispersion coefficients similar to those of densely packed ice-washing columns.

The concept of minimum reflux as related to column-crystallizer operation is presented by Brodie (op. cit.) and is applicable to

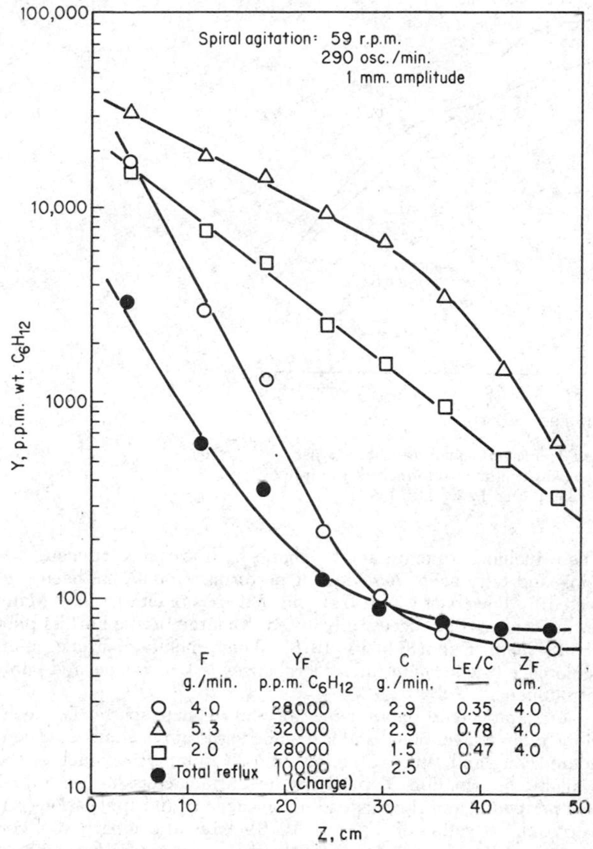

FIG. 17-11 Steady-state separation of benzene-cyclohexane in a center-fed column. To convert centimeters to inches, multiply by 0.3937; to convert grams to pounds, multiply by 0.002205. [*Data of Albertins*, Am. Inst. Chem. Eng. J., ***15**, 554 (1969); and Henry et al.*, Am. Inst. Chem. Eng. J., ***16**, 1055 (1970)*.]

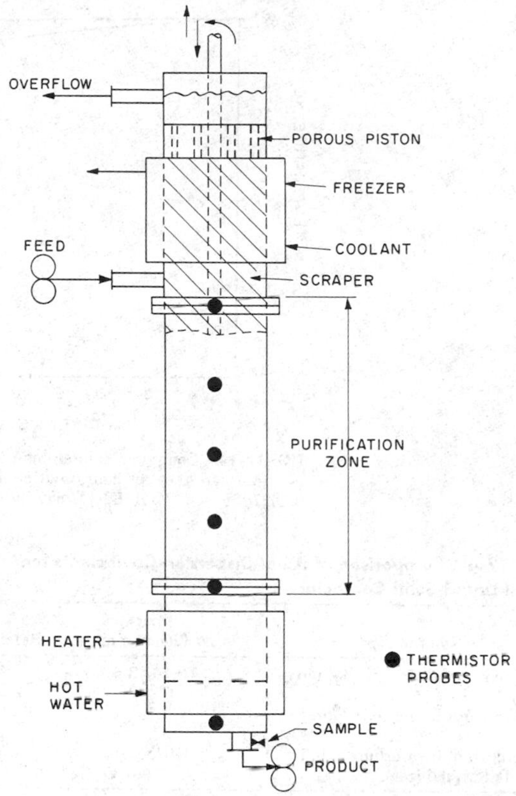

FIG. 17-12 Dense-bed center-fed column crystallizer. [*Moyers et al.*, Am. Inst. Chem. Eng. J., ***20**, 1121 (1974)*.]

all types of column crystallizers, including end-fed units. In order to stabilize column operation the sensible heat of subcooled solids entering the melting zone should be balanced or exceeded by the heat of fusion of the refluxed melt. The relationship in Eq. (17-11) describes the minimum reflux requirement for proper column operation.

$$R = (T_P - T_F)\, C_P/\lambda \tag{17-11}$$

R = reflux ratio, g reflux/g product; T_P = product temperature, °C; T_F = saturated-feed temperature, °C; C_P = specific heat of solid crystals, cal/(g·°C); and λ = heat of fusion, cal/g.

All refluxed melt will refreeze if reflux supplied equals that com-

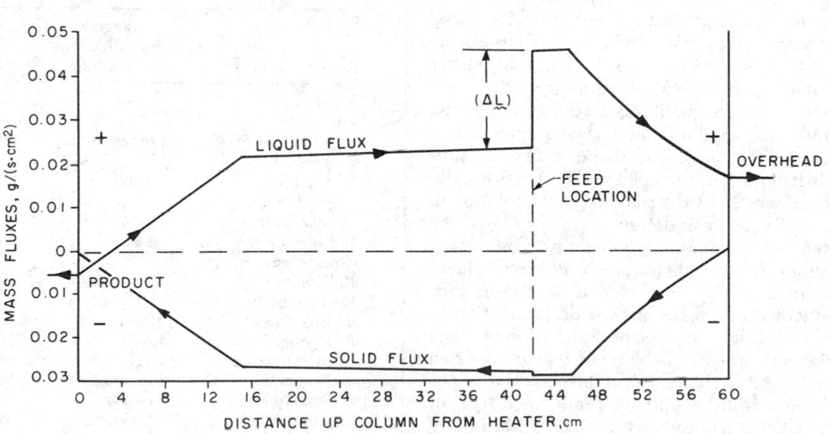

FIG. 17-13 Computed liquid and solid mass fluxes in a dense-bed column crystallizer. To convert centimeters to inches, multiply by 0.3937; to convert grams per second-square centimeter to pounds per hour-square foot, multiply by 7373. [*Moyers et al.*, Am. Inst. Chem. Eng. J., ***20**, 1121 (1974)*.]

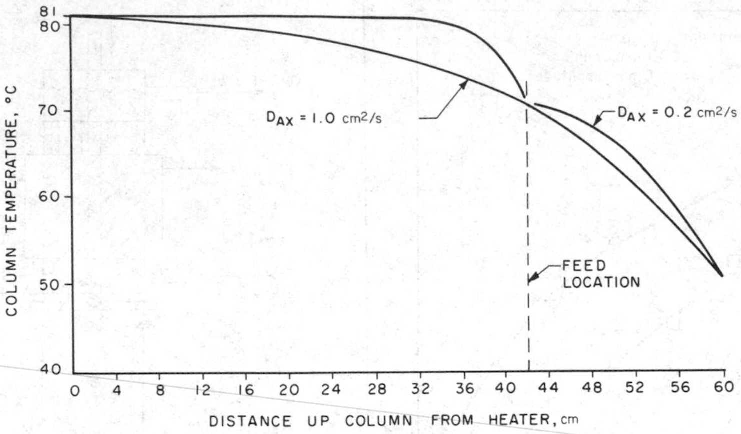

FIG. 17-14 Computed column-temperature profiles. To convert square centimeters per second to square feet per hour, multiply by 3.875; to convert centimeters to inches, multiply by 0.3937; °F = % °C + 32. [*Moyers et al.*, Am. Inst. Chem. Eng. J., *20, 1121 (1974).*]

TABLE 17-4 Comparison of Axial-Dispersion Coefficients for Several Liquid-Solid Contactors

Column type	Dispersion coefficient, cm^2/s	Reference
Center-fed crystallizer (oscillating spiral)	1.6–3.5	1
Center-fed crystallizer (oscillating spiral)	1.3–1.7	2
Countercurrent ice-washing column	0.025–0.17	3
Center-fed crystallizer	0.12–0.30	4

References:
1. Albertins et al., *Am. Inst. Chem. Eng. J.*, **15**, 554 (1969).
2. Gates et al., *Am. Inst. Chem. Eng. J.*, **16**, 648 (1970).
3. Ritter, Ph.D. thesis, Massachusetts Institute of Technology, 1969.
4. Moyers et al., *Am. Inst. Chem. Eng. J.*, **20**, 1119 (1974).

puted by Eq. (17-11). When reflux supplied is greater than the minimum, jacket cooling in the refining zone or additional cooling in the recovery zone is required to maintain product recovery. Since high-purity melts are fed near their pure-component freezing temperatures, little refreezing takes place unless jacket cooling is added.

Significant Variables To utilize a column-crystallizer design or rating model, a large number of parameters must be identified. Many of these are empirical in nature and must be determined experimentally in equipment identical to the specific device being evaluated. Hence macroscopic evaluation of systems by large-scale piloting is the rule rather than the exception. Included in this rather long list of critical parameters are factors such as impurity level trapped in the solid phase, product quality as a function of reflux ratio, degree of liquid and solids axial mixing in the equipment as a function of solids-conveyor design, size and shape of crystals produced, and ease of solids handling in the column. Heat is normally removed through metal surfaces; thus, the stability of the solution to subcooling can also be a major factor in design.

Applications The center-fed column crystallizer with oscillating spiral conveyor (Fig. 17-9) has been used on a preparative scale; however, the horizontal column (Fig. 17-8) marketed as the **Brodie Purifier,** has achieved commercialization for large-scale production of *para*-dichlorobenzene and naphthalene. A plant designed to produce 6000 tons per year of 99.9 percent *para*-dichlorobenzene from a rich 95 percent feedstock and a unit designed to produce 7200 tons per year of refined naphthalene from a phthalic-grade feedstock of approximately 78°C crystallizing point were reported by Muir (79th nat. meet., AIChE, Houston, 1975).

The spiral column has been used to purify binary and multicomponent mixtures of both the eutectic and the solid-solution types. These include aromatic and aliphatic hydrocarbons, aqueous systems, and fatty acids. A review of performance data has been presented by Powers et al. (in Zief and Wilcox, op. cit., p. 355). Many products have been successfully tested in a large Brodie Purifier pilot plant [*Chem. Eng.* (Feb. 13, 1978)]. These include benzene, monochloroacetic acid, *para*-nitrochlorobenzene, *para*-xylene, and *para*-nitrotoluene.

Comparative-cost data for melt crystallizers are sparse. The technique is less conventional and is specific to the application being considered. Zief and Wilcox (op. cit., p. 603) summarize much of the available information. Typical Brodie Purifier costs start around 1 cent per pound and rise, depending on capacity and application (private communication, R. F. Muir, 1979). A broad comparison of column-crystallizer cost factors with those of other separation methods is given in Table 17-5.

End-Fed Column Crystallizer End-fed columns were developed and successfully commercialized by the Phillips Petroleum Company in the 1950s. The sections of a typical end-fed column, often referred to as a Phillips column, are shown on Fig. 17-7. Impure liquor is removed through filters located between the prod-

TABLE 17-5 Advantages and Disadvantages of Column Crystallization Compared with Other Separating and Purifying Methods*

Selection criteria	Staged crystallization or washing	Distillation	Column crystallization
1. Field of application			
a. Number organic products	Moderate (50)	Very high (1000)	Moderate (50)
b. Maximum production, tons/year	50,000	100,000	50,000
2. Industrial experience	Moderate	Very large	Small
3. Reliability	Moderate	Good	Good
4. Relative costs			
a. Investment	100–120	60–120	100
b. Energy	Low	High	Very low
c. Maintenance	High	Small	Small
d. Personnel	Moderate	Small	Small
e. Startup losses	Moderate	None	None
5. Product purity	Moderate	Moderate to high	Very high
6. Corrosion	Moderate	High	Little

*Personal communication with the C. W. Nofsinger Co.

uct-freezing zone and the melter rather than at the end of the freezing zone, as occurs in center-fed units. The purification mechanism for end-fed units is basically the same as for center-fed devices. However, there are reflux restrictions in an end-fed column, and a high degree of solids compaction exists near the melter of an end-fed device. It has been observed that the free-liquid composition and the fraction of solids are relatively constant throughout most of the purification section but exhibit a sharp discontinuity near the melting section [McKay et al., *Ind. Eng. Chem.*, **52**, 197 (1969)]. Investigators of end-fed column behavior are listed in Table 17-6. Note that end-fed columns are adaptable only for eutectic-system purification and cannot be operated at total reflux.

TABLE 17-6 End-Fed-Crystallizer Investigations

	Treatments	
Eutectic systems	Theoretical	Experimental
Continuous—steady state	1, 2, 4	1, 4
Batch	3	3

1. McKay et al., *Ind. Eng. Chem. Process Des. Dev.*, **6**, 16 (1967).
2. Player, *Ind. Eng. Chem. Process Des. Dev.*, **8**, 210 (1969).
3. Yagi et al., *Kagaku Kogaku*, **72**, 415 (1963).
4. Shen and Meyer, Prepr. 19F, AIChE Symp., Chicago, 1970.

Theory Heat transfer between the crystals and free liquid plays a dominant role in the purification section. Because relatively impure feed is often processed, axial-temperature differences of 40 to 50°C are common. The crystals in the feed slurry are therefore substantially subcooled. The heat transfer between the subcooled crystals and the free liquid results in refreezing a portion of the liquid stream. This leads to an increase in the fraction of solids in the region of the column just above the melting section. This refreezing, which results in a dense bed of solids in the lower region of the column, is necessary for proper operation (McKay, loc. cit.). Most of the purification takes place in this region of the bed, as evidenced by the composition discontinuity cited earlier. It is emphasized that the high solids content which occurs in the purification section of the end-fed column is in contrast with the solids content in the purification section of a dilute-phase center-fed column. The crystal-liquid slurry entering an end-fed column is typically up to 50 percent solids as it enters the column (McKay, loc. cit.) and forms a dense immobile bed at the bottom of the purification section (the weight-fraction solids at the bottom of the column can exceed 95 percent).

Player (op. cit.) has developed a model for the end-fed column which incorporates the refreezing of free liquid and miscible displacement of the occluded liquid associated with the crystal phase. He views the upper section of the purification section (see Fig. 17-7) as a zone of compaction for the crystals while a relatively small portion of the volume where a dense bed of crystals has formed is actually serving as the purification section. His method predicts a discontinuity in both the liquid composition and the solids content near the melting section and therefore should be of value in interpreting performance data from end-fed column crystallizers.

The sensitivity of product purity to crystal-phase quality and refrozen-reflux quantity is expressed by Player (op. cit.).

$$X_1 = \frac{S_2Y_2 + (W_1 - W_2)}{S_2 + (W_1 - W_2)} \tag{17-12}$$

X_1 = product composition, mass fraction; S_2 = solid flux to compaction zone, kg/(s/m^2); Y_2 = composition of solids to compaction zone; W_1 = liquid flux entering purification zone (i.e., reflux from melter); and W_2 = liquid flux leaving compaction zone. If $S_2 = 100$ flux units and the change in liquid flux across the compaction-purification zone is 24.59, then product composition X_1 can be computed as a function of crystal quality ($Y_2 = 0.99$, $X_1 = 0.992$; $Y_2 = 0.995$, $X_1 = 0.996$; $Y_2 = 0.999$, $X_1 = 0.9992$).

For the example shown, product quality is only slightly enhanced by the refreezing of reflux on subcooled crystals. Player's analysis indicates that refreezing of high-purity melt simply enriches the solid phase by dilution of the impurity content and, at least for reflux levels attainable in end-fed columns, does not significantly enrich product. It is suggested that the high degree of solids compaction that occurs in an end-fed column crystallizer (some of which can be attributed to refreezing of reflux) minimizes axial mixing. Hence, excellent crystal washing is achieved with small quantities of reflux. If liquid flux leaving the compaction zone is slightly positive (i.e., greater than that required to heat subcooled solids), then high purification via a crystal-washing mechanism is achieved.

Performance information for the purification of *p*-xylene indicates that nearly 100 percent of the crystals in the feed stream are removed as product. This suggests that the liquid which is refluxed from the melting section is effectively refrozen by the countercurrent stream of subcooled crystals. A high-melting product of 99.0 to 99.8 weight percent *p*-xylene has been obtained from a 65 weight percent *p*-xylene feed. The major impurity was *m*-xylene. Figure 17-15 illustrates the column-cross-section-area–capacity relationship for various product purities.

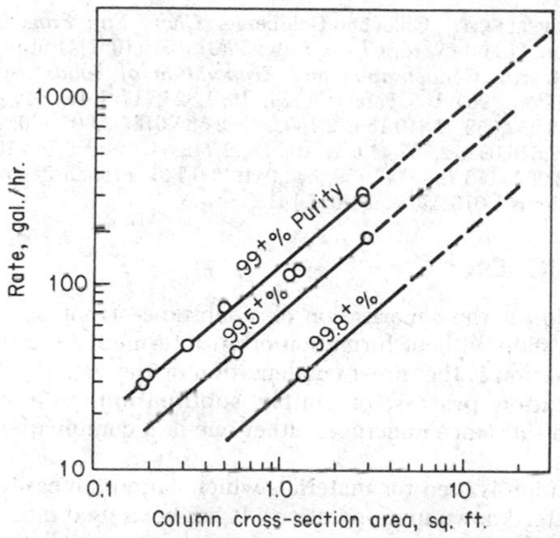

FIG. 17-15 Pulsed-column capacity versus column size for 65 percent *p*-xylene feed. To convert gallons per hour to cubic meters per hour, multiply by 0.9396; to convert square feet to square meters, multiply by 0.0929. *(McKay et al., prepr., 59th nat. meet., AIChE, East Columbus, Ohio.)*

Significant Variables A pulsing device is commonly used to obtain efficient crystal transport countercurrently to the free-liquid stream. The required pulse displacement is a function of the cross-sectional area of the column. Proper pulse displacements have been determined by McKay (in Zief and Wilcox, op. cit., p. 438). The nature of the crystals and their particle-size distribution affect column performance. These variables play a dominant role in the case of liquid removal and solids containment in the filtration region of the column. Also, the porosity of the dense bed of crystals at the bottom of the purification section is affected by the type of crystal involved. McKay (loc. cit.) has reviewed the effect of these variables on column performance.

The level of impurities in the column feed critically affects column performance. The mechanism of purification is strongly coupled to the degree of refreezing which occurs in the purification section. Refreezing will occur, however, only if the crystals are significantly subcooled in the column feed. Thus, the end-fed column may not be applicable to the processing of relatively pure feedstocks (0.1 to 3 weight percent impurities) because of the inability to obtain appropriate refreezing.

Applications Column crystallizers of the end-fed type can be used for purification of many eutectic-type systems and for aqueous as well as organic systems (McKay, loc. cit.). Column crystallizers have been used for xylene isomer separation, but recently other separation technologies have tended to supplant crystallization.

Other Techniques With dilute feedstock it is often advisable from an economic and particle-size-control viewpoint to cool the feed and form slurry externally in separate equipment and then charge the crystal slurry directly into the purification zone of an end-fed column. Conducting solidification and purification in separate equipment does not basically change the end-fed nature of the operation.

Ice-washing-type columns developed for freeze desalinization (Zief and Wilcox, op. cit., p. 451) have provision for melting and refluxing the frozen product, in this case water. In conjunction with appropriate external crystal formation equipment the ice-washing columns can be adapted for either concentrating solute or for the recovery and refining of organic or aqueous products (Concentration Specialists, Inc., Andover, Mass.). For very dilute systems, columns can be staged. A continuous purifier classified as end-fed in which the crystal slurry is formed in separate external equipment, either drum flaker or vessel, has been disclosed [*Chem. Eng.* (April 1979)]. The slurry is charged from the crystallizing equipment to a vertical column having a melter at the top, a spent-liquor drain at the bottom, and a screw conveyor to transport solids upward into the melting zone. Two plants are reportedly in commercial use in Japan; one produces *para*-dichlorobenzene and the other 3,5-dichloroaniline.

End-fed column crystallization, whether practiced in a single column or in several pieces of equipment, is not a conventional separation technique and tends to be very specific to the application being considered. One compilation of cost data (Zief and Wilcox, op. cit., p. 603) summarizes much of the information that is available.

SUBLIMATION

General References: Gillot and Goldberger, *Chem. Eng. Prog. Symp. Ser.*, **65**(91), 36–42 (1969). Nord, *Chem. Eng.*, **58**(9), 157 (1951). Rutner, Goldfinger, and Hearth, *Condensation and Evaporation of Solids*, Gordon and Breach, London, 1964. U.S. Patents 1,324,716; 1,324,717; 1,464,844; 1,987,301; 2,214,838; 2,252,052; 2,310,188; 2,499,255; 2,583,013; 2,607,440; 2,608,472; 2,628,892; 2,676,092; 2,737,439; 2,740,527; 2,742,342; and 2,743,169. British Patents 142,902; 173,789; 447,759; 644,941; 700,143. French Patent 948,039. German Patents 1,016,236 and 1,017,141.

INTRODUCTION

Sublimation is the vaporization of a substance from the solid into the vapor state without formation of an intermediate liquid phase. **Desublimation** is the direct condensation of the vapor to the solid. A **sublimation process,** or simply **sublimation,** is a procedure whereby a substance undergoes either one or a combination of these transitions.

Sublimation is used for materials which cannot be easily purified by the better-known unit operations. It has been used most often for separation of a volatile component from other essentially nonvolatile components, for example, for the separation of sulfur from impurities or for purification of benzoic acid. There has been interest in separating mixtures of volatile components by sublimation methods [Gillot and Goldberger, loc cit.; and Holden and Bryant, *Sep. Sci.*, **4**(1), 1–13 (1969)]. Fractional sublimation is analogous to distillation except that the volatile components are separated from the solid phase rather than from the liquid phase.

Sublimation as a separation tool should be considered when, according to Nord (loc. cit.), (1) the material is unstable or is temperature- or oxidation-sensitive; (2) it is desirable to produce a solid product directly from the vapor, i.e., for a certain crystal type, size, or appearance of the product; (3) the product to be recovered is nonvolatile and heat-sensitive and is to be separated from a volatile material, e.g., in freeze drying; (4) the material to be recovered has a high melting point and, if processed at elevated temperatures, presents problems such as corrosion of equipment; (5) the volatile material is mixed with a high percentage of nonvolatile materials; and (6) a mixture of volatile materials is to be separated.

MECHANISM AND THEORY OF SUBLIMATION PROCESSES

For vaporization to proceed, the vapor pressure of the subliming components must be greater than their partial pressures in the gas phase in contact with the solid. Relatively few substances have sufficiently high vapor pressures to sublime under atmospheric conditions. Therefore, sublimation must be accomplished by heating the solid or controlling the gaseous environment in contact with the solid, or both. The environment can be controlled either by vacuum operation, in which case the total pressure is lowered and the gas phase contains primarily the subliming components, or by using a nonreactive gaseous diluent to lower the partial pressures of the subliming components. The latter is known as **carrier** or **entrainer sublimation.** Vacuum sublimation is inherently a batch operation, whereas entrainer sublimation can be conducted as a continuous process.

Figure 17-16 is a generalized schematic for simple sublimation processes. For vacuum sublimation, the entrainer and quench gas lines, as well as any of the recycle streams shown, would be omitted. Entrainer sublimation can be operated with or without recycle of the entrainer gas.

Thermodynamic Limitations for Sublimation Processes A phase diagram for a pure substance may be used to represent completely the phase relationships during a sublimation process involving a pure substance. For a pure substance or a mechanical mixture of

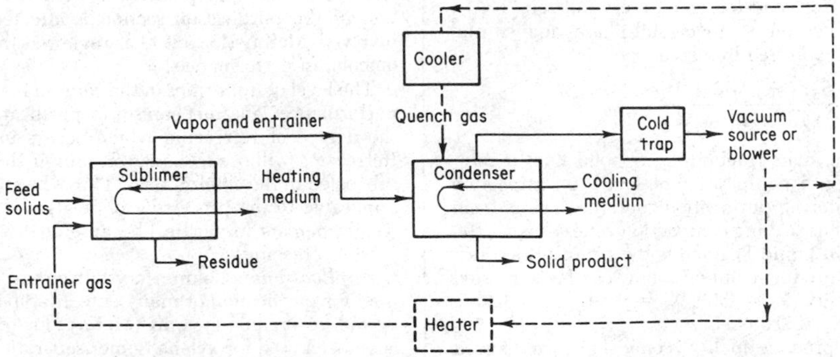

FIG. 17-16 Generalized schematic for simple sublimation.

solids containing a single volatile component there is no theoretical limitation with regard to the purity of the product obtained in a simple sublimation operation.

When more than one volatile component is involved, the system acquires an additional degree of freedom for each added component and the phase relationships during a sublimation process can no longer be completely represented on a single plane. For a system of two volatile solids in which there is no mutual solubility in the solid phase, a simple eutectic phase diagram (see Fig. 17-2 and replace "liquid" by "vapor" in the diagram) typically represents the phase relations at a fixed pressure. If the two volatile solids form a solid solution over the entire range of compositions, a figure like Fig. 17-1 with the term "liquid" replaced by the term "vapor" applies, also at fixed pressure. For a solid substance consisting of more than one volatile component, no pure component can be recovered in a simple sublimation operation if a total condenser is used. Pure components can be recovered only by a fractionating technique. For a two-component system in which solid solutions are not formed, fractional desublimation techniques may be successful for separation into pure components if there are relatively large vapor-pressure differences between the sublimable components. When the vapor-pressure difference is small or when a solid solution is formed, fractional sublimation must be used to recover any pure components. For the two-component case in which solid solutions are not formed, only one pure component can be obtained from a single column. When a solid solution is formed from two volatile components throughout the entire composition range, it is theoretically possible to separate the two components completely in a single column.

For multicomponent mixtures of similar volatility one or more pure components can be isolated, depending upon the solubilities in the solid phase. If mutual solubility does not exist in the solid phase, it will be impossible to isolate more than one pure component regardless of the number of columns used. If homogeneous solid solutions are formed throughout the entire composition range of the various binary pairs and in the multicomponent system, it is theoretically possible to isolate each of the individual pure components by using a number of columns that is one less than the number of components involved.

Rate-Limiting Factors for Sublimation Processes The production capacity of a sublimation process will usually be determined by one mechanistic step in the overall process. One of the following will normally control:

1. *Heat flow to the subliming solids.* Heat transfer to the solids is difficult to accomplish. A number of methods for improving heat transfer are used. Often the solids are fed in a finely divided form and are continuously agitated in the sublimer. For entrainer sublimation the entrainer gas may be preheated. Direct firing in kettles may be used for materials which are not heat-sensitive.

2. *Diffusion of sublimable components to the vapor-solid interface.* An important difference between distillation and sublimation is the way in which the volatile components arrive at the vapor–condensed-phase interface. Renewal of the interface is aided by convection in liquids, but this mechanism is absent for solids. Surface depletion of the volatile components can occur in solids, resulting in a solid-state diffusion-limited process.

Mass transfer in the solids is enhanced by some of the same provisions made to improve heat transfer.

3. *Change of solids to vapor.* This step is not usually of practical interest as a rate-limiting step. Discussion of the theory of vaporization and condensation of solids is available in Rutner et al. (loc. cit.).

4. *Mass transport from the vaporizing to the condensing zone.* For vacuum sublimation, mass transport from the subliming zone to the condensing zone may be rate-limiting. In simple sublimation mass transport is improved by continuous pumping and can be improved significantly by use of an entrainer gas.

5. *Change of vapors to solids.* This step is not usually rate-controlling.

6. *Heat flow from the condensed solids.* Unless means are provided to avoid the buildup of a condensed layer of solids on the cooling surface of the condenser in simple sublimation, with passage of time the rate-limiting mechanism is likely to become conduction from the condensed solid.

EQUIPMENT AND DESIGN EQUATIONS

Equipment for sublimation is not standardized to any extent. The patent literature noted in the general references is the best source of information for equipment designed or adapted for sublimation.

Equipment Typical equipment consists of jacketed pan dryers, vacuum rotary dryers, vacuum shelf dryers, direct-fired retorts, and Herreshoff roasting furnaces. The state of development of sublimers is much more advanced than that of condensers. Solids-condensing equipment often consists of nothing more than tanks furnished with mechanical scrapers, brushes, or vibrators for removing the condensed solids.

Two devices suitable for continuous fractional sublimation are those described by Gillot and Goldberger (loc. cit.) and by Jacque and Dumes (U.S. Patent 2,944,878). The former provide reflux by the use of circulating inert solid particles upon which the sublimable components are deposited as a thin film. The latter obtain reflux by mechanical transportation or free fall of the sublimable solids.

Design Equations: Simple Sublimation Simple sublimation can be conducted as either a vacuum or an entrainer operation. If the solids to be processed under equilibrium conditions (i.e., equilibrium between solid and vapor) in the sublimer and condenser do not form solid solutions, the theoretical separation obtained depends solely on the ratios of the saturated-vapor pressures of the components at the sublimer and condenser temperatures. If the ratios of the vapor pressures are small or solid solutions are formed, these limitations normally remove all interest in simple sublimation, and a fractional-sublimation process must be used.

Calculation of the yield per pass of sublimable solids is essentially the same for operation under vacuum or with an entrainer, provided all rate-controlling steps are equalized. For a system consisting of two sublimable components, the percent loss per pass is

$$\text{Percent loss} = \frac{r(P_{AC} + P_{BC})/(P_{AS} + P_{BS})}{(1 + r) - [(P_{AC} + P_{BC} - \Delta P)/(P_{AS} + P_{BS})]} \times 100 \qquad (17\text{-}13)$$

where r is the ratio of the moles of inert gas (either unavoidable as in vacuum operation or intentional as with an entrainer) to the moles of solids sublimed, *i.e.*, $r = P_1/(P_{AS} + P_{BS}) = (P - P_{AS} - P_{BS})/(P_{AS} + P_{BS})$. P_A and P_B are the vapor pressures of components A and B, subscripts S and C refer to the sublimer and condenser, P_1 is the partial pressure of inert gas, ΔP is the total pressure drop between sublimer and condenser, and P is the total pressure in the sublimer.

To calculate the percent yield per pass, a material balance must be made on the sublimer and condenser. In the actual case in which equilibrium may not be obtained and the gas in the sublimer and condenser is not saturated with sublimable components, the vapor pressures in the sublimer should be replaced with $E_S P_S$ and those in the condenser with $P_C E_C$, where E_S and E_C are relative-saturation or efficiency values.

For simple vacuum sublimation a recycle of vapors is not possible. Therefore, the ultimate yield of condensed solids is determined by subtracting the loss calculated by Eq. (17-13) from 100. For simple entrainer sublimation the entrainer can be recycled to increase the ultimate yield of products over that shown by Eq. (17-13). Since the yield loss increases with increasing r, air leakage should be kept small in vacuum sublimation. The advantages of entrainer sublimation usually offset the disadvantage of a high r or low yield per pass.

Example 1: Purification by Simple Entrainer Sublimation An intimate mechanical mixture consisting of 15 mole percent of component A and 85 mole percent of component B is to be purified by simple entrainer sublimation. Assume equilibrium conditions obtained in both the sublimer and the condenser, constant sublimer and condenser temperatures of 65.6°C (150°F) and 4.4°C (40°F) respectively, solids to be fed at a rate of 200 mol/h, operation at atmospheric pressure, and no recycle of the entrainer. It is known that pure A under these conditions will sublime at a rate of 20 mol/h and the pressure drop between the sublimer and the condenser is 2068 N/m^2 (0.3 lbf/in^2). The following vapor-pressure data are available:

Temperature, °C	Vapor pressure, N/m^3 absolute	
	P_A	P_B
65.6	9333	1167
4.4	1067	66.6

Estimate the percentage of sublimed material lost and the concentration of the condensed product.

If the vaporization rates are assumed to be proportional to the vapor pressures, $P_{AS}/P_{BS} = N_{AS}/N_{BS} = 9333/1167 = 8.0$. Since the amount of A vaporized is 20 mol/h, B sublimes at a rate of 2.5 mol/h; r is calculated from the vapor-pressure data as 8.65, and the percent yield loss is determined from Eq. (17-13) as 9.6 percent. The ratio of moles of A to moles of B leaving the condenser is given by the ratio of the vapor pressures at the condenser temperature, i.e., $N_{AC}/N_{BC} = 16$. This relationship and the equation $N_{AC} + N_{BC} = 22.5\ (0.096)$ are solved simultaneously to yield the moles of A and B leaving the condenser in the vapor phase. The composition of the condensed product is then determined from material balance as 88.3 mole percent A and 11.7 mole percent B.

In Example 1 the effluent vapors from the condenser could be condensed at a temperature of less than 4.4°C (40°F) to yield a minor product of higher concentration in A.

Example 2: Purification by Vacuum Sublimation Assume that component B in Example 1 has a negligible vapor pressure and that simple vacuum sublimation is to be used. If the loss of product A is limited to 0.1 percent, what is the allowable air leakage into the system in cubic meters per minute at standard temperature and pressure? Assume $E_C = 0.98$, $E_S = 0.90$, and ΔP is negligible.

Note that since only one volatile component is present, $P_B = 0$. Therefore, rearranging Eq. (17-13) and substituting the partial pressure yield

$$r = \frac{0.001\,[(0.90)(9333) - (1067/0.98)]}{(1067/0.98) - (0.001)(0.90)(9333)}$$
$$= 0.00677 \text{ mol of air/mol of product vaporized}$$

Since the moles of product vaporized = $^{20}/_{60}$ = 0.333 mol/min, the allowable air leakage will be (0.333)(0.00677) = 0.00225 mol/min, or (0.00225)(22.4) = 0.05 m^3/min.

Design Equations: Fractional Sublimation The use of reflux distinguishes fractional sublimation from simple sublimation. When fractional sublimation is conducted as a vacuum operation, the partial pressure differential between the bottom and the top of the column acts as the driving force to cause vapor flow up the column. A major advantage to operation using entrainer gas is that (from the phase rule) an additional degree of freedom is allowed, and thus the operating pressure and separation desired can be set and the column operated isothermally or with a selected temperature profile.

In fractional sublimation there is a tendency for the more volatile components to become concentrated in the upper region of the column and for the less volatile components to be concentrated in the lower region. Thus, a separation can be made.

For fractional sublimation of mixtures which form solid solutions the calculation procedures are similar to those for fraction distillation presented in Sec. 13.

MEMBRANE PROCESSES

The membrane permeation of liquids and gases, dialysis, reverse osmosis, and ultrafiltration are discussed in this subsection. Other membrane processes not covered in this section are listed below with pertinent references.

1. *Electrodialysis.* See the discussion of electrodialysis in the following subsection, "Separation Processes Based Primarily on Action in a Field."
2. *Helium separation through glass.* Hwang and Kammermeyer, *Membranes in Separations*, vol. VII in Weissberger (ed.), *Techniques of Chemistry*, Wiley-Interscience, New York, 1975. Kammermeyer, "Gas and Vapor Separations by Means of Membranes," in Perry (ed.), *Progress in Separation and Purification*, vol. I, Interscience, New York, 1968, p. 335.
3. *Hydrogen Separation through Palladium and Alloy Membranes.* Hwang and Kammermeyer, loc. cit. Kammermeyer, loc. cit. McBride and McKinley, *Chem. Eng. Prog.*, **61**, 81 (1965).
4. *Liquid-surfactant membranes.* Cahn, Frankenfeld, Li, Naden, and Subramanian, in Li (ed.), *Recent Developments in Separation Science*, vol. VI, CRC Press, Boca Raton, Fla., 1981. Cahn and Li, "Hydrocarbon Separation by Liquid Membrane Processes," in Meares (ed.), *Membrane Separation Processes*, Elsevier, Amsterdam and New York, 1976, chap. 9. Frankenfeld and Li, in Li (ed.), *Recent Developments in Separation Science*, vol. III, CRC Press, Boca Raton, Fla., 1977, p. 285. Ho, Hatton, Lightfoot, and Li, "Extraction with Liquid Membranes: A Diffusion-Controlled Model," presented at 2d World Congress of Chemical Engineering and World Chemical Exposition, Montreal, Oct. 4–9, 1981, published in *Am. Inst. Chem. Eng. J.*, *28*, 662 (1982). Li, *Am. Inst. Chem. Eng. J.*, **17**, 459 (1971).
5. *Immobilized-solvent membranes.* Baker, Tuttle, Kelly, and Lonsdale, *J. Membr. Sci.*, **2**, 213 (1977). Ward, in Li (ed.), *Recent Developments in Separation Science*, vol. I, CRC Press, Cleveland, 1972, p. 153.

MEMBRANE PERMEATION OF LIQUIDS AND GASES

GENERAL REFERENCES: Li, Long, and Henley, *Ind. Eng. Chem.*, **57**, 18 (1965). Michaels and Bixler, in Perry (ed.), *Progress in Separation and Purification*, vol. I, Interscience, New York, 1968, p. 143. Rogers, Fels, and Li, in Li (ed.), *Recent Developments in Separation Science*, vol. II, CRC Press, Cleveland, 1972, p. 107. Stannett, Koros, Paul, Lonsdale, and Baker, *Adv. Polym. Sci.*, **32**, 69 (1979).

Definitions of Membrane and Membrane Permeation This subsection covers permeation of liquids and gases through polymeric membranes when the driving force for transport is either concentration or pressure. A membrane is a thin barrier separating two fluids. The barrier prevents hydrodynamic flow so that transport through the membrane is by sorption and diffusion. The property of the membrane describing the rate of transport is its permeability. A membrane is semipermeable if, under identical conditions, it transports different molecular species at different rates.

Polymeric films in general can be regarded as interspersed crystalline and amorphous regions. The crystalline region having regular structures is generally assumed to be impermeable to liquids and gases. The polymer segments in the amorphous phase can have thermal motion and can be pushed aside to make room for permeating molecules.

Gas permeation involves gases on the high-pressure side of the membrane permeating through the membrane to its low-pressure side. Liquid permeation involves permeation of feed components from a liquid phase on one side of the membrane to another liquid phase or a vapor phase on the other side.

For transport through microporous membranes, the reader is referred to the literature [Hwang and Kammermeyer, loc. cit.; Quinn, Anderson, Ho, and Petzny, *Biophys. J.*, **12**, 990 (1972); and Bean, in Eisenman (ed.), *Membranes*, Marcel Dekker, New York, 1972, p. 1].

Theory The permeation process through a polymeric membrane involves three steps: (1) solution of the permeating molecules at the upstream side of the membrane, (2) diffusion of these molecules through the membrane, and (3) desorption at the downstream side of the membrane.

Fickian Diffusion in Rubbery Polymers The permeation of molecules in rubbery polymers, such as rubber and hydrocarbon polymers above their glass-transition temperatures (i.e., essentially nonpolar structure of elastomers), can be described by a diffusion process following Fick's first law:

$$N = -D(dC/dx) \tag{17-14}$$

where N = permeation flux, $[\text{cm}^3(\text{STP})]/(\text{s}\cdot\text{cm}^2)$; D = diffusion coefficient or diffusivity of the permeating molecules, cm^2/s; C = concentration, $[\text{cm}^3(\text{STP})]/\text{cm}^3$; and x = distance of permeation in polymer film, cm. If D is independent of concentration, Fick's law can be integrated to give

$$N = (D/\ell)(C_1 - C_2) \tag{17-15}$$

where ℓ = membrane thickness, cm; and the subscripts 1 and 2 refer to the upstream and downstream sides of the membrane respectively.

Equation (17-15) is used to describe liquid permeation when D is constant. Henry's law, $C = Hp$, is generally assumed to apply, where H = Henry's-law constant or solubility coefficient, $[\text{cm}^3(\text{STP})]/(\text{cm}^3\cdot\text{cmHg})$; and p = pressure, cmHg. For gas permeation, Henry's law can be combined with Eq. (17-15) to give

$$N = \frac{D(H_1p_1 - H_2p_2)}{\ell} \tag{17-16}$$

If the Henry's-law constant is assumed to be a function only of temperature and both membrane surfaces are at the same temperature, then $H_1 = H_2 = H$ and

$$N = \frac{P(p_1 - p_2)}{\ell} \tag{17-17}$$

where $P = DH$ = the permeability, $[\text{cm}^3(\text{STP})\cdot\text{cm}]/(\text{cm}^2\cdot\text{s}\cdot\text{cmHg})$.

For unsteady-state permeation, the amount of permeates retained per unit volume of film is equal to the rate of change of centration with time, t, s.

$$-(dN/dx) = (dC/dt) \tag{17-18}$$

Combining Eq. (17-18) with Eq. (17-14) leads to Fick's second law:

$$\frac{dC}{dt} = \frac{d}{dx}\left(D\frac{dC}{dx}\right) \tag{17-19}$$

A solution for the case of a finite solid with constant diffusion coefficient can be used to evaluate the total amount of permeates Q $[\text{cm}^3(\text{STP})/\text{cm}^2]$ passing through the film from $t = 0$ until $t = t$.

$$Q = (DC_1/\ell)t - (C_1/6)\ell \tag{17-20}$$

Equation (17-20) shows that Q increases linearly with t. When extrapolating back the linear portion of the Q versus t plot to the time axis, one obtains an intercept $t = \theta$, where $Q = 0$ and

$$D = \ell^2/6\theta \tag{17-21}$$

θ is known as the "diffusion time lag" and provides an experimental method for the determination of D [Daynes, *Proc. R. Soc. (London)*, **A97**, 286 (1920)].

The diffusion coefficient for gases has an Arrhenius form [Barrer, *Trans. Faraday Soc.*, **35**, 628 (1939); and Stannett et al., loc. cit.].

$$D = D_0 \exp(-E_D/RT) \tag{17-22}$$

where D_0 = constant, cm^2/s; E_D = activation energy for diffusion, $\text{cal}/(\text{g}\cdot\text{mol})$; R = gas constant = 1.987 $\text{cal}/(\text{g}\cdot\text{mol}\cdot\text{K})$, and T = absolute temperature, K. Similarly, the permeability and solubility coefficient can have the Arrhenius forms.

$$P = P_0 \exp[-E_p/RT] \tag{17-23}$$

$$H = H_0 \exp[-\Delta H_s/RT] \tag{17-24}$$

where P_0 = constant, $[\text{cm}^3(\text{STP})\cdot\text{cm}]/(\text{cm}^2\cdot\text{s}\cdot\text{cmHg})$; E_p = activation energy for permeation, $\text{cal}/(\text{g}\cdot\text{mol})$; H_0 = constant, $[\text{cm}^3(\text{STP})]/(\text{cm}^3\cdot\text{cmHg})$; and ΔH_s = heat of solution, $\text{cal}/(\text{g}\cdot\text{mol})$. Some permeability data and activation energy for permeation are presented in Table 17-7.

Dual-Mode Sorption Model for Glassy Polymers For glassy polymers the correlation between the diffusivity D and the physical property E_D traditionally invoked in rubbery polymers tends to break down. There exists an inflection in the Arrhenius plot of D near the glass-transition temperature. To account for this, the dual-mode-sorption model is proposed in such a way that the sorption isotherm is described by the combination of a Henry's-law "dissolved" component, C_D $[\text{cm}^3(\text{STP})/\text{cm}^3]$, and a Langmuir "hole-filling" term, C_h $[\text{cm}^3(\text{STP})/\text{cm}^3]$, i.e.,

$$C = C_D + C_H = Hp + \frac{C'_Hbp}{1 + bp} \tag{17-25}$$

[Stannett et al., loc. cit.; and Vieth, Howell, and Hsieh, *J. Membr. Sci.*, **1**, 177 (1976)], where C'_H $[\text{cm}^3(\text{STP})/\text{cm}^3]$ and b (1/cmHg) are the Langmuir capacity constant and the affinity constant respectively.

The permeation flux is expressed as follows:

$$N = -D_{\text{eff}}(dC/dx) \tag{17-26}$$

where D_{eff} = effective diffusion coefficient which is dependent on concentration. The effective diffusivity determined by low-pressure time-lag measurements corresponds to the following equation:

$$\lim_{p1\to 0}\left(\frac{\ell^2}{6\theta}\right) = \lim_{p1\to 0}(D_{\text{eff}}) = D\left(\frac{1 + FB}{1 + B}\right) \tag{17-27}$$

where D = diffusion coefficient which is due to "dissolved" concentration component and is the only applicable diffusion coefficient above the glass-transition temperature, F = ratio of the diffusion coefficient of the permeate in "holes" to the diffusion coefficient of the permeate in the true solution D, and $B = C'_Hb/H$. The permeability in this dual-mode-sorption model is given by the following equation:

$$P = HD\left(1 + \frac{FB}{1 + bp_1}\right) \tag{17-28}$$

Effects of Concentration and Pressure In liquid permeation, diffusivity often depends strongly on the concentration of solvents in the plastic film. Many expressions have been proposed to relate D to the solubility of solvent in the film and to a diffusivity D_0 obtained at zero concentration of solvents. The equation quite widely accepted is

$$D = D_0e^{ac} \tag{17-29}$$

where D_0 and a are constants at a given temperature and the constant a $[\text{cm}^3/\text{cm}^3(\text{STP})]$ essentially describes the plasticizing action of solvent on the film.

Substituting Eq. (17-29) in Eq. (17-14) gives the steady-state permeation flux for a single component:

$$N = \frac{D_0}{a\ell}(e^{ac}1 - e^{ac}2) \tag{17-30}$$

The second term $e^{ac}2$ is usually negligible with respect to $e^{ac}1$.

In gas permeation the permeation of permanent gases through polymeric films is usually independent of pressure. However, the permeability of organic gases and vapors shows a dependence on pressure and concentration by virtue of the strong interaction between solute and membrane. At low pressures, where Henry's law holds, Eq. (17-31) can be used to relate the permeation constant to pressure [Li and Long, in *Progress in Separation and Purification*, vol. 3, Interscience, New York, 1970, p. 153; and Stern and Fang, *J. Polym. Sci., Part A-2*, **10**, 201 (1972)].

$$P = P'_0e^{Ap} \tag{17-31}$$

where P'_0 = permeability at zero pressure; and A = the plasticizing constant, 1/cmHg. In order to evaluate the characteristic constants P'_0 and A in Eq. (17-31), one carries out the integration

$$N = (P'_0/A\ell)(e^{Ap}1 - e^{Ap}2) \tag{17-32}$$

TABLE 17-7 Gas Permeabilities and Permeation Activation Energies of Polymer Membranes

		Hydrogen		Helium		Nitrogen		Oxygen		Carbon dioxide		
Polymer membrane	Temperature	P	E_p	P	E_p	P	E_p	P	E_p	P	E_p	Ref.
Cellulose acetate	25	8.4	5.2	...	...	...	...	...	...	8.7	7.1	1
	30	...	...	...	...	0.28	6.5	0.78	5.0			1
Polyamide (nylon)	25	1.0	8.1									1
	30	...	...	...	...	0.02	11.2	0.038	10.4	0.16	9.7	1
Polybutadiene	25–30	42.0	6.6	...	...	6.4	8.2	19.0	7.1	138.0	5.2	2
Polybutadiene (amorphous)	25	87.0	5.6	...	...	20.0	7.0	...	...	342.0	3.4	3
Poly[butadiene(80%)-acrylonitrile]	25–30	25.0	7.2	17.0	6.8	2.5	9.9	8.2	8.6	64.0	7.0	2
Poly[butadiene(61%)-acrylonitrile]	25–30	7.2	8.8	6.9	7.7	0.24	13.8	0.96	12.0	7.5	10.5	2
Polycarbonate	25–30	14.0	...	...	...	0.3	...	1.5	4.6	8.5	3.8	2
Polychloroprene	25–30	14.0	8.1	4.5	...	1.2	10.6	4.0	9.9	25.0	8.5	2
Polychlorotrifluoroethylene	25–30	0.95	7.1	...	...	0.13	12.5	0.3	10.9	1.2	7.4	2
Polydialkylsiloxane (silicone rubber)	25	...	...	...	...	...	...	500.0	25.0	2800.0		1
Polyethylene (density, 0.922 g/cm^3)	25	8.6	8.2	7.4	8.3							1
	30	...	...	...	...	2.0	11.7	5.5	10.3	26.5	8.2	1
Polyethylene (density, 0.96 g/cm^3)	25–30	...	...	...	...	...	...	1.1	8.8	4.3	7.4	2
Polyethylene terephthalate (Mylar)	25	0.6	5.5	1.1	4.6	0.005	7.5	0.03	6.4	0.10	6.2	1
Poly(isobutylene-isoprene) (butyl rubber)	25–30	7.3	8.7	6.6	7.6	0.32	12.5	1.3	10.7	5.2	9.9	2
Polyisoprene (natural rubber)	25	50.0	6.9	30.8	6.3	8.4	9.0	23.0	7.1	133.0	6.1	1
Polypropylene	25–30	...	...	...	...	0.6	...	2.3	11.4	9.2	9.1	2
Polystyrene	25–30	...	...	...	...	0.3–8.0	...	1.5–25.0	...	7.5–37.0		2
Poly(styrene-butadiene)	25–30	40.0	6.8	23.0	6.6	6.4	8.7	17.0	7.3	124.0	5.7	2
Poly(vinyl acetate)	25–30	...	...	...	...	0.13	...	0.5	13.4	4.6		2
Poly(vinyl chloride)	25–30	3.6	1.9	...	...	0.04	...	0.12	...	1.0	12.2	2
Poly(vinyl chloride-vinyl acetate)	25–30	9.2	7.4	8.5	7.2	0.6	12.3	2.4	9.7	16.0	8.2	2
Poly(vinylidene chloride)(Saran)	30	0.08	...	...	...	0.001	16.8	0.005	15.9	0.03	12.3	1
Teflon FEP (fluorinated ethylene-propylene copolymer)	25	14.0	6.3	40.0	4.9	2.15	7.3	5.9	5.8	1.7		1
	40	...	...	...	...	...	...	9.07	6.1			4

NOTE: Temperature, °C; P, $10^{-10} \times [\text{cm}^3(\text{STP})\cdot\text{cm}]/(\text{cm}^2\cdot\text{s}\cdot\text{cmHg})$; E_p, kcal/(g·mol).

References:

1. Rogers, "Permeability and Chemical Resistence," in *Engineering Design for Plastics*, Reinhold, New York, 1964.
2. Rogers, Fells, and Li, in Li (ed.), *Recent Developments in Separation Science*, vol. II, CRC Press, Cleveland, 1972, p. 107.
3. Cowling and Park, *J. Membr. Sci.*, **5**, 199 (1979).
4. Koros, Wang, and Felder, *J. Appl. Polym. Sci.*, **26**, 2805 (1981).

For carbon dioxide permeating through a glassy polyacrylonitrile [Huvard, Stannett, Koros, and Hopfenberg, *J. Membr. Sci.*, **6**, 185 (1980)], methane permeating through glassy aromatic polyimide, and oxygen permeating through glassy Mylar [Pye, Hoehn, and Panar, *J. Appl. Polym. Sci.*, **20**, 1921 (1976)], the permeabilities decrease with increasing upstream pressure.

Reciprocal permeability is an additive property when permeabilities are independent of pressure. For a laminate composed by n layers, the relationship is

$$\frac{1}{P_t} = \sum_{i=1}^{n} \frac{1}{\ell_t} \frac{\ell_i}{P_i} \tag{17-33}$$

where P_t = permeability of the laminate, ℓ_t = total thickness of the laminate, P_i = permeability of layer i, and ℓ_i = thickness of layer i. The case in which individual permeabilities in a laminate depend on pressure is discussed by Sobolev et al. [*Ind. Eng. Chem.*, **49**, 441 (1957)].

Effect of Temperature As shown in Eqs. (17-22), (17-23), and (17-24) the temperature dependence of the diffusion coefficient, permeability, and solubility coefficient is usually given by Arrhenius relationships using the activation energies for diffusion, permeation, and solution respectively. In gas permeation, the Arrhenius relationship is applicable with the same constants in both subcritical and supercritical pressures for carbon dioxide and ethylene gases through polyethylene (Li and Henley, loc. cit.; and Li and Long, loc. cit.) but not for ethane, propane, and propylene through polyethylene. In the latter case, the plots of P versus $1/T$ are not linear. The temperatures at which the Arrhenius plots change their slopes vary from −2 to −27°C for the different gases tested [Henley and dos Santos, *Am. Inst. Chem. Eng. J.*, **13**, 1117 (1967)]. Similar behavior of the Arrhenius relationship at low temperatures was observed in liquid permeation (Li and Long, loc. cit.). A break in the Arrhenius plots was also often found in the vicinity of the glass-transition temperatures of polymers (Stannett et al., loc. cit.).

Prediction of Permeability, Diffusion Coefficient, and Solubility Coefficient The permeabilities of pure organic liquids through polyethylene and polypropylene were correlated with a parameter (the Permachor) characteristic of the penetrant molecule [Solame and Pinsky, *J. Mod. Packag.*, **36**, 153 (1962)]. This technique makes use of additive Permachor values which are obtained for various chemical groups, lengths of carbon chains, types of unsaturation, and degree of branching in the permeating molecule (Table 17-8). For polyethylene, the correlation equation is

$$\log_{10} P = 16.55 - (3700/T) - 0.22\Psi \tag{17-34}$$

TABLE 17-8 Permachor Values for Some Atoms and Groups

	Number of carbon atoms in the molecule					
Atoms or groups	1	2	3	4	5	6
Carbon	1.0	1.0	1.0	1.0	1.0	1.0
Chlorine	1.2	1.2	1.2	1.2	1.2	1.2
Nitrate	15.4	13.4	13.4	13.4	13.4	9.6
Ether	...	2.4	2.4	2.4	2.4	1.4
Ketone	...	10.8	10.8	8.5	8.5	8.5
Alcohol aliphatic	16.5	16.5	15.5	14.0	14.0	11.0
Alcohol aromatic	...	...	...	...	...	13.0
Phenyl	...	...	...	...	...	5.4

Iso substituent: Add 2.0.
C = C: Subtract 0.2.
Cyclic: Add 1.0 (below 25°).

where P = permeability in the unit of g/(0.001-in membrane thickness · day · 100 in^2) and Ψ = Permachor. For gas permeation, the Permachor method may also be used to correlate polymeric structure with permeability. Three correlation equations for 25°C are available for O_2, N_2, and CO_2 gases respectively [Salame, *Am. Chem. Soc. Prepr.*, **8,** 137 (April 1967)].

Although the effects of temperature or symmetry on polar-group interactions are not included in this technique, it is still the most powerful prediction method available today for pure solvents in hydrocarbon polymers when permeabilities are not concentration- or pressure-dependent. The permeabilities of gases can also be correlated with the vapor pressure of a reference substance [Othmer and Frohlich, *Ind. Eng. Chem.*, **47,** 1034 (1955)].

The diffusion coefficient and the activation energy for diffusion were correlated with a term relating to the molecular diameter by Michaels and Bixler [*J. Polym. Sci.*, **50,** 413 (1961)].

The solubility coefficient H was correlated with the normal boiling point of a gas which provides a measure of the ease of condensing the gas (Fig. 17-17; Stannett et al., loc. cit.). The solubility coefficient can also be correlated with the Lennard-Jones force constants for the solubilized gas. The heats of solution of gases in polymers (ΔH_s) can be correlated with the Lennard-Jones force constants [Fig. 17-18; Michaels and Bixler, *J. Polym. Sci.*, **50,** 393 (1961)].

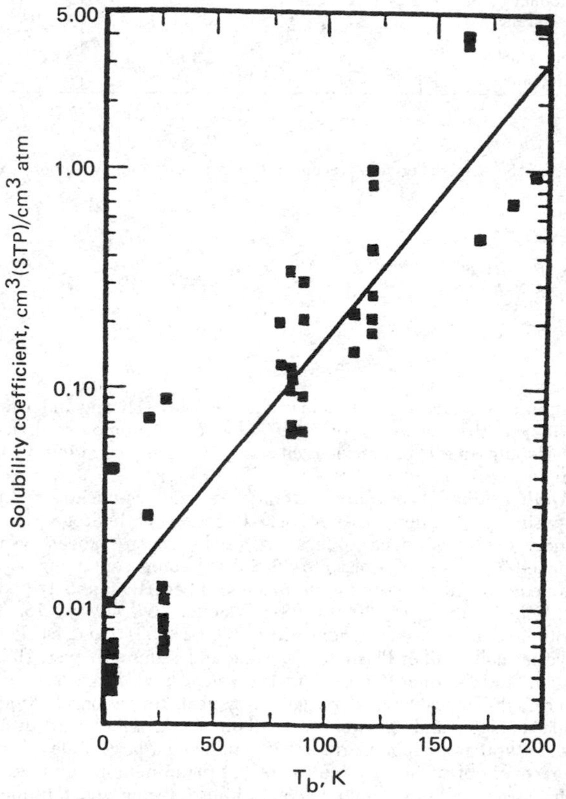

FIG. 17-17 Solubility coefficient versus normal boiling point in various polymers above their glass-transition temperatures. °R = % K. [*Stannett, Koros, Paul, Lonsdale, and Baker,* Adv. Polym. Sci., *32, 69 (1979).*]

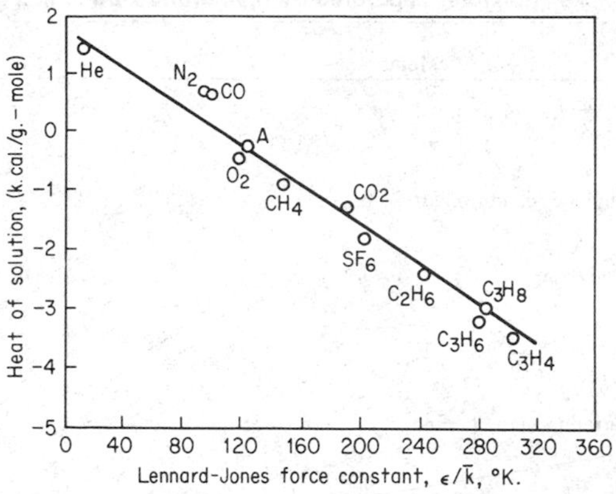

FIG. 17-18 Jolley-Hildebrand correlation of heats of solution of gases in amorphous polyethelene. To convert kilocalories per gram-mole to British thermal units per pound-mole, multiply by 1799; °R = % K.

Separation Processes and Equipment Design

Advantages of Membrane Separation Processes In general, the use of gas- or liquid-permeable membrane separation techniques is of special benefit for separating (1) mixtures of compounds of similar chemical and physical properties, (2) mixtures of structured or position isomers, and (3) mixtures containing thermally unstable components. The membrane process is particularly useful as a separation technique whenever conventional separation methods cannot be used economically to get reasonable separation. It can also be used as a unit operation in conjunction with a conventional separation unit. For example, a membrane permeation unit can be used to break an azeotropic mixture before feeding it to a distillation column. Large-scale membrane separation processes involve device and system design concepts which are unique in chemical engineering.

Definition of Separation Factor The degree of separation is commonly expressed in terms of separation factor, which is defined by the following concentration ratio:

$$\alpha = \frac{(C_A/C_B)_P}{(C_A/C_B)_R} \tag{17-35}$$

where α is the separation factor, and the subscripts A, B, P, and R refer to a permeate compound, a reference compound (also in the feed), permeate phase, and raffinate phase respectively. In general, if α is constant over a wide range of feed composition, it is likely that diffusivity is the process-controlling factor, whereas if the permeate composition is constant, solubility constitutes the process-controlling factor [Sanders and Choo, *Pet. Refiner*, **39,** 133 (1969)]. The separation factor in Eq. (17-35) can generally be approximated by the ratio of the permeabilities (Stannett et al., loc. cit.).

$$\alpha = P_A/P_B \tag{17-36}$$

Some separation data for various mixture-membrane systems are given in Table 17-9. Detailed permeation data for the separation of methane from carbon dioxide at various temperatures [Ellig, Althouse, and McCandless, *J. Membr. Sci.*, **6,** 259 (1980)] and for the separation of sulfur dioxide at various pressures [Kuehne and Friedlander, *Ind. Eng. Chem. Process Des. Dev.*, **19,** 609, 616 (1980)] have been reported. The permeabilities of polyethylene to ammonia at various temperatures and to methane have also been reported in the literature [Lee and Hart, *J. Appl. Polym. Sci.*, 955 (1980); and Matulevicius and Li, *Ind. Eng. Chem. Prod. Res. Dev.*, **11,** 312 (1972)].

General Theory of Separation The theory of separation in a single stage was developed for binary mixtures [Weller and Steiner, *Chem. Eng. Prog.*, **46,** 585 (1950); *J. Appl. Phys.*, **21,** 279 (1950); and Hwang and Kammermeyer, loc. cit.] and for ternary and quaternary mixtures [Brubaker and Kammermeyer, *Ind. Eng. Chem.*, **44,** 1465 (1952); Stern et al., ibid., **57,** 49 (1965); and Stern, "Gas Permeation Processes," in Lacey and Loeb (eds.), *Industrial Processing with Membranes*, repr. ed., Robert E. Krieger Publishing Co., Huntington, N.Y., 1979, chap. 13]. Principles of separations in cascade operation have been described in the literature [Naylor and Backer, *Am. Inst. Chem. Eng. J.*, **1,** 95 (1955); Hwang and Kam-

TABLE 17-9 Mixtures Separated by Membrane Permeation

Membrane	Temp., °C.	Components	Feed composition, %	α B/A	Ref.
Polyethylene	60	Methanol Benzene	39.1*w* 60.9*w*	6.5	1
	25	Ethylene Methane	50*w* 50*w*	3.0	5
Polyethylene (conditioned)	30	*o*-Xylene (*A*) *m*-Xylene (*B*) *p*-Xylene (*C*)	30 65 5	1.59 (*B/A*) 1.26 (*C/B*) 2.01 (*C/A*)	6
Polypropylene	25	Ethylene Methane	50*w* 50*w*	1.95	5
	60	*n*-Dodecane *n*-Heptane	46*w* 54*w*	84	2
Polystyrene	30	Nitrogen Helium	......	16*	3,8,9
Poly(tetrafluoroethylene)	25	Ethylene Methane	55*w* 45*w*	1.25	4
	30	Nitrogen Helium	....	25*	9
Polyethylene terephthalate	25	Methane Helium	....	170*	8
Nylon	40	Chloroform Trichloro-ethylene	37.5*w* 62.5*w*	1.34	7
Silicone rubber	30	Nitrogen Helium	....	1.5*	8
Cellulose acetate	...	*n*-Hexane Ethanol	....	17–130	10
	25	Ethylene Methane	45*w* 55*w*	1.66	4

*Calculated as ratio permeability coefficients.

References:
1. Binning and Lee, U.S. Patent 2,953,502, Sept. 20, 1960.
2. Eisenmann, "Purification of Organic Compounds by Membrane Permeation," U.S. Armed Services Technical Information Agency Report, No. AD 408670, 1963.
3. Jolley, U.S. Patent 3,172,741, Mar. 9, 1965.
4. Li, *Ind. Eng. Chem. Prod. Res. Develop.*, **8**, 281 (1969).
5. Li and Long, *Am. Inst. Chem. Engr. J.*, **15**, 73 (1969).
6. Michaels, Baddour, Bixler, and Choo, *Ind. Eng. Chem. Process Design Develop.*, **1**, 14 (1962).
7. Schrodt, Sweeny, and Rose, *Am. Chem. Soc. Div. Petrol. Chem. Preprints*, **6**, A-29 (April, 1961).
8. Stern, Sinclair, Gareis, Vahldieck, and Mohr, *Ind. Eng. Chem.*, **57**, 49 (1965).
9. Stern, Mohr, and Gareis, German Patent Ser. 1,139,474. Open Nov. 11, 1962.
10. Sweeny and Rose, *Ind. Eng. Chem. Prod. Res. Develop.*, **4**, 248 (1965).

mermeyer, *Can. J. Chem. Eng.*, **43**, 36 (1965); Blumkin, Oak Ridge Nat. Lab. Rep. No. K-OA-1559, January 1968; Stern, 1979, loc. cit..: and Hwang and Kammermeyer, 1975, loc. cit.]. Theoretical and experimental studies of gas separation in hollow-fiber permeators have been reported in the literature [Blaisdell and Kammermeyer, *Chem. Eng. Sci.*, **28**, 1249 (1973); and Thorman, Rhim, and Hwang, ibid., **30**, 751 (1975)]. The performance of hollow-fiber permeators for gas separation has been analyzed with respect to design parameters, operating variables, physical properties, flow patterns, and broken fibers [Antonson, Gardner, King, and Ko, *Ind. Eng. Chem. Process Des. Dev.*, **16**, 463 (1977)]. A continuous-membrane column, in which both the most and the least permeable components can be separated continuously to any degree without cascading, has been described and analyzed [Hwang and Thorman, *Am. Inst. Chem. Eng. J.*, **26**, 558 (1980); and Hwang, Yuen, and Thorman, *Sep. Sci. Technol.*, **15**, 1069 (1980)].

In separating organic gases and vapors, the plasticizing effect, which lowers membrane selectivity, can be characterized by the solubility differences of the permeates between their pure state and the state in which they are mixed with other compounds [Li, *Ind. Eng. Chem. Prod. Res. Dev.*, 8, 281 (1969)].

Design of a Membrane Separation Unit There is a variety of designs of membrane separation units (see Fig. 17-19). For more extensive general discussion of this topic, the reader should see the subsequent subsection "Reverse Osmosis" and the references (Michaels and Bixler, loc. cit.; and Stern, 1979, loc. cit.). For gas separation, three common designs of permeators are (1) the hollow-fiber module, (2) the spiral-wound module, and (3) the plate-and-frame module (Stern, 1976, loc. cit.). Some design equations specific for membrane modules are available in the literature (Hwang and Kammermeyer, 1975, loc. cit.; Stern, 1979, loc. cit.; Antonson et al., loc. cit.; Hwang and Thorman, loc. cit.; and Hwang, Yuen, and Thorman, loc. cit.).

Applications There are several industrial applications and promising permeation processes. (1) Recovery of hydrogen from refinery gases: Industrial applications on hydrogen recovery were achieved by the use of Monsanto PRISM separators containing polysulfone hollow fibers coated with silicone rubber [Henis and Tripodi, *Sep. Sci. Technol.*, **15**, 1059 (1980); Maciula, *Oil Gas J.*, **78**, 63 (1980); and Rosenzweig, *Chem. Eng.*, **88**, 62 (1981)] and Du Pont polyester hollow-fiber Permasep [Hwang and Kammermeyer, 1975, loc. cit.; and Gardner, Crane, and Hannan, *Chem. Eng. Prog.*, **73**, 76 (1977)]. The hollow fibers have a typical dimension of 20-μm inside diameter and 40-μm outside diameter. (2) Helium recovery from natural gas: Union Carbide has conducted a large-scale test on the recovery of helium by the use of flat-plate membrane modules with Eastman KP-98 cellulose acetate films (Hwang and Kammermeyer, 1975, loc. cit.). A considerable number of details on the process technology and operating data had already been reported (Litz and Smith, Union Carbide Res. Inst. Rep. UCRI-701, September 1972). A cost estimate of helium recovery in using Teflon FEP membranes is available (Stern, 1979, loc. cit.).

DIALYSIS

GENERAL REFERENCES: Lane and Riggle, *Chem. Eng. Prog. Symp. Ser.*, **55**, 127 (1959). Spriggs and Li, "Liquid Permeation through Polymeric Membranes," in Meares (ed.), *Membrane Separation Processes*, Elsevier, Amster-

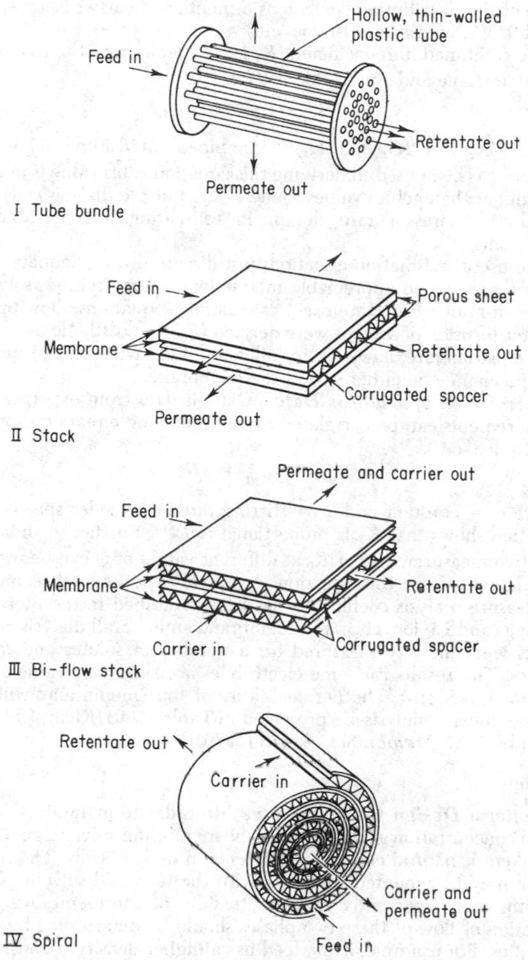

FIG. 17-19 Membrane-permeation module designs. (*Michaels, Bixler, and Rigopulos,* 7th World Pet. Congr. Proc., *4, 1967.*)

dam and New York, 1976, chap. 2. Tuwiner, *Diffusion and Membrane Technology,* Reinhold, New York, 1962. Vromen, *Ind. Eng. Chem.,* **54,** 20 (1962).

Definitions Dialysis is the transfer of solute molecules across a membrane by diffusion from a concentrated solution to a dilute solution. A simultaneous diffusion of solvent molecules through the membrane occurs in the opposite direction—a phenomenon called osmosis. The ratio of grams of water transported per gram of dissolved solute dialyzed is defined as the water-transport number. The membranes used can be either porous or nonporous. The solution on the feed side is called the dialyzate or retentate, and the solution on the other side is called the diffusate. Separation of solutes in dialysis is due to differences in their diffusion rates.

Theory

Mechanism of Dialysis The process of mass transfer from the dialyzate compartment to the diffusate compartment involves diffusion through a liquid film on each side of the membrane as well as through the membrane itself. Diffusion of solute molecules is governed by a number of factors, such as the inherent mobility of the molecules, the restrictive effect or drag exerted by the membrane pores on both solute and solvent molecules.

In describing the dialysis mechanism, Fick's first and second laws of diffusion, Eqs. (17-14) and (17-19), can serve as a starting point. In the case of dilute solutions, integrating Fick's first law over the diffusional path, with the assumptions that no volume changes take place on either side of the membrane and that the mass-transfer coefficients are constant along the diffusion path, results in the rate equation (Lane and Riggle, loc. cit.; and Spriggs and Li, loc. cit.):

$$W = K_0 A \, \Delta C_{\ell m} \tag{17-37}$$

where W = weight of solute passing through the membrane per unit time, g/s; K_0 = overall dialysis coefficient or mass-transfer coefficient, cm/s; A = membrane area, cm^2; $\Delta C_{\ell m}$ = logarithmic mean concentration difference across the membrane, based on inlet and outlet values in a continuous dialyzer and on initial and final values in a batch dialyzer, g/cm^3; and C = concentration of solute in the membrane, g/cm^3. In order to express Eq. (17-37) in terms of the concentrations in the solutions C_s, one obtains Eq. (17-38) by introducing the distribution coefficient β:

$$W = K_0 \beta A (\Delta C_s)_{\ell m} \tag{17-38}$$

$$\beta = C/C_s \tag{17-39}$$

For batch dialysis when mass transfer is not at steady state, the rate equation is

$$kt = -m \frac{\log (C_{so} - C_s)}{C_{so}} \tag{17-40}$$

where k = dialysis rate constant similar to the rate constant of a first-order reaction, t = time, m = volume ratio of feed to diffusate, C_{so} = concentration of solute in the feed, and C_s = concentration of solute in the dialyzate (Tuwiner, loc. cit.).

The overall dialysis coefficient is related to the coefficients of the two liquid films and to the membrane coefficient in a manner analogous to heat transfer:

$$1/K_0 = 1/K_m + 1/K_{\ell 1} + 1/K_{\ell 2} \tag{17-41}$$

where K_m = membrane permeability to solute and $K_{\ell 1}$ and $K_{\ell 2}$ = liquid-film mass-transfer coefficients on the dialyzate side and the diffusate side of a membrane respectively.

A significant amount of solvent may also transfer across the membrane. This may be calculated from Eqs. (17-38), (17-40), and (17-41) by substituting solvent concentration, diffusivity, and permeability for the corresponding solute properties.

For concentrated-feed solutions, the liquid-film resistance is a function of the pH of the feed because it affects the liquid-film thickness. And solvent transport by osmosis may be reduced or enhanced, depending on whether the dialyzing membrane exhibits positive or negative absorption of the solute. For instance, the liquid-film thickness decreased from 0.045 to 0.029 cm as the concentration of hydrochloric acid increased from 0.5 to 6.5*M* (Vromen and Chamberlin, Prepr. 59, 42d nat. meet., AIChE, Atlanta, Feb. 21–24, 1960). It has also been observed that when K_ℓ increases with increasing concentration, K_m may actually decrease. This indicates that the real relation between these two coefficients is complex (Vromen, loc. cit.). Furthermore, electrical interaction may also be important in affecting membrane permeability and the dialysis coefficient if the solution contains an electrolyte and the membrane contains fixed electrical charges [Carr, in Berl (ed.), *Physical Methods in Chemical Analysis,* vol. 4, Academic, New York, 1961, pp. 1–43].

For porous dialysis membranes, membrane selectivity is related to pore size as expressed mathematically by the following equation [Ferry, *J. Gen. Physiol.,* **20,** 95 (1936)]:

$$A_D = A_S \left(1 - \frac{s}{S}\right)^2 \tag{17-42}$$

where A_D = area available for diffusion, A_S = cross-sectional area of the pore, s = diameter of diffusing molecule or particle, and S = diameter of the pore. This equation shows that membrane selectivity becomes high when the membrane barely allows the solute to diffuse through it.

Estimation of Dialysis Coefficients The membrane dialysis coefficient can be calculated from the following equation (Lane and Riggle, loc. cit.):

$$K_m = D_s F V'/h\ell \tag{17-43}$$

where D_s = diffusion coefficient in the solution, cm^2/s; F = ratio of hindered diffusion rate to unhindered rate (drag factor); V' = fraction of membrane volume occupied by pores; h = tortuosity, or ratio of capillary length to membrane wet thickness; and ℓ = membrane wet thickness, cm.

The drag factor may be evaluated from the following equation [Bacon, *J. Franklin Inst.*, **221**, 251 (1936)]:

$$F = 1 - 2.104\frac{s}{S} + 2.09\left(\frac{s}{S}\right)^3 - 0.95\left(\frac{s}{S}\right)^5 \qquad (17\text{-}44)$$

where s = diameter of diffusing molecule or particle and S = average diameter of membrane pores. The diameter of diffusing molecule is found from its molar volume (Spriggs and Li, loc. cit.). For solids,

$$s = 1.465 \times 10^{-16}(M/\rho)^{1/3} \qquad (17\text{-}45)$$

where M = molecular weight, g/(g·mol); ρ = density, g/cm^3; and s is in cm. For liquids,

$$s = 10^{-16}V^{1/3} \qquad (17\text{-}46)$$

where V = molar volume at the boiling point, cm^3.

The diffusion coefficient in solution can be estimated (Reid, Prausnitz, and Sherwood, *The Properties of Gases and Liquids*, McGraw-Hill, New York, 1977). Approximate values of diffusion coefficients are given in Table 17-10.

The values of V', h, and ℓ for some commercial films are presented in Table 17-11. For better accuracy, they should be determined experimentally, because the swollen thickness may be affected by the dialyzing molecules and tortuosity may vary for the same kind of film but with different dry thicknesses. It should be noted that the value of the distribution coefficient β mentioned earlier is generally equal to V' (Spriggs and Li, loc. cit.).

The combined film coefficient K_ℓ is calculable from the following equation (Lane and Riggle, loc. cit.):

$$K_\ell = 16.7D_s \qquad (17\text{-}47)$$

where $1/K_\ell = 1/K_{\ell 1} + 1/K_{\ell 2}$. A combined liquid-film thickness of 0.06 cm has been used in deriving this equation. This value is in general higher than actual values found in most of the dialysis cells and therefore assures a safe design in estimating average dialysis coefficients.

The useful estimation equations just discussed are adequate only when there are no appreciable interactions among solutes, solvent, and membrane. For more exact calculations, equations allowing for counterdiffusion of solvent were derived (Lane and Riggle, loc. cit.). Such equations are, however, for dilute solutions when there is no net volume change on either side of the membrane.

Correlation of Dialysis Data Dialysis data from experimental measurements can be correlated by the following equation (Spriggs and Li, loc. cit.):

$$1/K_0 = 1/K_m + c/\omega^{2/3} \qquad (17\text{-}48)$$

where c = constant and ω = stirring rate or impeller speed. This equation shows that K_ℓ is proportional to $\omega^{2/3}$. Further, it indicates that two measurements of K_0 at different values of ω can be used to calculate K_m from this equation. A typical experimental technique to measure dialysis coefficients has been described in the literature (Spriggs and Li, loc. cit.). The membrane and overall dialysis coefficients were directly measured for a number of solutes and membranes. The results for some electrolytes are shown in Table 17-12 (Vromen, loc. cit.). The permeabilities of four membranes suitable for use in hemodialysis are presented in Table 17-13 [Klein, Holland, and Eberle, *J. Membr. Sci.*, **5**, 173 (1979)].

Design

General Design Considerations In order to maintain a maximum concentration gradient across the membrane, advantage should be taken of natural convection currents in dialysis cells. The membrane may be mounted vertically with the feed and solvent phases flowing countercurrently on opposite sides of the membrane. The direction of flow of these two phases should be determined by their densities. For example, if the feed has a higher density than the solvent, it should be introduced at the bottom of the dialyzate com-

TABLE 17-10 Diffusion Coefficients and Molecular Diameters of Nonelectrolytes

Molecular weight, M, g/(g·mol)	Diffusivity in solution, D_s, $10^{-5} \times (cm^2/s)$	Molecular diameter, s, $10^{-8} \times$ cm
10	2.20	2.9
100	0.70	6.2
1,000	0.25	13.2
10,000	0.11	28.5
100,000	0.05	62.0
1,000,000	0.025	132

TABLE 17-11 Properties of Commercial Films*

Film	Dry thickness, $10^{-3} \times$ cm	Wet thickness, ℓ, $10^{-3} \times$ cm	Relative volume occupied by pores, V'	Tortuosity, h	Pore diameter, S, $10^{-8} \times$ cm
Du Pont cellophane					
300 P.D.†	1.91	3.96	0.52	3.8	38
450 P.D.†	2.67	4.95	0.46	3.4	46
600 P.D.†	3.56	7.25	0.51	4.9	33
300 P.U.D.‡	1.78	4.32	0.59	3.9	48
450 P.U.D.‡	2.54	6.10	0.58	3.5	51
Avisco cellophane					
300 P.-1	2.54	5.07	0.50	4.1	31
450 P.-1	3.94	7.86	0.50	3.8	32
600 P.-1	9.90	17.80	0.44	2.4	29
Seamless tubing	1.91	3.56	0.46	4.3	33
Paterson parchment paper					
30-lb	5.08	6.85	0.26	2.0	36
40-lb	5.97	8.50	0.30	3.4	30
60-lb	9.00	12.70	0.29	3.0	22
Denitrated nitrocellulose					
Light	5.34	9.40	0.43	2.6	35
Heavy	9.15	16.50	0.44	2.2	33

*Do not use for caustic solutions.
†P.D. refers to sized materials.
‡P.U.D. refers to unsized materials.

TABLE 17-12 Comparing Liquid-Film and Membrane Coefficients for HI-Sep 70 Membrane*

Electrolyte	Concentration, mol/L	K_0†, 10^{-4} × (cm/s)	K_m†, 10^{-4} × (cm/s)	K_ℓ‡, 10^{-4} × (cm/s)
HCl	0.56	3.0	5.3	6.8
	2.45	4.5	6.5	11.3
	4.45	5.7	8.8	15.0
	6.5	6.5	9.7	19.3
H_2SO_4	0.212	2.1	4.2	3.8
	4.24	3.8	4.8	18.0
H_3PO_4	0.20	1.0	1.7	3.0
	3.20	1.2	1.5	5.8
N_aOH	0.5	2.2	3.3	5.7
	7.5	1.5	3.2	2.8

*Vromen, *Ind. Eng. Chem.*, **54**, 20 (1962).
†Measured.
‡Calculated from Eq. (17-41) with $1/K_\ell = 1/K_{\ell 1} + 1/K_{\ell 2}$.

partment and the solvent should be introduced at the top of the diffusate compartment. The solvent in the dialyzate compartment, entering by osmosis through the membrane, will produce a rising stream of low-density weak liquor, whereas the diffusing solute in the diffusate compartment will form a stream of high-density solution flowing down the membrane surface. This density streaming effect can reduce greatly the thickness of the liquid films at both sides of the membrane and thus lower the overall resistance to dialysis. The osmotic transfer of solvent is undesirable in a dialysis process because it tends to dilute the dialyzate and thereby reduces the dialysis performance.

Because the mass-transfer rate is directly proportional to membrane area and inversely proportional to membrane thickness, the dialyzing surface should be as large as possible and as thin as permitted by mechanical-strength requirements. Also, the dialyzing temperature should be as high as the thermal stability of the membrane and the solutions involved will allow.

In dialyzer design, the dialysis coefficients can usually be predicted quite accurately for dilute solutions. However, for concentrated solutions, dialyzer design should be based on data from experiments that duplicate the conditions of the application concerned.

Membranes A suitable membrane should have high permeability, proper pore size, and mechanical stability. The membranes originally employed for caustic recovery were of parchment or cellulose. When placed in water, they swell and form a microporous gel with a size of 30 to 50 Å (30 × 10^{-8} to 50 × 10^{-8} cm). Of these two materials, parchment is more stable physically and chemically. With the introduction of acid-stable microporous polyvinyl chloride (PVC) films in 1958, acid-recovery processes became commercially practical. This type of membrane has high permeabilities and long life (up to 2 years in acid solution). It is produced in thicknesses ranging from 2.5 to 4.5 mil (0.00635 to 0.0114 cm) and has estimated pore sizes in three ranges of 20 to 50, 30 to 60, and 60 to 100 Å. Other polymeric membranes for dialysis have been reported in the literature (Hwang and Kammermeyer, 1975, loc. cit.).

Ion-exchange membranes are capable of separating nonelectrolytes from electrolytes by hindering the passage of ions. Separations of considerable magnitude have been obtained with these membranes in laboratory experiments on mixtures of acetone and sulfuric acid, formaldehyde and sodium sulfate, formic acid and sodium chloride, and urea and calcium chloride (Carr, loc. cit.).

For hemodialysis, or artificial-kidney use, dialysis membranes are available in hollow-fiber form. Two types of commercially available hollow fibers for this use are regenerated cellulose material with 285-μm (0.0285-cm) outside diameter and 30-μm (0.0030-cm) wall thickness (Dow Chemical) and modified polyacrylonitrile material with 400-μm outside diameter and 50-μm wall thickness (Monsanto Company) (Hwang and Kammermeyer, 1975, loc. cit.).

Membrane porosity and pore size, which govern membrane selectivity and transfer rate, can be controlled to a certain degree by treatment with swelling agents, by stretching, and by chemical reaction, such as acetylation of cellophane, to reduce pore size. In general, the major mass-transfer resistance is the membrane. Therefore, future improvement of dialyzer performance will depend largely on finding better ways of controlling porosity and pore size and methods of making new membranes with high transfer coefficients.

Batch Dialyzers Stirred-batch dialyzers are generally used to evaluate the performance of different membranes. This includes the determination of membrane dialysis coefficients. The cell is markedly similar to that used for obtaining gas and liquid permeabilities, with the exception that, as a rule, one tries to adjust the hydrodynamics so that $K\ell \to \infty$. In a dialysis apparatus, this is done by inserting variable-speed stirrers in each compartment and increasing the speed until K_0 reaches a maximum.

Continuous Dialyzers Several continuous countercurrent laboratory models have been reported in the literature [Daniel, in *Encyclopedia of Chemical Technology*, vol. 5, Interscience, New York, 1955, pp. 1–20; Carr, loc. cit.; Vromen, loc. cit.; Marshall and Storrow, *Ind. Eng. Chem.*, **43**, 2934 (1951); Craig and Stewart, *Biochemistry*, **4**, 2712 (1965); Craig and Chen, *Anal. Chem.*, **41**, 590 (1969); Craig, Chen, and Taylor, *J. Macromol. Sci. Chem.*, **3**, 133 (1969);

TABLE 17-13 Membrane Permeabilities and Solution Diffusivities for Hemodialysis Solutes*

			Solute permeabilities in membranes, K_m			
Solute	Solute molecular weight	Solute diffusivity, D_s, 10^{-5} × (cm²/s)	Cuprophan 150 PM cellulosic membrane, 10^{-4} × (cm/s)	Celanese CA-2 cellulose acetate, 10^{-4} × (cm/s)	Rhône Poulenc RP AN-69 poly(acrylonitrile methallysuefonate), 10^{-4} × (cm/s)	Bard PCM polycarbonate, 10^{-4} × (cm/s)
Urea	60	1.81	11.5	9.22	15.5	17.7
Phosphate	96	2.78				
Creatinine	113	1.29	5.82	4.74	9.38	9.22
Uric acid	168	1.16				
Glucose	180	0.909	3.75	2.86	6.88	5.59
Sucrose	342	0.697	1.76	2.68	4.87	4.42
Raffinose	504	0.578	1.62	1.44	4.09	2.98
Sucrose dilaurate	715	0.438	0.541	1.11	1.27	1.09
Peptide A, foldable	778	0.494	0.650	1.43	2.73	2.89
Peptide B, foldable	820	0.478	0.412	0.988	1.49	1.78
Peptide C, stiff	1023	0.442	0.417	0.979	2.45	1.93
Peptide D, stiff	1003	0.442	0.270	0.596	1.43	0.944
Vitamin B_{12}	1355	0.379	0.594	0.85	2.66	1.80
Inulin	5200	0.215	0.120	0.25	0.877	0.296

*Klein, Holland, and Eberle, *J. Membr. Sci.*, **5**, 173 (1979).

Craig, in Li (ed.), "Recent Advances in Separation Techniques," *Am. Inst. Chem. Eng. Symp. Ser.*, **68,** 1 (1972); and Spriggs and Li, loc. cit.]. Dialysis coefficients obtained with laboratory units hold approximately for commercial units, provided that both are operated at the same liquor concentrations and flow rates per unit area of membrane surface.

Commercial Dialyzers There are three principal types of commercial dialyzers: (1) the tank type, (2) the filter-press type, and (3) the hollow-fiber type. In tank-type dialyzers, flat bags of membranes are simply immersed in a tank of liquor in such a way that the feed liquor is circulated through the tank outside the bags while the solvent is circulated inside the bags. The Cerini dialyzer (Cerini, U.S. Patent 1,719,754, 1929) is a typical example of this type, which consists of impregnated-membrane bags made of mercerized cotton. Its outstanding feature is the long life of the membrane—12 to 18 months in strong caustic solutions. However, this type of dialyzer has some disadvantages. It requires a heavy, thick membrane in order to ensure its strength, leading to a low dialyzing rate. Besides, it also requires a large space for the equipment.

The filter-press type is a newer model which can employ very thin membranes and has a dialysis coefficient 5 to 10 times larger than that in the tank-type Cerini dialyzer. Examples of this type are Asahi, Brosites, Kooij, and Graver Hi-Sep dialyzers [Daniel, loc. cit.; Chamberlin and Vromen, *Chem. Eng.*, **66,** 117 (May 4, 1959); Van Soye, ibid., **66,** 84 (Jan. 12, 1959); and Tuwiner, loc. cit.]. In this type of dialyzer, vertical membranes are sandwiched between alternate liquor and solvent frames, the liquor and solvent being fed to the bottom and top of these frames respectively. Dialyzate and diffusate are removed through channels located at the top and the bottom.

Hollow-fiber dialyzers arranged in the shell-and-tube configuration similar to that of heat exchangers have been developed recently for dialysis applications, particularly for hemodialysis. This type of dialyzer has been analyzed mathematically [Noda and Gryte, *Am. Inst. Chem. Eng. J.*, **25,** 113 (1979); Gostoli and Gatta, *J. Membr. Sci.*, **6,** 133 (1980); and Hermans, *Desalination*, **26,** 45 (1978)]. One of the models can predict the mass-transfer coefficient as a function of fiber-packing density, membrane thickness, membrane material, and solute type (Noda and Gryte, loc. cit.). Diffusion coefficients for hollow-fiber membranes have been reported as a function of solute molecular weight as shown in Fig. 17-20 [Klein, *J. Appl. Polym. Sci. Appl. Polym. Symp.*, **31,** 361 (1977)].

Applications Most industrial applications of dialysis are concerned with the treatment of waste streams and the recovery of acids from metallurgical liquors. The first large-scale commercial application was the recovery of caustic soda from colloidal hemicellulose in the viscose-rayon industry. Industrial dialysis units for recovery of spent acid from metallurgical liquors have been widely used since 1958, when acid-resistant vinyl-plastic membranes were introduced, such as in the recovery of hydrochloric and nitric acids from stainless-steel pickling liquor and the separation of chromic acid from anodizing, engraving, and etching liquors [Spriggs and Li, loc. cit.; and Dvorin, *Metal Finish.*, 52 (April 1959)]. This type of membrane has also been employed in the separation of mineral acids from their salts, such as the separation of sulfuric acid from copper and nickel sulfate, and in the purification of steel-pickling liquor by removing iron salts [Friedlander and Rickles, *Chem. Eng.*, 111 (Feb. 28, 1966)].

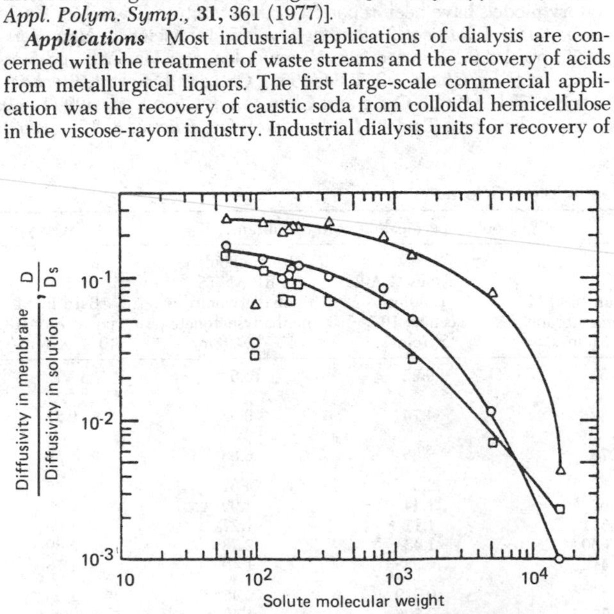

FIG. 17-20 Solute diffusivities at 37°C (98.6°F) for three hemodialysis hollow-fiber membranes: ○, Enka B2-AH hollow fiber; △, Amicon polysulfone hollow fiber; and □, Dow cellulosic hollow fiber. [*Klein,* J. Appl. Polym. Sci. Appl. Polym. Symp., *31, 361 (1977).*]

Dialysis has also found a recent application in the medical field: hemodialysis. The artificial kidney is a dialysis machine for purifying human blood. The principal solutes to be removed for chronic patients are urea, uric acid, creatinine, phosphates, and excess amounts of chloride [Chang, in Li (ed.), *Recent Developments in Separation Science*, vol. I, CRC Press, Cleveland, 1972, p. 203; and Lyman, in Bier (ed.), *Membrane Processes in Industry and Biomedicine*, Plenum Press, New York, 1971]. Membrane permeabilities and diffusion coefficients for some of these solutes are presented in Table 17-13 and Fig. 17-20 (Klein, Holland, and Eberle, loc. cit.; and Klein, loc. cit). The artificial-kidney dialyzer has been found useful for dialysis of protein solutions in biochemical preparative work in such a way that salt and low-molecular-weight constituents can be removed from the solutions [Vesterberg and Wadstrom, *Sep. Sci.*, **5,** 83 (1973); Wadstrom and Vesterberg, ibid., **5,** 91 (1973); and Hwang and Kammermeyer, 1975, loc. cit.].

Dialysis has many other interesting applications (Spriggs and Li, loc. cit.; Klein, loc. cit.; and Rickles, *Membranes: Technology and Economics*, Noyes Data Corp., Park Ridge, N.J., 1967). These include purification of high-cost materials in the pharmaceutical industry, separation of organic mixtures, and concentration of fruit juice. Combined with ultrafiltration, dialysis can also be used for desalting [Cross, in Li (ed.), "Recent Advances in Separation Techniques," *Am. Inst. Chem. Eng. Symp. Ser.*, **68,** 15 (1972)].

REVERSE OSMOSIS

GENERAL REFERENCES: Harris, Humphreys, and Spiegler, "Reverse Osmosis (Hyperfiltration) in Water Desalination," in Meares (ed.), *Membrane Separation Processes*, Elsevier, Amsterdam and New York, 1976, chap. 4. Merten (ed.), *Desalination by Reverse Osmosis*, M.I.T., Cambridge, Mass., 1966. Podall, "Reverse Osmosis," in Li (ed.) *Recent Developments in Separation Science*, vol. II, CRC Press, Cleveland, 1972, p. 171. Sourirajan (ed.), *Reverse Osmosis and Synthetic Membranes*, National Research Council Canada Publications, Ottawa, 1977.

Definitions Reverse osmosis separates a solute from a solution by forcing the solvent to flow through a membrane by applying a pressure greater than the normal osmotic pressure. In reverse osmosis the solute molecules are of about the same size as that of the solvent molecules. Such a separation process based on pressure difference across a membrane combines technical simplicity with versatility. Unlike distillation and freezing processes, it can operate at ambient temperature without phase change.

Theory

Solubility and Diffusion Model For a homogeneous diffusive-type membrane, the steady-state permeation of the solvent is

$$N_w = P_w[(\Delta p - \Delta\pi)/\ell] \qquad (17\text{-}49)$$

[Michaels, *Chem. Eng. Prog.*, **64,** 31 (1968)], where N_w = steady-state permeation flux of the solvent through the membrane, (g·mol)/(s·cm²); ℓ = membrane thickness, cm; P_w = specific permeability, (g·mol·s)/g,

$$P_w = (\overline{C}_{wm}\overline{D}_{wm}V_w)/RT \qquad (17\text{-}50)$$

$\overline{C}_{wm}$ = mean solvent concentration in the membrane, (g·mol)/cm³; $\overline{D}_{wm}$ = mean diffusivity of the solvent in the membrane, cm²/s; V_w = molar volume of the solvent, cm³/(g·mol); R = gas constant = 8.314×10^7 (g·cm²)/(s²·g·mol·K); T = absolute temperature, K;

Δp = applied pressure difference between the upstream and downstream sides of the membrane, $g/(s^2 \cdot cm)$ or dyn/cm^2; $\Delta\pi$ = osmotic pressure difference, $g/(s^2 \cdot cm)$ or dyn/cm^2;

$$\Delta\pi = -(RT/V_w) \ln (a_{w1}/a_{w2}) \tag{17-51}$$

and a_w = solvent activity. Subscripts 1 and 2 refer to the upstream and downstream sides of the membrane respectively. For dilute solutions Eq. (17-51) becomes

$$\Delta\pi = (C_{s1} - C_{s2})RT \tag{17-52}$$

where C_s = solute concentration in a solution, $(g \cdot mol)/cm^3$; and $\pi = C_sRT$ is the van't Hoff equation. Osmotic pressures for the sodium chloride solutions of various salinities at different temperatures are shown in Table 17-14. The osmotic pressures of fruit juices and sugar solutions at the ambient temperature are presented in Fig. 17-21. Osmotic pressures for alkaline earth-metal halides and some other salts are available in the literature [Goldberg and Nuttall, *J. Phys. Chem. Ref. Data,* **7**, 263 (1978); and Bonner, *J. Chem. Eng. Data,* **24**, 210, 211 (1979)]. For high-pressure dilute solutions or high-molecular-weight solutes, the osmotic correction is not as important, and Eq. (17-49) reduces to

$$N_w = P_w(\Delta p/\ell) \tag{17-53}$$

The solute transfer through the membrane is due to a combination of molecular diffusion and solvent drag as expressed by the following equation:

$$N_s = -D_{sm}\frac{dC_{sm}}{d\chi} + k_s\frac{C_{sm}}{C_{wm}}N_w \tag{17-54}$$

where N_s = solute flux through the membrane, $(g \cdot mol)/(s \cdot cm^2)$; D_{sm} = solute diffusion coefficient in the membrane, cm^2/s; C_{sm} = solute concentration in the membrane, $(g \cdot mol)/cm^3$; C_{wm} = solvent concentration in the membrane, $(g \cdot mol)/cm^3$; and k_s = coupling coefficient (between zero and unity). The solute rejection efficiency is defined as

$$R_s = 1 - (C_{s2}/C_{s1}) \tag{17-55}$$

Sieve Mechanism For a random isotropic microporous membrane, the steady-state solvent flux is

$$N_w = K_w \Delta p/\eta_w\ell \tag{17-56}$$

where K_w = hydraulic permeability:

$$K_w = \epsilon r^2/20 \tag{17-57}$$

η_w = viscosity of the solvent, r = hydraulic mean pore radius, and ϵ = porosity of the membrane.

The constant of 20 in Eq. (17-57) is an empirical number taking account of anomalies of pore shape, nonuniformity of pore cross section, and tortuosity of flow paths in the membrane [Michaels, in Perry (ed.), *Progress in Separation and Purification*, vol. I, Interscience, New York, 1968, p. 297].

Solute is transferred purely by convection with solvent through those pores of the membrane large enough to admit solute molecules. This can be represented by

$$N_s = \left[(1 - \phi)\frac{K_w \Delta p}{\eta_w \ell}\right]\frac{C_{s1}}{C_{w1}} \tag{17-58}$$

where ϕ = fraction of the pure-solvent flux which passes through solute-rejecting pores and C_{w1} = solvent concention in the solution at the upstream side.

Concentration Polarization Actual fluxes may be lower than predicted because of concentration polarization. This means, for example, that in desalination, when water transfers through a membrane, the salt left behind concentrates in a solution next to the membrane surface. The efficiency of membrane separation therefore decreases gradually as the concentrated-salt-solution layer gradually increases its thickness. Accompanying the increase in salt concentration at the interface is an increase in the osmotic pressure of the solution at the interface, which in turn decreases the pressure for driving water through the membrane. Concentration polarization could also lead to destruction of sensitive membrane surfaces.

The salt layer built up near the membrane surface may reach a constant thickness as a result of a balance of two opposing factors, i.e., the convective transport of salt toward the membrane by the bulk motion of the water and the back diffusion of salt away from the membrane surface due to the concentration gradient established near the phase boundary. This is shown by Eq. (17-59) [Sherwood et al., *Ind. Eng. Chem. Fundam.*, **4**, 113 (1965)].

$$N_w\frac{C_s}{C_{w1}} + (D_s + E)\frac{dC_s}{dy} = 0 \tag{17-59}$$

where D_s = solute diffusivity in the solution, E = eddy-diffusion coefficient, and y = distance normal to the phase boundary. The

TABLE 17-14 Osmotic Pressure of Sodium Chloride Solutions*

m_{NaCl}	Osmotic pressure π, atm. at 25°C.	40°C.	60°C.	100°C.
0.001	0.05	0.05	0.05	0.06
0.01	0.47	0.49	0.52	0.57
0.05	2.31	2.41	2.53	2.75
0.10	4.56	4.76	5.00	5.42
0.20	9.04	9.44	9.93	10.74
0.40	18.02	18.84	19.83	21.45
0.60	27.12	28.40	29.92	32.35
0.80	36.37	38.14	40.22	43.48
1.00	45.80	48.08	50.76	54.87
2.00	96.2	101.3	107.3	115.9
3.00	153.2	161.6	171.0	184.2
4.00	218.9	230.5	243.3	260.8
5.00	295.2	309.4	325.2	346.5
6.00	384.1	400.2	418.0	442.2

*Stoughton and Lietzke, *J. Chem. Eng. Data,* **10**, 254 (1965).

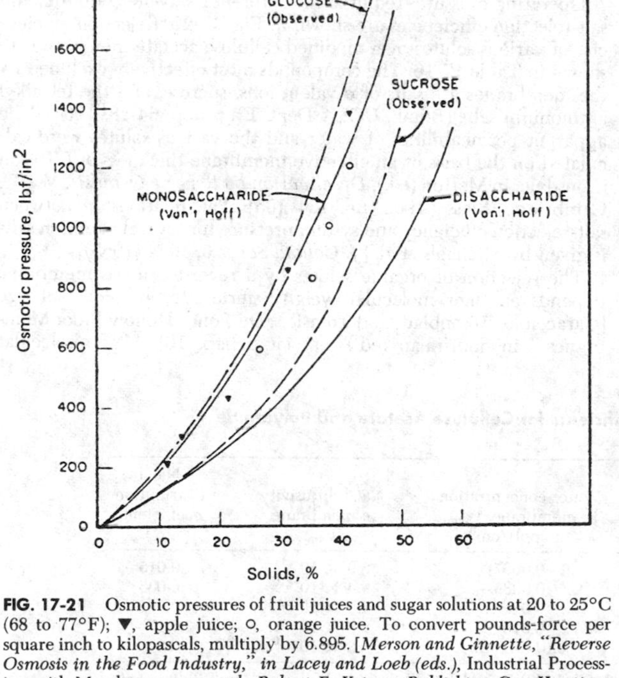

FIG. 17-21 Osmotic pressures of fruit juices and sugar solutions at 20 to 25°C (68 to 77°F); ▼, apple juice; ○, orange juice. To convert pounds-force per square inch to kilopascals, multiply by 6.895. [*Merson and Ginnette, "Reverse Osmosis in the Food Industry," in Lacey and Loeb (eds.),* Industrial Processing with Membranes, *repr. ed., Robert E. Krieger Publishing Co., Huntington, N.Y., 1979.*]

solution for the preceding equation for turbulent flow is

$$\frac{C_{sp}}{C_s} = \exp\left[\frac{2N_w Sc^{2/3}}{C_{w1} U_b f}\right] \qquad (17\text{-}60)$$

where C_{sp} = solute concentration in the upstream solution at the membrane surface, (g·mol)/cm^3; Sc = Schmidt number; U_b = bulk-fluid velocity, cm/s; and f = Fanning friction factor.

For laminar flow between sheet membranes, the salt diffusion is described by the equation

$$\frac{\partial}{\partial X}(UC) + \lambda\frac{\partial}{\partial Z}\left(VC - \gamma\frac{\partial C}{\partial Z}\right) = 0 \qquad (17\text{-}61)$$

where $X = x/h$, dimensionless axial distance; x = axial distance from channel entrance; h = half width of two-dimensional channel; $U = u/u(o)$, dimensionless axial velocity; u = velocity component in x direction; $u(o)$ = average fluid velocity at the channel entrance; $C = C_s/C_s(0)$, $C_s(0)$ = solute concentration in the solution at the channel entrance; $\lambda = v/u(o)$; v = withdrawal velocity through membrane; $Z = y/h$ = dimensionless transverse position; $V = v_y/v$ = dimensionless velocity normal to the membrane; v_y = velocity component in the y direction; and $\gamma = D_s/vh$. The solution is

$$C(X, Z) = \sum_{n=0}^{\infty} a_n Y_n(Z)(1 - \lambda X)^{2/3} a_{n-1} \qquad (17\text{-}62)$$

where a_n and $Y_n(Z)$ are the eigenvalues of the concentration polarization expressed as τ, where $\tau = (C_{sp} - \overline{C}_s)/\overline{C}_s$, then

$$\tau = (1 - \lambda X)C(X, 1) - 1 \qquad (17\text{-}63)$$

Similar mathematical analyses of concentration polarization in both laminar and turbulent flows in cylindrical and parallel-plate channels are available [Dresner, OSW Rep. 3621, Oak Ridge National Laboratory, 1964; and Gill et al., *Ind. Eng. Chem. Fundam.*, **5**, 36F (1966)].

The solute can transfer through the membrane by a solution-diffusion process as expressed in Eq. (17-64), but the driving force for the solute transport is the concentration difference rather than the net pressure difference (Stannett et al., loc. cit.).

$$N_s = \frac{D_{sm}\beta(C_{sp} - C_{s2})}{\ell} \qquad (17\text{-}64)$$

where β = the membrane-solution distribution coefficient for the solute. The diffusion coefficients and distribution coefficients for water and sodium chloride in cellulose acetate and polyamide membranes are listed in Table 17-15.

It should be noted that the concentration polarization for hollow fibers can be neglected because of the small filtrate flow rate as indicated from a mathematical analysis of a hollow-fiber reverse-osmosis system [Kabadi, Doshi, and Gill, *Chem. Eng. Commun.*, **3**, 339 (1979)].

Design

Membranes A membrane suitable for reverse-osmosis processes has to meet stringent requirements. For example, some important membrane properties in relation to the process economics of desalination are (1) membrane selectivity for water over ions, which determines the number of pressurized stages required to product potable water; (2) permeation rate of water per unit pressure gradient, which determines the size of the equipment per unit production rate of potable water; and (3) membrane durability, which determines how often the membrane must be replaced.

Two membrane materials are in common use today: cellulose acetate with a degree of acetylation ranging from 2.5 to 2.8 and a family of aromatic polyamide and polyamide-hydrazides (Stannett et al., loc. cit.). The properties of these materials are shown in Table 17-15. The dominant membranes in use presently are the Loeb-Sourirajan-type (asymmetric) cellulose acetate membranes in spiral-wound module form and the aromatic polyamide hollow-fiber membranes packaged into a multiple tube-in-shell design.

The asymmetric cellulose acetate membranes are made in a form of composite film in which a thin, dense layer estimated to be 0.1 to 10 μm thick is supported by a much thicker (2- to 5-mil) porous, spongy substrate with little or no resistance to permeation. Inperfection-free membranes of extreme thinness (effectively, about 0.1 μm or even less) have been developed to have water fluxes approaching 24.5 gal/(ft^2·day) [1 m^3/(m^2·day)] with high salt rejection. These membranes have been incorporated into inexpensive spiral-wound modules [Kremen, "Technology and Engineering of ROGA Spiral-Wound Reverse Osmosis Membrane Modules," in Sourirajan (ed.), op. cit., chap. 17].

In reverse osmosis, solute molecules whose dimensions are close to the pore size of membrane will block some of the pores, causing plugging and reducing permeability. To avoid plugging of microporous membranes, it is best to select a membrane with pore sizes below solute dimensions. The asymmetric microporous membranes exhibit superior resistance to plugging because the pores are of conical shape with their diameters increasing with distance through the membrane. Coke formation may also occur because the retained solute accumulates at the membrane surface. This may cause membrane deterioration. Such deterioration can be partially eliminated by periodic reversal of flow direction and by maintaining a high circulation velocity past the membrane (see previous discussion of concentration polarization).

Operating pressures required for different feedwater salinities and salt-rejection efficiencies are shown in Fig. 17-22. Rejection efficiencies of various solutes by a modified cellulose acetate membrane are shown in Table 17-16. The compounds most effectively excluded by the membranes are salts of divalent ions, sucrose, and the tetralkyl ammonium salts (Blunk, UCLA Dept. Eng. Rep. 64-28, 1964). The apparent permeabilities of water and the various solutes were calculated on the basis of an effective membrane thickness of 0.16 μm [Lonsdale, in Merten (ed.), *Desalination by Reverse Osmosis*, M.I.T., Cambridge, Mass., 1966, pp. 134–135]. The relationship between salt-rejection efficiency and system pressure for several common salts is given by Michaels et al. [*J. Colloid Sci.*, **20**, 1034 (1965)].

The rejection of organic solutes by a reverse-osmosis membrane depends on their molecular weight, steric size, and ionic charge [Caracciolo, Rosenblatt, and Tomsic, "Du Pont's Hollow Fiber Membranes," in Sourirajan (ed.), op. cit., chap. 16]. High molecular

TABLE 17-15 Diffusion Coefficients and Distribution Coefficients for Cellulose Acetate and Polyamide Membranes*

Membrane	Water diffusivity in membrane, D_{wm}, cm^2/s	Water concentration in membrane, C_{wm}, (g·mol)/cm^3	NaCl diffusivity in membrane, D_{sm}	NaCl distribution coefficient, β
Cellulose acetate, 43.2% acetyl	1.3×10^{-6}	0.0067	3.9×10^{-11}	0.015
Cellulose acetate, 39.8% acetyl	1.6×10^{-6}	0.0089	9.4×10^{-10}	0.035
Cellulose acetate, 37.6% acetyl	2.9×10^{-6}	0.011	4.3×10^{-9}	0.062
Cellulose acetate, 33.6% acetyl	5.7×10^{-6}	0.016	2.9×10^{-8}	0.17
Nomex® polyamide	1.5×10^{-6}	0.011–0.027	1.3×10^{-10}	0.2

*Stannett, Koros, Paul, Lonsdale, and Baker, *Adv. Polym. Sci.*, **32**, 69 (1979); Lonsdale, Merten, and Reiley, *J. Appl. Polym. Sci.*, **9**, 1341 (1965).

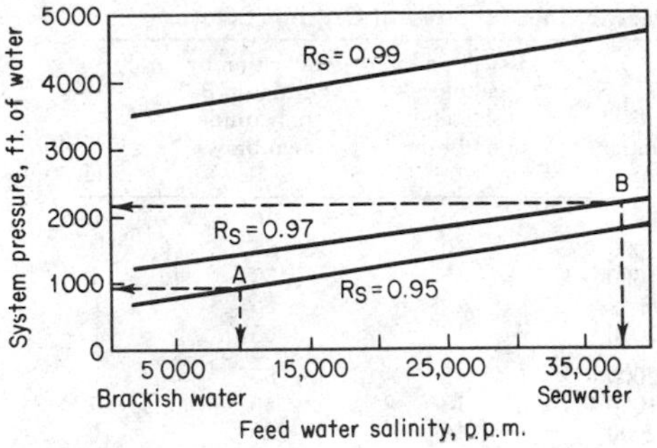

FIG. 17-22 Effect of salinity on salt-rejection efficiency and system pressure. To convert feet to meters, multiply by 0.3048.

weight, branching, and electrical charge with sterically bulky hydrated ions improve rejection. Higher pH increases dissociation and then enhances rejection. Rejection data of cellulose acetate and polyamide membranes for several organic solutes are presented in Table 17-17. Rejection results for some other organic species are available in the literature [Besik, "Reverse Osmosis in Treatment of Domestic and Municipal Waters," in Sourirajan (ed.), op. cit., chap. 24].

While cellulose acetate and polyamide membranes are used commonly in reverse osmosis today, research on the development of new and improved membranes is continuing. A number of different membranes have been found to possess interesting salt rejections and water permeation rates (Stannett et al., loc. cit.; and Podall, loc. cit.).

Membrane Modules There are four common designs: (1) spiral-wound, (2) hollow-fiber, (3) tubular, and (4) plate-and-frame. The spiral-wound module shown in Fig. 17-19 consists of an envelope of membrane around a porous backing matrix (such as a matrix of glass beads held together by a plastic resin) in such a way that two pieces of membranes sandwich the backing matrix and are glued around three edges of the matrix. The fourth edge of the membrane-envelope-matrix construction is connected to a perforated central tube. This membrane construction is then wound around the tube in a jelly-roll fashion, and the assembly is inserted into a canister-type pressure vessel. Feed solution flows over the membrane, whereas the purified solvent passing through the membrane flows to a collection system by way of the central tube. The canisters can be arranged in series so that feed solution can flow across a large number of membranes. Modules used in pilot-plant work are approximately 1 ft (0.3 m) long and 2½ to 3 in (6.35 to 7.62 cm) in diameter and contain 4 to 8 ft^2 (0.37 to 0.74 m^2) of membrane area (Brat and Merten, Off. Saline Water Res. Dev. Prog. Rep. 165, 1965). The larger modules are about 4 in (10 cm) in diameter (Hwang and Kammermeyer, 1975, loc. cit.).

The hollow-fiber module, as shown schematically in Fig. 17-23, consists of fibers with an outside diameter of 25 to 250 μm and a wall thickness of 5 to 50 μm. Input water, under high pressure, flows over the outside surface of the fibers. The permeated water then flows out through the base of the fibers and is collected as product.

Tubular devices shown in Fig. 17-19 consist of parallel bundles of rigid-walled porous or perforated tubes ½ to 1 in (1.3 to 2.5 cm) in diameter. The inside walls are lined with the membrane. Pressurized feed flows inside the tubes, and permeate drips off the outside surfaces and is collected in suitable troughs or vessels (Michaels, loc. cit.).

The plate-and-frame module has several variations used in the industry. The most common one consists of thin plastic plates covered on both sides by membranes which are sealed around the edge to prevent leaks (Fig. 17-24). These plates resemble phonograph records in that their flat surfaces contain small grooves through which the permeate flows after passing through the membrane. The permeate eventually flows into a central tube at the stack and is collected through this tube (Lonsdale, loc. cit.).

A comparison of the membrane modules is discussed later under "Ultrafiltration" (Table 17-24).

Applications Although reverse osmosis is most often used for the desalination of seawater and brackish water, it can also be used to fractionate mixtures of materials which are difficult to separate by other means, such as synthetic or natural polymers. The reverse-osmosis process is particularly appealing to manufacturers of thermally and chemically unstable products, e.g., biologicals, drugs, and food products, in which traditional purification and separation processes often lead to product loss or flavor deterioration.

Industrial applications include the following: (1) seawater desalination: Reverse-osmosis units have successfully desalted seawater to potable quality [Caracciolo et al., loc. cit.; Shields, "Five Years' Experience with Reverse Osmosis Systems Using Du Pont Permasep Permeators," *Proc. 6th Int. Symp. Fresh Water from Sea*, **3**, 395 (1978); and Tidball, Kuiper, Lange, and Kadaj, *Desalination*, **29**, 319 (1979)]. (2) Potable-water production: Reverse osmosis has been used to supply potable water to communities (such as Greenfield, Iowa, and Rotonda West, Florida) [Moore, *J. Am. Water Works Assoc.*, **64**, 781 (1972); and Caracciolo et al., loc. cit.] and to treat brackish or surface waters (Besik, loc. cit.). (3) Demineralization: Reverse-osmosis units have been integrated into the ultrapure-water systems that

TABLE 17-16 Selectivity of Modified Cellulose Acetate Membranes to Several Solutes*

Solute	Concentration, moles/liter	Water flux, g./sq. cm.-sec.	Rejection, %	Apparent water permeability P_w, g./cm.-sec.	Apparent solute permeability P_s, cm./sec.
NaCl	0.90	7.2×10^{-4}	98.1	2.6×10^{-7}	2.0×10^{-10}
NaBr	0.51	6.9×10^{-4}	98.0	1.9×10^{-7}	3.1×10^{-10}
KCl	0.71	7.4×10^{-4}	95.8	2.2×10^{-7}	5.3×10^{-10}
$NaNO_3$	0.029	1.2×10^{-3}	90.1	2.6×10^{-7}	1.9×10^{-9}
$NaClO_4$	0.43	1.1×10^{-3}	86.3	2.3×10^{-7}	2.9×10^{-7}
NH_4ClO_4	0.45	8.5×10^{-4}	77.4	2.1×10^{-7}	5.0×10^{-9}
NH_4NO_3	0.031	1.3×10^{-3}	80.3	2.8×10^{-7}	4.7×10^{-9}
$CaCl_2$	0.47	6.7×10^{-4}	99.1	2.1×10^{-7}	5.3×10^{-10}
Na_2SO_4	0.37	6.2×10^{-4}	99.3	1.6×10^{-7}	1.0×10^{-10}
HNO_3†	0.040	1.3×10^{-3}	51.3	2.7×10^{-7}	1.8×10^{-8}
NH_4OH†	0.050	1.4×10^{-3}	6.2	2.9×10^{-7}	3.0×10^{-7}
Sodium lauryl sulfate	0.0043	1.8×10^{-3}	98.2	3.8×10^{-7}	1.1×10^{-10}
Sucrose	0.15	1.5×10^{-3}	99.7	3.4×10^{-7}	5.0×10^{-11}
Tetramethyl ammonium chloride	0.48	8.9×10^{-4}	99.6	2.5×10^{-7}	5.0×10^{-11}
Tetraethyl ammonium chloride	0.32	1.1×10^{-3}	99.4	2.7×10^{-7}	1.0×10^{-10}

Pressure 102 atm.
Acetyl content of the membrane = 39.8 per cent.
Effective membrane thickness = 0.16

*Blunk, U.C.L.A., *Dept. Eng. Rept.* 64–28, 1964. Lonsdale, in Merten (ed.), "Desalination of Reverse Osmosis," pp. 134–135, M.I.T., Cambridge, Mass., 1966.

†This test was conducted at approximately 0°C. to reduce chemical attack. Nevertheless, some loss in selectivity to NaCl occurred during the test.

TABLE 17-17 Rejection Data of Cellulose Acetate and Polyamide Membranes for Several Organic Solutes

Solute	Molecular weight	Feed pH	Feed concentration, ppm	Rejection by cellulose acetate membrane, %	Rejection by Permasep B-9 polyamide membrane, %	Ref.
Acetic acid	60	3.7	500	...	40	1
	60	....	682–1000	10	31	2
Acetone	58	....	856–1000	5	53	2
Benzoic acid	122	3.7	500	...	83	1
n-Butanol	74	6–9	500–2000	17	65	1
iso-Butanol [$(CH_3)_2CHCH_2OH$]	74	4–10	500–2000	42	95	1
Diethyl ether	74	....	388–1000	13	58	2
Ethanol	46	5–10	500–2000	10	28	1
Ethylene glycol	62	6–7	2000	...	48	1
Formaldehyde	30	....	1000–1278	20	21	2
Formic acid	46	3.2	500	...	50	1
Glucose	180	5–8	500–2000	...	99	1
Glycerol	92	6–7	500–2000	...	91	1
	92	....	765–1000	80	88	2
Methanol	32	7	500–2000	5	0	1
n-Propanol	60	4–10	500–2000	25	62	1
iso-Propanol	60	5–11	500–2000	38	75	1
Phenol	94	7–9	500–2000	...	55	1
Raffinose ($C_{18}H_{32}O_{16}$)	504	7	2000	...	99.4	1
Sodium acetate	82	8.1	680	...	98	1
Sodium benzoate	144	8.1	590	...	99	1
Sodium formate	68	6.9	740	...	94	1
Sodium phenolate	116	10.7	2500	...	95	1
Sucrose	342	5–8	500–2000	...	99.8	1
Urea	60	....	1000–1188	18	34	2

References:
1. Caracciolo, Rosenblatt, and Tomsic, "Du Pont's Hollow Fiber Membranes," in Sourirajan (ed.), *Reverse Osmosis and Synthetic Membranes*, National Research Council Canada Publications, Ottawa, 1977, chap. 16.
2. Chian and Fang, *Am. Inst. Chem. Eng. Symp.*, **70**, ser. no. 136, 497 (1973).

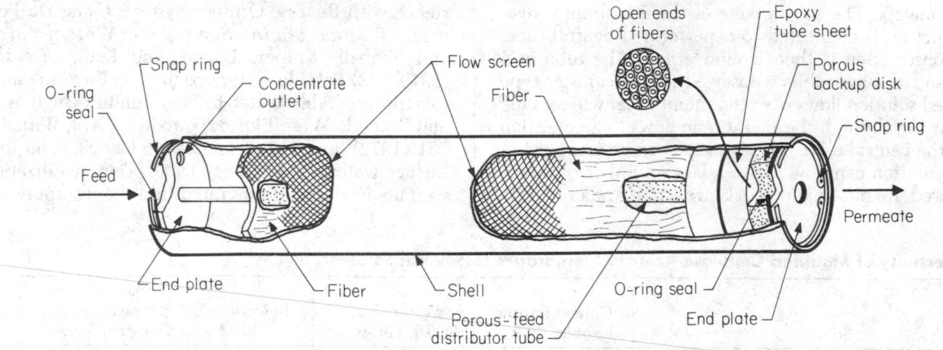

FIG. 17-23 Schematic of Du Pont hollow-fiber module, Permasep permeator. *(Courtesy E. I. du Pont de Nemours & Co.)*

supply water for electronic-manufacturing processes (Caracciolo et al., loc. cit.). (4) Recycle operations in electroplating and electrocoating plants: This application not only recovers valuable materials, such as nickel, chromium, and cyanide, but at the same time eliminates water-discharge problems [Caracciolo et al., loc. cit.; Golomb, "Application of Reverse Osmosis to Electroplating Water Treatment," in Sourirajan (ed.), op. cit., chap. 23; and Cohen and Loeb, "Industrial Water Treatment by Means of Reverse Osmosis Membranes," in Sourirajan (ed.), op. cit., chap. 25]. (5) Processing wheys, juices, and effluents in the food industry: Cheese and cottage-cheese wheys are processed by the use of reverse osmosis, which is combined with ultrafiltration to separate the solutes into fractions containing mostly protein, lactose, and lactic acid as well as to concentrate each of these fractions [Merson and Ginnette, "Reverse Osmosis in the Food Industry," in Lacey and Loeb (eds.), op. cit., chap. 10; and Delaney and Donnelly, "Applications of Reverse Osmosis in the Dairy Industry," in Sourirajan (ed.), op. cit., chap. 20]. A large reverse-osmosis plant in California removes solutes from the effluent from a food-fermentation process (Merson and Ginnette, loc. cit.). (6) Applications in the pulp and paper industry: Reverse osmosis can be employed to treat dilute processing flows in the pulp and paper industry for recycling water and reducing color, biochemical oxygen demand, and other objectionable components of paper-mill outfall [Bansal and Wiley, "Application of Reverse Osmosis in the Pulp and Paper Industry," in Sourirajan (ed.), op. cit., chap. 22]. Combined with ultrafiltration, reverse osmosis can be used to fractionate and concentrate marketable wood chemicals from spent sulfite liquors (Bansal and Wiley, loc. cit.). (7) Radioactive-laundry-

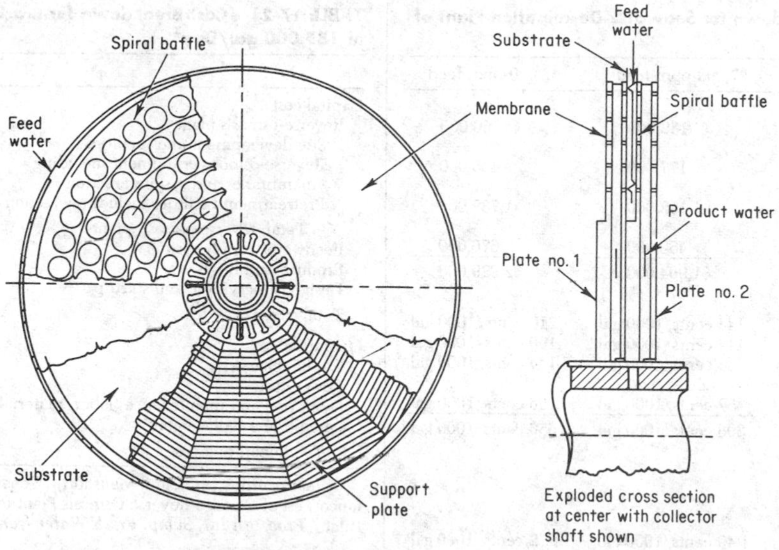

FIG. 17-24 Plate-and-frame system. *(Eastman Tech. Bull. TBM-1.)*

water concentration: Reverse osmosis removes both the radiochemicals and nonradioactive pollutants from the contaminated laundry water in nuclear power plants [Stana, "Westinghouse Membrane Systems," in Sourirajan (ed.), op. cit., chap. 18]. (8) Steam-generator blowdown: Steam-generator blowdown in the operation of a pressurized-water-reactor nuclear plant is processed via reverse osmosis to reduce by a factor of 10 or more the blowdown stream volume, which can then be handled with a radioactive-waste evaporator (Stana, loc. cit.). (9) Separation of oil-water emulsion: In metal-machining operations, an oil-water emulsion is used to lubricate and cool the tools and workplace. The emulsion picks up some of the metals. Reverse osmosis can be used to separate this emulsion to yield a water that can be discharged and an oil concentrate that can be either easily burned or further refined to produce a reusable oil (Stana, loc. cit.).

Cost Estimates Cost estimates for seawater-desalination plants with capacities of 528,000 gal (2000 m^3) per day and 1,057,000 gal (4000 m^3) per day are shown in Table 17-18. Both capital and operating costs decrease with increasing plant capacity and decreasing feedwater salinity. The cost breakdown for the seawater desalination plant of 528,000 gal/day is presented in Table 17-19.

Cost estimates for brackish-water-desalting plants with capacities of 185,000 gal (700 m^3) per day and 2,114,000 gal (8000 m^3) per day are listed in Table 17-20 for a product recovery of 60 percent. The cost breakdown for the brackish-water-desalting plant of 185,000 gal/day is shown in Table 17-21. Cost estimates as a function of product recovery for treating brackish or surface waters to produce potable water are presented in Fig. 17-25 for the capital cost and Fig. 17-26 for the operating cost.

In general, membrane cost per unit area is independent of plant capacity; therefore, the relative cost of the membrane increases with increasing plant size, while capital and operating costs decrease. In the near future, membrane life may be extended to 10 years. This, together with possible price lowering due to an expanded market and technical advances in mechanization and automation of membrane replacement, will decrease membrane-replacement costs.

TABLE 17-18 Cost Estimates for Seawater-Desalination Plants*

	Cost
Design and economic parameters	
Operating pressure	800 lb/in^2 gauge
Plant factor	85%
Membrane module	Polyamide hollow fiber
Membrane life	2 years
Fixed-charge rate	12%/year
Power cost	3.5 cents/kWh
Product salinity for 27,000-ppm feed	300 ppm
Product salinity for 42,000-ppm feed	500 ppm
Product recovery for 27,000-ppm feed	35%
Product recovery for 42,000-ppm feed	20%
Capital costs	
528,000-gal/day plant with 27,000-ppm feed	$1,960,000
528,000-gal/day plant with 42,000-ppm feed	$2,930,000
1,057,000-gal/day plant with 27,000-ppm feed	$3,330,000
1,057,000-gal/day plant with 42,000-ppm feed	$5,070,000
Operating costs	
528,000-gal/day plant with 27,000-ppm feed	398 cents/1000 gal
528,000-gal/day plant with 42,000-ppm feed	656 cents/1000 gal
1,057,000-gal/day plant with 27,000-ppm feed	364 cents/1000 gal
1,057,000-gal/day plant with 42,000-ppm feed	591 cents/1000 gal

*Glueckstern, Kantor, and Wilf, "Field Trials and Preliminary Evaluation of Reverse Osmosis Systems for Seawater Desalting at the Red-Sea Shore," *Proc. 6th Int. Symp. Fresh Water from Sea,* 3, 307 (1978).

ULTRAFILTRATION

GENERAL REFERENCES: See discussion of cross-flow filtration presented in subsection "Novel Solid-Liquid Separation Processes." Blatt, "Principles and Practice of Ultrafiltration," in Meares (ed.), *Membrane Separation Processes,* Elsevier, Amsterdam and New York, 1976, chap. 3. Porter, "Membrane Filtration," in Schweitzer (ed.), *Handbook of Separation Techniques for Chemical Engineers,* McGraw-Hill, New York, 1979, sec. 2.1. Porter and Nelson, "Ultrafiltration in the Chemical, Food Processing, Pharmaceutical and Medical Industries," in Li (ed.), *Recent Developments in Separation Science,* vol. II, CRC Press, Cleveland, 1972, chap. 10.

Definitions Similar to reverse osmosis, ultrafiltration is a pressure-driven membrane process being capable of separating solution components on the basis of molecular size and shape. Under an applied pressure difference across an ultrafiltration membrane, solvent and small solute species pass through the membrane and are collected as permeate while larger solute species are retained by the membrane and recovered as a concentrated retenate.

Ultrafiltration involves solutes whose molecular dimensions are 10 or more times larger than those of the solvent and are usually below

TABLE 17-19 Cost Breakdown for Seawater-Desalination Plant of 528,000 gal/Day*

Costs	27,000-ppm feed	42,000-ppm feed
Capital cost		
Site	$ 182,000	$ 69,000
Pretreatment and posttreatment	177,000	445,000
Reverse-osmosis plant including membranes	1,152,000	1,739,000
Design, erection, and other costs including contingency	453,000	676,000
Total capital cost	$1,964,000	$2,929,000
Operating cost		
Capital charges	144 cents/1000 gal	216 cents/1000 gal
Membrane replacement	114 cents/1000 gal	190 cents/1000 gal
Power cost	83 cents/1000 gal	144 cents/1000 gal
Operation and maintenance	57 cents/1000 gal	106 cents/1000 gal
Total operating cost	398 cents/1000 gal	656 cents/1000 gal
Effect of economic parameters on operating cost		
Fixed-charge rate		
Higher = 16%	+49 cents/1000 gal	+72 cents/1000 gal
Lower = 8%	−49 cents/1000 gal	−72 cents/1000 gal
Plant factor		
Lower = 80%	+19 cents/1000 gal	+30 cents/1000 gal
Higher = 90%	−19 cents/1000 gal	−30 cents/1000 gal
Membrane life		
Lower = 1 year	+114 cents/1000 gal	+190 cents/1000 gal
Higher = 3 year	−114 cents/1000 gal	−190 cents/1000 gal

*Glueckstern, Kantor, and Wilf, "Field Trials and Preliminary Evaluation of Reverse Osmosis Systems for Seawater Desalting at the Red-Sea Shore," *Proc. 6th Int. Symp. Fresh Water from Sea*, 3, 307 (1978).

TABLE 17-20 Cost Estimates for Brackish-Water-Desalting Plants*

	Cost
Design and economic parameters	
Operating pressure	400 lb/in^2 gauge
Plant factor	90%
Fixed-charge rate	12% per year
Power cost	3.5 cents/kWh
Feedwater salinity	5900 ppm
Product salinity	550 ppm
Product recovery	60%
Feedwater source	Well
Capital costs	
185,000-gal (700-m^3)/day plant	$301,000
2,114,000-gal (8000-m^3)/day plant	$3,000,000
Operating costs	
185,000-gal (700-m^3)/day plant	254 cents/1000 gal
2,114,000-gal (8000-m^3)/day plant	190 cents/1000 gal

*Glueckstern, Kantor, and Mansdorf, "General Design, Erection and Acceptance Test of a Large Reverse Osmosis Plant for Water Supply to the Town of Eilat," *Proc. 6th Int. Symp. Fresh Water from Sea*, 3, 297 (1978).

½-μm size. The solutes or the materials to be separated usually have molecular weights greater than 500, such as macromolecules (proteins, polymers, starches, natural gums, enzymes, etc.), colloidal dispersions (clays, pigments, minerals, latex particles, and microorganisms), and emulsions (grease-detergent and oil-water emulsions). Solutes of low molecular weight may be separated by first complexing with suitable macromolecules [Michaels, "Ultrafiltration," in Perry (ed.), *Advances in Separation and Purification*, Wiley, New York (1968); and Nguyen, Aptel, and Neel, *J. Membr. Sci.*, **6**, 71 (1980)]. As a consequence of relatively high molecular weight, the osmotic pressures of the solutes are usually low. Thus, the operating pressures of ultrafiltration range only from 3.4×10^5 to 68.9×10^5 g/(s^2·cm) (5 to 100 lbf/in^2). The lower pressures reduce the energy

TABLE 17-21 Cost Breakdown for Brackish-Water Desalting Plant of 185,000 gal/Day*

	Cost
Capital cost	
Reverse-osmosis plant	
Site development and facilities	36,600
Reverse-osmosis equipment including membranes and system erection	188,200
Pretreatment facilities including erection	10,700
Total, reverse-osmosis plant	$235,500
Feedwater-supply system	$20,400
Product-supply system	15,100
Design, supervision, and startup	30,300
Total capital cost	$301,300
Operating cost	
Capital charges	57 cents/1000 gal
Power cost	49 cents/1000 gal
Operation and maintenance (labor, materials, and membrane replacement)	148 cents/1000 gal
Total operating cost	254 cents/1000 gal

*Glueckstern, Kantor, and Mansdorf, "General Design, Erection and Acceptance Test of a Large Reverse Osmosis Plant for Water Supply to the Town of Eilat," *Proc. 6th Int. Symp. Fresh Water from Sea*, 3, 297 (1978).

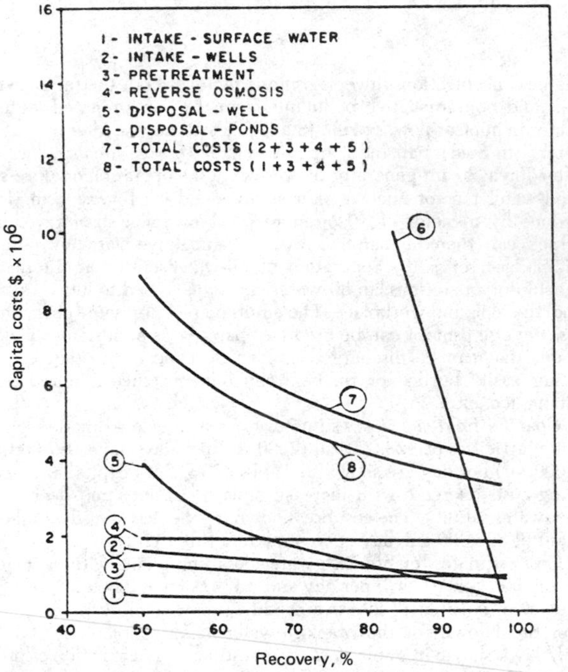

FIG. 17-25 Capital cost for treating brackish or surface waters to produce potable water at 790 m^3/h (5 million gal/day). [*Besik, "Reverse Osmosis in Treatment of Domestic and Municipal Waters," in Sourirajan (ed.),* Reverse Osmosis and Synthetic Membranes, *National Research Council Canada Publications, Ottawa, 1977, chap. 24.*]

requirement for pumping and compression and the equipment cost by a considerable margin over reverse osmosis. Ultrafiltration can be used to concentrate, purify, and fractionally separate macrosolutes or materials in the feed liquid.

The solute-rejection efficiency for ultrafiltration is defined in the same way as that for reverse osmosis, Eq. (17-55).

Theory

Sieve Model When solvent transports toward the membrane surface, it carries solute which is rejected at the membrane surface,

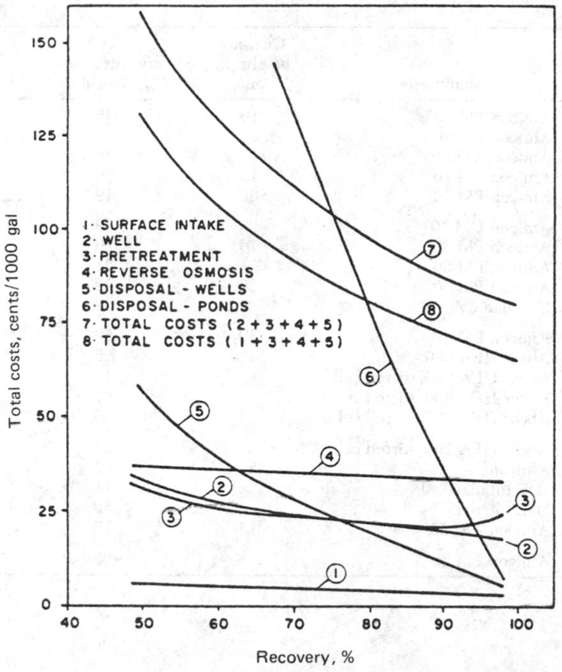

FIG. 17-26 Operating cost for treating brackish or surface waters to produce potable water at 790 m^3/h (5 million gal/day). To convert cents per 1000 gallons to dollars per cubic meter, multiply by 0.002642. [*Besik, "Reverse Osmosis in Treatment of Domestic and Municipal Waters," in Sourirajan (ed.),* Reverse Osmosis and Synthetic Membranes, *National Research Council Canada Publications, Ottawa, 1977, chap. 24.*]

resulting in an accumulation of solute on the membrane. This accumulation can lead to the formation of a gel layer, or secondary membrane. Thus, the transmembrane solvent flux may be expressed as

$$N_w = (\Delta p - \Delta\pi)/(R_g + R_m) \quad (17\text{-}65)$$

where R_g = resistance due to the gel layer, (g·cm)/(g·mol·s); R_m = resistance due to the membrane; and the rest of the symbols are the same as those described previously. (It should be noted that the symbols used in this section are the same as in the subsection "Reverse Osmosis.") The osmotic pressures for macrosolutes and colloidal dispersions are usually low. Thus, $\Delta\pi$ in Eq. (17-65) can be neglected. This equation therefore results in

$$N_w = \Delta p/(R_g + R_m) \quad (17\text{-}66)$$

[Porter and Michaels, *Chemtech,* 56 (January 1971); and Porter, in Shoemaker (ed.), *Am. Inst. Chem. Eng. Symp.,* **73,** ser. no. 173, 83 (1977)].

For the ultrafiltration of a dilute solution of macrosolute with a microporous membrane of relatively high retentivity (pore diameter of 1 μm or larger), R_g may be insignificant in comparison with R_m and Eq. (17-66) reduces to Eq. (17-53). This is the case in which there is no concentration polarization or the gel layer is mobile at infinite dilution. In this case, when the solute is transferred purely by connection with solvent through pores of the membrane large enough to admit solute molecules, the solute flux can be expressed by

$$N_s = N_w(1 - \phi)(C_{s1}/C_{w1}) \quad (17\text{-}67)$$

Concentration Polarization When solute is rejected by the membrane, it thus accumulates on the membrane surface. At high concentration it precipitates or forms a thixotropic gel. The gel layer can be considered as a secondary dynamic membrane which is hydraulically permeable to solvent. The resistance due to the gel layer, R_g, is often much greater than that of membrane, R_m. Thus, R_g is the limiting step, and the flux obtained becomes independent of membrane permeability. The gel layer will increase in thickness or become compacted. This results in increasing the resistance R_g to the solvent transport and slowing down the solvent flow. R_g increases until the net transport of solute to the membrane surface due to the connective transfer by solvent flow is equal to the back diffusion of solute away from the membrane into the bulk solution because of its polarized concentration gradient. Any further increase in the transmembrane pressure drop will cause the gel layer to thicken by an amount to render the transmembrane solvent flux unchanged with the pressure drop. The steady-state flow can be expressed by Eq. (17-68).

$$N_w\left(\frac{C_s}{C_w} - \frac{C_{s2}}{C_{w2}}\right) + D_s\frac{dC_s}{dy} = 0 \quad (17\text{-}68)$$

where C_w = solvent concentration in the solution, (g·mol)/cm^3; C_{w2} = C_w at the downstream side; and y = distance normal to the gel layer, cm. Equation (17-68) may be approximated as follows:

$$Q_w(C_s - C_{s2}) + D_s(dC_s/dy) = 0 \quad (17\text{-}69)$$

where Q_w = transmembrane solvent volumetric flux, cm/s. Integration of Eq. (17-69) results in Eq. (17-70) [Rautenbach and Rauch, *Int. Chem. Eng.,* **18,** 417 (1978)].

$$Q_w = \frac{D_s}{\delta}\ln\left(\frac{C_{sg} - C_{s2}}{C_{s1} - C_{s2}}\right) = K\ln\left(\frac{C_{sg} - C_{s2}}{C_{s1} - C_{s2}}\right) \quad (17\text{-}70)$$

where δ = boundary-layer thickness over the gel layer, cm; C_{sg} = gel concentration of the solute, (g·mol)/cm^3; and K = mass-transfer coefficient, cm/s. Usually, the solute concentration in ultrafiltrate, C_{s2}, is nearly zero or very small, particularly in comparison with C_{sg} and C_{s1}. Thus, Eq. (17-70) results in Eq. (17-71) [Porter, 1977, loc. cit.; Porter, 1979, loc. cit.; Porter and Michaels, loc. cit.; and Porter, *Ind. Eng. Chem. Prod. Res. Dev.,* **11,** 234 (1972)].

$$Q_w = K\ln(C_{sg}/C_{s1}) \quad (17\text{-}71)$$

It should be noted that solute concentrations can also be expressed in weight percent for the gel layer of constant density.

The validity of Eq. (17-71) has been demonstrated experimentally for a large number of macromolecular solutes and colloidal dispersions (Porter, 1972, loc. cit.; Blatt, 1976, loc. cit.; and Hwang and Kammermeyer, 1975, loc. cit.). The plot of Q_w versus ln C_{s1} is linear. The gel concentration C_{sg} is the concentration at which the solvent flux drops to zero. Normally, it is determined experimentally from the linear plot of Q_w versus ln C_{s1}. C_{sg} depends mainly on solute characteristics, i.e., the chemical and morphological properties of solute. It is virtually independent of bulk-solution concentration, fluid-flow conditions, operating pressure, and membrane characteristics. Some values of C_{sg} are listed in Table 17-22. The gel concentration for macromolecular solutes is about 25 percent (by weight) with a range of 5 to 50 percent. And the gel concentration for colloidal dispersions is about 65 percent (by weight) with a range of 50 to 75 percent. These results agree reasonably well with those that show that many protein solutions gel at around 25 percent solid and that colloidal dispersions are expected to have close packing with 65 to 75 percent solid (Porter, 1977, loc. cit.; Porter, 1972, loc. cit.; and Porter, 1979, loc. cit.).

The mass-transfer coefficient K is the measure of the mass transfer of solute away from the membrane surface. This is controlled mainly by fluid-flow conditions and operating temperature. For the laminar flow with a feed stream flowing over a membrane surface (cross-flow), K has been given (Porter, 1972, loc. cit.) as follows:

$$K = 0.816\left(\frac{\dot\gamma}{L}D_s^2\right)^{0.33} \quad (17\text{-}72)$$

where $\dot\gamma$ = fluid shear rate at the membrane surface, 1/s; $\dot\gamma = 8U_b/d$ for circular tubes of the tube diameter d, cm; $\dot\gamma = 6U_b/b$ for rectangular channels of the channel height b, cm; U_b = bulk fluid velocity, cm/s; and L = tube length or channel length, cm. For turbulent

TABLE 17-22 Experimental Values of Gel Concentrations

Solute	Diffusivity, D_s, cm²/s	Membrane	Channel height, h, mil	Gel concentration, C_{sg}, weight %	Ref.
Human albumin	6×10^{-7}	Amicon PM 30	10	45	1
	6×10^{-7}	Amicon PM 30	30	28	1
	6×10^{-7}	Amicon XM 50	15	45	1
γ-Globulin (whole bovine serum)	4×10^{-7}	Amicon PM 10	15	30	1
Collagen (gelatin) at 70°C	0.7×10^{-7}	Amicon PM 30	30	19	1, 2
Human blood plasma	....	Amicon UM 10	8	59	2
Bovine serum	....	Amicon PM 30	10	20	2
	....	Amicon PM 30	11	21	2
Egg albumen	....	Amicon PM 30	30	39	3
	....	Eastman CA		39	3
Skim-milk protein	....	Amicon PM 30	30	50	3
Polyethylene glycol (Carbowax 20 M)	5×10^{-7}	Abcor HFA 300	...	7.5	1, 4
Dextran 20	....	Abcor HFA 200, stirred cell	...	15	4
Dextran 40	....	Abcor HFA 200, stirred cell		21	4
Dextran 70	....	Abcor HFA 200, stirred cell	...	26	4
Dextran 110	....	Abcor HFA 200, stirred cell	...	34	4
Triton X-100 (isoctylphenoxy polyethoxy ethanol MW 628)	$\sim 8 \times 10^{-7}$	Amicon	30	5.2	5
Secondary effluents from municipal sewage treatment	....	IDE tubular	...	11	6
Cutting-oil emulsion	....	IDE tubular	...	50	6
Styrene-butadiene polymer latex	....	Amicon XM 50	15	75	1
Electrodeposition primer	....	Amicon PM 30	...	~65	1

References:
1. Porter, *Ind. Eng. Chem. Prod. Res. Dev.*, **11**, 234 (1972).
2. Porter and Michaels, *Chem. Technol.*, 440 (July 1971).
3. Porter and Michaels, *Chem. Technol.*, 248 (April 1971).
4. Goldsmith, *Ind. Eng. Chem. Fundam.*, **10**, 113 (1971).
5. Grieves et al., *Am. Inst. Chem. Eng. J.*, **19**, 766 (1973).
6. Matz and Meitlis, *Desalination*, **24**, 281 (1978).

flow with a cross-flow feed stream, K has been shown (Porter, 1972, loc. cit.) as follows:

$$K = 0.023(U_b^{0.8} D_s^{0.67} / d_h^{0.2} \nu^{0.47}) \tag{17-73}$$

where d_h = equivalent hydraulic diameter ($d_h = 2b$ for flat rectangular channels), cm; and ν = kinematic viscosity, cm²/s. The solute diffusivity may be calculated from the Stokes-Einstein relationship:

$$D_s = kT/6\pi\eta r_s \tag{17-74}$$

where k = Boltzmann constant = 1.380×10^{-16} (g·cm²)/(s²·K); T = absolute temperature, K; η = fluid viscosity, g/(s·cm) or P; and r_s = radius of the solute particle, cm. An estimation method for diffusion coefficients of high-molecular-weight solutes has been reported [Fedors, *Am. Inst. Chem. Eng. J.*, **25**, 883 (1979)].

The comparison between the experimental ultrafiltration flux and the theoretical one calculated from Eq. (17-72) or Eq. (17-73) has been made by Porter (Porter, 1972, loc. cit.) for both macromolecular solutions and colloidal dispersions. For macromolecular solutions, the agreement between experimental and theoretical fluxes is within 15 to 30 percent for both laminar and turbulent flows. However, for colloidal dispersions, the experimental flux is much higher than the theoretical one by a factor of 20 to 30 in laminar flow and by a factor of 8 to 10 in turbulent flow.

It should be noted that the concentration polarization for hollow fibers can be disregarded because of the small filtrate flow rate (Rautenbach and Rauch, loc. cit.). The solvent flux for hollow fibers is mainly controlled by membrane permeability. The flux is described by Eq. (17-66) with $R_g = 0$ or Eq. (17-53).

Polarization Control As described previously, concentration polarization is an unavoidable consequence of ultrafiltration. In order to maximize the ultrafiltration flux, the polarized-layer thickness has to be minimized. Cross-flow, i.e., the feed-stream flows over the membrane surface, can be employed to sweep away part of the polarized layer.

In laminar flow, Eq. (17-72) shows that the mass-transfer coefficient enhances with fluid shear rate, i.e., increases with fluid velocity and decreases with hydraulic diameter to the 0.33 power. Thus, ultrafiltration fluxes can be enhanced by increasing the fluid shear rate to decrease the boundary-layer thickness δ at the membrane surface and to promote the mass transport of solute away from the membrane surface. This principle has been utilized in laminar thin-channel ultrafiltration to control concentration polarization [Porter, 1972, loc. cit.; and Blatt, Dravid, Michaels, and Nelson, in Flinn (ed.), *Membrane Science and Technology*, Plenum Press, New York, 1970, p. 47]. In thin-channel ultrafiltration, the feed stream flows through a narrow channel, 0.025 to 0.076 cm (10 to 30 mil), at high velocity, 152 to 762 cm/s (5 to 25 ft/s). This usually gives an ultrafiltration flux 2 to 10 times higher than those obtained with conventional ultrafiltration devices.

For turbulent flow, Eq. (17-73) indicates that the mass-transfer coefficient increases with the fluid velocity to the 0.8 power and decreases with the hydraulic diameter to the 0.2 power. The velocity augments the mass-transfer coefficient more pronouncedly in turbulent flow than in laminar flow. Thus, turbulent flow has been utilized in tubular ultrafiltration to control concentration polarization and to enhance the mass-transfer coefficient [Goldsmith, *Ind. Eng. Chem. Fundam.*, **10**, 113 (1971)]. In contrast to laminar flow, the mass-transfer coefficient for turbulent flow does not depend on tube length (or channel length). In turbulent flow, both the velocity and concentration profiles are established rapidly in the entrance section of the tube.

Turbulent promotion in laminar channel flow to enhance the mass-transfer coefficient has been investigated by the use of detached strip-type turbulence promoters [Shen and Probstein, *Ind. Eng. Chem. Process Des. Dev.*, **18**, 547 (1979)]. Depending on the ratio of channel height to interpromoter spacing, a factor of 3 for the augmentation of the mass-transfer coefficient can be obtained by the use of turbulence promoters. Other types of inserts, such as static mixers, have been used to enhance the mass-transfer coefficient. For the tubular module with Kenics mixer inserts, mass transfer can be enhanced by a factor of 2.7 in comparison with that calculated from Eq. (17-72) [Pitera and Middleman, *Ind. Eng. Chem. Process Des. Dev.*, **12**, 52 (1973); and Rautenbach and Rauch, loc. cit.].

Most of the polarization-control techniques previously described—

laminar thin-channel flow, turbulent tubular flow, and inserted devices—involve a higher velocity or a narrower channel to increase fluid shear rate. It should be noted that high shear rate can induce denaturation of labile species [Charm and Lai, *Biotechnol. Bioeng.*, **13,** 185 (1971); and Blatt, loc. cit.]. Besides, an increase in mass-transfer coefficient and a subsequent decrease in gel layer will allow less inhibited transport of secondarily retained species (Blatt, loc. cit.). In addition to the control techniques mentioned, pulsed flow and mechanical scrubbing to keep the membrane clear (Charm and Matteo, *Methods in Enzymology*, vol. 22, Academic, New York, 1971, p. 476) can also be used to minimize concentration polarization.

Design

Membranes Ultrafiltration membranes may be classified into two types, anisotropic (asymmetric) and amorphous (homogeneous), with anisotropic membranes more commonly used. For more extensive discussion of this topic, the reader should see the subsection "Reverse Osmosis." In general, membranes for ultrafiltration have more open structures than those for reverse osmosis.

Anisotropic membranes exhibit higher fluxes because of the thinness of the skin and better resistance to plugging than homogeneous membranes. Anisotropic membranes are often cast on tough substructures of porous supports with extremely open porosity, such as coarse-woven polyethylene, to provide extra mechanical strength and durability. Most anisotropic membranes for industrial ultrafiltration applications are made of cellulose acetate–based material. Recently, the fabrication of anisotropic membranes has been extended to other polymers more resistant to bases, acids, solvents, and high temperatures. Some membrane materials in current industrial use and under development are shown in Table 17-23. A list of some commercially available membranes is available (Porter and Nelson, loc. cit.).

TABLE 17-23 Membrane Materials*

Material type	pH range	Maximum temperature at pH_7	Chlorine resistance	Solvent resistance
Cellulose acetate	4.5–9	55°C	Good	Poor
Polyamide	3–12	80°C	Poor	Good
Polysulfone	0–14	80°C	Good	Good
Polyacrylonitrile	2–12	60°C	Good	Poor
Polyfuran	2–12	90°C	Poor	Good

*Bailey, *Filtr. Sep.*, 213 (May–June 1977).

Anisotropic hollow-fiber membranes have also been made with the skin on either the outside or the inside of the fiber [Porter, *Biotechnol. Bioeng. Symp.*, no. 3, 125 (1972*a*)]. As indicated from Eq. (17-72), a mass-transfer coefficient increases with fluid velocity and decreases with fiber-lumen diameter. Thus, it is advantageous to pass the feed stream through the fiber lumen with the skin on the inside for better polarization control. In fact, as mentioned before, concentration polarization inside the hollow fiber can be disregarded because of small flux. Hollow fibers with a typical inside diameter of about 450 μm have a burst strength approaching 68.9 × 10^5 g/(s^2·cm)(100 lbf/in^2) (Porter, 1977, loc. cit.).

For repeated use, ultrafiltration membranes are generally cleaned with dilute detergent solutions to remove adsorbed solutes. Membranes treated with detergents often result in flux increase and retentivity decrease, which are most likely due to swelling of the membranes by detergent solutes [Swaminathan, Chaudhuri, and Sirkar, *J. Colloid Interface Sci.*, **76,** 573 (1980)].

Membrane Modules Membrane modules include tubular, plate-and-frame, spiral-wound, hollow-fiber, thin-channel, and parallel-leaf-type modules. The first four modules have been described under "Reverse Osmosis." Normally, in the ultrafiltration hollow-fiber module, the feed stream flows through the fiber lumen.

The thin-channel module has been developed for efficient fluid management in terms of flux per unit horsepower (Porter, 1972, loc. cit.; and Blatt, Dravid, Michaels, and Nelson, loc. cit.). As described earlier, in thin-channel ultrafiltration a feed stream flows through a narrow channel, 0.025 to 0.076 cm, at high velocity, 152 to 762 cm/s. Thin-channel plate-and-frame and tubular modules have been used. Figure 17-27 shows a Romicon thin-channel tubular module.

The parallel leaf-type module is shown in Fig. 17-28. Double sheets of membrane are sealed on three sides, and the membrane envelope formed is supported on a cardboardlike porous material.

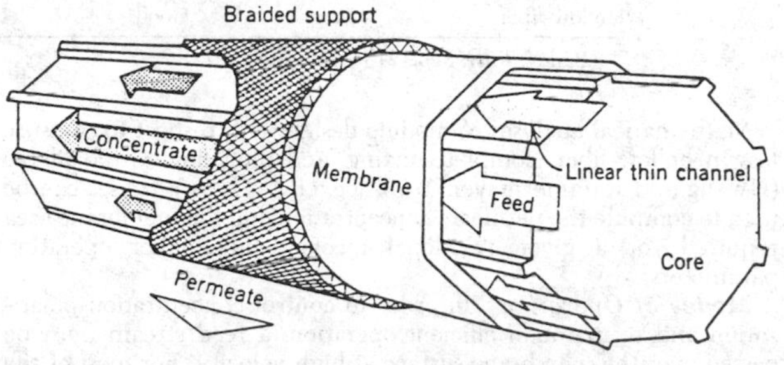

FIG. 17-27 Schematic diagram of a Romicon thin-channel tubular module.

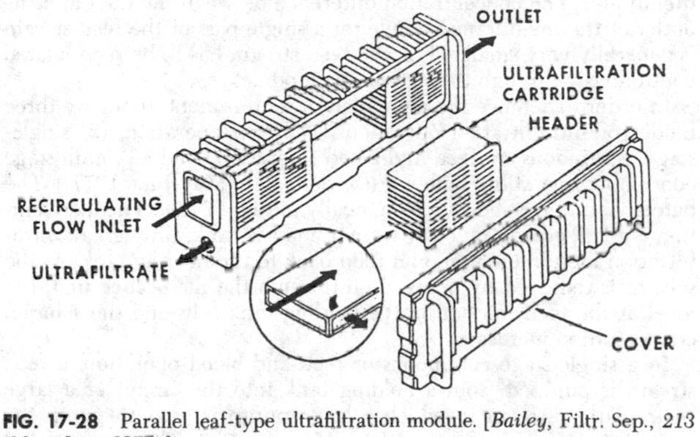

FIG. 17-28 Parallel leaf-type ultrafiltration module. [*Bailey,* Filtr. Sep., *213 (May–June 1977).*]

Several such membrane envelopes or leaves are attached in parallel to the same header to form a cartridge so that the ultrafiltrate can flow to the header. The feed stream is passed longitudinally through the cartridge and in parallel to the membrane envelopes. The module normally consists of a fiberglass-reinforced housing that can accommodate three replaceable cartridges. A comparison of major module characteristics is shown in Table 17-24.

For a tubular module, the pressure drop for pumping a feed stream through the module can be expressed by Eq. (17-75).

$$\Delta p = \left(K_H + 4f\frac{L}{d}\right)\left(\frac{\rho U_b}{2}\right)^2 \tag{17-75}$$

where K_H = hydraulic friction-loss coefficient for the two headers of the module and f = Fanning friction factor. The pressure drop for parallel single channels of relatively short length, such as in plate-and-frame, leaf-type, and spiral-wound modules, can be calculated from Eq. (17-76) (Blatt, 1976, loc. cit.).

$$\Delta p = 1.8(\eta Lq/nWb^3) = 1.8(\eta LU_b/b^2) \tag{17-76}$$

where q = pumping rate or circulation rate, cm^3/s; W = channel width, cm; η = solution viscosity, g/(s·cm) or P; and n = number of parallel channels ($n = 1$ for a single spiral channel). The pumping power consumption can be obtained from $q\cdot\Delta p$. Matz and Meitlis [*Desalination*, **24,** 281 (1978)] obtained 21 kWh/m^3 of feed treated for a batch tubular ultrafiltration system and 9.7 for a continuous tubular system (feed-and-bleed-operation mode).

TABLE 17-24 Comparison of Major Module Characteristics*

Module type	Simplicity of flow path	Resistance to mechanical damage	Lack of susceptibility to blockage	Holdup volume	Ease of mechanical cleaning	Ease of isolation of small volume	Power consumption based on membrane area
Tubular	Good	Good	Good	Poor	Good	Good	Poor
Plate-and-frame	Poor	Good	Poor	Good	Fair	Good	Good
Leaf-type	Fair	Poor	Poor	Good	Poor	Poor	Fair
Spiral-wound	Fair	Poor	Poor	Good	Poor	Poor	Good
Hollow-fiber	Fair	Good	Poor	Good	Poor	Poor	Good

*Bailey, *Filtr. Sep.*, 213 (May–June 1977).

Mathematical analyses of module design with respect to pressure loss in hollow-fiber, complete-mixing, and plug flow are available (Hwang and Kammermeyer, 1975, loc. cit.). These analyses can be used to compute the permeate concentration and the membrane area required for a given fractional recovery and other operating parameters.

Modes of Operation In order to control concentration polarization and to maintain efficient operation, a feed stream must be passed over the membrane surface at high velocity. For most of the membrane modules, this means that the flow rate of the feed stream is much higher than the ultrafiltration rate permeating through the membrane. The concentration difference between the inlet and the outlet of the membrane module for a single pass of the feed stream is generally very small. Thus, the feed stream has to be recirculated continuously through the membrane module.

In order to achieve the recirculation requirement, there are three modes of ultrafiltration operation: (1) batch operation, (2) single-stage continuous or feed-and-bleed operation, and (3) multistage continuous operation [Bailey, *Filtr. Sep.*, 213 (May-June 1977)]. The batch operation is shown schematically in Fig. 17-29*a*. The feed solution is pumped continuously from a holding tank, through an ultrafiltration membrane unit, and then back to the holding tank. As the solvent is removed by permeation through the membrane unit, the level of the feed solution in the holding tank falls and the solution concentration increases.

In a single-stage continuous or feed-and-bleed operation, a feed stream is pumped from a holding tank into the circuit of a large recirculation stream in which a large pump is used to pump the stream continuously through the membrane unit. The concentrated product is bled from this circuit slowly and continuously. The flow rates of the feed and the bleed streams are maintained to be equal. Figure 17-29*b* shows schematically the multistage continuous operation, which employs two or more feed-and-bleed stages. Each stage operates at a constant concentration, which increases from the first stage to the last. The concentration of the last stage is that of the concentrated product. A feed pump is needed to pump the feed stream from a holding tank to the first stage. Subsequent stages are fed from the preceding stages by small pressure differences.

Applications The types of applications by ultrafiltration may be classified into (1) concentration, (2) microsolute separation, and (3) macrosolute fractionation (Blatt, loc. cit.). Probably, concentration is the largest application. Most industrial applications belong to this category. As mentioned previously, microsolute separation can be achieved by the use of ligand binding or complexing with a macromolecule. The separation of free and protein-bound calcium is an example. Microsolute separations, such as salt removal and salt exchange, can be accomplished via ultrafiltration or its combination with dialysis. Macrosolute fractionation can be made by the use of membranes with various molecular-weight cutoff values or a tandem system in which the output of one cell is led into another with a progressively low molecular-weight cutoff membrane and so on.

Current large-scale industrial applications (Bailey, loc. cit.) include the following. (1) Electrophoretic paints: One of the most popular commercial uses of ultrafiltration has been the recovery of electrophoretic paints [Michaels, *Chemtech*, 36 (January 1981); and anonymous, *Chem. Eng.*, **87**, 57 (1980)]. Ultrafiltration is used to process the paint by retaining the polymer resins and pigment solids

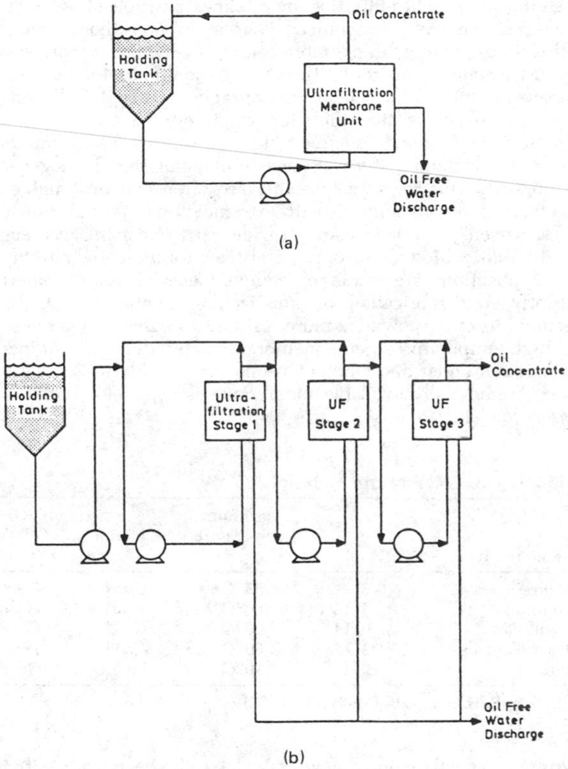

FIG. 17-29 Batch and continuous ultrafiltration operations. (*a*) Batch ultrafiltration. (*b*) Continuous ultrafiltration. [*Bailey,* Filtr. Sep., *213 (May–June 1977).*]

while allowing inorganic salts, water, and solvent to permeate through the membrane. The retained species are returned to the electropaint tank. The permeate is then used to rinse the freshly painted components as they emerge from the paint and to recover the drag-out excess paint. (2) Protein extraction in food and dairy industries: Large protein molecules from cheese, casein whey, or skim milk are concentrated via ultrafiltration (Michaels, 1981, loc. cit.). (3) Wastewaters containing starches and enzymes: Wastewaters from certain food processing, such as potato handling, contain dilute concentrations of starches, and certain effluents from the brewing industry contain enzymes, etc. Ultrafiltration has been employed to recover the starches and the enzymes and to produce acceptable effluents for discharge. (4) Textile-desizing water: Sizing materials such as starches and water-soluble polymers (polyvinyl alcohol) are often used to facilitate weaving processes. The woven cloth is later washed to remove the size, resulting in a dilute solution of the sizing material. Ultrafiltration can be used to recover this sizing material for reuse and to produce good-quality water permeate for discharge

TABLE 17-25 Applicability of Various Module Configurations for Ultrafiltration Applications*

Industrial application	Tubular module	Plate-and-frame module	Thin-channel module	Leaf-type module	Spiral-wound module	Hollow-fiber module
Electrodeposition paints	+	−	−	−	−	−
Protein extraction in food and dairy industries	+	+	+	+	−	−
Wastewater containing starches and enzymes	±	+	+	+	−	−
Textile-desizing water	−	+	+	+	±	−
Latex concentration	+	−	−	−	−	−
Oil-water emulsions	+	±	±	−	−	−
Waste-machining and metal-rolling emulsions	+	−	−	−	−	−
Wool-scouring effluent	+	−	−	−	−	−
Pulp-mill effluent	+	±	±	±	−	−
Alkaline cleaning of greasy or dirty metal parts	+	−	−	−	−	−

*Ouchiyama and Tanaka, *Int. Chem. Eng.*, **17,** 430 (1977); Bailey, *Filtr. Sep.*, 213 (May–June 1977). + = applicable; ± = limited applicability; − = not applicable.

or reuse. (5) Latex concentration: Concentrating dilute latex solutions from the wash-down during the manufacture and applications of various synthetic rubbers, containers, and reactor vessels, etc., has been proved to be successful with ultrafiltration. (6) Oil-water emulsions: Certain oil-water emulsions are used for repetitive drawing and metal-forming operations from which the components pass into the water rinse. Ultrafiltration is used here (*Chem. Eng.*, **87,** 57 (1980)] in a way similar to the electrophoretic-paint application described previously. (7) Waste-machining and metal-rolling emulsions: Ultrafiltration can be used to concentrate emulsified machining and metal-rolling oils from an initial 2 to 10 percent to a level of 25 to 50 percent. The concentrated emulsion can be disposed of by burning and the water permeate discharged to the drain. (8) Wool-scouring effluent: The wool-scouring effluent containing lanolin-type greases emulsified with detergent can be dewatered via ultrafiltration, often in conjunction with centrifugation. (9) Pulp-mill effluent: The high-molecular-weight lignosulfonates in pulp-mill effluents can be separated and concentrated by ultrafiltration. (10) Alkaline cleaning of greasy or dirty metal parts: Alkaline-solution baths are used to clean greasy or dirty metal parts. Ultrafiltration can be employed to remove the grease, oil, and dirty particles from the cleaning bath and to recover most of the cleaner as permeate. The applicability of various module configurations for these ultrafiltration uses is presented in Table 17-25.

Cost Estimate Ultrafiltration economics normally is expressed in terms of the cost for producing 1000 gal of permeate or for processing 1000 gal of feed. Klinkowski [*Chem. Eng.*, **85,** 164 (1978)] has developed an effective and simple estimate for the operating cost for 1000 gal of permeate, which is shown in Table 17-26. His estimates for capital and operating costs (less depreciation) are presented in Figs. 17-30 and 17-31 for whey-protein recovery and in Figs. 17-32 and 17-33 for the concentration of skim milk.

An economic analysis for concentrating latex wastewater has been reported by Bansal [*Ind. Water Eng.*, **13,** 6 (1976)]. For the recovery of sizing materials in textile industry, detailed ultrafiltration economics is available [Suchecki, *Text. Ind.*, **142,** 45 (1978)].

TABLE 17-26 Ultrafiltration Cost Estimate*

Assumptions	
1. Installed cost of membrane system	$70/ft^2
2. Membrane replacement cost	$10/ft^2
3. Membrane replacement period	1 year
4. Power cost	$0.02/kWh
5. Linear depreciation period	5 year
Installation cost (amortization) per 1000-gal permeate	$38.4/flux, in gal/(ft^2·day)
Membrane replacement per 1000-gal permeate	$27.4/flux, in gal/(ft^2·day)
Power per 1000-gal permeate	$12.2/flux, in gal/(ft^2·day)
Total operating cost per 1000-gal permeate	$78/flux, in gal/(ft^2·day)

*Klinkowski, *Chem. Eng.*, **85,** 164 (1978).

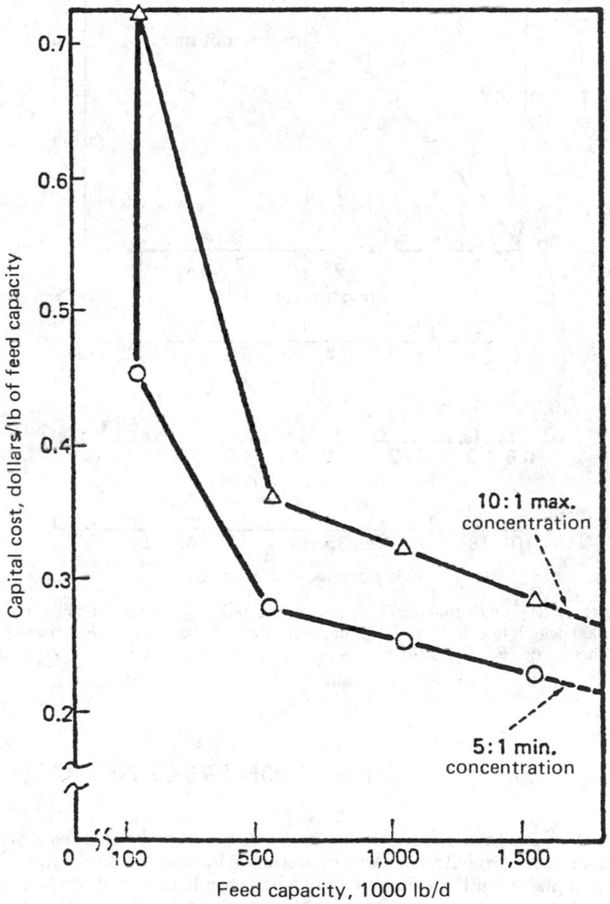

FIG. 17-30 Capital costs for whey-protein recovery. To convert dollars per pound to dollars per kilogram, multiply by 2.205; to convert 1000 pounds per day to kilograms per hour, multiply by 18.900. [*Klinkowski,* Chem. Eng., *85, 164 (1978).*]

Although a high circulation rate increases pumping energy consumption, it may reduce overall costs by increasing flux and so decreasing membrane replacement and fixed-cost components (Matz and Meitlis, loc. cit.). An optimization technique of flow conditions to minimize permeate product cost in laminar-channel ultrafiltration is available (Shen and Probstein, loc. cit.).

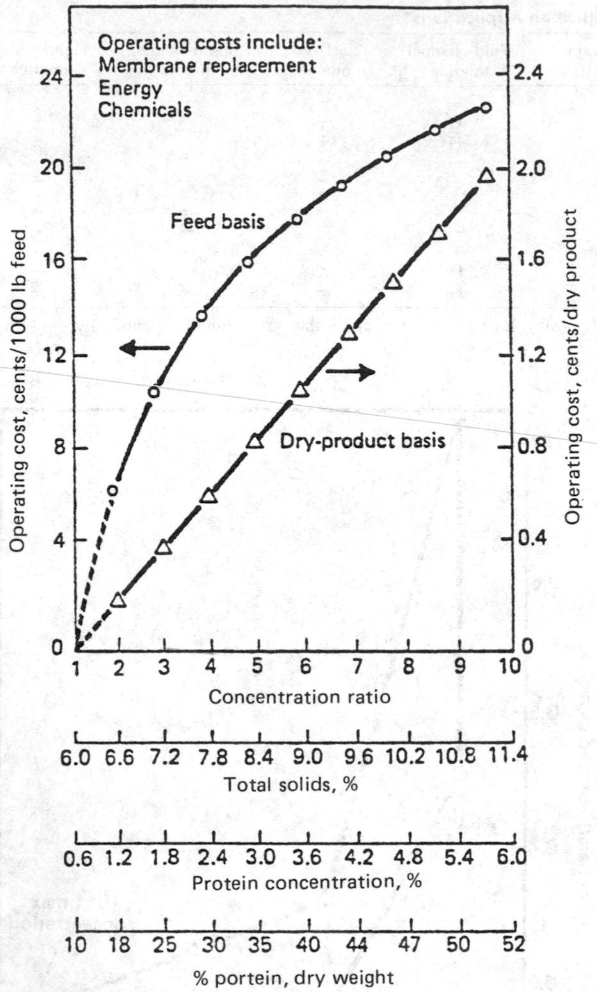

FIG. 17-31 Operating costs for whey-protein recovery. To convert dollars per 1000 pounds to dollars per kilogram, multiply by 2.205×10^{-3}. [*Klinkowski,* Chem. Eng., *85, 164 (1978).*]

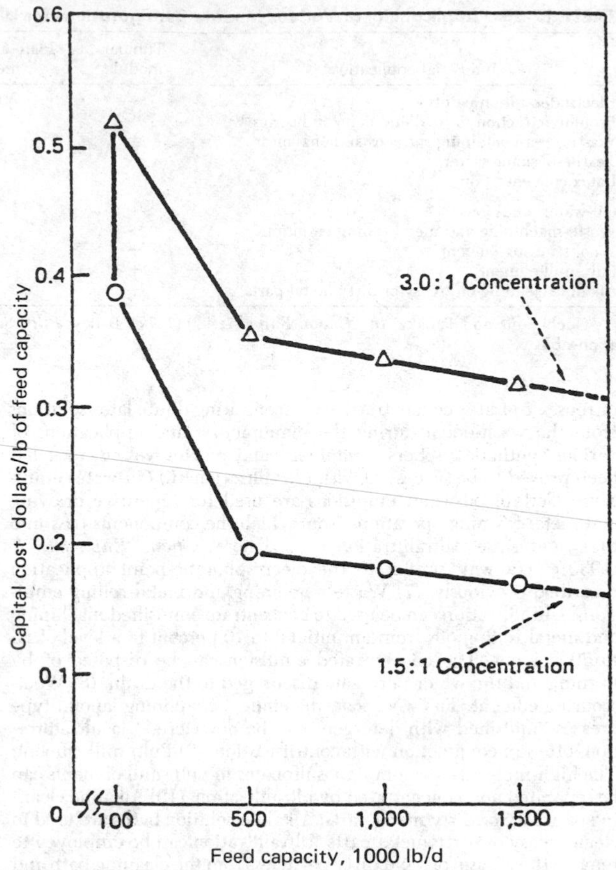

FIG. 17-32 Capital costs for concentrating skim milk. To convert dollars per pound to dollars per kilogram, multiply by 2.205; to convert 1000 pounds per day to kilograms per hour, multiply by 18.900. [*Klinkowski,* Chem. Eng., *85, 164 (1978).*]

SEPARATION PROCESSES BASED PRIMARILY ON ACTION IN A FIELD

Differences in mobilities of ions, molecules, or particles in an electric field can be exploited to perform useful separations. Primary emphasis is placed on the electrophoresis, dielectrophoresis, and electrodialysis processes. Analogous separation processes involving magnetic and centrifugal force fields are widely applied in the process industry (see Secs. 19 and 21).

THEORY OF ELECTRICAL SEPARATIONS

GENERAL REFERENCES: Newman, *Adv. Electrochem. Electrochem. Eng.*, **5**, 87 (1967); *Ind. Eng. Chem.*, **60**(4), 12 (1968).

For electrolytic solutions, migration of charged species in an electric field constitutes an additional mechanism of mass transfer. Thus the flux of an ionic species N_i in (g·mol)/(cm²·s) in dilute solutions can be expressed as

$$\mathrm{N}_i = -z_i u_i \mathcal{F} c_i \nabla E - D_i \nabla c_i + c_i v \qquad (17\text{-}77)$$

The ionic mobility u_i is the average velocity imparted to the species under the action of a unit force (per mole). v is the stream velocity, cm/s. In the present case, the electrical force is given by the product of the electric field ∇E in V/cm and the charge $z_i\mathcal{F}$ per mole, where $\mathcal{F}$ is the Faraday constant in C/g equivalent and z_i is the valence of the ith species. Multiplication of this force by the mobility and the concentration c_i [(g·mol)/cm³] yields the contribution of migration to the flux of the ith species.

The diffusive and convective terms in Eq. (17-77) are the same as in nonelectrolytic mass transfer. The ionic mobility u_i, (g·mol·cm²)/(J·s), can be related to the ionic-diffusion coefficient D_i, cm²/s, and the ionic conductance of the ith species λ_i, cm²/(Ω·g equivalent):

$$u_i = D_i/RT = \lambda_i/|z_i|\mathcal{F}^2 \qquad (17\text{-}78)$$

where T is the absolute temperature, K; and R is the gas constant, 8.3143 J/(K·mol). Ionic conductances are tabulated in the literature (Robinson and Stokes, *Electrolyte Solutions*, Academic, New York,

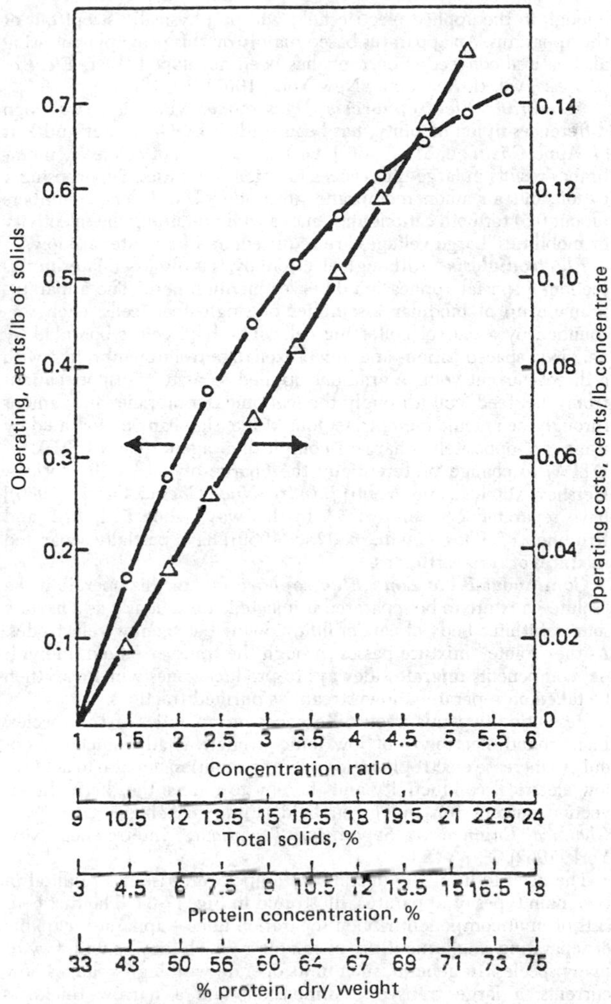

FIG. 17-33 Operating costs for concentrating skim milk. To convert cents per pound to cents per kilogram, multiply by 2.205. [*Klinkowski*, Chem. Eng., *85*, *164 (1978)*.]

1959). For practical purposes, a bulk electrolytic solution is electrically neutral.

$$\sum_i z_i c_i = 0 \qquad (17\text{-}79)$$

since the forces required to effect an appreciable separation of charge are prohibitively large.

The **current density** (A/cm^2) produced by movement of charged species is described by summing the terms in Eq. (17-80) for all species:

$$i = \mathcal{F} \sum_i z_i \mathbf{N}_i = -\kappa \, \nabla E - \mathcal{F} \sum_i z_i D_i \, \nabla c_i \qquad (17\text{-}80)$$

where the electrical conductivity κ in S/cm is given by

$$\kappa = \mathcal{F}^2 \sum_i z_i^2 u_i c_i \qquad (17\text{-}81)$$

In solutions of uniform composition, the diffusional terms vanish and Eq. (17-80) reduces to Ohm's law.

Conservation of each species is expressed by the relation

$$\partial c_i/\partial t = -\nabla \cdot \mathbf{N}_i \qquad (17\text{-}81a)$$

provided that the species is not produced or consumed in homogeneous chemical reactions. In two important cases, this conservation law reduces to the equation of convective diffusion:

$$(\partial c_i/\partial t) + \mathbf{v}\nabla \cdot c_i = D\,\nabla^2 c_i \qquad (17\text{-}82)$$

First, when a large excess of inert electrolyte is present, the electric field will be small and migration can be neglected for minor ionic components; Eq. (17-82) then applies to these minor components, where D is the ionic-diffusion coefficient. Second, Eq. (17-82) applies when the solution contains only one cationic and one anionic species. The electric field can be eliminated by means of the electroneutrality relation.

In the latter case the diffusion coefficient D of the electrolyte is given by

$$D = (z_+ u_+ D_- - z_- u_- D_+)/(z_+ u_+ - z_- u_-) \qquad (17\text{-}83)$$

which represents a compromise between the diffusion coefficients of the two ions. When Eq. (17-82) applies, many solutions can be obtained by analogy with heat transfer and nonelectrolytic mass transfer.

Because the solution is electrically neutral, conservation of charge is expressed by differentiating Eq. (17-80):

$$\nabla \cdot \mathbf{i} = 0 = -\kappa\,\nabla^2 E - \mathcal{F} \sum_i z_i D_i \,\nabla^2 c_i \qquad (17\text{-}84)$$

For solutions of uniform composition, Eq. (17-84) reduces to Laplace's equation for the potential:

$$\nabla^2 E = 0 \qquad (17\text{-}85)$$

This equation is the starting point for determination of the current-density distributions in many electrochemical cells.

Near an interface or at solution junctions, the solution departs from electroneutrality. Charges of one sign may be preferentially adsorbed at the interface, or the interface may be charged. In either case, the charge at the interface is counterbalanced by an equal and opposite charge composed of ions in the solution. Thermal motion prevents this countercharge from lying immediately adjacent to the interface, and the result is a "diffuse-charge layer" whose thickness is on the order of 10 to 100 Å.

A tangential electric field ∇E_t acting on these charges produces a relative motion between the interface and the solution just outside the diffuse layer. In view of the thinness of the diffuse layer, a balance of the tangential viscous and electrical forces can be written

$$\mu(\partial^2 v_t/\partial y^2) + \rho_e \,\nabla E_t = 0 \qquad (17\text{-}86)$$

where μ is the viscosity and ρ_e is the electric-charge density, C/cm^3. Furthermore, the variation of potential with the normal distance satisfies Poisson's equation:

$$\partial^2 E/\partial y^2 = -(\rho_e/\epsilon) \qquad (17\text{-}87)$$

with ϵ defined as the **permittivity** of the solution. [The relative dielectric constant is ϵ/ϵ_0, where ϵ_0 is the permittivity of free space; $\epsilon_0 = 8.8542 \times 10^{-14}$ C/(V·cm).] Elimination of the electric-charge density between Eqs. (17-86) and (17-87), with two integrations, gives a relation between ∇E_t and the velocity v_0 of the bulk solution relative to the interface.

$$\mu[v_t(\infty) - v_t(0)] = \epsilon\,\nabla E_t[E(\infty) - E(0)] \qquad (17\text{-}88)$$

or

$$v_0 = -(\epsilon\,\nabla E_t \zeta/\mu) \qquad (17\text{-}89)$$

The potential difference across the mobile part of the diffuse-charge layer is frequently called the **zeta potential,** $\zeta = E(0) - E(\infty)$. Its value depends on the composition of the electrolytic solution as well as on the nature of the particle-liquid interface.

The effects summarized by Eq. (17-89) form the basis of electrophoresis. For many particles, the diffuse-charge layer can be characterized adequately by the value of the zeta potential. For a spherical particle of radius r_0 which is large compared with the thickness of the diffuse-charge layer, an electric field uniform at a distance from the particle will produce a tangential electric field which varies with position on the particle. Laplace's equation [Eq. (17-85)] governs the distribution of potential outside the diffuse-charge layer; also, the Navier-Stokes equation for a creeping-flow regime can be applied to the velocity distribution. On account of the thinness of the diffuse-charge layer, Eq. (17-89) can be used as a local boundary condition, accounting for the effect of this charge in leading to movement of the particle relative to the solution. The result of this computation gives the velocity of the particle as

$$v = \epsilon\zeta \nabla E/\mu \qquad (17\text{-}90)$$

and it may be convenient to tabulate the mobility of the particle

$$U = v/\nabla E = \epsilon\zeta/\mu \qquad (17\text{-}91)$$

rather than its zeta potential. Note that this mobility gives the velocity of the particle for unit electric field rather than for unit force on the particle.

ELECTROPHORESIS

Electrophoretic Mobility Macromolecules move at speeds measured in tenths of micrometers per second in a field (gradient) of 1 V/cm. Larger particles such as bubbles or bacteria move up to 10 times as fast because U is usually higher. To achieve useful separations, therefore, voltage gradients of 10 to 100 V/cm are required. High voltage gradients are achieved only at the expense of power dissipation within the fluid, and the resulting heat tends to cause undesirable convection currents.

Several devices are available commercially to measure mobility. One of these (Zeta-Meter Inc., New York) allows direct microscopic measurement of individual particles. Another allows measurement in more concentrated suspensions (Numinco Instrument Corp., Monroeville, Pa.). The state of the charge can also be measured by a streaming-current detector (Waters Associates, Inc., Framingham, Mass.). For macromolecules, more elaborate devices such as the Tiselius moving-boundary apparatus are used.

Mobility is affected by the dielectric constant and viscosity of the suspending fluid, as indicated in Eq. (17-91). The ionic strength of the fluid has a strong effect on the thickness of the double layer and hence on ζ. As a rule, mobility varies inversely as the square root of ionic strength [Overbeek, *Adv. Colloid Sci.*, **3**, 97 (1950)].

Modes of Operation There is a close analogy between sedimentation of particles or macromolecules in a gravitational field and their electrophoretic movement in an electric field. Both types of separation have proved valuable not only for analysis of colloids but also for preparative work, at least in the laboratory. Electrophoresis is applicable also for separating mixtures of simple cations or anions in certain cases in which other separating methods are ineffectual.

Electrodecantation or Electroconvection This is one of several operations in which one mobile component (or several) is to be separated out from less mobile or immobile ones. The mixture is introduced between two vertical semipermeable membranes; for separating cations, anion membranes are used, and vice versa. When an electric field is applied, the charged component migrates to one or another of the membranes; but since it cannot penetrate the membrane, it accumulates at the surface to form a dense concentrated layer of particles which will sink toward the bottom of the apparatus. Near the top of the apparatus immobile components will be relatively pure. Murphy [*J. Electrochem. Soc.*, **97**(11), 405 (1950)] has used silver–silver chloride electrodes in place of membranes. Frilette [*J. Phys. Chem.*, **61**, 168 (1957)], using anion membranes, partially separated H^+ and Na^+, K^+ and Li^+, and K^+ and Na^+. Unfortunately no simple electrodecantation apparatus is available for bench-scale testing. A rather complex device described by Polson and Largier [in Alexander and Block (eds.), *Analytical Methods of Protein Chemistry*, vol. I, Pergamon, New York, 1960] is available commercially (Quickfit Reeve Angel, Inc., Clifton, N.J.).

Countercurrent Electrophoresis A mixture of mobile species can be split into two fractions by the electrical analog of elutriation. In such countercurrent electrophoresis, sometimes termed an ion still, a flow of the suspending fluid is maintained parallel to the direction of the voltage gradient. Species which do not migrate fast enough in the applied electric field will be physically swept out of the apparatus. An apparatus based mainly on this principle but using also natural convection currents has been developed (Bier, *Electrophoresis*, vol. II, Academic, New York, 1967).

Membrane Electrophoresis This mode, which is based upon differences in ion mobility, has been studied by Glueckauf and Kitt [*J. Appl. Chem.*, **6**, 511 (1956)]. Partial exclusion of coions by membranes results in large differences in coion mobilities. Superposing a cation and an anion membrane gives high transference numbers (about 0.5) for both cations and anions while retaining the selectivity of mobilities. Large voltages are required, and flow rates are low.

Electrodialysis Although electrodialysis is discussed later in this section, a special application deserves mention here. The apparatus is made up of modular assemblies of single feed cells, each surrounded by a pair of collecting cells. The feed cell is bounded by narrowly spaced cation- and anion-exchange resin membranes, with a direct-current voltage gradient applied so as to transport cations from the feed cell through the cationic membrane and anions through the anionic membrane. Ions of like sign can be separated by using an oppositely charged complexing agent (e.g., EDTA or DTPA) to change preferentially the charge of one of the species. Hershey, Mitchell, and Webb [*J. Inorg. Nucl. Chem.*, **28**, 645 (1966)] have separated CS^+ and SR^{++} in this way, while Bril, Bril, and Krumholz [*J. Phys. Chem.*, **63**, 256 (1959)] have partially separated mixtures of rare-earth ions.

Continuous-Flow Zone Electrophoresis In this operation the "solute" mixture to be separated is injected continuously as a narrow source within a body of carrier fluid flowing between two electrodes. As the "solute" mixture passes through the transverse field, individual components migrate sideways to produce zones which can then be taken off separately downstream as purified fractions.

Resolution depends upon differences in mobilities of the species. Background electrolyte of low ionic strength is advantageous, not only to increase electrophoretic (solute) mobilities, but also to achieve low electrical conductivity and thereby to reduce the thermal-convection current for any given field [Finn, in Schoen (ed.), *New Chemical Engineering Separation Techniques*, Interscience, New York, 1962].

The need to limit the maximum temperature rise has resulted in two main types of apparatus, illustrated in Fig. 17-34. The first consists of multicomponent ribbon separation units—apparatus capable of separating small quantities of mixtures which may contain few or many species. In general, such units operate with high voltages, low currents, a large transverse dimension, and a narrow thickness between cooling faces. Numerous units developed for analytic chemistry, generally with filter-paper curtains but sometimes with gran-

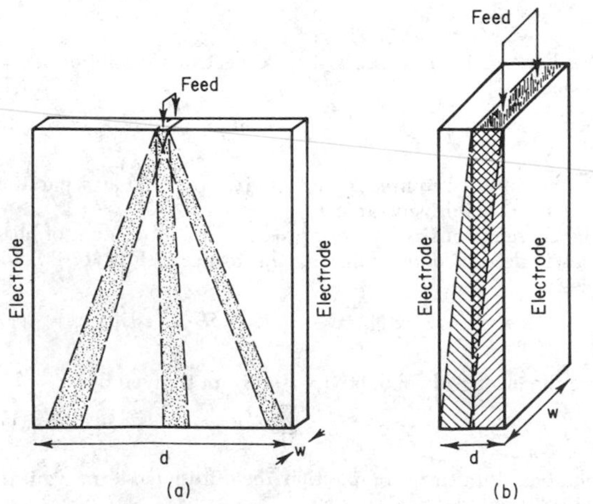

FIG. 17-34 Types of arrangement for zone electrophoresis or electrochromatography. (*a*) Ribbon unit, with $d > w$; cooling at side faces. (*b*) Block unit, with $w > d$; cooling at electrodes.

ular "anticonvectant" packing, are of this type. The second type consists of block separation units—apparatus designed to separate larger quantities of a mixture into two (or at most three) species or fractions. Such units generally use low to moderate voltages and high currents, with cooling by circulation of cold electrolyte through the electrode compartments. Scale-up can readily be accomplished by extending the thickness dimension *w*.

Both types of units have generally been operated in trace mode; that is, "background" or "elutant" electrolyte is fed to the unit along with the mixture to be separated. A desirable and possible means of operation for preparative applications is in bulk mode, in which one separated component follows the other without background electrolyte being present, except that other ions may be required to bracket the separated zones. Overlap regions between components should be recycled, and pure components collected as products.

For block units, the need to stabilize flow has given rise to a number of distinct techniques.

Free flow. Dobry and Finn [*Chem. Eng. Prog.*, **54**, 59 (1958)] used upward flow, stabilized by adding methyl cellulose, polyvinyl alcohol, or dextran to the background solution. Upward flow was also used in the electrode compartments, with cooling efficiency sufficient to keep the main solution within 1°C of entering temperature.

Density gradients to stabilize flow have been employed by Philpot [*Trans. Faraday Soc.*, **36**, 38 (1940)] and Mel [*J. Phys. Chem.*, **31**, 559 (1959)]. Mel's Staflo apparatus [*J. Phys. Chem.*, **31**, 559 (1959)] has liquid flow in the horizontal direction, with layers of increasing density downward produced by sucrose concentrations increasing to 7.5 percent. The solute mixture to be separated is introduced in one such layer. Operation at low electrolyte concentrations, low voltage gradients, and low flow rates presents no cooling problem.

Packed beds. A packed cylindrical electrochromatograph 9 in (23 cm) in diameter and 48 in (1.2 m) high, with operating voltages in the 25- to 100-V range, has been developed by Hybarger, Vermeulen, and coworkers [*Ind. Eng. Chem. Process Des. Dev.*, **10**, 91 (1971)]. The annular bed is separated from inner and outer electrodes by porous ceramic diaphragms. The unit is cooled by rapid circulation of cooled electrolyte between the diaphragms and the electrodes.

An interesting modification of zone electrophoresis resolves mixtures of ampholytes on the basis of **differing isoelectric points** rather than differing mobilities. Such **isoelectric spectra** develop when a pH gradient is established parallel to the electric field. Each species then migrates until it arrives at the region of pH where it possesses no net surface charge. A strong focusing effect is thereby achieved [Kolin, in Glick (ed.), *Methods of Biochemical Analysis*, vol. VI, Interscience, New York, 1958].

ELECTRODIALYSIS

In electrodialysis, the concentration and/or composition of electrolyte solutions is altered as a result of electromigration through membranes in contact with these solutions. Electrodialysis units are stacks of narrow compartments through which the feed solution is pumped (Fig. 17-35). These compartments are separated by alternating cation-exchange and anion-exchange membranes which are selectively permeable to positive and negative ions respectively. The modern development of an electrodialysis technology has led to the ready availability of mechanically sturdy, highly conductive cation- or anion-selective membranes, which in turn have opened many new possibilities for the design of processes requiring these types of separators. The terminal compartments are bounded by electrodes, for passing direct current through the whole stack. The compartments can be stacked horizontally or vertically; the assembly resembles a filter press. Plastic-mesh or other types of separators are inserted in the solution compartments to keep the cation- and anion-exchange membranes apart (compartment thickness is usually held to about 1 mm or less in order to reduce ohmic loss) and to promote mixing.

The situation before passage of current is schematically shown in Fig. 17-35*a*. When the electrodes are connected to a dc source, ion migration begins, as shown schematically in Fig. 17-35*b* for a group of compartments well within the stack. In each compartment posi-

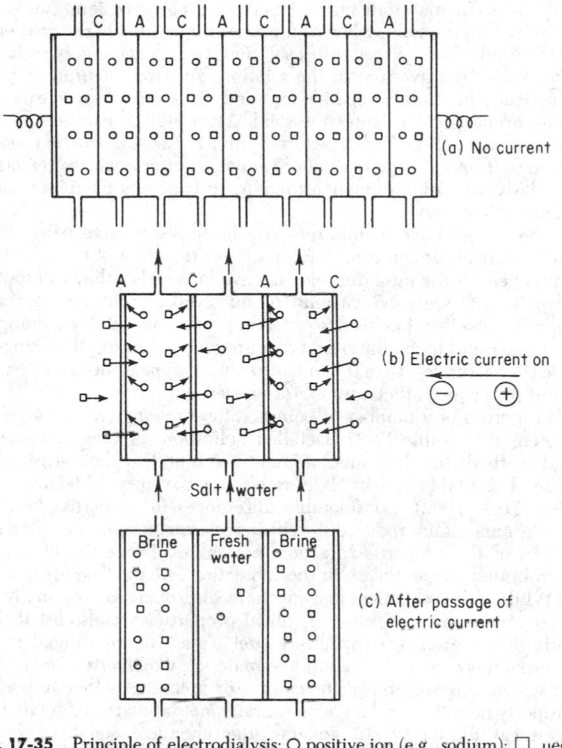

FIG. 17-35 Principle of electrodialysis; O positive ion (e.g., sodium); □, negative ion (e.g., chloride); *C* and *A* cation- and anion-permeable membrane respectively. Ion migration under the action of electric current causes salt depletion in alternate compartments and salt enrichment in adjacent ones. (*Spiegler*, Salt Water Purification, *2d ed., Plenum Press, New York, 1977.*)

tive ions (cations) travel from right to left; negative ones (anions), in the opposite direction. For the center compartment, the anion-permeable membrane on the right does not admit cations from the right, and the cation-permeable membrane on the left similarly acts as a barrier for the negative ions from the left. As a result, the electrolyte concentration decreases in the center compartment and increases in the neighboring compartments, as shown in Fig. 17-35*c*.

The most common application of electrodialysis has been the splitting of saline water into a more concentrated and more dilute portion [Wilson (ed.), *Demineralization by Electrodialysis*, Butterworth, London, 1960; Shaffer and Mintz in Spiegler and Laird (eds.), *Principles of Desalination*, 2d ed., Academic, New York, 1980]. In industrial electrodialysis installations, which contain from 10 to hundreds of compartments between one pair of electrodes, the passage of electric current thus creates fresh water ("diluate") and brine ("concentrate") in neighboring cells. In other words, half of the cells carry partly desalted water, and half carry brine. The solutions in the electrode compartments are contaminated with the products of the electrode reactions which take place as a result of the passage of the electric current. The largest desalting plant in operation in 1979 produced about 15,000 m^3 (4 million U.S. gal) per day of desalted water from brackish feedwater.

Electrodialysis is also being used to concentrate seawater prior to table-salt recovery and to demineralize seawater fully [Tsunoda and Kato, *Desalination*, **3**, 66 1967); and Nishiwaki, in Lacey and Loeb (eds.), *Industrial Processing with Membranes*, Wiley, New York, 1972, chap. 6]. High-temperature processing (up to 70°C) looks particularly promising for seawater treatment [McRae, Glass, Leitz, Clarke, and Alexander, *Desalination*, **4**, 236 (1968); and Forgacs, Koslowsky, and Rabinowitz, ibid., **5**, 349 (1968)].

Ion-Selective Membranes The selective permeability of membranes can be explained by the fact that they are ion-exchange materials. In such materials either positive or negative ions can easily move within the solid, whereas the ions of opposite charge are bound to the solid. When placed into a solution, cation exchangers exchange freely with positive ions in the solution, but from a *dilute* solution practically no mobile negative ions and no positive ions in excess of those exchanged can enter the solid. As a result, a cation-exchange membrane acts as a barrier for anions. Electric current passes through it only by the motion of cations from one face to the other. Similarly, an anion-exchange membrane conducts by migration of negative ions only.

The ion-selective membranes are the most sensitive parts of the unit. Their lifetime is considerably shorter than that of the rest of the unit, even for the most durable ones available. Also, they can foul by deposition of scale, which must be dissolved or removed mechanically. Much effort has been invested in producing suitable membrane materials, and many different types are on the market; they range in thickness from less than 0.1 to 1 mm and in appearance from parchment paper to synthetic upholstery covering.

Properties of a number of common ion-exchange membranes are presented in Table 17-27. Detailed definitions of these parameters and methods for their measurement are found in the source reference of the table and in Helfferich's *Ion Exchange* (McGraw-Hill, New York, 1962). Considerable differences in properties between batches manufactured at different times are common; the tabulated values of the properties are nominal values. Properties of neutral membranes (N) are given in the subsection "Membrane Processes."

While electrochemical and mechanical properties are usually the most important ones, some chemical properties are also listed. The ratio of ion-exchange capacity to gel water is proportional to the internal molality of the membrane material, which often determines its selective permeability for cations or anions. Another important property of an ion-exchange membrane, not tabulated in detail, is its chemical durability. In general, the chemical stability of ion-exchange membranes is similar to that of the more familiar granular ion-exchange resins. The common membranes have adequate resistance to acids and bases. Oxidation resistance is important in some applications, for example, for the membrane next to the anode in a stack, and for this reason some manufacturers have offered products that have particularly good oxidation resistance. Some oxidation-resistant membranes have rather poor resistance to alkalies. Some, but not all, anion-exchange membranes are damaged by strongly alkaline solutions, especially at high temperatures.

The resistance of ion-selective membranes is of considerable importance in determining ohmic losses, particularly when the solutions have relatively low resistance, e.g., in the electrodialytic concentration of seawater prior to table-salt manufacture, as practiced in Japan. The resistance is tabulated in terms of unit area of membrane sheet rather than specific resistance. Measurements are made with alternating current (60 or 1000 Hz) normal to the membrane surface; the apparent dc resistance is often considerably higher than the ac resistance. Different methods for the measurement of dc resistance have been developed [Berg, Brun, Schmitt, and Spiegler, in Liberti and Helfferich (eds.), *Mass Transfer and Kinetics of Ion Exchange*, M. Nijhoff, The Hague, 1983; and Guillou, Guillou, and Buvet, *Membranes à perméabilité sélective*, Centre National de la Recherche Scientifique, Paris, 1969, p. 131].

It is known that the resistances of membranes vary with the type of counterion. In *cation exchangers,* the mobile positive ions are the counterions, and the negative mobile ions are the coions. The opposite ionic charges apply to *anion-exchanger* nomenclature. The higher the charge of the counterion, the higher in general is the resistance. The hydrogen form of highly dissociated ("strong-acid") cation-exchange materials has, in general, a 5 to 10 times lower resistance than the sodium form; the hydroxyl form of highly dissociated ("strong-base") anion-exchange materials has usually a 2 to 5 times lower resistance than the chloride form. The resistance decreases with increasing concentration of the equilibrating solution because of "invasion" of the membrane by ions from that solution.

The **permselectivity** Ψ (also called "selectivity") of a membrane is a quantitative measure of the ability of cation- or anion-exchange materials to conduct selectively by positive or negative ions, respectively. It is given by

$$\Psi_c \equiv (\bar{t}_+ - t_+)/t_- \text{ for cation-exchange membranes} \quad (17\text{-}92a)$$

$$\Psi_a \equiv (\bar{t}_- - t_-)/t_+ \text{ for anion-exchange membranes} \quad (17\text{-}92b)$$

where $\bar{t}$ and t are the ionic transport numbers in the membrane and solution respectively, and subscripts + and − stand for cations and anions respectively in the solution of a single electrolyte. The transport number of an ion is the fraction of the total current carried by that ion. It can be determined by direct transport experiments (Berg, Brun, Schmitt, and Spiegler, loc. cit.) or from measurements of membrane potentials, i.e., potential differences E_m (volt), between two solutions of the same electrolyte on the right and left side of the membrane, from the relation

$$E_m = \frac{RT}{Z\mathcal{F}}(1 - 2\bar{t}) \ln \frac{a''}{a'} \quad (17\text{-}93)$$

where a'' and a' are the ionic activities of the counterions on the right and left side of the membrane respectively; $\bar{t}$ and Z the transport number and valence (positive for cations and negative for anions) of the counterions in the membrane respectively; R the universal gas constant, 8.31 (W·s)/(K·g·mol); T the temperature, K; and $\mathcal{F}$ the Faraday constant, 0.965×10^5 C/g equivalent. In both methods, corrections have to be made to account for the electroosmotic transport of water with the ions [Lee, Maskell, and Tye, in Meares (ed.), *Membrane Transport Processes*, Elsevier, Amsterdam, 1976]. Literature values for transport numbers obtained by the two methods do not always agree; values obtained by the direct-transfer method reflect mass transport in an electrodialysis stack more closely than those from the membrane-potential method. Transport numbers for coions are a measure of the nonselectivity of the membrane. For 1N solutions outside, these are highest where H^+ is the anion-exchanger coion or OH^- is the cation-exchanger coion. Next highest are those for a univalent coion with a polyvalent counterion, for instance, for Cl^- with Mg^{++} in a cation-exchange membrane. When the coion valence is the same as (or greater than) the counterion valence, its transport number is generally below 0.05 provided the equilibrating solutions are below 0.1 N.

Table 17-27 includes also some information on *mechanical properties* of the membranes. The strength is usually measured in terms of the **Mullen burst test** (ASTM D774-46). The tabulated data refer to new membranes; like many other synthetic polymers, ion-exchange membranes undergo aging effects which reduce their strength and often also their permselectivity and/or conductivity. Therefore, membrane attrition, necessitating replacement after periods of operation generally of several years (depending on membrane type and application), is an important factor in the cost of electrodialysis. Anion-permeable ("anion-exchange") membranes are usually more sensitive and also more expensive than cation-permeable membranes. Prices (per square foot) in 1979 for moderate quantities of cation-exchange membranes were of the order of $2, and $6 for anion-exchange membranes. This corresponds to about $22 and $65 per square meter respectively. Prices (per square foot) of the membranes vary from manufacturer to manufacturer. The prices of cation-exchange and anion-exchange membranes in quantities of 1000 membranes or more are frequently quoted.

Because of the higher attrition of anion-exchange membranes and their vulnerability to fouling and scaling, attempts have been made to replace them by neutral (nonselective) membranes, thus trading off the chemical and mechanical disadvantages of anion-exchange membranes for higher power costs. For instance, for solutions of KCl ($t_+ = 0.5 = t_-$), passage of 1 Faraday shifts 1 mol of salt from diluate to concentrate when alternating ideally selective cation- and anion-exchange membranes are used (Fig. 17-35), while only ½ mol is shifted with alternating ideal cation-exchange and neutral membranes. Therefore, to achieve the same production rate in cation-exchange–neutral-membrane stacks as in regular electrodialysis stacks (Fig. 17-35), twice the current is needed. The voltage is then also nearly twice; hence power costs quadruple. Information on cation-exchange–neutral-membrane stacks was obtained before anion-exchange membranes became available (Prausnitz and Reitstoetter, *Elektrophorese, Elektroosmose, Elektrodialyse,* Steinkopf, Dresden, 1931), but developments in electrodialysis equipment design since 1950 have again attracted attention to this process modification, often called "transport depletion" (Lacey and Loeb, *Industrial Processing with Membranes,* Wiley, New York, 1972).

Equipment Design The design of electrodialysis stacks bears some similarity to that of filter presses. Alternate membrane sheets are separated from each other by sheets of plastic mesh or other nonconductive spacer materials which provide a multiple-turn path for the solutions flowing past the membrane surfaces. This principle is illustrated in Fig. 17-36, which shows spacers for the freshwater and brine compartments in a plane parallel to the membranes. The thickness of these spacers, usually 1 mm or less, represents a compromise between expenditure of electric energy (Joule heat) and of the pumping energy. Four holes *A, B, C, D* punched in each spacer match holes punched into the membranes and into the electrodes. When the alternating layers of membranes and spacers are placed on top of each other and compressed tightly, these holes form liquid conduits. Feed enters the diluate compartments only through conduit *A* and leaves only through *C*. Similarly, feed for the concentrate compartments enters only through conduit *B* and leaves only through *D*. Diluate and concentrate leave the compartments through separate conduits *C* and *D* respectively. All practical units use this or similar distribution systems. In many units, however, the central part of the compartment contains a plastic network made of expanded PVC sheet which serves essentially the same purpose as the zigzag path shown in Fig. 17-36.

Assembly and disassembly of stacks is manual. For this reason, as well as considerations of the strength of the stack materials, the assembly, membranes, and spacers are placed between a pair of electrodes, and compressed by heavy end plates as in filter presses. The membrane dimensions are usually 2 by 1 m or less. Various materials for electrodes are in use, for example, graphite and stainless steel, which are gradually attacked and must be replaced. Platinum-coated metals (e.g., titanium, tantalum, or zirconium), with a life of several years, are now frequently used. Since the anodes are subject to greater danger of corrosion, platinum-coated materials are favored for the construction of anodes. Hastelloy or similar materials are sometimes used for cathodes, but since in many units polarity is reversed periodically to counteract the effects of polarization and to remove scale deposits, the use of platinum-coated metals for *both* electrodes is common. Often several small subassemblies ("packs") containing about 50 cell pairs (100 membranes) are used. As many as 10 of these packs have been placed into a single press. A single set of electrodes may be used for the entire assembly.

In order to decrease the mass-transfer resistances at the membrane-solution interfaces, liquid-flow velocities in the stacks are in general as high as 10 to 100 cm/s, depending on the specific application. In some modifications of electrodialysis flow velocities are low, e.g., in the "electrosorption" process in which *sealed* membrane sandwiches, each consisting of a cation, an anion-exchange membrane, and a spacer, are the fundamental desalting unit [Lacey and Lang, *Desalination*, **2**, 387 (1967); Kederu, Cohen, Warshawsky, and Cahana *Desalination,* **46**, 291 (1983)]. Ions from the external solution first migrate into the sandwich under the influence of a driving voltage; after some concentration has taken place, the voltage is reversed and the sandwich is "regenerated" in a cyclical process. This modification is similar to the process for preconcentration of seawater, used in Japan, in which the concentrate in the sandwiches is withdrawn. The slow electroosmotic water flow which accompanies the ion flow acts as a continuous flush. In the desalination of less concentrated solutions, however, the liquid-flow velocities have to be kept high in order to promote efficient mass transfer at the membrane-solution interfaces. At these velocities the degree of demineralization achieved in a single pass is usually less than required. In small installations, the diluate may be recirculated until it is sufficiently demineralized in a batch process. In large plants, it is more common to arrange two or more stages in series. Complete physical separation of these stages is not always necessary. Several groups of cells, each with its own set of electrodes, may be contained within the same clamping press. The flow of concentrate between stages may be either cocurrent or countercurrent with respect to the diluate; the concentrate may flow through the stages in series, or fresh raw water may be introduced separately into each stage. It is seen that many flow variations are possible. General methods for the design of multistage plants are available; the pros and cons of different flow schemes have been discussed in detail in the literature [Mason and Kirkham, *Chem. Eng. Prog. Ser.,* **55**(24), 173 (1959); Shaffer and Mintz, loc. cit.; Wilson, loc. cit.]. Single stacks producing up to 1000 m^3 (about 264,000 U.S. gal) diluate per day are in operation. For yet larger production rates, parallel arrangement of stacks rather than scale-up of single stacks is considered preferable because of considerations of the stacks' dimensional stability as well as ease of dismantling and cleaning the components.

Power Consumption and Process Optimization To optimize the electrodialysis process or to compare different methods of operation (e.g., continuous to batch recirculation methods), electrodialysis stacks are often simply represented as more or less complex networks of resistors: power consumption is balanced against investment and maintenance costs. In the development of these analyses, two stages can be distinguished which are designated as "ohmic" and "nonohmic." The ohmic analysis assumes proportionality between current and voltage in the electrodialysis stack. In fact, however, resistance increases considerably with increasing current densities. Therefore, attempts have been made to refine the ohmic analyses by inclusion of polarization and other factors leading to nonlinear current-voltage curves. This is important because some demineralization operations are performed at high current densities when resistance is nonohmic. For viscous solutions, this nonohmic approach is even more important than for saline water.

Using the method of ohmic analysis, the dc power dissipation in a given stack of constant resistance $\mathbf{R}$ at current I is $\mathbf{R}I^2$. The amount of salt shifted and hence the amount of fresh water produced are proportional to I. Therefore, the dc power dissipation *per gallon of diluate produced* is proportional to I. Fixed costs per gallon of diluate produced, on the other hand, are inversely proportional to I. Therefore, the total cost of unit amount of diluate produced C_t is

$$C_t = aI + (b/I) + g \tag{17-94}$$

where a, b, and g are taken as constants for a given stack if ohmic analysis applies. The first term represents electric power costs, the second the fixed charges, and the third unit costs which are independent of the total production, such as the cost of pretreatment chemicals. By differentiation of Eq. (17-94) to find minimum product cost, the optimum current is seen to be

$$I_{opt} = (b/a)^{1/2} \tag{17-95}$$

By substituting this value in the equation for the power cost C_t, it is found that for most economical operation the first two terms are to be equal (power costs = fixed costs). This conclusion is known as *Kelvin's law* (Grant, Ireson, and Leavenworth, *Principles of Engineering Economy,* 6th ed., Wiley, New York, 1976, p. 208).

This optimum condition can almost never be met, however, because polarization phenomena set an upper limit to permissible current densities. It is possible to increase the permissible current density by increasing the pumping rate, but mechanical-design considerations (e.g., membrane strength) have in the past set a limit to pumping rates also. Thus the polarization limitation, rather than the results of economic optimization, frequently controls the operating current density in practical electrodialysis installations. The current densities used in the desalting of brackish waters containing up to 5000 ppm dissolved solids generally lie between 6 and 20 mA/cm^2 (5.6 and 18.6 A/ft^2).

To calculate the stack resistance in the ohmic range from the solution concentration, it is necessary to integrate over the whole flow

TABLE 17-27 Properties of Commercial Ion-Exchange Membranes*

Manufacturer†	Chemical			Electrochemical		Mechanical				Remarks
	Membrane type	Ion-exchange capacity, meq/g	Gel water, %, dry basis	Area resistance, $\Omega \cdot cm^2$	Selectivity	Strength, wet	Reversible drying, dimensional stability	Nominal thickness, mil, wet	Size available	
AMF Incorporated										
Homogeneous, polyethylene base				1000 Hz ~, 0.6*N* KCl	Voltage ratio, 0.5/1.0*N* KCl	Mullen burst, lbf/in^2				
C-60	C	1.5 ± 0.2	40 ± 5	5 ± 2	80 ± 5	45 ± 5	Reversible, 10–13% linear expansion on rewetting	12	Rolls, 44 in wide	
C-100	C	1.3 ± 0.2	22 ± 7	7 ± 2	93 ± 2	60 ± 5		6		
A-60	A	1.6 ± 0.3	30 ± 5	6 ± 2	80 ± 4	45 ± 5	Reversible, 12–15% linear expansion on rewetting	12	Rolls, 44 in wide	
A-100	A	1.5 ± 0.3	20 ± 5	8 ± 2	90 ± 3	50 ± 5		7		
Fluorocarbon base										
C-311	C	0.65 ± 0.1	25 ± 8	3.1 ± 1	80 ± 5	100 ± 10	Reversible, 12–15% linear expansion on rewetting	12	Rolls, 44 in wide	Outstanding oxidation resistance
C-313	C	0.65 ± 0.1	25 ± 8	2.0 ± 1	85 ± 5	55 ± 5		6		
C-322	C	0.90 ± 0.1	35 ± 10	2.5 ± 1	75 ± 5	90 ± 5		0		
Asahi Chemical Industry Co., Ltd.				In seawater	Transport no. in seawater	kg/cm^2				
Reinforced	C			2.65	0.95	2.0	Reversible	7		
	A			2.25	0.95	2.0		7		
	N			5.71		2.5		6		
Asahi Glass Company, Ltd.										
Fabric-reinforced				1000 Hz ~, 0.5*N* NaCl	Voltage ratio, 0.5/1.0*N* NaCl	Mullen burst, kg/cm^2				
CMV	C	3.0–4.0	55–65	2.5–3.5	85–89	6–8	Reversible, dimensional stability better than 0.2%	5–6	1 × 2 m	Rejects multivalent positive ions
AMV	A	2.6–3.1	32–43	3.0–4.5	84–90	4–7		4–5.5	1 × 2 m	
ASV	A			3.5–5.0	87	4–7		4–6	1 × 2 m	
DMV	A			1.5–2.0		4–7		6–9	1 × 2 m	
Ionac Chemical Division, Sybron Corp.				ac, 1.0*N* NaCl	Corrected voltage ratio, 0.5/1.0*N* NaCl	lbf/in^2				
Heterogeneous, fabric reinforced										
MC-3142	C	1.06	~20	3.4	94	185	Reversible, dimensional stability good	6	Sheets, 40 × 120 in	Cut sheets, full width, also available; MC-3470 also as 30 × 96-in pieces
MC-3470	C	1.05		4.8	96	190		13–14		
MA-3148	A	0.96	~20	1.7	90	190		7		
MA-3475R	A	1.13		5.2	99	200		14–15		
DM-12	A		23	2–4	85–90	140–150		4.5		

Ionics, Incorporated				0.1*N* NaCl	Hittorf, 0.6*N* NaCl					
Fabric-reinforced										
CR-61 AZGG	C	2.1	47 (wet)	30	85	300	Cracks on drying; dimensional stability good	40	36 × 40 in	
CR-61 AZL	C	2.7	46 (wet)	11	90	115		23	18 × 40 in	
CR-61 CZL	C	2.7	40 (wet)	11	93	115		24	18 × 40 in	
CR-70	C	4.0	28 (wet)	5 (2*N* NaOH)	90	40		18	18 × 40 in	
AR-111 BZL	A	1.8	43 (wet)	11	90	125	Cracks on drying	24	18 × 40 in	
AR-111 EZL	A	1.7	36 (wet)	11	90	125		25	18 × 40 in	
AR-102	A	2.4	35 (wet)	6	97	130		24	18 × 40 in	
Negev Institute for Arid Zone Research					Voltage ratio, 0.05/0.10*N* KCl	Mullen burst, lbf/in^2				
Homogeneous, polyethylene-based										
PE-C-1	C	1.2–1.6	30–40	5–10	93			12		
PE-A-1	A	1.1–1.3	30–40	5–10	91			12		
Cellulosic										
CL-C-1	C	1.6	~50	3–9	85–92			8		
CL-A-1	A	3.5	~40	5–18	82–92			11		
Tokuyama Soda Co.				0.5*N* NaCl	Transport no. in seawater	kg/cm^2				
Fabric-reinforced										
CL-25T	C	1.5–1.8	30–40	2.7–3.2	>0.98	3–4		6–7	1 × 1.5 m	Electroosmotic water transference; CL-25T and AV-4T, 4.8 and 3.0 mol per Faraday respectively
CH-2T	C	1.8–2.3	40–50	1.6–2.0	>0.98	3–4		6–8		
C66-5T	C	2.2–2.6	35–45	1.3–1.8	>0.98	2–4		6–8		
AV-4T	A	1.5–2.0	20–30	3.0–4.0	>0.98	6–7		5.5–6.5		
AF-4T	A	1.8–2.5	25–35	1.8–2.5	>0.98	6–7		6–8		
AVS-4T	A	1.5–2.0	25–30	3.7–4.7	>0.98	4–6		6–7		
AFS-4T	A	1.3–2.5	25–35	2.5–3.2	>0.98	4–6		6–8		

*From Shaffer and Mintz, in Spiegler and Laird (eds), *Principles of Desalination*, 2d ed., Academic, New York, 1980, slightly revised.

†Manufacturers:
AMF Incorporated, B&D Division, 689 Hope Street, Springdale, Conn.
Asahi Chemical Industry Co., Ltd., 06907 Hibiya-Mitsui Building, 1-2 Yuraku-cho, Chiyoda-kse, Tokyo 100, Japan.
Asahi Glass Company, Ltd., 14, 2-chrome, Marunouchi, Chiyoda-ku, Tokyo, Japan.
Ionac Chemical Division, Sybron Corp., Birmingham, N.J.
Ionics, Incorporated, 65 Grove Street, Watertown, Mass. 02172.
Negev Institute for Arid Zone Research, Beersheva 84110, Israel.
Tokuyama Soda Co., Tokuyama City T745, Japan.

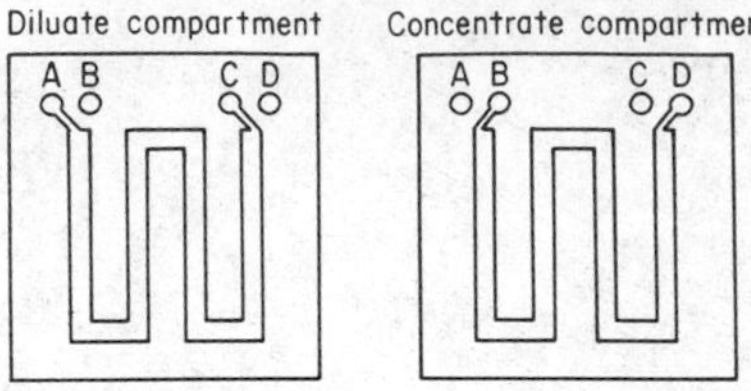

FIG. 17-36 Spacers for electrodialysis units. Principle of liquid-distribution system. *(Spiegler, Salt-Water Purification, 2d ed., Plenum Press, New York, 1977.)*

path. Mason and Kirkham (loc. cit.) have done this by using an approximate expression for the local resistance $\mathbf{R}_p$ ($\Omega \cdot cm^2$), of a cell pair consisting of unit area of cation-exchange membrane, diluate, anion-exchange membrane, and concentrate in series:

$$\mathbf{R}_p = (K_1/\bar{c}) + K_2 - K_3\bar{c} \qquad (17\text{-}96)$$

where K_1, K_2, and K_3 are empirical parameters for any spacer geometry, ionic composition, and membrane type and temperature, and $\bar{c}$ is the local "average" concentration calculated from

$$\frac{1}{\bar{c}} = \frac{1}{2}\left(\frac{1}{c_d} + \frac{r}{c_c}\right) \qquad (17\text{-}97)$$

Here c_d and c_c are the local diluate and concentrate concentrations, and r is the ratio of spacer thicknesses in the concentrate and in the diluate compartment (frequently r is unity). The rationale for the calculation of $\mathbf{R}_p$ is a model of dissipative resistance terms; viz., the first term in Eq. (17-96) represents the resistances of the bulk-solution streams, and the second those of the membranes. (The third term, which is negligible at low $\bar{c}$, allows for nonproportionality between solution concentration and conductance and for decrease in membrane resistances with increasing solution concentration.) Allowances for membrane potentials which oppose the applied voltage and thus represent an apparent resistance and for resistance increases due to concentration gradients in the diffusion layers are made in choosing the numerical values of K_1, K_2, and K_3. When using the Mason-Kirkham expression for $\mathbf{R}_p$, it is advisable to insert empirical constants whenever available from the manufacturer. In the absence of such data, the K's can be estimated from tabulated values of solution and membrane resistance (Table 17-27), but the resulting K's are usually somewhat low.

Nonohmic calculation takes into account the increase of stack resistance with increasing current density. When the voltage across the electrodialysis stack is raised, the current at first increases roughly in proportion to the voltage; eventually, further voltage increments cause only small current increases. At this point or near it, pH changes appear in concentrate and diluate streams. In electrodialysis technology these phenomena are collectively termed **polarization.** They are caused by mass-transfer limitations adjacent to the membrane-solution interfaces, electret effects in the membranes, and the undesirable participation of H^+ and OH^- ions in conduction ("acid-base generation"; this effect has frequently been called "water splitting," but the latter term is used for the production of hydrogen and oxygen *gas* in the energy literature). Polarization in electrodialysis sets an upper limit on practical current densities and production rates because of high power consumption and/or scale formation.

An analysis of the limiting-current phenomenon by classical electrochemical methods [Spiegler, *Desalination,* 9, 367 (1971)] reveals some analogy to polarographic situations at metal-solution interfaces. For example, when the Nernst-Planck equations of ionic flux [Eq. (17-77)] are applied to the simple case of a system comprising an anion-exchange membrane and a KCl solution (Fig. 17-37), the potential drop ΔE (V) between two identical probe electrodes placed into the solution streams on the right and left sides of the membrane is

$$\Delta E = -(2\mathbf{R}_s + \mathbf{R}_m)i \qquad (17\text{-}98)$$
$$+ \left[\frac{\mathcal{F}D}{(\bar{t}_- - t_-)\lambda} + \frac{RT}{\mathcal{F}}(2\bar{t}_- - 1)\right] \ln \frac{1 + (i/i_{lim})}{1 - (i/i_{lim})}$$

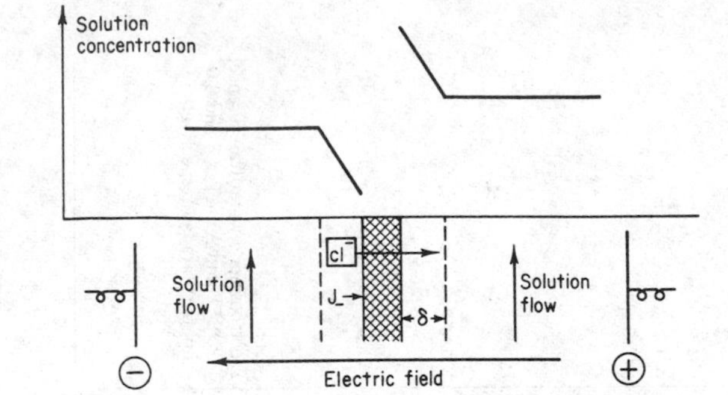

FIG. 17-37 Schematic of field and ion flow at anion-exchange membrane. [*Spiegler,* Desalination, **9,** *367 (1971).*]

(The negative sign appears because a system is considered in which positive current flows from right to left.) Here $\mathbf{R}_s$ is the resistance of the solution between probe electrode and diffusion layer, calculated for unit cross section, $\Omega \cdot cm^2$; $\mathbf{R}_m$, the membrane resistance, $\Omega \cdot cm^2$; $\mathcal{F}$, Faraday's constant, C/g equivalent; D, the diffusion coefficient of KCl in solution, cm^2/s; λ, the equivalent conductance of KCl, $\Omega^{-1}cm^2$ equivalent^{-1}; $\bar{t}_-$, t_-, the transport numbers of Cl^- in membrane and solution respectively; RT is in (W·s)/mol; and i_{lim} is the limiting current: $i_{lim} = -\mathcal{F}Dc_0/[\delta(\bar{t}_- - t_-)]$, where c_0 is the electrolyte concentration, equivalent cm^3 in the bulk solutions, and δ, cm, the local thickness of the diffusion layer near the probe electrodes. Equation (17-98) describes the potential drop for a small membrane element bracketed by identical solutions. It is of interest that the ohmic drops in the diffusion layers, which are dissipative in nature, are contained in the second term together with the membrane potential, which is at least partially "reversible." Figure 17-38 is a plot of this equation for a membrane of $\mathbf{R}_m = 11\ \Omega \cdot cm^2$, and $\bar{t}_- = 0.98$, in 0.03*M* KCl at 25°C. δ is taken as 0.02 cm, with the resistance $\mathbf{R}_s$ equivalent to a solution layer of 0.1 cm. Note that for $i \to 0$, the limiting slope contains, in addition to ohmic drop $-(\mathbf{R}_s + \mathbf{R}_m)\,i$, a contribution about as large from the membrane potential in the second term.

In principle, the diffusion-layer thickness depends on channel geometry, distance from solution inlet, and flow velocity. In practice, approximate average values of δ are used, which are determined from approximate design equations of the type

$$\text{Polarization parameter} \equiv \left|\frac{i_{lim}}{c_0}\right|$$
$$= \left|\frac{\mathcal{F}D}{\delta(\bar{t}_- - t_-)}\right| = \alpha u^{\beta} \qquad (17\text{-}99)$$

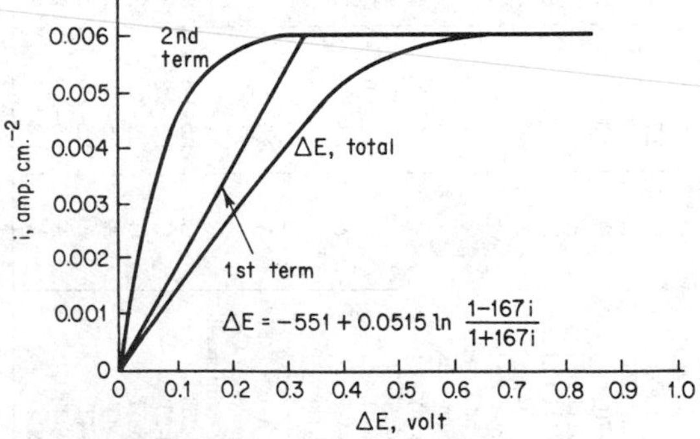

FIG. 17-38 Calculated current-voltage curve for system; KCl solution and anion-exchange membrane. Details of the conditions assumed and the numerical parameters used in this figure are listed in the original reference. [*Spiegler,* Desalination, **9,** *367 (1971).*]

where u is the flow velocity, and α, β are empirical constants characteristic of each spacer material [Smith, R&D Rep. 325, Office of Saline Water, U.S. Department of the Interior, 1968; Belfort and Guter, *Desalination*, **5**, 267 (1968); and Sonin and Probstein, *Desalination*, **5**, 293 (1968)]. For some spacers, β is close to 0.6.

Because of variation of concentration δ along the flow path, current-voltage curves in an electrodialysis stack cannot be expected to be identical with those shown in Fig. 17-38, but ion-flux plateaus are predicted by the Nernst-Planck treatment all the same [Forgacs, Ishibashi, Leibovitz, Sinkovic, and Spiegler, *Desalination*, **10**, 181 (1972)], as can be seen from Fig. 17-39, in which the electric current *due to the flow of chloride ions* through an anion-exchange membrane has been calculated from the *total* electric current including acid-base generation. It is seen that the chloride flow reaches a limiting-current plateau when plotted against terminal voltage, while the total current shows merely a gradual decrease of slope as the voltage is raised, followed by an increasing-slope region at yet higher voltages because of intensifying acid-base generation. The empirical practice has frequently been adopted to plot a graph of the apparent stack resistance versus the reciprocal of the current [Cowan and Brown, *Ind. Eng. Chem.*, **51**, 1445 (1959)] and to take the minimum of this "Cowan plot" as the "limiting current," although the occurrence of a genuine plateau of the electric current in a current-voltage graph (as routinely traced in conventional polarography) or of flow of salt ions (as opposed to H^+ and OH^-), as shown in Fig. 17-39, is seldom verified. The stack is usually operated at a current not higher than a fixed percentage (e.g., 70 percent) of this value.

A computer program for nonohmic optimization has been developed (Belfort and Guter, loc. cit.) and applied to a number of existing and projected electrodialytic desalination plants. In this type of optimization the total resistance of the stack is represented as a complex network made up of ohmic elements (e.g., concentrate and diluate bulk solutions; membranes proper) and nonohmic ones (e.g., the resistance of the diffusion layers and the resistance equivalent to the membrane potential).

Power Requirements The dc energy consumption W_c (Wh/kgal diluate) can be calculated from the resistance

$$W_e = \frac{I^2\mathbf{R}}{F} = \frac{I^2\mathbf{R}_p n}{A_p F} = \frac{V^2}{\mathbf{R}F} = \frac{V^2 A_p}{\mathbf{R}_p n F} \qquad (17\text{-}100)$$

where I is the total current through the stack, A; $\mathbf{R}$ the total stack resistance, Ω; n, the number of cell pairs; A_p, the active area of each membrane, cm^2; and F, the total diluate flow rate through the stack, kgal/h. The applied voltage is $I\mathbf{R} = V$; it is usually in the range of 1 to 5 V per membrane pair. $\mathbf{R}_p$ is the resistance of the unit area of a cell pair; it may be calculated from the Mason-Kirkham approximation described earlier [Eq. (17-96)], provided values of the parameters K_1, K_2, K_3 for the particular system envisaged are obtainable from the manufacturer, or it may be calculated by use of nonohmic analysis (Belfort and Guter, loc. cit.). The energy consumption (watthour *per equivalent salt shifted*) is

$$W_{es} = \frac{V\mathcal{F}}{3600\eta_I n} = \frac{26.8V}{\eta_I n} = \frac{26.8IR_p}{A_p \eta_I} \qquad (17\text{-}101)$$

where η_I is the current efficiency, viz., the ratio of the salt shifted to the theoretical requirement in a stack containing ideally permselective membranes. It is related to the permselectivity of both membrane types. While the maximum value $\eta_I = 1$ is never reached in practice, η_I is usually larger than 0.85 in electrodialyzers treating brackish water. These electric energy costs all refer to direct current. The converters for obtaining dc from available ac power represent an appreciable portion of the cost of electrodialysis installations. Pumping-power requirements, which vary with the channel, spacer, and stack geometry, must also be added, but these are in general less than the dc power requirements.

Detailed Operating and Design Data Tables 17-28 and 17-29 (Courtesy R. E. McDiarmid and D. I. Elyanow, Ionics, Incorporated, 1979) show characteristic design and operating data for automatic-periodic-reversal (EDR) units.

Membrane Scaling and Fouling The increased salt concentrations near the membrane-brine interfaces (Fig. 17-37) may induce scale precipitation if solubility limits are exceeded. pH changes in the solutions promote the formation of these scales: When the salt

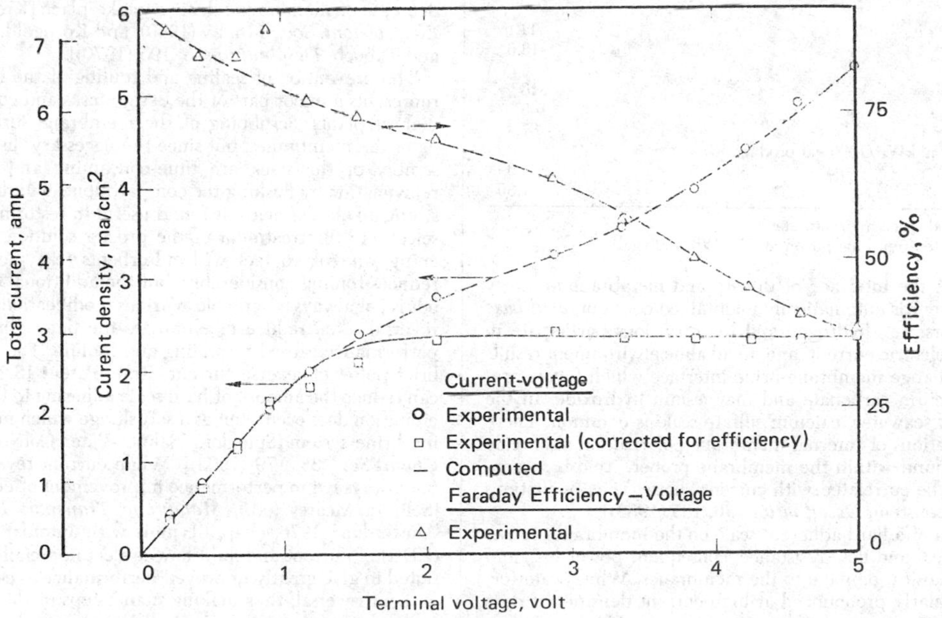

FIG. 17-39 Current-voltage curves for an anion-exchange membrane (111BZL 183; Ionics, Incorporated, Watertown, Mass.). Solutions: KCl, 5×10^{-5} (g·mol)/cm^3; linear flow velocities, 8.5 and 11.8 cm/s (0.279 and 0.387 ft/s) for concentrate and diluate respectively; 25°C. "Efficiency" refers to the percentage of electric current carried by Cl^-, as determined by chemical analyses. Active electrode area: 1246 cm^2 (1.341 ft^2). [*Forgacs*, Ishibashi, Leibovitz, Sinkovic, and Spiegler, Desalination, *10, 181 (1972).*]

TABLE 17-28 Design Characteristics of Typical EDR (Electrodialysis-Reversal) Membrane Stacks*

	Single stack, MK II, four stages	Single stack, MK III, three stages	Single stack, MK III, one stage	Three stacks in series, MK III
Typical hydraulic flow rate				
U.S. gal/24-h day	16,700	55,600	166,700	166,700
U.S. gal/min	11.6	38.6	116	116
Pressure drop at typical flow, lb/in^2	47	44	14	42
Number of membranes	540	900	900	2,700
Size of membranes, in × in	18 × 20	18 × 40	18 × 40	18 × 40
Total area of membranes, ft^2	1,350	4,500	4,500	13,500
% total area available for transfer	62	64	64	64
Approximate weight, lb	1,300	2,800	2,800	8,400
Approximate overall height, including legs	4′6″	6′10″	6′8″	6′8″
Demineralization per pass (25°C, high-Cl water, typical flow), %	88.5	88.3	52	90.0
Current required for 3000-ppm feed, A	Stages 1 and 2: 19 Stages 3 and 4: 8	Stages 1 and 2: 36 Stage 3: 12	46	Stage 1: 46 Stage 2: 24 Stage 3: 12
Voltage required for 3000-ppm feed†	Stages 1 and 2: 180 Stages 3 and 4: 150	Stages 1 and 2: 350 Stage 3: 150	640	Stage 1: 640 Stage 2: 500 Stage 3: 420
Direct-current kW/stack for 3000-ppm feed†	4.6	14.1	29	Stage 1: 29 Stage 2: 12 Stage 3: 5
Direct-current kWh/1000 gal product‡ for 3000-ppm feed†	7.4	6.8	4.7	7.4

*Ionics, Incorporated, Watertown, Mass, 1979. These units use the EDR process, in which polarity and fluid flow are periodically reversed. In general, addition of acid and antiprecipitant to the feed is not necessary in this process.

†For typical brackish water containing a high proportion of sodium chloride.

‡Approximately 10% of flow wasted during reversal.

TABLE 17-29 Performance Characteristics of a Two-Stack EDR Demineralizer on a Brackish Well in the Middle East*

Feedwater concentration, mg/L total dissolved solids (TDS)	3,585
Product-water concentration, mg/L TDS	450
Production per 24-h day, gal	90,072
Water wasted (including brine waste, electrode waste, and reversal losses), %	34.8
Feedwater temperature, °F (°C)	94 (34)
Current, A	
Stage 1	42.1
Stage 2	13.6
Voltage	
Stage 1	318
Stage 2	143
Demineralization, %	87.4
Energy consumption, kWh/1000 gal product	
Direct current	4.09
Total†	7.80

*Two electric and three hydraulic stages.

†Includes estimated pumping power at 3.5 kWh/1000 gal.

concentrations at the interface of diluate and membrane are very low and the current is automatically maintained constant, *acid-base generation* occurs; i.e., hydrogen and hydroxyl ions participate in the transport of electric current, and an alkaline environment results at the anion-exchange membrane-brine interface which favors precipitation of calcium carbonate and magnesium hydroxide. In the electrodialysis of seawater, calcium sulfate scale is common. There exist some indications of internal membrane polarization also; scale is occasionally found within the membrane proper. Acid-base generation seems to be correlated with current "noise," i.e., oscillations [Kedem and Rubenstein, *Desalination*, **46**, 185 (1983)].

The deposition of a hard adherent scale on the membranes causes increased electrical and flow resistance, consequent power loss, and frequently mechanical damage to the membranes. While scale formation is particularly pronounced at high current densities, it does also occur to a lesser extent at low current densities. To prevent calcium carbonate scale, hydrochloric or sulfuric acid is often added to the concentrate stream and also to the rinse liquid for the cathode compartment where hydroxyl ions are formed by the electrode reactions. The amounts of acid needed depend on the alkalinity of the raw water; for a raw-water alkalinity of 140 ppm (as $CaCO_3$), a dosage of 2-lb/kgal diluate produced has been used. When soft sludges (e.g., ferric hydroxide) deposit on membrane surfaces (Grossman and Sonin, *Desalination*, **10**, 157 (1972) or the adsorption of colloids and/or multivalent ions, often present in the raw water only in very low impurity concentrations, fouls the membranes (usually the anion-exchange membranes), acid-base generation at relatively low electric current densities and power losses due to excessive resistance and decrease of membrane selectivity take place [Kressman and Tye, *J. Electrochem. Soc.*, **116**, 25 (1969); and Korngold, de Körösy, Rehav, and Taboch, *Desalination*, **8**, 195 (1970)].

The prevention of scaling and fouling of the membrane surfaces represents a major part of the design tasks and costs of most electrodialysis plants. Scrubbing of the membrane surfaces often rejuvenates the membranes, but since the necessary disassembly and reassembly of the stack are time-consuming and expensive, in situ rejuvenation by flushing the compartments with leach solutions, e.g., strong alkalies or acids, is found useful in restoring conductivity and selectivity. Pretreatment of the process solutions by filtration, softening, and/or contact with adsorbents (e.g., active carbon) often reduces fouling considerably, but the additional cost of these methods is not always acceptable. Various modifications of electric current reversal often reduce or eliminate the deterioration of membrane performance caused by scaling and fouling. The application of very brief pulses of reverse current (Israel Patent 13,242, Mar. 23, 1961) can reduce the amount of hard scale adhering to the membrane. Precipitation does occur, but as a soft sludge which moves with the flowing brine system [Spiegler, "Saline-Water Conversion No. 2," *Adv. Chem. Ser.*, **38**, 179 (1963)]. While current-reversal methods have not always led to performance improvement of electrodialysis plants [Solt, in Meares (ed.), *Membrane Transport Processes*, Elsevier, Amsterdam, 1976, chap. 6], units with automatic periodic reversal (EDR) of current and fluid flow are commercially available and are stated to give greatly improved performance as compared with units without reversal, thus making many steps of the pretreatment process unnecessary (Aquamite bulletins, Ionics, Incorporated, Watertown, Mass.). Methods for keeping membrane surfaces clean and active are of particular importance when the liquids to be treated are colloid solutions (sols) and/or contain membrane contaminants. Wastewater treatment by membrane methods is sometimes feasible, provided proper precautions are taken [Belfort, in Shuval (ed.), *Water Renovation and Reuse*, Academic, New York, 1977, chap. 6).

Applications of Electrodialysis In principle, electrodialysis is applicable whenever removal, recovery, and/or concentration of ionic solutes from nonionic ones is required. Its major use to date has been for demineralization of brackish waters, but other applications have come to the fore and have been improved in the recent past.

Electrodialyzers which contain mixtures of granular cation- and anion-exchange resins in the diluate compartments have proved useful for the decontamination of dilute radioactive wastewater. The resins remove the last traces of radioactive ions; they also reduce the conductance of the diluate compartments and are being continuously regenerated by the passage of the current [Gittens and Glueckauf, *Am. Inst. Chem. Eng. Symp. Ser.*, **9,** 79 (1965)]. A fabric of ion-conducting spacers has advantages over mixed-bed packing [Kedem, *Desalination*, **16,** 105 (1975)].

The application of electrodialysis to the deashing of food products and fluids in pharmaceutical plants is growing because this relatively gentle separation method does not lead to thermal decomposition of the products. For instance, whey, which is too high in minerals to be useful for human consumption, is being converted to a digestible product by electrodialysis. Much work has been done by the paper industry on the recovery of constituents from pulping liquors (Lacey and Loeb, *Industrial Processing with Membranes*, Wiley, New York, 1972). Moreover, the chlorine-alkali industry has introduced a process in which a cation-exchange membrane separates between cathode and anode compartments to take the place of the classic mercury process, which is environmentally objectionable. Because these membranes must function in highly concentrated solutions and in a strongly alkaline milieu, they are based on more resistant (fluorinated) polymer lattices than the membranes used in water desalination. While these membrane cells are not complete electrodialysis cells, the recent use of this electrically driven membrane process is likely to influence the further development of electrodialysis in concentrated solutions and the production of membranes for use in devices containing such solutions, e.g., some batteries and fuel cells. Thus the development of electrodialysis technology has led to the ready availability of mechanically sturdy, highly conductive cation- or anion-selective membranes, which have opened many possibilities for the design of processes requiring these types of separators which had not been available before.

DIELECTROPHORESIS

GENERAL REFERENCES: Pohl, in Moore (ed.), *Electrostatics and Its Applications*, Wiley, New York, 1973, chap. 14 and chap. 15 (with Crane). Pohl, in Catsimpoolas (ed.), *Methods of Cell Separation*, vol. I, Plenum Press, New York, 1977, chap. 3. Pohl, *Dielectrophoresis: The Behavior of Matter in Nonuniform Electric Fields*, Cambridge, New York, 1978.

Introduction Dielectrophoresis (DEP) is defined as the motion of neutral, polarizable matter produced by a nonuniform electric (ac or dc) field. DEP should be distinguished from electrophoresis, which is the motion of charged particles in a uniform electric field (Fig. 17-40).

The DEP of numerous particle types has been studied, and many applications have been developed. Particles studied have included aerosols, glass, minerals, polymer molecules, living cells, and cell organelles. Applications developed include filtration, orientation, sorting or separation, characterization, and levitation and materials handling. Effects of DEP are easily exhibited, especially by large particles, and can be applied in many useful and desirable ways. DEP effects can, however, be observed on particles ranging in size even down to the molecular level in special cases. Since thermal effects tend to disrupt DEP with molecular-sized particles, they can be controlled only under special conditions such as in molecular beams.

Principle The principle of particle and cell separation, control, or characterization by the action of DEP lies in the fact that a net force can arise upon even neutral particles situated in a nonuniform electric field. The force can be thought of as rising from the imaginary two-step process of (1) induction or alignment of an electric dipole in a particle placed in an electric field followed by (2) unequal forces on the ends of that dipole. This arises from the fact that the force of an electric field upon a charge is equal to the amount of the charge and to the *local* field strength at that charge. Since the two (equal) charges of the (induced or oriented) dipole of the particle lie in unequal field strengths of the diverging field, a net force arises. If the particle is suspended in a fluid, then the polarizability of that medium enters, too. If, for example, the particle is more polarizable than the fluid, then the net force is such as to impel the particle to regions of greater field strength. Note that this statement implies that the effect is independent of the absolute sign of the field direction. This is found to be the case. Even rapidly alternating (ac) fields can be used to provide *unidirectional* motion of the suspended particles.

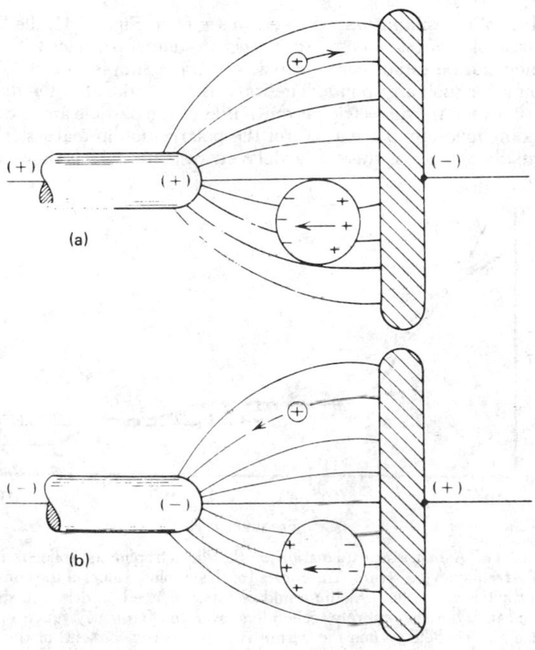

FIG. 17-40 Comparison of behaviors of neutral charged bodies in an alternating nonuniform electric field. (*a*) Positively charged body moves toward negative electrode. Neutral body is polarized, then is attracted toward point where field is strongest. Since the two charge regions on the neutral body are equal in amount of charge but the force is proportional to the local field, a net force toward the region of more intense field results. (*b*) Positively charged body moves toward the negative electrode. Again, the neutral body is polarized, but it does not reverse direction although the field is reversed. It still moves toward the region of highest field intensity.

Formal Theory A small neutral particle at equilibrium in a static electric field experiences a net force due to DEP that can be written as $\mathbf{F} = (\mathbf{p}\cdot\nabla)\mathbf{E}$, where $\mathbf{p}$ is the dipole moment vector and $\mathbf{E}$ is the external electric field. If the particle is a simple dielectric and is isotropically, linearly, and homogeneously polarizable, then the dipole moment can be written as $\mathbf{p} = \alpha v \mathbf{E}$, where α is the (scalar) polarizability, v is the volume of the particle, and $\mathbf{E}$ is the external field. The force can then be written as:

$$\mathbf{F} = \alpha v(\mathbf{E}\cdot\nabla)\mathbf{E} = \tfrac{1}{2}\alpha v \nabla|\mathbf{E}|^2 \qquad (17\text{-}102)$$

This force equation can now be used to find the force in model systems such as that of an ideal dielectric sphere (relative dielectric constant K_2) in an ideal perfectly insulating dielectric fluid (relative dielectric constant K_1). The force can now be written as

$$\mathbf{F} = 2\pi a^3 \epsilon_0 K_1 \left(\frac{K_2 - K_1}{K_2 + 2K_1}\right) \nabla|\mathbf{E}|^2 \qquad (17\text{-}103)$$

(ideal dielectric sphere in ideal fluid)

Heuristic Explanation As we can see from Fig. 17-41, the DEP response of *real* (as opposed to perfect insulator) particles with frequency can be rather complicated. We use a simple illustration to account for such a response. The force is proportional to the difference between the dielectric permittivities of the particle and the surrounding medium. Since a part of the polarization in real systems is thermally activated, there is a delayed response which shows as a

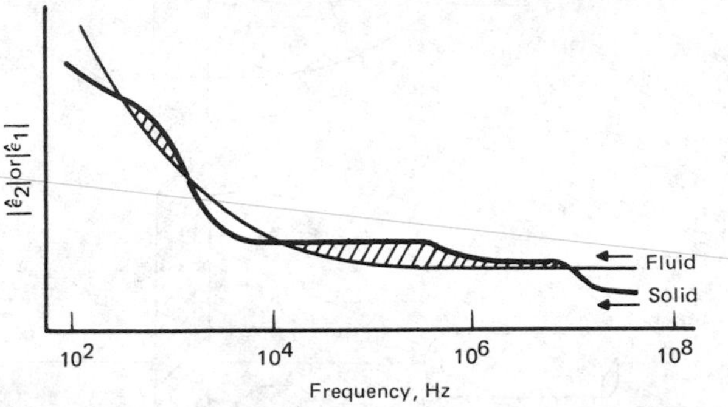

FIG. 17-41 A heuristic explanation of the dielectrophoretic-collection-rate (DCR)–frequency spectrum. The curves for the absolute values of the complex permittivities of the fluid medium and of the suspended particles are shown lying nearly, but not entirely, coincident over the frequency range of the applied electric field. When the permittivity (dielectric constant) of the particles exceeds that of the suspending medium, the collection, or "positive dielectrophoresis," occurs. In the frequency ranges in which the permittivity of the particles is less than that of the suspending medium no collection at the regions of higher field intensity occurs. Instead there is "negative dielectrophoresis," i.e., movement of the particles into regions of lower field intensity.

phase lag between **D**, the dielectric displacement, and **E**, the electric-field intensity. To take this into account we may replace the simple (absolute) dielectric constant ϵ by the complex (absolute) dielectric constant $\hat{\epsilon} = \epsilon' - i\epsilon'' = \epsilon' - i\sigma/w$, where ω is the angular frequency of the applied field. For treating spherical objects, for example, the replacement

$$F \propto \frac{\epsilon_1(\epsilon_2 - \epsilon_1)}{\epsilon_2 + 2\epsilon_1} \rightarrow \text{Re}\left\{\frac{\hat{\epsilon}_1^\circ(\hat{\epsilon}_2 - \hat{\epsilon}_1)}{\hat{\epsilon}_2 + 2\hat{\epsilon}_1}\right\} \qquad (17\text{-}104)$$

can be made, where $\hat{\epsilon}^\circ$ is the complex conjugate of $\hat{\epsilon}$.

With this force expression for real dielectrics, we can now explain the complicated DEP response with the help of Fig. 17-41.

A particle, such as a living cell, can be imagined as having a number of different frequency-dependent polarization mechanisms contributing to the total effective polarization of the particle $|\hat{\epsilon}_2|$. The heavy curve in Fig. 17-41 shows that the various mechanisms in the particle drop out stepwise as the frequency increases. The light curve in Fig. 17-41 shows the polarization for a simple homogeneous liquid that forms the surrounding medium. This curve is a smooth function which becomes constant at high frequency. As the curves cross each other (and hence $|\epsilon_2| = |\epsilon_1|$), various responses occur. The particle can thus be attracted to the strongest field region, be repelled from that region, or experience no force depending on the frequency. The net result will be like the curves of Fig. 17-53.

Limitations It is desirable to have an estimate for the smallest particle size that can be effectively influenced by DEP. To do this, we consider the force on a particle due to DEP and also due to the osmotic pressure. This latter diffusional force will randomize the particles and tend to destroy the control by DEP. Figure 17-42 shows a plot of these two forces, calculated for practical and representative conditions, as a function of particle radius. As we can see, the smallest particles that can be effectively handled by DEP appear to be in range of 0.01 to 0.1 μm (100 to 1000 Å).

Another limitation to be considered is the volume that the DEP force can affect. This factor can be controlled by the design of electrodes. As an example, consider electrodes of cylindrical geometry. A practical example of this would be a cylinder with a wire running

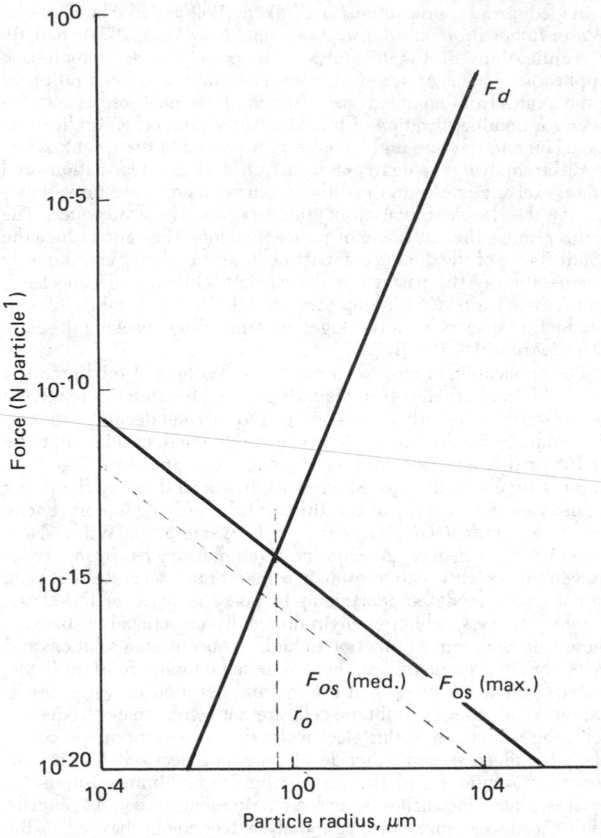

FIG. 17-42 Comparison of the dielectrophoretic (F_d) and osmotic (F_{os}) forces as functions of the particle size.

down the middle to provide the two electrodes. The field in such a system is proportional to $1/r$. The DEP force is then $F_{DEP} \propto \nabla|E^2| \propto 1/r^3$, so that any differences in particle polarization might well be masked merely by positional differences in the force. At the outer cylinder the DEP force may even be too small to affect the particles appreciably. The most desirable electrode shape is one in which the force is independent of position within the nonuniform field. This "isomotive" electrode system is shown in Fig. 17-43.

Applications of Dielectrophoresis Over the past 20 years the use of DEP has grown rapidly to a point at which it is in use for biological, colloidal, and mineral materials studies and handling. The effects of nonuniform electric fields are used for handling particulate matter far more often than is usually recognized. This includes the removal of particulate matter by "electrofiltration," the sorting of mixtures, or its converse, the act of mixing, as well as the coalescence of suspensions. In addition to these effects involving the translational motions of particles, some systems apply the orientational or torsional forces available in nonuniform fields. One well-known example of the latter is the placing of "tip-up" grit on emery papers commercially. Xerography and many other imaging processes are examples of multibillion-dollar industries which depend upon DEP for their success.

A clear distinction between **electrophoresis** (field action on an object carrying excess free charges) and **dielectrophoresis** (field gradient action on neutral objects) must be borne in mind at all times.

Dielectrofiltration A dielectrofilter is a device which uses the action of an electric field to aid the filtration and removal of particulates from fluid media. A dielectrofilter can have a very obvious advantage over a mechanical filter in that it can remove particles

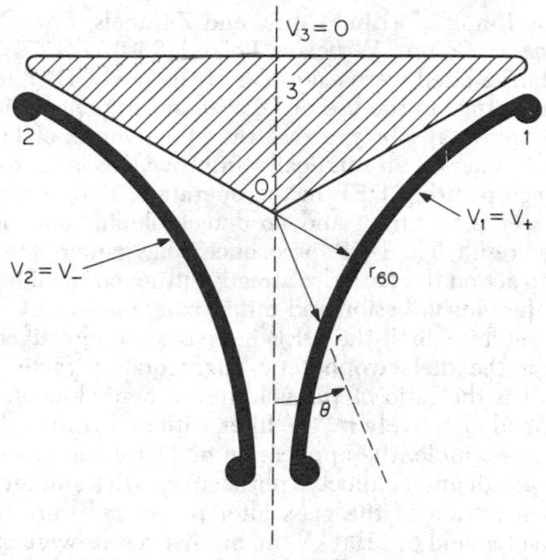

FIG. 17-43 A practical isomotive field geometry, showing r_{60}, the critical radius characterizing the isomotive electrodes. Electrode 3 is at ground potential, while electrodes 1 and 2 are at $V_1 = V_+$ and $V_2 = V_- = -V_+$ respectively. The inner faces of electrodes 1 and 2 follow $r = r_o\,[\sin(3\theta/2)]^{-2/3}$, while electrode 3 forms an angle of 120° about the midline.

which are much *smaller* than the flow channels in the filter. In contrast, the ideal mechanical filter must have all its passages smaller than the particles to be removed. The resultant flow resistance can be use-restrictive and energy-consuming unless a phenomenon such as dielectrofiltration is used.

Dielectrofiltration can (and often does) employ both electrophoresis and dielectrophoresis in its application. The precise physical process which dominates depends on a number of physical parameters of the system. Factors such as field intensity and frequency and the electrical conductivity and dielectric constants of the materials present determine this. Although these factors need constant attention for optimum operation of the dielectrofilter, this additional complication is often more than compensated for by the advantages of dielectrofiltration such as greater throughput and lesser sensitivity to viscosity problems, etc. To operate the dielectrofilter in the dominantly electrophoretic mode requires that excess free charges of one sign or the other reside on the particulate matter. The necessary charges can be those naturally present, as upon a charged sol; or they may need to be artificially implanted such as by passing the particles through a corona discharge. Dielectrofiltration by the corona-charging, electrophoresis-dominated Cottrell technique is now widely used.

The discussion of dielectrofiltration which follows deals primarily with the dielectrophoretic mode. See the later subsection "Novel Solid-Liquid Separation Processes" for a discussion of electrofiltration processes which involve electrophoresis.

To operate the dielectrofilter (dominantly dielectrophoretic mode), on the other hand, one must avoid the presence of free charge on the particles. If the particles can become charged during the operation, a cycle of alternate charging and discharging in which the particles dash to and from the electrodes can occur. This is most likely to occur if static or very low frequency fields are used. For this reason, corona and like effects may be troublesome and need often to be minimized. To be sure, the DEP force is proportional to the field applied [actually to $\nabla(E)^2$], but fields which are too intense can produce such troublesome charge injection. A compromise for optimal operation is necessary between having $\nabla(E)^2$ so low that DEP forces are insufficient for dependable operation, on the one hand, and having **E** so high that troublesome discharges (e.g., coronalike) interfere with dependable operation of the dielectrofilter. In insulative media such as air or hydrocarbon liquids, for example, one might prefer to operate with fields in the range of, say, 10 to 10,000 V/cm. In more conductive media such as water, acetone, or alcohol, for example, one would usually prefer rather lower fields in the range of 0.01 to 100 V/cm. The higher field ranges cited might become unsuitable if conductive sharp asperities are present.

Another factor of importance in dielectrofiltration is the need to have the DEP effect firmly operative upon *all* portions of the fluid passing through. Oversight of this factor is a most common cause of incomplete dielectrofiltration. Good dielectrofilter design will emphasize this crucial point. To put this numerically, let us consider the essential field factor for DEP force, namely $\nabla(\mathbf{E}_0)^2$. Near sharp points, e.g., **E**, the electric field varies with the radial distance r as $E \propto r^{-2}$; hence our DEP force factor will vary as $\nabla(E)^2 \propto r^{-5}$. In the neighborhood of sharp "line" sources such as at the edge of electrode plates, $E \propto r^{-1}$, hence, $\nabla(E)^2 \propto r^{-3}$. If, for instance, the distance is varied by a factor of 4 from the effective field source in these cases, the DEP force can be expected to weaken by a factor of 1024 or 64 respectively for the point source and the line source. The matter is even more keenly at issue when field-warping dielectrics (defined later) are used to effect maximal filtration. In this case the field-warping material is made to produce dipole fields as induced by the applied electric field. If we ask how the crucial factor, $\nabla(E)^2$, varies with distance away from such a dipole, we find that since the field E_d about a dipole varies approximately as r^{-3}, then $\nabla(E)^2$ can be expected to vary as r^{-7}. It then becomes critically important that the particles to be removed from the passing fluid do, indeed, pass very close to the surface of the field-warping material, or it will not be effectively handled. Clearly, it would be difficult to maintain successfully uniform dielectrofiltration treatment of fluid passing through such wildly variant regions. The problems can be minimized by ensuring that all the elements of the passing fluid go closely by such field sources in the dielectrofilter. In practice this is done by constructing the dielectrofilter from an assembly of highly comminuted electrodes or else by a set of relatively simple and widely spaced metallic electrodes between which is set an assembly of more or less finely divided solid dielectric material having a complex permittivity different from that of the fluid to be treated. The solid dielectric (fibers, spheres, chunks) serves to produce field nonuniformities or **field warpings** to which the particles to be filtered are to be attracted. In treating fluids of low dielectric constant such as air or hydrocarbon fluids, one sees field-warping materials such as sintered ceramic balls, glass-wool matting, open-mesh polyurethane foam, alumina, chunks, or $BaTiO_3$ particles.

An example of a practical dielectrofilter which uses both of the features described, namely, sharp electrodes *and* dielectric field-warping filler materials, is that described in Fig. 17-44 [H. J. Hall and R. F. Brown, *Lubric. Eng.*, **22**, 488 (1966)]. It is intended for use with hydraulic fluids, fuel oils, lubricating oils, transformer oils,

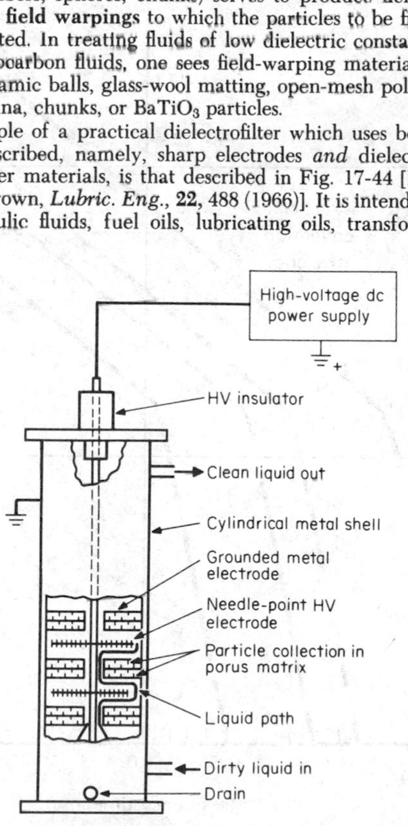

FIG. 17-44 Diagram illustrating the function of an electrostatic liquid cleaner.

lubricants, and various refinery streams. Performance data are cited in Fig. 17-45. It must be remarked that in the opinion of Hall and Brown the action of the dielectrofilter was "electrostatic" and due to free charge on the particles dispersed in the liquids. It is the present authors' opinion, however, that both electrophoresis and dielectrophoresis are operative here but that the dominant mechanism is that of DEP, in which neutral particles are polarized and attracted to the regions of highest field intensity.

Another example of the commercial use of DEP is in polymer clarification [A. N. Wennerberg, U.S. Patent 2,914,453, 1959; assignor to Standard Oil Co. (Indiana)]. Here, either ac or dc potentials were used while passing suspensions to be clarified through regions with an area-to-electrode-area ratio of 10:1 or 100:1 and with fields in the order of 10 kV/cm. Field warping by the presence of various solid dielectrics was observed to enhance filtration considerably, as expected for DEP. The filtration of molten or dissolved polymers to free them of objectionable quantities of catalyst residues, for example, was more effective if a solid dielectric material such as Attapulgus clay, silica gel, fuller's earth, alumina, or bauxite was present in the region between the electrodes. The effectiveness of percolation through such absorptive solids for removing color bodies is remarkably enhanced by the presence of an applied field. A given amount of clay is reported to remove from 4 to 10 times as much color as would be removed in the absence of DEP. Similar results are

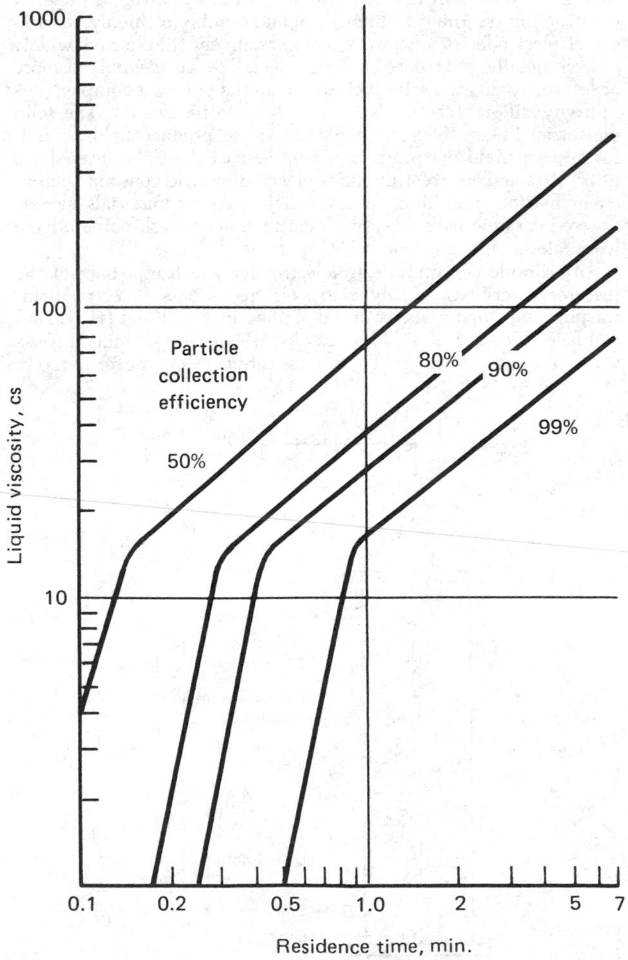

FIG. 17-45 Performance data for a typical high-efficiency electrostatic liquid cleaner.

reported by Lin et al. [Lin, Yaniv, and Zimmels, *Proc. XIIIth Int. Miner. Process. Congr.*, Wroclaw, Poland, 83–105 (1979)].

The instances cited were examples of the use of DEP to filter liquids. We now turn to the use of DEP to aid in dielectrofiltration of gases. Fielding et al. observe that the effectiveness of high-quality fiberglass air filters is dramatically improved by a factor of 10 or more by incorporating DEP in the operation. Extremely little current or power is required, and no detectable amounts of ozone or corona need result. The DEP force, once it has gathered the particles, continues to act on the particles already sitting on the filter medium, thereby improving adhesion and minimizing blowoff.

The degree by which the DEP increases the effectiveness of gas filtration, or the dielectrophoretic augmentation factor (DAF), is definable. It is the ratio of the volumes of aerosol-laden gas which can be cleaned effectively by the filter with and without the voltage applied. For example, the application of 11 kV/cm gave a DAF of 30 for 1.0-μm-diameter dioctyl phthalate particles in air, implying that the penetration of the glass filter is reduced thirtyfold by the application of a field of 1100 kV/m. Similar results were obtained by using "standard" fly ash supplied by the Air Pollution Control Office of the U.S. Environmental Protection Agency. The data obtained for several aerosols tested are shown in Table 17-30 and in Fig. 17-46. The relation DAF $= kV^2/v$ is observed to hold approximately for each aerosol. Here, the DEP augmentation factor DAF is observed to depend upon a constant K, a characteristic of the material, upon the square of the applied voltage, and upon the inverse of the volume flow rate v through the filter.

It is worth noting that in the case of the air filter described DEP serves as an augmenting rather than as an exclusive mechanism for the removal of particulate material. It is a unique feature of the dielectrophoretic gas filter that the DEP force is maximal when the particulates are at or on the fiber surface. This causes the deposits to be

TABLE 17-30 Dielectrophoretic Augmentation of Filtration of a Liquid Aerosol*

Air speed, cm/s	DAF at 2 kV	3.5 kV	5 kV	7 kV
	0.3-μm-diameter dioctyl phthalate aerosol			
3	8	19	95	330
6	3	13	39	120
9	3	11	28	100
15	2	6	13	42
20	2	5	9	27
28	2	4	6	14
39	2	3	4	9
50	1	2	3	6
	1.0-μm-diameter dioctyl phthalate aerosol			
3	30	110	300	1100
6	6	3	95	360
9	4	18	50	170
15	3	10	20	50
20	2	6	13	35
28	2	4	8	18
39	2	3	5	11
50	1	2	3	7
	Fly-ash aerosol			
6	10	30	80	
10	8	30	80	
14	5	20	40	
20	4	10	30	70
35	3	7	10	20
45	1	2	6	
53	1	2	7	10

*Experimentally measured dielectrophoretic augmentation factor DAF as a function of air speed and applied voltage for a glass-fiber filter (HP-100, Farr Co.). Cf. Fielding, Thompson, Bogardus, and Clark, *Dielectrophoretic Filtration of Solid and Liquid Aerosol Particulates*, Prepr. 75-32.2, 68th ann. meet., Air Pollut. Control Assoc., Boston, June 1975.

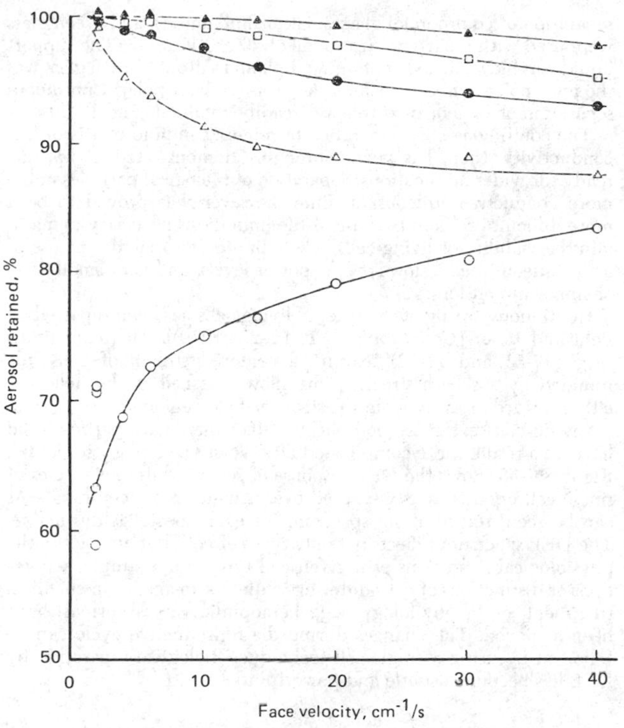

FIG. 17-46 Efficiency of an electrofilter as a function of gas flow rate at 5 different voltages. Experimental materials: 1-μm aerosol of dioctyl phthalate; glass-fiber filter. Symbols: O, no voltage applied; △, 2 kV; ●, 3.5 kV; □, 5 kV; ▲, 7 kV. *(After Fielting et al.,* Dielectrophoretic Filtration of Solid and Liquid Aerosol Particulates, *Prepr.* 75-32.2, *68th ann. meet., Air Pollut. Control Assoc., Boston, June 1975.)*

strongly retained by this particular filtration mechanism. It thus contrasts importantly with other types of gas filter in which the filtration mechanism no longer acts after the capture of the particle. In particular, in the case of the older electrostatic mechanisms involving only coulombic attraction, a simple charge alternation on the particle, such as caused by normal conduction, often evokes disruption of the filter operation because of particle repulsion from the contacting electrode. On the other hand, ordinary mechanical filtration depends upon the action of adventitious particle trapping or upon van der Waals forces, etc., to hold the particles. The high efficiency possible with electrofilters suggests their wider use.

Particle Sorting The techniques of DEP can also be applied to the sorting and separation of particles of many kinds. Separations based on DEP depend on differences in the polarizability of the particles. This can be of value when magnetic methods, mechanical filtration, or density-based separations are ineffective.

Black and Hammond used DEP to obtain separations of biological products. Separations were made in nonpolar organic solvents on a mixture of soybean phosphatides and soybean micella, also on a mixture of killed bacterial spores and killed bacterial vegetative cells. Theirs is a batch process in which an electrode chamber is filled with the mixture, the field applied, and the two separated portions drawn off.

An improved method of continuous particle separation was developed by Verschure and Ijlst [*Nature,* **211,** 619 (1966)], using screens to produce the nonuniform electric field. This separator, shown in Fig. 17-47, consists of two parallel electrodes spaced 3 mm apart. The electrodes are supplied with 0 to 2 kV at 50 Hz. The upper electrode is a metal plate insulated with Mylar [poly(ethylene terephthalate)] to prevent short circuits. The lower electrode is a screen which produces a nonuniform field and allows the particles to fall through it. While operating, the electrodes are immersed in a liquid medium of appropriate dielectric constant. The mineral mixture is introduced between the electrodes by a shaker. If the liquid is chosen so that its dielectric constant is between that of the minerals, the particles with the higher dielectric constant tend to form chains and stay between the electrodes to be carried down to a collector. The particles with the lower dielectric constant tend to fall through the screen electrode into another collector. This apparatus was used to separate mineral-powder mixtures: (1) nigerite, cassiterite, and columbite-tantalite; and (2) biotite and astrophyllite. The liquid medium for the first separation was di-*n*-butyl phthalate, while that for the second was a 2:3 mixture of butyl alcohol and kerosine. Good separations were obtained from both mixtures after several passes.

Pohl and Plymale [*J. Electrochem. Soc.,* **107,** 386 (1960)] developed a continuous separator by using isomotive-electrode geometry. In this separator, also shown in Figs. 17-48 and 17-49, the particles are shaken down between the electrodes. The electrodes are set at a rather large descending angle and twisted slightly (1 to 6°) about the long axis. There are two outlet ports at the lower end of the elec-

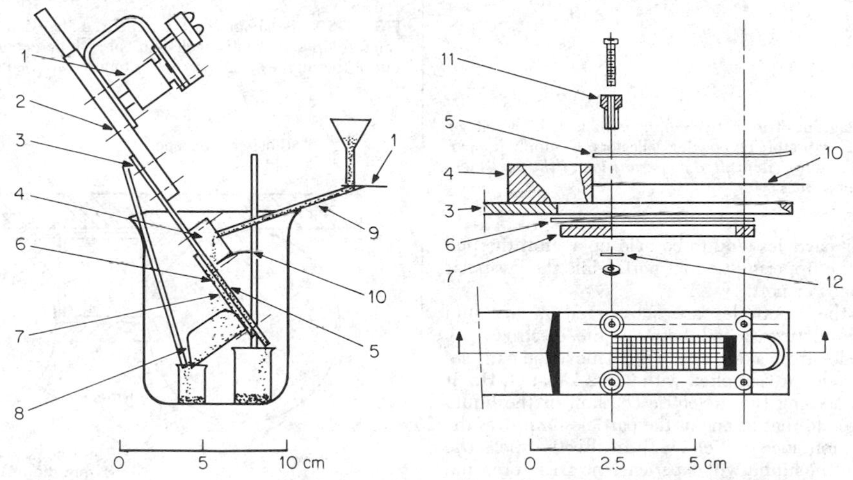

FIG. 17-47 Diagram of the separator devised by Verschure and Ilst: (1) variable vibrator, (2) vibrator base (rigid polyvinyl chloride), (3) separator body (Perspex), (4) separator funnel (Perspex), (5) upper electrode (copper), (6) sieve electrode (0.75-mm copper sieve), (7) clamping plate (Perspex), (8) chute (Mylar), (9) feeder chute, (10) insulating foil, (11) bushing (Teflon), and (12) ring (Teflon).

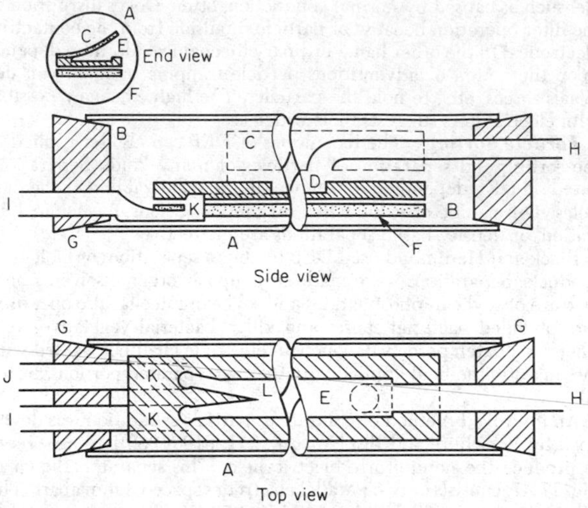

FIG. 17-48 Schematic diagram of an isomotive cell for continuous dielectrophoretic separations: *A*, glass outer cell wall; *B*, dielectric liquid; *C*, curved upper brass electrode; *D*, solid particles lying in flat groove of Teflon insulator plate; *E*, Teflon plate, grooved; *F*, lower flat brass electrode; *G*, neoprene stopper; *H*, feed inlet, copper; *I*, lower exit tube, copper; *J*, upper exit tube, copper; *K*, exit holes; and *L*, groove divider cut in Teflon. *(Reprinted by permission of the publisher, The Electrochemical Society, Inc.)*

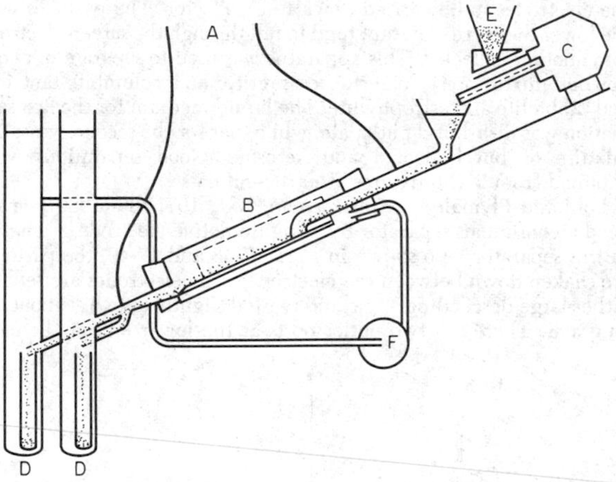

FIG. 17-49 Schematic diagram of isomotive-cell operation: *A*, high-voltage source; *B*, isomotive cell; *C*, vibrator; *D*, powder collectors; *E*, supply hopper for powder mixture; and *F*, pump. *(Reprinted by permission of the publisher, The Electrochemical Society, Inc.)*

trodes. The more polar particles tend to be held up against the pull of gravity and go out the uppermost outlet port, while the less polar particles exit from the lower port.

During operation, the electrodes are submerged in a liquid medium, the mineral mixture is fed between the electrodes by shaker, and the electrodes are shaken separately to move the particles down them. The electrodes are supplied with 0 to 2 kV at 60 Hz. It has been found that choosing the dielectric constant of the liquid medium to be very close to that of one of the particles improves the efficiency of the separation since particles with an effective dielectric constant close to that of the liquid will experience no DEP force and will not distort the field and, therefore, will not be affected by cooperative behavior (mutual DEP) of the other particles.

This apparatus was used to separate many mineral mixtures. More recently, Terry and Pohl used a similar apparatus to demonstrate the separation of commercial glasses. Clear flint, amber, and green glass were used with a particle size range of 20 to 140 mesh. The support liquid was bicyclohexyl containing diphenyl sulfone. Green glass was the most polar, while clear flint glass was the least polar. Continuous separation of various mixtures was readily obtained (Fig. 17-50.)

The continuous DEP separation of minerals in fluid media of low conductivity, then, has been thoroughly demonstrated. It appears ready for wider applications. Separation of biological particles in the more conductive aqueous medium, however, has proved to be a more difficult problem because of the limitations necessary to maintain the viability of living cells. These limitations include the use of lossy aqueous media, low specific power levels, and very small flow-chamber dimensions.

Continuous sorting of mixtures of living cells has been reported by Pohl and Kaler [*Cell Biophys.*, **1**, 15–28 (1979)]. Their apparatus (Figs. 17-51 and 17-52) features a centered stream of cells surrounded by a sheath stream. This allows the cells to be deflected either toward or away from the stronger-field region.

Living cells are more complicated than inanimate particles and have, as a result, more complicated DEP versus frequency responses. Figure 17-53 shows the DEP response of several different species of single-cell organisms as measured over a wide frequency range. As can be seen, the resulting spectrum for each species is distinctive. The DEP spectrum reflects not only types of cells but changes in the physiological conditions of a given cell type. For example, separations or distinctions of cells differing in diet, or in age, or in chemical treatment, or in physiology (e.g., hemophilic versus normal) have been achieved. The changes during the reproductive cycle can be followed by using a single-cell technique. Biological separation by DEP has become a subtle and powerful tool.

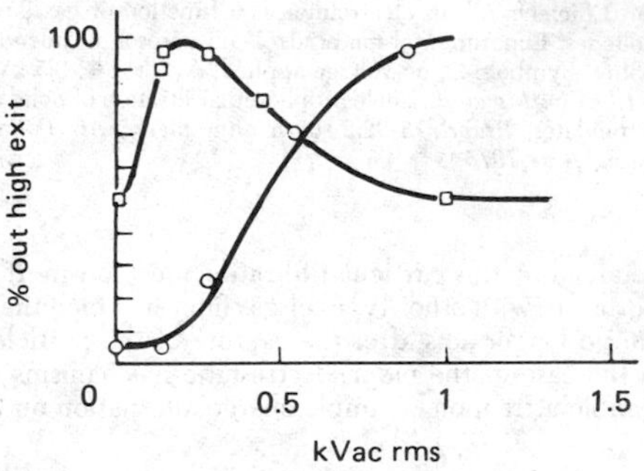

FIG. 17-50 Behavior of sodium dichromate–titanium dioxide (rutile) particle mixtures in an isomotive cell. Symbols: □, percent of rutile in powder coming out of the high exit; O, percent of total powder coming out of the high exit.

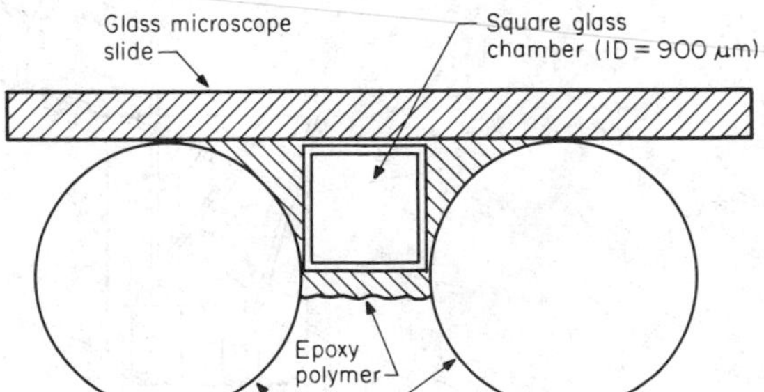

FIG. 17-51 Cross-sectional view of a continuous-flow chamber, showing the arrangement of the square chamber and wire electrodes. To convert micrometers to feet, multiply by 3.281×10^{-6}; to convert millimeters to feet, multiply by 3.281×10^{-3}.

NOVEL SOLID-LIQUID SEPARATION PROCESSES

CROSS-FLOW FILTRATION

GENERAL REFERENCE: Henry, "Cross Flow Filtration," in Li (ed.), *Recent Developments in Separation Science*, vol. 2, CRC Press, Cleveland, 1972.

Process Concept Filtration is a widely used solid-liquid separation process. Conventional filtration processes operate with the slurry flow "dead end" into the filtration media. As such, if the driving pressure differential is held constant, the filtration rate will decrease with time. This decrease in rate is directly attributable to the increase in flow resistance through the growing filter cake. This effect will often be more pronounced when filtering colloidal or micrometer-sized particles as the filtration media which retain these particles are often susceptable to plugging. In order to avoid this problem, fine-particle filtration must often be preceded by a pretreatment step such as flocculation or coagulation in order to increase the particle size of the solids that must be retained.

Limitation of filter-cake growth can reduce the decrease of filtration rate with time. This can be achieved if the slurry flow is tangential to, rather than directly into, the filter medium. This mode of filter operation is termed cross-flow filtration (CFF). Cross-flow filtration employs a high fluid circulation rate tangential to the filter medium in order to minimize the accumulation of particles on the filter surface. This permits CFF to operate as a steady-state process. Conventional filtration, conversely, employs a dead-end flow into the filter medium which produces particle accumulation and reduction of the filtration rate. This dead-end-type flow makes conventional filtration an inherently batch process. These two types of filter operation are compared in Fig. 17-54.

The increased filtration rates obtained with CFF, compared with those of conventional filtration, permit the use of commercially available microporous or ultrafiltration membranes for the retention of colloidal and/or micrometer-sized particles. Increases in filtration rate of two orders of magnitude have been observed with microporous membranes (Henry, op. cit., p. 213).

Advantages of Cross-Flow Filtration Cross-flow filtration has the following potential advantages over conventional filtration.

- Filtration rate is not affected significantly by the particle–suspending-medium density difference.
- Particle accumulation at the filter surface is minimized; i.e., high filtration rates are obtained.
- Feed additives such as flocculating agents are not required.
- The addition of filter aid is not required. This is especially important when contamination of the solid product must be minimized.

These advantages are obtained at an energy cost. The pumping energy required in order to move the slurry tangentially to the filter

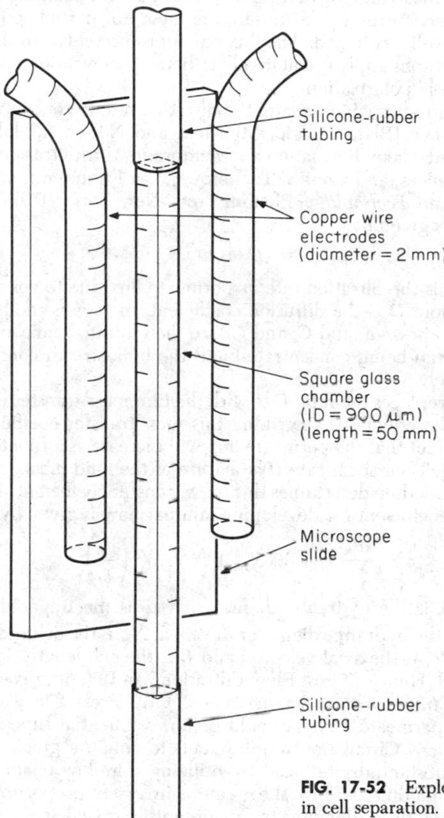

FIG. 17-52 Exploded view of a continuous dielectrophoretic chamber used in cell separation. To convert millimeters to feet, multiply by 3.281×10^{-3}. [*From Pohl and Kaler,* Cell Biophys., *1, 15–28 (1979).*]

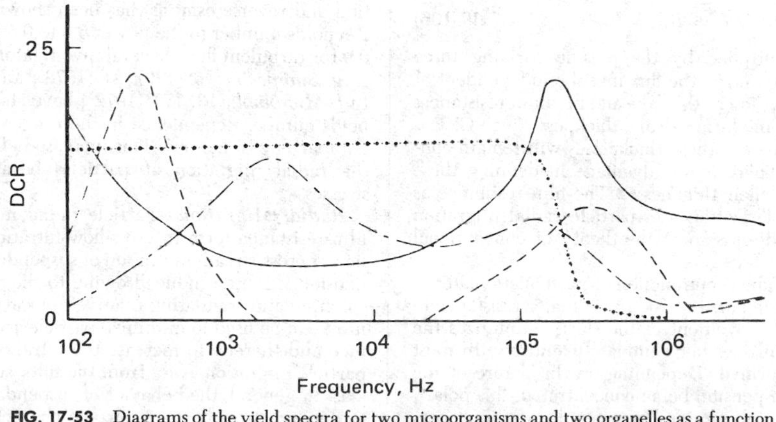

FIG. 17-53 Diagrams of the yield spectra for two microorganisms and two organelles as a function of frequency of the electric field. Symbols: . . . . , *Bacillus subtilis;* ——, *Pseudomonas aeruginosa;* – – – –, chloroplasts; —·—·—, mitochondria.

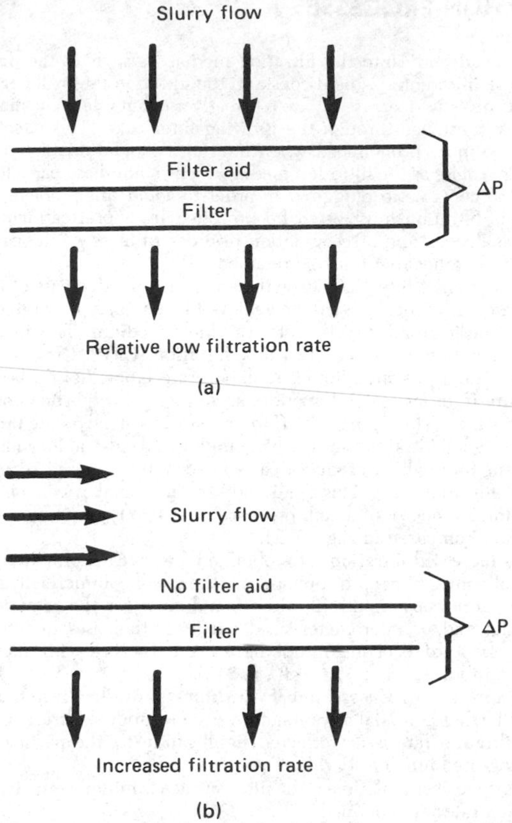

FIG. 17-54 Comparison of conventional and cross-flow filtration. (*a*) Conventional filtration. (*b*) Cross-flow-filtration arrangement to minimize particle accumulations at filter surface.

medium must be supplied in addition to the energy required to supply the pressure differential driving force for filtration.

Theory In general, the filtration flux J, [m/s (gal/ft^2·day)] can be expressed by the following resistance equation:

$$J = \Delta P/R_t \tag{17-105}$$

where the total resistance R_t, (N·s)/m^3, is expressed as a function of the film resistance R_f, the medium resistance R_m, and the cake resistance R_c:

$$R_t = R_f\,\Delta P + R_m + R_c \tag{17-106}$$

The film resistance is multiplied by the pressure driving force because, in the case of film control, the flux must be independent of this driving force. The definition of the cake and medium resistances includes the corresponding medium or cake thickness. Since CFF is inherently a steady-state process, these thicknesses will remain constant for a given system. There is no advantage in defining these resistances separately from their thicknesses. The film resistance as represented by particle polarization and particle radial migration will be discussed later. A discussion of the theory of conventional filtration can be found in Sec. 19.

Particle Polarization The accumulation of particulate material at the surface of the filtration barrier produces a major resistance to filtration. While tangential movement of the slurry minimizes the accumulation of particles, it does not eliminate it entirely. In most cases, a layer of solids is formed. Depending on the nature of the filtration barrier and the suspension being concentrated, this polarization can reduce the filtration rate and/or alter the apparent particle-rejection characteristics of the barrier.

The polarization of the filter occurs because particles are transported to the surface by the liquid and are retained while the liquid passes through the barrier. A gradient in the retained-solid concentration is established in the direction perpendicular to the filter medium. The retained solid is transported back to the bulk stream by a diffusional (brownian-motion) mechanism. This particle polarization can be controlled by varying the rate of flow tangential to the filter medium since this flow promotes the diffusive transport back to the bulk stream. At steady state, the net flux of material toward the membrane is exactly balanced by the flux of material due to diffusion back into the bulk stream (Fig. 17-55). The macrosolute analog, concentration polarization, has been studied widely for the cases of reverse osmosis and ultrafiltration. While particle polarization is similar to concentration polarization, the theory of particle polarization is not as well developed. The flux equations presented in the following paragraphs apply qualitatively to both concentration polarization and particle polarization.

A relationship which relates the filtration rate to concentration has been defined [Blatt, Dravid, Michaels, and Nelson, "Solute Polarization and Cake Formation in Membrane Ultrafiltration: Causes, Consequences, and Control Techniques," in Flinn (ed.), *Membrane Science and Technology,* Plenum Press, New York 1970]. This relationship is given by

$$J_c = (D/\delta)\ \ln\,(C_s/C_b) \tag{17-107}$$

where J_c is the filtration rate or permeate flux due to concentration polarization; D is the diffusion coefficient, m^2/s; δ is the boundary-layer thickness, m; and C_s and C_b are the saturated concentration of the material being concentrated and the bulk-stream concentration respectively.

If one replaces the ratio D/δ with the film mass-transfer coefficient k, then the equations describing this mass-transfer coefficient illustrate the fact that the permeate flux will increase as a function of the fluid velocity or shear rate. The nature of the fluid mechanics in the filtration module determines this mass-transfer coefficient. The mass-transfer coefficient for developing laminar flow is given by

$$\frac{kd_h}{D} = 1.86(N_{\text{Red}})^{1/3}(N_{\text{Sc}})^{1/3}\left(\frac{d_h}{L}\right)^{1/3} \tag{17-108}$$

where d_h is the hydraulic diameter; N_{Red} is the Reynolds number based on the hydraulic diameter $d_h V_1 \rho/\mu$; N_{Sc} is the Schmidt number $(\mu/\rho D)$; V_1 is the axial velocity; and L is the cell length. It has been suggested [Henry, "Cross Flow Filtration," in Li (ed.), *Recent Developments in Separation Science*, vol. 2, CRC Press, Cleveland, 1972] that the permeate flux J_c should be an exponential function of the shear rate $\dot{\gamma}$. Circulation requirements to obtain a given shear rate can be substantially reduced by utilizing thin-flow channels in the filtration module because shear rate is inversely proportional to the channel depth or diameter for a constant fluid velocity.

The degree of mass-transfer-coefficient dependence on the circulation rate for macrosolute and dissolved-solute systems (ultrafiltration and reverse osmosis) has been shown to be proportional to the Reynolds number to the power 0.3 to 0.5 for laminar flow and 0.8 to 0.9 for turbulent flow. Several investigators [Porter, *Am. Inst. Chem. Eng. Symp. Ser.*, **68**(120), 21 (1972); and Henry and Allred, *Dev. Ind. Microbiol.*, **13**, 177 (1972)] have shown a much stronger Reynolds-number dependence in the range of exponent 1 to 2 even in laminar-flow situations. This increased effect has been attributed to the radial migration of particles because of nonuniform shear stresses.

Radial Migration Particle radial migration often reduces the film-resistance term in cross-flow-filtration systems. This phenomenon of cross-stream migration of suspended particles in a flowing suspension is inertia-induced owing to the resultant forces of particles spinning and translating relative to local fluid velocity. This mechanism can be used to minimize particle accumulation at the filter surface and thereby to increase the filtration rate, as it tends to cause particle migration away from the filter surface and toward the duct axis. In general, the behavior of suspended particles has been found to depend strongly upon particle characteristics, bulk-flow configuration, and density difference between the particle and the suspending medium. Radial migration is more significant for a flexible par-

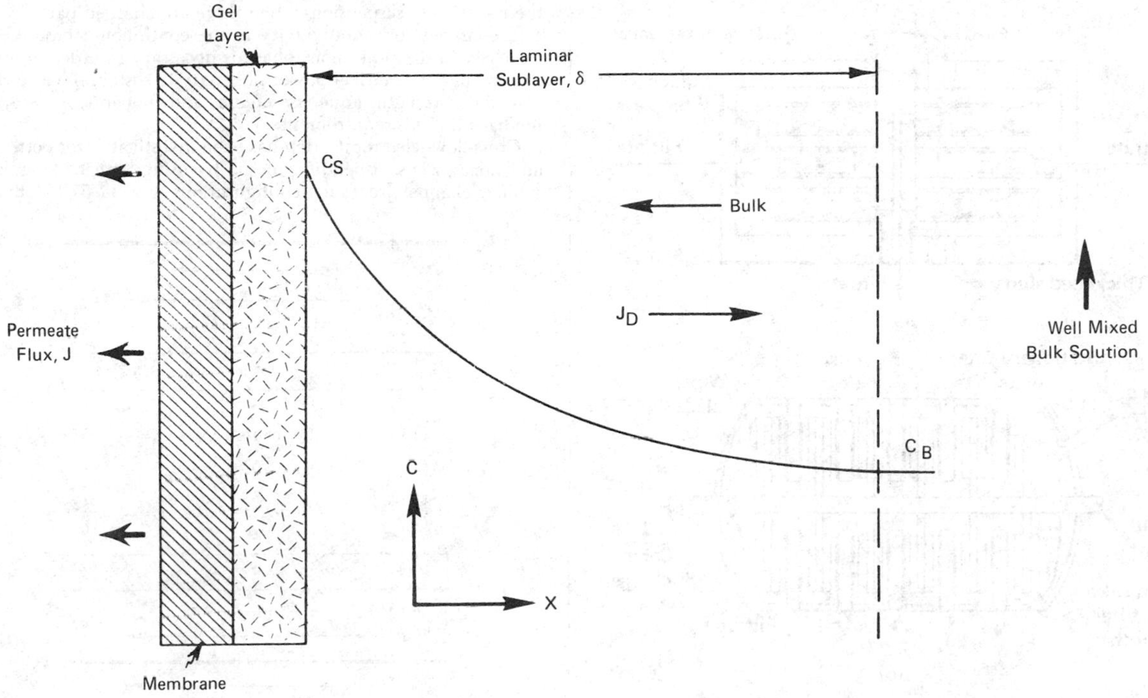

FIG. 17-55 Macrosolute-concentration-polarization model.

ticle than for a rigid particle. An excellent summary of work on radial migration has been presented by Brenner ("Hydrodynamic Resistance of Particles at Small Reynolds Numbers," in *Advances in Chemical Engineering*, vol. 6, Academic, New York, 1966).

The contribution of radial migration to the overall permeate flux in a cross-flow–electrofiltration system has been investigated by Kuo (Ph.D. dissertation, West Virginia University, 1978). The conclusions of this study can be directly applied to standard CFF. It was determined that the slip-shear radial-migration force based on the inertial lift force [Saffman, *J. Fluid Mech.*, **22**, 385 (1965)] dominates any other radial-migration forces by at least one order of magnitude for the experimental systems used in this study. This conclusion led to the use of the following equation to describe the particle radial-migration velocity U_r(m/s):

$$U_r = \left(\frac{81.2}{6\pi}\right) U_s \left(\frac{a}{b}\right) \left(\frac{6U_m b}{\nu}\right)^{0.5} \tag{17-109}$$

where U_s is the slip velocity of the particle relative to the fluid; U_m is the mean axial velocity of the feed flow; ν is the kinematic viscosity, m^2/s; a is the particle radius; and b is the channel height normal to the filter. Kuo also performed a simple order-of-magnitude analysis based on a clay particle in an aqueous suspension in order to demonstrate the effect of particle size on the relative importance of the diffusional and radial-migration velocities. This analysis showed that radial migration will dominate systems in which large particles ($\geq$100 μm) are present, while particle convective mass transfer will dominate systems containing small particles ($\leq$1 μm). These conclusions have been confirmed experimentally.

Since radial migration and particle polarization are both primary sources of resistance across the boundary layer, they can both be combined in order to define a relationship which describes the film resistance R_f:

$$R_f = \frac{\Delta P}{k \ln\left(\dfrac{C_s}{C_b}\right) + U_r} \tag{17-110}$$

It can be seen by examining Eqs. (17-108), (17-109), and (17-110) that increasing the slurry circulation rate (all other things being held constant) reduces film resistance by increasing both the diffusional flux and the radial-migration velocity.

Cross-Flow-Filtration Equipment There are two design considerations for CFF equipment. The first is efficient contacting of the flowing suspension with the surface of the filtration barrier, and the second is the provision of a high shear rate at the wall in order to control cake formation. Both of these design considerations can be satisfied through the use of filter configurations which include narrow flow channels.

Conventional ultrafiltration-equipment configurations can be adapted to use in cross-flow filtration. This is particularly appropriate for tubular turbulent-flow systems. This type of system can be described as slurry flow through a porous pipe. Filtration modules using this configuration often resemble shell-and-tube heat exchangers with the slurry being introduced tube-side and the filtrate being removed shell-side. Several porous substrates have been proposed for the filtration of suspensions of micrometer- and submicrometer-sized particles. These include porous ceramic tubes, porous metals, and fire-hose jackets [Zhevnovatyi, *Int. J. Chem. Eng.*, 4(1), 125 (1964); and Dahlheimer, Thomas, and Kraus, *Ind. Eng. Chem. Process Des. Dev.*, **9**(4), 566 (1970)]. The average pore size of these materials ranges from 0.2 to 50 μm. The porous metal and ceramic substrates offer very good chemical, temperature, and pressure stability compared with most polymer membranes. They are, however, more susceptible to plugging and depth filtration. The two-directional structural stability of many porous substrates permits backwashing of the filtration medium. In general, this is not possible with either ultrafiltration or homogeneous microporous membranes.

Another possible configuration for CFF involves the use of rotating elements in a stationary vessel. The slurry flows across a rotating element, while the filtrate flows through the filter medium and out through the hollow shaft of the rotating element. Centrifugal action, as well as fluid shear, helps in the cake removal. Svarovsky [*Chem. Eng.*, **86**(16), 73 (1979)], in a review article comments on several designs of the rotating "dynamic" filter (Fig. 17-56). The filter con-

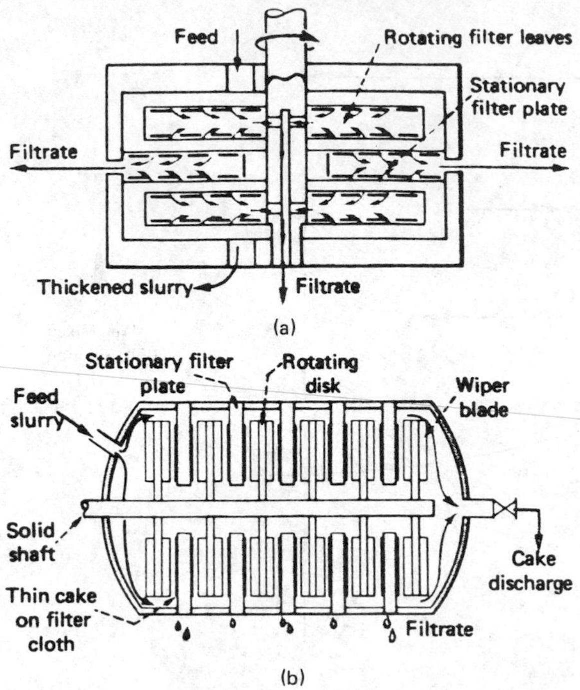

FIG. 17-56 Two versions of the dynamic filter, in which cross-flow filtration is performed with rotating elements. (*a*) European design (only two stages shown). (*b*) United States design. [*From Svarovsky*, Chem. Eng., *86(16)*, 72 *(1979)*.]

figuration shown in Fig. 17-56*a* allows filter permeate to pass through both rotating and stationary filter surfaces. Productivity increases of 5 to 25 times, relative to conventional filtration, for a given moisture content in the final slurry, have been claimed for these filters. This productivity is gauged in terms of the mass of dry cake produced per unit area and time. Another version of the dynamic filter, in which the rotors act only to provide the shear forces necessary for cake removal (Fig. 17-56*b*), is available.

CROSS-FLOW–ELECTROFILTRATION

General References: Henry, Lawler, and Kuo, *Am. Inst. Chem. Eng. J.*, 23(6), 851 (1977). Kuo, Ph.D. dissertation, West Virginia University, 1978.

Process Concept The application of a direct electric field of appropriate polarity when filtering should cause a net charged-particle migration away from the filter medium. This electrophoretic migration will prevent filter-cake formation and the subsequent reduction of filter performance. An additional benefit derived from the imposed electric field is an electroosmotic flux. The presence of this flux in the membrane and in any particulate accumulation may further enhance the filtration rate. The concept of utilizing an applied electric field in this manner is known as electrofiltration (Bier, *Electrophoresis*, vol. I, Academic, New York, 1959). (Note: The cross-flow–electrofiltration process described here involves particle transfer by electrophoresis rather than dielectrophoresis. See the earlier discussion on dielectrofiltration which is included in the subsection "Separation Processes Based Primarily on Action in a Field.")

Cross-flow–electrofiltration (CF-EF) is the multifunctional separation process which combines the electrophoretic migration present in electrofiltration with the particle diffusion and radial-migration forces present in CFF in order to reduce further the formation of filter cake. Cross-flow–electrofiltration can even eliminate the formation of filter cake entirely. This process should find application in the filtration of suspensions when there are charged particles as well as a relatively low conductivity in the continuous phase. Low conductivity in the continuous phase is necessary in order to minimize the amount of electrical power necessary to sustain the electric field. Low-ionic-strength aqueous media and nonaqueous suspending media fulfill this requirement.

Cross-flow–electrofiltration has been investigated for both aqueous and nonaqueous suspending media by using both rectangular- and tubular-channel processing configurations (Fig. 17-57). Henry, Lawler, and Kuo (op. cit.), using a rectangular-channel system with a 0.6-μ-pore-size polycarbonate Nuclepore filtration membrane, investigated CF-EF for 2.5-μm kaolin-water and 0.5- to 2-μm oil-in-water emulsion systems. Kuo (op. cit.), using similar equipment, studied 5-μm kaolin-water, ~ 100-μm Cr_2O_3-water, and ~ 6-μm Al_2O_3-methanol and/or -butanol systems. For both studies electrical fields of 0 to 60 V/cm were used for aqueous systems, and to 5000 V/cm were used for nonaqueous systems. The studies covered a wide range of processing variables in order to gain a better understanding of CF-EF fundamentals. Lee, Gidaspow, and Wasan [*Ind. Eng. Chem. Fundam.*, **19**(2), 166 (1980)] studied CF-EF by using a porous stainless-steel tube (pore size = 5 μm) as the filtration medium. A platinum wire running down the center of the tube acted as one electrode, while the porous steel tube itself acted as the other electrode. Nonaqueous suspensions of 0.3- to 2-μm Al_2O_3-tetralin and a coal-derived liquid diluted with xylene and tetralin were studied. By operating with applied electric fields (1000 to 10,000 V/cm) above the critical voltage, clear particle-free filtrates were produced. It should be noted that the pore size of the stainless-steel filter medium (5 μm) was greater than the particle size of the suspended Al_2O_3 solids (0.3 to 2 μm).

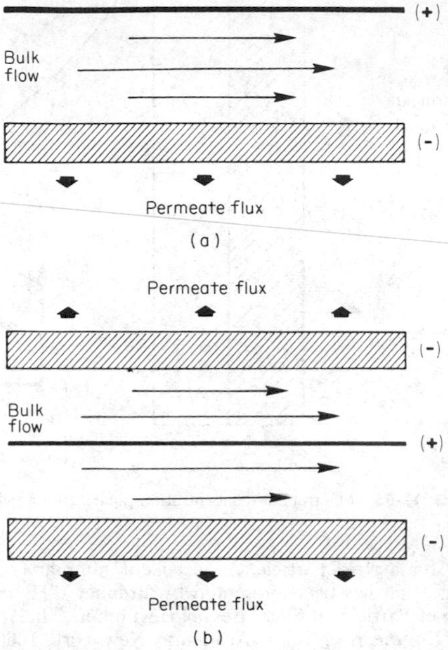

FIG. 17-57 Alternative electrode configurations for cross-flow–electrofiltration.

Theory Cross-flow–electrofiltration can theoretically be treated as if it were cross-flow filtration with superimposed electrical effects. These electrical effects include electroosmosis in the filter medium and cake and electrophoresis of the particles in the slurry. Equations (17-105) and (17-106) still apply to this system, with only the definition of the terms R_f, R_m, and R_c being changed. The addition of the applied electric field can, however, result in some qualitative differences in permeate-flux-parameter dependences.

The membrane resistance for CF-EF can be defined by specifying two permeate fluxes as

$$J_{om} = \Delta P/R_{om} \qquad (17\text{-}111)$$

$$J_m = \Delta P/R_m \qquad (17\text{-}112)$$

where J_{om} is the flux through the membrane in the absence of an electric field and any other resistance, m/s; J_m is the same flux in the presence of an electric field; and R_{om} is the membrane resistance in the absence of an electric field, $(N\cdot s)/m^3$. When electroosmotic effects do occur,

$$J_m = J_{om} + K_m E \qquad (17\text{-}113)$$

where K_m is the electroosmotic coefficient of the membrane, $m^2/(V\cdot s)$; and E is the applied-electric-field strength, V/m. Equations (17-111), (17-112), and (17-113) can be combined and rearranged to give Eq. (17-114), the membrane resistance in the presence of an electric field.

$$R_m = \frac{R_{om}}{1 + \left(\dfrac{K_m E}{J_{om}}\right)} \qquad (17\text{-}114)$$

Similarly, cake resistance can be represented as

$$R_c = \frac{R_{oc}}{1 + \left(\dfrac{K_c E}{J_{oc}}\right)} \qquad (17\text{-}115)$$

where J_{oc} is the flux through the cake in the absence of an electric field or any other resistance, R_{oc} is the cake resistance in the absence of an electric field, and K_o is the electroosmotic coefficient of the cake. The cake resistance is not a constant but is dependent upon the cake thickness, which is in turn a function of the transmembrane pressure drop and electrical-field strength.

Particulate systems require the addition of the term $\mu_e E$ in order to account for the electrophoretic migration of the particle. The constant μ_e is the electrophoretic mobility of the particle, $m^2/(V\cdot s)$. For the case of the CF-EF, the film resistance R_f can be represented as

$$R_f = \frac{\Delta P}{k \ln\left(\dfrac{C_s}{C_b}\right) + U_r + \mu_e E} \qquad (17\text{-}116)$$

The resistances, when incorporated into Eqs. (17-106) and (17-105), yield the general expression for the permeate flux for particulate suspensions in cross-flow-electrofiltration systems.

There are three distinct regimes of operation in CF-EF. These regimes (Fig. 17-58) are defined by the magnitude of the applied electric field with respect to the critical voltage E_c. The critical voltage is defined as the voltage at which the net particle migration velocity toward the filtration medium is zero. At the critical voltage, there is a balance between the electrical-migration and radial-migration velocities away from the filter and the velocity at which the particles are swept toward the filter by bulk flow. There is no diffusive transport at $E = E_c$ (Fig. 17-58*b*) because there is no gradient in the particle concentration normal to the filter surface. At field strengths below the critical voltage (Fig. 17-58*a*), all migration velocities occur in the same direction as in the cross-flow-filtration systems discussed earlier. At values of applied voltage above the critical voltage (Fig. 17-58*c*) qualitative differences are observed. In this case, the electrophoretic-migration velocity away from the filter medium is greater than the velocity caused by bulk flow toward the filtration medium. Particles concentrate away from the filter medium. This implies that particle concentration is lowest next to the filter medium (in actuality, a clear boundary layer has been observed). The influence of fluid shear still improves the transfer of particles down the concentration gradient, but in this case it is toward the filtration medium. When the particles are small and diffusive transport dominates radial migration, increasing the circulation velocity will decrease the permeate flux rate in this regime. When the particles are large and radial migration dominates, the increase in circulation velocity will still improve the filtration rate. These effects are illustrated qualitatively in Fig. 17-59*a*. The solid lines represent systems in which the particle diffusive effect dominates the radial-migration effect, while the dashed lines represent the inverse. Figure 17-59*b* illustrates the increase in filtration rate with increasing electric field strength. For field strengths $E > E_c$, increases in permeate flux rate are due only to electroosmosis in the filtration medium.

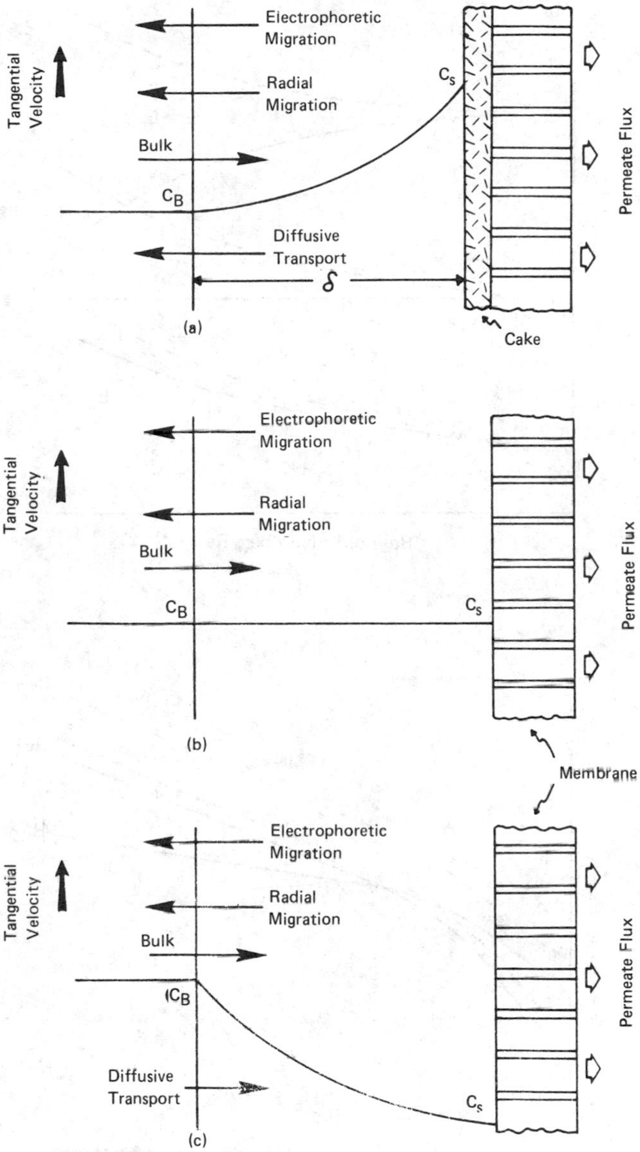

FIG. 17-58 Regimes of operation of cross-flow–electrofiltration: (*a*) voltage less than critical, (*b*) voltage equal to the critical voltage, (*c*) voltage greater than critical.

One potential difficulty with CF-EF is the electrodeposition of the particles at the electrode away from the filtration medium. This phenomenon, if allowed to persist, will result in performance decay of CF-EF with respect to maintenance of the electric field. Several approaches such as momentary reverses in polarity, protection of the

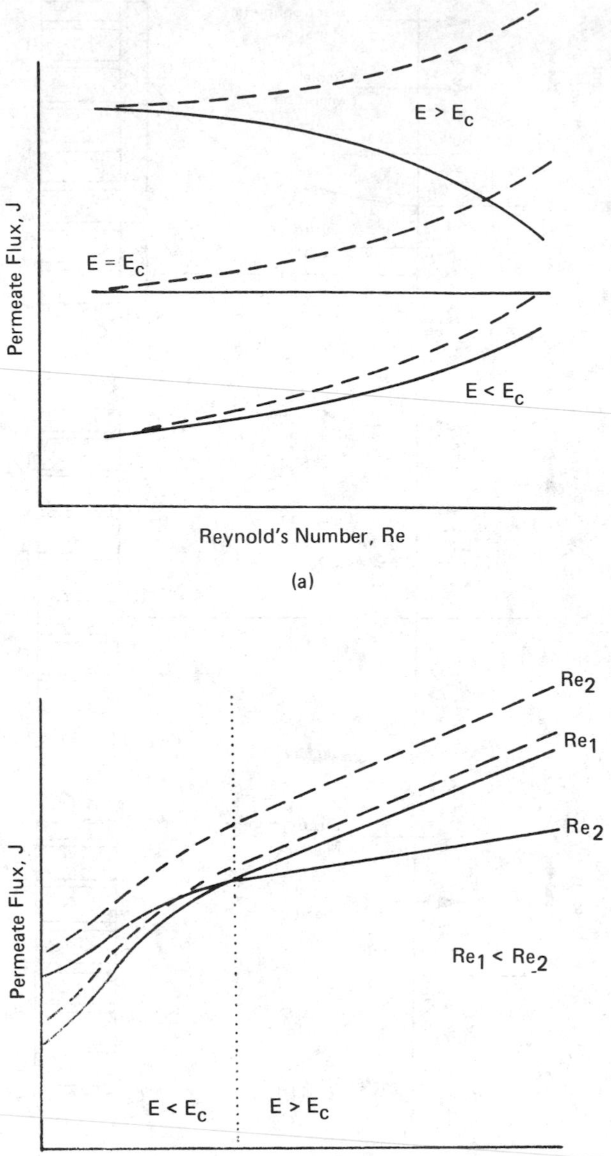

FIG. 17-59 Qualitative effects of Reynolds number and applied-electric-field strength on the filtration permeate flux J. Dashed lines indicate large particles (radial migration dominates); solid lines, small particles (particle diffusion dominates).

electrode with a porous membrane or filter medium, and/or utilization of a high fluid shear rate can minimize electrodeposition.

Related Processes There have been developed several electrofiltration systems which are suitable for processes in concentrated suspensions. These processes often rely primarily on the electroosmosis mechanism. The electrofiltration process developed by Freeman (U.S. Patent 4,107,026, 1978) is an example.

DUAL-FUNCTIONAL FILTER

The dual-functional filter, also called the Uni-Flow filter (Comirand and Popper, U.S. Patent 3,523,077, 1970), is a solid-liquid separation process which is capable of dewatering slurries of micrometer-sized, gelatinous materials that are difficult to dewater by other methods.

The dual-functional filter (Fig. 17-60) utilizes vertical, collapsible porous-fiber hoses (nylon, polypropylene, cotton, etc.) as a filter medium. The operation of this system is cyclic in nature. First, the feed slurry enters the top of the hose, the bottom dumping valve being closed. Filtrate is removed through the porous walls of the hose as filter cake builds up inside. The filtration phase is terminated before solids fill the hose. After the predetermined cycle time, the feed line is closed while the dumping valve opens simultaneously. The flexible hose collapses as a result of the associated pressure release, ejecting the filter cake and some unfiltered feed slurry into the settling receiver. The cake, of which only a fraction will redisperse, settles out into a product sludge, leaving the unfiltered slurry to be recycled. Rapid cycling of the filter, causing the formation of only very thin filter cakes, results in the achievement of high filtration rates.

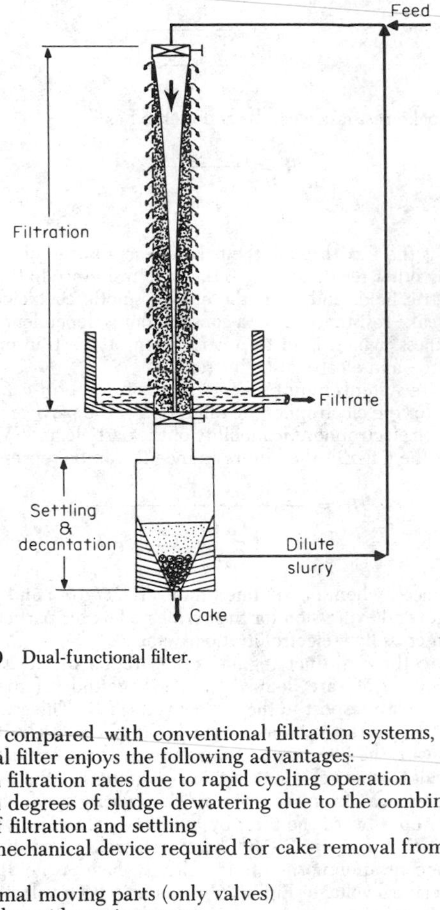

FIG. 17-60 Dual-functional filter.

When compared with conventional filtration systems, the dual-functional filter enjoys the following advantages:

- High filtration rates due to rapid cycling operation
- High degrees of sludge dewatering due to the combined mechanisms of filtration and settling
- No mechanical device required for cake removal from the filter hose
- Minimal moving parts (only valves)
- No filter-aid requirements
- Applicability to gelatinous suspensions of submicrometer particles
- Relative ease of scale-up

It should be noted that an empirical selection of the proper hose material must be made for a given slurry in order to obtain a filter medium which will simultaneously contain the submicrometer particles being separated and effectively dump the cake at the end of the filtration cycle.

Only a few studies of the dual-functional filter have been made. Most of them aim at the dewatering of neutralized-acid mine drainage, an environmental problem caused by coal mining. Page, Copely, and Shackelford ("A New Filtration Device for Concentrating Neutralized AMD Sludge," 4th Symposium on Coal Mine Drainage, Pittsburgh, 1972), tested a bank of two five-hose filter units piped in series. With hose lengths of 6 m (20 ft) they were able to filter a feed slurry containing 0.1 weight percent solids at an average rate of 0.28 gal/(min·ft^2). The product sludge contained 25 weight percent solids. The variables determined to be important to the design of a dual-functional filter system were the number and length of the hoses, the material and weave of the hoses, and the pressure applied to the system. Nawrocki (Contr. 68-01-0743, Office of Research and Development, U.S. EPA, 1974) assembled preliminary cost estimates for the dual-functional filter. Henry, Lui, and Kuo [*Am. Inst. Chem. Eng. J.*, **22**(3), 433 (1976)] used data derived from the filtering of neutralized-acid mine drainage in order to develop a mathematical model to describe the filter. This model includes the effects of both the axial variation of the pressure driving force and filter-cake compressibility.

SURFACE-BASED SOLID-LIQUID SEPARATIONS INVOLVING A SECOND LIQUID PHASE

GENERAL EFERENCE: Fuerstenau, "Fine Particle Flotation," in Somasundaran (ed.), *Fine Particles Processing*, vol. 1, American Institute of Mining, Metallurgical, and Petroleum Engineers, New York, 1980. Henry, Prudich, and Lau, *Colloids Surf.*, **1**, 335 (1980). Henry, Prudich, and Vaidyanathan, *Sep. Purif. Methods*, 8(2), 31 (1979). Jacques, Hovarongkura, and Henry, *Am. Inst. Chem. Eng. J.*, **25**(1), 160 (1979). Stratton-Crawley, "Oil Flotation: Two Liquid Flotation Techniques," in Somasundaran and Arbiter (eds.), *Beneficiation of Mineral Fines*, American Institute of Mining, Metallurgical, and Petroleum Engineers, New York, 1979.

Process Concept Three potential surface-based regimes of separation exist when a second, immiscible liquid phase is added to another, solids-containing liquid in order to effect the removal of solids. These regimes (Fig. 17-61) are:

1. Distribution of the solids into the bulk second liquid phase
2. Collection of the solids at the liquid-liquid interface
3. Bridging or clumping of the solids by the added fluid in order to form an agglomerate followed by settling or filtration

These separation techniques should find particular application in systems containing fine particles. The surface chemical differences involved among these separation regimes are only a matter of degree; i.e., all three regimes require the wetting of the solid by the second liquid phase. The addition of a surface-active agent is sometimes needed in order to achieve the required solids wettability. In spite of this similarity, applied processing (equipment configuration, operating conditions, etc.) can vary widely. Collection at the interface would normally be treated as a flotation process (see also Sec. 17: "Adsorptive-Bubble Separation Methods"; and Sec. 21: "Flotation"), distribution to the bulk liquid as a liquid-liquid extraction analog, and particle bridging as a settling (sedimentation) or filtration process.

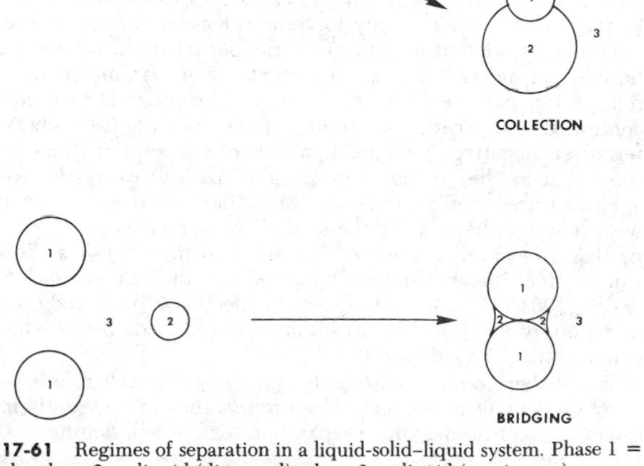

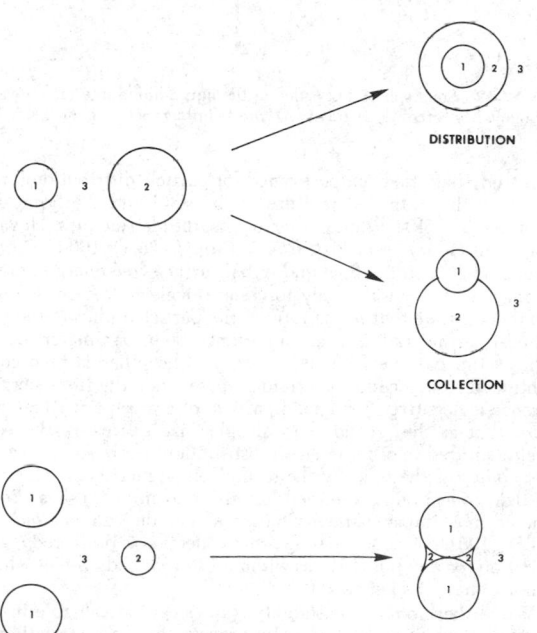

FIG. 17-61 Regimes of separation in a liquid-solid–liquid system. Phase 1 = particle; phase 2 = liquid (dispersed); phase 3 = liquid (continuous).

Even though surface-property-based liquid-solid-liquid separation techniques have yet to be widely used in significant industrial applications, several studies which demonstrate their effectiveness have appeared in literature.

Albertsson (*Partition of Cell Particles and Marcromolecules*, 2d ed., Wiley-Interscience, New York, 1971) has extensively used particle distribution to fractionate mixtures of biological products. In order to demonstrate the versatility of particle distribution, he has cited the example shown in Table 17-31. The feed mixture consisted of polystyrene particles, red blood cells, starch, and cellulose. Liquid-liquid particle distribution has also been studied by using mineral-matter particles (average diameter = 5.5 μm) extracted from a coal liquid as the solid in a xylene-water system [Prudich and Henry, *Am. Inst. Chem. Eng. J.*, **24**(5), 788 (1978)]. By using surface-active agents in order to enhance the water wettability of the solid particles, recoveries of better than 95 percent of the particles to the water phase were observed. All particles remained in the xylene when no surfactant was added.

TABLE 17-31 Separations of Particles between Two Phases

System	Top phase	Bottom phase
Polyethylene glycol / salt	Polystyrene	All others
PEG / Dextran; 20,000 MW	Algae	All others
PEG / Dextran; 200,000 MW	Red cells	All others
Methyl cellulose / Dextran	Cellulose particles	Starch

Particle collection at a liquid-liquid interface is a particularly favorable separation process when applied to fine-particle systems. Advantages of this type of processing include:

- Decreased liquid-liquid interfacial tension (when compared with a gas-liquid system) results in higher liquid-liquid interfacial areas, which favor solid-particle droplet collisions.
- Liquid-solid interactions due to long-range intermolecular forces are much larger than are gas-solid interactions. This means that it is easier to collect fine particles at a liquid-liquid interface than at a gas-liquid interface.
- The increased momentum of liquid droplets (when compared with gas) should favor solid-particle collection.

Fuerstenau [Lai and Fuerstenau, *Trans. Am. Inst. Min. Metall. Pet. Eng.*, **241**, 549 (1968); Raghavan and Fuerstenau, *Am. Inst. Chem. Eng.* Symp. Ser., **71**(150), 59 (1975)] has studied this process with respect to the removal of alumina particles (0.1 μm) and hematite particles (0.2 μm) from an aqueous solution by using isooctane. The use of isooctane as the collecting phase for the hematite particles resulted in an increase in particle recovery of about 50 percent over that measured when air was used as the collecting phase under the same conditions. The effect of the wettability of the solid particles (as measured by the three-phase contact angle) on the recovery of hematite in the water-isooctane system is shown in Fig. 17-62. This behavior is typical of particle collection. Particle collection at an oil-water interface has also been studied with respect to particle removal from a coal liquid. Particle removals averaging about 80 percent

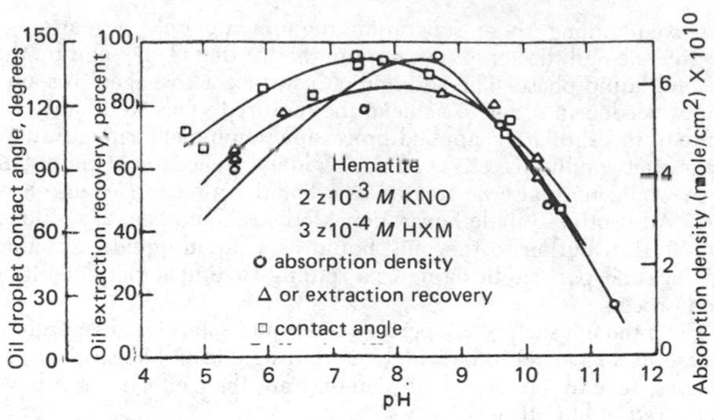

FIG. 17-62 The variation of adsorption density, oil-droplet contact angle, and oil-extraction recovery of hematite as a function of pH. To convert gram-moles per square centimeter to pound-moles per square foot, multiply by 2.048. [*From Raghavan and Fuerstenau,* Am. Inst. Chem. Eng. Symp. Ser., *71(150), 59 (1975).*]

have been observed when water is used as the collecting phase (Lau, master's thesis, West Virginia University, 1979). Surfactant addition was necessary in order to control the wettability of the solids.

Particle bridging has been chiefly investigated with respect to spherical agglomeration. Spherical agglomeration involves the collecting or transferring of the fine particles from suspension in a liquid phase into spherical aggregates held together by a second liquid phase. The aggregates are then removed from the slurry by filtration or settling. Like the other liquid-solid-liquid separation techniques, the solid must be wet by the second liquid phase. The spherical agglomeration process has resulted in the development of a pilot unit called the Shell Pelletizing Separator [Zuiderweg and Van Lookeren Campagne, *Chem. Eng. (London),* **220,** CE223 (1968)]. A detailed discussion of spherical agglomeration can be found in Sec. 8: "Size Enlargement."

The ability to determine in advance which of the separation regimes is most advantageous for a given liquid-solid-liquid system would be desirable. No set of criteria with which to make this determination presently exists. Work has been done with respect to the identification of system parameters which make these processes technically feasible. The results of these studies can be used to guide the selection of the second liquid phase as well as to suggest approximate operating conditions (dispersed-liquid droplet size, degree and type of mixing, surface-active-chemical addition, etc.).

Theory Theoretical analyses of spherical particles suspended in a planar liquid-liquid interface have appeared in literature for some time, the most commonly presented forms being those of a free energy and/or force balance made in the absence of all external body forces. These analyses are generally used to define the boundary criteria for the shift between the collection and distribution regimes, the bridging regime not being considered. This type of analysis shows that for a spherical particle possessing a three-phase contact angle between 0 and 180°, as measured through the receiving or collecting phase, collection at the interface is favored over residence in either bulk phase. These equations are summarized, using a derivation of Young's equation, as

$$\frac{\gamma_{s2} - \gamma_{s1}}{\gamma_{12}} > 1 \text{ particles wet to phase 1} \qquad (17\text{-}117)$$

$$\frac{\gamma_{s2} - \gamma_{s1}}{\gamma_{12}} < -1 \text{ particles wet to phase 2} \qquad (17\text{-}118)$$

$$\left|\frac{\gamma_{s2} - \gamma_{s1}}{\gamma_{12}}\right| \leq 1 \text{ particle at interface} \qquad (17\text{-}119)$$

where γ_{ij} is the surface tension between phases i and j, N/m (dyn/cm); s indicates the solid phase; and subscripts 1 and 2 indicate the two liquid phases.

Several additional studies [Winitzer, *Sep. Sci.,* 8(1), 45 (1973); ibid., 8(6), 647 (1973); Maru, Wasan, and Kintner, *Chem. Eng. Sci.,* **26,** 1615 (1971); and Rapacchietta and Neumann, *J. Colloid Interface Sci.,* **59**(3), 555 (1977)] which include body forces such as gravitational acceleration and buoyancy have been made. A typical example of a force balance describing such a system (Fig. 17-63) is summarized in Eq. (17-120).

$$[(\gamma_{s1} - \gamma_{s2}) \cos \delta + \gamma_{12} \cos B]L = g\,[V_{\text{total}}\,\rho_s - V_1\rho_1 - V_2\,\rho_2] \qquad (17\text{-}120)$$

where V_1 is the volume of the particle in fluid phase 1, V_2 is the volume in fluid phase 2, L is the particle circumference at the interface between the two liquid phases, ρ_i is the density of phase i, and g is the gravitational constant. The left-hand side of the equation represents the surface forces acting on the solid particle, while the right-hand side includes the gravitational and buoyancy forces. This example illustrates the fact that body forces can have a significant effect on system behavior. The solid-particle size as well as the densities of the solid and both liquid phases are introduced as important system parameters.

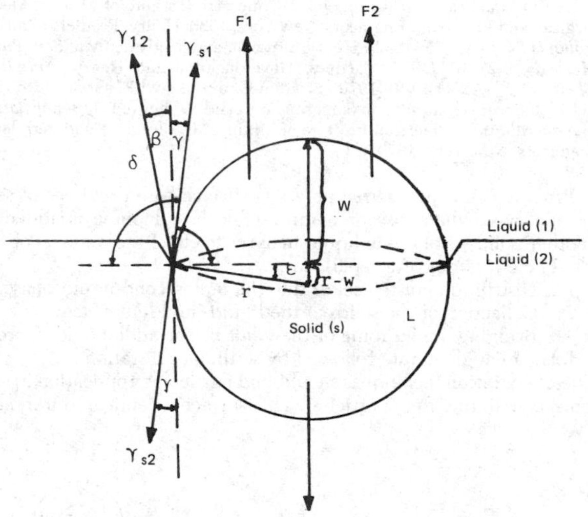

FIG. 17-63 Solid sphere suspended at the liquid-liquid interface. F_1 and F_2 are buoyancy forces; F_s is gravity. [*From Winitzer,* Sep. Sci., *8(1), 45 (1973).*]

A study has also been performed for particle distribution for cases in which the radii of curvature of the solid and the liquid-liquid interface are of the same order of magnitude [Jacques, Hovarongkura, and Henry, *Am. Inst. Chem. Eng. J.,* **25**(1), 160 (1979)]. Differences between the final and initial surface free energies are used to analyze this system. Body forces are neglected. Results (Fig. 17-64) demonstrate that n, the ratio of the particle radius to the liquid-liquid-interface radius, is an important system parameter. Distribution of the particle from one phase to the other is favored over continued residence in the original phase when the free-energy difference is negative. For a solid particle of a given size, these results show that as the second-phase droplet size decreases, the contact angle required in order to effect distribution decreases (the required wettability of the solid by the second phase increases). The case of particle collection at a curved liquid-liquid interface has also been studied in a similar manner [Smith and Van de Ven, *Colloids Surf.,* **2,** 387 (1981)]. This study shows that collection is preferred over distribution for any n in systems without external body forces when the contact angle lies between 0 and 180°.

While thermodynamic-stability studies can be valuable in evaluating the technical feasibility of a process, they are presently inadequate in determining which separation regime will dominate a par-

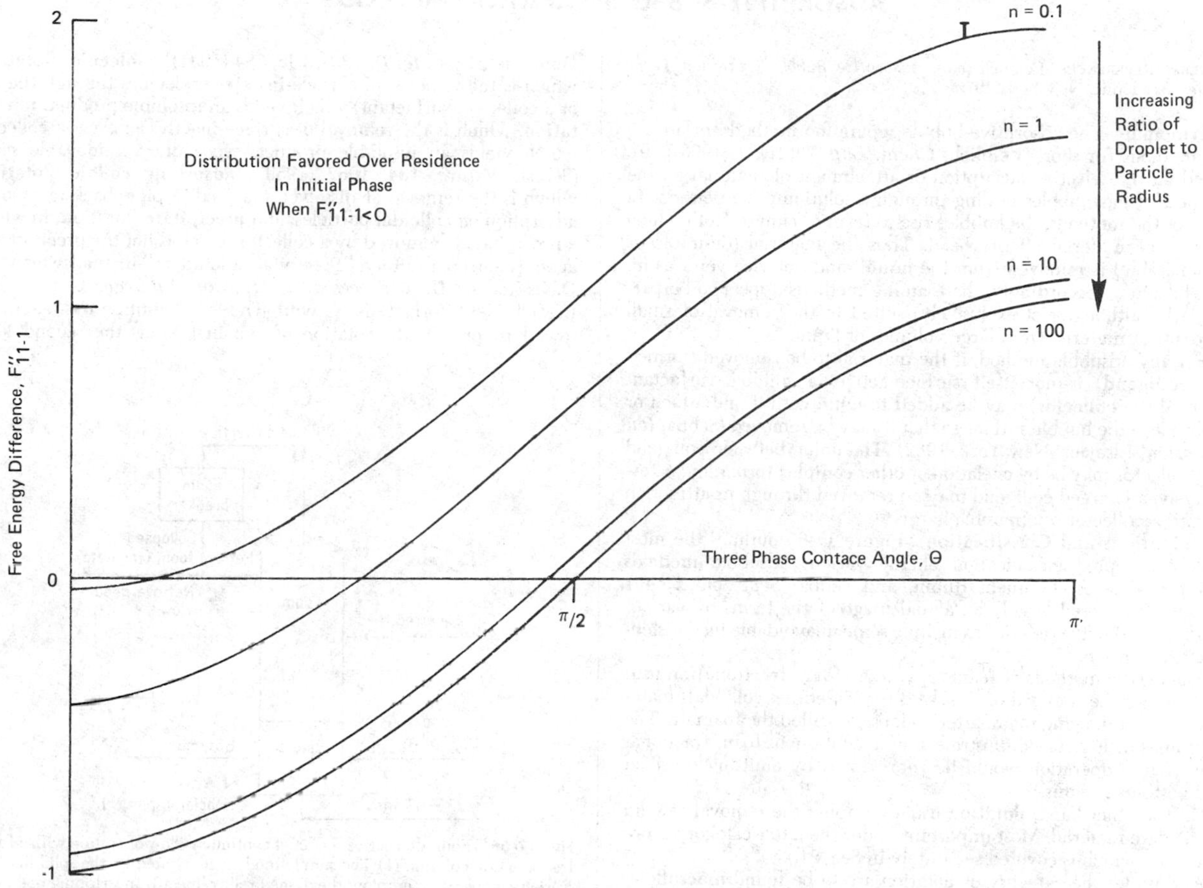

FIG. 17-64 Normalized free-energy difference between distributed (II) and nondistributed (I) states of the solid particles versus three-phase contact angle (collection at the interface is not considered). A negative free-energy difference implies that the distributed state is preferred over the nondistributed state. Note especially the significant effect of n, the ratio of the liquid droplet to solid-particle radius. [*From Jacques, Hovarongkura, and Henry,* Am. Inst. Chem. Eng. J., *25(1), 160 (1979)*].

ticular liquid-solid-liquid system. These analyses ignore important processing phenomena such as the mechanism of encounter of the dispersed-phase liquid with the solid particles, the strength of particle attachment, and the mixing-energy input necessary to effect the separation. No models of good predictive value which take all these variables into account have yet been offered. Until the effects of these and other system variables can be adequately understood, quantified, and combined into such a predictive model, no a priori method of performance prediction will be possible.

SIROFLOC PROCESS

The heterocoagulation of wastewater impurities with small magnetite particles has recently been developed into a commercial solid-liquid separation process. The novel aspect of this process is its ability to recycle the magnetite particles because of the reversibility of their surface chemistry. This reversibility is accomplished through pH control of the bulk aqueous phase. At values of pH below their isoelectric point, the magnetite particles exhibit a positive surface charge which helps them to collect and bind the negatively charged water contaminants. Upon adjusting the pH to values above the isoelectric point, the magnetite's surface charge is reversed and the contaminants are released. This surface chemistry has been described by Kolarik [*Chem. Aust.*, **47** (6), 234 (1980)]. The overall solid-liquid separation process works as follows:

1. The untreated water is acidified and mixed with the magnetite particles.
2. A coagulating agent is added to aid in the collection of the impurities by the magnetite particles.
3. The coagulated contaminant-magnetite particles are removed, leaving a treated, clean water.
4. The stream containing the magnetite particles is treated with caustic soda, resulting in the destruction of the coagulate.
5. The magnetite is washed clean and added back to the acidified, untreated water.

Step 3 is effected by a conventional drum magnetic separator. The pilot process utilizes sedimentation in an applied magnetic field. The applied magnetic field causes the magnetite particles to flocculate and thereby further accelerate the sedimentation rate.

The Sirofloc process, which involves heterocoagulation, has proved to be much faster and simpler than conventional alum flocculation processes. The total processing time for a water, including clarification, has been reduced from 2 to 3 h to 15 to 20 min through the use of this process. Furthermore, preliminary economics indicate that a 20 to 40 percent savings in capital costs can be realized by using the Sirofloc process (Anderson, Blesing, Bolto, Kolarik, Priestly, and Raper, 8th Federal Convention of the Australian Water and Wastewater Association, 1979). A 35-ML/day treatment facility is presently being operated near Perth, Western Australia.

ADSORPTIVE-BUBBLE SEPARATION METHODS

GENERAL REFERENCE: Lemlich (ed.), *Adsorptive Bubble Separation Techniques*, Academic, New York, 1972.

Principle The adsorptive-bubble separation methods, or adsubble methods for short [Lemlich, *Chem. Eng.* **73**(21), 7 (1966)], are based on the selective adsorption or attachment of material on the surfaces of gas bubbles passing through a solution or suspension. In most of the methods, the bubbles rise to form a foam or froth which carries the material off overhead. Thus the material (desirable or undersirable) is removed from the liquid, and not vice versa as in, say, filtration. Accordingly, the foaming methods appear to be particularly (although not exclusively) suited to the removal of small amounts of material from large volumes of liquid.

For any adsubble method, if the material to be removed (termed the **colligend**) is not itself surface-active, a suitable surfactant (termed the **collector**) may be added to unite with it and attach or adsorb it to the bubble surface so that it may be removed (Sebba, *Ion Flotation*, Elsevier, New York, 1962). The union between colligend and collector may be by chelation or other complex formation. Alternatively, a charged colligend may be removed through its attraction toward a collector of opposite charge.

Definitions and Classification Figure 17-65 outlines the most widely accepted classification of the various adsubble methods [Karger, Grieves, Lemlich, Rubin, and Sebba, *Sep. Sci.*, **2**, 401 (1967)]. It is based largely on actual usage of the terms by various workers, and so the definitions include some unavoidable inconsistencies and overlap.

Among the methods of foam separation, **foam fractionation** usually implies the removal of dissolved (or sometimes colloidal) material. The overflowing foam, after collapse, is called the foamate. The solid lines of Fig. 17-66 illustrate simple continuous foam fractionation. (Batch operation would be represented by omitting the feed and bottoms streams.)

On the other hand, **flotation** usually implies the removal of solid particulate material. Most important under the latter category is **ore flotation,** which is covered separately in Sec. 21.

Also under the category of flotation are to be found **macroflotation,** which is the removal of macroscopic particles; **microflotation** (also called **colloid flotation**), which is the removal of microscopic particles, particularly colloids or microorganisms [Dognon and Dumontet, *Comptes Rendus*, **135**, 884 (1941)]; **molecular flotation,** which is the removal of surface-inactive molecules through the use of a collector (surfactant) which yields an insoluble product; **ion flotation,** which is the removal of surface-inactive ions via a collector which yields an insoluble product, especially a removable scum [Sebba, *Nature*, **184**, 1062 (1959)]; **adsorbing colloid flotation,** which is the removal of dissolved material in piggyback fashion by adsorption on colloidal particles; and **precipitate flotation,** in which a precipitate is removed by a collector which is not the precipitating agent [Baarson and Ray, "Precipitate Flotation," in Wadsworth and Davis (eds.), *Unit Processes in Hydrometallurgy*, Gordon and Breach, New York, 1964, p. 656]. The last definition has been narrowed to precipitate flotation of the first kind, the second kind

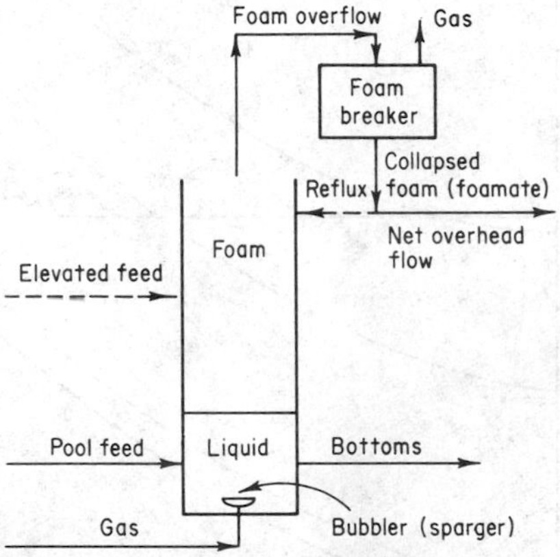

FIG. 17-66 Four alternative modes of continuous-flow operation with a foam-fractionation column: (1) The simple mode is illustrated by the solid lines. (2) Enriching operation employs the dashed reflux line. (3) In stripping operation, the elevated dashed feed line to the foam replaces the solid feed line to the pool. (4) For combined operation, reflux and elevated feed to the foam are both employed.

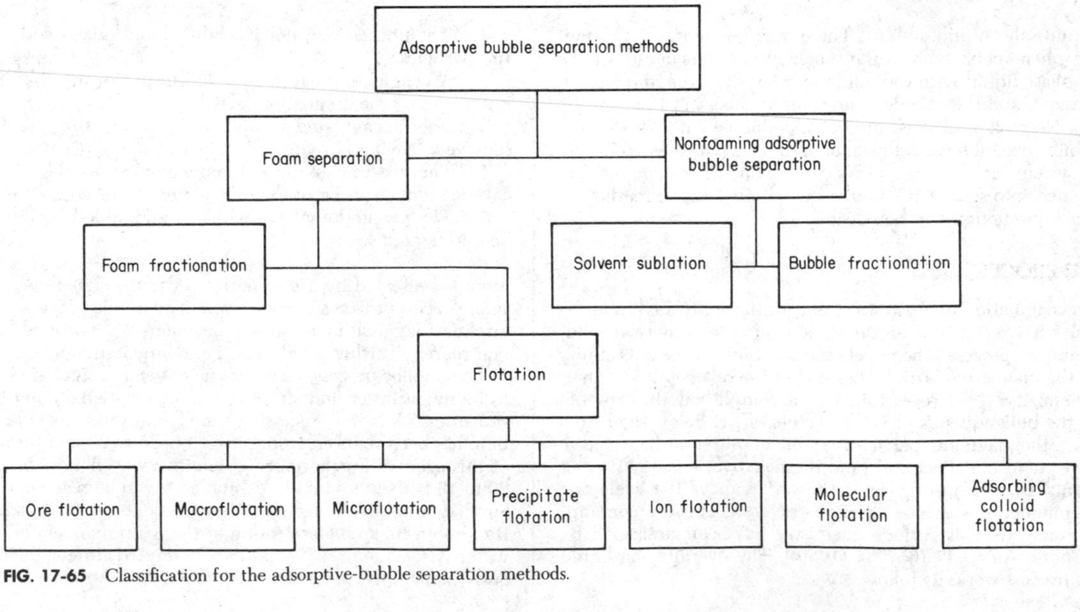

FIG. 17-65 Classification for the adsorptive-bubble separation methods.

requiring no separate collector at all [Mahne and Pinfold, *J. Appl. Chem.*, **18**, 52 (1968)].

A separation can sometimes be obtained even in the absence of any foam (or any floated floc or other surrogate). In **bubble fractionation** this is achieved simply by lengthening the bubbled pool to form a vertical column [Dorman and Lemlich, *Nature*, **207**, 145 (1965)]. The ascending bubbles then deposit their adsorbed or attached material at the top of the pool as they exit. This results in a concentration gradient which can serve as a basis for separation. Bubble fractionation can operate either alone or as a booster section below a foam fractionator, perhaps to raise the concentration up to the foaming threshold.

In **solvent sublation** an immiscible liquid is placed atop the main liquid to trap the material deposited by the bubbles as they exit (Sebba, *Ion Flotation*, Elsevier, New York, 1962). The upper liquid should dissolve or at least wet the material. With appropriate selectivity, the separation so achieved can sometimes be much greater than that with bubble fractionation alone.

The droplet analogs to the adsubble methods have been termed the **adsoplet methods** (from adsorptive droplet separation methods) [Lemlich, "Adsorptive Bubble Separation Methods," *Ind. Eng. Chem.*, **60**(10), 16 (1968)]. They are omitted from Fig. 17-65, since they involve adsorption or attachment at *liquid-liquid* interfaces. Among them are **emulsion fractionation** [Eldib, "Foam and Emulsion Fractionation," in Kobe and McKetta (eds.), *Advances in Petroleum Chemistry and Refining*, vol. 7, Interscience, New York, 1963, p. 66], which is the analog of foam fractionation; and **droplet fractionation** [Lemlich, loc. cit.; and Strain, *J. Phys. Chem.*, **57**, 638 (1953)], which is the analog of bubble fractionation. Similarly, the old beneficiation operation called bulk oil flotation (Gaudin, *Flotation*, 2d ed., McGraw-Hill, New York, 1957) is the analog of modern ore flotation. By and large, the adsoplet methods have not attracted the attention accorded to the adsubble methods.

Of all the adsubble methods, foam fractionation is the one for which chemical engineering theory is the most advanced. Fortunately, some of this theory also applies to other adsubble methods.

Adsorption The separation achieved depends in part on the selectivity of adsorption at the bubble surface. At equilibrium, the adsorption of dissolved material follows the Gibbs equation (Gibbs, *Collected Works*, Longmans Green, New York, 1928).

$$d\gamma = -\mathbf{R}T\Sigma\Gamma_i d \ln a_i \qquad (17\text{-}121)$$

Γ_i is the surface excess (Davies and Rideal, *Interfacial Phenomena*, 2d ed., Academic, New York, 1963). For most purposes, it is sufficient to view Γ_i as the concentration of adsorbed component i at the surface in units of, say, (g · mol)/cm^2. **R** is the gas constant, T is the absolute temperature, γ is the surface tension, and a_i is the activity of component i. The minus sign shows that material which concentrates at the surface generally lowers the surface tension, and vice versa. This can sometimes be a guide in determining preliminarily what materials can be separated.

When applied to a nonionic surfactant in pure water at concentrations below the critical micelle concentration, Eq. (17-121) simplifies into Eq. (17-122).

$$\Gamma_s = -\frac{1}{\mathbf{R}T}\frac{d\gamma}{d \ln C_s} \qquad (17\text{-}122)$$

C is the concentration in the bulk, and subscript s refers to the surfactant. Under some conditions, Eq. (17-122) may apply to an ionic surfactant as well (Lemlich, loc. cit.).

The major surfactant in the foam may usually be considered to be present at the bubble surfaces in the form of an adsorbed monolayer with a substantially constant Γ_s, often of the order of 3×10^{-10} (g · mol)/cm^2, for a molecular weight of several hundred. On the other hand, trace materials follow the linear-adsorption isotherm $\Gamma_i = K_i C_i$ if their concentration is low enough. For a wider range of concentration a Langmuir or other type of isotherm may be applicable (Davies and Rideal, loc. cit.).

Factors Affecting Adsorption K_i for a colligend can be adversely affected (reduced) through an insufficiency of collector. It can also be reduced through an excess of collector, which competes for the available surface against the collector-colligend complex [Schnepf, Gaden, Mirocznik, and Schonfeld, *Chem. Eng. Prog.*, **55**(5), 42 (1959)].

Excess collector can also reduce the separation by forming micelles in the bulk which adsorb some of the colligend, thus keeping it from the surface. This effect of the micelles on K_i for the colligend is given theoretically [Lemlich, "Principles of Foam Fractionation," in Perry (ed.), *Progress in Separation and Purification*, vol. 1, Interscience, New York, 1968, chap. 1] by Eq. (17-123) [Lemlich (ed.), *Adsorptive Bubble Separation Techniques*, Academic, New York, 1972] if Γ_s is constant when $C_s > C_{sc}$:

$$\frac{1}{K_2} = \frac{1}{K_1} + \frac{C_s - C_{sc}}{\Gamma_s E} \qquad (17\text{-}123)$$

K_1 is K_i just below the collector's critical micelle concentration, C_{sc}. K_2 is K_i at some higher collector concentration, C_s. E is the relative effectiveness, in adsorbing colligend, of surface collector versus micellar collector. Generally, $E > 1$. Γ_s is the surface excess of collector. More about each K is available [Lemlich, "Adsubble Methods," in Li (ed.), *Recent Developments in Separation Science*, vol. 1, CRC Press, Cleveland, 1972, pp. 113–127; Jashnani and Lemlich, *Ind. Eng. Chem. Process Des. Dev.*, **12**, 312 (1973)].

The controlling effect of various ions can be expressed in terms of thermodynamic equilibria [Karger and DeVivo, *Sep. Sci.*, **3**, 393 1968)]. Similarities with ion exchange have been noted. The selectivity of counterionic adsorption increases with ionic charge and decreases with hydration number [Jorne and Rubin, *Sep. Sci.*, **4**, 313 (1969); and Kato and Nakamori, *J. Chem. Eng. Japan*, **9**, 378 (1976)]. By analogy with other separation processes, the relative distribution in multicomponent systems can be analyzed in terms of a selectivity coefficient $\alpha_{mn} = \Gamma_m C_n/\Gamma_n C_m$ [Rubin and Jorne, *Ind. Eng. Chem. Fundam.*, **8**, 474 (1969); *J. Colloid Interface Sci.*, **33**, 208 (1970)].

Operation in the Simple Mode If there is no concentration gradient within the liquid pool and if there is no coalescence within the rising foam, then the operation shown by the solid lines of Fig. 17-66 is truly in the simple mode, i.e., a single theoretical stage of separation. Equations (17-124) and (17-125) will then apply to the steady-flow operation.

$$C_Q = C_W + (GS\Gamma_W/Q) \qquad (17\text{-}124)$$

$$C_W = C_F - (GS\Gamma_W/F) \qquad (17\text{-}125)$$

C_F, C_W, and C_Q are the concentrations of the substance in question (which may be a colligend or a surfactant) in the feed stream, bottoms stream, and foamate (collapsed foam) respectively. G, F, and Q are the volumetric flow rates of gas, feed, and foamate respectively. Γ_W is the surface excess in equilibrium with C_W. S is the surface-to-volume ratio for a bubble. For a spherical bubble, $S = 6/d$, where d is the bubble diameter. For variation in bubble sizes, d should be taken as $\Sigma n_i d_i^3/\Sigma n_i d_i^2$, where n_i is the number of bubbles with diameter d_i in a representative region of foam.

Finding Γ Either Eq. (17-124) or Eq. (17-125) can be used to find the surface excess indirectly from experimental measurements. To assure a close approach to operation as a single theoretical stage, coalescence in the rising foam should be minimized by maintaining a proper gas rate and a low foam height [Brunner and Lemlich, *Ind. Eng. Chem. Fundam.* **2**, 297 (1963)]. These precautions apply particularly with Eq. (17-124).

For laboratory purposes it is sometimes convenient to recycle the foamate directly to the pool in a manner analogous to an equilibrium still. This eliminates the feed and bottoms streams and makes for a more reliable approach to steady-state operation. However, this recycling may not be advisable for colligend measurements in the presence of slowly dissociating collector micelles.

To avoid spurious effects in the laboratory, it is advisable to employ a prehumidified chemically inert gas.

Bubble Sizes Subject to certain errors (de Vries, *Foam Stability*, Rubber-Stichting, Delft, 1957), foam bubble diameters can be measured photographically. Some of these errors can be minimized by taking pains to generate bubbles of fairly uniform size, say, by using a bubbler with identical orifices or by just using a bubbler with a single orifice (gas rate permitting). Otherwise, a correction for planar statistical sampling bias in the foam should be incorporated with actual diameters [de Vries, op. cit.] or truncated diameters [Lemlich, *Chem. Eng. Commun.* **16**, 153 (1982)]. Also, size segregation can reduce mean mural bubble diameter by roughly half the standard deviation [Cheng and Lemlich, *Ind. Eng. Chem. Fundam.* **22**, 105 (1983)]. Bubble diameters can also be measured in the liquid pool, either photographically or indirectly via measurement of the gas flow rate and stroboscopic determination of bubble frequency [Leonard and Lemlich, *Am. Inst. Chem. Eng. J.*, **11**, 25 (1965)].

Bubble sizes at formation generally increase with surface tension and orifice diameter. Prediction of sizes in swarms from multiple orifices is difficult. In aqueous solutions of low surface tension, bubble diameters of the order of 1 mm are common. Bubbles produced by the more complicated techniques of pressure flotation or vacuum flotation are usually smaller, with diameters of the order of 0.1 mm or less.

Enriching and Stripping Unlike truly simple foam fractionation without significant changes in bubble diameter, coalescence in a foam column destroys some bubble surface and so releases adsorbed material to trickle down through the rising foam. This downflow constitutes internal reflux, which enriches the rising foam by countercurrent action. The result is a richer foamate, i.e., higher C_Q than that obtainable from the single theoretical stage of the corresponding simple mode. Significant coalescence is often present in rising foam, but the effect on bubble diameter and enrichment is frequently overlooked.

External reflux can be furnished by returning some of the externally broken foam to the top of the column. The concentrating effect of reflux, even for a substance which saturates the surface, has been verified [Lemlich and Lavi, *Science*, **134**, 191 (1961)].

Introducing the feed into the foam some distance above the pool makes for stripping operation. The resulting countercurrent flow in the foam further purifies the bottoms, i.e., lowers C_W.

Enriching, stripping, and combined operations are shown in Fig. 17-66.

Foam-Column Theory The counterflowing streams within the foam are viewed as consisting *effectively* of a descending stream of interstitial liquid (equal to zero for the simple mode) and an ascending stream of interstitial liquid plus bubble surface. (By considering this ascending surface as analogous to a vapor, the overall operation becomes analogous in a way to distillation *with entrainment*.)

An effective concentration [$\bar{C}$] in the ascending stream at any level in the column is defined by Eq. (17-126):

$$\bar{C} = C + (GS\Gamma/U) \tag{17-126}$$

where U is the volumetric rate of interstitial liquid upflow, C is the concentration in this ascending liquid at that level, and Γ is the surface excess in equilibrium with C. Any effect of micelles should be included.

For simplicity, U can usually be equated to Q. An effective equilibrium curve can now be plotted from Eq. (17-126) in terms of $\bar{C}$ (or rather $\bar{C}^\circ$) versus C.

Operating lines can be found in the usual way from material balances. The slope of each such line is $\Delta\bar{C}/\Delta C = L/U$, where L is the downflow rate in the particular column section and C is now the concentration in the descending stream.

The number of theoretical stages can then be found in one of the usual ways. Figure 17-67 illustrates a graphical calculation for a stripper.

Alternatively, the number of transfer units (NTU) in the foam based on, say, the ascending stream can be found from Eq. (17-127):

$$\text{NTU} = \int_{\bar{C}^{*}_{W}}^{C_Q} \frac{d\bar{C}}{\bar{C}^\circ - \bar{C}} \tag{17-127}$$

$\bar{C}^\circ$ is related to C by the effective equilibrium curve, and $\bar{C}^\circ_W$ is similarly related to C_W. $\bar{C}$ is related to C by the operating line.

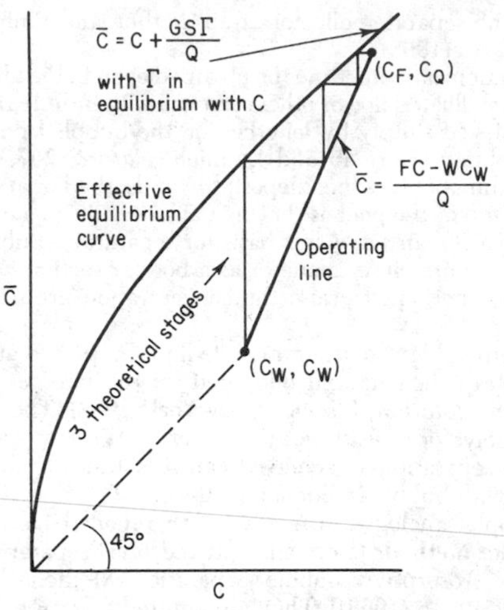

FIG. 17-67 Graphical determination of theoretical stages for a foam-fractionation stripping column.

To illustrate this integration analytically, Eq. (17-127) becomes Eq. (17-128) for the case of a stripping column removing a colligend which is subject to the linear-equilibrium isotherm $\Gamma = KC$.

$$\text{NTU} = \frac{F}{GSK - W} \ln \frac{FW + F(GSK - W)C_F/C_W}{GSK(GSK + F - W)} \tag{17-128}$$

As another illustration, Eq. (17-127) becomes Eq. (17-129) for an enriching column which is concentrating a surfactant with a constant Γ:

$$\text{NTU} = R \ln \frac{RGS\Gamma(F - D)}{(R + 1)GS\Gamma(F - D) - (R + 1)FD(C_D - C_F)} \tag{17-129}$$

Unless the liquid pool is purposely lengthened vertically in order to give additional separation via bubble fractionation, it is usually taken to represent one theoretical stage. A bubbler submergence of 30 cm or so is usually ample for a solute with a molecular weight that does not exceed several hundred.

In a colligend stripper, it may be necessary to add some collector to the pool as well as the feed because the collector is also stripped off.

Limiting Equations If the height of a foam-fractionation column is increased sufficiently, a concentration pinch will develop between the counterflowing interstitial streams (Brunner and Lemlich, loc. cit.). For an enricher, the separation attained will then approach the predictions of Eq. (17-130) and, interestingly enough, Eq. (17-125).

$$C_D = C_W + (GS\Gamma_W/D) \tag{17-130}$$

D is the volumetric rate at which net foamate (net overhead liquid product) is withdrawn. $D = Q/(R + 1)$. The concentration in the net foamate is C_D. In the usual case of total foam breakage (no dephlegmation), $C_D = C_Q$.

If the tall column is a stripper, the separation will approach that of Eqs. (17-131) and (17-132):

$$C_Q = C_F + (GS\Gamma_F/Q) \tag{17-131}$$

$$C_W = C_F - (GS\Gamma_F/W) \tag{17-132}$$

For a sufficiently tall combined column, the separation will approach that of Eqs. (17-133) and (17-132):

$$C_D = C_F + (GS\Gamma_F/D) \tag{17-133}$$

The formation of micelles in the foam breaker does not affect the limiting equations because of the theoretically unlimited opportunity in a sufficiently tall column for their transfer from the reflux to the ascending stream [Lemlich, "Principles of Foam Fractionation," in Perry (ed.), *Progress in Separation and Purification*, vol. 1, Interscience, New York, 1968, chap. 1].

In practice, the performance of a well-operated foam column several feet tall may actually approximate the limiting equations, provided there is little channeling in the foam and provided that reflux is either absent or is present at a low ratio.

Column Operation To assure intimate contact between the counterflowing interstitial streams, the volume fraction of liquid in the foam should be kept below about 10 percent—and the lower the better. Also, rather uniform bubble sizes are desirable. The foam bubbles will thus pack together as blunted polyhedra rather than as spheres, and the suction in the capillaries (Plateau borders) so formed will promote good liquid distribution and contact. To allow for this desirable deviation from sphericity, $S = 6.3/d$ in the equations for enriching, stripping, and combined column operation [Lemlich, *Chem. Eng.*, **75**(27), 95 (1968); **76**(6), 5 (1969)]. Diameter d still refers to the sphere.

Visible channeling or significant deviations from plug flow of the foam should be avoided, if necessary by widening the column or lowering the gas and/or liquid rates. The superficial gas velocity should probably not exceed 1 or 2 cm/s. Under proper conditions, HTU values of several cm have been reported [Hastings, Ph.D. dissertation, Michigan State University, East Lansing, 1967; and Jashnani and Lemlich, *Ind. Eng. Chem. Process Des. Dev.*, **12**, 312 (1973)]. The foam column height equals NTU × HTU.

For columns that are wider than several centimeters, reflux and feed distributors should be used, particularly for wet foam [Haas and Johnson, *Am. Inst. Chem. Eng. J.*, **11**, 319 (1965)]. Liquid content within the foam can be monitored conductimeterically [Chang and Lemlich, *J. Colloid Interface Sci.*, **73**, 224 (1980)]. See Fig. 17-68. Theoretically, as the limit $\mathcal{D} = K = 0$ is very closely approached, $\mathcal{D} = 3K$ [Lemlich, *J. Colloid Interface Sci.*, **64**, 107 (1978)].

Wet foam can be handled in a bubble-cap column (Wace and Banfield, *Chem. Process Eng.*, **47**(10), 70 (1966)] or in a sieve plate column [Aguayo and Lemlich, *Ind. Eng. Chem. Process Des. Dev.*, **13**, 153 (1974)]. Alternatively, individual short columns can be connected in countercurrent array [Banfield, Newson, and Alder, *Am. Inst. Chem. Eng. Symp. Ser.*, **1**, 3 (1965); Leonard and Blacyki, *Ind. Eng. Chem. Process Des. Dev.*, **17**, 358 (1978)].

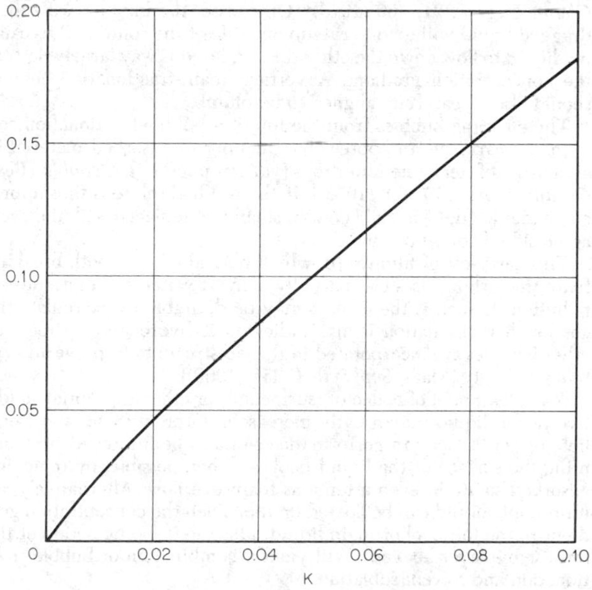

FIG. 17-68 Empirical relationship between $\mathcal{D}$, the volumetric fraction of liquid in common polydisperse foam, and K, the electrical conductivity of the foam divided by the electrical conductivity of the liquid. [*Chang and Lemlich, J. Colloid Interface Sci.*, 73, 224 (1980).]

A high gas rate can be used to achieve maximum throughput in the simple mode (Wace, Alder, and Banfield, AERE-R5920, U.K. Atomic Energy Authority, 1968) because channeling is not a factor in that mode. A horizontal drainage section can be used overhead [Haas and Johnson, *Ind. Eng. Chem. Fundam.*, **6**, 225 (1967)]. The highly mobile dispersion produced by a very high gas rate is not a true foam but is rather a so-called **gas emulsion** [Bikerman, *Ind. Eng. Chem.*, **57**(1), 56 (1965)].

A very low gas rate in a column several feet tall with internal reflux can sometimes be used to effect difficult multicomponent separations in batch operation [Lemlich, "Principles of Foam Fractionation," in Perry (ed.), *Progress in Separation and Purification*, vol. 1, Interscience, New York, 1968, chap. 1].

The same end may be achieved by continuous operation at total external reflux with a small U bend in the reflux line for foamate holdup [Rubin and Melech, *Can. J. Chem. Eng.*, **50**, 748 (1972)].

The slowly rising foam in a tall column can be employed as the sorbent for continuous chromatographic separations [Talman and Rubin, *Sep. Sci.*, **11**, 509 (1976)]. Low gas rates are also employed in short columns to produce the scumlike froth of batch-operated ion flotation, microflotation, and precipitate flotation.

Foam Drainage and Overflow The rate of foam overflow on a gas-free basis (i.e., the total volumetric foamate rate Q) can be estimated from a detailed theory for foam drainage [Leonard and Lemlich, *Am. Inst. Chem. Eng. J.*, **11**, 18 (1965)]. From the resulting relationship for overflow [Fanlo and Lemlich, *Am. Inst. Chem. Eng. Symp. Ser.*, **9**, 75, 85 (1965)], Eq. (17-134) can be employed as a convenient approximation to the theory so as to avoid trial and error over the usual range of interest for foam of low liquid content ascending in plug flow:

$$\frac{Q}{G} = 22\left(\frac{v_G^3\,\mu\,\mu_s^2}{g^3\rho^3 d^8}\right)^{1/4} \tag{17-134}$$

The superficial gas velocity v_g is G/A, where A is the horizontal cross-sectional area of the empty vertical foam column. Also, g is the acceleration of gravity, ρ is the liquid density, μ is the ordinary liquid viscosity, and μ_s is the effective surface viscosity.

To account for inhomogeneity in bubble sizes, d in Eq. (17-134) should be taken as $\sqrt{\Sigma n_i d_i^3/\Sigma n_i d_i}$ and evaluated at the top of the vertical column if coalescence is significant in the rising foam. Note that this average d for overflow differs from that employed earlier for S. Also, see "Bubble Sizes" regarding the correction for planar statistical sampling bias and the presence of size segregation at a wall.

For theoretical reasons, Q determined from Eq. (17-134) should be multiplied by the factor $(1 + 3Q/G)$ to give a final Q. However, for foam of sufficiently low liquid content this multiplication can be omitted with little error.

The effective surface viscosity is best found by experiment with the system in question, followed by back calculation through Eq. (17-134). From the precursors to Eq. (17-134), such experiments have yielded values of μ_s on the order of 10^{-4} (dyn·s)/cm for common surfactants in water at room temperature, which agrees with independent measurements [Lemlich, *Chem. Eng. Sci.*, **23**, 932 (1968); and Shih and Lemlich, *Am. Inst. Chem. Eng. J.*, **13**, 751 (1967)]. However, the expected high μ_s for aqueous solutions of such skin-forming substances as saponin and albumin was not attained, perhaps because of their nonnewtonian surface behavior [Shih and Lemlich, *Ind. Eng. Chem. Fundam.*, **10**, 254 (1971); and Jashnani and Lemlich, *J. Colloid Interface Sci.*, **46**, 13 (1974)].

The drainage theory breaks down for columns with tortuous cross section, large slugs of gas, or heavy coalescence in the rising foam.

Foam Coalescence Coalescence is of two types. The first is the growth of the larger foam bubbles at the expense of the smaller bubbles due to interbubble gas diffusion, which results from the smaller bubbles having somewhat higher internal pressures [Adamson, *The Physical Chemistry of Surfaces*, 3d ed., Wiley, New York, 1976)]. Small bubbles can even disappear entirely. In principle, the rate at which this type of coalescence proceeds can be estimated [Ranadive and Lemlich, *J. Colloid Interface Sci.*, **70**, 392 (1979)].

The second type of coalescence arises from the rupture of films between adjacent bubbles [Vrij and Overbeek, *J. Am. Chem. Soc.*, **90**, 3074 (1968)]. Its rate appears to follow first-order reaction kinetics with respect to the number of bubbles [New, *Proc. 4th Int. Congr. Surf. Active Substances*, Brussels, 1964, **2**, 1167 (1967)] and to decrease with film thickness [Steiner, Hunkeler, and Hartland, *Trans. Inst. Chem. Eng.*, **55**, 153 (1977)]. Many factors are involved [Bikerman, *Foams*, Springer-Verlag, New York, 1973; and Akers (ed.), *Foams*, Academic, New York, 1976].

Both types of coalescence can be important in the foam separations characterized by low gas flow rate, such as batchwise ion flotation producing a scum-bearing froth of comparatively long residence time. On the other hand, with the relatively higher gas flow rate of foam fractionation, the residence time may be too short for the first type to be important, and *if* the foam is sufficiently stable, even the second type of coalescence may be unimportant.

Unlike the case for Eq. (17-134), when coalescence is significant, it is better to find S from d evaluated at the feed level for Eqs. (17-131) to (17-133) and at the pool surface for Eqs. (17-125) and (17-130).

Foam Breaking It is usually desirable to collapse the overflowing foam. This can be accomplished by chemical means (Bikerman, op. cit.) if external reflux is not employed or by thermal means [Kishimoto, *Kolloid Z.*, **192**, 66 (1963)] if degradation of the overhead product is not a factor.

Foam can also be broken with a rotating perforated basket [Lemlich, "Principles of Foam Fractionation," in Perry (ed.), *Progress in Separation and Purification*, vol. 1, Interscience, New York, 1968, chap. 1]. If the foamate is aqueous (as it usually is), the operation can be improved by discharging onto Teflon instead of glass [Haas and Johnson, *Am. Inst. Chem. Eng. J.*, **11**, 319 (1965)]. A turbine can be used to break foam [Ng, Mueller, and Walden, *Can. J. Chem. Eng.*, **55**, 439 (1977)]. Foam which is not overly stable has been broken by running foamate onto it [Brunner and Stephan, *Ind. Eng. Chem.*, **57**(5), 40 (1965)]. Foam can also be broken by sound or ultrasound, a rotating disk, and other means [Ohkawa, Sakagama, Sakai, Futai, and Takahara, *J. Ferment. Technol.*, **56**, 428, 532 (1978)].

If desired, dephlegmation (partial collapse of the foam to give reflux) can be accomplished by simply widening the top of the column, provided the foam is not too stable. Otherwise, one of the more positive methods of foam breaking can be employed to achieve dephlegmation.

Bubble Fractionation Figure 17-69 shows continuous bubble fractionation. This operation can be analyzed in a simplified way in terms of the adsorbed carry-up, which furthers the concentration gradient, and the dispersion in the liquid, which reduces the gradient [Lemlich, *Am. Inst. Chem. Eng. J.*, **12**, 802 (1966); **13**, 1017 (1967)].

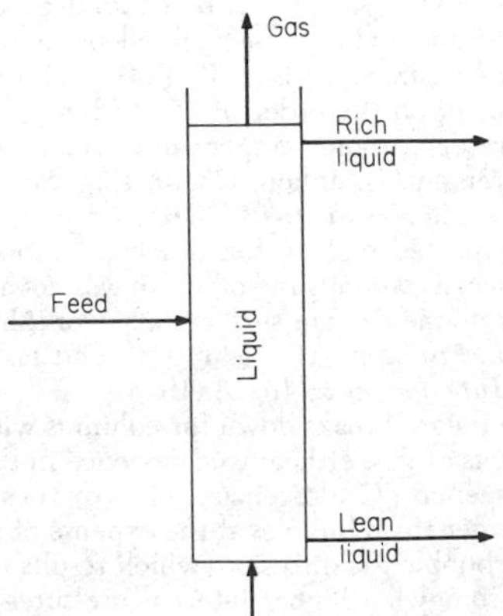

FIG. 17-69 Continuous bubble fractionation.

To illustrate, consider the limiting case in which the feed stream and the two liquid takeoff streams of Fig. 17-69 are each zero, thus resulting in batch operation. At steady state the rate of adsorbed carry-up will equal the rate of downward dispersion, or $af\Gamma = \overline{D}A dC/dh$. Here a is the surface area of a bubble, f is the frequency of bubble formation, $\overline{D}$ is the dispersion (effective diffusion) coefficient based on the column cross-sectional area A, and C is the concentration at height h within the column.

There are several possible alternative relationships for Γ [Lemlich, op. cit.]. For simplicity, consider $\Gamma = K'C$, where K' is not necessarily the same as the equilibrium constant K. Substituting and integrating from the boundary condition of $C = C_B$ at $h = 0$ yield

$$C/C_B = \exp(Jh) \qquad (17\text{-}135)$$

C_B is the concentration at the bottom of the column, and parameter $J = K'af/\overline{D}A$. Combining Eq. (17-135) with a material balance against the solute in the initial charge of liquid gives

$$\frac{C}{C_i} = \frac{JH\exp(Jh)}{\exp(JH) - 1} \qquad (17\text{-}136)$$

C_i is the concentration in the initial charge, and H is the total height of the column.

The foregoing approach has been extended to steady continuous flow as illustrated in Fig. 17-69 [Cannon and Lemlich, *Chem. Eng. Prog. Symp. Ser.*, **68**(124), 180 (1972); Bruin, Hudson, and Morgan, *Ind. Eng. Chem. Fundam.*, **11**, 175 (1972); and Wang, Granstrom, and Kown, *Environ. Lett.*, **3**, 251 (1972), **4**, 233 (1973), **5**, 71 (1973)]. The extension includes a rough method for estimating the optimum feed location as well as a very detailed analysis of column performance which takes into account the various local phenomena around each rising bubble [Cannon and Lemlich, op. cit.].

In agreement with experiment [Shah and Lemlich, *Ind. Eng. Chem. Fundam.*, **9**, 350 (1970); and Garmendia, Perez, and Katz, *J. Chem. Educ.*, **50**, 864 (1973)], theory shows that the degree of separation that is obtained increases as the liquid column is made taller. But unfortunately it decreases as the column is made wider. In simple terms, the latter effect can be attributed to the increase in the dispersion coefficient as the column is widened.

In this last connection it is important that the column be aligned precisely vertically (Valdes-Krieg, King, and Sephton, *Am. Inst. Chem. Eng. J.*, **21**, 400 (1975)]. Otherwise, the bubbles with their dragged liquid will tend to rise up one side of the column, thus causing liquid to flow down the other side, and in this way largely destroy the concentration gradient. A vertical foam-fractionation column should also be carefully aligned to be plumb.

The escaping bubbles from the top of a bubble-fractionation column can carry off an appreciable quantity of adsorbed material in an aerosol of very fine film drops [various papers, *J. Geophys. Res., Oceans Atmos.*, **77**(27), (1972)]. If the residual solute is thus appreciably depleted, C_i in Eq. (17-136) should be replaced with the average residual concentration.

This carry-off of film drops, which may also occur with breaking foam, in certain cases can partially convert water pollution into air pollution. If such is the case, it may be desirable to recirculate the gas. Such recirculation is also indicated if hydrocarbon vapors or other volatiles are incorporated in the gas stream to improve adsorptive selectivity [Maas, *Sep. Sci.*, **4**, 457 (1969)].

A small amount of collector (surfactant) or other appropriate additive in the liquid may greatly increase adsorption (Shah and Lemlich, op. cit.). Column performance can also be improved by skimming the surface of the liquid pool or, when possible, by removing adsorbed solute in even a tenuous foam overflow. Alternatively, an immiscible liquid can be floated on top. Then the concentration gradient in the tall pool of main liquid, plus the trapping action of the immiscible layer above it, will yield a combination of bubble fractionation and solvent sublation.

Systems Separated Some of the various separations reported in the literature are listed in Rubin and Gaden, "Foam Separation," in Schoen (ed.), *New Chemical Engineering Separation Techniques*, Interscience, New York, 1962, chap. 5; Lemlich, *Ind. Eng. Chem.*,

60(10), 16 (1968); Pushkarev, Egorov, and Khrustalev, *Clarification and Deactivation of Waste Waters by Frothing Flotation*, in Russian, Atomizdat, Moscow, 1969; Kuskin and Golman, *Flotation of Ions and Molecules*, in Russian, Nedra, Moscow, 1971; Lemlich (ed.), *Adsorptive Bubble Separation Technqiues*, Academic, New York, 1972; Lemlich, "Adsubble Methods," in Li (ed.), *Recent Developments in Separation Science*, vol. 1, CRC Press, Cleveland, 1972, chap. 5; Grieves, *Chem. Eng. J.*, **9**, 93 (1975); Valdes-Krieg, King, and Sephton, *Sep. Purif. Methods*, **6**, 221 (1977); and Clarke and Wilson, *Foam Flotation*, Marcel Dekker, New York, 1983.

Of the numerous separations reported, only a few can be listed here. Except for minerals beneficiation [ore flotation] which is covered in Sec. 21, the most important industrial applications are usually in the area of pollution control.

A pilot-sized foaming unit reduced the alkyl benzene sulfonate concentration of 500,000 gal of sewage per day to nearly 1 mg/L, using a G/F of 5 and producing a Q/F of no more than 0.03 [Brunner and Stephan, *Ind. Eng. Chem.*, **57**(5), 40 (1965); and Stephan, *Civ. Eng.*, **35**(9), 46 (1965). A full-scale unit handling over 45,420 m^3/day (12 million gal/day) performed nearly as well. The foam also carried off some other pollutants. However, with the widespread advent of biodegradable detergents, large-scale foam fractionation of municipal sewage has been discontinued.

Other plant-scale applications to pollution control include the flotation of suspended sewage particles by depressurizing so as to release dissolved air [Jenkins, Scherfig, and Eckhoff, "Applications of Adsorptive Bubble Separation Techniques to Wastewater Treatment," in Lemlich (ed.), *Adsorptive Bubble Separation Techniques*, Academic, New York, 1972, chap. 14; and Richter, *Internat. Chem. Eng.*, **16**, 614 (1976)]. Dissolved-air flotation is also employed in treating wastewater from pulp and paper mills [Coertze, *Prog. Water Technol.*, **10**, 449 (1978); and Severeid, *TAPPI* **62**(2), 61, 1979]. In addition, there is the flotation, with electrolytically released bubbles [Chambers and Cottrell, *Chem. Eng.*, **83**(16), 95 (1976)], of oily iron dust [Ellwood, *Chem. Eng.*, **75**(16), 82 (1968)] and of a variety of wastes from surface-treatment processes at the maintenance and overhaul base of an airline [Roth and Ferguson, *Desalination*, **23**, 49 (1977)].

Fats and, through the use of lignosulfonic acid, proteins can be flotated from the wastewaters of slaughterhouses and other food-processing installations [Hopwood, *Inst. Chem. Eng. Symp. Ser.*, **41**, M1 (1975)]. After further treatment, the floated sludge has been fed to swine.

A report of the recovery of protein from potato-juice wastewater by foaming [Weijenberg, Mulder, Drinkenberg, and Stemerding, *Ind. Eng. Chem. Process Des. Dev.*, **17**, 209 (1978)] is reminiscent of the classical recovery of protein from potato and sugar-beet juices [Ostwald and Siehr, *Kolloid Z.*, **79**, 11 (1937)]. The isoelectric pH is often a good choice for the foam fractionation of protein (Rubin and Gaden, loc. cit.). Adding a salt to lower solubility may also help.

With the addition of appropriate additives as needed, the flotation of refinery wastewaters reduced their oil content to less than 10 mg/L in pilot-plant operation [Steiner, Bennett, Mohler, and Clere, *Chem. Eng. Prog.*, **74**(12), 39 (1978)] and full-scale operation (Simonsen, *Hydrocarb. Process. Pet. Refiner*, **41**(5), 145, 1962]. Experiments with a cationic collector to remove oils reportedly confirmed theory [Angelidon, Keskavarz, Richardson, and Jameson, *Ind. Eng. Chem. Process Des. Dev.*, **16**, 436 (1977)].

Pilot-plant [Hyde, Miller, Packham, and Richards, *J. Am. Water Works Assoc.*, **69**, 369 (1977)] and full-scale [Ward, *Water Serv.*, **81**, 499 (1977)] flotation in the preparation of potable water is described.

Overflow at the rate of 2700 m^3 (713,000 gal) per day from a zinc-concentrate thickener is treated by ion flotation, precipitate flotation, and untrafine-particle flotation [Nagahama, *Can. Min. Metall. Bull.*, **67**, 79 (1974)]. In precipitate flotation only the surface of the particles need be coated with collector. Therefore, in principle less collector is required than for the equivalent removal of ions by foam fractionation or ion flotation.

By using an anionic collector and external reflux in a combined (enriching and stripping) column of 3.8-cm (1.5-in) diameter with a feed rate of 1.63 m/h [40 gal/(h · ft^2)] based on column cross section, D/F was reduced to 0.00027 with C_w/C_F for Sr^{2+} below 0.001 [Shonfeld and Kibbey, *Nucl. Appl.*, **3**, 353 (1967)]. Reports of the adsubble separation of 29 heavy metals, radioactive and otherwise, have been tabulated [Lemlich, "The Adsorptive Bubble Separation Techniques," in Sabadell (ed.), *Proc. Conf. Traces Heavy Met. Water*, 211–223, Princeton University, 1973, EPA 902/9-74-001, U.S. EPA, Reg. II, 1974). Some separation of ^{15}N from ^{14}N by foam fractionation has been reported [Hitchcock, Ph.D. dissertation, University of Missouri, Rolla, 1982].

The numerous separations reported in the literature include surfactants, inorganic ions, enzymes, other proteins, other organics, biological cells, and various other particles and substances. The scale of the systems ranges from the simple Crits test for the presence of surfactants in water, which has been shown to operate by virtue of transient foam fractionation [Lemlich, *J. Colloid Interface Sci.*, **37**, 497 (1971)], to the natural adsubble processes that occur on a grand scale in the ocean [Wallace and Duce, *Deep Sea Res.*, **25**, 827 (1978)]. For further information see the reviews cited earlier.

Example 3: Foam Fractionation
[Lemlich, *Chem. Eng.*, **75**(27), 95 (1968); **76**(5), 5 (1969).] Excerpted by special permission from *Chemical Engineering*, Dec. 16, 1968, copyright 1968, by McGraw-Hill, Inc., New York, N.Y. 10020, and modified.

An aqueous waste stream flowing at the rate of 3.785 m^3/h (1000 gal/h) contains a trace of a certain toxic ion (the colligend). Using continuous foam fractionation in the simple mode—with bubble diameters of 0.1 cm in the liquid pool—what air rate and column diameter are required to reduce the concentration of this ion by a factor of 10 while producing no more than 0.189 m^3/h (50 gal/h) of collapsed foam? The adsorption of the ion by the deliberately added surfactant (the collector) is governed by a linear isotherm with an equilibrium constant of 0.09 cm.

Also, if the foregoing requires a surfactant concentration of 2×10^{-4} molar in the pool, at what rate must the surfactant be supplied to the system?

Solution. The linear adsorption isotherm can be written as $\Gamma_W = KC_W$. Combining with Eq. (17-125) and substituting $S = 6/d$ yields $G = [(C_F - C_W)Fd]/(6KC_W)$. But in this instance, $C_F = 10\ C_W$. Substituting the given information and canceling C_W yields

$$G = \frac{(10 - 1) \times 3.785 \times 0.1}{6 \times 0.09} = 6.31 \text{ m}^3 \text{ of air/h } (3.71 \text{ ft}^3/\text{min})$$

The surfactant that is furnished must leave in the foam and bottom streams. The amount leaving in the former is $GS\Gamma_W + QC_W$ by Eq. (17-124); the amount in the latter is simply WC_W. (All concentrations now refer to the surfactant, not to the ion.) Summing, factoring, and noting that $F = Q + W$ give $GS\Gamma_W + FC_W$ as the surfactant rate. In the absence of further information, Γ_W for the surfactant monolayer will be taken at the typical value of 3×10^{-10} (g · mol)/cm^2. Substituting

$$\text{Surfactant rate} = \frac{6.31 \times 6 \times 3 \times 10^{-10}}{0.1 \times 10^{-6}} + \frac{3.785 \times 2 \times 10^{-4}}{10^{-3}}$$
$$= 0.871 \text{ (g} \cdot \text{mol)/h } [1.92 \times 10^{-3} \text{ (lb} \cdot \text{mol)/h]}$$

If the molecular weight of the surfactant is, say, 500, this corresponds to a surfactant input of roughly 435 g/h (1 lb/h). Of course, if the stream already contains surface-active substances, some or all of the added surfactant can be dispensed with.

Equation 17-134 is now used to find the column diameter. In the absence of physical-property data, the viscosity μ is estimated to be equal to that of water at room temperature, namely, 1 cP, which is 10^{-2} (dyn·s)/cm^2. Similarly, $\rho = 1$ g/cm^3. The surface viscosity μ_s is taken as the typical value 10^{-4} (dyn·s)/cm.

In view of the above uncertainties, the small wetness correction of $1 + 3Q/G$ is omitted.

Since $v_G = G/A$, rearranging Eq. (17-134) gives

$$A = \frac{61.6}{g\rho}\left(\frac{G^7 \mu \mu_s^2}{Q^4 d^8}\right)^{1/3}$$

Substituting,

$$A = \frac{61.6}{980.6 \times 1}\left(\frac{(6.31)^7 \times 10^{-2} \times (10^{-4})^2}{(0.189)^4 \times (0.1)^8}\right)^{1/3} \times \frac{100}{3600}$$
$$= 0.255 \text{ m}^2 \text{ (2.75 ft}^2\text{)}$$

This corresponds to a column diameter of $(4A/\pi)^{1/2}$, or about 0.57 m (1.87 ft).

Comments. In the overflowing foam, the volumetric ratio of liquid to gas is $Q/G = 0.189/6.31 = 0.03$. Neglecting the wetness correction is thus justified, especially in view of the approximations made for some of the physical-property data.

Q/G of 0.03 corresponds to an overflowing foam of less than 3 percent liquid fraction by volume. Since this implies a moderately dry foam in the column proper, such a foam may be rather unstable and may coalesce to a significant degree as it rises. In other words, d at the top of the foam may be appreciably larger than d lower down.

This means that the required A can be significantly different from 0.255 m^2, or conversely, if 0.255 m^2 is maintained, then the collapsed-foam overflow rate can be other than 0.189 m^3/h. Of course, since such internal coalescence would provide internal reflux, the operation would not strictly speaking be in the simple mode.

For further comments on this particular problem, see the original source. For additional illustrative problems, see Lemlich, "Principles of Foam Fractionation," in Perry (ed.), *Progress in Separation and Purification*, vol. 1, Interscience, New York, 1968, chap. 1. For further information, see Lemlich (ed.), *Adsorptive Bubble Separation Techniques*, Academic, New York, 1972; and Grieves, "Adsorptive Bubble Separation Methods," in Elving (ed)., *Treatise on Analytical Chemistry*, part 1, vol. 5, Wiley, New York, 1982, chap. 9.

DIFFUSIONAL SEPARATION PROCESSES

GENERAL REFERENCES: Benedict (ed.), "Developments in Uranium Enrichment," *Am. Inst. Chem. Eng. Symp. Ser.*, 73, no. 169, 1977. Benedict and Pigford, *Nuclear Chemical Engineering*, 2d ed., McGraw-Hill, New York, 1981; London ed., *Separation of Isotopes*, George Newnes, London, 1961. Cohen, *The Theory of Isotope Separation as Applied to the Large-Scale Production of U-235*, National Nuclear Energy Ser., Div. III, vol. 1B, McGraw-Hill, New York, 1951. Pratt, *Countercurrent Separation Processes*, Elsevier, Amsterdam, 1967. Shacter, Von Halle, and Hoglund, "Diffusion Separation Methods," in Standen (ed.), *Encyclopedia of Chemical Technology*, Wiley, New York, 1979. Villani, *Isotope Separation*, American Nuclear Society Monograph, ANS Publications, 1976.

The less well-known diffusional separation methods, gaseous diffusion, thermal diffusion, some pressure-diffusion processes (gas centrifugation and the separation nozzle), and mass diffusion, are treated here. No attempt is made to present mechanical details of the equipment, since these can be obtained from the references. The objective is to outline the ranges of applicability of these processes, to describe briefly the principles upon which their operation depends, and to recommend sources for additional information.

These processes are often little known to the chemical engineer. All of them except gas centrifugation are thermodynamically irreversible. Thus, although they can accomplish the separations commonly encountered in chemical engineering practice, they usually cannot compete economically with the more conventional, thermodynamically reversible processes. The conventional processes (e.g., distillation, chemical exchange) show sharply decreasing separation factors as the constituents of the process mixture become more similar. The separation becomes most difficult in the case of isotopes, and it is here that diffusional separation methods find application.

When the atomic weight of the isotopes is less than about 40, there are significant enough differences in their physical and chemical properties that the reversible distillation and chemical-exchange methods are applicable. These processes are used to separate the isotopes of hydrogen, lithium, boron, carbon, nitrogen, and oxygen. Above atomic weight 40 these conventional processes have unexploitably low separation factors, and the diffusional separation methods are preferred. An exception to this general rule was reported by the French Commissariat à L'Energie Atomique in 1977, when it announced that it had achieved a breakthrough in chemical-exchange technology and had developed an economical chemical-exchange process for the enrichment of the isotopes of uranium.

The choice of process is also influenced by the scale of production. Thermal diffusion is used for small-scale separations for which its relatively high power requirement (operating cost) is less important than ease of fabrication and plant construction (capital cost). Stagewise processes such as gaseous diffusion are not used for small-scale production because many stages, each with its own complement of components, would be required. Since stage capacity normally increases faster than stage cost, stagewise processes are used only for very large production rates in which full advantage can be taken of cost scaling.

The applicability of common isotope-separating processes as compared with isotopic weight and scale of production is shown in Table 17-32. The electromagnetic separator, which is a production mass spectrometer, is indicated for completeness, as are distillation and chemical exchange.

The stage-separation factor for isotope separation tends to be very small. Countercurrent-flow processes have the advantage of combining the equivalent of many equilibrium stages of separation in a single unit. Even here, quite frequently many units must be connected in series and/or parallel to span the required concentration interval and obtain the required throughput. This interconnecting of separating units is called cascading.

A concept widely used in isotope separation is that of separative capacity. Separative capacity is defined and discussed extensively in the general references cited. Briefly and qualitatively, separative capacity is useful because it combines the two important characteristics in any separation, throughput and enrichment. A separating unit in an isotope-separation process is characterized by its separation effect, $\alpha - 1$, and its throughput, or capacity, of which the upflow L is a measure, and in the case of a continuous process also by the stage length. Intuitive reasoning leads to the conclusion that large separations obtained at large throughputs are desirable. Analysis (see references) shows that for isotope-separation processes, which have relatively small separation factors, the separative capacity of a unit is directly proportional to the throughput and to the square of the separation effect and, in the case of a continuous process, inversely proportional to the stage length. A useful form for the separative capacity of a stagewise process is given in Eq. (17-142). Its use can be illustrated by an example. For a separation task of interest, the required separative capacity can be calculated. For a given process, then, the separative capacity of a separating element can be calculated and divided into the required separative capacity to determine the minimum number of separating elements which will be required.

Table 17-32 Applicability of Several Isotope-Separation Processes

Scale of production \ Isotopic weight	Light (atomic weights $<$ 40)	Medium (atomic weights 40–150)	Heavy (atomic weights $>$ 150)
Small, mg./day	Thermal diffusion, electromagnetic	Thermal diffusion, electromagnetic	Electromagnetic
Medium, g./day	Distillation, chemical exchange	Gas centrifuge, mass diffusion	Gas centrifuge
Large, kg./day	Distillation, chemical exchange	Gaseous diffusion, gas centrifuge	Gaseous diffusion, gas centrifuge, separation nozzle

The following treatment of the individual processes should permit a rough estimate of the separative capacity of their separating elements.

GASEOUS DIFFUSION

Separation Mechanism Gaseous diffusion is a popular name for a gas-phase separation process depending, for its separation effect, upon flow of the process-gas mixture through extremely small holes in a barrier or membrane. When the diameter of these holes is less than the mean free path of the process-gas molecules, individual molecules can pass through the barrier, but bulk flow of gas is effectively prevented. Since at a given temperature lighter molecules move with greater velocity than heavy molecules, the light molecules will strike the walls (barrier) more frequently, relative to their concentration, and thus they tend to pass through the barrier preferentially. Enrichment thereby occurs. Flow through holes sufficiently small is called "molecular effusion" if their length is negligible, or "molecular streaming" (Knudsen flow) in the case of a practical barrier of finite thickness. If the holes are larger in diameter, bulk laminar flow (Poiseuille flow) takes place. Poiseuille flow is nonseparative and must be minimized if the process is to operate efficiently. Furthermore, in order to process large quantities of gas in equipment of reasonable size, the barrier must have a significant fraction of its surface open to flow. From this short description of the separation effect, it is clear that a well-designed gaseous-diffusion barrier not only must be strong enough mechanically to withstand the required pressure drop across it and be chemically inert to the process gas but also must have a large number of very small-diameter holes.

Process Description Gaseous diffusion is a stagewise process. In order to obtain large enrichments, these stages must be connected in series to form cascades. Experience with the gaseous-diffusion process indicates its region of economic applicability to be the large-scale separation of heavy isotopes. The process is thermodynamically irreversible because of the irreversible flow of gas from a region of high pressure to a region of low pressure.

In general, a gaseous-diffusion stage consists of a converter which contains the barrier, a motor, a gas compressor to move the gas from stage to stage and to provide the required pressure drop across the barrier, a gas cooler to remove the heat of compression from the process gas, a control valve to balance flows and lend hydrodynamic stability to the stage, interconnecting piping, and instrumentation. A section of a cascade of gaseous-diffusion stages is shown schematically in Fig. 17-70.

The only known large-scale application of the gaseous-diffusion process is the separation of the isotopes of uranium. United States gaseous-diffusion plants contain 10,276 stages and were built at a cost of \$2.33 billion. Descriptions of the United States plants have been published (U.S. AEC Rep. ORO-658, 1968; ORO-668, 1969; ORO-684, 1972; ORO-685, 1972). The United States gaseous-diffusion-plant complex has a nominal separative capacity of 17,200,000 separative work units per year when powered to 6100 MW. Improvement programs have added 1300 MW to the cascades and increased the separative capacity of these plants to 28,000,000 separative work units per year. Eurodif, a multinational consortium organized under French leadership, has built at Tricastin, France, a large gaseous-diffusion plant with a separative capacity of about 10,800,000 separative work units per year.

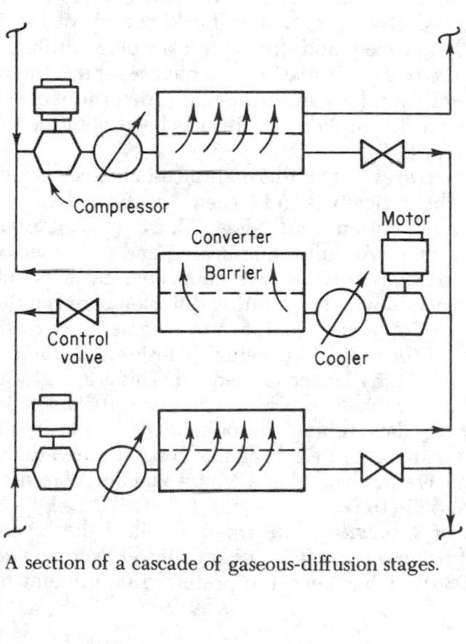

FIG. 17-70 A section of a cascade of gaseous-diffusion stages.

Stage Design ***Flow through the Barrier*** An ideal gaseous-diffusion barrier is one which permits only Knudsen flow. This condition is met when the hole diameter is small compared with the mean free path of the gas molecules at prevailing temperatures and pressure. If the holes are treated as a collection of long, straight circular capillaries, the flow rate is given by Knudsen's law:

$$N = \frac{4}{3}\frac{\sigma d}{l}(p_f - p_b)(2\pi MRT)^{-1/2} \tag{17-137}$$

where N is the molar flow rate of gas per unit area of barrier, σ the fractional open area of the barrier, d the hole diameter, l the length of the hole (barrier thickness), M the molecular weight of the gas, T the absolute temperature of the gas, p_f and p_b the high- and low-side pressures on the barrier, and R the gas constant. If the barrier has some larger holes, nonseparative Poiseuille flow occurs. Poiseuille flow is directly proportional to $(p_f^2 - p_b^2)$ and inversely proportional to gas viscosity μ. The total flow per unit area through the barrier N_t can be taken as the sum of the Knudsen and Poiseuille flows:

$$N_t = \frac{a}{\sqrt{M}}(p_f - p_b) + \frac{b}{\mu}(p_f^2 - p_b^2) \tag{17-138}$$

where a and b are functions of temperature and barrier properties.

Separation Effect The ideal binary separation factor across the barrier at a given point is evaluated for an ideal barrier (pure separative flow) and zero low-side pressure. The flow of light component is proportional to $p_f\, x/\sqrt{M_1}$, and the flow of heavy component is proportional to $p_f(1 - x)/\sqrt{M_2}$, where x is the mole fraction of light component in the uneffused gas and M_1 and M_2 are the molecular weights of the light and heavy gas respectively. The concentration of light component in the effused gas y is the ratio of the light-component flow to the total flow

$$y = \frac{x/\sqrt{M_1}}{x/\sqrt{M_1} + (1 - x)/\sqrt{M_2}} \tag{17-139}$$

It follows from the definition that the ideal separation factor is given by

$$\alpha^\circ \equiv \frac{y/(1 - y)}{x/(1 - x)} = \sqrt{\frac{M_2}{M_1}} \tag{17-140}$$

Stage Separation Factor The observed separation factor for a stage will differ from α°. This results from the existence of four efficiency terms whose effects must be included:

1. *Barrier efficiency* E_b. A portion of the flow through a real barrier will be of the Poiseuille (nonseparative) type. The barrier efficiency is given approximately by the ratio of separative flow (Knudsen flow) to the total flow through the barrier.

2. *Back-pressure efficiency* E_p. Since material flows through the barrier into a region whose pressure p_b is not negligible with respect to p_f, light material on the low-pressure side tends to effuse preferentially back toward the high-pressure side, thereby undoing the desired separation. To a first approximation, this efficiency factor will be equal to $(1 - p_b/p_f)$.

3. *Mixing efficiency* E_m. As a result of the preferential effusion of light component through the barrier, the gas layer at the surface of the barrier on the high-pressure side is depleted in the light component. Because the barrier enriches this depleted material immediately adjacent to it, the low-pressure-side concentration will be

lower than that expected from measurements of the average light-component concentration in the bulk of the gas on the high-pressure side. High mixing efficiency depends on reducing the thickness of the boundary layer.

4. *Cut correction* E_c. The stage separation is defined by relating the concentrations of streams at the stage exits, whereas the point separation factor relates the concentrations across the barrier, which change continuously as the gas flows through the stage. If the uneffused gas flows through the stage without appreciable mixing in the direction of flow and if the effused portion is removed as soon as it passes through the barrier, it follows that the exiting effused stream represents a stage average concentration while the concentration of the exiting uneffused gas is at its terminal or minimum value. The cut correction accounts for this effect. It can be shown that $E_c = \{\ln[1/(1-\theta)]\}/\theta$, where θ, the cut, is the ratio of the flow rate of the effused stream to the stage feed stream. Stages normally operate at a cut of ½; thus $E_c = 1.386$.

By applying these corrections, the actual stage separation factor can be calculated from the ideal point separation factor by

$$\alpha - 1 = E_b E_p E_m E_c (\alpha^\circ - 1) \qquad (17\text{-}141)$$

For a very high-quality barrier in which Poiseuille flow can be ignored, the barrier efficiency E_b will equal 1. If, in addition, mixing efficiency is assumed to be 100 percent, the separative capacity of a stage operating at a cut of one-half would be

$$\Delta U = \frac{1}{4} L(\alpha - 1)^2 = 0.480L\left(1 - \frac{p_b}{p_f}\right)^2 (\alpha^\circ - 1)^2 \qquad (17\text{-}142)$$

where L is the molar flow rate of the effused stream.

Power Requirement If the power requirement of the stage is assumed to be that associated with the isothermal compression of the process gas from p_b to p_f, the power requirement can be calculated with the aid of Eq. (17-142). For a stage of known separative capacity the stage power P is given by

$$P = LRT \ln \frac{p_f}{p_b} = \frac{2.082\,\Delta U\,RT}{(\alpha^\circ - 1)^2} \frac{\ln(1/r)}{(1 - r)^2} \qquad (17\text{-}143)$$

where r is the pressure ratio p_b/p_f. The power requirement is minimized when $r = 0.285$.

Barrier Area The barrier area A can be calculated from the ratio of the total molar-stage throughput to the separative (Knudsen) flow rate given in Eq. (17-137). Combination of Eqs. (17-137) and (17-142) gives the desired expression for the barrier area in terms of the stage separative capacity in the form

$$A = \frac{L}{N} = \frac{3.914\Delta U l(\mathrm{MRT})^{1/2}}{(\alpha^\circ - 1)^2 \sigma d} \frac{1}{p_f(1 - r)^3} \qquad (17\text{-}144)$$

which is a minimum at the impractical condition of $r = 0$ (zero back pressure). Equation (17-144) shows the advantage of having thin barriers with a high fraction of their surface area open to flow.

Compressor Capacity The volumetric pumping capacity of the process-gas compressors may also be estimated for a stage of known separative capacity. The volumetric capacity V is simply the molar flow L divided by the molar density of the process gas c. By using Eq. (17-142) and the ideal-gas law, V is obtained in the form

$$V = \frac{LRT}{p_b} = \frac{2.082\Delta RT}{(\alpha^\circ - 1)^2} \frac{1}{p_f r(1 - r)^3} \qquad (17\text{-}145)$$

which minimizes at $r = ⅓$.

Barrier area and compressor capacity are large contributors to the capital cost of a gaseous-diffusion cascade, and the power requirement comprises the bulk of operating cost. These three important parameters have conflicting individual optima. A more complete treatment of the design and optimization of gaseous-diffusion stages and cascades is given by Pratt (op. cit., chap. 7).

THERMAL DIFFUSION

General References: Grew and Ibbs, *Thermal Diffusion in Gases*, Cambridge, England, 1952. Jones and Furry, *Rev. Mod. Phys.*, **18**, 151 (1946). Powers, "Thermal Diffusion," in Schoen (ed.), *New Chemical Engineering Separation Techniques*, Interscience, New York, 1962, chap. 1. Vasaru et al., *Thermal Diffusion Column Theory and Practice*, VEB Deutscher Verlag, Berlin, 1969.

Thermal diffusion occurs whenever a mixture is subjected to a temperature gradient; it leads to a partial separation of the components of the mixture. Thus, thermal-diffusion phenomena can be applied to the separation of materials. However, since thermal diffusion in solids is extremely slow, as are all diffusional processes in solids, and since no practical method of separating solid mixtures by thermal diffusion has yet been devised, the following discussion will be limited to thermal diffusion in gases and liquids.

Literature Review Over 1000 references to the separation of mixtures by thermal diffusion appear in the scientific literature, too many to acknowledge individually the contributions of the many workers in this field. The majority of the work published can be categorized under the following headings.

Single-Stage Experiments Since single-stage thermal-diffusion cells yield only very small separations, these experiments are conducted primarily for the purpose of determining the value of the thermal-diffusion constant α.

1. *Liquid mixtures.* The thermal-diffusion effect was first observed in a liquid solution in 1856. Following its discovery, many investigators measured the thermal-diffusion effect in a variety of liquid systems. In general, they reported their results in terms of the Soret coefficient [after Soret, *Arch. Sci. (Geneva)*, **2**, 48 (1879)] σ, which is related to the thermal-diffusion constant by $\sigma = \alpha/T$, where T is the absolute temperature. For this purpose equipment of the type shown schematically in Fig. 17-71*a* and *b* has been customarily

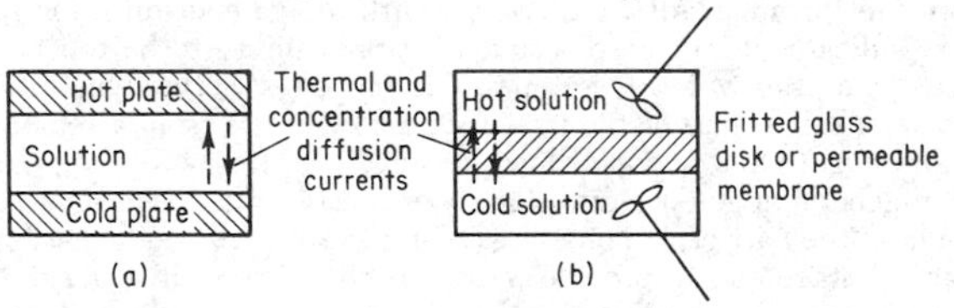

FIG. 17-71 Single-stage thermal-diffusion cells.

used. In the former [Tanner, *Trans. Faraday Soc.*, **23**, 75 (1927); **49**, 611 (1953); and others] the liquid mixture is confined between two horizontal parallel plates and subjected to a vertical temperature gradient, the higher temperature being applied at the top of the cell in order to minimize thermal-convection currents. In the latter [Riehl, Z. *Elektrochem.*, **49**, 306 (1943); Alexander, Z. *Phys. Chem.*, **195**, 175 (1950); and others] the hot and cold regions of the solution are mechanically agitated, and diffusion takes place through a permeable membrane or fritted-glass disk which separates them. A tabulation of several hundred experimentally determined values of the Soret coefficient for liquid mixtures has been published (Von Halle, U.S. AEC Rep. K-1420, 1959).

2. *Gas mixtures.* The thermal-diffusion effect in gases was first predicted theoretically and then confirmed experimentally [Chapman and Dootson, *Phil. Mag.*, **33**, 248 (1917)] by using a two-bulb device. The two bulbs, one heated and the other cooled, containing the gaseous mixture are connected by a capillary which inhibits convection while permitting diffusion between the chambers to take place. For more accurate measurements of α, the use of a Trennschaukel (or separation swing) [Clusius and Bühler, Z. *Naturforsch.*, **9a**, 775 (1954)] is recommended. This is actually a multistage device in that it consists of several thermal-diffusion cells connected in series by capillary tubing. A double-ended bellows is used to circulate a small amount of gas back and forth through the system. The theory of the Trennschaukel is given by Van der Waerden [Z. *Naturforsch.*, **12a**, 583 (1957)].

Column or Cascade Experiments Since the invention of the thermal-diffusion column [Clusius and Dickel, *Naturwissenschaften*, **26**, 546 (1938)], it has been the preferred equipment for effecting separations by thermal diffusion. In this apparatus, sometimes called a Clusius-Dickel column or a thermogravitational column, a fluid mixture is subjected to a horizontal temperature gradient. Natural laminar thermal-convection currents are thereby established, upward in the neighborhood of the hot wall and downward in the neighborhood of the cold wall. This countercurrent flow produces the equivalent of several hundred individual separation stages in a single column. Experiments of this type have been carried out either for the purpose of verifying column theory or for the production of a desired material.

1. *Liquid mixtures.* The separation of liquid mixtures has frequently been carried out in parallel-plate thermal-diffusion columns. This is due to the fact that in the case of liquids the desired spacing between the hot and cold walls is of the order of 0.25 mm (0.01 in). Although a variety of liquid organic mixtures has been separated by this means [Korsching and Wirtz, *Naturwissenschaften*, **27,** 110 (1939); Prigogine et al., *Physica*, **16,** 851 (1950); Powers, U.S. AEC Rep. UCRL-2618, 1954; and others] and in cylindrical concentric-tube columns (Fig. 17-72) [Jones and Milberger, *Ind. Eng. Chem.*, **45,** 2689 (1953); and others], liquid-phase thermal diffusion appears to remain a laboratory tool with little industrial-scale application. One noteworthy exception was the thermal-diffusion plant consisting of 2100 columns, each about 14.6 m (48 ft) in length, built at Oak Ridge, Tenn., during World War II for the separation of uranium isotopes (Abelson and Hoover, *Proc. Int. Sym. Isotope Sep. Amsterdam*, 1958, p. 483).

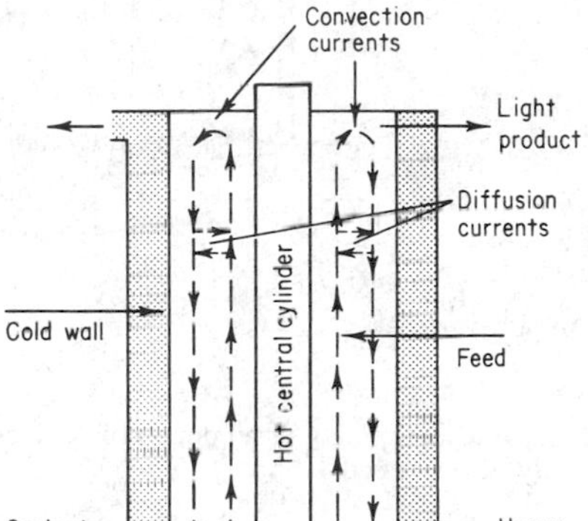

FIG. 17-72 Thermal-diffusion column.

Pioneering work on liquid-phase thermal diffusion in concentric-tube columns is currently being conducted at the Mound Laboratory, Miamisburg, Ohio [Rutherford, *Ind. Eng. Chem. Process Des. Dev.*, **17,** 77 (1978)].

2. *Gas mixtures.* Separations in the gas phase are usually carried out in concentric-tube-type columns (Fig. 17-72) in which the central tube is an electrically heated General Electric Calrod heating element or its equivalent, or in hot-wire columns, similar to the original Clusius-Dickel column, in which the central tube is replaced by an electrically heated wire which runs the length of the column. The column is surrounded by a water jacket which cools the outer wall. A modification used in constructing thermal-diffusion column cascades employs a common water jacket for a large number of columns (Rathkamp, U.S. AEC Rep. ORNL-TM-1047, 1965). Although thermal diffusion has been used to separate dissimilar gases [Thomas and Watkins, *Chem. Eng. Sci.*, **5,** 34 (1956); **6,** 26 (1956)], it has found its widest application in the separation of isotopes of gaseous elements. Thermal-diffusion cascades are currently in use at Mound Laboratory for the separation of the isotopes of helium, neon, argon, krypton, xenon, and carbon.

Theory Theoretical papers on thermal diffusion may be divided into two main categories.

1. Those which deal with the prediction of the thermal-diffusion constant α. Estimates of α for simple gaseous molecules based on kinetic theory and a suitable intermolecular-force law (Hirschfelder, Curtiss, and Bird, *Molecular Theory of Gases and Liquids*, Wiley, New York, 1954, chap. 8) especially for isotopic mixtures are in good agreement with the experimental values. On the other hand, estimates for liquids based on thermodynamic properties of the solution [Dougherty and Drickamer, *J. Phys. Chem.*, **59,** 443 (1955)] yield only qualitative agreement with experiment.

2. Those which deal with the phenomenological theory of thermal-diffusion-column performance. The subject is covered in detail in the references for this section.

Theory The degree of separation that can be attained in either a single-stage thermal-diffusion cell or a multistage column depends directly on the value of the thermal-diffusion constant α. For gases α can be estimated from

$$\alpha = \frac{105}{118}\,\frac{M_2 - M_1}{M_2 + M_1}\,k_T^\circ \tag{17-146}$$

where M_1 and M_2 are the molecular weights of the components. The quantity k_T° has been evaluated for several intermolecular-force models (Hirschfelder, Curtiss, and Bird, loc. cit.) (for rigid elastic spheres $k_T^\circ = 1$). The basic transport equation for the diffusive current of one component of a binary mixture in the presence of a temperature gradient is

$$J_1 = cD\left[\,ax(1-x)\frac{1}{T}\frac{dT}{dr} - \frac{dx}{dr}\right] \tag{17-147}$$

where J_1 is the current of component 1 in moles per unit time per unit area, c is the molar density and D the coefficient of diffusion of the gas mixture, x is the mole fraction of component 1, T the absolute temperature, and r the coordinate of distance. It follows directly from Eq. (17-147) that the separation which can be obtained in a single-stage cell is limited to

$$\ln\frac{x_h/(1-x_h)}{x_c/(1-x_c)} = \alpha \ln\frac{T_h}{T_c} \tag{17-148}$$

where the subscripts h and c refer to the hot and cold regions of the cell respectively. Equation (17-148) is also used to calculate values of α from experimental data.

The separating behavior of thermal-diffusion columns can be predicted from the column-transport equation

$$\tau = Hx(1-x) - (K_c + K_d)\,(dx/dz) \tag{17-149}$$

where τ is the axial transport of the desired component in the column in mass per unit time and z is the coordinate of column length. The derivation of Eq. (17-149) is treated in detail in the general references and also in those for this section. The quantities H, K_c, and K_d, called column-transport coefficients, depend on the physical properties of the mixture and the geometry and temperature of the column:

$$H = \frac{2\pi g}{6!}\,\frac{\alpha\rho^2}{\mu}\,r^4h \tag{17-150a}$$

$$K_c = \frac{2\pi g^2}{9!}\,\frac{\rho^3}{\mu^2 D}\,r^8k_c \tag{17-150b}$$

$$K_d = 2\pi(\rho D)r^2k_d \tag{17-150c}$$

where g is the acceleration of gravity, ρ is the mass density, r is the radius of the cold wall, and μ is the viscosity of the gas. The quantities h, k_c, and k_d are termed shape factors. Values have been tabulated for the inverse-power repulsion-force law (Von Halle, Greene, and Hoglund, U.S. Rep. K-1469, 1965) and for the Lennard-Jones (6-12) force law (Von Halle and Hoglund, U.S. AEC Rep. K-1679, 1966). It follows from Eq. (17-149) that the maximum enrichment attainable in the thermal-diffusion column is given by

$$\ln\frac{x_t/(1-x_t)}{x_b/(1-x_b)} = \frac{HZ}{K_c + K_d} \tag{17-151}$$

where the subscripts t and b refer to the top and bottom of the column respectively and Z is the column length. The theoretical separative capacity of a column is given by

$$\Delta U = \frac{H^2 Z}{4(K_c + K_d)} \qquad (17\text{-}152)$$

By using carefully constructed thermal-diffusion columns, good agreement has been obtained between theory and experiment [Rutherford et al., *J. Chem. Phys.*, **50**, 5359 (1969)].

PRESSURE DIFFUSION

The kinetic theory of gases predicts that a partial separation of the components of a gaseous mixture will occur when the gas is subjected to a pressure gradient. Although industrial processes for separating mixtures based on pressure-diffusion phenomena are relatively rare today, research and development work in this area is widespread. Relatively steep pressure gradients are required in order to obtain a significant separation effect. The best-known device for producing large pressure gradients in gas mixtures is the gas centrifuge; a second device utilizing pressure diffusion is the separation nozzle being developed in Germany for the enrichment of the isotopes of uranium.

Gas-Centrifuge Process President Carter announced plans to build a gas centrifuge plant at Portsmouth, Ohio. The plant was estimated to cost about $4.4 billion to construct and was expected to add about 8,800,000 separative work units per year to United States separative capacity. Great Britain, the Netherlands, and Germany are operating centrifuge demonstration plants at Almelo in the Netherlands and at Capenhurst, England. Projected overall capacity is 10,000,000 separative work units per year by 1990. Japan has also announced tentative plans for a 4,000,000-separative-work-unit-per-year centrifuge plant to be operational by 1990.

Gas Centrifuges The gas centrifuge consists essentially of a vertical cylinder, containing the gas mixture to be separated, which is rotated about its axis at high angular velocity. The effect of the rotation is to enrich the lighter components of the mixture in the vicinity of the axis and the heavier components in the vicinity of the wall. Countercurrent gas centrifuges, in which an axial convective circulation of the gas is induced in order to multiply the basic separation effect and provide multistage enrichment in a single machine, are of most interest commercially. Countercurrent flow may be provided either by external pumps (Beams, Hagg, and Murphree, *Developments in the Centrifuge Separation Project*, NNES, vol. X-1, 1951), by an axial temperature gradient (Groth, in London ed., *Separation of Isotopes*, George Newnes, London, 1961, chap. 6), or by the insertion of a stationary member in the rotating bowl (Zippe, Univ. Virginia EP-4420-101-60V, 1960). In addition, provision must be made for feeding and withdrawing gas from the bowl. An early-model gas centrifuge is shown schematically in Fig. 17-73. The scoop at the top of the bowl, in addition to removing the heavy product gas, also induces the countercurrent flow. The bottom scoop, shielded from the bowl by a rotating baffle, removes the light gas. The bowl is held in the vertical position by a magnetic suspension, rotates on a needle bearing, and is driven from below by either an electric motor or a gas turbine.

A survey of the theory of the countercurrent gas centrifuge has been published [Olander, *Adv. Nucl. Sci. Technol.*, **6**, 105 (1972)]. According to theory, the maximum separative work which a gas centrifuge can produce per unit time is given by

$$\Delta U_{max} = \frac{\pi Z c D}{2}\left(\frac{\Delta M V^2}{2RT}\right)^2 \qquad (17\text{-}153)$$

where ΔU is the separative capacity in moles per unit time, Z the length of the centrifuge, c the molar density and D the diffusivity of the process gas, ΔM the mass difference between the components being separated, V the peripheral velocity of the centrifuge, T the absolute temperature, and R the gas constant. This expression indicates the desirability of long centrifuge bowls rotating at very high speeds.

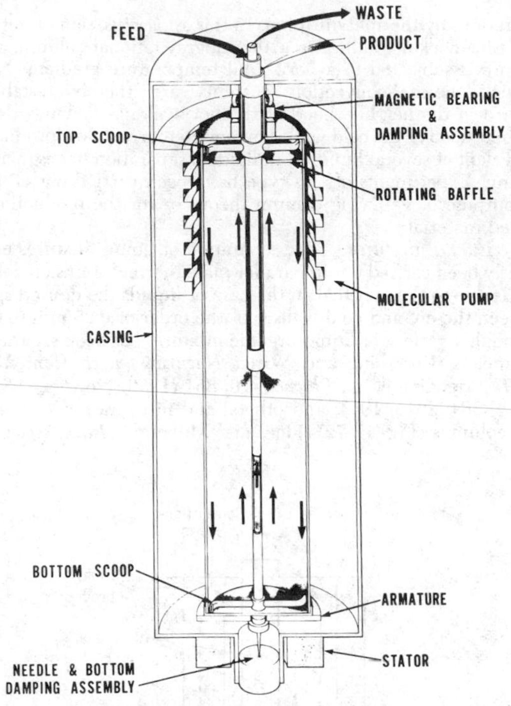

FIG. 17-73 An early-model gas centrifuge.

The actual separative capacity of a countercurrent gas centrifuge can conveniently be expressed

$$\delta U = e_I \cdot e_c \cdot e_f \cdot \delta U(\max) \qquad (17\text{-}154)$$

where e_I is the ideality efficiency, which is equal to about 0.815 for a well-designed centrifuge with nondecaying axial flows, e_c is the circulation efficiency, equal to $m^2/(1 + m^2)$, and e_f is the flow-profile efficiency, which depends on details of the internal flow in the centrifuge (Von Halle, "The Countercurrent Gas Centrifuge for the Enrichment of U-235," A.I.Ch.E. Symposium Series No. 192, Vol. 76, pp. 82-88, 1980.

A model of the internal flow yields a flow-profile efficiency at high peripheral velocities equal to $7.2/(MV^2/2RT)$.

Gas centrifuges have been tested on the separation of the isotopes of argon, krypton, xenon, and uranium. Zippe, with a centrifuge 30 cm in length, rotating with a peripheral speed of 360 m/s, obtained 35 percent of the maximum theoretical separative work with UF_6 as the process gas.

Separation Nozzle The separation nozzle (Becker et al., *Proc. Conf. Uranium Isotope Sep.*, London, May 1975) utilizes the pressure gradient in a curved, expanding, supersonic jet to achieve separation of a gas mixture. A separation-nozzle stage is shown schematically in Fig. 17-74. A light gas (helium) is added to the feed (UF_6) in order to increase the velocity of the jet. As the jet traverses the curved path, the heavy components are concentrated in the vicinity of the wall. A knife-edge divides the jet into two fractions, one enriched in the light components, the other enriched in the heavy components, which are pumped off separately. To obtain a desired separation, many separation-nozzle stages must be connected in a manner similar to gaseous-diffusion stages to form a cascade. This process avoids the problems associated with the fine-pored membranes required for the gaseous-diffusion process and with the high-speed rotating parts of the centrifuge process. It suffers from the disadvantage of a relatively high power requirement.

Figure 17-75 illustrates the design of a commercial separation-nozzle element manufactured by the Messerschmitt-Bölkow-Blohm

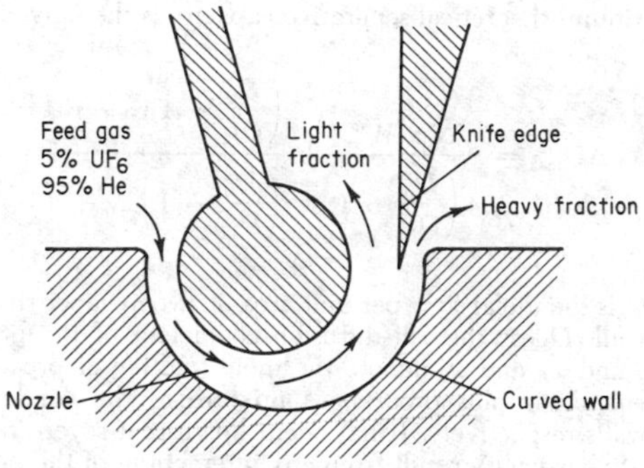

FIG. 17-74 Separation nozzle.

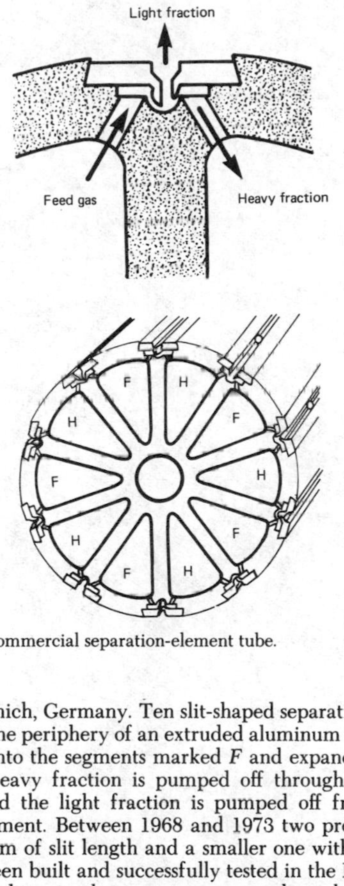

FIG. 17-75 A commercial separation-element tube.

Company, Munich, Germany. Ten slit-shaped separation nozzles are fabricated on the periphery of an extruded aluminum tube. Feed gas is introduced into the segments marked *F* and expands through the nozzles. The heavy fraction is pumped off through the segments marked *H*, and the light fraction is pumped off from the space around the element. Between 1968 and 1973 two prototype stages, one with 1620 m of slit length and a smaller one with 540 m of slit length, have been built and successfully tested in the Karlsruhe center. Technical data on the prototype stages have been published (Geppert et al., *Proc. Int. Conf. Uranium Isotope Sep.*, London, May 1975).

MASS DIFFUSION

Mass diffusion, in its general sense, refers to the gas-phase process in which the separation of a gas mixture is effected by its diffusion into a third gaseous component. In practical applications the third component, termed the sweep vapor or the separating agent, is a vapor which moves through the mixture to be separated by virtue of its partial-pressure gradient. This partial-pressure gradient is maintained by introducing the separating agent as a vapor into the process region through a porous wall and removing it as a liquid which drains from the face of a condenser on the opposite side of the process region. The flux of separating agent tends to sweep preferentially along with it the less diffusible component of the process-gas mixtures, thus effecting the desired separation. This action of the separating agent has led to its being called the sweep vapor and has given the process the alternative name of sweep diffusion. Since the sweep vapor receives its latent heat of vaporization at high temperature and gives it up at a reduced temperature, the process is thermodynamically irreversible.

Mass diffusion can be carried out either in stages or in continuous-countercurrent-flow columns which have an overall enrichment equivalent to many stages in series. Here only the column applications will be discussed. A simple schematic drawing of a mass-diffusion column is given in Fig. 17-76 to show its essential features. The sweep vapor moves in a closed cycle. It is boiled to produce a vapor, delivered to the column, distributed along the process region through a porous wall, diffused across the process region and mass-diffusion screen, condensed on the cold wall, and returned as a liquid to the boiler. The process-gas mixture is at the same time circulated through the column in countercurrent flow by a pump. The flow normally is downward near the condenser wall and upward near the porous (vapor-distributing) wall. The sweep vapor preferentially carries the less diffusible (usually the heavier) component of the process-gas mixture toward the condenser wall and the region of the downflow. Thus, in the operating column, the less diffusible component is enriched at the bottom of the column and the more diffusible (usually the lighter) component is enriched at the top of the column.

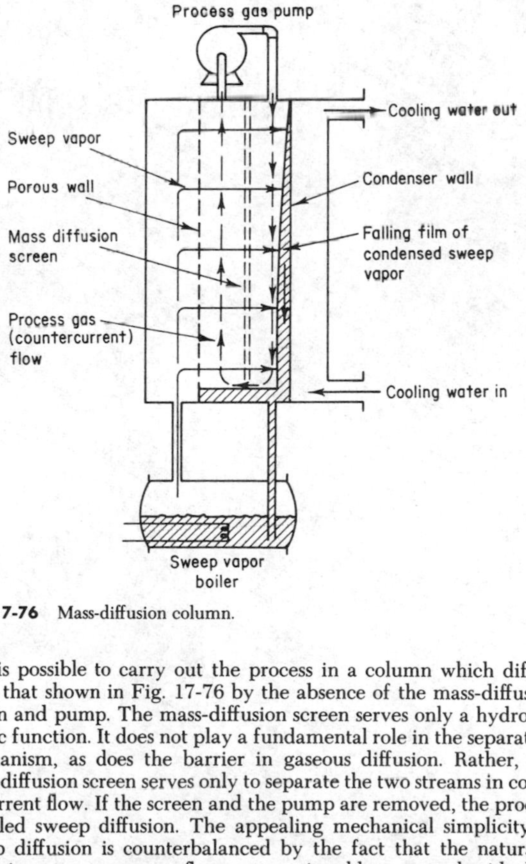

FIG. 17-76 Mass-diffusion column.

It is possible to carry out the process in a column which differs from that shown in Fig. 17-76 by the absence of the mass-diffusion screen and pump. The mass-diffusion screen serves only a hydrodynamic function. It does not play a fundamental role in the separation mechanism, as does the barrier in gaseous diffusion. Rather, the mass-diffusion screen serves only to separate the two streams in countercurrent flow. If the screen and the pump are removed, the process is called sweep diffusion. The appealing mechanical simplicity of sweep diffusion is counterbalanced by the fact that the naturally occurring countercurrent-flow pattern is seldom near the ideal in terms of the flow-velocity profile or adequate in terms of the gas-circulation rate.

Detailed theory of the process and descriptions of the equipment used are available in the following references: the theories of mass-diffusion stages and continuous-countercurrent-flow columns are treated by Benedict and Boas [*Chem. Eng. Prog.*, **47**, 51, 111 (1951)]; an excellent treatment of mass-diffusion-column theory and a comparison with experimental results are given by de Wet and Los [*Z. Naturforsch.*, **19a**, 740 (1964)]; and Pratt (op. cit. chap. 7) gives a thorough treatment of stage and column theory.

Theory Since no significant commercial use is known to have been made of this process, the presentation of column theory is limited to giving the expression for the maximum theoretical separative capacity of a column of length Z and width W.

The process depends on the differing diffusivities of the two isotopic components of the process gas into the sweep vapor. Let the subscripts 0, 1, and 2 denote, respectively, the sweep vapor, the light, and the heavy component. Then, if the diffusion coefficient D_0 is the arithmetic average of the diffusivities D_{01} and D_{02} of the isotopic components into the sweep vapor, the separability γ can be given by

$$\gamma = \frac{D_{01} - D_{02}}{D_0} = \left(\frac{M_2 - M_1}{M_2 + M_1}\right)\left(\frac{2M_0}{2M_0 + M_1 + M_2}\right) \qquad (17\text{-}155)$$

The maximum theoretical separative capacity is then given by

$$\Delta U_{\max} = \frac{N_0\gamma^2 WZ}{4\left(\frac{D_0}{D_{12}} - 1\right)} \ln \frac{\left(\frac{D_0}{D_{12}} - 1\right)\omega_R + 1}{\left(\frac{D_0}{D_{12}} - 1\right)\omega_0 + 1} \qquad (17\text{-}156)$$

where N_0 is the molar flow per unit area of sweep vapor through the porous wall, D_{12} is the self-diffusion coefficient of the process-gas mixture, and ω_R and ω_0 denote the mole fraction of process gas in the sweep vapor plus process-gas mixture at the condenser and porous walls respectively. It has been shown that severe reductions in separative capacity result from any interactions of the process gas with the sweep vapor. Ideally, the process gas should be insoluble in the condensed sweep vapor. Actual column separative capacities are likely to be no more than 25 percent of the value calculated from Eq. (17-156).

Section 18

Liquid-Gas Systems

J. R. Fair, Ph.D., P.E., *Professor of Chemical Engineering, University of Texas; Fellow, American Institute of Chemical Engineers; Member, American Chemical Society, American Society for Engineering Education. (Section Editor, Gas-Liquid Contacting)*

D. E. Steinmeyer, M.A., M.S., P.E., *Manager, Engineering, Corporate Engineering Department, Monsanto Company; Member, American Institute of Chemical Engineers, American Chemical Society. (Liquid-in-Gas Dispersions)*

W. R. Penney, Ph.D., P.E., *Director of Corporate Process Engineering, E. A. Staley Manufacturing Company; Member, American Institute of Chemical Engineers. (Gas-in-Liquid Dispersions)*

B. B. Crocker, S.M., P.E., *Distinguished Fellow—Engineering, Corporate Engineering Department, Monsanto Company; Fellow, American Institute of Chemical Engineers; Member, Air Pollution Control Association. (Phase Separation)*

INTRODUCTION

GAS-LIQUID CONTACTING

PHASE DISPERSION

PHASE SEPARATION

Nomenclature and Units

In this listing, symbols used in the section are defined in a general way and appropriate SI and U.S. customary units are given. The units are particularly applicable for dimensionless groups of variables and for dimensionally consistent equations (those not having specialized dimensional constants). Specific variable definitions, as denoted by subscripts, for example, are given at the point of application in the section. Some symbols used in the section are defined only at the point of application. *In all cases in which units are defined in the text those units should be used in place of the ones given in this table.*

Symbol	Definition	SI units	U.S. customary units
A	Area	m^2	ft^2
a	Interfacial area per unit volume	m^2/m^3	ft^2/ft^3
C	Specific heat	$J/(kg \cdot K)$	$Btu/(lb \cdot °F)$
c	Specific heat	$J/(kg \cdot K)$	$Btu/(lb \cdot °F)$
D	Diameter	m	ft
D	Diffusion coefficient or diffusivity	m^2/s	ft^2/h
d	Diameter, size	m	ft
E	Efficiency	Dimensionless	Dimensionless
G	Gas mass rate	$kg/(s \cdot m^2)$	$lb/(h \cdot ft^2)$
g	Gravitational constant	$9.807\ m/s^2$	$32.18\ ft/s^2$
g_c	Conversion factor	$1.0\ (kg \cdot m)/(N \cdot s^2)$	$32.22\ (lb \cdot ft)/(lbf \cdot s^2)$
H	height	m	ft
H	Height of a transfer unit	m	ft
h	Length, height	m	ft
h	Heat-transfer coefficient	$J/(s \cdot m^2 \cdot K)$	$Btu/(h \cdot ft^2 \cdot °F)$
k	Thermal conductivity	$J/(m \cdot s \cdot K)$	$Btu/(ft \cdot h \cdot °F)$
k	Mass-transfer coefficient	Defined in text	Defined in text
L	Liquid mass rate	$kg/(s \cdot m^2)$	$lb/(h \cdot ft^2)$
L	Length	m	ft
M	Molecular weight	$kg/(kg \cdot mol)$	$lb/(lb \cdot mol)$
N	Number of stages or plates	Dimensionless	Dimensionless
N	Speed	1/s	1/s
N_{Ca}	Capillary number	Dimensionless	Dimensionless
N_{Fr}	Froude number	Dimensionless	Dimensionless
N_{Pe}	Péclet number	Dimensionless	Dimensionless
N_{Pr}	Prandtl number	Dimensionless	Dimensionless
N_{Re}	Reynolds number	Dimensionless	Dimensionless
N_{Sc}	Schmidt number	Dimensionless	Dimensionless
N_{We}	Weber number	Dimensionless	Dimensionless
P	Pressure	kPa	lbf/in^2
Q	Volumetric flow rate	m^3/s	ft^3/s
q	Flow rate	m^3/s	ft^3/s
T	Temperature	°C, K	°F, °R
t	Time	s	s, h
U	Gas or liquid velocity	m/s	ft/s
u	Gas or liquid velociy	m/s	ft/s
V	Velocity	m/s	ft/s
V	Volume	m^3	ft^3
x	Liquid-phase fraction, mass or mole	Dimensionless	Dimensionless
y	Vapor-phase fraction, mass or mole	Dimensionless	Dimensionless
Z	Length, height	m	ft
	Greek symbols		
α	Thermal diffusivity	m^2/s	ft^2/h
Γ	Flow rate per unit of circumference	$kg/(s \cdot m)$	$lb/(s \cdot ft)$
μ	Viscosity	$Pa \cdot s$	$lb/(ft \cdot s)$
γ	Kinematic viscosity	m^2/s	ft^2/s
ρ	Density	kg/m^3	lb/ft^3
σ	Surface or interfacial tension	mN/m	dyn/cm

INTRODUCTION

Liquid-gas contacting systems are utilized for transferring mass, heat, and momentum between the phases, subject to constraints of physical and chemical equilibrium. Process equipment for such systems is designed to achieve the appropriate transfer operations with a minimum expenditure of energy and capital investment.

In this section emphasis is placed on the transfer of mass. Typical liquid-gas mass-transfer systems are:

Distillation	Evaporation
Flashing	Humidification
Rectification	Dehumidification
Absorption	Dephlegmation
Stripping	Spray drying

Distillation is the separation of the constituents of a liquid mixture via partial vaporization of the mixture and separate recovery of vapor and residue. The process of vaporization is generally of a differential nature.

Flashing is a distillation process in which the total vapor removed approaches phase equilibrium with the residue liquid.

Rectification is the separation of the constituents of a liquid mixture by successive distillations (partial vaporizations and condensations) and is obtained via the use of an integral or differential process. Separations into effectively pure components may be obtained through this procedure.

Stripping or desorption is the transfer of gas, dissolved in a liquid, into a gas stream. The term is also applied to that section of a fractionating column below the feed plate.

Absorption is the transfer of a soluble component in a gas-phase mixture into a liquid absorbent whose volatility is low under process conditions.

Evaporation generally refers to the removal of water, by vaporization, from aqueous solutions of nonvolatile substances.

Humidification and dehumidification refer to the transfer of water between a gas stream and a water stream.

Dephlegmation, or partial condensation, refers to the process in which a vapor stream is cooled to a desired temperature such that a portion of the less volatile components of the stream is removed from the vapor by condensation.

Spray drying is an extension of the evaporative process in which almost all the liquid is removed from a solution of a nonvolatile solid in the liquid.

All these processes are, in common, liquid-gas mass-transfer operations and thus require similar treatment from the aspects of phase equilibrium and kinetics of mass transfer. The fluid-dynamic analysis of the equipment utilized for the transfer also is similar for many types of liquid-gas process systems.

Process equipment utilized for liquid-gas contacting is based on a combination of operating principles of the three categories:

Mode of flow of streams
- Countercurrent
- Cocurrent
- Cross-flow

Gross mechanism of transfer
- Differential
- Integral

Continuous phase
- Gas*
- Liquid

The combination of these characteristics utilized in the various types of process equipment is indicated in Table 18-1.

*In this section the terms "gas" and "vapor" are used interchangeably. The latter is often used in distillation, in which the gas phase is represented by an equilibrium vapor.

Table 18-1 Characteristics of Liquid-Gas Systems

Equipment designation	Mode of flow	Gross mechanism	Continuous phase	Primary process applications
Plate column	Cross-flow, countercurrent	Integral	Liquid and/or gas	Absorption, rectification, stripping
Packed column	Countercurrent, cocurrent	Differential	Liquid and/or gas	Absorption, rectification, stripping, humidification, dehumidification
Wetted-wall (falling-film) column	Countercurrent, cocurrent	Differential	Liquid and/or gas	Absorption, rectification, stripping, evaporation
Spray chamber	Cocurrent, cross-flow, countercurrent	Differential	Gas	Absorption, stripping, humidification, dehumidification
Heat exchanger	Cocurrent, countercurrent	Differential	Gas	Evaporation, dephlegmation
Agitated vessel	Complete mixing	Integral	Liquid	Absorption
Line mixer	Cocurrent	Differential	Liquid or gas	Absorption, stripping

GAS-LIQUID CONTACTING

PLATE COLUMNS

Plate Types Plate columns utilized for liquid-gas contacting may be classified according to mode of flow in their internal contacting devices:

1. Cross-flow plates
2. Counterflow plates

The cross-flow plate (Fig. 18-1*a*) utilizes a liquid downcomer and is more generally used than the counterflow plate (Fig. 18-1*b*) because of transfer-efficiency advantages and greater operating range. The liquid-flow pattern on a cross-flow plate can be controlled by placement of downcomers in order to achieve desired stability and transfer efficiency. Commonly used flow arrangements are shown in Fig. 18-2. A guide for the tentative selection of flow pattern is given in Table 18-2.

It should be noted that the fraction of column cross-sectional area available for gas dispersers (perforations, bubble caps) decreases when more than one downcomer is used. Thus, optimum design of

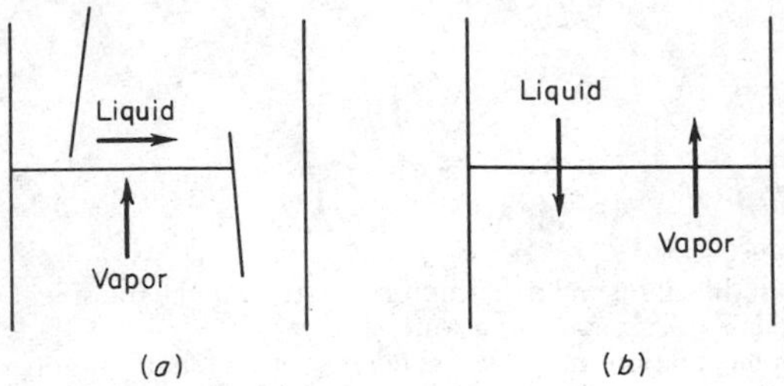

FIG. 18-1 (a) Cross-flow plate (side view). (b) Countercurrent plate (side view).

the plate involves a balance between liquid-flow accommodation and effective use of cross section for gas flow.

Most new designs of cross-flow plates employ perforations for dispersing gas into liquid on the plate. These perforations may be simple round orifices, or they may contain movable "valves" that provide variable orifices of noncircular shape. These perforated plates are called sieve plates (Fig. 18-3) or valve plates (Fig. 18-4). For sieve plates, liquid is prevented from flowing through the perforations by the flowing action of the gas; thus, when the gas flow is low, it is possible for some or all of the liquid to drain through the perforations and in effect bypass portions of the contacting zone. The valve plate is designed to minimize this drainage, or "weeping," since the valve tends to close as the gas flow becomes lower, the total orifice area varying to maintain a dynamic-pressure balance across the plate.

Historically the most common gas disperser for cross-flow plates has been the bubble cap. This device has a built-in seal which prevents liquid drainage at low gas-flow rates. Typical bubble caps are shown in Fig. 18-5. Gas flows up through a center riser, reverses flow under the cap, passes downward through the annulus between riser and cap, and finally passes into the liquid through a series of openings, or "slots," in the lower side of the cap.

Bubble caps were used almost exclusively as cross-flow-plate dispersers until about 1950, when they were largely displaced by simple or valve-type perforations. Many varieties of bubble-cap design were used (and therefore are extant in many operating columns), but in most cases bell caps of 75- to 150-mm (3- to 6-in) diameter were utilized.

In counterflow plates, liquid and gas utilize the same openings for flow. Thus, there are no downcomers. Openings are usually simple

Table 18-2 Selection of Cross-Flow-Plate Flow Pattern*

Estimated tower diam., ft.	Range of liquid capacity, gal./min.			
	Reverse flow	Cross-flow	Double pass	Cascade double pass
3	0–30	30–200		
4	0–40	40–300		
6	0–50	50–400	400– 700	
8	0–50	50–500	500– 800	
10	0–50	50–500	500– 900	900–1400
12	0–50	50–500	500–1000	1000–1600
15	0–50	50–500	500–1100	1100–1800
20	0–50	50–500	500–1100	1100–2000

*Bolles, chap. 14 in Smith, *Design of Equilibrium Stage Processes*, McGraw-Hill, New York, 1963. To convert feet to meters, multiply by 0.3048; to convert gallons per minute to decimeters per second (liters per second), multiply by 0.06309; and to convert gallons per minute to cubic meters per second, multiply by 6.309×10^{-5}.

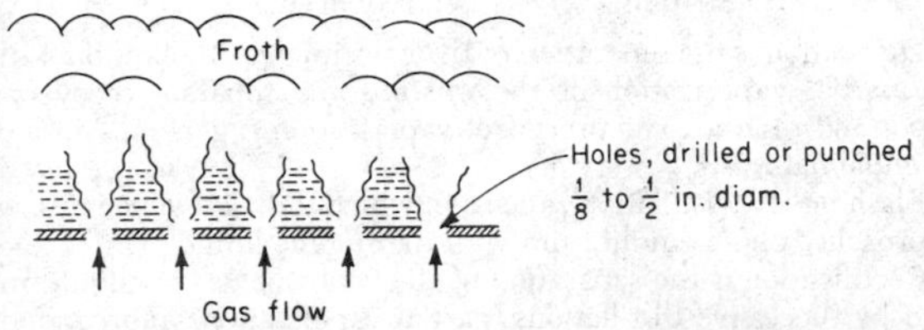

FIG. 18-3 Sieve-plate dispersers. To convert inches to millimeters, multiply by 25.4.

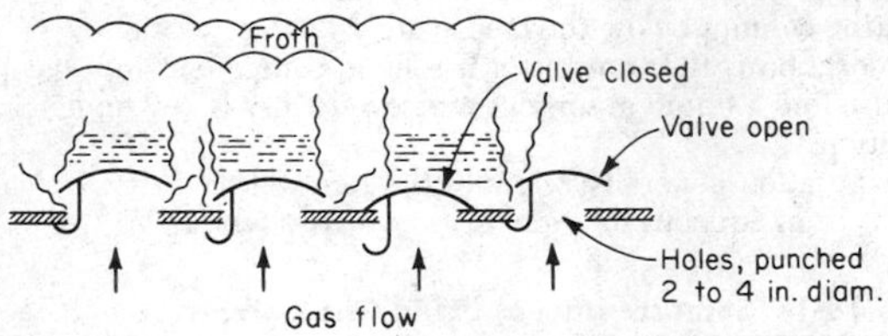

FIG. 18-4 Valve-plate dispersers. To convert inches to millimeters, multiply by 25.4.

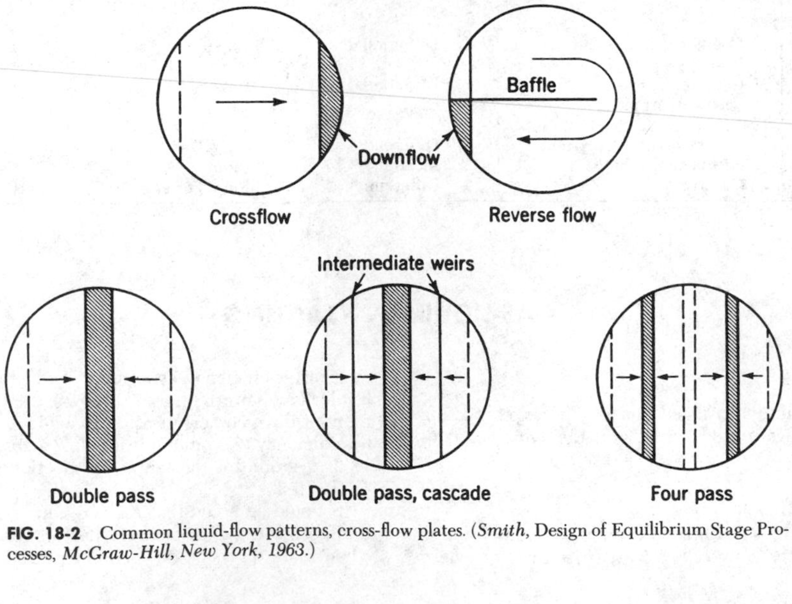

FIG. 18-2 Common liquid-flow patterns, cross-flow plates. (*Smith,* Design of Equilibrium Stage Processes, *McGraw-Hill, New York, 1963.*)

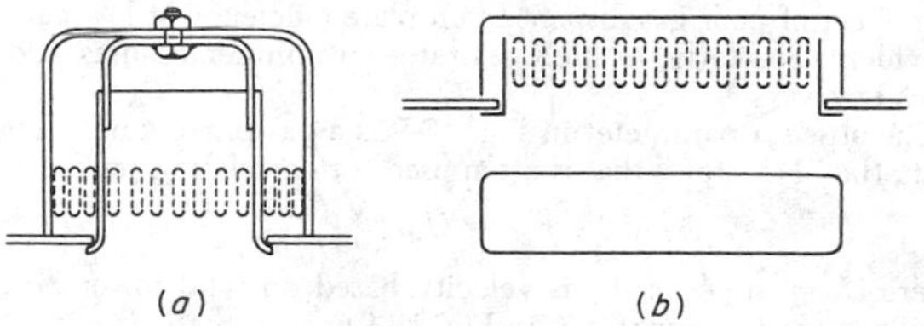

FIG. 18-5 (*a*) Circular or bell cap. (*b*) Tunnel cap.

round perforations in the 3- to 13-mm (⅛- to ½-in) range (dual-flow plate) or long slots with widths of 6 to 13 mm (¼ to ½ in) (Turbogrid tray). The plate material can be corrugated (Ripple tray) to segregate partially gas and liquid flow. In general, gas and liquid flow in a pulsating fashion with a particular opening passing both gas and liquid in an intermittent fashion.

A counterflow plate often used for contacting gases with liquids containing solids is the baffle plate, or "shower deck" (Fig. 18-6). Typically the plate is half-moon in shape and is sloped slightly in the direction of liquid flow. Gas contacts the liquid as it showers from the plate, and a serrated lip or weir at the edge of the plate can be used to improve the distribution of liquid in the shower.

The baffle plate operates with liquid dispersed and gas as the continuous phase and is used primarily in heat-transfer applications.

In summary, the perforated plate with liquid cross-flow (the "sieve plate") is the most common type specified for new designs. Schematic diagrams of such a plate are shown in Fig. 18-7. Nomenclature items are shown, with heights $h_{\ell i}$, h_f, $h_{\ell o}$, and h_ℓ referring to liquid entering, froth, liquid + froth leaving, and equivalent clear liquid averaged across the plate. For the plan view, area terms are as follows: A_t = tower total cross section; A_a = active area; A_d = area of one downcomer; A_n = net area for vapor flow (usually total cross section minus blocking downcomers); and A_h = area of holes or perforations. For the single cross-flow plate shown,

$$A_t = A_a + 2A_d$$

$$A_n = A_a + A_d = A_t - A_d$$

When downcomers are sloped or when perforations do not occupy essentially all the area between the downcomers, these simple relations do not apply. However, their adaptation should be obvious from the geometry involved.

The term "froth" in Fig. 18-7 suggests aeration in which the liquid phase is continuous. Under certain conditions there can be an inversion to a gas-continuous regime, or "spray." The spray has its phase boundaries equivalent to the boundaries for froth shown in Fig. 18-7.

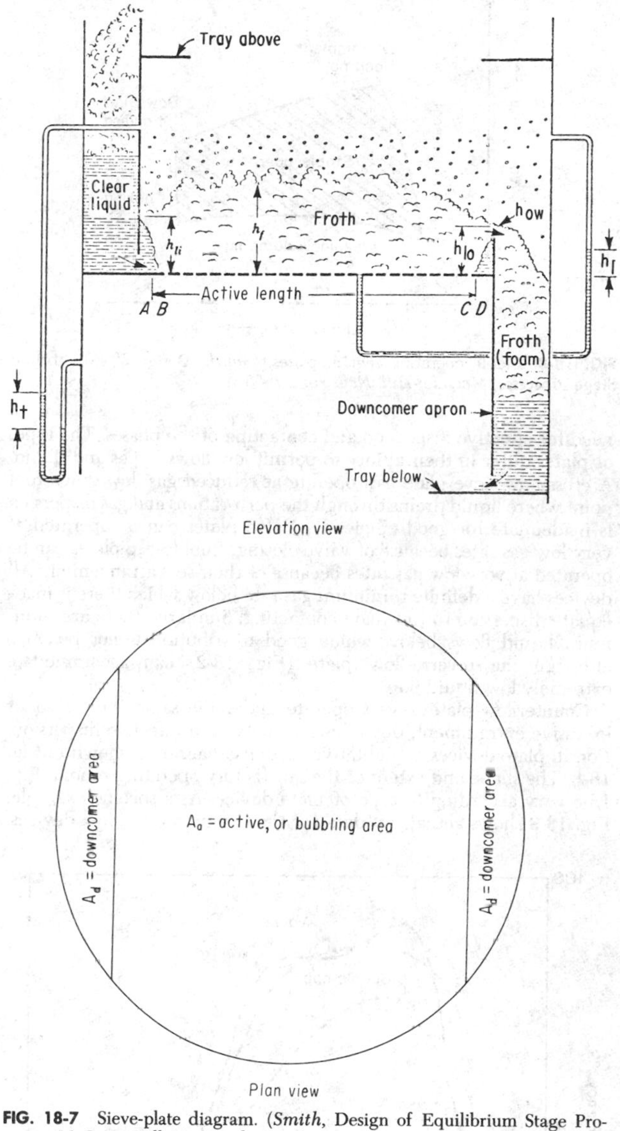

FIG. 18-7 Sieve-plate diagram. (*Smith,* Design of Equilibrium Stage Processes. *McGraw-Hill, New York, 1963.*)

Plate-Column Capacity

Introduction The maximum allowable capacity of a plate for handling gas and liquid flow is of primary importance because it fixes the minimum possible diameter of the column. For a constant liquid rate, increasing the gas rate results eventually in excessive entrainment and **flooding.** At the flood point it is difficult to obtain net downward flow of liquid, and any liquid fed to the column is carried out with the overhead gas. Furthermore, the column inventory of liquid increases, pressure drop across the column becomes quite large, and control becomes difficult. Rational design calls for operation at a safe margin below this maximum allowable condition.

Flooding may also be brought on by increasing the liquid rate while holding the gas rate constant. Excessive liquid flow can overtax the capacity of downcomers or other passages, with the ultimate result of increased liquid inventory, increased pressure drop, and the other characteristics of a flooded column.

These two types of flooding are usually considered separately when a plate column is being rated for capacity. For identification purposes they are called **entrainment flooding** (or "priming") and **downflow flooding.** When counterflow action is destroyed by either type, transfer efficiency is lost and reasonable design limits have been exceeded.

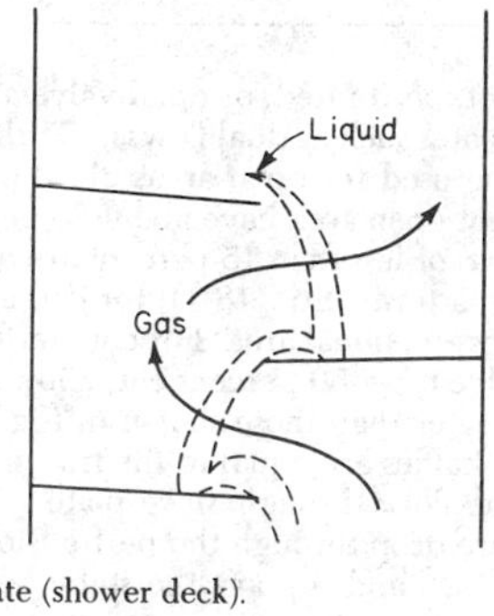

FIG. 18-6 Baffle plate (shower deck).

Minimum allowable capacity of a column is determined by the

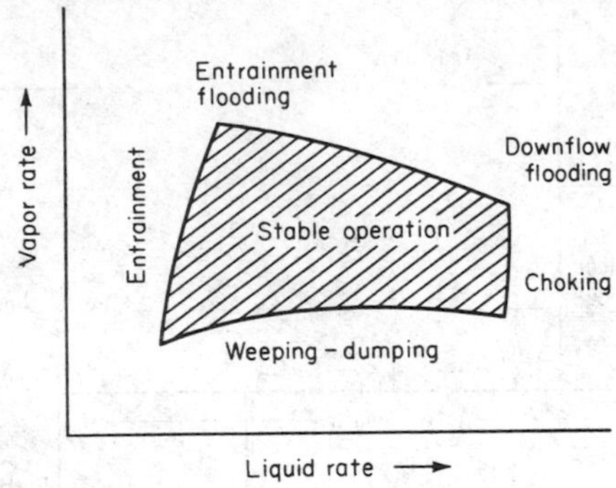

FIG. 18-8 Stable operating region, plates. (*Smith,* Design of Equilibrium Stage Processes, *McGraw-Hill, New York, 1963.*)

need for effective dispersion and contacting of the phases. The types of plates differ in their ability to permit low flows of gas and liquid. A cross-flow sieve plate can operate at reduced gas flow down to a point where liquid drains through the perforations and gas dispersion is inadequate for good efficiency. Valve plates can be operated at very low gas rates because of valve closing. Bubble-cap plates can be operated at very low gas rates because of their seal arrangement. All devices have a definite minimum gas rate below which there is inadequate dispersion for intimate contacting. Similarly, there are minimum liquid flows below which good distribution is not possible, although the reverse-flow plate (Fig. 18-2) can accommodate extremely low liquid flows.

Counterflow-plate towers operate under the same constraints of excessive entrainment, downflow capacity, and effective dispersion. For all plate devices a qualitative capacity diagram is shown in Fig. 18-8. The shape and extent of the satisfactory operating zone in Fig. 18-8 vary according to type of plate device. As a specific example, Fig. 18-9 shows actual test data for the common cross-flow devices.

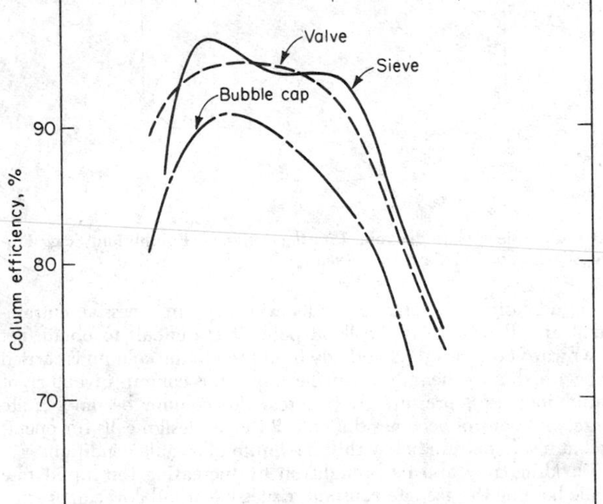

FIG. 18-9 Performance profiles of plates. Methanol-water, 3.2-ft column diameter; U_t = superficial gas velocity; and ρ_g = gas density. To convert (feet per second) (pounds per cubic foot)$^{1/2}$ to (meters per second) (kilograms per cubic meter)$^{1/2}$, multiply by 1.2199; to convert feet to meters, multiply by 0.3048. (*From Kastanek et al.,* Proc. Int. Symp. Distill., *Brighton, England, 1969; Institution of Chemical Engineers, London, 1969.*)

The effect of poor gas dispersion on plate efficiency at low gas rates is evident. Similarly, at high gas rates entrainment causes a loss of efficiency.

The abscissa parameter in Fig. 18-9 is a gas-phase kinetic-energy term, the "*F* factor," that is often used for correlating purposes:

$$F_{gt} = U_t \rho_g^{0.5}$$

where U_t = superficial gas velocity based on total tower area A_t, m/s (or ft/s as in Fig. 18-9)
ρ_g = gas density, kg/m³ (or lb/ft³ as in Fig. 18-9)

The flood point of the devices covered in Fig. 18-9 is noted to be in the range of F_{gt} = 1.4 to 1.6 (ft/s)(lb/ft³)$^{1/2}$.

Entrainment Flooding The early work of Souders and Brown [*Ind. Eng. Chem.*, **26,** 98 (1934)], based on a force balance on an "average" suspended droplet of liquid, led to the definition of a capacity parameter C_{sb}:

$$C_{sb} = U_n \sqrt{\rho_g/(\rho_\ell - \rho_g)} \qquad (18\text{-}1)$$

where U_n = linear gas velocity based on net area A_n, m/s (or ft/s as in Fig. 18-10)
ρ_g = gas density, kg/m³ (or lb/ft³)
ρ_ℓ = liquid density, kg/m³ (or lb/ft³)

For cross-flow plates net area is the column cross section less that area blocked by the downcomer or downcomers. For a single-pass plate with splash baffle (Fig. 18-7) the area of both downcomers should be deducted from the tower cross section to obtain net area. For counterflow plates, net area is the column cross section.

Maximum allowable values of the capacity parameter $C_{sb,\text{flood}}$ have been correlated against a flow parameter $F_{\ell g}$ as shown in Fig. 18-10. Figure 18-10 may be used for sieve plates, valve plates, and bubble-cap plates to determine the flooding-gas velocity:

$$U_{nf} = C_{sb,\text{flood}} \left(\frac{\sigma}{20}\right)^{0.2} \left(\frac{\rho_\ell - \rho_g}{\rho_g}\right)^{0.5} \qquad (18\text{-}2)$$

where U_{nf} = gas velocity through net area at flood, m/s (or ft/s as in Fig. 18-10)
$C_{sb,\text{flood}}$ = capacity parameter, m/s (or ft/s as in Fig. 18-10)
σ = liquid surface tension, mN/m (dyn/cm)
ρ_ℓ = liquid density, kg/m³ (or lb/ft³)
ρ_g = gas density, kg/m³ (or lb/ft³)

Figure 18-10 gives flooding-gas velocities to ±10 percent subject to the following restrictions:

1. System is low or nonfoaming.
2. Weir height is less than 15 percent of plate spacing.
3. Sieve-plate perforations are 13 mm (½ in) or less in diameter.
4. Ratio of slot (bubble cap), perforation (sieve), or full valve opening (valve plate) area A_h to active area A_a is 0.1 or greater. Otherwise, the value of U_{nf} obtained from Fig. 18-10 should be corrected:

A_h/A_a	U_{nf}/U_{nf}
0.10	1.00
0.08	0.90
0.06	0.80

where A_h = total slot, perforated, or open-valve area on plate.

For counterflow plates such as dual flows or Turbogrids, the curves of Fig. 18-10 may be used for open areas of 20 percent or greater. Plates with 15 percent open area have about 85 percent of the curve values, and open areas of less than 15 percent are not recommended. Note that the net-area term [Eq. (18-1)] for counterflow columns is equal to the total cross-sectional area. For counterflow-plate columns of the segmental-baffle type, 50 percent cut, allowable C_{sb} values are about 15 percent greater than those shown in Fig. 18-10, when vertical spacings of the baffles are equal to the tray spacings shown.

Weeping Liquid flow through sieve-plate perforations occurs when the gas-pressure drop through the perforations is not sufficient to create bubble surface and support the static head of froth above

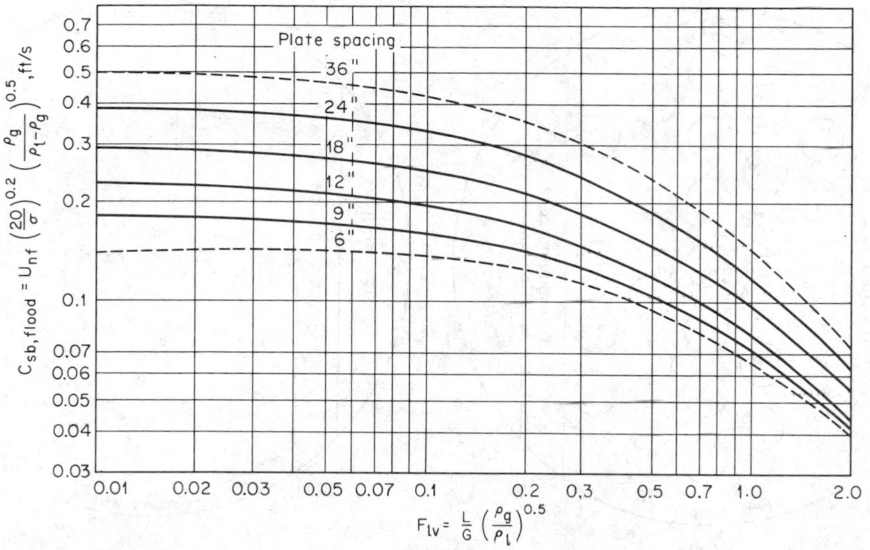

FIG. 18-10 Flooding limits for bubble-cap and perforated plates. L/G = liquid-gas mass ratio at point of consideration. To convert feet per second to meters per second, multiply by 0.3048; to convert inches (symbol ″) to meters, multiply by 0.0254. [*Fair*, Pet./Chem. Eng., *33(10), 45 (September 1969).*]

the perforations. Weeping can be deleterious in that liquid tends to short-circuit the primary contacting zones. On the other hand, some mass transfer to and from the weeping liquid occurs. Usual practice is to design so that deleterious weeping does not occur, based on a correlation such as that shown in Fig. 18-11.

In Fig. 18-11, h_d = head loss to gas flow through perforations, mm liquid [see Eq. (18-6)], and h_σ = head loss due to bubble formation, mm liquid. The latter loss is based on the energy required for bubble formation,

$$\Delta P = 4\sigma/d_h$$

with a convenient dimensional form for use in Fig. 18-11 being

$$h_\sigma = 409(\sigma/\rho_\ell d_h) \tag{18-2a}$$

where σ = surface tension, mN/m
ρ_ℓ = liquid density, kg/m^3
d_h = diameter of a perforation, mm
h_σ = head loss due to bubble formation, mm liquid

If design shows a condition *above* the appropriate curve of Fig. 18-11, weeping will not be deleterious to plate performance as measured by a sharp drop in plate efficiency (Fig. 18-9).

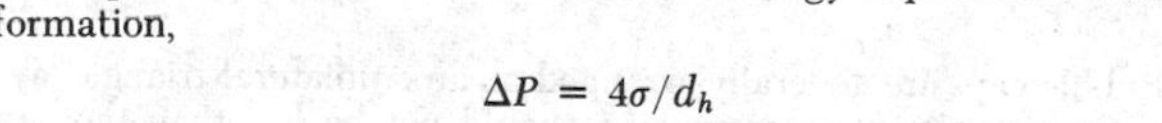

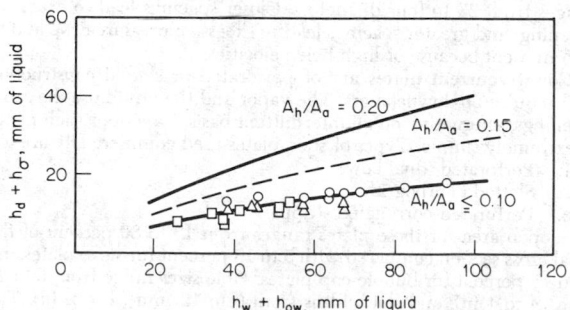

FIG. 18-11 Weeping sieve plates. To convert millimeters to inches, multiply by 0.0394. (*Smith*, Design of Equilibrium Stage Processes, *McGraw-Hill, New York, 1963.*)

Downflow Flooding Columns can flood because of their inability to handle large quantities of liquid. For cross-flow plates this limit on liquid rate is evidenced by **downcomer backup** to the plate above. To avoid downflow flooding, one must size the column downcomers such that excessive backup does not occur.

Downcomer backup is calculated from the pressure-balance equation

$$h_{dc} = h_t + h_w + h_{ow} + h_{da} + h_{hg} \tag{18-3}$$

where h_{dc} = height in downcomer, mm liquid
h_t = total pressure drop across the plate, mm liquid
h_w = height of weir at plate outlet, mm liquid
h_{ow} = height of crest over weir, mm liquid
h_{da} = head loss due to liquid flow under downcomer apron, mm liquid
h_{ha} = liquid gradient across plate, mm liquid.

The heights or head losses in Eq. (18-3) should be in consistent units, e.g., millimeters or inches of liquid under operating conditions on the plate.

As noted, h_{dc} is calculated in terms of equivalent clear liquid. Actually, the liquid in the downcomer may be aerated, and actual backup is

$$h'_{dc} = h_{dc}/\phi_{dc} \tag{18-4}$$

where ϕ_{dc} is an average relative froth density (ratio of froth density to liquid density) in the downcomer. Design must not permit h'_{dc} to exceed the value of plate spacing; otherwise, flooding can be precipitated. In fact, plate spacing may be determined by some safe approach to the calculated value of h'_{dc}.

The value of ϕ_{dc} depends upon the tendency for gas and liquid to disengage (froth to collapse) in the downcomer. For cases favoring rapid bubble rise (low gas density, low liquid viscosity, low system foamability) collapse is rapid and clear liquid fills the bottom of the downcomer (Fig. 18-7). For such cases it is usual practice to employ a value of ϕ_{dc} = 0.5. For cases favoring slow bubble rise (high gas density, high liquid viscosity, high system foamability), values of ϕ_{dc} = 0.2 to 0.3 should be used. As the critical point is approached in high-pressure distillations and absorptions, special precautions with downcomer sizing are mandatory, and sloping of the downcomer apron may be used to provide additional disengaging surface (but at

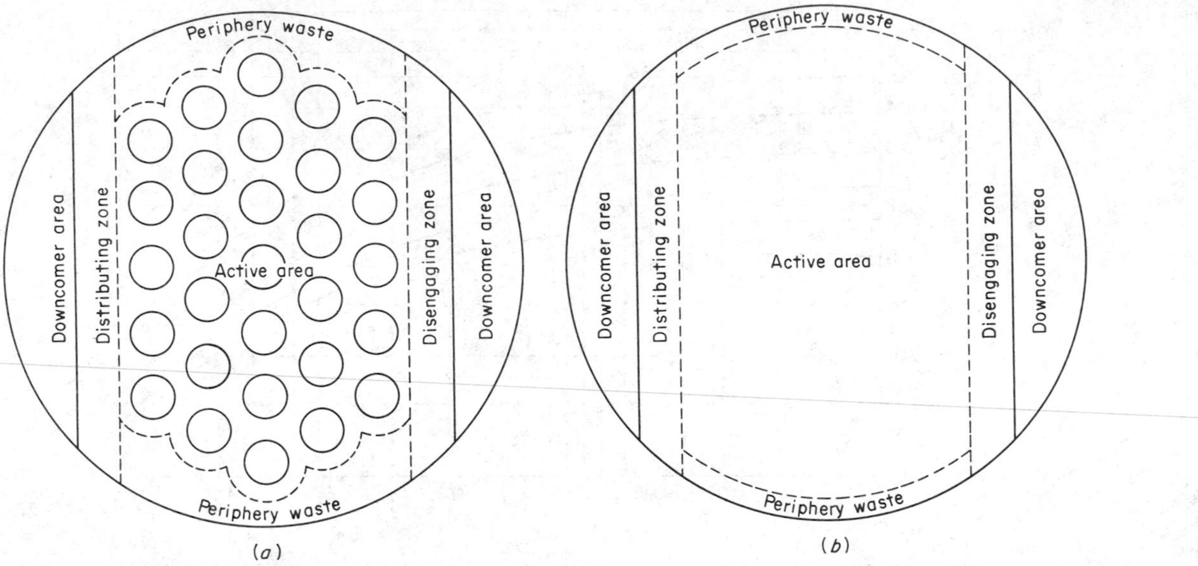

FIG. 18-12 Zone distribution. (*a*) Bubble-cap plate. (*b*) Sieve or valve plate.

the expense of cross-sectional area for perforations). Even so, some gas can be expected to recycle under the downcomer apron.

Plate Layouts **Cross-flow plates,** whether bubble-cap, sieve, or valve, are similar in layout (Fig. 18-12*a* and *b*). The plate consists of:

1. Active vapor-dispersion zone
2. Peripheral stiffening and support zone
3. Disengaging zone
4. Distributing zone
5. Downcomer zone

The downcomer zones generally occupy 10 to 30 percent of the total cross section.

The peripheral stiffening zone is generally 25 to 50 mm (1 to 2 in) wide and occupies 2 to 5 percent of the cross section, the fraction decreasing with increase in plate diameter.

The fraction of plate area occupied by disengaging and distributing zones ranges from 5 to 20 percent of the cross section. For some sieve-plate designs, these zones are eliminated completely.

For segmental downcomers, weir length ranges from 60 to 80 percent of the column diameter, so that the downcomer zone on each end of the plate occupies from 5 to 15 percent of the total cross section.

Periphery waste (Fig. 18-12) occurs primarily with bubble-cap trays and is due to the inability to fit the cap layout to the circular form of the plate. Valves and perforations can be located close to the wall, and little dead area results. Typical values of the fraction of the total cross-sectional area available for vapor dispersion and contact with the liquid for cross-flow plates with a chord weir equal to 75 percent of the column diameter are given in Table 18-3.

The plate thickness of a bubble-cap plate is generally established

Table 18-3 Typical Active Plate Areas

Column diameter, m	Cap diameter, mm	Active area, fraction of total cross section: Bubble cap	Sieve and valve
0.9	75	0.60	0.65
1.2	100	.57	.70
1.8	100	.66	.74
2.4	100	.70	.76
3.0	150	.74	.78

NOTE: To convert meters to feet, multiply by 3.281; to convert millimeters to inches, multiply by 0.0394.

by mechanical design factors and does not affect pressure drop. For a sieve plate, however, the plate is an integral component of the vapor-dispersion system, and thickness is significant.

For sieve plates, thickness is usually in the 10 to 14 U.S. Standard gauge range of 3.58 to 1.98 mm, or 0.141 to 0.078 in. Hardness of metal, size of die, and limits on hole size (for process reasons) lead to the following thickness criterion:

$$0.4 < \frac{\text{plate thickness}}{\text{hole diameter}} < 0.7$$

Bubble caps are generally arranged on an equilateral-triangle layout. Cap center-to-center spacing should not be less than the cap diameter plus 25 mm (1 in) in order to avoid the impact of vapor streams from adjacent caps. In practice, spacing varies from cap diameter plus 25 mm (1 in) to cap diameter plus 50 mm (2 in). Cap diameters vary from 50 to 150 mm (2 to 6 in) except in special applications such as cryogenic distillations, for which caps as small as 15-mm (0.6-in) diameter are used.

Hole sizes for sieve plates range from 1 mm to 25 mm (0.04 to 1 in) diameter, with sizes in the 4- to 6-mm (0.16- to 0.25-in) range being popular when fouling is not a serious problem and low entrainment rates are desired. The smaller hole sizes lead to punching problems, and the larger hole sizes lead to weeping and poor dispersion. The spacing of the holes, usually on an equilateral-triangle basis, ranges from 2½ to four diameters. Closer spacings lead to excessive weeping, and greater spacings lead to excessive pressure drop and to entrainment because of high hole velocities.

Countercurrent plates are of perforated or slotted construction and require no downcomers. The vapor and the liquid use the same openings, alternating on an intermittent basis. Layout of such plates is extremely simple. Types of such plates used commercially are

1. Perforated (dual flow)
2. Slotted (Turbogrid)
3. Perforated-corrugated (Ripple)

The open area for these plates ranges from 15 to 30 percent of the total cross section compared with 5 to 15 percent for sieve plates and 8 to 15 percent for bubble-cap plates. Hole sizes range from 6 to 25 mm (¼ to 1 in), and slot widths from 6 to 12 mm (¼ to ½ in). The Turbogrid and Ripple plates are proprietary devices.

Pressure Drop Methods for estimating fluid-dynamic behavior of cross-flow plates are analogous, whether the plates be bubble-cap,

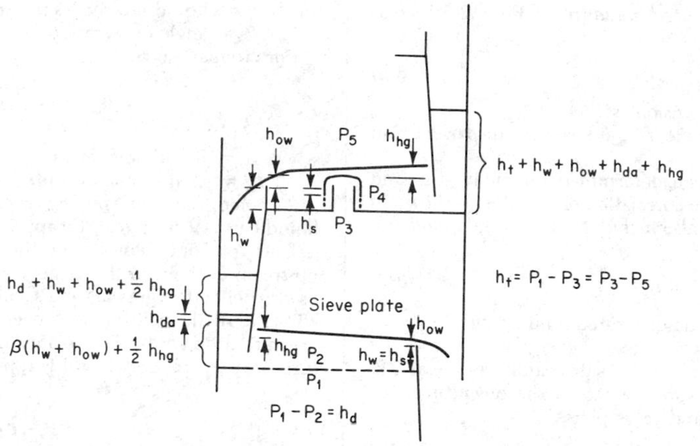

FIG. 18-13a Pressure-drop contributions for cross-flow plates. h_d = pressure drop through cap or sieve, equivalent height of plate liquid; h_w = height of weir; h_{ow} = weir crest; h_s = static liquid seal; h_{hg} = hydraulic gradient; and h_{da} = loss under downcomer.

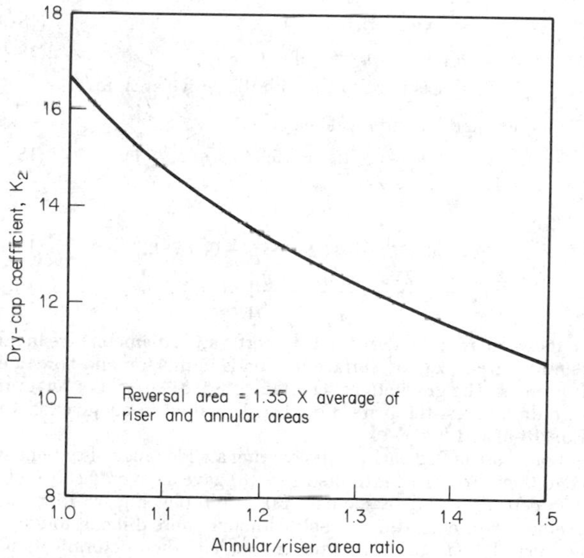

FIG. 18-13b Head-loss coefficient for dry bubble caps. (*Smith,* Design of Equilibrium Stage Processes, *McGraw-Hill, New York, 1963.*)

sieve, or valve. The total pressure drop across a plate is defined by the general equation (see Fig. 18-13a)

$$h_t = h_d + h'_\ell \tag{18-5}$$

where h_t = total pressure drop, mm liquid
h_d = pressure drop across the dispersion unit (dry cap + slot drop for bubble caps; dry hole for sieve plates; dry valve for valve plates), mm liquid
h'_ℓ = pressure drop through aerated mass over and around the disperser, mm liquid

It is convenient and consistent to relate all of these pressure-drop terms to height of equivalent clear liquid (deaerated basis) on the plate, either millimeters or inches of liquid.

Pressure drop across the disperser is calculated by variations of the standard orifice equation:

$$h_d = K_1 + K_2 \frac{\rho_g}{\rho_\ell} U_h^2 \tag{18-6}$$

where U_h = linear gas velocity through risers (bubble caps) or perforations (sieve plate), m/s

For bubble caps, K_1 is the drop through the slots which, according to Bolles (in Smith, *Equilibrium Stage Processes*, McGraw-Hill, New York, 1963, chap. 14), is

$$K_1 = 3.73 \left(\frac{\rho_g}{\rho_\ell - \rho_g}\right)^{1/5} h_{sh}^{4/5} U_s^{2/5} \tag{18-7}$$

where h_{sh} = cap slot height, mm
U_s = linear gas velocity through slots, m/s

and values of K_2 are obtained from Fig. 18-13b.

For sieve plates, $K_1 = 0$ and $K_2 = 50.8/C_v^2$. Values of C_v are taken from Fig. 18-14. For valve plates, values of K_1 and K_2 depend on whether the valves are fully open. They also depend on the shape and weight of the valves. Vendors of valve plates make K_1 and K_2 data (or their equivalent) readily available.

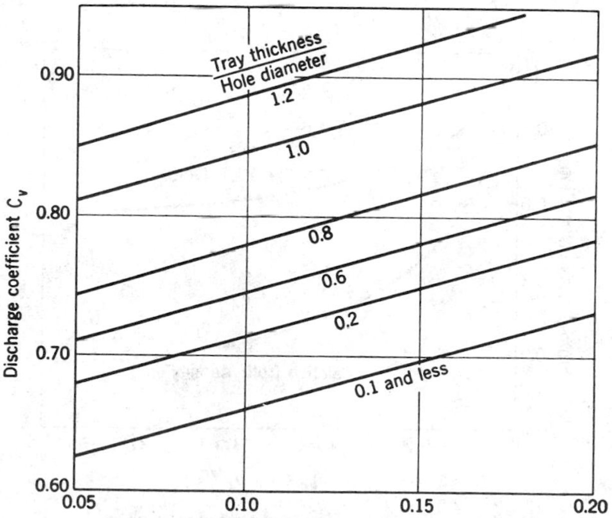

FIG. 18-14 Discharge coefficients for gas flow, sieve plates. [*Liebson, Kelley, and Bullington,* Pet. Refiner, *36(3), 288 (1957).*]

Pressure drop through the aerated liquid [h'_ℓ in Eq. (18-5)] is calculated by

$$h'_\ell = \beta h_{ds} \tag{18-8}$$

where β = aeration factor, dimensionless
h_{ds} = calculated height of clear liquid over the dispersers, mm (dynamic seal)

The aeration factor β has been determined for bubble-cap and sieve plates, and a representative correlation of its values is shown in Fig. 18-15. Note that the graph also includes values of *relative froth density* on the plate:

$$\phi_t = h'_\ell / h_f \tag{18-9}$$

where h'_ℓ = pressure drop through aerated liquid, mm
h_f = actual height of froth, mm

As can be seen, relative froth density on plates can be as low as 0.2, as contrasted with an average value of 0.5 in the downcomer, as noted for Eq. (18-4). For sieve and valve plates,

$$h_{ds} = h_w + h_{ow} + h_{hg}/2 \tag{18-10}$$

For bubble-cap plates,

$$h_{ds} = h_s + h_{ow} + h_{hg}/2 \tag{18-11}$$

where h_w = weir height, mm
h_s = static slot seal (weir height minus height of top of slot above plate floor), height of equivalent clear liquid, mm
h_{ow} = height of crest over weir, equivalent clear liquid, mm
h_{hg} = hydraulic gradient across plate, height of equivalent clear liquid, mm

The value of h_s is fixed by the selection of weir height and cap geometry. The value of weir crest h_{ow} may be calculated from the Francis weir equation and its modifications for various weir types. For a segmental weir and for height in *millimeters of clear liquid,*

$$h_{ow} = 664[(q/L_w)]^{2/3} \tag{18-12a}$$

where q = liquid flow, m^3/s
L_w = weir length, m

For serrated weirs,

$$h_{ow} = 851\left(\frac{q}{\tan \theta/2}\right)^{0.4} \tag{18-12b}$$

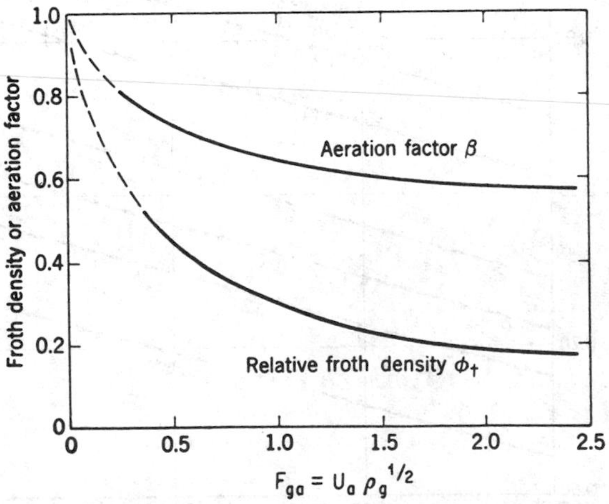

FIG. 18-15 Aeration factor, bubble-cap, sieve, and valve plates. U_a = linear gas velocity through active area, ft/s; and ρ_g = gas density, lb/ft^3. To convert F_{ga} in (feet per second) (pounds per cubic foot)$^{1/2}$ to (meters per second) (kilograms per cubic meter)$^{1/2}$, multiply by 1.2199. (*Smith,* Design of Equilibrium Stage Processes, *McGraw-Hill, New York, 1963.*)

where q = liquid flow, m^3/s per serration
θ = angle of serration, °

For circular weirs,

$$h_{ow} = 44{,}300\left(\frac{q}{d_w}\right)^{0.704} \tag{18-12c}$$

where q = liquid flow, m^3/s
d_w = weir diameter, mm

As noted, the weir crest h_{ow} is calculated on an equivalent clear-liquid basis. A more realistic approach is to recognize that in general a froth or spray flows over the outlet weir (settling can occur upstream of the weir if a large "calming zone" with no dispersers is used). Bennett, Agrawal, and Cook [*Am. Inst. Chem. Eng. J.*, **29**, 434 (1983)] allowed for froth overflow in a comprehensive study of pressure drop across sieve plates; their correlation for residual pressure drop [h'_ℓ in Eq. (18-5)] is represented by Eqs. (18-13) through (18-18):

$$h'_\ell = h_\ell + h'_\sigma \tag{18-13}$$

where h'_ℓ = pressure drop through the aerated liquid (= $h_t - h_d$), mm
h_ℓ = effective clear-liquid height (liquid holdup), mm
h'_σ = pressure drop for surface generation, mm

First, an effective froth density ϕ_e (dimensionless) is calculated:

$$\phi_e = \exp(-12.55 K_s^{0.91}) \tag{18-14}$$

where $K_s = U_a[\rho_g/(\rho_\ell - \rho_g)]^{0.5}$, m/s (18-15)
U_a = gas velocity through the active area, m/s

Then the liquid holdup is calculated:

$$h_\ell = \phi_e[h_w + 15{,}330C(q/\phi_e)^{2/3}] \tag{18-16}$$

where

$$C = 0.0327 + 0.0286 \exp(-0.1378 h_w) \tag{18-17}$$

$$h'_\sigma = \frac{472.4\sigma}{g\rho_\ell}\left[\frac{g(\rho_\ell - \rho_g)}{d_h \sigma}\right]^{1/3} \tag{18-18}$$

In these equations, h terms and d_h (perforation diameter) are in mm, densities are in kg/m^3, surface tension is in mN/m, and flow rate q is in m^3/s. The gravitational constant g is 9.81 m/s^2. For total pressure drop across the plate, Eq. (18-13) is used in conjunction with Eq. (18-6) and Fig. 8-14.

For a base of 302 data points covering a wide range of systems and conditions, Eqs. (18-14) through (18-18) gave an average error of ± 0.35 percent. For the same data Eq. (18-8) plus Fig. 18-15 gave an average error of ±10.7 percent. Although more difficult to use, the method of Bennett et al. is recommended when determination of pressure drop is of critical importance.

When straight or serrated segmental weirs are used in a column of circular cross section, a correction must be made for the distorted pattern of flow at the ends of the weirs. The correction factor F_w from Fig. 18-16 is used directly in Eq. (18-12*a*) or Eq. (18-12*b*). Even when circular downcomers are utilized, they are often preceded by a segmental weir. When the weir crest over a straight segmental weir is less than 6 mm (¼ in), it is desirable to use a serrated (notched) weir to provide good liquid distribution. Inasmuch as fabrication standards permit the tray to be 3 mm (⅛ in) out of level, weir crests less than 6 mm (¼ in) can result in maldistribution of liquid flow.

Loss under Downcomer The head loss under the downcomer apron, as millimeters of liquid, may be estimated from

$$h_{da} = 165.2(q/A_{da})^2 \tag{18-19}$$

where A_{da} = minimum area of flow under the downcomer apron, m^2. Although the loss under the downcomer is small, the clearance is significant from the aspect of tray stability and liquid distribution. The seal between the top of the liquid on the plate and the bottom of the downcomer should range between 13 and 38 mm (½ and 1½ in).

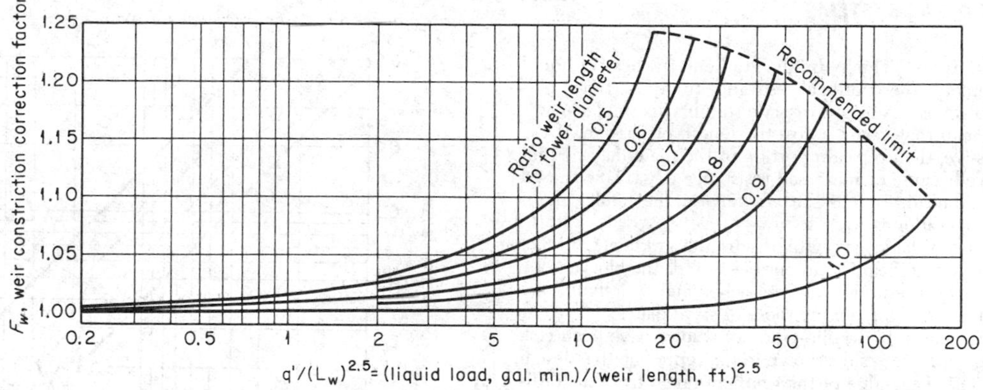

FIG. 18-16 Correction for effective weir length. To convert gallons per minute to cubic meters per second, multiply by 6.309×10^{-5}; to convert feet to meters, multiply by 0.3048. [*Bolles*, Pet. Refiner, **25**, *613 (1946)*.]

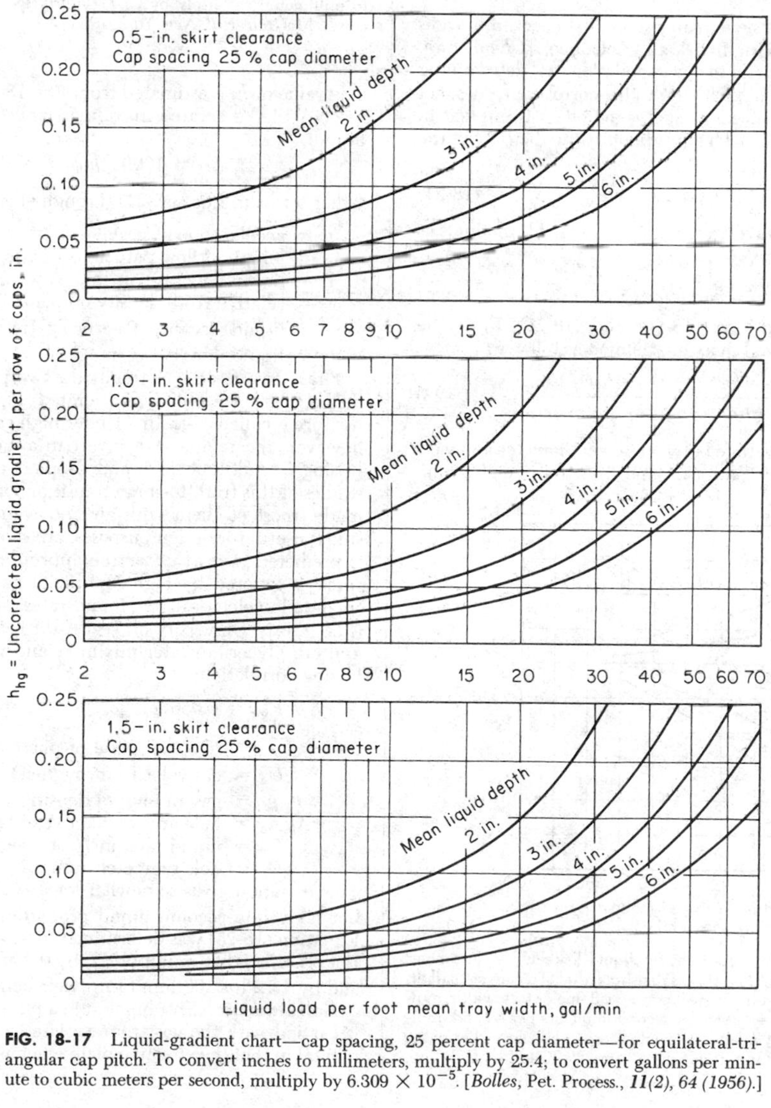

FIG. 18-17 Liquid-gradient chart—cap spacing, 25 percent cap diameter—for equilateral-triangular cap pitch. To convert inches to millimeters, multiply by 25.4; to convert gallons per minute to cubic meters per second, multiply by 6.309×10^{-5}. [*Bolles*, Pet. Process., ***11****(2), 64 (1956)*.]

Hydraulic Gradient The hydraulic gradient, the head of liquid necessary to overcome the frictional resistance to liquid (froth) passage across the plate, is significant to plate stability inasmuch as it is the only liquid head that varies across the length of passage. If the gradient is excessive, the upstream portion of the plate may be rendered inoperative because of increased resistance to gas flow caused by increased liquid head (Fig. 18-13*a*). In general, the empirical criterion for stable operation is $h_d > 2.5h_{hg}$.

Sieve plates usually have negligible hydraulic gradient. Bubble-cap plates can have significant gradient because of the blockage by the caps. Valve plates presumably are intermediate, with hydraulic-gradient characteristics approaching those of sieve plates.

For bubble-cap plates, hydraulic-gradient studies have been correlated by Bolles, with the results shown in the representative design charts of Fig. 18-17. The value of the gradient taken from the ordinate scale must be corrected for gas-flow effects by use of Fig. 18-18, which gives correction factor C_{vf}. Thus, for bubble-cap plates hydraulic gradient is obtained from

$$h_{hg} = C_{vf}h'_{hg} \tag{18-20}$$

with C_{vf} from Fig. 18-18 and h'_{hg} from Fig. 18-17. For cap-spacing–diameter ratios different from those in Fig. 18-17, the original reference should be consulted.

The hydraulic gradient on sieve plates should be checked in cases of long flow path of liquid. Hughmark and O'Connell [*Chem. Eng. Prog.*, **53**(3), 127 (1957)] presented a correlation for determining sieve-plate hydraulic gradient. Although the correlation does not explicitly indicate an effect of gas velocity, the effect is implicit in the choice of friction factor. The gradient is predicted by the relationship

$$h_{hg} = (1000fU_f^2L_f/gR_h) \tag{18-21}$$

where f = friction factor, correlated against a Reynolds modulus as in pipe flow

$$N_{Reh} = R_hU_f\rho_\ell/\mu_\ell \tag{18-22}$$

as shown in Fig. 18-19. In Eqs. (18-20) and (18-21), R_h is the hydraulic radius of the aerated mass, m, defined as follows:

$$R_h = \frac{\text{cross section}}{\text{wetted perimeter}} = \frac{h_fD_f}{2h_f + 1000D_f} \tag{18-23}$$

where D_f is the arithmetic average between tower diameter and weir length (average width of flow path), m, and h_f is froth height, mm.

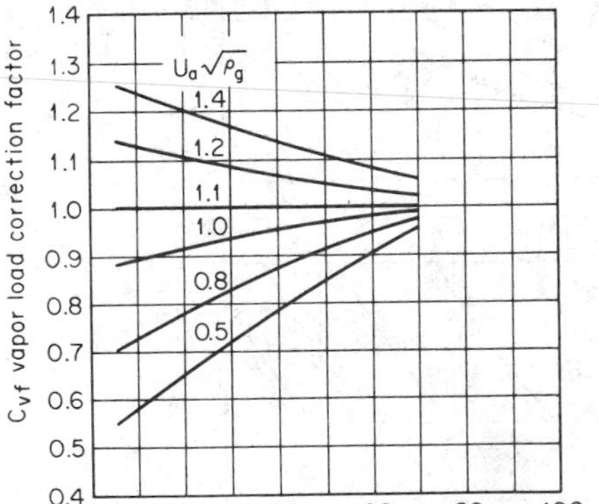

FIG. 18-18 Correction of liquid gradient for liquid load. U_a = linear gas velocity through active area, ft/s; ρ_g = gas density, lb/ft^3. To convert gallons per minute–foot tray width to cubic meters per second–meter tray width, multiply by 2.069×10^{-4}; to convert feet per second to meters per second, multiply by 0.3048; and to convert pounds per cubic foot to kilograms per cubic meter, multiply by 16.018. [*Davies*, Ind. Eng. Chem., *39, 774 (1947).*]

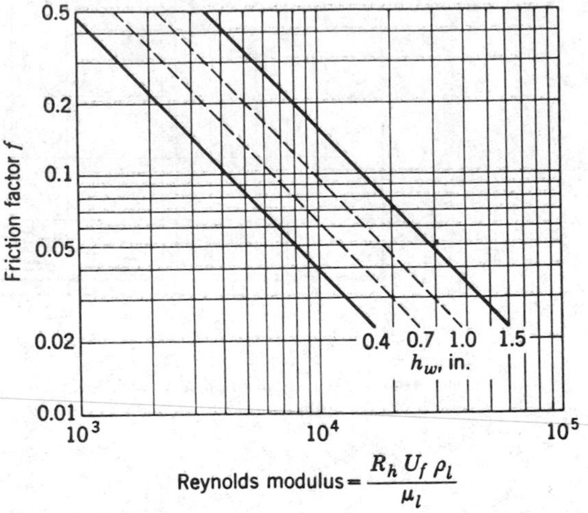

FIG. 18-19 Friction factor for froth cross-flow, sieve plates. To convert inches to millimeters, multiply by 25.4. (*Smith,* Design of Equilibrium Stage Processes, *McGraw-Hill, New York, 1963.*)

The value of h_f is estimated from Fig. 18-15 and Eq. (18-9). U_f is the velocity of the aerated mass, m/s, and is the same as for the clear liquid:

$$U_f = 1000q/h_f\phi_tD_f = 1000q/h_\ell D_f \tag{18-24}$$

Other terms in Eqs. (18-21) through (18-24) are:

g = acceleration of gravity, m/s^2
L_f = length of flow path across plate, m
q = liquid-flow rate, m^3/s
ϕ_t = relative froth density on plate, dimensionless
μ_ℓ = liquid viscosity, Pa·s or kg/(m·s)
ρ_ℓ = liquid density, kg/m^3

Phase Inversion Normally the two-phase mixture on the plate is in the form of a bubbly, or aerated, liquid. This liquid-continuous mixture is called a "froth." Under high gas rates and low liquid rates, however, the regime can invert to a gas-continuous "spray" comprising a multitude of liquid droplets of varying diameter. Many studies of this froth-to-spray transition, or phase inversion, have been made, most of them with air and water. The results of one such study, useful for design purposes, are shown in Fig. 18-20. The spray is predicted to exist *above* the appropriate curve. It appears that a spray is favored by high superficial gas velocities, low hole areas (high hole velocities), low liquid rates, and large hole sizes. Johnson (M.S. thesis, University of Texas, 1981) studied air aeration of mineral oil, glycerine-water mixtures, and water and obtained the following correlation:

$$F^\circ_{ga} = 0.0567\rho_\ell^{0.692}\sigma^{0.06}(A_h/A_a)^{0.25}(q/L_w)^{0.05}D_h^{-0.10} \tag{18-25}$$

where F°_{ga} = $U_a\rho_g^{1/2}$ at point of phase inversion
U_a = gas velocity through active area of plate, m/s
ρ_g,ρ_ℓ = gas and liquid densities, kg/m^3
A_h,A_a = hole area and active area, m^2
q/L_w = liquid rate, m^3/(s·m weir length)
D_h = hole diameter, mm

This equation gives somewhat lower values of F°_{ga} than Fig. 18-20 but takes into account liquid properties other than those of water. Equation (18-25) was determined for a weir height of 50 mm (2 in). For other heights, multiply F°_{ga} by 0.92 for 25-mm (1-in) weir height and by 1.12 for 100-mm (4-in) weir height.

Entrainment Entrainment in a plate column is that liquid which is carried with the gas from a plate to the plate above. It is detrimental in that the effective plate efficiency is lowered because liquid

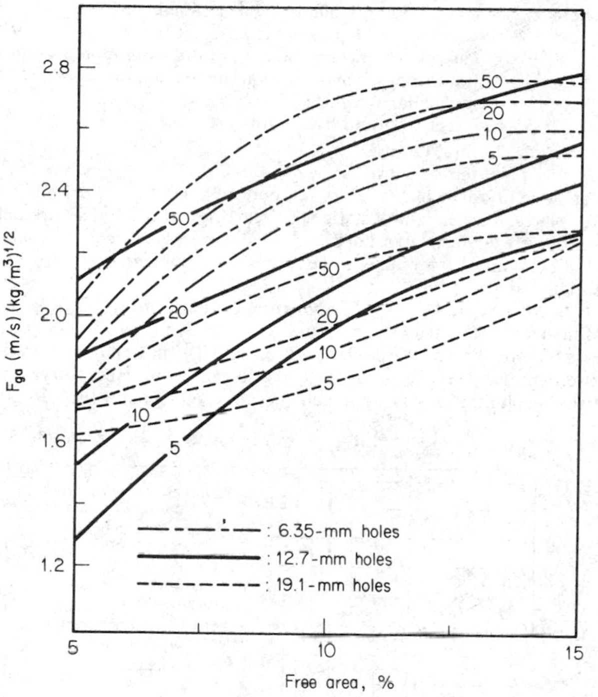

FIG. 18-20 Transition from froth to spray regime for holes of various diameters. Values on curves are liquid loadings, $m^3/(h \cdot m$ weir length). To convert cubic meters per hour-meter to cubic feet per hour-foot, multiply by 10.764; to convert (meters per second) (kilograms per cubic meter)$^{1/2}$ to (feet per second) (pounds per cubic foot)$^{1/2}$, multiply by 0.8197; and to convert millimeters to inches, multiply by 0.0394. [*Loon, Pinczewski, and Fell,* Trans. Inst. Chem. Eng., *51, 374 (1973).*]

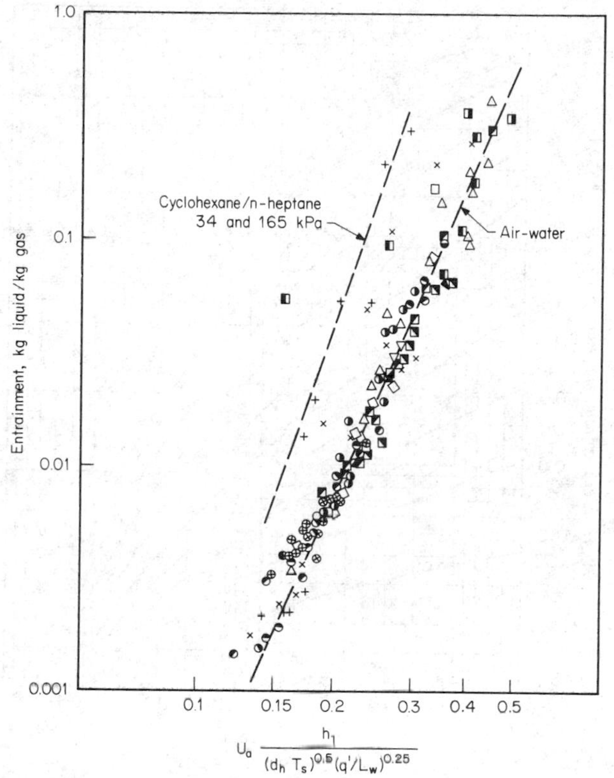

FIG. 18-21 Spray-regime entrainment correlation. T_s = tray spacing, mm; L_w = weir length, m; q' = liquid rate, m^3/h. To convert millimeters to inches, multiply by 0.0394; to convert meters to feet, multiply by 3.281; and to convert cubic meters per hour to cubic feet per hour, multiply by 35.31. [*Kister, Pinczewski, and Fell,* Ind. Eng. Chem. Process Des. Dev., *20, 528 (1981).*]

from a plate of lower volatility is carried to a plate of higher volatility, thereby diluting distillation or absorption effects. Entrainment is also detrimental when nonvolatile impurities are carried upward to contaminate the overhead product from the column.

Many studies of entrainment have been reported, most of them with small simulators using the air-water system. The dominant variable affecting entrainment is the gas velocity through the zone containing the froth or spray. Mechanisms of entrainment generation are discussed in the subsection "Liquid-in-Gas Dispersions."

Entrainment is expected to be more significant when flow conditions approach those favoring the spray regime; such conditions are common in vacuum distillations and in absorptions at low pressure. Figure 18-21 shows a correlation of observed water entrainment in air; the dashed line is based on a very limited amount of data for the distillation of cyclohexane–*n*-heptane mixtures (Sakata and Yanagi, *Distillation 1979,* Institution of Chemical Engineers, London, 1979, p. 3.2/21). More work on entrainment modeling will be required before a completely reliable predictive correlation will be available.

For distillations, it is often of more interest to ascertain the effect of entrainment on efficiency than to predict the quantitative amount of liquid entrained. For this purpose the correlation shown in Fig. 18-22 is useful. The parametric curves in the figure represent approach to the entrainment-flood point as predicted by Fig. 18-10 and the ratio of design gas velocity to flooding gas velocity. The abscissa values of Fig. 18-22 are the same as those of Fig. 18-10. The ordinate values ψ are fractions of gross liquid downflow, defined as follows:

$$\psi = e/(L + e) \tag{18-26}$$

where e = absolute entrainment of liquid
L = liquid downflow rate without entrainment

Figure 18-22 also accepts the validity of the Colburn equation [*Ind. Eng. Chem.,* **28**, 526 (1936)] for the effect of entrainment on efficiency:

$$\frac{E_a}{E_{mv}} = \frac{1}{1 + E_{mv}[\psi/(1 - \psi)]} \tag{18-27}$$

where E_{mv} = Murphree vapor efficiency [see Eq. (18-32)]
E_a = Murphree vapor efficiency, corrected for recycle effect of liquid entrainment

Equation (18-27) plus Fig. 18-22 may be used to estimate the effect of entrainment on plate efficiency. The figure alone may be used to estimate absolute values of entrainment. For the former situation, accuracy of ±15 percent may be expected. For the latter, poorer accuracy may be expected, especially in nondistillation applications.

Plate Efficiency The efficiency of a plate for mass transfer depends upon three sets of design parameters:

1. The system—composition and properties
2. Flow conditions—rates of throughput
3. Geometry—plate type and dimensions

The designer has little control over the first set but can deal effectively with the other two. Ultimate concern is with *overall column efficiency:*

$$E_{oc} = N_t/N_a \tag{18-28}$$

or the ratio of *theoretical plates* to *actual plates* required to make the separation. In arriving at a value of E_{oc} for design, the designer may rely on plant test data or on judicious use of pilot-plant-effi-

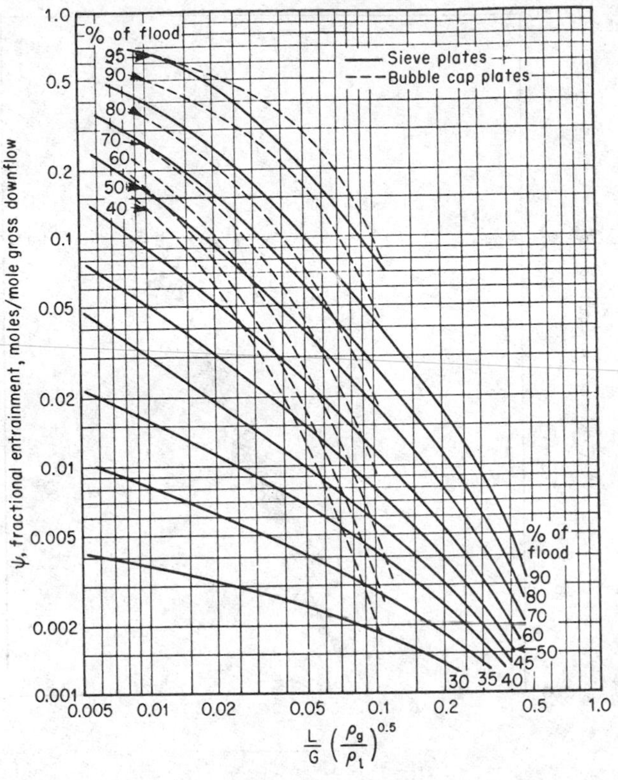

FIG. 18-22 Entrainment correlation. L/G = liquid-gass mass ratio; and ρ_ℓ and ρ_g = liquid and gas densities. [*Fair*, Pet./Chem. Eng., *33(10), 45 (September 1961).*]

ciency measurements. If such direct information is not available, the designer must resort to predictive methods.

Methods for predicting plate efficiency are of three general types:

1. Empirical methods
2. Direct scale-up from laboratory measurements
3. Theoretical or semitheoretical mass-transfer methods

The first of these gives E_{oc} directly. The second gives a point efficiency [Eq. (18-30)]. The third involves the prediction of individual phase efficiencies.

Empirical Predictive Methods Two empirical correlations which have found wide use are the one of Drickamer and Bradford [*Trans. Am. Inst. Chem. Eng.*, **39**, 319 (1943)] and a modification of it by O'Connell [*Trans. Am. Inst. Chem. Eng.*, **42**, 741 (1946)]. The latter is shown in Fig. 18-23; the Drickamer-Bradford data are included in the distillation plot. The absorber-efficiency correlation utilizes a solubility function: HP/μ_ℓ, where H is the Henry's-law constant, (lb·mol)/(ft^3·atm), P is pressure, atm, and μ_ℓ is liquid viscosity, cP. For both plots, properties are evaluated at average conditions in the column. Care should be taken not to use this correlation outside its intended range of application.

A semitheoretical method which gives overall efficiency is that of Bakowski [*Br. Chem. Eng.*, **8**, 384, 472 (1963); **14**, 945 (1969)]. It is based on the assumption that the mass-transfer rate for a component moving to the vapor phase is proportional to the concentration of the component in the liquid and to its vapor pressure. Also, the interfacial area is assumed proportional to liquid depth, and surface renewal rate is assumed proportional to gas velocity. The resulting general equation for binary distillation is

$$E_{oc} = \frac{1}{1 + 3.7(10^4)\dfrac{KM}{h'\rho_\ell T}} \tag{18-29}$$

where E_{oc} = overall column efficiency, fractional
K = vapor-liquid equilibrium ratio, y°/x
y° = gas-phase concentration at equilibrium, mole fraction
x = liquid-phase concentration, mole fraction
M = molecular weight
h' = effective liquid depth, mm
ρ_ℓ = liquid density, kg/m^3
T = temperature, K

For sieve or valve plates, $h' = h_w$, outlet weir height. For bubble-cap plates, h' = height of static seal. The original references present validations against laboratory and small-commercial-column data. Modifications of the efficiency equation for absorption-stripping are also included.

Direct Scale-Up of Laboratory Distillation Efficiency Measurements It has been found by Fair, Null, and Bolles [*Ind. Eng. Chem. Process Des. Dev.*, **22**, 53 (1983)] that efficiency measurements in 25- and 50-mm- (1- and 2-in-) diameter laboratory Oldershaw columns closely approach the point efficiencies (Eq. 18-30)

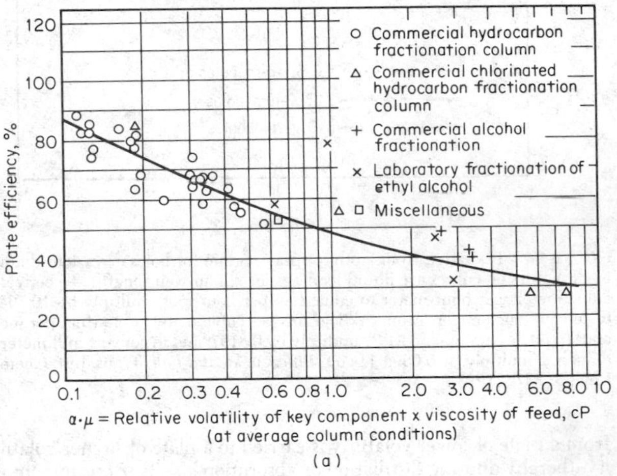

FIG. 18-23*a* O'Connell correlation for overall column efficiency E_{oc} for distillation. To convert centipoises to pascal-seconds, multiply by 10^{-3}. [*O'Connell*, Trans. Am. Inst. Chem. Eng., *42, 741 (1946).*]

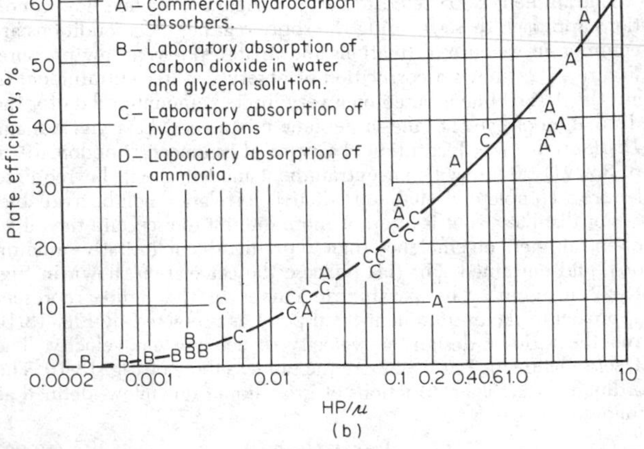

FIG. 18-23*b* O'Connell correlation for overall column efficiency E_{oc} for absorption. To convert HP/μ in pound-moles per cubic foot–centipoise to kilogram-moles per cubic meter–pascal-second, multiply by 1.60×10^4. [*O'Connell*, Trans. Am. Inst. Chem. Eng., **42**, *741 (1946).*]

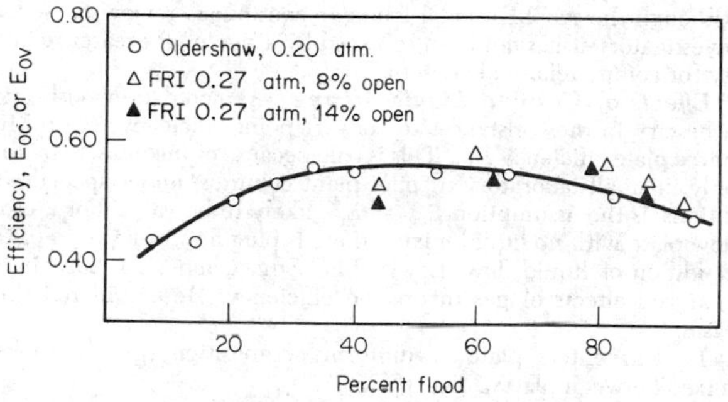

FIG. 18-24 Overall column efficiency of 25-mm Oldershaw column compared with point efficiency of 1.22-m-diameter-sieve sieve-plate column of Fractionation Research, Inc. System = cyclohexane–n-heptane. [(*Fair, Null, and Bolles*, Ind. Eng. Chem. Process Des. Dev., *22*, 53 (1982).]

measured in large sieve-plate columns. A representative comparison of scales of operation is shown in Fig. 18-24. Note that in order to achieve agreement between efficiencies it is necessary to ensure that (1) the systems being distilled are the same, (2) comparison is made at the same relative approach to the flood point, (3) operation is at total reflux, and (4) a standard Oldershaw device (a small perforated-plate column with downcomers) is used in the laboratory experimentation. Fair et al. made careful comparisons for several systems, utilizing as large-scale information the published efficiency studies of Fractionation Research, Inc.

Theoretical Predictive Methods The approach to equilibrium on a plate may be defined as the ratio of the actual change in gas composition as it passes through the plate to the change that would have occurred if the gas had reached a state of equilibrium with the liquid. If a *point* on plate n is considered, this definition leads to the **point efficiency**:

$$E_{og} = \left(\frac{y_n - y_{n-1}}{y_n^\circ - y_{n-1}}\right)_{\text{point}} \tag{18-30}$$

where y_n° is the gas concentration in equilibrium with liquid concentration at the point. This efficiency cannot exceed 1.0 (100 percent). If there are liquid-concentration gradients on the plate (i.e., plate liquid is not completely mixed), then y° will vary and E_{og} may vary from point to point on the plate. It should be noted that an analogous efficiency definition could be expressed on the basis of liquid concentrations. It should be noted also that **vaporization efficiency** (Holland, *Fundamentals of Multicomponent Distillation*, McGraw-Hill, New York, 1981) could be used:

$$E_v = y_n/y_n^\circ \tag{18-31}$$

For the entire plate and for gas concentrations, the **Murphree vapor efficiency** is used:

$$E_{mv} = \left(\frac{y_n - y_{n-1}}{y_n^\circ - y_{n-1}}\right)_{\text{plate}} \tag{18-32}$$

where y_n° is gas concentration in equilibrium with the concentration of the liquid leaving the plate (flowing into the downcomer, for a cross-flow plate). Because of concentration gradients in the liquid, E_{mv} can exceed 100 percent.

The best-established theoretical method for predicting E_{oc} is that of the AIChE (*Bubble-Tray Design Manual*, American Institute of Chemical Engineers, New York, 1958). It is based on the sequential prediction of point efficiency, Murphree efficiency, and overall column efficiency:

$$E_{og} \rightarrow E_{mv} \rightarrow E_{oc}$$

with suitable correction of E_{mv} for entrainment. The AIChE model is the basis for the development which follows.

On the basis of the two-film model for mass transfer, and relating all efficiencies to gas-phase concentrations (for convenience only; a similar development can be made on the basis of liquid concentrations), point efficiency can be expressed in terms of transfer units:

$$E_{og} = 1 - e^{-N_{og}} \tag{18-33}$$

where N_{og} = overall transfer units calculated from Eq. (18-34):

$$N_{og} = \frac{1}{1/N_g + \lambda/N_\ell} \tag{18-34}$$

where N_g = gas-phase transfer units
N_ℓ = liquid-phase transfer units
λ = mG_m/L_m (stripping factor)
m = slope of equilibrium curve
G_m = gas rate, mol/s
L_m = liquid rate, mol/s

Transfer units are dimensionless and are defined further in Sec. 14. According to Eq. (18-34), the evaluation of point efficiencies reduces to the prediction of point values of N_g and N_ℓ plus the evaluation of m, G_m, and L_m for the particular conditions under investigation.

Gas-phase transfer units are obtained from Eq. (18-35):

$$N_g = k_g a \theta_g \tag{18-35}$$

where k_g = gas-phase mass-transfer coefficient, $\dfrac{\text{kg}\cdot\text{mol}}{(\text{s}\cdot\text{m}^2)(\text{kg}\cdot\text{mol}/\text{m}^3)}$ or m/s
a = effective interfacial area for mass transfer, m^2/m^3 froth on plate
θ_g = residence time of gas in froth zone, s

The effect of increasing gas rate is to increase k_g and decrease θ_g, with the result that N_g tends to be constant over a range of gas rates. The AIChE work correlated N_g directly:

$$N_g = \frac{0.776 + 0.00457h_w - 0.238U_a\rho_g^{0.5} + 0.0712W}{N_{Scg}^{0.5}} \tag{18-36}$$

where h_w = weir height, mm
U_a = gas velocity through active area, m/s
W = liquid-flow rate, $m^3/(s\cdot m)$ of width of flow path on the plate
N_{Scg} = gas-phase Schmidt number $\mu_g/\rho_g D_g$, dimensionless

Equation (18-36) is based on experimental work with small columns and with bubble-cap plates. The range of conditions studied is discussed in the AIChE *Manual* and in supporting research reports that are referenced in the AIChE *Manual*. The equation appears to be equally applicable to bubble-cap, sieve, and valve plates. For the evaluation of W, an arithmetic-average flow-path width is adequate. For the evaluation of Schmidt number, see Sec. 3 for prediction of gas-phase diffusion coefficient D_g.

For distillation separations, most of the resistance to mass transfer occurs in the gas phase, and Eq. (18-35) assumes primary importance in efficiency prediction. Continuing effort is being directed toward breaking Eq. (18-35) down into its components for correlational purposes.

Liquid-phase transfer units are obtained from

$$N_\ell = k_\ell a \theta_\ell \tag{18-36a}$$

where k_ℓ = liquid-phase transfer coefficient, $\dfrac{\text{kg}\cdot\text{mol}}{(\text{s}\cdot\text{m}^2)(\text{kg}\cdot\text{mol}/\text{m}^3)}$ or m/s
a = effective interfacial area for mass transfer, m^2/m^3 froth or spray on the plate
θ_ℓ = residence time of liquid in the froth or spray zone, s

The froth or spray zone represents a volume with a base equal to the active area of the plate A_a and height h_f (see Fig. 18-7) equal to the level that would be observed through a window for viewing contacting action on the plate. In the absence of visual observations, h_f (mm) can be estimated from the AIChE equation:

$$h_f = 43.2U_a^2\rho_g + 1.89h_w - 40.6 \tag{18-37}$$

where U_a = linear gas velocity through the active area, m/s; ρ_g = gas density, kg/m^3; and h_w = weir height, mm. Thus, liquid residence time in seconds, s, is

$$\theta_\ell = h_f A_a \phi_t / 1000q \quad (18\text{-}38)$$

and since average froth or spray density on the plate ϕ_t is equal to h_ℓ / h_f,

$$\theta_\ell = h_\ell A_a / 1000q \quad (18\text{-}39)$$

where h_ℓ is obtained from Eq. (18-16) or, approximately, from Eq. (18-8) and q is the liquid flow rate, m^3/s.

The mass-transfer coefficient of Eq. (18-36*a*) is carried as a product with interfacial area (giving a volumetric mass-transfer coefficient):

$$k_\ell a = (3.875 \times 10^8 D_\ell)^{0.5}(0.40 U_a \rho_g^{0.5} + 0.17) \quad (18\text{-}40a)$$

(sieve plates)

$$k_\ell a = (4.127 \times 10^8 D_\ell)^{0.5}(0.21 U_a \rho_g^{0.5} + 0.15) \quad (18\text{-}40b)$$

(bubble-cap plates)

where D_ℓ = liquid-phase diffusion coefficient, m^2/s (see Sec. 3).

In summary, the point efficiency E_{og} is computed from Eq. (18-33) by using N_{og} from Eq. (18-34), N_g from Eq. (18-35), N_ℓ from Eq. (18-36*a*), and m based on the relative volatility of the system. For a binary mixture with constant relative volatility,

$$m = \frac{\alpha_{ij}}{[1 + (\alpha_{ij} - 1)x_i]^2} \quad (18\text{-}41)$$

where α_{ij} = relative volatility, component i (lighter material) relative to component j, and x_i = mole fraction of i in the liquid.

Although the AIChE model is under continuous review by several investigators, it has not been replaced by a model of greater reliability for commercial-scale design.

Effects of Gas and Liquid Mixing As noted previously, it is necessary in most instances to convert point efficiency E_{og} to Murphree plate efficiency E_{mv}. This is true because of incomplete mixing; only in small laboratory or pilot-plant columns, under special conditions, is the assumption $E_{og} = E_{mv}$ likely to be valid. For a cross-flow plate with *no* liquid mixing there is plug flow of liquid. For this condition of liquid flow, Lewis [*Ind. Eng. Chem.*, **28**, 399 (1936)] analyzed effects of gas mixing on efficiency. He considered three cases:

1. Gas enters plate at uniform composition (gas completely mixed between plates), Fig. 18-25.
2. Gas unmixed; liquid flows in the same direction on successive plates, Fig. 18-26.
3. Gas unmixed; liquid flows in alternate direction on successive plates, Fig. 18-27.

Case 1 has found the widest application in practice and is represented by the relationship

$$E_{mv} = 1/\lambda[\exp(\lambda E_{og}) - 1] \quad (18\text{-}42)$$

λ is defined as for Eq. (18-34). Equation (18-42) assumes the following in addition to the base conditions:

1. L/V is constant.
2. Slope of equilibrium curve m is constant.
3. Point efficiency is constant across the tray.

Most plate columns operate under conditions such that gas is completely mixed as it flows between the plates, but few operate with pure plug flow of liquid. Thus, Figs. 18-25 to 18-27 represent extreme cases and are important chiefly for evaluating geometric effects on mixing. Departure from plug flow of liquid has been stud-

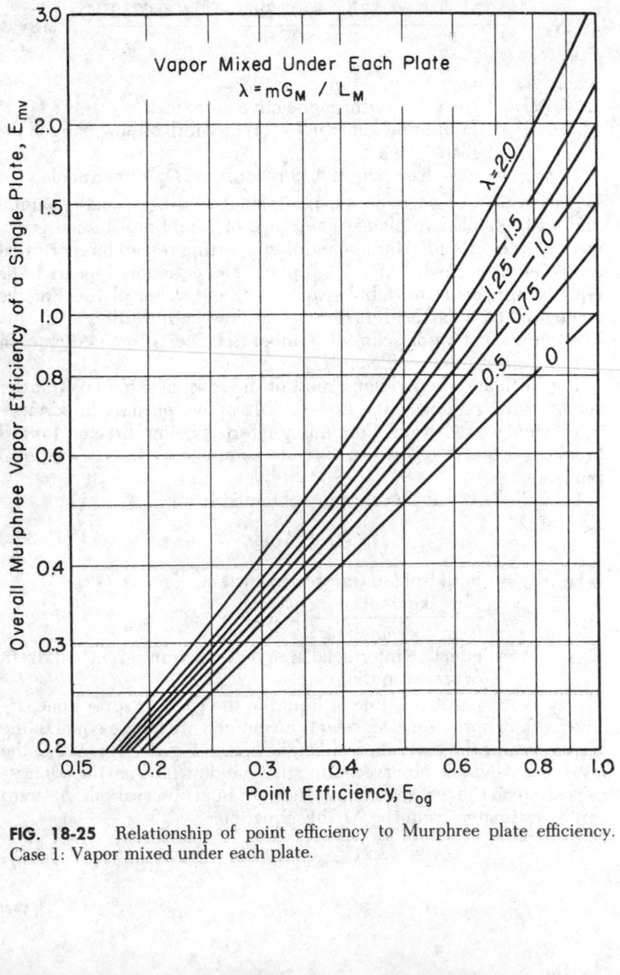

FIG. 18-25 Relationship of point efficiency to Murphree plate efficiency. Case 1: Vapor mixed under each plate.

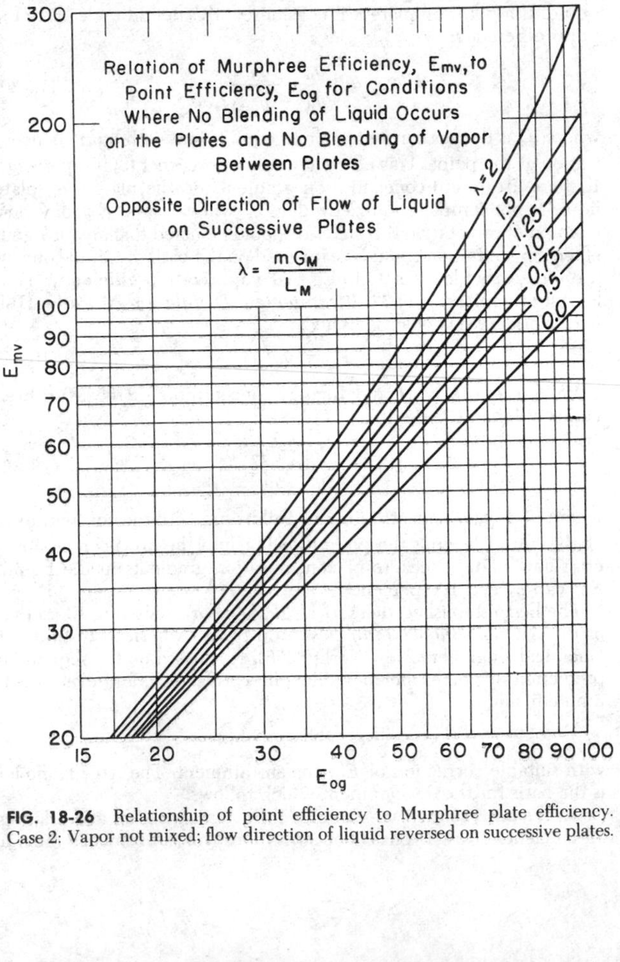

FIG. 18-26 Relationship of point efficiency to Murphree plate efficiency. Case 2: Vapor not mixed; flow direction of liquid reversed on successive plates.

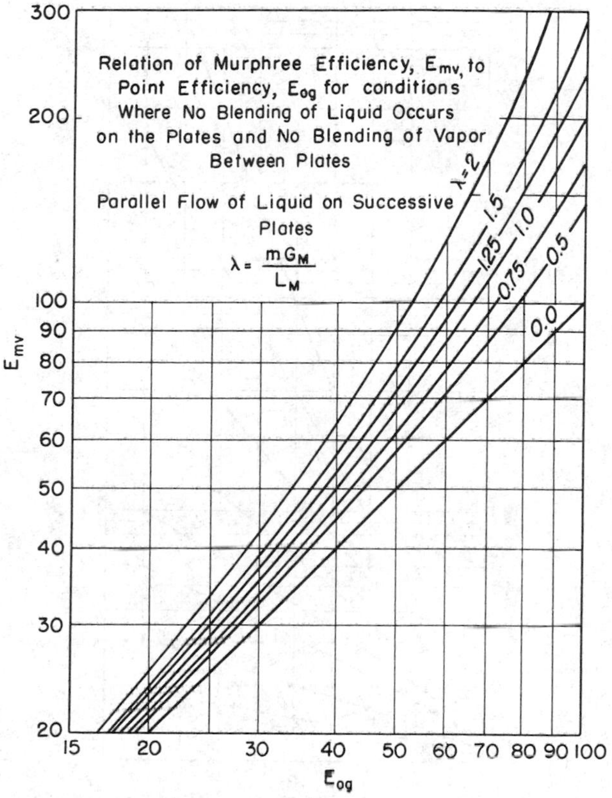

FIG. 18-27 Relationship of point efficiency to Murphree plate efficiency. Case 3: Vapor not mixed; liquid flow in same direction on successive plates.

ied by Gautreaux and O'Connell [*Chem. Eng. Prog.*, **51**, 232 (1955)] by assuming that liquid mixing can be represented as occurring in a series of stages of completely mixed liquid. For this model,

$$E_{mv} = \frac{1}{\lambda}\left[\left(1 + \lambda\frac{E_{og}}{n}\right)^n - 1\right] \tag{18-43}$$

where n = number of stages occurring on the tray.

An approximation of the number of stages can be obtained from Fig. 18-28, using the following criteria:

1. Increased liquid rate favors plug flow.
2. Sieve plates have less back mixing than bubble-cap plates because of less obstruction to flow.
3. Increased gas rate increases turbulence and the degree of back mixing of liquid.

An alternative approach is presented in the AIChE *Bubble-Tray Design Manual* and is based on an eddy-diffusion model. According to this model,

$$\frac{E_{mv}}{E_{og}} = \frac{1 - e^{-(n'+N_{Pe})}}{(n' + N_{Pe})\{1 + [(n' + N_{Pe})/n']\}} + \frac{e^n - 1}{n'\{1 + [n'/(n' + N_{Pe})]\}} \tag{18-44}$$

where $n' = \frac{N_{Pe}}{2}\left(\sqrt{1 - 4\lambda\frac{E_{og}}{N_{Pe}}} - 1\right)$

N_{Pe} = Péclet number (dimensionless) $= \frac{Z_\ell^2}{D_E\theta_l}$

Z_ℓ = length of liquid travel, m

λ = stripping factor [see Eq. (18-34)]

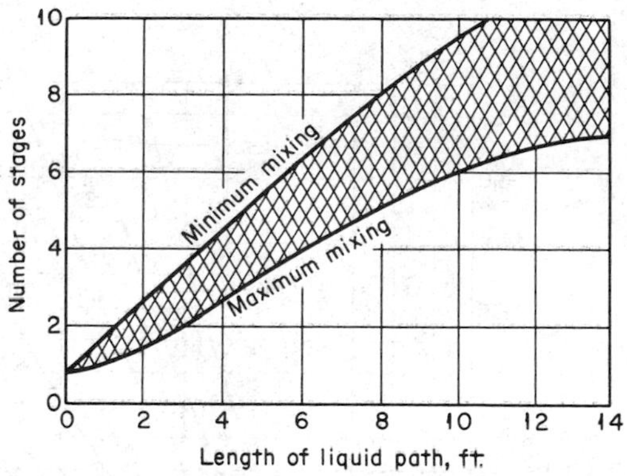

FIG. 18-28 Effect of length of liquid path on number of stages. To convert feet to meters, multiply by 0.3048. [*O'Connell and Gautreaux,* Chem. Eng. Prog., *51, 236 (1955).*]

The value of θ_l is calculated from Eq. (18-38). The term D_E is an eddy-diffusion coefficient that is obtained from experimental measurements. For sieve plates, Barker and Self [*Chem. Eng. Sci.*, **17**, 541 (1962)] obtained the following correlation:

$$D_E = 6.675(10^{-3})U_a^{1.44} + 0.922(10^{-4})h_\ell - 0.00562 \tag{18-45}$$

where D_E = eddy-diffusion coefficient, m^2/s
U_a = gas velocity through active area of plate, m/s
h_ℓ = liquid holdup on plate (Eq. 18-16), mm

For bubble-cap plates, the eddy-diffusion correlation in the AIChE *Bubble-Tray Design Manual* should be used.

The graphical representation of Eq. (18-44) is indicated in Fig. 18-29*a*, where as usual $\lambda = mG_m/L_m$.

Overall Column Efficiency Calculated values of E_{mv} must be corrected for entrainment, if any, by the Colburn equation [Eq. (18-27)]. The resulting corrected efficiency E_a is then converted to column efficiency by the relationship of Lewis [*Ind. Eng. Chem.*, **28**, 399 (1936)]:

$$E_{oc} = \frac{N}{N_t} = \frac{\log[1 + E_a(\lambda - 1)]}{\log\lambda} \tag{18-46}$$

Comparison of Efficiency of Various Plates Several studies of various plates have been carried out under conditions such that direct and meaningful comparisons are possible. Required conditions include identical system, same pressure, same column diameter, and equivalent submergence. Standart and coworkers [*Br. Chem. Eng.*, **11** (11), 1370 (1966); *Sep. Sci.*, **2**, 439 (1967)] used the methanol-water system at atmospheric pressure in a 1.0-m (3.3-ft) column. For a plate spacing of 0.4 m (15.7 in) they studied the following:

1. Bubble-cap plate—70-mm (2.75-in) round caps, 25.4-mm (1.0-in) submergence
2. Sieve plate—4.0-mm (5/32-in) holes, hole–active area = 0.048, 40-mm (1.57-in) outlet weir
3. Turbogrid plate—4.6-mm (0.18-in) slot width, 14.7 percent open area
4. Ripple plate—2.85-mm (7/64-in) holes, 10.8 percent open area

Efficiency data from the work are summarized in Fig. 18-29*b*.

Kirschbaum (*Distillier-Rektifiziertechnik*, 4th ed., Springer-Verlag, Berlin and Heidelberg, 1969) reported on studies of the ethanol-water system at atmospheric pressure, using several columns. For a 0.75-m (2.46-ft) column and 0.35-m (14-in) plate spacing, the following were covered:

1. Bubble-cap plate—90-mm (3.5-in) round caps, 30-mm (1.2-in) static submergence
2. Sieve plate—10-mm (25/64-in) holes, hole/active area = 0.104, 25.4-mm (1.0-in) outlet weir
3. Valve plate—40-mm (1.57-in) holes, 45 valves per plate, 25.4-mm (1.0-in) outlet weir.

Efficiency data are given in Fig. 18-30.

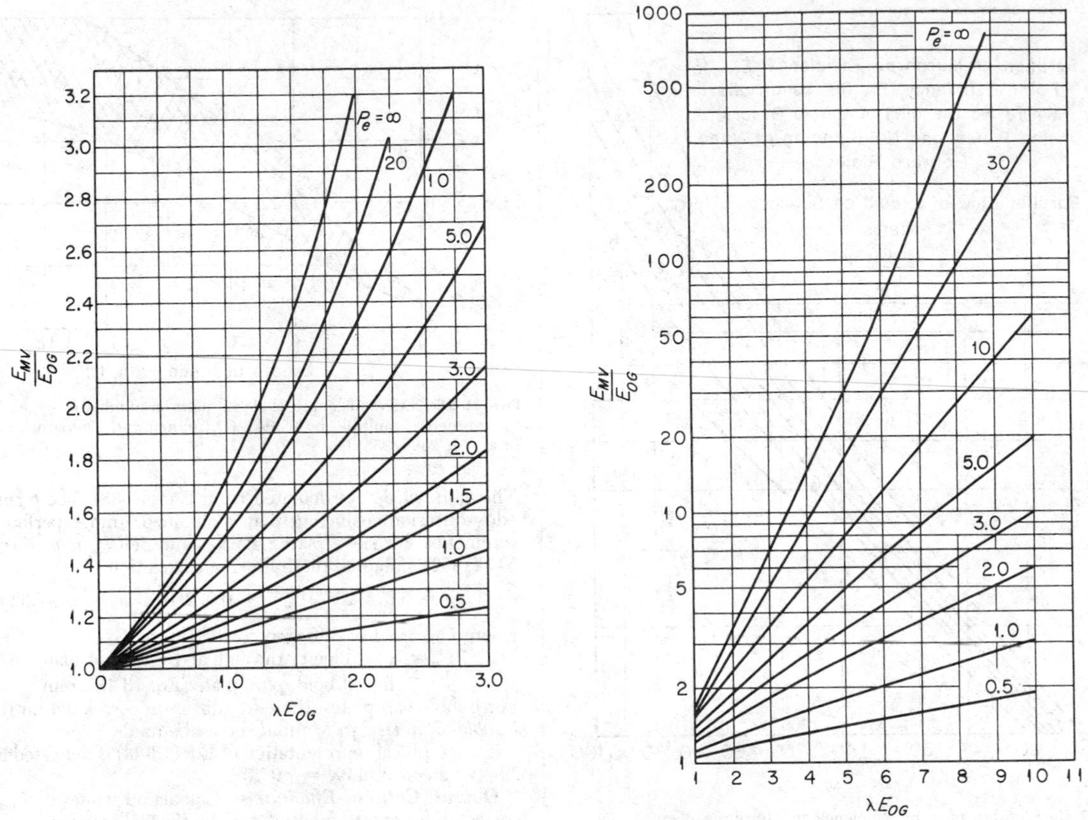

FIG. 18-29a Mixing curves. Pe = Péclet number (N_{Pe}). (Bubble-Tray Design Manual, *American Institute of Chemical Engineers, New York, 1958.*)

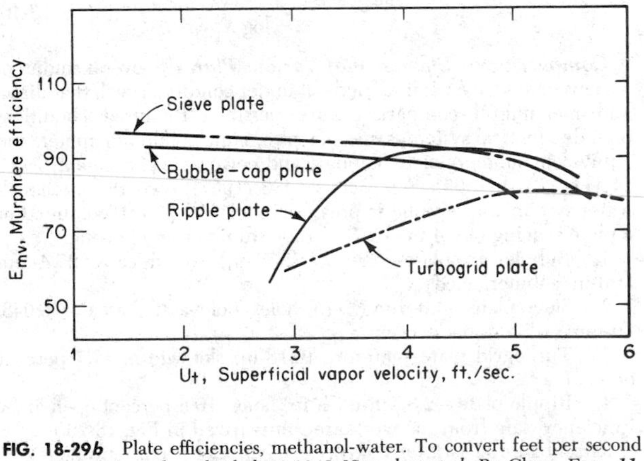

FIG. 18-29b Plate efficiencies, methanol-water. To convert feet per second to meters per second, multiply by 0.3048. [*Standart et al.*, Br. Chem. Eng., *11*, *1370 (1966);* Sep. Sci., *2*, *439 (1967)*.]

Billet and Raichle [*Chem. Ing. Tech.*, **38**, 825 (1966); **40**, 377 (1968)] used the ethylbenzene-styrene system at 100 mmHg absolute and a 0.80-m (2.6-ft) column with 0.5-m (19.7-in) plate spacing. Their studies included the following:

1. Bubble-cap plate—75-mm (3-in) round caps, 5-mm (0.2-in) static submergence
2. Valve plate—39-mm (1.52-in) holes, 64 valves per plate, 19-mm (0.75-in) outlet weir
3. Valve-sieve plate—49 valves with 39-mm (1.52-in) holes, 140 perforations with 9.5-mm (⅜-in) diameter, total open area 12.3 percent

Efficiency data are given in Fig. 18-31.

Testing of plates is carried out by Fractionation Research, Inc., for industrial sponsors. Some of the test data have been published for the cyclohexane–*n*-heptane system at 165 kPa (24 psia) in a 1.2-m- (4.0-ft-) diameter column with plate spacing of 0.61 m (24 in). Devices tested were:

1. Bubble-cap plate—thirty-seven 102-mm (4-in) round caps, 6.4-mm (0.25-in) submergence, 50.8-mm (2.0-in) outlet weir (AIChE Research Committee, final report, University of Delaware, Newark, Dec. 1, 1958)
2. Valve plate—136 valves (dimensions not given), 50.8-mm (2.0-in) outlet weir (*Glitsch V-1 Ballast Tray*, Bull. 160, Glitsch, Inc., Dallas, 1967)
3. Sieve plate—12.7-mm- (0.5-in-) diameter holes, hole/active area = 0.083, 50.8-mm (2.0-in) outlet weir [Sakata and Yanagi, *Inst. Chem. Eng. Symp. Ser.*, no. 56, 3.2/21 (1979)].

Efficiency data are given in Fig. 18-32. It should be noted that the submergence of the bubble-cap plate was less than that of the valve plate or the sieve plate.

All the foregoing test programs involve distillation under total-reflux conditions with well-defined binary systems. The primary value of the results is in the comparative data, but it should be emphasized that the design of each device was not necessarily optimized for the service.

Additional plate-efficiency data are listed in Table 18-4.

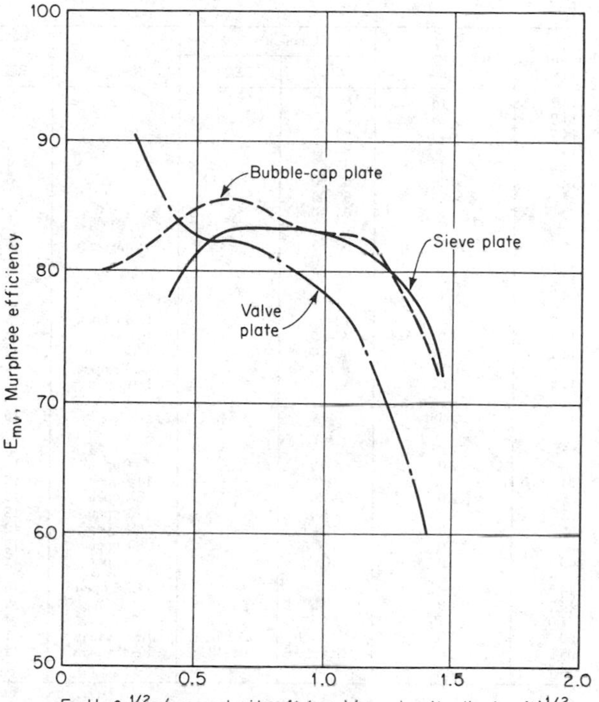

FIG. 18-30 Plate efficiencies, ethanol-water. To convert (feet per second) (pounds per cubic foot)$^{1/2}$ to (meters per second) (kilograms per cubic meter)$^{1/2}$, multiply by 1.2199. (*Kirschbaum,* Destillier-Rektifiziertechnik, 4th ed., *Springer-Verlag, Berlin and Heidelberg, 1969.*)

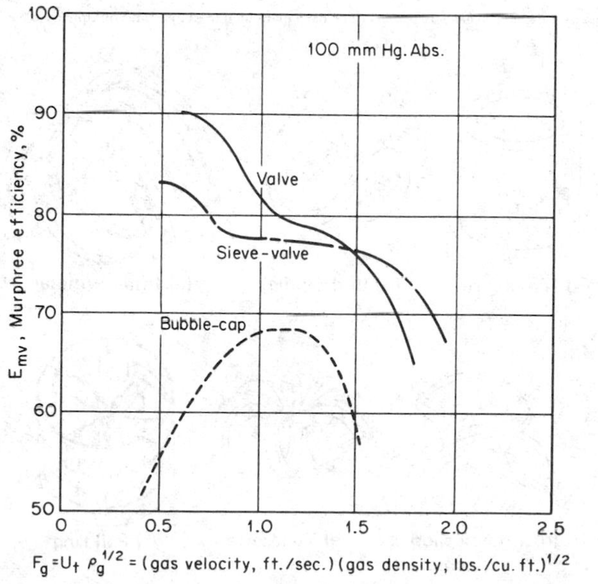

FIG. 18-31 Plate efficiencies, styrene–ethyl benzene. To convert (feet per second) (pounds per cubic foot)$^{1/2}$ to (meters per second) (kilograms per cubic meter)$^{1/2}$, multiply by 1.2199. [*Billet and Raichle,* Chem. Ing. Tech., *38,* 825 *(1966);* **40**, 377 *(1968).*]

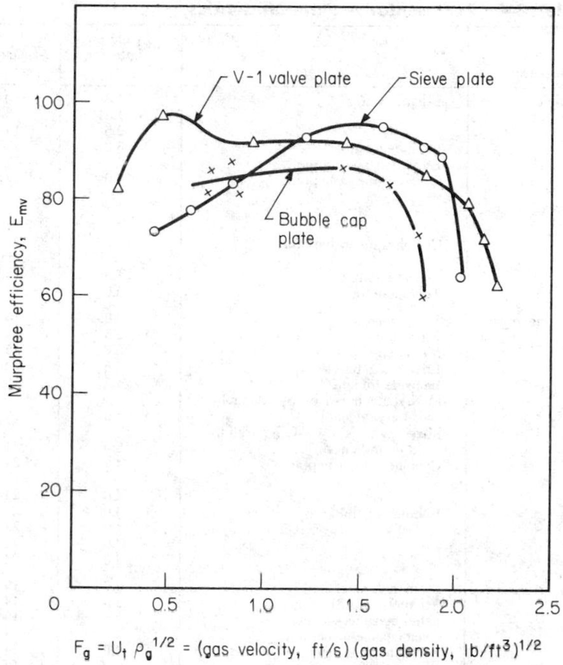

FIG. 18-32 Plate efficiencies, cyclohexane–*n*-heptane. To convert (feet per second) (pounds per cubic foot)$^{1/2}$ to (meters per second) (kilograms per cubic meter)$^{1/2}$, multiply by 1.2199.

PACKED COLUMNS

Introduction Packed columns for gas-liquid contacting are used extensively for absorption operations and, to a limited extent, for distillations. Usually the columns are filled with randomly oriented packing material, but in some cases the packing may be carefully positioned. The packed column is characteristically operated with counterflow of the phases.

The packed column is a simple device compared with plate columns (Fig. 18-33). A typical column consists of a cylindrical shell containing a support plate for the packing material and a liquid-distributing device designed to provide effective irrigation of the packing. Devices may be added to the packed bed to provide redistribution of liquid that might channel down the wall. Several beds may be used in the same column shell.

Many packings are commercially available, each possessing specific advantages for liquid-gas contacting from the aspects of cost, surface availability, interface regeneration, pressure drop, weight, and corrosion resistance. The packed bed is usually formed by dumping packing elements into the column ("dumped packings") and allowing them to form a random arrangement. Alternatively, larger sections of rigid, or "arranged," packing materials may be inserted carefully into the column. For most applications, the less expensive dumped packings are used, and typical packing elements are shown in Fig. 18-34.

Packed Columns versus Plate Columns Packed columns are usually specified when plate devices would not be feasible because of undesirable fluid characteristics or some special design requirement. Conditions favoring packed columns are:

1. For columns less than 0.6-m (2.0-ft) diameter, packings are usually cheaper than plates unless alloy-metal packings are required.
2. Acids and many other corrosive materials can be handled in packed columns because construction can be of ceramic, carbon, or other resistant materials.
3. Packings often exhibit desirable efficiency–pressure-drop characteristics for critical vacuum distillations.

Table 18-4 Representative Plate Efficiencies

Disperser	System	Column diameter, ft	Tray spacing, in	Pressure, psia	Static submergence, in	Efficiency, % E_{mv}*	Efficiency, % E_{oc}†	Remarks	Ref.
Bubble cap	Ethanol-water	1.31	10.6	14.7	1.18	83–87	......	......	1
		1.31	16.3	14.7	1.18	84–97			
		2.5	14	14.7	1.2	80–85	......	......	2
	Methanol-water	3.2	15.7	14.7	1.0	90–95	......	......	3
	Ethyl benzene–styrene	2.6	19.7	1.9	0.2	55–68	......	......	4
	Cyclohexane–n-heptane	4.0	24	14.7	0.25	65–90	......	......	5
				24	4.25	65–90	......		
				50		65–90			
	Cyclohexane–n-heptane	4.0	24	5	0.6	65–85	......	Tunnel caps	6
				24		75–100	......		
	Benzene-toluene	1.5	15.7	14.7	1.5	70–80	......	......	7
	Toluene-isooctane	5.0	24	14.7	0.4	......	60–80	......	8
Ripple sieve	Methanol-water	3.2	15.7	14.7	...	70–90	......	10.8% open	3
	Ethanol-water	2.5	14	14.7	1.0	75–85		10.4%	2
	Methanol-water	3.2	15.7	14.7	1.57	90–100		4.8% open	3
	Ethyl benzene–styrene	2.6	19.7	1.9	0.75	70		12.3% open	9
	Benzene-toluene	1.5	15.7	14.7	3.0	60–80		18% open	7
	Methyl alcohol–n-propyl alcohol–*sec*-butyl alcohol	6.0	18	18	1.38	......	64	......	10
	Mixed xylenes + C_8-C_{10} paraffins and naphthenes	13.0	21	25	1.25	......	86	......	5
	Cyclohexane–n-heptane	4.0	24	5	2.0	60–70	......	14% open	13
				24		80	......	14% open	13
		4.0	24	5	2.0	70–80	......	8% open	12
	Isobutane-n-butane	4.0	24	165	2.0	110	......	14% open	13
		4.0	24	165	2.0	120	......	8% open	12
		4.0	24	300	2.0	110	......	8% open	12
		4.0	24	400	2.0	100	......	8% open	12
Turbogrid valve	Methanol-water	3.2	15.7	14.7	...	70–80	......	14.7% open	3
	Ethanol-water	2.5	14	14.7	1.0	75–85	......	......	2
	Ethyl benzene–styrene	2.6	19.7	1.9	0.75	75–85	......	......	4
	Cyclohexane–n-heptane	4.0	24	20	3.0	......	50–96	Rect. valves	11
	n-Butane-isobutene	4.0	24	165	3.0	......	104–121	Rect. valves	11
	Benzene-toluene	1.5	15.7	14.7	3.0	75–80	......	......	7

References
1. Kirschbaum, *Z. Ver. Dtsch. Ing. Beih. Verfahrenstech.*, (5), 131 (1938); (3), 69 (1940).
2. Kirschbaum, *Distillier-Rektifiziertechnik*, 4th ed., Springer-Verlag, Berlin and Heidelberg, 1969.
3. Kastanek and Standart, *Sep. Sci*, **2,** 439 (1967).
4. Billet and Raichle, *Chem. Ing. Tech.*, **38,** 825 (1966); **40,** 377 (1968).
5. AIChE Research Committee, *Tray Efficiency in Distillation Columns*, final report, University of Delaware, Newark, 1958.
6. Raichle and Billet, *Chem. Ing. Tech.*, **35,** 831 (1963).
7. Zuiderweg, Verburg, and Gilissen, *Proc. Intn. Symp.*, Brighton, England, 1960.
8. Manning, Marple, and Hinds, *Ind. Eng. Chem.*, **49,** 2051 (1957).
9. Billet, *Proc. Intn. Symp.*, Brighton, England, 1970.
10. Mayfield, Church, Green, Lee, and Rasmussen, *Ind. Eng. Chem.*, **44,** 2238 (1952).
11. Fractionation Research, Inc., "Report of Tests of Nutter Type B Float Valve Tray," July 2, 1964, from Nutter Engineering Co., Tulsa.
12. Sakata and Yanagi, *Inst. Chem. Eng. Symp. Ser.*, no. 56, 3.2/21 (1979).
13. Yanagi and Sakata, *Ind. Eng. Chem. Process Des. Dev.* **21,** 712 (1982).

*See Eq. (18-32).
†See Eq. (18-28).
NOTE: To convert feet to meters, multiply by 0.3048; to convert inches to centimeters, multiply by 2.54; and to convert pounds-force per square inch to kilopascals, multiply by 6.895.

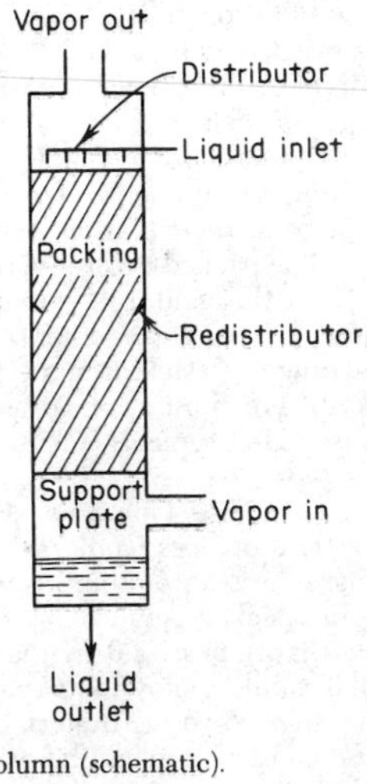

FIG. 18-33 Packed column (schematic).

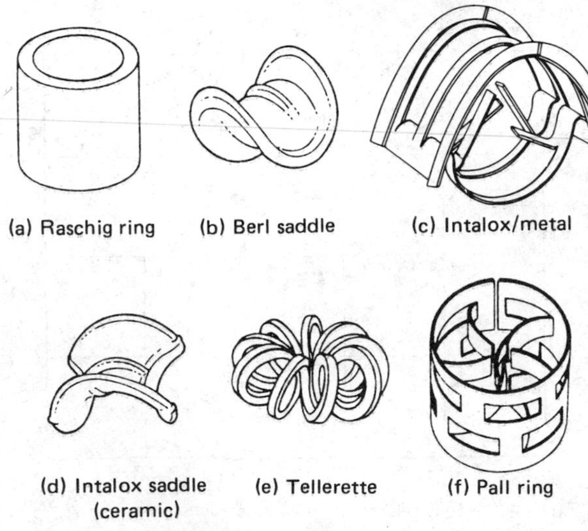

FIG. 18-34 Typical dumped-type packing elements.

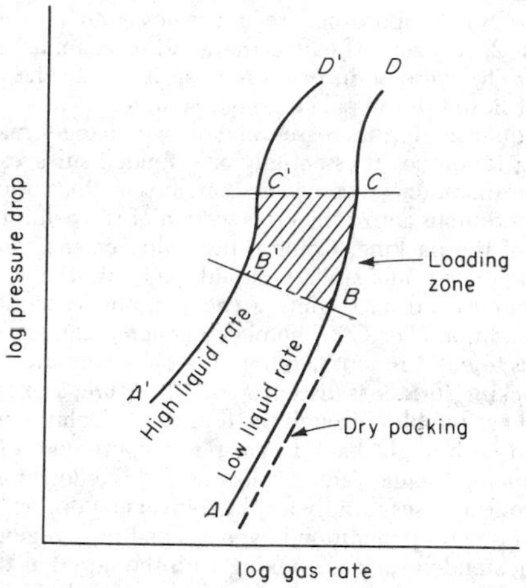

FIG. 18-35 Pressure-drop characteristics of packed columns.

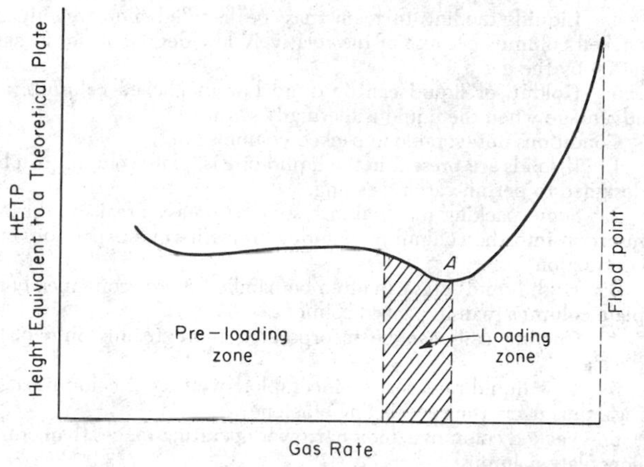

FIG. 18-36 Efficiency characteristics of packed columns (total-reflux distillation.)

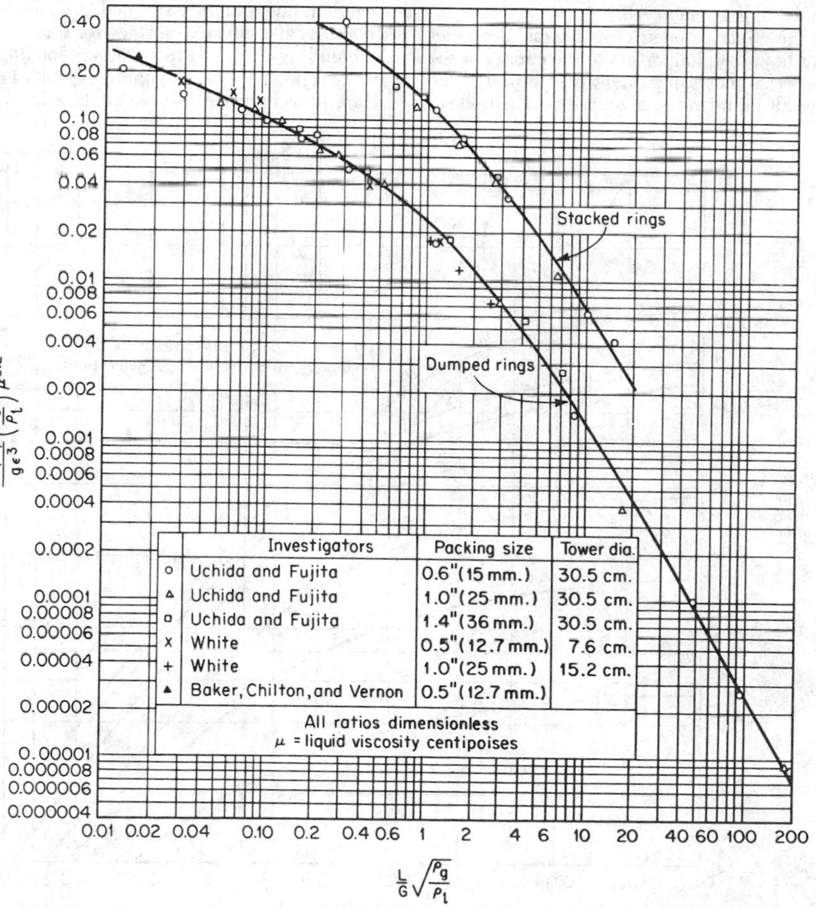

FIG. 18-37 Generalized correlation of flood points, packed columns. [*Sherwood et al.*, Ind. Eng. Chem., *30, 768 (1938).*]

4. Liquids tending to foam may be handled more readily in packed columns because of the relatively low degree of liquid agitation by the gas.

5. Holdup of liquid can be quite low in packed columns, an advantage when the liquid is thermally sensitive.

Conditions unfavorable to packed columns are:

1. If solids are present in the liquid or gas, plate columns can be designed to permit easier cleaning.

2. Some packing materials are subject to easy breakage during insertion into the column or resulting from thermal expansion and contraction.

3. High liquid rates can often be handled more economically in plate columns than in packed columns.

4. Cooling coils can be incorporated more readily into plate devices.

5. Low liquid rates lead to incomplete wetting of column packings, thus decreasing contacting efficiency.

6. Packed columns exhibit narrower operating ranges than crossflow plate columns.

Packed-Column Hydraulics Pressure drop of a gas flowing upward through a packing countercurrently to liquid flow, is characterized graphically in Fig. 18-35. At very low liquid rates, the effective open cross section of the packing is not appreciably different from that of dry packing, and pressure drop is due to flow through a series of randomly sized and located openings in the bed. Thus, pressure drop is proportional approximately to the square of the gas velocity, as indicated in the region *AB*.

At higher liquid rates the effective open cross section is smaller because of the presence of liquid, and a portion of the energy of the gas stream is used to support an increasing quantity of liquid in the column (region *A′B′*). For all liquid rates, a zone is reached where pressure drop is proportional to a gas-flow-rate power distinctly higher than 2; this zone is called the **loading zone,** as indicated in Fig. 18-35. The increase in pressure drop is due to the rapid accumulation of liquid in the packing-void volume.

As the liquid holdup increases, one of two changes may occur. If the packing is composed essentially of extended surfaces, the effective orifice diameter becomes so small that the liquid surface becomes continuous across the cross section of the column, generally at the top of the packing. Column instability occurs concomitantly with a rising continuous-phase liquid body in the column. The change in pressure drop is quite great with only a slight change in gas rate (condition *C* or *C′*). The phenomenon is called *flooding* and is analogous to entrainment flooding in a plate column.

If the packing surface is discontinuous in nature, a phase inversion occurs, and gas bubbles through the liquid. The column is not unstable and can be brought back to gas-phase continuous operation by merely reducing the gas rate. Analogously to the flooding condition, the pressure drop rises rapidly as phase inversion occurs.

A stable operating condition beyond "flooding" (region *CD* or *C′ D′*) for nonextended surface packing with the liquid as the continuous phase and the gas as the dispersed phase has been reported by Lerner and Grove [*Ind. Eng. Chem.*, **43**, 216 (1951)] and Teller [*Chem. Eng.*, **61**(9), 168 (1954)].

For total-reflux distillations carried out in packed columns, regions of loading and flooding are identified by their effects on mass-transfer efficiency, as shown in Fig. 18-36. Gas and liquid rate increase together, and a point is reached at which liquid accumulates rapidly (point *A*) and effective surface for mass transfer decreases rapidly.

Flooding and Loading Since flooding or phase inversion normally represents the maximum capacity condition for a packed column, it is desirable to predict its value for new designs. The first

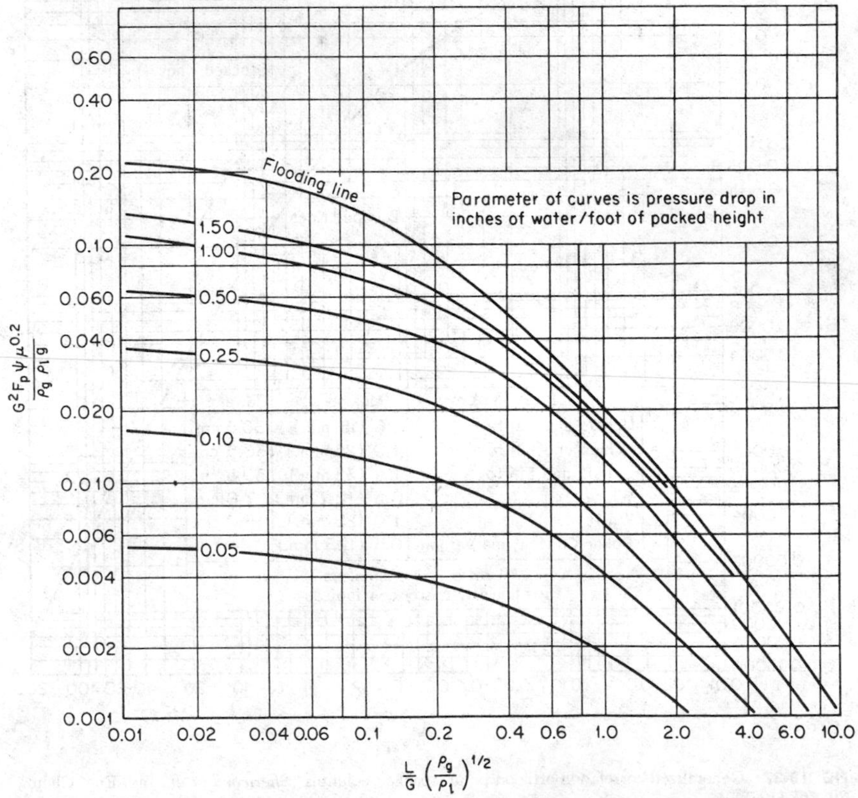

FIG. 18-38 Generalized flooding and pressure-drop correlation for packings. To convert inches of water per foot of packed height to millimeters of water per meter of packed height, multiply by 83.31. [*Eckert*, Chem. Eng. Prog., *66(3), 39 (1970).*]

generalized correlation of packed-column flood points was developed by Sherwood, Shipley, and Holloway [*Ind. Eng. Chem.*, **30**, 768 (1938)] on the basis of laboratory measurements primarily on the air-water system. It resulted in the relationship

$$\left(\frac{U_t^2 a_p \rho_g}{g \epsilon^3 \rho_\ell}\right) \mu_\ell^{0.2} = \text{function}\left(\frac{L}{G}\sqrt{\frac{\rho_g}{\rho_\ell}}\right) \tag{18-47}$$

where the first group on the left is dimensionless and viscosity of the liquid μ_ℓ is in cP (mPa·s). The group on the right is the dimensionless flow parameter introduced earlier in connection with the prediction of flooding and entrainment in plate columns (Figs. 18-10 and 18-22). Convenient units for Eq. (18-47) are:

U_t = superficial gas velocity, m/s
a_p = total area of packing, $m^2/(m^3$ bed)
ϵ = fractional voids in dry packing
g = gravitational constant, 9.8067 m/s^2
ρ_ℓ and ρ_g = liquid and gas densities, kg/m^3
L = liquid mass rate, $kg/(s \cdot m^2)$
G = gas mass rate, $kg/(s \cdot m^2)$
μ_ℓ = liquid viscosity, mPa·s (cP)

Table 18-5 Characteristics of Dumped Tower Packings*

Packing type	Nominal size, mm	Wall thickness, mm	Outside diameter and length, mm	Approximate no. elements per m^3	Approximate weight per m^3, kg	Approximate surface area, m^2/m^3	Percent void space	Packing factor F_p, m^{-1}
Berl saddles, ceramic	6	...	...	3.78×10^6	900	900	60	2950
	13	...	...	590,000	865	465	62	790
	25	...	...	77,000	720	250	68	360
	38	...	...	22,800	640	150	71	215
	50	...	...	8,800	625	105	72	150
Intalox saddles, ceramic	6	...	...	4.15×10^6	865	984	75	2380
	13	...	...	730,000	720	625	78	660
	25	...	...	84,000	705	255	77	320
	38	...	...	25,000	670	195	80	170
	50	...	...	9,400	760	118	79	130
	75	...	...	1,870	590	92	80	70
Intalox saddles, metal	(No. 25)	...	...	168,400	...	...	97	135
	(No. 40)	...	...	50,100	...	...	97	82
	(No. 50)	...	...	14,700	...	...	98	52
	(No. 70)	...	...	4,630	...	...	98	43
Intalox saddles, plastic (polypropylene)	25	...	...	55,800	76	206	91	105
	50	...	...	7,760	64	108	93	69
	75	...	...	1,520	60	88	94	50
Pall rings, metal	16	26 gauge	16				92	230
	25	24	25	49,600	480	205	94	157
	38	22	38	13,000	415	130	95	92
	50	20	50	6,040	385	115	96	66
	90		90	1,170	270	92	97	53
Pall rings, plastic (polypropylene)	16	...	16	214,000	116	340	87	310
	25	...	25	50,100	88	205	90	170
	38	...	38	13,600	76	130	91	105
	50	...	50	6,360	72	100	92	82
	90	...	90	1,170	68	85	92	52
Raschig rings, ceramic	6	1.6	6	3.02×10^6	960	710	62	5250
	13	2.4	13	378,000	880	370	64	2000
	19	2.4	19	109,000	800	240	72	840
	25	3.2	25	47,700	670	190	74	510
	38	6.4	38	13,500	740	120	68	310
	50	6.4	50	5,800	660	92	74	215
	75	9.5	75	1,700	590	62	75	120
	100	9.5	100	700	580	46	80	
Raschig rings, steel	19	1.6	19	111,000	1500	245	80	730
	25	1.6	25	46,300	1140	185	86	450
	38	1.6	38	14,100	785	130	90	270
	50	1.6	50	5,900	590	95	92	187
	75	1.6	75	1,800	400	66	95	105
Hy-Pac, steel	(No. 1)	...	30	30,000	300	177	96	141
	(No. 2)	...	60	3,780	225	95	97	59
Levapacking†	(No. 1)	...	...	34,000	270	164		
	(No. 2)	...	...	10,500	210	118		
Low-density polyethylene Tellerettes	1	...	...	39,700	160	250	83	

*Data are representative but vary slightly with vendor source of packing. Pall rings are sold under the alternate names of Flexirings and Ballast rings; Intalox saddles are sold under the alternate name of Flexisaddles. Number of elements per packed-bed volume can vary with the method of dumping the elements; see Billet, *Chem. Eng. Prog.*, **63**(9), 53 (1967). To convert millimeters to inches, multiply by 0.0394; to convert elements per cubic meter to elements per cubic foot, multiply by 0.0283; to convert square meters per cubic meter to square feet per cubic foot, multiply by 0.3048; and to convert $meters^{-1}$ to $feet^{-1}$, multiply by 0.3048.

†Leva, *Chem. Eng. Prog.*, **76**(9),73 (1980).

Table 18-6 Experimental Packing Factors

System	Column diameter, in	Pressure, psia	Packing	Flooding value, $L/G\sqrt{\rho_g/\rho_\ell}$	Packing factor F_p, ft^{-1} Measured	Table 18-5	Ref.
Pentanes	10	58.2	½-in Raschig rings	0.141	478	580	1
		73.2	½-in Raschig rings	0.161	424	580	1
	10	73.2	½-in Berl saddles	0.161	216	240	1
Butenes	16	143	½-in Intalox saddles	0.205	305	200	1
	16	313	½-in Intalox saddles	0.340	300	200	1
Hexenes	16	38	½-in Intalox saddles	0.093	381	200	1
Cyclohexane–n-heptane	48	24	¾-in Raschig rings	0.09–2.0	225–113	255	2
			1½-in Raschig rings	0.09–1.05	95–80	95	2
			3-in Raschig rings	0.1–0.7	63–44	37	2

References:
1. Clay, Clark, and Munro, *Chem. Eng. Prog.*, **62**(1), 51 (1966).
2. Silvey and Keller, *Proc. Intn. Symp. Distill.*, Brighton, England, 1970.

NOTE: To convert inches to millimeters, multiply by 25.4; to convert pounds-force per square inch absolute to kilopascals, multiply by 6.895; and to convert feet^{-1} to meters^{-1}, multiply by 3.28.

The functional relationship of Eq. (18-47) is shown graphically in Fig. 18-37. Later work with air and liquids other than water led to modifications of the Sherwood correlation, first by Leva [*Chem. Eng. Prog. Symp. Ser.*, **50**(10), 51 (1954)] and later in a series of papers by Eckert. The most recent modification by Eckert is shown in Fig. 18-38. It should be noted that the ordinate group in Fig. 18-38 includes ψ, the ratio of the density of water to the density of the liquid and also that the ratio a_p/ϵ^3, characteristic for a particular packing material, has been replaced by the **packing factor** F_p, also characteristic of a given packing (units = m^{-1}) but obtained experimentally instead of being calculated from packing geometry. It is common practice to retain the correlating parameters of Fig. 18-38 and adjust the values of the packing factor as additional experimental data become available.

Table 18-5 gives general characteristics of common dumped packings, including packing factors. The data have been obtained from various packing manufacturers and must be used with caution. Bolles and Fair [*Inst. Chem. Eng. Symp. Ser.*, no. 56, 3.3/35 (1979)] found that, with the use of packing factors from Table 18-5 plus the design chart Fig. 18-38 for ceramic Raschig rings, ceramic Berl saddles, metal Raschig rings, and metal pall rings, it would be necessary to multiply the predicted flooding-gas velocity by 0.62 in order to have 95 percent confidence that there would not be an unexpected flooding condition during operation at high gas rates. Table 18-6 shows some comparisons of measured and predicted packing factors. Part of the difficulty in generalizing the reported flood data lies in the variation of flood-point definition; Silvey and Keller [*Chem. Eng. Prog.*, **62**(1), 68 (1966)] list 10 different definitions that have been presented by investigators. In general, the design throughput for a packed column should allow for a ±30 percent error in the predicted flood point.

The **loading point** is a more nebulous quantity than the flood point. It has been described as the gas velocity at which a break in the slope of the pressure-drop–gas-rate curve occurs (Fig. 18-35) or a change in the mass-transfer efficiency occurs (Fig. 18-36). In view of the fact that the transition from preloading to loading conditions may be very gradual, the load point can only be approximated. For a rough estimate, the pressure-drop curve of Fig. 18-38 representing 0.50 in H_2O/ft (42 mm H_2O/m) may be taken as the lower limit of loading.

Pressure Drop For gas flow through dry packings, pressure drop may be estimated by use of the orifice equation, with suitable correction for the presence of liquid. On this basis, Leva [*Chem. Eng. Prog. Symp. Ser*, **50**(10), 51 (1954)] developed the following correlation for pressure drop in irrigated packed beds (parameters expressed in U.S. customary units):

$$\Delta P = C_2 10^{C_3 u_t} \rho_g U_t^2 \qquad (18\text{-}48)$$

where ΔP = in H_2O/ft packing (to convert to mm H_2O/m packing, multiply by 83.31)
ρ_g = gas density, lb/ft^3
U_t and u_t = superficial velocities of gas and liquid respectively, ft/s
C_2 and C_3 = constants, given in Table 18-7

The Leva correlation was developed from test data for the air-water system operating below the flood point. Although it has not been tested extensively, it was found by Silvey and Keller (*Proc. Intn. Symp. Distill.*, Brighton, England, 1970) to work well for the cyclohexane–n-heptane system at 165 kPa (24 psia), using 19-, 38-, and 75-mm (¾-, 1½-, and 3-in) ceramic Raschig rings in a 1.2-m (4.0-ft) column (Fig. 18-39).

Inasmuch as the development of flooding proceeds for each packing by the same mechanism, it is expected and confirmed that the pressure drop at flooding is independent of the liquid-gas ratio and is dependent only on the physical properties of the system. Pressure drops at flooding for various packings using the air-water system are indicated in Table 18-8.

The equality of pressure drops at the flood point led to the generalized pressure-drop correlation of Leva (op. cit.) and Eckert and coworkers as shown in Fig. 18-38. Packing factors for pressure-drop estimation are different from those for flood-point estimation and vary somewhat with flow conditions. Pressure-drop packing factors are given in Table 18-5. For the distillation system of Silvey and Keller (op. cit.), the Eckert correlation (using flooding packing factors) represented measured pressure drops as shown in Fig. 18-40.

Pressure-drop data for various packings are indicated in Figs. 18-41 to 18-44*a*. All data are for the air-water system, and use of the term G/ϕ permits comparison of other gases with air, since ϕ =

Table 18-7 Coefficients for Leva Pressure-Drop Correlation [Eq. (18-48)]

Type of packing	Nominal size, in.	Wall thickness, in.	C_2	C_3	Range of L, lb./(hr.)(sq. ft.)
Raschig rings	½	3/32	3.5	0.0577	300– 8,600
	¾	3/32	0.82	0.0361	1,800–10,800
	1	⅛	0.80	0.0348	360–27,000
	1½	¼	0.30	0.0320	720–18,000
	2	¼	0.28	0.0236	720–21,600
Berl saddles	½	. . .	1.5	0.0272	300–14,100
	¾	. . .	0.60	0.0236	360–14,400
	1	. . .	0.40	0.0236	720–28,800
	1½	. . .	0.20	0.0181	720–21,600
Intalox saddles . . .	1	. . .	0.31	0.0222	2,520–14,400
	1½	. . .	0.14	0.0181	2,520–14,400

NOTE: To convert inches to millimeters, multiply by 25.4; to convert pounds per hour–square foot to kilograms per second–square meter, multiply by 0.001356.

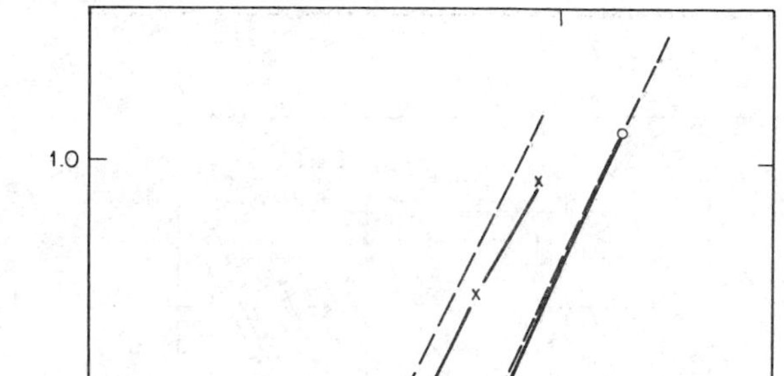

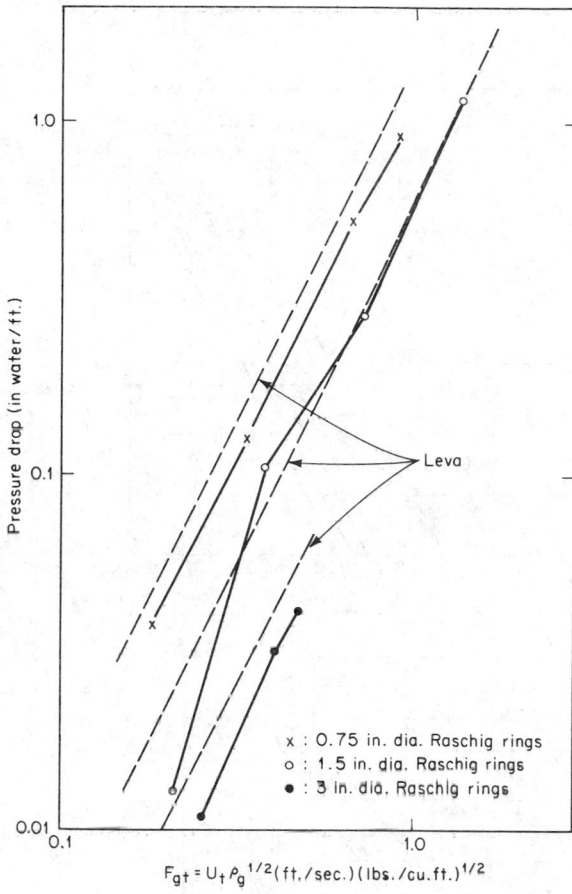

FIG. 18-39 Dry-bed pressure drop, ceramic Raschig rings. Cyclohexane–*n*-heptane system. To convert inches of water per foot to millimeters of water per meter, multiply by 83.31; to convert inches to millimeters, multiply by 25.4; and to convert (feet per second) (pounds per cubic foot)$^{1/2}$ to (meters per second) (kilograms per cubic meter)$^{1/2}$, multiply by 1.2199. (*Silvey and Keller*, Proc. Int. Symp. Distill., *Brighton, England, 1969; Leva*, Tower Packings and Packed Tower Design, *U.S. Stoneware Co., Akron, Ohio, 1951.*)

Table 18-8 Flooding Pressure Drops

Packing	Size	Pressure drop at flooding, in. water/ft. packing	Source
Raschig rings	¼	4	Zenz, *Chem. Eng.*, **60** (8), **176** (1953)
	½	3.5	
	1	4	
	1½	2.5	
	2	2.5	
Berl saddles	½	2.5	
	1	2.5	
	1½	2.2	
Tellerettes (phase inversion)	1	2.5	Teller and Ford, *Ind. Eng. Chem.*, **50**, 1201 (1958)

NOTE: To convert inches to millimeters, multiply by 25.4; to convert inches of water per foot to millimeters of water per meter, multiply by 83.31.

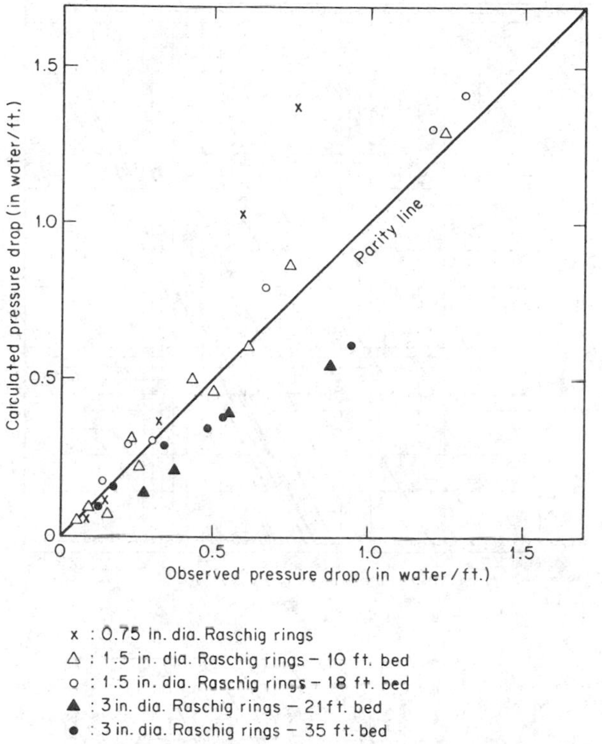

FIG. 18-40 Comparison of pressure-drop data with Eckert correlation. Cyclohexane–*n*-heptane system. To convert inches of water per foot to millimeters of water per meter, multiply by 83.31; to convert feet to meters, multiply by 0.3048. [*Silvey and Keller*, Proc. Int. Symp. Distill., *Brighton, England, 1969; Eckert*, Chem. Eng. Prog., *57*(9), 54 (1961).]

$\sqrt{\rho_g/\rho_{air}}$. For most of the experiments, the density of air was 1.20 kg/m^3 (0.075 lb/ft^3).

Arranged Packings Special packing materials are fabricated in larger sections that can be inserted into the column. Typical packings of this type are shown in Fig. 18-44*b;* the Koch Sulzer material is made from woven gauze, corrugated and formed to provide narrow channels having a high surface area, whereas the Flexipac (also known as Mellapak) is made from corrugated sheet metal, which may or may not be perforated with small holes. Other packings of this general type include Goodloe, Hyperfil, Kloss, neo-Kloss, Glitsch-grid, and Spraypak; descriptions and names of vendors may be found in the article by Nygren and Connolly [*Chem. Eng. Prog.*, **67**(3), 49 (1971)]. Arranged packings are generally more expensive than dumped packings, on an equivalent-tower-volume basis, but can exhibit favorable pressure-drop–efficiency characteristics that assume great importance in high-vacuum distillations.

Pressure-drop characteristics for Flexipac packing are shown in Fig. 18-45. Packing factors for this material are F_p = 108.2 m^{-1} (32.98 ft^{-1}) for No. 1 size and 52.5 m^{-1} (16.00 ft^{-1}) for No. 3 size. For flooding–pressure-drop characteristics of arranged packings contact should be made with the vendors.

Support Plates While the primary purpose of a packing support is to retain a bed of packing without excessive restriction to gas and liquid flow, it also serves to distribute both streams. Unless carefully designed, the support plate can also cause premature column flooding. Thus, design of the support plate significantly affects column pressure drop and stable operating range.

Two basic types of support plates may be utilized:

1. Countercurrent
2. Separate flow passages for liquid and gas

The two types are indicated in Figs. 18-46, 18-47, and 18-48.

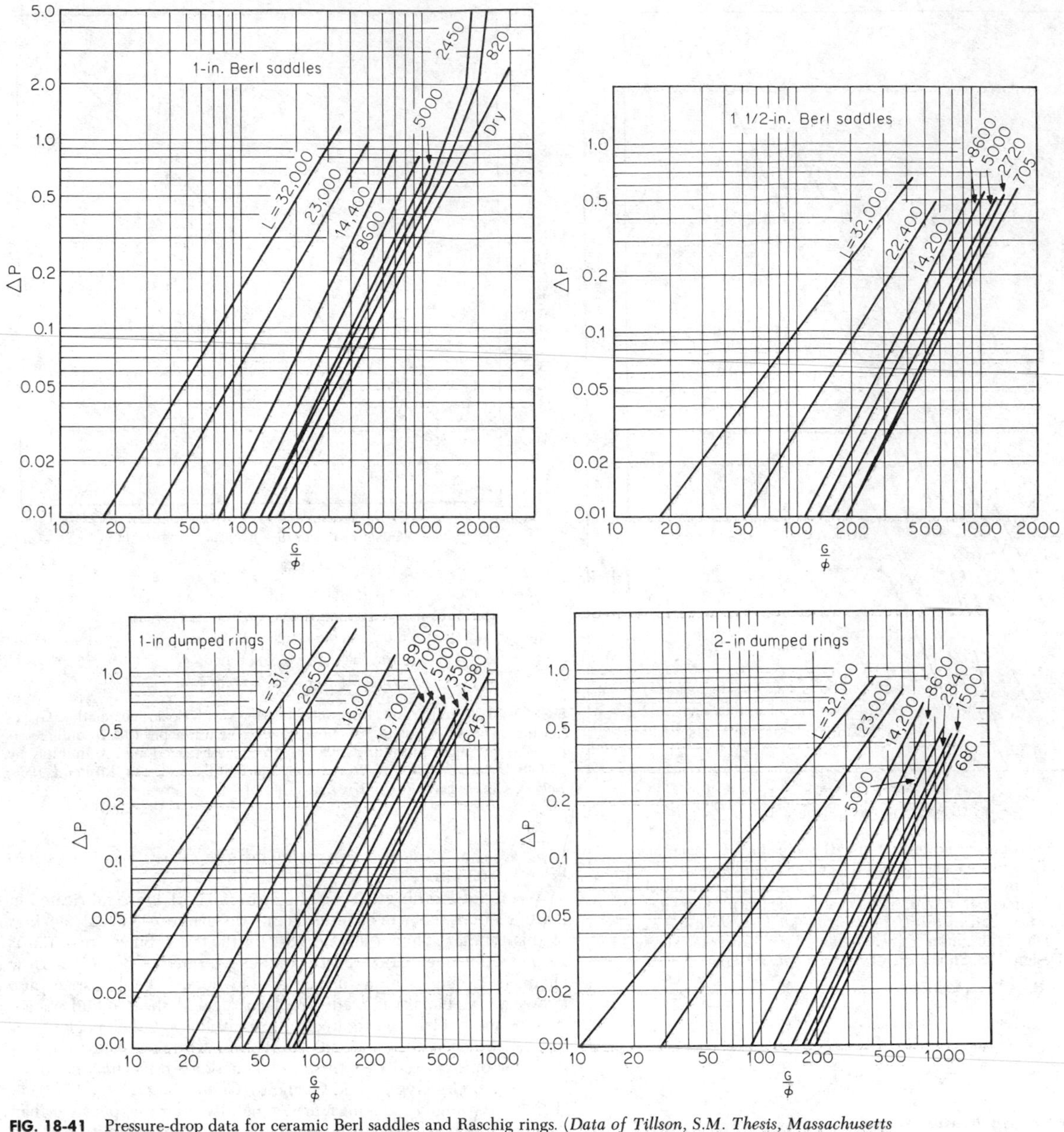

FIG. 18-41 Pressure-drop data for ceramic Berl saddles and Raschig rings. (*Data of Tillson, S.M. Thesis, Massachusetts Institute of Technology, 1939.*) Using air and water in a 508-mm (20-in) tower except for 1-in saddles at liquor rates below 5000 lb/(h·ft^2) from Mach [DECHEMA Monogr., *6, 38 (1933);* Z. Ver. Dtsch. Ing., *375 (1935).*] ΔP = pressure drop, in H_2O/ft depth; G = gas velocity, lb/(h·ft^2); L = liquid rate, lb/(h·ft^2); $\phi = (\rho_g/\rho_{air})^{1/2}$; ρ_g = gas density, lb/ft^3; and ρ_{air} = 0.075 lb/ft^3. To convert inches to millimeters, multiply by 25.4; to convert pounds per hour-square foot to kilograms per second-square meter, multiply by 0.001356; to convert inches of water per foot to millimeters of water per meter, multiply by 83.31; and to convert pounds per cubic foot to kilograms per cubic meter, multiply by 16.019.

The degree of open area on a support plate is the fraction of void inherent in the design of the plate minus that portion of the open area occluded by the packing. To avoid premature flooding, the net open area of the plate must be greater than that of the packing itself. With the countercurrent type of support plate the free area for gas flow can range up to 90 percent of the column cross-sectional area. However, such a plate is easily occluded by the packing pieces resting directly on it.

The separate flow passage devices can be designed for free areas up to 90 percent, and because of their geometry they will have very little occlusion by the packing.

Liquid Holdup Three modes of liquid holdup in packed columns are recognized:

1. Static, h_s
2. Total, h_t
3. Operating, h_o

Static holdup is the amount of liquid remaining on packing that has been fully wetted and then drained. Total holdup is the amount of

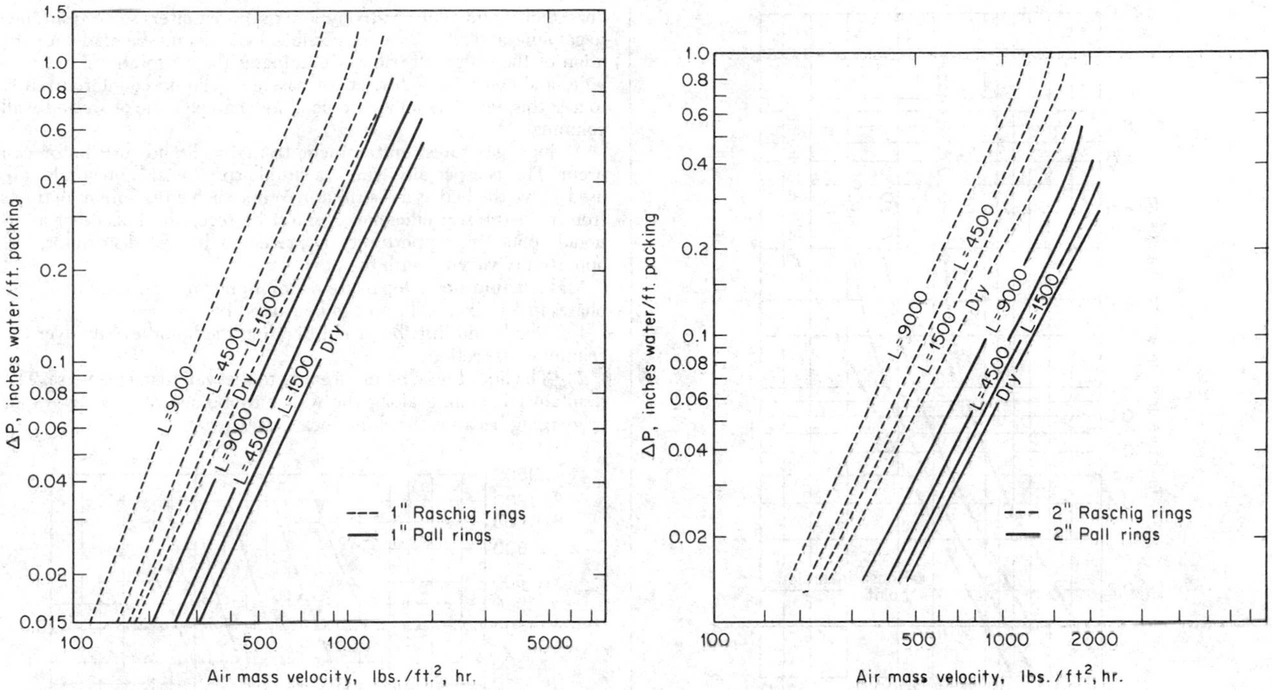

FIG. 18-42 Pressure drop for metal pall rings, 0.024-in wall thickness (1-in size) and 0.036-in wall (2-in size). Metal Raschig rings have 1/16-in wall. L = lb liquid/(h·ft²). To convert inches of water per foot to millimeters of water per meter, multiply by 83.31; to convert inches to millimeters, multiply by 25.4; and to convert pounds per hour-square foot to kilograms per second-square meter, multiply by 0.001356. [*Eckert, Foote, and Huntington,* Chem. Eng. Prog., *54(1), 70 (1958).*]

liquid on the packing under dynamic conditions. Operating holdup is the amount of liquid attributed to operation and is measured experimentally as the difference between total and static holdup. Thus,

$$h_t = h_o + h_s \tag{18-49}$$

where h values are in volumes of liquid per total volume of bed. The effective void fraction under operating conditions is

$$\epsilon' = \epsilon - h_t \tag{18-50}$$

Static holdup depends upon the balance between surface-tension forces tending to hold liquid in the bed and gravity or other forces that tend to displace the liquid out of the bed. Estimates of static holdup (for gravity drainage) may be made from the following relationship of Shulman et al. [*Am. Inst. Chem. Eng. J.*, **1**, 259 (1955)]:

$$h_s = 2.79\,\frac{C_1\,\mu^{C_2}\sigma^{C_3}}{\rho_\ell^{0.37}} \tag{18-51}$$

where μ_ℓ = liquid viscosity, mPa·s
σ = surface tension, mN/m
ρ_ℓ = liquid density, kg/m³

and constants are

Packing	C_1	C_2	C_3
1.0-in carbon Raschig rings	0.086	0.02	0.23
1.0-in ceramic Raschig rings	0.00092	0.02	0.99
1.0-in ceramic Berl saddles	0.0055	0.04	0.55

For other packings and for the case in which static holdup is changed by gas flowing through the bed, the method of Dombrowski and Brownell [*Ind. Eng. Chem.*, **46**, 1207 (1954)], which correlates static holdup with a dimensionless capillary number, should be used.

Typical total holdup data for packings are shown in Fig. 18-49. It should be noted that over much of the preloading range gas rate has little effect on holdup.

Operating holdup may be estimated by the dimensionless equation of Buchanan [*Ind. Eng. Chem. Fundam.*, **6**, 400 (1967)]:

$$h_o = 2.2\left(\frac{\mu'_\ell u_\ell}{g\rho_\ell d_p^2}\right)^{1/3} + 1.8\left(\frac{u_\ell^2}{gd_p}\right)^{1/2} \tag{18-52}$$

where μ'_ℓ = liquid viscosity, Pa·s
u_ℓ = liquid superficial velocity, m/s
g = gravitational constant, m/s²
ρ_ℓ = liquid density, kg/m³
d_p = nominal packing size, m

The first term is a "film number"; the second is the Froude number. The equation applies to ring packings only operating below the load point and correlates all literature data to about ±20 percent.

Operating holdup contributes effectively to mass-transfer rate, since it provides residence time for phase contact and surface regeneration via agglomeration and dispersion. Static holdup is limited in its contribution to mass-transfer rates, as indicated by Thoenes and Kramers [*Chem. Eng. Sci.*, **8**, 271 (1958)]. In laminar regions holdup in general has a negative effect on the efficiency of separation.

Liquid Distribution Uniform initial distribution of liquid at the top of the packed bed is essential for efficient column operation. This is accomplished by a device that spreads the liquid uniformly across the top of the packing. Baker, Chilton, and Vernon [*Trans. Am. Inst. Chem. Eng.*, **31**, 296 (1935)] studied the influence of the bed itself as a distributor and found that a single-point distribution in a 305-mm (12-in) column with 19-mm (¾-in) packing required 3.05m (10 ft) of bed before achieving uniform distribution across the bed. They

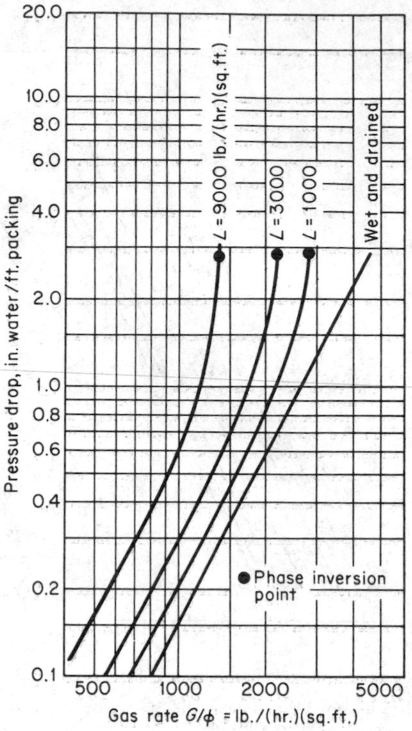

FIG. 18-43 Pressure drop for ¾- by 2-in Tellerettes (1-in nominal size). System: air–15 percent monoethanolamine (viscosity, 1.88 cS). To convert inches of water per foot to millimeters of water per meter, multiply by 83.31; to convert inches to millimeters, multiply by 25.4; and to convert pounds per hour–square foot to kilograms per second–square meter, multiply by 0.001356. [*Teller and Ford,* Ind. Eng. Chem., ***50**, 1201 (1958).*]

found also a tendency for liquid to migrate toward the column wall (Fig. 18-50), especially for ratios of column diameter to packing size less than 8. For a multipoint distribution, their recommendation was one liquid stream for each 194-cm² (30-in²) column area. Eckert [*Chem. Eng. Prog.*, **57,** 54 (1961)] recommends the following:

Column diameter, m	Streams/m²
1.2 or greater	40
0.75	170
0.40	340

Pratt [*Trans. Inst. Chem. Eng. (London)*, **29,** 226 (1951)] reported the effect of overhead liquid-distributor design for stacked 75-mm (3-in) rings in a 46-cm (18-in) square tower. Relative mass-transfer rates are shown in Table 18-9.

Several types of liquid distributors are used (Fig. 18-51). The perforated pipe distributor is satisfactory for use with clean liquids. It offers minimum restriction to gas flow and can be used for high liquid flows. Operating pressure drop for liquid ranges from 35 to 140 kPa (5 to 20 lbf/in²).

Trough-type distributors are often used in columns of 1.2-m (4-ft) diameter and larger. They can be fabricated from sheet metal, plastics, or ceramics, and consist of a series of troughs containing side notches for liquid overflow. Design can allow for positioning of notches, and flow rates through the notches can be calculated by weir formulas (Sec. 5). The distributor is not subject to plugging, does not restrict gas flow excessively, and has a wide operating range. For efficient operation at low liquid rates, accurate leveling is required.

The orifice distributor consists of a flat tray equipped with a number of risers for gas flow and perforations in the tray floor for discharge of liquid. The relatively low riser area offers some resistance to gas flow, and allowance for possible flooding must be made. A variation of the orifice distributor eliminates the perforations and provides a V notch in each riser for passage of liquid countercurrently to gas; this device is subject to flooding characteristic of wetted-wall columns.

At high gas rates, entrainment from the liquid distributor can occur. This is especially true if a simple spray nozzle or nozzles are used above the bed as a distributor. While such entrainment detracts from mass-transfer efficiency, it must be recognized also that additional contacting opportunity is presented by the distributor, the opportunity varying with the device used.

Maldistribution Departure from uniform distribution of the phases in a packed column can be caused by:

1. The liquid distributor not dividing the liquid evenly over the column cross section.
2. The liquid moving more easily to the wall than vice versa. The resultant channeling along the wall may be accentuated by vapor condensing because of column heat losses.

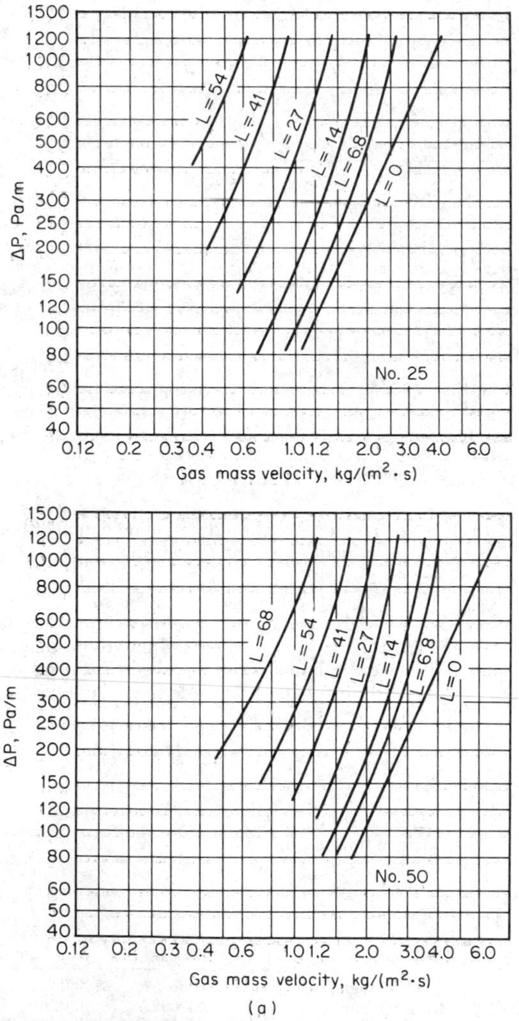

FIG. 18-44a Pressure for metal Intalox saddles, sizes No. 25 (nominal 25 mm) and No. 50 (nominal 50 mm). Air-water system at atmospheric pressure, 760-mm (30-in) column, bed height, 3.05 m (10 ft). L = liquid rate, kg/(s·m²). To convert kilograms per second–square meter to pounds per hour–square foot, multiply by 151.7; to convert pascals per meter to inches of water per foot, multiply by 0.1225. *(Courtesy Norton Company, Akron, Ohio.)*

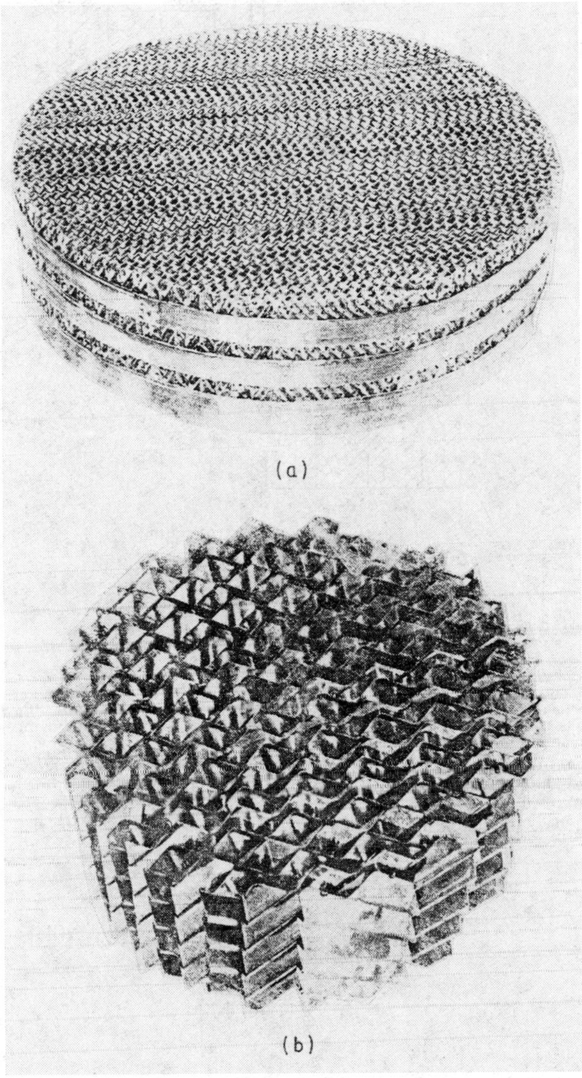

FIG. 18-44*b* Representative arranged-type packings: (*a*) Koch Sulzer, (*b*) Flexipac. *(Courtesy Koch Engineering Co., Inc.)*

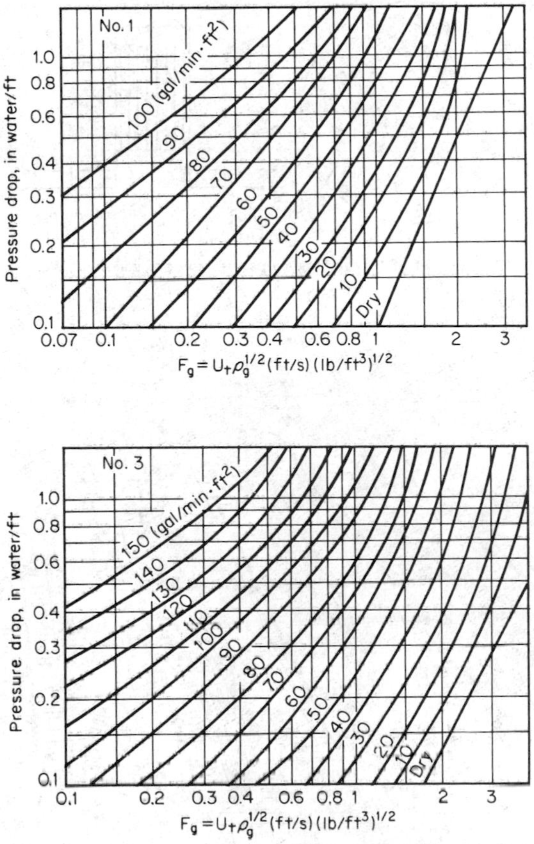

FIG. 18-45 Pressure drop for Flexipac packing, sizes No. 1 and No. 3. Air-water system at atmospheric pressure. Liquid rate in gallons per minute-square foot. To convert (feet per second) (pounds per cubic foot)$^{1/2}$ to (meters per second) (kilograms per cubic meter)$^{1/2}$, multiply by 1.2199; to convert gallons per minute–square foot to pounds per hour–square foot, multiply by 500; to convert inches of water per foot to millimeters of water per meter, multiply by 83.31; and to convert pounds per hour–square foot to kilograms per second–square meter, multiply by 0.001356. *(Courtesy Koch Engineering Co.)*

3. The packing geometry inhibiting lateral distribution.
4. Void variations due to packing being improperly installed.
5. The column being out of vertical alignment.

Cause 1 has been covered in the preceding subsection. Causes 4 and 5 can be handled through careful design and installation. Causes 2 and 3 bear additional discussion at this point.

The effect of channeling on mass-transfer efficiency has been studied theoretically by Manning and Cannon [*Ind. Eng. Chem.*, **49,** 347 (1957)], Huber and Hiltbrunner [*Chem. Eng. Sci.*, **21,** 819 (1966)], and Meier and Huber (*Proc. Intn. Symp. Distill.*, Brighton, England, 1970). Typical results are shown in Fig. 18-52. The effect of maldistribution on mass-transfer efficiency can be severe, but methods for predicting the amount of channeling are not available. Lateral-dispersion effects (cause 3) are the same as those of channeling.

Liquid migration to the wall appears to be favored by small column-diameter–packing-diameter ratios (e.g., less than 10) and can be corrected by the use of side wipers or redistributors (Fig. 18-53). Inhibition of lateral dispersion can be caused by the geometry of certain rigid packing elements, and according to Huber and Hiltbrunner (op. cit.) it is favored by column-diameter–packing-diameter ratios greater than 30. It is significant that Silvey and Keller (*Proc. Intn. Symp. Distill.*, Brighton, England, 1970) found essentially no maldistribution for packed heights up to 10.7 m (35 ft). The distillation of cyclohexane–*n*-heptane was used, with 19-, 38-, and 75-mm (¾-, 1½, and 3-in) ceramic Raschig rings in a 1.2-m (4-ft) column. The initial liquid distribution was by a 40-point trough-type device.

With careful attention given to the five causes of maldistribution,

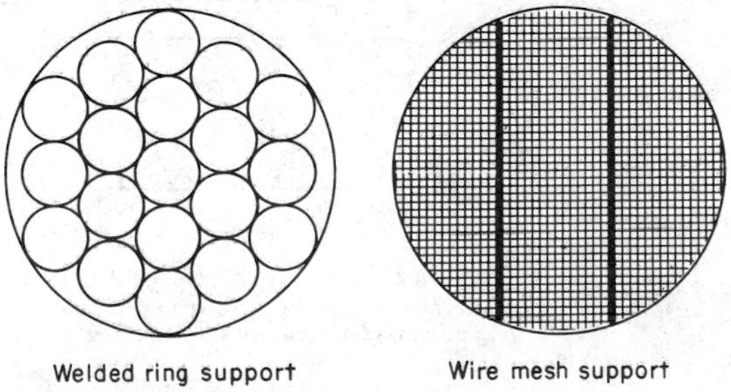

FIG. 18-46 Packing supports (countercurrent).

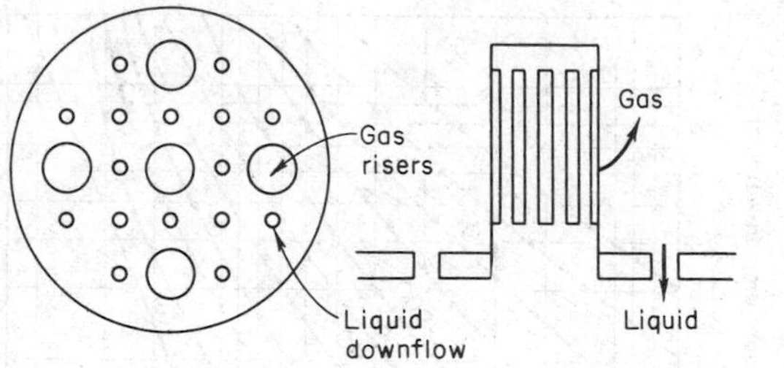

FIG. 18-47 Support plate, cap type (separate flow or "gas-injection" type).

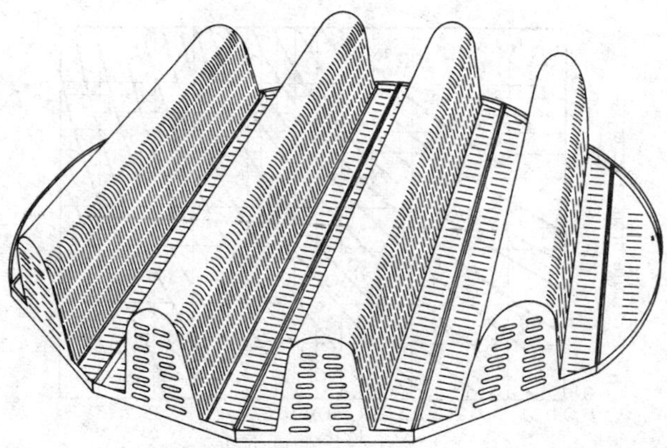

FIG. 18-48 Beam-type "gas-injection" support plate for large columns.

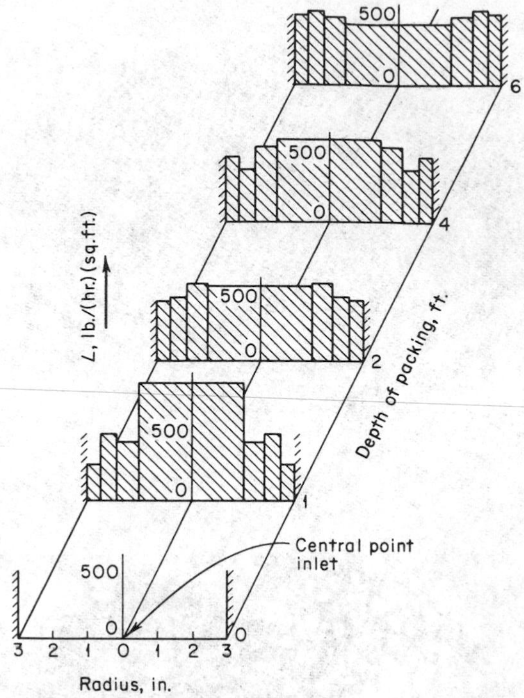

FIG. 18-50 Liquid distribution in a 6-in column packed with ½-in broken-stone packing. Increments of radius represent equal-annual-area segments of tower cross section. Central-point inlet. Water rate = 500 lb/(h·ft^2). Air rate = 810 lb/(h·ft^2). To convert pounds per hour–square foot to kilograms per second–square meter, multiply by 0.001356; to convert inches to centimeters, multiply by 2.54. (*Data from Baker, Chilton, and Vernon, in Sherwood and Pigford,* Absorption and Extraction, *2d ed., McGraw-Hill, New York, 1952.*)

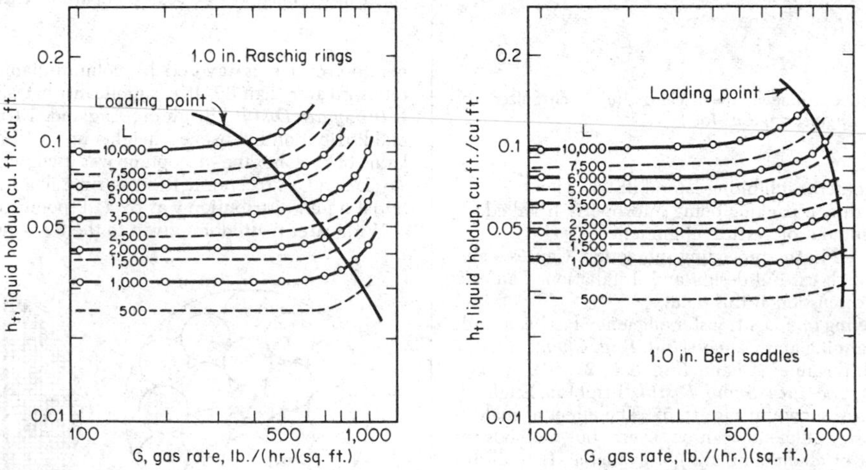

FIG. 18-49 Typical total holdup data for the air-water system. To convert pounds per hour–square foot to kilograms per second–square meter, multiply by 0.001356; to convert inches to millimeters, multiply by 25.4. [*Shulman et al.,* Am. Inst. Chem. Eng. J., *1, 247 (1955).*]

TABLE 18-9 Relative Transfer Rates as Function of Overhead Liquid Distributor

Distributor	Relative performance of packing
Serrated edge troughs 1 in. wide	100
Single nozzle entry	16
Splash plate below single nozzle	76
Single nozzle + 18-in. dumped 1-in. rings above stacked rings	63
Multiple nozzles + 18-in. dumped 1-in. rings above stacked rings	120

NOTE: To convert inches to millimeters, multiply by 25.4.

it is possible to design commercial packed columns for heights of 8 to 9 m (25 to 30 ft) between redistributors.

End Effects Analysis of the mass-transfer efficiency of a packed column should take into account that transfer which takes place outside the bed, i.e., at the ends of the packed sections. Inlet gas may very well contact exit liquid below the bottom support plate, and exit gas can contact liquid from some types of distributors (e.g., spray nozzles). The bottom of the column is the more likely place for transfer, and Silvey and Keller [*Chem. Eng. Prog.*, **62**(1), 68 (1966)] found that the reboiler plus the end effect could give up to two or more theoretical plates (Fig. 18-54).

Interfacial Area The effective area of contact between gas and liquid is that area which participates in the gas-liquid mass-exchange process. This area may be less than the actual interfacial area because of stagnant pools where liquid reaches saturation and no longer participates in the transfer process.

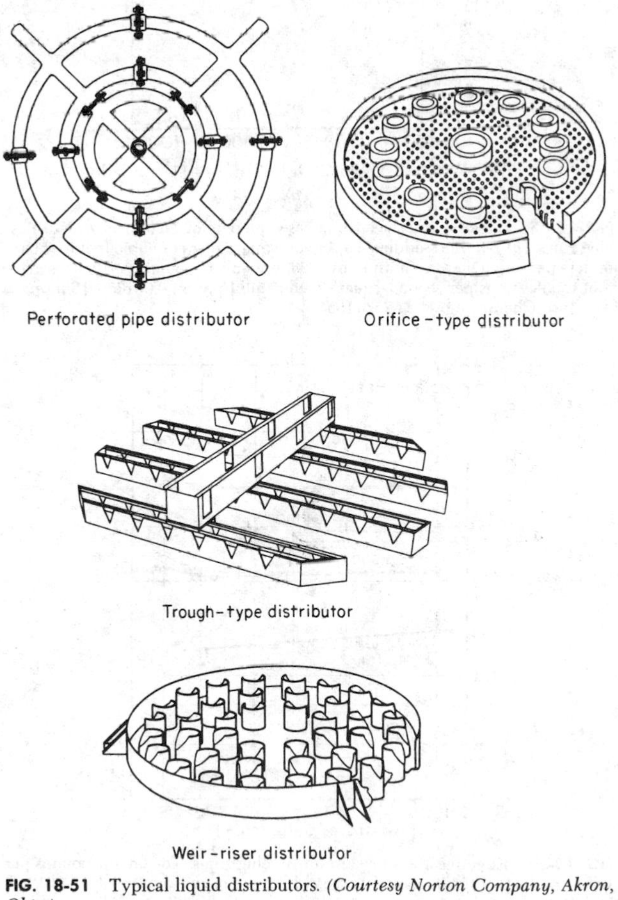

FIG. 18-51 Typical liquid distributors. *(Courtesy Norton Company, Akron, Ohio.)*

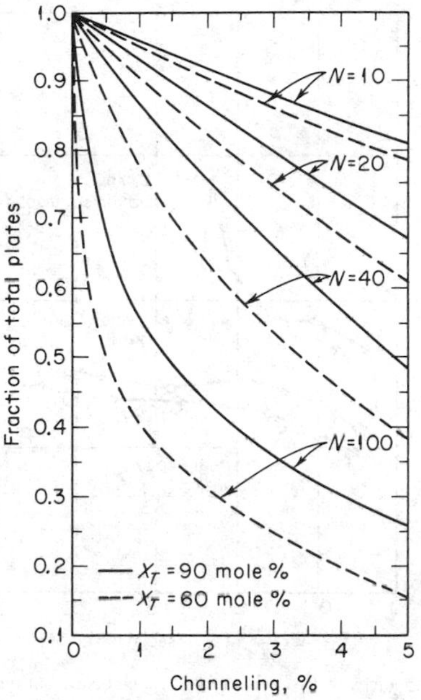

FIG. 18-52 Effect of liquid channeling on column efficiency for a system with a relative volatility of 1.07. Total number of theoretical plates N of 10, 20, 40, and 100 at top liquid composition X of 90 and 60 mole percent. [*Manning and Cannon*, Ind. Eng. Chem., *49, 347 (1957).*]

Effective area should not be confused with "wetted area." While film flow of liquid across the packing surface is a contributor, effective area includes also contributions from rivulets, drippings, and gas bubbles. Because of this complex physical picture, effective interfacial area is difficult to measure directly.

Weisman and Bonilla [*Ind. Eng. Chem.*, **42**, 1099 (1950)] determined effective area a_i of 25-mm (1-in) Raschig rings indirectly through the relationship $a_i = (k_g a_i)/k_g$. The k_g data were obtained via evaporation of water from presaturated rings by Taecker and Hougen [*Chem. Eng. Prog.*, **45**, 188 (1949)] and the vaporization $k_g a_i$ data from McAdams et al. [*Chem. Eng. Prog.*, **45**, 241 (1949)] for the air–water irrigated system. The authors proposed that

$$a_i/a_t = 0.54\ G^{0.31}L^{0.07} \tag{18-53}$$

for the range of liquid rates from 4 to 17 kg/(s·m^2) [2950 to 12,537 lb/(h·ft^2)], and where a_t = external surface area of the packing

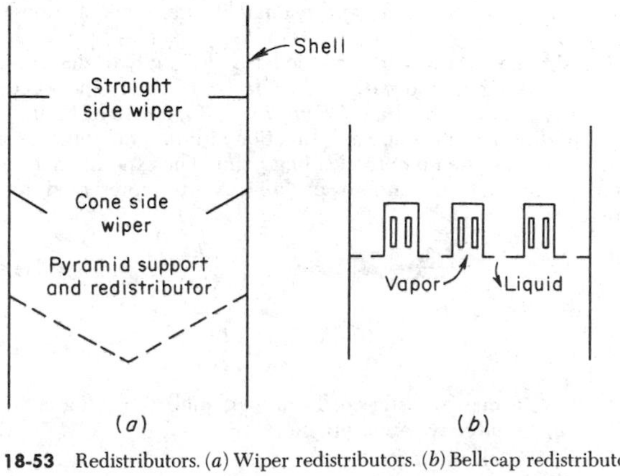

FIG. 18-53 Redistributors. (*a*) Wiper redistributors. (*b*) Bell-cap redistributor.

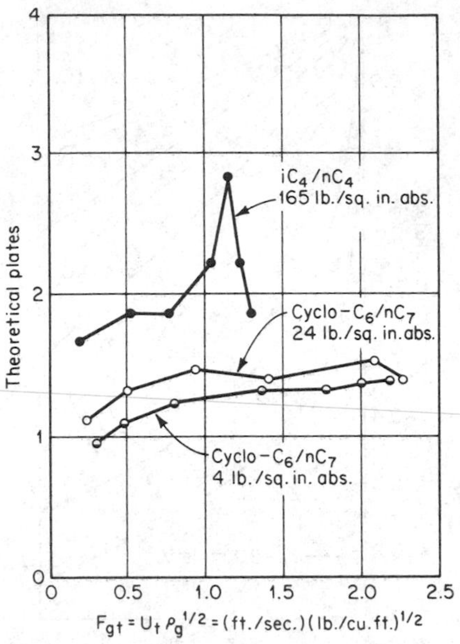

FIG. 18-54 Efficiency of FRI reboiler and space below bottom support plate. To convert pounds per square inch absolute to kilopascals, multiply by 6.8947; to convert (feet per second) (pounds per cubic foot)$^{1/2}$ to (meters per second) (kilograms per cubic meter)$^{1/2}$, multiply by 1.2199. [*Silvey and Keller,* Chem. Eng. Prog., *62(1), 68 (1966).*]

(Table 18-5). In Eq. (18-53), both gas and liquid rates, G and L, are in kg/(s·m^2). Areas are in consistent units.

A greater dependency on liquid rate was reported by Shulman et al. [*Am. Inst. Chem. Eng. J.*, **1**, 253 (1955)], who obtained the effective area via vaporization of packing constructed of naphthalene and from calculated ammonia absorption data of Fellinger (Sc.D. thesis, Massachusetts Institute of Technology, 1941), taking account of liquid-phase resistance. On the basis of gross system conditions, the values obtained are indicated in Fig. 18-55 for 25-mm (1-in) Raschig rings and Berl saddles. These packing types in the 12-, 38-, and 50-mm (0.5-, 1.5-, and 2.0-in) sizes were also studied.

Yoshida and Koyanagi [*Am. Inst. Chem. Eng. J.*, **8**, 309 (1962)] used the Weisman-Bonilla approach, accepting the Taecker-Hougen k_g data and making their own $k_g a_i$ measurements under vaporization, absorption, and distillation conditions. They found that the effective area differs between vaporization and absorption, as shown in Figs. 18-56 and 18-57. For distillation, they found effective areas to be different for systems with different surface tensions (Fig. 18-58) but, upon making the surface-tension correction, concluded that areas for distillation are approximately the same as those for absorption.

Liquid-Phase Transfer On the basis of a study of desorption of oxygen, hydrogen, and carbon dioxide from water, Sherwood and Holloway [*Trans. Am. Inst. Chem. Eng.*, **36**, 39 (1940)] found that the liquid-film coefficient was a function of liquid rate but was independent of gas rate up to the loading point. The experimental results from a number of random packings were represented by the equations

$$\frac{k_\ell a_e}{D_\ell} = \alpha \left(305 \frac{L}{\mu_\ell}\right)^{1-n} \left(\frac{\mu_\ell}{\rho_\ell D_\ell}\right)^{0.5} \quad (18\text{-}54)$$

$$H_\ell = \frac{1}{\alpha}\left(305 \frac{L}{\mu_\ell}\right)^{n} \left(\frac{\mu_\ell}{\rho_\ell D_\ell}\right)^{0.5} \quad (18\text{-}55)$$

where k_ℓ = mass-transfer coefficient, kg·mol/[(s·m^2)(kg·mol/m^3)]
a_e = effective area, m^2/m^3
D_ℓ = diffusion coefficient, m^2/s

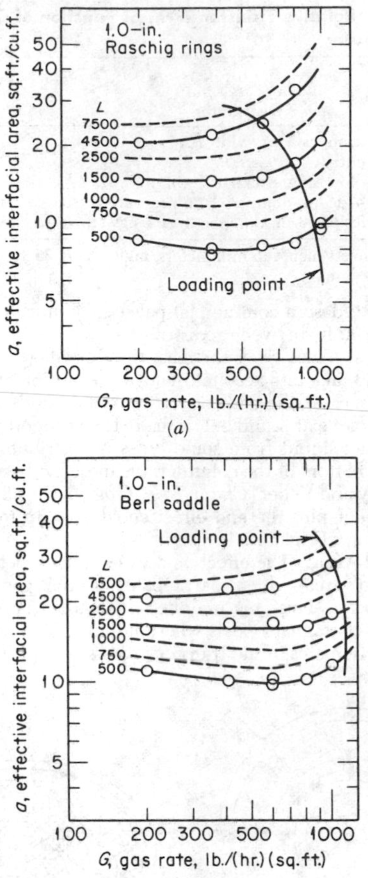

FIG. 18-55 Effective interfacial area based on data of Fellinger. (*a*) 1-in Raschig rings. (*b*) 1-in Berl saddles. To convert square feet per cubic foot to square meters per cubic meter, multiply by 3.28; to convert pounds per hour–square foot to kilograms per second–square meter, multiply by 0.001356. [*Shulman,* Am. Inst. Chem. Eng. J., *1, 257 (1955).*]

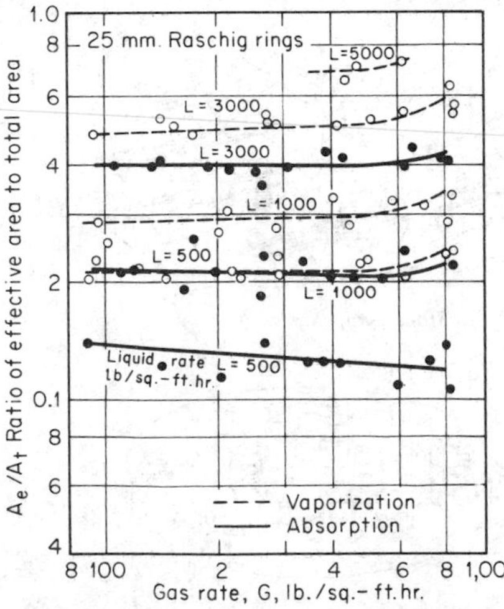

FIG. 18-56 Effective areas for 25-mm Raschig rings. To convert pounds per hour–square foot to kilograms per second–square meter, multiply by 0.001356. [*Yoshida and Koyanagi,* Am. Inst. Chem. Eng. J., *8, 309 (1962).*]

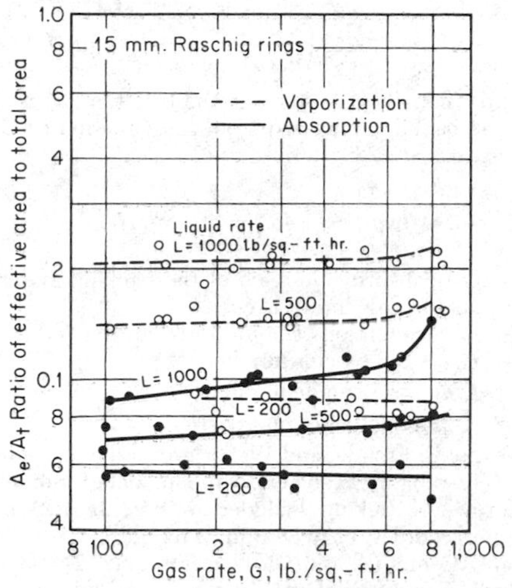

FIG. 18-57 Effective areas for 15-mm Raschig rings. To convert pounds per hour-square foot to kilograms per second–square meter, multiply by 0.001356. [*Yoshida and Koyanagi*, Am. Inst. Chem. Eng. J., *8, 309 (1962).*]

L = liquid rate, kg/(s·m²)
μ_ℓ = liquid viscosity, Pa·s
ρ_ℓ = liquid density, kg/m³
H_ℓ = height of a liquid-phase transfer unit, m [$= L/k_\ell a_e \rho_\ell$)]
α, n = constants as shown in Table 18-10

The Sherwood and Holloway results were generally confirmed by Molstad et al. [*Trans. Am. Inst. Chem. Eng.*, **38**, 410 (1942)], Deed et al. [*Ind. Eng. Chem.*, **39**, 766 (1947)], Vivian and Whitney [*Chem. Eng. Prog.*, **43**, 691 (1947)], Whitney and Vivian [*Chem. Eng. Prog.*, **45**, 323 (1949)], and Shulman and de Gouff [*Ind. Eng. Chem.*, **44**, 1915 (1952)], as outlined by Cornell et al. [*Chem. Eng. Prog.*, **56**(8), 68 (1960)].

The generalized equation of Cornell et al. (op cit.) for liquid-phase transfer units is

$$H_\ell = \frac{\Phi C}{3.28}\left(\frac{\mu_\ell}{\rho_\ell D_\ell}\right)^{0.5}\left(\frac{Z}{3.05}\right)^{0.15} \tag{18-56}$$

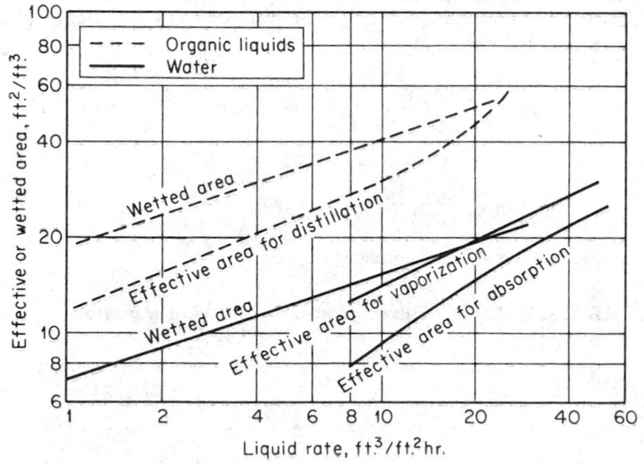

FIG. 18-58 Comparison of effective and wetted areas, 25-mm Raschig rings. To convert cubic feet per square foot–hour to cubic meters per square meter–second, multiply by 8.47 × 10⁻⁵; to convert square feet per cubic foot to square meters per cubic meter, multiply by 3.28. [*Yoshida and Koyanagi*, Am. Inst. Chem. Eng. J., *8, 309 (1962).*]

TABLE 18-10 Values of Constants for Eqs. (18-54) and (18-55)

	Packing, mm	α	n
Raschig rings	12	920	0.35
	25	330	0.22
	38	295	0.22
	50	260	0.22
Berl saddles	12	490	0.28
	25	560	0.28
	38	525	0.28
Tile	75	360	0.28

NOTE: To convert millimeters to inches, multiply by 0.0394.

where Φ = correlation parameter for a given packing, m
C = correction factor for high gas rates (Fig. 18-59)
μ_ℓ = liquid viscosity, Pa·s
ρ_ℓ = liquid density, kg/m³
D_ℓ = liquid-diffusion coefficient, m²/s
Z = height of packing, m

The correlation parameter Φ was adjusted by Bolles and Fair [*Inst. Chem. Eng. Symp. Ser.*, no. 56, 3.3/35 (1979); *Chem. Eng.*, **89**(14), 109 (July 12, 1982)] to give an improved fit of a large body of data covering additional packing types. Values of Φ may be obtained from Fig. 18-60.

It would be desirable to separate the effects of design variables on k_ℓ and on a_e. To this end, Shulman et al. [*Am. Inst. Chem. Eng. J.*, **1**, 253 (1955)] suggested use of the equation

$$\frac{k_\ell D_p}{D_\ell} = 25.1\left(\frac{D_p L}{\mu_\ell}\right)^{0.45}\left(\frac{\mu_\ell}{\rho_\ell D_\ell}\right)^{0.50} \tag{18-57}$$

for Raschig rings and Berl saddles, with effective area to be obtained from available data. In Eq. (18-57) the groups are dimensionless, and D_p is the diameter of a sphere possessing the same surface area as a unit of packing. See Eq. (18-54) for nomenclature and typical units.

In accordance with the j-factor analogy, Onda et al. [*Am. Inst. Chem. Eng. J.*, **5**, 235 (1959)] found a dimensionless relationship similar to those of van Krevelen and Hoftijzer [*Chem. Eng. Prog.*, **44**, 529 (1948)] and Yoshida and Koyanagi [*Ind. Eng. Chem.*, **50**, 365 (1958)]:

$$k_\ell\left(\frac{\rho_\ell}{\mu_\ell g}\right)^{1/3} = 0.021\left(\frac{L}{a_t\mu_\ell}\right)^{0.49}\left(\frac{\mu_\ell}{\rho_\ell D_\ell}\right)^{-0.50} \tag{18-58}$$

which gives

$$H_\ell = \frac{1}{0.021}\left(\frac{\mu_\ell^2}{\rho_\ell^2 g}\right)^{1/3}\left(\frac{L}{a_t\mu_\ell}\right)^{0.51}\left(\frac{\mu_\ell}{\rho_\ell D_\ell}\right)^{0.50} \tag{18-59}$$

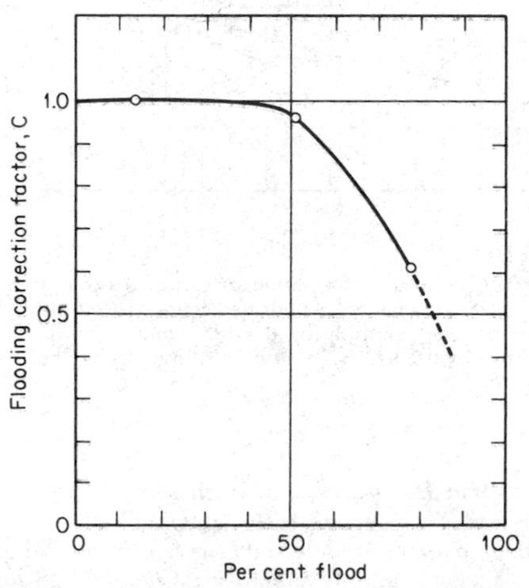

FIG. 18-59 Liquid-film correction factor for operation at high percent of flood. [*Cornell et al.*, Chem. Eng. Prog., *56(8), 68 (1960).*]

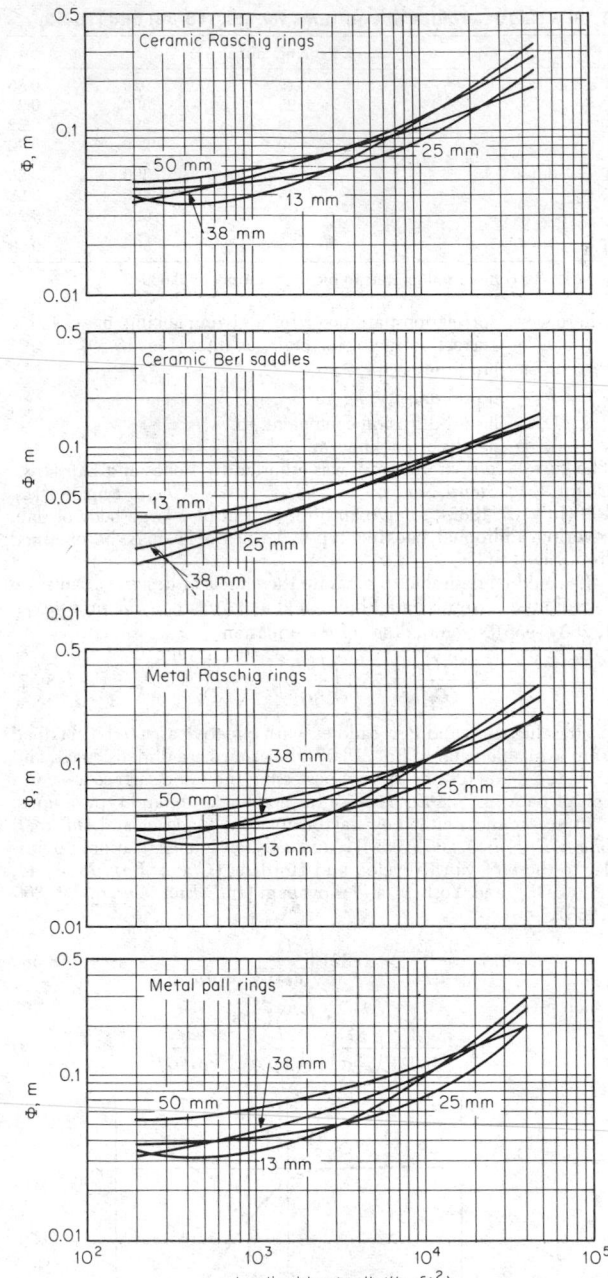

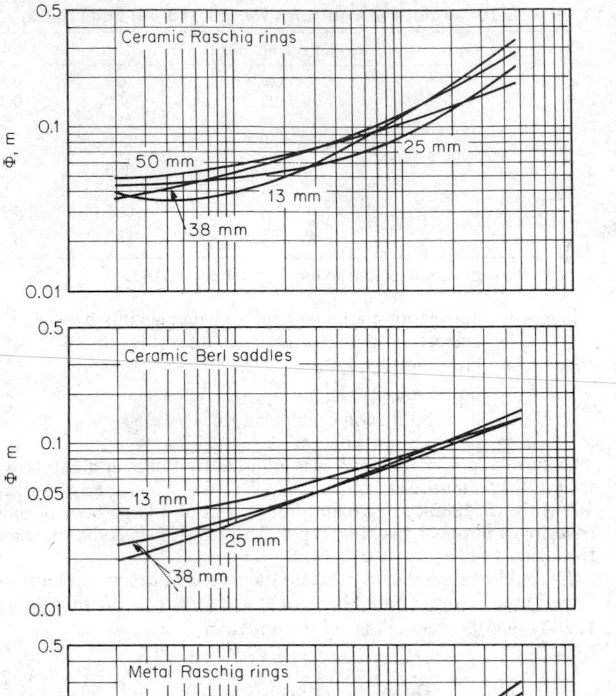

FIG. 18-60 H_ℓ correlation for various packings. To convert meters to feet, multiply by 3.281; to convert pounds per hour–square foot to kilograms per second–square meter, multiply by 0.001356; and to convert millimeters to inches, multiply by 0.0394. [*Bolles and Fair,* Inst. Chem. Eng. Symp. Ser., *no. 56, 3.3/35 (1969).*]

on the basis that $H_\ell = L/k_\ell a_t \rho_\ell$. In these equations, a_t is the total surface area of the packing, easily obtainable (Table 18-5). Other terms in the equations are selected to give dimensional consistency.

Onda, Takeuchi, and Okumoto [*J. Chem. Eng. Japan,* **1,** 56 (1968)] correlated a large amount of liquid-phase transfer data to ± 20 percent by the equation

$$k_\ell \left(\frac{\rho_\ell}{\mu_\ell g}\right)^{1/3} = 0.0051 \left(\frac{L}{a_w \mu_\ell}\right)^{2/3} \left(\frac{\mu_\ell}{\rho_\ell D_\ell}\right)^{0.50} (a_t D_p)^{0.4} \qquad (18\text{-}60)$$

which is dimensionally consistent. In the equation, a_w is the wetted surface of the packing, obtained from the equation of Onda et al. [*Kagaku Kogaku,* **31,** 126 (1967)]:

$$\frac{a_w}{a_t} = 1 - \exp\left[-1.45\left(\frac{\sigma_c}{\sigma}\right)^{0.75} N_{\mathrm{Re}}^{0.1} N_{\mathrm{Fr}}^{-0.05} N_{\mathrm{We}}^{0.2}\right] \qquad (18\text{-}61)$$

where σ_c = critical surface tension of the packing material (Table 18-11), mN/m (dyn/cm)
σ = surface tension of liquid, mN/m (dyn/cm)
N_{Re} = $L/a_t\mu_\ell$ (Reynolds number)
N_{Fr} = $L^2 a_t/\rho_\ell^2 g$ (Froude number)
N_{We} = $L^2/\rho_\ell \sigma a_t$ (Weber number)

with the Reynolds, Froude, and Weber groups being dimensionless.

Equation (18-60) is based on experimental data for organic liquids as well as for water. Packings included are Raschig rings, 6 to 50 mm (¼ to 2 in); Berl saddles, 12 to 38 mm (½ to 1½ in); 25-mm (1-in) pall rings; 12- and 25-mm (½- and 1-in) spheres; and 12- and 25-mm (½- and 1-in) rods. The range of conditions covered by the experiments is indicated in Figs. 18-61 and 18-62.

Gas-Phase Transfer Generalization of mass-transfer data for the gas phase has been much less successful than it has been for the liquid phase. Principal deterrents to the development of a general relationship have been (1) the variation of effective area as a function of flow rates and surface tension and (2) the lack of a gas-phase controlling test system unencumbered by side effects of heat development or chemical reaction.

Investigations of the behavior of gas films have been conducted with

1. Vaporization of pure liquids into a gas stream
2. Absorption of a solute gas into a liquid offering a high degree of solubility
3. Absorption of a solute gas into a liquid where the equilibrium partial pressure is zero because of an irreversible chemical reaction

In cases 1 and 2, heat-of-vaporization and heat-of-solution effects result in (1) localized temperature changes that are reflected in varying equilibrium constants not typical of the bulk conditions and (2) local changes in surface tension that may result in film splitting and change of effective area of mass transfer as indicated by Bond and Donald [*Chem. Eng. Sci.,* **6,** 237 (1957)]. In case 3, the situation is complicated by the fact that the chemical reaction rate or diffusion of the reacting component of the solvent may be rate-controlling, thus shifting a large portion of transfer resistance to the liquid phase.

Gamson, Thodos, and Hougen [*Trans. Am. Inst. Chem. Eng.,* **39,** 1 (1943)] evaluated the behavior in columns packed with spheres and saddles. It was found that

$$J_D = \frac{k_g p_{gf} M_m}{G}\left(\frac{\mu}{\rho D_g}\right)^{2/3} = 0.99\left(\frac{D_p G}{\mu}\right)^{-0.41} \qquad (18\text{-}62)$$

and

$$J_H = \frac{h_g}{CG}\left(\frac{C\mu}{k}\right)^{2/3} = 1.064\left(\frac{D_p G}{\mu}\right)^{-0.41} \qquad (18\text{-}63)$$

for values of N_{Re} exceeding 350.

TABLE 18-11 Critical Surface Tension of Packing Materials*

	σ_c, dyn/cm
Carbon	56
Ceramic	61
Glass	73
Paraffin	20
Polyethylene	33
Polyvinylchloride	40
Steel	75

*Onda, Takeuchi, and Koyama, *Kagaku Kogaku,* **31,** 126 (1967).
NOTE: dyn/cm = mN/m.

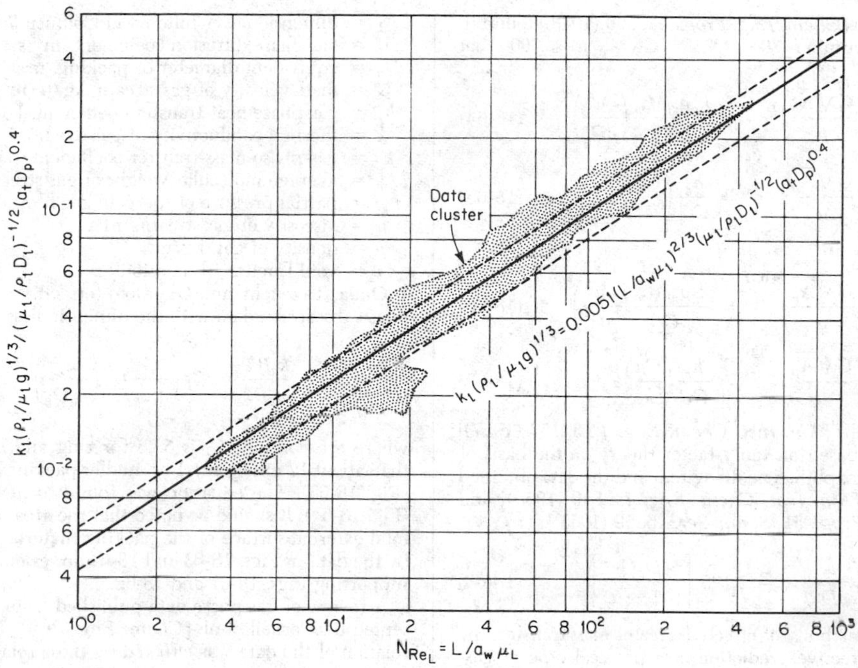

FIG. 18-61 Correlation of liquid-phase data for gas absorption and desorption by using water. [*Onda et al.*, J. Chem. Eng. Japan, *1, 56 (1968).*]

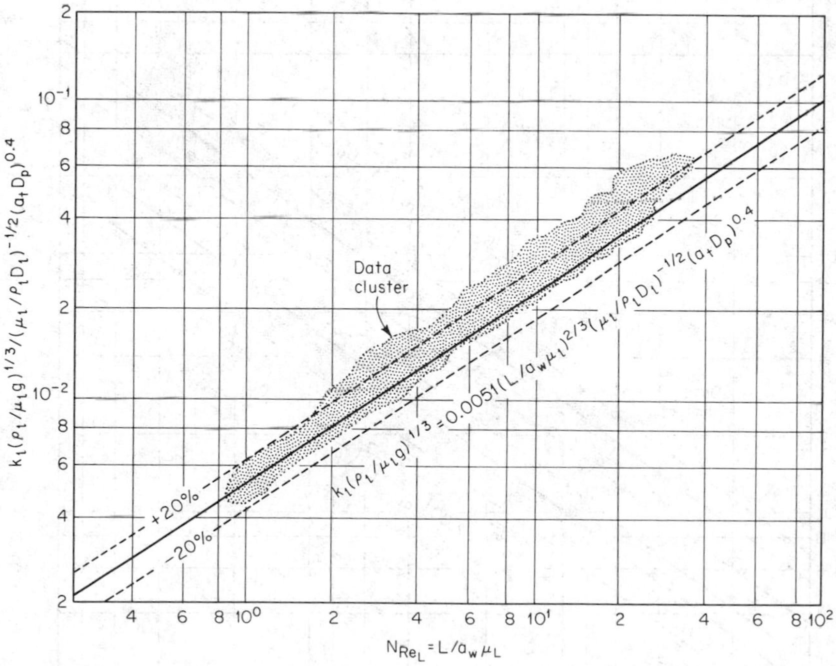

FIG. 18-62 Correlation of absorption data by using organic solvents. [Onda et al., *J. Chem. Eng. Japan, 1, 56 (1968).*]

Taecker and Hougen [*Chem. Eng. Prog.*, **44,** 529 (1948)] found in the range of Reynolds numbers $70 < (N_{Re} = G\sqrt{A_p}/\mu) < 3000$ that for Berl saddles

$$J_D = 0.855\left(\frac{G\sqrt{A_p}}{\mu}\right)^{-0.34} = \frac{k_g p_{gf} M_m}{G}\left(\frac{\mu}{\rho D_g}\right)^{2/3} \quad (18\text{-}64)$$

$$J_H = 0.920\left(\frac{G\sqrt{A_p}}{\mu}\right)^{-0.34} = \frac{h_g}{CG}\left(\frac{C\mu}{k}\right)^{2/3} \quad (18\text{-}65)$$

and for Raschig rings

$$J_D = 1.070\left(\frac{G\sqrt{A_p}}{\mu}\right)^{-0.41} = \frac{k_g p_{gf} M_m}{G}\left(\frac{\mu}{\rho D_g}\right)^{2/3} \quad (18\text{-}66)$$

$$J_H = 1.148\left(\frac{G\sqrt{A_p}}{\mu}\right)^{-0.41} = \frac{h_g}{CG}\left(\frac{C\mu}{k}\right)^{2/3} \quad (18\text{-}67)$$

Shulman and Margolis [*Am. Inst. Chem. Eng. J.*, **3,** 157 (1957)] proposed an equation based on the *J*-factor theory on the basis of data obtained with the naphthalene-air system and the data obtained by Lynch and Wilke [*Am. Inst. Chem. Eng. J.*, **1,** 9 (1955)] and Yoshida [*Chem. Eng. Prog.*, **51,** *Symp. Ser.* 16, 59 (1955)].

$$J_D = \frac{k_g M_m p_{gf}}{G}\left(\frac{\mu}{\rho D_g}\right)^{2/3} = 1.195\left[\frac{D_p G}{\mu(1-\epsilon)}\right]^{-0.36} \quad (18\text{-}68)$$

Combined with the availability of effective area of mass transfer, the authors believe that effective prediction of a physical type of gas-film mass-transfer coefficient can be made. The nomenclature for Eqs. (18-62) to (18-68) is as follows:

A_p = external surface of packing unit, m^2
C = specific heat of gas stream, J/(kg·K)
J = dimensionless Chilton and Colburn J factor
D_g = gas-phase diffusion coefficient, m^2/s
D_p = equivalent diameter of packing, m
G = mass velocity of gas stream, kg/(s·m^2)
h_g = gas-phase heat-transfer coefficient, J/(m^2·s·K)
k = thermal conductivity of gas stream, J/(m^2·s·K)
k_g = gas-phase mass-transfer coefficient, (kg·mol)/(s·m^2·Pa)
M_m = average molecular weight of gas stream kg/(kg·mol)
p_{gf} = partial pressure of inerts in gas film, Pa
μ = viscosity of gas stream, mPa · s
ρ = density of gas, kg/m^3
ϵ = void fraction, dimensionless

Onda, Takeuchi, and Okumoto (op. cit.) correlated available gas-phase absorption data with the dimensionless equation

$$\frac{k_g RT}{a_t D_g} = C_1\left(\frac{G}{a_t\mu_g}\right)^{0.7}\left(\frac{\mu_g}{\rho_g D_g}\right)^{1/3}(a_t D_p)^{-2.0} \quad (18\text{-}69)$$

where the constant C_1 is 5.23 for ring and saddle packings larger than about 12 mm (½ in). For smaller packings the value of C_1 is 2.00 (Fig. 18-63). The equation was found to fit vaporization data also (Fig. 18-64). It should be noted that the area term used is that of the total external surface of the packing material. Packings represented by the data in Figs. 18-63 and 18-64 are essentially the same as those supporting Figs. 18-61 and 18-62.

A review of gas-phase data published from 1940 to 1955 was presented by Cornell et al. [*Chem. Eng. Prog.*, **56**(8), 68 (1960)]. Correlation of the data was effected by these relationships:

Ring-type packings,

$$H_g = \frac{0.017\psi D^{1.24} Z^{0.33} Sc_g^{0.5}}{(L f_1 f_2 f_3)^{0.6}} \quad (18\text{-}70)$$

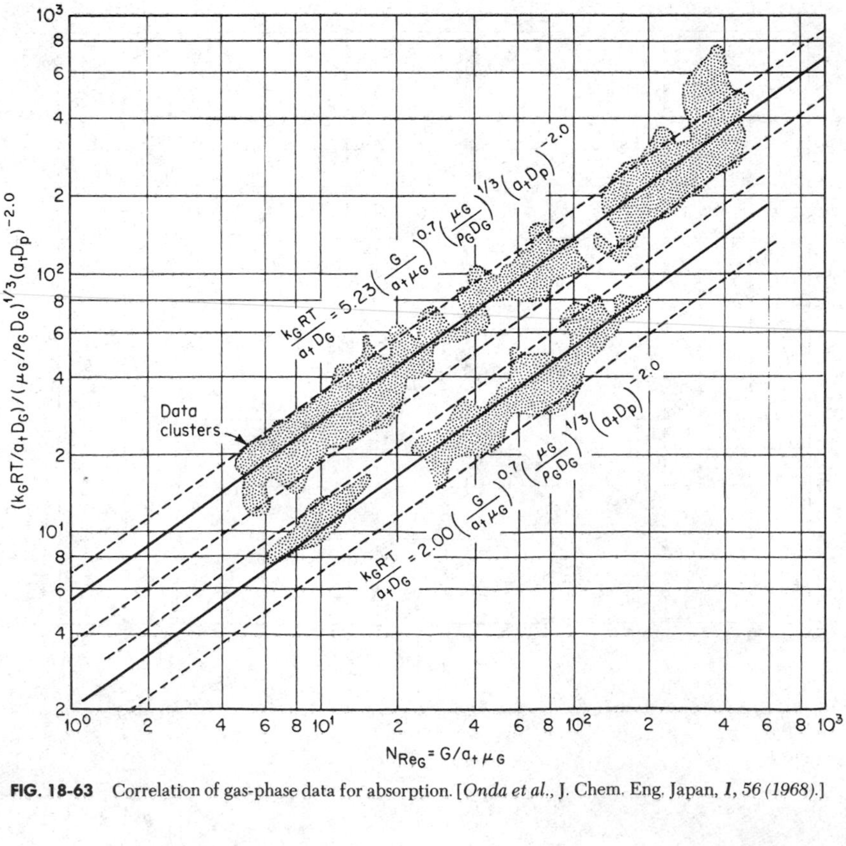

FIG. 18-63 Correlation of gas-phase data for absorption. [*Onda et al.*, J. Chem. Eng. Japan, *1, 56 (1968).*]

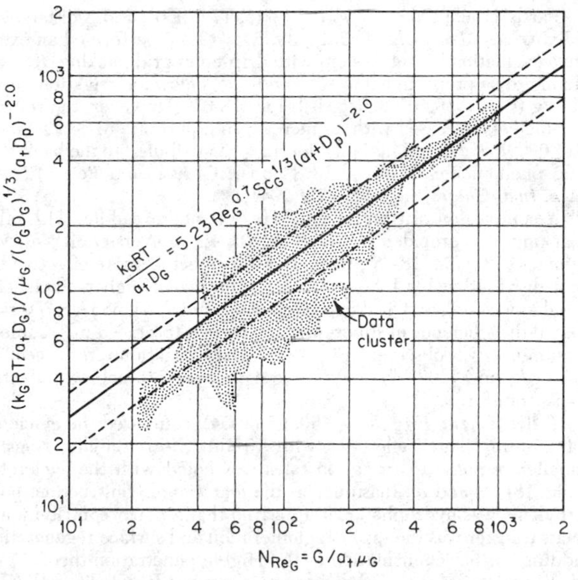

FIG. 18-64 Correlation of vaporization data with Eq. (18-69). Re = Reynolds number (N_{Re}). Sc = Schmidt number (N_{Sc}). [*Onda et al.*, J. Chem. Eng. Japan, *1, 56 (1968).*]

Saddle-type packings,

$$H_g = \frac{0.029\psi D^{1.11} Z^{0.33} Sc_g^{0.5}}{(L f_1 f_2 f_3)^{0.5}} \tag{18-70a}$$

where Sc_g = gas-phase Schmidt number (dimensionless) $= \mu_g/\rho_g D_g$
D = column diameter, m
Z = packed height, m
L = liquid rate, $kg/(s \cdot m^2)$
$f_1 = (\mu_\ell/\mu_w)^{0.16}$, with $\mu_w = 1.0$ mPa·s
$f_2 = (\rho_w/\rho_\ell)^{1.25}$, with $\rho_w = 1000\ kg/m^3$
$f_3 = (\sigma_w/\sigma_\ell)^{0.8}$, with $\sigma_w = 72.8$ mN/m (72.8 dyn/cm)

The correlation parameter ψ was adjusted by Bolles and Fair [*Inst. Chem. Eng. Symp. Ser.*, no. 56, 3.3/35 (1979); *Chem. Eng.*, **89**(14), 109 (July 12, 1982)] to give an improved fit of a large body of data covering additional packing types. Values of the adjusted parameters may be obtained from Fig. 18-65. In the use of Eq. (18-70) there is the following restriction: for column diameters larger than 0.6 m (2.0 ft), retain the diameter correction for 0.6 m.

An alternative approach that gives somewhat better fit of the distillation data contained in the data bank of Bolles and Fair (ibid.) is that of Bravo and Fair [*Ind. Eng. Chem. Process. Des. Dev.*, **21**(1), 162 (1982)], in which an effective area is obtained by the following correlating equation:

$$a_e/a_t = 0.310 \frac{\sigma^{0.5}}{Z^{0.4}} (Ca_\ell \cdot Re_g)^{0.392} \tag{18-71}$$

where a_e = effective area for mass transfer, m^2/m^3
a_t = total area of packing, m^2/m^3 (Table 18-5)
σ = surface tension, mN/m (dyn/cm)
Z = packed height, m
Ca_ℓ = liquid capillary number, dimensionless $(= \mu_\ell L/\rho_\ell \sigma)$
Re_g = gas Reynolds number, dimensionless $(= 6\ G/a_t \mu_g)$
μ_g, μ_ℓ = gas and liquid viscosities, Pa·s
L = liquid rate, $kg/(s \cdot m^2)$
G = gas rate, $kg/(s \cdot m^2)$
ρ_ℓ = liquid density, kg/m^3

The effective area obtained from Eq. (18-71) is combined with gas and liquid mass-transfer coefficients calculated by the Onda equations [Eqs. (18-60), (18-61), and (18-69)] to obtain heights of transfer units:

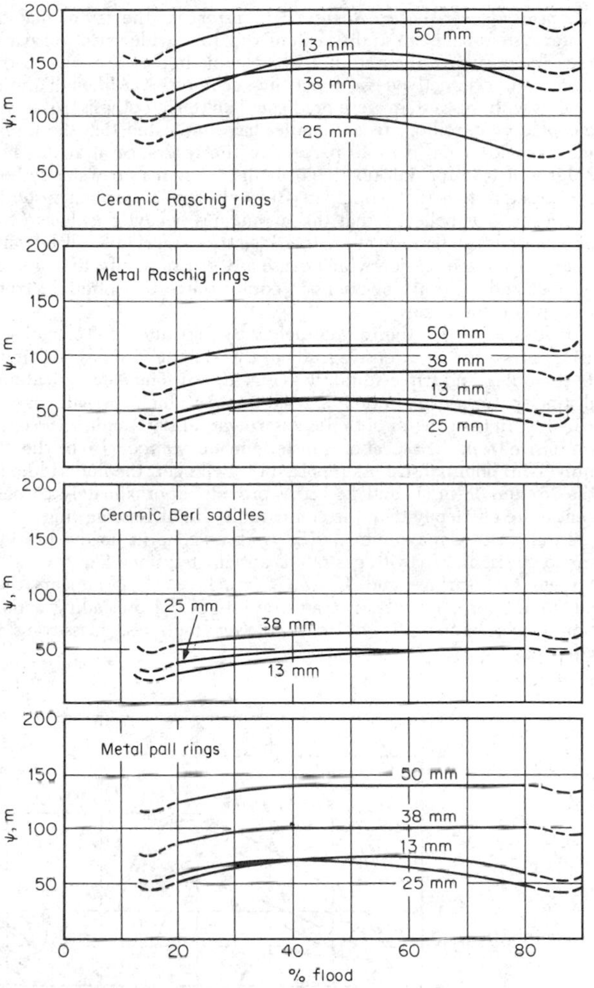

FIG. 18-65 H_g correlation for various packings. To convert meters to feet, multiply by 3.281; to convert millimeters to inches, multiply by 0.03937. [*Bolles and Fair*, Inst. Chem. Eng. Symp. Ser., *no. 56, 3.3/35 (1979).*]

$$H_g = G/(k_g a_e P\ M_g) \tag{18-72}$$

$$H_\ell = L/(k_\ell a_e \rho_\ell) \tag{18-73}$$

where P = total pressure, Pa
M_g = molecular weight of gas

and other terms are as defined earlier, with units consistent to give transfer-unit heights in meters.

Behavior of Various Systems and Packings In view of the incomplete development of the theory of mass-transfer behavior, much of the estimation of rate of transfer is based on comparison with systems already evaluated. Comparison is generally based on equivalent geometry and flow characteristics, the variation being related primarily to the physical properties of the systems.

The following system behaviors are reported:

Ammonia–air–water	Liquid- and gas-phase contributing—chemical reaction contributing
Air–water	Gas-phase controlling
Sulfur dioxide–air–water	Liquid- and gas-phase contributing
Carbon dioxide–air–water	Liquid-phase controlling
Distillation systems	Either phase may control—both contribute

Ammonia-Air-Water System Transfer in the ammonia-air-water system has been studied extensively for a wide variety of packings, and results of these studies are valuable in the estimation of the comparative effectiveness of packing geometries. Although initial studies with this system were predicated on the hypothesis that it was gas-phase controlling, recent studies have indicated that the liquid may provide from 5 to 40 percent of the resistance at 25°C. The heights of transfer unit obtained for the ammonia-air-water system are greater than that anticipated for physical-absorption systems. However, it is believed that this anomalous behavior is due to the effect of heat generation at the interface that results in localized variations in surface tension which cause a decrease in effective area of contact and to local deviations in equilibrium concentrations from bulk-phase conditions.

Fellinger (Sc.D. thesis, Massachusetts Institute of Technology, 1941) presented an extensive study of this system with Raschig rings, Berl saddles, and triple-spiral tiles (Fig. 18-66). The data indicate no distinct pattern of variation of mass-transfer efficiency with size of packing. In the cases of both Raschig rings and Berl saddles, the optimum size from the aspect of transfer efficiency appears to be the 25-mm (1-in) nominal size. As reported by Fellinger, the Berl saddle in the 25- and 38-m (1- and 1½-) sizes provides approximately 25 percent more efficiency than the comparable size of Raschig ring.

In all cases a maximum in H_{og} is observed prior to the loading range. A rise in H_{og} with gas rate is anticipated since $K_g a \; \alpha \; G^{0.5-0.8}$ for constant surface and $H_g \; \alpha \; G_m/k_g a(1 - y)$. It appears that effective interfacial area increases rapidly at the preloading conditions, overcoming the decrease of constant-area mass-transfer efficiency.

Parsly et al. [*Chem. Eng. Prog.*, **46,** 17 (1950)] and Molstad et al. [*Trans. Am. Inst. Chem. Eng.*, **39,** 605 (1943)] performed an extensive evaluation for this system with drip-point grid packing. $K_g a$ was found to be proportional to $G^{0.83}$ and $L^{0.17}$ below an irrigation rate of 14 kg (s·m²) [10,500 lb/(ft²·h)] (Fig. 18-67). However, the transfer coefficient decreased with an increase in liquid rate to 20 kg/(s·m²) [15,000 lb/(ft²·h)]. This behavior may be attributed to the back-mixing phenomenon observed by Cooper, Christl, and Perry [*Trans. Am. Inst. Chem. Eng.*, **37,** 979 (1941)].

A comparison of transfer efficiency of Intalox saddles with other packings was provided by Wen (M.S. thesis, University of West Virginia, 1953) (Fig. 18-68), indicating the effect of more effective liquid distribution for Intalox saddles resulting in higher efficiencies. Williams, Akell, and Talbot [*Chem. Eng. Prog.*, **43,** 585 (1947)] evaluated the efficiency of Fiberglas packing in a 15-cm- (6-in-) diameter column. It was observed (Fig. 18-69) that irrigation rates exceeding 13 kg/(s·m²) [2000 lb/(h·ft²)] were necessary to achieve effective mass transfer.

Teller [*Chem. Eng. Prog.*, **50,** 65 (1954)] compared the efficiency of 25-mm (1-in) Tellerettes with 19-mm (¾-in) Raschig rings. A smaller response to irrigation rates was noted with the Tellerettes (Fig. 18-70*a* and *b*). Inasmuch as the material of construction, polyethylene, was hydrophobic, it appeared that a major contribution to mass transfer was the rapid agglomeration and surface regeneration, adding further confirmation to the Higbie penetration theory. Further verification of this phenomenon was made by Khalif et al. (*Ref. Zh. Khim.*, no. 8843, 1955), who observed that the coefficient of mass transfer to droplets is 10 to 13 times larger than from a gas to a flat surface.

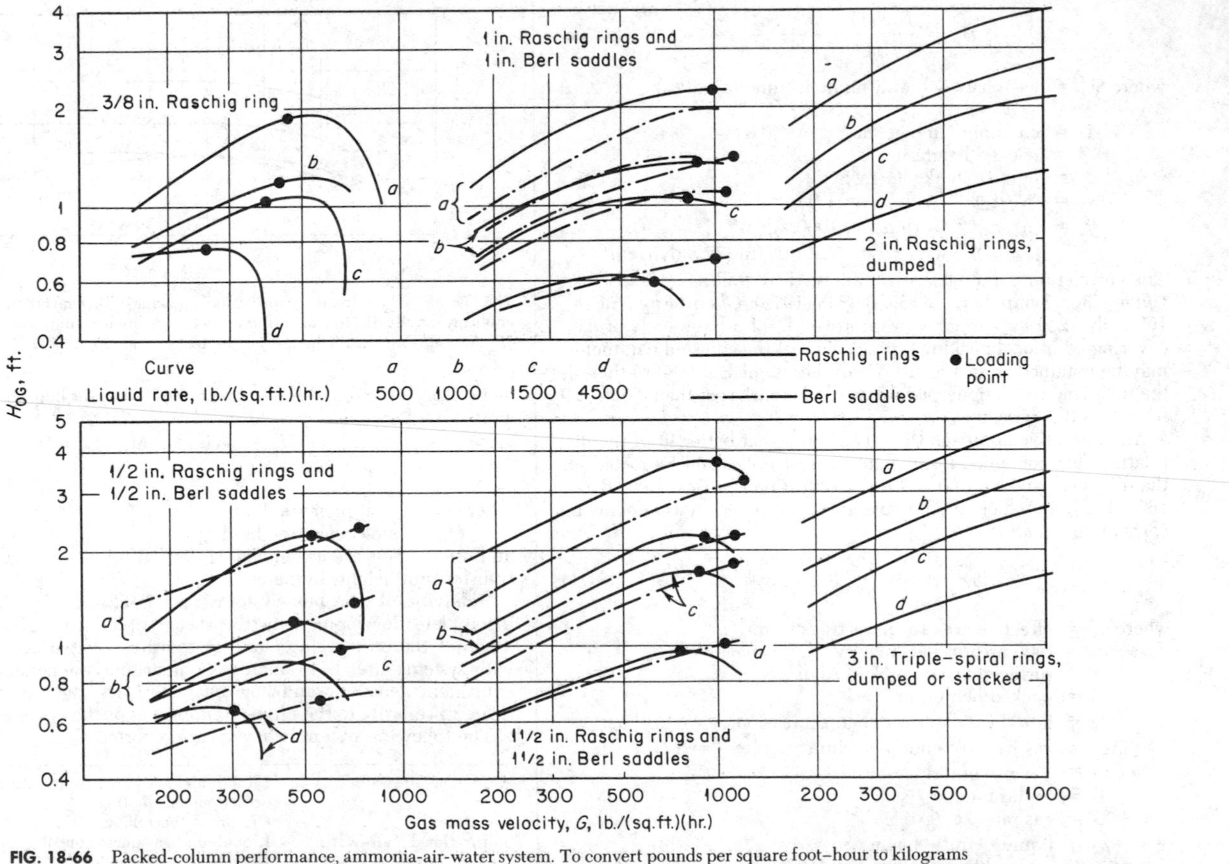

FIG. 18-66 Packed-column performance, ammonia-air-water system. To convert pounds per square foot–hour to kilograms per second–square meter, multiply by 0.001356; to convert feet to meters, multiply by 0.305; and to convert inches to millimeters, multiply by 25.4. *(Fellinger, Sc.D. thesis, Massachusetts Institute of Technology, 1941.)*

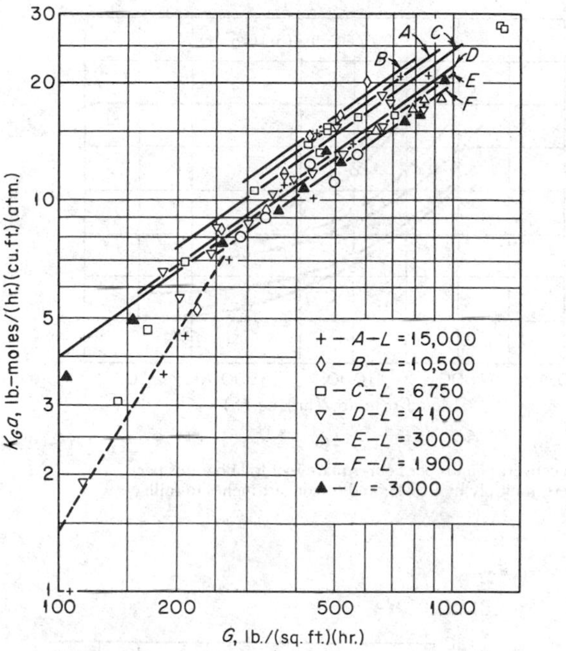

FIG. 18-67 $K_g a$ versus G for ammonia absorption with 2.25 ft of No. 6295 drip-point grid packing. To convert pounds per square foot–hour to kilograms per second–square meter, multiply by 0.001356; to convert pound-moles per hour–cubic foot–atmosphere to kilogram-moles per second–cubic meter–kilopascal, multiply by 4.390 × 10^{-5}. [*Parsly et al.*, Chem. Eng. Prog., *46, 17 (1950).*]

Air-Water System The air-water-system behavior for various packings has been extensively investigated from both the aspect of a vapor-phase controlling system and for developing design information for cooling towers and humidification systems.

Norman [*Trans. Inst. Chem. Eng.*, **29**(2), 226 (1951)] compared the data for performance of a variety of packings with that obtained for Paragrid packing. The packings reported were Raschig rings, Berl saddles, wood grid, and the Paragrid. A variation in the effect of the mass velocity of the gas is observed as a function of packing geometry. For the range of flow studied an increase in irrigation rates was reflected in an increase in efficiency of mass transfer (Fig. 18-71).

Parsly, Molstad, Cress, and Bauer [*Chem. Eng. Prog.*, **46**, 17 (1950)] evaluated the performance of drip-point grid tiles in the air-water system for the range $1.4 < L < 20$ and $0.14 < G < 1.4$ where flows are in kg/(s·m²). It was found that, for this packing, $k_g a$ was independent of liquor rate and proportional to $G^{0.839}$. The performance of 25-mm (1-in) carbon Raschig rings in this system was studied by McAdams, Pohlenz, and St. John [*Chem. Eng. Prog.*, **45**, 241 (1949)]. Using a 200-mm- (8-in-) diameter tower for the flow range $0.5 < G < 1.4$ and $0.7 < L < 3.5$, where flows are in kg(s·m²), it was found that end effects were equivalent to 18.2 cm (7.2 in) of packing. The small effect of liquid rate on the gas-phase mass and heat-transfer coefficient was unusual, as indicated by the equations

$$(h_g a_H)_{tf} = 1.78 G^{0.70} L^{0.07} e^{0.0023 t_f} \qquad (18\text{-}74a)$$

$$(k_g a)_m = 0.89 G^{0.70} L^{0.07} \qquad (18\text{-}74b)$$

$$h_\ell a = 0.82 G^{0.7} L^{0.5} \qquad (18\text{-}74c)$$

where the terms, expressed in *U.S. customary units*, are

$h_g a_H$ = heat-transfer coefficient, gas phase Btu/(h·ft³·°F)
$h_\ell a$ = heat-transfer coefficient, liquid phase Btu/(h·ft³·°F)
$(k_g a)_m$ = mass-transfer coefficient, gas phase lb/(h·ft³) (unit enthalpy difference); unit enthalpy difference = Btu/(lb·mol bone-dry air)

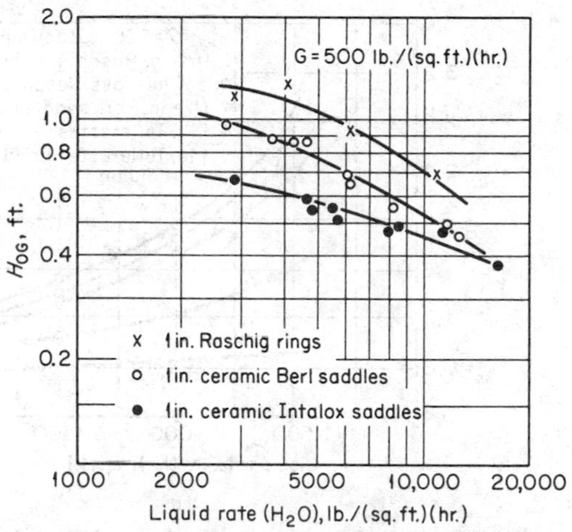

FIG. 18-68 Comparative absorption data in beds of Raschig rings, Berl saddles, and Intalox saddles, system NH_3-air-H_2O. To convert pounds per square foot–hour to kilograms per second–square meter, multiply by 0.001356; to convert inches to millimeters, multiply by 25.4. *(Wen, M. S. thesis, University of West Virginia, 1953.)*

L = liquid flow, lb/(h·ft²)
G = gas flow, lb/(h·ft²)
t_f = film temperature, °F

In addition, the data indicate a considerable liquid-film enthalpy-transfer resistance equivalent to 27 to 46 percent of the total resistance.

Cribb [*Br. Chem. Eng.*, 4(5), 264 (1959)] questioned the relative magnitude of the liquid-phase resistance, indicating that it was less than 20 percent. It was further indicated that in tests with 100-mm Raschig rings it was found that $h_\ell a \propto L^{1.5}$ was in good agreement with the data of Inazumi [*Chem. Eng. (Japan)*, **19**, 579 (1955)], who found that $h_\ell a \propto L^{1.36}$. Cribb (op. cit.) stated that the liquid rates used by McAdams et al. were below the MELR (minimum effective liquid rate) established by Pratt and thus the data were not reflective of the commercial operating range where a higher sensitivity to liquid rate exists.

Sulfur Dioxide-Air-Water Sulfur dioxide absorption by water represents a system in which both the liquid and the vapor resistances are significant. Whitney and Vivian [*Chem. Eng. Prog.*, **45**, 323 (1949)] evaluated the performance of 25-mm Raschig rings for the range $1.2 < L < 15.8$ kg/(s·m²) and $0.09 < G < 1.1$ kg/(s·m²). On the assumption that the gas rate did not affect the liquid-film transfer rate, the following relationship was developed:

$$\frac{1}{K_\ell a} = \frac{1}{0.44 L^{0.82}} + \frac{H'}{0.028 G^{0.7} L^{0.5}} \qquad (18\text{-}75)$$

$$H_{o\ell} = \frac{L^{0.18}}{27.4} + \frac{H' L^{0.5}}{1.75 G^{0.7}} \qquad (18\text{-}76)$$

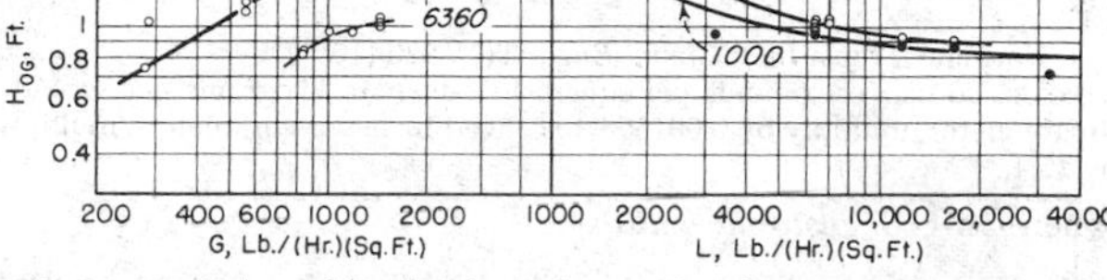

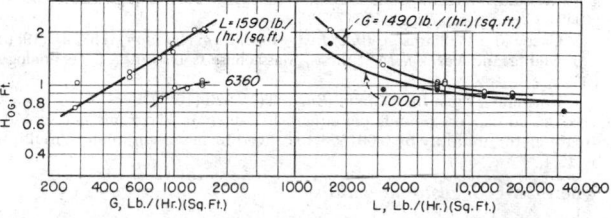

FIG. 18-69 Values of height of transfer unit for absorption of NH_3 in water from air. Fiberglas pads, fibers vertical, bulk density of 75.3 kg/m³ (4.7 lb/ft³). To convert pounds per hour–square foot to kilograms per second–square meter, multiply by 0.001356. [*Data of Williams, Akell, and Talbot*, Chem. Eng. Prog., *43, 585 (1947).*]

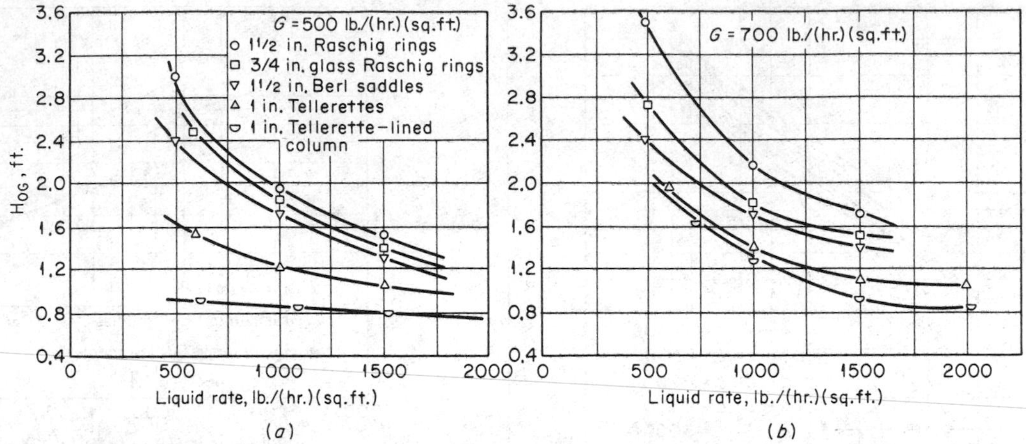

FIG. 18-70 Column-packing comparison, NH_3-air-H_2O system. To convert pounds per hour-square foot to kilograms per second-square meter, multiply by 0.001356; to convert feet to meters, multiply by 0.3048; and to convert inches to millimeters, multiply by 25.4. [*Teller*, Chem. Eng. Prog., *50, 70 (1954).*]

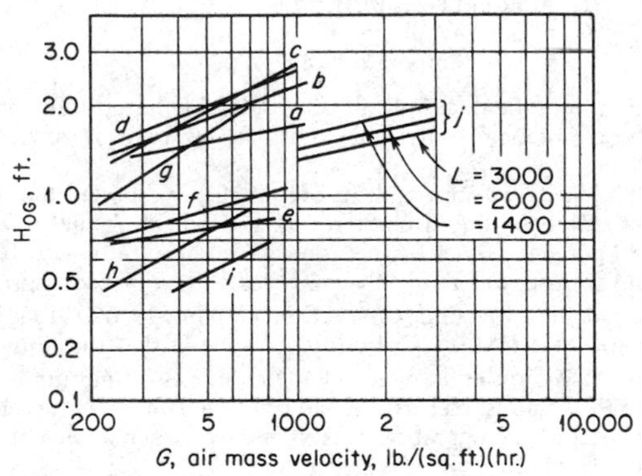

FIG. 18-71 Column-packing comparison, air-H_2O system.

Curve packing	Ref.	Liquid rate, (lb/(h·ft²)
a. 6295 grid	1	3000
b. 6146 grid	1	3000
c. 6897 grid	1	3000
d. Wood grid	2	3000
e. 1-in rings	1	3000
f. 1-in saddles	1	3000
g. 1-in rings	2	500
h. 1-in rings	3	1500
i. 1-in rings	3	3000
j. Carbon grid	4	3000

1. Molstad, McKinney, and Abbey, *Trans. Am. Inst. Chem. Eng.*, **39**, 605 (1943); *Chem. Eng. Prog.*, **46**, 17 (1950).
2. Carey and Williamson, *Proc. Inst. Mech. Eng. (London)*, **163**, 41 (1950).
3. Mehta and Parekh, M.S. thesis, Massachusetts Institute of Technology, 1939.
4. Norman, *Trans. Inst. Chem. Eng.*, **29**(2), 226 (1951).

NOTE: To convert pounds per square foot-hour to kilograms per second-square meter, multiply by 0.001356; to convert inches to millimeters, multiply by 25.4.

where, in *U.S. customary units,*

$K_\ell a$ = overall mass-transfer coefficient, (lb·mol)/[(h·ft³)(lb·mol)/(ft³)]
H' = modified Henry's-law constant = 0.107 at 21°C
L = liquid rate, lb/(h·ft²)
G = gas rate, lb/(h·ft²)

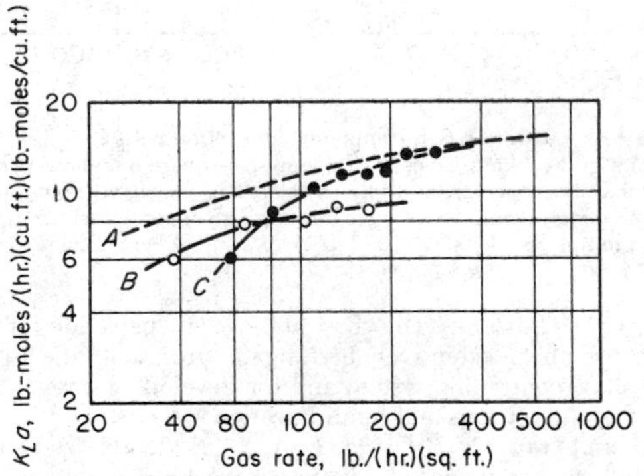

FIG. 18-72 Comparison of overall coefficients absorption of sulfur dioxide in water. (*A*) 1-in rings. [*Whitney and Vivian*, Chem. Eng. Prog., *45, 335 (1949).*] (*B*) 3-in spiral tile. [*Haslam, Ryan, and Weber*, Trans. Am. Inst. Chem. Eng., *15, 177 (1923).*] (*C*) 1-in coke. [*Haslam, Ryan, and Weber*, Trans. Am. Inst. Chem. Eng., *15, 177 (1923).*] To convert pounds per hour-square foot to kilograms per second-square meter, multiply by 0.001356; to convert pound-moles per [(hour-cubic foot) (pound-mole)/(cubic foot)] to kilogram-moles per [(second-cubic meter) (kilogram-mole)/(cubic meter)], divide by 3600.

The first term in the relationship represents the liquid-film behavior; the second term, the gas-film behavior.

The performance data obtained by the investigators are compared with those obtained by Haslam, Ryan, and Weber [*Trans. Am. Inst. Chem. Eng.*, **15**, 177 (1923)] for 25-mm (1-in) coke and 75-mm (3-in) spiral tile packings and are indicated in Fig. 18-72.

Pearson et al. [*Chem. Eng. Prog.*, **47**, 257 (1951)] evaluated the performance of 7.3 m (24 ft) of 25-mm (1-in) Raschig rings in a 300-mm- (12-in-) diameter tower for this system. The results were in good agreement with those obtained by Whitney and Vivian (op cit.), indicating that, with good distribution and accounting for end effects, laboratory data can be used for estimation of plant performance.

Carbon Dioxide-Air-Water System Performance of packed columns for this system has been reported by Sherwood and Holloway (op. cit.); Koch, Stutzman, Blum, and Hutchings [*Chem. Eng. Prog.*, **45**, 677 (1949)]; and Cooper, Christl, and Perry (op cit.). It was observed that H_ℓ was independent of gas rate for values of $U_\ell/U_g < 1$. At values of $U_\ell/U_g > 1$, backflow of the gas caused a decrease in effective rate of mass transfer. Sherwood and Holloway indicated that H_ℓ was proportional to $L^{0.22-0.46}$ as a function of the packing geometry and size.

Koch et al. (op. cit.) investigated the behavior of Raschig rings (9, 13, 19, and 25 mm, or ⅜, ½, ¾, and 1 in) in 150- and 250-mm- (6- and 10-in-) diameter columns packed to a height of 1.2 m (4 ft). In contradiction to the Sherwood and Holloway data, the authors proposed that H_ℓ was independent of packing size and that

$$k_\ell a = 0.25L^{0.96} \tag{18-77}$$

or

$$H_\ell \propto L^{0.04} \tag{18-77a}$$

where, in *U.S. customary units,*

$k_\ell a$ = mass-transfer coefficient, liquid phase (lb·mol)/[(h·ft³)(lb·mol) (ft³)]
L = liquid flow, (lb·mol)/(h·ft²)
H_ℓ = height of transfer unit, ft

A comparison of the data of these investigators with those of others is indicated in Fig. 18-73.

Distillation Applications Although formerly limited to column sizes of 1 m (3 ft) or less, packings of both the dumped and the arranged types are now being applied in much larger columns. A particular application is for vacuum distillations, for which packings can have theoretical plates–pressure-drop ratios that are more favorable than those of conventional plates. Eckert [*Chem. Eng. Prog.*, **59** (5), 76 (1963)] and Dolan and Strigle [*Chem. Eng. Prog.*, **76**(11), 78 (1980)] discuss the use of packings in columns with diameters up to 4 m (12 ft) or more.

The development of general methods for predicting packed-column distillation efficiency have been only moderately successful. Several correlations to give HETP (height of packing equivalent to one theoretical plate) have been proposed [Perry and Chilton (eds.), *Chemical Engineers' Handbook*, 5th ed., McGraw-Hill, New York, 1973, p. 18-49], but they have not been applied to larger column designs and are thought to be generally unreliable because the HETP concept has no fundamental mass-transfer basis. The transfer-unit concept has been more useful for generalized correlations, with the

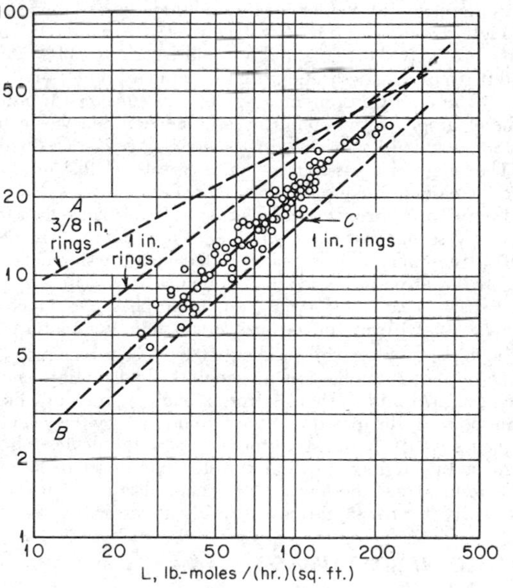

FIG. 18-73 $K_\ell a$ versus L for carbon dioxide absorption. (*A*) Sherwood and Holloway, *Trans. Am. Inst. Chem. Eng.*, **36**, 39 (1940). (*B*) Koch et al., *Chem. Eng. Prog.*, **45**, 681 (1949). (*C*) Draemel and Ruckman et al., *Chem. Eng. Prog.*, **45**, 677 (1949). To convert pound-moles per hour–square foot to kilogram-moles per second–square meter, multiply by 0.001356.

work of Cornell et al. [Eqs. (18-56) and (18-70)], Onda et al. [Eqs. (18-60) and (18-69)], and Bravo and Fair [Eqs. (18-71), (18-72), and (18-73)] providing a good basis for predicting the heights of transfer units for distillations using dumped packings. Comparisons of transfer units from these works with those measured in larger columns are provided by Bravo and Fair [*Ind. Eng. Chem. Process. Des. Dev.*, **21**(1), (1982)] and Bolles and Fair [*Ind. Eng. Chem. Symp. Ser.*, no. 56, 3.3/35 (1979)]. For each case, overall transfer units are obtained from

$$H_{og} = H_g + m(G_m/L_m)\,H_\ell \tag{18-78}$$

where H_{og} = height of an overall transfer unit based on gas-phase-concentration driving force, m
H_g = height of a gas-phase transfer unit, m
m = slope of equilibrium curve
G_m, L_m = gas and liquid rates, (kg·mol)/(s·m²)
H_ℓ = height of a liquid phase transfer unit, m

For a section of the packed column in which the operating and equilibrium lines can be considered straight, theoretical stages can be converted to numbers of transfer units as follows (see Sec. 14):

$$N_{og} = N_t \frac{\ln \lambda}{\lambda - 1} \tag{18-79}$$

where N_t = number of theoretical stages
$\lambda = m(G_m/L_m)$

Finally, the required packed height Z_P is obtained from

$$Z_P = (\text{HETP})\,(N_t) = (H_{og})(N_{og}) \tag{18-80}$$

LIQUID-DISPERSED CONTACTORS

Introduction Spray devices are the most common type of liquid-dispersed gas-liquid contactors. A second type is the baffle-plate or shower-deck column (Fig. 18-6). Spray columns are also discussed in the spray-drying discussion in Sec. 20.

Spray contactors find application for mass transfer when the gas is highly soluble or when the pressure drop must be very low. But their normal application is in heat-transfer service, and the following discussion will be in terms of thermal properties and measures of performance.

Heat-Transfer Applications Heat-transfer analogs of common mass-transfer terms are

$$\text{No. of gas-phase transfer units} = \frac{T_{g,\text{out}} - T_{g,\text{in}}}{(T_g - T_i)_{\text{mean}}} = N_g \tag{18-81}$$

$$\text{No. of liquid-phase transfer units} = \frac{T_{\ell,\text{out}} - T_{\ell,\text{in}}}{(T_\ell - T_i)_{\text{mean}}} = N_\ell \tag{18-82}$$

$$\text{Gas-phase volumetric transfer rate} = \frac{Gc_p}{H_g} = h_g a \tag{18-83}$$

$$\text{Diffusivity} = \frac{k}{\rho c_p} = \alpha_T \tag{18-84}$$

where T_g = gas temperature
T_ℓ = liquid temperature
T_i = interface temperature
H_g = height of a gas-phase transfer unit
G = weight gas flow/area
c_p = specific heat
ρ = density
k = thermal conductivity
α_T = thermal diffusivity

Note that the relative performance of a device can be converted from a mass-transfer basis to a heat-transfer basis by introducing these analogies together with the rate equations.

This conversion is simplest when $K_g a$ and $K_\ell a$ are defined in terms of mole-fraction driving force:

$$\frac{h_g a}{K_g a} \text{ or } \frac{h_\ell a}{K_\ell a} = \frac{c_p}{M}\left(\frac{\alpha_T}{D}\right)^{0.5} \tag{18-85}$$

where $h_\ell a$, $h_g a$ = volumetric heat-transfer rate
$K_\ell a$, $K_g a$ = volumetric mass-transfer rate
c_p = heat capacity
M = molecular weight
α_T/D = ratio of thermal diffusivity to molecular diffusivity, dimensionless

For gases, α_T/D is usually close to 1, since the same basic transfer mechanism exists. For liquids α_T/D is invariably much greater than 1. A simplified model yields the relation

$$\alpha_T/D = 1.9 \times 10^7(\mu/T) \qquad (18\text{-}86)$$

where μ is in Pa·s and T is in K. For a 1-cP (10^{-3}·Pa·s) liquid at 310 K, this gives a ratio of 61. This means that enormously high volumetric heat-transfer rates can be obtained for devices that are liquid-limited, as in a condenser of a pure vapor.

Theoretical Transfer Model Transfer from single droplets is theoretically well defined for the gas side. For a droplet moving counter to a gas, interfacial area is (in consistent units)

$$a = \frac{G_\ell}{\rho_\ell(U_\ell - U_g)}\frac{6}{D_d} \qquad (18\text{-}87)$$

where a = interfacial area/volume
G_ℓ = liquid-flow rate (weight/cross-sectional area)
ρ_ℓ = liquid density
U_ℓ = liquid velocity relative to the gas, often approximately the terminal velocity of droplets (see Sec. 5 for estimation)
U_g = superficial gas velocity
D_d = droplet diameter

The transfer coefficient is defined by (in consistent units)

$$h_g = (k/D_d)\,(2 + 0.6N_{Re}^{0.5}N_{Pr}^{0.33}) \qquad (18\text{-}88)$$

where h_g = transfer coefficient
k = thermal conductivity of gas
N_{Re} = Reynolds number = $D_d U_\ell \rho_g/\mu_g$
N_{Pr} = Prandtl number = $(c_p\mu/k)_g$

The volumetric coefficient $h_g a$ from the combination of Eqs. (18-87) and (18-88) is useful in defining the effect of variable changes but is limited in value because of its dependence on D_d. The product of area and coefficient obtained from a given mass of liquid is proportional to $1/D_d^2$ for small diameters. Similarly, the time to evaporate a small drop to dryness is proportional to $1/D_d^3$ The main problem is that droplet-size estimating procedures are often no better than ±50 percent for spray nozzles and are nonexistent for devices like shower-deck trays. A secondary problem is that there is no D_d that truly characterizes either the motion or transfer process for the whole spectrum of particle sizes present.

The corresponding theory for transfer in the liquid phase is even less certain. If one had static drops, the transfer would be

$$N_\ell = 0.5 + \frac{4(k/\rho c_p)\pi^2 t}{D_d^2} \text{ (dimensionless)} \qquad (18\text{-}89)$$

where N_ℓ = liquid-phase transfer units [Eq. (18-82)]
k = liquid thermal conductivity
ρ = liquid density
c_p = liquid specific heat
D_d = droplet diameter
t = time of contact

However, the static-drop assumption is rarely justified. For example, the high interfacial velocity in the spray from nozzles yields a high degree of internal mixing and much higher transfer.

A further weakness in the fundamental approach to estimating spray performance is deviation from countercurrent flow. Back mixing is hard to prevent, harder to define, and often limiting. The result of these limits is that the idealized countercurrent calculation of volumetric transfer rates rarely can sharply define performance.

For the case of the cocurrent operation such as line quenching, the liquid is assumed to accelerate quickly to gas velocity. As a result Eqs. (18-87) and (18-88) simplify to

$$a = \frac{G_\ell}{\rho_\ell}\frac{1}{U_g}\frac{6}{D_d} \qquad (18\text{-}90)$$

$$h_g = 2k/D_d \qquad (18\text{-}91)$$

Furthermore, since this type of operation usually involves little temperature rise in the liquid (because of evaporative cooling) liquid-phase resistance is seldom important; and with high-velocity cocurrent operation back mixing is not a problem. However, the wider variation of predictions shown for two fluid-atomization drop-size correlations (Table 18-14) again undercuts the utility of these equations. The one key point is that all correlations agree on the paramount importance of high velocity for obtaining fine atomization. This is reinforced by the common rule of thumb for this type of operation requiring a 60-m/s (200-ft/s) minimum relative injection velocity.

Venturi scrubbers are similar in that they need high velocity to achieve small droplets. They are primarily employed for mist and dust collection and are discussed further in the mist-collection portion of this section.

Empirical Approach: Sprays Back mixing in a single spray is so severe that many designers simply limit spray-chamber performance to a single equilibrium stage regardless of height. For a direct-contact heat-transfer device this means that the temperatures of the exiting gas and liquid would be equal.

The main cause of the high degree of back mixing is that there is no stabilizing pressure drop because of packing, plates, etc. Consequently the chief resistance to gas flow is the rain of drops. Anything less than perfect liquid distribution will induce a dodging action in the opposed vapor flow. The result is the development of large eddies and bypass streams. Other sources of back mixing are:

1. Large drops falling faster than small ones
2. Liquid striking the walls

At present about all that can be done is to take special care to obtain a uniform spray pattern with minimum collection at the walls and then to limit performance to one equilibrium stage. Liquid-flow rate and viscosity should be such that the spray nozzles are operating with their intended pattern.

For transfer-unit height of spray columns the best guide is the data of Pigford and Pyle [*Ind. Eng. Chem.*, **43**, 1949 (1951)]. These data show relatively short height of a liquid-transfer unit [0.5 to 1 m (1.5 to 3 ft)] in oxygen desorption from water in a 1.3-m- (4-ft-) tall spray zone. The high transfer rates result from enhanced transfer at the time of droplet formation. The same general behavior is indicated by Simpson and Lynn [*Am. Inst. Chem. Eng. J.*, **23**(5), 666 (1977)].

The height of a gas-transfer unit is greater [1 to 3 m (3 to 12 ft)] and increases as the gas flow rate increases.

Empirical Approach: Baffle Trays Performance data for baffles are very scant. Fair's [*Chem. Eng. Prog. Symp. Ser.*, **68**(118), 1 (1972)] summary of gas-phase-limited heat-transfer data indicates slightly lower performance than bubbling trays (1 to 2 N_g per tray). The sparse heat-transfer data on liquid-phase resistance indicate $h_\ell a/h_g a < 3{:}1$ even for the air-water system. For contrast, the ratio of bubbling trays typically is about 100:1. The explanation is the low level of internal droplet mixing generated by flow from one baffle to the next. (This differs sharply from a bubbling tray, in which significant power is dissipated into the liquid phase by the gas.) One consequence is that in mass-transfer operations this device is liquid-limited and the number of overall transfer units developed is much less. This results from the low ratio of molecular diffusivity to thermal diffusivity (for the liquid phase), as discussed earlier.

WETTED-WALL COLUMNS

Wetted-wall or falling-film columns have found application in mass-transfer problems when high-heat-transfer-rate requirements are concomitant with the absorption process. Large areas of open surface are available for heat transfer for a given rate of mass transfer in this

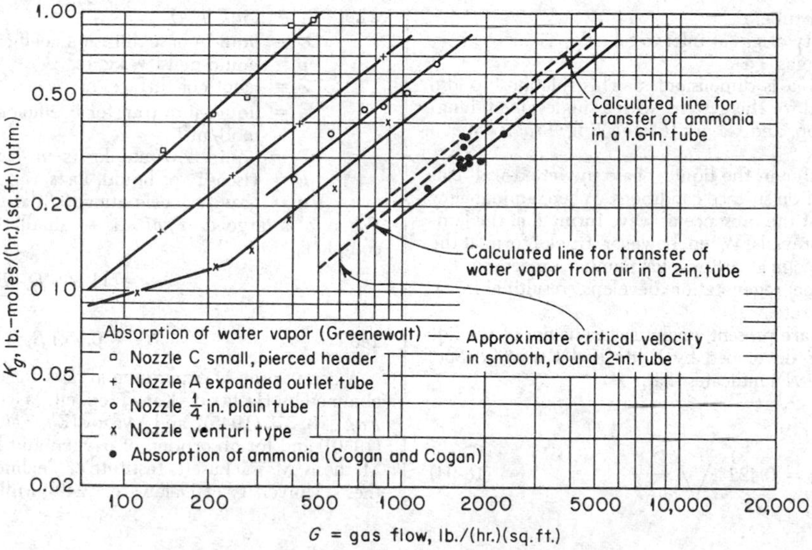

FIG. 18-74 Mass-transfer rates in wetted-wall columns having turbulence promoters. To convert pound-moles per hour–square foot–atmosphere to kilogram-moles per second–square meter–atmosphere, multiply by 0.00136; to convert pounds per hour–square foot to kilograms per second–square meter, multiply by 0.00136; and to convert inches to millimeters, multiply by 25.4. (*Data of Greenewalt and Cogan and Cogan, Sherwood, and Pigford,* Absorption and Extraction, *2d ed., McGraw-Hill, New York, 1952.*)

type of equipment because of the low mass-transfer rate inherent in wetted-wall equipment. In addition, this type of equipment lends itself to annular-type cooling devices.

Gilliland and Sherwood [*Ind. Eng. Chem.*, **26**, 516 (1934)] found that, for vaporization of pure liquids in air streams for streamline flow,

$$\frac{k_g D}{D_g}\frac{P_{BM}}{\pi} = 0.023 N_{Re}^{0.83} N_{Sc}^{0.44} \tag{18-92}$$

where D_g = diffusion coefficient
D = inside diameter of tube
k_g = mass-transfer coefficient, gas phase

Note that the group on the left side of Eq. (18-92) is dimensionless. When turbulence promoters are used at the inlet-gas section, an improvement in gas mass-transfer coefficient for absorption of water vapor by sulfuric acid was observed by Greenewalt [*Ind. Eng. Chem.*, **18**, 1291 (1926)]. A falling off of the rate of mass transfer below that indicated in Eq. (18-92) was observed by Cogan and Cogan (thesis, Massachusetts Institute of Technology, 1932) when a calming zone preceded the gas inlet in ammonia absorption (Fig. 18-74).

In work with the hydrogen chloride–air–water system, Dobratz, Moore, Barnard, and Meyer [*Chem. Eng. Prog.*, **49**, 611 (1953)] using a cocurrent-flow system found that $k_g \propto G^{1.8}$ (Fig. 18-75) instead of the 0.8 power as indicated by the Gilliland equation. Heat-transfer coefficients were also determined in this study. The radical increase in heat-transfer rate in the range of G = 30 kg/(s·m²) [20,000 lb/(h·ft²)] was similar to that observed by Tepe and Mueller [*Chem. Eng. Prog.*, **43**, 267 (1947)] in condensation inside tubes.

Gaylord and Miranda [*Chem. Eng. Prog.*, **53**, 139M (1957)] using a multitube cocurrent-flow falling-film hydrochloric acid absorber for hydrogen chloride absorption found

$$K_g = \frac{1.66(10^{-5})}{M_m^{1.75}}\left(\frac{DG}{\mu}\right) \tag{18-93}$$

where K_g = overall mass-transfer coefficient, (kg·mol)/(s·m²·atm)
M_m = mean molecular weight of gas stream at inlet to tube

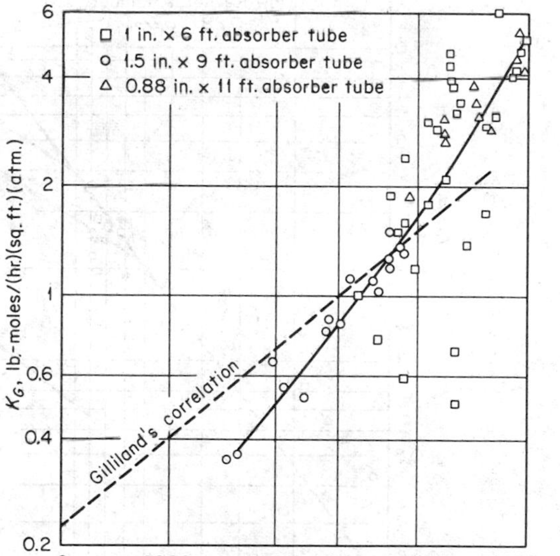

FIG. 18-75 Mass-transfer coefficients versus average gas velocity—HCl absorption, wetted-wall column. To convert pound-moles per hour–square foot–atmosphere to kilogram-moles per second–square meter–atmosphere, multiply by 0.00136; to convert pounds per hour–square foot to kilograms per second–square meter, multiply by 0.00136; to convert feet to meters, multiply by 0.305; and to convert inches to millimeters, multiply by 25.4. [*Dobratz et al.*, Chem Eng. Prog., *49, 611 (1953).*]

D = diameter of tube, m
G = mass velocity of gas at inlet to tube, kg/(s·m²)
μ = viscosity of gas, Pa·s

Note that the group DG/μ is dimensionless. This relationship also satisfied the data obtained for this system, with a single-tube falling-film unit, by Coull, Bishop, and Gaylor [*Chem. Eng. Prog.*, **45**, 506 (1949)].

The rate of mass transfer in the liquid phase in wetted-wall columns is highly dependent on surface conditions. When laminar-flow conditions prevail without the presence of wave formation, the laminar-penetration theory prevails. When, however, ripples form at the surface, and they may occur at a Reynolds number exceeding 4, a significant rate of surface regeneration develops, resulting in an increase in mass-transfer rate.

If no wave formations are present, analysis of behavior of the liquid-film mass transfer as developed by Hatta and Katori [*J. Soc. Chem. Ind.*, **37**, 280B (1934)] indicates that

$$k_\ell = 0.422\sqrt{\frac{D_\ell \Gamma}{\rho B_F^2}} \tag{18-94}$$

where B_F = $(3u\Gamma/\rho^2 g)^{1/3}$
D_ℓ = liquid-phase diffusion coefficient, m²/s
ρ = liquid density, kg/m³
Z = length of surface, m
k_ℓ = liquid-film-transfer coefficient, (kg·mol)/[(s·m²) (kg·mol)/m³]
Γ = liquid-flow rate, kg/(s·m) based on wetted perimeter
μ = viscosity of liquid, Pa·s
g = gravity acceleration, 9.81 m/s²

When Z is large or $\Gamma/\rho B_F$ is so small that liquid penetration is complete,

$$k_\ell = 11.800\, D_\ell/B_F \tag{18-95}$$

(18-96)

and $$H_\ell = 0.95\, \Gamma B_F/D_\ell$$

A comparison of experimental data for carbon dioxide absorption obtained by Hatta and Katori (op. cit.), Grimley [*Trans. Inst. Chem. Eng.*, **23**, 228 (1945)], and Vyazov [*Zh. Tekh. Fiz. U.S.S.R.*), **10**, 1519 (1940)] and for absorption of oxygen and hydrogen by Hodgson (S.M. thesis, Massachusetts Institute of Technology, 1949), Henley (B.S. thesis, University of Delaware, 1949), Miller (B.S. thesis, University

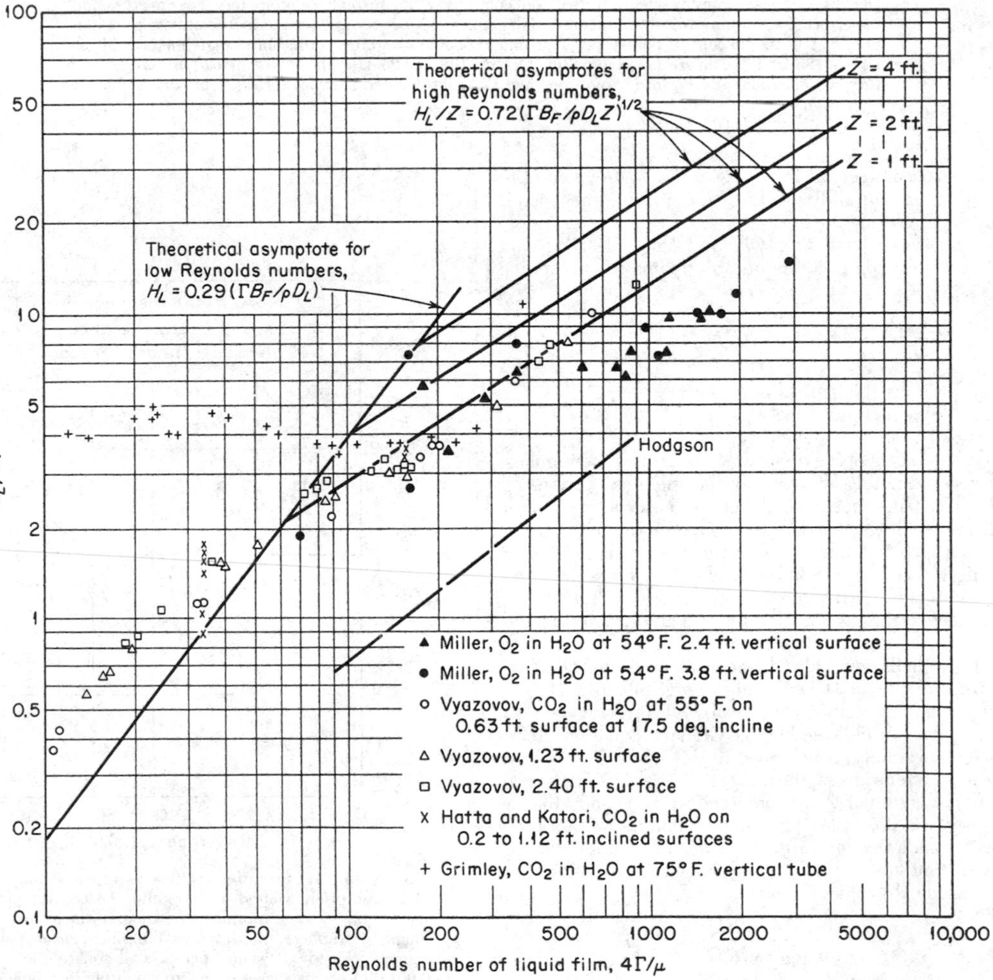

FIG. 18-76 Liquid-film resistance in absorption of gases in wetted-wall columns. Theoretical lines are calculated for oxygen absorption in water at 55°F. To convert feet to meters, multiply by 0.3048; C = ⅝ (°F − 32). (*Sherwood and Pigford,* Absorption and Extraction, *2d ed., McGraw-Hill, New York, 1952.*)

of Delaware, 1949), and Richards (B.S. thesis, University of Delaware, 1950) was made by Sherwood and Pigford (*Absorption and Extraction*, McGraw-Hill, New York, 1952) and is indicated in Fig. 18-76.

In general, the observed mass-transfer rates are greater than those predicted by theory and may be related to the development of surface rippling, a phenomenon which increases in intensity with increasing liquid path.

Vivian and Peaceman [*Am. Inst. Chem. Eng. J.*, **2**, 437 (1956)] investigated the characteristics of the CO_2-H_2O and Cl_2-HCl, H_2O system in a wetted-wall column and found that gas rate had no effect on the liquid-phase coefficient at Reynolds numbers below 2200. Beyond this rate, the effect of the resulting rippling was to increase significantly the liquid-phase transfer rate. The authors proposed a behavior relationship based on a dimensional analysis but suggested caution in its application concomitant with the use of this type of relationship. Cognizance was taken by the authors of the effects of column length, one to induce rippling and increase of rate of transfer, one to increase time of exposure which via the penetration theory decreases the average rate of mass transfer in the liquid phase. The equation is

$$\frac{k_\ell h}{D_\ell} = 0.433 \left(\frac{\mu_\ell}{\rho_\ell D_\ell}\right)^{1/2} \left(\frac{\rho_\ell^2 g h^3}{\mu_\ell^2}\right)^{1/6} \left(\frac{4\Gamma}{\mu_\ell}\right)^{0.4} \tag{18-97}$$

where D_ℓ = diffusion coefficient of solute in liquid
g = gravity-acceleration constant
h = length of wetted wall
k_ℓ = mass-transfer coefficient, liquid phase
Γ = mass rate of flow of liquid
μ_ℓ = viscosity of liquid
ρ_ℓ = density of liquid

The equation is dimensionless.

The effect of chemical reaction in reducing the effect of variation of the liquid rate on the rate of absorption in the laminar-flow regime was illustrated by the evaluation of the rate of absorption of chlorine in ferrous chloride solutions in a wetted-wall column by Gilliland, Baddour, and White [*Am. Inst. Chem. Eng. J.*, **4**, 323 (1958)].

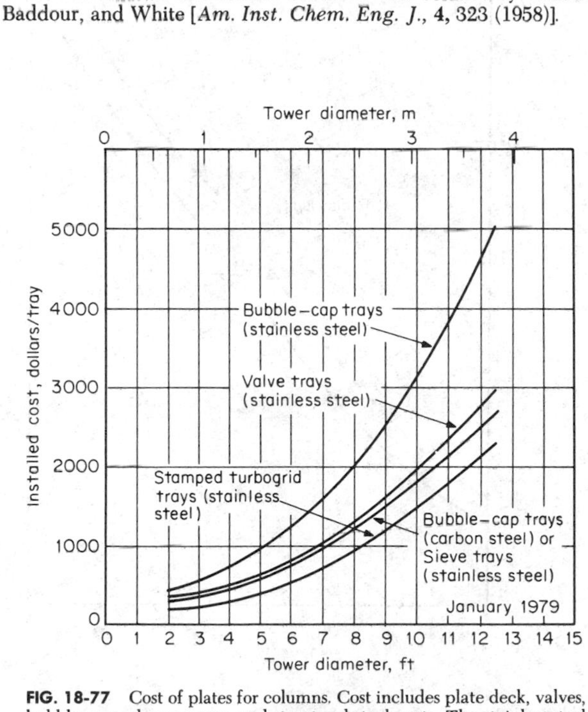

FIG. 18-77 Cost of plates for columns. Cost includes plate deck, valves, bubble caps, downcomers, and structural-steel parts. The stainless steel designated is type 410. (*Peters and Timmerhaus,* Plant Design and Economics for Chemical Engineers, *3d ed., McGraw-Hill, New York, 1980.*)

TABLE 18-12 Relative Fabricated Cost for Metals Used in Plate-Tower Construction*

Materials of construction	Relative cost per ft² of plate area (based on carbon steel = 1)
Sheet-metal plates	
Steel	1
4–6% chrome—½ molydenum-alloy steel	2
11–13% chrome-type 410 alloy steel	2.5
Red brass	3
Stainless-steel type 304	4
Stainless-steel type 347	4.8
Monel	7.0
Stainless-steel type 316	5.2
Inconel	8.2

*Peters and Timmerhause, *Plant Design and Economics for Chemical Engineers*, 3d ed., McGraw-Hill, New York, 1980. To convert cost per square foot to cost per square meter, multiply by 10.76.

Flooding in Wetted-Wall Columns When gas and liquid are in counterflow in wetted-wall columns, flooding can occur at high gas rates. Diehl and Koppany [*Chem. Eng. Prog.*, **65**, *Symp. Ser.* 42, 77 (1969)] correlated flooding data from a number of sources, including their own work, and developed the following expression:

$$U_f = F_1 F_2 \left(\frac{\sigma}{\rho_g}\right)^{1/2} \tag{18-98}$$

where U_f = flooding gas velocity, m/s
F_1 = 1.22 when $3.2\, d_i/\sigma > 1.0$
F_1 = $1.22\,(3.2\, d_i/\sigma)^{0.4}$ when $3.2\, d_i/\sigma < 1.0$
F_2 = $(G/L)^{0.25}$
G/L = gas-liquid mass ratio
d_i = inside diameter of column, mm
σ = surface tension, mN/m (dyn/cm)
ρ_g = gas density, kg/m³

The data covered column sizes up to 50-mm (2-in) diameter; the correlation should be used with caution for larger columns.

GAS-LIQUID-COLUMN ECONOMICS

Estimation of column costs for preliminary process evaluations requires consideration not only of the basic type of internals but also

TABLE 18-13 Costs of Tower Packings, Uninstalled, January 1979*

	Size, in			
	1	1½	2	3
Raschig rings				
Chemical porcelain	10.3	8.3	7.6	6.3
Carbon steel	23.7	15.5	13.3	10.9
Stainless steel	85.5	64.6	57.0	
Carbon	32.3	28.5	20.9	19.0
Intalox saddles				
Chemical stoneware	14.2	10.4	9.5	8.6
Chemical porcelain	15.2	11.4	10.4	9.5
Polypropylene	17.1	. . .	10.6	5.5
Berl saddles				
Chemical stoneware	22.0	17.0		
Chemical porcelain	27.0	20.0		
Pall rings				
Carbon steel	19.0	12.9	11.8	
Stainless steel	72.0	55.0	47.5	
Polypropylene	17.1	11.6	10.6	

*Peters and Timmerhaus, *Plant Design and Economics for Chemical Engineers*, 3d ed., McGraw-Hill, New York, 1980. Prices in dollars per cubic foot, 100-ft³ orders, FOB manufacturing plant. To convert cubic feet to cubic meters, multiply by 0.0283; to convert inches to millimeters, multiply by 25.4; and to convert dollars per cubic foot to dollars per cubic meter, multiply by 35.3.

of their effect on overall-system cost. For a distillation system, for example, the overall system can include the vessel (column); attendant structures, supports, and foundations; auxiliaries such as reboiler, condenser, feed heater, and control instruments; and connecting piping. The choice of internals influences all these costs, but other factors influence them as well. A complete optimization of the system requires a full-process-simulation model that can cover all pertinent variables influencing economics.

Cost of Internals Installed costs of plates (trays) may be estimated from Fig. 18-77, with corrections for plate material taken from Table 18-12. For two-pass plates the cost is 15 to 20 percent higher. Approximate costs of dumped packing materials may be obtained from Table 18-13, but it should be recognized that because of competition there can be significant variations in these costs from vendor to vendor. Also, packings sold in very large quantities carry discounts. Note that for Fig. 18-77 and Table 18-12 the effective cost date is 1979, with the Marshall and Swift cost index being taken as 560.

Packed-column internals include liquid distributors, packing support plates, redistributors (as needed), and holddown plates (to prevent movement of packing under flow conditions). Costs of these internals are given in Fig. 18-78, based on early 1976 prices and a Marshall and Swift cost index of 460.

Cost of Column The cost of the vessel, including heads, skirt, nozzles, and ladderways, is usually estimated on the basis of weight. Figure 18-79 provides 1979 cost data for the shell and heads, and Fig. 18-80 provides 1979 cost data for connections. For very approximate estimates of complete columns, including internals, Fig. 18-81 may be used. As for Figs. 18-79 and 18-80, the cost index is 560.

Plates versus Packings Bases for using packings instead of plates have been given (see subsection "Packed Columns"). In some circumstances either type of device may be chosen, e.g., for vacuum distillations. Fair [*Chem. Eng. Prog.*, **66**(3), 45 (1970)] has investigated the vacuum-distillation case and concludes that pall rings are superior among the packings but still suffer a capital-cost disadvantage compared with plates. Thus, packings must be justified on the basis of their operating-cost advantages, since they can provide lower bottoms temperatures and pressures; such justification is often possible. Billet (*Proc. Intn. Symp. Distill.*, Brighton, England, 1970) shows pall rings to be superior to a wide variety of devices when evaluated for the ethyl benzene–styrene system at 100-mmHg absolute pressure.

Optimization As stated previously, optimization studies should include the entire system. Such a study was made by Fair and Bolles [*Chem. Eng.*, **75**(9), 156 (1968)], using a light-hydrocarbon system and with the objective of defining optimum reflux ratio. Coolants used were at −87, −40, and +30°C (−125, −40, and +85°F), corresponding to different pressures of operation and associated different condensing temperatures. The results are shown in Fig. 18-82; the optimum reflux ratio is quite close to the calculated minimum reflux ratio.

Colburn (chemical engineering lecture notes, University of Delaware, 1943) proposed that the optimum reflux ratio is

$$R_{opt} = \frac{N + [(C_2/hG_b + C_3)/C_1] + (dN/dR)}{dN/dR} \tag{18-99}$$

where R = external reflux ratio = L/D
N = number of theoretical plates

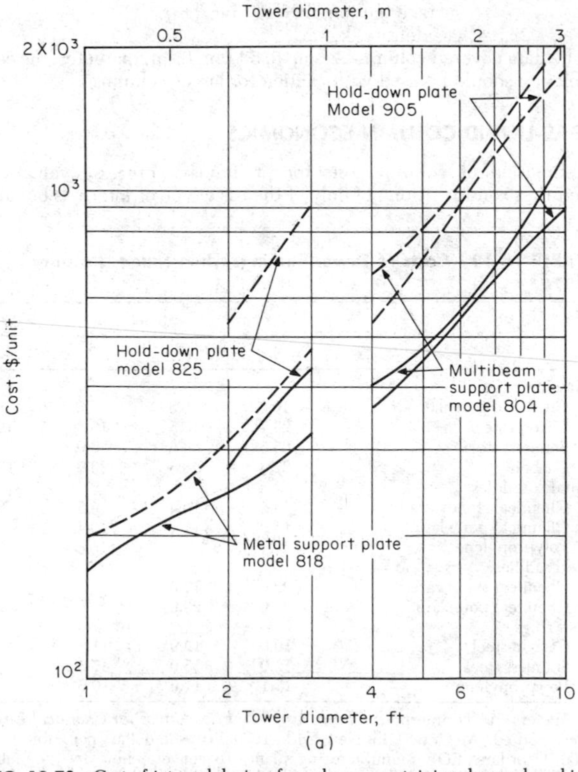

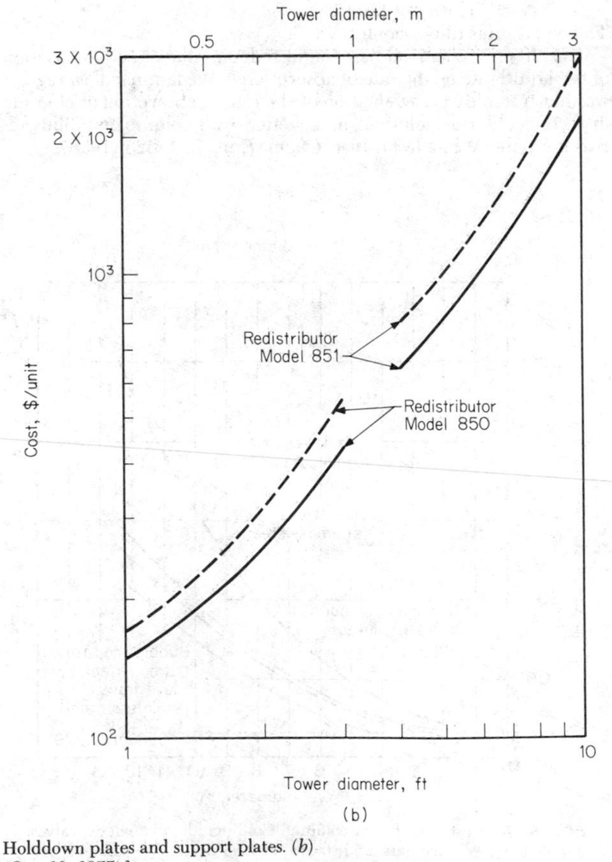

FIG. 18-78 Cost of internal devices for columns containing dumped packings. (*a*) Holddown plates and support plates. (*b*) Redistributors. (*c*) Liquid distributors. [*Pikulik and Diaz,* Chem. Eng., *84(21), 106 (Oct. 10, 1977).*]

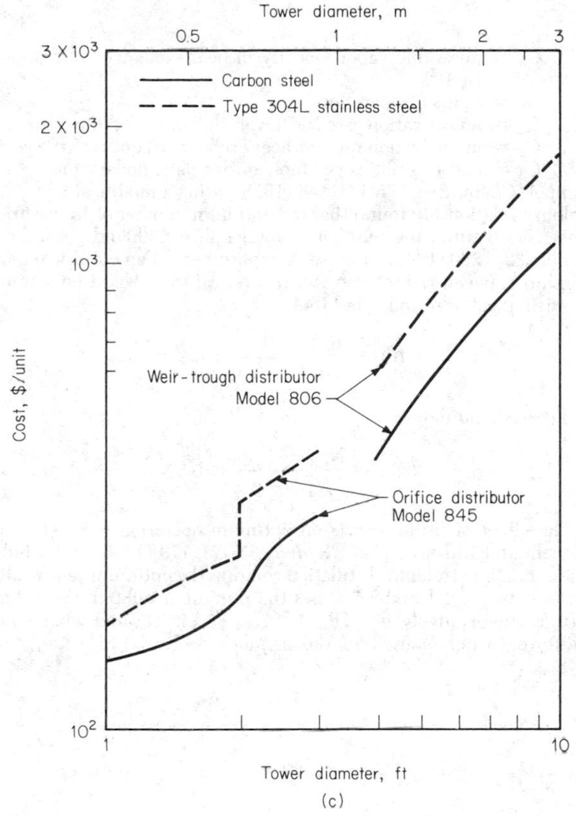

FIG. 18-78 *(Continued)*

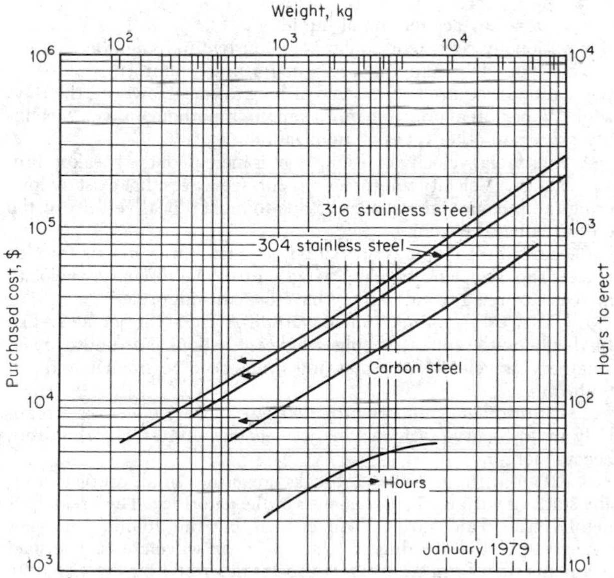

FIG. 18-79 Fabricated costs and installation time of towers. Costs are for shell with two heads and skirt but without trays, packing, or connections. (*Peters and Timmerhaus,* Plant Design and Economics for Chemical Engineers, *3d ed., McGraw-Hill, New York, 1980.*)

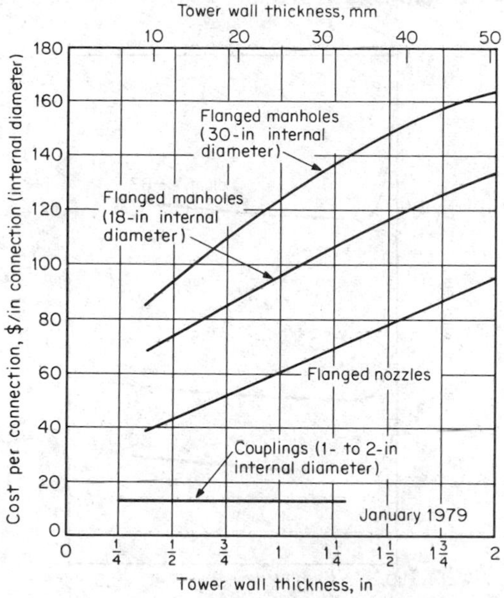

FIG. 18-80 Approxiamte installed cost of steel-tower connections. Values apply to 2070-kPa connections. Multiply costs by 0.9 for 1035-kPa (150-lb) connections and by 1.2 for 4140-kPa (600-lb) connections. To convert inches to millimeters, multiply by 25.4; to convert dollars per inch to dollars per centimeter, multiply by 0.394. (*Peters and Timmerhaus,* Plant Design and Economics for Chemical Engineers, *3d ed., New York, McGraw-Hill, 1980.*)

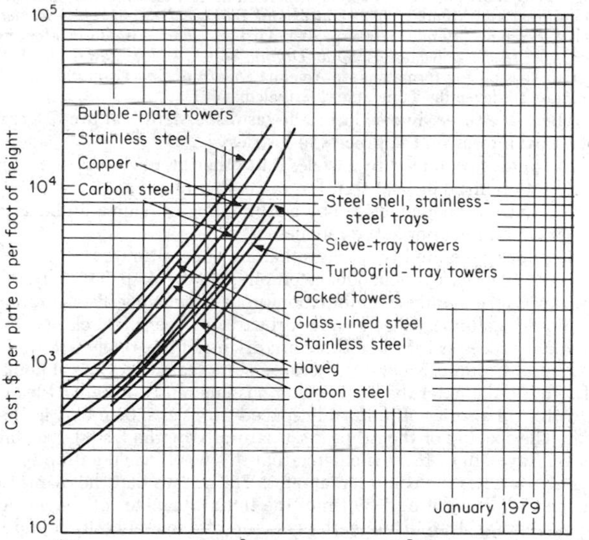

FIG. 18-81 Cost of towers including installation and auxiliaries. To convert inches to millimeters, multiply by 25.4; to convert feet to meters, multiply by 0.305; and to convert dollars per foot to dollars per meter, multiply by 3.28. (*Peters and Timmerhaus,* Plant Design and Economics for Chemical Engineers, *3d ed., McGraw-Hill, New York, 1980.*)

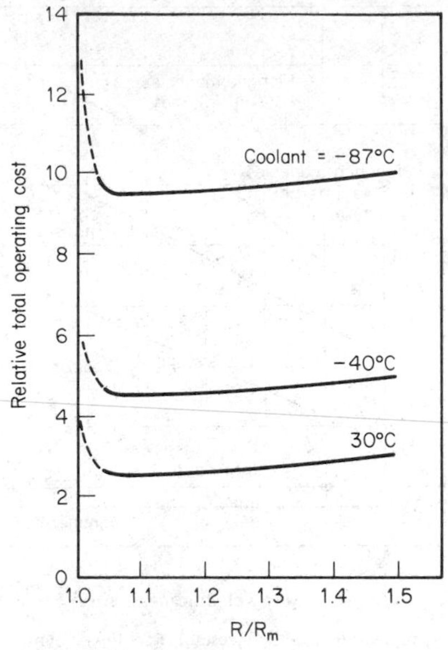

FIG. 18-82 Optimum-reflux-ratio determination. °F = 9/5 °C + 32. [*Fair and Bolles*, Chem. Eng., *75(4), 156 (1968).*]

G_b = allowable vapor velocity in heat exchangers, (lb·mol)/(h·ft²)
h = hours of operation
C_1 = amortization rate for tower, dollars/(ft²·plate·year)
C_2 = amortization rate for heat exchangers, dollars/(ft²·year)
C_3 = cost of utilities per mole of distillate, dollars/mol

Happel [*Chem. Eng.*, **65**(14), 144 (1958)] using a modification of the Colburn relationship found that the optimum number of trays varies from 2 to 3 times the number at total reflux. Gilliland [*Ind. Eng. Chem.*, **32**, 1220 (1940)] from the establishment of an empirical relationship between reflux ratio and theoretical trays based on a study of existing columns indicated that

$$0.1 < \frac{R_{opt} - R_{min}}{R_{min} + 1} < 0.3$$

and correspondingly

$$0.35 < \frac{N_{opt} - N_{min}}{N_{min} + 1} < 0.52$$

The effect of utilities costs on optimum operation was noted by Kiguchi and Ridgway [*Pet. Refiner*, **35**(12), 179 (1956)], who indicated that in petroleum-distillation columns the optimum reflux ratio varies between 1.1 and 1.5 times the minimum reflux ratio. When refrigeration is involved, $1.1R_{min} < R_{opt} < 1.2R_{min}$, and when cooling-tower water is used in the condensers, $1.2R_{mn;}in < R_{opt} < 1.4R_{min}$.

PHASE DISPERSION

LIQUID-IN-GAS DISPERSIONS

General References: For a general discussion of gas-liquid breakup processes see Brodkey, *The Phenomena of Fluid Motions*, Addison-Wesley, Reading, Mass., 1967. For a discussion of dispersion devices and how they work see Dombroski and Munday, *Biochemical and Biological Engineering Science*, vol. 2, Academic, London, 1968, p. 209; Marshall, *Chem. Eng. Prog. Monogr. Ser.*, **50**(2), 1954; and Masters, *Spray Drying*, 3d ed., Wiley, New York, 1979. For surveys on fog formation see Amelin, *Theory of Fog Formation*, Israel Program for Scientific Translations, Jerusalem, 1967.

There are three diverse liquid-in-gas dispersions that are of great practical interest in the process industries:

1. Sprays produced by nozzles and other atomizing systems for combustion, mass or heat transfer, or coating of surfaces
2. Entrainment generated by gas bubbling through a liquid as in a distillation tower or kettle reboiler
3. Fogs generated from gases that are supersaturated

The first two share a dependence on physical breakup. This physical breakup is frequently a combination process involving disintegration of liquid columns, sheets, and drops. However, for clarity the breakup processes for these three are discussed separately below.

Liquid-Column Breakup Because of increased pressure at points of reduced diameter the liquid column is inherently unstable. Ideally it will form a series of uniformly spaced drops as shown by Fig. 18-83*a*. The spacing of the drops is set naturally by the fastest-growing wave. Rayleigh in 1879 calculated that this would yield a drop spacing of 4.5 times the column diameter. This means that the diameter of the droplets will be 1.89 times the initial diameter of the jet. As shown by Fig. 18-83*b* and *c*, the breakup of a low-viscosity liquid is quite close to Rayleigh's prediction, although smaller satellite drops are also formed. (The natural frequency can be overriden by feeding external pulses to the column.)

For high-viscosity liquids the drops are larger. Weber in 1931 modified the relation for liquids of higher viscosity:

$$D = 1.89D_j\left[1 + \frac{3\,\mu_\ell}{(\sigma\rho_\ell D_j)^{1/2}}\right] \qquad (18\text{-}100)$$

where D = diameter of droplet
D_j = diameter of jet
μ_ℓ = viscosity of liquid
ρ_ℓ = density of liquid
σ = surface tension of liquid

(dimensionally consistent; any set of consistent units can be used)

This type of breakup is used when a nearly uniform-sized drop population is required, as in some prilling towers. However, the Rayleigh-Weber breakup mechanism's main importance is as a secondary process in other types of atomization.

As interfacial velocity of a simple jet is increased, the breakup process changes. Velocity reinforces any surface waves that exist by lowering the pressure on wave crests due to higher local velocity at the crests than at the trough.

Because of the complex breakup process no simple criteria for ultimate drop size and breakup length exist. As discharge velocity increases, three general regimes have been distinguished:

1. *Varicose (symmetrical oscillation).* Here the jet looks like Rayleigh's with symmetric bulges and contractions. The unbroken jet lengthens as velocity increases and drops become smaller and less uniform.
2. *Sinusoidal (transverse oscillation).* The jet weaves irregularly in an S-curve fashion. The jet becomes shorter and the drops become larger.
3. *Atomization.* The jet breaks down into small droplets, usually starting within 15 jet diameters of the jet orifice. The breakup is highly chaotic and not well understood, but the dominant process appears to be the shedding of ligaments from wave crests. The shed ligaments then form drops via a process described by Eq. (18-100). This stage corresponds to the normal condition employed when a simple orifice is used for atomization. Ohnesorge in 1936 gave the following criterion for the transition to atomization:

$$\frac{D_j U \rho_\ell}{\mu_\ell} > 2.8 \times 10^2 \left[\frac{\mu_\ell}{(\sigma\rho_\ell D_j)^{1/2}}\right]^{-0.82} \qquad (18\text{-}101)$$

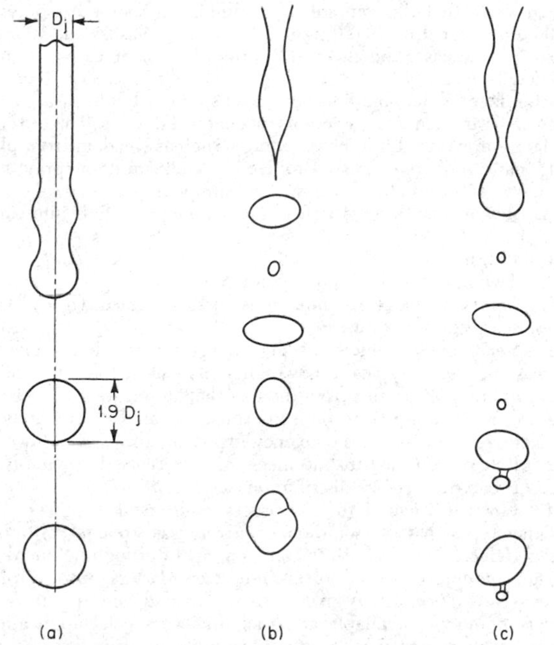

FIG. 18-83 (*a*) Idealized jet breakup suggesting uniform drop diameter and no satellites. (*b*) and (*c*) Actual breakup of a water jet as shown by high-speed photographs. [*From W. R. Marshall*, "Atomization and Spray Drying," *Chem. Eng. Prog. Monogr. Ser.*, no. 2 (1954).]

where U is the velocity of the jet relative to the surrounding gas. Any set of consistent units can be used.

The "plain" jet is seldom used in process applications because finer atomization can be obtained with other devices for equal energy input. It does have the advantages of yielding a very narrow spray angle (5 to 20°) and high penetration. For guidance on the effect of relative velocity and physical properties on droplet size the values given later under "Two-Fluid Atomizers" can be used.

Violent eruption of vapor within the jet and hence finer atomization can occur when a liquid is injected into a system below its vapor pressure. This is discussed by Brown and York [*Am. Inst. Chem. Eng. J.*, 8, 149 (1962)].

Liquid-Sheet Breakup The basic principle of most atomizers is to form a thin sheet which can break via a variety of mechanisms to form threads of liquid which in turn yield chains of droplets via the process described by Eq. (18-100). Because of the random nature of the thread formation a wide spectrum of droplet sizes results.

Unlike the liquid column, the sheet is basically stable. However, breakup can occur by four mechanisms.

1. *Rim disintegration.* The free edge of the sheet contracts into a cylinder, which then breaks from the surface as larger drops followed by thin liquid fingers. The position of the rim is fixed by the balance of surface-tension forces and the kinetic energy of the expanding sheet.

2. *Perforated-sheet disintegration.* Impingement of small drops or depressions due to turbulence puncture the sheet. The holes grow rapidly and their rims join to form threads, which then break (see Fig. 18-84*a*).

3. *Wave disintegration.* Reinforcement of surface waves by drag forces builds up until the sheet ruptures at the nodes. This ideally yields ribbons of liquid parallel to the leading edge. These in turn draw up into cylinders and pinch off droplets.

4. *Thick-sheet disintegration.* Here the crests of waves are shed as ligaments directly from the sheet. The process is closely analogous to atomization of jets as discussed previously and to the "stripping" type of droplet breakup discussed under "Droplet Breakup."

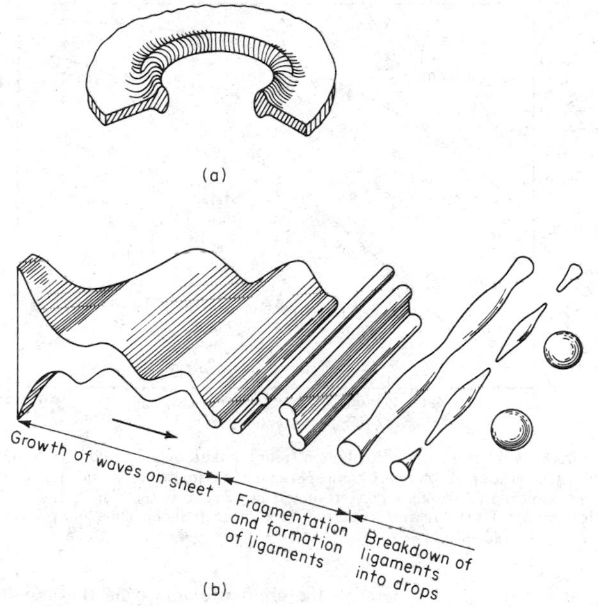

FIG. 18-84 Sheet breakup. (*a*) By perforation. [*After Fraser et al.*, Am. Inst. Chem. Eng. J., *8(5), 672 (1962).*] (*b*) By sinusoidal wave growth. [*After Dombrowski and Johns*, Chem. Eng. Sci., *18, 203 (1963).*]

Mayer's [*ARS J.*, **31** (12), 1783 (1963)] analysis of sheet breakup for high velocity (> 200 m/s) gave

$$D = 21.4\left[\frac{\mu_\ell(\sigma/\rho_\ell)^{1/2}}{\rho_g U_g^2}\right]^{2/3} \qquad (18\text{-}102)$$

(dimensionally consistent) where U_g = velocity of gas relative to the liquid.

Droplet Breakup Droplets themselves break into finer drops. Many studies of droplet breakup have centered on defining the conditions under which a drop will break (the largest stable drop size). The most important stability criterion is the ratio of aerodynamic forces to surface-tension forces defined by the Weber number (dimensionless):

$$N_{We} = U^2\rho_g D_{max}/\sigma g_c$$

However, the critical Weber number N_{Wecrit} also depends on contact time, liquid viscosity, and the relative-velocity history.

Commonly the N_{Wecrit} (for low-viscosity fluids) ranges from 10 to 20. Hinze [*Am. Inst. Chem. Eng. J.*, **1**, 289–295 (1953)] suggests that for a shock process the lower number applies and for a slow acceleration, e.g., a falling drop, the higher number can be used. Liquid viscosity as measured by a viscosity group also increases the critical Weber number. Brodkey (*The Phenomena of Fluid Motions*, Addison-Wesley, Reading, Mass., 1967) gives an empirical relation showing this effect

$$N_{Wecrit} = (N_{Wecrit})_{\mu\ell=0} + 14N_{Vi}^{1.6} \qquad (18\text{-}103)$$

where N_{Vi} = viscosity group = $\dfrac{\mu_\ell}{(\sigma\rho_\ell D_{max})^{1/2}}$

$(N_{Wecrit})_{\mu\ell=0}$ = 10 to 20 depending on contact time as discussed earlier.

Breakup normally yields a wide spectrum of droplet sizes and a great reduction in mean size. Figure 18-85 shows the droplet distribution from two different short-contact modes. The "bag" breakup mentioned in Fig. 18-85 occurs at relatively low differential velocity. Here the drop is first flattened and then blown up like a balloon which becomes unstable and bursts. The largest drops (one-fifth to

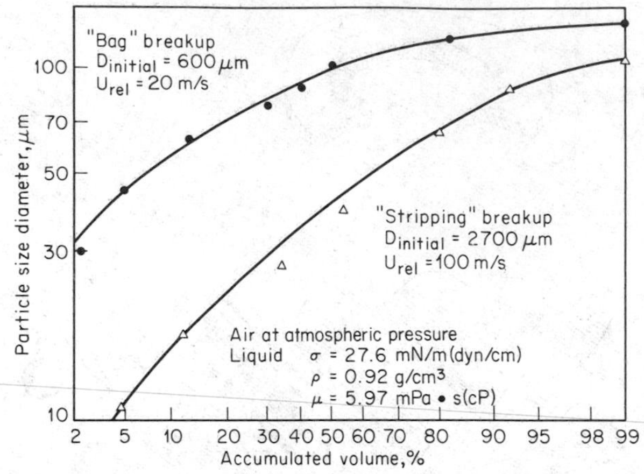

FIG. 18-85 Droplet-size distribution from breakup of drop undergoing sudden acceleration. To convert grams per cubic centimeter to pounds per cubic foot, multiply by 62.43; to convert meters per second to feet per second, multiply by 3.281. (*Wolfe and Andersen,* Proc. 5th Int. Shock Tube Symp., *April 1965, pp. 1145–1169.*)

one-tenth of the diameter of the original) come from the annular ring that forms the base of the balloon. "Stripping" breakup occurs at higher differential velocities. For stripping the process is essentially reversed, fine drops are stripped from the edges, and the drop does not inflate. Wolfe and Andersen (*Proc. 5th Int. Shock Tube Symp.*, April 1965, pp. 1145–1169) have developed a theoretical equation for this stripping mode. The equation correlates their data for the mass median diameter for both stripping and bag modes of breakup:

$$D_{vm} = \left(\frac{136 \mu_\ell \sigma^{3/2} D_I^{1/2}}{\rho_g^2 \rho_\ell^{1/2} U^4} \right)^{1/3} \qquad (18\text{-}104)$$

where D_{vm} = volume median droplet diameter
D_I = initial diameter of droplet
U = relative velocity

The equation is dimensionally consistent.

Simple equations like these offer a useful guide but fall short of complete description. The actual breakup process is complex, and the interrelation of variables such as contact time and turbulence also enters.

For example, at zero relative velocity, breakup can still occur in regions of high turbulence. Here the breakup is set by local variations in pressure causing droplet distortion. Hinze [*Am. Inst. Chem. Eng. J.*, **1**, 289–295 (1953)] applied turbulence theory to obtain Eq. (18-105):

$$D_{max} \alpha \left(\frac{\sigma}{\rho_g} \right)^{3/5} \epsilon^{-2/5} \qquad (18\text{-}105)$$

where ϵ = power dissipated/mass.

A further type of breakup can result from impingement on equipment walls as in pipe bends or compressor blades. Wachters and Westerling [*Chem. Eng. Sci.*, **21**, 1047 (1966)] studied 2000-µm droplets impinging on heated walls. They found two breakup modes for these nonwetted surfaces. For Weber numbers (defined with ρ_ℓ) of 30 to 80 the drops spread out on impact, then pulled back together because of surface tension, and rebounded in a columnar fashion. The column then followed a Rayleigh-type breakup and yielded two to three droplets. Above a N_{We} of roughly 80 the drops shatter on impact. Harvey (Ph.D. thesis, McMaster University, 1967) observed similar phenomena but at lower Weber numbers for 500-µm drops. Virtually no quantitative guidelines exist for impingement on wetted surfaces. In general, there is less tendency to shatter on wetted surfaces, although Hobbs and Kezweeny [*Science*, **155**, 112 (1967)] demonstrate that this can still be a significant source of spray. (A 3000-µm water droplet falling 46 cm (18 in) splashes out 10 spray particles; increased height yields proportional increases in spray particles.)

Atomizers The common need to disperse a liquid in a gas has spawned a tremendous variety of mechanical devices. The different designs emphasize different advantages such as freedom from plugging, pattern of spray, small droplet size, uniformity of spray, high turndown ratio, and/or low power consumption.

As shown in Table 18-13*a*, most atomizers fall into three categories:

1. Pressure nozzles ("hydraulic")
2. Two-fluid nozzles ("pneumatic")
3. Rotary devices ("spinning cups, disks, or vaned wheels")

These share certain features:

1. Relatively low efficiency. The energy required to produce the increase in area is typically less than 1 percent of the total energy consumption. This results from the fact that atomization is a secondary process resulting from high interfacial shear. Unfortunately, as droplet sizes decrease, this efficiency invariably drops even lower.
2. Reliance on the breakup mechanisms discussed previously.
3. Broad droplet-size distribution (see Fig. 18-87).
4. Low cost (relative to most process equipment).

Other types that use sonic energy (from gas streams), ultrasonic energy (electronic), and electrostatic energy offer high potential but are not commonly used in process industries. Masters (*Spray Drying*, Wiley, New York, 1979) gives a more detailed description of the variety of nozzles available and their diverse special applications. Special requirements such as size uniformity in prilling towers can dictate other approaches to dispersion. Here plates or spinning cylinders are drilled with many holes to develop nearly uniform columns which yield drops by the mechanism described by Eq. (18-100).

Usually the most important feature of a nozzle is the size of droplet it produces. As shown earlier, the heat or mass transfer that a given dispersion can produce is often proportional to $(1/D)^2$. As a result fine drops are favored. On the other extreme, drops that are too fine will not settle, and a common concern is the amount of liquid that will be entrained from a given spray operation. For example, if sprays are used to contact atmospheric air flowing at 1.5 m/s, drops smaller than 350 µm [terminal velocity = 1.5 m/s (4.92 ft/s)] will be entrained. Even for the relative coarse spray of the hollow-cone nozzle shown in Fig. 18-86, 7.5 percent of the total liquid mass will be entrained.

Pressure Nozzles Manufacturers' data such as shown by Fig. 18-87 are available for most nozzles for the air-water system but rarely for other systems. The dependence of drop size on geometry, velocity, and properties changes as the nature of the breakup process changes. This explains the wide divergence of findings on the importance of surface tension, viscosity, and densities. All the findings are presumably valid for the test conditions but fail on extrapolation to other systems and geometries. In Fig. 18-87 note the much coarser solid-cone spray; this causes slower mass and heat transfer.

Effect of Physical Properties Because of the extreme variety of available geometries, no attempt to encompass this variable is made here. The suggested predictive route starts with air-water data from the manufacturer at the desired flow rate. This drop size is then corrected by Eq. (18-106) for different physical properties:

$$\frac{D_{vm,\ system}}{D_{vm,\ water}} = \left(\frac{\sigma_{system}}{73} \right)^{0.5} \left(\frac{\mu_\ell}{1.0} \right)^{0.2} \left(\frac{1.0}{\rho_\ell} \right)^{0.3} \qquad (18\text{-}106)$$

where D_{vm} = volume median droplet diameter
σ = surface tension, mN/m (dyn/cm)
μ_ℓ = liquid viscosity, mPa·s (cP)
ρ_ℓ = liquid density, g/cm^3

The exponential dependences in Eq. (18-106) simply represent averages of values reported by a number of studies. This equation is intended only as a rough guide, since simple exponential relation-

TABLE 18-13a Atomizer Summary

Types of atomizer	Design features	Advantages	Disadvantages
Pressure.	Flow $\alpha(\Delta P/\rho_\ell)^{1/2}$. Only source of energy is from fluid being atomized.	Simplicity and low cost.	Limited tolerance for solids; uncertain spray with high-viscosity liquids; susceptible to erosion. Need for special designs (e.g., bypass, spill, or dual-orifice) to achieve turndown.
1. Hollow cone.	Liquid leaves as conical sheet as a result of centrifugal motion of liquid. Air core extends into nozzle.	High atomization efficiency.	Concentrated spray pattern at cone boundaries.
a. Whirl chamber (see Fig. 18-86*a*).	Centrifugal motion developed by tangential inlet in chamber upstream of orifice.	Minimum opportunity for plugging.	
b. Grooved core.	Centrifugal motion developed by inserts in chamber.	Smaller spray angle than 1*a* and ability to handle flows smaller than 1*a*.	
2. Solid cone (see Fig. 18.86*b*).	Similar to hollow cone but with insert to provide even distribution.	More uniform spatial pattern than hollow cone.	Coarser drops for comparable flows and pressure drops. Failure to yield same pattern with different fluids.
3. Fan (flat) spray.	Liquid leaves as a flat sheet or flattened ellipse.	Flat pattern is useful for coating surfaces and for injection into streams.	
a. Oval or rectangular orifice (see Fig. 18-86*c*). Numerous variants on cavity and groove exist.	Combination of cavity and orifice produces two streams that impinge within the nozzle.	. . . .	Small clearances.
b. Deflector (see Fig. 18-86*d*).	Liquid from plain circular orifice impinges on curved deflector.	Minimal plugging.	Coarser drops.
c. Impinging jets (see Fig. 18-86*e*).	Two jets collide outside nozzle and produce a sheet perpendicular to their plane.	Different liquids are isolated until they mix outside of orifice. Can produce a flat circular sheet when jets impinge at 180°.	Extreme care needed to align jets.
4. Nozzles with wider range of turndown.			
a. Spill (bypass) (see Fig. 18-86*f*).	A portion of the liquid is recirculated after going through the swirl chamber.	Achieves uniform hollow cone atomization pattern with very high turndown (50:1).	Waste of energy in bypass stream. Added piping for spill flow.
b. Poppet (see Fig. 18-86*g*).	Conical sheet is developed by flow between orifice and poppet. Increased pressure causes poppet to move out and increase flow area.	Simplest control over broad range.	Difficult to maintain proper clearances.
c. Dual-orifice.	Two concentric orifices, each with its own liquid-supply system. The conical sheets impinge so that the high-velocity stream provides atomization energy.	Uniform spray angle throughout range.	Small flow passages. Nonuniformity of droplet sizes.
Two-fluid (see Fig. 18-86*h*).	Gas impinges coaxially and supplies energy for breakup.	High velocities can be achieved at lower pressures because the gas is the high-velocity stream. Liquid-flow passages can be large, and hence plugging can be minimized.	Because gas is also accelerated, efficiency is inherently lower than pressure nozzles.
Sonic.	Gas generates an intense sound field into which liquid is directed.	Similar to two-fluid but with greater tolerance for solids.	Similar to two-fluid.
Rotary wheels (see Fig. 86*i*), disks, and cups.	Liquid is fed to a rotating surface and spreads in a uniform film. Flat disks, disks with vanes, and bowl-shaped cups are used. Liquid is thrown out at 90° to the axis.	The velocity that determines drop size is independent of flow. Hence these can handle a wide range of rates. They can also tolerate very viscous materials as well as slurries. Can achieve very high capacity in a single unit; does not require a high-pressure pump.	Mechanical complexity of rotating equipment. Radial discharge.

ships such as (18-106) do not define the real world. For example, Ford and Furmidge [*Br. J. Appl. Phys.*, **18**, 335 (1967)] show that in the fan nozzle there are four regions in which viscosity exerts different effects. Similarly, while all correlations show that drop size increases with surface tension, its effect is not simple. It resists deformation of a sheet but assists drop formation after the sheet has been broken. Increasing vapor density can both increase and decrease drop size depending on the regime of operation. Because of this uncertainty gas density was excluded from Eq. (18-106).

In practice viscosity exerts the greatest impact of the physical properties [despite the low exponent in Eq. (18-106)] because it can vary over a broader range. For example, it is not uncommon to find an oil with 1000 times the viscosity of water while most liquids fall within a factor of 2 of water density and a factor of 3 of its surface tension.

At some low flow, pressure nozzles do not develop their normal pattern but tend to approach solid streams. The required flow to achieve the normal pattern increases with viscosity.

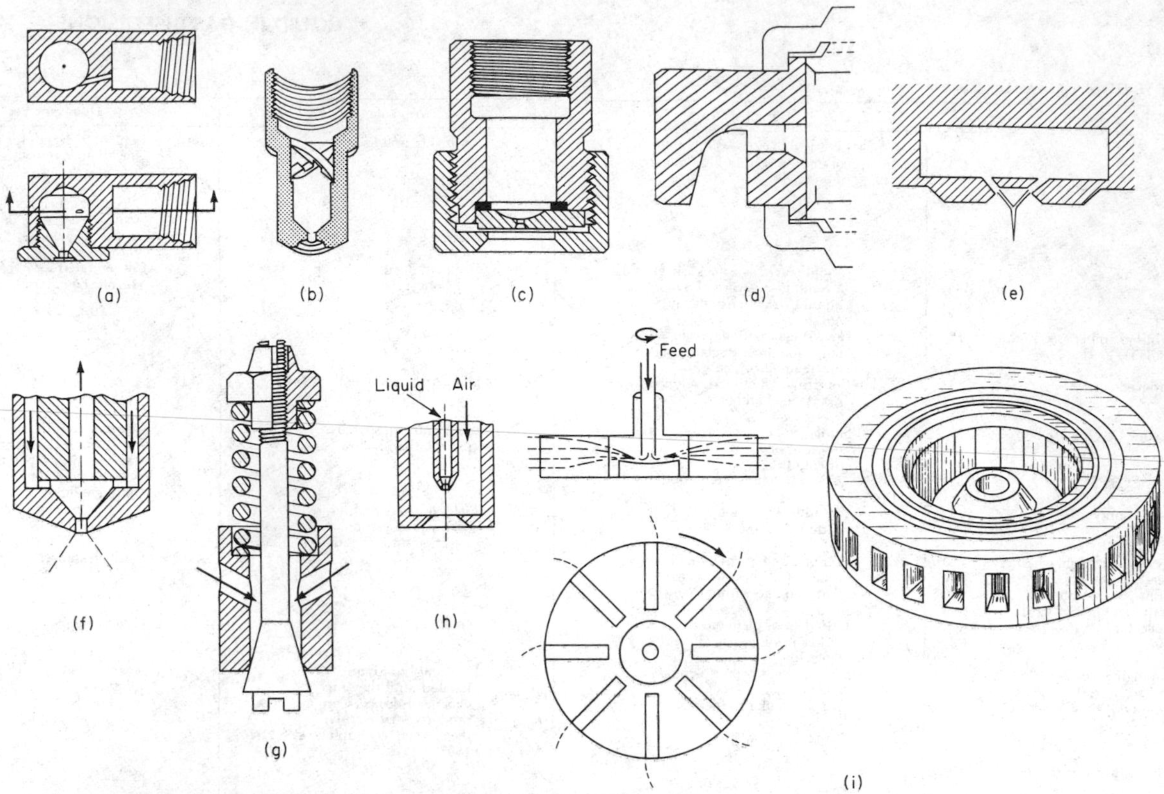

FIG. 18-86 Charactersitic spray nozzles. (*a*) Whirl-chamber hollow cone. (*b*) Solid cone. (*c*) Oval-orifice fan. (*d*) Deflector jet. (*e*) Impinging jet. (*f*) Bypass. (*g*) Poppet. (*h*) Two-fluid. (*i*) Vaned rotating wheel. (*a*), (*b*), (*e*), (*f*), (*h*), and (*i*) From Dombrowski and Munday, *Biochem. Biol. Eng. Sci.*, 1968. (*d*) and (*g*) Delavan Corporation. (*c*) Shutte & Koerting Division, Ametek, Inc.

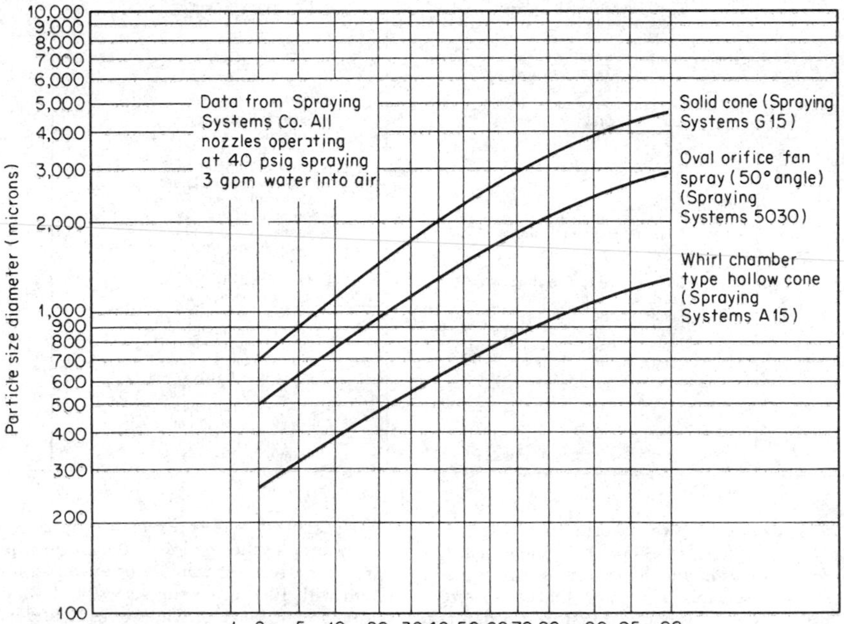

FIG. 18-87 Droplet-size distribution for three different types of nozzles. To convert pounds per square inch gauge to kilopascals, multiply by 6.89; to convert gallons per minute to cubic meters per hour, multiply by 0.227. *(Spraying Systems Inc.)*

Effect of Flow Rate, Pressure Drop, and Orifice Size For a nozzle with a developed pattern the volume median drop size can be estimated to fall with rising ΔP (pressure drop) by Eq. (18-107):

$$\frac{D_{vm}, \Delta P_1}{D_{vm}, \Delta P_2} = \left(\frac{\Delta P_2}{\Delta P_1}\right)^{1/3} \tag{18-107}$$

For similar nozzles and constant ΔP, the drop size will increase with nozzle size as indicated by Eq. (18-108):

$$\frac{D_{vm1}}{D_{vm2}} = \left(\frac{\text{orifice diameter}_1}{\text{orifice diameter}_2}\right)^{2/3} \tag{18-108}$$

The effect of ΔP and orifice size can also be recast in terms of volumetric flow and velocity. For similar nozzles

$$\frac{D_{vm1}}{D_{vm2}} = \left(\frac{\text{volumetric flow}_1}{\text{volumetric flow}_2}\right)^{1/3} \left(\frac{\text{velocity}_2}{\text{velocity}_1}\right) \tag{18-109}$$

These relationships are supported by data from Dietrich (*Proc. 1st Intn. Conf. Liq. Atomization Spray Syst.*, Tokyo, 1978) and Dombrowski and Wolfson [*Trans. Inst. Chem. Eng.*, **50**, 259 (1972)].

Effect of Spray Angle A shift to a smaller-angle nozzle gives slightly larger drops for a given type of nozzle because of the reduced tendency of the sheet to thin. Dietrich's data for fan sprays show the following:

Angle	25°	50°	65°	80°	95°
D_{vm}, μm	1459	1226	988	808	771

The spray angle itself can be important. The quoted nozzle angles can be expected to be sharply reduced as liquid viscosity goes up and to a lesser extent as gas density increases. Further, entrainment of gas causes the spray angle to close in as distance away from the nozzle increases.

Rotary Atomizers For rotating wheels, vaneless disks, and cups there are three regimes of operation. At low rates the liquid is shed directly as drops from the rim. At intermediate rates the liquid leaves the rim as threads of liquid, and at the highest rate (normal operating condition) the liquid extends from the edge as a thin sheet which breaks down similarly to a fan or hollow-cone spray nozzle.

For the transition to sheet formation Dombrowski [*Biochem. Biol. Eng. Sci.*, **2**, 209 (1968)] gives

$$\frac{(\sin \alpha)^{1/3} \rho_\ell^{0.81} n^{0.67} Q_V^{1.14} \mu_\ell^{0.19}}{\sigma D_{\text{cup}}^{0.81}} < 0.286 \tag{18-110}$$

(dimensionless) where Q_V is the liquid volumetric flow rate, n is the rotational speed, α is the angle of the lip to the axis, and D_{cup} is the cup diameter. The effect of physical properties on drop sizes formed from these sheets is not clear. They appear less sensitive to surface tension and viscosity than indicated by Eq. (18-106).

For the vaned-wheel atomizer, Masters and Moeller (*Proc. 1st Intn. Conf. Liq. Atomization Spray Syst.*, Tokyo, August 1978) suggest

$$D_{vm} \propto \frac{Q_v^a}{n^{0.8} \, (\text{wheel diameter})^{0.6}}$$

where the exponent a decreases from 0.24 to 0.12 as tip speed increases to 150 m/s (490 ft/s). Vaneless disks or cups give a coarser spray than wheels for a given set of conditions.

Two-Fluid Atomizers This general category includes such diverse applications as venturi atomizers and reactor-effluent quench systems in addition to two-fluid spray nozzles. Depending on the manner in which the two fluids meet, several of the breakup mechanisms discussed earlier may be applicable. The breakup process is a combination of ligament stripping, ballooning of fragments, and to some extent high-level turbulent rupture. The unifying feature of this category is dependence on high relative gas velocity to supply atomizing energy. Conceptually this type of breakup is close to that of the liquid jet.

In its simplest form, gas impinges on a solid jet of liquid. A number of experimental studies such as Weiss and Worsham [*ARS J.*, **29**, 252 (1959)] and Ingebo (NASA Tech. Note D 4640, July 1968) have been made of this system and have been supplemented by theoretical models of Mayer [*ARS J.*, **31**, 12, 1783 (1963)] and Adelberg [*AJAA J.*, **6**(6) (1968)]. These are summarized in Table 18-14 for the case of very high ratios of gas to liquid. Investigators agree on assigning a greater importance to relative velocity than suggested for other atomizers. Also differing from other atomizers, two-fluid atomizers generally find a strong decrease in drop size with increasing gas density. In addition, several of the drop-size correlations show a dependence on the ratio of gas to liquid, and most are much more complicated than simple exponential dependence. For example, the often-cited study by Nukiyama and Tanasawa [Defence Res. Board, Dep. Nat. Defence, Ottawa (Canada), Rep. 4, Mar. 18, 1950, transl. from *Trans. Soc. Mech. Eng. (Japan)*, **5**, 18, 68–75 (February 1939)] (made on small nozzles with a maximum liquid orifice of 2 mm) gives Sauter mean diameter:

$$D_{32} = \frac{585}{U}\left(\frac{\sigma}{\rho_\ell}\right)^{0.50} + 1683\left(\frac{\mu_\ell}{\sqrt{\sigma\rho_\ell}}\right)^{0.45}\left(1000\,\frac{L}{G}\frac{\rho_g}{\rho_\ell}\right)^{1.5} \tag{18-111}$$

where D_{32} = Sauter mean diameter, μm
σ = surface tension, mN/m (dyn/cm)
ρ_ℓ, ρ_g = liquid and gas density, g/cm^3
μ_ℓ = liquid viscosity, Pa·s
U = relative velocity, m/s
L/G = liquid/gas mass ratio, dimensionless

Jasuja [*Trans. Am. Soc. Mech. Eng.*, **103**, 514 (1981)] gives a relationship for a prefilming atomizer

$$D_{32} = 0.17\left[\left(\frac{\sigma}{\rho_g U^2}\right)^{0.45} D^{0.55}\left(1 + \frac{L}{G}\right)^{0.5} + 0.1\left(\frac{\mu_\ell^2}{\rho_\ell\sigma}\right)^{0.375} D^{0.625}\left(1 + \frac{L}{G}\right)^{0.8}\right] \tag{18-112}$$

TABLE 18-14 Dependence of Drop Size on Different Parameters in Two-Fluid Atomization as Predicted by Various Studies—Effective Exponents of Simplified Equations

	Relative velocity	Surface tension	Liquid viscosity	Gas density
Weiss and Worsham (jet in large duct)	−1.33	0.41	0.33	−0.7 to −1.0
Ingebo (jet in cocurrent flow—empirical for nonaccelerating case)	−1.6	0.66	−0.1	−0.3
Mayer (theoretical for sheets)	−1.33	0.33	0.67	−0.67
Adelberg (theoretical for jets)	−1.33 to −0.67	0.17 to 0.33	0.33 to 0.67	−0.67 to −0.33
Nukiyama and Tanasawa (empirical for small nozzle)	−1	0.5	0	0
Jasuja (empirical for small nozzle)	−0.9	0.45	0	−0.45
El-Shanawany and Lefebvre (empirical for small nozzle)	−1.2	0.6	0	−0.7
Rizkalla and Lefebvre (empirical for small nozzle)	−1.0	0.5	0	−1.0
Kim and Marshall (empirical for small nozzle)	−1.14	0.41	0.32	−0.57
Tatterson, Dallman, and Hanratty (pipe flow)	−0.9	0.5	0	−0.4
Hinze (turbulence theory)	−1.2	0.6	0	−0.6

The additional term D is the diameter, μm, of the prefilmer. El-Shanawany and Lefebvre [*J. Energy*, **4**, 184 (1980)], Rizkally and Lefebvre [*Trans. ASME J. Fluids*, **316** (1975)], and Kim and Marshall [*Am. Inst. Chem. Eng. J.*, **17**(3), 575 (1971)] develop similar correlations. The L/G impact on drop size, noted by all, is small as long as L/G is below 0.2.

Atomization in pipe flow is another variant on this type of atomizer. Tatterson, Dallman, and Hanratty [*Am. Inst. Chem. Eng. J.*, **23**(1), 68 (1977)] suggest

$$D_{vm} = 0.023 \left[\frac{D_{pipe}\,\sigma}{U^2 \rho_g f} \right]^{1.2} \tag{18-113}$$

The relationship is dimensionally consistent. Any consistent set of units on the right-hand side will lead length units on the left-hand side.

D_{vm} = volume median droplet diameter
D_{pipe} = pipe diameter
f = friction factor = 0.046/(Reynolds no.)$^{0.2}$
U = gas velocity

The spread shown by Table 18-14 is typical of that for other types of atomization. There are good reasons for variation such as subtle changes in geometry and dramatic differences in test conditions. However, any correlation should be extrapolated with caution and, when possible, tempered with data as in Eq. (18-106).

Droplet-Size Distribution Instead of the single droplet size implied by most of the equations given here, a spectrum of droplet sizes is produced. One has a number of possible ways to characterize the droplet size. The most common are:

1. *Volume median (mass median)* D_{vm}. This has no fundamental meaning but is easy to determine since it is at the midpoint of a cumulative-volume plot. It gives the largest size of the various "average" drop sizes.

2. *Sauter mean* D_{32}. This has the same ratio of surface to mass as the total drop population. It is typically 70 to 90 percent of D_{vm}. It is frequently used in transport processes.

3. *Maximum* D_{max}. This is the largest-sized particle in the population. It is typically 2 to 3 times D_{vm}. Its use yields the most conservative calculation of transport processes.

There are a number of other possible combinations that weight the length, surface, volume, or ratios of these as desired. However, any average drop size is fictitious, and none is completely satisfactory. For example, there is no way in which the high surface and transfer coefficients in small drops can be made available to the larger drops. Hence a process calculation based on a given droplet size describes only what happens to that size and gives at best an approximation to the total mass. Further, the residence time of large and small drops is grossly different because of different settling velocities.

The only proper approach is to break the population into several smaller groupings and calculate their behavior. High-speed computers now make this feasible.

In addition to the other choices there are also a variety of ways to describe the droplet population versus size curve. Figures 18-85 and 18-87 illustrate one of the most common methods, the plot of cumulative volume against droplet size on log-normal graph paper. This satisfies the restraint of not extrapolating to a negative drop size. Its other advantages are that it is easy to plot, the results are easy to visualize, and it yields a nearly straight line at lower drop sizes. Its chief failing is that it deviates strongly from a linear relation above D_{vm}. All plots tend to bend over because of physical limits on attainable drop sizes.

A similar method that overcomes this objection is to use $D/(D_{max} - D)$ in place of D on log-normal paper. Mugele [*Am. Inst. Chem. Eng. J.*, **6**, 3–8 (1960)] discusses the pitfalls of applying some other analytical descriptions to droplet populations.

Entrainment Due to Gas Bubbling through a Liquid Entrainment is a common problem in vaporizers and generally limits the capacity of distillation trays. It can stem from a variety of sources.

1. Excessive foaming. This is a case of a gas-in-liquid dispersion (covered in the next subsection).

2. Droplets formed from the collapse of the bubble dome (see Fig. 18-88*a*). These are virtually unavoidable. They are generally under 25 μm, which means that their terminal velocities are low and they will be entrained. Fortunately, because of their small size they contribute little on a weight basis (<0.001 kg liquid/kg vapor), although they dominate on a number basis.

3. Droplets from the jet caused by liquid rushing to fill the cavity left by the bubble (see Fig. 18-88*d*). These droplets range up to 1000 μm, their size depending on bubble size. A number of studies have found that bubbles greater than about 0.5 cm (0.2 in) yield a jet that is too short and fat to break. This is important at modest loadings. Once foam forms over the surface, drop ejection by this mode decreases sharply.

4. At higher vapor loads, the kinetic energy of the vapor rather than the bubble burst supplies the thrust for jets and sheets of liquid that are thrown up in an uneven pattern. This "splashing" mode can yield much higher levels of entrainment and ultimately causes flooding in distillation trays.

There are two distinct groupings of entrainment data. The first group deals with evaporators and steam generation. It covers a very wide range of vapor velocities but a limited physical-properties range. Typical data were summarized by Standiford (Fig. 11-31, p. 11-35, in *Chemical Engineers' Handbook*, 4th ed., McGraw-Hill, New York, 1963).

In the lowest velocity regime, E (kg liquid entrained/kg vapor) is proportional to velocity. At values of $E \approx 0.0001$, there is a shift to a region where the dependence is with (velocity)$^{3-5}$.

The transition between these two zones probably corresponds to a shift in the mechanism of droplet production. A second major contributor to the increasing exponent on velocity is the change in the drop-size–terminal-velocity relationship. Terminal velocity is proportional to drop size in the Stokes'-law region and to (drop size)$^{1/2}$ in the Newton's-law region. And since the cumulative fraction below a given drop size varies as drop size (see Fig. 18-89), one predicts that the fraction (drops entrained/drops produced) rises with (velocity)^1 at low velocity and with (velocity)^4 at high velocity. This effect combines with the increased droplet production to give a very steep velocity dependence near flooding.

The second grouping of data deals with distillation trays of assorted geometry. It spans a more limited velocity range but generally yields a variation of E as (velocity)$^{3-4}$. Near the flood point higher velocity dependence is sometimes shown. The data also show a very steep dependence of entrainment on tray spacing. This results from the sharp decrease in height to which larger drops are ejected (or splashed). It is unlikely that this strong dependence continues much beyond a 1.0-m (3.3-ft) spacing. This second data group also spans a wider range of physical properties. In addition to depen-

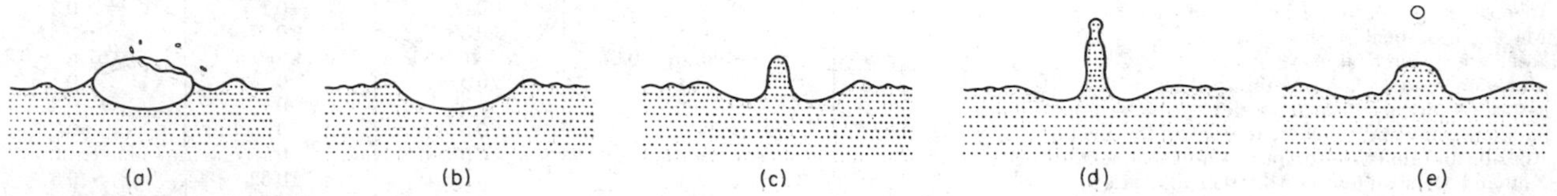

FIG. 18-88 Mechanism of the burst of an air bubble on the surface of water. [*Newitt, Dombrowski, and Knellman,* Trans. Inst. Chem. Eng., **32**, *244 (1954).*]

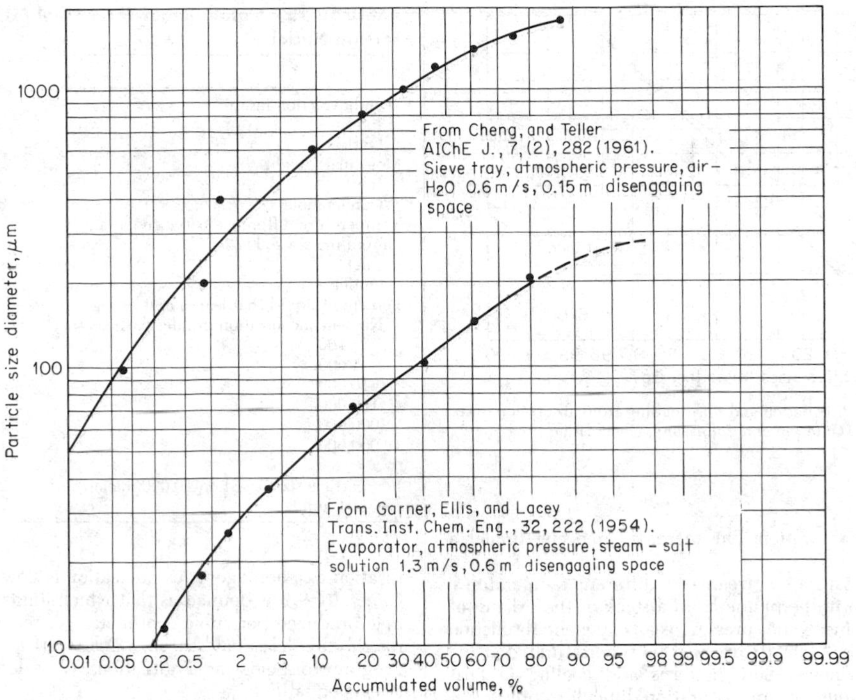

FIG. 18-89 Entrainment droplet-size distribution. To convert meters per second to feet per second, multiply by 3.281; to convert meters to feet multiply by 3.281.

dence on the parameter $\rho_g/(\rho_\ell - \rho_g)$, which one would expect from a force balance on the entrained particles, the data show that entrainment decreases with increased surface tension. This is consistent with the finding, discussed previously, that increased surface tension yields larger droplets (which are more difficult to entrain).

Manning [*Ind. Eng. Chem.*, **56**(4), 14 (1964)] combined the concepts of droplet force balance and maximum stable drop size to obtain a relation for maximum superficial gas velocity (where E goes to 1):

$$U_{max} = 0.9 \frac{[g\,(\rho_\ell - \rho_g)\,\sigma]^{1/4}}{\rho_g^{1/2}} \qquad (18\text{-}114)$$

(The relation is dimensionally consistent; any set of consistent units on the right-hand side yields velocity units for the left-hand side.)

Fair's correlation for sieve and bubble-cap trays shown in Fig. 18-22 implicitly considers these factors in that the main parameter is percent of flood, which in turn depends on tray spacing, vapor velocity, and physical properties (see Fig. 18-10). Note that Fig. 18-22 implicitly incorporates a higher velocity dependence than V^{3-4} above 90 percent of flood, particularly for high-density systems. This correlation can be extended to evaporators by using the sieve-tray curves and setting L/G to 1. This means that the ordinate of Fig. 18-22 reads as "kg liquid/kg vapor" or E as defined earlier.

For velocities that give less than 30 percent of flood,

$$E = E_{30\%}\left(\frac{\%\text{ flood}}{30}\right)^3 \qquad (18\text{-}115)$$

If E from Eq. (18-115) is less than 0.0001,

$$E = 0.0001\left[\frac{\%\text{ flood}}{30}\right]\left[\frac{E_{30\%}}{0.0001}\right]^{1/3} \qquad (18\text{-}116)$$

The detailed geometry of the vapor-liquid contact is also important as shown by the much higher entrainment of bubble-cap than sieve trays in Fig. 18-22. Pinczewski and Fell [*Trans. Inst. Chem. Eng.*, **55**, 46 (1977)] show, for example, that the key parameter in setting droplet distribution is the velocity at which vapor jets onto the tray rather than the superficial tray velocity. Kister, Pinczewski, and Fell [*Ind. Eng. Chem. Process Des. Dev.*, **20**, 528 (1981)] further show that even with identical jet velocity and superficial velocity a tripling of hole diameter can cause a fivefold increase in entrainment. This suggests that a correlation such as that of Fig. 18-22 can be relied on only for an order-of-magnitude estimate.

Accurate size prediction for entrainment is not possible beyond defining the maximum. This maximum can be set by either critical-Weber-number criteria or by the settling velocity matching the gas superficial velocity. For the settling velocity to limit size, at least 0.8 m (30 in) disengaging space is usually required. As shown by Fig. 18-89, actual entrainment spans a broad range. The reason for the much larger drop sizes of the upper curve is the short disengaging space. For this curve over 99 percent of the entrainment has a terminal velocity greater than the vapor velocity. For contrast, in the lower curve the terminal velocity of the largest particle reported is the same as the vapor velocity. Note that even for the lower curve less than 10 percent of the entrainment is in drops of less than 50 μm. This means that it is relatively easy to remove. Over 50 percent is typically captured by the underside of the next higher tray or by a turn in the piping leaving an evaporator. (Figure 18-89 shows "free" entrainment rather than that which actually passes through the next tray or outlet piping.) Conversely, though small on a mass basis, the smaller drops are extremely numerous. On a number basis more than one-half of the drops in the lower curve are under 5 μm. These can serve as nuclei for fog condensation in downstream equipment in addition to their nuisance value.

Fog Condensation This is an entirely different way of forming dispersions. Here the dispersion results from condensation of a vapor rather than mechanical breakup. The particle sizes are usually much finer (0.1 to 30 μm) and are designated as mist or fog.

Fog particles grow because of excess saturation in the gas. Sometimes this means that the gas is supersaturated (i.e., it is below its dew point). At other times fog can grow on soluble foreign nuclei at partial pressures below saturation. In either case fogging results from an

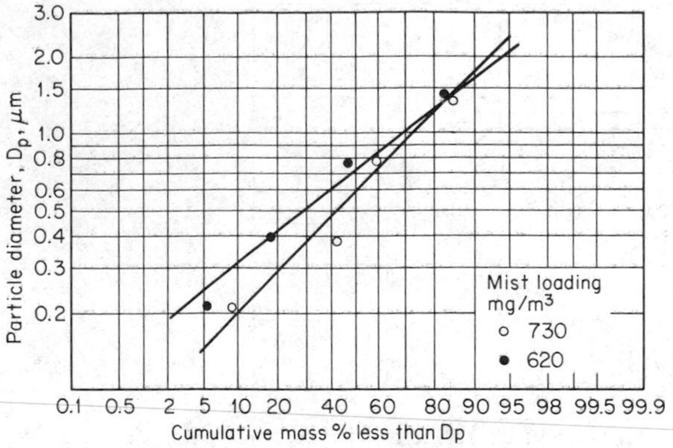

FIG. 18-90 Particle-size distribution and mist loading from absorption tower in a contact H_2SO_4 plant. [*Gillespie and Johnstone,* Chem. Eng. Prog., *51(2), 74 (1955).*]

increase in the relative saturation. This increase can occur through a variety of routes:

1. Mixing of two saturated streams at different temperatures. This is commonly seen in the plume from a stack or the exhaust of breath on a cold day. Since vapor pressure is an exponential function of temperature, the resultant mixture will be supersaturated at the mixed temperature. Uneven flow patterns and cooling in heat exchangers make this route to supersaturation difficult to prevent.
2. Increased partial pressure due to reaction. An example is the reaction of SO_3 and H_2O to yield H_2SO_4, which has a much lower vapor pressure than its components.
3. Isoentropic expansion (cooling) of a gas, as in a steam nozzle.
4. Cooling of a gas containing a condensable vapor. Here the problem is that the gas cools faster than condensable vapor can be removed by mass transfer. The problem arises for two reasons. First, the condensable vapor is generally dilute and of higher molecular weight than the noncondensable gas. This means that the molecular diffusivity of the vapor will be much less than the thermal diffusivity of the gas. Restated, the ratio of N_{Sc}/N_{Pr} is greater than 1. The result is that a cooler yields more heat-transfer units $\int dT_g/(T_g - T_i)$ than mass-transfer units $\int dY/(Y_g - Y_i)$. Second, both transfer processes derive their driving force from the temperature difference between the gas T_g and the interface T_i. Each incremental decrease in interface temperature yields the same relative increase in temperature driving force. However, the interface vapor pressure can only approach the limit of zero. Because of this, for equal molecular and thermal diffusivities a saturated mixture will supersaturate when cooled. The tendency to form supersaturated gases generally increases with increased molecular weight of the condensable, with increased temperature differences, with reduced temperature, and with reduced initial superheating. However, to evaluate whether a given condensing step yields fog requires rigorous treatment of the coupled heat- and mass-transfer processes through the entire condensation. (Again high-speed computers make this practical.) Steinmeyer [*Chem. Eng. Prog.*, **68**(7), 64 (1972)] illustrates this, showing the impact of foreign-nuclei concentration on calculated fog formation. See Table 18-15.

Fortunately supersaturation is not a sufficient criterion for fog formation. If it were, many and perhaps most partial condensers would be plagued with fog.

In order to form a fog, nuclei must first be found. Homogeneous nucleation is quite difficult because of the energy barrier associated with creation of the interface. It can be treated as a kinetic process, with the rate a very steep function of the supersaturation ratio (S = partial pressure of condensable per vapor pressure at gas temperature). For water an increase in S from 3.4 to 3.8 causes a 10,000-fold increase in the nucleation rate. As a result, below a critical supersaturation S_{crit} homogeneous nucleation is slow enough to be ignored. Generally S_{crit} is defined as that which limits nucleation to one particle produced per cubic centimeter per second. It can be estimated roughly by "classical" theory (*Theory of Fog Condensation,* Israel Program for Scientific Translations, Jerusalem, 1967), using the following equation:

TABLE 18-15 Simulation of Three Heat Exchangers with Varying Foreign Nuclei

	1	2	3
Weight fraction insert			
Inlet	0.51	0.42	0.02
Outlet	0.80	0.80	0.32
Molecular weight			
Inert	28	29	29
Condensable	86	99	210
Temperature difference between gas and liquid interface, K			
Inlet	14	24	67
Outlet	4	10	4
Percent of liquid that leaves unit as fog			
Nuclei concentration in inlet particles/cm³			
100	0.05	1.1	2.2
1,000	0.44	5.6	3.9
10,000	3.2	9.8	4.9
100,000	9.6	11.4	5.1
1,000,000	13.3	11.6	
10,000,000	14.7		
∞	14.7	11.8	5.1
Fog particle size based on 10,000 nuclei/cm³ at inlet, μm	28	25	4

$$S_{crit} = \exp\left[0.56\,\frac{M}{\rho_\ell}\left(\frac{\sigma}{T}\right)^{3/2}\right] \tag{18-117}$$

where σ = surface tension, mN/m (dyn/cm)
ρ_ℓ = liquid density, g/cm³
T = temperature, K
M = molecular weight of condensable

Table 18-16 shows typical experimental values of S_{crit}, taken from the work of Russell [*J. Chem. Phys.*, **50**, 1809 (1969)].

Since the critical supersaturation ratio for homogeneous nucleation is typically greater than 3, it is not often reached in process equipment. Unfortunately foreign nuclei are often present in abundance and permit nucleation at much lower supersaturations. For example,

1. *Solids.* Surveys have shown that air contains thousands of particles per cubic centimeter in the 0.1- to 1-μm range suitable for nuclei. The sources range through ocean-generated salt spray to combustion processes. The concentration is highest in large cities and industrial regions. When the foreign nuclei are soluble in the fog, nucleation occurs at S values very close to 1. This is the mechanism controlling atmospheric water condensation. Even when not soluble, a foreign particle is an effective nucleus if wet by the liquid. Thus a 1-μm insoluble particle with a 0° contact angle requires an S of only 1.001 in order to serve as a condensation site for water.
2. *Ions.* Amelin (*Theory of Fog Condensation,* Israel Program for Scientific Translations, Jerusalem, 1967) reports that ordinary air contains even higher concentrations of ions. These ions also reduce

TABLE 18-16 Experimental Critical Supersaturation Ratios

	Temperature, K°	S_{crit}
H_2O	264	4.91
C_2H_5OH	275	2.13
CH_4OH	264	3.55
C_6H_6	253	5.32
CCl_4	247	6.5
$CHCl_3$	258	3.73
C_6H_5Cl	250	9.5

°°R = 9/5 K.

the required critical supersaturation, but by only about 10 to 20 percent unless multiple charges are present. Russell [*J. Chem. Phys.*, **50**, 4, 1809 (1969)] compares theory with data for ion nucleation.

3. *Entrained liquids.* Production of small droplets is inherent in the bubbling process as shown by Fig. 18-88. Values range from near zero to 10,000/cm^3 of vapor, depending on how the vapor breaks through the liquid and on the opportunity for evaporation of the small drops after entrainment.

Hence, although the severe supersaturation needed to support homogeneous nucleation is seldom encountered, most process streams contain enough foreign nuclei to cause some fogging. While fogging has been reported in only a relatively low percent of process partial condensers, it is rarely looked for and volunteers its presence only when yield losses or pollution is intolerable.

Although monodisperse (nearly uniform droplet size) fogs can be grown by limiting nucleation and providing a long retention time for growth on these nuclei, more commonly industrial fogs show a broad distribution as in Fig. 18-90. Note that the sizes are several orders of magnitude smaller than those shown earlier for entrainment, for atomizers, and even for high-speed droplet rupture. The result, as discussed in a later subsection, is a demand for efficient removal devices for micrometer-sized particles.

While generally fog formation is a nuisance, it can also be a useful route to a high-surface-area liquid-in-gas dispersion, as in insecticide application.

GAS-IN-LIQUID DISPERSIONS

GENERAL REFERENCES: Comprehenisve treatments of bubbles or foams are given by Akers, *Foams: Symposium 1975,* Academic, New York, 1973; Bendure, *TAPPI*, **58**, 83 (1975); Benfratello, *Energ. Elettr.*, **30**, 80, 486 (1953); Berkman and Egloff, *Emulsions and Foams,* Reinhold, New York, 1941, pp. 112–152; Bikerman, *Foams,* Springer-Verlag, New York, 1975; *Encyclopedia of Chemical Technology,* vol. 11, secs. "Foamed Plastics" and "Foams," Wiley, New York, 1980, pp. 82–145; Haberman and Morton, Rep. 802, David W. Taylor Model Basin, Washington, 1953; Levich, *Physicochemical Hydrodynamics,* Prentice-Hall, Englewood Cliffs, N.J., 1962; and Soo, *Fluid Dynamics of Multiphase Systems,* Blaisdell, Waltham, Mass., 1967. The formation of bubbles is comprehensively treated by Clift, Grace, and Weber, *Bubbles, Drops and Particles,* Academic, New York, 1978; and Kumar and Kuloor, *Adv. Chem. Eng.*, **8**, 255–368 (1970). Design methods for units operation in bubble columns and stirred vessels are covered by Atika and Yoshida, *Ind. Eng. Chem. Process Des. Dev.*, **13**, 84 (1974); Calderbank, *Chem. Eng.*, CE209 (October 1967); and *Mixing,* vol. II, Academic, New York, 1967, pp. 1–111; Fair, *Chem. Eng.*, **74**, 67 (July 3, 1967); Jackson, *Chem. Eng.*, CE107 (May 1964); Jordan, *Chemical Process Development,* Interscience, New York, 1968, part 1, pp. 111–175; Mersmann, *Ger. Chem. Eng.*, **1**, 1 (1978); Resnick and Gal-Or, *Adv. Chem. Eng.*, **7**, 295–395 (1968); and Valentin, *Absorption in Gas-Liquid Dispersions,* E. & F. N. Spon, London, 1967. The influence of surface-active agents on bubbles and foams is summarized in selected passages from Schwartz and Perry, *Surface Active Agents,* vol. I, Interscience, New York, 1949; and from Schwartz, Perry, and Berch, *Surface Active Agents and Detergents,* vol. II, Interscience, New York, 1958. See also Elenkov, *Theor. Found. Chem. Eng.*, **1**, 1, 117 (1967); and Rubel, *Antifoaming and Defoaming Agents,* Noyes Data Corp., Park Ridge, N.J., 1972. A review of foam stability also is given by de Vries, *Meded. Rubber Sticht. Delft,* no. 328, 1957. Foam-separation methodology is discussed by Aguoyo and Lemlich, *Ind. Eng. Chem. Process Des. Dev.*, **13**, 153 (1974); and Lemlich, *Ind. Eng. Chem.*, **60**, 16 (1968). The following reviews of specific applications of gas-to-liquid dispersions are recommended: Industrial fermentations: Aiba, Humphrey, and Millis, *Biochemical Engineering,* Academic, New York, 1965. Finn, *Bacteriol. Rev.*, **18**, 254 (1954). Oldshue, "Fermentation Mixing Scale-Up Techniques," in *Biotechnology and Bioengineering,* vol. VIII, 1966, pp. 3–24. Aerobic oxidation of wastes: Dickey, "Turbine Agitated Gas Dispersion—Power, Flooding and Hold-Up," Pap. No. 116d, AIChE 72d annual meeting, San Francisco, 1979. Eckenfelder and McCabe, *Advances in Biological Waste Treatment,* Macmillan, New York, 1963. Eckenfelder and O'Connor, *Biological Waste Treatment,* Pergamon, New York, 1961. McCabe and Eckenfelder, *Biological Treatment of Sewage and Industrial Wastes,* vol. I, Reinhold, New York, 1955, part 2. *Proceedings of Industrial Waste Treatment Conference,* Purdue University, annually. Zlokarnik, *Adv. Biochem. Eng.*, **11**, 158–180 (1979). Cellular elastomers: Fling, *Natural Rubber Latex and Its Applications,* no. 4: *The Preparation of Latex Foam Products,* British Rubber Development Board, London, 1954. Gould, in *Symposium on Application of Synthetic Rubbers,* American Society for Testing and Materials, Philadelphia, 1944, pp. 90–103. Skochdopole, "Foamed Plastics," *Encyclopedia of Chemical Technology,* vol. 9, Wiley, New York, 1966, pp. 847–888. Fire-fighting foams: Perri, in Bikerman, op. cit., chap. 12. Ratzer, *Ind. Eng. Chem.*, **48**, 2013 (1956). Froth-flotation methods and equipment: Booth, in Bikerman, op. cit., chap. 13. Gaudin, *Flotation,* McGraw-Hill, New York, 1957. Taggart, *Handbook of Mineral Dressing,* Wiley, New York, 1945, sec. 12, pp. 52–81.

Objectives of Gas Dispersion The dispersion of gas as bubbles in a liquid or in a plastic mass is effected for one of the following purposes: (1) gas-liquid contacting (to promote absorption or stripping, with or without chemical reaction), (2) agitation of the liquid phase, or (3) foam or froth production. Gas-in-liquid dispersions also may be produced or encountered inadvertently, sometimes undesirably.

Gas-Liquid Contacting Usually this is accomplished with conventional columns or with spray absorbers (see preceding subsection "Gas-Liquid Contacting."). For systems containing solids or tar likely to plug columns, for absorptions accomplished by strongly exothermic reactions, or for treatments involving a readily soluble gas or a condensable vapor, however, gas dispersers may be used to advantage.

Agitation Agitation by a stream of gas bubbles (usually air) rising through a liquid is employed in tanks of such large volume or of such unsymmetrical shape as to make mechanical agitation ineffective or expensive. Gas spargers may replace mechanical agitators also for simple blending operations involving a liquid of low volatility or for applications in which it is difficult to seal around an agitator shaft.

Foam Production This is important in froth-flotation separations, in the manufacture of cellular elastomers, plastics, and glass, and in certain special applications (e.g., food products, fire extinguishers). Unwanted foam can occur in process columns, in agitated vessels, and in reactors in which a gaseous product is formed; it must be avoided, destroyed, or controlled. Berkman and Egloff (*Emulsions and Foams,* Reinhold, New York, 1941, pp. 112–152) have pointed out that foam is produced only in systems possessing the proper combination of interfacial tension, viscosity, volatility, and concentration of solute or suspended solids. From the standpoint of gas comminution, foam production requires the creation of small bubbles in a liquid capable of sustaining foam.

Theory of Bubble and Foam Formation A **bubble** is a globule of gas or vapor surrounded by a mass or thin film of liquid. By extension, globular voids in a solid are sometimes called bubbles. **Foam** is a group of bubbles separated from one another by thin films, the aggregation having a finite static life. Although nontechnical dictionaries do not distinguish between foam and froth, a technical distinction is often made. A highly concentrated dispersion of bubbles in a liquid is considered a **froth** even if its static life is substantially nil (i.e., it must be dynamically maintained); thus all foams are also froth, whereas the reverse is not true. The term **lather** implies a froth that is worked up on a solid surface by mechanical agitation; it is seldom used in technical discussions. The thin walls of bubbles comprising a foam are called **laminae** or **lamellae.**

Bubbles in a liquid originate from one of three general sources: (1) they may be formed by desupersaturation of a solution of the gas or by the decomposition of a component in the liquid; (2) they may be introduced directly into the liquid by a bubbler or sparger, or by mechanical entrainment; (3) they may result from the disintegration of larger bubbles already in the liquid.

Generation Spontaneous generation of bubbles of gas or vapor from a homogeneous liquid is theoretically impossible (Bikerman, *Foams: Theory and Industrial Applications,* Reinhold, New York, 1953, p. 10). The appearance of a bubble requires a gas nucleus as a void in the liquid. The nucleus may be in the form of a small bubble or of a solid carrying adsorbed gas, examples of the latter being dust particles, boiling chips, and a solid wall. A void can result from cavitation, mechanically or acoustically induced. Blander and Katz [*Am. Inst. Chem. Eng. J.*, **21**, 833 (1975)] have thoroughly reviewed bubble nucleation in liquids.

Theory permits the approximation of the maximum size of a bubble that can adhere to a submerged horizontal surface if the contact angle between bubble and solid (angle formed by solid-liquid and liquid-gas interfaces) is known [Wark, *J. Phys. Chem.*, **37**, 623

(1933); Jakob, *Mech. Eng.*, **58,** 643 (1936)]. Inasmuch as the bubbles that actually rise from a surface are always considerably smaller than those so calculated and inasmuch as the contact angle is seldom known, the theory is not directly useful.

Formation at a Single Orifice The formation of bubbles at an orifice or capillary immersed in a liquid has been the subject of much study, experimental and theoretical. Bikerman (op. cit., secs. 3 to 7), Valentin (op. cit., chap. 2), Jackson (op. cit.), Soo (op. cit., chap. 3), Fair (op. cit.), Kumar et. al. (op. cit.), and Clift et al. (op. cit.) have presented reviews and analyses of this subject.

There are three regimes of bubble production (Silberman in *Proceedings of the Fifth Midwestern Conference on Fluid Mechanics,* University of Michigan Press, Ann Arbor, 1957, pp. 263–284): (1) single-bubble, (2) intermediate, and (3) jet.

Single-Bubble Regime Bubbles are produced one at a time, their size being determined primarily by the orifice diameter D, the interfacial tension of the gas-liquid film σ, the densities of the liquid ρ_ℓ and gas ρ_g, and the gravitational acceleration g according to the relation

$$D_b/D = [6\sigma/D^2\,(\rho_\ell - \rho_g)]^{1/3} \tag{18-118}$$

where D_b is the bubble diameter. The bubble size is independent of gas flow rate; the frequency, therefore, is directly proportional to the rate. Equation (18-118) leads to

$$f = \frac{Qg(\rho_\ell - \rho_g)}{\pi D\sigma} \tag{18-119}$$

where f is the frequency of bubble formation and Q is the volumetric rate of gas flow, in consistent units.

Equations (18-118) and (18-119) result from a balance of bubble buoyancy against interfacial tension. They recognize no inertia or viscosity effects. At low bubbling rates, they are quite satisfactory. Van Krevelen and Hoftijzer [*Chem. Eng. Prog.*, **46,** 29 (1950)] and Guyer and Peterhaus [*Helv. Chim. Acta,* **26,** 1099 (1943)], for example, reported air-bubble diameters 0.84 to 1.02 times those calculated from Eq. (18-118) when the bubbles were formed at a frequency in the range 0.3 to 1.0 per second in water, transformer oil, ether, and carbon tetrachloride. Orifice diameters ranged from 0.004 to 0.95 cm (0.0016 to 0.37 in), and the orifices discharged vertically upward. If the orifice diameter becomes too large, the bubble diameter will be smaller than the orifice diameter, as predicted by Eq. (18-118), and instability results; consequently stable, stationary bubbles cannot be produced.

For the case of bubbles being formed in water, the orifice diameter which permits bubbles of about its own size is calculated as 0.66 cm (0.25 in). Davidson and Amick [*Am. Inst. Chem. Eng. J.*, **2,** 337 (1956)] confirmed this estimate in their observation that stable bubbles in water were formed at a 0.64-cm, orifice but could not be formed at a 0.79-cm one.

For very thin liquids, Eqs. (18-118) and (18-119) are expected to be valid up to a gas-flow Reynolds number of 200 (Valentin, op. cit., p. 8). For liquid viscosities up to 100 cP, Datta, Napier, and Newitt [*Trans. Inst. Chem. Eng.*, **28,** 14 (1950)] and Siems and Kauffman [*Chem. Eng. Sci.*, **5,** 127 (1956)] have shown that liquid viscosity has very little effect on the bubble volume, but Davidson and Schuler [*Trans. Inst. Chem. Eng.*, **38,** 144 (1960)] and Krishnamurthi et al. [*Ind. Eng. Chem. Fundam.*, **7,** 549 (1968)] have shown that the liquid viscosity can cause the bubble size to increase considerably over that predicted by Eq. (18-118) for liquid viscosities above 1000 cP. In fact, Davidson et al. (op. cit.) found that their data agreed very well with a theoretical equation obtained by equating the buoyant force to drag based on Stokes' law and the velocity of the bubble equator at break-off.

$$D_b^3 = \frac{6}{\pi}\left(\frac{4\pi}{3}\right)^{1/4}\left(\frac{15\nu Q}{2g}\right)^{3/4} \tag{18-120}$$

where ν is the liquid kinematic viscosity and Q is the gas volumetric flow rate. This equation is dimensionally consistent. The relative effect of liquid viscosity can be obtained by comparing the bubble diameters calculated from Eqs. (18-118) and (18-120). If liquid viscosity appears significant, one might want to use the long and tedious method developed by Krishnamurthi et al. (op. cit.), which considers both surface-tension forces and viscous-drag forces.

Intermediate Regime This regime extends approximately from a Reynolds number of 200 to one of 2100. As the gas flow through a submerged orifice increases beyond the limit of the single-bubble regime, the frequency of bubble formation increases more slowly and the bubbles begin to grow in size. Between the two regimes there may indeed be a range of gas rates over which the bubble size decreases with increasing rate, owing to the establishment of liquid currents that nip the bubbles off prematurely. The net result can be the occurrence of a minimum bubble diameter at some particular gas rate (Maier, U.S. Bur. Mines Bull. 260, 1927; and Bikerman, op. cit., p. 4). At the upper portion of this region, the frequency becomes very nearly constant with respect to gas rate and the bubble size correspondingly increases with gas rate. The bubble size is affected primarily by (1) orifice diameter, (2) liquid-inertia effects, (3) liquid viscosity, (4) liquid density, and (5) the relationship between the constancy of gas flow and the constancy of pressure at the orifice.

Kumar et al. have done extensive experimental and theoretical work reported in *Ind. Eng. Chem. Fundam.*, **7,** 549 (1968); *Chem. Eng. Sci.*, **24,** part I, 731; part II, 749; part III, 1711 (1969) and summarized in *Adv. Chem. Eng.*, **8,** 255 (1970). They along with other investigators—Swope [*Can. J. Chem. Eng.*, **44,** 169 (1972)], Tsuge and Hibino [*J. Chem. Eng. Japan,* **11,** 307 (1972)], Pinczewski [*Chem. Eng. Sci.*, **36,** 405 (1981)], Tsuge and Hibino [*Int. Chem. Eng.*, **21,** 66 (1981)], and Takahashi and Miyahara [ibid., p. 224]—have solved the equations resulting from a force balance on the forming bubble, taking into account buoyancy, surface tension, inertia, and viscous-drag forces for both conditions of constant flow through the orifice and constant pressure in the gas chamber. The results of the theoretical work are extremely complex and tedious algebraic and differential equations. Although Mersmann [*Ger. Chem. Eng.*, **1,** 1 (1978)] claims that the results of Kumar et al. (loc. cit.) well fit experimental data, Lanauze and Harn [*Chem. Eng. Sci.*, **29,** 1663 (1974)] claim differently: "Further, it has been shown that the mathematical formulation of Kumar's model, including the condition of detachment, do not adequately describe the experimental situation—Kumar's model has several fundamental weaknesses, the computational simplicity being achieved at the expense of physical reality." In lieu of careful independent checks of predictive accuracy, the results of the comprehensive theoretical work will not be presented here. Simpler, more easily understood predictive methods, for certain important limiting cases, will be presented. As a check on the accuracy of these simpler methods, it will perhaps be prudent to calculate the bubble diameter from the graphical representation by Mersmann (loc. cit.) of the results of Kumar et al. (loc. cit.).

For conditions approaching constant flow through the orifice, a relationship derived by equating the buoyant force to the inertia force of the liquid [Davidson et al., *Trans. Inst. Chem. Eng.*, **38,** 335 (1960)] (dimensionally consistent),

$$D_b^3 = 1.378\,\frac{6Q^{6/5}}{\pi g^{3/5}} \tag{18-121}$$

fits experimental data reasonably well. Surface tension and liquid viscosity tend to increase the bubble size at low Reynolds number. The effect of surface tension is greater for large orifice diameters. The magnitude of the diameter increase due to high liquid viscosity can be obtained from Eq. (18-120).

For conditions approaching constant pressure at the orifice entrance, which probably simulates most industrial applications, there is no independently verified predictive method. For air at near atmospheric pressure sparged into relatively inviscid liquids [$\mu \leq$ 100 mPa·s (cP)] the correlation of Kumar, Degaleesan, Ladda, and Hoelscher [*Can. J. Chem. Eng.*, **54,** 503 (1976)] fits experimental data well. Their correlation is presented here as Fig. 18-91.

Caution is recommended for the use of the correlation for gas densities and viscosities very different from atmospheric air because gas properties were not varied by Kumar et al. and no investigators have found or suspected that the bubble diameter will change significantly as gas viscosity changes and the effect of gas density is undefined.

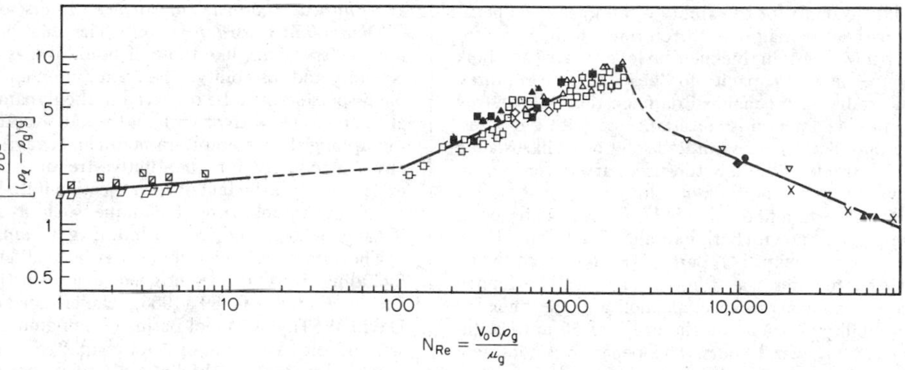

FIG. 18-91 Bubble-diameter correlation for air sparged into relatively inviscid liquids. D_b = bubble diameter, D = orifice diameter, V_o = gas velocity through sparging orifice, P = fluid density, and μ = fluid viscosity. [*From* Can. J. Chem. Eng., *54, 503 (1976).*]

Jet Regime With further rate increases, turbulence occurs at the orifice, and the gas stream approaches the appearance of a continuous jet which breaks up 7.6 to 10.2 cm (3 to 4 in) above the orifice. Actually, the stream consists of large, closely spaced, irregular bubbles with a rapid swirling motion. These bubbles disintegrate into a cloud of smaller ones of random size distribution between 0.025 cm (0.01 in) or smaller and about 1.27 cm (0.5 in), with a mean size for air and water of about 0.4 cm (0.16 in) (Leibson et al., loc. cit.). There are many contradictory reports about this regime, and theory, although helpful (see, for example, Siberman, loc. cit.), is as yet unable to describe the phenomena observed. The correlation of Kumar et al. (Fig. 18-91) is recommended for air-liquid systems.

Formation at Multiple Orifices At high velocities coalescence of bubbles formed at individual orifices occurs; Helsby and Tuson [*Research* (London), **8,** 270 (1955)], for example, observed the frequent coalescence of bubbles formed in pairs or in quartets at an orifice. Multiple orifices spaced by the order of magnitude of the orifice diameter increase the probability of coalescence, and when the magnitude is small (as in a sintered plate), there is invariably some. The broken lines of Fig. 18-95 presumably represent zones of increased coalescence and relatively less effective dispersion as the gas rate through porous-carbon tubes is increased. Savitskaya [*Kolloidn. Zh.*, **13,** 309 (1951)] found that the average bubble size formed at the surface of a porous plate was such as to maintain constancy of the product of bubble specific surface and interfacial tension as the latter was varied by addition of a surfactant. Konig, Buchholz, Lucke, and Schugerl [*Ger. Chem. Eng.*, **1,** 199 (1978)] produced bubble sizes varying from 0.5 to 4 mm by the use of two porous-plate spargers and one perforated-plate sparger with superficial gas velocities from 1 to 8 cm/s (0.033 to 0.26 ft/s). The small bubble sizes were stabilized by adding up to 0.5 percent of various alcohols to water.

At high flow rates through perforated plates such as occur in distillation columns, Calderbank and Rennie [*Trans. Inst. Chem. Eng.*, **40,** T3 (1962)]; Porter, Davies, and Wong [ibid., **45,** T265 (1967)]; Rennie and Evans [*Br. Chem. Eng.*, **7,** 498 (1962)]; and Valentin (op. cit., chap. 3) have investigated and discussed the effect of the flow conditions through the multiple orifices on the froths and foams which occur above perforated plates.

Entrainment and Mechanical Disintegration Gas can be entrained into a liquid by a solid or a stream of liquid falling from the gas phase into the liquid, by surface ripples or waves, and by the vertical swirl of a mass of agitated liquid about the axis of a rotating agitator. Small bubbles probably form near the surface of the liquid and are caught into the path of turbulent eddies whose velocity exceeds the terminal velocity of the bubbles. The disintegration of a submerged mass of gas takes place by the turbulent tearing of smaller bubbles away from the exterior of the larger mass or by the influence of surface tension on the mass when it is attenuated by inertial or shear forces into a cylindrical or disk form. A fluid cylinder that is greater in length than in circumference is unstable and tends to break spontaneously into two or more spheres. These effects account for the action of fluid attrition and of an agitator in the disintegration of suspended gas. Quantitative correlations for gas entrainment by liquid jets and in agitated vessels will be given later.

Foams Two excellent reviews (Shedlovsky, op. cit.; and Lemlich, op. cit.) covering the literature pertinent to foams have been published. A foam is formed when bubbles rise to the surface of a liquid and persist for a while without coalescence with one another or without rupture into the vapor space. The formation of foam, then, consists simply of the formation, rise, and aggregation of bubbles in a liquid in which foam can exist. The life of foams varies over many magnitudes—from seconds to years—but in general is finite. Maintenance of a foam, therefore, is a dynamic phenomenon.

Gravitational force favors the separation of gas from liquid in a disperse system, causing the bubbles to rise to the liquid surface and the liquid contained in the bubble walls to drain downward to the main body of the liquid. Interfacial tension favors the coalescence and ultimate disappearance of bubbles; indeed, it is the cause of bubble destruction upon the rupture of the laminae.

The viscosity of the liquid in a film opposes the drainage of the film and its displacement by the approach of coalescing bubbles. The higher the viscosity, the slower will be the film-thinning process; furthermore, if viscosity increases as the film grows thinner, the process becomes self-retarding. The viscosity of films appears to be greater than that of the main body of the parent liquid in many cases. Sometimes this is a simple temperature effect, the film being cooler because of evaporation; sometimes it is a concentration effect, dissolved or fine suspended solids migrating to the interface to produce classical or anomalous increases in viscosity; at yet other times, the effect seems to occur without explanation.

If the liquid laminae of a foam system can be converted to impermeable solid membranes, the film viscosity can be regarded as having become infinite and the resulting "solid foam" will be permanent. Likewise, if the laminae are composed of a Bingham plastic or a thixotrope, the foam will be permanently stable for bubbles whose buoyancy does not permit exceeding the yield stress. For other nonnewtonian fluids, however, and for all newtonian ones, no matter how viscous, the viscosity can only delay but never prevent foam disappearance. The popular theory, held since the days of Plateau, that foam life is proportional to surface viscosity and inversely proportional to interfacial tension, is not correct, according to Bikerman (op. cit., p. 161), who points out that it is contradicted by experiment.

Bikerman also rejects the idea that foam films drain to a critical thickness at which they spontaneously burst. Foam stability rather is keyed to the existence of a surface skin of low interfacial tension immediately overlying a solution bulk of higher tension, latent until it is exposed by rupture of the superficial layer [Maragoni, *Nuovo cimento*, no. 2, **5–6,** 239 (1871)]. Such a phenomenon of surface elasticity, resulting from concentration differences between bulk and surface of the liquid, accounts for the ability of bubbles to be penetrated by missiles without damage. With reference to it, it is conceivable that films below a certain thickness no longer carry any bulk of solution and hence have no capacity to "heal" surface ruptures, thus becoming vulnerable to mechanical damage that will destroy them. The Maragoni phenomenon is consistent also with the observation that neither pure liquids nor saturated solutions will sustain a foam, since neither extreme will allow the necessary differences in concentration between surface and bulk of solution.

The specific ability of certain finely divided, insoluble solids to stabilize foam has long been known (Berkman and Egloff, op. cit., p. 133; and Bikerman, op. cit., chap. 11). Bartsch [*Kolloidchem. Beih.*, **20**, 1 (1925)] found that the presence of fine galena greatly extended the life of air foam in aqueous isoamyl alcohol, and the finer the solids, the greater stability. Particles on the order of 50 μm lengthened the life from 17 s to several hours. This behavior is consistent with theory, which indicates that a solid particle of medium contact angle with the liquid will prevent the coalescence of two bubbles with which it is in simultaneous contact. Quantitative observations of this phenomenon are scanty.

Berkman and Egloff explain that some additives increase the flexibility or toughness of bubble walls, rather than their viscosity, to render them more durable. They cite as illustrations the addition of small quantities of soap to saponin solutions or of glycerin to soap solution to yield much more stable foam. The increased stability with ionic additives is probably due to electrostatic repulsion between charged, nearly parallel surfaces of the liquid film which acts to retard draining and hence rupture.

Characteristics of Dispersion

Properties of Component Phases As discussed in the preceding subsection, dispersions of gases in liquids are affected by the viscosity of the liquid, the density of the liquid and of the gas, and the interfacial tension between the two phases. They also may be affected directly by the composition of the liquid phase. Both the formation of bubbles and their behavior during their lifetime are influenced by these quantities as well as by the mechanical aspects of their environment.

Viscosity and density of the component phases can be measured with confidence by conventional methods, as can the interfacial tension between a pure liquid and a gas. The interfacial tension of a system involving a solution or micellar dispersion becomes less satisfactory, inasmuch as the interfacial free energy depends on the concentration of solute at the interface. Dynamic methods and even some of the so-called static methods involve the creation of new surfaces. Since the establishment of equilibrium between this surface and the solute in the body of the solution requires a finite amount of time, the value measured will be in error if the measurement is made more rapidly than the solute can diffuse to the fresh surface. Eckenfelder and Barnhart (AIChE 42d national meeting, Repr. 30, Atlanta, 1960) found that measurements of the surface tension of sodium lauryl sulfate solutions by maximum bubble pressure were higher than those by DuNuoy tensiometer by 40 to 90 percent, the larger factor corresponding to a concentration of about 100 ppm, and the smaller to a concentration of 2500 ppm of sulfate.

Even if the interfacial tension is measured accurately, there may be doubt about its applicability to the surface of bubbles being rapidly formed in a solution of a surface-active agent, for the bubble surface may not have time to become equilibrated with the solution. Coppock and Meiklejohn [*Trans. Inst. Chem. Eng. (London)*, **29**, 75 (1951)] reported that bubbles formed in single-bubble regime at an orifice in a solution of a commercial detergent had a diameter larger than that calculated in terms of the measured surface tension of the solution [Eq. (18-118)]. The disparity probably is a reflection of unequilibrated bubble laminae.

One concerned with the measurement of gas-liquid interfacial tension should consult the useful reviews of methods prepared by Harkins (in Weissberger, *Techniques of Organic Chemistry*, vol. I, part II, 2d ed., Interscience, New York, 1949, chap. 9), Schwartz and coauthors (*Surface Active Agents*, vol. I, Interscience, New York, 1949, pp. 263–271; *Surface Active Agents and Detergents*, vol. II, Interscience, New York, 1958, pp. 389–391, 417–418), and by Adamson (*Physical Chemistry of Surfaces*, Interscience, New York, 1960).

Dispersion Characteristics The chief characteristics of gas-in-liquid dispersions, like those of liquid-in-gas suspensions, are heterogeneity and instability. The composition and structure of an unstable dispersion must be observed in the dynamic situation by looking at the mixture, with or without the aid of optical devices, or by photographing it, preferably in nominal steady state; photographs usually are required for quantitative treatment. Stable foams may be examined after the fact of their creation if they are sufficiently robust or if an immobilizing technique such as freezing is employed [Chang, Schoen, and Grove, *Ind. Eng. Chem.*, **48**, 2035 (1956)].

The rate of rise of bubbles has been discussed in many papers, including two that present good reviews of the subject [Benfratello, *Energ. Elettr.*, **30**, 80 (1953); Haberman and Morton, Rep. 802, David W. Taylor Model Basin, Washington, September 1953]; Jackson, loc. cit.; Valentin, op. cit., chap. 2; Soo, op. cit., chap. 3; Calderbank, loc. cit., p. CE220; and Levich, op. cit., chap. VIII). Small bubbles (below 0.2 mm in diameter) are essentially rigid spheres and rise through water at terminal velocities that place them clearly in the laminar-flow region; hence their rising velocity may be calculated from Stokes' law. As bubble size increases to about 2 mm, the spherical shape is retained, and the Reynolds number is still sufficiently small (<10) that Stokes' law should be nearly obeyed.

Two effects set in, however, that alter the velocity. At about $N_{Re} = 100$, a wobble begins that can develop into a helical path if the bubbles are not liberated too closely to one another [Houghton, McLean, and Ritchie, *Chem. Eng. Sci.*, **7**, 40 (1957); and Houghton, Ritchie and Thompson, ibid., p. 111]. Furthermore, for bubbles in the range of 1 mm and larger (until distortion becomes serious) internal circulation can set in [Garner and Hammerton, *Chem. Eng. Sci.*, **3**, (1954); and Haberman and Morton, loc. cit.], and according to theoretical analyses by Hadamard and Rybczynski and given by Levich (op. cit.) the drag coefficient for a low-viscosity dispersed phase and a high-viscosity continuous phase will approach two-thirds of the drag coefficient for rigid spheres, namely, $C_D = 16/N_{Re}$. The rise velocity of a circulating drop will thus be 1.5 times that of a rigid sphere. Redfield and Houghton [*Chem. Eng. Sci.*, **20**, 131 (1965)] have found that CO_2 bubbles rising in pure water agree with the theoretical solution for a circulating drop below $N_{Re} = 1$. Many investigators (see Valentin, op. cit.) have found that extremely small quantities of impurities can retard or stop this internal circulation. In this behavior may lie the explanation of the fact that the addition of long-chain fatty acids to water to produce a concentration of 1.5×10^{-4} molar markedly reduces the rate of rise of bubbles [Stuke, *Naturwissenschaften*, **39**, 325 (1952)].

Above about 2 mm bubbles begin to change to ellipsoids, and above 1 cm they become lens-shaped, according to Davies and Taylor [*Proc. R. Soc. (London)*, **A200**, 379 (1950)]. The rising velocity in thin liquids for the size range 1 mm $< D_B <$ 20 mm has been reported as 20 to 30 cm/s by Haberman and Morton (op. cit.); and Davenport, Richardson, and Bradshaw [*Chem. Eng. Sci.*, **22**, 1221 (1967)]. Schwerdtfeger [ibid., **23**, 937 (1968)] even found the same for argon bubbles rising in mercury. Surface-active agents have no effect on the rise velocity of bubbles larger than 4 mm in thin liquids (Davenport et al., loc. cit.).

Above a Reynolds number of the order of magnitude of 1000, the bubbles assume a helmet shape, with a flat bottom (Eckenfelder and Barnhart, loc. cit.; and Leibson et al., loc. cit.). After bubbles become large enough to depart from Stokes' law at their terminal velocity, behavior is generally complicated and erratic, and the data reported scatter considerably. The rise can be slowed, furthermore, by a wall effect if the diameter of the container is not greater than 10 times the diameter of the bubbles, as shown by Uno and Kintner [*Am. Inst. Chem. Eng. J.*, **2**, 420 (1956); and Collins, *J. Fluid Mech.*, **28**(1), 97 (1967)]. Work has been done to predict the rise velocity of large bubbles [Rippin and Davidson, *Chem. Eng. Sci.*, **22**, 217(1967); Grace and Harrison, ibid., 1337; Mendelson, *Am. Inst. Chem. Eng. J.*, **13**, 250 (1967); Cole, ibid., Lehrer, *J. Chem. Eng. Japan*, **9**, 237; (1976) and Lehrer, *Am. Inst. Chem. Eng. J.*, **26**, 170 (1980)]. The works of Lehrer present correlations which accurately predict rise velocities for a wide range of system properties. An excellent review of the technical literature concerning the rise of single bubbles and drops has been published by Clift, Grace, and Weber (*Bubbles, Drops and Particles*, Academic, New York, 1978). Mendelson has used a wave theory to predict the terminal velocity, and Cole has checked the theory with additional data. The other authors listed solved some simplified form of the Navier-Stokes equations.

When bubbles are produced in clouds, as by a porous disperser, their behavior during rising is further complicated by interaction among themselves. In addition to the tendency for small bubbles to coalesce and large ones to disintegrate, there are two additional opposing influences on the rate of rise of bubbles of any particular size: (1) a "chimney" effect can develop in which a massive current upward appears at the axis of the bubble stream, leading to increased net bubble velocity; and (2) the proximity of the bubbles to one another can result in a hindered-settling condition, leading to reduced average bubble velocity. Figure 18-92 shows the data of Houghton, McLean, and Ritchie (op. cit.) for clouds of bubbles compared with the single-bubble data of Houghton, Ritchie, and Thompson (op. cit.) for pure water and seawater and of Peebles and Garber [*Chem. Eng. Prog.*, **49,** 88 (1953)] for acetic acid and ethyl acetate. The bubble clouds were produced with a sintered-glass plate of mean pore size (inferred from air wet-permeability data) of 81 μm.

The difference between the curves for pure water and seawater again illustrates the significance of small concentrations of solute with respect to bubble behavior. In commercial bubble columns and agitated vessels coalescence and breakup are so rapid and violent that the rise velocity of a single bubble is meaningless. The average rise velocity can, however, be readily calculated from holdup correlations which will be given later.

The quantitative examination of bubble systems is aided by the use of proper illumination and photography. The formation of bubbles at single sources often is sufficiently periodic to be "stopped" by stroboscopic light. Clouds of rising bubbles are more difficult to assess and require careful technique. Satisfactory photographic methods have been developed by Vermeulen, Williams, and Langlois [*Chem. Eng. Prog.*, **51,** 85 (1955)] and by Calderbank [*Trans. Inst. Chem. Eng.*, **36,** 443 (1958)] and are described by these authors. Calderbank's technique resulted in particularly precise measurements that permitted a good estimation of the surface area of the dispersed bubbles.

Methods of Gas Dispersion The problem of dispersing a gas in a liquid may be attacked in several ways: (1) the gas bubbles, of the desired size or which grow to the desired size, may be introduced

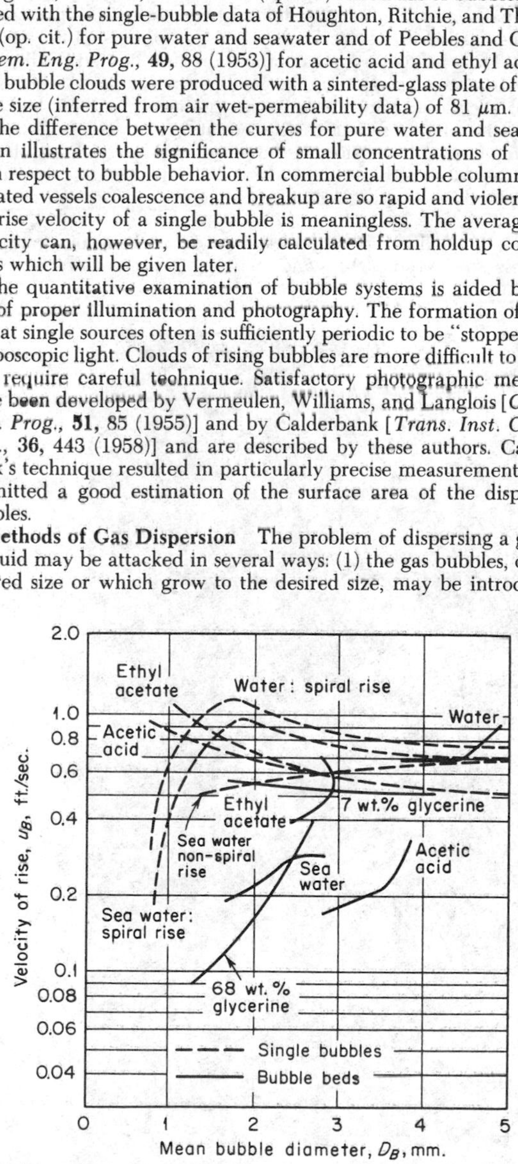

FIG. 18-92 Velocity of rising bubbles, singly and in clouds. To convert feet per second to meters per second, multiply by 0.305. [*From* Chem. Eng. Sci., *7, 48 (1957).*]

directly into the liquid; (2) a volatile liquid may be vaporized by either decreasing the system pressure or increasing its temperature; (3) a chemical reaction may produce a gas; or (4) a massive bubble or stream of gas is disintegrated by fluid shear and/or turbulence in the liquid.

Spargers: Simple Bubblers The simplest method of dispersing gas in a liquid contained in a tank is to introduce the gas through an open-end standpipe, a horizontal perforated pipe, or a perforated plate at the bottom of the tank. At ordinary gassing rates (corresponding to the jet regime) relatively large bubbles will be produced regardless of the size of the orifices.

Perforated-pipe or -plate spargers usually have orifices 3 to 12 mm (⅛ to ½ in) in diameter. Effective design methods to minimize maldistribution are presented in the *Chemical Engineers' Handbook* (5th ed., 1973, p. 5–47) and by Knaebel [*Chem. Eng.*, 116 (Mar. 9, 1981)]. In the author's experience, for turbulent flow conditions into the sparger, the following relationship will allow design of a perforated-pipe sparger for a given degree of maldistribution provided $N > 5$ and $L/D < 300$.

$$D_p = 0.95(NC)^{1/2} D_h/(\Delta V_h/V_h)^{1/4} \qquad (18\text{-}122)$$

where D_p = pipe diameter, D_h = sparging hole diameter, N = number of holes in sparger, C = orifice coefficient for sparger hole (see *Chemical Engineers' Handbook*, 5th ed., pp. 5–13, 5–34), V_h = average velocity through sparger holes, ΔV_h = difference between maximum and minimum velocities through sparger holes, and $\Delta V_h/V_h$ = fractional maldistribution of flow through sparger holes.

Simple spargers are used as agitators for large tanks, principally in the cement and oil industries. Kauffman [*Chem. Metall. Eng.*, **37,** 178–180 (1930)] reported the following air rates for various degrees of agitation in a tank containing 2.7 m (9 ft) of liquid:

Degree of agitation	Air rate, $ft^3/(ft^2$ tank cross section · min)
Moderate	0.65
Complete	1.3
Violent	3.1

NOTE: To convert feet per minute to meters per second, multiply by 0.0051.

For a liquid depth of 0.9 m (3 ft) he recommended that the listed rates be doubled.

An air lift consisting of a sparger jetting into a draft tube with ports discharging at several heights has been recommended by Heiser [*Chem. Eng.*, **55**(1), 135 (1948)] for maintaining agitation in a heavy, coarse slurry the level of which varies widely. The design is illustrated in Fig. 18-93.

The ability of a sparger to blend miscible liquids might be described in terms of a fictitious diffusivity. Siemes did so, reporting that the agitation produced by a stream of bubbles rising in a tube with a superficial velocity of about 8.24 cm/s (0.27 ft/s) corresponded to an apparent diffusion coefficient as large as 75 cm^2/s [*Chem. Ing. Tech.*, **29,** 727 (1957)]. The blending rate thus is several orders of magnitude higher than it would be by natural diffusive action.

These results are typical of later investigations on back mixing, which will be discussed in more detail later.

Lehrer [*Ind. Eng. Chem. Process Des. Dev.*, **7,** 226 (1968)] conducted liquid-blending tests with air sparging in a 0.61-m- (2-ft-) diameter by 0.61-m (2-ft-) tall vessel and found that an air volume equal to about one-half of the vessel volume gave thorough blending of inviscid liquids of equal viscosities. Using an analogy to mechanically agitated vessels in which equal tank turnovers give equal blend times, one would expect this criterion to be applicable to other vessel sizes. Liquids of unequal density would probably require somewhat more air.

Open-end pipes, perforated plates, and ring- or cross-style perforated-pipe spargers are used without mechanical agitation to promote mass transfer, as in chlorinators and biological sewage treatment. In the "quiescent regime" [superficial gas velocity less than 4.57 to 6.1 cm/s (0.15 to 0.2 ft/s)] the previously mentioned spargers

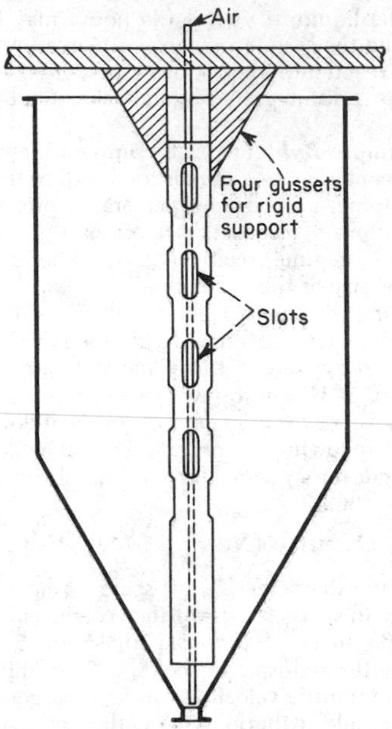

FIG. 18-93 Slotted air lift for agitation of a variable-level charge. [*From* Chem. Eng., *55(1), 135 (1948).*]

are usually operated at orifice Reynolds numbers in excess of 6000 in order to get small bubbles so as to increase the interfacial area and thus increase mass transfer. In the "turbulent regime" [superficial gas velocity greater than 4.57 to 6.1 cm/s (0.15 to 0.2 ft/s)] sparger design is not critical because a balance between coalescence and breakup is established very quickly according to Towell, Strand, and Ackerman [*Am. Inst. Chem. Eng. Symp. Ser.*, **10,** 97 (1965)]. However, a reasonably uniform orifice distribution over the column cross section is desirable, and according to Fair [*Chem. Eng.*, **74,** 67 (July 3, 1967); 207 (July 17, 1967)] the orifice velocity should be less than 75 to 90 m/s (250 to 300 ft/s).

Porous Septa In the quiescent regime porous plates, tubes, disks, or other shapes which are made by bonding or sintering together carefully sized particles of carbon, ceramic, polymer, or metal are frequently used for gas dispersion, particularly in foam fractionators. The resulting septa may be used as spargers to produce much smaller bubbles than will result from a simple bubbler. Figure 18-94 shows a comparison of the bubbles emitted by a perforated-pipe sparger [0.16-cm (0.062-in) orifices] and a porous carbon septum (120-μm pores). The gas flux through a porous septum is limited on the lower side by the requirement that for good performance the whole sparger surface should bubble more or less uniformly, and on the higher side by the onset of serious coalescence at the surface of the septum, resulting in poor dispersion. In the practical range of fluxes, the size of the bubbles produced depends on both the size of pores in the septum and the pressure drop imposed across it, being a direct function of both.

Table 18-17 lists typical grades of porous carbon, silica, alumina, stainless steel (type 316), and polymer commercially available.

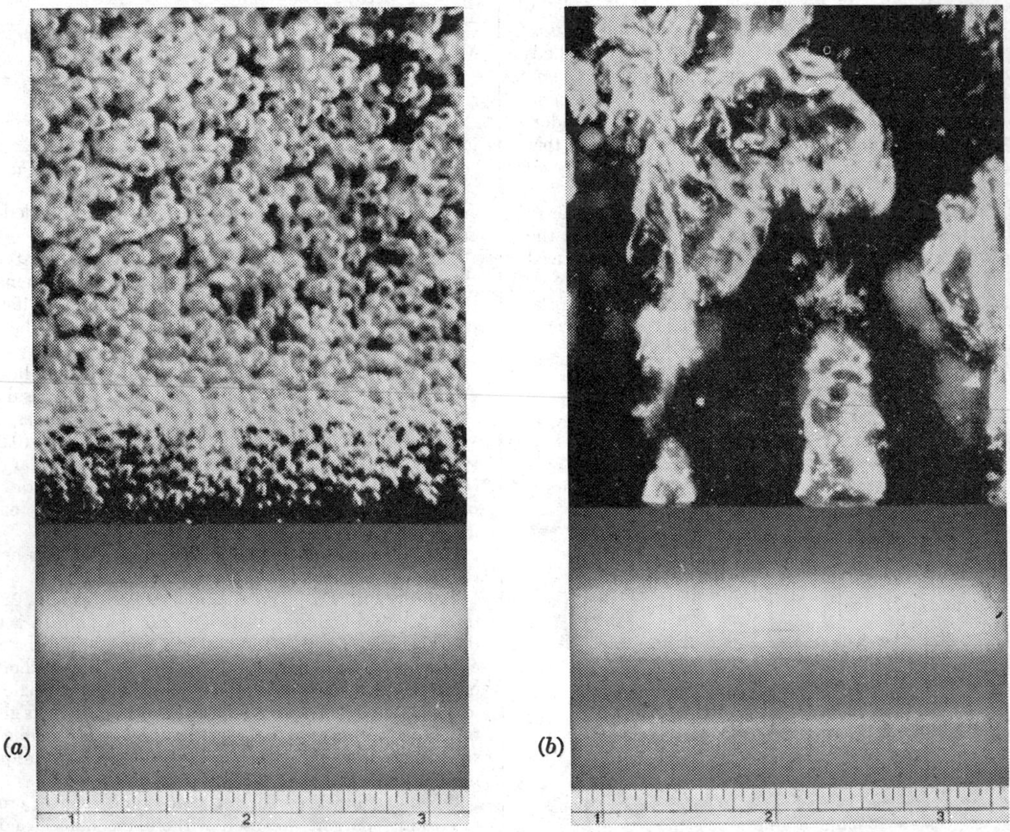

FIG. 18-94 Comparison of bubbles from a porous septum and from a perforated-pipe sparger. Air in water at 70°F. (*a*) Grade 25 porous-carbon diffuser operating under a pressure differential of 13.7 in of water. (*b*) Karbate pipe perforated with 1/16-in holes on 1-in centers. *(National Carbon Co.)* To convert inches to centimeters, multiply by 2.54; °C = 5/9 (°F−32).

TABLE 18-17 Characteristics of Porous Septa

Grade	Avg. % porosity	Avg. pore diam.	Air-permeability data: Diaphragm thickness, in.	Pressure differential, in. water	Air flow, cu. ft./(sq. ft.)(min.)
Alundum porous alumina*					
P2220	...	25	1	2	0.35
P2120	36	60	1	2	2
P260	35	164	1	2	15
P236	34	240	1	2	40
P216	...	720	1	2	110
National porous carbon†					
60	48	33	1	2	
45	48	58	1	2	2
25	48	120	1	2	13
Filtros porous silica‡					
Extra fine	26.0	55	1.5	2	1–3
Fine	28.8	110	1.5	2	4–8
Medium fine	31.1	130	1.5	2	9–12
Medium	33.7	150	1.5	2	13–20
Medium coarse	33.8	200	1.5	2	21–30
Coarse	34.5	250	1.5	2	31–59
Extra coarse	36.5	300	1.5	2	60–100
Porous plastic§					
Teflon	....	9	0.125	1.38	5
Kel-F	....	15	0.125	1.38	13
Micro Metallic porous stainless steel§,¶					
H	45	5	0.125	1.38	1.8
G	50	10	0.125	1.38	3
F	50	20	0.125	1.38	5
E	50	35	0.125	1.38	18
D	50	65	0.125	1.38	60
C	55	165	0.125	27.7	990

*Data by courtesy of Norton Co., Worcester, Mass. A number of other grades between the extremes listed are available.
†Data by courtesy of National Carbon Co., Cleveland, Ohio.
‡Data by courtesy of Filtros Inc., East Rochester, N.Y.
§Data by courtesy of Pall Corp., Glen Cove, N.Y.
¶Similar septa made from other metals are available.

Porous media are manufactured also from porcelain, glass, silicon carbide, and a number of metals: Monel, Inconel, nickel, bronze, Hastelloy C, Stellite L-605, gold, platinum, and many types of stainless steel. The air permeabilities of Table 18-17 indicate the relative flow resistances of the various grades to homogeneous fluid but may not be used in designing a disperser for submerged operation, for the resistance of a septum to the flow of gas increases when it is wet. The air permeabilities for water-submerged porous carbon of some of the grades listed in the table are shown in Fig. 18-95. The data were determined with septa 0.625 in thick in water at 70°F. Comparable wet-permeability data for 1-in Alundum plates of two grades of fineness are given in Table 18-18.

The gas rate at which coalescence begins to reduce the effectiveness of dispersion appears to depend not only on the pore size and pore structure of the dispersing medium but also on the liquid properties, liquid depth, agitation, and other features of the sparging environment; coalescence is strongly dependent on the concentration of surfactants capable of forming an electrical double layer and thus produce ionic bubbles, long-chain alcohols in water being excellent examples. For porous-carbon media, the manufacturer suggests that the best dispersion performance will result if the broken-line regions of Fig. 18-95 are avoided. For porous stainless-steel spargers, which extend to a lower pore size than carbon, Micro Metallic Division, Pall Corp., recommends (Release 120A, 1959) a working limit of 8 ft/min (0.044 m/s) to avoid serious coalescence. This agrees with the data reported by Konig et al. (loc. cit.), in which 0.08 m/s (0.26 ft/s) was used and bubbles as small as 1 mm (0.04 in) were produced from a 5-μm porous sparger.

Slabs of porous material are installed by grouting or welding together to form a diaphragm, usually horizontal. Tubes are prone to produce coalesced gas at rates high enough to cause bubbling from their lower faces, but they have the advantage of being demountable for cleaning or replacement (U.S. Patent 2,328,655). Roe [*Sewage Works J.*, **18**, 878 (1945)] claimed that silicon carbide tubes are superior to horizontal plates, principally because of the wiping action of the liquid circulating past the tube. He reported respective maximum capacities of 2.5 and 1.5 cm^3/gas ($cm^2 \cdot s$) [5 and 3 ft^3 gas/($ft^2 \cdot$ min)] for a horizontal tube and a horizontal plate of the same material (unspecified grade). Mounting a flat-plate porous sparger vertically instead of horizontally seriously reduces the effectiveness of the sparger for three reasons: (1) the gas is distributed over a reduced cross section; (2) at normal rates, the lower portion of the sparger may not operate because of difference in hydrostatic head; and (3) there is a marked tendency for bubbles to coalesce along the sparger surface. Bone (M.S. thesis in chemical engineering, University of Kansas, 1948) found that the oxygen sulfite solution coefficient identified with a 3.2- by 10-cm (1¼- by 4-in) rectangular porous carbon sparger was 26 to 41 percent lower for vertical than for horizontal operation of the sparger, the greatest reduction occurring when the long dimension was vertical.

Precipitation and Generation Methods For a thorough understanding of the phenomena involved, bubble nucleation should be considered. A discussion of nucleation phenomena is beyond the scope of this *Handbook;* however, excellent coverages are presented by Blander and Katz [*Am. Inst. Chem. Eng. J.*, **21**, 833 (1975)] and Attar [(ibid., **24**, 106 (1978)]. Precipitation of a gas from a supersaturated solution generally results in a fine dispersion of bubbles throughout the liquid [Bateman and Lang, *Can. J. Res.*, **23E**, 22

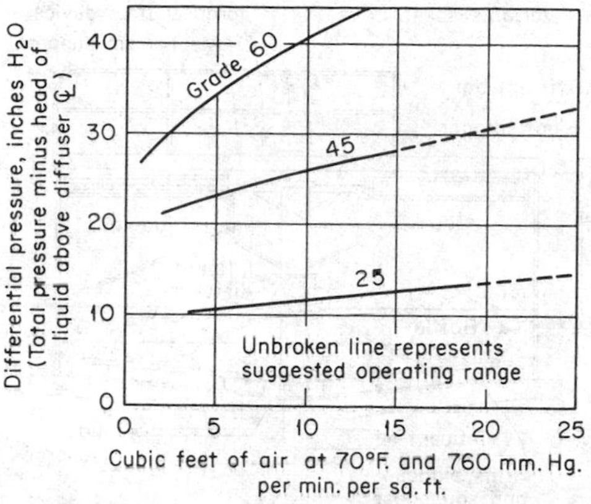

FIG. 18-95 Pressure drop across porous-carbon diffusers submerged in water at 70°F. To convert feet per minute to meters per second, multiply by 0.0051; to convert inches to millimeters, multiply by 25.4; °C = ⅝ (°F − 32). *(National Carbon Co.)*

TABLE 18-18 Wet Permeability of Alundum Porous Plates 1 in Thick*

Dry permeability at 2 in. of water differential, cu. ft./(min.)(sq. ft.)	Pressure differential across wet plate, in. of water	Air flow through wet plate, cu. ft./(min.)(sq. ft.)
4.3	20.67	2.0
	21.77	3.0
	22.86	4.0
	23.90	5.0
55.0	4.02	1.0
	4.14	2.0
	4.22	3.0
	4.27	4.0
	4.30	5.0

*Data by courtesy of Norton Company, Worcester, Mass. To convert inches to centimeters, multiply by 2.54; to convert feet per minute to meters per second, multiply by 0.0051.

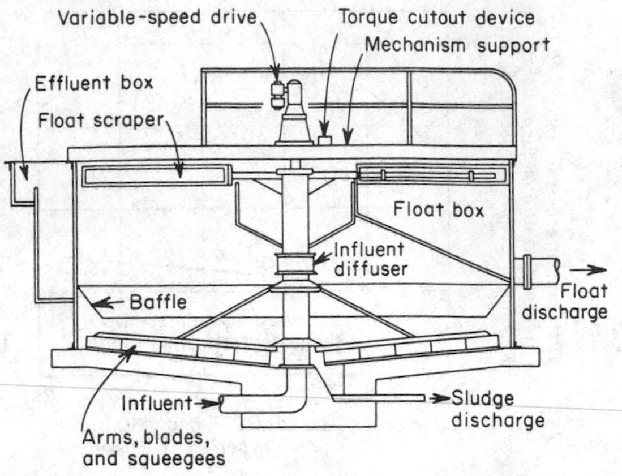

FIG. 18-96 The Flotator dissolved-air flotation thickener. *(Process Engineers, Inc., a division of Eimco Corp., now Envirotech Corporation.)*

(1945)]. Precipitation finds its widest use in the manufacture of cellular polymers. Skochdopole (op. cit.) has given a good general discussion of foamed polymers with many pertinent references. The polymer to be foamed is heated to elevated temperatures, raised to high pressure, usually by an extruder, and simultaneously saturated with an inert gas. The foam which results from gas precipitation upon expansion to atmospheric pressure is stabilized either totally by the physical process of cooling, which is common for thermoplastics, or partially by cooling and chemical reaction, which is common for vulcanizing rubbers and thermosetting plastics. Gould [*Rubber Age (N.Y.)*, **54**, 526; **55**, 65 (1944)] explains how foam rubber is made by saturating with an inert gas at elevated temperatures and 3.1×10^4 kPa (4500 lbf/in^2), and Taylor (U.S. Patent 2,372,695) made foamed thermoplastics by saturating with a volatile solvent at 210°C and 2.1×10^4 kPa (3000 lbf/in^2) pressure. Foamy household products such as shaving creams are formed by precipitation of the dissolved propellent upon expansion to atmospheric pressure (see "Foams and Aerosols" in *Encyclopedia of Chemical Technology*, op. cit.).

Precipation is the principle underlying two designs of a froth-flotation cell. In one, preaerated pulp is caused to froth by the application of a vacuum. An example is the Clemens cell. In another, the pulp is saturated with air at superatmospheric pressure and is flashed into the cell, which operates at atmospheric pressure. An example is the Juell cell. Neither of the flotation cells is extensively used in the mineral industries today.

Pressure saturation followed by flashing and bubble precipitation is the active principle in a combined flotation and thickening unit, the Eimco-Process Flotator, used by the process industries particularly for waste pretreatment and recovery (Fig. 18-96). The unit is a circular tank with the thickener and scraper mechanism supported on beams spanning the tank. The tank may be 2 to 12 m (6 to 40 ft) in diameter. The feed mixture is saturated with air at elevated pressure, usually about 308 kPa (30 psig). After passing through a back-pressure valve, the influent is introduced to the unit below the liquid level, where a mass of fine bubbles is released, carrying fine solids upward to a level at which they can be skimmed to the scum trough. Coarse, heavy solids meanwhile settle as in a conventional thickener. The method of influent introduction is said to be critical to the success of the Flotator's operation and must be specified for the particular problem being attacked. Examples of applications of the Flotator are sewage-sludge thickening, fat recovery from meat-packing waste, oil removal from refinery wastewaters, and fiber recovery in the pulp and paper industry.

Generation Fine, well-dispersed bubbles are produced if a dissolved or finely divided suspended material is decomposed to yield a gas.

Generation methods are employed to prepare cellular elastomers or thermoplastics to which the resulting products of decomposition are not harmful. A number of "blowing agents" are used, the most common being sodium and ammonium bicarbonate, calcium carbonate, ammonium nitrate, diazo derivatives, and diisocyanates. Colin-Russ [*Chem. Trade J.*, **115**, 631 (1944)] suggested gas-saturated leather charcoal as a blowing agent. In the Talaly process for foam rubber, hydrogen peroxide is the source of gas, the oxygen being liberated by the action of catalase from yeast (Winspear and Waterman, in Morton, *Introduction to Rubber Technology*, Reinhold, New York, 1959, chap. 18). In every case, the blowing agent is compounded into a latex before it is gelled or into an elastomeric mass before it is cured, the material being uniformly dispersed throughout the plastic mass before gas generation occurs.

Leavening agents are gas-generation sources used to produce the light cellular structure of breads and pastries. Examples are baking powder and yeast, the latter effecting the decomposition of carbohydrate.

Polyurethane and epoxy foams are made by intimately blending (several examples from the patent literature of pertinent blending equipment are discussed by Randolph, *Mixing in the Chemical and Allied Industries*, Noyes Data Corp., Park Ridge, N.J., 1967) the uncross-linked liquid polymer, a cross-linking catalyst, and a volatile liquid (usually one of the Freons): the heat generated by the chemical reaction vaporizes the volatile liquid, thereby producing foam. The thermosetting reaction also renders the foam rigid.

Fluid-Attrition Systems: Nozzles and Pipe-Line Contactors The turbulence developed during the rapid flow of fluid through a nozzle or a pipe sometimes is utilized to disperse a gas in a liquid. Steam-water mixers of the venturi-nozzle type are manufactured by several companies. Excellent dispersions can be obtained with such devices, although for a gas-in-water dispersion the gas-to-liquid ratio is relatively low. In one design of nozzle, both air and steam are dispersed into the water to reduce further the vibration and noise resulting from collapsing steam bubbles.

Jet diffusers (Fig. 18-97) and impingement aerators (Fig. 18-98) are used to aerate waste-treatment lagoons. Both devices are installed in multiple on adjacent air and liquor headers. In the jet diffuser the air is aspirated into the liquid, which is under 239 to 308 kPa (20 to 30 psig) in the liquor header. For the impingement aerators the

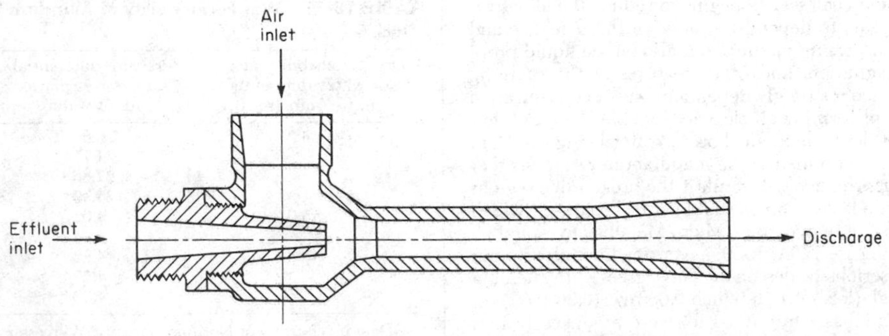

FIG. 18-97 Aeration ejector. *(Penberthy, a division of Houdaille Industries, Inc.)*

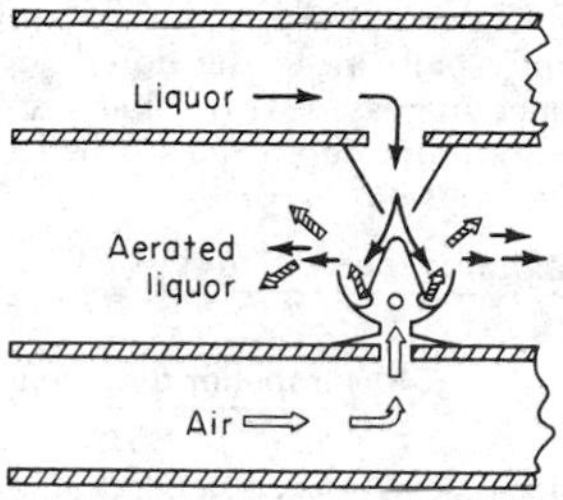

FIG. 18-98 Impingement aerator.

liquor is generally lifted to the liquor header by an air lift. Jet diffusers and pipe-line contactors are used extensively to produce fire-fighting foams [Ratzer, *Ind. Eng. Chem.*, **48**, 2013 (1956); Rivikind and Myerson, ibid., 2017; and "Foams," *Encyclopedia of Chemical Technology*, op. cit.].

Pipe-line contactors of gas-liquid mixtures usually involve orifices or baffles to redistribute the gas periodically. Pfirrmann (German Patent 740,674) described a pipe-line disperser with occasional short constrictions of such cross section that the fluid velocity through them would exceed 1 m/s (3 ft/s). Tell [*Chem. Metall. Eng.*, **52**(6), 115 (1945)] recommended an orifice of unstated dimension for dispersing continuously small quantities of gas in a hydrocarbon.

A downflow pipe-line disperser for air and sewage was developed by Nordell (U.S. Patent 2,374,722). Initial dispersion was accomplished by a vortex above the downpipe into which the air was entrained as small bubbles.

Nozzles and flow mixers may be used for gas-liquid contacting only when cocurrent flow is permissible.

Cascade Systems A stream of liquid falling through a gas into a pool will entrain, under the proper conditions, approximately its own volume of gas and will disperse the entrained gas into the pool. This principle was first employed in cascade-type froth-flotation machines.

Cooper (U.S. Patent 2,398,345) described a gravity cascade system designed for scrubbing a gas with the cascading liquid. Mertes (U.S. Patent 2,128,311) reported that a solution containing a ferrous compound, when discharged vertically downward with a velocity of 12 m/s (40 ft/s) through air from a nozzle into a pool less than 7½ cm (3 in) below it, entrained sufficient air for rapid oxidation of the ferrous salt.

De Frate and Rush (AIChE Prepr. 39D, 64th national meeting, New Orleans, March 1969) have measured gas entrainment due to liquid jets entering the free surface. The volumetric ratio of gas entrained to liquid-flow rate varied from 1 to 20. Their data were correlated approximately by

$$\frac{Q_g}{Q_\ell} = 0.0316\left(\frac{V^2\rho_\ell L}{\sigma g_c}\right)^{1/2} \tag{18-123}$$

where Q_g/Q_ℓ is volumetric entrainment ratio, V is jet velocity, ρ_ℓ is liquid density, L is the length of the jet, and σ is the interfacial tension.

Mechanical Agitators These are used primarily (1) to disperse gas into a liquid for the purpose of promoting mass transfer and (2) to create a foam. For producing foam many different agitators are available, especially from manufacturers catering to the food industry. Wire-whip agitators, which come in several styles, one of which is illustrated in Fig. 18-99, are used quite successfully to aerate a liquid through the liquid-free surface and produce a uniform foam. Several other agitator types (serrated disk, disk, turbine, etc.) were tested by Fundy and Bates [*Am. Inst. Chem. Eng. J.*, **9**, 338 (1963)] in producing a liquid-liquid dispersion. Their work is of fundamental importance for producing the more stable gas-liquid dispersions. They found for a noncoalescing system that the long-time drop (bubble) size was dependent only on the agitator-tip speed but that the time dependence of the average drop (bubble) size was strongly dependent on the agitator pumping rate. High-pumping-rate impellers approached the long-time drop size much more quickly than low-pumping-rate impellers. The findings of Fondy and Bates are quite important for scale-up purposes, because it is well known that scaling up on the basis of constant tip speed will give much lower vessel-turnover time in the large system, resulting in a much slower approach to equilibrium drop (bubble) size.

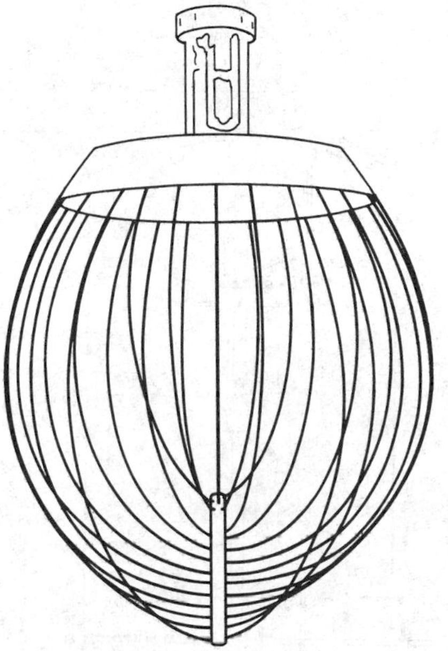

FIG. 18-99 Wire whip.

For gas dispersion in pot-type vessels for the purpose of enhancing mass transfer the four- or six-bladed disk-type turbine is now used almost exclusively. The disk prevents gas from escaping upward without passing through the high-shear zone at the impeller blade tips. A simple open-pipe sparger directly underneath the impeller gives the same performance as more expensive spargers [see Valentin, *Br. Chem. Eng.*, **12**, 1213 (1967)]. Particularly in hydrogenation reactors a second impeller, often an axial-flow turbine or a propeller, is placed near the free surface in order to reintroduce gas from the freeboard space back into the liquid. The Turbo-gas-absorber, illustrated in Fig. 18-100, is an example of equipment which is used in closed vessels to entrain gas from the vessel head space into the liquid. Gas-entraining capacities of the Turbo-gas-absorber are given in Table 18-19.

Tsao and Cramer (AIChE Prepr. 37D, 61st annual meeting, Los Angeles, 1968), Zlokarnik [*Chem. Ing. Tech.*, **38**, 357 (1966)], Martin [*Ind. Eng. Chem. Process Des. Dev.*, **11**, 397 (1972)], and Topiwala and Hamer [*Trans. Inst. Chem. Eng.*, **52**, 113 (1974)] have measured the gas-pumping rate of a hollow-shaft "Waldhof-type" impeller. Zlokarnik (op. cit.) has presented a comprehensive correlation for the gas-entraining capacity of a hollow-shaft agitator. The entire correlation is not presented here; however, with careful design, at least as much gas can be entrained as predicted by the following correlation:

$$Q/ND^3 = 0.075 \tag{18-124}$$

for an impeller with four arms, each arm being constructed of a hollow tube, when $d/D = 7$ and $(ND)^2/gH > 2$, where Q = gas-pumping capacity of the hollow-shaft agitator, N = agitator speed, D = impeller diameter, d = inside diameter of the hollow arm, g = gravitational acceleration, and H = liquid submergence of the impeller.

Boerma and Lankester [*Chem. Eng. Sci.*, **23**, 799 (1968) have measured the surface aeration capacity of a six-bladed disk-type turbine. In a fully baffled vessel the optimum depth to obtain maximum

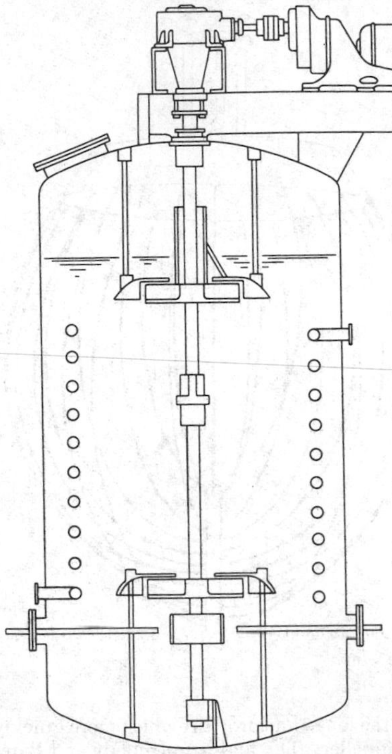

FIG. 18-100 Turbo-gas-absorber with combined pressure and self-induction impellers in hydrogenation vessel. The upper impeller is the gas entrainer. *(Turbo-Mixer Division, General American Transportation Corp.)*

TABLE 18-19 Turbo-Gas-Absorber Gas-Entraining Capacity*

Impeller diam., in.	Maximum gas-entraining capacity, cu. ft./min.	Maximum gas rate for pressure-fed operation, cu. ft./min.
4	2	3
6	10	15
9	15	20
12	25	40
18	60	100
22	100	180
27	160	300
34	...	600
42	...	1200
60	...	2500
72	...	3800
84	...	5000+

*Data by courtesy of Turbo-Mixer Division, General American Transportation Corp. To convert inches to centimeters, multiply by 2.54; to convert cubic feet per minute to cubic meters per second, multiply by 4.72×10^{-4}.

gas dispersion was 15 percent of the liquid depth. In a vessel with baffles extending only halfway to the liquid surface the optimum impeller submergence increased with agitator speed because of the vortex formed. At optimum depth the following correlation is recommended for larger vessels:

$$V(\text{m}^3/\text{s}) = 0.30\left(\frac{\text{agitator speed, r/min}}{500}\right)^{2.5} \times \left(\frac{\text{impeller diameter, cm}}{25.4}\right)^{4.5} \quad (18\text{-}125)$$

Gas dispersion through the free surface by mechanical aerators is becoming very prevalent in aerobic waste-treatment lagoons. Surface aerators are generally of two types: (1) large-diameter flow-speed turbines operating just below the free surface of the liquid, often pontoon-mounted; and (2) small-diameter high-speed (normally motor-speed) propellers operating in draft tubes, which units are always pontoon-mounted. An example of the turbine type is illustrated in Fig. 18-101, and the propeller type is illustrated in Fig. 18-102. There are several other styles of the turbine type; e.g., Mixing Equipment Co., Inc., uses an unshrouded 45° axial-flow turbine [see Dykman and Michel, *Chem. Eng.*, **117** (Mar. 10, 1969)], Infilco makes a unit which has a large-diameter vaned disk operating just below the free surface with a smaller-diameter submerged-disk turbine for additional solids suspension, and Yoemans Brothers, Inc., makes a Hi-Cone unit which has six separately mounted triangular-shaped blades operating over a draft tube.

Equipment Selection Ideally, selection of equipment to produce a gas-in-liquid dispersion should be made on the basis of a complete economic analysis. The design engineer and especially the pilot-plant engineer seldom have sufficient information or time to do a complete economic analysis. In the following discussion some guidelines are given as to what equipment might be feasible and what equipment migh prove most economical.

For producing foam for foam-separation processes usually perforated-plate or porous-plate spargers are used. Mechanical agitators are often not effective in the light foams needed in foam fractionation. Dissolved-air flotation, based on the release of a pressurized flow in which oxygen was dissolved, has been shown to be effective sometimes for particulate removal when sparged air failed because the bubbles formed upon precipitation are smaller—down to 80 μm—than bubbles possible with sparging, typically 1000 μm [Grieves and Ettelt, *Am. Inst. Chem. Eng. J.*, **13**, 1167 (1967)]. Mechanically agitated surface aerators such as the Wemco-Fagergren flotation unit are used extensively for ore flotation.

To produce foam by batch processes mechanical agitators are used almost exclusively. The gas can be either introduced through the free surface by the entraining action of the impeller or alternatively sparged beneath the impeller. In such batch operation the liquid level gradually rises as the foam is generated; thus squatty impellers such as turbines are rapidly covered with foam and must almost always be sparged from below. Tall impellers such as the previously mentioned wire whips are especially well suited to entrain gas from the vapor space. For a new application generally some experimentation with different impellers is necessary in order to get the desired fine final bubble size without getting frothing over initially. For producing foams continually an aspirating venturi nozzle and restric-

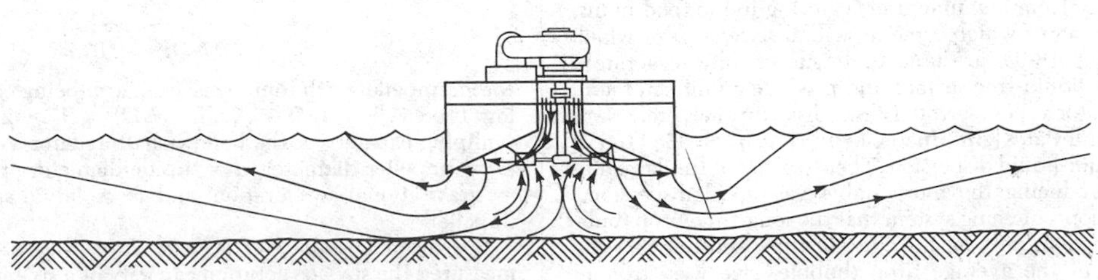

FIG. 18-101 The Cyclox surface aerator. *(Cleveland Mixer Co.)*

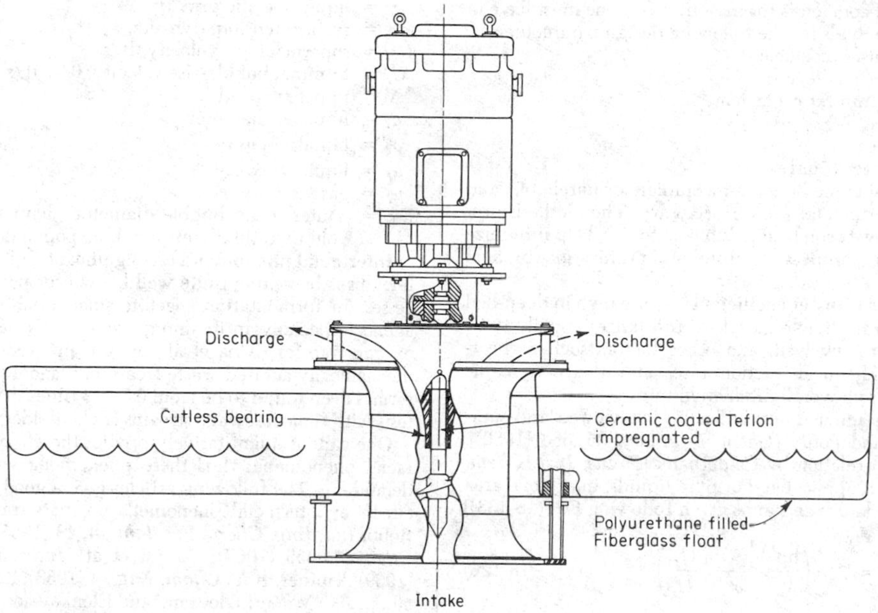

FIG. 18-102 Propeller-type surface aerator. *(Ashbrook-Simon-Hartley Corp.)*

tions in pipes such as baffles and metal gauzes are generally most economical.

For gas absorption the equipment possibilities are generally packed columns; plate distillation towers, possibly with mechanical agitation on every plate; deep-bed contactors (bubble columns or sparged lagoons); and mechanically agitated vessels or lagoons. Packed towers and plate distillation columns are discussed elsewhere. Generally these devices are used when a relatively large number of stages (more than two or three) is required to achieve the desired result practically.

The volumetric mass-transfer coefficients and heights of transfer units for bubble columns and packed towers have been compared for absorption of CO_2 into water by Houghton et al. [*Chem. Eng. Sci.*, **7**, 26 (1957)]. The bubble column will tolerate much higher vapor velocities, and in the overlapping region [superficial gas velocities of 0.9 to 1.8 cm/s (0.03 to 0.06 ft/s)] the bubble column has about 3 times higher mass-transfer coefficient and about 3 times greater height of transfer unit. The liquid in a bubble column is, for practical purposes, quite well mixed; thus, chemical reactions and component separations requiring significant plug flow of the liquid cannot be carried out with bubble columns. Bubble columns and agitated vessels are the ideal equipment for processes in which the fraction of gas absorbed need not be great, possibly the gas can be recycled, and the liquid is required to be well mixed. The gas phase in bubble columns is not nearly so well back-mixed as the liquid, and often plug flow of the gas is a logical assumption, but in agitated vessels the gas phase is also well mixed.

The choice of a bubble column or an agitated vessel depends primarily on the solubility of the gas in the liquid, the corrosiveness of the liquid (often a gas compressor can be made of inexpensive material, whereas a mechanical agitator may have to be made of exotic, expensive materials), and the rate of chemical reaction as compared with the mass-transfer rate. Bubble columns and agitated vessels seldom are used for gas absorption except in chemical reactors. As a general rule, if the overall reaction rate is 5 times greater than the mass-transfer rate in a simple bubble column, a mechanical agitator will be most economical unless the mechanical agitator would have to be made from considerably more expensive material than the gas compressor.

In bubble columns and simply sparged lagoons, selecting the sparger is a very important consideration. In the turbulent regime [superficial gas velocity greater than 4.6 to 6 cm/s (0.15 to 0.2 ft/s)] inexpensive perforated-pipe spargers should be used. Often the holes must be placed on the pipe bottom in order to make the sparger free-draining during operation. In the quiescent regime porous septa will often give considerably higher overall mass-transfer coefficients than perforated plates or pipes because of the formation of tiny bubbles which do not coalesce. Chain and coworkers (*First International Symposium on Chemical Microbiology*, World Health Organization, Monogr. Ser. 10, Geneva, 1952) claimed that porous disks are about twice as effective as open-pipe and ring spargers for the air oxidation of sodium sulfite. Eckenfelder [*Chem. Eng. Prog.* **52**(7), 290 (1956) has compared the oxygen-transfer capabilities of various equipment on the basis of the operating power required to absorb a given quantity of O_2.

The installed cost of the various pieces of equipment probably would not vary sufficiently to warrant including it in an economic analysis. Surface mechanical aerators are not included in this comparison. Of the units compared it appears that porous tubes give the most efficient power usage. Kalinske (*Adv. Biol. Waste Treatment*, 1963, p. 157) has compared submerged sparged aerators with mechanical surface aerators. He has summarized this comparison in *Water Sewage Works*, 33 (January 1968). He indicates that surface areators are significantly more efficient than subsurface aeration, both for oxygen-absorption and for gas-stripping operations.

Zlokarnik and Mann (paper at Mixing Conf., Rindge, N.H., August 1975) have found the opposite of Kalinske; i.e., subsurface diffusers, subsurface sparged turbines, and surface aerators compare approximately as 4:2:1 respectively in terms of O_2 transfer efficiency; however, Zlokarnik [*Adv. Biochem. Eng.*, **11**, 157 (1979)] later indicates that the scale-up correlation used earlier might be somewhat inaccurate. When all available information is considered, it appears that with near-optimum design any of the aeration systems (diffusers, submerged turbine, or surface impeller) should give a transfer efficiency of at least 2.25 kg O_2/kWh (5 lb O_2/kWh). Thus, the final selection should probably be made primarily on the basis of operational reliability, maintenance, and capital costs.

Mass Transfer Mass transfer in plate and packed gas-liquid contactors has been covered earlier in this subsection. Attention here will be limited to deep-bed contactors (bubble columns and agitated vessels). Theory underlying mass transfer was developed in Sec. 14.

To design deep-bed contactors for mass transfer, one must have, in general, predictive methods for the following design parameters:

Gas-phase mass-transfer coefficient
Interfacial area
Liquid-phase mass-transfer coefficient
Interfacial resistance
Holdup of gas phase
Mean driving force for transfer

In most cases available methods are incomplete or unreliable, and some supporting experimental work is necessary. The methods proposed here should allow theoretical feasibility studies, help minimize experimentation, and permit a measure of optimization in final design.

The **gas-phase mass-transfer coefficient** is quite high in deep-bed contactors, leading to negligible gas-phase resistance. Usually sparingly soluble gases are involved, and other factors such as heat-removal capability, chemical-reaction rate, and thermodynamic equilibrium tend to limit overall production rates.

Interfacial area in agitated vessels has been reviewed and summarized by Sridhar and Potter [*Chem. Eng. Sci.*, **35,** 683 (1980)]. They found that a correlation by Calderbank [*Trans. Inst. Chem. Eng.*, **36,** 443 (1958)] is applicable: For pure liquids, interfacial area a is given by [nomenclature and units given following Eq. (18-133)]

$$a = 215\left(\frac{\text{hp}}{\text{V}}\right)^{0.4}\frac{\gamma^{0.2}}{\sigma^{0.6}}\left(\frac{U_{gt}}{U_r}\right)^{0.5} \qquad (18\text{-}126)$$

where surface aeration is negligible, generally when

$$\left(\frac{Nd_i^2\rho_\ell}{\mu_\ell}\right)^{0.7}\left(\frac{Nd_i}{U_{gt}}\right)^{0.3} < 25{,}000$$

when surface aeration is significant,

$$a' = a \times 10^{-4}\left[\left(\frac{Nd_i^2\rho_\ell}{\mu_\ell}\right)^{0.7}\left(\frac{Nd_i}{U_{gt}}\right)^{0.2} - 25{,}000\right] \qquad (18\text{-}127)$$

Sridhar and Potter [*Ind. Eng. Chem. Fundam.*, **19,** 21 (1980)] have also recently studied **gas holdup** and **bubble diameters** in agitated vessels. The correlations of Calderbank (loc. cit.) are recommended. Fractional gas holdup ϵ and Sauter mean bubble diameter D_{32} are given by

$$\epsilon = \left(\frac{U_{gt}}{U_r}\right)^{1/2}\epsilon^{1/2} + 0.015a \qquad (18\text{-}128)$$

$$D_{32} = 0.0279\frac{\sigma^{0.6}}{(\text{hp}/\text{V})^{0.4}\gamma^{0.2}}\epsilon^{1/2} + 0.09 \qquad (18\text{-}129)$$

For solutions of electrolytes,

$$\epsilon = 470\left(\frac{\text{hp}}{\text{V}}\right)^{0.4}\frac{U_{gt}^{0.6}}{\gamma^{0.4}} \qquad (18\text{-}130)$$

$$D_{32} = 0.0153\frac{\sigma^{0.6}}{(\text{hp}/\text{V})^{0.4}\gamma^{0.2}}\epsilon^{0.4}\left(\frac{\mu_g}{\mu_\ell}\right)^{0.25} \qquad (18\text{-}131)$$

For aqueous solutions of alcohols,

$$D_{32} = 0.013\frac{\sigma^{0.6}}{(\text{hp}/\text{V})^{0.4}\gamma^{0.2}}\epsilon^{0.65}\left(\frac{\mu_g}{\mu_\ell}\right)^{0.25} \qquad (18\text{-}132)$$

In certain cases one may wish to estimate the specific interfacial area from measured or calculated values of gas holdup and Sauter mean bubble diameter. For **spherical** bubbles,

$$a = 6(\epsilon/D_{32}) \qquad (18\text{-}133)$$

Nomenclature for Eqs. (18-120) through (18-133) is:

a = specific interfacial area, cm^{-1}, in the absence of significant surface aeration
hp/V = agitator horsepower per vessel volume, hp/ft^3
γ = liquid specific gravity
σ = surface tension, dyn/cm
U_{gt} = superficial gas velocity, ft/s
U_r = terminal bubble-rise velocity, 0.87 ft/s for practical purposes
N = impeller speed
d_i = impeller diameter
ρ_ℓ = liquid viscosity
μ_ℓ = liquid viscosity
μ_2 = gas viscosity
D_{32} = Sauter mean bubble diameter (diameter such that surface-volume ratio of entire bubble population is represented), cm

Interfacial phenomena can significantly affect overall mass transfer; this is becoming quite well known for particular industrial processes. In fermentation reactors small quantities of surface-active agents (especially antifoaming agents) can drastically reduce overall oxygen transfer (Aiba et al., op. cit. pp. 153, 154), and in aerobic mechanically aerated waste-treatment lagoons overall oxygen transfer has been found to be from 0.5 to 3 times that for pure water from tests with typical sewage streams (Eckenfelder et al., op. cit., p. 105).

One cannot quantitatively predict the effect of the various interfacial phenomena; thus these phenomena will not be covered in detail here. The following articles give a good general review of the effects of interfacial phenomena on mass transfer: Goodridge and Robb, *Ind. Eng. Chem. Fundam.*, **4,** 49 (1965); Calderbank, *Chem. Eng.*, **CE 205** (1967); Gal-Or et al., *Ind. Eng. Chem.*, **61**(2), 22 (1969); Kintner, *Adv. Chem. Eng.*, **4** (1963); Resnick and Gal-Or, op. cit., p. 295; Valentin, loc. cit.; and Elenkov, loc. cit., *Ind. Eng. Chem. Ann. Rev. Mass Transfer*, **60**(1), 67 (1968); **60**(12), 53 (1968); **62**(2), 41 (1970). In the following outline the effects of the various interfacial phenomena on the factors which influence overall mass transfer are given.

Possible Effects of Interfacial Phenomena

1. Effect on continuous-phase mass-transfer coefficient
   a. Impurities concentrate at interface. Bubble motion produces circumferential surface-tension gradients which act to retard circulation and vibration, thereby decreasing the mass-transfer coefficient.
   b. Large concentration gradients and large heat effects (very soluble gases) can cause interfacial turbulence (Marangoni effect), which increases the mass-transfer coefficient.
2. Effect on interfacial area
   a. Impurities lower static surface tension and give smaller bubbles.
   b. Surfactants can electrically charge the bubble surface (produce ionic bubbles) and retard coalescence (soap stabilization of an oil-water emulsion is an excellent example of this phenomenon), thereby increasing the interfacial area.
   c. Large concentration gradients and large heat effects can cause bubble breakup.
3. Effect on mean mass-transfer driving force
   a. Relatively insoluble impurities concentrate at the interface, giving an interfacial resistance. This phenomenon has been used in retarding evaporation from water reservoirs.
   b. The axial concentration variation can be changed by changes in coalescence. The mean driving force for mass transfer is therefore changed.

Gas holdup (ϵ) **in bubble columns,** with coalescing systems, may be estimated from a correlation by Hughmark [*Ind. Eng. Chem. Process Des. Dev.*, **6,** 218–220 (1967)] reproduced here as Fig. 18-103. For noncoalescing systems, with considerably smaller bubbles, ϵ can be as great as 0.6 at U_{SG} = 0.05 m/s (0.16 ft/s) according to Mersmann [*Germ Chem. Eng.*, **1,** 1–11 (1978)].

It is often helpful to use the relationship between ϵ and superficial gas velocity (U_{SG}) and the rise velocity of a gas bubble relative to the liquid velocity ($U_r + U_\ell$) (with U_ℓ defined as positive upward):

$$\epsilon = U_{SG}/(U_r + U_\ell) \qquad (18\text{-}134)$$

Rise velocities of bubbles through liquids were discussed previously.

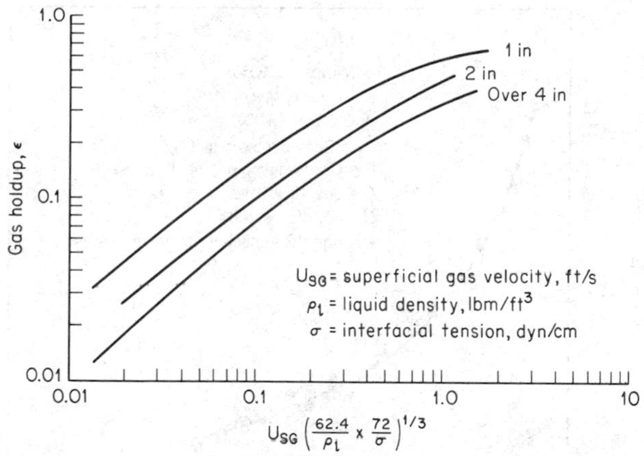

FIG. 18-103 Gas holdup correlation. [Ind. Eng. Chem. Process Des. Dev., *6*, *218 (1967).*]

For a better understanding of the interactions between parameters, it is often helpful to calculate the effective bubble rise velocity U_r from measured valves of ϵ; e.g., the data of Mersmann (loc. cit.) indicated $\epsilon = 0.6$ for $U_{SG} = 0.05$ m/s, giving $U_r = 0.083$ m/s (0.27 ft/s), which agrees with the data reported in Fig. 18-92 for the rise velocity of bubble clouds. The rise velocity of single bubbles, for $d_b \gtrsim 1$ mm, is about 0.3 m/s (0.8 ft/s), for liquids with viscosities not too different from water. Using this value in Eq. (18-125) and comparing with Fig. 18-103 one finds that at low values of U_{SG} the rise velocity of the bubbles is less than the rise velocity of a single bubble, especially for small-diameter tubes, but that the opposite occurs for large values of U_{SG}.

Liquid-phase mass-transfer coefficients in bubble columns have been reviewed by Calderbank ["Mixing," loc. cit.], Fair [*Chem. Eng.*, loc. cit.], Mersmann [*Ger. Chem. Eng.*, **1**, 1 (1978)], Dakwar, Hallensleben, and Popovic [*Can. J. Chem. Eng.*, **58**, 190 (1980)], and Hikita, Asai, Tanigawa, Segawa, and Kito [*Chem. Eng. J.*, **22**, 61 (1981)].

For prediction of the overall mass-transfer rate with negligible gas-phase resistance, the *product* of the *local liquid phase mass-transfer coefficient* and the *interfacial area per unit volume* ($k_\ell a$) is needed. Two approaches have been used: (1) separate correlations are obtained for k_ℓ and for a, and (2) a correlation is obtained for the overall volumetric coefficient $k_\ell a$. Dimensionless correlations have been proposed for the local coefficient by Calderbank (*Mixing*, loc. cit.). For small bubbles with $d_b \leq 0.5$ mm,

$$\frac{k_\ell D_b}{D_\ell} = 2 + 0.31\left[D_b \frac{3g\Delta\rho}{D_\ell \rho_\ell}\right]^{1/3} \quad (18\text{-}135)$$

and for $d_b \geq 2.5$ mm,

$$\frac{k_\ell D_b}{D_\ell} = 0.42\left[D_b \frac{3g\Delta\rho}{D_\ell \rho_\ell}\right]^{1/3}\left[\frac{\nu}{D_\ell}\right]^{1/6} \quad (18\text{-}136)$$

where k_ℓ = liquid-phase local mass-transfer coefficient, D_b = bubble diameter, D_ℓ = liquid-phase mass diffusivity, g = gravitational acceleration, ρ_ℓ = liquid density, $\Delta\rho$ = liquid density minus gas density, and ν kinematic viscosity of liquid.

A linear interpolation based on D_b is appropriate for $0.5 \leq D_b \leq 2.5$ mm. Prediction of D_b was discussed earlier for various sparger types. For pure liquids and otherwise coalescing systems, the sparger type does not significantly affect equilibrium bubble size; for such systems the correlation of Hughmark [*Ind. Eng. Chem. Process Des. Dev.*, **6**, 218–220 (1967)] is recommended:

$$D_b/0.63 = (\sigma/72)^{0.6}\,(1000/\rho_\ell)^{0.2} \quad (18\text{-}137)$$

where D_b = diameter of equivalent spherical bubble, cm; σ = interfacial tension, dyn/cm; and ρ_ℓ = liquid density, kg/m³.

With gas holdup ϵ and bubble diameter D_b known, the interfacial area per unit of volume a can be calculated:

$$a = 6\epsilon/D_b \quad (18\text{-}138)$$

Methods for predicting ϵ were previously discussed.

Hikita et al. (loc. cit.) have presented a dimensionless correlation for the volumetric liquid-phase mass-transfer coefficient $k_\ell a$

$$\frac{k_\ell a}{g} = 14.9\left(\frac{U_{SG}\mu_\ell}{\sigma}\right)^{1.76}\left(\frac{\mu_\ell g}{\rho_\ell \sigma^3}\right)^{-0.248}\left(\frac{\mu_g}{\mu_\ell}\right)^{0.243}\left(\frac{\mu_\ell}{\rho_\ell D_\ell}\right)^{-0.604} \quad (18\text{-}139)$$

for pure liquids and otherwise coalescing systems. They also investigated the effect of electrolytes on $k_\ell a$, and they found that salt concentrations < 1 weight percent increased $k_\ell a$ by up to 50 percent. They attributed the increase to the increase in interfacial area a caused by the occurrence of smaller gas bubbles owing to the coalescing hindering properties of the electrolyte solutions.

As mentioned earlier, two approaches have been presented for obtaining $k_\ell a$. The recommended procedure is to use both approaches and carefully evaluate the reasons for differences.

Axial dispersion in bubble columns has been extensively studied. An excellent review article by Shah, Stiegel, and Sharma [*Am. Inst. Chem. Eng. J.*, **24**, 369 (1978)] has summarized the literature prior to 1978. Works by Konig, Buchholz, Lucke, and Schugerl [*Ger. Chem. Eng.*, **1**, 199 (1978)], Field and Davidson [*Trans. Inst. Chem. Eng.*, **58**, 228 (1980)], and Riquarts [*Ger. Chem. Eng.*, **4**, 18 (1981)] are particularly useful references which postdate the work of Shah et al.

Axial dispersion occurs in both the liquid and the gas phases. The degree of axial dispersion is greatly affected by vessel diameter, vessel internals, gas superficial velocity, and surface-active agents which retard coalescence. For systems with coalescence-retarding surfactants the initial bubble size produced by the gas sparger is also very significant. The gas and liquid physical properties have only a slight effect on the degree of axial dispersion, except that liquid viscosity becomes important as the flow regime becomes laminar. With pure liquids, in the absence of coalescence-inhibiting, surface-active agents, the nature of the sparger has little effect on the axial dispersion, and experimental results are reasonably well correlated by the dispersion model. For the liquid phase Requarts (loc. cit.) has given the following dimensionless correlation for vessels without internals:

$$D_\ell/U_g d_t = 0.068[(U_g d_t/\nu_\ell)/(U_g^2/d_t g)^3] \quad (18\text{-}140)$$

where D_ℓ = liquid-phase axial dispersion coefficient, U_g = superficial velocity of the gas phase, d_t = vessel diameter, ν_ℓ = liquid-phase kinematic viscosity, and g = acceleration of gravity. The recommended correlation for the gas-phase axial-dispersion coefficient is given by Field and Davidson (loc. cit.):

$$D_g = 56.4\, d_t^{1.33}\,(U_g/\epsilon)^{3.56} \quad (18\text{-}141)$$

where D_g = gas-phase axial-dispersion coefficient, m²/s; d_t = vessel diameter, m; U_g = superficial gas velocity, m/s; and ϵ = fractional gas holdup, volume fraction.

The correlations given in the preceding paragraphs are applicable to vertical cylindrical vessels with pure liquids without coalescence inhibitors. For other vessel geometries such as columns of rectangular cross section, packed columns, and coiled tubes, the work of Shah et al. (loc. cit.) should be consulted. For systems containing coalescence-inhibiting surfactants, axial dispersion can be vastly different from that in systems in which coalescence is negligible. Konig et al. (loc. cit.) have well demonstrated the effects of surfactants and sparger type by conducting tests with weak alcohol solutions using three different porous spargers. With pure water the sparger—and, consequently, initial bubble size—had little effect on back mixing because coalescence produced a dynamic-equilibrium bubble size not far above the sparger. With surfactants the average bubble size was smaller than the dynamic-equilibrium bubble size. Small bubbles produced minimal back mixing up to $\epsilon \approx 40$ percent; however, above $\epsilon > 40$ percent back mixing increased very rapidly as U_g

increased. The rapid increase in back mixing as ϵ exceeds 40 percent was postulated to occur indirectly because a bubble carries upward with it a volume of liquid equal to about 70 percent of the bubble volume, and, for $\epsilon \gtrsim 40$ percent, the bubbles carry so much liquid upward that steady, uniform bubble rise can no longer be maintained and an oscillating, slugging flow develops, which produces fluctuating pressure at the gas distributor and the formation of large eddies. The large eddies greatly increase back mixing. For the air-alcohol-water system the minimum bubble size to prevent unsteady conditions was about 1, 1.5, and 2 mm for U_g = 1, 3, and 5 cm/s respectively. Any smaller bubble size produced increased back mixing. The results of Konig et al. (loc. cit.) clearly indicate that the interaction of surfactants and sparger can be very complex; thus, one should proceed very cautiously in designing systems in which surfactants significantly retard coalescence. Caution is particularly important because surfactants can produce either much more or much less back mixing than surfactant-free systems, depending on the bubble size, which, in turn, depends on the sparger utilized.

The **liquid-phase mass-transfer coefficient** for agitated vessels has been measured and correlated by several investigators. Sideman, Hortacsu, and Fulton [*Ind. Eng. Chem.*, **58**(7), 32 (1966)] and Valentin [*Br. Chem. Eng.*, **12**, 1213 (1967)] have presented reviews of early work. Yagi and Yoshida [*Ind. Eng. Chem. Process Des. Dev.*, **14**, 488 (1975)], Zlokarnik [*Adv. Biochem. Eng.*, **8**, 133 (1978)], Van't Riet [*Ind. Eng. Chem. Process Des. Dev.*, **18**, 357 (1979)], and Hocker, Langer, and Udo [*Ger. Chem. Eng.*, **4**, 51 (1981)] have published the most comprehensive and useful recent works. Van't Riet has recommended the following correlations for the air-water system.

For pure air-water systems,

$$k_\ell a = 0.026\,(P/V_\ell)^{0.4}\,V_g^{0.5} \qquad (18\text{-}142)$$

and for ionic air-aqueous solutions,

$$k_\ell a = 0.002\,(P/V_\ell)^{0.7}V_g^{0.2} \qquad (18\text{-}143)$$

where $k_\ell a$ = overall volumetric liquid-phase mass-transfer coefficient, s^{-1}; P = impeller power consumption, W; V_ℓ = liquid volume, m^3; and V_g = superficial gas velocity, m/s. Equations (18-142) and (18-143) were developed by using data taken in the range $500 < P/V < 10{,}000$ W/m^3.

For low-viscosity systems other than air-water, the method recommended by Fair [*Chem. Eng.*, **74**, 67–74 (July 3, 1967)] is suggested:

$$Ka_{\text{system}} = Ka_{O_2\text{-}H_2O}\left(\frac{D_{\ell\text{system}}}{D_{\ell O_2\text{-}H_2O}}\right)^{1/2} \qquad (18\text{-}144)$$

Hocker et al. (loc. cit.) have developed a dimensionless correlation for the liquid-phase mass-transfer coefficient:

$$k_\ell a/(Q/V) = 0.105\,[(P/Q)\rho_\ell(\nu/g)^{2/3}]^{0.59}\text{Sc}^{0.3} \qquad (18\text{-}145)$$

where $k_\ell a$ = mass-transfer coefficient, s^{-1}; Q = gas rate, m^3/s; V = liquid volume, m^3; P = agitator power input to gassed liquid, W [i.e., (kg·m)/s(m/s^2)]; ρ_ℓ = liquid density, kg/m^3; ν = liquid kinematic viscosity, m^2/s; g = gravitation acceleration, m/s^2; and Sc = Schmidt number. The recommended procedure for obtaining a reasonable estimate for the mass transfer is to use both methods presented here and then check carefully with literature data. One should carefully consider the possible effects of surfactants.

For horizontal vessels with multiple impellers on horizontal shafts, the works of Joshi and Sharma [*Can. J. Chem. Eng.*, **54**, 460 (1976)] and Fukuda, Idogawa, Ikeda, and Endoh [*J. Chem. Eng. Japan*, **13**, 298 (1980)] should be consulted. The *agitator power requirement is less for an aerated liquid* than for an unaerated liquid. Yung, Wong, and Chang [*Can. J. Chem. Eng.*, **59**, 672 (1979)] and Dickey (Paper 116d, *Am. Inst. Chem. Eng.*, 72d annual meeting, San Francisco, Nov. 25–29, 1979) have published works which include comprehensive literature surveys. Dickey found that all mathematical correlations have serious inadequacies. Figure 18-104, from the paper by Luong and Volesky [*Am. Inst. Chem. Eng. J.*, **25**, 893 (1979)] is recommended for the estimation of gassed power. Based on the results of Dickey's work (op. cit.), this correlation should be reasonably accurate for coalescing systems with $P_g/V > 0.5$ kW/m^3. For lower power inputs and gassed power can be significantly less than that predicted by Fig. 18-104. For noncoalescing systems with small bubbles and large gas holdup, the gassed power could be significantly less than predicted by Fig. 18-104. For more accurate estimations the work of Dickey (loc. cit.) should be consulted.

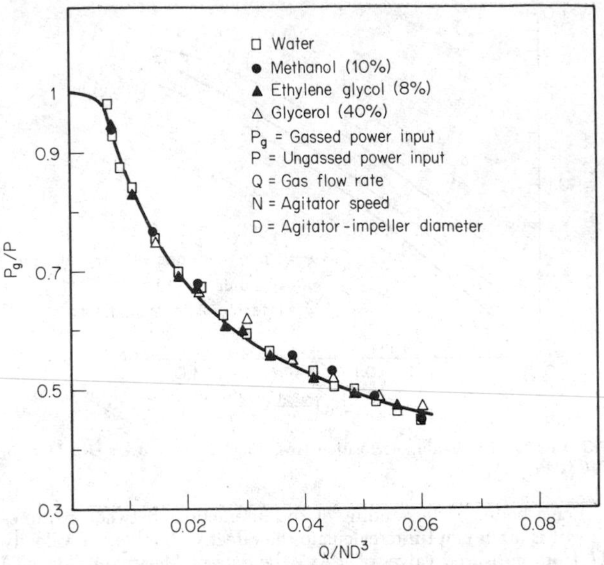

FIG. 18-104 Effect of gas flow on agitator power input for six-bladed disk turbine. [Am. Inst. Chem. Eng. J., **25**, *893 (1979)*.]

At sufficiently high gas rates a significant portion of the gas effectively bypasses the impeller without being dispersed, and the impeller is said to be "flooded." Dickey (loc. cit.) reviews the work relating to this important phenomenon, and he presents the following correlation (in dimensionless form): $Q/NH^3 \leq 1.9\,(N^2/D/g)^{1/2}(D/T)^{1.5}$, where Q = gas flow rate, N = agitator speed, D = agitator diameter, g = gravitational acceleration, and T = tank diameter. This correlation predicts higher gas flows than are often used in practical applications. $Q/ND^3 \leq 0.06$ is a more conservative correlation. One should proceed carefully when designing for gas flow rates exceeding that predicted by the latter correlation; in particular, the references presented by Van't Riet (loc. cit.) should be carefully considered.

PHASE SEPARATION

Gases and liquids may be intentionally contacted as in absorption and distillation, or a mixture of phases may occur unintentionally as in vapor condensation from inadvertent cooling or liquid entrainment from a film. Regardless of the origin, it is usually desirable or necessary ultimately to separate gas-liquid dispersions. While separation will usually occur naturally, the rate is often economically intolerable and separation processes are employed to accelerate the step.

GAS-PHASE CONTINUOUS SYSTEMS

Practical separation techniques for liquid particles in gases are discussed. Since gas-borne particulates include both liquid and solid particles, many devices used for dry-dust collection (discussed in Sec. 20 under "Gas-Solids Separation)" can be adapted to liquid-particle separation. Also, the basic subject of particle mechanics is covered in Sec. 5. Separation of liquid particulates is frequently desirable in chemical processes such as in countercurrent-stage contacting because liquid entrainment with the gas partially reduces true countercurrency. Separation before entering another process step may be needed to prevent corrosion, to prevent yield loss, or to prevent equipment damage or malfunction. Separation before the atmospheric release of gases may be necessary to prevent environmental problems and for regulatory compliance.

REFERENCES: *General:* Hesketh, *Fine Particles in Gaseous Media,* 1977; *Air Pollution Control,* Ann Arbor Science Pubs., Ann Arbor, Mich., 1979. Nonhebel, *Gas Purification Processes for Air Pollution Control,* Newnes-Butterworth, London, 1972. Stern, *Air Pollution,* 3d ed., vol. IV, Academic, New York, 1977. Strauss, *Industrial Gas Cleaning,* 2d ed., Pergamon, New York, 1975; *Air Pollution Control,* 2 vols., Wiley, New York, 1971–1972. Theodore and Buonicore, *Industrial Air Pollution Control Equipment for Particulates,* CRC Press, Cleveland, 1976.

Sampling: (R-1) *Federal Register,* **41,** 23059 (1976). (R-2) Cheremisinoff and Young, *Air Pollution Control and Design Handbook,* Marcel Dekker, New York, 1977, part 1, pp. 65–121. (R-3) Cooper and Rossano, *Source Testing for Air Pollution Control,* Environmental Science Services, Wilton, Conn., 1970. (R-4) *Industrial Guide for Air Pollution Control,* U.S. EPA Publ. EPA 625/6-78-004, 1978. (R-5) Stern, *Air Pollution,* 3d ed., vol III, Academic, New York, 1976, pp. 525–587. (R-6) *Methods of Air Sampling and Analysis,* 2d ed., American Public Health Association, Washington 1977. (R-8) Stockham and Fochtman, *Particle Size Analysis,* Ann Arbor Science Pubs., Ann Arbor, Mich., 1977.

Specific: (R-9) Calvert, Goldshmid, Leith, and Mehta, NTIS Publ. PB-213016, 213017, 1972. (R-10) Calvert, *J. Air Pollut. Control Assoc.,* **24,** 929 (1974). (R-11) Calvert, *Chem. Eng.,* **84**(18), 54 (1977). (R-12) Calvert, Yung, and Leung, NTIS Publ. PB-248050, 1975. (R-13) Calvert and Lundgren, *J. Air Pollut. Control Assoc.,* **18,** 677 (1968). (R-14) Calvert, Lundgren, and Mehta, *J. Air Pollut. Control Assoc.,* **22,** 529 (1972). (R-15) Yung, Barbarika, and Calvert, *J. Air Pollut. Control Assoc.,* **27,** 348 (1977). (R-16) Katz, M.S. thesis, Pennsylvania State University, 1958. (R-17) York and Poppele, *Chem. Eng. Prog.,* **59**(6), 45 (1963). (R-18) York, *Chem. Eng. Prog.,* **50,** 421 (1954). References with the notation (R-) are cited in the text.

Definitions: Mist and Spray Little standardization has been adopted in defining gas-borne liquid particles, and this frequently leads to confusion in the selection, design, and operation of collection equipment. **Aerosol** applies to suspended particulate, either solid or liquid, which is slow to settle by gravity and to particles from the submicrometer range up to 10 to 20 μm. **Mists** are fine suspended liquid dispersions usually resulting from condensation and ranging upward in particle size from around 0.1 μm. **Spray** refers to entrained liquid droplets. The droplets may be entrained from atomizing processes previously discussed under "Liquid-in-Gas Dispersions" in this section. In such instances, size will range from the finest particles produced up to a particle whose terminal settling velocity is equal to the entraining gas velocity if some settling volume is provided. Process spray is often created unintentionally, such as by the condensation of vapors on cold duct walls and its subsequent reentrainment, or from two-phase flow in pipes, gas bubbling through liquids, and entrainment from boiling liquids. Entrainment size distribution from sieve trays has been given by Cheng and Teller [*Am. Inst. Chem. Eng. J.,* **7**(2), 282 (1961)] and evaporator spray by Garner et al. [*Trans. Inst. Chem. Eng.,* **32,** 222 (1954)]. In general, spray can range downward in particle size from 5000 μm. There can be overlapping in size between the coarsest mist particles and the finest spray particles, but some authorities have found it convenient arbitrarily to set a boundary of 10 μm between the two. Actually, considerable overlap exists in the region of 5 to 40 μm. Table 18-19*a* lists typical ranges of particle size created by different mechanisms. The sizes actually entrained can be influenced by the local gas velocity. Figure 18-105 compares the approximate size range of liquid particles with other particulate material and the approximate applicable

TABLE 18-19*a* Particle Sizes Produced by Various Mechanisms

Mechanism or process	Particle-size range, μm
Liquid pressure spray nozzle	100 –5000
Gas-atomizing spray nozzle	1 – 100
Gas bubbling through liquid or boiling liquid	20 –1000
Condensation processes with fogging	0.1– 30
Annular two-phase flow in pipe or duct	10 –2000

size range of collection devices. Figure 20-102 gives an expanded chart by Lapple for solid particles. Mist and fog formation has been discussed previously.

Gas Sampling The sampling of gases containing mists and sprays may be necessary to obtain data for collection-device design, in which case particle-size distribution, total mass loading, and gas volume, temperature, pressure, and composition may all be needed. Other reasons for sampling may be to determine equipment performance, measure yield loss, or determine compliance with regulations.

Location of a sample probe in the process stream is critical especially when larger particles must be sampled. Mass loading in one portion of a duct may be severalfold greater than in another portion as affected by flow patterns. Therefore, the stream should be sampled at a number of points. The U.S. Environmental Protection Agency (R-1) has specified 8 points for ducts between 0.3 and 0.6 m (12 and 24 in) and 12 points for larger ducts, provided there are no flow disturbances for eight pipe diameters upstream and two downstream from the sampling point. When only particles smaller than 3 μm are to be sampled, location and number of sample points are less critical since such particles remain reasonably well dispersed by brownian motion. However, some gravity settling of such particles and even gases of high density have been observed in long horizontal breeching. Isokinetic sampling (velocity at the probe inlet is equal to local duct velocity) is required to get a representative sample of particles larger than 3 μm (error is small for 4- to 5-μm particles). Sampling methods and procedures for mass loading have been developed (R-1 through R-8).

Particle-Size Analysis Many particle-size-analysis methods suitable for dry-dust measurement are unsuitable for liquids because of coalescence and drainage after collection. Measurement of particle

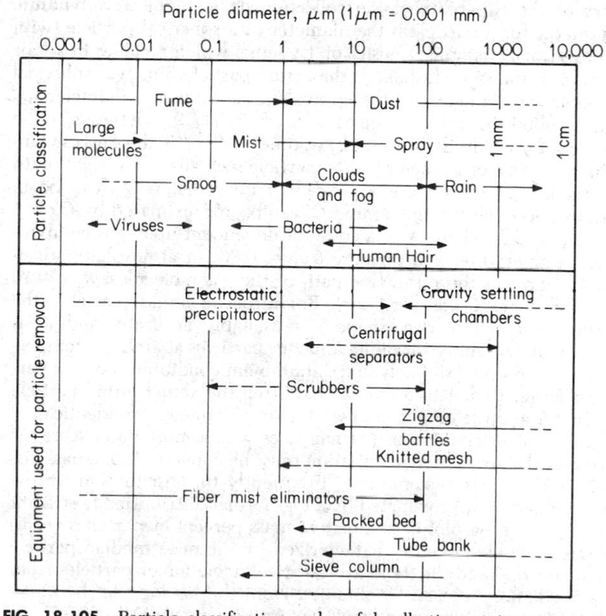

FIG. 18-105 Particle classification and useful collection equipment versus particle size.

sizes in the flowing aerosol stream by using a **cascade impactor** is one of the better means. The impacting principle has been described by Ranz and Wong [*Ind. Eng. Chem.*, **44,** 1371 (1952)] and Gillespie and Johnstone [*Chem. Eng. Prog.*, **51,** 75F (1955)]. The Andersen, Sierra, and University of Washington impactors may be used if the sampling period is kept short so as not to saturate the collection substrate. An impactor designed specifically for collecting liquids has been described by Brink, Kennedy, and Yu [*Am. Inst. Chem. Eng. Symp. Ser.*, **70**(137), 333 (1974)].

Collection Mechanisms Mechanisms which may be used for separating liquid particles from gases are (1) gravity settling, (2) inertial (including centrifugal) impaction, (3) flow-line interception, (4) diffusional (brownian) deposition, (5) electrostatic attraction, (6) thermal precipitation, (7) flux forces (thermophoresis, diffusiophoresis, Stefan flow), and (8) particle agglomeration (nucleation) techniques. Equations and parameters for these mechanisms are given in Table 20-23. Most collection devices rarely operate solely with a single mechanism, although one mechanism may so predominate that it may be referred to, for instance, as an **inertial-impaction device.**

After collection, liquid particles coalesce and must be drained from the unit, preferably without reentrainment. Calvert (R-12) has studied the mechanism of reentrainment in a number of liquid-particle collectors. Four types of reentrainment were typically observed: (1) transition from separated flow of gas and liquid to a two-phase region of separated-entrained flow, (2) rupture of bubbles, (3) liquid creep on the separator surface, and (4) shattering of liquid droplets and splashing. Generally, reentrainment increased with increasing gas velocity. Unfortunately, in devices collecting primarily by centrifugal and inertial impaction, primary collection efficiency increases with gas velocity; thus overall efficiency may go through a maximum as reentrainment overtakes the incremental increase in efficiency. Prediction of collection efficiency must consider both primary collection and reentrainment.

Procedures for Design and Selection of Collection Devices Calvert and coworkers (R-9 to R-12) have suggested useful design and selection procedures for particulate-collection devices in which direct impingement and inertial impaction are the most significant mechanisms. The concept is based on the premise that the mass median aerodynamic particle diameter d_{p50} is a significant measure of the difficulty of collection of the liquid particles and that the collection device cut size d_{pc} (defined as the aerodynamic particle diameter collected with 50 percent efficiency) is a significant measure of the capability of the collection device. The **aerodynamic diameter** for a particle is the diameter of a spherical particle (with an arbitrarily assigned density of 1 g/cm^3) which behaves in an air stream in the same fashion as the actual particle. For real spherical particles of diameter d_p, the equivalent aerodynamic diameter d_{pa} can be obtained from the equation $d_{pa} = dp(\rho_p C')^{1/2}$, where ρ_p is the apparent particle density (mass/volume) and C' is the Stokes-Cunningham correction factor for the particle size, all in consistent units. If particle diameters are expressed in micrometers, ρ_p can be in grams per cubic centimeter and C' can be approximated by $C' = 1 + A_c(2\lambda/D_p)$, where A_c is a constant dependent upon gas composition, temperature, and pressure ($A_c = 0.86$ for atmospheric air at 20°C) and λ is the mean free path of the gas molecules ($\lambda = 0.10$ μm for 20°C atmospheric air). For airborne liquid particles, the assumption of spherical shape is reasonably accurate, and ρ_p is approximately unity for dilute aqueous particles at ambient temperatures. C' is approximately unity at ambient conditions for such particles larger than 1 to 5 μm, so that often the actual liquid particle diameter and the equivalent aerodynamic diameter are identical.

When a distribution of particle sizes which must be collected is present, the actual size distribution must be converted to a mass distribution by aerodynamic size. Frequently the distribution can be represented or approximated by a log-normal distribution (a straight line on a log-log plot of cumulative mass percent of particles versus diameter) which can be characterized by the mass median particle diameter d_{p50} and the standard statistical deviation of particles from the median σ_g. σ_g can be obtained from the log-log plot by $\sigma_g = D_{pa50}/D_{pa}$ at 15.87 percent $= D_{pa}$ at 84.13 percent$/D_{pa50}$.

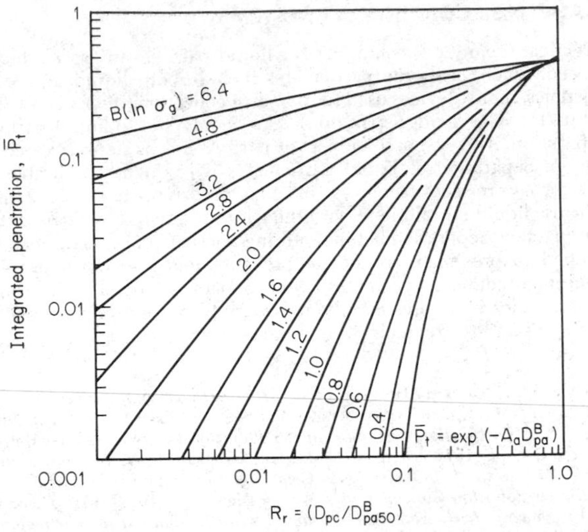

FIG. 18-106 Overall integrated penetration as a function of particle-size distribution and collector parameters. *(Calvert, Yung, and Leung, NTIS Publ. PB-248050, 1975.)*

The grade efficiency η of most collectors can be expressed as a function of the aerodynamic particle size in the form of an exponential equation. It is simpler to write the equation in terms of the particle penetration P_t (those particles not collected), where the fractional penetration $P_t = 1 - \eta$, when η is the fractional efficiency. The typical collection equation is

$$P_t = e^{(-A_a D p a B)} \tag{18-146}$$

where A_a and B are functions of the collection device. Calvert (R-12) has determined that for many devices in which the primary collection mechanism is direct interception and inertial impaction, such as packed beds, knitted-mesh collectors, zigzag baffles, target collectors such as tube banks, sieve-plate columns, and venturi scrubbers, the value of B is approximately 2.0. For cyclonic collectors, the value

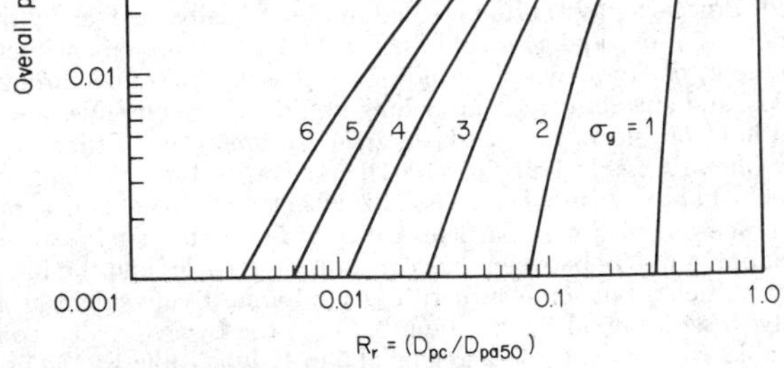

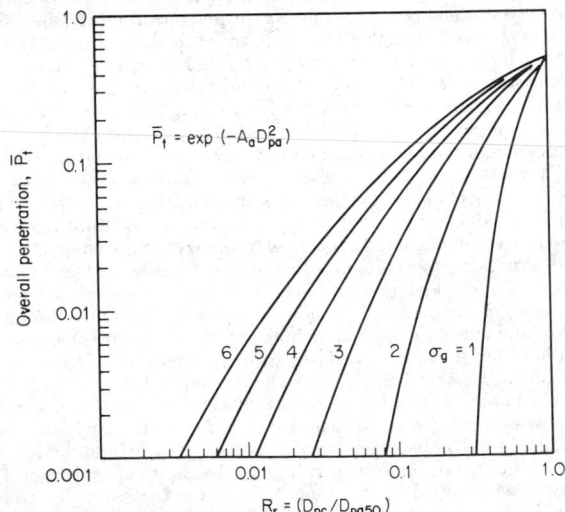

FIG. 18-107 Overall integrated penetration as a function of particle-size distribution and collector cut diameter when $B = 2$ in Eq. (18-146). *(Calvert, Goldshmid, Leith, and Mehta, NTIS Publ. PB-213016, 213017, 1972.)*

of B is approximately 0.67. The overall integrated penetration $\overline{P}_t$ for a device handling a distribution of particle sizes can be obtained by

$$\overline{P}_t = \int_0^W \left(\frac{dW}{W}\right) P_t \qquad (18\text{-}147)$$

where (dW/W) is the mass of particles in a given narrow size distribution and P_t is the average penetration for that size range. When the particles to be collected are log-normally distributed and the collection device efficiency can be expressed by Eq. (18-146), the required overall integrated collection efficiency $\overline{P}_t$ can be related to the ratio of the device aerodynamic cut size D_{pc} to the mass median aerodynamic particle size D_{pa50}. This required ratio for a given distribution and collection is designated R_r, and these relationships are illustrated graphically in Fig. 18-106. For the many devices for which B is approximately 2.0, a simplified plot (Fig. 18-107) is obtained. From these figures, by knowing the desired overall collection efficiency and particle distribution, the value of R_r can be read. Substituting the mass median particle diameter gives the aerodynamic cut size required from the collection device being considered. Therefore, an experimental plot of aerodynamic cut size for each collection device versus operating parameters can be used to determine the device suitability.

Collection Equipment

Gravity Settlers Gravity can act to remove larger droplets. Settling or disengaging space above aerated or boiling liquids in a tank or spray zone in a tower can be very useful. If gas velocity is kept low, all particles with terminal settling velocities (see Fig. 5-80) above the gas velocity will eventually settle. Increasing vessel cross section in the settling zone is helpful. Terminal velocities for particles smaller than 50 μm are very low and generally not attractive for particle removal. Laminar flow of gas in long horizontal paths between trays or shelves on which the droplets settle is another effective means of employing gravity. Design equations are given in Sec. 20 under "Gas-Solids Separations." Settler pressure drop is very low, usually being limited to entrance and exit losses.

Centrifugal Separation Centrifugal force can be utilized to enhance particle collection to several hundredfold that of gravity. The design of cyclone separators for dust removal is treated in detail in Sec. 20 under "Gas-Solids Separations," and typical cyclone designs are shown in Fig. 20-108. Dimension ratios for one family of cyclones are given in Fig. 20-106. Cyclones, if carefully designed, can be more efficient on liquids than on solids since liquids coalesce on capture and are easy to drain from the unit. However, some precautions not needed for solid cyclones are necessary to prevent reentrainment.

Tests by Calvert (R-12) show high primary collection efficiency on droplets down to 10 μm and in accordance with the efficiency equations of Leith and Licht [*Am. Inst. Chem. Eng. Symp. Ser.*, **68**(126), 196–206 (1972)] for the specific cyclone geometry tested if entrainment is avoided. Typical entrainment points are (1) creep along the gas outlet pipe, (2) entrainment by shearing of the liquid film from the walls, and (3) vortex pickup from accumulated liquid in the bottom (Fig. 18-108*a*). Reentrainment from creep of liquid along the top of the cyclone and down the outlet pipe can be prevented by providing the outlet pipe with a flared conical skirt (Fig. 18-108*b*) which provides a point from which the liquid can drip without being caught in the outlet gas. The skirt should be slightly shorter than the gas outlet pipe but extend below the bottom of the gas inlet. The cyclone inlet gas should not impinge on this skirt. Often the bottom edge of the skirt is V-notched or serrated.

Reentrainment is generally reduced by lower inlet gas velocities. Calvert (R-12) reviewed the literature on predicting the onset of entrainment and found that of Chien and Ibele (ASME Pap. 62-WA170) to be the most reliable. Calvert applies their correlation to a liquid Reynolds number on the wall of the cyclone, $N_{Re,L} = 4Q_L/h_i v_L$, where Q_L is the volumetric liquid flow rate, cm^3/s; h_i is the cyclone inlet height, cm; and v_L is the kinematic liquid viscosity,

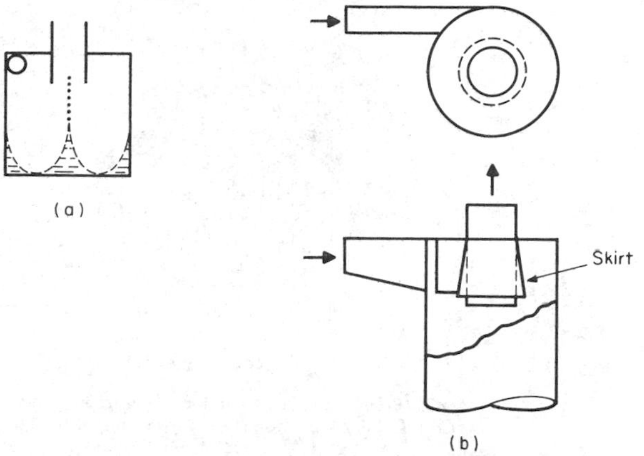

FIG. 18-108 (*a*) Liquid entrainment from the bottom of a vessel by centrifugal flow. (*Rietema and Verver*, Cyclones in Industry, Elsevier, Amsterdam, 1961.) (*b*) Gas-outlet skirt for liquid cyclones. (*Stern et al.*, Cyclone Dust Collectors, American Petroleum Institute, New York, 1955.)

cm^2/s. He finds that the onset of entrainment occurs at a cyclone inlet gas velocity V_{ci}, m/s, in accordance with the relationship $\ln V_{ci} = 6.516 - 0.2865 \ln N_{Re,L}$.

Reentrainment from the bottom of the cyclone can be prevented in several ways. If a typical long-cone dry cyclone is used and liquid is kept continually drained, vortex entrainment is unlikely. However, a vortex breaker baffle in the outlet is desirable, and perhaps a flat disk on top extending to within 2 to 5 cm (0.8 to 2 in) of the walls may be beneficial. Often liquid cyclones are built without cones and have dished bottoms. The modifications described earlier are definitely needed in such situations. Stern, Caplan, and Bush (*Cyclone Dust Collectors*, American Petroleum Institute, New York, 1955) and Rietema and Verver (in Tengbergen, *Cyclones in Industry*, Elsevier, Amsterdam, 1961, chap. 7) have discussed liquid-collecting cyclones.

As with dust cyclones, no reliable pressure-drop equations exist (see Sec. 20), although many have been published. A part of the problem is that there is no standard cyclone geometry. Calvert (R-12) experimentally obtained $\Delta P = 0.000513\, \rho_g (Q_g/h_i W_i)^2 (2.8 h_i w_i/d_o^2)$, where ΔP is in cm of water; ρ_g is the gas density, g/cm^3; Q_g is the gas volumetric flow rate, cm^3/s; h_i and w_i are cyclone inlet height and width respectively, cm; and d_o is the gas outlet diameter, cm. This equation is in the same form as that proposed by Shepherd and Lapple [*Ind. Eng. Chem.*, **31**, 1246 (1940)] but gives only 37 percent as much pressure drop.

Liquid cyclone efficiency can be improved somewhat by introducing a coarse spray of liquid in the cyclone inlet. Large droplets which are easily collected collide with finer particles as they sweep the gas stream in their travel to the wall. (See subsection "Wet Scrubbers" regarding optimum spray size.) Cyclones may also be operated wet to improve their operation on dry dust. Efficiency can be improved through reduction in entrainment losses since the dust particles become trapped in the water film. Collision between droplets and dust particles aids collection, and adequate irrigation can eliminate problems of wall buildup and fouling. The most effective operation is obtained by spraying countercurrently to the gas flow in the cyclone inlet duct at liquid rates of 0.7 to 2.0 L/m^3 of gas. There are also many proprietary designs of liquid separators using centrifugal force, some of which are illustrated in Fig. 18-109. Many of these were originally developed as steam separators to remove entrained condensate. In some designs, impingement on swirl baffles aids separation.

Impingement Separation Impingement separation employs direct impact and inertial forces between particles, the gas streamlines, and target bodies to provide capture. The mechanism is dis-

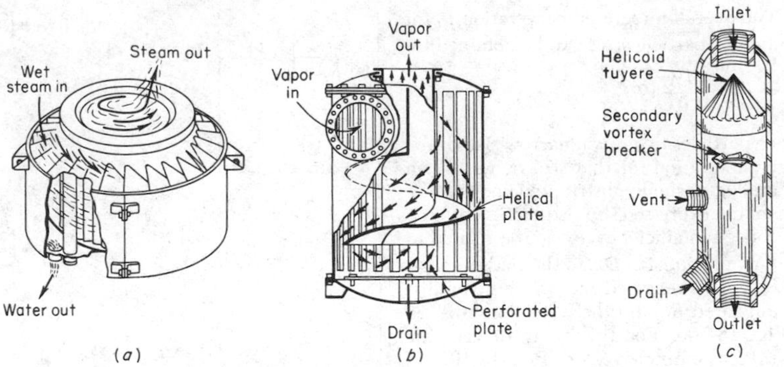

FIG. 18-109 Typical separators using impingement in addition to centrifugal force. (*a*) Hi-eF purifier. *(V. D. Anderson Co.)* (*b*) Flick separator. *(Wurster & Sanger, Inc.)* (*c*) Type RA line separator. *(Centrifix Corp., Bull. 220.)*

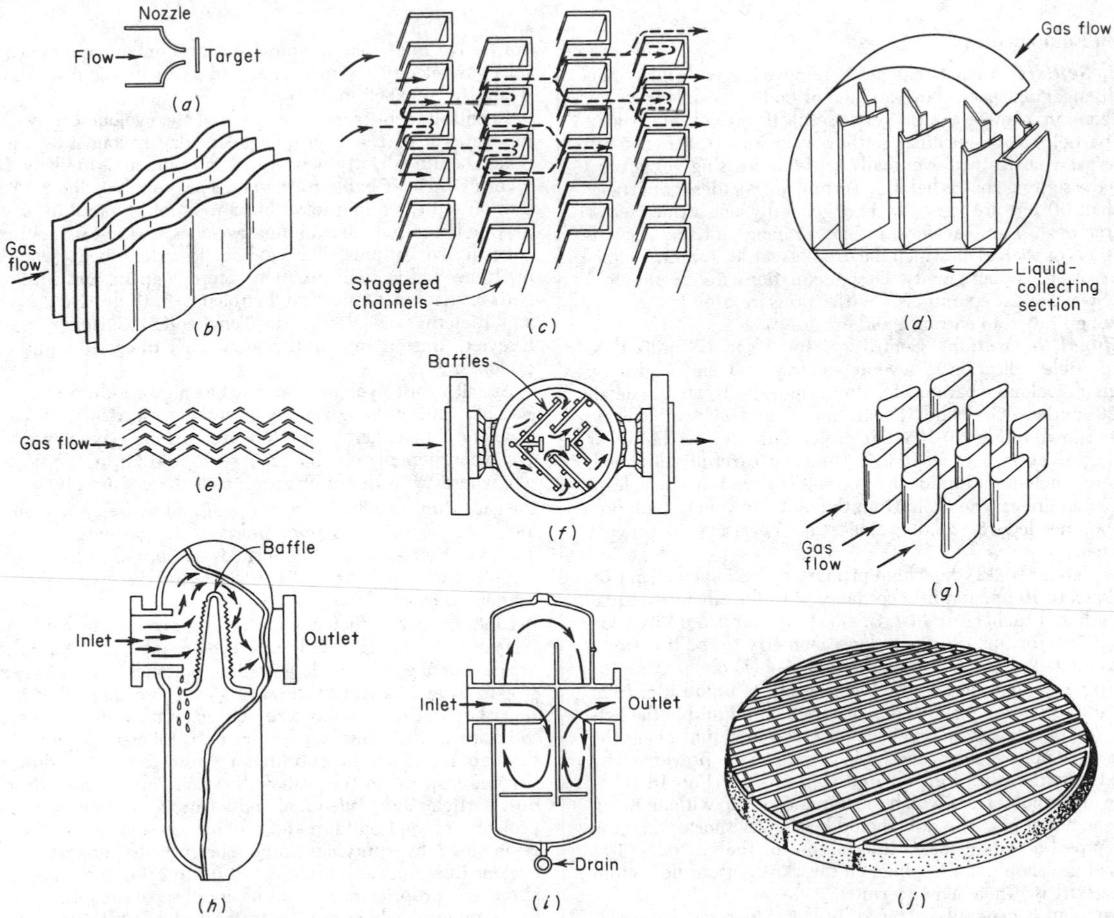

FIG. 18-110 Typical impingement separators. (*a*) Jet impactor. (*b*) Wave plate. (*c*) Staggered channels. *(Blaw-Knox Food & Chemical Equipment, Inc.)* (*d*) Vane-type mist extractor. *(Maloney-Crawford Tank and Mfg. Co.)* (*e*) Peerless line separator. *(Peerless Mfg. Co.)* (*f*) Strong separator. *(Strong Carlisle and Hammond.)* (*g*) Karbate line separator. *(Union Carbide Corporation)* (*h*) Type E horizontal separator. *(Wright-Austin Co.)* (*i*) PL separator. *(Ingersoll Rand.)* (*j*) Wire-mesh demister. *(Otto H. York Co.)*

cussed in Sec. 20 under "Gas-Solids Separations." With liquids, droplet coalescence occurs on the target surface, and provision must be made for drainage without reentrainment. Calvert (R-12) has studied droplet collection by impingement on targets consisting of banks of tubes, zigzag baffles, and packed and mesh beds. Figure 18-110 illustrates some other types of impingement-separator designs.

In its simplest form, an impingement separator may be nothing more than a target placed in front of a flow channel such as a disk at the end of a tube. To improve collection efficiency, the gas velocity may be increased by forming the end into a nozzle (Fig. 18-110*a*). Particle collection as a function of size may be estimated by using the target-efficiency correlation in Fig. 20-105. Since target efficiency will be low for systems with separation numbers below 5 to 10 (small particles, low gas velocities), the mist will frequently be subjected to a number of targets in series as in Fig. 18-110*c*, *d*, and *g*.

The overall droplet penetration is the product of penetration for each set of targets in series. Obviously, for a distribution of particle sizes, an integration procedure is required to give overall collection efficiency. This target-efficiency method is suitable for predicting efficiency when the design effectively prevents the bypassing or short-circuiting of targets by the gas stream and provides adequate time to accelerate the liquid droplets to gas velocity. Katz (R-16) investigated a jet and target-plate entrainment separator design and found the pressure drop less than would be expected to supply the kinetic energy both for droplet acceleration and gas friction. An estimate based on his results indicates that the liquid particles on the average were being accelerated to only about 60 percent of the gas velocity. The largest droplets, which are the easiest to collect, will be accelerated less than the smaller particles. This factor has a leveling effect on collection efficiency as a function of particle size so that experimental results on such devices may not show as sharp a decrease in efficiency with particle size as predicted by calculation. Such results indicate that in many cases our lack of predicting ability results, not from imperfections in the theoretical treatment, but from our lack of knowledge of velocity distributions within the system.

Katz (R-16) also studied *wave-plate impingement separators* (Fig. 18-110*b*) made up of 90° formed arcs with an 11.1-mm (0.44-in) radius and a 3.8-mm (0.15-in) clearance between sheets. The pressure drop is a function of system geometry. The pressure drop for Katz's system and collection efficiency for seven waves are shown in Fig. 18-111. Katz used the Souders-Brown expression to define a design velocity for the gas between the waves:

$$U = K\sqrt{(\rho_\ell - \rho_g)/\rho_g} \qquad (18\text{-}148)$$

K is 0.12 to give U in ms^{-1} (0.4 for ft/s), and ρ_ℓ and ρ_g are liquid and gas densities in any consistent set of units. Katz found no change in efficiency at gas velocities from one-half to 3 times that given by the equation.

Calvert (R-12) investigated *zigzag baffles* of a design more like Fig. 18-110*e*. The baffles may have spaces between the changes in direction or be connected as shown. He found close to 100 per collection for water droplets of 10 μm and larger. Some designs had

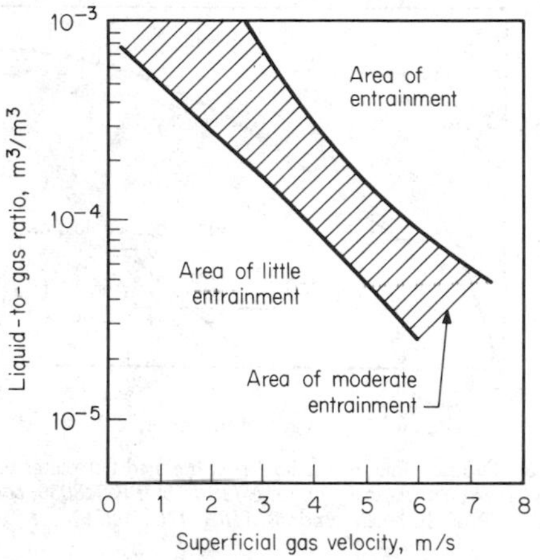

FIG. 18-112 Safe operating region to prevent reentrainment from vertical zigzag baffles with horizontal gas flow. *(Calvert, Yung, and Leung, NTIS Publ. PB-248050, 1975.)*

high efficiencies down to 5 or 8 μm. Desirable gas velocities were 2 to 3.5 m/s (6.6 to 11.5 ft/s), with a pressure drop for a six-pass baffle of 2 to 2.5 cm (0.8 to 1.0 in) of water. On the basis of turbulent mixing, an equation was developed for predicting primary collection efficiency as a function of particle size and collector geometry:

$$\eta = 1 - \exp\left[-\frac{u_{tc}nW\theta}{57.3\,U_g b \tan\theta}\right] \qquad (18\text{-}149)$$

where η is the fractional primary collection efficiency; u_{tc} is the drop terminal centrifugal velocity in the normal direction, cm/s; U_g is the superficial gas velocity, cm/s; n is the number of rows of baffles or bends; θ is the angle of inclination of the baffle to the flow path, °; W is the width of the baffle, cm; and b is the spacing between baffles in the same row, cm. For conditions of low Reynolds number ($N_{Re,D} < 0.1$) where Stokes' law applies, Calvert obtains the value for drop terminal centrifugal velocity of $u_{tc} = d_p^2\,\rho_p a/18\,\mu_g$, where d_p and ρ_p are the drop particle diameter, cm, and particle density, g/cm³, respectively; μ_g is the gas viscosity, P; and a is the acceleration due to centrifugal force. It is defined by the equation $a = 2U_g^2 \sin\theta/W\cos^3\theta$. For situations in which Stokes' law does not apply, Calvert recommends substitution in the derivation of Eq. (18-149) for u_{tc} of drag coefficients from drag-coefficient data of Foust et al. (*Principles of Unit Operations*, Toppan Co., Tokyo, 1959).

Calvert found that reentrainment from the baffles was affected by the gas velocity, the liquid-to-gas ratio, and the orientation of the baffles. Horizontal gas flow past vertical baffles provided the best drainage and lowest reentrainment. Safe operating regions with vertical baffles are shown in Fig. 18-112. Horizontal baffles gave the

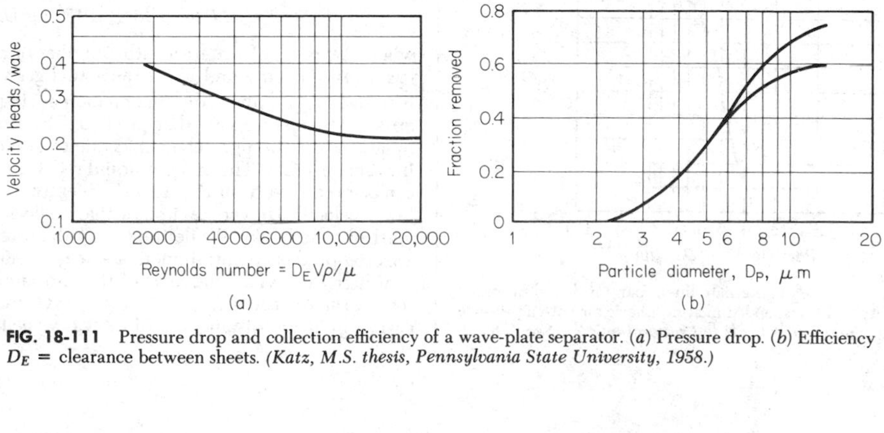

FIG. 18-111 Pressure drop and collection efficiency of a wave-plate separator. (*a*) Pressure drop. (*b*) Efficiency D_E = clearance between sheets. *(Katz, M.S. thesis, Pennsylvania State University, 1958.)*

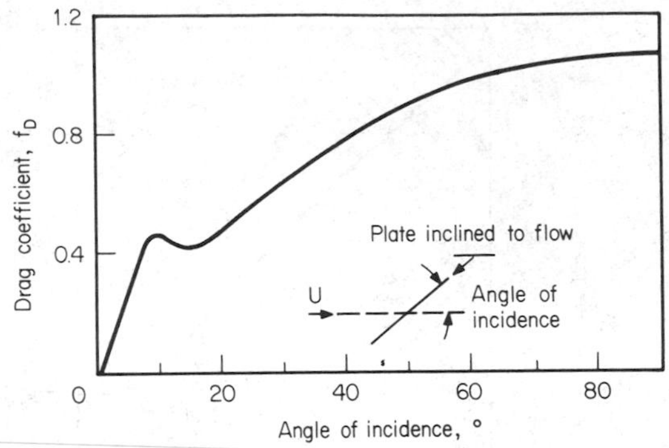

FIG. 18-113 Drag coefficient for flow past inclined flat plates for use in Eq. (18-150). [*Calvert, Yung, and Leung, NTIS Publ. PB-248050; based on Fage and Johansen,* Proc. R. Soc. (London), ***116A,*** *170 (1927).*]

poorest drainage and the highest reentrainment, with inclined baffles intermediate in performance. Equation (18-150), developed by Calvert, predicts pressure drop across zigzag baffles. The indicated summation must be made over the number of rows of baffles present.

$$\Delta P = \sum_{i=1}^{i=n} 1.02 \times 10^{-3} f_D \rho_g \frac{U'^2_g A_p}{2A_t} \qquad (18\text{-}150)$$

ΔP is the pressure drop, cm of water; ρ_g is the gas density, g/cm^3; A_p is the total projected area of an entire row of baffles in the direction of inlet gas flow, cm^2; and A_t is the duct cross-sectional area, cm^2. The value f_D is a drag coefficient for gas flow past inclined flat plates taken from Fig. 18-113, while U'_g is the actual gas velocity, cm/s, which is related to the superficial gas velocity U_g by $U'_g = U_g/\cos\theta$. It must be noted that the angle of incidence θ for the second and successive rows of baffles is twice the angle of incidence for the first row. Most of Calvert's work was with 30° baffles, but the method correlates well with other data on 45° baffles.

The **Karbate line separator** (Fig. 18-110*g*) is composed of several layers of teardrop-shaped target rods of Karbate. A design flow constant K in Eq. (18-148) of 0.035 m/s (1.0 ft/s) is recommended by the manufacturer. Pressure drop is said to be 5½ velocity heads on the basis of the superficial gas velocity. This value would probably increase at high liquid loads. Figure 18-114 gives the manufacturer's reported grade efficiency curve at the design air velocity.

The use of **multiple tube banks** as a droplet collector has also been studied by Calvert (R-12). He reports that collection efficiency for closely packed tubes follows equations for rectangular jet impaction which can be obtained graphically from Fig. 18-115 by using a dimensional parameter β which is based on the tube geometry; β =

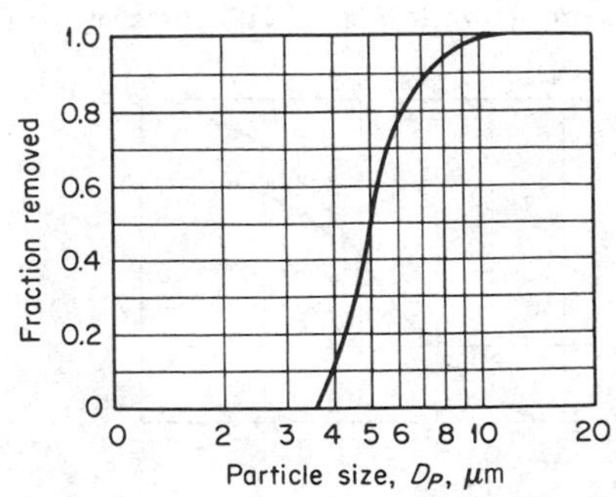

FIG. 18-114 Collection efficiency of Karbate line separator, based on particles with a specific gravity of 1.0 suspended in atmospheric air with a pressure drop of 2.5 cm water gauge. (*Union Carbide Corporation Cat. Sec. S-6900, 1960.*)

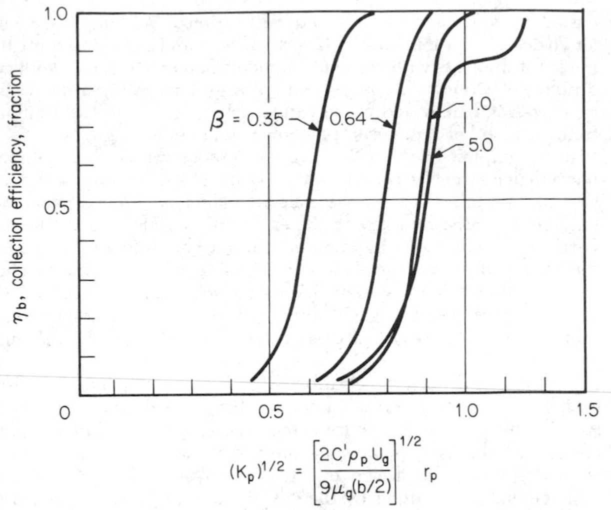

FIG. 18-115 Experimental collection efficiencies of rectangular impactors. C' is the Stokes-Cunningham correction factor; ρ_p, particle density, g/cm^3; U_g, superficial gas velocity, approaching the impactor openings, cm/s; and μ_g, gas viscosity, P. [*Calvert, Yung, and Leung, NTIS Publ. PB-248050; based on Mercer and Chow,* J. Coll. Interface Sci., ***27,*** *75 (1968).*]

$2l_i/b$, where b is the open distance between adjacent tubes in the row (orifice width) and l_i is the impaction length (distance between orifice and impingement plane), or approximately the distance between centerlines of successive tube rows. Note that the impaction parameter K_p is plotted to the one-half power in Fig. 18-115 and that the radius of the droplet is used rather than the diameter. Collection efficiency overall for a given size of particle is predicted for the entire tube bank by

$$\eta = 1 - (1 - \eta_b)^N \qquad (18\text{-}151)$$

where η_b is the collection efficiency for a given size of particle in one stage of a rectangular jet impactor (Fig. 18-115) and N is the number of stages in the tube bank (equal to one less than the number of rows). For widely spaced tubes, the target efficiency η_g can be calculated from Fig. 20-105 or from the impaction data of Golovin and Putnam [*Ind. Eng. Chem. Fundam.,* **1,** 264 (1962)]. The efficiency of the overall tube banks for a specific particle size can then be calculated from the equation $\eta = 1 - (1 - \eta_t a'/A)^n$, where a' is the cross-sectional area of all tubes in one row, A is the total flow area, and n is the number of rows of tubes.

Calvert reports pressure drop through tube banks to be largely unaffected by liquid loading and indicates that Grimison's correlations in Sec. 5 ("Tube Banks") for gas flow normal to tube banks or data for gas flow through heat-exchanger bundles can be used. However, the following equation is suggested:

$$\Delta P = 8.48 \times 10^{-5} n \rho_g U'^2_g \qquad (18\text{-}152)$$

where ΔP is cm of water; n is the number of rows of tubes; ρ_g is the gas density, g/cm^3; and U'_g is the actual gas velocity between tubes in a row, cm/s. Calvert did find an increase in pressure drop of about 80 to 85 percent above that predicted by Eq. (18-152) in vertical upflow of gas through tube banks due to liquid holdup at gas velocities above 4 m/s. The onset of liquid reentrainment from tube banks can be predicted from Fig. 18-116. Reentrainment occurred at much lower velocities in vertical upflow than in horizontal gas flow through vertical tube banks. While the top of the cross-hatched line of Fig. 18-116*a* predicts reentrainment above gas velocities of 3 m/s (9.8 ft/s) at high liquid loading, most of the entrainment settled to the bottom of the duct in 1 to 2 m (3.3 to 6.6 ft), and entrainment did not carry significant distances until the gas velocity exceeded 7 m/s (23 ft/s).

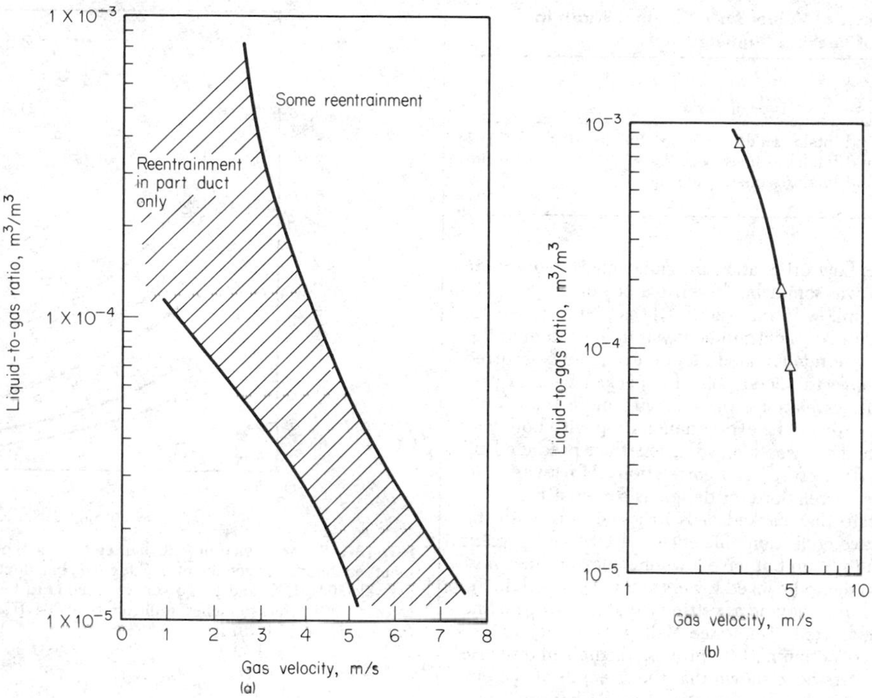

FIG. 18-116 Experimental results showing effect of gas velocity and liquid load on entrainment from (*a*) vertical tube banks with horizontal gas flow and (*b*) horizontal tube banks with upflow. To convert meters per second to feet per second, multiply by 3.281. *(Calvert, Yung, and Leung, NTIS Publ. PB-248050.)*

Packed-Bed Collectors Many different materials, including coal, coke, broken solids of various types such as brick, tile, rock, and stone, as well as normal types of tower-packing rings, saddles, and special plastic shapes, have been used over the years in packed beds to remove entrained liquids through impaction and filtration. Separators using natural materials are not available as standard commercial units but are designed for specific applications. Coke boxes were used extensively in the years 1920 to 1940 as sulfuric acid entrainment separators (see *Chemical Engineers' Handbook,* 5th ed., p. 18–87) but have now been largely superseded by more sophisticated and efficient devices.

Jackson and Calvert [*Am. Inst. Chem. Eng. J.,* **12,** 1075 (1966)] studied the collection of fine fuel-oil-mist particles in beds of ½-in glass spheres, Raschig rings, and Berl and Intalox saddles. The mist

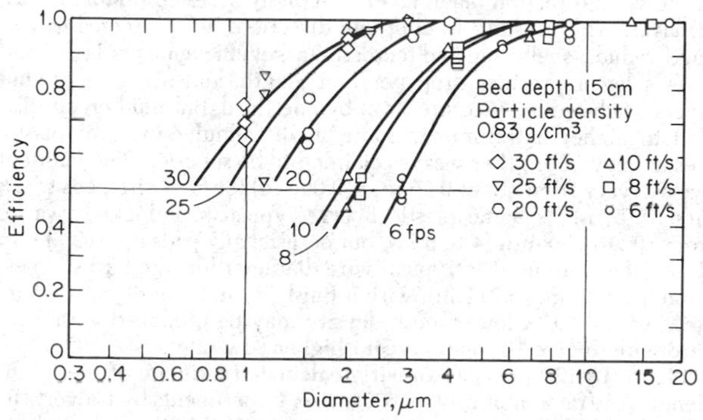

FIG. 18-117 Experimental collection efficiency. ½-in Intalox saddles. To convert feet per second to meters per second, multiply by 0.3048; to convert centimeters to inches, multiply by 0.394; and to convert grams per cubic centimeter to pounds per cubic foot, multiply by 62.43. [*Jackson and Calvert,* Am. Inst. Chem. Eng. J., *12, 1075 (1966).*]

had a mass median particle diameter of 6 μm and a standard deviation of 2.0. The collection efficiency as a function of particle size and gas velocity in a 355-mm- (14-in-) diameter by 152-mm- (6-in-) thick bed of Intalox saddles is given in Fig. 18-117. This and additional work have been generalized by Calvert (R-12) to predict collection efficiencies of liquid particles in any packed bed. Assumptions in the theoretical development are that the drag force on the drop is given by Stokes' law and that the number of semicircular bends to which the gas is subjected, η_1, is related to the length of the bed, Z (cm), in the direction of gas flow, the packing diameter, d_c (cm), and the gas-flow channel width, b (cm), such that $\eta_1 = Z/(d_c + b)$. The gas velocity through the channels, U_{gb} (cm/s), is inversely proportional to the bed free volume for gas flow such that $U_{gb} = U_g[1/(\varepsilon - h_b)]$, where U_g is the gas superficial velocity, cm/s, approaching the bed, ε is the bed void fraction, and h_b is the fraction of the total bed volume taken up with liquid which can be obtained from data on liquid holdup in packed beds. The width of the semicircular channels b can be expressed as a fraction j of the diameter of the packing elements, such that $b = j\,d_c$. These assumptions lead to an equation for predicting the penetration of a given size of liquid particle through a packed bed:

$$P_t = 1 - \exp\left[\frac{-\pi}{2(j+j^2)(\varepsilon-h_b)}\left(\frac{Z}{d_c}\right)K_p\right] \tag{18-153}$$

where

$$K_p = \frac{\rho_p d_p^2 U_g}{9\,\mu_g d_c} \tag{18-154}$$

Values of ρ_p and d_p are droplet density, g/cm³, and droplet diameter, cm; μ_g is the gas viscosity, P. All other terms were defined previously. Table 18-20 gives values of j calculated from experimental data of Jackson and Calvert. Values of j for most manufactured packing appear to fall in the range from 0.16 to 0.19. The low value of 0.03 for coke may be due to the porosity of the coke itself.

Calvert (R-12) has tested the correlation in cross-flow packed beds, which tend to give better drainage than countercurrent beds, and has

TABLE 18-20 Experimental Values for *j*, Channel Width in Packing as a Fraction of Packing Diameter

Packing size			
cm	in	Type of packing	*j*
1.27	0.5	Berl and Intalox saddles, marbles, Raschig rings	0.192
2.54	1.0	Berl and Intalox saddles, pall rings	0.190
3.8	1.5	Berl and Intalox saddles, pall rings	0.165
7.6–12.7	3–5	Coke	0.03

found the effect of gas-flow orientation insignificant. However, the onset of reentrainment was somewhat lower in a bed of 2.5-cm (1.0-in) pall rings with gas upflow [6 m/s (20 ft/s)] than with horizontal cross-flow of gas. The onset of reentrainment was independent of liquid loading (all beds were nonirrigated), and entrainment occurred at values somewhat above the flood point for packed beds as predicted by conventional correlations. In beds with more than 3 cm (1.2 in) of water pressure drop, the experimental drop with both vertical and horizontal gas flow was somewhat less than predicted by generalized packed-bed pressure-drop correlations. However, Calvert recommends these correlations for design as conservative.

Calvert's data indicate that packed beds irrigated only with the collected liquid can have collection efficiencies of 80 to 90 percent on mist particles down to 3 μm but have low efficiency on finer mist particles. Frequently, irrigated packed towers and towers with internals will be used with liquid having a wetting capability for the fine mist which must be collected. Tennessee Valley Authority (TVA) experiments with the collection of 1.0-μm mass median phosphoric acid mist in packed towers have shown that the strength of the circulating phosphoric acid is highly important [see Baskerville, *Am. Inst. Chem. Eng. J.*, **37**, 79 (1941); and p. 18–87, 5th ed. of the *Handbook*]. Hesketh (*J. Air Pollut. Control Assoc.*, **24**, 942 (1974)] has reported up to 50 percent improvement in collection efficiency in venturi scrubbers on fine particles with the addition of only 0.10 percent of a low-foaming nonionic surfactant to the scrubbing liquid, and others have experienced similar results in other gas-liquid-contacting devices. Calvert (R-9 and R-10) has reported on the efficiency of various gas-liquid-contacting devices for fine particles. Figure 18-118 gives the particle aerodynamic cut size for a single-sieve-plate gas scrubber as a function of sieve hole size d_h, cm; hole gas velocity u_h, m/s; and froth or foam density on the plate F, g/cm^3. This curve is based on standard air and water properties and wettable (hydrophilic) particles. The cut diameter decreases with an increase in froth density, which must be predicted from correlations for sieve-plate behavior (see Fig. 18-15). Equation (18-153) can be used to calculate generalized design curves for collection in packed columns in the same fashion by finding parameters of packing size, bed length, and gas velocity which give collection efficiencies of 50 percent for various size particles. Figure 18-119 illustrates such a plot for three unidentified gas velocities and two sizes of packing.

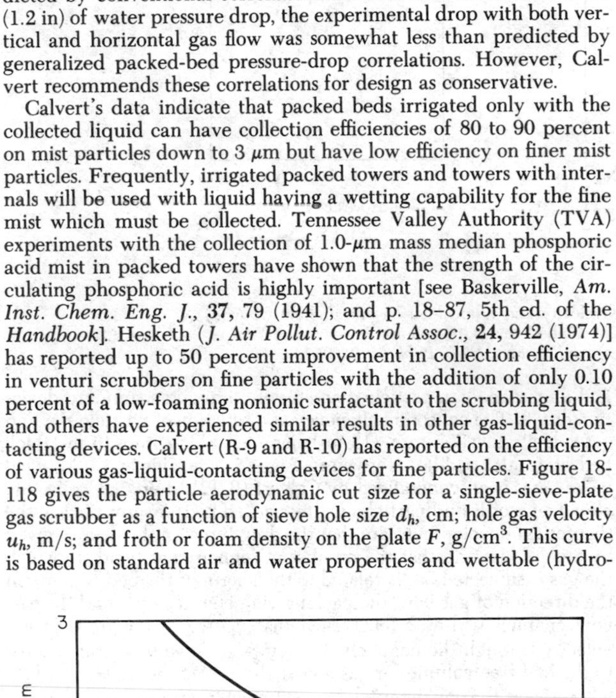

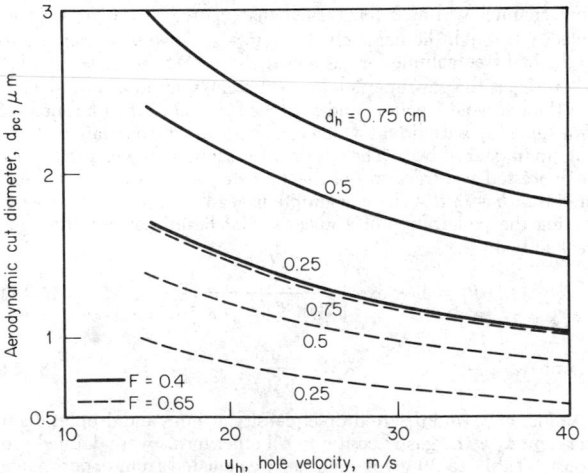

FIG. 18-118 Aerodynamic cut diameter for a single-sieve-plate scrubber as a function of hole size, hole-gas velocity, and froth density, F, g/cm^3. To convert meters per second to feet per second, multiply by 3.281; to convert grams per cubic centimeter to pounds per cubic foot, multiply by 62.43. [*Calvert*, J. Air Pollut. Control Assoc., *24, 929 (1974).*]

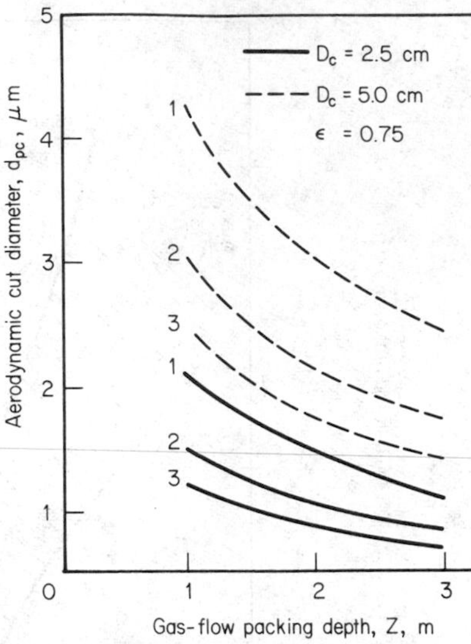

FIG. 18-119 Aerodynamic cut diameter for a typical packed-bed entrainment separator as a function of packing size, bed depth, and three unidentified gas velocities, 1, 2, and 3. To convert meters to feet, multiply by 3.281; to convert centimeters to inches, multiply by 0.394. [*Calvert*, J. Air Pollut. Control Assoc., *24, 929 (1974).*]

Wire-Mesh Mist Collectors Knitted mesh of varying density and voidage is widely used for entrainment separators. Its advantage is close to 100 percent removal of drops larger than 5 μm at superficial gas velocities from about 0.2 ms/s (0.6 ft/s) to 5 m/s (16.4 ft/s), depending somewhat on the design of the mesh. Pressure drop is usually no more than 2.5 cm (1 in) of water. A major disadvantage is the ease with which tars and insoluble solids plug the mesh. The separator can be made to fit vessels of any shape and can be made of any material which can be drawn into a wire. Stainless-steel and plastic fibers are most common, but other metals are sometimes used. Generally three basic types of mesh are used: (1) layers with a crimp in the same direction (each layer is actually a nested double layer); (2) layers with a crimp in alternate directions, which increases voidage, reduces sheltering and increases target efficiency per layer, and gives a lower pressure drop per unit length; and (3) spiral-wound layers whch reduce pressure drop by one-third, but fluid creep may lead to higher entrainment. Some small manufacturers of plastic meshes may offer other weaves claimed to be superior. The filament size can vary from about 0.15 mm (0.006 in) for fine-wire pads to 3.8 mm (0.15 in) for some plastic fibers. Typical pad thickness varies from 100 to 150 mm (4 to 6 in), but occasionally pads up to 300 mm (12 in) thick are used. A typical wire diameter for standard stainless mesh is 0.28 mm (0.011 in), with a finished mesh density of 0.15 g/cm^3 (9.4 lb/ft^3). A lower mesh density may be produced with standard wire to give 10 to 20 percent higher flow rates.

Figure 18-120 presents an early calculated estimate of mesh efficiency as a fraction of mist-particle size. Experiments by Calvert (R-12) confirm the accuracy of the equation of Bradie and Dickson (*Joint Symp. Proc. Inst. Mech. Eng./Yorkshire Br. Inst. Chem. Eng.*, 1969, pp. 24–25) for primary efficiency in mesh separators:

$$\eta = 1 - \exp\,(-2/3)\pi a\, l\eta_t) \qquad (18\text{-}155)$$

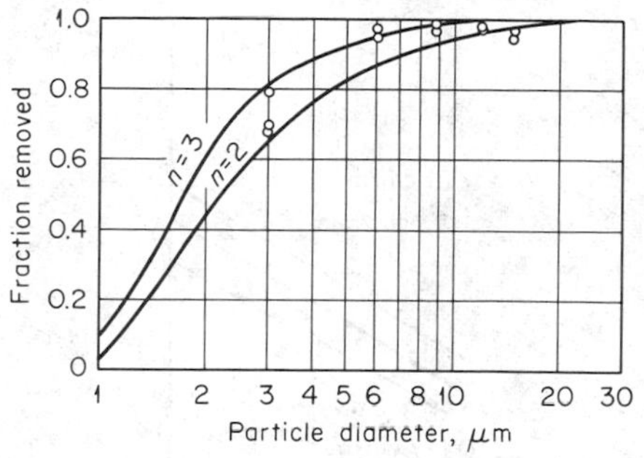

FIG. 18-120 Collection efficiency of wire-mesh separator; 6-in thickness, 98.6 percent free space, 0.006-in-diameter wire used for experiment points. Curves calculated for target area equal to 2 and 3 times the solids volume of packing. To convert inches to millimeters, multiply by 25.4.

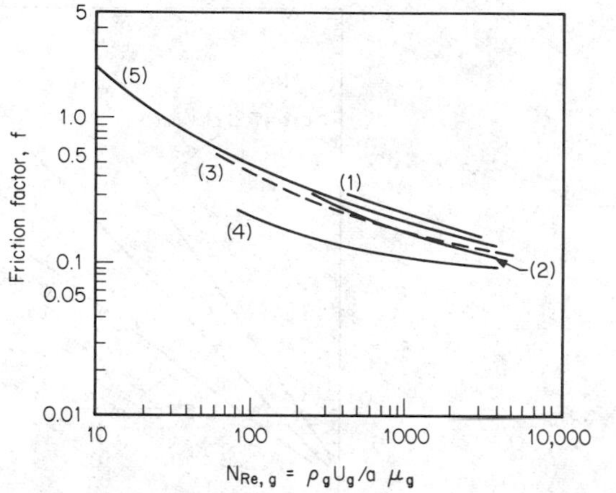

FIG. 18-121 Value of friction factor f for dry knitted mesh for Eq. (18-156). Values of York and Poppele [*Chem. Eng. Prog., 50, 421 (1954)*] are given in curve 1 for mesh crimped in the alternating direction and curve 2 for mesh crimped in the same direction. Data of Bradie and Dickson (*Joint Symp. Proc. Inst. Mech. Eng./Yorkshire Br. Inst. Chem. Eng.*, 1969, pp. 24–25) are given in curve 3 for layered mesh and curve 4 for spiral-wound mesh. Curve 5 is data of Satsangee (M.S. thesis, Brooklyn Polytechnic Institute, 1948) and Schurig (D.Ch.E. dissertation, Brooklyn Polytechnic Institute, 1946). *(From Calvert, Yung, and Leung, NTIS Publ. PB-248050, 1975.)*

where η is the overall collection efficiency for a given-size particle; l is the thickness of the mesh, cm, in the direction of gas flow; a is the surface area of the wires per unit volume of mesh pad, cm^2/cm^3; and η_t, the target collection efficiency for cylindrical wire, can be calculated from Fig. 20-105 or the impaction data of Golovin and Putnam [*Ind. Eng. Chem.*, 1, 264 (1962)]. The factor 2/3, introduced by Carpenter and Othmer [*Am. Inst. Chem. Eng. J.*, 1, 549 (1955)], corrects for the fact that not all the wires are perpendicular to the gas flow and gives the projected perpendicular area. If the specific mesh surface area a is not available, it can be calculated from the mesh void area ε and the mesh wire diameter d_w in cm, $a = 4(1-\varepsilon)/d_w$.

York and Poppele (R-17) have stated that factors governing maximum allowable gas velocity through the mesh are (1) gas and liquid density, (2) liquid surface tension, (3) liquid viscosity, (4) specific wire surface area, (5) entering-liquid loading, and (6) suspended-solids content. York (R-18) has proposed application of the Souders-Brown equation [Eq. (18-148)] for correlation of maximum allowable gas velocity with values of K for most cases of 0.1067 to give U in m/s (0.35 for ft/s). When liquid viscosity or inlet loading is high or the liquid is dirty, the value of K must be reduced. Schroeder (M.S. thesis, Newark College of Engineering, 1962) found lower values for K necessary when liquid surface tension is reduced such as by the presence of surfactants in water. Ludwig (*Applied Process Design for Chemical and Petrochemical Plants*, 2d ed., vol. I, Gulf, Houston, 1977, p. 157) recommends reduced K values under vacuum of 0.061 at an absolute pressure of 6.77 kPa (0.98 lbf/in^2) and 0.082 at 54 kPa (7.83 lbf/in^2) absolute to give U in m/s. Most manufacturers suggest setting the design velocity at three-fourths of the maximum velocity to allow for surges in gas flow.

York and Popple (R-17) have suggested that total pressure drop through the mesh is equal to the sum of the mesh dry pressure drop plus an increment due to the presence of liquid. They considered the mesh to be equivalent to numerous small circular channels and used the D'Arcy formula with a modified Reynolds number to correlate friction factor (see Fig. 18-121) for Eq. (18-156) giving dry pressure drop.

$$\Delta P_{dry} = fla\rho_g U_g^2/981\,\varepsilon^3 \qquad (18\text{-}156)$$

where ΔP is in cm of water; f is from Fig. 18-121; ρ_g is the gas density, g/cm^3; U_g is the superficial gas velocity, cm/s; and ε is the mesh porosity or void fraction; l and a are as defined in Eq. (18-155). Figure 18-121 gives data of York and Poppele for mesh crimped in the same and alternating directions and also includes the data of Satsangee, of Schuring, and of Bradie and Dickson.

The incremental pressure drop for wet mesh is not available for all operating conditions or for mesh of different styles. The data of York and Poppele for wet-mesh incremental pressure drop, ΔP_L in cm of water, are shown in Fig. 18-122 for parameters of liquid velocity L/A, defined as liquid volumetric flow rate, cm^3/min per unit of mesh cross-sectional area in cm^2; liquid density ρ_L is in g/cm^3.

York generally recommends the installation of the mesh horizontally with upflow of gas as in Fig. 18-110*j*; Calvert (R-12) tested the mesh horizontally with upflow and vertically with horizontal gas flow. He reports better drainage with the mesh vertical and somewhat higher permissible gas velocities without reentrainment, which is contrary to past practice. With horizontal flow through vertical mesh, he found collection efficiency to follow the predictions of Eq. (18-155) up to 4 m/s (13 ft/s) with air and water. Some reentrainment was encountered at higher velocities, but it did not appear serious until velocities exceeded 6.0 m/s (20 ft/s). With vertical upflow of gas, entrainment was encountered at velocities above and below 4.0 m/s (13 ft/s), depending on inlet liquid quantity (see Fig. 18-123). Figure 18-124 illustrates the onset of entrainment from mesh as a function of liquid loading and gas velocity and the safe operating area recommended by Calvert. Measurements of dry pressure drop by Calvert gave values only about one-third of those predicted from Eq. (18-156). He found the pressure drop to be highly affected by liquid load. The pressure drop of wet mesh could be correlated as a function of $U_g^{1.65}$ and parameters of liquid loading L/A, as shown in Fig. 18-125.

As indicated previously, mesh efficiency drops rapidly as particles decrease in size below 5 μm. An alternative is to use two mesh pads in series. The first mesh is made of fine wires and is operated beyond the flood point. It results in droplet coalescence, and the second mesh, using standard wire and operated below flooding, catches entrainment from the first mesh. Coalescence and flooding in the first mesh may be assisted with water sprays or irrigation. Massey [*Chem. Eng. Prog.*, **55**(5), 114 (1959)] and Coykendall et al. [*J. Air Pollut. Control Assoc.*, **18**, 315 (1968)] have discussed such applications. Calvert (R-12) presents data on the particle size of entrained drops from mesh as a function of gas velocity which can be used for sizing the secondary collector. A major disadvantage of this approach is high pressure drop, which can be in the range from 25 cm (10 in) of water to as high as 85 cm (33 in) of water if the mist is mainly submicrometer.

Wet Scrubbers Scrubbers have not been widely used for the collection of purely liquid particulate, probably because they are generally more complex and expensive than impaction devices of the types previously discussed. Further, scrubbers are no more efficient than the former devices for the same energy consumption. However, scrubbers of the types discussed in Sec. 20 and illustrated in Figs. 20-

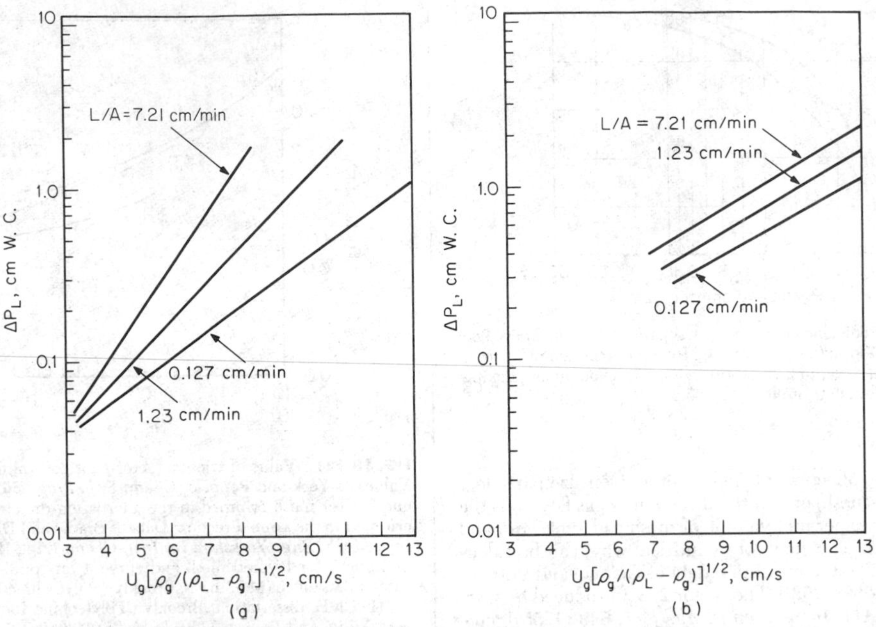

FIG. 18-122 Incremental pressure drop in knitted mesh due to the presence of liquid (a) with the mesh crimps in the same direction and (b) with crimps in the alternating direction, based on the data of York and Poppele [*Chem. Eng. Prog.*, **50,** 421 (1954)]. To convert centimeters per minute to feet per minute, multiply by 0.0328; to convert centimeters per second to feet per second, multiply by 0.0328. *(From Calvert, Yung, and Leung, NTIS Publ. PB-248050, 1975)*

120 to 20-122 can be used to capture liquid particles efficiently. Their use is primarily indicated when it is desired to accomplish simultaneously another task such as gas absorption or the collection of solid and liquid particulate mixtures.

Table 20-41 [*Chemical Engineers' Handbook*, 5th ed.)], showing the minimum size of particles collectible in different types of scrubbers at reasonably high efficiencies, is a good selection guide. Cyclonic spray towers (Fig. 20-120) can effectively remove liquid particles down to around 2 to 3 μm. Figures 20-112 and 20-113 (*Chemical Engineers' Handbook*, 5th ed.), giving target efficiency between spray drop size and particle size as calculated by Stairmand or Johnstone and Roberts, should be considered in selecting spray atomization for the most efficient tower operation. Figure 18-126 gives calculated particle cut size as a function of tower height (or length) for vertical countercurrent spray towers and for horizontal-gas-flow, vertical-liquid-flow cross-current spray towers with parameters for liquid drop size. These curves are based on physical properties of standard air and water and should be used under conditions in which these are reasonable approximations. Lack of uniform liquid distribution or liquid flowing down the walls can affect the performance, requiring empirical correction factors. Calvert (R-10) sug-

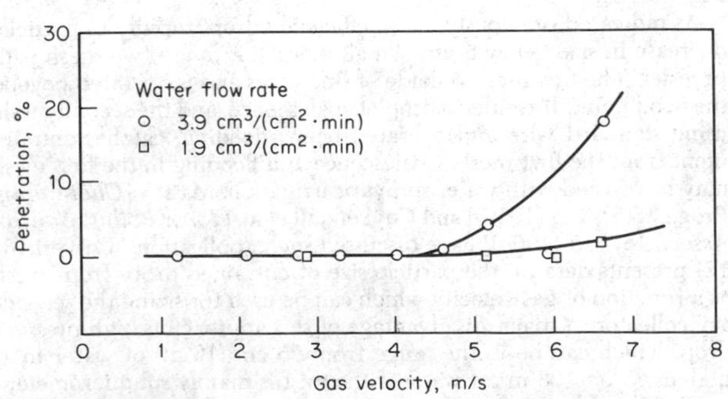

FIG. 18-123 Experimental data of Calvert with air and water in mesh with vertical upflow, showing the effect of liquid loading on efficiency and reentrainment. To convert meters per second to feet per second, multiply by 3.281; to convert cubic centimeters per square centimeter–minute to cubic feet per square foot–minute, multiply by 0.0328. (*Calvert, Yung, and Leung, NTIS Publ. PB-248050, 1975.*)

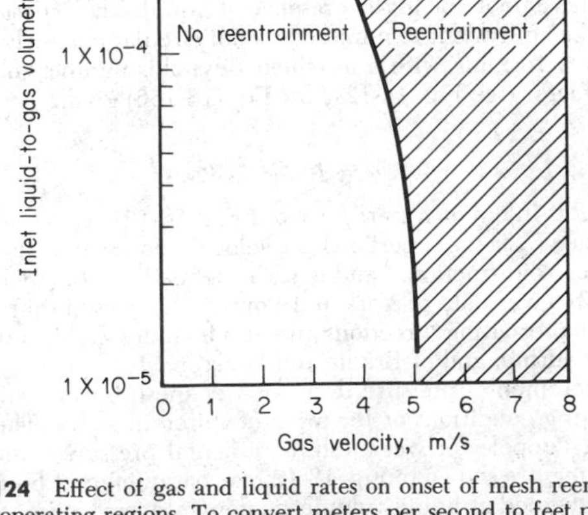

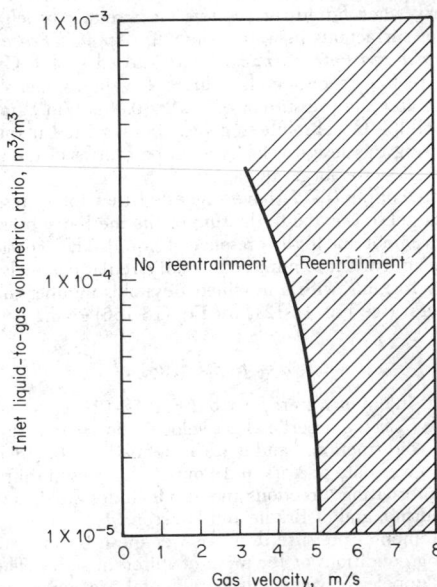

FIG. 18-124 Effect of gas and liquid rates on onset of mesh reentrainment and safe operating regions. To convert meters per second to feet per second, multiply by 3.281. (*Calvert, Yung, and Leung, NTIS Publ. PB-248050, 1975.*)

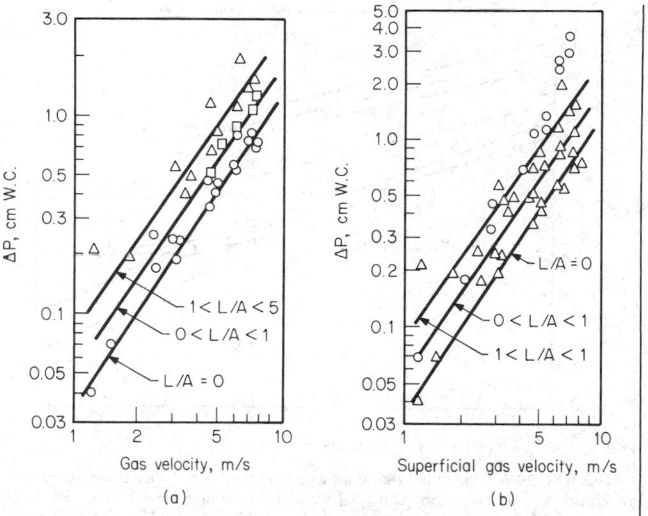

FIG. 18-125 Experimental pressure measured by Calvert as a function of gas velocity and liquid loading for (*a*) horizontal gas flow through vertical mesh and (*b*) gas upflow through horizontal mesh. Mesh thickness was 10 cm with 2.8-mm wire and void fraction of 98.2 percent, crimped in alternating directions. To convert meters per second to feet per second, multiply by 3.281; to convert centimeters to inches, multiply by 0.394. (*Calvert, Yung, and Leung, NTIS Publ. PB-248050, 1975.*)

gests that a correction factor of 0.2 be used in small-diameter scrubbers to account for the liquid on the walls, i.e., let $Q_L/Q_g = 0.2\ (Q_L/Q_g)_{actual}$. Many more complicated wet scrubbers employ a combination of sprays or liquid atomization, cyclonic action, baffles, and targets. These combinations are not likely to be more efficient than similar devices previously discussed that operate at equivalent pressure drop. The vast majority of wet scrubbers operate at moderate pressure drop [8 to 15 cm (3 to 6 in) of water or 18 to 30 cm (7 to 12 in) of water) and cannot be expected to have high efficiency on particles smaller than 10 μm or 3 to 5 μm. Fine and submicrometer particles can be captured efficiently only in wet scrubbers having high energy input such as venturi scrubbers, two-phase eductor scrubbers, and flux-force-condensation scrubbers.

Venturi Scrubbers One type of venturi scrubber is illustrated in Fig. 20-121. Venturi scrubbers have been used extensively for collecting fine and submicrometer solid particulate, condensing tars and mists, and mixtures of liquids and solids. To a lesser extent, they have also been used for simultaneous gas absorption, although Lundy [*Ind. Eng. Chem.*, **50**, 293 (1958)] indicates that they are generally limited to three transfer units. They have been used to collect submicrometer chemical incinerator fume and mist as well as sulfuric and phosphoric acid mists. The collection efficiency of a venturi scrubber is highly dependent on the throat velocity or pressure drop, the liquid-to-gas ratio, and the chemical nature of wettability of the particulate. Throat velocities may range from 60 to 150 m/s (200 to 500 ft/s). Liquid injection rates are typically 0.67 to 1.4 $m^3/1000\ m^3$ of gas. A liquid rate of 1.0 m^3 per 1000 m^3 of gas is usually close to optimum, but liquid rates as high as 2.7 m^3 (95 ft^3) have been used. Efficiency improves with increased liquid rate but only at the expense of higher pressure drop and energy consumption. Pressure-drop predictions for a given efficiency are hazardous without determining the nature of the particulate and the liquid-to-gas ratio. In general, particles coarser than 1 μm can be collected efficiently with pressure drops of 25 to 50 cm of water. For appreciable collection of submicrometer particles, pressure drops of 75 to 100 cm (30 to 40 in) of water are usually required. When particles are appreciably finer thatn 0.5 μm, pressure drops of 175 to 250 cm (70 to 100 in) of water have been used.

One of the problems in predicting efficiency and required pressure drop of a venturi is the chemical nature or wettability of the particulate, which on 0.5-μm-size particles can make up to a threefold difference in required pressure drop for its efficient collection. Calvert (R-9, R-10) has represented this effect by an empirical factor *f*, which is based on the hydrophobic ($f = 0.25$) or hydrophilic ($f = 0.50$) nature of the particles. Figure 18-127 gives the cut diameter of a venturi scrubber as a function of its operating parameters (throat velocity, pressure drop, and liquid-to-gas ratio) for hydrophobic particles. Figure 18-129 compares cut diameter as a function of pressure drop for an otherwise identically operating venturi on hydrophobic and hydrophilic particles. Calvert (R-9) gives equations which can be used for constructing cut-size curves similar to those of Fig. 18-127 for other values of the empirical factor *f*. Most real particles are neither completely hydrophobic nor completely hydrophilic but have *f* values lying between the two extremes. Phosphoric acid mist,

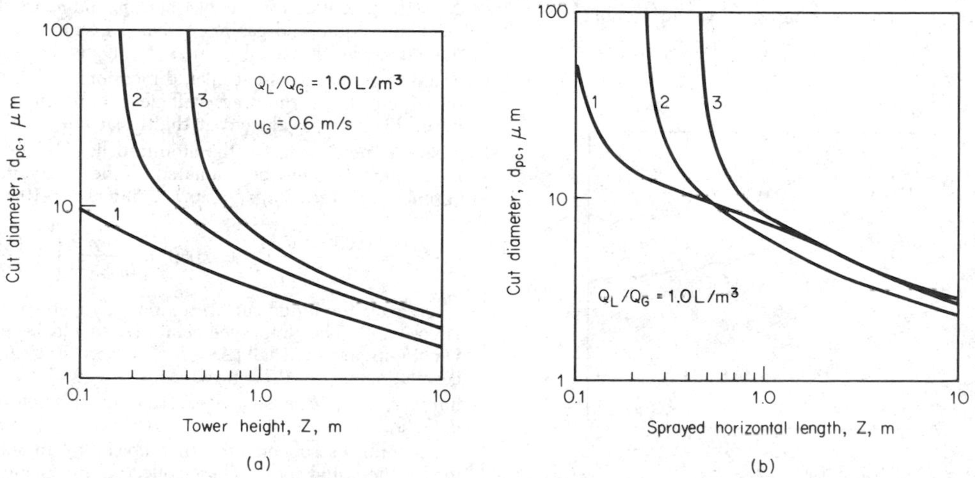

FIG. 18-126 Predicted spray-tower cut diameter as a function of sprayed length and spray droplet size for (*a*) vertical-countercurrent towers and (*b*) horizontal-cross-flow towers per Calvert [*J. Air Pollut. Control Assoc.*, *24*, 929 (1974)]. Curve 1 is for 200-μm spray droplets, curve 2 for 500-μm spray, and curve 3 for 1000-μm spray. Q_L/Q_G is the volumetric liquid-to-gas ratio, L liquid/m^3 gas, and u_G is the superficial gas velocity in the tower. To convert liters per cubic meter to cubic feet per cubic foot, multiply by 10^{-3}.

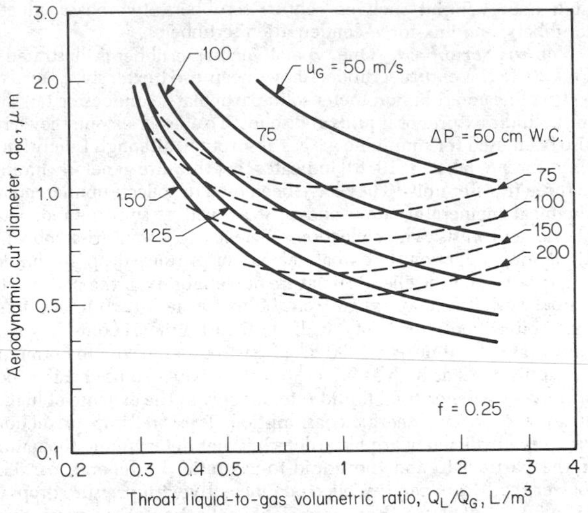

FIG. 18-127 Prediction of venturi-scrubber cut diameter for hydrophobic particles as functions of operating parameters as measured by Calvert [Calvert, Goldshmid, Leith, and Mehta, NTIS Publ. PB-213016, 213017, 1972; and Calvert, *J. Air Pollut. Control Assoc.*, **24**, 929 (1974)]. u_G is the superficial throat velocity, and ΔP is the pressure drop from converging to diverging section. To convert meters per second to feet per second, multiply by 3.281; to convert liters per cubic meter to cubic feet per cubic foot, multiply by 10^{-3}; and to convert centimeters to inches, multiply by 0.394.

on the basis of data of Brink and Contant [*Ind. Eng. Chem.*, **50**, 1157 (1958)] appears to have a value of $f = 0.46$. Unfortunately, no chemical-test methods have yet been devised for determining appropriate f values for a particulate in the laboratory.

Pressure drop in a venturi scrubber is controlled by throat velocity. While some venturis have fixed throats, many are designed with variable louvers to change throat dimensions and control performance for changes in gas flow. Pressure-drop equations have been developed by Calvert (R-13, R-14, R-15), Boll [*Ind. Eng. Chem. Fundam.*, **12**, 40 (1973)], and Hesketh [*J. Air Pollut. Control Assoc.*, **24**, 939 (1974)]. Hollands and Goel [*Ind. Eng. Chem. Fundam.*, **14**, 16 (1975)] have developed a generalized pressure-drop equation.

The Hesketh equation is empirical and is based upon a regression analysis of data from a number of industrial venturi scrubbers:

$$\Delta P = U_{gt}^2\,\rho_g\,A_t^{0.133}\,L^{0.78}/1270 \qquad (18\text{-}157)$$

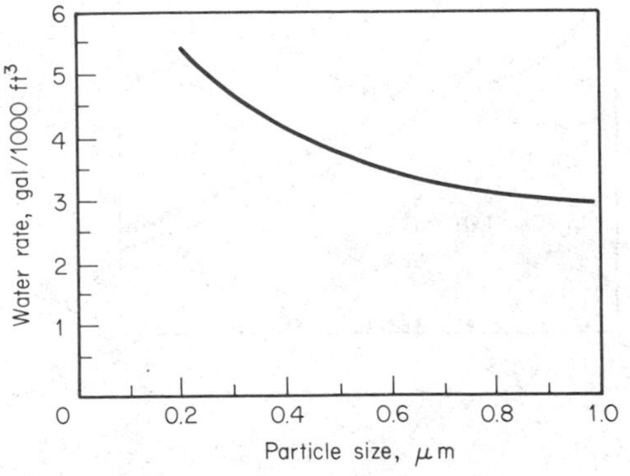

FIG. 18-128 Superheated high-pressure hot-water requirements for 99 percent collection as a function of particle size in a two-phase eductor jet scrubber. To convert gallons per 1000 cubic feet to cubic meters per 1000 cubic meters, multiply by 0.134. [*Gardenier*, J. Air Pollut. Control Assoc., *24, 954 (1974).*]

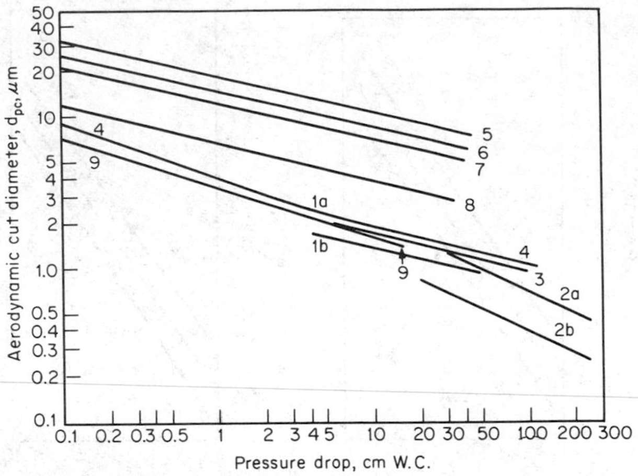

FIG. 18-129 Typical cut diameter as a function of pressure drop for various liquid-particle collectors. Curves 1*a* and *b* are single-sieve plates with froth density of 0.4 g/cm³; 1*a* has sieve holes of 0.5 cm and 1*b* holes of 0.3 cm. Curves 2*a* and *b* are for a venturi scrubber with hydrophobic particles (2*a*) and hydrophilic particles (2*b*). Curve 3 is an impingement plate, and curve 4 is a packed column with 2.5-cm-diameter packing. Curve 5 is a zigzag baffle collector with six baffles at $\Theta = 30°$. Curve 7 is for six rows of staggered tubes with 1-cm spacing between adjacent tube walls in a row. Curve 8 is similar, except that tube-wall spacing in the row is 0.3 cm. Curve 9 is for wire-mesh pads. To convert grams per cubic centimeter to pounds per cubic foot, multiply by 62.43; to convert centimeters to inches, multiply by 0.394. [*Calvert,* J. Air Pollut. Control Assoc., *24, 929 (1974); and Calvert, Yung, and Leung, NTIS Publ. PB-248050, 1975.*]

where ΔP is the pressure drop, in of water; U_{gt} is the gas velocity in the throat, ft/s; ρ_g is the gas density, lb/ft³; A_t is the throat area, ft²; and L is the liquid-to-gas ratio, gal/1000 acf.

Calvert (R-15) critiqued the many pressure-drop equations and suggested the following simplified equation as accurate to ±10 percent:

$$\Delta P = \frac{2\,\rho_\ell U_g^2}{981\,g_c}\left(\frac{Q_\ell}{Q_g}\right)[1 - x^2 + \sqrt{(x^4 - x^2)^{0.5}}] \qquad (18\text{-}158)$$

where
$$x = (3\,l_t C_{Di}\rho_g/16\,d_\ell\,\rho_\ell) + 1 \qquad (18\text{-}159)$$

ΔP is the pressure drop, cm of water; ρ_ℓ and ρ_g are the density of the scrubbing liquid and gas respectively, g/cm³; U_g is the velocity of the gas at the throat inlet, cm/s; Q_ℓ/Q_g is the volumetric ratio of liquid to gas at the throat inlet, dimensionless; l_t is the length of the throat, cm; C_{Di} is the drag coefficient, dimensionless, for the mean liquid diameter, evaluated at the throat inlet; and d_ℓ is the Sauter mean diameter, cm, for the atomized liquid. The atomized-liquid mean diameter must be evaluated by the Nukiyama and Tanasawa [*Trans. Soc. Mech Eng. (Japan)*, **4, 5, 6** (1937–1940)] equation:

$$d_\ell = \frac{0.0585}{U_g}\left(\frac{\sigma_\ell}{\rho_\ell}\right)^{0.5} + 0.0597\left[\frac{\mu_\ell}{(\sigma_\ell\rho_\ell)^{0.5}}\right]^{0.45}\left(\frac{Q_\ell}{Q_g}\right)^{1.5} \qquad (18\text{-}160)$$

where σ_ℓ is the liquid surface tension, dyn/cm; and μ_ℓ is the liquid viscosity, P. The drag coefficient C_{Di} should be evaluated by the Dickinson and Marshall [*Am. Inst. Chem. Eng. J.*, **14**, 541 (1968)] correlation $C_{Di} = 0.22 + (24/N_{\mathrm{Re}i})(1 + 0.15\,N_{\mathrm{Re}i}^{0.6})$. The Reynolds number, $N_{\mathrm{Re}i}$, is evaluated at the throat inlet considerations as $d_\ell G_g/\mu_g$.

All venturi scrubbers must be followed by an entrainment collector for the liquid spray. These collectors are usually centrifugal and will have an additional pressure drop of several centimeters of water, which must be added to that of the venturi itself.

Other Scrubbers A liquid-ejector venturi (Fig. 20-121), in which high-pressure water from a jet induces the flow of gas, has been used to collect mist particles in the 1- to 2-μm range, but submicrometer particles will generally pass through an eductor. Power costs for liquid pumping are high if appreciable motive force must be imparted to the gas because jet-pump efficiency is usually less than 10 percent. Harris [*Chem. Eng. Prog.*, **42**(4), 55 (1966)] has described their application. Two-phase eductors have been considerably more successful on capture of submicrometer mist particles and could be attractive in situations in which large quantities of waste thermal energy are available. However, the equivalent energy consumption is equal to that required for high-energy venturi scrubbers, and such devices are likely to be no more attractive than venturi scrubbers when the thermal energy is priced at its proper value. Sparks [*J. Air Pollut. Control Assoc.*, **24**, 958 (1974)] has discussed steam ejectors giving 99 percent collection of particles 0.3 to 10 μm. Energy requirements were 311,000 J/m^3(8.25 Btu/scf). Gardenier [*J. Air Pollut. Control Assoc.*, **24**, 954 (1974)] operated a liquid eductor with high-pressure (6900- to 27,600-kPa) (1000- to 4000-lbf/in^2) hot water heated to 200°C (392°F) which flashed into two phases as it issued from the jet. He obtained 95 to 99 percent collection of submicrometer particulate. Figure 18-128 shows the water-to-gas ratio required as a function of particle size to achieve 99 percent collection.

Effect of Gas Saturation in Scrubbing If hot unsaturated gas is introduced into a wet scrubber, spray particles will evaporate to cool and saturate the gas. The evaporating liquid molecules moving away from the target droplets will repel particles which might collide with them. This results in the forces of diffusiophoresis opposing particle collection. Semrau and Witham (Air Pollut. Control Assoc. Prepr. 75-30.1) investigated temperature parameters in wet scrubbing and found a definite decrease in the efficiency of evaporative scrubbers and an enhancement of efficiency when a hot saturated gas is scrubbed with cold water rather than recirculated hot water. Little improvement was experienced in cooling a hot saturated gas below a 50°C dew point.

Energy Requirements for Inertial-Impaction Efficiency Semrau [*J. Air Pollut. Control Assoc.*, **13**, 587 (1963)] proposed a "contacting-power" principle which states that the collecting efficiency of a given size of particle is proportional to the power expended and that the smaller the particle, the greater the power required. Mathematically expressed, $N_T = \propto P_T^\gamma$, where N_T is the number of particulate transfer units achieved and P_T is the total energy expended within the collection device, including gas and liquid pressure drop and thermal and mechanical energy added in atomizers. N_T is further defined as $N_T = \ln [1/(1 - \eta)]$, where η is the overall fractional collection efficiency. This was intended as a universal principle, but the constants $\propto$ and γ have been found to be functions of the chemical nature of the system. Others have pointed out that the principle is applicable only when the primary collection mechanism is impaction and direct interception. Calvert (R-10, R-12) has found that plotting particle cut size versus pressure drop (or power expended) as in Fig. 18-129 is a more suitable way to develop a generalized energy-requirement curve for impaction devices. The various curves fall close together and outline an imaginary curve that indicates the magnitude of pressure drop required as particle size decreases bound by the two limits of hydrophilic and hydrophobic particles. By calculating the required cut size for a given collection efficiency, Fig. 18-129 can also be used as a guide to deciding between different collection devices.

Collection of Fine Mists Inertial-impaction devices previously discussed give high efficiency on particles above 5 μm in size and often reasonable efficiency on particles down to 3 μm in size at moderate pressure drops. However, this mechanism becomes ineffective for particles smaller than 3 μm because of the particle gaslike mobility. Only impaction devices having extremely high energy input such as venturi scrubbers and a flooded mesh pad (the pad interstices really become miniature venturi scrubbers in parallel and in series) can give high collection efficiency on fine particles, defined as 2.5 or 3 μm and smaller, including the submicrometer range. Fine particles are subjected to brownian motion in gases, and diffusional deposition can be employed for their collection. Diffusional deposition becomes

TABLE 18-21 Brownian Movement of Particles*

Particle diameter, μm	Brownian displacement of particle, μm/s
0.1	29.4
0.25	14.2
0.5	8.92
1.0	5.91
2.5	3.58
5.0	2.49
10.0	1.75

*Brink, *Can. J. Chem. Eng.*, **41**, 134 (1963). Based on spherical water particles in air at 21°C and 1 atm.

highly efficient as particles become smaller, especially below 0.2 to 0.3 μm. Table 18-21 shows typical displacement velocity of particles. Randomly oriented fiber beds having tortuous and narrow gas passages are suitable devices for utilizing this collection mechanism. (The diffusional collection mechanism is discussed in Sec. 20 under "Gas-Solids Separations.") Other collection mechanisms which are efficient for fine particles are electrostatic forces and flux forces such as thermophoresis and diffusiophoresis. Particle growth and nucleation methods are also applicable. Efficient collection of fine particles is important because particles in the range of 2.0 to around 0.2 μm are the ones which penetrate and are deposited in the lung most efficiently. Hence, particles in this range constitute the largest health hazard.

Fiber Mist Eliminators These devices are produced in various configurations. Generally, randomly oriented glass or polypropylene fibers are densely packed between reinforcing screens, producing fiber beds varying in thickness usually from 25 to 75 mm (1 to 3 in), although thicker beds can be produced. Units with efficiencies as high as 99.9 percent on fine particles have been developed (see *Chemical Engineers' Handbook*, 5th ed., p. 18-88). A combination of mechanisms interacts to provide high overall collection efficiency. Particles larger than 2 to 3 μm are collected on the fibers by inertial impaction and direct interception, while small particles are collected by brownian diffusion. When the device is designed to use this latter mechanism as the primary means, efficiency turndown problems are eliminated as collection efficiency by diffusion increases with residence time. Pressure drop through the beds increases with velocity to the first power since the gas flow is laminar. This leads to design capability trade-offs. As pressure drop is reduced and energy is conserved, capital increases because more filtering area is required for the same efficiency.

Three series of fiber mist eliminators are typically available. A spray-catcher series is designed primarily for essentially 100 percent

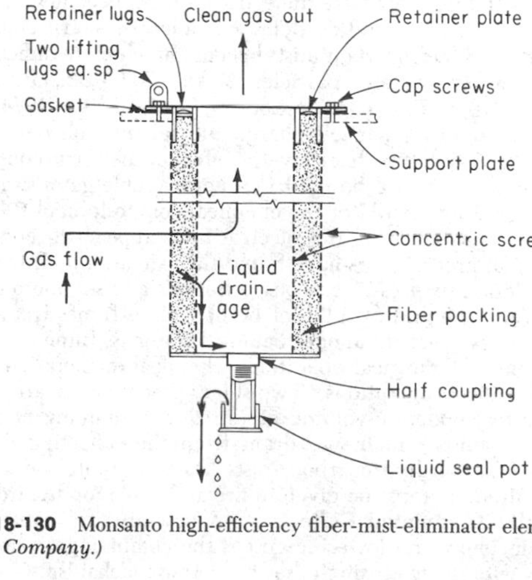

FIG. 18-130 Monsanto high-efficiency fiber-mist-eliminator element. *(Monsanto Company.)*

Table 18-22 Operating Characteristics of Various Types of Fiber Mist Eliminators as Used on Sulfuric Acid Plants*

	High efficiency	High velocity	Spray catcher
Controlling mechanism for mist collection	Brownian movement	Impaction	Impaction
Superficial velocity, m/s	0.075–0.20	2.0–2.5	2.0–2.5
Efficiency on particles greater than 3 μm, %	Essentially 100	Essentially 100	Essentially 100
Efficiency on particles 3 μm and smaller, %	95–99+	90–98	15–30
Pressure drop, cm H_2O	12–38	15–20	1.0–2.5

*Brink, Burggrabe, and Greenwell, *Chem. Eng. Prog.*, **64**(11), 82 (1968). To convert centimeters to inches, multiply by 0.394.

capture of droplets larger than 3 μm. The high-velocity type is designed to give moderately high efficiency on particles down to 1.0 μm as well. Both of these types are usually produced in the form of flat panels of 25- to 50-mm (1- to 2-in) thickness. The high-efficiency type is illustrated in Fig. 18-130. As mist particles are collected, they coalesce into a liquid film which wets the fibers. Liquid is moved horizontally through the bed by the gas drag force and downward by gravity. It drains down the downstream retaining screen to the bottom of the element and is returned to the process through a liquid seal. Table 18-22 gives typical operating characteristics of the three types of collectors. The application of these devices to sulfuric acid plants and other process gases has been discussed by Brink (see *Chemical Engineers' Handbook*, 5th ed., pp. 18-89, 18-90).

Solid particulates are captured as readily as liquids in fiber beds but can rapidly plug the bed if they are insoluble. Fiber beds have frequently been used for mixtures of liquids and soluble solids and with soluble solids in condensing situations. Sufficient solvent (usually water) is atomized into the gas stream entering the collector to irrigate the fiber elements and dissolve the collected particulate. Such fiber beds have been used to collect fine fumes such as ammonium nitrate and ammonium chloride smokes.

Electrostatic Precipitators The principles and operation of electrical precipitators are discussed in Sec. 20 under "Gas-Solids Separations." Precipitators are admirably suited to the collection of fine mists and mixtures of mists and solid particulates. Tube-type precipitators have been used for many years for the collection of acid mists and the removal of tar from coke-oven gas. The first practical installation of a precipitator by Cottrell was made on sulfuric acid mist in 1907. Most older installations of precipitators were tube-type rather than plate-type. However, recently two plate-type wet precipitators employing water sprays or overflowing weirs have been introduced by Mikropul Corporation [Bakke, *J. Air Pollut. Control Assoc.*, **25**, 163 (1975)] and by Fluid Ionics. Such precipitators operate on the principle of making all particles conductive when possible, which increases the particle migration velocity and collection efficiency. Under these conditions, particle dielectric strength becomes a much more important variable, and particles with a low dielectric constant such as condensed hydrocarbon mists become much more difficult to collect than water-wettable particles. Bakke (U.S.–U.S.S.R. Joint Work. Group Symp.: Fine Particle Control, San Francisco, 1974) has developed equations for particle charge and relative collection efficiency in wet precipitators that show the effect of dielectric constant. Wet precipitators can also be used to absorb soluble gases simultaneously by adjusting the pH or the chemical composition of the liquid spray. The presence of the electric field appears to enhance absorption. Wet precipitators have found their greatest usefulness to date in handling mixtures of gaseous pollutants and submicrometer particulate (either liquid or solid, or both) such as fumes from aluminum-pot lines, carbon anode baking, fiberglass-fume control, coke-oven and metallurgical operations, chemical incineration, and phosphate-fertilizer operations. Two-stage precipitators are used increasingly for moderate-volume gas streams containing nonconductive liquid mists which will drain from the collecting plates. Their application on hydrocarbon mists has been quite successful, but careful attention must be given to fire and explosion hazards.

Electrically Augmented Collectors A new area for enhancing collection efficiency and lowering cost is the combining of electrostatic forces with devices using other collecting mechanisms such as impaction and diffusion. Cooper (Air Pollut. Control Assoc. Prepr. 75-02.1) evaluated the magnitude of forces operating between charged and uncharged particles and concluded that electrostatic attraction is the strongest collecting force operating on particles finer than 2 μm. Nielsen and Hill [*Ind. Eng. Chem. Fundam.*, **15**, 149 (1976)] have quantified these relationships, and a number of practical devices have been demonstrated. Pilat and Meyer (NTIS Publ. PB-252653, 1976) have demonstrated up to 99 percent collection of fine particles in a two-stage spray tower in which the inlet particles and water spray are charged with opposite polarity. The principle has been applied to retrofitting existing spray towers to enhance collection.

Klugman and Sheppard (Air Pollut. Control Assoc. Prepr. 75-30.3) have developed an ionizing wet scrubber in which the charged mist particles are collected in a grounded, irrigated cross-flow bed of Tellerette packing. Particles smaller than 1 μm have been collected with 98 percent efficiency by using two units in series. Dembinsky and Vicard (Air Pollut. Control Assoc. Prepr. 78-17.6) have used an electrically augmented low-pressure [5 to 10 cm (2 to 4 in) of water] venturi scrubber to give 95 to 98 percent collection efficiency on submicrometer particles.

Particle Growth and Nucleation Fine particles may be subjected to conditions favoring the growth of particles either through condensation or through coalescence. Saturation of a hot gas stream with water, followed by condensation on the particles acting as nuclei when the gas is cooled, can increase particle size and ease of collection. Addition of steam can produce the same results. Scrubbing of the humid gas with a cold liquid can bring diffusiophoresis into play. The introduction of cold liquid drops causes a reduction in water-vapor pressure at the surface of the cold drop. The resulting vapor-pressure gradient causes a hydrodynamic flow toward the drop known as Stefan flow which enhances the movement of mist particles toward the spray drop. If the molecular mass of the diffusing vapor is different from the carrier gas, this density difference also produces a driving force, and the sum of these forces is known as diffusiophoresis. A mathematical description of these forces has been presented by Calvert (R-9) and by Sparks and Pilat [*Atmos. Environ.*, **4**, 651 (1970)]. Thermal differences between the carrier gas and the cold scrubbing droplets can further enhance collection through thermophoresis. Calvert and Jhaseri [*J. Air Pollut. Control Assoc.*, **24**, 946 (1974)]; and NTIS Publ. PB-227307, 1973)] have investigated condensation scrubbing in multiple-sieve plate towers.

Submicrometer droplets can be coagulated through brownian diffusion if given ample time. The introduction of particles 50 to 100 times larger in diameter can enhance coagulation, but the addition of a broad range of particle sizes is discouraged. Increasing turbulence will aid coagulation, so fans to stir the gas or narrow, tortuous passages such as those of a packed bed can be beneficial. Sonic energy can also produce coagulation, especially the production of standing waves in the confines of long, narrow tubes. Addition of water and oil mists can sometimes aid sonic coagulation. Sulfuric acid mist [Danser, *Chem. Eng.*, **57**(5), 158 (1950)] and carbon black [Stokes, *Chem. Eng. Prog.*, **46**, 423 (1950)] have been successfully agglomerated with sonic energy. Frequently sonic agglomeration has been unsuccessful because of the high energy requirement. Most sonic generators have very poor energy-transformation efficiency. Wegrzyn et al. (U.S. EPA Publ. EPA-600/7-79-004C, 1979, p. 233) have reviewed acoustic agglomerators. Mednikov (U.S.S.R. Akad.

Soc. Moscow, 1963) suggested that the incorporation of sonic agglomeration with electrostatic precipitation could greatly reduce precipitator size.

Other Collectors Tarry particulates and other difficult-to-handle liquids have been collected on a dry, expendable phenol formaldehyde–bonded glass-fiber mat [Goldfield, *J. Air Pollut. Control Assoc.*, **20**, 466 (1970)] in roll form which is advanced intermittently into a filter frame. Superficial gas velocities are 2.5 to 3.5 m/s (8.2 to 11.5 ft/s), and pressure drop is typically 41 to 46 cm (16 to 18 in) of water. Collection efficiencies of 99 percent have been obtained on submicrometer particles. Brady [*Chem. Eng. Prog.*, **73**(8), 45 (1977)] has discussed a cleanable modification of this approach in which the gas is passed through a reticulated foam filter that is slowly rotated and solvent-cleaned.

LIQUID-PHASE CONTINUOUS SYSTEMS

Practical separation techniques for gases dispersed in liquids are discussed. Processes and methods for dispersing gas in liquid have been discussed earlier in this section, together with information for predicting the bubble size produced. Gas-in-liquid dispersions are also produced in chemical reactions and electrochemical cells in which a gas is liberated. Such dispersions are likely to be much finer than those produced by the dispersion of a gas. Dispersions may also be unintentionally created in the vaporization of a liquid.

General References: Adamson, *Physical Chemistry of Surfaces*, 2d ed., Wiley-Interscience, New York, 1967. Akers, *Foams*, Academic, New York, 1976. Bikerman, *Foams*, Springer-Verlag, New York, 1973. Bikerman et al., *Foams: Theory and Industrial Applications*, Reinhold, New York, 1953. Kerner, *Foam Control Agents*, Noyes Data Corp., Park Ridge, N.J., 1976. Kitchener, in Danielli, Pankhurst, and Riddiford (eds.,) *Recent Progress in Surface Science*, vol. 1, Academic, New York, 1964. Rubel, *Antifoaming and Defoaming Agents*, Noyes Data Corp., Park Ridge, N.J., 1972. Sonntag and Strenge, *Coagulation and Stability of Disperse Systems*, Halsted-Wiley, New York, 1972.

Types of Gas-in-Liquid Dispersions Two types of dispersions exist. In one, gas bubbles produce an **unstable dispersion** which separates readily under the influence of gravity once the mixture has been removed from the influence of the dispersing force. Gas-liquid contacting means such as bubble towers and gas-dispersing agitators are typical examples of equipment producing such dispersions. More difficulties may result in separation when the gas is dispersed in the form of bubbles only a few micrometers in size. An example is the evolution of gas from a liquid in which it has been dissolved or released through chemical reaction such as electrolysis. Coalescence of the dispersed phase can be helpful in such circumstances.

The second type is a **stable dispersion,** or foam. Separation can be extremely difficult in some cases. A pure two-component system of gas and liquid cannot produce dispersions of the second type. Stable foams can be produced only when an additional substance is adsorbed at the liquid-surface interface. The substance adsorbed may be in true solution but with a chemical tendency to concentrate in the interface such as that of a surface-active agent, or it may be a finely divided solid which concentrates in the interface because it is only poorly wetted by the liquid. Surfactants and proteins are examples of soluble materials, while dust particles and extraneous dirt including traces of nonmiscible liquids can be examples of poorly wetted materials.

Separation of gases and liquids always involves coalescence, but enhancement of the rate of coalescence may be required only in difficult separations.

Separation of Unstable Systems The buoyancy of bubbles suspended in liquid can frequently be depended upon to cause the bubbles to rise to the surface and separate. This is a special case of gravity settling. The mixture is allowed to stand at rest or is moved along a flow path in laminar flow until the bubbles have surfaced. Table 18-23 shows the calculated rate of rise of air bubbles at atmospheric pressure in water at 20°C (68°F) as a function of diameter. It will be observed that the velocity of rise for 10-μm bubbles is very low, so that long separating times would be required for gas which is more finely dispersed.

For liquids other than water, the rise velocity can be approximated from Table 18-23 by multiplying by the liquid's specific gravity and the reciprocal of its viscosity (in centipoises). For bubbles larger than 100 μm, this procedure is erroneous, but the error is less than 15 percent for bubbles up to 1000 μm. More serious is the underlying assumption of Table 18-23 that the bubbles are rigid spheres. Circulation within the bubble causes notable increases in velocity in the range of 100 μm to 1 mm, and the flattening of bubbles 1 cm and larger appreciably decreases their velocity. However, in this latter size range the velocity is so high as to make separation a trivial problem.

In design of separating chambers, static vessels or continuous-flow tanks may be used. Care must be taken to protect the flow from turbulence, which could cause back mixing of partially separated fluids or which could carry unseparated liquids rapidly to the separated-liquid outlet. Vertical baffles to protect rising bubbles from flow currents are sometimes employed. Unseparated fluids should be distributed to the separating region as uniformly and with as little velocity as possible. When the bubble rise velocity is quite low, shallow tanks or flow channels should be used to minimize the residence time required.

Quite low velocity rise of bubbles due either to small bubble size or to high liquid viscosity can cause difficult situations. With low-viscosity liquids, separation-enhancing possibilities in addition to those previously enumerated are to sparge the liquid with large-diameter gas bubbles or to atomize the mixture as a spray into a tower. Large gas bubbles rising rapidly through the liquid collide with small bubbles and aid their coalescence through capture. Atomizing of the continuous phase reduces the distance that small gas bubbles must travel to reach a gas interface. Evacuation of the spray space can also be beneficial in promoting small-bubble growth and especially in promoting gas evolution when the gas has appreciable liquid solubility. Liquid heating will also reduce solubility.

Surfaces in the settling zone for bubble coalescence such as closely spaced vertical or inclined plates or tubes are beneficial. When clean low-viscosity fluids are involved, passage of the undegassed liquid through a tightly packed pad of mesh or fine fibers at low velocity will result in efficient bubble coalescence. Spielman and Goren [*Ind. Eng. Chem.*, **62**(10), (1970)] reviewed the literature on coalescence with porous media and reported their own experimental results [*Ind. Eng. Chem. Fundam.*, **11**(1), 73 (1972)] on the coalescence of oil-water liquid emulsions. The principles are applicable to a gas-in-liquid system. Glass-fiber mats composed of 3.5-, 6-, or 12-μm diameter fibers, varying in thickness from 1.3 to 3.3 mm, successfully coalesced and separated 1- to 7-μm oil droplets at superficial bed velocities of 0.02 to 1.5 cm/s (0.00067 to 0.049 ft/s).

In the deaeration of high-viscosity fluids such as polymers, the material is flowed in thin sheets along solid surfaces. Vacuum is applied to increase bubble size and hasten separation. The Versator (Cornell Machine Co.) degasses viscous liquids by spreading them into a thin film by centrifugal action as the liquids flow through an evacuated rotating bowl.

Table 18-23 Terminal Velocity of Standard Air Bubbles Rising in Water at 20°C*

Bubble diameter, μm	10	30	50	100	200	300
Terminal velocity, mm/s	0.061	0.488	1.433	5.486	21.95	49.38

*Calculated from Stokes' law. To convert millimeters per second to feet per second, multiply by 0.003281.

Separation of Foam Foam is a colloidal system containing relatively large volumes of dispersed gas in a relatively small volume of liquid. Foams are thermodynamically unstable with respect to separation into their components of gas and vapor, and appreciable surface energy is released in the bursting of foam bubbles. Foams are dynamic systems in which a third component produces a surface layer that is different in composition from the bulk of the liquid phase. The stabilizing effect of such components (often present only in trace amounts) can produce foams of troubling persistence in many operations. (Foams which have lasted for years when left undisturbed have been produced.) Bendure [TAPPI, **58**(2), 83 (1975)], Keszthelyi [*J. Paint Technol.*, **46**(11), 31 (1974)], Ahmad [*Sep. Sci*, **10**, 649 (1975)], and Shedlovsky ("Foams," *Encyclopedia of Chemical Technology*, 2d ed., Wiley, New York, 1966) have presented concise articles on the characteristics and properties of foams in addition to the general references cited at the beginning of this subsection.

Foams can be a severe problem in chemical-processing steps involving gas-liquid interaction such as distillation, absorption, evaporation, chemical reaction, and particle separation and settling. It can also be a major problem in pulp and paper manufacture, oil-well drilling fluids, production of water-based paints, utilization of lubricants and hydraulic fluids, dyeing and sizing of textiles, operation of steam boilers, fermentation operations, polymerization, wet-process phosphoric acid concentration, adhesive production, and foam control in products such as detergents, waxes, printing inks, instant coffee, and glycol antifreeze.

Foams, as freshly generated, are gas emulsions with spherical bubbles separated by liquid films up to a few millimeters in thickness. They age rapidly by liquid drainage and form polyhedrals in which three bubbles intersect at corners with angles of approximately 120°. During drainage, the lamellae become increasingly thinner, especially in the center (only a few micrometers thickness), and more brittle. This feature indicates that with some foams if a foam layer can be tolerated, it may be self-limiting, as fresh foam is added to the bottom of the layer with drained foam collapsing on the top. (A quick-breaking foam may reach its maximum life cycle in 6 s. A moderately stable foam can persist for 140 s.) During drainage, gas from small foam bubbles, which is at a high pressure, will diffuse into large bubbles so that foam micelles increase with time. As drainage proceeds, weak areas in the lamella may develop. However, the presence of a higher concentration of surfactants in the surface produces a lower surface tension. As the lamella starts to fail, exposing bulk liquid with higher surface tension, the surface is renewed and healed. This is known as the *Marangoni effect*. If drainage can occur faster than Marangoni healing, a hole may develop in the lamella. The forces involved are such that collapse will occur in milliseconds without concern for rupture propagation. However, in very stable foams, electrostatic surface forces (zeta potential) prevent complete drainage and collapse. In some cases, stable lamella thicknesses of only two molecules have been measured.

Drainage rate is influenced by surface viscosity, which is very temperature-sensitive. At a critical temperature, which is a function of the system, a temperature change of only a few degrees can change a slow-draining foam to a fast-draining foam. This change in drainage rate can be a factor of 100 or more; thus increasing the temperature of foam can cause its destruction. An increase in temperature may also cause liquid evaporation and lamella thinning. As the lamellae become thinner, they become more brittle and fragile. Thus, mechanical deformation or pressure changes, which cause a change in gas-bubble volume, can also cause rupture.

Bendure indicates 10 ways to increase foam stability: (1) increase bulk liquid viscosity, (2) increase surface viscosity, (3) maintain thick walls (higher liquid-to-gas ratio), (4) reduce liquid surface tension, (5) increase surface elasticity, (6) increase surface concentration, (7) reduce surfactant-adsorption rate, (8) prevent liquid evaporation, (9) avoid mechanical stresses, and (10) eliminate foam inhibitors. Obviously, the reverse of each of these actions, when possible, is a way to control and break foam.

Physical Defoaming Techniques Typical physical defoaming techniques include mechanical methods for producing foam stress, thermal methods involving heating or cooling, and electrical methods. Combinations of these methods may also be employed, or they may be used in conjunction with chemical defoamers. Some methods are only moderately successful when conditions are present to re-form the foam such as breaking foam on the surface of boiling liquids. In some cases it may be desirable to draw the foam off and treat it separately. Foam can always be stopped by removing the energy source creating it, but this is often impractical.

Thermal Methods Heating is often a suitable means of destroying foam. As indicated previously, raising the foam above a critical temperature (which must be determined experimentally) can greatly decrease the surface viscosity of the film and change the foam from a slow-draining to a fast-draining foam. Coupling such heating with a mechanical force such as a revolving paddle to cause foam deformation is frequently successful. Other effects of heating are expansion of the gas in the foam bubbles, which increases strain on the lamella walls as well as requiring their movement and flexing. Evaporation of solvent may occur causing thinning of the walls. At sufficiently high temperatures, desorption or decomposition of stabilizing substances may occur. Placing a high-temperature bank of steam coils at the maximum foam level is one control method. As the foam approaches or touches the coil, it collapses. The designer should consider the fact that the coil will frequently become coated with solute.

Application of radiant heat to a foam surface is also practiced. Depending on the situation, the radiant source may be electric lamps, Glowbar units, or gas-fired radiant burners. Hot gases from burners will enhance film drying of the foam. Heat may also be applied by jetting or spraying hot water on the foam. This is a combination of methods since the jetting produces mechanical shear, and the water itself provides dilution and change in foam-film composition.

Cooling can also destroy foam if it is carried to the point of freezing since the formation of solvent crystals destroys the foam structure. Less drastic cooling such as spraying a hot foam with cold water may be effective. Cooling will reduce the gas pressure in the foam bubbles and may cause them to shrink. This is coupled with the effects of shear and dilution mentioned earlier. In general, moderate cooling will be less effective than heating since the surface viscosity is being modified in the direction of a more stable foam.

Mechanical Methods Static or rotating breaker bars or slowly revolving paddles are sometimes successful. Their application in conjunction with other methods is frequently better. As indicated in the theory of foams, they will work better if installed at a level at which the foam has had some time to age and drain. A rotating breaker works by deforming the foam, which causes rupture of the lamella walls. Rapidly moving slingers will throw the foam against the vessel wall and may cause impact on other foam outside the envelope of the slinger. In some instances, stationary bars or closely spaced plates will limit the rise of foam. The action here is primarily one of providing surface for coalescence of the foam. Wettability of the surface, whether moving or stationary, is frequently important. Usually a surface not wetted by the liquid is superior, just as is frequently the case of porous media for foam coalescence. However, in both cases there are exceptions for which wettable surfaces are preferred. Shkodin [*Kolloidn. Zh.*, **14**, 213 (1952)] found molasses foam to be destroyed by contact with a wax-coated rod and unaffected by a clean glass rod.

Goldberg and Rubin [*Ind. Eng. Chem. Process Des. Dev.*, **6** 195 (1967)] showed in tests with a disk spinning vertically to the foam layer that most mechanical procedures, whether centrifugation, mixing, or blowing through nozzles, consist basically of the application of shear stress. Subjecting foam to an air-jet impact can also provide a source of drying and evaporation from the film, especially if the air is heated. Other effective means of destroying bubbles are to lower a frame of metal points periodically into the foam or to shower the foam with falling solid particles.

Pressure and Acoustic Vibrations These methods for rupturing foam are really special forms of mechanical treatment. Change in pressure in the vessel containing the foam stresses the lamella walls by expanding or contracting the gas inside the foam bubbles. Oscillation of the vessel pressure subjects the foam to repeated film flexing.

Parlow [*Zucker*, **3**, 468 (1950)] controlled foam in sugar-sirup evaporators with high-frequency air pulses. It is by no means certain that high-frequency pulsing is necessary in all cases. Lower frequency and higher amplitude could be equally beneficial. Acoustic vibration is a similar phenomenon causing localized pressure oscillation by using sound waves. Impulses at 6 kHz have been found to break froth from coal flotation [Sun, *Min. Eng.*, **3**, 865 (1958)]. Sonntag and Strenge (*Coagulation and Stability of Disperse Systems*, Halsted-Wiley, New York, 1972, p. 121) report foam suppression with high-intensity sound waves (11 kHz, 150 dB) but indicate that the procedure is too expensive for large-scale application. The Sontrifuge (Teknika Inc., a subsidiary of Chemineer, Inc.) is a commercially available low-speed centrifuge employing sonic energy to break the foam. Walsh [*Chem. Process.*, **29**, 91 (1966)], Carlson [*Pap. Trade J.*, **151**, 38 (1967)], and Thorhildsen and Rich [*TAPPI*, **49**, 95A (1966)] have described the unit.

Electrical Methods As colloids, most foams typically have electrical double layers of charged ions which contribute to foam stability. Accordingly, foams can be broken by the influence of an external electric field. While few commercial applications have been developed, Sonntag and Strenge (op. cit., p. 114) indicate that foams can be broken by passage through devices much like electrostatic precipitators for dusts. Devices similar to two-stage precipitators having closely spaced plates of opposite polarity should be especially useful. Sonntag and Strenge, in experiments with liquid-liquid emulsions, indicate that the colloid structure can be broken at a field strength of the order of 8 to 9 $\times$ 10^5 V/cm.

Chemical Defoaming Techniques Sonntag and Strenge (op. cit., p. 111) indicate two chemical methods for foam breaking. One method is causing the stabilizing substances to be desorbed from the interface, such as by displacement with other more surface-active but nonstabilizing compounds. Heat may also cause desorption. The second method is to carry on chemical changes in the adsorption layer, leading to a new structure. Some defoamers may act purely by mechanical means but will be discussed in this subsection since their action is generally considered to be chemical in nature. Often chemical defoamers act in more than one way.

Chemical Defoamers The addition of chemical foam breakers is the most elegant way to break a foam. Effective defoamers cause very rapid disintegration of the foam and frequently need be present only in parts per million. The great diversity of compounds used for defoamers and the many different systems in which they are applied make a brief and orderly discussion of their selection difficult. Compounds needed to break aqueous foams may be different from those needed for aqueous-free systems. The majority of defoamers are insoluble or nonmiscible in the foam continuous phase, but some work best because of their ready solubility. Lichtman (*Defoamers*, 3d ed., Wiley, New York, 1979) has presented a concise summary of the application and use of defoamers. Rubel (*Antifoaming and Defoaming Agents*, Noyes Data Corp., Park Ridge, N.J., 1972) has reviewed the extensive patent literature on defoamers. Defoamers are also discussed extensively in the general references at the beginning of this subsection.

One useful method of aqueous defoaming is to add a nonfoam stabilizing surfactant which is more surface-active than the stabilizing substance in the foam. Thus a foam stabilized with an ionic surfactant can be broken by the addition of a very surface-active but nonstabilizing silicone oil. The silicone displaces the foam stabilizer from the interface by virtue of its insolubility. However, it does not stabilize the foam because its foam films have poor elasticity and rupture easily.

A major requirement for a defoamer is cost-effectiveness. Accordingly, some useful characteristics are low volatility (to prevent stripping from the system before it is dispersed and does its work), ease of dispersion and strong spreading power, and surface attraction-orientation. Chemical defoamers must also be selected in regard to their possible effect on product quality and their environmental and health suitability. For instance, silicone antifoam agents are effective in textile jet dyeing but reduce the fire retardancy of the fabric. Mineral-oil defoamers in sugar evaporation have been replaced by specifically approved materials. The tendency is no longer to use a single defoamer compound but to use a formulation specially tailored for the application comprising carriers, secondary antifoam agents, emulsifiers, and stabilizing agents in addition to the primary defoamer. **Carriers,** usually hydrocarbon oils or water, serve as the vehicle to support the release and spread of the primary defoamer. **Secondary defoamers** may provide a synergistic effect for the primary defoamer or modify its properties such as spreadability or solubility. **Emulsifiers** may enhance the speed of dispersion, while **stabilizing agents** may enhance defoamer stability or shelf life.

Hydrophobic silica defoamers work on a basis which may not be chemical at all. They are basically finely divided solid silica particles dispersed in a hydrocarbon or silicone oil which serves as a spreading vehicle. Kulkarni [*Ind. Eng. Chem. Fundam.*, **16**, 472 (1977)] theo-

TABLE 18-24 Major Types and Applications of Defoamers

Classification	Examples	Applications
Silicones	Dimethyl silicone, trialkyl and tetraalkyl silanes	Lubricating oils; distillation; fermentation; jam and wine making; food processing
Aliphatic acids or esters	Mostly high-molecular-weight compounds; diethyl phthalate; lauric acid	Papermaking; wood-pulp suspensions; water-based paints; food processing
Alcohols	Moderate- to high-molecular-weight monohydric and polyhydric alcohols; octyl alcohol; C-12 to C-20 alcohols; lauryl alcohol	Distillation; fermentation; papermaking; glues and adhesives
Sulfates or sulfonates	Alkali metal salts of sulfated alcohols, sulfonic acid salts; alkyl-aryl sulfonates; sodium lauryl sulfate	Nonaqueous systems; mixed aqueous and nonaqueous systems; oil-well drilling muds; spent H_2SO_4 recovery; deep- fat frying
Amines or amides	Alkyl amines (undecyloctyl and diamyl methyl amine); polyamides (acyl derivatives of piperazine)	Boiler foam; sewage foam; fermentation; dye baths
Halogenated compounds	Fluochloro hydrocarbons with 5 to 50 C atoms; chlorinated hydrocarbons	Lubrication-oil and grease distillation; vegetable-protein glues
Natural products	Vegetable oils; waxes, mineral oils plus their sulfated derivatives (including those of animal oils and fats)	Sugar extraction; glue manufacture; cutting oils
Fatty-acid soaps	Alkali, alkaline earth, and other metal soaps; sodium stearate; aluminum stearate	Gear oils; paper stock; paper sizing; glue solutions
Inorganic compounds	Monosodium phosphate mixed with boric acid and ethyl carbonate, disodium phosphate; sodium aluminate, bentonite and other solids	Distillation; instant coffee; boiler feedwater; sugar extraction
Phosphates	Alkyl-alkalene diphosphates; tributyl phosphate in isopropanol	Petroleum-oil systems; foam control in soap solutions
Hydrophobic silica	Finely divided silica in polydimethyl siloxane	Aqueous foaming systems
Sulfides or thio derivatives	Metallic derivatives of thio ethers and disulfides, usually mixed with organic phosphite esters; long-chain alkyl thienyl ketones	Lubricating oils; boiler water

rizes that this mixture defoams by the penetration of the silica particle into the bubble and the rupture of the wall. Table 18-24 lists major types of defoamers and typical applications.

Other Chemical Methods These methods rely chiefly on destroying the foam stabilizer or neutralizing its effect through methods other than displacement and are applicable when the process will permit changing the chemical environment. Forms stabilized with alkali esters can be broken by acidification since the equilavent free acids do not stabilize foam. Foams containing sulfated and sulfonated ionic detergents can be broken with the addition of fatty-acid soaps and calcium salts. Several theories have been proposed. One suggests that the surfactant is tied up in the foam as double calcium salts of both the sulfonate and the soap. Another suggests that calcium soaps oriented in the film render it inelastic.

Ionic surfactants adsorb at the foam interface and orient with the charged group immersed in the lamellae and their uncharged tails pointed into the gas stream. As the film drains, the charged groups, which repel each other, tend to be moved more closely together. The repulsive force between like charges hinders drainage and stabilizes the film. Addition of a salt or an electrolyte to the foam screens the repulsive effect, permits additional drainage, and can reduce foam stability.

Foam Prevention Chemical prevention of foam differs from defoaming only in that compounds or mixtures are added to a stream prior to processing to prevent the formation of foam either during processing or during customer use. Such additives, sometimes distinguished as **antifoam agents,** are usually in the same chemical class of materials as defoamers. However, they are usually specifically formulated for the application. Typical examples of products formulated with antifoam agents are laundry detergents (to control excess foaming), automotive antifreeze, instant coffee, and jet-aircraft fuel. Foaming in some chemical processes such as distillation or evaporation may be due to trace impurities such as surface-active agents. An alternative to antifoam agents is their removal before processing such as by treatment with activated carbon [Pool, *Chem. Process.*, **21**(9), 56 (1958)].

Automatic Foam Control In processing materials when foam can accumulate, it is often desirable to measure the height of the foam layer continuously and to dispense defoamer automatically as required to control the foam. Other corrective action can also be taken automatically. Methods of sensing the foam level have included electrodes in which the electrical circuit is completed when the foam touches the electrode [Nelson, *Ind. Eng. Chem.*, **48**, 2183 (1956); and Browne, U.S. Patent 2,981,693, 1961], floats designed to rise in a foam layer (Carter, U.S. Patent 3,154,577, 1964), and change in power input required to turn a foam-breaking impeller as the foam level rises (Yamashita, U.S. Patent 3,317,435, 1967). Timers to control the duration of defoamer addition have also been used. Browne has suggested automatic addition of defoamer through a porous wick when the foam level reaches the level of the wick. Foam control has also been discussed by Kroll [*Ind. Eng. Chem.*, **48**, 2190 (1956)].

Section 19

Liquid-Solid Systems

Shelby A. Miller, Ph.D., *Senior Chemical Engineer, Argonne National Laboratory; Member, American Association for the Advancement of Science (Fellow), American Chemical Society, American Institute of Chemical Engineers (Fellow), American Institute of Chemists (Fellow), Filtration Society, New York Academy of Sciences, Society of Chemical Industry; Registered Professional Engineer, New York. (Section Editor, Leaching)*

Charles M. Ambler, B.S.Ch.E. (deceased), *Consultant; former Director of Chemical Engineering, Sharples Division, Pennwalt Corporation. (Centrifuges)*

Richard C. Bennett, B.S.Ch.E., *Division Manager, Swenson Division, Whiting Corp.; Member, American Chemical Society, American Institute of Chemical Engineers; Registered Professional Engineer, Illinois. (Crystallization from Solution)*

Donald A. Dahlstrom, Ph.D., *Senior Vice President, Research and Development, Eimco Process Equipment Company; Member, American Chemical Society, American Institute of Chemical Engineers, American Institute of Mining, Metallurgical and Petroleum Engineers, Canadian Institute of Mining and Metallurgy, Mining and Metallurgical Society of America, Filtration Society, Air Pollution Control Association, Water Pollution Control Federation. (Gravity Sedimentation Operations)*

J. D. Darji, M.S.Ch.E., *Senior Technical Consultant, Infilco Degremont Inc., Member, American Institute of Chemical Engineers, American Society of Mechanical Engineers. (Ion-Exchange and Adsorption Equipment)*

Robert C. Emmett, Jr., B.S.Ch.E., *Senior Process Consultant, Technology and Development, Eimco Process Equipment Company; Member, American Institute of Chemical Engineers, American Institute of Mining, Metallurgical and Petroleum Engineers. (Gravity Sedimentation Operations)*

Joseph B. Gray, Ph.D., *Senior Consultant (Retired), Engineering Department, E. I. du Pont de Nemours & Co.; Member, American Chemical Society, American Institute of Chemical Engineers; Registered Professional Engineer, Delaware. (Agitation of Low-Viscosity Particle Suspensions)*

C. Fred Gurnham, D.Eng.Sc., *Consultant; Member, American Academy of Environmental Engineers, American Chemical Society, American Institute of Chemical Engineers (Fellow), American Society of Civil Engineers, Water Pollution Control Federation; Registered Professional Engineer, Illinois, Indiana. (Expression)*

Louis J. Jacobs, Jr., M.S.Ch.E., *Director of Corporate Engineering Division, A. E. Staley Manufacturing Company; Member, American Institute of Chemical Engineers; Registered Professional Engineer, Missouri. (Filtration)*

Ronald P. Klepper, B.S.Ch.E., *Manager, Process Technology, Eimco Process Equipment Company; Member, American Institute of Chemical Engineers; Registered Professional Engineer, Utah. (Gravity Sedimentation Operations)*

A. W. Michalson, B.S.Min.E., *President, A. W. Michalson Company; Member, American Water Works Association. (Ion-Exchange and Adsorption Equipment)*

James Y. Oldshue, Ph.D., *Vice President, Mixing Technology, Mixing Equipment Company, Inc.; Member, American Chemical Society, American Institute of Chemical Engineers (Fellow), American Institute of Chemists (Fellow); Registered Professional Engineer, New York. (Agitation of Low-Viscosity Particle Suspensions)*

Charles E. Silverblatt, M.S.Ch.E., *Vice President, Technology and Development, Eimco Process Equipment Company; Member, American Institute of Chemical Engineers, American Institute of Mining, Metallurgical and Petroleum Engineers, Filtration Society, Technical Association of the Pulp and Paper Industry. (Gravity Sedimentation Operations)*

Julian C. Smith, B.Chem., Ch.E., *Professor of Chemical Engineering, Cornell University; Member, American Chemical Society, American Institute of Chemical Engineers (Fellow). (Selection of a Solids-Liquid Separator)*

David B. Todd, Ph.D., *Technical Director, Plastics Equipment Division, Baker Perkins Inc.; Member, American Association for the Advancement of Science, American Chemical Society, American Institute of Chemical Engineers (Fellow), American Institute of Chemists (Fellow), American Oil Chemists Society, Society of Plastics Engineers, Society of the Plastics Industry; Registered Professional Engineer, Michigan. (Paste and Viscous-Material Mixing)*

PHASE CONTACTING AND LIQUID-SOLID PROCESSING

AGITATION OF LOW-VISCOSITY PARTICLE SUSPENSIONS

PASTE AND VISCOUS-MATERIAL MIXING

CRYSTALLIZATION FROM SOLUTION

ION-EXCHANGE AND ADSORPTION EQUIPMENT

LEACHING

PHASE SEPARATION

GRAVITY SEDIMENTATION OPERATIONS

FILTRATION

CENTRIFUGES

EXPRESSION

SELECTION OF A SOLIDS-LIQUID SEPARATOR

Nomenclature and Units

Symbol	Definition	SI units	U.S. customary units
A	Area	m^2	ft^2
A	Amplitude of sine-wave variation of a property or characteristic	Units of property being varied, m	Units of property being varied, in
a	Rate of acceleration	m/s^2	ft/s^2
a	Exponent defining a power function	Dimensionless	Dimensionless
B	Specific rate of crystal formation (number of crystals formed in a unit volume of solution per unit time)	$m^{-3}\cdot s^{-1}$	$mL^{-1}\cdot s^{-1}$
b	Length or height	mm	in
C	Clearance interval	cm	in
C	Concentration of suspended solids	g/L	Weight percent, g/L
C	Concentration of dissolved solids	g/L	g/L
C	Coefficient, general		
c	Composition, mass fraction	Dimensionless	Dimensionless
c	Coefficient, general		
D	Diameter	m, mm, μm	in, μm, mesh
d	Diameter	μm	mesh, μm
E	Energy	J	ft·lbf
F	Force	N	lbf
G	Crystal growth rate	m/s	mm/h
G	Relative centrifugal force	Dimensionless	Dimensionless
g	Acceleration	m/s^2	ft/s^2
H	Height or head	m	in, cm
H	Mass	g	lb
I	Characteristic dimension	m, mm	in, mm
I	Velocity	m/s	ft/s
I	Moment of inertia	$kg\cdot m^2$	$lb\cdot ft^2$
i	Number of surfaces	Dimensionless	Dimensionless
i	Exponent	Dimensionless	Dimensionless
j	Exponent	Dimensionless	Dimensionless
K	Coefficient, general		
k	Coefficient, general		
L	Thickness or characteristic dimension	m	in, mm
M	Density	kg/m^3	lb/ft^3, g/mL
m	Mass	kg	lb
m	Mass ratio	Dimensionless	Dimensionless
m	Torque	N·m	lb·ft
m	Fourth or higher moment of distribution		
m	Exponent	Dimensionless	Dimensionless
N	Rotation rate	s^{-1}	min^{-1} (r/min)
N	Number, including a numeric (dimensionless group)	Dimensionless	Dimensionless
n	Rotation rate	s^{-1}	min^{-1} (r/min), s^{-1} (r/s)
n	Crystal population density (number per increment of characteristic dimension)	m^{-1}	in^{-1}
n	Number of units (e.g., stages in a series)	Dimensionless	Dimensionless
n	Exponent	Dimensionless	Dimensionless
O	Volume of liquid per unit of mass of solids (overflow)	m^3/kg	ft^3/lb
P	Pressure	Pa	lbf/in^2 (psi), atm
P	Power	W	hp
P	Mass of crystals	kg	lb
P	Turnover (ratio of crystallizer volume to volumetric circulation rate)	s	min, s
p	Pressure	Pa	lbf/in^2 (psi)
PD	Particle diameter	mm, μm	in, μm
Q	Volumetric flow rate	L/s, m^3/s	gal/min
q	Volumetric ratio of liquid to dry solid	Dimensionless	Dimensionless
R	Mass fraction	Dimensionless	Dimensionless
R	Mole ratio	Dimensionless	Dimensionless
R	Filter-medium resistance	m^{-1}	ft^{-1}
r	Radius	m	in, ft
r	Filter-medium resistance	m^{-1}	ft^{-1}
RCF	Relative centrifugal force (multiples of standard gravitational acceleration)	Dimensionless	Dimensionless
S	Solubility (mass ratio)	Dimensionless	Dimensionless
S	Supersaturation coefficient (ratio of solute concentrations at supersaturation and at saturation)	Dimensionless	Dimensionless
S	Rotation rate	s^{-1}	min^{-1} (r/min)
S	Stress	Pa	lbf/in^2 (psi)
s	Thickness	m	in
s	Supersaturation (solute concentration in excess of solubility)	g/L	g/L
s	Specific gravity	Dimensionless	Dimensionless
s	Compressibility index	Dimensionless	Dimensionless
T	Torque	N·m	lbf·in
t	Time	s	s, min, h
Unit area	Plan area per unit mass settling rate	$m^2/(Mg\cdot day)$	$ft^2/(ton\cdot day)$
U	Volume of liquid per unit mass of solids (underflow)	m^3/kg	ft^3/lb
U	Consolidation (length fraction)	Dimensionless	Dimensionless

Nomenclature and Units (*Continued*)

Symbol	Definition	SI units	U.S. customary units
V	Volume	m^3, L	gal, in^3
V	Volume per unit of mass settling rate	$m^3/(Mg \cdot day)$	$ft^3/(ton \cdot day)$
v	Velocity	m/s, cm/s	ft/s
W	Mass ("weight")	g	lb
w	Mass of solids per unit volume of filtrate	g/L	g/L, lb/gal
X	Length	m, mm	in
x	Mole fraction	Dimensionless	Dimensionless
x	Concentration	g/m^3	Optional
x	Exponent	Dimensionless	Dimensionless
y	Mole fraction	Dimensionless	Dimensionless
y	Exponent	Dimensionless	Dimensionless

Symbol	Definition	SI units	U.S. customary units
	Greek symbols		
α	Angle	rad	°
α	Specific resistance of filter cake	m/g	ft/lb
α	Coefficient, general		
γ	Rate of shear	s^{-1}	s^{-1}
γ	Exponent	Dimensionless	Dimensionless
δ	Thickness	m, mm	in, mm
ΔL	Change in length	m, mm	in, mm
θ	Time	s	Optional
θ	Period, frequency	s^{-1}	s^{-1}
θ	Angle	rad	°
ν	Viscosity	$Pa \cdot s$	P
ω	Kinematic viscosity	m^2/s	St
ρ	Density	g/m^3	lb/ft^3
Σ	Area	m^2	ft^2
ω	Angular velocity	s^{-1}	s^{-1}

NOTE: Some highly specialized terms found in the text are not listed.

PHASE CONTACTING AND LIQUID-SOLID PROCESSING

In chemical engineering literature, it is customary to treat agitation, paste mixing, two-phase flow (including slurry transportation), and spraying as discrete, multipurpose operations for which the principles of equipment design and operation can be stated generally and then adapted to different specific process ends. They are so presented in this *Handbook:* agitation and paste mixing in this section, two-phase flow in Sec. 5, and spraying in Sec. 18. It is also customary to deal with some process goals dependent on solid-liquid contactors in terms of single-purpose operations that may employ a variety of equipment options. Such operations include adsorption, colloiding, crystallization, flocculation, ion exchange, and leaching. Again, the custom is followed in the *Handbook,* the equipment for each of these areas being treated in an individual subsection of Sec. 19, except for two that are omitted. Colloiding has been left out because its special, narrow character makes it of less wide interest to chemical engineers than the others and because in the equipment sense it concerns liquid-liquid emulsions more often than it does liquid-solid suspensions. The interested reader is referred to the many reference texts and monographs on colloid chemistry and colloiding. Flocculation has not been included because the emphasis is generally less on equipment than on implementation of principles by selection of flocculating agents and by procedure. Gravity settlers, described later in this section, are in fact often simultaneous flocculators and separators. In this connection flocculation is discussed briefly later in the subsection "Flocculation"; it is also considered by Gale (in Purchas, *Solid/Liquid Separation Equipment Scale-Up,* Uplands Press, Croydon, England, 1977, pp. 46 ff.) and by Stevenson (ibid., pp. 127 ff.). Some of the chemical engineering implications of flocculation are summarized by Porter, Flood, and Rennier, *Chem. Eng.,* **73**(13), 141 (1966).

AGITATION OF LOW-VISCOSITY PARTICLE SUSPENSIONS

GENERAL REFERENCES: Holland and Chapman, *Liquid Mixing and Processing in Stirred Tanks,* Reinhold, New York, 1966. Jordon, *Chemical Process Development,* part 1, Interscience, New York, 1968, p. 111. Nagata, *Mixing Principles and Applications,* Wiley, New York, 1975. Oldshue and Todd, in *Kirk-Othmer Encyclopedia of Chemical Technology,* 3d ed., vol. 15, Wiley, New York, 1981, p. 604. Parker, *Chem. Eng.,* **71**(12), 165 (1964). Quillen, *Chem. Eng.,* **61**(12), 179 (1954). Uhl and Gray (eds.), *Mixing: Theory and Practice,* vol. 1, Academic, New York, 1966; vol. 2, 1967. Sterbacek and Tausk, *Mixing in the Chemical Industry,* trans. by Mayer and ed. by Bourne, Pergamon, London, 1965. Zlokarnik, in *Ullmann's Encyklopädie der technischen Chemie,* 4th ed., vol. 2, Verlag Chemie, Weinheim, Germany, 1972, p. 259.

A variety of process functions are carried out in vessels stirred by rotating impellers. Some examples are (1) blending miscible liquids; (2) contacting or dispersing immiscible liquids; (3) dispersing a gas in a liquid; (4) promoting heat transfer between the agitated liquid and a heat-exchange surface; (5) suspending or dispersing particulate solids in a liquid to produce uniformity, to promote mass transfer (such as dissolution), or to initiate and assist chemical reaction; and (6) reducing particle agglomerate size. Only the latter two of these are treated in this section, but material on some of the others will be found in Secs. 10, 18, and 21. Stirred vessels are emphasized in this discussion, but some mixing operations may be carried out continuously by turbulence or by mechanical devices in pipes with very little back circulation when the time for mixing can be short.

MIXING EQUIPMENT

Impellers may be roughly divided into two broad classes: axial-flow impellers and radial-flow impellers. The classification depends on the angle that the blade makes with the plane of impeller rotation.

Axial-Flow Impellers Axial-flow impellers include all impellers in which the blade makes an angle of less than 90° with the plane of rotation. Propellers and pitched-blade turbines or paddles, as illustrated in Figs. 19-1 and 19-2, are representative axial-flow impellers.

Propellers are often used for agitation in tanks smaller than 3.8 m^3 (1000 gal) or less than 1.8 m (6 ft) in diameter when less than 2.2 kW (3 hp) is satisfactory for obtaining the desired process results.

Propeller mixers may be clamped on the side of an open vessel in the angular, off-center position shown in Fig. 19-10 or bolted to a flange or plate on the top of a closed vessel with the shaft in the same angular, off-center position. This mounting results in a strong top-to-bottom circulation.

Two basic speed ranges are available: 1150 or 1750 r/min with direct drive and 350 or 420 r/min with a gear drive. The high-speed units produce higher velocities and shear rates in the propeller discharge stream and a lower circulation rate throughout the vessel than the low-speed units. For suspension of solids, it is common to use the gear-driven units, while for rapid dispersion or fast reactions the high-speed units are more appropriate.

Propellers may also be mounted near the bottom of the cylindrical wall of a vessel as shown in Fig. 19-3. Such side-entering agitators are used to blend low-viscosity fluids [<0.1 Pa·s (100 cP)] or to keep slowly settling sediment suspended in tanks as large as some 4000 m^3 (10^6 gal). Mixing of paper pulp is often carried out by side-entering propellers.

Pitched-blade turbines (Fig. 19-2) are used on top-entering agitator shafts instead of propellers when a high axial circulation rate is desired and the power consumption is more than 2.2 kW (3 hp). A pitched-blade turbine near the upper surface of liquid in a vessel is effective for rapid submergence of floating particulate solids.

Radial-Flow Impellers Radial-flow impellers have blades which are parallel to the axis of the drive shaft. The smaller multiblade ones are known as "turbines"; larger, slower-speed impellers, with two or four blades, are often called "paddles." The diameter of a turbine is normally between 0.3 and 0.6 of the tank diameter. Turbine impel-

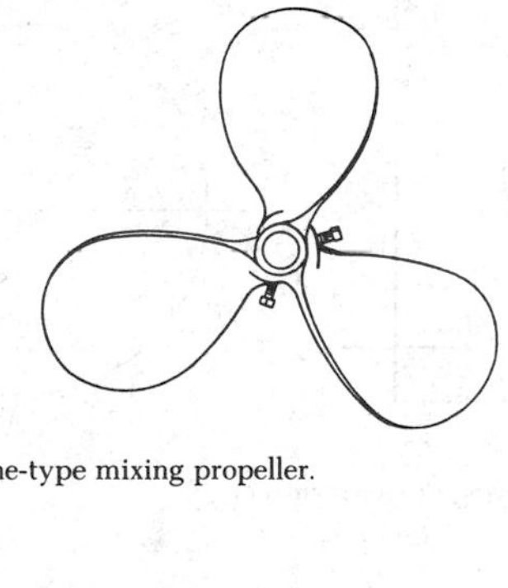

FIG. 19-1 Marine-type mixing propeller.

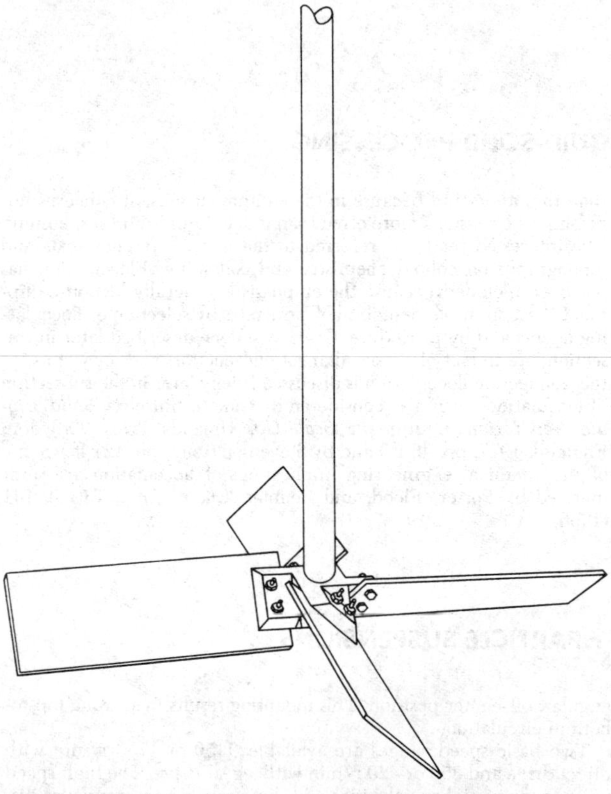

FIG. 19-2 Pitched-blade turbine.

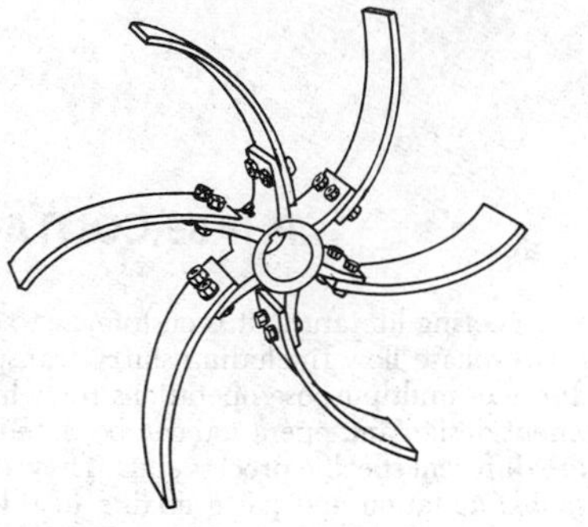

FIG. 19-4 Curved-blade turbine.

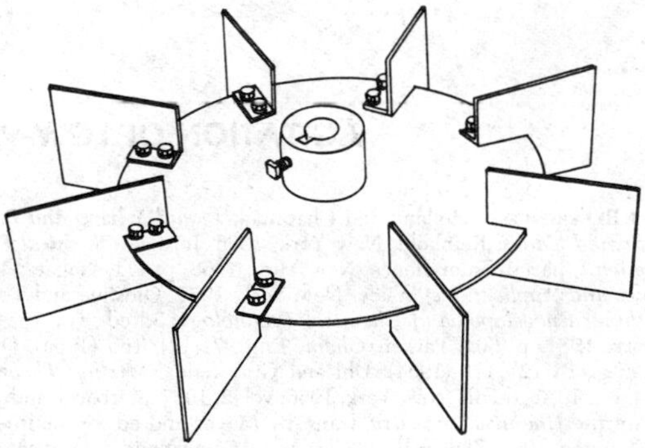

FIG. 19-5 Flat-blade turbine.

lers come in a variety of types, such as curved-blade and flat-blade, as illustrated in Figs. 19-4 and 19-5. Curved blades aid in starting an impeller in settled solids. A paddle agitator has a diameter usually greater than 0.6 of the tank diameter and turns at a slow speed. Construction is often similar to that shown in Fig. 19-4, but with two or four straight blades and with a relatively smaller hub.

Most large-scale agitation of solid-liquid suspensions is done with top-entering turbines or paddles. Power may range from 750 W (1 hp) to as high as 750 kW (1000 hp). The impeller speed is typically between 50 and 150 r/min; but, depending on process conditions, it may go as high as 400 or as low as 15 r/min.

For processes in which corrosion of commonly used metals is a

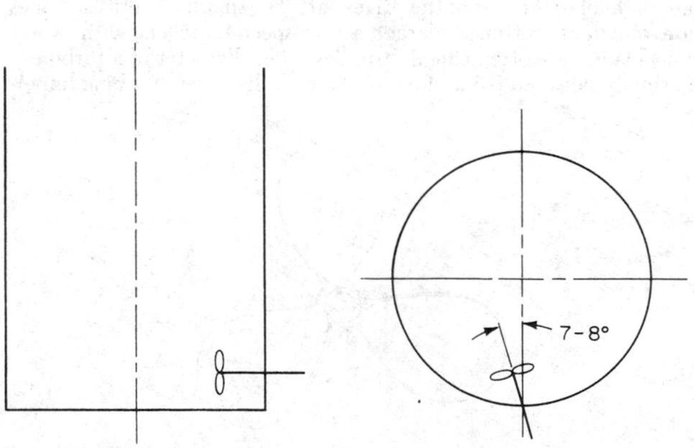

FIG. 19-3 Side-entering propeller mixer.

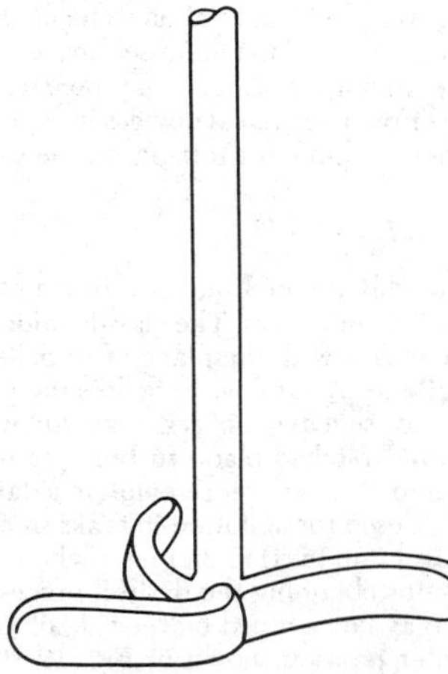

FIG. 19-6 Glassed-steel impeller. (*The Pfaudler Company.*)

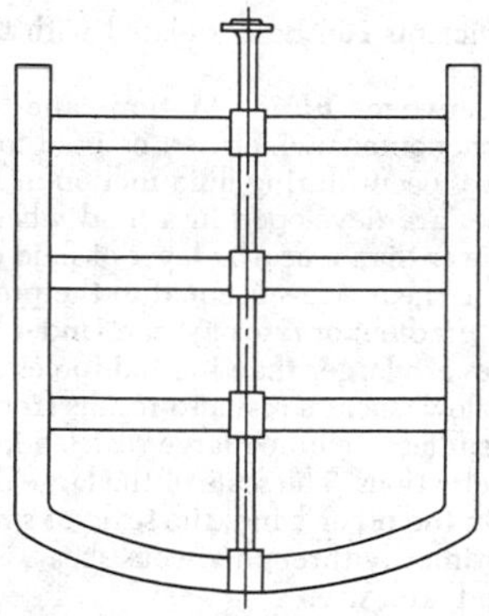

FIG. 19-7 Anchor impeller.

problem, glass-coated impellers may be economical. A typical modified curved-blade turbine of this type is shown in Fig. 19-6.

Close-Clearance Stirrers For some pseudoplastic fluid systems stagnant fluid may be found next to the vessel walls in parts remote from propeller or turbine impellers. In such cases, an "anchor" impeller may be used (Fig. 19-7). The fluid flow is principally circular in the direction of rotation of the anchor. Whether substantial axial or radial fluid motion also occurs depends on the fluid viscosity and the design of the upper blade-supporting spokes. Anchor agitators are used particularly to obtain improved heat transfer in high-consistency fluids.

Unbaffled Tanks If a low-viscosity liquid is stirred in an unbaffled tank by an axially mounted agitator, there is a tendency for a swirling flow pattern to develop regardless of the type of impeller. Figure 19-8 shows a typical flow pattern. A vortex is produced owing to centrifugal force acting on the rotating liquid. In spite of the presence of a vortex, satisfactory process results often can be obtained in an unbaffled vessel. However, there is a limit to the rotational speed that may be used, since once the vortex reaches the impeller, severe air entrainment may occur. In addition, the swirling mass of liquid often generates an oscillating surge in the tank, which coupled with the deep vortex may create a large fluctuating force acting on the mixer shaft.

Vertical velocities in a vortexing low-viscosity liquid are low relative to circumferential velocities in the vessel. Increased vertical circulation rates may be obtained by mounting the impeller off center, as illustrated in Fig. 19-9. This position may be used with either turbines or propellers. The position is critical, since too far or too little off center in one direction or the other will cause greater swirling, erratic vortexing, and dangerously high shaft stresses. Changes in viscosity and tank size also affect the flow pattern in such vessels. Off-center mountings have been particularly effective in the suspension of paper pulp.

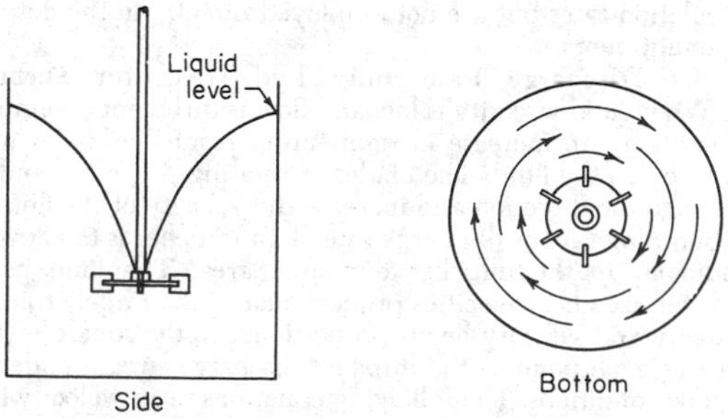

FIG. 19-8 Typical flow pattern for either axial- or radial-flow impellers in an unbaffled tank.

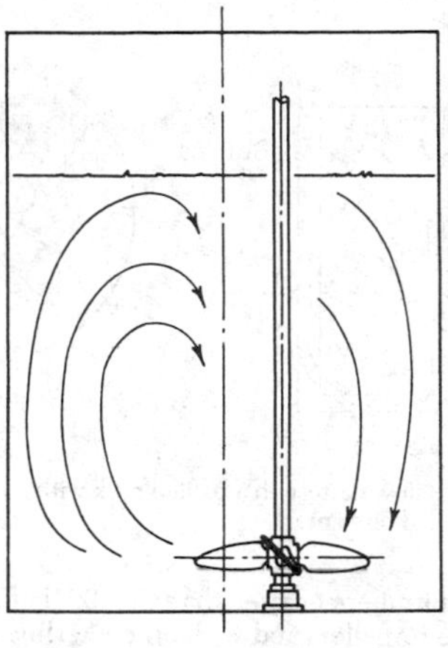

FIG. 19-9 Flow pattern with a paper-stock propeller, unbaffled; vertical off-center position.

With axial-flow impellers, an angular off-center position may be used. The impeller is mounted approximately 15° from the vertical, as shown in Fig. 19-10.

The angular off-center position used with propeller units is usually limited to propellers delivering 2.2 kW (3 hp) or less. The unbalanced fluid forces generated by this mounting can become severe with higher power.

Paddles and anchors normally operate coaxially within unbaffled tanks, since they may have a close clearance with the tank wall.

Baffled Tanks For vigorous agitation of thin suspensions, the tank is provided with baffles which are flat vertical strips set radially along the tank wall, as illustrated in Figs. 19-11 and 19-12. Four baffles are almost always adequate. A common baffle width is one-tenth to one-twelfth of the tank diameter (radial dimension). For agitating slurries, the baffles often are located one-half of their width from the vessel wall to minimize accumulation of solids on or behind them.

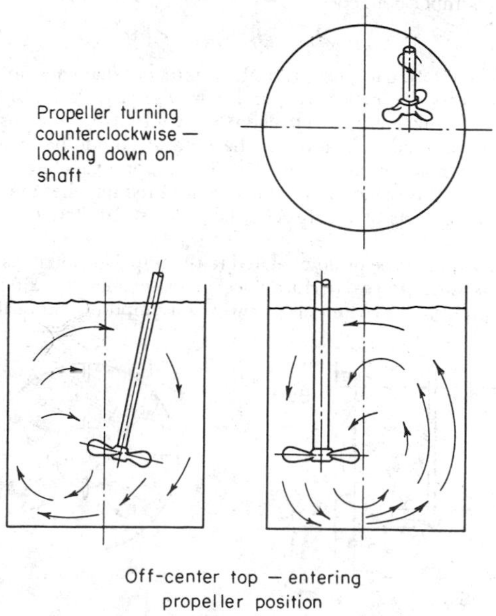

FIG. 19-10 Flow pattern for a propeller in angular off-center position without baffles.

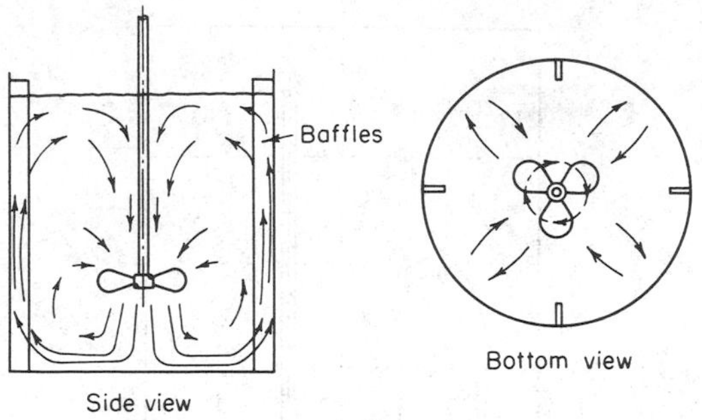

FIG. 19-11 Typical flow pattern in a baffled tank with a propeller or an axial-flow turbine positioned on center.

For Reynolds numbers greater than 10,000, baffles are commonly used with turbine impellers and with on-centerline axial-flow impellers. The flow patterns illustrated in Figs. 19-11 and 19-12 are quite different, but in both cases the use of baffles results in a large top-to-bottom circulation without vortexing or severely unbalanced fluid forces on the impeller shaft.

In the transition region [Reynolds numbers, Eq. (19-1), from 10 to 10,000], the width of the baffle may be reduced, often to one-half of standard width. If the circulation pattern is satisfactory when the tank is unbaffled but a vortex creates a problem, partial-length baffles may be used. These are standard-width and extend downward from the surface into about one-third of the liquid volume.

In the region of laminar flow ($N_{Re} < 10$), the same power is consumed by the impeller whether baffles are present or not, and they are seldom required. The flow pattern may be affected by the baffles, but not always advantageously. When they are needed, the baffles are usually placed one or two widths radially off the tank wall, to allow fluid to circulate behind them and at the same time produce some axial deflection of flow.

FLUID BEHAVIOR IN MIXING VESSELS

Impeller Reynolds Number The presence or absence of turbulence in an impeller-stirred vessel can be correlated with an impeller Reynolds number defined

$$N_{Re} = D_a^2 N\rho/\mu \tag{19-1}$$

where N = rotational speed, r/s; D_a = impeller diameter, m (ft); ρ = fluid density, kg/m^3 (lb/ft^3); and μ = viscosity, Pa·s [lb/(ft·s)]. Flow in the tank is turbulent when $N_{Re} > 10{,}000$. Thus viscosity alone is not a valid indication of the type of flow to be expected. Between Reynolds numbers of 10,000 and approximately 10 is a transition range in which flow is turbulent at the impeller and laminar in remote parts of the vessel; when $N_{Re} < 10$, flow is laminar only.

Not only is the type of flow related to the impeller Reynolds number, but also such process performance characteristics as mixing time, impeller pumping rate, impeller power consumption, and heat- and mass-transfer coefficients can be correlated with this dimensionless group.

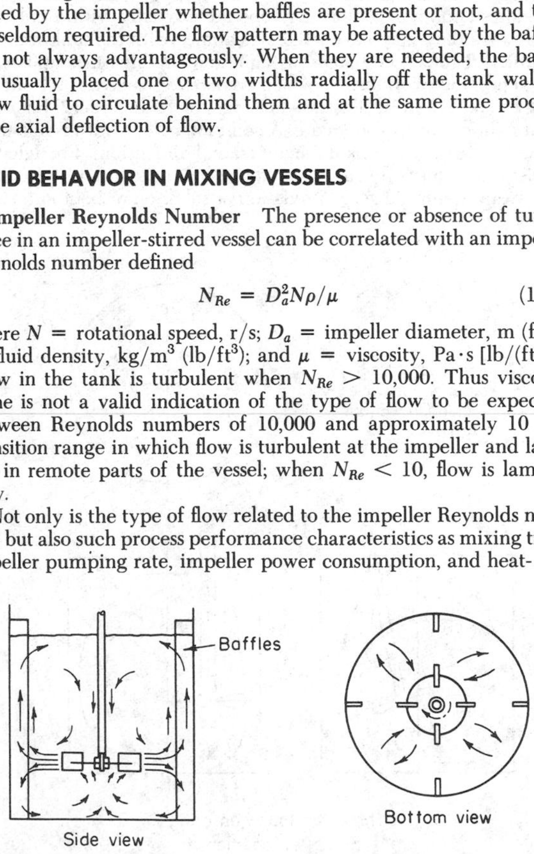

FIG. 19-12 Typical flow pattern in a baffled tank with a turbine positioned on center.

Relationship between Fluid Motion and Process Performance Several phenomena which can be used to promote various processing objectives occur during fluid motion in a vessel.

1. Shear stresses are developed in a fluid when a layer of fluid moves faster or slower than a nearby layer of fluid or a solid surface. In laminar flow, the shear stress is equal to the product of fluid viscosity and velocity gradient or rate of shear. Under laminar-flow conditions, shear forces are larger than inertial forces in the fluid.

With turbulent flow, shear stress also results from the behavior of transient random eddies, including large-scale eddies which decay to small eddies or fluctuations. The scale of the large eddies depends on equipment size. On the other hand, the scale of small eddies, which dissipate energy primarily through viscous shear, is almost independent of agitator and tank size.

The shear stress in the fluid is much higher near the impeller than it is near the tank wall. The difference is greater in large tanks than in small ones.

2. Inertial forces are developed when the velocity of a fluid changes direction or magnitude. In turbulent flow, inertia forces are larger than viscous forces. Fluid in motion tends to continue in motion until it meets a solid surface or other fluid moving in a different direction. Forces are developed during the momentum transfer that takes place. The forces acting on the impeller blades fluctuate in a random manner related to the scale and intensity of turbulence at the impeller.

3. The interfacial area between gases and liquids, immiscible liquids, and solids and liquids may be enlarged or reduced by these viscous and inertia forces when interacting with interfacial forces such as surface tension.

4. Concentration and temperature differences are reduced by bulk flow or circulation in a vessel. Fluid regions of different composition or temperature are reduced in thickness by bulk motion in which velocity gradients exist. This process is called bulk diffusion or Taylor diffusion (Brodkey, in Uhl and Gray, op. cit., vol. 1, p. 48). The turbulent and molecular diffusion reduces the difference between these regions. In laminar flow, Taylor diffusion and molecular diffusion are the mechanisms of concentration- and temperature-difference reduction.

5. Equilibrium concentrations which tend to develop at solid-liquid, gas-liquid, or liquid-liquid interfaces are displaced or changed by molecular and turbulent diffusion between bulk fluid and fluid adjacent to the interface. Bulk motion (Taylor diffusion) aids in this mass-transfer mechanism also.

Turbulent Flow in Stirred Vessels Turbulence parameters such as intensity and scale of turbulence, correlation coefficients, and energy spectra have been measured in stirred vessels. However, these characteristics are not used directly in the design of stirred vessels. For further details see Cutter, *Am. Inst. Chem. Eng. J.*, **12**, 35 (1966).

Fluid Velocities in Mixing Equipment Fluid velocities have been measured for various turbines in baffled and unbaffled vessels. Typical data are summarized in Uhl and Gray, op. cit., vol. 1, chap. 4. Velocity data have been used for calculating impeller discharge and circulation rates but are not employed directly in the design of mixing equipment.

Impeller Discharge Rate and Fluid Head for Turbulent Flow When fluid viscosity is low and flow is turbulent, an impeller moves fluids by an increase in momentum from the blades which exert a force on the fluid. The blades of rotating propellers and turbines change the direction and increase the velocity of the fluids.

The pumping rate or discharge rate of an impeller is the flow rate perpendicular to the impeller discharge area. The fluid passing through this area has velocities proportional to the impeller peripheral velocity and velocity heads proportional to the square of these velocities at each point in the impeller discharge stream under turbulent-flow conditions. The following equations relate velocity head, pumping rate, and power for geometrically similar impellers under turbulent-flow conditions:

$$Q = N_Q N D_a^3 \tag{19-2}$$

$$H = N_p N^2 D_a^2 / N_Q g \tag{19-3}$$

$$P = N_p \rho N^3 (D_a^5 / g_c) \tag{19-4}$$

$$P = \rho H Q g / g_c \tag{19-5}$$

where Q = impeller discharge rate, m^3/s (ft^3/s); N_Q = discharge coefficient, dimensionless; H = velocity head, m (ft); N_p = power number, dimensionless; P = power, (N·m)/s [(ft·lbf)/s]; g_c = dimensional constant, 32.2 (ft·lb)/(lbf·s^2)(g_c = 1 when using SI units); and g = gravitational acceleration, m/s^2 (ft/s^2).

The discharge rate Q has been measured for several types of impellers, and discharge coefficients have been calculated. The data of a number of investigators are reviewed by Uhl and Gray (op. cit., vol. 1, chap. 4). N_Q is 0.4 to 0.5 for a propeller with pitch equal to diameter at $N_{Re} = 10^5$. For turbines, N_Q ranges from 0.7 to 2.9, depending on the number of blades, blade-height-to-impeller-diameter ratio, and impeller-to-vessel-diameter ratio. The effects of these geometric variables are not well defined.

Power consumption has also been measured and correlated with impeller Reynolds number. The velocity head for a mixing impeller can be calculated, then, from flow and power data, by Eq. (19-3) or Eq. (19-5).

The velocity head of the impeller discharge stream is a measure of the maximum force that this fluid can exert when its velocity is changed. Such inertia forces are higher in streams with higher discharge velocities. Shear rates and shear stresses are also higher under these conditions in the smallest eddies. If a higher discharge velocity is desired at the same power consumption, a smaller-diameter impeller must be used at a higher rotational speed. According to Eq. (19-4), at a given power level $N \propto D_a^{-5/3}$ and $ND_a \propto D_a^{-2/3}$. Then, $H \propto D_a^{-4/3}$ and $Q \propto D_a^{4/3}$.

An impeller with a high fluid head is one with high peripheral velocity and discharge velocity. Such impellers are useful for (1) rapid reduction of concentration differences in the impeller discharge stream (rapid mixing), (2) production of large interfacial area and small droplets in gas-liquid and immiscible-liquid systems, (3) solids deagglomeration, and (4) promotion of mass transfer between phases.

The impeller discharge rate can be increased at the same power consumption by increasing impeller diameter and decreasing rotational speed and peripheral velocity so that $N^3 D_a^5$ is a constant (Eq. 19-4)]. Flow goes up, velocity head and peripheral velocity go down, but impeller torque T_Q goes up. At the same torque, $N^2 D_a^5$ is constant, $P \propto D_a^{-5/2}$, and $Q \propto D_a^{1/2}$. Therefore, increasing impeller diameter at constant torque increases discharge rate at lower power consumption. At the same discharge rate, $N D_a^3$ is constant, $P \propto D_a^{-4}$, and $T_Q \propto D_a^{-1}$. Therefore, power and torque decrease as impeller diameter is increased at constant Q.

A large-diameter impeller with a high discharge rate is used for (1) short times to complete mixing of miscible liquid throughout a vessel, (2) promotion of heat transfer, (3) reduction of concentration and temperature differences in all parts of vessels used for constant-environment reactors and continuous averaging, and (4) suspension of particles of relatively low settling rate.

Laminar Fluid Motion in Vessels When the impeller Reynolds number is less than 10, the flow induced by the impeller is laminar. Under these conditions, the impeller drags fluid with it in a predominantly circular pattern. If the impeller blades curve back, there is a viscous drag flow toward the tips of these blades. Under moderate-viscosity conditions in laminar flow, centrifugal force acting on the fluid layer dragged in a circular path by the rotating impeller will move fluid in a radial direction. This centrifugal effect causes any gas accumulated behind a rotating blade to move to the axis of impeller rotation. Such radial-velocity components are small relative to tangential velocity.

For turbines at Reynolds numbers less than 100, toroidal stagnant zones exist above and below the turbine periphery. Interchange of liquid between these regions and the rest of the vessel is principally by molecular diffusion.

Suspensions of fine solids may have pseudoplastic or plastic-flow properties. When they are in laminar flow in a stirred vessel, motion in remote parts of the vessel where shear rates are low may become negligible or cease completely. To compensate for this behavior of slurries, large-diameter impellers or paddles are used, with $(D_a/D_T) > 0.6$, where D_T is the tank diameter. In some cases, for example, with some anchors, $D_a > 0.95\, D_T$. Two or more paddles may be used in deep tanks to avoid stagnant regions in slurries.

In laminar flow ($N_{Re} < 10$), $N_P \propto 1/N_{Re}$ and $P \propto \mu N^2 D_a^3$. Since shear stress is proportional to rotational speed, shear stress can be increased at the same power consumption by increasing N proportionally to $D_a^{-3/2}$ as impeller diameter D_a is decreased.

Fluid circulation probably can be increased at the same power consumption and viscosity in laminar flow by increasing impeller diameter and decreasing rotational speed, but the relationship between Q, N, and D_a for laminar flow from turbines has not been determined.

As in the case of turbulent flow, then, small-diameter impellers ($D_a < D_T/3$) are useful for (1) rapid mixing of dry particles into liquids, (2) gas dispersion in slurries, (3) solid-particle deagglomeration, and (4) promoting mass transfer between solid and liquid phases. If stagnant regions are a problem, large impellers must be used and rotational speed and power increased to obtain the required results. Small continuous-processing equipment may be more economical than batch equipment in such cases.

Likewise, large-diameter impellers ($D_a > D_T/2$) are useful for (1) avoiding stagnant regions in slurries, (2) short mixing times to obtain uniformity throughout a vessel, (3) promotion of heat transfer, and (4) laminar continuous averaging of slurries.

Vortex Depth In an unbaffled vessel with an impeller rotating in the center, centrifugal force acting on the fluid raises the fluid level at the wall and lowers the level at the shaft. The depth and shape of such a vortex (Rieger, Ditl, and Novak, *Chem. Eng. Sci.*, **34**, 397 (1978)] depend on impeller and vessel dimensions as well as rotational speed.

Power Consumption of Impellers Power consumption is related to fluid density, fluid viscosity, rotational speed, and impeller diameter by plots of power number ($g_c P/\rho N^3 D_a^5$) versus Reynolds number ($D_a^2 N \rho/\mu$). Typical correlation lines for frequently used impellers operating in newtonian liquids contained in baffled cylindrical vessels are presented in Fig. 19-13. These curves may be used also for operation of the respective impellers in unbaffled tanks when the Reynolds number is 300 or less. When N_{Re} is greater than 300, however, the power consumption is lower in an unbaffled vessel than indicated in Fig. 19-13. For example, for a six-blade disk turbine with $D_T/D_a = 3$ and $D_a/W_i = 5$, $N_P = 1.2$ when $N_{Re} = 10^4$. This is only about one-fifth of the value of N_P when baffles are present.

Additional power data for other impeller types such as anchors, curved-blade turbines, and paddles in baffled and unbaffled vessels are available in the following references: Holland and Chapman, op. cit., chaps. 2, 4, Reinhold, New York, 1966; and Bates, Fondy, and Fenic, in Uhl and Gray, op. cit., vol. 1, chap. 3.

Power consumption for impellers in pseudoplastic, Bingham plastic, and dilatant nonnewtonian fluids may be calculated by using the correlating lines of Fig. 19-13 if viscosity is obtained from viscosity-shear rate curves as described here. For a pseudoplastic fluid, viscosity decreases as shear rate increases. A Bingham plastic is similar to a pseudoplastic fluid but requires that a minimum shear stress be exceeded for any flow to occur. For a dilatant fluid, viscosity increases as shear rate increases.

The appropriate shear rate to use in calculating viscosity is given by one of the following equations when a propeller or a turbine is used (Bates et al., in Uhl and Gray, op. cit., vol. 1, p. 149):

For dilatant liquids,

$$\dot{\gamma} = 13N(D_a/D_T)^{0.5} \tag{19-6}$$

For pseudoplastic and Bingham plastic fluids,

$$\dot{\gamma} = 10N \tag{19-7}$$

where $\dot{\gamma}$ = average shear rate, s^{-1}.

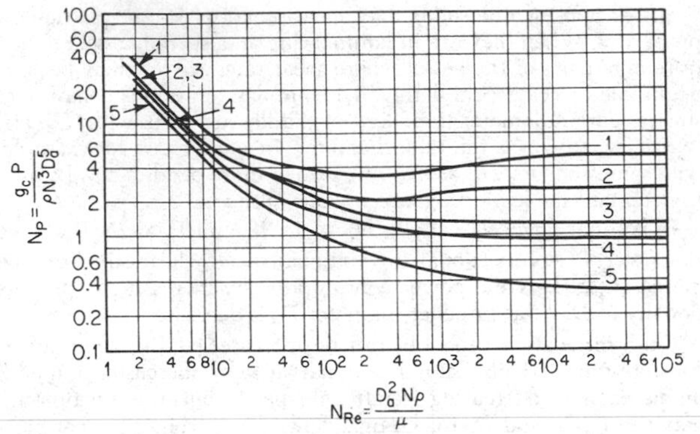

FIG. 19-13 Impeller power correlations: curve 1, six-blade turbine, $D_a/W_i = 5$, like Fig. 19-5 but with six blades, four baffles, each $D_T/12$; curve 2, vertical-blade, open turbine like Fig. 19-4 but with six straight blades, $D_a/W_i = 8$, four baffles each $D_T/12$; curve 3, 45° pitched-blade turbine like Fig. 19-2 but with six blades, $D_a/W_i = 8$, four baffles, each $D_T/12$; curve 4, propeller, pitch equal to $2D_a$, four baffles, each $0.1D_T$, also same propeller in angular off-center position with no baffles; curve 5, propeller, pitch equal to D_a, four baffles each $0.1D_T$, also same propeller in angular off-center position as in Fig. 19-10 with no baffles. D_a = impeller diameter, D_T = tank diameter, g_c = gravitational conversion factor, N = impeller rotational speed, P = power transmitted by impeller shaft, W_i = impeller-blade height, μ = viscosity of stirred liquid, and ρ = density of stirred mixture. Any set of consistent units may be used, but N must be rotations (rather than radians) per unit time. In the SI system, g_c is dimensionless and unity. [*Curves 4 and 5 from Rushton, Costich, and Everett,* Chem. Eng. Prog., *46, 395, 467 (1950), by permission; curves 2 and 3 from Bates, Fondy, and Corpstein,* Ind. Eng. Chem. Process Des. Dev., *2, 310 (1963), by permission of the copyright owner, the American Chemical Society.*]

The shear rate calculated from impeller rotational speed is used to identify a viscosity from a plot of viscosity versus shear rate determined with a capillary or rotational viscometer. Next N_{Re} is calculated, and N_P is read from a plot like Fig. 19-13.

DESIGN OF AGITATION EQUIPMENT

Selection of Equipment The principal factors which influence mixing-equipment choice are (1) the process requirements, (2) the flow properties of the process fluids, (3) equipment costs, and (4) construction materials required.

Ideally, the equipment chosen should be that of the lowest total cost which meets all process requirements. The total cost includes depreciation on investment, operating cost such as power, and maintenance costs. Rarely is any more than a superficial evaluation based on this principle justified, however, because the cost of such an evaluation often exceeds the potential savings that can be realized. Usually optimization is based on experience with similar mixing operations. Often the process requirements can be matched with those of a similar operation, but sometimes tests are necessary to identify a satisfactory design and to find the minimum rotational speed and power.

There are no satisfactory specific guides for selecting mixing equipment because the ranges of application of the various types of equipment overlap and the effects of flow properties on process performance have not been adequately defined. Nevertheless, what is frequently done in selecting equipment is described in the following paragraphs.

Top-Entering Propellers For vessels less than 1.8 m (6 ft) in diameter, a clamp- or flange-mounted, angular, off-center propeller with no baffles should be the initial choice for meeting a wide range of process requirements (Fig. 19-10). The vessel straight-side-height-to-diameter ratio should be 0.75 to 1.5, and the volume of stirred liquid should not exceed 4 m^3 (about 1000 gal).

For suspension of free-settling particles, circulation of pseudoplastic slurries, and heat transfer or mixing of miscible liquids to obtain uniformity, a speed of 350 or 420 r/min. should be stipulated. For dispersion of dry particles in liquids or for rapid initial mixing of liquid reactants in a vessel, a 1750- r/min propeller should be used at a distance $D_T/4$ above the vessel bottom. A second propeller can be added to the shaft at a depth D_a below the liquid surface if the submergence of floating liquids or particulate solids is otherwise inadequate. Such propeller mixers are readily available up to 2.2 kW (3 hp) for off-center sloped-shaft mounting.

Propeller size, pitch, and rotational speed may be selected by model tests, by experience with similar operations, or, in a few cases, by published correlations of performance data such as mixing time or heat transfer. The propeller diameter and motor power should be the minimum which meet process requirements.

If agitation is required for a vessel less than 1.8 m (6 ft) in diameter and the same operations will be scaled up to a larger vessel ultimately, the equipment type should be the same as that expected in the larger vessel.

Turbines For vessel volumes of 4 to 200 m^3 (1000 to 50,000 gal), a turbine mixer mounted coaxially within the vessel with four or more baffles should be the initial choice. Here also the vessel straight-side-height-to-diameter ratio should be 0.75 to 1.5. Four vertical baffles should be fastened perpendicularly to the vessel wall with a gap between baffle and wall equal to $D_T/24$ and a radial baffle width equal to $D_T/12$. A four- or six-blade turbine impeller like that shown in Fig. 19-5 (or in Fig. 19-4 but with straight blades) should be used. The blade-height-to-turbine-diameter ratio should be from 1 to 5 to 1 to 8.

For suspension of rapidly settling particles, the turbine diameter should be $D_T/3$ to $D_T/2$. A clearance of less than one-seventh of the fluid depth in the vessel should be used between the lower edge of the turbine blade tips and the vessel bottom. As the viscosity of a suspension increases, the impeller diameter should be increased. This diameter may be increased to 0.7 D_T and a second impeller added to avoid stagnant regions in pseudoplastic slurries. Moving the baffles halfway between the impeller periphery and the vessel wall will also help avoid stagnant fluid near the baffles.

As has been shown, power consumption is decreased and turbine discharge rate is increased as turbine diameter is increased at constant torque (in the completely turbulent regime). This means that for a stipulated discharge rate, more efficient operation is obtained (lower power and torque) with a relatively large turbine operating at a relatively low speed ($N \propto D_a^{-3}$). Conversely, if power is held constant, decreasing turbine diameter results in increasing peripheral velocity and decreasing torque. Thus at a stipulated power level the rapid, efficient initial mixing of reactants identified with high peripheral velocity can be achieved by a relatively small turbine operating at a relatively high speed ($N \propto D_a^{-5/3}$).

For circulation and mixing to obtain uniformity, the impeller should be located at one-third of the liquid depth above the vessel bottom unless rapidly settling material or a need to stir a nearly empty vessel requires a lower impeller location.

Side-Entering Propellers For vessels greater than 4 m^3 (1000 gal), a side-entering propeller agitator (Fig. 19-3) may be more economical than a top-mounted turbine on a centered vertical shaft. For vessels greater than 38 m^3 (10,000 gal), the economic attractiveness of side-entering propellers increases. For vessels larger than 380 m^3 (100,000 gal), units may be as large as 56 kW (75 hp), and two or even three may be installed in one tank. For the suspension of slow-settling particles or the maintenance of uniformity in a viscous slurry of small particles, the diameter and rotational speed of a side-entering agitator must be selected on the basis of model tests or experience with similar operations.

When abrasive solid particles must be suspended, maintenance costs for the submerged shaft seal of a side-entering propeller may become high enough to make this type of mixer an uneconomical choice.

Jet Mixers Continuous recycle of the contents of a tank through an external pump so arranged that the pump discharge stream appropriately reenters the vessel can result in a flow pattern in the tank which will produce a slow mixing action [Fossett, *Trans. Inst.*

Chem. Eng., **29,** 322 (1951)]. Jet agitation of vessels is restricted to mixing miscible liquids or keeping very slow-settling sediments in suspension. The method should not be chosen for suspending rapidly settling particles or for mixing pseudoplastic slurries.

Selection of Impeller Rotational Speed and Power The rotational speed for an impeller is usually based on tests or correlations of test results. Examples of correlations of a process performance criterion such as batch-mixing time or a mass-transfer coefficient are given later in this section.

Model theory using dimensionless groups is applicable to stirred vessels (Johnstone and Thring, *Pilot Plants, Models and Scale-Up Methods in Chemical Engineering*, McGraw-Hill, New York, 1957). Rarely can all pertinent dimensionless groups be kept the same when equipment dimensions are changed; nevertheless, correlations of dimensionless groups are often successfully used to predict within satisfactory engineering limits the performance of large-scale equipment from tests in models. A notable example is the power-number correlation (Fig. 19-13).

In some cases, the same power per unit volume for a similar mixing operation in equipment of different sizes is used as a basis for selecting power, and rotational speed then is calculated from a correlation like Fig. 19-13. This method should be used with caution, for it often leads to higher power than needed as scale of operation increases, and it occasionally predicts too little power.

Mechanical Design An agitator and its surroundings are a mechanical system of moving and fixed parts in which bending, torsional, shear, and fatigue stresses are encountered. Gears, bearing supports, shafts, couplings, keys, impeller hubs, impeller blades, baffles, thermometer wells, and supporting structured members—each of these parts experiences its own variety of these stresses and must be designed accordingly. Satisfactory design means not only that each member is strong enough but that deflections are sufficiently small to permit successful behavior of seals, close-clearance impellers, and protective coatings. The problem of an adequately corrosion-resistant material of construction may heighten the mechanical-design difficulty, as in the case of glass-enameled steel.

The rotational speed of an agitator shaft must be sufficiently far removed from the natural vibration frequencies of the agitator assembly and other parts of the system to avoid excitation of these natural frequencies. Usually only the first critical speed is of concern, although some agitators operate beyond their second critical. Methods of estimating the critical speeds of shaft-and-impeller assemblies are summarized by Leedom and Parker in Uhl and Gray, op. cit., vol. 2, chap. 11.

Leedom and Parker also treat the general subject of the mechanical design of impeller-type agitator systems. They deal with impellers, shafts, gear reducers, seals and stuffing boxes, and baffles.

MIXING TO OBTAIN UNIFORMITY

Allowable limits are often placed on composition or property values for products made by batch and by continuous processes and operations involving mixing. Statistical methods are available for describing the uniformity of these values and for controlling such operations. Methods of sampling, calculation, and interpretation of uniformity data in mixing operations (including particulate solids, liquids, and slurries) are presented in an AIChE equipment-testing procedure (*Dry Solids, Paste and Dough Mixing Equipment Testing Procedure,* American Institute of Chemical Engineers, New York, 1979).

These complex methods are often simplified in practice. For example, a range or spread in composition or properties may be used as a measure of uniformity instead of standard deviation. In a batch operation, the time to bring composition or properties within a specified range or spread in values is often used as a measure of mixing performance.

Batch Mixing Batch-mixing times have been studied for miscible liquids in several types of commonly used impeller-stirred vessels. The results of these studies have been summarized by Brennan and Lehrer [*Trans. Inst. Chem. Eng.*, **54,** 139 (1976)] and Gray (Uhl and Gray, op. cit., vol. 1, chap. 4). However, these correlations are not applicable to suspensions of rapidly settling particles or to pseudoplastic slurries. In the latter case, viscosities are higher in remote parts of the circulated fluid than they are at the impeller, and mixing times are longer [Blasinski and Rzyski, *Int. Chem. Eng.*, **12,** 24 (1972)]. For pseudoplastic fluids, shorter mixing times may be obtained in unbaffled than in baffled vessels.

Averaging in Continuous-Flow Stirred Vessels When a fluid stream passes into and out of a stirred vessel, each increment of the feed is mixed to some degree with the fluid in the vessel. If the composition or properties of the feed stream change, the vessel will average the variations which occur within intervals shorter than the average residence time of fluid within the vessel. The reduction in amplitude of variations (averaging) which occurs depends on the flow pattern and fluid velocities in the vessel, the residence time (volume divided by feed flow rate), and the manner in which the properties or composition change with time.

For a constant feed rate to a cascade series of equal-volume vessels, each of constant volume and each sufficiently well stirred that the content is uniform (although its properties may vary in time), the ratio of exit to inlet property-variation amplitudes for a sine-wave input variation is given by the following equation (Buckley, *Techniques of Process Control,* Wiley, New York, 1964, p. 24):

$$\frac{A_o}{A_i} = \left[\left(\frac{2\pi\theta_r}{n\theta_s}\right)^2 + 1\right]^{-n/2} \tag{19-8}$$

where θ_r = total average residence time of n equal-size vessels, min; and θ_s = sine-wave period, min. When $2\pi\theta_r/n\theta_s \gg 1$, a log-log plot of A_o/A_i versus θ_r/θ_s is a straight line of slope $-n$.

The effluent-composition variations can be calculated from feed compositions for a perfectly mixed vessel by transient material balances in the form of a differential equation:

$$V\,dc = Qc'\,d\theta - Qc\,d\theta \tag{19-9}$$

where V = vessel volume, m^3 (ft^3); c = concentration, composition, or property of fluid in vessel and effluent stream; Q = feed and effluent stream flow rate, m^3/s (ft^3/s); and c' = concentration, composition, or property of fluid in feed stream. For time variations of c' such as sine waves, square waves, or a single pulse, this equation can be solved by classical analytical methods, by Laplace transforms [Walker and Cholette, *Pulp Pap. Mag. Can.*, **59,** 113 (1958)], by an analog computer, or by analog simulation in a digital computer.

For random or other input variations that are difficult to describe mathematically, a finite-difference form of Eq. (19-9) can be used to calculate effluent variations.

Statistical methods can be used to calculate the effluent-stream variance from an autocorrelation function and variance of the input stream [Danckwerts and Sellers, *Ind. Chem.*, **27,** 395 (1951)].

SOLIDS SUSPENSION

Definitions of Degree of Suspension Definitions of solid suspension have not been standardized, and care should be exercised to distinguish in literature references the authors' definitions of suspension and the method in which samples were taken. In this subsection, definitions which are useful for suspension in opaque tanks are presented. Some definitions based on observations in transparent tanks are not useful with opaque slurries and opaque tanks.

1. *Percent suspension.* This is calculated as follows:

$$100 \times \frac{\text{weight percent solids at sample point}}{\text{weight percent solids in tank}} = \text{percent suspension}$$

Percent suspension can have values greater than, equal to, or less than 100 percent. It may also be based on a specific particle-size fraction.

2. *Complete uniformity.* This implies that the percent suspension at every point is 100 percent. The uppermost region of stirred liquid is, of course, the most resistant to uniformity. It is very difficult to cause particles with settling velocities above 0.03 m/s (0.1 ft/s) to

be suspended uniformly in the topmost 2 percent of the tank volume, since the primarily horizontal flow pattern in this layer cannot keep solids of high settling velocity in suspension.

3. *Complete off-bottom suspension.* This is defined as all particles moving and elevated from the tank bottom. It implies nothing about the value of percent suspension anywhere in the tank. Sometimes *complete motion on the tank bottom* (all unsuspended particles in motion on or at the tank bottom) is an adequate condition. For solids of high settling velocity it is much more realistic than complete off-bottom suspension, with which it may be confused.

4. *Height of suspension.* The liquid height in the tank to which solids are suspended may be used to describe the degree of suspension. Height of suspension is most commonly expressed as the percent solids of each of the various particle-size fractions at various liquid heights off bottom.

5. *Concentration profiles: mixtures of particle sizes.* Virtually no experimental data are reported on concentration profiles of mixtures of particle sizes. On the other hand, practically all industrial applications have a mixture of particle sizes.

Continuously Fed Tank A method of describing and analyzing the condition of an agitated suspension of particles of distributed size, one that is particularly applicable to a continuous-flow tank, has been proposed by Oldshue [in Wadsworth and Davis, *Unit Processes in Hydrometallurgy* (Metall. Soc. Conf. 24), Gordon and Breach, New York, 1964, p. 33]. A tank containing a suspension is divided into zones for sampling. The suspension performance may then be described by vertical concentration profiles of the particle-size fractions and of the total solids in the tank, the parameter of location being the zone number.

The particles occupying any particular region in the tank need not have the size distribution of the feed, with one exception: in a steady-state continuous-flow tank, the suspension at the point of effluent withdrawal must have the same average composition with respect to both total solids and size distribution as the feed,* since the effluent and feed streams must be identical on average. The composition at all other points in the tank will, in fact, adjust itself so that this is true as steady state is established. If the agitator is not capable of maintaining inlet composition at the draw-off point, surging composition will occur throughout the tank and in the effluent stream so that the average output is the same as the input.

Sampling Sampling of a stirred suspension of solids of a distribution of sizes and shapes is difficult and unreliable [Randolph, paper presented before AIChE, Los Angeles, December 1968; Rushton, *Am. Inst. Chem. Eng.-Inst. Chem. Eng. Symp. Ser.*, **10,** 3 (1965)]. Every measuring instrument or draw-off tube interferes with the velocity pattern in the stirred liquid and makes accurate sampling virtually impossible.

Batch Suspension of Closely Sized Particles The rotational speed to meet a desired particle-suspension criterion is a complex function of impeller, baffle, and vessel dimensions; particle-size distribution; particle density and concentration; and liquid density and viscosity. Published correlations of these variables are limited in scope and are not always satisfactory for predicting the rotational speed required to meet process requirements [Bohnet and Niesmak, *Ger. Chem. Eng. (Engl. transl.),* **3,** 57 (1980); and Herringe, *Proc. 3d Eur. Conf. Mixing,* Univ. of York, England (Apr. 4–6, 1979), British Hydromechanics Research Association, Cranfield, Bedford, 1979, pp. 199–216].

The general approach to the design of a slurry-suspending agitator is to select the type and geometry of impeller and tank, specify the rotational speed required for acceptable performance, and determine the shaft power demanded to drive the impeller at that speed. Down-pumping axial-flow impellers are suggested for most particle-suspension operations. Values of D_a/D_T between 0.3 and 0.5 and values of C/D_T between 0.25 and 0.50 are appropriate, where C is the clearance between the impeller and the tank bottom. The procedure of Gates, Morton, and Fondy [*Chem. Eng.*, **83**(11), 144 (1976)] is tentatively recommended for the selection of rotational speed, which will depend on the proportions of the system (especially D_a/D_T) as well as on the absolute magnitude of D_a. The shaft power then can be calculated from a power-function curve for the impeller-tank combination (e.g., Fig. 19-13). The reader is cautioned that the drive-selection tables of Gates et al. are illustrative rather than general, inasmuch as specific installation geometry and size are implicit parameters that affect the listed values of both speed and power.

A radial-flow impeller may be used to stir solids-liquid suspensions, but the power required is likely to be higher than that for an axial-flow impeller meeting the same suspension criterion (Oldshue, *Proc. 1st Pacific Chem. Eng. Congr.*, Kyoto, Society of Chemical Engineers of Japan and American Institute of Chemical Engineers, Oct. 10–14, 1972, pp. 294–300).

When tanks and impellers are geometrically similar, the power required to produce the same state of particle suspension in a tank of any diameter greater than 0.5 m is proportional to the D_T^x. The value of x approaches 3 as an upper limit (implying constant power per unit volume) when low percentages of discrete, free-settling particles are involved. For the mixing of hindered-settling solids of low settling velocity at high solids percentages, x can approach 2 as a lower limit.

The power for the same suspension performance of impellers of different sizes in the same tank (different D_a/D_T values at constant D_T) is proportional to D_a^y; y ranges from zero to -2.5 for typical applications. A value of -1.0 can be used for estimating purposes if actual experimental data are not available.

Free-Settling Velocity of Particles For some solid-suspension correlations, the free-settling velocity of the solid particles is needed. If complete suspension of a mixed population of particles is desired, the agitator should be designed for the most rapidly settling particles; if the mixture is composed of particles of the same shape and density but different sizes, this means designing for the largest ones. In general, settling velocities must be measured, but if there is sufficient information about the size and shape of a typical particle, the free-settling velocity can be calculated. Methods of doing so are summarized in Sec. 5. A useful method of inferring shape factors from settling velocities of similar particles of two sizes is outlined by Oldshue [*Ind. Eng. Chem.*, **61**(9), 83 (1969)].

Hindered Settling For the higher solids concentrations often encountered in processing practice, hindered-settling conditions exist and correlations based on free settling do not apply. At sufficiently high concentrations suspension behavior is similar to that of a homogeneous pseudoplastic fluid. The percentage of fine particles, on the order of 100 mesh and smaller, determines the approach to pseudoplasticity. In many natural ores, the presence of clay also makes the slurry similar to a homogeneous pseudoplastic fluid.

It may take hours or even days for the completion of hindered settling and for the solids to pack to their maximum consistency. In high-density storage applications in which such settling might occur, the concern is with the uniformity of feed materials in or out of the system as well as with how much volume of the tank is active storage in the case of a shutdown of feed supply. Since the level of settled solids is much higher than the impeller level, what will happen if a power failure occurs and the impeller has "sanded in" should be considered. Restarting buried impellers in these cases depends upon the length of time during which settling has occurred and the characteristics of the solids, and it can be one of the most important practical design considerations in an actual installation.

Height of Suspension In large slurry-storage tanks involving 5 to 70 percent by weight solids, it is sometimes not practical or necessary to have uniform, complete suspension of all the particles throughout the vessel.

At low mixer power there can be a zone in the tank where the solids concentration is quite uniform but solids-free liquid will be on top. The prediction of performance can be based on the volume of the uniform-solids portion, and the top layer of liquid does not affect the mixing action in the bottom part of the system. Under other conditions, such as lower solids concentration or lower slurry levels, supernatant liquid can be incorporated into the slurry and the entire tank volume can be utilized.

*With adjustments, of course, for any pertinent process occurring within the tank: dissolution, crystal growth, or chemical reaction, for example.

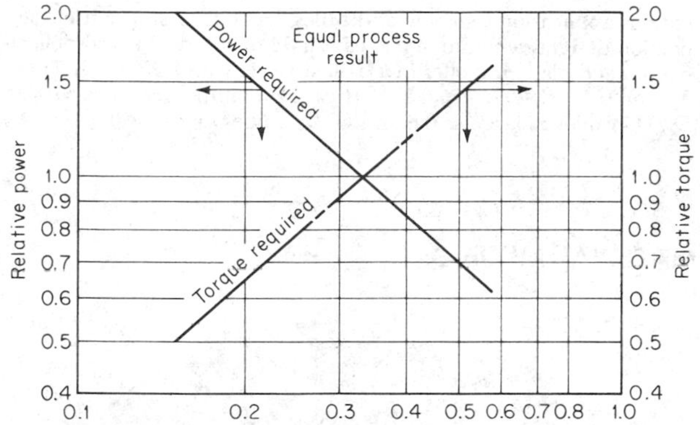

FIG. 19-14 Effect of varying impeller size, in a tank of fixed diameter, on power and torque requirements of a typical low-viscosity solids-suspension application. [*Oldshue,* Ind. Eng. Chem., *61, 79 (1969), by permission of the copyright owner, the American Chemical Society.*]

Effect of Impeller-to-Vessel-Diameter Ratio To provide complete uniformity through a slurry, it is usually found that the larger the impeller in a given tank, the less the power that is required. Large, slow-speed impellers require a lower horsepower for a given pumping capacity, and solids suspension is typically governed by circulation rate in the tank.

Figure 19-14 shows a typical curve for illustration purposes only. The actual slopes and boundaries of these curves vary depending upon the concentration and settling velocity of the solids and also on the particular definition of uniformity required.

At first glance it might be assumed that the bigger the impeller and the lower the horsepower, the more economical will be the mixer. It turns out, however, that beyond a certain point the increasing torque required dominates the economics, inasmuch as the initial cost of the agitator is governed largely by the torque required for the drive.

Table 19-1 illustrates the results of a typical cost analysis. For a short period of time, the 56-kW (75-hp), less expensive mixer is more desirable, but for longer periods of time, the 45-kW (60-hp) unit, even though it is initially more expensive, requires less total cost.

Air Agitation Air or another suitable gas is sometimes used for blending liquids or for the agitation of solid-liquid systems. Simple spargers have been used for years in the cement and oil industries. Kauffman [*Chem. Metall. Eng.*, **37**, 178 (1930)] reported that the degree of agitation in a large air-sparged tank containing 2.7 m (9 ft) of liquid increases from moderate to violent as the air flux increases from 0.2 to 1 $m^3/(m^2$ of tank cross section·min) [0.65 to 3.1 ft/min].

For solids suspension, a gas is generally ineffective if the particles are free-settling unless the settling velocity is very low. It can be used, however, for high-density slurries in which settling is hindered. A common technique is to inject air in the center of a draft tube so that the pulp rises through the tube to overflow and return downward in the outer annular space. Vessels employing this principle, called Pachuca tanks, are described in this section under "Leaching."

Flotation Cells Flotation cells are special agitated vessels common in the mining industry in conjunction with mechanical separations. Their job is to disperse air through the slurry and promote preferential attachment of air bubbles to the solid particles so that selected solids will rise into the froth at the top of the tank. Flotation is discussed in Sec. 21.

AIChE Equipment-Test Procedures The American Institute of Chemical Engineers has published *Equipment Test Procedures: Impeller Type Mixers* (1959), for testing an agitator in place in a process. These procedures cover equipment for liquid-solids contacting and describe the proper techniques for measuring power as well as process performance.

TABLE 19-1 Typical Cost Analysis: Mixer Size and Length of Service*

Costs of 45-kW (60-hp) and 56-kW (75-hp) mixers compared on the basis of savings achieved with the 56-kW installation.

	Savings for length of service indicated			
Source of savings	1 year, 8 h/day	1 year, 24 h/day	5 years, 24 h/day	10 years, 24 h/day
Purchase price†	$3,900	$3,900	$3,900	$3,900
Power cost‡	−800	−2,400	−9,400	−14,000
Total	$3,100	$1,500	−$5,500	−$10,100
Net effect for 56-kW mixer	Saving	Saving	Loss	Loss

*Adapted from Oldshue, *Ind. Eng. Chem.*, **61**(9), 82 (1969).

†A 56-kW mixer has smaller turbines and a smaller-sized speed reducer than a 45-kW unit. Therefore, the 56-kW mixer of this example costs $3,900 less than a 45-kW mixer.

‡Electricity is assumed to cost $254 per kilowatt-year. Discounted at 15%, this is equivalent to $170 per kilowatt-year for 5 years and $128 per kilowatt-year for 10 years.

PARTICLE-SIZE REDUCTION IN DISPERSIONS

Size reduction is achieved either by reducing the basic particle size of the solid phase or by breaking up agglomerates of particles. In either case, a major factor is the magnitude of fluid shear required. The higher the peripheral velocity of the impeller, the higher the shear rate and shear stress in the fluid. Not uncommonly tests are needed to determine the equipment geometry and impeller speed required. Turbine impellers with a low D_a/D_T ratio or special impellers such as flat or toothed disks may be used at high rotational speed. In some processes, size reduction is undesirable, and it is necessary to make sure that it does not proceed too far.

Typical operations involving size reduction are the dispersion of pigment in paints, the production of clays and coating materials, and the dispersion of lithium and sodium metals for chemical reaction.

SOLID-LIQUID MASS TRANSFER

Solid-liquid mass transfer between particles and agitated liquid in which they are free to move occurs in tanks used for such operations as leaching, adsorption, crystallization, and dissolution (with or without chemical reaction). This discussion deals principally with dissolution. Crystallizers, stirred-tank adsorbers, and leaching equipment are treated elsewhere in this section.

Most of the reported research on agitation-promoted dissolution has been done with closely sized, simple-geometry particles of slowly dissolving solute. Many correlations of dissolution-rate coefficients with physicochemical properties and mechanical parameters (geometry and agitator speed or power), usually involving such numerics as Reynolds number, Sherwood number, Schmidt number, and shape factors, have been proposed. Such general correlations are not reliable for design use. Furthermore, in industrial problems the liquids and solids are prevalently impure and diffusivities and solubilities are uncertain or unknown.

Mass-transfer correlation methods have been described by Batchelor, *J. Fluid Mech.*, **98**, 609 (1978); Boon-Long et al., *Chem. Eng. Sci.*, **33**, 816 (1978); Levins and Glastonbury, *Trans. Inst. Chem. Eng.*, **50**, 132 (1972); Miller, *Ind. Eng. Chem. Process Des. Dev.*, **10**, 365 (1971); Nienow, *Chem. Eng. J.*, **9**, 153 (1975); and Nienow and Miles, *Chem. Eng. J.*, **15**, 13 (1978).

Constant power per unit of volume is recommended tentatively as the basis for scale-up of mass-transfer coefficients between solids and liquids in stirred tanks. Tests in a geometrically similar model should be used as a basis for scale-up. The impeller rotational speeds used in model and scaled-up design should not be less than those needed for complete off-bottom suspension. At higher speeds than these, the mass-transfer rate is proportional to impeller power with an exponent of 0.1 to 0.15 [Kneule, *Chem. Ing. Tech,* **28**, 221 (1956); and Harriott, *Am. Inst. Chem. Eng. J.*, **8**, 93 (1962)].

The types of equipment suitable for suspending particles in liquids are suitable also for solid-liquid mass-transfer operations. However, optimum impeller and vessel geometries have not been well defined for this application (Nienow and Miles, loc. cit.). As in particle-suspension designs, a common choice is a 30 to 45° downward-deflecting pitched-blade impeller in a baffled vessel with $0.25 < D_a/D_T < 0.5$ and $0.25 < C/D_T < 0.5$. Vertical-blade impellers are used with D_a/D_T values as large as 0.6 and C/D_T values as low as 0.05.

PASTE AND VISCOUS-MATERIAL MIXING

General References: *Dry Solids, Paste and Dough Mixing Equipment Testing Procedure,* American Institute of Chemical Engineers, New York, 1979. Fischer, *Chem. Eng.*, **69**(3), 52 (1962). Irving and Saxton, "Mixing of High Viscosity Materials," in Uhl and Gray, *Mixing Theory and Practice*, vol. 2, Academic, New York, 1967, chap. 8. Mohr, in Bernhardt, *Processing of Thermoplastic Materials,* Reinhold, New York, 1967, chap. 3. Parker, *Chem. Eng.*, **71**(12), 166 (1964).

BATCH MIXERS

Change-Can Mixers Change-can mixers are vertical batch mixers in which the container is a separate unit easily placed in or removed from the frame of the machine. They are available in capacities of about 4 to 1200 L (1 to 300 gal). The commonest type is the pony mixer. Separate cans allow the batch to be carefully measured or weighed before being brought to the mixer itself. The mixer also may serve to transport the finished batch to the next operation or to storage. The identity of each batch is preserved, and weight checks are easily made.

Change cans are relatively inexpensive. A good supply of cans allows cleaning to be done in a separate department, arranged for efficient cleaning. In paint and ink plants, where mixing precedes milling or grinding and where there may be a long run of the same formulation and color, the cans may be used for an extended period without cleaning, as long as no drying out or surface oxidation occurs.

In most change-can mixers, the mixing elements are raised from the can by either a vertical lift or a tilting head; in others, the can is dropped away from the mixing elements. After separation the mixing elements drain into the can, and the blades can be wiped down. With the can out of the way, complete cleaning in the blades and their supports is simple. If necessary, blades may be cleaned by rotating then in solvent.

Intimate mixing is accomplished in change-can mixers in two ways. One method is to have the mixing-unit assembly revolve in a planetary motion so that the rotating blades sweep the entire circumference of the can (Fig. 19-15). The other is to mount the can on a rotating turntable so that all parts of the can wall pass fixed scraper blades or the agitator blades at a point of minimum clearance.

A type of heavy-duty planetary change-can mixer has been used extensively for processing critical solid-propellant materials. In a design offered by Baker Perkins Inc., the mixer blades pass through all portions of the can volume, and the two blades wipe each other, thus assuring no unmixed portions. A dead spot under the mixer blades is avoided by having both blades off the centerline of the can. The cans fit tightly enough so that mixing can be achieved under vacuum or pressure. There is no contact between the glands and the material being mixed. Charging ports located in the housing directly above the mixing can make it possible to charge materials to the mixer with the can in the operating position. This type of mixer is available in sizes of 0.5 L (about 1 pt) to 1.6 m^3 (420 gal), involving power input of 0.2 to 75 kW (0.25 to 100 hp).

Stationary-Tank Mixers Stationary-tank mixers are recommended when there is no advantage in having the change can for conveying or storage, when the batch size is above 600 L (approximately 150 gal), and when feed and product may be conveniently handled by permanent piping or chutes. The same type of mixing actions must take place in tank mixers as in change cans. The contents of the tank must be moved progressively into the active zone of intensive action at close clearances. However, as the tank and the bearings holding the mixing elements are part of the same structure, very close clearances can be maintained to give intense shear.

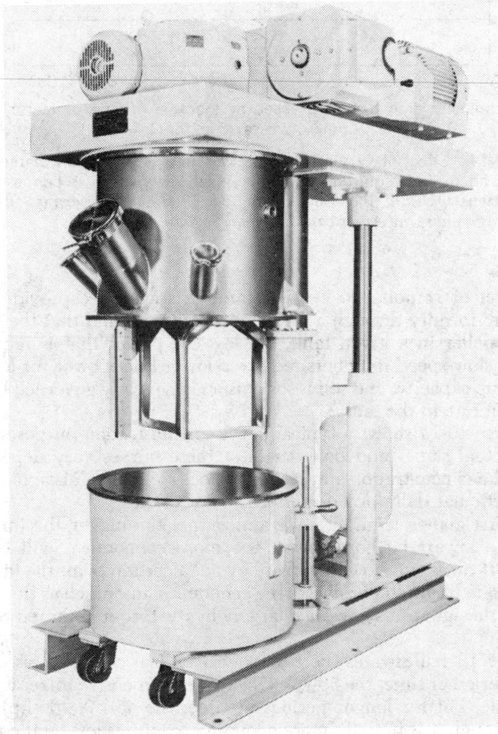

FIG. 19-15 Change-can mixer. (*Charles Ross & Son Co.*)

Stationary tanks are used with a wide variety of agitators for work with thin fluids, as discussed under "Agitation of Low-Viscosity Particle Suspensions." Paste-processing equipment, however, is limited to the following types.

Gate Mixers One of the oldest stationary-tank mixers is the gate mixer. A flat rotating structure of horizontal and vertical bars cuts the paste at different levels and at the tank wall where stationary bars may be fastened to give points of intensive shear. The speed is kept low to avoid rotating the entire mass in the tank. Slow mass mixing is produced by the mild centrifugal action of the rotating blades. The motion may be increased by sloping the blades. Paints, starch pastes, coatings, and sizes are effectively processed by gate mixers.

Instead of a gate, a close-fitting **anchor** or **horseshoe agitator** may be used (Fig. 19-7). The outer sweep assembly may be fitted with scraper blades to clean the container wall and improve heat transfer. Adhesives, greases, cosmetics, and pastes which require quick cooling or heating during mixing are handled successfully in these units.

Shear-Bar Mixers A modified gate mixer is the shear-bar mixer, which contains a series of vertical paddles passing between vertical stationary fingers. This construction increases the shear surface and produces more circulation. Similar intermeshing of moving and stationary bars can be obtained in a horizontal mixer. With this design the rotor blades may be uniformly staggered around the shaft to give a more uniform power load and better mixing. The end blades are shaped to move the material toward the center of the tank.

Helical-Blade Mixers Helical mixers are now available in a variety of configurations. The mixing element may be in the form of a conical or a cylindrical helix. It may be a ribbon spaced radially from the shaft by spokes or a screw consisting of a helical surface that is continuous from the shaft to the periphery of the helix. A venerable example of the latter type is the soap crutcher, in which the screw is mounted in a draft tube. Close screw-tube clearance and a high rotational speed result in rapid motion of the material and high shear. The screw lifts the material through the tube, and gravity returns it to the bottom of the tank. If the tank has well-rounded corners, this kind of mixer may be used for fibrous materials. Heavy paper pulp containing 16 to 18 percent solids is uniformly bleached in large mixers of this type.

A double helix shortens mixing time but requires more power. The disadvantages of the higher torque requirement are frequently offset by the better mixing and heat transfer.

A vertical helical ribbon blender can be combined with an axial screw of smaller diameter (Fig. 19-16). Such mixers are used in polymerization reactions in which uniform blending is required but in which high-shear dispersion is not a factor. Addition of the inner flight contributes little more turnover in mixing newtonian fluids but significantly shortens the mixing time in nonnewtonian systems and adds negligibly to the impeller power [Coyle et al., *Am. Inst. Chem. Eng. J.*, **15**, 903 (1970)].

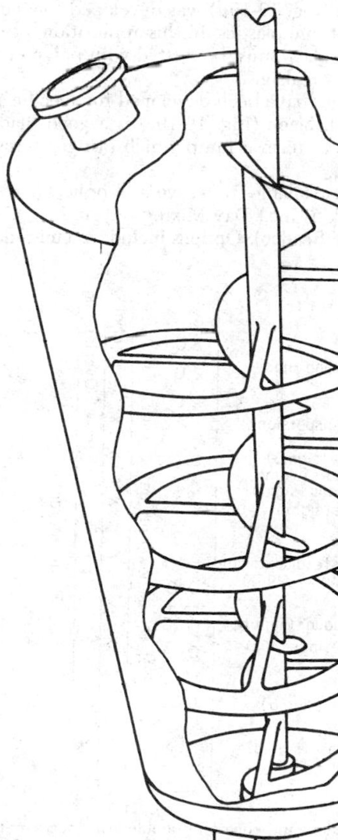

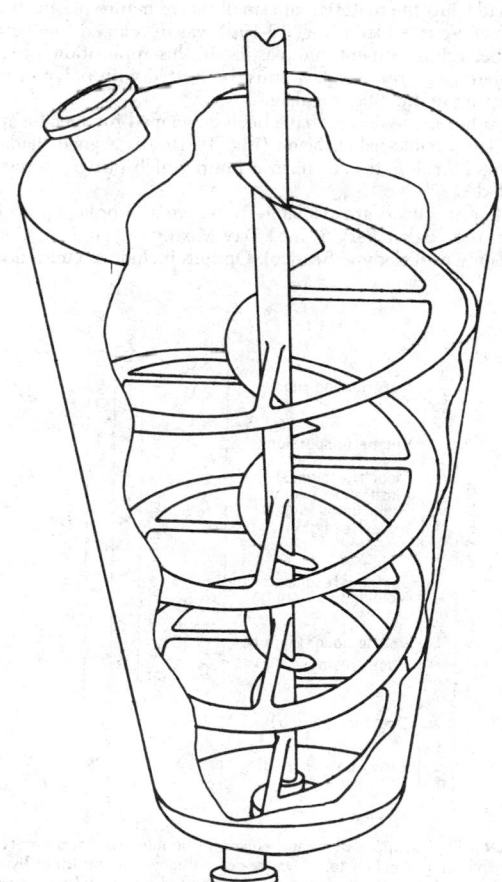

FIG. 19-16 Helical ribbon mixer.

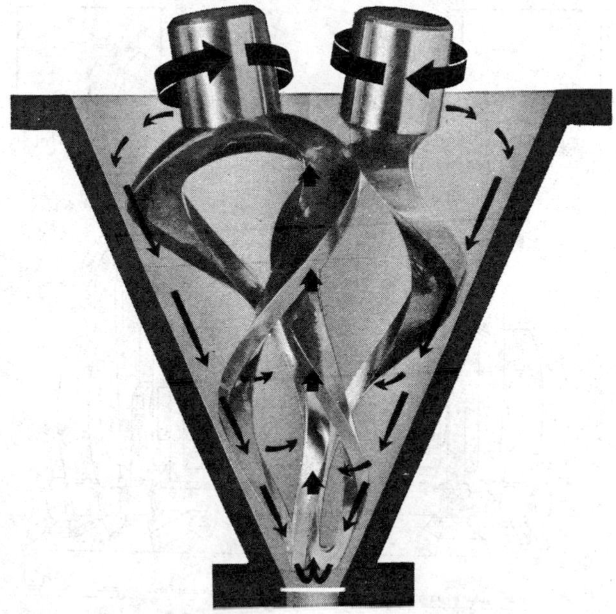

FIG. 19-17 Twin-blade conical mixer. *(Atlantic Research Corp.)*

Another variant of the helical mixer is the twin-blade conical unit shown in Fig. 19-17. This mixer has the advantage of adjustable blade-to-blade and blade-to-wall clearances.

Bourne and Butler [*Trans. Inst. Chem. Eng.*, **47**, T11 (1969)] proposed a scale-up method for the circulation rate and vortex depth produced by helical ribbon impellers based on an analysis of flow patterns. The ribbon causes all the fluid to circulate, using only 5 percent of the power needed by a turbine. It also provides a region of high shear stress near the wall, thereby achieving both bulk blending and dispersive mixing. Correlations of the power requirement and the wall heat-transfer coefficients associated with helical ribbon impellers were developed by Nagata et al. (Nagata, *Mixing Principles and Applications*, Kodansha, Tokyo, 1975, p. 99). Such scale-up methods and correlations are useful guides but must be employed with caution.

Double-Arm Kneading Mixers The universal mixing and kneading machine consists of two counterrotating blades in a rectangular trough curved at the bottom to form two longitudinal half cylinders and a saddle section (Fig. 19-18). The blades are driven by gearing at either or both ends. The oldest style empties through a bottom door or valve and is still in use when 100 percent discharge or thorough cleaning between batches is not an essential requirement. More commonly, however, double-arm mixers are tilted for discharge. The tilting mechanism may be manual, mechanical, or hydraulic.

A variety of blade shapes has evolved. The mixing action is a combination of bulk movement, smearing, stretching, folding, dividing, and recombining as the material is pulled and squeezed against blades, saddle, and sidewalls. The blades are pitched to achieve end-to-end circulation. Rotation is usually such that material is drawn down over the saddle. Clearances are as close as 1 mm (0.04 in).

The blades may be tangential or overlapping. Tangential blades are run at different speeds, with the advantages of faster mixing from constant change of relative position, greater wiped heat-transfer area per unit volume, and less riding of material above the blades. Overlapping blades can be designed to avoid buildup of sticky material on the blades.

The agitator design most widely used is the sigma blade (Fig. 19-19*a*). The sigma-blade mixer is capable of starting and operating with either liquids or solids or a combination of both. Modifications in blade-face design have been introduced to increase particular

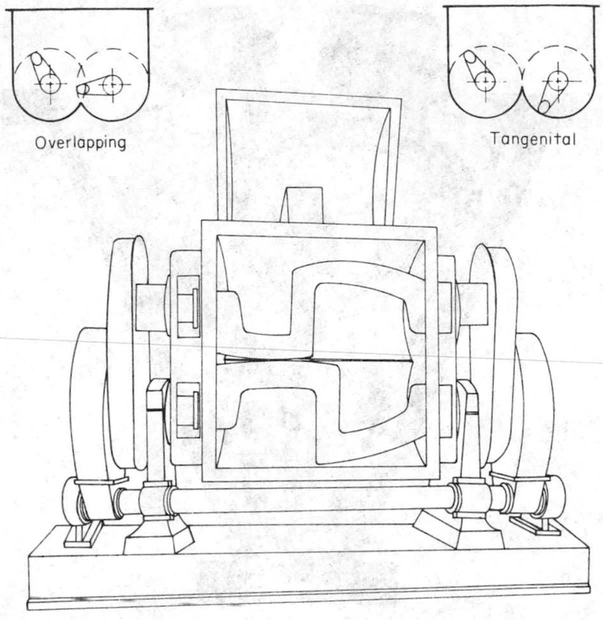

FIG. 19-18 Double-arm kneader mixer. *(Baker Perkins Inc.)*

effects, such as shreadding or wiping. The sigma blade has good mixing action, readily discharges materials which do not stick to the blades, and is relatively easy to clean when sticky materials are being processed.

The dispersion blade (Fig. 19-19*b*) was developed particularly to provide compressive shear higher than that achieved with standard

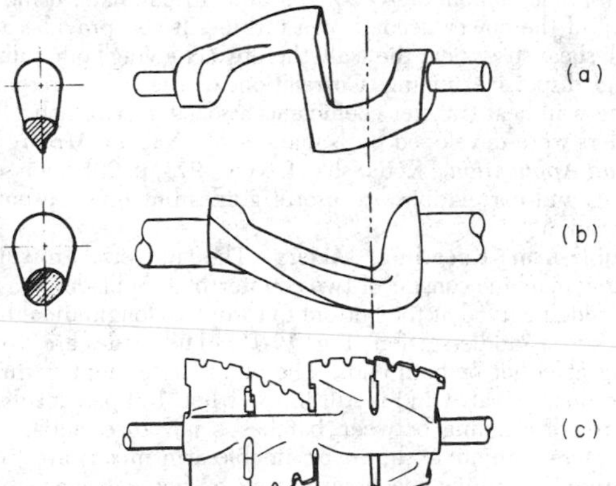

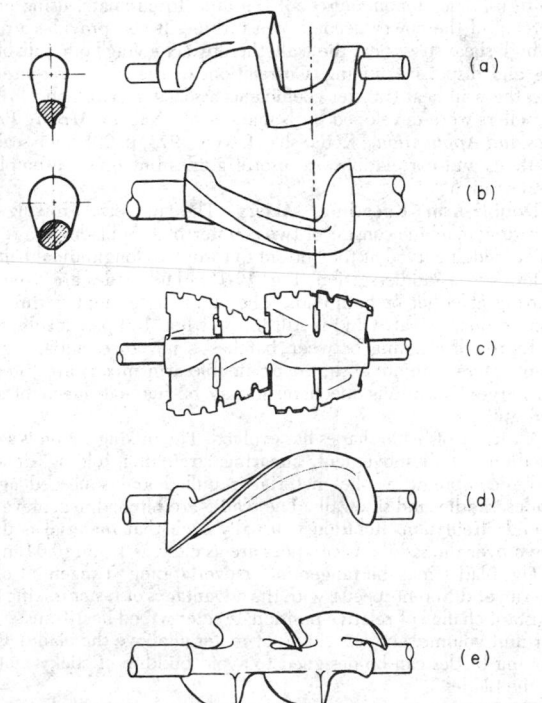

FIG. 19-19 Agitator blades for double-arm kneaders. (*a*) Sigma. (*b*) Dispersion. (*c*) Multiwiping overlap. (*d*) Single-curve. (*e*) Double-naben. *(Baker Perkins Inc.)*

TABLE 19-2 Characteristics of Double-Arm Kneading Mixers*

Size number	Capacity, U.S. gal.		Typical supplied horsepower		Floor space, ft.
	Working	Maximum	Sigma blade, MWOL blade	Dispersion blade	
4	0.7	1	1	2	1 × 3
6	2.3	3.5	2	5	2 × 3
8	4.5	7	5	7.5	3 × 4
11	10	15	15	20	5 × 6
12	20	30	25	40	6 × 6
14	50	75	30	60	6 × 8
15	100	150	50	100	8 × 10
16	150	225	60	150	9 × 11
17	200	300	75	200	9 × 13
18	300	450	100	—	10 × 14
20	500	750	150	—	11 × 16
21	600	900	175	—	12 × 16
22	750	1125	225	—	12 × 17
23	1000	1500	300	—	14 × 18

*Data from Baker Perkins Inc. To convert feet to meters, multiply by 0.3048; to convert gallons to cubic meters, multiply by 3.78×10^{-3}; and to convert horsepower to kilowatts, multiply by 0.746.

sigma blades. The blade face wedges material between itself and the trough, rather than scraping the trough, and is particularly suited for dispersing fine particles in a viscous mass. Rubbery materials have a tendency to ride the blades, and a ram is frequently used to keep the material in the mixing zone.

Multiwiping overlapping (MWOL) blades (Fig. 19-19*c*) are commonly used for mixtures which start tough and rubberlike, inasmuch as the blade cuts the material into small pieces before plasticating it.

The single-curve blade (Fig. 19-19*d*) was developed for incorporating fiber reinforcement into plastics. In this application, the individual fibers (e.g., sisal or glass) must be wetted with polymer without incurring undue fiber breakage.

Many other blade designs have been developed for specific applications. The double-naben blade (Fig. 19-19*e*) is a good blade for mixes which "ride," that is, form a lump which bridges across the sigma blade.

Double-arm mixers are available from several suppliers (e.g., Paul O. Abbe, Inc.; Baker Perkins Inc.; Day Mixing; Jaygo, Inc.; Charles Ross & Son Co.; Teledyne Readco). Options include vacuum design,

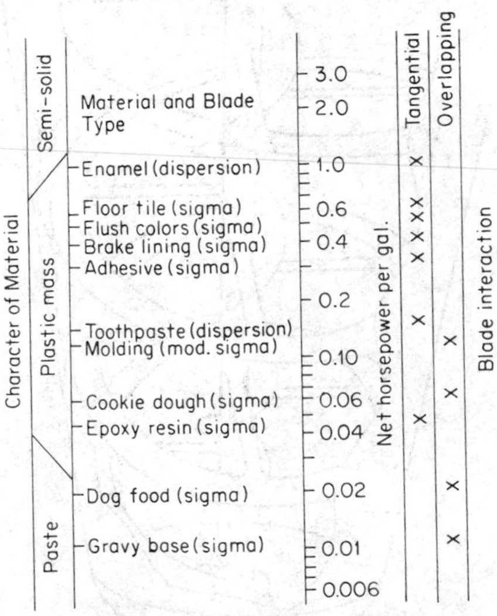

FIG. 19-20 Typical applications and power for double-arm kneaders. To convert horsepower per gallon to kilowatts per cubic meter, multiply by 197.3. [*Parker*, Chem. Eng., *72(18), 125 (1965); excerpted by special permission of the copyright owner, McGraw-Hill, Inc.*]

cored blades, jacketed trough, choice of cover design, and a variety of seals and packing glands. Power requirements vary from ⅙ to 2 hp/gal of capacity. Table 19-2 lists specifications and space requirements for typical tilting-type, double-ended-drive, double-arm mixers. The working capacity is generally at or near the top of the blades, and the total capacity is the volume contained when the mixer is filled level with the top of the trough.

Figure 19-20 provides a guide for typical applications. Individual formulation changes may require more power than indicated in the figure. Parker [*Chem. Eng.*, **72**(18), 121 (1965)] has described in greater detail how to select double-arm mixers.

Screw-Discharge Batch Mixers A variant of the sigma-blade mixer is now available with an extrusion-discharge screw located in the saddle section. During the mixing cycle the screw moves the material within the reach of the mixing blades, thereby accelerating the mixing process. At discharge time, the direction of rotation of the screw is reversed and the mixed material is extruded through suitable die openings in the side of the machine. The discharge screw is driven independently of the mixer blades by a separate drive.

Working capacities range from 4 to 3800 L (1 to 1000 gal), with up to 300 kW (400 hp). This type of kneader is offered by most of the double-arm-kneader manufacturers. It is particularly suitable when a heel from the prior batch can be left without detriment to succeeding batches.

Intensive Mixers

Banbury Mixer Preeminent in the field of high-intensity mixers, with power input up to 6000 kW/m^3 (30 hp/gal), is the Banbury mixer, made by the Farrel Co. (Fig. 19-21). It is used mainly in the plastics and rubber industries. The top of the charge is confined by an air-operated ram cover mounted so that it can be forced down on the charge. The clearance between the rotors and the walls is extremely small, and it is here that the mixing action takes place. The operation of the rotors of a Banbury at different speeds enables one rotor to drag the stock against the rear of the other and thus help clean ingredients from this area.

The extremely high power consumption of the machines operating at speeds of 40 r/min or lower calls for rotor shafts of large diameter. The combination of heavy shafts, stubby blades, close clearances, and the confined charge limits the Banbury mixer to small batches. The production rate is increased as much as possible by using powerful drives and rotating the blades at the highest speed that the material will stand. The friction produced in the confined space is great, and with heat-sensitive materials cooling may be the limiting factor. Recent innovations include a drop door, four-wing rotors which can provide 30 percent greater power than the older two-wing rotors, and separable gear housings. Equipment is available from laboratory size to a mixer capable of handling a 450-kg (1000-lb) charge and applying 2240 kW (3000 hp).

Prodex-Henschel and Welex-Papenmeir Mixers These mixers, of Purnell, Inc., and Welex, Inc., respectively, are high-intensity mixers combining vortex flow and high shear. Blades at the bottom of the vessel scoop the batch upward at peripheral speeds of about 40 m/s (130 ft/s). The high shear stress (to 20,000 s^{-1}) and blade impact easily reduce agglomerates and aid intimate dispersion. Since the energy input is high [200 kW/m^3 (about 8 hp/ft^3)], even powdery material is heated rapidly.

Mixers of this type are available in sizes from 6 to 800 L (1.5 to 200 gal), consuming from 1.5 to 150 kW (2 to 200 hp).

The Welex-Papenmeier impeller (Fig. 19-22) includes a secondary airfoil higher in the vessel, which imparts additional energy to the rising particles near the wall and also forces returning material to the eye of the bottom first-stage blades.

These mixers are particularly suited for rapid mixing of powders and granules with liquids, for dissolving resins or solids in liquids, or for removal of volatiles from pastes under vacuum. Scale-up is usually on the basis of maintaining constant peripheral velocity of the impeller.

Roll Mills Roll mills can provide exceedingly high localized shear while retaining extended surface for temperature control.

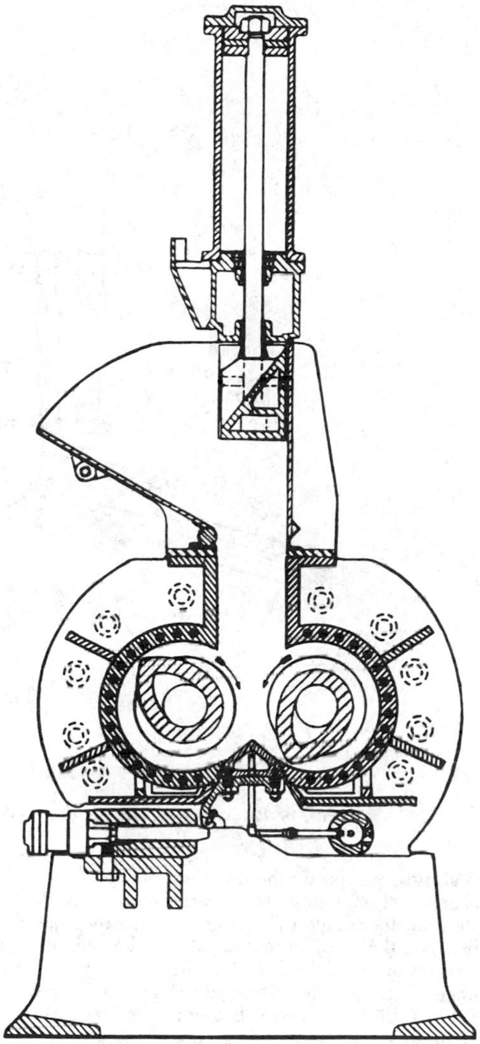

FIG. 19-21 Banbury mixer. *(Farrel Co.)*

Two-Roll Mills These mills contain two parallel rolls mounted in a heavy frame with provision for accurately regulating the pressure and distance between the rolls. As one pass between the rolls does little blending and only a small amount of work, the mills are practically always used as batch mixers. Only a small amount of material is in the high-shear zone at any one time.

To increase the wiping action, the rolls are usually operated at different speeds. The material passing between the rolls is returned to the feed point by the rotation of the rolls. If the rolls are at different temperatures, the material usually will stick to the hotter roll and return to the feed point as a thick layer.

At the end of the period of batch mixing, heavy materials may be discharged by dropping between the rolls, while thin mixes may be removed by a scraper bar pressing against the descending surface of one of the rolls.

Two-roll mills are used mainly for preparing color pastes for the ink, paint, and coating industries. There are a few applications in heavy-duty blending of rubber stocks, for which corrugated and masticating rolls are often used.

Three-Roll Mills These mills are continuous units containing parallel rolls of equal diameter mounted in a rigid framework. The

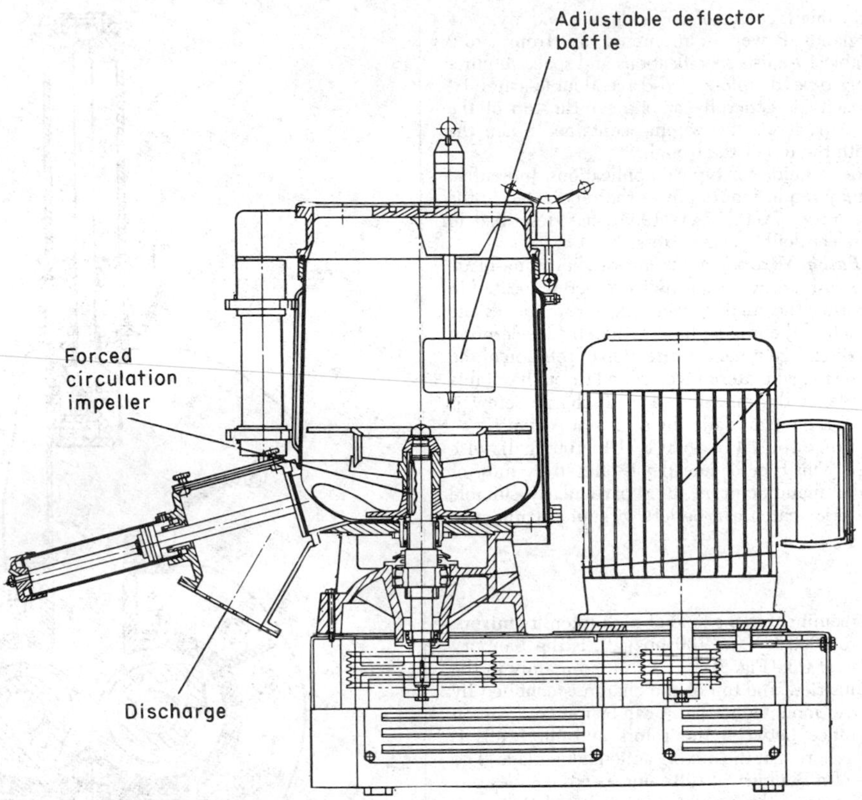

FIG. 19-22 Welex-Papenmeier mixer. (*Welex, Inc.*)

rolls run at different speeds, the receiving roll being the slowest and the discharge roll the fastest. The clearances between the front and back rolls and the middle rolls are independently adjustable. Feed enters between the first and second rolls, and a film of appreciable thickness is produced at the first nip, where aggregates and particles are crushed. These are then abraded by the rubbing action of the rolls turning at different speeds, but since the film is thick, there is probably no hydraulic shear. In the second nip a smaller clearance produces a thinner film and the speed of the takeoff roll is increased to compensate for the reduction of cross-sectional area. In the thinner film there is more crushing, less internal abrasion, and, because of the higher speed, more external abrasion against the rolls. The higher speed produces some hydraulic shear in the thin film. At both nips there is a rotary motion of the stock lying in the nip, which produces some mass mixing. The finished product is removed from the last roll by a tapered delivery chute fitted with a scraper bar. The chief application of three-roll mills is in dispersing and grinding inks and pigment pastes, but they may be used whenever an extremely uniform dispersion is desired.

Miscellaneous Batch Mixers

Bulk Blenders Many of the mixers used for solids blending (Sec. 21) are also suitable for some liquid-solids blending. **Ribbon blenders** can be used for such tasks as wetting out or coating a powder. When the final paste product is not too fluid, other solids-handling equipment finds frequent use.

Plow Mixers Plow mixers such as the **Littleford** (Littleford Bros., Inc.) and the **Marion** (Rapids Machinery Co.) machines can be used for either batch or continuous mixing. Plow-shaped heads arranged on the horizontal shaft rotate at high speed, hurling the material throughout the free space of the vessel. Additional intermixing and blending occur as the impellers plow through the solids bed. Special high-speed choppers (3600 r/min) can be installed to break up lumps and aid liquid incorporation. The choppers also disperse fine particles throughout viscous materials to provide a uniform suspension. The mixer is available in sizes of 40 L to 40 m^3 (10 to 10,000 gal) of working capacity.

Cone and Screw Mixers The **Nauta mixer** of Fig. 19-23 (Day Mixing) utilizes an orbiting action of a helical screw rotating on its own axis to carry material upward, while revolving about the centerline of the cone-shaped shell near the wall for top-to-bottom circulation. Reversing the direction of screw rotation aids discharge of pasty materials. A further variant, the **Vert-O-Mix** (Tower Iron Works, Inc.) uses an epicyclic action to provide more thorough coverage of the entire volume of the shell. Partial batches are mixed in the Nauta and Vert-O-Mix types of mixer as efficiently as full loads. These mixers, available in sizes of 40 L to 40 m^3 (10 to 10,000 gal), achieve excellent low-energy blending, with some hydraulic shear dispersion. At constant speed, both mixing time and power scale up with the square root of volume.

Pan Muller Mixers These mixers can be used if the paste is not too fluid or too sticky. The main application of muller mixers is now in the foundry industry, in mixing small amounts of moisture and binder materials with sand particles for both core and molding sand. In paste processing, pan-and-plow mixers are used principally for mixing putty and clay pastes, while muller mixers handle such diversified materials as clay, storage-battery paste, welding-rod coatings, and chocolate coatings.

In muller mixers the rotation of the circular pan or of the plows brings the material progressively into the path of the mullers, where the intensive action takes place. Figure 19-24 shows one type of mixer, in which the mullers and plows revolve around a stationary turret in a stationary pan. The outside plow moves material from the crib wall to the path of the following muller; the inside plow moves it from the central turret to the path of the other muller. The mullers crush the material, breaking down lumps and aggregates.

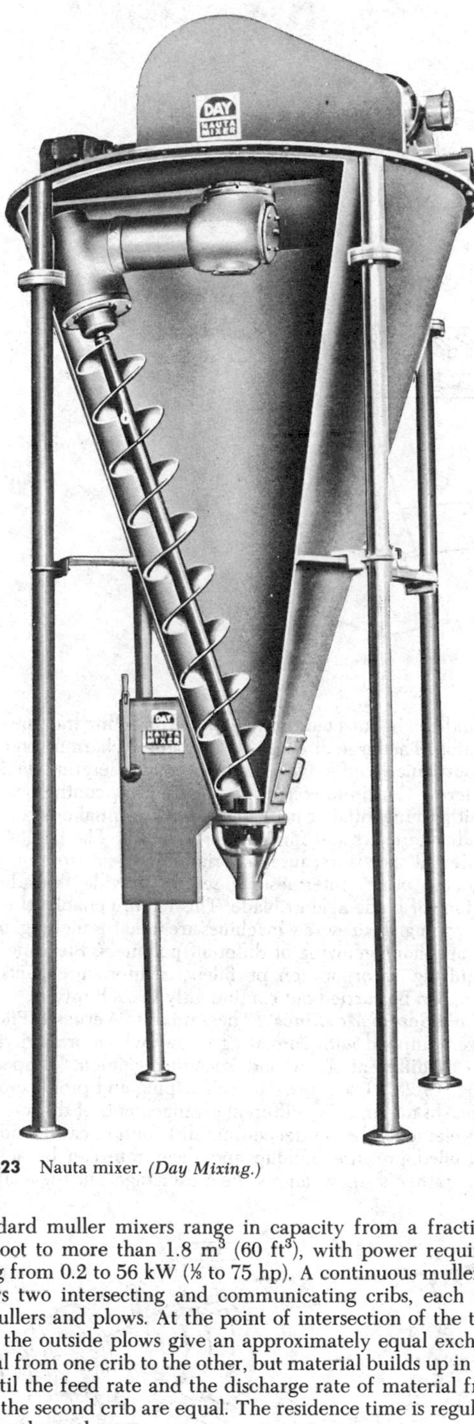

FIG. 19-23 Nauta mixer. *(Day Mixing.)*

Standard muller mixers range in capacity from a fraction of a cubic foot to more than 1.8 m^3 (60 ft^3), with power requirements ranging from 0.2 to 56 kW (⅛ to 75 hp). A continuous muller design employs two intersecting and communicating cribs, each with its own mullers and plows. At the point of intersection of the two crib bodies, the outside plows give an approximately equal exchange of material from one crib to the other, but material builds up in the first crib until the feed rate and the discharge rate of material from the gate in the second crib are equal. The residence time is regulated by adjusting the outlet gate.

CONTINUOUS MIXERS

Single-Screw Extruders The single-screw extruder is frequently used as a mixing device in the plastics industry. Stabilizers, color concentrates, etc., may be compounded with granular raw polymer, melted, and extruded into pellets, sheet, or rod. Detailed descriptions of extruders and procedures for calculating the degree of mixing attainable are available elsewhere (Irving and Saxton, op. cit.; and Paton et al., in Bernhardt, *Processing of Thermoplastic Materials*, Reinhold, New York, 1967, chap. 4). In essence a "circulating" movement is achieved by working against a discharge pressure such that there is a pressure flow opposite to the forwarding drag flow of the screw. Single-screw extruders can be equipped with large gears and thrust bearings to operate with high torque and high power input to the material.

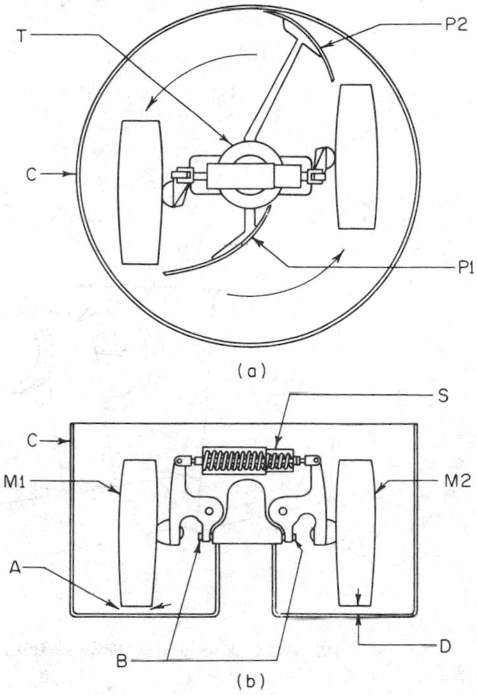

FIG. 19-24 Pan muller. (*a*) Plan view. (*b*) Sectional elevation. [*Bullock*, Chem. Eng. Prog., *51*, *243 (1955)*, *by permission*.]

Rietz Extructor This extruder, shown schematically in Fig. 19-25, has orifice plates and baffles along the vessel. The rotor carries multiple blades with a forward pitch, generating the head for extrusion through the orifice plates as well as battering the material to break up agglomerates between the baffles. Typical applications include wet granulation of pharmaceuticals, blending color in bar soap, and mixing and extruding cellulose materials. The Extructor is available in rotor diameters up to 600 mm (24 in) and in a power range of 5 to 112 kW (7 to 150 hp).

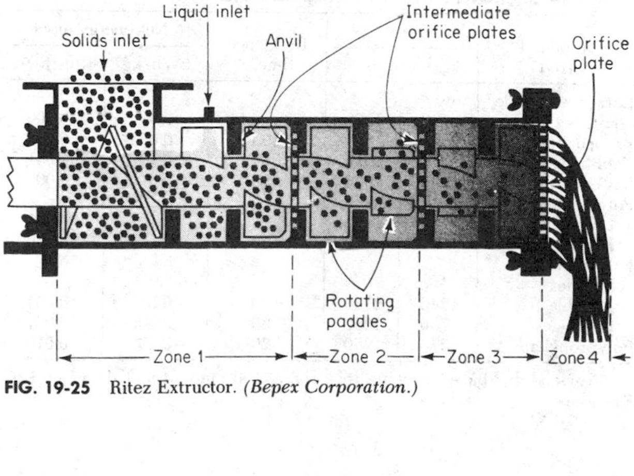

FIG. 19-25 Ritez Extructor. *(Bepex Corporation.)*

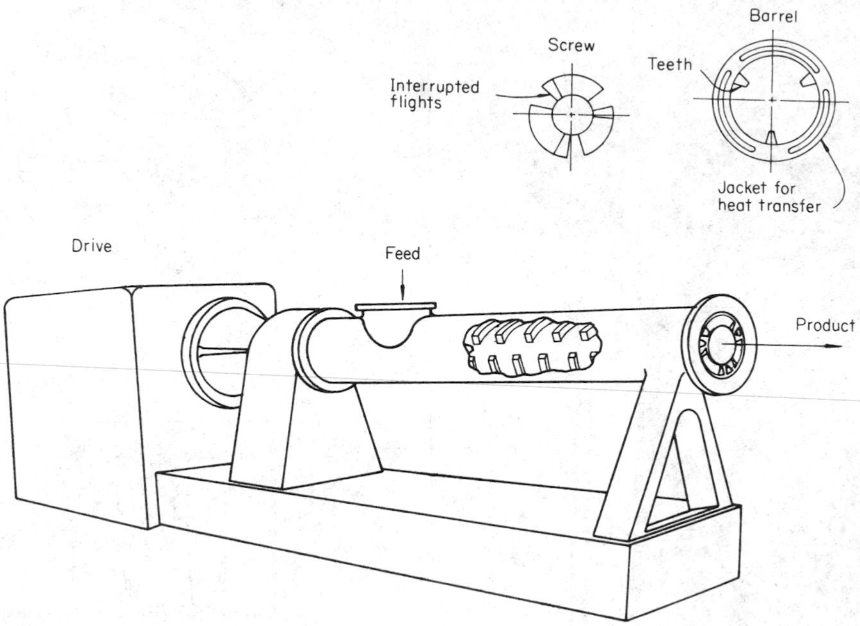

FIG. 19-26 Ko-Kneader. *(Baker Perkins Inc.)*

Baker Perkins Ko-Kneader Depicted in Fig. 19-26, this extruder is a single-screw mixer with an interrupted flight and with three rows of teeth protruding inward from the barrel wall. The screw is both rotated and reciprocated, with the stationary teeth passing through the interruptions in the thread of the screw. In essence, each tooth serves as a mixer to stir the material in the flight channel once each rotation. Thus it is possible to achieve a high degree of mixing in a relatively short retention time. Ko-Kneaders are available in nominal diameters ranging from 50 to 600 mm (2 to 24 in), with power up to 1100 kW (1500 hp). Table 19-3 shows typical uses and performance data for a 100-mm (4-in) Ko-Kneader.

Transfer-Mix This mixer (Sterling Extruder Corp.) is similar to a single-screw extruder except that both the screw and the barrel are divided into frustoconical sections, and both have helical channels. The helical channels are of opposite hand. As the screw turns, material moves forward in both helices but is also partially peeled off one into the other. This exchange circumvents the poor mixing occurring within the flights of a conventional single-screw extruder.

Baker Perkins Rotofeed This extruder (Fig. 19-27) is a light-duty mixer useful for forming pastes and slurries or for preblending doughs or resinous materials. Powdery material enters the top port, while liquid can be injected through teeth projecting into the conical screw section. The large disengaging area at the charging end makes the unit particularly effective as a continuous deaerating device.

Twin-Screw Continuous Mixers Twin-screw continuous mixers may be either tangential or intermeshing. Tangential designs permit larger shaft diameters and higher energy inputs. The blades can be run at different speeds to cause material movement from one barrel section to the other. Intermeshing screws provide the additional shear surface of blade against blade. This feature enables the blades to be self-wiping. Twin-screw machines are used for melting, mixing, coloring, and homogenizing of different polymers. Blending operations requiring incorporation of fillers, reinforcing agents, glass fibers, etc., can be carried out continuously in such mixers.

ZSK Twin-Screw Machines These mixers (Werner & Pfleiderer Corp.) are equipped with corotating screws which are individually made up of different screw and kneading elements slipped onto shafts (Fig. 19-28). The screws are self-wiping and produce positive conveyance of material. By different arrangements of the screws and kneading elements the residence-time distribution can be adjusted, and controlled pressure buildup and shear rate can be achieved. Owing to rather shallow flights, heat-exchange and devolatilizing

TABLE 19-3 Ko-Kneader Performance Data*

Compound mixed	Output		Residence time, s	Net energy input	
	kg/h	lb/h		kWh/kg	(hp·h)/lb
Carbon electrode paste	500	1100	50	0.01	0.006
Propellants	400	880	100	0.02	0.012
Kaolin clay	750	1650	60	0.10	0.061
Battery paste	1000	2200	60	0.01	0.006
Polyethylene					
Low-density	750	1650	20	0.10	0.061
High-density	600	1320	20	0.13	0.079
Polyvinyl chloride					
Flexible	1240	2730	20	0.06	0.037
Rigid	750	1650	30	0.10	0.061
Polypropylene	570	1250	20	0.13	0.079

*Provided by Baker Perkins Inc. for a 100-mm- (4-in-) diameter Ko-Kneader.

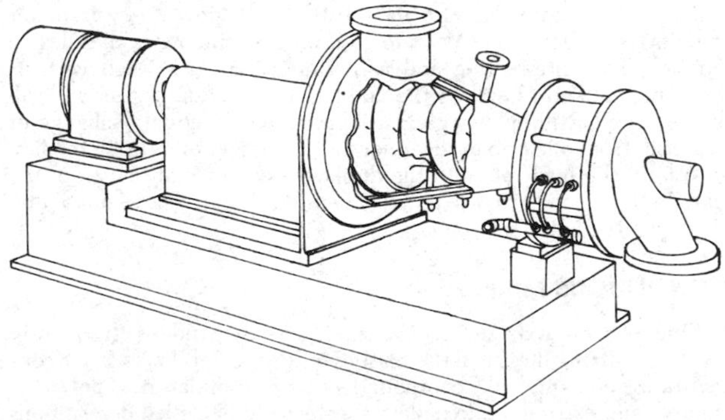

FIG. 19-27 Rotofeed mixer. *(Baker Perkins Inc.)*

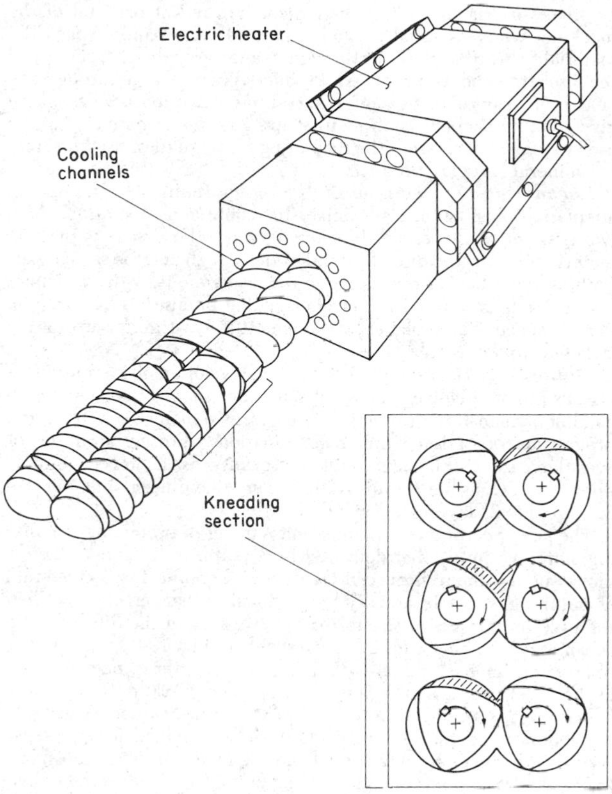

FIG. 19-28 ZSK twin-screw compounding extruder. *(Werner & Pfleiderer Corp.)*

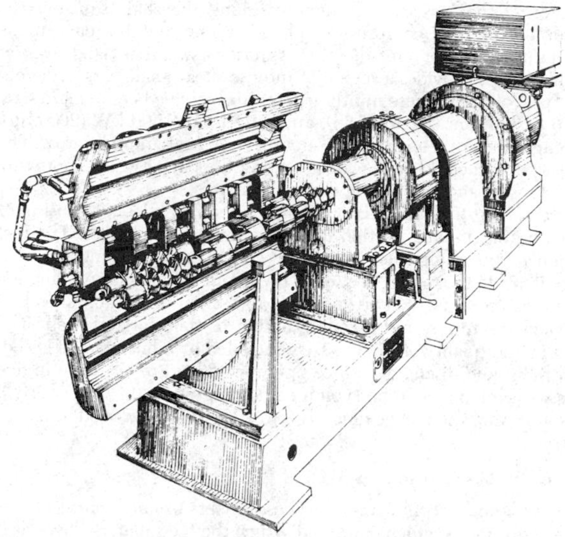

FIG. 19-29 M-P multipurpose mixer. *(Baker Perkins Inc.)*

processes can be carried out, made feasible by the continuous renewal of surfaces of the material stock. The housing of the processing section is made up of different barrel sections, which can be arranged in different numbers according to the process to be performed. The barrel sections are jacketed; they can be electrically heated, vapor-heated, or cooled by water or oil. These mixers are available with a maximum length-to-diameter ratio of 36 in sizes of 28 to 300 mm (1.1 to 12 in). They operate at speeds up to 300 r/min, requiring up to 3000 kW (4000 hp).

Multipurpose (M-P) Mixer This somewhat similar mixer (Baker Perkins Inc.) is shown in Fig. 19-29. Each pair of agitator elements causes an alternate compression and expansion twice each revolution.

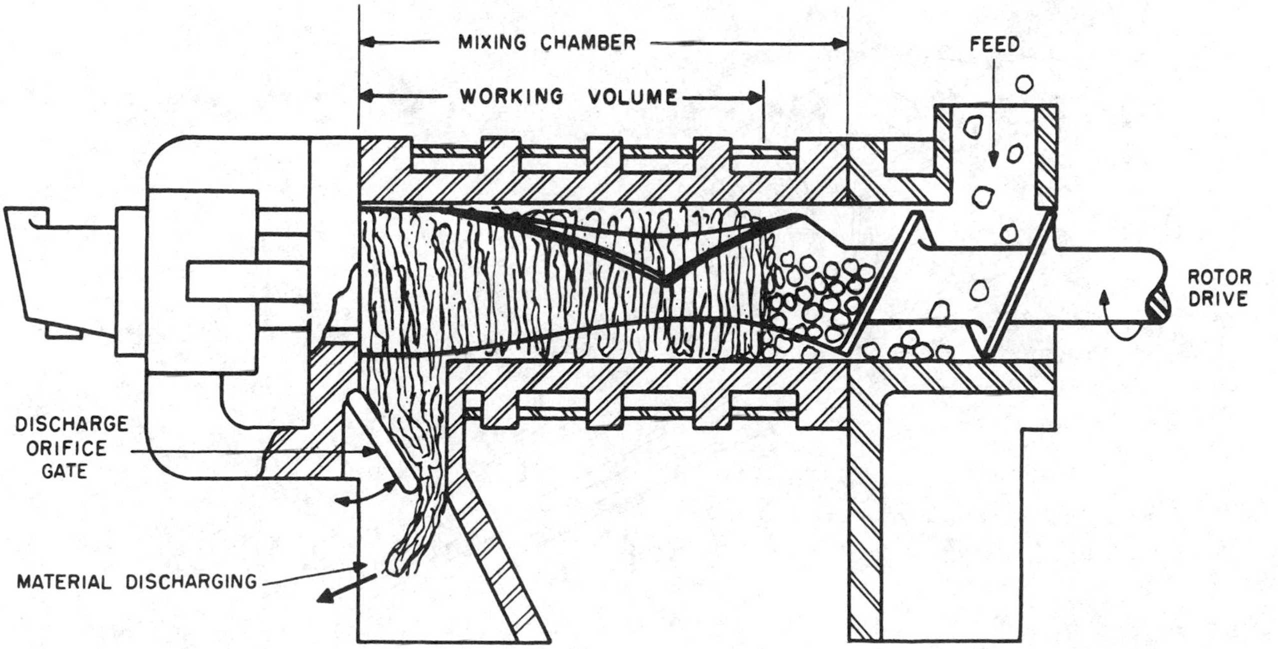

FIG. 19-30 Farrel continuous mixer. *(Farrel Co.)*

Staggering the lens-shaped elements along the shaft squeezes the material from the compression phase of one agitator pair to the expansion phase of an adjoining agitator pair. The land area at the blade tips provides a region of intense shear analogous to the nip of a two-roll mill. These multipurpose mixers have been built in sizes of 30 to 890 mm (1.2 to 35 in), applying up to 1500 kW (2000 hp). The agitators can be cored to double the heat-transfer area. The blades wipe each other as well as wiping the barrel walls, improving heat transfer and preventing any dead spots.

Farrel Continuous Mixer This mixer (Fig. 19-30) consists of rotors similar in cross section to the Banbury batch mixer. The first section of the rotor acts as a screw conveyor, propelling the feed ingredients to the mixing section. The mixing action is a combination of intensive shear, between rotor and chamber wall, kneading between the rotors, and a rolling action of the material itself. The amount and quality of mixing are controlled by adjustment of speed, feed rates, and discharge-orifice opening. Units are available in five sizes with mixing-chamber volumes ranging up to 0.12 m^3 (4.2 ft^3). At 200 r/min, the power range is 5 to 2200 kW (7 to 3000 hp).

Miscellaneous Continuous Mixers

Trough-and-Screw Mixers These mixers usually consist of single or twin rotors which continually turn the feed material over as it progresses toward the discharge end. Some have been designed with extensive heat-transfer area. The continuous-screw **Holo-Flite Processor** (Western Precipitation Division, Joy Manufacturing Company) is used primarily for heat transfer, since the hollow screws present extended surface without contributing much shear. Two or four screws may be used. Bethlehem Corp.'s **Porcupine Processor** (Fig. 19-31) also has heat-transfer media going through the flights of the rotor, but the agitator flights are cut to provide a folding action on the process mass. Breaker-bar assemblies, consisting of fingers extending toward the shaft, are frequently used to improve agitation.

Pug Mills A pug mill contains one or two shafts fitted with short, heavy paddles, mounted in a cylinder or trough which holds the material being processed. In two-shaft mills the shafts are parallel and may be horizontal or vertical. The paddles may or may not intermesh. Clearances are wide so that there is considerable mass mixing. Unmixed or partially mixed ingredients are fed at one end of the machine, which is usually totally enclosed. The paddles push the material forward as they cut through it, and carry the charge toward the discharge end as it is mixed. Product may discharge through one or two open ports or through one or more extrusion nozzles which give roughly shaped, continuous strips. Automatic cutters may be used to make blocks from the strips. Pug mills are most used for mixing mineral and clay products.

Kneadermaster This mixer (Patterson Industries, Inc.) is an adaptation of a sigma-blade mixer for continuous operation. Each two pairs of blades establish a mixing zone, the first pair pushing materials toward the discharge end of the trough and the second pair pushing them back. Forwarding to the next zone is by displacement with more feed material. Control of mixing intensity is by variation in rotor speed. Cored blades supplement the heat-transfer area of the jacketed trough.

Motionless Mixers A fairly recent development in continuous viscous mixing involves the use of stationary shaped diverters inside conduits which force the fluid media to mix themselves through a progression of divisions and recombinations, forming striations of ever-decreasing thickness until uniformity is achieved. Simple diverters, such as the **Kenics static mixer** (Chemineer, Inc.; Fig. 19-32), provide 2^n layerings per n diverters.

The power consumed by a motionless mixer in producing the mixing action is simply that delivered by a pump to the fluid which it moves against the resistance of the diverter conduit. For a given rate of pumping, it is substantially proportional to that resistance. When the diverter consists of several passageways, as in the **Sulzer static mixer** (Koch Engineering Co., Inc.) shown in Fig. 19-33, the number of layerings (hence, the rate of mixing) per diverter is increased, but at the expense of a higher pressure drop. The pressure drop, usually expressed as a multiple K of that of the empty duct, is strongly dependent upon the hydraulic radius of the divided flow passageway. The value of K, obtainable from the mixer supplier, can range from 6 to several hundred, depending on the Reynolds number and the geometry of the mixer.

Motionless mixers continuously interchange fluid elements between the walls and the center of the conduit, thereby providing enhanced heat transfer and relatively uniform residence times.

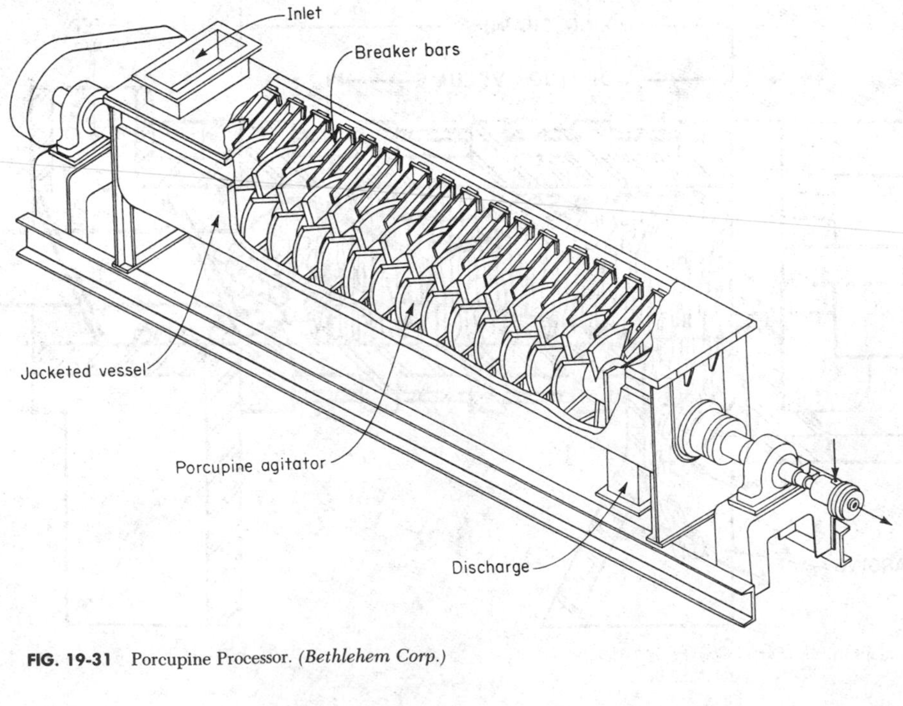

FIG. 19-31 Porcupine Processor. *(Bethlehem Corp.)*

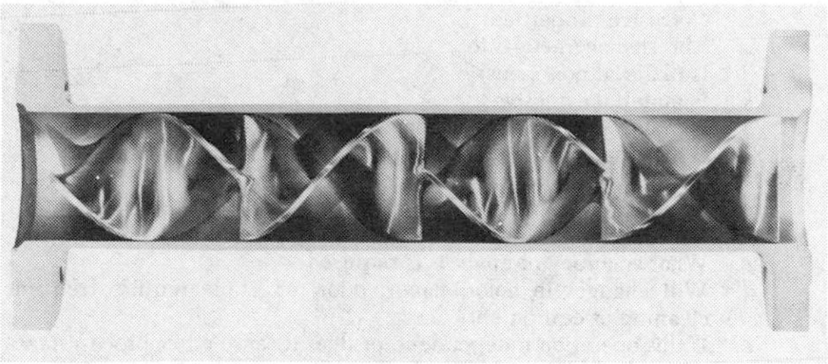

FIG. 19-32 Kenics static mixer. *(Chemineer, Inc.)*

PROCESS DESIGN CONSIDERATIONS

Scaling Up Mixing Performance

Scale-Up of Batch Mixers The prime basis of scale-up of batch mixers has been equal power per unit volume, although the most desirable practical criterion is equal blending per unit time. As size is increased, mechanical-design requirements may limit the larger mixer to lower agitator speeds; if so, blend times will be longer in the larger mixer than in the smaller prototype. If the power is high, the lower surface-to-volume ratio as size is increased may make temperature buildup a limiting factor. Since the impeller in a paste mixer generally comes close to the vessel wall, it is not possible to add cooling coils. In some instances, the impeller blades can be cored for additional heat-transfer area.

Experience with double-arm mixers indicates that power is proportional to the product of blade radius, blade-wing depth, trough length, and average of the speeds of the two blades (Irving and Saxton, loc. cit.). The mixing time scales up inversely with blade speed. Goodness of mixing is dependent primarily on the number of revolutions that the blades have made. As indicated previously, the minimum possible mixing time may become dependent on heat-transfer rate.

Frequently, the physical properties of a paste vary considerably during the mixing cycle. Even if one knew exactly how power depended upon density and viscosity, it might be better to predict the requirements for a large paste mixer from the power-time curve observed in the prototype mixer rather than to try to calculate or measure all intermediate properties during the processing sequence (i.e., the prototype mixer may be the best instrument to use to measure the effective viscosity).

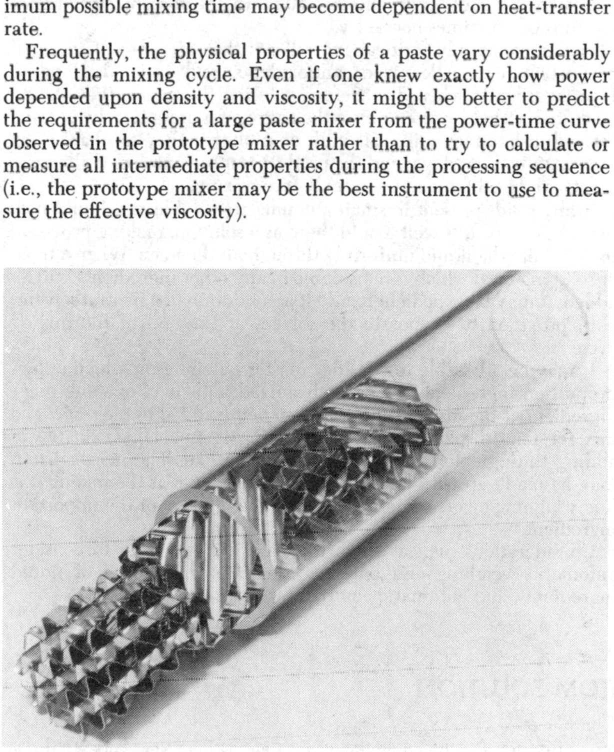

FIG. 19-33 Sulzer static mixer. *(Koch Engineering Co., Inc.)*

Scale-Up of Continuous Mixers Although scaling up on the basis of constant power per unit feed rate [kWh/kg or (hp·h)/lb] is usually a good first estimate, several other factors may have to be considered. As the equipment scale is increased, geometric similarity being at least approximated, there is a loss in surface-to-volume ratio. As size is increased, changing shear rate or length-to-diameter ratio may be required because of equipment-fabrication limitations. Furthermore, even if a reliable method of scaling up power exists, the determination of net power in small-scale test equipment is frequently difficult and inaccurate because of fairly large no-load power.

As a matter of fact, geometrical similarity usually cannot be maintained exactly as the size of the model is increased. In single-screw extruders, for example, channel depth in the flights cannot be increased in proportion to screw diameter because the distribution of heat generated by friction at the barrel wall requires more time as channel depth becomes greater. With constant retention time, therefore, nonhomogeneous product would be discharged from the scaled-up model. As the result of the departure from geometrical similarity, the throughput rate of single-screw extruders scales up with diameter to the power 2.0 to 2.5 (instead of diameter cubed) at constant length-to-diameter ratio and screw speed. The Ko-Kneader (Fig. 19-26) can be held geometrically similar, however, and its throughput rate is scaled up with the cube of diameter at constant speed.

The throughput rate of intermeshing twin-screw extruders (Fig. 19-28) and the Farrel continuous mixer (Fig. 19-30) is scaled up with diameter to about the 2.6 power. The production capacity of the M-P mixer (Fig. 19-29) is scaled up as the cube of diameter since geometry, shear rate, residence time, and power input per unit volume all can be held constant.

Residence-time distributions. For flow through a conduit, the extent of axial dispersion can be characterized either by an axial-diffusion coefficient or by analogy to a number of well-mixed stages in series. Retention time can control the performance of a mixing system. As the number of apparent stages increases, there is greater assurance that all the material will have the required residence time. Under conditions requiring uniform retention time, it is imperative that the feed streams be fed in the correct ratio on a time scale much shorter than the average residence time of the mixer; otherwise, a perturbation in the feed will produce a comparable perturbation in the product. The mixing impellers in continuous mixers can be designed to cover the full range from minimum axial mixing (plug flow) to maximum (to damp out feed irregularities). Residence-time distributions and effective Peclet numbers have been determined for a wide variety of twin-screw configurations [Todd and Irving, *Chem. Eng. Prog.*, **65**(9), 84 (1969)]. Conventional single-screw extruder mixers have Peclet numbers about equal to the length-to-diameter ratio, or an equivalent number of stages equal to one-half of that.

Heating and Cooling Mixers

Heat Transfer Pastes are often heated or cooled by heat transfer through the walls of the container or hollow mixing arms. Good agitation, a large ratio of transfer surface to mixer volume, and frequent removal of material from the surface are essential for high rates of heat transfer. Sometimes evaporation of part of the mix is used for cooling.

In most mixers, the metal wall has a negligible thermal resistance. The paste film, however, usually has high resistance. It is important, therefore, while minimizing the resistance of the heating or cooling medium, to move the paste up to and away from the smooth wall surface as steadily and rapidly as possible. This is best achieved by having the paste flow so as to follow a close-fitting scraper which wipes the film from the wall with each rotation. Typical overall heat-transfer coefficients are between 25 and 200 J/(m^2·s·K) [4 to 35 Btu/(h·ft^2·°F)].

Heating Methods The most economical heating method varies with plant location and available facilities. Direct firing is rarely used, since it does not permit good surface-temperature control and may cause scorching of the material on the vessel walls. Steam heating is the most widely used method. It is economical, safe, and easily controlled. With thin-wall mixers there must be automatic release of the vacuum that results when the pressure is reduced and the steam in the jacket condenses; otherwise, weak sections will collapse. Transfer-liquid heating using water, oil, special organic liquids, or molten inorganic salts permits good temperature control and provides insurance against overheating the processed material. Jackets for transfer-liquid heating usually must be baffled to provide good circulation. Higher temperatures can be achieved without requiring the heavy vessel construction otherwise required by steam.

Electrical heating is accomplished with resistance bands or ribbons which must be electrically insulated from the machine body but in good thermal contact with it. The heaters must be carefully spaced to avoid a succession of hot and cold areas. Sometimes they are mounted in aluminum blocks shaped to conform to the container walls. Their effective temperature range is 150 to 500°C (about 300 to 930°F). Temperature control is precise, maintenance and supervision costs are low, and conversion of electrical energy to useful heat is almost 100 percent. The cost of electrical energy is usually large, however, and may be prohibitive.

Frictional heat develops rapidly in some units such as a Banbury mixer. The first temperature rise may be beneficial in softening the materials and accelerating chemical reactions. High temperatures detrimental to the product may easily be reached, however, and provision for cooling or frequent stopping of the machine must be made. Frictional heating may be lessened by reducing the number of working elements, their area, and their speed. Cooling thus is facilitated, but at the expense of increased mixing time.

Cooling Methods In air cooling, air may be blown over the machine surfaces, the area of which is best extended with fins. Air or cooled inert gas may also be blown over the exposed surface of the mix, provided contamination or oxidation of the charge does not result. Evaporation of excess water or solvent under vacuum or at atmospheric pressure provides good cooling. A small amount of evaporation produces a large amount of cooling. Removing too much solvent, however, may damage the charge. Direct addition of ice to the mixer provides rapid, convenient cooling, but the resulting dilution of the mix must be permissible. Addition of dry ice is more expensive but results in lower temperatures, the mix is not diluted, and the CO_2 gas evolved provides a good inert atmosphere. Many mixers are cooled by circulation of water or refrigerants through jackets or hollow agitators. In general, this is the least expensive method, but it is limited by the magnitude of heat-transfer coefficient obtainable.

Selection of Equipment If a new product is being considered, the preliminary study must be highly detailed. Laboratory or pilot-plant work must be done to establish the controlling factors. The problem is then to select and install equipment which will operate for quantity production at minimum overall cost. Most equipment vendors have pilot equipment available on a rental basis or can conduct test runs in their own customer-demonstration facilities.

One approach to proper equipment specification is by analogy. What current product is most similar to the new one? How is this material produced? What difficulties are being experienced?

In other situations the following procedure is recommended:

1. List carefully all materials to be handled at the processing point and describe their characteristics, such as:
   a. How received at the processing unit: in bags, barrels, or drums, in bulk, by pipe line, etc.?
   b. Must storage and/or weighing be done at the site?
   c. Physical form.
   d. Specific gravity and bulk characteristics.
   e. Particle size or size range.
   f. Viscosity.
   g. Melting or boiling point.
   h. Corrosive properties.
   i. Abrasive characteristics.
   j. Is material poisonous?
   k. Is material explosive?
   l. Is material an irritant to skin, eyes, or lungs?
   m. Is material sensitive to exposure of air, moisture, or heat?
2. List pertinent data covering production:
   a. Quantity to be produced per 8-h shift.
   b. Formulation of finished product.
   c. What accuracy of analysis is required?
   d. Will changes in color, flavor, odor, or grade require frequent cleaning of equipment?
   e. Is this operation independent, or does it serve other process stages with which it must be synchronized?
   f. Is there a change in physical state during processing?
   g. Is there a chemical reaction? Is it endothermic or exothermic?
   h. What are the temperature requirements?
   i. What is the form of the finished product?
   j. How must the material be removed from the apparatus (by pumping, free flow through pipe or chute, dumping, etc.)?
3. Describe in detail the controlling characteristics of the finished product:
   a. Permanence of the emulsion or dispersion.
   b. Degree of blending of aggregates or of ultimate particles.
   c. Ultimate color development required.
   d. Uniformity of the dispersion of active ingredients, as in a drug product.
   e. Degree of control of moisture content for pumping extrusion, etc.

Preparation and Addition of Materials To ensure maximum production of high-grade mixed material, the preliminary preparation of the ingredients must be correct and they must be added in the proper order. There are equipment implications to these considerations.

Some finely powdered materials, such as carbon black, contain much air. If possible they should be compacted or wet out before being added to the mix. If a sufficient quantity of light solvent is a part of the formula, it may be used to wet the powder and drive out the air. If the powder cannot be wet, it may be possible to densify it somewhat by mechanical means. Removal of adsorbed gas under vacuum is sometimes necessary.

Not uncommonly, critical ingredients that are present in small proportion (e.g., vulcanizers, antioxidants, and antiacids) tend to form aggregates when dry. Before entering the mixer, they should be fluffed, either by screening if the aggregates are soft or by passage through a hammer mill, roll mill, or muller if they are hard. The mixing time is cut down and the product is more uniform if all ingredients are freed from aggregates before mixing.

If any solids present in small amounts are soluble in a liquid portion of the mix, it is well to add them as a solution, making provision to distribute the liquid uniformly throughout the mass. When a trace of solid material which is not soluble in any other ingredients is to be added, it may be expedient to add it as a solution in a neutral solvent, with provision to evaporate the solvent at the end of the mixing cycle.

It may be advisable to consider master batching, in which a low-proportion ingredient is separately mixed with part of some other ingredient of the mix, this premix then being added to the rest of the mix for final dispersion. Master batching is especially valuable in adding tinting colors, antioxidants, and the like. The master batch may be made up with laboratory accuracy, while at the mixing station weighing errors are minimized by the dilution of the important ingredient.

Considerations such as these may make it desirable to consider automatic weighing and batch accumulation, metering of liquid ingredients, and automatic control of various time cycles.

CRYSTALLIZATION FROM SOLUTION

General References: *AIChE Testing Procedures: Crystallizers*, American Institute of Chemical Engineers, New York, 1970; *Evaporators*, 1961. Bennett, *Chem. Eng. Prog.*, 58(9), 76 (1962). Buckley, *Crystal Growth*, Wiley, New York, 1951. Campbell and Smith, *Phase Rule*, Dover, New York, 1951. De Jong and Jancic (eds.), *Industrial Crystallization*, North-Holland Publishing Company, Amsterdam, 1979. Faraday Society, *Crystal Growth*, Butterworth,

London, 1949. Larson (ed.), "Crystallization from Solution: Factors Influencing Size Distribution," *Chem. Eng. Prog. Symp. Ser.*, **67**(110), (1971). Mullin (ed.), *Industrial Crystallization*, Plenum, New York, 1976. Mullin, in *Kirk-Othmer Encyclopedia of Chemical Technology*, 3d ed., vol. 7, Wiley, New York, 1979, p. 243. Newman and Bennett, *Chem. Eng. Prog.*, **55**(3), 65 (1959). Palermo and Larson (eds.), "Crystallization from Solutions and Melts," *Chem. Eng. Prog. Symp. Ser.*, **65**(95), (1969). Randolph (ed.), "Design, Control and Analysis of Crystallization Processes," *Am. Inst. Chem. Eng. Symp. Ser.*, **76**(193), (1980). Randolph and Larson, *Theory of Particulate Processes*, Academic, New York, 1971. Reed, *Dislocation in Crystals*, McGraw-Hill, New York, 1953. Rousseau and Larson (eds.), "Analysis and Design of Crystallization Processes," *Am. Inst. Chem. Eng. Symp. Ser.*, **72**(153), (1976). Seidell, *Solubilities of Inorganic and Metal Organic Compounds*, American Chemical Society, Washington, 1965. Van Hook, *Crystallization*, Reinhold, New York, 1961.

Crystallization is important as an industrial process because of the number of materials that are and can be marketed in the form of crystals. Its wide use is probably due to the highly purified and attractive form of a chemical solid which can be obtained from relatively impure solutions in a single processing step. In terms of energy requirements, crystallization requires much less energy for separation than do distillation and other commonly used methods of purification. In addition, it can be performed at relatively low temperatures and on a scale which varies from a few grams up to thousands of tons per day.

Crystallization may be carried out from a vapor, from a melt, or from a solution. Most of the industrial applications of the operation involve crystallization from solutions. Nevertheless, crystal solidification of metals is basically a crystallization process, and much theory has been developed in relation to metal crystallization. This topic is so specialized, however, that it is outside the scope of this subsection, which is limited to crystallization from solution.

PRINCIPLES OF CRYSTALLIZATION

Crystals A crystal may be defined as a solid composed of atoms arranged in an orderly, repetitive array. The interatomic distances in a crystal of any definite material are constant and are characteristic of that material. Because the pattern or arrangement of the atoms is repeated in all directions, there are definite restrictions on the kinds of symmetry that crystals can possess.

There are five main types of crystals, and these types have been arranged into seven crystallographic systems based on the crystal interfacial angles and the relative length of its axes. The treatment of the description and arrangement of the atomic structure of crystals is the science of **crystallography.** The material in this discussion will be limited to a treatment of the growth and production of crystals as a unit operation.

Solubility and Phase Diagrams Equilibrium relations for crystallization systems are expressed in the form of solubility data which are plotted as phase diagrams or solubility curves. Solubility data are ordinarily given as parts by weight of anhydrous material per 100 parts by weight of total solvent. In some cases these data are reported as parts by weight of anhydrous material per 100 parts of solution. If water of crystallization is present in the crystals, this is indicated as a separate phase. The concentration is normally plotted as a function of temperature and has no general shape or slope. It can also be reported as a function of pressure, but for most materials the change in solubility with change in pressure is very small. If there are two components in solution, it is common to plot the concentration of these two components on the *X* and *Y* axes and represent the solubility by isotherms. When three or more components are present, there are various techniques for depicting the solubility and phase relations in both three-dimension and two-dimension models. For a description of these techniques, refer to Campbell and Smith (loc. cit.). Shown in Fig. 19-34 is a phase diagram for magnesium sulfate in water. The line *p–a* represents the freezing points of ice (water) from solutions of magnesium sulfate. Point *a* is the eutectic, and the line *a–b–c–d–q* is the solubility curve of the various hydrates. Line *a–b* is the solubility curve for $MgSO_4 \cdot 12H_2O$, *b–c* is the solubility curve for $MgSO_4 \cdot 7H_2O$, *c–d* is the solubility curve for $MgSO_4 \cdot 6H_2O$, and *d–q* is the portion of the solubility curve for $MgSO_4 \cdot H_2O$.

As shown in Fig. 19-35, the mutual solubility of two salts can be

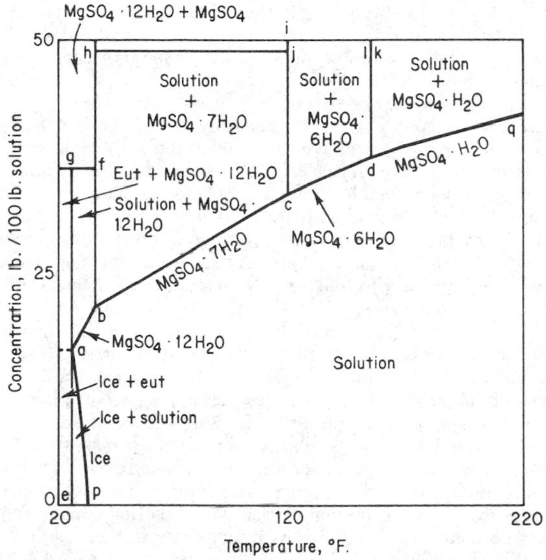

FIG. 19-34 Phase diagram, $MgSO_4 \cdot H_2O$. To convert pounds to kilograms, divide by 2.2; K = (°F + 459.7)/1.8.

plotted on the *X* and *Y* axes with temperatures as isotherm lines. In the example shown, all the solution compositions corresponding to 100°C with solid-phase sodium chloride present are shown on the line *DE*. All the solution compositions at equilibrium with solid-phase KCl at 100°C are shown by the line *EF*. If both solid-phase KCl and NaCl are present, the solution composition at equilibrium can only be represented by point *E*, which is the invariant point (at constant pressure). Connecting all the invariant points results in the mixed-salt line. The locus of this line is an important consideration in making phase separations.

There are numerous solubility data in the literature; the standard reference is by Seidell (loc. cit.). Valuable as they are, they nevertheless must be used with caution because the solubility of compounds is often influenced by pH and/or the presence of other soluble impurities which usually tend to depress the solubility of the major constituents. While exact values for any system are frequently best determined by actual composition measurements, the difficulty of reproducing these solubility diagrams should not be underestimated. To obtain data which are readily reproducible, elaborate pains must be taken to be sure the system sampled is at equilibrium, and often this means holding a sample at constant temperature for a

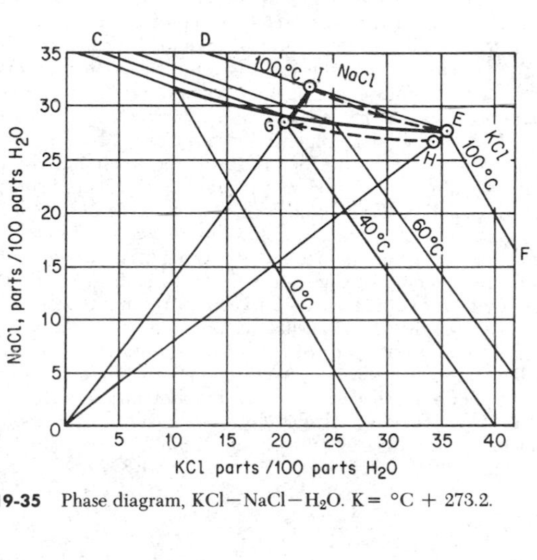

FIG. 19-35 Phase diagram, KCl—NaCl—H_2O. K = °C + 273.2.

period of from 1 to 100 h. While the published curves may not be exact for actual solutions of interest, they generally will be indicative of the shape of the solubility curve and will show the presence of hydrates or double salts.

Heat Effects in a Crystallization Process The heat effects in a crystallization process can be computed by two methods: (1) a heat balance can be made in which individual heat effects such as sensible heats, latent heats, and the heat of crystallization can be combined into an equation for total heat effects; or (2) an enthalpy balance can be made in which the total enthalpy of all leaving streams minus the total enthalpy of all entering streams is equal to the heat absorbed from external sources by the process. In using the heat-balance method, it is necessary to make a corresponding mass balance, since the heat effects are related to the quantities of solids produced through the heat of crystallization. The advantage of the enthalpy-concentration-diagram method is that both heat and mass effects are taken into account simultaneously. This method has limited use because of the difficulty in obtaining enthalpy-concentration data. This information has been published for only a few systems.

With compounds whose solubility increases with increasing temperature there is an absorption of heat when the compound dissolves. In compounds with decreasing solubility as the temperature increases, there is an evolution of heat when solution occurs. When there is no change in solubility with temperature, there is no heat effect. The solubility curve will be continuous as long as the solid substance of a given phase is in contact with the solution, and any sudden change in the slope of the curve will be accompanied by a change in the heat of solution and a change in the solid phase. Heats of solution are generally reported as the change in enthalpy associated with the dissolution of a large quantity of solute in an excess of pure solvent. Tables showing the heats of solution for various compounds are given in Sec. 3.

At equilibrium the heat of crystallization is equal and opposite in sign to the heat of solution. Using the heat of solution at infinite dilution as equal but opposite in sign to the heat of crystallization is equivalent to neglecting the heat of dilution. With many materials the heat of dilution is small in comparison with the heat of solution and the approximation is justified; however, there are exceptions. Relatively large heat effects are usually found in the crystallization of hydrated salts. In such cases the total heat released by this effect may be a substantial portion of the total heat effects in a cooling-type crystallizer. In evaporative-type crystallizers the heat of crystallization is usually negligible when compared with the heat of vaporizing the solvent.

Yield of a Crystallization Process In most cases the process of crystallization is slow, and the final mother liquor is in contact with a sufficiently large crystal surface so that the concentration of the mother liquor is substantially that of a saturated solution at the final temperature in the process. In such cases it is normal to calculate the yield from the initial solution composition and the solubility of the material at the final temperature. If evaporative crystallization is involved, the solvent removed must be taken into account in determining the final yield. If the crystals removed from solution are hydrated, account must be taken of the water of crystallization in the crystals, since this water is not available for retaining the solute in solution. The yield is also influenced in most plants by the removal of some mother liquor with the crystals being separated from the process. Typically, with a product separated on a centrifuge or filter, the adhering mother liquor would be in the range of 2 to 10 percent of the weight of the crystals.

The actual yield may be obtained from algebraic calculations or trial-and-error calculations when the heat effects in the process and any resultant evaporation are used to correct the initial assumptions on calculated yield. When calculations are made by hand, it is generally preferable to use the trial-and-error system, since it permits easy adjustments for relatively small deviations found in practice, such as the addition of wash water, or instrument and purge water additions. The following calculations are typical of an evaporative crystallizer precipitating a hydrated salt. If SI units are desired, kilograms = pounds × 0.454; K = (°F + 459.7)/1.8.

Example 1 A 10,000-lb batch of a 32.5 percent $MgSO_4$ solution at 120°F is cooled without appreciable evaporation to 70°F. What weight of $MgSO_4 \cdot 7H_2O$ crystals will be formed (if it is assumed that the mother liquor leaving is saturated)?

From the solubility diagram in Fig. 19-34, at 70°F the concentration of solids is 26.3 lb $MgSO_4$ per 100-lb solution.

The mole weight of $MgSO_4$ is 120.38.

The mole weight of $MgSO_4 \cdot 7H_2O$ is 246.49.

For calculations involving hydrated salts, it is convenient to make the calculations based on the hydrated solute and the "free water."

$$0.325 \text{ weight fraction} \times \frac{246.94}{120.38} = 0.662\ MgSO_4 \cdot 7H_2O \text{ in the feed solution}$$

$$0.263 \times \frac{246.94}{120.38} = 0.538\ MgSO_4 \cdot 7H_2O \text{ in the mother liquor}$$

Since the free water remains constant (except when there is evaporation), the final amount of soluble $MgSO_4 \cdot 7H_2O$ is calculated by the ratio of $\frac{0.538 \text{ lb } MgSO_4 \cdot 7H_2O}{(1 - 0.538) \text{ lb free water}}$

	Total	$MgSO_4 \cdot 7H_2O$	Free water	$MgSO_4 \cdot 7H_2O$ / Free water
Feed	10,000	6620	3380	0.662/0.338
Mother liquor	7,280	3900*	3380	0.538/0.562
Yield	2,720	2720		

*3380 × (0.538/0.462) = 3900.

A formula method for calculation is sometimes used where

$$P = R\frac{100W_0 - S(H_0 - E)}{100 - S(R - 1)}$$

P = weight of crystals in final magma, lb
R = mole weight of hydrate/mole weight of anhydrous = 2.04
S = solubility at mother-liquor temperature (anhydrous basis) in lb per 100 lb solvent. [0.263/(1 − 0.263)] × 100 = 35.7
W_0 = weight of anhydrous solute in the original batch. 10,000(0.325) = 3250 lb
H_0 = total weight of solvent at the beginning of the batch. 10,000 − 3250 = 6750 lb
E = evaporation = 0

$$P = 2.04\frac{(100)(3250) - 35.7(6750)}{100 - 35.7(2.04 - 1)} = 2700 \text{ lb}$$

Note that taking the difference between large numbers in this method can increase the chance for error.

Fractional Crystallization When two or more solutes are dissolved in a solvent, it is often possible to (1) separate these into the pure components or (2) separate one and leave the other in the solution. Whether or not this can be done depends on the solubility and phase relations of the system under consideration. Normally alternative 2 is successful only when one of the components has a much more rapid change in solubility with temperature than does the other. A typical example which is practiced on a large scale is the separation of KCl and NaCl from water solution. A phase diagram for this system is shown in Fig. 19-35. In this case the solubility of NaCl is plotted on the Y axis in parts per 100 parts of water, and the solubility of KCl is plotted on the X axis. The isotherms show a marked decrease in solubility for each component as the amount of the other is increased. This is typical for most inorganic salts. As explained earlier, the mixed-salt line is *CE*, and to make a separation of the solutes into the pure components it is necessary to be on one side of this line or the other. Normally a 95 to 98 percent approach to this line is possible. When evaporation occurs during a cooling or concentration process, this can be represented by movement away from the origin on a straight line through the origin. Dilution by water is represented by movement in the opposite direction.

A typical separation might be represented as follows: Starting at *E* with a saturated brine at 100°C a small amount of water is added to dissolve any traces of solid phase present and to make sure the solids precipitated initially are KCl. Evaporative cooling along line *HG* results in the precipitation of KCl. During this evaporative cooling, part of the water evaporated must be added back to the solution to prevent the coprecipitation of NaCl. The final composition at *G* can be calculated by the $NaCl/KCl/H_2O$ ratios and the known amount of NaCl in the incoming solution at *E*. The solution at point *G* may be concentrated by evaporation at 100°C. During this process the solution will increase in concentration with respect to both components until point *I* is reached. Then NaCl will precipitate, and the solution will become more concentrated in KCl, as indicated by the line *IE*, until the original point *E* is reached. If concentration is carried beyond point *E*, a mixture of KCl and NaCl will precipitate.

Example 2 Starting with 1000 lb of water in a solution at *H* on the solubility diagram in Fig. 19-35, calculate the yield on evaporative cooling and concentrate the solution back to point *H* so the cycle can be repeated, indicating the amount of NaCl precipitated and the evaporation and dilution required at the different steps in the process.

In solving problems of this type, it is convenient to list the material balance and the solubility ratios. The various points on the material balance are calculated by multiplying the quantity of the component which does not precipitate from solution during the transition from one point to another (normally the NaCl in cooling or the KCl in the evaporative step) by the solubility ratio at the next step, illustrated as follows:

Basis. 1000 lb of water at the initial conditions.

				Solubility ratios		
Solution component	KCl	NaCl	Water	KCl	NaCl	Water
H	343	270	1000	34.3	27.0	100
G(*a*)	194	270	950	20.4	28.4	100
KCl yield	149					
Net evaporation	. . .	. . .	50			
I(*b*)	194	270	860	22.6	31.4	100
E(*c*)	194	153	554	35.0	27.5	100
NaCl yield	. . .	117				
Evaporation	. . .	. . .	306			
Dilution	. . .	. . .	11			
H'	194	153	565	34.3	27.0	-100

The calculations for these steps are:

a. 270 lb NaCl (100 lb water/28.4 lb NaCl) = 950 lb water
 950 lb water (20.4 lb KCl/100 lb water) = 194 lb KCl

b. 270 lb NaCl (100 lb water/31.4 lb NaCl) = 860 lb water
 860 lb water (22.6 lb KCl/100 lb water) = 194 lb KCl

c. 194 lb KCl (100 lb water/35.0 lb KCl = 554 lb water
 554 lb water (27.5 lb NaCl/100 lb water) = 153 lb NaCl

Note that during the cooling step the maximum amount of evaporation which is permitted by the material balance is 50 lb for the step shown. In an evaporative-cooling step, however, the actual evaporation which results from adiabatic cooling is more than this. Therefore, water must be added back to prevent the NaCl concentration from rising too high; otherwise, coprecipitation of NaCl will occur.

Inasmuch as only mass ratios are involved in these calculations, kilograms or any other unit of mass may be substituted for pounds without affecting the validity of the example.

Although the figures given are for a step-by-step process, it is obvious that the same techniques will apply to a continuous system if the fresh feed containing KCl and NaCl is added at an appropriate part of the cycle, such as between steps *G* and *I* for the case of dilute feed solutions.

Another method of fractional crystallization, in which advantage is taken of different crystallization rates, is sometimes used. Thus, a solution saturated with borax and potassium chloride will, in the absence of borax seed crystals, precipitate only potassium chloride on rapid cooling. The borax remains behind as a supersaturated solution, and the potassium chloride crystals can be removed before the slower borax crystallization starts.

Crystal Formation There are obviously two steps involved in the preparation of crystal matter from a solution. The crystals must first form and then grow. The formation of a new solid phase either on an inert particle in the solution or in the solution itself is called **nucleation.** The increase in size of this nucleus with a layer-by-layer addition of solute is called **growth.** Both nucleation and crystal growth have supersaturation as a common driving force. Unless a solution is supersaturated, crystals can neither form nor grow. Supersaturation refers to the quantity of solute present in solution compared with the quantity which would be present if the solution were kept for a very long period of time with solid phase in contact with the solution. The latter value is the equilibrium solubility at the temperature and pressure under consideration. The supersaturation coefficient can be expressed

$$S = \frac{\text{parts solute/100 parts solvent}}{\text{parts solute at equilibrium/100 parts solvent}} \geqq 1.0 \qquad (19\text{-}10)$$

Solutions vary greatly in their ability to sustain measurable amounts of supersaturation. With some materials, such as sucrose, it is possible to develop a supersaturation coefficient of 1.4 to 2.0 with little danger of nucleation. With some common inorganic solutions such as sodium chloride in water, the amount of supersaturation which can be generated stably is so small that it is difficult or impossible to measure.

Certain qualitative facts in connection with supersaturation, growth, and the yield in a crystallization process are readily apparent. If the concentration of the initial solution and the final mother liquor are fixed, the total weight of the crystalline crop is also fixed if equilibrium is obtained. The particle-size distribution of this weight, however, will depend on the relationship between the two processes of nucleation and growth. Considering a given quantity of solution cooled through a fixed range, if there is considerable nucleation initially during the cooling process, the yield will consist of many small crystals. If only a few nuclei form at the start of the precipitation and the resulting yield occurs uniformly on these nuclei without secondary nucleation, a crop of large uniform crystals will result. Obviously, many intermediate cases of varying nucleation rates and growth rates can also occur, depending on the nature of the materials being handled, the rate of cooling, agitation, and other factors.

When a process is continuous, nucleation frequently occurs in the presence of a seeded solution by the combined effects of mechanical stimulus and nucleation caused by supersaturation (heterogeneous nucleation). If such a system is completely and uniformly mixed (i.e., the product stream represents the typical magma circulated within the system) and if the system is operating at steady state, the particle-size distribution has definite limits which can be predicted mathematically with a high degree of accuracy, as will be shown later in this section.

Geometry of Crystal Growth Geometrically a crystal is a solid bounded by planes. The shape and size of such a solid are functions of the interfacial angles and of the linear dimension of the faces. As the result of the constancy of its interfacial angles, each face of a growing or dissolving crystal, as it moves away from or toward the center of the crystal, is always parallel to its original position. This concept is known as the "principle of the parallel displacement of faces." The rate at which a face moves in a direction perpendicular to its original position is called the translation velocity of that face or the rate of growth of that face.

From the industrial point of view, the term "crystal habit" refers to the relative sizes of the faces of a crystal. No general law controlling crystal habit has been discovered. Nevertheless the habit of various products is of great importance. Long, needlelike crystals tend to be easily broken during centrifugation and drying. Flat, platelike crystals are very difficult to wash during filtration or centrifugation and result in relatively low filtration rates. Complex or twinned crystals tend to be more easily broken in transport than chunky, compact crystal habits. Spherical crystals (caused generally by attrition during growth) tend to give considerably less difficulty with caking than do cubical or other compact sizes.

Very small amounts of foreign substances will often completely change the crystal habit. The selective adsorption of dyes by different faces of a crystal or the change from an alkaline to an acidic environment will often produce pronounced changes in the crystal habit. The presence of other soluble anions and cations often has a similar influence. In the crystallization of ammonium sulfate the reduction in soluble iron to below 50 ppm of ferric ion is sufficient to cause a significant change in the habit of an ammonium sulfate crystal from a long, narrow form to a relatively chunky and compact form. Buckley (loc. cit.) lists additives having an influence on various crystallization systems. Additional information is available in the patent literature. Table 19-4 lists some of the better-known additives and their effects.

Since the relative sizes of the individual faces of a crystal vary between wide limits, it follows that different faces must have different translational velocities. A geometric law of crystal growth known as the **overlapping principle** is based on those velocity differences: in growing a crystal, only those faces having the lowest translational velocities survive; and in dissolving a crystal, only those faces having the highest translational velocities survive.

For example, consider the cross sections of a growing crystal as in Fig. 19-36. The polygons shown in the figure represent varying stages in the growth of the crystal. The faces marked *A* are slow-growing faces (low translational velocities), and the faces marked *B* are fast-growing (high translational velocities). It is apparent from Fig. 19-36 that the faster *B* faces tend to disappear as they are overlapped by the slower *A* faces.

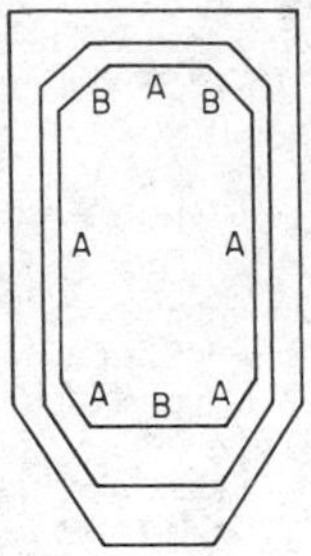

FIG. 19-36 Overlapping principle.

Purity of the Product If a crystal is produced in a region of the phase diagram where a single-crystal composition precipitates, the crystal itself will normally be pure provided that it is grown at relatively low rates and constant conditions. With many products these purities approach a value of about 99.5 to 99.8 percent. The difference between this and a purity of 100 percent is generally the result of small pockets of mother liquor called occlusions trapped within the crystal. Although frequently large enough to be seen with an ordinary microscope, these occlusions can be submicroscopic and represent dislocations within the structure of the crystal. They can be caused by either attrition or breakage during the growth process or by slip planes within the crystal structure caused by interference between screw-type dislocations and the remainder of the crystal faces. To increase the purity of the crystal beyond the point where such occlusions are normally expected (about 0.1 to 0.5 percent by volume), it is generally necessary to reduce the impurities in the mother liquor itself to an acceptably low level so that the mother liquor contained within these occlusions will not contain sufficient impurities to cause an impure product to be formed. It is normally necessary to recrystallize material from a solution which is relatively pure to surmount this type of purity problem.

In addition to the impurities within the crystal structure itself, there is normally an adhering mother-liquid film left on the surface of the crystal after separation in a centrifuge or on a filter. Typically a centrifuge may leave about 2 to 10 percent of the weight of the crystals as adhering mother liquor on the surface. This varies greatly with the size and shape or habit of the crystals. Large, uniform crystals precipitated from low-viscosity mother liquors will retain a min-

TABLE 19-4 Some Impurities Known to Be Habit Modifiers

Material crystallized	Additive or additives	Effect	Concentration	References
$Ba(NO_2)^2$	Mg, Te^{+4}	Help growth	....	1
$LiCl \cdot H_2O$	Cr, Mn^{+2}, Sn^{+2}, Co, Ni, Fe^{+3}	Help growth	Small	1
NaCl	Pb, Mn^{+2}, Bi, Sn^{+2}, Ti, Fe, Hg	Help growth	Small	1
	Urea	Forms octahedra	Small	2
	Tetraalkyl ammonium salts	Help growth and hardness	1–100 ppm	U.S. Patent 3,095,281
	Polyethylene-oxy compounds	Help growth and hardness	....	U.S. Patent 3,000,708
$NaClO_3$	Na_2SO_4, $NaClO_4$	Tetrahedrons	....	3
$Na_2CO_3 \cdot H_2O$	$SO_4^=$	Reduces L/D ratio	0.1–1.0%	Canadian Patent 812,685
	Ca^{+2} and Mg^{+2}	Increase bulk density	400 ppm	U.S. Patent 3,459,497
$Na_2B_4O_7$	Casein, gelatin	Flat crystals	....	2
Na_2SO_4	Alkyl aryl sulfonates	Aid growth	....	2
NH_4Cl	Mn, Fe, Cu, Co, Ni, Cr	Aid growth	Small	1
$(NH_4)_2HPO_4$	H_2SO_4	Reduces L/D ratio	7%	
$(NH_4)_2SO_4$	Cr^{+3}, Fe^{+3}, Al^{+3}	Needles	50 ppm	
	H_2SO_4	Needles	2–6%	
	Oxalic acid, citric acid	Chunky crystals	1000 ppm	U.S. Patent 2,092,073
	H_3PO_4, SO_2	Chunky crystals	1000 ppm	U.S. Patent 2,228,742
$MgSO_4 \cdot 7H_2O$	$Na_2B_4O_7$	Aids growth	5%	1
$NaCO_3 \cdot NaHCO_3 \cdot 2H_2O$	D-40 detergent	Aids growth	20 ppm	U.S. Patent 3,233,983
KH_2PO_4	$Na_2B_4O_7$	Aids growth	....	1
$NH_4H_2PO_4$	Fe^{+3}, Cr, Al, Sn	Help growth	Traces	1
NF_4F	Ca	Helps growth	Small	1
$ZnSO_4 \cdot 7H_2O$	Borax	Aids growth	....	1
Adipic acid	Surfactant-SDBS	Aids growth	50–100 ppm	2
Pentaerythritol	Sucrose	Aids growth	....	1
	Acetone solvent	Forms plates	....	2
Urea	Biuret	Reduces L/D and aids growth	2–7%	
	NH_4Cl	Reduces L/D and aids growth	5–10%	
Naphthalene	From cyclohexane	Forms needles	....	2
	From methanol	Forms plates		
KCl	Pb, Bi, Sn^{+2}, Ti, Zr, Th, Cd, Fe, Hg, Mg	Help growth	Small	1
KNO_3	Pb, Th, Bi	Help growth	Small	1
KNO_2	Fe	Helps growth	Small	1
K_2SO_4	Cl, Mn, Fe, Fe, Cu, Al, Mg, Bi	Help growth	Small	1

1. Gillman, *The Art and Science of Growing Crystals*, Wiley, New York, 1963.
2. Mullin, *Crystallization*, Butterworth, London, 1961.
3. Buckley, *Crystal Growth*, Wiley, New York, 1951.

imum of mother liquor, while nonuniform or small crystals precipitated from viscous solutions will retain a considerably larger proportion. Comparable statements apply to the filtration of crystals, although normally the amounts of mother liquor adhering to the crystals are considerably larger. It is common practice when precipitating materials from solutions which contain appreciable quantities of impurities to wash the crystals on the centrifuge or filter with either fresh solvent or feed solution. In principle, such washing can reduce the impurities quite substantially. It is also possible in many cases to reslurry the crystals in fresh solvent and recentrifuge the product in an effort to obtain a longer residence time during the washing operation and better mixing of the wash liquors with the crystals.

Coefficient of Variation One of the problems confronting any user or designer of crystallization equipment is the expected particle-size distribution of the solids leaving the system and how this distribution may be adequately described. Most crystalline-product distributions plotted on arithmetic-probability paper will exhibit a straight line for a considerable portion of the plotted distribution. In this type of plot the particle diameter should be plotted as the ordinate and the cumulative percent on the log-probability scale as the abscissa.

It is common practice to use a parameter characterizing crystal-size distribution called the coefficient of variation. This is defined as follows:

$$CV = 100 \frac{PD_{16\%} - PD_{84\%}}{2PD_{50\%}} \tag{19-11}$$

where CV = coefficient of variation, as a percentage
PD = particle diameter from intercept on ordinate axis at percent indicated

In order to be consistent with normal usage, the particle-size distribution when this parameter is used should be a straight line between approximately 10 percent cumulative weight and 90 percent cumulative weight. By giving the coefficient of variation and the mean particle diameter, a description of the particle-size distribution is obtained which is normally satisfactory for most industrial purposes. If the product is removed from a mixed-suspension crystallizer, this coefficient of variation should have a value of approximately 50 percent (Randolph and Larson, op. cit., chap. 2).

Crystal Nucleation and Growth

Rate of Growth Crystal growth is a layer-by-layer process, and since growth can occur only at the face of the crystal, material must be transported to that face from the bulk of the solution. Diffusional resistance to the movement of molecules (or ions) to the growing crystal face, as well as the resistance to integration of those molecules into the face, must be considered. As discussed earlier, different faces can have different rates of growth, and these can be selectively altered by the addition or elimination of impurities.

If L is a characteristic dimension of a crystal of selected material and shape, the rate of growth of a crystal face that is perpendicular to L is, by definition,

$$G \equiv \lim_{\Delta L \to O} \frac{\Delta L}{\Delta t} = \frac{dL}{dt} \tag{19-12}$$

where G is the growth rate over time internal t. It is customary to measure G in the practical units of millimeters per hour. It should be noted that growth rates so measured are actually twice the facial growth rate.

The delta L law. It has been shown by McCabe [*Ind. Eng. Chem.*, **21**, 30, 112 (1929)] that all geometrically similar crystals of the same material suspended in the same solution grow at the same rate if growth rate is defined as in Eq. (19-12). The rate is independent of crystal size, provided that all crystals in the suspension are treated alike. This generalization is known as the delta L law. Although there are some well-known exceptions, they usually occur when the crystals are very large or when movement of the crystals in the solution is so rapid that substantial changes occur in diffusion-limited growth of the faces.

It is emphasized that the delta L law does not apply when similar crystals are given preferential treatment based on size. It fails also when surface defects or dislocations significantly alter the growth rate of a crystal face. Nevertheless, it is a reasonably accurate generalization for a surprising number of industrial cases. When it is, it is important because it simplifies the mathematical treatment in modeling real crystallizers and is useful in predicting crystal-size distribution in many types of industrial crystallization equipment.

Important exceptions to McCabe's growth-rate model have been noted by Bramson, by Randolph, and by Abegg. These are discussed by Canning and Randolph, *Am. Inst. Chem. Eng. J.*, **13**, 5 (1967).

Nucleation The mechanism of crystal nucleation from solution has been studied by many scientists, and recent work suggests that—in commercial crystallization equipment, at least—the nucleation rate is the sum of contributions by (1) homogeneous nucleation and (2) nucleation due to contact between crystals and (*a*) other crystals, (*b*) the walls of the container, and (*c*) the pump impeller. If B^0 is the net number of new crystals formed in a unit volume of solution per unit of time,

$$B^0 = B_{ss} + B_e + B_c \tag{19-13}$$

where B_e is the rate of nucleation due to crystal-impeller contacts, B_c is that due to crystal-crystal contacts, and B_{ss} is the homogeneous nucleation rate due to the supersaturation driving force. The mechanism of the last-named is not precisely known, although it is obvious that molecules forming a nucleus not only have to coagulate, resisting the tendency to redissolve, but also must become oriented into a fixed lattice. The number of molecules required to form a stable crystal nucleus has been variously estimated at from 80 to 100 (with ice), and the probability that a stable nucleus will result from the simultaneous collision of that large number is extremely low unless the supersaturation level is very high or the solution is supersaturated in the absence of agitation. In commercial crystallization equipment, in which supersaturation is low and agitation is employed to keep the growing crystals suspended, the predominant mechanism is contact nucleation or, in extreme cases, attrition.

In order to treat crystallization systems both dynamically and continuously, a mathematical model has been developed which can correlate the nucleation rate to the level of supersaturation and/or the growth rate. Because the growth rate is more easily determined and because nucleation is sharply nonlinear in the regions normally encountered in industrial crystallization, it has been common to assume

$$B^0 = ks^i \tag{19-14}$$

where s, the supersaturation, is defined as $(C - C_s)$, C being the concentration of the solute and C_s its saturation concentration; and the exponent i and dimensional coefficient k are values characteristic of the material.

While Eq. (19-14) has been popular among those attempting correlations between nucleation rate and supersaturation, recently it has become commoner to use a derived relationship between nucleation rate and growth rate by assuming that

$$G = k's \tag{19-15}$$

whence, in consideration of Eq. (19-14),

$$B^0 = k''G^i \tag{19-16}$$

where the dimensional coefficient k' is characteristic of the material and the conditions of crystallization and $k'' = k/(k')^i$. Feeling that a model in which nucleation depends only on supersaturation or growth rate is simplistically deficient, some have proposed that contact nucleation rate is also a power function of slurry density and that

$$B^0 = k_n G^i M_T^j \tag{19-17}$$

where M_t is the density of the crystal slurry, g/L.

Although Eqs. (19-16) and (19-17) have been adopted by many as a matter of convenience, they are oversimplifications of the very complex relationship that is suggested by Eq. (19-13); Eq. (19-17) implicitly and quite arbitrarily combines the effects of homogeneous nucleation and those due to contact nucleation. They should be used only with caution.

In work pioneered by Clontz and McCabe [*Chem. Eng. Prog. Symp. Ser.*, **67**(110), 6 (1971)] and subsequently extended by others, contact nucleation rate was found to be proportional to the input of energy of contact, as well as being a function of contact area and supersaturation. This observation is important to the scaling up of crystallizers: at laboratory or bench scale, contact energy level is relatively low and homogeneous nucleation can contribute significantly to the total rate of nucleation; in commercial equipment, on the other hand, contact energy input is intense and contact nucleation is the predominant mechanism. Scale-up modeling of a crystallizer, therefore, must include its mechanical characteristics as well as the physiochemical driving force.

Nucleation and Growth From the preceding, it is clear that no analysis of a crystallizing system can be truly meaningful unless the simultaneous effects of nucleation rate, growth rate, heat balance, and material balance are considered. The most comprehensive treatment of this subject is by Randolph and Larson (op. cit), who developed a mathematical model for continuous crystallizers of the mixed-suspension or circulating-magma type [*Am. Inst. Chem. Eng. J.*, **8**, 639 (1962)] and subsequently examined variations of this model that include most of the aberrations found in commercial equipment. Randolph and Larson showed that when the total number of crystals in a given volume of suspension from a crystallizer is plotted as a function of the characteristic length as in Fig. 19-37, the slope of the line is usefully identified as the crystal population density, n:

$$n = \lim_{\Delta L \to 0} \frac{\Delta N}{\Delta L} = \frac{dN}{dL} \tag{19-18}$$

where N = total number of crystals up to size L per unit volume of magma. The population density thus defined is useful because it characterizes the nucleation-growth performance of a particular crystallization process or crystallizer.

The data for a plot like Fig. 19-37 are easily obtained from a screen analysis of the total crystal content of a known volume (e.g., a liter) of magma. The analysis is made with a closely spaced set of testing sieves, as listed in Table 21-6, the cumulative number of particles smaller than each sieve in the nest being plotted against the aperture dimension of that sieve. The fraction retained on each sieve is weighed, and the mass is converted to the equivalent number of particles by dividing by the calculated mass of a particle whose dimension is the arithmetic mean of the mesh sizes of the sieve on which it is retained and the sieve immediately above it.

In industrial practice, the size-distribution curve usually is not actually constructed. Instead, a mean value of the population density for any sieve fraction of interest (in essence, the population density of the particle of average dimension in that fraction) is determined directly as $\Delta N/\Delta L$, ΔN being the number of particles retained on the sieve and ΔL being the difference between the mesh sizes of the retaining sieve and its immediate predecessor. It is common to employ the units of $(\text{mm}\cdot\text{L})^{-1}$ for n.

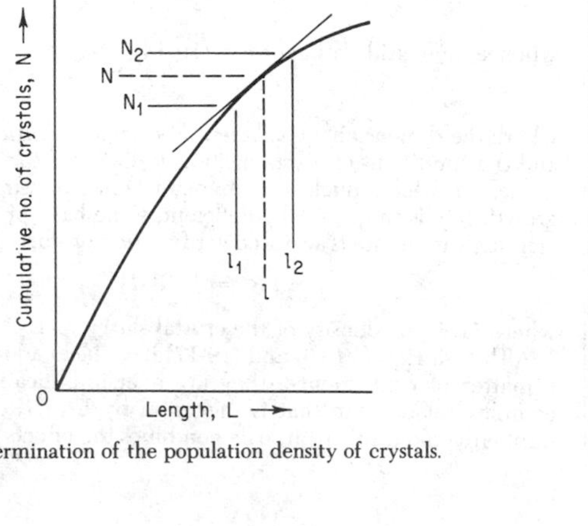

FIG. 19-37 Determination of the population density of crystals.

For a steady-state crystallizer receiving solids-free feed and containing a well-mixed suspension of crystals experiencing negligible breakage, a material-balance statement degenerates to a particle balance (the Randolph-Larson general-population balance); in turn, it simplifies to

$$dn/dL + n/Gt = 0 \tag{19-19}$$

if the delta L law applies (i.e., G is independent of L) and the drawdown (or retention) time is assumed to be invariant and calculated as $t = V/Q$. Integrated between the limits n^0, the population density of nuclei (for which L is assumed to be zero), and n, that of any chosen crystal size L, Eq. (19-19) becomes

$$\int_{n^0}^{n} \frac{dn}{n} = -\int_0^L \frac{dL}{Gt} \tag{19-20}$$

$$\ln n = \frac{-L}{Gt} + \ln n^0 \tag{19-21a}$$

or

$$n = n^0 e^{-L/Gt} \tag{19-21b}$$

A plot of ln n versus L is a straight line whose intercept is ln n^0 and whose slope is $-1/Gt$. (For plots on base-10 log paper, the appropriate slope correction must be made.) Thus, from a given product sample of known slurry density and retention time it is possible to obtain the nucleation rate and growth rate for the conditions tested if the sample satisfies the assumptions of the derivation and yields a straight line. A number of derived relations which describe the nucleation rate, size distribution, and average properties are summarized in Table 19-5.

If a straight line does *not* result (Fig. 19-38), at least part of the explanation may be violation of the delta L law (Canning and Randolph, loc. cit.). The best current theory about what causes size-dependent growth suggests what has been called growth dispersion or "Bujacian behavior" [Mullen (ed), op. cit., p. 23]. In the same environment different crystals of the same size can grow at different rates owing to differences in dislocations or other surface effects. The graphs of "slow" growers (Fig. 19-38, curve A) and "fast" growers (curve B) sum to a resultant line (curve C), concave upward, that is described by Eq. (19-22) (Randolph, in deJong and Jancic, op. cit., p. 295):

$$n = \sum \frac{B^0{}_i}{G_i} e^{(-L/G_i t)} \tag{19-22}$$

Equation (19-19) contains no information about the crystallizer's influence on the nucleation rate. If the crystallizer is of a mixed-suspension, mixed-product-removal (MSMPR) type, satisfying the criteria for Eq. (19-19), and if the model of Clontz and McCabe is valid, the contribution to the nucleation rate by the circulating pump can be calculated [Bennett, Fiedelman, and Randolph, *Chem. Eng. Prog.*, **69**(7), 86 (1973)]:

$$B_e = K_e \left(\frac{I^2}{P}\right) \rho G \int_0^\infty nL^4 dL \tag{19-23}$$

where I = tip speed of the propeller or impeller, m/s
ρ = crystal density, g/cm^3
P = volume of crystallizer/circulation rate (turnover), m^3/(m^3/s) = s

Since the integral term is the fourth moment of the distribution (m_4), Eq. (19-23) becomes

$$B_e = K_e \rho G \left(\frac{I^2}{P}\right) m_4 \tag{19-24}$$

Equation (19-24) is the general expression for impeller-induced mucleation. In a fixed-geometry system in which only the speed of the circulating pump is changed and in which the flow is roughly proportional to the pump speed, Eq. (19-24) may be satisfactorily replaced with

$$B_e = K_e'' \rho G (S_R)^3 m_4 \tag{19-25}$$

TABLE 19-5 Common Equations for Population-Balance Calculations

Name	Symbol	Units	Systems without fines removal	Systems with fines removal: Fines stream	Systems with fines removal: Product stream	References
Drawdown time (retention time)	t	h	$t = V/Q$	$t_F = V_{liquid}/Q_F$	$t = V/Q$	
Growth rate	G	mm/h	$G = dL/dt$	$G = dL/dt$	$G = dL/dt$	
Volume coefficient	K_v	1/no. (crystals)	$K_v = \frac{\text{volume of one crystal}}{L^3}$	$K_v = \frac{\text{volume of one crystal}}{L^3}$	$K_v = \frac{\text{volume of one crystal}}{L^3}$	
Population density	n	No. (crystals)/mm	$n = dN/dL$	$n = dN/dL$	$n = dN/dL$	1
Nuclei population density	n^o	No. (crystals/mm)	$n^o = K_M M^j G^{i-1}$			2
Population density	n	No. (crystals)/mm	$n = n^o e^{-L/Gt}$	$n_F = n^o e^{-L/Gt_F}$	$n = n^{o-Le/Gt_F} e^{-L/Gt}$	1, 3
Nucleation rate	B_0	No. (crystals)/h	$B_0 = Gn^o = K_M M^j G^i$	$B_0 = G_n^o$		4
Dimensionless length	x	None	$x = \frac{L}{Gt}$	$x_F = \frac{L}{Gt_F}, L_0 \to L_f$	$x = \frac{L}{Gt}, L_f \to L$	1
Mass/unit volume (slurry density)	M_T	g/L	$M_t = K_v\rho \int_0^\infty nL^3\, dL$ $M_t = K_v\rho 6\, n^o(Gt)^4$	$M_{T_F} = K_v\rho \int_0^{L_f} n^o e^{-L/Gt_F} L^3 dL$	$M_T = K_v\rho \int_{L_f}^\infty n^o e^{-L/Gt_F} e^{-L/Gt} L^3 dL$	1
Cumulative mass to x / Total mass	W_x	None	$W_x = 1 - e^{-x}\left(\frac{x^3}{6} + \frac{x^2}{2} + x + 1\right)$	$W_F = \frac{e^{-x}(x^3 + 3x^2 + 6x + 6) - 6}{e^{-xC}(x_c^3 + 3x_c^2 + 6x_c + 6) - 6}$	$W = \frac{6K_v\rho n^o e^{-Le/GtF}Gt)^4 \left[1 - e^{-x}\left(\frac{x^3}{6} + \frac{x^2}{2} + x + 1\right)\right]}{\text{Slurry density } M, g/L}$ when $L_c \approx 0$, compared with L_a	5
Dominant particle	L_d	mm	$l_d = 3Gt$			
Average particle, weight	L_a	mm	$l_a = 3.67\ Gt$			6
Total number of crystals	N_T	No./L	$N_T = \int_0^\infty n\ dl$	$N_F = \int_0^{L_f} n_F dL$	$N_T = \int_{L_f}^\infty n\ dL$	1, 3

1. Randlolph and Larson, *Am. Inst. Chem. Eng. J.*, **8**, 639 (1962).
2. Timm and Larson, *Am. Inst. Chem. Eng. J.*, **14**, 452 (1968).
3. Larson, private communication.
4. Larson, Timm, and Wolff, *Am. Inst. Chem. Eng. J.*, **14**, 448 (1968).
5. Larson and Randolph, *Chem. Eng. Prog. Symp. Ser.* **65**(95), 1 (1969).
6. Schoen, *Ind. Eng. Chem.*, **53**, 607 (1961).

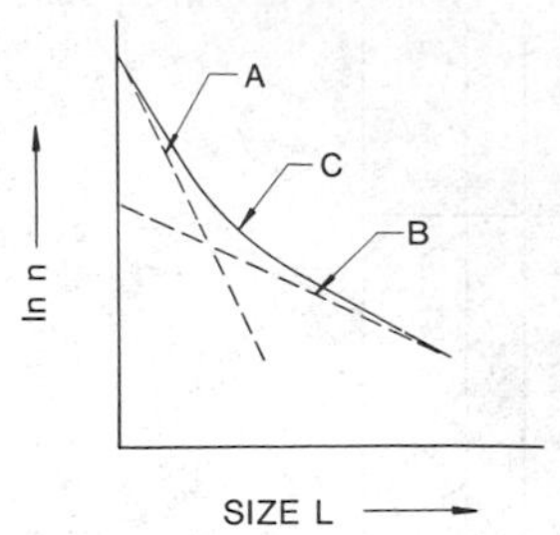

FIG. 19-38 Population density of crystals resulting from Bujacian behavior.

where S_R = rotation rate of impeller, r/min. If the maximum crystal-impeller impact stress is a nonlinear function of the kinetic energy, shown to be the case in at least some systems, Eq. (19-25) no longer applies.

In the specific case of an MSMPR exponential distribution, the fourth moment of the distribution may be calculated as

$$m_4 = 4!n^0(Gt)^5 \tag{19-26}$$

Substitution of this expression into Eq. (19-24) gives

$$B_e = kn^0\ G(S_R)^3L_D^5 \tag{19-27}$$

where $L_D = 3Gt$, the dominant crystal (mode) size.

Equation (19-27) displays the competing factors that stabilize secondary nucleation in an operating crystallizer when nucleation is due mostly to impeller-crystal contact. Any increase in particle size produces a fifth-power increase in nucleation rate, tending to counteract the direction of the change and thereby stabilizing the crystal-size distribution. From dimensional argument alone the size produced in a mixed crystallizer for a (fixed) nucleation rate varies as $(B^0)^{1/3}$. Thus, this fifth-order response of contact nucleation does not wildly upset the crystal size distribution but instead acts as a stabilizing feedback effect.

Nucleation due to crystal-to-crystal contact is greater for equal striking energies than crystal-to-metal contact. However, the viscous drag of the liquid on particle sizes normally encountered limits the velocity of impact to extremely low values. The assumption that only the largest crystal sizes contribute significantly to the nucleation rate by crystal-to-crystal contact permits a simple computation of the rate:

$$B_c = K_c\ \rho G m_j^2 \tag{19-28}$$

where m_j = the fourth, fifth, sixth, or higher moments of the distribution.

A number of different crystallizing systems have been investigated by using the Randolph-Larson technique, and some of the published growth rates and nucleation rates are included in Table 19-6. Although the usefulness of these data is limited to the conditions tested, the table gives a range of values which may be expected, and it permits resolution of the information gained from a simple screen analysis into the fundamental factors of growth rate and nucleation rate. Experiments may then be conducted to determine the independent effects of operation and equipment design on these parameters.

Although this procedure requires laborious calculations because of the number of samples normally needed, these computations and the determination of the best straight-line fit to the data are readily programmed for digital computers.

Example 3 Calculate the population density, growth, and nucleation rates for a crystal sample of urea for which there is the following information. These data are from Bennett and Van Buren [*Chem. Eng. Prog. Symp. Ser.*, **65**(95), 44 (1969)].

Slurry density = 450 g/L
Crystal density = 1.335 g/cm^3
Drawdown time t = 3.38 h
Shape factor k_v = 1.00

TABLE 19-6 Growth Rates and Kinetic Equations for Some Industrial Crystallized Products*

Material crystallized	G, m/s × 10^8	Range t, h	Range M_T, g/L	Temperature, °C	Scale†	Kinetic equation for B_0, no./(L·s)	References
Ammonium sulfate	1.67	3.83	150	70	P	$B^0 = 6.62 \times 10^{-25}\ G^{0.82}\ p^{-0.92}\ m_2^{2.05}$	Bennett and Wolf, American Institutie of Chemical Engineers, SFC, 1979
Ammonium sulfate	0.20	0.25	38	18	B	$B^0 = 2.94(10^{10})G^{1.03}$	Larsen and Mullen, *J. Cryst. Growth*, **20**, 183 (1973)
Ammonium sulfate	......	0.20	....	34	B	$B^0 = 6.14(10^{-11})\ S_R^{7.84}\ M_T^{0.98}G^{1.22}$	Youngquist and Randolph, *Am. Inst. Chem. Eng. J.*, **18**, 421 (1972)
Citric acid	1.1–3.7	....	....	16–24	B	$B^0 = 1.09(10^{10})m_4^{0.084}G^{0.84}$	Sikdar and Randolph, *Am. Inst. Chem. Eng. J.*, **22**, 110 (1976)
Magnesium sulfate	3.0–7.0	....	....	25	B	$B^0 = 9.65(10^{12})M_T^{0.67}\ G^{1.24}$	Sikdar and Randolph, *Am. Inst. Chem. Eng. J.*, **22**, 110 (1976)
Potassium chloride	2–12	....	200	32	P	$B^0 = 7.12(10^{39})M_T^{0.14}\ G^{4.99}$	Randolph et al., *Am. Inst. Chem. Eng. J.*, **23**, 500 (1977)
Potassium dichromate	1.2–9.1	0.25–1	14–42	.....	B	$B^0 = 7.33(10^4)M_T^{0.6}\ G^{0.5}$	Desari et al., *Am. Inst. Chem. Eng. J.*, **20**, 43 (1974)
Potassium dichromate	2.6–10	0.15–0.5	20–100	26–40	B	$B^0 = 1.59(10^{-3})S_R^3M_TG^{0.48}$	Janse, Ph.D. thesis, Delft Technical University, 1977
Potassium nitrate	8–13	0.25–0.50	10–40	20	B	$B^0 = 3.85(10^{16})M_T^{0.5}\ G^{2.06}$	Juzaszek and Larson, *Am. Inst. Chem. Eng. J.*, **23**, 460 (1977)
Potassium sulphate		0.03–0.17	1–7	30	B	$B^0 = 2.62(10^3)S_R^{2.5}M_T^{0.5}G^{0.54}$	Randolph and Sikdar, *Ind. Eng. Chem. Fundam.*, **15**, 64 (1976)
Potassium chloride	3.3	1–2	100	37	B	$B^0 = 5.16(10^{22})M_T^{0.91}G^{2.77}$	Randolph et al., *Ind. Eng. Chem. Process Des. Dev.*, **20**, 496 (1981)
Sodium chloride	4–13	0.2–1	25–200	50	B	$B^0 = 1.92(10^{10})S_R^2M_TG^2$	Asselbergs, Ph.D. thesis, Delft Technical University, 1978
Sodium chloride	0.5	1–2.5	70–190	72	P	$B^0 = 1.47(10^2)\left(\frac{I^2}{P}\right)m_4^{0.84}\ G^{0.98}$	Bennett et al., *Chem. Eng. Prog.*, **69**(7), 86 (1973)
Urea	0.4–4.2	2.5–6.8	350–510	55	P	$B^0 = 5.48(10^{-1})\ M_T^{-3.87}G^{-1.66}$	Bennett and Van Buren, *Chem. Eng. Prog. Symp. Ser.*, **95**(7), 65 (1973)
Urea	......	....	....	3–16	B	$B^0 = 1.49(10^{-31})\ S_R^{2.3}\ M_T^{1.07}\ G^{-3.54}$	Lodaya et al., *Ind. Eng. Chem. Process Des. Dev.*, **16**, 294 (1977)

*Additional data on many components are in Garside and Shah, *Ind. Eng. Chem. Process Des. Dev.*, **19**, 509 (1980).
†B = bench scale; P = pilot plant.

Product size:

−14 mesh, + 20 mesh	4.4 percent
−20 mesh, + 28 mesh	14.4 percent
−28 mesh, + 35 mesh	24.2 percent
−35 mesh, + 48 mesh	31.6 percent
−48 mesh, + 65 mesh	15.5 percent
−65 mesh, + 100 mesh	7.4 percent
−100 mesh	2.5 percent

n = number of particles per liter of volume
14 mesh = 1.168 mm, 20 mesh = 0.833 mm, average opening 1.00 mm
Size span = 0.335 mm = ΔL

$$n_{20} = \frac{(450\text{ g/L})(0.044)}{(1.335/1000)\text{ g/mm}^3(1.00^3\text{ mm}^3/\text{particle})(0.335\text{ mm})(1.0)}$$

$$n_{20} = 44{,}270$$

$$\ln n_{20} = 10{,}698$$

Repeating for each screen increment:

Screen size	Weight, %	k_v	ln n	L, average diameter, mm
100	7.4	1.0	18.099	0.178
65	15.5	1.0	17.452	0.251
48	31.6	1.0	16.778	0.356
35	24.2	1.0	15.131	0.503
28	14.4	1.0	13.224	0.711
20	4.4	1.0	10.698	1.000

Plotting ln n versus L as shown in Fig. 19-39, a straight line having an intercept at zero length of 19.781 and a slope of −9.127 results. As mentioned in discussing Eq. (19-12), the growth rate can then be found.

$$\text{Slope} = -1/Gt \text{ or } -9.127 = -1/[G(3.38)]$$

or $$G = 0.0324\text{ mm/h}$$

and $$B_0 = Gn^0 = (0.0324)(e^{19.781}) = 12.65 \times 10^6 \frac{n^0}{\text{L}\cdot\text{h}}$$

and $$L_a = 3.67(0.0324)(3.38) = 0.40\text{ mm}$$

An additional check can be made of the accuracy of the data by the relation

$$M_T = 6k_v\rho n^0(Gt)^4 = 450\text{ g/L}$$

$$M_T = (6)(1.0)\frac{1.335\text{ g/cm}^3}{1000\text{ mm}^3/\text{cm}^3}e^{19.78}[(0.0324)(3.38)]^4$$

$$M_T = 455\text{ g/L} \approx 450\text{ g/L}$$

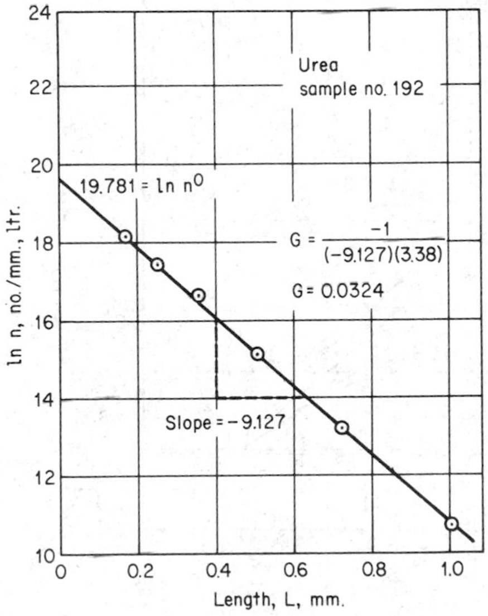

FIG. 19-39 Population-density plot for Example 3.

Had only the growth rate been known, the size distribution of the solids could have been calculated from the equation

$$W_f = 1 - e^{-x}\left(\frac{x^3}{6} + \frac{x^2}{2} + x + 1\right)$$

where W_f is the weight fraction *up to* size L and $x = L/Gt$.

$$x = \frac{L}{(0.0324)(3.38)} = \frac{L}{0.1095}$$

Screen size	L, mm	x	W_f*	Cumulative % retained 100 $(1 - W_f)$	Measured cumulative % retained
20	0.833	7.70	0.944	5.6	4.4
28	.589	5.38	.784	21.6	18.8
35	.417	3.80	.526	47.4	43.0
48	.295	2.70	.286	71.4	74.6
65	.208	1.90	.125	87.5	90.1
100	.147	1.34	.048	95.2	97.5

*Values of W_f as a function of x may be obtained from a table of Wick's functions.

Note that the calculated distribution shows some deviation from the measured values because of the small departure of the actual sample from the theoretical coefficient of variation (i.e., 47.5 versus 52 percent).

From several other samples taken from the same machine, a table of different values may be constructed:

Sample No.	ln n^0	G	ln G
191	18.81	0.0330	−3.41
192	19.78	.0324	−3.43
193	18.70	.0317	−3.45
194	20.51	.0200	−3.91

As shown in the graph in Fig. 19-40, the slope of the best line drawn through these points is −4.45. From the equation

$$n^0 = K_m M^j G^{i-1}$$

the quantity $(i - 1) = -4.45$, or $i = -3.45$; therefore,

$$n^0 = K_m M^j G^{-4.45} \text{ (nuclei population density)}$$

and $$B_0 = K_m M^j G^{-3.45} \text{ (nucleation rate)}$$

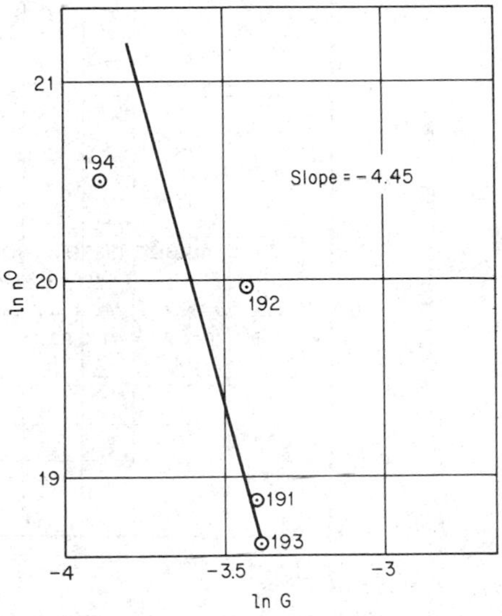

FIG. 19-40 Growth rate versus nucleation rate, Example 3.

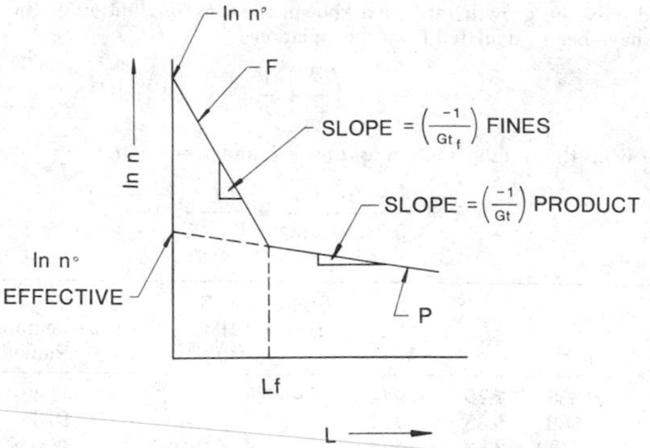

FIG. 19-41 Plot of ln n against L for a crystallizer with fines removal.

Here it can be seen that the nucleation rate is a decreasing function of growth rate (and supersaturation). The physical explanation is believed to be the mechanical influence of the crystallizer on the growing suspension and/or the effect of Bujacian behavior.

Had sufficient data indicating a change in n^0 for various values of M at constant G been available, a plot of ln n^0 versus ln M at corresponding G's would permit determination of the power j.

Crystallizers with Fines Removal In Example 3, the product was from a forced-circulation crystallizer of the MSMPR type. In many cases, the product produced by such machines is too small for commercial use; therefore, a separation baffle is added within the crystallizer to permit the removal of unwanted fine crystalline material from the magma, thereby controlling the population density in the machine so as to produce a coarser crystal product. When this is done, the product sample plots on a graph of ln n versus L as shown in line P, Fig. 19-41. The line of steepest slope, line F, represents the particle-size distribution of the fine material, and samples which show this distribution can be taken from the liquid leaving the fines-separation baffle. The product crystals have a slope of lower value, and typically there should be little or no material present smaller than L_f, the size which the baffle is designed to separate. The effective nucleation rate for the product material is the intersection of the extension of line P to zero size.

As long as the largest particle separated by the fines-destruction baffle is small compared with the mean particle size of the product, the seed for the product may be thought of as the particle-size distribution corresponding to the fine material which ranges in length from zero to L_f, the largest size separated by the baffle.

The product discharged from the crystallizer is characterized by the integral of the distribution from size L_f to infinity:

$$M_T = k_v \rho \int_{Lf}^{\infty} n^0 \exp(-L_f/Gt_f) \exp(L/GT) L^3 dL \qquad (19\text{-}29)$$

The integrated form of this equation is shown in Table 19-5.

For a given set of assumptions it is possible to calculate the characteristic curves for the product from the crystallizer when it is operated at various levels of fines removal as characterized by L_f. This has been done for an ammonium sulfate crystallizer in Fig. 19-42. Also shown in that figure is the actual size distribution obtained. In calculating theoretical size distributions in accordance with the Eq. (19-29), it is assumed that the growth rate is a constant, whereas in fact larger values of L_f will interact with the system driving force to raise the growth rate and the nucleation rate. Nevertheless, Fig. 19-42 illustrates clearly the empirical result of the operation of such equipment, demonstrating that the most significant variable in changing the particle-size distribution of the product is the size removed by the baffle. Conversely, changes in retention time for a given particle-removal size L_f make a relatively small change in the product-size distribution.

It is implicit that increasing the value of L_f will raise the supersaturation and growth rate to levels at which mass homogeneous nucleation can occur, thereby leading to periodic upsets of the sys-

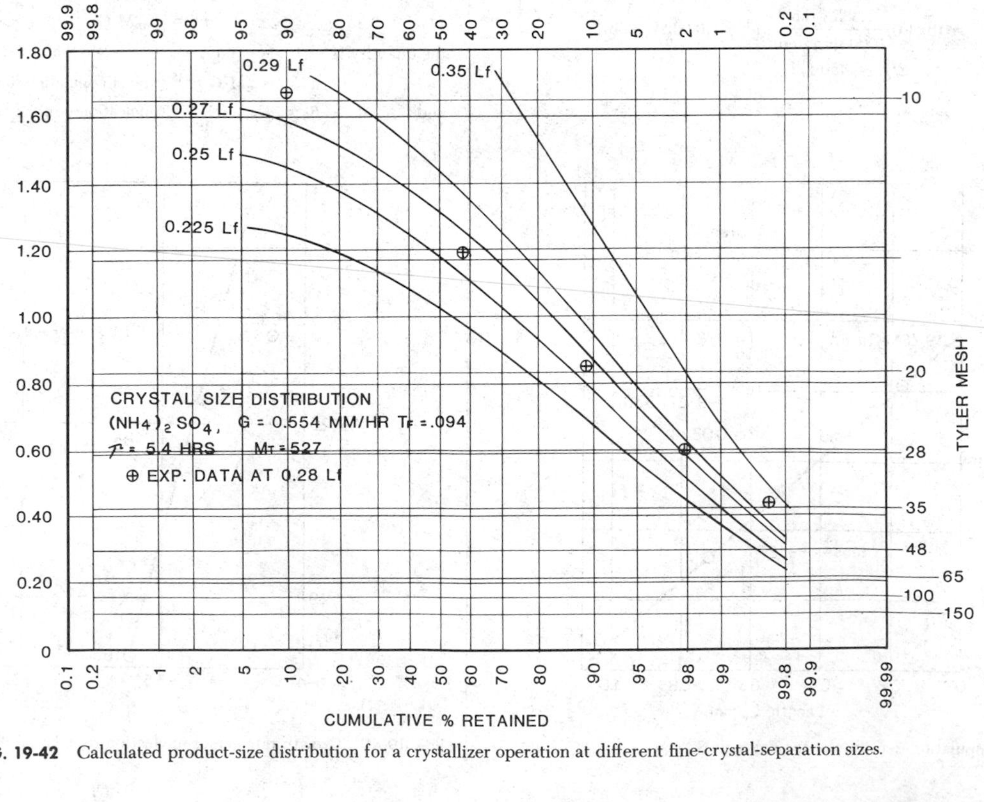

FIG. 19-42 Calculated product-size distribution for a crystallizer operation at different fine-crystal-separation sizes.

tem or cycling [Randolph, Beer, and Keener, *Am. Inst. Chem. Eng. J.*, **19**, 1140 (1973)]. That this could actually happen was demonstrated experimentally by Randolph, Beckman, and Kraljevich [*Am. Inst. Chem. Eng. J.*, **23**, 500 (1977)], and that it could be controlled dynamically by regulating the fines-destruction system was shown by Beckman and Randolph [ibid., (1977)]. Dynamic control of a crystallizer with a fines-destruction baffle and fine-particle-detection equipment employing a light-scattering (laser) particle-size-measurement instrument is described in U.S. Patent 4,263,010.

CRYSTALLIZATION EQUIPMENT

Whether a vessel is called an evaporator or a crystallizer depends primarily on the criteria used in arriving at its sizing. In an evaporator of the salting-out type, sizing is done on the basis of vapor release. In a crystallizer, sizing is normally done on the basis of the volume required for crystallization or for special features required to obtain the proper product size. In external appearance, the vessels could be identical. Evaporators are discussed in Sec. 11.

In the discussion which follows, crystallization equipment has been classified according to the means of suspending the growing product. This technique reduces the number of major classifications and segregates those to which Eq. (19-19) applies.

Mixed-Suspension, Mixed-Product-Removal Crystallizers This type of equipment, sometimes called the circulating-magma crystallizer, is by far the most important in use today. In most commercial equipment of this type, the uniformity of suspension of product solids within the crystallizer body is sufficient for the theory [Eqs. (19-19) to (19-21*b*)] to apply. Although a number of different varieties and features are included within this classification, the equipment operating with the highest capacity is the kind in which the vaporization of a solvent, usually water, occurs.

Although surface-cooled types of MSMPR crystallizers are available, most users prefer crystallizers employing vaporization of solvents or of refrigerants. The primary reason for this preference is that heat transferred through the critical supersaturating step is through a boiling-liquid–gas surface, avoiding the troublesome solid deposits that can form on a metal heat-transfer surface.

Forced-Circulation Evaporation Crystallizer This crystallizer is shown in Fig. 19-43. Slurry leaving the body is pumped through a circulating pipe and through a tube-and-shell heat exchanger, where its temperature increases by about 2 to 6°C (3 to 10°F). Since this heating is done without vaporization, materials of normal solubility should produce no deposition on the tubes. The heated slurry, returned to the body by a recirculation line, mixes with the body slurry and raises its temperature locally near the point of entry, which causes boiling at the liquid surface. During the consequent cooling and vaporization to achieve equilibrium between liquid and vapor, the supersaturation which is created causes deposits on the swirling body of suspended crystals until they again leave via the circulating pipe. The quantity and the velocity of the recirculation, the size of the body, and the type and speed of the circulating pump are critical design items if predictable results are to be achieved. A further discussion of the parameters affecting this type of equipment is given by Bennett, Newman, and Van Buren [*Chem. Eng. Prog.*, **55**(3), 65, (1959); *Chem. Eng. Prog. Symp. Ser.*, **65**(95), 34, 44 (1969)].

If the crystallizer is not of the evaporative type but relies only on *adiabatic evaporative cooling* to achieve the yield, the heating element is omitted. The feed is admitted into the circulating line after withdrawal of the slurry, at a point sufficiently below the free-liquid surface to prevent flashing during the mixing process.

Conispherical Magma Crystallizer This crystallizer is shown in Fig. 19-44. Although the same basic type as the crystallizer shown in Fig. 19-43, this unit employs a radial rather than tangential recirculation inlet. The inlet *LI* is designed in such a way as to favor the distribution of the incoming solution uniformly across the liquid surface in the vessel. Crystals precipitating during the boiling action are removed at the product outlet *P*, and the liquor is removed from outlet *LO* behind the baffle in the lower portion of the conisphere. As shown, this vessel would operate at or below the natural slurry

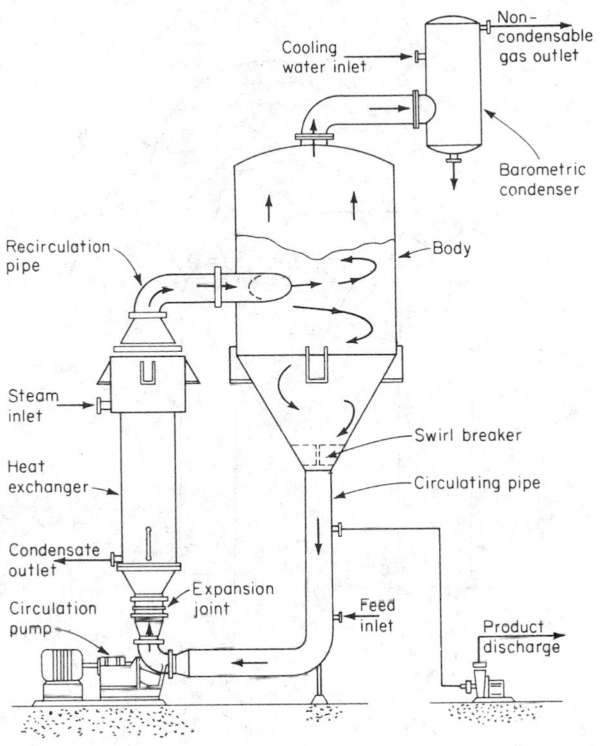

FIG. 19-43 Forced-circulation (evaporative) crystallizer. *(Swenson Division, Whiting Corporation.)*

density owing to the action of the conical settler located below the recirculating slurry outlet *LO*.

The size of the crystalline material produced in forced-circulation crystallizers varies with the operating parameters and the characteristics of the material being handled. With sodium chloride, urea, citric acid, and many similar inorganic and organic chemicals, the crystalline product is in the range 20 to 150 mesh.

Draft-Tube-Baffle (DTB) Evaporator-Crystallizer Because mechanical circulation greatly influences the level of nucleation within the crystallizer, a number of designs have been developed that use circulators located within the body of the crystallizer, thereby reducing the head against which the circulator must pump. This technique reduces the power input and circulator tip speed and therefore the rate of nucleation. A typical example is the draft-tube-baffle (DTB) evaporator-crystallizer (Swenson Division, Whiting Corporation) shown in Fig. 19-45. The suspension of product crystals in maintained by a large, slow-moving propeller surrounded by a draft tube within the body. The propeller directs the slurry to the liquid surface so as to prevent solids from short-circuiting the zone of the most intense supersaturation. Slurry which has been cooled is returned to the bottom of the vessel and recirculated through the propeller. At the propeller, heated solution is mixed with the recirculating slurry.

The design of Fig. 19-45 contains a fines-destruction feature comprising the settling zone surrounding the crystallizer body, the circulating pump, and the heating element. The heating element supplies sufficient heat to meet the evaporation requirements and to raise the temperature of the solution removed from the settler so as to destroy any small crystalline particles withdrawn. Coarse crystals are separated from the fines in the settling zone by gravitational sedimentation, and therefore this fines-destruction feature is applicable only to systems in which there is a substantial density difference between crystals and mother liquor.

This type of equipment can also be used for applications in which

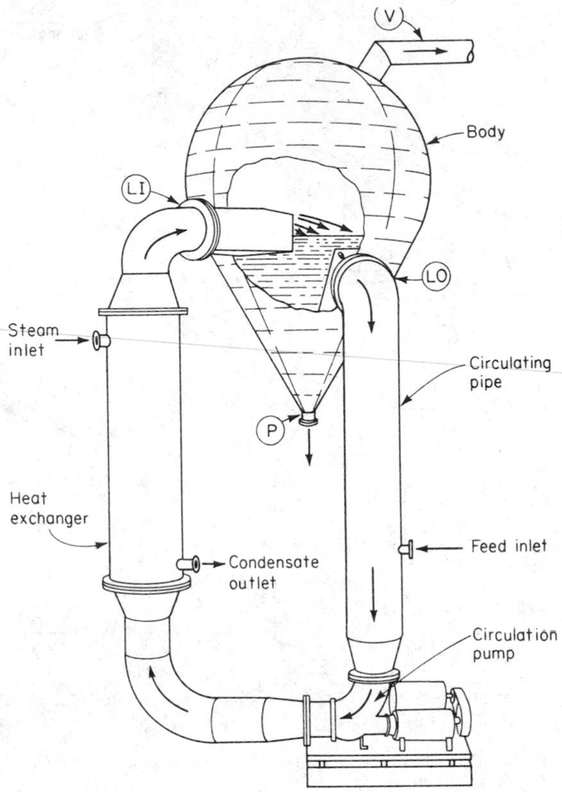

FIG. 19-44 Conispherical magma crystallizer.

the only heat removed is that required for adiabatic cooling of the incoming feed solution. When this is done and the fines-destruction feature is to be employed, a stream of liquid must be withdrawn from the settling zone of the crystallizer and the fine crystals must be separated or destroyed by some means other than heat addition—for example, either dilution or thickening and physical separation.

In some crystallization applications it is desirable to increase the solids content of the slurry within the body above the natural consistency, which is that developed by equilibrium cooling of the incoming feed solution to the final temperature. This can be done by withdrawing a stream of mother liquor from the baffle zone, thereby thickening the slurry within the growing zone of the crystallizer. This mother liquor is also available for removal of fine crystals for size control of the product.

Draft-Tube (DT) Crystallizer This crystallizer may be employed in systems in which fines destruction is not needed or wanted. In such cases the baffle is omitted, and the internal circulator is sized to have the minimum nucleating influence on the suspension.

In DTB and DT crystallizers the circulation rate achieved is generally much greater than that available in a similar forced-circulation crystallizer. The equipment therefore finds application when it is necessary to circulate large quantities of slurry to minimize supersaturation levels within the equipment. In general, this approach is required to obtain long operating cycles with material capable of growing on the walls of the crystallizer. The draft-tube and draft-tube-baffle designs are commonly used in the production of granular materials such as ammonium sulfate, potassium chloride, photographic hypo, and other inorganic and organic crystals for which product in the range 8 to 30 mesh is required.

Surface-Cooled Crystallizers For some materials, such as sodium chlorate, it is possible to use a forced-circulation tube-and-shell exchanger in direct combination with a draft-tube-crystallizer body, as shown in Fig. 19-46. Careful attention must be paid to the

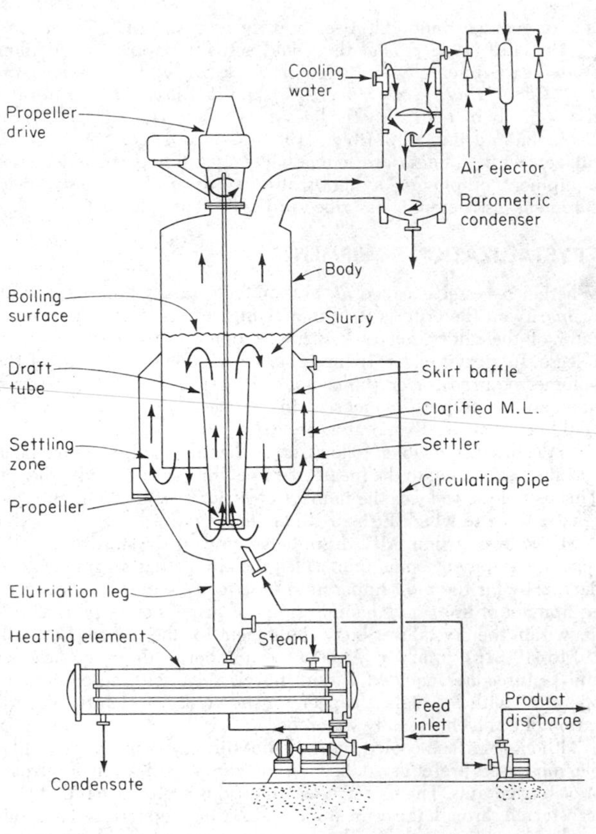

FIG. 19-45 Draft-tube-baffle (DTB) crystallizer. *(Swenson Division, Whiting Corporation.)*

temperature difference between the cooling medium and the slurry circulated through the exchanger tubes. In addition, the path and rate of slurry flow within the crystallizer body must be such that the volume contained in the body is "active." That is to say, crystals must be so suspended within the body by the turbulence that they are effective in relieving supersaturation created by the reduction in

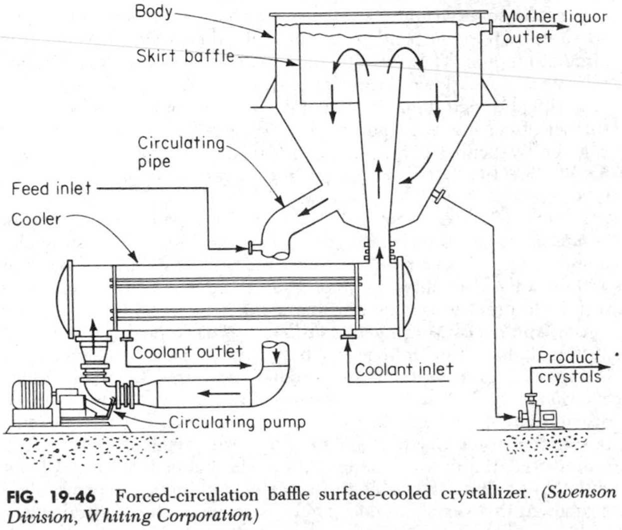

FIG. 19-46 Forced-circulation baffle surface-cooled crystallizer. *(Swenson Division, Whiting Corporation)*

temperature of the slurry as it passes through the exchanger. Obviously, the circulating pump is part of the crystallizing system, and careful attention must be paid to its type and its operating parameters to avoid undue nucleating influences.

The use of the internal baffle permits operation of the crystallizer at a slurry consistency other than that naturally obtained by the cooling of the feed from the initial temperature to the final mother-liquor temperature. The baffle also permits fines removal and destruction.

With most inorganic materials this type of equipment produces crystals in the range 30 to 100 mesh. The design is based on the allowable rates of heat exchange and the retention required to grow the product crystals.

Direct-Contact-Refrigeration Crystallizer For some applications, such as the freezing of ice from seawater, it is necessary to go to such low temperatures that cooling by the use of refrigerants is the only economical solution. In such systems it is sometimes impractical to employ surface-cooled equipment because the allowable temperature difference is so small (under 3°C) that the heat-exchanger surface becomes excessive or because the viscosity is so high that the mechanical energy put in by the circulation system requires a heat-removal rate greater than can be obtained at reasonable temperature differences. In such systems, it is convenient to admix the refrigerant with the slurry being cooled in the crystallizer, as shown in Fig. 19-47, so that the heat of vaporization of the refrigerant cools the slurry by direct contact. The successful application of such systems requires that the refrigerant be relatively immiscible with the mother liquor and be capable of separation, compression, condensation, and subsequent recycle into the crystallizing system. The operating pressures and temperatures chosen have a large bearing on power consumption.

This technique has been very successful in reducing the problems associated with buildup of solids on a cooling surface. The use of direct-contact refrigeration also reduces overall process-energy requirements, since in a refrigeration process involving two fluids a greater temperature difference is required on an overall basis when the refrigerant must first cool some intermediate solution, such as calcium chloride brine, and that solution in turn cools the mother liquor in the crystallizer.

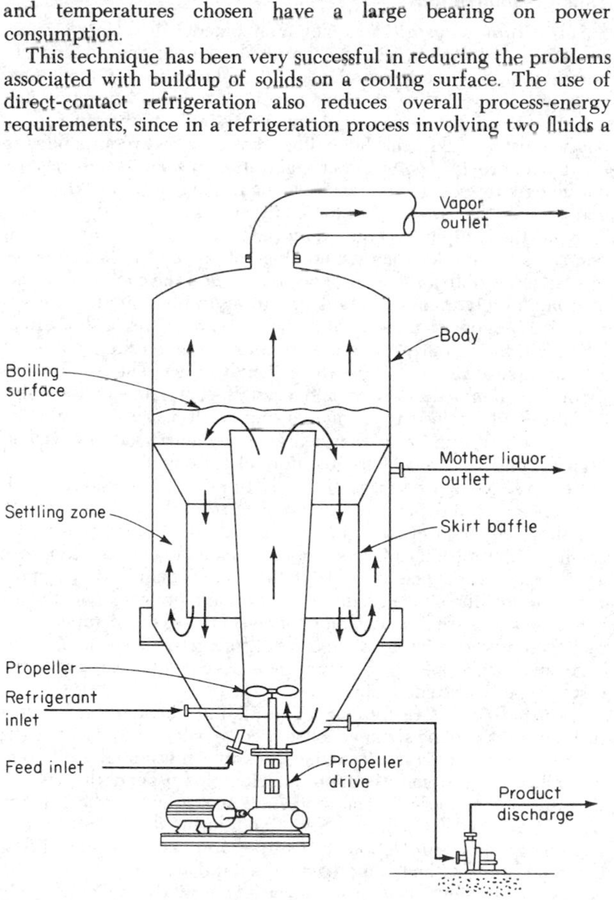

FIG. 19-47 Direct-contact-refrigeration crystallizer (DTB type). *(Swenson Division, Whiting Corporation.)*

Equipment of this type has been successfully operated at temperatures as low as −59°C (−75°F).

Reaction-Type Crystallizers In chemical reactions in which the end product is a solid-phase material such as a crystal or an amorphous solid the type of equipment described in the preceding subsections or shown in Fig. 19-47 may be used. By mixing the reactants in a large circulated stream of mother liquor containing suspended solids of the equilibrium phase, it is possible to minimize the driving force created during their reaction and remove the heat of reaction through the vaporization of a solvent, normally water. Depending on the final particle size required, it is possible to incorporate a fines-destruction baffle as shown in Fig. 19-47 and take advantage of the control over particle size afforded by this technique. In the case of ammonium sulfate crystallization from ammonia gas and concentrated sulfuric acid, it is necessary to vaporize water to remove the heat of reaction, and this water so removed can be reinjected after condensation into the fines-destruction stream to afford a very large amount of dissolving capability.

Other examples of this technique are the absorption of CO_2 and the neutralization of sulfuric acid to precipitate calcium sulfate dihydrate (gypsum).

Mixed-Suspension, Classified-Product-Removal Crystallizers Many of the crystallizers just described can be designed for classified-product discharge. Classification of the product is normally done by means of an elutriation leg suspended beneath the crystallizing body as shown in Fig. 19-45. Introduction of clarified mother liquor to the lower portion of the leg fluidizes the particles prior to discharge and selectively returns the finest crystals to the body for further growth. A relatively wide distribution of material is usually produced unless the elutriation leg is extremely long. Inlet conditions at the leg are critical if good classifying action or washing action is to be achieved.

If an elutriation leg or other product-classifying device is added to a crystallizer of the MSMPR type, the plot of the population density versus L is distorted in the region of largest sizes. Also the incorporation of an elutriation leg destabilizes the crystal-size distribution and under some conditions can lead to cycling. The theoretical treatment of both the crystallizer model and the cycling relations is discussed by Randolph, Beer, and Keener (loc. cit.). Although such a feature can be included on many types of classified-suspension or mixed-suspension crystallizers, it is most common to use this feature with the forced-circulation evaporative-crystallizer and the DTB crystallizer.

Classified-Suspension Crystallizer This equipment is also known as the **growth** or **Oslo crystallizer** and is characterized by the production of supersaturation in a circulating stream of liquor. Supersaturation is developed in one part of the system by evaporative cooling or by cooling in a heat exchanger, and it is relieved by passing the liquor through a fluidized bed of crystals. The fluidized bed may be contained in a simple tank or in a more sophisticated vessel arranged for a pronounced classification of the crystal sizes. Ideally this equipment operates within the metastable supersaturation field described by Miers and Isaac, *J. Chem. Soc.*, **1906**, 413.

In the **evaporative crystallizer** of Fig. 19-48, solution leaving the vaporization chamber at B is supersaturated slightly within the metastable zone so that new nuclei will not form. The liquor contacting the bed at E relieves its supersaturation on the growing crystals and leaves through the circulating pipe F. In a cooling-type crystallization hot feed is introduced at G, and the mixed liquor flashes when it reaches the vaporization chamber at A. If further evaporation is required to produce the driving force, a heat exchanger is installed between the circulating pump and the vaporization changer to supply the heat for the required rate of vaporization.

The transfer of supersaturated liquor from the vaporizer (point B, Fig. 19-48) often causes salt buildup in the piping and reduction of the operating cycle in equipment of this type. The rate of buildup can be reduced by circulating a thin suspension of solids through the vaporizing chamber; however, the presence of such small seed crys-

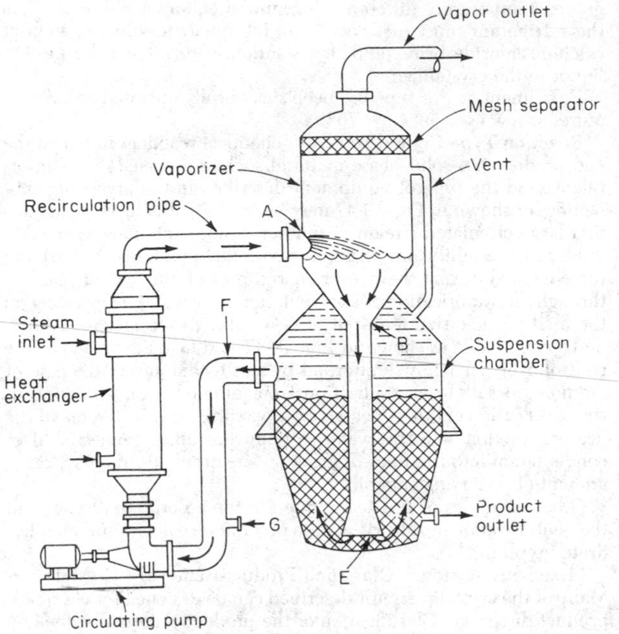

FIG. 19-48 Oslo evaporative crystallizer. *(Unitech Division, Union Tank Car Co.)*

tals tends to rob the supersaturation developed in the vaporizer, thereby lowering the efficiency of the recirculation system.

An **Oslo surface-cooled crystallizer** is illustrated in Fig. 19-49. Supersaturation is developed in the circulated liquor by chilling in the cooler *H*. This supersaturated liquor is contacted with the suspension of crystals in the suspension chamber at *E*. At the top of the suspension chamber a stream of mother liquor *D* can be removed to be used for fines removal and destruction. This feature can be added on either type of equipment. Fine crystals withdrawn from the top

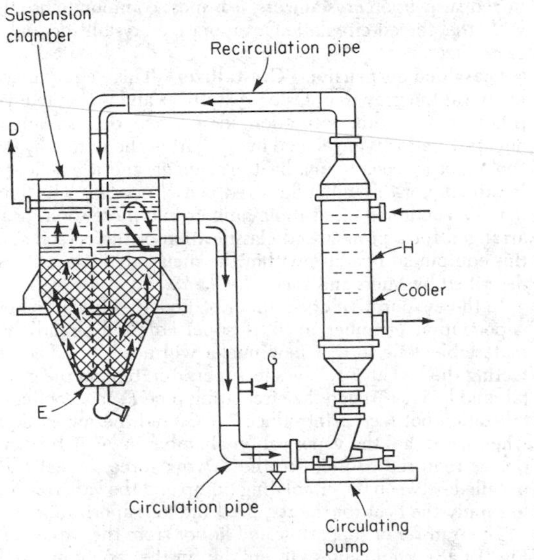

FIG. 19-49 Oslo surface-cooled crystallizer. *(Unitech Division, Union Tank Car Co.)*

of the suspension are destroyed, thereby reducing the overall number of crystals in the system and increasing the particle size of the remaining product crystals.

Scraped-Surface Crystallizer For relatively small-scale applications a number of crystallizer designs employing direct heat exchange between the slurry and a jacket or double wall containing a cooling medium have been developed. The heat-transfer surface is scraped or agitated in such a way that the deposits cannot build up. The scraped-surface crystallizer provides an effective and inexpensive method of producing slurry in equipment which does not require expensive installation or supporting structures.

Double-Pipe Scraped-Surface Crystallizer This type of equipment consists of a double-pipe heat exchanger with an internal agitator fitted with spring-loaded scrapers that wipe the wall of the inner pipe. The cooling liquid passes between the pipes, this annulus being dimensioned to permit reasonable shell-side velocities. The scrapers prevent the buildup of solids and maintain a good film coefficient of heat transfer. The equipment can be operated in a continuous or in a recirculating batch manner.

Such units are generally built in lengths to above 12 m (40 ft). They can be arranged in parallel or in series to give the necessary liquid velocities for various capacities. Heat-transfer coefficients have been reported in the range of 170 to 850 W/($m^2 \cdot K$) [30 to 150 Btu/($h \cdot ft^2 \cdot °F$)] at temperature differentials of 17°C (30°F) and higher [Garrett and Rosenbaum, *Chem. Eng.*, **65**(16), 127 (1958)]. Equipment of this type is marketed as the **Votator** and the **Armstrong crystallizer.**

Tank Crystallization

Static-Tank Crystallizer Common practice in plants of very small capacity dealing with concentrated solutions and materials of normal solubility is to pump the hot feed solution into an unagitated tank and allow cooling to occur either by natural convection and radiation or by surface cooling through coils in the tank or in a jacket on the outside of the tank. Sometimes rods or wires are suspended in the tank to provide centers for crystallization. Insulation may be used on the tank to reduce the rate of cooling in the first part of the cycle and prevent the formation of too many nuclei.

While the equipment required for such a system is extremely inexpensive and simple, there is nothing simple about its operation. Nucleation is difficult to control or predict, and the cooling rate varies considerably in an open tank according to the humidity and air velocity. Because of the lack of agitation there is only a slow circulation within the system caused by differences in density, and supersaturation levels normally rise to very high values. The result is formation of dendritic crystals and crystals containing considerable quantities of occlusions of mother liquor. It is also common to observe the formation of very large singular crystals as well as "slush" consisting of copious quantities of extreme fines.

Removal of the crystals is generally time-consuming and expensive, since crystal deposits must be broken from the sides of the tank and shoveled out after the liquor has been drained. Sometimes plate-cooling coils on which a heavy crust of crystalline material can form are inserted. At the end of the batch the coils are removed by a crane so that unloading of the crystals can be done more conveniently.

Such systems are now used only for certain specialized applications such as the production of Glauber's salt for synthetic sponges, or for very small-scale operations, or in underdeveloped areas where the cost of labor is extremely low.

Agitated-Tank Crystallizer The addition of a propeller or a turbine and a cooling system (jacket or coils) will greatly increase the capacity of a tank crystallizer but leads to a substantially different crystallization system. Agitation sufficient to suspend the crystals introduces the possibility of increased nucleation as the solution reaches its initial crystallization point and during the cooling process from mechanical stimulation of the supersaturated solution and contact nucleation. Although more widely used than the static tank, this type of equipment has limited capacities and finds application only for the relatively low-rate production of such items as fine chemicals and some pharmaceutical products. The limitation preventing more widespread application is generally the buildup of crystallized material on the cooling coils. While vigorous agitation tends to reduce this tendency, it usually does not eliminate it. For maximum production capacity, the coolant should be tempered by recirculation so that the temperature difference across the coils is as small as feasible. For materials which adhere readily to the crystallizer walls and cooling surfaces, the temperature difference should be in the range 3 to 6°C (5 to 10°F). This imposes a severe limitation on the capacity of the equipment. The overall heat-transfer coefficient will generally be in the range 285 to 1135 W/(m^2·K) [50 to 200 Btu/(h·ft^2·°F)].

The application of **Teflon-tube heat exchangers** to cooling-type tank crystallizers appears promising. Teflon-tube bundles are installed so that the tubes are relaxed and the movement of the tubing due to agitation within the system causes a continual descaling of the heat-transfer surface. Transfer coefficients in such systems are in the range 115 (organics) to 370 (inorganics) W/(m^2·K) [20 to 65 Btu/(h·ft^2·°F)].

INFORMATION REQUIRED TO SPECIFY A CRYSTALLIZER

The following information regarding the product, properties of the feed solution, and required materials of construction must be available before a crystallizer application can be properly evaluated and the appropriate equipment options identified. Is the crystalline material being produced a hydrated or an anhydrous material? What is the solubility of the compound in water or in other solvents under consideration, and how does this change with temperature? Are other compounds in solution which coprecipitate with the product being crystallized, or do these remain in solution, increasing in concentration until some change in product phase occurs? What will be the influence of impurities in the solution on the crystal habit, growth, and nucleation rates? What are the physical properties of the solution and its tendency to foam? What is the heat of crystallization of the product crystal? What is the production rate, and what is the basis on which this production rate is computed? What is the tendency of the material to grow on the walls of the crystallizer? What materials of construction can be used in contact with the solution at various temperatures? What utilities will be available at the crystallizer location, and what are the costs associated with the use of these utilities? Is the final product to be blended or mixed with other crystalline materials or solids? What size of product and what shape of product are required to meet these requirements? How can the crystalline material be separated from the mother liquor and dried? Are there temperature requirements or wash requirements which must be met? How can these solids or mixtures of solids be handled and stored without undue breakage and caking?

Another basic consideration is whether crystallization is best carried out on a batch basis or on a continuous basis. The present tendency in most processing plants is to use continuous equipment whenever possible. Continuous equipment permits adjusting of the operating variables to a relatively fine degree in order to achieve the best results in terms of energy usage and product characteristics. It allows the use of a smaller labor force and results in a continuous utility demand, which minimizes the size of boilers, cooling towers, and power-generation facilities. It also minimizes the capital investment required in the crystallizer and in the feed-storage and product-liquor-storage facilities.

Materials that have a tendency to grow readily on the walls of the crytallizer require periodic washout, and therefore an otherwise continuous operation would be interrupted once or even twice a week for the removal of these deposits. The impact that this contingency may have on the processing-equipment train ahead of the crystallizer must be considered.

The batch handling of wet or semidry crystalline materials is substantially more difficult than the storing and handling of dry crystalline materials. A batch operation has economic application only on a relatively small scale or when temperature or product characteristics require unusual precautions.

CRYSTALLIZER OPERATION

Crystal growth is a layer-by-layer process, and the retention time required in most commercial equipment to produce crystals of the size normally desired is on the order of 2 to 6 h. On the other hand, nucleation in a supersaturated solution can be generated in a fraction of a second. The influence of any upsets in operating conditions, in terms of the excess nuclei produced, is very short-term in comparison with the total growth period of the product removed from the crystallizer. In a practical sense, this means that steadiness of operation is much more important in crystallization equipment than it is in many other types of process equipment.

It is to be expected that four to six retention periods will pass before the effects of an upset will be damped out. Thus, the recovery period may last from 8 to 36 h.

The **rate of nuclei formation** required to sustain a given product size decreases exponentially with increasing size of the product. Although when crystals in the range of 100 to 50 mesh are produced, the system may react quickly, the system response when generating large crystals in the 14-mesh size range is quite slow. This is because a single pound of 150-mesh seed crystals is sufficient to provide the total number of particles in a ton of 14-mesh product crystals. In any system producing relatively large crystals, nucleation must be carefully controlled with respect to all internal and external sources. Particular attention must be paid to preventing seed crystals from entering with the incoming feed stream or being returned to the crystallizer with recycle streams of mother liquor coming back from the filter or centrifuge.

Experience has shown that in any given body operating at a given production rate, control of the magma (slurry) density is important to the control of crystal size. Although in some systems a change in slurry density does not result in a change in the rate nucleation, the more general case is that an increase in the magma density increases the product size through reduction in nucleation and increased retention time of the crystals in the growing bed. The reduction in supersaturation at longer retention times together with the smaller distance between growing crystals, which lowers the driving force that is required to transport material from the liquid phase to the growing solids (**propinquity effect**), appears to be responsible for the larger product.

A reduction in the magma density will generally increase nucleation and decrease the particle size. This technique has the disadvantage that crystal formation on the equipment surfaces increases because lower slurry densities create higher levels of supersaturation within the equipment, particularly at the critical boiling surface in a vaporization-type crystallizer.

High levels of supersaturation at the liquid surface or at the tube walls in a surface-cooled crystallizer are the dominant cause of wall salting. Although some types of crystallizers can operate for several months continuously when crystallizing KCl or $(NH_4)_2SO_4$, most machines have much shorter operating cycles. Second only to control of particle size, the extension of operating cycles is the most difficult operating problem to be solved in most installations.

In the forced-circulation-type crystallizer (Fig. 19-43) primary control over particle size is exercised by the designer in selecting the circulating system and volume of the body. From the operating standpoint there is little that can be done to an existing unit other than supply external seed, classify the discharge crystals, or control the slurry density. Nevertheless, machines of this type are frequently carefully controlled by these techniques and produce a predictable and desirable product-size distribution.

When crystals cannot be grown sufficiently large in forced-circulation equipment to meet product-size requirements, it is common to employ one of the designs that allow some influence to be exercised over the population density of the finer crystals. In the DTB design (Fig. 19-45) this is done by regulating the flow in the circulating pipe so as to withdraw a portion of the fines in the body in the amount of about 0.05 to 0.5 percent by settled volume. The exact quantity of solids depends on the size of the product crystals and on the capacity of the fines-dissolving system. If the machine is not operating stably, this quantity of solids will appear and then disappear, indicating changes in the nucleation rate within the circuit. At steady-state operation, the quantity of solids overflowing will remain relatively constant, with some solids appearing at all times. Should the slurry density of product crystals circulated within the machine rise to a value higher than about 50 percent settled volume, large quantities of product crystals will appear in the overflow system, disabling the fines-destruction equipment. Too high a circulating rate through the fines trap will produce this same result. Too low a flow through the fines circuit will remove insufficient particles and result in a smaller product-size crystal. To operate effectively, a crystallizer of the type employing fines-destruction techniques requires more sophisticated control than does operation of the simpler forced-circulation equipment.

The classifying crystallizer (Fig. 19-48) requires approximately the same control of the fines-removal stream and, in addition, requires control of the fluidizing flow circulated by the main pump. This flow must be adjusted to achieve the proper degree of fluidization in the suspension chamber, and this quantity of flow varies as the crystal size varies between start-up operation and normal operation. As with the draft-tube-baffle machine, a considerably higher degree of skill is required for operation of this equipment than of the forced-circulation type.

While most of the industrial designs in use today are built to reduce the problems due to excess nucleation, it is true that in some crystallizing systems a deficiency of seed crystals is produced and the product crystals are larger than are wanted or required. In such systems nucleation can be increased by increasing the mechanical stimulus created by the circulating devices or by seeding through the addition of fine crystals from some external source.

CRYSTALLIZER COSTS

Because crystallizers can come with such a wide variety of attachments, capacities, materials of construction, and designs, it is very difficult to present an accurate picture of the costs for any except certain specific types of equipment, crystallizing specific compounds. This is illustrated in Fig. 19-50, which shows the prices of equipment for crystallizing two different compounds at various production rates, one of the compounds being produced in two alternative crystallizer modes. Installed cost (including cost of equipment and accessories, foundations and supporting steel, utility piping, process piping and pumps, electrical switchgear, instrumentation, and labor, but excluding cost of a building) will be approximately twice these price figures.

It should be ever present in the reader's mind that for every par-

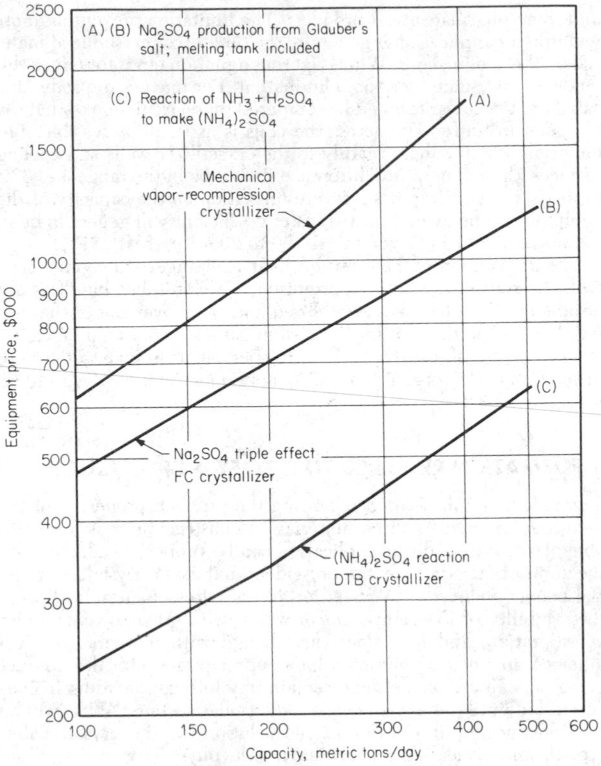

FIG. 19-50 Equipment prices, FOB point of fabrication, for typical crystallizer systems. Prices are for crystallizer plus accessories, including vacuum equipment. *(Data supplied by Swenson Division, Whiting Corporation, effective August 1983.)*

ticular case the appropriate crystallizer manufacturers should be consulted for reliable price estimates. Most crystallization equipment is custom-designed, and costs for a particular application may vary greatly from those illustrated in Fig. 19-50. Realistic estimation of installation costs also requires reference to local labor rates, site-specific factors, and other case specifics.

ION-EXCHANGE AND ADSORPTION EQUIPMENT

General References: Abrams, "Color Removal from Sugar Solutions by Adsorbent Resins," paper presented at 157th national meeting, American Chemical Society, Minneapolis, Apr. 15, 1969. Applebaum, *Demineralization by Ion Exchange*, Academic, New York, 1968. Arden, *Water Purification by Ion Exchange*, Plenum, New York, 1968. Calmon and Golds, *Ion Exchange for Pollution Control*, CRC Press, Boca Raton, Fla., 1979. Frazer, *Proc. Am. Power Conf.*, **29**, 815 (1967). Gilwood, *Chem. Eng.*, **74**(26), 83 (1967). Hiester, Cohen, and Phillips, *Chem. Eng. Prog. Symp. Ser.*, **50**(14), 23, 51, 63 (1954). Merritt, *The Extractive Metallurgy of Uranium*, Colorado School of Mines Research Institute, Golden, 1971. Michalson and Reents, *Proc. 28th Int. Water Conf., Eng. Soc. West. Pa.*, December 1967. Nachod and Schubert, *Ion Exchange Technology*, Academic, New York, 1956. Nordell, *Water Treatment for Industrial and Other Uses*, Reinhold, New York, 1961. Roberts, *Developments in Continuous Ion Exchange Equipment for AEC Applications*, Oak Ridge Nat. Lab. Rep. ORNL-2504, May 21, 1958. Thompson and Reents, *Proc. Am. Power Conf.*, **21**, 699 (1959). Wheaton and Lefevre, "Ion Exchange," *Kirk-Othmer Encyclopedia of Chemical Technology*, 3d ed., vol. 13, 1981, p. 678. Wirth, *Proc. Int. Water Conf., Eng. Soc. West. Pa.*, **41**, 1 (1980).

Operations that involve sorptive mass transfer between a liquid and the active surface of a solid are an important means of liquid purification, dissolved-solute recovery, and solute separation. In general, there are two broad classes of such operations: (1) ion exchange, in which a layer of free ions, held on a resinous sorbent by bound groups in the sorbent carrying the opposite charge, can be displaced by other ions of the same charge; and (2) adsorption, in which solute molecules are attracted to vacant active sites on the surface of a microcrystalline or resinous sorbent, condensing on these sites by virtue of the action of physical forces or chemical bonding. The apparatus in which these two operations are carried out is substantially the same. Because of differences in methods of regeneration, however, and characteristic differences between the processes to which they are respectively applied, the equipment for each class of operation will be treated individually.

ION-EXCHANGE OPERATIONS

General Design Ion-exchange resins are used in fixed beds, intermittent (called continuous) countercurrent columns, and slurry units. Of these, fixed beds are the commonest by far. They consist simply of pressure columns equipped with piping, valves, and accessory equipment to permit regeneration of the resins in place. The design of such columns is based upon the required flow rate, the size of batch to be treated between regenerations, and the capacity of the resin under the conditions of operation. In general, the flow rate determines the minimum and maximum allowable diameter of the column. The depth of resin bed must be sufficient to include the necessary volume of resin to provide sufficient contact time (minimum bed depth) for the exchange of ions to be completed. Required minimum bed depths have been empirically determined. If a bed is too shallow, distribution of the liquid is difficult, and complete exchange may not be achieved throughout a cycle. The lower 5 to 10 cm of resin will not be utilized fully because of uneven distribution, which represents an appreciable waste in a shallow bed. Distribution is particularly difficult if the equipment must operate over a wide range of flow rates. Usually, in water-treatment units a range of 4 to 1 is permissible. A smaller range is better when treating valuable products or when exchanging relatively high concentrations of ions.

If the volume of resin required is much larger than can be accommodated in a column suitable for the desired flow rate, storage of the product may be desirable to permit operation at an acceptable minimum rate. If the desired flow rate requires a larger column than is needed for the necessary resin volume, there must be a compromise. This problem and that which arises because ion exchange, an inherently batch operation, usually must satisfy a product demand that is continuous, can be solved by adroit process design. For example, smaller multiple columns equipped for automatic regeneration can be sized so that when one column is in regeneration the remaining unit can handle the total flow rate. When product storage is available, ion-exchange units can be designed to exhaust at higher than required flows and the surplus stored. The stored product can then be used during the regeneration cycle.

Before the sizes of ion-exchange equipment may be estimated, it is necessary to have a complete analysis of the water or other solution to be treated. The capacity of the resins for the ions to be exchanged must be known. Most currently available ion-exchange resins are relatively nonselective, so that all similar ions will be exchanged by a given resin whether the process requires this or not. Large amounts of data are available in published literature and sales brochures on capacities and applications of the various resins, but these are generally not specific to a given practical application and should be used for design purposes only by specialists. For any process that is not already thoroughly proved in other plants under similar conditions and for all major projects, laboratory and pilot-plant work is advisable to determine usable resin capacities, regenerant quantities, resin life, and quality of product. Firms which manufacture ion-exchange equipment will cooperate in such tests; they have experience in scaling up equipment from small laboratory columns. It is usually feasible to scale up to almost any size of plant from results obtained in columns 12.7 or 25.4 mm (1 or 2 in) in diameter, but such scale-up should not be attempted except by specialists.

Table 19-7 gives typical design data. These should be used for preliminary evaluation purposes only. Characteristic design calculations are presented and illustrated by Applebaum (op. cit., p. 226).

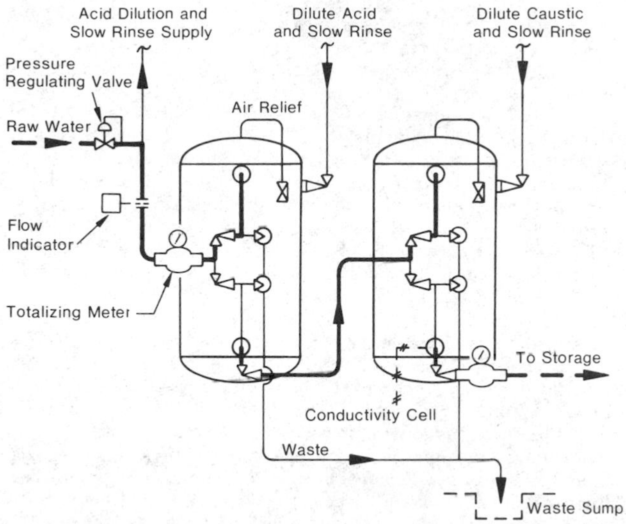

FIG. 19-51 Typical two-bed deionizing system. *(Infilco Degremont Inc.)*

Typical Ion-Exchange Column A typical fixed-bed ion exchanger consists of a vertical cylindrical pressure vessel of lined steel or stainless steel. Linings are usually of natural or synthetic rubber. Spargers are provided at the top and bottom, and frequently a separate distributor is used for regenerant solution. The resin bed, consisting of a meter or more of ion-exchange resin beads or particles, is supported by the screen of the bottom distributor. A distributor may be designed to prevent escape of resin particles by wrapping perforated pipe laterals with stainless steel or Saran screen or by placing a similar screen between perforated plates mounted in the top or bottom of the column. Externally, the unit is provided with a valve manifold to permit downflow operation, upflow backwashing, injection of regenerant, and rinsing of excess regenerant. A two-exchanger assembly comprising a cation and an anion bed in series, each contained in its own vessel, is a common configuration (Fig. 19-51).

If the unit is to treat a product of greater value than water, provision is made to remove the product as thoroughly as possible before regeneration and to displace the rinse water thoroughly before refilling with the valuable fluid. These steps are commonly known as "sweetening off" and "sweetening on" respectively. Because the resin beads are porous, some dilution of the product with water is

TABLE 19-7 Design Data for Fixed-Bed Ion Exchanger*

Type of resin	Maximum and minimum flow, m/h [gal/(min·ft²)]	Minimum bed depth, m (in)	Maximum operating temperatures, °C (°F)	Usable capacity, g-equivalent/L†	Regenerant, g/L resin‡
Weak acid cation	20 max. (8) 3 min. (1)	0.6 (24)	120 (248)	0.5–2.0	110% theoretical (HCl or H_2SO_4)
Strong acid cation	30 max. (12) 3 min. (1)	0.6 (24)	120 (248)	0.8–1.5 0.5–1.0 0.7–1.4	80–250 NaCl 35–200 66° Bé. H_2SO_4 80–500 20° Bé. HCl
Weak and intermediate base anions	17 max. (7) 3 min. (1)	0.75 (30)	40 (104)	0.8–1.4	35–70 NaOH
Strong-base anions	17 max. (7) 3 min. (1)	0.75 (30)	50 (122)	0.35–0.7	70–140 NaOH
Mixed cation and strong-base anion (chemical-equivalent mixture)	40 max. (16)	1.2 (47)	50 (122)	0.2–0.35 (based on mixture)	Same as cation and anion individually

*These figures represent the usual ranges of design for water-treatment applications. For chemical-process applications, allowable flow rates are generally somewhat lower than the maximums shown, and bed depths are usually somewhat greater.

†To convert to capacity in terms of kilograins of $CaCO_3$ per cubic foot of resin, multiply by 21.8.

‡To convert to pounds of regenerate per cubic foot of resin, multiply by 0.0625.

unavoidable. This can be minimized by using air to help displace water or product from the resin or by collecting dilute product obtained during the sweetening-off step and using it for sweetening on.

Freeboard is required above the resin bed to permit bed expansion during backwashing. Usually, the freeboard depth ranges from 50 to 100 percent of the resin depth, depending upon the type of resin.

Counterflow Regeneration Many ion exchangers operate downflow and are regenerated in the same direction (Fig. 19-52*a*). Generally, however, a better regeneration can be achieved by regenerating countercurrently to the service flow. For example, the monovalent ions (chiefly sodium) are most difficult to elute and are the cause of ion leakage in cation exchangers, and these ions concentrate in the lower part of the bed. In upflow regeneration, they are contacted by the fresh, more concentrated regenerant and are displaced upward so that after regeneration the lower part of the bed is most thoroughly regenerated. The result is a practical reduction of two-thirds of the subsequent leakage during the service run that would be encountered with downflow regeneration. However, counterflow regeneration requires that the resin bed be in a packed condition; this cannot be achieved in conventional equipment, in which there is space for expansion above the bed and the regenerant must pass upward to the top distributor. In order to overcome the problem of resin expansion during counterflow regeneration, some means must be provided to prevent rising of the bed. Arden (op. cit.) describes several methods that have been used to do this in Europe. The method which best prevents movement of resin bed during counterflow regeneration makes use of a dry layer of resin on top of the regular bed of resin. The dry layer is produced by applying vacuum through the collector in the top portion of the resin (Fig. 19-52*b*). The internals for such a unit are shown in Fig. 19-53.

Counterflow regeneration is essential to the efficient operation of "layered beds" consisting of two resins of differing ionic dissociation. The use of such beds has been recommended by some, but undesirable mixing of the two resins during backwash and regeneration and difficulty of achieving optimum regenerant concentration for both resins (problems often overlooked) have resulted in many unsatisfactory installations.

Mixed-Bed Ion Exchangers The typical ion exchanger contains only one ionic type of resin, either cation or anion, and regeneration is normally accomplished with one chemical solution, either a strong acid or a strong base. In a "mixed-bed" ion exchanger, however, a cation and an anion resin are contained in the same column. During the service (loading) step they are intimately mixed. For regeneration, backwashing separates the lighter anion resin from the denser cation resin. The unit has a screened distributor at the plane or interface between the two resins, so that they may be separately regenerated without removing them from the column.

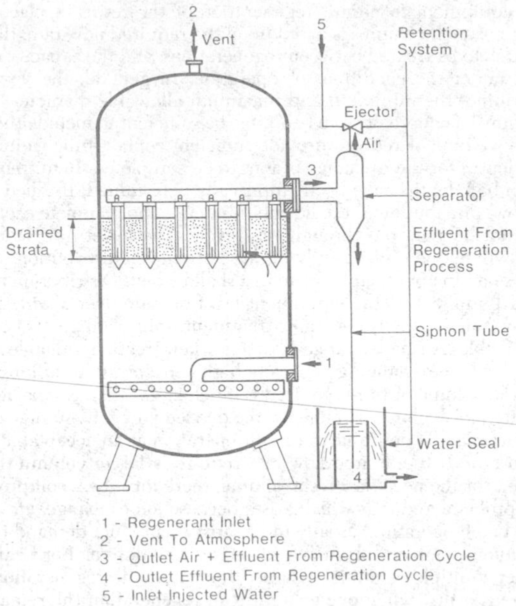

FIG. 19-53 Internals of an upflow regenerated unit. *(Infilco Degremont Inc.)*

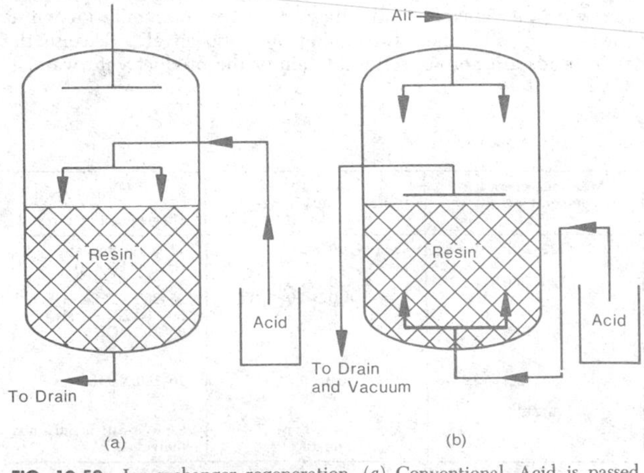

FIG. 19-52 Ion-exchanger regeneration. (*a*) Conventional. Acid is passed downflow through the cation-exchange resin bed. (*b*) Counterflow. Regenerant solution is introduced upflow with the resin bed held in place by a dry layer of resin.

The commonest method of regeneration (Fig. 19-54) permits sequential regeneration of the two resins, with first alkali flowing downward through the anion resin to the interface distributor and then acid flowing downward from the interface distributor through the cation resin. After regeneration and rinsing, the resins are remixed by compressed air. The bottom distributor is ordinarily of the screened type, in recognition of the difficulty of introducing compressed air into a support bed without upsetting graded layers and with the aim of providing product contamination. To permit uniform columnar exhaustion of the resin, the tank often has a flat bottom; in large-diameter columns this is accomplished by welding a flat false bottom in a conventional dished-head tank.

It is difficult to prevent intermixing of resins and some chemical penetration through the wrong one. This problem is alleviated by the use of inert resins of a density between the densities of cation and anion resins. During backwash and separation, the inert material forms a buffer zone between cation and anion layers.

Mixed-bed deionizers are often the choice to produce high-purity process water. Commonly the water to be purified is passed first through successive cation and anion beds and then through a mixed bed. Inasmuch as the influent to the mixed bed is relatively low in dissolved solids, the effluent can approach complete purity. To provide water of the high quality required by the semiconductor industry, the effluent of a mixed-bed unit may be polished further at the point of use by smaller cartridge-type mixed-bed exchangers. Numerous other examples of instances of the use of mixed beds to improve deionized water for process application can be found in the chemical, electronic, and metal-finishing industries.

Mixed-bed units are also used to polish condensate (boiler feedwater) in central power stations. Barring abnormal operation, the condensate contains very few impurities (on the order of micrograms per liter). However, even this amount of contaminant may be unacceptable for feedwater for supercritical boilers. In such applications, the superficial flow velocity may be as high as 100 m/h [41 gal/(min · ft^2)], with a resin depth of no more than 0.9 m (about 3 ft). Recent improvements in low-level ion detection (below 1 μg/L) indicate that such high velocity may not allow sufficient contact time for required removal of monovalent anionic species. This concern has

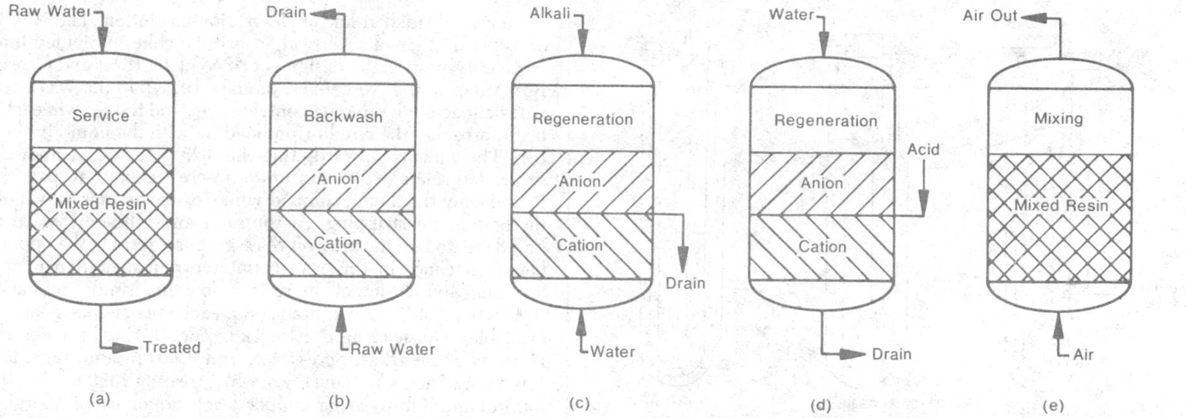

FIG. 19-54 Principles of mixed-bed ion exchange. (*a*) Service period. (*b*) Backwash period. (*c*) Caustic regeneration. (*d*) Acid regeneration. (*e*) Resin mixing.

led to increased resin bed depth (1.2 m) and flow rates in the neighborhood of 40 m/h [16 gal/(min·ft^2)].

The resins are flushed out of the service columns into separate regeneration equipment, since the relatively low regeneration flow rates would greatly complicate the design if the proper distributors were housed in the same vessel with the high-service flow distributors. Also, these installations may operate at 1.4 to 5.0 MPa (14 to 50 atm), whereas separate regeneration equipment, including valve manifolds, can be designed for relatively low pressures.

The effluent quality of a condensate polisher system depends critically on the mechanical design of the system—much more so than for other ion-exchange units. Although difficult to detect, uneven distribution of condensate through the resin beds should be avoided at all costs. Header-lateral distributors should be provided both for influent and effluent. Any resin left behind in the service columns (after flushing out for regeneration) may degrade the effluent quality and cause shortened service cycle. Resin selection is very important, and special resins that have abnormally large beads and are low in fines are used for this purpose.

Automation of Ion Exchangers As in other types of equipment, the term "automatic" is subject to various interpretations. A fully automatic deionizer has the following features: (1) bulk storage of regenerants; (2) automatic pumping and dilution of regenerants from bulk storage to ion-exchange units; (3) conductivity controllers or other method for sensing the quality of effluent and taking corrective procedures when needed; (4) a method for distinguishing between the need for regeneration and a transitory drop in effluent quality; (5) a fully automatic regeneration procedure, including automatic control of flows and dilution of regenerants; (6) fully automatic return to service or to standby condition, as required; (7) fail-safe protection to eliminate any risk of contamination of effluent with regenerant solutions, drain-down of ion-exchange columns or backup of water or product into regenerant storage; and (8) alarms to notify the operator of any malfunction and provision to advise the operator of the nature and location of the malfunction.

Since regeneration is accomplished by sequential operation of a number of valves, it is particularly adaptable to automatic operation. Multiple units are commonly used to permit uninterrupted flow and to allow individual units to be taken out of service and regenerated fully automatically. The units are commonly controlled by measurements of the conductivity and the silica content of the effluent. When high-purity water is essential, other ions (e.g., sodium and chloride) are also measured. Programmable controllers are uniquely suitable for controlling demineralizer-plant operation. These solid-state devices replace much of the mechanical hardware (such as timers and relays) of older control panels. Although microprocessors and computers can effectively control sophisticated processes, they do not appear to be particularly suitable for demineralizer automation.

Continuous and Quasi-Continuous Ion Exchange: Countercurrent Columns No packed beds of a truly continuous countercurrent design are presently in successful commercial operation, but the general term "continuous" is applied to ion-exchange equipment in which the resin and the contacting solutions flow countercurrently but intermittently. Typically, solution flow proceeds through a packed bed of resin for several minutes or longer and is stopped for less than a minute to permit the flow of a measured slug of resin. This period of interruption is short enough for the economical provision of surge tanks that permit the close approximation of continuous operation. Furthermore, resin flow is in increments which represent a small fraction of the total column of resin in the loading or stripping sections, so that the chemical advantages of true countercurrent operation can be closely approximated. When the resin-loading rate is known, the other sections can be independently sized to match that rate. This is in contrast to the fixed bed, in which the loading rate affects the size of column and amount of resin but in which the length of service run between regenerations and the time required for regeneration limit the daily capacity of the unit and dictate the number of parallel units needed to assure that one is always operating. If the loading rate is very high, multiple fixed beds are necessary in order to maintain *one* in service; a single continuous unit may then be designed to provide the same daily capacity, with a saving in capital cost and space requirements.

These advantages notwithstanding, it is apparent that a single continuous unit provides no emergency standby for maintenance, whereas duplex or multiple fixed beds automatically incorporate some standby capacity, the amount being subject to the designer's intent. This difference, combined with the fact that fixed-bed technology is the better established and employs equipment requiring little maintenance, has led the cautious water-treatment market to the choice of fixed-bed systems for some extremely large installations for which overall consideration appeared to favor continuous operation. Even in the classic environment of sophisticated continuous processing, that of the large petroleum or petrochemical complex, water-treatment plants tend to have intermittent fixed-bed designs. Yet, it is by the chemical industry that continuous exchangers can expect the likeliest increasing use.

As it turns out, a number of continuous designs have been marketed since 1960, and they were mainly for chemical-processing applications. The **Higgins** design was originally developed to recover uranium from reduction-residue sulfate leach slurries at the Oak Ridge Y-12 plant (Roberts, op. cit.), plutonium from nitrate solutions at the Hanford Purex plant, and uranium in chloride solution at the Japan Atomic Fuel Corporation uranium-metal production plant. More recently it has been adapted to a wide variety of applications, including large-volume water softening. The design is shown diagrammatically in Fig. 19-55. Larger units have relatively large-

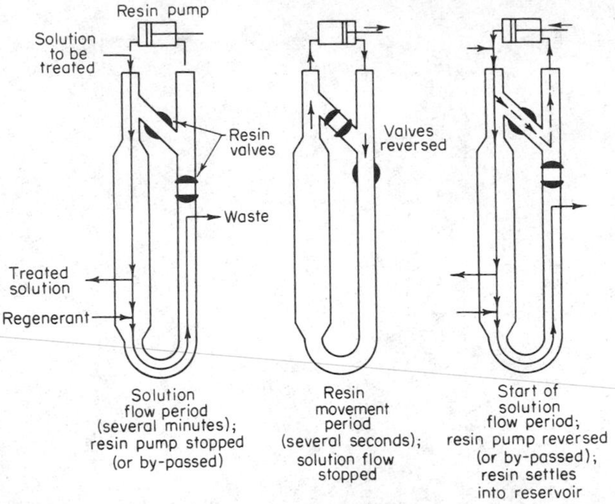

FIG. 19-55 Mode of operation of the Higgins contractor. *(ORNL-LR-Dwg. 27857R.)*

diameter loading and stripping sections, with resin valves and resin-transfer sections scaled down to a cross-sectional area of as little as 5 percent of the major contactor sections. To provide uniform resin flow, conical sections are used between these extremes. Resin is moved by a hydraulic impulse combined with valve operation.

The **Asahi** process (Fig. 19-56), a Japanese development, is used principally for high-volume water treatment [Gilwood, loc. cit.; and Newman, *Chem. Eng.*, **74**(26), 72 (1967)]. The liquid to be treated is passed upward through a column of resin in the adsorption tank. The upward flow at 30 to 40 m/h [12 to 16 gal/(min·ft^2)] keeps the bed packed against the top. After a preset time—10 to 60 min—the flow is interrupted for about 30 s, and the entire bed drops. A small portion (10 percent or less) of the resin is removed from the bottom of the adsorption tank and transferred hydraulically to the hopper feeding the regeneration tank. Simultaneously an equal quantity of regenerated resin is dropped into the adsorption tank. The process is then resumed. Meanwhile, regeneration is occurring by a similar flow system in the regeneration tank, from which the regenerated and partly reused resin is transferred periodically to the hopper above the wash-rinse tank. In the latter, the resin beds are fluidized by rinse water to flush away fines and foreign matter before the resin is returned to the adsorption-tank hopper.

Various schemes have been developed to provide contact between the resin and solution by agitating resin beads in the slurry and separating the loaded resin for desorption or elution. The U.S. Bureau of Mines developed a vertical adsorption column divided into compartments by specially designed orifice plates (Ross and George, U.S. Bur. Mines Rep. Invest. 7471, January 1971)]. In this way each compartment acts as a separate contact stage, and the resin in each chamber approaches the equilibrium loading with the liquid in the chamber. The flow is upward, thus fluidizing the resin and allowing suspended solids (e.g., as in uranium-ore leachate) to pass through. Periodically the flow is stopped while loaded resin is drawn off from the bottom for stripping. At the same time the downward flow of liquid created by the drawoff of resin causes resin in the upper chamber to descend. In this way countercurrent operation is obtained. Stripping and elution of the resin is done in a similar manner.

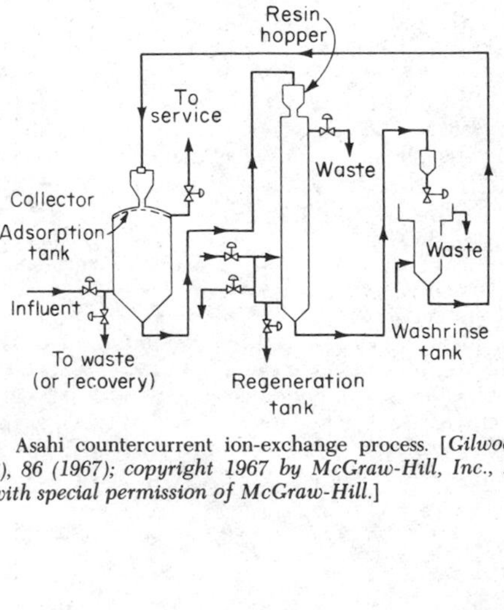

FIG. 19-56 Asahi countercurrent ion-exchange process. [*Gilwood,* Chem. Eng., *74(26), 86 (1967); copyright 1967 by McGraw-Hill, Inc., New York. Excerpted with special permission of McGraw-Hill.*]

Another fluidized continuous ion-exchange system commercially available (Himsley Engineering Ltd., Toronto) has been described by **Himsley and Farkas** (*Soc. Chem. Ind. Conf.*, Cambridge, England, July 1976). Such a system (Fig. 19-57), treating 1590 m^3/h (7000 gal/min) of uranium-bearing copper leach liquor using fiberglass-construction columns 3.7 m (12 ft) in diameter, has been described by Krist and Brost (paper presented at AIME annual meeting, Chicago, February 1981). The column is divided vertically into stages but differs from others in that it operates truly continuously, transferring measured batches of resin from stage to stage down the column without any interruption of feed flow. This is accomplished by pumping feed solution from one stage (A) to the stage (B) immediately above by means of external piping in such manner that the net flow through stage B is downward, carrying with it all of the resin in that stage B. When resin transfer is complete, the resin from stage C above is transferred downward in a similar manner. The process continues until the last stage (F) is empty. The regenerated resin is then transferred from the elution column to the empty stage (F).

The stripping of the loaded resin in the Himsley system employs a moving packed bed of resin. Each batch is introduced into the bottom of a tall column containing several batches of resin, and eluant or stripping solution passes down through the entire resin bed. This countercurrent-series elution results in a strong eluate with a minimum consumption of eluant.

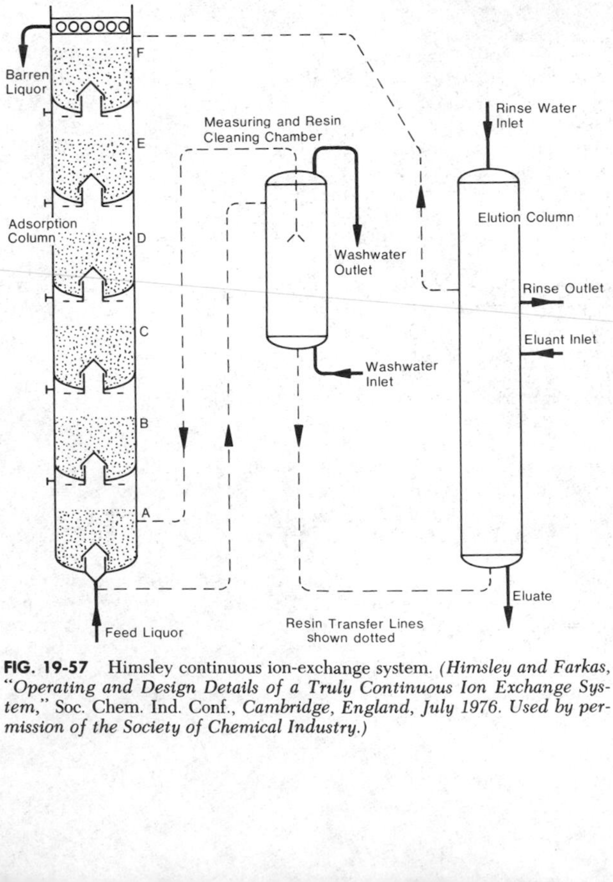

FIG. 19-57 Himsley continuous ion-exchange system. *(Himsley and Farkas, "Operating and Design Details of a Truly Continuous Ion Exchange System,"* Soc. Chem. Ind. Conf., *Cambridge, England, July 1976. Used by permission of the Society of Chemical Industry.)*

Such systems may be provided with rinse chambers to treat chemically eluted resin in order to change its ionic form or to remove silica or other unwanted material. The resin inventory in these continuous systems treating low-suspended solid-feed liquors is appreciably less than would be needed by conventional ion-exchange columns treating clarified liquor of the same ionic composition.

Disadvantages of countercurrent columns are essentially the same as those for continuous-sorption equipment, discussed in the following passage. Since ion-exchange resins are costly, the attrition resulting from resin transport is especially significant and must be evaluated. Some relatively fragile resins probably should not be used in this equipment. Processes requiring a series of resins, as, for example, weakly acid, strongly acid, weakly basic, and strongly basic mixed-bed, so far seem better adapted to fixed-bed designs.

Ion-Exchanger Costs As aids in estimation, Downing [*Chem. Eng.*, **72**(25), 170 (1965)] and Prater [*Pet. Refiner*, **39**(11), 261 1960] offer useful formulas, nomograms, and graphs for sizing and pricing ion exchangers. Obviously allowances must be made for subsequent price inflation. A number of economic indices are available for updating and bringing to a consistent basis such obsolete prices: for chemical-process equipment one of the best is the Marshall and Stevens index (published biweekly in *Chemical Engineering*).

As a general rule, fixed beds are more economical for flow rates below 20 L/s (317 gal/min) [Dallman, *Combustion*, **40**(7), 17 (January 1969)] and for circumstances in which duplex units will provide continuous product flow. Continuous units are more economical for flows above 20 L/s, when the loading rate is such that three or more parallel fixed beds are required to maintain continuous service. In the capacity range 50 to 100 L/s (800 to 1600 gal/min), for example, continuous processes like the Asahi required 25 to 30 percent less capital investment and 25 to 40 percent less expenditure for regenerating chemicals than fixed-bed processes did in 1967, according to Gilwood (op. cit.).

Typical costs for regenerating chemicals for both fixed-bed and continuous countercurrent water demineralizers are shown in Fig. 19-58. Allowances must be made for subsequent price inflation. A factor of 3 is suggested to bring the cost to 1983.

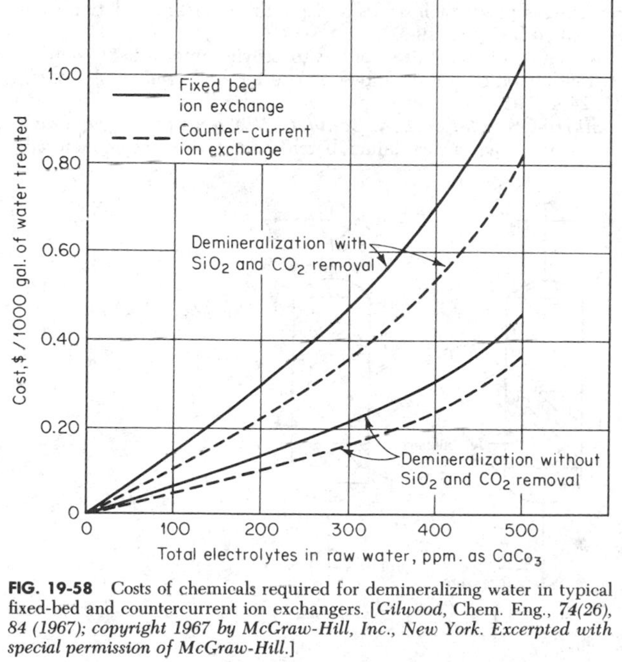

FIG. 19-58 Costs of chemicals required for demineralizing water in typical fixed-bed and countercurrent ion exchangers. [*Gilwood*, Chem. Eng., *74(26), 84 (1967); copyright 1967 by McGraw-Hill, Inc., New York. Excerpted with special permission of McGraw-Hill.*]

SORPTION EQUIPMENT

Ion-exchange resins, as well as some similar adsorbents without significant ion-exchange capabilities, are frequently used in sorption operations, in which no ion exchange takes place but in which the sequence of operation, and therefore the equipment design, is essentially identical to ion-exchange operations. Suitable adsorbents may be used, for example, for the removal of organic color from sugar solutions, glycerin, or water; they can be regenerated with caustic soda, acid, or in some cases organic solvents such as methanol. Color adsorbers are frequently incorporated in a train of ion exchangers for such applications, and are indistinguishable from the ion-exchange equipment in appearance and design.

Granular activated carbon is also used in adsorption applications like those mentioned previously, but the system is quite different from ion-exchange equipment because the carbon must be thermally regenerated. Such a system is illustrated in Fig. 19-59. Here, the choice of regeneration in place is not applicable, since regeneration must be accomplished in a furnace at approximately 925°C (1700°F).

Three major types of solid-fluid sorption equipment are in use: (1) batch units; (2) fixed beds of adsorbent through which the process fluid passes, with periodic interruption for regeneration; and (3) systems which provide for countercurrent or cocurrent movement of adsorbent and fluid in a continuous or quasi-continuous (intermittent) operation.

Batch Operations It is often advantageous to carry out sorbent-liquid contact in batch equipment; such methods are less frequently employed for the treatment of gases. Batch methods are well adapted to laboratory use and have also been applied on a larger scale in several specific instances.

Laboratory Methods In purifying the products of organic-chemical synthesis, decolorizing carbons and clays are frequently used as **contact adsorbents** (i.e., they are stirred directly into a liquid-phase mixture or solution) and are subsequently separated by filtration.

Batch tests or measurements are often conducted on portions of adsorbent or ion-exchange material intended for larger-scale use. For example, either the equilibrium sorptive uptake of a solid or its ultimate sorption capacity can be determined in this way. The solid is presaturated with solution that contains the solute at a specified concentration level; or preequilibrated with a pair of solutes, whose final solution-phase concentrations are determined separately. The solid can then be drained or centrifuged or rinsed so as to be wholly free of the initial solution. After this it is suspended in a quantity of a different liquid or solution, for titration or other similar analysis.

Contact Filtration of Lubricating Oils This process is used to remove colored and carbon-forming materials from lubricant stocks, as well as the traces of products formed in sulfuric acid treatment.

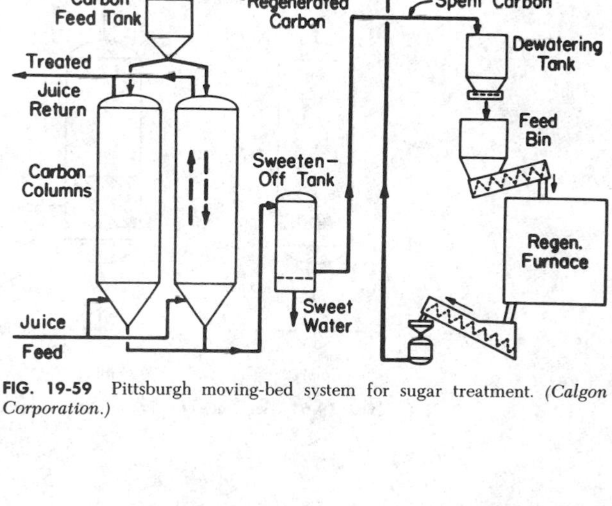

FIG. 19-59 Pittsburgh moving-bed system for sugar treatment. *(Calgon Corporation.)*

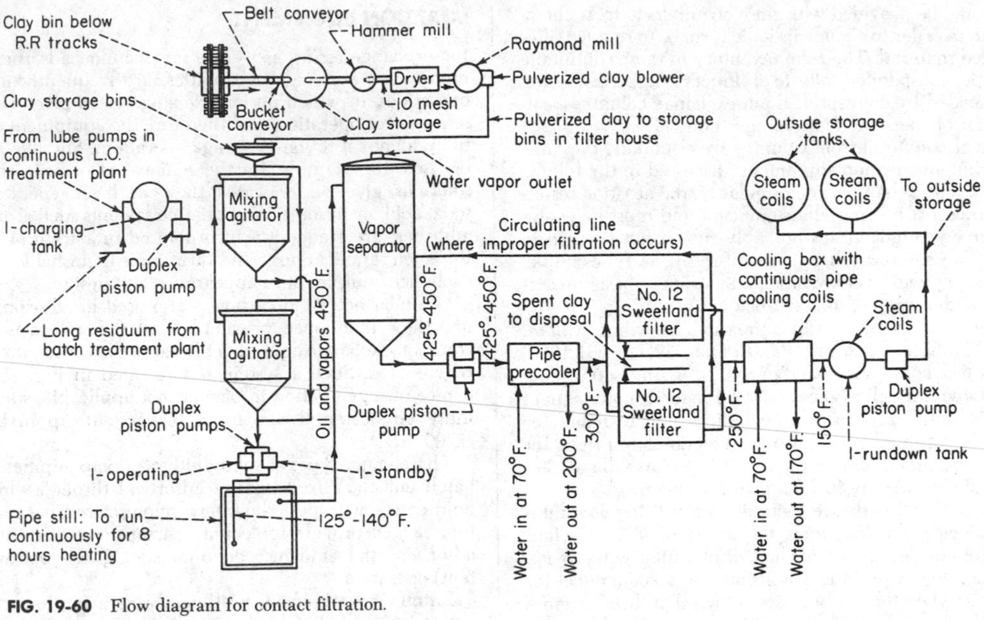

FIG. 19-60 Flow diagram for contact filtration.

In some cases, either vacuum distillates or residuum fractions from crude petroleum may require only acid and clay treatment in order to meet product specifications. Elsewhere, solvent extraction will be the principal means of treatment, followed by refining with acid and clay.

Figure 19-60 shows a contact-filter plant used for a "long-residuum" or "cylinder stock" having a flash point of 227 to 232°C (440 to 450°F) and a kinematic viscosity of 15.4 to 16.6 mm^2/s (Saybolt Universal viscosity of 80 to 85 s) at 99°C (210°F) [Kauffman, *Chem. Metall. Eng.*, 34, 155 (1927)]. This residuum is first treated with sulfuric acid in the proportion of 115 to 130 g/L (40 to 45 lb/bbl) of oil, at 60 to 66°C (140 to 150°F). After this treatment, the sludge is settled at the bottom of the agitator and drawn off. The oil, with a small amount of added clay, is filtered to remove emulsified acid. It is then mixed with about 60 g of clay/L (0.5 lb/gal) and pumped to a pipe still or similar heating unit, where it is brought to about 232°C (450°F). It is held at this temperature for several minutes, then cooled to 150°C (302°F) and filtered.

Particle size of the clay is usually 80 to 100 mesh, substantially finer than can be used in the alternative method of percolation treatment in fixed beds. Diatomaceous earth may be used as a filter precoat or may be mixed with the slurry to improve its filtering properties. The clay cake from the filter is usually washed with naphtha and blown with inert (flue) gas before it is discarded; it can be revivified by heating in air at between 540 and 750°C (about 1000 and 1400°F).

Mixer-Settler Operations Mixing and filtering can be accomplished in the same vessel by using large-particle sorbent material (50 mesh and coarser) through which a liquid can drain quite readily. The solid sorbent is reused a substantial number of times before it is regenerated or replaced. With the vessel filled with a charge of process liquor, gentle agitation is usually obtained by sparging the slurry with air. Draining of the liquor is subsequently hastened by use of a reverse gas flow which serves to expel the interstitial solution from the settled bed (Fig. 19-61).

A combination of mixer-settlers to provide intermittent countercurrent operation was described by Hiester et al. [*Ind. Eng. Chem.*, 45, 2402 (1953)].

Slurry-Granular Sorbent Contact When the process liquor is a slurry rather than a clear liquid, batch contact with sorbent materials

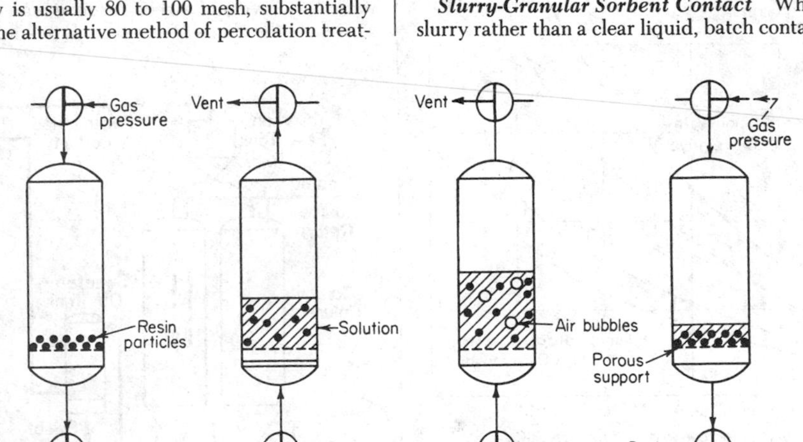

FIG. 19-61 Sequence of operations in a batch mixer-settler.

is generally preferred to percolation-type contact. The latter involves less agitation and is more likely to give a progressive accumulation of slurry particles in the body of sorbent. If the slurry particles have a different size or settling velocity from those of the sorbent, the sorbent may be incorporated in the slurry. After sufficient contact, the process liquor may be centrifuged or filtered. Subsequently, the sorbent is recovered by hydraulic classification or by wet screening.

Much of the ion-exchange development work in recent years has been done in connection with U.S. AEC-ERDA-DOE chemical-processing problems. One of the first and largest examples of such development is the Anaconda "resin-in-pulp" uranium mill at Grants, N.M., which produces about 1000 metric tons (1100 short tons) of ore per day.

The resin-in-pulp method was developed to utilize relatively coarse anion-exchange resin with a slurry of finely ground uranium ore in acid-leach liquor [Hollis and McArthur, *Proc. Int. Conf. Peaceful Uses At. Energy*, 8, 54 (1955)]. To avoid filtration, which would be necessary with conventional ion-exchange beds, large-mesh resin is loosely contained in stainless-steel screen "baskets" which oscillate up and down in troughs through which the pulp flows. This vertical action keeps the resin fluidized and prevents the slimes from fouling the resin. The pulp is fed through 10 series-connected troughs, each 15 m by 15 cm by 15 cm (50 ft by 6 in by 6 in). Each trough is fitted with 10 baskets, each containing 425 L (15 ft^3) of resin. Desorption of the uranium is carried out simultaneously by 4 additional troughs in series. The baskets oscillate at 5 cycles per minute through a 38-cm (15-in) stroke. The resin does not actually move countercurrently to the solutions, but the effect of continuous countercurrent flow is approximated by changing the feed and takeoff points from one trough to the next in a 14-trough time cycle.

A variation of this type of operation involves semicontinuous feed of fluid to an agitated vessel with fluid discharge through a screen that retains the resin (Davis, Ward, and Klinger, 1st Inter-American Congress on Chemical Engineering, Puerto Rico, 1961).

Fixed Beds The most frequently used method of fluid-solid contact for sorption operations is in columnar units, with the solid particles closely packed in a relatively fixed arrangement. Adsorbent particle sizes in such equipment usually lie in a narrow range but may average from as large as 4 mesh to as small as 250 mesh. **Pressure drop** is often taken as a determining factor in particle-size specification; on the other hand, performance of adsorbent columns often improves rapidly with decreasing particle size. Thus in liquid-phase processing, total cost of the adsorption step can sometimes be reduced by designing for overall pressure drops as large as 3 to 6 bar (300 to 600 kPa, or 50 to 100 lbf/in^2). Calculation of pressure drop in packed beds is described in Sec. 5.

For regenerating an adsorbent, a change in the equilibrium is usually desirable in order to make solute removal more rapid and thorough. This may involve a change in temperature, a change in carrier fluid, or frequently both. The regenerant may be a gas instead of a liquid—steam or hot air, for example—that vaporizes the adsorbate and sweeps it away. If so, the off gas passes through a condenser, where the solvent is recovered; any remaining fixed gas will not be completely free of solvent and may be sent to a gaseous-adsorption unit before being vented.

Joint Use of Adsorbents When an expensive adsorbent is needed to bring about the full extent of purification that is required, it is sometimes possible to reduce costs by combining its use with that of less expensive material. For example:

1. Suspended colloidal materials in an aqueous feed should be coalesced onto gravel or an inexpensive adsorbent, rather than being allowed to foul the surface of high-cost ion-exchange resin particles. In some cases, however, recently developed resins may be used as adsorbents for colloids and may be regenerated successfully and used for many cycles (Kunin and Hetherington, *Proc. 30th Int. Water Conf., Eng. Soc. West. Pa.*, October 1969).
2. For a solvent-rich air stream that also contains large proportions of water vapor, it may be advantageous to dry the air with a separate selective adsorbent (e.g., silica gel or alumina) ahead of the solvent-adsorption step.
3. Often as much as 90 percent of the total desired removal can be accomplished with a less selective (and less expensive) adsorbent, either in a single bed or in a cyclically operated cascade of beds. Use of a more expensive material thus is limited to final purification in a relatively small trimmer bed.

Cyclic Operations Often two fixed-bed adsorber or exchanger units are provided, so that one is on stream while the other is being regenerated. If the breakthrough curve is quite shallow and thus leaves a large proportion of unused capacity when the break point is reached, it may be desirable to use two or more beds in series, introducing a new bed at the downstream end each time that a completely spent bed upstream is removed from service. In this way, by a valve arrangement that provides the proper succession of saturant and regenerant streams to each bed, a fixed-bed assembly can simulate a completely countercurrent system.

Figure 19-62 shows the sequence of operating conditions in a **three-stage cascade,** or "merry-go-round" [Bulkeley, *Chem. Metall. Eng.*, **45**, 300 (1938)]. In any one step the first adsorber in the line is the partially charged unit which was second in the line in the preceding step, while the second adsorber is the unit which was being regenerated in the preceding step. Meanwhile, the third adsorber is undergoing regeneration (and cooling, if necessary). By this method each unit is rotated in turn through the three steps of regeneration, second-stage, and first-stage adsorption.

While this illustration deals only with vapor-phase adsorption and with a three-stage cascade, the principles it encompasses can be extended readily to other fixed-bed operations and to cascades containing a larger number of stages.

Continuous and Quasi-Continuous Operations Continuous-flow units involving the transport of solid particles have been proposed for both gas and liquid sorption operations. Although the operability of several of the available designs has been demonstrated, their relative economic advantage (compared with fixed-bed operation) is not well established. The primary problems to be overcome in continuous countercurrent sorption operations are the following:

1. Mechanical complexity of equipment
2. Gradual attrition of the solid sorbent
3. Limitations in particle-size range to avoid either classification or excessive pressure drop
4. Channeling (nonuniform flow) of either fluid or solid
5. Contamination between functional sections of the equipment, due to granular and porous structure of the solid

An analysis of the technical and economic factors involved shows that there is generally no overwhelming basis for preferring a countercurrent system over fixed-bed units [Hiester et al., *Chem Eng. Prog. Symp. Ser.*, **50**(14), 23, (1954)]. This conclusion changes, however, if there is an exceptionally favorable equilibrium for the removal process (as in water softening); and even more so if the regeneration step can be made favorable by the use of relatively concentrated eludent. An additional consideration is that continuous units generally require more headroom but less floor area.

Carbon-Treatment Systems The typical granular activated-carbon adsorber, incorporated in a system designed to permit thermal regeneration of the carbon, may be classified as a continuous countercurrent unit on the same basis as the so-called continuous ion exchangers described previously. Figure 19-59 illustrates such a system.

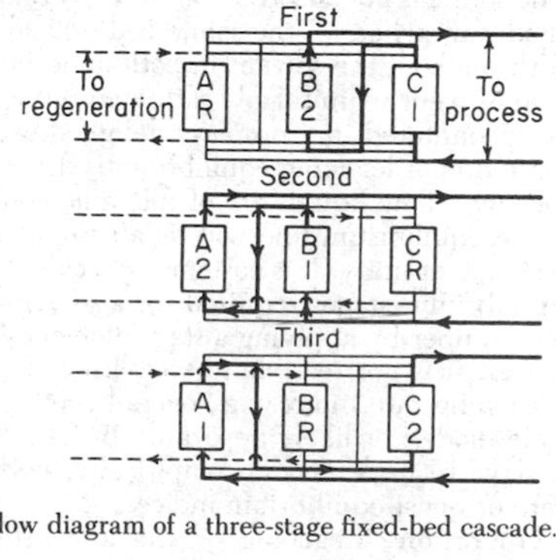

FIG. 19-62 Flow diagram of a three-stage fixed-bed cascade.

The adsorber tank is usually a vertical cylindrical pressure vessel, with fluid distributors at top and bottom, similar to the arrangement of an ion exchanger. The column is filled with granular carbon. Fluid flow is upward, and carbon is intermittently displaced downward by opening a valve at the bottom and injecting a measured slug of carbon into the top of the vessel. The exhausted slug (a small fraction of the total charge, in order to approximate fully countercurrent operation very closely) is transferred to the sweeten-off tank, where residual product is displaced. It next is dewatered and then is fed to the regeneration furnace, whence it eventually returns to the adsorber.

Fluid flow is interrupted briefly to permit carbon transfer, as in continuous ion exchangers.

LEACHING

GENERAL REFERENCES: Brunische-Olsen, *Solid Liquid Extraction*, NYT Nordisk Forlag, Arnold Busch, Copenhagen, 1962. Hamm, in *Kirk-Othmer Encyclopedia of Chemical Technology*, 3d ed., vol. 9, Wiley, New York, 1980, p. 721. Lerman, in Kirk and Othmer, *Encyclopedia of Chemical Technology*, 1st ed., vol. 6, Interscience Encyclopedia, New York, 1951, p. 91. McCabe and Smith, *Unit Operations of Chemical Engineering*, 3d ed., McGraw-Hill, New York, 1976, p. 607. Prabhudesai, in Schweitzer, *Handbook of Separation Techniques*, McGraw-Hill, New York, 1979, p. 5–3. Rickles, *Chem. Eng.*, **72** (6), 157 (1965).

DEFINITION

Leaching is the removal of a soluble fraction, in the form of a solution, from an insoluble, permeable solid phase with which it is associated. The separation usually involves selective dissolution, with or without diffusion, but in the extreme case of simple washing it consists merely of the displacement (with some mixing) of one interstitial liquid by another with which it is miscible. The soluble constituent may be solid or liquid; and it may be incorporated within, chemically combined with, adsorbed upon, or held mechanically in the pore structure of the insoluble material. The insoluble solid may be massive and porous; more often it is particulate, and the particles may be openly porous, cellular with selectively permeable cell walls, or surface-activated.

It is common practice to exclude from consideration as leaching the elution of surface-adsorbed solute. This process is treated instead as a special case of the reverse operation, adsorption (q.v.). Also usually excluded is the washing of filter cakes, whether in situ or by reslurrying and refiltration.

Because of its variety of applications and its importance to several ancient industries, leaching is known by a number of other names. Among those encountered in chemical engineering practice are extraction, solid-liquid extraction, lixiviation, percolation, infusion, washing, and decantation-settling.

Mechanism The mechanism of leaching may involve simple physical solution or dissolution made possible by chemical reaction. The rate of transport of solvent into the mass to be leached, or of soluble fraction into the solvent, or of extract solution out of the insoluble material, or some combination of these rates may be significant. A membranous resistance may be involved. A chemical-reaction rate may also affect the rate of leaching.

Inasmuch as the overflow and underflow streams are not immiscible phases but streams based on the same solvent, the concept of equilibrium for leaching is not the one applied in other mass-transfer separations. If the solute is not adsorbed on the inert solid, true equilibrium is reached only when all the solute is dissolved and distributed uniformly throughout the solvent in both underflow and overflow (or when the solvent is uniformly saturated with the solute, a condition never encountered in a properly designed extractor). The practical interpretation of leaching equilibrium is the state in which the overflow and underflow liquids are of the same composition; on a y-x diagram, the equilibrium line will be a straight line through the origin with a slope of unity. It is customary to calculate the number of ideal (equilibrium) stages required for a given leaching task and to adjust the number by applying a stage efficiency factor.

Usually, however, it is not feasible to establish a stage or overall efficiency or a leaching rate index (e.g., overall coefficient) without testing small-scale models of likely apparatus. In fact, the results of such tests may have to be sealed up empirically, without explicit evaluation of rate or quasi-equilibrium indices.

Methods of Operation Leaching systems are distinguished by operating cycle (batch, continuous, or multibatch intermittent); by direction of streams (cocurrent, countercurrent, or hybrid flow); by staging (single-stage, multistage, or differential-stage); and by method of contacting (sprayed percolation, immersed percolation, or solids dispersion). In general, descriptors from all four categories must be assigned to stipulate a leaching system completely (e.g., the Bollman-type extractor is a continuous hybrid-flow multistage sprayed percolator).

Whatever the mechanism and the method of operation, it is clear that the leaching process will be favored by increased surface per unit volume of solids to be leached and by decreased radial distances that must be traversed within the solids, both of which are favored by decreased particle size. Fine solids, on the other hand, cause slow percolation rate, difficult solids separation, and possible poor quality of solid product. The basis for an optimum particle size is established by these characteristics.

LEACHING EQUIPMENT

It is classification by contacting method that provides the two principal categories into which leaching equipment is divided: (1) that in which the leaching is accomplished by percolation and (2) that in which particulate solids are dispersed into a liquid and subsequently separated from it. Each includes batch and continuous units. Materials which disintegrate during leaching are treated in equipment of the second class.

A few designs of continuous machines fall in neither of these major classes.

Percolation In addition to being applied to ores and rock in place and by the simple technique of heap leaching, percolation is carried out in batch tanks and in continuous extractors.

Batch Percolators The **batch tank** is not unlike a big nutsche filter; it is a large circular or rectangular tank with a false bottom. The solids to be leached are dumped into the tank to a uniform depth. They are sprayed with solvent until their solute content is reduced to an economic minimum and are then excavated. Countercurrent flow of the solvent through a series of tanks is common, with fresh solvent entering the tank containing most nearly exhausted material. In a typical ore-dressing operation the tanks are 53 by 20 by 5.5 m (175 by 67 by 18 ft) and extract about 8200 Mg (9000 short tons) of ore on a 13-day cycle. Some tanks operate under presure, to contain volatile solvents or increase the percolation rate. A series of pressure tanks operating with countercurrent solvent flow is called a **diffusion battery.**

Continuous Percolators Coarse solids are also leached by percolation in moving-bed equipment, including single-deck and multideck rake classifiers, bucket-elevator contactors, and horizontal-belt conveyors.

The Bollman-type extractor shown in Fig. 19-63 is a bucket-elevator unit designed to handle about 2000 to 20,000 kg/h (50 to 500 short tons/day) of flaky solids (e.g., soybeans). Buckets with perforated bottoms are held on an endless moving belt. Dry flakes, fed into the descending buckets, are sprayed with partially enriched solvent. As the buckets rise on the other side of the unit, the solids are sprayed with a countercurrent stream of pure solvent. Exhausted flakes are dumped from the buckets at the top of the unit into a paddle conveyor; enriched solvent, the "full miscella," is pumped from the bottom of the casing. Because the solids are unagitated and because the final miscella moves cocurrently, the Bollman extractor permits the use of thin flakes while producing extract of good clarity. It is only

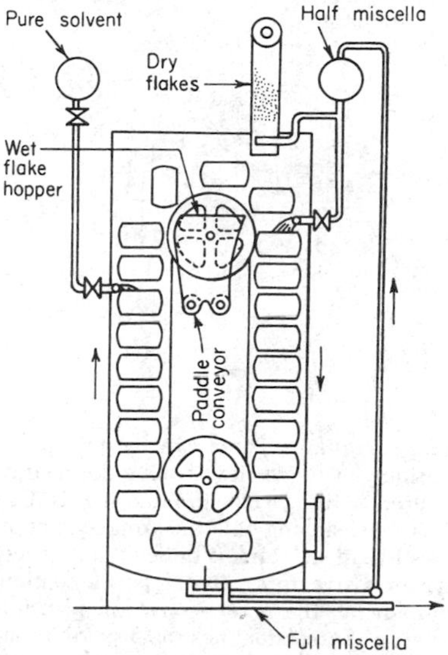

FIG. 19-63 Bollman-type extractor. *(McCabe and Smith,* Unit Operations of Chemical Engineering, *3d ed., p. 609. Copyright 1976 by McGraw-Hill, Inc., New York. Used with permission of McGraw-Hill Book Company.)*

partially a countercurrent device, however, and it sometimes permits channeling and consequent low stage efficiency.

In the **horizontal-basket design,** illustrated by the **Rotocel extractor** (Fig. 19-64), walled compartments in the form of annular sectors with liquid-permeable floors are rotated about a central axis. The compartments successively pass a feed point, a number of solvent sprays, a drainage section, and a discharge station (where the floor opens to discharge the extracted solids). The discharge station is circumferentially contiguous to the feed point. Countercurrent extraction is achieved by feeding fresh solvent only to the last compartment before dumping occurs and by washing the solids in each preceding compartment with the effluent from the succeeding one. The Rotocel is simple and inexpensive, and it requires little headroom. This type of equipment is made by a number of manufacturers. Horizontal table and tilting-pan vacuum filters (Figs. 19-104 and 19-105), of which it is the gravity counterpart, are used as extractors for leaching processes involving difficult solution-residue separation.

The **endless-belt percolator** (Hamm, loc. cit.) is similar in principle, but the successive feed, solvent spray, drainage, and dumping stations are linearly rather than circularly disposed. Examples are the **de Smet belt extractor** (uncompartmented) and the **Lurgi frame belt** (compartmented), the latter being a kind of linear equivalent of the Rotocel. Horizontal-belt vacuum filters (Fig. 19-106), which resemble endless-belt extractors, are sometimes used for leaching.

The **Kennedy extractor** (Fig. 19-65), also requiring little headroom, operates substantially as a percolator that moves the bed of solids through the solvent rather than the conventional opposite. It comprises a nearly horizontal line of chambers through each of which in succession the solids being leached are moved by a slow impeller enclosed in that section. There is an opportunity for drainage between stages when the impeller lifts solids above the liquid level before dumping them into the next chamber. Solvent flows by gravity from chamber to chamber countercurrently to the solids movement. The solids are subjected to mechanical action somewhat more intense than in other types of continuous percolator.

Dispersed-Solids Leaching Equipment for leaching fine solids by dispersion and separation includes batch tanks agitated by rotating impellers or by air and a variety of continuous devices.

Batch Stirred Tanks Tanks agitated by coaxial impellers (turbines, paddles, or propellers) are commonly used for batch dissolution of solids in liquids and may be used for leaching fine solids. Insofar as the controlling rate in the mass transfer is the rate of transfer of material into or from the interior of the solid particles rather than the rate of transfer to or from the surface of particles, the main function of the agitator is to supply unexhausted solvent to the particles while they reside in the tank long enough for the diffusive process to be completed. The agitator does this most efficiently if it just gently circulates the solids across the tank bottom or barely suspends them above the bottom. The design of solids-suspending agitators is discussed in the subsection "Batch Suspension of Closely Sized Particles."

The leached solids must be separated from the extract by settling and decantation or by external filters, centrifuges, or thickeners, all of which are treated elsewhere in Sec. 19. The difficulty of solids-extract separation and the fact that batch stirred tanks provide only a single equilibrium stage are their major disadvantages.

Pachuca tanks. Ores of gold, uranium, and other metals are

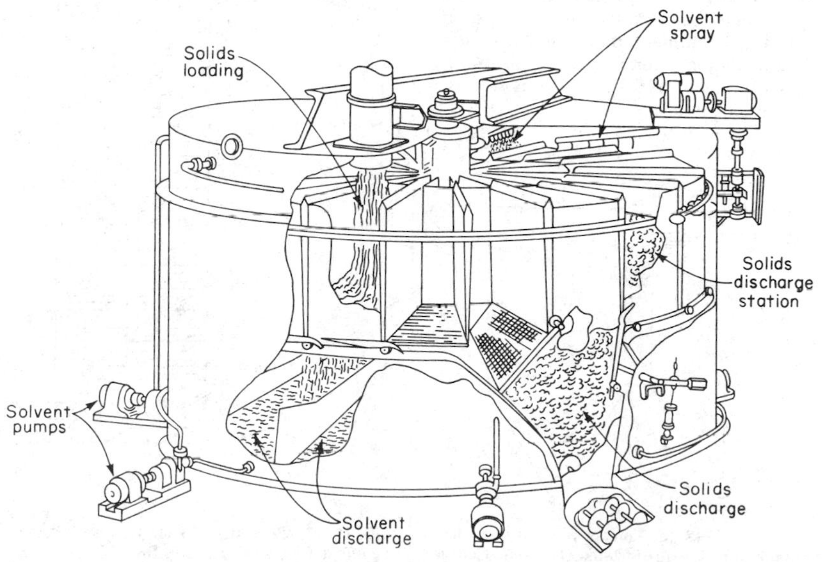

FIG. 19-64 Rotocel extractor. [*Rickles,* Chem. Eng., *72(6), 164 (1965). Used with permission of McGraw-Hill, Inc.*]

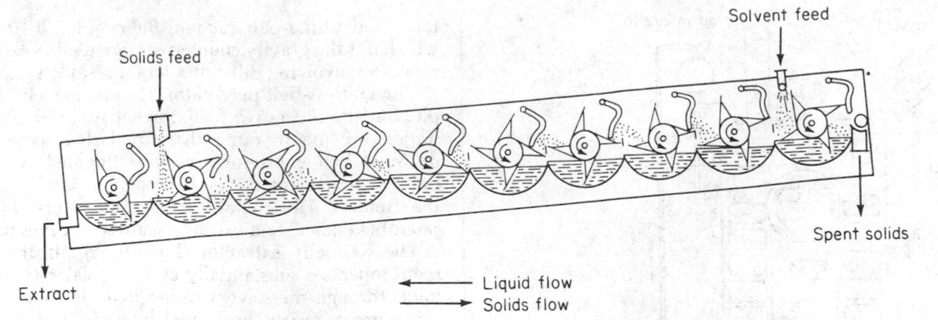

FIG. 19-65 Kennedy extractor. *(Vulcan Cincinnati, Inc.)*

commonly batch-leached in large air-agitated vessels known as Pachuca tanks. A typical tank is a vertical cylinder with a conical bottom section usually with a 60° included angle, 7 m (23 ft) in diameter and 14 m (46 ft) in overall height. In some designs air is admitted from an open pipe in the bottom of the cone and rises freely through the tank; more commonly, however, it enters through a central vertical tube, characteristically about 46 cm (18 in) in diameter, that extends from the bottom of the tank to a level above the conical section—in some cases, almost to the liquid surface (Fig. 19-66). Before it disengages at the liquid surface, the air induces in and above the axial tube substantial flow of pulp, which then finds its way down the outer part of the tank, eventually reentering the riser. The circulation rate in Pachuca tanks is discussed by Lamont [*Can. J. Chem. Eng.*, **36**, 153 (1958)].

Continuous Dispersed-Solids Leaching

Vertical-plate extractor. Exemplified by the **Bonotto extractor** (Fig. 19-67), this consists of a column divided into cylindrical compartments by equispaced horizontal plates. Each plate has a radial opening staggered 180° from the openings of the plates immediately above and below it, and each is wiped by a rotating radial blade. The solids, fed to the top plate, thus are caused to fall to each lower plate in succession. The solids fall as a curtain into solvent which flows upward through the tower. They are discharged by a screw conveyor and compactor.

Gravity sedimentation tanks. Operated as thickeners, these tanks can serve as continuous contacting and separating devices in which fine solids may be leached continuously. A series of such units properly connected permit true continuous countercurrent washing of fine solids. If appropriate, a mixing tank may be associated with each thickener to improve the contact between the solids and liquid being fed to that stage. Gravity sedimentation thickeners are described under "Gravity Sedimentation Operations." Of all continuous leaching equipment, gravity thickeners require the most area, and they are limited to relatively fine solids.

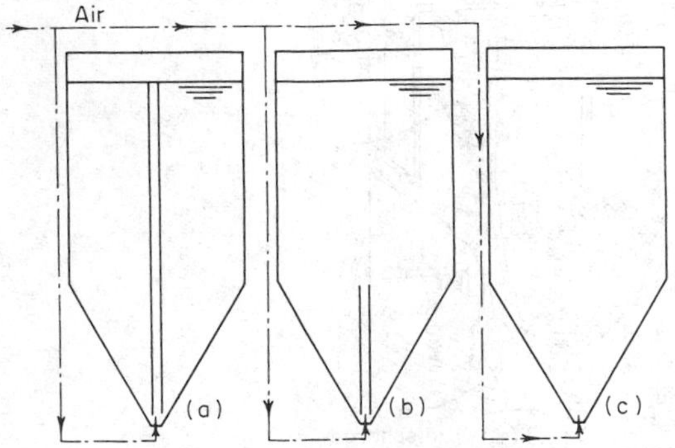

FIG. 19-66 Three designs of Pachuca tanks. (*a*) Komata-Reefs (full-center-column tank). (*b*) Stub-column tank. (*c*) Free-air-lift tank. [*Lamont,* Can. J. Chem. Eng., *36, 156 (1958); used with permission.*]

Impeller-agitated tanks. These can be operated as continuous leaching tanks, singly or in a series. If the solids feed is a mixture of particles of different settling velocities and if it is desirable that all particles reside in the leaching tank the same lengths of time, design of a continuous stirred leach tank is difficult and uncertain.

Screw-Conveyor Extractors One type of continuous leaching equipment, employing the screw-conveyor principle, is strictly speaking neither a percolator nor a dispersed-solids extractor. Although it is often classed with percolators, there can be sufficient agitation of the solids during their conveyance by the screw that the action differs from an orthodox percolation.

The **Hildebrant total-immersion extractor** is shown schematically in Fig. 19-68. The helix surface is perforated so that solvent can pass

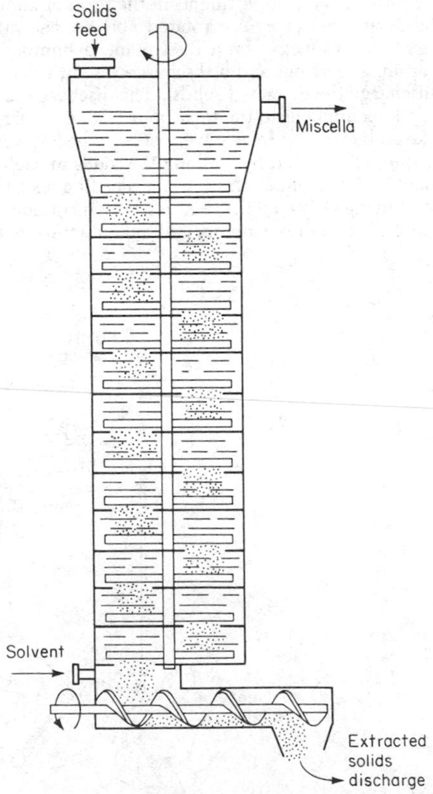

FIG. 19-67 Bonotto extractor. [*Rickles,* Chem. Eng., *72(6), 163 (1965); copyright 1965 by McGraw-Hill, Inc., New York. Excerpted with special permission of McGraw-Hill.*]

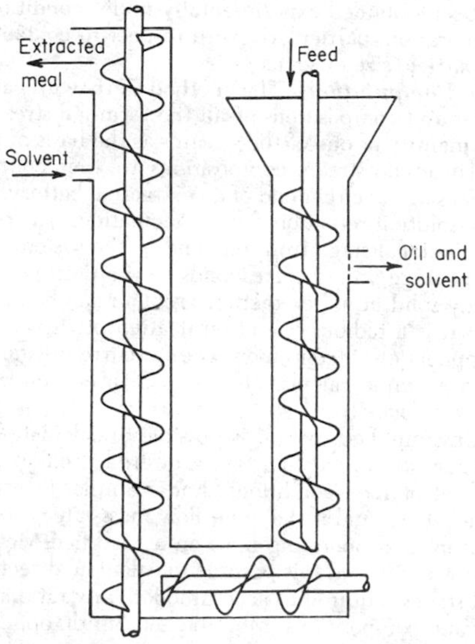

FIG. 19-68 Hildebrandt extractor. *(McCabe and Smith,* Unit Operations of Chemical Engineering, *3d ed., p. 609. Copyright 1976 by McGraw-Hill, Inc., New York. Used with permission of McGraw-Hill Book Company.)*

through countercurrently. The screws are so designed to compact the solids during their passage through the unit. The design offers the obvious advantages of countercurrent action and continuous solids compaction, but there are possibilities of some solvent loss and feed overflow, and successful operation is limited to light, permeable solids.

A somewhat similar but simpler design uses a horizontal screw section for leaching and a second screw in an inclined section for washing, draining, and discharging the extracted solids.

In the **De Danske Sukkerfabriker,** the axis of the extractor is tilted to about 10° from the horizontal, eliminating the necessity of two screws at different angles of inclination.

Sugar-beet cossettes are successfully transported upward in a vertical tower by an arrangement of inclined plates or wings attached to an axial shaft, the action being assisted by staggered guide plates on the tower wall. The shell is filled with water that passes downward as the beets travel upward. This configuration is employed in the **BMA diffusion tower** (Hamm, loc. cit.).

SELECTION OR DESIGN OF A LEACHING PROCESS*

At the heart of a leaching plant design at any level—conceptual, preliminary, firm engineering, or whatever—is unit-operations and process design of the extraction unit or line. The major aspects that are particular for the leaching operation are the selection of process and operating conditions and the sizing of the extraction equipment.

Process and Operating Conditions The major parameters that must be fixed or identified are the solvent to be used, the temperature, the terminal stream compositions and quantities, leaching cycle (batch or continuous), contact method, and specific extractor choice.

Choice of Solvent The solvent selected will offer the best balance of a number of desirable characteristics: high saturation limit and selectivity for the solute to be extracted, capability to produce extracted material of quality unimpaired by the solvent, chemical stability under process conditions, low viscosity, low vapor pressure, low toxicity and flammability, low density, low surface tension, ease and economy of recovery from the extract stream, and price. These factors are listed in an approximate order of decreasing importance, but the specifics of each application determine their interaction and relative significance, and any one can control the decision under the right combination of process conditions.

Temperature The temperature of the extraction should be chosen for the best balance of solubility, solvent-vapor pressure, solute diffusivity, solvent selectivity, and sensitivity of product. In some cases, temperature sensitivity of materials of construction to corrosion or erosion attack may be significant.

Terminal Stream Compositions and Quantities These are basically linked to an arbitrary given: the production capacity of the leaching plant (rate of extract production or rate of raw-material purification by extraction). When options are permitted, the degree of solute removal and the concentration of the extract stream chosen are those that maximize process economy while sustaining conformance to regulatory standards.

Leaching Cycle and Contact Method As is true generally, the choice between continuous and intermittent operation is largely a matter of the size and nature of the process of which the extraction is a part. The choice of a percolation or solids-dispersion technique depends principally on the amenability of the extraction to effective, sufficiently rapid percolation.

Type of Reactor The specific type of reactor that is most compatible (or least incompatible) with the chosen combination of the preceding parameters seldom is clearly and unequivocally perceived without difficulty, if at all. In the end, however, that remains the objective. As is always true, the ultimate criteria are reliability and profitability.

Extractor-Sizing Calculations For any given throughput rate (which fixes the cross-sectional area and/or the number of extractors), the size of the units boils down to the number of stages required, actual or equivalent. In calculation, this resolves into determination of the number of ideal stages required and application of appropriate stage efficiencies. The methods of calculation resemble those for other mass-transfer operations (see Secs. 13, 14, and 15), involving equilibrium data and contact conditions, and based on material balances. They are discussed briefly here with reference to countercurrent contacting.

Composition Diagrams In its elemental form, a leaching system consists of three components: inert, insoluble solids; a single nonadsorbed solute, which may be liquid or solid; and a single solvent.† Thus, it is a ternary system, albeit an unusual one, as already mentioned, by virtue of the total mutual "insolubility" of two of the phases and the simple nature of equilibrium.

The composition of a typical system is satisfactorily presented in the form of a diagram. Those diagrams most frequently employed are a right-triangular plot of mass fraction of solvent against mass fraction of solute (Fig. 19-69*a*) and a plot suggestive of a Ponchon-Savarit diagram, with inerts taking the place of enthalpy (Fig. 19-69*b*). A third diagram, less frequently used, is a modified McCabe-Thiele plot in which the overflow solution (inerts-free) and the underflow solution (traveling out of a stage with the inerts) are treated as pseudo phases, the mass fraction of solute in overflow, y, being plotted against the mass fraction of solute in underflow, x. (An additional representation, the equilateral-triangular diagram frequently employed for liquid-liquid ternary systems, is seldom used because the field of leaching data is confined to a small portion of the triangle.)

With reference to Fig. 19-69 (both graphs), EF represents the locus of overflow compositions for the case in which the overflow stream contains no inert solids. $E'F'$ represents the overflow streams containing some inert solids, either by entrainment or by partial solubility in the overflow solution. Lines GF, GL, and GM represent the loci of underflow compositions for the three different conditions indicated on the diagram. In Fig. 19-69*a*, the constant underflow line

*Portions of the passage on "Design Calculation Methods" in Sec. 17 of *Chemical Engineers' Handbook,* 5th ed., were incorporated into this section without substantial change. Authorship of them by R. K. Prabhudesai is gratefully acknowledged.

†The solubility of the inert, adsorption of solute on the inert, and complexity of solvent and extracted material can be taken into account if necessary. Their consideration is beyond the scope of this treatment.

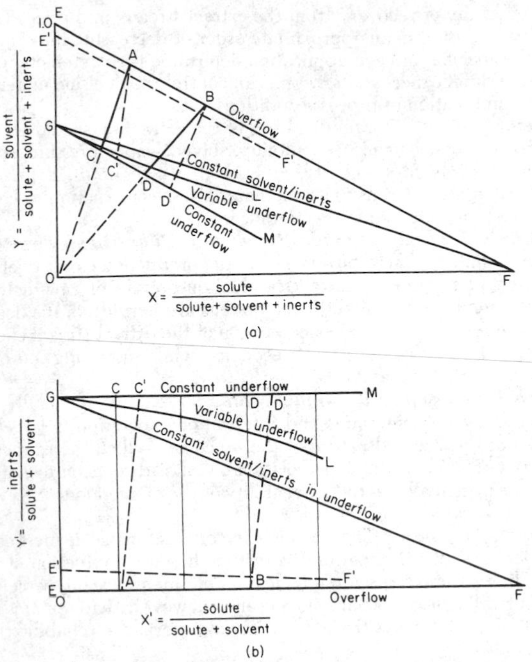

FIG. 19-69 Composition diagrams for leaching calculations. (*a*) Right-triangular diagram. (*b*) Modified Ponchon-Savarit diagram.

GM is parallel to *EF*, the hypotenuse of the triangle, whereas *GF* passes through the right-hand vertex representing 100 percent solute. In Fig. 19-69*b*, underflow line *GM* is parallel to the abscissa, and *GF* passes through the point on the abscissa representing the composition of the clear solution adhering to the inert solids.

Compositions of overflow and underflow streams leaving the same stage are represented by the intersection of the composition lines for those streams with a tie line (*AC*, *AC'*, *BD*, *BD'*). Equilibrium tie lines (*AC*, *BD*) pass through the origin (representing 100 percent inerts) in Fig. 19-69*a*, and are vertical (representing the same inert-free solution composition in both streams) in Fig. 19-69*b*. For nonequilibrium conditions with or without adsorption or for equilibrium conditions with selective adsorption, the tie lines are displaced, such as *AC'* and *BD'*. Point *C'* is to the right of *C* if the solute concentration in the overflow solution is less than that in the underflow solution adhering to the solids. Unequal concentrations in the two solutions indicate insufficient contact time and/or preferential adsorption of one of the components on the inert solids. Tie lines such as *AC'* may be considered as "practical tie lines" (i.e., they represent actual rather than ideal stages) if data on underflow and overflow composition have been obtained experimentally under conditions simulating actual operation, particularly with respect to contact time, agitation, and particle size of solids.

Algebraic Computation This method starts with calculation of the quantities and compositions of all the terminal streams, using a convenient quantity of one of the streams as the basis of calculation. Material balance and stream compositions are then computed for a terminal ideal stage at either end of an extraction battery, using equilibrium and solution-retention data. Calculations are repeated for each successive ideal stage from one end of the system to the other until an ideal stage which corresponds to the desired conditions is obtained. Any solid-liquid extraction problem can be solved by this method, but it is a tedious operation if attempted by a hand computer. Its application to continuous countercurrent leaching is illustrated by an example calculation in the fifth edition of *Chemical Engineers' Handbook* (p. 17-4).

For certain simplified cases it is possible to calculate directly the number of stages required to attain a desired product composition for a given set of feed conditions. For example, if equilibrium is attained in all stages and if the underflow mass rate is constant, both the equilibrium and operating lines on a modified McCabe-Thiele diagram are straight, and it is possible to calculate directly the number of ideal stages required to accommodate any rational set of terminal flows and compositions (McCabe and Smith, op. cit.):

$$N = \frac{\log [(y_b - x_b)/(y_a - x_a)]}{\log [(y_b - y_a)/(x_b - x_a)]} \tag{19-30}$$

Even when the conditions of equilibrium in each stage and constant underflow obtain, Eq. (19-30) normally is not valid for the first stage because the unextracted solids entering that stage usually are not premixed with solution to produce the underflow mass that will leave. This is easily rectified by calculating the exit streams for the first stage and using those values in Eq. (19-30) to calculate the number of stages required after stage 1.

Graphical Method This method of calculation is simply a diagrammatic representation of all the possible compositions in a leaching system, including equilibrium values, on which material balances across ideal (or, in some cases, nonideal) stages can be evaluated in the graphical equivalent of the stage-by-stage algebraic computation. It normally is simpler than the hand calculation of the algebraic solution, and it is viewed by many as helpful because it permits visualization of the process variables and their effect on the operation. Any of the four types of composition diagrams described above can be used, but modified Ponchon-Savarit or right-triangular plots (Fig. 19-69) are most convenient for leaching calculations.

The techniques of graphical solution are not unlike those for distillation and absorption (binary) problems using McCabe-Thiele, Ponchon-Savarit, and right-triangular diagrams and are very similar to those described in Sec. 15 for solvent-extraction (ternary) systems, using Janecke coordinates. For more detailed exposition and illustrations, see Brown et al., *Unit Operations*, Wiley, New York, 1950; McCabe and Smith, loc. cit., also chap. 18; and Treybal, *Mass Transfer Operations*, 2d ed., McGraw-Hill, New York, 1968.

PHASE SEPARATION

Equipment by which solids-liquid separations are performed is classified by the driving force invoked for the separation. Thus the subject is usually divided into the fields of gravity sedimentation, magnetic separation, filtration (excluding centrifugal filtration), centrifugation (both filtering and sedimentary), and expression. These subjects are treated in the balance of Sec. 19; only a few lines are devoted to magnetic separation, however, since it is primarily of importance to the mineralogical and metallurgical industries rather than the chemical. A summary passage on the competitive selection of a solids-liquid separator also is included.

Not dealt with are liquid elutriation and classification by settling-velocity methods. Although these operations are carried out in liquid-solid systems, they are separations achieved among groups of particles rather than of particles from suspending liquid. They are presented, therefore, in Sec. 21, and the principles underlying them are reviewed in Sec. 5.

GRAVITY SEDIMENTATION OPERATIONS

General References: Coe and Clevenger, *Trans. Am. Inst. Min. Eng.*, **55**, 356 (1916). Comings, Pruiss, and DeBord, *Ind. Eng. Chem.*, **46**, 1164 (1954). Counselmann, *Trans. Am. Inst. Min. Eng.*, **187**, 223 (1950). Fitch, *Ind. Eng. Chem.*, **58**(10), 18 (1966). *Trans. Soc. Min. Eng. AIME*, **223**, 129 (1962). Kynch, *Trans. Faraday Soc.*, **48**, 166 (1952). Pearse, *Gravity Thickening Theories: A Review*, Warren Spring Laboratory, Hertfordshire, 1977. Purchas, *Solid/Liquid Separation Equipment Scale-Up*, Uplands Press, Croydon, England, 1977. Talmage and Fitch, *Ind. Eng. Chem.*, **47**, 38 (1955). Wilhelm and Naide, *Min. Eng. (Littleton, Colo.)*, **33**, 1710 (1981).

Sedimentation is the partial separation or concentration of suspended solid particles from a liquid by gravity settling. This field may be divided into the functional operations of thickening and clarification. The primary purpose of thickening is to increase the concentration of suspended solids in a feed stream, while that of clarification is to remove a relatively small quantity of suspended particles and produce a clear effluent. These two functions are similar and occur simultaneously, and the terminology merely makes a distinction between the primary process results desired. Generally, thickener mechanisms are designed for the heavier-duty requirements imposed by a large quantity of relatively concentrated pulp, while clarifiers usually will include features that ensure essentially complete suspended-solids removal, such as greater depth, special provision for coagulation or flocculation of the feed suspension, and greater overflow-weir length.

CLASSIFICATION OF SETTLEABLE SOLIDS; SEDIMENTATION TESTS

The types of sedimentation encountered in process technology will be greatly affected not only by the obvious factors—particle size, liquid viscosity, solid and solution densities—but by the characteristics of the particles within the slurry. These properties, as well as the process requirements, will help determine both the type of equipment which will achieve the desired ends most effectively and the testing methods to be used to select the equipment.

Figure 19-70 illustrates the relationship between solids concentration, interparticle cohesiveness, and the type of sedimentation that may exist. "Totally discrete" particles include many mineral particles (usually greater in diameter than 20 μm), salt crystals, and similar substances that have little tendency to cohere. "Flocculent" particles generally will include those smaller than 20 μm (unless present in a dispersed state owing to surface charges), metal hydroxides, many chemical precipitates, and most organic substances other than true colloids.

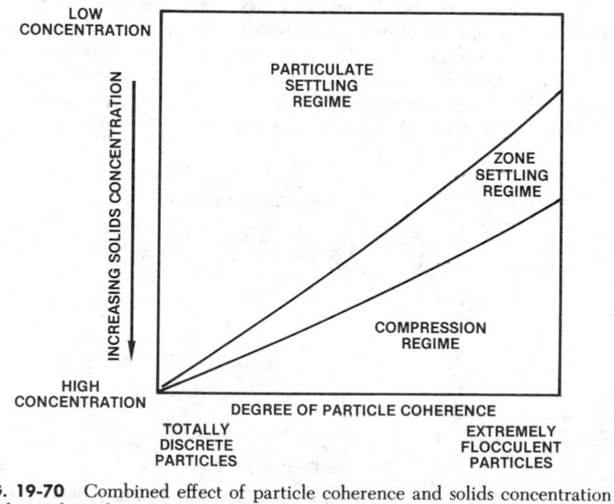

FIG. 19-70 Combined effect of particle coherence and solids concentration on the settling characteristics of a suspension.

At low concentrations, the type of sedimentation encountered is called particulate settling. Regardless of their nature, particles are sufficiently far apart to settle freely. Faster-settling particles may collide with slower-settling ones and, if they do not cohere, continue downward at their own specific rate. Those that do cohere will form floccules of a larger diameter that will settle at a rate greater than that of the individual particles.

There is a gradual transition from particulate settling into the zone-settling regime, where the particles are constrained to settle as a mass. The principal characteristic of this zone is that the settling rate of the mass, as observed in batch tests, will be a function of its solids concentration (for any particular condition of flocculation, particle density, etc.).

The solids concentration ultimately will reach a level at which particle descent is restrained not only by hydrodynamic forces but partially by mechanical support from the particles below; therefore, the weight of particles in mutual contact can influence the rate of sedimentation of those at lower levels. This compression, as it is termed, will result in further solids concentration because of compaction of the individual floccules and partial filling of the interfloc voids by the deformed floccules. Accordingly, the rate of sedimentation in the compression regime is a function of both the solids concentration and the depth of pulp in this particular zone. As indicated in Fig. 19-70, granular, nonflocculent particles may reach their ultimate solids concentration without passing through this regime.

As an illustration, the aluminum oxide trihydrate particles produced in the Bayer process would be located near the extreme left of Fig. 19-70. These solids settle in a particulate manner, passing through a zone-settling regime only briefly, and reach a terminal density or ultimate solids concentration without any significant compressive effects. At this point, the solids concentration may be as much as 80 percent by weight. The same compound, but of the gelatinous nature it has when precipitated in water treatment as aluminum hydroxide, would be on the extreme right-hand side of the figure. This flocculent material enters into a zone-settling regime at a low concentration (relative to the ultimate concentration it can reach) and gradually thickens. With sufficient pulp depth present, preferably aided by gentle stirring or vibration, the compression-zone effect will occur; this is essential for the sludge to attain its maximum solids concentration, around 10 percent. Certain fine-size (1- to 2-μm) precipitates of this compound will possess characteristics intermediate between the two extremes.

A feed stream to be clarified or thickened can exist at any state represented within this diagram. As it becomes concentrated owing to sedimentation, it may pass through all the regimes, and the settling rate in any one may be the size-determining factor for the required equipment.

Sedimentation-Test Procedures

Determination of Clarification-Zone Requirements In the treatment of solids suspensions which are in the particulate-settling regime, the usual objective will be the production of a clear effluent, and test methods limited to this type of settling will be the normal sizing procedure, although the area demand for thickening should be verified. With particulate or slightly flocculent matter, any method that measures the rate of particle subsidence will be suitable, and either long-tube or short-tube procedures (described later) may be used. If the solids are strongly flocculent and particles cohere easily during sedimentation, the long tube will yield erratic data, with better clarities being observed in samples taken from the lower taps (i.e., clarity appears to improve at higher settling rates). In these instances, time alone usually is the principal variable in clarification, and a simple detention test is recommended.

Long-Tube Method A transparent tube 2 to 4 m long and at least 100 mm in diameter (preferably larger), fitted with sampling taps every 200 to 300 mm, is used in this test. The tube is mounted vertically and filled with a representative sample of feed suspension. At timed intervals approximately 100-mL samples are withdrawn from successive taps, beginning with the uppermost one. The time intervals will be determined largely by the settling rate of the particles, and they will be so chosen that a series of at least four will produce samples that bracket the desired solids-removal target. Also, this procedure will indicate whether or not detention time is a factor in the rate of clarification. Typically, intervals may be 30 min long, the last series of samples representing the results obtainable with 2-h detention. The samples are analyzed for suspended-solids concentration by any suitable means, such as filtration through Gooch crucibles or centrifugation with calibrated tubes.

A plot is made of the suspended-solids concentration in each of the samples as a function of the nominal settling velocity of the sample, which is determined from the corresponding sample-tap depth divided by the elapsed time between the start of the test and the time of sampling. For each sampling series after the first, the depth will have changed because of the removal of the preceding samples, and this must be taken into account. With particulate solids having little or no tendency to cohere, the data points generally will fall on one line irrespective of the detention time. This indicates that the settling areas available in the basin will determine the degree of solids removal, and the depth will have little bearing on the results.

Since the solids concentration in the 100-mL samples withdrawn at different depths does not correspond to the average concentration in the fluid above the sample point (which would be equivalent to the overflow from the clarifier at that particular design rate), an adjustment must be made in calculating the required area. Accordingly, an approximate average of the solids concentration throughout the column of liquid above a given tap can be obtained by summing the values obtained from all the higher taps and the one which is being sampled and dividing by the total number of taps sampled. This value is then used in a plot of overflow suspended-solids concentration versus nominal settling velocity.

Areal efficiencies for properly designed clarifiers in which detention time is not a serious factor range from 65 to 80 percent, and the surface area should be increased accordingly to reduce the overflow rate.

Should the particles have a tendency to cohere slightly during sedimentation, each sampling time, representing a different nominal detention time in the clarifier, will produce different suspended-solids concentrations at similar rates. These data can be plotted as sets of curves of concentration versus settling rate for each detention time by the means just described. Scale-up will be similar, except that detention time will be a factor, and both depth and area of the clarifier will influence the results. In most cases, more than one combination of diameter and depth will be capable of producing the same clarification result.

These data may be conveniently evaluated by selecting different nominal overflow rates (equivalent to settling rates) for each of the detention-time values, and then plotting the suspended-solids concentrations for each nominal overflow rate (as a parameter) against the detention time. For a specified suspended-solids concentration in the effluent, a curve of overflow rate versus detention time can be prepared from this plot.

Short-Tube Method This test is suitable in cases in which detention time does not change the degree of particle flocculation and hence has no significant influence on particle-settling rates. It is also useful for hydroseparator tests when the sedimentation device is to be used for classification (see Sec. 21). A tube 50 to 75 mm in diameter and 300 to 500 cm long is employed. A sample placed in the tube is mixed to ensure uniformity, and settling is allowed to occur for a measured interval. At the end of this time, the supernatant liquid is siphoned off quickly down to a chosen level, and the collected sample is analyzed for suspended-solids concentration. The depth selected usually is based on the relative expected volumes of overflow and underflow. The suspended-solids concentration measured in the siphoned sample will be equivalent to the "averaged" values obtained in the long-tube test, at a corresponding settling rate.

In hydroseparator tests, it is necessary to measure solids concentrations and size distributions on both the supernatant sample withdrawn and the fraction remaining in the cylinder. The volume of the latter sample should be such as to produce a solids concentration that would be typical of a readily pumped underflow slurry.

Detention Test This test utilizes a 1- to 4-L beaker or similar vessel. The sample is placed in the container, flocculated by suitable means if required, and allowed to settle. Small samples for suspended-solids analysis are withdrawn from the vessel, usually in the region immediately below the midpoint, taken with sufficient care that settled solids are not resuspended. Sampling times may be at consecutively longer intervals, such as 5, 10, 20, 40, and 80 min.

The suspended-solids concentration can be plotted on log-log paper as a function of the sampling (detention) time. A straight line usually will result, and the required static detention time t to achieve a certain suspended-solids concentration C in the overflow of an ideal basin can be taken directly from the graph. If the plot is a straight line, the data are described by the equation

$$C = Kt^m \tag{19-31}$$

where the coefficient K and exponent m are characteristic of the particular suspension.

Should the suspension contain a fraction of solids which can be considered "unsettleable," the data are more easily represented by using the so-called second-order procedure. This depends on the data being reasonably represented by the equation

$$Kt = [1/(C - C_\infty)] - [1/C_0 - C_\infty)] \tag{19-32}$$

where C_∞ is the unsettleable-solids concentration and C_0 is the concentration of suspended solids in the unsettled (feed) sample. The residual-solids concentration remaining in suspension after a sufficiently long detention time (C_∞) must be determined first, and the data then plotted on linear paper as the reciprocal concentration function $1/(C - C_\infty)$ versus time.

Bulk-settling test. In cases involving detention time only, the overflow rate must be considered by other means. This is done by carrying out a settling test in which the solids are first concentrated to a level at which zone settling just begins. This is usually marked by a very diffuse, almost indiscernible interface during initial settling. Its rate of descent is measured with a graduated cylinder of suitable size, preferably at least 1 L, and the initial straight-line portion of the settling curve is used for specifying a bulk-settling rate. The design overflow rate should not exceed half of the bulk-settling rate.

Detention efficiency. Conversion from the ideal basin sized by detention-time procedures to an actual clarifier requires the inclusion of an efficiency factor to account for the effects of turbulence and nonuniform flow. Efficiencies vary greatly, being dependent not only on the relative dimensions of the clarifier and the means of feeding but also on the characteristics of the particles. The curve shown in

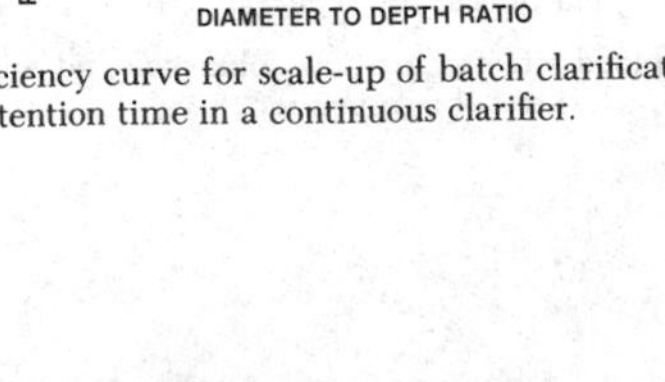

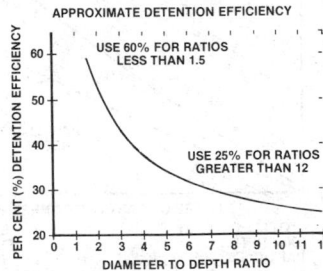

FIG. 19-71 Efficiency curve for scale-up of batch clarification data to determine nominal detention time in a continuous clarifier.

Fig. 19-71 can be used to scale up laboratory data in sizing circular clarifiers. The static detention time determined from a test to produce a specific effluent solids concentration is divided by the efficiency (expressed as a fraction) to determine the nominal detention time, which represents the volume of the clarifier above the settled pulp interface divided by the overflow rate. Different diameter-depth combinations are considered by using the corresponding efficiency factor. In most cases, area may be determined by factors other than the bulk-settling rate, such as practical tank-depth limitations.

Clarity in a thickener for which the feed concentration is in the zone-settling regime can depend on the mode of operation of the thickener. If the feed well is submerged in the pulp (i.e., below the pulp-water interface), overflow clarity generally will be improved over that produced when the pulp level is well below the feed-well outlet.

Thickener-Basin Area The area requirements for thickeners frequently are based on the solids flux rates measured in the zone-settling regime. Theory holds that, for any specific sedimentation condition, a critical concentration which will limit the solids throughput rate will exist in the thickener. This critical concentration will be evidenced as a pulp bed of variable depth in which the solids concentration is fairly uniform from top to bottom. Since the underflow concentration usually will be higher than this, gradually increasing concentrations will be found with progressing depth in the region beneath this constant-concentration zone. As the concentration within this critical zone represents a steady-state condition, its vertical extent may vary continually, responding to minor changes in the feed or underflow rate of concentration. In thickeners operating at relatively high underflow concentrations, with long solids-detention times and lower throughput, this zone may not be present.

Many batch-test methods which are based on determining the solids flux rate at this critical concentration have been developed. Most methods recognize that as the solids enter compression, thickening behavior is no longer a function only of solids concentration. Hence, these methods attempt to utilize the "critical" point dividing these two zones and size the area on the basis of the settling rate of a layer of pulp at this concentration. The difficulty lies in discerning where this point is located on the settling curve.

Many procedures have been developed, but two in particular have been more widely used: the Coe and Clevenger approach and the Kynch method as defined by Talmage and Fitch (references are cited at the beginning of this subsection).

The former requires measurement of the initial settling rate of a pulp at different solids concentrations varying from feed to final underflow value. The area requirement for each solids concentration tested is calculated by dividing the overflow-liquor-volume rate by the corresponding interface settling rate. A maximum value will be reached for a particular underflow concentration, and this will determine the basin area, generally expressed as the so-called unit area, a quantity with the dimensions of thickener plan area per unit of mass rate of settling of solids. The equation for determining the unit area at various solids concentrations is

$$\text{Unit area} = (1/C_i - 1/C_u)/s_L v_i \qquad (19\text{-}33)$$

where C_i is the solids concentration at the interfacial settling velocity v_i, C_u is the underflow concentration, and s_L is the specific gravity of the solids-free liquid in the thickener. The dimensions of C_i and C_u are the mass of suspended solids per unit volume of solids-free liquid in the sludge. Any consistent set of units may be used in Eq. (19-33), but the usual practical units ascribed to unit area are square meters per metric ton (i.e., per megagram) per day or square feet per short ton per day.

In the Kynch method a single settling curve generally is used, which is preferable when flocculation with polymers is employed. The floccule size will be greatly affected by the initial solids concentration, and this could lead to erroneous results if the Coe-Clevenger procedure were used. The curve is a plot of pulp level in a graduated cylinder versus settling time; the test should be continued until the desired underflow concentration is attained. In calculating the unit area based on this theory, the Talmage and Fitch approach is to solve the equation

$$\text{Unit area} = t_u/C_0H_0 \qquad (19\text{-}34)$$

where t_u is the time, days; C_0 is the initial solids concentration in the feed, Mg/m^3; and H_0 is the initial height, m, of the slurry in the test cylinder. The term t_u is taken from the intersection of a tangent to the curve at the critical point and a horizontal line representing the depth of pulp at underflow concentration. There are various means for selecting this critical point, all of them empirical, and the unit-area value determined cannot be considered precise.

Thickener-Basin Depth The pulp depth required in the thickener will be greatly affected by the role that compression plays in determining the rate of sedimentation. If the zone-settling conditions define the area needed, then depth of pulp will be unimportant and can be largely ignored, as the "normal" depth found in the thickener will be sufficient. On the other hand, with the compression zone controlling, depth of pulp will be significant, and it is essential to measure the sedimentation rate under these conditions.

To determine the compression-zone requirement in a thickener, a test should be run in a deep cylinder in which the average settling-pulp depth approximates the depth anticipated in the full-scale basin. The average density of the pulp in compression is calculated and used in Eq. (19-35) to determine the required compression-zone volume:

$$V = \Theta_c(\rho_s - \rho_L)/\rho_s(\rho_{sL} - \rho_L) \qquad (19\text{-}35)$$

where V is the volume, m^3, required per metric ton (or Mg) per day; Θ_c is the compression time, days, required in the test to reach underflow concentration; and ρ_s, ρ_L, and ρ_{sL} are the densities of the solids, liquid, and slurry (average) respectively, metric ton/m^3. This value divided by the average depth of the pulp during the period represents the unit area defined by compression requirements. If it exceeds the value determined from the zone-settling tests, it is the quantity to be used.

The **side depth** of the thickener is determined as the sum of the depth needed for the compression zone and for the clear zone. Normally, 1 to 2 m of clear-zone depth above the expected pulp level in a thickener will be sufficient for stable, effective operation. When the location of the pulp level cannot be predicted in advance or it is expected to be relatively low, a thickener-sidewall depth of 2 to 3 m is usually safe. Greater depth may be used in order to provide better clarity, although in most thickener applications the improvement obtained by this means will be marginal.

Two other approaches avoid using the critical point by sizing the area requirements from the settling conditions existing at the underflow concentration. One, an extension of zone-settling theory, is based on the settling rate of the layer of pulp at underflow concentration; the other employs Eq. (19-34), replacing t_u with t_x, the time value on the settling curve represented by the height of pulp at underflow concentration. Both give equivalent values, and, to be applied properly, both should be based on data obtained in cylinders at expected full-scale thickener-pulp depth. If a shorter laboratory cylinder is used, a depth-correction factor must be applied.

When compression is likely to be the controlling factor, it is desirable that tests be carried out in cylinders at least 1 m deep, although 1- and 2-L cylinders have been used with success. When a deep tube is used, the diameter should be at least 150 mm to reduce wall effect. It is essential to use a slow-speed (approximately 0.1-r/min) picket rake in all cylinder tests to prevent particle bridging and allow the sample to approach the underflow density obtainable in a full-scale thickener.

Scale-up factors used in thickening will vary, but typically a 1.2 to 1.3 multiplier applied to the unit area calculated from laboratory data is sufficient if proper testing procedures have been followed and the samples are representative.

Flocculation The use of flocculants has become quite common because of the considerable reduction in equipment size and capital cost that can be effected with very nominal reagent dosages. In general, flocculants will result in overflows of improved clarity and underflows of higher concentration from sedimentation units smaller than would be required for minimum acceptable performance without their use. Selection of the reagents usually involves simple bench-scale comparison tests on small samples of pulp for rough screening, followed by larger-scale tests in cylinders or in continuous pilot-plant thickeners with promising flocculants and dosages. Determination of the optimum dosage is complicated; it involves a number of economic factors such as reagent and capital costs, cost of a shutdown due to island formation, and influence of underflow concentration on cost.

Polymeric flocculants are available in various molecular-weight ranges, and they may be nonionic in character or be modified to have predominately cationic or anionic charges. The range of application varies; but, in general, nonionics are well suited to acidic suspensions, anionic flocculants work well in neutral or alkaline environments, and cationics are most effective on organic material and colloidal matter.

Colloidal solids (e.g., as encountered in waste treatment) may require initial treatment with a chemical having strong ionic properties, such as acid, lime, alum, or ferric sulfate. The latter two will precipitate at neutral pH and produce a gelatinous, flocculent structure which further helps collect extremely small particles. Some cationic polymers also may be effective in flocculation of particles of this type. (This action is commonly termed "coagulation.") Prolonged, gentle agitation improves the degree and rate of flocculation under these conditions. The solids concentration often should be increased artificially by recirculation of partially settled material for further improvement of the floc growth rate.

With polymer flocculation of pulps, however, extended agitation after the addition of the polymer may be detrimental. The reagent should be added to a sample under conditions which promote rapid dispersion and uniform, complete mixing with continuous polymer addition typically over a span of 3 to 15 s. In cylinder tests this can be accomplished by injecting flocculant into the pulp with an apparatus consisting of a syringe, a tube, and an inverted rubber stopper. The stopper, which should cover approximately three-fourths of the cylinder diameter, provides sufficient turbulence as it is moved gently up and down though the sample to cause good blending of reagent and pulp.

This difference in behavior is characteristic of these two types of flocculent structure. The typically smaller floccules resulting from coagulation re-form readily following mild shear conditions. On the other hand, the large floccules formed by polymers are easily ruptured in an irreversible manner and can be re-formed only by adding more polymer.

Torque Requirements Sufficient torque must be available in the raking mechanism of a full-scale thickener to allow it to move through the slurry and, when required, to rake the solids to the underflow outlet. Granular particulate solids that settle rapidly and reach a terminal solids concentration without going through any apparent compression or zone-settling regime require a maximum raking load, as they must be moved to the outlet solely by the mechanism. At the other end of the spectrum, extremely fine materials, such as clays and precipitates, require a minimum of raking, for most of the solids may reach the underflow outlet hydrodynamically. The rakes prevent a gradual buildup of some solids on the bottom, however, and a gentle stirring action often aids the thickening process. As the underflow concentration approaches its ultimate limit the consistency will increase greatly, resulting in a greater raking requirement and an increase in torque demand.

Most materials lie somewhere between these two extremes, but the torques provided in two properly designed thickeners of the same size but in different applications can differ greatly. Unfortunately, test methods to specify torque from small-scale tests are of questionable value, since it is difficult to duplicate actual conditions. Manufacturers of sedimentation equipment select torque ratings from experience with similar substances and will recommend a torque capability on this basis. Normally, thickeners and clarifiers should operate in the torque range that does not exceed 20 percent of design, to minimize shutdown when upsets occur.

Underflow Pump Requirements Many suspensions will thicken to a concentration higher than that which can be handled by conventional slurry pumps. Thickening tests should be performed with this in mind, for, in general, the unit area to produce the maximum concentration that can be pumped is the maximum needed. Determination of this ultimate "pumpable" concentration is largely a judgmental decision requiring some experience with slurry pumping; however, the behavior of the thickened suspension can be used as an approximate guide to pumpability. The supernatant should be decanted following a test and the settled solids repulped in the cylinder to a uniform consistency. Repulping is done easily with a rubber stopper fastened to the end of a rigid rod. If the bulk of the repulped slurry can be poured from the cylinder when it is tilted 10 to 30° above the horizontal, the corresponding thickener underflow can be handled by most types of slurry pumps. But if the slurry requires cylinder shaking or other mechanical means for its removal, it should be diluted to a more fluid condition.

THICKENERS

The primary function of a continuous thickener is to concentrate suspended solids by gravity settling so that a steady-state material balance is achieved, solids being withdrawn continuously, principally in the underflow, at the rate at which they are supplied in the feed. Thus, once an inventory of pulp has been accumulated, it will remain constant if the feed does. Continuous thickeners are used in most applications in which large quantities of solids must be concentrated or removed from large volumes of solid-liquid slurries.

A thickener has several basic components: a tank to contain the slurry, feed piping and a feed well to allow the feed stream to enter the tank, a rotating rake mechanism to assist in moving the concentrated solids to the withdrawal points, an underflow solids-withdrawal system, and an overflow launder.

Continuous thickeners have undergone several modifications that have resulted from the development of a wide variety of organic polymeric flocculents. As a result, there are now two basic types of continuous thickeners: **conventional** and **high-rate.**

Conventional Thickeners These are divided into three classes, differentiated by drive mechanism: (1) bridge-supported, (2) center-column-supported, and (3) traction-drive. These classes are described later in the subsection "Components and Accessories for Sedimentation Units." As indicated there, the diameter of the tank chosen, which for a conventional thickener may be as small as 3 m or as large as 120 m (10 to 400 ft), is related to the drive-support structure and vice versa.

The basic design of the conventional thickener is illustrated in Fig. 19-72, which shows the bridge-supported type. The feed stream is brought to the center of the thickener in a pipe or an open launder and enters the feed well, which is designed to minimize turbulence resulting from the feed entry velocity and to force the entering slurry below the clear-liquid surface. The thickened solids flow and/or are raked to the center of the thickener and removed. Overflow liquid is removed from the thickener in a peripheral launder.

The conventional thickener can be used with or without flocculants, depending upon the application. If employed, flocculants are normally added to the feed launder or in the feed well, and flocculation occurs in resulting turbulence.

High-Rate Thickeners High-rate thickeners are designed specifically to maximize the flocculation efficiency of flocculants. They differ from conventional thickeners in feed-well design, size, and control. Unlike conventional units, high-rate thickeners *must* use flocculants.

High-rate-thickener feed wells are designed to disperse flocculants thoroughly into the feed and to admit the flocculated slurry into the settling zone of the thickener without destruction of the newly formed floccules. The feed may enter into the sludge bed if flocculation must be completed by solids-contact means, or it may enter above the pulp level if it is sufficiently flocculated to produce the desired clarity and underflow density. Generally, both modes of

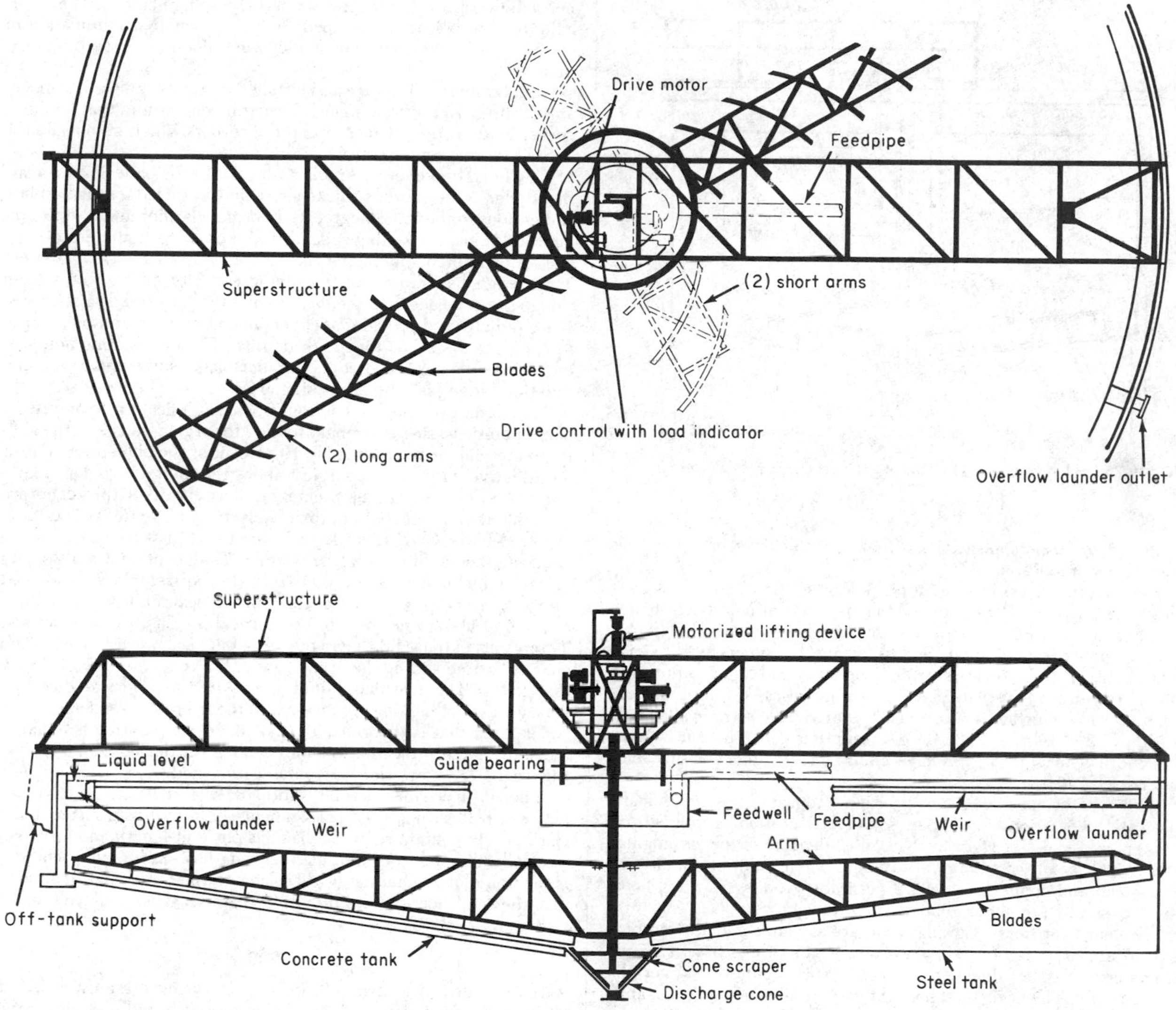

FIG. 19-72 Unit thickener with bridge-supported mechanism. *(Eimco Process Equipment Co.,)*

operation can be tested on the full-scale unit and the one which produces the better results employed subsequently.

Figure 19-73 illustrates a high-rate thickener, the Eimco **Hi-Capacity** design. A deaeration step must be used to remove entrained air from the feed stream so that it does not create turbulence in the thickener or cause flotation of solids. The feed stream enters the feed well, where flocculant is added at various locations to optimize mixing with the feed stream. The design of Fig. 19-73 uses a mechanical mixer to disperse the flocculant, whereas other designs depend on turbulence from the feed stream. Flocculation may occur both in the feed well and in the pulp bed, the amount of feed-well flocculation depending on the residence time and turbulence in the feed well.

The increase in flocculation efficiency realized in high-rate thickeners may increase the bulk-settling rate 2 to 10 times over that obtained in a conventional thickener, thus reducing the unit-area requirement by a similar factor. The corresponding volume may be 4 to 15 times smaller than with conventional thickeners in the same application. As a consequence, change in the solids feed rate or the settling characteristics of the solids will change the sludge level much more rapidly; therefore, high-rate thickeners require some form of automatic control. The control parameters used to maintain stable operation in high-rate thickeners are underflow solids density, underflow withdrawal rate, flocculant dosage, solids inventory in the tank, solids feed rate, and mechanical-torque-limit restriction of the rotating rake-arm mechanism.

CLARIFIERS

Continuous clarifiers generally are employed with dilute suspensions, principally industrial-process streams and domestic industrial wastes, and their primary purpose is to produce a relatively clear overflow. They are basically identical to thickeners in design and layout except that they employ a mechanism of lighter construction and a drive-head with a lower torque capability. These differences are permitted because in clarification applications the thickened pulp produced is smaller in volume and appreciably lower in suspended-solids concentration owing to the large percentage of relatively fine (smaller than 10-μm) solids. The installed cost of a clarifier therefore is approxi-

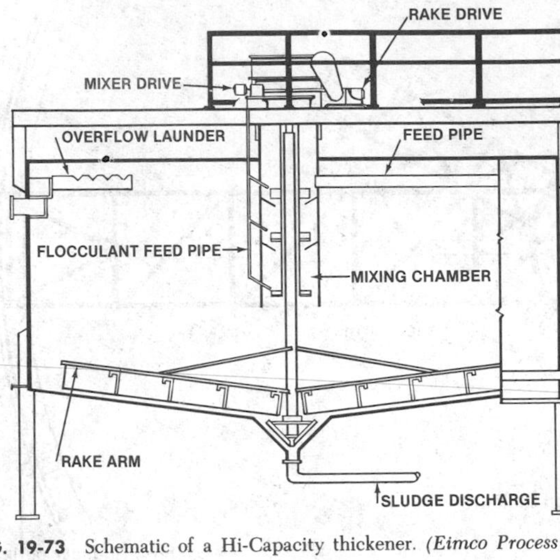

FIG. 19-73 Schematic of a Hi-Capacity thickener. *(Eimco Process Equipment Co.)*

mately 5 to 10 percent less than that of a thickener of equal tank size, as given in Table 19-9.

Rectangular Clarifiers A typical rectangular clarifier design is illustrated in Fig. 19-74. The raking mechanism employed in this design is a chain-type drag. The drag moves the deposited pulp to a sludge hopper located at one end by means of scrapers fixed to endless chains. During their return to the sludge-raking position, the flights travel near the water level and thus act as skimming devices for removal of surface scum. Rectangular clarifiers are available in widths of 2 to 10 m (6 to 33 ft). The length is generally 3 to 5 times the width. The larger widths have multiple raking mechanisms, each with a separate drive.

This type of clarifier is used in such applications as preliminary oil-water separations in refineries and clarification of rolling-mill effluent and scale recovery in steel mills. Because of the rectangular tank, it has an economic advantage when multiple units are employed and common walls are possible. Overflow clarities, however, generally are not as good as with other types.

Circular Clarifiers Circular units are available in diameters of 3 to 130 m (10 to 425 ft). They are of the same three basic types as single-compartment thickeners: bridge, center-column, and peripheral-traction. The bridge-supported type is limited generally to tanks less than 20 m in diameter because of economic considerations.

A circular clarifier often is equipped with a surface-skimming device, which includes a rotating skimmer, scum baffle, and scum-box assembly. In sewage and organic-waste applications, squeegees normally are provided for the rake-arm scraping blades, as it is desirable that the bottom be scraped clean to preclude accumulation of organic solids, with resultant septicity and flotation of decomposing material.

Circular mechanisms are also installed in square tanks. This mechanism differs from the standard circular mechanism in that a hinged corner blade is provided to sweep the corners which lie outside the path of the main mechanism.

Clarifier-Thickeners Another variation of the basic circular clarifier design is the clarifier-thickener, which is utilized to clarify dilute suspensions to the same degree as a normal clarifier and also to provide additional thickening capability and produce an underflow concentration appreciably higher than available from a normal clarifier. This design has additional thickening capacity because of a deep sludge sump adjacent to the center column that extends a short distance radially and provides adequate detention time and pulp depth to compact the solids to a high density. The compacting action is enhanced by extended-post and scraper-blade construction reaching into the sludge pit from the raking arms.

The clarifier-thickener has found wide application to industrial and domestic wastes. A primary feature for organic wastes is the ability to recirculate approximately 10 percent of the effluent volume as an oxidative liquor to a point just above the sludge bed. This maintains a fresh sludge and eliminates septicity. Clarifier-thickeners are available in diameters of approximately 10 to 60 m (30 to 200 ft).

Tilted-Plate Clarifiers Lamella or tilted-plate separators have achieved increased use for clarification. They contain a multiplicity of plates inclined at 45 to 60° from the horizontal. Various feed methods are employed so that the influent passes into each inclined channel at about one-third of the vertical height from the bottom. This results in the solids having to settle only a short distance in each channel before sliding down the base to the collection zone beneath the plates. The clarified liquid passes in the opposite direction beneath the ceiling of each channel to the overflow connection.

The area that is theoretically available for separation is equal to the sum of the projections of all channels on the horizontal plane. Figure 19-75 shows the horizontally projected area A_s of a single channel in a clarifier of unit width. If X is the uniform distance between plates (measured perpendicularly to the plate surface), the clarifier will contain $\sin \alpha/X$ channels per unit length and an effective collection area per unit clarifier length of $A_s \sin \alpha/X$, where α is the angle of inclination of the plates to the horizontal. It follows that the total horizontally projected plate area per unit volume of sludge in the clarifier A_{st} is

$$A_{st} = \cos \alpha/X \tag{19-36}$$

As α and X are decreased, A_s is increased. However, α must be larger than the angle of repose of the sludge so that it will slide down the plate, and the most common range is 55 to 60°. Plate spacing must be large enough to accommodate the opposite flows of liquid and sludge while reducing interference and preventing plugging and to

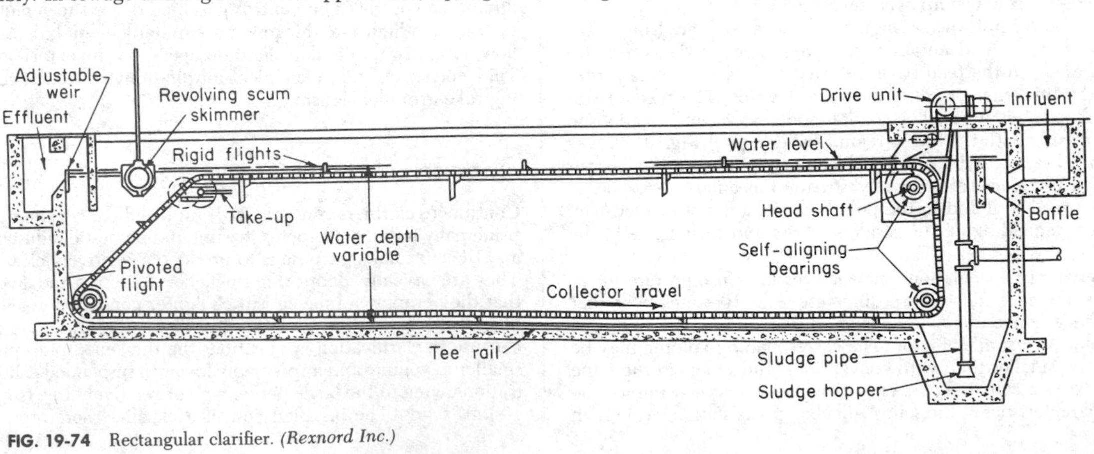

FIG. 19-74 Rectangular clarifier. *(Rexnord Inc.)*

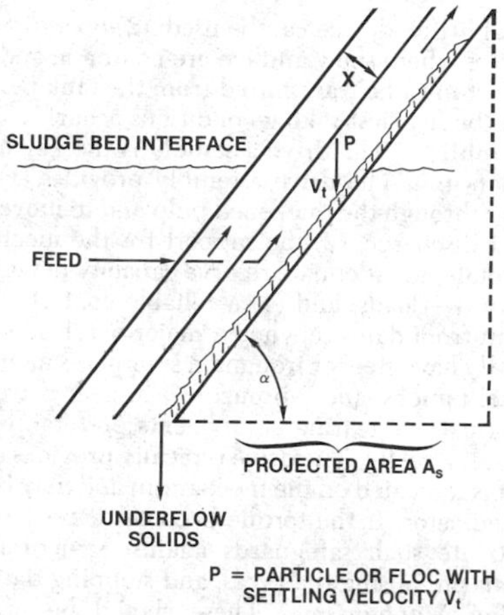

FIG. 19-75 Basic concept of the lamella-type clarifier.

provide enough residence time for the solids to settle to the bottom plate. Usual X values are 50 to 75 mm (2 to 3 in).

A wide variety of units are manufactured, the major difference among them being in feed-distribution methods. Operating capacities range from 1 to 3 m^3 of feed/(h·m^2) of projected horizontal area [0.4 to 1.2 gal/(min·ft^2)].

The principal advantage of the tilted-plate clarifier is the increased capacity per unit of plane area. Major disadvantages are a varying underflow solids concentration that is lower than in other gravity clarifiers and difficulty of cleaning when scaling or deposition occurs. The variable underflow composition is due primarily to the reduced compression-zone volume relative to the large settling area. When flocculants are employed, flocculating equipment and tankage preceding the separator are required, as geometry does not permit internal flocculation.

Solids-Contact Clarifiers When desirable, mixing, flocculation, and sedimentation all may be accomplished in a single tank. Of the various designs available, those employing mechanically assisted mixing in the reaction zone are the most efficient. They generally permit the highest overflow rate at a minimum chemical dosage while producing the best effluent quality. The unit illustrated in Fig.

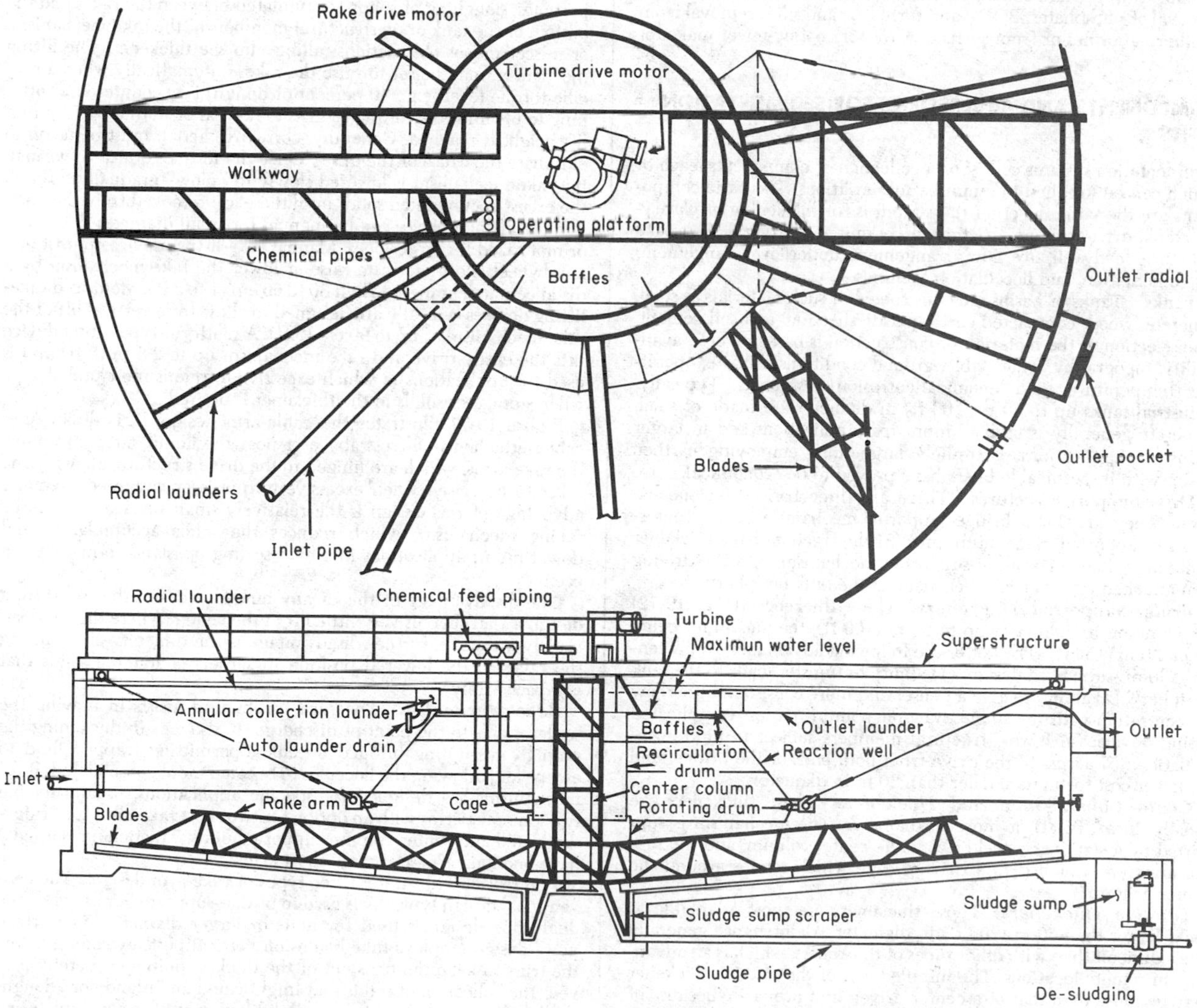

FIG. 19-76 Reactor-clarifier of the high-rate solids-contact type. *(Eimco Process Equipment Co.)*

19-76 consists of a combination dual drive which moves the rake mechanism at a very slow speed as it rotates a high-pumping-rate, low-shear turbine located in the top portion of a center reaction well at a very much higher speed. The influent, dosed with chemicals as it enters, is contacted with previously settled solids in a recirculation draft tube within the reaction well by means of the pumping action of the turbine, resulting in a thorough mixing of these streams. Owing to the higher concentration of solids being recirculated, all chemical reactions are more rapid and more nearly complete, and flocculation is improved. The mixture passes out of the contacting and reaction well into the clarification area, where the flocculated particles settle out. They are raked to the center to be used again in the recirculation process, with a small amount being discharged through the sludge pump. When floccules are too heavy to be circulated up through the draft tube (as in the case of metallurgical pulps), a modified design using external recirculation of a portion of the thickened underflow is chosen. These units employ a special mixing impeller in a large feed well with a small-diameter central outlet.

Solids-contact clarifiers are advantageous for clarifying turbid waters or slurries that require coagulation and flocculation for the removal of bacteria, suspended solids, or color. Applications include softening water by lime addition; clarifying industrial-process streams, sewage, and industrial wastewaters; tertiary treatment for removal of phosphates, BOD_5, and turbidity; and silica removal from geothermal brines or from surface water for cooling-tower makeup.

COMPONENTS AND ACCESSORIES FOR SEDIMENTATION UNITS

Sedimentation systems consist of a collection of components, each of which can be supplied in a number of variations. The basic components are the same, whether the system is for thickening or clarifying: tank, drive-support structure, drive unit and lifting device, rake structure, feed well, overflow arrangement, underflow arrangement, instrumentation, and flocculation facilities.

Tanks Tanks or basins are constructed of such materials as steel, concrete, wood, compacted earth, plastic sheeting, and soil cement. The selection of the materials of construction is based on cost, availability, topography, water table, ground conditions, climate, operating temperature, and chemical-corrosion resistance. Typically, industrial tanks up to 30 m (100 ft) in diameter are made of steel. Concrete generally is used in municipal applications and in larger industrial applications. Extremely large units employing earthen basins with impermeable liners have proved to be economical.

Drive-Support Structures There are three basic drive mechanisms. These are (1) the bridge-supported mechanism, (2) the center-column-supported mechanism, and (3) the traction-drive thickener containing a center-column-supported mechanism with the driving arm attached to a motorized carriage at the tank periphery.

Bridge-Supported Thickeners These thickeners (Fig. 19-72) are common in diameters up to 30 m (100 ft), the maximum being about 45 m (150 ft). They offer the following advantages over a center-column-supported design: (1) ability to transfer loads to the tank periphery; (2) ability to give a denser and more consistent underflow concentration with the single draw-off point; (3) a less complicated lifting device; (4) fewer structural members subject to mud accumulation; (5) access to the drive from both ends of the bridge; and (6) lower cost for units smaller than 30 m in diameter.

Center-Column-Supported Thickeners These thickeners are usually 20 m (65 ft) or more in diameter. The mechanism is supported by a stationary steel or concrete center column, and the raking arms are attached to a driving cage which rotates around the center column.

Traction Thickeners These thickeners are most adaptable to tanks larger than 60 m (200 ft) in diameter. Maintenance generally is less difficult than with other types of thickeners, which is an advantage in remote locations. The installed cost of the traction thickener may be more than that of a center-driven unit primarily because of the cost of constructing the heavy concrete wall required to support the drive carriage. Disadvantages of the traction thickener are that (1) no practical lifting device can be used; (2) operation may be difficult in climates where snow and ice are common; and (3) the driving-torque effort must be transmitted from the tank periphery to the center, where the heaviest raking conditions occur.

Drive Assemblies The drive assembly is the key component of a sedimentation unit. The drive assembly provides (1) the force to move the rakes through the thickened pulp and to move settled solids to the point of discharge, (2) the support for the mechanism which permits it to rotate, (3) adequate reserve capacity to withstand upsets and temporary overloads, and (4) a reliable control which protects the mechanism from damage when a major overload occurs.

Drives usually have steel or iron main spur gears mounted on bearings; alloy-steel pinions; and a bronze worm gear driven by a hardened-steel worm. The gearing components preferably are enclosed for maximum service life. The drive typically provides a torque measurement that is indicated on the mechanism and may be transmitted to a remote indicator. If the torque becomes excessive, it can automatically activate such safeguards against structural damage as sounding an alarm, raising the rakes, and stopping the drive.

Rake-Lifting Mechanisms These should be provided when abnormal thickener operation is probable. Abnormal thickener operation or excessive torque may result from insufficient underflow pumping, surges in the solids feed rate, excessive amounts of large particles, sloughing of solids accumulated between the rakes and the bottom of the tank or on structural members of the rake mechanism, or miscellaneous obstructions falling into the thickener. The lifting mechanism may be set to raise the rakes automatically when a specific torque level (e.g., 40 percent of design) is encountered, continuing to lift until the torque returns to normal or until the maximum lift height is reached. Generally, corrective action must be taken to eliminate the cause of the upset. Once the torque returns to normal, the rake mechanism is lowered slowly to "plow" gradually through the excess accumulated solids until these are removed from the tank.

Rake-lifting devices can be manual for small-diameter thickeners or motorized for larger ones. Manual rake-lifting devices consist of a handwheel and a worm to raise or lower the rake mechanism by a distance usually ranging from 30 to 60 cm (1 to 2 ft). Motorized rake-lifting devices typically are designed to allow for a vertical lift of the rake mechanism of up to 90 cm (3 ft). A platform-type lifting device lifts the entire drive and rake mechanism up to 2.5 m (8 ft) and is used for applications in which excessive torque is most probable or when storage of solids in the thickener is desired.

Figure 19-77 illustrates the cable-arm design. This design uses cables attached to a truss above or near the liquid surface to move the rake arms, which are hinged to the drive structure, allowing the rakes to be raised when excessive torque is encountered. A major advantage of this design is the relatively small surface area of the raking mechanism, which reduces the solids accumulation and downtime in applications in which scaling or island formation can occur.

One disadvantage of this or any hinged-arm or other self-lifting design is that there is very little lift at the center, where the overload usually occurs. A further disadvantage is the difficulty of returning the rakes to the lowered position in settlers containing solids that compact firmly.

Rake Mechanism The rake mechanism assists in moving the settled solids to the point of discharge. It also aids in thickening the pulp by disrupting bridged floccules, permitting trapped fluid to escape and allowing the floccules to become more consolidated. Rake mechanisms are designed for specific applications, usually having two long rake arms with an option for two short rake arms for bridge-supported and center-column-supported units. Traction units usually have one long arm and three short arms.

Figure 19-78 illustrates three types of rake-arm designs. The conventional design typically is used in bridge-supported units, while the dual-slope design is used for units of larger diameter. The "thixo post" design employs rake blades on vertical posts extending below the truss to keep the truss out of the thickest pulp and thereby prevent the collection of solids that might cause an "island" or "doughnut" to form. An island is a semisolidified mass that accumulates on the rake mechanism, rotating with it. Commonly, it is caused by use

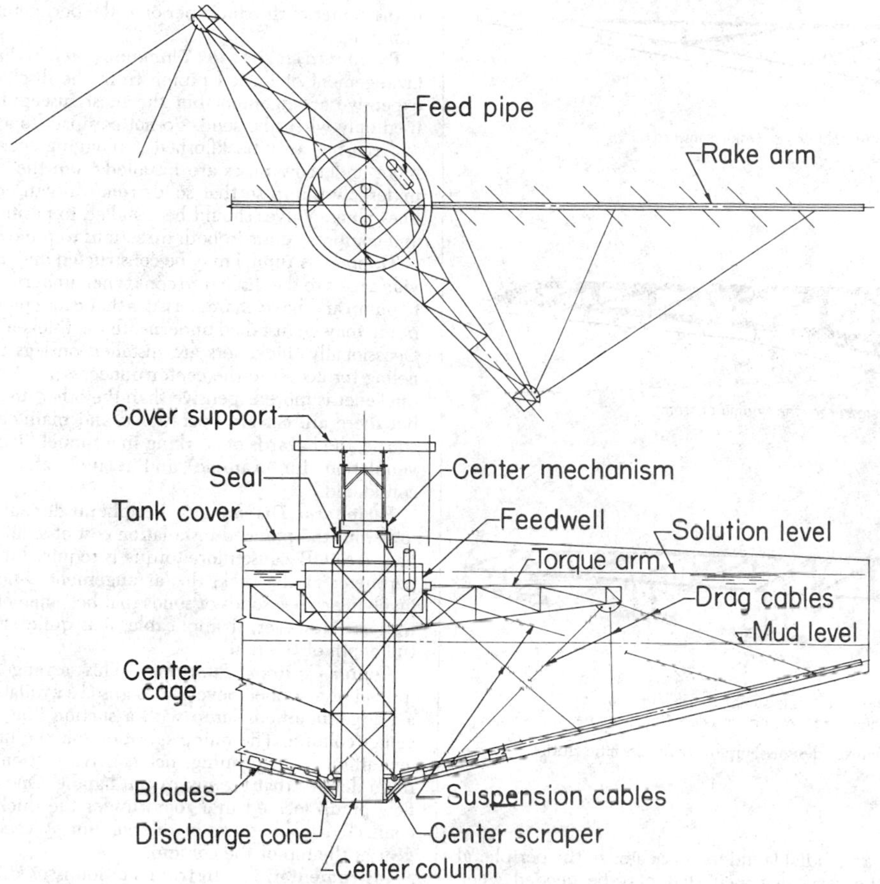

FIG. 19-77 Cable-arm design. *(Dorr-Oliver Inc.)*

of excessive flocculant or excessive detention time. An island reduces the capacity of a thickener and increases the torque load on the drive. If unchecked, it can cause a thickener to stall, requiring the accumulated contents to be washed out. Islands can be prevented in most cases by lifting the rakes periodically to slough off accumulations.

Rake blades can have attached spikes or serrated bottoms to cut into solids that have a tendency to compact. Lifting devices typically are used for these applications.

Rake-speed requirements depend on the type of solids entering the thickener. Peripheral speed ranges used are, for slow-settling solids, 3 to 8 m/min (10 to 26 ft/min), for fast-settling solids, 8 to 12 m/min (26 to 40 ft/min), and for coarse solids or crystalline materials, 12 to 30 m/min (40 to 100 ft/min).

Feed Well The feed well is designed to allow the feed to enter the conventional thickener with minimum turbulence and uniform distribution while dissipating most of its kinetic energy. Feed slurry enters the feed well, which is usually located in the center of the thickener, through a pipe or launder suspended from the bridge. To avoid excess velocity, an open launder normally has a slope no greater than ⅝₀₀. Pulp should enter the launder at a velocity that prevents sanding at the inlet. With nonsanding pulps, the feed may also enter upward through the center column.

The standard feed well for conventional thickeners is designed for a maximum nominal vertical flow velocity of 1.5 m/min (5 ft/min). High turbidity caused by short-circuiting the feed to the overflow can be reduced by increasing the depth of the feed well. When overflow clarity is important or the solids specific gravity is close to the liquid specific gravity, deep feed wells of large diameter are used.

Shallow feed wells may be used when overflow clarity is not important, the overflow rate is low, and/or the solids density is appreciably greater than water. Some special feed-well designs used to dissipate entrance velocity and create quiescent settling conditions split the feed stream and allow it to enter the feed well tangentially on opposite sides. With an intermediate annular shelf, the two streams shear one another to dissipate kinetic energy.

Overflow Arrangements Clarified effluent typically is removed in a peripheral launder located inside or outside the tank. The effluent enters the launder by overflowing a V-notch or level flat weir, or through submerged orifices in the bottom of the launder. Uneven overflow rates caused by wind blowing across the liquid surface in large thickeners can be better controlled when submerged orifices or V-notch weirs are used. Radial launders are used when uniform upward liquid flow is desired. This arrangement provides an additional benefit in reducing the effect of wind, which can seriously impair clarity in applications that employ basins of large diameter.

The hydraulic capacity of a launder must be sufficient to prevent flooding, which can cause short circuiting of the feed and deterioration of overflow clarity. Standards are occasionally imposed on weir overflow rates for clarifiers used in municipal applications; typical rates are 3.5 to 15 $m^3/(h \cdot m)$ [7000 to 30,000 gal/(day · ft)], and they are highly dependent on clarifier side-water depth. Industrial clarifiers may have higher overflow rates, depending on the application and the desired overflow clarity. Launders can be arranged in a variety of configurations to achieve the desired overflow rate. Two alternatives in the direction of increased weir length per unit of clarifier surface area are an annular launder inside the tank (the liquid

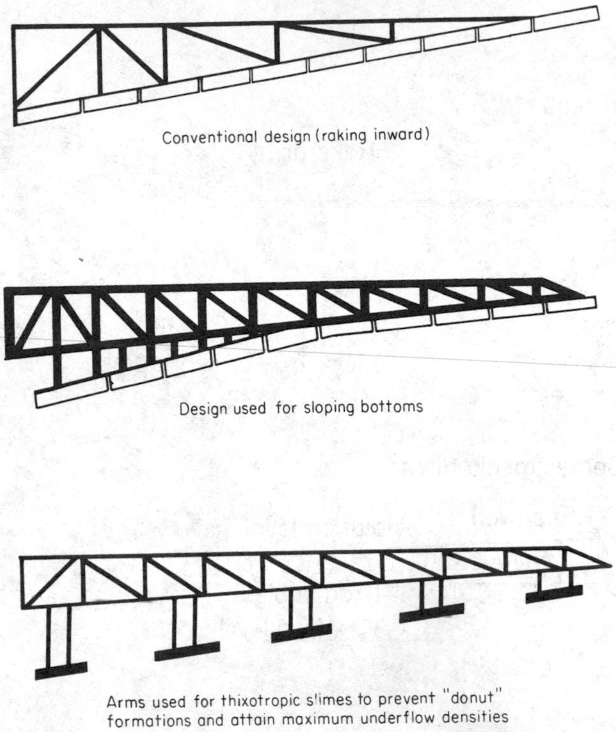

FIG. 19-78 Rake-mechanism designs employed for specific duties. *(Eimco Process Equipment Co.)*

flows over both sides) and radial launders connected to the peripheral launder (providing the very long weir that may be needed when abnormally high overflow rates are encountered and overflow clarity is important).

In many thickener applications, on the other hand, complete peripheral launders are not required, and no difference in either overflow clarity or underflow concentration will be obtained through their use. These systems utilize the available area for concentration of the solids, an operation which is unaffected by the launder configuration. For design purposes, a weir-loading rate in the range of 5 to 25 $m^3/(h \cdot m)$ [10,000 to 50,000 gal/(h·ft)] can be used, the higher values being employed with well-flocculated, rapidly settling slurries. The overflow launder required may occupy only a portion of the perimeter and be confined to one segment rather than being spaced uniformly around the tank.

Underflow Arrangements Concentrated solids are removed from the thickener by use of a centrifugal slurry pump or a positive-displacement pump (occasionally by gravity discharge through a flow-control valve suitable for slurry applications). The underflow arrangement must be designed to remove solids, without plugging upsets, at the maximum rate at which they will enter the thickener. Provisions *must* be made to unplug or to bypass plugged piping so that solids can always be removed from the thickener to preclude their filling the thickener and stalling the mechanism. Underflow recycle back to the feed well is used in some applications to aid in flocculation, to store solids in the thickener, and to maintain proper functioning of the unit during periods when the feed is reduced or interrupted.

There are four basic underflow arrangements: (1) the underflow pump adjacent to the thickener with buried piping from the discharge cone, (2) the underflow pump under or adjacent to the thickener with the piping from the discharge cone in a tunnel, (3) the underflow pump adjacent to the thickener with a peripheral discharge from the tank sidewall, and (4) the underflow pump located in the center of the thickener or at the perimeter with center-column pumping.

Pump Adjacent to Thickener with Buried Piping This arrangement of buried piping from the discharge cone is the least expensive arrangement but the most susceptible to plugging. It is used only when the solids do not compact to an unpumpable slurry and can be easily backflushed if plugging occurs. Typically, two or more underflow pipes are installed from the discharge cone to the underflow pump so that solids removal can continue if one of the lines plugs. Valves should be installed to permit flushing with water and compressed air in both directions to remove plugs.

Tunnel A tunnel may be constructed under the thickener to provide access to the discharge cone when underflow slurries are difficult to pump and have characteristics that cause plugging. The underflow pump may be installed underneath the thickener or at the perimeter. Occasionally thickeners are installed on legs or piers, making tunneling for access to the center unnecessary. A tunnel or an elevated thickener is more expensive than the other underflow arrangements, but there are certain operational and maintenance advantages. Of course, the hazards of working in a tunnel (flooding and interrupted ventilation, for example) and related safety regulations must be considered.

Peripheral Discharge Peripheral discharge sometimes is used to permit the reduced installation cost of a flat-bottom tank on compacted soil. Because more torque is required to rake the solids to the perimeter of the tank, this arrangement is not suitable for service involving coarse solids or solids that become nonfluid at high concentrations. However, if applicable, it is quite efficient, for it uses the full depth of the tank.

Center-Column Pumping This arrangement may be used instead of a tunnel. Several designs are available. The commonest is a bridge-mounted pump with a suction line through a wet or dry center column. The pump selection may be limiting, requiring special attention to priming, net positive suction head, and the maximum density that the pump can handle. One design has the underflow pump located in a room under the thickener mechanism and connected to openings in the column. Access is through the drive gear at the top of the column.

Instrumentation Instrumentation is used on most thickeners to measure torque in order to prevent mechanical damage to the drive or rake mechanism by actuating a rake-lifting device or shutting off power to the drive in the event of an overload. Some applications also require automatic control of thickener feed or discharge to maintain acceptable performance when departures from the norm occur. An example is the high-rate thickener, which must have some degree of automation for efficient normal operation. This is not difficult to provide, for the high-rate thickener is characterized by a relatively short time lag between a process change and the effect of the change; control by conventional instruments or minicomputers therefore is feasible. Automatic control of conventional thickeners, on the other hand, is more difficult because of the longer lag between process change and effect.

In addition to drive torque and rates of feed and underflow, other process variables that can be controlled to advantage in some applications are flocculant dosage, underflow-solids concentration, solids-liquid-interface height, and overflow clarity. Underflow pumping control is sometimes used to control two of these: pulp level and underflow solids concentration. An approximately steady-state material balance must be maintained around the thickener. Although the pulp volume in the thickener can be allowed to increase or decrease within harmless limits, it must be controlled to prevent, on the one hand, solids from overflowing the thickener and, on the other hand, the pulp level from falling so low that the underflow becomes too dilute. The concentration of solids in the underflow also must be maintained at the desired level—particularly at such a level as to avert compaction to a nonfluid condition that could lead to excessive torque on the rake mechanism. Variable pumping rates can be obtained by use of variable-speed pumps or flow-control valves. Underflow-density measurements can be made by using gamma-radiation-sensing or sonic devices, but since all such devices tend imperceptibly to lose accuracy, it is highly desirable to employ procedurally a simple direct-reading device like a Marcy pulp-density scale for routine use by the thickener operator.

Flocculation: Plant Practice and Facilities Flocculants are manufactured as dry solids, liquids, emulsions, and suspensions. The dry solids require properly designed systems for thorough dissolution to a typical concentration of about 0.5 to 1.0 weight percent. The fluid forms of flocculants are easier to disperse. The prepared solutions usually are diluted to 0.1 to 0.01 percent prior to addition to the thickener feed pulps. In some cases it may be necessary to dilute still further to the equivalent of 1 to 2 percent of feed volume for efficient flocculation.

In conventional thickeners, the flocculant typically is added in the feed launder, in the feed well, or at various other locations that will accommodate relatively uniform distribution of the reagent. In many applications, however, more carefully controlled mixing of flocculant with the feed stream is necessary to maintain high flocculation efficiency and rate of sedimentation. This is particularly true for a high-rate thickener.

Flocculant dosage relative to the concentration of suspended solids in the feed controls the size of the flocs and, thereby, the settling rate of the solids. Increasing the amount of flocculant usually increases the settling rate. Excessive use of flocculant, however, can cause island formation and should be avoided because of the resulting operational problems and the increased operation expense.

CONTINUOUS COUNTERCURRENT DECANTATION

The system of separation of solid-phase material from an associated solution by repeated stages of dilution and gravity sedimentation is adapted for most industrial-processing applications through an operation known as continuous countercurrent decantation (CCD). The flow of solids proceeds in a direction countercurrent to the flow of solution diluent (water, usually), with each stage composed of a mixing step followed by settling of the solids from the suspension. The number of stages ranges from 2 to more than 10, depending on the degree of separation required, the amount of wash fluid added (which influences the final solute concentration in the first-stage overflow), and the underflow solids concentration attainable. Most designs use 3 to 7 stages. Applications include processes in which the solution is the valuable component (as in alumina extraction), or in which purified solids are sought (magnesium hydroxide from seawater), or both (as frequently encountered in the chemical-processing industry and in base-metal hydrometallurgy).

Application Other alternatives to settler CCD for the separation of solids from solution include continuous filters, batch filters, batch-decanting systems, centrifuges, and special adsorption systems (e.g., carbon-in-pulp). The factors which may make CCD a preferred choice include the following: rapidly settling solids, assisted by flocculation; relatively high ratio of solids concentration between underflow and feed; moderately high wash ratios allowable (2 to 4 times the volume of liquor in the thickened underflows); large quantity of solids to be processed; and the presence of significant quantities of fine-size solids that are difficult to concentrate by other means. A technical feasibility and economic study is desirable in order to make the optimum choice.

Flow-Sheet Design Thickener-sizing tests, as described earlier, will determine unit areas, flocculant dosages, and underflow densities for the various stages. For most cases, unit areas will not vary significantly throughout the circuit; similarly, underflow concentrations should be relatively constant. In practice, the maximum unit area for the circuit is generally used for all thickeners to simplify construction.

Equipment The equipment selected for CCD circuits may consist of multiple-compartment washing-tray thickeners or a train of unit thickeners. The washing-tray thickener consists of a vertical array of coaxial trays connected in series, contained in a single tank. The advantages of this design are smaller floor-area requirements, less pumping equipment and piping, and reduced heat losses in circuits operating at elevated temperatures. However, there are several disadvantages: generally the design is more difficult to operate and experiences more operational upsets, desired underflow density is not obtained consistently and is lower than in unit thickeners, and problems with one compartment of the system can shut down the entire washing circuit. User preference has shifted almost entirely toward unit thickeners despite the larger floor-area requirement and greater initial cost, disadvantages that are substantially offset by improved performance and greater product recovery.

Underflow Pumping This is one of the most important influences on the entire system performance, and pump selection is a critical decision. Diaphragm pumps with open discharge are preferred in many cases, primarily because underflow densities are readily controlled with these units. Disadvantages include the relatively high cost in terms of volumetric capacity, generally higher maintenance and initial costs than for other types, and inability to transfer the slurry any great distance. Pressure-type diaphragm pumps and other positive-displacement pumps overcome the last-named objection. Large flows often are best handled with low-speed, large-diameter rubber-lined centrifugal pumps. Automatic control of the flow rate on these units is necessary.

Discharge of the underflow through an automatically controlled, variable-opening slurry valve, preferably located at or near the lowest level of the thickener, is an effective means of underflow control. The slurry should discharge into a sump, where it may be mixed with overflow prior to transfer to the next stage.

Overflow Pumps These can be omitted if the thickeners are located at increasing elevations from first to last so that overflows are transferred by gravity or if the mixture of underflow and overflow is to be pumped. Pumps are necessary, however, when maximum flexibility and control are sought. Usually, standard centrifugal pumps are employed.

Uniform Feed Rate One of the essential requirements in any CCD circuit is a relatively uniform feed rate. If the rate of solids feed to the first unit of the circuit varies greatly, it will be difficult to maintain the proper wash ratio in each thickener. If possible, surge storage ahead of the circuit should be provided to damp feed fluctuations. A less preferable alternative approach is a CCD circuit having overflow liquor surge capacity in each tank made possible by increased freeboard. This technique allows control of the wash ratio as underflow volumes vary, but it requires very close operator attention or even total automatic control of the system.

Interstage Mixing Efficiencies These efficiencies rarely approach the ideal 100 percent, representing solute concentrations in overflow and underflow liquor from each thickener that are identical. Part of the deficiency is due to insufficient blending of the two streams, and their equilibration will be hampered further by heavily flocculated solids. In systems in which flocculants are used, interstage efficiencies often will drop gradually from first to last thickener, and typical values will range from 98 percent down to as low as 70 percent. Many instances of poor efficiency can be improved by the means described in the following paragraphs.

In some cases, operators will add the flocculant to an overflow solution which is to be blended with the corresponding underflow. This results in reflocculation of the solids before the entrained solution has had a chance to blend completely with the overflow liquor. If this very effective means of flocculation is to be used, the preferable procedure is to recycle a portion of the overflow back to the feed line of the same thickener, adding the reagent to this liquor. The increased volume will not decrease the capacity of the thickener, and the added pumping cost probably will be more than offset by improved performance and a reduction in the consumption of flocculant.

The usual method of interstage mixing consists of a relatively simple arrangement in which the flows from preceding and succeeding stages are added to a feed box at the thickener periphery. The turbulence in the feed box and in the pipe or launder leading to the feed well usually provides enough mixing to obtain near equilibrium between underflow and overflow streams. However, viscous underflows or a durable floc structure may require a properly designed interstage mechanical mixer. Energy input from the mixer should be sufficient to disrupt the floccules partially, but not so great as to destroy the flocculant structure completely, lest additional reagent be required to restore the pulp to the design settling rate. A nominal detention time in the mixing tank of 30 to 60 s and an energy input equivalent to 0.2 to 0.4 kW/m^3 (1 to 2 hp/kgal) of tank volume should yield interstage efficiencies greater than 95 percent.

In CCD circuits employing flocculants, generally about one-half of the total flocculant consumption will be needed in the first unit, as the slurry in later stages is flocculated more easily. Overflocculation is to be avoided, since the underflows will become more difficult to repulp, and stage efficiency may drop significantly.

Instrumentation Instrumentation and partial automatic control of CCD circuits are now commonplace. Simultaneous recording of such data as feed rate, wash-water rate, underflow rate and density, thickener-drive-mechanism torque output, and flocculant dosage provides a means for maximizing the performance of the system.

The performance of a CCD circuit can be estimated through use of the following equations:

$$R = O/U[(O/U)^N - 1]/[(O/U)^{N+1} - 1] \qquad (19\text{-}37a)$$

$$R = 1 - (U/O')^N \qquad (19\text{-}37b)$$

for $O/U \neq 1$ and $U/O' \neq 1$. R is the fraction of dissolved value in the feed which is recovered in the overflow liquor from the first thickener, O and U are the overflow and underflow liquor volumes

TABLE 19-8 Typical Thickener and Clarifier Design Criteria and Operating Conditions

	Percent solids		Unit area, (m^2·day)/Mg°	Overflow rate, m^3/(m^2·h)°
	Feed	Underflow		
Alumina Bayer process				
Red mud, primary	3 - 4	10 -25	2 - 5	
Red mud, washers	6 - 8	15 -35	1 - 4	
Hydrate, fine or seed	1 -10	20 -50	1.2 - 3	0.07–0.12
Brine purification	0.2 - 2.0	8 -15	....	0.5 -1.2
Coal, refuse	0.5 - 6	20 -40	0.5 - 1†	0.7 -1.7
Coal, heavy-media (magnetic)	20 -30	60 -70	0.05 – 0.1†	
Cyanide, leached-ore	16 -33	40 -60	0.3 - 1.3†	
Flue dust, blast-furnace	0.2 - 2.0	40 -60	....	1.5 -3.7
Flue dust, BOF	0.2 - 2.0	30 -70	....	1 -3.7
Flue-gas desulfurization sludge	3 -12	20 -45	0.3 - 3†	
Magnesium hydroxide from brine	8 -10	25 -40	5 -10	
Magnesium hydroxide from seawater	1 - 4	15 -20	3 -10	0.5 -0.8
Metallurgical				
Copper concentrates	14 -50	40 -75	0.2 - 2	
Copper tailings	10 -30	45 -65	0.4 - 1	
Iron ore				
Concentrate (magnetic)	20 -35	50 -70	0.01- 0.08	
Concentrate (nonmagnetic), coarse: 40–65% -325	25 -40	60 -75	0.02- 0.1	
Concentrate (nonmagnetic), fine: 65–100% -325	15 -30	60 -70	0.15- 0.4	
Tailings (magnetic)	2 - 5	45 -60	0.6 - 1.5	1.2 -2.4
Tailings (nonmagnetic)	2 -10	45 -50	0.8 - 3	0.7 -1.2
Lead concentrates	20 -25	60 -80	0.5 - 1	
Molybdenum concentrates	10 -35	50 -60	0.2 - 0.4	
Nickel, $(NH_4)_2CO_3$ leach residue	15 -25	45 -60	0.3 - 0.5	
Nickel, acid leach residue	20	60	0.8	
Zinc concentrates	10 -20	50 -60	0.3 - 0.7	
Zinc leach residue	5 -10	25 -40	0.8 - 1.5	
Municipal waste				
Primary clarifier	0.02–0.05	0.5- 1.5	....	1 -1.7
Thickening				
Primary sludge	1 - 3	5 -10	8	
Waste activated sludge	0.2 - 1.5	2 - 3	33	
Anaerobically digested sludge	4 - 8	6 -12	10	
Phosphate slimes	1 - 3	5 -15	1.2 -18†	
Pickle liquor and rinse water	1 - 8	9 -18	3.5 - 5	
Plating waste	2 - 5	5 -30	....	1.2
Potash slimes	1 - 5	6 -25	4 -12	
Potato-processing waste	0.3 - 0.5	5 - 6	....	1
Pulp and paper				
Green-liquor clarifier	0.2	5	....	0.8
White-liquor clarifier	8	35 -45	0.8- 1.6	
Kraft waste	0.01- 0.05	2 - 5	....	0.8 -1.2
Deinking waste	0.01- 0.05	4 - 7	....	1 -1.2
Paper-mill waste	0.01- 0.05	2 - 8	....	1.2 -2.2
Sugarcane defecation	....	....	0.05‡	
Sugar-beet carbonation	2 - 5	15 -20	0.03- 0.07†‡	
Uranium				
Acid-leached ore	10 -30	25 -65	0.02- 1	
Alkaline-leached ore	20	60	1†	
Uranium precipitate	1 - 2	10 -25	5 -12.5	
Water treatment				
Clarification (after 30-min flocculation)	....	....	....	1 -1.3
Softening lime-soda (high-rate, solids-contact clarifiers)	....	....	....	3.7
Softening lime-sludge	5 -10	20 -45	0.6 - 2.5	

°m^2/(Mg·day) × 9.79 = ft^2/(short ton·day); m^3/(m^2·h) × 0.41 = gal/(ft^2·min); 1 Mg = 1 metric ton.

†High-rate thickeners using required flocculant dosages operate at 10 to 50 percent of these unit areas.

‡Basis: 1 Mg of cane or beets.

per unit weight of underflow solids, and N is the number of stages. Equation (19-37*b*) applies to a system in which the circuit receives dry solids with which second overflow is mixed to extract the soluble component. In this instance, O' refers to the overflow volume from any thickener except the first.

For more precise values, programmable calculators or computers can be used to determine soluble recovery as well as solution compositions for conditions that are typical of a CCD circuit, with varying underflow concentrations, stage efficiencies, and solution densities in each of the stages. The calculation sequence is easily performed by utilizing material-balance equations around each thickener, starting with a final thickener underflow liquid solute content of x and proceeding forward to the amount of solute in the feed entering the first unit expressed in x units. To accommodate interstage mixing efficiencies other than 100 percent, the amount of solute contributing toward a higher concentration in the underflow liquor can be considered as a bypass of this component directly from one underflow to the following one. Mixing efficiency is defined here as the ratio of solute concentration in the overflow and underflow liquors from the same thickener. For example, if the soluble concentration in the overflow from one stage were 9 g/L and mixing efficiency were 90 percent, the underflow liquor soluble concentration from the same stage would be 10 g/L.

DESIGN OR STIPULATION OF A SEDIMENTATION UNIT

Selection Selection of the type of unit thickener or clarifier depends primarily on installation and operating costs. Most manufacturers have overlapping sizes in the bridge-supported, center-column-supported, and traction types even though an economical size range exists for each type. For example, if a unit must be covered to conserve heat, the bridge-supported type may be more economical up to about 45 m (150 ft) in diameter, although 30 m (100 ft) may be the economic limit for an uncovered unit. Traction units often are least expensive in sizes over 75 m (250 ft) if ground conditions permit installing proper supporting walls to carry the loads.

Materials of Construction A wide variety of materials is available for tanks, as indicated earlier. Most mechanisms are made of steel; however, submerged parts may be made of wood, stainless steel, rubber-covered or coated steel, or special alloys.

Design Sizing Criteria Table 19-8, which lists typical design sizing criteria and operating conditions for a number of thickener and clarifier applications, is presented for purposes of illustration or preliminary estimate. Final designs should be based on bench-scale tests involving the methods previously discussed.

If a solids-contact clarifier is required, the surface-area requirement must exclude the area taken up by the reaction chamber. The reaction chamber itself is normally sized for a detention time of 15 to 45 min, depending on the type of treatment and the design of the unit. High-rate solids-contact designs normally require only 15 min in the reaction chamber.

Torque Rating The choice of torque rating has been discussed earlier. It is a function of such factors as quantity and quality of underflow (therefore of such parameters as particle characteristics and flocculant dosage that affect underflow character), unit area, and rake speed; but in the final analysis, as pointed out, torque must be specified on the basis of experience modified by these factors. Unless one is experienced in a given application, it is wise to consult a thickener or clarifier manufacturer.

TABLE 19-9 Approximate Installed Cost of Single-Compartment Thickeners*

Diameter		Cost	
m	ft	\$/m²	\$/ft²
3	10	3400–5500	315–510
10	30	1100–1700	100–160
15	50	800–1000	74–93
20	65	700–780	65–72
30	100	430–520	40–48
45	150	320–380	30–35
60	200	250–260	23–24
75	250	240–250	22–23
90	300	215–225	20–21
105	350	215	20
120	400	205	19

*Cost complete including installation; based on steel mechanisms, steel tank and bottom to 30 m, and steel tank and concrete bottom from 30 to 120 m. Consult manufacturer for accurate costs and for sizes above 120 m. Costs are based on 1981 dollars.

Thickener Costs

Equipment Costs vary widely for a given diameter because of the many types of construction. As a general rule, total installed cost will be about 2 to 4 times the cost of the raking mechanism (including drivehead and lift), plus walkways and bridge or centerpier cage, railings, and overflow launders. Table 19-9 lists the approximate installed costs of thickeners up to 120 m (400 ft) in diameter. These costs are to be used only as a guide and do not include costs of special design modifications. They include the erection of mechanism and tank plus normal uncomplicated site preparation, excavation, reinforcing-bar placement, underflow tunnel, backfill, and surveying. The price does not include any electrical work, pumps, piping, or instrumentation.

Operating Costs Power cost for a continuous thickener is an almost insignificant item. For example, a unit thickener 60 m (200 ft) in diameter with a torque rating of 1.0 MN·m (8.8 Mlbf·in) will normally require 12 kW (16 hp). The low power consumption is due to the very slow rotative speeds. Normally, a mechanism will be designed for a peripheral speed of about 9 m/min (0.5 ft/s), which corresponds to only 3 r/h for a 60-m (200-ft) unit. This low speed also means very low maintenance costs. Operating labor is low because little attention is normally required after initial operation has balanced the feed and underflow. If chemicals are required for flocculation, the chemical cost frequently dwarfs all other operating costs.

FILTRATION

GENERAL REFERENCES: Moir, *Chem. Eng.*, **89**(15), 46 (1982); Brown, ibid., 58; also published as McGraw-Hill Repr. A078. Cheremisinoff and Azbel, *Liquid Filtration*, Ann Arbor Science, Woburn, Mass., 1983. Orr (ed.), *Filtration: Principles and Practice*, part I, Marcel Dekker, New York, 1977; part II, 1979. Purchas (ed.), *Solid/Liquid Separation Equipment Scale-Up*, Uplands Press, Croydon, England, 1977. Schweitzer (ed.), *Handbook of Separation Techniques for Chemical Engineers*, part 4, McGraw-Hill, New York, 1979. Shoemaker (ed.), "What the Filter Man Needs to Know about Filtration," *Am. Inst. Chem. Eng. Symp. Ser.*, **73**(171), (1977). Talcott et al., in *Kirk-Othmer Encyclopedia of Chemical Technology*, 3d ed., vol. 10, Wiley, New York, 1980, p. 284. Tiller et al., *Chem. Eng.*, **81**(9), 116–136 (1974); also published as McGraw-Hill Repr. R203.

DEFINITIONS AND CLASSIFICATION

Filtration is the separation of a fluid-solids mixture involving passage of most of the fluid through a porous barrier which retains most of the solid particulates contained in the mixture. This subsection deals only with the filtration of solids from liquids; gas filtration is treated in Sec. 20. **Filtration** is the term for the unit operation. A filter is a piece of unit-operations equipment by which filtration is performed. The **filter medium** or **septum** is the barrier that lets the liquid pass while retaining most of the solids; it may be a screen, cloth, paper, or bed of solids. The liquid that passes through the filter medium is called the **filtrate.**

Filtration and filters can be classified several ways:

1. *By driving force.* The filtrate is induced to flow through the filter medium by hydrostatic head (gravity), pressure applied upstream of the filter medium, vacuum or reduced pressure applied downstream of the filter medium, or centrifugal force across the medium. Centrifugal filtration is closely related to centrifugal sedimentation, and both are discussed later under "Centrifuges."

2. *By filtration mechanism.* Although the mechanism for separation and accumulation of solids is not clearly understood, two models are generally considered and are the basis for the application of theory to the filtration process. When solids are stopped at the surface of a filter medium and pile upon one another to form a cake of increasing thickness, the separation is called **cake filtration.** When solids are trapped within the pores or body of the medium, it is termed **depth, filter-medium,** or **clarifying filtration.**

3. *By objective.* The process goal of filtration may be dry solids (the cake is the product of value), clarified liquid (the filtrate is the product of value), or both. Good solids recovery is best obtained by cake filtration, while clarification of the liquid is accomplished by either depth or cake filtration.

4. *By operating cycle.* Filtration may be intermittent (batch) or continuous. Batch filters may be operated with constant-pressure driving force, at constant rate, or in cycles that are variable with respect to both pressure and rate. Batch cycle can vary greatly, depending on filter area and solids loading.

5. *By nature of the solids.* Cake filtration may involve an accumulation of solids that is compressible or substantially incompressible, corresponding roughly in filter-medium filtration to particles that are deformable and to those that are rigid. The particle or particle-aggregate size may be of the same order of magnitude as the minimum pore size of most filter media (1 to 10 μm and greater), or may be smaller (1 μm down to the dimension of bacteria and even large molecules). Most filtrations involve solids of the former size range; those of the latter range can be filtered, if at all, only by filter-medium-type filtration or by ultrafiltration unless they are converted to the former range by aggregation prior to filtration.

These methods of classification are not mutually exclusive. Thus filters usually are divided first into the two groups of cake and clarifying equipment, then into groups of machines using the same kind of driving force, then further into batch and continuous classes. This is the scheme of classification underlying the discussion of filters of this subsection. Within it, the other aspects of operating cycle, the nature of the solids, and additional factors (e.g., types and classification of filter media) will be treated explicitly or implicitly.

THEORY OF FILTRATION

Filtration has evolved as a practical art rather than as a theoretical science; nevertheless, filtration theory has received steady attention since the pioneering work of Carman and Ruth half a century ago. In a very simple sense, the theoretical effort has consisted principally of trying to quantify the common engineering rate relationship

$$\text{Rate} = \text{driving force}/\text{resistance}$$

where, for the general filtration case, the resistance is the sum of that of the filter medium and that of the cake. In recent years, application of the theory to industrial problems has increased under the encouragement of several leaders, notably Tiller [*Filtr. Sep.*, **12,** 386 (1975)] and Purchas (op. cit.).

The theory is never used as the sole basis for the design of a filter, but it is valuable in interpreting laboratory tests, in seeking optimum conditions for filtration, and in predicting effects of changes in operating conditions. The use of filtration theory is limited by the fact that filtering characteristics must always be determined on the actual slurry in question, data obtained on one slurry being inapplicable to another. This is true because of the role of the nature and history of the solid particles and their interaction, a complex matter just beginning to get the required attention.

In cake filtration, once a layer of solid particles has formed on the filtering medium, its surface becomes the de facto filter medium, solids being deposited and adding to the thickness of the cake while the clear liquor passes through. The cake is therefore composed of a bulky mass of particles of irregular shape, among which run small channels. The flow of liquor through the channels is always streamline and may therefore be represented by Poiseuille's equation, which may be adapted in the following form:

$$\frac{dV}{A\,d\theta} = \frac{P}{\mu[\alpha(W/A) + r]} \tag{19-38}$$

[Carman, *Trans. Inst. Chem. Eng. (London)*, **16,** 174 (1938); also McCabe and Smith, *Unit Operations of Chemical Engineering*, 3d ed., McGraw-Hill, New York, 1976, p. 937], expressing the differential or instantaneous rate of filtration per unit area as the ratio of a driving force, pressure, to the product of viscosity by the sum of cake resistance and filter medium resistance.

The rate of filtration is expressed conveniently in terms of volume of filtrate collected V, area of filtering surface A, and time θ. P is the total pressure drop across the filter medium and the cake deposited on it. The viscosity μ is that of the filtrate.

The rate may also be expressed in terms of W, the mass of accumulated dry-cake solids corresponding to V. W is related to V by a simple material balance,* thus:

$$W = wV = \frac{\rho c}{1 - mc} V \tag{19-39}$$

where w is the mass of dry-cake solids per unit volume of filtrate, ρ is the density of the filtrate, c is the mass fraction of cake solids in the slurry, and m is the mass ratio of wet cake to dry.

The symbol α represents the average specific cake resistance, which is a constant for the particular cake in its immediate condition. In the usual range of operating conditions it is related to the pressure by the expression

$$\alpha = \alpha' P^s \tag{19-40}$$

where α' is a constant determined largely by the size of the particles forming the cake; s is the cake compressibility, varying from 0 for rigid, incompressible cakes, such as fine sand and diatomite, to 1.0 for very highly compressible cakes. For most industrial slurries, s lies between 0.1 and 0.8. The symbol r represents the resistance of unit area of filter medium but includes other losses (besides those across the cake and the medium) in the system across which P is the pressure drop.

To use Eq. (19-38) one must know the pattern of the filtration process, i.e., the variation of the flow rate and pressure with time. Generally the pumping mechanism determines the filtration flow characteristics and serves as a basis for the following three categories† [Tiller and Crump, *Chem. Eng. Prog.*, **73**(10), 65 (1977)]:

1. *Constant-pressure filtration.* The actuating mechanism is compressed gas maintained at a constant pressure.

2. *Constant-rate filtration.* Positive-displacement pumps of various types are employed.

3. *Variable-pressure, variable-rate filtration.* The use of a centrifugal pump results in this pattern: the discharge rate decreases with increasing back pressure.

Flow rate and pressure behavior for the three types of filtration are shown in Fig. 19-79. Depending on the characteristics of the centrifugal pump, widely differing curves may be encountered, as suggested by the figure.

Constant-Pressure Filtration For constant-pressure filtration Eq. (19-38) can be integrated to give the following relationships between total time and filtrate measurements:

*This material-balance statement depends for its validity on the premise that the cake either is incompressible or is compacted uniformly throughout its thickness. Compressible cakes are not compacted uniformly, in fact, and for them the balance, the assumption of constant average cake resistance, and the resulting dependent equations are in error. The error is negligibly small, however, for most materials encountered in normal plant practice. For very-high-consistency slurries, the more exact analysis of Tiller and Huang [*Ind. Eng. Chem.*, 53, 529 (1961)] should be followed.

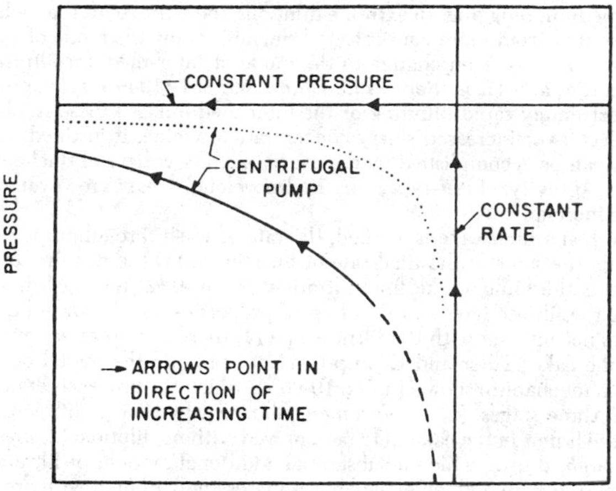

FIG. 19-79 Typical filtration cycles. [*Tiller and Crump*, Chem. Eng. Prog., *73(10), 72 (1977), by permission.*]

$$\frac{\theta}{V/A} = \frac{\mu\alpha}{2P}\frac{W}{A} + \frac{\mu r}{P} \qquad (19\text{-}41a)$$

$$\frac{\theta}{V/A} = \frac{\mu\alpha w}{2P}\frac{V}{A} + \frac{\mu r}{P} \qquad (19\text{-}41b)$$

$$\frac{\theta}{V/A} = \frac{\mu\alpha\rho c}{2P(1-mc)}\frac{V}{A} + \frac{\mu r}{P} \qquad (19\text{-}41c)$$

For a given constant-pressure filtration, these may be simplified to

$$\frac{\theta}{V/A} = K_p\frac{W}{A} + C = K_p'\frac{V}{A} + C \qquad (19\text{-}41d)$$

where K_p, K_p', and C are constants for the conditions employed. It should be noted that K_p, K_p', and C depend on filtering pressure not only in the obvious explicit way but also in the implicit sense that α, m, and r are generally dependent on P.

Constant-Rate Filtration For substantially incompressible cakes, Eq. (19-38) may be integrated for a constant rate of slurry feed to the filter to give the following equations, in which filter-medium resistance is treated as the equivalent constant-pressure component to be deducted from the rising total pressure drop to give the variable pressure through the filter cake [Ruth, *Ind. Eng. Chem.*, **27,** 717 (1935)]:

$$\frac{\theta}{V/A} = \frac{1}{\text{rate per unit area}} = \frac{\mu\alpha}{P-P_1}\frac{W}{A} \qquad (19\text{-}42a)$$

which may also be written

$$\frac{\theta}{V/A} = \frac{\mu\alpha w}{P-P_1}\frac{V}{A} = \frac{\mu\alpha\rho c}{(P-P_1)(1-mc)}\frac{V}{A} \qquad (19\text{-}42b)$$

In these equations P_1 is the pressure drop through the filter medium.

$$P_1 = \mu r(V/A\theta)$$

For a given constant-rate run, the equations may be simplified to

$$V/A = P/K_r + C' \qquad (19\text{-}42c)$$

where K_r and C' are constants for the given conditions.

Variable-Pressure, Variable-Rate Filtration The pattern of this category complicates the use of the basic rate equation. The method of Tiller and Crump (loc. cit.) can be used to integrate the equation when the characteristic curve of the feed pump is available.

In the filtration of small amounts of fine particles from liquid by means of bulky filter media (such as absorbent cotton or felt) it has been found that the preceding equations based upon the resistance of a cake of solids do not hold, since no cake is formed. For these cases, in which filtration takes place on the surface or within the interstices of a medium, analogous equations have been developed [Hermans and Bredée, *J. Soc. Chem. Ind.*, **55T,** 1 (1936)]. These are usefully summarized, for both constant-pressure and constant-rate conditions, by Grace [*Am. Inst. Chem. Eng. J.*, **2,** 323 (1956)]. These equations often apply to the clarification of such materials as sugar solutions, viscose and other spinning solutions, and film-casting dopes.

Practical Significance of the Filtration Equations The differential form [Eq. (19-38)] of the filtration equation yields interesting information about the mutual effects of the operating variables.

When the cake is composed of hard granular particles that make it rigid and incompressible, an increase in pressure results in no deformation of the particles or their interstices, whereby $s = 0$, and, if filter-medium resistance is neglected, Eq. (19-38) becomes

$$\frac{dV}{d\theta} = \frac{AP}{\mu\alpha'(W/A)} \qquad (19\text{-}38a)$$

For incompressible cakes, therefore, the flow rate is directly proportional to the area and pressure and inversely proportional to the viscosity, to the total amount of cake (or filtrate), and to α'.

When the cake consists of extremely soft, easily deformed particles, such as ferric and other metal hydroxides, s approaches 1.0, whereby Eq. (19-38), with the filter medium again neglected, reduces to

$$\frac{dV}{d\theta} = \frac{A}{\mu\alpha'(W/A)} \qquad (19\text{-}38b)$$

For very compressible cakes, therefore, the rate is independent of pressure.

The **effect of pressure** shown here is modified in most industrial filtrations, in which cake compressibility usually lies between 0.1 and 0.8. Furthermore, the resistance of the filter medium reduces the effects of the respective variables. It has been found true, however, that in the filtration of granular or crystalline solids an increase in pressure causes a nearly proportionate increase in flow rate. Flocculent or slimy precipitates, on the other hand, have their filtration rates increased only slightly by an increase in pressure. Some materials have a critical pressure above which a further increase results in an actual decrease in flow rate.

In the filtration of certain nonhomogeneous sludges, such as those of slimy solids to which filter aids have been added, it has been found that a constant flow rate during filtration is more satisfactory than a constant pressure, for the latter results in poor initial clarity of the filtrate and a rapid buildup of cake resistance. As a matter of fact, filtration of any but the most incompressible sludges is more satisfactory when a low pressure is used at the beginning of the run. This is especially important in filtering slurries of low solid content.

Most pressure filters are fed by centrifugal pumps, and the operation is by the variable-rate, variable-pressure mode illustrated in Fig. 19-79. It is usually advisable to throttle the pump somewhat until the cake is fully formed, as indicated by the broken portion of the lower centrifugal pump curve.

Cake thickness is an important factor in determining the capacity and design of a filter, and upon it the cycle of operation depends. Filtration theory shows that, with cloth resistance neglected, the average flow rate during a filtration is inversely proportional to the amount of cake deposited. It is also directly proportional to the square of the filtering area. It should be noted, as a consequence of these two relationships, that the average filtration rate for a given quantity of filtrate or cake is inversely proportional to the square of the thickness of the cake at the end of the filtration.

If the specific cake resistance is so high that even a very thin cake

†A combination of category 2 followed by category 1 as parts of the same filtration cycle is considered by some as a fourth category. For a method of combining the constant-rate and constant-pressure equations for such a cycle, see Brown, loc. cit.

offers a high flow resistance relative to that of the filter medium, maximum filter productivity ($W/A\theta$) is obtained with a cake of infinitesimal thickness. This argument neglects several factors, however: the resistance of the filter medium, the time required to remove the product cake and prepare the medium for the next cycle, the difficulty of discharging extremely thin cakes, and the increased investment required for a filter of greater area. In practice, therefore, the economic choice is a cake of appreciable thickness. The greater the resistance of the filter medium and the longer the preparation downtime, the thicker is the optimum cake.

The **effect of viscosity** is as indicated by the rate equations: the filtrate flow rate at any instant is inversely proportional to the filtrate viscosity. The high viscosity of some filtrates (for example, oils or concentrated solutions) can be reduced by the dilution of the prefilt with low-viscosity solvent, sometimes with a net gain in the filtration rate in spite of the increased volume of filtrate [Reeves, *Ind. Eng. Chem.*, **39,** 203 (1947); *Pet. Process.*, **4,** 885 (1949); and Göttner, *Erdöl Kohle*, **7,** 287 (1954)]. If the filtrate is required in high concentration for subsequent treatment or as a product, dilution will be feasible only if the cost of reconcentration does not make the economics of the filtration unfavorable.

The **effect of temperature** on the filtration rate of incompressible solids is evident principally through its effect on viscosity. The viscosity of most liquids decreases markedly with increasing temperature. Higher temperatures thus permit higher filtration rates; if the filtrate were water, for example, an increase from 20 to 60°C would double the rate of flow. Compressible sludges are affected in more complicated ways by temperature increases, but the general effect is apt to be increased filtration rate. Limits to the extent to which a prefilt may be heated are imposed by the cost of heating and, in vacuum filtration, by the vapor pressure of the filtrate.

The **effect of particle size** on cake and cloth resistances is marked. Even small changes in particle size affect the coefficient α' in the equation for cake resistance [Eq. (19-40)], and larger changes affect the compressibility s. Decreased particle size results in lower filtration rates and higher moisture content of the cake but sometimes in better washing efficiency. It is important, therefore, that close control be kept of the particle size in the feed to the filter. Degradation of particle size by violent pump action or agitation must be avoided. Pretreatment of a slurry by both chemical and physical means can make separation easier by increasing particle size. Several techniques are possible. Addition of certain electrolytes such as alum, lime, and iron salts may effect particle-surface changes and act as coagulants or flocculants. Control of pH may affect the surface charge or zeta potential of particles. The addition of high-molecular-weight polyelectrolytes is often a successful pretreatment to entrap fine particles forming a flocculation structure. Changing the temperature, recrystallization, aging, and freezing are sometimes effective physical techniques. More detailed overviews of pretreatment are provided by Tiller et al. [*Chem. Eng.*, **81**(9), 123 (1974)] and by Thomas [*Am. Inst. Chem. Eng. Symp. Ser.*, **73**(171), 18 1977)].

The **effect of the type of filter medium** often is not fully recognized. In the selection of the medium for a given filtration, a balance must be struck between as open a weave as is feasible to reduce plugging and as tight a weave as is necessary to prevent excessive "bleeding" of fine particles. After a small thickness of cake has formed on the medium, bleeding usually stops, fine particles being caught in the cake.

Of the weaves of filter cloths described under a following subsection, the number-duck weaves have the greatest ability to retain fine solids, followed in decreasing ability by chains (broken twills), twills, and satins. The tendency to plug, however, is in the reverse order. Thick, stiff cloths tend to plug more readily than thin, pliable ones. The effect of cloth plugging on filtration rate is so appreciable that it will usually be the cause of replacement of the cloth. It also results in a need for using a safety factor in predicting filter capacities.

The explicit **effect of solids concentration** in the slurry is shown in Eqs. (19-41*b*), (19-41*c*), and (19-42*b*). These equations indicate that, with negligible r, the time to deposit a stated mass of solids is inverse to the mass ratio of solids to filtrate. There may be implicit effects also. Change in slurry concentration may affect α and the rate of medium plugging. In extreme dilutions, the same solids that, when deposited from more concentrated slurries, follow the mode of cake filtration, may even change to the mode of filter-medium filtration [Heertjes and Haas, *Rec. Trav. Chim.*, **68,** 361 (1949)], tantamount to extremely rapid blinding of the filter medium. All these implicit effects favor increased slurry concentration, which, if justified, usually can be accomplished by the use of one of a variety of thickeners (e.g., gravity, filter-type, or hydrocyclone) that are available commercially.

When a filter cake is washed, the rate of wash throughput is generally the same as the final rate of filtration if (1) the washing pressure is the same as the final filtration pressure, (2) the wash liquid and the filtrate have similar physical properties, (3) the wash liquor does not interact with the filtrate, and (4) there is no rearrangement of the cake. Tiller and Crump (loc. cit.) provide the model of the wash mechanism shown in Fig. 19-80, in which the process is divided into three stages: (1) *displacement,* during which up to 50 percent of the liquor in the cake may be removed with no dilution; (2) *breakthrough,* during which a substantial additional amount of filtrate is displaced with high efficiency but is increasingly diluted with breakthrough wash; and (3) *diffusional,* during which the displacement of filtrate has practically ended, solute being removed only by diffusion from pockets. Normally washing should not be continued beyond the breakthrough stage, which usually ends when the wash volume equals about twice the cake void volume. If further solute removal is required, repulping and filtering probably will be more economical than diffusion washing. Washing effectiveness, unpredictable by theory, must be determined experimentally. A useful way of reducing experimental data is as in Fig. 19-81, a plot of fraction of original solute content remaining in the cake against wash ratio (ratio of wash volume to cake void volume) [Silverblatt et al., *Chem. Eng.*, **81**(9), 127 (1974)]. The broken line of maximum possible efficiency represents perfect, complete displacement. Curve *A* represents the excellent, near-ideal washing performance seldom encountered but sometimes characteristic of dense, low-porosity cakes of relatively high resistance. Curve C indicates poor washing; data falling on or above it may point to wash bypass because of a cracked or malformed cake. Curve B is typical of most cakes, the practical limit of washing being the removal of about 90 percent of the filtrate.

Further mechanical removal of intrapore liquid (filtrate or wash liquor) from filter cakes is accomplished by air or gas displacement through the cake ("blowing" or "drying") or by some form of mechanical compaction.

Application of Filtration Theory to the Interpretation of Data The filtration equations are useful in predicting the effect of a change in any variable if the constants are determined from data taken on the slurry in question. For example, vacuum test data can be extrapolated to show the approximate filtering rates that could be obtained if the slurry were filtered under pressure. Another problem often of interest is the effect of cake thickness or time cycle on the

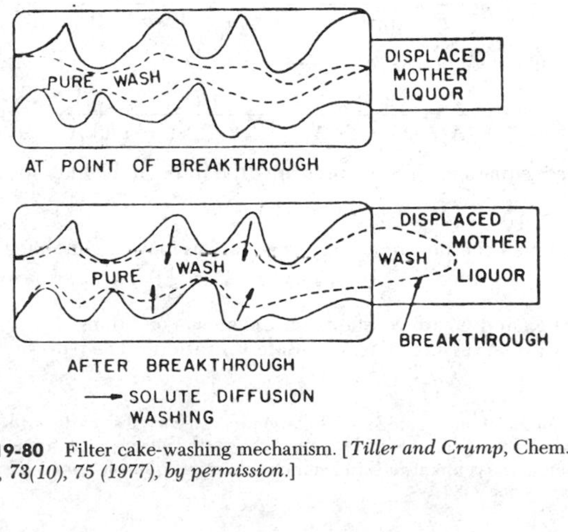

FIG. 19-80 Filter cake-washing mechanism. [*Tiller and Crump,* Chem. Eng. Prog., *73(10), 75 (1977), by permission.*]

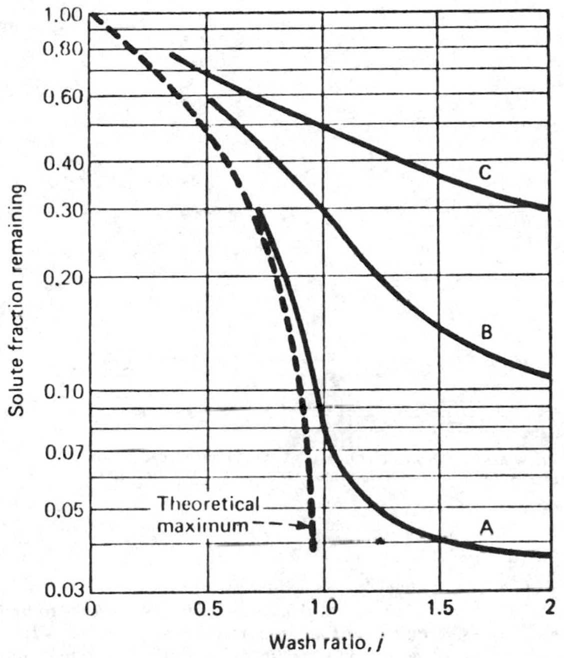

FIG. 19-81 Cake-washing effectiveness. [*Silverblatt et al.*, Chem. Eng., *81(9), 131 (1974), by permission.*]

overall filtration rate. Of greatest importance is the use of theory in interpreting the results of bench-scale or pilot-plant filtrations to estimate the size, operating cycle, and type of filter to be employed in a process under design or change. Scale-up tests are discussed in the next subsection.

If a **constant-pressure test** is run on a slurry, care being taken that not only the pressure but also the temperature and the solid content remain constant throughout the run and that time readings begin at the exact start of filtration, one can observe values of filtrate volume or weight and corresponding elapsed time. With the use of the known filtering area, values of $\theta/(V/A)$ can be calculated for various values of V/A which, when plotted with $\theta/(V/A)$ as the ordinate and V/A as the abscissa (Fig. 19-82*a*), result in a straight line having the slope $\mu\alpha w/2P$ and an intercept on the vertical axis of $\mu r/P$. Since μ, w, and P are known, α and r can be calculated from

$$\alpha = 2P/\mu w \times \text{(slope)}$$

and

$$r = P/\mu \times \text{(vertical intercept)}$$

The effect of the change of any variable not affecting α or r can now be estimated. It should be remembered that α and r usually depend on P, and may be affected by w.

It should be noted also that the intercept is difficult to determine accurately because of large potential experimental error in observing the time of the start of filtration and the time-volume correspon-

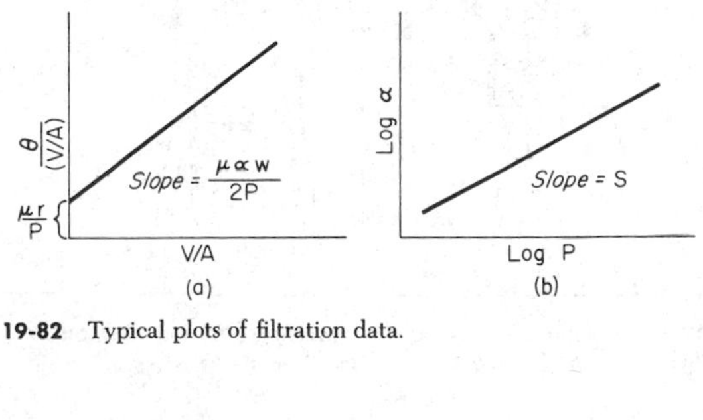

FIG. 19-82 Typical plots of filtration data.

dence during the first moments when the filtration rate is high. The value of r calculated from the intercept may vary appreciably from test to test, and will almost always be different from the value measured with clean medium in a permeability test.

To determine the effect of a change in pressure, it is necessary to run tests at three or more pressures, preferably spanning the range of interest. Plotting α or r against P on log-log paper (or log α or log r against the log P on cartesian coordinates) results in an approximate straight line (Fig. 19-82*b*) from which one may estimate values of α or r at interpolated or reasonably extrapolated magnitudes of P. The slope of the line is the index of a power relationship between α and P or r and P.

Not uncommonly r is found to be only slightly dependent on pressure. When this is true and especially when the filter-medium resistance is, as it should be, relatively small, an average value may be used for all pressures.

It is advisable to start a constant-pressure filtration test, like a comparable plant operation, at a low pressure, and smoothly increase the pressure to the desired operating level. In such cases, time and filtrate-quantity data should not be taken until the constant operating pressure is realized. The value of r calculated from the extrapolated intercept then reflects the resistance of both the filter medium and that part of the cake deposited during the pressure-buildup period. When only the total mass of dry cake is measured for the total cycle time, as is usually true in vacuum leaf tests, at least three runs of different lengths should be made to permit a reliable plot of θ/V against W. If rectification of the resulting three points is dubious, additional runs should be made.

Constant-rate cake-filtration tests, should be avoided because the data are awkward to interpret. When such data must be treated, it is suggested that the experimental procedure and correlation method of Bonilla [*Trans. Am. Inst. Chem. Eng.,* 34, 243 (1938)] be used, involving the determination of P_1 and α_0 in the equation

$$\alpha = \alpha_0 + \alpha'(P - P_1)^s \qquad (19\text{-}43)$$

and plotting $(\alpha - \alpha_0)$ versus $(P - P_1)$ to determine α' and s.

SMALL-SCALE TESTS

Filters cannot be sized strictly from theory owing to the undiscovered nature and apparent vagaries of the particles in the slurry. Therefore, unless exact data have already been established, small-scale tests must be performed to aid in selection of the proper filter and to determine sizing. The commonest test methods are the vacuum leaf test, funnel tests, and pressure-bomb filter tests. These devices are relatively simple and can be purchased, rented, or borrowed from major filter manufacturers. Some filter vendors may even perform the test work for the prospective customer. Great care is required to assure that the test slurry is formed and handled in the same way that the production slurry will be. Samples that have been stored for several days or that have been shipped have been known to filter quite differently from fresh samples. All tests should be made under operating conditions as close as possible to those expected for the large-scale operation being modeled.

Vacuum Tests

Leaf Tests For leaf tests equipment like that shown in Fig. 19-83 is used. A small needle valve, such as is used in valve-in-base bunsen burners, is a useful addition to the receiver to permit fine adjustment of the vacuum level by a controlled air leak. A specimen of selected filter medium is clamped to the leak, which is normally about 11 cm (4.3 in) in face diameter [93 cm^2 (0.1 ft^2) of filtering area]. Different media may be used in successive tests for comparison.

Leaf tests are used to predict the performance of rotary drum or disk filters, and they should be conducted with the operation of such equipment in mind. The simplest continuous cycle is divided into periods: pickup (cake formation), drying, and discharge. Sometimes pickup is followed by a period of displacement washing, with or without intermediate drying; sometimes also the cake may be com-

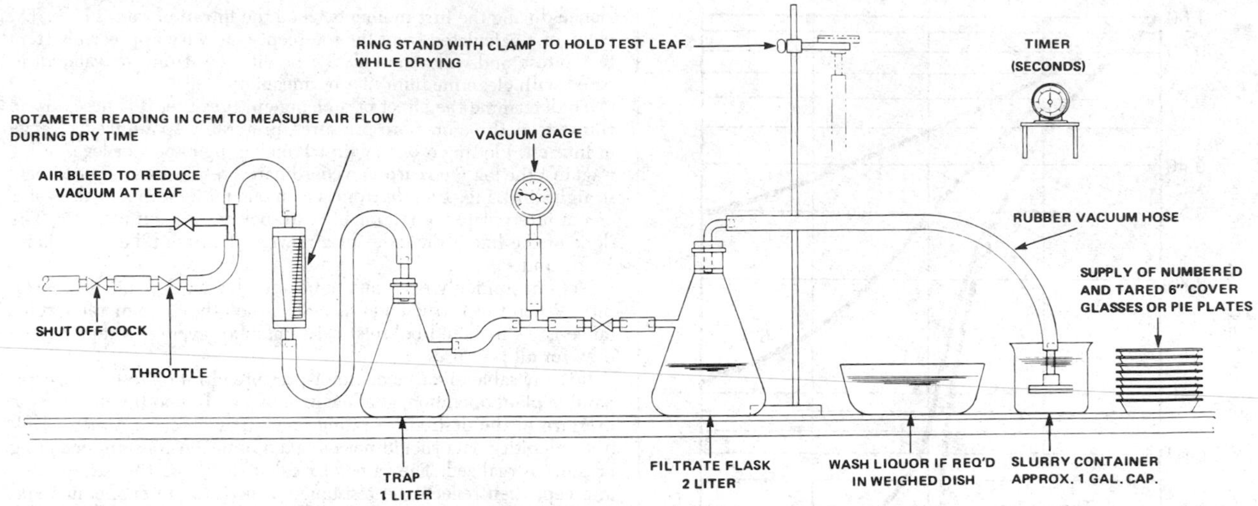

FIG. 19-83 Small-scale vacuum leaf filtration testing setup. *(Dorr-Oliver Inc.)*

pacted mechanically during drying. Counterpart operations can be carried out during leaf tests, and a plan of the whole cycle to be followed should be drafted before the tests are started.

While under vacuum, the test leaf is slid into the material to be filtered and is held totally immersed in the gently agitated slurry for a predetermined pickup period (not longer than 4 min or shorter than time enough to form 6 mm of cake). The leaf is then removed and held with the drain pipe down for the drying time allotted. With the vacuum still on, it may be immersed in a tank of water or other wash liquor if displacement washing is anticipated, and redried. Discharge can be tested by disconnecting the leaf and mouth-blowing through the drainpipe; a spatula may be used in the cake removal.

Usually a few rough preliminary tests at a selected vacuum level will indicate the appropriate time range over which three cakes of different thickness should be formed. Careful tests may then be made, exploring variations in temperature, slurry concentration and preconditioning, vacuum level, washing, or whatever conditions are of interest. During these tests, the primary data are total filtering time, volume of filtrate, thickness and uniformity of cake, mass of cake formed (both wet and dry), and vacuum reading. The slurry temperature and concentration must always be known. Other observations of importance are washing rate and effectiveness, drying behavior, whether and when the cake cracks, discharge characteristics, pH, and foaming tendency of the filtrate.

Funnel Tests For funnel tests, the test leaf of Fig. 19-83 is replaced by a small Büchner funnel in which a piece of appropriate filter medium has been placed so as to minimize leakage around it. With the vacuum on, slurry in the amount predetermined to produce a cake of desired thickness is poured rapidly into the funnel. The slurry should be so added as to allow the cake neither to become dry during the filtration nor to be damaged mechanically by the cascade of liquid. Washing, drying, and discharge observations may be made, as in leaf tests.

Funnel tests simulate the performance of horizontal-surface continuous vacuum filters (belts, tables, and pans).

Pressure Tests

Leaf Tests A bomb filter is used for small-scale leaf tests to simulate the performance of pressure-leaf (leaf-in-shell) filters. The equipment used is a small [50.8- by 50.8-mm (2- by 2-in)] leaf, covered with appropriate filter medium, suspended in a cell large enough to contain sufficient slurry to form the desired cake (Fig. 19-84). The slurry may be agitated gently, for example, by an air sparger.

Although incremental time and filtrate volume may be taken during a cake-forming cycle at a selected pressure to permit a plot like Fig. 19-82*a* from a single run, it may be more satisfactory to make several successive quick runs at the same pressure but for different lengths of time, recording only the terminal values of filtrate volume, time, and cake mass. Operation of the commercial unit should be

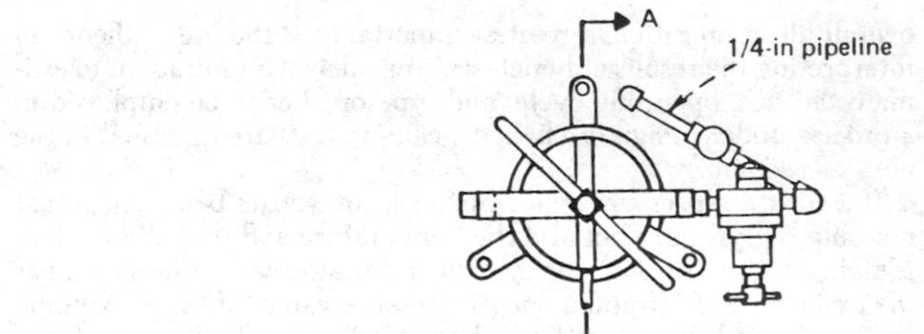

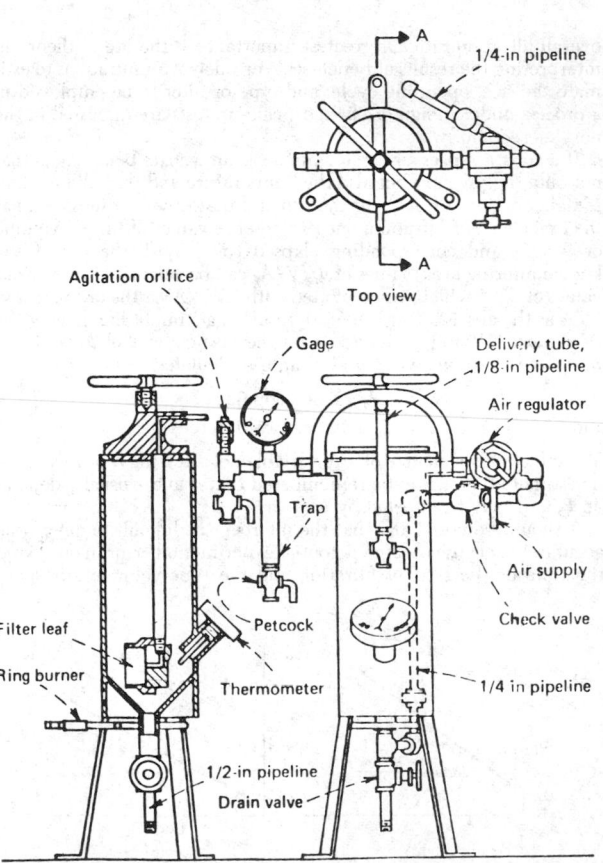

FIG. 19-84 Bomb filter for small-scale pressure filtration tests. [*Silverblatt et al.,* Chem. Eng., *81(9), 132 (1974), by permission.*]

kept in mind when the test cycles are planned. Displacement washing and air blowing of the cake should be tried if appropriate. Wet discharge can be simulated by opening the cell and playing a jet of water on the cake; dry discharge, by applying a gentle air blast to the filtrate-discharge tube. Tests at several pressures must be conducted to determine the compressibility of the cake solids.

Plate-and-Frame Tests These tests should be conducted if the use of a filter press in the plant is anticipated; at least a few confirming tests are advisable after preliminary leaf tests, unless the slurry is very rapidly filtering. A laboratory-size filter press consisting of two plates and a single frame may be used. It will permit the observation of solids-settling, cake-packing, and washing behavior, which may be quite different for a frame than for a leaf.

Compression-Permeability Tests Instead of model leaf tests, compression-permeability experiments may be substituted with advantage for appreciably compressible solids. As in the case of constant-rate filtration, a single run provides data equivalent to those obtained from a series of constant-pressure runs, but it avoids the data-treatment complexity of constant-rate tests.

The equipment consists of a cylindrical cell with a permeable bottom and an open top, into which is fitted a close-clearance, hollow, cylindrical piston with a permeable bottom. Slurry is poured into the cell, and a cake is formed by applying gentle vacuum to the filtrate discharge line. The cell is then filled with filtrate, and the counterweighted piston is allowed to descend to the cake level. Successive increments of mechanical stress are applied to the solids, at each of which the permeability of the cake is determined by passing filtrate through the piston under low head.

The experimental procedure and method of treatment of compression-permeability data have been explained by Grace [*Chem. Eng. Prog.*, **49,** 303, 427 (1953)], who showed that the values of α measured in such a cell and in a pressure filter were the same, and by Tiller [*Filtr. Sep.*, **12,** 386 (1975)].

Scaling Up Test Results The results of small-scale tests are determined as dry weight of solids or volume of filtrate per unit of area per cycle. This quantity multiplied by the number of cycles per day permits the calculation of either the filter area required for a stipulated daily capacity or the daily capacity of a specified plant filter. The scaled-up filtration area should be increased by 25 percent as a factor of uncertainty. In the calculation of cycle length, proper account must be made of the downtime of a batch filter or the "dead area" (between cake discharge and new cake deposit) of a continuous one.

Results obtained by leaf tests for capacity are irregular and unreliable for extremely free-filtering materials, like a magma of large crystals. For such a slurry, it is better to conduct tests with pilot-plant equipment. In any event, before undertaking filtration tests—whatever the prefilt material—it is advisable to consult one or more filter manufacturers, providing as much information as possible about the slurry and the objectives of filtration. The assistance of a reputable vendor can be invaluable to the inexperienced who must carry out a campaign of small-scale filtration tests.

FILTER MEDIA

All filters require a filter medium to retain solids, whether the filter is for cake filtration or for filter-medium or depth filtration. Specification of a medium is based on retention of some minimum particle size at good removal efficiency and on acceptable life of the medium in the environment of the filter. The selection of the type of filter medium is often the most important decision in success of the operation. For cake filtration, medium selection involves an optimization of the following factors:

1. Ability to bridge solids across its pores quickly after the feed is started (i.e., minimum propensity to bleed)
2. Low rate of entrapment of solids within its interstices (i.e., minimum propensity to blind)
3. Minimum resistance to filtrate flow (i.e., high production rate)
4. Resistance to chemical attack
5. Sufficient strength to support the filtering pressure
6. Acceptable resistance to mechanical wear
7. Ability to discharge cake easily and cleanly
8. Ability to conform mechanically to the kind of filter with which it will be used
9. Minimum cost

For filter-medium filtration, attributes 3, 4, 5, 8, and 9 of the preceding list apply and must have added to them (*a*) ability to retain the solids required, (*b*) freedom from discharge of lint or other adulterant into the filtrate, and (*c*) ability to plug slowly (i.e., long life).

Filter-medium selection embraces many types of construction: fabrics of woven fibers, felts, and nonwoven fibers, porous or sintered solids, polymer membranes, or particulate solids in the form of a permeable bed. Media of all types are available in a wide choice of materials.

Fabrics of Woven Fibers For cake filtration these fabrics are the most common type of medium. A wide variety of materials are available; some popular examples are listed in Table 19-10, with ratings for chemical and temperature resistance. In addition to the material of the fibers, a number of construction characteristics describe the filter cloth: (1) weave, (2) style number, (3) weight, (4) count, (5) ply, and (6) yarn number. Of the many types of weaves available, only four are extensively used as filter media: plain (square) weave, twill, chain weave, and satin.

All these weaves may be made from any textile fiber, natural or synthetic. They may be woven from spun staple yarns, multifilament continuous yarns, or monofilament yarns. The performance of the filter cloth depends on the weave and the type of yarn, with ratings indicated in Table 19-11.

Metal Fabrics or Screens These are available in several types of weave in nickel, copper, brass, bronze, aluminum, steel, stainless steel, Monel, and other alloys. In the plain weave, 400 mesh is the closest wire spacing available, thus limiting use to coarse crystalline slurries, pulps, and the like. The "Dutch weaves" employing relatively large, widely spaced, straight warp wires and relatively small crimped filling wires can be woven much more closely, providing a good medium for filtering fine crystals and pulps. This type of weave tends to plug readily when soft or amorphous particles are filtered and makes the use of filter aid desirable. Good corrosion and high temperature resistance of properly selected metals makes filtrations with metal media desirable for long-life applications. This is attractive for handling toxic materials in closed filters to which minimum exposure by maintenance personnel is desirable.

Pressed Felts and Cotton Batting These materials are used to filter gelatinous particles from paints, spinning solutions, and other viscous liquids. Filtration occurs by deposition of the particles in and on the fibers throughout the mat.

Nonwoven Fabrics Made of such synthetic fibers as polyester, nylon, or polyolefin, these fabrics are self-bonded in sheets just after melt extrusion. They are lighter and thinner than felts and are often used in multilayers. Weight of fabric may vary from 14 to 369 g/m^2 (0.4 to 10.9 oz/yd^2). In the low-weight range they are used for the gravity filtration of light oils, while the higher weights are used for highly viscous fluids to remove particles as small as 5 μm in diameter.

Filter Papers These papers come in a wide range of permeability, thickness, and strength. As a class of material, they have low strength, however, and require a perforated backup plate for support.

Rigid Porous Media These are available in sheets or plates and tubes. Materials used include sintered stainless steel and other metals, graphite, aluminum oxide, silica, porcelain, and some plastics—a gamut that allows a wide range of chemical and temperature resistance. Most applications are for clarification.

Polymer Membranes These are used in filtration applications for fine-particle separations such as microfiltration and ultrafiltration (clarification involving the removal of 1-μm and smaller particles). The membranes are made from a variety of materials, the commonest being cellulose acetates and polyamides. Membrane filtration, discussed in Sec. 17, has been well covered by Porter (in Schweitzer, op. cit., sec. 2.1).

Granular Beds of Particulate Solids Beds of solids like sand or coal are used as filter media to clarify water or chemical solutions containing small quantities of suspended particles. Filter-grade grains of desired particle size can be purchased. Frequently beds will

TABLE 19-10 Characteristics of Filter-Fabric Materials*

Generic name and description	Breaking tenacity, g/denier	Abrasion resistance	Resistance to acids	Resistance to alkalies	Resistance to oxidizing agents	Resistance to solvents	Specific gravity	Maximum operating temperature, °F†
Acetate—cellulose acetate. When not less than 92% of the hydroxyl groups are acetylated, "triacetate" may be used as a generic description.	1.2–1.5	G	F	P	G	G	1.33	210
Acrylic—any long-chain synthetic polymer composed of at least 85% by weight of acrylonitrile units.	2.0–4.8	G	G	F	G	E	1.18	300
Glass—fiber-forming substance is glass.	3.0–7.2	P	E	P	E	E	2.54	600
Metallic—composed of metal, metal-coated plastic, plastic-coated metal, or a core completely covered by metal.	......	G						
Modacrylic—fiber-forming substance is any long-chain synthetic polymer composed of less than 85% but at least 35% by weight of acrylonitrile units.	2.5–3.0	G	G	G	G	G	1.30	180
Nylon—any long-chain synthetic polyamide having recurring amide groups as an integral part of the polymer chain.	3.8–9.2	E	F–P	G	F–P	G	1.14	225
Polyester—any long-chain synthetic polymer composed of at least 85% by weight of an ester of a dihydric alcohol and terephthalic acid (*p*—HOOC—C_6H_4—COOH).	2.2–7.8	E–G	G	G–F	G	G	1.38	300
Polyethylene—long-chain synthetic polymer composed of at least 85% weight of ethylene.	1.0–7.0	G	G	G	F	G	0.92	165‡
Polypropylene—long-chain synthetic polymer composed of at least 85% by weight of propylene.	4.8–8.5	G	E	E	G	G	0.90	250§
Rayon—composed of regenerated cellulose and regenerated cellulose manufactured fibers in which substituents have replaced not more than 15% of the hydrogens of the hydroxyl groups.	1.5–5.7	G	P	F–P	F	G	1.52	210
Saran—fiber-forming substance is any long-chain synthetic polymer composed of at least 80% by weight of vinylidene chloride units (—CH_2—CCl_2—).	1.0–2.3	G	G	G	F	G	1.70	160
Cotton—natural fibers.	3.3–6.4	G	P	F	G	E–G	1.55	210
Fluorocarbon—long-chain synthetic polymer composed of tetrafluoroethylene units.	1.0–2.0	F	E	E	E	G	2.30	550¶

*Adapted from Mais, *Chem. Eng.*, 78(4), 51(1971). Symbols have the following meaning: E = excellent, G = good, F = fair, P = poor.
†°C = (°F − 32)/1.8; K = (°F + 459.7)/1.8.
‡Low-density polymer. Up to 230°F for high-density.
§Heat-set fabric; otherwise lower.
¶Requires ventilation because of release of toxic gases above 400°F.

be constructed of layers of different materials and different particle sizes.

Various types of filter media and the materials of which they are constructed are surveyed extensively by Purchas (*Industrial Filtration of Liquids*, CRC Press, Cleveland, 1967, chap. 3), and characterizing measurements (e.g., pore size, permeability) are reviewed in detail by Rushton and Griffiths (in Orr, op. cit., chap. 3). Briefer summaries of classification of media and of practical criteria for the selection of a filter medium are presented by Shoemaker (op. cit., p. 26) and Purchas [*Filtr. Sep.*, **17**, 253, 372 (1980)].

FILTER AIDS

Use of filter aids is a technique frequently applied for filtrations in which problems of slow filtration rate, rapid medium blinding, or unsatisfactory filtrate clarity arise. Filter aids are granular or fibrous solids capable of forming a highly permeable filter cake in which very fine solids or slimy, deformable flocs may be trapped. Application of filter aids may allow the use of a much more permeable filter medium than the clarification would require to produce filtrate of the same quality by depth filtration.

Filter aids should have low bulk density to minimize settling and aid good distribution on a filter-medium surface that may not be horizontal. They should also be porous and capable of forming a porous cake to minimize flow resistance, and they must be chemically inert to the filtrate. These characteristics are all found in the two most popular commercial filter aids: diatomaceous silica (also called diatomite, or diatomaceous earth), which is an almost pure silica prepared from deposits of diatom skeletons; and expanded perlite, particles of "puffed" lava that are principally aluminum alkali silicate.

TABLE 19-11 Effect of Types of Weave and Yarn on Filter Media*

	Highest flow rate	Greatest retention	Best cake discharge	Least moisture in cake	Resistance to blinding
Weave:					
Plain	4	1	4	4	4
Twill	2	3	2	2	2
Chain	3	2	3	3	3
Satin	1	4	1	1	1
Yarn:					
Monofilament	1	3	1	1	1
Multifilament	2	2	2	2	2
Spun-staple	3	1	3	3	3

* Adapted from French, *Chem. Eng.*, **70**(21), 189 (1963). In the table, 1 is best, 4 is poorest.

Cellulosic fibers (ground wood pulp) are sometimes used when siliceous materials cannot be used but are much more compressible. The use of other less effective aids (e.g., carbon and gypsum) may be justified in special cases. Sometimes a combination of carbon and diatomaceous silica permits adsorption in addition to filter-aid performance.

Diatomaceous Silica Filter aids of diatomaceous silica have a dry bulk density of 128 to 320 kg/m^3 (8 to 20 lb/ft^3), contain particles mostly smaller than 50 μm, and produce a cake with porosity in the range of 0.9 (volume of voids/total filter-cake volume). The high porosity (compared with a porosity of 0.38 for randomly packed uniform spheres and 0.2 to 0.3 for a typical filter cake) is indicative of its filter-aid ability. Different methods of processing the crude diatomite result in a series of filter aids having a wide range of permeability.

Perlite Perlite filter aids are somewhat lower in bulk density (48 to 96 kg/m^3, or 3 to 6 lb/ft^3) than diatomaceous silica and contain a higher fraction of particles in the 50- to 150-μm range. Perlite is also available in a number of grades of differing permeability and cost, the grades being roughly comparable to those of diatomaceous silica. Diatomaceous silica will withstand slightly more extreme pH levels than perlite, and it is said to be somewhat less compressible.

Filter aids are used in two ways: (1) as a precoat and (2) mixed with the slurry as a "body feed." Precoat filtration, employing a thin layer of about 0.5 to 1.0 kg/m^2 (0.1 to 0.2 lb/ft^2) deposited on the filter medium prior to beginning feed to the filter, is in wide use to protect the filter medium from fouling by trapping solids before they reach the medium. It also provides a finer matrix to trap fine solids and assure filtrate clarity. Body-feed application is the continuous addition of filter aid to the filter feed to increase the porosity of the cake. The amount of addition must be determined by trial, but a general rule of thumb calls for twice the amount of solids to be removed. For solids loadings greater than 1000 ppm this may become a significant cost factor. An acceptable alternative might be to use a rotary vacuum precoat filter [Smith, *Chem. Eng.*, **83**(4), 84 (1976)]. Further details of filter-aid filtration are set forth by Cain (in Schweitzer, op. cit., sec. 4.2) and Hutto [*Am. Inst. Chem. Eng. Symp. Ser.*, **73**(171), 50 (1977)]. Figure 19-85 shows a flow sheet indicating arrangements for both precoat and body-feed applications. Most filter aid is used on a one-time basis, although some techniques have been demonstrated to reuse precoat filter aid on vertical-tube pressure filters.

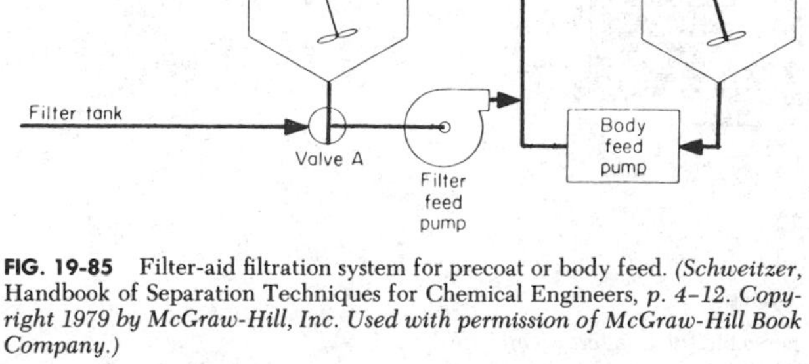

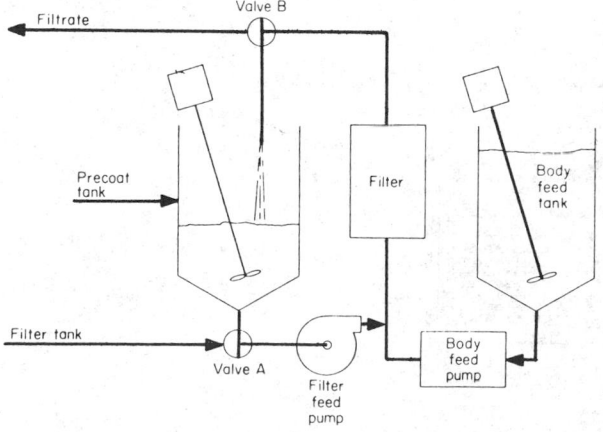

FIG. 19-85 Filter-aid filtration system for precoat or body feed. *(Schweitzer, Handbook of Separation Techniques for Chemical Engineers, p. 4–12. Copyright 1979 by McGraw-Hill, Inc. Used with permission of McGraw-Hill Book Company.)*

FILTRATION EQUIPMENT

Cake Filters Filters that accumulate appreciable visible quantities of solids on the surface of a filter medium are called cake filters. The slurry feed may have a solids concentration from about 1 percent to greater than 40 percent. The filter medium on which the cake forms is relatively open to minimize flow resistance, since once the cake forms, it becomes the effective filter medium. The initial filtrate therefore may contain unacceptable solids concentration until the cake is formed. This situation may be made tolerable by recycling the filtrate until acceptable clarity is obtained or by using a downstream polishing filter (clarifying type).

Cake filters are used when the desired product of the operation is the solids, the filtrate, or both. When the filtrate is the product, the degree of removal from the cake by washing or blowing with air or gas becomes an economic optimization. When the cake is the desired product, the incentive is to obtain the desired degree of cake purity by washing, blowing, and sometimes mechanical expression of residual liquid.

Implicit in cake filtration is the removal and handling of solids, since the cake is usually relatively dry and compacted. Cakes can be sticky and difficult to handle; therefore, the ability of a filter to discharge the cake cleanly is an important equipment-selection criterion.

In the operational sense, some filters are batch devices, whereas others are continuous. This difference provides the principal basis for classifying cake filters in the discussion that follows. The driving force by which the filter functions—hydrostatic head ("gravity"), pressure imposed by a pump or a gas blanket, or atmospheric pressure ("vacuum")—will be used as a secondary criterion.

Batch Cake Filters

Nutsche Filters A nutsche is one of the simplest batch filters. It is a tank with a false bottom, perforated or porous, which may either support a filter medium or act as the filter medium. The slurry is fed into the filter vessel, and separation occurs by gravity flow, gas pressure, vacuum, or a combination of these forces. The term "nutsche" comes from the German term for sucking, and vacuum is the common operating mode.

The design of most nutsche filters is very simple, and they are often fabricated by the user at low cost. The filter is very frequently used in laboratory, pilot-plant, or small-plant operation. For large-scale processing, however, the excessive floor area encumbered per unit of filtration area and the awkwardness of cake removal are strong deterrents. For small-scale operations, cake is manually removed. For large-scale applications, cake may be further processed by reslurrying or redissolving; or it may be removed manually (by shovel) or by mechanical discharge arrangements that sometimes are complicated.

Thorough displacement washing is possible in a nutsche if the wash solvent is added before the cake begins to be exposed to air displacement of filtrate. If washing needs to be more effective, an agitator can be provided in the nutsche vessel to reslurry the cake to allow adequate diffusion of solute from the solids.

Rosenmund filter. This is an example of a commercially available large-scale nutsche filter. The filter is offered in sizes of 1 to 10 m² (11 to 108 ft²). It is designed for totally closed operation, which allows the handling of hazardous solvents. The Rosenmund filter is equipped with a cake spreader and smoother, an automatic discharge screw, and an agitator to reslurry cake for washing, all of which can be sequence-controlled. The filter medium can be a filter cloth, a sintered plate, a porous ceramic structure, a wire screen, or a perforated plate.

Nutrex. A recent novel variation of a nutsche filter is the **Nutrex,** made by Bertrams Ltd. of Switzerland and available in the United States through Rosenmund, Inc. The Nutrex combines in one vessel a reactor, a filter, and a dryer. It consists of a cylindrical pressure vessel, dished at one end and fitted with a flat single-surface filter leaf at the other. The vessel is jacketed over most of its surface and can be rotated about a central diametric axis on trunnions. Mounted at the dished end is an adjustable-clearance agitator, anchor-shaped but with a straight top bar, driven from the underside and sealed with a double mechanical seal. Reaction or crystallization is done at any desired pressure up to 448 kPa (50 psig) with the dish head down. When this step is complete, the entire vessel is rotated 180° so that the filter surface is down and filtration is accomplished with either pressure or vacuum. The filter cake can be smoothed, squeezed, reslurried, and washed by varying the clearance between the straight member of the agitator and the filter medium. Drying can be done by using the agitator to scrape the cake off the filter medium and by tumbling the vessel on its axis as with a double cone dryer.

The Nutrex is particularly well suited for operations involving toxic materials or requiring sterile conditions. It is available with filtration surface area between 0.5 and 6 m² (5.4 and 64.6 ft²) and a reactor volume of 0.4 to 16.8 m³ (14 to 593 ft³).

Horizontal Plate Filter The horizontal multiple-plate pressure filter consists of a number of horizontal circular drainage plates and guides placed in a stack in a cylindrical shell (Fig. 19-86). In normal practice the filtering pressure is limited to 448 kPa (50 psig), although special filters have been designed for shell pressures of 2.2 MPa (300 psig) or higher.

A filter cloth or paper is placed on each plate, and a precoat of filter aid is applied if needed. Slurry is distributed to the plates via a central or an annular feed manifold. Filtration is continued until the cake capacity of the unit is reached or until the filtrate rate becomes too slow owing to cake resistance. The filter assembly may include a scavenger plate at the bottom to clear up the heel of a cycle. A wash or an air blow can be applied to the cake. The cake is removed manually by opening the vessel, removing the plates, and scraping or hosing off the solids.

The filter is compact, it has good cake distribution and effective cake washing, and the unit is amenable to easy cleaning and sterilization. Limitations are its small size, high labor requirements, and the requirement to open the unit to remove cake.

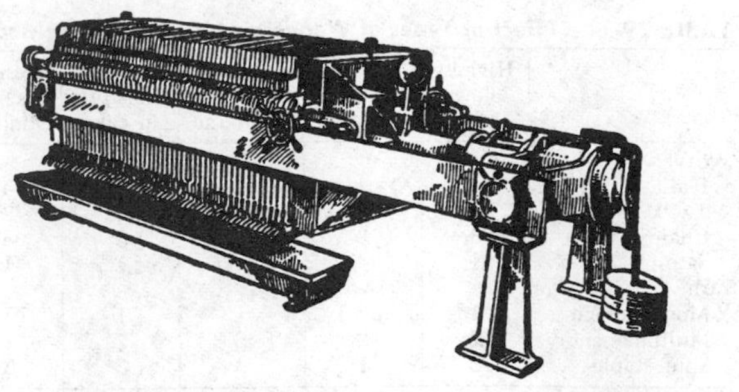

FIG. 19-87 Plate-and-frame filter press. *(Eimco Process Equipment Co.)*

The horizontal filter is best for applications in which small quantities of cake or intermittent flow rates are involved and in which cleanliness or sterile conditions are essential, as in the food and pharmaceutical industries. It is particularly useful in pilot plants and small specialty plants, where equipment must be flexible enough to accommodate a succession of process streams and products (implying, among other characteristics, equipment that can be cleaned quickly and dependably between campaigns). Sizes range from 0.2 to 0.84 m (8 to 33 in) in plate diameter, with up to 42 plates in a stack.

Filter Press The filter press, one of the most frequently used filters in the early years of the chemical industry, is still widely employed. Often referred to generically (in error) as the plate-and-frame filter, it has probably over 100 design variations. Two basic popular designs are the flush-plate, or plate-and-frame, design and the recessed-plate press. Both are available in a wide range of materials: metals, coated metals, plastics, or wood.

Plate-and-frame press. This press is an alternate assembly of plates covered on both sides with a filter medium, usually a cloth, and hollow frames that provide space for cake accumulation during filtration (Fig. 19-87). The frames have feed and wash manifold ports, while the plates have filtrate drainage ports. The plates and frames usually are rectangular, although circles and other shapes also are used (Fig. 19-88). They are hung on a pair of horizontal support bars and pressed together during filtration to form a watertight closure between two end plates, one of which is stationary. The press may be closed manually, hydraulically, or by a motor drive. Several

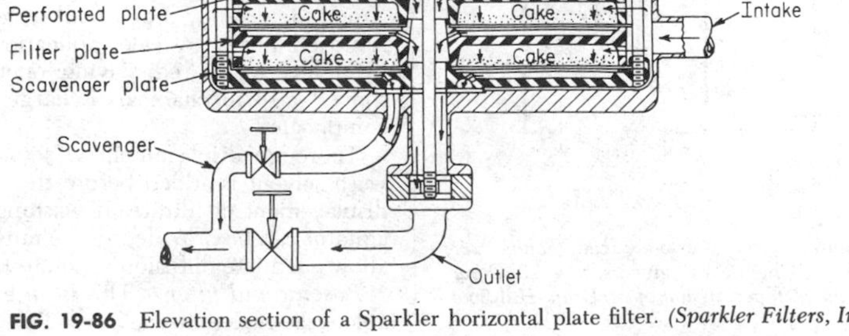

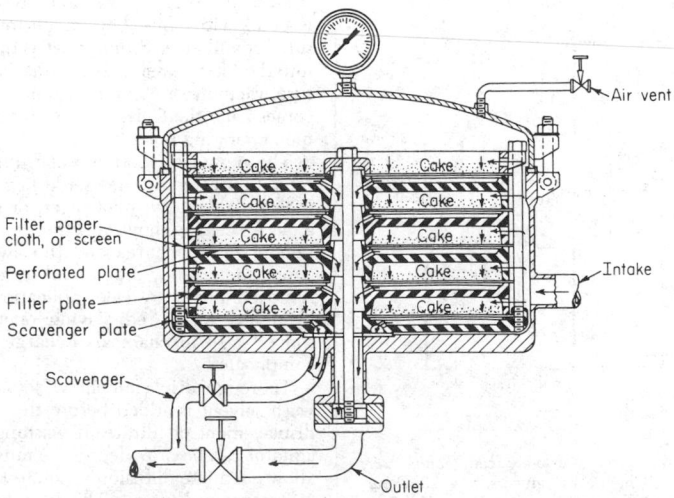

FIG. 19-86 Elevation section of a Sparkler horizontal plate filter. *(Sparkler Filters, Inc.)*

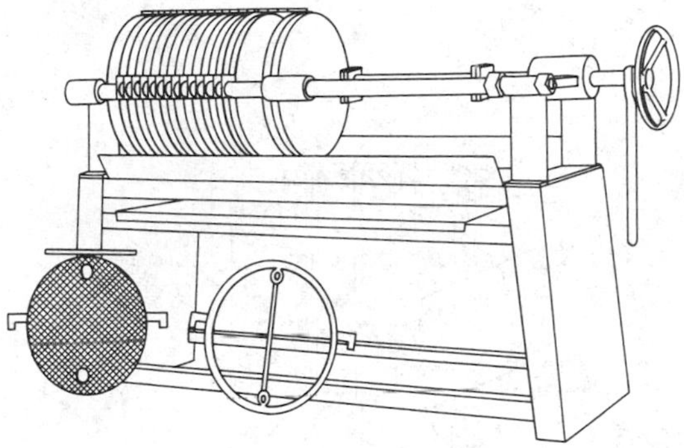

FIG. 19-88 Circular-plate fabricated-metal filter press. *(Star Systems Filtration Division.)*

feed and filtrate discharge arrangements are possible. In the most popular, the feed and discharge of the several elements of the press are manifolded via some of the holes that are in the four corners of each plate and frame (and filter cloth) to form continuous longitudinal channels from the stationary end plate to the other end of the press. Alternatively, the filtrate may be drained from each plate by an individual valve and spigot (for open discharge) or tubing (for closed). Top feed to and bottom discharge from the chambers provide maximum recovery of filtrate and maximum mean cake dryness. This arrangement is especially suitable for heavy fast-settling solids. For most slurries, bottom feed and top filtrate discharge allow quick air displacement and produce a more uniform cake.

Two wash techniques are used in plate-and-frame filter presses, illustrated in Fig. 19-89. In simple washing, the wash liquor follows the same path as the filtrate. If the cake is not extremely uniform and highly permeable, this type of washing is ineffective in a well-filled press. A better technique is thorough washing, in which the wash is introduced to the faces of alternate plates (with their discharge channels valved off). The wash passes through the entire cake and exits through the faces of the other plates. This improved technique requires a special design and the assembly of the plates in proper order. Thorough washing should be used only when the frames are well filled, since an incomplete fill of cake will allow cake collapse during the wash entry. The remainder of the wash flow will bypass through cracks or channels opened in the cake.

Filter presses are made in plate sizes from 10 by 10 cm (4 by 4 in) to 1.5 by 1.8 m (61 by 71 in). Frame thickness ranges from 0.3 to 20 cm (0.125 to 8 in). Operating pressures up to 780 kPa (100 psig) are common, with some presses designed for 7.8 MPa (1000 psig). Some metal units have cored plates for steam or refrigerant. Maximum pressure for wood or plastic frames is 500 to 600 kPa (60 to 70 psig).

The filter press has the advantage of simplicity, low capital cost, flexibility, and ability to operate at high pressure in either a cake-filter or a clarifying-filter application. Floor-space and headroom needs per unit of filter area are small, and capacity can be adjusted by adding or removing plates and frames. Filter presses are cleaned easily, and the filter medium is easily replaced. With proper operation a denser, drier cake compared with that of most other filters is obtained.

There are several serious disadvantages, including imperfect washing due to variable cake density, relatively short filter-cloth life due to the mechanical wear of emptying and cleaning the press (often involving scraping the cloth), and high labor requirements. Presses frequently drip or leak and thereby create housekeeping problems, but the biggest problem arises from the requirement to open the filter for cake discharge. The operator is thus exposed routinely to the contents of the filter, and this is becoming an increasingly severe disadvantage as more and more materials once believed safe are given restricted exposure limits.

Recessed-plate filter press. This press is similar to the plate-and-frame press in appearance but consists only of plates. Both faces of each plate are hollowed to form a chamber for cake accumulation between adjacent plates. This design has the advantage of about half as many joints as a plate-and-frame press, making a tight closure more certain. Figure 19-90 shows some of the features of one type of recessed-plate filter which has a gasket to further minimize leaks. Air can be introduced behind the cloth on both sides of each plate to assist cake removal.

Some interesting variations of standard designs include the ability to roll the filter to change from a bottom to a top inlet or outlet and the ability to add blank dividers to convert a press to a multistage press for further clarification of the filtrate or to do two separate filtrations simultaneously in the same press. Some designs have diaphragms between plates which can be expanded when filtration is finished to squeeze out additional moisture. Some designs feature automated opening and cake-discharge operations to reduce labor requirements.

Industrial Tubular Filter This product of the Industrial Filter & Pump Mfg. Co. consists of one or more perforated tubes supported horizontally or vertically by a transverse tube sheet within a cylindrical shell the axis of which is parallel to those of the tubes. A horizontal unit is illustrated in Fig. 19-91. The end of each tube farthest from the header is closed by an inverted dome. A sheet of filter paper

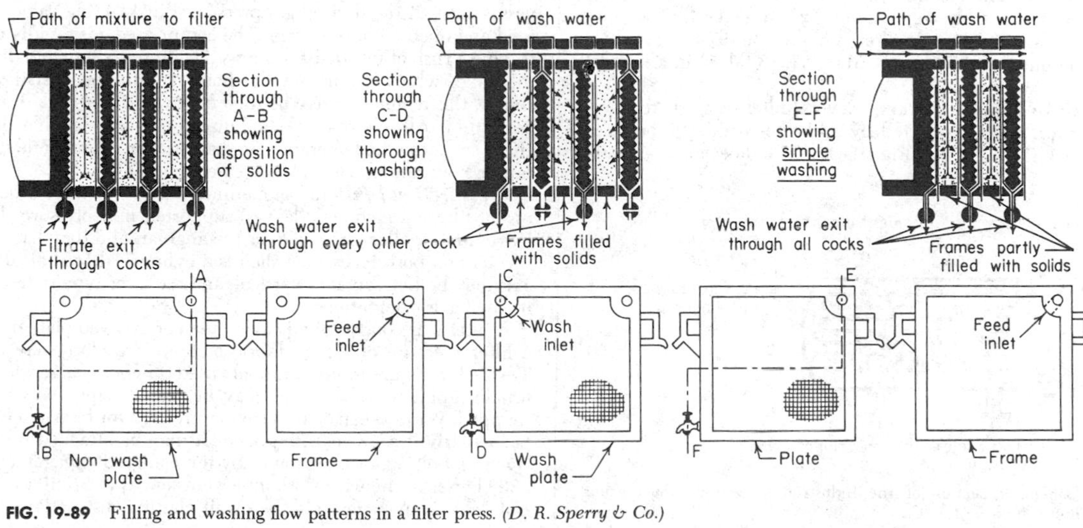

FIG. 19-89 Filling and washing flow patterns in a filter press. *(D. R. Sperry & Co.)*

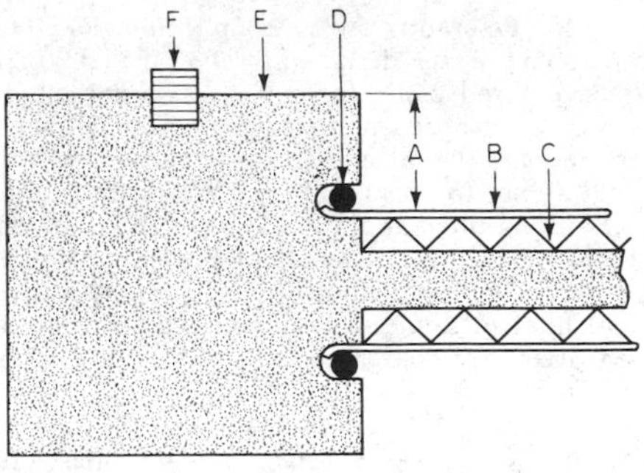

FIG. 19-90 Section detail of a caulked-gasketed-recessed filter plate. (*a*) Cake recess. (*b*) Filter cloth. (*c*) Drainage surface of plate. (*d*) Caulking strip. (*e*) Plate joint. (*f*) Sealing gasket. *(Eimco Process Equipment Co.)*

is rolled and inserted into each tube to form a filter-medium liner that is held against the wall of the tube by means of the inverted dome of the dead end and a tapered sealing ring inserted at the entry end.

Slurry under pressure is admitted to the chamber between the head of the shell and the tube sheet, whence it enters and fills the tubes. Filtration occurs as the filtrate passes radially outward through the paper liner and the wall of each tube into the shell and on out the filtrate discharge line, depositing cake on the liner. The filtration cycle is ended when the tubes have filled with cake or when the media have become plugged. The cake can be washed (if it has not been allowed to fill the tubes completely) and air-blown. The filter has a hinged head to provide easy access to the tube sheet and mouth of the tubes; thus "sausages" of cake can be removed by taking out the sealing rings and withdrawing the filter paper from each tube. The tubes themselves are easily removed for inspection and cleaning, each being sealed into the header by an O ring.

The advantages of the tubular filter are that it uses an easily replaced and inexpensive disposable filter medium, its filtration cycle can be interrupted and the shell can be emptied of prefilt at any time without loss of the cake, the cake is readily recoverable in dry form, and the inside of the filter is conveniently accessible. Disadvantages are the necessity and attendant labor requirements of emptying by hand and replacing the filter media and the tendency for heavy solids to settle out in the header chamber. Applications are as a scavenger filter to remove fines not removed in a prior filtration stage with a different kind of equipment, to handle the runoff from other filters, and in semiworks and small-plant operations in which the filter's size, versatility, and cleanliness recommend it.

Each tube provides a filtering area of 0.22 m^2 (2.4 ft^2). The maximum area available in a single filter is 8.3 m^2 or 197 ft^2 (82 tubes) in a horizontal unit and 47.5 m^2 or 511 ft^2 (213 tubes) in a vertical one.

External-Cake Tubular Filters Several filter designs are available with vertical tubes supported by a filtrate-chamber tube sheet in a vertical cylindrical vessel (Fig. 19-92). The tubes may be made

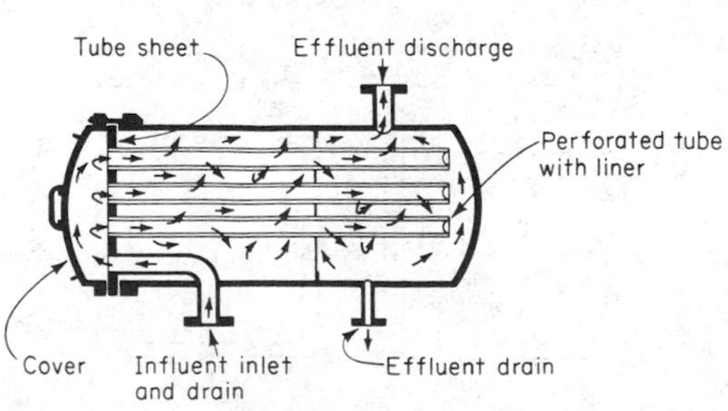

FIG. 19-91 Schematic section of the Industrial horizontal tubular filter. *(Industrial Filter & Pump Mfg. Co.)*

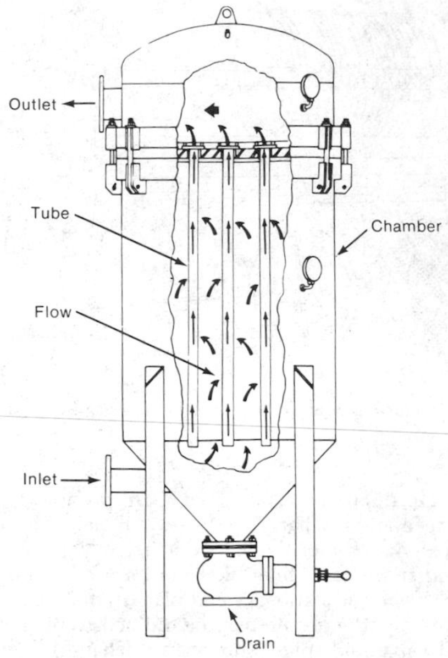

FIG. 19-92 Top-outlet tubular filter. *(Industrial Filter & Pump Mfg. Co.)*

of wire cloth; porous ceramic, carbon, plastic, or metal; or closely wound wire. The tubes may have a filter cloth on the outside. Frequently a filter-aid precoat will be applied to the tubes. The prefilt slurry is fed near the bottom of the vertical vessel. The filtrate passes from the outside to the inside of the tubes and into a filtrate chamber at the top or the bottom of the vessel. The solids form a cake on the outside of the tubes with the filter area actually increasing as the cake builds up, partially compensating for the increased flow resistance of the thicker cake. The filtration cycle continues until the differential pressure reaches a specified level, or until about 25 mm (1 in) of cake thickness is obtained.

Cake-discharge methods are the chief distinguishing feature among the various designs. That of the Industrial Filter & Pump **Hydro-Shoc,** for example, removes cake from the tubes by filtrate backflushing assisted by the "shocking" action of a compressed-gas pocket formed in the filtrate chamber at the top of the vertical vessel. Closing the filtrate outlet valve while continuing to feed the filter causes compression of the gas volume trapped in the dome of the vessel until, at the desired gas pressure, quick-acting valves stop the feed and open a bottom drain. The compressed gas rapidly expands, forcing a rush of filtrate back across the filter medium and dislodging the cake, which drains out the bottom with the flush liquid. Of course, this technique may be used only when wet-cake discharge is permitted. All external-cake tubular filters are capable of totally enclosed, automated operation, so that strict environmental standards can be maintained with relative ease.

Pressure Leaf Filters Sometimes called tank filters, they consist of flat filtering elements (leaves) supported in a pressure shell. The leaves are circular, arc-sided, or rectangular, and they have filtering surfaces on both faces. The shell is a cylindrical or conical tank. Its axis may be horizontal or vertical, and the filter type is described by its shell axis orientation.

A filter leaf consists of a heavy screen or grooved plate over which a filter medium of woven fabric or fine wire cloth may be fitted. Textile fabrics are more commonly used for chemical service and are usually applied as bags that may be sewed, zippered, stapled, or snapped. Wire-screen cloth is frequently used for filter-aid filtrations, particularly if a precoat is applied. It may be attached by welding, riveting, bolting, or caulking or by the clamped engagement of two 180° bends in the wire cloth under tension, as in Multi Metal's **Rim-Lok** leaf. Leaves may also be of all-plastic construction. The filter medium, regardless of material, should be as taut as possible to minimize sagging when it is loaded with a cake; excessive sag can cause cake cracking or dropping. Leaves may be supported at top, bottom, or center and may discharge filtrate from any of these locations. Figure 19-93 shows the elevation section of a precoated bottom-support wire leaf.

Pressure leaf filters are operated batchwise. The shell is locked, and the prefilt slurry is admitted from a pressure source (pump or monte-jus). The slurry enters in such a way as to minimize settling of the suspended solids. The shell is filled, and filtration occurs on the leaf surfaces, the filtrate discharging through an individual delivery line or into an internal manifold, as the filter design dictates. Filtration is allowed to proceed only until a cake of the desired thickness has formed, since to overfill will cause cake consolidation with consequent difficulty in washing and discharge. The decision of when to end the filtering cycle is largely a matter of experience, guided roughly by the rate in a constant-pressure filter or pressure drop in a constant-rate filter. This judgment may be supplanted by the use of a detector which "feels" the thickness of cake on a representative leaf.

If the cake is to be washed, the slurry heel can be blown from the filter and wash liquor can be introduced to refill the shell. If the cake tends to crack during air blowing, it may be necessary to displace the slurry heel with wash gradually so as never to allow the cake to dry. Upon the completion of filtration and washing, the cake is discharged by one of several methods, depending on the shell and leaf configuration.

Horizontal pressure leaf filters. In these filters the leaves may be rectangular leaves which run parallel to the axis and are of varying sizes since they form chords of the shell; or they may be circular or square elements parallel to the head of the shell, and all of the same dimension. The leaves may be supported in the shell from an independent rack, individually from the shell, or from a filtrate manifold. Horizontal filters are particularly suited to dry-cake discharge.

Most of the currently available commercial horizontal pressure filters have leaves parallel to the shell head. Cake discharge may be wet or dry; it can be accomplished by sluicing with liquid sprays, vibration of the leaves, or leaf rotation against a knife, wire, or brush. If a wet-cake discharge is allowable, the filters will probably be sluiced with high-pressure liquid. If the filter has a top or bottom

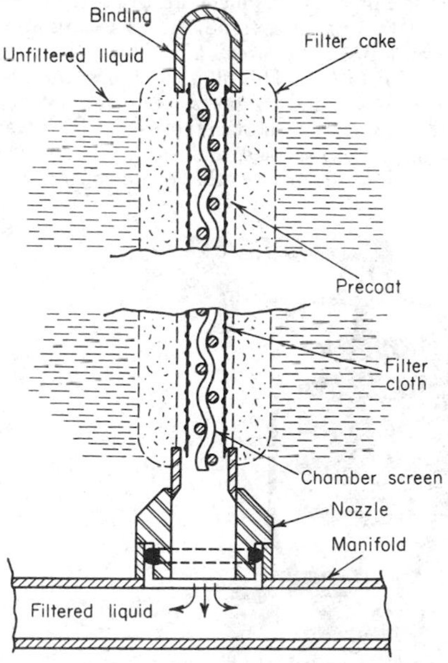

FIG. 19-93 Section of precoated wire filter leaf. *(Multi Metal Wire Cloth Inc.)*

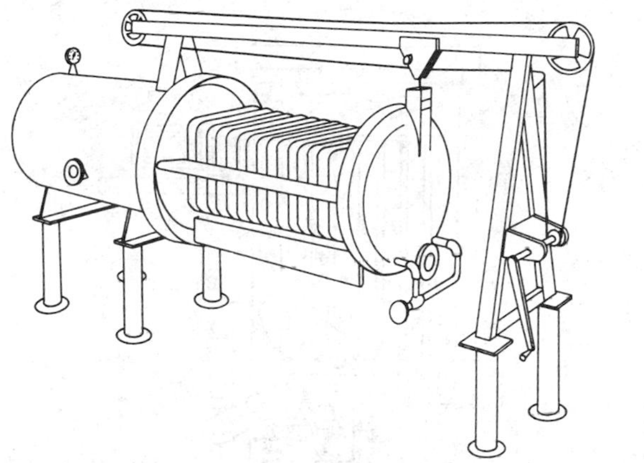

FIG. 19-94 Horizontal-tank pressure leaf filter. *(Ametek, Inc.: Process Equipment Division.)*

filtrate manifold, the leaves are usually in a fixed position, and the spray header is rotated to contact all filter surfaces. If the filtrate header is center-mounted, the leaves are generally rotated at about 3 r/min and the spray header is fixed. Some units may be wet-cake-discharged by mechanical vibration of the leaves with the filter filled with liquid. Dry-cake discharge normally will be accomplished by vibration if leaves are top- or bottom-manifolded and by rotation of the leaves against a cutting knife, wire, or brush if they are center-manifolded.

In many designs the filter is opened for cake discharge, and the leaf assembly is separated from the shell by moving one or the other on rails (Fig. 19-94). For processes involving toxic or flammable materials, a closed filter system can be maintained by sloping the bottom of the horizontal cylinder to the drain nozzle for wet discharge or by using a screw conveyor in the bottom of the shell for dry discharge (Fig. 19-95).

Vertical pressure leaf filters. These filters have vertical, parallel, rectangular leaves mounted in an upright cylindrical pressure tank. The leaves usually are of such different widths as to allow them to conform to the curvature of the tank and to fill it without waste space. The leaves often rest on a filtrate manifold, the connection being sealed by an O ring, so that they can be lifted individually from the top of the filter for inspection and repair. A scavenger leaf frequently is installed in the bottom of the shell to allow virtually complete filtration of the slurry heel at the end of a cycle.

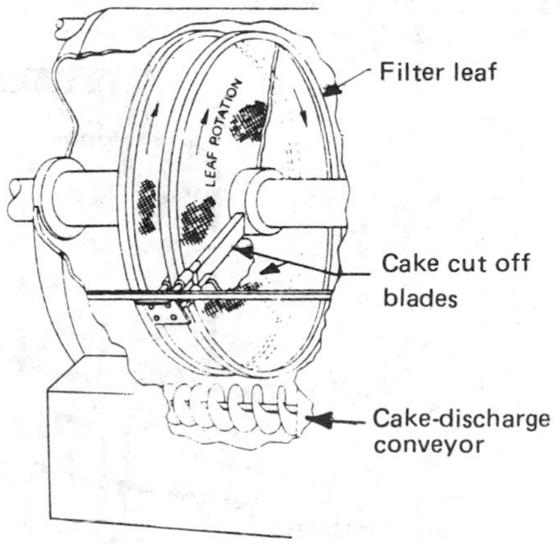

FIG. 19-95 Cutaway of a horizontal tank filter with dry-cake-discharge features. *(United States Filter: Fluid Systems Corp.)*

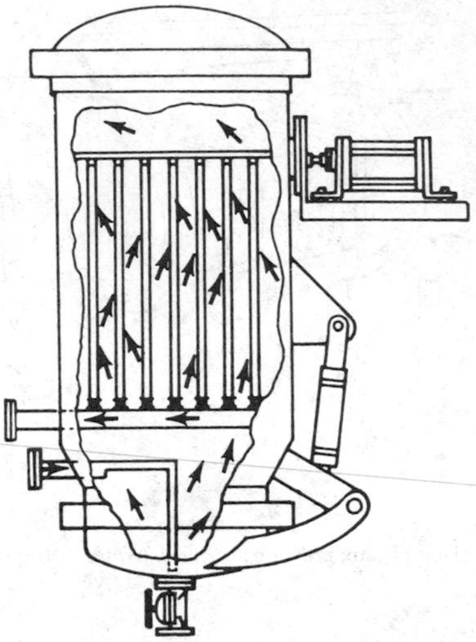

FIG. 19-96 Vertical leaf "dry-discharge" filter with a mechanical leaf vibrator. *(Industrial Filter & Pump Mfg. Co.)*

Vertical filters are not convenient for the removal of dry cake, although they can be used in this service if they have a bottom that can be retracted to permit the cake to fall into a bin or hopper below (Fig. 19-96). They are adapted rather to wet-solids discharge, a process that may be assisted by leaf vibration, air or steam sparging of a filter full of water, sluicing from fixed, oscillating, or traveling nozzles, and blowback. They are made by many companies, and they enjoy their widest use for filter-aid precoat filtration.

Advantages and uses. The advantages of pressure leaf filters are their considerable flexibility (up to the permissible maximum, cakes of various thickness can be formed successfully), their low labor charges, particularly when the cake may be sluiced off or the dry cake discharged cleanly by blowback, the basic simplicity of many of the designs, and their adaptability to quite effective displacement washing. Their disadvantages are the requirement of exceptionally intelligent and watchful supervision to avoid cake consolidation or dropping, their inability to form as dry a cake as a filter press, their tendency to classify vertically during filtration and to form misshapen nonuniform cakes unless the leaves rotate, and the restriction of most models to 610 kPa (75 psig) or less.

Pressure leaf filters are used to separate much the same kinds of slurries as are filter presses and are used much more extensively than filter presses for filter-aid filtrations. They should be seriously considered whenever uniformity of production permits long-time operation under essentially constant filtration conditions, when thorough washing with a minimum of liquor is desired, or when vapors or fumes make closed construction desirable. Under such conditions, if the filter medium does not require frequent changing, they may show a considerable advantage in cycle and labor economy over a filter press, which has a lower initial cost, and advantages of economy and flexibility over continuous vacuum filters, which have a higher first cost.

Pressure leaf filters are available with filtering areas of 930 cm^2 (1 ft^2) (laboratory size) up to about 79 m^2 (850 ft^2) for vertical filters and 158 m^2 (1700 ft^2) for horizontal ones. Leaf spacings range from 5 to 15 cm (2 to 6 in) but are seldom less than 7.5 cm (3 in) since 1.3 to 2.5 cm (0.5 to 1 in) should be left open between surfaces.

Centrifugal-Discharge Filter Horizontal top-surface filter plates may be mounted on a hollow motor-connected shaft that serves both as a filtrate-discharge manifold and as a drive shaft to permit centrifugal removal of the cake. An example is the **Funda filter** (manufactured in Switzerland and marketed in the United States by Chemapec, Inc.), illustrated schematically in Fig. 19-97. The filtering surface may be a textile fabric or a wire screen, and the use of a precoat is optional. The Funda filter is driven from the top, leaving the bottom unobstructed for inlet and drainage lines; a somewhat similar machine that employs a bottom drive, providing a lower center of mass and ground-level access to the drive system, is the German-made **Schenk filter** (marketed in the United States by the Votator Division, Chemetron Process Equipment Co.).

During filtration, the vessel that coaxially contains the assembly of filter plates is filled with prefilt under pressure, the filtrate passes through the plates and out the hollow shaft, and cake is formed on the top surfaces of the plates. After filtration, the vessel is drained, or the heel may be filtered by recirculation through a cascade ring at the top of the filter. The cake may be washed—or it may be extracted, steamed, air-blown, or dried by hot gas. It is discharged, wet or dry, by rotation of the shaft at sufficiently high speed to sling away the solids. If flushing is permitted, the discharge is assisted by a backwash of appropriate liquid.

The operating advantages of the centifugal-discharge filter are those of a horizontal-plate filter and, further, its ability to discharge cake without being opened. It is characterized by low labor demand, easy adaptability to automatic control, and amenability to the processing of hazardous, noxious, or sterile materials. Its disadvantages are its complexity and maintenance (stuffing boxes, high-speed drive) and its cost. The Funda filter is made in sizes that cover the filtering area range of 1 to 50 m^3 (11 to 537 ft^2). The largest Schenk filter provides 100 m^2 (1075 ft^2) of area.

The Idrex HLF Filter This filter combines features of the centrifugal-discharge filter and the horizontal plate filter in a unit having a vertical axis shell with horizontal circular filter leaves. Cake is formed only on the top surface, and filtrate is collected through a center shaft. Washing, drying, or other related operations can be per-

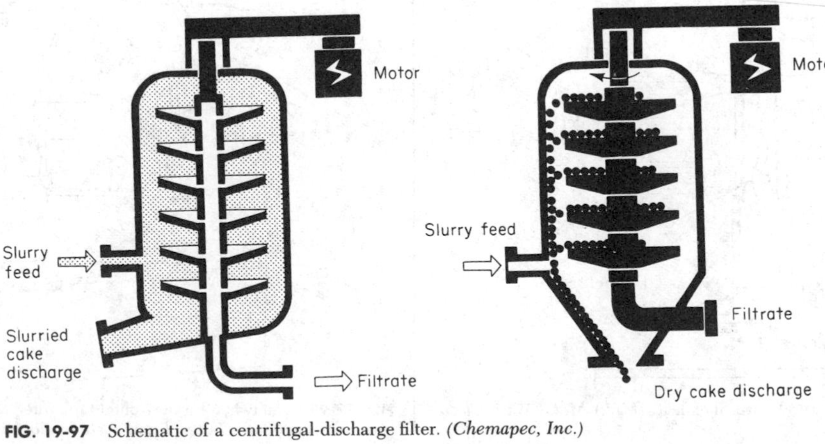

FIG. 19-97 Schematic of a centrifugal-discharge filter. *(Chemapec, Inc.)*

formed on the cake. Cake removal is accomplished by rotating the shaft-mounted leaves at 1.5 r/min past curved brushes, which force the cake to the outer edge of the leaves and cause it to drop to the bottom of the vessel for auger discharge. The brushing action provides complete cake removal without progressive plugging of the filter medium. The Idrex unit has all the advantages of the centrifugal discharge filter except centrifugal discharge, which it trades for low rotation speed to minimize mechanical maintenance problems.

Continuous Cake Filters Continuous cake filters are applicable when cake formation is fairly rapid, as in situations in which slurry flow is greater than about 5 L/min (1 to 2 gal/min), slurry concentration is greater than 1 percent, and particles are greater than 100 μm in diameter. Liquid viscosity below 0.1 Pa·s (100 cP) is usually required for maintaining rapid liquid flow through the cake. Some designs of continuous filters can compromise some of these guidelines by sacrificial use of filter aid when the cake is not the desired product.

Rotary Drum Filters The rotary drum filter is the most widely used of the continuous filters. There are many design variations, including operation as either a pressure filter or a vacuum filter. The major difference between designs is in the technique for cake discharge, to be discussed later. All the alternatives are characterized by a horizontal-axis drum covered on the cylindrical portion by filter medium over a grid support structure to allow drainage to manifolds. Basic materials of construction may be metals or plastics. Sizes (in terms of filter areas) range from 0.37 to 93 m^2 (4 to 1000 ft^2).

All drum filters (except the single-compartment filter) utilize a rotary-valve arrangement in the drum-axis support trunnion to facilitate removal of filtrate and wash liquid and to allow introduction of air or gas for cake blowback if needed. The valve controls the relative duration of each cycle as well as providing "dead" portions of the cycle through the use of bridge blocks. A typical valve design is shown in Fig. 19-98. Internal piping manifolds connect the valve with various sections of the drum.

Most drum filters are fed by operating the drum with about 35 percent of its circumference submerged in a slurry trough, although submergence can be set for any desired amount between zero and almost total. Some units contain an oscillating rake agitator in the trough to aid solids suspension. Others use propellers, paddles, or no agitator.

Slurries of free-filtering solids that are difficult to suspend are sometimes filtered on a top-feed drum filter or filter-dryer. An example application is in the production of table salt. An alternative for slurries of extremely coarse, dense solids is the internal drum filter. In the chemical-process industry both top-feed and internal drums (which are described briefly by Emmett in Schweitzer, op. cit., p. 4-41) have largely been displaced by the horizontal vacuum filter (q.v.).

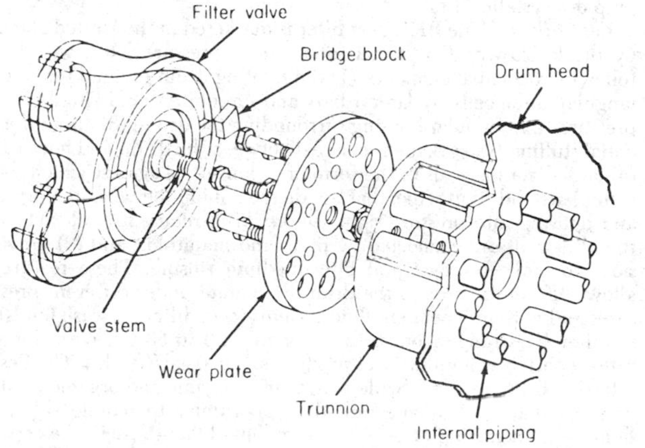

FIG. 19-98 Component arrangement of a continuous-filter valve. *(Emico Process Equipment Co.)*

Most drum filters operate at a rotation speed in the range of 0.1 to 10 r/min. Variable-speed drives are usually provided to allow adjustment for changing cake-formation and drainage rates.

Drum filters commonly are classified according to the feeding arrangement and the cake-discharge technique. They are so treated in this subsection. The characteristics of the slurry and the filter cake usually dictate the cake-discharge method.

Scraper-Discharge Filter The filter medium is usually caulked into grooves in the drum grid, with cake removal facilitated by a scraper blade just prior to the resubmergence of the drum (Fig. 19-99). The scraper serves mainly as a deflector to direct the cake, dislodged by an air blowback, into the discharge chute, since actual contact with the medium would cause rapid wear. In some cases the filter medium is held by circumferentially wound wires spaced 50 mm (4 in) apart, and a flexible scraper blade may rest lightly against the wire winding. A taut wire in place of the scraper blade may be used in some applications in which physical dislodging of sticky, cohesive cakes is needed.

For a given slurry, the maximum filtration rate is determined by the minimum cake thickness which can be removed—the thinner the cake, the less the flow resistance and the higher the rate. The minimum thickness is about 6 mm (0.25 in) for relatively rigid or cohesive cakes of materials such as mineral concentrates or coarse precipitates like gypsum or calcium citrate. Solids that form friable cakes composed of less cohesive materials such as salts or coal will usually require a cake thickness of 13 mm (0.5 in) or more. Filter cakes composed of fine precipitates such as pigments and magnesium hydroxide, which often produce cakes that crack or adhere to the medium, usually need a thickness of at least 10 mm (0.38 in).

String-Discharge Filter A system of endless strings or wires spaced about 13 mm (0.5 in) apart pass around the filter drum but are separated tangentially from the drum at the point of cake discharge, lifting the cake off as they leave contact with the drum. The strings return to the drum surface guided by two rollers, the cake separating from the strings as they pass over the rollers. If it has the required body, a thinner cake (5 mm or about $\frac{3}{16}$ in) than can be handled by drum filters is feasible, allowing more difficult materials to be filtered. This is done at the expense of greater dead area on the drum. Success depends on the ability of the cake to be removed with the strings and must be determined experimentally. Applications are mainly in the starch and pharmaceutical industries, with some in the metallurgical field.

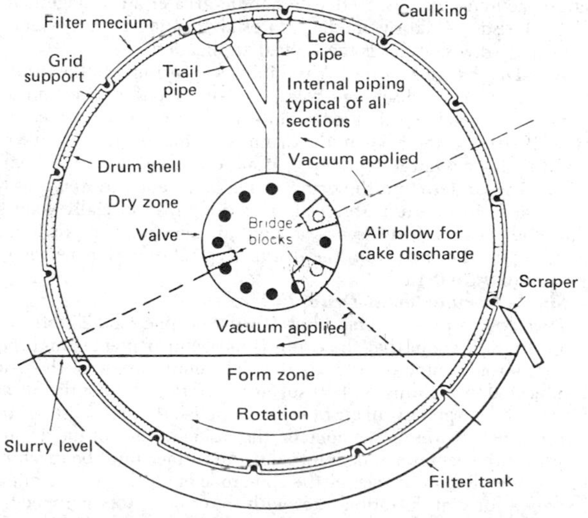

FIG. 19-99 Schematic of a rotary-drum vacuum filter with scraper discharge, showing operating zones. *(Schweitzer,* Handbook of Separation Techniques for Chemical Engineers, *p. 4-38. Copyright 1979 by McGraw-Hill, Inc. Used with the permission of McGraw-Hill Book Company.)*

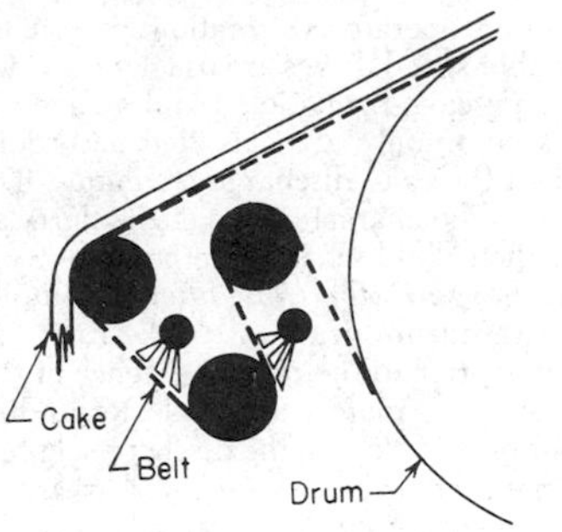

FIG. 19-100 Cake discharge and medium washing on an Eimco-Belt filter. *(Eimco Process Equipment Co.)*

Removable-Medium Filters Some drum filters provide for the filter medium to be removed and reapplied as the drum rotates. This feature permits the complete discharge of thin or sticky cake and provides the regenerative washing of the medium to reduce blinding. Higher filtration rates are possible because of the thinner cake and clean medium, but this is compromised by a less pure filtrate than normally produced by a nonremovable medium.

Belt-discharge filter. This is a drum filter carrying a fabric that is removed, passed over rollers, washed, and returned to the drum. Figure 19-100 shows the path of the medium while it is off the drum. A special aligning device keeps the medium wrinkle-free and in proper line during its travel. Thin cakes of difficult solids which may be slightly soluble are good applications. When acceptable, a sluice discharge makes cakes as thin as 1.5 to 2 mm (about 1⁄16 in) feasible. Several manufacturers offer belt-discharge filters.

Coilfilter. The **Coilfilter** (Komline-Sanderson Engineering Corp.) is a drum filter with a medium consisting of one or two layers of stainless-steel helically coiled springs, about 10 mm (0.4 in) in diameter, placed in a corduroy pattern around the drum. The springs follow the drum during filtration with cake forming the coils. They are separated from the drum to discharge the cake and undergo washing; if two layers are used, the coils of each layer are further separated from those of the other, passing over different sets of rolls. The use of stainless steel in spring form provides a relatively permanent medium that is readily cleaned by washing and flexing. Filtrate clarity is poorer than with most other media, and a relatively large vacuum pump is needed to handle greater air leakage than is characteristic of fabric media. Material forming a slimy, matlike cake (e.g., raw sewage) is the typical application.

Roll-Discharge Filters A roll in close proximity to the drum at the point of cake discharge rotates in the opposite direction at a peripheral speed equal to or slightly faster than that of the drum (Fig. 19-101). If the cake on the drum is adequately tacky and cohesive for this discharge technique, it adheres to cake on the smaller roll and separates from the drum. A blade or taut wire removes the material from the discharge roll. This design is especially good for thin, sticky cakes. If necessary, a slight air blow may be provided to help release the cake from the drum. Typical cake thickness is 1 to 10 mm (0.04 to 0.4 in).

Single-Compartment Drum Filter

Bird-Young filter. This filter (Bird Machine Co.) differs from most drum filters in that the drum is not compartmented and there is no internal piping or rotary valve. The entire inside of the drum is subjected to vacuum, with its surface perforated to pass the filtrate. Cake is discharged by an air blowback applied through a "shoe" that covers a narrow discharge zone on the inside surface of the drum to interrupt the vacuum. The internal drum surface must be machined to provide close clearance of the shoe to avoid leakage. The filter is designed for high filtration rates with thin cakes. Rotation speeds to 40 r/min are possible with cakes typically 3 to 6 mm (0.12 to 0.24 in) thick. Filter sizes range from 930 cm^2 to 13 m^2 (1 to 140 ft^2) with 93 percent of the area active. The slurry is fed into a conical feed tank designed to prevent solids from settling without the use of mechanical agitators. The proper liquid level is maintained by overflow, and submergence ranges from 5 to 70 percent of the drum circumference.

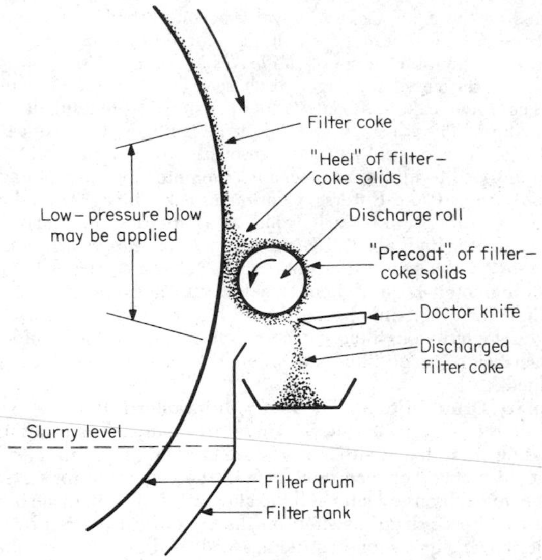

FIG. 19-101 Operating principle of a roll-discharge mechanism. *(Schweitzer,* Handbook of Separation Techniques for Chemical Engineers, *p. 4-40. Copyright 1979 by McGraw-Hill, Inc. Used with the permission of McGraw-Hill Book Company.)*

The perforated drum cylinder is divided into sections about 50 to 60 mm (2 to 2.5 in) wide. The filter medium is positioned into tubes between the sections and locked into place by round rods. No caulking, wires, or other fasteners are needed.

Wash sprays may be applied to the cake, with collection troughs or pans inserted inside the drum to keep the wash separate from the filtrate. Filtrate is removed from the lower section of the drum by a pipe passing through the trunnions.

The major advantages of the Bird-Young filter are its ability to handle thin cakes and operate at high speeds, its washing effectiveness, and its low internal resistance to air and filtrate flow. An additional advantage is the possibility of construction as a pressure filter with up to 1.14-MPa (150-psig) operating pressure to handle volatile liquids. The chief disadvantages are its high cost and the limited flexibility imposed by not having an adjustable rotary valve. Best applications are on free-draining nonblinding materials such as paper pulp or crystallized salts.

Fest Filter The **BHS-Fest filter** is marketed in the United States by Black Clawson Co. It is a rotary-drum pressure filter, with the following principal elements: (1) the rotating drum divided into rectangular filter cells by lateral bars and circumferential rings; (2) a pressure-tight annular housing surrounding the drum; (3) circumferential stuffing boxes that effect a seal between the drum and housing; (4) lateral zoning seals in the form of nylon shoes, against which the drum bars and rings wipe, that divide the annulus between the drum and housing into pressure-tight sectors; (5) a rotary valve to which the filter cells are connected by pipes and manifolds; and (6) provisions for cake removal and filter-medium rinsing. These features allow different sectors of the drum to operate under different pressures and with independent fluid systems (e.g., different wash fluids).

Filter sizes range from 0.12 to 7.7 m^2 (1.3 to 83 ft^2) with slurry rates reported as high as 0.011 $m^3/(m^2 \cdot s)$ [1000 gal/($ft^2 \cdot h$)]. The Fest filter has been used on a wide variety of inorganic and organic products. Its major advantages are the opportunity to handle volatile fluids with operating pressures to about 405 kPa (45 psig), the capability of excellent washing and drying, and atmospheric cake dis-

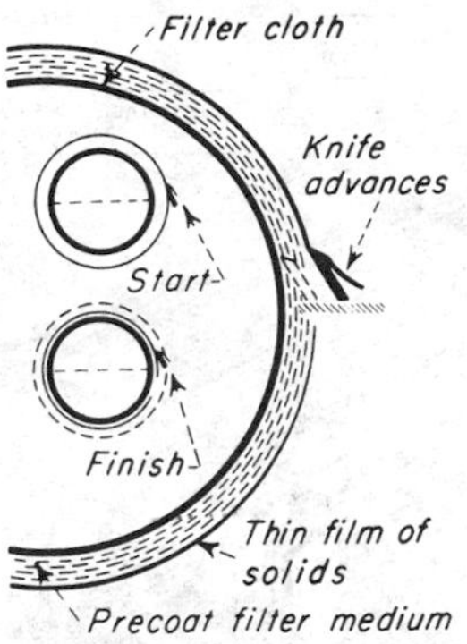

FIG. 19-102 Operating method of a vacuum precoat filter. *(Dorr-Oliver Inc.)*

charge. Its disadvantages are its complexity, the operating and maintenance problems associated with pressure seals in the presence of solids, and a very high capital cost.

Continuous Precoat Filters These filters may be operated as either pressure or vacuum filters, although vacuum operation is the prevailing one. The filters are really not continuous but have an extremely long batch cycle (3 to 4 days). Applications are for continuous clarification of liquids from slurries containing 50 to 5000 ppm of solids when only very thin unacceptable cakes would form on other filters.

Construction is similar to that of other drum filters, except that vacuum is applied to the entire rotation. Before feeding slurry a precoat layer of filter aid or other suitable solids, 75 to 125 mm (3 to 5 in) thick, is applied. The feed slurry is introduced and trapped in the outer surface of the precoat, where it is removed by a progressively advancing doctor knife which trims a thin layer of solids plus precoat (Fig. 19-102). The blade advances 0.05 to 0.2 mm (0.002 to 0.008 in) per revolution of the drum. When the precoat has been cut to a predefined minimum thickness, the filter is taken out of service, washed, and freshly precoated. This turnaround time may be 1 to 3 h.

Disk Filters A disk filter is a vacuum filter consisting of a number of vertical disks attached at intervals on a continuously rotating horizontal hollow central shaft (Fig. 19-103). Rotation is by a gear drive. Each disk consists of 10 to 30 sectors of metal, plastic, or wood, ribbed on both sides to support a filter cloth and provide drainage via an outlet nipple into the central shaft. Each sector may be replaced individually. The filter medium is usually a cloth bag slipped over the sectors and sealed to the discharge nipple. For some heavy-duty applications on ores, stainless-steel screens may be used.

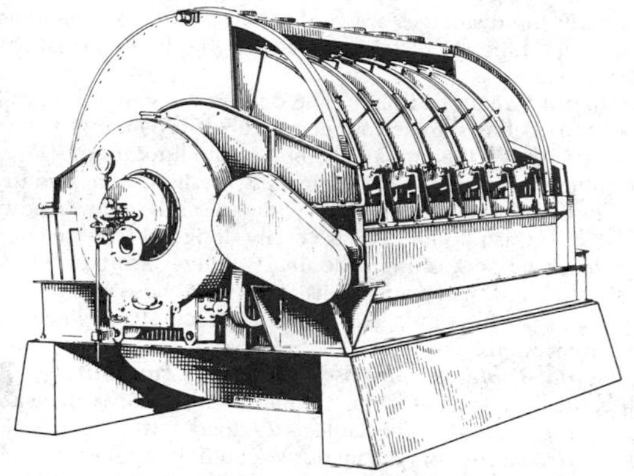

FIG. 19-103 Rotary disk filter. *(Dorr-Oliver Inc.)*

The disks are typically 40 to 45 percent submerged in a troughlike vessel containing the slurry. Another horizontal shaft running beneath the disks may contain agitator paddles to maintain suspension of the solids, as in the Eimco **Agidisc filter.** In some designs, feed is distributed through nozzles below each disk. Vacuum is supplied to the sectors as they rotate into the liquid to allow cake formation. Vacuum is maintained as the sectors emerge from the liquid and are exposed to air. Wash may be applied with sprays, but most applications are for dewatering only. As the sectors rotate to the discharge point, the vacuum is cut off, and a slight air blast is used to loosen the cake. This allows scraper blades to direct the cake into discharge chutes positioned between the disks. Vacuum and air blowback is controlled by an automatic valve as in rotary-drum filters.

Of all continuous filters, the vacuum disk is the lowest in cost per unit area of filter when mild steel, cast iron, or similar materials of construction may be used. It provides a large filtering area with minimum floor space, and it is used mostly in high-tonnage dewatering applications in sizes up to about 300 m^2 (3300 ft^2) of filter area.

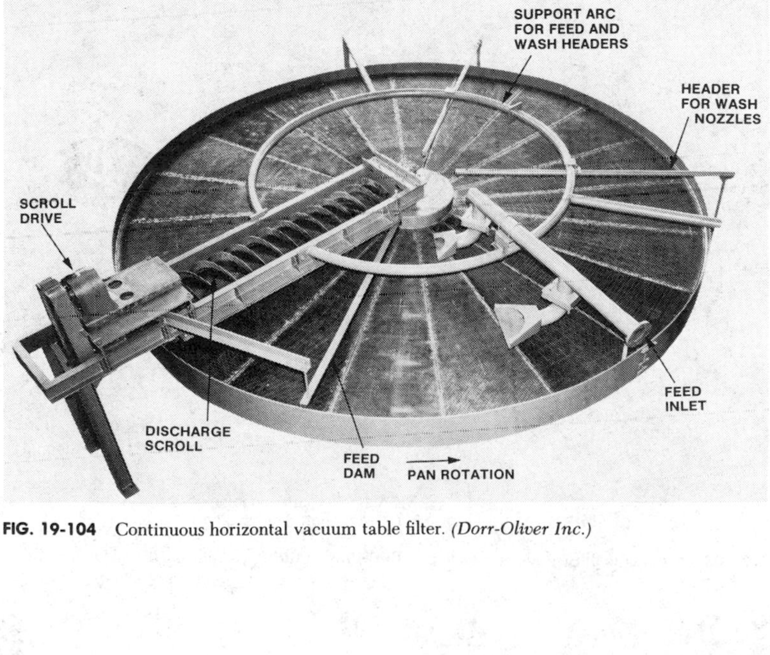

FIG. 19-104 Continuous horizontal vacuum table filter. *(Dorr-Oliver Inc.)*

The main disadvantages are the inadaptability to have effective wash and the difficulty of totally enclosing the filter for hazardous-material operations.

Horizontal Vacuum Filters These filters are generally classified into two broad classes: rotary circular and belt-type units. Regardless of geometry, they have similar advantages and limitations. They provide flexibility of choice of cake thickness, washing time, and drying cycle. They effectively handle heavy, dense solids, allow flooding of the cake with wash liquor, and are easily designed for true countercurrent leaching or washing. The disadvantages are they are more expensive to build than drum or disk filters, they use a large amount of floor space per filter area, and they are difficult to enclose for hazardous applications.

Horizontal-Table, Scroll-Discharge, and Pan Filters These are all basically revolving annular tables with the top surface a filter medium (Fig. 19-104). The table is divided into sectors, each of which is a separate compartment. Vacuum is applied through a drainage chamber beneath the table that leads to a large rotary valve. Slurry is fed at one point, and cake is removed after completing more than three-fourths of the circle, by a horizontal scroll conveyor which elevates the cake over the rim of the filter. A clearance of about 10 mm (0.4 in) is maintained between the scroll and the filter medium to prevent damage to the medium. Residual cake on the medium may be loosened by an air blow from below or with high-velocity liquid sprays from above. This residual cake is a disadvantage peculiar to this type of filter. With material that can cause blinding, frequent shutdowns for thorough cleaning may be needed. Unit sizes range from about 0.9 to 7 m (3 to 24 ft) in diameter, with about 80 percent of the surface available for filtration.

Tilting-Pan Filter This is a modification of the table or pan filter in which each of the sectors is an individual pan pivoted on a radial axis to allow its inversion for cake discharge, usually assisted by an air blast. Figure 19-105 shows the typical operating sequence. Filter-cake thicknesses of 50 to 100 mm (2 to 4 in) are common. Most applications involve free-draining inorganic-salt dewatering. In addition to the advantages and disadvantages common to all horizontal continuous filters, tilting-pan filters have the relative advantages of complete wash containment per sector, good cake discharge, filter-medium washing, and feasibility of construction in very large sizes, up to about 25 m (80 ft) in diameter, with about 75 percent of the area usable. Relative disadvantages are high capital cost (especially in smaller sizes) and mechanical complexity leading to higher maintenance costs.

Horizontal-Belt Filter This filter consists of a slotted or perforated elastomer belt driven as a conveyor belt carrying a filter fabric belt (Fig. 19-106). Both belts are supported by and pass across a support deck, which is sectioned to form vacuum chambers to collect filtrate and multiple wash zones if desired. Several manufacturers provide horizontal-belt filters, the major differences among which lie in the construction of the drainage belt, the method of retaining cake on the belt, and the method of alignment for the filter medium. The **Pannevis filter** replaces the drainage belt with a series of vacuum pans which avoids a problem of hydrocarbon attack on the elastomeric belts. The filters are rated according to the available drainage-belt area exposed to vacuum. Slurry is fed at one end by overflow weirs or a fantail chute; wash liquor is applied by sprays or weirs at one or more points as the belts travel along the deck. The cake is dumped as the belt passes over the end pulley, where the filter-medium belt is separated from the elastomeric drainage belt; from this point each is returned to the feed point over a separate pulley. This procedure allows the filter-medium belt to be washed thoroughly with sprays prior to its rejoining the drainage belt. Some units have wiping dams which ride on the cake to separate filtration and wash zones. Belt speeds up to 0.5 m/s (1.6 ft/s) can be used, with some operating-speed variation allowed, depending on the cake-

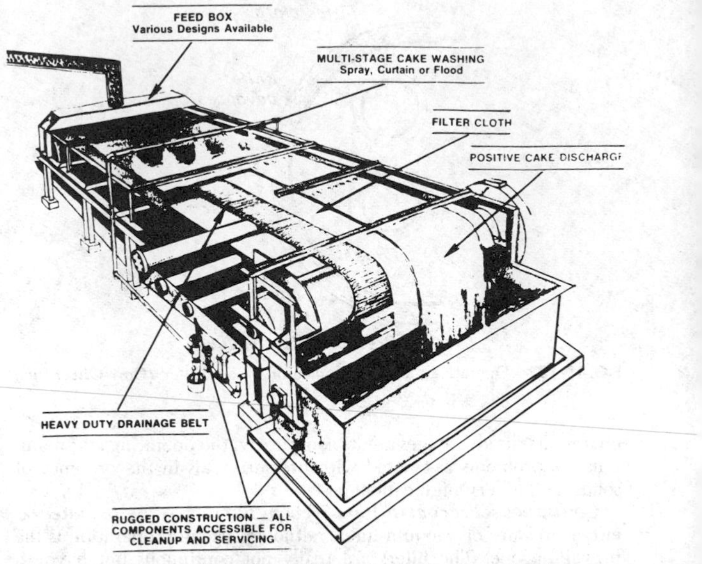

FIG. 19-106 Horizontal-belt filter. *(Eimco Process Equipment Co.)*

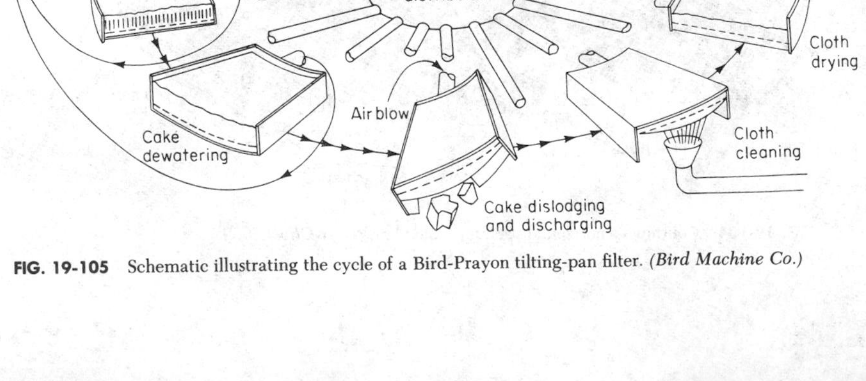

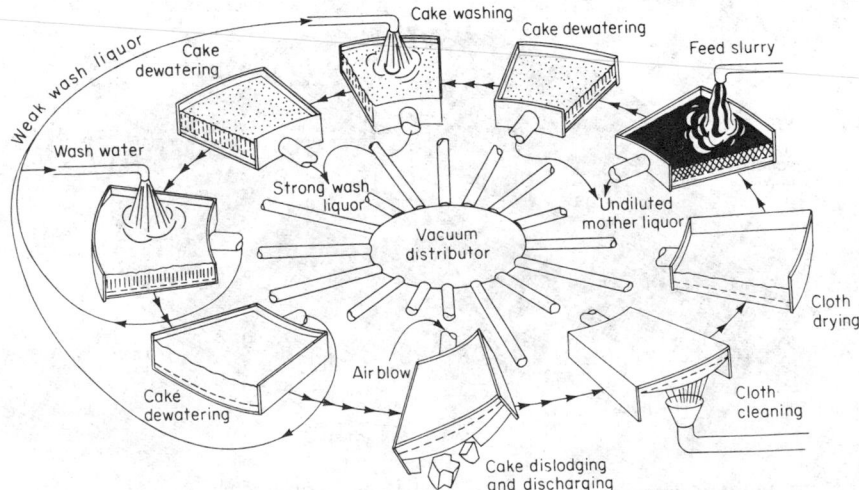

FIG. 19-105 Schematic illustrating the cycle of a Bird-Prayon tilting-pan filter. *(Bird Machine Co.)*

drainage rate. Cakes up to 100 to 150 mm (4 to 6 in) thick are possible with some fast-draining materials. Horizontal-belt filters have the advantages of complete cake removal and effective filter-medium washing. Their particular disadvantage is that at least half of the filter area is always idle on the return loop. Horizontal-belt filters are available with filtering areas from 0.18 to 61 m^2 (2 to 655 ft^2).

Some horizontal-belt filters are integrated with additional belt configurations to obtain further dewatering by pressing the cake between two belts. This type of equipment is described later in the subsection "Expression."

Filter Thickeners Thickeners are devices which remove a portion of the liquid from a slurry to increase the concentration of solids in suspension. Thickening is done to prepare a dilute slurry for more economical filtration or to change the consistency or concentration of the slurry for process reasons. The commonest method of thickening is by gravity sedimentation, discussed earlier in this section. Occasions may arise, however, in which a filter may be called upon for thickening service. Many of the filters previously discussed as cake filters can be operated as thickeners: the rotating disk with total submergence, cake being blown back into the slurry trough; the filter press with special plates containing flow channels that keep velocity high enough to prevent cake buildup; and cycled leaf filters with the cake discharge into the filter tank.

Newer approaches to filtration thickening involve various cross-flow filters, some of which are discussed in Sec. 17. The **Artisan continuous filter** is an example of a mechanical cross-flow filter consisting of a horizontal-axis vessel with doughnut-shaped filter-medium plates attached to the vessel wall. A series of rotating disks with wiper blades maintains a velocity across the filter medium to minimize cake buildup. Clarified filtrate passes through the medium, while the slurry thickens as it passes the various doughnuts before exiting the vessel at the end opposite the feed.

Various high-velocity tubular devices concentrate slurry while producing a clarified filtrate. These units consist of various ultrafiltration membranes such as hollow fibers or spiral-wound sheets, or porous tubes as in the **Giha Corp. Dyna-Sep** system. Each of these approaches depends on fluid velocity to minimize the boundary layer of solids on the filter medium and to maintain a good filtration flux. Some filtrate backflush may be applied as needed to maintain flux. These devices also are treated in Sec. 17.

Clarifying Filters Clarifying filters are used to separate liquid mixtures which contain only very small quantities of solids. When the solids are finely divided enough to be observed only as a haze, the filter which removes them is sometimes called a polishing filter. The prefilt slurry generally contains no more than 0.10 percent solids, the size of which may vary widely (0.01 to 100 μm). The filter usually produces no visible cake, sometimes because the amount of solids removed is so small, sometimes because the particles are removed by being entrapped within rather than upon the filter medium. Compared with cake filters, clarifying filters are of minor importance to pure chemical-process work, their greatest use being in the fields of beverage and water polishing, pharmaceutical filtration, fuel- and lubricating-oil clarification, electroplating-solution conditioning, and dry-cleaning-solvent recovery. They are essential, however, to the processes of fiber spinning and film extrusion; the spinning solution or dope must be free of particles above a certain size to maintain product quality and to prevent the clogging of spinnerets.

Most cake filters can be so operated as to function as clarifiers, although not necessarily with efficiency. On the other hand, a number of clarifying filters which can be used for no purpose other than clarifying or straining have been developed. In general, clarifying filters are less expensive than cake filters. Clarifying filters may be classified as disk and plate presses, cartridge clarifiers, precoat pressure filters, deep-bed filters, and miscellaneous types. Membrane filters constitute a special class of plate presses and cartridge filters. Simple strainers sometimes are used as clarifiers of liquids containing very large particles. Because they more closely resemble wet screens than filters and because they have little primary process application, they are not discussed here.

Disk Filters and Plate Presses Filters employing asbestos-pulp disks, cakes of cotton fibers (filtermasse), or sheets of paper or other media are used widely for the polishing of beverages, plating solutions, and other low-viscosity liquids containing small quantities of suspended matter. The term **disk filter** is applied to assemblies of pulp disks made of asbestos and cellulose fibers and sealed into a pressure case. The disks may be preassembled into a self-supporting unit (Fig. 19-107), or each disk may rest on an individual screen or plate against which it is sealed as the filter is closed (Fig. 19-108). The liquid flows through the disks, and into a central or peripheral discharge manifold. Flow rates are on the order of 122 L/(min·m^2) [3 gal/(min·ft^2)], and the operating pressure does not normally exceed 446 kPa (50 psig) (usually it is less). Disk filters are almost always operated as pressure filters. Individual units deliver up to 378 L/min (6000 gal/h) of low-viscosity liquid.

Disk-and-plate assemblies somewhat resemble horizontal-plate pressure filters, which, in fact, may be used for polishing. In one design (**Sparkler VR filter**) both sides of each plate are used as filtering surfaces, having paper or other media clamped against them.

Pulp filters. These filters employ one or more packs of filtermasse (cellulose fibers compressed to a compact cylinder) stacked into a pressure case. The packs are sometimes supported in individual trays which provide drainage channels and sometimes rest on one another with a loose spacer plate between each two packs and with a drainage screen buried in the center of each pack. The liquid being clarified flows under a pressure of 440 kPa (50 psig) or less through the pulp packs and into a drainage manifold. Flow rates are somewhat less than for disk filters, on the order of 20 L/(min·m^2) [0.5 gal/(min·ft^2)]. Pulp filters are used chiefly to polish beverages. The filtermasse may be washed in special washers and re-formed into new cakes.

Plate presses. Sometimes called sheet filters, these are assemblies of plates, sheets of filter media, and sometimes screens or frames. They are essentially modified filter presses with practically no cake-holding capacity. A press may consist of many plates or of a single filter sheet between two plates, the plates may be rectangular or circular, and the sheets may lie in a horizontal or vertical plane. The operation is similar to that of a filter press, and the flow rates are about the same as for disk filters. The operating pressure usually does not exceed 235 kPa (20 psig). The presses are used most frequently for low-viscosity liquids, but an ordinary filter press with thin frames is commonly used as a clarifier for 100-Pa·s (1000-P) rayon-spinning solution. Here the filtration pressure may be 6900 kPa (1000 psig).

Disk, pulp, and sheet filters accomplish extreme clarification. Not

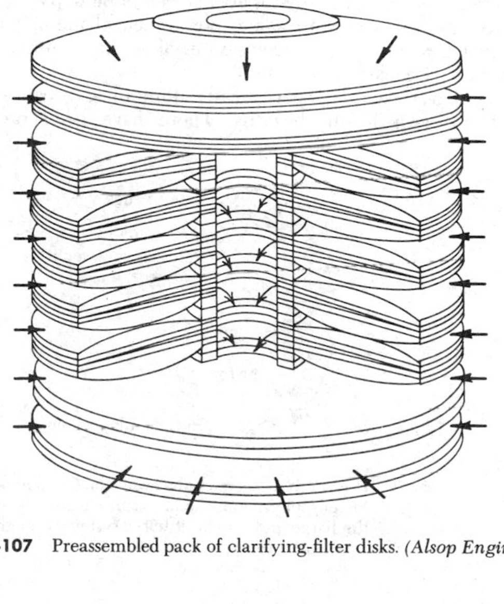

FIG. 19-107 Preassembled pack of clarifying-filter disks. *(Alsop Engineering Co.)*

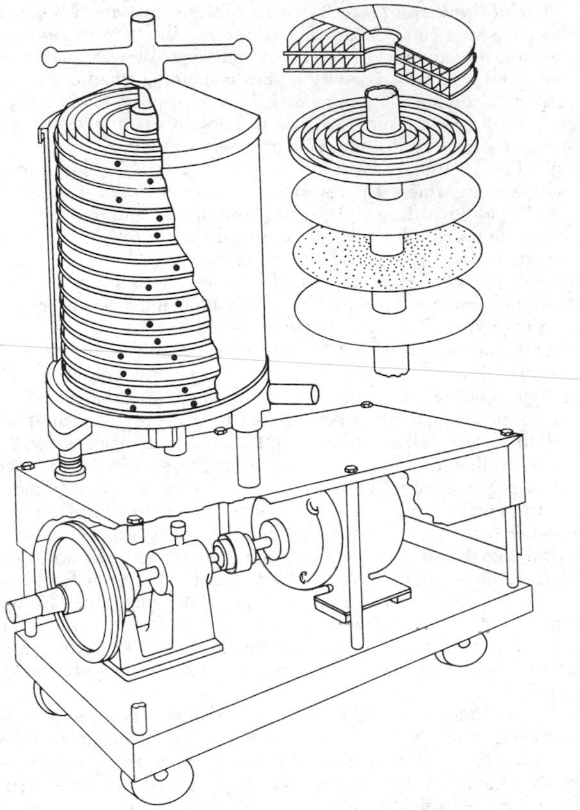

FIG. 19-108 Disk-and-plate clarifying-filter assembly. *(Alsop Engineering Co.)*

infrequently their mission is complete removal of particles above a stipulated cut size, which may be much less than 1 μm. They operate over a particle-size range of four to five orders of magnitude, contrasting with two orders of magnitude for most other filters. It is not surprising, therefore, that they involve a variety of kinds and grades of filter media, often in successive stages. In addition to packs or disks of cellulosic, polymeric, or asbestos fiber, sheets of pulp, paper, asbestos, carded fiber, woven fabrics, and porous cellophane or polymer are employed. Sandwich-pack composites of several materials have been used for viscous-dope filtration.

The use of asbestos has been greatly diminished because of its identification with health hazards. There have been proposed replacement materials such as the **Zeta Plus** filter media from the AMF Cuno Division, consisting of a composite of cellulose and inorganic filter aids that have a positive charge and provide an electrokinetic attraction to hold colloids (usually negatively charged). These media therefore provide both mechanical straining and electrokinetic adsorption.

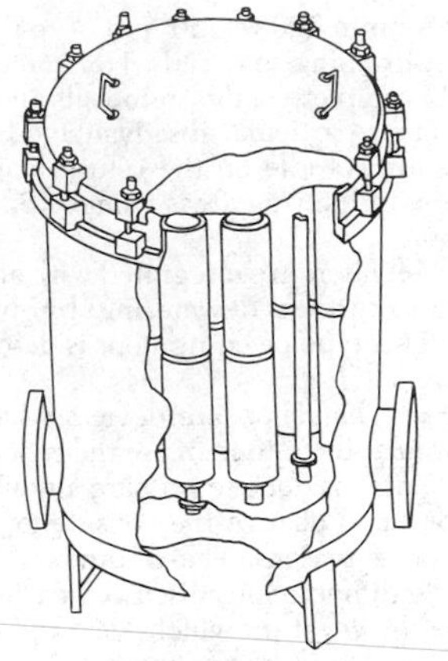

FIG. 19-110 Multielement cartridge clarifier. *(Commercial Filters Division, Kennecott Corp.)*

Cartridge Clarifiers Cartridge clarifiers are units which consist of or use one or more replaceable or renewable cartridges containing the active filter element. The unit usually is placed in a line carrying the liquid to be clarified; clarification thus occurs while the liquid is in transit.

Mechanical or edge filters. These consist of stacks of disks separated to precise intervals by spacer plates, or a wire wound on a cage in grooves of a precise pitch, or a combination of the two. The liquid to be filtered flows radially between the disks, wires, or layers of paper, and particles larger than the spacing are screened out. Edge filters can remove particles down to 0.001 in (25 μm) but more often have a minimum spacing of twice this value. They have small solids-retaining capacity and hence must be cleaned often to avoid plugging. Continuous cleaning is provided in some filters. For example, the Cuno **Flo-Klean** (Fig. 19-109), a wire-wound unit, employs a slowly rotating nozzle which backwashes the element with filtered liquid; and the Cuno **Auto-Klean** is equipped with a scraper that fits into the interdisk slots to comb away accumulated solids. In either case, the dislodged solids fall into a sump that may be drained at intervals.

Micronic clarifiers. The greatest number of cartridge clarifiers are of the micronic class, with elements of fiber, resin-impregnated filter paper, porous stone, or porous stainless steel of controlled porosity. Other rustless metals are also available. The elements may be

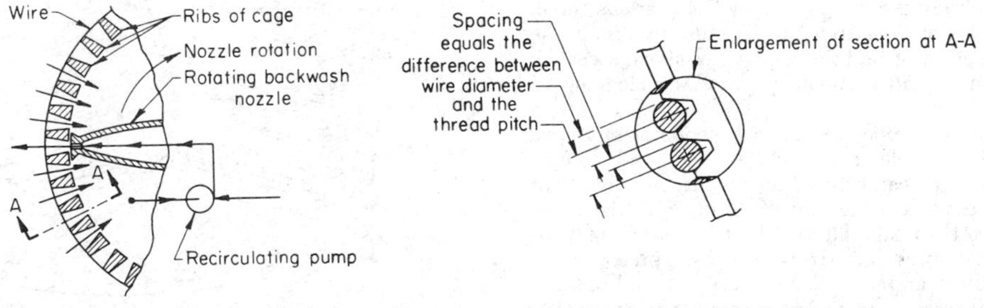

FIG. 19-109 Sectional view of the Cuno Flo-Klean backwashing edge filter. Fluid pumped through the nozzle loosens solids from the filter surface and clears the filtering area. The pump draws filtered fluid from the filter discharge and returns it to the system through the nozzle. Thus there is no loss of backwash fluid. *(Cuno Division, AMF Incorporated.)*

chosen to remove particles larger than a fraction of a micrometer, although many are made to pass 10-μm solids and smaller. By proper choice of multiple-cylinder cartridges or multiple cartridges in parallel any desired flow rate can be obtained at a reasonable pressure drop, often less than 235 kPa (20 psig). A multielement filter using wound-yarn tubes is shown in Fig. 19-110.

When the pressure rises to the permissible maximum, the cartridge must be opened and the element replaced. Micronic elements of the fiber type cannot be cleaned and are so priced that they can be discarded or the filter medium replaced economically. Stone elements usually must be cleaned, a process best accomplished by the manufacturer of the porous ceramic or in accordance with the manufacturer's directions. The user can clean stainless-steel elements by chemical treatment.

Flexibility. Cartridge filters are flexible: cartridges of different ratings and materials of construction can be interchanged, permitting ready accommodation to shifting conditions. They have the disadvantage of very limited solid-handling capability so that the maximum solid concentrations in the feed are limited to about 0.01 percent solids. The biggest limitation for modern process-plant operation is the need to open the filter to replace cartridges, which makes their use for the processing of hazardous materials undesirable. Some manufacturers—for example, the Hydraulic Research Division of Textron Inc. and the Fluid Dynamics Division of Brunswick Corp.—have designed cartridges of bonded metal fibers that can be backflushed or chemically cleaned without opening the unit. These filters, which can operate at temperatures to 482°C (900°F) and at pressures of 33 MPa (325 atm) or greater, are particularly useful for filtering polymers.

Cyclamatic pressure filter. This filter (Industrial Filter & Pump Mfg. Co.) is a semicontinuous renewable-surface cartridge clarifier. Its principle is indicated in Fig. 19-111. A vertical pressure tank contains a number of vertical cartridges arranged on a circle concentric with the tank and a windup spool coaxial with the tank. Each cartridge is a roll of paper or nonwoven sheet wound about a perforated metal or plastic core and feeding onto the windup spool. Filtration is radially inward through the cartridges, the filtrate passing through all the layers of the paper roll and outward through a bottom manifold to which the cartridge cores are connected. The principal solids removed are collected on the outside of the cartridges, although very fine particles may be trapped within the paper roll, usually in the topmost few layers. When the pressure drop across the cartridges attains a predetermined level, an external drive to the winder spool is activated automatically and sufficient filter medium is unwound from the cartridges onto the spool to provide each cartridge with a clean, unplugged surface. Cartridge life is usually 1 to 4 weeks, at the end of which an alarm sounds and the cartridges and windup spool must be replaced. This operation requires about 30 min.

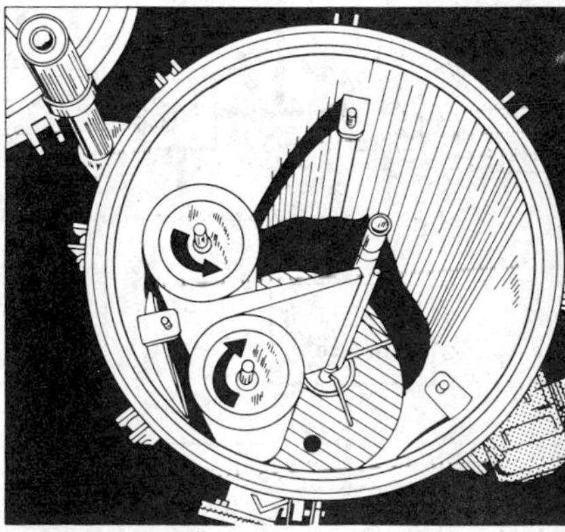

FIG. 19-111 Principle of the Cyclamatic clarifying filter. *(Industrial Filter & Pump Mfg. Co.)*

The Cyclamatic filter is 61 cm (24 in) in diameter and carries a maximum of four cartridges, each of which is 91 cm (36 in) long and has an average effective filtering area of about 0.37 m^2 (4 ft^2). Larger sizes, up to 122 cm (48 in) in diameter with up to 12 tubes, can be built on special demand. A process-water filtration rate of 163 L/($min \cdot m^2$) [4 gal/($min \cdot ft^2$)] can be maintained with a pressure 21 kPa (3 lbf/in^2).

Precoat Pressure Filters Precoat pressure filters consist of one or more leaves, plates, or tubes on which a coat of diatomaceous earth or other filter aid is deposited to form a filtering surface for clarification. Filter paper may be substituted for a precoat. Additional filter aid may be mixed with the liquid to be filtered (body feed), particularly if solids are gelatinous or sticky, in order to maintain a higher average filtration rate. Precoat pressure clarifying filters are essentially no different from pressure cake filters, except for the purpose to which they are put. When they are operating as "polishing filters," the precoat can be considered part of the filter medium rather than the cake.

In operating precoat filters care must be taken that a complete uniform layer of filter aid is deposited on the element before filtration starts. Otherwise the capacity of the filter can soon be seriously reduced or bleed-through of solids be allowed.

Filter aid, while inexpensive on a per weight basis, can add a significant cost to clarifying operations. Techniques have been developed, such as that of Xodar Corporation's **Ultra-kleen regenerative DE filters,** to reuse filter aid for several cycles by backwashing filter aid and debris from the outside of a tubular filter medium, reslurrying the solids into a homogeneous mix, and redepositing the mix as a precoat for further filtration. This process can be repeated until the solids filtered from the dirty process stream become a significant fraction of the precoat mix; then a new filter-aid charge is needed.

Deep-Bed Filters The typical deep-bed filter consists of a medium of sand or anthracite at least 1 m deep in a vertical tank. The bed is usually graded with decreasing solid-particle size from the bottom to the top of the bed, which is sometimes composed of more than one material (e.g., sand and anthracite). Variations in design introduce the feed above, below, or in the middle of the bed. A false perforated vessel bottom or perforated pipes near the bottom of the bed serve as inlet or filtrate-removal devices. Upflow designs have greater solids-handling capacity than downflow designs because larger particles in the feed are removed by the coarser solids of the bed, only the finest solids in the feed being polished out by the top fine-grain layers. In downflow designs these upper-layer solids tend to load rapidly, requiring more frequent backwash. Backflush with a liquid and sometimes an air sparge lifts the filtered solids from the bed. This concentrated backflush must then be handled; if sewering is not permitted, it may be filtered on a more conventional filter.

Deep-bed filters are strictly for clarification, usually in high-flow applications such as water treatment. The feed to a deep-bed filter typically carries less than 1000 ppm of solids if flocculation and gravity settling have been used as pretreatment routine. Equipment may be large open vessels made of concrete, as in water-treatment plants, or closed vertical process vessels that can operate under pressure.

United States Filter Corp. Maxi-Flo filter. The **Maxi-Flo Filter** is an example of the upflow closed-vessel design. Filtration rates to 0.0081 $m^3/(m^2 \cdot s)$ [12 gal/($ft^2 \cdot min$)] and filter cross-section areas up to 10.5 m^2 (113 ft^2) are possible. Deep-bed filtration has been reviewed by Tien and Payatakes [*Am. Inst. Chem. Eng. J.*, **25,** 737 (1979)] and by Oulman and Baumann [*Am. Inst. Chem. Eng. Symp. Ser.*, 73(171), 76 (1977)].

Dyna Sand Filter. A recently developed novel filter that avoids batch backwashing for cleaning, the **Dyna Sand Filter** is available from Parkson Corporation. The bed is continuously cleaned and regenerated by recycling solids internally through an air-lift pipe and a sand washer. Thus a constant pressure drop is maintained across the bed, and the need for parallel filters to allow continued on-stream operation, as with conventional designs, is avoided.

Ultrafilters These are membrane filters, usually of cellulose acetate or other polymers, in the form of hollow fibers or spiral-wound sheets. Ultrafilters represent the ultimate in particle sieving and liquid clarification. Solids as small as 0.001 μm can be removed. Ultrafiltration is treated in Sec. 17 of this volume; see also Porter (in Schweitzer, op. cit., sec. 2.1).

Miscellaneous Clarifiers

Bag filters. Filter bags usually made out of felt fabrics of various natural or synthetic fibers are used either in the open, attached with a snap ring to the end of a pipe, or in a housing much in the manner of cartridge filters. The felt serves as a depth medium to trap solids generally in the 20- to 200-μm range with typical applications filtering paints or lubricating oils. Wrotnowski (in Schweitzer, op. cit. p. 4–95) provides a detailed discussion of bag filters. The disadvantage of having to replace bags and expose personnel to process material results in very restrictive use of bag filters.

Magnetic filters. For selective removal of iron or other magnetic particles from liquid, several magnetic separators are available. The **Frantz Ferro-Filter** consists of a stack of soft-steel grids strongly magnetized by a direct-current coil or by a permanent magnet. The liquid to be filtered flows over the grids, which collect any magnetic solids present. Particles as small as 1 μm can be removed. The grids are cleaned by demagnetizing and flushing. The **Stearns magnetic-screen filter** and the **Eriez ferrous filter** use the same general principle.

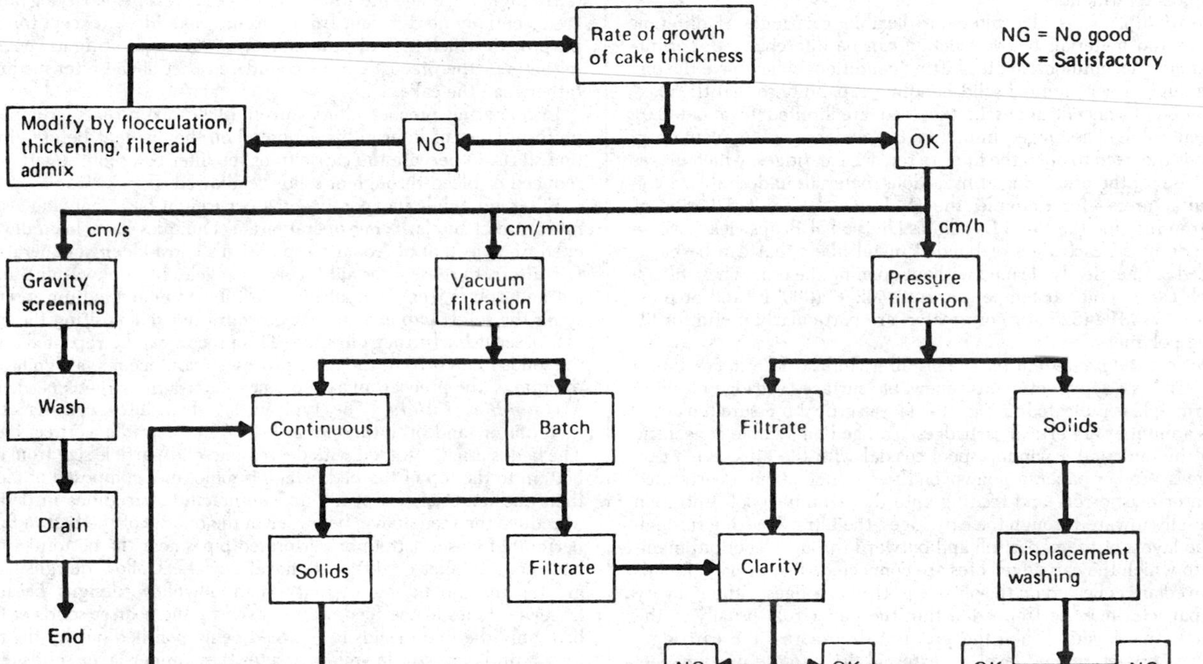

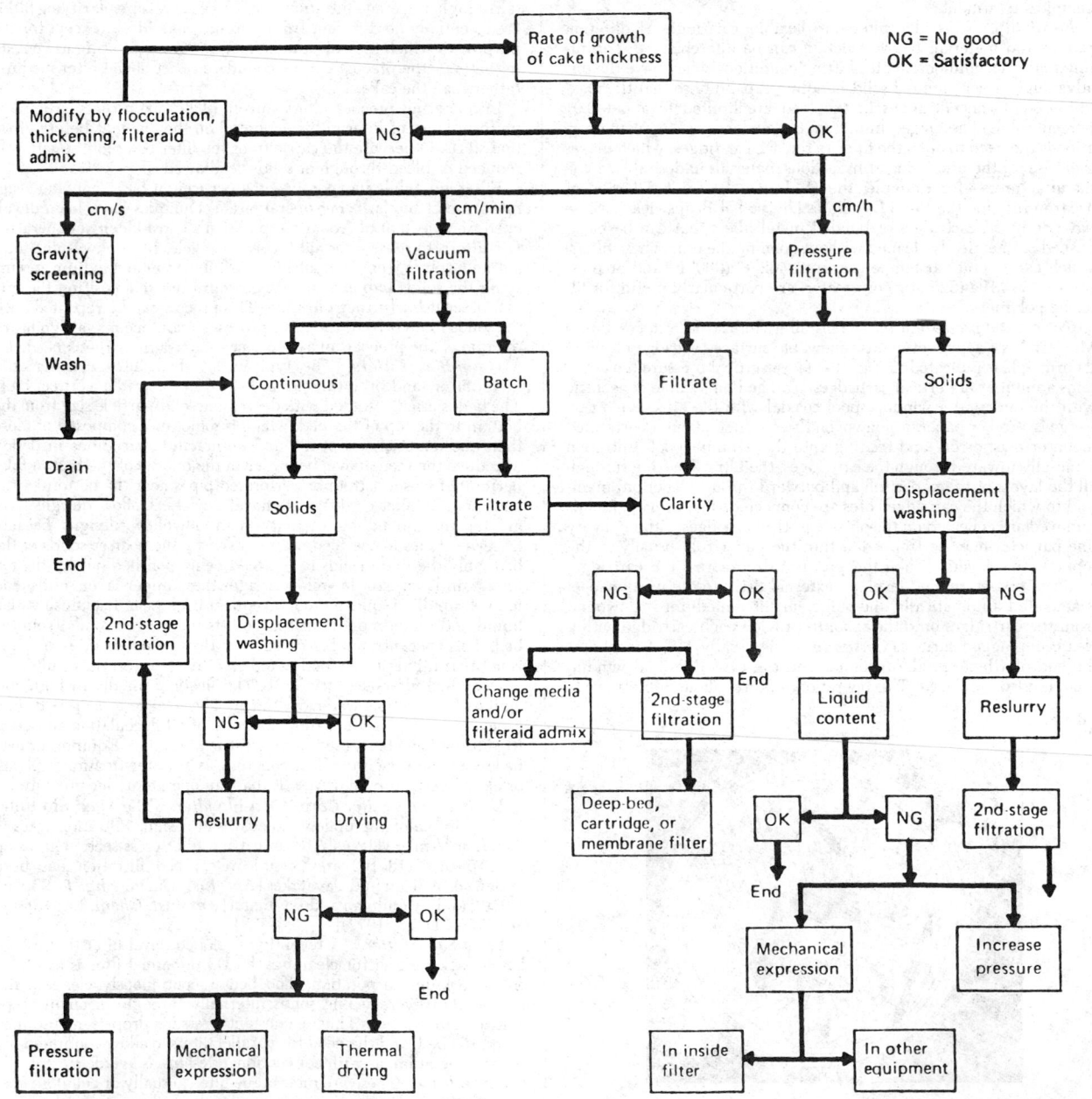

FIG. 19-112 Decision pattern for solving a filtration problem. [*Tiller*, Chem. Eng., *81(9), 118 (1974), by permission.*]

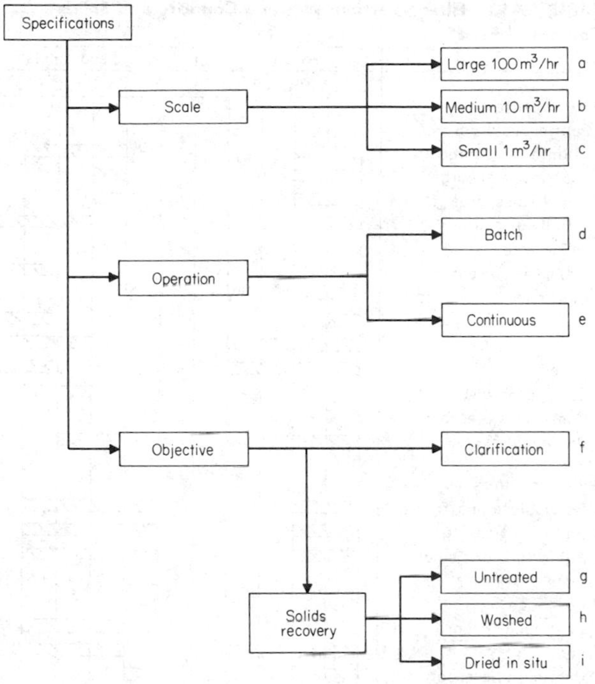

FIG. 19-113 Coding the problem specification. (Purchas, Solid/Liquid Separation Equipment Scale-Up, *Uplands Press, Croydon, England, 1977, p. 10, by permission.)*

Electro-filter separator. This filter (Petreco Division of Petrolite Corp.) is a specialized clarifying filter. A high-voltage (usually 460-V) electrostatic field polarizes a filter medium (bed of solids), causing it to attract suspended solids from the liquid. Because essentially total removal of fine solids is achieved, the device is attractive for the recovery of high-value particulates like some catalysts and other metals. However, its application is limited to nonaqueous hydrocarbon streams. When the filter becomes loaded, the power supply is deenergized and the flow of fluid through the unit resuspends the collected solids, thus simultaneously regenerating the bed and providing a concentrated slurry of the solids from which they can be retrieved if their recovery is desired. This filter is totally enclosed, and it can operate over a wide range of temperature and pressure.

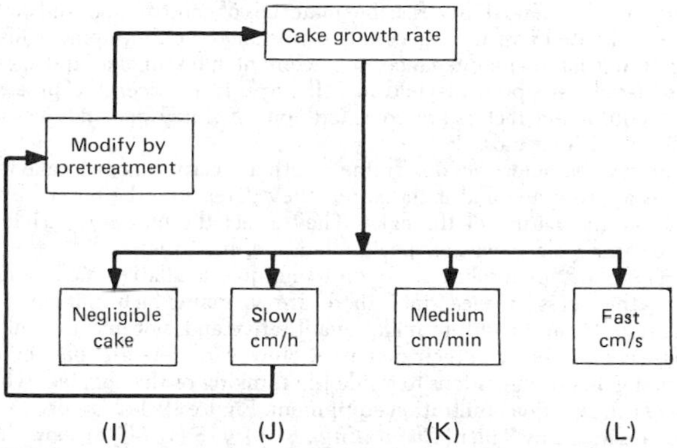

FIG. 19-115 Coding the filtration characteristics of a slurry. *(Purchas,* Solid/Liquid Separation Equipment Scale-Up, *Uplands Press, Croydon, England, 1977, p. 12, by permission.)*

SELECTION OF FILTRATION EQUIPMENT

If a process developer who must provide the mechanical separation of solids from a liquid has cleared the first decision hurdle by determining that filtration is the way to get the job done (see the final subsection of Sec. 19, "Selection of a Solids-Liquid Separator")—or that it must remain in the running until some of the details of equipment choice have been settled—choosing the right filter and right filtration conditions may still be difficult. Much as in the broader determination of which unit operation to employ, the selection of filtration equipment involves the balancing of process specifications and objectives against capabilities and characteristics of the various equipment choices (including filter media) available. The important process-related factors are slurry character, production throughput, process conditions, performance requirements, and permissible materials of construction. The important equipment-related factors are type of cycle (batch or continuous), driving force, production rates of the largest and smallest units, separation sharpness, washing

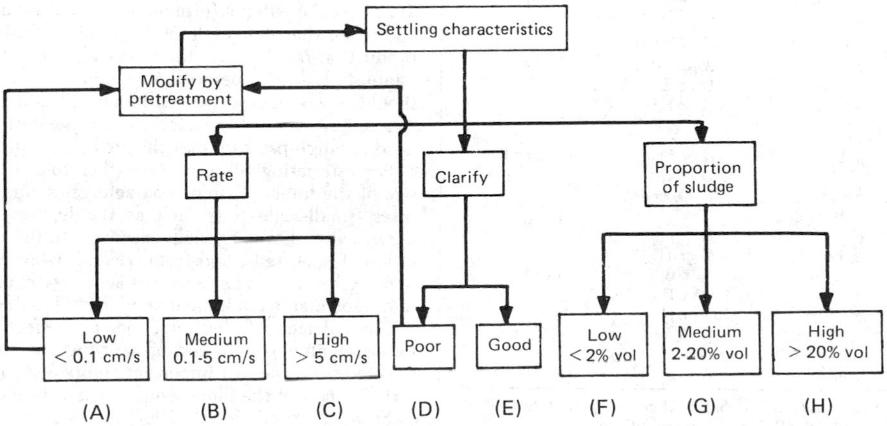

FIG. 19-114 Coding the settling characteristics of a slurry. *(Purchas,* Solid/Liquid Separation Equipment Scale-Up, *Uplands Press, Croydon, England, 1977, p. 11, by permission.)*

capability, dependability, feasible materials of construction, and cost. The estimated cost must account for installed cost, equipment life, operating labor, maintenance, replacement filter media, and costs associated with product-yield loss (if any). In between the process and equipment factors are considerations of slurry preconditioning and use of filter aids.

Slurry characteristics determine whether a clarifying or a cake filter is appropriate; and if the latter, they determine the rate of formation and nature of the cake. They affect the choice of driving force and cycle as well as specific design of machine.

There are no absolute selection techniques available to come up with the "best" choice since there are so many factors involved, many of them difficult to make quantitative and, not uncommonly, some contradictory in their demands. However, there are some published general suggestions to guide the thinking of the engineer who faces the selection of filtration equipment. Figure 19-112 is a decision tree designed by Tiller [*Chem. Eng.*, **81**(9), 118 (1974)] to show the steps to be followed in solving a filtration problem. It is erected on the premise that rate of cake formation is the most important guide to equipment selection. A filter-selection process proposed by Purchas (op. cit., p. 10–14) employs additional criteria and is based on a combination of process specifications and the results of simple tests. The application is coded by use of Figs. 19-113, 19-114, and 19-115, and the resulting codes are matched against Table 19-12 to identify possible filters. Information needed for Fig. 19-114 can be obtained by observing the settling of a slurry sample (Purchas suggests 1 L) in a graduated cylinder. Filter-cake-growth rate (Fig. 19-115) is determined by small-scale leaf or funnel tests as described earlier.

TABLE 19-12 Classification of Filters According to Duty and Slurry-Separation Characteristics*

Type of equipment	Suitable for duty specification†	Required slurry-separation characteristics‡: Slurry-settling characteristics	Slurry-filtering characteristics
Deep-bed filters	a or b	A	*T*
	e	D	
	f	F	
Cartridges	b or c	A or B	
	d	D or E	
	f	F	
Batch filters			
Pressure vessel with vertical elements	a, b, or c	A or B	I or J
	d	D or E	
	f, g, h, or i	F or G	
Pressure vessel with horizontal elements	b or c	A or B	J or K
	d	D or E	
	g or h	F or G	
Filter presses	a, b, or c	A (or B)	I or J
	d	D or E	
	f, g, h, or i	F, G, or H	
Variable-volume filters	a, b, or c	A (or B)	J or K
	d or e	D or E	
	g (or h)	G or H	
Continuous filters			
Bottom-fed drum or belt drum	a, b, or c	A or B	I, J, or K
	e	D or E	
	f, g, h, or i	F, G, or H	
Top-fed drum	a, b, or c	C	L
	e	E	
	g, i (or h)	G or H	
Disk	a, b, or c	A or B	J or K
	e	D or E	
	g	G or H	
Horizontal belt, pan or table	a, b, or c	A, B, or C	J, K, or L
	d or e	D or E	
	g or h	F, G, or H	

*Adapted from Purchas, *Solid/Liquid Separation Equipment Scale-Up*, Uplands Press, Croydon, England, 1977, p. 13, by permission.
†Symbols are identified in Fig. 19-113.
‡Symbols are identified in Figs. 19-114 and 19-115.

TABLE 19-13 Filter Selection: Washing Capacity and Solids Content of Feed*

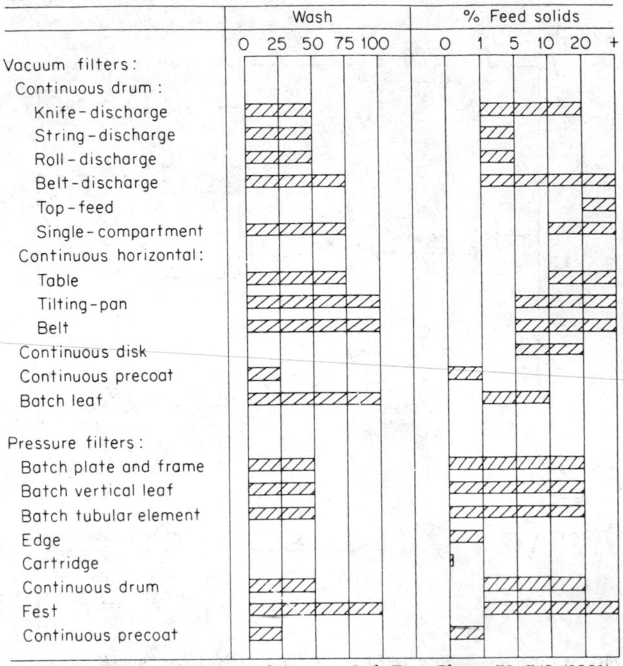

*Adapted from Van Note and Weems, *Ind. Eng. Chem.*, **53**, 549 (1961), by permission of the copyright owner, the American Chemical Society.

Cake-washing requirements also may significantly affect equipment choice, as capability varies greatly with filter type. The relative washing effectiveness of a number of filter types is indicated in Table 19-13.

Continuous filters are most attractive when the process application is a steady-state continuous one, but the rate at which cake forms and the magnitude of production rate are sometimes overriding factors. A rotary vacuum filter, for example, is a dubious choice if a 3-mm (0.12-in) cake will not form under normal vacuum in less than 5 min and if less than 1.4 m^3/h (50 ft^3/h) of wet cake is produced. Upper production-rate limits to the practicality of batch units are harder to establish, but any operation above 5.7 m^3/h (200 ft^3/h) of wet cake should be considered for continuous filtration if it is at all feasible. Again, however, other factors such as the desire for flexibility or the need for high pressure may dictate batch equipment.

For estimating filtration rate (therefore, operating pressure and size of the filter), washing characteristics, and other important features, small-scale tests such as the leaf or pressure bomb tests described earlier are usually essential. In the conduct and interpretation of such tests, and for advice on labor requirements, maintenance schedule, and selection of accessory equipment, the assistance of a dependable equipment vendor is advisable.

Filter Prices As indicated, one of the factors affecting the selection of a filter is total cost of carrying out the separation with the selected machine. An important component of this cost item is the installed cost of the filter, which starts with the purchase price.

From a survey of early 1982, prices of a number of widely used types of process filter were collated by Hall and coworkers [*Chem. Eng.*, **89**(7), 80 (1982)]. These data are drawn together in Fig. 19-

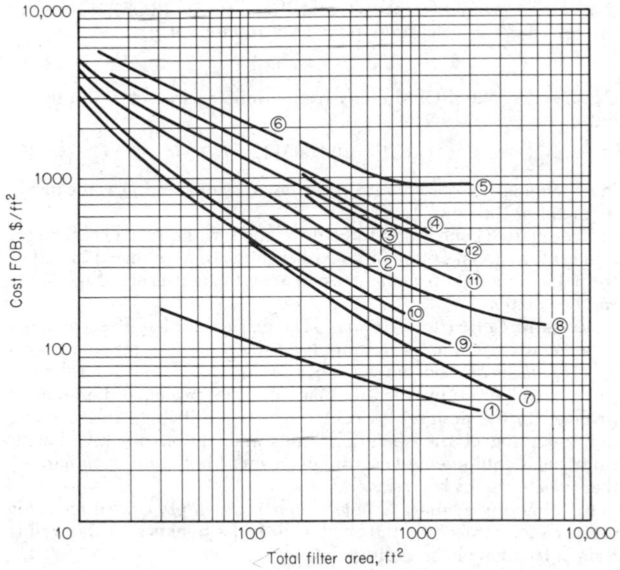

FIG. 19-116 Price of filters, uninstalled. Figures are for filters without accessories (except as indicated), FOB point of manufacture. (1) Batch pressure leaf and tubular filters, horizontal and vertical; carbon steel with stainless filter elements. (2) Horizontal table vacuum filter; carbon steel. (3) Horizontal table vacuum filter; 316 stainless steel. (4) Corrosion-resistant horizontal vacuum belt filter. (5) Tilting-pan vacuum filter; 316 stainless steel. (6) Rotary vacuum disk filter with agitator; carbon steel; heavy duty, for metallurgical and general industrial service. (7) Rotary vacuum disk filter with feed box, hood, and repulper but without agitator; 304 stainless steel; for pulp and paper applications. For 316 stainless steel, multiply by 1.19 (the smallest unit) to 1.24 (the largest), depending on filter size. (8) Single-compartment rotary vacuum drum filter; 316 stainless steel. (9) Multicompartment rotary vacuum drum filter with scraper discharge; carbon steel; for chemical, metallurgical, and general industrial service. For 316 stainless steel, multiply by 1.15 to 1.7, depending on filter size. (10) Multicompartment rotary vacuum drum with belt discharge and continuous vacuum precoat filter; carbon steel. For 316 stainless steel, use the same multipliers as for curve 9. (11) Multicompartment rotary vacuum drum with feed box, wash showers, and repulper; carbon steel; for pulp and paper service. (12) Same as curve 11, except that it is made of 304 stainless steel. For 316 stainless steel, multiply by 1.1 to 1.16 depending on filter size. *(Adapted from Hall et al.,* Chem. Eng., *89(7), 108, 109, 114, 115 (1982); copyright 1982 by McGraw-Hill, Inc., New York. Used with special permission of McGraw-Hill.)*

116. They have a claimed accuracy of ±10 percent, but they should be used confidently only with study-level cost estimations (±25 percent) at best. Cost of delivery to the plant can be approximated as 3 percent of the FOB price [Pikulik and Diaz, *Chem. Eng.*, **84**(21), 106 (1977)].

The cost of the filter station includes not only the installed cost of the filter itself but also that of all the accessories dedicated to the filtration operation. Examples are feed pumps and storage facilities, precoat tanks, vacuum systems (often a major cost factor for a vacuum filter station), and compressed-air systems. The delivered cost of the accessories plus the cost of installation of filter and accessories generally is of the same order of magnitude as the delivered filter cost and commonly is several times as large. Installation costs, of course, must be estimated with reference to local labor costs and site-specific considerations.

The relatively high prices of pulp and paper filters reflect the construction features that accommodate the very high hydraulic capacity that is required. The absence of data for some common types of filters, in particular the filter press, is explained by Hall as due to the complex variety of individual features and materials of construction. For information about missing filters and for firmer estimates for those types presented, vendors should be consulted. In all cases of serious interest, consultation should take place early in the evaluation procedure so that it can yield timely advice on testing, selection, and price.

CENTRIFUGES

GENERAL REFERENCES: Ambler, in McKetta, *Encyclopedia of Chemical Processing and Design,* vol. 7, Marcel Dekker, New York, 1978; also in Schweitzer, *Handbook of Separation Techniques for Chemical Engineers,* McGraw-Hill, New York, 1979, sec. 4. Ambler and Keith, in Perry and Weissberger, *Separation and Purification Techniques of Chemistry,* 3d ed., vol. 12, Wiley, New York, 1978. Flood, Porter, and Rennie, *Chem. Eng.,* 73(13), 190 (1966). Hultsch and Wilkesmann, in Purchas, *Solid/Liquid Separation Equipment Scale-Up,* Uplands Press, Croydon, England, 1977, chap. 12. Lavanchy and Keith, in *Kirk-Othmer Encyclopedia of Chemical Technology,* 3d ed., vol. 5, Wiley, New York, 1979, p. 194. Moyers, *Chem. Eng.,* 73(13), 182 (1966). Records, in Purchase, op. cit., chap. 6. Smith, *Ind. Eng. Chem.,* **43**, 439 (1961). Sullivan and Erikson, ibid., p. 434.

Centrifuges for the separation of solids from liquids are of two general types: (1) sedimentation centrifuges, which require a difference between the densities of the two phases; and (2) centrifugal filters, in which the solid phase is supported and retained on a permeable membrane through which the liquid phase is free to pass. Liquid-liquid centrifugal separators may be considered as an extension of the first type. The particles of a disperse liquid phase exhibit the same characteristics as particles of dispersed solids.

The use of centrifuges covers a broad range of applications, from the separation of gases of different molecular weights to the dewatering of 0.25-in (approximately 6-mm) coal.

GENERAL PRINCIPLES

When a body of mass m is acted on by a force F, it is accelerated in the direction of that force at a rate a that is inversely proportional to the mass:

$$a = F/m \tag{19-44}$$

When the force is removed, the body continues its motion in the same direction at constant velocity v. Its acceleration is zero until it is acted on again by a force. If the body is constrained to move in a circular path, its scalar velocity remains v, but its vector velocity changes continuously. The change in vector velocity is the **centrifugal acceleration,** a_c:

$$a_c = v \cdot v/r = \omega^2 r \tag{19-45}$$

in which r is the radius of the circular path and ω is the angular velocity. If this body is whirled on the end of a string, the pull on the string is **centrifugal force** F_c calculated

$$F_c = m\omega^2 r \tag{19-46}$$

The pull that must be exerted on the other end of the string to keep the body on its circular path and to prevent it from flying away on a tangent to this path is **centripetal force** F_{cp}:

$$F_{cp} = -m\omega^2 r \tag{19-47}$$

All components of a system comprising one or more particles suspended in a continuous liquid phase that is enclosed in a rotating cylindrical container experience centrifugal force. It is this force that causes solid particles denser than the liquid to migrate radially outward toward the wall of the rotating cylinder (sedimentation), particles less dense than the liquid to migrate radially inward toward the axis of rotation until they reach the air-liquid interface (flotation), and the liquid to pass through the sedimented solids and through the container wall if it is perforated or permeable.

These principles are illustrated in Fig. 19-117. In Fig. 19-117*a*, a stationary cylindrical bowl contains a slurry of liquid and solids denser than the liquid. The liquid surface is horizontal, and in time the solids come to rest on the floor of the bowl. In Fig. 19-117*b*, the bowl is rotating about its vertical axis. Liquid and solids are now being acted on by two forces: gravitational force downward and centrifugal force horizontally. In commercial centrifuges the centrifugal component is so much greater than the gravitational that the latter practically may be neglected. The liquid assumes the position shown, with an almost vertical inner surface. The solids settle horizontally to form a compact layer on the bowl wall. In Fig. 19-117*c*, the bowl wall is perforated and lined with a filter medium capable of retaining the particles. Nearly all the liquid flows out, leaving behind the relatively dry cake of solids.

Any rotating machine in which centrifugal force is applied for a useful purpose (e.g., phase separation) is called a **centrifuge** or, less commonly, a **centrifugal.** A centrifuge normally consists of (1) a bowl or rotor in which material to be treated is centrifugally accelerated, (2) a feed tube to introduce the material, (3) a drive shaft, (4) drive-shaft bearings, (5) a drive mechanism (usually an electric motor) to rotate the shaft and rotor, (6) a casing or cover to segregate the separated products, (7) a frame to support and align these elements, and (8) casing-to-shaft seals (when pressure containment is desired).

Magnitude of Centrifugal Force It is convenient to reference centrifugal acceleration to the standard acceleration of gravity g by

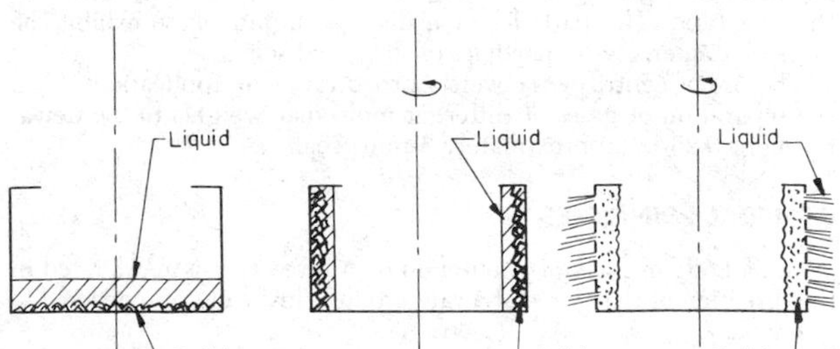

FIG. 19-117 Principles of centrifugal separation and filtration. (*a*) Bowl stationary. (*b*) Sedimentation in rotating imperforate bowl. (*c*) Filtration in rotating perforate basket.

use of the concept of "relative centrifugal force" (RCF), describing the centrifugal acceleration in terms of number of g's:

$$\text{RCF} = \omega^2 r/g \tag{19-48}$$

One can compute RCF directly from familiar parameters in English units:

$$\text{RCF} = 0.0000142 n^2 D_i \tag{19-49}$$

in which n is the rotation rate of the bowl, r/min; and D_i is the inside diameter of the bowl, in.

The relative centrifugal force developed in industrial centrifuges ranges from about 200 in large basket units to more than 125,000 in special sedimentation centrifuges. In analytical ultracentrifuges RCF can be 250,000.

Stress in Centrifuge Rotors The exact complete stress analysis of a centrifuge rotor is a complex subject of concern mainly to experts that is beyond the scope of this *Handbook*. But the relationships (or their approximation) among the principal parameters involved in the design of a centrifuge can contribute to the user's understanding of the operating limits and options identified with centrifugal equipment. Accordingly, a simplified, brief summary of these relationships is presented.

Rotation of a cylindrical object such as an empty centrifuge rotor creates a self-stress S_s in the bowl wall. If the thickness of the wall is a small fraction of the radius,

$$S_s = \omega^2 r_i^2 \rho_m \tag{19-50}$$

in which ρ_m is the density of the material of construction of the bowl shell and $r_i = D_i/2$.

The material being centrifuged exerts pressure on the inner wall of the bowl that produces a stress increment S_c in the bowl wall:

$$S_c = \omega^2 r_2 (r_2^2 - r_1^2)\rho_c/4\delta \tag{19-51}$$

in which r_1 and r_2 are respectively the inner and outer radii of the bowl contents (solids and liquid), ρ_c is the average density of these contents, and δ is the thickness of the bowl wall. Normally it is allowable to assume that $r_2 = r_i$.

The total stress experienced by the bowl wall is the sum of S_c and S_s:

$$S_t = S_s + S_c = \omega^2 r_2 \left[r_2 \rho_m + \frac{(r_2^2 - r_1^2)\rho_c}{4\delta} \right] \tag{19-52}$$

or, in common engineering terms and units,

$$S_t = 4.11 \times 10^{-9} n^2 D_i \left[D_i \rho_m + \frac{D_i^2 - D_1^2)\rho_c}{4\delta} \right] \tag{19-53}$$

With few exceptions industrial centrifuges are so designed that S_s is 45 to 65 percent of S_t.

In Eq. (19-53), the units that *must be used* are D_i, D_1, and δ, in; n, r/min; S_t, lbf/in^2; and ρ_c and ρ_m, lb/ft^3. *Use of any other units requires a different coefficient*. In Eqs. (19-50) through (19-52), however, *consistent units are required*. Any set of consistent units may be used.

Increased centrifugal separating effectiveness (i.e., increased centrifugal acceleration) can be realized by increasing either the diameter or the rate of rotation of the centrifuge bowl [Eq. (19-49)]. Increasing either of these parameters also increases both the self-stress and the stress caused by the process load; therefore, it is the maximum allowable stress of the material of which the centrifuge rotor is made that ultimately limits the centrifugal acceleration that the machine can produce. It is clear, however, that for a given bowl stress, centrifugal acceleration is an inverse function of bowl diameter—for example, doubling rotor speed and halving rotor diameter double the acceleration with no change in empty-bowl stress and some decrease in loaded-bowl stress. The highest attainable centrifugal field is achieved in rotors of the smallest diameter practical; and, conversely, large-bowl machines can be used only when a relatively low centrifugal field is required.

Figure 19-118 shows the range of diameter of commercial centrifuges and the range of maximum RCF developed in each type.

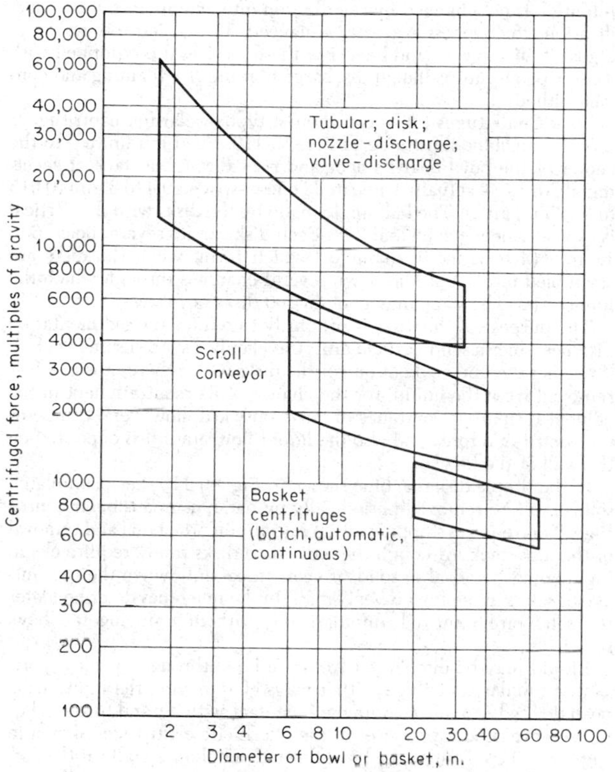

FIG. 19-118 Variation of centrifugal force with diameter in industrial centrifuges.

Materials of Construction Centrifuge bowls are made of almost every machinable alloy of reasonably high strength. Preference is given to those alloys having at least 1 percent elongation to minimize the risk of cracking at stress-concentration points. Typically, the list includes carbon steel, types 304, 316, and 317 stainless steels, alloy 20 (Carpenter) stainless steel, Monel Metal, Inconel, nickel, Hastelloy B, titanium, and alloyed aluminum. Vertical-basket centrifuges are frequently constructed of carbon steel or stainless steel coated with rubber, neoprene, Penton, or Kynar. Casings and feed, rinse, and discharge lines that are stationary and lightly stressed may be constructed of any suitable rigid corrosion-resistant material.

Critical Speed In the design of any high-speed rotating machinery attention must be paid to the phenomenon of critical speed. This is the speed at which the frequency of rotation matches the natural frequency of the rotating part. At this speed any vibration induced by slight unbalance in the rotor is strongly reinforced, resulting in large deflections, high stresses, and even failure of the equipment. Speeds corresponding to harmonics of the natural frequency are also critical speeds but give relatively small deflections and are much less troublesome than the fundamental frequency. Critical speed of simple shapes may be calculated from the moment of inertia; with complex elements such as a loaded centrifuge bowl it is best found by experiment.

Nearly all centrifuges operate at speeds well above the primary critical speed and therefore must pass through this speed during acceleration and deceleration. To permit them to do so safely, some degree of damping in their mounting must be provided. This may result from the design of the spindle or drive shaft alone, spring loading of the spindle bearing nearest the rotor, elastic loading of the suspension, or a combination of these. Smaller and medium-sized centrifuges of the cream-separator and bottle-centrifuge design are frequently mounted on elastic cushions.

SEDIMENTATION CENTRIFUGES

Sedimentation centrifuges remove or concentrate particles of solids in a liquid by causing the particles to migrate through the fluid radially toward or away from the axis of rotation, depending on the density difference between particles and liquid. If there is no difference in the density of the phases, the centrifuge cannot achieve separation. The discharge of the liquid may be intermittent, as in the case of the bottle or laboratory centrifuge. In commercial centrifuges, the liquid-phase discharge is usually continuous. The heavy-solid phase is deposited against the bowl wall for intermittent removal, either manually or by the action of a cutter knife; for continuous removal, by a differential screw conveyor; or for intermittent or continuous discharge with a portion of the continuous liquid phase through appropriate openings in the periphery of the bowl (Table 19-14). The time required for solids removal may be as much as an hour for completely manual operation or as little as a few seconds for fully automated intermittent operation. When the separated solids have less density than the continuous phase, they can be removed from the surface of the liquid with a skimming tube.

Batch Laboratory Equipment Sedimentation centrifuges handling discrete batches of liquid are nearly always for laboratory or analytical use, rarely for production. Representative types are test-tube centrifuges and ultracentrifuges.

Tubular-Bowl Centrifuges The tubular-bowl centrifuge is widely employed for purifying used lubricating and other industrial oils and in the food, biochemical, and pharmaceutical industries. Commercial models have bowls 102 to 127 mm (4 to 5 in) in diameter and 762 mm (30 in) long (Table 19-14). The smallest size (44- by 229-mm or 1.75- by 9-in bowl) is a laboratory model, also used for harvesting virus.

The bowl is suspended from an upper bearing and drive assembly through a flexible-drive spindle. It hangs freely with only a loose guide in a controlled damping assembly at the bottom. Thus, it can find its natural axis of rotation if it becomes slightly unbalanced because of its process load.

TABLE 19-14 Specifications and Performance Characteristics of Typical Sedimentation Centrifuges

Type	Bowl diameter, in	Speed, r/min	Maximum centrifugal force, × gravity	Throughput: Liquid, gal/min	Throughput: Solids, tons/h	Typical motor size, hp
Tubular	1¾	50,000*	62,400	0.05–0.25	. . . .	*
	4⅛	15,000	13,200	0.1–10	. . . .	2
	5	15,000	15,900	0.2–20	. . . .	3
Disk	7	12,000	14,300	0.1–10	. . . .	⅓
	13	7,500	10,400	5–50	. . . .	6
	24	4,000	5,500	20–200	. . . .	7½
Nozzle discharge	10	10,000	14,200	10–40	0.1–1	20
	16	6,250	8,900	25–150	0.4–4	40
	27	4,200	6,750	40–400	1–11	125
	30	3,300	4,600	40–400	1–11	125
Helical conveyor	6	8,000	5,500	To 20	0.03–0.25	5
	14	4,000	3,180	To 75	0.5–1.5	20
	18	3,500	3,130	To 50	0.5–1.5	15
	25	3,000	3,190	To 250	2.5–12	150
	32	1,800	1,470	To 250	3–10	60
	40	1,600	1,450	To 375	10–18	100
	54	1,000	770	To 750	20–60	150
Knife discharge	20	1,800	920	†	1.0‡	20
	36	1,200	740	†	4.1‡	30
	68	900	780	†	20.5‡	40

*Turbine drive, 100 lb/h (45 kg/h) of steam at 40 lbf/in^2 gauge (372 kPa) or equivalent compressed air.

†Widely variable.

‡Maximum volume of solids that the bowl can contain, ft^3.

NOTE: To convert inches to millimeters, multiply by 25.4; to convert revolutions per minute to radians per second, multiply by 0.105; to convert gallons per minute to liters per second, multiply by 0.063; to convert tons per hour to kilograms per second, multiply by 0.253; and to convert horsepower to kilowatts, multiply by 0.746.

Feed enters the bottom of the bowl through a stationary feed nozzle under pressure. The pressure and nozzle size are selected to give a clean jet upward into the bowl at the desired flow rate. The incoming liquid is accelerated to rotor speed, moves upward through the bowl as an annulus, and discharges at the top. Solids travel upward with the liquid and, at the same time, receive a radial velocity based on their size and weight in the centrifugal-force field. If the trajectory of a given particle intersects the bowl wall, the particle is removed from the fluid; if it does not, the particle appears in the effluent. Successful process performance depends on the correct balance of many factors, as discussed later under "Theory of Centrifugal Sedimentation."

The depth of the liquid layer is controlled by the radial position of the overflow port at the top of the bowl. To accelerate and maintain the liquid at the rotational speed of the bowl, an internal vane set, frequently of Y or trefoil form, is provided. It also serves to damp surge waves and minimize out-of-balance during deceleration.

The centrifuged liquid leaves the top of the bowl at the peripheral velocity of the overflow port. Not only is it subject to high shearing forces at this point, but it also strikes the casing or collection cover with considerable force and breaks into small droplets. If its surface tension is low, foaming may result. If this is deleterious to the product, as in the case of fruit juices, it may be desirable to remove the liquid from the top of the bowl by a skimmer or "centripetal" pump. In another configuration, the "full-bowl" principle, the feed and discharge openings are equipped with radial or face seals and no air-liquid interface is present in either the bowl or the discharge from it.

By providing two liquid outlets at different radii and different elevations at the top of the bowl, it is possible to separate continuously two immiscible liquids of different density while solid particles are being collected from either or both phases. The positioning of the interface inside the bowl between the separated liquids is usually controlled for optimum performance by adjustment of the heavy-phase outlet with an interchangeable washer or "ring dam" of selected inside diameter. Tubular-bowl centrifuges are available with a variety of casings and covers for the containment of noxious vapors and of volatile or flammable fluids.

Solids that have sedimented against the bowl wall are removed manually from this type of centrifuge when the quantity collected is sufficient to impair the quality of the clarification or separation. Their removal can be facilitated by lining the bowl with a layer of parchment paper so that solids, accelerator vanes, and paper are taken out as a cylindrical package, leaving only traces in the bowl to be removed by washing.

The liquid-handling capacity of the tubular-bowl centrifuge varies widely with the specific application. In the commercial sizes, it ranges from 38 to 57 L/h (10 to 15 gal/h) for stripping small bacteria from a culture medium to 4500 L/h (1200 gal/h) for purifying transformer oil and restoring its dielectric value. The solids-holding capacity of this centrifuge is small, usually limited to 4.5 kg (10 lb) or less. Since bowl cleaning usually takes 0.25 person-hour, the principal utility is for systems containing 1 percent or less of sedimentable solids.

Multichamber Centrifuges This design, a modification of the tubular-bowl centrifuge, is driven from below in the same manner as a conventional disk separator. It has a large-diameter, relatively short rotor. The bowl consists of a series of short tubular sections of increasing diameter nested to form a continuous tubular passage of stepwise increasing diameter for the flow of liquid. The feed is introduced into the smallest-diameter tube, the zone of least centrifugal force, and is subjected to successive zones of greater centrifugal force as it passes into and through the larger-diameter tubes. The heaviest particles are deposited in the smallest-diameter tube and smaller, lighter particles in the larger-diameter zones of higher centrifugal force. Performance may be improved by spacing the outer tubes more closely together to reduce the distance that the smaller particles need to travel to the tube wall. This also serves to maintain a constant velocity of flow between adjacent tubes. This type of bowl may contain up to six annular chambers and have a holding capacity of up to 0.064 m^3 (17 gal) of solids. The total holding volume for solids retention is not changed by the division into chambers, but the clarification effectiveness is greatly enhanced. The largest use is clarifying fruit juices, wort, and beer. For these services it is equipped with a centripetal-pump effluent discharge to minimize foaming and contact with air.

Disk Centrifuges The commonest type of clarifier centrifuge is the disk machine illustrated in Fig. 19-119. Feed is admitted to the center of the bowl near its floor and rises through a stack of sheet-metal "disks"—actually truncated cones—spaced 0.4 to 3 mm (0.015 to 0.125 in) apart. The half angle made by the disks with the vertical is typically between 35 and 50°. Each disk carries several holes, 6 to 13 mm (¼ to ½ in) in diameter, which form, when the disks are assembled in place in the bowl, several channels through which the liquid rises. The stack may contain 100 disks or more.

The purpose of the disks is primarily to reduce the sedimentation distance, since a solid particle must travel only a short distance before it reaches the underside of one of the disks. Once there, it is in effect removed from the liquid, for the chance of its reentrainment in the effluent is small. It continues to move outward, however, because of the centrifugal force and also the liquid flow, until it is deposited on the wall of the bowl.

In the simple disk machine shown in Fig. 19-119 the accumulated solids must be removed periodically by hand, as in a tubular centrifuge. This requires stopping and disassembling the bowl and removal of the disk stack. Although the individual disks rarely require cleaning, manual removal of solids is economical only when the percentage of solids in the feed is very small. In the machines described later the solids are removed automatically without reducing the bowl speed.

Liquid may be discharged from the bowl through overflow ports as in a tubular centrifuge. Other ways of removing the liquid have been devised to avoid foaming and contact with air and to allow liquid removal under pressure. In the "hermetic centrifuge" shown in Fig. 19-119 feed enters the bowl through a hollow spindle at the bottom, and clarified liquid leaves through the central pipe at the top. Rotary seals between the bowl and the stationary feed and discharge pipes permit operation under pressures of 644 to 780 kPa (80 to 100 psig). A valve in the discharge line keeps this pressure constant. Other designs contain a centripetal pump or "paring ring," with a vaned pump impeller mounted on the stationary feed pipe. Rotating liquid enters the outside of the impeller and flows inward between the stationary vanes to the annular chamber surrounding the feed pipe, and from there to the discharge. In the impeller the kinetic

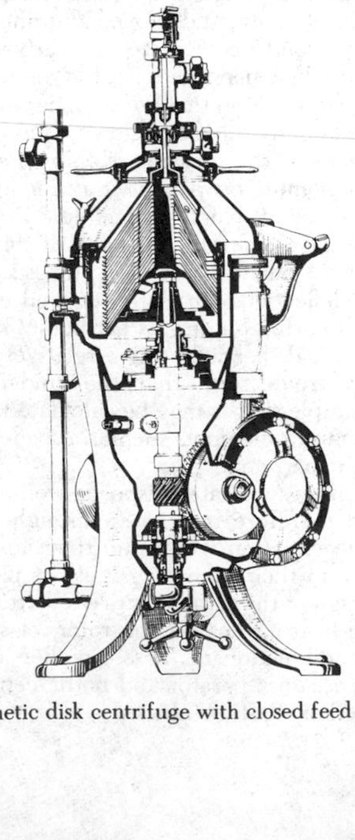

FIG. 19-119 Hermetic disk centrifuge with closed feed and discharge.

energy of the liquid is largely converted into pressure energy, so that the liquid leaves the bowl at pressures up to 780 kPa (100 psig).

Disk centrifuges range in diameter from 102 to 762 mm (4 to 30 in) and develop 4000 to 14,000 times the force of gravity. Their sedimenting effectiveness is very nearly the same as that of a tubular centrifuge, despite the lower centrifugal force. With some solids they are slightly less effective, on others slightly more effective, than a tubular machine. They are effective also as liquid-liquid separators, and especially so for the concentration of emulsions. By far the largest number of centrifuges sold for one purpose are cream separators for the concentration of butterfat in milk. These machines are all of the disk type.

Most disk-centrifuge rotors, particularly in the smaller sizes, are mounted on top of a stiff spindle and driven from below. A thrust bearing at the bottom carries the weight of the rotating assembly and provides a pivot point for the spindle. The main bearing is just below the rotor. It is preloaded radially with springs or elastic cushioning to provide the required degree of flexibility and damping. The drive train is almost unique. It provides a speed step-up from a horizontal prime-mover shaft to the vertical spindle through a gear wheel on the low-speed shaft to a worm shell on the high-speed spindle. Conventionally, a bearing-grade-bronze gear wheel is mated to a steel worm, and the system is splash-lubricated. This type of drive is usually limited to the transmission of approximately 19 kW (25 hp) maximum. For larger sizes consuming more power, the rotor may be either suspended and driven by belts from an offset motor or underdriven through belts or through a more conventional gear train.

Peripheral-Discharge Disk Centrifuges As shown in Fig. 19-120, the relatively small vertical height of the disk-centrifuge bowl makes it feasible to slope the bowl walls to direct the sedimented solids to a narrow annulus at the periphery and to accomplish this without undue increase in the stress level in the bowl shell. From

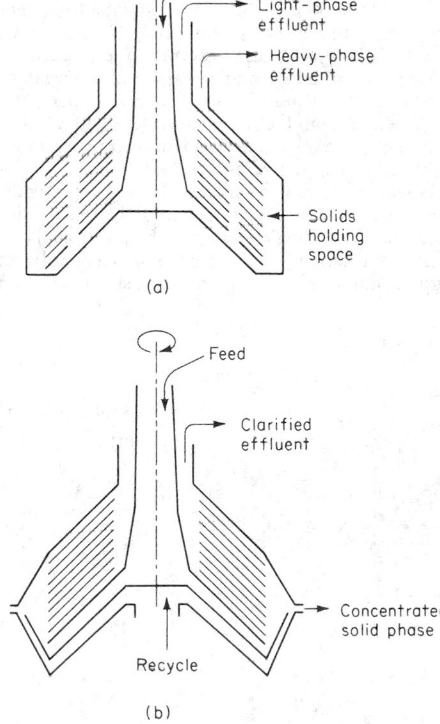

FIG. 19-120 Disk-centrifuge bowls. (*a*) Separator, solid wall. (*b*) Recycle clarifier, nozzle discharge.

here the solids can be discharged continuously through nozzles or intermittently by a variety of devices. It can be shown that when the disk stack height and diameter are controlled by the slope of the bowl walls, optimum centrifuge performance is obtained when the disk stack diameter is three-fourths of the maximum inside diameter of the bowl shell.

Nozzle-Discharge Centrifuges Solids are discharged continuously along with a portion of the liquid phase from the centrifuge shown in Fig. 19-120*b* through nozzles spaced around the periphery of the bowl. The angle of repose of the sedimented solids determines the slope of the bowl walls for satisfactory operation. This angle also controls the maximum permissible distance between nozzles and, therefore, the number of nozzles that must be used in a given size of bowl. Clarification efficiency is seriously impaired if the buildup of solids between nozzles reaches into the disk stack. The nozzle diameter should be at least twice the diameter of the largest particle normally present in the system to be centrifuged, and prescreening is recommended to remove larger extraneous solids. Nozzles typically range from 0.6 to 3 mm (0.025 to 0.125 in) in diameter, and the larger bowls accommodate up to 24 of them. The nozzles are almost always directed tangentially backward to the direction of rotation to recover the kinetic energy of the flow through them. This reduces the power demand of such a centrifuge by as much as 50 percent.

Many nozzle-discharge centrifuges are designed for only a single passage of the solids through the centrifugal field. The resultant solids concentration then depends on the feed rate and concentration and on the number and size of nozzles required for steady-state operation. Typically, each 1.3-mm (0.050-in) nozzle will discharge about 8.3 L/min (2.2 gal/min). Such once-through centrifuges are used for the separation of oil-water emulsions when the concentration of solids in the waste aqueous phase is of little importance, as well as for a number of straightforward clarification operations.

For the clarification of a single liquid phase with controlled concentration of the discharged slurry, a centrifuge that provides recirculation is used. The usual method is shown in Fig. 19-120*b*. A portion of the sludge discharge is returned to the bowl through a recycle system that directs it to an area immediately adjacent to the nozzles. This has the effect of preloading the nozzles with sludge that has already been separated and reduces the net flow of liquid with the newly sedimented solids from the feed. Increased concentration can be obtained alternatively by recycling a portion of the sludge to the feed, but this increases the solids loading in the disk stack, with a corresponding sacrifice in the clarity of the effluent at a given feed rate. The recycle loop may be built into the centrifuge piping system or it may be external with an intervening surge tank. The former gives, not always desirably, a rapid response to changes in feed rate and feed concentration; the latter responds to these changes more slowly and allows more time for necessary adjustments.

The maximum concentration of solids that can be obtained in the underflow depends strongly on their nature. It approaches 50 percent by weight for kaolin clay and 40 percent for cornstarch, and it may be as low as 15 percent for hydrophilic solids like yeast and corn gluten.

The effluent discharge may be over an annular dam or through a centripetal pump. The kinetic energy of the underflow nozzle discharge from some models is enough to raise it a few feet. The casing can be enclosed and fitted with mechanical shaft-to-casing seals for pressure containment to 1.1 MPa (150 psig) or higher. Operating temperatures to 315°C (about 600°F) are possible with some maximum limit to rotor speed.

Type 316/317 stainless steel is the commonest material of construction for disks and bowls, and tungsten carbide for nozzles. The bowls may be underdriven or suspended and range from a few inches to over 1 m (about 40 in) in outside diameter. The largest size, capable of clarifying up to 32 L/s (500 gal/min), requires 112 kW (150 hp).

"Self-Opening" Centrifuges In the self-opening-bowl configuration an annular seal is provided around the periphery of the rotor by contact between an elastomeric ring in the bowl top and a movable sleeve in the bottom section of the bowl. In the several available designs the closure is maintained by (1) spring pressure, (2) the pressure of a constant flow of auxiliary hydraulic fluid that drains off through a leak hole, or (3) a residual pool of the hydraulic operating fluid. To discharge the accumulated solids, the sleeve is caused to move away from the seal ring, exposing an annular slot through which centrifugal force causes solids of appropriate consistency to discharge. This is accomplished in design 1 by introducing operating fluid to a lower chamber, where, under centrifugal force, it overcomes the spring pressure; in design 2 by interrupting the flow of operating fluid; and in design 3 by introducing the operating fluid to a compartment that has a smaller inner radius than that of the residual pool. In all three designs a leak hole provides for drainage of the operating fluid. The unloading operation may be under manual or automatic-timer control, or the unload signal may be triggered by a reduction in light transmission through the effluent or by the resistance, due to cake buildup, to a recycled portion of the effluent.

The use of these centrifuges is limited to solids having the degree of fluidity or plasticity required to move through the exposed peripheral openings. They are available in sizes to handle up to 14 L/s (220 gal/min), with sludge-holding volume up to 0.023 m^3 (6 gal). It is recommended that the intervals of feeding between desludging be at least 60 s. The bowls are underdriven, with right-angle gearing to 15 kW (20 hp) and V belts from a vertical motor to 45 kW (60 hp).

Continuous Decanter Centrifuges (with Helical Conveyor) The continuous decanter centrifuge consists of a solid-wall bowl with a horizontal or a vertical axis of rotation. The bowl may be conical or cylindrical or, most often, a combination of the two (Fig. 19-121). Centrifugal force causes the liquid surface to be essentially parallel to the axis of rotation, whether horizontal or vertical. The solids-discharge ports at one end of the bowl are conventionally at a smaller radius than the liquid-discharge ports at the other end. The latter can be adjusted to control the level or depth of the pond and correspondingly the length of the dry "beach" section. Feed is introduced through a concentric tube to an appropriate point in the bowl. The liquid phase seeks the level of the ports at the larger radius and continuously discharges through them. The heavy solids that have sedimented against the bowl wall are transported continuously to the other end of the bowl by a helical screw conveyor that extends the full length of the bowl. If the sedimented solids are of appropriate consistency, the screw conveyor will transport them along the bowl; up the conical end section and onto the dry beach, where they have an opportunity to drain; and onward to the solids-discharge ports. The necessary differential speed between the bowl and the conveyor is usually accomplished by a two-stage planetary gearbox the housing of which rotates at the speed of the bowl, with a fixed first-stage pinion shaft. This provides transmission of the torque necessary to turn the conveyor at a speed lower by 20 to 80 r/min than that of the bowl itself. If the pinion shaft is allowed to turn in the direction of bowl rotation at an appropriate intermediate speed, the conveyor differential can be reduced to a magnitude below that resulting from the gear ratio. This can be accomplished by driving the pinion shaft from a variable-speed motor or by permitting it to rotate against a restraining brake. The torque at the pinion shaft can be used to control the feed rate or to signal an overload condition.

Soft solids that cannot be conveyed or are resuspended by higher conveyor differential speeds can often be separated by using lower differential speed. Preflocculation of the feed is seldom effective; the change in rotational energy disintegrates the flocs. However, the controlled addition of a flocculating agent (for example, a polyelectrolyte) directly into the bowl downstream from the feed zone frequently improves both effluent clarity and discharged-solids dryness, as in the case of secondary-waste-treatment sludges and the separation of finely divided clay from potash brines.

The horizontal continuous decanter centrifuge is operated below its critical speed. The bowl is mounted between fixed bearings that, in turn, are fixed to a rigid frame. The gearbox is cantilevered outboard of one of these bearings, and the feed tube enters the bowl through the other. Particularly in the larger sizes, the frame is connected to ground through vibration isolators. In the vertical configuration, the bowl and gearbox are suspended from the drive head, which, in turn, is connected to the frame and casing through vibration isolators. A clearance bushing at the bottom limits the excursion of the bowl during start-up and shutdown but does not provide the radial constraint of a bearing under normal operating conditions.

Screw-conveyor centrifuges with mechanical shaft-to-casing seals are available for pressure containment up to 1.1 MPa (150 psig). They can be built to operate at temperatures from −87 to +260°C (−125 to +500°F).

When abrasive solids are processed, the points of wear are protected with replaceable hard surfacing such as Colmonoy, Hastelloy, or tungsten carbide. These wear points include the feed zone, the conveyor leading face and edge, the beach, the solids-discharge ports, and, less frequently, the solids-discharge end of the casing and the cylindrical section of the bowl itself. Axial transport of solids is encouraged in some applications by ribs or grooves in the shell, particularly in the conical beach section, and by polished conveyor flights.

Washing in a continuous decanter is fairly effective on solid particles no smaller than 80 μm (200 mesh) provided the centrifugally compacted cake is reasonably porous. If it is not, the wash flows across the cake surface with little penetration. The wash liquid flows down the beach and into the pond. It cannot be segregated from the effluent mother liquor. Rinsing efficiency, the proportion of soluble impurities displaced from the solids, is in the range of 50 to 80 percent, depending on cake porosity and mass-transfer rate. A wash-solids weight ratio of up to 0.25 is required.

Various bowl configurations with a wide range of effective length-to-diameter ratios, from less than 1 to over 3.5, are available for specific applications, depending on whether the major goal is maximum clarification, classification, or solids dryness. Generally, the movement of liquid and solids is in opposite directions, but one design

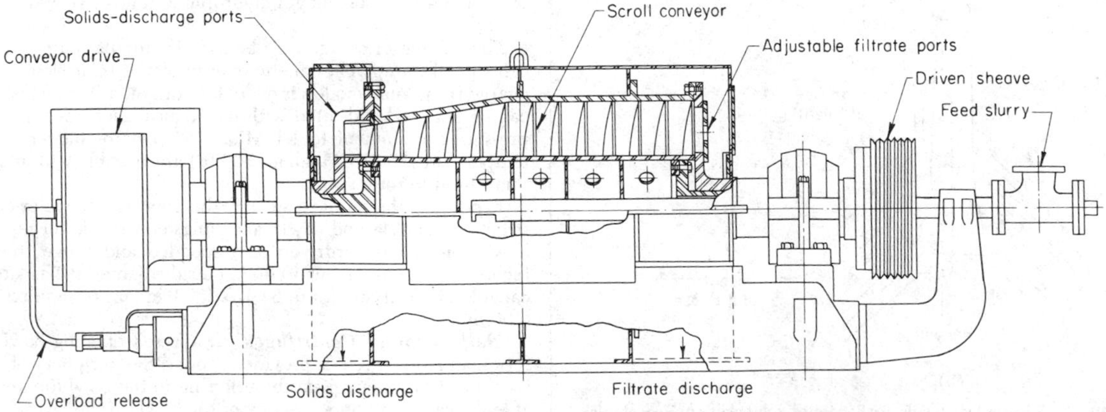

FIG. 19-121 Cylindrical-conical helical-conveyor centrifuge. *(Bird Machine Co.)*

variation provides concurrent flow from one end with skimmer take-off of the clarified liquid near the center of the bowl. Compound-angle beaches are used in specific applications such as the washing and drying of polystyrene beads. The pond level is held at the intersection of the two beach angles, the steeper angle being under the liquid and the long, shallow angle providing additional time for the solids to drain. The wash is applied on the pond side of the beach-angle intersection and functions as a continuously replenished annulus of wash liquid through which the solids are conveyed.

Centrifugal force, bowl radius, and effective length are the controlling parameters for clarification. The quantity of solids handled is a function of bowl diameter, conveyor pitch, and differential speed.

Helical-conveyor continuous decanters are available in a wide range of sizes (Table 19-14).

Screen-Bowl Continuous Decanter Centrifuges The screen-bowl decanter consists of a solid-bowl decanter to which has been added a cylindrical screen section on the small end. The helical screw conveyor is continuous through both sections. The feed is introduced in the cylindrical solid section, and the effluent discharges over plate dams or weirs at the large end. The solids are conveyed up the beach and then across the screen section before being discharged at the small end. The screen is conventionally of wedge-bar construction to minimize blinding. This construction also reduces friction to the movement of the solids across the screen, thus minimizing torque on the conveyor and gearbox. The screen section provides additional time for the solids to drain under centrifugal force and a more effective washing environment than is possible on the beach section of the conventional decanter.

Knife-Discharge Centrifugal Clarifiers Knife-discharge centrifuges with solid instead of perforated bowls are used as sedimentation centrifuges. The liquid flow is usually continuous, until a quantity of solids that interferes with further clarification has been accumulated. The feed enters the bowl at the hub end, where accelerators bring it up to bowl speed. The clarified effluent either overflows the lip ring or discharges through a skimmer pipe. In some designs, internal baffling is necessary to prevent the formation of surge waves and resultant out-of-balance. When an appropriate load of solids has built up in the bowl, the supernatant liquid is skimmed off by moving the skimmer pipe radially toward the bowl shell. The solids are then knifed out, as from centrifugal filters. Such centrifuges are usually used on systems containing coarse, easily sedimentable solids. When greater clarification effectiveness is required, the operation may be totally batchwise, with prolonged spinning of each batch. If the solids loading of the feed is low, several batches may be successively charged and the resulting supernatant liquor skimmed off before unloading of the collected solids is necessary.

Commercial centrifuges of this type range in bowl diameter from 0.3 to 2.4 m (12 to 96 in). The very large sizes are used on heavy-duty applications such as coal dewatering and are limited by stress considerations to the development of about 300 g. The intermediate sizes for chemical process service develop up to 1000 g, as indicated in Table 19-14.

Performance Characteristics of Sedimentation Centrifuges Centrifuges are often rated in terms of liquid- or solids-handling capacity, but an essential implicit consideration is the quality requirements of the separated streams produced. Centrifuge performance indeed depends on the characteristics of the system to be centrifuged as well as on the parameters of the centrifuge itself. Table 19-14 lists operating characteristics of some typical centrifuges. The capacity ranges indicated are broad but still do not cover the complete operating range, particularly with respect to specialty applications. The usual economic operation range of a 102-mm (4-in) tubular bowl is about 4 to 20 L/min (1 to 5 gal/min) of liquid, but this may be extended to from 0.75 L/min (0.2 gal/min), for harvesting small bacteria, to 75 L/min (20 gal/min), for an easy clarification job. Similarly, a bowl equipped with a 318-mm- (12.5-in-) diameter disk stack, in the nozzle discharge configuration, will satisfactorily concentrate starch at 530 L/min (140 gal/min), while the same disk stack at an even higher rotative speed will concentrate only a few gallons of natural-rubber latex per minute. The power required also may vary by a factor of 2 from that indicated in Table 19-14, depending on liquid and solids capacity and required pressure containment.

Successful applications of continuous sedimentation centrifuges have been made at temperatures from −73 to 316°C (−100 to 600°F) and at pressures from less than 132 Pa (1 mmHg absolute) to 1.1 MPa (150 psig). Obviously, the feed must be fluid enough to flow into the centrifuge, and the separated products must have characteristics that will permit them to be discharged from the centrifuge. The viscosity of the liquid phase has an important bearing on centrifuge capacity, and heating of the feed to reduce high viscosity will frequently give marked improvement in performance. Tubular centrifuges maintain their efficiency on highly viscous systems such as molten chicle and nitrocellulose "dopes" better than disk or helical-conveyor centrifuges. The upper practical limit of viscosity is about 100 Pa·s (1000 P).

Good performance on high-speed disk and tubular-bowl centrifuges has been obtained with specific-gravity differences as low as 0.02 between the phases. Commercial centrifuges successfully separate 1-μm particles when a specific-gravity difference of approximately 0.1 exists. The relationship of particle size, density difference, and viscosity to centrifuge performance is indicated in the next subsection.

Theory of Centrifugal Sedimentation A solid particle settling through a liquid in a centrifugal-force field is subjected to a constantly increasing force as it travels away from the axis of rotation. It therefore never reaches a true "terminal" velocity. However, at any given radial distance r the settling velocity u_t of a sufficiently small particle is very nearly given by the Stokes-law relation

$$v_t = \frac{\omega^2 r(\rho_p - \rho)D_p^2}{18\mu} \tag{19-54}$$

where ω is the angular velocity, ρ_p and ρ are the densities of solid and liquid respectively, D_p is the equivalent diameter of a spherical solid particle, and μ is the liquid viscosity.

If Stokes settling of a dilute suspension of uniform particles occurs in a tubular bowl of radius r, containing a thin layer of liquid of thickness s, the flow rate at which half the solid particles will be removed from the liquid is given by

$$Q_c = \frac{(\rho_p - \rho)D_p^2 V\omega^2 r}{9\mu s} \tag{19-55}$$

where Q_c is the volumetric flow rate of liquid through the bowl and V is the volume of liquid held in the bowl. With a given flow rate Q, the **critical diameter**, or "cut point," D_{pc} is given by

$$D_{pc} = \sqrt{\frac{9Q\mu s}{(\rho_p - \rho)V\omega^2 r}} \tag{19-56}$$

Most particles with diameters larger than D_{pc} will be eliminated by the centrifuge, most particles with smaller diameters will appear in the effluent, and particles with diameter D_{pc} will be divided equally between effluent stream and settled-solids phase.

When the space for sedimentation is not cylindrical or the liquid layer is thick, Eq. (19-55) must be written

$$Q_c = \frac{(\rho_p - \rho)D_p^2 V\omega^2 r_e}{9\mu s_e} \tag{19-57}$$

where r_e and s_e are the appropriate averaged values of radius and layer thickness for the given conditions. Equation (19-57) may be written

$$Q_c = \frac{2(\rho_p - \rho)D_p^2 g}{18\mu} \frac{V\omega^2 r_e}{g s_e} = 2v_g\Sigma \tag{19-57a}$$

in which v_g is the terminal settling velocity of a dispersed particle in the gravitational field and Σ is the area of a settling tank in the same field of equivalent settling characteristics to the centrifuge. For a bottle centrifuge

$$\Sigma = \omega^2 V/2g \ln [2r_3^2/(r_3^2 - r_1^2)] \tag{19-58}$$

in which r_3 is the final radius of the interface between the liquid and the sedimented solids and r_1 is the radius of the liquid surface. For a tubular-bowl centrifuge of length b, a similarly complicated precise expression for Σ simplifies with ± 4 percent accuracy to the approximation

$$\Sigma = \pi b\omega^2(3r_2^2 + r_1^2)/2g \qquad (19\text{-}59)$$

where r_2 is the radius of the inner surface of the bowl. For the disk centrifuge of Fig. 19-120,

$$\Sigma = \frac{2\pi(N-1)(r_b^3 - r_a^3)\omega^2}{3g \tan\theta} \qquad (19\text{-}60)$$

where N is the number of disks in the stack, r_a and r_b are the inner and outer radii of the disk stack, and θ is the conical half angle.

Typical Σ values for three types of sedimentation centrifuges are given in Table 19-15. In scaling up from laboratory tests, sedimentation performance should be the same if the value of Q/Σ is the same for the two machines.

This is a dependable criterion for the comparison of centrifuges of similar geometry and liquid-flow patterns developing approximately the same centrifugal acceleration. It should be used only with extreme caution when comparing centrifuges of different configurations or when the acceleration developed differs by a factor of more than 2. The reason for this is that the Σ concept assumes (1) an idealized liquid-flow pattern and (2) sedimentation of the particles that is unhindered and at the Stokes settling velocity.

A more accurate comparison is obtained by multiplying Σ by an efficiency factor that is characteristic of the type of centrifuge. The factor is almost 100 percent for the bottle centrifuge, about 80 percent for tubular-bowl centrifuges at low to moderate rates, and less than 55 percent for disk centrifuges. It varies widely in the helical-conveyor centrifuge, depending on the ease with which the solids compact to "conveyable" consistency. The performance of any centrifuge may deviate from the theoretical because of such factors as particle-size distribution, particle shape, deagglomeration of flocs in the feed system, reagglomeration of flocs within the centrifuge, maldistribution of flow in the centrifuge, and hindered-settling effects due to the changing concentration of the settling solids. Adjustments of scale-up indices for these factors can be achieved only through experience or by testing the particular application on a centrifuge of the selected configuration.

TABLE 19-15 Scale-Up Factors for Sedimentation Centrifuges

Type of centrifuge	Inside diameter, in	Disk diameter, in/no. of disks	Speed, r/min	Σ value, units of 10^4 ft^2	Recommended scale-up factors*
Tubular	1.75	....	23,000	0.32	1†
Tubular	4.125	....	15,000	2.7	21
Tubular	4.90	....	15,000	4.2	33
Disk	....	4.1/33	10,000	1.1	1
Disk	....	9.5/107	6,500	21.5	15
Disk	....	12.4/98	6,250	42.5	30
Disk	....	13.7/132	4,650	39.3	25
Disk	....	19.5/144	4,240	105	73
Helical conveyor	6	....	6,000	0.27	1
Helical conveyor	14	....	4,000	1.34	5
Helical conveyor	14‡	....	4,000	3.0	10
Helical conveyor	20	....	3,350	4.0	13.3
Helical conveyor	25	....	3,000	6.1	22
Helical conveyor	25‡	....	2,700	8.6	31

*These scale-up factors are relative capacities of centrifuges of the same type but different sizes when performing at the same level of separation achievement (e.g., same degree of clarification). These factors must not be used to compare the capacities of different types of centrifuges.

†Approaches 2.5 at rates below 100 mL/min.

‡Long-bowl configuration.

NOTE: To convert inches to millimeters, multiply by 25.4; to convert revolutions per minute to radians per second, multiply by 0.105; and to convert 10^4 square feet to square meters, multiply by 929.

CENTRIFUGAL FILTERS

Centrifuges which filter—that is, cause the liquid to flow through a bed of solids held on a screen—are commonly called **centrifugals** or **centrifugal filters.** They are sometimes known by other names: "wringers," "extractors," or "dryers." They include both batch machines (often designed to fit into continuous processes) and continuous machines. In principle they are all the same: in each type a cake of granular solids is deposited on a filter medium held in a rotating basket, washed, and spun "dry." They differ in whether the feed is batch, intermittent, or continuous and in the way in which the solids are removed from the basket.

Variable-Speed Basket Centrifuges Variable-speed basket centrifuges rotate on a vertical axis. The basket can be driven from below or be suspended from the drive shaft. The basket shell is usually perforated and cylindrical in shape. It is connected to the drive shaft through a hub and topped with a lip ring. The hub may be solid (as in Fig. 19-122) or open. If it is solid, the basket can be fed while rotating at any appropriate speed from zero upward and must be brought to rest for the centrifuged solids to be unloaded at the top. If the hub is open, the basket must be rotated during loading at sufficient speed to keep the feed from running out the bottom. Unloading is through the bottom either manually while the basket is at rest or by an unloader knife while the basket is rotating at low speed (less than 100 r/min). The filter medium can be wire mesh or cloth on a backing.

Solid-Bottom Batch Basket Centrifuges. These centrifuges are used for small-scale operations, when it is desired to preserve the integrity of a basket-size batch of solids, when the dewatered solids cannot tolerate mechanical handling, or when the traces of solids remaining in a more automated centrifuge would be subject to decomposition or spoilage.

Base-Bearing Centrifuges In the smaller sizes, the casing and base are connected to each other and to ground with no flexibility (Fig. 19-122). The spindle is supported from below on a thrust bearing often held in a ball joint that provides a pivot point. The spindle is centered by radial springs or rubber in compression. This provides damped freedom of motion of the axis of rotation, to compensate partially for the out-of-balance of a basket load.

The drive motor is vertical, and its axis of rotation is fixed with respect to the base. The centrifuge shaft is pivoted on its thrust bearing or ball joint. The drive and driven sheaves are aligned with this pivot point to minimize variation in the dimension between their centerlines, which would result in excessive belt wear as drive-shaft perturbations occur from normal unbalanced loading.

This construction is used extensively in "chip wringers" that recover excess oil from metal chips and turnings. The basket wall is usually solid and tapered outward toward the top. A loose-fitting lid confines the chips but provides adequate opening for the relatively small amount of oil to discharge freely. The bowl itself or a formed liner may be used as a tote box to transport the oily chips from the machine tool to the centrifuge. A typical chip-wringer basket is 660 mm (26 in) in diameter at the top and 584 mm (23 in) at the bottom, holds up to 0.15 m^3 (5 ft^3) or about 225 kg (500 lb) of crushed steel chips, and is driven at 1025 r/min by a 7.5-kW (10-hp) motor.

Link-Suspended Basket Centrifuges In 762-mm- (30-in-) diameter and larger sizes, the basket, curb, curb cover, and drive form a rigid assembly flexibly suspended from three fixed posts. In one variation the curb and curb cover are fixed to the posts and only the basket and drive assembly are suspended. The three suspension members may be either chain links or stiff rods in ball-and-socket joints, spring-loaded. The suspended assembly has restrained freedom to oscillate to compensate for normal out-of-balance, only a fraction of the out-of-balance forces being conveyed to the supporting structure. The motor shaft is vertical and the centerline distance

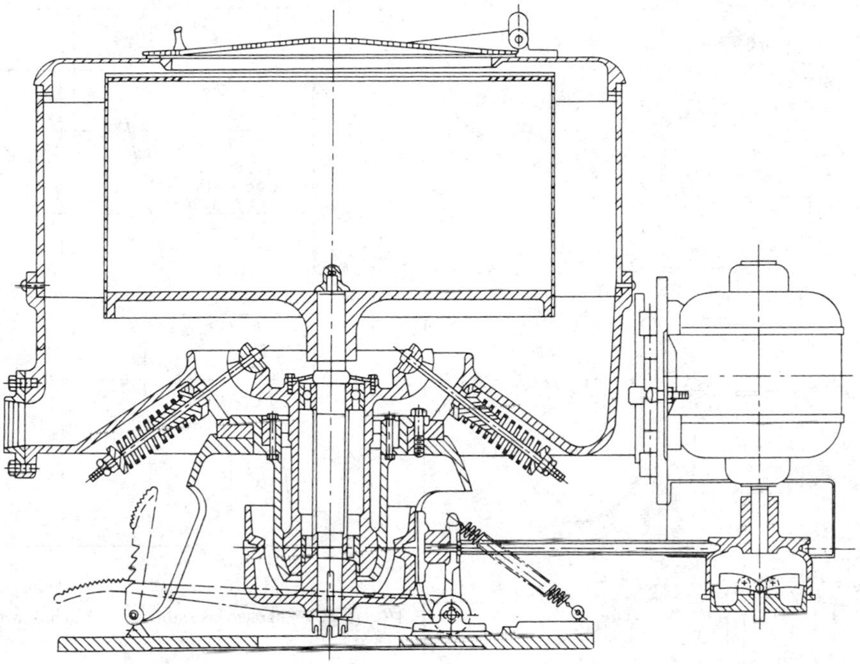

FIG. 19-122 Base-bearing, solid-bottom centrifugal filter.

between it and the drive shaft is fixed, providing a more efficient power train than that of the base-bearing type.

These centrifuges are usually loaded at zero speed. The cover lid is raised to expose an opening in the curb cover at least as large as the inside diameter of the basket lip ring. A safety interlock prevents raising the lid while the basket is rotating or starting rotation while the lid is raised. The load may be prepackaged for ease of handling, or it simply may be dumped in the basket. In either case the solids should be evenly distributed to minimize out-of-balance. The cover is closed, the basket is brought to full speed and maintained there until the free liquid has drained off through an opening in the bottom of the curb, the basket is brought to rest, and the lid is opened for unloading. Unloading of bulk solids may be facilitated if the filter medium is in the form of a bag contoured to fit the inside of the basket.

Link-suspended centrifuges with solid bottoms are available in sizes from 305- to 2743-mm (12- to 108-in) inside basket diameter.

By confining the perforations to a small area near the top of the shell, the solid-bottom basket can also be used for the successive extraction of a single batch of solids. The solids are not unloaded until the final extraction stage has been completed.

Open-Bottom Basket Centrifuges These centrifuges are made in top-suspended and link-suspended configurations. In both, the bottom hub consists of three functional components contained in a single casting or fabrication: (1) the central nave by which the basket is attached to the drive shaft, (2) an outer ring to which the cylindrical shell is attached and whose inside diameter is less than that of the lip ring, and (3) spokes or ribs connecting the nave to the outer ring. The space between the spokes is available for the gravity discharge of the centrifuged solids. The steps of the operating cycle are as follows: (1) accelerate to loading speed, (2) rinse screen, (3) load, (4) accelerate to purging speed, (5) wash cake (once or several times), (6) spin to dryness, (7) decelerate to unloading speed, and (8) unload. In many cases the screen rinse and cake wash may be superimposed on the preceding acceleration period. The cycle may be under manual, semiautomatic, or fully automated control. The drive may be a variable-speed electric motor, either direct or through V belts; a high-pressure, fixed-volume hydraulic motor receiving its energy from a constant-speed variable-volume pump; or, infrequently in modern practice, a steam or water turbine.

The unloading may be accomplished manually with a hand-driven plow or knife mounted on the curb cover or automatically with a double- or single-acting knife with hydraulic or pneumatic piston actuators. Safety interlocks prevent unloader-knife actuation except when the basket has decelerated to a safe unloading speed. In some designs the low unloading speed is controlled by an auxiliary gear-head motor. In at least one design, the direction of bowl rotation is reversed during unloading so that the plow that is trailing during normal rotation cannot be forced into the cake except at design unloading speed and direction. In some designs the opening through the basket bottom is covered with a plate that is lifted during unloading.

Top-Suspended Centrifuge This centrifuge (Fig. 19-123) is widely used for purging molasses from crystallized sugar as well as for many other applications. Conventionally, it consists of two A frames connected by a crossbar at the top. The crossbar supports the drive motor and the drive head. The drive head, which is connected to the motor or a driven pulley through a flexible coupling, carries the thrust and radial bearings that support the shaft and its load. The shaft has a controlled degree of flexibility built into it. The bowl shell is cylindrical with a lip ring at the top. The casing that receives the filtrate and channels it to a suitable outlet is fixed to the foundation between the A-frame uprights.

The entire weight of the motor is carried on the frame with no component other than its rotary motion reacting on the centrifuge proper. This permits the use of very large special motors for rapid cycling. On white-sugar service, up to 24 cycles/h with 364 kg (800 lb) of sugar per load are provided by wound-rotor variable-speed ac motors with a torque rating equivalent to over 112 kW (150 hp).

Typical pilot-plant top-suspended baskets are 305 mm (12 in) in diameter by 127 mm (5 in) deep; commercial machines are available in sizes from 508-mm (20-in) diameter by 305-mm (12-in) depth to 1524-mm (60-in) diameter by 1016-mm (40-in) depth and develop up to 1800 g in the smaller and intermediate sizes. Except for the

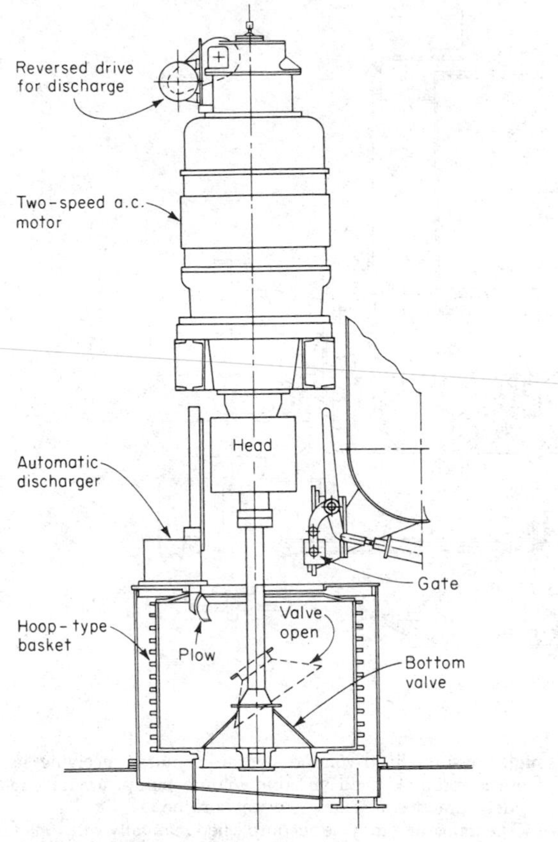

FIG. 19-123 Top-suspended centrifugal filter. *(Western States Machine Co.)*

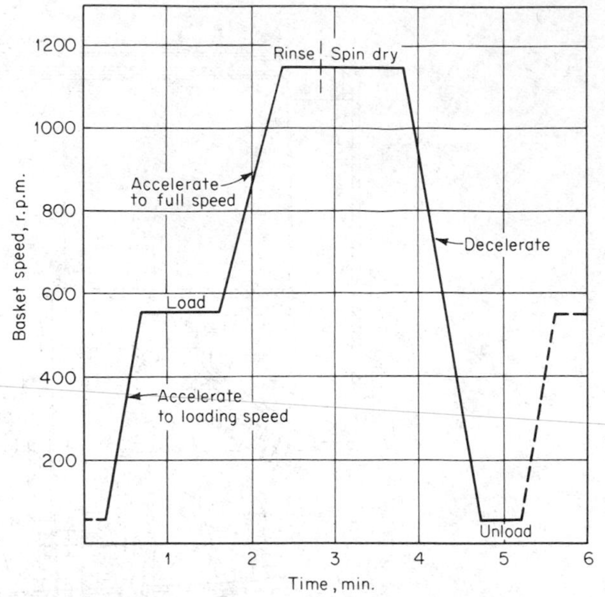

FIG. 19-124 Typical operating cycle, batch centrifugal filter.

sugar application, operation with a two-speed motor (half speed for loading and full speed for purging) is conventional. Hydraulic drives with infinite speed variation are commonly used in the chemical industry.

To maximize the number of cycles per hour, a combination of electrical and mechanical braking is employed during the deceleration period. This period has no processing value but is necessary for the transition from spinning speed to unloading speed.

Link-Suspended Bottom-Discharge Centrifuge With the top-suspended design, the only limit to the size and weight of the drive motor is the strength of the supporting structure, and this can be made as strong as is necessary. In the link-suspended bottom-discharge centrifuge, the weight of the side-mounted motor constitutes an overturning moment, and this weight is limited to a proportion of remaining suspended components. This limit corresponds to an especially lightweight 45-kW (60-hp) electric motor on a heavy-duty 1219-mm- (48-in-) diameter, 762-mm- (30-in-) depth basket. Conventionally, link-suspended centrifuge drive motors are of the two-speed type, half speed for loading and full speed for purging or drying. Greater torque transmission and more flexibility in speed control are possible, however, with hydraulic drives.

Operating Cycle Figure 19-124 illustrates a typical operating cycle for a variable-speed automatic basket centrifuge. The slope of the acceleration portion of the curve is a function of the moment of inertia of the basket and its contents and of the driving torque available from the prime mover. The slope of the deceleration portion is a function of the rate of energy absorption from the rotating assembly by the prime mover and by a braking system. The unloading time depends on the torque available for turning the basket at unloading speed, the characteristics of the cake to be unloaded, and the design of the unloader knife. The remainder of the cycle depends on the processing characteristics of the system to be centrifuged, slurry concentration and mother-liquor drain rate during loading, wash liquor required and its drain rate, and spin time necessary to reach required terminal dryness.

The operating cycle may be under manual control or fully automated through a sequence of automatic reset timers, speed sensors, and limit switches.

Constant-Speed Basket Centrifuges Some batch centrifugal filters operate at a constant bowl speed during the entire sequence of loading, deliquoring, and unloading so that no process time is lost in acceleration and deceleration. Centrifugals of this type are almost invariably operated completely automatically on a preprogrammed cycle, and are known as **batch-automatic centrifugal filters** or "peeler" centrifuges. The operating cycle is screen rinse, loading, wet spin (if required), cake wash, dry spin, and unloading.

Constant-speed batch-automatic centrifuges practically always operate on a horizontal axis of rotation with the drive shaft supported by fixed bearings. The basket may be cantilevered at one end of the drive shaft with the driven pulley at the other end. In another design the drive shaft extends through the basket and is also supported by an outboard bearing. The through-shaft design may carry two baskets with a common hub to permit feeding one while the other is spinning to smooth out the power demand. In still another variation, the basket hub is large enough (like that of the link-suspended centrifugal filter) to position the inboard bearing near the center of gravity of the loaded basket.

A cake distributor may be provided to level the load during feeding, to indicate cake depth, and to activate the feed-valve closure when the desired depth of cake has been reached. After the load has been spun dry, the cake is peeled out by rotating an unloader knife into it to discharge the solids down a chute extending through the opening in the lip ring at the front of the basket. In other designs, the unloader knife is smaller and double-acting, and in very large sizes, the discharge of the cake through the front of the basket may be facilitated with a horizontal screw conveyor. Since the unloader knife cannot be allowed to contact the filter medium, a heel of product remains in the basket after each unloading. This serves as a precoat to prevent loss of fines through the screen during the next cycle. It may become glazed and impervious from the rubbing action of the knife, and a screen rinse is frequently required to restore its permeability.

Although such centrifuges are used to dewater insoluble solids like coal and starch, their widest application is for dewatering and washing crystals having medium to fast drain rates, typically 50 to 200 mesh (−297 to 74 μm) in size. The prime mover, usually an electric motor, must be large enough to bring the empty basket to operating speed and also to accelerate the feed slurry. Usually, for optimum performance the feed rate should be matched to the drain rate so that a minimum of free mother liquor is left on the surface of the cake when the feed valve closes.

Continuous Filtering Centrifuges The trend toward continuous processing and reduction in operating labor has resulted in an increasing requirement for continuous processing equipment, including centrifuges. There are several types of truly continuous filtering centrifuges, the principal distinction among which is the method of solids movement. As in the case of other unit-operations equipment, the continuous filtering centrifuge is less flexible than its batch counterpart, and it must be specified more precisely for its particular process application.

Conical-Screen Centrifugal Filters When a conical screen is rotated about its axis, centrifugal force will impel a liquid through the openings in the screen and also will provide a component to encourage the advance of retained solids from the small to the large diameter. The sliding of the solids on the cone is favored by smooth perforated plates or wedge-wire sections with the slots parallel to the axis of rotation, rather than woven wire mesh.

Wide-Angle Centrifugal Filters If the half angle of the cone screen is greater than the angle of repose of the solids, the solids will slide across it with a velocity that is independent of feed rate. If the angle is too great, this velocity will be high and the retention time under centrifugal force will be too short for effective filtration. The angle selected is therefore highly critical with respect to performance on a specific application. Wide-angle and compound-angle centrifuges are used to dewater coal and rubber crumb and to dewater and wash crude sugar and vegetable fibers such as from corn and potatoes.

Shallow-Angle Centrifugal Filters By selecting a half angle for the cone screen that is less than the angle of repose of the solid fraction and providing supplementary means for the controlled advancement of the solids from the small diameter to the large diameter, longer retention time for drainage of the liquid fraction is made possible. Three methods are in common use for advancing the solids:

1. *Vibrational advance.* A relatively high-frequency force is superimposed on the rotating assembly. This can be either in line with the axis of rotation or torsional, around the drive shaft. In either case the solids are at least partly fluidized and shaken away from the screen. Even though the applied vibrational force is symmetrical, centrifugal force combined with the screen angle provides a steady advance of the solids being dewatered toward the large end of the cone from which they discharge.

2. *Oscillating or "tumbling" advance.* The drive shaft is supported at its lower end on a pivot point. A supplementary power source causes the shaft and the rotating bracket it carries to oscillate or gyrate about the pivot at a controlled amplitude and at a frequency lower than the rate of rotation of the basket. This also provides partial fluidization of the bed of solids in the basket, causing their controlled advance across the screen toward its large diameter, as in the vibrational-advance type.

3. *Metered advance.* The movement of the solids from the small diameter to the large diameter of the filter member is controlled by a conveyor, usually in the form of a modified helix or screw whose pitch decreases in the direction of the solids motion from the small end to the large end of the screen.

The several types of conical-screen centrifugal filters are constructed with both vertical and horizontal axis of rotation for various applications and installation requirements. Their energy requirement is low; for example, as little as 792 J/kg (0.2 kWh/ton) to dewater −19 + 6 mm (−¾ + ¼ in) stoker coal at the rate of 38 kg/s (150 tons/h) to 6 percent surface moisture on a large oscillating-advance type. Their liquid-handling capacity is limited, and 50 percent feed-slurry concentrations are advisable to obtain best performance and capacity. For this reason, the amount of cake wash that can be applied is restricted and rinsing efficiency is relatively low. As with any centrifugal filter, performance is also optimized with operation on large and uniformly sized particles. Screen thickness, and hence screen life, is a function of the size of the openings that will support solids of the size being centrifuged.

The metered-advance type is somewhat more flexible than the others in terms of ability to handle lower-concentration feed slurries and provide fair to good rinsing efficiency. Both the wide-angle and the metered-advance types are used for the dewatering of cellulosic fibers.

"Pusher" (Reciprocating) Centrifugal Filters

Single-Stage In conventional form these consist of a rotating perforated cylindrical rotor lined with wedge-wire screen with the slots parallel to the axis of rotation. The cylinder is open with no lip ring at the solids-discharge end and cantilevered at the other end through its hub to a hollow drive shaft (Fig. 19-125). Inside the basket, and fitting closely to the cylindrical screen, is a circular pusher plate. This is mounted on its own shaft inside the drive shaft. It rotates at the same speed as the cylindrical rotor and also receives a reciprocating motion from a concentric piston that is usually powered hydraulically. The feed enters at the centerline and is distributed and accelerated in the rotating feed cone. As the pusher plate retracts, a clean surface of screen is exposed for bulk drainage of the incoming feed slurry. As the pusher plate advances, the incremental annulus of cake thus formed transmits its pressure to the annulus of cake already in the basket, causing an equivalent amount to dis-

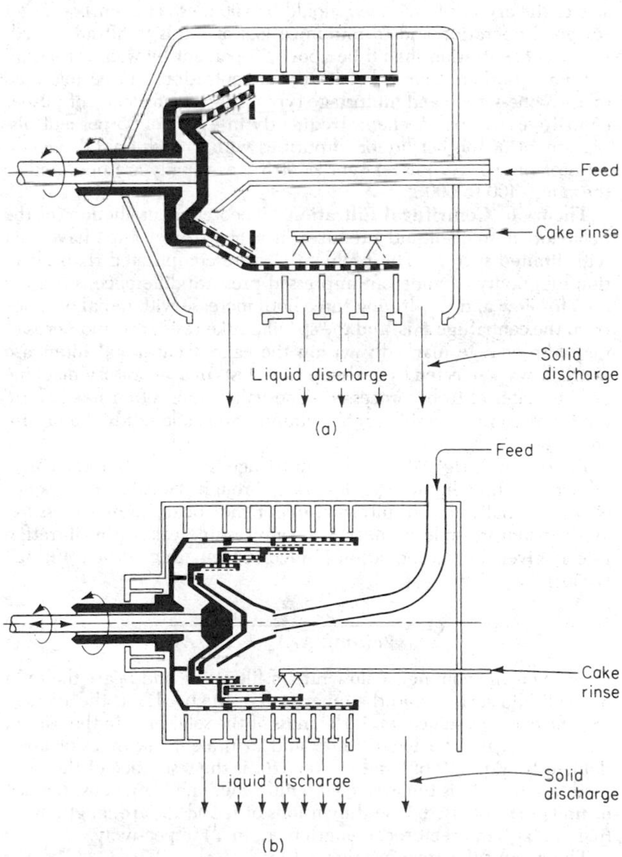

FIG. 19-125 Pusher centrifuges (rotor and casing). (*a*) Single-stage with conical pusher screen. (*b*) Multistage. (*From* Kirk-Othmer Encyclopedia of Chemical Technology, *2d ed., vol. 4, Wiley, New York, 1964, p. 230; also 3d ed., vol. 5, 1979, p. 751; by permission.*)

charge off the rotor at the open end. Cake wash is applied as a spray through which the cake advances stepwise. Pusher amplitude and frequency are usually controllable externally to optimize performance, with pusher amplitude up to 0.1 times basket diameter and pusher frequency in cycles per second up to about 20/in basket diameter [8 cycles/(cm·s)].

The wedge-bar screen construction minimizes friction between the screen and the advancing cake to permit the use of a relatively long screen, and correspondingly long retention time for the cake, without buckling.

In one modification, the pusher plate consists of a conical screen with an angle slightly greater than the angle of repose of the solids. This accelerates the feed slurry and provides extra area for bulk drainage, permitting operation over a wider range of feed concentrations.

Multistage In a multistage variation, the basket consists of a series of concentric cylindrical stages that increase in diameter in the direction of solids discharge (Fig. 19-125*b*). The first (smallest-diameter) stage and each alternate stage are fixed to the inner shaft and rotate and reciprocate. The final stage and alternate smaller stages are fixed to the outer shaft and rotate but do not reciprocate. Each screen section is relatively short with its own pusher action from the preceding stage so that there is less tendency for cake buckling with even a relatively long bowl. Bulk drainage takes place in the smaller-diameter section with minimum power consumption, and film drainage takes place in the larger-diameter sections where maximum centrifugal force is developed.

Uses Pusher-type centrifuges are used for dewatering and washing crystals and other particulate solids, including short fibers. The size of the crystals in the feed should be 100 mesh (149 μm) or larger for good operation and to minimize loss of solids to filtrate. Feed-slurry concentration should be above 35 percent by weight for conventional pushers. Somewhat lower concentrations can be tolerated on the cone-screen and multistage types. Rinsing efficiency of pusher centrifuges can be excellent, frequently in excess of 95 percent displacement of mother-liquor impurities with a wash-crystals weight ratio of only 1:10. Pusher-type centrifuges usually are operated in the range 400 to 600 *g*.

Theory of Centrifugal Filtration Theoretical predictions of the behavior of solids-liquid mixtures in a centrifugal filter have met with limited success. The problem is more complicated than filtration by gravity or under an impressed pressure difference, since the area for flow and the driving force both increase with radial distance from the centrifuge axis, and the specific cake resistance and porosity may also change markedly within the cake. Centrifugal filters are nearly always selected by scale-up from tests in a laboratory machine on the material to be processed. The values most often needed are the filtration rate, washing rate, spinning time, and residual moisture content.

Filtration Rate When the centrifuge cake is submerged in a known depth of liquid, the flow rate through the cake corresponds closely to that found in filtration under the corresponding pressure. With incompressible or nearly incompressible cakes, the filtration rate is given by the equation [Grace, *Chem. Eng. Prog.*, **49**, 427 (1953)]

$$Q = \frac{\rho\omega^2(r_2^2 - r_1^2)}{2\mu(\alpha m_c/\overline{A}_L\overline{A}_a + R_m/A_2)} \qquad (19\text{-}61)$$

where Q is the volumetric flow rate of filtrate, r_1 and r_2 are the radii of the liquid surface and the inner wall of the bowl, α is the average specific cake resistance, m_c is the mass of the solid cake in the basket, $\overline{A}_L$ and $\overline{A}_a$ are the logarithmic and arithmetic mean cake areas defined by Eqs. (19-62) and (19-63), R_m is the resistance of the filter medium, and A_2 is the area of the filter medium. Any consistent set of units may be used. The dimensions of α and R_m are length/mass (e.g., m/kg) and reciprocal length (e.g., m^{-1}) respectively.

The mean cake areas are defined by

$$\overline{A}_L = \frac{2\pi b(r_2 - r_3)}{\ln r_2/r_3} \qquad (19\text{-}62)$$

$$\overline{A}_a = (r_2 + r_3)\pi b \qquad (19\text{-}63)$$

where b is the height of the basket and r_3 is the radius of the inside surface of the cake.

Even with noncompressible cakes, Eq. (19-61) applies only when the cake is of uniform thickness. With compressible cakes or deformable particles, the relative filtration rate in a centrifuge is lower than in a filter; the specific cake resistance in the centrifuge may be larger than that predicted from pressure filtration tests by as much as an order of magnitude. For a given application the variation in specific resistance with cake thickness must be determined in order to scale up accurately centrifuge performance.

Film Drainage and Residual Moisture Content The prediction of the relationship between spinning time and residual moisture content of centrifuge cakes is even more complex and subject to error than the prediction of bulk filtration rate. Estimation of porosity from filter data is uncertain, since the compressive force in the centrifuge increases with radius through the cake depth and since the specific weight of an individual particle, which is proportional to $\rho_s - \rho$ at the start of filtration, increases to approach ρ_s during film drainage.

As a first approximation, the following formulas have some utility:

$$q \propto \frac{1}{d^{0.5}G^{0.5}\rho^{0.25}} \qquad (19\text{-}64)$$

where q is the ratio of the volume of liquid/volume of solids after spinning for an infinite time under conditions in which air drying does not occur, d is the equivalent diameter of the solid particle, and G is the relative centrifugal force; and

$$q_t - q \propto \frac{\nu^{0.5}(r_2 - r_3)^{0.5}}{dG^{0.5}t^{0.3}} \qquad (19\text{-}65)$$

where q_t is volume of liquid/volume of solid ratio at time t and ν is kinematic viscosity of the mother liquor. Experimentally it has been found that the exponent of $(r_2 - r_3)$ approaches 0.8 with compressible cakes and the exponent of t ranges from 0.3 to 0.5 with different types of solids.

In centrifugal fields of the order of 1000 *g* coarse solids can be spun substantially to condition q in a few seconds, while finely divided solids may require an hour or longer. The residual moisture content in a centrifuged cake is considerably lower than that in a filter cake. Values between one-third and two-thirds that of blown filter cake are normal.

For predicting commercial centrifuge performance with accuracy, it is best to obtain data on a centrifuge of similar geometry and large enough size to evaluate Q, q_t, and q over a reasonable range of G and t for the system to be centrifuged.

COSTS

Neither the investment cost (including installation) nor the operating cost of a centrifuge can be accurately correlated with any single characteristic of a given type. These costs also depend on the physical and chemical nature of the materials being separated, the environment in which the centrifuge is located, the auxiliary equipment needed to complete the installation, and many other factors, including the difficulty of making the separation to the required degree. The cost figures presented in this section are only representative of centrifuges for use in the chemical-process industries as of 1981. In any particular installations, the costs may be somewhat less or much greater than those presented here.

The useful parameter for value analysis is the installed cost of the number of centrifuges required to produce the demanded separative effect at the specified capacity of the plant. The possible benefits of adjustments in the upstream and downstream components of the plant and the process should be carefully examined in order to minimize total overall plant costs; the systems approach should be used.

Purchase Price Typical purchase prices, including drive motors, of tubular and disk sedimentation centrifuges are given in Table 19-16. The purchase price will vary upward with the use of more exotic materials of construction, the need for explosion-proof electrical gear, the type of enclosure required for vapor containment, and the

TABLE 19-16 Costs of High-Speed Centrifuges (1981 Base)

Type	Bowl diameter, in (mm)	Approximate Σ value, units of 10^4 ft^2 (10^3 m^2)	Designation	Purchase price, $	Remarks
Tubular	4 (102)	2.7 (2.5)	Oil purifier	5,000–12,000	Steel
	4 (102)	2.7 (2.5)	Chemical separation	13,800–22,000	Stainless steel
	5 (127)	4.2 (3.9)	Blood fractionation	22,000	Stainless steel
Disk	13.5 (343)	21 (20)	Hermetic	26,500–31,000	Stainless steel
	24 (610)	95 (88)	Centripetal pump	62,000–77,000	Stainless steel
Nozzle-discharge	12 (305)	12 (11)	Clarifier	24,500–28,700	Stainless steel
	18 (457)	25 (23)	Separator	44,000–48,500	Stainless steel
	30 (762)	100 (93)	Recycle clarifier	88,400–97,000	Stainless steel
Self-opening	14 (356)	13 (12)	Centripetal pump	28,700–33,200	Stainless steel
	18 (457)	22 (20)	Centripetal pump	46,500–51,000	Stainless steel
	24 (610)	38 (35)	Centripetal pump	71,000–77,500	Stainless steel

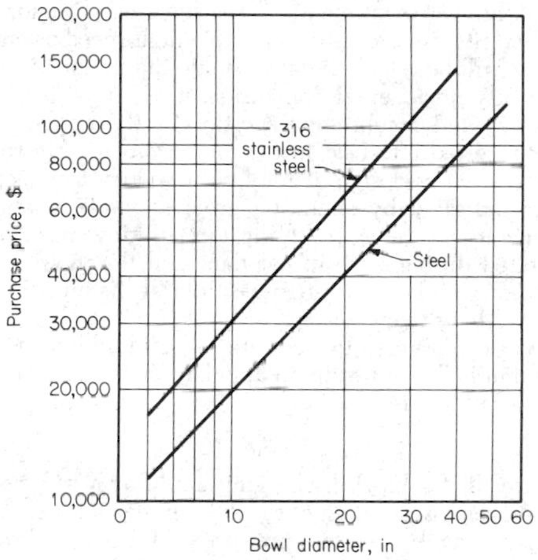

FIG. 19-126 Costs of helical-conveyor centrifuges.

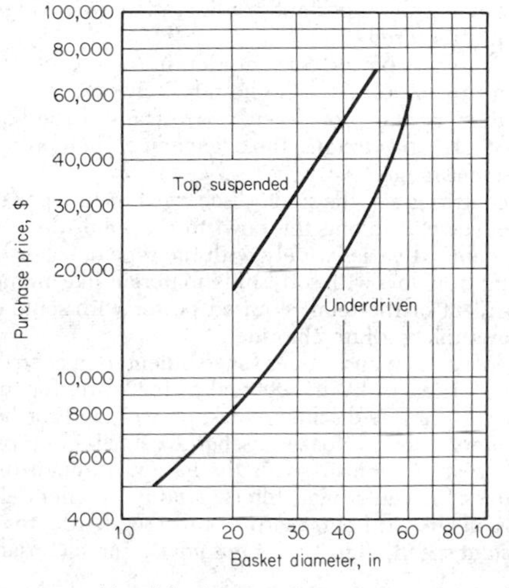

FIG. 19-127 Costs of batch basket centrifuges (316 stainless steel).

degree of portability, and this is true of all types of centrifuges. The Σ values are included to give an approximation of the relative amount of useful separative work available from each type. In comparing different types of centrifuges, these Σ values must be corrected for the efficiency of each type under its operating conditions, as was discussed under "Theory of Centrifugal Sedimentation."

The average purchase prices of scroll-conveyor sedimentation centrifuges with minimum add-ons are shown in Fig. 19-126; of top-suspended filtering centrifuges and bottom-driven basket centrifuges in Fig. 19-127; of batch automatic centrifuges in Fig. 19-128; and of pusher and oscillating filtering centrifuges in Fig. 19-129. These prices do not include drive trains and motors, which may add 20 to 30 percent to the stated price—less in the case of constant-speed centrifuges and more in the case of high-performance automatic vertical variable-speed centrifuges.

Installation Costs Installation costs of centrifuges will vary over an extremely wide range depending on the type of centrifuge, on the area and kind of structure in which it is to be installed, and on exactly

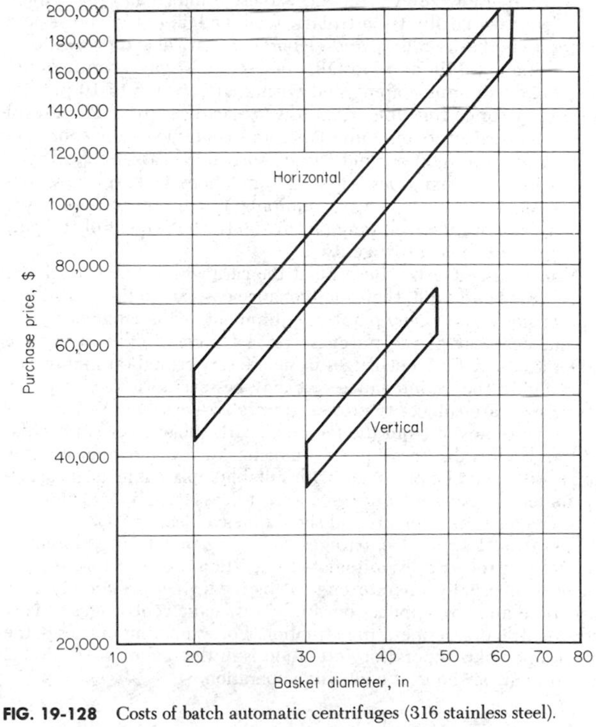

FIG. 19-128 Costs of batch automatic centrifuges (316 stainless steel).

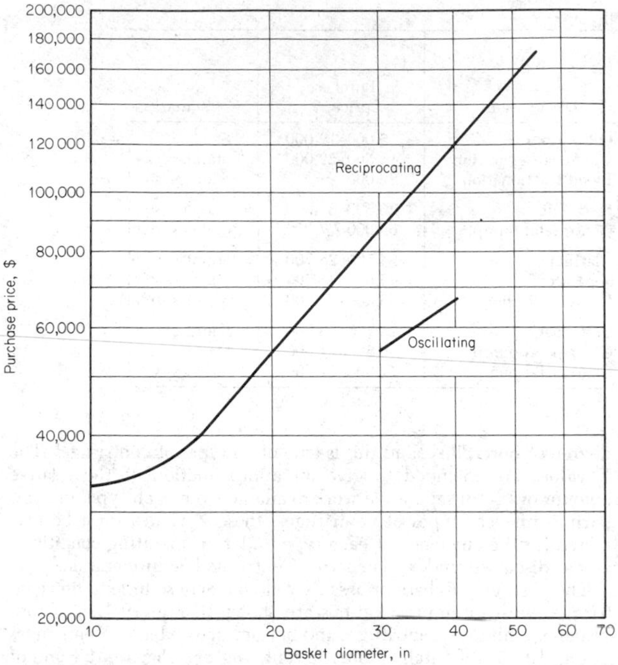

FIG. 19-129 Costs of reciprocating (pusher) and oscillating centrifuges (316 stainless steel).

what is meant by "installation costs." Some centrifuges, such as portable tubular and disk oil purifiers, are shipped as package units and require no foundation and a minimum of connecting piping and electrical wiring. Others, such as large batch automatic and continuous conveyor-type centrifuges, may require substantial foundations and even building reinforcement, extensive interconnecting piping with required flexibility, auxiliary feed and discharge tanks and pumps and other facilities, and elaborate electrical and process-control equipment. Minimum installation costs, covering a simple foundation and minimum piping and wiring, are about 5 to 10 percent of purchase price for tubular and disk centrifuges; 10 to 25 percent for bottom-driven, batch automatic, and continuous conveyor centrifuges; and up to 30 percent for top-suspended basket centrifuges. If the cost of all auxiliaries—special foundations, tanks, pumps, conveyors, electrical and control equipment, etc.—is included, the installation cost may well range from 100 to 400 percent of the purchase price of the centrifuge itself.

Maintenance Costs Because of the care with which centrifuges are designed and built, their maintenance costs are in line with those of other slower-moving separative equipment, in the range of 5 to 10 percent per year of the purchase price for centrifuges in light to moderate duty. For centrifuges in severe service and on highly corrosive fluids, the maintenance cost may be several times this value. Centrifuges are subject to erosion from abrasive solids. When these cannot be removed from the feed, the parts subject to wear such as feed and solids-discharge ports, unloader knives, and helical-screw conveyors should be protected with suitable replaceable hard-surfacing material. Excessive out-of-balance forces strongly contribute to maintenance requirements and should be avoided.

Operating Labor Centrifuges run the gamut from completely manual control to fully automated operation. For the former, one person can usually operate one to five centrifuges, depending on their type and the application. Fully automatic centrifuges usually require little direct operating attention. For estimating purposes, the direct labor plus supervisory attention is in the range of 0.05 to 0.2 person-hour per hour of centrifuge operation.

Energy and Power Requirements The energy required to bring the rotating centrifuge up to operating speed or to change its speed and the power required to overcome windage and friction and to accelerate the process stream fed to the centrifuge all must be considered in selecting the proper size of motor to drive it and in evaluating its operating costs.

The energy that is required to change the angular velocity of a rotating body from ω_1 to ω_2 is

$$E = 0.5I(\omega_2^2 - \omega_1^2) \tag{19-66}$$

where I is the moment of inertia of the rotating assembly about its axis of rotation. If the centrifuge is being started from rest, $\omega_1 = 0$ and

$$E = 0.5I\omega^2 \tag{19-67}$$

The energy supplied by the motor to bring the assembly to its operating speed is

$$E_m = \int_0^t m\omega_m \, dt \tag{19-68}$$

where t is the time required to bring the assembly to speed and m is the torque transmitted to it at the instantaneous angular velocity ω_m. The relationship between m and ω_m is a function of motor design; if known, it permits the calculation of t. In a high-speed or high-inertia centrifuge, t will be relatively long. While the motor is reaching its rated speed, its efficiency of conversion of electrical to mechanical energy is low, and a substantial portion of the applied electrical energy is converted into heat in the motor. Since t is a function of motor size as well as design, it may be necessary to select a larger motor than indicated by operating process, windage, and friction requirements to get the assembly up to operating speed before the heat generated during start-up has damaged the motor. It is commonly necessary on larger centrifuges to limit the number of allowable starts per hour to one or two.

The power P consumed by accelerating a liquid stream of Q flow rate and ρ density from rest to an angular velocity ω and being discharged at a radius r is

$$P = Q\rho\omega^2r^2 \tag{19-69}$$

This indicates that a bowl discharging liquid at $r = 0$ requires no power. However, viscous drag of the fluid in the passages of the bowl has to be overcome by external pressure, as from a feed pump, and energy-consuming seals between the stationary feed tube and the bowl are necessary to permit transfer of this pressure.

Bearings, belts or gears, and seals all consume power that is converted to heat and must be considered in the selection of motor size.

A large source of power consumption in a centrifuge is windage, the viscous drag of the atmosphere surrounding the rotor. Since this power is proportional to the density of the surrounding atmosphere, windage becomes an important consideration when high-pressure containment is required.

Another source of power consumption in some centrifuges is the use of a centripetal pump on the liquid-discharge side to minimize vapor entrainment and provide energy to transfer the liquid. Even with the best available designs, the efficiency of such pumps is in the 25 to 30 percent range.

Finally, centrifuges of the scroll-discharge type use power to move the sedimented solids along the bowl to the solids-discharge areas. The power demand varies widely with the type of solids being handled, ranging from low with soft, slimy materials like animal protein, up to 50 percent of the total required power with some crystalline solids (for example, sodium chloride).

Typical energy-demand values for sedimentation centrifuges handling dilute slurries, in kWh/1000 gal of feed, are, for tubular and disk, 1 to 10, for nozzle-discharge disk, 2 to 12, and for helical-conveyor decanters, 3 to 15. Nozzle-discharge centrifuges typically consume 15 to 40 kWh/ton of solids discharged through the nozzles. Typical values for centrifugal filters handling moderately concentrated feeds, in kilowatthours per ton of dry solids, are, for automatic batch (constant speed), 3 to 10, for automatic batch (variable speed),

TABLE 19-17 Characteristics of Commercial Centrifuges

Method of separation	Rotor type	Centrifuge type	Manner of liquid discharge	Manner of solids discharge or removal	Centrifuge speed for solids discharge	Capacity*
Sedimentation	Batch	Ultracentrifuge	....	....	....	1 mL
		Laboratory, clinical	Batch	Batch manual	Zero	To 6 L
	Tubular	Supercentrifuge	Continuous†	Batch manual	Zero	To 1200 gal/h
		Multipass clarifier	Continuous†	Batch manual	Zero	To 3000 gal/h
	Disk	Solid wall	Continuous†	Batch manual	Zero	To 30,000 gal/h
		Light-phase skimmer	Continuous	Continuous for light-phase solids	Full	To 1200 gal/h
		Peripheral nozzles	Continuous	Continuous	Full	To 24,000 gal/h
		Peripheral valves	Continuous	Intermittent	Full	To 3000 gal/h
		Peripheral annulus	Continuous	Intermittent	Full	To 12,000 gal/h
	Solid bowl	Constant-speed horizontal	Continuous†	Cyclic	Full (usually)	To 60 ft^3
		Variable-speed vertical	Continuous†	Cyclic	Zero or reduced	To 16 ft^3
		Continuous decanter	Continuous	Continuous screw conveyor	Full	To 18,000 gal/h To 75 tons/h solids
Sedimentation and filtration	....	Screen bowl decanter	Continuous	Continuous	Full	To 16,000 gal/h To 75 tons/h solids
Filtration	Conical screen	Wide-angle screen	Continuous	Continuous	Full	To 40 tons/h solids
		Differential conveyor	Continuous	Continuous	Full	To 40 tons/h solids
		Vibrating, oscillating, and tumbling screens	Continuous	Essentially continuous	Full	To 100 tons/h solids
		Reciprocating pusher	Continuous	Essentially continuous	Full	Limited data
	Cylindrical screen	Reciprocating pusher, single and multistage	Continuous	Essentially continuous	Full	To 30 tons/h solids
		Horizontal	Cyclic	Intermittent, automatic	Full (usually)	To 25 tons/h solids
		Vertical, underdriven	Cyclic	Intermittent, automatic or manual	Zero or reduced	To 6 tons/h solids
		Vertical, suspended	Cyclic	Intermittent, automatic or manual	Zero or reduced	To 10 tons/h solids

*To convert gallons per hour to liters per second, multiply by 0.00105; to convert tons per hour to kilograms per second, multiply by 0.253; and to convert cubic feet to cubic meters, multiply by 0.0283.

†Feed and liquid discharge interrupted while solids are unloaded.

5 to 25, for "pusher" centrifuges, 2 to 7.5, and for vibrating and oscillating conical-screen machines, 0.3 to 1.0. To convert these energy requirements to units of joules per liter, multiply by 952.

SELECTION OF CENTRIFUGES

Table 19-17 summarizes the several types of commercial centrifuges, their manner of liquid and solid discharge, their unloading speed, and their relative maximum (pumping) capacity. When either the liquid or the solid discharge is not continuous, the operation is said to be cyclic. Cyclic or batch centrifuges are often used in continuous processes by providing appropriate upstream and downstream surge capacity.

Sedimentation Centrifuges These centrifuges frequently are selected on the basis of tests on tubular, disk, or helical-conveyor centrifuges of small size. The test centrifuge should be of a configuration similar to that of the commercial centrifuge it is proposed to use. The results in terms of capacity for a given performance (effluent clarity and solids concentration) may be scaled up using the Σ concept of Eq. (19-57) and Eq. (19-57*a*). Bottle centrifuge tests [see Eq. (19-58)] may be used for information on systems containing well-dispersed solids. Such tests are totally unreliable on systems containing a dispersed phase that agglomerates or flocculates during the time of centrifuging.

Centrifugal Filters These filters often can be selected on the basis of batch tests on a laboratory unit, preferably one at least 12 in (305 mm) in diameter. Several methods are described in detail by Moyers [*Chem. Eng.*, **73**(13), 182 (1966)] and Ambler [*Ind. Eng. Chem.*, **53**, 430 (1961)].

Unless operating data on similar material are available from other sources, continuous centrifuges should be selected and sized only after tests on a centrifuge of identical configuration.

It seems needless to state, but is frequently overlooked, that test results are valid only to the extent that the slurry and the test conditions duplicate those that will exist in the operating plant. This may involve testing on a small scale (or even on a large one) with a slipstream from an existing unit, but the dependability of the data is often worth the extra effort involved. Most centrifuge manufacturers provide testing services and demonstration facilities in their own plants and maintain a supply of equipment for field testing in the customer's plant.

EXPRESSION

Definition Expression is the separation of liquid from a two-phase solid-liquid system by compression under conditions that permit the liquid to escape while the solid is retained between the compressing surfaces. Expression is distinguished from filtration in that pressure is applied by movement of the retaining walls instead of by pumping the material into a fixed space.

Purpose Expression has the same purpose as filtration: to separate liquid and solid phases from a mechanical mixture of the two. In filtration, the original mixture is sufficiently fluid to be pumpable; in expression, this is usually not true, and the material may appear entirely solid. Expression is therefore employed to separate systems that are not readily pumpable. It is also used instead of filtration or as a postfiltration operation when a more thorough removal of liquid from the cake is desired.

In some applications, such as the dewatering of paper, expression is competitive with drying. Mechanical removal of water is usually far cheaper than any thermal method; hence expression is almost universally employed for one stage of water removal. In vegetable-oil production, expression has generally yielded to solvent extraction. In the recovery of juice from sugarcane, expression on three-roll mills is combined with solvent extraction using water, in a series of alternating or simultaneous operations. The use of belt presses for the dewatering of sludges and slurries is increasing.

EXPRESSION EQUIPMENT

Hydraulic presses of the batch type, employed with little fundamental change for centuries, are gradually being displaced. Nevertheless, they still find extensive use in small or traditional applications, for which their simplicity, relatively low installed cost, and familiarity make them attractive. The principal batch presses are the box, platen, pot, curb, and cage. For continuous operation, the screw press and various types of roller mills are in common use. A more recent development is the belt press, a machine that can combine the operations of continuous filtration and expression.

Batch Presses

Box Press The material to be expressed is wrapped in canvas cloths and placed in a series of steel boxes fitting between the fixed and movable heads of a vertical hydraulic press. Each bag lies on a perforated mat over a grid of drainage channels and is covered and enclosed by the next higher box. The series of loaded boxes is compressed as a unit under hydraulic pressure.

A 15-box press will handle about 7000 kg (8 tons) of conditioned cottonseed meats in a 24-h period, reducing the oil content from 30 to about 6 percent. The operating cycle is from 20 to 30 min per batch. The press is first closed rapidly under low pressure until flow of oil starts at about 1.4-MPa (200-lbf/in^2) pressure on the cake; then high-pressure fluid is used to close the press slowly to the maximum of 11-MPa (1600-lbf/in^2) pressure on the cake, with 27.6-MPa (4000-lbf/in^2) on the hydraulic fluid. The maximum pressure is continued for a few minutes to permit drainage.

Platen Press The platen press is similar to the box press, but the cloth bags are not enclosed on the sides during pressing. The platens or plates are sometimes cored for heating and usually have gutters to collect the expressed liquid. The whole press may be tilted backward slightly to provide better drainage. This type of press is also built in horizontal form. With steam-heated platens, a cold-pressed oil of superior quality can be obtained first, followed by a further yield of hot-pressed oil of poorer quality.

Pot Press Material to be pressed is enclosed in a cylindrical pot, with filter pads or screens beneath and on top, and is compressed by a ram entering from above. The filter medium is flat and covers only the top and bottom of the material; hence it is not subject to stretching or tearing as in box or platen presses. Because the material is entirely enclosed, it may be more fluid than in other types of press. In practice a series of pots is used in each press, the bottom of each pot serving as the ram for the pot below.

The largest use of pot presses is in the chocolate industry, but they also find application for pressing olives and palm and similar nuts and for separating liquid from slushy materials such as chemical products. In a typical cycle, from 225 to 275 kg (500 to 600 lb) of chocolate liquor, produced by grinding cocoa nibs, is pumped to the press chambers, which are then closed under a pressure of 41.4 MPa (6000 lbf/in^2). Cocoa butter is expressed, leaving a cake of cocoa powder.

A development of the pot press is the Carver combined filter and hydraulic press, used as a conventional filter press until the chambers are full, and then closed under hydraulic pressure to obtain a further yield of liquor and a drier cake. This press is used in making cocoa and cocoa butter, in recovering crystals from other liquors, and for separating chemical precipitates. The feed to the press must be pumpable, but part of the filtrate may be recycled if necessary to obtain this condition.

Curb Press In the curb press, material to be expressed is enclosed in a cylinder of wooden slats or beveled steel bars or even of perforated steel plate. Compression by a ram causes the liquid to escape through the walls of the cylinder and flow to collecting channels at the base. Because no filter cloths are used, this type of press is best suited to the expression of fibrous nonoily materials, and the expressed liquid may contain some solids. Curb presses are used in the production of cider and other fruit and vegetable juices, sometimes with a screw mechanism instead of hydraulic pressure. They have been used for expressing olive oil, fish oils, and other oils that do not require high pressures and for dewatering and recovering grease from garbage before incineration.

Cage Press The cage press is similar to the curb press except that the inside of the cylinder has fine longitudinal grooves leading through the cylinder walls to larger drainage channels. It is suitable for oilier and less fibrous materials than those handled by the curb press. Intermediate drain plates and cloths are sometimes used within the cake. Cage presses are not extensively used in the United States but are employed in Europe for expressing castor beans and copra.

Continuous Presses

Screw Press The continuous screw press, typified by the Anderson Expeller (Fig. 19-130), consists of a rotating screw fitting closely inside a slotted or perforated curb. The curb and screw may be tapered toward the discharge end to increase the pressure on the material. This may also be achieved by varying the pitch on the screw or the diameter of the screw spindle in a uniform cylinder. The discharge end of the curb is partly closed by an adjustable cone or other device to change the size of the opening and thus to vary the pressure on the material. Rotation of the screw moves the material forward; and, as the pressure increases, liquid is expelled and escapes through the openings in the curb. The operation is continuous, and labor and other operating costs are lower than for hydraulic pressing. Screw presses of both horizontal and vertical designs are available.

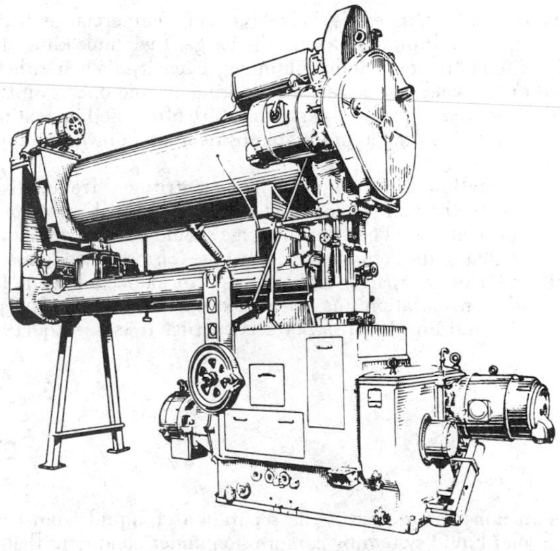

FIG. 19-130 Anderson Expeller press with oilseed-tempering unit. *(V. D. Anderson Co.)*

This type of equipment is extensively used in the vegetable- and animal-oil industries and is applied to the dewatering of such materials as paper pulp, plastics, synthetic rubber, garbage, and paunch manure. During dewatering, washing or dilution water may be injected at one or more points in the cage. Shear action in the mass effectively disintegrates the solid without damaging individual fibers. The capacity of commercial screw presses for oils ranges from 2.7 to 910 metric tons (3 to 1000 short tons) of raw material per 24 h. The residual oil content may vary from 2 to 18 percent, depending on the oilseed pressed and the type of press.

Cost of Screw Presses Continuous screw presses are categorized as high-, medium-, or low-pressure. Their costs vary with the pressure, capacity, and materials of construction. High-pressure, low-capacity units (about 225 to 450 kg, or 500 to 1000 lb, of press cake/h) for the animal- and vegetable-oil industry sell for about $50,000; a unit of double this capacity and operating at somewhat higher pressure (lower residual oil content) costs $60,000 to $100,000. Presses of still higher capacity, up to 5.5 metric tons/h (12,000 lb/h), range upward to $140,000. A two-stage dewatering press that handles up to 5.5 metric tons/h of dry fiber weight costs from $100,000 to $120,000, depending on materials of construction. Rubber- and polymer-drying presses are usually made of stainless steel; they cost from $40,000 for a pilot or semiworks model to $125,000 for a popular commercial size and $220,000 for the high-capacity model. Prices for vertical continuous screw presses are comparable; for vegetable juicing and similar food-industry applications (stainless steel in contact with the product), prices are quoted from $50,000 for a laboratory unit to $750,000 for a 4.5-metric-ton/h (5-ton/h) two-machine dewatering and drying line. The prices given here are as of September 1980.

Roller Mills Continuous roller mills (Fig. 19-131), as used in the cane-sugar industry, combine a mechanical breaking and crushing action with the application of pressure to express juice. Three-roll mills are common, with the top roll above and between the other two, and pressed against them by hydraulic rams at each end. Material is squeezed between the top and first rolls and is then directed by a turnplate into the nip of the top and second rolls for a second pressing. The rolls are made of cast iron, corrugated or grooved in various patterns. A feed roll is sometimes used to force-feed the first pair of rolls, permitting the use of a smaller mill opening or a higher rate of feed.

In the cane-sugar industry, trains of four to seven of these mills are used, with the blanket of crushed cane carried between them by apron conveyors. The cane is first crushed dry, but at selected points

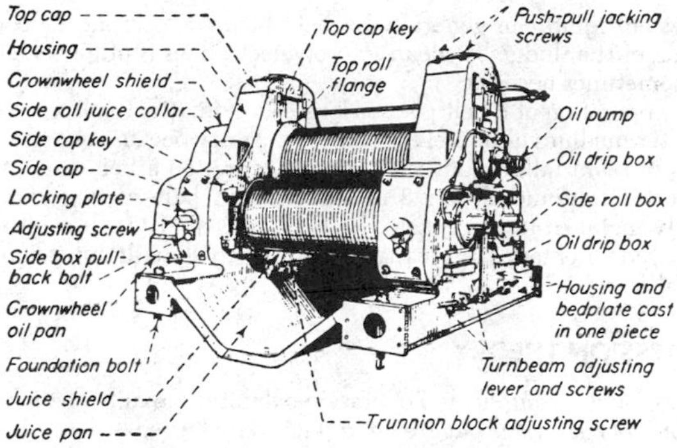

FIG. 19-131 Three-roll sugar mill. *(Farrel Co.)*

in the later mills water or weak liquor is added as a spray or bath to improve sugar recovery. This process is known as maceration and is equivalent to leaching of the cane combined with expression in the mills.

Two-roll mills cause expression of liquid without the crushing and tearing action of three-roll units. Double rolls are widely used for dewatering paper, usually supported on a felt, in papermaking machines. Similar two-roll units, often with padded surfaces instead of steel, squeeze water or process liquids from textiles after kiering, dyeing, bleaching, and related operations. In the process of dyeing, the cloth is thoroughly impregnated with small quantities of dye by successive squeezings with padded rolls.

Belt Presses The belt filter press encloses slurry or sludge between two endless moving belts and expresses the liquid under pressure applied by rollers. Commercial application of belt presses was developed during the 1970s; and they are currently used in the pulp and paper and the mineral industries as well as for the dewatering of many types of municipal and industrial waste sludges.

A dewatering belt filter may include three regions: a gravity drainage or straining zone for the removal of free water; a low-pressure zone, with gradually increasing pressure, to express surface and interstitial water; and a high-pressure zone to continue these actions and perhaps to induce the separation of cellular or bound water (Fig. 19-132). Feed slurries containing 2 to 8 percent solids can be dewatered to 12 to 40 percent solids or even further in the case of mineral materials that contain little cellular water. Loss of solids into the

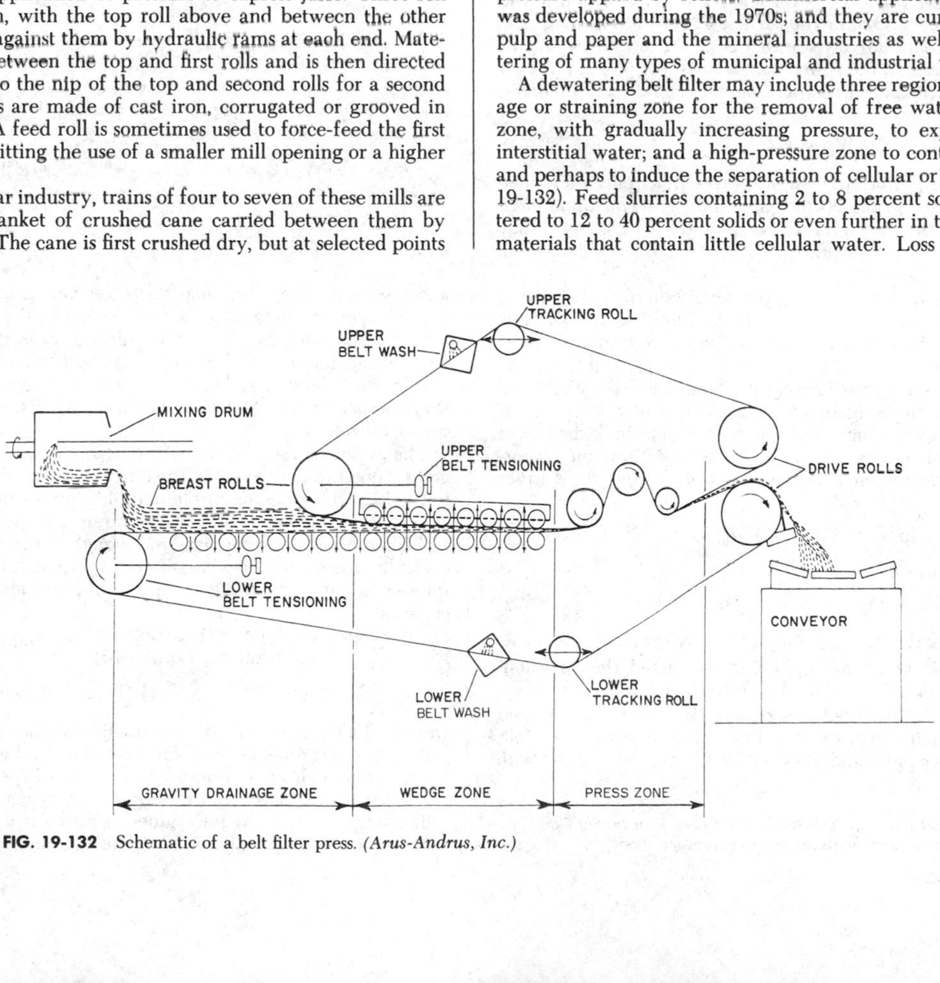

FIG. 19-132 Schematic of a belt filter press. *(Arus-Andrus, Inc.)*

expressed liquid can generally be held below 2 percent. Preconditioning of the sludge by means of polyelectrolytes is often beneficial and sometimes necessary.

The operation of a belt press is largely automatic. Replacement of the filter medium ultimately becomes necessary because of puncture, wear, or seam failure, but with careful operation a belt can remain in service continuously for 3 or 4 months. A belt-washing spray is usually included in the system. Belts up to 3 m (10 ft) in width are employed to obtain large capacity with such slow-draining slurries as coal washings.

EXPRESSION THEORY*

Expression is a complicated operation exhibiting mainly phenomena of the mechanics of aggregated and semiaggregated solids but involving also the mechanics of the liquid being separated from the solids. It is not surprising, then, that the theory of expression is far from complete. It consists largely of semitheoretical hypotheses intended to rationalize plant and laboratory observations. Much of the reported experimental work has been done on specific materials studied for their own sake and has led to empirical equations unsuitable for general application. (The research of Gurnham and Masson and that of Shirato and coworkers, summarized later, are exceptions to this pattern.)

Because empirical relationships formulated by the major investigators of expression have usefully rationalized specific cases and because they offer the only known quantitative insight to the operation, they are reviewed here briefly.

Equilibrium Conditions Knowledge of the equilibrium obtaining in a press—i.e., the shape and composition of the compressed solid-liquid mass after it has been subjected to constant pressure until no further deformation and flow occur—presumably is fundamental to understanding the process of expression and investigation of the rate at which it occurs. One of the earliest attempts to describe the equilibrium state quantitatively was that of Deerr for the pressing of sugarcane and bagasse in the equivalent of small pot presses [*Hawaii Sugar Plant. Assoc. Exp. Sta., Agric. Chem. Ser.*, Bull. 22 (1908); 30 (1910); 38 (1912)]:

$$V_c = C/p^n \tag{19-70}$$

in which V_c is the equilibrium volume of cake (fiber plus unexpelled juice) under pressure p, the coefficient C has whatever dimensions are required by those chosen for V_c and p, and the exponent n may vary with p. Entirely empirical, Eq. (19-70) fits Deerr's data quite well.

In the first definitive study of expression equilibrium, the classic experiments of Gurnham and Masson [*Ind. Eng. Chem.*, **38**, 1309 (1946)] were carried out in a test cylinder [equivalent to a pot press 6.5 cm² (1 in²) in area] and in a cage press 64.5 cm² (10 in²) in area, at pressures between approximately 1.7 and 136 MPa (250 and 20,000 lbf/in²). The investigators reasoned that as compacting pressure on a fibrous mass is increased, the equilibrium bulk density of the solids portion of the mass should increase by an amount proportional to the *fractional* increase in pressure causing the further compaction:

$$dp/p = K d\rho_c' = K' d(1/V_c) \tag{19-71}$$

or, in integrated form,

$$\log p = k + k'/V_c \tag{19-72}$$

where ρ_c' is the bulk density of the liquid-free cake. The values of K, K', k, and k' depend on the material being compacted, the conditions of expression, and the dimensions chosen for p, V_c, and ρ_c'.

Equation (19-72) correlated the observations of Gurnham and Masson for such materials as cotton fiber, woolen yarn, wool felt, asbestos fiber, paper pulp, and wood sawdust—dry and wetted with water, oil, or other liquid. It also correlated most of Deerr's data (loc. cit.).

Rate of Expression For the design of an expression plant, equilibrium data are not enough: the size or capacity of the presses depends on the rate at which equilibrium is approached—that is, the rate of liquid expression from the cake.

In a series of laboratory tests, Koo and coworkers [*Ind. Res. (China)*, **6**, 9 (1937); *J. Chem. Eng. China*, **4**, 15, 107 (1937); **5**, 47, 69 (1938); **7**, 1, 23 (1940); *Ind. Eng. Chem.*, **34**, 342 (1942)] observed that the compression time θ to produce a particular fractional recovery of seed oil W/W_0 (where W is the mass of oil expressed in time θ and W_0 is mass of oil in the original material) depends principally on the compacting pressure and the viscosity of the expelled oil:

$$W/W_0 = C' p^{1/2} \theta^{1/6} / \nu^a \tag{19-73}$$

where ν is the kinematic viscosity of the oil at press temperature, the exponent a depends on the type and condition of the oilseed, and the coefficient C' is a so-called press constant that depends on the type and condition of material being pressed and on the units chosen for the variables. Koo developed Eq. (19-73) from experiments in which pressure ranged from 6.8 to 30.6 MPa (1000 to 4500 lbf/in²); temperature, from 288 to 398 K (15 to 125°C), corresponding to a fifteenfold variation in ν; and pressing time, from 0.5 to 9 h.

In the most fundamental and extensive studies of expression rate yet conducted, Shirato and his coworkers cantilevered the rationale of soil compaction (e.g., Terzaghi and Peck, *Soil Mechanics in Engineering Practice*, Wiley, New York, 1947) and that of filtration [e.g., Tiller and Huang, *Ind. Eng. Chem.*, **53**, 529 (1961)] toward one another to bridge, at least tentatively, the gap of expression as applied to slurries and filter cakes.

Defining a consolidation ratio U_c as

$$U_c = (L_0 - L)/(L_0 - L_\infty) \tag{19-74}$$

where L_0, L, and L_∞ are the cake thicknesses at the respective compression times of 0, θ_c, and ∞ (i.e., L_∞ is the equilibrium thickness), Shirato et al. (in Orr, *Filtration: Principles and Practices*, part I, Marcel Dekker, New York, 1977, p. 433) showed that Terzaghi's model leads to the evaluation of U_c as

$$U_c = 1 - \exp[-\pi^2 i^2 C_e \theta_c / 4w_c^2] \tag{19-75}$$

where

$$C_e = \rho_s p / \mu c_c \alpha \tag{19-76}$$

and i is the number of drainage surfaces per cake, w_c is the mass of dry cake per unit area, ρ_s is the true density of the solids in the cake, μ is the absolute viscosity of the filtrate, c_c is the Terzaghi-Peck compression index (Terzaghi and Peck, *Soil Mechanics in Engineering Practice*, 2d ed., Wiley, New York, 1967, art. 13), and α is the average specific cake filtration resistance (see discussion under "Theory of Filtration").

The application of Eq. (19-75) confronts the user with two difficulties, one fundamental and the other a matter of practicability. As to the first, the equation implies equilibrium expression response in U_c for θ_c. In actual fact, there is often a delayed, slow response (called "creep") not recognized in Terzaghi's model. The practicable difficulty arises from the necessity of experimental facility for determining α and c_c (i.e., compression-permeability-measurement capability).

An escape from both of these obstacles is suggested in the use of a semiempirical substitute for Eq. (19-75):

$$U_c = (4i^2 C_e \theta_c / \pi w_0^2)^{1/2} / [1 + (4i^2 C_e \theta_c / \pi w_0^2)^{\gamma}]^{1/(2\gamma)} \tag{19-77}$$

In this equation, γ is called the consolidation behavior index, and it reflects the creep behavior of the cake. The evaluation of C_e and γ, furthermore, can be accomplished from the expression experiment results without recourse to the supplementary compression-permeability determination. At low values of θ_c, U_c is approximately proportional to $\theta_c^{0.5}$; thus a plot of U_c against $\theta_c^{0.5}$ allows an adequate eval-

*The editor is grateful to J. C. Yingst, Eimco Process Equipment Company, for his constructive comments with respect to expression theory.

uation of $4i^2C_e/\pi w_0^2$, hence of C_e, at the linear end of the curve. Upon substitution of C_e in Eq. (19-77), γ can be evaluated by curve fitting. Shirato et al. found values of γ ranging from 2.85 for cakes exhibiting no creep [*J. Chem. Eng. Japan*, **13**, 397 (1980)] to as low as 0.4 for strongly creeping cakes (*Proc. World Filtr. Congr. III*, vol. I, Sept. 13–17, 1982, p. 280).

Shirato's correlations have not been reported as substantiated outside the laboratory or for materials other than filterable slurries and the cakes deposited therefrom. Meanwhile, as pointed out by Purchas (*Solid/Liquid Separation Equipment Scale-Up*, Uplands Press, Croydon, England, 1977, p. 568), filter cakes that are literally incompressible (s, the compressibility exponent, is zero) present no opportunity for deliquoring by postfiltration expression. As compressibility increases, so does the practicality of expression, and at some threshold value of s the operation becomes viably attractive. What this value is may depend on the material and application involved, but a conservative value suggested by Purchas is 0.2.

Continuous Expression No comprehensive theory has been developed for expression in a continuous screw press, although there is some similarity to feed screws and extrusion presses. Worm efficiency, or forward movement of solids referred to a projected worm area, may range from 25 to 80 percent. The discrepancy in displacement appears to be due to shear motion within the mass and perhaps to a rolling movement near the periphery or cage wall.

Whereas a cake possessing internal tensile strength (e.g., paper pulp or coal) is dewatered easily in a screw press, mud or clay slurry may slip ineffectually past the screw. To deliquor such a slurry in a screw press, it is necessary to run the screw at a speed lower than a limiting value that depends on the nature of the feed material and the press dimensions. The provision of a screw-type die and selection of a suitable pitch of the worm can be important to favorable press performance [Shirato et al., *Kagaku Kogaku Rombunshu*, **3**, 303 (1977); *Proc. World Filtr. Congr. III*, vol. I, Sept. 13–17, 1982, p. 286].

Pressures inside a screw press may reach 136 to 272 MPa (20,000 to 40,000 lbf/in^2) at some points but are far lower in many applications, e.g., 6.8 to 13.6 MPa (1000 to 2000 lbf/in^2) in pulp dewatering.

Expression of liquids from continuously moving webs in a roller press has been studied with paper and papermakers' felts. The principal variables have been defined, although quantitative conclusions are largely empirical. Total pressure applied by means of the upper roll is an important variable, but its effect is not simple; increased pressure increases the force, causing dewatering, but may more than counteract this effect by compacting and lengthening the flow channels. Roll hardness is a factor, as soft rolls cause an increase in nip area, hence a diminished force per unit area on the material, and also an increased distance of flow; soft rolls thus perform a less thorough dewatering than hard rolls. Similarly, rolls of large diameter distribute the pressing force and increase the flow distance, so are less effective than small rolls. At high peripheral roll speed, time in the nip is diminished, and dewatering is less thorough.

An important but not-well-understood variable is resistance to flow within the material. This property varies manyfold as the material advances into the roll nip and is compressed. Transverse flow is necessary to remove liquid from the interior of the mass; flow parallel to the web but opposite in direction is the only possible liquid movement in the nip itself. The structure of the material and the variation of structure with pressure are controlling factors. Use of a perforated roll, sometimes with suction from the inside, permits transverse flow even at the nip and greatly increases the degree of liquid removal. Additional research is needed on all these variables. Three-roll mills presumably act like two successive pressings on two-roll mills; further dewatering occurs in the second nip because equilibrium is never reached in a single pass.

SELECTION OF A SOLIDS-LIQUID SEPARATOR

Selection of a separator of solids from liquids begins with a preliminary choice of possible devices and ends with the purchase of a particular machine of specific size, type, and material of construction. Successful operation and reliability over time are the goals: the machine must meet performance specifications, day after day, and not require constant maintenance and repair.

In selecting a solids-liquid separator, it is important to keep in mind the capabilities and limitations of commercially available devices. Among the multiplicity of types on the market many are designed for fairly specific applications, and unthinking attempts to apply them to other situations are likely to meet with failure. The danger is the more insidious because failure often is not of the clean no-go type; rather it is likely to be in the character of underproduction, subspecification product, or excessively costly operation—the kinds of limping failure that may be slowly detected and difficult to analyze for cause. In addition, it should be recognized that the performance of mechanical separators—more, perhaps, than most chemical-processing equipment—strongly depends on preceding steps in the process. A relatively minor upstream process change, one that might be inadvertent, can alter the optimal separator choice.

PRELIMINARY DEFINITION AND SELECTION

The steps in solving a solids-liquid separation problem, in general, are:

1. Define the overall problem.
2. Establish process conditions.
3. Make preliminary selections.
4. Take representative samples.
5. Make simple tests.
6. Modify process conditions if necessary.
7. Consult equipment manufacturers.
8. Make final selection.

Problem Definition Intelligent selection of a separator requires a careful and complete statement of the nature of the separation problem. Focusing narrowly on the specific problem, however, is not sufficient, especially if the separation is to be one of the steps in a new process. Instead the problem must be defined as broadly as possible, beginning with the chemical reactor or other source of material to be separated and ending with the separated materials in their desired final form. In this way the influence of preceding and subsequent process steps on the separation step will be illuminated. Sometimes, of course, the new separator is proposed to replace an existing unit; the new separator must then fit into the current process and accept feed materials of more or less fixed characteristics. At other times the separator is only one item in a train of new equipment, all parts of which must work in harmony if the separator is to be effective.

Assistance in problem definition and in developing a test program should be sought from persons experienced in the field. If your organization has a consultant in separations of this kind, by all means make use of the expertise available. If not, it may be wise to employ an outside consultant, whose special knowledge can save much time, headaches, and money by appropriate guidance. Again, it is important to do this early; after the separation equipment has been installed, there is little a consultant can do to remedy the sometimes disastrous effects of a poor selection.

Establishing Process Conditions Step 2 is taken by defining the problem in detail. Properties of the materials to be separated, the quantities of feed and products required, the range of operating variables, and any restrictions on materials of construction must be accurately fixed, or reasonable assumptions must be made. Accurate data

TABLE 19-18 Data for Selecting a Solids-Liquid Separator*

1. Process
   a. Describe the process briefly. Make up a flow sheet showing places where liquid-solid separators are needed.
   b. What are the objections to the present process?
   c. Briefly, what results are expected of the separator?
   d. Is the process batch or continuous?
   e. Number the following objectives in order of importance in your problem: (*a*) Separation of two different solids............; (*b*) Removal of solids to recover valuable liquor as overflow............; (*c*) Removal of solids to recover the solids as thickened underflow........... or as "dry" cake............; (*d*) Washing of solids............; (*e*) Classification of solids ....; (*f*) Clarification or "polishing" of liquid............; (*g*) Concentration of solids............
   f. List the available power and current characteristics.
2. Feed
   a. Quantity of feed:
   Continuous process: ....gal./min.; ....hr./day; .... lb./hr. of dry solids.
   Batch process: volume of batch: ....; total batch cycle: ....hr.
   b. Feed properties: Temp............; pH............; viscosity............
   c. What maximum feed temperature is allowable?
   d. Chemical analysis and specific gravity of carrying liquid.
   e. Chemical analysis and specific gravity of solids.
   f. Percentage of solids in feed slurry.
   g. Screen analysis of solids: Wet............ dry............
   h. Chemical analysis and concentration of solubles in feed.
   i. Impurities: Form and probable effect on separation.
   j. Is there a volatile component in the feed?............ Should the separator be vapor-tight?............ Must it be under pressure?............ If so, how much?............
3. Filtration and settling rates
   a. Filtration rate on Büchner funnel: ....gal./(min.)(sq. ft.) of filter area under a vacuum of ....in. Hg. Time required to form a cake ....in. thick: ....sec.
   b. At what rate do the solids settle by gravity?
   c. What percentage of the total feed volume do the settled solids occupy after settling is complete? After how long?
4. Feed preparation
   a. If the feed tends to foam, can antifoaming agents be used? If so, what type?
   b. Can flocculating agents be used? If so, what agent?
   c. Can a filter aid be used?
   d. What are the process steps immediately preceding the separation? Can they be modified to make the separation easier?
   e. Could another carrying liquid be used?
5. Washing
   a. Is washing necessary?
   b. What are the chemical analysis and specific gravity of wash liquid?
   c. Purpose of wash liquid: To displace residual mother liquor; or to dissolve soluble material from the solids?
   d. Temperature of wash liquid.
   e. Quantity of wash allowable, in lb./lb. of solids.
6. Separated solids
   a. What percentage of solids is desired in the cake or thickened underflow?
   b. Is particle breakage important?
   c. Amount of residual solubles allowable in solids.
   d. What further processing will have to be carried out on the solids?
7. Separated liquids
   a. Clarity of liquor: what percentage of solids is permissible?
   b. Must the filtrate and spent wash liquid be kept separate?
   c. What further processing will be carried out on the filtrate and/or spent wash?
8. Materials of construction
   a. What metals look most promising?
   b. What metals must not be used?
   c. What gasket and packing materials are suitable?

*U.S. customary engineering units have been retained in this data form. The following SI or modified-SI units might be used instead: centimeters = inches × 2.54; kilograms per kilogram = pounds per pound × 1.0; kilograms per hour = pounds per hour × 0.454; liters per minute = gallons per minute × 3.785; liters per second–square meter = gallons per minute–square foot × 0.679; and pascals = inches mercury × 3377.

on particle size, size distribution, densities, viscosity, and other physical properties should be obtained *before* selection is made, not after an installed separator fails to perform. Qualitative considerations influence separator selection also. Is the liquid the valuable phase, or the solid (or perhaps both)? Is partial or incomplete separation acceptable, or must the liquid be sparkling clear? Are crystal size and appearance important? Is good separation a key step in the process, or can occasional small lapses be tolerated?

Table 19-18 lists the pertinent background information which should be assembled. It is typical of data requested by manufacturers when they are asked to recommend and quote on a liquid-solids separator. The more accurately and thoroughly these questions can be answered, the better the final choice is likely to be.

Preliminary Selections Assembling background information permits tentative selection of promising equipment and rules out clearly unsuitable types. Recognition that a separation is mechanical rather than diffusive or chemical is usually immediate, as is recognition of the requirement of expression as opposed to the simpler sorting of particulate solids from an associated fluid. In the latter class nearly all separators operate either by sedimentation or by filtration, and choices must be made among the types of equipment available in each category. Figure 19-133 illustrates the nature of the choices available. Sedimentation, generally, does not give quite the separation of solids from liquid that filtration does, for it generally yields concentrated solids which contain appreciable amounts of liquid in comparison with the "dryer" cake yielded by filters. It can, however, be used for classifying solids particles by size or by density, whereas filtration usually cannot.

SAMPLES AND TESTS

Once the initial choice of promising separator types is made, representative liquid-solids samples should be collected for preliminary tests.

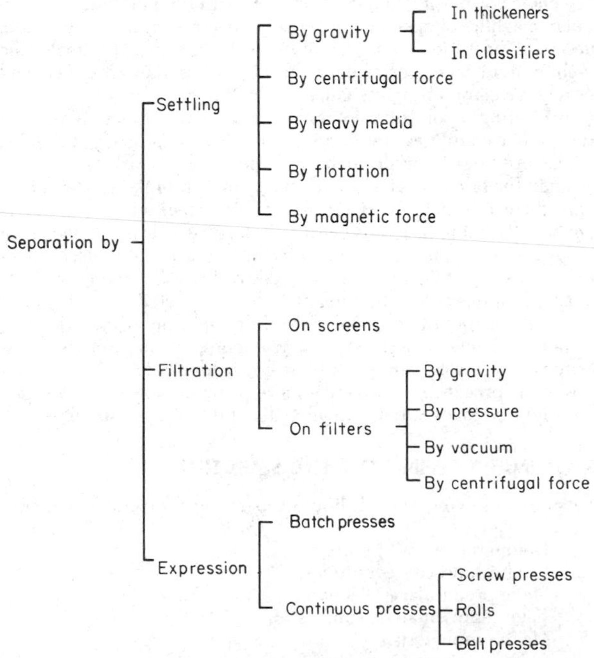

FIG. 19-133 Main paths to solids-liquid separation.

Representative Samples For meaningful results, tests must be run on representative samples. In liquid-solids systems good samples are hard to get. Frequently a liquid-solid mixture from a chemical process varies significantly from hour to hour, from batch to batch, or from week to week. A well-thought-out sampling program over a prolonged period, with samples spaced randomly and sufficiently far apart, under the most widely varying process conditions possible, should be formulated. Samples should be taken from all shifts in a continuous process and from many successive batches in a batch process. The influence of variations in raw materials on the separating characteristics should be investigated, as should the effect of reactor or crystallizer temperature, intensity of agitation, or other process variables.

Once samples are taken, they must be preserved unchanged until tested. Unfortunately cooling or heating the samples or the addition of preservatives may markedly change the ease with which solids may be separated from the liquid. Sometimes they make the separation easier, sometimes harder; in either case, tests made on deteriorated samples give a false picture of the capabilities of separation equipment. Even shipping of the samples can have a significant effect. Often it is so difficult to preserve liquid-solids samples without deterioration that accurate results can be obtained only by incorporating a test separation unit directly in the process stream.

Simple Tests It is usually profitable, however, to make simple preliminary tests, recognizing that the results may require confirmation through subsequent large-scale studies.

Preliminary gravity settling tests are made in a large graduated cylinder in which a well-stirred sample of slurry is allowed to settle, the height of the interface between clear supernatant liquid and concentrated slurry being recorded as a function of settling time. Centrifugal settling tests are normally made in a bottle centrifuge in which the slurry sample is spun at various speeds for various periods of time and the volume and consistency of the settled solids are noted. In gravity settling tests in particular, it is important to evaluate the effects of flocculating agents on settling rates.

Preliminary filtration tests may be made with a Büchner funnel or a small filter leaf, covered with canvas or other appropriate medium and connected to a vacuum system. Usually the suspension is poured carefully into the vacuum-connected funnel, whereas the leaf is immersed in a sample of the slurry and vacuum is applied to pull filtrate into a collecting flask. The time required to form each of several cakes in the range of 3 to 25 mm (⅛ to 1 in) thick under a given vacuum is noted, as is the volume of the collected filtrate. Properly conducted tests with a Büchner or a vacuum leaf closely simulate the action of rotary vacuum filters of the top- and bottom-feed variety, respectively, and may give the experienced observer enough information for complete specification of a plant-size filter. Alternatively, they may point to pressure-filter tests or, indeed, to a search for an alternative to filtration. Centrifugal filter tests are made in a perforated basket centrifugal 254 or 305 mm (10 or 12 in) in diameter lined with a suitable filter medium. Slurry is poured into the rotating basket until an appropriately thick cake—say, 25 mm (1 in)— is formed. Filtrate is recycled to the basket at such a rate that a thin layer of liquid is just visible on the surface of the cake. The discharge rate of the liquor under these conditions is the draining rate. The test is repeated with cakes of other thicknesses to establish the productive capacity of the centrifugal filter.

More detailed descriptions of small-scale sedimentation and filtration tests are presented in other parts of this section. Interpretation of the results and their conversion into preliminary estimates of such quantities as thickener size, centrifuge capacity, filter area, sludge density, cake dryness, and wash requirements also are discussed. Both the tests and the data treatment must be in experienced hands if error is to be avoided.

Modification of Process Conditions Relatively small changes in process conditions often markedly affect the performance of specific solids-liquid separators, making possible their application when initial test results indicated otherwise or vice versa. Flocculating agents are an example; many gravity settling operations are economically feasible only when flocculants are added to the process stream. Changes in precipitation or crystallization steps may greatly enhance or diminish filtration rates and hence filter capacity. Changes in the temperature of the process stream, the solute content, or the chemical nature of the suspending liquid also influence solids-settling rates. Occasionally it is desirable to add a heavy, finely divided solid to form a pseudo-liquid suspending medium in which the particles of the desired solid will rise to the surface. Attachment of air bubbles to solid particles in a flotation cell, using a suitable flotation agent, is another way of changing the relative densities of liquid and solid.

Consulting the Manufacturer Early in the selection campaign—certainly no later than the time at which the preliminary tests are completed—manufacturers of the more promising separators should be asked for assistance. Additional tests may be made at a manufacturer's test center; again a major problem is to obtain and preserve representative samples. As much process information as tolerable should be shared with the manufacturers to make full use of their experience with their particular equipment. Full-scale plant tests, although expensive, may well be justified before final selection is made. Such tests demonstrate operation on truly representative feed, show up long-term operating problems, and give valuable operating experience.

In summary, separator selection calls for clear problem definition, in broad terms; thorough cataloging of process information; and preliminary and tentative equipment selection, followed by refinement of the initial selections through tests on an increasingly larger scale. Reliability, flexibility of operation, and ease of maintenance should be weighed heavily in the final economic evaluation; rarely is purchase price, by itself, a governing factor in determining the suitability of a liquid-solids separator.

Solids Drying and Gas-Solid Systems

Harold F. Porter, B.S., *Principal Division Consultant, E. I. du Pont de Nemours & Co. (Retired); Fellow, American Institute of Chemical Engineers; Registered Professional Engineer (Delaware). (Section Editor)*

George A. Schurr, Ph.D., *Consultant, E. I. du Pont de Nemours & Co.; Member, American Institute of Chemical Engineers. (Solids Drying)*

David F. Wells, B.S., *President, David F. Wells & Associates; Member, American Institute of Chemical Engineers; Registered Professional Engineer (Delaware). (Fluidized-Bed Systems)*

Konrad T. Semrau, M.S., *Senior Chemical Engineer, SRI International; Member, American Chemical Society, American Institute of Chemical Engineers, Air Pollution Control Association. (Gas-Solids Separations)*

INTRODUCTION

SOLIDS-DRYING FUNDAMENTALS

SOLIDS-DRYING EQUIPMENT

Nomenclature and Units

In this table, symbols used in this section are defined in a general way. SI and customary U.S. units are listed. Specialized symbols are either defined at the point of application or in a separate table of nomenclature.

Symbol	Definition	SI units	U.S. customary units
A	Area	m^2	ft^2
a	Area	m^2	ft^2
C	Heat capacity	J/kg	Btu/lbm
D	Diameter	m	ft
D	Diffusivity	m^2/s	ft^2/s
d	Thickness	m	ft
E	Entrainment	kg/kg	lbm/lbm
E	Potential difference	V	V
F	Void fraction	Dimensionless	Dimensionless
G	Mass velocity	$k/(m^2 \cdot s)$	$lbm/(ft^2 \cdot s)$
H	Humidity	kg/kg	lbm/lbm
h	Heat-transfer coefficient	$J/(m^2 \cdot s \cdot K)$	$Btu/(ft^2 \cdot h \cdot °F)$
k	Mass-transfer coefficient	$kg/(m^2 \cdot s \cdot atm)$	$lbm/(h \cdot ft^2 \cdot atm)$
K	Thermal conductivity	$W/(m \cdot K)$	$Btu/(ft \cdot h \cdot °F)$
L	Length	m	ft
M	Molecular weight	kg/mol	lbm/mol
N	Rotational speed	1/s	1/min
P	Pressure	Pa	lbf/in^2
P	Pressure drop	Pa	lbf/ft^2
Q	Flow rate	m^3/s	ft^3/s
Q	Total heat flow	J/s	Btu/h
S	Slope	m/m	ft/ft
t	Time	s	s, h
t	Temperature	°C	°F
T	Absolute temperature	K	°R
U	Heat-transfer coefficient	$J/(m^2 \cdot s \cdot K)$	$Btu/(ft^2 \cdot h \cdot °F)$
u	Velocity	m/s	ft/s
V	Velocity	m/s	ft/s
V	Volume	m^3	ft^3
W	Mass	kg	lbm
w	Loading of fabric	kg/m^2	lbm/ft^2
	Greek symbols		
ϵ	Void fraction	Dimensionless	Dimensionless
θ	Time	s	s, h
λ	Latent heat	J/kg	Btu/lbm
μ	Viscosity	$Pa \cdot s$	$lbm/(s \cdot ft)$
ρ	Density	kg/m^3	lbm/ft^3
σ	Surface tension	N/m	dyn/cm

INTRODUCTION

This section is divided into four main subsections, "Solids-Drying Fundamentals," "Solids-Drying Equipment," "Fluidized-Bed Systems," and Gas-Solids Separation." In this introductory part some elementary definitions are given. In solids-gas contacting equipment, the solids bed can exist in any of the following four conditions.

Static This is a dense bed of solids in which each particle rests upon another at essentially the settled bulk density of the solids phase. Specifically, *there is no relative motion among solids particles* (Fig. 20-1).

Moving This is a slightly expanded bed of solids in which the particles are separated only enough to flow one over another. Usually the flow is downward under the force of gravity, but upward motion by mechanical lifting or agitation may also occur within the process vessel. In some cases, lifting of the solids is accomplished in separate equipment, and solids flow in the presence of the gas phase is downward only. The latter is a moving bed as usually defined in the petroleum industry. In this definition, *solids motion is achieved by either mechanical agitation or gravity force* (Fig. 20-2).

Fluidized This is an expanded condition in which the solids particles are supported by drag forces caused by the gas phase passing through the interstices among the particles at some critical velocity. It is an unstable condition in that the superficial gas velocity upward is less than the terminal setting velocity of the solids particles; the gas velocity is not sufficient to entrain and convey continuously all the solids. At the same time, there exist, within the stream of gas, eddies traveling at high enough velocities to lift the particles temporarily. Particle motion is continually upward and falling back. Specifically, the solids phase and the gas phase are intermixed and *together behave like a boiling fluid* (Fig. 20-3).

Dilute This is a fully expanded condition in which the solids particles are so widely separated that they exert essentially no influence upon each other. Specifically, the solids phase is so fully dispersed in the gas that *the density of the suspension is essentially that of the gas phase alone* (Fig. 20-4). Commonly, this situation exists when the gas velocity at all points in the system exceeds the terminal settling velocity of the solids and the particles can be lifted and continuously conveyed by the gas; however, this is not always true. Gravity settling chambers such as prilling towers and countercurrent-flow spray dryers are two exceptions in which gas velocity is insufficient to entrain the solids completely.

Gas-Solids Contacting Terms used in this section to describe the method by which gas may contact a bed of solids are the following:

1. *Parallel flow.* The direction of gas flow is parallel to the surface of the solids phase. Contacting is primarily at the interface between phases, with possibly some penetration of gas into the voids among the solids near the surface. The solids bed is usually in a static condition (Fig. 20-5).
2. *Perpendicular flow.* The direction of gas flow is normal to the phase interface. The gas impinges on the solids bed. Again the solids bed is usually in a static condition (Fig. 20-6).
3. *Through circulation.* The gas penetrates and flows through interstices among the solids, circulating more or less freely around the individual particles (Fig. 20-7). This may occur when solids are in static, moving, fluidized, or dilute conditions.

Three additional terms require definition.

1. *Cocurrent gas flow.* The gas phase and solids particles both flow in the same direction (Fig. 20-8).
2. *Countercurrent gas flow.* The direction of gas flow is exactly opposite to the direction of solids movement.
3. *Cross-flow of gas.* The direction of gas flow is at a right angle to that of solids movement, across the solids bed (Fig. 20-9).

Because in a gas-solids-contacting operation heat transfer and mass

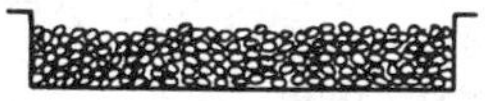

FIG. 20-1 Solids bed in static condition (tray dryer).

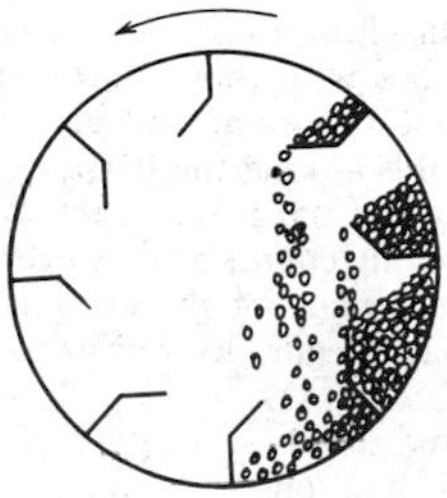

FIG. 20-2 Moving solids bed in a rotary dryer with lifters.

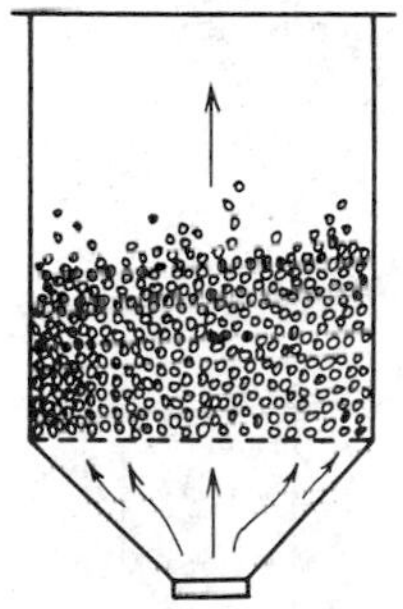

FIG. 20-3 Fluidized solids bed.

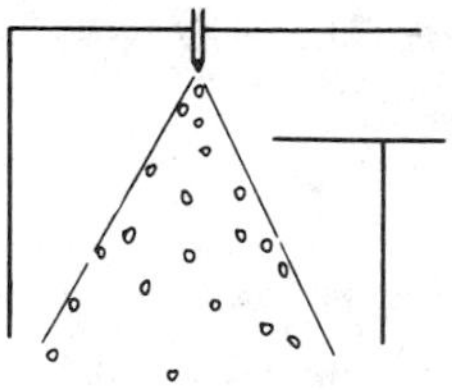

FIG. 20-4 Solids in a dilute condition near the top of a spray dryer.

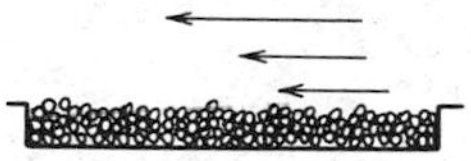

FIG. 20-5 Parallel gas flow over a static bed of solids.

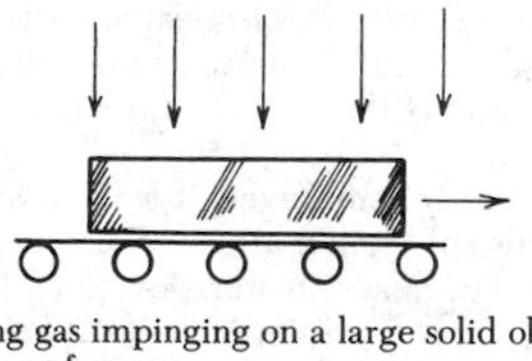

FIG. 20-6 Circulating gas impinging on a large solid object in perpendicular flow, in a roller-conveyor furnace.

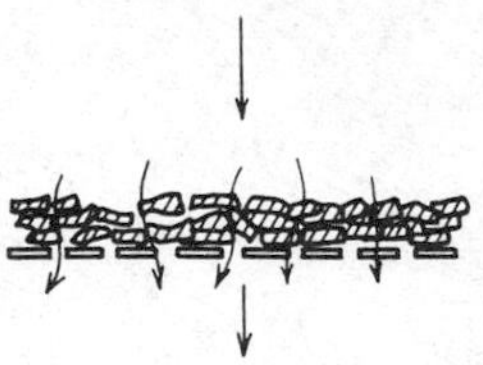

FIG. 20-7 Gas passing through a bed of preformed solids, in through circulation on a perforated-apron conveyor.

transfer take place at the solids' surfaces, maximum process efficiency can be expected with a maximum exposure of solids surface to the gas phase, together with thorough mixing of gas and solids. Both are important. Within any arrangement of particulate solids, gas is present in the voids among the particles and contacts all surfaces except at the points of particle contact. When the solids bed is in a static or slightly moving condition, however, gas within the voids is cut off from the main body of the gas phase. Some transfer of energy and mass may occur by diffusion, but it is usually insignificant.

Equipment design and selection are governed by two factors:

1. Mechanical considerations
2. Solids flow and surface characteristics.

The former usually involves process temperature or isolation. Solids surface characteristics are important in that they control the extent to which an operation is diffusion-limited, i.e., diffusion into

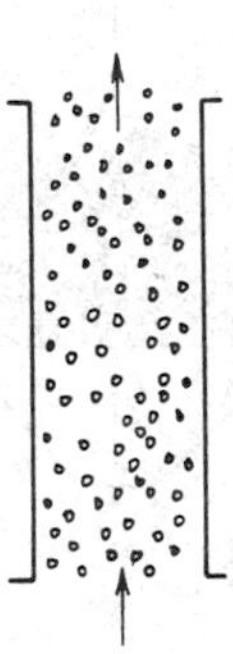

FIG. 20-8 Cocurrent gas-solids flow in a vertical-lift dilute-phase pneumatic conveyor.

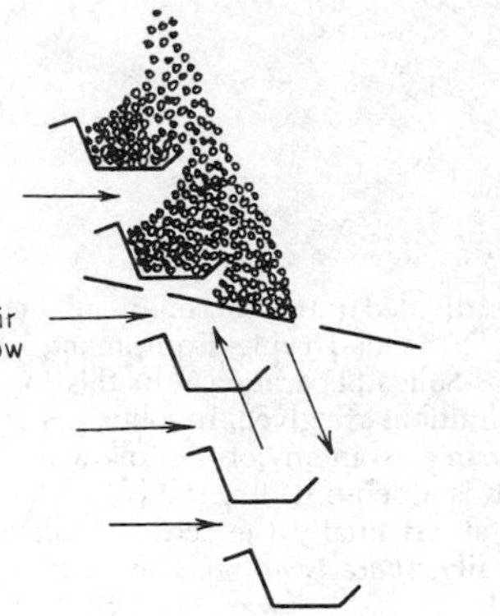

FIG. 20-9 Cross-flow of gas and solids in a cascade-type gravity dryer. *(Link-Belt Co., Multi-Louvre principle.)*

and out of the pores of a given solids particle, not through the voids among separate particles. The size of the solids particles, the surface-to-mass ratio, is also important in the evaluation of surface characteristics and the diffusion problem.

Gas-Solids Separations After the solids and gas have been brought together and mixed in a gas-solids contactor, it becomes necessary to separate the two phases. If the solids are sufficiently coarse and the gas velocity sufficiently low, it is possible to effect a complete gravitational separation in the primary contactor. Applications of this type are rare, however, and supplementary dust-collection equipment is commonly required. The recovery step may even dictate the type of primary contacting device selected. For example, when treating an extremely friable solid material, a deep fluidized-solids contactor might overload the collection system with fines, whereas the more gentle contacting of a traveling-screen contactor would be expected to produce a minimum of fines by attrition. Therefore, although gas-solids separation is usually considered as separate and distinct from the primary contacting operation, it is usually desirable to evaluate the separation problem at the same time that contacting methods are evaluated.

This section will consider methods and equipment employed to recover entrained solids from a gas stream and means for removing fine dispersoids from atmospheric air, as well as wet-dust collectors, scrubbers, and equipment for removing entrained liquid mist from gases.

SOLIDS-DRYING FUNDAMENTALS

GENERAL REFERENCES: Hall, *Dictionary of Drying*, Marcel Dekker, New York. 1979. Keey, *Drying Principles and Practice*, Pergamon, New York, 1972.

DEFINITIONS

Drying generally refers to the removal of a liquid from a solid by **evaporation.** Mechanical methods for separating a liquid from a solid are not generally considered drying, although they often precede a drying operation, since it is less expensive and frequently easier to use mechanical methods than to use thermal methods.

This subsection presents the theory and fundamental concepts of the drying of solids.

Equipment commonly employed for the drying of solids is described both in this subsection and in Sec. 11. The latter contains information on indirect-heat-transfer devices. This subsection contains mainly descriptions of direct-heat-transfer equipment. It also includes some indirect units; e.g., vacuum dryers, furnaces, steam-tube dryers, and rotary calciners. For the drying of gases, see Sec. 16.

Terminology Generally accepted definitions are given alphabetically in the following paragraphs.

Bound moisture in a solid is that liquid which exerts a vapor pressure less than that of the pure liquid at the given temperature. Liquid may become bound by retention in small capillaries, by solution in cell or fiber walls, by homogeneous solution throughout the solid, and by chemical or physical adsorption on solid surfaces.

Capillary flow is the flow of liquid through the interstices and over the surface of a solid, caused by liquid-solid molecular attraction.

Constant-rate period is that drying period during which the rate of water removal per unit of drying surface is constant.

Critical moisture content is the average moisture content when the constant-rate period ends.

Dry-weight basis expresses the moisture content of wet solid as kilograms of water per kilogram of bone-dry solid.

Equilibrium moisture content is the limiting moisture to which a given material can be dried under specific conditions of air temperature and humidity.

Falling-rate period is a drying period during which the instantaneous drying rate continually decreases.

Fiber-saturation point is the moisture content of cellular materials (e.g., wood) at which the cell walls are completely saturated while the cavities are liquid-free. It may be defined as the equilibrium moisture content as the humidity of the surrounding atmosphere approaches saturation.

Free-moisture content is that liquid which is removable at a given temperature and humidity. It may include bound and unbound moisture.

Funicular state is that condition in drying a porous body when capillary suction results in air being sucked into the pores.

Hygroscopic material is material that may contain bound moisture.

Initial moisture distribution refers to the moisture distribution throughout a solid at the start of drying.

Internal diffusion may be defined as the movement of liquid or vapor through a solid as the result of a concentration difference.

Moisture content of a solid is usually expressed as moisture quantity per unit weight of the dry or wet solid.

Moisture gradient refers to the distribution of water in a solid at a given moment in the drying process.

Nonhygroscopic material is material that can contain no bound moisture.

Pendular state is that state of a liquid in a porous solid when a continuous film of liquid no longer exists around and between discrete particles so that flow by capillary cannot occur. This state succeeds the funicular state.

Unaccomplished moisture change is the ratio of the free moisture present at any time to that initially present.

Unbound moisture in a hygroscopic material is that moisture in excess of the equilibrium moisture content corresponding to saturation humidity. All water in a nonhygroscopic material is unbound water.

Wet-weight basis expresses the moisture in a material as a percentage of the weight of the wet solid. Use of a dry-weight basis is recommended since the percentage change of moisture is constant for all moisture levels. When the wet-weight basis is used to express moisture content, a 2 or 3 percent change at high moisture contents (above 70 percent) actually represents a 15 to 20 percent change in evaporative load. See Fig. 20-10 for the relationship between the dry- and wet-weight bases.

APPLICATION OF PSYCHROMETRY TO DRYING

In any drying process, if an adequate supply of heat is assumed, the temperature and rate at which liquid vaporization occurs will depend on the vapor concentration in the surrounding atmosphere.

In vacuum drying or other processes containing atmospheres of 100 percent vapor, the temperature of liquid vaporization will equal or exceed the saturation temperature of the liquid at the system pressure. (When a free liquid or wetted surface is present, drying will occur at the saturation temperature, just as free water at 101.325 kPa vaporizes in a 100 percent steam atmosphere at 100°C.)

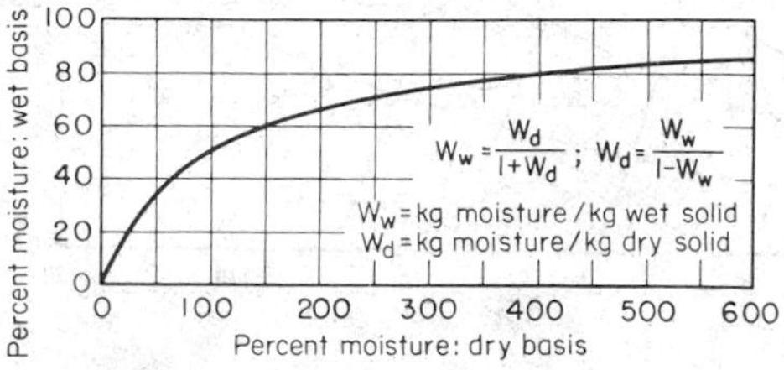

FIG. 20-10 Relationship between wet-weight and dry-weight bases.

On the other hand, when evolved vapor is purged from the dryer environment by using a second (inert) gas, the temperature at which vaporization occurs will depend on the concentration of vapor in the surrounding gas. In effect, the liquid must be heated to a temperature at which its vapor pressure equals or exceeds the partial pressure of vapor in the purge gas. In the reverse situation, condensation will occur.

In most drying operations, water is the liquid evaporated and air is the normally employed purge gas. For drying purposes, a psychrometric chart found very useful is that reproduced in Fig. 20-11.

1. The **wet-bulb or saturation temperature line** gives the maximum weight of water vapor that 1 kg of dry air can carry at the intersecting **dry-bulb temperature** shown on the abscissa at saturation humidity. The partial pressure of water in air equals the water-vapor pressure at that temperature. The saturation humidity is defined by

$$H_s = p_s/(P - p_s)18/28.9 \qquad (20\text{-}1)$$

where H_s = saturation humidity (kg/kg dry air), p_s = vapor pressure of water at temperature t_s, P = absolute pressure, and 18/28.9 = ratio of molecular weights of water (18) and air (28.9). Similarly, the humidity at any condition less than saturation is given by

$$H = p/(P - p)18/28.9 \qquad (20\text{-}2)$$

2. The percent relative humidity is defined by

$$H_R = 100(p/p_s) \qquad (20\text{-}3)$$

where p = partial pressure of water vapor in the air, p_s = vapor pressure of water at the same temperature, and H_R = percent relative humidity.

3. Humid volumes are given by the curves entitled "Volume m^3/kg dry air." The volumes are plotted as functions of absolute humidity and temperature. The difference between dry-air specific volume and humid-air volume at a given temperature is the volume of water vapor.

4. Enthalpy data are given on the basis of kilojoules per kilogram of dry air. **Enthalpy-at-saturation data** are accurate only at the saturation temperature and humidity. Enthalpy deviation curves permit enthalpy corrections for humidities less than saturation and show how the wet-bulb-temperature lines do not precisely coincide with constant-enthalpy, adiabatic cooling lines.

5. There are no lines for humid heats on Fig. 20-11. These may be calculated by

$$C_s = 1.0 + 1.87H \qquad (20\text{-}4)$$

where C_s = humid heat of moist air, kJ/(kg·K); 1.0 = specific heat of dry air, kJ/(kg·K), 1.87 = specific heat of water vapor, kJ/(kg·K); and H = absolute humidity, kg/kg dry air.

6. The wet-bulb-temperature lines represent also the adiabatic-saturation lines for air and water vapor only. These are based on the relationship

$$H_s - H = (C_s/\lambda)(t - t_s) \qquad (20\text{-}5)$$

where H_s and t_s = adiabatic saturation humidity and temperature respectively, corresponding to the air conditions represented by H and t, and C_s = humid heat for humidity H. The slope of the adiabatic-saturation curve is C_s/λ, where λ = latent heat of evaporation at t_s. These lines show the relationship between the temperature and humidity of air passing through a continuous dryer operating adiabatically.

The wet-bulb temperature is established by a dynamic equilibrium between heat and mass transfer when liquid evaporates from a small mass, such as the wet bulb of a thermometer, into a very large mass of gas such that the latter undergoes no temperature or humidity change. It is expressed by the relationship

$$h_c(t - t_w) = k_g'\lambda(H_w - H_a) \qquad (20\text{-}6)$$

where h_c = heat-transfer coefficient by convection, J/(m^2·s·K) [Btu/(h·ft^2·°F)]; t = air temperature, K; t_w = wet-bulb temperature of air, K; k_g' = mass-transfer coefficient, kg/(s·m^2) (kg/kg) [lb/

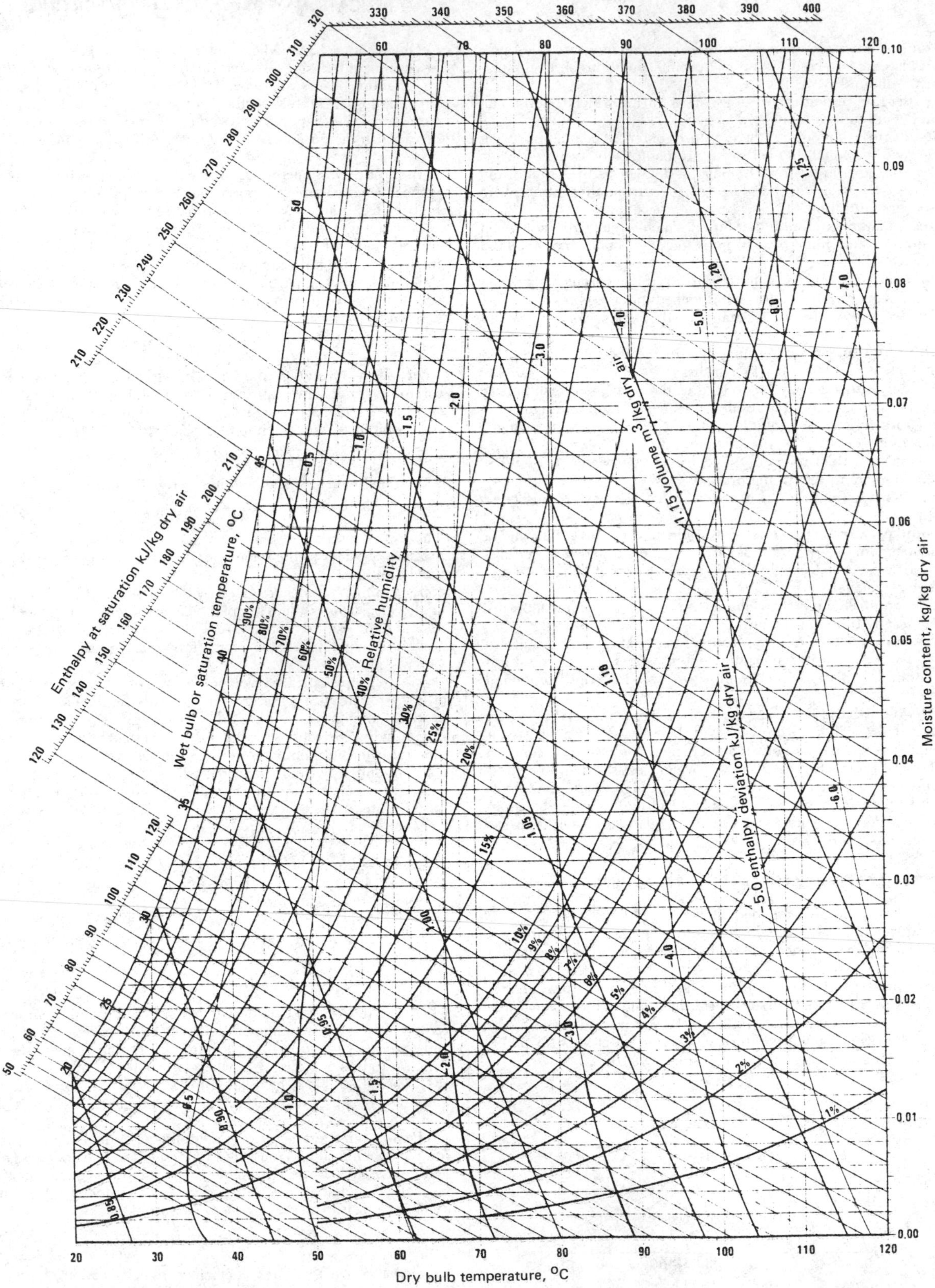

FIG. 20-11 Psychrometric chart: properties of air and water-vapor mixtures from 20 to 120°C. *(Carrier Corporation.)*

(h·ft²)(lb/lb)]; λ = latent heat of evaporation at t_w, J/kg (Btu/lb); H_w = saturated humidity at t_w = kg/kg of dry air; and H_a = humidity of the surrounding air, kg/kg of dry air.

For air–water-vapor mixtures, it so happens that $h_c/k'_g = C_s$ approximately, although there is no theoretical reason for this. Hence, since the ratio $(H_w - H_a)/(t_w - t)$ equals $h_c/k'_g/\lambda$, which represents the slope of the wet-bulb-temperature lines, it is also equal to C_s/λ, the slope of the adiabatic-saturation lines as shown previously.

A given humidity chart is precise only at the pressure for which it is evaluated. Most air–water-vapor charts are based on a pressure of 1 atm. Humidities read from these charts for given values of wet- and dry-bulb temperature apply only at an atmospheric pressure of 760 mmHg. If the total pressure is different from 760 mmHg, the humidity at a given wet-bulb and dry-bulb temperature must be corrected according to the following relationship.

$$H_a = H_o + 0.622p_w\left(\frac{1}{P - p_w} - \frac{1}{760 - p_w}\right) \tag{20-7}$$

where H_a = humidity of air at pressure P, kg/kg of dry air; H_o = humidity of air as read from a humidity chart based on 760-mm pressure at the observed wet- and dry-bulb temperatures, kg/kg dry air, p_w = vapor pressure of water at the observed wet-bulb temperature, mmHg; and P = the pressure at which the wet- and dry-bulb readings were taken. Similar corrections can be derived to correct specific volume, the saturation-humidity curve, and the relative-humidity curves.

HUMIDITY CHARTS FOR SOLVENT VAPORS

Humidity charts for other solvent vapors may be prepared in an analogous manner. There is one important difference involved, however, in that the wet-bulb temperature differs considerably from the adiabatic-saturation temperatures for vapors other than water.

Figures 20-12 to 20-14 show humidity charts for carbon tetrachloride, benzene, and toluene. The lines on these charts have been calculated in the manner outlined for air–water vapor except for the wet-bulb-temperature lines. The determination of these lines depends on data for the psychrometric ratio h_c/k'_g, as indicated by Eq. (20-6). For the charts shown, the wet-bulb-temperature lines are based on the following equation:

$$H_w - H = (\alpha h_c/\lambda_w k'_g)(t - t_w) \tag{20-8}$$

where α = radiation correction factor, a value of 1.06 having been used for these charts. Values of h_c/k'_g, obtained from values of h_c/k'_gC_s as presented by Walker, Lewis, McAdams, and Gilliland (*Principles of Chemical Engineering*, 3d ed., McGraw-Hill, New York, 1937), where C_s = humid heat of air with respect to the vapor involved, are as follows:

Material	Carbon tetrachloride	Benzene	Toluene
h_c/k'_gC_s	0.51	0.54	0.47

A discussion of the theory of the relationship between h_c and k'_g may be found in Sec. 12. Because both theoretical and experimental values of h_c/k'_g apply only to dilute gas mixtures, the wet-bulb lines at high concentrations have been omitted. For a discussion of the precautions to be taken in making psychrometric determinations of solvent vapors at low solvent wet-bulb temperatures in the presence of water vapor, see the paper by Sherwood and Comings [*Trans. Am. Inst. Chem. Eng.*, **28**, 88 (1932)].

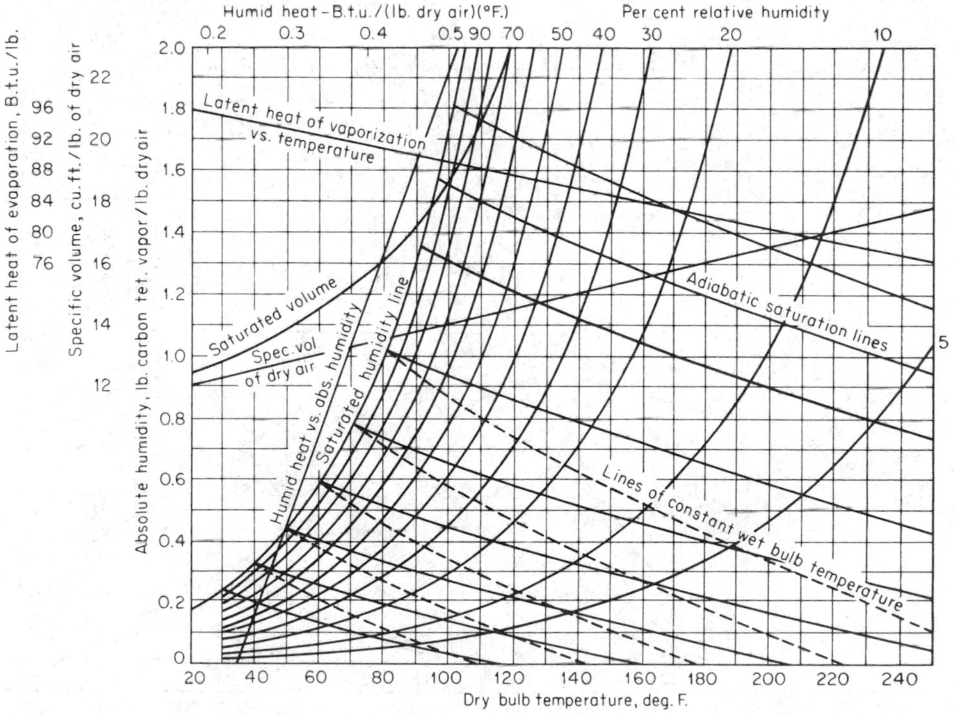

FIG. 20-12 Humidity chart for air–carbon tetrachloride vapor mixture. To convert British thermal units per pound to joules per kilogram, multiply by 2326; to convert British thermal units per pound dry air–degree Fahrenheit to joules per kilogram-kelvin, multiply by 4186.8; and to convert cubic feet per pound to cubic meters per kilogram, multiply by 0.0624.

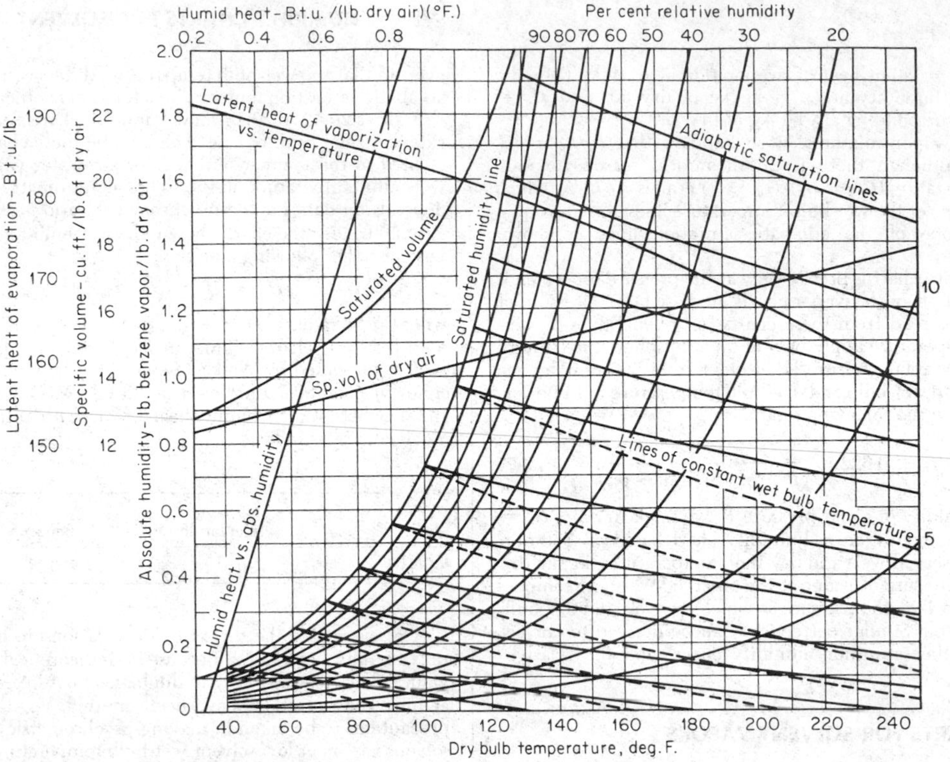

FIG. 20-13 Humidity chart for air–benzene-vapor mixture. To convert British thermal units per pound to joules per kilogram, multiply by 2326; to convert British thermal units per pound dry air–degree Fahrenheit to joules per kilogram-kelvin, multiply by 4186.8; and to convert cubic feet per pound to cubic meters per kilogram, multiply by 0.0624.

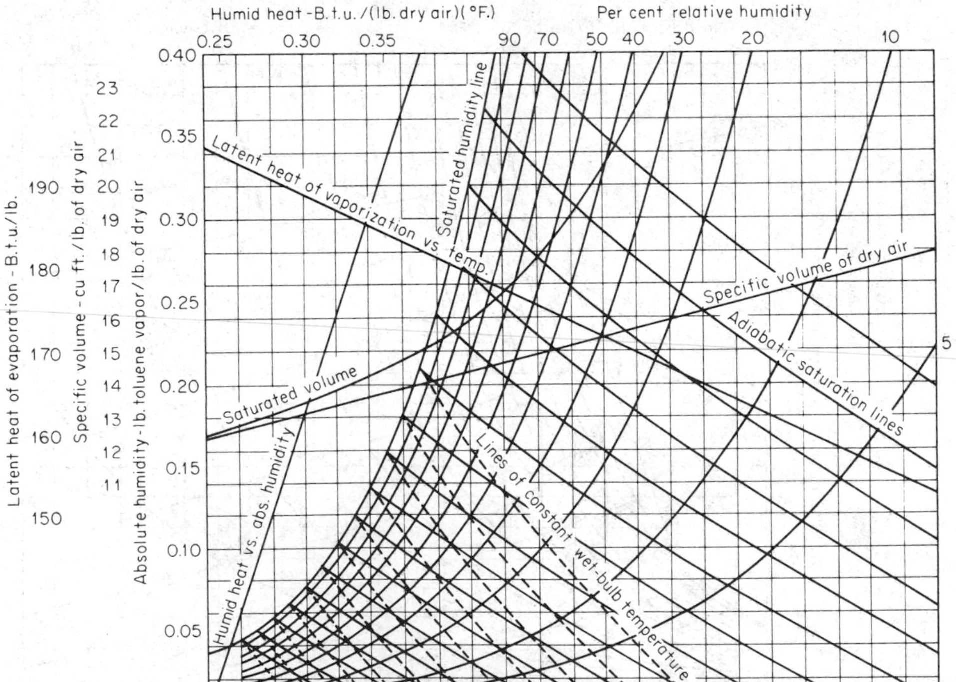

FIG. 20-14 Humidity chart for air–toluene-vapor mixture. To convert British thermal units per pound to joules per kilogram, multiply by 2326; to convert British thermal units per pound dry air–degree Fahrenheit to joules per kilogram-kelvin, multiply by 4186.8; and to convert cubic feet per pound to cubic meters per kilogram, multiply by 0.0624.

GENERAL CONDITIONS FOR DRYING

Solids drying encompasses two fundamental and simultaneous processes: (1) heat is transferred to evaporate liquid, and (2) mass is transferred as a liquid or vapor within the solid and as a vapor from the surface. The factors governing the rates of these processes determine the drying rate.

Commercial dryers differ fundamentally by the methods of heat transfer employed (see classification of dryers, Fig. 20-20). These industrial-dryer operations may utilize heat transfer by convection, conduction, radiation, or a combination of these. In each case, however, heat must flow to the outer surface and then into the interior of the solid. The single exception is dielectric and microwave drying, in which high-frequency electricity generates heat internally and produces a high temperature within the material and on its surface.

Mass is transferred in drying as a liquid and vapor within the solid and as vapor from the exposed surfaces. Movement within the solid results from a concentration gradient which is dependent on the characteristics of the solid. A solid to be dried may be porous or nonporous. It can also be hygroscopic or nonhygroscopic. Many solids fall intermediately between these two extremes, but it is generally convenient to consider the solid to be one or the other.

A study of how a solid dries may be based on the **internal** mechanism of liquid flow or on the effect of the **external** conditions of temperature, humidity, air flow, state of subdivision, etc., on the drying rate of the solids. The former procedure generally requires a fundamental study of the internal condition. The latter procedure, although less fundamental, is more generally used because the results have greater immediate application in equipment design and evaluation.

INTERNAL MECHANISM OF LIQUID FLOW

The structure of the solid determines the mechanism for which internal liquid flow may occur. These mechanisms can include (1) **diffusion** in continuous, homogeneous solids, (2) **capillary flow** in granular and porous solids, (3) flow caused by **shrinkage** and **pressure** gradients, (4) flow caused by **gravity,** and (5) flow caused by a **vaporization-condensation** sequence.

In general, one mechanism predominates at any given time in a solid during drying, but it is not uncommon to find different mechanisms predominating at different times during the drying cycle.

The study of internal moisture gradients establishes the particular mechanism which controls during the drying of a solid. The experimental determination of reliable moisture gradients is extremely difficult.

Hougen, McCauley, and Marshall [*Trans. Am. Inst. Chem. Eng.*, **36,** 183 (1940)] discussed the conditions under which capillary and diffusional flow may be expected in a drying solid and analyzed the published experimental moisture-gradient data for the two cases. Their curves indicate that capillary flow is typified by a moisture gradient involving a double curvature and point of inflection (Fig. 20-15*a*) while diffusional flow is a smooth curve, concave downward (Fig. 20-15*b*), as would be predicted from the diffusion equations. They also showed that the liquid-diffusion coefficient is usually a function of moisture content which decreases with decreasing moisture. The effect of variable diffusivity is illustrated in Fig. 20-15*b*, where the dashed line is calculated for constant diffusivity and the solid line is experimental for the case in which the diffusion coefficient is moisture-dependent. Thus, the integrated diffusion equations assuming constant diffusivity only approximate the actual behavior.

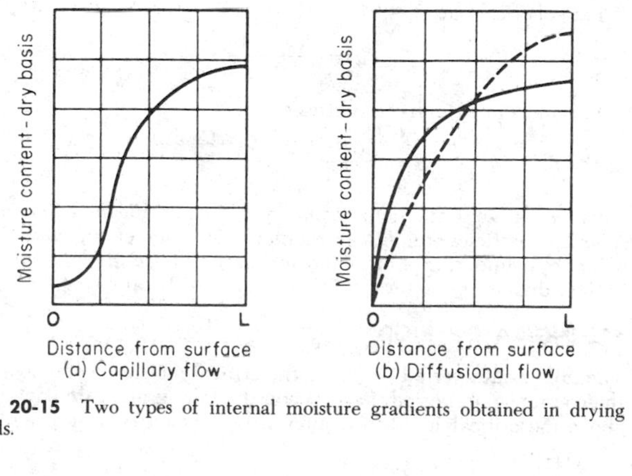

FIG. 20-15 Two types of internal moisture gradients obtained in drying solids.

These authors classified solids on the basis of capillary and diffusional flow:

Capillary Flow Moisture which is held in the interstices of solids, as liquid on the surface, or as free moisture in cell cavities, moves by gravity and capillarity, provided that passageways for continuous flow are present. In drying, liquid flow resulting from capillarity applies to liquids not held in solution and to all moisture above the fiber-saturation point, as in textiles, paper, and leather, and to all moisture above the equilibrium moisture content at atmospheric saturations, as in fine powders and granular solids, such as paint pigments, minerals, clays, soil, and sand.

Vapor Diffusion Moisture may move by vapor diffusion through the solid, provided that a temperature gradient is established by heating, thus creating a vapor-pressure gradient. Vaporization and vapor diffusion may occur in any solid in which heating takes place at one surface and drying from the other and in which liquid is isolated between granules of solid.

Liquid Diffusion The movement of liquids by diffusion in solids is restricted to the equilibrium moisture content below the point of atmospheric saturation and to systems in which moisture and solid are mutually soluble. The first class applies to the last stages in the drying of clays, starches, flour, textiles, paper, and wood; the second class includes the drying of soaps, glues, gelatins, and pastes.

External Conditions The principal external variables involved in any drying study are temperature, humidity, air flow, state of subdivision of the solid, agitation of the solid, method of supporting the solid, and contact between hot surfaces and wet solid. All these variables will not necessarily occur in one problem.

PERIODS OF DRYING

When a solid is dried experimentally, data relating moisture content to time are usually obtained. These data are then plotted as moisture content (dry basis) W versus time θ, as shown in Fig. 20-16*a*. This curve represents the general case when a wet solid loses moisture first by evaporation from a saturated surface on the solid, followed in turn by a period of evaporation from a saturated surface of gradually decreasing area, and, finally, when the latter evaporates in the interior of the solid.

Figure 20-16*a* indicates that the drying rate is subject to variation with time or moisture content. This variation is better illustrated by graphically or numerically differentiating the curve and plotting $dW/d\theta$ versus W, as shown in Fig. 20-16*b*, or as $dW/d\theta$ versus θ, as shown in Fig. 20-16*c*. These **rate curves** illustrate that the drying process is not a smooth, continuous one in which a single mechanism controls throughout. Figure 20-16*c* has the advantage of showing how long each drying period lasts.

The section AB on each curve represents a **warming-up** period of the solids. Section BC on each curve represents the **constant-rate** period. Point C, where the constant rate ends and the drying rate begins falling, is termed the **critical-moisture** content. The curved portion CD on Fig. 20-16*a* is termed the **falling-rate** period and, as shown in Fig. 20-16*b* and *c*, is typified by a continuously changing rate throughout the remainder of the drying cycle. Point E (Fig. 20-16*b*) represents the point at which all the exposed surface becomes completely unsaturated and marks the start of that portion of the drying cycle during which the rate of internal moisture movement controls the drying rate. Portion CE in Fig. 20-16*b* is usually defined as the first falling-rate drying period; portion DE, as the second falling-rate period.

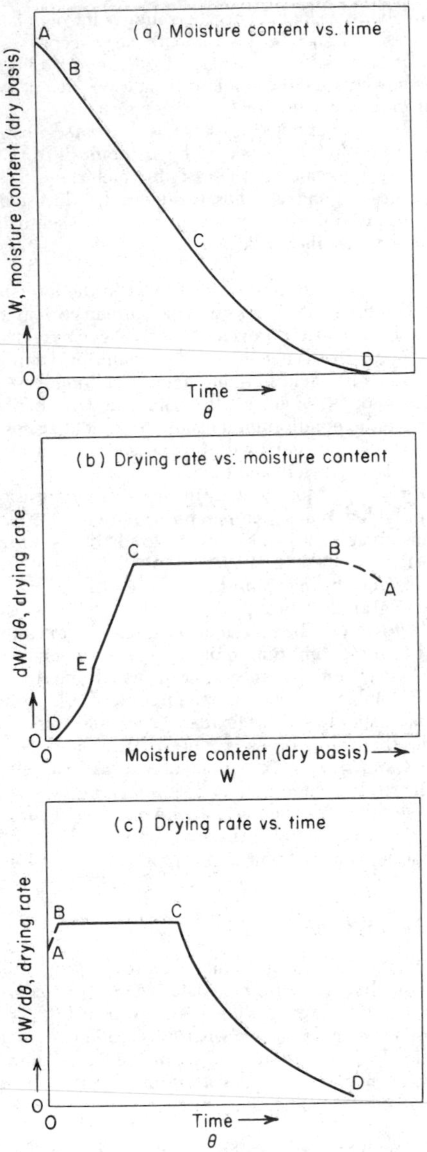

FIG. 20-16 The periods of drying.

CONSTANT-RATE PERIOD

In the constant-rate period moisture movement within the solid is rapid enough to maintain a saturated condition at the surface, and the rate of drying is controlled by the rate of heat transferred to the evaporating surface. Drying proceeds by diffusion of vapor from the saturated surface of the material across a stagnant air film into the environment. The rate of mass transfer balances the rate of heat transfer, and the temperature of the saturated surface remains constant. The mechanism of moisture removal is equivalent to evaporation from a body of water* and is essentially independent of the nature of the solids.

If heat is transferred solely by convection and in the absence of other heat effects, the surface temperature approaches the wet-bulb temperature. However, when heat is transferred by radiation, convection, or a combination of these and convection, the temperature at the saturated surface is between the wet-bulb temperature and the boiling point of water. Under these conditions, the rate of heat transfer is increased and a higher drying rate results.

When heat is transferred to a wet solid by convection to hot surfaces and heat transfer by convection is negligible, the solids approach the boiling-point temperature rather than the wet-bulb temperature. This method of heat transfer is utilized in indirect dryers (see classification of dryers in Fig. 20-20). Radiation is also effective in increasing the constant rate by augmenting the convection heat transfer and raising the surface temperature above the wet-bulb temperature.

When the heat for evaporation in the constant-rate period is supplied by a hot gas, a dynamic equilibrium establishes the rate of heat transfer to the material and the rate of vapor removal from the surface:

$$dw/d\theta = h_t A \, \Delta t/\lambda = k_g A \, \Delta p \qquad (20\text{-}9)$$

where $dw/d\theta$ = drying rate, kg water/s; h_t = total heat-transfer coefficient, J/(m^2·s·k) [Btu/(h·ft^2·°F)]; A = area for heat transfer and evaporation, m^2; λ = latent heat of evaporation at t'_s, J/kg (Btu/lb); k_g = mass-transfer coefficient, kg/(s·m^2·atm) [(lb/(h·ft^2·atm)]; $\Delta t = t - t'_s$, where t = gas (dry-bulb) temperature, K; $p = p_s - p$, where p_s = vapor pressure of water at surface temperature t'_s, atm; and p = partial pressure of water vapor in the gas, atm.

The magnitude of the constant rate depends upon three factors:

1. The heat- or mass-transfer coefficient
2. The area exposed to the drying medium
3. The difference in temperature or humidity between the gas stream and the wet surface of the solid

All these factors are the external variables. The internal mechanism of liquid flow does not affect the constant rate.

For drying calculations, it is convenient to express Eq. (20-9) in terms of the decrease in moisture content rather than in the quantity of water evaporated. For evaporation from a tray of wet material, if no change in volume during drying is assumed, Eq. (20-9) becomes

$$dw/d\theta = (h/\rho_s d\lambda)(t - t'_s) \qquad (20\text{-}10)$$

where $dw/d\theta$ = drying rate, kg water/(s·kg dry solids); h_t = total heat-transfer coefficient, J/(m^2·s·K) [Btu/(h·ft^2·°F)]; ρ_s = bulk density dry material, kg/m^3; d = thickness of bed, m; λ = latent heat of vaporization, J/kg (Btu/lb); t = air temperature, K; and t'_s = evaporating surface temperature, K. (Note that $dw/d\theta$ is inherently negative.)

A similar equation can be written for the through-circulation case:

$$dW/d\theta = (h_t a/\rho_s \lambda(t - t'_s) \qquad (20\text{-}11)$$

where a = m^2 of heat-transfer area/m^3 of bed, 1/m; and other symbols are as in Eq. (20-10).

The values of ρ_s and/or a must be known in order to use Eqs. (20-10) and (20-11). The value of a is difficult to estimate without experimental data. When the void fraction is known, a can sometimes be estimated from the following relationships:

For spherical particles,

$$a = \frac{6(1 - F)}{(D_p)_m} \qquad (20\text{-}12)$$

For uniform cylindrical particles,

$$a = \frac{4(0.5D_0 + Z)(1 - F)}{D_0 Z} \qquad (20\text{-}13)$$

where F = void fraction; $(D_p)_m$ = harmonic mean diameter of spherical particles, m; D_0 = diameter of cylinder, m; and Z = height of cylinder, m. For cylindrical particles that are long relative to their diameter, the term $0.5D_0$ in Eq. (20-13) can be neglected.

FALLING-RATE PERIOD

The falling-rate period begins at the critical moisture content when the constant-rate period ends. When the falling moisture content is above the critical moisture content, the whole drying process will

*The term "water" is used for convenience; this discussion applies equally to other liquids.

occur under constant-rate conditions. If, on the other hand, the initial moisture content is below the critical moisture content, the entire drying process will occur in the falling-rate period. This period is usually divided into two zones: (1) the zone of **unsaturated surface drying** and (2) the zone where **internal moisture movement** controls. In the first zone, the entire evaporating surface can no longer be maintained and saturated by moisture movement within the solid. The drying rate decreases from the unsaturated portion, and hence the rate for the total surface decreases. Generally, the drying rate depends on factors affecting the diffusion of moisture away from the evaporating surface and those affecting the rate of internal moisture movement.

As drying proceeds, the point is reached where the evaporating surface is unsaturated. The point of evaporation moves into the solid, and the dry process enters the second falling-rate period. The drying rate is now governed by the rate of internal moisture movement; the influence of external variables diminishes. This period usually predominates in determining the overall drying time to lower moisture content.

LIQUID DIFFUSION

Diffusion-controlled mass transfer is assumed when the vapor or liquid flow conforms to Fick's second law of diffusion. This is stated in the unsteady-state-diffusion equation using mass-transfer notation as

$$\delta c/\delta\theta = D_{AB}(\delta^2 C/dx) \qquad (20\text{-}14)$$

where c = concentration of one component in a two-component phase of A and B, θ = diffusion time, x = distance in the direction of diffusion, and D_{AB} = binary diffusivity of the phase AB. This equation applies to diffusion in solids, stationary liquids, and stagnant gases.

The diffusion equation for the falling-rate drying period for a slab can be derived from the diffusion equation if one assumes that the surface is dry or at an equilibrium moisture content and that the initial moisture distribution is uniform. For these conditions, the following equation is obtained:

$$\frac{W - W_e}{W_c - W_e} = \frac{8}{\pi^2}\left[\sum_{n=0}^{n=\infty} \frac{1}{(2n+1)} e^{-(2n+1)2D_\ell\theta(\pi/2d)2}\right] \qquad (20\text{-}15)$$

where W, W_e, and W_c = average moisture content (dry basis) at any time, θ, at the start of the falling-rate period and in equilibrium with the environment respectively, kg/kg; D_ℓ = liquid diffusivity, m^2/s; θ = time from start of falling-rate period, s; and d = one-half of the thickness of the solid layer through which diffusion occurs, m. When evaporation occurs from only one face, d = total thickness, m.

Equation (20-15) assumes that D_ℓ is constant; however, D_ℓ is rarely constant but varies with moisture content, temperature, and humidity. For long drying times, Eq. (20-15) simplifies to a limiting form of the diffusion equation as

$$\frac{W - W_e}{W_c - W_e} = \frac{8}{\pi^2}[e^{-D_\ell\theta(\pi/2d)2}] \qquad (20\text{-}16)$$

Equation (20-16) may be differentiated to give the drying rate as

$$-\frac{dW}{d\theta} = \frac{\pi^2 D_\ell}{4d^2}(W - W_e) \qquad (20\text{-}17)$$

where $dW/d\theta$ = drying rate, kg/s.

When Eq. (20-15) is plotted on semilogarithmic graph paper, a straight line is obtained for values of $(W - W_e)/(W_c - W_e) < 0.6$. It is in the straight-line portion that the approximate form [Eq. (20-17)] applies.

Equations (20-15), (20-16), and (20-17) hold only for a slab-sheet solid whose thickness is small relative to the other two dimensions. For other shapes, reference should be made to Crank (*The Mathematics of Diffusion*, Oxford, London, 1956).

An approximate equation for the falling-rate period may be obtained by integration of Eq. (20-17). This gives an equation for materials in which moisture movement is controlled by diffusion:

$$\theta_f = \frac{4d^2}{D_\ell\pi^2}\ln\frac{W_c - W_e}{W - W_e} \qquad (20\text{-}18)$$

where θ_f = drying time in the falling-rate period.

Diffusion equations may also be used to study vapor diffusion in porous materials. It should be clear that all estimates based on relationships that assume constant diffusivity are approximations. Liquid diffusivity in solids usually decreases with moisture concentration. Liquid and vapor diffusivity also change, and material shrinks during drying.

CAPILLARY THEORY

If the porous size of a granular material is suitable, moisture may move from a region of high to one of low concentration as the result of capillary action rather than by diffusion. The capillary theory assumes that a bed of nonporous spheres is composed of particles surrounding a space called a pore. These pores are connected by passages of various sizes. As water is progressively removed from the bed, the curvature of the water surface in the interstices of the top layer of spheres increases and a suction pressure resulting from curvature is set up. As the removal of water continues, the suction pressure attains a value in which air is drawn into the pore spaces between successive layers of spheres.

This entry suction or suction potential is a measure of the resultant forces tending to draw water from the interior of the bed to the surface. For a pore formed by regularly packed nonporous spheres, the suction potential is given by

$$P_s = x\sigma/r\rho g \qquad (20\text{-}19)$$

where P_s = suction potential, m of water; σ = surface tension; dyn/m; ρ = density of water, kg/m^3; g = 9.8 m/s^2; r = sphere radius, m; and x is a packing factor equal to 12.9 for rhombohedral and 4.8 for cubical packing.

As drying proceeds, the surface moisture evaporates, causing retreat of the surface menisci until the suction potential reaches a value given by Eq. (20-19). At this point, the pores of the surface will open, air will enter, and the moisture will redistribute itself with a slight lowering of the suction potential. As evaporation proceeds, the suction potential again increases until a slightly higher entry value is reached, when a further redistribution occurs.

The drying rate curve (Fig. 20-16*b*) can be analyzed in terms of capillary theory. In region *BC*, there is a loss of moisture with a gradual increase in suction and emptying of the bulk of the larger pores in the solid. In region *CE*, there is an increase in suction as the moisture content decreases and finer pores are opened. Section *ED* represents a condition in which moisture is being removed by vapor diffusion from the interior of the body, although there is still sufficient water in the bed to give rise to capillary forces.

An approximate equation for use for materials in which moisture movement is controlled by capillary flow is given as

$$\theta_f = \frac{\rho_s d\lambda(W_c - W_e)}{h_t(t - t_s')}\ln\frac{W_c - W_e}{W - W_e} \qquad (20\text{-}20)$$

where θ_f = drying time in the falling-rate period.

TABLE 20-1 Materials Obeying Eqs. (20-18) and (20-22)

Materials obeying Eq. (20-18)	Materials obeying (Eq. (20-22)
1. Single-phase solid systems such as soap, gelatin, and glue	1. Coarse granular solids such as sand, paint pigments, and minerals
2. Wood and similar solids below the fiber-saturation point	2. Materials in which moisture flow occurs at concentrations above the equilibrium moisture content at atmospheric saturation or above the fiber-saturation point
3. Last stages of drying starches, textiles, paper, clay, hydrophilic solids, and other materials when bound water is being removed	

Table 20-1 gives an approximate classification of materials that obey Eqs. (20-18) and (20-20).

CRITICAL MOISTURE CONTENT

To use the preceding equations for estimating drying times in the falling-rate period, it is necessary to know values of critical moisture content W_c. Such values are difficult to obtain without making actual drying tests, which in themselves would give the required drying time and thereby obviate solving the equations. However, when drying tests are not feasible, some estimate of critical moisture content must be made.

Values of critical moisture contents for some representative materials are given in Table 20-2 for drying by cross circulation and in Table 20-10 for drying by through circulation. The tabulated values are only approximate, since critical moisture content depends on the drying history. It appears that the constant-rate period ends when the moisture content at the surface reaches a specific value. Since the critical moisture content is the average moisture through the material, its value depends on the rate of drying, the thickness of the material, and the factors influencing moisture movement and resulting gradients within the solid. As a result, the critical moisture content increases with increased drying rate and with increased thickness of the mass of material being dried.

EQUILIBRIUM MOISTURE CONTENT

In drying solids it is important to distinguish between hygroscopic and nonhygroscopic materials. If a hygroscopic material is maintained in contact with air at constant temperature and humidity until equilibrium is reached, the material will attain a definite moisture content. This moisture is termed the equilibrium moisture content for the specified conditions. Equilibrium moisture may be adsorbed as a surface film or condensed in the fine capillaries of the solid at reduced pressure, and its concentration will vary with the temperature and humidity of the surrounding air. However, at low temperatures, e.g., 15 to 50°C, a plot of equilibrium moisture content versus percent relative humidity is essentially independent of temperature. At zero humidity the equilibrium moisture content of all materials is zero.

Equilibrium moisture content depends greatly on the nature of the solid. For nonporous, i.e., nonhygroscopic, materials, the equilibrium moisture content is essentially zero at all temperatures and humidities. For organic materials such as wood, paper, and soap, equilibrium moisture contents vary regularly over wide ranges as temperature and humidity change. In the special case of the dehydration of hydrated inorganic salts such as copper sulfate, sodium sulfate, or barium chloride, temperature and humidity control is very important in obtaining the desired degree of moisture removal, and the proper conditions must be determined from data on the water of hydration or crystallization as a function of air temperature and humidity.

Equilibrium moisture content of a solid is particularly important in drying because it represents the limiting moisture content for given conditions of humidity and temperature. If the material is dried to a moisture content less than it normally possesses in equilibrium with atmospheric air, it will return to its equilibrium value on storage unless special precautions are taken.

Equilibrium moisture content of a hygroscopic material may be determined in a number of ways, the only requirement being a source of constant-temperature and constant-humidity air. Determination may be made under static or dynamic conditions, although the latter case is preferred. A simple static procedure is to place a

TABLE 20-2 Approximate Critical Moisture Contents Obtained on the Air Drying of Various Materials, Expressed as Percentage Water on the Dry Basis

Material	Thickness, in.	Critical moisture, % water, dry basis
Barium nitrate crystals, on trays	1.0	7
Beaverboard	0.17	Above 120
Brick clay	.62	14
Carbon pigment	1	40
Celotex	0.44	160
Chrome leather	.04	125
Copper carbonate (on trays)	1–1.5	60
English china clay	1	16
Flint clay refractory brick mix	2.0	13
Gelatin, initially 400% water	0.1–0.2 (wet)	300
Iron blue pigment (on trays)	0.25–0.75	110
Kaolin	....	14
Lithol red	1	50
Lithopone press cake (in trays)	0.25	6.4
	.50	8.0
	.75	12.0
	1.0	16.0
Niter cake fines, on trays	....	Above 16
Paper, white eggshell	0.0075	41
Fine book	.005	33
Coated	.004	34
Newsprint	....	60–70
Plastic clay brick mix	2.0	19
Poplar wood	0.165	120
Prussian blue	....	40
Pulp lead, initially 140% water	....	Below 15
Rock salt (in trays)	1.0	7
Sand, 50–150 mesh	2.0	5
Sand, 200–325 mesh	2.0	10
Sand, through 325 mesh	2.0	21
Sea sand (on trays)	0.25	3
	.5	4.7
	.75	5.5
	1.0	5.9
	2.0	6.0
Silica brick mix	2.0	8
Sole leather	0.25	Above 90
Stannic tetrachloride sludge	1	180
Subsoil, clay fraction 55.4%	....	21
Subsoil, much higher clay content	....	35
Sulfite pulp	0.25–0.75	60–80
Sulfite pulp (pulp lap)	0.039	110
White lead	....	11
Whiting	0.25–1.5	6.9
Wool fabric, worsted	....	31
Wool, undyed serge	....	8

TABLE 20-3 Maintenance of Constant Humidity

Solid phase	Max. temp., °C.	% humidity
$H_3PO_4.\frac{1}{2}H_2O$	24.5	9
$ZnCl_2.\frac{1}{2}H_2O$	20	10
$KC_2H_3O_2$	168	13
$LiCl.H_2O$	20	15
$KC_2H_3O_2$	20	20
KF	100	22.9
NaBr	100	22.9
$CaCl_2.6H_2O$	24.5	31
$CaCl_2.6H_2O$	20	32.3
$CaCl_2.6H_2O$	18.5	35
CrO_3	20	35
$CaCl_2.6H_2O$	10	38
$CaCl_2.6H_2O$	5	39.8
$K_2CO_3.2H_2O$	24.5	43
$K_2CO_3.2H_2O$	18.5	44
$Ca(NO_3)_2.4H_2O$	24.5	51
$NaHSO_4.H_2O$	20	52
$Mg(NO_3)_2.6H_2O$	24.5	52
$NaClO_3$	100	54
$Ca(NO_3)_2.4H_2O$	18.5	56
$Mg(NO_3)_2.6H_2O$	18.5	56
$NaBr.2H_2O$	20	58
$Mg(C_2H_3O_2).4H_2O$	20	65
$NaNO_2$	20	66
$(NH_4)_2SO_4$	108.2	75
$(NH_4)_2SO_4$	20	81
$NaC_2H_3O_2.3H_2O$	20	76
$Na_2S_2O_3.5H_2O$	20	78
NH_4Cl	20	79.2
NH_4Cl	25	79.3
NH_4Cl	30	79.5
KBr	20	84
Tl_2SO_4	104.7	84.8
$KHSO_4$	20	86
$Na_2CO_3.10H_2O$	24.5	87
K_2CrO_4	20	88
$NaBrO_3$	20	92
$Na_2CO_3.10H_2O$	18.5	92
$Na_2SO_4.10H_2O$	20	93
$Na_2HPO_4.12H_2O$	20	95
NaF	100	96.6
$Pb(NO_3)_2$	20	98
$TlNO_3$	100.3	98.7
TlCl	100.1	99.7

For a more complete list of salts, and for references to the literature see "International Critical Tables," vol. 1, p. 68.

number of samples in ordinary laboratory desiccators containing sulfuric acid solutions of known concentrations which produce atmospheres of known relative humidity. The sample in each desiccator is weighed periodically until a constant weight is obtained. Moisture content at this final weight represents the equilibrium moisture content for the particular conditions.

The value of equilibrium moisture content, for many materials, depends on the direction in which equilibrium is approached. A different value is reached when a wet material loses moisture by desorption, as in drying, from that obtained when a dry material gains it by adsorption. For drying calculations the desorption values are preferred. In the general case, the equilibrum moisture content reached by losing moisture is higher than that reached by adsorbing it.

Equilibrium moisture content can be measured dynamically by placing the sample in a U tube through which is drawn a continuous flow of controlled-humidity air. Again the sample is weighed periodically until a constant weight is reached. Properly humidified air for such a procedure can be obtained by bubbling dry air through a large volume of a saturated salt solution which produces a definite degree of saturation of the air. Care must be taken to ensure that the air and salt solution reach equilibrium. Values of the humidity over various salt solutions may be found in Table 20-3.

ESTIMATIONS FOR TOTAL DRYING TIME

Estimates of both the constant-rate and the falling-rate periods are needed to estimate the total drying time for a given drying operation. If estimates for these periods are available, the total drying time is estimated by summing as

$$\theta_t = \theta_c + \theta_f \qquad (20\text{-}21)$$

where θ_t = total drying time, h; θ_c = drying time for constant-rate period, h; and θ_f = drying time for falling-rate period, h. The difficulty in estimating critical moisture content greatly reduces the number of drying cases in which calculation of a good estimate is possible.

ANALYSIS OF DATA

When experiments are carried out to select a suitable dryer and to obtain design data, the effect of changes in various external variables is studied. These experiments should be conducted in an experimental unit that simulates the large-scale dryer from both the thermal and the material-handling aspects, and only material which is truly representative of full-scale production should be used.

Data expressing moisture content in terms of elapsed time should be obtained and the results plotted as shown in Fig. 20-17. For purposes of analysis, the moisture-time curve must be differentiated graphically or numerically and the drying rates so obtained plotted to determine the nature and extent of the drying periods in the cycle. It is customary to plot drying rate versus moisture content as in Fig. 20-16*b*. Although instructive, this type of plot gives no information on duration of the drying periods. These are better shown by plots similar to Fig. 20-16*c*, in which drying rate is plotted as a function of time on either arithmetic or logarithmic coordinates. Logarithmic plots permit easy reading at low moisture contents or long times.

In order to determine whether a simple relationship exists in the falling-rate period, the unaccomplished moisture change, defined as ratio of free moisture in the solid at time θ to total free moisture present at start of the falling-rate period $(W - W_e)/(W_c - W_e)$, is plotted as a function of time on semilogarithmic paper. If a straight line is obtained such as curve *B* of Fig. 20-18 by using the upper scale of abscissa, either Eq. (20-18), for materials in which the moisture moves by diffusion, or Eq. (20-21), for materials in which the moisture movement is by capillary flow, may be applicable. If Eq. (20-21) applies, K_1, the slope of the falling-rate drying curve, is related to the constant drying rate. The latter is calculated from Eq. (20-22) and can be compared with the measured value. If the slopes agree, the moisture movement is by capillary flow. If the slopes do not agree, the moisture movement is by diffusion and the slope of the line should equal $\pi^2 D_\ell/4d^2$.

The dependency of drying rate on material thickness must be established experimentally. With the effect of material thickness established, liquid diffusivity can be calculated as indicated here. For this calculation, the theoretical values for an infinite slab are required. These are:

$\frac{D_\ell \theta}{d^2}$	0.02	0.05	0.10	0.15	0.20	0.30	0.50	1.0
$\frac{W - W_e}{W_c - W_e}$	0.84	0.75	0.642	0.568	0.496	0.387	0.238	0.069

A plot of these values is shown as curve *A* in Fig. 20-18.

If a straight line such as curve *B* of Fig. 20-18 represents the experimental data and if it has been established that the drying time varies inversely as the square of the thickness, the average liquid diffusivity can be obtained as follows. At a given value of $(W - W_e)/(W_c - W_e)$, read the corresponding value of $D_\ell\theta/d^2$ from curve *A*, Eq. (20-16), Fig. 20-18. At the same value of $(W - W_e)/(W_c - W_e)$, read

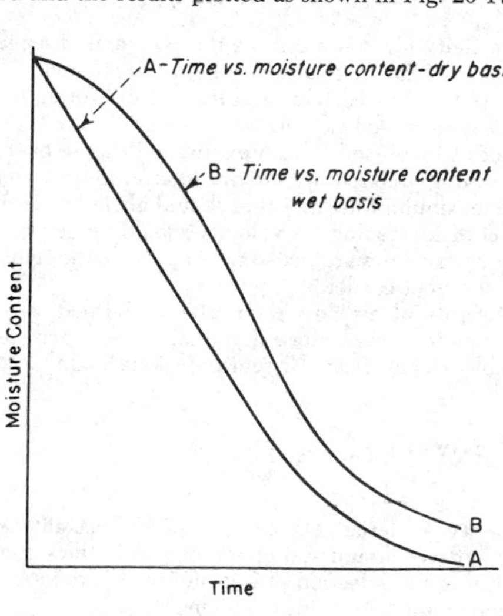

FIG. 20-17 Drying-time curves.

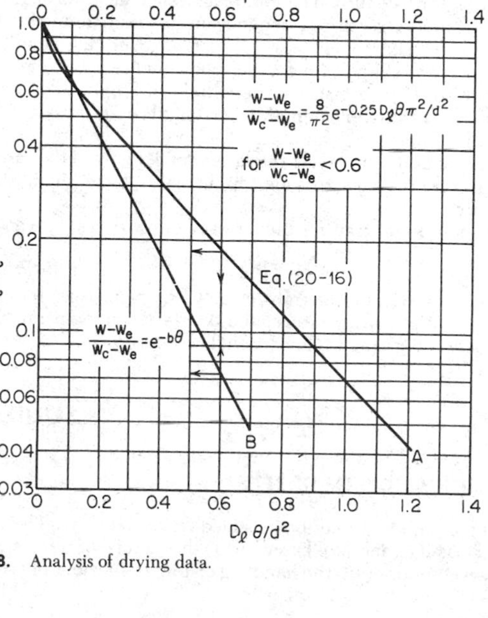

FIG. 20-18. Analysis of drying data.

the corresponding experimental value of θ from curve B (upper scale). Then

$$D_{\ell_{avg}} = \frac{(D_\ell \theta/d^2)\text{ theoretical}}{(\theta/d^2)\text{ experimental}} \tag{20-22}$$

where $D_{\ell_{avg}}$ = average experimental value of liquid diffusivity, m^2/h.

The value of diffusivity calculated from Eq. (20-22) must be recognized as an average value over the entire range of moisture change from $(W - W_e)/(W_c - W_e) = 1$ to the value $(W - W_e)/(W_c - W_e)$ at which θ/d^2 was evaluated. Further, Eq. (20-22) assumes that the theoretical curve is a straight line for all values of time. This is not true for values $(W - W_e)/(W_c - W_e)$ less than 0.6.

A more accurate value of $D_{\ell_{avg}}$ can be obtained by taking a ratio of slopes of the curves of Fig. 20-18. Thus the ratio of the slope of the experimental curve of unaccomplished moisture change versus drying time on a semilogarithmic plot [Eq. (20-16)] to the slope of the theoretical curve at the same unaccomplished moisture change, again on a semilogaritmic plot, equals the quantity D_ℓ/d^2. If d is known, D_ℓ can be evaluated.

Tests on Plant Dryers Tests on plant-scale dryers are usually carried out to obtain design data for a specific material, to select a suitable dryer type, or to check present performance of an existing dryer with the objective of determining its capacity potential. In these tests overall performance data are obtained and the results used to make heat and material balances and to estimate overall drying rates or heat-transfer coefficients.

Generally, the minimum data to be taken in order to calculate the performance of a dryer are:

1. Inlet and outlet moisture contents
2. Inlet and outlet gas temperatures
3. Inlet and outlet material temperatures
4. Feed rate
5. Gas rate
6. Inlet and outlet humidities
7. Retention time or time of passage through the dryer
8. Fuel consumption

Whenever possible, moisture contents and temperatures should be measured at various points within the dryer.

Typical experimental and calculated results of a drying test for a continuous adiabatic convection dryer are shown in Fig. 20-19. Test data as elaborate as those shown are not usually justified economically except when basic studies, aimed at clarifying the effect of operating variables, are being carried out in order to arrive at a reliable design procedure. The completeness of the information which is sought in any given test depends on the ultimate use of the data. In any case data for at least two sets of operating conditions are needed if a good analysis of dryer performance is to be made.

Results of drying tests can be correlated empirically in terms of **overall heat-transfer coefficient** or **length of a transfer unit** as a function of operating variables. The former is generally applicable to all types of dryers, while the latter applies only in the case of continuous dryers. The relationship between these quantities is as follows.

The number of transfer units in any direct dryer is given by

$$N_t = (t_1 - t_2)/\Delta t_m \tag{20-23}$$

where N_t = number of transfer units; t_1 = inlet gas temperature, K; t_2 = exit gas temperature; and Δt_m = mean temperature difference between gas and solids through the dryer, K.

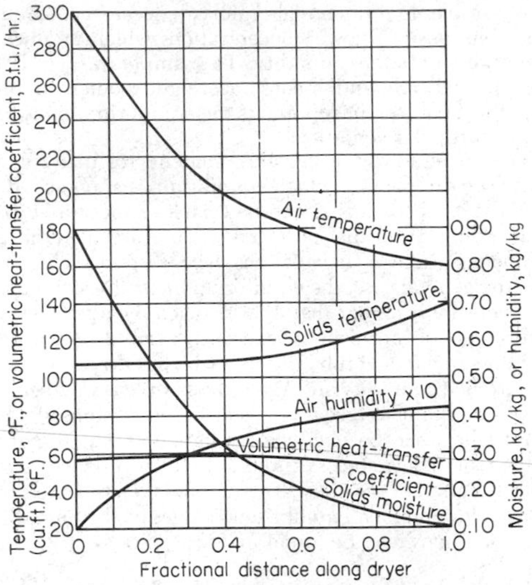

FIG. 20-19 Typical results of dryer-performance tests. To convert British thermal units per hour–cubic foot–degree Fahrenheit to joules per cubic meter–second–kelvin, multiply by 1.73.

The volumetric heat-transfer coefficient is given by

$$U_v = \frac{q_d}{V_d\,\Delta t_m} = \frac{wc_s(t_2 - t_1)}{A_d L_d\,\Delta t_m} \tag{20-24}$$

where U_v = volumetric heat-transfer coefficient, $J/(m^3 \cdot s \cdot K)$ [$Btu/(h \cdot ft^3 \cdot °F)$]; q = cross-sectional area of dryer, m^2; and L_α = dryer length, m.

The volumetric heat-transfer coefficients along the dryer are lower at the discharge end (Fig. 20-19) because of the internal resistance to moisture movement in the later stages of drying. When drying data are expressed in terms of overall performance, care and judgment should be exercised in extrapolating the results to other conditions, particularly conditions of different feed and product moisture. If, for example, the overall heat-transfer coefficients, from the data of Fig. 20-19, were used to predict a dryer design for reducing the product moisture below 10 percent, the design would be in error. Obviously, this problem can be circumvented by making sure that the final moisture in the experiments is below that desired in the product.

In any capacity test to determine the potential of a plant dryer, the effects of the following variables should be studied:

1. **Effect of increased temperature.** This is often the simplest way to achieve increased capacity.
2. **Effect of increased final moisture.** Because of the marked increase in drying time required to dry to low moisture contents, the permissible maximum final moisture should always be established.
3. **Effect of increasing air velocity** should be determined. Frequently, higher air rates are necessary to provide the required additional heat at higher capacities.
4. **Uniformity of air flow** should be established. Air-flow maldistribution can seriously reduce dryer capacity and efficiency.
5. Possible benefits from **air recirculation** should be considered.

SOLIDS-DRYING EQUIPMENT

CLASSIFICATION OF DRYERS

Drying equipment may be classified in several ways. The two most useful classifications are based on (1) the method of transferring heat to the wet solids or (2) the handling characteristics and physical properties of the wet material. The first method of classification reveals differences in dryer design and operation, while the second method is most useful in the selection of a group of dryers for preliminary consideration in a given drying problem.

A classification chart of drying equipment on the basis of heat

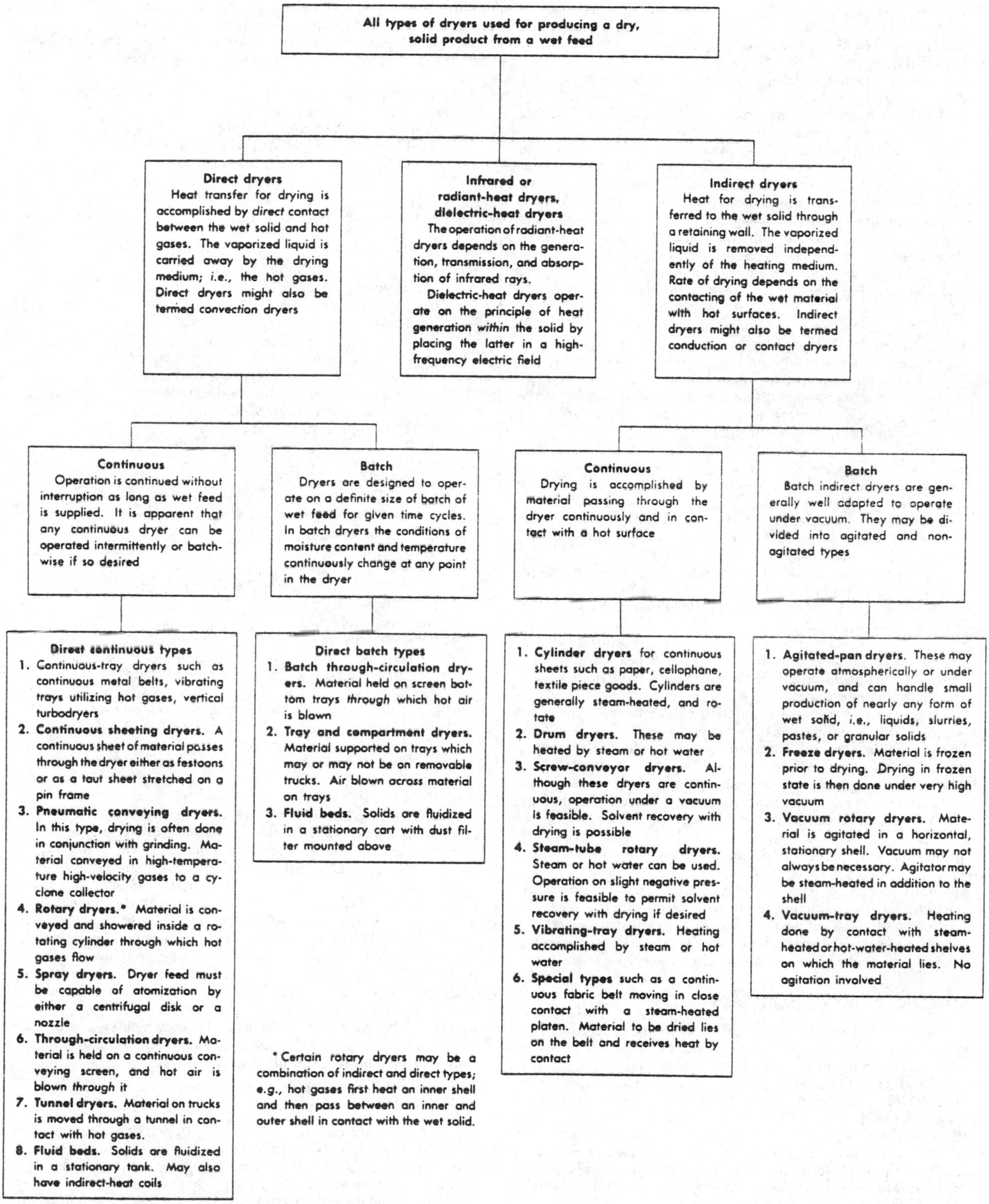

FIG. 20-20 Classification of dryers, based on method of heat transfer. [*Revised from Marshall,* Heat. Piping Air Cond., *18,* 71 (1946).]

transfer is shown in Fig. 20-20. This chart classifies dryers as direct or indirect, with subclasses of continuous or batchwise operation.

Direct Dryers The general operating characteristics of direct dryers are these:

1. Direct contacting of hot gases with the solids is employed for solids heating and vapor removal.
2. Drying temperatures may range up to 1000 K, the limiting temperature for most common structural metals. At the higher temperatures, radiation becomes an important heat-transfer mechanism.
3. At gas temperatures below the boiling point, the vapor content of gas influences the rate of drying and the final moisture content of the solid. With gas temperatures above the boiling point throughout, the vapor content of the gas has only a slight retarding effect on the drying rate and final moisture content. Thus, superheated vapors of the liquid being removed can be used for drying.
4. For low-temperature drying, dehumidification of the drying air may be required when atmospheric humidities are excessively high.
5. A direct dryer consumes more fuel per pound of water evaporated, the lower the final moisture content.
6. Efficiency increases with an increase in the inlet-gas temperature for a constant exhaust temperature.
7. Because large amounts of gas are required to supply all the heat for drying, dust-recovery equipment may be very large and expensive when drying very small particles.

Indirect dryers differ from direct dryers with respect to heat transfer and vapor removal:

1. Heat is transferred to the wet material by conduction through a solid retaining wall, usually metallic.
2. Surface temperatures may range from below freezing in the case of freeze dryers to above 800 K in the case of indirect dryers heated by combustion products.
3. Indirect dryers are suited to drying under reduced pressures and inert atmospheres to permit the recovery of solvents and to prevent the occurrence of explosive mixtures or the oxidation of easily decomposed materials.
4. Indirect dryers using condensing fluids as the heating medium are generally economical from the standpoint of heat consumption, since they furnish heat only in accordance with the demand made by the material being dried.
5. Dust recovery and dusty materials can be handled more satisfactorily in indirect dryers than in direct dryers.

Miscellaneous Dryers **Infrared dryers** depend on the transfer of radiant energy to evaporate moisture. The radiant energy is supplied electrically by infrared lamps, by electric resistance elements, or by incandescent refractories heated by gas. The latter method has the added advantage of convection heating. Infrared heating is not widely used in the chemical industries for the removal of moisture. Its principal use is in baking or drying paint films and in heating thin layers of materials.

Dielectric dryers have not as yet found a wide field of application. Their fundamental characteristic of generating heat within the solid indicates potentialities for drying massive geometrical objects such as wood, sponge-rubber shapes, and ceramics. Power costs may range to 10 times the fuel costs of conventional methods.

SELECTION OF DRYING EQUIPMENT

1. *Initial selection of dryers.* Select those dryers which appear best suited to handling the wet material and the dry product, which fit into the continuity of the process as a whole, and which will produce a product of the desired physical properties. This preliminary selection can be made with the aid of Table 20-4, which classifies the various types of dryers on the basis of the materials handled.
2. *Initial comparison of dryers.* The dryers so selected should be evaluated approximately from available cost and performance data. From this evaluation, those dryers which appear to be uneconomical or unsuitable from the standpoint of performance should be eliminated from further consideration.
3. *Drying tests.* Drying tests should be conducted in those dryers still under consideration. These tests will determine the optimum operating conditions and the product characteristics and will form the basis for firm quotations from equipment vendors.
4. *Final selection of dryer.* From the results of the drying tests and quotations, the final selection of the most suitable dryer can be made.

The important factors to consider in the preliminary selection of a dryer are the following:

1. Properties of the material being handled
   a. Physical characteristics when wet
   b. Physical characteristics when dry
   c. Corrosiveness
   d. Toxicity
   e. Flammability
   f. Particle size
   g. Abrasiveness
2. Drying characteristics of the material
   a. Type of moisture (bound, unbound, or both)
   b. Initial moisture content
   c. Final moisture content (maximum)
   d. Permissible drying temperature
   e. Probable drying time for different dryers
3. Flow of material to and from the dryer
   a. Quantity to be handled per hour
   b. Continuous or batch operation
   c. Process prior to drying
   d. Process subsequent to drying
4. Product qualities
   a. Shrinkage
   b. Contamination
   c. Uniformity of final moisture content
   d. Decomposition of product
   e. Overdrying
   f. State of subdivision
   g. Product temperature
   h. Bulk density
5. Recovery problems
   a. Dust recovery
   b. Solvent recovery
6. Facilities available at site of proposed installation
   a. Space
   b. Temperature, humidity, and cleanliness of air
   c. Available fuels
   d. Available electric power
   e. Permissible noise, vibration, dust, or heat losses
   f. Source of wet feed
   g. Exhaust-gas outlets

The physical nature of the material to be handled is the primary item for consideration. A slurry will demand a different type of dryer from that required by a coarse crystalline solid, which, in turn, will be different from that required by a sheet material (Table 20-4).

Following preliminary selection of suitable types of dryers, a high-spot evaluation of the size and cost should be made to eliminate those which are obviously uneconomical. Information for this evaluation can be obtained from material presented under discussion of the various dryer types. When data are inadequate, preliminary cost and performance data can usually be obtained from the equipment manufacturer. In comparing dryer performance, the factors in the preceding list which affect dryer performance should be properly weighed. The possibility of eliminating or simplifying processing steps whch precede or follow drying, such as filtration, grinding, or conveying, should be carefully considered.

DRYING TESTS

These tests should establish the optimum operating conditions, the ability of the dryer to handle the material physically, product quality and characteristics, and dryer size. The principal manufacturers of drying equipment are usually prepared to perform the required tests on dryers simulating their equipment. Occasionally, simple laboratory experiments can serve to reduce further the number of dryers under consideration.

TABLE 20-4 Classification of Commercial Dryers Based on Materials Handled

Type of dryer	Liquids	Slurries	Pastes and sludges	Free-flowing powders	Granular, crystalline, or fibrous solids	Large solids, special forms and shapes	Continuous sheets	Discontinuous sheets
	True and colloidal solutions; emulsions. Examples: inorganic salt solutions, extracts, milk, blood, waste liquors, rubber latex, etc.	Pumpable suspensions. Examples: pigment slurries, soap and detergents, calcium carbonate, bentonite, clay slip, lead concentrates, etc.	Examples: filter-press cakes, sedimentation sludges, centrifuged solids, starch, etc.	100 mesh or less. Relatively free flowing in wet state. Dusty when dry. Examples: centrifuged precipitates, pigments, clay, cement.	Larger than 100 mesh. Examples: rayon staple, salt crystals, sand, ores, potato strips, synthetic rubber.	Examples: pottery, brick, rayon cakes, shotgun shells, hats, painted objects, rayon skeins, lumber.	Examples: paper, impregnated fabrics, cloth, cellophane, plastic sheets.	Examples: veneer, wallboard, photograph prints, leather, foam rubber sheets.
Tray and compartment. Direct type, batch operation	Not applicable	For very small batch production. Laboratory drying	Suited to batch operation. At large capacities, investment and operating costs are high. Long drying times	Dusting may be a problem. See comments under Pastes and Sludges	Suited to batch operation. At large capacities, investment and operating costs are high. Long drying times	See comments under Granular solids	Not applicable	See comments under Granular solids
Batch through-circulation. Direct type, batch operation	Not applicable	Not applicable	Suitable only if material can be preformed. Suited to batch operation. Shorter drying time than tray dryers	Not applicable	Usually not suited for materials smaller than 30 mesh. Suited to small capacities and batch operation	Primarily useful for small objects	Not applicable	Not applicable
Tunnel. Continuous Tray. Direct type, continuous operation	Not applicable	Not applicable	Suitable for small and large-scale production.	See comments under Pastes and Sludges. Vertical-turbo applicable	Essentially large-scale, semicontinuous tray drying.	Suited to a wide variety of shapes and forms. Operation can be made continuous. Widely used	Not applicable	Suited for leather, wallboard, veneer.
Continuous through-circulation. Direct type, continuous operation	Not applicable	Only crystal filter dryer may be suited	Suitable for materials that can be preformed. Will handle large capacities. Roto-louvre requires dry-product recirculation	Not generally applicable, except Roto-louvre in certain cases	Usually not suited for materials smaller than 30 mesh. Material does not tumble, except in Roto-louvre dryer. Latter operates at higher temperatures	Suited to smaller objects that can be loaded on each other. Can be used to convey materials through heated zones. Roto-louvre not suited.	Not applicable	Special designs are required. Suited to veneers. Roto-louvre not applicable
Direct rotary. Direct type, continuous operation	Applicable with dry-product recirculation	Applicable with dry-product recirculation	Suitable only if product does not stick to walls and does not dust. Recirculation of product may prevent sticking	Suitable for most materials and capacities, provided that dusting is not too severe	Suitable for most materials at most capacities. Dusting or crystal abrasion will limit its use	Not applicable	Not applicable	Not applicable
Pneumatic conveying. Direct type, continuous operation	See comments under Slurries	Can be used only if product is recirculated to make feed suitable for handling	Usually requires recirculation of dry product to make suitable feed. Well suited to high capacities. Disintegration usually required	Suitable for materials that are easily suspended in a gas stream and lose moisture readily. Well suited to high capacities	Suitable for materials that are easily suspended in a gas stream. Well suited to high capacities. Product may suffer physical degradation	Not applicable	Not applicable	Not applicable
Spray. Direct type, continuous operation	Suited for large capacities. Product is usually powdery, spherical, and free-flowing. High temperatures can be used with heat-sensitive materials. Product may have low bulk density	See comments under Liquids. Pressure-nozzle atomizers subject to erosion	Requires special pumping equipment to feed the atomizer. See comments under Liquids	Not applicable	Not applicable	Not applicable	Not applicable	Not applicable
Continuous sheeting. Direct type, continuous operation	Not applicable	Not applicable	Not applicable	Not applicable	Not applicable	Not applicable	Different types are available for different requirements. Suitable for drying without contacting hot surfaces	Not applicable
Vacuum shelf. Indirect type, batch operation	Not applicable	Applicable for small-batch production	Suitable for batch operation, small capacities. Useful for heat-sensitive or readily oxidizable materials. Solvents can be recovered	See comments under Pastes and Sludges	Suitable for batch operation, small capacities. Useful for heat-sensitive or readily oxidizable materials. Solvents can be recovered	See comments under Granular solids	Not applicable	See comments under Granular solids

TABLE 20-4 Classification of Commercial Dryers Based on Materials Handled

Type of dryer	Liquids	Slurries	Pastes and sludges	Free-flowing powders	Granular, crystalline, or fibrous solids	Large solids, special forms and shapes	Continuous sheets	Discontinuous sheets
	True and colloidal solutions; emulsions. Examples: inorganic salt solutions, extracts, milk, blood, waste liquors, rubber latex, etc.	Pumpable suspensions. Examples: pigment slurries, soap and detergents, calcium carbonate, bentonite, clay slip, lead concentrates, etc.	Examples: filter-press cakes, sedimentation sludges, centrifuged solids, starch, etc.	100 mesh or less. Relatively free flowing in wet state. Dusty when dry. Examples: centrifuged precipitates, pigments, clay, cement.	Larger than 100 mesh. Examples: rayon staple, salt crystals, sand, ores, potato strips, synthetic rubber.	Examples: pottery, brick, rayon cakes, shotgun shells, hats, painted objects, rayon skeins, lumber.	Examples: paper, impregnated fabrics, cloth, cellophane, plastic sheets.	Examples: veneer, wallboard, photograph prints, leather, foam rubber sheets.
Vacuum freeze. Indirect type, batch or continuous operation	Usually used only for pharmaceuticals such as penicillin and blood plasma. Expensive. Used on heat-sensitive and readily oxidized materials	See comments under Liquids	See comments under Liquids	See comments under Liquids	Expensive. Usually used on pharmaceuticals and related products which cannot be dried successfully by other means. Applicable to fine chemicals.	See comments under Granular solids	Applicable in special cases such as emulsion-coated films	See comments under Granular solids
Pan. Indirect type, batch operation	Atmospheric or vacuum. Suitable for small batches. Easily cleaned. Solvents can be recovered. Material agitated while dried	See comments under Liquids	See comments under Liquids	See comments under Liquids	Suitable for small batches. Easily cleaned. Material is agitated during drying, causing some degradation	Not applicable	Not applicable	Not applicable
Vacuum rotary. Indirect type, batch operation	Not applicable, except when pumping slowly on dry "heel"	May have application in special cases when pumping onto dry "heel"	Use is questionable. Material usually cakes to dryer walls and agitator. Solvents can be recovered	Suitable for non-sticking materials. Useful for large batches of heat-sensitive materials and for solvent recovery	Useful for large batches of heat-sensitive materials or where solvent is to be recovered. Product will suffer some grinding action. Dust collectors may be required	Not applicable	Not applicable	Not applicable
Screw conveyor and indirect rotary. Indirect type, continuous operation	Applicable with dry-product recirculation	Applicable with dry-product recirculation	Generally requires recirculation of dry product. Little dusting occurs	Chief advantage is low dust loss. Well suited to most materials and capacities, particularly those requiring drying at steam temperature	Low dust loss. Material must not stick or be temperature-sensitive	Not applicable	Not applicable	Not applicable
Fluid beds. Batch, continuous, direct, and indirect	Applicable only with inert bed or dry-solids recirculator	See comments under Liquids	See comments under Liquids	Suitable, if not too dusty	Suitable for crystals, granules, and short fibers	Not applicable	Use hot inert particles for contacting	Use hot inert particles for contacting
Vibrating tray. Indirect type, continuous operation	Not applicable	Not applicable	Not applicable	Suitable for free-flowing materials	Suitable for free-flowing materials that can be conveyed on a vibrating tray	Not applicable	Not applicable	Not applicable
Drum. Indirect type, continuous operation	Single, double or twin. Atm. or vacuum operation. Product flaky and usually dusty. Maintenance costs may be high	See comments under Liquids. Twin-drum dryers are widely used	Can be used only when paste or sludge can be made to flow. See comments under Liquids	Not applicable	Not applicable	Not applicable	Not applicable	Not applicable
Cylinder. Indirect type, continuous operation	Not applicable	Not applicable	Not applicable	Not applicable	Not applicable	Not applicable	Suitable for thin or mechanically weak sheets which can be dried in contact with a heated surface. Special surface effects obtainable	Suitable for materials which need not be dried flat and which will not be injured by contact with hot drum
Infrared. Batch or continuous operation	Only for thin films	See comments under Liquids	See comments under Liquids (only for thin layers)	Only for thin layers	Primarily suited to drying surface moisture. Not suited for thick layers	Specially suited for drying and baking paint and enamels	Usually used in conjunction with other methods. Useful when there are space limitations	Useful for laboratory work or in conjunction with other methods
Dielectric. Batch or continuous operation	Very expensive	See comments under Liquids	See comments under Liquids	Very expensive	Very expensive	Rapid drying of large objects suited to this method	Applications for final stages of paper dryers	Successful on foam rubber. Not fully developed on other materials

Once a given type and size of dryer has been installed, the product characteristics and drying capacity can be changed only within relatively narrow limits. Thus it is more economical and far more satisfactory to experiment in small-scale units than on the dryer that is finally installed.

On the basis of the results of the drying tests that establish size and operating characteristics, formal quotations and guarantees should be obtained from dryer manufacturers. Initial costs, installation costs, operating costs, product quality, dryer operability, and dryer flexibility can then be given proper weight in final evaluation and selection.

BATCH TRAYS AND COMPARTMENTS

Description A tray or compartment dryer is an enclosed, insulated housing in which solids are placed upon tiers of trays in the case of particulate solids or stacked in piles or upon shelves in the case of large objects. Heat transfer may be *direct* from gas to solids by circulation of large volumes of hot gas or *indirect* by use of heated shelves, radiator coils, or refractory walls inside the housing. In indirect-heat units, excepting vacuum-shelf equipment, circulation of a small quantity of gas is usually necessary to sweep moisture vapor from the compartment and prevent gas saturation and condensation. Compartment units are employed for the heating and drying of lumber, ceramics, sheet materials (supported on poles), painted and metal objects, and all forms of particulate solids.

Field of Application Because of the high labor requirements usually associated with loading or unloading the compartments, batch compartment equipment is rarely economical except in the following situations:

1. A long heating cycle is necessary because the size of the solid objects or permissible heating temperature requires a long holdup for internal diffusion of heat or moisture. This case may apply when the cycle will exceed 12 to 24 h.
2. The production of several different products requires strict batch identity and thorough cleaning of equipment between batches. This is a situation existing in many small color-pigment-drying plants.
3. The quantity of material to be processed does not justify investment in more expensive, continuous equipment. This case would apply in many pharmaceutical-drying operations.

Further, because of the nature of solids-gas contacting, which is usually by parallel flow and rarely by through circulation, heat transfer and mass transfer are comparatively inefficient. For this reason, use of tray and compartment equipment is restricted primarily to ordinary drying and heat-treating operations. Despite these harsh limitations, when the listed situations do exist, economical alternatives are difficult to develop.

Auxiliary Equipment If noxious gases, fumes, or dust are given off during the operation, dust- or fume-recovery equipment will be necessary in the exhaust-gas system. Wet scrubbers are employed for the recovery of valuable solvents from dryers. In order to minimize heat losses, thorough insulation of the compartment with brick, asbestos, or other insulating compounds is necessary. Modern fabricated dryer-compartment panels usually have 7.5 to 15 cm of blanket insulation placed between the internal and external sheet-metal walls. Doors and other access openings should be gasketed and tight. In the case of tray and truck equipment, it is usually desirable to have available extra trays and trucks so that they can be preloaded for rapid emptying and loading of the compartment between cycles. Air filters and gas dryers are occasionally employed on the inlet-air system for direct-heat units.

Vacuum-shelf dryers require auxiliary stream jets or other vacuum-producing devices, intercondensers for vapor removal, and occasionally wet scrubbers or (heated) bag-type dust collectors.

Uniform depth of loading in dryers and furnaces handling particulate solids is essential to consistent operation, minimum heating cycles, or control of final moisture. After a tray has been loaded, the bed should be leveled to a uniform depth. Special preform devices, noodle extruders, pelletizers, etc., are employed occasionally for preparing pastes and filter cakes so that screen bottom trays can be used and the advantages of through circulation approached.

Control of tray and compartment equipment is usually maintained by control of the circulating-air temperature (and humidity) and rarely by solids temperature. On vacuum units, control of the absolute pressure and heating-medium temperature is utilized. In direct dryers, cycle controllers are frequently employed to vary the air temperature or velocity across the solids during the cycle, e.g., high air temperatures may be employed during a constant-rate drying period while the solids surface remains close to the air wet-bulb temperature. During the falling-rate periods, this temperature may be reduced to prevent casehardening or other degrading effects caused by overheating the solids surfaces. In addition, higher air velocities may be employed during early drying stages to improve heat transfer; however, after surface drying has been completed, this velocity may need to be reduced to prevent dusting. Two-speed circulating fans are employed commonly for this purpose.

Direct-Heat Tray Dryers Satisfactory operation of tray-type dryers depends on maintaining a constant temperature and a uniform air velocity over all the material being dried.

Circulation of air at velocities of 1 to 10 m/s is desirable to improve the surface heat-transfer coefficient and to eliminate stagnant air pockets. Proper air flow in tray dryers depends on sufficient fan capacity, on the design of ductwork to modify sudden changes in direction, and on properly placed baffles. *Nonuniform air flow is one of the most serious problems in the operation of tray dryers.*

Tray dryers may be of the tray-truck or the stationary-tray type. In the former, the trays are loaded on trucks which are pushed into the dryer; in the latter, the trays are loaded directly into stationary racks within the dryer. Trucks may be fitted with flanged wheels to run on tracks or with flat swivel wheels. They may also be suspended from and moved on monorails. Trucks usually contain two tiers of trays, with 18 to 48 trays per tier, depending upon the tray dimensions.

Trays may be square or rectangular, with 0.5 to 1 m^2 per tray, and may be fabricated from any material compatible with corrosion and temperature conditions. When the trays are stacked in the truck, there should be a clearance of not less than 4 cm between the material in one tray and the bottom of the tray immediately above. When material characteristics and handling permit, the trays should have screen bottoms for additional drying area. Metal trays are preferable to nonmetallic trays, since they conduct heat more readily. Tray loadings range usually from 1 to 10 cm deep.

Steam is the usual heating medium, and a standard heater arrangement consists of a main heater before the circulating fan. When steam is not available or the drying load is small, electrical heat can be used. For temperatures above 450 K, products of combustion can be used, or indirect-fired air heaters.

Air is circulated by propeller or centrifugal fans; the fan is usually mounted within or directly above the dryer. Above 450 K, external or water-cooled bearings become necessary. Total pressure drop through the trays, heaters, and ductwork is usually in the range of 2.5 to 5 cm of water. Air recirculation is generally in the order of 80 to 95 percent except during the initial drying stage of rapid evaporation. Fresh air is drawn in by the circulating fan, frequently through dust filters. In most installations, air is exhausted by a separate small exhaust fan with a damper to control air-recirculation rates.

PREDICTION OF HEAT- AND MASS-TRANSFER COEFFICIENTS

In convection phenomena, heat-transfer coefficients depend on the geometry of the system, the gas velocity past the evaporating surface, and the physical properties of the drying gas. In estimating drying rates, the use of heat-transfer coefficients is preferred because they are usually more reliable than mass-transfer coefficients. In calculating mass-transfer coefficients from drying experiments, the partial pressure at the surface is usually inferred from the measured or calculated temperature of the evaporating surface. Small errors in temperature have negligible effect on the heat-transfer coefficient but introduce relatively large errors in the partial pressure and hence in the mass-transfer coefficient.

For many cases in drying, the heat-transfer coefficient can be expressed as

$$h_c = \alpha G^n / D_c p \qquad (20\text{-}25)$$

where h_c = heat-transfer coefficient, J/(m^2·s·K) [Btu/(h·ft^2·°F)]; G = mass velocity of drying gas, kg/(s·m^2) [lb/(h·ft^2)]; D_c = characteristic dimension of the system, m; and α, n, and p are empirical constants. When radiation and conduction effects are negligible, the constant rate of drying from a surface is thus given by the following heat-transfer expression derived from Eqs. (20-10) and (20-25):

$$dw/d\theta = (\alpha G^n A/\lambda D_c^m)(t - t_s') \qquad (20\text{-}26)$$

When the liquid is water and the drying gas is air, t_s' is the wet-bulb temperature.

In order to estimate drying rates from Eq. (20-26) values of the empirical constants are required for the particular geometry under consideration. For flow parallel to plane plates, exponent n has been reported to range from 0.35 to 0.8 [Chu, Lane, and Conklin, *Ind. Eng. Chem.*, **45**, 1856 (1953); Wenzel and White, *Ind. Eng. Chem.*, **51**, 275 (1958)]. The differences in exponent have been attributed to differences in flow pattern in the space above the evaporating surface. In the absence of applicable specific data, the heat-transfer coefficient for the parallel-flow case can be taken, for estimating purposes, as

$$h = 8.8G^{0.8}/D_c^{0.2} \qquad (20\text{-}27)$$

where the experimental data have been weighted in favor of an exponent of 0.8 in conformity with the usual Colburn j factor and average values of the properties of air at 370 K have been incorporated.

Experimental data for drying from flat surfaces have been correlated by using the equivalent diameter of the flow channel or the length of the evaporating surface as the characteristic length dimension in the Reynolds number. However, the validity of one versus the other has not been established. The proper equivalent diameter probably depends at least on the geometry of the system, the roughness of the surface, and the flow conditions upstream of the evaporating surface. For most tray-drying calculations, the equivalent diameter (4 times the cross-sectional area divided by the perimeter of the flow channel) should be used.

For air flow impinging normally to the surface from slots, nozzles, or perforated plates, the heat-transfer coefficient can be obtained from the data of Friedman and Mueller (*Proceedings of the General Discussion on Heat Transfer*, Institution of Mechanical Engineers, London, and American Society of Mechanical Engineers, New York, 1951, pp. 138–142). These investigators give

$$h_c = \alpha G^{0.78} \qquad (20\text{-}28)$$

where the gas mass velocity G is based on the total heat-transfer area and α is dependent on the plate open area, hole or slot size, and spacing between the plate, nozzle, or slot and the heat-transfer surface.

Most efficient performance is obtained with plates having open areas equal to 2 to 3 percent of the total heat-transfer area. The plate should be located at a distance equal to four to six hole (or equivalent) diameters from the heat-transfer surface.

Data from tests employing multiple slots, with a correction calculated for slot width, were reported by Korger and Kizek [*Int. J. Heat Mass Transfer*, London, **9**, 337 (1966)].

Air impingement is commonly employed for drying sheets, film, thin slabs, and coatings. Another application in which it is used is as a secondary heat source on drum and can dryers (see Fig. 20-21).

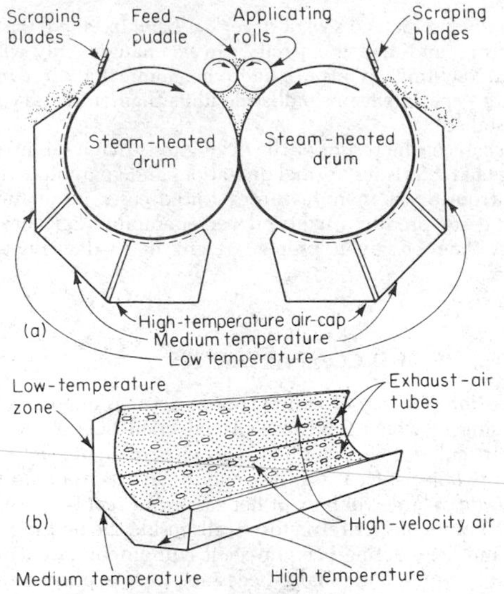

FIG. 20-21 Example of the use of air impingement in drying as a secondary heat source on a double-drum dryer. [Chem. Eng., *197 (June 19, 1967).*]

DETERMINATION OF THE TEMPERATURE OF THE EVAPORATING SURFACE

When radiation and conduction are negligible, the temperature of the evaporating surface approaches the wet-bulb temperature and is readily obtained from the humidity and dry-bulb temperature. Frequently, however, radiation and conduction cause the temperature of the evaporating surface to exceed the wet-bulb temperature. When this occurs, the true surface temperature must be estimated.

Under steady-state conditions the temperature of the evaporating surface increases until the rate of sensible heat transfer to the surface equals the rate of heat removed by evaporation from the surface. To calculate this temperature, it is convenient to modify Eq. (20-10) in terms of humidity rather than partial-pressure difference, as follows:

$$k_g(p_s - p) = k'(H_s - H) \qquad (20\text{-}29)$$

where k' = mass-transfer coefficient, kg/(s·m^2)(unit humidity difference), and $k' = pk_g(M_a/M_w)$ is a suitable approximation at low humidities; k_g = mass-transfer coefficient, kg/(s·m^2·atm); M_a = molecular weight of air; M_w = molecular weight of the diffusing vapor; p_s = vapor pressure of the liquid at the temperature of the evaporating surface, atm; p = partial pressure of vapor in air, atm; H_s = saturation humidity of the air at the temperature of the drying surface, kg/kg dry air; H = humidity of the drying air, kg/kg dry air; and p = total pressure, atm. For air-water mixtures k' is approximately 1.6 k_g at atmospheric pressure.

A rate balance between evaporation and heat transfer when radiation occurs may be modified by means of the psychrometric ratio for air-water vapor mixtures to give:

$$\frac{\lambda}{C_s}(H_s - H) = (t - t_s') + \frac{h_t\epsilon}{h_c}(t_r - t_s') \qquad (20\text{-}30)$$

where λ = latent heat of evaporation, J/kg at t_s'; $C_s = h_c/k'$, where h_c = convection heat-transfer coefficient, J/(s·m^2·K), and C_s = heat capacity of humid air, J/(kg dry air·K), as defined in the subsection "Application of Psychrometry to Drying"; t = temperature of drying gases, K; t_s' = temperature of the wet surface, K; h_c = radiation heat-transfer coefficient, J/(s·m^2·K); h_c = convection heat-transfer coefficient, J/(s·m^2·K); t_r = temperature of source radiating heat to the wet surface, K; and ϵ = emissivity of surface receiving radiation.

Equation (20-30) may be solved by trial and error or graphically to estimate the true values of H_s and t_s' and, hence, the actual drying rate. The values of λ and h_r depend on the value of t_s' but can generally be considered constant over the range of temperatures usually encountered in air drying.

Frequently, particularly in tray drying, heat arrives at the evaporating surface from the tray walls by conduction through the wet material. For this case, in which both radiation and conduction are significant, the total heat-transfer coefficient is given by Shepherd, Brewer, and Hadlock [*Ind. Eng. Chem.*, **30**, 388 (1938)] as

$$h_t = (h_c + h_r)\left[1 + \frac{A_u}{1 + d(h_c + h_r)/k}\right] \qquad (20\text{-}31)$$

where h_t = total heat-transfer coefficient, J/(s·m²·K); A_u = ratio of outside unwetted surface to evaporating surface area; d = depth of material in tray, m; and k = thermal conductivity of the wet material, J/[(s·m²)(K/m)]. Note that h_c must be corrected for emissivity of the surface. For insulated trays, the arithmetic average of inside and outside unwetted area should be used.

Equation (20-31) assumes that all heat sources are at the same temperature and that the convection coefficients to the evaporating surface and to the unwetted portions of the tray are equal. When radiation occurs from a source at a different temperature, the radiation coefficient can be corrected to the same basis by multiplying by the ratio $(t - t'_s)/(t_r - t'_s)$, where t, t'_s, and t_r are the drying-gas, evaporating-surface, and radiator temperatures respectively.

A relationship for estimating the surface temperature t'_s, based on the use of Eq. (20-31) to determine h_t, is as follows:

$$(H_s - H) = (h_t C_s/\lambda h_c)(t - t'_s) \qquad (20\text{-}32)$$

Equation (20-32) can be solved numerically or graphically. Figure 20-22 indicates how H_s and t'_s may be determined graphically on a humidity chart by the point of intersection on the saturation-humidity curve of a straight line of slope $h_t C_s/\lambda h_c$ passing through point (H_s, t_s).

PERFORMANCE DATA

A standard two-truck dryer is illustrated in Fig. 20-23. Adjustable baffles or a perforated distribution plate is normally employed to develop 0.3 to 1.3 cm of water-pressure drop at the wall through which air enters the truck enclosure. This will enhance the uniformity of air distribution, from top to bottom, among the trays. In three (or more) truck ovens, air-reheat coils may be placed between trucks if the evaporative load is high. Means for reversing air-flow direction may also be provided in multiple-truck units.

Performance data on some typical tray and compartment dryers are tabulated in Table 20-5. These indicate that an overall rate of evaporation of 0.0025 to 0.025 kg water/(s·m²) of tray area may be expected from tray and tray-truck dryers. The thermal efficiency of this type of dryer will vary from 20 to 50 percent, depending on the drying temperature used and the humidity of the exhaust air. In drying to very low moisture contents under temperature restrictions, the thermal efficiency may be in the order of 10 percent. The major operating cost for a tray dryer is the labor involved in loading and unloading the trays. About two labor-hours are required to load and unload a standard two-truck tray dryer. In addition, about one-third

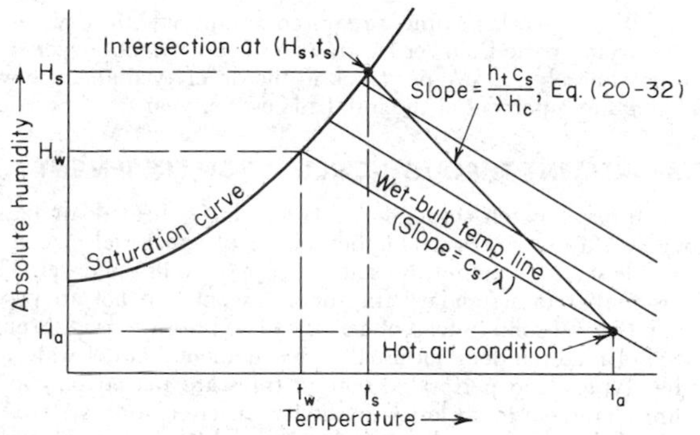

FIG. 20-22 Graphical estimation of surface temperature during constant-rate period.

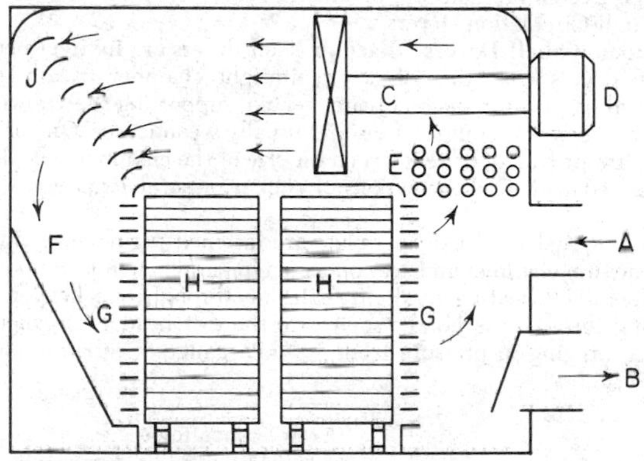

FIG. 20-23 Double-truck dryer. (*A*) Air-inlet duct. (*B*) Air-exhaust duct with damper. (*C*) Adjustable-pitch fan 1 to 15 hp. (*D*) Fan motor. (*E*) Fin heaters. (*F*) Plenum chamber. (*G*) Adjustable air-blast nozzles. (*H*) Trucks and trays. (*J*) Turning vanes. (*National Drying Machinery Co.*)

TABLE 20-5 Manufacturer's Performance Data for Tray and Tray-Truck Dryers*

Material	Color	Chrome yellow	Toluidine red	Half-finished Titone	Color
Type of dryer	2-truck	16-tray dryer	16-tray	3-truck	2-truck
Capacity, kg product/h	11.2	16.1	1.9	56.7	4.8
Number of trays	80	16	16	180	120
Tray spacing, cm	10	10	10	7.5	9
Tray size, cm	60 × 75 × 4	65 × 100 × 2.2	65 × 100 × 2	60 × 70 × 3.8	60 × 70 ×2.5
Depth of loading, cm	2.5 to 5	3	3.5	3	
Initial moisture, % bone-dry basis	207	46	220	223	116
Final moisture, % bone-dry basis	4.5	0.25	0.1	25	0.5
Air temperature, °C	85–74	100	50	95	99
Loading, kg product/m²	10.0	33.7	7.8	14.9	9.28
Drying time, h	33	21	41	20	96
Air velocity, m/s	1.0	2.3	2.3	3.0	2.5
Drying, kg water evaporated/(h·m²)	0.59	65	0.41	1.17	0.11
Steam consumption, kg/kg water evaporated	2.5	3.0	. . .	2.75	
Total installed power, kW	1.5	0.75	0.75	2.25	1.5

*Courtesy of Proctor & Schwartz, Inc.

to one-fifth of a worker's time is required to supervise the dryer during the drying period. Power for tray and compartment dryers will be approximately 1.1 kW per truck in the dryer. Maintenance will run from 3 to 5 percent of the installed cost per year.

COMPARTMENT THROUGH-CIRCULATION EQUIPMENT

In one type of batch through-circulation dryer, heated air passes through a stationary permeable bed of the wet material placed on removable screen-bottom trays suitably supported in the dryer. This type is similar to a standard tray dryer except that hot air passes through the wet solid instead of across it. The pressure drop through the bed of material does not usually exceed about 2 cm of water. In another type, deep perforated-bottom trays are placed on top of plenum chambers in a closed-circuit hot-air-circulating system. In some food-drying plants, the material is placed in finishing bins with perforated bottoms; heated air passes up through the material and is removed from the top of the bin, reheated, and recirculated. The latter types involve a pressure drop through the bed of material of 1 to 8 cm of water at relatively low air rates. Table 20-6 gives performance data on three applications of batch through-circulation dryers. Batch through-circulation dryers are restricted in application to granular materials that permit free flow-through circulation of air. Drying times are usually much shorter than in parallel-flow tray dryers. Design methods are included in the subsection "Continuous Through-Circulation Dryers."

TABLE 20-6 Performance Data for Batch Through-Circulation Dryers*

Kind of material	Granular polymer	Vegetable	Vegetable seeds
Capacity, kg product/h	122	42.5	27.7
Number of trays	16	24	24
Tray spacing, cm	43	43	43
Tray size, cm	91.4 × 104	91.4 × 104	85 × 98
Depth of loading, cm	7.0	6	4
Physical form of product	Crumbs	0.6-cm diced cubes	Washed seeds
Initial moisture content, % dry basis	11.1	669.0	100.0
Final moisture content, % dry basis	0.1	5.0	9.9
Air temperature, °C	88	77 dry-bulb	36
Air velocity, superficial, m/s	1.0	0.6 to 1.0	1.0
Tray loading, kg product/m²	16.1	5.2	6.7
Drying time, h	2.0	8.5	5.5
Overall drying rate, kg water evaporated/(h·m²)	0.89	11.86	1.14
Steam consumption, kg/kg water evaporated	4.0	2.42	6.8
Installed power, kW	7.5	19	19

*Courtesy of Proctor & Schwartz, Inc.

Vacuum-Shelf Dryers Vacuum-shelf dryers are indirect-heated batch dryers consisting of a vacuumtight chamber usually constructed of cast iron or steel plate, heated, supporting shelves within the chamber, a vacuum source, and usually a condenser. One or two doors are provided, depending on the size of the chamber. The doors are sealed with resilient gaskets of rubber or similar material (Fig. 20-24).

Hollow shelves of flat steel plate are fastened permanently inside the vacuum chamber and are connected in parallel to inlet and outlet headers. The heating medium, entering through one header and passing through the hollow shelves to the exit header, is generally steam, ranging in pressure from 700 kPa gauge to subatmospheric pressure for low-temperature operations. Low temperatures can be provided by circulating hot water, and high temperatures can be obtained by circulating hot oil or Dowtherm. Some small dryers employ electrically heated shelves. The material to be dried is placed in pans or trays on the heated shelves. The trays are generally of metal to ensure good heat transfer between the shelf and the tray.

Vacuum-shelf dryers may vary in size from 1 to 24 shelves, the largest chambers having overall dimensions of 6 m wide, 3 m long, and 2.5 m high.

Vacuum is applied to the chamber and vapor is removed through a large pipe which is connected to the chamber in a manner such

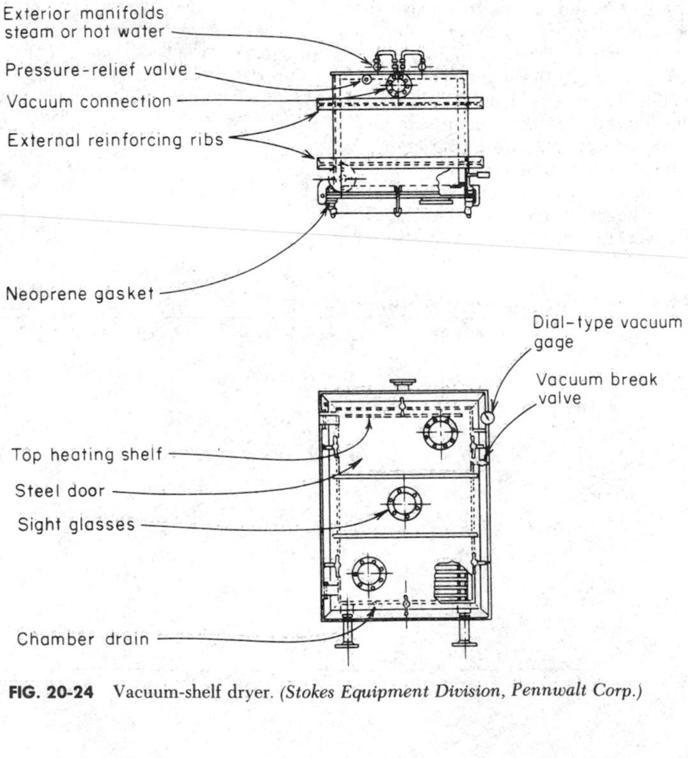

FIG. 20-24 Vacuum-shelf dryer. *(Stokes Equipment Division, Pennwalt Corp.)*

that, if the vacuum is broken suddenly, the inrushing air will not greatly disturb the bed of material being dried. This line leads to a condenser where moisture or solvent that has been vaporized is condensed. The noncondensable exhaust gas goes to the vacuum source, which may be a wet or dry vacuum pump or a steam-jet ejector.

Vacuum-shelf dryers are used extensively for drying pharmaceuticals, temperature-sensitive or easily oxidizable materials, and materials so valuable that labor cost is insignificant. They are particularly useful for handling small batches of materials wet with toxic or valuable solvents. Recovery of the solvent is easily accomplished without danger of passing through an explosive range. Dusty materials may be dried with negligible dust loss. Hygroscopic materials may be completely dried at temperatures below that required in atmospheric dryers. The equipment is employed also for freeze-drying processes (Sec. 17), for metallizing-furnace operations, and for the manufacture of semiconductor parts in controlled atmospheres. All these latter processes demand much lower operating pressures than do ordinary drying operations.

Design Methods Heat is transferred to the wet material by conduction through the shelf and bottom of the tray and by radiation from the shelf above. The critical moisture content will not be necessarily the same as for atmospheric tray drying [Ernst, Ridgway, and Tiller, *Ind. Eng. Chem.*, **30**, 1122 (1938)].

During the constant-rate period, moisture is rapidly removed. Often 50 percent of the moisture will evaporate in the first hour of a 6- to 8-h cycle. The drying time has been found to be proportional to between the first and second power of the depth of loading. Shelf vacuum dryers operate in the range of 1- to 25-mmHg pressure. For size-estimating purposes, a heat-transfer coefficient of 20 J/(m^2·s·K) may be used. The area employed in this case should be the shelf area in direct contact with the trays. Trays should be maintained as flatly as possible to obtain maximum area of contact with the heated shelves. For the same reason, the shelves should be kept free from scale and rust. Air vents should be installed on steam-heated shelves to vent noncondensable gases. The heating medium should not be applied to the shelves until after the air has been evacuated from the chamber in order to reduce the possibility of the material's overheating or boiling at the start of drying. Casehardening can sometimes be avoided by retarding the rate of drying in the early part of the cycle.

Performance Data The purchase price of a vacuum-shelf dryer depends upon the cabinet size and number of shelves per cabinet. For estimating purposes, typical prices and auxiliary-equipment requirements are given in Table 20-7. Installed cost of the equipment will be roughly 100 percent of the carbon steel purchase cost.

The thermal efficiency of a vacuum-shelf dryer is usually on the order of 60 to 60 percent. Table 20-8 gives operating data for one organic color and two inorganic compounds. Labor may constitute 50 percent of the operating cost; maintenance, 20 percent. Annual maintenance costs amount to 5 to 10 percent of the total installed cost. Actual labor costs will depend on drying time, facilities for loading and unloading trays, etc. The power required for these dryers is only that for the vacuum system; for vacuums of 680 to 735 mmHg the power requirements are in the order of 0.06 to 0.12 kW/m^2 tray surface.

Batch Furnaces These are employed mainly for the heat treating of metals, such as annealing, normalizing, and "drawing" (tempering), and for the drying and calcination of ceramic articles. Many specialized furnaces have been designed for these purposes and may be either batch or continuous in operation. Batch furnaces are used in chemical processing for the same purposes as batch tray and truck dryers, when the drying or process temperature exceeds that which can be tolerated by unlined metal walls; ordinary tray and truck dryers are rarely employed when the circulating-gas temperature will exceed 600 to 700 K. They are employed for small-batch calcination, thermal decompositions, and other chemical reactions; these are the same as those reactions performed on a larger scale in rotary kilns, hearth furnaces, and shaft furnaces.

Design procedures and information on heat release in furnaces are given in Sec. 11. Tables indicating normal operating temperatures in various heating furnaces and the more common process furnaces are also included. Specialized designs of batch furnaces are shown in Figs. 20-25 to 20-28 and are described briefly in the following paragraphs. All may be heated by gas, oil, or electricity. **Standard oven furnaces** are similar in design to the small muffle furnace depicted in Fig. 20-28, but with the muffle housing eliminated.

Description **Forced-convection pit furnaces** are employed for heat-treating small metal parts in bulk. Small pieces are suspended in a mesh-bottom basket, while larger pieces are placed on racks. Air heating is by means of Nichrome electric coils set in refractory walls around the periphery of the pit. A high-velocity fan beneath the basket circulates heated air up past the coils and then down through the basket. Some heat is radiated to the outer basket shell, but most is transferred by direct convection from the circulating gas to the solids.

Car-bottom furnaces differ from standard types in that the charge is placed upon movable cars for running into the furnace enclosure. The top of the car is refractory-lined and forms the furnace hearth. The top only is exposed to heat, the lower metal structure being protected by the hearth brick, sand, and water seals at the sides and ends and by the circulation of cooling air around the car structure below the hearth. For use where floor space is limited **elevator furnaces** serve similar purposes.

The **rotary-hearth furnace** consists of a heating chamber lined with refractory brick within which is an annular-shaped refractory-lined rotating hearth. Around the periphery of the rotating hearth, sand or circulating liquid seals are employed to prevent air infiltra-

TABLE 20-8 Performance Data of Vacuum-Shelf Dryers

Material	Sulfur black	Calcium carbonate	Calcium phosphate
Loading, kg dry material/m^2	25	17	33
Steam pressure, kPa gauge	410	410	205
Vacuum, mmHg	685–710	685–710	685–710
Initial moisture content, % (wet basis)	50	50.3	30.6
Final moisture content, % (wet basis)	1	1.15	4.3
Drying time, h	8	7	6
Evaporation rate, kg/(s·m^2)	8.9×10^{-4}	7.9×10^{-4}	6.6×10^{-4}

TABLE 20-7 Standard Vacuum-Shelf Dryers*

Shelf area, m^2	Floor space, m^2	Weight average, kg	Pump capacity, m^3/s	Pump motor, kW	Condenser area, m^2	Price/m^2 Carbon steel	Price/m^2 304 stainless steel
0.4–1.1	4.5	540	0.024	1.12	1	$60	$95
1.1–2.2	4.5	680	0.024	1.12	1	40	60
2.2–5.0	4.6	1130	0.038	1.49	4	25	35
5.0–6.7	5.0	1630	0.038	1.49	4	20	35
6.7–14.9	6.4	3900	0.071	2.24	9	15	25
16.7–21.1	6.9	5220	0.071	2.24	9	15	20

*Stokes Equipment Division, Pennwalt Corp.

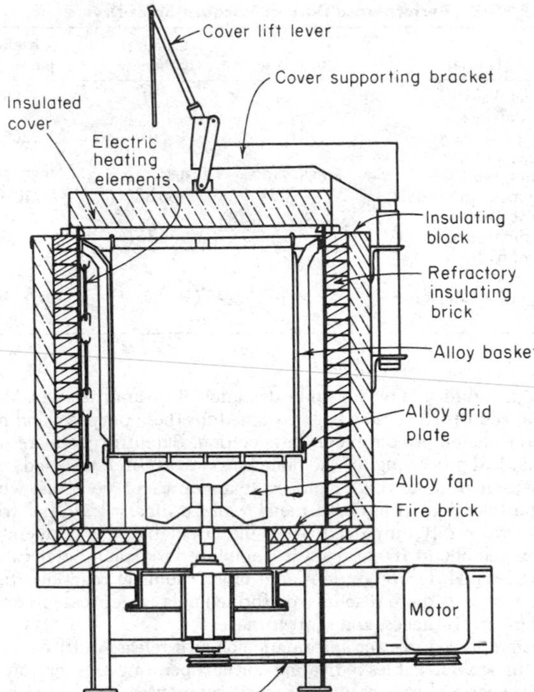

FIG. 20-25 Pit furnace. *(W. S. Rockwell Co.)*

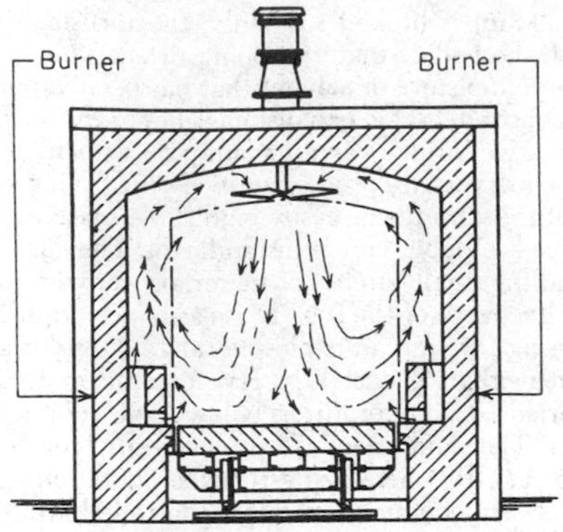

FIG. 20-26 Car-bottom furnace. *(W. S. Rockwell Co.)*

tion. It can be made semicontinuous in operation. The hearth speed can be varied to meet changing requirements in size, weight, and load of the charge. For gas and oil heating, the burners fire from the sides of the chamber, tangentially to the hearth.

Standard furnaces are usually direct-heated, in that the burner combustion gases circulate directly over the charge; occasionally the flame may be permitted to impinge on the charge. For bright annealing, tool hardening, powdered-metal sintering, and other work requiring protection of the charge by special atmospheres, **muffle-type furnaces** are frequently employed. In these, the charge is separated from the burners and combustion gases by a refractory arch. Heat is transferred by hot-gas radiation and convection to the arch and by radiation from the arch to the charge.

When used for **ceramic heating,** furnaces are called kilns. Operations include drying, oxidation, calcination, and vitrification. These kilns employ horizontal space burners with gaseous, liquid, or solid fuels. If product quality is not injured, ceramic ware may be exposed to flame and combustion gases; otherwise, muffle kilns are employed. Dutch ovens are used frequently for heat generation.

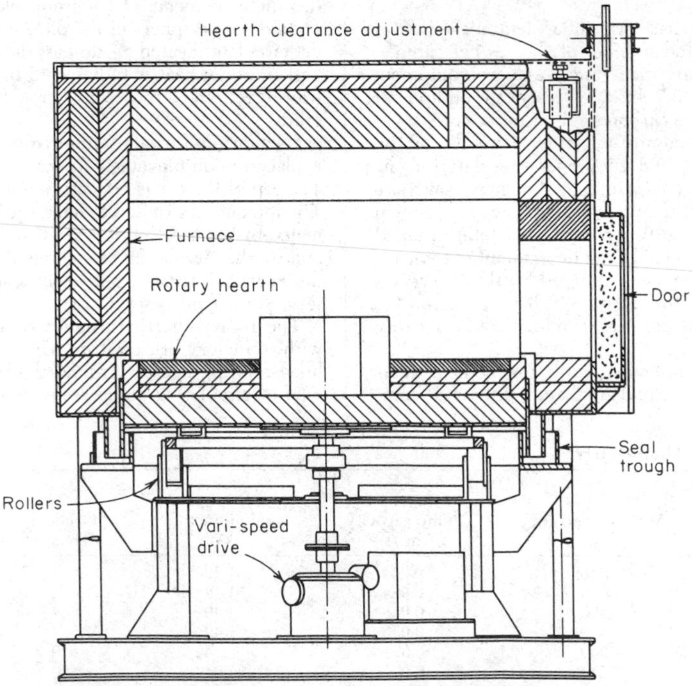

FIG. 20-27 Rotary-hearth furnace. *(W. S. Rockwell Co.)*

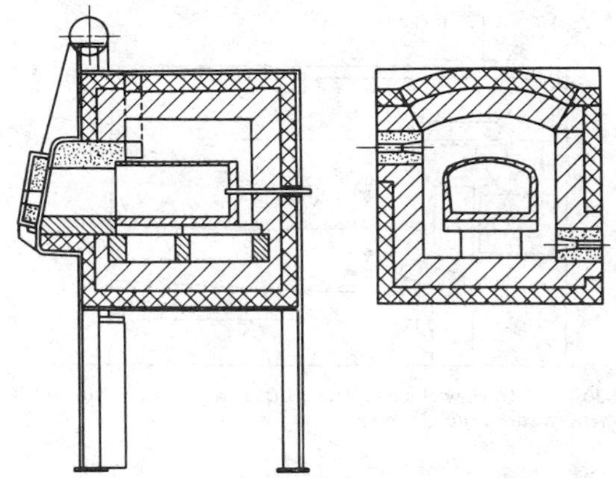

FIG. 20-28 Small muffle furnace. *(W. S. Rockwell Co.)*

Downdraft kilns are the most common type, being used for brick, pipe, tile, and stoneware. The name is derived from the direction of combustion-gas flow when contacting the charge. The gases then flow up inside the walls to the top of the kiln and chimney. **Updraft kilns** are similar except in direction of gas flow, which is upward past the charge. They are employed commonly for pottery burning. **Stove kilns** are variations of updraft kilns used for burning common brick. The kiln is built of green brick and covered with a layer of burned brick. It is completely dismantled after each burning. **Clamp kilns** are another variation of updraft kilns used for common brick and temporary in nature. They have no tops or flue systems but consist only of sidewalls with arched spaces for combustion.

CONTINUOUS TUNNELS

Continuous tunnels are in many cases **batch truck or tray compartments, operated in series.** The solids to be processed are placed in trays or on trucks which move progressively through the tunnel in contact with hot gases. Operation is **semicontinuous;** when the tunnel is filled, one truck is removed from the discharge end as each new truck is fed into the inlet end. In some cases, the trucks move on tracks or monorails, and they are usually conveyed mechanically, employing chain drives connecting to the bottom of each truck. Schematic diagrams of three typical tunnel arrangements are shown in Fig. 20-29. **Belt-conveyor and screen-conveyor tunnels are truly continuous** in operation, carrying a layer of solids on an endless conveyor.

Air flow can be totally **cocurrent, countercurrent,** or a combination of both as shown in Fig. 20-29. In addition, **cross-flow** designs are employed frequently, with the heating air flowing back and forth across the trucks in series. Reheat coils may be installed after each cross-flow pass to maintain constant-temperature operation; large propeller-type circulating fans are installed at each stage, and air may be introduced or exhausted at any desirable points. Tunnel equipment possesses maximum flexibility for any combination of air flow and temperature staging. When handling granular, particulate solids which do not offer high resistance to air flow, perforated or screen-type belt conveyors are employed with **through circulation** of gas to improve heat- and mass-transfer rates.

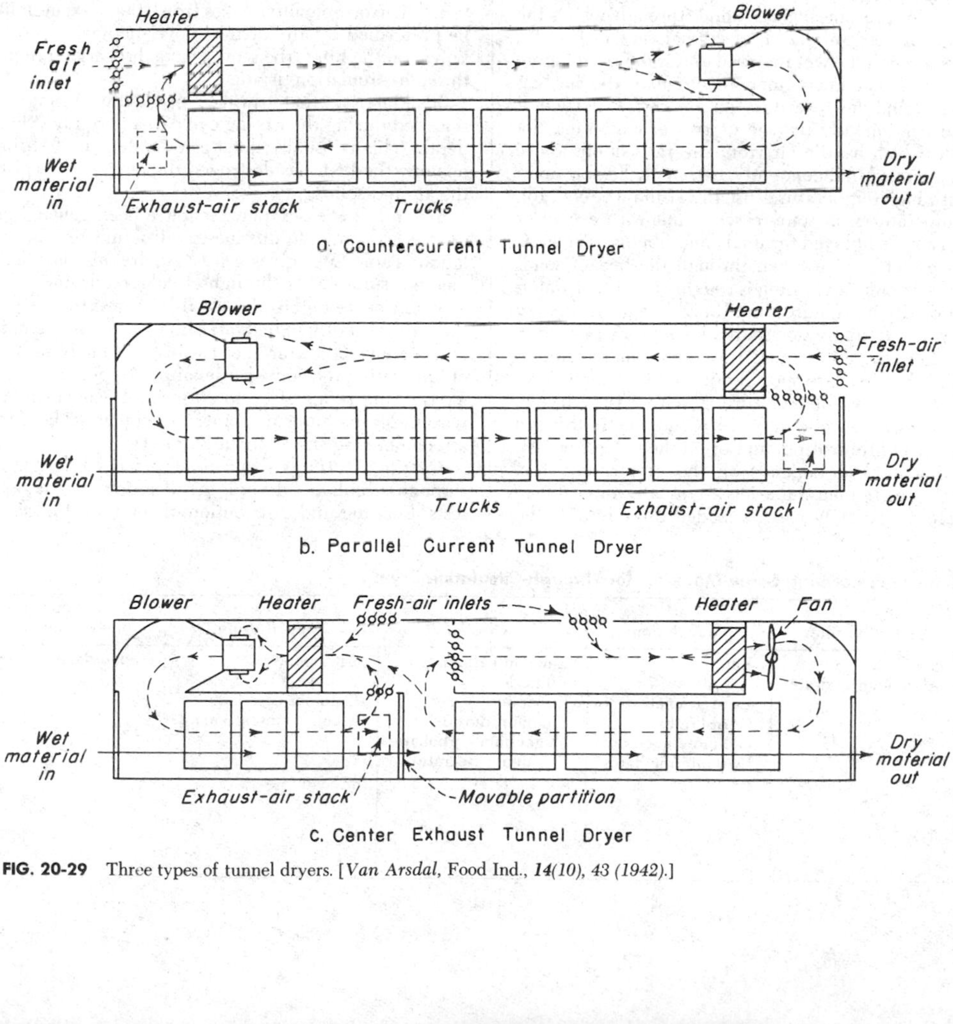

FIG. 20-29 Three types of tunnel dryers. [*Van Arsdal,* Food Ind., *14(10), 43 (1942).*]

In tunnel equipment, the solids are usually heated by direct contact with hot gases. In high-temperature operations, radiation from walls and refractory lining may be significant also. The air in a direct-heat unit may be heated directly or indirectly by combustion or, at temperature below 475 K, by finned steam coils.

Applications of tunnel equipment are essentially the same as for batch tray and compartment units previously described, namely, practically all forms of particulate solids and large solid objects. In operation, they are more suitable for large-quantity production, usually representing investment and installation savings over (multiple) batch compartments. In the case of truck and tray tunnels, labor savings for loading and unloading are not significant compared with batch equipment. Belt and screen conveyors which are truly continuous represent major labor savings over batch operations but require additional investment for automatic feeding and unloading devices.

Auxiliary equipment and the special design considerations discussed for batch trays and compartments apply also to tunnel equipment. For size-estimating purposes, tray and truck tunnels and furnaces can be treated in the same manner as discussed for batch equipment.

Continuous Through-Circulation Dryers Continuous through-circulation dryers operate on the principle of blowing hot air through a permeable bed of wet material passing continuously through the dryer. Dryer rates are high because of the large area of contact and short distance of travel for the internal moisture.

The most widely used type is the **horizontal conveying-screen dryer** in which wet material is conveyed as a layer, 2 to 15 cm deep, on a horizontal mesh screen or perforated apron, while heated air is blown either upward or downward through the bed of material. Its drying characteristics were studied by Marshall and Hougen [*Trans. Am. Inst. Chem. Eng.*, **38**, 91 (1942)]. This dryer consists usually of a number of individual sections, complete with fan and heating coils, arranged in series to form a housing or tunnel through which the conveying screen travels. As shown in the sectional view in Fig. 20-30, the air circulates through the wet material and is reheated before reentering the bed. It is not uncommon to circulate the hot gas upward in the wet end and downward in the dry end. A portion of the air is exhausted continuously by one or two exhaust fans, not shown in the sketch, which handle air from several sections. Since each section can be operated independently, extremely flexible operation is possible, with high temperatures usually at the wet end, followed by lower temperatures; in some cases a unit with cooled or specially humidified air is employed for final conditioning. The **maximum pressure drop** that can be taken through the bed of solids without developing leaks or air bypassing is roughly 50 mm of water.

Through-circulation drying requires that the wet material be in a state of granular or pelleted subdivision so that hot air may be readily blown through it. Many materials meet this requirement without special preparation. Others require special and often elaborate pretreatment to render them suitable for through-circulation drying. The process of converting a wet solid into a form suitable for through circulation of air is called **preforming,** and often the success or failure of this contacting method depends on the preforming step. Fibrous, flaky, and coarse granular materials are usually amenable to drying without preforming. They can be loaded directly onto the conveying screen by suitable spreading feeders of the oscillating-belt or vibrating type or by spiked drums or belts feeding from bins. When materials must be preformed, several methods are available, depending on the physical state of the wet solid.

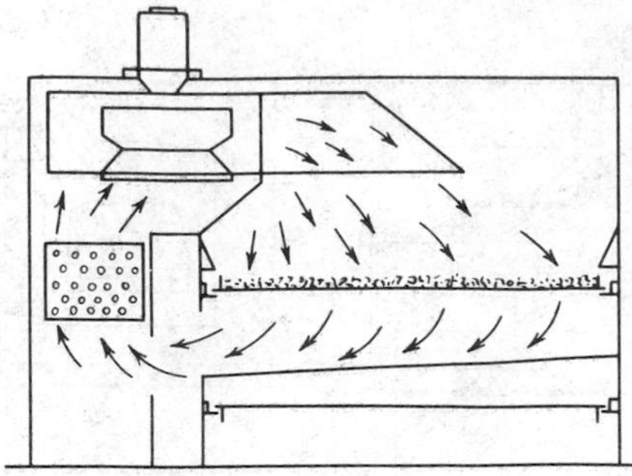

FIG. 20-30 Section view of a continuous through-circulation conveyor dryer. *(Proctor & Schwartz, Inc.)*

1. Relatively dry materials such as centrifuge cakes can sometimes be granulated to give a suitably porous bed on the conveying screen.
2. Pasty materials can often be preformed by extrusion to form sphaghetti-like pieces, about 6 mm in diameter and several centimeters long.
3. Wet pastes that cannot be granulated or extruded may be predried and preformed on a steam-heated finned drum. Preforming on a finned drum may be desirable also in that some predrying is accomplished.
4. Thixotropic filter cakes from rotary vacuum filters that cannot be preformed by any of the above methods can often be scored by knives on the filter, the scored cake discharging in pieces suitable for through-circulation drying.
5. Material that shrinks markedly during drying is often reloaded during the drying cycle to 2 to 6 times the original loading depth. This is usually done after a degree of shrinkage which, by opening the bed, has destroyed the effectiveness of contact between the air and solids.
6. In a few cases, powders have been pelleted or formed in briquettes to eliminate dustiness and permit drying by through circulation. Table 20-9 gives a list of materials classified by preforming methods suitable for through-circulation drying.

Steam-heated air is the usual heat-transfer medium employed in these dryers, although combustion gases may be used also. Temperatures above 600 K are not usually feasible because of the problems of lubricating the conveyor, chain, and roller drives. Recirculation of air is in the range of 60 to 90 percent. Conveyors may be made of wire-mesh screen or perforated-steel plate. The minimum practical screen opening size is about 30 mesh.

Design Methods In actual practice, design of a continuous through-circulation dryer is best based upon data taken in pilot-plant tests. Loading and distribution of solids on the screen are rarely as

TABLE 20-9 Methods of Preforming Some Materials for Through-Circulation Drying

No preforming required	Scored on filter	Granulation	Extrusion	Finned drum	Flaking on chilled drum	Briquetting and squeezing
Cellulose acetate	Starch	Kaolin	Calcium carbonate	Lithopone	Soap flakes	Soda ash
Silica gel	Aluminum hydrate	Cryolite	White lead	Zinc yellow		Cornstarch
Scoured wool		Lead arsenate	Lithopone	Calcium carbonate		Synthetic rubber
Sawdust		Cornstarch	Titanium dioxide	Magnesium carbonate		
Rayon waste		Cellulose acetate	Magnesium carbonate			
Fluorspar		Dye intermediates	Aluminum stearate			
Tapioca			Zinc stearate			
Breakfast food						
Asbestos fiber						
Cotton linters						
Rayon staple						

nearly uniform in commercial installations as in test dryers; 50 to 100 percent may be added to the test drying time for commercial design.

A mathematical method of a through-circulation dryer has been developed by Thygeson [*Am. Inst. Chem. Eng. J.*, **16**(5), 749 (1970)]. Results obtained by Gamson, Thodos, and Hougen [*Trans. Am. Inst. Chem. Eng.*, **39**, I (1943)] and Wilke and Hougen [ibid., **41**, 444 (1945)] for the rates of adiabatic evaporation of water from packed beds of porous solids are applicable when drying gases flow upward to downward. Use of average additive properties of the drying gas leads to

$$h_c = 0.11 \frac{G^{0.59}}{D_p^{0.41}} \quad \text{for} \quad \frac{D_p G}{\mu} > 350 \qquad (20\text{-}33)$$

$$\text{and } h_c = 0.15 \frac{G^{0.49}}{D_p^{0.51}} \quad \text{for} \quad \frac{D_p G}{\mu} < 350 \qquad (20\text{-}34)$$

where μ = gas viscosity, lb/(ft·h); D_p = diameter of sphere having the same surface area as particle, ft; and G = mass velocity of drying gas, lb/(h·ft²) [to convert from pounds per foot-hour to newtons per second–square meter, multiply by 4.133×10^{-4}; to convert from feet to meters, multiply by 0.3048; and to convert from pounds per hour–square foot to kilograms per second–square meter, multiply by 1.3562×10^{-3}].

Performance and Cost Data Experimental performance data are given in Table 20-10 for numerous common materials. Performance data from several commercial through-circulation conveyor dryers are given in Table 20-11. Labor requirements vary depending on the time required for feed adjustments, inspection, etc. These dryers will consume from 0.9 to 1.1 kg of steam/kg of water evaporated. Thermal efficiency is a function of final moisture required and percent air recirculation.

Conveying-screen dryers are fabricated with conveyor widths from 0.3- to 4.4-m sections 1.6 to 2.5 m long. Each section consists of a sheet-metal enclosure, insulated sidewalls and roof, heating coils, a circulating fan, inlet-air distributor baffles, a fines catch pan under the conveyor, and a conveyor screeen (Fig. 20-31). Table 20-12 gives

TABLE 20-10 Experimental Through-Circulation Drying Data for Miscellaneous Materials

Material	Physical form	Moisture contents, kg/kg dry solid			Inlet-air temperature, K	Depth of bed, cm	Loading, kg product/m²	Air velocity, m/s × 10¹	Experimental drying time, s × 10⁻²
		Initial	Critical	Final					
Alumina hydrate	Briquettes	0.105	0.06	0.00	453	6.4	60.0	6.0	18.0
Alumina hydrate	Scored filter cake	9.60	4.50	1.15	333	3.8	1.6	11.0	90.0
Alumina hydrate	Scored filter cake	5.56	2.25	0.42	333	7.0	4.6	11.0	108.0
Aluminum stearate	0.7-cm extrusions	4.20	2.60	0.003	350	7.6	6.5	13.0	36.0
Asbestos fiber	Flakes from squeeze rolls	0.47	0.11	0.008	410	7.6	13.6	9.0	5.6
Asbestos fiber	Flakes from squeeze rolls	0.46	0.10	0.0	410	5.1	6.3	9.0	3.6
Asbestos fiber	Flakes from squeeze rolls	0.46	0.075	0.0	410	3.8	4.5	11.0	2.7
Calcium carbonate	Preformed on finned drum	0.85	0.30	0.003	410	3.8	16.0	11.5	12.0
Calcium carbonate	Preformed on finned drum	0.84	0.35	0.0	410	8.9	25.7	11.7	18.0
Calcium carbonate	Extruded	1.69	0.98	0.255	410	1.3	4.9	14.3	9.0
Calcium carbonate	Extruded	1.41	0.45	0.05	410	1.9	5.8	10.2	12.0
Calcium stearate	Extruded	2.74	0.90	0.0026	350	7.6	8.8	5.6	57.0
Calcium stearate	Extruded	2.76	0.90	0.007	350	5.1	5.9	6.0	42.0
Calcium stearate	Extruded	2.52	1.00	0.0	350	3.8	4.4	10.2	24.0
Cellulose acetate	Granulated	1.14	0.40	0.09	400	1.3	1.4	12.7	1.8
Cellulose acetate	Granulated	1.09	0.35	0.0027	400	1.9	2.7	8.6	7.2
Cellulose acetate	Granulated	1.09	0.30	0.0041	400	2.5	4.1	5.6	10.8
Cellulose acetate	Granulated	1.10	0.45	0.004	400	3.8	6.1	5.1	18.0
Clay	Granulated	0.277	0.175	0.0	375	7.0	46.2	10.2	19.2
Clay	1.5-cm extrusions	0.28	0.18	0.0	375	12.7	100.0	10.7	43.8
Cryolite	Granulated	0.456	0.25	0.0026	380	5.1	34.2	9.1	24.0
Fluorspar	Pellets	0.13	0.066	0.0	425	5.1	51.4	11.6	7.8
Lead arsenate	Granulated	1.23	0.45	0.043	405	5.1	18.1	11.6	18.0
Lead arsenate	Granulated	1.25	0.55	0.054	405	6.4	22.0	10.2	24.0
Lead arsenate	Extruded	1.34	0.64	0.024	405	5.1	18.1	9.4	36.0
Lead arsenate	Extruded	1.31	0.60	0.0006	405	8.4	26.9	9.2	42.0
Kaolin	Formed on finned drum	0.28	0.17	0.0009	375	7.6	44.0	9.2	21.0
Kaolin	Formed on finned drum	0.297	0.20	0.005	375	11.4	56.3	12.2	15.0
Kaolin	Extruded	0.443	0.20	0.008	375	7.0	45.0	10.16	18.0
Kaolin	Extruded	0.36	0.14	0.0033	400	9.6	40.6	15.2	12.0
Kaolin	Extruded	0.36	0.21	0.0037	400	19.0	80.7	10.6	30.0
Lithopone (finished)	Extruded	0.35	0.065	0.0004	408	8.2	63.6	10.2	18.0
Lithopone (crude)	Extruded	0.67	0.26	0.0007	400	7.6	41.1	9.1	51.0
Lithopone	Extruded	0.72	0.28	0.0013	400	5.7	28.9	11.7	18.0
Magnesium carbonate	Extruded	2.57	0.87	0.001	415	7.6	11.0	11.4	17.4
Magnesium carbonate	Formed on finned drum	2.23	1.44	0.0019	418	7.6	13.2	8.6	24.0
Mercuric oxide	Extruded	0.163	0.07	0.004	365	3.8	66.5	11.2	24.0
Silica gel	Granular	4.51	1.85	0.15	400	3.8–0.6	3.2	8.6	15.0
Silica gel	Granular	4.49	1.50	0.215	340	3.8–0.6	3.4	9.1	63.0
Silica gel	Granular	4.50	1.60	0.218	325	3.8–0.6	3.5	9.1	66.0
Soda salt	Extruded	0.36	0.24	0.008	410	3.8	22.8	5.1	51.0
Starch (potato)	Scored filter cake	0.866	0.55	0.069	400	7.0	26.3	10.2	27.0
Starch (potato)	Scored filter cake	0.857	0.42	0.082	400	5.1	17.7	9.4	15.0
Starch (corn)	Scored filter cake	0.776	0.48	0.084	345	7.0	26.4	7.4	54.0
Starch (corn)	Scored filter cake	0.78	0.56	0.098	380	7.0	27.4	7.6	24.0
Starch (corn)	Scored filter cake	0.76	0.30	0.10	345	1.9	7.7	6.7	15.0
Titanium dioxide	Extruded	1.2	0.60	0.10	425	3.0	6.8	13.7	6.3
Titanium dioxide	Extruded	1.07	0.65	0.29	425	8.2	16.0	8.6	6.0
White lead	Formed on finned drum	0.238	0.07	0.001	355	6.4	76.8	11.2	30.0
White lead	Extruded	0.49	0.17	0.0	365	3.8	33.8	10.2	27.0
Zinc stearate	Extruded	4.63	1.50	0.005	360	4.4	4.2	8.6	36.0

TABLE 20-11 Performance Data for Continuous Though-Circulation Dryers*

Kind of material	Inorganic pigment	Cornstarch	Fiber staple	Charcoal briquettes	Gelatin	Inorganic chemical
Capacity, kg dry product/h	712	4536	1724	5443	295	862
Approximate dryer area, m^2	22.11	66.42	Stage A, 57.04; Stage B, 35.12	52.02	104.05	30.19
Depth of loading, cm	3	4	. . . ; . . .	16	5	4
Air temperature, °C	120	115 to 140	130 to 100; 100	135 to 120	32 to 52	121 to 82
Loading, kg product/m^2	18.8	27.3	3.5; 3.3	182.0	9.1	33
Type of conveyor, mm	1.59 by 6.35 slots	1.19 by 4.76 slots	2.57-diameter holes, perforated plate	8.5 × 8.5 mesh screen	4.23 × 4.23 mesh screen	1.59 × 6.35 slot
Preforming method or feed	Rolling extruder	Filtered and scored	Fiber feed	Pressed	Extrusion	Rolling extruder
Type and size of preformed particle, mm	6.35-diameter extrusions	Scored filter cake	Cut fiber	64 × 51 × 25	2-diameter extrusions	6.35-diameter extrusions
Initial moisture content, % bone-dry basis	120	85.2	110	37.3	300	111.2
Final moisture content, % bone-dry basis	0.5	13.6	9	5.3	11.1	1.0
Drying time, min	35	24	11	105	192	70
Drying rate, kg water evaporated/(h·m^2)	38.39	42.97	17.09	22.95	9.91	31.25
Air velocity (superficial), m/s	1.27	1.12	0.66	1.12	1.27	1.27
Heat source per kg water evaporated, steam kg/kg gas (m^3/kg)	Gas 0.11	Steam 2.0	Steam 1.73	Waste heat	Steam 2.83	Gas 0.13
Installed power, kW	29.8	119.3	194.0	82.06	179.0	41.03

*Courtesy of Proctor & Schwartz, Inc.

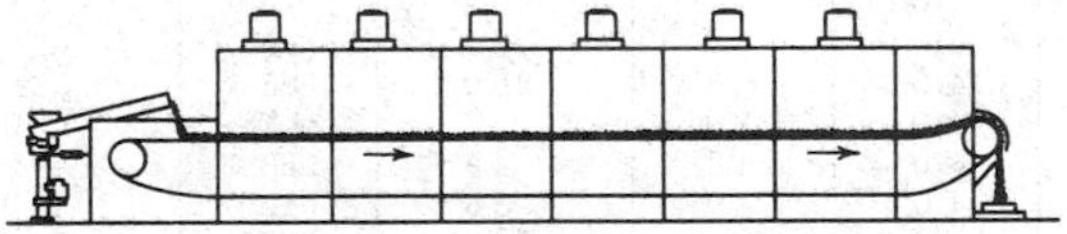

FIG. 20-31 Section assembly of a continuous through-circulation conveyor dryer. *(Proctor & Schwartz, Inc.)*

TABLE 20-12 Conveyor-Screen-Dryer Costs*

2.4-m-wide conveyor		3.0-m-wide conveyor
\$5200/m^2	(6–9 m long)	\$4400/m^2
\$4800/m^2	(9–12 m long)	\$4000/m^2
\$4400/m^2	(12–15 m long)	\$3600/m^2
\$4000/m^2	(15–18 m long)	\$3200/m^2

*National Drying Machinery Company, October 1981.

approximate purchase costs for equipment with type 304 stainless-steel hinged conveyor screens and includes steam-coil heaters, fans, motors, and a variable-speed conveyor drive. Cabinet- and auxiliary-equipment fabrication is of aluminized steel or stainless-steel materials. Prices do not include temperature controllers, motor starters, preform equipment, or auxiliary feed and discharge conveyors. These may add \$30,000 to \$75,000 to the dryer purchase cost (1981 costs).

A continuous conveyor dryer employing a combination of air impingement and through circulation is shown in Fig. 20-32.

Continuous Furnaces Continuous furnaces are employed for the same general duties cited for batch furnaces. Units are gas, oil, or electrically heated and utilize direct circulation of combustion gases or muffles for heat transfer. Continuous furnaces frequently have an extension added for **cooling** the charge before exposure to atmospheric air.

Conveyors may be of parallel-chain, mat, slat, woven wire-mesh belt, or cast-alloy type. Automatic tensioning devices are used to maintain belt tension during heating and cooling. The product may rest directly on the conveyor or on special supports built into it. Roller-conveyors are used for large pieces. **Flame curtains** are provided for sealing the ends and for protection of special treating atmospheres.

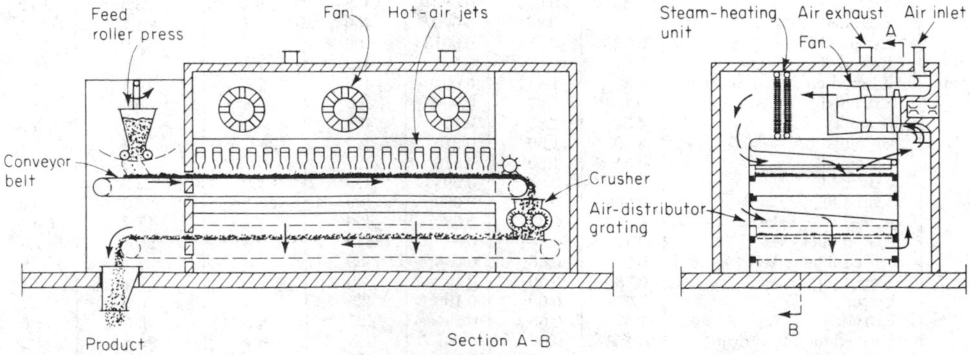

FIG. 20-32 Special conveyor dryer with air jets impinging on surface of bed on first pass. Dried material is crushed and passed again through dryer, with air going through the now-permeable bed. [Chem. Eng., *192 (June 19, 1967).*]

The pusher-type furnace is relatively free from mechanical problems because all mechanical parts are located outside the hot zone. It employs a roller-conveyor usually and will handle charges weighing considerably more per square meter than a belt-conveyor furnace. Pushers are driven by electric motors, compressed air, or hydraulic systems and can be automatically timed and synchronized with door-opening timers. For small solids, trays of perforated metal alloys are used to carry the product. These carriers ride through the tunnel on rollers, skid rails, and occasionally refractory skids, one tray pushing the next ahead. The charge may travel in a straight line or in counterflow movement in single or multiple chambers.

In counterflow movement, heat from the outgoing solids is transferred directly to cold incoming solids, reducing heat losses and fuel requirements. Continuous conveyor ovens are employed also for drying refractory shapes and for drying and baking enameled pieces. In many of these latter, the parts are suspended from overhead chain conveyors.

Ceramic tunnel kilns handling large irregular-shaped objects must be equipped for precise control of temperature and humidity conditions to prevent cracking and condensation on the product. The internal mechanism causing cracking when drying clay and ceramics has been studied extensively. Information on ceramic tunnel-kiln operation and design is reported fully in publications such as *The American Ceramic Society Bulletin*, *Ceramic Industry*, and *Transactions of the British Ceramic Society*.

ROTARY DRYERS

Description A rotary dryer consists of a cylinder, rotated upon suitable bearings and usually slightly inclined to the horizontal. The length of the cylinder may range from 4 to more than 10 times its diameter, which may vary from less than 0.3 to more than 3 m. Feed solids fed into one end of the cylinder progress through it by virtue of rotation, head effect, and slope of the cylinder and discharge as finished product at the other end. Gases flowing through the cylinder may retard or increase the rate of solids flow, depending upon whether gas flow is countercurrent or cocurrent with solids flow.

Rotary dryers have been classified as **direct, indirect-direct, indirect,** and **special types.** The terms refer to the method of heat transfer, being "direct" when heat is added to or removed from the solids by direct exchange between flowing gas and solids and being "indirect" when the heating medium is separated from physical contact with the solids by a metal wall or tube.

Only totally direct and totally indirect types will be discussed extensively here, as it must be recognized that an infinite number of variations between the two are possible. Their operating characteristics when performing heat- and mass-transfer operations make them suitable for the accomplishment of drying, chemical reactions, solvent recovery, thermal decompositions, mixing, sintering, and agglomeration of solids. The specific types included are the following:

Direct rotary dryer (cooler). This is usually a bare metal cylinder, with or without flights. It is suitable for low- and medium-temperature operations, the operating temperature being limited primarily by the strength characteristics of the metal employed in fabrication.

Direct rotary kiln. This is a metal cylinder lined on the interior with insulating block and/or refractory brick. It is suitable for high-temperature operations.

Indirect steam-tube dryer. This is a bare metal cylinder provided with one or more rows of metal tubes installed longitudinally in the shell. It is suitable for operation up to available steam temperatures or in processes requiring water cooling of the tubes.

Indirect rotary calciner. This is a bare metal cylinder surrounded on the outside by a fired or electrally heated furnace. It is suitable for operation at medium temperatures up to the maximum which can be tolerated by the metal wall of the cylinder, usually 650 to 700 K for carbon steel and 800 to 1025 K for stainless steel.

Direct Roto-Louvre dryer. This is one of the more important special types, differing from the direct rotary unit in that true *through circulation* of gas through the solids bed is provided. Like the direct rotary, it is suitable for low- and medium-temperature operation.

Field of Application Rotating equipment is applicable to batch or continuous processing solids which are relatively free-flowing and granular when discharged as product. Materials which are not completely free-flowing in their feed condition are handled in a special manner, either by recycling a portion of final product and premixing with the feed in an external mixer to form a uniform granular feed to the process or by maintaining a bed of free-flowing product in the cylinder at the feed end and, in essence, performing a premixing operation in the cylinder itself. A properly designed recycle process will permit processing of many forms of slurry and solution feeds in rotating vessels. Direct rotary kilns and indirect calciners without internal flights or other obstructions are often provided with **hanging link chains.** These may serve as surfaces upon which material can accumulate until it is no longer sticky, at which time it will break off as a granular solid and continue its movement through the cylinder. **Scraper chains** may also be provided on indirect calciners to maintain clean internal walls.

As a general rule, the direct-heat units are the simplest and most economical in construction and are employed when direct contact between the solids and flue gases or air can be tolerated. Because the total heat load must be introduced or removed in the gas stream, large gas volumes and high gas velocities are usually required. The latter will be rarely less than 0.5 m/s in an economical design. Therefore, employment of direct rotating equipment with solids containing extremely fine particles is likely to result in excessive entrainment losses in the exit-gas stream.

The indirect forms require only sufficient gas flow through the cylinder to remove vapors or otherwise complete the internal process. In addition, these can be sealed for processes requiring special gas atmospheres and exclusion of outside air.

Auxiliary Equipment On direct-heat rotating equipment, a combustion chamber is required for high temperatures and finned steam coils are used for low temperatures. If contamination of the product with combustion gases is undesirable on direct-heat units, indirect gas- or oil-fired air heaters may be employed to achieve temperatures in excess of available steam.

The *method of feeding* rotating equipment depends upon material characteristics and the location and type of upstream processing equipment. When the feed comes from above, a chute extending into the cylinder may be employed. For sealing purposes or if gravity feed is not convenient, a screw feeder is normally used. On cocurrent direct-heat units, cold-water jacketing of the feed chute or conveyor may be desirable if it is contacted by the inlet hot-gas stream. This will prevent overheating of the metal wall with resultant scaling or overheating of heat-sensitive feed materials.

Any type of solids conveyor may be suitabile for **recycle** mixing; however, the most universally applicable is the double-shaft pugmill-type paddle mixer (see Fig. 20-67). This conveyor or mixer should be insulated to prevent excessive heat losses from the hot, dry recycle product. To ensure uniformity in the recycle operation, a **surge storage** reserve of recycle solids should be installed for startup purposes and in the event of interruption of product discharge from the cylinder. In recycle operations, 50 to 60 percent product recirculation is found economical in many instances.

One method of feeding direct cocurrent drying equipment utilizes dryer exhaust gases to convey, mix, and predry wet feed. The latter is added to the exhaust gases, at high velocity, from the dryer. The wet feed, mixed with dust entrained from the dryer, separates from the exhaust gases in a cyclone and drops into the feed end of the cylinder. The technique combines pneumatic and rotary drying. High thermal efficiency results from two cocurrent-flow stages operating countercurrently.

Pneumatic conveyors are frequently employed as both dry-product conveyors and coolers. Other types of cooling equipment often used are screw conveyors, vibrating conveyors, and direct or indirect rotating coolers.

Dust entrained in the exit-gas stream is customarily removed in cyclone collectors. This dust may be discharged back into the process or separately collected. For expensive materials or extremely fine particles, bag collectors may follow a cyclone collector, provided fabric temperature stability is not limiting. When toxic gases or solids are present, the exit gas is at a high temperature, the gas is close to saturation as from a steam-tube dryer, or gas recirculation in a sealed system is involved, wet scrubbers may be used independently or following a cyclone. Cyclones and bag collectors in drying applications frequently require insulation and steam tracing. The exhaust fan should be located downstream from the collection system.

Rotating equipment, except brick-lined vessels, operated above ambient temperatures is usually insulated to reduce heat losses. Exceptions are direct-heat units of bare metal construction operating at high temperatures, on which heat losses from the shell are necessary to prevent overheating of the metal. Insulation is particularly necessary on cocurrent direct-heat units. It is not unusual for product cooling or condensation on the shell to occur in the last 10 to 50 percent of the cylinder length if it is not well insulated.

For best operation, the feed rate to rotating equipment should be closely controlled and uniform in quantity and quality. Because solids temperatures are difficult to measure and changes slowly detected, most rotating-equipment operations are controlled by indirect means. Inlet and exit gas temperatures are measured and controlled on direct-heat units such as direct dryers and kilns, steam temperature and pressure and exit-gas temperature and humidity are controlled on steam-tube units, and direct shell temperature measurements are taken on indirect calciners. Product temperature measurements are taken for secondary control purposes only in most instances.

Equipment which is electrically driven and operated with metal temperatures exceeding 425 K should be provided with auxiliary drives or power sources. Loss of rotation of a heated calciner or high-temperature dryer carrying a heavy bed of hot solids will quickly result in sagging of the cylinder due to nonuniform cooling.

Direct-Heat Rotary Dryers The direct-heat rotary dryer is usually equipped with **flights** on the interior for lifting and showering the solids through the gas stream during passage through the cylinder. These flights are usually offset every 0.6 to 2 m to ensure more continuous and uniform curtains of solids in the gas. The shape of the flights depends upon the handling characteristics of the solids. For free-flowing materials, a radial flight with a 90° lip is employed. For sticky materials, a flat radial flight without any lip is used. When materials change characteristics during drying, the flight design is changed along the dryer length. Many standard dryer designs employ flat flights with no lips in the first one-third of the dryer measured from the feed end, flights with 45° lips in the middle one-third, and flights with 90° lips in the final one-third of the cylinder. **Spiral flights** are usually provided in the first meter or so at the feed end to accelerate forward flow from under the feed chute or conveyor and to prevent leakage over the feed-end retainer ring into the gas seals.

When cocurrent gas-solids flow is used, flights may be left out to the final meter or so at the exit end to reduce entrainment of dry product in the exit gas. Showering of wet feed at the feed end of a countercurrent dryer will, on the other hand, frequently serve as an effective means for scrubbing dry entrained solids from the gas stream before it leaves the cylinder. Some dryers are provided with sawtooth flights to obtain uniform showering, while others use lengths of chain, attached to the underside of the flights, to scrape over and knock the walls of the cylinder, thereby removing sticky solids which might normally adhere to it. In kilns, the chains may contribute significantly to heat transfer; however, their use contributes to maintenance costs when flights are present in direct dryers. Solids sticking on flights and walls are usually removed more efficiently by **external shell knockers.** In dryers of large cross section, internal elements or partitions are sometimes used to increase the effectiveness of material distribution and reduce dusting and impact grinding. Use of internal members increases the difficulty of cleaning and maintenance unless sufficient free area is left between partitions for easy access of a person. Some examples of the more common flight arrangements are shown in Fig. 20-33. Component arrangements of countercurrent direct rotary dryers are shown in Fig. 20-34, and those of a cocurrent unit in Fig. 20-35.

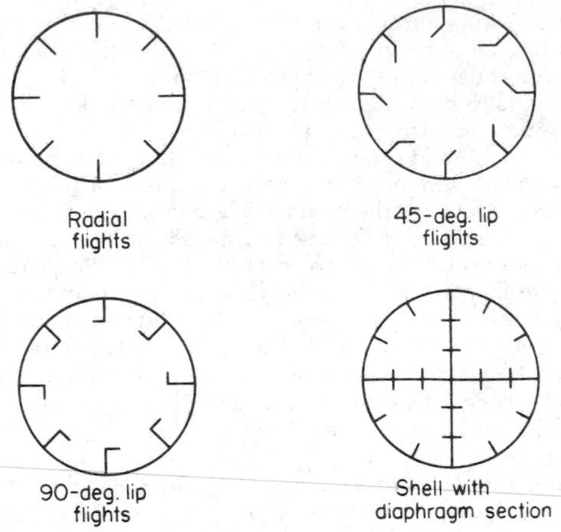

FIG. 20-33 Alternative direct-heat rotary-dryer flight arrangements.

Countercurrent flow of gas and solids gives greater **heat-transfer efficiency** with a given inlet-gas temperature, but cocurrent flow can be used more frequently to dry heat-sensitive materials at higher inlet-gas temperatures because of the rapid cooling of the gas during initial evaporation of surface moisture.

A number of different methods are employed to seal the rotating cylinder and prevent gas leakage through the annular opening between the rotating cylinder and the stationary throat pieces. *None is an effective solids seal*, nor will any function satisfactorily as a gas seal if solids leakage over the retaining ring on the cylinder is permitted. Three examples of ordinary gas seals are shown in Fig. 20-36. On direct rotary dryers, few gas seals are intended to be completely gastight, but by careful control of the internal pressure, generally between 0.25 and 2.5 mm of water below atmosphere, dusting to the outside is prevented and in-leakage of outside air is minimized.

Figure 20-36 also illustrates three basic types of **trunnion roll-bearing assemblies.** Antifriction pillow blocks are the most common on modern dryers; however, when the dryer load requires larger than a 12.7- to 15.2-cm-diameter bearing on the trunnion shaft, the dead-shaft antifriction bearing is substituted. This represents a considerable cost saving compared with the larger pillow blocks. They are completely sealed and continuously bathed in lubricant. Pillow-block bushings are less often used. The thrust washers are difficult to seal against dust, and they draw more power. Thrust roll mountings are depicted also in Fig. 20-36. These are usually dead-shaft.

Gases are forced through the cylinder by either an exhauster or an exhauster-blower combination. With the latter arrangement it is possible to maintain very precise control of internal pressure even when the total system pressure drop is high. When a low-pressure-drop air heater is employed, however, the exhauster alone is usually sufficient, as the major gas pressure losses are found in the exit-air ductwork and dust collectors. Use of a blower by itself to force gas through the cylinder is an unusual practice, because the internal pressure is above atmospheric and hot air and dust may be blown into the gas seals or out into the surrounding working areas.

Special designs of direct rotary dryers, such as the Renneburg DehydrO-Mat (Edward Renneburg & Sons Co.), are constructed especially to provide lower retention during the falling-rate drying period for the escape of internal moisture from the solids. The DehydrO-Mat is a cocurrent dryer employing a small-diameter shell at the feed end, where rapid evaporation of surface moisture in the stream of initially hot gas is accomplished with low holdup. At the solids- and gas-exit end, the shell diameter is increased to reduce gas velocities and provide increased holdup for the solids while they are exposed to the partially cooled gas stream.

The Louisville type P dryer is a cocurrent dryer developed for the drying of heat-sensitive polymers. It is designed for use on rather finely divided and bulky materials which are easily airborne since its

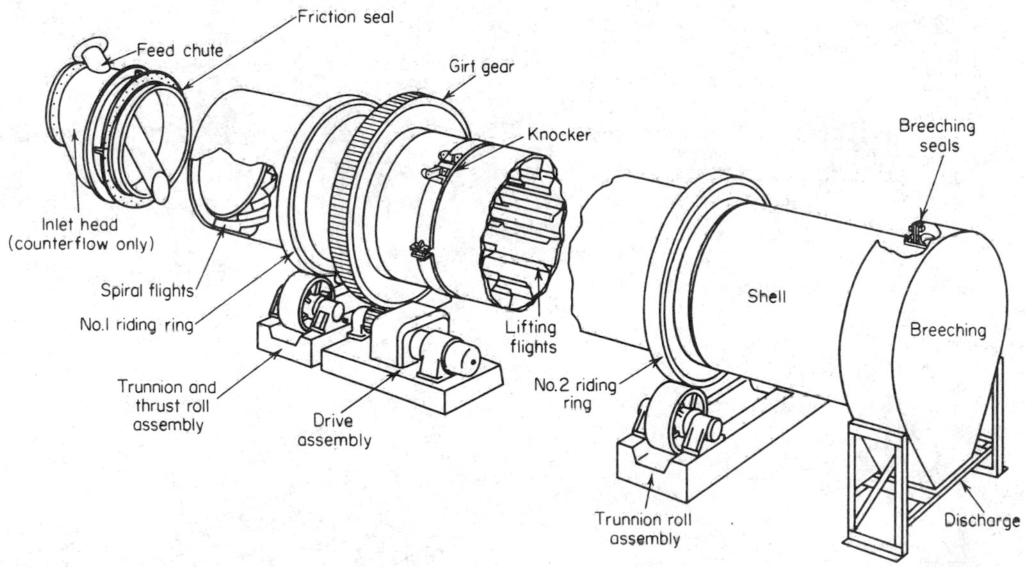

FIG. 20-34 Component arrangements of a countercurrent direct-heat rotary dryer. *(CE Raymond/Bartlett-Snow Co.)*

basic design utilizes a discharge cone permitting pneumatic conveying of dried solids from the dryer. Its internal design provides additional retention time by slowing the progress of the material through the dryer cylinder, permitting a comparatively high velocity of the drying medium without excessive blow-through.

The Louisville type H dryer is a modified cocurrent dryer with "flash-drying" characteristics. Its internal arrangements consist of alternating disks and doughnuts which give high differential velocities between the drying medium and the solids being processed to increase the heat transfer and hence the rate of moisture removal.

Design Methods Direct drying in a direct-heat rotary dryer is best expressed as a heat-transfer mechanism as follows:

$$Q_t = UaV(\Delta t)_m \tag{20-35}$$

where Q_t = total heat transferred, J/s; Ua = volumetric heat-transfer coefficient, J/(s·m³·K); V = dryer volume, m³; and $(\Delta t)_m$ = true mean temperature difference between the hot gases and material, K. When a considerable quantity of surface moisture is removed from the solids and the solids temperatures are unknown, a good approximation of $(\Delta t)_m$ is the logarithmic mean between the wet-bulb depressions of the drying air at the inlet and exit of the dryer.

Data for evaluating Ua were developed by Miller [*Trans. Am. Inst. Chem. Eng.*, **38**, 841 (1942)], Friedman and Marshall [*Chem. Eng. Prog.*, **45**, 482, 573 (1949)], and Seaman and Mitchell [*Chem. Eng. Prog.*, **50**, 467 (1954)]. These authors all employed relationships that could be reduced to the form $Ua = KG^n/D$, where K = a proportionality constant, G = gas mass velocity, D = diameter, and n = a constant.

P. Y. McCormick [*Chem. Eng. Prog.*, **58**(6), 57 (1962)] compared all available data. The comparisons showed that flight geometry and shell speed should be accounted for in the value of K. He suggested that shell rotational speed and flight number and shape must affect the overall balance; however, data for evaluating these variables separately are not available. Also, it is not believed that the effect of gas velocity is toward reducing the surface film thickness around each particle; rather, an increase in gas velocity more likely breaks up more effectively the showering curtains of solids and thereby exposes more solids surface; i.e., the effect is to increase the a in Ua rather than U.

The following relationship is recommended for commercial dryers now manufactured in the United States, which usually have a flight count per circle of 2.4 to 3.0 D and operate at shell peripheral speeds of 60 to 75 ft/min:

$$Q_t = (0.5G^{0.67}/D)V\,\Delta t_m \tag{20-36}$$

$$= 0.4LDG^{0.67}\,\Delta t_m \tag{20-37}$$

where Q_t = total heat transferred, Btu/h; L = dryer length, ft; D = dryer diameter, ft; G = gas mass velocity, lb/(h·ft² of cross sec-

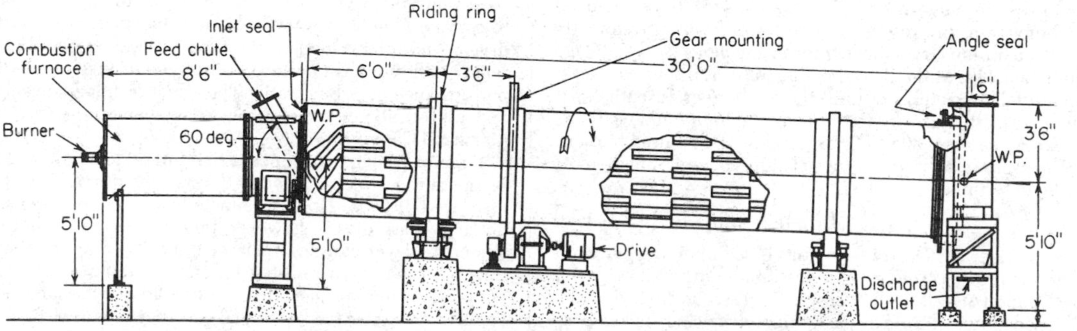

FIG. 20-35 Elevation of a 60-in-diameter by 30-ft-long direct-heat cocurrent rotary dryer. To convert inches to centimeters, multiply by 2.54; to convert feet to meters, multiply by 0.3048. *(CE Raymond/Bartlett-Snow Co.)*

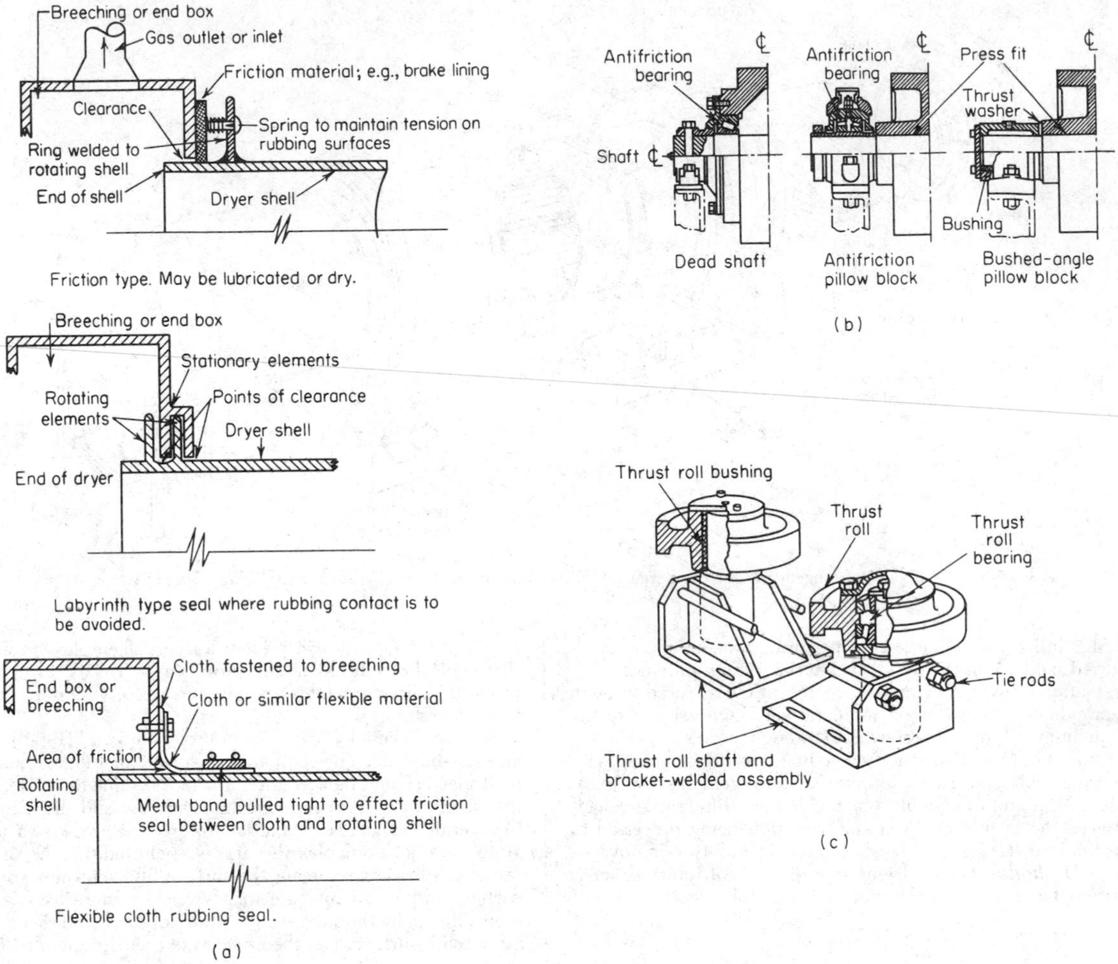

FIG. 20-36 Rotary-dryer components. (*a*) Alternative rotary gas seals. (*b*) Alternative trunnion roll bearings. (*c*) Thrust roll assembly. *(CE Raymond/Bartlett-Snow Co.)*

tion); and Δt_m = log mean of the drying-gas wet-bulb depression at the inlet end and exit end of the dryer shell.

Typical operating data for cocurrent rotary dryers are given in Table 20-13. (Note that the driving force ΔT_m must be based on wet-bulb depression and not on material temperatures. Use of material temperatures, particularly when the dry solids are superheated after drying, will yield conservative results.)

In most cases, direct-heat rotary dryers are still sized on the basis of pilot-plant tests, because rarely is all the moisture to be removed truly "free" moisture, and residence time for diffusion is frequently needed. For maximum dryer heat-transfer efficiency, dryer fillage must be sufficient to load the lifting flights fully, as discussed later.

Unless material characteristics limit the gas temperature, the inlet temperature is usually fixed by the heating medium employed; i.e., 400 to 450 K for steam or 800 to 1100 K for gas- and oil-fired burners. The proper exit-gas temperature is largely an economic function. Its value may be determined as follows:

$$N_t = (t_1 - t_2)/(\Delta t)_m \qquad (20\text{-}38)$$

where N_t = number of heat-transfer units based upon the gas; t_1 = initial-gas temperature, K; t_2 = exit-gas temperature allowing for heat losses, K; and $(\Delta t)_m$ is as defined for Eq. (20-35). Equation (20-38) can be used to select an exit-gas temperature since it has been found (empirically) that rotary dryers are most economically operated between N_t = 1.5 and N_t = 2.5.

The L/D (length-diameter) ratio found most efficient in commercial practice lies between 4 and 10. If the length calculated previously does not fall within these limits, another value of N_t which will place L/D in the proper range may be computed.

Rotary dryers usually operate with 10 to 15 percent of *their volume filled with material*. Lower fillage will be insufficient to utilize the lifters fully, while greater fillage creates the possibility of short-circuiting feed solids across the top of the bed. Under normal fillage conditions, the dryer usually can be made to hold solids long enough to complete the removal of *internal moisture*. If the holdup in the dryer is not great enough, the time of passage may be too short to remove all internal moisture, or because of incomplete flight fillage performance may be erratic. The effect of fillage on retention time and uniformity in rotary dryers has been studied by Miskell and Marshall [*Chem. Eng. Prog.*, **52**, 1 (1956)].

Time of passage is defined as holdup divided by feed rate. It can be measured directly in rotary dryers if holdup and feed rate can be measured directly. Holdup cannot always be measured conveniently on large plant dryers, however, unless a period of shutdown occurs when the dryer can be discharged and its contents weighed. Other methods have been resorted to, one of which consists of adding a pound or two of an inert detectable solid or a radioisotope of a feed constituent to the feed and analyzing for it in the discharged product. The time required for the maximum concentration to occur represents the average time of passage.

TABLE 20-13 Warm-Air Direct-Heat Cocurrent Rotary Dryers: Typical Performance Data*

Material: heat-sensitive solid
Maximum solids temperature: 65°C
Feed conditions: 25 percent moisture, 27°C
Product conditions: 0.5 percent moisture, 65°C
Inlet-air temperature: 165°C
Exit-air temperature: 71°C

Dryer size, m × m	1.219 × 7.62	1.372 × 7.621	1.524 × 9.144	1.839 × 10.668	2.134 × 12.192	2.438 × 13.716	3.048 × 16.767
Evaporation, kg/h	136.1	181.4	226.8	317.5	408.2	544.3	861.8
Work, 10^8 J/h	3.61	4.60	5.70	8.23	1.12	1.46	2.28
Steam, kg/h at kg/m^2 gauge	317.5	408.2	521.6	725.7	997.9	1315	2041
Discharge, kg/h	408	522	635	953	1270	1633	2586
Exhaust velocity, m/min	70	70	70	70	70	70	70
Exhaust volume, m^3/min	63.7	80.7	100.5	144.4	196.8	257.7	399.3
Exhaust fan, kW	3.7	3.7	5.6	7.5	11.2	18.6	22.4
Dryer drive, kW	2.2	5.6	5.6	7.5	14.9	18.6	37.3
Shipping weight, kg	7700	10,900	14,500	19,100	35,800	39,900	59,900
Price, FOB Chicago	$100,000	$125,000	$140,000	$170,000	$210,000	$245,000	$320,000

*Courtesy of Swenson Division, Whiting Corporation.
NOTE:
Assumed pressure drop in system: 200 mm.
System includes finned air heaters, transition piece, dryer, drive, product collector, duct, and fan.
Prices are for carbon steel construction and include entire dryer system (January 1982).
For 304 stainless-steel fabrication, multiply the prices given by 1.5.

The time of passage in rotary dryers can be estimated by the relationships developed by Friedman and Marshall (op. cit.), as given here:

$$\theta = \frac{0.23L}{SN^{0.9}D} \pm 0.6\frac{BLG}{F} \qquad (20\text{-}39)$$

$$B = 5(D_p)^{-0.5} \qquad (20\text{-}40)$$

where B = a constant depending upon the material being handled and approximately defined by Eq. (20-40); D_p = weight average particle size of material being handled, μm; F = feed rate to dryer, lb dry material/(h·ft^2 of dryer cross section); θ = time of passage, min; S = slope, ft/ft; N = speed, r/min; L = dryer length, ft; G = air-mass velocity, lb/(h·ft^2); and D = dryer diameter, ft. The plus sign refers to countercurrent flow and the negative sign to cocurrent flow. To convert from British thermal units per hour to watts, multiply by 0.293; to convert from pounds per hour–square foot of cross section to kilograms per second–square meter, multiply by 0.00135.

Air-mass velocities in rotary dryers usually range from 0.5 to 5.0 kg/(s·m^2). It is customary to employ the highest air velocity possible without serious dusting. The amount of dusting occurring during operation is a complex function of the material being dried, its physical state, the air velocity employed, the holdup in the dryer, the number of flights, the rate of rotation, and the construction of the breeching at the end of the dryer. It can be predicted accurately only by experimental tests. An air rate of 1.4 kg/(s·m^2) can usually be safely used with 35-mesh solids. Information on the dusting of a number of materials in a 0.3- by 2-m rotary dryer has been presented by Friedman and Marshall (ibid.). Rotary dryers operate at peripheral speeds of 0.25 to 0.5 m/s. Slopes of rotary-dryer shells vary from 0 to 8 cm/m. In some cases of cocurrent-flow operations, negative slopes have been used. The radial flight heights in a direct dryer will range from one-twelfth to one-eighth of the dryer diameter. The number of flights will range from 0.6 D to D, where D = diameter, m, for dryers larger than 0.6 m in diameter, and should be designed to carry and shower all the holdup and minimize any kiln action.

Performance and Cost Data Table 20-13 gives estimating-price data for direct rotary dryers employing steam-heated air. Higher-temperature operations requiring combustion chambers and fuel burners will cost more. The total installed cost of rotary dryers including instrumentation, auxiliaries, allocated building space, etc., will run from 150 to 300 percent of the purchase cost. Simple erection costs average 10 to 20 percent of the purchase cost.

Operating costs will include 5 to 10 percent of one worker's time, plus power and fuel required. Yearly maintenance costs will range from 50 to 10 percent of total installed costs. Total power for fans, dryer drive, and feed and product conveyors will be in the range of 0.5 D^2 to 1.0 D^2. Thermal efficiency of a high-temperature direct-heat rotary dryer will range from 55 to 75 percent and, with steam-heated air, from 30 to 55 percent.

A representative list of materials dried in direct-heat rotary dryers is given in Table 20-14.

Direct-Heat Rotary Kilns One of the most important of the high-temperature process furnaces is the direct-fired rotary kiln. It replaces the ordinary rotary dryer when the wall temperature exceeds that which can be tolerated by a bare metal shell (650 to 700 K for carbon steel). Rotary-kiln shells are lined in part or for their entire length with a **refractory brick** to prevent overheating of the steel with resulting weakening. Occasionally two linings are used, the one next to the shell being an **insulating brick.** Insulation is infrequently used on the outside of the shell, and caution must be observed not to overheat the shell metal by this confinement. When wet feeds are applied to a kiln lining at the cold end, there may be leakage of liquid through the lining to the shell, which will cause trouble if the liquid is corrosive.

TABLE 20-14 Representative Materials Dried in Direct-Heat Rotary Dryers*

Material dried	Moisture content, % (wet basis) Initial	Final	Heat efficiency, %
High-temperature:			
Sand	10	0.5	61
Stone	6	0.5	65
Fluorspar	6	0.5	59
Sodium chloride (vacuum salt)	3	0.04	70–80
Sodium sulfate	6	0.1	60
Ilmenite ore	6	0.2	60–65
Medium-temperature:			
Copperas	7	1 (moles)	55
Ammonium sulfate	3	0.10	50–60
Cellulose acetate	60	0.5	51
Sodium chloride (grainer salt)	25	0.06	35
Cast-iron borings	6	0.5	50–60
Styrene	5	0.1	45
Low-temperature:			
Oxalic acid	5	0.2	29
Vinyl resins	30	1	50–55
Ammonium nitrate prills	4	0.25	30–35
Urea prills	2	0.2	20–30
Urea crystals	3	0.1	50–55

*Taken from *Chem. Eng.*, June 19, 1967, p. 190, Table III.

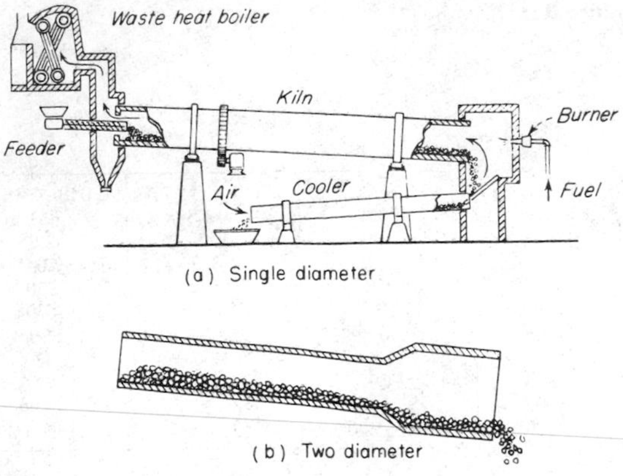

FIG. 20-37 Rotary kilns.

The feed is introduced into the upper end of the kiln by various methods, i.e., inclined chutes, overhung screw conveyors, slurry pipes, etc. Sometimes **ring dams** or chokes of a refractory material are installed within the kiln to build a deeper bed at one or more points, thus changing the flow pattern. The hot product is discharged from the lower end of the kiln into quench tanks, onto conveyors, or into cooling devices which may or may not recover its heat content. These cooling and heat-recovery devices include rotating inclined cylinders, inclined slow-moving grates, shaking grates, etc.

Some kilns have two or three diameters, part of the length being one diameter and the remainder being another diameter. It is claimed that this arrangement increases kiln capacity, decreases fuel consumption, and improves product quality. Two types of kilns are depicted in Fig. 20-37. An enlarged cross section near the discharge end (and hot-gas inlet) reduces the gas velocity and provides increased holdup for a "soaking" period at high temperature.

The first rotary kilns used in the United States were very small, 2 by 20 m. Sizes gradually increased and seemed to stop for a period at a maximum size of 4 by 150 m. A few much larger units have been installed for cement production.

Modern rotary-kiln shells are of all-welded construction. Riding rings are forged or cast steel; support rollers are forged or cast steel and, on rare occasions, tool steel. Main bearings are sleeve-type, normally bronze. Antifriction bearings are frequently used on very small kilns but never on large units. However, bearings on the pinion shafts are normally of the antifriction type.

Gearing is single helical or spur; gear lubrication usually is an automatic spray type. Single drives are used up to 150 kW. Kilns requiring more than 150 kW may be equipped with dual drives, i.e., two driving pinions and two motors, both driving one bull gear. In this manner, the power load is split through two separate driving mechanisms, meshing with one and the same gear.

Kiln inclination varies wtih processes from 2 to 6 cm/m. *Speed of rotation* also varies from very slow, i.e., a peripheral speed of 0.15 m/s for a TiO_2 pigment kiln, or 0.22 m/s for a cement kiln, to 0.64 m/s for a unit calcining phosphate materials.

Special features include the discharge end designed for air cooling or kilns that operate at high temperatures, such as cement, dead-burned dolomite, and magnesia. Firing hoods are designed with retractable fronts and large side doors and are mounted on wheels. *Internal heat recuperators* are of numerous designs and are becoming more popular as fuel prices increase. *Thermocouple collector rings* are placed at various points on the shell for indicating and recording internal temperatures.

Scoop systems are provided for introducing collected dust or, in some cases, a feed component through the shell at some intermediate point or points. Ports are installed in the shell for admitting combustion air at points beyond the hot zone; these are used in reducing kilns for burning carbon monoxide and volatiles from materials being processed.

Firing may be accomplished at either end, depending on whether cocurrent or countercurrent flow of the charge and gases is desired. Sometimes a solid fuel is mixed with the charge and burned as it moves down the kiln. Gaseous, liquid, or powdered fuels may be used. The burner may be installed directly at the end of the kiln with combustion occurring inside it. In this case, the discharge-end housing usually consists of a fixed or movable kiln hood through which the fuel pipe enters the kiln. A center position for the fuel pipe is used when the flame is wanted off the charge. Some users prefer an off-center position toward the trough between the charge chord and the descending kiln lining. The kiln and the hood (combustion chamber) usually have open ends which coincide with each other with the gap being closed by a sliding seal (Fig. 20-38). Sometimes a special offset chamber for the introduction of secondary tempering air is provided on dryers and kilns (Fig. 20-39).

The exhaust gases are generally discharged into dust and fume knockdown equipment to avoid contamination of the atmosphere. Gas-cleaning equipment includes cyclones, settling chambers, scrubbing towers, and electrical precipitators. Heat-recovery devices are utilized both within and outside the kiln. These result in an increase in kiln capacity or a decrease in fuel consumption. Waste-heat boilers, grates, coil systems, and chains are used for this purpose.

The feed end of a rotary kiln is partially closed by a ring-shaped **feed head** which retains the end brick and dams back flow of solids. On the discharge end, a brick retaining-ring casting is made up to suit the application. For low temperatures, segmental alloy-iron rings may be employed. For high-temperature processes, either segmental alloy-steel rings or kiln ends of the air-cooled type are employed; the latter provides longer life for both kiln end and the brick ring.

Efficient **air seals** are essential for the controlled and economical operation of kilns. They reduce outside air entrance; certain types

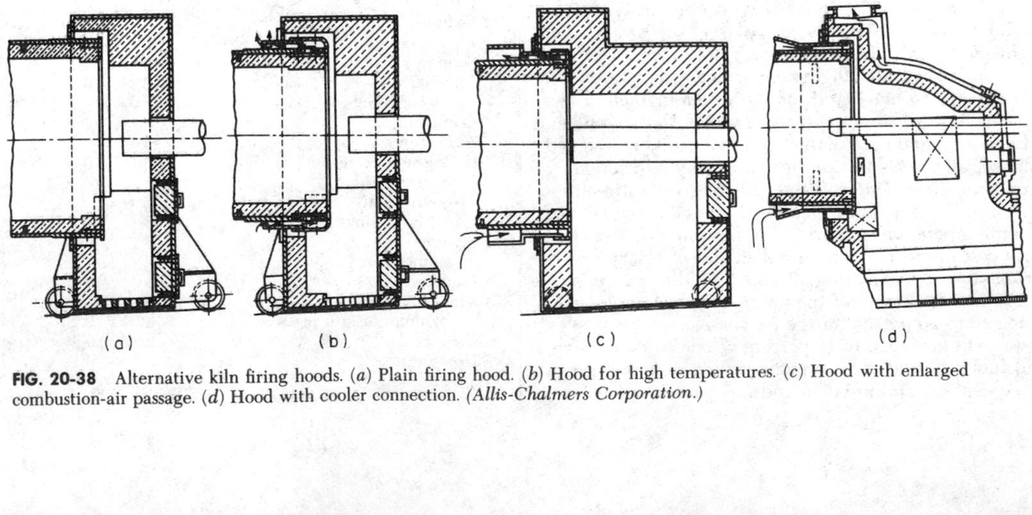

FIG. 20-38 Alternative kiln firing hoods. (*a*) Plain firing hood. (*b*) Hood for high temperatures. (*c*) Hood with enlarged combustion-air passage. (*d*) Hood with cooler connection. *(Allis-Chalmers Corporation.)*

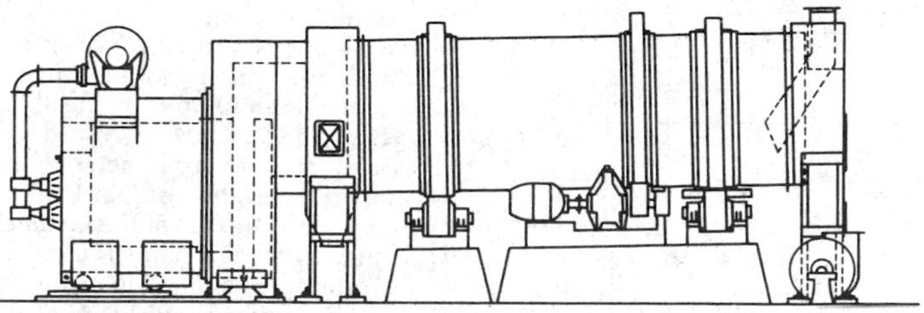

FIG. 20-39 Dryer firing hood with air-tempering chamber. *(CE Raymond/Bartlett-Snow Co.)*

effectively prevent entrance of all outside air. The simplest type of air seal is a floating T-section ring mounted on a wearing pad around the feed end of the kiln and free to slide with expansion of the kiln shell. The web of the T ring is confined within circular retainer plates. Figure 20-40 shows two arrangements. The floating-type discharge-end air seal consists of a circular bar which floats on a wearing pad and which can be moved to provide the desired operating clearance between air seal and firing hood. The floating ring and the fixed portion of these seals can be furnished with renewable wearing surfaces. Air infiltration through this type of seal is usually less than 10 percent (Allis-Chalmers Corporation). For further reduction of air infiltration, lantern-ring-type floating seals, pressurized with inert gas or stack gases, are employed. Accelerated drying of slurries in the feed end of rotary kilns in wet-process operations is achieved by installation of *hanging chains*. Conveying spirals support suspended lengths of chain which are arranged in such a way that they form an effective pattern for drying. With the chain system, slurry is heated in three ways: by direct transfer from chains after suspension in hot gases, by lifting material into the path of hot gases, and by directing the flow of hot gas over the slurry bed in the space formed under the suspended chains. Frequently, the product forms into uniformly sized pellets which progress through the rest of the kiln in that form, resulting in improved heat transfer and reduced dust losses (Fig. 20-41).

Design Methods In rotary kilns, the material is not showered through the air stream but is retained in the lower part of the cylinder. Gas-solids contacting is much less efficient than in flighted units. Heat transfer is by radiation and convection from the flowing gas to the kiln brick and exposed bed surface and by radiation from the brick to the bed. For units employing separate combustion chambers, it can be assumed that at high temperatures the wall-film resistance to convection heat transfer from the gas to the brick is limiting and that at any point the bed temperature approaches the wall temperature. Hence, the effective heat-transfer area is the inner kiln surface. For kilns under these conditions, the following empirical relationship is recommended for the convection heat-transfer coefficient from gas to brick:

$$Us = 23.7G^{0.67} \qquad (20\text{-}41)$$

where Us = heat-transfer coefficient, J/(m^2·s·K) [(Btu/(h·ft^2 kiln surface·°F)]; and G = gas mass flow rate, kg/(s·m^2 kiln cross section) [lb/(h·ft^2)].

Equation (20-41) does not account for gas radiation at high temperature when the kiln charge can "see" the burner flame; hence, the method will yield a conservative design. When a kiln is fired internally, the major source of heat transfer is radiation from the flame and hot gases. This occurs directly to both the solids surface and the wall, and from the latter to the product by reradiation (with some conduction).

Generally, a dry-feed kiln will have three zones of heating, and a wet-feed kiln will have four:

1. Drying zone at feed end, when moisture is removed
2. Heating zone, where the charge is heated to the reaction temperature, i.e., the decomposition temperature for limestone or "burning" temperature for cement
3. Reaction zone, in which the charge is burned, decomposed, reduced, oxidized, etc.
4. Soaking zone, where the reacted charge is superheated or "soaked" at temperature or, if desired, cooled before discharge

The rates of heat transfer in each zone will be different.

Rotary kilns operate at various temperatures throughout their length. A graph of approximate gas and charge temperatures for wet-process cement is shown in Fig. 20-42. The maximum charge temperature is 1700 to 1800 K; for the gases, 1800 to 1925 K. Overall heat-transfer rates have been estimated to be in the range of 25 to 60 KJ/(s·m^3) on the basis of total kiln volume.

Some commercial performance data for cement and lime kilns are shown in Table 20-15.

Some of the other major uses of direct rotary kilns are in the following processes:

Roasting. Rotary kilns are used for oxidizing and driving off sulfur and arsenic from various ores, including gold, silver, iron, etc. Temperatures employed will vary from 800 to 1600 K.

Chloridizing. Silver ores are chloridized successfully in rotary kilns. Temperatures must be closely controlled between 1030 and 1090 K.

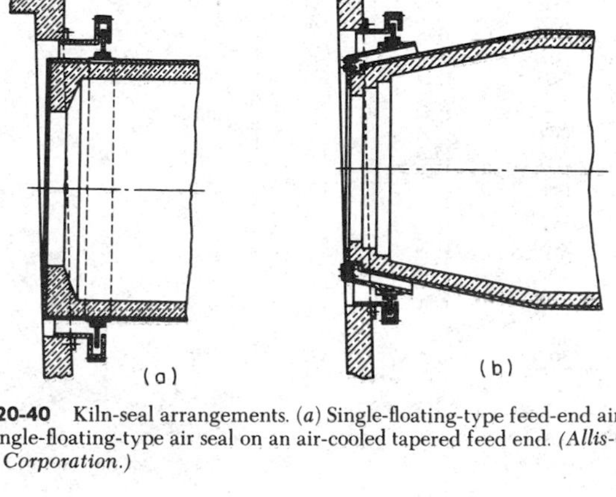

FIG. 20-40 Kiln-seal arrangements. (*a*) Single-floating-type feed-end air seal. (*b*) Single-floating-type air seal on an air-cooled tapered feed end. *(Allis-Chalmers Corporation.)*

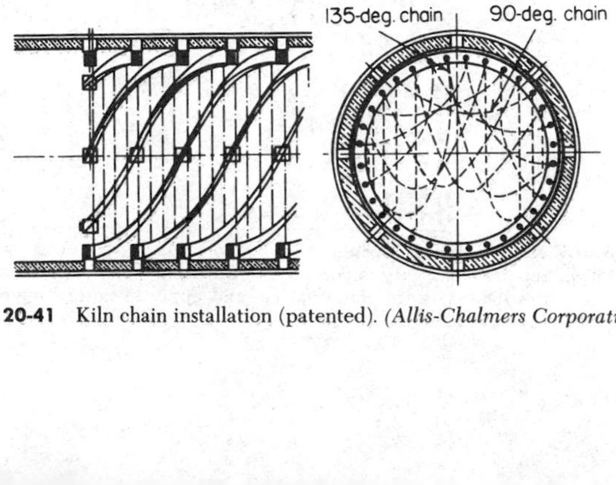

FIG. 20-41 Kiln chain installation (patented). *(Allis-Chalmers Corporation.)*

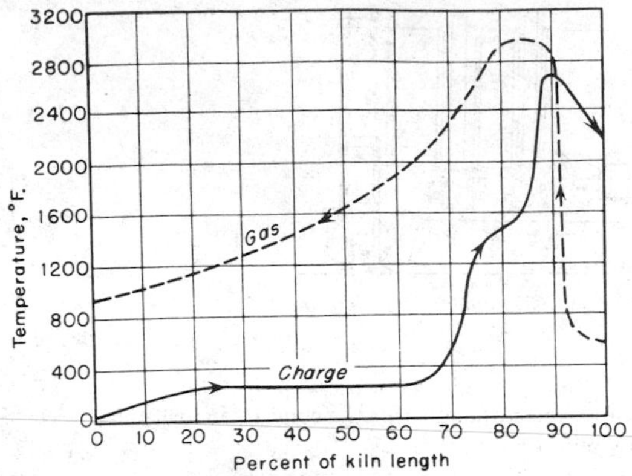

FIG. 20-42 Temperatures in rotary kiln on wet-process cement.

Black ash. Barium sulfide (BaS) is produced by calcining a mixture of barite ($BaSO_4$) and carbon at a temperature of 1350 K in continuous rotary kilns.

Spodumene. A mixture of quartz, feldspar, and spodumene is being calcined in rotary kilns at 1475 K to produce lithium aluminum silicate.

Vermiculite. A micaceous mineral is roasted to cause exfoliation for use as an insulating material.

Revivification. Temperatures of 800 to 1030 K are used to revivify fuller's and diatomaceous earth, although for some earths lower temperatures are employed.

Zinc. Oxidized ores are calcined to drive off water of hydration and carbon dioxide. The sulfide ore is always roasted before smelting.

Titanium oxide (TiO_2). This is produced from ilmenite ore by mixing ore with carbon and heating in a rotary kiln. Also, the rotary kiln is used in the process of recovery of titanium oxide from hydrated titanium precipitate at about 1250 K.

Roofing granules. Crushed quartz or sand of definite size is treated with various minerals, borax, soda ash, etc., and calcined at temperatures ranging from 1250 to 1600 K. Glass of different colors forms on the surface of the granules at various temperatures. An oxidizing or reducing flame is used to influence the final coloring.

Alumina (Al_2O_3). Alumina is produced by calcining either bauxite or aluminum hydroxide in rotary kilns at temperatures from 1250 to 1600 K. In obtaining the highest-purity alumina, the bauxite is digested with alkali to remove impurities; the resultant aluminum hydroxide [$Al_2(OH)_3$], of approximately 200-mesh size, is then calcined in rotary kilns at 1350 K.

Potassium salts. In this operation, potassium chloride (KCl) is introduced to the rotary kiln at a fineness of minus 100 mesh and containing 9 percent water. The salt is brought to the fusion temperature of 1048 K.

Magnesium oxide. The natural minerals, i.e., magnesite ($MgCO_3$), brucite [$Mg(OH)_2$], etc., after being crushed to predetermined size, are calcined at temperatures varying from 1055 to 2000 K, depending upon whether a caustic or a dead-burned product (periclase) is being produced. Magnesium hydroxide, recovered from seawater or salt brine, is also being treated in a similar manner except that it is added in the form of a sludge.

Sodium aluminum sulfate. This product is now being successfully calcined in rotary kilns. In this process, the salt cake is broken up just before it enters the kiln. Calcination is for the purpose of driving off the combined water (45 percent) and sulfuric acid (3 percent). Temperatures employed are approximately 800 K.

Phosphate rock. In this application, the rotary kiln is used to nodulize the fines in the ore and prepare them for electric-furnace

TABLE 20-15 Typical Rotary-Kiln Installations*

Size, diam. × length	Usual No. of supports	Range of motor hp. to operate†	Nominal 24-hr. capacities‡			
			Portland cement, 376-lb. bbl.		Lime, net tons	
			Dry process	Wet process	Lime sludge	Limestone
5 × 80 ft.	2	5–7.5	140	100	10	16
6 × 70 ft.	2	7.5–15	190	135	15	24
7 × 70 ft.	2	15–20	275	200	20	35
5 ft. 6 in. × 180 ft.	4	15–20	285	250	30	45
7 × 120 ft.	2	15–25	475	340	35	55
7 ft. 6 in. × 125 ft.	2	20–30	575	415	40	70
6 × 220 ft.	4	20–30	420	375	45	65
8 × 140 ft.	2	25–30	750	540	55	90
9 × 160 ft.	2	30–50	1100	800	80	130
8 ft. 6 in. × 185 ft.	4	30–50	1125	810	80	135
10 × 150 ft.	2	40–75	1300	950	...	145
10 × 175 ft.	2	50–75	1500	1100	...	155
8 × 300 ft.	5	50–75	1150	1000	110	160
7 ft. 6 in. × 8 ft. 6 in. × 320 ft.	5	50–75	1175	1020	115	165
7 ft. 6 in. × 10 ft. × 8 ft. 6 in. × 300 ft.	5	50–75	1175	1020		
10 × 11 × 175 ft.	2	50–75	1650	1200	120	180
10 ft. 6 in. × 185 ft.	2	50–75	1800	1300	130	190
11 × 175 ft.	2	60–100	1850	1375	...	205
8 ft. 6 in. × 10 ft. × 8 ft. 6 in. × 300 ft.	5	50–75	1400	1200		
8 × 10 × 300 ft.	5	50–75	1425	1225	140	200
9 ft. 6 in. × 265 ft.	4	60–100	1500	1300	150	215
9 × 10 ft. 6 in. × 9 ft. × 325 ft.	5	60–100	1700	1500		
10 ft. 6 in. × 250 ft.	4	60–100	1750	1525	175	240
9 ft. 6 in. × 11 ft. × 9 ft. 6 in. × 300 ft.	5	60–100	1800	1550		
10 × 300 ft.	5	75–125	1900	1650	190	250
9 ft. 6 in. × 11 ft. × 9 ft. 6 in. × 375 ft.	6	75–125	2025	1800		
11 × 300 ft.	5	75–125	2400	2100	225	300
11 ft. 6 in. × 300 ft.	4	100–150	2600	2250	240	320
10 ft. 6 in. × 375 ft.	5	100–150	2700	2400	250	325
11 ft. 3 in. × 360 ft.	5	125–175	2900	2500	275	350
11 ft. 6 in. × 475 ft.	7	150–250	4000	3500	375	450
12 × 500 ft.	8	200–300	4600	4000	425	500

*Allis-Chalmers Manufacturing Co.

†Power requirements vary according to size of kiln, character of material handled, and method of operation.

‡Capacities indicated are conservative, and apply to normal operation at sea level. Corrections would apply at increased altitudes, and for differing methods of operation.

operation. Ore under 5 cm in size and containing 50 percent or more minus 100 mesh is calcined. Ore nodulizes at approximately 1475 to 1500 K.

Mercury. In recovering mercury from cinnabar ores, the ore is crushed to minus 1.5 cm and fed to rotary kilns, where it is calcined to over 800 K. Since the mercury exists as mercuric sulfide (HgS), the sulfur is oxidized to SO_2 and the mercury vaporized. The gases are passed through cooling chambers, where the mercury condenses and is collected. Mercury vaporizes at 625 K.

Gypsum. The rotary kiln is rapidly replacing the kettle in producing plaster of paris. Great care is required, as the temperatures for reaction are low and within narrow limits, 382 to 403 K. Gypsum ($CaSO_4 \cdot 2H_2O$) is heated to drive off three-fourths of the water of crystallization to produce plaster of paris [$(CaSO_2) \cdot H_2O$]. Any overheating drives off all the water, producing gypsite ($CaSO_4$), which is unsatisfactory.

Clay. To produce lightweight aggregate for concrete, clay is calcined in rotary kilns. Temperatures employed vary from 1350 to 1600 K. The apparent density of the clay is reduced by 50 to 75 percent.

Iron ores. Crushed iron ores are partially reduced in rotary kilns to obtain nodules which are used in blast-furnace charges.

Manganese. Manganese ore, rhodochrosite, or manganese carbonate ($MnCO_2$) is calcined at about 1525 K to produce the oxide (Mn_3O_4). When the oxide ore is available but is in a finely divided state, the rotary kiln is used only for nodulizing.

Petroleum coke. In order to eliminate excess volatile matter, petroleum coke is calcined at temperatures of 1475 to 1525 K. This is a sensitive material, and temperature control is difficult to maintain.

When it is desired to increase the capacity of an existing kiln installation, consideration should be given to the following changes:

1. Increase charge volume held in kiln.
2. Increase temperature and quantity of combustion gases.
3. Decrease quantity of air in excess of combustion needs.
4. Increase speed of rotation of kiln.
5. Install ring dams at intermediate and discharge points.
6. Increase capacity of feeding and discharge mechanisms.
7. Decrease moisture content of feed material.
8. Increase temperature of feed material.
9. Install chains or flights, etc., in feed end.
10. Preheat all combustion air.
11. Reduce leakage of cold air into kiln at hot end.
12. Increase stack draft by increasing height or by use of jets.
13. Install instrumentation to control the kiln at maximum-capacity conditions.

The time of passage in rotary kilns (from which holdup can be calculated) can be estimated by the following formula (U.S. Bur. Mines Tech. Pap. 384, 1927):

$$\theta = 0.19L/NDS \qquad (20\text{-}42)$$

where θ = time of passage in the kiln, min; L = kiln length, ft; N = rotational speed, r/min; S = slope of kiln, ft/ft; and D = diameter inside brick, ft. Other equations for estimating the time of passage employing internal dams and a discharge dam are given by Bayard [*Chem. Metall. Eng.*, **52**(3), 100–102 (1945)].

The total power required to drive a rotary kiln or a dryer with lifters can be calculated by the following formulas (courtesy of CE Raymond Division, Combustion Engineering Inc.). For a rotary kiln or calciner without lifters,

$$\text{bhp} = \frac{N\,[18.85y\,(\sin B)w + 0.1925DW + 0.33W]}{100{,}000} \qquad (20\text{-}43)$$

For a rotary dryer or section of a kiln with lifters,

$$\text{bhp} = \frac{N\,(4.75dw + 0.1925DW + 0.33W)}{100{,}000} \qquad (20\text{-}44)$$

where bhp = brake horsepower required (1 bhp = 0.75 kW); N = rotational speed, r/min; y = distance between centerline of kiln and the center of gravity of material bed, ft; B = angle of repose of material; W = total rotating load (equipment plus material), lb; w = live load (material), lb; D = riding-ring diameter, ft; and d = shell diameter, ft. For estimating purposes, let $D = (d + 2)$.

Drive motors should be of the high-starting-torque type and selected for 1.33 times maximum rotational speed. For two- or three-diameter kilns, the brake horsepower for the several diameters should be calculated separately and summed. Auxiliary drives should be provided to maintain shell rotation in the event of power failure. These are usually gasoline or diesel engines.

Thermal Efficiency of Rotary Kilns Kiln length is a major factor in determining thermal efficiency, and kilns with a high ratio of length to diameter have a greater thermal efficiency than those with a low ratio. The use of chains inside the kiln and of heat-recovery equipment on the gases and product leaving the kiln can increase substantially the thermal efficiency of a kiln installation. Efficiencies ranging from 45 to more than 80 percent have been reported. A reasonably satisfactory range based on present fuel prices and construction costs would be 65 to 75 percent utilization and recovery of the heat content of the fuel plus any heat of reaction of the charge. No distinction is made from an efficiency-calculation standpoint between the heat utilized in the kiln and that recovered (or utilized) outside the kiln. With countercurrent flow of the combustion gases and the charge material, an excepionally long kiln will give high efficiencies within itself. However, good economics may dictate that a shorter kiln be installed with a waste-heat boiler on the hot gases to obtain an equivalent thermal efficiency at a lower investment. The heat in the hot product usually is recovered as preheat in the combustion air.

Size Segregation in Kilns When an assemblage of solid particles, not very closely screened, is rotated within a cylinder, the solids assume a lunar shape, as shown in Fig. 20-43. This causes serious size segregation. The finest sizes remain at the bottom, in contact with the hot brick. The coarser particles form the upper layer of the agitated mass. As the kiln completes a revolution, the exposed brick, in an upper position, absorbs sensible heat from the gas mass. As the heated brick completes its circuit, it passes under and is in conductive contact with the fine particles. These fines are thus effectively heated by direct solid-to-solid transfer. The larger particles are heated by direct radiation from gas and brick, and become adequately calcined. The particles of size intermediate between the fine and coarse remain, throughout a complete revolution, "sandwiched" between the coarse and fine layers and are protected from heat by the excellent insulation properties of these layers, thus perhaps escaping complete calcination. This factor of segregation is offset by some kiln operators who classify or screen the kiln feed so that only a narrow range of particle size is fed at one time. Also, faster kiln speeds which give a better agitation of the charge are used.

Rotary kilns are usually operated with between 3 and 12 percent of their volume filled with material; 7 percent is considered normal.

Cost Data Purchase prices, weights, and horsepower requirements of typical units are given in Table 20-16. Installed costs will run to from 300 to 500 percent of purchase cost. Maintenance will average 5 to 10 percent of the total installed cost per year but is dependent largely on the life of the refractory lining.

For estimating purposes refractory lining for a 2.7- to 3.4-m-diameter kiln costs \$6000 to \$15,000 per meter of kiln length installed (50 percent material, 50 percent labor).

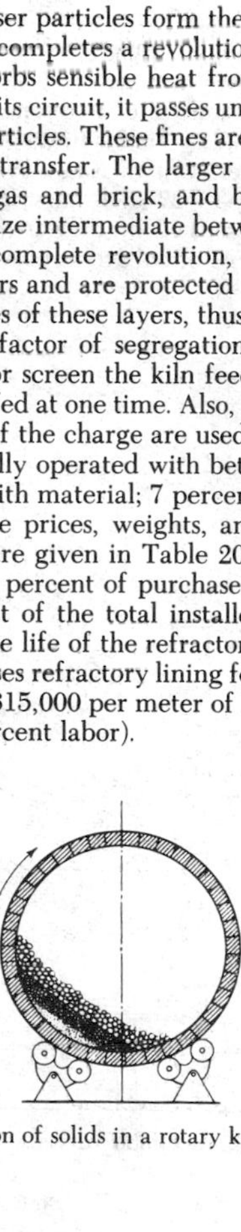

FIG. 20-43 Size segregation of solids in a rotary kiln.

TABLE 20-16 Approximate Purchase Costs and Weight of Rotary Kilns*

Kiln size, diameter × length	Total purchase price includes drive, burner, and controls (not including brick)
8′0″ × 80′0″	$ 448,000
8′0″ × 140′0″	600,000
8′0″ × 200′0″	960,000
8′0″ × 300′0″	1,240,000
9′0″ × 250′0″	1,373,000
9′0″ × 300′0″	1,545,000
10′0″ × 100′0″	682,000
10′0″ × 150′0″	965,000
10′0″ × 250′0″	1,502,000
10′0″ × 300′0″	1,682,000
10′0″ × 350′0″	1,779,000
10′6″ × 175′0″	1,182,000
10′6″ × 250′0″	1,677,000
10′6″ × 350′0″	1,942,000
11′0″ × 160′0″	1,213,000
11′0″ × 250′0″	1,670,000
11′0″ × 300′0″	1,858,000
11′0″ × 350′0″	2,344,000
11′0″ × 400′0″	2,544,000
11′6″ × 160′0″	1,251,000
11′6″ × 250′0″	1,768,000
11′6″ × 350′0″	2,393,000
11′6″ × 425′0″	2,676,000
12′0″ × 250′0″	1,837,000
12′0″ × 325′0″	2,598,000
12′0″ × 400′0″	2,645,000
12′0″ × 450′0″	3,570,000
13′0″ × 500′0″	4,388,000
14′0″ × 400′0″	4,155,000
16′6″ × 600′0″	8,190,000

*Courtesy of Fuller Co.; General American Transportation Corp.

A discussion of retention time in rotary kilns is given in *Brit. Chem. Eng.*, 27–29 (January 1966). Rotary-kiln heat control is discussed in detail by Bauer [*Chem. Eng.*, 193–200 (May 1954)] and Zubrzycki [*Chem. Can.*, 33–37 (February 1957)]. Reduction of iron ore in rotary kilns is described by Stewart [*Min. Congr. J.*, 34–38 (December 1958)]. The use of balls to improve solids flow is discussed in [*Chem. Eng.*, 120–222 (March 1956)]. Brisbane examined problems of shell deformation [(*Min. Eng.*, 210–212 (February 1956)]. Instrumentation is discussed by Dixon [*Ind. Eng. Chem. Process Des. Dev.*, 1436–1441 (July 1954)], and a mathematical simulation of a rotary kiln was developed by Sass [*Ind. Eng. Chem. Process Des. Dev.*, 532–535 (October 1967)]. This last paper employed the empirical convection heat-transfer coefficient given previously, and its use is discussed in later correspondence [ibid., 318–319 (April 1968)].

Indirect-Heat Rotary Steam-Tube Dryers Probably the most common type of indirect-heat rotary dryer is the steam-tube dryer (Fig. 20-44). Steam-heated tubes running the full length of the cylinder are fastened symmetrically in one, two, or three concentric rows inside the cylinder and rotate with it. Tubes may be simple pipe with condensate draining by gravity into the discharge manifold or bayonet-type. Bayonet-type tubes are also employed when units are used as water-tube coolers. When handling sticky materials, one row of tubes is preferred. These are occasionally shielded at the feed end of the dryer to prevent buildup of solids behind them. Lifting flights are usually inserted behind the tubes to promote solids agitation.

Wet feed enters the dryer through a chute or screw feeder. The product discharges through peripheral openings in the shell in ordinary dryers. These openings also serve to admit purge air to sweep moisture or other evolved gases from the shell. In practically all cases, gas flow is countercurrent to solids flow. To retain a deep bed of material within the dryer, normally 10 to 20 percent fillage, the discharge openings are supplied with removable chutes extending radially into the dryer. These, on removal, permit complete emptying of the dryer.

Steam is admitted to the tubes through a revolving steam joint into the steam side of the manifold (Fig. 20-45). Condensate is removed continuously, by gravity through the steam joint to a condensate receiver and my means of lifters in the condensate side of the manifold. By employing simple tubes, noncondensables are continuously vented at the other ends of the tubes through Sarco-type vent valves mounted on an auxiliary manifold ring, also revolving with the cylinder.

Vapors (from drying) are removed at the feed end of the dryer to the atmosphere through a natural-draft stack and settling chamber or wet scrubber. When employed in simple drying operations with 3.5×10^5 to 10×10^5 Pa steam, draft is controlled by a damper to admit only sufficient outside air to sweep moisture from the cylinder,

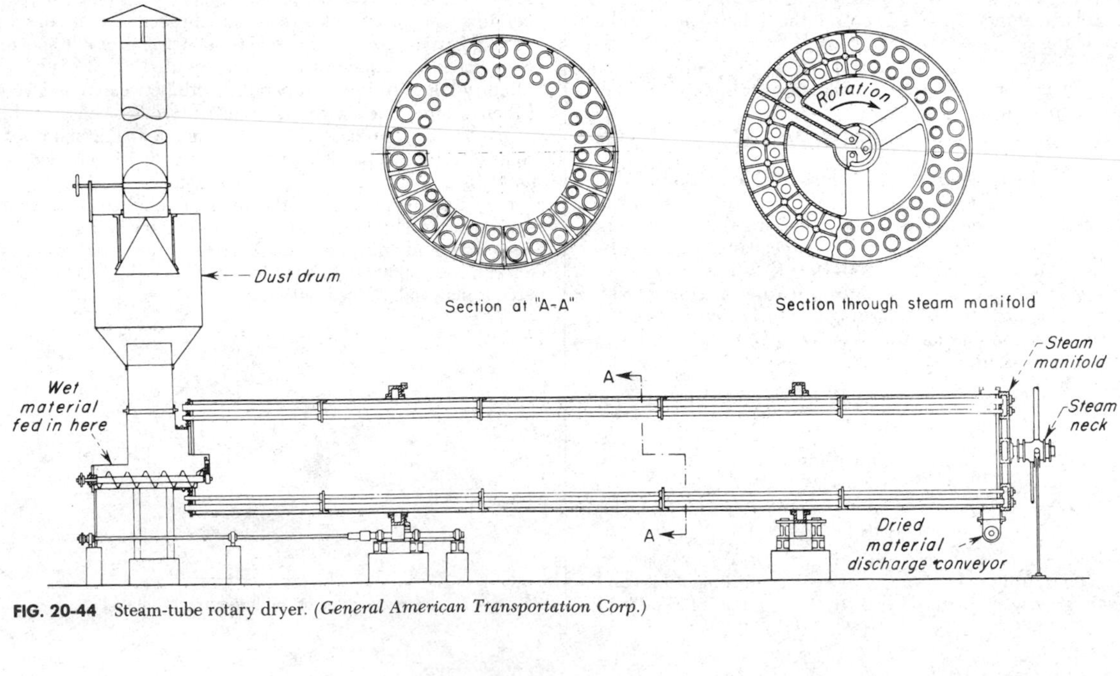

FIG. 20-44 Steam-tube rotary dryer. *(General American Transportation Corp.)*

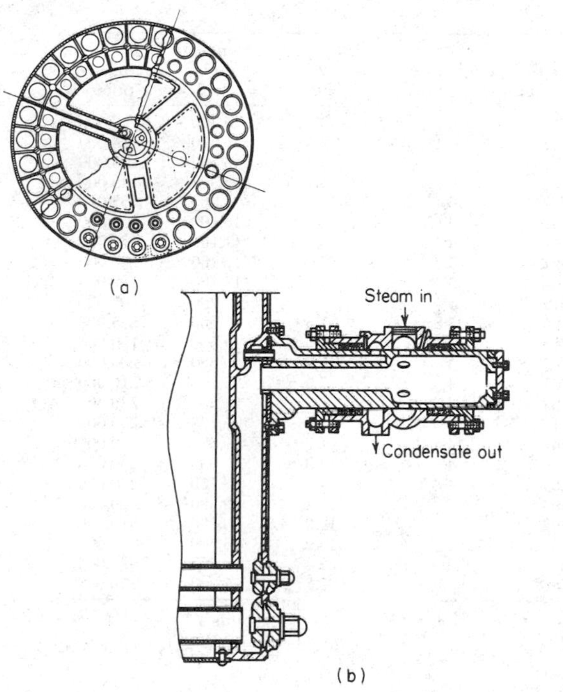

FIG. 20-45 Rotary steam joint for a standard steam-tube dryer. (*a*) Section of cast steam manifold. (*b*) Section of manifold and steam joint. *(Patented, General American Transportation Corp.)*

discharging the air at 340 to 365 K and 80 to 90 percent saturation. In this way, shell gas velocities and dusting are minimized. When used for solvent recovery or other processes requiring a sealed system, sweep gas is recirculated through a scrubber-gas cooler and blower.

Steam manifolds for pressures up to 10×10^5 Pa are of cast iron. For higher pressures, the manifold is fabricated from plate steel, stay-bolted, and welded. The tubes are fastened rigidly to the manifold face plate and are supported in a close-fitting annular plate at the other end to permit expansion. Packing on the steam neck is normally graphite-asbestos. Ordinary rotating seals are similar in design to those depicted in Fig. 20-36, with allowance for the admission of small quantities of outside air when the dryer is operated under a slight negative internal pressure.

Steam-tube dryers are used for the continuous drying, heating, or cooling of granular or powdery solids which cannot be exposed to ordinary atmospheric or combustion gases. They are especially suitable for fine dusty particles because of the low gas velocities required for purging of the cylinder. Tube sticking is avoided or reduced by employing recycle, shell knockers, etc., as previously described; tube scaling by sticky solids is one of the major hazards to efficient operation. The dryers are suitable for drying, solvent recovery, and chemical reactions. Steam-tube units have found effective employment in soda-ash production, replacing more expensive indirect-heat rotary calciners.

Special types of steam-tube dryers employ packed and purged seals on all rotating joints, with a central solids-discharge manifold through the steam neck to reduce the seal diameter. This manifold contains the product discharge conveyor and a passage for the admission of sweep gas. Solids are removed from the shell by special volute lifters and dropped into the discharge conveyor. Units have been fabricated for operation at 76 mm of water, internal shell pressure, with no detectable air leakage.

Design Methods Heat-transfer coefficients in steam-tube dryers range from 30 to 85 $J/(m^2 \cdot s \cdot K)$. Coefficients will increase with increasing steam temperature because of increased heat transfer by

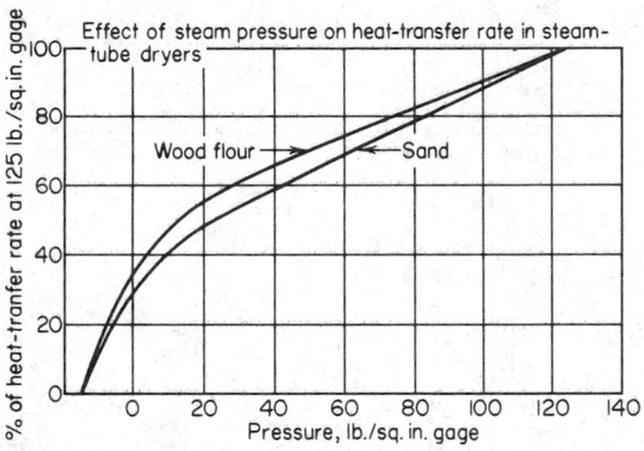

FIG. 20-46 Effect of steam pressure on the heat-transfer rate in steam-tube dryers. *(General American Transportation Corp.)*

radiation. In units carrying saturated steam at 420 to 450 K, the heat flux $U\Delta T$ will range from 3400 to 6800 $J/(m^2 \cdot s \cdot K)$ for difficult-to-dry and organic solids and to 11,350 $J/(m^2 \cdot s \cdot K)$ for finely divided inorganic materials. The effect of steam pressure on heat-transfer rates up to 8.6×10^5 Pa is illustrated in Fig. 20-46.

Performance and Cost Data Table 20-17 contains data for a number of standard sizes of steam-tube dryers. Prices tabulated are for ordinary carbon steel construction. Installed costs will run from 150 to 300 percent of purchase cost.

The thermal efficiency of steam-tube units will range from 70 to 90 percent, if a well-insulated cylinder is assumed. This does not allow for boiler efficiency, however, and is therefore not directly comparable with direct-heat units such as the direct-heat rotary dryer or indirect-heat calciner.

Operating costs for these dryers include 5 to 10 percent of one person's time. Maintenance will average 5 to 10 percent of total installed cost per year.

Table 20-18 outlines typical performance data from three drying applications in steam-tube dryers.

Indirect-Heat Calciners Indirect-heat calciners, either batch or continuous, are employed for heat treating and drying at higher temperatures than can be obtained in steam-heated rotating equipment. They require a minimum flow of gas to purge the cylinder which, when handling granular solids, reduces dusting; they are suitable for gas-sealed operation with oxidizing, inert, or reducing atmospheres. Indirect calciners are widely and successfully used in the following specific applications:

1. Activating wood charcoal
2. Reducing mineral high oxides to low oxides
3. Drying fluoride precipitates in a hydrogen fluoride atmosphere
4. Calcination of silica gel
5. Drying and removal of sulfur from cobalt, copper, and nickel
6. Reduction of metal oxides
7. Oxidizing and "burning off" of organic impurities
8. Reclamation of foundry sand from the shell-molding process

This unit consists essentially of a cylindrical retort, rotating within a stationary refractory-lined cylindrical furnace. The latter is arranged so that fuel combustion occurs within the annular ring between the retort and the furnace. The retort cylinder extends at both ends beyond the furnace. These end extensions carry the riding rings and drive gear. Material may be fed continuously at one end and discharged continuously at the other. Feeding and solids discharging are usually accomplished with screw feeders or other positive feeders to prevent leakage of gases into or out of the retort with the solids.

In some cases in which it is desirable to cool the product before removal to the outside atmosphere, the discharge end of the cylinder is provided with an additional extension, the exterior of which is water-spray-cooled. In cocurrent-flow calciners, hot gases from the

TABLE 20-17 Standard Steam-Tube Dryers*

Size, diameter × length, m	Tubes No. OD (mm)	Tubes No. OD (mm)	m^2 of free area	Dryer speed, r/min	Motor size, hp	Shipping weight, kg	Estimated price
0.965 × 4.572	14 (114)		21.4	6	2.2	5,500	$120,000
0.965 × 6.096	14 (114)		29.3	6	2.2	5,900	130,000
0.965 × 7.620	14 (114)		36.7	6	3.7	6,500	138,000
0.965 × 9.144	14 (114)		44.6	6	3.7	6,900	145,000
0.965 × 10.668	14 (114)		52.0	6	3.7	7,500	155,000
1.372 × 6.096	18 (114)	18 (63.5)	58.1	4.4	3.7	10,200	160,000
1,372 × 7.620	18 (114)	18 (63.5)	73.4	4.4	3.7	11,100	170,000
1.372 × 9.144	18 (114)	18 (63.5)	88.7	5	5.6	12,100	180,000
1.372 × 10.668	18 (114)	18 (63.5)	104	5	5.6	13,100	192,000
1.372 × 12.192	18 (114)	18 (63.5)	119	5	5.6	14,200	205,000
1.372 × 13.716	18 (114)	18 (63.5)	135	5.5	7.5	15,000	215,000
1.829 × 7.62	27 (114)	27 (76.2)	118	4	5.6	19,300	190,000
1.829 × 9.144	27 (114)	27 (76.2)	143	4	5.6	20,600	200,000
1.829 × 10.668	27 (114)	27 (76.2)	167	4	7.5	22,100	210,000
1.829 × 12.192	27 (114)	27 (76.2)	192	4	7.5	23,800	220,000
1.829 × 13.716	27 (114)	27 (76.2)	217	4	11.2	25,700	230,000
1.829 × 15.240	27 (114)	27 (76.2)	242	4	11.2	27,500	240,000
1.829 × 16.764	27 (114)	27 (76.2)	266	4	14.9	29,300	250,000
1.829 × 18.288	27 (114)	27 (76.2)	291	4	14.9	30,700	260,000
2.438 × 12.192	90 (114)		394	3	11.2	49,900	430,000
2.438 × 15.240	90 (114)		492	3	14.9	56,300	510,000
2.438 × 18.288	90 (114)		590	3	14.9	63,500	580,000
2.438 × 21.336	90 (114)		689	3	22.4	69,900	660,000
2.438 × 24.387	90 (114)		786	3	29.8	75,300	730,000

*Courtesy of Swenson Division, Whiting Corporation (prices of January 1982). Carbon steel fabrication; multiply by 1.75 for 304 stainless steel.

interior of the heated portion of the cylinder are withdrawn through a special exit tube. This tube extends centrally through the cooled section to prevent flow of gas near the cooled-shell surfaces and possible condensation. Frequently a separate cooler is used, isolated from the calciner by an air lock.

Operating temperatures in indirect-heat calciners are limited only by structural considerations, normally 700 K for carbon steel and 920 K for stainless steels. Use of special metals may permit operation up to 1365 K.

To prevent sliding of solids over the smooth interior of the shell, lifting bars running longitudinally and welded to the inside wall are frequently provided. These normally do not shower the solids as in a direct-heat vessel but merely prevent sliding so that the bed will turn over and constantly expose new surface for heat and mass transfer. To prevent scaling of the shell interior by sticky solids, a scraper "chain" is occasionally employed. This may, for example, consist of a series of I-beam sections, pinned together. These will be fastened at each end to rigid swivels by link chain to permit turning and prevent wrapping of the beam chain upon itself. The beam sections must be sufficiently heavy to sink through the solids bed, so that they ride directly on and scrape the shell. In this instance, lifting bars would not be used, agitation being provided by chain motion. The use of a scraper chain is a fairly common practice in, for example, indirect-heat soda-ash calciners. For precise control of retention, approaching plug flow, continuous spiral flights may be attached to the inside of the shell.

Because indirect-heat calciners frequently require close-fitting gas seals, it is customary to support all parts on a self-contained steel base, for sizes up to approximately 1.25 m in diameter by 9.5 m long. Electric, gas, or oil heating is used, with multiple-burner arrangements beneath the shell to ensure uniform heating. Process control is normally by shell temperature, measured by radiation pyrometers.

TABLE 20-18 Steam-Tube Dryer Performance Data

	Class 1	Class 2	Class 3
Class of materials handled	High-moisture organic, distillers' grains, brewers' grains, citrus pulp	Pigment filter cakes, blanc fixe, barium carbonate, precipitated chalk	Finely divided inorganic solids, water-ground mica, water-ground silica, flotation concentrates
Description of class	Wet feed is granular and damp but not sticky or muddy and dries to granular meal	Wet feed is pasty, muddy, or sloppy; product is mostly hard pellets	Wet feed is crumbly and friable; product is powder with very few lumps
Normal moisture content of wet feed, % dry basis	233	100	54
Normal moisture content of product, % dry basis	11	0.15	0.5
Normal temperature of wet feed, K	310–320	280–290	280–290
Normal temperature of product, K	350–355	380–410	365–375
Evaporation per product, kg	2	1	0.53
Heat load per lb product, kJ	2250	1190	625
Steam pressure normally used, kPa gauge	860	860	860
Heating surface required per kg product, m^2	0.34	0.4	0.072
Steam consumption per kg product, kg	3.33	1.72	0.85

TABLE 20-19 Indirect-Heat Rotary Calciners: Sizes and Purchase Costs*

Diameter, ft	Overall cylinder length	Heated cylinder length	Motor hp based on 2 r/min of cylinder, 10% loading 100-lb/ft^3 material	Shipping weight, lb exclusive of refractories	Approximate sale price in carbon steel construction	Approximate sale price in No. 316 stainless construction
4	37 ft 6 in	31 ft 0 in	5	25,000	$145,000	$185,000
5	44 ft 4 in	36 ft 8 in	7.5	50,000	180,000	240,000
6	45 ft 0 in	37 ft 4 in	10	55,000	225,000	285,000
7	60 ft 0 in	52 ft 3 in	25	85,000	290,000	360,000

*CE Raymond Division, Combustion Engineering Inc. (Bartlett-Snow).

When a special gas atmosphere must be maintained inside the cylinder, positive rotary gas seals, with one or more pressurized and purged annular chambers, are employed. The diaphragm-type seal CE Raymond (Bartlett-Snow) is suitable for pressures up to 5 cm of water, with no detectable leakage.

In general, the temperature range of operation for indirect-heated calciners can vary over a wide range, from 425 to 475 K at the low end to approximately 1400 K at the high end. All types of carbon steel, stainless, and alloy construction are used, depending upon temperature, process, and corrosion requirements. Fabricated-alloy cylinders can be used over the greater part of the temperature spectrum; however, the greater creep-stress abilities of cast alloys makes their use desirable for the highest calciner-cylinder temperature applications.

Design Methods In indirect-heated calciners, heat transfer is primarily by radiation from the cylinder wall to the solids bed. The thermal efficiency ranges from 35 to 65 percent. The limiting factors in heat transmission lie in the conductivity and radiation constants of the shell metal and solids bed. If the characteristics of these are known, equipment may be accurately sized by employing the Stefan-Boltzmann radiation equation. Apparent heat-transfer coefficients will range from 17 J/(m^2·s·K) in low-temperature operations to 85 J/(m^2·s·K) in high-temperature processes.

Cost Data Power, operating, and maintenance costs are similar to those previously outlined for direct- and indirect-heat rotary dryers. Estimating purchase costs for continuous rotating calciners with carbon steel and type 316 stainless-steel cylinders are given in Table 20-19 together with size, weight, and motor requirements. Sale price includes the cylinder, ordinary angle seals, furnace, drive, feed and discharge conveyors, burners, etc. Installed cost may be estimated at 200 to 300 percent of the purchase cost. A layout of a typical continuous calciner with an extended cooler section is illustrated in Fig. 20-47.

Small batch retorts, heated electrically or by combustion, are widely used as carburizing furnaces and are applicable also to chemical processes involving the heat treating of particulate solids. These are mounted on a structural-steel base, complete with cylinder, furnace, drive motor, burner, etc. Units are commercially available in diameters from 0.24 to 1.25 m and lengths of 1 to 2 m. Continuous retorts with helical internal spirals are employed for metal-heat-treating purposes. Precise retention control is maintained in these operations. Standard diameters are 0.33, 0.5, and 0.67 m with effective lengths up to 3 m. These vessels are employed in many small-scale chemical-process operations which require accurate control of retention. Their operating characteristics and applications are identical to those of the larger indirect-heat calciners.

Direct-Heat Roto-Louvre Dryer One of the more important special types of rotating equipment is the Roto-Louvre dryer. As illustrated in Fig. 20-48, hot air (or cooling air) is blown through louvers in a double-wall rotating cylinder and up through the bed of solids. The latter moves continuously through the cylinder as it rotates. Constant turnover of the bed ensures uniform gas contacting for heat and mass transfer. The annular gas passage behind the louvers is partitioned so that contacting air enters the cylinder only beneath the solids bed. The number of louvers covered at any one time is roughly 30 percent. Because air circulates through the bed, fillages of 13 to 15 percent or greater are employed.

Roto-Louvre dryers range in size from 0.8 to 3.6 m in diameter and from 2.5 to 11 m long. The largest unit is reported capable of evaporating 5500 kg/h of water. Hot gases from 400 to 865 K may be employed. Because gas flow is through the bed of solids, high pressure drop, from 7 to 50 cm of water, may be encountered within the shell. For this reason, both a pressure inlet fan and an exhaust fan are provided in most applications to maintain the static pressure within the equipment as closely as possible to atmospheric. This prevents excessive in-leakage or blowing of hot gas and dust to the outside. For pressure control, one fan is usually operated under fixed conditions, with an automatic damper control on the other, regulated by a pressure detector-controller.

In heating or drying applications, when cooling of the product is desired before discharge to the atmosphere, cool air is blown through a second annular space, outside the inlet hot-air annulus, and released through the louvers at the solids-discharge end of the shell.

Roto-Louvre dryers are suitable for processing coarse granular solids which do not offer high resistance to air flow, require intimate gas contacting, and do not contain significant quantities of dust.

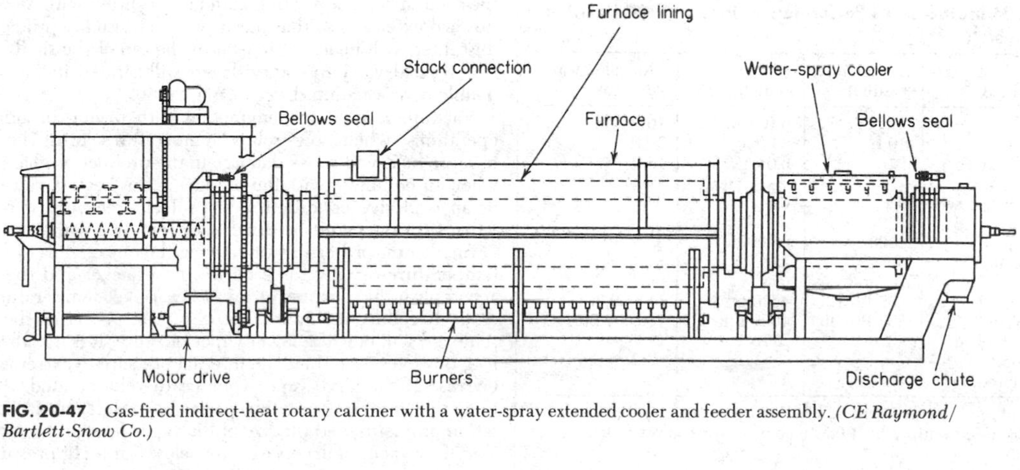

FIG. 20-47 Gas-fired indirect-heat rotary calciner with a water-spray extended cooler and feeder assembly. *(CE Raymond/Bartlett-Snow Co.)*

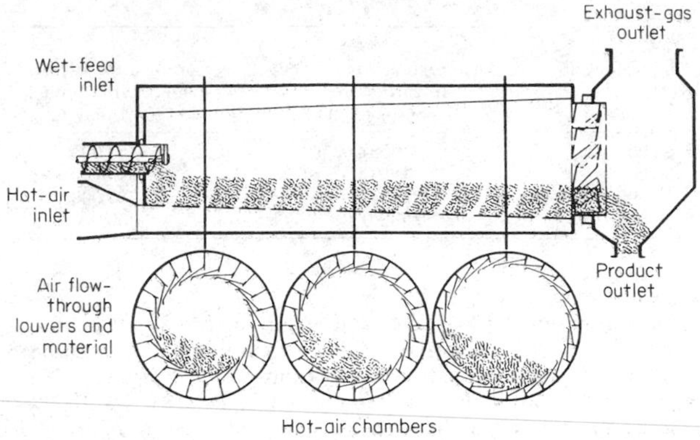

FIG. 20-48 Link-Belt Roto-Louvre dryer. *(Material Handling Systems Division, FMC Corporation.)*

Heat and mass transfer from the gas to the surface of the solids is extremely efficient; hence the equipment size required for a given duty is frequently less than required when an ordinary direct-heat rotary vessel with lifting flights is used. Purchase-price savings are partially balanced, however, by the more complex construction of the Roto-Louvre unit. A Roto-Louvre dryer will have a capacity roughly 1.5 times that of a single-shell rotary dryer of the same size under equivalent operating conditions. Because of the cross-flow method of heat exchange, the average Δt is not a simple function of inlet and outlet Δt's. There are currently no published data which permit the sizing of equipment without pilot tests as recommended by the manufacturer. Three applications of Roto-Louvre dryers are outlined in Table 20-20. Installation, operating, power, and maintenance costs will be similar to those experienced with ordinary direct-heat rotary dryers. *Thermal efficiency* will range from 30 to 70 percent.

AGITATED DRYERS

Description An agitated dryer is defined as one on which the housing enclosing the process is stationary while solids movement is accomplished by an internal mechanical agitator. Many forms are in use; however, this discussion covers only three basic types.

Field of Application Agitated dryers are applicable to processing solids which are relatively free-flowing and granular when discharged as product. Materials which are not free-flowing in their feed condition can be treated by recycle methods as described in the subsection "Rotary Dryers." In general, agitated dryers have applications similar to those of rotating vessels. Their chief advantages compared with the latter lie in the fact that (1) large-diameter rotary seals are not required at the solids and gas feed and exit points because the housing is stationary, and for this reason gas-leakage problems are minimized. Rotary seals are required only at the points of entrance of the mechanical agitator shaft. (2) Use of a mechanical agitator for solids mixing introduces shear forces which are helpful for breaking up lumps and agglomerates. Balling and pelleting of sticky solids, an occasional occurrence in rotating vessels, can be prevented by special agitator design. The problems concerning dusting of fine particles in direct-heat units are identical to those discussed under "Rotary Dryers."

TABLE 20-20 Manufacturer's Performance Data for FMC Link-Belt Roto-Louvre Dryers*

Material dried	Ammonium sulfate	Foundry sand	Metallurgical coke
Dryer diameter	2 ft 7 in	6 ft 4 in	10 ft 3 in
Dryer length	10 ft	24 ft	30 ft
Moisture in feed, % wet basis	2.0	6.0	18.0
Moisture in product, % wet basis	0.1	0.5	0.5
Production rate, lb/h	2500	32,000	38,000
Evaporation rate, lb/h	50	2130	8110
Type of fuel	Steam	Gas	Oil
Fuel consumption	255 lb/h	4630 ft^3/h	115 gal/h
Calorific value of fuel	837 Btu/lb	1000 Btu/ft^3	150,000 Btu/gal
Efficiency, Btu, supplied per lb evaporation	4370	2170	2135
Total power required, hp	4	41	78

*Material Handling Systems Division, FMC Corp. To convert British thermal units to kilojoules, multiply by 1.06; to convert horsepower to kilowatts, multiply by 0.746.

Vacuum Rotary Dryers Vacuum rotary dryers are batch dryers, at least in currently available commercial forms. Design of continuous equipment awaits development of continuous solids-discharging (and feeding) devices which will continuously convey particulate solids across a 100-kPa pressure barrier, with no back leakage of air into the vessel. So-called continuous equipment now available is continuous in the drying stage but requires two or more batch hoppers to serve as air locks; the product output remains batch.

The more common type of vacuum rotary dryer consists of a stationary cylindrical shell, mounted horizontally, in which a set of agitator blades mounted on a revolving central shaft stirs the solids being treated. Heat is supplied by circulation of hot water, steam, or Dowtherm through a jacket surrounding the shell and, in larger units, through the hollow central shaft. The agitator is either a single discontinuous spiral or a double continuous spiral. The outer blades are set as closely as possible to the wall without touching, usually leaving a gap of 0.3 to 0.6 cm. Modern units occasionally employ spring-loaded shell scrapers mounted on the blades. The dryer is charged through a port at the top and emptied through one or more discharge nozzles at the bottom. Vacuum is applied and maintained by any of the conventional methods, i.e., steam jets, vacuum pumps, etc.

Another type of vacuum rotary dryer consists of a rotating horizontal cylindrical shell, suitably jacketed. Vacuum is applied to this unit through hollow trunnions with suitable packing glands. Rotary glands must be used also for admitting and removing the heating medium from the jacket. The inside of the shell may have lifting bars, welded longitudinally, to assist agitation of the solids.

The double-cone rotating vacuum dryer is a more common design. Although it is identical in operating design, the sloping walls of the cones permit more rapid emptying of solids when the dryer is in a stationary position. The older cylinder shape required continuous rotation during emptying to convey product to the discharge nozzles. As a result, a circular dust hood was frequently necessary to enclose the discharge-nozzle turning circle and prevent serious dust losses to the atmosphere during unloading. Several new designs of the double-cone type employ internal tubes or plate coils to provide additional heating surface.

On all rotating dryers, the vapor-outlet tube is stationary; it enters the shell through a rotating gland and is fitted with an elbow and an upward extension so that the vapor inlet, usually protected by a felt dust filter, will be at all times near the top of the shell.

A typical vacuum rotary dryer is illustrated in Fig. 20-49 and a double-cone vacuum dryer in Fig. 20-50.

Vacuum is used in conjunction with drying or other chemical operations when low solids temperatures must be maintained because heat will cause damage to the product or change its nature, when air combines with the product as it is heated, causing oxidation or an explosive condition, when solvent recovery is required, and when materials must be dried to extremely low moisture levels.

In vacuum processing and drying the objective is to create a large temperature-driving force between the jacket and the product. To accomplish this purpose at fairly low jacket temperatures, it is necessary to reduce the internal process pressure so that the liquid being removed will boil at a lower vapor pressure. It is not always economical, however, to reduce the internal pressure to extremely low levels because of the large vapor volumes thereby created. It is necessary to compromise on operating pressure, considering leakage, condensation problems, and the size of the vapor lines and pumping system. Very few vacuum dryers operate below 5-mmHg pressure on a com-

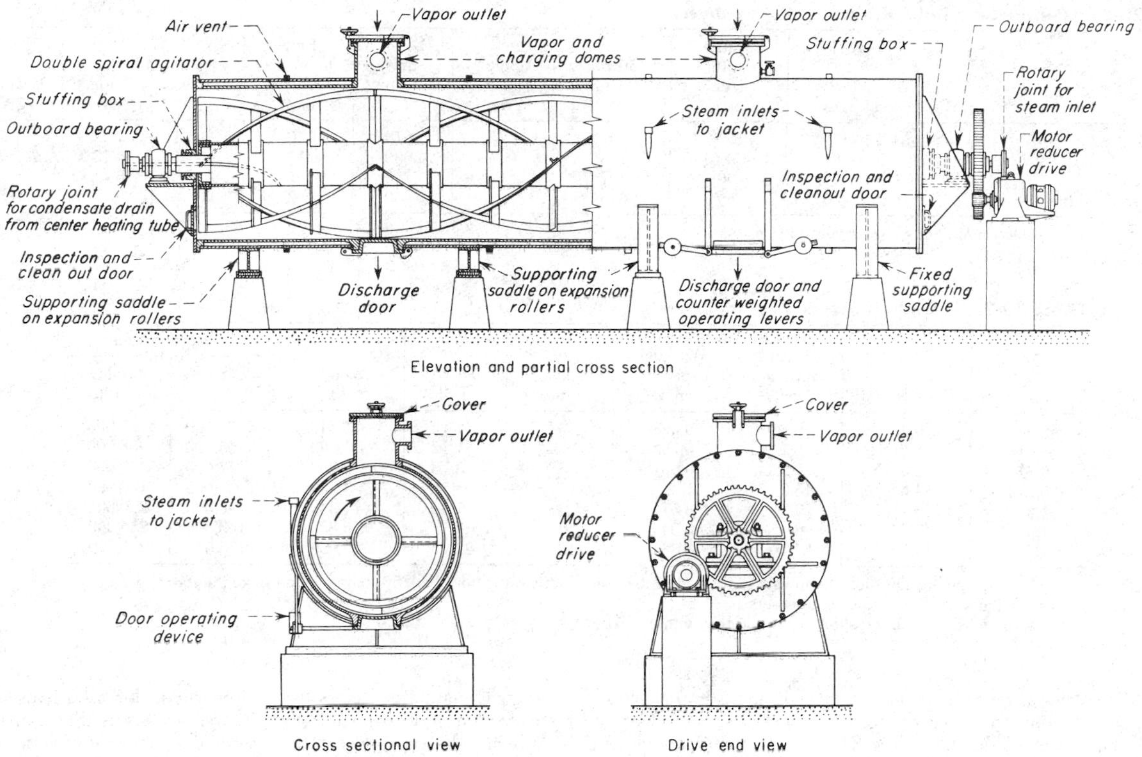

FIG. 20-49 A typical vacuum dryer. *(Blaw-Knox Food & Chemical Equipment, Inc.)*

mercial scale. Air in-leakage through gasket surfaces will be in the range of 0.2 kg/(h·linear m of gasketed surface) under these conditions.

Design Methods The rate of heat transfer from the heating medium through the dryer wall to the solids can be expressed by

$$Q = UA\Delta t_m \qquad (20\text{-}45)$$

where Q = heat flux, J/s (Btu/h); U = overall heat-transfer coefficient, J/(m^2·s·K) [Btu/(h·ft^2 jacket area·°F)]; A = total jacket area, m^2; and Δt_m = log-mean-temperature driving force from heating medium to the solids, K.

The overall heat-transfer rate is almost entirely dependent upon the film coefficient between the inner jacket wall and the solids, which depends to a large extent on the solids characteristics. Overall coefficients may range from 30 to 200 J/(m^2·s·K), based upon total area if the dryer walls are kept reasonably clean. Coefficients as low as 5 or 10 may be encountered if caking on the walls occurs.

For estimating purposes without tests, a reasonable coefficient for ordinary drying, and without taking the product to absolute dryness, may be assumed at U = 50 J/(m^2·s·K) for rotary agitator dryers and 35 J/(m^2·s·K) for rotating units.

Vacuum dryers are usually filled to 50 to 65 percent of their total shell volume. Agitator speeds range from 3 to 8 r/min. Faster speeds yield a slight improvement in heat transfer but consume more power.

Performance and Cost Data Typical performance data for vacuum rotary dryers are given in Table 20-21. Size and cost data for rotary agitator units are given in Table 20-22. Data for double-cone units are in Table 20-23.

Turbo-Tray Dryers The turbo-tray dryer is a continuous dryer consisting of a stack of rotating annular shelves in the center of which turbo-type fans revolve to circulate the air over the shelves. Wet material enters through the roof, falling onto the top shelf as it rotates beneath the feed opening. After completing one revolution, the material is wiped by a stationary wiper through radial slots onto the shelf below, where it is spread into a uniform pile by a stationary

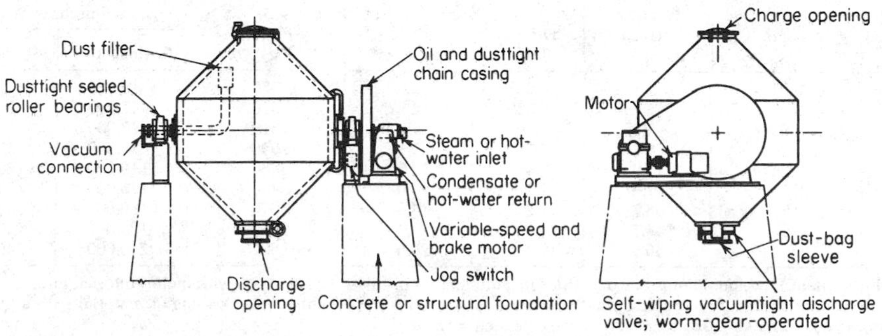

FIG. 20-50 Rotating (double-cone) vacuum dryer. *(Stokes Equipment Division, Pennwalt Corp.)*

TABLE 20-21 Performance Data of Vacuum Rotary Dryers*

Material	Diameter × length, m	Initial moisture, % dry basis	Steam pressure, Pa × 10³	Agitator speed, r/min	Batch dry weight, kg	Final moisture, % dry basis	Pa × 10³	Time, h	Evaporation, kg/(h·m²)
Cellulose acetate	1.5 × 9.1	87.5	97	5.25	610	6	90–91	7	1.5
Starch	1.5 × 9.1	45–48	103	4	3630	12	88–91	4.75	7.3
Sulfur black	1.5 × 9.1	50	207	4	3180	1	91	6	4.4
Fuller's earth/mineral spirit	0.9 × 3.0	50	345	6	450	2	95	8	5.4

*Stokes Equipment Division, Pennwalt Corporation.

TABLE 20-22 Standard Rotary Vacuum Dryers*

Diameter, m	Length, m	Heating surface, m²	Working capacity, m³†	Agitator speed, r/min	Drive, kW	Weight, kg	Purchase price: Carbon steel	Purchase price: Stainless steel (304)
0.46	0.49	0.836	0.028	7½	1.12	540	$ 22,500	$ 27,900
0.61	1.8	3.72	0.283	7½	1.12	1,680	55,000	68,000
0.91	3.0	10.2	0.991	6	3.73	3,860	77,000	95,000
0.91	4.6	15.3	1.42	6	3.73	5,530	95,000	108,000
1.2	6.1	29.2	3.57	6	7.46	11,340	140,000	200,000
1.5	7.6	48.1	6.94	6	18.7	15,880	160,000	230,000
1.5	9.1	57.7	8.33	6	22.4	19,050	175,000	245,000

*Stokes Equipment Division, Pennwalt Corp. (December 1981). Prices include shell, 50-lb/in²-gauge jacket, agitator, drive, and motor; auxiliary dust collectors, condensers.

†Loading with product level on or around the agitator shaft.

leveler. The action is repeated on each shelf, with transfers occurring once in each revolution. From the last shelf, material is discharged through the bottom of the dryer (Fig. 20-51). The steel-frame housing consists of removable insulated panels for access to the interior. All bearings and lubricated parts are exterior to the unit with the drives located under the housing. Parts in contact with the product may be of steel or special alloy. The trays can be of any sheet material, such as enameled steel, asbestos-cement composition board, or plastic-glass laminates.

The rate at which each fan circulates air can be varied by changing the pitch of the fan blades. In final drying stages, in which diffusion controls or the product is light and powdery, the circulation rate is considerably lower than in the initial stage, in which high evaporation rates prevail. In the majority of applications, air flows through the dryer upward in counterflow to the material. In special cases, required drying conditions dictate that air flow be cocurrent or both countercurrent and cocurrent with the exhaust leaving at some level between solids inlet and discharge. A separate cold-air-supply fan is provided if the product is to be cooled before being discharged.

By virtue of its vertical construction, the turbo-type tray dryer has a stack effect, the resulting draft being frequently sufficient to operate the dryer with natural draft. Pressure at all points within the dryer is maintained close to atmospheric, as low as 0.1, usually less than 0.5 mm of water. Most of the roof area is used as a breeching, lowering the exhaust velocity to settle dust back into the dryer.

Heaters can be located in the space between the trays and the dryer housing, where they are not in direct contact with the product, and thermal efficiencies up to 3500 kJ/kg (1500 Btu/lb) of water evaporated can be obtained by reheating the air within the dryer. Steam is the usual heating medium. The high cost of heating electrically generally restricts its use to relatively small equipment. For materials which have a tendency to foul internal heating surfaces, an external heating system is employed.

The turbo-tray dryer can handle materials from thick slurries [1 million (N·s)/m² (100,000 cP) and over] to fine powders. It is not suitable for fibrous materials which mat or for doughy or tacky materials. Thin slurries can often be handled by recycle of dry product. Filter-press cakes are granulated before feeding. Thixotropic materials are fed directly from a rotary filter by scoring the cake as it leaves the drum. Pastes can be extruded onto the top shelf and subjected to a hot blast of air to make them firm and free-flowing after one revolution.

The turbo-tray dryer is manufactured in sizes from package units 2 m in height and 2 m in diameter to large outdoor installations 20 m in height and 11 m in diameter. Tray areas range from 5.5 to 1675

TABLE 20-23 Standard (Double-Cone) Rotating Vacuum Dryers*

Working capacity, m³	Total volume, m³	Heating surface, m²	Drive, kW	Floor space, m²	Weight, kg	Purchase cost: Carbon steel	Purchase cost: Stainless steel
0.085	0.130	1.11	.373	2.60	730	$ 18,000	$ 21,000.00
0.283	0.436	2.79	.560	2.97	910	21,000	24,000.00
0.708	1.09	5.30	1.49	5.57	1810	28,000	31,500.00
1.42	2.18	8.45	3.73	7.15	2040	54,000	59,000.00
2.83	4.36	13.9	7.46	13.9	3860	110,000	120,000.00
4.25	6.51	17.5	11.2	14.9	5440	125,000	135,000.00
7.08	10.5	*38.7	11.2	15.8	9070	180,000	195,000.00
9.20	13.9	*46.7	11.2	20.4	9980	195,000	215,000.00
11.3	16.0	*56.0	11.2	26.0	10,890	210,000	245,000.00

*Stokes Equipment Division, Pennwalt Corp. Price includes dryer, 15-lb/in² jacket, drive with motor, internal filter, and trunnion supports for concrete or steel foundations. Horsepower is established on 65 percent volume loading of material with a bulk density of 50 lb/ft³. Models of 250 ft³, 325 ft³, 400 ft³ have extended surface area.

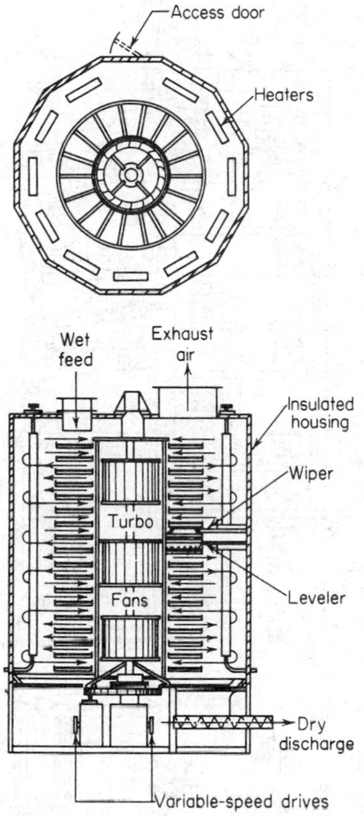

FIG. 20-51 Turbo-tray dryer. *(Wyssmont Company, Inc.)*

m^2 in a single unit. The number of shelves in a tray rotor varies according to space available and minimum rate of transfer required, from as few as 12 shelves to as many as 58 in the largest units. Standard construction permits operating temperatures up to 615 K.

Design Methods The heat- and mass-transfer mechanisms are similar to those in batch tray dryers, except that constant turning over and mixing of the solids significantly improves drying rates. Design must usually be based on previous installations or pilot tests by the manufacturer; apparent heat-transfer coefficients will range

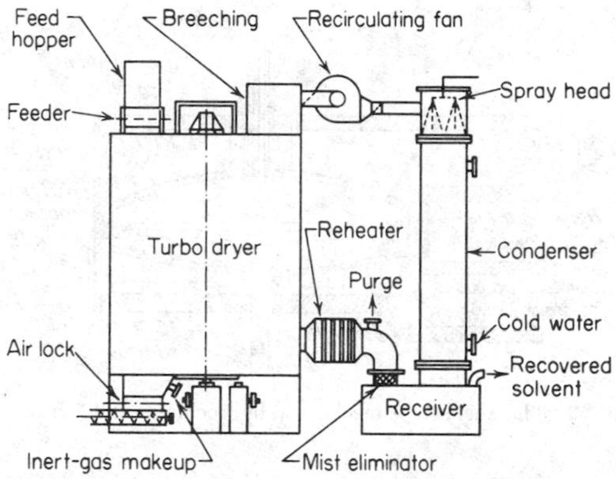

FIG. 20-52 Turbo-tray dryer in closed circuit for continuous drying with solvent recovery. *(Wyssmont Company, Inc.)*

from 28 to 55 $J/(m^2 \cdot s \cdot K)$ for dry solids to 68 to 115 $J/(m^2 \cdot s \cdot K)$ for wet solids. Turbo-tray dryers have been employed successfully for the drying and cooling of calcium hypochlorite, urea crystals, calcium chloride flakes, and sodium chloride crystals. The Wyssmont "closed-circuit" system, as shown in Fig. 20-52, consists of the turbo-tray dryer with or without internal heaters, recirculation fan, condenser with receiver and mist eliminators, and reheater. Feed and discharge are through a sealed wet feeder and lock respectively. This method is used for continuous drying without leakage of fumes, vapors, or dust to the atmosphere.

Performance and Cost Data Performance data for three applications of closed-circuit drying are included in Table 20-24. Operating, labor, and maintenance costs compare favorably with those of direct-heat rotating equipment.

Hearth Furnaces A special design of a circular hearth furnace is the **Mannheim furnace,** in which sulfuric acid is reacted with sodium chloride to produce salt cake and hydrochloric acid. It consists of a refractory hearth, up to 6 m in diameter, with a silicon carbide arch. Hot flue gases are circulated around the muffle. The major portion of heat is transmitted through the arch and radiated to the product on the hearth. Feed materials are mixed and charged continuously to the center of the hearth, where they are stirred by underdriven rabble arms. The charge is gradually worked toward the periphery as the reaction generates hydrogen chloride gas. The gas is withdrawn through a separate duct to an absorption system. The salt cake is discharged at the periphery. Figure 20-53 shows a diagrammatic cross section of a Mannheim furnace. Combustion-cham-

TABLE 20-24 Turbo-Tray Dryer Performance Data in Wyssmont Closed-Circuit Operations*

Material dried	Antioxidant	Water-soluble polymer	Antibiotic filter cake	Petroleum coke
Dried product, kg/h	500	85	2400	227
Volatiles composition	Methanol and water	Xylene and water	Alcohol and water	Methanol
Feed volatiles, % wet basis	10	20	30	30
Product volatiles, % wet basis	0.5	4.8	3.5	0.2
Evaporation rate, kg/h	53	16	910	302
Type of heating system	External	External	External	External
Heating medium	Steam	Steam	Steam	Steam
Drying medium	Inert gas	Inert gas	Inert gas	Inert gas
Heat consumption, J/kg	0.56×10^6	2.2×10^6	1.42×10^6	1.74×10^6
Power, dryer, kW	1.8	0.75	12.4	6.4
Power, recirculation fan, kW	5.6	5.6	37.5	15
Materials of construction	Stainless-steel interior	Stainless-steel interior	Stainless-steel interior	Carbon steel
Dryer height, m	4.4	3.2	7.6	6.5
Dryer diameter, m	2.9	1.8	6.0	4.5
Recovery system	Shell-and-tube condenser	Shell-and-tube condenser	Direct-contact condenser	Shell-and-tube condenser
Condenser cooling medium	Brine	Chilled water	Tower water	Chilled water
Location	Outdoor	Indoor	Indoor	Indoor
Approximate cost of dryer (1980)	$140,000	$70,000	$390,000	$140,000
Dryer assembly	Packaged unit	Packaged unit	Field-erected unit	Field-erected unit

*Courtesy of Wyssmont Company, Inc.

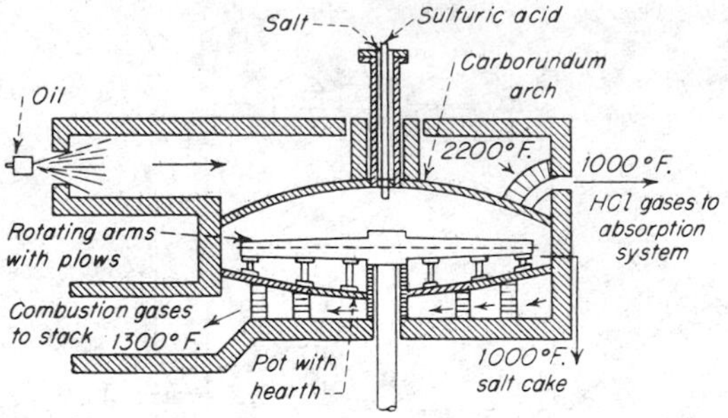

FIG. 20-53 Mannheim-type mechanical hydrochloric acid furnace.

ber temperatures of about 1475 K are used for heating. The salt cake is discharged from the hearth at about 800 K.

Multiple-Hearth Furnaces Multiple-hearth furnaces are known under various names: the Herreshoff, McDougall, Wedge, Pacific, etc. Figure 20-54 shows a general design. It consists of a number of annular-shaped hearths mounted one above the other. There are rabble arms on each hearth driven from a common center shaft. The feed is charged at the center of the upper hearth. The arms move the charge outward to the periphery, where it falls to the next hearth. Here it is moved again to the center, from which it falls to the next hearth. This continues down the furnace. The hollow center shaft is cooled internally by forced-air circulation.

Burners may be mounted at any of the hearths, and the circulated air is used for combustion. These furnaces handle granular materials and provide a long countercurrent path between the flue gases and the charge material. Industrial sizes are built from 2 to 7 m in diameter and include 4 to 16 hearths. Total hearth areas range from 6.5 to 335 m^2. The furnaces are used for roasting ores, drying and calcining lime, magnesite, and carbonate sludges, reactivation of decolorizing earths, and burning of sulfides to produce sulfur dioxide. The following is a partial list of applications:

1. Lime (*a*) from crushed limestone, (*b*) from oyster or sea shell, and (*c*) from dolomitic limestone
2. Lead and zinc; roasting of sulfides
3. Mercury from cinnabar ores by volatilization
4. Gold and silver: (*a*) chloridizing roast of gold-silver ore, and (*b*) removal of arsenic
5. Sulfuric acid from iron pyrites
6. Paint pigments; roasting of metallic oxides
7. Refractory clays; calcination of refractory clay to reduce shrinkage
8. Foundry sand; removal of carbon from used foundry sand
9. Fuller's earth; calcination of fuller's-earth material
10. Sewage disposal; calcination of sewage slurry

Table 20-25 lists three specific applications with a brief description of the furnaces as to design and operating conditions.

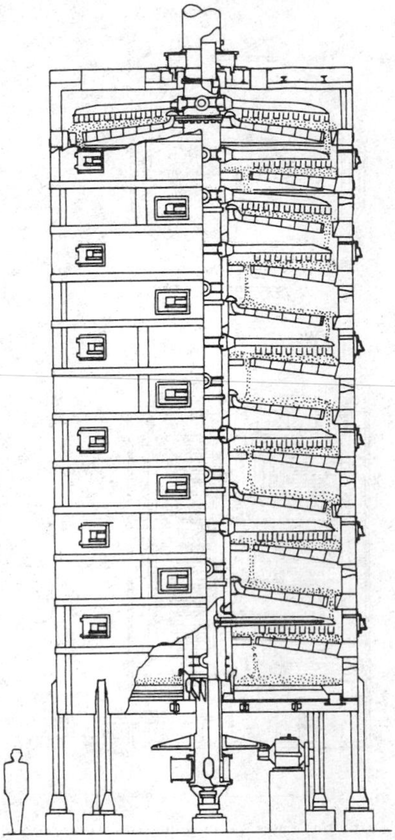

FIG. 20-54 Pacific multiple-hearth furnace.

GRAVITY DRYERS

Description A body of solids in which the particles, consisting of granules, pellets, beads, or briquettes, flow downward by gravity at substantially their normal settled bulk density through a vessel in contact with gases is defined frequently as a **moving bed.** Moving-bed equipment finds application in blast furnaces, shaft furnaces, and petroleum refining.

A gravity dryer consists of a stationary vertical, usually cylindrical housing with openings for the introduction of solids (at the top) and removal of solids (at the bottom). Gas flow is through the solids bed and may be cocurrent or countercurrent and, in some instances,

TABLE 20-25 Applications of Multiple-Hearth Furnaces*

Product	Production rate	Furnace size	Special features
Mercury from cinnabar ore . . .	225 tons ore/day (95% recovery)	(2) 18.0 ft. diam., 8 hearth furnaces	Furnaces fired on hearths 3 to 7, inclusive; retention time of 1.0 hr.; furnaces are oil-fired with low-pressure atomizing air burners; all air, both primary and secondary, introduced through the burners; draft control by Monel · cold-gas fans downstream from mercury condensers.
Lime from oyster shell	240 tons/day, shell (120 tons/day, lime)	(1) 22 ft., 3 in. diam., 12 hearth furnaces	
Magnesium oxide from magnesium hydroxide	100 tons/day, 50% magnesium hydroxide slurry; yields 50 tons/day magnesium oxide	(2) 22 ft., 3 in., 10 hearth furnaces	Furnace walls of 4.5-in. firebrick, 9-in. insulation for 1550°F. operating temp. Furnace fired on hearths 4 to 10, inclusive

*BSP, Envirotech.

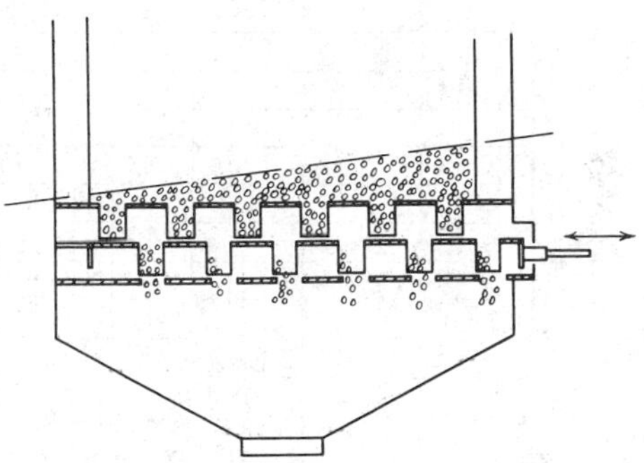

FIG. 20-55 Gravity-bed reactor; solids-discharge mechanism.

cross-flow. By definition, the rate of gas flow upward must be less than that required for fluidization.

Fields of Application One of the major advantages of the gravity-bed technique is that it lends itself well to true intimate countercurrent contacting of solids and gases. This provides for efficient heat transfer and mass transfer. Gravity-bed contacting also permits the use of the solid as a heat-transfer medium, as in pebble heaters.

Gravity vessels are applicable to coarse granular free-flowing solids which are comparatively dust-free. The solids must possess physical properties in size and surface characteristics so that they will not stick together, bridge, or segregate during passage through the vessel. The presence of significant quantities of fines or dust will close the passages among the larger particles through which the gas must penetrate, increasing pressure drop. Fines may also segregate near the sides of the bed or in other areas where gas velocities are low, ultimately completely sealing off these portions of the vessel. The high efficiency of gas-solids contacting in gravity beds is due to the uniform distribution of gas throughout the solids bed; hence choice of feed and its preparation are important factors to successful operation. Preforming techniques such as pelleting and briquetting are employed frequently for the preparation of suitable feed materials.

Gravity vessels are suitable for low-, medium-, and high-temperature operation; in the last case, the housing will be lined completely with refractory brick. Dust-recovery equipment is minimized in this type of operation since the bed actually performs as a dust collector itself, and dust in the bed will not, in a successful application, exist in large quantities.

Other advantages of gravity beds include flexibility in gas and solids flow rates and capacities, variable retention times from minutes to several hours, space economy, ease of startup and shutdown, the potentially large number of contacting stages, and ease of control by using the inlet- and exit-gas temperatures.

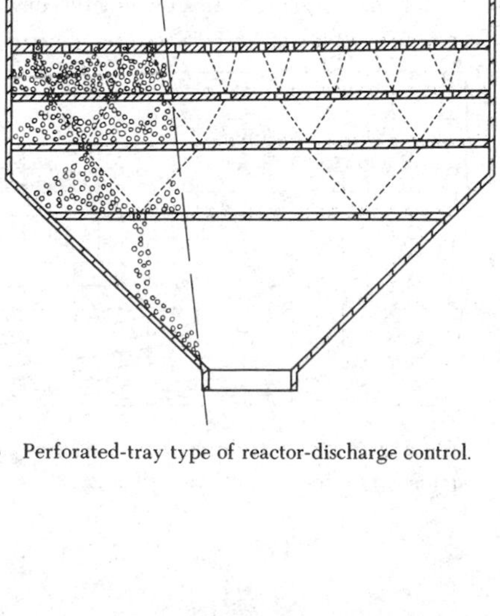

FIG. 20-56 Perforated-tray type of reactor-discharge control.

Maintenance of a uniform rate of solids movement downward over the entire cross section of the bed is one of the most critical operating problems encountered. For this reason gravity beds are designed to be as high and narrow as practical. In a vessel of large cross section, discharge through a conical bottom and center outlet will usually result in some degree of "ratholing" through the center of the bed. Flow through the center will be rapid while essentially stagnant pockets are left around the sides. To overcome this problem, multiple outlets are provided in the center and around the periphery; table unloaders, rotating plows, wide moving grates, and multiple-screw unloaders are employed; insertion of inverted cone baffles in the lower section of the bed, spaced so that flushing at the center is retarded, is also a successful method for improving uniformity of solids movement. Figure 20-55 illustrates a moving tray with multiple downspouts used to remove a precise amount of solids from each increment of area across the base of a gravity-bed reactor. The various pockets are filled at one extremity of its motion and emptied at the other. It is suitable primarily for fine nonabrasive solids. Figure 20-56 depicts a perforated-plate design, taking advantage of the flow characteristics and angle of repose of the solids to control the unloading rate. Still another design of this general type involves the use of a nest of inclined pipes, discharging into a common header, and placed to draw solids at geometrically spaced points across the base of the reactor.

Gas disengaging from the solids may represent another serious operating problem in a gravity bed. One method employs downspouts at the top for solids feeding while leaving an open space in the vessel above the downspout outlet for gas disengaging, as illustrated in Fig. 20-57*a*. Another uses a series of inverted V-shaped channels inserted into the top of the solids bed. Gas and vapor are collected and removed from under the V's, while the solids flow over the top and around the channels (Fig. 20-57*b*). These methods for both gas and solids removal were developed originally for petroleum-refining catalytic reactors.

Shaft Furnaces The oldest and most important application of the shaft furnace is the **blast furnace** used for the production of pig iron. Another use is in the manufacture of phosphorus from phos-

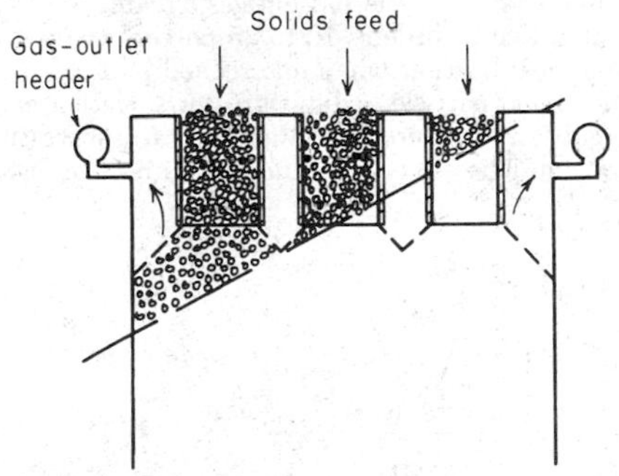

FIG. 20-57*a* Countercurrent gas-solids flow at the top disengaging section of a moving-bed catalytic reactor.

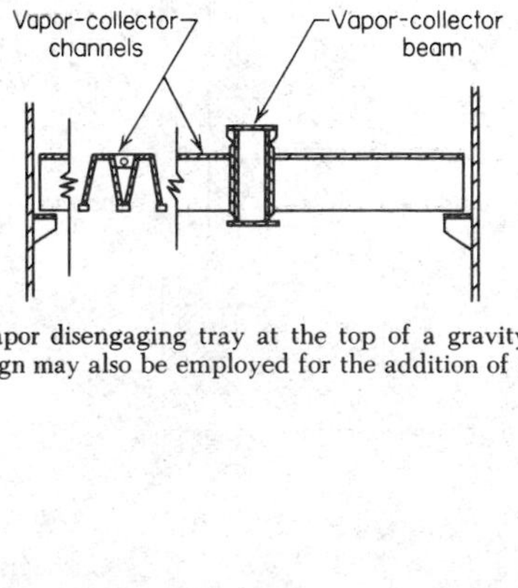

FIG. 20-57*b* Vapor disengaging tray at the top of a gravity-bed catalytic reactor. This design may also be employed for the addition of gas to a bed of solids.

phate rock. Formerly lime was calcined exclusively in this type of furnace. Shaft furnaces are widely used also as gas producers. Chemicals are manufactured in shaft furnaces from briquetted mixtures of the reacting components.

A shaft furnace is a vertical refractory-lined cylinder in which a stationary or descending column of solids is maintained and through which an ascending stream of hot gas is forced. Three methods of fuel application may be employed: (1) one in which a solid fuel is added alone or mixed with the reacting solids, (2) one in which the fuel is burned in a separate combustion chamber with the hot gases being blown into the furnace at some level of the column, (3) one in which the fuel is introduced and burned in the bottom of the shaft.

For maximum heat economy, recovered exhaust heat is employed for preheating of the incoming solids and combustion air. The fuels used may be gas, oil, or pulverized coal.

Bucket elevators, skip hoists, and cranes are used for top feeding of the furnace. Retention and downward flow are controlled by timing of the bottom discharge. Gases are propelled by a blower or by induced draft from a stack or discharge fan. In normal operation, the downward flow of solids and upward flow of gas are constant with time, maintaining ideal steady-state conditions.

Figure 20-58 illustrates a shaft lime kiln.

Design Methods The size and shape of the charge particles control the amount of surface over which heat may be transmitted to the particle and also the depth of penetration through which the heat must pass to reach the center of each particle. Also, this size and shape control the nature of the random packing in the shaft and the extent of voids for gas passage. As particle size is decreased, the surface area of the particles increases. At the same time, the depth of heat penetration decreases. Both these factors tend to improve performance. With small particle size, however, the charge column presents high resistance to the passage of gas.

With closely screened material, the percentage of voids (usually 37 percent) is independent of particle size. With unscreened particles showing a wide variation in size, the void volume is decreased; irregularity in gas flow results.

There is a large difference between the total surface of the particles (as determined by their size and shape) and the "effective surface" actually exposed to the passing gas stream. In practice, it has been estimated that as little as 10 to 25 percent of the total surface is effective in heat transfer when unscreened particles are treated.

Irregular-shaped particles exhibit greater surface area than regular-shaped cubes and spheres, the amount of this increase being possibly 25 percent. The effect of particle size and size distribution on

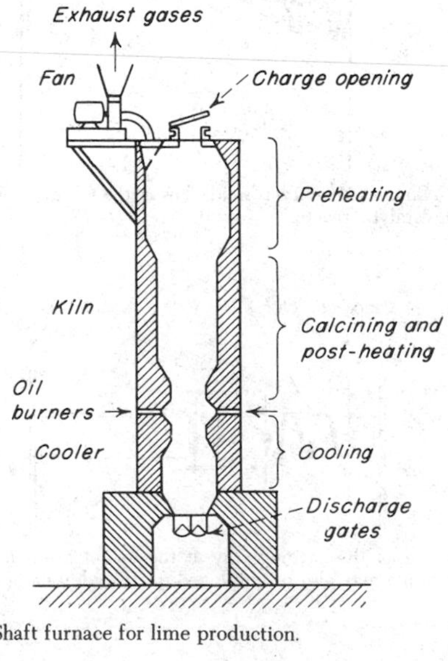

FIG. 20-58 Shaft furnace for lime production.

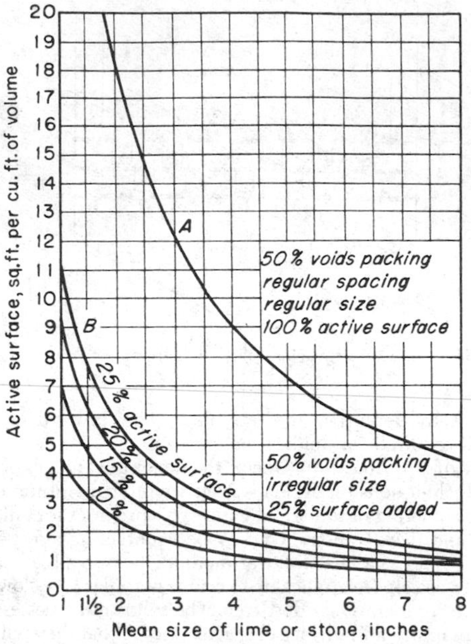

FIG. 20-59 Curve *A* shows surface variations with stone size, 100 percent active surfaces. Curves in group *B* show the effect of irregular stone size.

effective surface, in a shaft employed for calcination of limestone, is shown in Fig. 20-59. Curve *A* shows the calculated surface based on an assumed 50 percent void volume and cubical-shaped particles. The *B* set of curves applies to such unscreened irregularly shaped particles as are usually encountered in practice.

The laws governing the flow of fluids through packed beds given in Sec. 5 are applicable to shaft furnaces. Since the pressure drop in a bed is affected by the size and shape of the interstitial voids, the horizontal and vertical nonuniformity of the bed, the changes in gas composition during passage, and other operating factors, test data for a given material are necessary for proper design. In the case of limestone, Fig. 20-60 shows the effect of particle size on the gas-flow friction through the bed, assuming that the friction varies as the square of the gas mass velocity and inversely with the particle size, and utilizing base points established during actual kiln operations. Information on the mathematical treatment of heat transfer in packed beds is included in Sec. 10.

Pellet Coolers and Dryers Gravity beds are employed for the cooling and drying of extruded pellets and briquettes from size-enlargement processes. The rotary cooler illustrated in Fig. 20-61 consists of a stationary steel tank having a wear cylinder at the top for entry of gas and solids (usually from a pneumatic conveyor), with

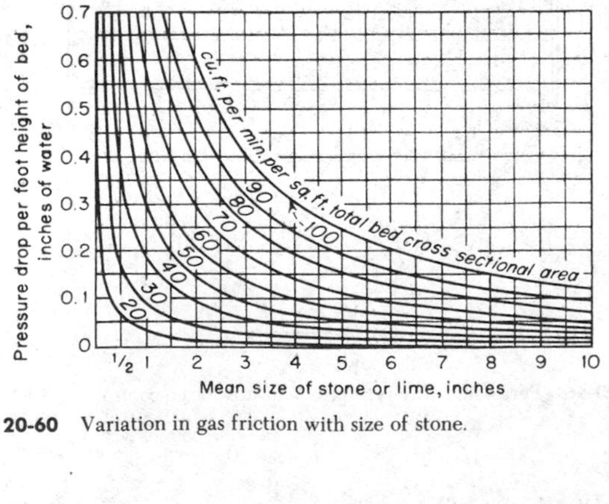

FIG. 20-60 Variation in gas friction with size of stone.

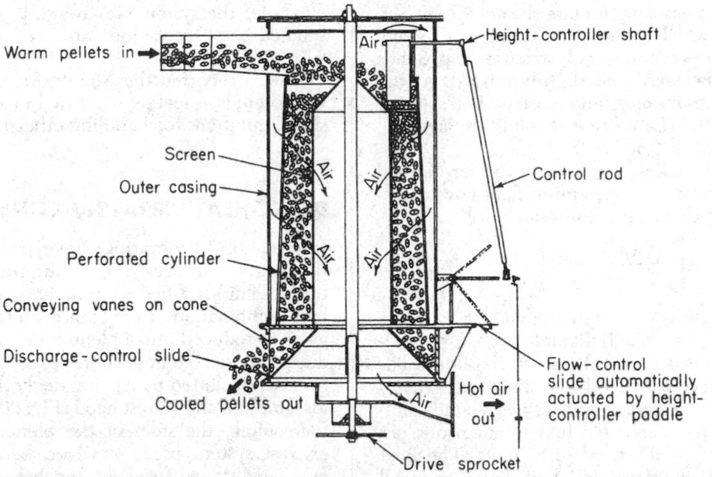

FIG. 20-61 Sprout, Waldron rotary cooler. *(Sprout, Waldron Companies.)*

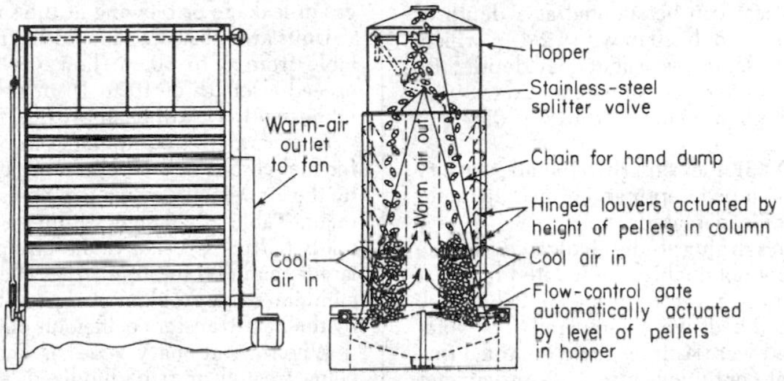

Fig. 20-62 Vertical gravity-bed cooler with louvers. *(Sprout, Waldron Companies.)*

air holes staggered around the outer wall to admit additional air for optimum circulation. The tank encloses a rotating cage for retention of the solids bed. This cage consists of an inner cylinder of wire mesh and an outer perforated shell. Air entering the tank is circulated in cross-flow through the pellet bed and discharges through the center column. Usually a rotary unloading gate and air lock are located underneath the cage.

Another cross-flow design employs a rectangular housing, partitioned into three vertical sections (Fig. 20-62). Solids move downward in the two outer sections, while cooling or drying air is drawn through the louvered outer walls and through the solids bed and is discharged through the center section. Solids are discharged over a baffled shaking shoe. Units of this general type are used for drying wheat and other grain products and numerous forms of pelleted feeds. Gravity-bed dryers are most suitable for drying of granular heat-sensitive products employing moderate air temperatures. These require extended holdup during the falling-rate drying period.

Spouted Beds The spouted-bed technique was developed primarily for solids which are too coarse to be handled in fluidized beds. Although their applications overlap, the methods of gas-solids mixing are completely different. A schematic view of a spouted bed is given in Fig. 20-63. Mixing and gas-solids contacting are achieved first in a fluid "spout," flowing upward through the center of a loosely packed bed of solids. Particles are entrained by the fluid and conveyed to the top of the bed. They then flow downward in the surrounding annulus as in an ordinary gravity bed, countercurrently to gas flow. The mechanisms of gas flow and solids flow in spouted beds were first described by Mathur and Gishler [*Am. Inst. Chem. Eng. J.*, **1**, 2, 157–164 (1955)]. Drying studies have been carried out by Cowan [*Eng. J.*, **41**, 5, 60–64 (1958)], and a theoretical equation for predicting the minimum fluid velocity necessary to initiate spouting was developed by Madonna and Lama [*Am. Inst. Chem. Eng. J.*, **4**, 4, 497 (1958)]. Investigations to determine maximum spoutable depths and to develop theoretical relationships based on vessel geometry and operating variables have been carried out by Lefroy [*Trans. Inst. Chem. Eng.*, **47**(5), T120–128 (1969)] and Reddy [*Can. J. Chem. Eng.*, **46**(5), 329–334 (1968)].

Gas flow in a spouted bed is partially through the spout and partially through the annulus. About 30 percent of the gas entering the system immediately diffuses into the downward-flowing annulus. Near the top of the bed, the quantity in the annulus approaches 66

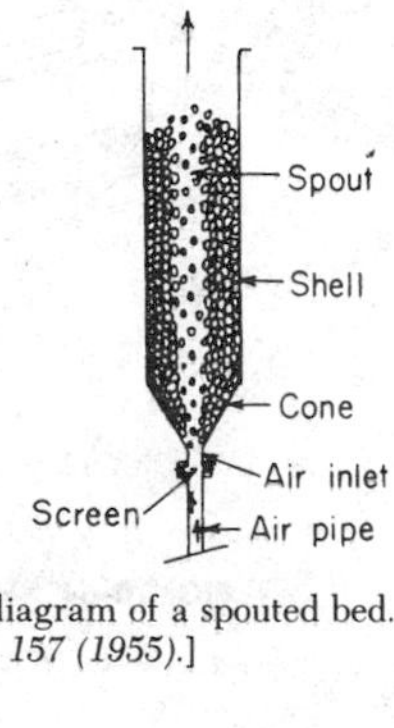

FIG. 20-63 Schematic diagram of a spouted bed. [*Mathur and Gishler,* Am. Inst. Chem. Eng. J., *1, 2, 157 (1955).*]

percent of the total gas flow; the gas flow through the annulus at any point in the bed equals that which would flow through a loosely packed solids bed under the same conditions of pressure drop. Solids flow in the annulus is both downward and slightly inward. As the fluid spout rises in the bed, it entrains more and more particles, losing velocity and gas into the annulus. The volume of solids displaced by the spout is roughly 6 percent of the total bed.

On the basis of experimental studies, Mathur and Gishler derived an empirical correlation to describe the minimum fluid flow necessary for spouting, in 3- to 12-in-diameter columns:

$$u = \frac{D_p}{D_c}\left(\frac{D_o}{D_c}\right)^{0.33}\left[\frac{2gL(p_s - p_f)}{p_f}\right]^{0.5} \qquad (20\text{-}46)$$

where u = superficial fluid velocity through the bed, ft/s; D_p = particle diameter, ft; D_c = column (or bed) diameter, ft; D_o = fluid-inlet orifice diameter, ft; L = bed height, ft; p_s = absolute solids density, lb/ft^3; p_f = fluid density, lb/ft^3; and g = 32.2 ft/s^2, gravity acceleration. To convert feet per second to meters per second, multiply by 0.305; to convert pounds per cubic foot to kilograms per cubic meter, multiply by 16. g = 9.8 m/s^2 in SI units. The inlet orifice diameter, air rate, bed diameter, and bed depth were all found to be critical and interdependent:

1. In a given-diameter bed, deeper beds can be spouted as the gas-inlet orifice size is decreased. Using air, a 12-in-diameter bed containing 0.125- by 0.250-in wheat can be spouted at a depth of over 100 in with a 0.8-in orifice, but at only 20 in with a 2.4-in orifice.

2. Increasing bed diameter increases spoutable depth. By employing a bed-orifice diameter ratio of 12 for air spouting, a 9-in-diameter bed was spouted at a depth of 65 in while a 12-in-diameter bed was spouted at 95 in.

3. As indicated by Eq. (20-46) the superficial fluid velocity required for spouting increases with bed depth and orifice diameter and decreases as the bed diameter is increased.

Employing wood chips, Cowan's drying studies indicated that the volumetric heat-transfer coefficient obtainable in a spouted bed is at least twice that in a direct-heat rotary dryer. By using 20- to 30-mesh Ottawa sand, fluidized and spouted beds were compared. The volumetric coefficients in the fluid bed were 4 times those obtained in a spouted bed. Mathur dried wheat continuously in a 12-in-diameter spouted bed, followed by a 9-in-diameter spouted-bed cooler. A drying rate of roughly 100 lb/h of water was obtained by using 450 K inlet air. Six hundred pounds per hour of wheat was reduced from 16 to 26 percent to 4 percent moisture. Evaporation occurred also in the cooler by using sensible heat present in the wheat. The maximum drying-bed temperature was 118°F, and the overall thermal efficiency of the system was roughly 65 percent. Some aspects of the spouted-bed technique are covered by patent (U.S. Patent 2,786,280).

Cowan reported that significant size reduction of solids occurred when cellulose acetate was dried in a spouted bed, indicating its possible limitations for handling other friable particles.

DIRECT-HEAT VIBRATING-CONVEYOR DRYERS

Information on vibrating conveyors and their mechanical construction is given in Sec. 7. The vibrating-conveyor dryer is a modified form of fluidized-bed equipment, in which fluidization is maintained by a combination of pneumatic and mechanical forces. The heating gas is introduced into a plenum beneath the conveying deck through ducts and flexible hose connections, passes up through a screen, perforated, or slotted conveying deck, through the fluidized bed of solids, and into an exhaust hood (Fig. 20-64). If ambient air is employed for cooling, the sides of the plenum may be open and a simple exhaust system used; however, because the gas-distribution plate may be designed for several inches of water-pressure drop to ensure a uniform velocity distribution through the bed of solids, a combination pressure-blower–exhaust-fan system is desirable to balance the pressure above the deck with the outside atmosphere and prevent gas in-leakage or blowing at the solids feed and exit points.

Units are fabricated in widths from 0.3 to 1.5 m. Lengths are variable from 3 to 50 m; however, most commercial units will not exceed a length of 10 to 16 m per section. Power required for the vibrating drive will be approximately 0.4 kW/m^2 of deck.

In general, this equipment offers an economical heat-transfer area for first cost as well as operating cost. Capacity is limited primarily by the air velocity which can be used without excessive dust entrainment. Table 20-26 shows limiting air velocities suitable for various solids particles. Usually, the equipment is satisfactory for particles larger than 100 mesh in size. [The use of indirect-heated conveyors eliminates the problem of dust entrainment, but capacity is limited by the heat-transfer coefficients obtainable on the deck (see Sec. 11)].

When a stationary vessel is employed for fluidization, all solids being treated must be fluidized; nonfluidizable fractions fall to the bottom of the bed and may eventually block the gas distributor. The addition of mechanical vibration to a fluidized system offers the following advantages:

1. Equipment can handle nonfluidizable solids fractions. Although these fractions may drop through the bed to the screen, directional-throw vibration will cause them to be conveyed to the

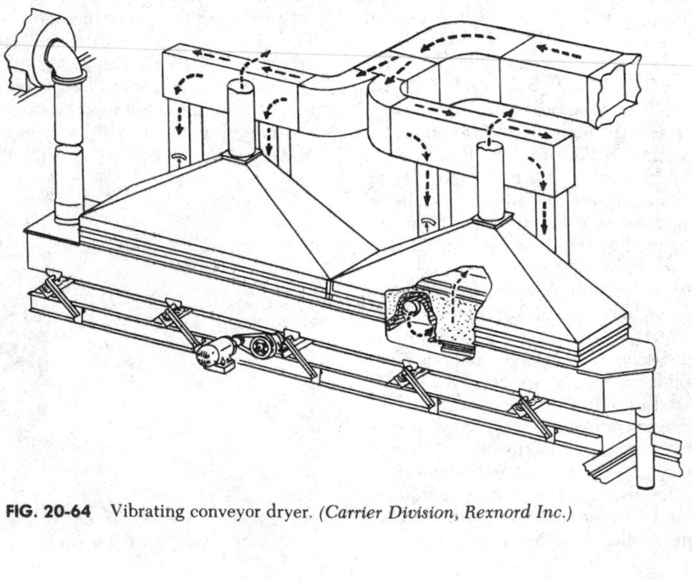

FIG. 20-64 Vibrating conveyor dryer. *(Carrier Division, Rexnord Inc.)*

TABLE 20-26 Table for Estimating Maximum Superficial Air Velocities through Vibrating-Conveyor Screens*

Mesh size	Velocity, m/s	
	2.0 specific gravity	1.0 specific gravity
200	0.22	0.13
100	0.69	0.38
50	1.4	0.89
30	2.6	1.8
20	3.2	2.5
10	6.9	4.6
5	11.4	7.9

*Carrier Division, Rexnord Inc.

discharge end of the conveyor. Prescreening or sizing of the feed is less critical than in a stationary fluidized bed.

2. Because of mechanical vibration, incipient channeling is reduced.

3. Fluidization may be accomplished with lower pressures and gas velocities. This has been evidenced on vibratory units by the fact that fluidization stops when the vibrating drive is stopped.

Vibrating-conveyor dryers are suitable for free-flowing solids containing mainly surface moisture. Retention is limited by conveying speeds which range from 0.02 to 0.12 m/s. Bed depth rarely exceeds 7 cm, although units are fabricated to carry 30- to 46-cm-deep beds; these also employ plate and pipe coils suspended in the bed to provide additional heat-transfer area. Vibrating dryers are not suitable for fibrous materials which mat or for sticky solids which may ball or adhere to the deck.

For estimating purposes for direct-heat drying applications, it can be assumed that the average exit-gas temperature leaving the solids bed will approach the final solids discharge temperature on an ordinary unit carrying a 5- to 15-cm-deep bed. Calculation of the heat load and selection of an inlet-air temperature and superficial velocity (Table 20-26) will then permit approximate sizing, provided an approximation of the minimum required retention time can be made.

Vibrating conveyors employing direct contacting of solids with hot, humid air have also been employed for the agglomeration of fine powders, chiefly for the preparation of agglomerated water-dispersible food products. Control of inlet-air temperature and dew point permits the uniform addition of small quantities of liquids to solids by condensation on the cool incoming-particle surfaces. The wetting section of the conveyor is followed immediately by a warm-air-drying section and particle screening.

PNEUMATIC-CONVEYOR DRYERS AND SPRAY DRYERS

A gas-solids contacting operation in which the solids phase exists in a dilute condition is termed a **pneumatic system.** It is called pneumatic because, in most cases, the quantity and velocity of the gas are sufficient to lift and convey the solids against the force of gravity. Pneumatic systems may be distinguished by two characteristics:

1. Retention of a given solids particle in the system is on the average very short, usually no more than a few seconds. This means that any process conducted in a pneumatic system cannot be diffusion-controlled. The reaction must be mainly a surface phenomenon, or the solids particles must be very small so that heat transfer and mass transfer from the interiors are essentially instantaneous.

2. On an energy-content basis, the system is balanced at all times; i.e., there is sufficient energy in the gas (or solids) present in the system at any time to complete the work on all the solids (or gas) present at the same time. This is significant in that there is no lag in response to control changes or in starting up and shutting down the system; no partially processed residual solids or gas need be retained between runs.

It is for these reasons that pneumatic equipment is especially suitable for processing *heat-sensitive, easily oxidized, explosive,* or *flammable* materials which cannot be exposed to process conditions for extended periods. Further, pneumatic installations may be operated satisfactorily in either batch- or continuous-process installations.

Gas flow and solids flow are usually cocurrent, one exception being a countercurrent-flow spray dryer. The method of gas-solids contacting is best described as through circulation; however, in the dilute condition solids particles are so widely dispersed in the gas that they exhibit apparently no effect upon one another, and they offer essentially no resistance to the passage of gas among them.

Pneumatic-Conveyor Dryers

Description A pneumatic-conveyor dryer consists of a long tube or duct carrying a gas at high velocity, a fan to propel the gas, a suitable feeder for addition and dispersion of particulate solids in the gas stream, and a cyclone collector or other separation equipment for final recovery of solids from the gas.

The solids feeder may be of any type; screw feeders, venturi sections, high-speed grinders, and dispersion mills are employed. For pneumatic conveyors, selection of the correct feeder to obtain thorough initial dispersion of solids in the gas is of major importance. For example, by employing an air-swept hammer mill in a drying operation, 65 to 95 percent of the total heat may be transferred within the mill itself if all the drying gas is passed through it. Fans may be of the induced-draft or the forced-draft type. The former is usually preferred because the system can then be operated under a slight negative pressure. Dust and hot gas will not be blown out through leaks in the equipment. Cyclone separators are preferred for low investment. If maximum recovery of dust or noxious fumes is required, the cyclone may be followed by a wet scrubber or bag collector.

In ordinary heating and cooling operations, during which there is no moisture pickup, continuous recirculation of the conveying gas is frequently employed. Also, solvent-recovery operations employing continuously recirculated inert gas with intercondensers and gas reheaters are carried out in pneumatic conveyors.

Pneumatic conveyors are suitable for materials which are granular and free-flowing when dispersed in the gas stream, so they do not stick on the conveyor walls or agglomerate. Sticky materials such as filter cakes may be dispersed and partially dried by an air-swept disintegrator in many cases. Otherwise, dry product may be recycled and mixed with fresh feed, and then the two dispersed together in a disintegrator. Coarse material containing internal moisture may be subjected to fine grinding in a hammer mill. The main requirement in all applications is that the operation be instantaneously completed; internal diffusion of moisture must not be limiting in drying operations, and particle sizes must be small enough so that the thermal conductivity of the solids does not control during heating and cooling operations. Pneumatic conveyors are rarely suitable for abrasive solids. Pneumatic conveying can result in significant particle-size reduction, particularly when crystalline or other friable materials are being handled. This may or may not be desirable but must be recognized if the system is selected. The action is similar to that of a fluid-energy grinder.

Pneumatic conveyors may be single-stage or multistage. The former is employed for evaporation of small quantities of surface moisture. Multistage installations are used for difficult drying processes, e.g., drying heat-sensitive products containing large quantities of moisture and drying materials initially containing internal as well as surface moisture. Typical single- and two-stage drying systems are illustrated in Figs. 20-65, 20-66*a* and *b*, and 20-67. Figure 20-65 incorporates a single-stage dryer with a second stage containing a cage-mill disintegrator. The second stage ensures complete drying after thorough dispersion of lumps and agglomerates. If disintegration is required to disperse the wet feed, the stages can be reversed, or disintegration can be employed in both stages. Systems of the type illustrated are employed for drying synthetic resins, of which low-pressure polyethylene and polypropylene are examples.

Figure 20-67 illustrates a single-stage dryer employing a paddle mixer, recycle, and a CE-Raymond cage mill for fine grinding and dispersion of the mixed feed in the air stream. These units are

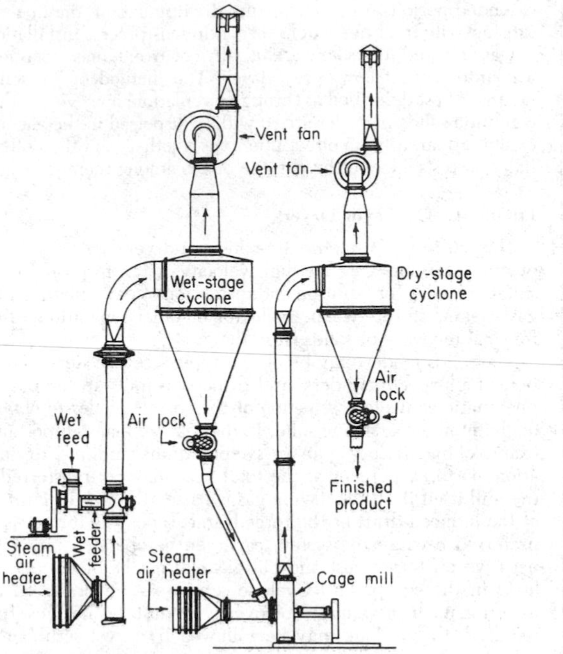

FIG. 20-65 Two-stage air-stream and cage mill, pneumatic-conveyor dryer. *(Raymond Division, Combustion Engineering Inc.)*

designed to handle filter and centrifuge cakes and other sticky or pasty feeds. Capacities are given in Table 20-27.

Several typical products dried in pneumatic conveyors are described in Table 20-28. The air-stream type referred to is an ordinary single-stage dryer like the first stage of Fig. 20-65.

Design Methods Depending upon the temperature sensitivity of the product, inlet-air temperatures between 400 and 1000 K are employed. With a heat-sensitive solid, a high initial moisture content should permit use of a high inlet-air temperature. Evaporation of surface moisture takes place at essentially the wet-bulb air temperature. Until this has been completed, by which time the air will have cooled significantly, the surface-moisture film prevents the solids temperature from exceeding the wet-bulb temperature of the air. Pneumatic conveyors are used for solids having initial moisture contents ranging from 3 to 90 percent, wet basis. The air quantity required and solids-to-gas loading are fixed by the moisture load, the inlet-air temperature, and, frequently, the exit-air humidity. If the last is too great to permit complete drying, i.e., if the exit-air humidity is above that in equilibrium with the product at required dryness, the solids-to-gas loading must be reduced together with the inlet-air temperature.

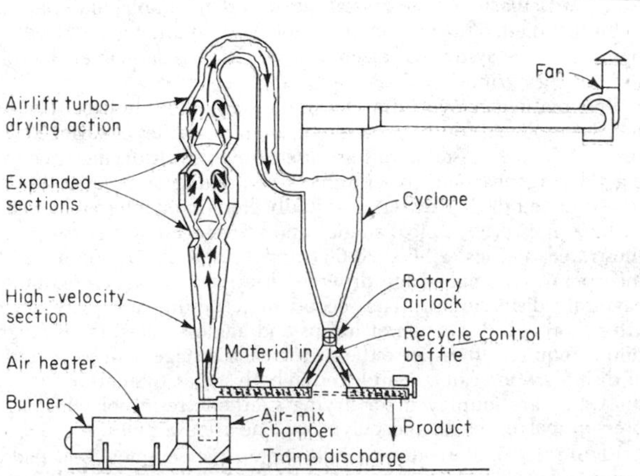

FIG. 20-66a Air-lift pneumatic-conveyor dryer; includes partial recycle of dry product and expanding tube and cone sections to provide longer holdup for coarse particles. *(Bepex Corporation.)*

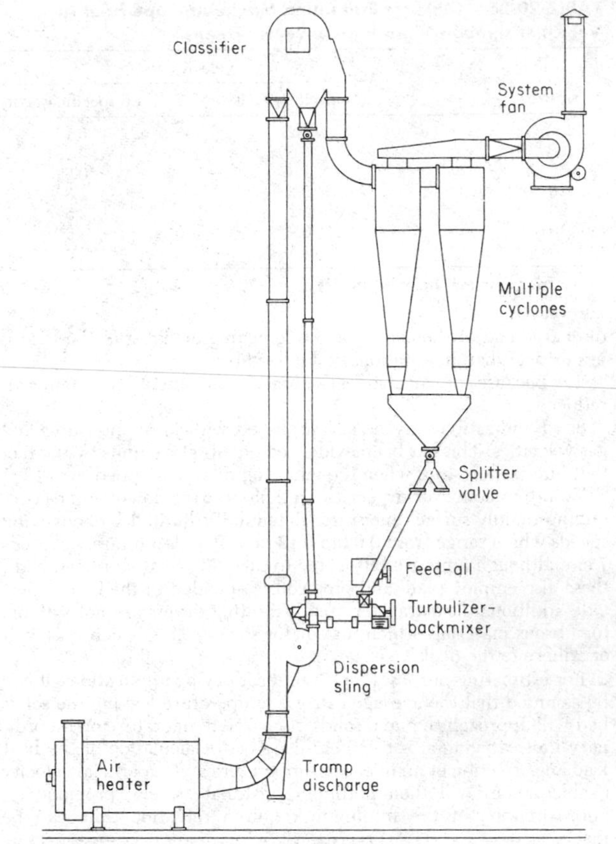

FIG. 20-66b Strong-Scott flash dryer with integral coarse-fraction classifier to separate undried particles for recycle. *(Bepex Corporation.)*

The gas velocity in the conveying duct must be sufficient to convey the largest particle. This may be calculated accurately by methods given in Sec. 5. For estimating purposes, a velocity of 25 m/s, calculated at the exit-air temperature, is frequently employed. If mainly surface moisture is present, the temperature driving force for drying will approach the log mean of the inlet- and exit-gas wet-bulb depressions. (The exit solids temperature will approach the exit-gas dry-bulb temperature.)

Observation of operating conveyors indicates that the solids are rarely uniformly dispersed in the gas phase. With infrequent exceptions, the particles move in a laminar pattern, following a streamline along the duct wall where the flow velocity is at a minimum. Complete or even partial diffusion in the gas phase is rarely experienced even with low-specific-gravity particles. Air velocities may approach 20 to 30 m/s. It is doubtful, however, that even finer and lighter materials reach more than 80 percent of this speed, while heavier and larger fractions may travel at much slower rates [Fischer, *Mech. Eng.*, **81**, 11, 67–69 (1959)]. Very little information and operating data on pneumatic-conveyor dryers which would permit a true theoretical basis for design have been published. Therefore, firm design always requires pilot tests. It is believed, however, that the significant velocity effect in a pneumatic conveyor is the difference in velocities between gas and solids, which is the reason why a major part of the total drying actually occurs in the feed section.

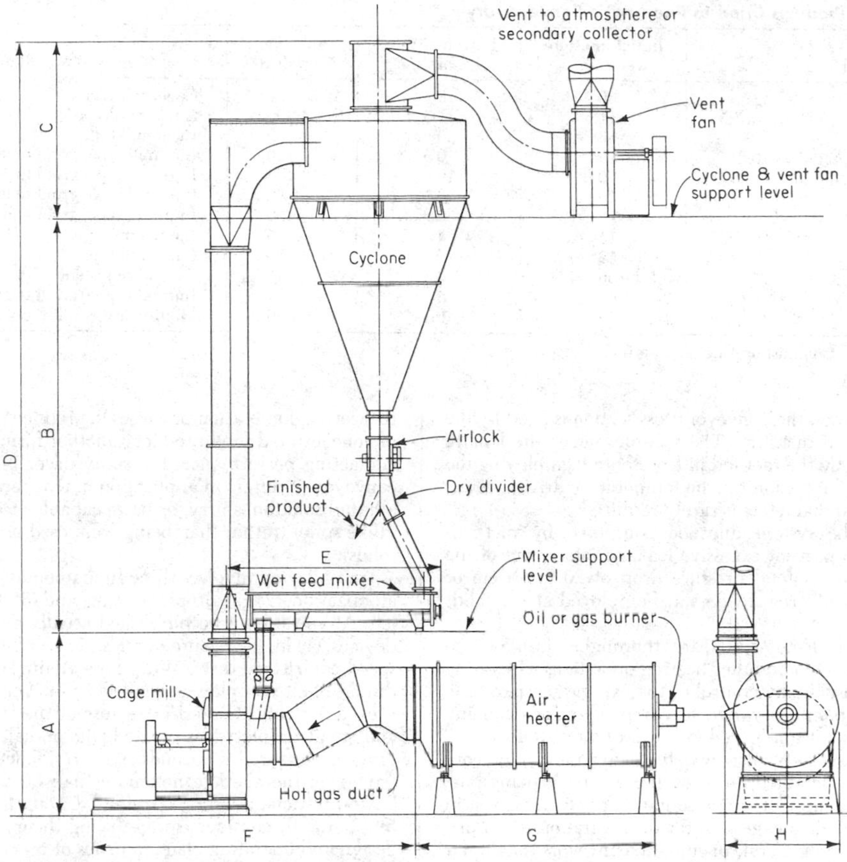

FIG. 20-67 Single-stage pneumatic-conveyor dryer. *(Raymond Division, Combustion Engineering Inc.)*

One manner in which size may be computed, for estimating purposes, is by employing a volumetric heat-transfer concept as used for rotary dryers. If it is assumed that contacting efficiency is in the same order as that provided by efficient lifters in a rotary dryer and that the velocity difference between gas and solids controls, Eq. (20-36) may be employed to estimate a volumetric heat-transfer coefficient. By assuming a duct diameter of 0.3 m (D) and a gas velocity of 23 m/s, if the solids velocity is taken as 80 percent of this speed, the velocity difference between the two would be 4.6 m/s. If the exit gas has a density of 1 kg/m^3, the relative mass flow rate of the gas G becomes 4.8 kg/(s·m^2); the volumetric heat-transfer coefficient is 2235 J/(m^3·s·K). This is not far different from many coefficients found in commercial installations; however, it is usually not possible to predict accurately the actual difference in velocity between gas and solids. Furthermore, the coefficient is influenced by the solids-to-gas loading and particle size, which control the total solids surface exposed to the gas. Therefore, the figure given is only an approximation.

TABLE 20-27 Sizes and Capacities of CE-Raymond Cage-Mill Flash-Drying Systems*

Based upon 1200°F inlet temperatures†

Cyclone	*A*	*B*	*C*	*D*	*E*	*F*	*G*	*H*	Evaporation capacity, lb water/h	Price,‡ FOB shops
18 ft 0 in	18 ft 0 in	38 ft 0 in	20 ft 0 in	76 ft 0 in	19 ft 6 in	27 ft 0 in	22 ft 6 in	10 ft 0 in	20,000	$365,000
16 ft 0 in	15 ft 0 in	34 ft 0 in	17 ft 4 in	66 ft 4 in	17 ft 0 in	25 ft 0 in	20 ft 6 in	9 ft 9 in	15,700	311,000
14 ft 0 in	14 ft 0 in	31 ft 0 in	14 ft 4 in	59 ft 4 in	16 ft 6 in	24 ft 0 in	19 ft 4 in	9 ft 6 in	12,000	270,000
12 ft 0 in	13 ft 0 in	29 ft 0 in	12 ft 6 in	54 ft 6 in	15 ft 3 in	23 ft 0 in	17 ft 4 in	8 ft 2 in	8,850	246,000
10 ft 0 in	11 ft 6 in	26 ft 0 in	10 ft 9 in	48 ft 3 in	13 ft 3 in	22 ft 0 in	15 ft 8 in	7 ft 1 in	6,150	204,000
9 ft 0 in	11 ft 0 in	24 ft 0 in	10 ft 3 in	45 ft 3 in	13 ft 3 in	21 ft 0 in	15 ft 8 in	6 ft 11 in	5,000	190,000
8 ft 0 in	10 ft 6 in	24 ft 0 in	9 ft 6 in	44 ft 0 in	12 ft 6 in	20 ft 6 in	13 ft 5 in	6 ft 6 in	3,940	179,000
7 ft 0 in	10 ft 0 in	23 ft 0 in	8 ft 9 in	41 ft 9 in	12 ft 6 in	20 ft 0 in	12 ft 3 in	6 ft 8 in	3,000	171,000
6 ft 0 in	9 ft 6 in	22 ft 0 in	8 ft 0 in	39 ft 6 in	11 ft 0 in	19 ft 6 in	11 ft 3 in	5 ft 7 in	2,220	160,000
5 ft 0 in	9 ft 0 in	20 ft 0 in	7 ft 3 in	36 ft 3 in	10 ft 9 in	18 ft 0 in	10 ft 4 in	5 ft 5 in	1,540	144,000
4 ft 0 in	8 ft 6 in	19 ft 0 in	6 ft 3 in	33 ft 9 in	10 ft 9 in	18 ft 0 in	9 ft 1 in	4 ft 0 in	985	135,000

*CE Raymond Division, Combustion Engineering Inc.

†With inlet temperature of 600 to 700°F, consider the water evaporation to be one-half of that listed in the table. Considerably lower inlet temperatures are also frequently used for many materials.

‡Price based upon carbon steel construction (motors and secondary dust collectors by others).

TABLE 20-28 Typical Products Dried in Pneumatic-Conveyor Dryers*

Material	Initial moisture, wet basis, %	Final moisture, wet basis, %	Rate, kg/s product	Remarks
Clay, acid-treated	60	18	64	Cage-mill type
Coal, 1 cm × 0	11.5	1.5	1,200	Air-stream type
Corn-gluten feed	65	20	90	Cage-mill type
Kaolin (H_2O washed and partially dried)	10	0.5	120	Imp-mill type; grind to 99.9% −325 mesh
Gluten (vital wheat)	70	10	4	Imp-mill type; grind to −80 mesh
Clay, ball	25	0.5	60	Imp-mill type; grind to 95% −100 mesh
Gypsum, raw	25 total	5	53	Imp-mill type; grind and calcine to stucco
Pharmaceuticals	15	4	8	Air-stream type
Silica-gel catalyst	53	10	80	Cage-mill type
Synthetic resin	50	0.5	16	Two-stage system
Carboxymethyl cellulose	40	3	7	Imp-mill type; grind to 80% −200 mesh
Sewage sludge	82	0	140	Multiple-cage-mill-type systems; dry and incinerate

*CE Raymond Division, Combustion Engineering Inc.

For estimating purposes, the conveyor cross section is fixed by the assumed air velocity and quantity. The volume, hence the length, can then be calculated by the method just presented, employing the log-mean air wet-bulb depression for the temperature driving force. A conveyor length >50 diameters is rarely required.

Pressure drop in the system may be computed by methods described in Sec. 5. To prevent excessive leakage into or out of the system, which may have a total pressure drop of 20 to 38 cm of water, rotary air locks or screw feeders are employed at the solids inlet and discharge.

The conveyor and collector parts are thoroughly insulated to reduce heat losses in drying and other heating operations. Operating control is maintained usually by control of the exit-gas temperature, with the inlet-gas temperature varied to compensate for changing feed conditions. A constant solids feed rate must be maintained.

Cost Data Purchase costs vary widely; many pneumatic-conveyor installations are assembled units, each component being purchased from a different supplier. Representative prices are given in Table 20-27. These include a cage mill for disintegration and a primary cyclone collector. In general, pneumatic conveyors for similar duties will compete in cost with cocurrent rotary dryers. Space economics may reduce the total installed investment slightly below that of the rotary unit. Operating costs, thermal efficiency, etc., are similar to those of cocurrent rotary units sized for the same duty. When other operations, such as conveying, grinding, or classifying, are simultaneously performed, operating and investment costs may be reduced for the pneumatic-drying process itself by being partially written off on the secondary function. In this situation, a pneumatic conveyor becomes particularly attractive.

Spray Dryers

Description A spray dryer consists of a large cylindrical and usually vertical chamber into which material to be dried is sprayed in the form of small droplets and into which is fed a large volume of hot gas sufficient to supply the heat necessary to complete evaporation of the liquid. Heat transfer and mass transfer are accomplished by direct contact of the hot gas with the dispersed droplets. After completion of drying, the cooled gas and solids are separated. This may be accomplished partially at the bottom of the drying chamber by classification and separation of the coarse dried particles. Fine particles are separated from the gas in external cyclones or bag collectors. When only the coarse-particle fraction is desired for finished product, fines may be recovered in wet scrubbers; the scrubber liquid is concentrated and returned as feed to the dryer. Horizontal spray chambers are manufactured with a longitudinal screw conveyor in the bottom of the drying chamber for continuous removal of settled coarse particles.

The principal use of spray dryers is for ordinary drying of water solutions and slurries. They are used also in combined drying and heat-treating operations, and for melt fusion and cooling of molten materials, e.g., ammonium nitrate "prilling." The latter may be considered a solids size-enlargement process. Spray dryers are employed for wet-agglomeration processes to produce rapidly dispersible forms of concentrated food products, another form of size enlargement. In contacting performance, the spray dryer is similar to a pneumatic conveyor. It differs in application in that the feed material is usually a liquid solution, slurry, or paste capable of being dispersed in a fluidlike spray (rather than being composed of free-flowing particulate solids).

Spray drying involves three fundamental unit processes: (1) liquid atomization, (2) gas-droplet mixing, and (3) drying from liquid droplets. Atomization is accomplished usually by one of three atomizing devices: (1) high-pressure nozzles, (2) two-fluid nozzles, and (3) high-speed centrifugal disks. With these atomizers, thin solutions may be dispersed into droplets as small as 2 μm. The largest drop sizes rarely exceed 500 μm (35 mesh). Because of the large total drying surface and small droplet sizes created, the actual drying time in a spray dryer is measured in seconds. Total residence of a particle in the system is on the average not more than 30 s. A review by Marshall ["Atomization and Spray Drying," *Chem. Eng. Prog. Monogr. Ser.*, **50**, 2 (1954)] considers spray-drying theory in detail as well as the design and operating characteristics of modern spray dryers. A later survey of spray drying, which constitutes a good supplement to Marshall, was published by Masters [*Ind. Eng. Chem.*, **60**(10), 53–63 (1968)]. Liquid atomization and dispersion are discussed in detail in Sec. 18. Atomizers commonly employed on spray dryers are described briefly in the following paragraphs.

Special designs of spray dryers may provide for cooling air to enter around the chamber, closed systems for the recovery of solvents, and air sweepers or mechanical rakes to remove dry product from the walls and bottom of the chamber. Some are followed by pneumatic conveyors as depicted in Fig. 20-68, in which drying air is diluted with cool air for product cooling before separation. Spray dryers may operate with cocurrent, mixed, or countercurrent flow of gas and solids. Inlet-gas temperatures may range from 425 to 1100 K.

1. Pressure nozzles effect atomization by forcing the liquid under high pressure and with a high degree of spin through a small orifice. Pressures may range from 2700 to 69,000 kPa/m^2, depending on the degree of atomization, capacity, and physical properties. Nozzle orifices may range in size from 0.25- to 0.4-mm diameter, depending on the pressure desired for a given capacity and the degree of atomization required. For high pressures and when solids are in suspension in the liquid, the nozzle orifice will be subject to wear by erosion, and the orifice should be made of a hard alloy such as tungsten carbide or stellite. Maintenance on pressure nozzles is always a problem since erosion occurs with even the hardest inserts, and once the orifice has become scratched and nonuniform, good atomization is no longer possible. Likewise, incrustation and plugging by particles of foreign matter cause trouble. Piston pumps furnish the liquids at high pressure; erosion of the valves in these pumps is another maintenance problem.

Spray characteristics of pressure nozzles depend on the pressure and nozzle-orifice size. Pressure affects not only the spray characteristics but also the capacity. If it is desired to reduce the amount of liquid sprayed by lowering the pressure, then the spray may become

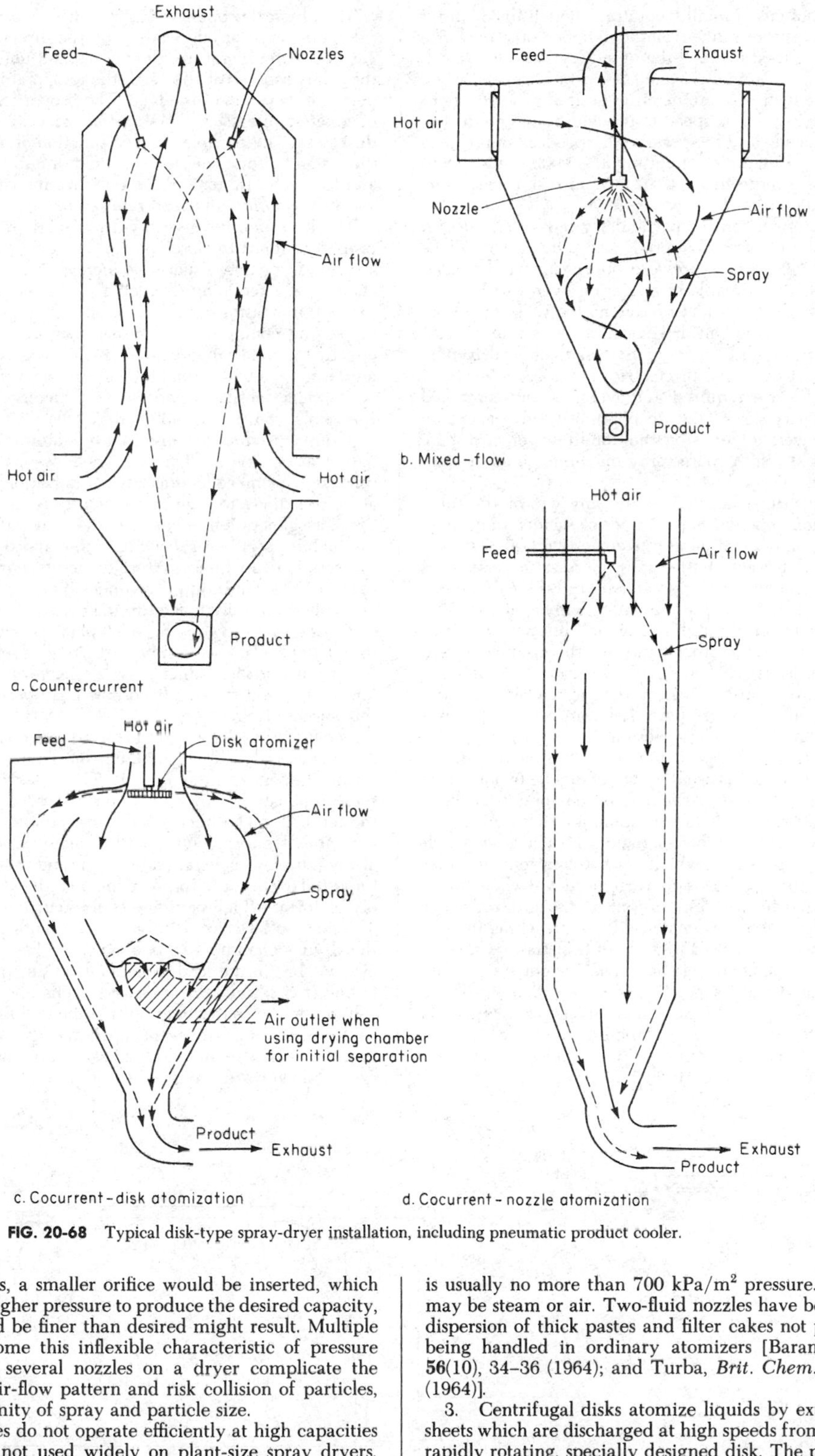

FIG. 20-68 Typical disk-type spray-dryer installation, including pneumatic product cooler.

coarser. To correct this, a smaller orifice would be inserted, which might then require a higher pressure to produce the desired capacity, and a spray that would be finer than desired might result. Multiple nozzles tend to overcome this inflexible characteristic of pressure atomization, although several nozzles on a dryer complicate the chamber design and air-flow pattern and risk collision of particles, resulting in nonuniformity of spray and particle size.

2. Two-fluid nozzles do not operate efficiently at high capacities and consequently are not used widely on plant-size spray dryers. Their chief advantage is that they operate at relatively low pressure, the liquid being 0 to 400 kPa/m^2 pressure, while the atomizing fluid is usually no more than 700 kPa/m^2 pressure. The atomizing fluid may be steam or air. Two-fluid nozzles have been employed for the dispersion of thick pastes and filter cakes not previously capable of being handled in ordinary atomizers [Baran, *Ind. Eng. Chem.*, **56**(10), 34–36 (1964); and Turba, *Brit. Chem. Eng.*, 9(7), 457–460 (1964)].

3. Centrifugal disks atomize liquids by extending them in thin sheets which are discharged at high speeds from the periphery of the rapidly rotating, specially designed disk. The principal objectives in disk design are to ensure bringing the liquid to disk speed and to obtain a uniform drop-size distribution in the atomized liquid. Disk

diameters range from 5 cm in small laboratory models to 35 cm for plant-size dryers. Disk speeds range from 3000 to 50,000 r/min. The high speed is generally used in small-diameter dryers. Usual speeds on plant-size dryers range from 4000 to 20,000 r/min, depending on disk diameter and the degree of atomization desired. The degree of atomization as a function of disk speed is affected by the product of disk diameter and speed, i.e., by peripheral speed as opposed to angular speed. Thus, a 13-cm disk operating at 30,000 r/min would be expected to atomize more finely than a 5-cm disk of the same design running at 50,000 r/min.

Centrifugal-disk atomization is particularly advantageous for atomizing suspensions and pastes that erode and plug nozzles. Thick pastes can be handled if positive-pressure pumps are used to feed them to the disk. Disks are capable of operating over a wide range of feed rates and disk speeds without producing too variable a product. Centrifugal disks may be belt-driven, direct-driven by a high-speed electric motor powered by a frequency changer, or driven by a steam turbine. Direct drive by an electric motor has advantages when very high speeds are required and when closely controlled speed variations are necessary. The life of high-speed bearings in centrifugal-disk atomizers depends on the conditions of operation. Average life may be 2000 h. A spare spray machine should be standard equipment.

The particle-size distribution obtained by any one of the three methods of atomization depends on a number of factors. In general, the size distribution will depend on atomizer design, liquid properties, and degree of atomization. If the finest atomization possible is attempted, a limiting condition is approached, and the particle-size range, regardless of the method of atomization, will be narrow. This is particularly true of pressure nozzles, in which uniformity of size increases with pressure. On the other hand, for the production of a coarse product with a high percentage of large particles, the method of atomization will have a large effect on the particle-size distribution. Production of uniform coarse particles from centrifugal disks frequently can be obtained by careful design.

One of the principal advantages of spray drying is the production of **a spherical particle,** which is usually not obtainable by any other drying method. This spherical particle may be solid or hollow, depending on the material, the feed condition, and the drying conditions. In general, aqueous solutions of materials such as soap, gelatin, and water-soluble polymers which form tough tenuous outer skins on drying will form hollow spherical particles when spray-dried. This is attributed to the formation of a casehardened outer surface on the particle which prevents liquid from reaching the surface from the particle interior. Because of high heat-transfer rates to the drops, the liquid at the center of the particle vaporizes, causing the outer shell to expand and form a hollow sphere. Sometimes the rate of vapor generation within the particle is sufficient to blow a hole through the wall of the spherical shell. Spherical particles may be obtained from true solutions or from slurries and may be produced by any of the previously described atomizers.

The physical properties of spray-dried materials are subject to considerable variation, depending on the direction of flow of the inlet gas and its temperature, the degree and uniformity of atomization, the solids content of the feed, the temperature of the feed, and the degree of aeration of the feed. The properties of the product usually of greatest interest are (1) **particle size,** (2) **bulk density,** and (3) **dustiness**. The particle size is a function of atomizer-operating conditions and also of the solids content, liquid viscosity, liquid density, and feed rate. In general, particle size increases with solids content, viscosity, density, and feed rate.

The bulk density of spray-dried solids is frequently the critical property subject to close control. The bulk density of material from a spray dryer may usually be increased by the following operating changes: (1) reducing droplet size, (2) reducing inlet-air temperature, (3) increasing air throughput, (4) increasing air turbulence, (5) employing countercurrent rather than cocurrent gas flow, and (6) effecting a wide range of size distribution from the atomizer. Chaloud et al. evaluated qualitatively the effects of operating variables on the bulk density of particles from detergent spray dryers [*Chem. Eng. Prog.*, **53,** 12, 593–596 (1957)].

A dusty product is caused by fine atomization or particle degradation after drying. Thin-wall hollow particles are susceptible to breakage during collection. Fine atomization and a high gas temperature contribute to high production rates in small drying chambers; they also generate fine particles and thin-wall spheres. Spray-drying installations yielding exceedingly fine and dusty products are often the result of an honest effort to design equipment for maximum capacity at a minimum investment. Large solids particles or heavy-wall spheres require longer drying cycles, hence larger drying chambers. Careful study in the pilot plant is necessary. In commercial installations, classification of particles and separation of a fine fraction from coarse product may be accomplished by countercurrent flow of gas and solids. Mixed-flow chambers are also adaptable for this purpose (Fig. 20-69).

The majority of spray dryers in commercial use employ cocurrent flow of gas and solids. Countercurrent-flow dryers are used primarily for drying soaps and detergents. Their classifying ability is useful in these applications. Air flow is upward, carrying entrained fines from the top of the chamber. The coarse product settles and is removed separately from the bottom of the chamber. Horizontal spray dryers always employ cocurrent flow of gas and solids. A swirling motion is imparted to the air to improve mixing. Mixed-flow dryers take a variety of forms which combine countercurrent and cocurrent drying. The flow patterns are complex with a high degree of turbulence in the drying chamber. In one type, air flow is similar to that in a cyclone. It is introduced tangentially at the top of a conical chamber, travels in a spiral pattern down the chamber wall, and returns in a column up the center to exhaust at the top. Feed is introduced at the center of the top, travels outward and downward countercurrently to the exit-gas stream, and then as it nears the wall is picked up and carried downward cocurrently with the inlet-gas stream. Many vari-

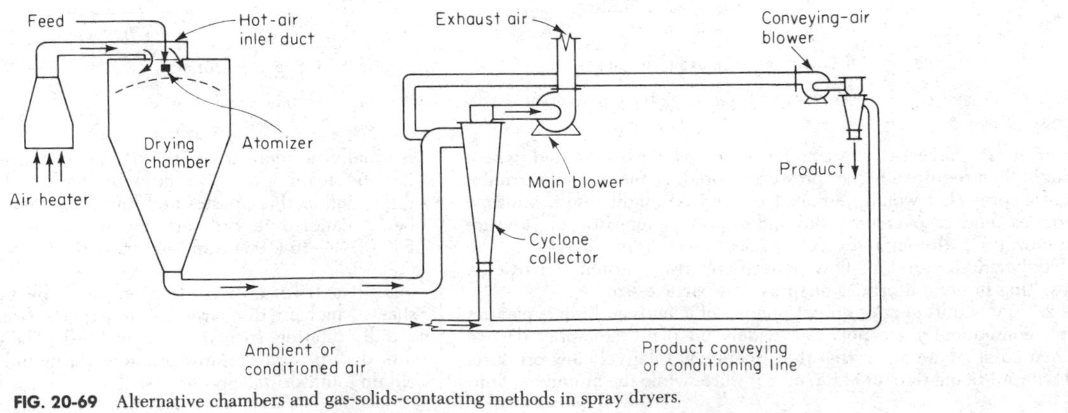

FIG. 20-69 Alternative chambers and gas-solids-contacting methods in spray dryers.

TABLE 20-29 Some Materials That Have Been Successfully Spray-Dried in a 6-m-Diameter by 6-m-High Chamber with a Centrifugal-Disk Atomizer*

Material	Air temperature, K: In	Air temperature, K: Out	% water in feed	Evaporation rate, kg/s
Blood, animal	440	345	65	5.9
Yeast	500	335	86	8.2
Zinc sulfate	600	380	55	10.0
Lignin	475	365	63	6.9
Aluminum hydroxide	590	325	93	19.4
Silica gel	590	350	95	16.9
Magnesium carbonate	590	320	92	18.2
Tanning extract	440	340	46	5.2
Coffee extract	420	355	70	3.8
Detergent *A*	505	395	50	5.0
Detergent *B*	510	390	63	6.2
Detergent *C*	505	395	40	2.6
Manganese sulfate	590	415	50	5.5
Aluminum sulfate	415	350	70	1.7
Urea resin *A*	535	355	60	3.8
Urea resin *B*	505	360	70	1.9
Sodium sulfide	500	340	50	2.0
Pigment	515	335	73	13.2

*Courtesy of Bowen Engineering, Inc.

NOTE: The fan on this dryer handles about 5.2 m^3/s at outlet conditions. The outlet-air temperature includes cold air in-leakage, and the true temperature drop caused by evaporation must therefore be estimated from a heat balance.

ations of air-flow patterns are employed commercially; most are intended primarily to produce turbulence and thorough mixing of gas and droplets and to achieve the most effective use of the chamber volume.

Applications The major and most successful drying applications of spray dryers are for solutions, slurries, and pastes which (1) cannot be dewatered mechanically, (2) are heat-sensitive and cannot be exposed to high-temperature atmospheres for long periods, or (3) contain ultrafine particles which will agglomerate and fuse if dried in other than a dilute condition. In other applications, spray drying is rarely competitive on a cost basis with two-step dewatering and solids-drying processes. The cost of bag collectors for solids recovery from large volumes of exit gas may double the cost of a spray-dryer installation. Additional costs must usually be justified on the basis of some improvement in product quality, such as particle form, size, flavor, color, or heat stability. Spray drying is applicable to heat-sensitive products such as milk powders and other foods and pharmaceuticals because of the short contact time in the dryer hot zone. Further, the water film on the liquid drop protects the solids from high gas temperatures. Drying is carried out at essentially the drying-air wet-bulb temperature. Color pigments are examples of the class of products for which it is desired to maintain as closely as possible the original solids particle size. Table 20-29 lists typical materials which have been successfully spray-dried. One other class of products particularly applicable to spray dryers is solids slurries, containing extremely fine particles, which is nonnewtonian in flow characteristics and remains fluid at very low moisture content. Certain classes of clays are found in this category. Also, spray dryers have been developed for encapsulation processes to convert liquid volatile flavors and perfumes to particulate solids forms [Maleeny, *Soap Chem. Spec.*, **34**, 1, 135–141 (1958)].

If the product in no way adheres to the dryer parts and simple cyclone collectors are sufficient for gas-solids separation, batch operation of a spray dryer may be considered. Otherwise, the time and costs for cleaning the large equipment parts make them rarely economical for other than continuous processing of a single material.

A standard cocurrent-flow spray dryer is illustrated in Fig. 20-68. It includes a primary cyclone for separation of fines from the dryer exit gas and a pneumatic conveyor following the dryer used for product cooling. Large-diameter drying chambers are required when disk atomization is employed. Small-diameter high vertical chambers are used with two-fluid and pressure atomizers. The chamber shape must conform to the atomizer spray pattern so that sprayed particles will not contact the walls before they are completely dry.

Design Methods Design variables must be established by experimental tests before final design of a chamber can be carried out. In general, chamber size, atomizer selection, and separation auxiliaries will be determined by the desired physical characteristics of the product. Drying by itself is rarely a problem. An installed spray dryer is relatively inflexible in meeting changing operating requirements while maintaining a constant production rate. Important variables which must be fixed before design of a commercial dryer are the following:

1. The form and particle size of product required
2. The physical properties of the feed: moisture, viscosity, density, etc.
3. The maximum inlet-gas and product temperatures

Theoretical correlations of spray-dryer performance published by Gluckert [*Am. Inst. Chem. Eng. J.*, 8(4), 460–466 (1962)] may be employed for the scale-up of laboratory dryers and, in some instances, for estimating dryer requirements in the absence of any tests.

Several assumptions are necessary.

1. The largest droplets, which dry most slowly, are the limiting portion of the spray. They determine ultimate chamber dimensions and are employed for the evaluation.
2. The largest droplet in a spray population is 3 times the diameter of the average drop size [see Eq. (20-50)].
3. A droplet Nusselt number = 2, corresponding to pure conduction (Reynolds number = 0) to infinity, is employed for evaluating the coefficient of heat transfer.
4. Drying conditions, because of turbulence and gas mixing, are uniform throughout the chamber; i.e., the entire chamber is at the gas exit temperature—this fact has been well established in many chambers except in the immediate zone of gas inlet and spray atomization.
5. The temperature driving force for drying is the difference between the drying-gas outlet temperature and, in the case of pure water, the gas wet-bulb temperature. In the case of a solution, the adiabatic saturation temperature of the pure saturated solution is employed rather than the wet-bulb temperature.

Methods for calculating average and maximum drop sizes from various atomizers are given by Marshall (op. cit.). For pneumatic nozzles, an expression developed by Nukiyama and Tanasawa is recommended:

$$\overline{X}_{vs} = \frac{1920\sqrt{\alpha}}{V_a\sqrt{\rho_l}} + 597\left(\frac{\mu}{\sqrt{\alpha\rho_l}}\right)^{0.45}\left(\frac{1000Q_L}{Q_a}\right)^{1.5} \tag{20-47}$$

where $\overline{X}_{vs}$ = average drop diameter, μm (a drop with the same volume-surface ratio as the total sum of all drops formed)
α = surface tension, dyn/cm
μ = liquid viscosity, P
V_a = relative velocity between air and liquid, ft/s
ρ_l = liquid density, g/cm^3
Q_L = liquid volumetric flow rate
Q_a = air volumetric flow rate

For single-fluid pressure nozzles, a rule of thumb is employed:

$$\overline{X}_{vs} = 500/\sqrt[3]{\Delta P} \tag{20-48}$$

where ΔP = pressure drop across nozzle, lb/in^2.

For centrifugal disks, the relation of Friedman, Gluckert, and Marshall is employed [*Chem. Eng. Prog.*, **48**, 181 (1952)]:

$$\frac{D_{vs}}{r} = 0.4\left(\frac{\Gamma}{\rho_l N r^2}\right)^{0.6}\left(\frac{\mu}{\Gamma}\right)^{0.2}\left(\frac{\alpha\rho_l L_w}{\Gamma^2}\right)^{0.1} \tag{20-49}$$

where D_{vs} = average drop diameter, ft
r = disk radius, ft
Γ = spray mass velocity, lb/(min·ft of wetted disk periphery)
ρ_l = liquid density, lb/ft^3

N = disk speed, r/min
μ = liquid viscosity, lb/(ft·min)
α = surface tension, lb/min^2
L_w = wetted disk periphery, ft

NOTE: All groups are dimensionless. To convert dynes per square centimeter to joules per square meter, multiply by 10^{-3}; to convert poises to newton-seconds per square meter, multiply by 10^{-1}; to convert feet per second to meters per second, multiply by 0.3048; to convert feet to meters, multiply by 0.3048; to convert pounds per minute-foot to kilograms per second-meter, multiply by 0.025; to convert pounds per cubic foot to kilograms per cubic meter, multiply by 16.019; to convert pounds per minute squared to kilograms per second squared, multiply by 1.26×10^{-4}; to convert British thermal units per hour to kilojoules per second, multiply by 2.63×10^{-4}; and to convert British thermal units per hour–square foot–degree Fahrenheit per foot to joules per square meter–second–kelvin per meter, multiply by 1.7307.

Inspection of these relationships will show that the variables are difficult to specify in the absence of tests except when handling pure liquids—which in spray drying is rare indeed. The most useful method for employing these equations is to conduct small-scale drying tests in a chamber under conditions in which wall impingement and sticking are incipient. The maximum particle size can then be back-calculated by using the relationships given in the following paragraphs, and the effects of changing atomizing variables evaluated by using the preceding equations:

$$\overline{X}_m = 3\overline{X}_{vs} \qquad (20\text{-}50)$$

where $\overline{X}_m$ = maximum drop diameter, μm.

Gluckert gives the following relationships for calculating heat transfer under various conditions of atomization:

Two-fluid pneumatic nozzles:

$$Q = \frac{6.38K_f v^{2/3}\,\Delta t\; w_s}{D_m^2\;\;\;\;\;\;\;\rho_s}\sqrt{\frac{\rho_a}{w_a V_a}\,\frac{w_a + w_s}{w_a}} \qquad (20\text{-}51)$$

Single-fluid pressure nozzles:

$$Q = \frac{10.98K_f v^{2/3}\,\Delta t}{D_m^2}\,D_s\sqrt{\frac{\rho_t}{\rho_s}} \qquad (20\text{-}52)$$

Centrifugal-disk atomizers:

$$Q = \frac{4.19K_f(R_c - r/2)^2\,\Delta t}{D_m^2\rho_s}\sqrt{\frac{w_s\rho_t}{rN}} \qquad (20\text{-}53)$$

where Q = rate of heat transfer to spray, Btu/h
K_f = thermal conductivity of gas film surrounding the droplet, Btu/(h·ft^2)(°F·ft), evaluated at the average between dryer gas and drop temperature
v = volume of dryer chamber, ft^3
Δt = temperature driving force (under terminal conditions described above), °F
D_m = maximum drop diameter, ft
w_s = weight rate of liquid flow, lb/h
ρ_s = density of liquid, lb/ft^3
w_a = weight rate of atomizing air flow, lb/h
ρ_a = density of atomizing air, lb/ft^3
V_a = velocity of atomizing air at atomizer, ft/h
D_s = diameter of pressure-nozzle discharge orifice, ft
ρ_t = density of dryer gas at exit conditions, lb/ft^3
R_c = radius of drying chamber with centrifugal disk, ft
r = radius of disk, ft
N = rate of disk rotation, r/h

For proper use of the equations, the chamber shape must conform to the spray pattern. With cocurrent gas-spray flow, the angle of spread of single-fluid pressure nozzles and two-fluid pneumatic nozzles is such that wall impingement will occur at a distance approximately four chamber diameters below the nozzle; therefore, chambers employing these atomizers should have vertical height-to-diameter ratios of at least 4 and, more usually, 5. The discharge cone below the vertical portion should have a slope of at least 60°, to minimize settling accumulations, and is used entirely to accelerate gas and solids for entry into the exit duct.

The critical dimension of a centrifugal-disk chamber is the diameter. Vertical height is usually 0.5 to 1.0 times the diameter; the large cone is needed mainly to accelerate to the discharge duct and prevent settling; it contributes little to drying capacity.

Cost Data Drying chambers, ductwork, and cyclone separators are usually constructed of stainless steel. Savings of roughly 20 percent may be achieved on the total purchase cost by using carbon steel; the increasing tendency toward the use of heat-resistant and corrosion-resistant plastic coatings (epoxy resins) makes the future appear promising for greater use of carbon steel construction. Wide differences in cost may be experienced in the selection of basic equipment. Air heaters vary in price range according to the selection of steam, electricity, direct-fired, and indirect-fired oil or gas heaters. Dust-collection equipment may consist of cyclone collectors or bag-type filters and may include a wet scrubber. Costs of nozzle and centrifugal atomizers are usually comparable. While the centrifugal atomizer requires mechanical gearing and motor drive, a high-pressure nozzle requires a high-pressure pump, which will usually more than offset the cost of gearing and motor for the centrifugal atomizer. Auxiliary equipment which may be included comprises air filters, drying-chamber insulation, and mechanical or pneumatic cooling conveyors. A minimum of instrumentation consists of indicating and recording thermometers for inlet-air and outlet temperatures, an ammeter for atomizer motor drive (or a pressure gauge for nozzle atomization), a flowmeter, manometers, a high-temperature alarm, and a panelboard with push-button stations for all equipment. The drying process may be completely controlled automatically with some additional instrumentation.

Spray dryers may operate under positive, negative, or neutral pressures. In general, pressure drop in a complete system will range from 15 to 50 cm of water, depending on duct size and separation equipment employed.

FLUIDIZED-BED SYSTEMS

Fluidization, or fluidizing, converts a bed of solid particles into an expanded, suspended mass that has many properties of a liquid. This mass has zero angle of repose, seeks its own level, and assumes the shape of the containing vessel.

Fluidized beds are used successfully in a multitude of processes both catalytic and noncatalytic. Among the catalytic uses are hydrocarbon cracking and re-forming, oxidation of naphthalene to phthalic anhydride, and ammoxidation of propylene to acrylonitrile. A few of the noncatalytic uses are roasting of sulfide ores, coking of petroleum residues, calcination of limestone, aluminum hydroxide, and phosphate ores, drying, and classification. Considerable effort and interest are now centered in the areas of coal and waste combustion to raise steam and the gasification of coal.

The size of solid particles which can be fluidized varies greatly from less than 1 μm to 6 cm (2½ in). It is generally concluded that particles distributed in sizes between 150 μm and 10 μm are the best for smooth fluidization (least formation of large bubbles). Large particles cause instability and result in slugging or massive surges. Small particles (less than 20 μm) frequently, even though dry, act as if damp, forming agglomerates or fissures in the bed, or spouting. Adding finer-sized particles to a coarse bed or coarser-sized particles to a bed of fines usually results in better fluidization.

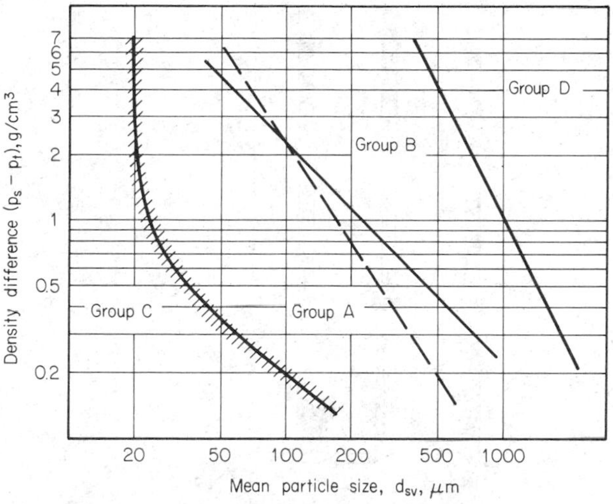

FIG. 20-70 Powder-classification diagram for fluidization by air (ambient conditions). [*From Geldart,* Powder Technol., *7, 285–292 (1973).*]

The upward velocity of the gas is usually between 0.15 m/s (0.5 ft/s) and 6 m/s (20 ft/s). This velocity is based upon the flow through the empty vessel and is referred to as the **superficial velocity**.

For details beyond the scope of this subsection, reference should be made to Zenz and Othmer, *Fluidization and Fluid Particle Systems,* Reinhold, New York, 1960; Kunii and Levenspiel, *Fluidization Engineering,* Krieger, Huntington, N.Y., 1977; Vanecek, Markvart, and Drbohlav, *Fluidized Bed Drying,* Leonard Hill, London 1966; Davidson and Harrison (eds.), *Fluidization,* Academic, London and New York, 1971; and the vast number of papers published in periodicals, transcripts of symposia, and the American Institute of Chemical Engineers symposium series.

GAS-SOLID SYSTEMS

Several workers in the field have systemized the various types of fluidization. Several of these types are discussed in the following subsections because each adds another dimension to the understanding of the phenomena.

Types of Solids Geldart [*Powder Technol.,* **7,** 285–292 (1973)] has characterized four groups of solids that exhibit different properties when fluidized with a gas. Figure 20-70 shows the division of the classes as a function of mean particle size, $\bar{d}_{sv}$, μm, and density difference, $(\rho_s - \rho_f)$, g/cm^3, where ρ_s = particle density and ρ_f = fluid density, $\bar{d}_{sv} = 1/\Sigma(x/d_{sv})$, d_{sv} = surface-volume diameter of particle, and x = weight fraction of particles in each size range.

When gas is passed upward through a bed of particles of groups *A, B,* or *D,* friction causes a pressure drop expressed by the Carman-Kozeny fixed-bed correlation. As the gas velocity is increased, the pressure drop increases until it equals the weight of the bed divided by the cross-sectional area. This velocity is called minimum fluidizing velocity, U_{mf}. When this point is reached, the bed of group *A* particles will expand uniformly until at some higher velocity gas bubbles will form (minimum bubbling velocity, U_{mb}). For group *B* and group *D* particles U_{mf} and U_{mb} are essentially equal. Group *C* particles exhibit cohesive tendencies, and as the gas flow is further increased, usually "rathole," the gas opens channels that extend from the gas distributor to the surface. If channels are not formed, the whole bed will lift as a piston. At higher velocities or with mechanical agitation or vibration, this type of particle will fluidize but with the appearance of clumps or clusters of particles. For all groups of powder (*A, B, C,* and *D*) as the gas velocity is further increased, bed density is decreased and turbulence increased. In smaller-diameter beds, especially with group *C* and *D* powders, slugging will occur as the bubbles increase in size to greater than half of the bed diameter. Bubbles grow by vertical and lateral merging. Bubbles also increase in size as the gas velocity is increased [Whitehead, in Davidson and Harrison (eds.), *Fluidization,* Academic, London and New York, 1971.] As the gas velocity is increased further and bubbles tend to disappear and streamers of solids and gas prevail, pressure fluctuations in the bed are greatly reduced. Further increase in velocity results in dilute-phase pneumatic transport.

Phase Diagram (Zenz and Othmer) Zenz and Othmer (op. cit.) have graphically represented (Fig. 20-71) all gas-solid systems in which the gas is flowing counter to gravity as a function of pressure drop per unit of height versus velocity. Note that line *OAB* in Fig. 20-71 is the pressure-drop versus gas-velocity curve for a packed bed and *BD* the curve for a fluid bed. Zenz indicates an instability between *D* and *H* because with no solids flow all the particles will be entrained from the bed; however, if solids are added to replace those entrained, system *IJ* prevails. The area *DHIJ* will be discussed further.

Phase Diagram (Reh) Reh [*Ger. Chem. Eng.,* **1,** 319–329 (1978); Fig. 20-72] has correlated the various types of gas-solid systems in which the gas is flowing counter to gravity in a status graph using the parameters of particle Reynolds number (Re_p), reciprocal of the drag-force coefficient ($1/C_D$), the Archimedes number (Ar), and the similarity number (*M*). By means of this plot, the regime of fluidization can be predicted.

Phase Diagram (Yerushalmi, Turner, and Squires) Yerushalmi, Turner, and Squires [*Ind. Eng. Chem Process Des. Dev.,* **15,** 1, 47–53 (1976); Fig. 20-73] have characterized the various fluidization regimes in a plot similar to that of Zenz and Othmer, log pressure gradient versus log velocity. They have distinguished between the bubbling and the turbulent fluidized-bed regimes and have given considerable attention to the area that Zenz and Othmer consider to be discontinuous, calling this the fast fluid-bed regime. This latter regime is similar to or the same as Reh's turbulent fluid bed.

Solids Concentration versus Height From the foregoing it is apparent that there are several regimes of fluidization. These are, in order of increasing gas velocity, particulate fluidization (Geldert group *A*), bubbling (aggregative), turbulent, fast, and transport. Each of these regimes has characteristic solids concentration profiles as shown in Fig. 20-74.

Equipment Types Fluidized-bed systems take many forms. Figure 20-75 shows some of the more prevalent concepts with approximate ranges of gas velocities.

Minimum Fluidizing Velocity U_{mf}, the minimum fluidizing velocity, is frequently used in fluid-bed calculations and in quantifying one of the particle properties. This parameter is best measured in small-scale equipment. However, if this is inconvenient, there are several correlations that can be used to predict U_{mf}. The relationship established by Baeyens and Geldart ["Fluidization and Its Applications," *Proc. Int. Symp. Toulouse,* 263 (1973)] is one of the better correlations:

$$U_{mf} = \frac{0.0009(\rho_s - \rho_f)^{0.934} g^{0.934} \bar{d}^{1.8}}{\mu 0.87 \rho_f 0.006} \tag{20-54}$$

where $\bar{d}$
= $1/[\Sigma(xi/dp_i)]$, m
U_{mf} = m/s
ρ_s = density of particles, kg/m^3
ρ_f = density of gas, kg/m^3
g = gravitational constant, 9.81 m/s^2
μ = gas viscosity, kg/(m·s)

The flow required to maintain a complete homogeneous bed of solids in which coarse or heavy particles will not segregate from the fluidized portion is very different from the minimum fluidizing velocity. See Rowe, Nienow, and Aghim, *Trans. Inst. Chem. Eng.,* **50**(4), 310–323, 324–333 (1972), for a discussion of segregation or mixing mechanism as well as the means of predicting this; also see Geldart and Abrahamson, *Am. Inst. Chem. Eng. Symp. Ser.,* **205,** 77 (1981).

Particulate Fluidization Fluid beds of Geldart class *A* powders that are operated at gas velocities above the minimum fluidizing velocity (U_{mf}) but below the minimum bubbling velocity (U_{mb}) are said to be particularly fluidized. As the gas velocity is increased above

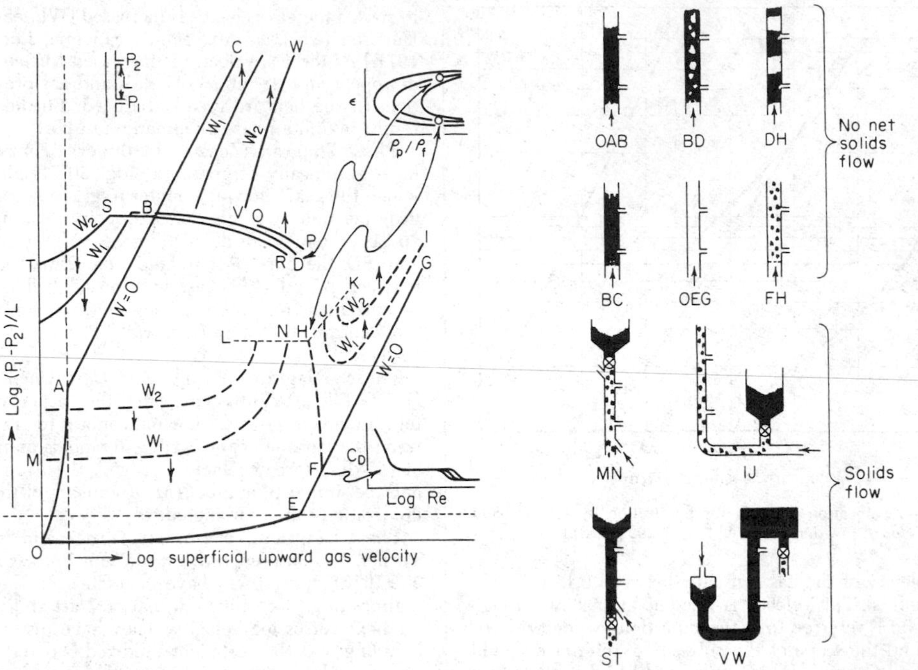

FIG. 20-71 Schematic phase diagram in the region of upward gas flow. W = mass flow solids, lb/(h·ft^2); ϵ = fraction voids; ρ_p = particle density, lb/ft^3; ρ_f = fluid density, lb/ft^3; Cd = drag coefficient; Re = modified Reynolds number. (*Zenz and Othmer,* Fluidization and Fluid Particle Systems, *Reinhold, New York, 1960.*)

Key:

OAB = packed bed
BD = fluidized bed
DH = slugging bed
IJ = cocurrent flow (dilute phase)
ST = countercurrent flow (dense phase)
AC = packed bed (restrained at top)
OEG = fluid only (no solids)
FH = dilute phase
MN = countercurrent flow (dilute phase)
VW = cocurrent flow (dense phase)

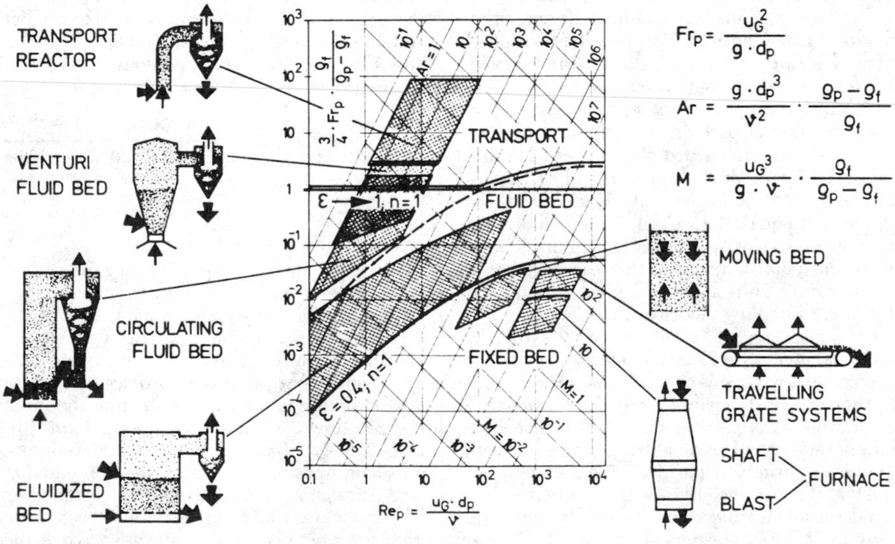

FIG. 20-72 Simplified fluid-bed status graph. [*From Reh,* Ger. Chem. Eng., *1, 319–329 (1978).*]

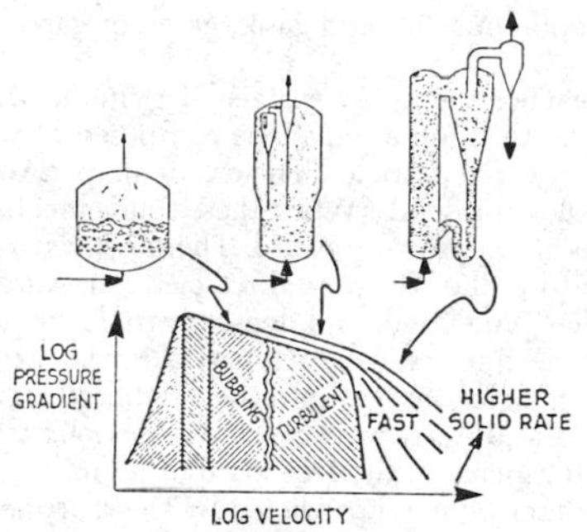

FIG. 20-73 Fluidization phase diagram for a fine powder, showing schematic diagrams of equipment for use in bubbling, turbulent, and fast fluidization regimes. [*From Yerushalmi, Turner, and Squires,* Ind. Eng. Chem. Process Des. Dev., *15, 1, 47–53 (1976).*]

U_{mf}, the bed further expands. Decreasing $(\rho_s - \rho_f)$, d_p and/or increasing μ_f increases the spread between U_{mf} and U_{mb} until at some point, usually at high pressure, the bed is fully particularly fluidized. Richardson and Zakis [*Trans. Inst. Chem. Eng.*, **32**, 35 (1954)] showed that $U/U_i = \epsilon^n$, where n is a function of system properties, ϵ = void fraction, U = superficial fluid velocity, and U_i = theoretical superficial velocity from the Richardson and Zakis plot when $\epsilon = 1$.

DESIGN OF FLUIDIZED-BED SYSTEMS

The use of the fluidization technique requires in almost all cases the employment of a fluidized-bed system rather than an isolated piece of equipment. Figure 20-76 illustrates the arrangement of components of a system.

The major parts of a fluidized-bed system can be listed as follows:

1. Fluidization vessel
   a. Fluidized-bed portion
   b. Disengaging space or freeboard
   c. Gas distributor
2. Solids feeder or flow control
3. Solids discharge
4. Dust separator for the exit gases
5. Instrumentation
6. Gas supply

Fluidization Vessel The most common shape is a vertical cylinder. Just as for a vessel designed for boiling a liquid, space must be provided for vertical expansion of the solids and for disengaging splashed and entrained material. The volume above the bed is called the disengaging space. The cross-sectional area is determined by the volumetric flow of gas and the allowable or required fluidizing velocity of the gas at operating conditions. In some cases the lowest permissible velocity of gas is used, and in others the greatest permissible

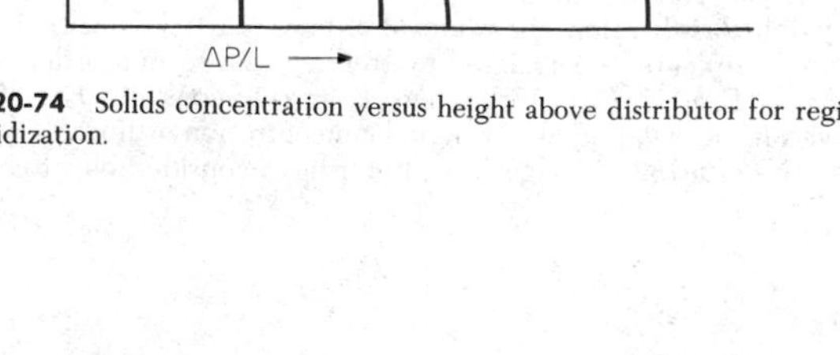

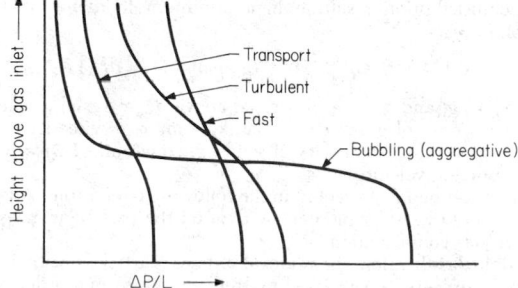

FIG. 20-74 Solids concentration versus height above distributor for regimes of fluidization.

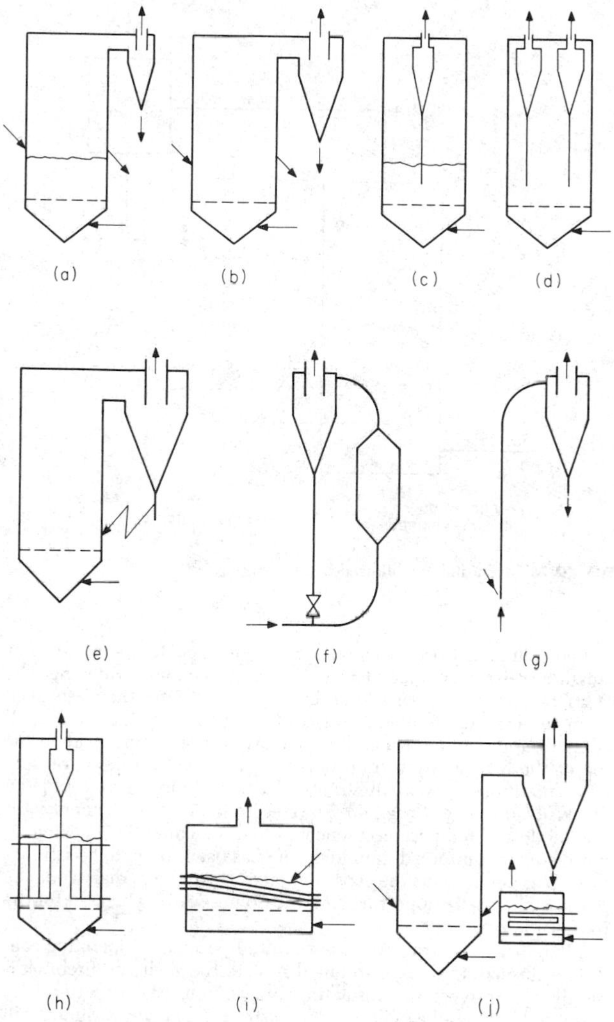

FIG. 20-75 Fluidized-bed systems. (*a*) Bubbling bed, external cyclone, $U < 20 \times U_{mf}$. (*b*) Turbulent bed, external cyclone, $20 \times U_{mf} < U < 200 \times U_{mf}$. (*c*) Bubbling bed, internal cyclones, $U < 20 \times U_{mf}$. (*d*) Turbulent bed, internal cyclones, $20 \times U_{mf} < U < 200 \times U_{mf}$. (*e*) Circulating (fast) bed, external cyclones, $U > 200 \times U_{mf}$. (*f*) Circulating bed, $U > 200 \times U_{mf}$. (*g*) Transport, $U > U_T$. (*h*) Bubbling or turbulent bed with internal heat transfer, $2 \times U_{mf} < U < 200 \times U_{mf}$. (*i*) Bubbling or turbulent bed with internal heat transfer, $2 \times U_{mf} < U < 100 \times U_{mf}$. (*j*) Circulating bed with external heat transfer, $U > 200 \times U_{mf}$.

velocity is used. The maximum flow is generally determined by the carry-over or entrainment of solids, and this is related to the dimensions of the disengaging space (cross-sectional area and height).

Bed Bed height is determined by a number of factors, either individually or collectively, such as:

1. Gas-contact time
2. L/D ratio required to provide staging
3. Space required for internal heat exchangers
4. Solids-retention time

Generally, bed heights are not less than 0.3 m (12 in) or more than 15 m (50 ft).

Although the reactor is usually a vertical cylinder, there is no real limitation on shape. The specific design features vary with operating conditions, available space, and use. The lack of moving parts lends toward simple, clean design.

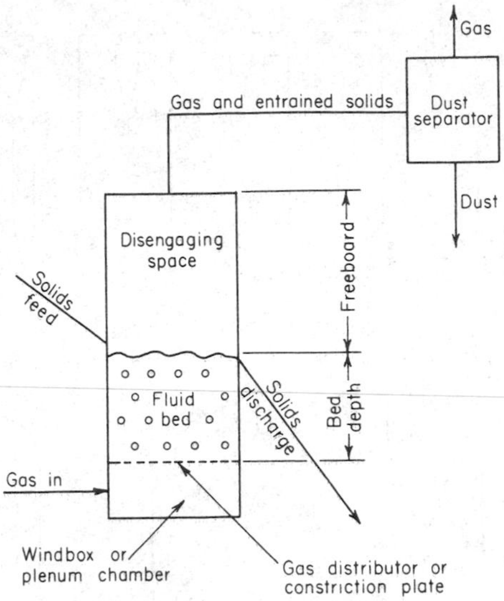

FIG. 20-76 Noncatalytic fluidized-bed system.

Many fluidized-bed units operate at elevated temperatures. For this use, refractory-lined steel is the most economical design. The refractory serves two main purposes: (1) it insulates the metal shell from the elevated temperatures, and (2) it protects the metal shell from abrasion by the bed and particularly the splashing solids at the top of the bed resulting from bursting bubbles. Depending on specific conditions, several different refractory linings are used [Van Dyck, *Chem. Eng. Prog.*, 46–51 (December 1979)]. Generally, for the moderate temperatures encountered in catalytic cracking of petroleum, a reinforced-gunnite lining has been found to be satisfactory. This also permits the construction of larger units than would be permissible if self-supporting ceramic domes were to be used for the roof of the reactor.

When heavier refractories are required because of operating conditions, insulating brick is installed next to the shell and firebrick is installed to protect the insulating brick. Industrial experience in many fields of application has demonstrated that such a lining will successfully withstand the abrasive conditions for many years without replacement. Most serious refractory wear occurs with coarse particles at high gas velocities and is usually most pronounced near the operating level of the fluidized bed.

Gas leakage behind the refractory has plagued a number of units. Care should be taken in the design and installation of the refractory to reduce the possibility of the formation of "chimneys" in the refractories. A small flow of solids and gas can quickly erode large passages in soft insulating brick or even in dense refractory. Gas stops are frequently attached to the shell and project into the refractory lining. Care in design and installation of openings in shell and lining is also required.

In many cases, cold spots on the reactor shell will result in condensation and high corrosion rates. Sufficient insulation to maintain the shell and appurtenances above the dew point of the reaction gases is necessary.

The violent motion of a fluidized bed requires ample foundations and sturdy supporting structure for the reactor. Even a relatively small differential movement of the reactor shell with the lining will materially shorten refractory life. The lining and shell must be designed as a unit.

Freeboard The freeboard or disengaging height is the distance between the top of the fluid bed and the gas-exit nozzle in bubbling- or turbulent-bed units. The distinction between bed and freeboard is difficult to determine in fast and transport units (see Fig. 20-74).

At least two actions can take place in the freeboard (classification of solids and reaction of solids and gases, gases, or gases catalyzed by the solids).

As a bubble reaches the upper surface of a fluidized bed, the gas breaks through the thin upper envelope composed of solid particles entraining some of these particles. The crater-shaped void formed is rapidly filled by flowing solids. When these solids meet at the center of the void, solids are geysered upward. The downward pull of gravity and the upward pull of the drag force of the upward-flowing gas act on the particles. The larger and denser particles return to the top of the bed, and the finer and lighter particles are carried upward. Apparently the particles thrown into the freeboard have a random distribution of initial upward velocity and direction. As a result, the classification taking place in the freeboard of a fluidized bed can be mathematically described and predicted by the sharpness-index correlation developed and described by W. F. Carey and C. H. Bosanquet of Imperial Chemical Industries, Ltd., which can be used for all classifiers. $C/F = (D_p/D_{pc})^{SI}$, where C = weight ratio of particle size D_p remaining in bed, F = weight ratio of particle size D_p entrained, SI = sharpness index, and D_{pc} = cut size or particle size, equivalent to the particle whose terminal velocity equals the gas velocity.

The sharpness index is proportional to $(U^2/gz)^{-0.594}(\mu_c/\mu_0)$, or SI $= 0.30(U^2/gz)^{-0.594}(\mu_c/\mu_0)$, where U = superficial velocity, m/s; g = acceleration of gravity, m/s^2; z = disengaging height, m; μ_c = viscosity of gas at conditions; and μ_0 = viscosity of gas at 25°C. This relationship does not hold for particles smaller that about 80 μm. These smaller particles are not entrained as readily as expected.

Several methods of predicting entrainment have been presented in the literature for bubbling and turbulent beds [Zenz and Othmer, op. cit.; Gugnoni and Zenz, in Grace and Matson (eds.), *Fluidization*, Plenum, New York and London, 1980; S. T. Pemberton, Ph.D. thesis, Cambridge University, 1982; Wen and Hashinger, *Am. Inst. Chem. Eng. J.*, **6**, 220 (July 1960); Yagi and Aochi, *Soc. Chem. Eng. (Japan)*, spring meeting, 1955; and Merrick and Highley, *Am. Inst. Chem. Eng. Symp. Ser.* **70**(137), 366–378 (1974)]. Entrainment even when carefully measured is nonconsistent over short periods of time (minutes). Several conclusions regarding entrainment found in the literature are erroneous because of the paucity of the data used and the lack of reproducibility of such data. Furthermore, most disregard the effect of disengaging height.

Entrainment increases as gas velocity increases, as gas viscosity increases, as vessel diameter increases, and as "fines" concentration increases. Entrainment decreases as disengaging height increases, as particle density increases, and as gravity increases.

Entrainment (mass of solids/mass of gas) is proportional to $(U^2/gz)^n$ (U = superficial gas velocity, m/s; g = acceleration of gravity, m/s^2; z = disengaging height, m; n = exponent, usually about 1.5). Entrainment from beds of solids of different types (A, B, or D) as defined by Geldart (loc. cit.) differ. The same holds true if the particle-size distribution is bimodal. Entrainment from beds of Geldart (loc. cit.) Class A powers with a broad and continuous size distribution can be predicted by Eq. (20-55). The term (U^2/gz) has been corrected to the base case of gas at atmospheric conditions, a 1.0-m-diameter vessel, particles with a density of 1.0 gm/cm^3, and a particle-size distribution resulting in a minimum fluidizing velocity (U^{mf}) of 1.0 m/s.

$$E = 0.154(U^2/gz)^{1.53}(D)^{0.75}(\mu_c/\mu_0)^{1.78}(1/\rho_s)^{2.5}(1/U_{mf})^{1.5}$$

where E, V, g, and z are as described above; D = vessel diameter, m; μ_c = viscosity of gas at conditions, kg/s.m; μ_0 = viscosity of gas at 25°C, kg/s·m; ρ_s = density of solids, gm/cm^3; and U_{mf} = minimum fluidizing velocity, m/s.

Another method is presented in the following paragraphs. All predictions should be substantiated by data on the particular gas-solid system under consideration.

In batch classification, the removal of fines (particles less than any arbitrary size) can be correlated by treating as a second-order reaction $K = (F/\theta)[1/x(x - F)]$, where K = rate constant, F = fines removed in time θ, and x = original concentration of fines.

Gas Distributor The gas distributor has a considerable effect on

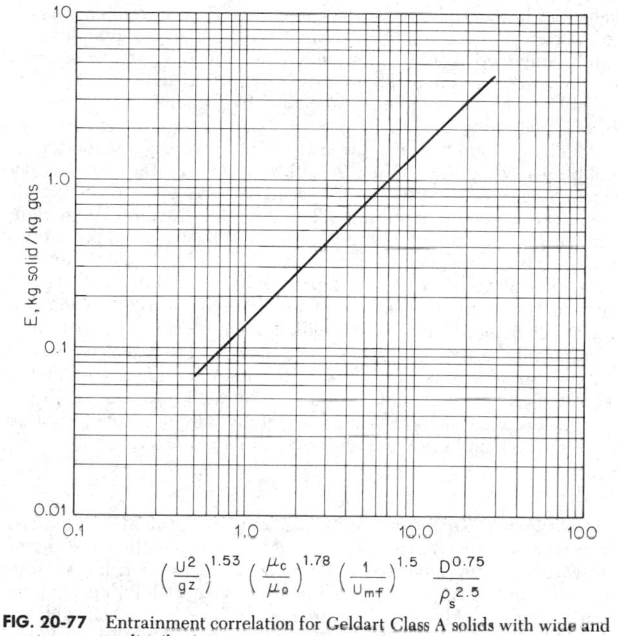

FIG. 20-77 Entrainment correlation for Geldart Class A solids with wide and continuous size distribution.

proper operation of the fluidized bed. Basically there are two types: (1) for use when the inlet gas contains solids and (2) for use when the inlet gas is clean. In the latter case, the distributor is designed to prevent back flow of solids during normal operation, and in many cases it is designed to prevent back flow during shutdown. In order to provide distribution, it is necessary to restrict the gas or gas and solids flow so that pressure drops across the restriction amount to from 0.5 kPa (2 in of water) to 20 kPa (3 lbf/in^2). As a general rule, pressure drops in excess of 15 kPa (2 lbf/in^2) are not used.

One school of thought contends that the pressure drop across the distributor should be at least 30 percent of the bed pressure drop. Another school maintains the former is true but that a maximum of 2.5 kPa (10 in of water) is required. Probably, the actual design of the distributor, whether upshot, downshot, slotted orifice, tuyered, etc., has a tremendous amount to do with the pressure drop required to achieve good distribution of the gas.

When both solids and gases pass through the distributor, such as in catalytic-cracking units, a number of variations are or have been used, such as concentric rings in the same plane, with the annuli open (Fig. 20-78*a*); concentric rings in the form of a cone (Fig. 20-78*b*); grids of T bars or other structural shapes (Fig. 20-78*c*); flat metal perforated plates supported or reinforced with structural members (Fig. 20-78*d*); dished and perforated plates concave both upward and downward (Fig. 20-78*e* and *f*). The last two forms are generally more economical.

Experience has shown that the concave-upward type is a better arrangement than the concave-downward type, as it tends to increase the flow of gases in the outer portion of the bed. This counteracts the normal tendency of higher gas flows in the center of the bed.

Structurally, distributors must withstand the differential pressure across the restriction during normal and abnormal flow. In addition, during a shutdown all or a portion of the bed will be supported by the distributor until sufficient back flow of the solids has occurred both to reduce the weight of solids above the distributor and to support some of this remaining weight by transmitting the force to the walls and bottom of the reactor. During startup considerable upward thrust can be exerted against the distributor as the settled solids under the distributor are carried up into the normal reactor bed.

When the feed gas is devoid of or contains only small quantities of fine solids, more sophisticated designs of gas distributors can be

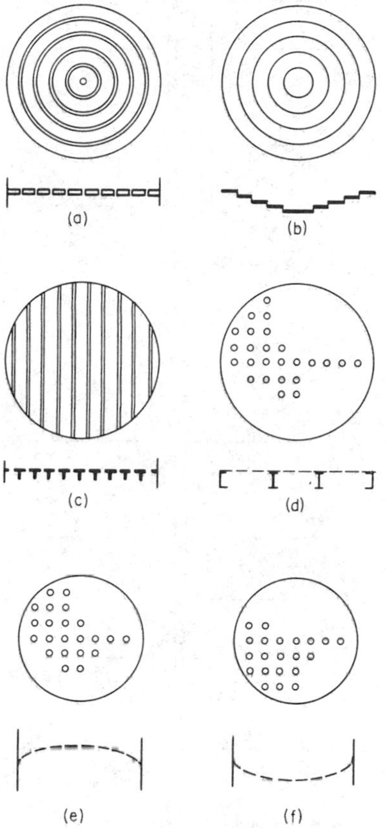

FIG. 20-78 Gas distribution for gases containing solids.

used to effect economies in initial cost and maintenance. This is most pronounced when the inlet gas is cold and noncorrosive. When this is the case, the plenum chamber gas distributor, and distributor supports can be fabricated of mild steel by using normal temperature design factors. The first commercial fluidized-bed ore roaster [Mathews, *Trans. Can. Inst. Min. Metall.*, **L11**, 97 (1949)], supplied by the Dorr Co. (now Dorr-Oliver Inc.) in 1947 to Cochenour-Willans, Red Lake, Ontario, was designed with a mild-steel constriction plate covered with castable refractory to insulate the plate from the calcine and also to provide cones in which refractory balls were placed to act as ball checks. The balls eroded unevenly, and the castable cracked. However, when the unit was shut down by closing the air-control valve, the runback of solids was negligible because of bridging. If, however, the unit was shut down by deenergizing the centrifugal blower motor, the higher pressure in the reactor would relieve through the blower and fluidizing gas plus solids would run back through the constriction plate. Figure 20-79 illustrates two designs of gas inlets which have been successfully used to prevent flowback of solids. For best results, irrespective of the design, the gas flow should be stopped and pressure-relieved from the bottom upward through the bed.

Some units have been built and successfully operated with simple slot-type distributors made of heat-resistant steel. This requires a heat-resistant plenum chamber but eliminates the frequently encountered problem of corrosion caused by condensation of acids and water vapor on the cold metal of the distributor.

When the inlet gas is hot, such as in dryers or in the upper distributors of multibed units, ceramic arches or heat-resistant metal grates are generally used.

Self-supporting ceramic domes have been in successful use for many years as gas distributors when temperatures range up to 1400

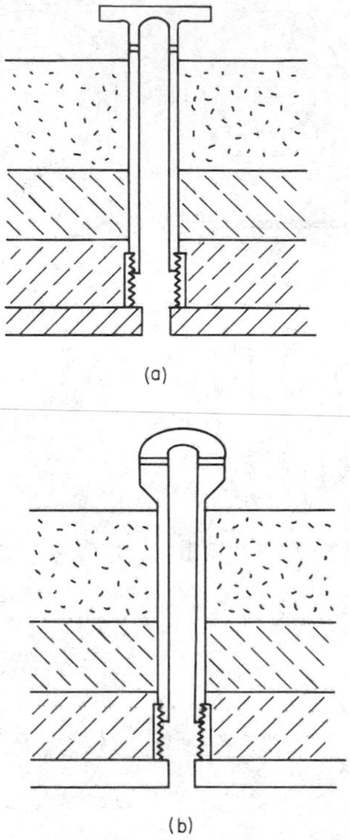

FIG. 20-79 Gas inlets designed to prevent backflow of solids. (*a*) Mortarboard. (*b*) Club head. *(Dorr-Oliver Inc.)*

K (1100°C). Some of these domes are fitted with alloy-steel orifices to regulate air distribution. However, the ceramic arch presents the same problem as the dished head positioned concave downward. Either the holes in the center must be smaller so that the sum of the pressure drops through the distributor plus the bed is constant across the whole cross section, or the top of the arch must be flattened so that the bed depth in the center and outside is equal. This is especially important when shallow beds are used.

When the atmosphere in the bed is sufficiently benign, a sparger-type distributor may be used. See Fig. 20-80.

In some cases, it is impractical to use a plenum chamber under the constriction plate. This condition arises when a flammable or explosive mixture of gases is being introduced to the reactor. One solution is to pipe the gases to a multitude of individual gas inlets in the floor of the reactor. In this way it may be possible to maintain the gas

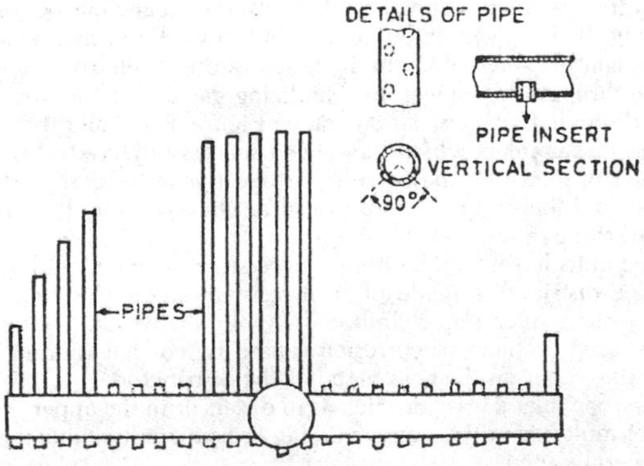

FIG. 20-80 Multiple-pipe gas distributor. [*From Stemerding, de Groot, and Kuypers,* Soc. Chem. Ind. J. Symp. Fluidization Proc., *35–46, London (1963).*]

velocities in the pipes above the flame velocity or to reduce the volume of gas in each pipe to the point at which an explosion can be safely contained. Another solution is to provide separate inlets for the different gases and depend on mixing in the fluidized bed. The inlets should be fairly close to one another, as lateral gas mixing in fluidized beds is poor.

Much attention has been given to the effect of gas distribution on bubble growth in the bed and the effect of this on catalyst utilization, space-time yield, etc. It would appear that the best gas distributor would be a porous membrane. This type of distributor is seldom practical for commercial units because of both structural limitations and the need for absolutely clean gas. Practically, the limitations on hole spacing are dependent on particle size of solids, materials of construction, and type of distributor. If easily worked metals are used, punching, drilling, and welding are not expensive operations and permit the use of large numbers of holes. The use of tuyeres or bubble caps permits horizontal distribution of the gas so that a smaller number of gas-inlet ports can still achieve good gas distribution. If a ceramic arch is used, generally only one hole per brick is permissible and brick dimensions must be reasonable.

Scale-Up

Bubbling or Turbulent Beds Scale-up of noncatalytic fluidized beds when the reaction is fast, as in roasting or calcination, is straightforward and is usually carried out on an area basis. Small-scale tests are made to determine physical limitations such as sintering, agglomeration, solids-holdup time required, etc. Slower ($k < 1/s$) catalytic or more complex reactions in which several gas interchanges are required are usually scaled up in several steps, from laboratory to commercial size. The hydrodynamics of gas-solids flow and contacting is quite different in small-diameter high-L/D fluid beds as compared with large-diameter moderate-L/D beds. In small-diameter beds the bubbles cannot grow larger than the vessel diameter; when the bubble size approaches the vessel diameter, the bubble changes from a freely rising bubble to a slug which moves upward at a lower velocity than a bubble of the same volume. The gas-solids contacting and gas holdup time in slug flow are significantly greater than in bubbling flow. Furthermore, the solids and gas back mixing is much less in high-L/D beds, either slugging or bubbling, as compared with low-L/D beds. Thus, the conversion or yield in large unstaged reactors is sometimes considerably lower than in small high-L/D units. To overcome some of the problems of scale-up, staged units are used (see Fig. 20-81). It is generally concluded than an unstaged 1-m- (40-in-) diameter unit will achieve about the same conversion as a large industrial unit. The validity of this conclusion is dependent on many variables, including bed depth, particle size, and size distribution, temperature, and pressure. A brief history of fluidization, fluidized-bed scale-up, and modeling will illustrate the problems.

Fluidized beds were used in Europe in the 1920s to gasify coal. Scale-up problems either were insignificant or were not publicized. During World War II, catalytic cracking of oil to produce gasoline was successfully commercialized by scaling up from pilot-plant size (a few centimeters in diameter) to commercial size (several meters in diameter). It is fortunate that the kinetics of the cracking reactions are fast, that the ratio of crude oil to catalyst is determined by thermal balance and the required catalyst circulation rates, and that the crude feed point was in the plug-flow riser. The first experience of problems with scale-up was associated with the production of gasoline from natural gas by using the Fischer-Tropsch process. Some 0.10-m- (4-in-), 0.20-m- (8-in-), and 0.30-m- (12-in-) diameter pilot-plant results were scaled to a 7-m-diameter commercial unit, where the yield was only about 50 percent of that achieved in the pilot units. The Fischer-Tropsch synthesis is a relatively slow reaction; therefore, gas-solid contacting is very important. Since this unfortunate experience or perhaps because of it, much effort has been given to the scale-up of fluidized beds. Many models have been developed; these basically are of two types, the two-phase model [May, *Chem. Eng. Prog.*, **55,** 12, 5, 49–55 (1959); and Van Deemter, *Chem. Eng. Sci.*, **13,** 143–154 (1961)] and the bubble model (Kunii and Levenspiel, *Fluidization Engineering*, Wiley, New York, 1969). The two-

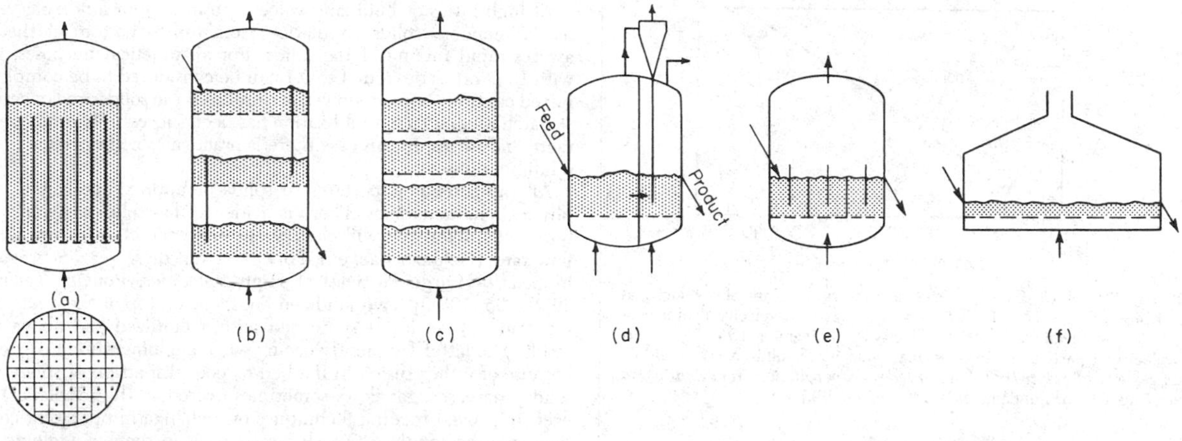

FIG. 20-81 Methods of providing staging in fluidized beds.

phase model according to May and Van Deemter is shown in Fig. 20-82. In these models all or most of the gas passes through the bed in plug flow in the bubbles which do not contain solids (catalyst). The solids form a dense suspension-emulsion phase in which gas and solids mix according to an axial dispersion coefficient (E). Cross flow between the two phases is predicted by a mass-transfer coefficient. Refinements of the basic model and predictions of mass-transfer and axial-dispersion coefficients are the subject of many papers [Van Deemter, *Proc. Symp. Fluidization*, Eindhoven (1967); de Groot, ibid.; Van Swaaij; and Zuidiweg, *Proc. 5th Eur. Symp. React. Eng., Amsterdam*, B9-25 (1972); DeVries, Van Swaaij, Mantovani, and Heijkoop, ibid., B9-59 (1972); and Werther, *Ger. Chem. Eng.*, **1**, 243-251 (1978)].

The bubble model (Kunii and Levenspiel, *Fluidization Engineering*, Wiley, New York, 1969; Fig. 20-83) assumes constant-sized bubbles (effective bubble size d_b) rising through the suspension phase. Gas is transferred from the bubble void to the mantle and wake at mass-transfer coefficient K_{bc} and from the mantle and wake to the emulsion phase at mass-transfer coefficient K_{ce}. Experimental results have been fitted to theory by means of adjusting the effective bubble size. As mentioned previously, bubble size changes from the bottom to the top of the bed, and thus this model is not realistic though of considerable use in evaluating reactor performance. Several bubble models using bubbles of increasing size from the distributor to the top of the bed and gas interchange between the bubbles and the emulsion phase according to Kunii and Levenspiel have been proposed [Kato and Wen, *Chem. Eng. Sci.*, **24**, 1351-1369 (1969); and Fryer and Potter, in Keairns (ed.), *Fluidization Technology*, vol. I, Hemisphere, Washington, 1975, pp. 171-178].

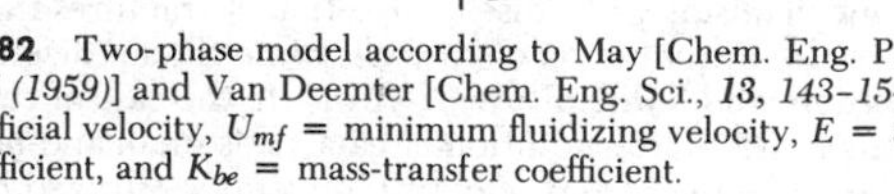

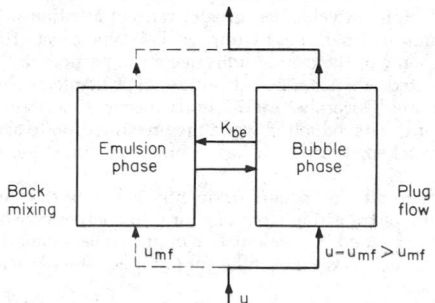

FIG. 20-82 Two-phase model according to May [Chem. Eng. Prog., *55, 12, 5, 49-55 (1959)*] and Van Deemter [Chem. Eng. Sci., *13, 143-154 (1961)*]. U = superficial velocity, U_{mf} = minimum fluidizing velocity, E = axial dispersion coefficient, and K_{be} = mass-transfer coefficient.

The use of models to interpret and understand data from small, intermediate, or large fluidized beds is extremely helpful. To build a full-scale plant on the basis of results from a laboratory-size unit by using a model for scale-up could invite problems such as occurred in the scale-up of the dense-fluid-bed Fischer-Tropsch synthesis. Experience with similar reactions and gas-solid systems would be helpful in understanding the hydrodynamics, which are so very important.

Circulating or Fast Beds Mass transfer in the circulating or fast regime is believed to be greater than in a bubbling or turbulent bed. It can be predicted by assuming dispersed particles with a slip velocity equal to the gas velocity. The solids concentration versus the height above the gas inlet (Fig. 20-74) is not accurately predictable (Yerushalmi, Turner, and Squires, op. cit.), as data from larger-diameter units have not been published to date.

Entrainment from fast, circulating, or transport units is based upon steady-state material balances. The feed or circulating rate controls the solids holdup because of the effect of solids concentration on slip velocity [Fig. 20-84; Reddy and Pei, *Ind. Eng. Chem. Fundam.*, 8(3), 420 (August 1969)]. The solids concentration as measured by $\Delta P/L$ (see Fig. 20-74) is different from the entrainment or circulating rate. $\Delta P/L$ is a measure of the solids in a given cross-sectional slice of the unit, not a measure of the rate of entrance and exit from this slice.

Transport Transport units can be scaled up on the principles of pneumatic conveying. Mass and heat transfer can be predicted on

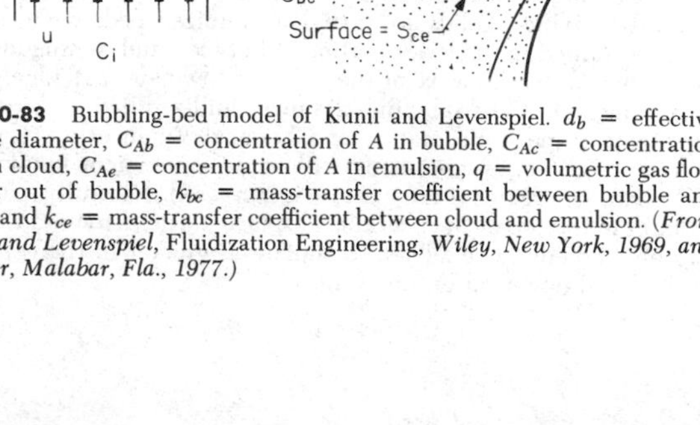

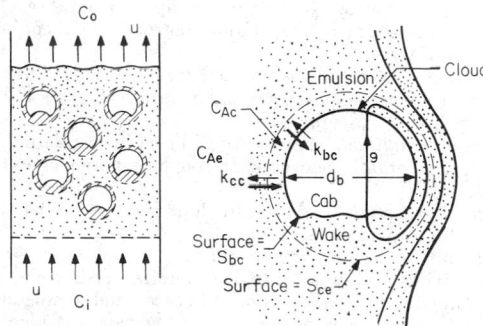

FIG. 20-83 Bubbling-bed model of Kunii and Levenspiel. d_b = effective bubble diameter, C_{Ab} = concentration of A in bubble, C_{Ac} = concentration of A in cloud, C_{Ae} = concentration of A in emulsion, q = volumetric gas flow into or out of bubble, k_{bc} = mass-transfer coefficient between bubble and cloud, and k_{ce} = mass-transfer coefficient between cloud and emulsion. (*From Kunii and Levenspiel,* Fluidization Engineering, *Wiley, New York, 1969, and Krieger, Malabar, Fla., 1977.*)

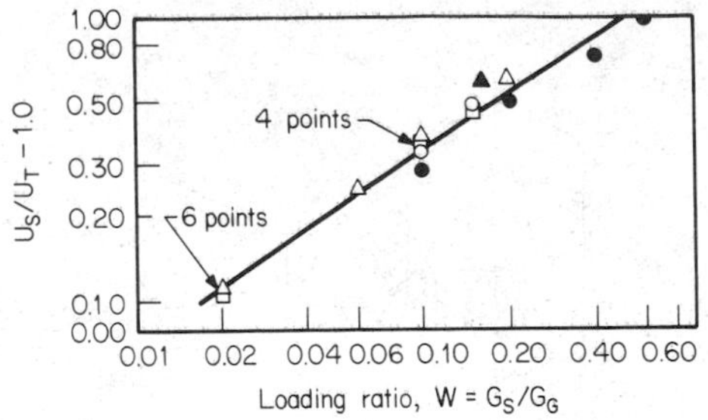

FIG. 20-84 Plot of $U_s/U_T - 1$ versus loading ratio. U_s = slip velocity at center of pipe, ($U_{go} - U_{po}$), L/T; U_{go} = mean axial gas velocity at pipe center, L/T; U_{po} = mean axial particle velocity at pipe center, L/T; U_T = terminal velocity of particle, L/T; G_g = mass velocity of solids, m/L^{2}T; and G_s = mass velocity of gas, m/L^{2}T. [*Reprinted with permission from Reddy and Pei,* Ind. Eng. Chem. Fundam., *8(3), 490–497 (1969).*]

both the slip velocity during acceleration and the slip velocity at full acceleration. The slip velocity is increased as the solids concentration is increased.

Heat Transfer Heat-exchange surfaces have been used to provide means of removing or adding heat to fluidized beds. Usually, these surfaces are provided in the form of vertical tubes manifolded at top and bottom or in trombone shape manifolded exterior to the vessel.

Other shapes such as horizontal bayonets have been used. In any such installations adequate provision must be made for abrasion of the exchanger surface by the bed.

The prediction of the heat-transfer coefficient is covered in Secs. 10 and 11. Normally, the transfer rate is between 5 and 25 times that for the gas alone. See Davidson and Harrison (loc. cit.).

Heat transfer from solids to gas and gas to solids usually results in a coefficient of about 6 to 20 J/(m^2·s·K) [3 to 10 Btu/(h·ft^2·°F)]. However, the large area of the solids per cubic foot of bed, 5000 m^2/m^3 (15,000 ft^2/ft^3) for 60-μm particles of 600 kg/m^3 (40 lb/ft^3) bulk density, results in the rapid approach of gas and solids temperatures. With a fairly good distributor, essential equalization of temperatures occurs within 2 to 6 cm (1 to 3 in) of the top of the distributor.

Bed thermal conductivities in the vertical direction have been measured in the laboratory in the range of 40 to 60 kJ/(m^2·s·K) [20,000 to 30,000 Btu/(h·ft^2·°F·ft)]. Horizontal conductivities for 3-mm (⅛-in) particles in the range of 2 kJ/(m^2·s·K) [1000 Btu/(h·ft^2·°F·ft)] have been measured in large-scale experiments.

Except in extreme L/D ratios, the temperature in the fluidized bed is uniform—generally the temperature at any point being within 5 K (10°F) of any other point.

Temperature Control Because of the rapid equalization of temperatures in fluidized beds, temperature control can be accomplished in a number of ways.

1. *Adiabatic.* Control gas flow and/or solids feed rate so that the heat of reaction is removed as sensible heat in off gases and solids or heat supplied by gases or solids.
2. *Solids circulation.* Remove or add heat by circulating solids.
3. *Gas circulation.* Recycle gas through heat exchangers to cool or heat.
4. *Liquid injection.* Add volatile liquid so that the latent heat of vaporization equals excess energy.
5. *Cooling or heating surfaces in bed.*

Solids Mixing Solids are mixed in fluidized beds by means of solids entrained in the lower portion of bubbles, and the shedding of these solids from the wake of the bubble (Rowe and Patridge, "Particle Movement Caused by Bubbles in a Fluidized Bed," Third Congress of European Federation of Chemical Engineering, London, June 2, 1962). Thus, no mixing will occur at incipient fluidization, and mixing increases as the gas rate is increased. Naturally, particles brought to the top of the bed must displace particles toward the bottom of the bed. Generally, solids upflow is greater in the center of the bed and downward at the wall.

At high ratios of fluidizing velocity to minimum fluidizing velocity, tremendous solids circulation from top to bottom of the bed assures rapid mixing of the solids. For all practical purposes, beds with L/D ratios of from 4 to 0.1 can be considered to be completely mixed continuous-reaction vessels insofar as the solids are concerned.

Batch mixing using fluidization has been successfully employed in many industries. In this case there is practically no limitation to vessel dimensions.

All the foregoing pertains to solids of approximately the same physical characteristics. There is evidence that solids of widely different characteristics will classify one from the other at certain gas flow rates [Geldart, Baeyens, Pope, and van de Wijer, "Segregation in Beds of Large Particles at High Velocities, *Powder Technol.*, **30**(2), 195 (1981)]. Two fluidized beds, one on top of the other, may be formed, or a lower static bed with a fluidized bed above may result. The latter frequently occurs when agglomeration takes place because of either fusion in the bed or poor dispersion of sticky feed solids. Increased gas flows sometimes overcome the problem; however, improved feeding techniques or a change in operating conditions may be required. Another solution is to remove agglomerates either continuously or periodically from the bottom of the bed.

Gas Mixing The mixing of gases as they pass vertically up through the bed has never been considered a problem. However, horizontal mixing is very poor and requires effective distributors if two gases are to be mixed in the fluidized bed.

In bubbling beds operated at velocities of less than about 5 to 11 times U_{mf} the gases will flow upward in both the emulsion and the bubble phases. At velocities greater than about 5 to 11 times U_{mf} the downward velocity of the emulsion phase is sufficient to carry the contained gas downward. The back mixing of gases increases as U/U_{mf} is increased until the circulating or fast regime is reached where the back mixing decreases as the velocity is further increased.

Size Enlargement Under proper conditions, particles of solids can be caused to grow. That is sometimes advantageous and at other times disadvantageous. Growth is associated with the liquefaction or softening of some portion of the bed material (i.e., addition of soda ash to calcium carbonate feed in lime reburning, tars in fluidized-bed coking, or lead or zinc roasting cause agglomeration of dry particles in much the same way as binders act in rotary pelletizers). The motion of the particles, one against the other, in the bed results in spherical pellets. If the size of these particles is not controlled, segregation of the large particles from the bed will occur.

In drying solutions or slurries of solutions, the location of the feed-injection nozzle (spray nozzle) has a great effect on the size of particle formed in the bed. Also of importance are the operating temperature, relative humidity of the off gas, and gas velocity. Particle growth can occur as agglomeration or as an "onion skinning."

Size Reduction Three major size-reduction mechanisms occur in the fluidized bed. These are attrition, impact, and thermal decrepitation.

Because of the random motion of the solids, some abrasion of the surface occurs. This is generally quite small, usually amounting to about 0.25 to 1 percent of the solids per day.

In areas of high gas velocities, greater rates of attrition will occur as well as fracture of particles by impact. This type of jet grinding is employed in some of the coking units to control particle size (Griffin et al., presented at American Institute of Chemical Engineers national meeting, December 1957). It also occurs to a lesser degree at the point of gas introduction when the pressure drop to assure gas distribution is taken across an orifice or pipe that discharges directly into the bed.

Thermal decrepitation occurs frequently when crystals are rearranged because of transition from one form to another or when new compounds are formed (i.e., calcination of limestone). Sometimes the strains in cases such as this are sufficient to reduce the particle to the basic crystal size.

All these mechanisms will cause completion of fractures that were started before the introduction of the solids into the fluidized bed.

Solids Feeders and Solids Flow Control In the case of catalytic-cracking units in which the addition of catalyst is small and need not be steady, the makeup catalyst may be fed from pressurized hoppers

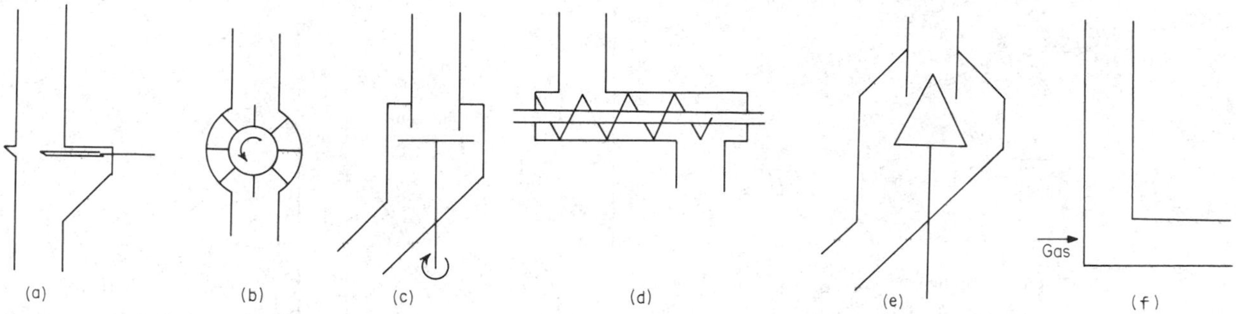

FIG. 20-85 Solids-flow-control devices. (*a*) Slide valve. (*b*) Star valve. (*c*) Table feeder. (*d*) Screw feeder. (*e*) Cone valve. (*f*) L Valve.

into one of the conveying lines. The main solids-flow-control problem is to maintain balanced inventories of catalyst in and controlled flow from and to the reactor and regenerator. This flow of solids from an oxidizing atmosphere to a reducing one, or vice versa, usually necessitates stripping gases from the interstices of the solids as well as gases adsorbed by the particles. Steam is usually used for this purpose. The point of removal of the solids from the fluidized bed is usually under a lower pressure than the point of feed introduction into the carrier gas. The pressure is increased at the bottom of the solids draw-off pipe or standpipe by introducing gas at the bottom or at intervals along the length so that the pressure drop of the gas flowing upward counter to the solids flow downward results in a higher pressure at the bottom as compared with the top. This standpipe may be fluidized, or the solids may be flowing with no appreciable expansion. In any event, the pressure above the solids control valve must be maintained at the same or greater pressure than at the point of solids introduction into the carrier gas stream. See Judd and Rowe, in Davidson and Keairns (eds.), *Fluidization*, Cambridge, New York, 1978, p. 110; and Leung, Jones, and Knowlton, *Powder Technol.*, **19**, 7–15 (1976).

Several designs of valves for solids flow control are used. These should be chosen with care to suit the specific conditions. Usually, block valves are used in conjunction with the control valves. Figure 20-85 shows schematically some of the devices used for solids flow control. Not shown in Fig. 20-85 is the flow-control arrangement used in the Exxon Research & Engineering Co. model IV catalytic-cracking units. This device consists of a U bend. A variable portion of regenerating air is injected into the riser leg. Changes in air-injection rate change the fluid density in the riser and thereby achieve control of the solids flow rate. Catalyst circulation rates of 1200 kg/s (70 tons/min) have been reported.

When the solid is one of the reactants, such as in ore roasting, the flow must be continuous and precise in order to maintain constant conditions in the reactor. Feeding of free-flowing granular solids into a fluidized bed is not difficult. Standard commercially available solids-weighing and -conveying equipment can be used to control the rate and deliver the solids to the feeder. Screw conveyors, dip pipes, seal legs, and injectors are used to introduce the solids into the reactor proper (Fig. 20-85). Difficulties arise and special techniques must be used when the solids are not free-flowing, such as is the case with most filter cakes. One solution to this problem was developed at Cochenour-Willans. After much difficulty in attempting to feed a wet and sometimes frozen filter cake into the reactor by means of a screw feeder, experimental feeding of a water slurry of flotation concentrates was attempted. This trial was successful, and this method has been used in almost all cases in which the heat balance, particle size of solids, and other considerations have permitted. Gilfillan et al. (*J. Chem. Metall. Min. Soc. S. Afr.*, May 1954) and Soloman and Beal (*Uranium in South Africa, 1946–56*) present complete details on the use of this system for feeding.

When slurry feeding is impractical, **recycling** of solids product to mix with the feed, both to dry and to achieve a better-handling material, has been used successfully. Also, the use of a rotary table feeder mounted on top of the reactor, discharging through a mechanical disintegrator, has been successful. The wet solids generally must be broken up into discrete particles of very fine agglomerates either by mechanical action before entering the bed or by rapidly vaporizing water. If lumps of dry or semidry solids are fed, the agglomerates do not break up but tend to fuse together. As the size of the agglomerate is many times the size of the largest individual particle, these agglomerates will segregate out of the bed, and in time the whole of the fluidized bed may be replaced with a static bed of agglomerates.

Solids Discharge The type of discharge mechanism utilized is dependent upon the necessity of sealing the atmosphere inside the fluidized-bed reactor and the subsequent treatment of the solids. The simplest solids discharge is an overflow weir. This can be used only when the escape of fluidizing gas does not present any hazards due to nature or dust content or when the leakage of gas into the fluidized-bed chamber from the atmosphere into which the bed is discharged is permitted. Solids will overflow from a fluidized bed through a port even though the pressure above the bed is maintained at a slightly lower pressure than the exterior pressure. When it is necessary to restrict the flow of gas through the opening, a simple flapper valve is frequently used. Overflow to combination seal and quench tanks (Fig. 20-86) is used when it is permissible to wet the solids and when disposal or subsequent treatment of the solids in slurry form is desirable. The FluoSeal is a simple and effective way of sealing and purging gas from the solids when an overflow-type discharge is used (Fig. 20-87).

Seal legs are frequently used in conjunction with solids-flow-control valves to equalize pressures and to strip trapped or adsorbed gases from the solids. The operation of a seal leg is shown schemati-

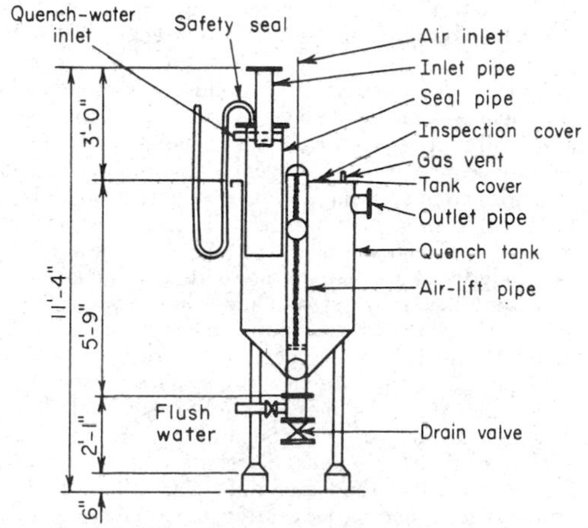

FIG. 20-86 Quench tank for overflow or cyclone solids discharge. [*Gilfillan et al., "The FluoSolids Reactor as a Source of Sulphur Dioxide,"* J. Chem. Metall. Min. Soc. S. Afr. *(May 1954).*]

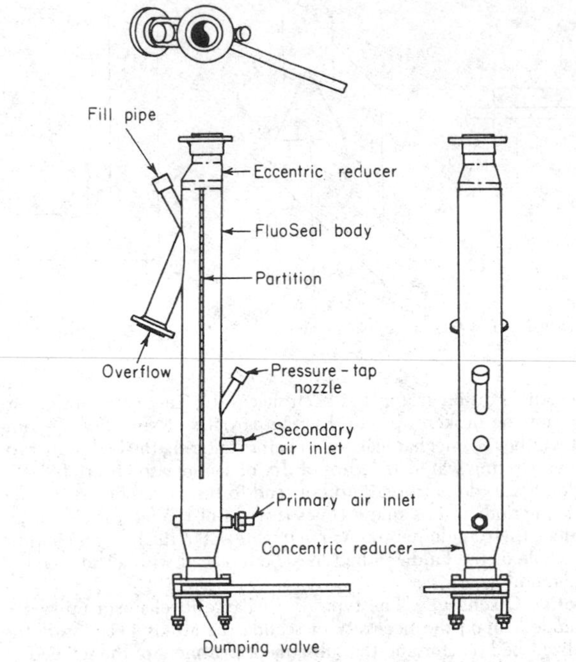

FIG. 20-87 Dorrco FluoSeal, type UA. *(Dorr-Oliver Inc.)*

cally in Fig. 20-88. The solids settle by gravity from the fluidized bed into the seal leg or standpipe. Seal and/or stripping gas is introduced near the bottom of the leg. This gas flows both upward and downward. Pressures indicated in the illustration have no absolute value but are only relative. The legs are designed for either fluidized or settled solids.

The ICI valve shown schematically in Fig. 20-89 serves better as a seal device than as a solids-flow-control valve. Gas introduced below the normal solids level and above the discharge port will flow upward and downward. The relative flow in each direction is self-adjusting, depending upon the differential pressure between the point of solids feed and discharge and the level of solids in the leg. The length and diameter of the discharge spout are selected so that the undisturbed angle of repose of the solids will prevent discharge of the solids. As solids are fed into the leg, height H of solids increases. This in turn reduces the flow of gas in an upward direction and increases the flow of gas in a downward direction. When the flow of gas downward and through the solids-discharge port reaches a given rate, the angle of repose of the solids is upset and solids discharge commences. Usually, the level of solids above the point of gas introduction will float between H and H'. Changes in conditions such as temperature or pressure can frequently result in uncontrolled discharge of solids and loss of seal. The fixed opening also makes it subject to plugging from stray lumps, etc. Bottom draw-offs are used when a possibility of segregation in the bed is encountered.

In most catalytic-reactor systems, no solids removal is necessary as the catalyst is retained in the system and solids loss is in the form of fines that are not collected by the dust-recovery system.

Dust Separation It is usually necessary to recover the solids carried by the gas leaving the disengaging space or freeboard of the fluidized bed. Generally, cyclones are used to remove the major portion of these solids (see "Gas-Solids Separation"). However, in a few cases, usually on small-scale units, filters are employed without the use of cyclones to reduce the loading of solids in the gas. For high-temperature usage, either porous ceramic or sintered metal has been employed. Multiple units must be provided so that one unit can be blown back with clean gas while one or more are filtering.

Cyclones are arranged generally in any one of the arrangements

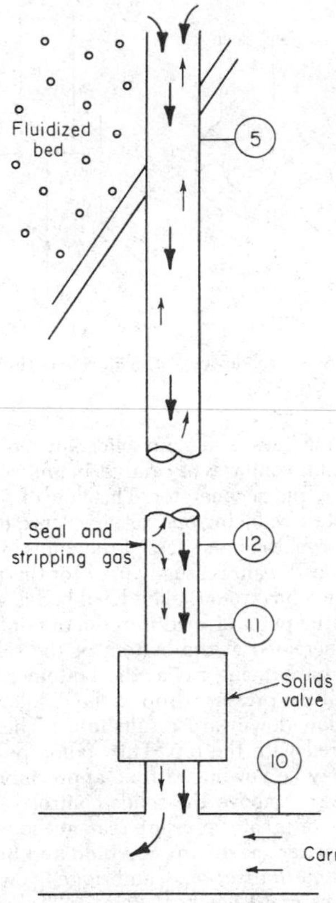

FIG. 20-88 Fluidized-bed seal leg.

shown in Fig. 20-90. The effect of cyclone arrangement on the height of the vessel and the overall height of the system is apparent. Details regarding cyclone design and collection efficiencies are to be found in another portion of this section.

Discharging of the cyclone into the fluidized bed requires some care. It is necessary to seal the bottom of the cyclone so that the collection efficiency of the cyclone will not be impaired by the passage of appreciable quantities of gas up through the solids-discharge port. This is usually done by sealing the dip leg in the fluid bed. Experience has shown, particularly in the case of deep beds, that the bottom of the dip pipe must be protected from the action of large gas bubbles which, if allowed to pass up the leg, would carry quantities of fine solids up into the cyclone and cause momentarily high losses. This can be done by attaching a plate larger in diameter than the

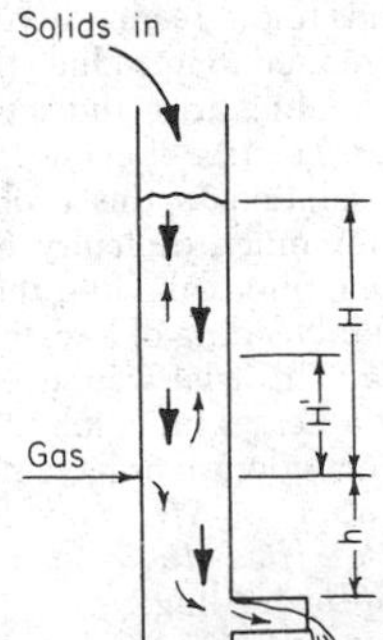

Fig. 20-89 ICI value. *(Imperial Chemical Industries, Ltd., British Patent 607,723.)*

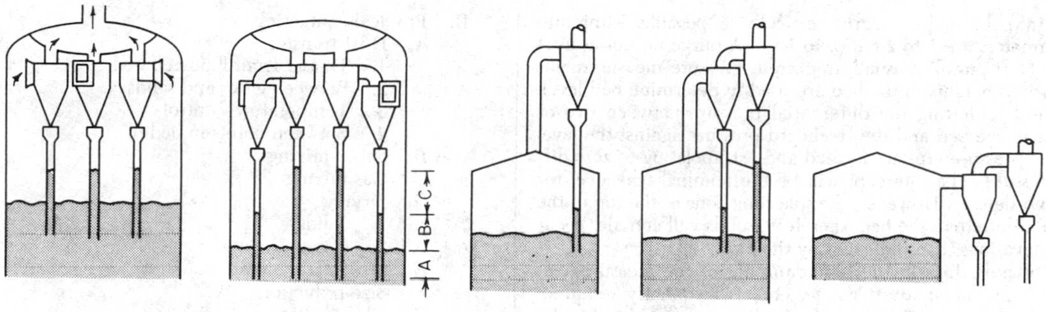

FIG. 20-90 Fluidized-bed cyclone arrangements. (*a*) Single-stage internal cyclone. (*b*) Two-stage internal cyclone. (*c*) Single-stage external cyclone; dust returned to bed. (*d*) Two-stage external cyclone; dust returned to bed. (*e*) Two-stage external cyclone; dust collected externally.

pipe to the bottom (see Fig. 20-91*e*). The length of the seal leg can be estimated as shown in the following example.

Given: Fluid density of bed at 0.3-m/s (1-ft/s) superficial gas velocity = 1100 kg/m^3 (70 lb/ft^3).

Fluid density of cyclone product at 0.15 m/s (0.5 ft/s) = 650 kg/m^3 (40 lb/ft^3).

Settled bed depth = 1.8 m (6 ft)

Fluidized-bed depth = 2.4 m (8 ft)

Pressure drop through cyclone = 1.4 kPa (0.2 lbf/in^2)

In order to assure seal at startup, the bottom of the seal leg is 1.5 m (5 ft) above the constriction plate or submerged 0.9 m (3 ft) in the fluidized bed.

The pressure at the solids outlet of a gas cyclone is usually about 0.7 kPa (0.1 lbf/in^2) lower than the pressure at the discharge of the leg. Total pressure to be balanced by the fluid leg in the cyclone dip leg is

$$(0.9 \times 1100 \times 9.81)/1000 = 1.9 \text{ m}$$
$$[(3 \times 70)/144 + 0.2 + 0.1 = 1.7 \text{ lb/in}^2]$$

Height of solids in dip leg = (11.8 × 1000)/(650 × 9.81) = 1.9 m [(1.7 × 144)/40 = 6.1 ft]; therefore, the bottom of the separator pot on the cyclone must be at least 1.9 + 1.5 or 3.4 m (6.1 + 5 or 11.1 ft) above the gas distributor. To allow for upsets, changes in size distribution, etc., use 4.6 m (15 ft).

In addition to the simple dip leg, various other devices have been used to seal cyclone solids returns, especially for second-stage cyclones. A number of these are shown in Fig. 20-91. One of the most frequently used is the flapper valve (20-91*a*). There is no general agreement as to whether this valve should discharge below the bed level or in the freeboard. All the others are discharged above the bed level. In any event, the legs must be large enough to carry momentarily high rates of solids and must provide seals to overcome cyclone pressure drops as well as to allow for differences in fluid density of bed and cyclone products. It has been reported that, in the case of catalytic-cracking catalysts, the fluid density of the solids collected by the primary cyclone is essentially the same as that in the fluidized bed. However, as a general rule the fluidized density of solids collected by the first cyclone is less than the fluidized density of the bed. Each succeeding cyclone collects finer and less dense solids. The velocity of gas up the tailpipe of a cyclone is less than the velocity in the bed, frequently being one-half of that of the bed.

As cyclones are less effective as the particle size decreases, secondary collection units are frequently required, i.e., filters, electrostatic precipitators, and scrubbers. When dry collection is not required, elimination of cyclones is possible if allowance is made for heavy solids loads in the scrubber (see "Gas-Solids Separations"; see also Sec. 18: "Scrubbers").

Instrumentation

Temperature Measurement This is usually simple, and standard temperature-sensing elements are adequate for continuous use. Because of the high abrasion wear on horizontal protection tubes, vertical installations are frequently used. In highly corrosive atmospheres in which metallic protection tubes cannot be used, short, heavy ceramic tubes have been used successfully.

Pressure Measurement Although successful pressure-measurement probes or taps have been fabricated by using porous materials, the most universally accepted pressure tap consists of a purged tube

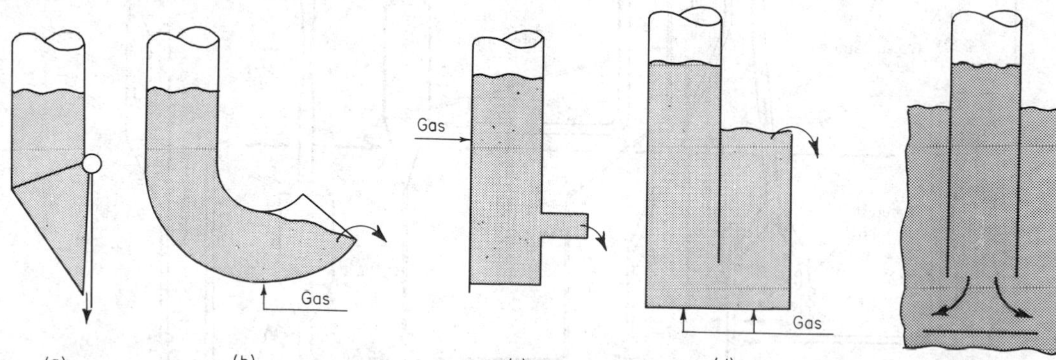

FIG. 20-91 Cyclone solids-return seals. (*a*) Flapper valve. *(Ducon Co., Inc.)* (*b*) J valve. (*c*) ICI valve. *(Imperial Chemical Industries, Ltd.; British Patent 607,723.)* (*d*) Fluid-seal pot (see Fig. 20-62). (*e*) "Dollar" plate. *a*, *b*, *c*, and *d* may be used above the bed; *a* and *e* are used below the bed.

projecting into the bed as nearly vertically as possible. Minimum internal diameters are 1 to 2 cm (½ to 1 in). A purge rate of at least 0.005 m^3/s (1 ft^3/min) is usually required. Pressure measurements taken at various heights in the bed are used to determine bed level. This is done by plotting the differential pressure between two or more points in the bed and the freeboard pressure against the level at which the pressure taps are located and extrapolating to zero differential pressure. The intercept will be the nominal bed level for constant fluid density. However, the splashing zone at the top of the bed is more dilute than the bed; therefore, solids will actually occur at greater elevations than indicated by the plot.

Because of pulsations in the bed, damping of the measurement instrument (manometer, draft gauge, etc.) is frequently required. Experienced operators can frequently predict operating troubles by observing the motion of the measuring instrument.

Flow Measurement Measurement of flow rates of clean gases presents no problem. Flow measurement of dirty gases is usually avoided. The flow of solids is usually controlled but not measured except externally to the system. Solids flows in the system are usually adjusted on an inferential basis (temperature, pressure level, catalyst activity, gas analysis, etc.). In many roasting operations the color of the calcine indicates solids feed rate.

USES OF FLUIDIZED BEDS

The possible uses of fluidized beds are manifold. A number of applications have become commercial successes; others are in the pilot-plant stage, and others in bench-scale stage. Generally, the fluidized bed is used for gas-solids contacting; however, in some instances the presence of the gas or solid is used only to provide a fluidized bed to accomplish the end result. Uses or special characteristics follow:

I. Chemical reactions
   A. Catalytic
   B. Noncatalytic
      1. Homogeneous
      2. Heterogeneous
II. Physical contacting
   A. Heat transfer
      1. To and from fluidized bed
      2. Between gases and solids
      3. Temperature control
      4. Between points in bed
   B. Solids mixing
   C. Gas mixing
   D. Drying
      1. Solids
      2. Gases
   E. Size enlargement
   F. Size reduction
   G. Classification
      1. Removal of fines from solids
      2. Removal of fines from gas
   H. Adsorption-desorption
   I. Heat treatment
   J. Coating

Chemical Reactions

Catalytic Reactions This use has provided the greatest impetus for use, development, and research in the field of fluidized solids. Some of the details pertaining to this use are to be found in the preceding pages of this section. Reference should also be made to Sec. 4.

Cracking. The evolution of fluidized catalytic cracking (FCC) since the early 1940s has resulted in several configurations depending upon the particular use and designer. One example is shown in Fig. 20-92.

The high rate of transfer of solids between the regenerator and the reactor permits a balancing of the exothermic burning of carbon and tars in the regenerator and the endothermic cracking of petroleum in the reactor, so that temperature in both units can usually be controlled without resorting to auxiliary heat-control mechanisms. The high rate of catalyst circulation also permits the maintenance of

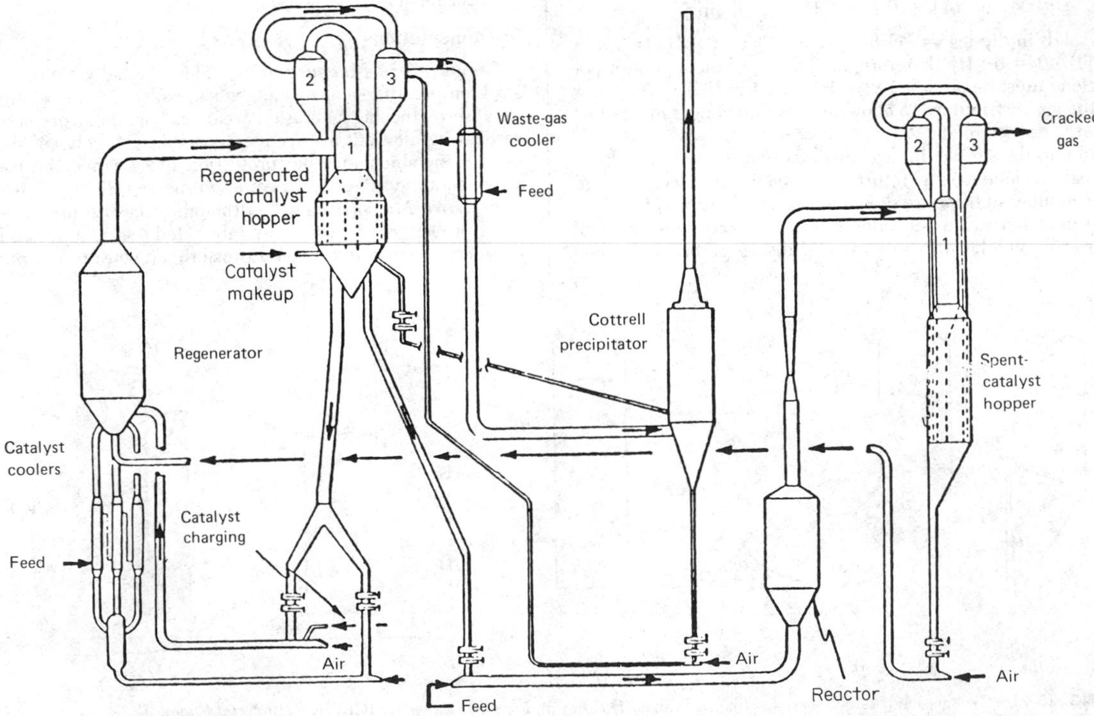

FIG. 20-92 Model I catalytic-cracking unit. [*Reprinted from Luckenbach, Reichle, Gladrow, and Worley, "Catalytic Cracking," in McKetta (ed.),* Encyclopedia of Chemical Processing and Design, *vol. 13, Marcel Dekker, New York, 1981, pp. 1–132, by courtesy Marcel Dekker, Inc.*]

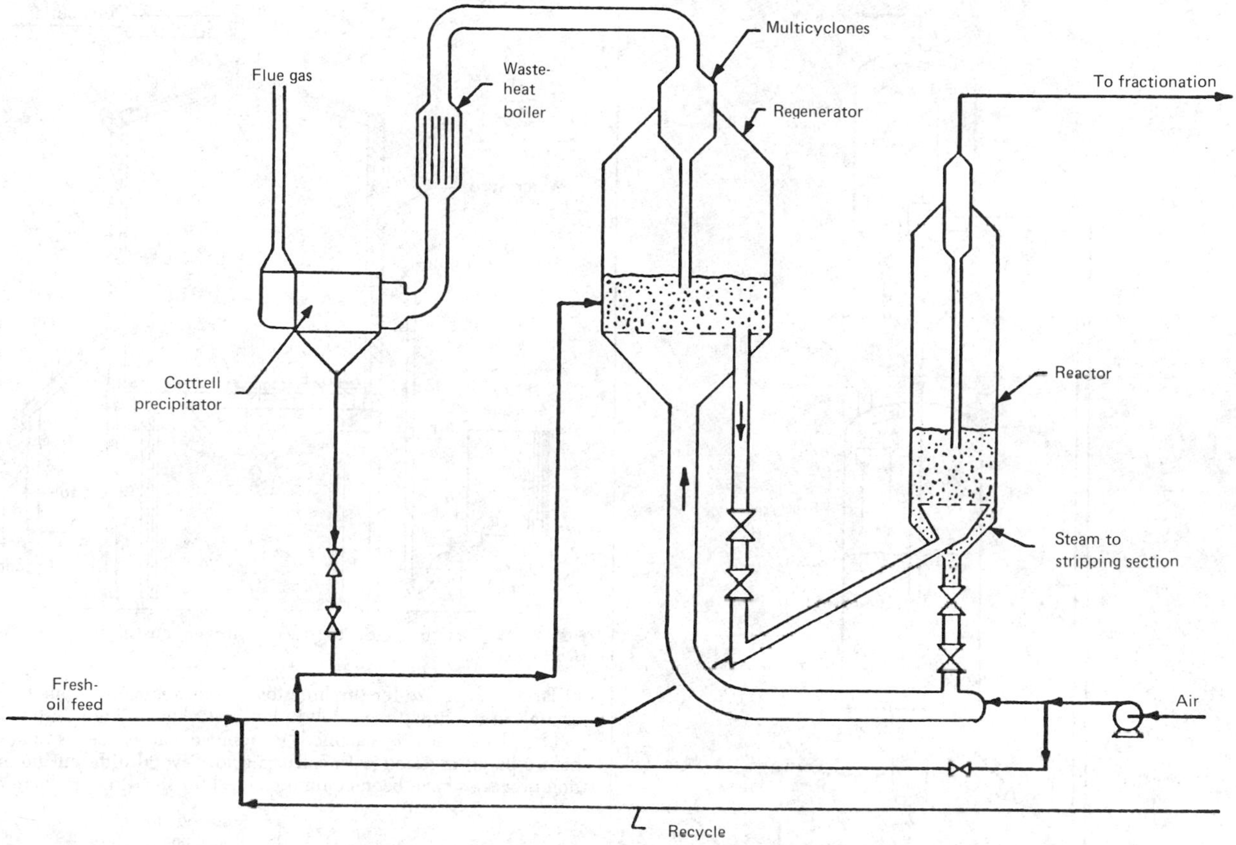

FIG. 20-93 Downflow Model II catalytic-cracking unit. [*Reprinted from Luckenbach, Reichle, Gladrow, and Worley, "Catalytic Cracking," in McKetta (ed.),* Encyclopedia of Chemical Processing and Design, *vol. 13, Marcel Dekker, New York, 1981, pp. 1–132, by courtesy Marcel Dekker, Inc.*]

the catalyst at a constantly high activity. It should be noted that the regenerator of catalytic-cracking units is generally considered to give results which agree fairly closely with those expected for a completely mixed reactor. The use of the riser reactor (transport or fast fluid bed) results in much lower gas and solids back mixing and an approach to plug flow.

The first fluid catalytic-cracking unit was placed in operation in Baytown, Texas, in 1942. This was a low-pressure, 115- to 120-kPa (2- to 3-psig) unit operating in what is now called the turbulent fluidization mode, 1.2 to 1.8 m/s (4 to 6 ft/s) (see Fig. 20-92). Even before the startup of the first model I, it was realized that by lowering the velocity, a dense, aggregative fluidized bed, 300 to 400 kg/m^3 (20 to 25 lb/ft^3), would be formed, allowing completion of reaction and regeneration. Pressure was increased to 240 to 320 kPa (20 to 30 psig) (Fig. 20-93). In the 1960s more active catalysts resulted in the use of riser cracking as shown in Fig. 20-94. Many companies participated in the development of the cat cracker, including Exxon Research & Engineering Co., Universal Oil Products Companies, Kellogg Co., Texaco Development Corp., Gulf Research Development Co., and Shell Oil Company. Many of the companies provide designs and/or licenses to operate to others. For further details, see Luckenbach, Reichle, Gladrow, and Worley, "Cracking, Catalytic," in McKetta (ed.), *Encyclopedia of Chemical Processing and Design,* vol. 13, Marcel Dekker, New York, 1981, pp. 1–132.

Alkyl chlorides. Olefins are chlorinated to alkyl chlorides in a single fluidized bed. HCl reacts with O_2 over a copper chloride catalyst to form chlorine. The chlorine reacts with the olefin to form the alkyl chloide. The process developed by the Shell Development Co. uses a recycle of catalyst fines in aqueous HCl to control the temperature [*Chem. Proc.,* **16,** 42 (1953)].

Phthalic anhydride. Naphthalene is oxidized by air to phthalic anhydride in a bubbling fluidized reactor. Even though the naphthalene feed is in liquid form, the reaction is highly exothermic. Temperature control is achieved by removing heat through vertical tubes in the bed to raise steam [Graham and Way, *Chem. Eng. Prog.,* **58,** 96 (January 1962)].

Acrylonitrile. Acrylonitrile is produced by reacting propylene, ammonia, and oxygen (air) in a single fluidized bed of a complex catalyst. Known as the SOHIO process, this process was first operated commercially in 1960. In addition to acrylonitrile, significant quantities of HCN and acetonitrile are also produced. This process is also exothermic. Temperature control is achieved by raising steam inside vertical tubes immersed in the bed [Veatch, *Hydrocarbon Process. Pet. Refiner,* **41,** 18 (November 1962)].

Fischer-Tropsch synthesis. The scale-up of a bubbling-bed reactor to produce gasoline from CO and H_2 was unsuccessful (see "Design of Fluidized-Bed Systems: Scale-Up"). However, Kellogg Co. developed a successful Fischer-Tropsch synthesis reactor based on a dilute-phase or transport-reactor concept. Kellogg, in its design, prevented gas bypassing by using the transport reactor and maintained temperature control of the exothermic reaction by inserting heat exchangers in the transport line. This process has been very successful and repeatedly expanded at the South African Synthetic Oil Limited (SASOL) plant in the Republic of South Africa, where politics and economics favor the conversion of coal to gasoline and other hydrocarbons. Refer to Jewell and Johnson, U.S. Patent 2,543,974, Mar. 6, 1951.

Additional catalytic processes. Nitrobenzene is hydrogenated to aniline (U.S. Patent 2,891,094). Melamine and isophthalonitrile are produced in catalytic fluidized-bed reactors. Badger has announced

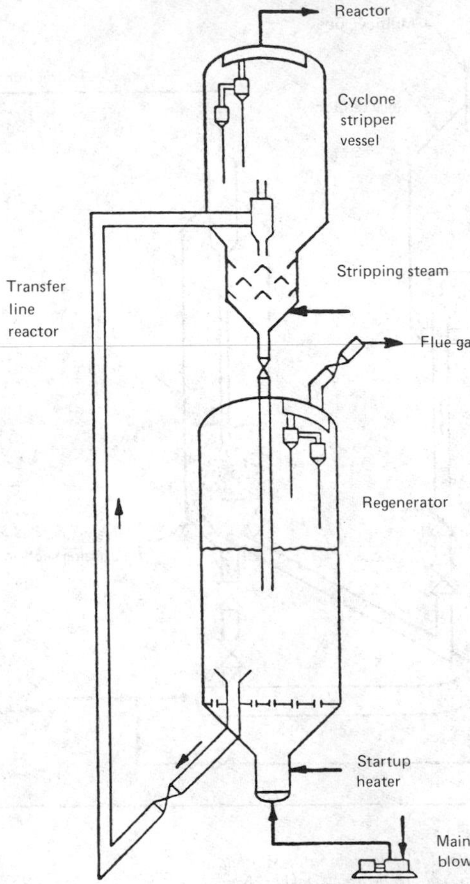

FIG. 20-94 Exxon transfer-line catalytic-cracking unit. [*Reprinted from Luckenbach, Reichle, Gladrow, and Worley, "Catalytic Cracking," in McKetta (ed.),* Encyclopedia of Chemical Processing and Design, *vol. 13, Marcel Dekker, New York, 1981, pp. 1–132, by courtesy Marcel Dekker, Inc.*]

a process to produce maleic anhydride by the partial oxidation of butane (Schaffel, Chen, and Graham, "Fluidized Bed Catalytic Oxidation of Butane to Maleic Anhydride," presented at Chemical Engineering World Congress, Montreal, 1981). Mobil Oil Corp. has announced the construction of a pilot plant to convert methanol to gasoline.

Noncatalytic Reactions

Homogeneous reactions. Homogeneous noncatalytic reactions are normally carried out in a fluidized bed to achieve mixing of the gases and temperature control. The solids of the bed act as a heat sink or source and facilitate heat transfer from or to the gas or from or to heat-exchange surfaces. Reactions of this type include chlorination of hydrocarbons or oxidation of gaseous fuels.

Heterogeneous reactions. This category covers the greatest commercial use of fluidized beds other than petroleum cracking. The **roasting of sulfide, arsenical, and/or antimonial ores** to facilitate the release of gold or silver values; the roasting of pyrite, pyrrhotite, or naturally occurring sulfur ores to provide SO_2 for sulfuric acid manufacture; and the roasting of copper, cobalt, and zinc sulfide ores to solubilize the metal values are the major metallurgical uses. Figure 20-95 shows basic items in the system.

Calcination of lime and dolomite and clay in a commercial unit has been successfully demonstrated (Fig. 20-96). Fuels are burned in a fluidized bed of the product to produce the required heat. Bunker C oil, natural gas, and coal are used in commercial units. Temperature control is accurate enough to permit production of lime of very high availability with close control of slaking characteristics. Also, half calcination of dolomite is an accepted practice. The requirement

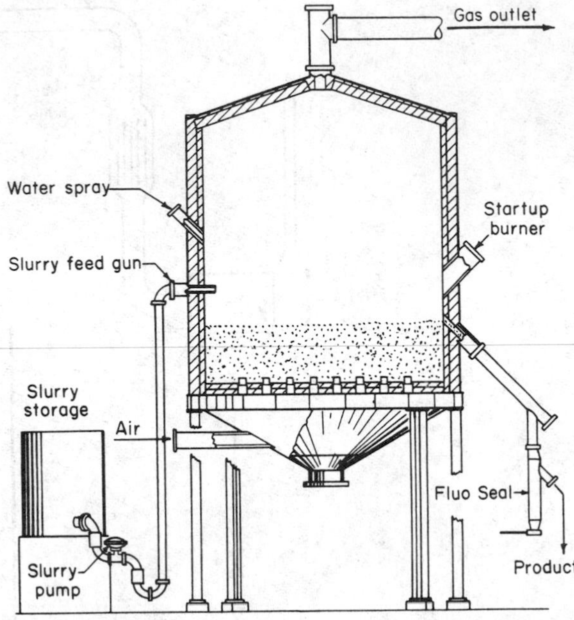

FIG. 20-95 Dorrco FluoSolids reactor; single-compartment, slurry feed. *(Dorr-Oliver Inc.)*

of large crystal size for the limestone limits application. Small-sized crystals in the limestone result in low yields due to high dust losses.

Phosphate rock is calcined to remove carbonaceous material before being digested with sulfuric acid. Several different fluidization processes have been commercialized for the direct reduction of

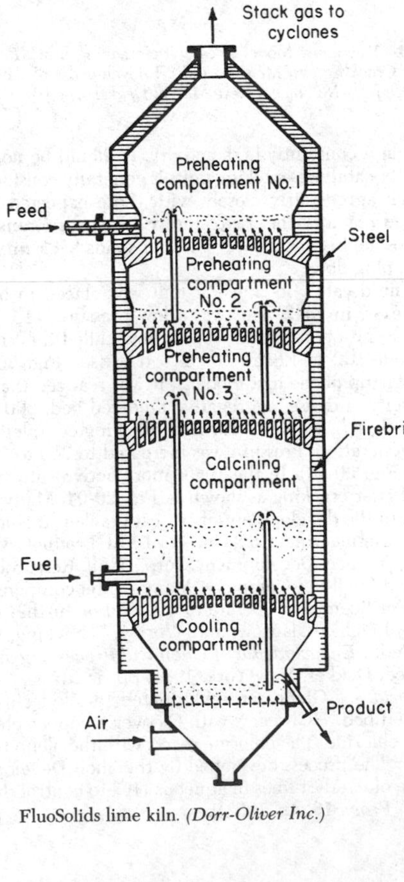

FIG. 20-96 FluoSolids lime kiln. *(Dorr-Oliver Inc.)*

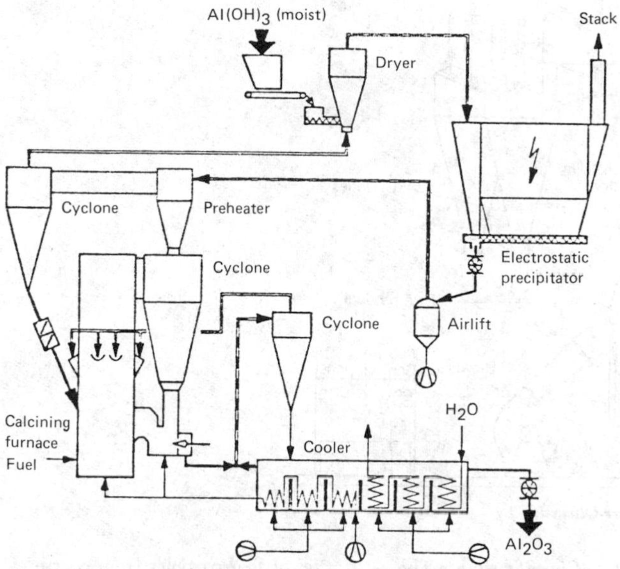

FIG. 20-97 Circulating-fluid-bed calciner. *(Lurgi Corp.)*

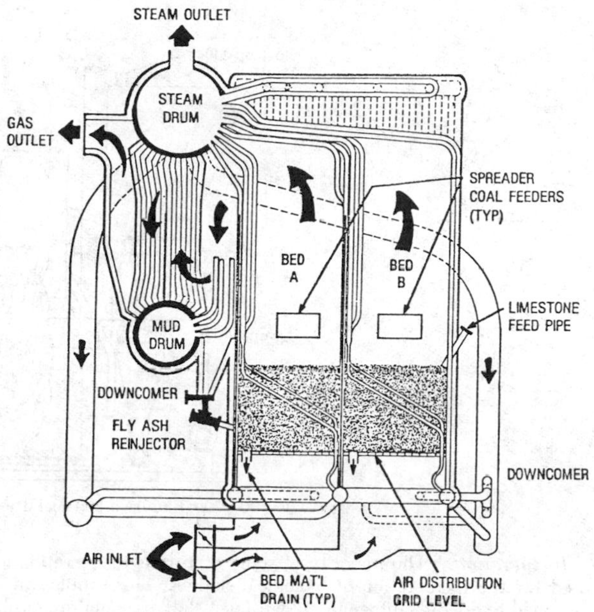

FIG. 20-98 Fluidized-bed steam generator at Georgetown University; 12.6-kg/s (100,000-lb/h) steam at 4.75-MPa (675-psig) pressure. *(From Georgetown Univ. Q. Tech. Prog. Rep. METC/DOE/10381/135, July–September 1980.)*

hematite to high-iron, low-oxide products. Foundry sand is also calcined to remove organic binders and release fines.

The calcination of $Al(OH)_3$ to Al_2O_3 in a circulating fluidized process produces a high-grade product. The process combines the use of circulating, bubbling, and transport beds to achieve high thermal efficiency. See Fig. 20-97.

An interesting feature of these high-temperature-calcination applications is the direct injection of either heavy oil, natural gas, or fine coal into the fluidized bed. Combustion takes place at well below flame temperatures without atomization. Considerable care in the design of the fuel- and air-supply system is necessary to take full advantage of the fluidized bed, which serves to mix the air and fuel.

Coal can be burned in fluidized beds in an environmentally acceptable manner by adding limestone or dolomite to the bed to react with the SO_2 to form $CaSO_4$. Because of moderate combustion temperature, about 800 to 900°C, NO_x, which results from the oxidation of nitrogen compounds contained in the coal, is kept at a low level. NO_x is increased by higher temperatures and higher excess oxygen. Two-stage air addition reduces NO_x.

Several concepts of fluidized-bed combustion have been or are being developed. **Atmospheric fluidized-bed combustion (AFBC),** in which most of the heat-exchange tubes are located in the bed, is illustrated in Fig. 20-98. **Pressurized fluidized-bed combustion (PFBC)** is, as the name implies, operated at above atmospheric pressures. The beds and heat-transfer surface are stacked to conserve space and to reduce the size of the pressure vessel. This type of unit is usually conceived as a cogeneration unit. Steam raised in the boilers would be employed to drive turbines or for other uses, and the hot pressurized gases after cleaning would be let down through an expander coupled to a compressor to supply the compressed combustion air and/or electric generator. **Circulating fluidized-bed combustors** have many advantages, the main one being the absence of heat-transfer surface in the combustion zone. This concept is being commercialized in several forms as shown in Fig. 20-99.

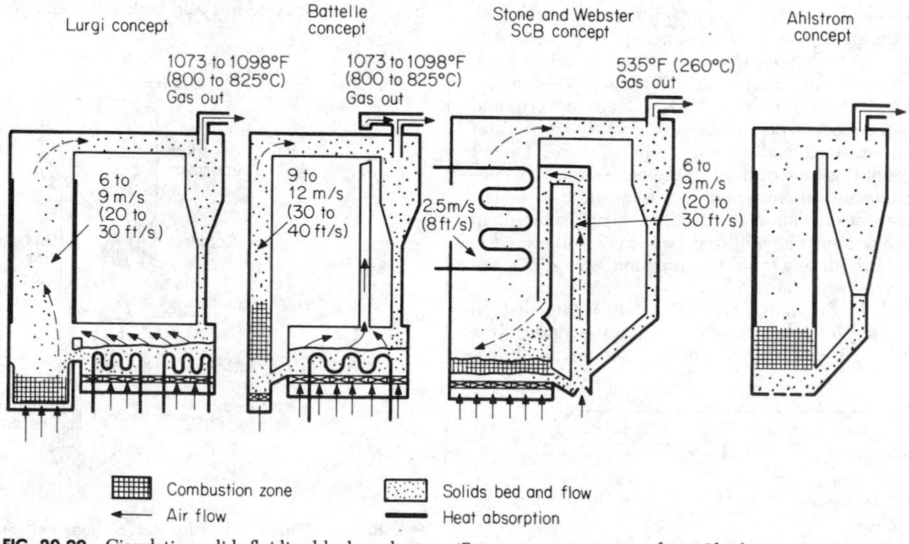

FIG. 20-99 Circulating solids fluidized-bed combustors. *(Private communication from Charles H. Rice, Conoco Inc.)*

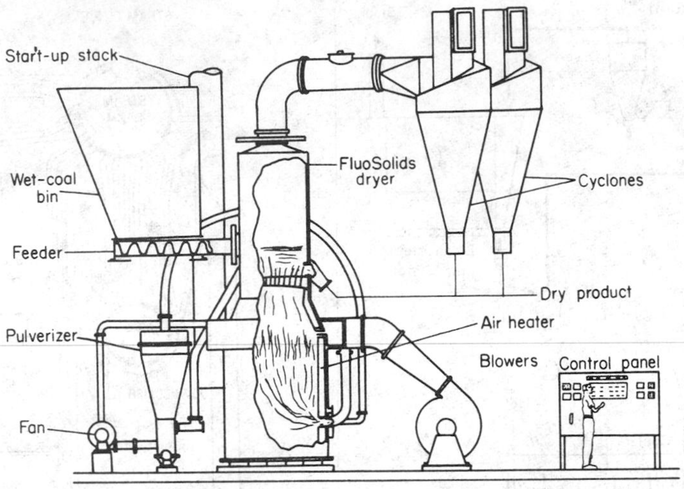

FIG. 20-100 Fluidized-bed coal dryer. *(Dorr-Oliver Inc.)*

Incineration. The majority of some 300 units in operation are used for the incineration of biological sludges. These units can be designed to operate autogenously with wet sludges containing as little as 6 MJ/kg (2600 Btu/lb) heating value. Depending on the calorific value of the feed, heat can be recovered as steam either by means of waste-heat boilers or by a combination of waste-heat boilers and the heat-exchange surface in the fluid bed.

Several units are used for sulfite-paper-mill waste-liquor disposal. At least six units are used for oil-refinery wastes, which sometimes include a mixture of liquid sludges, emulsions, and caustic waste [Flood and Kernel, *Chem. Proc.* (Sept. 8, 1973)]. Miscellaneous uses include the incineration of sawdust, carbon-black waste, pharmaceutical waste, grease from domestic sewage, spent coffee grounds, and domestic garbage.

Toxic or hazardous wastes can be disposed of in fluidized beds by either chemical capture or complete destruction. In the former case, bed material, such as limestone, will react with halides, sulfides, metals, etc., to form stable compounds which can be landfilled. Contact times of up to 5 or 10 s at 1200 K (900°C) to 1300 K (1000°C) assure complete destruction of most compounds.

Physical Contacting

Drying Fluidized-bed units for drying solids, particularly coal, cement, rock, and limestone, are in general acceptance. Economic considerations make these units particularly attractive when large tonnages of solids are to be handled (Fig. 20-100). Fuel requirements are 3.3 to 4.2 MJ/kg (1500 to 1900 Btu/lb of water removed), and total power for blowers, feeders, etc., is about 0.08 kWh/kg of water removed. The maximum-sized feed is 6 cm (1½ in) × 0 coal. One of the major advantages of this type of dryer is the close control of conditions so that a predetermined amount of free moisture may be left with the solids to prevent dusting of the product during subsequent material-handling operations. The fluidized-bed dryer is also used as a classifier so that both drying and classification operations are accomplished simultaneously.

Wall and Ash [*Ind. Eng. Chem.*, **41**, 1247 (1949)] state that, in drying 4.8-mm (−4 mesh) dolomite with combustion gases at a superficial velocity of 1.2 m/s (4 ft/s), the following removals of fines were achieved:

Particle size	% removed
− 65 + 100 mesh	60
−100 + 150 mesh	79
−150 + 200 mesh	85
−200 + 325 mesh	89
−325 mesh	89

Classification The separation of fine particles from coarse can be effected by use of a fluidized bed (see "Drying"). However, for economic reasons (**i.e.**, initial cost, power requirements for compression of fluidizing gas, etc.), it is doubtful except in special cases if a fluidized-bed classifier would be built for this purpose alone.

It has been proposed that fluidized beds be used to remove fine solids from a gas stream. This is possible under special conditions.

Adsorption-Desorption An arrangement for gas fractionation is shown in Fig. 20-101.

The effects of adsorption and desorption on the performance of fluidized beds are discussed under "Catalytic Reactions." Adsorption of carbon disulfide vapors from air streams as great as 300 m^3/s (540,000 ft^3/min) in a 17-m- (53-ft-) diameter unit has been reported by Avery and Tracey ("The Application of Fluidized Beds of Activated Carbon to Recover Solvent from Air or Gas Streams," Tripartate Chemical Engineering Conference, Montreal, Sept. 24, 1968).

Heat Treatment Heat treatment can be divided into two types, treatment of fluidizable solids and treatment of large, usually metallic objects in a fluid bed. The former is generally accomplished in multicompartment units to conserve heat (Fig. 20-96). The heat treatment of large metallic objects is accomplished in long, narrow heated beds. The objects are conveyed through the beds by an overhead conveyor system. Fluid beds are used because of the high heat-

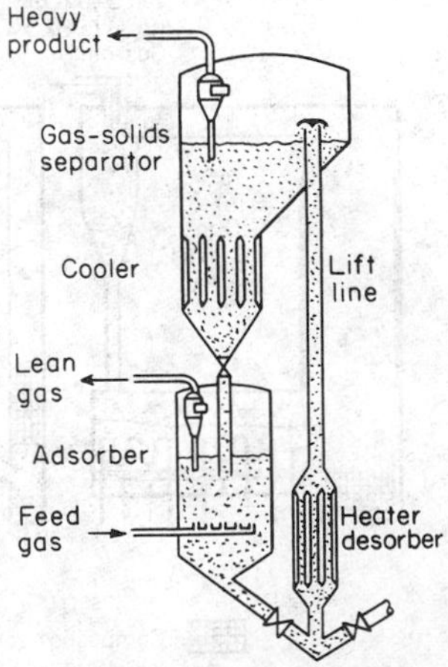

FIG. 20-101 Fluidized bed for gas fractionation. [*Sittig,* Chem. Eng. *(May 1953).*]

transfer rate and uniform temperature. See Reindl, "Fluid Bed Technology," *American Society for Metals*, Cincinnati, Sept. 23, 1981; Fennell, *Ind. Heat.*, **48**, 9, 36 (September 1981).

Coating Fluidized beds of thermoplastic resins have been used to facilitate the coating of metallic parts. A properly prepared, heated metal part is dipped into the fluidized bed, which permits complete immersion in the dry solids. The heated metal fuses the thermoplastic, forming a continuous uniform coating.

GAS-SOLIDS SEPARATIONS

This subsection is concerned with the application of particle mechanics (see Sec. 5, "Fluid and Particle Mechanics") to the design and application of dust-collection systems. It includes wet collectors, or scrubbers, for particle collection. Scrubbers designed for purposes of mass transfer are discussed in Secs. 14 and 18. Equipment for removing entrained liquid mist from gases is described in Sec. 18.

Nomenclature

Except where otherwise noted here or in the text, either consistent system of units (SI or U.S. customary) may be used. Only SI units may be used for electrical quantities, since no comparable electrical units exist in the U.S. customary system. When special units are used, they are noted at the point of use.

Symbols	Definition	SI units	U.S. customary units	Special units
A_c	Cyclone inlet area = B_cH_c for cyclone with rectangular inlet	m^2	ft^2	
A_e	Area of collecting electrode (side on which particles collect, only)	m^2	ft^2	
B_c	Width of rectangular cyclone inlet duct	m	ft	
B_e	Spacing between wire and plate, or between rod and curtain, or between parallel plates in electrical precipitators	m	ft	
B_s	Width of gravity settling chamber	m	ft	
c_d	Dust concentration in gas stream	g/m^3		gr/ft^3
c_h	Specific heat of gas	J/(kg·K)	Btu/(lbm·°F)	
c_{hb}	Specific heat of collecting body	J/(kg·K)	Btu/(lbm·°F)	
c_{hp}	Specific heat of particle	J/(kg·K)	Btu/(lbm·°F)	
D_b	Diameter or other representative dimension of collector body or device	m	ft	
D_{b1}, D_{b2}	Other characteristic dimensions of collector body or device	m	ft	
D_c	Cyclone diameter	m	ft	
D_d	Outside diameter of wire or discharge electrode of concentric-cylinder type of electrical precipitator	m	ft	
D_e	Diameter of cyclone gas exit duct	m	ft	
D_o	Volume/surface-mean-drop diameter			μm
D_p	Diameter of particle	m	ft	μm
D_{pc}	Cut diameter, diameter of particles of which 50% of those present are collected	m	ft	μm
$D_{p,min}$	Minimum diameter of particle that is completely collected	m	ft	μm
D_t	Inside diameter of collecting tube of concentric-cylinder type of electrical precipitator	m	ft	
D_v	Diffusion coefficient for particle	m^2/s	ft^2/s	

Symbols	Definition	SI units	U.S. customary units	Special units
DF	Decontamination factor = $1/(1 - \eta)$	Dimensionless	Dimensionless	
DI	Decontamination index = $\log_{10} [1/(1 - \eta)]$	Dimensionless	Dimensionless	
e	Natural (napierian) logarithmic base	2.718 . . .	2.718 . . .	
E	Potential difference	V		
E_c	Potential difference required for corona discharge to commence	V		
E_d	Voltage across dust layer	V		
E_s	Potential difference required for sparking to commence	V		
F_{cv}	Cyclone friction loss, expressed as number of cyclone-inlet-velocity heads, based on area A_c	Dimensionless	Dimensionless	
F_E	Effective friction loss across wetted equipment in scrubber	kPa		in water
g_c	Conversion factor		32.17 (lbm/lbf)(ft/s^2)	
g_L	Local acceleration due to gravity	m/s^2	ft/s^2	
h_{vi}	Cyclone inlet velocity head			in water
H_c	Height of rectangular cyclone inlet duct	m	ft	
H_s	Height of gravity settling chamber	m	ft	
I	Electrical current per unit of electrode length	A/m		
j	Corona current density at dust layer	A/m^2		
k_ρ	Density of gas relative to its density at 0°C, 1 atm	Dimensionless	Dimensionless	Dimensionless
k_t	Thermal conductivity of gas	W/(m·K)	Btu/(s·ft·°F)	
k_{tb}	Thermal conductivity of collecting body	W/(m·K)	Btu/(s·ft·°F)	
k_{tp}	Thermal conductivity of particle	W/(m·K)	Btu/(s·ft·°F)	
K	Empirical proportionality constant for cyclone pressure drop or friction loss	Dimensionless	Dimensionless	
K_1	Resistance coefficient of "conditioned" filter fabric	kPa/(m/min)		in water/(ft/min)
K_2	Resistance coefficient of dust cake on filter fabric	$\frac{kPa}{(m/min)(g/m^2)}$	$\frac{in\ water}{(ft/min)(lbm/ft^2)}$	
K_α	Proportionality constant, for target efficiency of a single fiber in a bed of fibers	Dimensionless	Dimensionless	
K_c	Resistance coefficient for "conditioned" filter fabric			$\frac{in\ water}{(ft/min)(cP)}$
K_d	Resistance coefficient for dust cake on filter fabric			$\frac{in\ water}{(ft/min)(gr/ft^2)(cP)}$
K_e	Electrical-precipitator constant	s/m	s/ft	

Symbols	Definition	SI units	U.S. customary units	Special units
K_F	Resistance coefficient for clean filter cloth			in water / (ft/min)(cP)
K_o	"Energy-distance" constant for electrical discharge in gases	m		
K_m	Stokes-Cunningham correction factor	Dimensionless	Dimensionless	Dimensionless
K_s	Proportionality factor in "slip-flow" correction factor	Dimensionless	Dimensionless	Dimensionless
L	Thickness of fibrous filter or of dust layer on surface filter	m	ft	
L_e	Length of collecting electrode in direction of gas flow	m	ft	
L_s	Length of gravity settling chamber in direction of gas flow	m	ft	
ln	Natural logarithm (logarithm to the base e)	Dimensionless	Dimensionless	Dimensionless
M	Molecular weight	kg/mol	lbm/mol	
n	Exponent	Dimensionless	Dimensionless	Dimensionless
N	Number of gas molecules in a mole	6.06×10^{26} Molecules/(kg·mol)	2.76×10^{26} Molecules/(lb·mol)	
N_e	"Effective" number of turns made by a gas stream in a cyclone separator	Dimensionless	Dimensionless	Dimensionless
N_{Kn}	Knudsen number = λ_m/D_b	Dimensionless	Dimensionless	
N_{Ma}	Mach number	Dimensionless	Dimensionless	
N_o	Number of elementary electrical charges acquired by a particle	Dimensionless	Dimensionless	
N_{Pr}	Prandtl number $= c_h\mu/k_t$	Dimensionless	Dimensionless	
N_{Re}	Reynolds number = $(D_p\rho V_o/\mu)$ or $(D_p\rho u_t/\mu)$	Dimensionless	Dimensionless	
N_{sc}	Interaction number = $18\ \mu/K_m\ \rho_p\ D_v$	Dimensionless	Dimensionless	
N_{sd}	Diffusional separation number	Dimensionless	Dimensionless	
N_{sec}	Electrostatic-attraction separation number	Dimensionless	Dimensionless	
N_{sei}	Electrostatic-induction separation number	Dimensionless	Dimensionless	
N_{sf}	Flow-line separation number	Dimensionless	Dimensionless	
N_{sg}	Gravitational separation number	Dimensionless	Dimensionless	
N_{si}	Inertial separation number	Dimensionless	Dimensionless	
N_{sic}	Modified inertial separation number for centrifugal field in cyclone	Dimensionless	Dimensionless	
N_{st}	Thermal separation number	Dimensionless	Dimensionless	
N_t	Number of transfer units = $\ln[1/(1-\eta)]$	Dimensionless	Dimensionless	
N_{tc}	Number of turns made by gas stream in a cyclone separator	Dimensionless	Dimensionless	
Δp	Gas pressure drop	kPa	lbf/ft^2	in water
Δp_i	Gas pressure drop in cyclone or filter			in water
Δp_{cv}	Pressure drop, expressed as number of cyclone inlet velocity heads, based on area A_c	Dimensionless	Dimensionless	Dimensionless
p_F	Gauge pressure of water fed to scrubber	kPa		lbf/in^2
P_G	Gas-phase contacting power	MJ/1000 m^3		hp/(1000 ft^3/min)
P_L	Liquid-phase contacting power	MJ/1000 m^3		hp/(1000 ft^3/min)

Symbols	Definition	SI units	U.S. customary units	Special units
P_M	Mechanical contacting power	MJ/1000 m^3		hp/(1000 ft^3/min)
P_T	Total contacting power	MJ/1000 m^3		hp/(1000 ft^3/min)
q	Gas flow rate	m^3/s	ft^3/s	
Q_G	Gas flow rate		ft^3/s	ft^3/min
Q_L	Liquid flow rate		ft^3/s	gal/min
Q_p	Electrical charge on particle	C		
r	Radius; distance from centerline of cyclone separator; distance from centerline of concentric-cylinder electrical precipitator	m	ft	
R	Gas constant	8.3143 J/(mol·K)	$1546 \dfrac{(\text{ft})(\text{lbf})}{(\text{lb·mol})(°\text{F})}$	
Time		s	s	
t_m	Time			min
T	Absolute gas temperature	K	°R	
T_b	Absolute temperature of collecting body	K	°R	
T_p	Absolute temperature of particle	K	°R	
u_e	Velocity of migration of particle toward collecting electrode	m/s	ft/s	
u_t	Terminal settling velocity of particle under action of gravity	m/s	ft/s	ft/s
V_c	Average cyclone inlet velocity, based on area A_c	m/s	ft/s	ft/s
V_e	Average velocity of gas flowing through electrical precipitator	m/s	ft/s	
V_f	Filtration velocity (superficial gas velocity through filter)	m/min		ft/min
V_o	Gas velocity	m/s	ft/s	
V_s	Average gas velocity in gravity settling	m/s	ft/s	
V_{ct}	Tangential component of gas velocity in cyclone	m/s	ft/s	
w	Loading of collected dust on filter	g/m^2	lbm/ft^2	gr/ft^2
	Greek symbols			
α	Empirical constant in equation of scrubber performance curve	$\left[\dfrac{\text{MJ}}{1000\ \text{m}^3}\right]^{-\gamma}$		$\left[\dfrac{\text{hp}}{(100\ \text{ft}^3/\text{min})}\right]^{-\gamma}$
γ	Empirical constant in equation of scrubber performance curve	Dimensionless		Dimensionless
δ	Dielectric constant	Dimensionless		
δ_g	Dielectric constant at 0°C, 1 atm	Dimensionless		
δ_o	Permittivity of free space	F/m		
δ_b	Dielectric constant of collecting body	Dimensionless		
δ_p	Dielectric constant of particle	Dimensionless		
Δ	Fractional free area (for screens, perforated plates, grids)	Dimensionless	Dimensionless	
ϵ	Elementary electrical charge	1.60210×10^{-19} C		
ϵ_b	Characteristics potential gradient at collecting surface	V/m		
ϵ_v	Fraction voids in bed of solids	Dimensionless	Dimensionless	Dimensionless
ζ	$= 1 + 2\dfrac{(\delta-1)}{(\delta+2)}$; ranges from a value of 1 for materials with a dielectric constant of 1 to 3 for conductors	Dimensionless		

Symbols	Definition	SI units	U.S. customary units	Special units
η	Collection efficiency, weight fraction of entering dispersoid collected	Dimensionless	Dimensionless	Dimensionless
η_o	Target efficiency of an isolated collecting body, fraction of dispersoid in swept volume collected on body	Dimensionless	Dimensionless	Dimensionless
η_t	Target efficiency of a single collecting body in an array of collecting bodies, fraction of dispersoid in swept volume collected on body	Dimensionless	Dimensionless	Dimensionless
λ_i	Ionic mobility of gas	(m/s)/(V/m)		
λ_m	Mean free path of gas molecules	m	ft	
λ_p	Particle mobility = $u_e/\mathcal{E}$	(m/s)/(V/m)		
μ	Gas viscosity	Pa·s	lbm/(s·ft)	cP
μ_L	Liquid viscosity			cP
ρ_d	Resistivity of dust layer	Ω · m		
ρ_L	Liquid density		lbm/ft^3	lbm/ft^3
ρ_s	True (*not* bulk) density of solids or liquid drops	kg/m^3	lbm/ft^3	lbm/ft^3
ρ'	Density of gas relative to its density at 25°C, 1 atm	Dimensionless	Dimensionless	Dimensionless
σ	Ion density	Number/m^3		
σ_{avg}	Average ion density	Number/m^3		
σ_L	Liquid surface tension			dyn/cm
ϕ	Cumulative weight fraction larger than size	Dimensionless	Dimensionless	Dimensionless
ϕ_s	Particle shape factor = (surface of sphere)/(surface of particle of same volume)	Dimensionless	Dimensionless	Dimensionless
	Script symbols			
$\mathcal{E}$	Potential gradient	V/m		
$\mathcal{E}_c$	Potential gradient required for corona discharge to commence	V/m		
$\mathcal{E}_i$	Average potential gradient in ionization stage	V/m		
$\mathcal{E}_o$	Electrical breakdown constant for gas	V/m		
$\mathcal{E}_p$	Average potential gradient in collection stage	V/m		
$\mathcal{E}_s$	Potential gradient required for sparking to commence	V/m		

GENERAL REFERENCES: Burchsted, Kahn, and Fuller, *Nuclear Air Cleaning Handbook*, ERDA 76-21, Oak Ridge, Tenn., 1976. Cadle, *The Measurement of Airborne Particles*, Wiley, New York, 1975. Davies, *Aerosol Science*, Academic, New York, 1966. Davies, *Air Filtration*, Academic, New York, 1973. Dennis, *Handbook on Aerosols*, ERDA TID-26608, Oak Ridge, Tenn., 1976. Drinker and Hatch, *Industrial Dust*, 2d ed., McGraw-Hill, New York, 1954. Friedlander, *Smoke, Dust, and Haze*, Wiley, New York, 1977. Fuchs, *The Mechanics of Aerosols*, Pergamon, Oxford, 1964. Green and Lane, *Particulate Clouds: Dusts, Smokes, and Mists*, Van Nostrand, New York, 1964. Lapple, *Fluid and Particle Mechanics*, University of Delaware, Newark, 1951. Licht, *Air Pollution Control Engineering—Basic Calculations for Particle Collection*, Marcel Dekker, New York, 1980. Liu, *Fine Particles—Aerosol Generation, Measurement, Sampling, and Analysis*, Academic, New York, 1976. Lunde and Lapple, *Chem. Eng. Prog.*, **53**, 385 (1957). Lundgren et al., *Aerosol Measurement*, University of Florida, Gainesville, 1979. Mercer, *Aerosol Technology in Hazard Evaluation*, Academic, New York, 1973. Nonhebel, *Processes for Air Pollution Control*, CRC Press, Cleveland, 1972. Shaw, *Fundamentals of Aerosol Science*, Wiley, New York, 1978. Stern, *Air Pollution: A Comprehensive Treatise*, vols. 3 and 4, Academic, New York, 1977. Strauss, *Industrial Gas Cleaning*, 2d ed., Pergamon, New York, 1975. Theodore and Buonicore, *Air Pollution Control Equipment: Selection, Design, Operation, and Maintenance*, Prentice-Hall, Englewood Cliffs, N.J., 1982. White, *Industrial Electrostatic Precipitation*, Addison-Wesley, Reading, Mass., 1963. White and Smith, *High-Efficiency Air Filtration*, Butterworth, Washington, 1964.

PURPOSE OF DUST COLLECTION

Dust collection is concerned with the removal or collection of solid dispersoids in gases for purposes of:

1. Air-pollution control, as in fly-ash removal from power-plant flue gases
2. Equipment-maintenance reduction, as in filtration of engine-intake air or pyrites furnace-gas treatment prior to its entry to a contact sulfuric acid plant
3. Safety- or health-hazard elimination, as in collection of siliceous and metallic dusts around grinding and drilling equipment and in some metallurgical operations and flour dusts from milling or bagging operations
4. Product-quality improvement, as in air cleaning in the production of pharmaceutical products and photographic film
5. Recovery of a valuable product, as in collection of dusts from dryers and smelters
6. Powdered-product collection, as in pneumatic conveying; the spray drying of milk, eggs, and soap; and the manufacture of high-purity zinc oxide and carbon black

PROPERTIES OF PARTICLE DISPERSOIDS

An understanding of the fundamental properties and characteristics of gas dispersoids is essential to the design of industrial dust-control equipment. Figure 20-102 shows characteristics of dispersoids and other particles together with the types of gas-cleaning equipment that are applicable to their control. Two types of solid dispersoids are shown: (1) dust, which is composed of particles larger than 1 μm; and (2) fume, which consists of particles generally smaller than 1 μm. Dusts usually result from mechanical disintegration of matter. They may be redispersed from the settled, or bulk, condition by an air blast. Fumes are submicrometer dispersoids formed by processes such as combustion, sublimation, and condensation. Once collected, they cannot be redispersed from the settled condition to their original state of dispersion by air blasts or mechanical dispersion equipment.

The primary distinguishing characteristic of gas dispersoids is particle size. The generally accepted unit of particle size is the micrometer, μm. (Prior to the adoption of the SI system, the same unit was known as the micron and was designated by μ.) The particle size of a gas dispersoid is usually taken as the diameter of a sphere having the same mass and density as the particle in question. Some writers occasionally specify particle size by the radius, particularly in the older European literature. Another common method is to designate the screen mesh that has an aperture corresponding to the particle diameter; the screen scale used must also be specified to avoid confusion.

From the standpoint of collector design and performance, the most important size-related property of a dust particle is its dynamic behavior. Particles larger than 100 μm are readily collectible by simple inertial or gravitational methods. For particles under 100 μm, the range of principal difficulty in dust collection, the resistance to motion in a gas is viscous (see Sec. 5: "Particle Dynamics"), and for such particles, the most useful size specification is commonly the Stokes settling diameter, which is the diameter of the spherical particle of the same density that has the same terminal velocity in viscous flow as the particle in question. It is yet more convenient in many circumstances to use the "aerodynamic diameter," which is the diameter of the particle of unit density (1 g/cm^3) that has the same terminal settling velocity. Use of the aerodynamic diameter permits direct comparisons of the dynamic behavior of particles that are actually of different sizes, shapes, and densities [Raabe, *J. Air Pollut. Control Assoc.*, **26**, 856 (1976)].

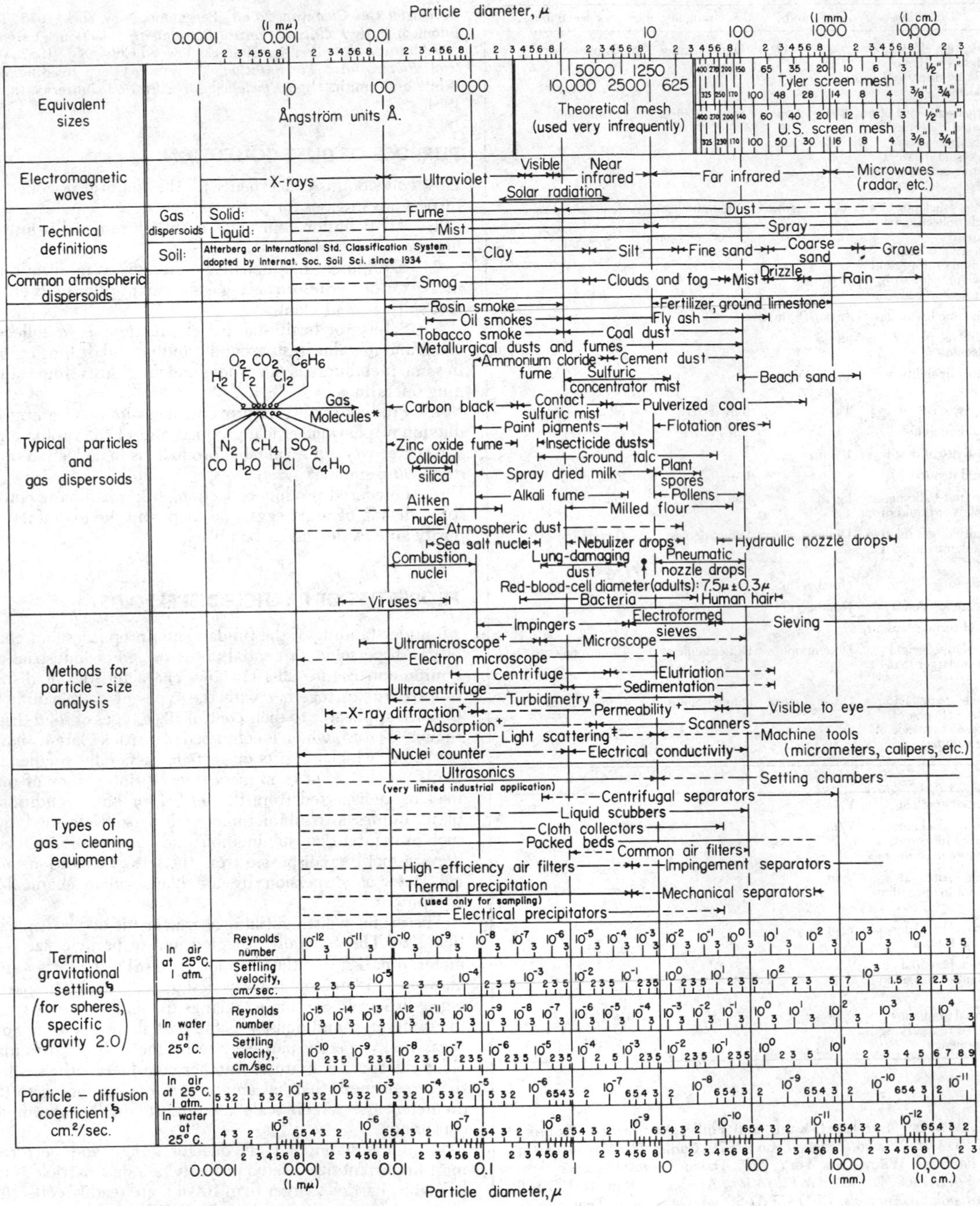

* Molecular diameters calculated from viscosity data at 0°C.

+ Furnishes average particle diameter but no size distribution.

‡ Size distribution may be obtained by special calibration.

§ Stokes – Cunningham factor included in values given for air but not included for water.

FIG. 20-102 Characteristics of particles and particle dispersoids. *(Courtesy of the Stanford Research Institute; prepared by C. E. Lapple.)*

When the size of a particle approaches the same order of magnitude as the mean free path of the gas molecules, the settling velocity is greater than predicted by Stokes' law because of molecular slip. The slip-flow correction is appreciable for particles smaller than 1 μm and is allowed for by the Cunningham correction for Stokes' law (Lapple, op. cit.; Licht, op. cit.). The Cunningham correction is applied in calculations of the aerodynamic diameters of particles that are in the appropriate size range.

Although solid fume particles may range in size down to perhaps 0.001 μm, fine particles effectively smaller than about 0.1 μm are not of much significance in industrial dust and fume sources because their aggregate mass is only a very small fraction of the total mass emission. At the concentrations present in such sources (e.g., production of carbon black) the coagulation, or flocculation, rate of the ultrafine particles is extremely high, and the particles speedily grow to sizes of 0.1 μm or greater. The most difficult collection problems are thus concerned with particles in the range of about 0.1 to 2 μm, in which forces for deposition by inertia are small. For collection of particles under 0.1 μm, diffusional deposition becomes increasingly important as the particle size decreases.

In a gas stream carrying dust or fume, some degree of particle flocculation will exist, so that both discrete particles and clusters of adhering particles will be present. The discrete particles composing the clusters may be only loosely attached to each other, as by van der Waals forces [Lapple, *Chem. Eng.*, **75** (11), 149 (1968)]. Flocculation tends to increase with increases in particle concentration and may strongly influence collector performance.

PARTICLE MEASUREMENTS

Measurements of the concentrations and characteristics of dust dispersed in air or other gases may be necessary (1) to determine the need for control measures, (2) to establish compliance with legal requirements, (3) to obtain information for collector design, and (4) to determine collector performance.

Atmospheric-Pollution Measurements The dust-fall measurement is one of the common methods for obtaining a relative long-period evaluation of particulate air pollution. Stack-smoke densities are often graded visually by means of the Ringelmann chart. Plume opacity may be continuously monitored and recorded by a photoelectric device which measures the amount of light transmitted through a stack plume. Equipment for local atmospheric-dust-concentration measurements fall into five general types: (1) the impinger, (2) the hot-wire or thermal precipitator, (3) the electrostatic precipitator, (4) the filter, and (5) impactors and cyclones. The filter is the most widely used, in the form of either a continuous tape, or a number of filter disks arranged in an automatic sequencing device, or a single, short-term, high-volume sampler. Samplers such as these are commonly used to obtain mass emission and particle-size distribution. Impactors and small cyclones are commonly used as size-discriminating samplers and are usually followed by filters for the determination of the finest fraction of the dust (Lundgren et al., *Aerosol Measurement*, University of Florida, Gainesville, 1979; and Dennis, *Handbook on Aerosols*, U.S. ERDA TID-26608, Oak Ridge, Tenn., 1976).

Process-Gas Sampling In sampling process gases either to determine dust concentration or to obtain a representative dust sample, it is necessary to take special precautions to avoid inertial segregation of the particles. To prevent such classification, a traverse of the duct may be required, and at each point the sampling nozzle must face directly into the gas stream with the velocity in the mouth of the nozzle equal to the local gas velocity at that point. This is called "isokinetic sampling." If the sampling velocity is too high, the dust sample will contain a lower concentration of dust than the mainstream, with a greater percentage of fine particles; if the sampling velocity is too low, the dust sample will contain a higher concentration of dust with a greater percentage of coarse particles [Lapple, *Heat. Piping Air Cond.*, **16**, 578 (1944); *Manual of Disposal of Refinery Wastes*, vol. V, American Petroleum Institute, New York, 1954; and Dennis, op. cit.].

TABLE 20-30 Examples of Size-Analysis Methods and Equipment*

Particle size, μ	General method	Examples† of specific instruments
37 and larger	Dry-sieve analysis	Tyler Ro-Tap, Alpine Jet sieve
10 and larger	Wet-sieve analysis	Buckbee-Mears sieves
1–100	Optical microscope	Zeiss, Bausch & Lomb, Nikon microscopes
	Microscope with scanner and counter	Millipore IIMC system
	Dry gravity sedimentation	Roller analyzer, Sharples Micromerograph
	Wet gravity sedimentation	Andreasen pipet
	Electrolyte resistivity change	Coulter counter
0.2–20	Light scattering	Royco
	Cascade impactor	Brink, Anderson, Casella, Lundgren impactors
	Wet centrifugal sedimentation	M.S.A.-Whitby analyzer
0.01–10	Ultracentrifuge	Goetz aerosol spectrometer
	Transmission electron microscope	Phillips, RCA, Hitachi, Zeiss, Metropolitan-Vickers, Siemens microscopes
	Scanning electron microscope	Reist & Burgess‡ system

*Krockta and Lucas, *Air Pollution Control Assoc. J.*, **22**, 461 (1972).

†This table gives examples of specific equipment. It is not intended to be a complete listing, nor is it intended as an endorsement of any instrument.

‡Reist and Burgess, "Development of an Automatic Particle Assaying Instrument Utilizing a Scanning Microscope," Paper No. 69–124, Air Pollution Control Association Meeting, New York, 1969.

Particle-Size Analysis Methods for particle-size analysis are shown in Fig. 20-102, and examples of size-analysis methods are given in Table 20-30. More detailed information may be found in the following references: Irani and Callis, *Particle Size: Measurement, Interpretation, and Application*, Wiley, New York, 1963; Dallavalle and Orr, *Fine Particle Measurement: Size, Surface and Pore Volume*, Macmillan, New York, 1959; Green and Lane, *Particulate Clouds: Dusts, Smokes, and Mists*, 2d ed., Van Nostrand, New York, 1964; Lapple, *Chem. Eng.*, **75**(11), 140 (1968); Lapple, "Particle-Size Analysis," in *Encyclopedia of Science and Technology*, 5th ed., McGraw-Hill, New York, 1982; and Cadle, *The Measurement of Airborne Particles*, Wiley, New York, 1975. Particle-size distribution may be presented on either a frequency or a cumulative basis; the various methods are discussed in the references just cited. The most common method presents a plot of particle size versus the cumulative weight percent of material larger or smaller than the indicated size, on logarithmic-probability graph paper.

For determination of the aerodynamic diameters of particles, the most commonly applicable methods for particle-size analysis are those based on inertia: aerosol centrifuges, cyclones, and inertial impactors (Lundgren et al., *Aerosol Measurement*, University of Florida, Gainesville, 1979; and Liu, *Fine Particles—Aerosol Generation, Measurement, Sampling, and Analysis*, Academic, New York, 1976). Impactors are the most commonly used. Nevertheless, impactor measurements are subject to numerous errors [Rao and Whitby, *Am. Ind. Hyg. Assoc. J.*, **38**, 174 (1977); Marple and Willeke, "Inertial Impactors," in Lundgren et al., *Aerosol Measurement;* and Fuchs, "Aerosol Impactors," in Shaw, *Fundamentals of Aerosol Science*, Wiley, New York, 1978]. Reentrainment due to particle bouncing and blowoff of deposited particles makes a dust appear finer than it actually is, as does the breakup of flocculated particles. Processing cascade-impactor data also presents possibilities for substantial errors (Fuchs, *The Mechanics of Aerosols*, Pergamon, Oxford, 1964) and is laborious as well. Lawless (Rep. No. EPA-600/7-78-189, U.S. EPA, 1978) discusses problems in analyzing and fitting

TABLE 20-31 Summary of Mechanisms and Parameters in Aerosol Deposition

Deposition	Origin of force field	Deposition mechanism measureable in terms of: Basic parameter	Deposition mechanism measureable in terms of: Specific modifying parameters	System parameters
Flow-line interception*	Physical gradient*	$N_{sf} = \left(\frac{D_p}{D_b}\right)$	$N_{sc} = \left(\frac{N_{sf}^2}{N_{st}N_{sd}}\right) = \left(\frac{18\mu}{K_m\rho_p D_v}\right)$†	Geometry: (D_{b1}/D_b), (D_{b2}/D_b), etc. ϵ_v Δ
Inertial deposition	Velocity gradient	$N_{st} = \left(\frac{K_m \rho s D_p^2 V_o}{18\mu D_b}\right)$		
Diffusional deposition	Concentration gradient	$N_{sd} = \left(\frac{D_v}{V_o D_b}\right)$		
Gravity settling	Elevation gradient	$N_{sg} = \left(\frac{u_t}{V_o}\right)$		Flow pattern: N_{Re}‡ N_{Ma} N_{Kn} Surface accommodation
Electrostatic precipitation	Electric-field gradient§ *a.* Attraction *b.* Induction	$N_{sec} = \left(\frac{K_m Q_p \epsilon_b}{\mu D_p V_o}\right)$; $N_{sei} = \left(\frac{\delta_p - 1}{\delta_p + 2}\right)\left(\frac{K_m D_p^2 \delta_o \epsilon_b 2}{\mu D_b V_o}\right)$	δ_p, δ_b¶	
Thermal precipitation	Temperature gradient	$N_{st} = \left(\frac{T - T_b}{T}\right)\left(\frac{\mu}{K_m \rho D_b V_o}\right)\left(\frac{k_t}{2k_t + k_{tp}}\right)$	(T_b/T), (T_p/T),† (N_{Pr}), (k_{tp}/k_t), (k_{tb}/k_t),¶ (c_{hp}/c_h), (c_{hb}/c_h)¶	

SOURCE: Lunde and Lapple, *Chem. Eng. Prog.*, **53**, 385 (1957).

*This has also commonly been termed "direct interception" and in conventional analysis would constitute a physical boundary condition imposed upon the particle path induced by action of other forces. By itself it reflects deposition that might result with a hypothetical particle having finite size but no mass or elasticity.

†This parameter is an alternative to N_{sf}, N_{st}, or N_{ad} and is useful as a measure of the interactive effect of one of these on the other two. It is comparable with the Schmidt number.

‡When applied to the inertial deposition mechanism, a convenient alternative is $(K_m\rho_s/18\rho) = N_{st}/(N_{sf}^2 N_{Re})$.

§In cases in which the body charge distribution is fixed and known, ϵ_b may be replaced with Q_{bs}/δ_o.

¶Not likely to be significant contributions.

cascade-impactor data to obtain dust-collector efficiencies for discrete particle sizes.

The measured diameters of particles should as nearly as possible represent the effective particle size of a dust as it exists in the gas stream. When significant flocculation exists, it is sometimes possible to use measurement methods based on gravity settling.

For dust-control work, it is recommended that a preliminary qualitative examination of the dust first be made without a detailed particle count. A visual estimate of particle-size distribution will often provide sufficient guidance for a preliminary assessment of requirements for collection equipment.

MECHANISMS OF DUST COLLECTION

The basic operations in dust collection by any device are (1) separation of the gas-borne particles from the gas stream by deposition on a collecting surface; (2) retention of the deposit on the surface; and (3) removal of the deposit from the surface for recovery or disposal. The separation step requires (1) application of a force that produces a differential motion of a particle relative to the gas and (2) a gas retention time sufficient for the particle to migrate to the collecting surface. The principal mechanisms of aerosol deposition that are applied in dust collectors are (1) gravitational deposition, (2) flow-line interception, (3) inertial deposition, (4) diffusional deposition, and (5) electrostatic deposition. Thermal deposition is only a minor factor in practical dust-collection equipment because the thermophoretic force is small. Table 20-31 lists these six mechanisms and presents the characteristic parameters of their operation [Lunde and Lapple, *Chem. Eng. Prog.*, **53**, 385 (1957)]. The actions of the inertial-deposition, flow-line-interception, and diffusional-deposition mechanisms are illustrated in Fig. 20-103 for the case of a collecting body immersed in a particle-laden gas stream.

Two other deposition mechanisms, in addition to the six listed, may be in operation under particular circumstances. Some dust particles may be collected on filters by sieving when the pore diameter is less than the particle diameter. Except in small membrane filters, the sieving mechanism is probably limited to surface-type filters, in which a layer of collected dust is itself the principal filter medium.

The other mechanism appears in scrubbers. When water vapor diffuses from a gas stream to a cold surface and condenses, there is a net hydrodynamic flow of the noncondensable gas directed toward the surface. This flow, termed the Stefan flow, carries aerosol particles to the condensing surface (Goldsmith and May, in Davies, *Aerosol Science*, Academic, New York, 1966) and can substantially improve the performance of a scrubber. However, there is a corresponding Stefan flow directed away from a surface at which water is evaporating, and this will tend to repel aerosol particles from the surface.

In addition to the deposition mechanisms themselves, methods for preliminary conditioning of aerosols may be used to increase the effectiveness of the deposition mechanisms subsequently applied. One such conditioning method consists of imposing on the gas high-intensity acoustic vibrations to cause collisions and flocculation of the aerosol particles, producing large particles that can be separated by simple inertial devices such as cyclones. This process, termed "sonic (or acoustic) agglomeration," has not attained commercial acceptance.

Another conditioning method, adaptable to scrubber systems, consists of inducing condensation of water vapor on the aerosol particles as nuclei, increasing the size of the particles and making them more susceptible to collection by inertial deposition.

Most forms of dust-collection equipment use more than one of the collection mechanisms, and in some instances the controlling mechanism may change when the collector is operated over a wide range of conditions. Consequently, collectors are most conveniently classified by type rather than according to the underlying mechanisms that may be operating.

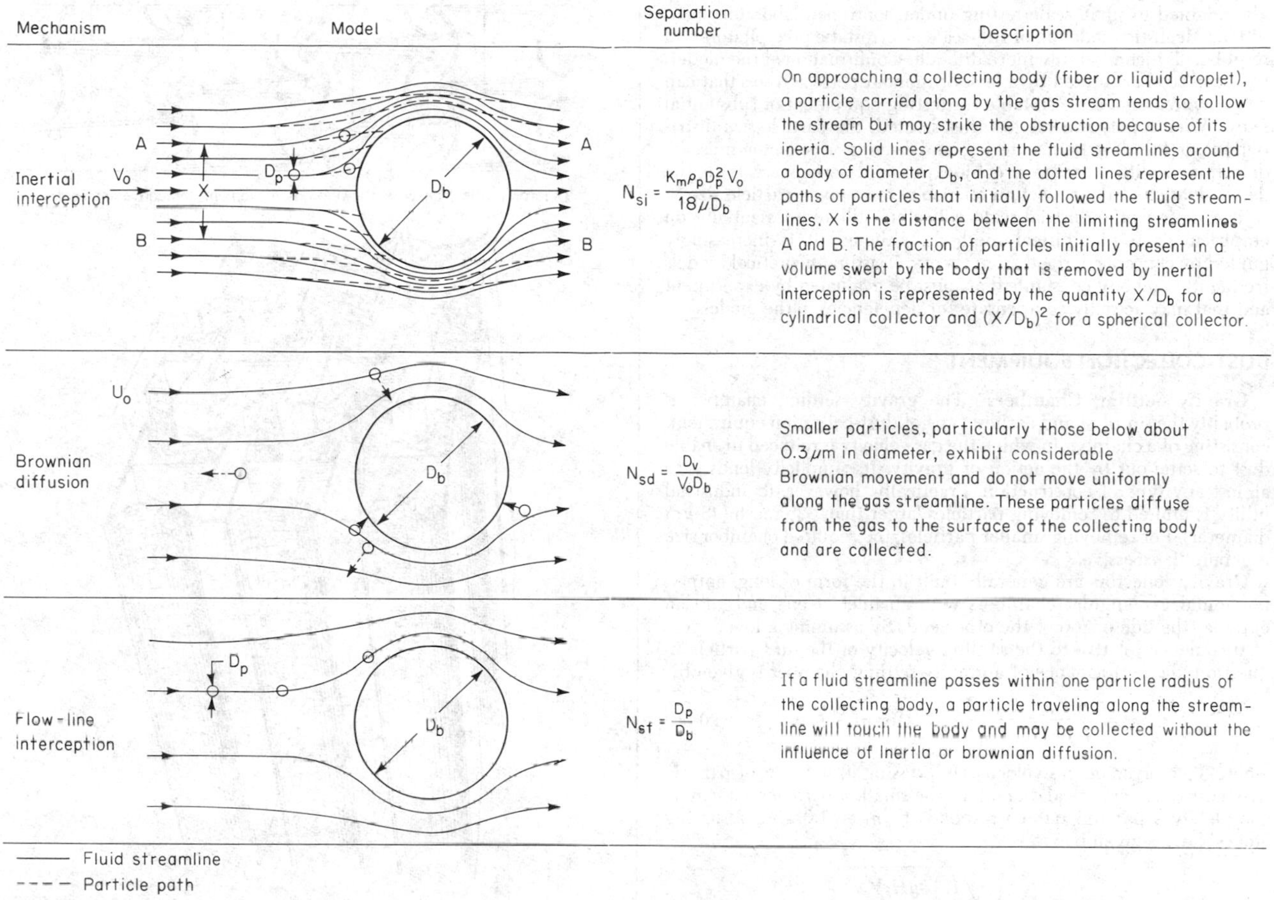
Mechanism
Model
Separation number
Description
Inertial interception
A
B
V_o
D_p
X
D_b
$N_{si} = \frac{K_m \rho_p D_p^2 V_o}{18 \mu D_b}$
On approaching a collecting body (fiber or liquid droplet), a particle carried along by the gas stream tends to follow the stream but may strike the obstruction because of its inertia. Solid lines represent the fluid streamlines around a body of diameter D_b, and the dotted lines represent the paths of particles that initially followed the fluid streamlines. X is the distance between the limiting streamlines A and B. The fraction of particles initially present in a volume swept by the body that is removed by inertial interception is represented by the quantity X/D_b for a cylindrical collector and $(X/D_b)^2$ for a spherical collector.
Brownian diffusion
U_o
D_b
$N_{sd} = \frac{D_v}{V_o D_b}$
Smaller particles, particularly those below about 0.3 μm in diameter, exhibit considerable Brownian movement and do not move uniformly along the gas streamline. These particles diffuse from the gas to the surface of the collecting body and are collected.
Flow-line interception
D_p
D_b
$N_{sf} = \frac{D_p}{D_b}$
If a fluid streamline passes within one particle radius of the collecting body, a particle traveling along the stream-line will touch the body and may be collected without the influence of inertia or brownian diffusion.
Fluid streamline
Particle path
FIG. 20-103 Particle deposition on collector bodies.

PERFORMANCE OF DUST COLLECTORS

The performance of a dust collector is most commonly expressed as the collection efficiency η, the weight ratio of the dust collected to the dust entering the apparatus. However, the collection efficiency is usually related exponentially to the properties of the dust and gas and the operating conditions of most types of collectors and hence is an insensitive function of the collector operating conditions as its value approaches 1.0. Performance in the high-efficiency range is better expressed by the penetration $1 - \eta$, the weight ratio of the dust escaping to the dust entering. Particularly in reference to collection of radioactive aerosols, it is common to express performance in terms of the reciprocal of the penetration $1/(1 - \eta)$, which is termed the "decontamination factor" and is designated by DF. Blasewitz and Judson [*Chem. Eng. Prog.*, **51**, 6-J (1955)] used the quantity $\log_{10} [1/(1 - \eta)]$, which they also termed the decontamination factor. Licht (*Air Pollution Control Engineering*, Marcel Dekker, New York, 1980) proposes the term "decontamination index," symbol DI, for $\log_{10} [1/(1 - \eta)]$ to avoid confusion with $1/(1 - \eta)$. The number of transfer units N_t, which is equal to $\ln [1/(1 - \eta)]$ in the case of dust collection, was first proposed for use by Lapple (Wright, Stasny, and Lapple, "High Velocity Air Filters," WADC Tech. Rep. 55-457, ASTIA No. AD-142075, October 1957) and is more commonly used than the DI. Because of the exponential form of the relationship between efficiency and process variables for most dust collectors, the use of N_t (or DI) is particularly suitable for correlating collector performance data.

In comparing alternative collectors for a given service, a figure of merit is desirable for ranking the different devices. Since power consumption is one of the most important characteristics of a collector, the ratio of N_t to power consumption is a useful criterion. Another is the ratio of N_t to capital investment.

DUST-COLLECTOR DESIGN

In dust-collection equipment, most or all of the collection mechanisms may be operating simultaneously, their relative importance being determined by the particle and gas characteristics, the geometry of the equipment, and the fluid-flow pattern. Although the general case is exceedingly complex, it is usually possible in specific instances to determine which mechanism or mechanisms may be controlling. Nevertheless, the difficulty of theoretical treatment of dust-collection phenomena has made necessary simplifying assumptions, with the introduction of corresponding uncertainties. Theoretical studies have been hampered by a lack of adequate experimental techniques for verification of predictions. Although theoretical treatment of collector performance has been greatly expanded in the period since 1960, few of the resulting performance models have received adequate experimental confirmation because of experimental limitations.

The best-established models of collector performance are those for fibrous filters and fixed-bed granular filters, in which the structures and fluid-flow patterns are reasonably well defined. These devices are also adapted to small-scale testing under controlled laboratory conditions. Realistic modeling of full-scale electrostatic precipitators and scrubbers is incomparably more difficult. Confirmation of the models has been further limited by a lack of monodisperse aerosols that can be generated on a scale suitable for testing equipment of substantial sizes. When a polydisperse test dust is used, the particle-size distributions of the dust both entering and leaving a collector must be determined with extreme precision to avoid serious errors in the determination of the collection efficiency for a given particle size.

The design of industrial-scale collectors still rests essentially on empirical or semiempirical methods, although it is increasingly guided by concepts derived from theory. Existing theoretical models frequently embody constants that must be evaluated by experiment and that may actually compensate for deficiencies in the models.

DUST-COLLECTION EQUIPMENT

Gravity Settling Chambers The gravity settling chamber is probably the simplest and earliest type of dust-collection equipment, consisting of a chamber in which the gas velocity is reduced to enable dust to settle out by the action of gravity. Its simplicity lends it to almost any type of construction. Practically, however, its industrial utility is limited to removing particles larger than 325 mesh (43-μm diameter). For removing smaller particles, the required chamber size is generally excessive.

Gravity collectors are generally built in the form of long, empty, horizontal, rectangular chambers with an inlet at one end and an outlet at the side or top of the other end. By assuming a low degree of turbulence relative to the settling velocity of the dust particle in question, the performance of a gravity settling chamber is given by

$$\eta = \frac{u_t L_s}{H_s V_s} = \frac{u_t B_s L_s}{q} \qquad (\text{for } \eta \leqq 1.0) \qquad (20\text{-}56)$$

where V_s = average gas velocity. Expressing μ_t in terms of particle size (equivalent spherical diameter), the smallest particle that can be completely separated out corresponds to η = 1.0 and, assuming Stokes' law, is given by

$$D_{p,\min} = \sqrt{\frac{18\mu H_s V_s}{g_L L_s(\rho_s - \rho)}} \qquad (20\text{-}57)$$

$$= \sqrt{\frac{18\mu q}{g_L B_s L_s(\rho_s - \rho)}}$$

where ρ = gas density and ρ_s = particle density. For a given volumetric air-flow rate, collection efficiency depends on the total plan cross section of the chamber and is independent of the height. The height need be made only large enough so that the gas velocity V_s in the chamber is not so high as to cause reentrainment of separated dust. Generally V_s should not exceed about 3 m/s (10 ft/s).

Horizontal plates arranged as shelves within the chamber will give a marked improvement in collection. This arrangement is known as the Howard dust chamber (Fume Arrester, U.S. Patent 896,111, 1908). The disadvantage of the unit is the difficulty of cleaning owing to the close shelf spacing and warpage at elevated temperatures.

The pressure drop through a settling chamber is small, consisting primarily of entrance and exit losses. Because low gas velocities are used, the chamber is not subject to abrasion and may therefore be used as a precleaner to remove very coarse particles and thus minimize abrasion on subsequent equipment.

Impingement Separators Impingement separators are a class of inertial separators in which particles are separated from the gas by impingement on collecting bodies arrayed across the path of the gas stream. In general, impingement separators are designed for pressure drops in the range of 0.25 to 0.40 kPa (0.1 to 1.5 in water), depending on the type and application, and are limited to removing dusts that are predominantly larger than 10- to 20-μm diameter. The chief advantage of such units over other types of collectors is that they are more adaptable to existing flues or ducts. Rappers are sometimes provided to shake the collected dust off the collecting bodies at selected intervals. Further description will be found in Powers [*Rock Prod.*, **46,** 70, 72 (1943)] and Roberts [*Power,* **83,** 345, 392 (1939)]. A typical impingement separator (no longer manufactured commercially) is shown in Fig. 20-104.

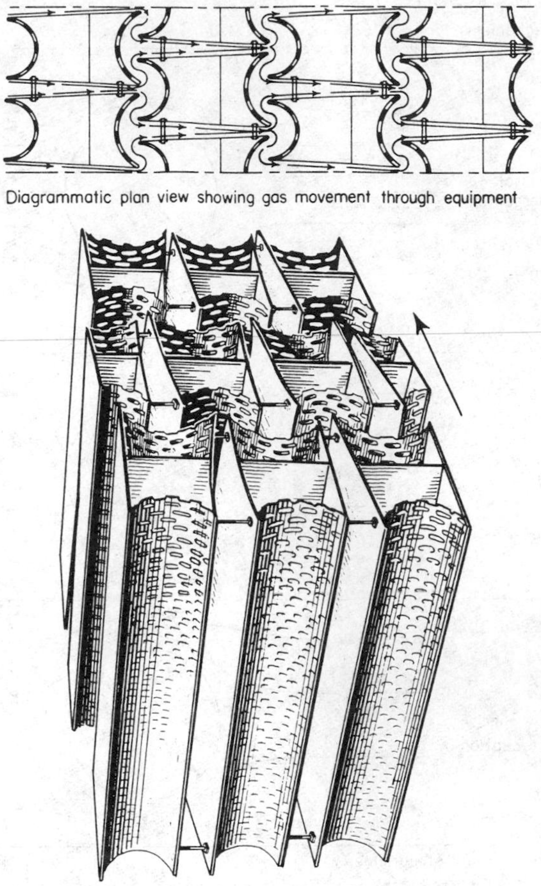

Diagrammatic plan view showing gas movement through equipment

FIG. 20-104 Reverse-nozzle impingement separator. *(By-Products Recoveries, Inc.)*

Impingement separators function according to the principles illustrated in Fig. 20-103. However, over the range of separator constructions and operating conditions used, the influence of the diffusional deposition mechanism is negligible, and the effective collection mechanisms are inertial deposition and flow-line interception. Collection by a body such as the cylinder shown in Fig. 20-103 can be presented in terms of a so-called target efficiency. Thus, all particles that are initially carried in the fluid between streamlines *A* and *B* will be collected on the body, and the target efficiency will be equal to X/D_b. The target efficiency η_0 is a function of the inertial separation number. For simple collector-body shapes and fluid-flow conditions it is possible to compute the relationship from classical hydrodynamics. Fig. 20-105 presents the target efficiencies for ribbons, spheres, and cylinders as computed by Langmuir and Blodgett [U.S. Army Air Forces Tech. Rep. 5418, Feb. 19, 1946 (U.S. Dept. of Commerce, Office of Technical Services PB-27565); General Electric Res. Lab. Rep. No. RL-225, December 1944–July 1945, reissued June 1949]. The curves apply for conditions under which Stokes' law holds for the motion of the particle. The relationships are derived for conditions of potential (streamline) flow around the bodies but should hold closely even if the flow around the body is turbulent, since conditions on the upstream side of the body should approach those of potential flow in any case. It should be noted that the relationships

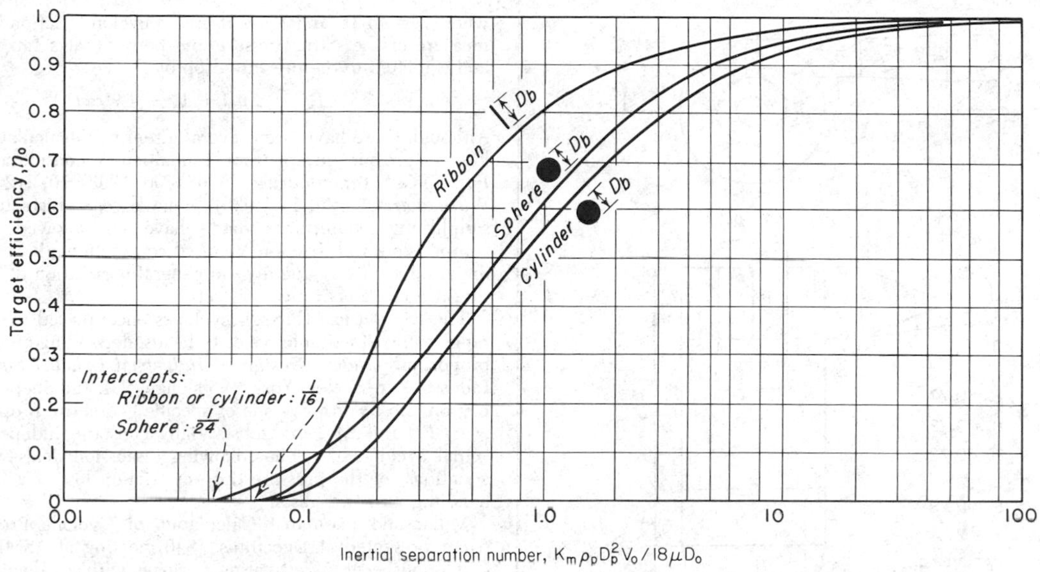

FIG. 20-105 Target efficiency of spheres, cylinders, and ribbons. The curves apply for conditions in which Stokes' law holds for the motion of the particle. [*Langmuir and Blodgett, U.S. Army Air Forces Tech. Rep. 5418, Feb. 19, 1946 (U.S. Department of Commerce, Office of Technical Services PB 27565).*]

should apply whether the fluid moves around the body or the body moves through the fluid as long as V_0 is taken as the relative velocity between the body and the bulk of the fluid.

Although the curves of Fig. 20-105 apply to collecting bodies in an infinite fluid, they may be applied for the direct calculation of collection efficiency of units employing such bodies in parallel and in series, provided that adjacent collecting members are not so close as to cause an appreciable distortion of the flow pattern. This is substantially the case in many types of air filters. When collecting members are relatively close, these curves would give a conservative approximation of collection efficiency. For collecting members having shapes widely different from those shown in Fig. 20-105, experimental determinations of the target-efficiency relationship are generally required, although order-of-magnitude estimates can be made by judicious interpolation of the curves shown. Ranz (Penn. State Univ. Eng. Res. Bull. B-66, December 1956) presents target efficiencies for a recessed body, or cup, and for a rectangular half body as well as additional curves for ribbons, cylinders, and spheres. He also gives methods for estimating the efficiencies of some typical impingement separators.

Cyclone Separators The most widely used type of dust-collection equipment is the cyclone, in which dust-laden gas enters a cylindrical or conical chamber tangentially at one or more points and leaves through a central opening (Fig. 20-106). The dust particles, by virtue of their inertia, will tend to move toward the outside separator wall, from which they are led into a receiver. A cyclone is essentially a settling chamber in which gravitational acceleration is replaced by centrifugal acceleration. At operating conditions commonly employed, the centrifugal separating force or acceleration may range from 5 times gravity in very large diameter, low-resistance cyclones, to 2500 times gravity in very small, high-resistance units. The immediate entrance to a cyclone is usually rectangular.

Fields of Application Within the range of their performance capabilities, cyclone collectors offer one of the least expensive means of dust collection from the standpoint of both investment and operation. Their major limitation is that unless very small units are used, their efficiency is low for collection of particles smaller than 5 μm and, particularly, for particles smaller than 2 to 3 μm. Although cyclones may be used to collect particles larger than 200 μm, gravity settling chambers or simple inertial separators (such as gas-reversal chambers) are usually satisfactory and less subject to abrasion. In special cases in which the dust is highly flocculated or high dust concentrations (over 230 g/m^3, or 100 gr/ft^3) are encountered, cyclones will remove dusts having small particle sizes. In certain instances efficiencies as high as 98 percent have been attained on dusts having ultimate particle sizes of 0.1 to 2.0 μm because of the predominant effect of flocculation.

Cyclones are used to remove both solids and liquids from gases and have been operated at temperatures as high as 1000°C and pressures as high as 50,700 kPa (500 atm).

Flow Pattern In a cyclone the gas path involves a double vortex with the gas spiraling downward at the outside and upward at the inside. When the gas enters the cyclone, its velocity undergoes a redistribution so that the tangential component of velocity increases with decreasing radius as expressed by $V_{ct} \sim r^{-n}$. The spiral velocity in a cyclone may reach a value several times the average inlet-gas velocity. Theoretical considerations indicate that n should be equal to 1.0 in the absence of wall friction. Actual measurements [Shepherd and Lapple, *Ind. Eng. Chem.*, **31,** 972 (1939); **32,** 1246 (1940)], however, indicate that n may range from 0.5 to 0.7 over a large portion of the cyclone radius. Ter Linden [*Inst. Mech. Eng. J.*, **160,** 235 (1949)] found n to be 0.52 for tangential velocities measured in the cylindrical portion of the cyclone at positions ranging from the radius of the gas-outlet pipe to the radius of the collector. Although the velocity approaches zero at the wall, the boundary layer is sufficiently thin that pitot-tube measurements show relatively high tangential velocities there, as shown in Fig. 20-107. The radial velocity V_r is directed toward the center throughout most of the cyclone, except at the center, where it is directed outward.

Superimposed on the "double spiral," there may be a "double eddy" [Van Tongeran, *Mech. Eng.*, **57,** 753 (1935); and Wellmann, *Feuerungstechnik*, **26,** 137 (1938)] similar to that encountered in pipe coils. Measurements on cyclones of the type shown in Fig. 20-106 indicate, however, that such double-eddy velocities are small compared with the spiral velocity (Shepherd and Lapple, op. cit.).

Pressure Drop The pressure drop through a cyclone as well as the friction loss is most conveniently expressed in terms of the velocity head based on the immediate cyclone inlet area. The inlet velocity head, expressed in inches of water, is related to the average inlet-gas velocity and density by

$$h_{vi} = 0.0030 \rho V_c^2 \tag{20-58}$$

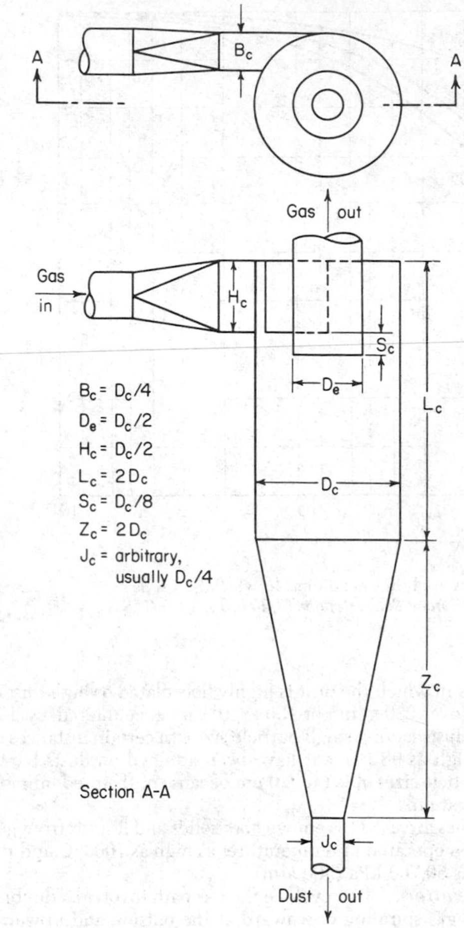

FIG. 20-106 Cyclone-separator proportions.

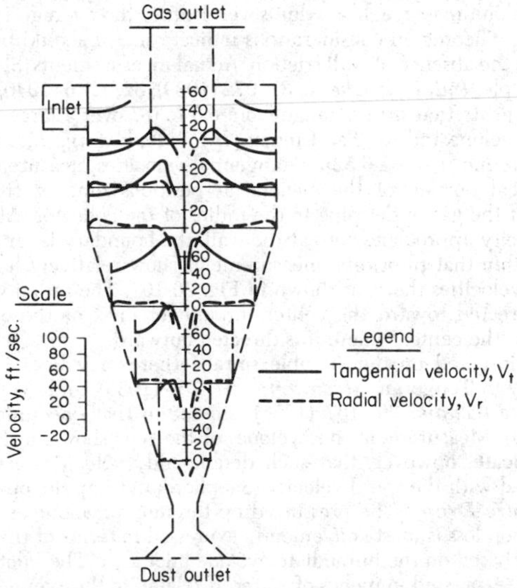

FIG. 20-107 Variation of tangential velocity and radial velocity at different points in a cyclone. [*Ter Linden*, Inst. Mech. Eng. J., *160, 235 (1949)*.]

where ρ = lb/ft^3 and V_c = ft/s. The cyclone friction loss is a direct measure of the static pressure and power that a fan must develop and is related to the pressure drop by

$$F_{cv} = \Delta p_{cv} + 1 - (4A_c/\pi D_e^2)^2 \qquad (20\text{-}59)$$

Although there have been several attempts to calculate the friction loss or pressure drop from fundamental considerations [Feifel, *Forsch. Geb. Ingenieurwes.*, **9**, 68, 306 (1938); **10**, 212 (1939); *Arch. Wärmewirtsch.*, **20**, 15 (1939)], none is very satisfactory, since the simplifying assumptions made have not allowed for entrance compression, wall friction, and exit contraction, all of which have a major effect. Consequently, no general correlation of cyclone pressure-drop data is available as yet.

The friction loss through cyclones encountered in practice may range from 1 to 20 inlet velocity heads, depending on the geometric proportions (Alden, *Design of Industrial Exhaust System*, 3d ed., Industrial Press, New York, 1959, chap. VII; and Shepherd and Lapple, op. cit.). For a cyclone of specific geometric proportions, however, F_{cv} and Δp_{cv} are substantially constant, independent of the actual cyclone size. The following discussion deals with reported equations for the pressure drop or friction loss of a cyclone when handling dust-free gases.

Miller and Lissman ["Calculation of Cyclone Pressure Drop," paper presented at December 1940 meeting of ASME, New York (not published)], investigating cyclones with an involute entrance, obtained the following empirical expression:

$$\Delta p_{cv} = K(D_c/D_e)^2 \qquad (20\text{-}60)$$

The value of K was found to be substantially constant with a value of 3.2 over the following range in proportions: (B_c/D_c) = ⅛ to ⅜; $(H_c/D_c) \cong 1.0$; (D_e/D_c) = ¼ to ¾. For smaller values of (D_e/D_c) the value of K increased, while for smaller values of (B_c/D_c) it decreased. In these tests D_c, D_e, and B_c were varied but not H_c.

Shepherd and Lapple (op. cit.), investigating cyclones of the general type shown in Fig. 20-106, obtained the following empirical expression:

$$F_{cv} = KB_cH_c/D_e^2 \qquad (20\text{-}61)$$

These tests covered the following range in proportions: (B_c/D_c) = $\frac{1}{12}$ to ¼; (H_c/D_c) = ¼ to ½; (D_e/D_c) = ¼ to ½. With the normal arrangement in which the rectangular inlet terminates at the outer elements of the cyclone body or cylinder, K was found to have a value of 16.0. If the inner side of the inlet duct was extended past the cyclone cylinder wall and into the annular space halfway to the opposite wall to form an "inlet vane," the friction loss was reduced by over 50 percent and K was found to be 7.5. Pressure-drop values calculated by means of Eq. (20-61) for a value of K of 13.0 will check the Miller and Lissman data within ±30 percent for the most part. For the specific proportions shown in Fig. 20-106, F_{cv} = 8.0.

Data reported in the trade literature for the type of cyclone shown in Fig. 20-108*b* can be closely represented by Eq. (20-61) for a value of K of 18.4. The proportions covered by this design are approximately: (B_c/D_c) = ⅜; (H_c/D_c) = ⅝; (D_e/D_c) = ½ to 1. The term D_c, as specified here, is the diameter of the main or upper cylinder. The large diameter of the upper cone is 1⅜ times as large. The cyclone shown in Fig. 20-108*a* has a reported (trade literature) pressure drop as follows:

$$\Delta p_i = 0.013\rho V_c^2 \qquad (20\text{-}62)$$

where Δp_i = in of water. Iinoya [*Mem. Fac. Eng. Nagoya Univ.*, 5(2), (September 1953)] investigated the effect of wall roughness on pressure drop and found that pressure drop *decreased* with increased surface roughness. He pasted sand particles of various sizes on the wall of the cyclone and obtained pressure-drop data as shown in Table 20-32. First ("Fundamental Factors in the Design of Cyclone Dust Collectors," doctoral thesis, Harvard University, May 1950) independently concluded that wall friction was a negligible part of the total pressure drop. Wall friction apparently reduces the vortex intensity, thereby decreasing pressure drop.

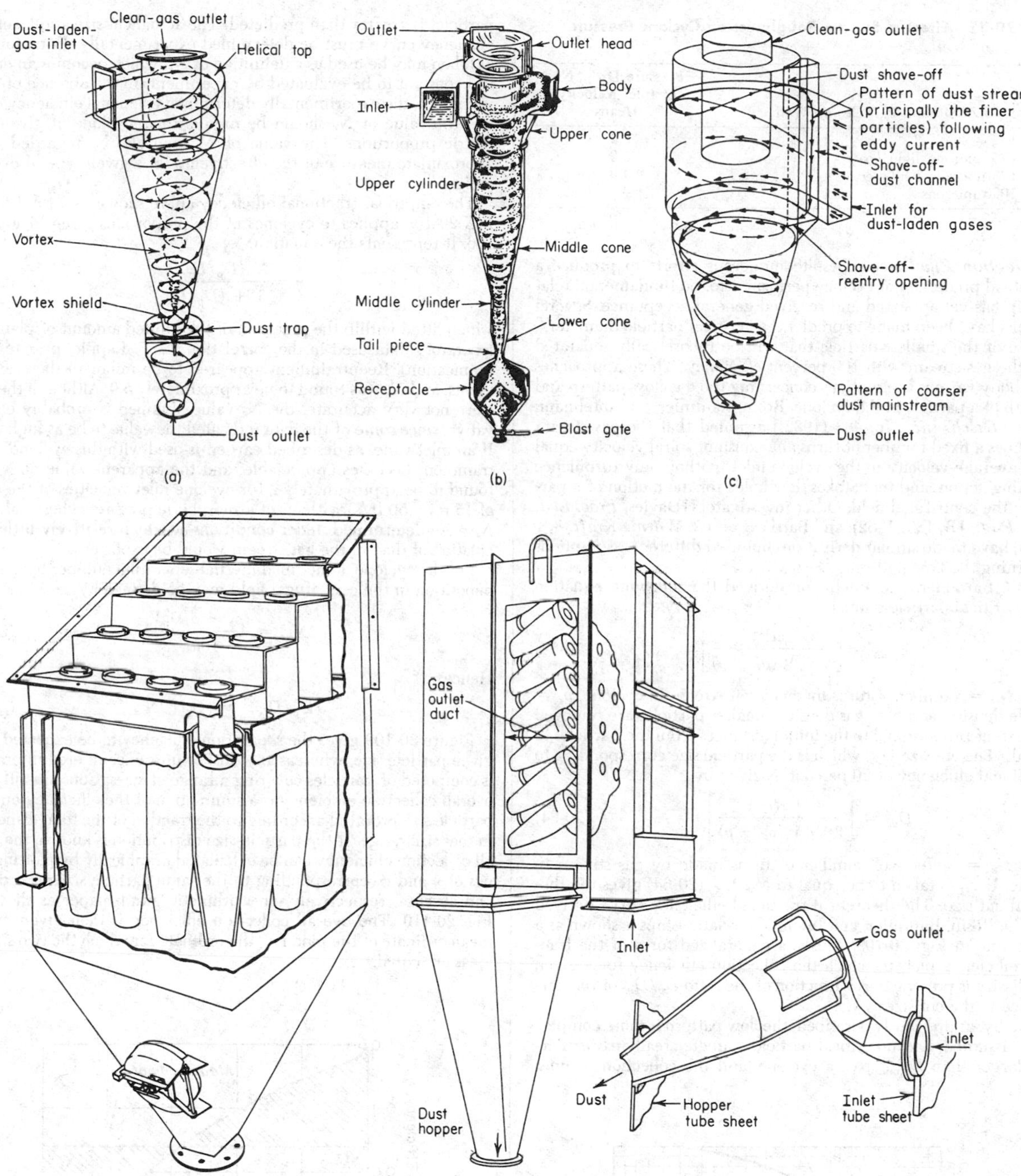

FIG. 20-108 Typical commercial cyclones. (*a*) Duclone collector. *(Ducon Co., Inc.)* (*b*) Sirocco type D collector. *(American Blower Corp.)* (*c*) Van Tongeren cyclone. *(Buell Engineering Co.)* (*d*) Multiclone collector. *(Western Precipitation Division, Joy Manufacturing Company.)* (*e*) Dustex miniature collector assembly. *(Dustex Corp.)* (*f*) Cutaway of the Dustex cyclone tube.

TABLE 20-32 Effect of Surface Roughness on Cyclone Pressure Drop

Size of Sand Particles on Wall	Pressure Drop, No. of Inlet Velocity Heads
None	8.0
147 to 175 microns, light coat	5.8
147 to 175 microns, heavy coat	4.9
500 to 1000 microns	4.1

Collection Efficiency Despite numerous efforts to produce a theoretical prediction of cyclone performance, no fundamental relationship has yet appeared and received general acceptance. Several attempts have been made to predict the "critical particle diameter," the size of the smallest particle that will be theoretically separated from the gas stream with 100 percent efficiency. These approaches all embody various assumptions concerning the gas-flow pattern and the path of a particle in the cyclone. Rosin, Rammler, and Intelmann [*Z. Ver. Dtsch. Ing.*, **76**, 433 (1932)] assumed that the gas stream undergoes a fixed number of turns at a constant spiral velocity equal to the average velocity in the cyclone inlet, without any turbulence or mixing action, and that Stokes' law holds for the motion of a particle in the centrifugal field. Other investigators [Davies, *Proc. Inst. Mech. Eng.*, **1B**, 185 (1952); and Barth, *Brennst. Wärme Kraft*, **8**, 1 (1956)] have made similar derivations but used different assumptions concerning the flow pattern.

Rosin, Rammler, and Intelmann derived the following equation for the critical particle diameter:

$$D_{p,\min} = \left[\frac{9\mu B_c}{\pi N_{tc} V_c(\rho_s - \rho)}\right]^{0.5} \tag{20-63}$$

where N_{tc} = number of turns made by gas stream in cyclone, ρ_s = particle density, and ρ = gas density. Smaller particles are removed to an extent proportional to the initial distances from the particles to the wall. The cut size D_{pc}, which is the particle size corresponding to a fractional efficiency of 50 percent, is given by

$$D_{pc} = \left[\frac{9\mu B_c}{2\pi N_e V_c(\rho_s - \rho)}\right]^{0.5} \tag{20-64}$$

where N_e = "effective" number of turns made by gas stream in cyclone. If N_e is taken to be equal to N_{tc}, Eq. (20-64) gives the theoretical cut size. The theoretical fractional efficiency curve derived from the Rosin, Rammler, and Intelmann relationships is shown as a broken line in Fig. 20-109. This is a generalized form of the fractional efficiency plot, in which the collection efficiency for a given particle size is presented as a function of the ratio D_p/D_{pc} of the particle size to the cut size.

Actually, as previously described, the flow pattern is more complex than is assumed, and no critical particle diameter really exists. Particles larger than $D_{p,\min}$ pass a cyclone, and the collection of finer particles is greater than predicted. The actual cut size and fractional efficiency curve must be determined experimentally. Equation (20-64) then may be used as a definition of N_e, which becomes an empirical constant to be evaluated by experiment. In the absence of reentrainment, the experimentally determined fractional efficiency curve and the value of N_e should be unique for a cyclone of given geometric proportions. The value of N_e may also be regarded as an approximate measure of the effectiveness of a given type of cyclone design.

The empirical fractional efficiency curve shown as a solid line in Fig. 20-109 applies to cyclones of the proportions given in Fig. 20-106. It represents the equation

$$\eta = \frac{(D_p/D_{pc})^2}{1 + (D_p/D_{pc})^2} \tag{20-65}$$

which fitted within the precision of the limited amount of plant and laboratory data used in the correlation (C. E. Lapple, private communication). Reentrainment appeared to be minor in these experiments, and N_e was found to be approximately 5.0. Although the data were not very accurate, the N_e value obtained is probably conservative, since some of the data indicated the value to be as high as 10. If an inlet vane, as described earlier, is used with this cyclone, reentrainment becomes appreciable, and the apparent value of N_e was found to be approximately 2 for cyclone inlet velocities of the order of 15 m/s (50 ft/s) with air at atmospheric pressure. These values of N_e were determined under conditions at which relatively little flocculation of dust in the gas stream would be expected.

For the cyclone, a modified inertial separation number for particle deposition in the centrifugal field may be defined by

$$N_{stc} = \frac{(\rho_s - \rho)D_p^2 V_c}{18\mu B_c} \tag{20-66}$$

Hence,

$$(D_p/D_{pc})^2 = (4\pi N_e)N_{stc} \tag{20-67}$$

Figure 20-109 gives the separation efficiency to be expected for a given particle size, whereas any dust or mist encountered in practice is composed of particles covering a range of sizes. Consequently, the overall collection efficiency is a summation of the efficiency on each particle size prorated according to the fraction of the total dispersoid in that size range. If the particle-size distribution is known, the overall collection efficiency can be calculated graphically by plotting values of η and ϕ, corresponding to the same particle size, as ordinate and abscissa, respectively, on arithmetic graph paper as shown in Fig. 20-110. The overall collection efficiency is then given by the mean ordinate of the plot, i.e., the ordinate for which the two shaded areas are equal.

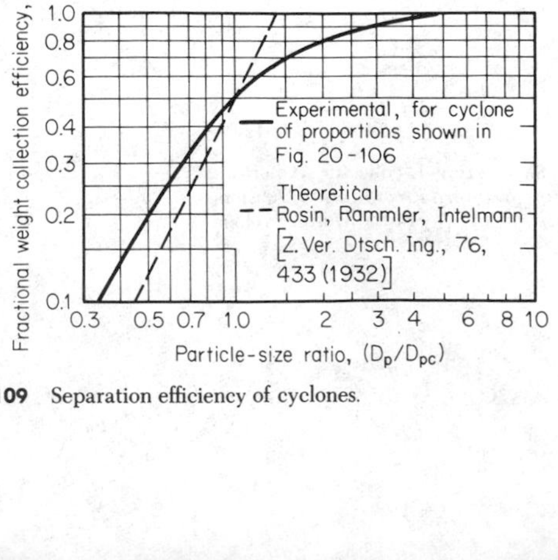

FIG. 20-109 Separation efficiency of cyclones.

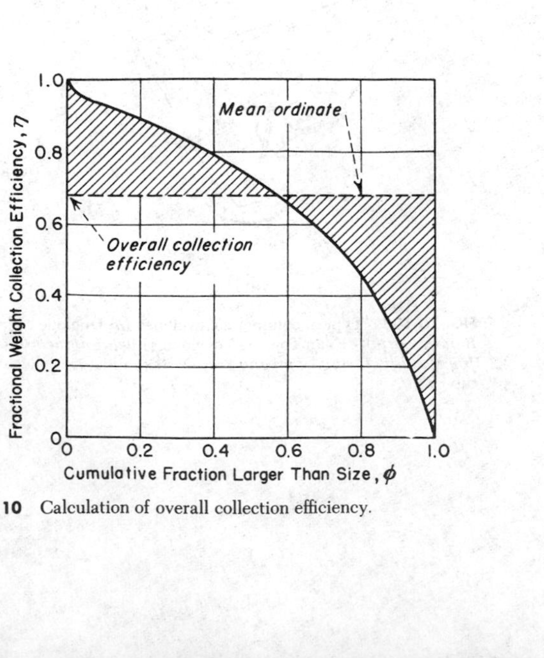

FIG. 20-110 Calculation of overall collection efficiency.

TABLE 20-33 Experimental Cyclone Collection Efficiencies*

Cyclone inlet velocity: 44 ft/s (13.4 m/s)
Cyclone pressure drop: 4 in water (1.0 kPa)
Inlet dust concentration: 2–5 gr/ft^3 (4.6–11.5 g/m^3)
Specific gravity of dust: 3.0
Cyclone proportions: $B_e \cong D_e/6$
Gas: atmospheric air

Inlet Dust Particle-Size Analysis

Particle diameter, μm	Cumulative % larger than size
5	74
10	64
20	43

Cyclone diameter, in	Dust collected, %		
6	Total 90	− 5 μm 66	+ 5 μm 98
9	Total 83	−10 μm 60	+10 μm 99
24	Total 70	−20 μm 47	+20 μm 98

*Data reported by Anderson in Perry, *Chemical Engineers' Handbook*, 2d ed., McGraw-Hill, New York, 1941, p. 1860.

In many cases a good approximation is obtained if the overall collection efficiency is taken as equal to the cumulative percentage of material ϕ larger than the cut size D_{pc} in the dust fed to the cyclone. Equation (20-64) allows for operating temperature in the viscosity term, which means that, for a given inlet velocity, increased temperature results in a larger cut size, corresponding to a lower efficiency. As another good approximation, it should be noted that a given size of cyclone will have substantially the same collection efficiency at any temperature, provided that the pressure drop is the same, because of counterbalancing effects of gas density and viscosity.

Lapple [*Chem. Eng.*, **58**, 144 (May 1951)] presents design curves for a cyclone of the proportions shown in Fig. 20-106, based on data of Fig. 20-109. Table 20-33 gives experimental collection-efficiency data reported by Anderson for geometrically similar cyclones of the type shown in Fig. 20-108*d*. These will serve to illustrate the order of magnitude of the collection efficiency to be expected for various particle sizes.

Leith and Licht [*Am. Inst. Chem. Eng. Symp. Ser.*, **68**(126), 196 (1972)] have attempted to produce a theoretical model of cyclone performance that will be generally applicable to tangential-inlet cyclones of the general type illustrated by Fig. 20-106. In the derived relationship, the collection efficiency is an exponential function of a modified inertial separation number, a dimensionless "cyclone design number" calculated from the geometry of the cyclone, and the vortex exponent in the equation relating the tangential component of the gas velocity to the radial position. The Leith-Licht model is based on an assumption of continual radial back mixing of the uncollected particles together with calculation of an average residence time for the gas. Various simplifying assumptions are also made concerning the flow pattern. The deficiencies in the model and some of the assumptions are discussed briefly by Dietz [*Am. Inst. Chem. Eng. J.*, **27**, 888 (1981)]. The computations of the cyclone design number and the vortex exponent depend on empirical relationships given by Alexander [*Proc. Australas. Inst. Min. Metall.* (N.S.), **152–153**, 203 (1949)], but values of these factors might also be calculated from experimentally determined fractional efficiency data. The Leith-Licht model has been fitted to some experimental data of Stairmand [*Trans. Inst. Chem. Eng.*, **29**, 356 (1951)] and Peterson and Whitby [*ASHRAE J.*, **7** (5), 42 (1965)] but has received no extensive independent experimental confirmation.

With few exception [Beeckmans and Kim, *Can J. Chem. Eng.*, **55**, 640 (1977)], testing of cyclone efficiencies has been conducted with redispersed polydisperse test dusts or with industrial dusts. The actual state of dispersion of the entering dusts has been uncertain. Flocculation of dusts is probably the chief cause of numerous inconsistencies and discrepancies in reported data on cyclone performance.

Cyclone Design Factors Cyclones are generally designed to meet specified pressure-drop limitations. For ordinary installations, operating at approximately atmospheric pressure, fan limitations generally dictate a maximum allowable pressure drop corresponding to a cyclone inlet velocity in the range of 6 to 21 m/s (20 to 70 ft/s). Consequently, cyclones are usually designed for an inlet velocity of 15 m/s (50 ft/s), though this need not be strictly adhered to.

In the removal of dusts, collection efficiency can be changed by only a relatively small amount by a variation in operating conditions. The primary design factor that can be utilized to control collection efficiency is the cyclone diameter, a smaller-diameter unit operating at a fixed pressure drop having the higher efficiency [Anderson, *Chem. Metall.*, **40,** 525 (1933); Drijver, *Wärme*, **60,** 333 (1937); and Whiton, *Power,* **75,** 344 (1932); *Chem. Metall.*, **39,** 150 (1932)]. Small-diameter cyclones, however, will require a multiple of units in parallel for a specified capacity. In such cases the individual cyclones can discharge the dust into a common receiving hopper [Whiton, *Trans. Am. Soc. Mech. Eng.*, **63,** 213 (1941)]. The final design involves a compromise between collection efficiency and complexity of equipment. It is customary to design a single cyclone for a given capacity, resorting to multiple parallel units only if the predicted collection efficiency is inadequate for a single unit.

Reducing the gas-outlet duct diameter will increase both collection efficiency and pressure drop. Increasing the length of a cyclone is generally conceded to increase collection efficiency, though there are no reliable supporting data. There is also no reliable information on the effect of inlet proportions. It is essential that the inlet transition be relatively gradual in order to avoid excessive pressure drop due to gas jetting into the cyclone chamber. There is disagreement among cyclone designers regarding the optimum cone angle, but most "high-efficiency" cyclones have cone lengths in the range of 1.6 to 3.0 cyclone diameters.

Collection efficiency is normally increased by increasing the gas throughput (Drijver, op. cit.). However, if the entering dust is flocculated, increased gas velocities may cause deflocculation in the cyclone, so that efficiency remains the same or actually decreases. Also, variations in design proportions that result in increased collection efficiency with dispersed dusts may be detrimental with flocculated dusts. Kalen and Zenz [*Am. Inst. Chem. Eng. Symp. Ser.*, **70** (137), 388 (1974)] report that collection efficiency increases with increasing gas inlet velocity up to a minimum tangential velocity at which dust is either reentrained or not deposited because of saltation. Koch and Licht [*Chem. Eng.*, **84**(24), 80 (1977)] estimate that for typical cyclones the saltation velocity is consistent with cyclone inlet velocities in the range of 15 to 27 m/s (50 to 90 ft/s). The conditions of the confirming experiments by Kalen and Zenz are not well described, but the data reported strongly suggest that the incoming test dust was flocculated and that it was deflocculated to a greater and greater extent as the gas inlet velocity was increased. C. E. Lapple (private communication) reports that in cyclone tests with talc dust collection efficiency increased steadily as the inlet velocity was increased up to the maximum of 52 m/s (170 ft/s). With ilmenite dust, which was much more strongly flocculated, efficiency decreased over the same inlet-velocity range. In later experiments with well-dispersed talc dust, collection efficiency continued to increase at inlet velocities up to the maximum used, 82 m/s (270 ft/s).

Cyclones in series may be justified under some circumstances:

1. The dust has a broad size distribution, including particles under 10 to 15 μm as well as larger and possibly abrasive particles. A large low-velocity cyclone may be used to remove the coarse particles ahead of a unit with small-diameter multiple tubes.
2. The dust is composed of fine particles but is highly flocculated or tends to flocculate in preceding equipment and in the cyclones themselves. Efficiencies predicted on the basis of ultimate particle size will be highly conservative.
3. The dust is relatively uniform, and the efficiency of the second-stage cyclone is not greatly lower than that of the first stage.
4. Dependable operation is critical. Second-stage or even third-stage cyclones may be used as backup.

A cyclone will operate equally well on the suction or pressure side of a fan if the dust receiver is airtight. Probably the greatest single cause for poor cyclone performance, however, is the leakage of air into the dust outlet of the cyclone. A slight air leak at this point can result in a tremendous drop in collection efficiency, particularly with fine dusts. For a cyclone under pressure, air leakage at this point is objectionable primarily because of the local dust nuisance created. For batch operation, an airtight hopper or receiver may be used. For continuous withdrawal of collected dust, a rotary star valve, a double-lock valve, or a screw conveyor may be used, the latter only with fine dusts. A collapsible open-ended rubber tube can be used for cyclones operating under slight negative pressure; mechanical flapgate valves and fluidized seal legs can be used (see "Fluidized-Bed Systems: Solids Discharge"). Special pneumatic unloading devices can also be used with dusts. In any case it is essential that sufficient unloading and receiving capacity be provided to prevent collected material from accumulating in the cyclone.

Generally cone-and-disk baffles, helical guide vanes, etc., placed inside a cyclone, will have a detrimental effect on performance. A few of these devices do have some merit, however, under special circumstances. Although an inlet vane will reduce pressure drop, it causes a correspondingly greater reduction in collection efficiency. Its use is recommended only when collection efficiency is normally so high as to be a secondary consideration and when it is desired to decrease the resistance of an existing cyclone system for purposes of increased air-handling capacity or when floor-space or headroom requirements are controlling factors. If an inlet vane is used, it is advantageous to increase the gas-exit-duct length inside the cyclone chamber. A disk or cone baffle located beneath the gas-outlet duct may be beneficial if air in-leakage at the dust outlet cannot be avoided. A heavy chain suspended from the gas-outlet duct has been found beneficial to minimize dust buildup on the cyclone walls. Such a chain should be suspended from a swivel so that it is free to rotate without twisting. At present there are no known devices that will recover the gas spiral-velocity energy in the gas-outlet duct. Substantially all devices that have been reported to reduce pressure drop do so by reducing spiral velocities in the cyclone chamber and consequently result in reduced collection efficiency.

At low dust loadings the pressure in the dust receiver of a single cyclone will generally be lower than in the gas-outlet duct. Increased dust loadings will increase the pressure in the dust receiver. Such devices as cones, disks, and inlet vanes will generally cause the pressure in the dust receiver to exceed that in the gas-outlet duct. A cyclone will operate as well in a horizontal position as in a vertical position. However, departure from the normal vertical position results in an increasing tendency to plug the dust outlet. If the dust outlet becomes plugged, collection efficiency will, of course, be low. If the cyclone exit duct must be reduced to tie in with proposed duct sizes, the transition should be made at least five diameters downstream from the cyclone and preferably after a bend. In the event that the transition must be made closer to the cyclone, a Greek cross should be installed in the transition piece in order to avoid excessive pressure drop.

Increased dust loadings will result in both decreased pressure drop and increased collection efficiency (Drijver, op. cit.; and Shepherd and Lapple, op. cit.). At dust loadings of over 460 g/m^3 (200 gr/ft^3), the pressure drop may be as low as half of that calculated in the absence of dust.

Commercial Equipment Simple cyclones are available in a wide variety of shapes ranging from long, slender units similar to that shown in Fig. 20-106 to short, large-diameter units. The body may be conical or cylindrical, and entrances may be involute or tangential and round or rectangular.

In Fig. 20-108 are shown some of the special types of commercial cyclones. In the Multiclone a spiral motion is imparted to the gas by annular vanes, and it is furnished in multiple units of 15.2- and 22.9-cm (6- and 9-in) diameter. Its largest field of application has been in the collection of fly ash from steam boilers. The tubes are commonly constructed of cast iron and other abrasion-resistant alloys. The Ducon cyclone utlizes a scroll inlet and a helical roof, both of which should reduce pressure drop. A major field of application has been in series and parallel installations in connection with fluidized-solids contactors in the petroleum and metallurgical industries. The cyclone is commonly fabricated of welded steel or stainless steel and may be lined with ceramic materials. The manufacturer claims an improvement in efficiency resulting from a conical baffle, which is sometimes placed in the bottom of the cone. This device is used only on free-flowing dusts, since it has been found to cause plugging when handling sticky materials. The Van Tongeran cyclone claims to utilize the double eddy for increased collection efficiency by providing a bypass from the top to the conical portion of the cyclone. It is made of welded steel and alloy plates and has had broad application. The Sirocco type D cyclone has an exit-duct collar that can be changed to increase or decrease collection efficiency with a corresponding increase or decrease in pressure drop. It is of welded-steel or alloy construction and may be furnished with a cast-iron cone for abrasive applications. A wide range of sizes and types are offered for applications requiring various cut sizes. The Dustex unit employs multiple 12.7-cm- (5-in-) diameter cyclones in parallel for applications in which the dust loading does not exceed 23 g/m^3 (10 gr/ft^3). Efforts to avoid dust buildup, plugging, and subsequent maldistribution of gas flow in this collector include the use of vertical tube sheets, individual tangential inlets, and a design which permits circulation of the cleaned gases around the outside of the cyclones so as to minimize condensation from gases at temperatures close to the dew point. Standard construction materials are cast iron, cast aluminum, and welded stainless steels.

In addition to the conventional reverse-flow cyclones typified by Figs. 20-106 and 20-108, some use is made of uniflow, or straight-through, cyclones, in which the gas and solids discharge at the same end (Fig. 20-111). These devices act as concentrators; the concentrated dust, together with 5 to 20 percent of the inlet gas, is discharged at the periphery, while the clean gas passes out axially. The purge gas and concentrated dust enter a conventional cyclone for final separation. The straight-through cyclones are usually multiple-tube units.

The rotary-flow cyclone consists of a cylindrical body into which two separate gas streams are introduced. The primary flow of dirty gas being cleaned enters near the bottom through a spin-vane element and travels upward in a spiral along the cylinder axis to the

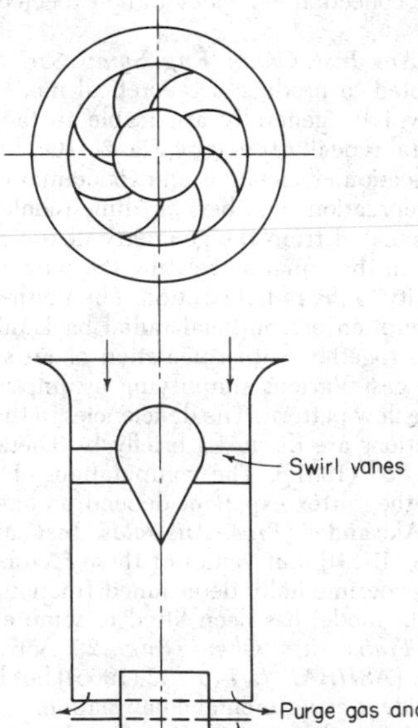

FIG. 20-111 Uniflow cyclone. [*Ter Linden*, Inst. Mech. Eng. J., *160*, 233 *(1949)*.]

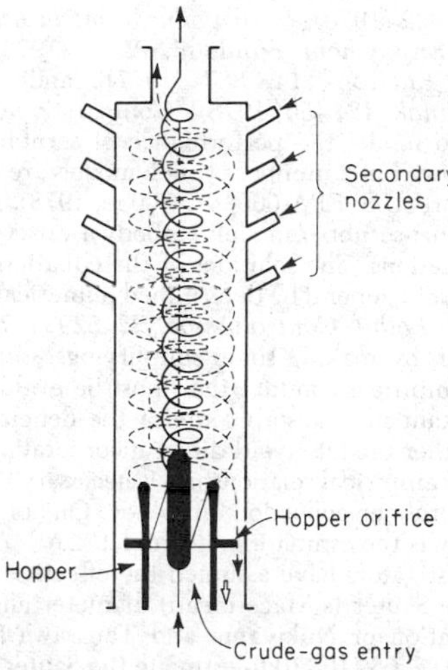

FIG. 20-112 Rotary-flow cyclone. [*Teske*, Staub, *English ed., 30, 35 (1970).*]

clean-gas outlet. The secondary gas enters at the top of the unit and travels downward in a spiral along the wall of the cylinder, augmenting the angular momentum of the total gas stream and carrying separated dust down to the hopper below. In the original form of the collector (Fig. 20-112), the secondary gas was introduced through a series of downward-angled tangential jets in the upper part of the cylinder wall. In later developments [Alt and Schmidt, *Staub*, English trans., **29** (7), 1 (1969)], the secondary gas is introduced through a single tangential inlet or through annular turning vanes located between the outer cylinder and the clean-gas outlet. The secondary gas may be clean air, partially cleaned gas, or part of the dirty gas being cleaned.

The rotary-flow cyclone was developed by Siemens AG and licensed to Aerodyne Development Corp. in the United States and Clarke Chapman Co., Ltd., in Great Britain. Early reports from the developer issued by the licensees indicated extremely high collection efficiencies for particles smaller than 5 μm. These showed a cut size D_{pc} of approximately 0.5 μm for a dust of specific gravity 2.6, which would be comparable to the performance of a high-energy scrubber. However, tests by Dunson ("Dust Collector Performance Evaluation by Midget Impinger and Coulter Counter," Pap. No. 5E, AIChE annual meeting, Chicago, 1970), with a collector of 200-ft^3/min (340-m^3/h) capacity, gave measured cut sizes of 2 to 3 μm, depending on the powder tested. This is consistent with the performance of a conventional cyclone (Fig. 20-106) of comparable capacity and pressure drop. Later analyses by Ciliberti and Lancaster [*Chem. Eng. Sci.*, **31**, 499 (1976); *Am. Inst. Chem. Eng. J.*, **22**, 394 (1976)] generally confirmed the observations of Dunson. The optimistic early reports probably resulted from the use of flocculated test dusts.

Mechanical Centrifugal Separators A number of collectors in which the centrifugal field is supplied by a rotating member are commercially available. In the typical unit shown in Fig. 20-113, the exhauster or fan and dust collector are combined as a single unit. The blades are especially shaped to direct the separated dust into an annular slot leading to the collection hopper while the cleaned gas continues to the scroll.

Although no comparative data are available, the collection efficiency of units of this type is probably comparable with that of the single-unit high-pressure-drop-cyclone installation. The clearances are smaller and the centrifugal fields higher than in a cyclone, but these advantages are probably compensated for by the shorter gas path and the greater degree of turbulence with its inherent reentrainment tendency. The chief advantage of these units lies in their compactness, which may be a prime consideration for large installations or plants requiring a large number of individual collectors. Caution should be exercised when attempting to apply this type of unit to a dust that shows a marked tendency to build up on solid surfaces, because of the high maintenance costs that may be encountered from plugging and rotor unbalancing.

Particulate Scrubbers Wet collectors, or scrubbers, form a class of devices in which a liquid (usually water) is used to assist or accomplish the collection of dusts or mists. Such devices have been in use for well over 100 years, and innumerable designs have been or are offered commercially or constructed by users. Wet-film collectors logically form a separate subcategory of devices. They comprise inertial collectors in which a film of liquid flows over the interior surfaces, preventing reentrainment of dust particles and flushing away the deposited dust. Wetted-wall cyclones are an example [Stairmand, *Trans. Inst. Chem. Eng.*, **29**, 356 (1951)]. Wet-film collectors have not been studied systematically but can probably be expected to perform much as do equivalent dry inertial collectors, except for the benefit of reduced reentrainment.

In particulate scrubbers, the liquid is dispersed into the gas as a spray, and the liquid droplets are the principal collectors for the dust particles. Depending on their design and operating conditions, particulate scrubbers can be adapted to collecting fine as well as coarse particles. Collection of particles by the drops follows the same principles illustrated in Fig. 20-103. Various investigations of the relative contributions of the various mechanisms have led to the conclusion that the predominant mechanism is inertial deposition. Flow-line interception is only a minor mechanism in the collection of the finer dust particles by liquid droplets of the sizes encountered in scrubbers. Diffusion is indicated to be a relatively minor mechanism for the particles larger than 0.1 μm that are of principal concern. Thermal deposition is negligible. Gravitational settling is ineffective because of the high gas velocities and short residence times used in scrubbers. Electrostatic deposition is unlikely to be important except in cases in which the dust particles or the water, or both, are being deliberately charged from an external power source to enhance collection. Deposition produced by Stefan flow can be significant when water vapor is condensing in a scrubber.

Despite numerous claims or speculations that wetting of dust particles by the scrubbing liquid plays a major role in the collection process, there is no unequivocal evidence that this is the case. The issue is whether wetting is an important factor in the adherence of a particle to a collecting droplet upon impact. From the body of general experience, it can be inferred that wettable particles probably are not collected much, if any, more readily than nonwettable particles of the same size. However, the available experimental techniques have not been adequate to permit any direct test to resolve the question. Changing from a wettable to a nonwettable test aerosol or from one scrubbing liquid to another is virtually certain to introduce other (and possibly unknown) factors into the scrubbing process. The most

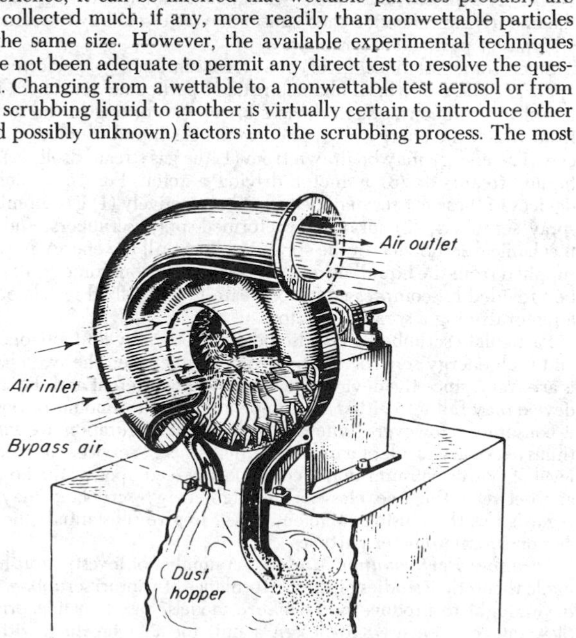

FIG. 20-113 Typical mechanical centrifugal separator. Type D Rotoclone (cutaway view). *(American Air Filter Co., Inc.)*

informative experimental studies appear to be some by Weber [*Staub*, English trans., **28**, 37 (November 1968); **29**, 12 (July 1969)], who bombarded single drops of various liquids with dust particles at different velocities and studied the behavior at impact by means of high-speed photography. Dust particles hitting the drops were invariably retained by the latter, regardless of their wettability by the liquid used.

The use of wetting agents in scrubbing water is equally controversial, and there has been no clear demonstration that it is beneficial.

A particulate scrubber may be considered as consisting of two parts: (1) a contactor stage, in which a spray is generated and the dust-laden gas stream is brought into contact with it; and (2) an entrainment separation stage, in which the spray and deposited dust particles are separated from the cleaned gas. These two stages may be separate or physically combined. The contactor stage may be of any form intended to bring about effective contacting of the gas and spray. The spray may be generated by the flow of the gas itself in contact with the liquid, by spray nozzles (pressure-atomizing or pneumatic-atomizing), by a motor-driven mechanical spray generator, or by a motor-driven rotor through which both gas and liquid pass.

Entrainment separation is accomplished with inertial separators, which are usually cyclones or impingement separators of various forms. If properly designed, these devices can remove virtually all droplets of the sizes produced in scrubbers. However, reentrainment of liquid can take place in poorly designed or overloaded separators.

Scrubber Types and Performance The diversity of particulate scrubber designs is so great as to defy any detailed and self-consistent system of classification based on configuration or principle of operation. However, it is convenient to characterize scrubbers loosely according to prominent constructional features, even though the modes of operation of different devices in a group may vary widely.

A relationship of power consumption to collection efficiency is characteristic of all particulate scrubbers. Attaining increased efficiency requires increased power consumption, and the power consumption required to attain a given efficiency increases as the particle size of the dust decreases. Experience generally indicates that the power consumption required to provide a specific efficiency on a given dust does not vary widely even with markedly different devices. The extent to which this generalization holds true has not been fully explored, but the known extent is sufficient to suggest that the underlying collection mechanism may be essentially the same in all types of particulate scrubbers.

Since some relationship of power consumption to performance appears to be a universal characteristic of particulate scrubbers, it is useful to characterize such devices broadly according to the source from which the energy is supplied to the gas-liquid contacting process. The energy may be drawn from (1) the gas stream itself, (2) the liquid stream, or (3) a motor driving a rotor. For convenience, devices in these classes may be termed respectively (1) gas-atomized spray scrubbers, (2) spray, or preformed-spray, scrubbers, and (3) mechanical scrubbers. In the spray scrubbers, all the energy may be supplied from the liquid, using a pressure nozzle, but some or all may be provided by compressed air or steam in a two-fluid nozzle or by a motor driving a spray generator.

Particulate scrubbers may also be classed broadly into low-energy and high-energy scrubbers. The distinction between the two classes is arbitrary, since the devices are not basically different and the same device may fall into either class depending on the amount of power it consumes. However, some differences in configuration are sometimes necessary to adapt a device for high-energy service. No specific level of power consumption is commonly agreed upon as the boundary between the two classes, but high-energy scrubbers may be regarded as those using sufficient power to give substantial efficiencies on submicrometer particles.

Scrubber Performance Models A number of investigators have made theoretical studies of the performance of venturi scrubbers and have sought to produce performance models, based on first principles, that can be used to design a unit for a given duty without recourse to experimental data other than the particle size and size distribution of the dust. Among these workers are Calvert [*Am. Inst. Chem. Eng. J.*, **16**, 392 (1970); *J. Air Pollut. Control Assoc.*, **24**, 929 (1974)], Boll [*Ind. Eng. Chem. Fundam.*, **12**, 40 (1973)], Goel and Hollands [*Atmos. Environ.*, **11**, 837 (1977)], and Yung et al. [*Environ. Sci. Technol.*, **12**, 456 (1978)]. Comparatively few efforts have been made to model the performance of scrubbers of types other than the venturi, but a number of such models are summarized by Yung and Calvert (U.S. EPA-600/8-78-005b, 1978).

The various venturi-scrubber models embody a variety of assumptions and approximations. The solutions of the equations for particulate collection must in general be determined numerically, although Calvert et al. [*J. Air Pollut. Control Assoc.*, **22**, 529 (1972)] obtained an explicit equation by making some simplifying assumptions and incorporating an empirical constant that must be evaluated experimentally; the constant may absorb some of the deficiencies in the model. Although other models avoid direct incorporation of empirical constants, use of empirical relationships is necessary to obtain specific estimates of scrubber collection efficiency. One of the areas of greatest uncertainty is the estimation of droplet size.

Most of the investigators have assumed the effective drop size of the spray to be the Sauter (surface-mean) diameter and have used the empirical equation of Nukiyama and Tanasawa [*Trans. Soc. Mech. Eng., Japan*, **5**, 63 (1939)] to estimate the Sauter diameter:

$$D_o = \frac{1920\sqrt{\sigma_L}}{V_o\sqrt{\rho_L/62.3}} + 75.4\left(\frac{\mu_L}{\sqrt{\sigma_L\rho_L/62.3}}\right)^{0.45}\left(\frac{1000Q_L}{Q_G}\right)^{1.5} \tag{20-68}$$

where D_o = drop diameter, μm; V_o = gas velocity, ft/s; σ_L = liquid surface tension, dyn/cm; ρ_L = liquid density, lb/ft^3; μ_L = liquid viscosity, cP; Q_L = liquid flow rate, ft^3/s; and Q_G = gas flow rate, ft^3/s.

The Nukiyama-Tanasawa equation, which is not dimensionally homogeneous, was derived from experiments with small, internal-mix pneumatic atomizing nozzles with concentric feed of air and liquid (Lapple et al., "Atomization: A Survey and Critique of the Literature," Stanford Res. Inst. Tech. Rep. No. 6, AD 821-314, 1967; Lapple, "Atomization," in *McGraw-Hill Encyclopedia of Science and Technology*, 5th ed., vol. 1, McGraw-Hill, New York, 1982, p. 858). The effect of nozzle size on the drop size is undefined. Even within the range of parameters for which the relationship was derived, the drop sizes reported by various investigators have varied by twofold to threefold from those predicted by the equation (Boll, op. cit.). The Nukiyama-Tanasawa equation has, nevertheless, been applied to large venturi and orifice scrubbers with configurations radically different from those of the atomizing nozzles for which the equation was originally developed.

Primarily because of the lack of adequate experimental techniques (particularly, the production of appropriate monodisperse aerosols), there has been no comprehensive experimental test of any of the venturi-scrubber models over wide ranges of design and operating variables. The models for other types of scrubbers appear to be essentially untested.

Contacting Power Correlation A scrubber design method that has achieved wide acceptance and use is based on correlation of the collection efficiency with the power dissipated in the gas-liquid contacting process, which is termed "contacting power." The method originated from an investigation by Lapple and Kamack [*Chem. Eng. Prog.*, **51**, 110 (1955)] and has been extended and refined in a series of papers by Semrau and coworkers [*Ind. Eng. Chem.*, **50**, 1615 (1958); *J. Air Pollut. Control Assoc.*, **10**, 200 (1960); **13**, 587 (1963); U.S. EPA-650/2-74-108, 1974; U.S. EPA-600/2-77-234, 1977; *Chem. Eng.*, **84**(20), 87 (1977); and "Performance of Particulate Scrubbers as Influenced by Gas-Liquid Contactor Design and by Dust Flocculation," EPA-600/9-82-005c, 1982, p. 43]. Other workers have made extensive independent studies of the correlation method [Walker and Hall, *J. Air Pollut. Control Assoc.*, **18**, 319 (1968)], and numerous studies of narrower scope have been made. The major conclusion from these studies is that the collection efficiency of a scrubber on a given dust is essentially dependent only on the contacting power and is affected to only a minor degree by the size or geometry of the scrubber or by the way in which the contacting power is applied. This contacting-power rule is strictly empirical, and the full extent of its validity has still not been explored. It has been best verified for the class of gas-atomized spray scrubbers, in which the contacting power is derived from the gas stream and takes the form of gas pressure drop. Tests of the equivalence of contacting power supplied from the liquid stream in pressure spray nozzles have been far less extensive and are strongly indicative but not yet conclusive. Evidence for the equivalence of contacting power from mechanically driven devices is also indicative but extremely limited in quantity.

Contacting power is defined as the power per unit of volumetric gas flow rate that is dissipated in gas-liquid contacting and is ultimately converted to heat. In the simplest case, in which all the energy is obtained from the gas stream in the form of pressure drop, the contacting power is equivalent to the friction loss across the wetted equipment, which is termed "effective friction loss," F_E. The pressure drop may reflect kinetic-energy changes rather than energy dissipation, and pressure drops that result solely from kinetic-energy changes in the gas stream do not correlate with performance. Likewise, any friction losses taking place across equipment that is operating dry do not contribute to gas-liquid contacting and do not correlate with performance. The gross power input to a scrubber includes losses in motors, drive shafts, fans, and pumps that obviously should be unrelated to scrubber performance.

The effective friction loss, or "gas-phase contacting power," is easily determined by direct measurements. However, the "liquid-phase contacting power," supplied from the stream of scrubbing liquid, and the "mechanical contacting power," supplied by a mechanically driven rotor, are not directly measurable; the theoretical power inputs can be estimated, but the portions of these quantities effectively converted to contacting power can only be inferred from comparison with gas-phase contacting power. Such data as are available indicate that the contributions of contacting power from different sources are directly additive in their relation to scrubber performance.

Contacting power is variously expressed in units of MJ/1000 m^3 (SI), kWh/1000 m^3 (meter-kilogram-second system), and hp/(1000 ft^3/min) (U.S. customary). Relationships for conversion to SI units are

$$1.0 \text{ kWh}/1000 \text{ m}^3 = 3.60 \text{ MJ}/1000 \text{ m}^3$$
$$1.0 \text{ hp}/(1000 \text{ ft}^3/\text{min}) = 1.58 \text{ MJ}/1000 \text{ m}^3$$

The gas-phase contacting power P_G may be calculated from the effective friction loss by the following relationships:

SI units:

$$P_G = 1.0F_E \tag{20-69}$$

where F_E = kPa.

U.S. customary units:

$$P_G = 0.1575F_E \tag{20-70}$$

where F_E = in of water.

The power input from a liquid stream injected with a hydraulic spray nozzle may usually be taken as approximately equal to the product of the nozzle feed pressure p_F and the volumetric liquid rate. The liquid-phase contacting power P_L may then be calculated from the following formulas:

SI units:

$$P_L = 1.0p_F(Q_L/Q_G) \tag{20-71}$$

where p_F = kPa gauge, and Q_L and Q_G = m^3/s.

U.S. customary units:

$$P_L = 0.583p_F(Q_L/Q_G) \tag{20-72}$$

where p_G = lbf/in^2 gauge, Q_L = gal/min, and Q_G = ft^3/min.

The correlation of efficiency data is based on the total contacting power P_T, which is the sum of P_G, P_L, and any power P_M that may be supplied mechanically by a power-driven rotor.

In general, the liquid-to-gas ratio does not have an influence independent of contacting power on the collection efficiency of scrubbers of the venturi type. This is true at least of operation with liquid-to-gas ratios above some critical lower value. However, several investigations [Semrau and Lunn, "Performance of Particulate Scrubbers as Influenced by Gas-Liquid Contactor Design and by Dust Flocculation," EPA-600/9-82-005c, 1982, p. 43; and Muir et al., *Filtr. Sep.*, **15**, 332 (1978)] have shown that at low liquid-to-gas ratios relatively poor efficiencies may be obtained at a given contacting power. Such regions of operation are obviously to be avoided.

It has sometimes been asserted that multiple gas-liquid contactors in series will give higher efficiencies at a given contacting power than will a single contacting stage. However, there is little experimental evidence to support this contention. Lapple and Kamack (op. cit.) obtained slightly higher efficiencies with a venturi and an orifice in series than they did with a venturi alone. Muir and Mihisei [*Atmos. Environ.*, **13**, 1187 (1979)] obtained somewhat higher efficiencies on two redispersed dusts when using two venturis in series rather than one. The improvement obtained with two-stage scrubbing was greatest with the coarser of the two dusts and was relatively small with the finer dust. Flocculation or deflocculation of the dusts may have been responsible for some of the behavior encountered. Semrau et al. (EPA-600/2-77-234, 1977) compared the performance of a four-stage multiple-orifice contactor with that of a single-orifice contactor, using well-dispersed aerosols generated from ammonium fluorescein, The multiple-orifice contactor gave about the same efficiency as the single-orifice in the upper range of contacting power but lower efficiencies in the lower range. The deviations in performance in this case were probably characteristic of the particular multiple-orifice contactor rather than of multistage contacting as such.

Most scrubbers actually incorporate more than one stage of gas-liquid contacting even though these may not be identical (e.g., the contactor and the entrainment separator). The preponderance of evidence indicates that multiple-stage contacting is not inherently either more or less efficient than single-stage contacting. However, two-stage contacting may have practical benefits in dealing with abrasive or flocculated dusts.

Some investigators have proposed, mostly on the basis of mathematical modeling, to optimize the design of scrubbers to obtain a given efficiency with a minimum power consumption (e.g., Goel and Hollands, op. cit.). In fact, no optimum in performance appears to exist; apart from some avoidable regions of unfavorable operation, increased contacting power yields increased efficiency.

Scrubber Performance Curves The scrubber performance curve, which shows the relationship of scrubber efficiency to the contacting power, has been found to take the form

$$N_t = \alpha P_T^{\gamma} \tag{20-73}$$

where α and γ are empirical constants that depend primarily on the aerosol (dust or mist) collected. In a log-log plot of N_t versus P_T, γ is the slope of the performance curve and α is the intercept at $P_T = 1$. Figure 20-114 shows such a performance curve for the collection of coal fly ash by a pilot-plant venturi scrubber (Raben "Use of Scrubbers for Control of Emissions from Power Boilers," United States–U.S.S.R. Symposium on Control of Fine-Particulate Emissions from Industrial Sources, San Francisco, 1974). The scatter in the data reflects not merely experimental errors but actual variations in the particle-size characteristics of the dust. Because the characteristics of an industrial dust vary with time, the scrubber performance curve necessarily must represent an average material, and the scatter in the data is frequently greater than is shown in Fig. 20-114. For best definition, the curve should cover as wide a range of contacting power as possible. Obtaining the data thus requires pilot-plant equipment with the flexibility to operate over a wide range of conditions. Because scrubber performance is not greatly affected by the size of the unit, it is feasible to conduct the tests with a unit handling no more than 170 m^3/h (100 ft^3/min) of gas.

A clear interpretation of γ, the slope of the curve, is still lacking. Presumably, it should be related to the particle-size distribution of the dust. Because scrubbing preferentially removes the coarser particles, the fraction of the dust removed (or the increment of N_t) per unit of contacting power should decrease as the contacting power and efficiency increase, so that the value of γ should be less than unity. In fact, the value of γ has been less than unity for most dusts. Nevertheless, some data in the literature have displayed values of γ

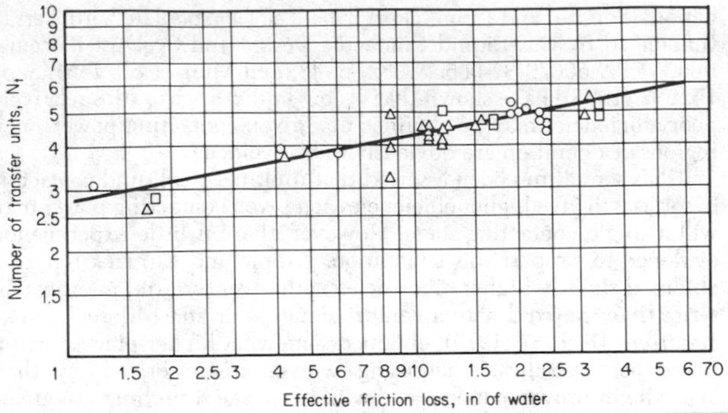

FIG. 20-114 Performance of pilot-plant venturi scrubber on fly ash. Liquid-to-gas ratio, gal/1000 ft^3: ○, 10; △, 15; □, 20. *(Raben, United States–U.S.S.R. Symposium on Fine-Particulate Emissions from Industrial Sources, San Francisco, 1974.)*

greater than unity when plotted on the transfer-unit basis, indicating that the residual fraction of dust became more readily collectible as contacting power and efficiency increased. More recent studies by Semrau et al. (EPA-650/2-74-108 and EPA-600/2-77-237) have revealed performance curves having two branches (typified by Fig. 20-115), the lower having a slope greater than unity and the upper a slope less than unity. This suggests that had the earlier tests been extended into higher contacting-power ranges, performance curves with flatter slopes might have appeared in those ranges.

Among the aerosols that gave performance curves with $\gamma > 1$, the only obvious common characteristic was that a large fraction of each was composed of submicrometer particles.

Cut-Power Correlation Another design method, also based on scrubber power consumption, is the cut-power method of Calvert [*J. Air Pollut. Control Assoc.*, **24,** 929 (1974); *Chem. Eng.*, **84**(18), 54 (1977)]. In this approach, the cut diameter (the particle diameter for which the collection efficiency is 50 percent) is given as a function of the gas pressure drop or of the power input per unit of volumetric gas flow rate. The functional relationship is presented as a log-log plot of the cut diameter versus the pressure drop (or power input). In principle, the function could be constructed by experimentally determining scrubber performance curves for discrete particle sizes and then plotting the particle sizes against the corresponding pressure drops necessary to give efficiencies of 50 percent. In practice, Calvert and coworkers evidently have in most cases constructed the cut-power functions for various scrubbers by modeling (Yung and Calvert, U.S. EPA-600/8-78-005b, 1978). They show a variety of curves, whereas empirical studies have indicated that different types of scrubbers generally have about the same performance at a given level of power consumption.

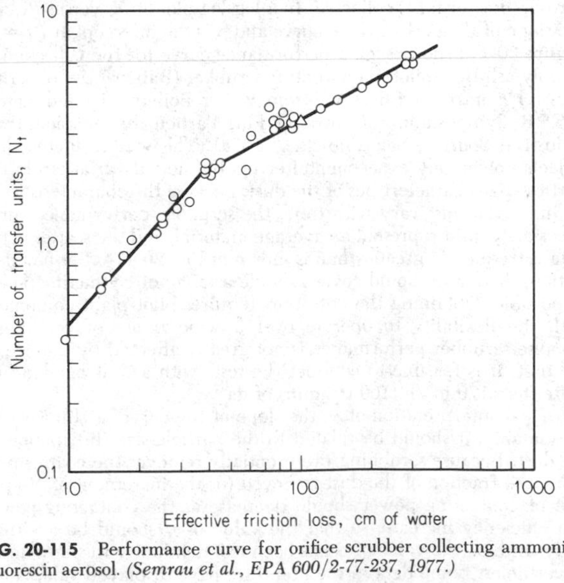

FIG. 20-115 Performance curve for orifice scrubber collecting ammonium fluorescin aerosol. *(Semrau et al., EPA 600/2-77-237, 1977.)*

Condensation Scrubbing The collection efficiency of scrubbing can be increased by the simultaneous condensation of water vapor from the gas stream. Water-vapor condensation assists in particle removal by two entirely different mechanisms. One is the deposition of particles on cold-water droplets or other surfaces as the result of Stefan flow. The other is the condensation of water vapor on particles as nuclei, which enlarges the particles and makes them more readily collected by inertial deposition on droplets. Both mechanisms can operate simultaneously. However, for the buildup of particles by condensation to be effective, there must be adequate time for the particles to grow substantially before the principal gas-liquid-contacting operation takes place. Hence, if particle buildup is to be sought, the scrubber should be preceded by an appropriate gas-conditioning section. On the other hand, particle collection by Stefan flow can be induced simply by scrubbing the hot, humid gas with sufficient cold water to bring the gas below its initial dew point. Any practical method of inducing condensation on the dust particles will incidentally afford opportunities for the operation of the Stefan-flow mechanism. The hot gas stream must, of course, have a high initial moisture content, since the magnitude of the effects obtained is related to the quantity of water vapor condensed.

Although there is a considerable body of literature on particle collection by condensation mechanisms, most of it is either theoretical or, if experimental, treats basic phenomena in simplified cases. Few studies have been made to determine what performance may be expected from condensation scrubbing under practical conditions in industrial applications. In a series of studies, Calvert and coworkers investigated several types of equipment for condensation scrubbing, generally emphasizing the use of the condensation center effect to build up the particles for collection by inertial deposition (Calvert and Parker, EPA-600/8-78-005c, 1978). From early estimates, they predicted that a condensation scrubber would require only about one-third or less of the power required by a conventional high-energy scrubber. A subsequent demonstration-plant scrubber system consisted of a direct-contact condensing tower fed with cold water followed by a venturi scrubber fed with recirculated water (Chmielewski and Calvert, EPA-600/7-81-148, 1981). The condensation and particle buildup took place in the cooling tower. In operation on humidified iron-foundry-cupola gas, this system still required about 65 percent as much power as for conventional high-energy scrubbing.

Semrau and coworkers [*Ind. Eng. Chem.*, **50,** 1615 (1958); *J. Air Pollut. Control Assoc.*, **13,** 587 (1963); EPA-650/2-74-108, 1974] investigated condensation scrubbing in pilot-plant studies in the field and, later, under laboratory conditions. Hot, humid gases were scrubbed directly with cold water under conditions that were favorable for the Stefan-flow mechanism but offered little or no opportunity for particle buildup. Some of the field studies indicated a contacting-power saving of as much as 50 percent for condensation scrubbing of Kraft-recovery-furnace fume. Laboratory tests on a predominantly submicrometer synthetic aerosol showed contacting-power savings of up to 40 percent with condensation scrubbing.

In the scrubbing of hot gases with high water content, condensation reduces contacting power and affords a direct power saving through the reduction of the gas volume by cooling and water-vapor condensation, but it incurs other costs for power and equipment for heat transfer and water cooling. However, condensation scrubbing may offer a net economic advantage if recovery of low-level heat is practical. It should also be advantageous when a hot gas must be not only cleaned but cooled and dehumidified as well; examples are the cleaning of blast-furnace gas for use as fuel and of SO_2-bearing waste gases for feed to a sulfuric acid plant.

Entrainment Separation The entrainment separator is a critical element of a scrubber, since the collection efficiency of the scrubber depends on essentially total removal of the spray from the gas stream. The sprays generated in scrubbers are generally large enough in droplet size that they can be readily removed by properly designed inertial separators. Primary collection of the spray is seldom the critical limitation on separator performance, but reentrainment is a common problem. In dust scrubbers it is essential that the entrainment separator not be of a form readily subject to blockage by solids deposits and that it be readily cleared of deposits if they should occur. Cyclone separators are advantageous in this respect and are widely used with venturi contactors. However, they cannot readily be made integral with scrubbers of some other configurations, which can be more conveniently fitted with various forms of impingement separators (see Sec. 18). Although separator design can be important, the most common cause of reentrainment is simply the use of excessive gas velocities, and few data are available on the gas-handling capacities of separators. In the absence of good data, there is a frequent tendency to underdesign separators in an effort to reduce costs.

Venturi Scrubbers The venturi scrubber is one of the most widely used types of particulate scrubbers. The designs have become generally standardized, and units are manufactured by a large number of companies. Venturi scrubbers may be used as either high- or low-energy devices but are most commonly employed as high-energy units. The units originally studied and used were designed to the proportions of the classical venturis used for metering, but since it was discovered that these proportions have no special merits, simpler and more practical designs have been adopted. Most "venturi" contactors in current use are in fact not venturis but variable orifices of one form or another. Any of a wide range of devices can be used, including a simple pipe-line contactor. Although the venturi scrubber is not inherently more efficient at a given contacting power than other types of devices, its simplicity and flexibility favor its use. It is also useful as a gas absorber for relatively soluble gases, but because it is a cocurrent contactor, it is not well suited to absorption of gases having low solubilities.

Current designs for venturi scrubbers generally use the vertical downflow of gas through the venturi contactor and incorporate three features: (1) a "wet-approach" or "flooded-wall" entry section, to avoid dust buildup at a wet-dry junction; (2) an adjustable throat for the venturi (or orifice), to provide for adjustment of the pressure drop; and (3) a "flooded elbow" located below the venturi and ahead of the entrainment separator, to reduce wear by abrasive particles. The venturi throat is sometimes fitted with a refractory lining to resist abrasion by dust particles. The entrainment separator is commonly, but not invariably, of the cyclone type. An example of the "standard form" of venturi scrubber is shown in Fig. 20-116. The wet-approach entry section has made practical the recirculation of slurries. Various forms of adjustable throats, which may be under manual or automatic control, permit maintaining a constant pressure drop and constant efficiency under conditions of varying gas flow.

Self-Induced Spray Scrubbers Self-induced spray scrubbers form a category of gas-atomized spray scrubbers in which a tube or a duct of some other shape forms the gas-liquid-contacting zone. The gas stream flowing at high velocity through the contactor atomizes the liquid in essentially the same manner as in a venturi scrubber. However, the liquid is fed into the contactor and later recirculated from the entrainment separator section by gravity instead of being circulated by a pump as in venturi scrubbers. The scheme is well illustrated in Fig. 20-117*a*. A great many such devices using contactor ducts of various shapes, as in Fig. 20-117*b*, are offered commercially. Although self-induced spray scrubbers can be built as high-energy units and sometimes are, most such devices are designed for only low-energy service.

The principal advantage of self-induced spray scrubbers is the elimination of a pump for recirculation of the scrubbing liquid. However, the designs for high-energy service are somewhat more complex and less flexible than those for venturi scrubbers.

Plate Towers Plate (tray) towers are countercurrent gas-atomized spray scrubbers using one or more plates for gas-liquid contacting. They are essentially the same as, if not identical to, the devices used for gas absorption and are frequently employed in applications

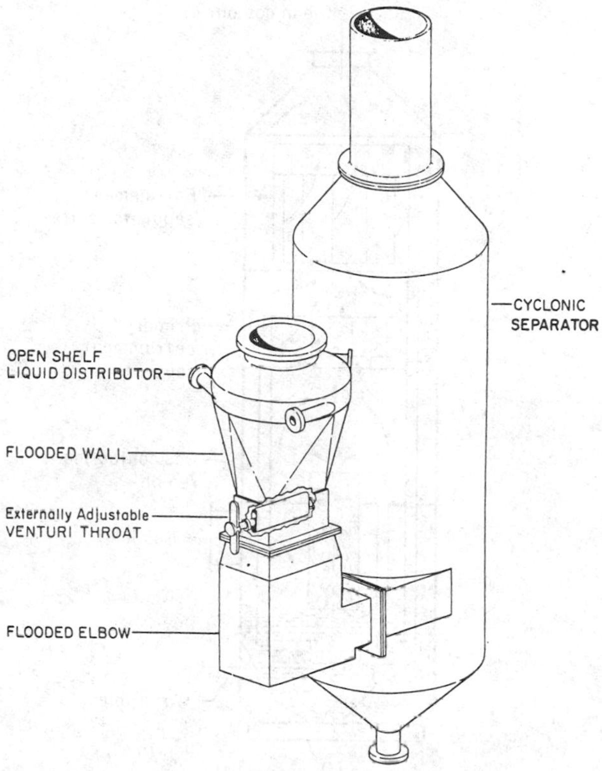

FIG. 20-116 Venturi scrubber. *(Neptune AirPol.)*

in which gases are to be absorbed simultaneously with the removal of dust. Except possibly in cases in which condensation effects are involved, countercurrent operation is not significantly beneficial in dust collection.

The plates may be any of several types, including sieve, bubble-cap, and valve trays. The impingement baffle plate (Fig. 20-118) is commonly used for dust collection applications. Impingement on the baffles is not the controlling mechanism of particle collection; the principal collecting bodies are the droplets produced from the liquid by the gas as it flows through the perforations and around the baffles. The slot stage (Fig. 20-118) is in effect a miniature venturi contactor. Valve trays constitute multiple self-adjusting orifices that provide nearly constant gas pressure drop over considerable ranges of variation in gas flow. The gas pressure drop that can be taken across a single plate is necessarily limited, so that units designed for high contacting power must use multiple plates.

Plate towers are more subject to plugging and fouling than venturi-type scrubbers that have large passages for gas and liquid.

Packed-Bed Scrubbers Packed-bed scrubbers of the types used for gas absorption (see Sec. 18) may also be used for dust collection but are subject to plugging by deposits of insoluble solids. Random packings, such as dumped Raschig rings and Berl saddles, are most seriously affected by plugging. Regular packings, such as stacked grids, are better in dust-collection service. When both a gas and particulate matter are to be collected, it is advisable to use a primary-stage scrubber of the venturi or similar type to collect the particulate matter ahead of a packed gas absorber.

Packed-bed scrubbers may be constructed for either vertical or horizontal gas flow. Vertical-flow units (packed towers) commonly use countercurrent flow of gas and liquid, although cocurrent flow is sometimes used. Packed scrubbers using horizontal gas flow usually employ cross-flow of liquid.

Scrubber packings are too large to serve as collecting bodies for

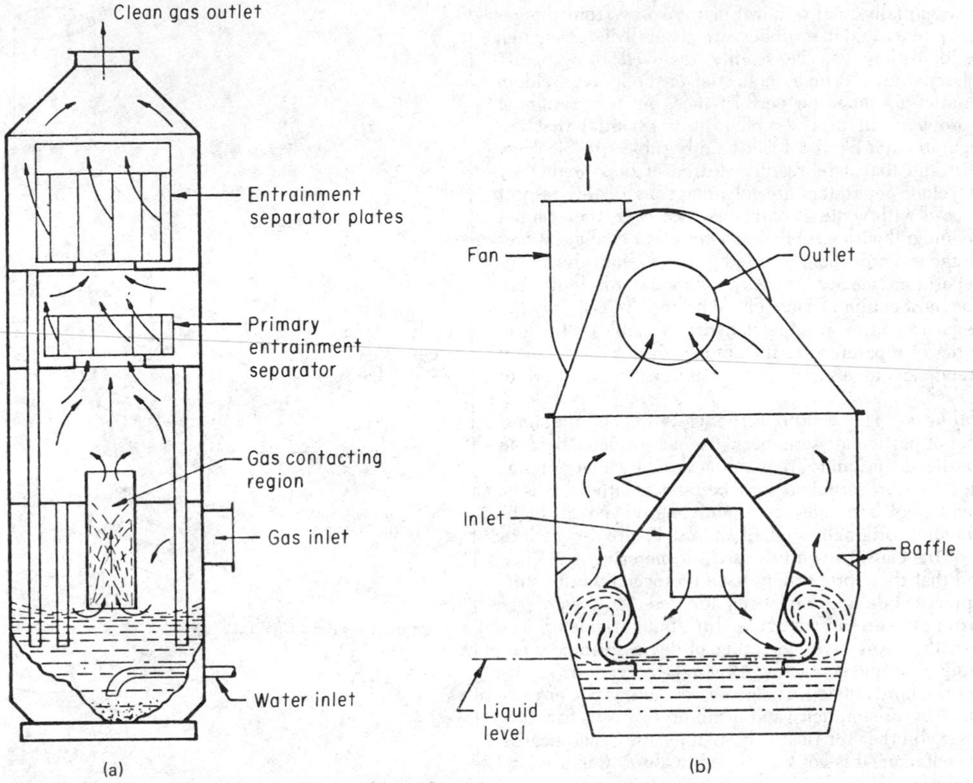

FIG. 20-117 Self-induced spray scrubbers. (*a*) Blaw-Knox Food & Chemical Equipment, Inc. (*b*) American Air Filter Co., Inc.

any except very large dust particles. In the collection of fine particles, the packings serve primarily to promote fluid turbulence that aids the deposition of the dust particles on droplets. In a packed tower operating below the flooding point, with most of the liquid flowing in films and little spray formation, the relative efficiency in collection of particles may possibly be lower than that of a venturi-type scrubber operating at the same contacting power. However, no data are available to resolve the question.

Mobile-Bed Scrubbers Mobile-bed scrubbers (Fig. 20-119) are constructed with one or more beds of low-density spheres that are free to move between upper and lower retaining grids. The spheres are commonly 1.0 in (2.5 cm) or more in diameter and made from rubber or a plastic such as polypropylene. The plastic spheres may be solid or hollow. Gas and liquid flows are countercurrent, and the spherical packings are fluidized by the upward-flowing gas. The movement of the packings is intended to minimize fouling and plugging of the bed. Mobile-bed scrubbers were first developed for absorbing gases from gas streams that also carry solid or semisolid particles.

The spherical packings are too large to serve as effective targets for the deposition of fine dust particles. In dust-collection service, the packings actually serve as turbulence promoters, while the dust particles are collected primarily by the liquid droplets.

The gas pressure drop through the scrubber may be increased by increasing the gas velocity, the liquid-to-gas ratio, the depth of the bed, the density of the packings, and the number of beds in series. In an experimental study, Yung et al. (EPA-600/7-79-071, 1979) determined that the collection efficiency of a mobile-bed scrubber was dependent only on the gas pressure drop and was not influenced independently by the gas velocity, the liquid-to-gas ratio, or the number of beds except as these factors affected the pressure drop. Yung et al. also reported that the mobile-bed scrubber was less efficient at a given pressure drop than scrubbers of the venturi type, but without offering comparable experimental supporting evidence.

Spray Scrubbers Spray scrubbers consist of empty chambers of some simple form in which the gas stream is contacted with liquid droplets generated by spray nozzles. A common form is a spray tower, in which the gas flows upward through a bank or successive banks of spray nozzles. Similar arrangements are sometimes used in spray chambers with horizontal gas flow. Such devices have very low gas pressure drops, and all but a small part of the contacting power is derived from the liquid stream. The required contacting power is obtained from an appropriate combination of liquid pressure and flow rate. Most spray scrubber are low-energy units. Collection of fine particles is possible but may require very high liquid-to-gas ratios, liquid feed pressures, or both. Plugging of the nozzles can be a persistent maintenance problem. Entrainment separators are necessary to prevent carry-over of spray into the exit gas.

Cyclone Scrubbers The vessels of cyclone scrubbers are all in the form of cyclones, which provide for entrainment separation. However, the gas-liquid-contacting devices may be of either the gas-atomized-spray or the preformed-spray type. The cyclone-spray scrubber shown in Fig. 20-120*a* has an axial spray tree, or manifold, equipped with hydraulic spray nozzles. Similar units are available with the spray nozzles mounted in the wall of the cyclone, discharging inward. This latter arrangement makes the nozzles more accessible for maintenance. In the cyclone scrubber shown in Fig. 20-120*b*, most of the gas-liquid contacting is accomplished in the swirl vanes, with the energy being supplied from the gas stream in the form of pressure drop. The swirl vanes serve the same function as do the trays in a tray tower. Higher contacting power can be provided by using additional sets of swirl vanes in series.

Ejector-Venturi Scrubbers In the ejector-venturi scrubber (Fig. 20-121) the cocurrent water jet from a spray nozzle serves both to

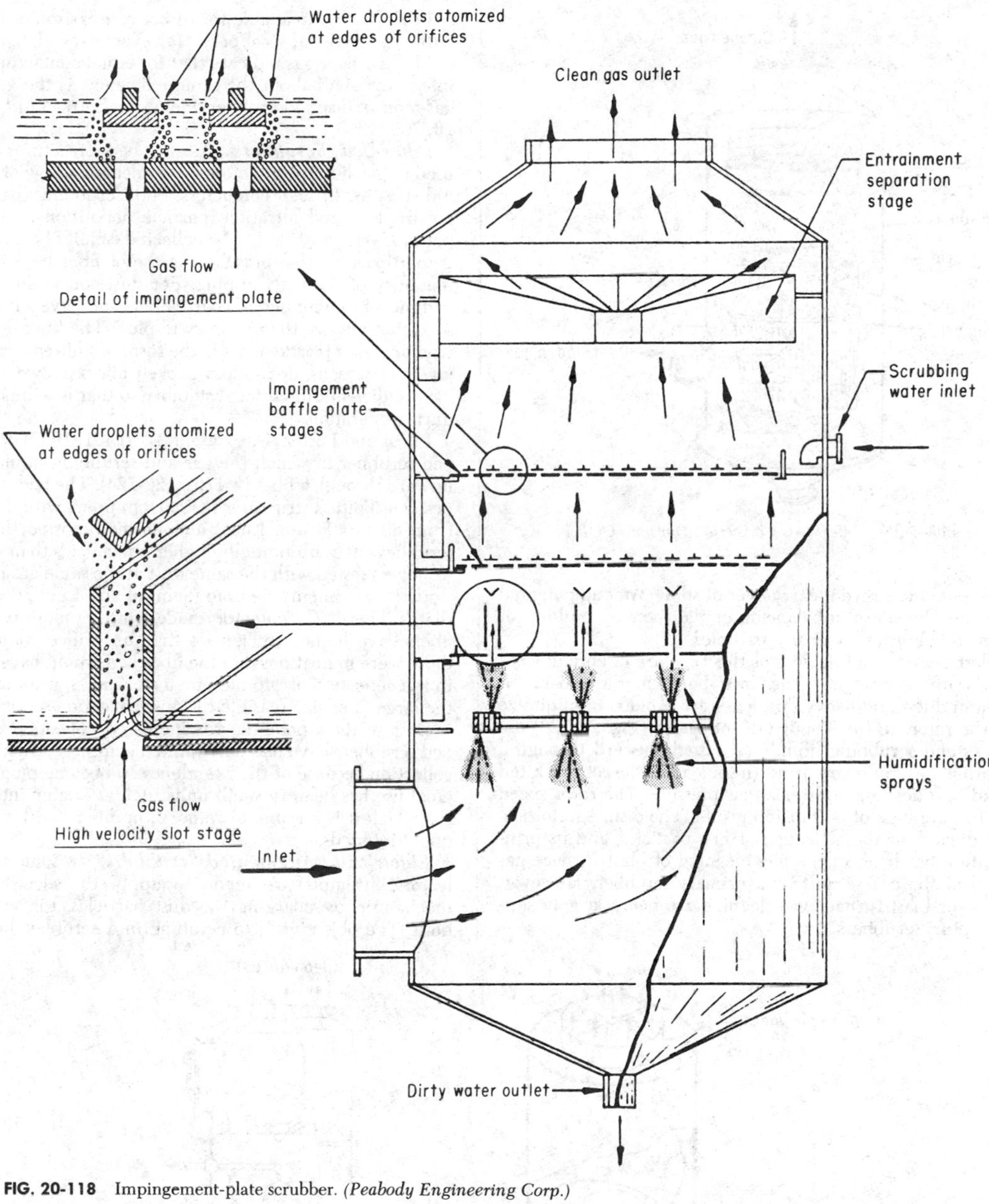

FIG. 20-118 Impingement-plate scrubber. *(Peabody Engineering Corp.)*

scrub the gas and to provide the draft for moving the gas. No fan is required, but the equivalent power must be supplied to the pump that delivers water to the ejector nozzle. The water must be supplied in sufficient volume and at high enough pressure to provide both adequate draft and enough contacting power for the required scrubbing operation. Considered as a gas pump, the ejector is not a very efficient device, but the dissipated energy that is not effective in pumping does serve in gas-liquid contacting. The energy equivalent to any gas pressure rise across the scrubber is not part of the contacting power (Semrau et al., EPA-600/2-77-234, 1974).

The ejector-venturi scrubber is widely used as a gas absorber, but the combinations of water pressure and flow rate that are sufficient to provide the required draft usually do not also yield enough contacting power to give high collection efficiency on submicrometer particles. Other types of ejectors have been employed to provide higher contacting-power levels. In one, superheated water is discharged through the nozzle, and part flashes to steam, increasing the mechanical energy available for scrubbing [Gardenier, *J. Air Pollut. Control Assoc.*, **24**, 954 (1974)]. Some units use two-fluid nozzles, with either compressed air or steam as the compressible fluid [Sparks, *J. Air Pollut. Control Assoc.*, **24**, 958 (1974)]. Most of the energy for gas movement and for atomizing the liquid and scrubbing the gas is derived from the compressed air or steam. In some ejector-venturi-scrubbers installations, part of the draft is supplied by a fan[Williams and Fuller, TAPPI, **60**(1), 108 (1977)].

Mechanical Scrubbers Mechanical scrubbers comprise those devices in which a power-driven rotor produces the fine spray and the contacting of gas and liquid. As in other types of scrubbers, it is the droplets that are the principal collecting bodies for the dust particles. The rotor acts as a turbulence producer. An entrainment sep-

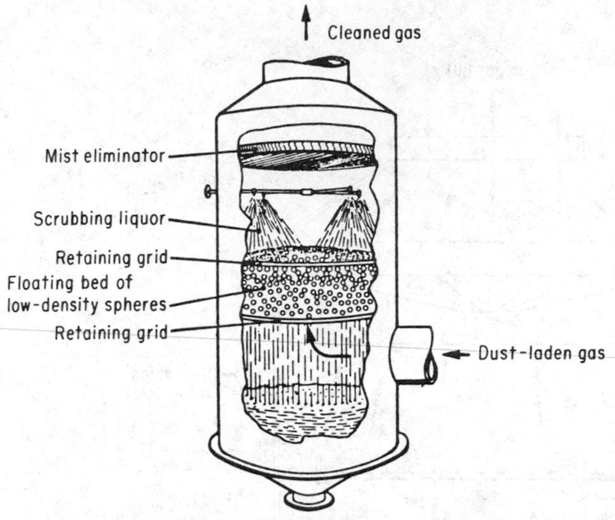

FIG. 20-119 Mobile-bed scrubber. *(Air Correction Division, UOP.)*

arator must be used to prevent carry-over of spray. Among potential maintenance problems are unbalancing of the rotor by buildup of dust deposits and abrasion by coarse particles.

The simplest commercial devices of this type are essentially fans upon which water is sprayed. The unit shown in Fig. 20-122 is adapted to light duty, and heavy dust loads are avoided to minimize buildup on the rotor. In the Theisen disintegrator (Fig. 20-123) the dust-laden gas and scrubbing liquid are passed outward through a series of rotating and stationary arms. In some units the rotor is fitted with fan blades to develop additional gas pressure. The rotor speed is generally in the range of 350 to 750 r/min. The disintegrator is a high-energy device for the collection of fine particles, and its principal application has been in the fine cleaning of blast-furnace gas after removal of the coarse dust by a primary scrubber. However, disintegrators for blast-furnace-gas cleaning are being largely supplanted by venturi scrubbers.

Few data are available for direct comparison of the disintegrator with other types of scrubbers. However, a few data on the scrubbing of blast-furnace gas indicate that for equal contacting power the disintegrator gives about the same efficiency as the venturi and other gas-atomized spray scrubbers [Semrau, *J. Air Pollut. Control Assoc.*, **10,** 200 (1960)].

Fiber-Bed Scrubbers Fibrous-bed structures are sometimes used as gas-liquid contactors, with cocurrent flow of the gas and liquid streams. In such contactors, both scrubbing (particle deposition on droplets) and filtration (particle deposition on fibers) may take place. If only mists are to be collected, small fibers may be used, but if solid particles are present, the use of fiber beds is limited by the tendency of the beds to plug. For dust-collection service, the fiber bed must be composed of coarse fibers and have a high void fraction, so as to minimize the tendency to plug. The fiber bed may be made from metal or plastic fibers in the form of knitted structures, multiple layers of screens, or random-packed fibers. However, the bed must have sufficient dimensional stability so that it will not be compacted during operation.

Lucas and Porter (U.S. Patent 3,370,401, 1967) developed a fiber-bed scrubber in which the gas and scrubbing liquid flow vertically upward through a fiber bed (Fig. 20-124). The beds tested were composed of knitted structures made from fibers with diameters ranging from 89 to 406 μm. Lucas and Porter reported that the fiber-bed scrubber gave substantially higher efficiencies than did venturi-type scrubbers tested with the same dust at the same gas pressure drop. In similar experiments, Semrau (Semrau and Lunn, op. cit.) also found that a fiber-bed contactor made with random-packed steel-wool fibers gave higher efficiencies than an orifice contactor. However, there were indications that the fiber bed would have little advantage in the collection of submicrometer particles, presumably because of the large fiber size feasible for dust-collection service.

Despite their potential for increased collection efficiency, fiber-bed scrubbers have had only limited commercial acceptance for dust collection because of their tendency to become plugged. Their principal use has been in small units such as engine-intake-air cleaners, for which it is feasible to remove the fiber bed for cleaning at frequent intervals.

Electrically Augmented Scrubbers In some types of wet collectors, attempts are made to apply the electrostatic-deposition mechanism by charging the dust particles, the water droplets, or both. The objective is to combine in a scrubber high efficiency in

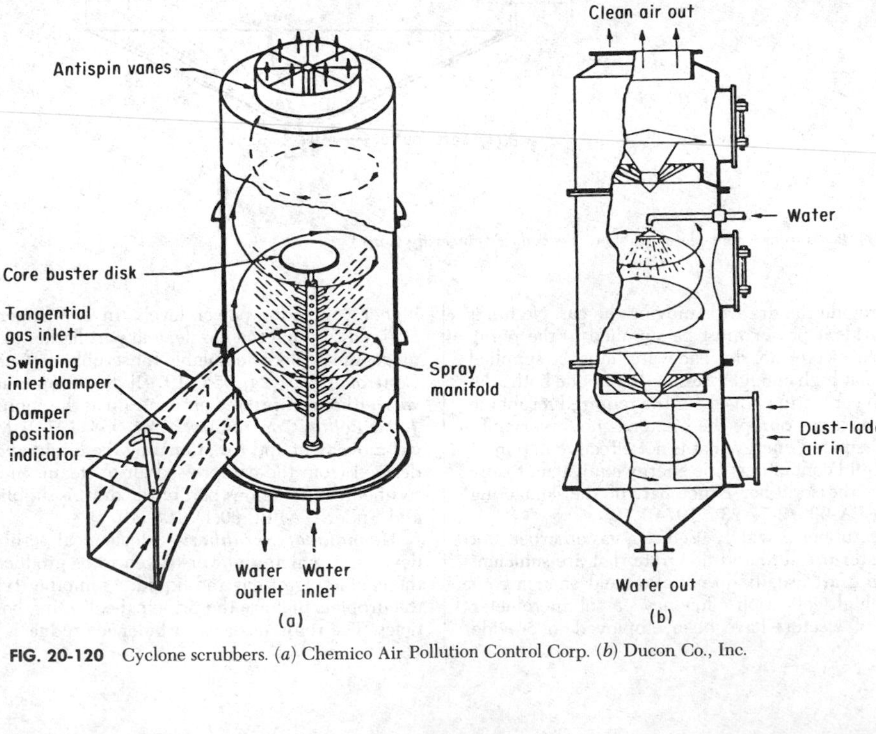

FIG. 20-120 Cyclone scrubbers. (*a*) Chemico Air Pollution Control Corp. (*b*) Ducon Co., Inc.

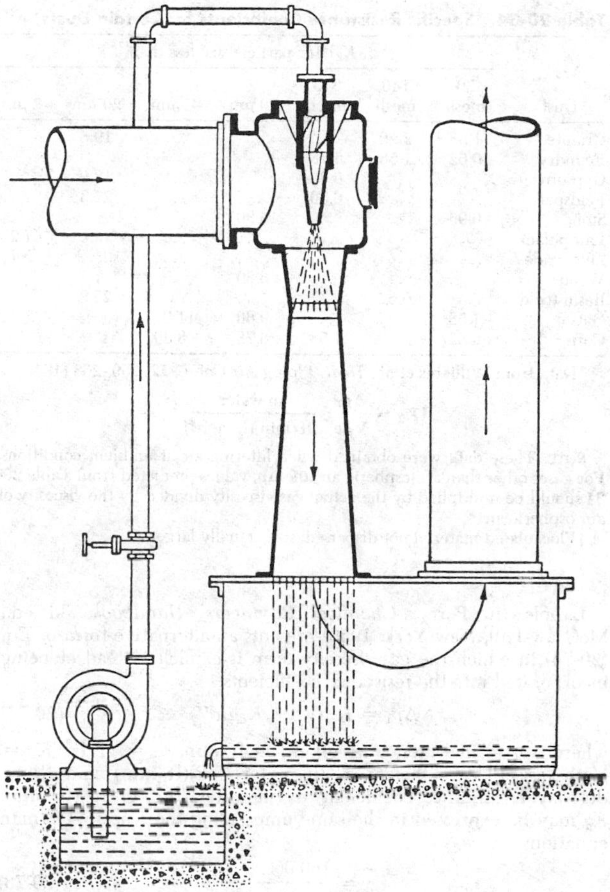

FIG. 20-121 Ejector-venturi scrubber. *(Schutte & Koerting Division, Amtek, Inc.)*

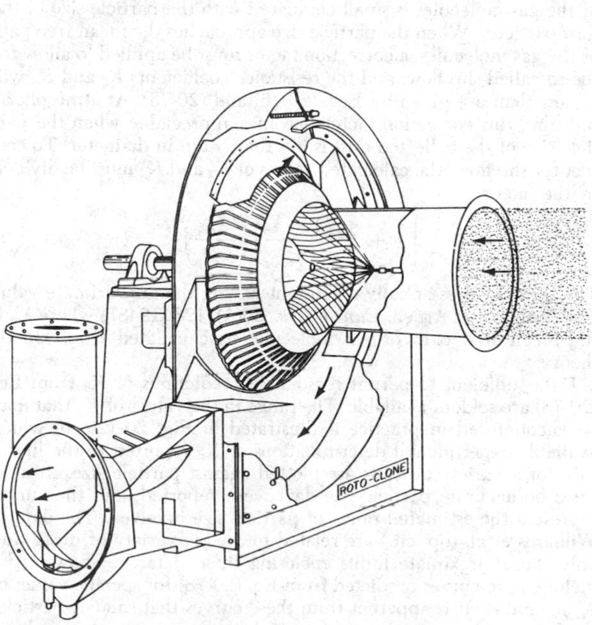

FIG. 20-122 Mechanical scrubber. *(American Air Filter Co., Inc.)*

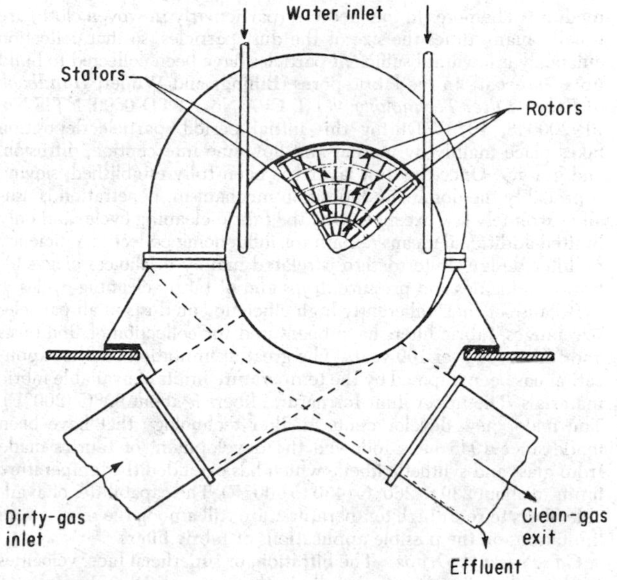

FIG. 20-123 Disintegrator scrubber. [*Stairmand*, J. Inst. Fuel., *29, 58 (1956).*]

collecting fine particles and the moderate power consumption characteristic of an electrical precipitator. Successful devices of this type have been essentially wet electrical precipitators and should properly be discussed in that category (see "Electrical Precipitators"). So far, there has been no clear demonstration of a device that combines the small size, compactness, and high efficiency of a high-energy scrubber with the relatively low power consumption of an electrical precipitator.

Fabric Filters Fabric filters, commonly termed "bag filters" or "baghouses," are collectors in which dust is removed from the gas stream by passing the dust-laden gas through a fabric of some type (e.g., woven cloth, felt, or porous membrane). These devices are "surface" filters in that dust collects in a layer on the surface of the filter medium, and the dust layer itself becomes the effective filter

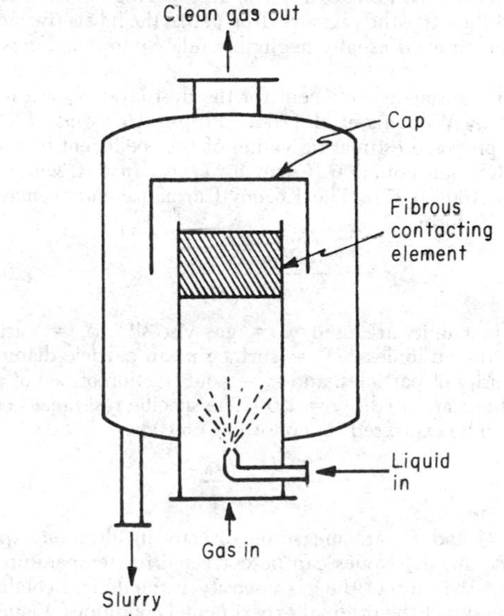

FIG. 20-124 Fibrous-bed scrubber. *(Lucas and Porter, U.S. Patent 3,370,401, 1967.)*

medium. The pores in the medium (particularly in woven cloth) are usually many times the size of the dust particles, so that collection efficiency is low until sufficient particles have been collected to build up a "precoat" in the fabric pores (Billings and Wilder, *Handbook of Fabric Filter Technology,* vol. I, EPA No. APTD-0690, NTIS No. PB-200648, 1979). During this initial period, particle deposition takes place mainly by inertial and flow-line interception, diffusion, and gravity. Once the dust layer has been fully established, sieving is probably the dominant deposition mechanism, penetration is usually extremely low except during the fabric-cleaning cycle, and only limited additional means remain for influencing collection efficiency by filter design. Filter design is related mainly to choices of gas filtration velocities and pressure drops and of fabric-cleaning cycles.

Because of their inherently high efficiency on dusts in all particle-size ranges, fabric filters have been used for collection of fine dusts and fumes for over 100 years. The greatest limitation on filter application has been imposed by the temperature limits of available fabric materials. The upper limit for natural fibers is about 90°C (200°F). The major new developments in filter technology that have been made since 1945 have followed the development of fabrics made from glass and synthetic fibers, which has extended the temperature limits to about 230 to 260°C (450 to 500°F). The capabilities of available fibers to resist high temperatures are still among the most severe limitations on the possible applications of fabric filters.

Gas Pressure Drops The filtration, or superficial face, velocities used in fabric filters are generally in the range of 0.3 to 3 m/min (1 to 10 ft/min), depending on the types of fabric, fabric supports, and cleaning methods used. In this range, gas pressure drops conform to Darcy's law for streamline flow in porous media, in which the pressure drop is directly proportional to the flow rate. The pressure drop across the fabric and the collected dust layer may be expressed (Billings and Wilder, op. cit.) by

$$\Delta p = K_1 V_f + K_2 w V_f \tag{20-74}$$

where Δp = kPa, or in of water; V_f = superficial velocity through filter, m/min, or ft/min; w = dust loading on filter, g/m^2, or lbm/ft^2; and K_1 and K_2 are resistance coefficients for the "conditioned" fabric and the dust layer respectively. The conditioned fabric is that fabric in which a relatively consistent dust load remains deposited in depth following cycles of filtration and cleaning. K_1, expressed in units of kPa/(m/min) or in water/(ft/min), may be more than 10 times the value of the resistance coefficient for the original clean fabric. If the depth of the dust layer on the fabric is greater than about 0.2 cm (1/16 in), corresponding to a fabric dust loading on the order of 500 g/m^2 (0.1 lbm/ft^2), the pressure drop across the fabric (including the dust in the pores) is usually negligible relative to that across the dust layer.

The specific resistance coefficient for the dust layer K_2 was originally defined by Williams et al. [*Heat. Piping Air Cond.*, **12,** 259 (1940)], who proposed estimating values of the coefficient by use of the Kozeny-Carman equation [Carman, *Trans. Inst. Chem. Eng. (London),* **15,** 150 (1937)]. The Kozeny-Carman equation may be expressed in the form

$$\Delta p = \frac{180 \mu V_f w (1 - \epsilon_v)}{g_c \phi_s^2 D_p^2 \rho_s \epsilon_v^3} \tag{20-75}$$

where consistent units are used; μ = gas viscosity; ϕ_s = particle shape factor, dimensionless; D_p = surface mean particle diameter; ρ_s = true density of particles; and ϵ_v = void fraction of bed of particles. (If SI units are used, g_c = 1.0.) The specific resistance coefficient may then be expressed (in consistent units) by

$$K_2 = \frac{180 \mu (1 - \epsilon_v)}{g_c \phi_s^2 D_p^2 \rho_s \epsilon_v^3} \tag{20-76}$$

In practice, K_1 and K_2 are measured directly in filtration experiments. The K_1 and K_2 values can be corrected for temperature by multiplying by the ratio of the gas viscosity at the desired condition to the gas viscosity at the original experimental conditions. Values of K_2 determined for certain dusts by Williams et al. (op. cit.) are presented in Table 20-34.

Table 20-34 Specific Resistance Coefficients for Certain Dusts*

	K_2† for particle size less than						
Dust	20 mesh	140 mesh	375 mesh	90 μm	45 μm	20 μm	2 μm
Granite	1.58	2.20	....	....	....	19.8	
Foundry	0.62	1.58	3.78				
Gypsum	....	....	6.30	....	....	18.9	
Feldspar	....	....	6.30	....	....	27.3	
Stone	0.96	....	....	6.30			
Lampblack	....	....	....	....	....	....	47.2
Zinc oxide	....	....	....	....	....	....	15.7‡
Wood	....	....	....	6.30			
Resin (cold)	....	0.62	....	....	....	25.2	
Oats	1.58	....	....	9.60	11.0		
Corn	0.62	....	1.58	3.78	8.80		

*Data from Williams et al., *Heat. Piping Air Cond.*, **12,** 259–263 (1940).

$$\dagger K_2 = \frac{\Delta p_i}{V_f w}, \frac{\text{in water}}{(\text{ft/min})(\text{lbm/ft}^2)}.$$

NOTE: These data were obtained when filtering air at ambient conditions. For gases other than atmospheric air, the Δp_i values predicted from Table 20-34 should be multiplied by the actual gas viscosity divided by the viscosity of atmospheric air.

‡Flocculated material not dispersed; size actually larger.

Lapple (in Perry, *Chemical Engineers' Handbook,* 3d ed., McGraw-Hill, New York, 1950) presents an alternative form of Eq. (20-74) in which the gas-viscosity term is explicit instead of being incorporated into the resistance coefficients:

$$\Delta p_i = K_c \mu V_f + K_d \mu w V_f \tag{20-77}$$

where Δp_i = in water, μ = cP, V_f = ft/min, w = gr/ft^2, K_c = cloth resistance coefficient = (in water)/(cP)(ft/min), and K_d = dust-layer resistance coefficient = (in water)/(cP)(gr/ft^2)(ft/min). K_d may be expressed in the same units, using the Kozeny-Carman equation:

$$K_d = \frac{160.0(1 - \epsilon_v)}{\phi_s^2 D_p^2 \rho_s \epsilon_v^3} \tag{20-78}$$

where D_p = μm and ρ_s = lbm/ft^3.

Equations (20-76) and (20-78) apply only when the mean free path of the gas molecules is small compared with the particle size of the dust particles. When the particle size approaches the mean free path of the gas molecules, a correction factor must be applied to allow for the so-called slip flow, and the resistance coefficients K_2 and K_d will be less than are given by Eqs. (20-76) and (20-78). At atmospheric pressure, this correction factor becomes appreciable when the particle size of the collected dust is less than 5 μm in diameter. To correct for slip flow, the calculated values of K_2 and K_d must be divided by the factor

$$\left[1 + K_s \left(\frac{1 - \epsilon_v}{\epsilon_v}\right)\left(\frac{\lambda_m}{\phi_s D_p}\right)\right]$$

The term K_s is essentially a constant having the approximate value 15 [Carman and Arnell, *Can. J. Res.*, **26(A),** 128 (1948)], where λ_m is the mean free path of gas molecules as calculated from kinetic theory.

Data sufficient to permit reasonable predictions of K_d from Eq. (20-78) are seldom available. The range in the values of K_d that may be encountered in practice is illustrated in Fig. 20-125, in which available experimental determinations of K_d reported in the literature for a variety of dusts are plotted against particle size. In most cases no accurate particle-size data were reported, and the curves represent the estimated range of particle size involved. The data of Williams et al. (op. cit.) are related to a wide variety of dusts, and only the approximate limits enclosing these data are shown. Also included are curves predicted from Eq. (20-78) for specific values of ϕ_s, ρ_s, and ϵ_v. It is apparent from these curves that smaller particles tend toward higher values of ϵ_v, which is consistent with the obser-

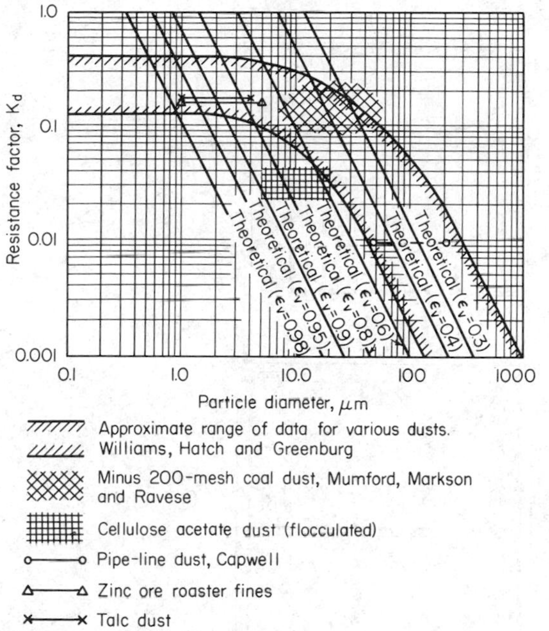

FIG. 20-125 Resistance factors for dust layers. Theoretical curves given are based on Eq. (20-78) for a shape factor of 0.5 and a true particle specific gravity of 2.0. [*Williams, Hatch, and Greenburg,* Heat. Piping Air. Cond., *12, 259 (1940); Mumford, Markson, and Ravese,* Trans. Am. Soc. Mech. Eng., *62, 271 (1940); Capwell,* Gas, *15, 31 (August 1939)*].

vation that fine dusts (particularly those smaller than 10 μm) have lower bulk densities than coarser fractions, apparently because of the effects of surface forces. For coarse dusts, K_d varies approximately inversely as the square of the particle diameter, which implies that the void fraction (or bulk density) does not change with particle size. However, for particles finer than 10 μm, the value of K_d appears to become constant, increased voids compensating for the reduction in size. In addition to the increased voidages encountered with small particles, slip flow also contributes to the relative constancy of the experimental values of K_d for particle sizes under 5 μm.

Because of the assumptions underlying its derivation, the Kozeny-Carman equation is not valid at void fractions greater than 0.7 to 0.8 (Billings and Wilder, op. cit.). In addition, in situ measurement of the void fraction of a dust layer on a filter fabric is extremely difficult and has seldom even been attempted. The structure of the layer is dependent on the character of the fabric surface as well as on the characteristics of the dust, whereas the derivations of Eqs. (20-74) and (20-77) and the application of Eq. (20-78) implicitly assume that K_2 is dependent only on the properties of the dust. A smooth fabric surface permits the dust to become closely packed, leading to a relatively high value of K_2. If the surface is napped or has numerous extended fibrils, the dust cake formed will be more porous and have a lower value of K_2 [Billings and Wilder, op. cit.; Snyder and Pring, *Ind. Eng. Chem.*, **47**, 960 (1955); and K. T. Semrau, unpublished data, SRI International, Menlo Park, Calif., 1952–1953].

Equation (20-74) indicates that for filtration at a given velocity the pressure drop is a linear function of the fabric dust loading w. In some cases, particularly with smooth-surfaced fabrics, this is at least approximately the case, but in other instances the function displays an upward curvature with increases in w, indicating compression of the dust layer, the fabric, or both, and a consequent increase in K_2 (Snyder and Pring, op. cit.; Semrau, op. cit.). Several investigations have shown K_2 to be increased by increases in the filtration velocity [Billings and Wilder, op. cit.; Spaite and Walsh, *Am. Ind. Hyg. Assoc. J.*, **24**, 357 (1968)]. However, the various investigators do not agree on the magnitude of the velocity effects. Billings and Wilder suggest assuming as an approximation that K_2 is directly proportional to the filtration velocity, but the actual relationship is probably dependent on the nature of the fabric and fabric surface, the characteristics of the dust, the dust loading on the fabric, and the pressure drop.

Clearly, the factors determining K_2 are far more complex than is indicated by a simple application of the Kozeny-Carman equation, and when possible, filter design should be based on experimental determinations made under conditions approximating those expected in the planned installation.

Types of Filters Current fabric-filter designs fall into three types, depending on the method of cleaning used: (1) shaker-cleaned, (2) reverse-flow-cleaned, and (3) reverse-pulse-cleaned. The shaker-cleaned filter is the earliest form of bag filter (Fig. 20-126). The open lower ends of the bags are fastened over openings in the tube sheet that separates the lower dirty-gas inlet chamber from the upper clean-gas chamber. The bag supports from which the bags are suspended are connected to a shaking mechanism. The dirty gas flows upward into the filter bags, and the dust collects on the inside surfaces of the bags. When the gas pressure drop rises to a chosen upper limit as the result of dust accumulation, the gas flow is stopped and the shaker is operated, giving a whipping motion to the bags. The dislodged dust falls into the dust hopper located below the tube sheet. If the filter is to be operated continuously, it must be constructed with multiple compartments, so that the individual compartments can be sequentially taken off line for cleaning while the other compartments continue in operation (Fig. 20-127).

Shaker-cleaned filters are available as standard commercial units, although large baghouses for heavy-duty service are commonly custom-designed and -fabricated. The oval or round bags used in the standard units are usually 12 to 20 cm (5 to 8 in) in diameter and 2.5 to 5 m (8 to 17 ft) long. The large, heavy-duty baghouses may use bags up to 30 cm (12 in) in diameter and 9 m (30 ft) long. The bags must be made of woven fabrics to withstand the flexing and stretching involved in shaking. The fabrics may be made from natural fibers (cotton or wool) or synthetic fibers. Fabrics of glass or mineral fibers are generally too fragile to be cleaned by shaking and are usually used in reverse-flow-cleaned filters.

In small units, the shaking of the bags may be carried out manually, but motor shaking is generally provided for the larger units. Completely automatic operation can be provided by fitting the filter with a timer, a shaker motor, and air- or motor-operated gas-discharge valves. The small sizes (under 1000 ft² of cloth area) are available completely assembled as "unit" filters. Large units (other than custom units) are usually built up of standardized rectangular sections in parallel. Each section contains on the order of 1000 to 2000 ft² of cloth, and the sections are assembled in the field to form a single filter housing. In this manner, the filter can be partitioned so that one or more sections at a time can be cut out of service for shaking or general maintenance. Additional capacity can be provided at a later date by adding more sections. Normally, such filters are furnished with rectangular housings for the sake of economy, but cylindrical housings are available when greater strength is required for pressure or vacuum service.

Ordinary shaker-cleaned filters may be shaken every ¼ to 8 h, depending on the service. A manometer connected across the filter is useful in determining when the filter should be shaken. Fully automatic filters may be shaken as frequently as every 2 min, but bag maintenance will be greatly reduced if the time between shakings can be increased to 15 or 20 min without developing excessive pressure drop. Cleaning may be actuated automatically by a differential-pressure switch. It is essential that the gas flow through the filter be stopped when shaking in order to permit the dust to fall off. With very fine dust, it may even be necessary to equalize the pressure across the cloth [Mumford, Markson, and Ravese, *Trans. Am. Soc. Mech. Eng.*, **62**, 271 (1940)]. In practice this can be accomplished without interrupting the operation by cutting one section out of service at a time, as shown in Fig. 20-127. In automatic filters this operation involves closing the dampers, shaking the filter units either pneumatically or mechanically, sometimes with the addition of a reverse flow of cleaned gas through the filter, and lastly reopening

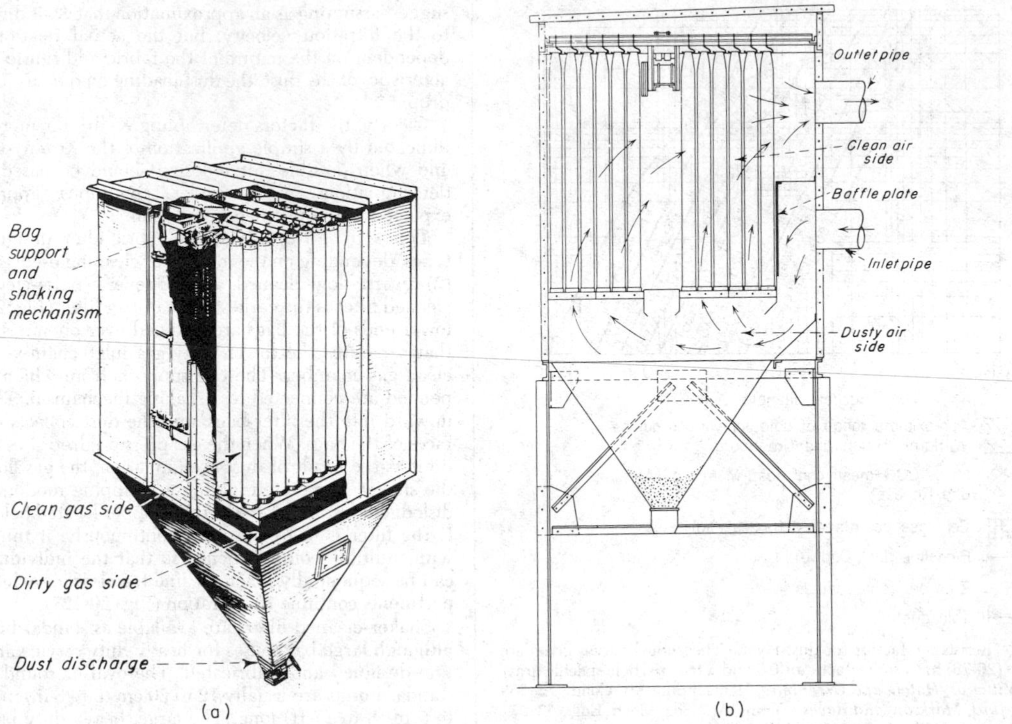

FIG. 20-126 Typical shaker-type fabric filters. (*a*) Buell Norblo (cutaway view). (*b*) Wheelabrator-Frye Inc. (sectional view).

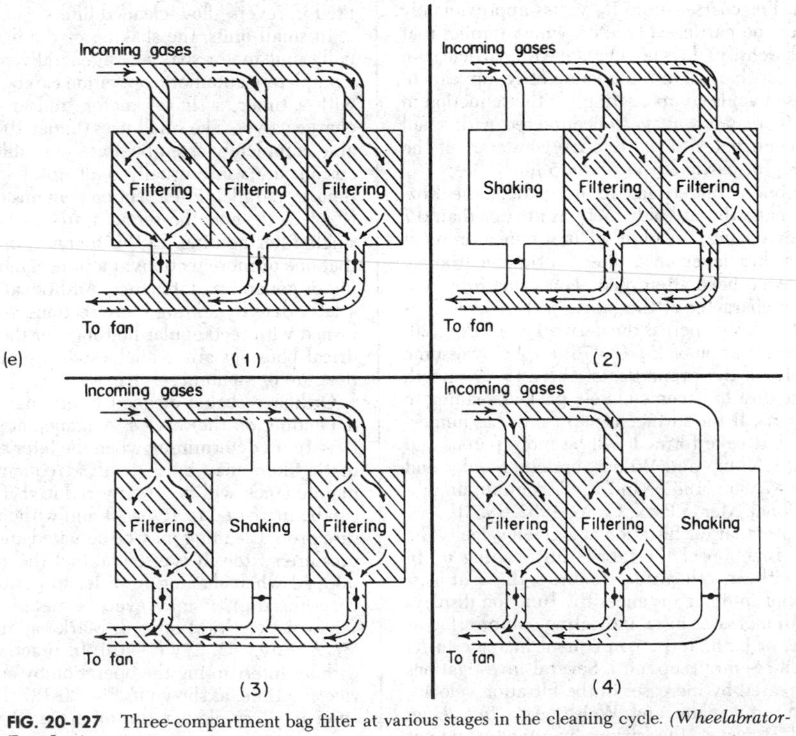

FIG. 20-127 Three-compartment bag filter at various stages in the cleaning cycle. (*Wheelabrator-Frye Inc.*)

the dampers. For compressed-air-operated automatic filters, this entire operation may take only 2 to 10 s. For ordinary mechanical filters equipped for automatic control, the operation may take as long as 3 min.

Equation (20-77) may be rewritten as

$$\Delta p_i = K_d \mu c_d V_f^2 t_m \tag{20-79}$$

where c_d = dust concentration in dirty gas, gr/ft^3; and t_m = filtration time, min. This shows that the pressure drop due to dust accumulation varies as the square of the gas velocity through the filter. (The actual effect of velocity on pressure drop may be even greater in some instances.) Greater cloth area and reduced filtration velocity therefore afford substantial reductions in shaking frequency and in bag wear. Consequently, it is generally economical to be conservative in specifying cloth area. Shaker-cleaned filters are generally operated at filtration velocities of 0.3 to 2.5 m/min (1 to 8 ft/min) and at pressure drops of 0.5 to 1.5 kPa (2 to 6 in water). For very fine dusts or high dust concentrations, filtration velocities should not exceed 1 m/min (3 ft/min). For fine fumes and dusts in heavy-duty installations, filtration velocities of 0.3 to 0.6 m/min (1 to 2 ft/min) have long been accepted on the basis of operating experience.

Cyclone precleaners are sometimes used to reduce the dust load on the filter or to remove large hot cinders or other materials that might damage the bags. However, reducing the dust load on the filter by this means may not reduce the pressure drop, since the increase in K_2 produced by the reduction in average particle size may compensate for the decrease in the fabric dust loading.

In filter operation, it is essential that the gas be kept above its dew point to avoid water-vapor condensation on the bags and resulting plugging of the bag pores. However, fabric filters have been used successfully in steam atmospheres, such as those encountered in vacuum dryers. In such cases, the housing is generally steam-chased.

Reverse-flow-cleaned filters are generally similar to the shaker-cleaned filters except for the elimination of the shaker. After the flow of dirty gas has stopped, a fan is used to force clean gas through the bags from the clean-gas side. This flow of gas partly collapses the bags and dislodges the collected dust, which falls to the dust hopper. Rings are usually sewn into the bags at intervals along the length to prevent complete collapse, which would obstruct the fall of the dislodged dust. The principal applications of reverse-flow cleaning are in units using fiberglass fabric bags for dust collection at temperatures above 150°C (300°F). Collapsing and reinflation of the bags can be made sufficiently gentle to avoid putting excessive stresses on the fiberglass fabrics [Perkins and Imbalzano, "Factors Affecting the Bag Life Performance in Coal-Fired Boilers," 3d APCA Specialty Conference on the User and Fabric Filtration Equipment, Niagara Falls, N.Y., 1978; and Miller, *Power*, **125**(8), 78 (1981)]. As with shaker-cleaned filters, compartments of the baghouse are taken off line sequentially for bag cleaning. The gas for reverse-flow cleaning is commonly supplied in an amount necessary to give a superficial velocity through the bags of 0.5 to 0.6 m/min (1.5 to 2.0 ft/min), which is the same range as the filtration velocities frequently used.

In the reverse-pulse filter (frequently termed a reverse-jet filter), the filter bag forms a sleeve that is drawn over a wire cage, which is usually cylindrical (Fig. 20-128). The cage supports the fabric on the clean-gas side, and the dust is collected on the outside of the bag. A venturi nozzle is located in the clean-gas outlet from the bag. For cleaning, a jet of high-velocity air is directed through the venturi nozzle and into the bag, inducing a flow of cleaned gas to enter the bag and flow through the fabric to the dirty-gas side. The high-velocity jet is released in a sudden, short pulse (typical duration 100 ms or less) from a compressed-air line by a solenoid valve. The pulse of air and clean gas expands the bag and dislodges the collected dust. Rows of bags are cleaned in a timed sequence by programmed operation of the solenoid valves. The pressure of the pulse is sufficient to dislodge the dust without cessation of the gas flow through the filter unit.

It has been a common practice to clean the bags on line (i.e., without stopping the flow of dirty gas into the filter), and reverse-pulse bag filters have been built without division into multiple compartments. However, investigations [Leith et al., *J. Air Pollut. Control Assoc.*, **27**, 636 (1977)] and experience have shown that, with online cleaning of reverse-pulse filters, a large fraction of the dust dislodged from the bag being cleaned may redeposit on neighboring bags rather than fall to the dust hopper. As a result, there is a growing trend to off-line cleaning of reverse-pulse filters. The baghouse is sectionalized so that the outlet-gas plenum serving the bags in a section can be closed off from the clean-gas exhaust, thereby stopping the flow of inlet gas through the bags. On the dirty-gas side of the tube sheet, the bags of the section are separated by partitions from the neighboring sections, where filtration is continuing. Sections of the filter are cleaned in rotation, as in shaker and reverse-flow filters.

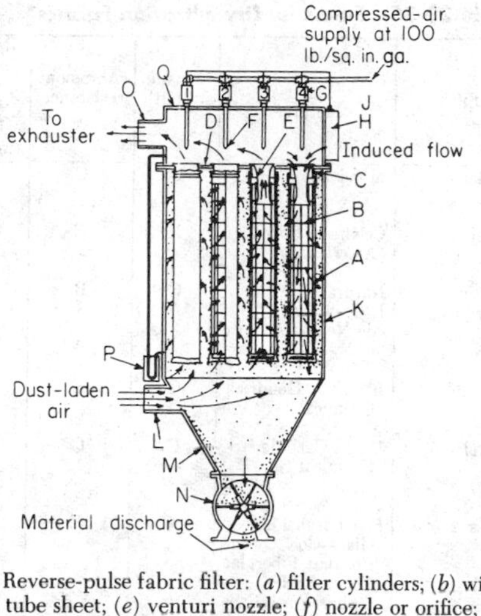

FIG. 20-128 Reverse-pulse fabric filter: (*a*) filter cylinders; (*b*) wire retainers; (*c*) collars; (*d*) tube sheet; (*e*) venturi nozzle; (*f*) nozzle or orifice; (*g*) solenoid valve; (*h*) timer; (*j*) air manifold; (*k*) collector housing; (*l*) inlet; (*m*) hopper; (*n*) air lock; (*o*) upper plenum. *(Mikropul Division, U.S. Filter Corp.)*

Some manufacturers are using relatively low-pressure air (100 kPa, or 15 lbf/in^2, instead of 690 kPa, or 100 lbf/in^2) and are eliminating the venturi tubes for clean-gas induction. Others have eliminated the separate jet nozzles located at the individual bags and use a single jet to inject a pulse into the outlet-gas plenum.

Reverse-pulse filters are typically operated at higher filtration velocities (air-to-cloth ratios) than shaker or reverse-flow filters designed for the same duty. Filtration velocities may range from 1 to 4.5 m/min (3 to 15 ft/min), depending on the dust being collected, but for most dusts the commonly used range is about 1.2 to 2.5 m/min (4 to 8 ft/min). The frequency of cleaning is also dependent on the nature and concentration of the dust, with the intervals between pulses varying from about 2 to 15 min.

The cleaning action of the pulse is so effective that the dust layer may be completely removed from the surface of the fabric. Consequently, the fabric itself must serve as the principal filter medium for at least a substantial part of the filtration cycle. Woven fabrics are unsuitable for such service, and felts of various types must be used. The bulk of the dust is still removed in a surface layer, but the felt ensures that an adequate collection efficiency is maintained until the dust layer has formed.

Filter Fabrics The cost of the filter bags represents a substantial part of the erected cost of a bag filter—typically 5 to 20 percent, depending on the bag material [Reigel and Bundy, *Power*, **121**(1), 68 (1977)]. The cost of bag repair and replacement is the largest component of the cost of bag-filter maintenance. Consequently, the proper choice of filter fabric is critical to both the technical performance and the economics of operating a filter. With the advent of synthetic fibers, it has become possible to produce fabrics having a wide range of properties (Table 20-35). However, demonstrating the acceptability of a fabric still depends on experience with prolonged operation under the actual or simulated conditions of the proposed

Table 20-35 Fibers for Dry-Filtration Fabrics*

Fiber	Manu-facturer	Tensile strength ‡	Abrasion resistance ‡	Recommended max. operating temp.; exposure time in degrees F.		Chemical resistance		Flammability; will support combustion?	Special properties	General chemical classi-fication
				Long (months)	Short (hours)	Acids	Alkalies			
Acrilan	Monsanto	C	C	250	300	D	D	Yes	....	Polyacrylo-nitrile (acrylic)
Arnel	Celanese Corp. of America	E	E	250	300	D	D	Yes	A modified cellulose, has improved heat and bacterial resistance	Triacetate
Cotton	Natural fiber	C	B	160	250	E	A	Yes	....	Cellulose
Dacron	E. I. du Pont de Nemours	A	A	275	350	B	C–D fair to good	Yes	More rapid de-gradation may occur in the presence of heat and moisture. Holds crease	Polyester
Darvan	B. F. Goodrich Chemical	C	E	310	320	B	D	Yes	Melts above 330°F. Has excellent dimensional stability at 300°F.	Nytril
Dynel	Union Carbide Chemical Co.	C	C	180	240	B	A	No	Will soften and distort if exposed to temp. above 180°F. unless heat-set	Copolymer of acrylonitrile and vinyl chloride
Glass	Pittsburgh Plate Glass Co.; Owens Corning Fiberglas Corp.; Libby Owens; Ford Glass Fiber Co.	A	E	500	650	C	E	No	Limited by poor flex-abrasion qualities. Finishes limit max. temp. range	Glass
Kodel	Eastman Chemical Products, Inc.	C	C	275	350	B–C	C–D	Yes	Excellent stability under heat	Polyester
Nylon 66	E. I. du Pont de Nemours; Monsanto	A	A	200	250	E	A	Yes	Stays soft and pliable when exposed to heat	Polyamide
Nylon 6	American Enka; Industrial Rayon; Allied Chemical	A	A	200	250	E	B	Yes	....	Polyamide
Orlon 42	E. I. du Pont de Nemours	C	B	260	300	C	D	Yes	Best all-around high-temperature fiber	Polyacrylo-nitrile (acrylic)
Polyethylene	Union Carbide Chemical Co.	A	A	150	212 (heat-set)	A	A	Yes	Affected by some organic solvents. Can be heat-set to operate at about 212°F. If subjected to load for a long time, it will continue to stretch. Lighter than water	Polyethylene
Polypro-pylene	Hercules, Alamo Polymer, National plastic products	A	B–C	190	190	A	A	Yes	Strong, excellent chemical resistance	Polyolefin
Q957	Dow Chemical	C	X	220	240	B	B	No	Fibers made from film. They are flat ribbons	Vinylidene chloride
Saran	Saran Yarn Co. and others	D	C	150	200	A	B (ammonia, E)	No	Outstanding chemical resistance, but severe temp. limitation	Vinylidene chloride
Teflon (multi-filament)	E. I. du Pont de Nemours	C	D	450	550	A	A	No	Expensive. Best chemical resistance, good heat resistance. When exposed to temp. in excess of 400°F. toxic fumes are given off. Strength decreases rapidly at high temp.	Polyfluoro-ethylene
Nomex	E. I. du Pont de Nemours	A–B	A–B	425	450	D	A	No	Outstanding temperature resistance	Nylon aromatic polyamide
Verel	Eastman Chemical Products, Inc.	C	E	200	250	C	D	No	Verel FR has better flame resistance than wool	Modified acrylic
Wool	Natural fiber	Wet, E Dry, D	Wet, C Dry, C	200	250	E	E	No	When wet has excellent elastic recovery. Can be felted	Protein
Zefran	Dow Chemical	C	C	220	270	C	D	Yes	....	Acrylic alloy

*Courtesy of the Globe Albany Corp. This company offers information as best currently available; no obligation or liability whatsoever is assumed in connection with its use. Data apply only to staple fibers, although continuous-filament yarns are also made; Teflon is excepted, as test fabric was made from filament yarn.

‡A = excellent; B = above average; C = good, average; D = fair; E = poor; X = unknown.

application. The choice of a fabric material for a given service is necessarily a compromise, since no single material possesses all the properties that may be desired. Following the choice of material, the type of fabric construction is critical.

Two principal types of fabric are adaptable to filter use: woven fabrics, which are used in shaker and reverse-flow filters; and felts, which are used in reverse-pulse filters. The felts made from synthetic fibers are needle felts (i.e., felted on a needle loom) and are normally reinforced with a woven insert. The physical properties and air permeabilities of some typical woven and felt filter fabrics are presented in Tables 20-36 and 20-37. The "air permeability" of a filter fabric is defined as the flow rate of air in cubic feet per minute (at 70°F, 1 atm) that will pass through 1 ft^2 of clean fabric under an applied differential pressure of ½ in water. The resistance coefficient K_F of the clean fabric is defined by the equation in Table 20-36, which may be used to calculate the value of K_F from the air permeability. If Δp_i is taken as 0.5 in water, μ as 0.0181 cP (the viscosity of air at 70°F and 1 atm), and V_f as the air permeability, then K_F = 27.8/air permeability.

Collection Efficiency The inherent collection efficiency of fabric filters is usually so high that, for practical purposes, the precise level has not commonly been the subject of much concern. Furthermore, for collection of a given dust, the efficiency is usually fixed by the choices of filter fabric, filtration velocity, method of cleaning,

Table 20-36 Resistance Factors and Air Permeabilities for Typical Woven Fabrics

Cloth	Pore size,* in	Threads/in	Weight, oz/yd²	Thread* diameter, in	K_F†	Air permeability, (ft³/min)/ft² at Δp_i = ½ inH_2O
Osnaburg cotton	0.01	32 × 28	....	0.02	0.51	55
Osnaburg cotton (soiled)‡	....	32 × 28	....	....	4.80	5.8
Drill cotton	0.01	68 × 40	5.28	0.01	0.093	300
Cotton§	....	46 × 56	....	....	1.39	20
Cotton§	....	104 × 68	....	....	1.54	18
Cotton sateen (unnapped)	0.007	96 × 56	6.88	0.009	0.27	103
Cotton sateen (unnapped)	0.005	96 × 64	8.23	0.01	0.88	32
Cotton sateen (unnapped)	....	96 × 60	....	0.012	1.63	17
Cotton sateen (unnapped)	0.004	96 × 56	10.2	0.011	1.12	25
Wool	....	....	....	....	0.25	111
Wool	....	40 × 50	11.5	0.014	0.33	84
Wool, white§	....	36 × 32	....	....	0.15	185
Wool, black§	....	28 × 30	....	....	0.25	110
Wool§	....	30 × 26	....	....	0.51	55
Vinyon§	....	37 × 37	....	....	0.12	23
Nylon tackle twill	....	72 × 196	....	0.010	0.66	42
Nylon sailcloth	....	130 × 130	....	0.007	1.66	17
Nylon§	....	37 × 37	....	....	1.74	16
Nylon§	....	....	....	....	3.71	7.5
Asbeston§	....	....	....	....	0.56	50
Orlon§	....	72 × 72	....	....	0.66	42
Orlon§	....	74 × 38	....	....	0.75	37
Orlon§	....	....	....	....	1.16	24
Orlon§	....	....	....	....	1.98	14
Smoothtex nickel screen	....	(300 mesh)	....	....	0.16	174
Glass	....	32 × 28	....	0.03	1.60	17
Dacron	....	60 × 40	5.8	....	0.84	33
Dacron	....	76 × 48	13.4	....	0.29	9.5
Teflon	....	76 × 70	8.7	....	1.39	20

*Estimates based on microscopic examination.

†Measured with atmospheric air. This value will be constant only for streamline flow, which is the case for values of $\rho V_f/\mu$ of less than approximately 100.

$$K_F = \Delta p_i/\mu V_f$$

where Δp_i = pressure drop, in water; μ = gas viscosity, cP; V_f = superficial gas velocity through cloth, ft/min; and ρ = gas density, lb/ft³.

‡Cloth, similar to previous one, that had been in service and contained dust in pores although free of surface accumulation.

§Data from Pring, *Air Pollution*, McGraw-Hill, New York, 1952, p. 280.

and cleaning cycle, leaving few if any controllable variables by which efficiency can be further influenced. Inefficiency usually results from bags that are poorly installed, torn, or stretched from excessive dust loading and pressure drop. Of course, certain types of fabrics may simply be unsuited for filtration of a particular dust, but usually this will soon become obvious.

Few basic studies of the efficiency of bag filters have been made. Increased dust penetration immediately following cleaning has been readily observed while the dust layer is being reestablished. However, field and laboratory studies have indicated that during the rest of the filtration cycle the effluent-dust concentration tends to remain constant regardless of the inlet concentration [Dennis, *J. Air Pollut. Control Assoc.*, **24**, 1156 (1974)]. In addition, there has been little indication that the penetration is strongly related to dust-particle size, except possibly in the low-submicrometer range. These observations appear to be generally consistent with sieving being the principal collection mechanism.

Leith and First [*J. Air Pollut. Control Assoc.*, **27**, 534 (1977); **27**,

Table 20-37 Physical Properties of Selected Felts for Reverse-Pulse Filters

Fiber	Weight, oz/yd²	Thickness, in	Breaking strength, lbf/in width	Elongation, % to rupture	Air permeability, (ft³/min)/ft² at Δp_i = ½ in water	K_F
Wool	23.1	0.135	...	...	27.1	1.03
Wool	21.2	0.129	...	...	29.8	0.93
Orlon*	10.9	0.045	65	18	20–25	1.11–1.39
Orlon*	17.9	0.088	85	18	15–20	1.39–1.85
Orlon*	24	0.125	110	60	10–20	1.39–2.78
Acrilan*	17.9	0.075	100	22	15–20	1.39–1.85
Dynel*	24	0.125	60	80	30–40	0.70–0.93
Dacron*	17.9	0.080	125	22	15–20	1.39–1.85
Dacron*	9.9	0.250	20	150	200–225	0.11–0.14
Dacron*	24	0.125	175	80	20–30	0.93–1.39
Nylon*	24	0.125	100	100	30–40	0.70–0.93
Arnel*	24	0.125	60	80	30–40	0.70–0.93
Teflon	15.6	0.053	...	...	82.5	0.34
Teflon	43.5	0.119	...	...	21.6	1.29

*These data courtesy of American Felt Co.

754 (1977)] studied the collection efficiency of reverse-pulse filters and concluded that once the dust cake has been established, "straight-through" penetration by dust particles that pass through the filter without being stopped is negligible by comparison with penetration by dust that actually deposits initially and then "seeps" through the fabric to be reentrained into the exit air stream. They also noted that "pinholes" may form in the dust cake, particularly over pores between yarns in a woven fabric, and that particles may subsequently penetrate straight through at the pinholes. The formation of pinholes, or "cake puncture," had been observed earlier by Stephan et al. [*Am. Ind. Hyg. Assoc. J.*, **21**, 1 (1960)], but without measurement of the associated loss of collection efficiency. When a supported flat filter medium with extremely fine pores (e.g., glass-fiber paper, membrane filter) was used, no cake puncture took place even with very high pressure differentials across the cake. However, puncture did occur when a cotton-sateen filter fabric was used as the cake support. The formation of pinholes with certain combinations of dusts, fabrics, and filtration conditions was also observed by Koscianowski et al. (EPA-600/7-78-056, 1978). Evidently puncture occurs when the local cake structure is not strong enough to maintain a bridge over the aperture represented by a large pore and the portion of the cake covering the pore is blown through the fabric. This suggests that formation of pinholes will be highly dependent on the strength of the surface forces between particles that produce flocculation of dusts. The seepage of a dust through a filter is probably also closely related to the strength of the surface forces.

Granular-Bed Filters Granular-bed filters may be classified as "depth" filters, since dust particles deposit in depth within the bed of granules. The granules themselves present targets for the deposition of particles by inertia, diffusion, flow-line interception, gravity, and electrostatic attraction, depending on the dust and filter characteristics and the operating conditions. Other deposition mechanisms are minor at most. Although it is physically possible under some circumstances for a dust layer to form on the inlet face of the filter, the practical limits of gas pressure drop will normally have been reached long before a surface dust layer can be established.

Granular-bed filters may be divided into three classes:

1. *Fixed-bed, or packed-bed, filters.* These units are not cleaned when they become plugged with deposited dust particles but are broken up for disposal or simply abandoned. If they are constructed from fine granules (e.g., sand particles), they may be designed to give high collection efficiencies on fine dust particles. However, if such a filter is to have a reasonable operating life, it can be used only on a gas containing a low concentration of dust particles.

2. *Cleanable granular-bed filters.* In these devices provisions are made to separate the collected dust from the granules either continuously or periodically, so that the units can operate continuously on gases containing moderate to high dust concentrations. The necessity for cleaning and recycling the granules generally restricts the practical lower granule size to about 3 to 10 mm. This in turn makes it difficult to attain high collection efficiencies on fine particles with granule beds of reasonable depth and gas pressure drop.

3. *Fluidized-bed filters.* Fluidized beds of granules have received considerable study on theoretical and experimental levels but have not been applied on a practical commercial scale.

Fixed Granular-Bed Filters Fixed-bed filters composed of granules have received considerable theoretical and experimental study [Thomas and Yoder, *AMA Arch. Ind. Health*, **13**, 545 (1956); **13**, 550 (1956); Knettig and Beeckmans, *Can. J. Chem. Eng.*, **52**, 703 (1974); Schmidt et al., *J. Air Pollut. Control Assoc.*, **28**, 143 (1978); Tardos et al., *J. Air Pollut. Control Assoc.*, **28**, 354 (1978); and Gutfinger and Tardos, *Atmos. Environ.*, **13**, 853 (1979)]. The theoretical approach is the same as that used in the treatment of deep-bed fibrous filters.

Fibers for filter applications can be produced with diameters smaller than it is practical to obtain with granules. Consequently, most concern with filtration of fine particles has been focused on fibrous-bed rather than granular-bed filters. However, for certain specialized applications granular beds have shown some superior properties, such as greater dimensional stability. Granular-bed filters of special design (deep-bed sand filters) have been used since 1948 for removing radioactive particles from waste air and gas streams in atomic energy plants (Lapple, "Interim Report—200 Area Stack Contamination," U.S. AEC Rep. HDC-743, Oct. 11, 1948; Juvinall et al., "Sand-Bed Filtration of Aerosols: A Review of Published Information," U.S. AEC Rep. ANL-7683, 1970; and Burchsted et al., *Nuclear Air Cleaning Handbook*, U.S. ERDA 76-21, 1976). The filter characteristics needed included high collection efficiency on fine particles, large dust-holding capacity to give long operating life, and low maintenance requirements. The sand filters are as much as 2.7 m (9 ft) in depth and are constructed in graded layers with about a 2:1 variation in the granule size from one layer to the next. The airflow direction is upward, and the granules decrease in size in the direction of the air flow. The bottom layer is composed of rocks about 5 to 7.5 cm (2 to 3 in) in diameter, and granule sizes in successive layers decrease to 0.3 to 0.6 mm (50 to 30 mesh) in the finest layer. With superficial face velocities of about 1.5 m/min (5 ft/min), gas pressure drops of clean filters have ranged from 1.7 to 2.8 kPa (7 to 11 in water). Collection efficiencies of up to 99.98 percent with a polydisperse dioctyl phthalate aerosol of 0.7-μm mean diameter have been reported (Juvinall et al., op. cit.). Operating lives of 5 years or more have been attained.

Cleanable Granular-Bed Filters The principal objective in the development of cleanable granular-bed filters is to produce a device that can operate at temperatures above the range that can be tolerated with fabric filters. In some of the devices, the granules are circulated continuously through the unit, then are cleaned of the collected dust and returned to the filter bed. In others, the granular bed remains in place but is periodically taken out of service and cleaned by some means, such as backflushing with air.

A number of moving-bed granular filters have used cross-flow designs. The Dorfan Impingo filter (Dorfan, U.S. Patent 2,604,187) shown in Fig. 20-129 is representative. It was offered commercially in the 1950s but is no longer marketed. The granules moved downward through two panels, which were tapered to help prevent hang-ups of the particles and collected dust. The granules flowed at a rate controlled by rotary valves below each panel and upon discharge were separated from the dust by screening or other means. The average panel thickness was 30 cm (12 in), and the granules were rocks of 1.3- to 3.8-cm (½- to 1½-in) diameter. The gas flowed through the two panels in series. The inventor claimed high collection efficiencies

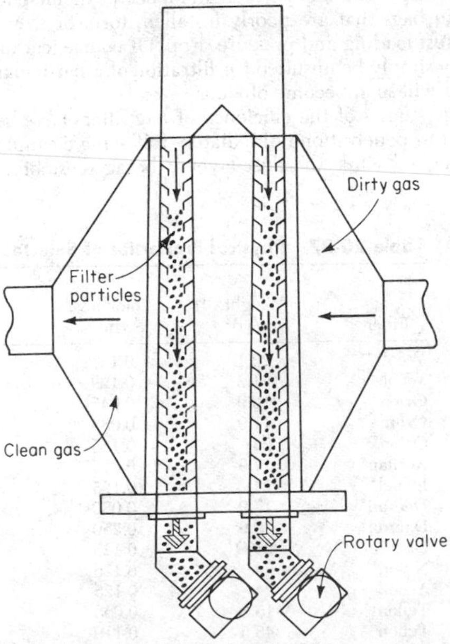

FIG. 20-129 Dorfan Impingo granular-bed filter. *(Dorfan, U.S. Patent 2,604,187.)*

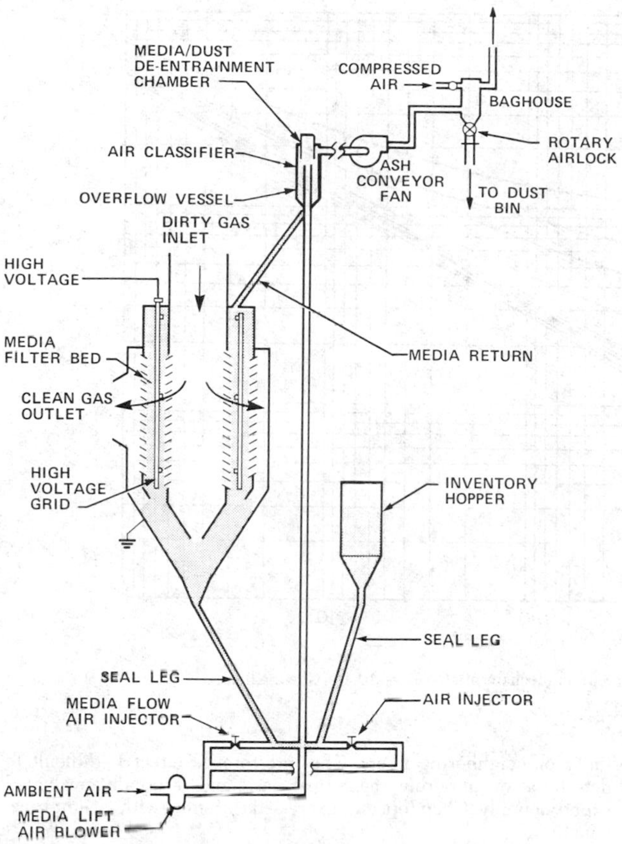

FIG. 20-130 Electrically augmented granular-bed filter. *(Combustion Power Company.)*

on 2- to 10-μm dust with a filter face velocity of 1.8 m/s (6 ft/s). However, in experiments with a similar cross-flow filter, Taub ("Filtration Phenomena in a Packed Bed Filter," Ph.D. thesis, Carnegie-Mellon University, Pittsburgh, 1979) found that once the bed had reached an equilibrium loading of dust, reentrainment produced a serious loss of efficiency.

Another form of cross-flow moving-granular-bed filter, produced by the Combustion Power Company (Fig. 20-130), is currently in commercial use in some applications. The granular filter medium consists of ⅛- to ¼-in (3- to 6-mm) pea gravel. Gas face velocities range from 30 to 46 m/min (100 to 150 ft/min), and reported gas pressure drops are in the range of 0.5 to 3 kPa (2 to 12 in water). The original form of the device [Reese, TAPPI, **60**(3), 109 (1977)] did not incorporate electrical augmentation. Collection efficiencies for submicrometer particles were low, and the electrical augmentation was added to correct the deficiency (Parquet, "The Electroscrubber Filter: Applications and Particulate Collection Performance," EPA-600/9-82-005c, 1982, p. 363). The electrostatic grid immersed in the bed of granules is charged to a potential of 20,000 to 30,000 V, producing an electric field between the grid and the inlet and outlet louvers that enclose the bed. No ionizing electrode is used to charge particles in the incoming gas; reliance is placed on the existence of natural charges on the dust particles. Individual dust particles commonly carry positive or negative charges even though the net charge on the dust as a whole is normally neutral. Depending on their charges, dust particles are attracted or repelled by the electrical field and are therefore caused to deposit on the rocks in the bed.

Self et al. ("Electrical Augmentation of Granular Bed Filters," EPA-600/9-80-039c, 1980, p. 309) demonstrated in theoretical studies and laboratory experiments that such an augmentation system should yield substantial increases in the collection efficiency for fine particles if the particles carry significant charges. Significant improvements in the performance of the Combustion Power units with electrical augmentation have been reported by the manufacturer (Parquet, op. cit.).

Another type of gravel-bed filter, developed by GFE in Germany, has had limited commercial application in the United States [Schueler, *Rock Prod.*, **76**(7), 66 (1973); **77**(11), 39 (1974)]. After precleaning in a cyclone, the gas flows downward through a stationary horizontal filter bed of gravel. When the bed becomes loaded with dust, the gas flow is cut off, and the bed is backflushed with air while being stirred with a double-armed rake that is rotated by a gear motor. The backflush air also flows backward through the cyclone, which then acts as a dropout chamber. Multiple filter units are constructed in parallel so that individual units can be taken off the line for cleaning. The dust dislodged from the bed and carried by the backflush air is flocculated, and part is collected in the cyclone. The backflush air with the remaining suspended dust is cleaned in the other gravel-bed filter units that are operating on line. Performance tests made on one installation for the U.S. Environmental Protection Agency (EPA-600/7-78-093, 1978) did not give clear results but indicated that collection efficiencies were low on particles under 2 μm and that some of the dust in the backflush air was redispersed sufficiently to penetrate the operating filter units.

Air Filters The types of equipment previously described are intended primarily for the collection of process dusts, whereas air filters comprise a variety of filtration devices designed for the collection of particulate matter at low concentrations, usually atmospheric dust. The difference in the two categories of equipment is not in the principles of operation but in the adaptations required to deal with the different quantities of dust. Process-dust concentrations may run as high as several hundred grams per cubic meter (or grains per cubic foot) but usually do not exceed 45 g/m³ (20 gr/ft³). Atmospheric-dust concentrations that may be expected in various types of locations are shown in Table 20-38 and are generally below 12 mg/m³ (5 gr/1000 ft³).

The most frequent application of air filters is in cleaning atmospheric air for building ventilation, which usually requires only moderately high collection-efficiency levels. However, a variety of industrial operations developed mostly since the 1940s require air of extreme cleanliness, sometimes for pressurizing enclosures such as clean rooms and sometimes for use in a process itself. Examples of applications include the manufacture of antibiotics and other pharmaceuticals, the production of photographic film, and the manufacture and assembly of semiconductors and other electronic devices. Air cleaning at the necessary efficiency levels is accomplished by the use of high-efficiency fibrous filters that have been developed since the 1940s.

Air filters are also used to protect internal-combustion engines and gas turbines by cleaning the intake air. In some locations and applications, the atmospheric-dust concentrations encountered are much higher than those normally encountered in air-conditioning service.

High-efficiency air filters are sometimes used for emission control when particulate contaminants are low in concentration but present special hazards; cleaning of ventilation air and other gas streams exhausted from nuclear plant operations is an example.

Air-Filtration Theory Current high-efficiency air- and gas-filtration methods and equipment have resulted largely from the devel-

Table 20-38 Average Atmospheric-Dust Concentrations*
1 gr/1000 ft³ = 2.3 mg/m³

Location	Dust concentration, gr/1000 ft³
Rural and suburban districts	0.02–0.2
Metropolitan districts	0.04–0.4
Industrial districts	0.1 –2.0
Ordinary factories or workrooms	0.2 –4.0
Excessive dusty factories or mines	4.0 –400

* *Heating Ventilating Air Conditioning Guide*, American Society of Heating, Refrigerating and Air-Conditioning Engineers, New York, 1960, p. 77.

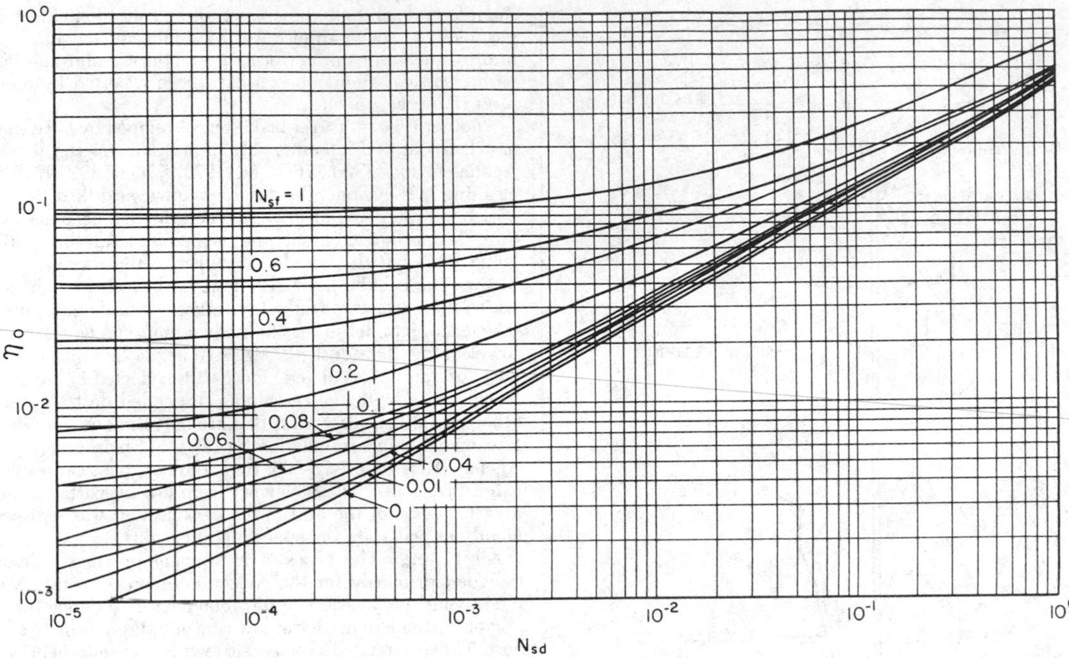

FIG. 20-131 Isolated fiber efficiency for combined diffusion and interception mechanism at $N_{Re} = 10^{-2}$. [*Chen,* Chem. Rev., *55, 595 (1955).*]

opment of filtration theory since about 1930 and particularly since the 1940s. Much of the theoretical advance was originally encouraged by the requirements of the military and atomic energy programs. The fibrous filter has served both as a practical device and as a model for theoretical and experimental investigation. Extensive reviews and new treatments of air-filtration theory and experience have been presented by Chen [*Chem. Rev.*, **55**, 595 (1955)], Dorman ("Filtration," in Davies, *Aerosol Science*, Academic, New York, 1966), Pich (*Theory of Aerosol Filtration by Fibrous and Membrane Filters*, in ibid.), Davies (*Air Filtration*, Academic, New York, 1973), and Kirsch and Stechkina ("The Theory of Aerosol Filtration with Fibrous Filters," in Shaw, *Fundamentals of Aerosol Science*, Wiley, New York, 1978). The theoretical treatment of filtration starts with the processes of dust-particle deposition on collecting bodies, as outlined in Fig. 20-103 and Table 20-31. All the mechanisms shown in Table 20-31 may come into play, but inertial deposition, flow-line interception, and diffusional deposition are usually dominant. Electrostatic precipitation may become a major mechanism if the collecting body, the dust particle, or both, are charged. Gravitational settling is a minor influence for particles in the size range of usual interest. Thermal precipitation is nil in the absence of significant temperature gradients. Sieving is a possible mechanism only when the pores in the filter medium are smaller than or approximately equal to the particle size and will not be encountered in fibrous filters unless they are loaded sufficiently for a surface dust layer to form.

The theoretical prediction of the efficiency of collection of dust particles by a fibrous filter consists of three steps (Chen, op. cit.):

1. Calculation of the target efficiency η_o of an isolated fiber in an air stream having a superficial velocity the same as that in the filter
2. Determining the difference between the target efficiency of the isolated fiber and that of an individual fiber in the filter array η_t
3. Determining the collection efficiency of the filter η from the target efficiency of the individual fibers

The results of computations of η_o for an isolated fiber are illustrated in Figs. 20-131 and 20-132. The target efficiency η_t of an individual fiber in a filter differs from η_o for two main reasons (Pich, op. cit.): (1) the average gas velocity is higher in the filter, and (2) the velocity field around the individual fibers is influenced by the proximity of neighboring fibers. The interference effect is difficult to determine on a purely theoretical basis and is usually evaluated experimentally. Chen (op. cit.) expressed the effect with an empirical equation:

$$\eta_t = \eta_o[1 + K_\alpha(1 - \epsilon_v)] \tag{20-80}$$

This indicates that the target efficiency of the fiber is increased by the proximity of other fibers. The value of K_α averaged 4.5 for values of the void fraction ϵ_v, ranging from 0.90 to 0.99. Extending use of the equation to values of ϵ_v lower than 0.90 may result in large errors.

The collection efficiency of the filter may be calculated from the fiber target efficiency and other physical characteristics of the filter (Chen, op. cit.):

$$N_t = \frac{4\eta_t L(1 - \epsilon_v)}{\pi D_b \epsilon_v} \tag{20-81}$$

where D_b = fiber diameter and L = filter thickness. The derivation of Eq. (20-81) assumes that (1) η_t is the same throughout the filter, (2) all fibers are of the same diameter D_b, are cylindrical and are normal to the direction of the gas flow, (3) the fraction of the particles deposited in any one layer of fiber is small, and (4) the gas passing through the filter is essentially completely remixed after it leaves one layer of the filter and before it enters the next. The first assumption requires that Eq. (20-81) apply only for particles of a single size for which there are corresponding values of η_t, η, and N_t.

For filters of high porosity, ϵ_v approaches unity and Eq. (20-81) reduces to the expression used by Wong et al. [*J. Appl. Phys.*, **27**, 161 (1956)] and Thomas and Lapple [*Am. Inst. Chem. Eng. J.*, **7**, 203 (1961)]:

$$N_t = \frac{4\eta_t L(1 - \epsilon_v)}{\pi D_b} \tag{20-82}$$

The foregoing procedure is commonly employed in reverse to determine or confirm fiber target efficiencies from the experimentally determined efficiencies of fibrous filter pads.

Filtration theory assumes that a dust particle that touches a collector body adheres to it. This assumption appears to be valid in most

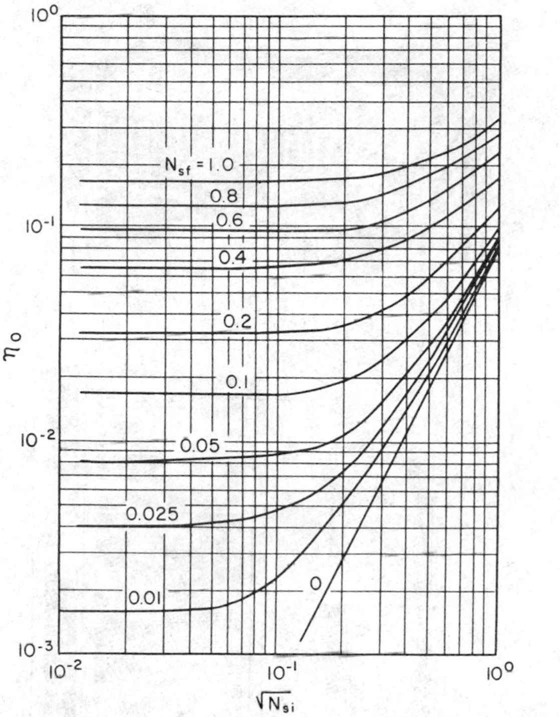

FIG. 20-132 Isolated fiber efficiency for combined inertia and interception mechanisms at N_{Re} = 0.2. [*Chen*, Chem. Rev., *55, 595 (1955).*]

cases, but evidence of nonadherence, or particle bouncing, has appeared in some instances. Wright et al. ("High Velocity Air Filters," WADC TR 55-457, ASTIA Doc. AD-142075, 1957) investigated the performance of fibrous filters at filtration velocities of 0.091 to 3.05 m/s (0.3 to 100 ft/s), using 0.3-μm and 1.4-μm supercooled liquid aerosols and a 1.2-μm solid aerosol. The collection efficiencies agreed well with theoretical predictions for the liquid aerosols and apparently also for the solid aerosol at filtration velocities under 0.3 m/s (1 ft/s). But at filtration velocities above 0.3 m/s some of the solid particles failed to adhere. With a filter composed of 30-μm glass fibers and a filtration velocity of 9.1 m/s (30 ft/s), there were indications that 90 percent of the solid aerosol particles striking a fiber bounced off.

Bouncing may be regarded as a defect in the particle-deposition process. However, particles that have been deposited in filters may subsequently be blown off and reentrained into the air stream (Corn, "Adhesion of Particles," in Davies, *Aerosol Science*, Academic, New York, 1966; and Davies, op. cit.).

The theories of filtration by a fibrous filter relate only to the initial efficiency of the clean filter in the "static" period of filtration before the deposition of any appreciable quantity of dust particles. The deposition of particles in a filter increases the number of targets available to intercept particles, so that collection efficiency increases as the filter loads. At the same time, the filter undergoes clogging and the pressure drop increases. No theory is available for dealing with the "dynamic" period of filtration in which collection efficiency and pressure drop vary with the loading of collected dust. The theoretical treatment of this filtration period is incomparably more complex than that for the "static" period. Investigators have noted that both the increase in collection efficiency and the increase in pressure drop are exponential functions of the loading of collected dust or are at least roughly so (Davies, op. cit.). Some empirical relationships have been derived for correlating data in particular instances.

The dust particles collected by a fibrous filter do not deposit in uniform layers on fibers but tend to deposit preferentially on previously deposited particles (Billings, "Effect of Particle Accumulation in Aerosol Filtration," Ph.D. dissertation, California Institute of Technology, Pasadena, 1966), forming chainlike agglomerates termed "dendrites." The growth of dendritic deposits on fibers has been studied experimentally [Billings, op. cit.; Bhutra and Payatakes, *J. Aerosol Sci.*, **10**, 445 (1979)], and Payatakes and coworkers [Payatakes and Tien, *J. Aerosol Sci.*, **7**, 85 (1976); Payatakes, *Am. Inst. Chem. Eng. J.*, **23**, 192 (1977); and Payatakes and Gradon, *Chem. Eng. Sci.*, **35**, 1083 (1980)] have attempted to model the growth of dendrites and its influence on filter efficiency and pressure drop.

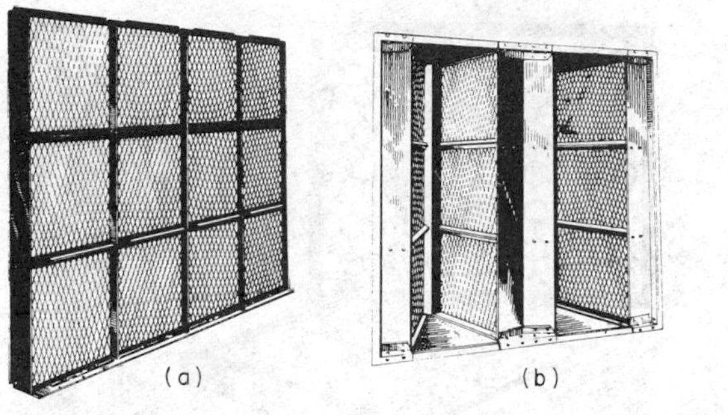

FIG. 20-133 Typical uses of filter-bank installation. (*a*) Flat or L-type installation. (*b*) V-type installation.

Air-Filter Types and Characteristics Air filters may be broadly divided into two classes: (1) panel, or unit, filters; and (2) automatic, or continuous, filters. Panel filters are constructed in units of convenient size (commonly 20- by 20-in or 24- by 24-in face area) to facilitate installation, maintenance, and cleaning. Each unit consists of a cleanable or replaceable cell or filter pad in a substantial frame that may be bolted to the frames of similar units to form an airtight partition between the source of the dusty air and the destination of the cleaned air (Fig. 20-133). Felt or other liners are sometimes used to make the assembly of individual cells airtight. For more critical installations, the individual panels may be fastened into a mounting frame that accommodates the entire filter bank. Depending on the air velocity through the filter medium that is allowable, the panels may be placed in a flat array, normal to the direction of the air stream, or in a V-type array. Filter banks in flat array may use cells with media in the form of deep pockets to provide the necessary filter area.

Panel filters may use either viscous or dry filter media. Viscous filters are so called because the filter medium is coated with a tacky liquid of high viscosity (e.g., mineral oil and adhesives) to retain the dust. The filter pad consists of an assembly of coarse fibers (now usually metal, glass, or plastic). Because the fibers are coarse and the media are highly porous, resistance to air flow is low and high filtration velocities can be used. Viscous filters are of two classes. In one, the filter medium or the entire panel is discarded when loaded (replaceable media and throwaway units). In the other, the medium is constructed of metal (e.g., crimped metal mesh) and is washed or steamed when loaded, then reoiled and reused. Typical viscous panel filters are shown in Fig. 20-134.

Dry filters are usually deeper than viscous filters. The dry filter media use finer fibers and have much smaller pores than the viscous media and need not rely on an oil coating to retain collected dust. Because of their greater resistance to air flow, dry filters must use lower filtration velocities to avoid excessive pressure drops. Hence, dry media must have larger surface areas and are usually pleated or arranged in the form of pockets (Fig. 20-135). The filter media are generally sheets of cellulose pulp, cotton, felt, or spun glass. Filters using felt or similar materials are usually reconditioned by vacuum or dry cleaning. With less expensive media, the medium is usually replaced. Mechanical loading devices are often used to replace the filter sheets in the frames. In some types of units, the complete unit cell is discarded when the maximum dust load has been reached.

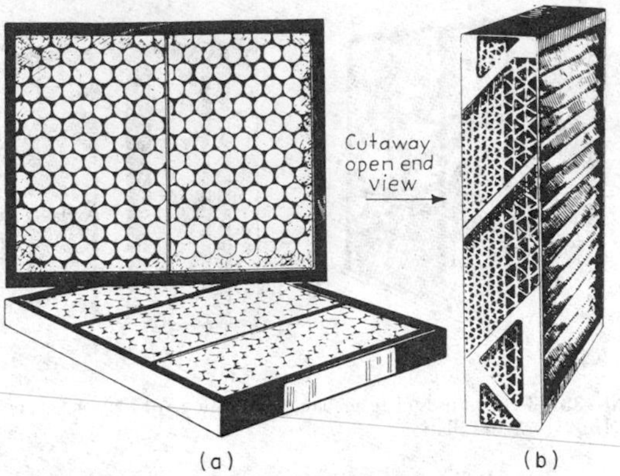

FIG. 20-134 Typical viscous unit filters. (*a*) Throwaway type, Dustop. (*Owens-Corning Fiberglas Corp.*) (*b*) Cleanable type. Air-Maze type B, cutaway open-end view. (*Air-Maze.*)

Automatic filters are made with either viscous-coated or dry filter media. However, the cleaning or disposal of the loaded medium is essentially continuous and automatic. In most such devices the air passes horizontally through a movable filter curtain. As the filter loads with dust, the curtain is continuously or intermittently advanced to expose clean media to the air flow and to clean or dispose of the loaded medium. Movement of the curtain can be provided by a hand crank or a motor drive. Movement of a motor-driven curtain can be actuated automatically by a differential-pressure switch connected across the filter. Viscous-type automatic filters (Fig. 20-136*b* and *c*) consist of perforated, crimped, or woven metallic screens in series. As the curtain moves through the oil bath, the panels are rinsed and coated with a fresh film of oil.

Dry-type automatic filters use a filter curtain consisting of a glass-fiber medium. Clean medium is dispensed from a roll, and the dust-laden medium is rerolled for disposal.

The Airmat dust arrester (Fig. 20-136*a*) is a dry-type air filter that is also used as a dust collector, since it can handle relatively heavy dust loads. It can be used for dusts that are not sticky and that can

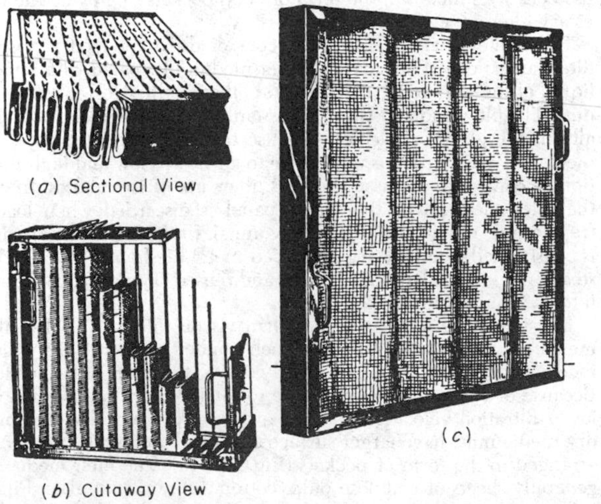

FIG. 20-135 Typical dry filters. (*a*) Throwaway type, Airplex. (*Davies Air Filter Corporation.*) (*b*) Replaceable medium type, Airmat PL-24, cutaway view. (*American Air Filter Co., Inc.*) (*c*) Cleanable type, Amirglass sawtooth. (*Amirton Company.*)

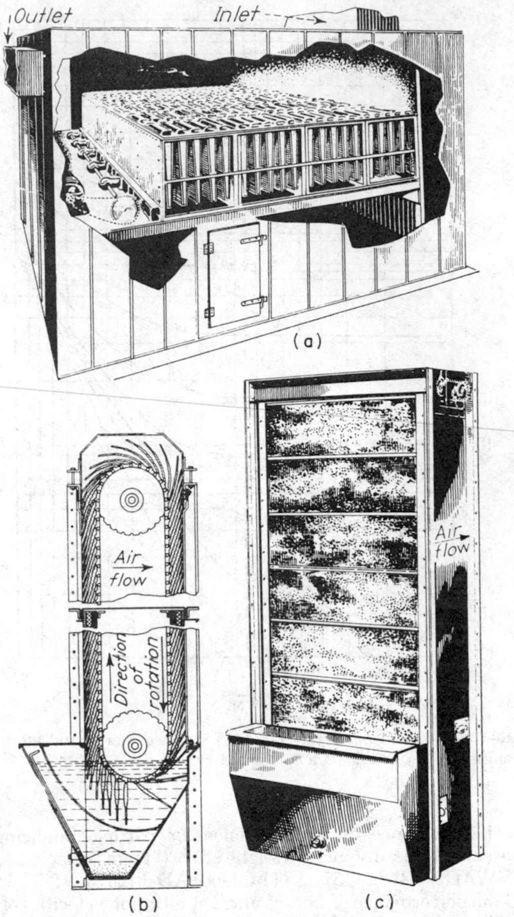

FIG. 20-136 Typical automatic air filters. (*a*) Dry-type Airmat dust arrester. (*American Air Filter Co., Inc.*) (*b*) Multipanel. (*American Air Filter Co., Inc.*) (*c*) Stay-new model A. (*Dollinger Corp.*)

be easily shaken off, but the air flow must be stopped or diverted while the filter is vibrated.

The characteristics of the various types of air filters are compared in Table 20-39. The pressure drop through a filter increases as dust accumulates. The filter should be replaced when the pressure drop starts to increase rapidly; otherwise, the air capacity will decrease. The maximum allowable pressure drop ranges from about 0.20 to 0.50 in water, depending on the type of filter medium. The cleanliness of the filtered air may also suffer if cleaning or replacement of dirty air filters is neglected. The dust loading of the air handled generally determines the life of unit filters. For this reason automatic filters become increasingly attractive as the dust concentration of the air to be cleaned increases, since dust capacity is not usually an important item with such filters. Dust capacities for unit filters generally range from 0.5 to 4.0 lb for a standard 20- by 20-in unit. In general, the life of viscous-type filters handling average city air may range from 2 to 5 months, while for dry-type filters it will range from 1 to 3 months. For the average chemical plant the life will be from one-half to one-third of that of the same filter handling "average" city air and may at times be considerably less because of the higher dust loadings involved.

The matter of a proper schedule for servicing filters cannot be too strongly stressed if satisfactory operation is to be obtained. The overall time cycle for reconditioning washable viscous filters is generally

Table 20-39 Comparative Air-Filter Characteristics

	Unit filters				Automatic filters
	Viscous type		Dry type		
	Cleanable	Throwaway	Throwaway	Cleanable	
Dust capacity	1. Well adapted for heavy dust loads (up to 2 grains/1000 cu. ft.) due to high dust capacity		1. Well adapted to light or moderate dust loads of less than 1 grain/1000 cu. ft.		1. Well adapted for heavy dust loads (> 2 grains/1000 cu. ft.) since it is serviced automatically
Filter size	1. Common size of unit filter is 20- × 20-in. face area handling 800 cu. ft./min. at rated capacity 2. Face velocity is generally 300–400 ft./min. for all types				1. Automatic viscous units supplied to handle 1000 cu. ft./min. and over 2. Face velocity is 350–750 ft./min.
Air velocity	1. Rated velocity is 300–400 ft./min. through the filter medium 2. Entrainment of oil may occur at very high velocities		1. Rated velocity is 10–50 ft./min. through the medium. (Some dry glass types run as high as 300 ft./min.) 2. Higher velocities may result in rupture of filter medium		1. Rated velocity is 350–750 ft./min. through the filter medium for viscous types. For dry types, it is 10–50 ft./min.
Resistance	1. Resistance ranges from 0.05–0.30 in. when clean to 0.4–0.5 in. when dirty 2. When the resistance exceeds a given value, the cells should be replaced or reconditioned 3. Cycling cells in large installations will serve to maintain a nearly constant resistance				1. Resistance runs about 0.3–0.4 in. water
	4. High resistance due to excessive dust loading results in channeling and poor efficiency		4. Excessive pressure drops resulting from high dust loading may result in rupture of filter medium		
Efficiency	1. Commercial makes are found in a variety of efficiencies, these depending roughly on filter resistance for similar types of medium 2. Efficiency decreases with increased dust load and increases with increased velocity up to certain limits		1. In general, give higher efficiency than viscous type, particularly on fine particles 2. Efficiency increases with increased dust load and decreases with increased velocity		
Operating cycle		1. Well adapted for short-period operations (less than 10 hr./day) due to relatively low investment cost			1. Well adapted for continuous operation
	2. Operating cycle is 1–2 months for general "average" industrial air conditioning		2. Operating cycle is 2–4 weeks for general "average" industrial air conditioning		
Method of cleaning	1. Washed with steam, hot water, or solvents and given fresh oil coating	1. Filter cell replaced. Life may in some cases be lengthened by shaking or vacuum cleaning, but this is not often successful		1. Vacuum cleaned, blown with compressed air, or dry cleaned	1. Automatic. Filter may clog in time and cleaning by blowing with compressed air may be necessary
Space requirement	1. Well adapted for low headroom requirements 2. Form of banks can be chosen to fit any shaped space 3. Space should be allowed for a man to remove filter cells for cleaning or replacement				1. Have a high headroom requirement 2. Take up less floor space than other types
	4. Requires space for washing, reoiling, and draining tanks			4. Requires space for mechanical loader in some cases	
Type of filter medium	1. Crimped, split, or woven metal, glass fibers, wood shavings, hair—all oil coated		1. Cellulose pulp, felt, cotton gauze, spun glass 2. Dry medium cannot stand direct wetting. Oil-impregnated mediums are available to resist humidity and prevent fluff entrainment		1. Metal screens, packing, or baffling. One type uses cellulose pulp
Character of dust	1. Not well suited for linty materials			1. Not well suited for handling oily dusts	1. Not suited for linty material if of viscous types
	2. Well adapted for make-up air and granular materials		2. Well adapted for linty material 3. Better adapted for fine dust than other types		
Temperature limitations	1. All metal types may be used up as high as 250°F. if suitable oil or grease is used. Those utilizing cellulosic materials are limited to 180°F.		1. Limited to 180°F. except for glass types which may be used up to 700°F. if suitable frames and gaskets are used		1. Viscous may be used up to 250°F. if suitable oil is used. Dry type limited to 180°F.

about 24 h. Unless the filters are allowed to drain sufficiently, an entrainment of oil by the filtered air may result.

In viscous air filters, dust collection is achieved by the impingement of particles on the filter surface, with the viscous coating serving to prevent reentrainment of separated dust. Collection efficiency generally increases as the gas velocity increases unless the velocity becomes so high as to reentrain the dust together with the viscous coating. Efficiency also tends to decrease with increasing dust accumulation because of a saturation of the viscous coating with dust.

High-Efficiency Air Cleaning Air-filter systems for nuclear facilities and for other applications demanding extremely high standards of air purity require filtration efficiencies well beyond those attainable with the equipment described above. The *Nuclear Air Cleaning Handbook* (Burchsted et al., op. cit.) presents an extensive treatment of the requirements for and the design of such air-cleaning facilities. Much of the material is pertinent to high-efficiency air-filter systems for applications to other than nuclear facilities.

HEPA (high-efficiency particulate air) filters were originally developed for nuclear and military applications but are now widely used and are manufactured by numerous companies. By definition, an HEPA filter is a "throwaway, extended-medium dry-type" filter having (1) a minimum particle-removal efficiency of not less than 99.97 percent for 0.3-μm particles, (2) a maximum resistance, when clean, of 1.0 in water when operated at rated air-flow capacity, and (3) a rigid casing extending the full depth of the medium (Burchsted et al., op. cit.). The filter medium is a paper made of submicrometer glass fibers in a matrix of larger-diameter (1- to 4-μm) glass fibers. An organic binder is added during the papermaking process to hold the fibers and give the paper added tensile strength. Filter units are made in several standard sizes (Table 20-40).

Table 20-40 Standard HEPA Filters*

Face dimensions, in	Depth, less gaskets, in	Design air-flow capacity at clean-filter resistance of 1.0 in water (standard ft^3/min)
24 × 24	11½	1000
24 × 24	5⅞	500
12 × 12	5⅞	125
8 × 8	5⅞	20
8 × 8	3 1/16	25

*Burchsted et al., *Nuclear Air Cleaning Handbook,* ERDA 76-21, Oak Ridge, Tenn., 1976.

Table 20-41 Classification of Common Air Filters*

Group	Efficiency	Filter type	Stain test efficiency, %	Arrestance, %
I	Low	Viscous impingement, panel type	<20†	40–80†
II	Moderate	Extended medium, dry type	20–60†	80–96†
III	High	Extended medium, dry type	60–98‡	96–99†
HEPA	Extreme	Extended medium, dry type	100§	100†

*Burchsted et al., *Nuclear Air Cleaning Handbook*, ERDA 76-21, Oak Ridge, Tenn., 1976.
†Test using synthetic dust.
‡Stain test using atmospheric dust.
§ASHRAE/52-68, American Society of Heating, Refrigerating and Air-Conditioning Engineers.

Because HEPA filters are designed primarily for high efficiency, their dust-loading capacities are limited, and it is common practice to use prefilters to extend their operating lives. In general, HEPA filters should be protected from (1) lint, (2) particles larger than 1 to 2 μm in diameter, and (3) dust concentrations greater than 23 mg/m^3 (10 gr/1000 ft^3). Air filters used in nuclear facilities as prefilters and building-supply air filters are classified as shown in Table 20-41. The standard of the American Society of Heating, Refrigerating and Air-Conditioning Engineers (*Method of Testing Air Cleaning Devices Used in General Ventilation for Removing Particulate Matter*, ASHRAE 52-68, 1968) requires both a dust-spot (dust-stain) efficiency test made with atmospheric dust and a weight-arrestance test made with a synthetic test dust. A more precise comparison of the different groups of filters, based on removal efficiencies for particles of specific sizes, is presented in Table 20-42.

Table 20-43 presents the relative performance of Group I, II, and III filters with respect to air-flow capacity, resistance, and dust-holding capacity. The dust-holding capacities correspond to the manufacturers' recommended maximum allowable increases in air-flow resistance. The values for dust-holding capacity are based on tests with a synthetic dust and hence are relative. The actual dust-holding capacity in a specific application will depend on the characteristics of the dust encountered. In some instances it may be appropriate to use two or more stages of precleaning in air-filter systems to achieve a desired combination of operating life and efficiency. In very dusty locations, inertial devices such as multiple small cyclones may be used as first-stage separators.

Electrical Precipitators When particles suspended in a gas are exposed to gas ions in an electrostatic field, they will become charged and migrate under the action of the field. The functional mechanisms of electrical precipitation may be listed as follows:

1. Gas ionization
2. Particle collection
   a. Production of electrostatic field to cause charging and migration of dust particles
   b. Gas retention to permit particle migration to a collection surface
   c. Prevention of reentrainment of collected particles
   d. Removal of collected particles from the equipment

Table 20-42 Comparison of Air Filters by Percent Removal Efficiency for Various Particle Sizes*

Group	Efficiency	Removal efficiency, %, for particle size of 0.3 μm	1.0 μm	5.0 μm	10.0 μm
I	Low	0–2	10–30	40–70	90–98
II	Moderate	10–40	40–70	85–95	98–99
III	High	45–85	75–99	99–99.9	99.9
HEPA	Extreme	99.97 min	99.99	100	100

*Burchsted et al., *Nuclear Air Cleaning Handbook*, ERDA 76-21, Oak Ridge, Tenn., 1976.

Table 20-43 Air-Flow Capacity, Resistance, and Dust-Holding Capacity of Air Filters*

Group	Efficiency	Air-flow capacity, $ft^3/(min \cdot ft^2$ of frontal area)	Resistance, in water: Clean filter	Used filter	Dust-holding capacity, g/(1000 $ft^3 \cdot$ min of air-flow capacity)
I	Low	300–500	0.05–0.1	0.3–0.5	50–1000
II	Moderate	250–750	0.1 –0.5	0.5–1.0	100– 500
III	High	250–750	0.20–0.5	0.6–1.4	50– 200

*Burchsted et al., *Nuclear Air Cleaning Handbook*, ERDA 76-21, Oak Ridge, Tenn., 1976.

There are two general classes of electrical precipitators: (1) single-stage, in which ionization and collection are combined; (2) two-stage, in which ionization is achieved in one portion of the equipment, followed by collection in another. Various types in each class differ essentially in the details by which each function is accomplished.

The underlying theory presented in the following paragraphs assumes that the dust concentration is small, since only very incomplete evaluations for conditions of high dust concentration have been made.

Field Strength Whereas the applied potential or voltage is the quantity commonly known, it is the field strength that determines behavior in an electrostatic field. When the current flow is low (i.e., before the onset of spark or corona discharge), these are related by the following equations for two common forms of electrodes:

Parallel plates:

$$\mathcal{E} = E/B_e \tag{20-83}$$

Concentric cylinders (wire-in-cylinder):

$$\mathcal{E} = \frac{E}{r \ln (D_t/D_d)} \tag{20-84}$$

The field strength is uniform between parallel plates, whereas it varies in the space between concentric cylinders, being highest at the surface of the central cylinder. After corona sets in, the current flow will become appreciable. The field strength near the center electrode will be less than given by Eq. (20-84), and that in the major portion of the clearance space will be greater and more uniform [see Eqs. (20-89) and (20-90)].

Potential and Ionization In order to obtain gas ionization it is necessary to exceed, at least locally, the electrical breakdown strength of the gas. Corona is the name applied to such a local discharge that fails to propagate itself. Sparking is essentially an advanced stage of corona in which complete breakdown of the gas occurs along a given path. Since corona represents a local breakdown, it can occur only in a nonuniform electrical field (Whitehead, *Dielectric Phenomena—Electrical Discharge in Gases*, Van Nostrand, Princeton, N.J., 1927, p. 40). Consequently, for parallel plates, only sparking occurs at a field strength or potential difference given by the empirical expressions

$$\mathcal{E}_s = \mathcal{E}_o k_p \left[1 + \left(\frac{K_o}{k_p B_e} \right) \right] \tag{20-85}$$

$$E_s = \mathcal{E}_o k_p B_e + K_o \mathcal{E}_o \tag{20-86}$$

For air in the range of $k_p B_e$ from 0.1 to 2, $\mathcal{E}_o = 111.2$ and $K_o = 0.048$. Thornton [*Phil. Mag.*, **28** (7), 666 (1939)] gives values for other gases. For concentric cylinders (Loeb, *Fundamental Processes of Electrical Discharge in Gases*, Wiley, New York, 1939; Peek, *Dielectric Phenomena in High-Voltage Engineering*, McGraw-Hill, New York, 1929; and Whitehead, op. cit.), corona sets in at the central wire when

$$\mathcal{E}_c = \mathcal{E}_o k_p \left(1 + \sqrt{\frac{K_o}{k_p D_d}} \right) \tag{20-87}$$

$$E_c = \left(\frac{\mathcal{E}_o k_p D_d}{2}\right)\left(1 + \sqrt{\frac{K_o}{k_p D_d}}\right)\ln\left(\frac{D_t}{D_d}\right) \tag{20-88}$$

For air approximate values are $\mathcal{E}_o = 110$, $K_o = 0.18$. Corona, however, will set in only if $(D_t/D_d) > 2.718$. If this ratio is less than 2.718, no corona occurs, and only sparking will result, following the laws given by Eqs. (20-87) and (20-88) (Peek, op. cit.).

In practice, precipitators are usually operated at the highest voltage practicable without sparking, since this increases both the particle charge and the electrical precipitating field. The sparking potential is generally higher with a negative charge on the discharge electrode and is less erratic in behavior than a positive corona discharge. It is the consensus, however, that ozone formation with a positive discharge is considerably less than with a negative discharge. For these reasons negative discharge is generally used in industrial precipitators, and a positive discharge is utilized in air-conditioning applications. In Table 20-44 are given some typical values for the sparking potential for the case of small wires in pipes of various sizes. The sparking potential varies approximately directly as the density of the gas but is very sensitive to the character of any material collected on the electrodes. Even small amounts of poorly conducting material on the electrodes may markedly lower the sparking voltage. For positive polarity of the discharge electrode, the sparking voltage will be very much lower. The sparking voltage is greatly affected by the temperature and humidity of the gas, as shown in Fig. 20-137.

Current Flow Corona discharge is accompanied by a relatively small flow of electric current, typically 0.1 to 0.5 mA/m^2 of collecting-electrode area (projected, rather than actual area). Sparking usually involves a considerably larger flow of current which cannot be tolerated except for occasional periods of a fraction of a second duration, and then only when suitable electrical controls are provided to limit the current. However, when suitable controls are provided, precipitators have been operated continuously with a small amount of sparking to ensure that the voltage is in the correct range to ensure corona. Besides disruptive effects on the electrical equipment and electrodes, sparking will result in low collection efficiency because of reduction in applied voltage, redispersion of collected dust, and current channeling. Although an exact calculation can be made for the current flow for a direct-current potential applied between concentric cylinders, the following simpler expression, based on the assumption of a constant space charge or ion density, gives a good approximation of corona current [Ladenburg, *Ann. Phys.*, 4(5), 863 (1930)]:

$$I = \frac{8\lambda_i E(E - E_c)}{D_t^2 \ln (D_t/D_d)} \tag{20-89}$$

and the average space charge is given by (Whitehead, op. cit.)

$$\sigma_{avg} = \frac{4(E - E_c)}{\pi D_t^2 \epsilon} \tag{20-90}$$

In the space outside the immediate vicinity of corona discharge, the field strength is sensibly constant, and an average value is given by

$$\mathcal{E} = \sqrt{2I/\lambda_i} \tag{20-91}$$

which applies if the potential difference is above the critical potential required for corona discharge so that an appreciable current flows.

Ionic mobilities are given by Loeb (*International Critical Tables*, vol. 6, McGraw-Hill, New York, 1929, p. 107). For air at 0°C, 760 mmHg, $\lambda_i = 624$ (cm/s)/(statV/cm) for negative ions. Positive ions usually have a slightly lower mobility. Loeb (*Basic Processes of Gaseous Electronics*, University of California Press, Berkeley and Los Angeles, 1955, p. 53) gives a theoretical expression for ionic mobility of gases which is probably good to within ±50 percent:

$$\lambda_i = \frac{100.0}{k_p \sqrt{(\delta_g - 1)M}} \tag{20-92}$$

In general, ionic mobilities are inversely proportional to gas density. Ionic velocities in the usual electrostatic precipitator are on the order of 30.5 m/s (100 ft/s).

Electric Wind By virtue of the momentum transfer from gas ions moving in the electrical field to the surrounding gas molecules, a gas circulation, known as the "electric" or "ionic" wind, is set up between the electrodes. For conditions encountered in electrical precipitators, the velocity of this circulation is on the order of 0.6 m/s (2 ft/s). Also, as a result of this momentum transfer, the pressure at the collecting electrode is slightly higher than at the discharge electrode (Whitehead, op. cit., p. 167).

Charging of Particles [Deutsch, *Ann. Phys.*, **68**(4), 335 (1922); **9**(5), 249 (1931); **10**(5), 847 (1931); Ladenburg, op. cit.; and Mierdel, *Z. Tech. Phys.*, **13**, 564 (1932).] Three forces act on a gas ion in the vicinity of a particle: attractive forces due to the field strength and the ionic image; and repulsive forces due to the Coulomb effect. For spherical particles larger than 1-μm diameter, the ionic image effect is negligible, and charging will continue until the other two forces balance according to the equation

$$N_o = \left(\frac{\zeta \mathcal{E} D_p^2}{4\epsilon}\right)\left(\frac{\pi\sigma\epsilon t\lambda_i}{1 + \pi\sigma\epsilon\lambda_i t}\right) \tag{20-93}$$

The ultimate charge acquired by the particle is given by

$$N_o = \zeta \mathcal{E} D_p^2/4\epsilon \tag{20-94}$$

and is very nearly attained in a fraction of a second. For particles smaller than 1-μm diameter, the initial charging will occur according to Eq. (20-93). However, owing to the ionic-image effect, the ultimate charge will be considerably greater because of penetration resulting from the kinetic energy of the gas ions. For charging times of the order encountered in electrical precipitation, the ultimate charge acquired by spherical particles smaller than about 1-μm diameter may be approximated (±30 percent) by the empirical expression

$$N_o = 3.4 \times 10^3 D_p T \tag{20-95}$$

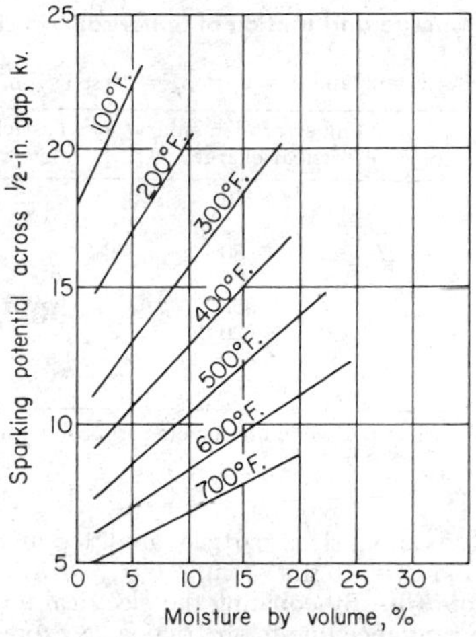

FIG. 20-137 Sparking potential for negative point-to-plane ½-in (1.3-cm) gap as a function of moisture content and temperature of air at 1-atm (101.3-kPa) pressure. [*Sproull and Nakada*, Ind. Eng. Chem., ***43**, 1356 (1951)*.]

Table 20-44 Sparking Potentials* (Small Wire Concentric in Pipe)

	Sparking potential,† volts	
Pipe diameter, in.	Peak	Root mean square
4	59,000	45,000
6	76,000	58,000
9	90,000	69,000
12	100,000	77,000

*Data reported by Anderson in Perry, "Chemical Engineers' Handbook," 2d ed., p. 1873, McGraw-Hill, New York, 1941.

†For gases at atmospheric pressure, 100°F., containing water vapor, air, CO_2, and mist, and negative-discharge–electrode polarity.

Table 20-45 Charge and Motion of Spherical Particles in an Electric Field

For $\zeta = 2$, and $\varepsilon = \varepsilon_i = \varepsilon_p = 10$ statV/cm

Particle diam., μ	Number of elementary electrical charges, N_0	Particle migration velocity,* u_e, ft./sec.
0.1	10	0.27
.25	25	.15
.5	50	.12
1.0	105	.11
2.5	655	.26
5.0	2,620	.50
10.0	10,470	.98
25.0	65,500	2.40

NOTE: To convert feet per second to meters per second, multiply by 0.3048.

Table 20-46 Performance Data on Typical Single-Stage Electrical Precipitator Installations*

Type of precipitator	Type of dust	Gas volume, cu. ft./min.	Average gas velocity, ft./sec.	Collecting electrode area, sq. ft.	Over-all collection efficiency, %	Average particle migration velocity, ft./sec.
Rod curtain	Smelter fume	180,000	6	44,400	85	0.13
Tulip type	Gypsum from kiln	25,000	3.5	3,800	99.7	.64
Perforated plate . .	Fly ash	108,000	6	10,900	91	.40
Rod curtain	Cement	204,000	9.5	26,000	91	.31

*Research-Cottrell, Inc. To convert cubic feet per minute to cubic meters per second, multiply by 0.00047; to convert feet per second to meters per second, multiply by 0.3048; and to convert square feet to square meters, multiply by 0.0929.

Values of N_o for various sized particles are listed in Table 20-45 for 70°F, $\zeta = 2$, and $\mathscr{E} = 10$ statV/cm.

Particle Mobility By equating the electrical force acting on a particle to the resistance due to air friction, as expressed by Stokes' law, the particle velocity or mobility may be expressed by

1. For particles larger than 1-μm diameter:

$$\lambda_p = \left(\frac{u_e}{\mathscr{E}_p}\right) = \frac{\zeta D_p \mathscr{E}_i K_m}{12\pi\mu} \tag{20-96}$$

2. For particles smaller than 1-μm diameter:

$$\lambda_p = \left(\frac{u_e}{\mathscr{E}_p}\right) = \frac{360 K_m \epsilon T}{\mu} \tag{20-97}$$

For single-stage precipitators, $\mathscr{E}_i$ and $\mathscr{E}_p$ may be considered as essentially equal. It is apparent from Eq. (20-97) that the mobility in an electric field will be almost the same for all particles smaller than about 1-μm diameter, and hence, in the absence of reentrainment, collection efficiency should be almost independent of particle size in this range. Very small particles will actually have a greater mobility because of the Stokes-Cunningham correction factor. Values of u_e are listed in Table 20-45 for 70°F, $\zeta = 2$, and $\mathscr{E} = \mathscr{E}_i = \mathscr{E}_p = 10$ statV/cm.

Collection Efficiency Although actual particle mobilities may be considerably greater than would be calculated on the basis given in the preceding paragraph because of the action of the electric wind in single-stage precipitators, the latter acts in a compensating fashion, and the overall effect of the electric wind is probably to provide an equalization of particle concentration between the electrodes similar to the action of normal turbulence (Mierdel, op. cit.). On this basis Deutsch (op. cit.) has derived the following equations for collection efficiency, the form of which had previously been suggested by Anderson on the basis of experimental data:

$$\eta = 1 - e^{-(u_e A_e/q)} = 1 - e^{-K_e u_e} \tag{20-98}$$

For the concentric-cylinder (or wire-in-cylinder) type of precipitator, $K_e = 4L_e/D_t V_e$; for rod-curtain or wire-plate types, $K_e = L_e/B_e V_e$. Strictly speaking, Eq. (20-98) applies only for a given particle

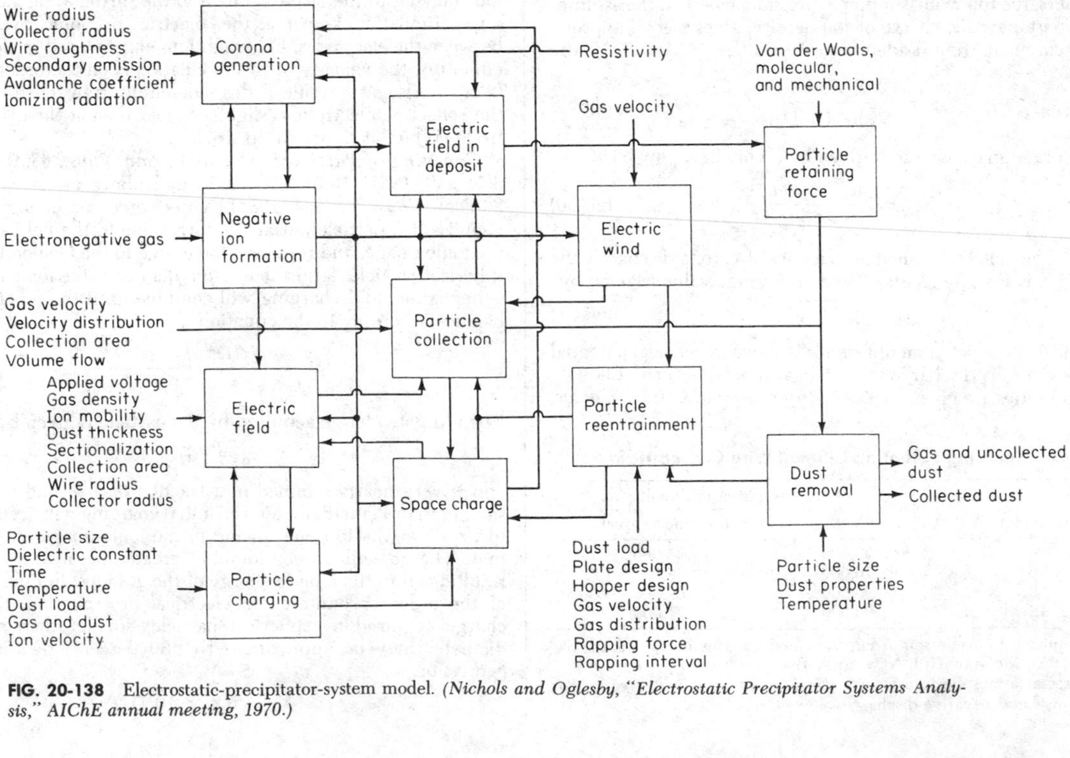

FIG. 20-138 Electrostatic-precipitator-system model. *(Nichols and Oglesby, "Electrostatic Precipitator Systems Analysis," AIChE annual meeting, 1970.)*

size, and the overall efficiency must be obtained by an integration process for a specific dust distribution, as described in the subsection "Cyclone Separators." However, over limited ranges of performance conditions, Eq. (20-98) has been found to give a good approximation of overall collection efficiency, with the term for particle migration velocity representing an empirical average value. Such values, calculated from overall collection-efficiency measurements, are given in Table 20-46 for specific installations.

For two-stage precipitators with close collecting-plate spacings (Figs. 20-152, 20-153), the gas flow is substantially streamline, and no electric wind exists. Consequently, with reentrainment neglected, collection efficiency may be expressed as [Penny, *Electr. Eng*, **56**, 159 (1937)]

$$\eta = u_e L_e / V_e B_e \tag{20-99}$$

which holds for values of $\eta \leqq 1.0$. In practice, however, extraneous factors may cause the actual efficiency to approach a relationship of the type given by Eq. (20-98).

Application The theoretical considerations that have been expounded should be used only for order-of-magnitude estimates, since a number of extraneous factors may enter into actual performance. In actual installations rectified alternating current is employed. Hence the electric field is not fixed but varies continuously, depending on the waveform of the rectifier, although Schmidt and Anderson [*Electr. Eng.*, **57**, 332 (1938)] report that the waveform is not a critical factor. Allowances for high dust concentrations have not been fully studied, although Deutsch (op. cit.) has presented a theoretical approach. In addition, irregularities on the discharge electrode will result in local discharges. Such irregularities can readily result from dust incrustation on the discharge electrodes due to charging of particles with opposite polarity within the thin but appreciable flow or ionization layer surrounding this electrode. Very high dust loadings increase the potential difference required for corona and reduce the current due to the space charge of the particles. This tends to reduce the average particle charge and reduces collection efficiency. This can be compensated for by increasing the potential difference when high dust loadings are involved.

Several investigators have attempted to modify the basic Deutsch equation so that it would more nearly describe precipitator performance. Cooperman ("A New Theory of Precipitator Efficiency," Pap. 69-4, APCA meeting, New York, 1969) introduced correction factors for diffusional forces arising from variations in particle concentration along the precipitator length and also perpendicular to the collecting surface. Robinson [*Atmos. Environ.* **1**(3), 193 (1967)] derived an equation for collection efficiency in which two erosion or reentrainment terms are introduced.

An analysis of precipitator performance based on theoretical considerations was undertaken by the Southern Research Institute for the National Air Pollution Control Administration (Nichols and Oglesby, "Electrostatic Precipitator Systems Analysis," AIChE annual meeting, 1970). A mathematical model was developed for calculating the particle charge, electric field, and collection efficiency based on the Deutsch-Anderson equation. The system diagram is shown in Fig. 20-138. This system-analysis method, using high-speed computers, makes it possible to analyze what takes place in each increment of precipitator length. Collection efficiency versus particle size is computed for each 1 ft (0.3 m) of gas travel, and the inlet particle-size distribution is modified accordingly. Computed overall efficiencies compare well with measured values on three precipitators. The model assumes that field charging is the only charging mechanism. The authors considered the addition of several refinements to the program: the influence of diffusion charging; reentrainment effects due to rapping and erosion; and loss of efficiency due to maldistribution of gas, dust resistivity, and gas-property effects. The modeling technique appeared promising, but much more work was needed before it could be used for design. The same authors prepared a general treatise (Oglesby and Nichols, *A Manual of Electrostatic Precipitator Technology*, parts I and II, Southern Research Institute, Birmingham, Ala., U.S. Government Publications PB196360,196381, 1970).

High-Pressure–High-Temperature Electrostatic Precipitation In general, increased pressure increases precipitation efficiency, although a somewhat higher potential is required, because it reduces ion mobility and hence increases the potential required for corona and sparking. Increased temperature reduces collection efficiency because ion mobility is increased, lowering critical potentials, and because gas viscosity is increased, reducing migration velocities.

Precipitators have been operated at pressures up to 5.5 MPa (800 psig) and temperatures to 800°C.

The effect of increasing gas density on sparkover voltage has been investigated by Robinson [*J. Appl. Phys.*, **40**, 5107 (1969); *Air Pollution Control*, part 1, Wiley-Interscience, New York, 1971, chap. 5]. Figure 20-139 shows the effect of gas density on corona-starting and sparkover voltages for positive and negative corona in a pipe precipitator. The sparkover voltages are experimental and are given by the solid points. Experimental corona-starting voltages are given by the hollow points. The solid lines are corona-starting voltage curves calculated from Eq. (20-99). This is an empirical relationship developed by Robinson.

$$\frac{E_c}{\rho'} = A \frac{B}{\sqrt{D_d \rho'/2}} \tag{20-100}$$

E_c is the corona-starting field, kV/cm. ρ' is the relative gas density, equal to the actual gas density divided by the density of air at 25°C, 1 atm. D_d is the diameter of the ionizing wire, cm. A and B are constants which are characteristics of the gas. In dry air, A = 32.2 kV/cm and B = 8.46 kV/cm$^{1/2}$. Agreement between experimental

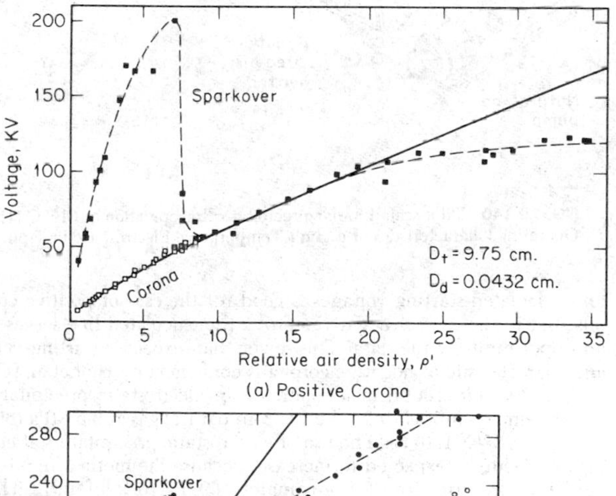

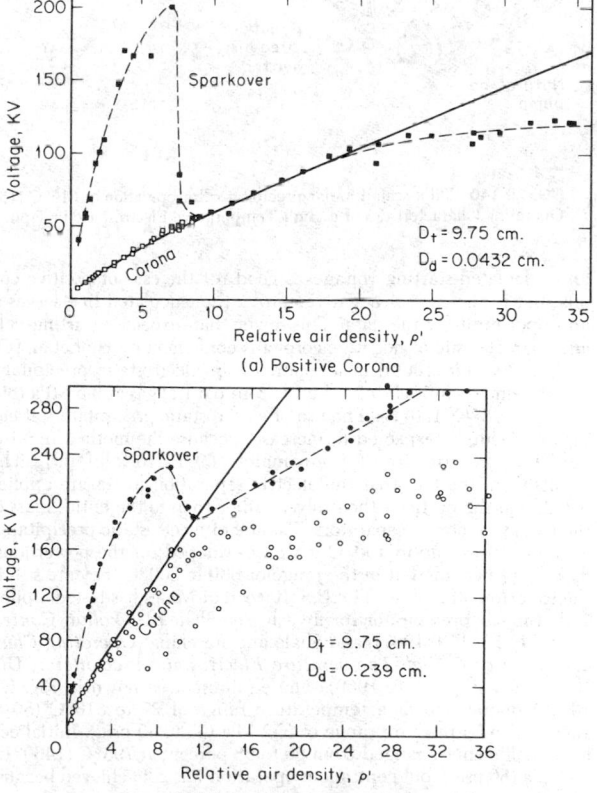

FIG. 20-139 Corona-starting and sparkover voltages for coaxial wire-pipe electrodes in air (25°C). D_t and D_d are the respective pipe and wire diameters. The voltage is unvarying direct current. (*Robinson,* Air Pollution Control, *part 1, Wiley-Interscience, New York, 1971, chap. 5.*)

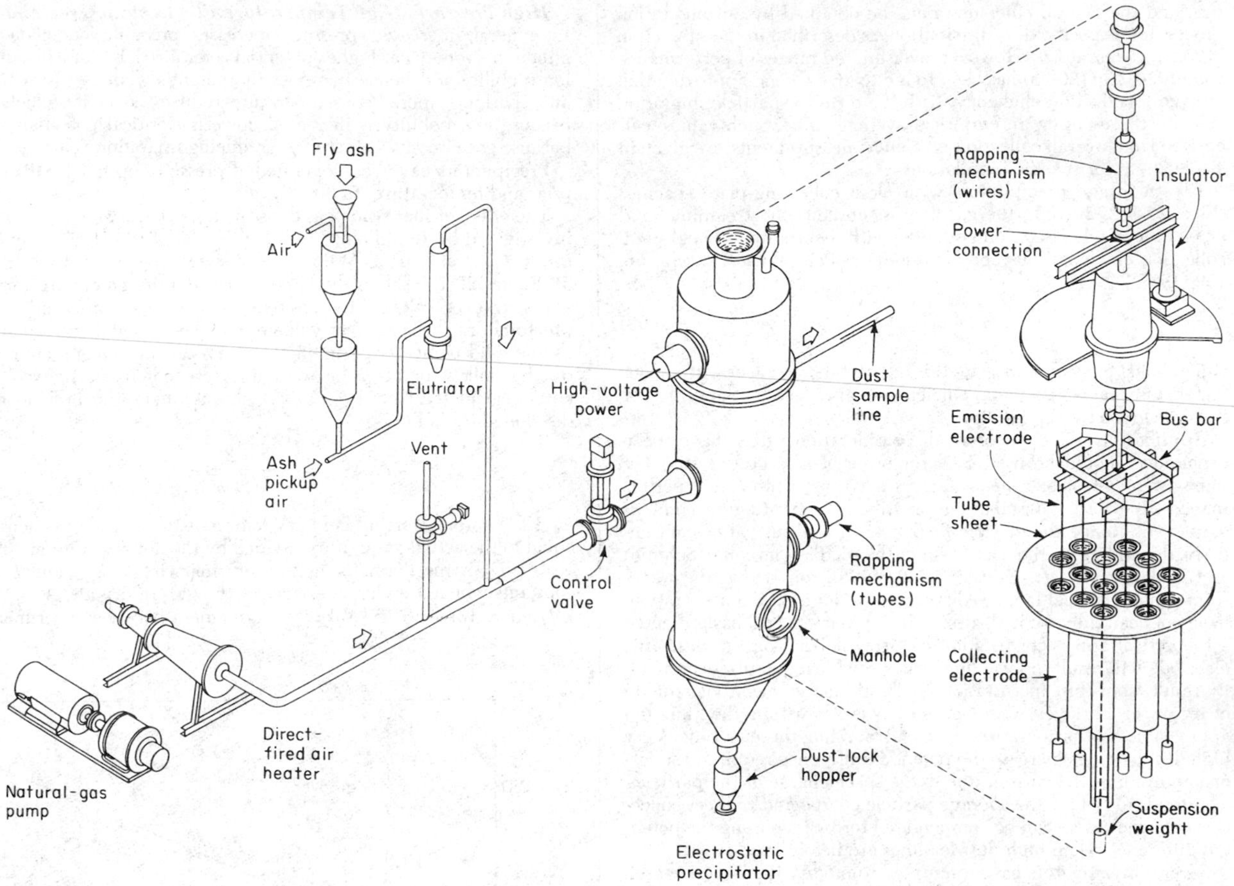

FIG. 20-140 Pilot-scale tubular precipitator for operation at 816°C (1500°F) and 552 kPa (80 psig). (*Shale and Faschig, Operating Characteristics of a High-Temperature Electrostatic Precipitator, U.S. Bur. Mines Rep. 7276, 1969.*)

and calculated starting voltages is good for the case of positive corona, but in the case of negative corona the calculated line serves as an upper limit for the data. This lower-than-expected starting-voltage characteristic of negative corona is confirmed by Hall et al. [*Oil Gas J.*, **66**, 109 (1968)] in a report of an electrostatic precipitator which removes lubricating-oil mist from natural gas at 5.5 MPa (800 psig) and 38°C (100°F). The use of electrostatic precipitators at elevated pressure is expected to increase, because the method requires very low pressure drop [approximately 69 Pa (0.1 lbf/in^2)]. This results from the fact that the electric separation forces are applied directly to the particles themselves rather than to the entire mass of the gas, as in inertial separators. The use of electrostatic precipitators at temperatures up to 400°C is well developed for the powerhouse fly-ash application, but in the range of 600 to 800°C they are still in the experimental phase. The U.S. Bureau of Mines has tested a pilot-scale tubular precipitator for fly ash. See Shale [*Air Pollut. Control Assoc. J.*, **17**, 159 (1967)] and Shale and Fasching (*Operating Characteristics of a High-Temperature Electrostatic Precipitator*, U.S. Bur. Mines Rep. 7276, 1969). The equipment is shown in Fig. 20-140. It operated over a temperature range of 27 to 816°C (80 to 1500°F) and a pressure range of 552 kPa (35 to 80 psig). Initial collection efficiencies ranged from 90 to 98 percent at 793°C (1460°F), 552 kPa (80 psig), but continuous operation was not achieved because of excessive thermal expansion of internal parts.

Resistivity Problems Optimum performance of electrostatic precipitators is achieved when the electrical resistivity of the collected dust is sufficiently high to result in electrostatic pinning of the particles to the collecting surface, but not so high that dielectric breakdown of the dust layer occurs as the corona current passes through it. The optimum resistivity range is generally considered to be from 10^8 to 10^{10} Ω·cm, measured at operating conditions. As the dust builds up on the collecting electrode, it impedes the flow of current, so that a voltage drop is developed across the dust layer:

$$E_d = j\rho_d L_d \tag{20-101}$$

If E_d/L_d exceeds the dielectric strength of the dust layer, sparks occur in the deposit and form back-corona craters. Ions of both polarities are formed. Positive ions formed in the craters are attracted to the negatively charged particles in the gas stream, whose charge level is reduced so that collection efficiency decreases. Some of the positive ions neutralize part of the negative-space-charge cloud normally present near the wire, thereby increasing total current. Collection efficiency under these conditions will not correlate with total power input (Owens, E. I. du Pont de Nemours & Co. internal communication, 1971). Under normal conditions, collection efficiency is an exponential function of corona power (White, *Industrial Electrostatic Precipitation*, Addison-Wesley, Reading, Mass., 1963). With typical ion density in the range of 10^9/cm^3, overall voltage gradient would be about 4000 V/cm, and current about 1 μA/cm^2. Dielectric breakdown of the dust layer (at about 10,000 V/cm) would therefore be expected for dusts with resistivities above 10^{10} Ω·cm.

Problems due to high resistivity are of great concern in **fly-ash** precipitation because air-pollution regulations require that coals have low (<1 percent) sulfur content. Figure 20-141 shows that the resis-

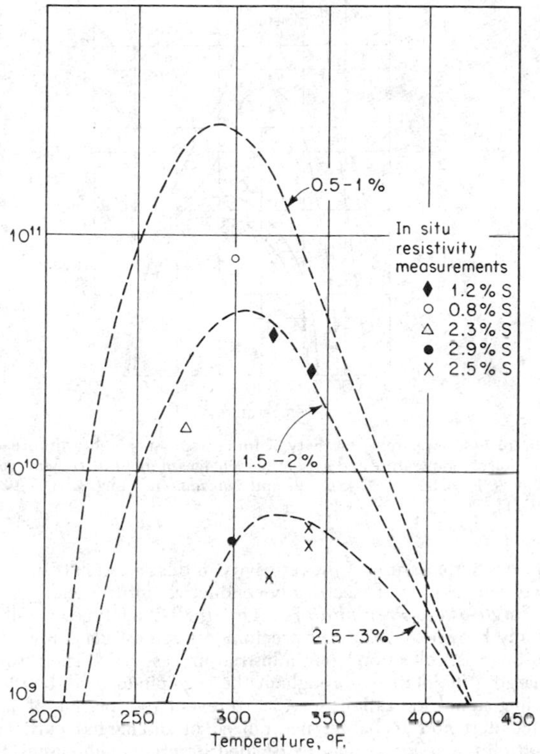

FIG. 20-141 Trends in resistivity of fly ash with variations in flue-gas temperature and coal sulfur content. °C = (°F − 32) × ⅝. (*Oglesby and Nichols,* A Manual of Electrostatic Precipitator Technology, *part II, Southern Research Institute, Birmingham, Ala., 1970.*)

tivity of low-sulfur coal ash exceeds the threshold of 10^{10} Ω·cm at common operating temperatures. This has resulted in the installation of a number of precipitators which have failed to meet guaranteed performance. This has occurred to an alarming extent in the United States but has also been encountered in Australia, where the sulfur content is typically 0.3 to 0.6 percent. Maartmann (Pap. EN-34F, 2d International Clean Air Congress, Washington, 1970) reports the installation of a number of precipitators which performed below guarantees, so that the Electricity Commission of New South Wales decided that each manufacturer wishing to bid on a new station must first make pilot tests to prove performance on the actual coal to be burned in that station. Problems of back corona and excessive sparking with low-sulfur coal usually require that the operating voltage be reduced. This reduces the migration velocity and leads to larger precipitators. Ramsdell (*Design Criteria for Precipitators for Modern Central Station Power Plants,* American Power Conference, Chicago, 1968) developed the curves in Fig. 20-142. They show the results of extensive field tests by the Consolidated Edison Co. In another paper ("Anti-pollution Program of Consolidated Edison Co. of New York," ASCE, May 13–17, 1968), Ramsdell traces the remarkable growth in the size of precipitators required for high efficiency on low-sulfur coals. The culmination of this work was the precipitator at boiler 30 at Ravenswood Station, New York. Resistivity problems were avoided by operating at high temperature [343°C (650°F)]. The mechanical (cyclone) collector was installed after the precipitator to clean up puffs due to rapping.

Maartmann (op. cit.) agrees that sulfur content is important but feels that it should not be the sole criterion for the determination of collecting surface. He points to specific collecting-surface requirements as high as 500 $ft^2/(1000\ ft^3\cdot min)$ for 95 percent collection efficiency with high-resistivity Australian ash.

Schmidt and Anderson (op. cit.) and Anderson [*Physics,* 3, 23 (July 1932)] claim that resistivity of the collected dust may be a controlling factor which is very sensitive to moisture. They state that an increase in relative humidity of 5 percent may double the precipitation rate because of its effect on the conductivity of the collected dust layer. This sensitivity to moisture makes realistic measurement of dust

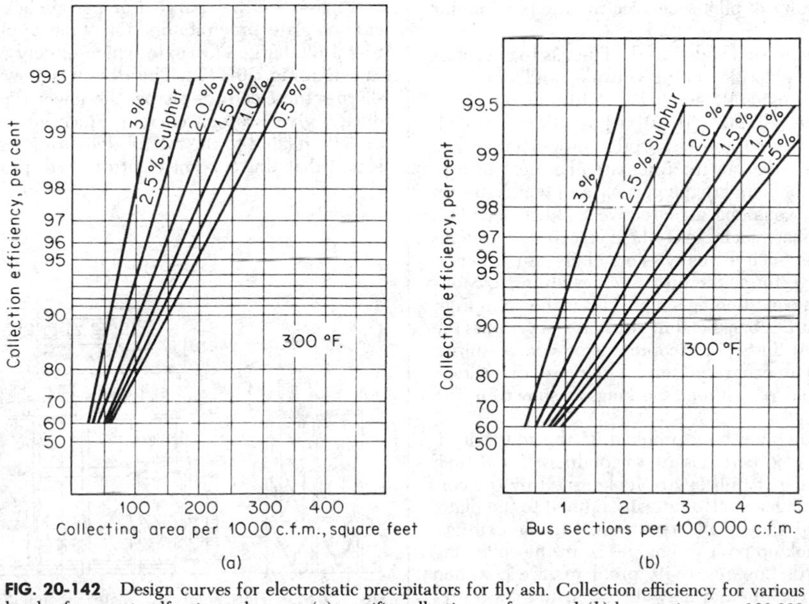

FIG. 20-142 Design curves for electrostatic precipitators for fly ash. Collection efficiency for various levels of percent sulfur in coal versus (*a*) specific collecting surface, and (*b*) bus sections per 100,000 ft^3/min (4.7 m^3/s). °C = (°F − 32) × ⅝. (*Ramsdell,* Design Criteria for Precipitators for Modern Central Station Power Plants, *American Power Conference, Chicago, Ill., 1968.*)

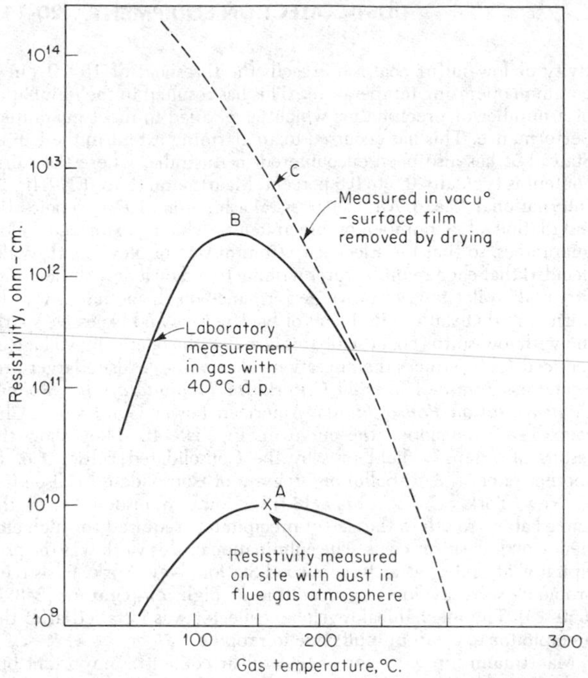

FIG. 20-143 Variation in resistivity of dust. (*A*) On site. (*B*) Same sample in laboratory. (*C*) Sample heated to 350°C in vacuum to destroy surface film. *(Busby, Whitehead, and Darby, Pap. EN-34H, 2d International Clean Air Congress, Washington, Dec. 6–11, 1970.)*

resistivity an extremely difficult task. Figure 20-143 shows that measurements made on site in the actual flue gas (in situ) can vary onehundredfold from laboratory measurements. Oglesby and Nichols (op. cit.) describe two devices for in situ resistivity measurements, but they emphasize that the techniques are not yet standardized and that the data should be interpreted cautiously and the method used in the measurements identified. The uncertain state of this art gives additional support to the value of pilot-scale electrostatic precipitator tests.

Conditioning agents have been added to the flue gas to alter dust resistivity. Steam, sodium chloride, sulfur trioxide, and ammonia have all been successfully used. Research by Chittum and others [Schmidt, *Ind. Eng. Chem.*, **41,** 2428 (1949)] led to a theory of conditioning by alteration of the moisture-adsorption properties of dust surfaces. Chittum proposed that an intermediate chemical-adsorption film, which was strongly bound to the particle and which in turn strongly adsorbed water, would be an effective conditioner. This explains how acid conditioners, such as SO_3, help resistivity problems associated with basic dusts, such as many types of fly ash, whereas ammonia is a good additive for acidic dusts such as alumina. Moisture alone can be used as a conditioning agent. This is shown in Figs. 20-137 and 20-144. Moisture is beneficial in two ways: it reduces the electrical resistivity of most dusts (an exception is powdered sulfur, which apparently does not absorb water), and it increases the voltage which may safely be employed without sparking, as shown in Fig. 20-137.

Low resistivity can sometimes be a problem. If the resistivity is below $10^4\ \Omega\cdot$cm, the collected particles are so conductive that their charges leak to ground faster than they are replenished by the corona. The particles are no longer electrostatically pinned to the plate, and they may then be swept away and reentrained in the exit gas. The particles may even pick up positive charges from the collecting plate and then be repelled. Low-resistivity problems are common with dusts of high carbon content and may also occur in fly-ash precipitators which handle the ash from high-sulfur coal and operate at

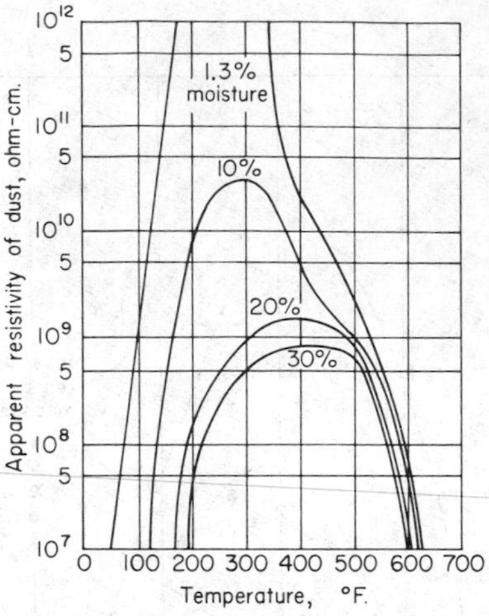

FIG. 20-144 Apparent resistivity of fume from an open-hearth furnace as a function of temperature and various percentages of moisture content (volume). °C = (°F − 32) × 5/9. [*Sproull and Nakada,* Ind. Eng. Chem., *43, 1355 (1951).*]

low gas temperatures. Low resistivity in this case results from excessive condensation of electrically conductive sulfuric acid.

Single-Stage Precipitators The single-stage type of unit, commonly known as a Cottrell precipitator, is most generally used for dust or mist collection from industrial-process gases. The corona discharge is maintained throughout the precipitator and, besides providing initial ionization, serves to prevent redispersion of precipitated dust and recharges neutralized or discharged particle ions. Cottrell precipitators may be divided into two main classes, the so-called plate type (Fig. 20-145), in which the collecting electrodes consist of parallel plates, screens, or rows of rods, chains, or wires; and the pipe type (Fig. 20-146), in which the collecting electrodes consist of a nest of parallel pipes which may be square, round, or of any other shape. The discharge or precipitating electrodes in each case are wires or rods, either round or edged, which are placed midway between the collecting electrodes or in the center of the pipes and may be either parallel or perpendicular to the gas flow in the case of plate precipitators. On some applications the Koppers Co. uses a discharge electrode which closely resembles barbed wire, as shown in Fig. 20-147*a*. This design is based on the principle that the sharper the electrode shape, the lower the voltage required to produce a given corona current. Since efficiency of dust collection is directly related to the rate of delivering electric energy to the gas, it is said that this type of electrode will produce the same collection

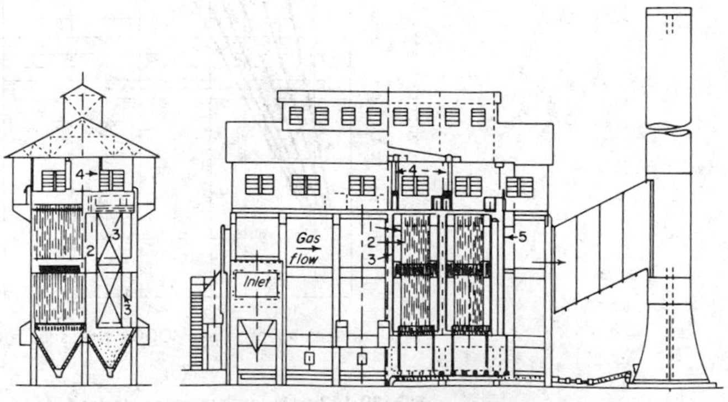

FIG. 20-145 Horizontal-flow plate precipitator used in a cement plant. *(Western Precipitation Division, Joy Manufacturing Company.)*

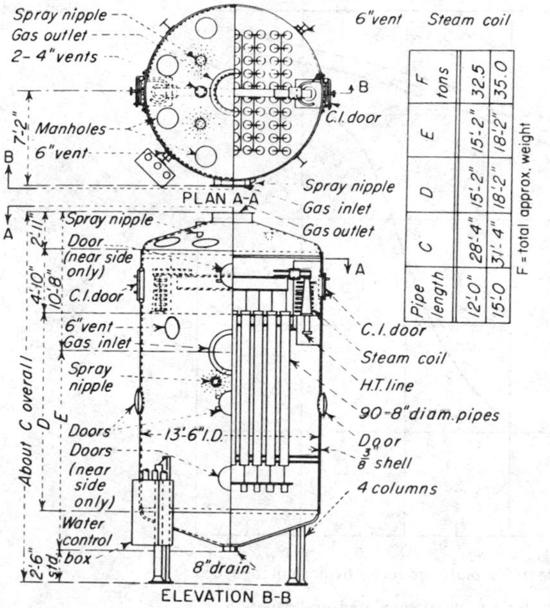

Pipe length	C	D	E	F tons
12'-0"	28'-4"	15'-2"	15'-2"	32.5
15'-0"	31'-4"	18'-2"	18'-2"	35.0

F = total approx. weight

FIG. 20-146 Blast-furnace pipe precipitator. *(Research-Cottrell, Inc.)*

efficiency at a lower voltage than other electrode types. Comparative data are plotted in Fig. 20-147*b*. The data from Fig. 20-147*b* reduce to a single curve when plotted in Fig. 20-147*c* as input energy in watt-seconds per cubic foot of gas treated. The manufacturer states that plant installations have demonstrated that insulating dusts may be permitted to build up on the main wire without changing the corona current or its distribution, since the barbs remain clean. When the collecting electrodes are screens or rows of rods or wires, the gases are usually passed parallel to the plane of each but may also be passed through it. In pipe precipitators, the gas flow is generally vertical up through the pipe, although downflow is not unusual. The pipe-type precipitator is usually used for the removal of liquid particles and volatilized fumes [Beaver, op. cit.; and Cree, *Am. Gas J.*, **162**, 27 (March 1945)], and the plate type is used mainly on dusts. In the pipe type, the discharge electrodes are usually suspended from an insulated support and kept taut by a weight at the bottom. Cree (op. cit.) discusses the application of electrical precipitators to tar removal in the gas industry.

Rapping Except when liquid dispersoids are being collected or, in the case of film precipitators, when a liquid is circulated over the collecting-electrode surface (Fig. 20-148), thus continuously removing the precipitated material, the collected dust is dislodged from the electrodes either periodically or continuously by mechanical rapping or scraping, which may be performed automatically or manually. Automatic rapping with either impact-type or vibrator-type rappers is common practice. White (op. cit.) recommends fairly continuous rapping with magnetic-impulse rappers. Rapping with excessive force leads to dust reentrainment and possible mechanical failure of the plates, while insufficient rapping leads to excessive dust buildup with poor electrical operation and reduced collection efficiency. Intermittent rapping at intervals of an hour or more causes heavy puffs of reentrained dust. Sproull [*Air Pollut. Control Assoc. J.*, **15**, 50 (1965)] reports the importance of electrode acceleration and shows that it varies with the type of dust, whether the electrode is rapped perpendicularly (normally) to the plate or parallel to it. Figure 20-149 shows the accelerations required for rapping normally to the plate. Difficult dusts may require as much as 100 G acceleration for 90 percent removal, and even higher accelerations are required when the vibrating force is applied in the plane of the plate.

Perforated-plate or rod-curtain precipitators are frequently

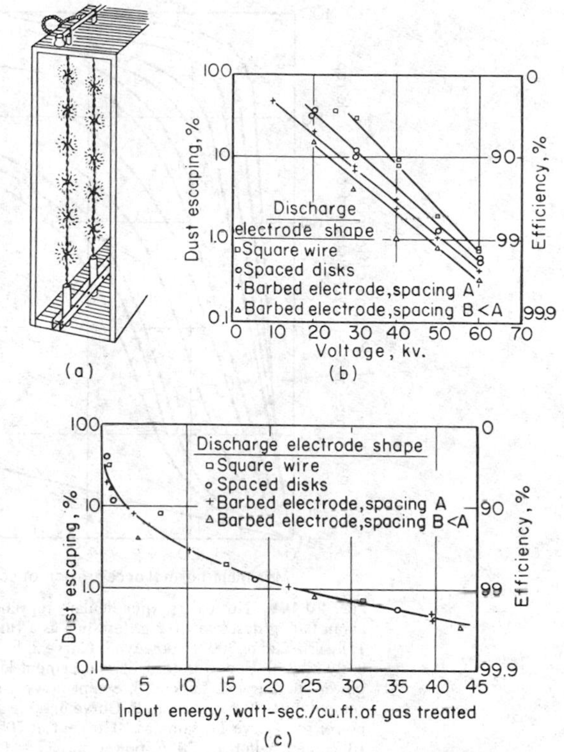

FIG. 20-147 Barbed-wire discharge electrode. (*a*) Illustration of corona location at barbs. (*b*) Efficiency of precipitator versus voltage for several discharge electrode types. (*c*) Efficiency versus input energy. *(Lagarias, Pap. 59-51, Air Pollution Control Association, June 21–26, 1959.)*

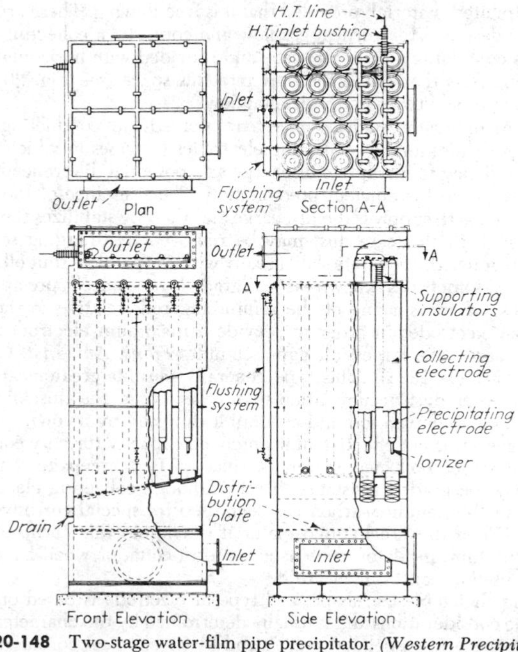

FIG. 20-148 Two-stage water-film pipe precipitator. *(Western Precipitation Division, Joy Manufacturing Company.)*

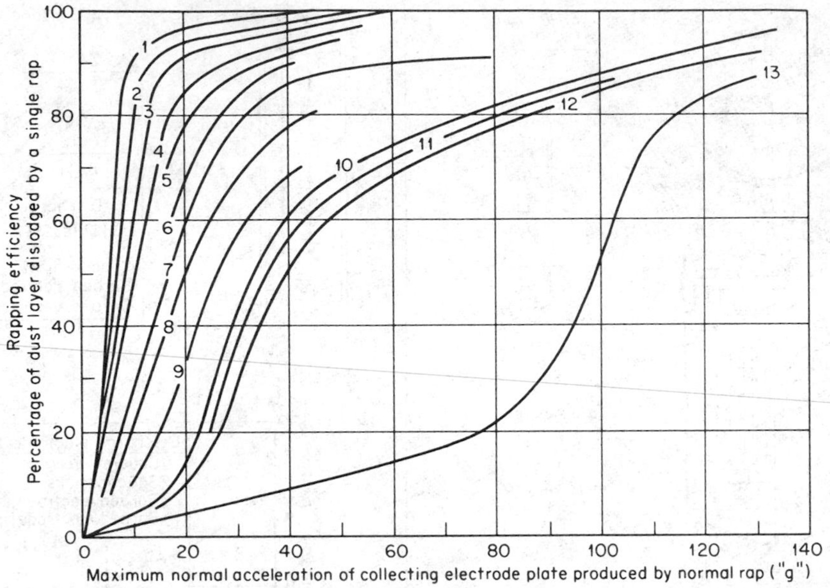

FIG. 20-149 Normal (perpendicular) rapping efficiency for various precipitated dust layers having about 0.03 g dust/cm^2 (0.2 g dust/in^2) as a function of maximum acceleration in multiples of g. Curve 1, fly ash, 200 or 300°F, power off. Curve 2, fly ash, 70°F, power off; also 200 or 300°F, power on. Curve 3, fly ash, 70°F, power on. Curve 4, cement-kiln feed, 300°F, power off. Curve 5, cement dust, 300°F, power off. Curve 6, same as 5, except power on. Curve 7, cement-kiln feed, 300°F, power on. Curve 8, cement dust, 200°F, power off. Curve 9, same as 8, except power on. Curve 10, cement-kiln feed, 200°F, power off. Curve 11, same as 10, except at 70°F. Curve 12, cement-kiln feed, 200°F, power on. Curve 13, cement-kiln feed, 70°F, power on. °C = (°F − 32) × 5/9. [*Sproull, Air Pollut. Control Assoc. J.*, ***15,*** *50 (1965).*]

rapped without shutting off the gas flow and with the electrodes energized. This procedure, however, results in a tendency for reentrainment of collected dust. Sectional or composite-plate collecting electrodes (sometimes known as hollow, pocket, or tulip electrodes) are used to minimize this tendency in the continuous removal of the precipitated material, provided that it is free-flowing. These are generally designed for vertical gas flow and comprise a collecting electrode containing a dead air space and provided with horizontal protruding slots that guide the dust into this space (see Fig. 20-150), although some types use horizontal flow.

Semiconductors, such as concrete reinforced with conducting rods, are sometimes used as collecting electrodes for gases in which there is a tendency to disruptive discharge at a potential difference below that required for efficient precipitation. The resistance of the electrode tends to suppress the discharge and thereby stabilizes the electric field. In this case dust may be removed by dragging scraper chains across the concrete slab, usually with the gas flow shut off. This type is sometimes known as a "graded resistance" precipitator because of the spacing of the reinforcing rods relative to the discharge electrodes in order to provide a maximum electrode resistance across the largest air gap [Schmidt, *Trans. Am. Inst. Chem. Eng.*, **21,** 11 (1928)]. This type generally permits greater capacity and greater dust accumulation than other types, the dust in some cases being allowed to build up until it drops off by its own weight. However, it is not effective in its intended capacity for very conductive gases or collected materials, since surface creepage tends to destroy the graded resistance effect. Although collecting electrodes are usually metallic, carbon has been used for special corrosive service. When the precipitated material is a liquid that forms a conducting film, insulators such as glass, terra-cotta, or wood have also been used as collecting electrodes.

The choice of size, shape, and type of electrode is based on economic considerations and is usually determined by the characteristics of the gas and suspended matter and by mechanical considerations such as flue arrangement, the available space, and previous experience with the electrodes on similar problems. The spacing between collecting electrodes in plate-type precipitators and the pipe diameter in pipe-type precipitators usually ranges from 15 to 38 cm (6 to 15 in). The smaller the spacing, the lower the necessary voltage and overall equipment size, but the greater the difficulties involved in

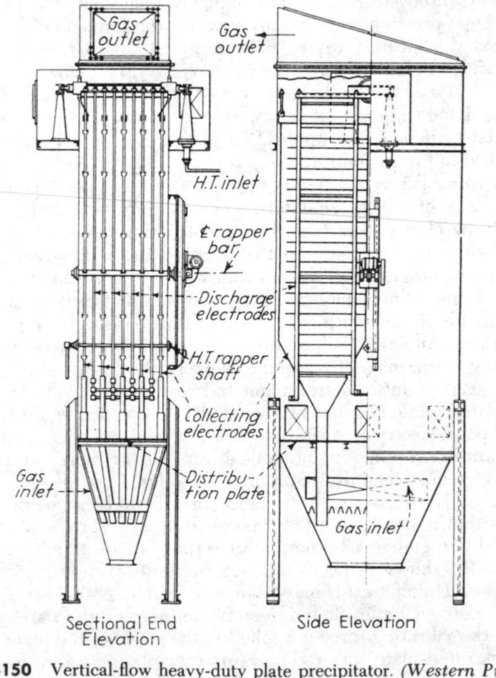

FIG. 20-150 Vertical-flow heavy-duty plate precipitator. *(Western Precipitation Division, Joy Manufacturing Company.)*

maintaining proper alignment and resulting from disturbances due to collected material. Large spacings are usually associated with high dust concentration in order to minimize sparkover due to dust buildup. For very high dust concentrations, such as those encountered in fluid-catalyst plants, it is advantageous to use greater spacings in the first half of the precipitator than in the second half. Precipitators, especially of the plate type, are frequently built with groups of collecting electrodes in series in a common housing. Collecting electrodes are generally on the order of 0.9 to 1.8 m (3 to 6 ft) wide and 3 to 5.5 m (10 to 18 ft) high in plate-type precipitators and 1.8 to 4.6 m (6 to 15 ft) high in pipe types. It is essential for good collection efficiency that the gas be evenly distributed across the various electrode elements. Although this can be achieved by proper gas-inlet transitions and guide vanes, perforated plates or screens located on the upstream side of the electrodes are generally used for distribution. Perforated plates or screens located on the downstream side may be used in special cases.

Electrical precipitators are generally designed for collection efficiency in the range of 90 to 99.9 percent. It is essential, however, that the units be properly maintained in order to achieve the required collection efficiency. Electric power consumption is generally 0.2 to 0.6 kW/(1000 $ft^3 \cdot min$) of gas handled, and the pressure drop across the precipitator unit is usually less than 124 Pa (0.5 in water), ranging from 62 to 248 Pa (¼ to 1 in) and representing primarily distributor and entrance-exit losses. Applied potentials range from 30,000 to 100,000 V. Gas velocities and retention times are generally in the range of 0.9 to 3 m/s (3 to 10 ft/s) and 1 to 15 s respectively. Velocities are kept low in conventional precipitators to avoid reentrainment of dust. There are, however, precipitator installations on carbon black in which the precipitator acts to flocculate the dust so that it may be subsequently collected in multiple small-diameter cyclone collectors. By not attempting to collect the particles in the precipitator, higher velocities may be used with a correspondingly lower investment cost.

Power Supply Electrical precipitators are generally energized by rectified alternating current of commercial frequency. The voltage is stepped up to the required value by means of a transformer and then rectified. The rectifying equipment has undergone an evolution which began with the synchronous mechanical rectifier in 1904 and was followed by mercury-vapor rectifiers in the 1920s; the first solid-state selenium rectifiers were introduced about 1939. Silicon rectifiers are the latest and most widely used type, since they provide high efficiency and reliability. Automatic controls commonly are tied to voltage, current, spark rate, or some combination of these parameters. Modern precipitators use control circuits similar to those shown on Fig. 20-151. A high-voltage silicon rectifier is used together with a saturable reactor and means for limiting current and controlling voltage and/or spark rate. One popular method adjusts the voltage to give a specified sparking frequency (typically 50 to 150 sparks per minute per bus section). Half-wave rectification is sometimes

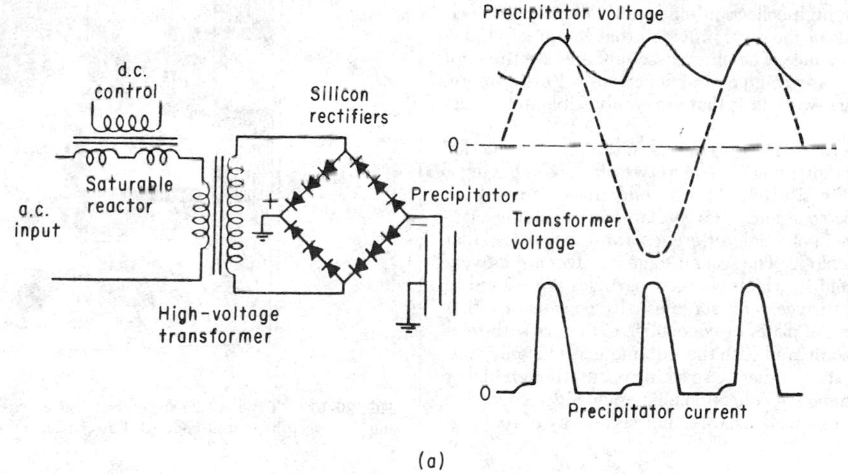

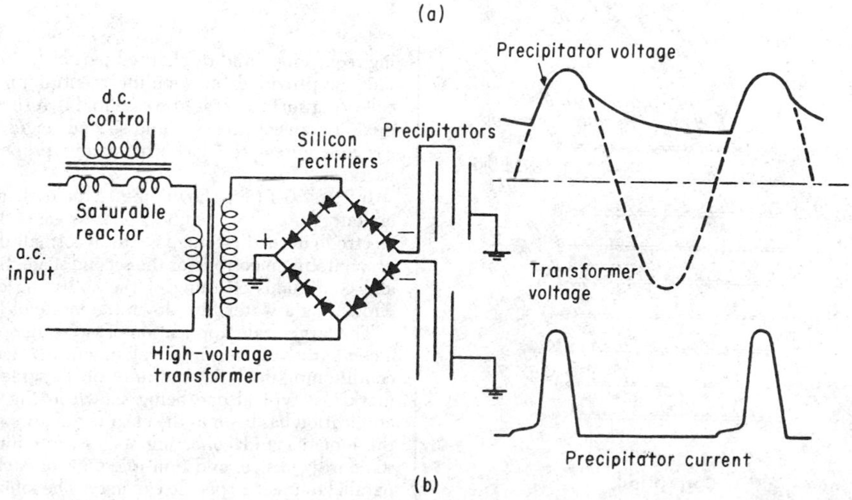

FIG. 20-151 Schematic circuits for silicon rectifier sets with saturable reactor control. (*a*) Full-wave silicon rectifier. (*b*) Half-wave silicon rectifier. (*White,* Industrial Electrostatic Precipitation, *Addison-Wesley, Reading, Mass., 1963.*)

used because of its lower equipment requirements and power consumption. It also has the advantage of longer decay periods for sparks to extinguish between current pulses.

Electrode **insulators** must also be designed for a particular service. The properties of the dust or mist and gas determine their design as well as the physical details of the installation. Conducting mists require special allowances such as oil seals, energized shielding cups, or air bleeds. With saturated gas, steam coils are frequently used to prevent condensation on the electrodes.

Typical applications in the chemical field (Beaver, op. cit.) include detarring of manufactured gas, removal of acid mist and impurities in contact sulfuric acid plants, recovery of phosphoric acid mists, removal of dusts in gases from roasters, sintering machines, calciners, cement and lime kilns, blast furnaces, carbon-black furnaces, regenerators on fluid-catalyst units, chemical-recovery furnaces in soda and sulfate pulp mills, and gypsum kettles. Figure 20-150 shows a vertical-flow steel-plate-type precipitator similar to a type used for catalyst-dust collection in certain fluid-catalyst plants.

A development of interest to the chemical industry is the tubular precipitator of reinforced-plastic construction (Wanner, *Gas Cleaning Plant after TiO_2 Rotary Kilns*, technical bulletin, Lurgi Corp., Frankfurt, Germany, 1971). Tubes made of polyvinyl chloride plastic are reinforced on the outside with polyester-fiber glass. The use of modern economical materials of construction to replace high-maintenance materials such as lead has been long awaited for corrosive applications.

Electrical precipitators are probably the most versatile of all types of dust collectors. Very high collection efficiencies can be obtained regardless of the fineness of the dust, provided that the precipitators are given proper maintenance. The chief disadvantages are the high initial cost and, in some cases, high maintenance costs. Furthermore, caution must be exercised with dusts that are combustible in the carrier gas.

Two-Stage Precipitators In two-stage precipitators, corona discharge takes place in the first stage between two electrodes having a nonuniform field (see Fig. 20-152). This is generally obtained by a fine-wire discharge electrode and a large-diameter receiving electrode. In this stage the potential difference must be above that required for corona discharge. The second stage involves a relatively uniform electrostatic field in which charged particles are caused to migrate to a collecting surface. This stage usually consists of either alternately charged parallel plates or concentric cylinders with relatively close clearances compared with their diameters. The only voltage requirement in this stage is that no sparking occur, though higher voltages will result in increased collection efficiency. Since collection occurs in the absence of corona discharge, there is no way of recharging reentrained and discharged particles. Consequently, some means must be provided for avoiding reentrainment of particles from the collecting surface. It is also essential that there be sufficient time and mixing between the first and second stages to secure distribution of gas ions across the gas stream and proper charging of the dust particles.

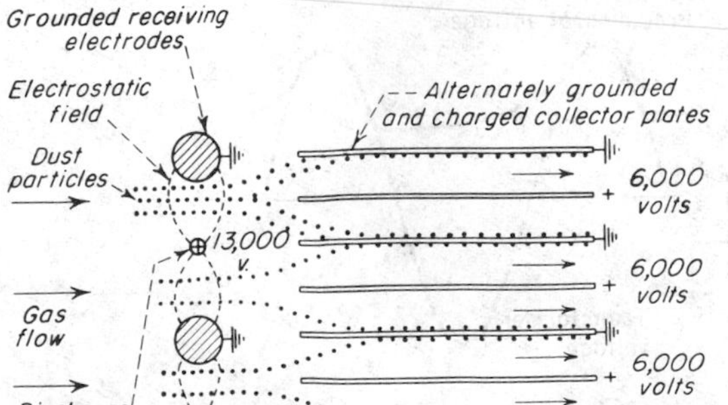

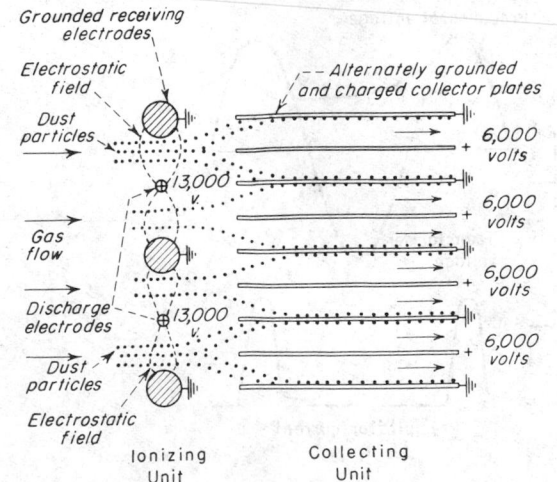

FIG. 20-152 Two-stage electrical-precipitation principle.

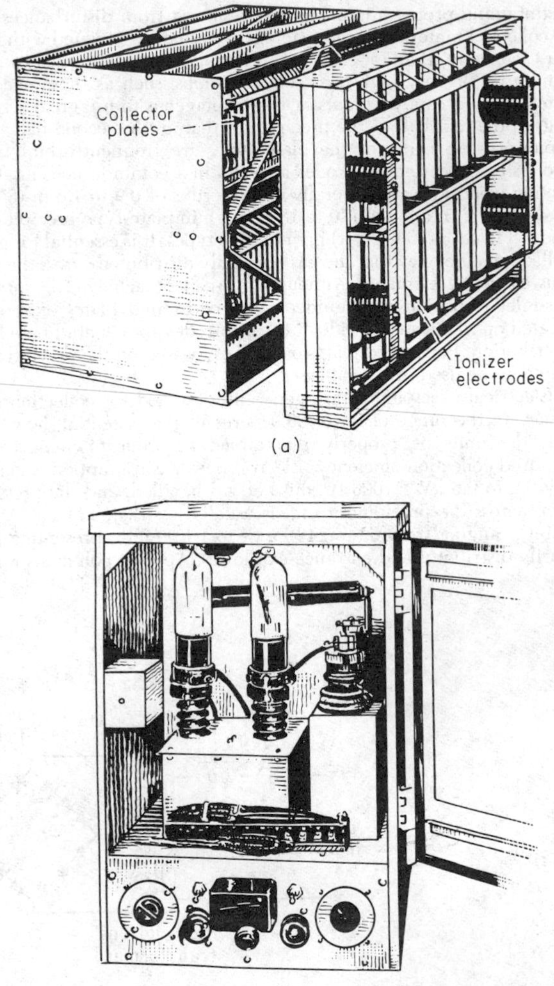

FIG. 20-153 Typical two-stage electrical precipitator used in air conditioning. (*a*) Precipitron unit cells. (*b*) Power pack. (*Westinghouse Electric Corp.*)

In Fig. 20-148 is shown one of the earlier types of two-stage precipitators used for cleaning process gases. In this unit the ionizing electrode in the first stage is simply a small-diameter extension of the precipitating electrode of the second stage. Reentrainment is avoided and continuous cleaning of the collecting electrode is achieved by circulating a water film down the inside of the collecting electrode.

The large-scale application of two-stage precipitators is, however, a comparatively recent development that has taken place in the air-conditioning field. Several units of the same general type are on the market, a typical one being shown in Fig. 20-153, whose primary application has been in the cleaning of atmospheric air. In these units the ionizing and collecting stages are built in separate sections of standardized size, and multiple units of each stage are assembled in parallel to meet a specific capacity. The ionizer unit is generally built up of vertical grounded tubes, circular or streamline in cross section and 1¼-in nominal diameter, spaced about 8.4 cm (3½ in) apart.

Between these tubes are stretched parallel discharge or ionizer electrodes, consisting of approximately 0.2032-mm- (0.008-in-) diameter tungsten wire. The collecting unit, located on the downstream side of the ionizer unit, consists of 20 BWG plates arranged parallel to the gas flow (see Fig. 20-152), usually vertical, with alternate plates grounded. These plates are spaced on approximately a 7.9-mm (5/16-in) pitch and are about 30 to 46 cm (12 to 18 in) deep.

Both ionizer and collector sections are generally made in 61- and 91-cm- (24- and 36-in-) wide units, the heights being variable depending on the specific type or make. The plates are generally either zinc-coated steel or aluminum. A dc potential of 13,000 V is applied between the ionizer wire and tubes, the wires being positive. A positive dc potential of 6000 V is applied between adjacent collector plates. The necessary voltages for the ionizer and collector plates are obtained through a compact, self-contained vacuum-tube rectifier unit operated directly off of a 110-V ac supply line.

The plates are coated with a viscous oil to avoid reentrainment of collected dust. When the dust buildup exceeds a depth of approximately 1.6 mm (1/16 in), the plate sections must be taken out, washed, and reoiled. Automatic means for cleaning and reoiling in place are also available. Depending on dust concentrations, cleaning may be required every 2 weeks to 3 months. Installations are usually provided with guarded doors that automatically cut off the power when the unit is entered. Where poor approach conditions are involved, perforated-plate air distributors may be employed. The units are rated at 85 to 90 percent efficiency (U.S. Bureau of Standards Discoloration Test method) at superficial velocities in the range of 1.5 to 2.8 m/s (300 to 500 ft/min). Electric power consumption is approximately 0.02 kW/(1000 $ft^3 \cdot$min), and pressure drops range from 24.9 to 49.8 Pa (0.1 to 0.2 in water).

A unit is available in which electrostatic precipitation is combined with a dry-air filter of the type shown in Fig. 20-135*b*. In another unit an electrostatic field is superimposed on an automatic filter of the type shown in Fig. 20-136*b*. In this case the ionizer wires are located on the leading face of the unit, and the collecting electrodes consist of alternate stationary and rotating parallel plates. Cleaning in this case is automatic and continuous.

Although intended primarily for air-conditioning applications, these units have been successfully applied to the collection of relatively nonconducting mists such as oil. However, other process applications have been limited largely to experimental installations. The large cost advantage of these units over the Cottrell precipitator lies in the smaller equipment size made possible by the close plate spacing, in the lower power consumption due to the two-stage operation, and primarily in the mass production of standardized units. In process applications, the close plate spacing is objectionable because of the relatively high dust concentrations involved. Special material or weight requirements for the structural members may eliminate the mass-production advantage except for individual wide applications. Consequently, application to process gases would appear to be limited. A possible outstanding field of application may be that of sulfuric acid mist collection, although there are no commercial installations. The chief difficulty encountered in this case is the short-circuiting of the insulation.

Alternating-Current Precipitators High-voltage alternating current may be employed for electrical precipitation. Corona discharge will result in a net rectification, provided that no spark gaps are used in series with the precipitator. However, the equipment capacity for a given efficiency is considerably lower than for direct current. In addition, difficulties due to induced high-frequency currents may be encountered. The simplicity of an ac system, on the other hand, has permitted very satisfactory adaptation for laboratory and sampling purposes [Drinker, Thomson, and Fitchet, *J. Ind. Hyg.*, **5**, 162 (September 1923)].

Some promising work with alternating current has been undertaken at the University of Karlsruhe. Lau [*Staub*, English ed., **29**, 10 (1969)] and coworkers found that ac precipitators operated at 50 Hz were more effective than dc precipitators for dusts with resistivities higher than 10^{11} Ω·cm. An insulating screen covering the collecting electrode permitted higher-voltage operation without sparkover.

Section 21

Solid-Solid and Liquid-Liquid Systems

W. M. Goldberger, D.Ch.E., P.E., *Director of Research and Development, Superior Graphite Company, Chicago. (Section Editor)*

Lanny A. Robbins, Ph.D., *Research Scientist, Dow Chemical Company; Member, American Institute of Chemical Engineers, Sigma Xi. (Co-Section Editor, Liquid-Liquid Systems)*

R. A. Fiedler, B.S., Ch.E., *Manager of Applications-Sedimentation Technology, Dorr-Oliver Inc., Stamford, Connecticut. (Wet Classification)*

T. L. B. Jepsen, M.Sc., Min. Proc., *Metallurgical Process Engineer, Basic, Inc., Gabbs, Nevada. (Dense-Media Separation)*

Frank S. Knoll, M.Sc., Min. Proc., *President, Carpco, Inc., Jacksonville, Florida. (Electrostatic Separation)*

James O. Maloney, Ph.D., P.E., *Professor of Chemical Engineering, University of Kansas; Member, American Institute of Chemical Engineers, American Chemical Society, American Society for Engineering Education. (Liquid-Liquid Systems)*

D. W. Mitchell, Ph.D., Met. Eng., P.E., *Technical Director, Carpco, Inc., Jacksonville, Florida. (Electrostatic Separation)*

Bhupendra K. Parekh, Ph.D., Min. Sci., *Research Specialist, Exxon Minerals Company, Houston. (Separation of Ultrafine Solids)*

Thomas C. Sorenson, M.B.A., Min. Eng., *President, Galigher Ash (Canada) Ltd. (Flotation)*

Paul L. Stavenger, M.S., Ch.E., *Director of Technology, Dorr-Oliver Inc., Stamford, Connecticut. (Wet Classification)*

Rich L. Thelen, B.S., M.E., *Vice President, EIE Company, Inc., Lisle, Illinois. (Solids Sampling)*

Robert E. Treybal, Ph.D., P.E. *(deceased), Profesor and Chairman, Department of Chemical Engineering, University of Rhode Island; Member, American Institute of Chemical Engineers, American Chemical Society, American Society for Engineering Education; Fellow, New York Academy of Sciences, American Institute of Chemists. (Liquid-Liquid Systems)*

Ionel Wechsler, M.S., Min. and Met., *Vice President, Sala Magnetics, Inc., Cambridge, Massachusetts. (Magnetic Separation)*

SOLID-SOLID SYSTEMS

PROCESSING BULK SOLIDS

SELECTIVE PHYSICAL SEPARATION OF MIXED SOLIDS

LIQUID-LIQUID SYSTEMS

SOLID-SOLID SYSTEMS

The processing of mixed particulate solids entails treatment of the bulk solids for mixing, sampling, sizing, and classification. Also important are operations for selective separation or concentration of certain solids from a mixture. The methods used for these operations as applied to solid-solid systems will depend on the size range of the system. A convenient guide to the application of various means for classification and solid-solid separation in relation to the size range of general application is presented in Fig. 21-1 after Roberts et al.

This subsection discusses the basic considerations involved in these various unit operations and describes present industrial practice and equipment in general use.

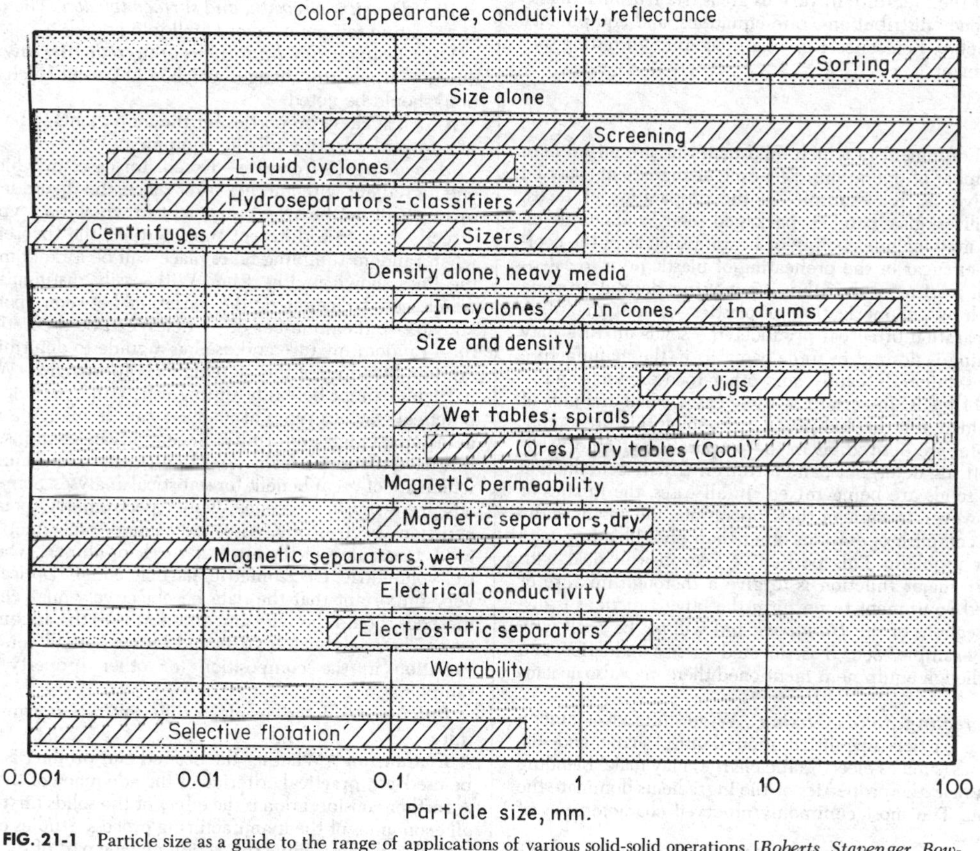

FIG. 21-1 Particle size as a guide to the range of applications of various solid-solid operations. [*Roberts, Stavenger, Bowersox, Walton, and Mehta,* Chem. Eng., *78(4), 89 (Feb. 15, 1971).*]

PROCESSING BULK SOLIDS

MIXING

General References

1. *AIChE Standard Testing Procedure for Solids Mixing Equipment,* American Institute of Chemical Engineers, New York.
2. Bullock, *Chem. Eng.*, **66**, 177 (Apr. 20, 1959).
3. Danckwerts, *Research*, **6**, 355–361 (1953).
4. Fischer, *Chem. Eng.*, **67**, 107 (Aug. 8, 1960).
5. Kirk and Othmer (eds), *Encyclopedia of Chemical Technology,* Wiley-Interscience, New York: Rushton, Boutros, and Selheimer, "Mixing and Agitating," vol. 9, 1st ed., pp. 133–166; Rushton and Boutros, "Mixing and Blending," vol. 13, 2d ed., pp. 577–613; Oldshue and Todd, "Mixing and Blending," vol. 15, 3d ed., pp. 604–637.
6. Lacey, *J. Appl. Chem.*, **4**, 257–268 (1954).
7. Quillen, *Chem. Eng.*, **61**, 178 (June 1954).
8. Scott, in Cremer and Davies (eds.), *Chemical Engineering Practice*, vol. 3, Butterworth, London, 1957, p. 362.
9. Weidenbaum, in Drew and Hoopes (eds.), *Advances in Chemical Engineering,* vol. II, Academic, New York, 1958, chap. on mixing of solids.
10. Work, *Chem. Eng. Prog.*, **50**(9), (September 1954).
11. Vance, "Statistical Properties of Dry Blends," *Ind. Eng. Chem.*, **58**, 37–44 (1966).
12. Gren, "Solids Mixing: A Review of Theory," *Br. Chem. Eng.*, **12**, 1733–1738 (1967).
13. Ashton and Valentin, "Mixing of Powders and Particles in Industrial Mixers," *Trans. Inst. Chem. Eng. (London)*, **44**, t165–t188 (1966).
14. Valentin, "Mixing of Powders and Pastes: Basic Concepts," *Chem. Eng. (London)*, **45**, CE99–CE106 (1967).

15. Uhl and Gray (eds.), *Mixing*, vols. I and II, Academic, New York, 1966–1967.
16. Bhatia and Cheremisinoff (eds.), *Solids Separation and Mixing*, vol. 1, Technomic Publishing Co., Westport, Conn., 1979.
17. Beddow, *Particulate Science and Technology*, Chemical Publishing, New York, 1980.

A comprehensive bibliography is available in Ref. 9. Equipment photographs and details are available in Refs. 2, 4, 5, 7, 8, and 10. References 3 and 6 give excellent theoretical work. Reference 5 gives a tabulation and summary of many mixer types and applications. References 8 and 9 are book chapters dealing with mixing of solids and cover both the theoretical and the equipment aspects. Interpretive summaries of the literature in various areas (state of mixedness, theoretical frequency distributions, rate equations, and equipment) are included in Ref. 9. Reference 1 gives a procedure for testing solids-mixing equipment.

Fundamentals

Objectives Equipment in which solid materials are mixed may be used for a number of operations. Blending of ingredients may be the main objective, as, for example, in the preparation of feeds, insecticides, fertilizer, glass batches, packaged foods, and cosmetics. Other objectives may include cooling or heating such as in the cooling of limestone or sugar or the preheating of plastic prior to calendering. Drying or roasting of the solids is sometimes desired. In some applications, such as polymerization of plastics, catalyst manufacture, or the preparation of cereal products, the solids mixture may be reacted. Coating is desired in some cases, as in the manufacture of pigments, dyes, minerals, candy, and other food products and in the preparation of feeds. In certain of these cases, small amounts of liquid may be added, but the end product is a solids mixture. Sometimes agglomerates are desired, as in the preparation of food products, pharmaceuticals, detergents, and fertilizer. Often size reduction is desired while solids are being mixed. In all cases, the mixing of solids occurs. However, in some of these operations, the details of the equipment to accomplish operations other than pure blending may become a major problem. This portion of Sec. 21 will deal with equipment whose major function is to give a thorough mixture of solids. Specialized equipment to perform the other functions is discussed in other sections of the *Handbook* and will not be dealt with here. Thus, for example, Sec. 8 is devoted to size reduction and enlargement, although equipment mentioned there may also accomplish mixing.

Properties Affecting Solids Mixing Wide differences among properties such as particle-size distribution, density, shape, and surface characteristics (such as electrostatic charge) may make blending very difficult. In fact, the properties of the ingredients dominate the mixing operation. The most commonly observed characteristics of solids are as follows:

1. *Particle-size distribution.* This tells the percentages of the material in different size ranges.
2. *Bulk density.* This is the weight per unit of volume of a quantity of solid particles, usually expressed in kilograms per cubic meter (pounds per cubic foot). It is not a constant and can be decreased by aeration and increased by vibration or mechanical packing.
3. *True density.* The true density of the solid material is usually expressed in kilograms per cubic meter (pounds per cubic foot). This, divided by the density of water, equals specific gravity.
4. *Particle shape.* Some types are pellets, egg shapes, blocks, spheres, flakes, chips, rods, filaments, crystals, or irregular shapes.
5. *Surface characteristics.* These include surface area and tendency to hold a static charge.
6. *Flow characteristics.* Angle of repose and flowability are measurable characteristics for which standard tests are available (e.g., ASTM Test B213-48, Flow Rate of Metal Powders, etc.). A steeper angle of repose would indicate less flowability. The term "lubricity" has sometimes been used for solid particles to correspond roughly to viscosity of a fluid.
7. *Friability.* (Also see "Grindability," Sec. 8.) This is the tendency of the material to break into smaller sizes in the course of handling. There are quantitative tests specially devised for certain materials such as coal which can be used to estimate this property. Abrasiveness of one ingredient upon another should also be considered.
8. *State of agglomeration.* This refers to whether the particles exist independently or adhere to one another in clusters. The kind and degree of energy employed during mixing and the friability of the agglomerates will affect the extent of agglomerate breakdown and particle dispersion.
9. *Moisture or liquid content of solids.* Often a small amount of liquid is added for dust reduction or special requirements (such as oils for cosmetics). The resultant material may still have the appearance of a dry solid rather than a paste.
10. *Density, viscosity, and surface tension.* These are properties at operating temperature of any liquid added.
11. *Temperature limitations of ingredients.* Any unusual effects due to temperature changes which might occur (such as heat of reaction) should be noted.

A look at these properties for the ingredients to be mixed is a first step toward selecting mixing equipment.

Measuring Uniformity Except for cases in which a coating of one ingredient with another takes place, the theoretical end result of mixing will not be an arrangement in which one type of particle is directly next to a different type. Rather, the theoretical end result when random tumbling takes place will be a random mixture along the lines shown in Fig. 21-2. With easily distinguishable particles which can be counted, the variation between spot samples of a known size (i.e., number of particles) can be theoretically predicted for a random mixture and used as a guide to determine how closely random blending of the ingredients is approached. When individual particles cannot be easily distinguished and particle counts are not practical, various types of analyses can be made on spot samples to determine batch uniformity. Recent advances in instrumental analysis have made it much easier to give rapid and numerous analyses which are of great benefit for statistical analyses. Some of these methods are x-ray fluorescence, flame spectrometry, polarography, and emission spectroscopy. Also, radioactive-tracer methods have been used. Regardless of the analytic methods chosen, whether gravimetric, volumetric, electrometric, particle counts, optical, or other, it is very important that the data be objectively analyzed via statistical methods should there be any question as to the adequacy of the mixture. The analytical error should be very small compared with the variation in the composition (or other property) between spot samples.

Reference 9 describes many different types of measuring uniformity.

Evaluation Whether the desired end product is satisfactory can be used as a practical criterion of the adequacy of the solids mixture. A further consideration is the effect of the solids mixture on the overall economics of the manufacturing process. Studies of the type mentioned in the preceding subsection *may* be part of such an evaluation.

FIG. 21-2 Random arrangement of black and white particles. [*Lacey*, Trans. Inst. Chem. Eng. (London), **21**, 52 *(1943)*.]

When the solids mixture is made directly into a product, as in the case of feed pellets or pharmaceutical tablets, uniformity tests on these items will speak for themselves. If the solids mixture must be further processed, as in the manufacture of glass or plastics, the efficiency and costs of the subsequent operations can often be related to the starting solids mixture. In such cases, knowledge of the homogeneity of the solids mixture is needed to determine its effect on the manufacturing process.

Regardless of the method of evaluating the solids mixture, the sampling procedure is vital. Often a sampling thief, or other special device, is used to remove samples from the mixture without excessive disturbance of the batch. If an easier method of sampling is obvious and will bring less contamination to the batch, it should be used.

Method of sampling, location, size and number of samples, method of sample analysis, and fraction of the batch removed for sampling all contribute to how well the sampling study reflects the actual conditions.

A standard testing procedure for solids-mixing equipment is available (Ref. 1). This contains details and references pertaining to sampling from solids mixtures for both batch and continuous mixing.

Segregation Problems Previously it was pointed out that wide differences among properties may make blending very difficult. For example, natural segregating tendencies will be observed with extreme differences in specific gravity, size, or shape. The heavier, smaller, or smoother and rounder particles tend to sink through the lighter, larger, or jagged ones respectively. In some cases, preparation of the materials to avoid extreme differences in such ingredient properties can avoid segregation problems.

There are also other factors which can cause segregation.

Electrostatic charges may cause particles to repel each other. When continued blending may cause such charges to build up, it is important to determine the precise blending time required and not to overblend.

Loss of material as dust must be considered as a possible means of segregation and should not be aggravated by too strong suction in the dust-collection apparatus.

If there are smeary particles which have an almost pastelike behavior and barely flow (high angle of repose), frictional anchorage of these onto the other particles in the mixture may be necessary in order to achieve good mixing.

If a batch ingredient is in agglomerate form, some device to break up the agglomerates should be used to prevent them from segregating from the rest of the mixture and to ensure the intimate dispersion of this ingredient throughout the mixture.

The use of a liquid such as water (possibly with a surface-active agent) can have remarkable effects in overcoming segregation which may appear inevitable otherwise.

Although these statements apply to the actual **solids-mixing** operation, thought must also be given to the subsequent processing steps. Thus, the solids-mixing operation must be checked from the point of view of delivering a well-mixed batch to a certain point. The system must be scrutinized for possible segregating points such as transfer points, long drops, flow through silos, and vibratory equipment. Where a liquid is used, the amount that can be added without getting into caking problems which may upset the later processing of the solids mixture should be determined.

Equipment

Mixing Mechanisms There are several basic mechanisms by which solid particles are mixed. These include small-scale random motion (diffusion), large-scale random motion (convection), and shear.

Motions which increase the mobility of the individual particles will promote diffusive mixing. If there are no opposing segregating effects, this diffusive mixing will in time lead to a high degree of homogeneity. Diffusive mixing occurs when particles are distributed over a freshly developed surface and when individual particles are given increased internal mobility. A plain tumbler gives the former, while an impact mill gives the latter.

For most rapid mixing, in addition to diffusive (fine-scale) mixing, there should be a means by which large groups of particles are intermixed. This can be accomplished by either the convective or the shear mechanism. A ribbon mixer illustrates the former, whereas a plain tumbler gives the latter.

Types of Solids-Mixing Machines There are several types of solids-mixing machines. In some machines the container moves. In others a device rotates within a stationary container. In some cases, a combination of rotating container and rotating internal device is used. Sometimes baffles or blades are present in the mixer. Table 21-1 classifies solids-mixing machines via the characteristics given in the column headings. Illustrations of several of the machines listed there are shown in Fig. 21-3. The various types listed in Table 21-1 will be briefly discussed, with paragraph numbers referring to the columns.

1. *Tumbler.* Suitable for gentle blending; capable of handling large volumes; easily cleaned; suitable for dense powders and abrasive materials. Not for breaking up agglomerates.

Figure 21-3*a* and *b* (without broken-line portions) shows some unbaffled tumblers.

Figure 21-3*c* and *d* shows some baffled tumblers.

2. *Tumbler with agglomerate breaker.* See Sec. 8: "Tumbling Mills," for ball mill, rod mill, and vibratory pebble mill which will accomplish mixing along with size reduction.

Several tumblers are available with separately driven internal rotating devices for breaking up agglomerates. The tumbler itself can be used for gentle blending if agglomerate breakdown is not required.

The broken-line portions of Fig. 21-3*a* and *b* show some types of agglomerate-breaking devices for tumblers.

Table 21-2 includes impact velocities for some internal rotating devices in tumblers as well as other mixers. Contamination and wear problems of internal rotating devices are discussed under "Performance Characteristics."

3. *Stationary shell or trough.* There are a number of different types of mixers in which the container is stationary and material displacement is accomplished by single or multiple rotating inner mixing devices.

a. Ribbon mixer (Fig. 21-3*e*). Within this subgroup there are several types. Ribbon cross section and pitch, clearances between outer ribbon and shell, and number of spirals on the ribbon are some features which can be varied to accommodate materials ranging from low-density finely divided materials that aerate rapidly to fibrous or sticky materials that require positive discharge aid. Other construction variations are center or end discharge and the mounting of paddles or cutting blades on the center shaft. A broad ribbon can be used for lifting as well as for conveying, while a narrow one will cut through the material while conveying. The ribbon is adaptable to batch or continuous mixing.

b. Vertical screw mixer. This subgroup also has several variations. One type is shown in Fig. 21-3*f*. In this type, the screw rotates about its own axis while also orbiting around the center axis of the conical tank. In another variation, the screw does not orbit but remains in the center of the conical tank and is tapered so that the swept area steadily increases with increasing height. In another type, the central screw is contained in an inner cylindrical casing. This type of mixer is primarily suitable for free-flowing dry solids.

c. Muller mixer. The stationary-pan muller with rotating turret is one of several types. Other muller types are the countercurrent type, in which the pan and muller turret rotate in opposite directions, and the rotating-pan type, in which the muller turret is stationary.

The heavy, wide roller rides over the material. There is some skidding action where the rollers engage the mass of materials. This gives local shearing plus coarse-scale mixing which is aided by the plows and scrapers.

The muller is useful for mixing problems requiring certain types of aggregate breakdown, frictional anchorage of particles to one another, and densification of the final mix. Materials which are excessively fluid or sticky should be avoided. The muller mixer is generally used for batch operations (Fig. 21-3*g*), although Fig. 21-3*h* shows a continuous muller.

d. Twin rotor (Fig. 21-3*i*). This consists of two shafts with either paddles or screws encased in a cylindrical shell. There are var-

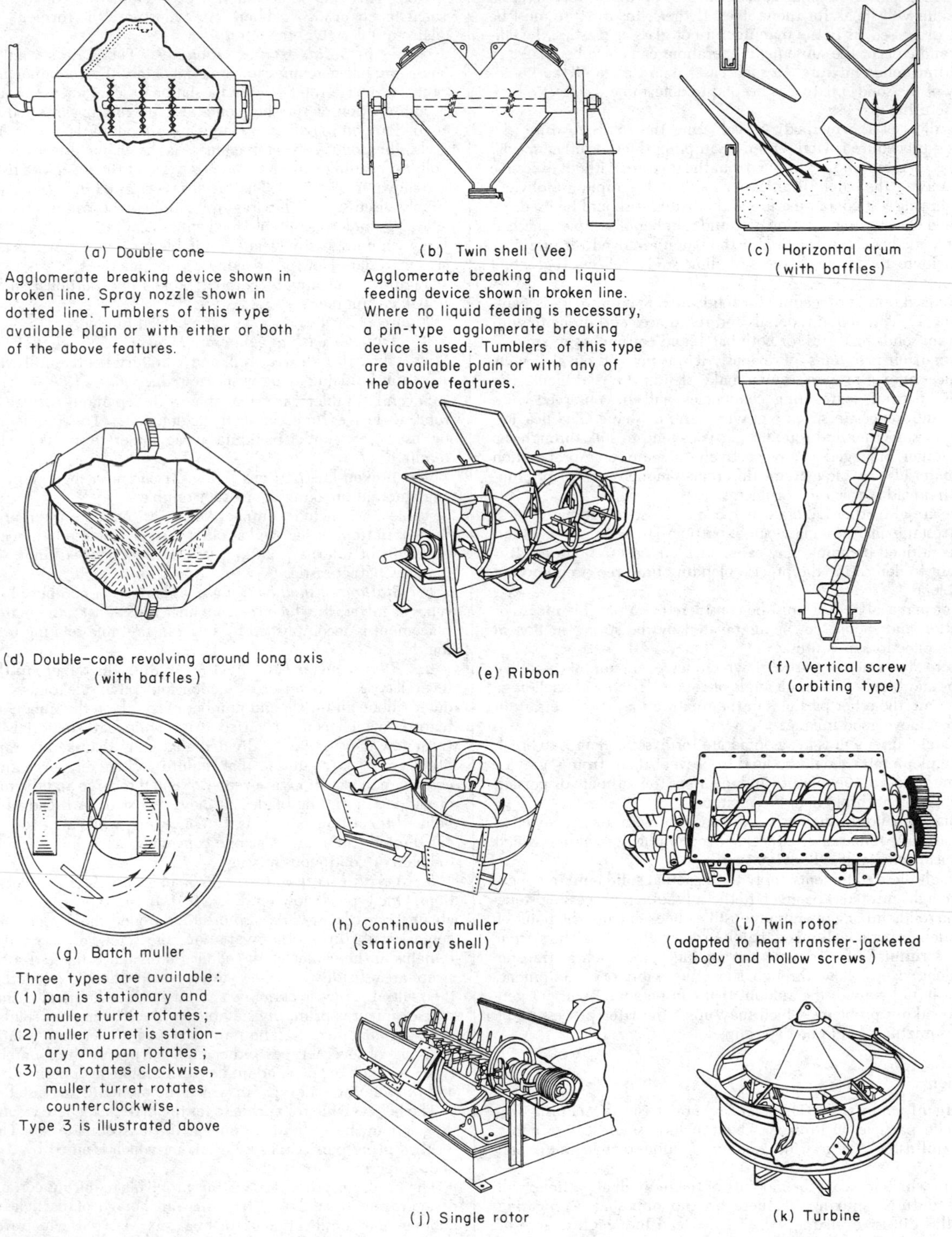

(a) Double cone
Agglomerate breaking device shown in broken line. Spray nozzle shown in dotted line. Tumblers of this type available plain or with either or both of the above features.
(b) Twin shell (Vee)
Agglomerate breaking and liquid feeding device shown in broken line. Where no liquid feeding is necessary, a pin-type agglomerate breaking device is used. Tumblers of this type are available plain or with any of the above features.
(c) Horizontal drum (with baffles)
(d) Double-cone revolving around long axis (with baffles)
(e) Ribbon
(f) Vertical screw (orbiting type)
(g) Batch muller
Three types are available:
(1) pan is stationary and muller turret rotates;
(2) muller turret is stationary and pan rotates;
(3) pan rotates clockwise, muller turret rotates counterclockwise.
Type 3 is illustrated above
(h) Continuous muller (stationary shell)
(i) Twin rotor (adapted to heat transfer-jacketed body and hollow screws)
(j) Single rotor
(k) Turbine

FIG. 21-3 Several types of solids-mixing machines. (See Table 21-1.)

TABLE 21-1 Types of Solids-Mixing Machines*

Tumbler (1)	Tumbler with internal agglomerate breaker (2)	Stationary shell or trough (3)	Both shell and internal device rotate (4)	Impact mixing (5)	Process steps which can affect solids mixing (6)
Without baffles:					
Drum, either horizontal or inclined	Ball mill Pebble mill	Ribbon	Countercurrent, muller turret and pan rotate in opposite directions	Hammer mill	Filling of hoppers
Double cone.	Rod mill	Stationary pan, rotating muller turret†		Impact mill	Fluidization
Twin shell. . . . . Cube	Vibratory pebble mill	Vertical screw	Planetary types	Cage mill	Screw feeders
Mushroom type.	Double cone Twin shell Cube	Single rotor Twin rotor		Jet mill Attrition mill	Conveyor-belt loading Elevator loading
		Turbine Paddle mixer			Pneumatic conveying Vibrating
With baffles: Horizontal drum.		Sifter (turbosifter)			
Double cone revolving around long axis					

*Diagrammatic sketches of many of these machines are shown in Fig. 21-3.
†There is also a muller in which the turret is stationary but the pan rotates.
‡Although these steps, when carefully selected, can aid mixing, caution must be exercised with pneumatic conveying and vibrating, as they may tend to separate materials.

ious types available with shaft speeds ranging from moderately low to relatively high (see Table 21-2). The twin rotor is useful for continuously mixing non-free-flowing solids; liquids can be added, there is minor product attrition, and materials can be added beyond the inlet. It is easily adaptable to heating or cooling. Some machines are specifically designed for heat transfer during mixing. The pug mill is one type of twin rotor.

e. Single rotor (Fig. 21-3*j*). This consists of a single shaft with paddles encased in a cylindrical shell. This type is available with relatively high speeds (see Table 21-2), although in certain cases lower speeds are used. A high-speed single rotor gives the maximum impact short of a grinding mill. It is used for intensive dispersion and disintegration. The type is available with split casing and is suitable for heating or cooling and for small amounts of liquid addition.

f. Turbine mixer (Fig. 21-3*k*). This is a circular trough with a housing in the center around which revolves a spider or a series of legs with plowshares or moldboards on each leg. The moldboards spin around through the circular trough. This mixer is suitable for free-flowing dry materials or semiwet materials which do not flow well and is also adaptable to liquid-solid mixing and coating problems.

4. *Shell and internal device rotate.* The countercurrent muller (Fig. 21-3*g*), which is in this category, is mentioned under "Muller mixer." This machine has a clockwise rotating mixing pan with a counterclockwise rotating mixing tool head mounted off center of the pan, thus providing a planetary mixing pattern. For the mixing of free-flowing solids not requiring the shearing and compressive action of mullers, plows are sometimes used alone. When used with mullers, plows deflect material into their path. Special mixing tools are also available.

5. *Impact mixing.* This process, which includes size reduction, is covered in Sec. 8.

The process steps listed in Table 21-1 can sometimes be used to promote mixing. However, they are primarily for functions other than solids mixing. (Note precautions for pneumatic conveying and vibrating in Table 21-1.)

Since paste mixing is not within the scope of this section, such widely used paste mixers as the sigma blade and banbury types will not be covered here but instead are discussed in Sec. 19: "Batch Mixers."

Performance Characteristics Before selecting solids-mixing equipment, a careful study should be made of various performance characteristics. These are given here.

Uniformity of Mixture. The proper type of mixer should be chosen to assure the desired degree of batch homogeneity. This cannot be compromised for other conveniences. Information is given under "Types of Solids-Mixing Machines" about the special abilities of various kinds of machines to blend different types of materials.

Care should be taken to avoid mixing too long, as in some cases this will result in a poorer blend. A graph of degree of mixing versus time should be made to select the proper mixing time quantitatively.

Mixing Time The actual time during which the batch is being mixed is usually less than 15 min if the proper type of machine and working capacity have been chosen. In some cases much more lengthy mixing times are tolerated so as to avoid the cost of purchasing more efficient equipment. However, there is usually a machine that can properly homogenize almost any type of mixture in less than 15 min provided one is willing to pay the price. In fact, proper mixer design in most instances will produce the desired blend in a few minutes.

Besides actual mixing time, however, the total cycle time should be optimized.

Charging and Discharging The total handling system must be considered in order to obtain optimum charging and discharging conditions. This includes the efficient use of weigh hoppers and surge bins, minor-ingredient premixing, location of discharge gates, etc.

Power In general, power requiremens are not a major consideration in choosing a solids mixer since other requirements usually predominate. However, sufficient power must be supplied to handle the maximum needs should there be changes during the mixing operation. Also, when a variety of mixes may be required, power must be sufficient for the heaviest bulk-density materials. If the loaded mixer is to be started from rest, there should be sufficient power for this. When speed variation may be desirable, this should be taken into account in planning power requirements.

Horsepower requirements of several types of mixers are listed in Table 21-3.

Cleaning The ease, frequency, and thoroughness of cleaning may be crucial considerations when incompatible batches are to be mixed at different times in the same machine. Plain tumbling vessels are easy to clean provided that adequate openings are available. Areas that may present cleaning problems are (1) seals or stuffing boxes, (2) crevices at baffle supports, (3) any corners, and (4) discharge arrangement. If cleaning between different batches may be

TABLE 21-2 Approximate Impact Velocities of Some Rotating Internal Devices in Mixers*

Type of mixer (see Table 21-1)	Tip speed, ft/min
Ribbon	280
Turbine	600
Twin-shell tumbler with	
Pin-type intensifier	1700
Liquid-feed bar	3300
Twin rotor	Up to 1300
Single rotor	6000–9000
Mills of various types	2500–20,000

*To convert feet per minute to meters per second, multiply by 0.00508.

TABLE 21-3 Horsepower Requirements and Speeds of Rotation for Some Commercial Solids Mixers [Approximately 1.5 m^3 (50 ft^3) Working Capacity]

Type of solids-mixing machine	Approximate working capacity, ft^3	Horsepower, hp		Rotational speed, r/min		Comments
		Shell	Internal device	Shell	Internal-device shaft speed	
1. Tumbler						
Without baffles						
Double cone	54	7½	...	18	...	Based on 100-lb/ft^3 material.
Twin shell	50	5	...	13.7	...	Maximum bulk density of material = 55 lb/ft^3.
With baffles						
Horizontal drum						
Manufacturer E	50	20	...	11.1	...	Heavy-duty (material 100 lb/ft^3). For extremely heavy duty (150–200-lb/ft^3 material), the maximum working capacity with 20-hp motor is 35 ft^3.
Manufacturer F	50	10	...	14	...	For material of 40-lb/ft^3 maximum bulk density.
Double cone revolving about horizontal axis	56	25	...	11.5	...	Mixer can be tilted. Rear end charger. Capacity based on mixed concrete.
2. Tumbler with agglomerate breaker						
Double cone	54	7½	See Comments.	18	See Comments.	Horsepower requirement for internal device depends on character of material, type and speed of agitator. These are to be determined by adequate testing.
Twin shell	50	5	5 (pin-type intensifier bar) 7½ (liquid-solids intensifier bar)	13.7	945 (1730-ft/min tip speed) 1055 (3320-ft/min tip speed)	Maximum bulk density of material = 55 lb/ft^3.
3. Stationary shell or trough						
Ribbon						
Manufacturer C	50	...	12	...	28	Horsepower required based on material of 50–60-lb/ft^3 bulk density, medium free-flowing, using 10 hp/ton for average mix cycle of 3–10 min (depending on material, range can be 3–18 hp/ton).
Manufacturer A	46	...	10	...	37	Based on material of 30-lb/ft^3 bulk density.
Manufacturer D	50	...	15	...	45	Based on material of 40–50-lb/ft^3 bulk density.
Three-shaft ribbon	50	...	Blender shaft 20 Feeder shaft 7½ (total)	...	Variable-speed drives on all shafts	This blender is rated at 300 ft^3/h on batch-mixing basis; 900 ft^3/h on continuous-mixing basis. Materials rated at 70-lb/ft^3 bulk density.
Vertical screw	52.9	...	5	...	Screw, 64.4 Orbit, 2.2	Horsepower based on 37-lb/ft^3 bulk density. This may vary with different materials. Maximum hp = 10, maximum weight = 4410 lb.
Muller:						
Batch; stationary pan, rotating turret	40	...	60	...	24 (turret speed)	Based on material of 60–75-lb/ft^3 bulk density.
Continuous; stationary pan, rotating turret	Basically, the continuous mullers are merely two-batch mullers joined together at the cribs, making a figure-8 design. Thus, the 40-ft^3 batch muller rated at 60 hp becomes an 80-ft^3 working-capacity continuous muller requiring 125 hp. This would give 125 tons/h with a 2½-min residence time. Tarret sperds are 24 r/min.					
Single rotor	...	See Comments.		...	...	In this *continuous* unit, output can range from 25–600 lb/min with hp from 5 to 100 and r/min of 500 to 4000, depending on the materials mixed.
Double rotor	...	See Comments.		...	...	In this *continuous* unit the output can range from 200–500 lb/min with hp from 5 to 40 and r/min from 200 to 300, depending on the materials mixed.

TABLE 21-3 Horsepower Requirements and Speeds of Rotation for Some Commercial Solids Mixers [Approximately 1.5 m³ (50 ft³) Working Capacity] *(Continued)*

Type of solids-mixing machine	Approximate working capacity, ft³	Horsepower, hp: Shell	Horsepower, hp: Internal device	Rotational speed, r/min: Shell	Rotational speed, r/min: Internal-device shaft speed	Comments
Twin-rotor heat-exchanger mixer	49.2	. . .	5–15	. . . .	20–100	Amount of conveying and mixing action affected by amount of pitch and type of ribbons mounted on exterior of hollow screws.
Turbine	50	. . .	50	. . . .	Peripheral speed of 600 ft/min	
4. Both shell and internal device rotate						
Countercurrent muller	45	°	°	6.75–8.75	28–35	
	60–90†	20	25	6.65	20	

°One 25-hp motor drives both the shell (mixing pan) and the internal device (mixing star).
†Batch-capacity range depends on nature of materials to be mixed.
NOTE: To convert cubic feet to cubic meters, multiply by 0.02832; to convert horsepower to kilowatts, multiply by 0.7457; to convert pounds per cubic foot to kilograms per cubic meter, multiply by 16.02; to convert tons per hour to kilograms per second, multiply by 0.252; to convert revolutions per minute to radians per second, multiply by 0.1047; to convert pounds per minute to kilograms per minute, multiply by 0.4535; and to convert horsepower per ton to kilowatts per metric ton, multiply by 0.8352.

time-consuming, several small mixers should be considered. Special sanitary construction can usually be provided at extra expense.

Agglomerate Breakdown and Attrition The two methods of producing agglomerate breakdown and attrition are as follows:

1. *Impact.* The major factor is the peripheral speed of the rotating internal device. Table 21-3 gives impact-velocity data for various mixers.

2. *Shearing and compressive action.* In mullers this depends upon the clearance between muller and pan and the muller weight or spring load respectively.

When an attrition device is necessary to break down aggregates but may also produce too much size reduction on other batch ingredients, tolerable attrition should be determined by tests.

Dust Formation Loss of dust can seriously affect batch composition, particularly when vital minor ingredients are lost. Methods of minimizing dust formation are: (1) Use of less dusty but equally satisfactory batch ingredients. Sometimes a pelletized form of an extremely dusty material is available. (2) Proper venting so as to enable filtering of displaced air rather than unregulated loss of dust-laden air. (3) Dust-tight arrangements for loading and unloading the mixer. (4) Addition of liquids if tolerable. Not only is water effective in minimizing dust upon discharging from the mixer, but if properly added it will also render the batch less dusty in subsequent handling steps. The addition of a small quantity of surface-active agent will improve the penetration of the water throughout the batch and enable it to wet even such materials as coal dust. The method of adding water is important (see "Method of Adding Liquids").

Care should be taken to avoid powerful suction or air flow on the mixer or the weigh hopper from which the ingredients feed into the mixer. If the dust-collection suction on the mixer is too strong, vital ingredients may be sucked out. If the dust-collection suction on the weighing system is too strong, errors in weighing may result.

Electrostatic Charge Certain batch materials such as plastics tend to accumulate a charge easily. Work input will affect the charge on the batch. Coating of the inside of the mixer shell or rotating elements may occasionally result because of electrostatic charge. This can present a cleaning problem. Possible aids in overcoming this are (1) addition of special solid materials with very high surface area to weight ratios, (2) addition of liquids (see "Dust Formation" and "Method of Adding Liquids"), (3) proper choice of material of construction of the mixer, (4) controlling humidity, (5) preparation of the batch ingredients so as to minimize accumulated charge.

Equipment Wear Simple tumbling mixers give the least wear. Attrition devices in tumblers may present serious abrasion problems with certain materials such as sand and abrasive grinding-wheel grains. Abrasion-resistant coating such as rubber coating, special alloys, or platings should be considered for these cases. An internal agitator device may wear even though its speed is low. Particularly when highly abrasive materials are to be mixed, the benefits of an agglomerate-breaking device must be weighed against potential contamination and replacement and maintenance costs.

Contamination of Product This has been partially covered under "Cleaning" and "Equipment Wear." Other sources of contamination are lubricants and repair materials. Types which are not compatible with the batches to be mixed should be avoided.

Heating or Cooling Nearly all commercial mixers can be heated or cooled. Some can be provided with heated or cooled agitators. If temperature rise during mixing is detrimental, cooling facilities should be provided. The various manufacturers can provide details on the means of heating their machines. Most common heating means are (1) water or steam in the jacket and in hollow-screw or paddle-type internal agitator, (2) hot oil, (3) Dowtherm liquid or vapor, (4) electric heaters, contact or radiant, (5) hot air in direct contact with product (suitable only for revolving-drum-type mixers), (6) exterior heating of drum by direct or indirect firing. For cooling, the most common means are (1) water or refrigerated fluid in the jacket and in hollow-screw or paddle-type internal agitator, (2) an evaporant such as liquid ammonia, (3) direct air contact (for rotating-shell mixers), and (4) oil or Dowtherm (or its equivalent) for cooling high-temperature materials.

Flexibility When batches of widely different size must be mixed, flexibility of operating capacity may enable use of fewer mixers. Certain features may necessitate a nonflexible capacity requirement. For example, ordinarily an internal agitating device in a tumbling mixer does not function effectively unless the batch is loaded to a certain level. The need for such features must be weighed against the limitations imposed by a narrow operating-capacity range when choosing equipment for an operation in which batch size will vary considerably.

In general, the effect of percentage of mixer volume occupied by the batch on the adequacy of mixing should be borne in mind, particularly when any change from the recommended volume percent is considered.

Vacuum or Pressure Most tumbling mixers can have provision for vacuum or pressure. Mixers which cannot be adapted to these conditions are mullers with rotating pans. Continuous mixers introduce problems of sealing the charge and discharge ends.

Method of Adding Liquids When the addition of liquids may be desirable (see "Dust Formation" and "Electrostatic Charge"), this should be considered when designing the mixing system rather than hastily improvised. The purpose of the liquid should be considered, whether for (1) dust suppression, (2) product, or (3) heating and cooling. If a viscous liquid must be well distributed, this requirement should be considered when choosing the mixer.

Liquid should be directed into the batch materials and not onto bare mixer surface since this could cause buildup. Nozzle spray pressure should be sufficient to penetrate the batch but not so high as to cause heavy splashing. The liquid should be added to the well-mixed batch. In particular, when premature addition of liquid could impair the adequacy of blending, both the time during which it is added in the mixing cycle and the time taken to add the liquid are important.

Automated equipment for the addition of liquids can be worked into the overall mixing plant when necessary. For dust-reduction purposes, a volumetric method of metering is satisfactory. However, should a critical batch ingredient be added in liquid form, a more precise method of metering may be necessary.

Other considerations are (1) proper ventilation and discharge enclosures, (2) provision for relief of internal explosion, (3) vibration isolation (shock mounts), (4) remote operation of charge and discharge, (5) noise during operation.

Equipment Selection Types of mixers and performance characteristics have been given. Segregating tendencies among solid materials have also been described. A sound approach to solids-mixer selection starts with a careful examination of these areas. However, mixer selection should also involve consideration of the mixer's place in the overall process. Possible consolidation of many solids-processing steps or the opposite (splitting one operation into several) deserves scrutiny at this time. If no one standard machine has all the necessary requirements, thought should be given to which machine can best be modified to achieve the most desirable combination of features. One should look at the overall process objectives as well as at equipment details when selecting a solids mixer.

Pilot Tests In some cases, it is possible to perform pilot tests on a small-scale version of the equipment to be used in production. Much useful information can be found here but the following must be borne in mind:

1. In general, the larger the pilot unit, the more reliable the prediction of large-scale performance. The pilot unit should be a prototype with all dimensions properly scaled down.
2. Published solids-mixing scale-up data are rare. Equipment suppliers can provide scale-up information for their particular types of equipment on the basis of experience. With geometrically similar tumblers, if the speeds are adjusted to give comparable motion and the mixer volume fraction occupied by the charge is the same, scale-up of results will be straightforward. The presence of a rotating internal device presents problems in the scaling up of clearances, blade area to mixture volume, and sizes and speeds of the rotating devices. For agglomerate breakers, the key factor in scaling up is impact velocity. Scale-up in cylinders is discussed on pages 290–292 of Ref. 9. Solids-processing scale-up is discussed in a paper by Sterret (*Chem. Eng.*, Sept. 21, 1959).
3. The actual process materials should be used if possible. If substitute materials must be used, they should have the same mixing characteristics. Tests with differently colored but otherwise identical beads can be misleading, and so can tracers. The reason is that the flow properties of the specific materials to be mixed in the plant may not be the same as these demonstration materials. Regardless of how the mixer contents appear to be moved around, the properties of the actual batch ingredients may cause segregation or other problems.
4. Differences in materials of construction between the pilot unit and the production unit should be considered. These may have a bearing on caking, abrasion, and electrostatic effects.

Continuous Mixing Although batch mixing has been the predominant method of mixing solids, consideration is being given to the use of continuous mixing in many industries. There are two types of continuous-mixing operations. The first type has a low holdup volume and will provide fine-scale blending of the particles via impact and shear elements such as are used in grinding machines. Some machines of this type are hammer, impact, cage, and jet mills. It is essential that the feed to these machines be properly proportioned and premixed to achieve a uniform product.

The second type of continuous mixer involves high holdup machines which contain agitating and conveying mechanisms. These rearrange the individual particles and also displace large volumes of material and move the batch through the machine. Mixers of this type can produce both fine-scale and coarse-scale blending. The ribbon-type mixer is frequently used for continuous mixing, although this is also used for batch mixing. A continuous muller mixer has been developed as shown in Fig. 21-3*h*.

The average composition of the stream leaving a continuous mixer is the same as the average of the added entering streams. Variations in proportions of the entering streams will be damped out by the mixing action of a continuous mixer. These effluent-stream variations will become smaller as average solids residence time is increased and the frequency of the variations increases.

Certain general criteria can be used to determine whether continuous flow will be beneficial. Continuous flow is worth consideration if (1) a single formulation can be run for an extended period, (2) the fluctuations of the outgoing product are within process requirements, (3) sufficiently accurate metering of ingredients can be achieved, (4) the rest of the process warrants continuous mixing. Continuous flow is of doubtful benefit if (1) frequent changes of formulations are anticipated, (2) fluctuations of product composition will be outside the permitted range, (3) the ingredients cannot be metered with the necessary level of accuracy, (4) complex temperature or pressure cycles are involved.

Sometimes a system of mixing and dispersing is composed of one or more batch units providing a feed to a continuous intensive dispersion unit. Another possibility would be a batch mixer and surge bin which provide a continuous feed to a final dispersion unit. Various combinations of this type with adequate sampling at the proper points may be used when continuous flow would be beneficial provided that certain features could be overcome.

SOLIDS SAMPLING

GENERAL REFERENCES: Beddow, *Particulate Science and Technology*, Chemical Publishing, New York, 1980, pp. 392–402. "Economics of Automatic Sampling," 3d Stevens Symposium on Statistical Methods in the Chemical Industry, American Society for Quality Control, Jan. 24, 1959. Gy, *Sampling of Particulate Materials*, Elsevier, New York, 1979. Mular and Bhappu (eds.), *Mineral Processing Plant Design*, 2d ed., Society of Mining Engineers, AIME, New York, 1980. Pryor, *Mineral Processing*, 3d ed., Elsevier, New York, 1965. Taggart, *Handbook of Mineral Dressing*, 2d ed., Wiley, New York, 1945.

The control of processes involving the treatment of solids generally requires means for sampling and analysis of the solids at various points in the operation. Unlike liquids, solids are not homogeneous. The composition of individual particles will vary with particle size and particle density. It follows that care must be exercised to take a sample that represents the entire solids mixture at the point of interest in the process. If the solids are not sampled in a representative manner, process and product control will not be reliable. Grab samples and the use of cutters that do not take a portion of the stream uniformly will lead to nonrepresentative results.

There are a number of reasons for the segregation of solids by both size and gravity within a solids-handling system. Vibration such as that which can occur on belts and conveyors is one cause of variation in sample composition. Segregation can also occur whenever granular solids are stacked on piles or in storage bins. Thus, the best technique involves moving the sample cutter at a uniform speed completely through the entire stream while the stream is in a state of free fall. The direction of movement of the cutter should be at right angles to the direction of flow of the stream.

The initial practice of manual sampling has been largely replaced by automatic mechanical sampling devices that require little maintenance and are relatively low in cost, impersonal, and statistically reliable. This discussion concerns the selection of sample size, type of sampling devices available, and their arrangement and cost.

Dry Sampling The sampling of dry granular solids is common. Dust seals make it possible to sample in pneumatic conveying sys-

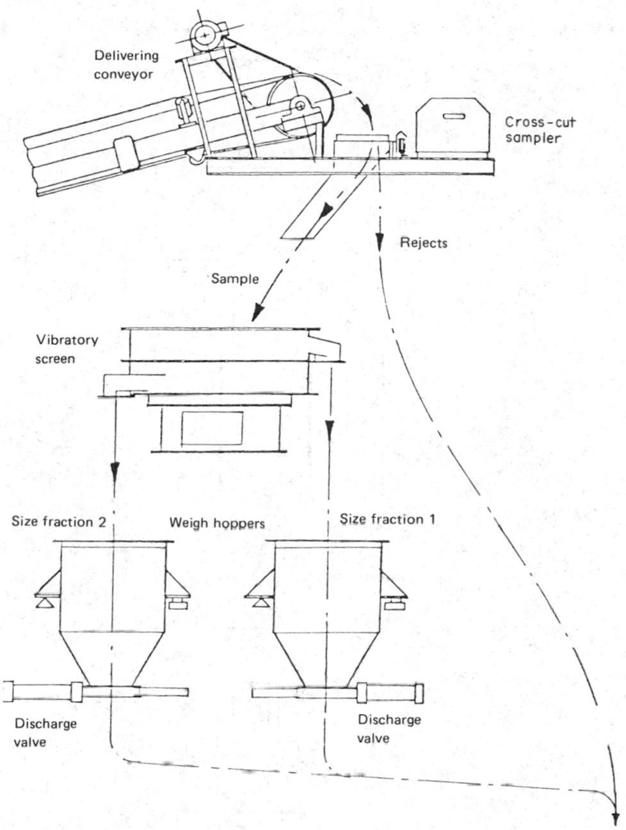

FIG. 21-4 Equipment arrangement for sampling for size analysis. (*Courtesy of Denver Equipment Company.*)

tems. Care should be taken to locate the sample cutter at a point in the free-fall trajectory where particle velocities are not so high that air entrainment could cause some segregation of fines.

Sampling for Size Analysis This may be accomplished manually or automatically by removing a sample increment periodically and discharging it to a screening device. If sampling is done automatically, screened fractions are discharged into bins mounted on scales (Fig. 21-4). Size distribution can be computed manually or electronically. This procedure is generally used for solids with large top sizes and when a large amount of sample is required to be representative. Such a sample could not be handled practically by hand screening and weighing.

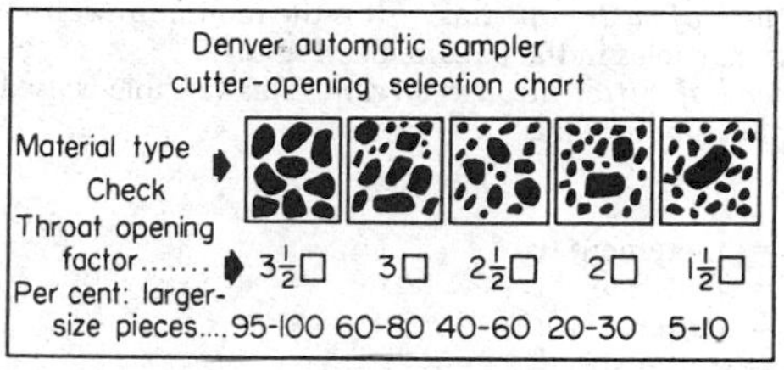

FIG. 21-5 Throat-opening selection chart for an automatic sampler. (*Courtesy of Denver Equipment Company.*)

Sampling for Moisture Analysis This must be done in a manner that avoids change in moisture content after the point of sampling and before the analysis is made. Closed chutework and sealed sample containers should be considered. One cause of moisture change may be the heating of solids in equipment such as a crusher. Also high-moisture-content materials can lose some moisture in sample chutes, conveyors, and long-free-fall arrangements.

Sampling for Chemical Analysis This procedure, accomplished either in the laboratory or online by x-ray or nuclear analyzers, requires that a sampling system be designed for self-cleaning. No areas where solids can adhere must exist.

Slurry Sampling The sampling of slurries is generally very reliable, although the handling of the sample after the cutter can cause problems. Solids may separate from the liquid by settling or centrifugal action, or material may accumulate at the dry-wall–liquid interface. If liquid analysis is not critical, it may be suitable to use liquid sprays to wash the cutter and key zones of the sampling system.

Sample Size There are no formulas to determine the amount of the sample, the percentage of material to be taken as a sample, or the frequency of the sampling interval. Some industries have established standards for sampling, but practices generally will differ from plant to plant. Factors that should be considered in selecting the sample size are:

1. *Lot size.* A larger gross sample must be taken from a very large lot than from a small lot.
2. *Analysis.* The sample taken for sizing analysis should be adequate to represent the actual size distribution. Moisture and composition analysis will require sufficient final sample for all tests to be performed.
3. *Variability.* The variability of analyzed components is an important factor. A smaller sample size can be taken for cases that involve components which are uniformly distributed than for cases of high variability.
4. *Quality-control tolerances.* For very close control of tolerances, a large gross sample should be taken to provide great precision. This procedure is required for costly products.

Increment Sample Size This is determined as follows:

1. *Process feed rate.* This is usually fixed by process and not varied for convenience of sampling.

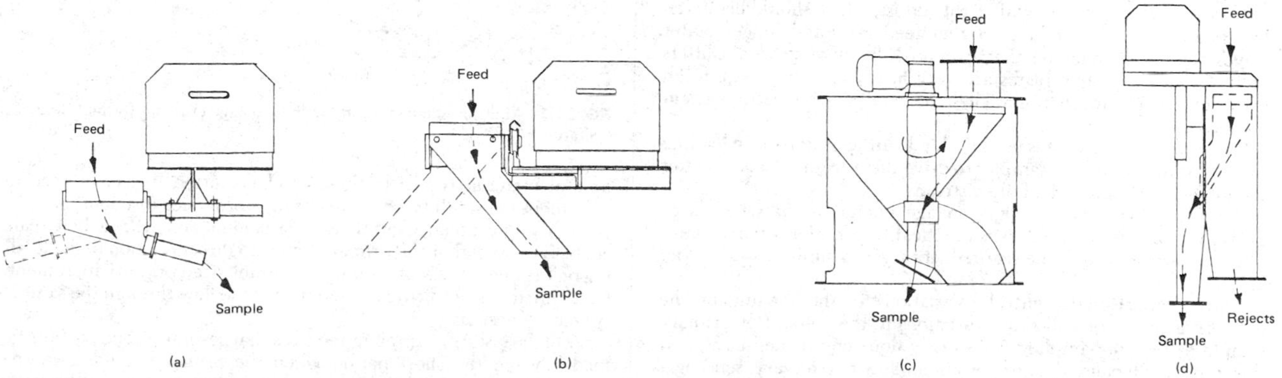

FIG. 21-6 Common types of mechanical samplers. (*a*) Automatic slurry. (*b*) Automatic dry solid. (*c*) Vezin. (*d*) Chute type.

2. *Width of cutter opening.* It is the minimum width based on the size of particles in the stream. See Fig. 21-5.

3. *Speed of cutter through stream.* This variable is used to control sample volume. Thus,

$$I = mW_c/S_c$$

where I = increment size
m = flow rate
W_c = cutter width
S_c = cutter speed

Units must be consistent.

Each increment must be sufficient to represent the size distribution of the lot. Larger particle sizes therefore require larger increments.

Gross Sample Size This can be determined by:

1. *Frequency of increments.* A variable timer may be used to establish the gross sample bulk.

2. *Lot size.* Large lots require more increments than small lots. A lot may consist of the quantity of material contained in a shipment, a bin or tank, a batch, or a time period during which a process is adjusted.

3. *Variability.* Given two equal-sized lots, the lot having the more variable component requires the greater number of increments. Variability may fluctuate as the process control of quality varies. The number of increments per lot is usually determined by the standard for the industry or by quality-control considerations.

Final Sample Size In many cases, the final sample size is a single increment. Often, however, a gross sample is much too large for final analysis and must be divided as follows:

1. Coning and quartering or riffling by hand is a long-established practice for reducing a sample to convenient size.

2. Automatic reduction by secondary (or tertiary) sampling may be required when the gross sample is too large for convenient handling at the laboratory. In multiple-stage solids sampling, each increment must be fed at a constant rate to the succeeding sampler over a time period equal to the period between original sample cuts.

3. Crushing prior to sample reduction permits the extraction of smaller sample increments (each representative of the crushed size distribution) and subsequent reduction to convenient sample size. This is, of course, not applicable for particle-size-analysis applications.

Types of Mechanical Samplers Some of the more common mechanical samplers are illustrated in Fig. 21-6. Automatic straight-line samplers offer both the lowest cost and the greatest flexibility of operation. Vezin-type samplers are arc-type samplers which normally operate continuously and can sample dry solids or slurries. They are inherently dust-tight.

Automatic straight-line samplers are available with electric or hydraulic drives. The electromechanical units are usually fixed-speed units. The hydraulic drives, either the cylinder or the hydraulic-motor type, allow in-field revisions with respect to cutter speed. Hydraulic drives are generally preferred over pneumatic types. They are better able to maintain a constant cutter speed despite changes that affect the resistance of the solids stream.

Low feed rates require small, light cutters, which should be driven by a standard-duty machine. Higher feed rates and longer cutter travel require a machine that can provide more power. Cutters mounted in dust baffle plates are very heavy and may require an extra-heavy-duty mechanism. These normally have electromechanical drives.

The enormous flow rates used in ship loading or unit-train loading may require a hydraulic sampler to drive the extremely heavy cutter and baffle plate across the falling stream.

Sampling Systems In high-tonnage operations or when large gross samples are required, it is practical to consider separate sampling towers with multiple size-reduction and sampling stages (Fig. 21-7).

Auto Proportional Control Variations in the loading of the delivering conveyor will cause variations in the amount of primary sample extracted. Sampling is normally done on a timed basis, and this in turn will cause the sample-system flow rate to vary, sending a variable volume of sample to the secondary cutter. If too small a volume of material is sent to the secondary cutter, it may extract an increment too small to be representative of its size distribution.

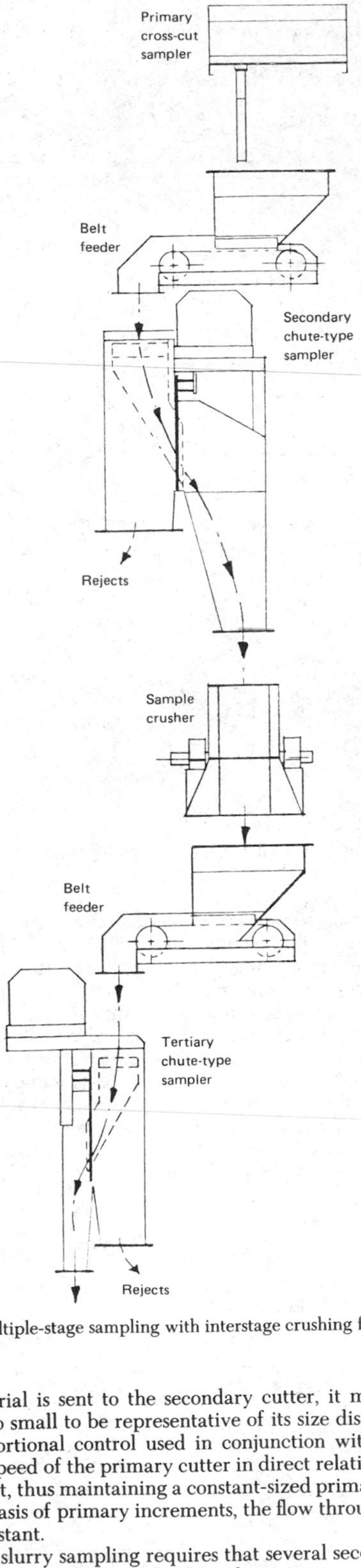

FIG. 21-7 Multiple-stage sampling with interstage crushing for high-tonnage operation.

Auto proportional control used in conjunction with a belt scale controls the speed of the primary cutter in direct relation to the loading on the belt, thus maintaining a constant-sized primary increment. On a timed basis of primary increments, the flow through the sample system is constant.

Secondary slurry sampling requires that several secondary cuts be made during the short period when the primary cutter is in the stream. This may be only 2 to 4 s, depending on the width of the

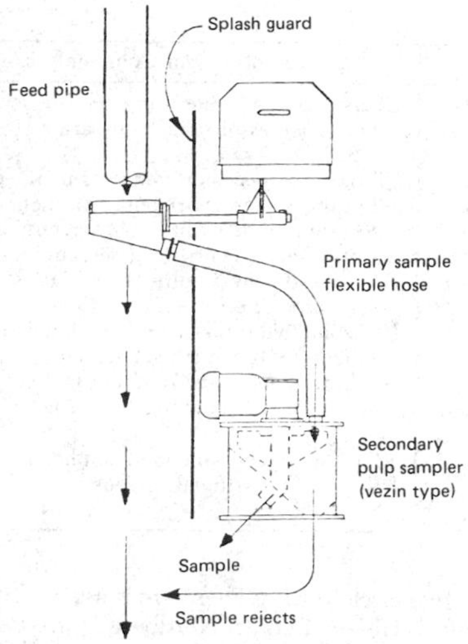

FIG. 21-8 Vezin-type secondary sampler with four sample cutters. *(Courtesy of Denver Equipment Company.)*

stream and the speed of the cutter. A Vezin-type sampler equipped with four cutters is often used (Fig. 21-8).

Costs Costs for various sizes and types of sampling equipment, as well as for sampling towers and systems, are given in Table 21-4.

INTRODUCTION TO SCREENING AND WET CLASSIFICATION

The **classification** of solids by particle size is carried out for a number of reasons. Size classification can facilitate subsequent processing steps. An example is the scalping of tramp oversize material to avoid clogging a piece of processing apparatus. Similarly, better efficiency is achieved by removing fines before size reduction in crushers or ball or rod mills. Finished products generally are required to meet particle-size limits. Size separation is accomplished either in the dry condition or with the solids in suspension as a slurry. Wet classification allows higher process rates, particularly for materials of very fine sizes. Classification often is an integral part of a unit operation, as in closed-circuit grinding. Air classification methods for dry size classification in conjunction with size-reduction operations is covered in Sec. 8. Dry size separation by **screening** and various methods of **wet classification** are discussed in this section.

TABLE 21-4 Cost of Sampling Equipment, 1981

Cutter travel, in*	Size and type of sampler	Approximate price, $	Addition for cutter, $
18	Straight-line automatic (standard duty)	2,875	550 (wet type)
30	Straight-line automatic (heavy duty)	4,000	1,100 (dry type, front travel)
42	Straight-line automatic (heavy duty)	4,200	1,200 (dry type, end travel)
55	Straight-line automatic (extra heavy duty)	11,400	6,800 (dry type, slide plate)
75	Straight-line automatic (extra heavy duty)	14,900	8,400 (dry type, slide plate)
20, diameter	Vezin-type simplex (continuous)	4,900	Cutter included
16½, diameter	Chute type (hydraulic)	8,200	Cutter included
36, diameter	Chute type (hydraulic)	11,900	Cutter included
. . .	Riffle type 18 × 18 opening	400	

*To convert inches to meters, multiply by 0.0254.

SCREENING

GENERAL REFERENCES: Beddow, "Dry Separation Techniques," *Chem. Eng.*, **88,** 70 (Aug. 10, 1981). Colman, "Selection Guidelines for Size and Type of Vibrating Screens in Ore Crushing Plants," in Mular and Bhappu (eds.), *Mineral Processing Plant Design*, 2d ed., Society of Mining Engineers, AIME, New York, 1980. Kuenhold, "Factors to Consider in Vibrating Screen Installations," *Min. Eng.*, 650–653 (June 1957). Matthews, *Chem. Eng.*, deskbook issue, 99 (Feb. 15, 1971). Matthews, "Screening," *Chem. Eng.*, **79,** 76 (July 10, 1972). Moir, "Recent Developments in Mineral Processing and Their Implications," *Economics of Mineral Engineering Mining Journal*, Books Ltd., London, 1976, p. 125. Mular, *Mineral Processing Equipment Costs and Preliminary Capital Cost Estimations*, spec. vol. 18, Canadian Institute of Mining and Metallurgy, Montreal, 1978. Pryor, *Mineral Processing*, 3d ed., Elsevier, New York, 1965. Reed, "The Story behind the New Sieve Specifications," *Test. World*, October 1959. Taggart, *Handbook of Mineral Dressing*, 2d ed., Wiley, New York, 1945.

Definitions

Screening Screening is the separation of a mixture of various sizes of grains into two or more portions by means of a screening surface, the screening surface acting as a multiple go–no-go gauge and the final portions consisting of grains of more uniform size than those of the original mixture.

Material that remains on a given screening surface is the oversize or plus material, material passing through the screening surface is the undersize or minus material, and material passing one screening surface and retained on a subsequent surface is the intermediate material.

The screening surface may consist of woven-wire, silk, or plastic cloth, perforated or punched plate, grizzly bars, or wedge wire sections.

Classification of screening operations and the range of separations that can be attained with various screens were given in concise form by Matthews (op. cit., 1971). See Table 21-5. Further details are given under "Equipment." Figure 21-9 indicates the size-range applicability of various screen types.

Mesh and Space Cloth Wire cloth is generally specified by "mesh," which is the number of openings per linear inch counting from the center of any wire to a point exactly 25.4 mm (1 in) distant, or by an opening specified in inches or millimeters, which is understood to be the clear opening or space between the wires. Mesh is generally favored for cloth 2 mesh and finer and clear opening for space cloth of 12.7-mm (½-in) opening and coarser.

Aperture Aperture, or screen-size opening, is the minimum clear space between the edges of the opening in the screening surface and is usually given in inches or millimeters.

Open Area The open area of a screen is the percentage of actual openings versus total screen area and can be determined by the formulas given in Fig. 21-16.

Particle-Size Distribution This is defined as the relative percentage by weight of grains of each of the different size fractions represented in the sample. It is one of the most important factors in evaluating a screening operation and is best determined by a complete size analysis using testing sieves.

Sieve Scale A sieve scale is a series of testing sieves having openings in a fixed succession; for example, in the original basic Tyler standard sieve scale the widths of the successive openings have a constant ratio of the square root of 2, or 1.414, while the areas of the successive openings have a constant ratio of 2. The Tyler scale has been enlarged to include intermediate openings so that the entire scale has successive openings according to the fourth root of 2, or 1.189. The sieve series adopted by the National Bureau of Standards, American Society for Testing and Materials, American National Standards Institute, and many countries applies the fourth-root-of-2 principle, and the openings are fully compatible with the Tyler standard scale even though the sieve designations may vary (Table 21-6).

TABLE 21-5 Types of Screening Operations

Operation and Description	Type of Screen Commonly Employed
Scalping—Strictly, the removing of a small amount of oversize from a feed which is predominantly fines. Typically, the removal of oversize from a feed with approximately a maximum of 5% oversize and a minimum of 50% half-size.	Coarse (grizzly); fine, same as fine separation; ultrafine, same as ultrafine separation.
Separation (coarse)—Making a size separation at 4 mesh and larger.	Vibrating screen, horizontal or inclined.
Separation (fine)—Making a size separation smaller than 4 mesh and larger than 48 mesh.	Vibrating screen, horizontal or inclined; high-speed low-amplitude vibrating screens; sifter screens; static sieves; centrifugal screens.
Separation (ultrafine)—Making a size separation smaller than 48 mesh.	High-speed low-amplitude vibrating screen; sifter screens; static sieves; centrifugal screens.
Dewatering—Removal of free water from a solids-water mixture. Generally limited to 4 mesh and above.	Horizontal vibrating screen; inclined vibrating screens (about 10°); centrifugal screen.
Trash removal—Removal of extraneous foreign matter from a processed material. Essentially a form of scalping operation. Type of screen employed will depend on size range of processed material—coarse, fine, or ultrafine.	Vibrating screen, horizontal or inclined; sifter screens; static sieves; centrifugal screen.
Other applications: Desliming—Removal of extremely fine particles from a wet material by passing it over a screening surface. *Conveying*—In some instances transport of the material may be as important as the operation. *Media recovery*—A combination washing and dewatering operation.	Vibrating screens, inclined and horizontal; oscillating screens; centrifugal screens.

Equipment Screening machines may be divided into five main classes: grizzlies, revolving screens, shaking screens, vibrating screens, and oscillating screens. Grizzlies are used primarily for scalping at 0.05 m (2 in) and coarser, while revolving screens and shaking screens are generally used for separations above 0.013 m (½ in). Vibrating screens cover this coarse range and also down into the fine meshes. Oscillating screens are confined in general to the finer meshes below 4 mesh.

Grizzly Screens These consist of a set of parallel bars held apart by spacers at some predetermined opening. Bars are frequently made of manganese steel to reduce wear. A grizzly is widely used before a primary crusher in rock- or ore-crushing plants to remove the fines before the ore or rock enters the crusher. It can be a stationary set of bars or a vibrating screen.

Stationary grizzlies. These are the simplest of all separating devices and the least expensive to install and maintain. They are normally limited to the scalping or rough screening of dry material at 0.05 m (2 in) and coarser and are not satisfactory for moist and sticky material. The slope, or angle with the horizontal, will vary between 20 and 50°. Stationary grizzlies require no power and little maintenance. It is, of course, difficult to change the opening between the bars, and the separation may not be sufficiently complete.

Flat grizzlies. These, in which the parallel bars are in a horizontal plane, are used on tops of ore and coal bins and under unloading trestles. This type of grizzly is used to retain occasional pieces too large for the following plant equipment. These lumps must then be broken up or removed manually.

Vibrating grizzlies. These are simply bar grizzlies mounted on eccentrics so that the entire assembly is given a back-and-forth movement or a positive circle throw. These are made by companies such as Allis-Chalmers, Hewitt Robins, Nordberg, Link-Belt, Simplicity, and Tyler.

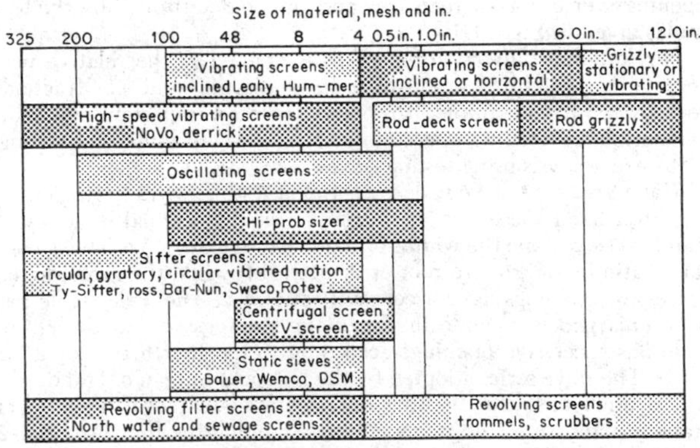

FIG. 21-9 Range of separations that can be obtained with various kinds of screens. To convert inches to meters, multiply by 0.0254. [*Matthews*, Chem. Eng. *(Feb. 15, 1971)*.]

Revolving Screens Revolving screens, or trommel screens, once widely used, are being largely replaced by vibrating screens. They consist of a cylindrical frame surrounded by wire cloth or perforated plate, open at both ends, and inclined at a slight angle. The material to be screened is delivered at the upper end, and the oversize is discharged at the lower end. The desired product falls through the wire-cloth openings. The screens revolve at relatively low speeds of 15 to 20 r/min. Their capacity is not great, and efficiency is relatively low.

Mechanical Shaking Screens These screens consist of a rectangular frame which holds wire cloth or perforated plate and is slightly inclined and suspended by loose rods or cables or supported from a base frame by flexible flat springs. The frame is driven with a reciprocating motion. The material to be screened is fed at the upper end and is advanced by the forward stroke of the screen while the finer particles pass through the openings. In many screening operations such devices have given way to vibrating screens.

Shaking screens, such as the mechanical-conveyor type made by Syntron Co., may be used for both screening and conveying.

The advantages of this type are low headroom and low power requirement. The disadvantages are the high cost of maintenance of the screen and the supporting structure owing to vibration and low capacity compared with inclined high-speed vibrating screens.

Vibrating Screens These screens are used as standard practice when large capacity and high efficiency are desired. The capacity, especially in the finer sizes, is so much greater than that of any of the other screens that they have practically replaced all other types when efficiency of the screen is an important factor. Advantages include accuracy of sizing, increased capacity per unit area, low maintenance cost per ton of material handled, and a saving in installation space and weight.

There are a great number of vibrating screens on the market, but basically they can be divided into two main classes: (1) mechanically vibrated screens and (2) electrically vibrated screens.

Mechanically Vibrated Screens The most versatile vibration for medium to coarse sizing is generally conceded to be the vertical circle produced by an eccentric or unbalanced shaft, but other types of vibration may be more suitable for certain screening operations, particularly in the finer sizes. One well-known *four-bearing mechanically vibrated screen,* installed in an inclined position, is the Ty-Rock (Fig. 21-10). This is a balanced circle-throw machine mounted on a base frame, having a full-floating body mounted on shear rubber mounting units which absorb the shocks of heavy material and allow the shaft to revolve around its own natural center of rotation.

Two-bearing screens, of which there are many types, have the same screen body as the four-bearing type but without the two outer bearings and the base frame. The gyrating motion is caused by eccentric weights on the shaft, and the screen itself is supported by overhead cables or springs on the floor.

TABLE 21-6 U.S. Sieve Series and Tyler Equivalents (ASTM—E-11-61)

Sieve designation		Sieve opening		Nominal wire diam.		Tyler equivalent designation
Standard	Alternate	mm.	in. (approx. equivalents)	mm.	in. (approx. equivalents)	
107.6 mm.	4.24 in.	107.6	4.24	6.40	0.2520	
101.6 mm.	4 in.†	101.6	4.00	6.30	.2480	
90.5 mm.	3½ in.	90.5	3.50	6.08	.2394	
76.1 mm.	3 in.	76.1	3.00	5.80	.2283	
64.0 mm.	2½ in.	64.0	2.50	5.50	.2165	
53.8 mm.	2.12 in.	53.8	2.12	5.15	.2028	
50.8 mm.	2 in.†	50.8	2.00	5.05	.1988	
45.3 mm.	1¾ in.	45.3	1.75	4.85	.1909	
38.1 mm.	1½ in.	38.1	1.50	4.59	.1807	
32.0 mm.	1¼ in.	32.0	1.25	4.23	.1665	
26.9 mm.	1.06 in.	26.9	1.06	3.90	.1535	1.050 in.
25.4 mm.	1 in.†	25.4	1.00	3.80	.1496	
22.6 mm.*	⅞ in.	22.6	0.875	3.50	.1378	0.883 in.
19.0 mm.	¾ in.	19.0	.750	3.30	.1299	.742 in.
16.0 mm.*	⅝ in.	16.0	.625	3.00	.1181	.624 in.
13.5 mm.	0.530 in.	13.5	.530	2.75	.1083	.525 in.
12.7 mm.	½ in.†	12.7	.500	2.67	.1051	
11.2 mm.*	7/16 in.	11.2	.438	2.45	.0965	.441 in.
9.51 mm.	⅜ in.	9.51	.375	2.27	.0894	.371 in.
8.00 mm.*	5/16 in.	8.00	.312	2.07	.0815	2½ mesh
6.73 mm.	0.265 in.	6.73	.265	1.87	.0736	3 mesh
6.35 mm.	¼ in.†	6.35	.250	1.82	.0717	
5.66 mm.*	No. 3½	5.66	.223	1.68	.0661	3½ mesh
4.76 mm.	No. 4	4.76	.187	1.54	.0606	4 mesh
4.00 mm.*	No. 5	4.00	.157	1.37	.0539	5 mesh
3.36 mm.	No. 6	3.36	.132	1.23	.0484	6 mesh
2.83 mm.*	No. 7	2.83	.111	1.10	.0430	7 mesh
2.38 mm.	No. 8	2.38	.0937	1.00	.0394	8 mesh
2.00 mm.*	No. 10	2.00	.0787	0.900	.0354	9 mesh
1.68 mm.	No. 12	1.68	.0661	.810	.0319	10 mesh
1.41 mm.*	No. 14	1.41	.0555	.725	.0285	12 mesh
1.19 mm.	No. 16	1.19	.0469	.650	.0256	14 mesh
1.00 mm.*	No. 18	1.00	.0394	.580	.0228	16 mesh
841 micron	No. 20	0.841	.0331	.510	.0201	20 mesh
707 micron*	No. 25	.707	.0278	.450	.0177	24 mesh
595 micron	No. 30	.595	.0234	.390	.0154	28 mesh
500 micron*	No. 35	.500	.0197	.340	.0134	32 mesh
420 micron	No. 40	.420	.0165	.290	.0114	35 mesh
354 micron*	No. 45	.354	.0139	.247	.0097	42 mesh
297 micron	No. 50	.297	.0117	.215	.0085	48 mesh
250 micron*	No. 60	.250	.0098	.180	.0071	60 mesh
210 micron	No. 70	.210	.0083	.152	.0060	65 mesh
177 micron*	No. 80	.177	.0070	.131	.0052	80 mesh
149 micron	No. 100	.149	.0059	.110	.0043	100 mesh
125 micron*	No. 120	.125	.0049	.091	.0036	115 mesh
105 micron	No. 140	.105	.0041	.076	.0030	150 mesh
88 micron*	No. 170	.088	.0035	.064	.0025	170 mesh
74 micron	No. 200	.074	.0029	.053	.0021	200 mesh
63 micron*	No. 230	.063	.0025	.044	.0017	250 mesh
53 micron	No. 270	.053	.0021	.037	.0015	270 mesh
44 micron*	No. 325	.044	.0017	.030	.0012	325 mesh
37 micron	No. 400	.037	.0015	.025	.0010	400 mesh

*These sieves correspond to those proposed as an international (I.S.O.) standard. It is recommended that wherever possible these sieves be included in all sieve analysis data or reports intended for international publication.

†These sieves are not in the fourth-root-of-2 series, but they have been included because they are in common usage.

Screening machines actuated by rotating unbalanced weights have a symmetrical shaft through the screen body with an unbalanced flywheel on each end. Counterweights on each flywheel, which may be moved in relation to the shaft, permit adjustment of the amplitude of vibration. On some makes of machines the complete shaft assembly is contained in a unit bolted to the top of the screen body.

The horizontal-type screen is actuated by an enclosed mechanism consisting of off-center weights geared together on short horizontal shafts. The mechanism is usually mounted between the side plates and above the screen body (Fig. 21-11).

Electrically Vibrated Screens These screens are particularly useful in the chemical industry. They handle very successfully many light, fine, dry materials and metal powders from approximately 4 mesh to as fine as 325 mesh. Most of these screens have an intense, high-speed (25 to 120 vibrations/s) low-amplitude vibration supplied by means of an electromagnet.

Typical of these is the Hum-mer screen used throughout the chemical industry. Figure 21-12 shows one used throughout the fertilizer industry for handling mixed chemical fertilizers.

Oscillating Screens These screens are characterized by low-speed oscillations [5 to 7 oscillations per second (300 to 400 r/min)] in a plane essentially parallel to the screen cloth.

Screens in this group are usually from 0.013 m (½ in) to 60 mesh. Some light free-flowing materials, however, can be separated at 200 to 300 mesh. Silk cloths are often used.

Reciprocating Screens These screens have many applications in chemical work. An eccentric under the screen supplies oscillation, ranging from gyratory [about 0.05-m (2-in) diameter] at the feed end to reciprocating motion at the discharge end. Frequency is 8 to 10 oscillations per second (500 to 600 r/min), and since the screen is inclined about 5°, a secondary high-amplitude normal vibration of about 0.0025 m (1/10 in) is also set up. Further vibration is caused by balls bouncing against the lower surface of the screen cloth.

These screens are used extensively in the United States and are standard equipment in many chemical and processing plants for handling fine separations even down to 300 mesh. They are used to handle a variety of chemicals, usually dry, light, or bulky materials, light metal powders, powdered foods, and granular materials. They are not designed for handling heavy tonnages of materials like rock or gravel. Machines of this type are exemplified by Fig. 21-13.

Gyratory Screens These are boxlike machines, either round or square, with a series of screen cloths nested atop one another. Oscillation, supplied by eccentrics or counterweights, is in a circular or near-circular orbit. In some machines a supplementary whipping action is set up. Most gyratory screens have an auxiliary vibration caused by balls bouncing against the lower surface of the screen cloth. A typical machine is shown in Fig. 21-14. Machines of this type are operated continuously and can be located in line in pneumatic

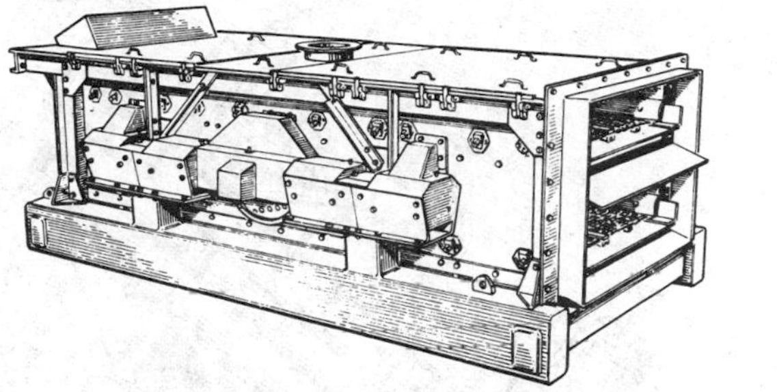

FIG. 21-10 Ty-Rock screen with air-seal enclosure. *(W. S. Tyler, Inc.)*

FIG. 21-11 Mechanically vibrated horizontal screen. *(Courtesy of Diester Concentrator Company, Inc.)*

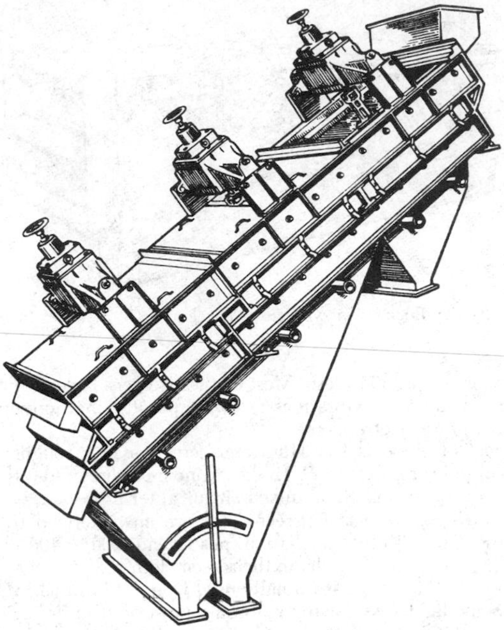

FIG. 21-12 Type 38 Hum-mer screen. *(W. S. Tyler, Inc.)*

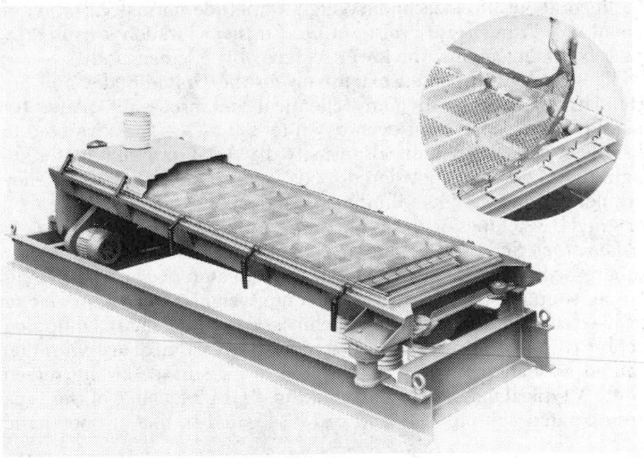

FIG. 21-13 Reciprocating screen. *(Courtesy of Rotex Corp.)*

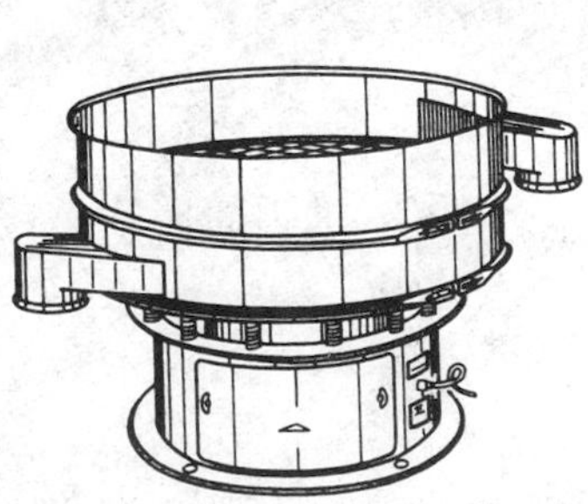

FIG. 21-14 Vibro-energy screen. *(Southwestern Engineering Company.)*

conveying systems as scalping screens. The size ranges from 0.6 to 1.5 m (24 to 60 in).

Gyratory Riddles These screens are driven in an oscillating path by a motor attached to the support shaft of the screen. The gyratory riddle is the least expensive screen on the market and is intended normally for batch screening.

Screen Surfaces The selection of the proper screening surface is very important, and the opening, wire diameter, and open area should all be carefully considered. The four general types of screening surfaces are woven-wire cloth, silk bolting cloth, punched plate, and bar or rod screens.

Woven-Wire Cloth This type has by far the greatest selection as to screen opening, wire diameter, and percentage of open area. Thousands of specifications are available from over 0.10 m (4 in) clear opening to 500 mesh. Woven-wire screens are obtainable in a variety of metals and alloys. Steel and high-carbon steel are generally favored for the coarser openings because of their abrasion-resistant qualities, and other materials, such as phosphor bronze, Monel, and stainless steel, are used for their corrosion-resisting or noncontamination qualities.

Square-mesh cloth is the conventional type of screen cloth, but there are many types of cloth with an oblong weave. This latter construction provides greater open area and capacity and in addition makes it possible to use stronger wire for the same size of screen opening and for the same percentage of open area.

In choosing a wire-cloth specification there must be a compromise between sharpness of separation, capacity, freedom from blinding, and life of the wire cloth. The square-mesh cloth will give the closest control of the maximum size particle in the undersize material; but the effective size of the openings will be reduced, because of the foreshortening when used at an angle of inclination, with consequent reduction in capacity. It should be realized that it is often necessary to use a cloth specification with an aperture larger than the smallest-size material acceptable in the oversize in order to ensure thorough removal of the undersize. A screen with a rectangular opening will increase the capacity with but little loss of sharpness when handling rounded or cubical grains. Slabby or flat material may also be handled on rectangular-opening cloth if the final-product specification will allow in the undersize a certain percentage of flat pieces having one dimension greater than the specified square-opening sieve. In other words pieces that might fall through a rectangular cloth and be allowed in the product might not go through the limiting square-mesh sieve on which the specification is based. If the through product is to be further ground or processed, a small amount of this material will not be objectionable.

Screen-cloth specifications having a relatively large length-to-width ratio are desirable when moisture or sticky material tends to cause blinding with square or short rectangular openings.

The finer the diameter of the wire from which a given specification is woven, the greater will be its screening capacity, although its screening life will be shorter. Since production capacity is generally more important than screen-surface cost, care should be taken to avoid using too heavy a specification which might restrict the capacity of the screening unit on which it is used and thus create a bottleneck in the flow.

Catalogs of wire-cloth manufacturers should be consulted for further study of the different types of wire-cloth specifications.

Silk Bolting Cloth This material originated in Switzerland and is generally woven from twisted multistrand natural silk. The system of numbers and grades for both bolting cloth and gritz gauze has been handed down from the original Swiss weavers. In recent years, nylon and similar synthetic materials woven largely from monofilaments have been introduced. The nylon grades are generally designated by their micrometer opening and are available in light, standard, and heavy weights.

Comparative Openings of Screening Cloths In screening any material, the size of the particles going through the screen is determined by the actual opening and not by the number of meshes per linear unit. As a rule, the lighter grades of wire-screen cloth, having greater percentages of open area, screen more freely and accurately and should be used whenever they will give satisfactory length of service. Tables of comparative openings are available for selecting a screen specification with a specific opening or for picking a specification having a heavier or lighter wire but the same opening.

Punched Plates These are available in a variety of perforations including round, square, hexagonal, and elongated openings. Punched metal will generally wear longer than wire cloth and has more rigidity, which is an advantage in certain applications. However, it usually does not give the capacity per unit area that wire cloth does and is generally heavier. Its use is normally limited to the coarser separations.

Bar Screens These screens are generally used in handling large and heavy pieces of material. They are formed from rails, rods, or bars, suitably shaped; made from rolled steel or castings; fixed in parallel position and held by crossbars and spacers. Bars, which taper in thickness from top to bottom and may also taper in width from one end to the other, are recommended because they tend to avoid blinding.

Rod decks, composed of spring-steel rods approximately 0.6 m (2 ft) long, sprung into position between molded-rubber blocks, and held in position by means of rubber spacers, are also available.

Factors in Selecting Screening Equipment In attempting to pick a screening machine for a specific screening problem it should be emphasized that generalized formulas and charts used to predict screen capacity can give only an approximation because of the many variables which may affect performance. Screen consultants will readily admit that they must depend largely on laboratory tests and field experience. However, two governing factors bear mention: generally width of screen relates to capacity and length of screen relates to efficiency. Width is necessary to reduce the bed thickness to a practical maximum and length to allow the undersize to be removed without an inordinate amount of fines in the oversize. In attempting to choose a screening machine for a particular screening application the customer and manufacturer should consider the following: (1) Full description of material involved, including the name and type of material, bulk density, and physical characteristics such as hardness, particle shape, flow characteristics (free-flowing, sluggish, or sticky), percent of moisture and temperature. (2) Normal and maximum total rate of feed to screen. (3) Complete sieve analysis of screen feed, including maximum lump size, and sieve analysis of desired product. (4) Separation or separations required and the purpose of screening. Can slotted or rectangular openings be used in place of square openings? (5) Is screening to be accomplished dry or wet, and what amount of water is available? (6) Other important factors include method of delivering feed to the screen, open or closed circuit, open or enclosed screens, previous screening experience with the material, flow sheet or description of related equipment, operating hours per day, power available, and space limitations.

Variables in Screening Operations It will readily be seen that many variables in a screening operation can easily be changed in the field, and practical operators will always be trying to improve their operations or adapt them to new products or processes. Capacity and efficiency in screening operations are closely related. Capacity may be large if low efficiency is not objectionable. Usually, as the tonnage to a screen is increased, efficiency is decreased.

Method of Feed The screening machine must be fed properly in order to obtain maximum capacity and efficiency. The feed should be spread evenly over the full width of the screen cloth and approach the screen surface in a direction parallel to the longitudinal axis of the screen and at as low a practical velocity as is possible.

Screening Surfaces It is generally agreed that the most efficient screening results when a series of single-deck screens is used. This is true because lower decks of multiple-deck screens are not fed so that their entire area is used and because each separation requires a different combination of angle, speed, and amplitude of vibration for maximum performance.

Angle of Slope The optimum slope of inclined vibrating screens is that which will handle the greatest volume of oversize and still remove the available undersize required by the standards of the particular operation. To separate a material into coarse and fine fractions, the bed thickness must be limited so that vibration can stratify the load and allow fines to work their way to the screen surface and pass through the opening. Increased slope naturally increases the rate of travel, and at a given rate it reduces the bed thickness.

In the oscillating screen the angle of inclination must be coordinated with the speed and stroke for best results.

Direction of Rotation In circle-throw screens somewhat greater efficiency can be obtained by counterflow rotation, that is, having the material move down the screen against the rotation. Screens rotating with the flow of material will handle greater tonnage and operate at a lower angle.

Vibration Amplitude and Frequency Speed and amplitude of vibration should be designed to convey the material properly and to prevent blinding of the cloth. They are somewhat dependent upon the size and weight of the material being handled and are related to the angle of installation and the type of screen surface. The object, of course, is to see that the feed is properly stratified for the most efficient separation.

Performance Formulas

Screen Efficiency There is confusion concerning the meaning of screen efficiency, as a uniform method for figuring efficiency has never been established. A sound method of evaluating screen performance is given by W. S. Tyler, Inc., Mentor, Ohio, in its *Sieve Handbook*, no. 53. In this formula, when material put through the screen is the desired product, "efficiency" is the ratio of the amount of undersize obtained to the amount of undersize in the feed.

$$E = (R \times d)/b \tag{21-1}$$

where E = efficiency, R = percent of fines through the screen, d = percent finer than the designated size in screen fines, and b = percent finer than the designated size in screen feed.

When the object is to recover an oversize product from the screen, efficiency may be expressed as a ratio of the amount of oversize obtained to the amount of true oversize:

$$E = (O \times c)/a \tag{21-2}$$

where O = percent of oversize over the screen, c = percent coarser than the designated size in screen oversize, and a = percent coarser than the designated size in screen feed.

Other formulas for the derivation of screen efficiency are used. Taggart *(Handbook of Mineral Dressing)* gives the formula

$$E = 100 \times \frac{100(e - v)}{e(100 - v)}$$

where E is the efficiency, e is the percentage of undersize in the feed, and v is the percentage of undersize in the screen oversize.

Graphical methods of evaluating efficiency, using sieve analyses, are also employed and are recommended when serious research on screening is done.

Estimating Screen Capacity Various methods of predicting screening capacity have been proposed, and each has its limitations. The throughflow method of Matthews uses the following equation:

$$A = 0.4C_t/C_uF_{oa}F_s \tag{21-3}$$

where A = screen area
C_t = throughflow rate
C_u = unit capacity
F_{oa} = open-area factor
F_s = slotted-area factor.

The unit capacity C_u can be determined from Fig. 21-15. Figure 21-16 can be used to determine the open-area factor F_{oa}, and the slotted-opening factor F_s for various screen types is given in Table 21-7.

Probability Screening Principle Probability screening uses the fact that particles moving almost at right angles to a screening surface are not likely to pass through when the particle size is greater than about half of the distance between the screen elements. Screens utilizing the probability principle are manufacturing by Dutch State Mines (DSM), Bartles (CTS), and Morgensen. The last-named incorporates multiple decks. Higher throughput, longer screen life, and

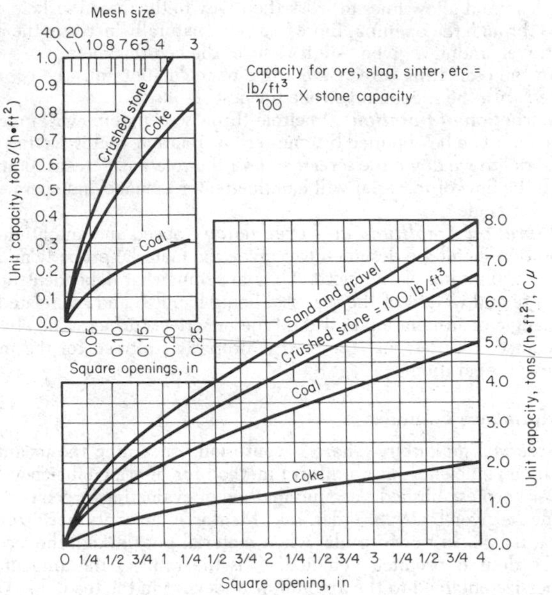

FIG. 21-15 Unit capacity (C_u) for square-opening screens. To convert inches to meters, multiply by 0.0254; to convert tons per hour–square foot to kilograms per second–square meter, multiply by 2.7182.

lower capital costs are claimed for these screening systems. The performance of several types of probability screens was reviewed by Moir (op cit.)

Constant Screen Bed Thickness Another screen design approach to improve screening performance was patented (British Patent 124,904, 1970) and is utilized in the HCC-Burstlein system. Rather than the gradually decreasing bed thickness from feed to discharge of the conventional screening system, the Burstlein system breaks up the screen surface into a number of inclined elements that are each inclined at different angles and vibrated differently to thin

Aperture	Formula	
Rectangular openings 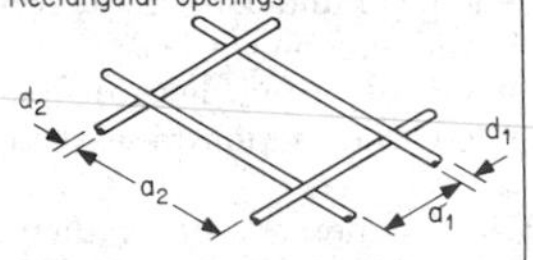	$F_{oa} = \frac{a_1 a_2}{(a_1 + d_1)(a_2 + d_2)} \times 100$; F_{oa} is open area, %; d is diameter of wire, or horizontal width of bar (for plate); a is clear opening dimension	(21-4)
Square openings: Specified by opening size	$F_{oa} = 100\left(\frac{a}{a+d}\right)^2$, $a_1 = a_2 = a$, $d_1 = d_2 = d$	(21-5)
Square openings: Specified in mesh, m	$F_{oa} = 100 a^2 m^2$, $m = \frac{1}{a+d}$	(21-6)
Parallel-rod decks 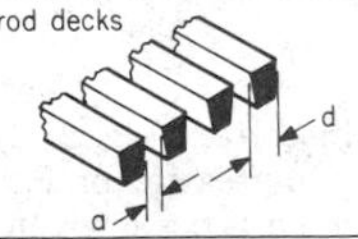	$F_{oa} = \frac{100a}{(a+d)}$	(21-7)
Special weaves 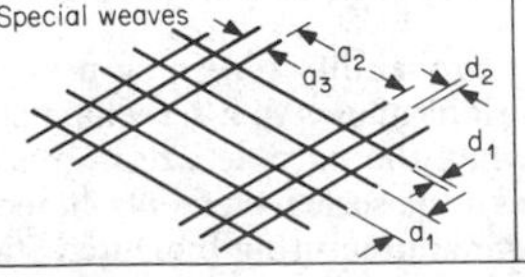	Assuming $a_3 = a_1$; $F_{oa} = 100\left[\frac{a_1(a_2 + 2a_1)}{(a_2 + 2a_1 + 3d_2)(a_1 + d_1)}\right]$	(21-8)

FIG. 21-16 Open-area factor (F_{oa}) for flow-through screen-capacity calculation.

TABLE 21-7 Slotted-Opening Factors

Screen type	Length-to-width ratio	Slotted-opening factor, E_s
Square and slightly rectangular openings	Less than 2	1.0
Rectangular openings	Equal to or greater than 2 but less than 4	1.1
Slotted openings	Equal to or greater than 4 but less than 25	1.2
Parallel-rod decks	Equal to or greater than 25	1.4

out the feed end and thicken the discharge end, thereby allowing the finer particles to penetrate the bed and reach the screening surface more uniformly. Advantages are claimed in capacity and in capital and operating costs.

Noise and Safety Noise is generated in screening plants because of the impact of the feed material on the screen surfaces. The drive mechanism also generates noise. Rubber and rubber-clad decks can reduce substantially feed-impact noise with the added benefit of longer life of the decks. Noise from the drive mechanism can be reduced by enclosing the mechanism in a box or by adding material to the side plates to dampen the noise.

Depending on the feed materials, the dust generated in a screening operation may be hazardous because of possible explosion and toxicity. These hazards must be carefully evaluated before fixing the design of the facility and selecting the apparatus.

Testing Sieves Many product specifications require definite sizes of material in terms of given percentages passing or retained on specified test sieves. Test sieves are also generally used to determine the efficiency of screening devices and the work of crushing and grinding machinery.

It is essential that standard sieves, with standard-size openings, be used for sieve analyses. The time of screening and the method of agitating the material on the sieve should also be standard, and many industries the practice of specifying the test-sieve designation and the time and method of sieving is followed. An excellent booklet on the theory and use of standard testing sieves is given in the *Testing Sieve Handbook*, No. 53, published by W. S. Tyler, Inc., Mentor, Ohio.

U.S. Sieve Series The American Society for Testing and Materials in cooperation with the National Bureau of Standards and the American National Standards Institute has further refined the U.S. sieve series, combining the former coarse and fine series into a single series with a fourth-root-of-2 ratio (Table 21-6). The openings in the individual sieves have remained unchanged except for minor adjustments in sieves coarser than 0.00673 m (6.73 mm). In the revised series, sieves 1 mm and coarser are identified by openings in millimeters, and those finer than 1 mm by their openings in micrometers.

Tyler Standard Sieve Series Many users have their standards and tests based on using Tyler standard screen scale testing sieves (Table 21-6). The only difference between the U.S. sieves and the Tyler screen scale sieves is the identification method. Tyler screen scale sieves are identified by nominal meshes per linear inch, while the U.S. sieves are identified by millimeters or micrometers or by an arbitrary number which does not necessarily mean the mesh count. The Tyler standard sieve scale series has as its base a 200-mesh screen in which the opening is 7.37×10^{-5} m (0.0029 in) and the wire diameter 5.35×10^{-5} m (0.0021 in).

International Test Sieve Series The International Organization for Standardization (ISO) has been intensifying its efforts to establish an international test sieve series. At a meeting held at The Hague in October 1959, the ISO provisionally recommended for adoption as an international standard 19 sieves as shown by the ° in Table 21-6. These sieves correspond to every alternating sieve in the fourth-root-of-2 U.S. sieve series from 0.022-m (⅞-in) opening to 325 mesh. The ISO has prepared a manual of sieving procedures that is available through the American Society for Testing and Materials.

The Ro-Tap testing **sieve shaker** (Fig. 21-17) manufactured by W.

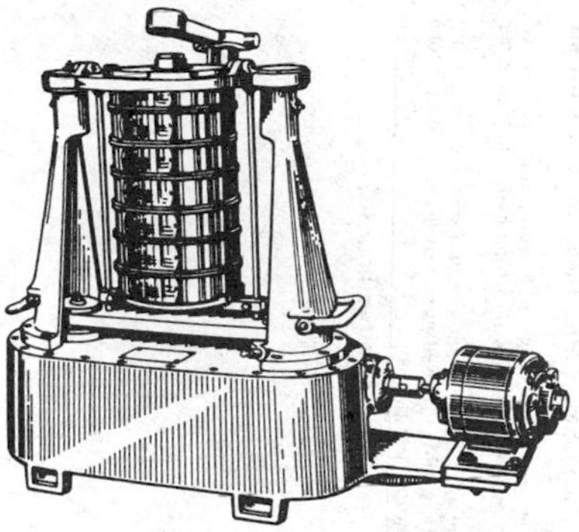

FIG. 21-17 Ro-Tap testing sieve shaker. *(W. S. Tyler, Inc.)*

S. Tyler, Inc., is the standard machine for automatically carrying out sieve-test procedures with accuracy and dependability. This device is built to hold a series of 0.203-m- (8-in-) diameter Tyler standard scale testing sieves and imparts to the sieves both a circular and a tapping motion. In effect, it reproduces the circular and tapping motion given testing sieves in hand sieving but does it with a uniform mchanical action. An important feature of the Ro-Tap is that both speed and stroke are fixed and not adjustable. This ensures the comparability between a number of sieve tests not only in a manufacturer's plant but between tests of a supplier and a customer.

The Ro-Tap is equipped to handle from 1 to 13 sieves at a time and is equipped with a timer that automatically terminates the test after any predetermined time.

Another mechanical shaker is the End-Shak, made by the Newark Wire Cloth Co. Sieves used are Newark test sieves, made to conform with the U.S. standard series.

A number of less expensive sieve shakers are on the market, such as the Dynamic, by Soiltest Inc., Chicago; the Cenco-Meinzer, by Central Scientific Co., Chicago; the Tyler portable, by W. S. Tyler, Inc., Mentor, Ohio; and also a number of electromagnetic vibratory shakers. The latter should be used only when strict comparability with other tests is not required, since it is difficult to be sure that identical intensity of vibration was present in the tests being compared.

WET CLASSIFICATION

General References: Dahlstrom, "Fundamentals and Application of the Liquid Cyclone," *Chem. Eng. Prog.*, **50,** *Symp. Ser.*, no. 15, 41 (1954). Fitch, "Calculating Terminal Settling Rates in Sedimentation," *Chem. Eng.*, **79**(17), 96 (Aug. 7, 1972). Fitch, "A Generalized Phenomenological Model for Classification," *Proc. Int. Conf. Particle Technol.*, August 1973. Gaudin, *Principles of Mineral Dressing*, McGraw-Hill, New York, 1939. Hitzrot, "A Guide to the Proper Application of Classifiers," *Min. Eng.*, **199,** 534 (May 1954). Richards and Locke, *Textbook of Ore Dressing*, 3d ed., McGraw-Hill, New York, 1940. Somasundaran (ed.), *Fine Particle Processing*, vol. 1, Society of Mining Engineers, AIME, New York, 1980. Taggart, *Handbook of Mineral Dressing*, 2d ed., Wiley, New York, 1945.

Introduction Wet classification is defined here as that art of separating the solid particles in a mixture of solids and liquid into fractions according to particle size or density by methods other than screening. In general, the products resulting are (1) a partially drained fraction containing the coarse material (called the sand) and (2) a fine fraction along with the remaining portion of the liquid medium (called the overflow).

The classifying operation is carried out in a pool of fluid pulp confined in a tank arranged to allow the coarse solids to settle out, whereupon they are removed by gravity, mechanical means, or induced pressure. Solids which do not settle report as overflow from the pool. Mesh of separation as used in this text is the screen size retaining 1½ percent of the overflow solids.

All wet classifiers depend on the difference in settling rate between coarse and fine or heavy- and light-gravity particles in the pool confined within the tank of the machine. Rates can be controlled to some extent by mild agitation, providing for hindered settling, and power versus gravity in centrifuging types of units.

Several fundamental laws on classification are:

1. Coarse particles have a relatively faster settling velocity than fine particles of the same specific gravity.
2. Heavy-gravity particles have a relatively faster settling velocity than light-gravity particles of the same size.
3. Settling rates of solid particles become progressively slower as the viscosity or density of the fluid medium increases.

a. There is a point (called critical dilution) where the lowering of density or viscosity in the pool by addition of more liquid creates a velocity effect which overcomes normal classification settling velocity, thereby coarsening the separation.

b. Conversely, at this point less liquid will cause a viscosity and buoyancy effect which will also coarsen the separation.

Typical problems to be solved by wet-classification means fall into several broad categories such as (1) to effect a simple sand-slime separation resulting in two products; (2) to effect a concentration of smaller heavy-gravity particles in a product containing larger light-gravity particles; (3) to obtain a washing effect by successive dewatering, repulping in weaker solution, and further dewatering; (4) to sort solids having a full range of screen sizes into a number of partials each having a short range of screen sizes; and (5) to achieve closed-circuit control of grinding mills.

Classification is by definition used preponderantly in the treatment of raw materials. However, these raw materials find their way into chemical processing per se and thus become of interest to the chemical engineer, particularly when the products to be treated react better when of a defined cleanliness, size, gravity, or moisture content.

Classifier types fall into three basic categories: (1) nonmechanical, (2) mechanical, and (3) hydraulic. Functionally types 1 and 2 are similar and differ only in the means of sand removal. In hydraulic types 3 the character of separation is different because of the hindered settling induced by the hydraulic water.

There are numerous machines and machine types to effect the separations and products under consideration, and there is much overlapping in the possibilities. Usually, one type will provide optimum economy for the specific problem involved.

The quick reference Table 21-8 will help by way of rapid elimination of poor possibilities. Following that the brief comments and illustrations will help pinpoint most probable selections. Further study of the more elaborate data in the references and contact with the usual suppliers are recommended, as there are many possible modifications of equipment which can improve operating results from any type of machine finally selected.

Nonmechanical Classifiers

Cone Type Cone classifiers are one of the oldest types but are still used for relatively crude work because of low cost of installation. They are limited in diameter because of high headroom requirements caused by the $\pm 60°$ sloping sides. Units are simple and are often fabricated locally with millwright ingenuity fashioning the apex opening arrangement for adjustment or control of the spigot (sand) product. Cones are not suitable for pulps having a tendency to hang up or build mudbanks. Operating attention is often necessary to a greater degree than for the more positive mechanical types. Cost figures are not available. See Fig. 21-18.

Liquid Cyclone The wet cyclone classifier has rapidly achieved prominence since the 1950s and continues to gain popularity throughout chemical and ore-dressing industries. Standout virtues are its low capital cost and ability to make extremely fine separations and to deliver a given separation at high overflow percent solids. See Fig. 21-19.

In simplest terms the unit has a top cylindrical section and a lower conical section terminating in an apex opening, often adjustable. The unit operates under pressure induced by a static hydraulic head or by means of a pump forcing new feed into the cylindrical portion tangentially, thus producing centrifuging action and vortexing. The cover has a downward-extending pipe to cut the vortex and remove

TABLE 21-8 Sizes, Limitations, and Major Applications of Wet Classification Machines

Figure	Type of classifier	Normal size range, ft.			Normal mesh of separation range*	Normal feed tonnage range	Max. oversize in feed	Normal overflow, % solids range	Normal feed density, % solids range	Normal sand product, % solids range	Motor range, hp.	Typical applications
		Width	Diam.	Max. length								
	Non-mechanical:											
21-60	Cone classifier	......	2-12	....	28-325	2-100 tons/hr.	1/4 in.	5-30	Not critical	35-60	None	For desliming and primary dewatering
21-61	Liquid cyclone	......	10 mm. to 4 ft.	9	35 mesh to 5 μ	1/2-1500 gal./min.	14-325 mesh	5-30	10-60	55-70	Power for pressure head 5-60 lb./sq. in.	For medium or fine separations and closed-circuit grinding
	Mechanical:											
21-62	Drag classifier	1-10	....	Not critical	28-200	5-350 tons/hr.	1 1/2 in.	5-30	Not critical	70-83	1-10	For desliming, conveying, and closed-circuit grinding
21-63 and 21-64	Rake and spiral classifiers	1-20	....	40	20-200	5-350 tons/hr.	1 in.	5-30	Not critical	75-83	1/2-25	Closed-circuit grinding, washing and dewatering, desliming, process feed control
21-65	Bowl classifier	1 1/2-20	4-28	40	100-325	5-200 tons/hr.	1/2 in.	5-25	Not critical	75-80	Bowl: 1-7 1/2 Rake: 1-25	Closed-circuit grinding usually in secondary circuits
21-66	Bowl desiltor	4-16	20-50	40	100-325	5-250 tons/hr.	1/2 in.	1-15	Not critical	75-83	Bowl: 1-10 Rake: 5-25	Recovery of fine sand, limestone, coal, and fine phosphate rock from large flow volumes
21-67	Hydroseparator	......	10-150	....	100-325	5-700 tons/hr.	1/4 in.	1-20	Not critical	30-50	1-15	For fine separation where large feed volumes are involved and drainage not critical
21-68	Solid-bowl centrifuge	......	18-54 in.	70 in.	200 mesh to 1 μ	10-600 gal./min.	1/4 in.	1-40	5-50	10-70	15-150	For fine-size fractionating
21-69	Countercurrent classifier	......	1 1/2-10	40	35-100	1-600 tons/hr.	3 in.	5-30	Not critical	75-83	1/4-25	Sand-slime separations, washing, closed-circuit grinding
	Hydraulic:											
21-70	Jet sizer	1 1/2-20	....	5-20	8-150	2-100 tons/hr.	3/16 in.	1-10	30-60	40-60	1-2 for air pressure	Multiproduct unit for exceptionally clean sands fractionated into narrow size ranges. Min. 3 tons hydraulic water per ton sand
21-71	SuperSorter†	6	....	40	8-150	40-150 tons/hr.	3/8 in.	1-10	30-60	40-60	1 to operate pincer valves	Multiproduct unit for exceptionally clean sands fractionated into narrow size ranges. Min. 3 tons hydraulic water per ton sand
21-72	SiphonSizer‡	......	3-30	....	14-150	1-100 tons/hr.	1 in.	1-10	30-60	40-60	None	Two-product unit efficient for desliming and exceptionally clean sands, washing, closed-circuit grinding. Min. 2 tons hydraulic water per ton sand

NOTE: Equipment can be provided in stainless steel and rubber-covered steel for corrosion resistance. To convert feet to meters, multiply by 0.3048; to convert tons per hour to kilograms per second, multiply by 0.2520; to convert inches to meters, multiply by 0.0254; and to convert horsepower to kilowatts, multiply by 0.7457.

*Size of screen retaining 1 1/2% of the overflow solids.

†Trademark of Deister Concentrator Company, Inc.

‡Trademark of Dorr-Oliver Inc.

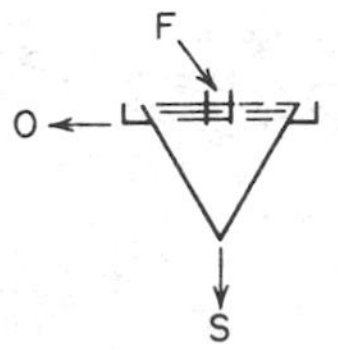

FIG. 21-18 Cone classifier. F = feed; O = overflow product; S = sand product.

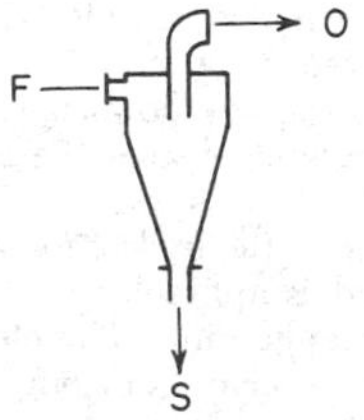

FIG. 21-19 Liquid cyclone.

FIG. 21-20 A cluster of liquid cyclones. *(Courtesy of Dorr-Oliver Inc.)*

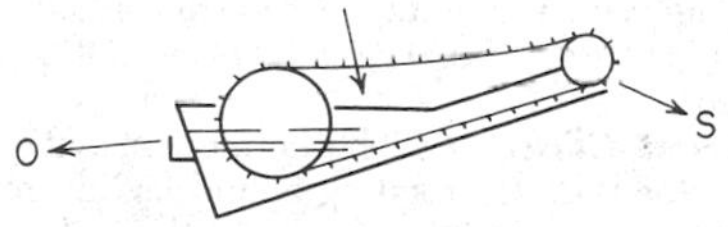

FIG. 21-21 Drag classifier.

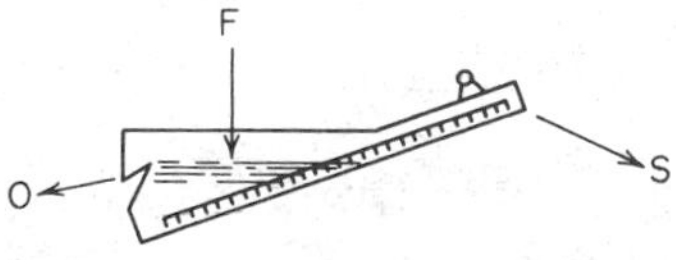

FIG. 21-22 Rake classifier.

the overflow product. Coarse solids travel down the sides of the steeply sided cone section and are removed in a partially dewatered form at the apex.

Liquid cyclones are available in numerous sizes and types ranging from pencil-sized 10-mm diameters of plastic or aluminum oxide to the 1.2-m (48-in) diameter of rubber-protected mild or stainless steel. Porcelain units 25 to 100 mm (1 to 4 in) in diameter are becoming popular, and in the 150-mm (6-in) size the starch industry has standardized on special molded nylon types. Small units for fine-size separations are usually manifolded in multiple units in parallel with up to 480 ten-mm cyclones in a single case. Larger sizes may be used singly or manifolded by outside piping as shown in the photograph of Fig. 21-20.

The liquid cyclone is largely replacing classifiers in closed-circuit grinding.

Typical uses more in line with chemical applications are degritting milk of lime and of red mud in alumina production, removal of carbonaceous material in upgrading gypsum produced in making phosphoric acid, open-circuit washing of fine uranium pulps, classification of crystal magma such as lactose and sodium bisulfite, and classifying pigment and plastic beads into size ranges.

Costs run from a few dollars each for single 10-mm units to many thousands for multiples in closed housings and from a few hundred dollars to \$6000 for the conventional 76-, 152-, 305-, and 610-mm- (3-, 6-, 12-, and 24-in-) diameter ore-dressing types. Unless static head is available the cost of the pumping system must be added.

Mechanical Classifiers

Drag Classifiers Single endless-belt or chain suspensions with cross flights running in an inclined trough have long been used for draining and classifying. Many styles, sizes, and shapes have resulted from locally built units, and operating results on a scientific basis are meager. In general, they have served their purpose consistent with the type of engineering and cost included. See Fig. 21-21.

The Hardinge Overdrain° classifier is of the belt type, but it embodies the innovation of allowing entrapped water and slimes to escape through holes in the belt just uphill of the cross flights in an upward direction and thence flow down on top of the belt into the pool without again intermingling with the sand product being advanced by the cross flights. Sands with lower moisture and fines content result from this action. Modern design and materials of construction permit sizes up to 3 m (10 ft) wide and 12.5 m (41 ft) long on steeper than average slopes and for very high tonnages.

Approximate FOB factory costs range from \$10,000 to \$15,000 per foot of belt width, including tanks but without drives or accessories.

Rake and Spiral Classifiers (Figs. 21-22 and 21-23) Rake-type classifiers such as the Dorr† classifier (shown in Fig. 21-24) and spiral types such as the Akins‡ have been the workhorses for general-classification problems for half a century, and their names describe the mechanisms installed in sloping-bottom tanks. Mechanically the machines are powerfully built, and functionally they are versatile and flexible. They were the first classifiers used successfully for closed-circuit grinding. Separations as fine as 325 mesh can be accomplished at reduced tonnage rates.

Control of water into the classifiers is important since separation into overflow and sand products is made largely by the buoyancy, viscosity, and degree of agitation in the pool.

°Trademark of Koppers Co., Inc.
†Trademark of Dorr-Oliver Inc.
‡Trademark of Mine & Smelter.

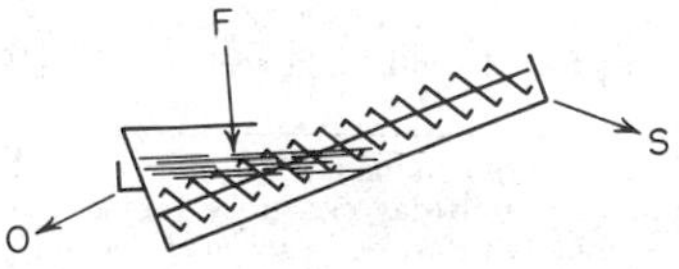

FIG. 21-23 Spiral classifier.

FIG. 21-24 Rake classifier. *(Courtesy of Dorr-Oliver Inc.)*

Both types of machines will produce rake products of consistent moisture content even with considerable variation in feed tonnage or volume. Operating costs including maintenance and labor run from 2 to 15 mil/t of sand raked depending upon the abrasiveness of the material. For additional operating and cost data see Table 21-9.

Bowl Classifier The bowl classifier (Fig. 21-25) was developed to provide more pool area necessary for fine separations consistent with high tonnage. In essence a shallow bowl with revolving plows is superimposed over a rake or screw dewatering section. Feed enters at the center of the bowl, and fine solids overflow at the periphery. Coarse solids collected on the bowl bottom are raked to the center for discharge into the dewatering compartment below where wash water may be added for counterflow.

Liquid cyclones are rapidly taking over the functions formerly handled by bowl classifiers because of lower capital costs and floor-area requirements. However, when drained sand products are wanted, there is still a place for bowls in ore-dressing practice.

Costs including tanks and drives range from $6000 to $9000 per foot of bowl diameter depending upon raking capacity needed.

Bowl Desilter The bowl desilter (Fig. 21-26) provides for pool surface areas well beyond areas possible in bowl classifiers, in which larger sizes are limited by mechanical design. Its use is in operations involving large flow volumes and fine separations. Rake tonnages can be great or small with a dewatering compartment to suit the conditions.

In the bowl desilter the rotating blades in the bowl plow outward and discharge settled coarse material at the periphery, where it drops into the drainage compartment. This configuration does away with the long cantilevered rake construction necessary in bowl classifiers.

Widest application has been for the recovery of and drainage of very fine material overflowing coarser washing units in glass sand, concrete sand, coal, and limestone processing plants.

Costs including steel tanks and drives range from $5000 to $7500 per foot of bowl diameter, depending upon the raking capacity needed.

TABLE 21-9 Approximate Operating and Cost Data for Rake and Spiral Classifiers*

Characteristic	Light duty	Heavy duty
Typical size, ft	4 by 20	8 by 30
Raking capacity, 2.7 sp gr solids, tons/day per stroke/min	25	320
Maximum speed, strokes/min	30	25
Approximate weight, lb	1250	3200
Approximate cost per foot of width, mild steel, FOB factory	$10,000	$15,000

*1981 costs. To convert feet to meters, multiply by 0.3048; to convert tons/day per stroke/minute to tons/day per stroke/second, multiply by 66.08; to convert strokes per minute to strokes per second, multiply by 0.0167; and to convert pounds to kilograms, multiply by 0.454.

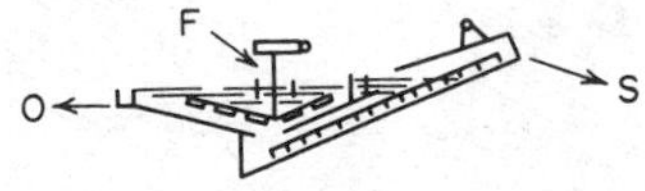

FIG. 21-25 Bowl classifier.

Hydroseparator The hydroseparator (Fig. 21-27) is merely a thickener-type machine receiving more flow than can be clarified in the area provided. Thus the overflow contains fine solids, and the greater the feed rate per unit of area the coarser the solids in the overflow.

Classification efficiency of the hydroseparator compares with that of the cone classifier and is appreciably lower than that obtained from mechanical or hydraulic units. The chief virtue of the hydroseparator is its ability to receive and slough off great quantities of water at low per-unit-volume cost.

Typical applications include primary dewatering of phosphate rock matrix and silica sand products following wet screening. In ore dressing it is used mainly to protect large-diameter thickeners by scalping out +65-mesh material.

Mechanism costs range from $110 to $270 per square meter ($1000 to $2500 per 100 ft^2) of pool area, not including tanks, which may be steel, wood, or concrete.

Solid-Bowl Centrifuge The Bird solid-bowl centrifuge (Fig. 21-28) uses power instead of gravity and can develop centrifugal forces up to 1800 times the force of gravity. It is therefore a unique type in classification practice.

The unit consists essentially of two rotating elements, the outer being a solid-shell conical-shaped bowl and the inner comprising a helical-screw conveyor revolving at a speed slightly lower than that of the bowl. Raw feed slurry is delivered through a stationary feed pipe to the conveyor, where, urged by centrifugal force, it is transferred to the revolving bowl. A circumferential classifying pool is formed and contained at the larger diameter of the cone shell. The ports for oversize material are located closer to the axis of rotation than the ports for the overflow to effect a beach line and drainage.

Centrifugal force deposits the oversize particles against the bowl wall, from which they are conveyed by the helix. The overflow fractions flow around the helix to the liquid-discharge ports. Size of separation is controlled by feed rate and degree of centrifugal force.

Several prime features of this totally enclosed unit are its high capacity per unit of floor area, small volume of material in process, high degree of separation, and shear action for dispersion of solids. Typical applications are desliming to upgrade cement rock, sizing of abrasives, fractionating for reagent control, and classification of pigments.

Costs and weights including motors run from $75,000 and 1500 kg (3300 lb) to $275,000 and 11,000 kg (22,000 lb) FOB factory in plain-steel construction.

Countercurrent Classifier The countercurrent classifier (Fig. 21-29) is an inclined, slowly rotating cylindrical drum with continuous spiral flights attached to the interior of the shell forming helical troughs. Direction of rotation is such that material in the troughs is impelled toward the higher end. The lower end of the shell is closed except for a central overflow opening. Attached to the upper end is

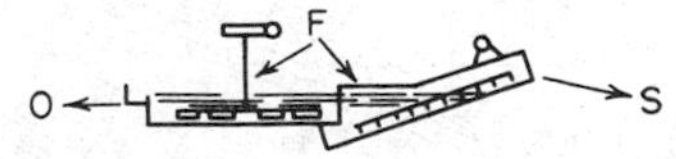

FIG. 21-26 Bowl desilter.

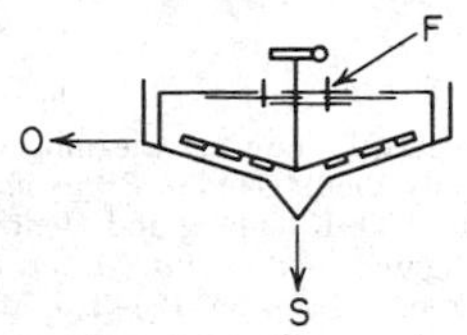

FIG. 21-27 Hydroseparator.

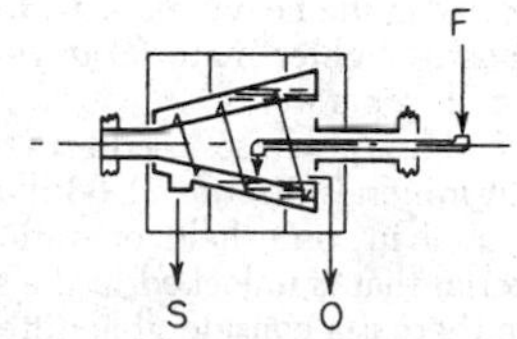

FIG. 21-28 Solid-bowl centrifuge.

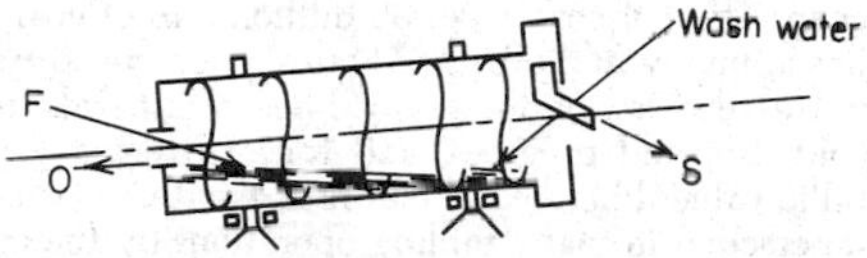

FIG. 21-29 Countercurrent classifier.

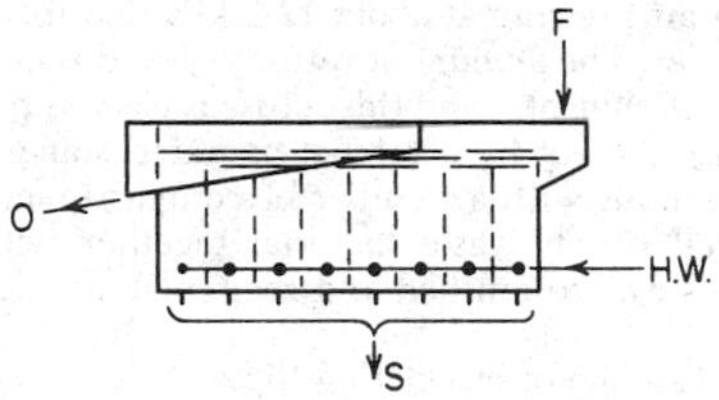

FIG. 21-30 Jet Sizer.

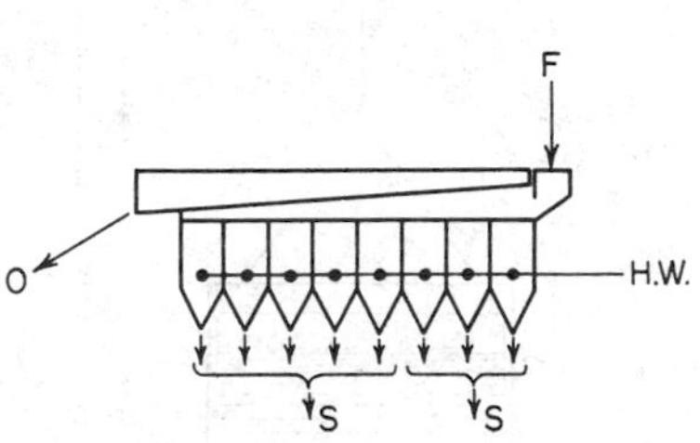

FIG. 21-31 SuperSorter.

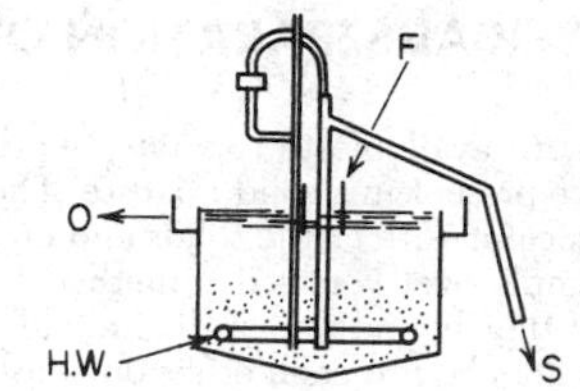

FIG. 21-32 D-O SiphonSizer.

a sand-dewatering elevator which rotates with the shell. Wash water introduced at the upper end drains from the lifting flights above the normal water level and progresses countercurrently to the sand toward the overflow.

Usual application is for sand-slime separations, washing and for closed construction restricting escape of heat and chemical fumes, easy start-up after shutdown, and general simplicity.

Weights range from 500 to 55,000 kg (1100 to 120,000 lb) and approximate costs FOB factory range from \$4.50 to \$1.60 per kilogram (\$10 to \$3.50 per pound), including drives.

Hydraulic Classifiers

***Jet Sizer** and SuperSorter.†** The Jet Sizer (Fig. 21-30) and SuperSorter (Fig. 21-31) are multicompartment and, therefore, multiproduct classifiers operating on the basis of hindered settling. The classification pockets are arranged in series for throughflow with parallel pockets to take care of high tonnage fractions in the range of sizes. Each compartment is served with low-pressure hydraulic water.

Hydraulic classification ensures the highest separating efficiency obtainable by wet-classification means. The amount of hydraulic water is controlled so that in each succeeding compartment the coarsest particles are maintained in hindered-settling condition and the finer fractions pass along for similar treatment. Two compartments will normally capture 90 percent of a two-screen-size fraction. Spigot discharge is controlled by air-actuated valves in the Jet Sizer and motor-driven pincer-type valves in the SuperSorter. Sand fractions can be taken from single or combinations of compartments as desired.

Typical applications include careful sizing of silica-glass sand, washing phosphate rock, sizing of abrasives, smokeless powder, sodium aluminate, etc.

Jet Sizer costs average approximately \$10,000 per compartment.

D-O SiphonSizer* The D-O SiphonSizer (Fig. 21-32) is a high-efficiency hydraulic classifier developed originally for the washing and sizing of phosphate rock. In ore-dressing work it is normally a two-product unit; but by use of an upper column sealed at the top and open at the bottom, three products are possible: coarse, intermediate, and fine fractions.

Feed to be sized is put into hindered-settling condition by hydraulic water in quantity only sufficient to teeter the smallest particle wanted in the sand product. Thus the sand will contain all solids coarser in size, and the finer fractions report to the overflow or pass into the upper column for removal in a three-product unit.

Sands are discharged by siphons extending to the bottom of the hindered-settling zone. Siphon control is obtained by a novel hydrostatically actuated valve which makes or breaks the siphon to flow only when the teeter zone is in correct condition. Discharge by an intermediate fraction from the upper column is by means of additional siphons. Hydraulic-water consumption is considerably lower than required for multipocket sizers.

SiphonSizers vary so widely in configuration that general cost data are not meaningful.

*Trademark of Dorr-Oliver Inc.
†Trademark of Deister Concentrator Company, Inc.

SELECTIVE PHYSICAL SEPARATION OF MIXED SOLIDS

Various techniques are available to separate the different types of particles that may be present in a solid mixture. The choice depends on the physicochemical nature of the solids and on site-specific considerations (for example, wet versus dry methods). A key consideration is the extent of the "liberation" of the individual particles to be separated. Particles attached to each other obviously cannot be separated by direct mechanical means except after the attachment has been broken. In ore processing, the mineral values are generally liberated by comminution. Rarely is liberation complete at any one size, and a physical-separation flow sheet will incorporate a sequence of operations that often are designed first to reject as much unwanted material as is possible at a coarse size and subsequently to recover the values after further size reduction. Any difference in physical properties of the individual solids can be used as the basis for separation. Differences in density, size, shape, color, and electrical and magnetic properties are used in successful commercial separation processes. An important factor in determining the techniques that can be practically applied is the particle-size range of the mixture. A guide to selection of the generally used separation methods was given earlier in Fig. 21-1. It can be noted that most physical separation methods lose their effectiveness as the particle size decreases. For particle sizes below about 0.10 mm (10 mesh) few conventional selective separation techniques are available. Surface chemical differences become the important basis for separation (flotation). Magnetic-property differences are also used effectively in the finer-particle-size solid systems.

Physical separation methods are most widely used for the processing of coal and ore materials, and their basic development was designed for that purpose. Tremendous tonnages of solids are processed routinely at costs often as low at $1 per ton of material separated. The methods are applicable for other than ore processing, and solid-separation technology has become a more integral part of chemical-process operations. Recent requirements to recover values from various solid wastes have emphasized the need to adapt the relatively low cost physical separation techniques of the ore processor, and as the needs to treat new types of materials and to improve recovery efficiency are constantly increasing, new designs are being developed. The methods most commonly used are presented first. A brief discussion of more specialized methods is also provided.

Gravity concentration is one of the oldest of the solids-separation techniques and the most important mineral-dressing method for obtaining ore concentrates. It is used mainly now for coal cleaning, yet Mills ["Process Design, Scale-Up and Plant Design for Gravity Concentration," in Mular and Bhappu (eds.), *Mineral Processing Plant Design*, 2d ed., Society of Mining Engineers, AIME, New York, 1980] notes that still more tonnage and greater values of materials are concentrated by gravity methods than by a method such as froth flotation. The major unit operations which comprise gravity separation are jigging, tabling, spiral concentration, and dense-media separation. For high-capacity treatment of finer-sized low-grade ore materials, particularly the heavy mineral sands, the Reichert cone is becoming an industry standard [Ferree, "An Expanded Role in Minerals Processing Is Seen for the Reichert Cone," *Min. Eng.*, **25**(3), 29 (1973)].

The key to the effective use of gravity methods is knowledge of the inherent association of the solid particles to be separated. Because the different methods and apparatus perform best in different size ranges, it is necessary to know the degree of liberation as a function of size for the feed material of interest. In addition, there are limitations due to specific-gravity differences. Laboratory sink-float-test work done with heavy liquids covering a range of specific gravities and using different-sized fractions is essential to determining the limitations of any proposed gravity separation system (see "Dense-Media Separation").

JIGGING

GENERAL REFERENCES: Aplan, "Gravity Concentration," in Kirk and Othmer (eds.), *Encyclopedia of Chemical Technology*, vol. 12, 3d ed., Wiley, New York, 1980, pp. 1–29. Bogert, "Fine Coal Cleaning with the Feldspar Jig," *Min. Congr. J.*, **46,** 42 (July 1960). Gaudin, *Principles of Mineral Dressing*, McGraw-Hill, New York, 1939. Kirchberg and Hentzschel, "A Study of the Behavior of Particles in Jigging," *Trans. Int. Miner. Dressing Congr.*, 1957, Almquiste, Wiksell, Stockholm, 1958, pp. 193–215. Marincel, "Use of Jigs in the Concentration of Iron Ores," 14th Annual Mining Symposium, University of Minnesota, Minneapolis, 1953. Mayer, "Fundamentals of Potential Theory of Jigging," 7th International Mineral Processing Congress, New York, 1964. Pryor, *Mineral Processing*, 3d ed., Elsevier, New York, 1965. Taggart, *Handbook of Mineral Dressing*, 2d ed., Wiley, New York, 1945.

Introduction A jig is a mechanical device used for separating materials of different specific gravities by the pulsation of a stream of liquid flowing through a bed of materials. The liquid pulsates, or "jigs" up and down, causing the heavy material to work down to the bottom of the bed and the lighter material to rise to the top. Each product is then drawn off separately.

Jigging is one of the oldest processes used for concentrating heavy mineral from the light minerals. Figure 21-33 illustrates a primitive hand jig device still used in some field operations. Jigging is best suited for coarse material that is unlocked in the size range 20 mesh and coarser and when there is a considerable difference between the effective specific gravity (sp gr mineral minus sp gr water) of the valuable and the waste material. Jigs are simple in operation. Water consumption is high, and the tailings losses on metallic ores are usually high. Also, because of the scarcity of still-available ore deposits having coarse mineralization, the jigs are used to a limited extent, mostly to treat iron ores, a few lead-zinc ores, and some heavy nonmetallic ores like barite and diamonds. Jigging is widely employed for the concentration of coal. Over 50 million tons of coal is concentrated by jigs annually in the United States. High-speed types of jigs are used for the recovery of fine-grained heavy minerals from placer deposits, gold, tin, and tungsten; and for recovering a portion of coarse metallic values liberated in ball-mill grinding circuits. Jigging has been superseded in many milling operations by the adoption of the dense-media process or by fine grinding followed by flotation.

Types of Jigs There are two principal types of jigs. In the first type, the sieve is stationary, and water is forced up through the screen.

A modern form of the fixed-sieve-type jig is the Jeffrey air-operated Baum jig shown in Fig. 21-34, which is used extensively in coal washing. In this jig the pulsations are caused by alternately applying and exhausting air pressure at about 17.2 kPa (2.5 lbf/in^2) from the pulsation chamber. The amount of refuse rejected is controlled automatically by a "flash float," and this refuse is ejected positively from the screen compartment by a ratchet-operated star gate. Such jigs customarily are built with a number of compartments. Each compartment or cell rejects waste material together with some coal. These middlings can be crushed and recirculated, and some of the coal recovered.

A further advantage of circulating these middlings is to raise the density of the material in the jig bed, giving the effect of a heavy-medium process and thus sharpening the separation. A mechanical jig that has come into widespread use since 1950 is the Wemco-

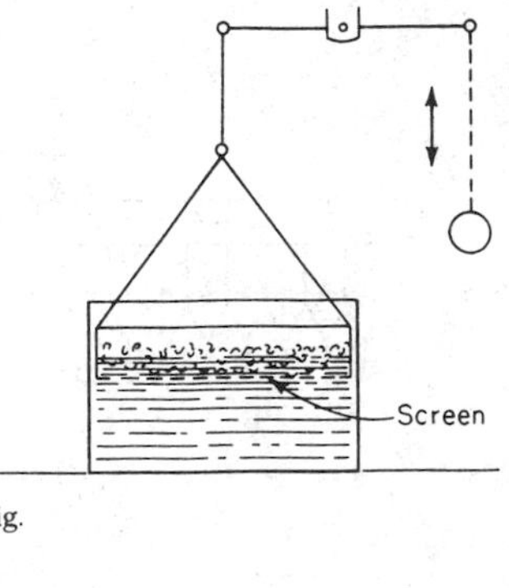

FIG. 21-33 Hand jig.

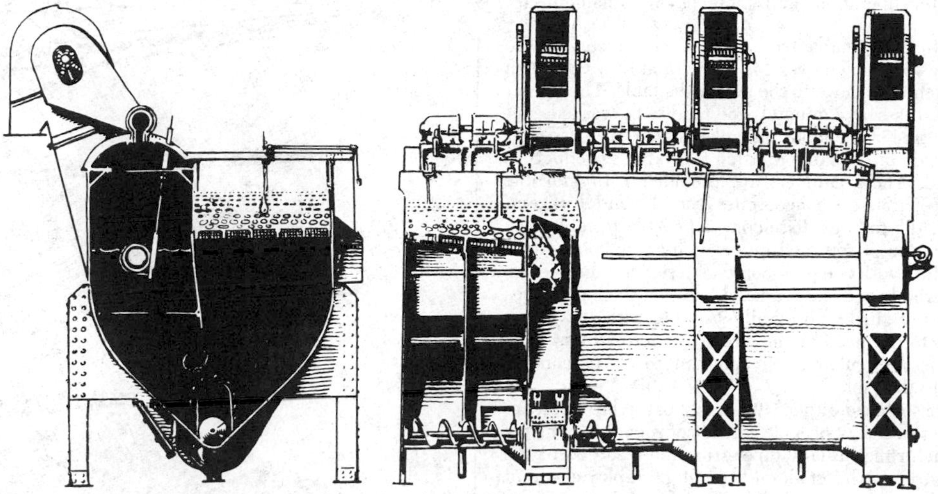

FIG. 21-34 Jeffrey (Baum-type) coal jig.

Remer jig. This jig is used mainly in concentrating materials such as iron and barite ores and in removing impurities such as wood, shale, and lignite from sand and gravel. These jigs are unusual in that they use a stroke of approximately 9.5 mm (⅜ in) at 2 Hz (150 r/min) with a secondary motion superimposed having a 1.6-mm (1/16-in) stroke with a frequency of 6 to 7 Hz (400 strokes/min). Jigs of the type in which the sieve moves up and down such as the Hancock jigs are now in little use. A type of high-speed jig in rather common use is the Denver mineral jig. This is usually a two-compartment all-steel diaphragm-actuated jig. This jig is sometimes used in grinding circuits to recover heavy minerals as soon as they have been liberated. These minerals are recovered as a hutch product. This jig is used mostly in the treatment of gold, tungsten, and chromite ores.

Jig Feed In coal washing jigging is practiced on unsized material as coarse as 175 mm (7 in). In metal-milling practice jigging is now seldom employed on material coarser than 20 mm (¾ in). Float-and-sink methods have largely superseded jigs as a way of concentrating metallic ores in the minus 75 to plus 10-mm (3 to plus ½-in) range. Shaking tables usually are considered more efficient than jigs for treating ores finer than 2 mm (10 mesh). Jigs are used in some plants to obtain flow-sheet simplicity since they can handle a wide range of sizes. Jigs, except when extremely heavy minerals such as gold, galena, cassiterite, or tungsten minerals are treated, recover only a small percentage of the sizes finer than 65 mesh (¼ mm).

Capacity The Jeffrey-Baum will treat minus 100-mm (4-in) coal at the rate of 8 kg/(s·m^2) [3 tons/(h·ft^2)] of active screen area. For fine sizes capacity decreases. A standard 1.52- by 4.87-m (5- by 16-ft) Wemco-Remer jig will treat minus 9.5-mm (⅜-in) iron ore at the rate of 7.5 to 11.3 kg/s (30 to 45 tons/h). A Cooley jig, a variation of the Harz jig consisting of six compartments 1.07 by 1.22 m (42 by 48 in), will handle 6 to 7.5 kg/s (25 to 30 tons/h) of minus 19-mm (¾-in) Mid-Continent zinc ore. The largest commercially available jig is the IHC Cleveland 25, a circular jig of 7.5-m (24.6-ft) diameter with a nominal capacity range of 30 to 60 kg/s (130 to 260 tons/h) of coal.

Power Requirements The power required in jigging depends on the screen area, the size of material treated, the percentage of opening in the jig screen, the depth of the bed, the length of stroke, and the choke frequency. The power required for plunger-type jigs treating 12.7-mm (½-in) material is about 7 W/m^2 (0.1 hp/ft^2) jig screen surface.

Water Consumption Jigs require much water. In most installations, the Harz-type jig uses 0.006 to 0.01 m^3 water/kg (1500 to 2500 gal/ton) material treated. Water requirements for treating minus 10-mm (⅜-in) iron ore in a Wemco-Remer rougher-cleaner jig circuit are approximately 0.005 m^3 water/kg (1200 gal/ton) of material processed.

Cost The direct operating cost of jigging depends on the nature and size of material to be treated, the number of jigging stages required, and the size of the plant. In large-tonnage plants treating coal or iron ore, this unit cost will vary from 10 to 50 cents per ton of feed.

TABLING

GENERAL REFERENCES: "Automation Keys Two Stage Precision Washing at Moss No. 3," *Coal Age*, **64**, 80 (July 1959). Burdick, "Beneficiation of Scheelite Ores by Gravity Concentration," *Min. Technol.*, **6**(6), 1 (1942). Coghill, DeVaney, Clemmer, and Cooke, *Concentration of Potash Ores of Carlsbad, N.M., by Ore Dressing Methods*, U.S. Bur. Mines Rep. Invest. 3271, 1935. Dickson, Trepp, and Nichols, "Virginia Plant Concentrates Sulphide Ore with Air Tables," *Eng. Min. J.*, **160**(4), (April 1959). Gaudin, *Principles of Mineral Dressing*, McGraw-Hill, New York, 1939. Kirchberg and Berger, *Trans. Int. Miner. Process. Congr.*, London, 1960, p. 537. "Linka Mill Added to Nevada WO_3 Output," *Min. World*, **18**, 52 (June 1956). McLeod, "Tungsten Milling and Current Metallurgy at Canadian Exploration Limited," *Can. Min. Metall. Bull.*, **50**, 137 (March 1957). Mitchell, "The Recovery of Pyrite from Coal Mine Refuse," *Min. Technol.*, **8**(4), 2 (1944). Norman and O'Meara, *Froth Flotation and Agglomerate Tabling of Mica*, U.S. Bur. Mines Rep. Invest. 3558, 1941. O'Meara, Norman, and Hammond, "Froth Flotation and Agglomerate Tabling of Feldspars," *Bull. Am. Ceram. Soc.*, **18**, 286 (1939). Stockett, "Milling Practice of the St. Joseph Lead Co.," *Min. Technol.*, **7**(3), 1 (1943). Taggart, *Handbook of Mineral Dressing*, 2d ed., Wiley, New York, 1945. "Upgrading Fragile Coal to Premium Metallurgical Product," *Coal Age*, **64**, 94 (July 1959).

Wet Tabling Tabling is a concentration process whereby a separation between two or more minerals is effected by flowing a pulp across a riffled plane surface inclined slightly from the horizontal, differentially shaken in the direction of the long axis, and washed with an even flow of water at right angles to the direction of motion. A separation between two or more minerals depends mainly on the difference in specific gravity between the minerals and to a lesser degree on the shape and size of the particles. The process is best suited for the concentration of ore and coal where there is a considerable difference between the effective specific gravity (sp gr mineral minus sp gr water) of the valuable and the waste material. Tables treat metallic ores effectively in the size range from 6 to 150 mesh but can be used to treat lighter materials such as coal of a considerably larger size.

Tabling is best suited for the treatment of material containing only one valuable mineral that is free at a granular size and when a considerable difference exists between the effective specific gravities of the mineral constituents. Flotation has been found to be best in treating complex ores containing several valuable minerals, those requiring fine grinding for liberation, and those having small gravity differentials.

The heaviest particles in a table feed are the least affected by the current of water washing down over the tables, and they collect in the riffles along which they move to the end of the table. The lighter materials ride above the heavy minerals and tend to be washed over the riffles to the low side of the table. Suitable launders are placed at the end of the low side of the table to catch the various products as they are discharged. These launders are provided with movable dividing devices to separate the concentrates from the middlings and the middlings from the tailings. It seldom is possible in tabling to make a sharp separation of the feed into a high-grade concentrate and a low-grade tailing with one pass. Some material of intermediate grade is almost invariably present as a band between these products, and it is customary to return such middlings either with or without additional grinding to the head of the circuit for retreatment. The amount of middling recirculated may amount to 25 percent of weight of the feed to the table.

Tables usually are surfaced either with heavy battleship linoleum or with rubber. The riffles may be a clear grade of sugar pine or may be rubber strips. Such riffles usually taper from the feed end of the table to the discharge end. Almost all mill operators employ different styles of riffling table, which they believe best for their particular separations. The usual method of riffling is shown in Fig. 21-35.

If the object of tabling is to produce as clean a concentrate as possible, a diagonal area in the upper discharge side corner is left unriffled. This area is known as the cleaning deck. If the table is to be used in making only a rough concentrate and a finished tailing, the riffling is extended by many operators. Tables are provided with adjustable tilting devices so that the transverse slope may be varied. The head motion is such that the deck reverses its direction with a maximum velocity at one end and a minimum velocity at the other end of the stroke. It is the quickness of the return that causes the material to migrate toward the discharge end. The length of stroke may be adjusted. This will vary from 0.03 m (1¼ in) for coarse material to 0.01 m (½ in) for fines. Modern tables operate from 4 Hz (270 strokes/min) for coarse to 6 Hz (350 strokes/min) for fines.

Present table practice is to use multiple decks. Multiple-deck tables consisting of from two to three decks effect space saving proportionate to the number of decks employed. They also have the advantage in that no heavy floor supports need be supplied since such tables are supported by suspended mountings. Multiple-deck installations reduce capital expenditures since a single motor and less piping and fewer launders are required than for a comparable number of single-deck installations. A two-deck configuration is shown in Fig. 21-36.

General information for standard-size tables operating on various-sized feeds is shown in Table 21-10. The No. 6 table of the Deister Concentrator Company, Fort Wayne, Indiana, has a diagonal deck approximately 1.83 m (6 ft) wide and 4.27 m (14 ft) long. The No. 7 table used primarily for coal work is approximately 2.44 m (8 ft) wide and 4.88 m (16 ft) long. The figures given apply to single-deck installations. In modern practice, each table, whether it be a single-deck or a multiple-deck installation, is driven by a single motor which is connected to the actuating mechanism by a V-belt drive. The installed horsepower for the large No. 7 deck is 1120 W (1.5 hp).

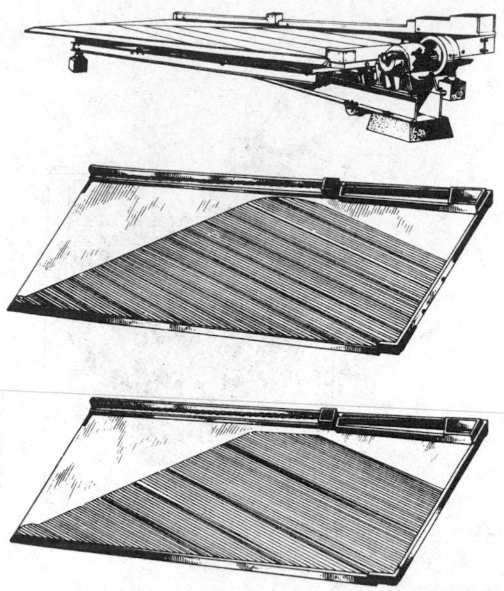

FIG. 21-35 Deister-Overstrom diagonal deck table. Center, diagonal deck with pool riffle system for sand; bottom, diagonal deck with pool riffle system for fine sand and slime.

FIG. 21-36 Two-deck concentrating table. *(Courtesy Diester Concentrator Company, Inc.)*

TABLE 21-10 Generalized Operating Data for Superduty Diagonal-Deck Concentrating Table

Table No.	Feed	Feed size	Feed capacity, tons/hr.	Speed, r.p.m.	Stroke, in.	Water with feed, gal./min.	Dressing water, gal./min.	Size of deck
6	Ore	¼ in.–35 mesh	2.0 –10.0	275	1.25	30–150	10–100	6′5″ × 14′1″
6	Ore	35–150 mesh	1.0 – 2.5	285	0.75	16– 40	5– 20	6′5″ × 14′1″
6	Ore	Minus 150 mesh	0.25– 1.0	300	.50	3– 12	3– 10	6′5″ × 14′1″
7	Coal	1½ in.	15.0–25.0	270	1.25	125–210	55– 90	8′¼″ × 16′9¼″
7	Coal	¾ in.	10.0–15.0	280	1.00	60– 85	20– 35	8′¼″ × 16′9¼″
7	Coal	½ in.	7.5–12.0	285	1.00	42– 65	18– 31	8′¼″ × 16′9¼″
7	Coal	⅛ in.	5.0– 7.5	290	0.75	28– 42	12– 18	8′¼″ × 16′9¼″
7	Coal	1/16 in.	3.0– 5.0	290	.75	15– 28	9– 12	8′¼″ × 16′9¼″

NOTE: To convert inches to meters, multiply by 0.0254; to convert tons per hour to kilograms per second, multiply by 0.252; to convert revolutions per minute to hertz, multiply by 0.0167; to convert gallons per minute to cubic meters per second, multiply 6.309×10^{-5}; and to convert feet to meters, multiply by 0.3048.

A comparable figure for the smaller No. 6 deck is 746 W (1 hp) per deck. The actual power consumed in operation is somewhat less.

An essential factor for good table operation is that the rate of feed must be uniform, both as to tonnage and as to physical properties. No one factor will cause more trouble to the table operator than to have a surging feed. The feed to tables may be unsized, or it may be either screened or hydraulically classified. For treating fine coals a common procedure is to use hydrocyclones both to deslime the material and to give a cyclone underflow of about 40 percent solids, which constitutes the table feed.

Tabling is a relatively cheap operation. If the feed is uniform, one operator can take care of many tables. In a modern coal plant with multiple-deck tables, a single operator can handle the tabling of as much as 300 kg/s (1200 tons/h). In an ore-tabling plant such as a lead or zinc operation, a table operator can watch as many as 50 tables with a total capacity in the order of 50 kg/s (200 tons/h). Labor is the principal item of cost. Power requirements and maintenance are both low. The installed cost of a table including supports and launders is from $8000 to $15,000 per deck. In the past, one of the disadvantages of table installation was the relatively large floor space required for the tonnage treated. This disadvantage has now largely been overcome by the use of multiple-deck tables. Their main advantage is that, in the size range for which they are suited, tabling is a cheap and effective method of concentrating simple ores and coal.

Dry Tabling Tabling may be done dry as well as wet, and for such use tables of special design are used. The Sutton, Steele and Steele table is an example of this type of equipment. It has a shaking motion somewhat similar to that of a wet table, except that the direction of motion is inclined upward from the horizontal, and instead of water acting as the medium of distribution, a blast of air is driven through a perforated deck. The table has application when it is desirable to treat material dry, either because of water shortage or because it is undesirable to wet the materials. An advantage of this table is the ability to handle material coarser than that treated on most wet tables. Ores as coarse as 0.006 m (¼ in) and coal as coarse as 0.076 m (3 in) can be treated.

Close sizing is necessary to give good results, and until recently this has militated against adoption of the table for fine sizes, owing to the difficulties of screening most ores dry below about 40 mesh. The development of improved dry methods for sizing fine material by the use of various cyclonelike devices has tended to increase the use of this apparatus on finer sizes.

Dry tables are used commercially in the separation of many types of minerals. Their greatest use is in the treatment of coal, but ilmenite, various tungsten ores, and even copper ores are so treated. Another important use is the cleaning of industrial materials such as seeds, cork, bagasse, fiber, nuts, wood chips, and coffee. One interesting use is in the sorting of silicon carbide by grain shapes. Flat and splintery grains are removed from others of more nearly equal dimensions.

Agglomeration Tabling Agglomeration tabling is a process whereby selective flocculation or agglomeration of grains of one mineral in an aggregate is caused by the addition of an agglomerating agent in a conditioning cell or in the ball-mill circuit, the slurry containing the agglomerated grains then being fed across gravity tables. The larger size, the oil-filmed surface, and the feathery texture of the floccules cause them to be washed over the side of the table by the current of cross water, while the unflocculated discrete particles remain on the table and are carried off the end in the position followed normally by the concentrate in the usual table feed. An oiled particle will tend to ride on the surface of the water and thus is more readily carried across the side of the table than an unoiled particle. Agglomeration tabling has had more application in the concentration of phosphate minerals than in any other field, although successful tests have been run on limestone, potash, mica, and other ores.

The process is limited to granular material in the size range from 10 to 100 mesh. In this respect it differs from flotation, which functions best on material 48 mesh and finer. For best results the material should be well deslimed and should be conditioned with the agglomerating reagents at a high percentage of solids, 65 percent or greater.

TABLE 21-11 Operating Data for Agglomerate Tabling of Phosphate and Potash Ore

Type of table	Feed size	Feed capacity, tons/h	Table speed, r/min	Table stroke, in	Water with feed, gal/min	Dressing water, gal/min	Size of deck
No. 6 superduty diagonal deck	10–48 mesh	2.5–3.5	295	1.0	20–40	8–15	6′5″ × 14′1″

NOTE: See Table 2-10 for conversion to SI units.

A collector is used that will selectively film the mineral to be agglomerated. In phosphate and limestone practice, this collector is usually a cheap fatty acid such as tall oil. In potash separation long-chain amines are used to film sylvite (KCl).

A bulk oil is always used in addition to the collector to give body to the film and to assist in forming agglomerates. In Florida practice, it is customary to use 0.14 to 0.23-kg/ton (0.3- to 0.5-lb/ton) tall oil and 1.8 to 23 kg/ton (4 to 5 lb/ton) of a 22°Bé fuel oil. Operating data for the agglomerate tabling of phosphate and potash ore are shown in Table 21-11.

Agglomerate tabling works best on simple ores consisting of two free minerals. It has several advantages over the usual tabling method in that it can be used to separate two minerals the difference in specific gravity of which is so small that an effective separation cannot be made by gravity separation alone. Tables treating an agglomerated feed have a considerably larger capacity than tables using untreated feeds, since the capacity of a table treating an agglomerated feed is limited only by the carrying capacity of the riffles. Disadvantages of the method that must be considered are the cost of the reagents used and the fact that if the mineral fraction filmed is the one to be sold, the oily film may be objectionable and must be removed.

SPIRAL CONCENTRATION

GENERAL REFERENCES: Adair, "New Method for Recovery of Flake Mica," *Min. Eng.*, **3**, 252 (1951). Brown, "Humphreys Spiral Concentration on Mesabi Range Ores," *Trans. Am. Inst. Min. Metall. Pet. Eng., Min. Branch*, **184**, 187 (1949). Gleeson, "Why the Humphreys Spiral Works," *Eng. Min. J.*, **146**(3), 85 (1945). Humphreys, "Where Spirals Replaced Tables, Flotation Cells," *Eng. Min. J.*, **146**(3), 82 (1945). Jacobs, "Long Range Iron Project," *Eng. Min. J.*, **168**(4), 100 (1967). Lenhart, "Spiral Concentrators for Gravity Separation of Minerals," *Rock Prod.*, **54**(12), 92, 131 (1951). Otto, "Preparation of Anthracite Silt for Boiler Fuel in a Humphreys Spiral Test Plant," *Trans. 5th Ann. Anthracite Conf.*, Lehigh University, May 1947. Roberts, "How New Highland Plant Recovers Titaniferous Minerals," *Min. World*, **17**(11), 52, 72 (1955). Roe, *Iron Ore Beneficiation*, Minerals Publishing Company, Lake Bluff, Ill., 1957. Thompson, "The Humphreys Spiral: Some Present and Potential Applications," *Eng. Min. J.*, **151**(8), 87 (1950). Thompson, "The Humphreys Spiral Concentrator: Its Place in Ore Dressing," *Min. Eng.*, **10**(1), 84 (1958).

Principle of Operation Spiral concentration of ores and industrial materials is based primarily on the specific-gravity differentials of the materials to be separated. The shape factor of the feed material is also important, and utilization of reagentized feed can change the apparent specific gravity of component minerals by forced attachment of air bubbles to mineral flocs. The best known spiral-type concentrator is the Humphreys spiral concentrator, which first proved its commercial feasibility in 1943. In that year an Oregon plant successfully demonstrated ability to recover chromium minerals from low-grade beach sand deposits.

The Humphreys spiral concentrator is a spirally shaped channel or launder with a modified semicircular cross section, as illustrated in Fig. 21-37. The standard spiral consists of five complete turns, but three-turn units are used in some instances when an unusually rapid and clean separation takes place, as in second-stage or cleaner spirals. There is a drop of 0.34 m (13.5 in)/turn as the flowing pulp progresses from the top to the bottom of the spiral. One spiral concen-

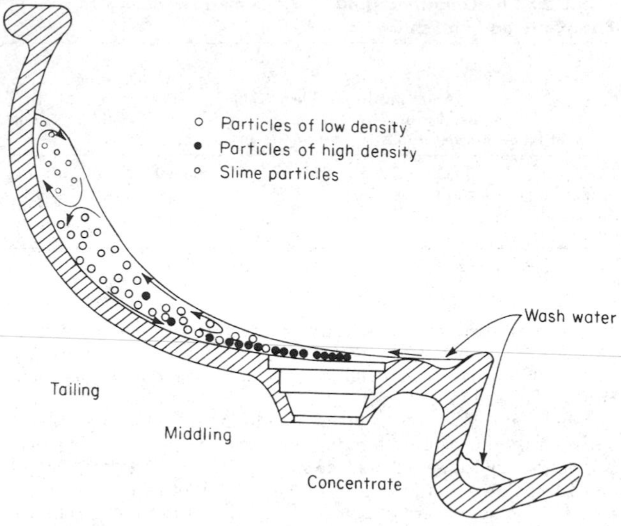

FIG. 21-37 Heavy-mineral separation in the Humphreys spiral concentrator.

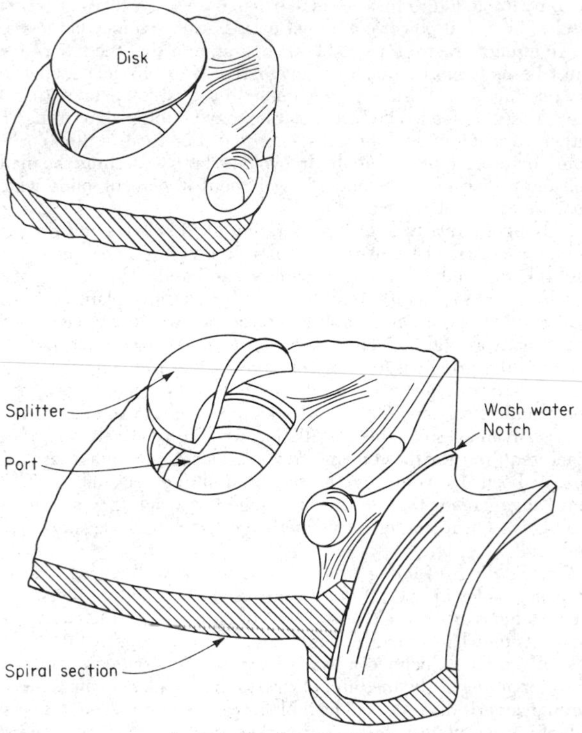

FIG. 21-38 Disks and splitters as used in the Humphreys spiral concentrator.

trator occupies about 0.37 m^2 (4 ft^2) of floor space and about 2.1 m (7 ft) of headroom measured from feed to discharge box. The optimum particle-size range of feed particles for spirals is about 10 to 200 mesh (2 to 0.074 mm).

As the feed slurry flows down the spiral channel, the particles with the highest specific gravity sink to the bottom and move inward toward the inside of the channel. The lighter-weight particles move to the outside and are carried away by the faster, more dilute pulp stream. At 120° intervals circular concentrate "ports," or openings, appear in the bottom of the channel near the inside edge, as illustrated in Fig. 21-38. There are 15 ports in a five-turn spiral, but usually more than half of them are blocked off with smooth stainless-steel disks in order to allow proper configuration of the concentrate stream and good washing of the concentrate. Wash water is available along the entire inside edge of the spiral, where it flows at the rate of 0.2 to 0.6 L/s (3 to 10 gal/min) in a separate wash-water channel. Thus the spiral provides repeated washing stages as the pulp flows down the channel. Generally the richest concentrate is withdrawn from the concentrate ports near the top end of the spiral. Concentrate ports are fitted with very simple stainless-steel "belt-disk splitters," which can split out the desired portion of the concentrate stream. As the gradually impoverished pulp flows down the spiral, wash water is proportioned from the wash-water channel by a series of notches and directed so as to wash repeatedly across the concentrate band and sweep out unwanted gangue particles. The lowest-specific-gravity solids wash outward, and the finest particles actually climb the sloping wall of pulp on the outside of the channel. The concentrate withdrawn from ports near the bottom end of the spiral is usually low-grade and, if liberated, may be recirculated to obtain additional recovery of values and a higher grade of concentrate.

Although the spiral concentrator is mechanically a very simple piece of equipment, the separating action taking place is complex. It involves centrifugal force, friction against the spiral surface, gravity, and the drag of the water.

Basic Requirements for Spiral Concentration Minerals or materials of different specific gravity can usually be concentrated on spirals if the heavy particles do not exceed 10 mesh (2 mm) or are not finer than 200 mesh (0.074 mm.). The size of the low-specific-gravity component is not critical when the values to be recovered are in the heavier-particle fraction. In this case the size of the light particles may range from 4 mesh (4.76 mm) to zero. The quantity of locked grain components (referred to as middlings in the mineral industry) that are present in a given pulp can be critical because this material is frequently recirculated and may eventually accumulate to a degree that will inhibit the separation operations. One solution to such a problem is continual removal of all or part of the middling stream to a grinding mill, followed by separate recovery of values in another spiral circuit or other concentrating machines.

Examples of good feed materials for spiral concentration are (1) beach sands that are processed for recovery of chromite, ilmenite, rutile, zircon, tin, and iron-ore minerals; (2) hard-rock iron ores in which good liberation of iron values occurs in the 10- to 200-mesh size range; (3) some mica and phosphate ores; (4) tailings from concentrating plants that contain heavy mineral components not recovered by flotation and other concentrating methods; and (5) some fractions of coal [minus 6-mm (¼-in) sizes] that can be upgraded by spiral concentration. The spiral used for coal cleaning has six complete turns with a more gradual slope [a 0.25-m (10-in) pitch]. The six turns require about the same headroom as the conventional five-turn spiral.

The spiral concentrator has shown unusual capability in the gravity processing of tailing streams from conventional magnetic and froth-flotation types of ore-processing plants. There are a number of minerals plants where some of the iron values are first recovered by spirals and the tailings are then sent to magnetic separators. There are also iron plants in which the reverse order of processing is used. In spirals the nonmagnetic iron minerals can be efficiently recovered as a high-grade product. An outstanding example of tailings processing in spirals is illustrated by a Colorado molybdenum-ore treatment plant. Spirals recover salable tungsten, pyrite, and tin concentrates from thousands of tons of flotation-plant waste every 24 h. There is no other known ore-processing method that can economically recover tin and tungsten values from this source. The crude ore contains only 0.03 percent tungstic oxide and a trace of tin. The tin occurs as the mineral cassiterite and the tungsten as the mineral hubnerite.

Operating Characteristics Spiral **capacity** can range from 0.12

to over 0.5 kg/s (0.5 to over 2 tons/h) of new feed. The grade of concentrate produced can be adversely affected by either too low or too high a feed rate. A good average feed rate for most spiral installations is 1.5 short tons of new feed per hour. The pulp density of spiral feed may range from 10 to 50 percent solids. If the values are contained in coarse heavy minerals, a high pulp density is preferred, whereas if the values are in fine-sized heavy minerals, it is better to use low pulp densities. Generally 20 to 30 percent solids by weight will constitute a suitable pulp feed.

Water Requirement The water requirement per spiral can range from 1.0 to 2.5 L/s (15 to 40 gal/min); this includes 0.2 to 0.6 L/s (3 to 10 gal/min) of water used in the wash-water channel. An attractive feature of the spiral is that reclaimed water can generally be used in all except the very final upgrading step.

Maintenance The only moving parts in spiral concentrators are those in the pumps that supply the feed and recirculate intermediate products. However, there are sometimes minor maintenance problems associated with the spiral trough itself. Some ores contain sharp particles of very abrasive minerals. The presence of these minerals in some ore causes rapid formation of deep grooves in the surface of cast-iron spirals. Wear grooves can be patched with a variety of plastic and metallic cements. Most spirals presently in service are made of cast iron with molded and vulcanized liners. These liners have successfully solved most wear problems.

Other than the wear problems, actual in-plant maintenance usually involves removal of wood, pieces of blasting wire, and other trash from the ports. When a reagentized feed is used, layers of oily reagents can build up on the spiral surface and sometimes require scrubbing for removal. With feeds containing oily reagents that attack rubber, abrasion-resistant alloy spiral sections are used.

Spirals have been manufactured from concrete, plastics, solid rubber, iron, and special iron alloys.

Operating Costs The operating cost of a spiral concentrator plant will be among the lowest costs of any ore-processing plant handling similar feed material. The only moving equipment parts involved are the pumps included in the flow sheet for the purpose of elevating feed and water to the spirals. One large pump can feed 100 or more spirals in a large plant. At a Canadian iron-ore plant twelve 0.14 m^3/s (2200-gal/min) pumps provide feed for 1152 rougher (or first-stage) spirals. In many plants the second- and third-stage spirals are gravity-fed. Thus maintenance is largely limited to pump repair. When unusually abrasive ores are processed, maintenance of worn sections of the spirals can be extensive unless rubber coating or other abrasion-resistant materials are used on the wearing surfaces.

Labor requirements for a spiral plant are low and are governed primarily by the type of material being fed to the spirals. For example, a phosphate ore containing roots, leaves, and other trash will contain sufficient fibrous material to block concentrate ports, and generally prevent good spiral operation. When such an ore is processed, considerably more labor is required for cleaning and adjustments. Generally metallic ores are quite free of fibrous material that will hang up in the spirals, and one person can operate 100 or more spirals.

Power requirements for spiral plants are low, consisting primarily of pumping energy and possibly a thickener or other pulp-handling equipment associated with the flow sheet.

A typical summary of an approximate range of spiral concentration plant direct costs is given in Table 21-12.

TABLE 21-12 Approximate Range of Direct Costs for Spiral Concentration

Cost element	Cents per short ton of spiral feed
Labor	3.0–5.0
Power	1.6–3.0
Maintenance	2.0–3.0
Depreciation	2.8–4.0
Total	9.4–15.0

DENSE-MEDIA SEPARATION

GENERAL REFERENCES: Aplan and Spedden, "Viscosity Control in Heavy Media Suspension," *Proc. 7th Int. Miner. Process. Congr.*, New York, Sept. 20, 1964, Gordon and Breach, New York, 1965, p. 103. Browning, *Heavy Liquids and Procedures for Laboratory Separation of Minerals*, U.S. Bur. Mines Inf. Circ. 8007. *Chemical Engineers' Handbook*, 5th ed., McGraw-Hill, New York, 1973, sec. 21. Deurbrouck and Hudy, *Performance Characteristics of Coal-Washing Equipment: Dense-Medium Cyclones*, U.S. Bur. Mines Rep. Invest. 7673. Doyle, "The Sink-Float Process in Lead-Zinc Concentration," *AIME Symp. Lead Zinc*, St. Louis, 1970. Gaudin, *Principles of Mineral Dressing*, McGraw-Hill, New York, 1939. *MBI Patented Sink and Float Process*, brochure, American Zinc, Lead and Smelting Co., 1940. "Mineral Engineering Techniques," *Chem. Eng. Prog. Symp. Ser.*, **50**(15), (1954). Mular and Bhappu (eds.), *Mineral Processing Plant Design*, 2d ed., Society of Mining Engineers, AIME, New York, 1980. Ore Dressing Notes, 11, 14, and 16, American Cyanamid Co. Oss and Erickson, "Instrumentation and Control of the Heavy Media Process," *Min. Eng.*, **14**, 41 (May 1962). Richards and Locke, *Textbook of Ore Dressing*, 3d ed., McGraw-Hill, New York, 1940. Rodis and Cremer, "Why an Atomized Ferrosilicon?" *Min. World*, **22**(3), 36 (March 1960). Taggart, *Handbook of Mineral Dressing*, 2d ed., Wiley, New York, 1945. Tippin and Browning, *Heavy Liquid Cyclone Concentration of Minerals*, U.S. Bur. Mines Rep. Invest. 6969 and 7134. Volin and Valentyik, "Control of Heavy Media Plants," *Pit Quarry*, **62**, 111 (December 1969). Walker and Allen, "Beneficiation of Industrial Minerals by Heavy Media Separation," *Trans. Am. Inst. Min. Metall. Pet. Eng., Min. Branch*, **184**, 17 (1949).

Dense-media separation, also known as heavy-media or sink-float processing, is an adaptation of the common laboratory procedure for separating solids of differing specific gravities by immersing them in a heavy liquid of specific gravity intermediary between those of the solids, thereby causing the lighter particles to float while the heavier sink. However, in dense-media separation, the parting liquid is produced by dispersing relatively fine-grained solids of a high specific gravity in water and maintaining this pulp in suspension by light agitation. The method is very effective and can be used to separate solids with differences in specific gravity of as little as 0.005. It is often the only process needed for the removal of deleterious wastes from coal. The method is used extensively for beneficiating ore minerals, and it is finding increasing use for the processing of shredded automobile scrap and for the recovery of values from solid municipal waste.

Dense-media separation may be used to produce either a finished concentrate or an upgraded feed for subsequent processing. In the latter case, it provides a low-cost means to reject a significant amount of essentially barren waste at a coarse size.

Sink-float plants are usually custom-designed for each individual application. However, for coal beneficiation modular units are available. For most large mineral-processing applications the plants will be permanently located for easy access of feed and disposal of waste, but for smaller coal and aggregate operations the plants are often constructed with the anticipation of relocation when the deposits have been depleted.

The response of any given feed to sink-float processing can be accurately established in the laboratory by testing with various heavy liquids. The liquids generally used for this purpose are listed in Table 21-13. These halogenated hydrocarbons are mutually miscible,

TABLE 21-13 Liquids Used to Test Feeds

Name	Specific gravity, 25°C
Methylene iodide	3.33
Tetrabromoethane	2.96
Bromoform	2.89
Tribromoethane	2.61
Methylene bromide	2.48
Ethylene dibromide	2.17
Methylene chlorobromide	1.92
Pentachloroethane	1.67
Carbon tetrachloride	1.59
Trichloroethylene	1.46
Ethylene dichloride	1.26

which enables the preparation of almost any pulp density attainable in a commercial plant. Heavy-liquid test work provides the basis for specifying the optimum screen size for the preparation of the feed.

Continuous pilot-plant test runs are generally recommended to verify the laboratory results and to establish criteria for plant design. Facilities for these runs are available at a number of minerals-processing research centers.

Feed Preparation and Feed Size The ability to achieve a separation of different solid particles on the basis of density, as in all physical separation, depends on the degree to which the particles are liberated (detached) from each other. Liberation can be achieved by breaking the material in a manner that causes it to fracture and free the individual grains of the constituents to be recovered. The degree of separation that can be realized by the dense-media process will depend on the degree of liberation of the individual grains.

There will be an optimum size reduction of the feed material for the dense-media process. This size range can depend on the overall objectives of its use. For example, if the process is to be used in conjunction with a subsequent separation method such as flotation, the intent may be more the rejection of barren waste material at a relatively coarse size and at high recovery of the values, although the resulting grade from the dense-media operation may still be low. On the other hand, if the concentrate grade must be high, a finer degree of liberation will be needed at some loss in recovery. The initial test work can establish the so-called grade versus recovery limitations of the dense-media operation for the specific material of interest. This testing should recognize that the dense-media process is not effective for treating material which contains any substantial amount of particles smaller in size than about 0.5 mm (20 to 28 Tyler mesh).

The largest size that can be treated depends mostly on the dimensions of the separating vessel; coal up to 0.3 m (12 in) has been successfully processed in a drum separator of the type illustrated in Fig. 21-41. Complete removal of fines is usually necessary to ensure proper viscosity of the media. Fines increase viscosity and slow the separation process.

The feed-preparation screen between crusher and separatory vessel may be of either the revolving or the vibrating type. Wash water is applied only to the feed end of the screen so that the process feed will enter the separator moist but without any free water, which would lower pulp density. In a few instances it has been found advantageous to provide for surge storage between screen and separator to drain off further excess water.

A typical flow sheet is shown on Fig. 21-39.

Preparation of the Media Various solid materials have been used to prepare the media. In the initial development of the process, a suspension of sand and also mixtures of barite and clay were used for separating coal from slate. Galena (lead sulfide mineral) was also used to achieve a higher pulp density. In present processing, iron-based particles such as magnetite and ferrosilicon are preferred because they offer suitable density, high resistance to attrition, and ease of recovery by magnetic methods. With magnetite, a pulp density 2.5 times that of water can be obtained. Ferrosilicon can provide a density factor of 3.3, which is effective for separating most gangue constituents from metallic ores. Both materials might be used to obtain intermediate media densities.

Media-Particle Size The size of the media particle is important. A relatively coarse medium (minus 100 mesh) is commonly used in larger-volume static-type separators such as cones. However, in dynamic separators, a much finer size is desirable. Ground magnetite or atomized ferrosilicon is advantageous in this application. The latter is produced by pouring molten ferrosilicon into an atomizing chamber, where a jet of steam forms the alloy into spherical particles. These particles are more resistant to wear than are the particles in a ground product and cause less abrasion to the equipment. Atomized ferrosilicon permits pulp densities above about 3.4-density factor.

Custom-ground natural magnetite is available in many size ranges.

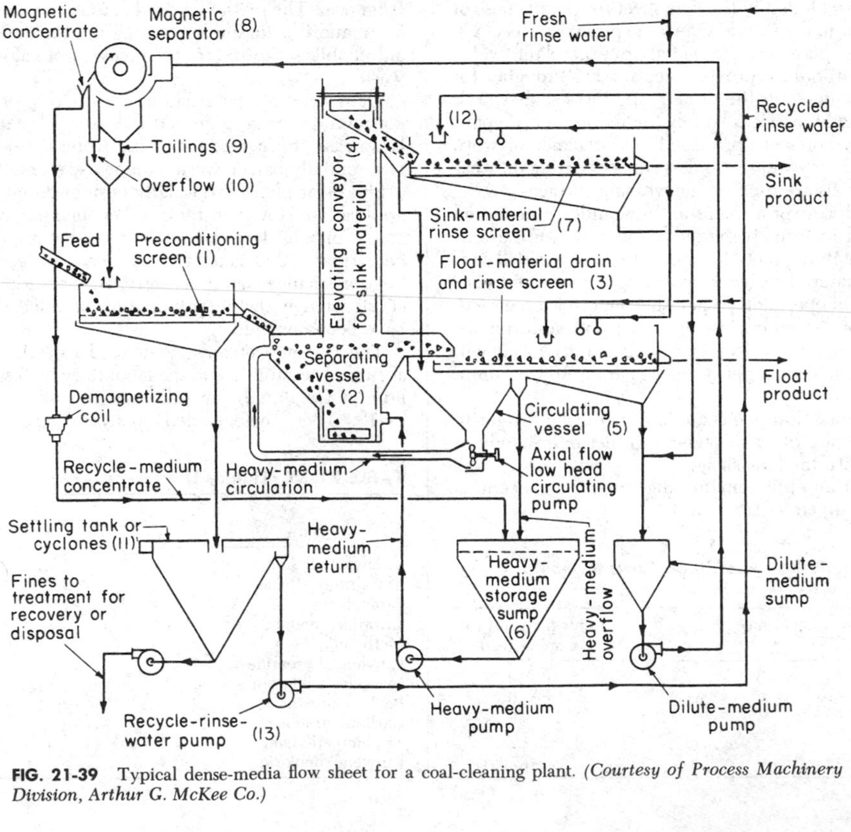

FIG. 21-39 Typical dense-media flow sheet for a coal-cleaning plant. *(Courtesy of Process Machinery Division, Arthur G. McKee Co.)*

TABLE 21-14 Typical Size Distribution of Ground Natural Magnetite*

Product grade	Percent retained by weight for mesh size 100	200	325	Less than 325	1978 cost
A	0.6	12.0	17.8	69.6	$72
B	0.1	1.0	7.6	91.4	$74
C	5.0	22.0	23.0	50.0	$72
D	6.0	29.5	22.9	36.3	$71
E	0.1	0.4	1.9	97.7	$85
G	0.2	6.2	15.5	78.1	$74

*Foote Mineral Company.

Table 21-14 gives a typical specification sheet. Cost is based on truckload quantities in 45.4-kg (100-lb) paper bags, FOB Frazer, Pennsylvania.

Pulverized ferrosilicon containing approximately 15 percent silicon is available from the Foote Mineral Company and from Carborundum Co. in the sizes and at the prices shown in Table 21-15. Cost is based on truckload quantities in 227-kg (500-lb) steel drums, FOB Keokuk, Iowa, and Niagara Falls, Ontario.

Atomized ferrosilicon is at present available only from West Germany through American Hoechst Corp. in the sizes shown on Table 21-16. Costs vary with the exchange of U.S. dollars to deutsche marks but will be around $770 per metric ton, FOB Germany (1978 estimate).

Chemical Additives The use of chemical additives in sink-float processing is not common except for the use of lime to prevent oxidation and decomposition of the medium. A small amount of clay is sometimes added to improve the kinetic stability of the suspension.

Considerable laboratory work has indicated that the use of a dispersant such as sodium hexametaphosphate may assist in the stabilization of the medium; more recent data report the beneficial effect of the addition of polymers that reduce media viscosity while simultaneously producing a very low settling rate of the ferrous compound. This should be of great value for difficult separations, but at present no data are available from commercial operations.

Separating Vessels Many different types of separating vessels have been proposed and used for sink-float separation. For applications at coarse sizes involving a high ratio of float to sink or a high gravity differential, as in the case of coal, trough-shaped vessels (as shown in Fig. 21-40) or rotating-wheel separators are commonly used. However, for the beneficiation of most ores, other types of separators have found general acceptance. The optimum design for any given ore will depend on such variables as the rate of feed, the size of feed particles, the ratio of float to sink, and the gravity differential between the solids to be separated. Separators are classified as either **static** or **dynamic,** depending on whether or not centrifugal force is applied.

Drum Separators Very coarse solids, up to 0.3 m (12 in), are often processed in a drum separator of the type shown in Fig. 21-41.

TABLE 21-15 Typical Size Distribution of Pulverized Ferrosilicon

Particle size; mesh size less than	1978 cost range
48	$245–279
65	$248–282
100	$252–291
200	$289–400

This is similar to a ball-mill shell with lifters permanently attached to the wall. Medium and feed enter at one end, and the float product flows out through the discharge trunnion, while the sink is lifted by the rotation of the drum to a stationary launder, through which it is flushed out. Modifications of this type include division of the shell into two compartments, which permits simultaneous operation at two different pulp densities resulting in various grades of products. The two-compartment revolving drum is illustrated in Fig. 21-41.

Drum separators have capacities up to 250 kg/s (900 t/h), cone separators to about 125 kg/s (500 t/h), and dynamic separators a maximum of 28 kg/s (100 t/h); however, these can readily be manifolded for any required tonnage. Economics will dictate the minimum tonnage for which a plant would be justified; several plants of 2.8-kg/s (10-t/h) capacity have been built.

Cone Separators Feed materials in the intermediate sizes, 0.1 to 0.01 m (4 to 0.5 in), may be processed in cone separators as shown in Fig. 21-42. These have a large surface area and increased volume pulp, which permit longer retention time than that of most other types of separators; this is a great advantage when separating solids of small gravity differential. The feed is introduced into the cone at a point below the pool surface and as far from the overflow baffle as possible. Slow-moving scrapers prevent a buildup of medium on the cone wall, provide the necessary agitation to prevent settling of the medium, and push the float particles toward the overflow weir. The sink product is removed from the cone bottom by rock pump, internal air lift, or external air lift. Mechanical elevators of the screw or bucket type have also been used but require more maintenance.

Cyclone Separators Finer feed solids, from 0.04 to 0.0005 m (1.5 in to 28 mesh), may be treated in dynamic separators of the Dutch State Mines cyclone type (Fig. 21-43). In cyclone separators, the medium and the feed enter the separator together tangentially at the feed inlet (1); the short cylindrical section (2) carries the central vortex finder (3), which prevents short circuiting within the cyclone. Separation is made in the cone-shaped part of the cyclone (4) by the action of centrifugal and centripetal forces. The heavier portion of the feed leaves the cyclone at the apex opening (5), and the lighter portion leaves at the overflow top orifice (6).

The sharpness of separation of the mineral from the gangue is dependent on (1) the stability of the suspension, which is influenced by the size of the medium; (2) the specific gravity of the medium; (3) the cleanliness of the medium; (4) the cone angle; (5) the size and ratios of the internal openings in the cyclone (inlet, apex, and vortex); and (6) the pressure at which the pulp is introduced into the cyclone.

TABLE 21-16 Size Distribution of Atomized Ferrosilicons

Particle size, greater than Tyler mesh	Manufacturers' grade distribution: Extra coarse	Coarse	Fine	Cyclone 60	Cyclone 40	Cyclone 20
48	15– 0	5– 0				
65	25– 7	8– 2	4– 0			
100	42–20	22– 7	8– 2	3–0		
150	55–35	35–17	22– 5	8–0	3–0	
200	70–50	50–35	30–15	20–5	8–3	2–0
Particle size, less than Tyler mesh						
200	30–50	50–65	70–85	80–95	92–97	98–100
325	.....	.....	.....	40–60	70–85	90–100
625	.....	.....	.....	10–20	40–55	70– 80

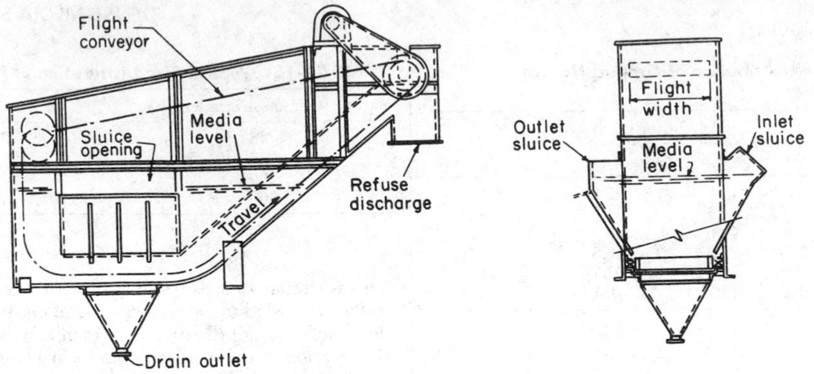

FIG. 21-40 Drag-tank-type dense-media separatory vessel. *(Courtesy of Link-Belt Co.)*

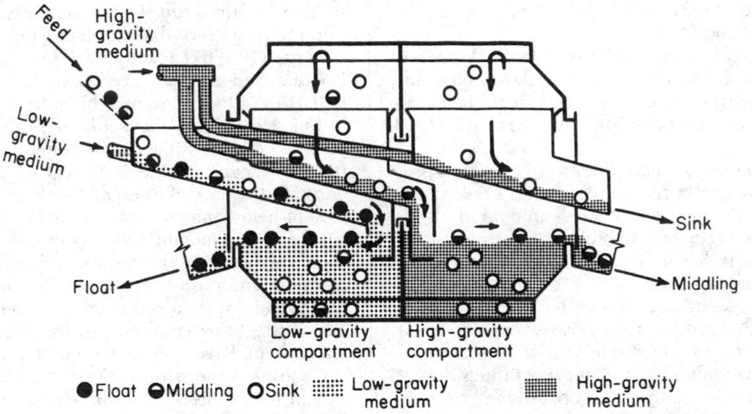

FIG. 21-41 Revolving-drum-type dense-media separatory vessel. *(Courtesy of Western Machinery Co.)*

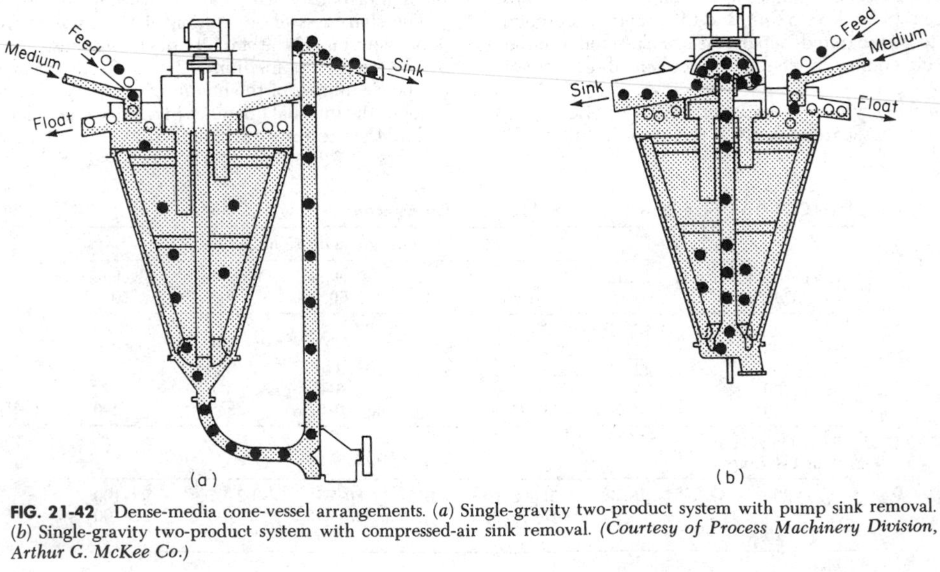

FIG. 21-42 Dense-media cone-vessel arrangements. (*a*) Single-gravity two-product system with pump sink removal. (*b*) Single-gravity two-product system with compressed-air sink removal. *(Courtesy of Process Machinery Division, Arthur G. McKee Co.)*

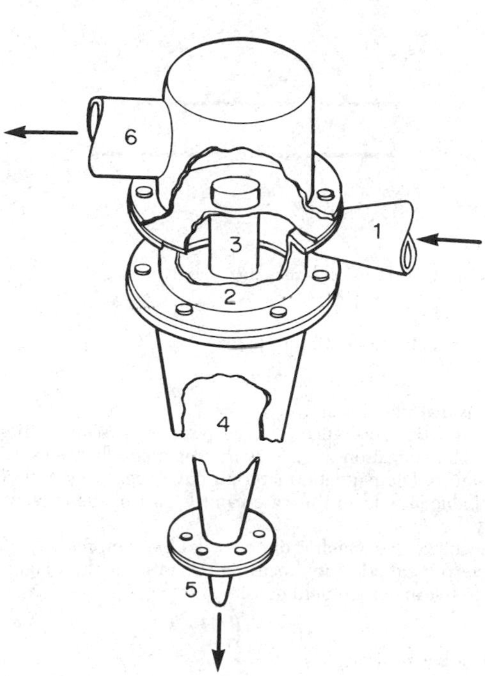

FIG. 21-43 Dutch State Mines cyclone separator.

A 20° cone angle is the most common. Cyclone diameter will be determined by the separation to be made as well as by the capacity required. The 0.5- and 0.6-m (20- and 24-in) cyclones are most common in coal plants, whereas multiple cones of 0.25- or 0.3-m (10- or 12-in) diameter are used in higher-gravity separations.

Dense-media cyclones are generally operated in the (0.7–1.0) × 10^6-Pa (10–15-lbf/in^2) range. It is not advisable to go below (0.4–0.56) × 10^6 Pa because the recovery of low-specific-gravity material and the rejection of impurity are improved at higher pressures, especially for the finer sizes. Pressures as high as 2.5 × 10^6 Pa (36 lbf/in^2) have been used, and they increase capacity but accelerate wear. Residence time of the ore particles is very short in the cyclone, and a large volume of medium is circulated for each ton of feed treated in the cone. Loss of media is higher in cyclone plants because of the finer media required and the additional volume encountered in these plants. Media loss may be 2 to 5 kg/ton (5 to 10 lb/ton) of ore treated in cyclone plants, as compared with 0.2 to 0.8 kg/ton (0.5 to 1.5 lb/ton) in coarse, static heavy-medium circuits. Cyclone-plant labor requirements are low, and efficiency is high. A 0.6-m (24-in) heavy-medium cyclone can handle 75 tons of coal per hour.

Dyna Whirlpool A unique vessel design for capacities up to 100 t/h has been developed by the American Zinc Co. The separation occurs in a cylindrical-shaped separatory vessel maintained in an inclined position from horizontal. This system, known as the Dyna Whirlpool (DWP) process, provides for separate entry of the medium and the feed solids, as illustrated in Fig. 21-44. A distinct feature of this separator is that the feed enters the separator via gravity flow. Feed size may range from 0.05 to 0.0002 m (2 in to 65 mesh). Magnetite or ferrosilicon is generally used.

Process Control As is the case in all concentration processes, optimum results will be obtained under steady operating conditions. Because of the simplicity of the dense-media process, these can readily be maintained.

Uniformity of the rate of feed will be ensured by a constant-weight feeder; density control may be automatically obtained through a measuring probe on the media-return line that adjusts delivery of the necessary volume of media from the densifier or media thickener; the viscosity can be controlled automatically by continuously testing a predetermined volume of return media and adjusting the divider under the drainage screen for media cleaning as needed; pH control can be automated by conventional methods.

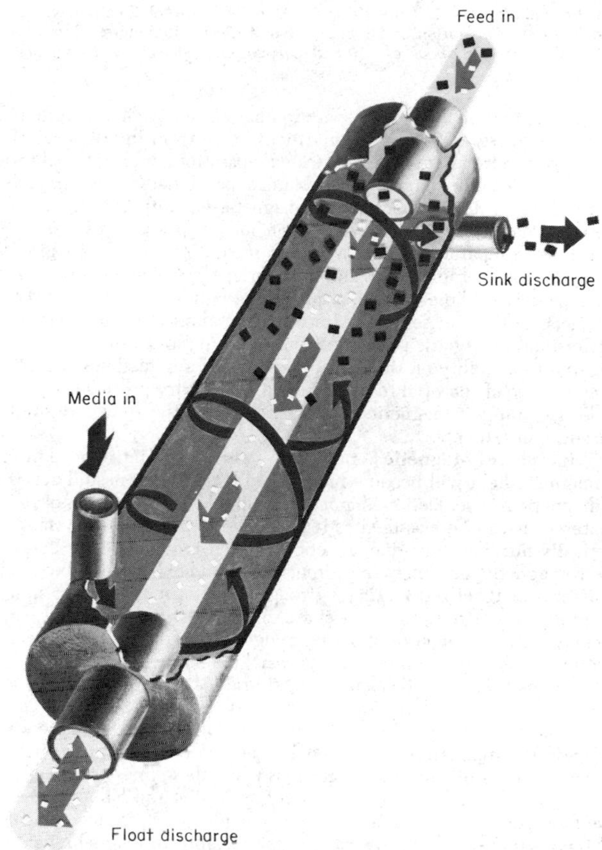

FIG. 21-44 Dyna Whirlpool separator. *(Courtesy of American Zinc Co.,)*

Notwithstanding the possibility of such automation, many successful operations depend almost entirely on manual sampling. Density determinations of the pulp on the media-return line and on each of the drainage screens are made at scheduled intervals, and the operator adjusts the media flow as needed.

Costs Because sink-float processing is applied to relatively coarse particles and is a single-pass operation, capital and operating costs are usually considerably lower than would be required for a flotation or a gravity mill of the same capacity. A large flow of water is required for feed preparation and for media recovery, but almost total recovery for recirculation is possible. A minimum of two job-trained operators per shift is generally required by law, but these would be able to attend several separators at almost any feed rate.

Estimates for a 30-kg/s (100-ton/h) plant using a dynamic separator is approximately $350,000 (1978), exclusive of power, water, compressed air, crushing, foundations, and housing; installed power for such a plant will be about 298 kW (400 hp).

Direct operating costs usually vary between 30 cents and $1 per ton of feed, depending on hourly tonnage and on media recovery. Media losses are usually higher in a dynamic than in a static separator, mainly because a finer particle size is required. Media loss may be from 2 to 20 kg/t (5 to 10 lb/ton) of mixture processed in a cyclone plant compared with a static plant, which generally operates with media losses of an order of magnitude lower.

MAGNETIC SEPARATION

General References: Kolm, Oberteuffer, and Kelland, "High-Gradient Magnetic Separation," *Sci. Am.*, **233**(5), 46–54 (November 1975). Lawyer and Hopstock, "Wet Magnetic Separation of Weakly Magnetic Materials," *Min.*

Sci. Eng., **6**(3), 154–172 (July 1974). Marston, "The Use of Electromagnetic Fields for the Separation of Minerals," World Electrical Congress, Moscow, 1977. Taggart, *Handbook of Mineral Dressing*, 2d ed., Wiley, New York, 1945.

The principles of magnetic separation have been applied commercially for nearly 100 years. Applications range from the removal of coarse tramp iron to more sophisticated separations, such as the elimination of weakly magnetic iron-stained particulates from paper-coating clays. The application of magnetic-separation methods to weakly magnetic particles has been made possible by recent advances in separator design. Magnetic separators now have a great many industrial applications and range in size from small laboratory-scale devices to those capable of processing hundreds of tons hourly.

Selecting the best separator for a specific application requires an understanding of basic principles of magnetism plus an evaluation of separator capability on the basis of design and application variables such as type of material to be processed, wet or dry processing, particle-size range, magnetic characteristics of the feed, desired throughput rate, etc.

Principles of Magnetic Separation Any particle introduced into a magnetic field will become magnetized to some extent and act as a magnetic dipole. Depending on the magnetic characteristics of the material, it can be classified as ferromagnetic, paramagnetic (magnetically attracted), or diamagnetic (repelled by a magnetic field). Ferromagnetic substances (e.g., iron, nickel, and cobalt) may be permanently magnetized and have strong magnetic moments per unit volume. Paramagnetic substances are further classified as strongly or weakly magnetic according to the strength of the magnetic moment produced per unit volume in the external magnetic field.

A magnetic field and magnetic-field gradients are produced in a variety of ways and vary in both field geometry and strength. The magnetic field of a magnet is the space through which its influence extends. It is mapped by the lines of magnetic force. A magnetic field is considered uniform or homogeneous when these lines are parallel and equally spaced. It can be noted in Fig. 21-45*a* and *b* that neither the bar (permanent) magnet nor the coils plus iron magnet typical of the C-frame magnet type can produce a uniform magnetic field.

The intensity of the magnetic field H is measured in amperes per meter. For a single-layer solenoid, at any point along its axis the magnetic field intensity is

$$H = \tfrac{1}{2}NI(\cos\theta_2 - \cos\theta_1) \quad (21\text{-}4)$$

where H is measured in amperes per meter, A/m
N is the number of turns per unit length or the number of turns per meter
I is the current per turn in amperes, A
θ_1 and θ_2 are the angles included between the axis and the lines drawn from the measured point to its near and far edges

The magnetic flux density is

$$B = \mu_0(H + M) \quad (21\text{-}5)$$

Flux density is calculated as the permeability of free space times the sum of the magnetic-field intensity and the induced magnetization observed whenever a magnetic material is placed in a magnetic field; it is measured in teslas, T.

The strength of the induced magnetization is equivalent to M dipoles per cubic meter where μ_0 is the permeability of free space, equal to $4\pi \times 10^{-7}$, N/A^2; and M is the magnetization, A/m.

Another method for calculating B is

$$B = \mu H \quad (21\text{-}6)$$

where μ is the permeability of the material.

The magnetic susceptibility of a material (χ, volume susceptibility) is dimensionless and is defined as the ratio of induced magnetization to magnetic field intensity. It is expressed as

$$\chi = M/H \quad (21\text{-}7)$$

Thus,

$$B = \mu_0 H(1 + \chi) \quad (21\text{-}8)$$

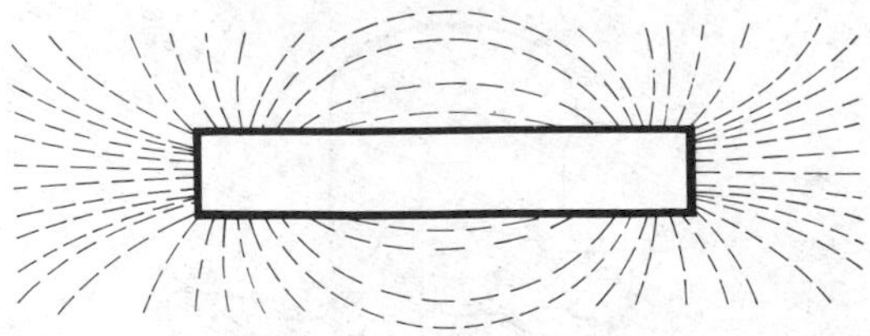

FIG. 21-45*a* Lines of force surrounding a bar-type magnet.

Specific magnetic susceptibility (ψ) is

$$\psi = \chi/\rho \quad (21\text{-}9)$$

where ρ is material density.

Paramagnetic substances have positive susceptibilities, and induced magnetization augments the magnetic-flux density within the substance. Diamagnetic materials have negative susceptibilities, and an induced field in this case cancels part of the magnetic-field intensity.

Permeability (μ), which is often used, albeit imprecisely, in referring to ferromagnetic substances, is the ratio of the magnetic-flux density to the magnetic-field density.

$$\mu = B/H \quad (21\text{-}10)$$

Relative permeability is μ/μ_0.

In cgs units,

$$B = H + 4\pi I \quad (21\text{-}11)$$

$$K = I/H \quad (21\text{-}12)$$

$$K = \chi/4\pi \quad (21\text{-}13)$$

Table 21-17 shows the magnetic susceptibility of minerals and elements. Magnetization of various materials is directly dependent on two factors: (1) the degree of magnetic susceptibility and (2) the applied magnetic-field intensity. It can be seen in Fig. 21-46 that ferromagnetic materials quickly become magnetically saturated and that an increase in magnetic-field intensity will have no effect after a certain point. For paramagnetic materials (e.g., hematite), which are more difficult to magnetize, the magnetic-flux density is directly proportional to the magnetic-field intensity, and some of these substances, practically speaking, cannot be saturated.

In addition, the magnetic characteristics of a material can change as a function of stress (e.g., unannealed series 316 stainless steel can be magnetic after machining), temperature, pressure, and physical and chemical treatment. Therefore, when two paramagnetic materials with similar magnetic susceptibilities are to be separated, the

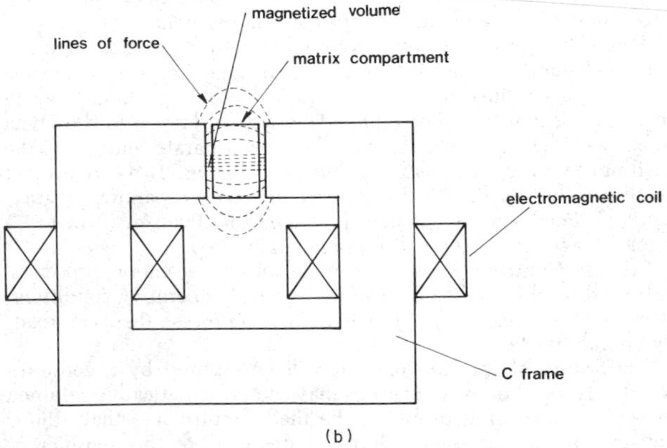

FIG. 21-45*b* Lines of force produced by a C-frame magnet (coils and iron-magnet surface).

TABLE 21-17 Magnetic Susceptibility of Elements and Minerals

Substance	Susceptibility, 10^{-6} cgs	Substance	Susceptibility, 10^{-6} cgs
Aluminum	+10.5	Ferberite	+39.3
Al_2O_3	−37.0	Galena	−0.4
Apatite	+1.0 to +18.0	Garnierite	+30.7
Aragonite	−0.4	Gold	−28.0
Asbolan	+150.0	Ilmenite	+15.45 to 70.0
Azurite	+12.2 to +19.0	Lead	−23
Anatase	+0.96 to +5.60	Malachite	+10.5 to +14.5
Beryl	+0.4	Millerite	+0.21 to +3.85
Braunite	+35.0 to +150.0	Molybdenite	+4.93 to +7.07
Biotite	+40.0	Molybdenum	+89.0
Barite	+10.0	Platinum	+201.9
Barite (pure)	−71.3	Rutile	+0.85 to +4.78
Brannerite	+3.5	Scheelite	+0.13 to +0.27
Chromium	+180.0	Siderite	+65.19 to +103.81
Chromite	+125.6 to +450.0	Titanium	+150.0
Cobalt	Ferromagnetic	Tungsten	+59.0
Cobaltine	+2.0	Uranium	+395.0
Cobaltite	+0.34 to +0.64	Vanadium	+255.0
Columbite	+32.55 to +37.20	Vanadinite	−0.2 to +0.27
Copper	−0.1	Wolframite	+42.2
Chalcopyrite	+1.0 to +5.0		

NOTE: Extensive listing of magnetic susceptibilities of elements and organic and inorganic compounds can be found in G. Foex, *Tables de constantes et donnes numériques*, Massou et Cie., Paris, 1957.

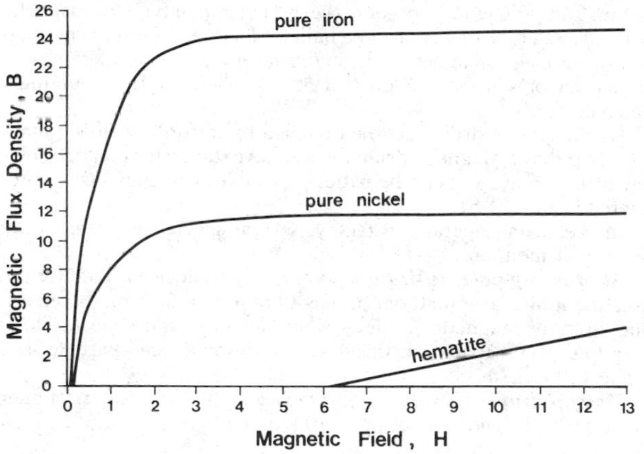

FIG. 21-46 Magnetization curves for ferromagnetic and paramagnetic materials.

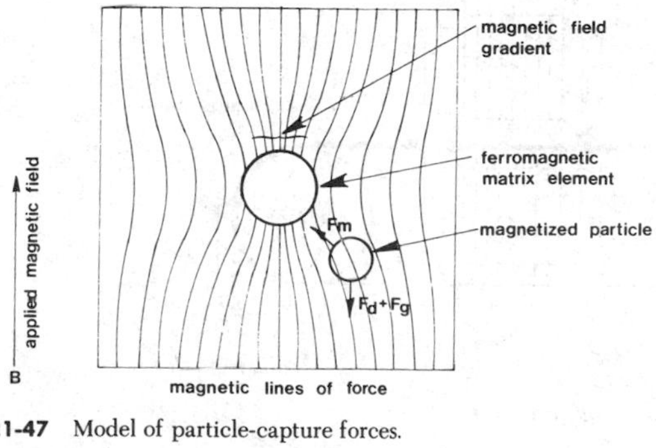

FIG. 21-47 Model of particle-capture forces.

possibility that pretreatment will facilitate subsequent separation should be studied.

A magnetic field exerts a force on each of the two poles of a dipole (particle), forcing it to align itself with the lines of magnetic force. These are exerted in opposite directions, and if the magnetic field is uniform, they will be equal. Therefore, the net force on the dipole will be zero. However, if the field varies in space (has a gradient), the force on the dipole will be greater in the direction of the higher field and will be proportional to the magnetic-dipole moment and the magnitude of the magnetic-field gradient.

A model of the forces operating in such a case is shown in Fig. 21-47, where

$$F_m = m\,dB/dz \tag{21-14}$$

$$F_d = 3\pi\eta b v \tag{21-15}$$

where F_m = magnetic tractive force
F_d = hydrodynamic drag force
F_g = gravitational force
m = magnetization characterization of the particle ($m = \chi HV$)
H = magnetic-field intensity
dB/dz = magnetic-field gradient
η = fluid viscosity
b = particle diameter
v = fluid velocity
V = particle volume
χ = magnetic permeability of particle.

In order to retain the magnetic fraction of the material in the collection volume of the separator it is necessary that

$$F_m \geq F_d + F_g \tag{21-16}$$

The preceding analysis shows that density is important and cannot be influenced and that particle size is of extreme importance, as the magnetic force F_m is directly dependent on the b^3 of the particle, while the drag force F_d is dependent on b. This means that separations between particles with close magnetic susceptibilities can be successfully performed only if the particle-size distribution is within a relatively narrow range. The influence of gravitational force is dependent on the relative direction of the slurry flow. [Buoyant forces are neglected in the relationship expressed in Eq. (21-16).]

It should be noted that effective magnetic separation requires that particles of different species be liberated from each other. It is also important that the finest possible matrices (filamentary type) be used because these produce the highest magnetic-field gradients. The use of high-gradient-producing matrices can substantially reduce the magnetic-field intensity (magnet strength) required to gain the same separation results, thus lowering both capital investment and process costs. The most economical results are achieved when the diameters of the matrix filaments are matched to the size of the particulates being processed.

Factors which adversely influence the separation of very fine particle systems are brownian motion and London forces. However, it is possible to counter these forces by the use of dispersants, temperature control, etc.

Equipment Separator designs differ for the various types of materials to be separated. In general, magnetic separation devices can be grouped as follows:

Grate-Type Magnets This type of device consists of a series of tubes (often of stainless steel) which are packed with ceramic magnets and installed in a trap perpendicular to the fluid-flow direction. Grate magnets are used for the wet or dry removal of tramp coarse or fine iron. The various available equipment designs include self-cleaning grates, wing-and-drawer-type magnetic grates, permanent magnetic grates, vibratory grates, and rota-grates. Each of the designs is manufactured in a range of sizes, with single or multiple rows (banks) of magnetic tubes. Applications include ferrous traps for slurries such as detergents (e.g., in chemical plants), sugar and candy (e.g., in food plants), ink recycling (e.g., in printing operations), or pulp in paper mills.

Grates may be installed in all circuits of dry, pulverized material

TABLE 21-18 Tramp-Iron Removal with Plate Magnet, 0.6 m from Top*

	Relative capacity, percent		
Particle size	Chute angle, 35°	Chute angle, 45°	Chute angle, 60°
Over 30 g (1 oz)	125	100	75
Over 8 mesh (2.38 mm) to 30 g	100	75	45
Under 8 mesh (2.38 mm)	33	25	10

*Courtesy of Eriez Magnetics.

where contamination or accidents may occur from tramp or fine iron.

Plate Magnets and Magnetic Humps These devices are used to remove tramp iron from materials being conveyed pneumatically or falling in gravity flow. Tramp iron is removed by being trapped against a magnetized plate. This type of magnet must be cleaned periodically. A chute angle of 45° is recommended. The plate magnet should be close to the feed point to eliminate the influence of velocity. Complete lines of plate magnets and magnetic humps are offered by many manufacturers.

Plate magnets, which are used in chutes, can be either permanent magnets or electromagnets. For the permanent type, magnet width extends to 1.23 m; there is a maximum width of 2.85 m for electromagnets. Capacities are approximately 250 m^3/h for each meter of width, at a 45° angle, with the magnet located in the bottom of the chute approximately 0.6 m from the top and material introduced at a slow velocity. Capacity varies with the size of the tramp-iron particles to be removed and with the angle of the chute (see Table 21-18).

Lifting Magnets These devices operate in either a continuous or a cyclic manner. Continuous devices usually have a belt which moves over the lifting magnetic poles to carry the magnetized particles into a region of low or zero magnetic field, where they are released. Depending on the design of the poles, these units can be either high- or low-intensity devices. Figure 21-48 shows the in-line and cross-belt methods for installing a lifting magnet above a conveyor belt.

Cross-belt magnetic separators are based on the same principle as lifting magnets. Although these units have relatively low capacities, the same unit can produce selective separations with different products by using different pole gaps and field strengths. (see Fig. 21-49.)

The magnet designs shown in Figs. 21-48 and 21-49 are used for tramp-iron removal. Suspended magnets are positioned from 5 to 10 cm above the highest point of the material on the conveyor and may be designed to be self-cleaning. Sizes for devices of this type range up to 2.8/1.6 m. The installation shown in Fig. 21-48*a* is often preferred because it requires a less powerful magnet and can clean material from a higher-speed conveyor belt (over 1.75 m/s). For self-cleaning units, the belt is run at up to 2.5 m/s.

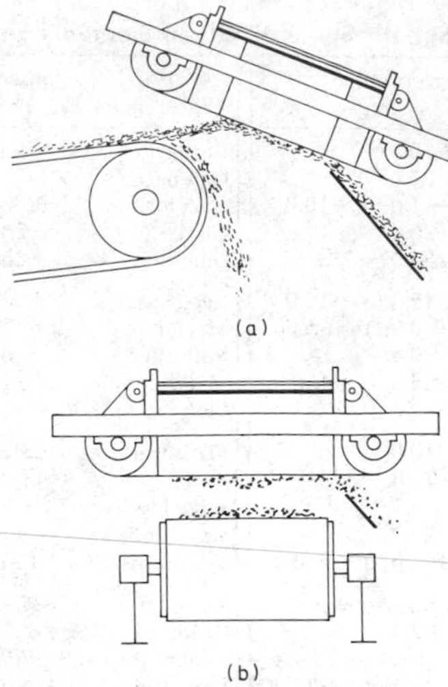

FIG. 21-48 Types of lifting magnets. (*a*) In-line lifting magnet. (*b*) Cross-belt lifting magnet. *(Courtesy of Eriez Magnetics.)*

Drum and Pulley Magnets Since Thomas Edison invented and developed the magnetic pulley for the concentration of nickel ore, drums and pulleys have become the most common types of magnetic separators. These devices can be built with either a permanent magnet or an electromagnet, and the drum separator can operate with either dry or wet feeds. Figure 21-50 is a schematic for mounting a magnetic pulley.

Dry magnetic drums can be designed to perform as lifting magnets or pulleys. Magnetic drum devices have stationary magnets; pulley drums rotate. Other schematics of possible arrangements are presented in Fig. 21-51.

In the drum-separator category, several specialized devices are worthy of mention.

Alternating-polarity drum separator. This device is used for the treatment of coarse material (minus 40 mm, plus 0.15 mm) containing strongly magnetic particles when a high-grade concentrate is required. The capacity of this device varies with feed-particle size, up to 100 t/(h·m).

Unigap drum separator. This device is used for materials finer than 6 mm at feed rates of up to 10 t/(h·m).

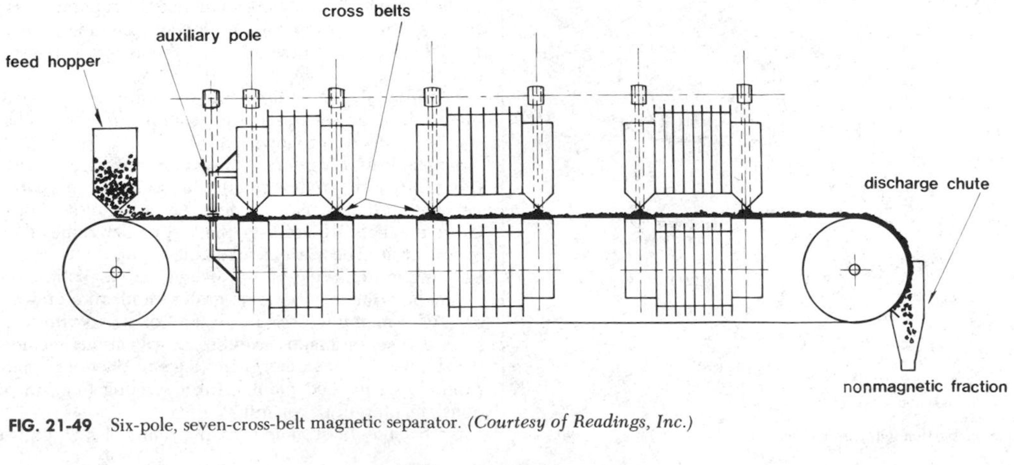

FIG. 21-49 Six-pole, seven-cross-belt magnetic separator. *(Courtesy of Readings, Inc.)*

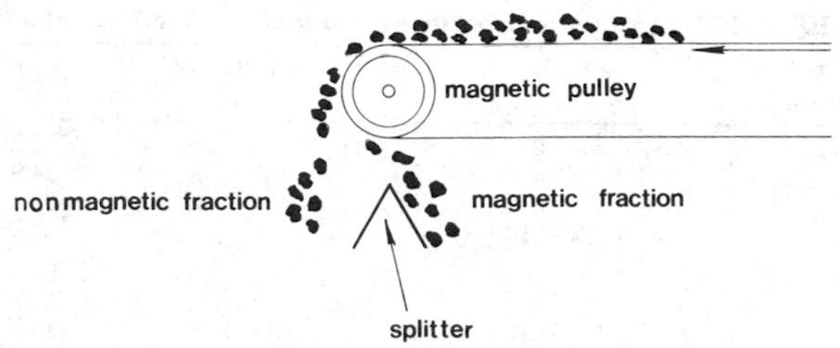

FIG. 21-50 Magnetic pulley.

High-speed, low-intensity drum magnetic separator. This device is designed to handle very fine material (minus 0.15 mm and finer) to produce a high-grade magnetic concentrate.

Depending on the required results—high recovery of magnetics or high-grade concentrates (clean magnetics)—wet drum separators are designed to work in concurrent, countercurrent, or counterrotating fashion by using one or more drums in any possible combination. Figure 21-52 presents schematics of these wet drum magnetic separators.

Magnetic pulleys. These vary in size from 0.203 to 1.219 m in diameter and from 2.03 to 1.526 m in width. The acceptable depth of the material on the conveyor belt depends on the diameter of the pulley and the linear velocity of the belt (see Table 21-19). Table 21-20 indicates the maximum capacity for such units. Depending on the application, the correction factors given in Table 21-21 should be applied. For sizing and maximum efficiency, multiply the actual volume of material to be handled by the correction factor shown and select the magnetic pulley having a capacity equal to or greater than the resultant volume.

Wet Drum Magnetic Separators These devices are used for the concentration of strongly magnetic coarse particles. The size of the separator is influenced by several variables: slurry volume, percent solids in the slurry, percent magnetics in the slurry, required recovery of magnetic particles, and required concentration of magnetic product. This type of separator is built by several manufacturers; drum sizes range from 0.023 to 1.2 m in diameter, with widths up to 3.0 m. The concurrent type can process slurries with 20 percent solids by weight for single-drum separators and with 35 to 45 percent for

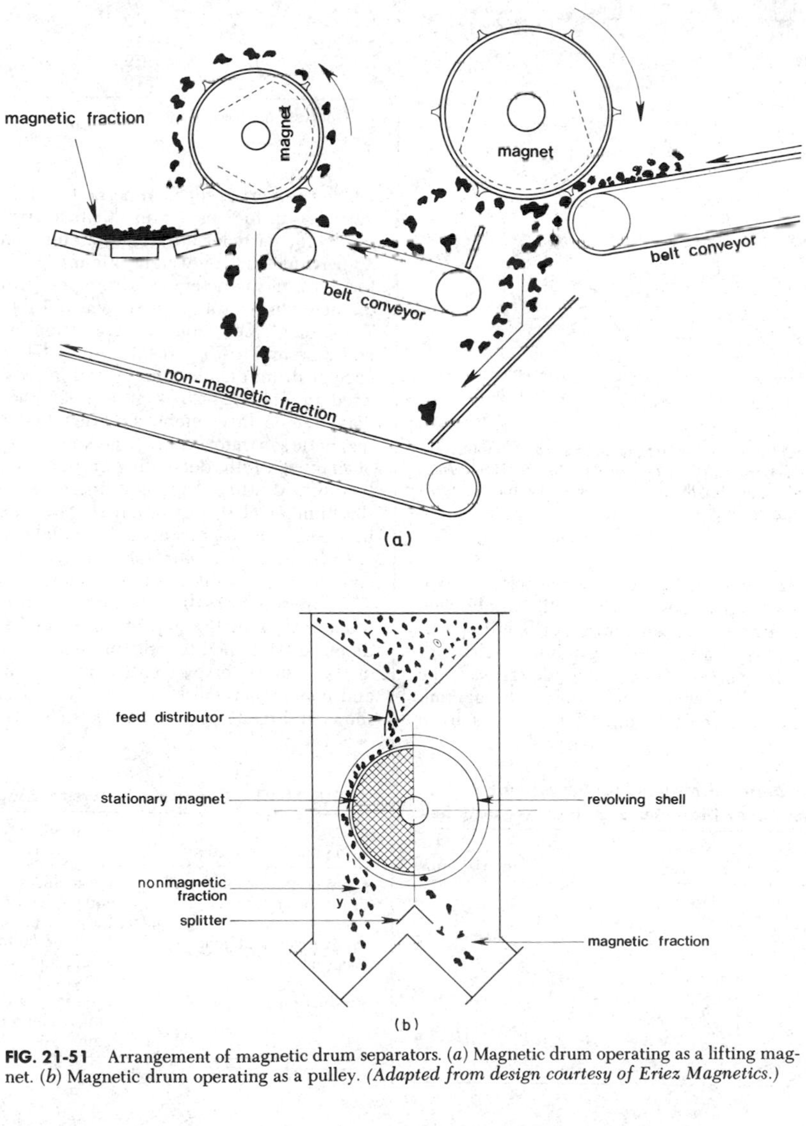

FIG. 21-51 Arrangement of magnetic drum separators. (*a*) Magnetic drum operating as a lifting magnet. (*b*) Magnetic drum operating as a pulley. *(Adapted from design courtesy of Eriez Magnetics.)*

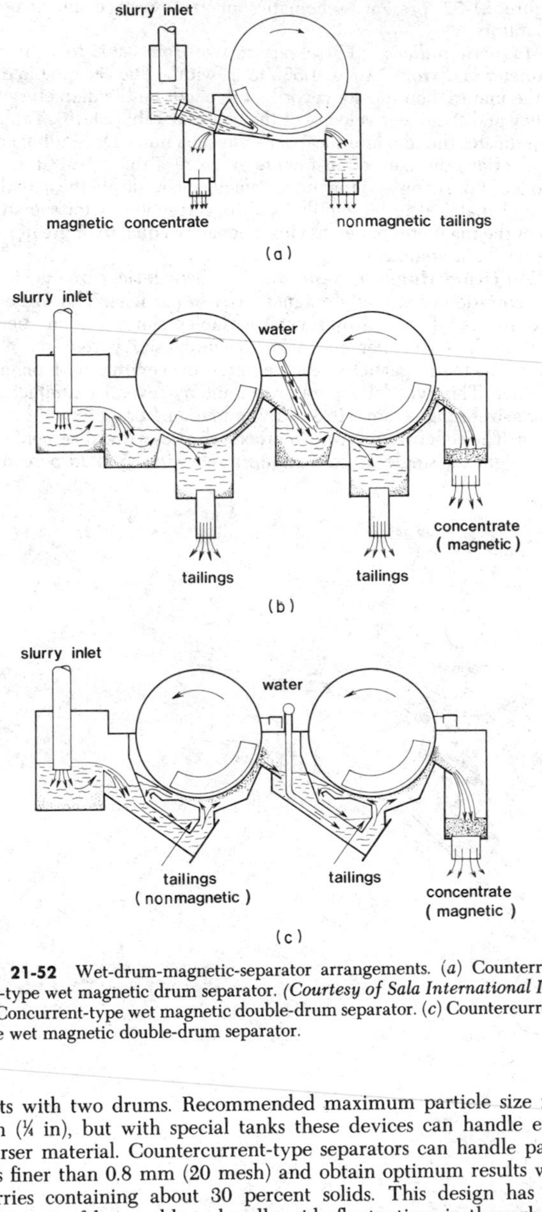

FIG. 21-52 Wet-drum-magnetic-separator arrangements. (*a*) Counterrotation-type wet magnetic drum separator. *(Courtesy of Sala International Inc.)* (*b*) Concurrent-type wet magnetic double-drum separator. (*c*) Countercurrent-type wet magnetic double-drum separator.

units with two drums. Recommended maximum particle size is 6 mm (¼ in), but with special tanks these devices can handle even coarser material. Countercurrent-type separators can handle particles finer than 0.8 mm (20 mesh) and obtain optimum results with slurries containing about 30 percent solids. This design has the advantage of being able to handle wide fluctuations in throughput. The counterrotating separator is recommended for applications in

TABLE 21-19 Maximum Depth of Material for Separator by Magnetic Pulley Based on Pulley Diameter and Linear-Velocity Belt

Diameter of pulley, mm	Belt linear velocity, m/s	Depth of mterial, mm
203	0.584	38
305	0.890	70
380	1.017	89
508	1.271	121
610	1.448	140
762	1.678	165
914	1.855	191
1067	2.033	210
1219	2.211	235

TABLE 21-20 Maximum Capacity for Magnetic Pulley Separator*

Pulley diameter, mm	Belt width, mm	Belt velocity, m/s	Capacity, m^3/h
203	203	0.585	12.2
	406		24.9
	610		62.3
	914		133.0
381	305	1.017	38.0
	406		50.0
	610		113.0
	914		255.0
	1219		515.0
457	305	1.143	47.0
	406		59.0
	610		130.0
	914		300.0
	1219		623.0
610	406	1.448	88.0
	610		170.0
	914		374.0
	1219		755.0
	1524		1133.0
914	457	1.855	153.0
	610		218.0
	914		481.0
	1219		935.0
	1529		1500.0

*Courtesy of Eriez Magnetics.

which recovery is more important than grade. This unit can handle particles up to 3 or 4 mm (⅛ in) in size, but with less satisfactory results for particles finer than 0.5 mm and slurries containing 30 to 40 percent solids by weight. Figure 21-53 shows a gauss (tesla) chart for a 1.2-m-diameter wet drum separator. The influence of drum diameter in separation is shown in Table 21-22 and in Fig. 21-54. These data result from the processing of a partially martised magnetite ground to 75 percent minus 0.044 mm (325 mesh). The influence of drum diameter and grind in separator capability is summarized in Table 21-23. Figure 21-55 shows the influence of drum diameter on investment. The **installed cost** of a wet single-drum magnetic separator can vary between $25,000 and $75,000 per meter of magnet width, depending on the design, dimensions, and manufacturer. Multiple-drum costs increase in about direct proportion to the number of drums required. **Maintenance costs** per year vary between 3 and 5 percent of the initial investment.

Induced-Roll Separators These devices, which have been in commercial use since 1890, handle only dry, granulated material. They are similar to drum separators, with the difference that the cylinder rotates in the gap of an electromagnet. Magnetic-field gradients are obtained by creating sharply edged ridges on the surface of the cylinder or by constructing a cylinder of alternate magnetic and nonmagnetic disks. A schematic of an induced-roll separator is shown in Fig. 21-56. The best particle-size distribution for separation

TABLE 21-21 Correction Factors for Magnetic-Pulley Capacities*

Type of application	Type of tramp iron to be removed	Correction factor
Crusher and primary-mill protection	Large and medium, over 30 g (1 oz)	1.0
Secondary-mill, pulverizer, and general separation	Large, over 30 g	1.0
	Medium, 30 to 240 g	1.3
	Small, 8 mesh (2.38 mm) to 30 g	2.0
Product purification	Fine ferrous contamination; finer than 8 mesh (2.38 mm)	4.0

*Courtesy of Eriez Magnetics.

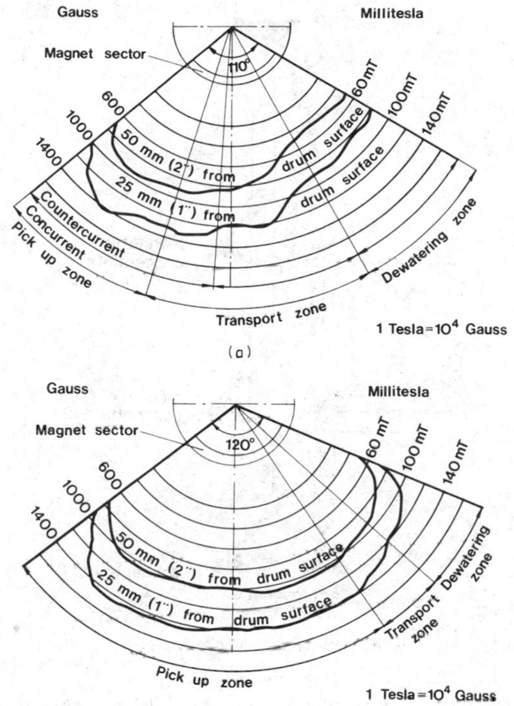

FIG. 21-53 Magnetic-field distribution charts. (*a*) Concurrent and countercurrent wet drum magnetic separator, 1.2-m diameter. (*b*) Counterrotation wet drum magnetic separator, 1.2-m diameter. *(Courtesy of Sala International Inc.)*

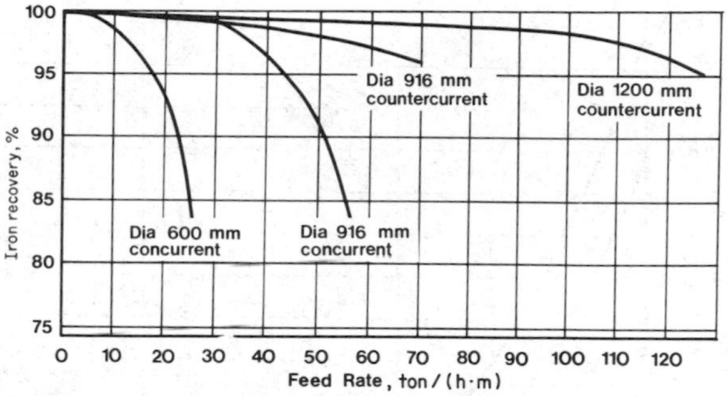

FIG. 21-54 Influence of drum diameter on separation. *(Courtesy of Sala International Inc.)*

TABLE 21-22 Influence of Magnetic-Drum Diameter on Separation*

Separator diameter, m	No. of stages	Composition of concentrates: SiO_2, %	Composition of concentrates: Fe, %	Fe recovery, %	Feed rate, t/(h·m)
0.600	6	1.1–1.3	70.0	92	10–12
0.916	6	0.9–1.0	70.0	98	28–33
1.200	4	0.9–1.0	70.0	98	62–85

*Courtesy of Sala International Inc.

is minus 2 mm, plus 0.074 mm (minus 10 mesh, plus 200 mesh). Industrial devices are built with multiple rolls, which operate either in a series or in parallel, and can be used as concentrators or as purifiers (see Fig. 21-57). Standard widths for the rolls are 0.25, 0.5, and 0.75 m. Capacities vary between 1.5 and 18 t/(h·m). Induced-roll separators are used only to process weakly magnetic materials. **Capital costs** for this type of device are relatively low compared with those of other high-intensity magnetic separators, but **total process costs** are high owing to moisture-free feed requirements. A wet-process induced-roll separator was developed in the U.S.S.R. during the early 1960s and is reported to have a capacity of up to 100 t/h. In 1964 an Australian manufacturer introduced a wet-type induced-roll separator designed with a laminated, grooved rotor that rotates around a vertical axis (pole). These devices are built with up to 10 poles and are used principally to concentrate ilmenite sands. Capacity is approximately 0.8 t/h per magnetic pole.

Separations similar to those obtained with dry induced-roll devices can be obtained with cross-belt separators (Fig. 21-49). These units are built with up to eight poles, each of which can operate at different magnetic-field intensities to allow simultaneous production of different concentrates. However, capacity is low, and installed costs per ton capacity are high compared with induced-roll units.

Induced-Pole Separators In devices of this type, magnetic-field gradients are produced by the application of background magnetic field to a ferromagnetic matrix, thereby inducing magnetic poles around matrix edges. The correlation of edge and field directions determines whether the separator is a parallel field-to-flow unit or a perpendicular field-to-flow unit. In this category there are only two practical types of separator constructions, C-frame and solenoid. Uniformity of background magnetic field depends on design.

1. *C-frame magnets.* As shown in Fig. 21-45*b*, these magnets employ a ferromagnetic matrix placed between the poles of an electromagnet. With this design, however, the background magnetic field is not uniform. Also, high magnetic fringe fields are usually noticed in the flush region of these separators, and these can cause possible matrix clogging when even relatively small amounts of ferromagnetic particles are present in the slurry. The ferromagnetic material used to transfer the magnetic-flux lines from pole to pole occupies between 40 and 80 percent of the magnetized volume.

TABLE 21-23 Influence of Drum Diameter and Grind on Separator Capability*

Feed: Description	Feed: Percent of feed minus 74 μm	Separator arrangement	Recommended capacities, t/(h·m), Diameter of drum, m: 0.60	0.90	1.20
Coarse	15–25	Concurrent	15–25	70–90	120–160
Medium	50	Concurrent or full countercurrent	10–15	35–50	60–90
Fine	75–95	Semicountercurrent	6–10	30–50	60–90

*Courtesy of Sala International Inc.

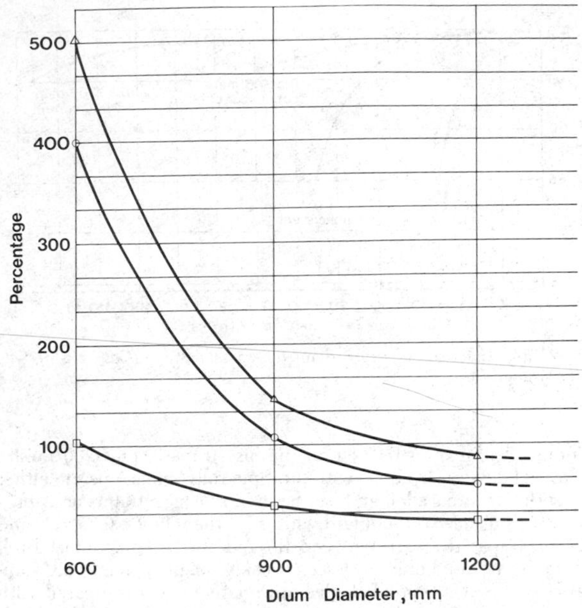

FIG. 21-55 Influence of drum diameter on separation cost. △, two stages coarse separation; ○, three stages fine separation; □, sum of coarse and fine separations. *(Courtesy of Sala International Inc.)*

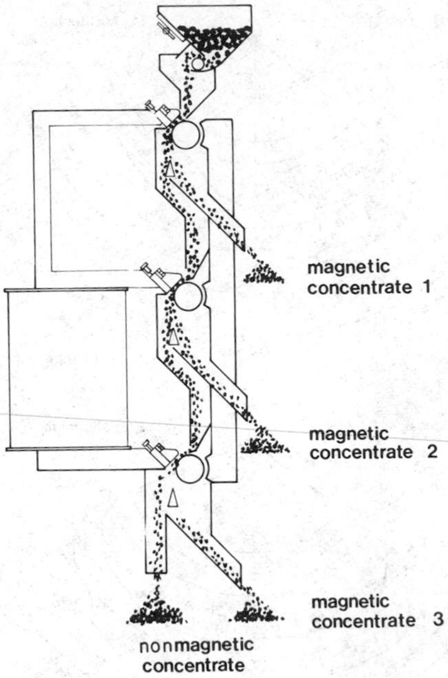

FIG. 21-57 Multiple induced-roll magnetic separator.

2. *Solenoid magnets*. These devices can be designed for wet or dry feeds. Depending on design, they can have a relatively uniform background magnetic field. It can be noted from Fig. 21-58 that the use of the return frame is important for generating a uniform magnetic field. Filamentary-type matrices, which occupy less than 10 percent of the magnetized volume yet still provide very high field gradients, can be used with these types of magnets.

The most familiar of the C-frame, matrix-type industrial magnetic separators are the Carpco, Eriez, Readings, and Jones devices. The Carpco separator employs steel balls as a matrix, Eriez uses a combination of expanded metal matrices, and the Readings and Jones separators have grooved-plate matrices. Capacities for this type of unit are reported to up to 180 t/h (in the case of Brazilian-hematite processing).

Solenoid magnetic separators are designed for batch-type, cyclic, and continuous operation. Devices which can use matrices of expanded metal, grooved plates, steel balls, or filamentary metals have been designed. Continuous separators with capacities to 600 t/h for iron ores (similar to the Brazilian hematite) are commercially available (Sala International Inc.). Selection of the method of operation is application-dependent, being based on variables such as temperature, pressure, volume of magnetics in the feed, etc.

A familiar type of cyclically operated solenoid electromagnet is the Franz separator, a well-known continuous type of solenoid separator manufactured by Krupp-Sol. An enclosed flux return-frame solenoid design for cyclic and continuous use is built by Sala International Inc. A schematic of a continuous Sala high-gradient magnetic separator is shown in Fig. 21-59.

Depending on the type of matrix used, induced-pole magnetic separators can be classified as either high-intensity magnetic separators, which utilize grooved plates or steel balls as the matrix material, or as high-gradient magnetic separators, which use filamentary matrices such as steel wool or expanded metal. Filamentary matrices have proved to be more advantageous.

The maximum magnetic field produced by a C-type device is 2 T.

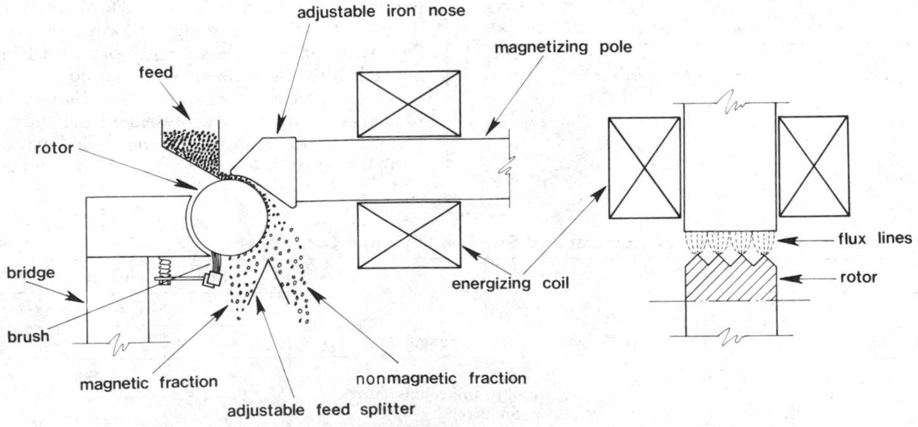

FIG. 21-56 Schematic diagram of an induced-roll separator.

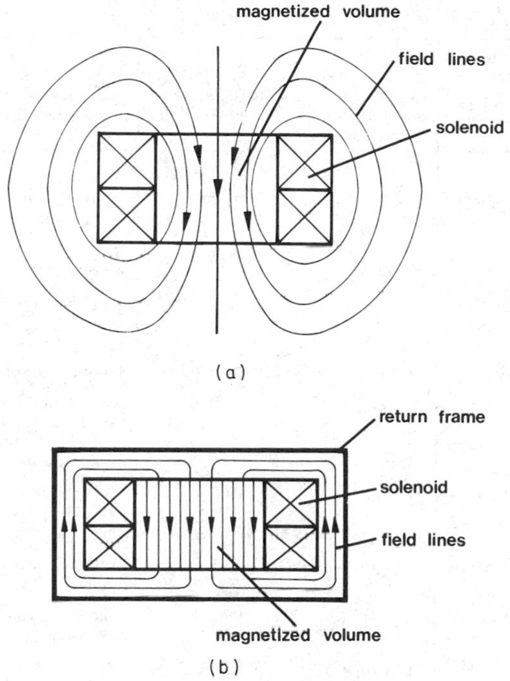

FIG. 21-58 Solenoid-type magnets. (*a*) Nonuniform field. (*b*) Uniform field by use of return frame.

For solenoids, conventional designs produce magnetic-field intensities up to 2 T, while superconducting units can be constructed with ratings up to 8 T.

In all induced-pole devices, the more magnetic particles are retained on the matrix while the less magnetic fraction is carried away in the slurry.

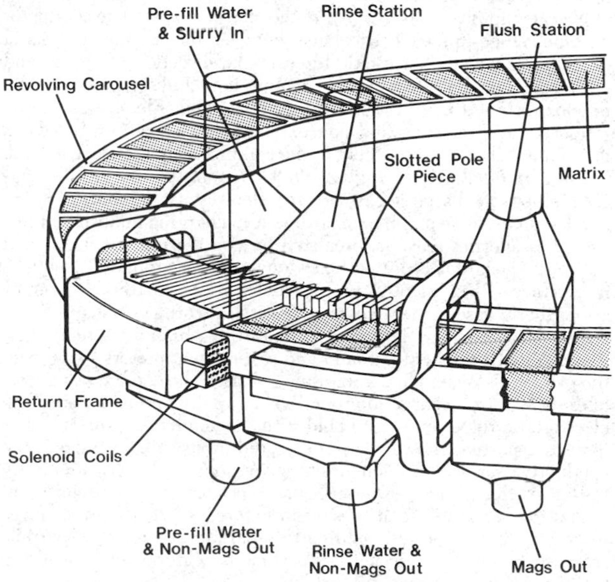

FIG. 21-59 Schematic of continuous high-gradient magnetic separator. *(Courtesy of Sala International Inc.)*

Dynamic (or Deflecting) Devices The oldest type of dynamic separation device is the Franz Isodynamic separator. This laboratory device has a dipole configuration with the poles shaped so that the value of $H\,dB/dz$ is constant throughout the working separation volume. Material to be separated can be fed through a vibratory chute or dropped between the poles, producing a separation based only on relative magnetic susceptibility.

There are a variety of new developments in magnetic separation which are of possible interest, but their commercial applicability is still not yet assured. These include quadripole separators and a spiral-flow device.

Table 21-24 lists potential applications for all types of magnetic separators.

ELECTROSTATIC SEPARATION

GENERAL REFERENCES: Fraas, *Electrostatic Separation of Granular Materials,* U.S. Bur. Mines Bull. 603, 1962. Gaudin, "The Principles of Electrical Processing with Particular Applications to Electrostatic Separation." *Min. Sci. Eng.,* 3(2), 46–57 (1961). Moore, *Electrostatics and Its Applications,* Wiley-Interscience, New York, 1973. Ralston, *Electrostatic Separation of Mixed Granular Solids,* Elsevier, Amsterdam, 1961.

General Principles Electrostatic separation (of particles), also commonly known as high-tension separation, is a method of separation based on the differential attraction or repulsion of charged particles under the influence of an electrical field. Applying an electrostatic charge to the particles is a necessary step before particle separation can be accomplished. Various techniques can be used for charging. These include contact electrification, conductive induction, and ion bombardment.

Regardless of the method of charging, the amount of charge that can be accumulated on a particle is limited by the maximum achievable charge density and the surface area of the particle. Electrostatic separation of mixed particles is possible when the electrostatic force acting on some particles is great enough to overcome gravity or inertial forces. Because the surface area of a solid varies as the square of a linear dimension whereas the mass varies as the cube of that dimension, gravity and inertial forces acting on solid particles increase faster with particle size than do electrostatic forces for charged particles in electric fields. Thus, there are upper size limits beyond which electrostatic separation of particles of a given shape is not feasible. For granular materials this upper size limit is about 1.5 mm; for thin pieces of large cross-sectional area and for long pieces of small cross-sectional area the limit can be greater than 25 mm.

The motion of fine particles immersed in a moving fluid is more greatly affected by fluid drag forces than that for similar large particles. For very small particles in a fluid, particle motion approximates the motion of the enveloping fluid. Industrial electrostatic separation of solid particles, which is universally conducted in air (or other easily ionizable gas), is difficult at particle sizes less than about 0.074 mm (200 mesh) and almost impossible at less than 0.044 mm (325 mesh).

Charging Mechanisms

Contact Electrification (Fig. 21-60*A*) When dissimilar materials touch each other, there is an opportunity for the transfer of electric charges. The extent of charge transfer can be such that a significant surface charge of opposite sign is developed when the materials are later separated. High temperatures and low humidity favor the development of high surface charges through the mechanism of contact electrification. Rubbing the materials together to increase the area of effective contact can also lead to high surface charges.

Particles carrying charges of opposite polarity due to contact electrification will be attracted to opposite electrodes when passing through an electric field and thus can be separated from each other.

Conductive Induction (Fig. 21-60*B*) The term "conductive induction" describes the process by which an initially uncharged particle that comes into contact with a charged surface assumes the polarity and, eventually, the potential of the surface. A particle of a good electrical conductor will assume the polarity and potential of the charged surface very rapidly. However, a dielectric particle will

TABLE 21-24 Potential Applications of Magnetic Separators

Device type	Type of construction	Maximum background magnetic field, Oe	Type of matrix which can be used	Maximum field gradient obtainable, G/cm	Required magnetic susceptibility for particulates	Particle size to be treated, 0 mm	Materials which can be treated; fields of use
Grate	Permanent magnet	500	Rods	500	Ferro	< 12	Tramp and fine iron
Pulley	Permanent magnet and electromagnet	100–200	....	100–1000	Ferro, strongly	< 50	Ferro and strongly magnetic
Belt	Electromagnet	100–1000	....	100–1000	Strongly	0.15–30	Strongly magnetic
Drum	Permanent magnet and electromagnet	500–1000	....	500–1000	Strongly	0.02–20	Magnetite processing
Franz Isodynamic	Electromagnet	10,000	....	2000	Strongly, weakly	> 0.01	Only for laboratory
Solenoid; Franz ferrofilter	Electromagnet	20,000	Steel ribbons, balls	200,000	Strongly, weakly	> 0.01	Tramp and fine iron, ceramic slurries, industrial minerals, chemical industry
Induced rolls	Electromagnet	20,000	....	200,000	Strongly	0.03–3	Dry, dedusted, weakly magnetic particles
C-frame type; Jones	Electromagnet	20,000	Grooved plates	200,000	Strongly, weakly	0.01–2	Iron ores, industrial minerals
Carpco	Electromagnet	20,000	Steel balls	45,000	Weakly	0.01–1	Iron ores, industrial minerals
Marston Sala high-gradient magnetic separator	Electromagnet, superconducting	20,000 50,000	Steel wool, expanded metal, steel balls	25×10^6	Strongly to very weakly	0.0001–2	Iron ores, industrial minerals, coal, liquefied coal, wastewaters, purifiers, catalyst recovery, chemical industry

become polarized so that the side of the particle away from the charged surface develops the same polarity as the surface. Particles of intermediate conductivity may be initially polarized but approach the potential of the charged surface at a rate depending on their conductivity.

If a good conductor particle and a good dielectric particle are just separated from contact with a charged plate, the conductor particle will be repelled by the charged plate and the dielectric particle will be neither repelled nor attracted by it.

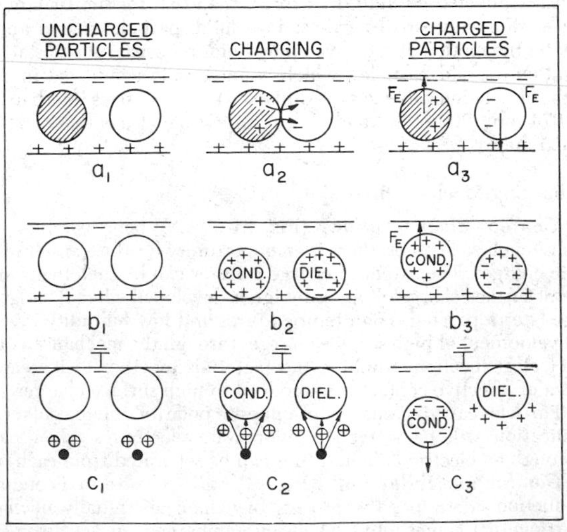

FIG. 21-60 Schematic representation of charging mechanisms. (*A*) Contact electrification. (*B*) Conductive induction. (*C*) Ion bombardment. Cond. = conductor particle; diel. = dielectric particle; ● = high-voltage dc electrode; ⊕ = ions from corona discharge at high-voltage electrode.

The charged plate must be balanced by other oppositely charged (or earthed) bodies to maintain overall neutrality. In electrostatic separation this is usually accomplished by means of a single electrode of charge opposite in sign to that of the charged plate. The conductor particle is then in the electrical field between the two electrodes and experiences a net electrostatic force in the direction of the second electrode. The dielectric particle, having no net charge, experiences no electrostatic force in a uniform electric field. Electrostatic separation of the conductor and nonconductor (dielectric) particles can be accomplished by movement of the conductors in the electric field.

Ion Bombardment (Fig. 21-60*C*) The most positive and strongest method of charging particles for electrostatic separation is ion bombardment. Use of ion bombardment in charging materials of dissimilar properties may be visualized by considering conductor and nonconductor (dielectric) particles touching the grounded conducting surface of Fig. 21-60*C*. Both particles are bombarded by ions of atmospheric gases generated by an electrical corona discharge from a high-voltage electrode (usually a fine tungsten-alloy wire at ±20 to 30 kV with respect to ground and several centimeters away from the particles). When ion bombardment ceases, the conductor particle loses its acquired charge to ground very rapidly and experiences no electrostatic force tending to hold it to the conducting surface. The dielectric particle, however, being coated on its side away from the conducting surface with ions of charge opposite in electrical polarity to that of the surface, experiences an electrostatic force tending to hold it to the surface. If the electrostatic force is larger than the force of gravity or other forces tending to separate the dielectric particle from the conducting surface, the particle is held in contact with the surface and is said to be "pinned."

Electrostatic-Separation Machines The first electrostatic machines to be used commercially employed the principle of contact

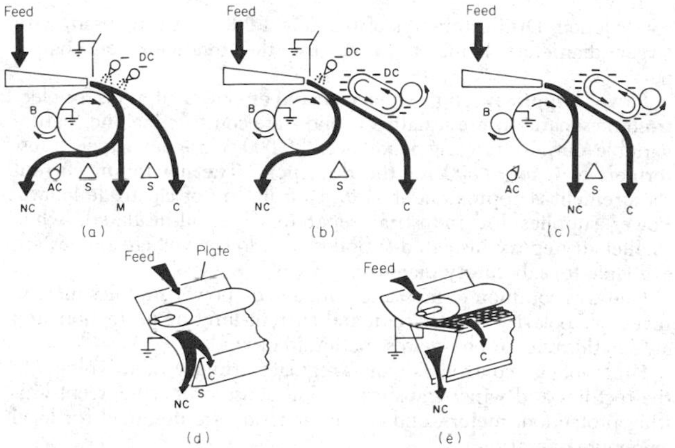

FIG. 21-61 Operating principles of electrostatic separators. C = conductors; NC = nonconductors; M = middling; DC = high-voltage dc electrodes; AC = high-voltage ac wiper (electrodes); B = brush; S = splitter.

electrification. These were free-fall devices incorporating large vertical plates between which an electrostatic field was maintained. These separators required elaborate feeding systems for charging by contact electrification. In addition, they needed internal humidity control and five or more stages to achieve an effective separation. Commercial application of free-fall separators was largely discontinued in the late 1940s because of these complexities and related high operating costs.

The common types of present industrial electrostatic separators employ charging by conductive induction and/or ion bombardment. Figure 21-61 illustrates the principles of application.

Conductive-Induction Machines Electrostatic separators exploiting the principle of conductive induction will generally use the following electrode designs:

Plate separators. These separators introduce material typically by gravity onto a grounded-metal slide in front of which is placed a static electrode of large surface area. The elaborate contact electrification used in earlier free-fall electrostatic separators was avoided in later devices by incorporating the slide principle. The inclined plate separator is illustrated in Figs. 21-61*b* and 21-62*a*. Separation occurs by particles selectively acquiring an induced charge from the grounded plate and then being attracted in the direction of the charged electrode.

Screen-plate separators. These separators represent a further improvement of the plate separator. They include a metal slide at ground potential that is extended with a conducting screen of suitable screen-opening size to allow easy passage of the largest grains being treated. A stationary electrode is placed above the slide and screen sections as shown in Figs. 21-61*e* and 21-62*b*. Particles capable of assuming an induced charge from the slide are attracted by the electrode and prevented from passing through the screen grid; the other particles pass through the screen unaffected.

Rotating-electrode separators. These separators are similar in principle to plate separators except that a metal slide or a large rotating grounded-metal drum may be used in conjunction with a rotating electrode of large surface area and opposite potential. This electrode can be either circular or elliptical in cross section and is self-cleaning. Figures 21-61*b* and 21-61*c* illustrate this type of device.

Electrostatic sieves. Electrostatic-sieve separation evolved from investigations into the dry treatment of particles smaller than 0.074 mm (200 mesh). The fine nature of material treated by this method dictates that vibratory or gravitational feeding methods be replaced by an electrostatic technique, referred to as "pumping," to overcome naturally occurring surface forces which tend to cause agglomeration of particles. Material is conveyed electrostatically into a screen chamber, which consists of electrified parallel sieves and matching attracting plates of opposite polarity. Multiple sets of sieves can be incorporated. The principle of operation of one such machine is shown in Fig. 21-63.

Ion Bombardment

Conductive roll (drum) separators. These separators are by far the most widely employed industrial-machine type. The electrostatic elements of these machines consist of a conducive rotating drum at ground potential coupled with one or more high-voltage ionizing electrodes. Suitably placed nondischarging (static) electrodes are often used in conjunction with an ionizing electrode to create a static

FIG. 21-62 Plate and screen-plate types of conductive-induction electrostatic separator. (*a*) Readings electrostatic plate separator. *(Courtesy of Readings, Inc.)* (*b*) Mineral Deposits Limited MK III electrostatic screen-plate separator. *(Courtesy of Mineral Deposits Limited.)*

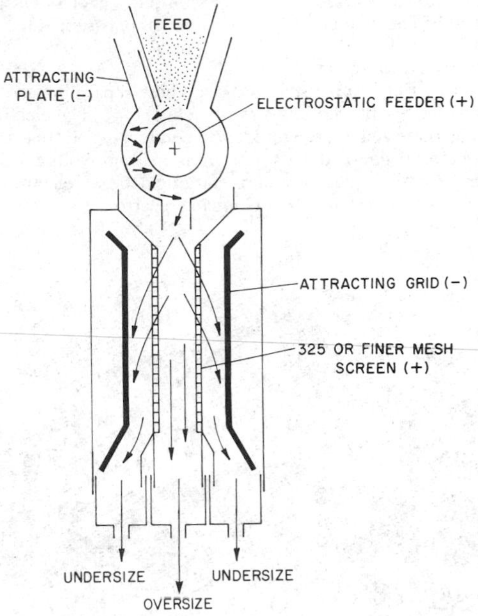

FIG. 21-63 Electrostatic sieve for fine metal powders.

field which aids centrifugal force in removing conductive particles from the drum surface.

Drum construction is typically of carbon or stainless steel when treating granular materials minus 1 mm in size. Recently, nonmetallic conducting materials such as rubber impregnated with carbon black have been used for drum construction when treating materials coarser than 1 mm. Conducting surfaces with a higher coefficient of friction are extending the treatment of materials by electrostatic separation up to 25 mm and, in some cases, to 50 mm, at which point favorable shape factors can be exploited. Drum diameter has ranged from 0.150 to 0.360 m, while drum length varies from 0.460 to 3.050 m in industrial ion-bombardment (high-tension) machines.

Feeding these separators is accomplished by vibratory, belt, rotary spline, or gravity methods, depending on the particle size being treated. Vibratory and belt feeding techniques are preferred for coarser sizes, and rotary spline and gravity methods are normally used for finer materials. Exceptions to this generalization can be observed in plant practice. The ionizing electrodes employed in these machines vary considerably in appearance, but all produce a corona.

An alternating-current electrode system referred to in the industry as a "wiper" is often installed in the nonconductor product-collection section behind each drum. The function of the wiper is to use an ac corona to neutralize the charge on the nonconductor particles pinned to the surface of the drum and thereby reduce the workload for mechanically operated brushing systems.

Ion-bombardment machines are available in horizontal and vertical (stacked) configurations. Horizontal units are preferred for large-tonnage applications, in which machines are arranged in rows for ease of maintenance and operation. Stacked units up to four rolls high have been used to reduce material-handling costs in multipass treatment schemes when the material is capable of being passed vertically from one roll to another by gravity. Large and small roll-type separators and internal details are shown in Fig. 21-64.

Nonconductive roll (drum) separators. These separators fall into the class of shape separators. The so-called shape separator does not depend on preferential leakage by conductors of the charge given to all entering particles. Separation is accomplished by selectively pinning high-surface-area particles in preference to other, more rounded particles. Machine construction is similar to that of the larger conducting-roll ion-bombardment separators except for drum construction. Drum dimeters of 0.250 to 1.200 m can be used, with larger diameters being required for the treatment of coarser particles.

Power Supplies High-voltage ac and dc power supplies for electrostatic separators are usually of solid-state construction and feature variable outputs ranging from 0 to 30,000 V for ac wiper transformers to 0 to 50,000 for the dc supply. The maximum current requirement is approximately 1.0 to 1.5 mA/m of electrode length. Power supplies for industrial separators are oil-insulated, while smaller dry-epoxy-insulated supplies with lower voltage ranges are available for laboratory machines.

Features common to most high-voltage dc power supplies include reversible polarity, short-circuit and current-limiting protection, and automatic residual-charge dissipation to ground.

High-voltage controllers which regulate primary input voltage to the rectifier and wiper transformer and house primary current-limiting protection, meters, and instrumentation are designed for local or remote operation.

Machine Capacities and Costs (1979) Table 21-25 presents capital-cost information as related to machine capacity for conductive-induction and ion-bombardment electrostatic separators.

Applications of Electrostatic Separation

Mineral Beneficiation Electrostatic methods are widely used in the processing of ores with mineral concentrates. Generally, electrostatic separation is used as a part of an overall flow sheet comprising various combinations of physical separation procedures. It is particularly well established in the processing of heavy-mineral beach sands from which are recovered ilmenite, rutile, zircon, monazite silicates, and quartz. High-grade specular hematite concentrates have been recovered at rates of 300 kg/s (1000 tons/h) in Labrador. Applications also include processing tin ores to separate cassiterite from columbite and ilmenite.

Charging by ion bombardment is the technique used in most mineral separations. The conductive-induction (nonionizing) types of separators have also been used. Applications of this device in the minerals industry include rutile-zircon, biotite-feldspar, quartz–phosphate pebble, carbon-silica, diamond-gangue minerals, quartz-kyanite and sillimanite, muscovite-feldspar, talc-magnesite, and asbestos from host rock.

Generally, separators of the conductive-induction type have a lower capacity per unit length of electrode than the ion-bombardment (ionizing) type of apparatus, and multipass operation is generally required. This disadvantage is offset by the ability of these separators to (1) produce high-grade concentrates from ore materials that are otherwise difficult to process and (2) process a coarser material than competitive processes such as froth flotation. Electrostatic separation is being tested as an alternative to the presently used process of flotation of pebble phosphates for coarser-size fractions. Advantages sought include reduced reagent costs, a lower water requirement, and fewer tailings-disposal problems when a part of the flotation circuit is eliminated.

Metal-Powder Processing Removal of nonmetallic impurities from metal powders by ion-bombardment electrostatic separation is readily accomplished in the minus 0.300- to plus 0.074-mm-size range. Examples of metal powders treated are high-purity metal or alloy powders used for making powder-metallurgy parts and "scrap" metal powder from such operations as grinding and sandblasting.

Food Processing Most food products contain sufficient water so that they behave as conductors in electrostatic separation. Stones can be removed from products such as beans and lentils, while stems and dry leafy materials can be separated from moist leafy ones provided the maximum particle size is in the range of about 15 to 30 mm. Separation of the shells of almonds and walnuts from the meat of the nuts can be accomplished electrostatically.

Waste Processing Industrial and municipal wastes can be processed for separation and recovery of resources. For example, many kinds of scrap wire can be separated into metal and plastic products of high purity by mechanical granulation to free the wire metal from the insulation, followed by electrostatic separation to concentrate the

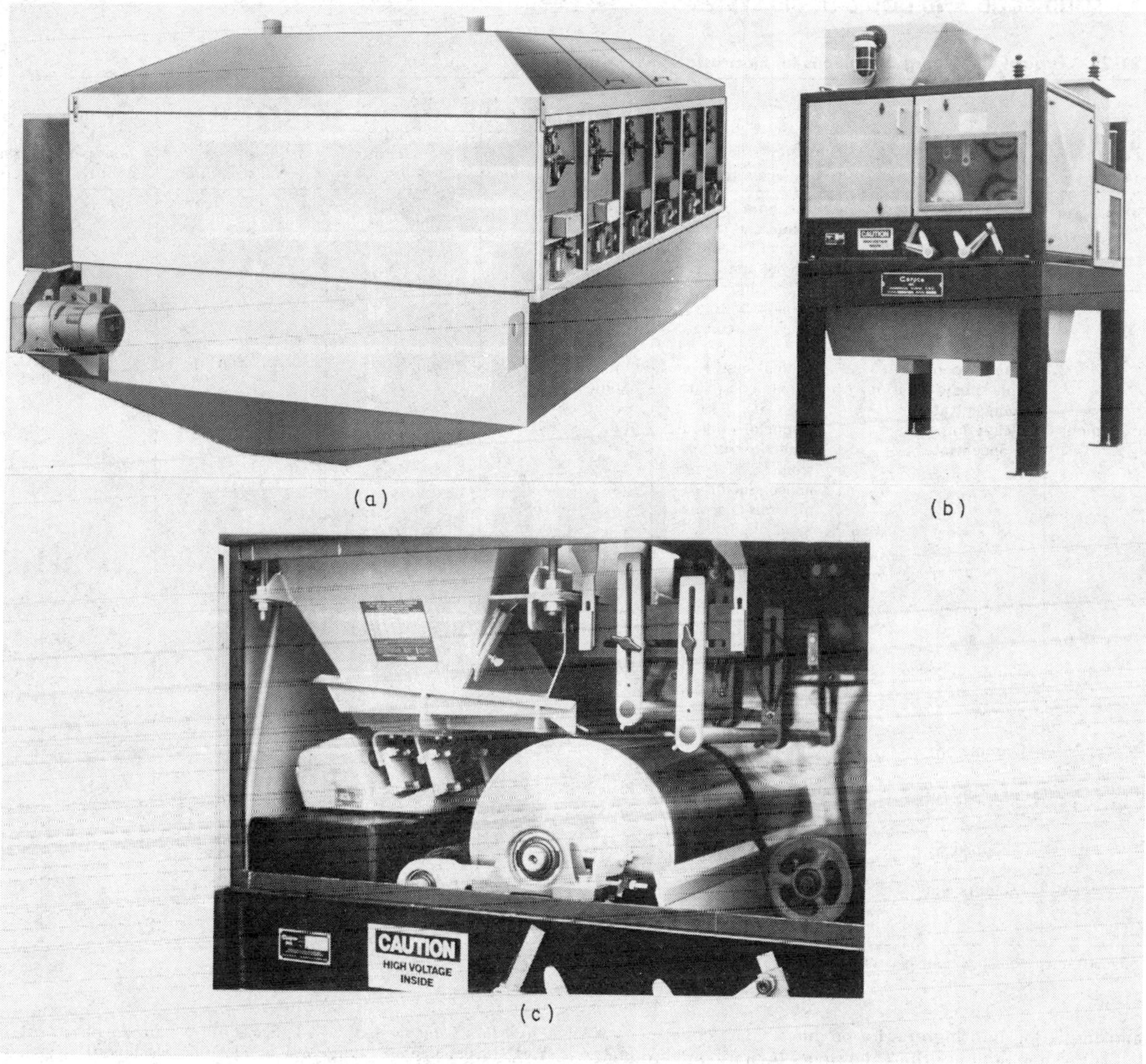

FIG. 21-64 Roll-type ion-bombardment electrostatic separators. (*a*) Carpco high-capacity electrostatic roll separator. (*b*) Carpco pilot-plant–industrial roll electrostatic separator. (*c*) International configuration of roll-type electrostatic separator, ionizing mode. *(Courtesy Carpco, Inc.)*

TABLE 21-25 Capital Costs of Electrostatic Separators

	Capacity, t/h	Capital cost includes power supplies, $	$/(t/h)
Conductive-induction electrostatic separators			
Plate separator: basis, 5 pass × 2 start; 1.42 m in length per start; capacity, 900 kg/(h·m)	2.6	19,500	7,500
versus			
Rotating electrode: basis, 2 pass × 2 start; 1.42 m in length per start; capacity, 1800 kg/(h·m)	5.1	30,500	6,000
Ion-bombardment electrostatic separators			
3-m roll units: basis, 1 pass × 6 or 12 start; 2.92 m in length per start; capacity, 2150 kg/(h·m)	37.6 for 6 start 75.2 for 12 start	94,500 for 6 start 144,000 for 12 start	2,500 1,900
1.5-m roll units: basis, 1 pass × 4 start; 1.42 m in length per start; capacity, 1600 kg/(h·m)	9.1	25,000	2,800
Basis, 3 pass × 8 start; 1.42 m in length per start; capacity, 1600 kg/(h·m)	18.3	79,000	4,300

NOTE: Capacities and 1979 costs based on the treatment of industrial minerals having a specific gravity of 2.6 to 4.0. High-voltage power supplies and controls *are included*. Special materials of constuction, automatic controls, and varying equipment options have not been considered. To convert kilograms per hour-meter to kilograms per second-meter, multiply by 2.778×10^{-4}; to convert metric tons per hour to kilograms per second, multiply by 0.2778.

TABLE 21-26 Typical Operating Conditions for Electrostatic Separations

Type of particle charging	Feed	Separation	Type of separator	Feed temperature, °C	Feed size, mm	Feed rate, metric tons per hour per start*	No. of stages of separation
Conductive induction	Zircon concentrate (eastern Australia)	Residual silica and conductor minerals from zircon; upgrading of concentrate from 98.95 to 99.35% zircon at 92% recovery	Screen plate	50–80	−0.21 + 10,074	0.6–0.7	2
	Florida-pebble-phosphate flotation concentrate	Residual silica from pebble phosphate	Roll with rotating static electrode	70–90	−1.0 + 0.10	2.7	2
Ion bombardment	Heavy-mineral concentrate	Conductor minerals (ilmenite, rutile) from nonconductor minerals (zircon, monazite, aluminum silicates, quartz and others)	Roll	120	−1.0 + 0.04	2.5	3
	Iron ore	Iron oxides from quartz and silicates	Roll	120	−1.0	6–7	2–4
	Tungsten concentrate	Scheelite from iron oxides and other conductor minerals	Roll	150	−0.6	1.0–1.5	3
	Chrome ore	Chromite from silica and silicates	Roll	120	−0.85		
	Foodstuffs	Removal of small amounts of impurities	Plate or roll	Ambient	−25	0.3–1.0	2
	Chopped wire	Metal from plastic insulation	Roll	Ambient	−12.5	1.5–2.0	2
	Metal powder	Removal of nonmetallic impurities	Roll	Ambient to 120	−0.20	1–2	2

*To convert metric tons per hour per start to kilograms per second per start, multiply by 0.2778.

metal and insulation into high-purity products. Glass cullet can be separated from metallic impurities and stones by ion-bombardment electrostatic separation.

Other Applications Electrostatic separators have been used to separate a number of different types of materials not only on the basis of differences in dielectric properties but also in combination with differences in surface conductivity and shape factors. Among these operations are seed sorting, cleaning of spices, separating of pill coatings from base materials, removal of textile from reclaimed plastics, and separation of paper and plastic. Electrostatic separation has also been adapted for use in classification and sizing when elongated particles or extremely fine sizes cause difficulty in conventional dry screening applications.

Typical Operating Conditions Table 21-26 presents some typical values of important operating conditions for the separation of several different types of feed materials. In considering candidate processes for a given separation job, the table can sometimes be helpful in showing that materials of similar properties and/or economic value can be treated by electrostatic separation.

FLOTATION

GENERAL REFERENCES: *Froth Flotation*, 50th anniversary vol., AIME, New York, 1962. Fuerstenau (ed.), *Flotation*, A. M. Gaudin Memorial Vols. I and II, Society of Mining Engineers, AIME, New York, 1976. Gaudin, *Principles of Mineral Dressing*, McGraw-Hill, New York, 1939. Gaudin, *Flotation*, 2d ed., McGraw-Hill, New York, 1957. Glembotskii, Klassen, and Plaksin, *Flotation*, Primary Sources, New York, 1972. Mular and Bhappu (eds.), *Mineral Processing Plant Design*, 2d ed., Society of Mining Engineers, AIME, New York, 1980. Sorensen, "Large Agitair Flotation Machines Design and Operation," paper presented at 14th International Mineral Processing Congress, Toronto, Oct. 17–23, 1982. Sutherland and Wark, *Principles of Flotation*, Australasian Institute of Mining and Metallurgy, Melbourne, 1955. Taggart, *Handbook of Mineral Dressing*, 2d ed., Wiley, New York, 1945. Taggart, *Elements of Ore Dressing*, Wiley, New York, 1951.

General Description Mixed liberated particles can be separated from each other by flotation if there is sufficient difference in their wettability. The flotation process operates by preparing a water suspension of a mixture of relatively finely sized solids. This is usually done in an agitated chamber open at the top. Fine bubbles of air are then dispersed through the agitated suspension to form a froth which rises to the top of the chamber (cell). Particles which are readily wetted by water (hydrophilic) tend to remain in the water suspension. Those particles that are not easily wetted (hydrophobic) tend to be collected at the air-bubble–water interface and rise to the surface attached to air bubbles. Thus differences in the surface chemical properties of the solids are the basis for separation. Surfaces which do not have strong surface chemical bonds that were broken tend to be nonpolar surfaces and are not readily wetted. Substances such as graphite and talc are examples of solids that can be broken easily along weakly bonded layer planes without rupturing strong chemical bonds. These solids are naturally floatable.

The relative wettability of the solids in a mixture can be enhanced by the addition of various agents which are adsorbed selectively on the surfaces of certain species in the mixture. Frothing agents are also added to generate a stable bubble and foam action. In addition, reagents are used to prevent flocculation and maintain an appropriate degree of dispersion of the suspended solids. Other reagents may be used to alter the selectivity of the collector agent and to depress (prevent the flotation of) certain species.

The flotation process is widely used in the mineral-process industry to concentrate mineral values in ores. The flotation of various sulfide minerals away from gangue constituents is common practice throughout the world, with millions of tons of mixed solids being processed daily. Although specific-gravity difference is not the basis for the separation, metallic minerals are generally ground to finer than 48 mesh to enable the particles to remain in active suspension. Nonmetallics, such as coal, that have lower specific gravity can be processed at a coarser grind (10 to 28 mesh).

Flotation Reagents

Promoters or Collectors These reagents provide minerals which are to be floated with a water-repellent air-avid coating that will adhere to an air bubble. Typical collectors for flotation of metallic sulfides and native metals are xanthates, R—O—C(=S)—SNa, and dithiophosphates, (R—O)₂P(=S)—SNa, where R is an alkyl group of two to six carbon atoms. The ionized collector is adsorbed on a sulfide mineral surface with bonding through the sulfur atoms. The alkyl group provides the water-repellent coating. Quantities of the order of (0.05 to 1.0) $\times 10^{-4}$ kg reagent/kg ore (0.01 to 0.2 lb reagent/ton ore) are generally used.

Crude or refined fatty acids and their soaps, petroleum sulfonates, and sulfonated fatty acids are widely used as collectors in flotation of fluorspar, phosphate rock, iron ore, and other nonmetallics. In these operations reagent dosages are much higher, of the order of (1.0 to 10.0) $\times 10^{-4}$ kg reagent/kg ore (0.2 to 2 lb reagent/ton ore).

Cationic collectors such as fatty amines and amine salts are widely used for the flotation of quartz, potash, and silicate minerals in quantities of (0.5 to 5) $\times 10^{-4}$ kg/kg (0.1 to 1 lb/ton).

Fuel oil and kerosine are used as collectors for coal, graphite, sulfur, and molybdenite since they are readily adsorbed by such naturally hydrophobic minerals. In fact, a frother alone can often be used to float these minerals. These hydrocarbons are also used as extenders or diluents in nonmetallic flotation with sulfonates, fatty acids, and fatty amines.

Frothers Commonly used frothers are pine oil, cresylic acid, polypropylene glycol ether, and 5- to 8-carbon aliphatic alcohols such as methyl isobutyl carbinol and methyl amyl alcohol. Quantities of frothers required are usually (0.05 to 1.0) $\times 10^{-4}$ kg/kg (0.01 to 0.2 lb/ton).

Modifiers Flotation modifiers include several classes of chemicals:

1. *Activators.* These are used to make a mineral surface amenable to collector coating. Copper ion is used, for example, to activate sphalerite (ZnS), rendering the sphalerite surface capable of adsorbing a xanthate or dithiophosphate collector. Sodium sulfide is used to coat oxidized copper and lead minerals so that they can be floated by a sulfide mineral collector.

2. *Alkalinity regulators.* Regulators such as lime, caustic soda, soda ash, and sulfuric acid are used to control or adjust pH, a very critical factor in many flotation separations.

3. *Depressants.* Depressants assist in selectivity (sharpness of separation) or stop unwanted minerals from floating. Typical are sodium or calcium cyanide to depress pyrite (FeS_2) while floating galena (PbS), sphalerite (ZnS), or copper sulfides; zinc sulfate to depress ZnS while floating PbS; sodium ferrocyanide to depress copper sulfides while floating molybdenite (MoS_2); lime to depress pyrite; sodium silicate to depress quartz; quebracho to depress calcite ($CaCO_3$) during fluorite (CaF_2) flotation; and lignin sulfonates and dextrins to depress graphite and talc during sulfide flotation.

4. *Dispersants or deflocculants.* These are important for the control of slimes which sometimes interfere with selectivity and increase reagent consumption. Soda ash, lime, sodium silicate, and lignin sulfonates are used for this purpose.

Quantities of modifying agents used vary widely, ranging from as little as 0.25 $\times 10^{-4}$ kg/kg (0.05 lb/ton) to as high as (25 or 50) $\times 10^{-4}$ kg/kg (5 or 10 lb/ton), depending upon the reagent and the metallurgical problem.

Mineral Applications A U.S. Bureau of Mines survey covering 202 froth flotation plants in the United States showed that 198 million tons of material was treated by flotation in 1960 to recover 20 million tons of concentrates which contained approximately $1 billion in recoverable products. Most of the world's copper, lead, zinc, molybdenum, and nickel are produced from ores that are concentrated first by flotation. Much of the phosphate and potash required for fertilizers is concentrated by flotation. In addition, flotation is commonly used for the recovery of fine coal and for the concentration of a wide range of mineral commodities including fluorspar, barite, glass sand, iron oxide, pyrite, manganese ore, clay, feldspar, mica, spodumene, bastnaesite, calcite, garnet, kyanite, and talc.

Other Applications In addition to its main application in separation of minerals from one another, froth flotation and modifications of the flotation process have many applications in other fields. These include recovery of bitumen from tar sands, deinking of wastepaper, flotation of solids from white water in papermaking, flotation of impurities from peas, flotation of peeling "beewing" from wheat, and removal of oil or organic contaminants from water or aqueous solutions.

Flotation Machines The machines that are most widely used today in sulfide, coal, and nonmetallic flotation operations in the western hemisphere are the Fagergren,* the D-R Denver,* and the Agitair* flotation machines. Often one type of machine will be used for roughing and another for cleaning.

These machines provide mechanical agitation and aeration by means of a rotating impeller on an upright shaft. In addition, the Agitair and Denver cells also utilize air from a blower to help aerate the pulp.

In the Fagergren machine (Fig. 21-65) pulp is drawn upward into a rotor *A* by the rotor's lower portion *B*. Simultaneously the rotor's

*Agitair flotation machines are sold by the Galigher Ash Company, D-R Denver machines by the Denver Equipment Company, division of Joy Manufacturing Company, and Fagergren machines by the WEMCO Division, Envirotech Corporation.

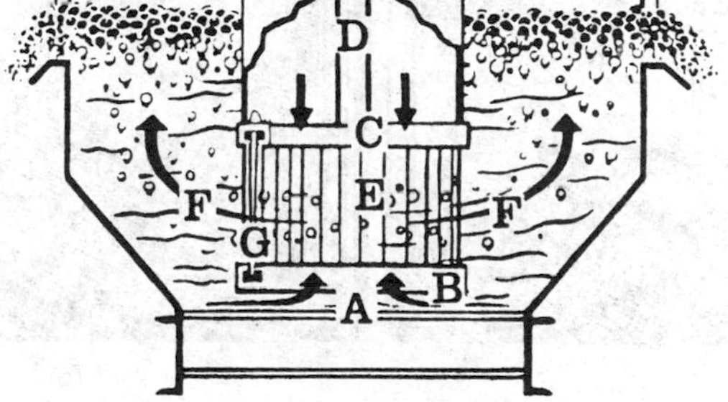

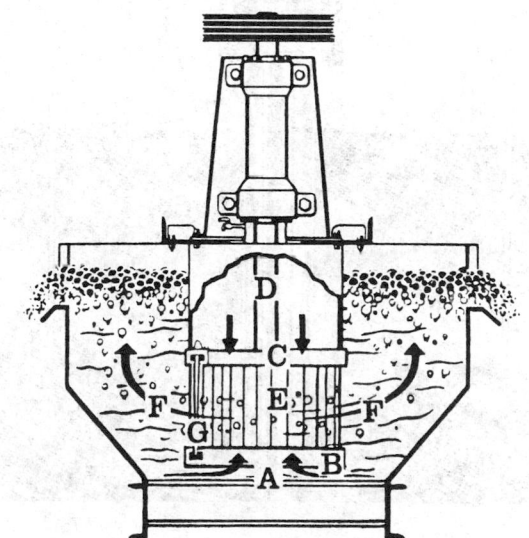

FIG. 21-65 Fagergren flotation machine.

upper end *C* draws air down the standpipe *D* for thorough mixing with the pulp inside the rotor *E*. The aerated pulp is then expelled by a strong centrifugal force *F*. The shearing action of the stator *G*, a stationary cage fitting closely around the rotor, breaks the air into minute bubbles. This action uniformly distributes a large volume of air in the form of minute bubbles in all parts of the cell.

In the D-R Denver machine (Fig. 21-66) the pulp enters the top of the recirculation well *A*, while low-pressure air enters through the air passage *B*. Pulp and air are intmately mixed and thrown outward by the rotating impeller *C* through the stationary diffuser *D*. The collector-coated mineral particles adhere to the rising bubbles and are carried to the top of the cell to be removed in the froth product.

In the Agitair flotation machine (Fig. 21-67) the impeller is a flat rubber-covered disk with steel fingers extending downward from the periphery. A rubber-covered stabilizer eliminates dead spots in the agitation zone and improves bubble-ore contact. Degree of aeration is controlled by regulating air volume on each cell with an individual air valve. Air is supplied at 10×10^3 Pa (1.5 lbf/in^2).

Many other types of flotation cells may still be found in older mills. The Callow cell has no mechanical parts. Ore pulp is suspended and aerated by air bubbles coming through a porous medium (usually a cloth mat) forming the bottom of the cell. In the MacIntosh cell air is introduced through a porous medium wrapped around a rotating pipe near the bottom of a V-shaped trough. The Forrester cell is a V-shaped trough with air introduced through 0.013- to 0.025-m (0.5- to 1.0-in) vertical pipes spaced at 0.10- to 0.15-m (4- to 6-in) intervals lengthwise.

Capacities Tonnage handled by flotation equipment will vary with the pulp density of feed and flotation time (residence time) required for roughing and cleaning. Number of cells required for a specific job can be calculated as follows:

$$\text{No. cells} = (T \times G \times P)/V \qquad (21\text{-}17)$$

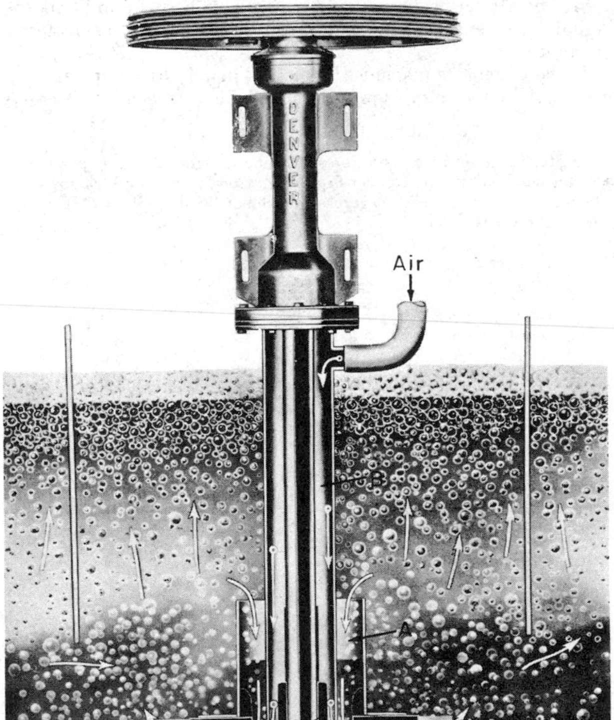

FIG. 21-66 D-R Denver flotation machine.

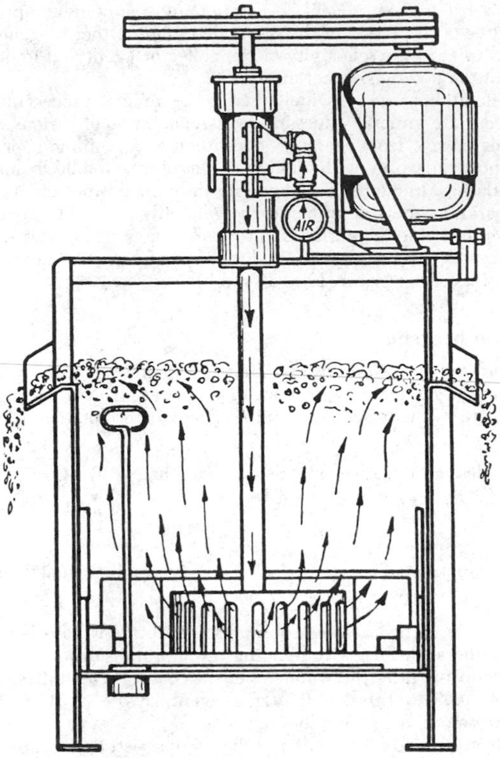

FIG. 21-67 Agitair flotation machine.

where T = flotation time
G = mass rate
P = pulp volume per unit mass of dry solids
V = cell volume
(all units to be consistent)

Typical data for several types of machines are given in Table 21-27.

Maximum Cell Size The use of the formula to calculate the numbers of cells requires knowledge of the required flotation time for the specific case; otherwise, substantial deficiencies may be encountered in actual performance. For each mineral and each machine function (roughing, scavenging, cleaning) there are a typical flotation time required and a typical number of cells to be operated in series to achieve economic recovery. Example flotation times and cells operating in series are given in Table 21-28. After the required time and number of cells to be used in series have been determined, the maximum cell size can be selected.

Mineral-processing plants are being designed with capacities on the order of 500 to 1000 kg/s (2000 to 4000 tons/h). The unit capacities of flotation machines now being manufactured are 10 times greater than those in common use 5 to 7 years ago (Fig. 21-69). Larger-scale flotation machines offer the advantages of lower installed cost, lower operating cost, and lower floor-space requirements (Fig. 21-68). However, it should be noted that large flotation cells do not permit a reduction in the number of cells in series. The use of larger cells does enable the number of parallel rows to be reduced and thereby permits a reduction in pumps, piping, and other auxiliaries.

Plant Operation Ores must be ground to a point of complete or nearly complete liberation. Even though this might possibly be accomplished by coarse crushing, grinding to finer than 10 mesh is necessary in all cases and to finer than 48 mesh in most cases prior to flotation. Grinding is done in closed circuit with classifiers.

In many instances, superior flotation results are obtained by con-

TABLE 21-27 Approximate Capacities of Flotation Cells, Tons of Dry Feed per Hour with Pulp at 33 Percent Solids, Specific-Gravity Ore at 3.0

Cell	Cell volume, ft³	Motor hp per cell	8 cells, flotation time, min: 4	8 cells: 8	8 cells: 12	12 cells, flotation time, min: 4	12 cells: 8	12 cells: 12
Agitair No. 120A, 120 by 120 in	375	30* or 40*	594	297	198	891	445	297
D-R Denver No. 300, 78 by 78 in	300	25*	475	238	158	713	357	238
Fagergren No. 120,† 120 by 120 in	400	30	634	317	211	950	475	317
Booth No. 120, 120 by 120 in	430	60	681	341	227	1022	512	341
Agitair No. 165A × 1500, 144 by 166 in	1500	75	2360	1180	790	1770	1180	890

*Excluding power for auxiliary air blower.
†Availability also in 120- by 90-in size, with volume of 300 ft³ for heavy-duty applications.
NOTE: To convert cubic feet to cubic meters, multiply by 0.0283; to convert minutes to seconds, multiply by 60; and to convert horsepower to watts, multiply by 746.

ditioning the ore with the reagents before the flotation step. Oily-type collectors are sometimes added to the grinding circuit to ensure dispersion. For proper selectivity, a definite contact time is sometimes required between reagent and ore, and this is usually secured by mixing the reagent and ore pulp in a "conditioner" consisting of a cylindrical tank with a vertical impeller.

Flotation machines are built in multiple units, and the flow of the pulp through various units is adjusted for the best results. Common practice is to feed the pulp to several cells known as "roughers," which produce a barren tailing and low-grade concentrate. The concentrate is treated, sometimes after regrinding, in "cleaner" cells and "recleaner" cells for final concentration. The tailings from the cleaner and recleaner cells are recirculated back through the system or concentrated separately in additional cells. Regrinding of these middlings is necessary in many ores.

Important auxiliary equipment in a flotation plant includes reagent feeders and controls, sampling and weighing devices, slurry

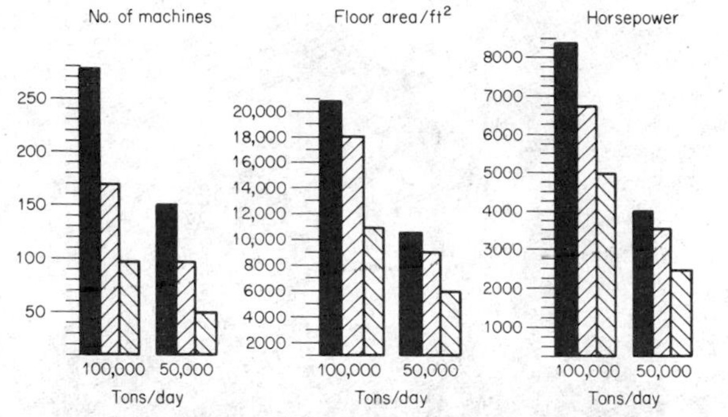

FIG. 21-68 Comparative number of machines, floor area (exclusive of walkways, connection boxes, and junction boxes), and power requirement based on average 100,000-tons/day and 50,000-tons/day copper concentrator. ■ = 300-ft³ machine; ▨ = 500-ft³ machine; ▧ = 1000-ft³ machine.

pumps, filters and thickeners for dewatering solids, reagent storage and makeup equipment, and analytical devices for process control.

Figure 21-70 is a flow sheet of a typical simple flotation plant.

Economics of Flotation

Equipment Costs The cost (1980) of flotation cells for the sizes listed in Table 21-27 is approximately \$880 to \$1060 per cubic meter (\$25 to \$30 per cubic foot), FOB the factory for the cells in the size from 28 to 42 m³ (1000 to 1500 ft³). Smaller cells are higher in unit cost. For example, cells of 2.8 m³ (100 ft³) entail approximately double the cost per unit of cell volume. This approximate cost is for mild-steel construction and rubber-covered wear parts and includes motors and drives but excludes feed boxes, junction boxes, discharge boxes, launders, and air blowers.

Plant Costs Froth-flotation plants generally are completely integrated concentrating operations, with equipment also installed for crushing, grinding, sizing, materials handling, and water recovery. Hence the actual cost of the flotation section of a milling plant is a small part of the overall capital cost. Large sulfide mills (10,000 tons/day and up) cost approximately \$2000 per ton per day for a single-product plant or \$2500 for a multiple-product plant (1980). Costs for smaller plants run higher than this. This approximation includes an allowance for a complete milling installation with wiring, piping, auxiliaries, ore bins, and mill water tank, but no shops or warehouses.

TABLE 21-28 Typical Flotation Time and Number of Cells in Series for Various Minerals

Mineral*	Usual percent solids for rougher feed†	Batch laboratory retention times, min	Usual roughing retention times,‡ min	Number of cells in series: Minimum	Usual range	Satisfactory
Barite	30–40	4–5	8–10	4–6	6–9	8–10
Coal	4–8	2–3	3–5	3–4	4–5	5–6
Copper	32–42	6–8	13–16	11–14	14–17	18–20
Industrial waste	As received	4–5	6–12	4–5	5–8	9–11
Fluorspar	25–32	4–5	8–10	5–6	6–10	10–12
Feldspar	25–35	3–4	8–10	4–6	6–8	8–10
Lead	25–35	3–5	6–8	4–6	6–8	8–10
Molybdenum	35–45	6–7	14–20	12–14	14–17	17–20
Nickel	28–32	6–7	10–14	10–12	12–16	16–20
Oily water	As received	2–3	4–6	.....	4–6	6–8
Phosphate	30–35	2–3	4–6	3–4	4–5	5–6
Potash	25–35	2–3	4–6	.....	4–6	6–8
Sand (impurity flot)	30–40	3–4	7–9	4–6	6–8	
Silica (iron ore)	40–50	3–5	8–10	8–10	10–14	
Silica (phosphate)	30–35	2–3	4–6	.....	4–6	
Tungsten	25–32	5–6	8–12	.....	6–10	10–12
Zinc	25–32	5–6	8–12	.....	6–9	9–12

*Floatable form.
†For cleaning applications, use 50 to 65% of normal roughing percent solids.
‡For cleaning applications, the retention time required is between 60 and 75% of the roughing time for a particular mineral.

FIG. 21-69 Large flotation cell No. 165 AX 1500 Agitair, 42.5 m^3 (1500 ft^3). *(Courtesy of Galigher Ash Company.)*

Of the $2000 per ton per day total, about $50 per ton per day is the cost of flotation equipment alone in larger mills. In small mills (up to 1000 tons/day), total cost may run as high as $4000 to $5000 per ton per day with flotation machines running to $100 per ton per day. The highest capital-cost item in a mill is usually the grinding equipment.

A capital-cost estimate for a hypothetical flotation mill for concentrating a porphyry copper ore of 9070 metric tons (10,000 short tons) per day capacity is given in Table 21-29, based on the analysis by Henderson and Crowell ["Comparison of Estimated vs. Actual Capital Cost and Operating Data for a Copper Concentrator," in Mular and Bhappu (eds.), *Mineral Processing Plant Design*, 2d ed., Society of Mining Engineers, AIME, New York, 1980]. It should be noted that the costs given in Table 21-29 are strictly estimates. Actual mill costs will vary depending on the following factors:

1. Hardness of ore and mineralogical associations (ease and fineness of grinding to be determined)
2. Metal content of ore (nonmetallics require much greater materials-handling investment for concentrates than do base-metal sulfides, for example)
3. Complexity of process (for example, selective flotation of several minerals will result in higher cost than flotation of a single mineral)
4. Location of mill
5. Topography of area
6. Soil conditions at mill site
7. Tailing-disposal considerations
8. Weather conditions
9. Estimated life of plant
10. Degree of instrumentation and control
11. Design factors for capacity

To obtain a rough estimated maximum cost of a sulfide-concentrating plant, a rule of thumb is to multiply the total mill-equipment cost (FOB factory) by 4. Less elaborate mills may cost 3 times the FOB factory cost of equipment.

TABLE 21-29 Approximate Capital-Cost Estimate*

Description	Cost
Grizzly feeder	$ 32,600
Jaw crusher	182,415
Vibrating feeders (5)	17,407
Tramp-iron magnet (2)	11,362
Weightometers (3)	16,201
Metal detector	9,950
Screens (3)	66,785
Cone crushers (2)	300,512
Belt feeders (6)	33,990
Grinding mills (3)	841,559
Horizontal slurry pumps	43,259
Vertical slurry pumps	33,919
Cyclones	30,279
Distributors	22,891
Flotation cells (all)	213,368
Thickeners (3)	143,255
Filters (2)	60,956
Railroad scale	25,930
Lime unloading	12,809
Lime pumps	2,128
Lime slaker	16,700
Air compressor	48,408
Installed air dryer	2,310
Water pumps	5,244
All electric motors	300,000
Total process-equipment cost	$2,474,237
Miscellaneous equipment	275,000
Spare parts	200,000
Total	$2,949,237 (or $2,950,000)
Concentrator cost (concentrator cost equals factor of 3.5 times major-equipment cost, or)	$10,325,000
Additional items	
Power supply	65,000
Water supply	800,000
Auxiliary facilities (shop, assay office, warehouse, main office)	1,000,000
Tailing dam	500,000
Subtotal	$12,690,000
Contingency at 15%	1,900,000
Total milling plant cost	$14,590,000

*Based on analysis by Henderson and Crowell, in Mular and Bhappu (eds.), *Mineral Processing Plant Design*, 2d ed., Society of Mining Engineers, AIME, New York, 1980.

Operating Costs

Power consumption. For most sulfide ores power consumption will range from 2 to 3 kWh/ton of feed (for flotation alone). Easy-to-float ores such as the well-deslimed Florida phosphate ores can be floated for about 1 kWh/ton. A typical copper sulfide ore will require 2½ to 3 kWh/ton.

Chief factors influencing power requirements are (1) flotation time, (2) pulp density (dilution) of feed, (3) number of cleaning stages, and (4) froth and middling pumping requirements.

Labor requirements. In modern sulfide flotation plants labor requirements are low; one operator can readily handle flotation sections treating up to 30,000 tons/day of feed, provided that the ore is not too complex and treatment is fairly simple.

Total direct milling cost (1980) for sulfide flotation plants ranges from $1 to $4 per ton of ore processed and the size of the operation. Of this total, generally 15 to 20 percent is for maintenance and the remainder for operating cost. An approximation of direct operating costs is given in Table 21-30.

TECHNIQUES FOR SELECTIVE SEPARATION OF ULTRAFINE SOLIDS

GENERAL REFERENCES: Aplan and Fuerstenau, "Principles of Nonmetallic Mineral Flotation," *Froth Flotation*, 50th anniversary vol., AIME, New York,

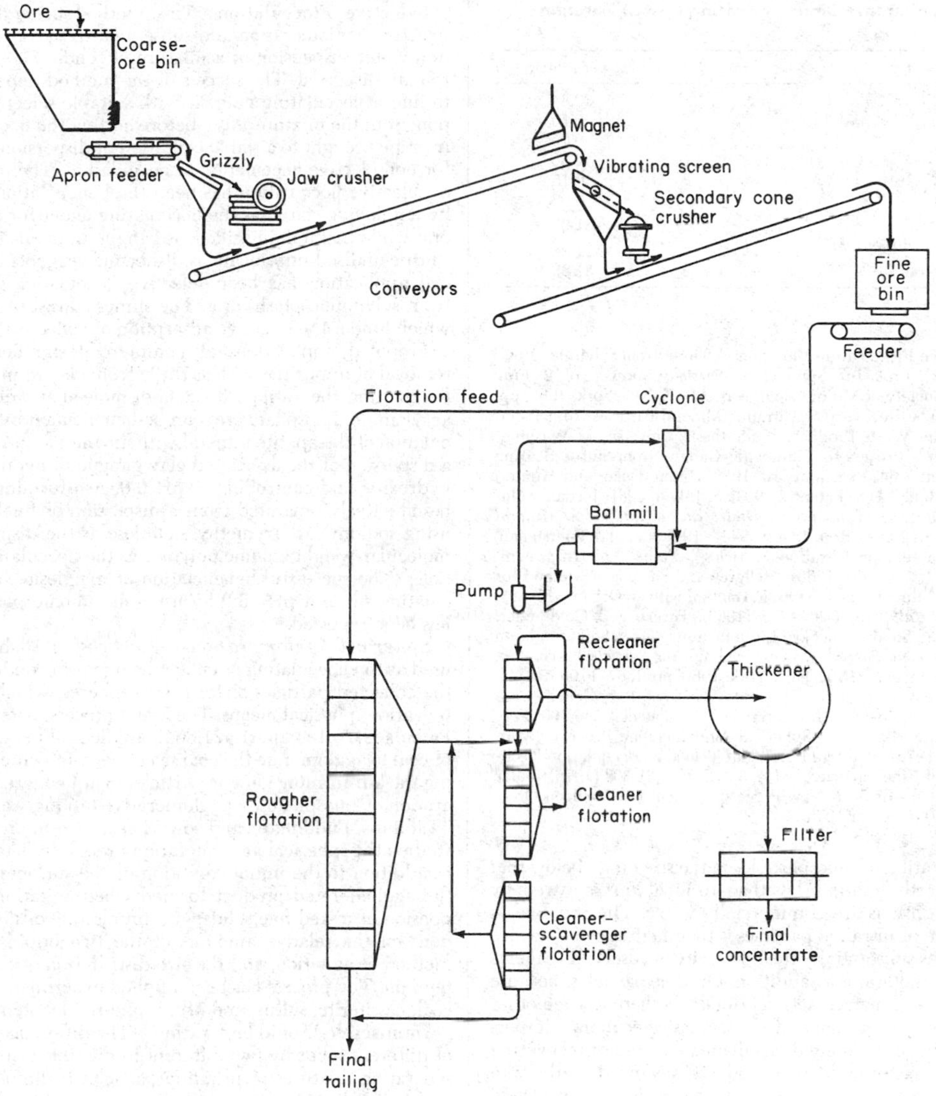

FIG. 21-70 Flow sheet of simple flotation plant.

TABLE 21-30 Approximate Direct Operating Cost of Flotation Plant

Cost item	Cost per metric ton
Power	$0.34
Operating labor	0.16
Maintenance labor	0.09
Operating supplies	0.30
Maintenance supplies	0.15
Supervision	0.03
Total	$1.07
Indirect costs at 30% of direct costs	.32
Total operating costs	$1.39

1962. Banks, "Selective Flocculation-Flotation of Slimes from Sylvinite Ores," in Somasundaran (ed.), *Proc. Int. Symp. Fine Particle Process.*, vol. 2: *Fine Particle Processing*, Society of Mining Engineers, AIME, New York, 1980, pp. 1104–1111. Buller, "The Recovery of Ultrafine Mineral Particles from Flotation Tailings and Other Waste Products," M.S. thesis, University of Washington, 1975. Goldberger, "Process for Removing Gangue from Sulfur Bearing Materials," U.S. Patent 3,647,398, Mar. 10, 1972. Green, Duke, and Hunter, "Froth Flotation Method," U.S. Patent 2,990,958, July 4, 1961. Jordan, Sullivan, Davis, and Weaver, *A Continuous Dielectric Separator for Mineral Beneficiation*, U.S. Bur. Mines Rep. Invest. 8437, 1980. Lai and Fuerstenau, "Liquid-Liquid Extraction of Ultrafine Particles," *Trans. Am. Inst. Min. Metall. Pet. Eng.*, **241**, 549–556 (1968). Mellgren and Shergold, "Method for Recovering Ultrafine Mineral Particles by Extraction with an Organic Phase," *Trans. Inst. Min. Metall.*, **75**, C267–268 (1966). Parekh and Goldberger, "Removal of Suspended Solids from Coal Liquefaction Oils," in Somasundaran (ed), *Proc. Int. Symp. Fine Particle Process.*, vol. 2: *Fine Particle Processing*, Society of Mining Engineers, AIME, New York, 1980, pp. 1687–1706. Parekh and Goldberger, "Recovery of Silicon Carbide Whiskers from Coked, Converted Rice Hulls by Liquid-Liquid Separation," U.S. Patent 4,249,700, Feb. 10, 1981. Puddington and Sparks, "Spherical Agglomeration Process," *Min. Sci. Eng.*, 7(4), 282 (1975). Read and Hollick, "Selective Flocculation Techniques for Recovery of Fine Particles," *Min. Sci. Eng.*, 8(3), 202 (1976). Villar and Dawe, "The Tilden Mine—A New Processing Technique for Iron Ore," *Min. Congr. J.*, **61**, 40 (October 1975).

The physical separation of solid particles in the size range below that for conventional froth flotation (less than 10 to 20 μm) is extremely difficult, and few options are commercially viable. The problem in selective separation of ultrafine particles is their high specific surface area, which reduces substantially the selectivity in adsorbed reagents of the type used in flotation. In addition, ultrafine particles, because of their high degree of surface activity, tend to adhere to each other nonselectively. However, because of the increasing economic importance of treating the more finely disseminated ores and of recovering values from slime wastes and fume products, advanced methods to separate ultrafine particles selectively are being developed. These methods are based on differences in the chemical and electrochemical character of particle surfaces and include selective flocculation, electrophoresis, colloid, foam, and ultraflotation procedures, and use of immiscible liquids.

Ultraflotation The tendency of ultrafine solid particles to adhere to coarser particles generally causes inefficiency in the conventional flotation process, and for this reason the pulp is deslimed to eliminate the ultrafines in advance of the flotation circuit. However, this same tendency is used to advantage in ultraflotation, a process applied commercially for improving the brightness of kaolin by removing micrometer-sized titaniferous impurities. The ultraflotation process employs an auxiliary solid to effect a selective adhesion of one ultrafine mineral type onto its surface. The auxiliary or carrier mineral is subsequently removed from the suspension by conventional flotation.

The carrier mineral used in the beneficiation of kaolin is limestone, which is cleaned of adsorbed reagents and particulates and recycled. Ultraflotation requires increased reagent use because of the higher specific surface. Pulp density is generally less to achieve effective dispersion, and more cells are required for the same capacity. Ultraflotation is reported to cost substantially more than conventional flotation because of the increased handling of solid materials and the cost of reagents.

Selective Flocculation This method generally comprises the selective agglomeration and subsequent removal of one component in a water suspension of a mixture of solids. The other components remain dispersed. The success of the method depends on the ability to find a flocculating reagent with suitable selectivity for one component in the mixture. Also, before adding the flocculant, the system must be brought to a stable condition of dispersion with no tendency for nonselective flocculation or coagulation. The main industrial use of selective flocculation has been the beneficiation of hematite ores by employing starch as the flocculating agent for the hematite mineral while maintaining siliceous gangue minerals in suspension with caustic and sodium silicate as dispersing reagents. Another commercial application has been selective flocculation-flotation of slimes from sylvinite (potash) ores. The slimes consist of fine clay minerals which hinder the effective adsorption of collectors onto the sylvinite.

Proposed applications of commercial significance include the removal of impurities such as rutile from clay minerals by selectively flocculating the rutile with a high-molecular-weight anionic polyacrylamide. Phosphate ores can be beneficiated by the selective flocculation of the apatite mineral with an anionic starch after achieving a dispersion of the associated clay gangue minerals by using sodium hydroxide and controlling to pH 9.0. Ash-forming components can be effectively separated from a dispersion of finely ground coal by using sodium carboxymethyl cellulose as the dispersant and a low-molecular-weight anionic polymer as the flocculant for the coal particles. The successful beneficiation of magnesite by selectively flocculating silica at pH 13.0 by using an anionic polymeric flocculant has been reported.

Spherical Agglomeration A low-cost oil such as fuel oil can be used as the flocculating or collecting agent in conditions which cause the collected particles to form into spheres, which can be removed by various physical means. The Trent process, patented in 1922 (U.S. Patents 1,420,164 and 1,421,862), applied oil to aqueous suspensions of coal to agglomerate the coal selectively into oil-coal granules, leaving the ash-forming gangue particles in aqueous suspension. The separation of phosphate by agglomerative tabling was described under "Tabling." Puddington and coworkers (see references) more recently studied the spherical-agglomeration process in detail and noted that, in addition to the initial condition of the surfaces, the character of the agglomerated product formed when agitating an aqueous suspension of mixed fine solids with immiscible oils is critically dependent on the relative amounts of the two liquids, the degree and method of agitation, and the size distribution of the particles in suspension. The process has been applied experimentally to beneficiate coal, cassiterite, sulfur, and other solid-solid systems.

Immiscible-Liquid Separation The difference in the wettability of different solids by two different immiscible liquids is the basis for separating mixtures of ultrafine particles in liquid suspension. The immiscible-liquid system is usually composed of water and a water-insoluble organic or hydrocarbon liquid (oil). When a sufficient quantity of oil has been dispersed as fine droplets through an aqueous suspension of a mixture of solids, the oil droplets will collect the hydrophobic particles selectively. The collected solids may then migrate from the droplet surface into the oil phase if the oil-wetting tendency is sufficient. This technique, referred to earlier as bulk-oil

TABLE 21-31 Contact-Angle Values of Isooctane on Alumina and Hematite Surfaces in Sodium Dodecyl Sulfonate Solutions at Various pH*

Solid	SDS concentration, molar	pH	Isooctane contact angle,°
Alumina	5×10^{-5}	3.0	125
	5×10^{-5}	6.0	90
	5×10^{-5}	9.0	40
Hematite	1×10^{-3}	3.0	120
	1×10^{-3}	6.0	48
	1×10^{-3}	9.0	18

*After Mellgren and Shergold, *Trans. Inst. Min. Metall.*, **75**, C267–C268 (1966); and Lai and Fuerstenau, *Trans. Am. Inst. Min. Metall. Pet. Eng.*, **241**, 549–556 (1968).

flotation, was described in the patents issued to Cattermole around the turn of the twentieth century (U.S. Patents 763,259, 763,260, 727,273, and 777,274).

The immiscible-liquid separation technique, which is analogous to liquid-liquid extraction (Sec. 15), requires two distinct bulk-liquid phases. The particles will selectively migrate to one or the other phase, depending on their surface affinity for that liquid. Theoretically, the direction of the migration can be predicted by Young's equation:

$$\sigma_{so} - \sigma_{sw} = \sigma_{wo} \cos \theta \qquad (21\text{-}18)$$

where θ is the contact angle of an oil droplet formed on a solid submerged in water. Contact-angle values of isooctane liquid on alumina and hematite surfaces submerged in aqueous solutions of surfactant are given in Table 21-31. The σ_{sw}, σ_{wo}, and σ_{so} are the interfacial tensions of solid-water, water-oil, and solid-oil respectively. One of three situations will occur for each solid:

1. If $\theta < 90°$ (or $\sigma_{sw} < \sigma_{so}$), the particles will tend to be drawn into the aqueous phase.
2. If $\theta = 90°$ (or $\sigma_{sw} = \sigma_{so}$), the particles will tend to remain concentrated at the oil-water interface.
3. If $\theta > 90°$ (or $\sigma_{sw} > \sigma_{so}$), the particles will tend to be drawn into the oil phase.

The effective application of the immiscible-liquid separation method would generally require that condition 3 be satisfied. The contact angle can be controlled by control of the pH and by the use of surfactants as indicated in Table 21-31. Other factors which determine the efficiency of the separation include the type of oil used, the oil concentration and the degree of dispersion which affects the size of the oil droplets, and the extent of the interfacial area between oil and water phases. The influence of surfactant concentration and the length of the surfactant molecule also influence separation efficiency, as shown in Fig. 21-71.

In other applications of immiscible-liquid separation, submicrometer-sized siliceous gangue particles present in native sulfur ores were removed by melting the ore concentrate and then introducing water as the second liquid phase to collect the hydrophilic minerals by the process of Goldberger. Ultrafine ash-forming mineral particles suspended in oils derived by liquefaction of coal also can be removed by immiscible-liquid separation using the process of Goldberger and Faulkner (U.S. Patent 4,012,314). In this case, the tendency for hydrophobic and hydrophilic particles to partition between oil and

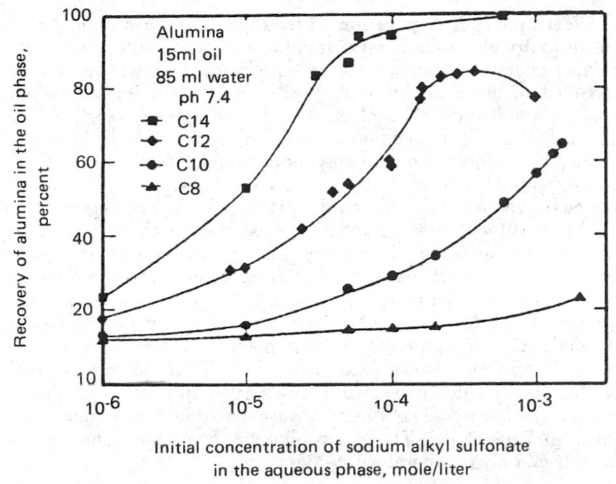

FIG. 21-71 The recovery of alumina in the oil phase as a function of the concentration of C_8, C_{10}, C_{12}, and C_{14} alkyl sulfonate at pH 7.4 and 15 volume percent isooctane. *(After Lai and Fuerstenau, 1968).*

water phases maintained in contact with each other is enhanced by the application of centrifugal force.

Dielectric Separation Solid particles can be separated by utilizing differences in their dielectric properties. A particle placed in a liquid of lower dielectric constant (K) will be attracted to one electrode in a high-gradient electric field, whereas a particle of lower K than the liquid will be repelled. The U.S. Bureau of Mines designed a continuous dielectric separator for mineral particles in the size range 65 by 400 mesh. The process was found less effective as the size of the particles decreased. Studies on a mixture of rutile (TiO_2) and quartz (SiO_2) indicated that the grade of the rutile concentrate decreased from 65 to 42 percent and that rutile recovery decreased from 97 to 83 percent as the particle size decreased from 65 mesh to 400 mesh. A two-stage separation resulted in higher grade and rutile recovery. Other mineral mixtures tried also showed adequate separation ability, doubling the grade of the product from the feed. A diagram of the separation apparatus is shown in Fig. 21-72.

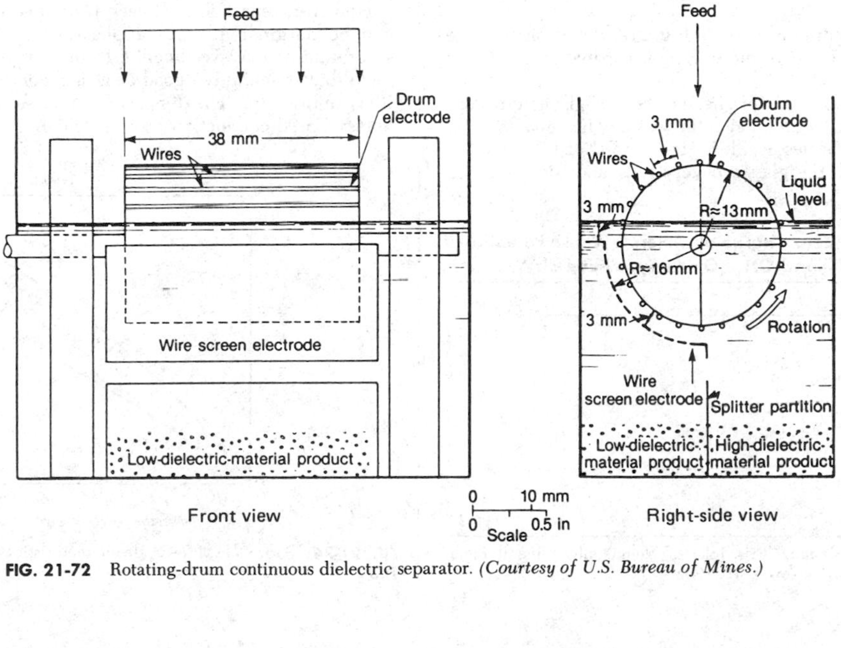

FIG. 21-72 Rotating-drum continuous dielectric separator. *(Courtesy of U.S. Bureau of Mines.)*

Electrophoretic Separation Electrophoretic separation is analogous to dry electrostatic separation; it is based on differences in the surface charge on solid particles in liquid suspension which cause the particles to move at different speeds through the liquid medium under the influence of an electric field. The sign and magnitude of the surface charge depend on the surface properties of the particular solid and on the pH of the suspending liquid. For salt-type solids having appreciable solubility (e.g., $BaSO_4$, $CaSO_4$, etc.), the sign of the charge depends notonly on the pH but also on the concentration of the constituent ions in soluton. For example, in the case of barite ($BaSO_4$) the surface of the mineral in aqueous suspension could be made positively charged by adding excess of Ba^{++} ions or could be made negatively charged by adding SO_4- ions to the suspension. For nonsoluble oxides (i.e., SiO_2, TiO_2, etc.) the sign of the charge depends only on the pH because the potential determining ions are the H^+ and OH^-. Every solid surface has a "point of zero charge" (PZC), above which it is negatively charged and below which it is positively charged. The PZCs of various oxide-type minerals are given in Table 21-32. The electrophoretic-separation technique has been tried only in laboratory studies.

AUTOMATIC SORTING

Difference in optical properties can be used as the basis to separate solids in a mixture. Optical properties include color, light reflectance, opacity, and fluorescence excited by ultraviolet rays or x-rays. Differences in electrical conductance can also be used for separation. With appropriate sensing, the particles in a moving stream can be sorted by using an air jet or other means to deflect certain particles away from the mainstream (Fig. 21-73). The lower limit of particle size is about 0.003 m (⅛ in); below this limit the process rate would be slow and equipment costs become exceedingly high.

In a typical **optical sorting** installation, the mixture of particles is fed from a hopper onto a vibrating feeder. Dust may be removed by dry screening or by water spray. The solids then enter a troughed conveyor belt and align the flow and cause the particles to be projected in a continuous stream along their free-fall trajectory. They are viewed in midair during fall through an optical chamber by a series of cameras arranged to view the entire surface of each particle. The color or reflectivity of the surface of each particle sets up a characteristic voltage pattern in the output circuit of a light-sensing photomultiplier. The patterns are analyzed electronically and compared with a preset reflectance level. When appropriate, a reject signal delayed electronically will activate on air jet to deflect a particle from the main stream.

An example of throughput is given in Fig. 21-74. For the machine referenced in Fig. 21-73, electronic sensing is capable of inspecting 80 particles per second.

Color sorting has been used for the recovery of glass from nontransparent materials in municipal solid waste. The separation of mixed glasses from opaques is illustrated in Fig. 21-75. As the throughput is increased, the glass recovery falls; and for any given installation there will be a breakeven point for the optimum number of machines against the amount of glass recovered. Color sorting can also separate mixed colored glasses to produce amber and green products that meet a specification of 10 percent contamination on one type of glass in the other.

When **electrical conductivity** is used as the basis of the sorting process, contact of the particles is made by a brush type of electrode to generate the signal for analysis. Materials having a resistance difference of 2000 kΩ can be readily separated from material of 100-kΩ resistance.

However, separation between resistance levels of 1000 to 300 kΩ may be marginal. A typical application of conductivity sorting is the separation of massive ilmenite from anorthosite. Both are compact rocks, but ilmenite is a good electrical conductor, whereas anorthosite is an insulator. The dimensions and operating information for the Sortex CS-03 conductivity sorter, which is capable of processing up

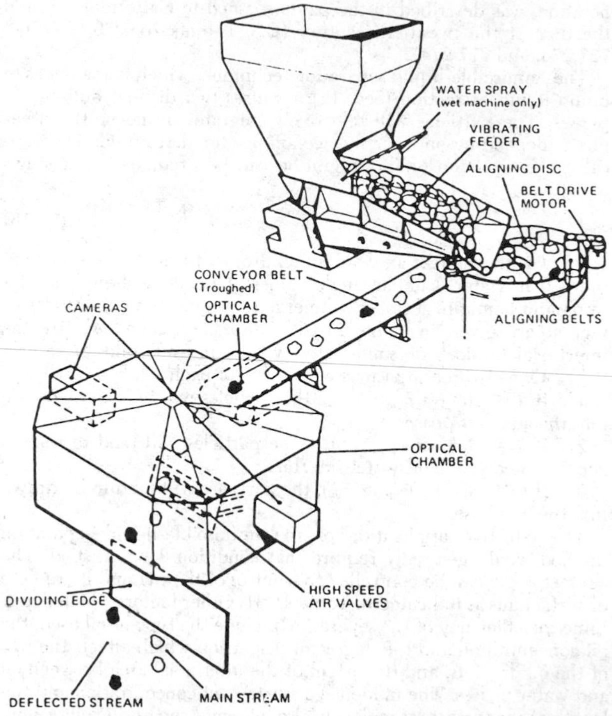

FIG. 21-73 Sortex 711M optical separator. *(Courtesy Gunson's Sortex Ltd.)*

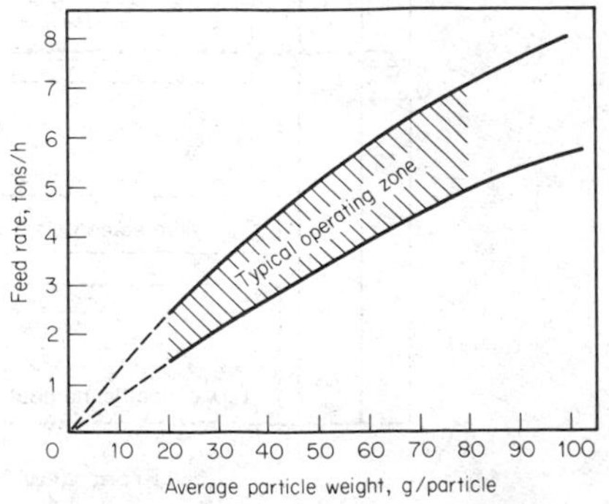

FIG. 21-74 Sortex 711M feed-rate characteristics. *(Courtesy Gunson's Sortex Ltd.)*

TABLE 21-32 Point of Zero Charge of Minerals Whose Potential Determining Ions are H^+ and OH^- (Oxide-Type Minerals)*

Mineral	pH
Quartz, SiO_2	2 to 3.7
Corundum, Al_2O_3	9.4
Rutile, TiO_2	6.0
Cassiterite, SnO_2	4.5
Magnetite, Fe_3O_4	6.5
Hematite, Fe_2O_3	6.7
Goethite, $Fe_2O_3 \cdot 2H_2O$	6.7
Cummingtonite, (Mg,Fe)SiO_3	5.2
Kaolinite, clay	3.4
Bentonite, clay	< 3.0

*After Aplan and Fuerstenau, "Principles of Nonmetallic Mineral Flotation," *Froth Flotation*, AIME, New York, 1962.

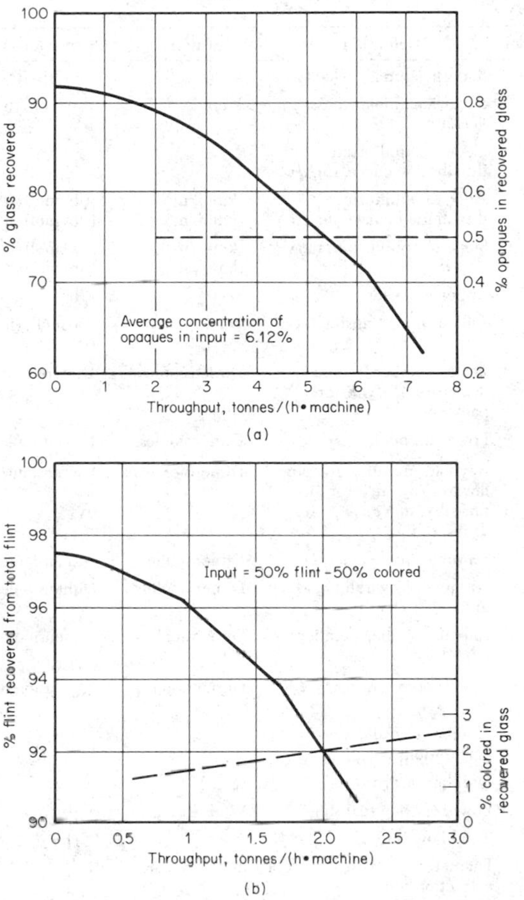

FIG. 21-75 Optical separation of mixed glasses. (*a*) Separation of opaques from glass. (*b*) Separation of flint from colored. *(Courtesy Gunson's Sortex Ltd.)*

TABLE 21-33 Specifications for Sortex Conductivity Sorter CS-03*

Dimensions	Height (excluding feeder), 1.7 m (70 in) Width, 1.9 m (76 in) Length, 2.5 m (100 in) Diameter of disk, 1.50 m (49 in)
Electric power	380–440 V, three-phase 50/60 Hz; consumption, 2 kW (excluding vibrating feeder)
Air consumption (depending on rejection rate)	7 m^3/min at 6 bar (250 ft^3/min at 80 lbf/in^2)
Water supply	4 L/min (1 gal/min) for cleansing the light source
Net weight	560 kg (22 cwt) approximately

*Courtesy of Gunson's Sortex Ltd.

to about 25,000 kg/h (27.5 tons/h) of 0.05- to 0.15-in mesh size (2 to 6 in), are given in Table 21-33.

The differences in absorptivity of radiant energy by different substances can also be used to separate materials. If a mixture of materials is exposed to radiant energy, for example, infrared, depending on the properties of the material involved, some particles will become heated more than others. The more opaque particles will be heated more by infrared heating than clear particles. Thus, the more opaque will become hotter. By spreading the irradiated material onto a surface coated with a low-melting thermoplastic or a heat-sensitive polymer, the higher-temperature particles will adhere, while the cooler particles will not. This is the basis for the **thermoadhesive-separation** method of Brison ("Separation of Materials," U.S. Patent 2,907,456, 1959) used by the International Salt Co. for removing impurities from mined rock salt. The heat-sensitive resin employed is a mixture of polymerized styrene resins, Piccolastic A-25 and Piccolastic A-50. The proportion of each was adjusted to give the required softening point to achieve the desired results. In practice, the resin can be continuously applied to a moving belt by brush or hot spray. Periodic scrapping and redressing are required to maintain belt performance.

Use of specific forms of radiant energy, infrared, ultraviolet, dielectric heating, etc., can allow specific separations to be made. The separation of clear and colored grains of glass and the separation of different metals are possible applications of the thermoadhesive method being considered in the field of solid-waste processing.

LIQUID-LIQUID SYSTEMS

GENERAL REFERENCES: *Books.* Astarita, *Mass Transfer with Chemical Reactions,* Elsevier, New York, 1967. Calderbank, "Mass Transfer," in Uhl and Gray (eds.), *Mixing,* vol. 2, Academic, New York, 1967, p. 2. Cremer and Davies, *Chemical Engineering Practice,* vol. 5, Academic, New York, 1958. Davies, "Mass Transfer and Interfacial Phenomena," in Drew, Hoopes, and Vermeulen (eds.), *Advances in Chemical Engineering,* vol. 4, Academic, New York, 1963, p. 3. Hanson (ed.), *Recent Advances in Liquid-Liquid Extraction,* Pergamon, New York, 1971. Hyman, "Mixing and Agitation," ibid., vol. 3, 1962, p. 120. Kalichevsky and Kobe, *Petroleum Refining with Chemicals,* Van Nostrand, Princeton, N.J., 1965. Kintner, "Drop Phenomena Affecting Liquid Extraction," in Drew et al. (eds.), *Advances in Chemical Engineering,* vol. 4, Academic, New York, 1963, p. 52. Lo, Baird, and Hanson (eds.), *Handbook of Solvent Extraction,* Wiley-Interscience, New York, 1983. Molyneux, *Chemical Plant Design,* vol. I, Butterworth, Washington, 1965. Olney and Miller, "Liquid Extraction," in Acrivos (ed.), *Modern Chemical Engineering,* vol. 1, Reinhold, New York, 1963, p. 89. Olson and Stout, "Mixing and Chemical Reactions," in Uhl and Gray (eds.), *Mixing,* vol. 2, Academic, New York, 1967, p. 115. Rietema, "Segregation in Liquid-Liquid Dispersions," in Drew et al. (eds.), *Advances in Chemical Engineering,* vol. 5, Academic, New York, 1964, p. 237. Rod, Misek, and Sterbacek, *Liquid Extraction,* Statne Nakladatelstor Techniki Literatury, Prague, 1964. Schweitzer (ed.), *Handbook of Separation Techniques for Chemical Engineers,* McGraw-Hill, New York, 1979. Sideman, "Direct-Contact Heat Transfer between Immiscible Liquids," in Drew et al. (eds.), *Advances in Chemical Engineering,* vol. 6, Academic, New York, 1966, p. 207. Treybal, "Mechanically Aided Liquid Extraction," ibid., vol. 1, 1956, p. 289. Treybal, *Liquid Extraction,* 2d ed., McGraw-Hill, New York, 1963. Ziolkowski, *Liquid Extraction in the Chemical Industry,* Gos. Nauchn. Tekln. Izd. Khim. Lit., Leningrad, 1963.

Journals. Akell, *Chem. Eng. Prog.,* **62**(9), 50 (1966). Brown and Hanson, *Br. Chem. Eng.,* **11,** 695 (1966). Davis, Hicks, and Vermeulen, *Chem. Eng. Prog.,* **50,** 188 (1954). Hanson, *Br. Chem. Eng.,* **10,** 34 (1965). Jackson, *Chem. Eng. Prog.,* **62**(9), 82 (1966). Miller, *Ind. Eng. Chem.,* **56**(10), 18 (1964). Morello and Poffenberger, ibid., **42,** 1021 (1950). Pratt, *Ind. Chem.,* **30,** 437, 475, 597 (1954); **31,** 63, 505, 552 (1955). Reman, *Chem. Eng. Prog.,* **62**(9), 56 (1966). Thornton, *Nucl. Eng.,* **1,** 156, 204 (1956). Treybal, *Chem. Eng. Prog.,* **62**(9), 67 (1966).

Annual reviews. Elgin, *Ind. Eng. Chem.,* **38,** 26 (1946); **39,** 23 (1947); **40,** 53 (1948); **41,** 35 (1949); **42,** 47 (1950). Teybal, ibid., **43,** 79 (1951); **44,** 53 (1952); **45,** 55 (1953); **46,** 91 (1954); **47,** 536 (1955); **48,** 518 (1956); **49,** 514 (1957); **50,** 463 (1958); **51,** 378 (1959); **52,** 262 (1960); **53,** 161 (1961). Beckmann et al., ibid., **57**(11), 103, 108 (1965); **58**(11), 97 (1966); **59**(11), 71 (1967); **60**(11), 43 (1968).

Nomenclature

Symbol	Definition	SI units	U.S. customary units
A_p	Area of a drop	m^2	ft^2
a	Specific interfacial surface between liquids	m^2/m^3	ft^2/ft^3
a_p	Specific packing surface	m^2/m^3	ft^2/ft^3
B	Ratio of total length to characteristic length		
b	Constant		
C	Constant		
C_O	Orifice coefficient	Dimensionless	Dimensionless
c	Concentration	$kmol/m^3$	$(lb \cdot mol)/ft^3$
D	Solute diffusivity	m^2/s	ft^2/h
D'	Enhanced diffusivity	m^2/s	ft^2/h
d	Differential operator		
d_F	Packing size	m	ft
d_{FC}	Critical packing size	m	ft
d_i	Impeller diameter	m	ft
d_O	Nozzle, perforation, orifice diameter	m	ft
d_p	Drop diameter, diameter of sphere of same volume per surface	m	ft
d_{pJ}	Drop diameter at jetting if no jet forms	m	ft
d_s	Diameter of stator-ring opening	m	ft
d_t	Tube or pipe diameter	m	ft
E	Fractional efficiency of a single stage (mixer-settlers)		
E	Longitudinal dispersion coefficient (differential extractors)	m^2/s	ft^2/h
E_{MD}	Murphree dispersed-phase stage efficiency, fractional		
E_O	Overall stage efficiency of a cascade, fractional		
e	2.7183 (napierian logarithm base)		
f_E	Weighting factor		
f_R	Weighting factor		
g	Local acceleration due to gravity	$9.83\ m/s^2$	$4.18\ E08\ ft/h^2$
g_c	Gravitational conversion factor	$1(kg \cdot m)/(N \cdot s)$	$4.18\ E08\ (lbm \cdot ft)/(lbf \cdot h^2)$
H_{to}	Overall height of a transfer unit	m	ft
HETS	Height equivalent to a theoretical stage	m	ft
h	Head loss due to friction	m	ft
h_C	Contribution to h due to continuous phase	m	ft
h_D	Contribution to h due to dispersed phase	m	ft
h_o	Contribution to h_D due to orifice	m	ft
h_σ	Contribution to h_D due to interfacial tension	m	ft
K	Overall mass-transfer coefficient	$kmol/(s \cdot m^2)(kmol/m^3)$	$(lb \cdot mol)/(h \cdot ft^2)[(lb \cdot mol)/ft^3]$
k	Individual-phase mass-transfer coefficient	$kmol/(s \cdot m^2)(kmol/m^3)$	$(lb \cdot mol)/(h \cdot ft^2)[(lb \cdot mol)/ft^3]$
k_t	Thermal conductivity	$W/(m \cdot K)$	$Btu/[h \cdot ft^2 \cdot {}^\circ F)/ft]$
L	Superficial mass velocity	$kg/(s \cdot m^2)$	$lbm/(h \cdot ft^2)$
L'	Superficial molar mass velocity	$kmol/(s \cdot m^2)$	$(lb \cdot mol)/(h \cdot ft^2)$
m	Slope of equilibrium distribution curve, dy/dx		
m'	Slope of equilibrium distribution curve, dc_E/dc_R	$(kmol/m^3)/(kmol/m^3)$	$[(lb \cdot mol)/ft^3]/[(lb \cdot mol)/ft^3]$
m'_{CD}	Slope of equilibrium curve, dc_C/dc_D	$(kmol/m^3)/(kmol/m^3)$	$[(lb \cdot mol)/ft^3]/[(lb \cdot mol)/ft^3]$
N	Impeller speed	r/s	r/h
N_f	Flux of mass transfer	$kmol/(s \cdot m^2)$	$(lb \cdot mol)/(h \cdot ft^2)$
N_{Pe}	Péclet number for axial dispersion, Vd_F/E for packing	Dimensionless	Dimensionless
N_{Po}	Power number, $Pg_c/\rho N^3 d_i^5$	Dimensionless	Dimensionless
N_{Re}	Reynolds number; for pipe flow, $d_t V \rho_{av}/u_{av}$; for an impeller, $d_i^2 N \rho_{av}/u_{av}$; for drops, $d_p V_t u_c/\rho c$	Dimensionless	Dimensionless
N_{Sc}	Schmidt number, $u/\rho D$	Dimensionless	Dimensionless
N_{to}	Number of overall transfer units	Dimensionless	Dimensionless
$N_{We,i}$	Impeller Weber number, $\rho d_i^3 N^2/\sigma g_c$	Dimensionless	Dimensionless
$N_{We,t}$	Pipe Weber number, $\rho d_t V^2/\sigma g_c$	Dimensionless	Dimensionless
n	Number of orifices or perforations per plate		
n	Number of drops		
P	Power for one real stage	W	$(ft \cdot lbf)/h$
Q	Total flow rate	m^3/s	ft^3/h
T	Diameter of mixing vessel or extraction tower	m	ft
U	Overall heat-transfer coefficient	$W/(m^2 \cdot K)$	$Btu/(h \cdot ft^2 \cdot {}^\circ F)$
V	Superficial velocity	$m/s = m^3/(s \cdot m^2)$	$ft/h = ft^3/(h \cdot ft^2)$
V_d	Velocity in a down spout	m/s	ft/h
V_K	Characteristic velocity	m/s	ft/h
V_O	Velocity through an orifice or nozzle	m/s	ft/h
V'_O	Velocity through an orifice or nozzle	m/s	ft/h
V'_{OJ}	Jetting velocity	m/s	ft/h
V_S	Slip velocity	m/s	ft/h
V_t	Terminal settling velocity	m/s	ft/h
v	Liquid volume	m^3	ft^3
v_p	Drop volume	m^3	ft^3
Z	Height of liquid in vessel or mixer; for towers, height of packed section	m	ft
Z_t	Distance between trays	m	ft
Z'_t	Distance between trays	m	in
z	Distance	m	ft
	Greek symbols		
Δp	Pressure drop	Pa	lbf/ft^2
$\Delta\rho$	Difference in density	kg/m^3	lbm/ft^3
δ	Dimensionless amplitude for oscillating drops		

Symbol	Definition	SI units	U.S. customary units
ϵ	Fraction void volume in packed section		
θ	Time of contact	s	h
θ_C	Time between coalescences	s	h
θ_F	Time of drop formation	s	h
λ	Eigenvalue		
μ	Viscosity	Pa·s	lbm/(ft·h)
μ'	Viscosity	Pa·s	cP
ν	Coalescence frequency, fraction of drops coalescing per time	L/s	L/h
π	3.1416		
ρ	Density	kg/m^3	lbm/ft^3
Σ	Summation		
σ	Interfacial tension	N/m	lbf/ft
σ'	Interfacial tension	N/m	dyn/cm
φ	Volume fraction of a liquid in a vessel or extractor's void volume		
ω	Vibration frequency for oscillating drops	L/s	L/h
	Additional subscripts		
av	Average		
C	Continuous phase		
D	Dispersed phase		
E	Extract		
F	Flooding		
H	Heavy liquid		
L	Light liquid		
max	Maximum		
o	Organic		
plug	Plug flow		
R	Raffinate		
w	Water or aqueous liquid		
1	Concentrated end		
2	Dilute end		

Introduction Insoluble liquids may be brought into direct contact to cause transfer of dissolved substances, to allow transfer of heat, and to promote chemical reaction. This subsection concerns the design and selection of equipment used for conducting this type of liquid-liquid contact operation.

Objectives There are four principal purposes of operations involving the direct contact of immiscible liquids. The purpose of a particular contact operation may involve any one or any combination of the following objectives:

1. *Separation of components in solution.* This includes the ordinary objectives of liquid extraction, in which the constituents of a solution are separated by causing their unequal distribution between two insoluble liquids, the washing of a liquid with another to remove small amounts of a dissolved impurity, and the like. The theoretical principles governing the phase relationships, material balances, and number of ideal stages or transfer units required to bring about the desired changes are to be found in Sec. 15. Design of equipment is based on the quantities of liquids and the efficiency and operating characteristics of the type of equipment selected.

2. *Chemical reaction.* The reactants may be the liquids themselves, or they may be dissolved in the insoluble liquids. The kinetics of this type of reaction is treated in Sec. 4.

3. *Cooling or heating a liquid by direct contact with another.* Although liquid-liquid-contact operations have not been used widely for heat transfer alone, this technique is one of increasing interest. Applications also include cases in which chemical reaction or liquid extraction occurs simultaneously.

4. *Creating permanent emulsions.* The objective is to disperse one liquid within another in such finely divided form that separation by settling either does not occur or occurs extremely slowly. The purpose is to prepare the emulsion. Neither extraction nor chemical reaction between the liquids is ordinarily sought.

Liquid-liquid contacting equipment may be generally classified into two categories: **stagewise** and **continuous (differential)** contact.

STAGEWISE EQUIPMENT: MIXER-SETTLERS

The function of a stage is to contact the liquids, allow equilibrium to be approached, and to make a mechanical separation of the liquids. The contacting and separating correspond to mixing the liquids, and settling the resulting dispersion; so these devices are usually called **mixer-settlers.** The operation may be carried out in batch fashion or with continuous flow. If batch, it is likely that the same vessel will serve for both mixing and settling, whereas if continuous, separate vessels are usually but not always used.

Mixer-Settler Equipment The equipment for extraction or chemical reaction may be classified as follows:

I. Mixers
  A. Flow or line mixers
    1. Mechanical agitation
    2. No mechanical agitation
  B. Agitated vessels
    1. Mechanical agitation
    2. Gas agitation
II. Settlers
  A. Nonmechanical
    1. Gravity
    2. Centrifugal (cyclones)
  B. Mechanical (centrifuges)
  C. Settler auxiliaries
    1. Coalescers
    2. Separator membranes
    3. Electrostatic equipment

In principle, at least, any mixer may be coupled with any settler to provide the complete stage. There are several combinations which are especially popular. Continuously operated devices usually, but not always, place the mixing and settling functions in separate vessels. Batch-operated devices may use the same vessel alternately for the separate functions.

Flow or Line Mixers

Definition Flow or line mixers are devices through which the liquids to be contacted are passed, characterized principally by the very small time of contact for the liquids. They are used only for continuous operations or semibatch (in which one liquid flows continuously and the other is continuously recycled). If holding time is required for extraction or reaction, it must be provided by passing the mixed liquids through a vessel of the necessary volume. This may be a long pipe of large diameter, sometimes fitted with segmental baffles, but frequently the settler which follows the mixer serves. The energy for mixing and dispersing usually comes from pressure drop resulting from flow.

There are many types, and only the most important can be mentioned here. [See also Hunter, in Dunstan (ed.), *Science of Petroleum,* vol. 3, Oxford, New York, 1938, pp. 1779–1797.] They are used fairly extensively in treating petroleum distillates, in vegetable-oil refining, in extraction of phenol-bearing coke-oven liquors, in some metal extractions, and the like. Kalichevsky and Kobe (*Petroleum Refining with Chemicals*, Elsevier, New York, 1956) discuss detailed application in the refining of petroleum.

Jet Mixers These depend upon impingement of one liquid on the other to obtain a dispersion, and one of the liquids is pumped through a small nozzle or orifice into a flowing stream of the other. Both liquids are pumped. They can be used successfully only for liquids of low interfacial tension. See Fig. 21-76 and also Hunter and Nash [*Ind. Chem.*, **9**, 245, 263, 317 (1933)]. Treybal (*Liquid Extraction*, 2d ed., McGraw-Hill, New York, 1963) describes a more elab-

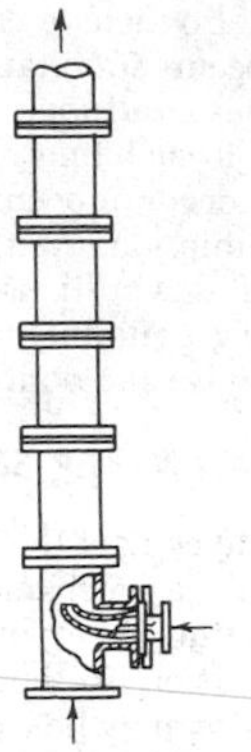

FIG. 21-76 Elbow jet mixer with orifice column. (*Treybal,* Liquid Extraction, *2d ed., McGraw-Hill, New York, 1963, with permission.*)

orate device. For a study of the extraction of antibiotics with jet mixers, see Anneskova and Boiko, *Med. Prom. SSSR,* **13**(5), 26 (1959). Insonation with ultrasound of a toluene-water mixture during methanol extraction with a simple jet mixer improves the rate of mass transfer, but the energy requirements for significant improvement are large [Woodle and Vilbrandt, *Am. Inst. Chem. Eng. J.,* **6,** 296 (1960)].

Trice (U.S. AEC ANL-5741, 1957) used two cylindrical vessels, $T = Z$, $T = 0.10$ and 0.15 m (0.333 and 0.5 ft), through which insoluble liquid pairs were pumped. The arrangement was a form of jet mixer. The droplet size was measured by a light transmittance scheme, and for two systems, $d_p V_D\, \rho_C/\mu_C$ was a function of $(TV_D\, \rho_C/\mu_C)(TV_D^2\, \rho_C/\sigma)^{2/3}\varphi_D^{0.5}$. The mass-transfer coefficient for the continuous phase in two systems k_C' was given by

$$k_C' T/D_C = 0.03(TV_D \rho_C/\mu_C)^{0.88} N_{ScC}^{0.5} \qquad (21\text{-}19)$$

Injectors The flow of one liquid is induced by the flow of the other, with only the majority liquid being pumped at relatively high velocity. Figure 21-77 shows a typical device used in semibatch fashion for washing oil with a recirculated wash liquid. It is installed directly in the settling drum. See also Hampton (U.S. Patent 2,091,709, 1933), Sheldon (U.S. Patent 2,009,347, 1935), and Ng

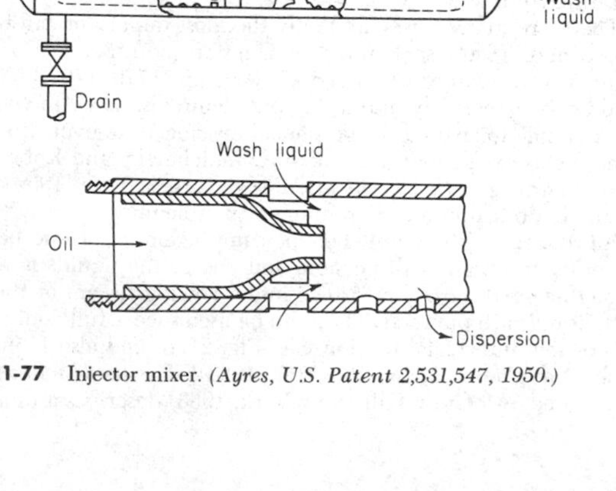

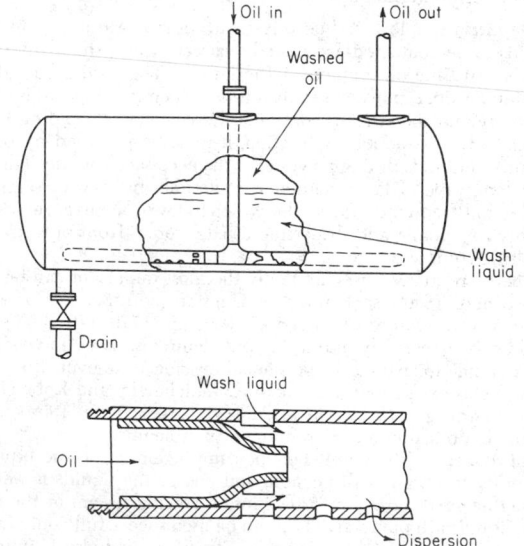

FIG. 21-77 Injector mixer. (*Ayres, U.S. Patent 2,531,547, 1950.*)

(U.S. Patent 2,665,975, 1954). Folsom [*Chem. Eng. Prog.,* **44,** 765 (1948)] gives a good review of basic principles. The most thorough study for extraction is provided by Kafarov and Zhukovskaya [*Zh. Prikl. Khim.,* **31,** 376 (1958)], who used very small injectors. With an injector measuring 73 mm from throat to exit, with 2.48-mm throat diameter, they extracted benzoic acid and acetic acid from water with carbon tetrachloride at the rate of 58 to 106 L/h, to obtain a stage efficiency $E = 0.8$ to 1.0. Data on flow characteristics are also given. Boyadzhiev and Elenkov [*Collect. Czech. Chem. Commun.,* **31,** 4072 (1966)] point out that the presence of surface-active agents exerts a profound influence on drop size in such devices.

Orifices and Mixing Nozzles Both liquids are pumped through constrictions in a pipe, the pressure drop of which is partly utilized to create the dispersion (see Fig. 21-78). Single nozzles or several in series may be used. For the orifice mixers, as many as 20 orifice plates each with 13.8-kPa (2-lb/in^2) pressure drop may be used in series [Morell and Bergman, *Chem. Metall. Eng.,* **35,** 211 (1928)]. In the Dualayer process for removal of mercaptans from gasoline, 258 m^3/h (39,000 bbl/day) of oil and treating solution are contacted with 68.9-kPa (10-lb/in^2) pressure drop per stage [Greek et al., *Ind. Eng. Chem.,* **49,** 1938 (1957)]. Holland et al. [*Am. Inst. Chem. Eng. J.,* **4,** 346 (1958); **6,** 615 (1960)] report on the interfacial area produced between two immiscible liquids entering a pipe (diameter 0.8 to 2.0 in) from an orifice, $\varphi_D = 0.02$ to 0.20, at flow rates of 0.23 to 4.1 m^3/h (1 to 18 gal/min). At a distance 17.8 cm (7 in) downstream from the orifice,

$$a_{av} = \frac{0.179}{\sigma g_c}(C_O^2\,\Delta p)^{0.75}\left(\frac{\sigma\sqrt{g_c\rho_{av}}}{\mu_D}\right)^{0.158}\left[\left(\frac{d_t}{d_O}\right)^4 - 1\right]^{0.117}\varphi_D^{0.878} \qquad (21\text{-}20)$$

where a_{av} = interfacial surface, cm^2/cm^3; C_O = orifice coefficient, dimensionless; d_t = pipe diameter, in; d_O = orifice diameter, in; g_c = gravitational conversion factor, (32.2 lbm·ft)/(lbf·s^2); Δp = pressure drop across orifice, lbf/ft^2; μ_D = viscosity of dispersed phase, lbm/(ft·s); ρ_{av} = density of dispersed phase, lbm/ft; and σ = interfacial tension, lbf/ft. See also Shirotsuka et al. [*Kagaku Kogaku,* **25,** 109 (1961)].

Valves Valves may be considered to be adjustable orifice mixers. In desalting crude petroleum by mixing with water, Hayes et al. [*Chem. Eng. Prog.,* **45,** 235 (1949)] used a globe-valve mixer operating at 110- to 221-kPa (16- to 32-lb/in^2) pressure drop for mixing 66 m^3/h (416 bbl/h) oil with 8 m^3/h (50 bbl/h) water, with best results at the lowest value. Simkin and Olney [*Am. Inst. Chem. Eng. J.,* **2,** 545 (1956)] mixed kerosine and white oil with water, using 0.35- to 0.62-kPa (0.05- to 0.09-lb/in^2) pressure drop across a 1-in gate valve, at 22-m^3/h (10-gal/min) flow rate for optimum separating conditions in a cyclone, but higher pressure drops were required to give good extractor efficiencies.

Pumps Centrifugal pumps, in which the two liquids are fed to the suction side of the pump, have been used fairly extensively, and they offer the advantage of providing interstage pumping at the same time. They have been commonly used in the extraction of

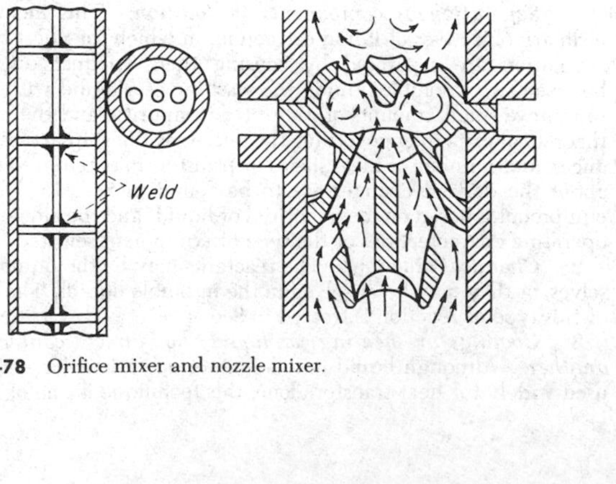

FIG. 21-78 Orifice mixer and nozzle mixer.

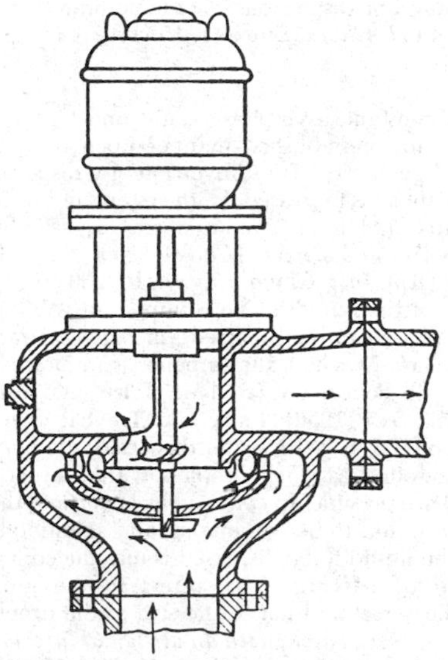

FIG. 21-79 Nettco Corp. Flomix. *(Chase, U.S. Patent 2,183,859, 1939.)*

phenols from coke-oven liquors with light oil [Gollmar, *Ind. Eng. Chem.*, **39**, 596 1947); Carbone, *Sewage Ind. Wastes*, **22**, 200 (1950)], but the intense shearing action causes emulsions with this low-interfacial-tension system. Modern plants use other types of extractors. Pumps are useful in the extraction of slurries, as in the extraction of uranyl nitrate from acid–uranium-ore slurries [*Chem. Eng.*, **66**, 80 (Nov. 2, 1959)]. Shaw and Long [*Chem. Eng.*, **64**(11), 251 (1957)] obtain a stage efficiency of 100 percent (E = 1.0) in a uranium-ore-slurry extraction with an open impeller pump. In order to avoid emulsification difficulties in these extractions, it is necessary to maintain the organic phase continuous, if necessary by recycling a portion of the settled organic liquid to the mixer.

Agitated Line Mixer See Fig. 21-79. This device, which combines the features of orifice mixers and agitators, is used extensively in treating petroleum and vegetable oils. It is available in sizes to fit ½- to 10-in pipe. The device of Fig. 21-80, with two impellers in separate stages, is available in sizes to fit 4- to 20-in pipe.

Packed Tubes Cocurrent flow of immiscible liquids through a packed tube produces a one-stage contact, characteristic of line mixers. For flow of isobutanol-water° through a 0.5-in diameter tube packed with 6 in of 3-mm glass beads, Leacock and Churchill [*Am. Inst. Chem. Eng. J.*, **7**, 196 (1961)] find

$$k_C a_{av} = c_1 L_C^{0.5} L_D \qquad (21\text{-}21)$$

$$k_D a_{av} = c_2 L_C^{0.75} L_D^{0.75} \qquad (21\text{-}22)$$

where c_1 = 0.00178 using SI units and 0.00032 using U.S. customary units; and c_2 = 0.0037 using SI units and 0.00057 using U.S. customary units. These indicate a stage efficiency approaching 100 percent. Organic-phase holdup and pressure drop for larger pipes similarly packed are also available [Rigg and Churchill, ibid., **10**, 810 (1964)].

Pipe Lines The principal interest here will be for flow in which one liquid is dispersed in another as they flow cocurrently through a pipe (stratified flow produces too little interfacial area for use in liquid extraction or chemical reaction between liquids). Drop size of dispersed phase, if initially very fine at high concentrations, increases as the distance downstream increases, owing to coalescence [see Holland, loc. cit.; Ward and Knudsen, *Am. Inst. Chem. Eng. J.*, **13**, 356 (1967)]; or if initially large, decreases by breakup in regions of high shear [Sleicher, ibid., **8**, 471 (1962); *Chem. Eng. Sci.*, **20**, 57 (1965)]. The maximum drop size is given by (Sleicher, loc. cit.)

$$\frac{d_{p,\max}\rho_C V^2}{\sigma g_c}\sqrt{\frac{\mu_C V}{\sigma g_c}} = C\left[1 + 0.7\left(\frac{\mu_D V}{\sigma g_c}\right)^{0.7}\right] \qquad (21\text{-}23)$$

where C = 43 (d_t = 0.013 m or 0.0417 ft) or 38 (d_t = 0.038 m or 0.125 ft), with $d_{p,av} = d_{p,\max}/4$ for high flow rates and $d_{p,\max}/13$ for low velocities.

Extensive measurements of the rate of mass transfer between n-butanol and water flowing in a 0.008-m (0.314-in) ID horizontal pipe are reported by Watkinson and Cavers [*Can. J. Chem. Eng.*, **45**, 258 (1967)] in a series of graphs not readily reproduced here. Length of a transfer unit for either phase is strongly dependent upon flow rate and passes through a pronounced maximum at an organic-water phase ratio of 0.5. In energy (pressure-drop) requirements and volume, the pipe line compared favorably with other types of extractors. Boyadzhiev and Elenkov [*Chem. Eng. Sci.*, **21**, 955 (1966)] concluded that, for the extraction of iodine between carbon tetrachloride and water in turbulent flow, drop coalescence and breakup did not influence the extraction rate. Yoshida et al. [*Coal Tar* (Japan), **8**, 107 (1956)] provide details of the treatment of crude benzene with sulfuric acid in a 1-in diameter pipe, N_{Re} = 37,000 to 50,000. Fernandes and Sharma [*Chem. Eng. Sci.*, **23**, 9 (1968)] used cocurrent flow downward of two liquids in a pipe, agitated with an upward current of air.

The pipe has also been used for the transfer of heat between two immiscible liquids in cocurrent flow. For hydrocarbon oil–water, the heat-transfer coefficient is given by

$$\frac{U a_{av} d_t^2}{v k_{to}} = \frac{\varphi_D N_{We,t}^{6/5}}{\dfrac{k_{to}}{0.415 k_{tC}} + \dfrac{k_{to}}{0.173 k_{tD}}} \qquad (21\text{-}24)$$

for φ_D = 0 to 0.2. Additional data for φ_D = 0.4 to 0.8 are also given. Data for stratified flow are given by Wilke et al. [*Chem. Eng. Prog.*, **59**, 69 (1963)] and Grover and Knudsen [*Chem. Eng. Prog.*, **51**, *Symp. Ser.* 17, 71 (1955)].

Mixing in Agitated Vessels Agitated vessels may frequently be used for either batch or continuous service and for the latter may be sized to provide any holding time desired. They are useful for liquids of any viscosity up to 750 Pa·s (750,000 cP), although in contacting two liquids for reaction or extraction purposes viscosities in excess of 0.1 Pa·s (100 cP) are only rarely encountered.

Mechanical Agitation This type of agitation utilizes a rotating impeller immersed in the liquid to accomplish the mixing and dispersion. There are literally hundreds of devices using this principle,

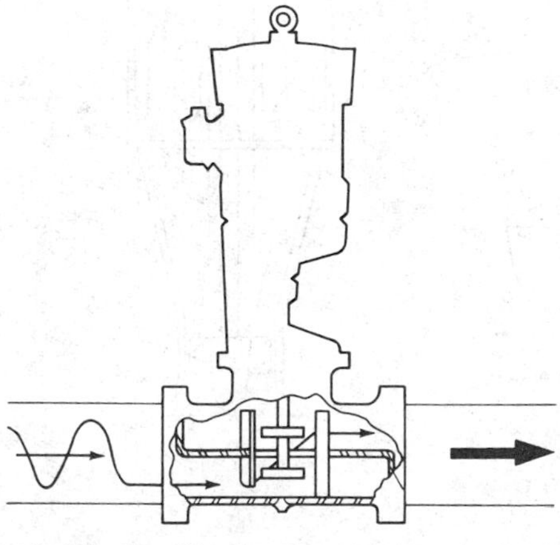

FIG. 21-80 Lightnin line blender. *(Mixing Equipment Co., Inc., with permission.)*

°Isobutanol dispersed: L_D = 3500 to 27,000; water continuous; L_C = 6000 to 32,000 in pounds-mass per hour–square foot (to convert to kilograms per second–square meter, multiply by 1.36×10^{-3}).

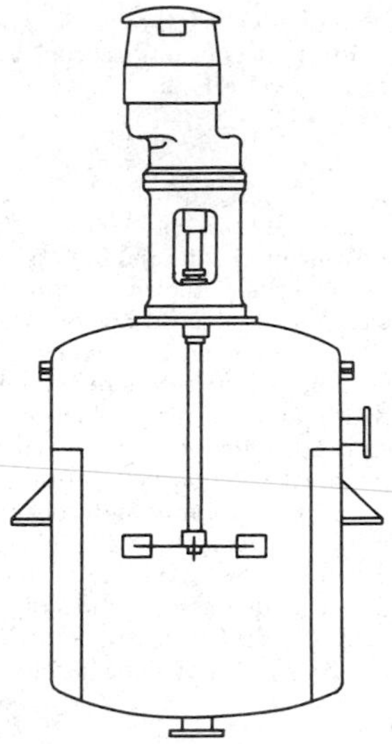

FIG. 21-81 Baffled mixing vessel.

the major variations being found when chemical reactions are being carried out. The basic requirements regarding shape and arrangement of the vessel, type and arrangement of the impeller, and the like are essentially the same as those for dispersing finely divided solids in liquids, which are fully discussed in Sec. 19. Figure 21-81 shows a typical single-compartment vessel for extraction or chemical reaction. Back mixing may be reduced, and the extraction stage efficiency thereby increased, by dividing the vessel into a number of compartments by horizontal separators, each compartment containing an agitator; or, for chemical reactions, in the manner of Fig. 21-82. Figure 21-83 shows an example of a reaction vessel in which extensive heat-transfer surface is also required.

The following summary of operating characteristics of mechanically agitated vessels is confined to the data available on liquid-liquid contacting.

Phase Dispersed There is an ill-defined upper limit to the volume fraction of dispersed liquid which may be maintained in an agitated dispersion. For dispersions of organic liquids in water [Quinn and Sigloh, *Can. J. Chem. Eng.*, **41,** 15 (1963)],

$$\varphi_{Do,\max} = \varphi' + (C/N^3) \qquad (21\text{-}25)$$

where φ' is a constant, asymptotic value, and C is a constant, both depending in an unestablished manner upon the system physical properties and geometry. Thus, inversion of a dispersion may occur if the agitator speed is increased. With systems of low interfacial tension ($\sigma' = 2$ to 3 mN/m or 2 to 3 dyn/cm), φ_D as high as 0.8 can be maintained. Selker and Sleicher [*Can. J. Chem. Eng.*, **43,** 298 (1965)] and Yeh et al. [*Am. Inst. Chem. Eng. J.*, **10,** 260 (1964)] feel that the viscosity ratio of the liquids alone is important. Within the limits in which either phase can be dispersed, for *batch operation* of baffled vessels, that phase in which the impeller is immersed when at rest will normally be continuous [Rodger, Trice, and Rushton, *Chem. Eng. Prog.*, **52,** 515 (1956); Laity and Treybal, *Am. Inst. Chem. Eng. J.*, **3,** 176 (1957)]. With water dispersed, *dual emulsions* (continuous phase found as small droplets within larger drops of dispersed phase) are possible. In *continuous operation*, the vessel is first filled with the liquid to be continuous, and agitation is then begun, after which the liquid to be dispersed is introduced.

Uniformity of Mixing This refers to the gross uniformity throughout the vessel and not to the size of the droplets produced. For *unbaffled vessels, batch, with an air-liquid interface*, Miller and Mann [*Trans. Am. Inst. Chem. Eng.*, **40,** 709 (1944)] mixed water with several organic liquids, measuring uniformity of mixing by sampling the tank at various places, comparing the percentage of dispersed phase found with that in the tank as a whole. A power application of 200 to 400 W/m^3 [(250 to 500 ft·lb)/(min·ft^3)] gave maximum and nearly uniform performance for all. See also Nagata et al. [*Chem. Eng. (Japan)*, **15,** 59 (1951)].

For *baffled vessels operated continuously, no air-liquid interface*, flow upward, light liquid dispersed [Treybal, *Am. Inst. Chem. Eng. J.*, **4,** 202 (1958)], the average fraction of dispersed phase in the vessel $\varphi_{D,av}$ is less than the fraction of the dispersed liquid in the feed mixture, unless the impeller speed is above a certain critical value which depends upon vessel geometry and liquid properties. Thornton and Bouyatiotis [*Ind. Chem.*, **39,** 298 (1963); *Inst. Chem. Eng. Symp. Liquid Extraction*, Newcastle-upon-Tyne, April 1967] have presented correlations of data for a 17.8-cm (7-in) vessel, but these do not agree with observations on 15.2- and 30.5-cm (6- and 12-in) vessels in Treybal's laboratory. See also Kovalev and Kagan [*Zh. Prikl. Khim.*, **39,** 1513 (1966)] and Trambouze [*Chem. Eng. Sci.*, **14,** 161 (1961)]. Stemerding et al. [*Can. J. Chem. Eng.*, **43,** 153 (1965)] pre-

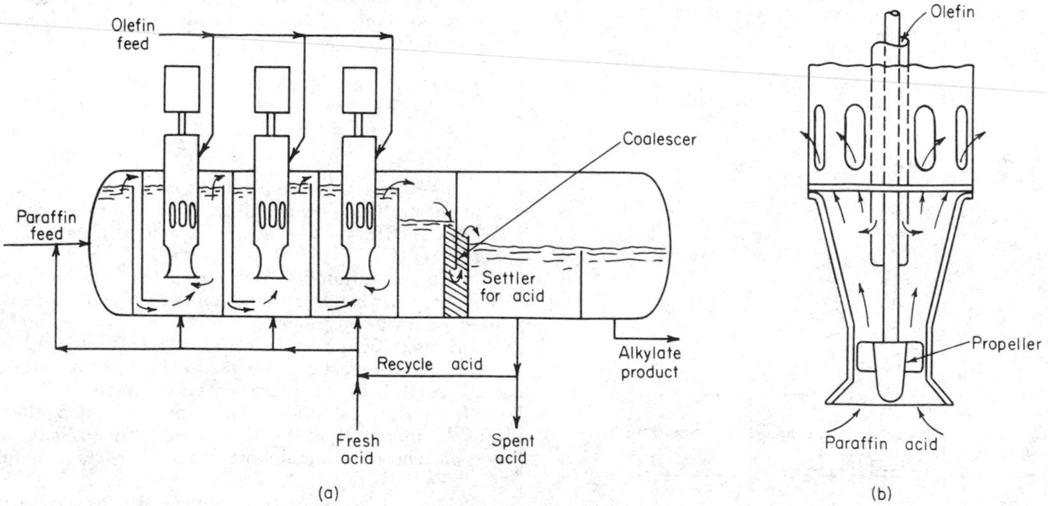

FIG. 21-82 (*a*) Kellogg Co. cascade alkylator for sulfuric acid alkylation of paraffins and olefins, simplified. (*b*) Mixer detail, simplified. *(Stiles et al., U.S. Patents 2,852,581, 1958, and 2,920,124, 1960.)*

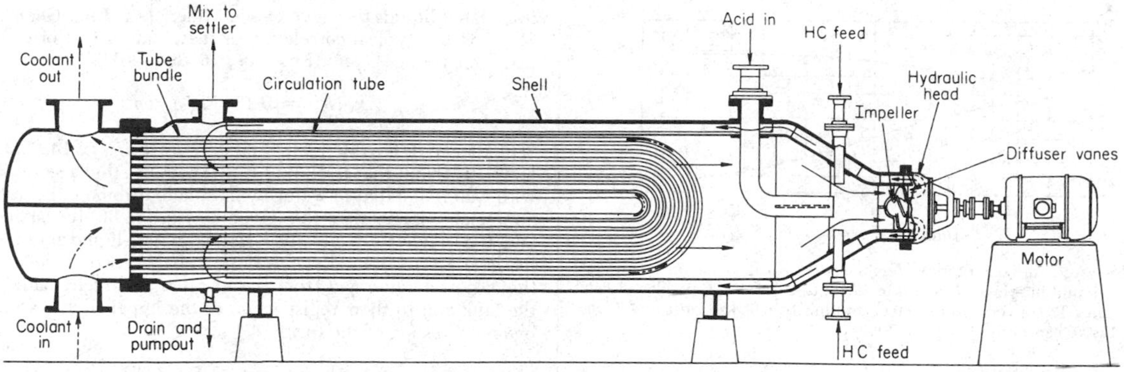

FIG. 21-83 Stratco contactor for HF alkylation of hydrocarbons. *(Courtesy of Stratford Engineering Co.)*

sent data on a large mixing tank [15 m^3 (530 ft^3)] fitted with a marine-type propeller and a draft tube.

Drop Size and Interfacial Area The drops produced have a size range [Sullivan and Lindsey, *Ind. Eng. Chem. Fundam.*, **1,** 87 (1962); Sprow, *Chem. Eng. Sci.*, **22,** 435 (1967); and Chen and Middleman, *Am. Inst. Chem. Eng. J.*, **13,** 989 (1967)]. The average drop size may be expressed as

$$d_{p,\text{av}} = \Sigma n_i d_{pi}^3 / \Sigma n_i d_{pi}^2 \tag{21-26}$$

and if the drops are spherical,

$$a_{\text{av}} = 6\varphi_{D,\text{av}}/d_{p,\text{av}} \tag{21-27}$$

The drop size varies locally with location in the vessel, being smallest at the impeller and largest in regions farthest removed from the impeller owing to coalescence in regions of relatively low turbulence intensity [Schindler and Treybal, *Am. Inst. Chem. Eng. J.*, **14,** 790 (1968); Vanderveen, U.S. AEC UCRL-8733, 1960]. Interfacial area and hence average drop size have been measured by light transmittance, light scattering, direct photography, and other means. Typical of the resulting correlations is that of Thornton and Bouyatiotis (*Inst. Chem. Eng. Symp. Liquid Extraction*, Newcastle-upon-Tyne, April 1967) for a 17.8-cm- (7-in-) diameter baffled vessel, six-bladed flat-blade turbine, d_i = 6.85 cm (0.225 ft), operated full, for organic liquids (σ' = 8.5 to 34, ρ_D = 43.1 to 56.4, μ_D = 1.18 to 1.81) dispersed in water, in the absence of mass transfer, and under conditions giving nearly the vessel-average $d_{p,\text{av}}$:

$$\frac{d_{p,\text{av}}}{d_p^0} = 1 + 1.18\phi_D \left(\frac{\sigma^2 g_c^2}{d_p^0 \mu_C^2 g}\right) \left(\frac{\mu_C^4 g}{\Delta\rho\ \sigma^3 g_c^3}\right)^{0.62} \left(\frac{\Delta\rho}{\rho_C}\right)^{0.05} \tag{21-28}$$

where d_p^0 is given by

$$\frac{(d_p^0)^3 \rho_C^2 g}{\mu_C^2} = 29.0 \left(\frac{P^3 g_c^3}{v^3 \rho_C^2 \mu_C g^4}\right)^{-0.32} \left(\frac{\rho_C \sigma^3 g_c^3}{\mu_C^4 g}\right)^{0.14} \tag{21-29}$$

Caution is needed in using such correlations, since those available do not generally agree with each other. For example, Eq. (21-28) gives $d_{p,\text{av}}$ = 4.78(10^{-4}) ft for a liquid pair of properties a' = 30, ρ_C = 62.0, ρ_D = 52.0, μ_C = 2.42, μ_D = 1.94, $\varphi_{D,\text{av}}$ = 0.20 in a vessel T = Z = 0.75, a turbine impeller d_i = 0.25 turning at 400 r/min. Other correlations provide 3.28(10^{-4}) [Thornton and Bouyatiotis, *Ind. Chem.*, **39,** 298 (1963)], 8.58(10^{-4}) [Calderbank, *Trans. Inst. Chem. Eng. (London)*, **36,** 443 (1958)], 6.1(10^{-4}) [Kafarov and Babinov, *Zh. Prikl. Khim*, **32,** 789 (1959)], and 2.68(10^{-3}) (Rushton and Love, paper at AIChE, Mexico City, September 1967). See also Vermeulen et al. [*Chem. Eng. Prog.*, **51,** 85F (1955)], Rodgers et al. [ibid., **52,** 515 (1956); U.S. AEC ANL-5575 (1956)], Rodrigues et al. [*Am. Inst. Chem. Eng. J.*, **7,** 663 (1961)], Sharma et al. [*Chem. Eng. Sci.*, **21,** 707 (1966); **22,** 1267 (1967)], and Kagan and Kovalev [*Khim. Prom.*, **42,** 192 (1966)]. For the effect of absence of baffles, see Fick et al. (U.S. AEC UCRL-2545, 1954) and Schindler and Treybal [*Am. Inst. Chem. Eng. J.*, **14,** 790 (1968)]. The latter have observations during mass transfer.

Coalescence Rates The droplets coalesce and redisperse at rates that depend upon the vessel geometry, N, $\varphi_{D,\text{av}}$, and liquid properties. The few measurements available, made with a variety of techniques, do not as yet permit quantitative estimates of the coalescence frequency ν. Madden and Damarell [*Am. Inst. Chem. Eng. J.*, **8,** 233 (1962)] found for baffled vessels that ν varied as $N^{2.2}\varphi_{D,\text{av}}^{0.5}$, and this has generally been confirmed by Groothius and Zuiderweg [*Chem. Eng. Sci.*, **19,** 63 (1964)], Miller et al. [*Am. Inst. Chem. Eng. J.*, **9,** 196 (1963)], and Howarth [ibid., **13,** 1007 (1967)], although absolute values of ν in the various studies are not well related. Hillestad and Rushton (paper at AIChE, Columbus, Ohio, May 1966), on the other hand, find ν to vary as $N^{0.73}\varphi_{D,\text{av}}$ for impeller Weber numbers $N_{\text{We},i}$ below a certain critical value and as $N^{-3.5}\varphi_{D,\text{av}}^{1.58}$ for higher Weber numbers. The influence of liquid properties is strong. There is clear evidence [Groothius and Zuiderweg, loc. cit.; *Chem. Eng. Sci.*, **12,** 288 (1960)] that coalescence rates are enhanced by mass transfer from a drop to the surrounding continuum and retarded by transfer in the reverse direction. See also Howarth [*Chem. Eng. Sci.*, **19,** 33 (1964)]. For a theoretical treatment of drop breakage and coalescence and their effects, see Valentas and Amundsen [*Ind. Eng. Chem. Fundam.*, **5,** 271, 533 (1966); **7,** 66 (1968)], Gal-Or and Walatka [*Am. Inst. Chem. Eng. J.*, **13,** 650 (1967)], and Curl [ibid., **9,** 175 (1963)].

Power for Agitation The data for single liquids in baffled and unbaffled vessels with an air-liquid interface are very extensive and are summarized by Rushton, Costich, and Everett [*Chem. Eng. Prog.*, **46,** 395, 467 (1950)]. See also Sec. 19. In general, for baffled vessels N_{Po} is a function of N_{Re}; whereas for unbaffled vessels operated with a vortex N_{Po} is a function of N_{Re} and N_{Fr}. The data for two-phase liquid mixtures are very limited.

Unbaffled vessels; air-liquid interface. Miller and Mann [*Trans. Am. Inst. Chem. Eng.*, **40,** 709 (1944)] worked with tanks, T = 0.15 to 0.46 m (0.5 to 1.5 ft), and flat- and pitched-blade turbines, a two-bladed propeller, and a spiral turbine. For each impeller in the absence of a vortex ($N_{\text{Re}} < 10^4$), power to agitate single liquids and two-phase mixtures (water continuous) was correlated by the same relation, $N_{\text{Po}} = f(N_{\text{Re}})$. This required that density and viscosity for two-phase mixtures be computed as

$$\rho_{\text{av}} = \rho_C\varphi_C + \rho_D\varphi_D \tag{21-30}$$

$$\mu_{\text{av}} = \mu_C^{\varphi_C}\mu_D^{\varphi_D} \tag{21-31}$$

Baffled vessels, with or without an air-liquid interface. Figure 21-84 shows a correlation of power for two-phase liquids and single liquids, using a six-bladed flat-blade turbine [Laity and Treybal, *Am. Inst. Chem. Eng. J.*, **3,** 176 (1957)]. The two-liquid data are correlated by the single-liquid curve provided the density is computed by Eq. (21-30) and the viscosity by [Vermeulen et al., *Chem. Eng. Prog.*, **51,** 85F (1955)]

$$\mu_{\text{av}} = \frac{\mu_C}{\varphi_C}\left(1 + \frac{1.5\mu_D\varphi_D}{\mu_D + \mu_C}\right) \tag{21-32}$$

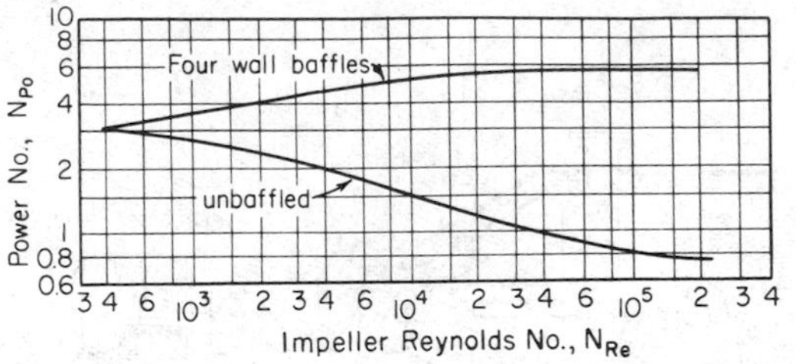

FIG. 21-84 Agitator power for two-liquid mixtures, baffled and unbaffled vessels, no air-liquid interface. T = 1.0 to 1.5 ft, d_i/T = 0.333, six-bladed flat-blade turbines. To convert feet to meters, multiply by 0.3048. [*Laity and Treybal,* Am. Inst. Chem. Eng. J., *3, 176 (1959).*]

There is essentially no effect of liquid flow rate when the vessel is operated continuously. Olney and Carlson [*Chem. Eng. Prog.*, **43**, 473 (1947)] report power data for an arrowhead disperser used with wall baffles and for a spiral turbine (Turbo-Mixer) with a stator-ring baffle but no wall baffles, with single- and two-phase liquids. These data may also be correlated through Eqs. (21-30) and (21-32).

At least until more data are obtained, it is therefore recommended that the general correlations for single liquids in baffled vessels (Sec. 19) be used for two-liquid mixtures, with density and viscosity computed through Eqs. (21-30) and (21-32) respectively.

Unbaffled vessels; no air-liquid interface. Figure 21-84 shows the results of batch agitation of one- and two-liquid mixtures in 0.305- and 0.457-m- (12- and 18-in-) diameter vessels with a six-bladed flat-blade turbine (Laity and Treybal, loc. cit.). The correlation shown requires that densities for two liquids be calculated by Eq. (21-30) and viscosities by

$$\mu_{av} = \frac{\mu_w}{\varphi_w}\left(1 + \frac{6\varphi_o\mu_o}{\mu_w + \mu_o}\right) \tag{21-33}$$

for $\varphi_w > 0.40$, and

$$\mu_{av} = \frac{\mu_o}{\varphi_o}\left(1 - \frac{1.5\varphi_w\mu_w}{\mu_w + \mu_o}\right) \tag{21-34}$$

for $\varphi_w < 0.40$. The viscosity of the mixture can exceed that of either constituent. (The μ_{av} is not the same when calculated at φ_w = 0.40. Ed. note.) Below N_{Re} = 10^4 the liquids were not uniformly mixed. There is no effect of impeller height from d_i to $2d_i$ from the vessel bottom and a small effect of liquid flow rate at very high liquid rates when operated continuously. Owing to the lack of data, no generalizations regarding other impeller types may be made. Additional power data are provided by Wingard et al. (Ala. Polytech. Inst. Eng. Exp. Sta. Bull. 17, p. 3, 1952).

Rates of Mass Transfer Measurements simply of the extent of extraction in an agitated vessel lead to the overall "volumetric" mass-transfer coefficients, $K_C a_{av}$ or $K_D a_{av}$, or the equivalent stage efficiency. The coefficients K_C and K_D are made up of the coefficients for the individual liquids, k_C and k_D:

$$\frac{1}{K_D} = \frac{1}{k_D} + \frac{1}{m'_{CD}k_C} \qquad \frac{1}{K_C} = \frac{1}{k_C} + \frac{m''_{CD}}{k_D} \tag{21-35}$$

The evidence is that the coefficients k_C and k_D and the interfacial area a_{av} depend differently upon operating variables. For purposes of design, therefore, it is ultimately necessary to have separate information on the quantities k_C, k_D, and a_{av}. The role of an additional surface resistance is emphasized by the studies of Kishinevski and Moehalova [*Zh. Prikl. Khim.*, **33**, 2049 (1960)].

Information on the coefficients is relatively undeveloped. They are evidently strongly influenced by rate of drop coalescence and breakup, presence of surface-active agents, "interfacial turbulence" (Marangoni effect), drop-size distribution, and the like, none of which can be effectively evaluated at this time.

Continuous-phase coefficients. There have been a large number of measurements of k_C for solid particles and gas bubbles suspended in agitated liquids [for review, see Miller, *Ind. Eng. Chem.*, **56**(10), 18 (1964)]. A typical correlation of these data is that of Calderbank and Moo-Young [*Chem. Eng. Sci.*, **16**, 39 (1961)]:

$$k_C N_{Sc}^{2/3} = 0.13\,(P\mu_C g_c/v\rho^2)^{1/4} \tag{21-36}$$

Schindler and Treybal [*Am. Inst. Chem. Eng. J.*, **14**, 790 (1968)], however, found that for liquid dispersions of ethyl acetate saturated with water, agitated in water by flat-blade turbine impellers, k_C was appreciably larger than that given by Eq. (21-36) for baffled vessels and even higher for unbaffled vessels (no air-liquid interface). The increase was attributed to the rate of coalescence of the droplets as the dispersion emerged from the impeller and recirculated through the tank and to their redispersion at the impeller. It was described by an expression of the form

$$k_C = k_S + C(D_C/\theta_C)^{0.5} \tag{21-37}$$

where k_S was calculated from Harriott's data for small-diameter solids [Harriott, *Am. Inst. Chem. Eng. J.*, **8**, 93 (1962)]. The continuous phase was found to be completely back-mixed and of uniform composition throughout for both baffled and unbaffled vessels.

Dispersed-phase coefficients. There have been no direct measurements of k_D for liquid dispersions in agitated vessels. If the drops are small (as they usually are), internal circulation causes them to behave like rigid spheres with an enhanced diffusivity D'_D. In stirred vessels, the ratio D'_D/D_D has been estimated to lie in the range of about 1:2 [Olney, *Am. Inst. Chem. Eng. J.*, **7**, 348 (1961); Treybal *Liquid Extraction*, 2d ed., McGraw-Hill, New York, 1963]. For a pump-mix impeller (Fig. 21-91), Coughlin and von Berg [*Chem. Eng. Sci.*, **21**, 3 (1966)], on the other hand, estimate k_D to be higher than that for circulating drops but not so large as that for oscillating drops (see below). These estimates do not take into account drop coalescence, interfacial turbulence, etc.; they are based on an assumed value for k_C and measured overall coefficients.

Overall coefficients and stage efficiency. If it is assumed that values of a_{av}, k_C, k_D (and therefore K_D) can somehow be estimated, the stage efficiency can be calculated through

$$E_{MD} = 1 - \exp\left(-\frac{K_D a_{av} Z}{V_D}\right) = 1 - \exp\left[-\frac{K_D a_{av}\theta(V_C + V_D)}{V_D}\right] \tag{21-38}$$

See also Treybal [*Am. Inst. Chem. Eng. J.*, **4**, 202 (1958); **6**, 5M (1960)] and Olander [*Chem. Eng. Sci.*, **18**, 47 (1963); **19**, 275 (1964)]. The remaining discussion is confined to measured values of stage efficiency or volumetric overall coefficients. These are largely of value only for the particular systems studied. For this reason, one fairly complete study will be described, and the others will only be mentioned.

Figure 21-85 summarizes the results for the extraction of n-butylamine from kerosine into water in a continuously operated mixer [T = 0.37 m (1.23 ft); Z = 0.48 m (1.562 ft)] fed cocurrently upward, with and without four wall baffles and with a variety of impellers [Overcashier, Kingsley, and Olney, *Am. Inst. Chem. Eng. J.*, **2**, 529 (1956)]. When unbaffled, the vessel was full and without an air-liquid interface. E_O represents the overall countercurrent efficiency of a single stage. E_O at zero agitator speed was 0.18 at a liquid residence time of 1.08 min. The improved performance in the absence of baffles may be attributed to the reduction in back mixing and to the reduced power requirement for a given impeller speed. In the absence of baffles, vertical location of the impeller is immaterial. With baffles, the best performance is given with the impeller at 0.667 Z from the bottom, the worst at 0.25 Z from the bottom. For the spiral turbine, wall baffles and stator-ring baffles produced the same power-efficiency relationship. Off-center unbaffled operation at a propeller was intermediate between centered baffled and centered unbaffled operation. The data for propellers, spiral turbines, and flat-blade turbines, d_i = 0.10 to 0.25 m (0.333 to 0.833 ft), in both unbaffled and baffled tanks, with a flow rate to produce a residence time θ

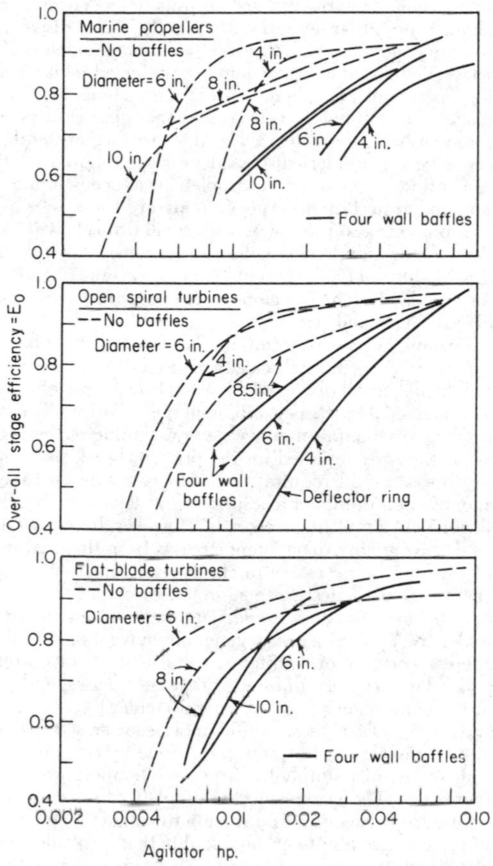

FIG. 21-85 Continuous extraction of n-butylamine from kerosine into water. T = 1.23 ft, Z = 1.56 ft, no air-liquid interface, impellers centered, V_R/V_E × 1.57, residence time × 1.08 min. To convert feet to meters, multiply by 0.3048; to convert inches to centimeters, multiply by 2.54; and to convert horsepower to kilowatts, multiply by 0.746. [*Overcashier, Kingsley, and Olney,* Am. Inst. Chem. Eng. J., *2, 529 (1956), with permission.*]

= 0.18 h, kerosine-water ratio = 1.57 by volume, are empirically correlated by

$$E_O = 1 - \frac{0.318(10^{15})(d_i/T)^{\alpha}}{N_{\text{Re}}^{3.2}N_{\text{Po}}^{1.37}} \qquad (21\text{-}39)$$

where α = 0 for baffled operation and 1.6 for unbaffled operation. For Eq. (21-39), the viscosity and density were computed through Eqs. (21-30) and (21-31).

Other detailed studies are the following:

1. Hixson and Smith [*Ind. Eng. Chem.*, **41,** 973 (1949)]. Batch extraction of iodine from water into carbon tetrachloride; unbaffled vessels, propeller agitated. Log $(1 - E)$ is linear with time.

2. Karr and Scheibel [*Chem. Eng. Prog.*, **50,** *Symp. Ser.* 10, 73 (1954)]. Continuous extraction of acetic acid between methyl isobutyl ketone and water, and xylene and water, and of acetone between xylene and water; unbaffled vessels consisting of the unpacked section of the extractor of Fig. 21-108. Rate of extraction is larger when organic liquid is dispersed in the extractant than with other arrangements.

3. Flynn and Treybal [*Am. Inst. Chem. Eng. J.*, **1,** 324 (1955)]. Continuous extraction of benzoic acid from toluene and kerosine into water; baffled vessels, turbine agitators. Stage efficiency is correlated with agitator energy per unit of liquid treated.

4. Mottel and Colvin (U.S. AEC DP-254, 1957). Continuous heat transfer between kerosine and water; vessel of Pump-Mix design (Fig. 21-91).

5. Ryon, Daley, and Lowrie [*Chem. Eng. Prog.*, **55**(10), 70, (1959); U.S. AFC ORNL-2951, 1960]. Continuous extraction of uranium from sulfate-ore-leach liquors and kerosine + tributyl phosphate and di(2-ethylhexyl)-phosphoric acid; baffled vessels, turbine agitated. There is strong evidence of the influence of a slow chemical reaction.

6. Ryon and Lowrie (U.S. AEC ORNL-3381, 1960). Batch and continuous extraction of uranium from aqueous sulfate solutions into kerosine + amines, stripping of extract with aqueous sodium carbonate; baffled vessels, turbine agitated. A detailed process study.

7. David and Colvin [*Am. Inst. Chem. Eng. J.*, **7,** 72 (1961)]. Continuous heat transfer between kerosine and water; unbaffled vessel. Open impellers (paddles and propellers) are better than closed (centrifugal and disk impellers) at the same tip speed.

8. Simard et al. [*Can. J. Chem. Eng.*, **39,** 229 (1961)]. Continuous extraction of uranium from aqueous nitrate solutions into kerosine + tributyl phosphate and from sulfate solutions containing tricaprylamine; unbaffled vessel, propeller agitated. Process details for high recovery and low reagent costs.

9. Rushton, Nagata, and Rooney [*Am. Inst. Chem. Eng. J.*, **10,** 298 (1964)]. Batch extraction of octanoic acid from water and corn syrup into xylene, paraffin oil, and their mixtures; baffled vessel, turbine impeller. $K_C a_{av}$ proportional to $N^{2.1}\mu_C^{-0.6}\mu_D^{-0.55}$.

10. Coughlin and von Berg [*Chem. Eng. Sci.*, **21,** 3 (1966)]. Continuous heat transfer and extraction of ethylbutyric acid between kerosine and water; unbaffled vessel, Pump-Mix design (Fig. 21-91). Interfacial area measured.

Scale-up. Treybal [*Chem. Eng. Prog.*, **62**(9), 67 (1966)] demonstrates that, for geometrically similar agitated vessels with equal holding time and power per unit volume on the two scales, the stage efficiency for simple extraction will likely increase on scale-up. This is not necessarily the most economical method of scale-up, however [see Treybal, *Am. Inst. Chem. Eng. J.*, **5,** 474 (1959)]. See also Hills [*Br. Chem. Eng.*, **6,** 104 (1961)].

Intrastage recycle. Recycling some of one of the settled liquids back to the agitated vessel sometimes improves settling of the dispersion. In addition, the stage efficiency of a stirred vessel can be considerably enhanced by recycling the liquid favored by solute distribution, whereas recycling the other liquid reduces the stage efficiency. When solute distribution favors the dispersed phase and mass-transfer rates are poor, recycling the settled dispersed phase can result in minimizing the volume of a cascade of extraction vessels [Treybal, *Ind. Eng. Chem. Fundam.*, **3,** 185 (1964)]. See also Gel'perin et al. (*Khim. Neft. Mashinostr.*, **1966,** 23).

Rates of Heterogeneous Chemical Reaction Detailed studies of the role of agitation are practically nonexistent. Most kinetic studies merely report that "efficient agitation" was used, with no details. In the case of one more detailed study [McKinley and White, *Trans. Am. Inst. Chem. Eng.*, **40,** 143 (1944)], the effect of agitator speed in the continuous nitration of toluene with mixed acids in a laboratory-size stirred vessel was measured. Below 800 r/min the degree of agitation was insufficient to produce appreciable interfacial surface for rapid mass transfer of the reagents to the reaction site, whereas above 1200 r/min mass-transfer equilibrium was reached so rapidly that further increase in agitator speed no longer appreciably influenced the reaction rate. The data are specific for the particular apparatus used, but presumably are typical of the sort to be obtained for most heterogeneous liquid reactions. For example, Otake and Komasawa [*Kagaku Kogaku*, **32,** 475 (1968)], for the same nitration, report a similar observation. This study also included varying which phase was dispersed and the nitric acid concentration. Engel and Hougen [*Am. Inst. Chem. Eng. J.*, **9,** 724 (1963)] hydrolyzed isoamyl acetate dispersed in a large volume of water (normally a two-phase reaction) catalyzed by an acidic ion-exchange resin in a baffled vessel. The overall resistance to hydrolysis, which is the sum of the resistances of the liquid surrounding the catalyst and of the catalyst itself, was influenced by degree of agitation only for catalyst particles below 100-μm diameter. Molony and Millrain [Benzole Prod. Ltd. (London) Res. Pap. 1-1967] report extensive data on the slow-reaction acid washing of thiophene from benzene in laboratory vessels and plant washers of up to 45.5-m^3 (10,000-U.K. gal) capacity.

Kircher, Miller, and Geiser [*Ind. Eng. Chem.*, **46,** 1925 (1954)] specify that agitated vessels for sulfonation reactions should be equipped with turbine agitators, d_i/T = 0.35 to 0.5, operated at a peripheral speed of 3.30 to 3.56 m/s (650 to 700 ft/min). For many practical details of the use of the apparatus of Fig. 21-83, see *Hydrofluoric Acid Alkylation,* Phillips Petroleum Company, Bartlesville, Okla., 1946. For details of heterogeneous reactions generally, see

Groggins, *Unit Processes in Organic Synthesis*, 5th ed., McGraw-Hill, New York, 1958.

There is an abundance of theoretical treatment in the literature [see, for example, Hofmann, *Chem. Eng. Sci.*, **8**, 113 (1958); Olander, ibid., **6**, 233 (1960); Scriven, *Am. Inst. Chem. Eng. J.*, **7**, 524 (1963); Spalding, *Chem. Ing. Tech.*, **32**, 91 (1960); Trambouze, *Chem. Eng. Sci.*, **14**, 161 (1961); and for the effect of coalescence and redispersion, Curl, *Am. Inst. Chem. Eng. J.*, **9**, 175 (1963); Shain, ibid., **12**, 806 (1966); and Rietema, in Drew et al. (eds.), *Advances in Chemical Engineering*, vol. 5, Academic, New York, 1964, p. 237]; but there have been remarkably few experimental studies. Extractive reaction, in which a solvent extracts one of the products to enhance the yield, is considered by Piret, Trambouze, et al. [*Am. Inst. Chem. Eng. J.*, **6**, 394, 574 (1960); **7**, 138 (1961)]. See also Schmitz and Amundsen [*Chem. Eng. Sci.*, **18**, 265, 415, 447 (1963)].

Gas Agitation The gas may be a vapor such as steam which is generated in place by boiling the liquids to be contacted or which may be admitted through spargers at the bottom of the vessel. Permanent gases such as air may also be used. Air, for example, has been used extensively for the mixing of reagents such as sulfuric acid with all but the most volatile of petroleum liquids. It can provide the gentlest of agitation, as in the washing of nitroglycerin with water, as well as vigorous mixing. There is danger of oxidation of product with air, and with any gas there will necessarily be some volatilization of the liquids being mixed. Gas agitation in the extraction of radioactive liquids offers the advantages of no maintenance-requiring moving parts, but it may require decontamination of the effluent air.

Although gas agitation has usually been considered an uneconomical method of applying mixing power, there is little in the way of quantitative data with which to judge its effectiveness in contacting immiscible liquids. Mathers and Winter [*Can. J. Chem. Eng.*, **37**, 99 (1959)] describe a mixer-settler in which air is used as an air-lift type of mixer. With a mixer of 5-L volume, aqueous acetic acid (3.48 L/min) was extracted with hexone (6.9 L/min), using 8.5 L/min (0.3 ft^3/min) of air. The average stage efficiency (including the effect of a 10-L settler) was $E_{MR} = 0.93$, and the power for air was 0.001 hp, corresponding to 90 ft·lb energy expended/ft^3 liquids treated. Thornton [*Nucl. Eng.*, **1**, 156, 204 (1956)] described a somewhat similar air-agitated mixer.

Settlers

Emulsions and Dispersions The mixture of liquids leaving a mixer is a cloudy dispersion which must be settled, coalesced, and separated into its liquid phases in order to be withdrawn as separate liquids from a stage. For a dispersion to "break" into separate phases, both sedimentation and coalescence of the drops of the dispersed phase must occur. Unstable dispersions usually have droplet diameters of about 1 mm or larger and settle rapidly. Stable dispersions, or emulsions, are generally characterized by droplet diameters of about 1 μm or less. The unstable dispersions are preferred in liquid-liquid-extraction operations and chemical-reaction systems involving two liquid phases that ultimately need to be separated. Dispersions and emulsions are usually characterized by the terms **water-in-oil** (meaning aqueous liquid droplets dispersed in organic liquid continuous phase) and **oil-in-water** (organic droplets in aqueous liquid). **Dual emulsions** and **liquid-membrane systems** are those in which the continuous phase is also present as very small droplets within larger drops of the other liquid. See Becher, *Emulsions: Theory and Practice*, ACS Monogr. 175, Reinhold, New York, 1957; and Li and Shrier, in Li (ed.), *Recent Developments in Separation Science*, vol. I, CRC Press, Cleveland, 1972, p. 163.

The "breaking" of a dispersion in a batch settler may be divided into two periods: (1) primary break, or rapid settling and coalescence of most of the dispersed phase, which often leaves a fog of very small droplets suspended as parts per million in the majority phase; and (2) secondary break, which represents the slow settling of the fog. Most industrial settlers are designed for the primary break since the slow secondary break would require much longer residence times. The small amount of entrainment to a subsequent stage seldom influences stage efficiency in a multistage cascade. However, for conserving solvent and desolventizing the effluent streams from the final stages of a cascade, it may be necessary to clarify as completely as possible, including the use of coalescers to eliminate secondary fog.

Sedimentation Isolated droplets, settling or rising in a stagnant liquid under the force of gravity, generally move more rapidly than solid spheres. The rate of settling or rising is more rapid for large droplet size, large density difference between phases, and low viscosity of the continuous phase. Felix and Holder [*Am. Inst. Chem. Eng. J.*, **1**, 296 (1955)] show considerably shorter settling time of petroleum-oil dispersions in water and phenol by reducing the continuous-phase viscosity simply by raising the temperature.

Coalescence The coalescence of droplets can occur whenever two or more droplets collide and remain in contact long enough for the continuous-phase film to become so thin that a hole develops and allows the liquid to become one body. A clean system with a high interfacial tension will generally coalesce quite rapidly. Particulates and polymeric films tend to accumulate at droplet surfaces and reduce the rate of coalescence. This can lead to the buildup of a "rag" layer at the liquid-liquid interface in an extractor. Rapid drop breakup and rapid coalescence can significantly enhance the rate of mass transfer between phases.

Gravity Settlers; Decanters These are tanks in which a liquid-liquid dispersion is continuously settled and coalesced and from which the settled liquids are continuously withdrawn. They can be

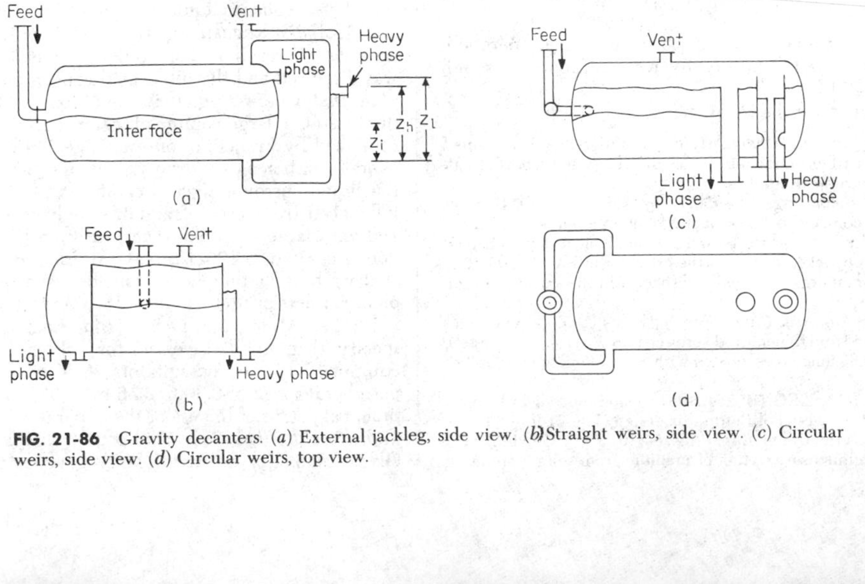

FIG. 21-86 Gravity decanters. (*a*) External jackleg, side view. (*b*) Straight weirs, side view. (*c*) Circular weirs, side view. (*d*) Circular weirs, top view.

either horizontal or vertical. Figure 21-86 shows some typical horizontal decanters. For an uninstrumented decanter the height of the heavy-phase-liquid leg above the interface is balanced against the height of the light-liquid phase above the interface, Eq. 21-40.

$$(Z_h - Z_i)\rho_h = (Z_L - Z_i)\rho_L \qquad (21\text{-}40)$$

The velocity of the liquid entering the decanter should be kept low to minimize disturbance of the interface. Sometimes an impingement baffle, or "picket fence," has been used. In other cases, opposing inlets as in Fig. 21-86*c* and *d* have been used. For an external jackleg shown in Fig. 21-86*a* the heavy-liquid takeoff requires a siphon break to prevent emptying the vessel by siphoning. Some problems can occur because of pressure drop through the outlet piping and variable levels under flow conditions. The horizontal weirs in Fig. 21-86*b* and the circular weirs in Fig. 21-86*c* can be designed for a very low crest height at maximum design flow rates. When rag builds up at the interface, sometimes it can be purged by withdrawing a small stream, filtering out the solids, and returning the liquids to the decanter. The decanter can also be instrumented with an interface detector and automatic control valve on the heavy-phase flow. The light phase can still overflow from the vessel.

For general reviews, see Ingersoll [*Pet. Refiner*, **30**(6), (1951)] and Hart [*Pet. Process.*, **2**, 282, 471, 513, 632 (1947)]. In the petroleum industry, settler volumes have frequently been sufficiently large so as to provide a holding time from 0.5 to 1.0 h, which in most cases is probably excessive and costly. For most thin liquids, in which unusual emulsification problems do not occur, 5 to 10 min is ample. The size of the settler seems to be set by the rate of flow per unit of horizontal cross-sectional area as well as holding time [Williams et al., *Trans. Inst. Chem. Eng. (London)*, **36**, 464 (1958)]; Ryon, Daley, and Lowrie [*Chem. Eng. Prog.*, **55**(10), 70 (1959)], for the settling of aqueous uranium solutions and kerosine–alkyl phosphate solvents, used decanters of the type shown in Fig. 21-86*b*. The depth of the decanter having been chosen, these authors recommend that the horizontal cross section for the prevailing flow rate be set at twice the value which would give a dispersion-band thickness equal to the depth of the tank. In this manner dispersions of 9.08 m^3/h aqueous + 15.90 m^3/h solvent (40 gal/min aqueous + 70 gal/min solvent) were successfully settled in a decanter of 1.4-min holding time.

Gravity settlers, basically of the type shown in Fig. 21-86*a*, were used by Wilke et al. (UCRL-10625, 1963; UCRL-11182, 1964) to settle water and Aroclor (specific gravity, 1.36). The dispersion may occupy a wedge-shaped volume in the region of the interface at low flow rates instead of covering the entire interface as for higher flow rates. Important variables influencing the performance are (1) the value of φ_D in the entering liquid mixture, (2) whether the dispersion is introduced above or below the interface, and (3) the distance of an impact baffle from the inlet pipe. The length of the dispersion wedge for kerosine-water (dispersed) is proportional to V_D/d_p^3 (Jeffreys and Pitt, paper at AIChE meeting, Salt Lake City, May 1967). Higher temperatures (to decrease viscosity) and longer residence times within each phase improve the settling of water (dispersed)-coconut fatty acids [Manchanda and Woods, *Ind. Eng. Chem. Process Des. Dev.*, **7**, 182 (1968)].

In the extraction of uranium from ore leach liquors with kerosine-reagent solvents, there is a saving in the cost of thickeners and filters if the aqueous liquors are not clarified before extraction. If such slurries are extracted, however, it is necessary to increase the solvent-aqueous ratio in the extractor in order to make the organic phase continuous; otherwise, unsettleable emulsions are produced. Table 21-34 gives the data of Shaw and Long [*Chem. Eng.*, **64**(11), 251 (1957)] for settling areas required for such extractions. The high organic-aqueous ratios are obtained by recycling settled organic phase from the settler to the mixer. Entrainment of organic solvent with the settled solids represents a serious problem in such operations.

Cyclones Cyclones have been suggested as simple means of enhancing by centrifugal force the rate of settling of liquid dispersions. Tepe and Woods (U.S. AEC AECD-2864, 1943) report a few data for the separation of isobutanol-water dispersions in such devices, but the results were poor. The most thorough studies are those of Simkin and Olney [*Am. Inst. Chem. Eng. J.*, **2**, 545 (1956)] and of Hitchon [At. Energy Res. Estab. (Gt. Brit.) CE/R-2777, 1959], who conclude that high extraction efficiencies (requiring high degrees of dispersion) and good clarification of both effluents cannot be obtained in one stage. Tepe and Woods (loc. cit.) also tried *helical coils of pipe* for separating isobutanol-water mixtures, with poor results.

TABLE 21-34 Settling of Aqueous Uranium Leach Liquors with Kerosine–Alkyl Phosphate Solvent*

Nature of aqueous feed	Organic-aqueous ratio required	Permissible settler flow rate, U.S. gal/(min·ft^2) horizontal area
Clear liquor	4	1.4–1.6
Slimes (5% solids)	8	0.6
Dense pulps (50–60% solids)	10	0.3

*Shaw and Long, *Chem. Eng.*, **64**(11), 251 (1957).

To convert gallons per minute–square foot to cubic meters per hour–square meter, multiply by 2.44.

Centrifuges Mechanical centrifuges, high-speed machines, have been used for many years for separating liquid-liquid dispersions, for example, in the separation of caustic solutions and oils in the soap-making process, more recently in uranium extractions, and in many others. By enhancing the settling rate (without, however, influencing coalescence), they reduce the settling time considerably. See, for example, Landis [*Chem. Eng. Prog.*, **61**(10), 58 (1965)]. For details see Sec. 19.

Settler Auxiliaries These include the use of coalescers, separating membranes, and electrical devices and the addition of emulsion-breaking reagents. These last are used for treating permanent emulsions and will not be discussed here.

Coalescers. The small drops of a fine dispersion may be caused to coalesce and thus become larger by passing the dispersion through a coalescer. The enlarged drops then settle more rapidly. Coalescers are mats, beds, or layers of porous or fibrous solids whose properties are especially suited for the purpose at hand. In an extensive study, Sareen et al. [*Am. Inst. Chem. Eng. J.*, **12**, 1045 (1966)] found, in part, that (1) coalescence is promoted by decreased fiber diameter, (2) a minimum bed density is required to achieve complete coalescence, dependent upon the system characteristics, (3) wetting of the fibers by droplets of dispersed phase is not necessary for good coalescence, (4) a fibrous bed of medical cotton can be made to coalesce almost any kind of liquid dispersed in another except if $\sigma' < 3$ mN/m (dyn/cm), (5) cotton fibers are best supported from collapse by mixing with fibers of glass or Dynel [see also Langdon et al., *Petro/Chem Eng.*, **1963**(11), 35], (6) the optimum bed thickness of a mixed bed depends on the ratio of cotton to support (0.75 in for 50 percent cotton), (7) the maximum velocity through the bed with effective coalescing increases with bed depth, but increased pressure drop causes redispersion, presumably at values depending upon the liquid system, and (8) some surfactants interfere with coalescence, but others do not. For tests on petroleum-brine emulsions and Fiberglas, see Burtis and Kirkbride [*Trans. Am. Inst. Chem. Eng.*, **42**, 413 (1946)] and Hayes et al. [*Chem. Eng. Prog.*, **45**, 235 (1949)]. Beds of granular solids such as sand, etc., and bats of excelsior, steel wool, and the like have also been used.

Separating membranes. If the capillary size of a porous substance is very small, the liquid which preferentially wets the solid may flow through the capillaries readily but strong interfacial films block the capillaries for flow of nonwetting liquid. Sufficient pressure will cause disruption of the films and permit passage of the nonwetting liquid, but regulation of the pressure commensurate with the pore size permits perfect phase separation. Separating membranes of this type are made of a variety of materials such as porcelain, paper which has been coated with special resins, and the like and may be either hydrophilic or hydrophobic in character. They are made thin so as to permit maximum passage of the wetting liquid [see Jordan, *Trans. Am. Soc. Mech. Eng.*, **77**, 393 (1955); and Belk, *Chem. Eng. Prog.*, **61**(10), 72 (1965)]. In practice, the dispersion is usually first passed through a coalescer so as to permit settling of the bulk of the

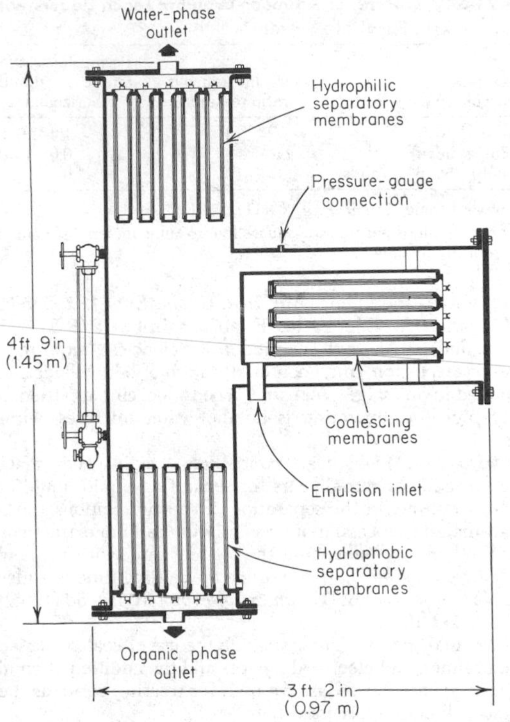

FIG. 21-87 Combination coalescer, settler, and membrane separator. *(Courtesy of Selas Corporation of America.)*

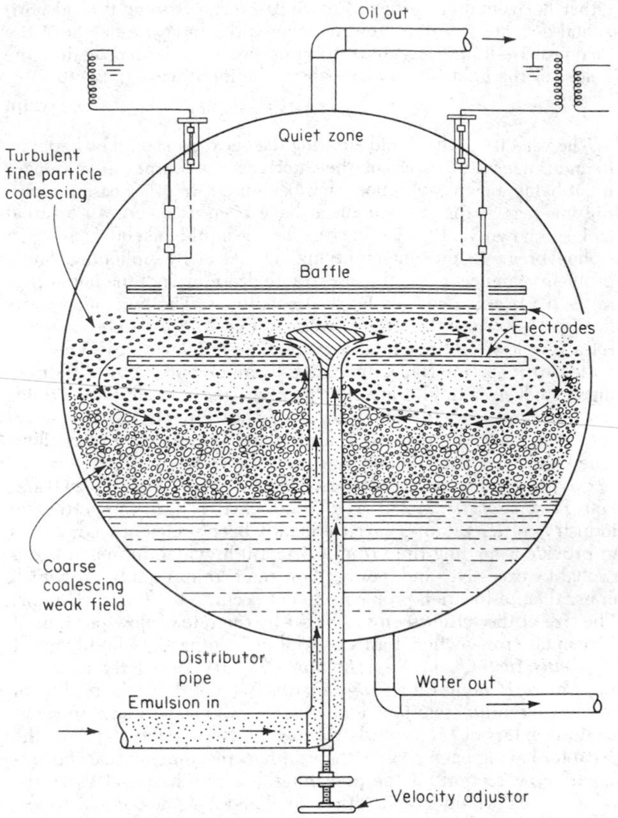

FIG. 21-89 Internal circulation and electric field, Petreco Cylectric coalescer (schematic). [*Waterman,* Chem. Eng. Prog., *61(10), 51 (1965), with permission.*]

dispersed phase before the mixture is presented to the separating membrane, thus relieving the load on the membrane.

Figure 21-87 shows a combination device containing coalescers and both hydrophobic and hydrophilic separating membranes. Coalescers and separating membranes are fashioned in the form of hollow cylinders, and flow is radially through the wall. After passing through the coalescers, the bulk of the liquids settles in the vertical member of the device, and then the settled phases are passed through their respective separating membranes. Devices of this type are designed to handle 0.57 to 6.81 m^3/h (150 to 1800 gal/h), delivering completely separated phases; and further settling is unnecessary. Figure 21-88 shows another design for removing dispersed water from jet fuel or gasoline, available in sizes to handle from 68 to 250 m^3/h (300 to 1100 gal/min) and delivering clear effluents. In this case only a hydrophobic membrane is required [Redmon, *Chem. Eng. Prog.*, **59**(9), 87 (1963)].

Electrical devices. Subjecting electrically conducting emulsions or dispersions to high-voltage electric fields may cause rupture of the protective film about a droplet and thus induce coalescence. Dispersions of low conductivity are subject, in an electric field, to forces between particles resulting from acquired induced dipoles, which induce coalescence. These phenomena have been used particularly for the desalting of petroleum emulsified with brine, and for similar applications. Devices have been built to handle 828 m^3/h (125,000 bbl/day) of crude oil, at costs of approximately 0.1 to 0.5 cent/bbl [Waterman, *Chem. Eng. Prog.*, **61**(10), 51 (1965)]. For a detailed study see Sjoblom and Goren [*Ind. Eng. Chem. Fundam.*, **5**, 519 (1966)] and Brown and Hanson [*Trans. Faraday Soc.*, **61**, 1754 (1965)]. Figure 21-89 shows schematically the flow through a typical device.

Mixer-Settler Combinations Any mixer and settler can be combined to produce a stage, and the stages in turn arranged in a multistage cascade. Figure 21-90 shows such an arrangement with compartmented vessels.

A great many commonly used other arrangements have been

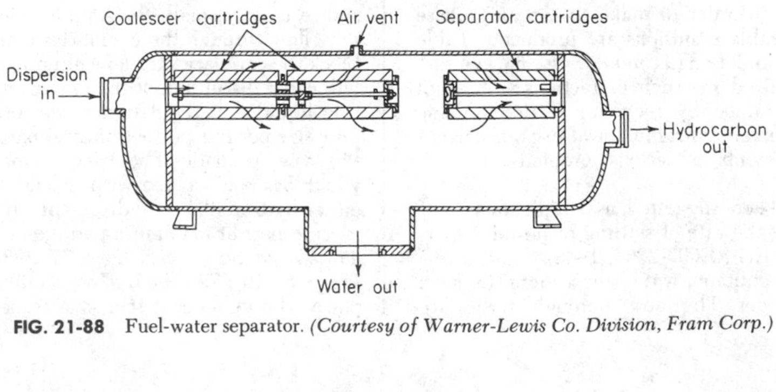

FIG. 21-88 Fuel-water separator. *(Courtesy of Warner-Lewis Co. Division, Fram Corp.)*

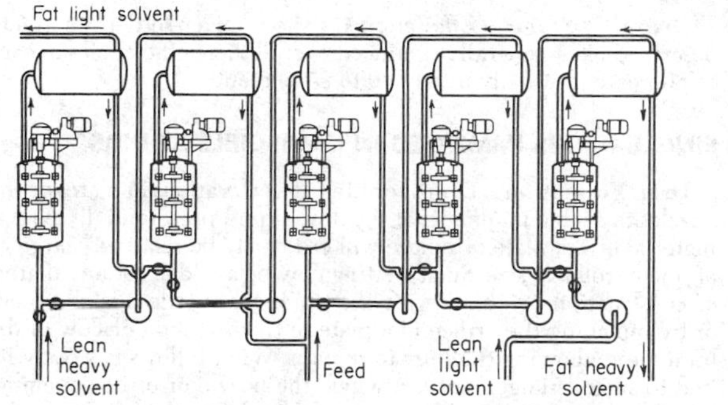

FIG. 21-90 Five-stage countercurrent cascade arranged for fractional extraction. Mixers compartmented, fitted with Turbo-Mixer agitators. *(Courtesy of Turbo-Mixer Division, General American Transportation Corp.)*

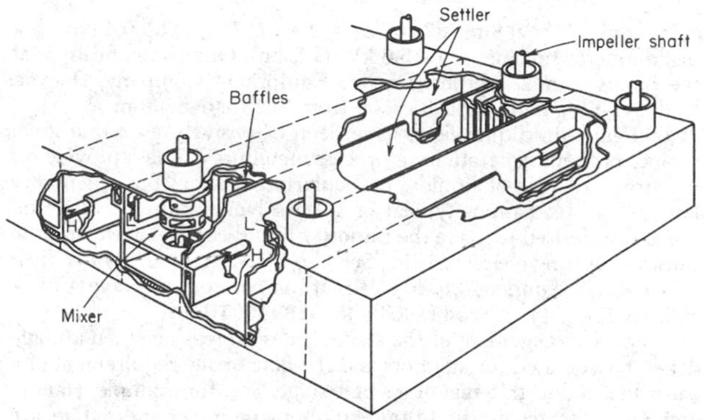

FIG. 21-91 Pump-Mix mixer-settler. [*Coplan, Davidson, and Zebroski,* Chem. Eng. Prog., *50, 403 (1954), with permission.*]

developed in an effort to reduce or eliminate interstage pumping and to reduce costs generally. Only a few of the more commonly used types are mentioned here.

A compact alternating arrangement of mixers and settlers has been adopted in many of the "box-type" extractors developed originally for processing radioactive solutions, but now used in principle for many processes, with literally dozens of modifications. An example is the Pump-Mix mixer-settler (Fig. 21-91), in which adjacent stages have common walls [Coplan, Davidson, and Zebroski, *Chem. Eng. Prog.*, **50**, 403 (1954)]. The impellers in this case pump as well as mix by drawing the heavy liquid upward through the hollow impeller shaft and discharging it at a higher level through the hollow impeller. These extractors or variants of them have been built not only in relatively large sizes but also in miniature for bench-scale work.

Figure 21-92 represents still further modification for low cost [Hazen and Henrickson, *Min. Eng.*, 994 (1957); Quinn, Trefoil (Denver Equipment Company) Bull. M4-B90, 1957]. At *a* and *b* in the figure, the settler is a circular tank T = 4.9 m (16 ft), Z = 2.1 m (7 ft), with the mixing vessel, 1.2 by 1.2 m (4 by 4 ft), contained inside. Agitators are turbines, d_i = 0.46 m (1.5 ft), operated at 150 r/min (1.12 kW) and 200 r/min (2.02 kW). The aqueous feed is 22.7-m^3/h (100-gal/min) uranium-bearing ore leach liquor, the organic solvent 4.5-m^3/h (20-gal/min) alkyl phosphate solutions in kerosine. Adjacent stages are at 0.3-m (1-ft) elevation difference, allowing gravity flow of the aqueous liquor, while the organic phase is pumped in countercurrent by air lifts. Provision is made for recycle of settled organic phase by overflow to the mixer, the amount of which can be adjusted by changing the height of the organic-overflow pipe. The vanadium extractor at *c* in the figure is a box type, built into a cir-

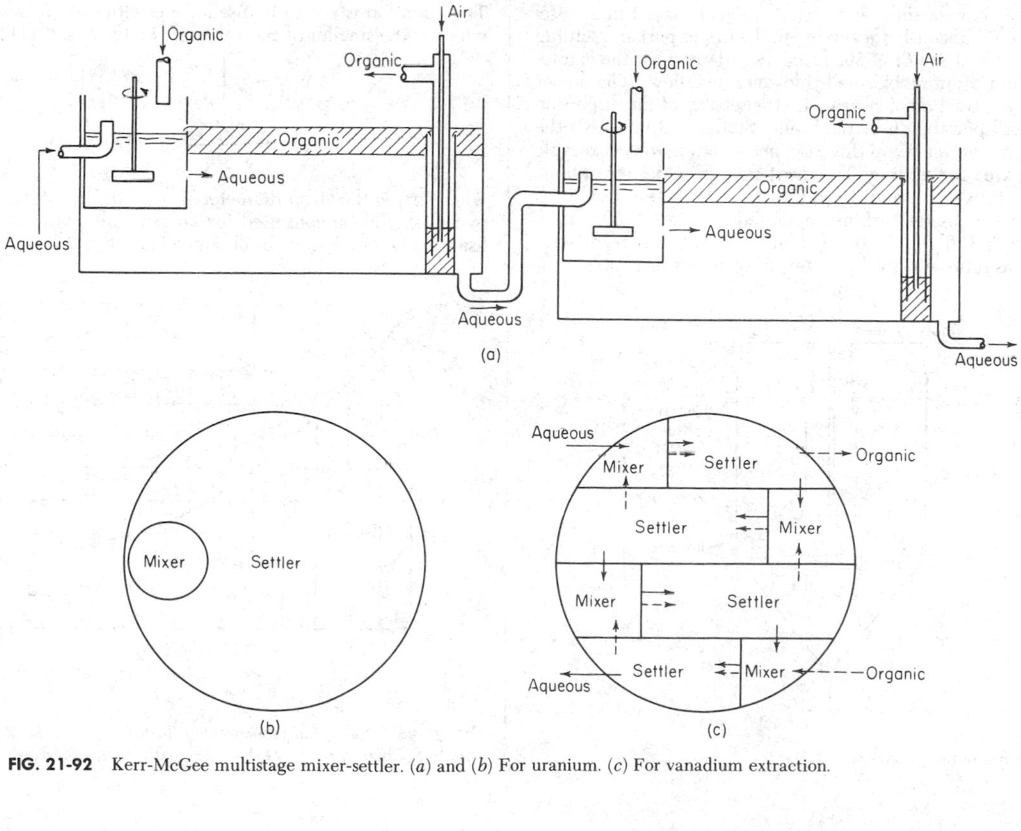

FIG. 21-92 Kerr-McGee multistage mixer-settler. (*a*) and (*b*) For uranium. (*c*) For vanadium extraction.

cular tank, T = 9.8 m (32 ft), Z = 2.1 m (7 ft). The 0.46-m- (18-in-) diameter turbines draw 5.6 kW (7.5 hp). Other modifications of the box-type mixer-settler (Denver Equipment Company, Denver, Bull. A1-B6), with capacities of from 0.23- to 5700-m^3/h (1- to 25,000-gal/min) liquid flow, have been extensively used in a great variety of metal separations in process metallurgy. These provide for intrastage recycle of liquids, particularly advantageous when very low solvent-feed ratios typical of good solvents must be used and when it is desired to make the minority liquid continuous in order to improve settling characteristics. See also Williams et al. [*Trans. Inst. Chem. Eng. (London)*, **36**, 464 (1958)] and Hanson and Kaye [*Chem. Process Eng.*, **44**, 27, 654 (1963); **45**, 413 (1964)].

Vertical arrangement of the stages is desirable, since then a single drive may be used for agitators and the floor-space requirement of a cascade is reduced to that of a single stage. See, for example, Hanson and Kaye, loc. cit. In the Lurgi extractor, the mixer and settlers are in separate vertical shells interconnected with piping [Guccione, *Chem. Eng.*, 78 (July 4, 1966)].

A great many other devices are known. The Fenske extractor [Fenske and Long, *Chem. Eng. Prog.*, **51**, 194 (1955); *Ind. Eng. Chem.*, **53**, 791 (1961); *Ind. Eng. Chem. Fundam.*, **1**, 152 (1962)] is a vertical stack of mixer-settler stages, with mixing done by a vertically moving reciprocating plate in each mixer. One very successful device, particularly in the extraction of radioactive solutions, uses a centrifuge instead of a settler to separate the mixed liquids, and the pump-mixer and centrifuge of each stage operate on a common shaft [Clark, U.S. AEC DP-752 (1962); Kisbaugh, ibid., DP-841 (1963)]. See also Goncharenko et al. (*Tr. Vses. Khemosorbtsii Nauchn-Tekhn. Sovesch. Protessy Zhidkostnoi Ekstraktsii Khemosorbtsii*, 2d, Leningrad, **1964**, 75) and Berestovoi et al. (ibid., 171). Still another uses a cyclone (hydroclone) for separating [Whatley and Woods, U.S. AEC ORNL-3533 (1964); Finsterwalder, ibid., ORNL-4088 (1967)]. The Graesser extractor (Coleby, U.S. Patent 3,017,253, 1962) is a horizontal shell filled with stratified settled liquids, with a series of buckets revolving around the inner periphery which rain droplets of one liquid through the other. It has been used primarily in Europe for easily emulsified liquids.

Figure 21-93 shows a shrouded mixer impeller placed in a vessel which serves simultaneously for mixing in the upper part and settling in the lower [Lash, *Min. Eng.* **10**, 1161 (1958)], used for the extraction of uranium from unclarified-thickener overflow. The major problems in the extraction of slurries are the settling of the dispersion and solvent loss upon the discharged solid. Settling requires that the organic phase be continuous in this case, necessitating solvent recycle and adequate area for settling. The draft tube provides for organic recycle. Plant-size vessels are 6.10 m (20 ft) in diameter, fitted with 0.46-m- (18-in-) diameter turbines operated at 100 to 150 r/min. Flow capacity is 100 m^3/h (440 gal/min) of aqueous feed, with organic-aqueous ratio = 3 in the mixer, fixed by settling rates.

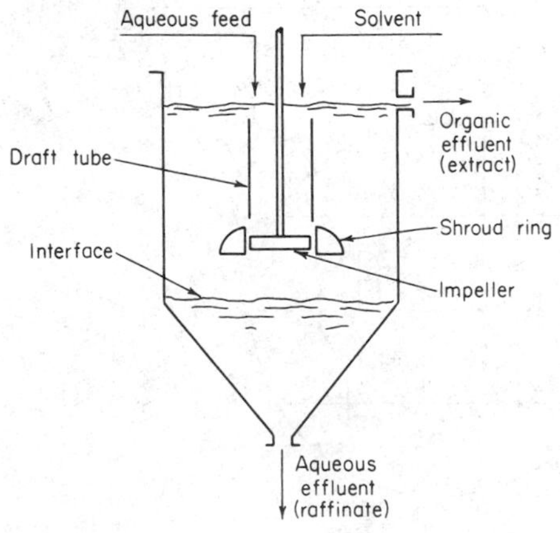

FIG. 21-93 Vitro uranium mixer-settler.

Overall Stage Efficiencies The mixer-settler extractors described have generally produced overall stage efficiencies in excess of 80 percent, usually nearly 90 to 95 percent.

SINGLE DROPS IMMERSED IN IMMISCIBLE LIQUIDS

Drop Formation Drops forming at perforations in plates or from nozzles in such a manner that the drop liquid preferentially wets the material of the plate or nozzle will ordinarily be relatively large and of uncontrollable size. Such wetting may be avoided for any material of construction by chamfering the tip of the nozzle to a sharp edge or by punching the orifice in a plate in the direction of flow of drop liquid and allowing the burr to remain. What follows is strictly limited to such arrangements, in which the nozzle or orifice opening is horizontal, and in the absence of mass transfer (the effect of which is not yet established).

At relatively low velocities through the opening, drops form at the opening and are of uniform size. At higher velocities, a jet of drop liquid projects from the opening, and drops form by breakup of the jet and are generally nonuniform. At very high velocities, "atomization" into very small, nonuniform drops occurs directly at the opening.

For *drops forming slowly at the opening*, the most recent work is that of Scheele and Meister [*Am. Inst. Chem. Eng. J.*, **14**, 9, 15 (1968)], covering the following experimental conditions: d_O = 0.081- to 0.78-cm (0.00266- to 0.0255-ft-) long nozzles to ensure fully developed laminar velocity profiles, ρ_D = 680 to 985 kg/m^3 (42.5 to 61.5 lbm/ft^3), $\Delta\rho$ = 10 to 570 kg/m^3 (0.623 to 35.6 lbm/ft^3), μ_C = 0.0001 to 5166 Pa·s [2.32 to 1250 lbm/(ft·h)], μ_D = 0.0004 to 0.12 Pa·s [0.95 to 2.93 lbm/(ft·h)], σ' = 1.79 mN/m or dyn/cm. The drop size is given by

$$v_p = \frac{\pi d_p^3}{6} = F\left[\frac{\pi\sigma g_c d_O}{g\,\Delta\rho} + \frac{20\mu_c Q d_O}{d_p^2 g\,\Delta\rho} - \frac{4\rho_D Q V_O'}{3g\,\Delta\rho} + 4.5\left(\frac{Q^2 d_O^2 \rho_D \sigma g_c}{(g\,\Delta\rho)^2}\right)^{1/3}\right] \tag{21-41}$$

The equation is not to be used at velocities above which jets form, which is the smaller of the values given by Eqs. (21-42) and (21-43).

$$V_{OJ}' = 2\left[\frac{\sigma g_c}{\rho d_O}\left(1 - \frac{d_O}{d_{pJ}}\right)\right]^{1/2} \tag{21-42}$$

$$V_{OJ}' = \frac{2d_{pJ}^2 V_t'}{3d_O^2} \tag{21-43}$$

where d_{pJ} is the drop diameter if a jet did not form. The original papers should be consulted for an estimate of the likely errors. To use Eq. (21-41), F must be obtained from Fig. 21-94. The calculation

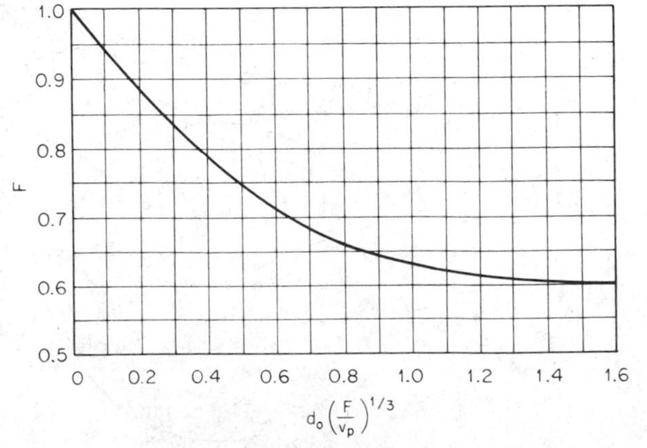

FIG. 21-94 Correction factor for use in Eq. (21-41). [*Scheele and Meister, Am. Inst. Chem. Eng. J., 14, 9 (1968), with permission.*]

is one of trial and error: as a first approximation, the second term in the brackets on the right of Eq. (21-41) is neglected, whereupon F/v_p and d_p may be estimated. Figure 21-94 then provides F, and the calculation may be repeated without neglecting the indicated term. The results are frequently in conflict with two earlier correlations [Hayworth and Treybal, *Ind. Eng. Chem.*, **42,** 1174 (1950); Null and Johnson, *Am. Inst. Chem. Eng. J.*, **4,** 273 (1958)]. It has also been observed that the lowest velocity at which a jet is formed may depend upon whether this is reached by increasing or decreasing the velocity. See also Rao et al. [*Chem. Eng. Sci.*, **21,** 867 (1966)] and Siemes [*Chem. Ing. Tech.*, **28,** 727 (1956)]. For *drops formed by breakup of the jet*, see Christiansen and Hixson [*Ind. Eng. Chem.*, **49,** 1017 (1957)].

Mass Transfer during Drop Formation The literature is very large and cannot be reviewed in its entirety here. References which provide a good review of modern thinking are Angelo, Lightfoot, and Howard [*Am. Inst. Chem. Eng. J.*, **12,** 751 (1966); **14,** 531 (1968)]; Heertjes and de Nie [*Chem. Eng. Sci.*, **21,** 755 (1966)]; Ilkovic [*Collect. Czech. Chem. Commun.*, **6,** 498 (1934)]; Popovich, Jevis, and Trass [*Chem. Eng. Sci.*, **19,** 357 (1964)]; and Zheleznyak [*Zh. Prikl. Khim.*, **40,** 870 (1967)]. The overall mass-transfer coefficient K_D is used as follows:

$$N_f = K_D\left(c_D - \frac{c_C}{m'_{CD}}\right) \tag{21-44}$$

where the flux N_f is a time-area-average value. For example, if K_D is based on the total time θ_F of drop formation and the area A_p of the drop at breakaway, the total moles of solute transferred is $N_f A_p \theta_F$. The concentration c_D is that of the solute in the entering drop liquid, c_C in the bulk continuous liquid. Transfer is out of the drop if N_f is positive, into the drop if negative. On the assumption that the two-resistance theory applies and that the mechanism of mass transfer is the same for both phases,

$$K_D = k_D\left[\frac{1}{1 + (1/m'_{CD})(D_D/D_C)^{0.5}}\right] \tag{21-45}$$

The various theories for treating the mass-transfer mechanisms generally result in an expression of the form

$$k_D = C(D_D/\pi\theta_F)^{0.5} \tag{21-46}$$

where C depends upon the assumed nature of the rate of growth of drop volume and interfacial surface and the mass-transfer process (penetration, surface-renewal, surface-stretch theory, etc.).

For k_D based on θ_F and A_p and an always spherical shape assumed for the growing drop, the various theories lead to values of C in the range 0.857 to 3.43 (Popovich et al., loc. cit.). Most of the available data seem to fit values of C in the range 1.3 to 1.8. There are notable exceptions, variously ascribed to the presence of surface-active agents and interfacial turbulence.

Angelo et al. (loc. cit.) have observed, however, that the surface of the drop grows linearly with time; and by application of the surface-stretch theory they have derived results that appear most promising. Unfortunately, the additionally required quantities (residual surface left at the orifice after breakaway, for example) are not yet predictable without specific measurements for each situation.

Terminal Velocity Liquid drops rising or falling in a liquid under the force of gravity generally move at rates different from those of rigid spheres of the same volume and density. For small diameters, although the drops are essentially spherical, internal circulation of the drop substance results in surface motion and produces larger terminal velocities. As the diameter increases, the shape departs from the spherical [Wellek et al., *Am. Inst. Chem. Eng. J.*, **12,** 854 (1966)]; and eventually oscillations of the shape occur, leading to velocities substantially lower than for rigid spheres and usually decreasing somewhat with increasing diameter. In both regions, the drop velocity can be altered strongly by the presence of surface-active agents. For a thorough review, see Kintner [in Drew et al. (eds.), *Advances in Chemical Engineering*, vol. 4, Academic, New York, 1963, p. 51]. Figure 21-95 is a general correlation giving the terminal velocity V_t of liquid drops immersed in a quiet continuous

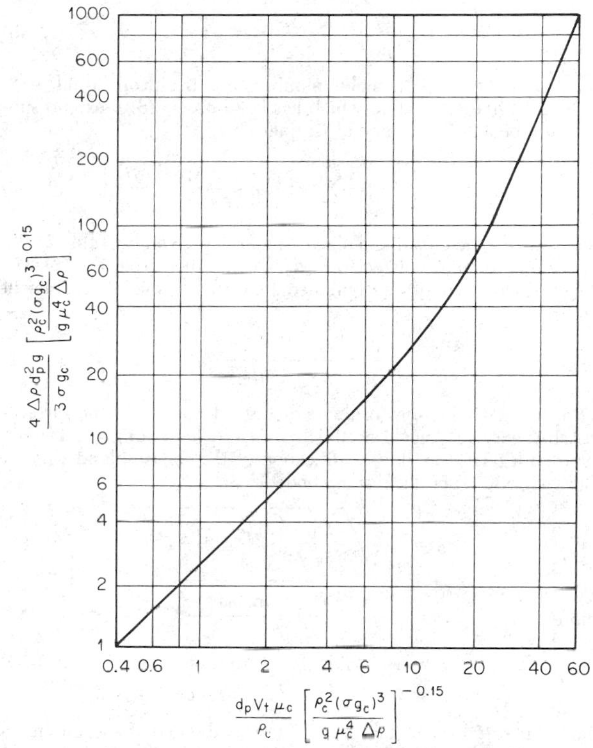

FIG. 21-95 Terminal velocity, liquid drops in low-viscosity liquids. [*Hu and Kintner*, Am. Inst. Chem. Eng. J., *1, 42 (1955), with permission.*]

liquid [$\mu'_C \leq 0.005$ Pa·s (5 cP)] in the absence of wall effects. The peak velocity occurs at an ordinate of the figure of about 70. See Klee and Treybal [*Am. Inst. Chem. Eng. J.*, **2,** 444 (1956)] for convenient equations. For $\mu'_C = 0.005$ to 0.020 Pa·s (5 to 20 cP), the ordinate of the figure should be multiplied by $(\mu_{H_2O}/\mu_C)^{0.14}$ [Johnson and Braida, *Can. J. Chem. Eng.*, **35,** 165 (1957)]. For higher values of μ'_C, see Warshay et al. [ibid., **37,** 29 (1959)]. For wall effects ($d_p/T > 0.18$), see Harmathy [*Am. Inst. Chem. Eng. J.*, **6,** 281 (1960)], Salami, Vignes, and Le Goff [*Genie Chim.*, **94,** 67 (1965)], and Strom and Kintner [*Am. Inst. Chem. Eng. J.*, **4,** 153 (1958)]. For influence of Marangoni instability, see Linde and Sehrt [*Z. Phys. Chem. (Leipzig)*, **231,** 151 (1966)].

Mass Transfer during Drop Rise or Fall There are a substantial number of data for rates of mass transfer, which are difficult to interpret because of the complications offered by the possible presence of interfacial turbulence (which leads to unusually high rates of mass transfer) and contamination of the drop interface with surface-active agents (which usually lowers the rates). The various regimes of motion produce different regimes of mass transfer, and these have been given much theoretical study. It will not be possible to review these in detail here. The following recommendations are considered reasonable estimates of the effects to be expected in the absence of interfacial turbulence and surface-active agents, but it must be emphasized that the available data occasionally show greatly different results.

For **circulating drops** (ordinate of Fig. 21-95 < 70), k_C may be estimated by applying an empirical correction F to the values for solid spheres to account for the circulation [Hughmark, *Ind. Eng. Chem. Fundam.*, **6,** 408 (1967)]:

$$\frac{k_C d_p}{D_C} = \left[2 + 0.463 N_{Re}^{0.484} N_{Sc,C}^{0.339}\left(\frac{d_p g^{1/3}}{D_C^{2/3}}\right)^{0.072}\right] F \tag{21-47}$$

where

$$F = 0.281 + 1.615\kappa + 3.73\kappa^2 - 1.874\kappa^3 \tag{21-48}$$

$$\kappa = N_{Re}^{1/8}\left(\frac{\mu_C}{\mu_D}\right)^{1/4}\left(\frac{\mu_C V_t}{\sigma g_c}\right)^{1/6} \tag{21-49}$$

and N_{Re} is the drop Reynolds number. For the drop liquid, with Hadamard-like circulation which has been observed to extend substantially beyond the region of laminar flow,

$$k_D = -\frac{d_p}{6\theta}\ln\left[\frac{3}{8}\sum_{n=1}^{\infty} B_n^2 \exp\left(-\frac{\lambda_n 64 D_D \theta}{d_p^2}\right)\right] \tag{21-50}$$

where the coefficients B and eigenvalues λ are given by Table 21-35. Values may be interpolated for $k_C d_p/D_C$ obtained from Eq. (21-47). The coefficients may be combined to form an overall coefficient through

$$\frac{1}{K_D} = \frac{1}{k_D} + \frac{1}{m'_{CD}k_C} \tag{21-51}$$

For **oscillating drops** (ordinate of Fig. 21-95 > 70), the surface-stretch theory of Angelo et al. [*Am. Inst. Chem. Eng. J.*, **12**, 751 (1966); **14**, 531 (1968)], a modification of that of Rose and Kintner [ibid., **12**, 530 (1966)] is recommended:

$$k_D = \sqrt{\frac{4D_D\omega}{\pi}\left(1 + \delta + \frac{3}{8}\delta^2\right)} \tag{21-52}$$

where

$$\omega = \frac{1}{2\pi}\sqrt{\frac{192\sigma g_c b}{d_p^3(3\rho_D + 2\rho_C)}} \tag{21-53}$$

$$b = 1.052 d_p^{0.225} \text{ for } d_p \text{ in meters} \tag{21-54}$$

$$b = 0.805 d_p^{0.225} \text{ for } d_p \text{ in feet}$$

The value of δ may be taken as 0.2 in the absence of specific information. Equation (21-57), when joined through Eq. (21-56) with a k_C given by the same mechanism, produces K_D as given by Eq. (21-50).

The above must be considered to be estimates only. For further details and other viewpoints, useful references are Cheh and Tobias [*Ind. Eng. Chem. Fundam.*, **7**, 48 (1968)], Griffith [*Chem. Eng. Sci.*, **12**, 198 (1960)], Handlos and Baron [*Am. Inst. Chem. Eng. J.*, **3**, 127 (1957)], Johns and Beckmann [ibid., **11**, 10 (1966)], Johnson and Hamielec [ibid., **6**, 145 (1960)], Kornienko [*Tr. Kishinev, Politekh. Inst.*, **5**, 16 (1966)], Olander [*Am. Inst. Chem. Eng. J.*, **12**, 1018 (1966)], Sideman and Shabtai [*Can. J. Chem. Eng.*, **42**, 107, 238 (1964)], Skelland and Wellek [*Am. Inst. Chem. Eng. J.*, **10**, 491, 798 (1964); **11**, 557 (1965)], and Thorsen and Terjesen [*Chem. Eng. Sci.*, **17**, 137 (1962)].

Coalescence. Coalescence of a single drop into a bulk of the same liquid through a flat interface is a complex phenomenon. The drop will normally rest on the surface for a time before coalescing with the bulk, and upon doing so a small droplet may be expelled, to go through the same process. The extent of extraction during such a process is roughly the same as during drop formation [Licht and Conway, *Ind. Eng. Chem.*, **42**, 1151 (1950)], but there are few data. See also Johnson and Hamielec [*Am. Inst. Chem. Eng. J.*, **6**, 145 (1960)].

TABLE 21-35 Eigenvalues and Coefficients for Circulating Drops*

$k_C d_p/D_C$	λ_1	λ_2	λ_3	B_1	B_2	B_3
3.20	0.262	4.24	....	1.49	0.107	
5.33	0.386					
8.00	0.534					
10.7	0.680	4.92	....	1.49	0.300	
16.0	0.860	5.26	....	1.48	0.382	
21.3	0.982	5.63	....	1.47	0.428	
26.7	1.082	5.90	15.7	1.49	0.495	0.205
53.3	1.324	7.04	17.5	1.43	0.603	0.298
107	1.484	7.88	19.5	1.39	0.603	0.384
213	1.560	8.50	20.8	1.31	0.588	0.396
320	1.600	8.62	21.3	1.31	0.583	0.391
∞	1.656	9.08	22.2	1.29	0.596	0.386

*Elzinga and Banchero, *Chem. Eng. Prog.*, **55**, *Symp. Ser.* 29, 149 (1959), with permission.

CONTINUOUS (DIFFERENTIAL) CONTACT EQUIPMENT

Equipment in this category is usually arranged for multistage countercurrent contact of the insoluble liquids, without repeated complete separation of the liquids from each other between stages or their equivalent. Instead, the liquids remain in continuous contact throughout their passage through the equipment.

General Characteristics Countercurrent flow is maintained by virtue of the difference in densities of the liquids and either the force of gravity (vertical towers) or centrifugal force (centrifugal extractors). Only one of the liquids may be pumped through the equipment at any desired velocity. The maximum velocity for the second is then fixed; if it is attempted to exceed this limit, the second liquid will be rejected and the extractor will be **flooded.**

It cannot be overemphasized that knowledge of the characteristics of such equipment is surprisingly underdeveloped. The number of quantities that influence the rate of extraction is very large, and many of them are not well understood. Most of the available data were taken from small laboratory devices, frequently only a few inches in diameter and a few feet high. For these reasons the generalizations given here should be used only for very rough estimates, with allowance for generous factors of safety.

Axial Dispersion The devices in this category are subject to axial (longitudinal) dispersion within both liquids or departure from strictly "plug," countercurrent flow. As a result, the towers must be taller than simple application of the plug-flow numbers of transfer units (see Sec. 15) would indicate. The problem has been extensively studied by Sleicher [*Am. Inst. Chem. Eng. J.*, **5**, 145 (1959)] and Vermeulen et al. [U.S. AEC UCRL-3911, 1958; suppl., 1958; 10928, 1963; *Ind. Eng. Chem. Fundam.*, **2**, 113, 304 (1963); *Chem. Eng. Prog.*, **62**(9), 95 (1966)]. The two studies lead to essentially the same results although they are expressed somewhat differently. For a review, see Li and Zeigler [*Ind. Eng. Chem.*, **59**(3), 30 (1967)]. It will not be possible to outline in detail here all the considerations taken into account; for these the original papers should be consulted. For present purposes the procedure to be used in design, as developed by Vermeulen et al., will be outlined. It is limited to cases in which flow rates, distribution coefficients, and mass-transfer coefficients are constant throughout the extractor.

1. Obtain $N_{tOR,plug}$ from Colburn's equation [*Trans. Am. Inst. Chem. Eng.*, **35**, 211 (1938)]:

$$N_{tOR,plug} = \frac{1}{1 - V_R/m'V_E}\ln\left[\left(\frac{c_{R1} - c_{E2}/m'}{c_{R2} - c_{E2}/m'}\right)\left(1 - \frac{V_R}{m'V_E}\right) + \frac{V_R}{m'V_E}\right] \tag{21-55}$$

2. Obtain H_{tOR} (from data correlations, etc.) and Z_{plug}:

$$Z_{plug} = N_{tOR,plug}H_{tOR} \tag{21-56}$$

3. Solve Eqs. (21-57) to (21-59) together with Fig. 21-96 simultaneously by trial and error to obtain N_{tOR}:

$$1/N_{tOR} = 1/N_{tOR,plug} - 1/N'_{tOR} \tag{21-57}$$

$$N'_{tOR} = (N_{Pe}B)_E + \frac{\ln(V_R/m'V_E)}{V_R/m'V_E - 1} \tag{21-58}$$

Equation (21-58) is applicable only for cases in which $N_{tOR}V_R/m'V_E$ and $(N_{Pe}B)_E \geq 1.0$.

$$(N_{Pe}B)_E = \left(\frac{V_R/m'V_E}{f_R N_{Pe,R}B} + \frac{1}{f_E N_{Pe,E}B}\right)^{-1} \tag{21-59}$$

4. The final height of the effective portion of the tower, Z, is then

$$Z = Z_{plug}(N_{tOR}/N_{tOR,plug}) \tag{21-60}$$

In these expressions, $B = Z/d$, $N_{Pe,E} = dV_E/E_E$, $N_{Pe,R} = dV_R/E_R$, where d = some characteristic length such as d_F for packed towers or T for spray towers. E_E and E_R are the longitudinal dispersion coefficients, which must ultimately be determined experimentally. They are usually reported as E_C, E_D, $N_{Pe,C}$, and $N_{Pe,D}$, since they are more characteristic of the continuous or dis-

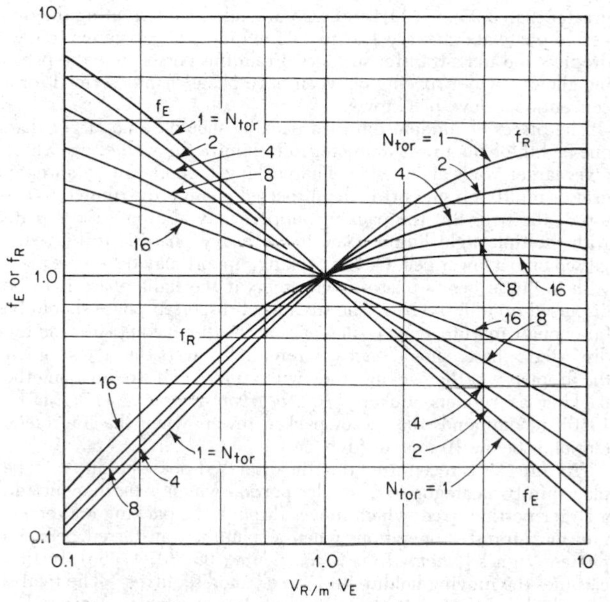

FIG. 21-96 Factors for Eq. (21-64). [*Vermeulen et al.*, Chem. Eng. Prog., *62(9), 95 (1966), with permission.*]

persed nature of the liquid than whether the liquid is extract or raffinate. For plug flow, $E = 0$: for complete mixing, $E = \infty$. In using these expressions, H_{tOR} should represent data that have been corrected for axial dispersion; unfortunately, very few data have been so corrected. Rod [*Br. Chem. Eng.*, **9**, 300 (1964)] presents a graphical calculation suitable even for curvilinear equilibrium curves.

Devices that are stagelike in character (sieve trays, compartmented extractors, etc.) are perhaps better treated by a somewhat different procedure which space does not permit outlining here. See Sleicher [*Am. Inst. Chem. Eng. J.*, **6**, 529 (1960)], Miyauchi and Vermeulen [*Ind. Eng. Chem. Fundam.*, **2**, 304 (1963)], and Van der Laan [*Chem. Eng. Sci.*, **7**, 187 (1958)].

Equipment Classification Equipment can be broadly classified into the following categories, generally in order of increasing complexity of internal construction. Those most generally used are:

I. Gravity-operated extractors
  A. No mechanical agitation
    1. Spray towers
    2. Packed towers
    3. Perforated-plate (sieve-plate) towers
  B. Mechanically agitated extractors
    1. Towers with rotating stirrers
    2. Pulsed towers
      a. Liquid contents pulsed
      b. Reciprocating plates
II. Centrifugal extractors

Spray Towers These are simple gravity extractors, consisting of empty towers with provisions for introducing and removing liquids at the ends (see Fig. 21-97). The interface can be run above the top distributor, below the bottom distributor, or in the middle, depending on where the best performance is achieved. Because of severe axial back mixing, it is difficult to achieve the equivalent of more than one or two theoretical stages or transfer units on one side of the interface.

Distributors The orifices or nozzles for introducing the dispersed phase are usually not smaller than 0.13 cm (0.05 in) in diameter in order to avoid clogging, nor larger than 0.64 cm (0.25 in) in order to avoid formation of excessively large drops. They should be designed to eliminate wetting by the dispersed liquid (see subsection "Drop Formation").

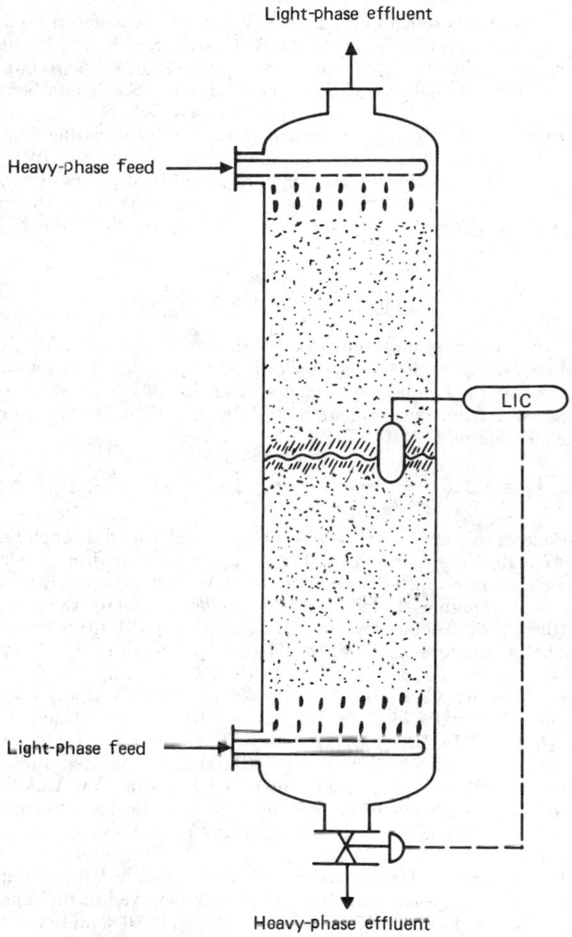

FIG. 21-97 Spray tower with both phases dispersed.

Dispersed-Phase Holdup Under conditions such that the rate of coalescence of the dispersed drops into the bulk dispersed liquid at the dispersed-liquid exit exceeds the rate of their arrival at this point, the dispersed-liquid holdup in the tower is relatively low. By restricting the area of the coalescence surface [Woodward, *Chem. Eng. Prog.*, **57**(1), 52 (1961); and Kehat and Letan, *Ind. Eng. Chem. Process Des. Dev.*, **7**, 385 (1968)] or by increasing the rate of flow of the dispersed liquid, the drops crowd to form a region of high holdup ($\varphi_D = 0.65$ or more) which extends into the column proper. Ultimately, if this region extends to the bottom of the tower, the drops are rejected at their inlet and the tower is flooded. Slip velocity V_S is the relative velocity of the two liquids, and for countercurrent flow,

$$V_S = \frac{V_D}{\varphi_D} + \frac{V_C}{1 - \varphi_D} \qquad (21\text{-}61)$$

For regions in the tower of constant low holdup and in the *absence of rapid interdroplet coalescence promoted by mass transfer from drops to the continuous liquid* (which can lower φ_D by as much as 50 percent), Elgin et al. [*Am. Inst. Chem. Eng. J.*, **3**, 63 (1957); **5**, 533 (1959); **11**, 158 (1965)] found that the ratio V_S/V_t for the average drop diameter can be estimated from Zenz's correlation for fluid-solid particulate systems [*Pet. Refiner*, **36**(8), 147 (1957)], which then permits an estimate of φ_D. Alternatively, Letan and Kehat [ibid., **13**, 443 (1967)] empirically correlate conditions of both low and high holdup by

$$V_S = V_{S,\varphi_D=0} \exp(-C\varphi_D) \qquad (21\text{-}62)$$

where $V_{S,\varphi D=0}$ is something less than V_t, and C depends on liquid properties. See also Harmathy [*Acta Tech. Acad. Sci. Hung.*, **12,** 209 (1955) (in English)] and Johnson and Lavergne [*Can. J. Chem. Eng.*, **39,** 37 (1961)]. Equation (21-27) then provides the specific interfacial area.

Flooding This can be estimated theoretically by setting $\partial V_C/\partial\varphi_D = \partial V_D/\partial\varphi_D = 0$ [Thornton, *Chem. Eng. Sci.*, **5,** 201 (1956)], using Eq. (21-61). On the basis of purely statistical comparison of observed and calculated data, the empirical correlation of Minard and Johnson [*Chem. Eng. Prog.*, **48,** 62 (1952)], slightly modified, is recommended:

$$V_{CF} = \frac{10{,}000\Delta\rho^{0.28}}{[0.453\mu_C^{0.075}\rho_C^{0.5} + d_p^{0.56}\rho_D^{0.5}(Q_D/Q_C)^{0.5}]^2} \qquad (21\text{-}63)$$

Use U.S. customary units only in this equation.

Mass Transfer In the absence of interdrop coalescence, for *circulating* drops, k_D may be estimated from Eq. (21-50) and k_C for $\mu_C/\mu_D < 1$ from the correlation of Ruby and Elgin [*Chem. Eng. Prog.*, **51,** *Symp. Ser.* 16, 17 (1955)]:

$$k_C = 0.725\left(\frac{d_pV_S\rho_C}{\mu_C}\right)^{-0.43}\left(\frac{\mu_C}{\rho_C D_C}\right)^{-0.58} V_S(1-\varphi_D) \qquad (21\text{-}64)$$

This agrees reasonably well with coefficients calculated through Eq. (21-47) with V_t replaced by V_S. For $\mu_C/\mu_D > 1$, k_C for drop swarms is smaller than Eq. (21-47) would indicate, by factors that may be as large as 3 (Hughmark, loc. cit.). For *oscillating* drops, there is a dearth of data. Tentatively, Eq. (21-52) with Eq. (21-45) is recommended. For other recent data, see Heertjes [*Chem. Eng. Sci.*, **3,** 122 (1954)], Smith and Beckmann [*Am. Inst. Chem. Eng. J.*, **5,** 533 (1959)], Astarita [*Chim. Ind. (Milan)*, **43,** 10 (1961)], Kuznetsov and Smirnov [*Zh. Prikl. Khim.*, **34,** 2111 (1961)], Rao and Rao [*J. Sci. Ind. Res. (India)*, **20D,** 101 (1961); *Indian J. Technol.*, **4,** 68 (1966); and *J. Appl. Chem.*, **16,** 218 (1966)]. For earlier data, see Treybal (*Liquid Extraction*, 1st ed., 1951; 2d ed., 1963, McGraw-Hill, New York).

Extraction with chemical reaction. A theoretical treatment is offered by Luss and Amundsen [*Ind. Eng. Chem. Fundam.*, **6,** 436 (1967)].

Heat Transfer Heat-transfer rates are generally large despite severe axial dispersion, with Ua_{av} frequently observed in the range 18.6 to 74.5 and even to 130 kW/(m^3·K) [1000 to 4000 and even to 7000 Btu/(h·ft^3·°F)] [see Bauerle and Ahlert, *Ind. Eng. Chem. Process Des. Dev.*, **4,** 225 (1965); and Greskovich et al., *Am. Inst. Chem. Eng. J.*, **13,** 1160 (1967); Sideman, in Drew et al. (eds.), *Advances in Chemical Engineering*, vol. 6, Academic, New York, 1966, p. 207, reviewed earlier work]. In the absence of specific heat-transfer correlations, it is suggested that rates be estimated from mass-transfer correlations via the heat-mass-transfer analogy.

Axial Dispersion For low values of φ_D and in the absence of interdrop coalescence, axial dispersion for the dispersed phase is evidently very small ($E_D \sim 0$). For the continuous phase, low μ_C and φ_D, Vermeulen et al. [*Chem. Eng. Prog.*, **62**(9), 95 (1966)] reviewed the available data and concluded that, for T = 3.6 to 15.2 cm (0.117 to 0.5 ft), E_C is given empirically by

$$E_C = c(V_DT)^{1/2} \qquad (21\text{-}65)$$

where c = 23.6 for U.S. customary units and 7.2 for SI units. For treatment of heat transfer, particularly for high values of φ_D when axial dispersion evidently is the controlling factor, see Letan and Kehat [*Am. Inst. Chem. Eng. J.*, **11,** 804 (1965); **14,** 398 (1968)] and Mixon et al. [ibid., **13,** 21 (1967)].

Packed Towers For a packed-tower liquid-liquid extractor the empty shell of a spray tower is filled with packing to reduce the vertical circulation of the continuous phase. The standard commercial packings used in vapor-liquid systems are also used in liquid-liquid systems. This includes Raschig and pall rings, Berl and Intalox saddles, and other random-dumped packings as well as the newer structured packings. The packing reduces the available free space for flow but also significantly reduces the height required for mass transfer. However, Nemunaitis, Eckert, Foote, and Rollinson [*Chem. Eng. Prog.*, **67**(11), 60 (1971)] reported little benefit from a packed height greater than 3.05 m (10 ft) and recommended redistributing the dispersed phase about every 1.52 to 3.05 m (5 to 10 ft) to generate new droplets and mass-transfer surfaces. From this perspective the packing allows a wider spacing between sieve plates than described for a conventional sieve-plate tower.

The pieces of random-dumped packing should be no larger than one-eighth of the tower diameter to minimize the wall effect which gives larger voids at the wall. The packing support can be an open grid or multiarch support if the dispersed phase is distributed to the top of the bed. But the packing support may also be a sieve plate with multiple light-liquid risers if the heavy phase is to be redispersed onto a lower bed. Or the packing support may be a sieve plate with multiple heavy-phase downcomers if the light phase is to be dispersed up into the bed. The streams of dispersed phase should be far enough apart to avoid coalescence at the dispersion plate, and the dispersed phase should not preferentially wet the packing. If the droplets wet the packing, they will coalesce and stream along the packing as rivulets. Eckert [*Hydrocarbon Process.*, 117 (March 1976)] recommends the use of packed towers when the interfacial tension is below 10 mN/m (dyn/cm).

Holdup It is recognized that the dispersed-phase holdup may be placed in two categories: a smaller portion which is permanent and a larger portion, free, which moves through the packing and enters into mass-transfer operations when a solute is transferred between phases. Vignes [*Chem. Ind. Genie Chim.*, **95,** 307 (1966)] further classifies the moving holdup into "free" and "semifree." The total is φ_D, which here refers to the volume of dispersed phase expressed as a fraction of the void space in the packed section. See Beckmann et al. [*Am. Inst. Chem. Eng. J.*, **1,** 426 (1955); **3,** 223 (1957)].

What follows is a very brief summary of the extensive work of Pratt and his coworkers, Dell, Gayler, Lewis, Jones, Roberts, and White [*Trans. Inst. Chem. Eng. (London)*, **29,** 89, 110, 126 (1951); **31,** 57, 69 (1953); *Chem. Ind. (London)*, 1952, p. 358]. For the standard commercial packings of 1.27-cm (½-in) size and larger, at low values of V_D, φ_D varies linearly with V_D up to values of $\varphi_D \doteq 0.10$. With further increase of V_D, φ_D increases sharply up to a "lower transition point," resembling "loading" in gas-liquid contact. At still higher values of V_D an upper transition point occurs, the drops of dispersed phase tend to coalesce, and V_D can increase without a corresponding increase in φ_D. This regime ends in flooding. Drops of the dispersed phase reach a characteristic size after leaving the distributor nozzles regardless of their initial size. For each system there is a critical packing size above which the mean drop size is a minimum. For smaller packing, the drop size is larger (and the interfacial area smaller). The critical size of packing, usually 1.27 cm (½ in) or more, is given by

$$d_{FC} = 2.42(\sigma g_c/\Delta\rho g)^{0.5} \qquad (21\text{-}66)$$

For packing larger than d_{FC}, the characteristic drop diameter, for liquids that are in concentration equilibrium, is given by

$$d_p = 0.92(\sigma g_c/\Delta\rho g)^{0.5}(V_K\epsilon\varphi_D/V_D) \qquad (21\text{-}67)$$

For liquids that are not in concentration equilibrium and when an unequilibrated solute is present, the characteristic drop size will generally be larger. If the drops formed at the distributor nozzle are smaller than this, there may be a tendency to flood until they grow to size. Thornton [*Ind. Chem.*, **39,** 632 (1963)] finds that large drops decay in exponential fashion to their final size. It is therefore best to design the nozzles to give drop sizes which are larger than that given by Eq. (21-67). V_K is a characteristic drop velocity (at $V_C = 0$, V_D approaching 0), and is given by Fig. 21-98. Below the upper transition point, the holdup is given by

$$\frac{V_D}{\varphi_D} + \frac{V_C}{1-\varphi_D} = \epsilon V_K(1-\varphi_D) \qquad (21\text{-}68)$$

Additional holdup correlations are offered by Sitarmayya and Laddha [*Chem. Eng. Sci.*, **13,** 263 (1961)] and Ghosal et al. [*Trans. Indian Inst. Chem. Eng.*, **11,** 23 (1958–1959)]. The interfacial area is given by

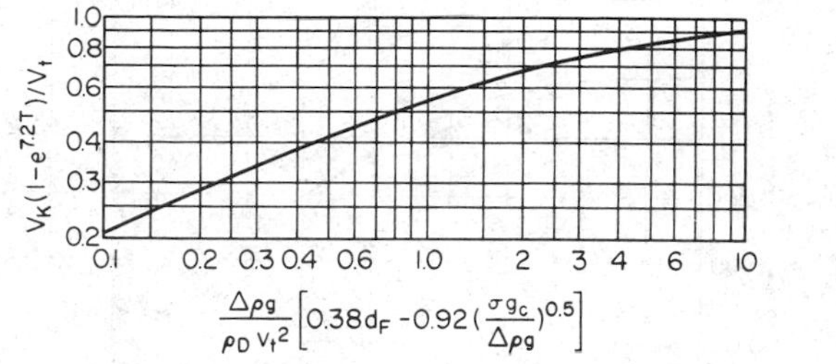

FIG. 21-98 Characteristic drop velocity for packed towers, for equilibrium liquids $d_F > d_{FC}$ and $T > 0.25$ ft. [*Pratt,* Ind. Chem., *31, 552 (1955), with permission.*]

$$a = 6\epsilon\varphi_D/d_p \tag{21-69}$$

It is generally desirable to design for φ_D in the range 0.15 to 0.25 (the lower value for $V_D/V_C < 0.5$).

Flooding Many correlations are available. By a comparison of the observed and calculated velocities at flooding for all available data, those of Crawford and Wilke [*Chem. Eng. Prog.*, **47**, 423 (1951)] and Hoffing and Lockhart [*Chem. Eng. Prog.*, **50**, 94 (1954)] are best and about equally effective. The Crawford-Wilke correlation is the simpler and is given in Fig. 21-99. Nemunaitis, Eckert, Foote, and Rollinson [*Chem. Eng. Prog.*, **67**(11), 60 (1971)] updated the correlation using packing factors. See also Dell and Pratt [*Trans. Inst. Chem. Eng. (London),* **29,** 89, 270 (1951)], Fujita et al. [*Chem. Eng. (Japan),* **17,** 230 (1957)], Sakiadis and Johnson [*Ind. Eng. Chem.,* **46,** 1229 (1954)], and Kafarov and Dytnerskii [*Zh. Prikl. Khim.,* **30,** 1698 (1957)]. For very small packings, see Rao and Rao [*Chem. Eng. Sci.,* **9,** 170 (1958)] and Venkatoramen and Laddha [*Am. Inst. Chem. Eng. J.,* **6,** 355 (1960)]. It is recommended that flow rates be set at no more than 50 percent of the flooding values, less if the interfacial tension of the liquids is high.

Mass Transfer Extraction rates for packed towers are usually excellently correlated for a given situation on the coordinate system of Fig. 21-100. Treybal [*Chem. Eng. Prog.*, **62**(9), 67 (1966)] has suggested means whereby overall H_{tO}'s may be resolved into constituent H_t's. In connection with the data on this figure, it should be noted that economical values of $m'V_E/V_R$ will usually lie in the range between 1 and 2, so that overall heights of transfer units are not too unreasonable even for this high-interfacial-tension system. For lower interfacial tensions, H_{tOC} will ordinarily be appreciably less.

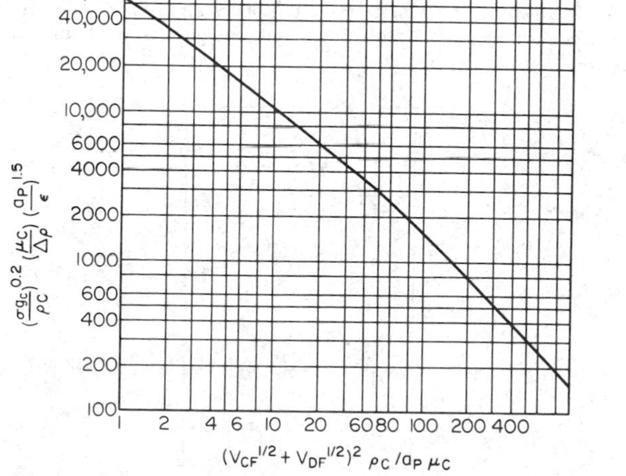

FIG. 21-99 Flooding in packed towers. Use only customary units in the variables. [*Crawford and Wilke,* Chem. Eng. Prog., *47, 423 (1951), with permission.*]

The number of variables that are known to influence the rate of extraction is exceedingly large, and includes at least the following:

Size, shape, and material of packing
Tower diameter
Packing depth
Dispersed-phase distributor design
Which liquid is dispersed
Direction of extraction, whether from dispersed to continuous, organic liquid to water, or the reverse
Dispersed-phase holdup
Flow rates and flow ratio of the liquids
Physical properties of the liquids
Presence or absence of surface-active agents

Although many attempts have been made to establish a method for estimating the extraction rates [see, for example, Ellis, *Ind. Chem.*, **28,** 483 (1952); Jeffreys and Ellis, *Congr. Chem. Eng. Des.*, **1962,** 65; and Treybal, *Liquid Extraction,* 2d ed., McGraw-Hill, New York, 1963], it is still most important to pilot-plant any new process. About the most that can be said is that, for a given system, packing, and method of operation, H_{tD} should be practically constant for all flow rates up to transition and that H_{tC} should vary roughly as $C(V_C/V_D)^n$, where C and n are constants, and to both H_t's correction must be applied on scale-up for axial dispersion [Treybal, *Chem. Eng. Prog.*, **62**(9), 67 (1966)]. Table 21-36 lists additional selected data sources.

Axial Dispersion Vermeulen et al. [*Chem. Eng. Prog.*, **62**(9), 95 (1966)] summarized many of the data for packings. Their correlation for the continuous phase is shown in Fig. 21-101. For the dispersed phase, their correlation is given by

1. Nonwetted carbon rings and wetted Berl saddles:

$$\log \frac{E_D}{V_D d_F} = 0.046 \frac{V_C}{V_D} + 0.301 \tag{21-70}$$

2. Wetted ceramic rings:

$$\log \frac{E_D}{V_D d_F} = 0.161 \frac{V_C}{V_D} + 0.347 \tag{21-71}$$

The measurements were made with kerosine or diisobutyl ketone dispersed in water. Additional work is reported by Komasawa et al. [*Kagaku Kogaku,* **30,** 237, 450, 928, 1103 (1966); English version, **4,**

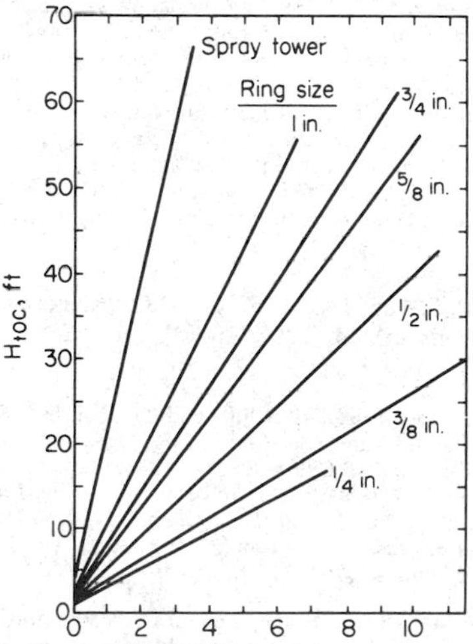

FIG. 21-100 Extraction of diethylamine from water into toluene (dispersed) in towers packed with unglazed porcelain Raschig rings. To convert feet to meters, multiply by 0.3048; to convert inches to centimeters, multiply by 2.54. [*Leibson and Beckmann,* Chem. Eng. Prog., *49, 405 (1953), with permission.*]

TABLE 21-36 Selected Sources of Packed-Tower Mass-Transfer Data

System	Tower diameter, in.	Packing	Ref.
Water–acetic acid–ethyl acetate, cyclohexane, methylcyclohexane, ethyl acetate + benzene	1	0.25-in. saddles	*b*
Water–acetic acid–methyl isobutyl ketone	1.95	0.23-in. rings	*g*
	3	0.375-in. plastic spheres	*j*
		0.375-in. plastic, ceramic rings	*k*
		0.5-in. plastic, ceramic saddles	*k*
Water-acetone-hydrocarbon . . .	1.88	0.25-, 0.375-in. rings, 6-mm. beads	*o*
	2–4	0.5-, 0.75-in. rings	*a*
Water–adipic acid–ethyl ether . .	6	0.5-, 0.75-in. rings, 0.375-in. spheres	*e*
Water–benzoic acid–carbon tetrachloride	1.95	0.25-in. rings	*f*
Water-diethylamine-toluene . . .	3, 4, 6	0.25–1-in. rings	*i*
	3	0.375-in. rings	*m*
Water-ethyl acetate	4	0.5-in. rings	*c*
Water-methylisobutyl-carbinol . .	4	0.5-in. rings	*n*
Water–methyl ethyl ketone	4	0.5-in. rings	*n*
Water–propionic acid–methyl isobutyl ketone	1.88	0.25-, 0.375-in. rings, 6-mm. beads	*o*
Acetone (aq.)–soybean oil, linseed oil	2	0.25-in. saddles, 0.5-in. rings	*p*
Petroleum-furfural	2	0.25-in. rings	*d*
	1.2	0.16-in. rings	*l*
Toluene–heptane–diethylene glycol	1.4, 2.25	Glass and brass rings	*h*

a Degaleesan and Laddha, *Chem. Eng. Sci.*, **21**, 199 (1966); *Indian Chem. Eng.*, 8(1), 6 (1966).
b Eaglesfield, Kelly, and Short, *Ind. Chem.*, **29**, 147, 243 (1953).
c Gaylor and Pratt, *Trans. Inst. Chem. Eng. (London)*, **31**, 78 (1953).
d Garwin and Barber, *Pet. Refiner*, **32**(1), 144 (1953).
e Gier and Hougen, *Ind. Eng. Chem.*, **45**, 1362 (1953).
f Guyer, Guyer, and Mauli, *Helv. Chim. Acta*, **38**, 790 (1955).
g Guyer, Guyer, and Mauli, *Helv. Chim. Acta*, **38**, 955 (1955).
h Kishinevskii and Mochalova, *Zh. Prikl. Khim.*, **33**, 2344 (1960).
i Liebson and Beckmann, *Chem. Eng. Prog.*, **49**, 405 (1953).
j Moorhead and Himmelblau, *Ind. Eng. Chem. Fundam.*, **1**, 68 (1962).
k Osmon and Himmelblau, *J. Chem. Eng. Data*, **6**, 551 (1961).
l Sef and Moretu, *Nafta (Zagreb)*, **5**, 125 (1954).
m Shih and Kraybill, *Ind. Eng. Chem. Process. Des. Dev.*, **5**, 260 (1966).
n Smith and Beckmann, *Am. Inst. Chem. Eng. J.*, **4**, 180 (1958).
o Rao and Rao, *J. Chem. Eng. Data*, **6**, 200 (1961).
p Young and Sullans, *J. Am. Oil Chem. Soc.*, **32**, 397 (155).
NOTE: To convert inches to centimeters, multiply by 2.54.

288, 363 (1966); **5**, 125, 182 (1967)], and Olbrich et al. [*Trans. Inst. Chem. Eng. (London)*, **44**, T207 (1966)].

GENERAL REFERENCES: Bussolari, Schiff, and Treybal, *Ind. Eng. Chem.*, **45**, 2413 (1953). Fujita and Tanizawa, *Chem. Eng. (Japan)*, **17**, 111 (1953). Garner, Ellis, and Hill, *Am. Inst. Chem. Eng. J.*, **1**, 185 (1955); *Trans. Inst. Chem. Eng. (London)*, **34**, 223 (1956). Major and Hertzog, *Chem. Eng. Prog.*, **51**, 17 (1955). Mayfield and Church, *Ind. Eng. Chem.*, **44**, 2253 (1952). Planovskii and Bulatov, *Khim. Mashinostr.*, **1960**(2), 10; (3), 9. Pyle, Duffey, and Colburn, *Ind. Eng. Chem.*, **42**, 1042 (1950).

Perforated-Plate (Sieve-Plate) Towers A schematic diagram for the most common design of perforated-plate, or sieve-plate, tower, arranged for light liquid dispersed, is shown in Fig. 21-102. The light liquid flows through the perforations of each plate and is thereby dispersed into drops which rise through the continuous phase. The continuous liquid flows horizontally across each plate and passes to the plate beneath through the down spout. For heavy liquid dispersed, the same design may be used, but turned upside down. The plates serve to eliminate essentially completely the vertical recirculation of continuous phase characteristic of the spray tower. Furthermore, extraction rates are enhanced by the repeated coalescence and redispersion into droplets of the dispersed phase. Towers of the simple design suggested by Fig. 21-102 have been used successfully in a great variety of services and for petroleum-refining processes have commonly been built to diameters of 3.66 m (12 ft). With careful design, these towers may have excellent flow capacities, and with systems of low interfacial tension equally excellent mass-transfer characteristics.

Many variations in design have been suggested and tried, for example, the use of tower packing in the down spouts to prevent entrainment of dispersed phase, arrangements in which both liquids must pass through perforations at each plate, arrangements with vertical perforated plates, etc. As examples of these, see Bradley (U.S. Patent 2,642,341, 1953), Williams (U.S. Patent 2,652,316, 1953), Maycock and Hartwig (U.S. Patent 2,729,550, 1956), and Pohlenz (U.S. Patent 2,872,295, 1959). Data are available only for arrangements of the sort shown in Fig. 21-102. In general, caplike sieve plates, bubble caps, and vertical perforated plates have not been as satisfactory as horizontal plates.

Sieve-Plate Design For best tray efficiency, it is well established that the dispersed phase must issue cleanly from the perforations. This requires that the material of the plates be preferentially wet by the continuous phase (requiring the use of plastics or plastic-coated

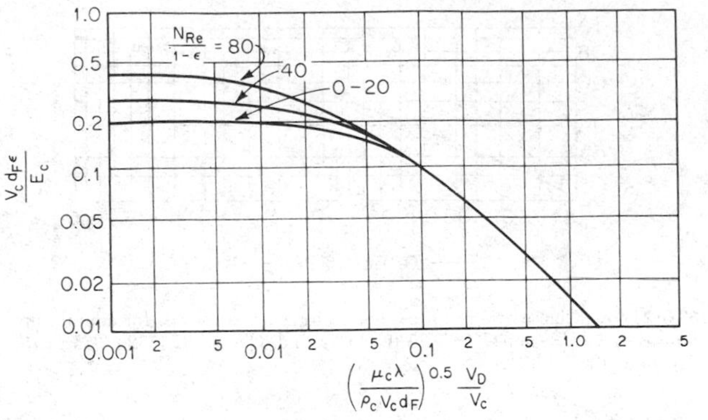

FIG. 21-101 Axial dispersion for the continuous phase in packed towers. Spheres (0.75-in, ϵ = 0.32 to 0.41; 0.50-in, ϵ = 0.62), Raschig rings (0.50-in, ϵ = 0.62; 0.75-in, ϵ = 0.65), Berl saddles (1.0-in, ϵ = 0.67). Use customary units in the variables. [*Vermeulen et al.*, Chem. Eng. Prog., ***62****(9), 95 (1966), with permission.*]

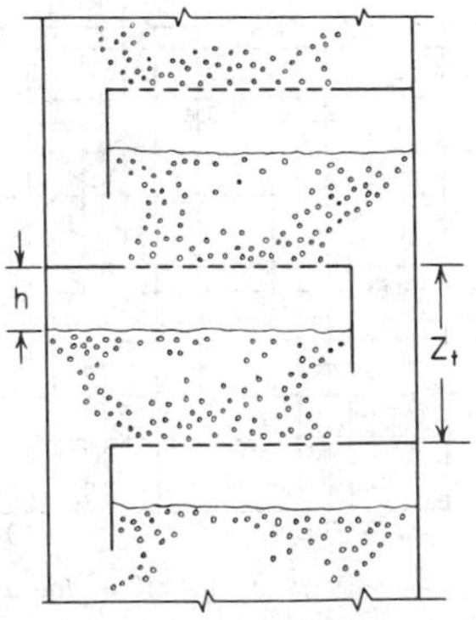

FIG. 21-102 Portion of a perforated-tray tower, arranged for light liquid dispersed.

plates in some instances) or that the dispersed phase issue from nozzles projecting beyond the plate surface. These may be formed by punching the holes and leaving the burr in place or otherwise forming the jets (see Mayfield and Church, loc. cit.). The liquid flowing at the larger volume rate should be dispersed.

Perforations are usually 0.32 to 0.64 cm (⅛ to ¼ in) in diameter, set 1.27 to 1.81 cm (½ to ¾ in) apart, on square or triangular pitch. There appears to be relatively little effect of hole size on extraction rate, except that with systems of high interfacial tension smaller holes will produce somewhat better rates. The entire hole area is suitably set at 15 to 25 percent of the column cross section, subject, however, to check through calculations as outlined below. The velocity through the holes should be such that drops do not form slowly at the holes, but rather that the dispersed phase streams through the openings to be broken up into droplets at a slight distance from the plate. This generally requires average linear velocities through the holes of from 15.2 to 30.5 cm/s (0.5 to 1.0 ft/s). The plate area directly opposite down spouts is kept free of perforations. A scum or "interface-rag" bypass can be incorporated in the trays (see Mayfield and Church, op. cit.) at the expense of tray efficiency, or provision may be made for periodic withdrawal of accumulations through the side of the tower between plates.

Down spouts (or up spouts) are best set flush with the plate from which they lead, with no weir as in gas-liquid contact. The velocity of the continous phase in the down spout V_d, which sets the downspout cross section, should be set at a value lower than the terminal velocity of some arbitrarily small droplet of dispersed phase, say, 0.08 or 0.16 cm (1/32 or 1/16 in) in diameter; otherwise, recirculation of

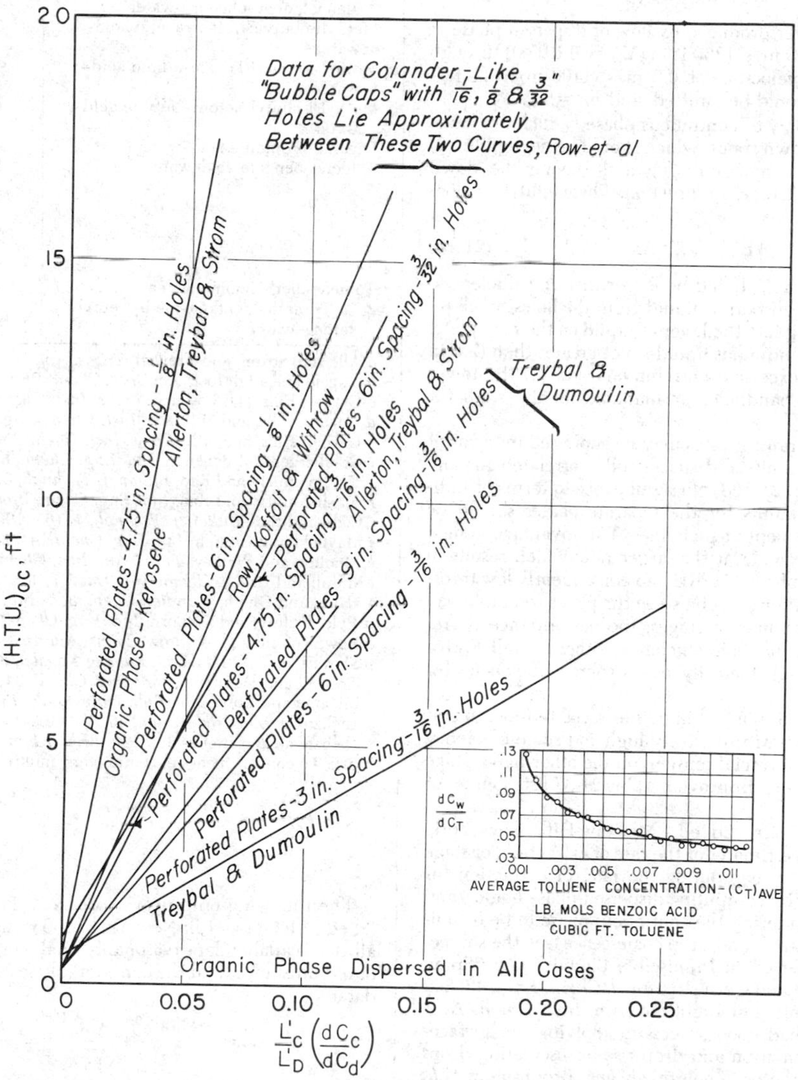

FIG. 21-103 Extraction rates for sieve-plate and modified bubble-plate columns. System: benzoic acid–water–toluene, except where noted. To convert feet to meters, multiply by 0.3048; to convert inches to centimers, multiply by 2.54. [*Allerton, Strom, and Treybal*, Trans. Am. Inst. Chem. Eng., *39, 361 (1943); Row, Koffolt, and Withrow, ibid., 37, 559 (1941); Treybal and Dumoulin*, Ind. Eng. Chem., *34, 709 (1942).*]

entrained dispersed phase around a plate will result in flooding. The down spouts should extend beyond the accumulated layer of dispersed phase on the plate.

The depth of dispersed liquid h accumulating on each plate is determined by the pressure drop required for flow of the liquids,

$$h = h_C + h_D \tag{21-72}$$

For the dispersed phase,

$$h_D = h_\sigma + h_O \tag{21-73}$$

The available data indicate that, for the orifice effect,

$$h_O = \frac{(V_O^2 - V_D^2)\rho_D}{2g(0.67)^2\,\Delta\rho} \doteq \frac{V_O'^2\rho_D}{28.9\,\Delta\rho} \tag{21-74}$$

and that h_σ to overcome interfacial-tension effects may be estimated for drop formation at a low velocity through the holes,

$$h_\sigma = 6\sigma g_c/d_{p0.1}\,\Delta\rho g \tag{21-75}$$

where $d_{p0.1}$ = drop diameter produced by flow of dispersed phase at V_O = 109 m/h (V_O' = 0.03 m/s) [360 ft/h (V_O' = 0.1 ft/s)] through the perforations. At hole velocities of 0.3 m/s (1100 m/h) [1 ft/s (3600 ft/h)] or more, h_σ should be omitted, and $h_D = h_O$.

The head required for flow of continuous phase h_C includes losses due to (1) friction in the down spout, which should be negligible, (2) contraction and expansion upon entering and leaving the down spout, and (3) two abrupt changes in direction. These total 4.5 velocity heads:

$$h_C = 4.5V_d^2\rho_C/2g\,\Delta\rho \tag{21-76}$$

The distance between trays Z_t should be larger than h, sufficient so that (1) the "streamers" of dispersed liquid from the holes break up into drops before coalescing into the layer of liquid on the next plate, (2) the linear velocity of continuous liquid is not greater than that in the down spout to avoid excessive entrainment, and (3) the tower may be entered through handholes or manholes in the sides for cleaning.

Mass Transfer Mass-transfer rates may be expressed in terms of overall heights of transfer units and successfully correlated for any tower and system as in Fig. 21-103. No significance in terms of individual heights of transfer units for the separate phases should be given to the slope and intercept of such lines. The advantage gained by dispersing the liquid flowing at the larger rate, which results in low values for the abscissa of Fig. 21-103 and consequently low transfer-unit heights, is clear. Alternatively, since the plates resemble and basically behave in the manner of stages, the performance is frequently expressed in terms of stage efficiency, either overall E_O for the entire tower or, more satisfactorily, as Murphree efficiencies for each tray.

The system of Fig. 21-103 is one of high interfacial tension, so that the heights of transfer units are relatively high and stage efficiency low. For systems of low interfacial tension, on the other hand, stage efficiencies may be very much improved. Table 21-37 lists sources of mass-transfer data.

Treybal (*Liquid Extraction*, 2d ed., McGraw-Hill, New York, 1963) has shown that good estimates of the rate of extraction, or stage efficiency, may be made by computing the rates of extraction for drop formation, drop rise (by computing dispersed-phase holdup and drop velocity and by considering the continuous phase to be of uniform concentration vertically), and drop coalescence (see the subsections "Single Drops Immersed in Immiscible Liquids" and "Spray Towers." See also Skelland and Cornish [*Can. J. Chem. Eng.*, **43**, 302 (1965)]. Specifically, Angelo and Lightfoot [*Am. Inst. Chem. Eng. J.*, **14**, 531 (1968)] have had good success in applying the surface-stretch theory to drop formation and drop rise for oscillating drops on a perforated-tray extractor. Zheleznyak and Brounshtein [*Zh. Prikl. Khim.*, **40**, 584, 689 (1967)] have shown that if the mass-transfer resistance lies within the drop phase, the approach to equilibrium of that phase produced by an extractor is simply related to the approach reached in one section.

TABLE 21-37 Mass-Transfer Data for Perforated-Tray Towers

System	Tower diameter, in.	Tray spacing, in.	Ref.
Benzene–acetic acid–water	1.97	3.9–6.3	*t*
	1.97	3.2–6.3	*s*
	2.2	2.8–6.3	*r*
	1.6 × 3.2	5.9	*p*
Benzene–acetone–water	3	4, 8	*m*
Benzene–benzoic acid–water	3	4	*m*
Benzene–monochloroacetic acid–water	1.97	3.9–6.3	*t*
Benzene–propionic acid–water	1.97	3.2–6.3	*s*
Carbon tetrachloride–propionic acid–water	1.97	3.9–6.3	*t*
Ethyl acetate–acetic acid–water	2	8–24	*j*
Ethyl ether–acetic acid–water	8.63	4–7.2	*n*
Gasoline–methyl ethyl ketone–water	3.75	4.5, 6	*k*
Kerosene-acetone-water	3	4, 8	*m*
Kerosene–benzoic acid–water	3.63	4.75	*a*
Isopar-H–benzyl alcohol, methyl benzyl alcohol, acetophenone-water	2 × 12	24	*b*
Methylisobutylcarbinol–acetic acid–water	3	6	*l*
Methyl isobutyl ketone–adipic acid–water	4.18	6	*e*
Methyl isobutyl ketone–butyric acid–water	4.8	6–23	*g*
Pegasol–propionic acid–water	4.8	6–11	*g*
Toluene–benzoic acid–water	8.75	6	*o*
	3.63	4.75	*a*
	3.56	3–9	*q*
	3	6	*l*
	2.72	9	*f*
	2	24	*j*
Toluene–diethylamine–water	4.18	6	*c, d*
2,2,4-Trimethylpentane–methyl ethyl ketone–water	3.75	4.5, 6	*k*

a Allerton, Strom, and Treybal, *Trans. Am. Inst. Chem. Eng.*, **39**, 361 (1943).
b Angelo and Lightfoot, *Am. Inst. Chem. Eng. J.*, **14**, 53 (1968).
c Garner, Ellis, and Fosbury, *Trans. Inst. Chem. Eng. (London)*, **31**, 348 (1953).
d Garner, Ellis, and Hill, *Am. Inst. Chem. Eng. J.*, **1**, 185 (1955).
e Garner, Ellis, and Hill, *Trans. Inst. Chem. Eng. (London)*, **34**, 223 (1956).
f Goldberger and Benenati, *Ind. Eng. Chem.*, **51**, 641 (1959).
g Krishnamurty and Rao, *Indian J. Technol.*, **5**, 205 (1967).
h Krishnamurty and Rao, *Ind. Eng. Chem. Process Des. Dev.*, **7**, 166 (1968).
i Lodh and Rao, *Indian J. Technol.*, **4**, 163 (1966).
j Mayfield and Church, *Ind. Eng. Chem.*, **44**, 2253 (1952).
k Moulton and Walkey, *Trans. Am. Inst. Chem. Eng.*, **40**, 695 (1944).
l Murali and Rao, *J. Chem. Eng. Data*, **7**, 468 (1962).
m Nandi and Ghosh, *J. Indian Chem. Soc., Ind. News Ed.*, **13**, 93, 103, 108 (1950).
n Pyle, Duffey, and Colburn, *Ind. Eng. Chem.*, **42**, 1042 (1950).
o Row, Koffolt, and Withrow, *Trans. Am. Inst. Chem. Eng.*, **37**, 559 (1941).
p Shirotsuka and Murakami, *Kagaku Kogaku*, **30**, 727 (1966).
q Treybal and Dumoulin, *Ind. Eng. Chem.*, **34**, 709 (1942).
r Ueyama and Koboyashi, *Bull. Univ. Osaka Prefect.*, **A7**, 113 (1959).
s Zheliznyak, *Zh. Prikl. Khim.*, **40**, 689 (1967).
t Zheliznyak and Brounshtein, *Zh. Prikl. Khim.*, **40**, 584 (1967).
NOTE: To convert inches to centimeters, multiply by 2.54.

The following empirical expression (Treybal, *Liquid Extraction*, 2d ed., McGraw-Hill, New York, 1963) has been found to represent all the available data reasonably well, considering the great variety of circumstances and the considerable scatter in many of the original data:

$$E_O = \frac{89{,}500Z_t^{0.5}}{\sigma g_c}\left(\frac{V_D}{V_C}\right)^{0.42} = \frac{0.9Z_t'^{0.5}}{\sigma'}\left(\frac{V_D}{V_C}\right)^{0.42} \tag{21-77}$$

Use only U.S. customary units in this equation. Krishnamurty and Rao [*Ind. Eng. Chem. Process Des. Dev.*, **7**, 166 (1968)] suggest that Eq. (21-77) is improved if the right-hand side is multiplied by $0.1123/d_O^{0.35}$.

Mechanically Agitated Gravity Devices Owing to the usual small density differences between the contacted liquids, the energy available from simple counterflow under the force of gravity is insufficient to disperse one liquid in the other and to establish turbulence levels to the extent necessary for rapid mass transfer, particularly for systems of high interfacial tension. Application of energy, mechanically applied through stirring devices, pulsations, etc., assists. The devices of major importance are considered below in order of increasing complexity of design.

Rotary-Disk Contactors (RDC)

GENERAL REFERENCES: Logsdail, Thornton, and Pratt, *Trans. Inst. Chem. Eng. (London),* **35,** 301 (1957). Misek, *Collect. Czech. Commun.,* **28,** 426, 570, 1631 (1963); **32,** 4018 (1967) (in English); *Ratacni Diskove Extraktory a Jejich Vypocty,* SNTL, Prague, 1964. Olney et al., *Am. Inst. Chem. Eng. J.,* **8,** 252 (1962); **10,** 827 (1964). Reman et al., U.S. Patent 2,601,674 (1952); *Chem. Eng. Prog.,* **51,** 141 (1955); **62**(9), 56 (1966); *Joint Symposium: Scaling-Up Chemical Plant and Processes,* London, 1957, p. 26.

Refer to Fig. 21-104. The tower is formed into compartments by horizontal doughnut-shaped or annular baffles, and within each compartment agitation is provided by a rotating, centrally located, horizontal disk. Somewhat similar devices have been known for some time. The features here are that the rotating disk is smooth and flat and of a diameter less than that of the opening in the stationary baffles, which facilitates fabrication and apparently improves extraction rates. Typical proportions are: tower diameter to rotating-disk diameter = 1.5:3; tower diameter to distance between disks = 2:8. The general proportions may be varied from one end of the tower to the other to accommodate changing liquid volumes and physical properties. These towers have been used in diameters ranging from a few inches for laboratory work up to 2.4 m (8 ft) in diameter by 12.2 m (40 ft) tall for purposes of deasphalting petroleum. Other commercial services include furfural extraction of lubricating oils, desulfurization of gasoline, phenol recovery from wastewaters, and many others.

A reliable design procedure for new systems, without the necessity for laboratory work, is not yet established. The data available show that the flow capacity increases with (1) decreased rotor speed, (2) decreased diameter of rotating disks, (3) increased diameter of opening in the stationary baffles, and (4) increased compartment height. Logsdail et al. (loc. cit.) have proposed that the slip velocity of Eq. (21-61), in the absence of mass transfer, can be set equal to $V_K(1 - \varphi_D)$, where V_K is a "characteristic" velocity which can be related to the liquid properties, speed of agitation, and tower geometry. Kung and Beckmann [*Am. Inst. Chem. Eng. J.,* **7,** 319 (1961)] and Olney et al. (loc. cit.) have also used this. Misek (loc. cit.), however, has had considerable success by setting the slip velocity equal to $V_K(1 - \varphi_D) \exp[\varphi_D(z - 4.1)]$, where z is a "coalescence coefficient" which depends on the liquid properties. Evidently mass transfer has a profound effect, as a result of drop coalescence; variation in the flooding rate from −15 to +200 percent has been noted in the extraction of acetone to and from water, respectively, with organic solvents. See also Kagan et al., *Izv. Vyssh. Uchebn. Zaved. Khim. Khim. Tekhnol.,* **9,** 836 (1966). Drop-size distribution which has an important influence on axial dispersion in the dispersed phase has been studied extensively by Misek and Olney (loc. cit.).

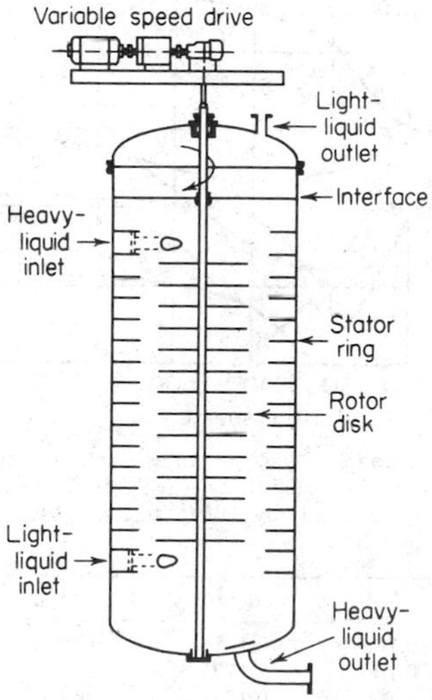

FIG. 21-104 Rotating-disk (RDC) extractor. *(Courtesy of General American Transportation Corp.)*

The value of HETS becomes smaller with (1) increased rotor speed but passes through a minimum, (2) increased diameter of rotating disks, (3) decreased diameter of stationary baffle opening, and (4) decreased compartment height. Reman and Olney [*Chem. Eng. Prog.,* **51,** 141 (1955)] show a correlation of stage height for two sizes of RDCs with the system water-kerosine-butylamine, as in Fig. 21-105. That such correlations cannot be general is indicated by these authors' data on caustic extraction of gasoline, which show quite different curves. Logsdail, Thornton, and Pratt (loc. cit.) tentatively suggest that data can be correlated through

$$\frac{H_{tOC}}{V_C}\left(\frac{g^2\rho_C}{\mu_C}\right)^{1/3}\varphi_D = C\left[\frac{\mu_C g}{V_K^3(1-\varphi_D)^3\rho_C}\right]^{2\beta/3}\left(\frac{\Delta\rho}{\rho_C}\right)^{2(\beta-1)/3} \tag{21-78}$$

the constants C and β to be determined for each system. For toluene-water-acetone, $\beta = 0.13$; for butyl acetate–water–acetone, $\beta = 0.4$; in both cases, transfer was from water to organic solvent. For transfer in the reverse direction, V_K could not be computed (see above).

A large number of studies of axial mixing have been made [Gel'perin et al., *Teor. Osn. Khim. Tekhnol.,* **1,** 666 (1967); Kagan et al., *Zh. Prikl. Khim.,* **39,** 88 (1966); Miyauchi et al., *Am. Inst. Chem. Eng. J.,* **12,** 508 (1966); Stainthorp and Sudall, *Trans. Inst. Chem. Eng. (London),* **42,** 198 (1964); Stemerding and Zuiderweg, *Chem. Ing. Tech.,* **35,** 844 (1963); and Strand et al., *Am. Inst. Chem. Eng. J.,* **8,** 252 (1962)]. Reman [*Chem. Eng. Prog.,* **62**(9), 56 (1966)] recommends, for the continuous phase in columns 0.08 to 2.13 m (3 in to 7 ft) in diameter,

$$E_C = 0.5Z_tV_C + 0.012d_iNZ_t(d_s/T)^2 \tag{21-79}$$

For the dispersed phase firm relationships have not been established, but at high rotor speeds, E_D may be 1 to 3 times E_C. In any event,

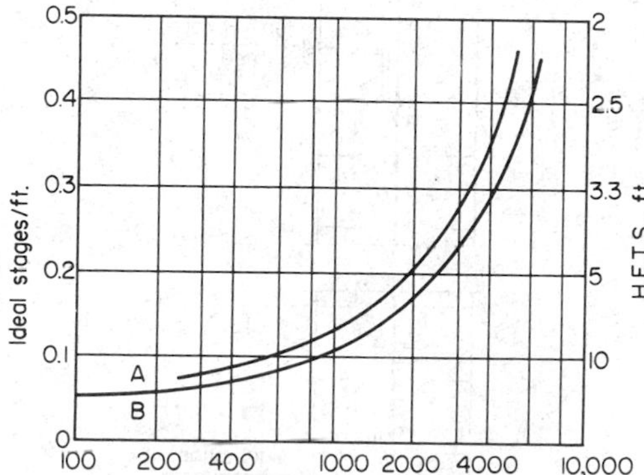

FIG. 21-105 Extraction in RDC columns, water-butylamine-kerosine (continuous). T = 0.33 and 1.33 ft. Curve *A*: V_D = 50.7, V_C = 78.9 ft/h. Curve *B*: V_D = 25.4, V_C = 78.9 ft/h. Use customary units in the variables. [*Data of Reman and Olney,* Chem. Eng. Prog., *51, 141 (1955).*]

axial mixing for the liquid flowing at the lower rate becomes very severe for extreme flow ratios (>10).

Costs are given by Clerk (*Chem. Eng.*, 232 (Oct. 12, 1964).

Several modifications of the design have appeared. Modifications of the rotors include perforation of the disk [Krishnara et al., *Br. Chem. Eng.*, **12**, 719 (1967)] and radially supported arc plates [Nakamura and Hiratsuka, *Kagaku Kogaku*, **30**, 1003 (1966)]. An "asymmetric" modification, with off-center rotors and arrangement of settling spaces for the liquids between dispersions (Misek, loc. cit.) is available in Europe.

Lightnin Mixer (Oldshue-Rushton) Tower

General References: Bibaud and Treybal, *Am. Inst. Chem. Eng. J.*, **12**, 472 (1966). Dykstra, Thompson, and Clouse, *Ind. Eng. Chem.*, **50**, 161 (1958). Gustison, Treybal, and Capps, *Chem. Eng. Prog.*, **58**, *Symp. Ser.* 39, 8 (1962). Gutoff, *Am. Inst. Chem. Eng. J.*, **11**, 712 (1965). Oldshue and Rushton, *Chem. Eng. Prog.*, **48**, 297 (1952). Miyauchi et al., *Am. Inst. Chem. Eng. J.*, **12**, 508 (1966).

Refer to Fig. 21-106. The extractor is an extension of the simple baffled mixing vessel into a multistage column. Although commercial application has been made, data are scarce and are limited to towers of small diameter. The preferred proportions are $Z_t = 0.5T$, $d_s > d_i$.

For water (continuous) and toluene or kerosine (dispersed), in a tower with $T = 0.152$ m (0.5 ft), $Z_t = 0.082$ m (0.27 ft), $d_i = 0.051$ m (0.1667 ft), dispersed-phase holdup is given by Eq. (21-66) with $V_S = V_K(1 - \varphi_D)$ and the following relationship by Wong (M.Ch.E. thesis, New York University, 1963):

$$V_K\mu_C/\sigma g_c = 1.77(10^{-4})(g/d_iN^2)(\Delta\rho/\rho_C)^{0.9} \quad (21\text{-}80)$$

For the same liquids axial mixing is described by (Bibaud and Treybal, loc. cit.)

$$E_C\varphi_C/V_CZ_t = -0.1400 + 0.0268(d_iN\varphi_C/V_C) \quad (21\text{-}81)$$

$$\frac{d_i^2N}{E_D} = 0.393(10^{-8})\left(\frac{d_i^3N^2\rho_C}{\sigma g_c}\right)^{1.54}\left(\frac{\rho_C}{\Delta\rho}\right)^{4.18}\left(\frac{d_i^2N\rho_C}{\mu_C}\right)^{0.61} \quad (21\text{-}82)$$

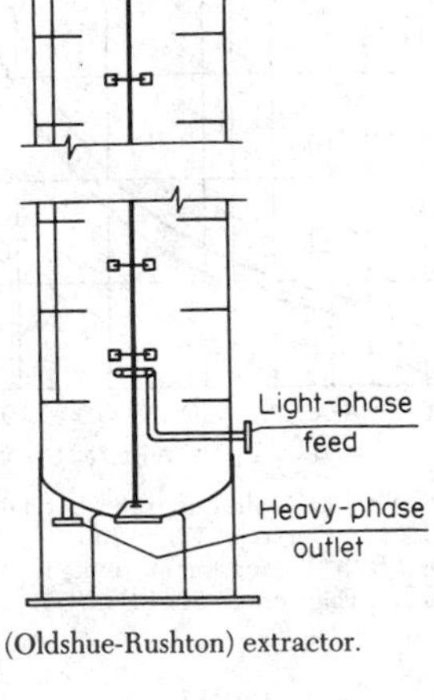

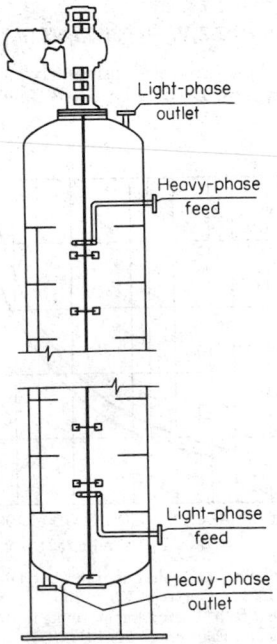

FIG. 21-106 Mixco (Oldshue-Rushton) extractor.

See also Miyauchi et al. (loc. cit.), who express the axial mixing in terms of interstage flow. For the continuous phase with no dispersed-phase flow, see Bibaud and Treybal, and Gutoff (loc. cit.).

Figure 21-107 presents some of the data of Oldshue and Rushton (loc. cit.) which show an optimum agitator speed for each configuration studied. The optimum would be expected to vary with physical properties of the liquids contacted. HETS is improved, although capacity is decreased, by smaller openings in the stationary baffles. In the more difficult (because of high interfacial tension) extraction of uranium between kerosine-diluted solvents and aqueous solutions, Dykstra et al. (loc. cit.) have also shown the development of an optimum impeller speed. Gustison et al. (loc. cit) have found it possible to correlate the stage efficiency with the ratio of flow rates (V_D/V_C) and the distribution coefficient, which varies considerably with concentration in the extraction of uranium. They also found it possible to scale up performance from 0.152- to 0.305-m (6- to 12-in) diameter geometrically, on the assumption that the continuous phase was thoroughly mixed in each compartment, by applying equal power per unit volume of liquids treated on the large and the small scale and using the same mass velocities of flow. Bibaud (loc. cit.) found that, for butylamine extracted from kerosine (dispersed) into water, the extraction rates corrected for axial mixing in either phase were described by assuming the drops to be rigid spheres, with Thornton's correlation [*Ind. Chem.*, **39**, 298 (1963] for drop size.

A somewhat related design has been studied by Nagata, Eguchi, and coworkers [*Chem. Eng. (Japan)*, **17**, 20 (1953); **20**, 2 (1956); *Mem. Fac. Eng., Kyoto Univ.*, **19**, 102 (1957); *Kagaku Kogaku*, **22**, 483 (1958)]. This column is characterized by the relatively small, separate openings between compartments for passage of liquids and the eccentric location of the impeller shaft. In a pilot-plant column, T = 0.3 m (0.983 ft), phenol was extracted from water [V_C = 11.6 m/h (38.1 ft/h)] into benzene [V_D = 6.4 m/h (21 ft/h)] at a stage efficiency of 0.618.

Scheibel Extraction Towers The original Scheibel tower design [*Chem. Eng. Prog.*, **44**, 681, 771 (1948); U.S. Patent 2,493,265, 1950] used knitted-mesh packed sections in a tower for coalescence with a centrally located impeller between the packed sections for drop breakup. Scheibel and Karr [*Ind. Eng. Chem.*, **42**, 1048 (1950)] presented data on a 0.305-m- (12-in-) diameter column of this design (Fig. 21-109) for systems which are difficult to extract because of high interfacial tension or easy because of low interfacial tension.

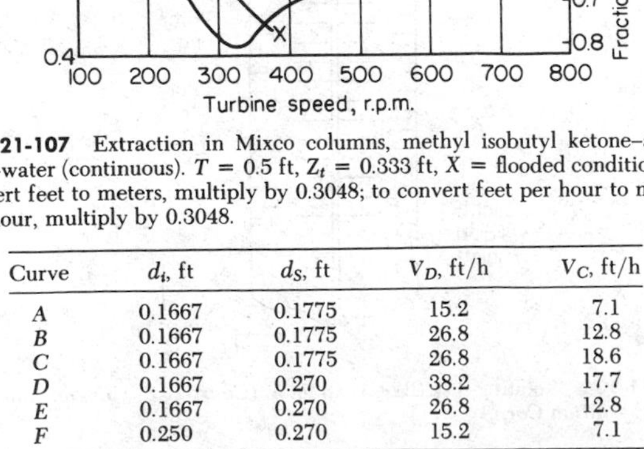

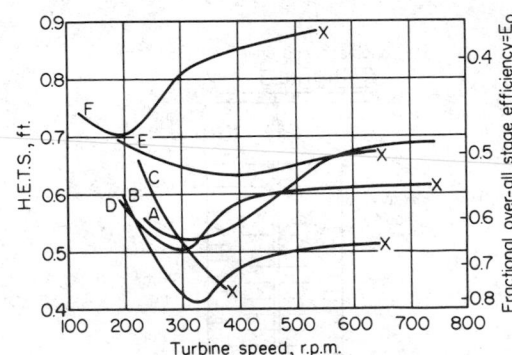

FIG. 21-107 Extraction in Mixco columns, methyl isobutyl ketone–acetic acid–water (continuous). T = 0.5 ft, Z_t = 0.333 ft, X = flooded condition. To convert feet to meters, multiply by 0.3048; to convert feet per hour to meters per hour, multiply by 0.3048.

Curve	d_i, ft	d_S, ft	V_D, ft/h	V_C, ft/h
A	0.1667	0.1775	15.2	7.1
B	0.1667	0.1775	26.8	12.8
C	0.1667	0.1775	26.8	18.6
D	0.1667	0.270	38.2	17.7
E	0.1667	0.270	26.8	12.8
F	0.250	0.270	15.2	7.1

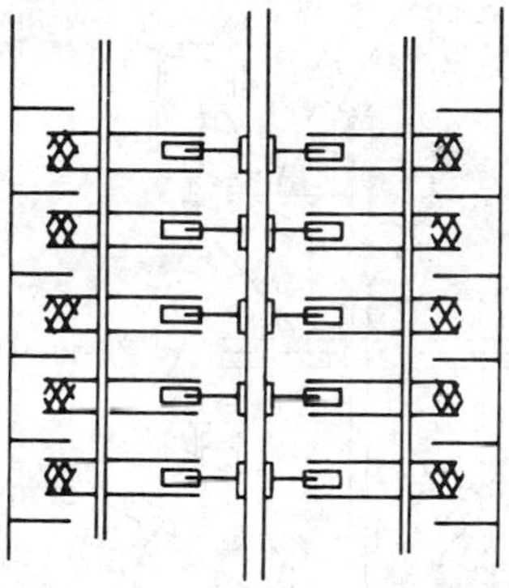

FIG. 21-108 Second Scheibel extractor with horizontal baffles and no wire-mesh packing between stages. [*Reprinted by permission of* Am. Inst. Chem. Eng. J., **2,** *74 (1956).*]

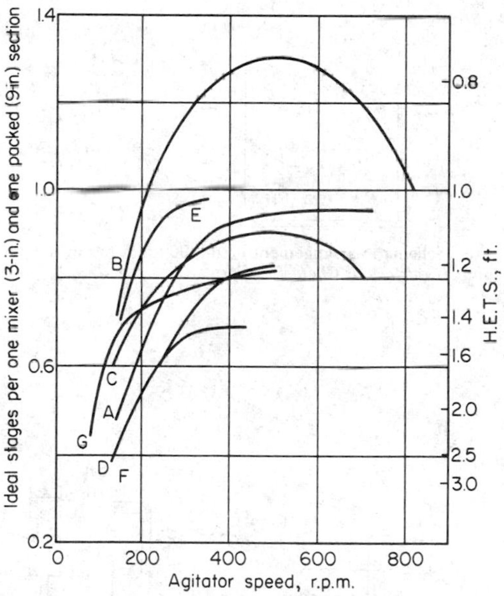

FIG. 21-109 Extraction in first Scheibel column. T = 0.94 ft, d_i = 0.333 ft, height of mixer section = 3 in, height of packed section = 9 in. To convert inches to centimeters, multiply by 2.54; to convert feet to meters, multiply by 0.3048; and to convert feet per hour to meters per hour, multiply by 0.3048. [*Data of Scheibel and Karr,* Ind. Eng. Chem., **42,** *1048 (1950).*]

Curve	System	V_D, ft/h	V_C, ft/h
A	MIBK(C)–water(D,E)–acetic acid	41.7	41.7
	MIBK(D)–water(C,E)–acetic acid		
B	MIBK(C,E)–water(D)–acetic acid	41.7	41.7
C	MIBK(C)–water(D,E)–acetic acid	23.2	23.2
	MIBK(C,E)–water(D)–acetic acid		
	MIBK(D)–water(C,E)–acetic acid		
D	o-Xylene(D)–water(C,E)–acetone	25.9	17.3
E	o-Xylene(D,E)–water(C)–acetone	22.1	21.2
F	o-Xylene(C)–water(D,E)–acetone	25.9	17.3
G	o-Xylene(C,E)–water(D)–acetone	21.1	22.1

MIBK = methyl isobutyl ketone; C = continuous; D = dispersed; E = extractant.

Excellent values of HETS were obtained with a wide variety of conditions. Low throughput and ratios of flow rates greatly different from unity required high agitator speeds for best results. Both direction of extraction and which phase was dispersed influenced the rates. The liquids of Fig. 21-109 were also used in tests involving the mixing sections alone (see operating characteristics of mechanically agitated vessels). Honekamp and Burkhart [*Ind. Eng. Chem. Process Des. Dev.,* **1,** 176 (1962)] found very little change in drop size to occur within the knitted-wire mesh and measured extraction rates in the mesh zone for one system.

A second Scheibel tower design [*Am. Inst. Chem. Eng. J.,* **2,** 74 (1956); U.S. Patent 2,850,362, 1958] reduced HETS and permitted more direct scale-up. The impellers are surrounded by stationary shroud baffles to direct the flow of droplets as they are discharged from the tips of the impellers. Data taken from a 0.305-m- (12-in-) diameter tower are shown in Fig. 21-111 and correlated in terms of the power applied per unit volume of liquids handled per compartment. For the impeller used, the power number at turbulent Reynolds numbers is N_{Po} = 1.85. The data show that while packing in alternate sections may increase mass-transfer rates, it decreases flow capacity. For many industrial systems, the knitted mesh was not used because of fouling (Fig. 21-108). Towers up to 2.13 m (7 ft) in diameter are in service. A third design by Scheibel (U.S. Patent 3,389,970, 1968) uses closed impellers plus horizontal baffles in the tower.

Kühni Tower The extraction towers designed at Kühni [see Mögli and Bühlmann, in Lo, Baird, and Hanson (eds.), *Handbook of Solvent Extraction,* Wiley-Interscience, New York, 1983, sec. 13.5] use shrouded (closed) impellers on a central shaft in the tower (Fig. 21-110). The droplet size can be controlled by the speed and diameter of the impeller, while the circulation rate can be controlled by the design of the width of the impeller. A perforated plate between each stage can control the droplet holdup by the percentage of open area in the plate.

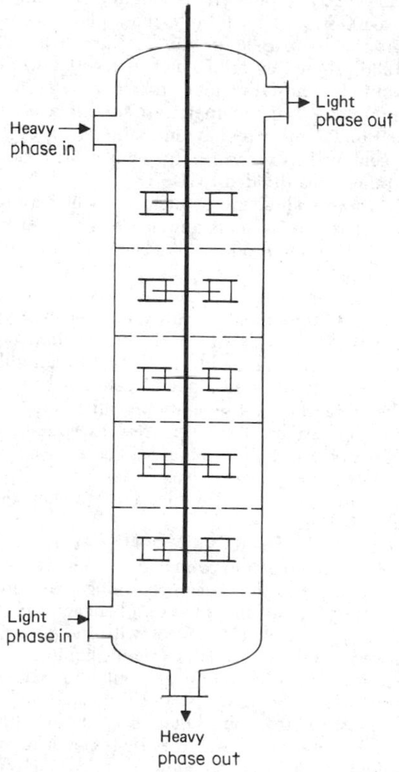

FIG. 21-110 Kühni tower.

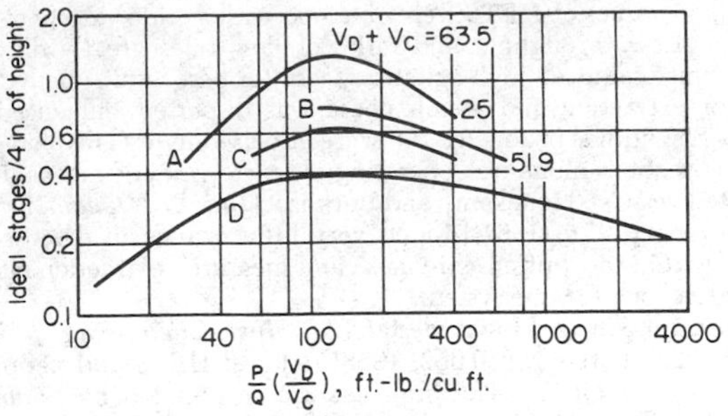

FIG. 21-111 Extraction in second Scheibel column. T = 0.94 ft, d_i = 0.333 ft, height of packed section = height of mixer section = 2 in. Use customary units in the variables. [*Data of Scheibel,* Am. Inst. Chem. Eng. J., *2, 74 (1956).*]

Curve	System
A, B*	Methyl isobutyl ketone–water–acetic acid
C*	o-Xylene–water–acetic acid
D†	o-Xylene–water–phenol
	Methyl isobutyl ketone–water–acetic acid
	o-Xylene–water–acetic acid

*Alternate mixing and packed sections.
†Packing omitted. Agitators in alternate and also every section.

Treybal Tower Treybal [U.S. Patent 3,325,255, 1967); *Chem. Eng. Prog.,* **60**(5), 77 (1964)] adapted a mixer-settler cascade in tower form in which the liquids are settled between stages.

Karr Reciprocating Plate Tower The reciprocating plate extractor developed by Karr [*Am. Inst. Chem. Eng. J.,* **5**, 446 (1959)] is a mechanically agitated tower using dual-flow plates with 50 to 60 percent open area, mounted on a central shaft and reciprocated vertically (Fig. 21-112). A typical stroke length is 2.54 cm (1 in) with a speed of 10 to 400 strokes per minute and a plate spacing of 5 to 15 cm (2 to 6 in). Scale-up relationships by Karr (*Sep. Sci. Technol.,* **15**(4), 877 (1980)] show that HETS increases with tower diameter to the 0.38 power in the most difficult case. Laboratory columns of 2.54- and 5.08-cm- (1- and 2-in-) diameter are used to scale up to towers as large as 0.9 to 1.2 m (3 to 4 ft) in diameter. A high volumetric efficiency is achieved as measured by total volumetric throughput per cross-sectional area divided by HETS.

Pulsed Columns These are extractors in which a rapid reciprocating motion of relatively short amplitude is applied to the liquid contents. The agitation so produced has been found to give improved rates of extraction. The principle originated with van Dijck (U.S. Patent 2,011,186, 1935). Because agitation was necessary to reduce tower heights and consequently the expense of massive shielding, and because pulsing provided a means of agitation not requiring moving parts, bearings, and the like in contact with highly corrosive, dangerously radioactive liquids, pulsed columns have been freely applied in the extraction and separation of metals from solutions of atomic energy operations. With very few exceptions, applications appear thus far to be limited to this area. There are two major types of columns: (1) ordinary (spray, packed, etc.) extractors on which pulsations are imposed and (2) a special sieve-plate design. Their characteristics are quite different.

Pulsing devices. Refer to Fig. 21-113. At *a*, a reciprocating plunger or piston pump from which the check valves have been removed is connected to the space containing continuous phase, as shown. This arrangement suffers the disadvantages that (1) the corrosive liquid may be in direct contact with the piston and (2) too rapid pulsing, especially with volatile organic liquids, may cause cavitation. The pipe connecting column and pulser may be of any length to pass through shielding, barriers, and the like, but high pressure drop in the transfer pipe contributes to cavitation difficulties. An alternative arrangement using an air pulse is shown at *b* in the figure [Thornton, *Chem. Eng. Prog.,* **50,** *Symp. Ser.* 13, 39 (1954); U.S. Patent 2,818,324, 1957]. This keeps corrosive liquids out of contact with the pulsing device and obviates the cavitation problem but because

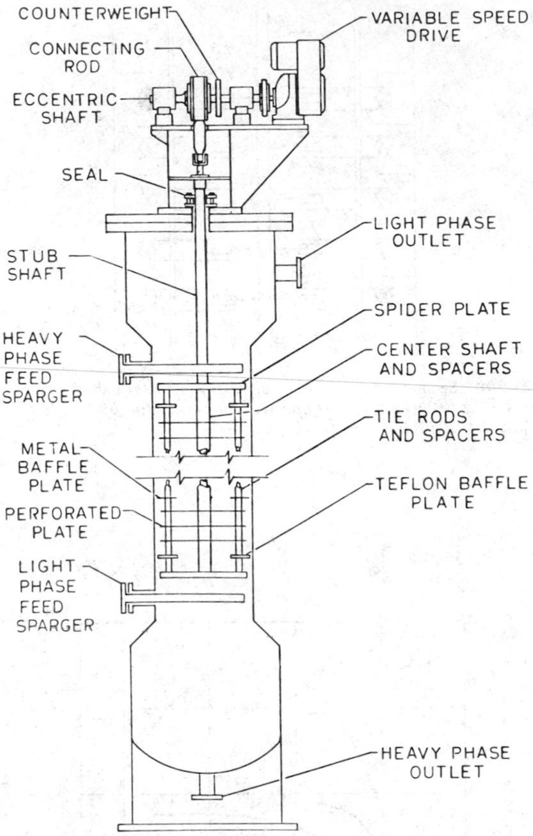

FIG. 21-112 Schematic arrangement of the 900-mm (36-in) reciprocating-plate column. (*Courtesy of Chem-Pro Corp.*)

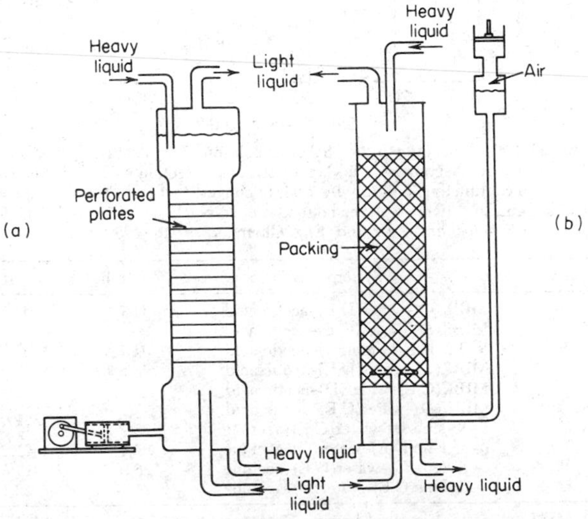

FIG. 21-113 Pulsed columns: (*a*) Perforated-plate column with pump pulse generator. (*b*) Packed column with air pulser.

of the compressibility of the gas requires greater application of pulsing power for the same results. For design, see Week and Knight [*Ind. Eng. Chem. Process Des. Dev.*, **6**, 480 (1967); **7**, 156 (1968)]. For pulsing at the natural frequency of the column, Baird [*Proc. Am. Inst. Chem. Eng.-Inst. Chem. Eng. Joint Meeting*, London, **1956**(6), 53] connected the liquid space to a volume of gas which acts as a spring. Flexible bellows or diaphragms of reinforced rubber, plastic, or metal in contact with the liquids may be flexed mechanically or by an electromagnetic transducer (Thornton, loc. cit.). If hydraulically activated, these may have a life of up to 30,000,000 cycles or more [Jealous and Johnson, *Ind. Eng. Chem.*, **47**, 1159 (1955)]. With suitable cam mechanisms, pulsations whose amplitude-time characteristics appear as sine, square, or sawtooth wave shapes are possible.

Pressure at the pulsing device and the conditions for cavitation and "water hammer" may be estimated by the methods of Williams and Little [*Trans. Inst. Chem. Eng. (London)*, **32**, 174 (1954)] provided the pressure-drop characteristics of the tower internals are known. Jealous and Johnson (loc. cit) have had good success in computing the power required for pulsing. Since power requirement alternates, the use of a flywheel on the pulse mechanism to act as an energy reservoir is suggested as a means of reducing power requirements. Alternatively, two columns could be pulsed 180° out of phase with one pulse generator (Griffith, Jasney, and Tupper, U.S. AEC AECD-3440, 1952). Irvine (U.S. AEC ORNL-2377, 1957) devised a pulse pump to utilize part of the pulse energy. Concatenated columns (long extractors built as several short columns, with liquids led from one to the other in strictly countercurrent fashion) may be pulsed by a single pulse generator to advantage, since less power is required owing to reduced static head [Jealous and Lieberman, *Chem. Eng. Prog.*, **52**, 366 (1956)].

The following terms are generally used to describe the pulse action: **Frequency** is the rate of application of the pulse action, cycles/time. **Amplitude** is the linear distance between extreme positions of the liquid in the column (not of the pulser) produced by pulsing. **Pulsed volume** = amplitude × frequency × column cross-sectional area = volumetric rate of movement of liquid, expressed as volume/time or volume/(time-area).

Pulsed spray columns. Billerbeck et al. [*Ind. Eng. Chem.*, **48**, 183 (1956)] applied pulsing to a laboratory [3.8-cm- (1.5-in-) diameter] column. At pulse amplitude 1.11 cm (7⁄16 in), rates of mass transfer improved slightly with increased frequency up to 400 cycles/min, but the effect was relatively small. Shirotsuka [*Kagaku Kogaku*, **22**, 687 (1958)] provides additional data. There is not believed to be commercial application.

Pulsed packed columns. Any of the ordinary packings may be used, although random packings tend to orient on pulsing, which may lead to channeling. For this reason, Thornton [*Chem. Eng. Prog.*, **50**, *Symp. Ser.* 13, 39 (1954); *Br. Chem. Eng.*, **3**, 247 (1958)] recommends fixed packing made from plates of corrugated expanded metal. Polyethylene packing, not wet by aqueous solutions, provides higher flow capacities and mass-transfer rates than ceramic (wetted) packing [Jackson, Holman, and Grove, *Am. Inst. Chem. Eng. J.*, **8**, 659 (1952)]. Pulsing reduces the size of dispersed-phase droplets, increases holdup, and increases interfacial area for mass transfer. There is a greater tendency toward emulsification, and maximum throughput is decreased, but HETS is reduced considerably, by the pulsing. Pulsing can be applied on existing nonpulsed packed towers to good mass-transfer advantage, provided limiting flow rates are not exceeded.

Figure 21-114 is perhaps typical of the results obtainable, although no generalizations have been devised for estimating the mass-transfer rates in the absence of experiment. For additional data, see Crico [*Genie Chim.*, **73**, 57 (1955)], Feich and Anderson [*Ind. Eng. Chem.*, **44**, 404 (1952)], Karpacheva et al. [*Khim. Masinostr.*, **1959**(3), 6; **1960**(2), 13; *Khim. Prom.*, **1960**, 469], Honda et al. [*Kagaku Kikai*, **21**, 645 (1957); *Kagaku Kogaku*, **22**, 97 (1958)], Oyama and Yamaguchi [*Kagaku Kogaku*, **22**, 668 (1958)], Potnis et al. [*Ind. Eng. Chem.*, **51**, 645 (1959)], Widmer [*Chem. Ing. Tech.*, **39**, 900 (1967)]; Worall and Thwaites [*Br. Chem. Eng.*, **10**, 158 (1965)], Ziolkowski and Naumowicz [*Chem. Stosow.*, **2**, 457 (1958); **3**, 475 (1959); **5**, 363 (1961)].

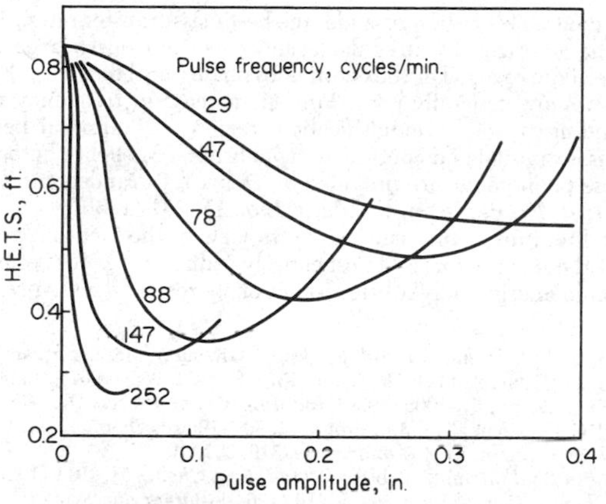

FIG. 21-114 Effect of pulsing on extraction in a packed column: methyl isobutyl ketone–acetic acid–water (continuous). Tower diameter = 1.58 in, 27-in depth of ¼-in Raschig rings. $V_D = V_C$ = 7.5 to 10. To convert inches to centimeters, multiply by 2.54. [*Data of Chantry, von Berg, and Wiegandt,* Ind. Eng. Chem., *47, 1153 (1955), with permission.*]

A small perforated-plate column of conventional design was pulsed by Goldberger and Benenati [*Ind. Eng. Chem.*, **51**, 641 (1959)] with marked improvement in mass-transfer rates.

Pulsed sieve-plate columns. The standard arrangement (see Fig. 21-113*a*) consists of a tower fitted with horizontal sieve plates which occupy the entire cross section of the columns. There are *no down spouts* as in ordinary sieve-plate columns. Typical arrangements use 0.32-cm- (⅛-in-) diameter perforations sufficient to provide 20 to 25 percent free space, with 5.08-cm (2-in) plate spacing, pulse amplitudes in the range 0.64 to 2.5 cm (0.25 to 1 in), and frequencies of 100 to 250 cycles/min, although the pulse characteristics will depend upon the system and flow rates under consideration. Plates are usually of metal, but Sobotik and Himmelblau [*Am. Inst. Chem. Eng. J.*, **6**, 619 (1960)] indicate that for certain services plates which are not wet by water (polyethylene) may be advantageous.

Sege and Woodfield [*Chem. Eng. Prog.*, **50**, *Symp. Ser.* 13, 179 (1954)] provide a good description of the operational characteristics. Refer to Fig. 21-115. Since in many cases the perforations are too small to permit flow owing to interfacial tension of the liquids, the total pulsed volume must ordinarily approximate the volumetric rate of flow of the liquids [Edwards and Beyer, *Am. Inst. Chem. Eng. J.*, **2**, 148 (1956), show that slightly higher rates than $V_D + V_C$ = pulsed volume may be obtained]. In region 1 of the figure, the column is flooded because of insufficient pulsed volume. In region 2, discrete layers of liquid appear between plates during the quiet portion of the pulse cycle. During upward pulsing, the light liquid is forced through the perforations and forms drops which rise to the plate above. During downward pulsing, the heavy liquid behaves similarly. Flow is stable, but mass-transfer rates are generally poor. In region 3 there is little change in phase dispersion throughout the pulse cycle, and a fairly uniform dispersion of small droplets persists

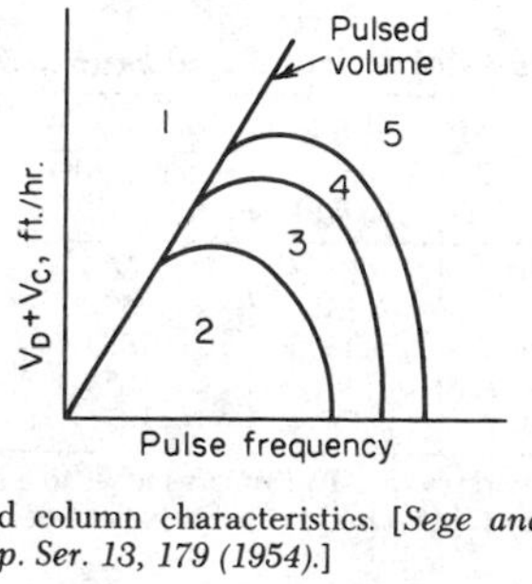

FIG. 21-115 Pulsed column characteristics. [*Sege and Woodfield,* Chem. Eng. Prog., *50, Symp. Ser. 13, 179 (1954).*]

throughout. This region provides the best mass-transfer rates. Region 4 is characterized by irregular coalescence into fairly large drops, and periodic reversal of the continuous phase (local flooding). Extraction rates are generally poor. Further increase in frequency results in flooding owing to emulsification, region 5. Transition between regions is gradual and continuous, not abrupt. Excellent photographs of these phenomena are provided by Defives, Durandet, and Gladel [*Rev. Inst. Fr. Pet. Ann. Combust. Liq.*, **11**, 231 (1956)].

The literature is unusually large. In view of the fact that application of these extractors is almost entirely confined to processes related to atomic energy, only a brief listing of sources of data is presented here.

Dispersed-phase holdup and flooding. Groenier, McAllister, and Ryon [U.S. AEC ORNL-3890, 1966; *Chem. Eng. Sci.*, **22**, 931 (1967)]; Babb et al. [*Ind. Eng. Chem.*, **51**, 1005 (1959); *Ind. Eng. Chem. Process Des. Dev.*, **2**, 38 (1963)]; Gel'perin et al. [*Khim. Prom.*, **42**, 607 (1966)]; Thornton and Logsdail [*Trans. Inst. Chem. Eng. (London)*, **35**, 316, 331 (1957)].

Longitudinal mixing. Babb et al. [*Ind. Eng. Chem.*, **51**, 1011 (1959); *Ind. Eng. Chem. Process Des. Dev.*, **3**, 210 (1964)]; Burger and Swift (U.S. AEC HW-29010, 1953); Miyauchi et al. [*Am. Inst. Chem. Eng. J.*, **11**, 395 (1965); *Kagaku Kogaku*, **30**, 895 (1966)]; Otake and Komasawa [ibid., **32**(6), 19 (1968)].

Mass-transfer rates. Correlations are offered by Smoot, Mar, and Babb [*Ind. Eng. Chem.*, **51**, 1005 (1959); *Ind. Eng. Chem. Fundam.*, **1**, 93 (1962)] and Zwolkowski and Kubica [*Chem. Stosow., Ser.* **B2**, 392 (1965)].

Controlled Cycling The compartmented character of sieve-plate columns described above lends itself particularly well to this technique, which is, however, not confined to these devices [Cannon, *Oil Gas J.*, **51**, 268 (1952); **55**, 68 (1956); Szabo et al., *Chem. Eng. Prog.*, **60**(1), 66 (1964); Belter and Speaker, *Ind. Eng. Chem. Process Des. Dev.*, **6**, 36 (1967); Horn, ibid., **6**, 30 (1967); Robinson and Engel, *Ind. Eng. Chem.*, **59**(3), 22 (1967); and Lövland, *Ind. Eng. Chem. Process Des. Dev.*, **7**, 65 (1968)]. A cycle is completed by the following sequence of events: (1) a light-phase flow period, during which the heavy phase does not flow; (2) a coalescing period, during which neither phase flows; (3) a heavy-phase flow period, during which the light phase does not flow; and (4) a repeat of the coalescing period. The net result can be an increased flow capacity (in the case of sieve-plate pulsed columns) and stage efficiency, such that the effect of $2N$ stages may be obtained with a column of N stages, provided the total holdup of each phase is displaced during each cycle.

Centrifugal Extractors The force of gravity for counterflow of liquids of different density may be replaced and in effect increased (many thousandfold if desired) by centrifugal machines. These then become especially useful for handling liquids of low density difference and those with tendencies to form emulsions.

Podbielniak Extractor (Podbielniak, U.S. Patent 2,044,996, 1935, and other patents) This is the most important of the group. Refer to Fig. 21-116. Rotation is about a horizontal shaft. The body of the extractor is a cylindrical drum containing concentric perforate cylinders. The liquids are introduced through the rotating shaft with the help of special mechanical seals; the light liquid is led internally to the drum periphery and the heavy liquid to the axis of the drum. Rapid rotation (up to several thousand revolutions per minute, depending on size) causes radial counterflow of the liquids, which are then led out through the shaft. Materials of construction include steel, stainless steel, Hastelloy, and other corrosion-resistant alloys.

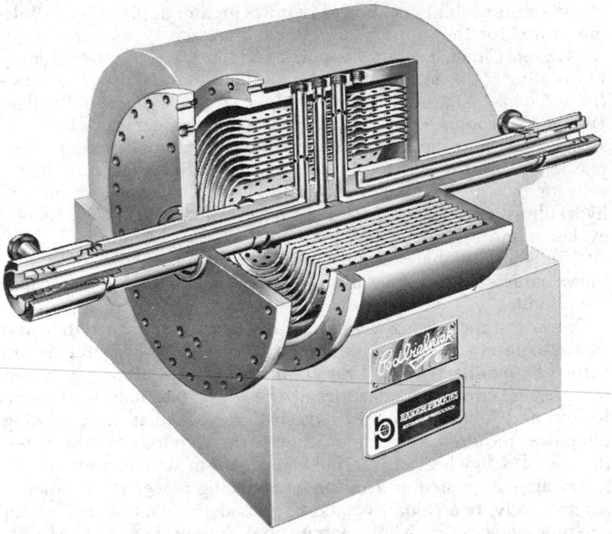

FIG. 21-116 Podbielniak centrifugal extractor. *(Courtesy of Baker Perkins Inc.)*

The machines are particularly characterized by extremely low holdup of liquid per stage, and this led to their extensive use in the extraction of antibiotics, such as penicillin and the like, for which multistage extraction and phase separation must be done rapidly to avoid chemical destruction of the product under conditions of extraction. They have been used extensively in all phases of pharmaceutical manufacture and are increasingly being used in other fields: petroleum processing, both solvent refining and acid treating, dephenolization of wastewaters, extraction of uranium from ore leach liquors, as well as for clarification and phase-separation work. See Kaiser, *Sewage Ind. Wastes*, **27**, 311 (1955); Podbielniak, Gavin, and Kaiser, *J. Am. Oil Chem. Soc.*, **36**, 238 (1959); Doyle and Rauch, *Pet. Eng.*, **27**(5), C-49 (1955); Anderson and Lau, *Chem. Eng. Prog.*, **51**, 507 (1955); Todd and Podbielniak, ibid., **61**(5), 69 (1965); and Todd, ibid., **62**(8), 119 (1966). The last contains data on interstage back mixing. Table 21-38 lists some of the characteristics of the machines.

With a laboratory model [0.55 m (18 in) in diameter, 5.08 cm (2 in) wide, 18 concentric cylinders slotted at 180° intervals], Barson and Beyer [*Chem. Eng. Prog.*, **49**, 243 (1953)] obtained from two to eight ideal stages with isoamyl alcohol–boric acid–water at 5000 r/min. The number of stages increased with ratio of light-to-heavy-liquid flow but with varying position of the interface and consequently varying fraction of the machine devoted to light-liquid-dispersed. At constant flow rate, the number of stages was essentially independent of rotational speed. Jacobson and Beyer [*Am. Inst. Chem. Eng. J.*, **2**, 283 (1956)] obtained about the same results. Alexandre and Gentilini [*Rev. Inst. Fr. Pet. Ann. Combust. Liq.*, **11**, 389 (1956)] similarly obtained five ideal stages with benzene–acetic acid–water, and 3.4 to 12.5 ideal stages with methyl isobutyl ketone–acetic

TABLE 21-38 Podbielniak Centrifugal Extractors*

Model number	Over-all dimensions, in.			Total wt., lb.	Horsepower		Flow capacity, gal./min.	
	Width	Height	Length (incl. drive)		Connected	Continuous	Multistage extraction	Neutralization, acid treating, extraction of fermentation broths
A-1	16	12	30	150	3.0	2.5	1.0	0.5
B-10	55.5	33	67.5	2,700	7.5	6.7	30	30
D-18	76	45	85	8,600	15	10	150	75
D-36	94	45	85	10,250	25	15	300	150
E-48	113	59	107	21,500	40	22	500	300

*Courtesy Baker Perkins Inc. To convert inches to centimeters, multiply by 2.54; to convert pounds to kilograms, multiply by 0.454; to convert horsepower to kilowatts, multiply by 0.746; and to convert gallons per minute to cubic meters per hour, multiply by 0.227.

acid–water. Anderson and Lau [*Chem. Eng. Prog.*, **51,** 507 (1955) describe a model handling 10 to 15 percent suspended solids in the liquids, and report a fraction to two ideal stages when extracting penicillin and chloromycetin, 7.04 to 8.71 m^3/h (1860 to 2300 gal/h) total flow rate.

Quadronics (Liquid Dynamics) Extractor (Doyle et al., U.S. Patent 3,114,707, 1963, and others; paper at AIChE meeting, St. Louis, February 1968) This is a horizontally rotated device, a variant of the Podbielniak extractor, in which either fixed or adjustable orifices may be inserted as a package radially. These permit control of the mixing intensity as the liquids pass radially through the extractor. Flow capacities, depending on machine size, range from 0.34 to 340 m^3/h (1.5 to 1500 gal/min).

Luwesta (Centriwesta) Extractor This is a development from Coutor (U.S. Patent 2,036,924, 1936). See also Eisenlohr [*Ind. Chem.*, **27,** 271 (1951); *Chem. Ing. Tech.*, **23,** 12 (1951); *Pharm. Ind.*, **17,** 207 (1955); *Trans. Indian Inst. Chem. Eng.*, **3,** 7 (1949–1950)] and Husain et al. [*Chim. Ind. (Milan)*, **82,** 435 (1959)]. This centrifuge revolves about a vertical axis and contains three actual stages. It operates at 3800 r/min and handles approximately 4.92 m^3/h (1300 gal/h) total liquid flow at 12-kW power requirement. Provision is made in the machine for the accumulation of solids separated from the liquids, for periodic removal. It is used, more extensively in Europe than in the United States, for the extraction of acetic acid, pharmaceuticals, and similar products.

De Laval Extractor (Palmqvist and Beskow, U.S. Patent 3,108,953, 1959) This machine contains a number of perforated cylinders revolving about a vertical shaft. The liquids follow a spiral path about 25 m (82 ft) long, in countercurrent fashion radially, and mix when passing through the perforations. There are no published performance data.

Section 22

Process Control*

T. C. Wherry, B.S.E.E., B.S.Ch.E., *Vice President (Retired), Director of Systems Research, Applied Automation, Inc.; Member, Board of Directors; Member, American Institute of Chemical Engineers, American Automatic Control Council; Fellow, Instrument Society of America; Fellow, American Institute of Chemists; Registered Professional Engineer (Oklahoma). (Section Editor)*

Jerry R. Peebles, M.S., P.E., *Engineering Associate, Cities Service Oil and Gas Company. (Assistant Section Editor)*

Patrick M. McNeese, B.S., *General Manager of Purchasing, Cities Service Oil and Gas Company*

Philip O. Teter, Jr., B.S., P.E., *Electrical Engineer, Corps of Engineers*

Richard E. Worsham, B.S., *Region Engineer, Cities Service Oil and Gas Company*

Roy M. Young, B.S., Ch.E., *Process Automation Consultant, R & M Associates, Inc.*

FUNDAMENTALS OF AUTOMATIC CONTROL

*The following persons are acknowledged for their contributions to the fifth edition: E. C. Miller, B. O. Ayers, E. N. Fuller, D. A. Jewett, D. W. Lane, R. D. McCoy, R. S. Moser, F. T. Ogle, T. J. Pemberton, W. G. Ragains, R. C. Richardson, O. E. Ririe, R. A. Sanford, H. W. Staten, W. S. Stewart, and E. D. Tolin.

COMPUTER PROCESS CONTROL

UNIT-OPERATIONS CONTROL

CONTROL-SYSTEM ANALYSIS

FUNDAMENTALS OF AUTOMATIC CONTROL

GENERAL REFERENCES: Caldwell, Coon, and Zoss, *Frequency Response for Process Control*. McGraw-Hill, New York, 1959. Considine, *Process Instruments and Control Handbook*, McGraw-Hill, New York, 1957. Considine and Ross, *Handbook of Applied Instrumentation*, McGraw-Hill, New York, 1964. Coughanouer and Koppel, *Process Systems Analysis and Control*, McGraw-Hill, New York, 1965. Eckman, *Automatic Process Control*, Wiley, New York, 1958. Gibson, *Nonlinear Automatic Control*, McGraw-Hill, New York, 1963. Harriott, *Process Control*, McGraw-Hill, New York, 1964. Liptak, *Instrument Engineers Handbook*, Chilton, Philadelphia, 1969. Murrill, *Automatic Control of Processes*, International Textbook, Scranton, Pa., 1967. Shinskey, *Process Control System*, McGraw-Hill, New York 1967. Williams and Lauher, *Automatic Control of Chemical and Petroleum Processes*, Gulf, Houston, 1961.

Processes may be controlled more precisely to give more uniform and higher-quality products by the application of automatic control, often leading to higher profits. Additionally, processes which respond too rapidly to be controlled by human operators can be controlled automatically. Automatic control is also beneficial in certain remote, hazardous, or routine operations. After a period of experimentation, computers are being used to operate and automatically control processing systems, many too large and too complex for effective direct human control.

Since process profit is usually the most important benefit to be obtained by applying automatic control, the quality of control and its cost should be compared with the economic return expected and the process technical objectives. The economic return includes reduced operating costs, maintenance, and off-specification product along with improved process operability and increased throughput.

Investment costs may be affected if the consequences of control are reflected in the selection of processing equipment. This requires active interaction between those people doing process design and those doing control-system design. Neglect of this interaction leads to the "hang-it-on" concept of control-system design in which the control system is added to the process design to make it operable.

THE GENERAL CONTROL SYSTEM

The various aspects of automatic control can best be described by use of an example. Consider a process such as that shown in Fig. 22-1. The flowing liquid is to be heated to a desired temperature by steam flowing through heating coils. The temperature of the exit flow is affected by factors (process variables) such as the temperature of the incoming liquid, the flow rate of the liquid, the temperature of steam, the flow rate of steam, heat capacities of the fluids, heat loss from the vessel, and mixer speed.

Open- and Closed-Loop Systems The system as shown in Fig.22-1 is normally classified as "open-loop." *Open-loop* control systems are those in which information about the controlled variable (in this case, temperature) is not used to adjust any of the system inputs to compensate for variations in the process variables. The term open-loop is often encountered in discussions of control systems to indicate that the uncontrolled process dynamics are being studied.

A *closed-loop* control system implies that the controlled variable is measured and that the result of this measurement is used to manipulate one of the process variables, such as steam flow.

Feedback Control In the closed-loop control system, information about the controlled variable is fed back as the basis for control of a process variable; hence the designation "closed-loop feedback control." This feedback can be accomplished by a human operator (manual control) or by use of instruments (automatic control).

For *manual control*, referring to Fig. 22-1, an operator periodically measures the temperature; if this temperature, for example, is below the desired value, the operator increases the steam flow by opening the valve slightly. For *automatic control*, a temperature-sensitive device is used to produce a signal (electrical, pneumatic, etc.) proportional to the measured temperature. This signal is fed to a controller, which compares it with a preset desired value, or set point. If a difference exists, the controller changes the opening of the steam-control valve to correct the temperature as in Fig. 22-2.

Feedforward Control Feedforward control is becoming widely used. Process disturbances are measured and compensated for without waiting for a change in the controlled variable to indicate that a disturbance has occurred. Feedforward control is also useful when the final controlled variable cannot be measured. In the example shown in Fig. 22-3, the feedforward controller has the computational ability, using the measured input-liquid flow rate and temperature, to compute the necessary steam flow rate to maintain the desired output-liquid temperature.

The equation solved by the controller relating input-liquid heat content, steam flow, and output-liquid temperature is usually designated as the *process model*. Perfect models and controllers are rare; so a combination of feedback and feedforward control is more desirable (see Fig. 22-4). The arrangement of a controller supplying the set point for another controller is known as *cascade control* and is commonly used in feedback control.

Block Diagrams These diagrams show the relationships among the system variables and are the standard method of representing systems for analysis or discussion. Conventions for block-diagram construction have been established. Lines represent signals which may be flows of information, material, or energy. A circular summing junction represents an algebraic summation of the input signals to that point. An algebraic sign, + or −, is placed beside the arrow to the summing junction to represent addition or subtraction. A branch point, or a line branching from another line, represents a division of a signal into more than one path without modification. Rectangles represent a modification of the entering signal and are used for system elements. The rectangles normally contain notations which describe the dynamic characteristics of the system they rep-

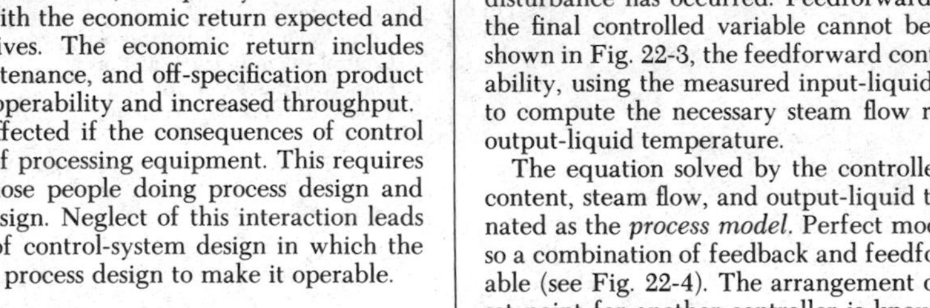

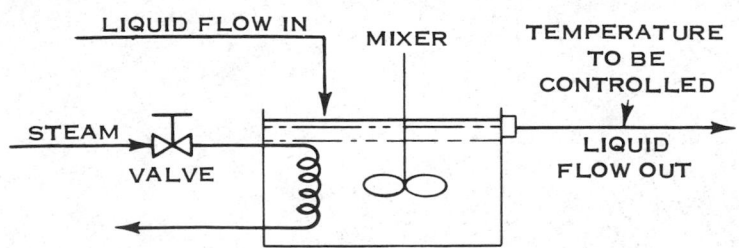

FIG. 22-1 Simple heat-exchange process.

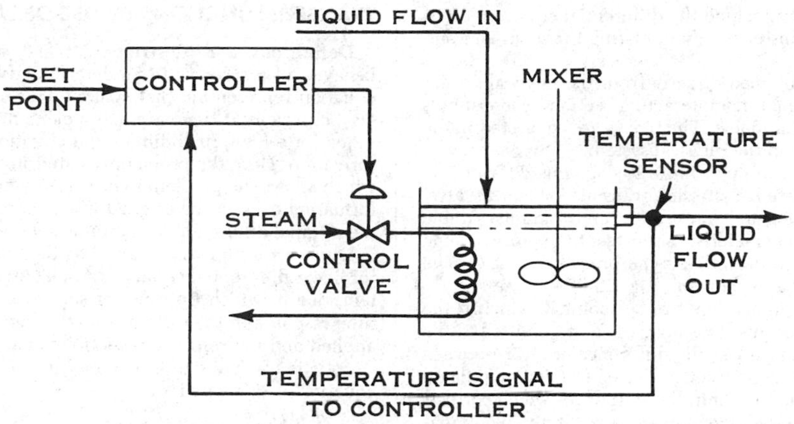

FIG. 22-2 Automatic feedback control of heat-exchange process.

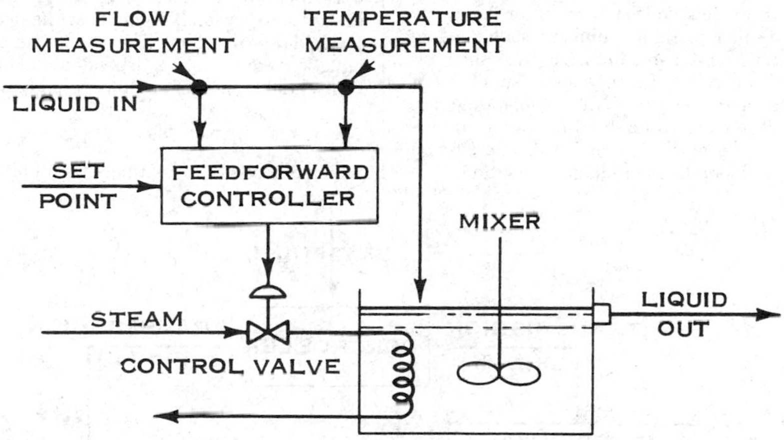

FIG. 22-3 Feedforward control of heat-exchange process.

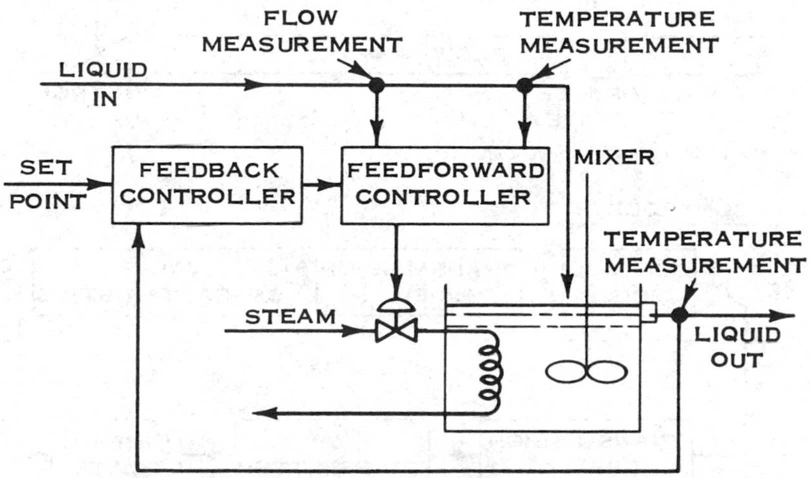

FIG. 22-4 Combined feedforward and feedback control of heat exchange.

resent. These notations may include the differential equation, units-conversion constant, or transfer function relating the input and output of the element.

The block diagram is obtained directly from the physical system by dividing it into functional, noninteracting-sections whose inputs and outputs are readily identifiable. The blocks are connected in the same order as they appear in the physical system.

The pneumatic-flow control loop illustrated in Fig. 22-5 has six main sections to be considered: controller, transmission line *a*, valve, orifice plate, differential-pressure transmitter, and transmission line *b*. For this case, the valve characteristics represent the process being controlled; the flow through the valve is the process output *C*. The block diagram of this system is shown in Fig. 22-6.

The controller has a reference input or set point *R*, which is the desired value for the process-measurement signal transmitted to the controller. The controller measures the difference or error between the set point and the measurement signal. The error *E* is manipulated by the controller to provide the controller output *M*, which corrects the valve position to drive the error toward zero. In the block diagram, the controller is represented by the summing junction and the control-modes block.

The controller output feeds transmission line *a*; the block representing transmission line *a* has *M*, the manipulated variable, as an input and *A* as an output. The transmission line may be very fast, as in the case of electronic instruments, so that $A = M$, or it may be slow, as in the case of some pneumatic-instrument installations, so that *A* lags behind *M* in time. The signal *A* in turn controls the position of the control valve; this position and the valve characteristics determine the flow rate through the valve. In a similar manner, blocks are added for the orifice plate, differential-pressure transmitter, and transmission line *b*. The block diagram of the closed-loop feedback system gives a closed loop in block-diagram notation.

TRANSFER FUNCTIONS BY USE OF LAPLACE TRANSFORMS

Definitions and Restrictions The steady-state and dynamic behavior of a system can be determined by solving the differential equation representing that system. This may be a long and tedious task, especially if there are many elements in the system. One technique for solving such differential equations uses the Laplace transformation. Here the problem is stated in terms of a second variable which allows the problem to be solved algebraically. Then, by transformation back to the original independent variable, the solution to the original differential equation is obtained (see Sec. 2).

Laplace transforms, as useful as they are, can be employed only for linear differential equations. Such equations describe a linear system, one in which the rules of superposition apply. That is, if the time response of the system is $y_1(t)$ when a forcing function $x_1(t)$ is applied and the response is $y_2(t)$ when $x_2(t)$ is applied, then if $x_1(t) + x_2(t)$ is applied, the response is $y_1(t) + y_2(t)$. A linear differential equation of this type is

$$p_n(t)\frac{d^n y(t)}{dt^n} + p_{n-1}(t)\frac{d^{n-1}y(t)}{dt^{n-1}} + \cdots + p_0(t)y(t) = x(t) \qquad (22\text{-}1)$$

where the coefficients $p_i(t)$ are not functions of the dependent variable $y(t)$ or any of its derivatives. Generally, the solution of such an equation with time-varying coefficients is difficult. Most chemical operations are nonlinear over a wide range, but the assumption of linearity over a small region near the operating point can usually be justified (linearization). The coefficients of this linearized differential equation are usually independent of time. Therefore, for process control the basic process differential equation is usually of the form

$$m_n\frac{d^n y(t)}{dt^n} + m_{n-1}\frac{d^{n-1}y(t)}{dt^{n-1}} + \cdots + m_0 y(t) = x(t)$$

which can be solved routinely with Laplace transforms.

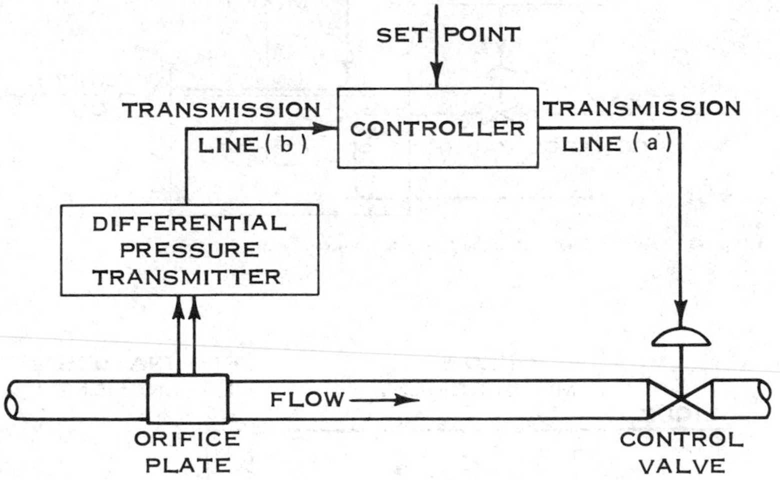

FIG. 22-5 Process-flow control loop.

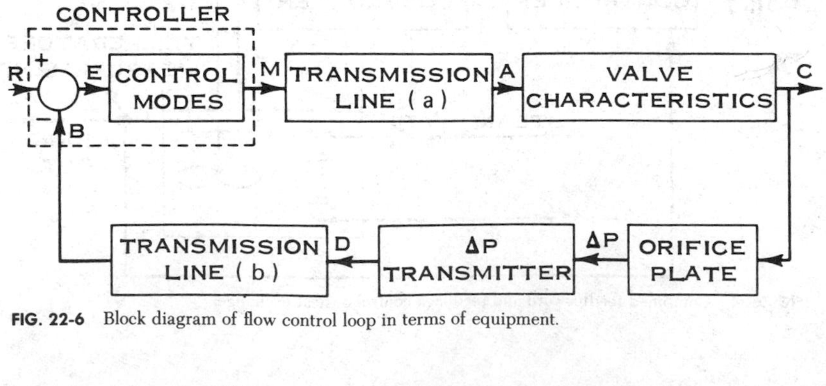

FIG. 22-6 Block diagram of flow control loop in terms of equipment.

Transfer Functions To illustrate the convenience of the Laplace transformation, consider a differential equation given by

$$T\frac{dc(t)}{dt} + c(t) = RMu(t) \qquad c(0) = 0 \tag{22-2}$$

T represents the time constant of the system, and $Mu(t)$ describes a step input to the system of magnitude M. We wish to find the response of the output $c(t)$. The Laplace transforms $\mathcal{L}$ needed are taken from tables.

$$\mathcal{L}\left[\frac{dc(t)}{dt}\right] = sC(s) - c(0+)$$

$c(0+) = 0$ since the initial conditions were zero.

$$\mathcal{L}[c(t)] = C(s)$$
$$\mathcal{L}[Mu(t)] = M(1/s)$$

The Laplace transform of Eq. (22-1) is then

$$TsC(s) + C(s) = R(M/s)$$
$$C(s) = \frac{RM}{s(TS + 1)} \tag{22-3}$$

To transform this equation back into the time domain, we look up the inverse transform $\mathcal{L}^{-1}$ and find that

$$c(t) = \mathcal{L}^{-1}[C(s)] = RM(1 - e^{-t/T})$$

In a system of many elements the transformed equation corresponding to Eq. (22-3) may be quite complicated. However, it can usually be manipulated into a form for which the inverse transform can be found in a table.

In process-control work Laplace transforms are used to determine responses to disturbances. Steady-state or constant terms will usually drop out of the solutions of the differential equations because initial conditions will usually be assumed to be zero.

The *transfer function* is defined as the ratio of the Laplace transform of the responding variable (output) to the Laplace transform of the disturbing variable (input). From Eq. (22-3), we get for the transfer function $KG(s)$

$$KG(s) = \frac{\text{output}}{\text{input}} = \frac{C(s)}{M(s)} = \frac{R}{Ts + 1} \tag{22-4}$$

The convention for designating transfer functions in a control diagram is the expression $KG(s)$. Capital letters are used when the functions are in the s domain, and small letters are used in the time domain. $G(s)$ represents the dynamic portion of transfer function, and K is related to the steady-state gain through an element. In the transfer function, Eq. (22-4), $K = R$ and $G(s) = 1/(Ts + 1)$.

It is common practice in drawing block diagrams in the s domain to omit the (s) from the $F(s)$'s. Instead only the capital-letter designation is used to represent the s-domain transforms. A block diagram in terms of transfer functions is illustrated by Fig. 22-7, where H is the transfer function of the feedback-measurement device.

Combining Transfer Functions Fluid- and thermal-process systems exhibit many different dynamic characteristics, but many systems may be described by combinations of five transfer functions:

K	proportional element
$1/Ts$	capacitance element
$1/(Ts + 1)$	first-order element
$1/(T^2s^2 + 2\xi Ts + 1)$	second-order element
e^{-Ls}	dead-time element

Transfer functions are important tools in the analysis of control systems. Each block or element of the control system has its own characteristic transfer function. If the s-domain transfer function notation $KG(s)$ is used for each block, the system elements can be combined by algebraic procedures into an overall expression for the entire control system as illustrated in Fig. 22-8. Then, by substitution of the detailed individual transfer functions, one obtains the overall transfer function from which the analysis of the control system can be made.

PROCESS CHARACTERISTICS

When using block diagrams for control-system analysis, it is necessary that the transfer function representing both the static or steady state and the dynamic behavior for each block be known. To describe the behavior of a physical process, the specific characteristics of that process must be defined. A process block will be defined as having a single input and a single output; in the s domain, the output transform is determined by multiplying the input transform by the transfer function of the process. Each block can be described in terms of the five elements listed previously or combinations of them.

Proportional Element The analogy of electrical resistance describes many of the characteristics of flow, diffusion, and thermal systems. Electrical resistance is given by Ohm's law.

$$R = \frac{e_1 - e_2}{i} = \frac{\Delta e}{i} = \frac{\text{potential difference}}{\text{current flow}}$$

where e = potential, V; i = current, A; and R = resistance, Ω.

In turbulent liquid flow, a small opening is equivalent to electrical resistance. Flow rate through the restriction is given by Bernoulli's law.

$$f = kA\sqrt{2g_c(h_1 - h_2)} \tag{22-5}$$

where f = flow rate, k = flow coefficient, A = area of the restriction, g_c = gravitational constant, and h = liquid head (potential). This is a nonlinear relationship between flow and potential which is unlike the linear electrical system. It is possible, however, to "linear-

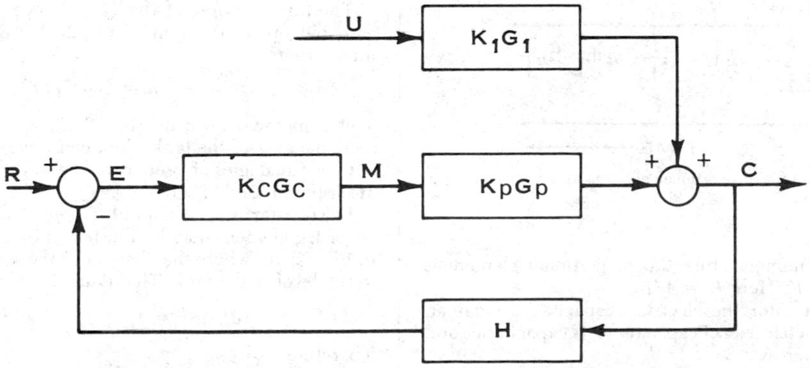

FIG. 22-7 Generalized block diagram of a feedback control system in terms of transfer function for a process subject to a load disturbance.

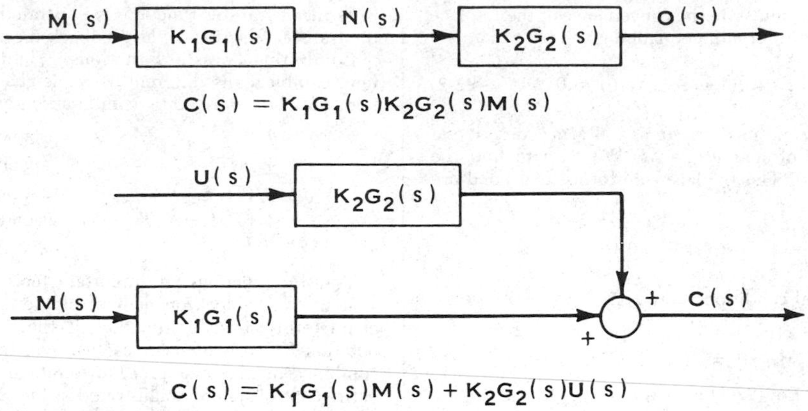

FIG. 22-8 Block diagram of process elements in transfer-function notation and the respective composite transfer functions.

ize" the system at a particular operating point by assuming that $h_1 - h_2$ is large compared with the change in $h_1 - h_2$ to be considered. That is, the resistance can be defined by

$$R_{f0} = \left(\frac{\text{change of head difference}}{\text{change of flow}}\right)_{f0} = \left.\frac{d(h_1 - h_2)}{df}\right|_{f0}$$

where R_{f0} is defined for a particular value of flow f_0. From this definition and Eq. (22-5), one gets

$$R_{f0} = f_0/(gk^2A^2)$$

Thus, at a particular flow rate f_0, a small change in potential difference $d(h_1 - h_2)$ will result in a proportional change in flow rate. Note that in this example the resistance is not a constant but depends on the flow rate.

The transient response of the output of a proportional element is identical to the input except for the magnitude. Consider a step change in head difference for the flow example. The linearized flow system is represented in the time domain by Fig. 22-9. The time response of f to a step change in $h_1 - h_2$ at $t = 0$ is shown in Fig. 22-10.

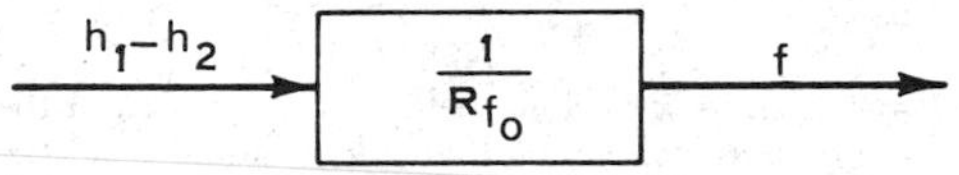

FIG. 22-9 Liquid-resistance-element block diagram.

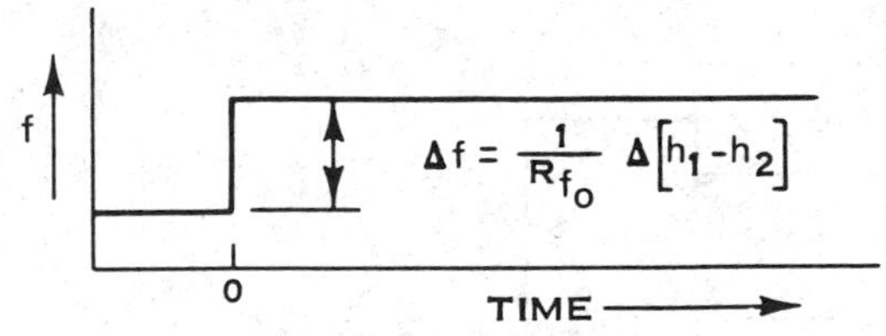

FIG. 22-10 Time response of a proportional element to a step change in potential at time equals zero.

In Laplace s-domain nomenclature the proportional element is frequently designated as K. Here $K = 1/R_{f0}$.

Capacitance Element For the electrical capacitance element, Faraday's law relates the charge on a capacitor to its capacitance and the potential difference across it.

$$C = q/e = \text{quantity/potential} \tag{22-6}$$

where q = charge, C; C = capacitance, F; and e = potential, V. From the definition of current and Eq. (22-6)

$$dq/dt = i = C\,(de/dt)$$

A capacitance is an energy-storing device. That is, current flow into the capacitor is the product of the capacitance and the rate of change of potential.

Liquid capacitance is directly analogous to electrical capacitance and may be represented by a surge tank used for temporary liquid storage.

$$C_L\,(dh/dt) = f \tag{22-7}$$

where C_L = liquid capacitance or cross-sectional area. In the Laplace s domain, Eq. (22-7) becomes

$$C_L sH(s) = F(s) \qquad \text{or} \qquad H(s) = \frac{1}{C_L s}\,F(s)$$

The block diagram is Fig. 22-11. The term $1/C_L s$ is the transfer function. In Laplace s-domain nomenclature this type of transfer function is usually designated as $1/Ts$ where $C_L = T$.

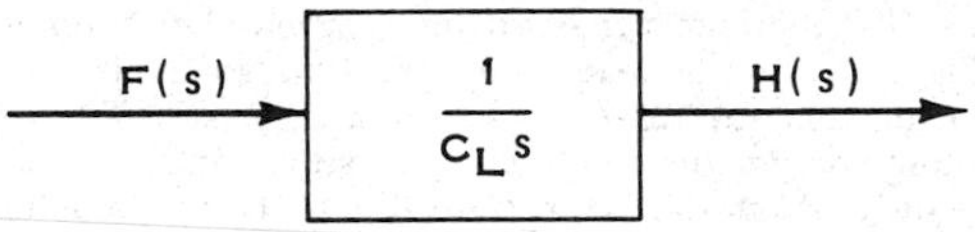

FIG. 22-11 Liquid-capacitance-element block diagram.

The time response of this capacitance element to a given flow f_0 can be obtained from Eq. (22-7). Separating variables and integrating,

$$h = (f_0/C_L)\,t + k$$

This response, shown in Fig. 22-12, will be nonlinear if the cross-sectional area of the tank is not constant.

Other analogies of proportional and capacitance process elements are listed in Table 22-1.

First-Order Lag Element (Time-Constant Element) A first-order lag element may be illustrated by the liquid level in the tank in Fig. 22-13, where the flow out of the tank is assumed proportional to the level in the tank. The dynamics are described by

$$C_L\,(dh/dt) = f_1 - f_2 \qquad \text{and} \qquad f_2 = h/R_L$$

Therefore,

$$R_L C_L\,(dh/dt) + h = R_L f_1 \tag{22-8}$$

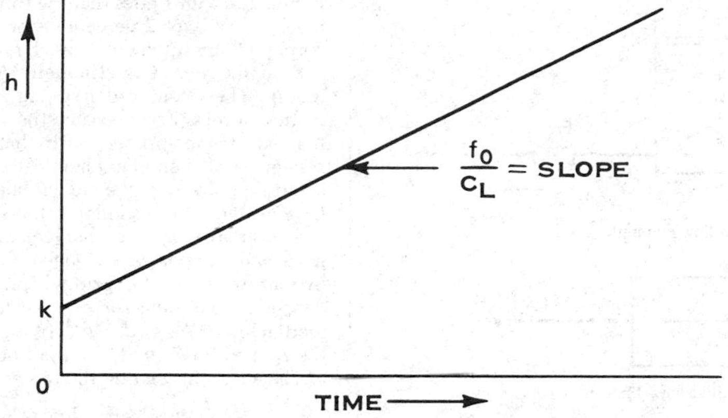

FIG. 22-12 Response of liquid-capacitance element for constant flow and constant cross section.

TABLE 22-1 Some Electrical, Flow, and Heat Analogies

	Electrical	Liquid flow	Gas flow	Heat flow
Quantity units Potential units Flow units	q, coulomb e, volt i, ampere, coulomb/sec.	V, volume, ft.^{3} h, liquid head, ft. f, volume flow rate, ft.^{3}/sec.	M, mass, lb.$_m$ p, pressure, lb.$_f$/ft.^{2} W, mass flow rate, lb.$_m$/sec.	Q, heat, B.t.u. T, temperature, degrees q, heat flow, B.t.u./sec.
Resistance to flow $R = \frac{\Delta(\text{potential})}{\Delta(\text{flow})}$	$R(\text{ohms}) = \frac{e_1 - e_2}{i}$	$R\left(\frac{\text{sec.}}{\text{ft.}^2}\right) = \frac{d(h_1 - h_2)}{df}$	$R\,\frac{1}{(\text{ft.})(\text{sec.})} = \frac{d(p_1 - p_2)}{dW}$	$R\left(\frac{(\text{deg.})(\text{sec.})}{\text{B.t.u.}}\right) = \frac{d(T_1 - T_2)}{dq}$
Symbolic notation	i →, e_1, e_2, R	f →, h_1, h_2, R	W →, p_1, p_2, R	q →, T_1, R, T_2
Governing equation	$i = \frac{e}{R}$ R = constant	$f = K\,S\,\sqrt{2g(h_1 - h_2)}$ $= f_{ss} + \frac{\Delta(h_1 - h_2)}{R}$ $R = \frac{f_{ss}}{K^2S^2g}$	$W = K_1S\,\sqrt{g_c\rho_{av}(p_1 - p_2)}$ $= W_{ss} + \frac{\Delta(p_1 - p_2)}{R}$ $R = \frac{2W_{ss}}{K_1^2S^2g_c\rho_{av}}$ where $p_1 - p_2 < 0.1p_1$	**Conduction** **Convection** $q = kA\,\frac{T_1 - T_2}{\Delta x} = HA(T_1 - T_2)$ $= q_{ss}\,\frac{\Delta(T_1 - T_2)}{R}$ $R = \frac{\Delta x}{kA} \qquad = \frac{1}{HA}$
Capacity $C = \frac{\Delta(\text{quantity})}{\Delta(\text{potential})}$	$C(\text{farad}) = \frac{q}{e} = \frac{i}{de/dt}$	$A(\text{ft.}^2) = \frac{\Delta V}{\Delta h}$	$C\left(\frac{(\text{lb.}_m)(\text{ft.}^2)}{\text{lb.}_f}\right) = \frac{\Delta M}{\Delta p}$	$C\left(\frac{\text{B.t.u.}}{\text{deg.}}\right) = MC_p = \frac{dQ}{dT}$
Symbolic notation	C, e, i →	f, A, ↑h	constant volume, W →, p	q →, T
Governing equation	$e = e_0 + \frac{1}{C}\int i\,dt$	$h = h_0 + \frac{1}{A}\int f\,dt$	$p = p_0 + \frac{1}{C_1}\int W\,dt$	$T = T_0 + \frac{1}{C}\int q\,dt$

Subscripts: ss indicates steady state and 0 indicates initial condition

t = time
S = area of restriction
A = area of liquid surface
Δx = thickness
g = 32.2 ft./sec.^{2}
g_c = 32.2 lb$_m$. ft./lb$_f$. sec.^{2}

ρ_{av} = density at average pressure
$C_1 = V/R_gT$ for ideal gas at constant volume and temperature
R_g = gas constant
C_p = specific heat at constant pressure
K, K_1, k, and H are the appropriate coefficients

The product RC appears often in analyses of automatic control systems. By analogy from electric-circuit theory, it is called the *time constant* and is usually designated as T. Time constants are probably the most used indicators of process-element characteristics. Substituting $T = R_LC_L$ and rewriting Eq. (22-8) in the s domain gives

$$(Ts + 1)H(s) = R_LF_1(s)$$

$$H(s) = RL/(Ts + 1)\,F_1(s) \qquad (22\text{-}9)$$

The transfer function for this first-order lag element is $R_L/(Ts + 1)$; see Fig. 22-14. R_L is a magnitude factor arising here because head or

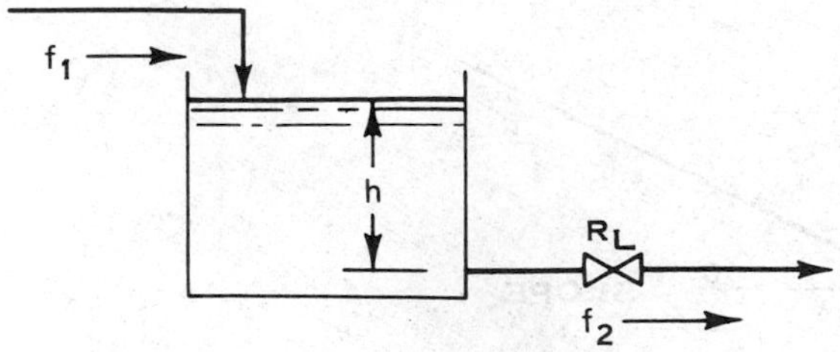

FIG. 22-13 First-order lag, liquid-flow element.

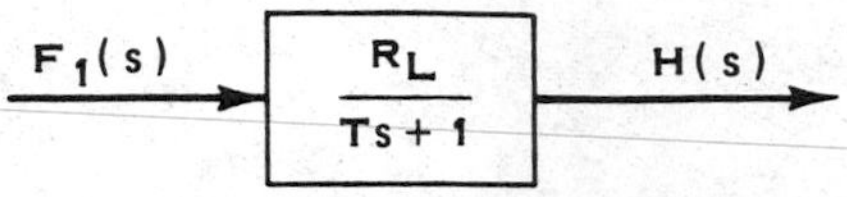

FIG. 22-14 Block diagram of a liquid-flow, first-order lag process.

potential was the process-element output. First-order lag elements are characterized by the expression $1/(Ts + 1)$.

The time response of a first-order lag can be obtained by solving Eq. (22-8) for a step change of inflow with the tank initially empty. Rearranging terms,

$$dh/(R_L f_1 - h) = dt/T \qquad f_1 = \text{constant}$$

Integrating between the limits $h_0 = 0$ gives

$$-\ln (R_L f_1 - h) \Big|_{h_0}^{h} = \frac{t}{T}$$

$$h(t) = R_L f_1[1 - e^{-t/T}]$$

A plot of the process-element output for a unit-step increase in input in Fig. 22-15 illustrates that the time constant is the time required to reach $1 - e^{-1}$, 63.2 percent of the final value. The final value of the output will be R, where $R = R_L f_1$.

Second-Order Lag Element (Quadratic or Oscillatory Element) The second-order system is important in the study of automatic control systems because the actual response of a controlled system is often compared with that of an oscillatory second-order system. Few open-loop chemical processes exhibit oscillatory characteristics; however, the closed-loop control system is often tuned to have characteristics similar to the second-order oscillatory system.

Second-order-system characteristics are illustrated by the spring, mass, and damper system shown in Fig. 22-16. The mass is acted on by four forces: (1) the spring-displacement force kc, (2) the velocity-dependent damping force $p\ dc/dt$, (3) the acceleration-dependent inertial force $(W/g)(d^2c/dt^2)$ (g is gravitational acceleration), and (4) the applied force m. The output of this element is the displacement c. The equation relating the forces is

$$(W/g)\,(d^2c/dt^2) + p\,(dc/dt) + kc = m$$

If the initial displacement and velocity are zero, the Laplace transformations are

$$\mathcal{L}[c(t)] = C(s) \qquad \mathcal{L}\left[\frac{dc}{dt}\right] = sC(s) \quad \text{and} \quad \mathcal{L}\left[\frac{d^2c}{dt^2}\right] = s^2C(s)$$

After rearranging, the transformed equation is

$$\frac{C(s)}{M(s)} = \frac{1/k}{(W/gk)\,s^2 + (p/k)\,s + 1}$$

where $M(s)$ and $C(s)$ are the system input and output respectively. This transfer function is characteristic of all quadratic lags. For dis-

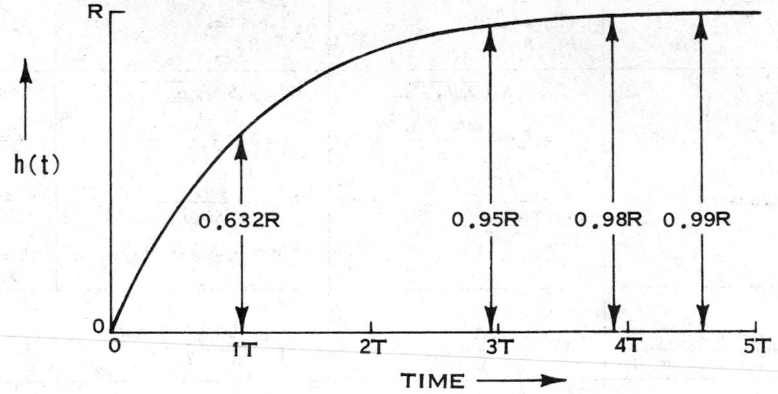

FIG. 22-15 Time response of a first-order lag resulting from a unit-step increase in input at time equals zero.

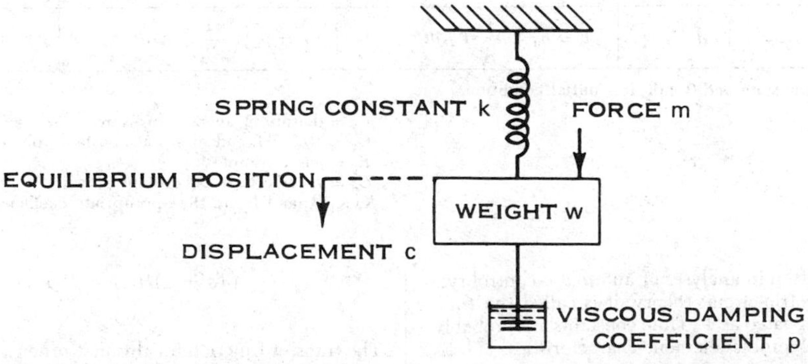

FIG. 22-16 Mechanical representation of the oscillatory element.

cussion, it is convenient to write this transfer function in a standard form.

$$KG(s) = \frac{C(s)}{M(s)} = \frac{K}{T_c^2 s^2 + 2\xi T_c s + 1} \qquad (22\text{-}10)$$

where K = system gain = $1/k$
T_c = characteristic time = $\sqrt{W/gk}$
ξ = damping factor = $(p/2k)(1/T_c)$

The significance of T_c and ξ can be seen from the response of the system to a unit-step change in the input shown in Fig. 22-17. The response depends on the value of ξ. For $\xi < 1$ it is oscillatory, while for $\xi > 1$ it is nonoscillatory. Systems for which $\xi > 1$ are termed *overdamped;* for $\xi < 1$, they are termed *underdamped;* and for $\xi = 1$, the systems are *critically damped.*

For $\xi < 1$:

$$c(t) = K\left[1 - \frac{1}{\sqrt{1-\xi^2}} e^{-\xi(t/T_c)} \sin(\omega_r t + \phi)\right] \qquad (22\text{-}11)$$

where $\omega_r = \dfrac{\sqrt{1-\xi^2}}{T_c}$ and $\phi = \tan^{-1}\dfrac{\sqrt{1-\xi^2}}{\xi} = \cos^{-1}\xi$

For $\xi = 1$:

$$c(t) = K\left[1 - \left(1 + \frac{t}{T_c}\right) e^{-(t/T_c)}\right]$$

For $\xi > 1$:

$$c(t) = K\left[1 - \frac{1}{\sqrt{\xi^2 - 1}} e^{-\xi(t/T_c)} \sinh(\omega_d t + \phi)\right]$$

where $\omega d = \dfrac{\sqrt{\xi^2 - 1}}{T_c}$ and $\phi = \cosh^{-1}\xi$

The underdamped response $\xi < 1$ is commonly used for describing control-system performance. From Eq. (22-11), some specific relationships can be noted. The frequency of oscillation f is a function of both ξ and T_c.

$$f = \omega_n/2\pi = \sqrt{1 - \xi^2/2\pi T_c}$$

The undamped natural frequency f_n for $\xi = 0$,

$$f_n = \omega_n/2\pi = 1/2\pi T_c \qquad (22\text{-}12)$$

defines the characteristic time T_c.

Chemical processes rarely exhibit oscillatory characteristics but are often represented by several first-order lags in series. Two first-order lags in series (Fig. 22-18) are characterized by an overdamped second-order system:

$$KG(s) = \frac{R_1}{T_1 s + 1}\,\frac{R_2}{T_2 s + 1} = \frac{R_1 R_2}{T_1 T_2 s^2 + (T_1 + T_2)s + 1}$$

Comparing this with Eq. (22-10) gives the characteristic values of $K = R_1R_2$, $T_c^2 = T_1T_2$, and $\xi = \frac{1}{2}(T_1 + T_2/\sqrt{T_1T_2}$, $\xi > 1$, except for the case $T_1 = T_2$, for which $\xi = 1$.

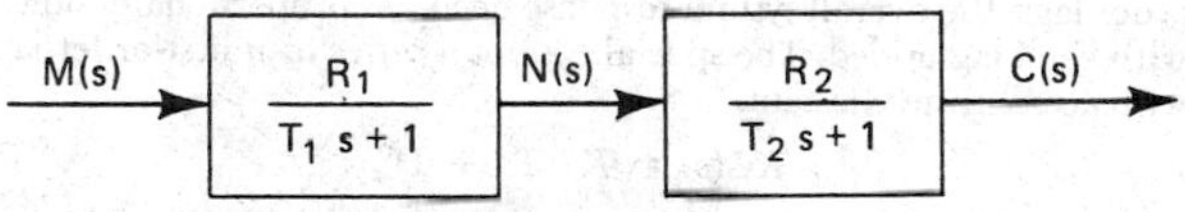

FIG. 22-18 Block diagram of two first-order lags forming a second-order lag.

Distance-Velocity Lag (Dead-Time Element) The dead-time element, commonly called distance-velocity lag or true time delay, is often encountered in process systems. For example, if a temperature-measuring element is located downstream from a heat exchan-

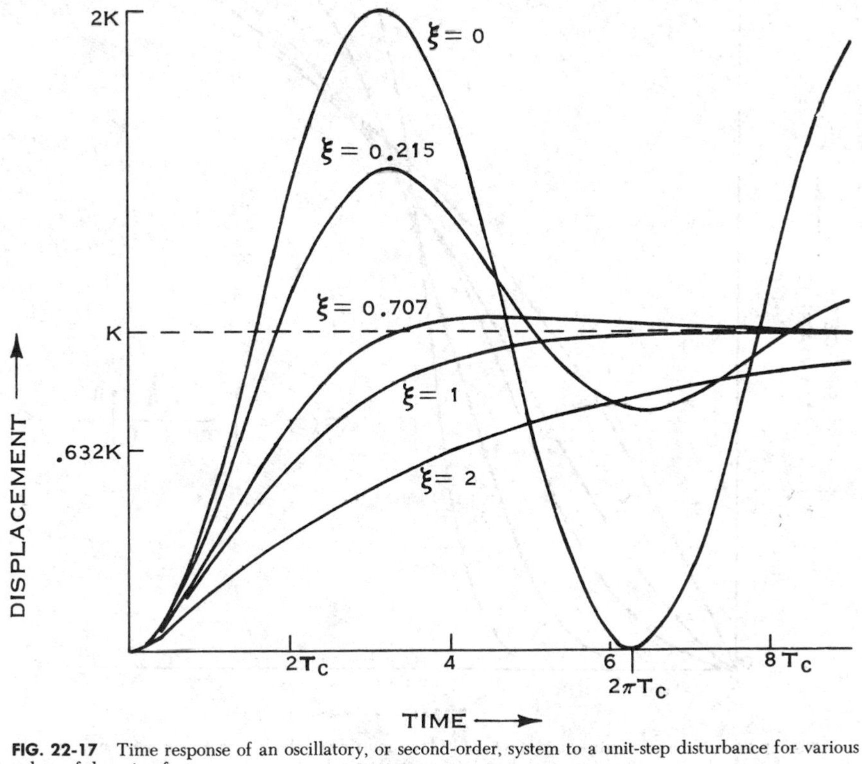

FIG. 22-17 Time response of an oscillatory, or second-order, system to a unit-step disturbance for various values of damping factor.

ger, a time delay occurs before the heated fluid leaving the exchanger arrives at the temperature-measurement point. If some element of a system produces a dead time of L time units, then any input $f(t)$ to that element will be reproduced at the output as $f(t - L)$. Transforming to the s domain gives

$$\mathcal{L}[f(t)] = F(s) = \text{input}$$

$$\mathcal{L}[f(t - L)] = e^{-Ls}F(s) = \text{output}$$

and $$KG(s) = \frac{\text{output}}{\text{input}} = \frac{e^{-Ls}F(s)}{F(s)} = e^{-Ls}$$

Putting this into block-diagram notation gives Fig. 22-19.

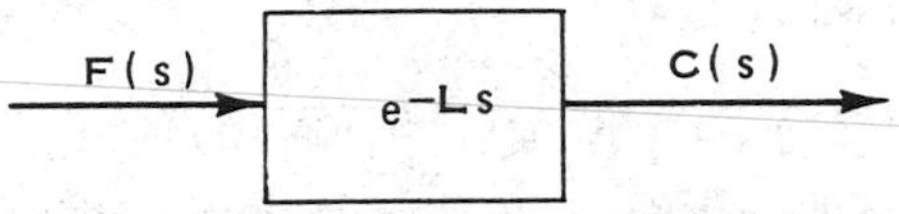

FIG. 22-19 Block diagram of a dead-time process element.

Higher-Order Lags If a process is described by a series of n first-order lags, the overall system response becomes proportionally slower with each lag added. The special case of a series of n first-order lags with equal time constant,

$$KG(s) = K/(Ts + 1)^n$$

is illustrated in Fig. 22-20. Note that all curves reach about 60 percent of their final value at $t = nT$.

Higher-order systems can be approximated by a first- or second-order plus dead-time system for control-system design. Consider the response of a fifth-order system to a unit-step input (Fig. 22-21). This system can be modeled by a first-order lag plus dead time,

$$KG(s) = Ke^{-Ls}/(Ts + 1)$$

where $T = t_1 - L$, t_1 being the apparent response time constant. The system gain K is the response magnitude at final value, and the dead time L is given by the point of intersection of the maximum slope with the time axis.

FEEDBACK-CONTROL-SYSTEM CHARACTERISTICS

When an automatic controller is added to a process, the performance depends on the nature of the process, the type of control equipment, and the care with which the controller is tuned. For example, for a set-point change, one would like the process to come to its new operating point as quickly as possible. Ideally, in a simple process this would mean to change the manipulated variable instantly to its maximum or minimum value and hold it there until the process reaches the new desired value, then instantly readjust the manipulated variable to its new equilibrium value. Figure 22-22 illustrates three degrees of performance of automatic control of a simple process: idealized conditions, attainable performance, and poor control. The use of computers as process controllers has made possible control performance on some processes very nearly equal to the idealized action illustrated in Fig. 22-22.

Within the economic and technical requirements of the process,

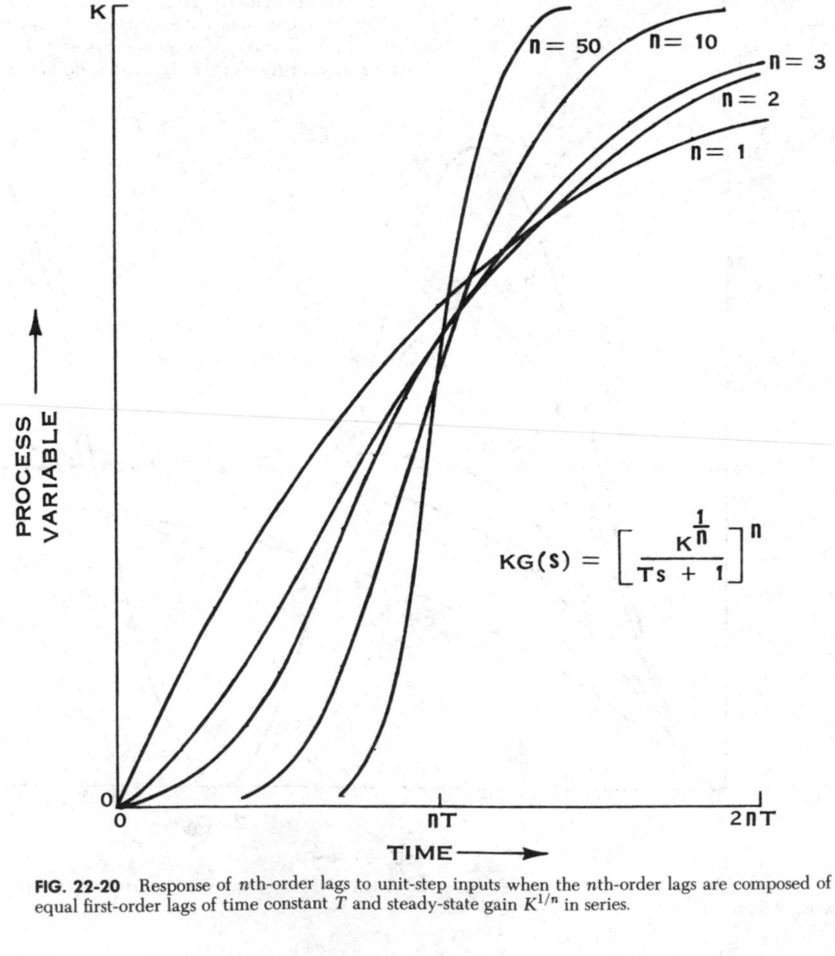

FIG. 22-20 Response of nth-order lags to unit-step inputs when the nth-order lags are composed of n equal first-order lags of time constant T and steady-state gain $K^{1/n}$ in series.

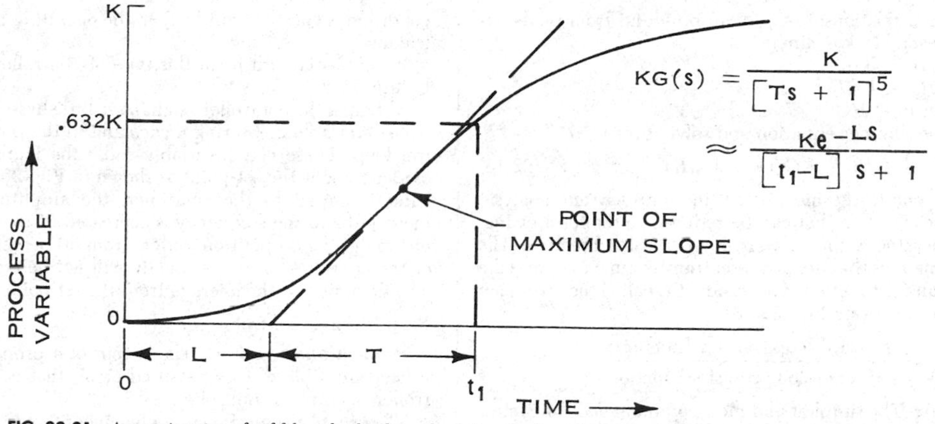

FIG. 22-21 Approximation of a fifth-order lag by a dead time L plus a first-order lag of time constant $T = t_1 - L$.

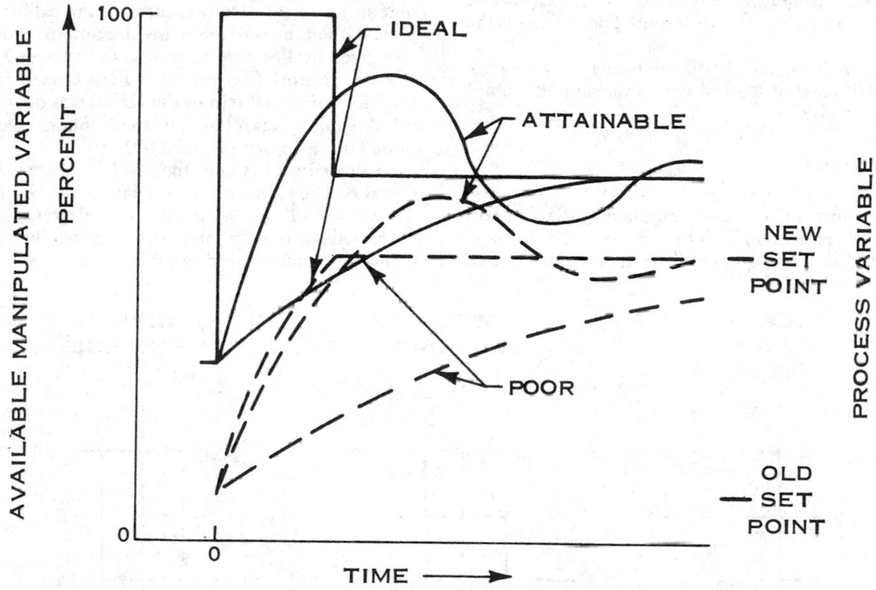

FIG. 22-22 Comparison of responses to a step set-point change for idealized control action, attainable conventional control action, and poor control action. Solid lines represent the portion of the available manipulated variable sent to the process, and dashed lines represent the response of the controlled variable.

two objectives of applying an automatic controller to a process are (1) to reduce the order of the controller-process system to the lowest practical order and (2) to reduce the time constant of the controller-process system to the smallest practical value.

Closing the Loop Many linear-feedback control-system block diagrams can be reduced to the form shown in Fig. 22-23. KG, the forward-loop transfer function, includes all elements between the error sensor and output C. H is the transfer function of the feedback path. The summer determines the difference between set point R and feedback signal B.

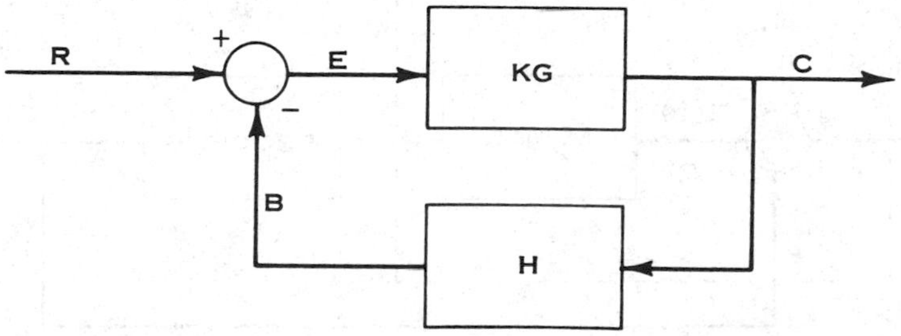

FIG. 22-23 Generalized block diagram of a feedback control system.

The input-output relationship C/R can be found from the equations describing each block. Namely,

Summer: $E = R - B$
Forward path: $C = KGE$
Feedback path: $B = HC$

Substituting in the summer equation and solving for C/R,

$$C/R = KG/(1 + KGH) \qquad (22\text{-}13)$$

This equation, one of the most useful in control-system analysis. describes any linear system that can be reduced to the form of Fig. 22-23. The numerator is the forward-loop transfer function. The denominator is one plus the forward-loop transfer function times the feedback-loop transfer function. This product is called the open-loop transfer function. Rewriting Eq. (22-13)

$$\frac{C}{R} = \frac{\text{forward-loop transfer function}}{1 + \text{open-loop transfer function}}$$

On-Off Control The simplest and most common mode of control is the on-off, or two-position, type such as the familiar thermostats used in space heating, refrigeration, and tank heating. Of the several variations of the on-off controller or relay, three are shown in Fig. 22-24. The controller with hysteresis or differential dead band is the most common.

For convenience of analysis, the on-off controller is frequently described as a three-position controller in which the output, manipulated variable M, is given by

$$M = M_0 \operatorname{sgn} E = \begin{cases} +M_0 \text{ for } E > 0 \\ 0 \text{ for } E = 0 \\ -M_0 \text{ for } E < 0 \end{cases}$$

where M_0 is the maximum value of the manipulated variable available. In many cases $+M_0$ corresponds to the process variable being turned on and $-M_0$ to that being turned off. In these cases the zero condition is chosen as midway in the operating range for mathematical analysis.

In block-diagram form this type of controller may be illustrated by Fig. 22-25.

Although the controller might have little hysteresis, lags in the controlled-variable measuring device and in the process are in the control loop. These lags invariably cause the manipulated variable to oscillate about the set point as shown in Fig. 22-26. If a differential band is added to the controller, the amplitude of oscillation is increased and the frequency is decreased. If the duty cycle, fraction of time in the on position, differs from 50 percent, the time average of the controlled variable usually will not be equal to the set point. The deviation of the average from the set point is termed *offset*.

Proportional Control

Proportional Action The output of a proportional controller is a fixed multiple of the measured error; that is, a proportional controller is simply a multiplier.

The two terms most commonly encountered in describing proportional controllers are proportional band PB and controller gain K_c. *Controller gain* is the amount by which the error is multiplied to obtain the output. Many controllers are calibrated in PB rather than gain. Although there are many definitions, *proportional band* is usually given by $PB = 100 \text{ percent}/K_c$. A lower PB gives a higher gain.

Proportional Control of a First-Order Process The transfer function for the system in Fig. 22-13 was developed earlier [Eq. (22-9)]. A simple control system to maintain a constant head c in this same tank is shown in Fig. 22-27.

In this control system the level is measured and transduced into signal B. The value of B is compared with the signal representing the desired level R, the set point. The difference is the error E, which is amplified K_c times by proportional controller gain. The output of the controller, manipulated variable M, drives a valve which changes the

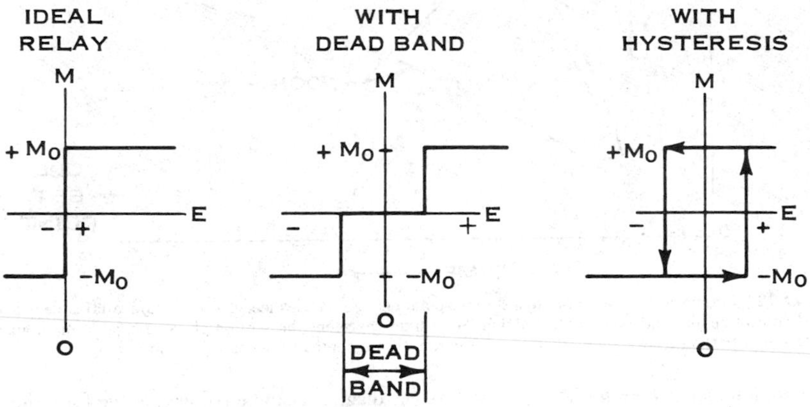

FIG. 22-24 Operational characteristics of on-off control relays or switches compared with the ideal relay.

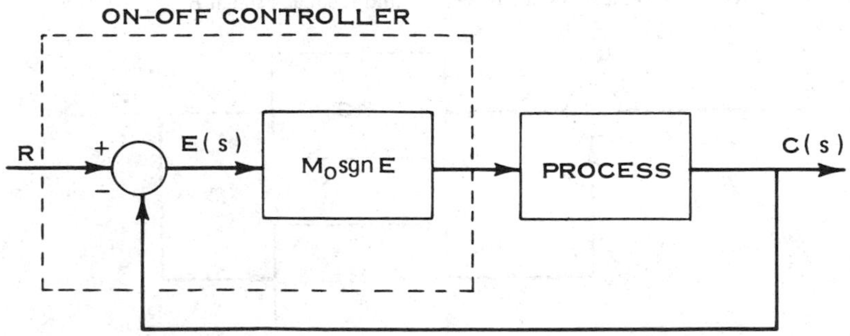

FIG. 22-25 Block diagram of an on-off controller applied to a process.

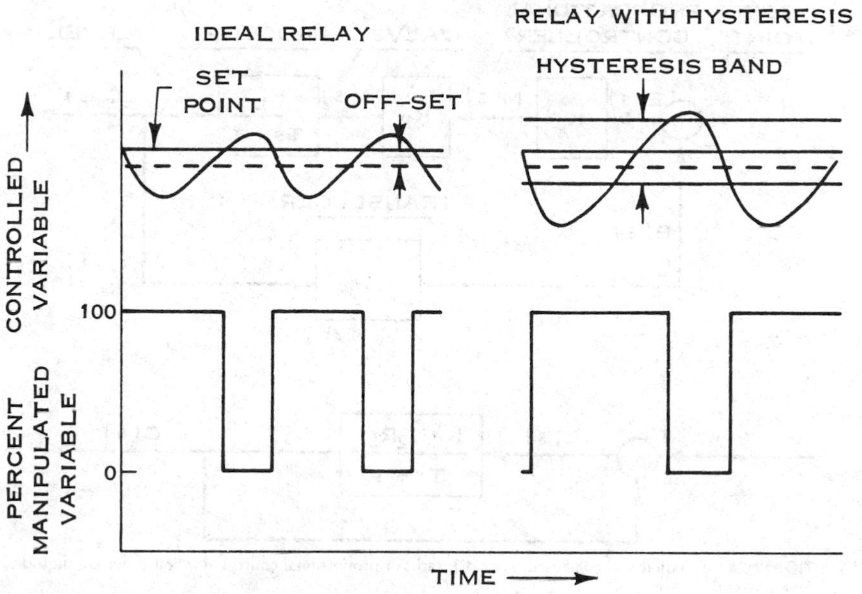

FIG. 22-26 Relationship between the controlled variable and the manipulated variable for an ideal relay and a relay with hysteresis in a process with lags in the control loop.

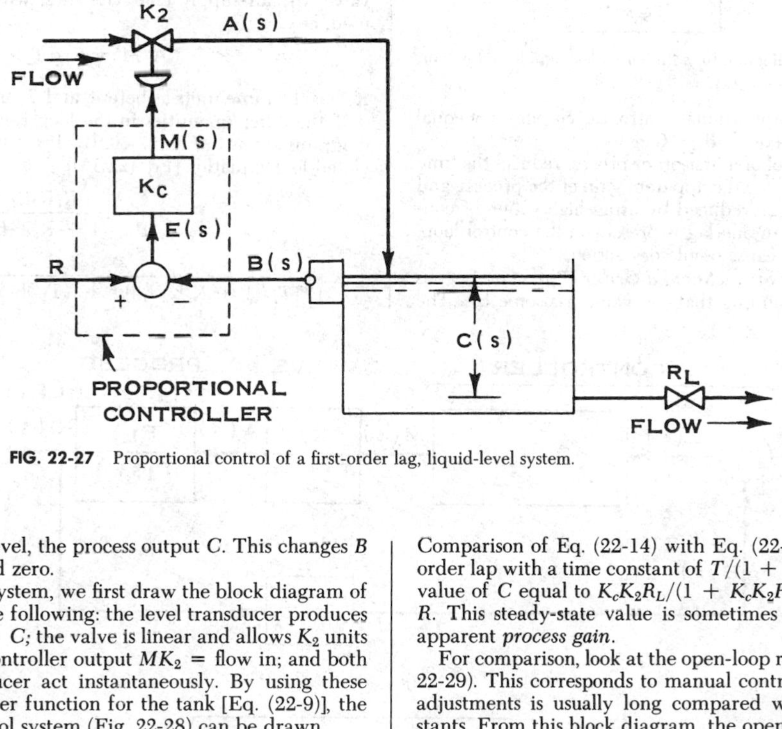

FIG. 22-27 Proportional control of a first-order lag, liquid-level system.

flow *A* and therefore the level, the process output *C*. This changes *B* and forces the error toward zero.

In order to analyze the system, we first draw the block diagram of the system by assuming the following: the level transducer produces a signal equal to head, $B = C$; the valve is linear and allows K_2 units of flow for every unit of controller output MK_2 = flow in; and both the valve and the transducer act instantaneously. By using these assumptions and the transfer function for the tank [Eq. (22-9)], the block diagram of the control system (Fig. 22-28) can be drawn.

By using the general closed-loop equation [Eq. (22-13)], the relationship between *C* and *R* can be written directly:

$$\frac{C}{R} = \frac{K_cK_2R_L/(Ts + 1)}{1 + [K_cK_2R_L/(Ts + 1)]} = \frac{K_cK_2R_L/(1 + K_CK_2R_L}{[T/(1 + K_cK_2R_L)]\, s + 1} \tag{22-14}$$

Comparison of Eq. (22-14) with Eq. (22-9) shows this to be a first-order lap with a time constant of $T/(1 + K_cK_2R_L)$ and a steady-state value of *C* equal to $K_cK_2R_L/(1 + K_cK_2R_L)$ following a unit step in *R*. This steady-state value is sometimes called the system gain or apparent *process gain*.

For comparison, look at the open-loop response of this system (Fig. 22-29). This corresponds to manual control when the time between adjustments is usually long compared with the process time constants. From this block diagram, the open-loop transfer function is

$$C/M = K_2R_L/(Ts + 1)$$

This is a first-order lag with a time constant *T*. The final value of *C* after a unit step in *M* is K_2R_L, the process gain. Comparing steady-state values of the output variable for open- and closed-loop operation, when $M = R$, shows that the open-loop value is larger by K_2R_L $[1 - K_c/(1 + K_cK_2R_L)]$. This difference in steady-state value is

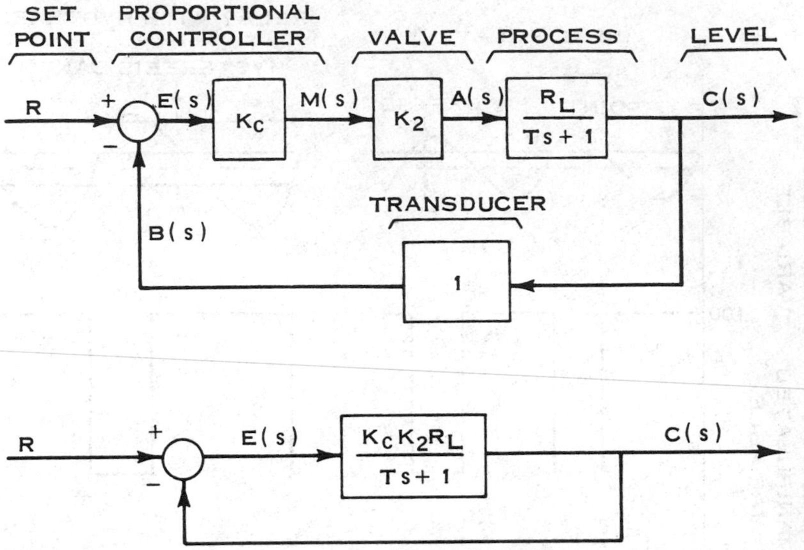

FIG. 22-28 Detailed and combined block diagrams of proportional control of a first-order lag, liquid-level system.

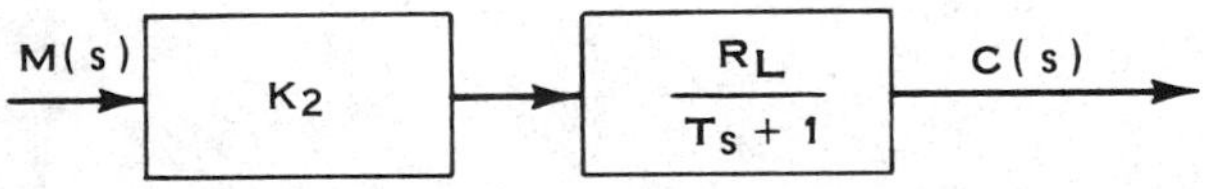

FIG. 22-29 Open-loop block diagram for a first-order lag, liquid-level system.

termed *offset*. Thus, with proportional control the output is not equal to the set point, and offset equals $R - C$.

Thus, proportional control of a first-order process reduces the time constant of the system, reduces the apparent gain of the process, and introduces offset that may be reduced by using high values of controller gain K_c. If more than one lag is present in the control loop, high proportional gain can cause oscillatory action.

Proportional Control of a Second-Order Process Let us reconsider liquid level, assuming that the valve has some lag. The dynamic response of a valve can be roughly approximated by a first-order lag. The time constant of the lag depends on the size of the valve, the air-supply characteristics, whether a valve positioner is used, etc.

$$A/M = K_2/(T_2s + 1)$$

K_2 has the same units as before, and T_2 is the valve time constant.

If the other quantities in the loop remain unchanged, the block diagrams are as in Fig. 22-30. By direct substitution into the general closed-loop equation [Eq. (22-13)].

$$\frac{C}{R} = \frac{K_cK_2R_L}{1 + K_cK_2R_L}$$

$$\times \frac{1}{[T_1T_2/(1 + K_cK_2R_L)]s^2 + [(T_1 + T_2)/(1 + K_cK_2R_L)]s + 1}$$

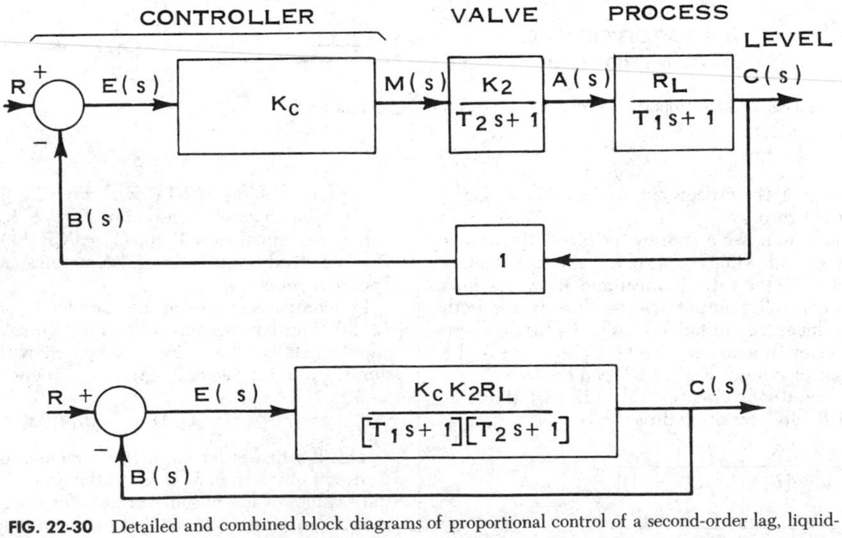

FIG. 22-30 Detailed and combined block diagrams of proportional control of a second-order lag, liquid-level system.

This equation is of the same form as Eq. (22-10), indicating that this controlled system has the same characteristics as the spring, mass, and damper system. The step response will be like one of the curves in Fig. 22-17. Terms K_c, K_2, R_L, T_1, and T_2 combine to describe the system as in Eqs. (22-11) and (22-13).

$$K = \text{system gain} = \frac{K_cK_2R_L}{1 + K_cK_2R_L}$$

$$T_c = \text{characteristic time} = \sqrt{\frac{T_1T_2}{1 + K_cK_2R_L}}$$

$$\xi = \text{damping factor} = \frac{T_1 + T_2}{2\sqrt{T_1T_2(1 + K_cK_2R_L)}}$$

$$f_n = \text{natural frequency} = \frac{1}{2\pi}\sqrt{\frac{1 + K_cK_2R_L}{T_1T_2}}$$

From Fig. 22-17, it is noted that speed of response is determined by characteristic time. The general shape of the response curve is determined by the damping factor. Therefore, if the system response is to be changed, T_c and/or ξ must be manipulated. K_c, the controller gain, is the only variable easily manipulated to modify the response.

For a unit-step change in R the change in the steady-state value of C is $K_cK_2R_L/(1 + K_cK_2R_L)$. The offset is still present and is exactly the same amount as before.

An increase in controller gain will speed up the response time, tend to make the output have more overshoot and oscillate following a step change in set point (or load disturbance), increase the frequency of oscillation, and decrease the amount of offset. If the gain is reduced to the point where ξ is equal to or greater than 1.0, the response will have no overshoot and will be quite slow. Thus, several of the controlled-system response characteristics are affected by the controller gain setting.

Proportional plus Reset Control (PI Control)

Proportional plus Reset Controller Action The transfer function for an ideal proportional plus reset controller is

$$\frac{M}{E} = K_c\left(1 + \frac{1}{T_i s}\right) \tag{22-15}$$

where T_i = reset time. Transforming Eq. (22-15) to the time domain,

$$m = K_c e + \frac{K_c}{T_i}\int e\,dt$$

The controller output consists of two parts, the first proportional to the error and the second proportional to the integral of the error. Thus the controller has proportional plus integral action, or proportional plus reset action, as it is usually called. Reset action causes the controller output to change as long as an error exists. Even small errors can eventually provide enough controller output to force the error to zero, the main purpose of reset action.

The controller open-loop response to a unit-step change in error is shown in Fig. 22-31 for an ideal 20.7- to 103.4-kPa (3- to 15-lbf/in^2) pneumatic controller where $K_c = 4$ and $T_i = 2$ min. T_i, called reset time or integral time, is the time required for the integral portion of the controller output to become equal to the proportional portion with a constant error.

There are two popular ways to designate reset in controllers. The controller reset knob may be calibrated in *reset time* and is usually referred to as *minutes per repeat*, minutes to repeat the initial proportional-action change in controller output. On some models the reset may be calibrated in *reset rate* $1/T_i$, or *repeats per minute*.

Proportional plus Reset Control of a First-Order Process If a proportional plus reset controller is placed in the level-control system in Fig. 22-27, we will have Fig. 22-32 and the block diagram in Fig. 22-33. If T_i is set equal to T (an optimum controller setting by several criteria), the block diagram reduces to Fig. 22-34.

Substituting into the general closed-loop equation,

$$\frac{C}{R} = \frac{K_cK_2R_L/Ts}{1 + K_cK_2R_L/Ts} = \frac{1}{(T/K_cK_2R_L)s + 1} \tag{22-16}$$

This transfer function represents a first-order lag with a time constant similar to that obtained with the proportional controller. However, the steady-state value of C/R following a unit-step change in R is 1. This means that there is no offset; the output will go to the set point desired.

The proportional plus reset action has slowed the response compared with proportional control. However, compared with manual

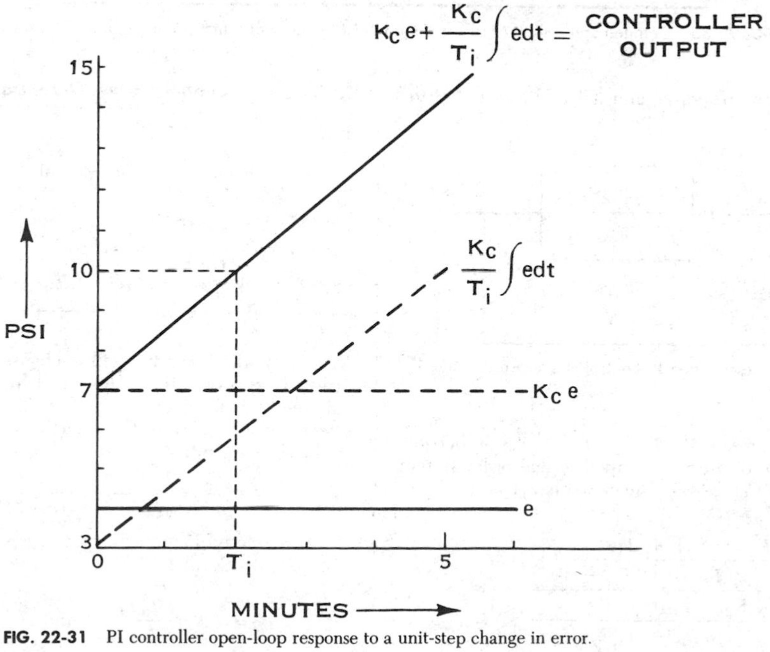

FIG. 22-31 PI controller open-loop response to a unit-step change in error.

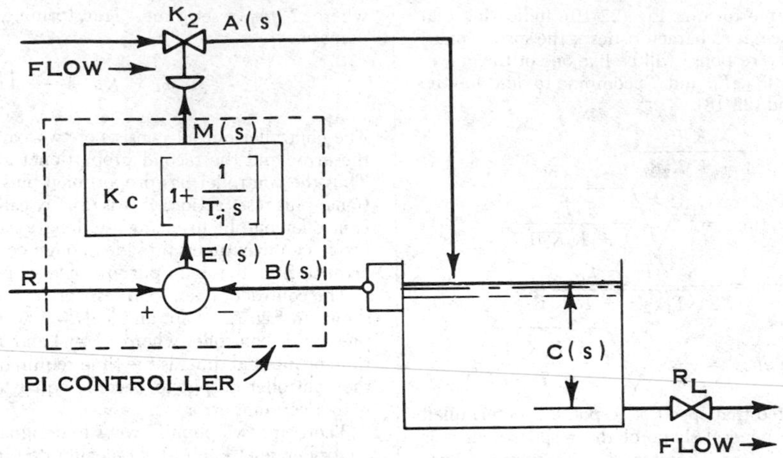

FIG. 22-32 Proportional plus reset control of a first-order lag, liquid-level system.

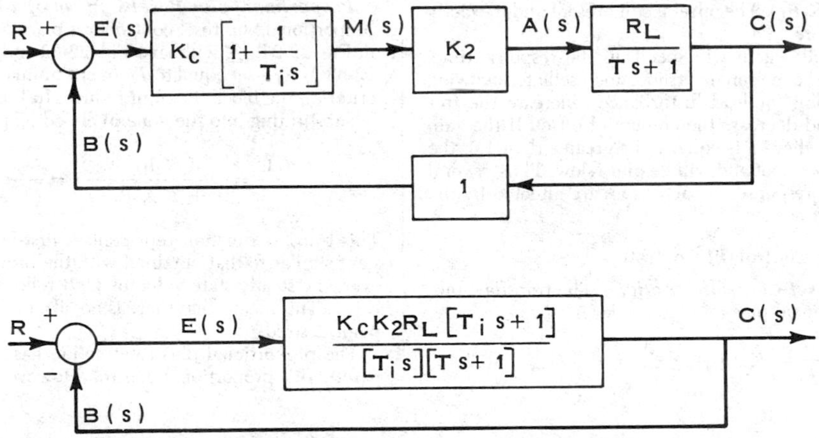

FIG. 22-33 Detailed and combined block diagram of PI control of a first-order lag, liquid-level system.

control, it has speeded up the response, and it has eliminated offset (see Table 22-2).

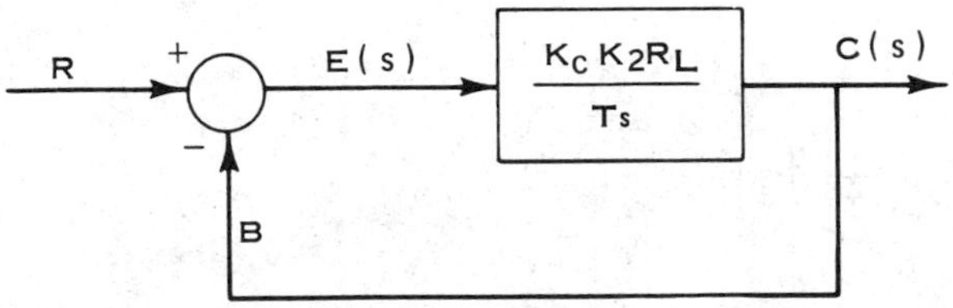

FIG. 22-34 Block diagram of reset-tuned PI control of a first-order lag, liquid-level system.

Proportional plus Reset Control of a Second-Order Process A block diagram of an overdamped second-order system with PI control is shown in Fig. 22-35. The closed-loop transfer function is

$$\frac{C}{R} = \frac{\dfrac{K_c K_2}{T_i} \dfrac{1 + T_i s}{s(T_1 s + 1)(T_2 s + 1)}}{1 + \dfrac{K_c K_2}{T_i} \dfrac{1 + T_i s}{s(T_1 s + 1)(T_2 s + 1)}}$$

TABLE 22-2 Control for First-Order Liquid-Level Process

	Type of control		
Variable	Manual	Proportional	Proportional plus reset
Time constant	T	$\frac{T}{1 + K_c K_2 R_L}$	$\frac{T}{K_c K_2 R_L}$
Final value of $\frac{C}{R}$	$K_2 R_L$	$\frac{K_c K_2 R_L}{1 + K_c K_2 R_L}$	1

Various methods can be used to choose K_c and T_i to obtain a lower-ordered expression for C/R. A feel for initial choices for K_c and T_i can be obtained by choosing $T_i = T_2$, where $T_2 > T_1$. This choice effectively provides for the cancellation of the largest time constant in the system, so that

$$\frac{C}{R} = \frac{1}{(T_i T_1/K_c K_2)s^2 + (T_i/K_c K_2)s + 1} \qquad (22\text{-}17)$$

Thus the closed-loop performance is second-order. Comparing Eq. (22-17) with Eq. (22-10) gives

$$T_c = \sqrt{\frac{T_1 T_i}{K_c K_2}} \qquad \text{and} \qquad \xi = \frac{1}{2}\sqrt{\frac{T_i}{K_c K_2 T_1}}$$

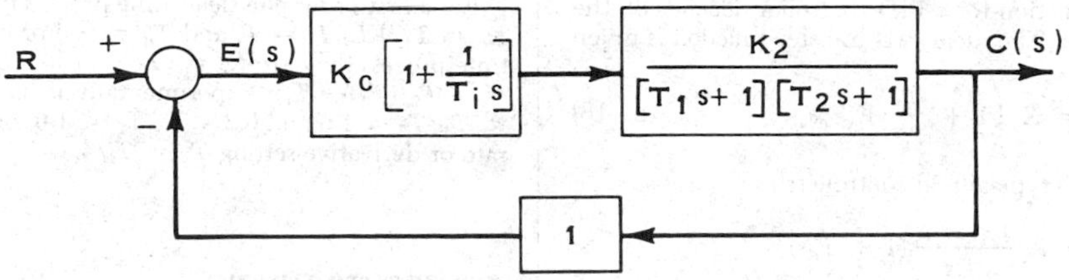

FIG. 22-35 Block diagram of an overdamped second-order process controlled by a PI controller; $T_1 \neq T_2$.

From these results, a value of K_c can be chosen to give a desired value of T_c or ξ. As illustrated by Eq. (22-17), the proportional plus reset controller can be used for the second-order system to eliminate offset. The addition of reset action into the second-order system will not alter the fact that the system can be oscillatory if the gain is high enough.

Proportional plus Reset plus Rate Control (PID)

Rate or Derivative Action The main purpose of rate or derivative action is to speed up the control action. It does this by anticipating where a process is going and applying correction to stop the change in error. It "anticipates" by measuring the rate of change of the error and applying a control action proportional to the rate of change. Derivative action is most useful for high-ordered processes with a large inertia, i.e., a slow-to-start response to a manipulated-variable change.

The ideal proportional plus rate controller has the transfer function

$$M/E = K_c(1 + T_d s)$$

where T_d = rate or derivative time (the time base over which the error change is measured). In the time domain,

$$m = K_c e + K_c T_d\,(de/dt)$$

Thus, the controller output is proportional to both size of error and rate of change of error.

The characteristics of an ideal proportional plus rate (PD) controller are shown in Fig. 22-36, where the error is assumed to be increasing at a constant rate. T_d is the time required for the controller output to become equal to twice the proportional response. That is, T_d is the time base over which the derivative action is computed.

In PD control the rate control action is added to the proportional control action. The size of the rate action depends on proportional gain, rate time, and the rate of change of the error. It is of no help in reducing offset.

Rate action is seldom used in control systems with large amounts of noise. Since noise is a fast-varying signal, the rate action it causes is large. Under certain conditions the controller may increase the peak-to-peak amplitude of the noise and therefore be undesirable. This possibility causes many process operators to avoid using rate action in places where it could be used to advantage, thus sacrificing process performance.

Three-Mode Controllers (PID) The three-mode controller combines the actions of the proportional, integral, and derivative elements into a single unit. The major effect of the various elements is as follows:

1. *Proportional (gain).* Generally shapes the response curve: higher gains generally give faster transient but more oscillatory responses.
2. *Integral (reset).* Eliminates steady-state offset.
3. *Derivative (rate).* Allows higher proportional gains for high-ordered systems.

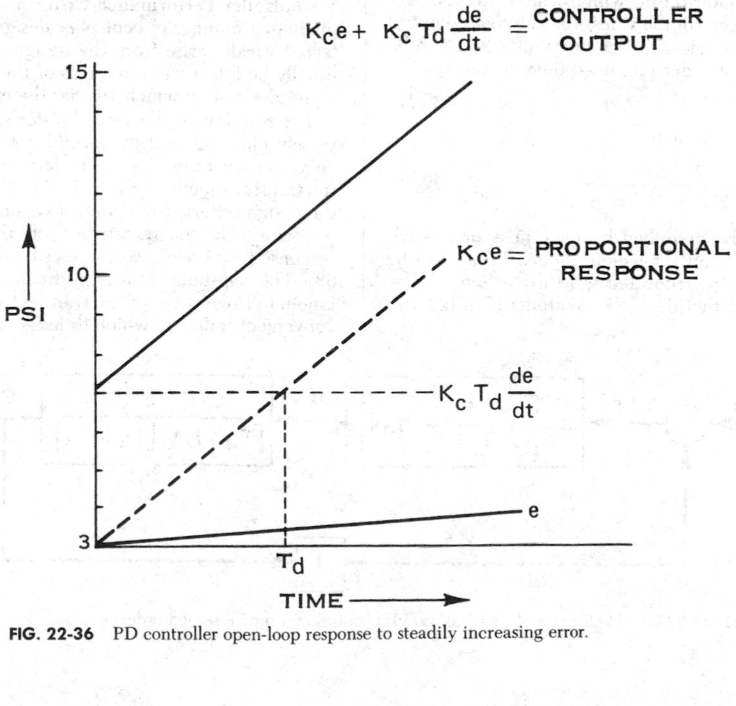

FIG. 22-36 PD controller open-loop response to steadily increasing error.

The precise transfer function for a PID controller depends on the manufacturer and its type. The ideal PID transfer function is given by

$$\frac{M(s)}{E(s)} = K_c\left(1 + \frac{1}{T_i s} + T_d s\right) \qquad (22\text{-}18)$$

Transfer functions for two typical real controllers are

$$\frac{M(s)}{E(s)} = K_c \frac{1 + T_i s}{T_i s}(1 + T_d s)$$

and

$$\frac{M(s)}{E(s)} = K_c \frac{1 + \alpha T_i s}{1 + T_i s}\,\frac{1 + T_d s}{1 + \beta T_d s}$$

where $\alpha, \beta \approx 0.1$

PID Control of a Second-Order Process A block diagram of a second-order system with a PID controller is illustrated in Fig. 22-37. Combining transfer functions and obtaining the closed-loop transfer function give

$$\frac{C}{R} = \frac{\dfrac{K_c K_2}{T_i}\,\dfrac{1 + T_i s + T_i T_d s^2}{s(T_1 s + 1)(T_2 s + 1)}}{1 + \dfrac{K_c K_2}{T_i}\,\dfrac{1 + T_i s + T_i T_d s^2}{s(T_1 s + 1)(T_2 s + 1)}}$$

This is a cumbersome transfer function, and design of a PID controller for systems of this high order would ordinarily be carried out by other techniques. A feed for the desired controller settings can be obtained if T_i and T_d are chosen so that $T_i = T_1 + T_2$ and $T_i T_d = T_1 T_2$. Under such conditions

$$\frac{C}{R} = \frac{1}{(T_i/K_c K_2)s + 1}$$

This transfer function represents a first-order lag similar to the one obtained for PI control of a first-order system [Eq. (22-16)].

Higher-Order Systems It has been shown that PI and PID controllers can be used to control first- and second-order systems effectively. When more than two lags are present in the control loop, analysis and design problems become more difficult, and other techniques are available. Some of these techniques are discussed under "Controller Tuning."

Most chemical processes have a number of lags, and an accurate knowledge of these lags may be difficult to obtain. Also, many chemical processes have an inherent dead time which complicates the control problem. In spite of these complexities, for automatic-control purposes the dynamics of the majority of chemical processes can be adequately described by a first-order plus dead-time model,

$$KG(s) = Ke^{-Ls}/(Ts + 1) \qquad (22\text{-}19)$$

or by a second-order plus dead-time model,

$$KG(s) = \frac{Ke^{-Ls}}{T_1 s + 1)(T_2 s + 1)} \qquad (22\text{-}20)$$

For processes which can be described by these first- or second-order plus dead-time models and for which the dead time can be approximated adequately by the truncated series expansion $e^{-Ls} = 1 - Ls$, it can be shown that the ideal PID controller [Eq. (22-18)] is the best controller.

For a first-order plus dead-time process the controller constants are $K_c = T_1/KL$, $T_i = T$, and $T_d = 0$. For a second-order plus dead-time process they are $K_c = (T_1 + T_2)/KL$, $T_i = T_1 + T_2$, and $T_d = T_1T_2/(T_1 + T_2)$. Experimentally it has been shown that for $0.1 < T_1/T_2 < 1.0$ and $0.2 < L/T_2 < 1.0$, one may use the controller rate or derivative setting $T_d = T_i/4$.

CONTROLLER TUNING

GENERAL REFERENCES: Cohen and Coon, "Theoretical Investigation of Retarded Control," *Trans. Am. Inst. Mech. Eng.*, **75,** 827–834 (1953). Graham and Lathrop, "The Synthesis of Optimum Transient Response," *Trans. Am. Inst. Electr. Eng.*, **72(II),** 273–288 (1953). Haalman, "Adjusting Controllers for a Deadtime Process," *Control Eng.*, **12**(7), 71–73 (1965). Laspe, "Estimating Controller Actions," *Instrum. Control Syst.*, **36**(10), 109–114 (1963). Liptak, "How to Set Process Controllers," *Chem. Eng.*, **71**(25), 129–134 (1964). Lopez, Miller, Smith, and Murrill, "Tuning Controllers with Error-Integral Criteria," *Instrum. Technol.*, **14**(11), 57–62 (1967). Miller, Lopez, Smith, and Murrill, "A Comparison of Controller Tuning Techniques," *Control Eng.*, **14**(12), 72–75 (1967). Ziegler and Nichols, "Optimum Settings for Automatic Controllers," *Trans. Am. Soc. Mech. Eng.*, **64,** 759–768 (1942); *Instrum. Soc. Am. J.*, **11**(6), 73–74 (1964); (7), 75–76 (1964); (8) 63–64 (1964).

This is an introduction to several methods which are used to design or tune proportional-integral-derivative (PID) controllers. PID controllers are widely used and are optimum for many design criteria. Many design or tuning methods do not require the controller transfer function to be of a PID form; however, limiting the choice to the PID controller usually makes the job more straightforward.

Tuning techniques can be classified into two categories, those used when the process model is unknown and those used when the process model is known. The methods in the first category require tests to be performed on the process which effectively supply the user with sufficient process-model information to proceed with controller tuning. These methods are widely used for rapid field adjustment of control loops. The second category includes a variety of analytical approaches to controller tuning.

This discussion is restricted to linear, continuous control systems. Techniques available for tuning nonlinear-response controllers, sampled-data controllers, and special-mode controllers are mainly used in process control in conjunction with computer control systems.

Controller Performance Criteria In this discussion reference is made to a number of controller design or tuning criteria. These criteria basically arise from the design technique being used and can usually be interpreted in terms of the closed-loop response of a second-order system which satisfies the criteria.

The response of the second-order system to a disturbance is completely characterized by specification of the time constants or the characteristic time T_c and the damping ratio or damping factor ξ in the transfer function in Eq. (22-11). These values can be specified as the design criteria for a control system.

Usually the characteristics of the underdamped closed-loop response $\xi < 1$ are used to describe the design criteria (see Fig. 22-38). The damping ratio is particularly useful since it specifies the amount of overshoot of the transient response. As such, it provides a convenient index by which to judge any controller criteria.

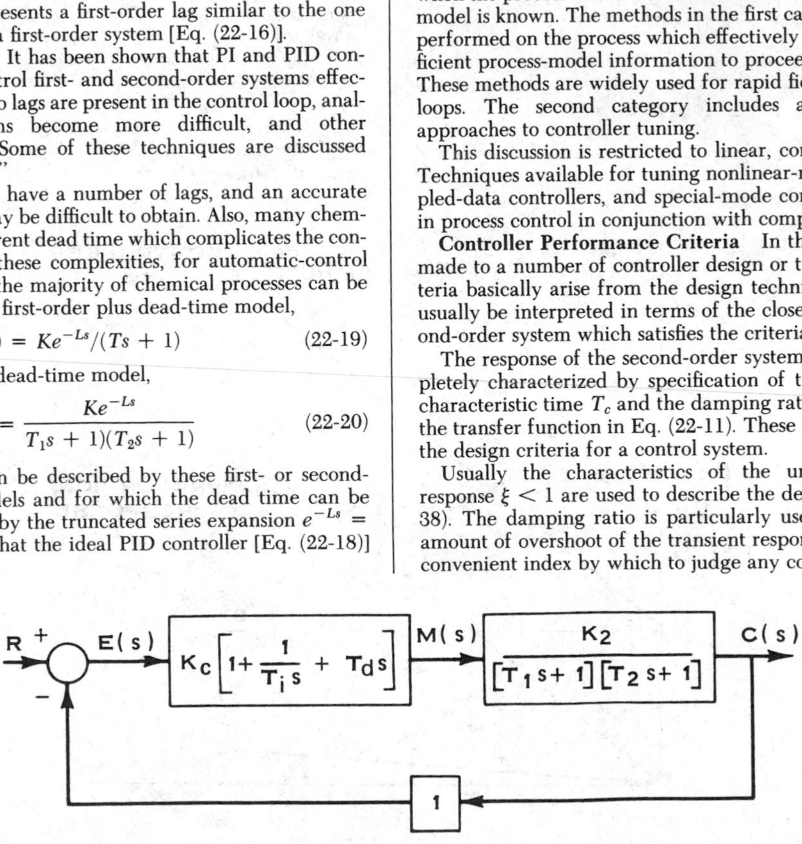

FIG. 22-37 Block diagram of an idealized PID control of a simple second-order process.

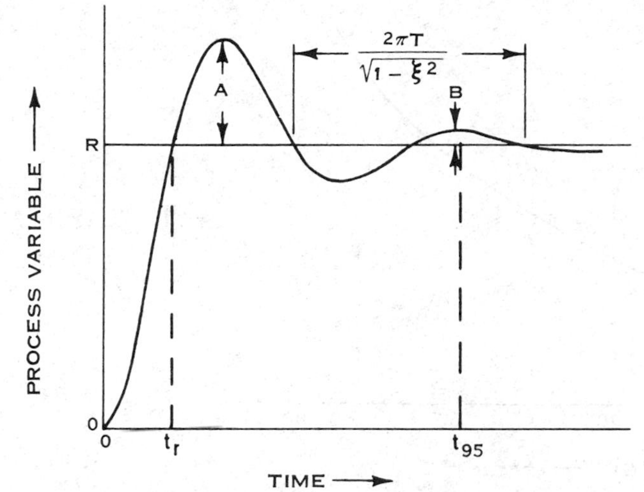

FIG. 22-38 Response of a second-order system to a step input where $\xi = 0.3$.

Overshoot Overshoot, A/R in Fig. 22-38 expresses the degree to which the initial response exceeds the final value. It is related to the damping ratio ξ by

$$\text{Overshoot} = e^{-(\pi\xi/\sqrt{1-\xi^2})}$$

for the second-order system described by Eq. (22-11).

Decay Ratio The decay ratio is a very common design criterion for process control. Often a quarter decay ratio is specified, $B/A = \frac{1}{4}$ in Fig. 22-38. The decay ratio for the second-order system described by Eq. (22-11) is given by decay ratio = (overshoot)^2. Specification of the decay ratio does not completely define the system response since the characteristic time T_c is not restricted.

Rise Time The rise time for an underdamped second-order system may be defined as the time required for the first crossing of the ultimate value t_r in Fig. 22-38. It is a function of both T_c and ξ.

Response Time This is the time required for the system to remain within 5 percent of its ultimate value following a step input t_{95} in Fig. 22-38.

Frequency of Oscillation The radian frequency of oscillation ω_r and natural frequency ω_n can be obtained from Eqs. (22-11) and (22-12).

$$\omega_r = \sqrt{1-\xi^2}/T_c \qquad \text{and} \qquad \omega_n = 1/T_c$$

The ratio ω_r/ω_n may thus be used to specify the damping ratio (see Fig. 22-39).

Phase and Gain Margin Phase and gain margins are commonly used as design criteria when Bode-plot or Nyquist-plot techniques are used. The phase margin is related to the damping ratio and thus to the decay ratio. Comparing Figs. 22-39 and 22-40 indicates that the commonly used decay ratio of 0.25, a phase margin of 30°, and a damping ratio of 0.225 are equivalent design criteria. A phase margin of 30° means that, in an open-loop control system with an oscillating manipulated variable, the manipulated variable is 150° out of phase with the output variable. This is a stability margin of 30°, since 180° out of phase gives rise to sustained undamped oscillation and the system is on the verge of instability.

The gain margin is the factor by which the controller gain may be increased before instability occurs and thus is a measure of relative stability. Gain margins of 1.7 to 3 are usually specified.

Error Integrals The time integral of the process deviation from the set point is called the error integral. Such error integrals can be expressed as functions of the damping ratio for a second-order system. The objective of using an error-integral criterion of control-system performance is to minimize the error integral and rise time by tuning the controller. Most integral criteria emphasize the effects of

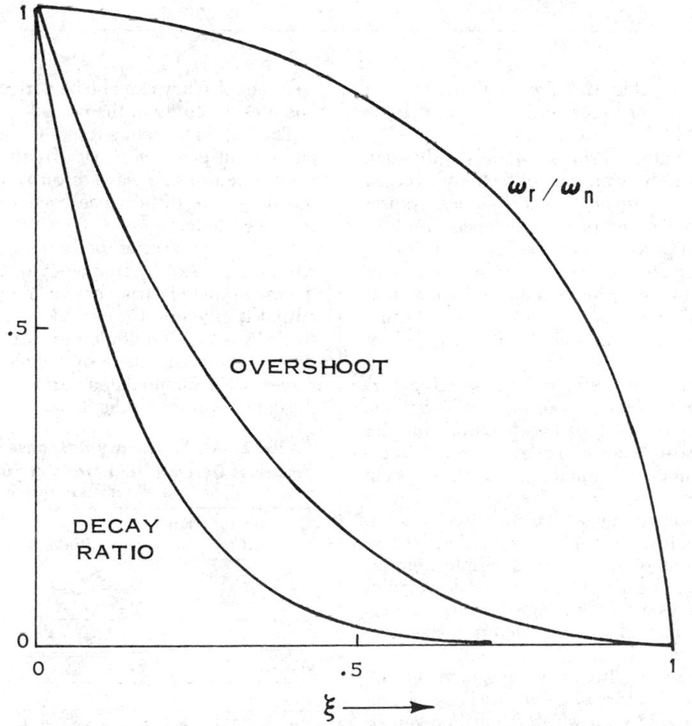

FIG. 22-39 Plots of overshoot, decay ratio, and ω_r/ω_n versus damping ratio.

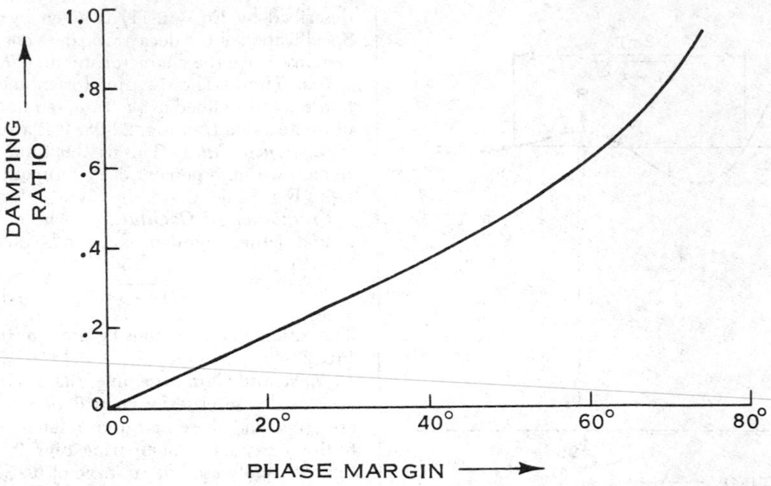

FIG. 22-40 Relationship between damping ratio and phase margin for a second-order system.

TABLE 22-3 Typical Error-Integral Criteria Compared with Other Performance Criteria of Underdamped Second-Order Control Systems

Criterion type		Damping ratio	Decay ratio	Phase margin
Integral squared error (ISE)	$\int e^2\,dt$	0.5	0.03	70 deg.
Integral of absolute error (IAE)	$\int \lvert e \rvert\,dt$	~0.65	<0.01	>90 deg.
Integral of (time) (absolute error) (ITAE)	$\int t\lvert e \rvert\,dt$	0.7	<0.01	>90 deg.
Integral of (time) (squared error) (ITSE)	$\int te^2\,dt$	0.6	0.01	>90 deg.
Mean square error (MSE)	$\lim_{T\to\infty} \frac{1}{2T} \int_{-T}^{T} e^2\,dt$	0.5	0.03	70 deg.
Quarter decay		0.225	0.25	30 deg.

offset. The comparisons made in Table 22-3 are for underdamped second-order systems only. Results for higher-order systems may be significantly different.

Tuning Methods Based on Known Process Models Although most controllers are tuned by search techniques or techniques based on simple on-site tests, the effective use of these techniques requires an understanding of the methods based on known process models. Additionally, the techniques employed when the process model is unknown may not provide adequate information for best controller tuning, and a more rigorous design may be desirable. Often an adequate model for the process to be controlled is available from testing the process, by feedback of experience, or from the development of a theoretical model.

There are several methods for determining process models experimentally. Generally, these methods require an input of a specific type and an accurate measurement of the process output for this input. Some of these methods are discussed under "Tuning Procedures When Process Model Is Unknown" and under "Control-System Analysis."

The determination of controller-tuning settings requires, in addition to the process model, a specification of the control criteria. These criteria vary and generally depend on the chosen design technique.

Usually, one is justified in assuming a first- or second-order lag plus dead-time model for a process for controller design. See Eqs. (22-19) and (22-20). The terms K, L, T, T_1, and T_2 can be determined experimentally or from a theoretical model.

Frequency-Response Bode Plots Frequency-response controller-tuning techniques are popular because of the relative ease of construction of Bode plots and the ease with which dead time can be handled. Frequency-response plots are usually the direct result of process testing and do not require forcing the system to fit an idealized model. They can also be derived from the transfer function that one uses to represent the process.

The basic frequency-response data are the process gain variation and output phase shift with frequency when the open-loop process is subjected to sine-wave disturbances. By varying the frequency of the sine-wave disturbance over a wide range, one can obtain the Bode-plot data.

For linear processes the frequency of the process output variation will be equal to the frequency of the input disturbance. That is, if the manipulated variable is given by $m = R_0 \sin \omega t$, then the process output is given by $C(s) = KG(s)M(s)$ or $c = R(\omega) \sin [\omega t - \phi(\omega)]$. $R(\omega)/R_0$ is the amplitude ratio AR, or process gain. $\phi(\omega)$, the phase shift, is the phase angle by which the process output lags the sine-wave-varied manipulated variable. The variables log AR and ϕ are the data required for the Bode plots illustrated in Table 22-4 and Fig.

TABLE 22-4 Frequency-Response Characteristics of Typical Processes Derived from Transfer Functions

ω = radian frequency of sine-wave input

Transfer function $KG(s)$	Amplitude ratio $AR(\omega)$	Phase shift $\phi(\omega)$
$\frac{K}{Ts + 1}$	$\frac{K}{\sqrt{1 + T^2\omega^2}}$	$-\tan^{-1}(\omega T)$
$\frac{K}{(T_1 s + 1)(T_2 s + 1)}$	$\frac{K}{\sqrt{(1 + T_1^2\omega^2)(1 + T_2^2\omega^2)}}$	$-\tan^{-1}(\omega T_1) - \tan^{-1}(\omega T_2)$
$\frac{K}{T_c^2 s^2 + \xi T_c s + 1}$	$\frac{K}{\sqrt{(1 - \omega^2 T_c^2)^2 + (2\xi\omega T_c)^2}}$	$-\tan^{-1} \frac{2\xi\,\omega\,T}{1 - \omega^2 T_c^2}$
Ke^{-Ls}	K	$-L\omega$

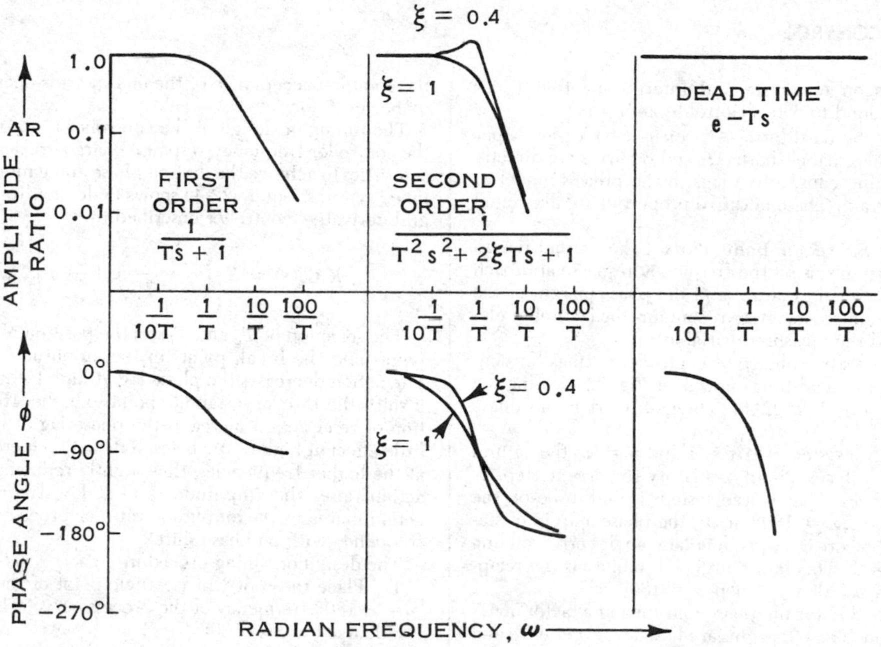

FIG. 22-41 Bode plots for typical first-order, second-order, and dead-time processes.

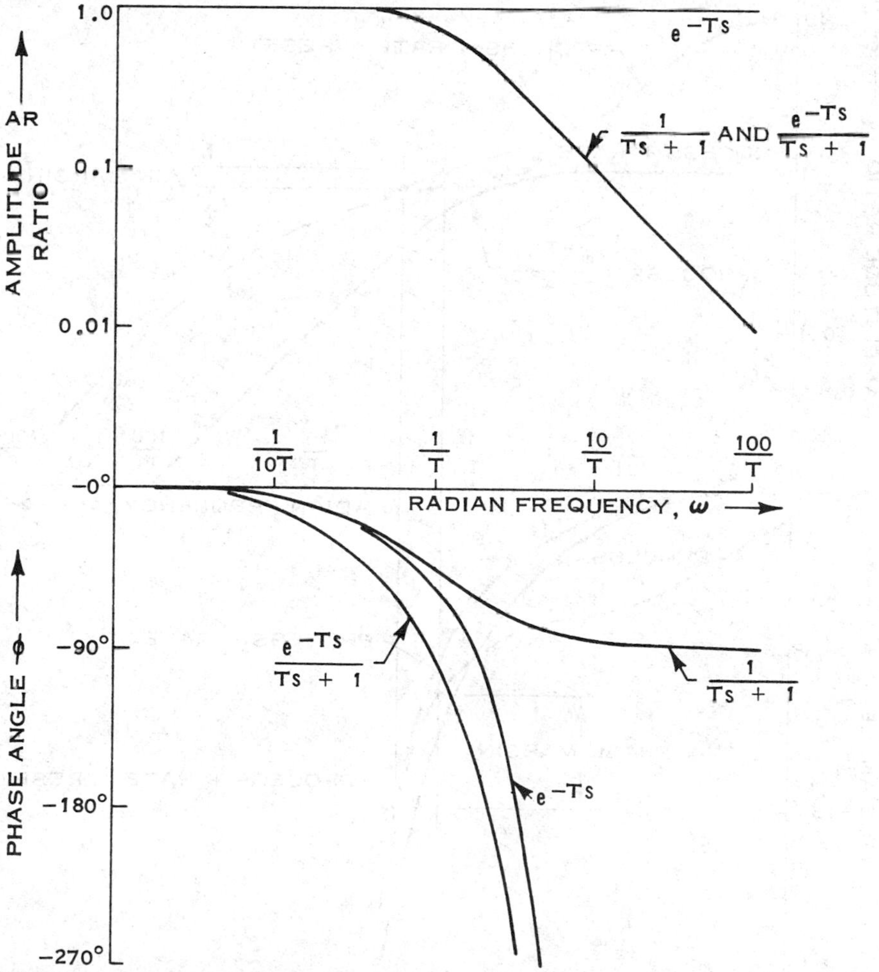

FIG. 22-42 Bode plots showing how a Bode plot for a process composed of a first-order lag and a dead time is obtained by adding the responses of the two component processes.

22-41. Further details on obtaining and interpreting frequency-response data will be found under "Control-Systems Analysis."

When plotted in the standard form, AR versus ω on log-log paper and ϕ versus ω on semilog paper, both AR and ϕ curves are directly additive when combining consecutive lags in the process model or units in the block diagram. These additive properties are illustrated in Fig. 22-42.

Controller Tuning Based on Bode Plots The normal design criterion for control systems using the frequency-response approach is the specification of the gain margin and the phase margin of the open-loop system. The open-loop system contains the controller elements, the process, and the measured variable.

Figure 22-43 represents one choice of controller settings for control of the first-order plus dead-time process of Fig. 22-42 using the controller characterized by Fig. 22-44. The gain margin and phase margin are indicated.

In a closed-loop control system if $AR = 1$ and $\omega = \omega_n$, the natural frequency, the system will sustain an oscillatory response to step- or impulse-type disturbances. That is, the system is said to be on the verge of instability. Since $\phi = 180°$ at ω_n, the phase margin of stability is $180° - \phi_{AR=1}$ where $\phi_{AR=1}$ is the phase angle corresponding to the condition $AR = 1$. The gain margin of stability is the reciprocal of AR when $\omega = \omega_n$, that is, when $\phi = 180°$.

Common criteria are 30° for the phase margin and a factor of 1.7 to 3 for the gain margin. These margins are based on experience and the common acceptance of the one-quarter-decay controller performance criterion.

The tuning or design of the controller is accomplished by adding the controller frequency-response characteristics to the system characteristics to achieve the desired phase and gain margins for the combined systems. Figure 22-44 shows Bode plots for a proportional-integral-derivative controller described by

$$K_cG_c(s) = K_c\left(1 + \frac{1}{T_is}\right)(1 + T_ds) \text{ with } K_c = 1$$

The selection of T_d and T_i sets the position of the 45° phase-angle points and the break points in the amplitude-ratio curves. Adding rate action decreases the phase lag at high frequencies; in particular it shifts the 180° or instability point to higher frequencies. The addition of reset action increases the phase lag at low frequencies with little effect at high frequencies. Rate action increases amplitude ratio at the higher frequencies, thus slightly reducing gain margin. Reset action raises the amplitude ratio at low frequencies. Proportional action increases the amplitude ratio or process gain uniformly with frequency with no phase shift.

The design or tuning procedure is:

1. Place the +50° phase-angle point of the rate curve over the $\phi = -180°$ frequency of the process curve. That is, $1/T_d$ is slightly less than ω for $\phi = 180°$.

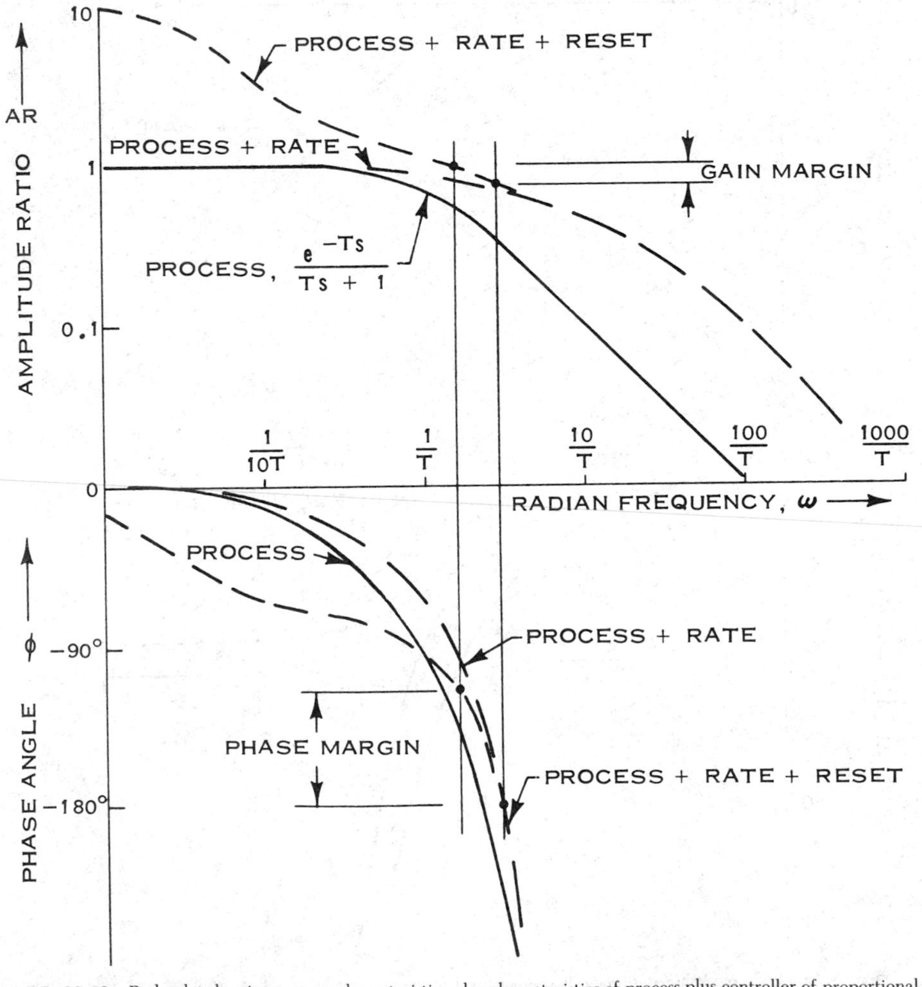

FIG. 22-43 Bode plot showing process characteristics plus characteristics of process plus controller of proportional gain $K_c = 10$. Phase margin and gain margin of a tuned controller-process system are indicated.

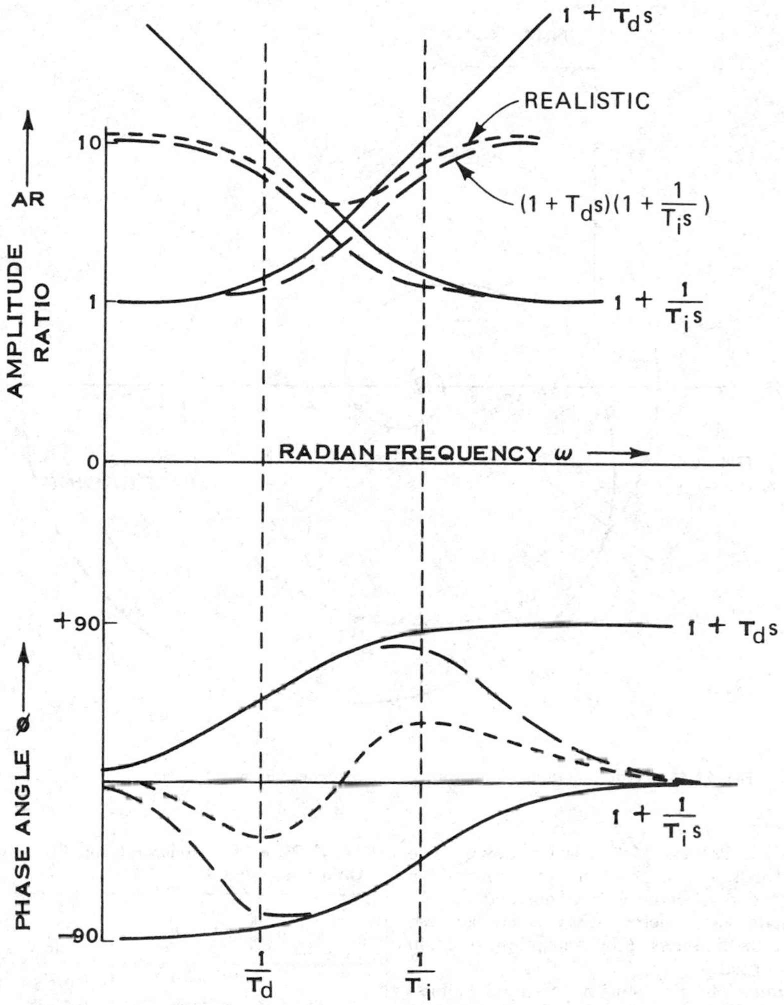

FIG. 22-44 Bode plots illustrating frequency characteristics of ideal noninteracting controllers, $K_c(1 + T_d s)$ and $K_c(1 + 1/T_i s)$, in terms of rate time and reset time for $K_c = 1$. Dashed curves (——)represent the characteristics of more realistic controllers when gains are limited to 10 at $K_c = 1$. Dashed curves (--) represent the composite PID controller.

2. Combine amplitude and phase curves for T_d controller and process.

3. Place the $-10°$ phase-angle point of the reset curve over the frequency corresponding to $\phi = -170°$ for the combined rate and process curves. This specifies $1/T_i$.

4. Combine amplitude and phase curves for T_i controller and combined process plus T_d controller.

5. Add sufficient proportional gain to slide composite process plus reset plus rate curve upward to satisfy the gain-margin criterion.

The upper curve, tuned controller plus process, of Fig. 22-43 represents one attempt at tuning by this procedure. Different gain- and phase-margin criteria will result in slightly different tunings.

An estimate of whether rate action will be useful may be obtained from the process Bode plots. If

$$2 < \frac{AR \text{ for } \phi = 0°}{AR \text{ for } \phi = -180°} < 10 \quad \text{or} \quad \frac{\omega \text{ for } \phi = -270°}{\omega \text{ for } \phi = -180°} > 5$$

rate action will probably be useful. Reset action will probably be useful in all processes where $d(AR)/dw \approx 0$ at low frequencies except when AR is very large.

Nyquist Diagram If one plots the amplitude-ratio versus phase-shift frequency-response data on polar-coordinate paper, one has the Nyquist diagram. Although the same information is available from the Nyquist diagram as from the Bode plot, it is not widely used for controller tuning. The phase and gain margins are more easily visualized in the Nyquist diagram, but the diagrams are more difficult to construct from small amounts of data and controller settings are more difficult to derive.

A typical Nyquist diagram is shown in Fig. 22-45. If $a < 1$ at $\phi = -180°$ (along the negative real axis), the process is stable; if $a = 1$, it is oscillatory; and if $a > 1$, the process is unstable. The gain and phase margins are indicated.

Root Locus The root-locus method of control-system analysis is perhaps the most educational technique available for describing the behavior of feedback control systems. An understanding of the root-locus method of analysis and the results obtainable provide the control engineer with a sound fundamental understanding of other design or tuning techniques commonly used. This technique, however, has limited usefulness in practical process control because of the difficulty of handling dead time.

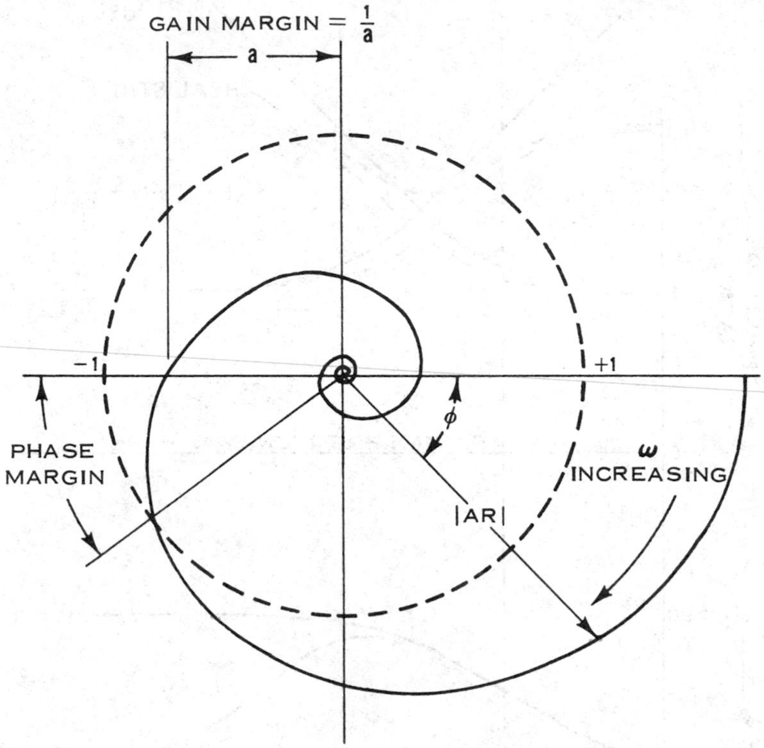

FIG. 22-45 Nyquist diagram of an overdamped low-order system with dead time.

Tuning Procedures When Process Model Is Unknown The most widely used field methods of tuning controllers incorporate the ultimate-period method, the reaction-curve method, and various search methods. The ultimate-period method uses a closed-loop test, whereas the reaction-curve method uses an open-loop test to obtain process dynamic-model information.

Ultimate-Period Method The first step in this method (also called the Ziegler-Nichols method) for tuning controllers is to obtain the required process and closed-loop control-system dynamic-response data. These data are obtained by tuning out the integral and derivative actions of the controller and using only the proportional-control mode. The proportional gain is gradually increased until the closed-loop system is forced to cycle continuously at the point of instability. A small upset such as a change in set point may be required to initiate the oscillation.

The proportional gain at the point of continuous cycling (ultimate gain) and the period of oscillation (ultimate period) identify the frequency response of the open-loop system at one point: the phase angle at the ultimate frequency (1/ultimate period) is −180°, and the open-loop gain is unity, including controller and feedback loop. Suggested controller settings are listed in Table 22-5.

Reaction-Curve Method This method of tuning is based on the open-loop response of the process to a step input. This response curve can be used to derive the dynamic characteristics of the process. Moreover, making the assumption that the process may be described by a first-order lag and a dead time, the controller settings can be calculated.

Assuming that a recorder-controller is connected to the process, to obtain the required process response data, place the controller on manual control and apply a step change ΔM to the controller signal to the control valve. Record the size of the step change and the time of application on the recorder chart. The reaction curve, the recorded response of the measured variable, should be similar to Fig. 22-46.

TABLE 22-5 Controller Settings Based on Ultimate Gain K_u and Ultimate Period T_u

Control action	Performance criterion	Gain	Integral time, min.	Derivative time, min.
P	¼ decay	$0.5\ K_u$		
PI	¼ decay	$0.45\ K_u$	$0.833\ T_u$	
PID	¼ decay	$0.6\ K_u$	$0.5\ T_u$	$0.125\ T_u$
PID	Some overshoot	$0.33\ K_u$	$0.5\ T_u$	$0.33\ T_u$
PID	No overshoot	$0.2\ K_u$	$0.33\ T_u$	$0.5\ T_u$

Referring to Fig. 22-46, determine the maximum rate of change R and the dead time L. Determine the rate of change per unit-step change in the manipulated variable, i.e., $R_1 = R/\Delta M$. Table 22-6 gives the controller tuning desired in terms of R_1 and L. These controller settings are based on the quarter-decay performance criterion.

Search Techniques The foregoing implies that controller tuning requires a detailed control-loop analysis; in practice many controllers are tuned by trial-and-error methods based on process experience. Such methods are generally successful if the individual doing the tuning understands the fundamentals of control and controller tuning or if the process does not require a high level of performance from the installed control equipment.

When controllers have been tuned by the procedures previously outlined, these tunings are based on the assumption that disturbances enter the process at one particular point. If the process characteristics have not been precisely determined, there may be a number of reasons why the more formalized techniques described do not give satisfactory control. In these cases final adjustments must be made by trial-and-error search methods. The quality of the results depends on

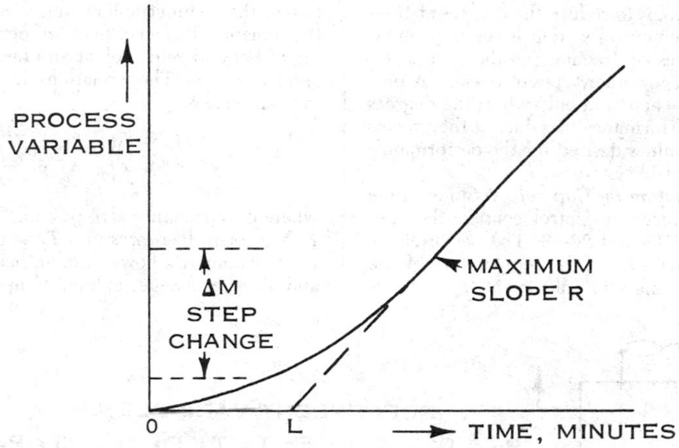

FIG. 22-46 Reaction curve of a typical controlled-process variable when the manipulated variable is changed by ΔM.

TABLE 22-6 Controller Settings Based on Reaction-Curve Process Test

Control action	Gain	Integral time, min.	Derivative time, min.
P	$\frac{1}{R_1 L}$		
PI	$\frac{0.9}{R_1 L}$	$3\,L$	
PID	$\frac{1.2}{R_1 L}$ to $\frac{2}{R_1 L}$	$2.5\,L$ to $2\,L$	$0.5\,L$ to $0.3\,L$

the degree of process knowledge available as well as on knowledge of control.

Formalized search techniques are used in computerized adaptive control systems. In these systems the effectiveness of the computerized controller settings are continuously evaluated quantitatively, and process tests are automatically conducted to determine when new settings are needed and what they should be. The applications for which these procedures are necessary are quite specialized and are beyond the scope of this *Handbook*.

Analytical Methods Analytical techniques for design or tuning of controllers are beyond the scope of this *Handbook* and are not widely used in process control. They are, however, instructive for the chemical and control engineer interested in the more difficult control problems.

With the increasing use of computers for process analysis, analytical methods may become more popular, particularly for nonlinear systems. Analytical techniques generally involve two areas: direct solution of the system differential equations in the time domain, usually by state variables, and optimization of a specific performance criterion. The criteria for optimization by analytical techniques usually involve minimum response time or an integral time cost function.

Analog-Computer Techniques The use of the analog computer for simulation has provided an effective method for designing control systems. By simulating both process system and controller on the analog computer, controller settings can be chosen by any method which might be applied to the plant. The reliability of the settings when applied to the existing process depend on the accuracy of the model used for simulation of the process. Analog-computer simulation is discussed in some detail under "Control-Systems Analysis."

Tuning of On-Off Control Systems The tuning of on-off control systems is a somewhat special case. In practice many on-off controllers do not possess adequate adjustment facilities for tuning the dead band or hysteresis loop width. Therefore, tuning consists of adjusting the process as well as the controller or, perhaps, instead of the controller.

Reducing the dead band reduces the amplitude and increases the frequency of oscillation when the controller dead band is not small compared with the apparent dead band imposed by the process and measurement lags (see Fig. 22-26). Increased frequency of oscillation increases wear on control equipment.

Reduction of offset is accomplished by increasing frequency of oscillation or by adjusting the duty cycle toward 50 percent. The latter adjustment must generally be made by the adjustment of process equipment. Such an adjustment might consist of changing the on-off difference in the manipulated variable or changing the fraction of the manipulated variable under control of the controller by bypassing the control equipment with a constant amount of the manipulated variable.

ADDITIONAL TOPICS ON CONTROL

General References: Cadman, Rothfus, and Kermode, "Linear Feedforward Control of a Continuous Flow Stirred Tank Reactor," 59th Annual Am. Inst. Chem. Eng. Meeting, Detroit, 1966. Johnson and Lupfer, "Distillation Column Models," *Chem. Eng. Prog.*, **62**(6), 75–79 (1966). Lupfer and Oglesby, "Applying Dead-Time Compensation for Linear Predictor Process Control, *Instrum. Soc. Am. J.*, **8**(11), 53–57 (1961). Luyben, "Distillation Decoupling," 64th National Am. Inst. Chem. Eng. Meeting, New Orleans, 1969. Tyner and May, *Process Engineering Control*, Ronald, New York, 1968. Webb, "Reducing Process Disturbances with Cascade Control," *Control Eng.*, **8**(8), 73–76 (1961).

Although the great majority of control problems can be solved by the application of a conventional feedback controller, the performance of many systems can be improved by special techniques. Some of these techniques have become sufficiently common to be discussed here.

Degrees of Freedom and Control

Degrees of Freedom and Purpose of Control In the design of a control system, one should use methods which define the result desired and which provide some indication of when this is achieved. The concept of degrees of freedom provides such a tool.

A degree of freedom is accounted for by any independent variable entering the process which, if changed, will in some way affect one or more of the output or performance variables. The state or condition of a process is completely described when each of its degrees of freedom is specified or controlled.

The purpose of automatic control is to reduce the degrees of freedom of the process. An automatic control system is the implementation of the reduction of degrees of freedom of the process by mechanical, pneumatic, electronic, or other types of devices. A process is on fully automatic control when, and only when, the degrees of freedom with respect to the performance variables of the process are identically zero. That is, the values desired for the performance variables are set points for the control system.

Reduction in Degrees of Freedom by Control As an example of how degrees of freedom are reduced by control, consider the coolant system illustrated in Figs. 22-47 and 22-48. The performance variables of interest are the mass flow and the temperature of the coolant flowing into the jacket. The mass flow of coolant M_c is a function of the volumetric flow and variables which affect the density of the coolant. The jacket inlet temperature T_i is a function of the mixing of hot and cold coolant and heat losses between point of mixing and jacket inlet. These relationships may be illustrated by the following expressions:

$$M_c = f_1(\phi, P_b, P_f, T_f)$$
$$T_i = f_2(\phi, P_b, P_f, T_f, F_1, T_1, F_2, T_2, A)$$

where ϕ = coolant-valve position, P_b = back pressure from jacket, P_f = coolant feed pressure, T_f = coolant feed temperature, F_1 and F_2 = volumetric flows through heating and cooling exchangers, T_1 and T_2 = exchanger effluent temperatures, and A = ambient tem-

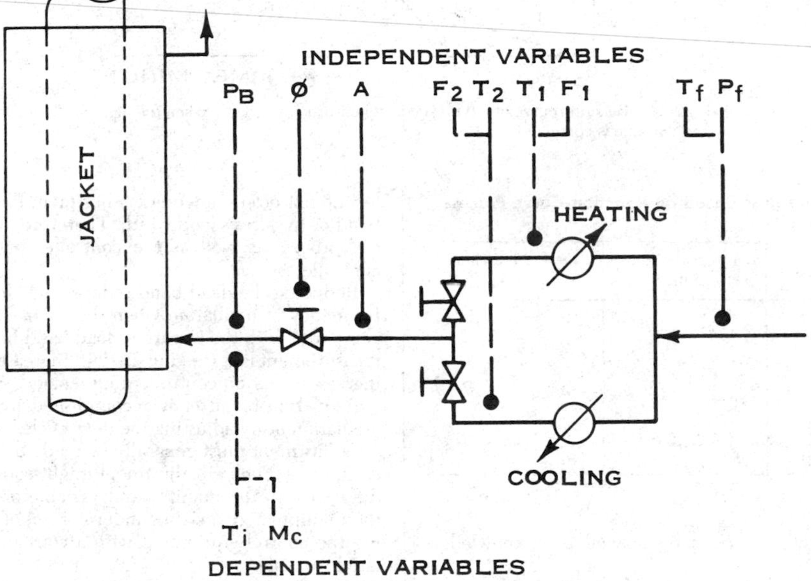

FIG. 22-47 Uncontrolled coolant system showing the performance of dependent variables and of independent variables or degrees of freedom.

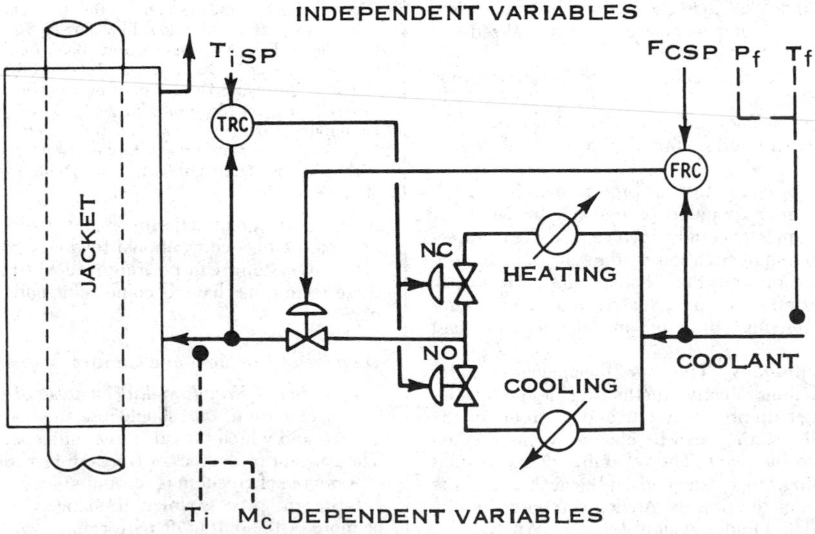

FIG. 22-48 Controlled coolant system showing the independent variables or degrees of freedom remaining after the addition of controls.

perature. There are nine degrees of freedom, or independent variables.

Control of coolant volumetric flow is illustrated in Fig. 22-48 by controller *FRC*. The mass flow is now given by

$$M_c = f_3(F_{cSP}, P_f, T_f)$$

where F_{cSP} = coolant volumetric-flow-rate set point. Since $F_1 = F_{cSP} - F_2$, the energy balance becomes

$$T_i = f_4(F_{cSP}, P_f, T_f, F_2, T_1, T_2, A)$$

There are still a total of seven degrees of freedom. However, now one, F_{cSP}, is under control, that is, manipulated or specified by the operator.

The addition of the temperature controller *TRC* (Fig. 22-48) further reduces the degrees of freedom. It regulates the split of the coolant flow through the two heat exchangers by means of two range valves, one direct-acting and the other reverse-acting. The performance variables are now

$$M_c = f_5(F_{cSP}, P_f, T_f)$$
$$F_i = f_6(T_{iSP})$$

There are now four degrees of freedom, of which two are independent variables P_f and T_f and two are manipulated or specified by the operator, F_{CSP} and T_{iSP}.

The reasons for picking these particular performance variables will be clarified later in the discussion associated with Fig. 22-191.

Dependent upon the design of the rest of the processing system, P_f and T_f may or may not be independent variables. One or both of them may be a fixed operating condition. One must be careful in determining that a particular variable is a fixed operating condition. There are numerous examples of control systems which operated unsatisfactorily because a so-called fixed operating condition was not, in reality, fixed.

Cascade Control An often-used method for minimizing disturbances entering a slow process is cascade, or multiloop, control. Cascade control can also speed the response of the control system by reducing the time constant of the process transfer function relating the manipulated variable and the process output. A block diagram of a cascade control system is shown in Fig. 22-49. Instead of adjusting a final control element, such as a control valve, the output of the primary controller is made the set point of the secondary control loop.

The secondary control loop encompassing only a portion of the total process is a lower-ordered system so that the controller can be tuned to give faster response. Consider Fig. 22-50. A flow controller has been cascaded with the temperature controller of Fig. 22-2. Time constants in the secondary loop are much shorter than for the total process, so good tuning of the secondary controller effectively eliminates or minimizes the flow disturbances which enter the process through the steam supply.

Elimination of one source of disturbances reduces the degrees of freedom of the process and also the order and time constants of the process. These reductions in degrees of freedom, order, and time

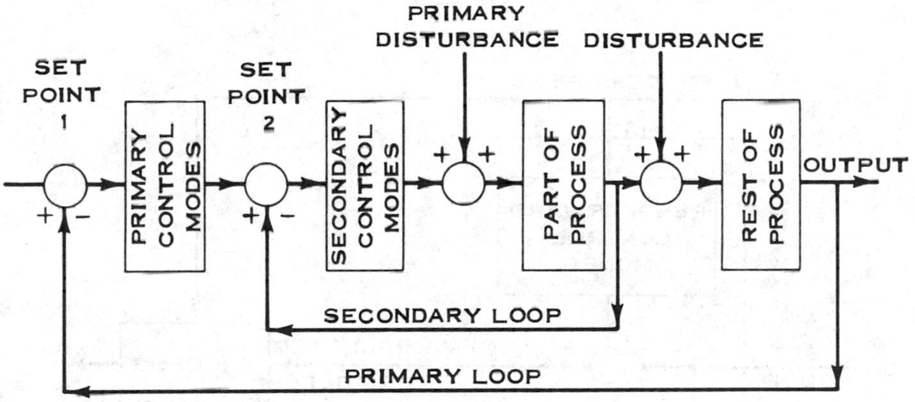

FIG. 22-49 Block diagram of cascaded feedback controllers.

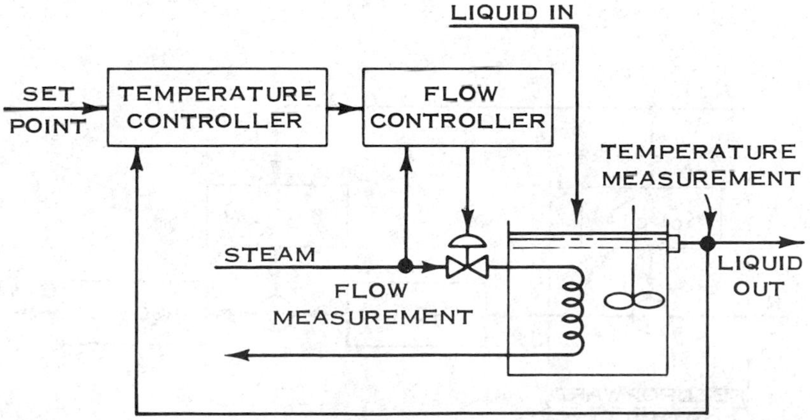

FIG. 22-50 Cascade control system in which disturbances originating in the steam supply are prevented from entering the heat-exchanger process.

constants increase the speed of response possible in the primary control loop. They also reduce the size of variations in the controlled variable beyond that which would be achievable by the increased speed of response of the primary control system.

Three major characteristics are usually present if cascade control is to be effective: the closed-loop time constant of the secondary loop should be less than one-third of the time constant of the primary loop, the secondary loop should include a source of a major process disturbance, and the process variable being manipulated must be able to move the primary controlled variable to its desired value.

Cascade control of a slightly different nature is the technique of using computer control systems to reduce the manual process-operating adjustments to those which represent setting process performance or quality parameters.

Feedforward Control Feedforward control (sometimes called predictive or anticipatory control) basically involves a multiple-input process in which the inputs can be measured. With reference to Fig. 22-3, feedforward control makes use of input measurements and relationships between inputs and outputs to adjust the process in order to minimize or eliminate the effects of input disturbances on the process output.

In Fig. 22-51 the process model in the feedforward control modes has dynamic characteristics very different from those represented by the process block. The dynamics that are of primary interest are those associated with the feed measurements and the manipulated variable. Frequently these dynamics may be considered as instantaneous, and the control modes may be very simple, only proportional-mode.

In the generalized block diagram in Fig. 22-52, the relationship between the output variable $C(s)$ and the disturbance variable $U(s)$ is given by

$$[C(s)/U(s)] = K_2G_2(s)[K_3G_3(s) + K_cG_c(s)K_1G_1(s)] \quad (22\text{-}21)$$

The purpose of the feedforward control is to make the process output dependent only on $R(s)$, the set point. That is,

$$[C(s)/R(s)] = K_1G_1(s)K_2G_2(s)$$

Equation (22-21) shows that the disturbance has no effect if

$$K_cG_c(s)K_1G_1(s) + K_3G_3(s) = 0$$

or the controller is described by

$$K_cG_c(s) = [K_3G_3(s)/K_1G_1(s)] \quad (22\text{-}22)$$

To illustrate with a simple example, if in the process $K_3G_3(s) = K_3$ and $K_1G_1(s) = K_1/(T_1s + 1)$, then from Eq. (22-22) the feedforward controller will be described by

$$K_cG_c(s) = -(K_3/K_1)(T_1s + 1)$$

which is the form of a proportional plus rate controller.

Combined Feedforward-Feedback Systems As was mentioned in conjunction with Fig. 22-4, feedforward control can rarely fulfill all the control requirements, so that feedback control is normally used in combination with it. Such an arrangement, illustrated in Fig. 22-53, reduces the accuracy and amount of process-knowledge detail required for specification of $K_cG_c(s)$.

A rather specialized application for combined feedforward-feedback controllers illustrates some of the potential of such systems. In a two-input, two-output process (Fig. 22-54) the outputs can be

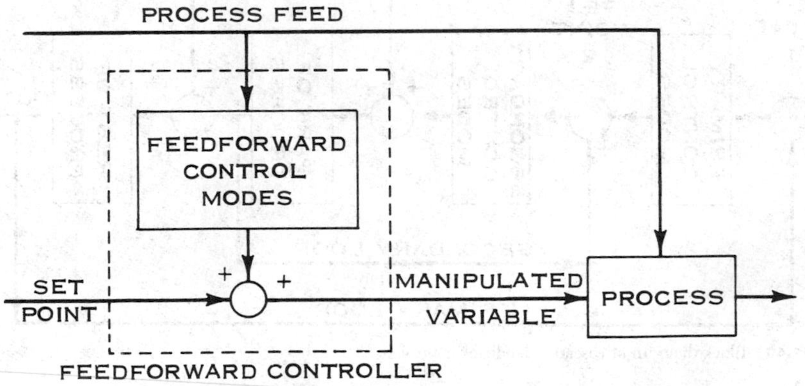

FIG. 22-51 Block diagram of a feedforward control system.

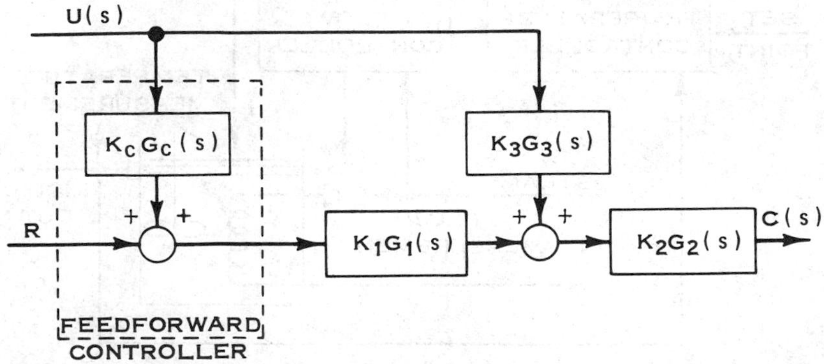

FIG. 22-52 Generalized block diagram of a feedforward control system.

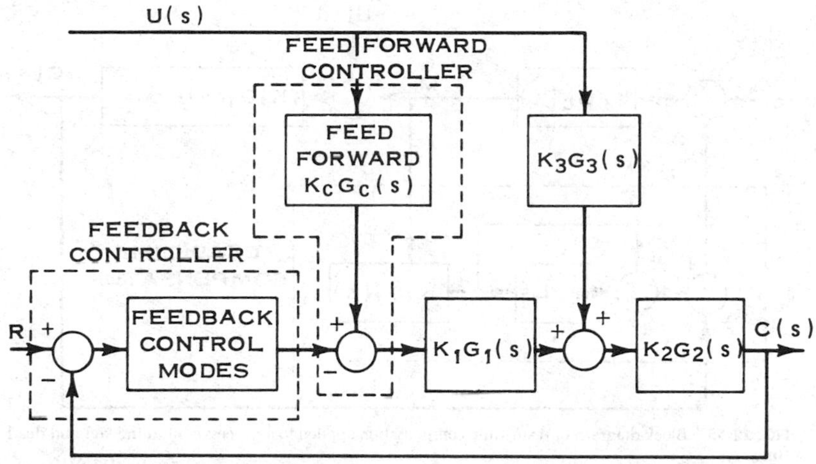

FIG. 22-53 Generalized block diagram of a combined feedforward-feedback control system.

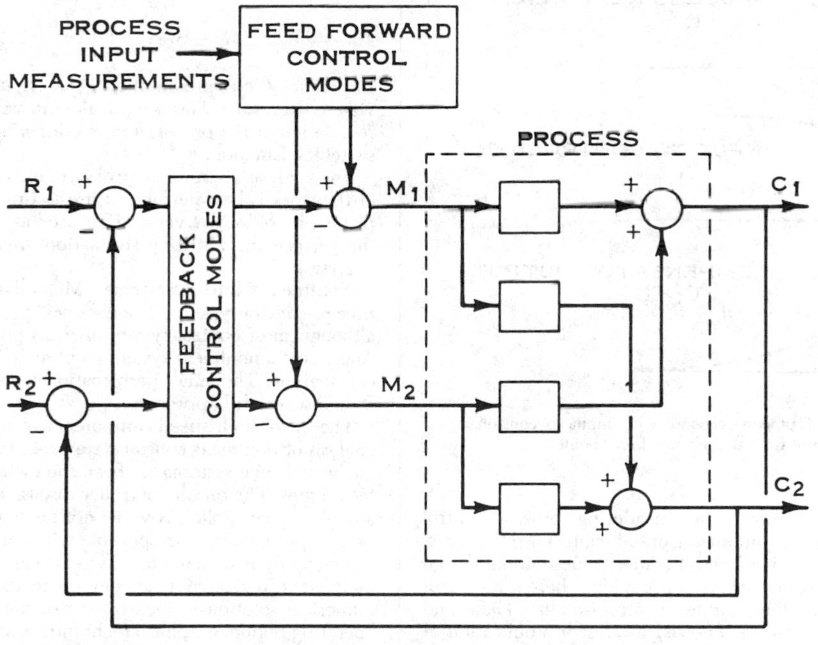

FIG. 22-54 Block diagram of a combined feedforward-feedback control system.

decoupled so that control action taken as a result of a measurement on one output variable will affect only that one output variable. Determination of the character of the feedback and feedforward controllers is generally a matrix-algebra job, and a computer is normally required to implement the controllers; see subsection "Fractionation System."

Dead-Time Compensation When a conventional feedback controller is used on a process which has an appreciable dead time, the control system is a slowly responding one. Controller tunings are characterized by small proportional gains and small reset times. An effective technique for speeding controller response is illustrated in Fig. 22-55. The objective of the compensation is to permit tuning the controller as if no dead time were present. The models of the process, a low-order lag and dead time, may be difficult to implement if the control system is not computerized.

The manipulated variable acts on the $K_1G_1(s)$ portion of the model and provides a low-order lag signal which is added to the process output feedback signal and compared immediately with the set point. This closed-loop control signal has no dead time.. As the model $K_1G_1(s)$ output is acted upon by $e^{-}{}_{Ls}$ the signal added to the process output feedback signal starts toward zero at the time when the dead time has elapsed. The compensator signal becomes zero by the time that the feedback process signal C reaches its full response and C_1 becomes equal to C as shown in Fig. 22-56. That is,

$$C_1(s) = \overbrace{K_1G_1(s)e^{-Ls}M(s)}^{\text{process}} + \overbrace{K_1G_1(s)(1 - e^{-Ls})M(s)}^{\text{model}}$$
$$= K_1G_1(s)M(s)$$

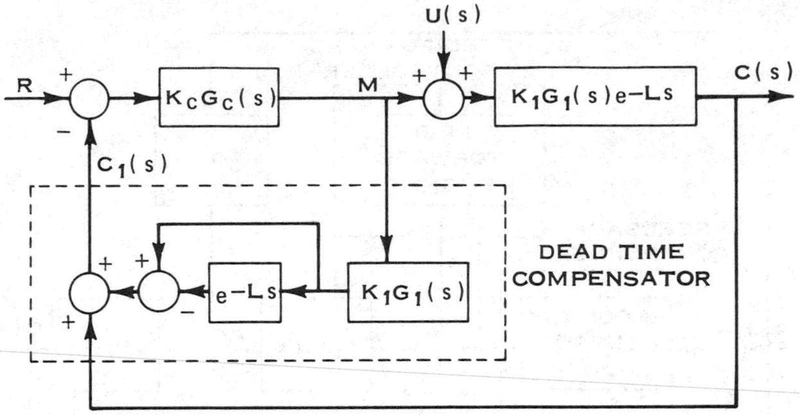

FIG. 22-55 Block diagram of dead-time compensation applied to a process containing lags and dead time.

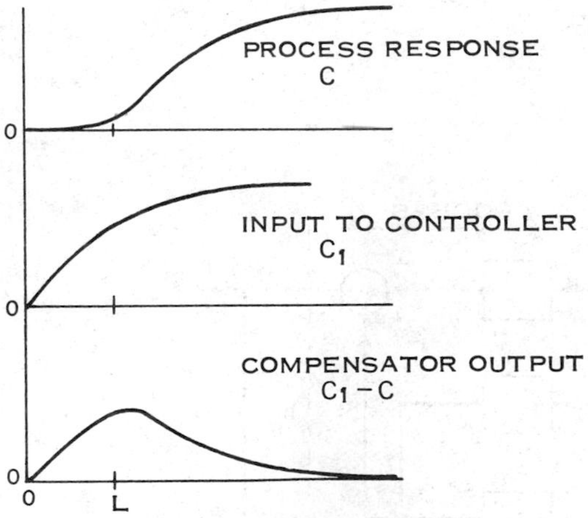

FIG. 22-56 Comparison of process response with input to controller and dead-time compensator output for a step change in set point.

On-Off (Bang-Bang) Control In considering process control when disturbances are being minimized, on-off control is rather drastic control action. On the other hand, for most time-optimal design criteria, the on-off controller is optimum provided there are no process constraints which prohibit sudden control actions. These and other considerations have led to renewed interest in application of the on-off controller to process control.

The optimum switching function for an n-order control system includes $n - 1$ derivations of E. That is,

$$M = M_0 \operatorname{sgn} F\left(E, \frac{dE}{dt}, \ldots, \frac{d^{n-1}E}{dt^{n-1}}\right)$$

In practice the approximation $F(E, dE/dt) = E + T(dE/dt)$ provides good results. This is equivalent to assuming a second-order representation of the process and making a linear approximation of the switching function.

The analysis of on-off control systems is not widely practiced. Most of the work is best performed in the time domain where the phase plane, plot of dE/dt versus E, is used as a visual aid in interpreting the performance. Analog simulation greatly simplifies this type of analysis.

Nonlinear Control Systems Many nonlinear elements appear in process-control systems. The chemical process is normally nonlinear, although most control-system analysis procedures assume linearity. Analysis of a nonlinear system is considerably more difficult than linear analysis. The small-perturbation technique of linearization has been a successful approach.

The use of high-speed computers has made possible more detailed analysis of nonlinear control systems. In particular, the time-domain analysis of such systems has been somewhat simplified by simulation techniques. The on-off controller discussed earlier is one example of a nonlinear controller. A variety of nonlinear actions which improve process performance are possible with computer controllers.

The analysis of nonlinear systems can be approached by several methods: direct analytical solution in the time domain (including numerical solutions), linearization of nonlinear terms over a small operating region, graphical techniques such as phase-plane plots, and simulation.

PROCESS MEASUREMENTS

General References: Benedict, *Fundamentals of Temperature, Pressure, and Flow Measurements*, Wiley, New York, 1969. Considine, *Process Instruments and Control Handbook*, McGraw-Hill, New York, 1957. Considine and Ross, *Handbook of Applied Instrumentation*, McGraw-Hill, New York, 1964. *ISA Transducer Compendium*, 2d ed., Plenum, New York, 1969. Liptak, *Instrument Engineers Handbook*, Chilton, Philadelphia, 1969. "Process Instrument Elements," *Chem. Eng.*, **76**(11), 137–164 (June 2, 1969).

TEMPERATURE MEASUREMENTS

General References: Benedict, "International Practical Temperature Scale of 1968," *Leeds Northrup Tech. J.*, no. 6, 1–12 (spring, 1969); *Instrum. Control Syst.*, **42**, 85–89 (October 1969).

International Temperature Scale To provide a consistent basis for precise and convenient temperature measurements the International Practical Temperature Scale has been established. This scale covers the range from the triple point of hydrogen to incandescent bodies and flames. The scale is based on reproducible equilibrium temperatures or fixed points to which numerical values have been assigned and on specified interpolation formulas which relate temperatures between or above these points to the indications of standard temperature-measuring instruments. The nature of these formulas and standard instruments is indicated in Table 22-7.

Thermocouples Temperature measurements using thermocouples are based on the discovery by Seebeck in 1821 that an electric current flows in a continuous circuit of two different metallic wires

TABLE 22-7 Basis for the International Practical Temperature Scale of 1968

Temperature range, °C.	Interpolation equation	Standard instrument
−259.34 to zero	A 20th-order reference function with the range broken into four parts for interpolation	Platinum resistance thermometer
Zero to 630.74	$R_t = R_0[1 + A(t - \Delta t) - B(t - \Delta t)^2]$ where Δt is a fourth-order function, R_t = resistance of thermometer resistor at temperature t, R_0 = resistance at 0°C., and A and B are constants	
630.74 to 1064.42	$E = a + bt + ct^2$ where E = e.m.f. of standard thermocouple and a, b, and c are constants*	Platinum vs. platinum–10% rhodium thermocouple
1064.43 and above	$\frac{J_T}{J_{Au}} = \frac{e^{C_2/\lambda T_{Au}} - 1}{e^{C_2/\lambda T} - 1}$ where J_T and J_{Au} are the radiant energies per unit wave-length interval at wave length λ emitted per unit time by a unit area of a black body at the temperature T, °K., and at the gold point T_{Au}, respectively; C_2 = 1.4388 cm. deg.	Optical pyrometer

*The temperature of the antimony point (630.74°K.) to be determined with a standard resistance thermometer.

if the two junctions are at different temperatures. The thermocouple may be represented diagrammatically as shown in Fig. 22-57. A and B are the two metals, and T_1 and T_2 are the temperatures of the junctions. If T_1 is the colder junction and the thermoelectric current i flows in the direction indicated in Fig. 22-57*a*, metal A is customarily referred to as thermoelectrically positive to metal B.

In electric circuits, the current is dependent on the electromotive force (emf) developed and the resistance of the circuit. For accurate temperature measurements, the measuring instrument is constructed so that a no-current emf is measured to eliminate the effects of circuit resistance.

The thermal emf, as indicated in Fig. 22-57*b*, is a measure of the difference in temperature between T_2 and T_1. In control systems the reference junction (cold junction) is usually located at the emf-measuring device. The reference junction may be held at constant temperature such as in an ice bath or a thermostated oven, or it may be at ambient temperature but electrically compensated so that it appears to be held at a constant temperature.

Series combinations of several identical thermocouples can be used to provide a larger temperature-measurement signal or to average the temperature at several locations. The cold junctions, as shown in Fig. 22-58, must all be at the same temperature, or there will also be an averaging of their temperatures.

Intermediate Metals and Temperatures In Fig. 22-57*a* variations of temperature along A or B do not affect the thermal emf generated by the temperature difference $T_2 - T_1$, provided such variations do not modify the junction temperatures and provided the wires A and B are homogeneous throughout their length.

When the measuring junction is located a long distance from the reference junction and measuring device, it may be more economical to use special *extension wires* as indicated by C and D in Fig. 22-59. The emf is unchanged when $T_3 = T_4 = T_5 = T_6$ even though C and D are metals different from A and B. However, in many pro-

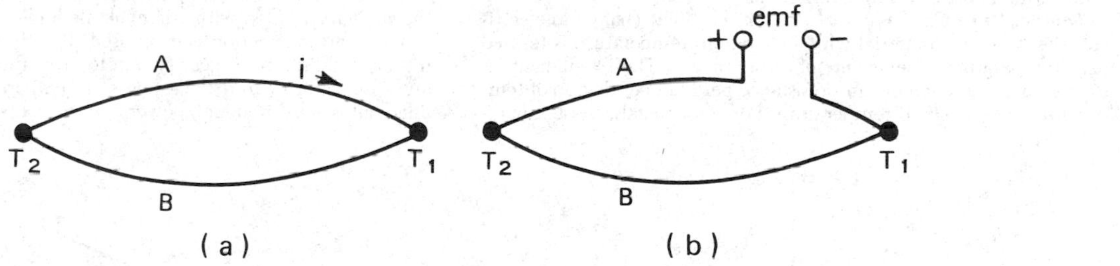

FIG. 22-57 Simple thermocouple circuit.

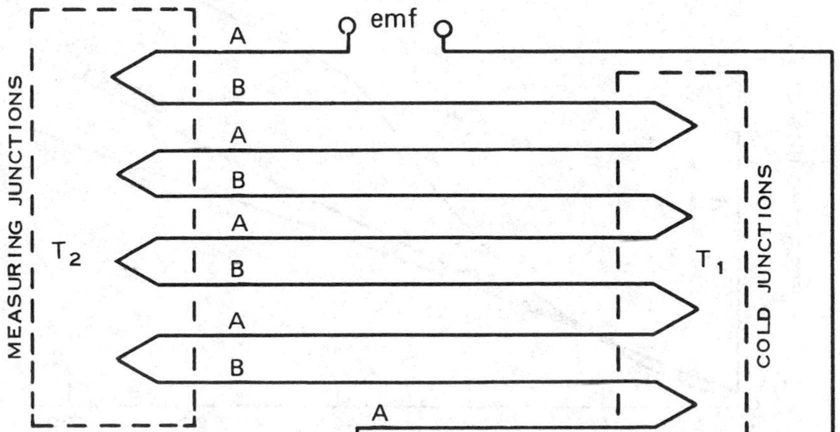

FIG. 22-58 Use of multiple thermocouples to increase the emf signal for small-temperature-differential measurements or to obtain average temperature.

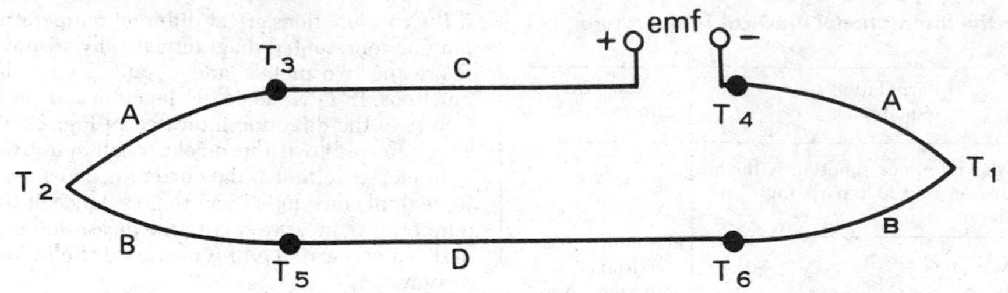

FIG. 22-59 Thermoelectric circuit with additional conductors C and D and junction temperatures T_3, T_4, T_5, and T_6 added to the simple circuit of Fig. 22-57.

cess applications not all these temperatures are equal. Since $T_3 - T_4$ and $T_5 - T_6$ are usually smaller than $T_2 - T_1$, the extension-wire combination C and D need only have the same thermoelectric properties as the combination A and B over the smaller temperature range encountered by T_3, T_4, T_5, and T_6 in order to maintain the accuracy of the measurement.

Resistance Thermometers The resistance thermometer depends upon the inherent characteristics of metals to change in electrical resistance when they undergo a change in temperature. Although industrial resistance thermometers are usually constructed of platinum, copper, or nickel, semiconducting materials such as thermistors are finding increased use.

Basically, a resistance thermometer is an instrument for measuring electrical resistance that is calibrated in units of temperature instead of in units of resistance. Several common forms of bridge circuits are employed in industrial resistance thermometry, the most common being the Wheatstone bridge. Such bridges may be excited with either direct- or alternating-current unbalance and may be indicated by null-balance-type or deflection-type instruments; see subsection "Indicating and Recording Instruments."

Temperature Coefficient of Resistance The change in electrical resistance of a material with a change in temperature is termed its "temperature coefficient of resistance." The coefficient is expressed as a change in ohms resistance per ohm per degree of temperature at a specified temperature. For most metals, the temperature coefficient is positive. For many pure metals, the coefficient is essentially constant over large portions of their useful range.

Typical resistance versus temperature curves for platinum, copper, and nickel are given in Fig. 22-60. Detailed resistance versus temperature tables are available from the National Bureau of Standards and suppliers of resistance thermometers.

Filled-System Thermometers The filled-system thermometer is designed to provide an indication of temperature some distance removed from the point of measurement as illustrated in Fig. 22-61. The sensitive or measuring element (bulb) contains a gas or liquid which changes in volume, pressure, or vapor pressure with temperature. This change is communicated through a capillary tube to the Bourdon tube or other pressure- or volume-sensitive device. The Bourdon tube responds so as to provide a motion related in a definite way to the bulb temperature. Those systems which respond to volume changes are completely filled with a liquid. Systems that respond to pressure changes either are filled with a gas or are partially filled with a volatile liquid. Changes in gas or vapor pressure with changes in bulb temperatures are carried through the capillary to the Bourdon. The latter bulbs are sometimes constructed so that the capillary is filled with a nonvolatile liquid.

The Bourdon motion can be used directly or be amplified by a mechanical linkage or gear system to drive the pointer of a temperature indicator or to drive a pen of a temperature recorder. Fluid-filled bulbs deliver enough power to drive controller mechanisms and

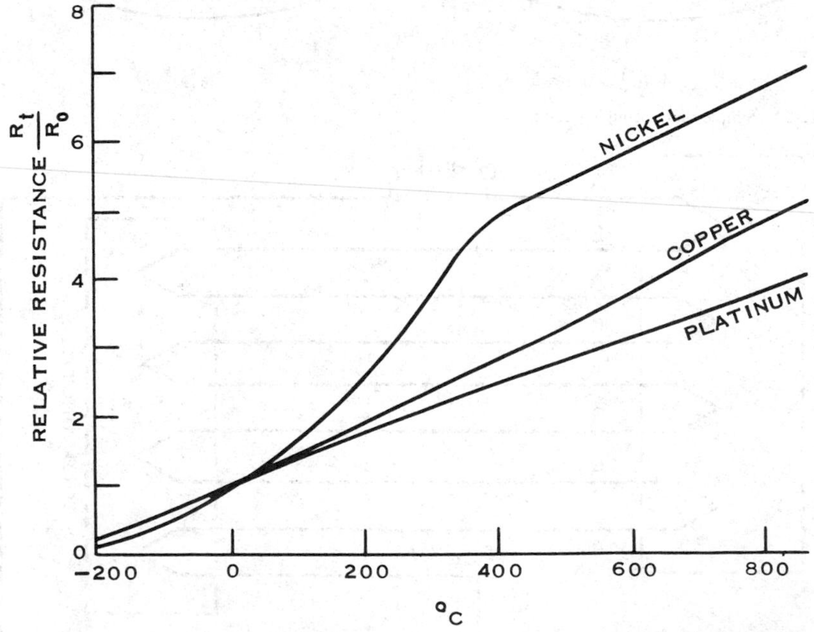

FIG. 22-60 Typical resistance-thermometer curves for platinum, copper, and nickel wire, where R_t = resistance at temperature t and R_0 = resistance at 0°C.

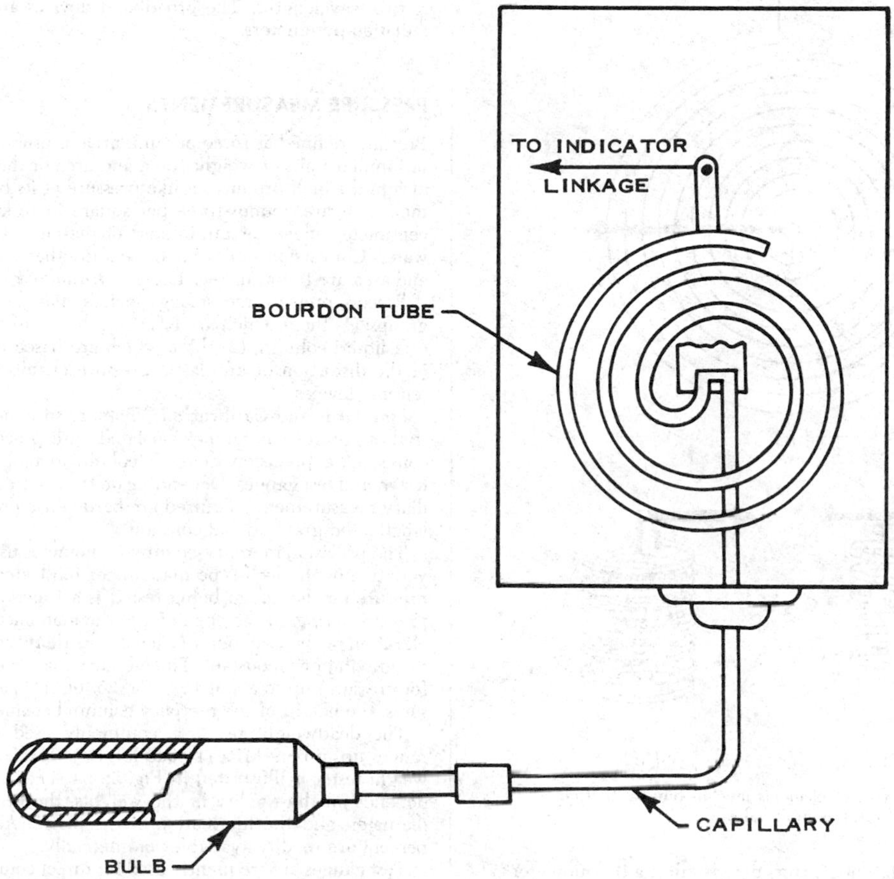

FIG. 22-61 Filled-system thermometer.

even directly actuate control valves. These devices are characterized by large thermal capacity, which sometimes leads to slow response, particularly when they are enclosed in a thermal well for process measurements.

Application Filled-system thermometers are used extensively in industrial processes for a number of reasons. The simplicity of these devices allows rugged construction, minimizing the possibility of damage failure with a generally minor amount of upkeep. In case of system failure, the entire unit must be replaced or repaired. Its simplicity may allow inexpensive overall design of control equipment.

As normally used in the process industries, the sensitivity and percentage of span accuracy of these thermometers are generally the equal of those of other quality industrial-type temperature-measuring instruments. Sensitivity and absolute accuracy are not the equal of those of short-span electrical instruments used in connection with resistance-thermometer bulbs. Also the maximum temperature is more limited than in some electrical measuring devices.

The capillary allows considerable separation between the point of measurement and the point of temperature indication but is usually limited to 76 m (250 ft). It is frequently more economical to employ transmitters for signal transmission beyond 30 m (100 ft).

These devices are self-contained and need no auxiliary power supply such as compressed air or electricity unless a signal-transmission system is desired.

Bimetal Thermometers Thermostatic bimetal can be defined as a composite material made up of strips of two or more metals fastened together. This composite, because of the different expansion rates of its components, tends to change curvature when subjected to a change in temperature.

With one end of a straight strip fixed, the other end deflects in proportion to the temperature change and the square of the length, and inversely as the thickness, throughout the linear portion of the deflection characteristic curve. If a bimetallic strip is wound into a helix or a spiral and one end is fixed, the other end will rotate when heat is applied. For a thermometer with uniform scale divisions, a bimetal must be designed to have linear deflection over the desired temperature range.

Types of Elements The three types of elements most commonly used in thermometers, the flat spiral, the single helix, and the multiple helix, are shown in Fig. 22-62.

Bimetal thermometers are made for use at temperatures ranging from 583°C (1000°F) down to −184°C (−300°F) and lower. However, at the low temperatures the rate of deflection drops off quite rapidly. Bimetal thermometers do not have long-time stability at temperatures above 427°C (800°F).

Both bimetal strips and spirals form the sensing basis for a wide variety of temperature switches and thermostatic controls. Differential expansion between two metals not held together along their length forms the basis for another group of temperature switches and thermostatic controls.

Liquid-in-Glass Thermometers The three forms of liquid-in-glass thermometers are (1) all glass (etched stem or enclosed scale), (2) tube and scale, and (3) industrial. These three classes are summarized in Table 22-8. While liquid-in-glass thermometers are not used in automatic process-control systems, they are widely used as measuring devices for manual control and in control laboratories. In using the common mercury-in-glass thermometer care should be exercised that the condition of usage corresponds to the calibration

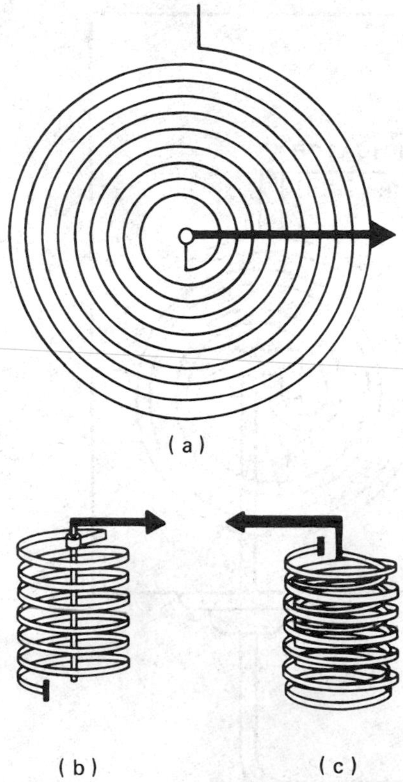

FIG. 22-62 Principal types of elements used in bimetal thermometers. (*a*) Flat spiral. (*b*) Single helix. (*c*) Multiple helix.

condition. That is, substantial errors may result if a thermometer calibrated for total immersion is used at partial immersion.

Pyrometers Measuring the temperature of an object by means of the quantity and character of the energy which it radiates has been designated "radiation pyrometry." This field of pyrometry has produced several different devices which may be broadly classified in two groups: (1) optical pyrometers, those instruments in which the brightness of a hot object is compared with that of a source of standard brightness; and (2) radiation pyrometers, those instruments which measure the rate of energy emission per unit area over a relatively broad range of wavelengths or which compare the radiation at two wavelengths. The broadband devices are classified as total-radiation pyrometers.

PRESSURE MEASUREMENTS

Pressure, defined as force per unit area, is usually expressed in terms of familiar units of weight-force and area or the height of a column of liquid which produces a like pressure at its base. The more common units are pounds-force per square inch, kilograms per square centimeter, inches or millimeters of mercury, and inches or feet of water. Conversion tables for these and other units of weight, force, and area are found in Sec. 1 of this *Handbook*.

Process pressure-measuring devices may be divided into three groups: (1) those which are based on the measurement of the height of a liquid column, (2) those which are based on the measurement of the distortion of an elastic pressure chamber, and (3) electrical sensing devices.

Standards and Calibration The pressure standards in the industrial or control laboratory associated with processing plants usually consist of a precision mercury-column manometer, a deadweight tester, and test gauges. Depending on the accuracy required, the auxiliary measurements required are barometric pressure, temperature, length, and gravitational constant.

The precision mercury-column manometer used is generally some variation of the well-type manometer illustrated in Fig. 22-63. The pressure on the gauge being tested is a function of the barometric pressure acting on the open end of the manometer, the difference in elevation of the two mercury levels, the density of the mercury, and the gravitational constant. The mercury manometer is normally used for pressures up to about 172 kPa (25 lbf/in^2) gauge. At higher pressures, the height of the mercury column becomes inconvenient.

The deadweight tester is commonly used for higher pressure ranges up to 68.9 MPa (10,000 lbf/in^2). The basic form of a deadweight tester is illustrated in Fig. 22-64. The accuracy of this tester depends on the quality of the weights, the accuracy of the piston diameter, and the lubrication of the piston. Accuracies within 0.1 percent are readily available commercially.

Test gauges are frequently used for direct comparison with process gauges. These test gauges can be certified for accuracy at prescribed temperatures and barometric pressures by commercial testing laboratories and the National Bureau of Standards. Typical available commercial test gauges have an accuracy within 0.25 percent. They can readily be checked with deadweight testers.

Liquid-Column Methods Liquid-column pressure-measuring devices are those in which the pressure being measured is balanced against the pressure exerted by a column of liquid. If the density of the liquid is known, the height of the liquid column is a measure of the pressure. Most forms of liquid-column pressure-measuring

TABLE 22-8 Characteristics of Liquid-in-Glass Thermometers

Class and type	Range °F.	Range °C.	Accuracy
All glass:			
Einchluss	−328 to +680	−201 to +360	Usually 1 scale division
Beckmann	−22 to +392	−30 to +200	±0.002° to ±0.005°C. Scale of 5° or 6°C. must be set to desired range
Clinical	+96 to +106	+35 to +41	±0.2°F. (0.1°C.)
Laboratory or chemical	−328 to +1200	−201 to +648	Usually 1 scale division
Max. or min. registering	−40 to +400	−40 to +204	1 to 2 scale divisions
Tube and scale:			
Cup case	−30 to +500	−22 to +260	Usually 1 scale division
Tin, copper, or stainless-steel case	−40 to +400	−40 to +204	Usually 1 scale division
Industrial:			
Spirit-filled	−150 to +120	−100 to +50	Usually 1 scale division
Mercury-filled	−40 to +1200	−40 to +648	Usually 1 scale division

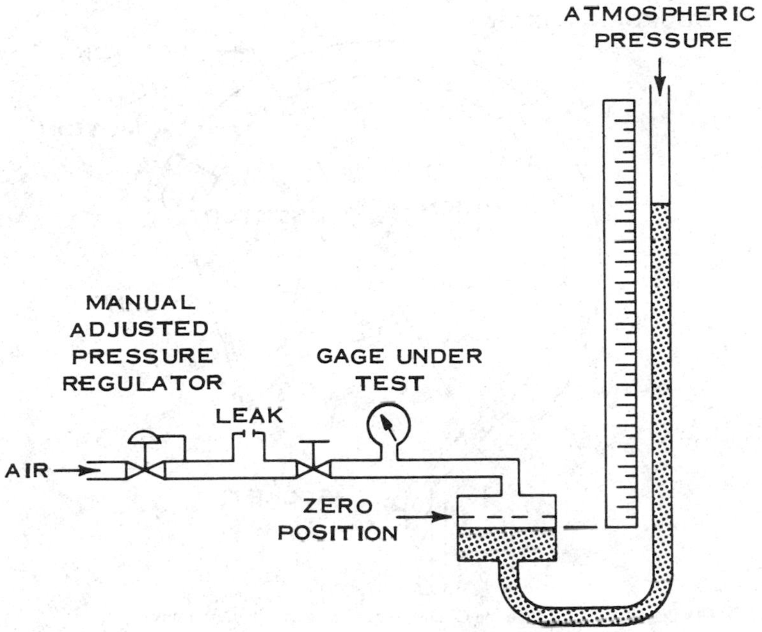

FIG. 22-63 Mercury-column pressure-gauge test stand using a well-type manometer.

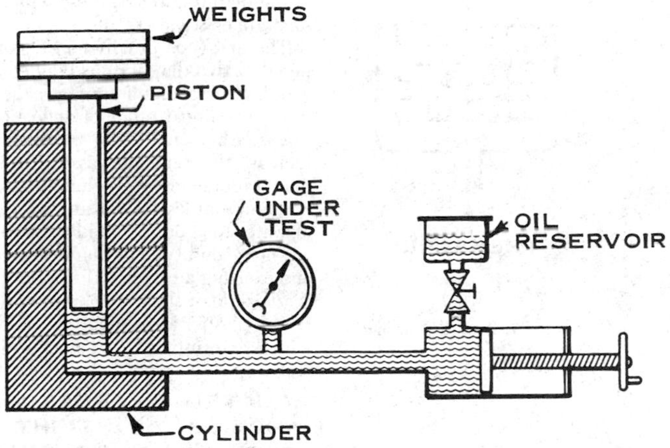

FIG. 22-64 Simple deadweight tester.

devices are commonly called *manometers*. When the height of the liquid is observed visually, the liquid columns are contained in glass or other transparent tubes. The height of the liquid column may be measured in length units or be calibrated in pressure units.

Depending on the pressure range, water and mercury are the liquids most frequently used. Since the density of the liquid used varies with temperature, the temperature must be taken into account for accurate pressure measurements. Several manometer variations are described in Sec. 5 of this *Handbook*.

Elastic-Element Methods Elastic-element pressure-measuring devices are those in which the measured pressure deforms some elastic material (usually metallic) within its elastic limit, the magnitude of the deformation being approximately proportional to the applied pressure. These devices may be loosely classified into three types: Bourdon tube, bellows, and diaphragm.

Bourdon-Tube Elements Probably the most frequently used process pressure-indicating device is the C-spring Bourdon-tube pressure gauge, illustrated in Fig. 22-65. Gauges of this general type are available in a wide variety of pressure ranges, case styles, dial sizes, and materials of construction. Materials are selected on the basis of pressure range, resistance to corrosion by the process materials, and effect of temperature on calibration. Gauges calibrated with pressure, vacuum, compound (combination pressure and vacuum), and suppressed-zero ranges are available.

Bellows Elements The bellows element is an axially elastic cylinder with deep folds or convolutions. The bellows may be used unopposed, or it may be restrained by an opposing spring as illustrated in Fig. 22-66. The pressure to be measured may be applied either to the inside or to the space outside the bellows, with the other side exposed to atmospheric pressure. For measurement of absolute pressure either the inside or the space outside of the bellows can be evacuated and sealed. Differential pressures may be measured by applying the pressures to opposite sides of a single bellows or to two opposing bellows.

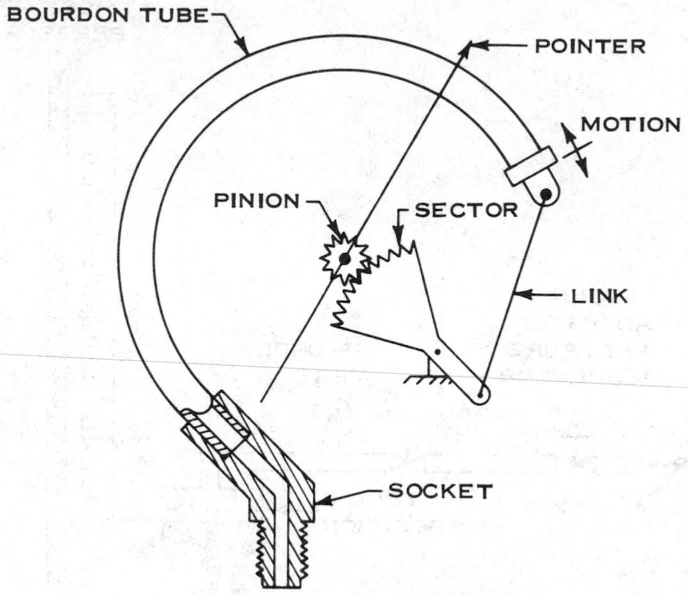

FIG. 22-65 Typical Bourdon C-tube pressure gauge and indicator linkage.

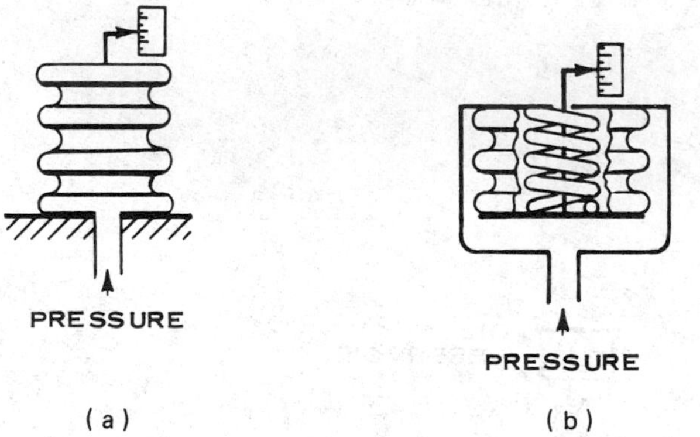

FIG. 22-66 Typical bellows elements. (*a*) Unopposed bellows. (*b*) Spring-loaded bellows.

Diaphragm Elements Diaphragm elements may be classified into two principal types: those which utilize the elastic characteristics of the diaphragm and those which are opposed by a spring or other separate elastic element.

The first type usually consists of one or more capsules each composed of two diaphragms bonded together by soldering, brazing, or welding. The diaphragms are flat or corrugated circular metallic disks. The linear range of deflection versus pressure for a single capsule is rather small; so several capsules may be connected together to increase the deflection as in Fig. 22-67. Metals commonly used in diaphragm elements include brass, phosphor bronze, beryllium copper, and stainless steel. Ranges are available from fractions of an inch of water to about 206.8 kPa (30 lbf/in^2) gauge.

The second type of diaphragm is used for containing the pressure and exerting a force on the opposing elastic element. The diaphragm is a flexible or slack diaphragm of rubber, leather, impregnated fabric, or plastic. Movement of the diaphragm is opposed by a spring which determines the deflection for a given pressure (see Fig. 22-

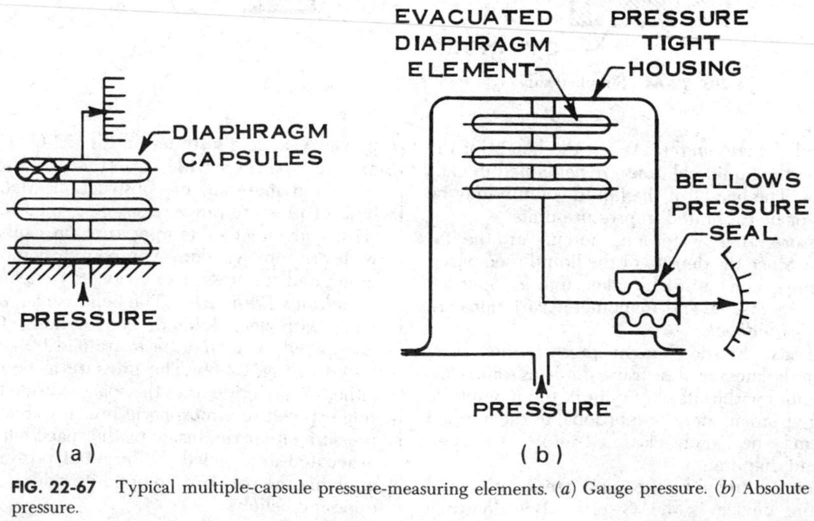

FIG. 22-67 Typical multiple-capsule pressure-measuring elements. (*a*) Gauge pressure. (*b*) Absolute pressure.

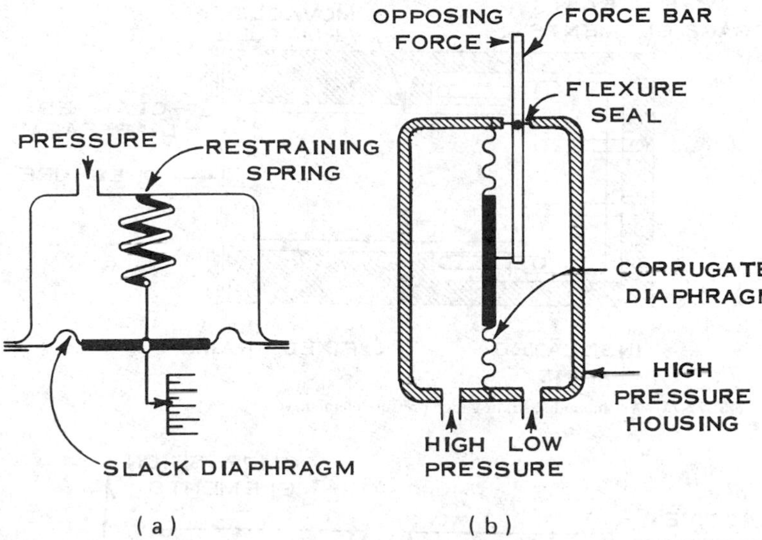

FIG. 22-68 Typical diaphragm pressure elements. (*a*) Slack diaphragm element. (*b*) Force-balance differential-pressure element.

68*a*). This type of diaphragm is used for the measurement of extremely low pressure, vacuum, or differential pressure.

Another opposed-type diaphragm element is illustrated in Fig. 22-68*b*. The diaphragm element is typically used to transmit the pressure force to a force-balance transmitter with little motion of the diaphragm (see also Fig. 22-95). Diaphragms of this type may be metallic (either flat or corrugated) or slack. Configurations for measurement of gauge, absolute, and differential pressures are available.

Electrical Methods

Strain Gauges When a wire or other electrical conductor is stretched elastically, its length is increased and its diameter is decreased. Both of these dimensional changes result in an increase in the electrical resistance of the conductor. Devices utilizing resistance-wire grids for measuring small distortions in elastically stressed materials are commonly called strain gauges. Pressure-measuring elements utilizing strain gauges are available in a wide variety of forms. They usually consist of one of the elastic elements described earlier to which one or more strain gauges have been attached to measure the deformation.

There are two basic strain-gauge forms: bonded and unbonded. Bonded strain gauges are those which are bonded directly to the surface of the elastic element whose strain is to be measured. The conductors may be either wire or etched foil, as illustrated in Fig. 22-69. The wire-type gauges are constructed from a wire with a diameter of 0.254 to 0.381 mm (1 to 1.5 mils). Both types are usually cemented to a carrier or substrate of plastic or paper which is in turn bonded to the elastic surface, as in Fig. 22-70.

The unbonded-strain-gauge transducer consists of a fixed frame and an armature which moves with respect to the frame in response to the measured pressure. The strain-gauge wire filaments are stretched between the armature and frame. A typical diaphragm-actuated unbonded-strain-gauge transducer is illustrated in Fig. 22-71.

The strain gauges are usually connected electrically in the familiar Wheatstone-bridge configuration illustrated in Fig. 22-72. In the two-active-arm bridges, the active element is mounted on the stressed elastic part, and the dummy element is mounted on an unstressed part. This configuration compensates for the thermal expansion of the supports and for the change in resistance of the strain-gauge elements with temperature. In the four-active-arm bridge, two elements are mounted so that they are placed in tension

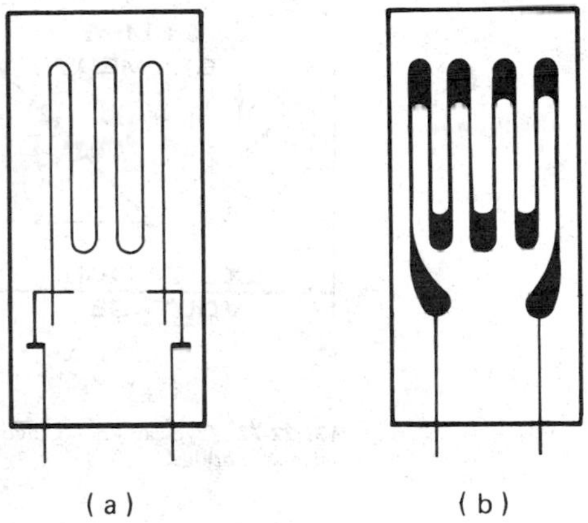

FIG. 22-69 Bonded strain gauges. (*a*) Wire type. (*b*) Foil type.

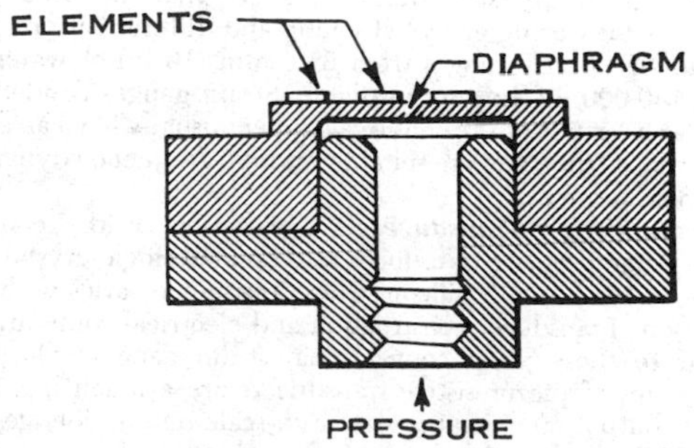

FIG. 22-70 A typical bonded strain-gauge pressure element.

by increased pressure, and two either are placed in compression or are unstressed. This configuration increases the sensitivity of the transducer and retains the temperature-compensation feature. Standard strain-gauge bridges are designed for an excitation voltage of 20 V or less, either ac or dc. The bridge output voltage is compara-

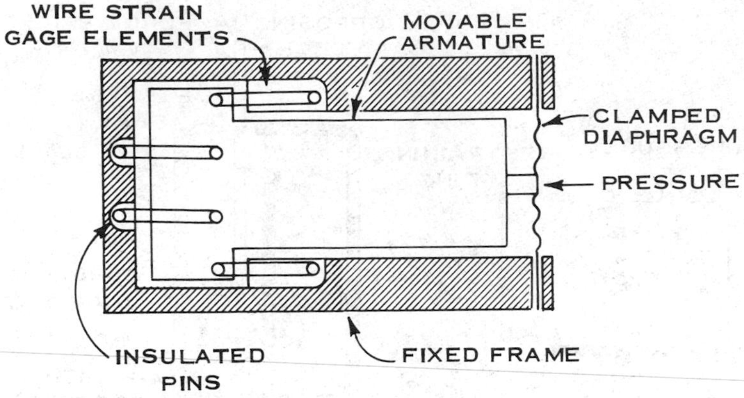

FIG. 22-71 An unbonded strain-gauge pressure element.

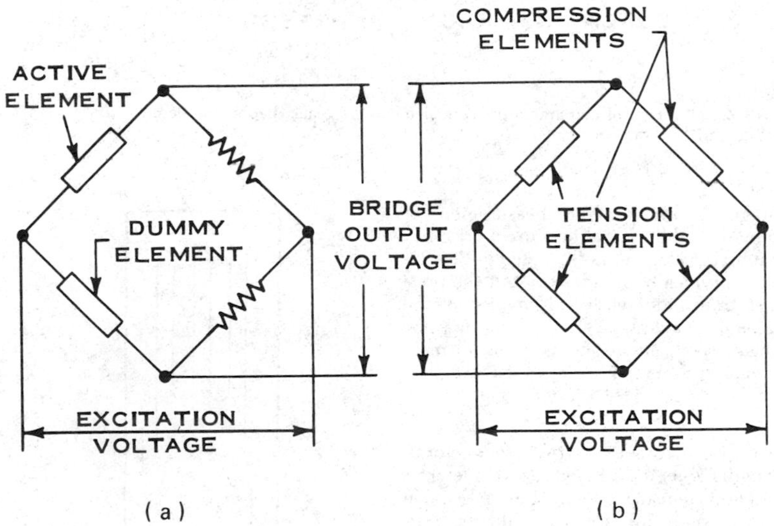

FIG. 22-72 Typical strain-gauge bridge configurations. (*a*) Two-active-arm bridge. (*b*) Four-active-arm bridge.

tively low, with full-scale values of up to 4 mV/V of excitation being typical.

Strain-gauge pressure transducers are manufactured in many forms for measuring gauge, absolute, and differential pressures and vacuum. Full-scale ranges from 25.4 mm (10 in) of water to 1034 MPa (150,000 lbf/in^2) are available. Strain gauges bonded directly to a diaphragm pressure-sensitive element usually have an extremely fast response time and are suitable for high-frequency dynamic-pressure measurements.

Piezoresistive Transducers A variation of the conventional strain-gauge pressure transducer uses bonded single-crystal semiconductor wafers, usually silicon, whose resistance varies with strain or distortion. Transducer construction and electrical configurations are similar to those using conventional strain gauges. The principal advantages of piezoresistive transducers are a much higher bridge voltage output, and smaller size. Full-scale output voltages of 50 to 100 mV/V of excitation are typical.

Piezoelectric Transducers Certain crystals produce a potential difference between their surfaces when stressed in appropriate directions. Piezoelectric pressure transducers generate a potential difference proportional to a pressure-generated stress. Because of the extremely high electrical impedance of piezoelectric crystals at low frequency, these transducers are usually not suitable for measurement of static process pressures.

FLOW MEASUREMENTS

General References: "Fluid Meters—Their Theory and Application," report of ASME Research Committee on Fluid Meters, 5th ed., ASME, New York, 1959. Spink, *Principles and Practice of Flow Meter Engineering*, 9th ed., Foxboro Co., Foxboro, Mass., 1967.

Flow, defined as volume per unit of time at specified temperature and pressure conditions, is for the most part measured by positive-displacement or rate meters. The term "positive-displacement meter" applies to a device in which the flow is divided into isolated measured volumes when the number of fillings of these volumes is counted in some manner.

The term "rate meter" applies to all types of flowmeters through which the material passes without being divided into isolated quantities. Movement of the material is usually sensed by a primary measuring element which activates a secondary device. The flow rate is then inferred from the response of the secondary device by means of known physical laws or from empirical relationships.

The principal classes of flow-measuring instruments used in the process industries are variable-head, variable-area, positive-displacement, and turbine instruments, mass flowmeters, wiers and flumes for flow measurement in open channels, and, more recently, vortex-shedding and ultrasonic flowmeters. Examples of such devices (except vortex-shedding and ultrasonic flowmeters) are presented and discussed in Sec. 5 along with principles governing their use. Vortex-shedding and ultrasonic flowmeters are presented in this subsection.

Vortex-Shedding Flowmeters These flowmeters take advantage of vortex shedding, which occurs when a fluid flows past a non-streamlined object (a blunt body). The flow cannot follow the shape of the object and separates from it, forming turbulent vortices or eddies at the object's side surfaces. As the vortices move downstream, they grow in size and are eventually shed or detached from the object. Shedding takes place alternately at either side of the object, and the rate of vortex formation and shedding is directly proportional to the volumetric flow rate.

Various vortex-shedding flowmeters are distinguished mainly by the methods used to detect the formed vortices. Four principal methods are in use:

1. One method utilizes a thermistor mounted in such a way that the thermistor is cyclically cooled by the changing flow of the fluid as a vortex passes. The thermistor's electrical resistance varies with temperature.
2. Since the vortices are shed alternately on each side of the blunt body, a pressure port through the body can be used to shift a lightweight metal sphere or disk alternately back and forth as the vortices pass. The second method then uses a magnetic pickup coil to count the sphere's or disk's oscillations.
3. In the third method shedded vortices are counted with a vane which extends behind the blunt body. The alternating vortices move the vane from side to side as they pass. The vane's motion or the induced mechanical stress is detected.
4. The last method uses ultrasonic transmitters and receivers to detect the vortices. Either the vortices are detected with a sonic beam directed across the stream from the blunt body, or the sonic beam is reflected off or refracted from the vortices downstream.

In all cases the vortices are counted and used to develop a signal linearly proportional to the flow rate. The signals are inherently digital and can easily be totaled; consequently they are directly compatible with digital control systems. Instrument ranges of 10:1 or 20:1 are claimed, with the restriction that the flow must remain turbulent. Accuracy is said to be maintained regardless of density, viscosity, temperature, or pressure when the Reynolds number is greater than 30,000.

Ultrasonic Flowmeters All ultrasonic flowmeters are based upon the variable time delays of received sound waves which arise when a flowing liquid's rate of flow is varied. Two fundamental measurement techniques, depending upon liquid cleanliness, are generally used.

In the first technique two opposing transducers are inserted in a pipe so that one transducer is downstream from the other. These transducers are then used to measure not the sound velocity but the difference between the velocity at which the sound travels with the direction of flow and the velocity at which it travels against the direction of flow.

The differential velocity is measured either by (1) direct time delays using sound-wave burst or (2) frequency shifts derived from beat-together, continuous signals. The frequency-measurement technique is usually preferred because of its simplicity and independence of the liquid static velocity. A relatively clean liquid is required to preserve the uniqueness of the measurement path.

In the second technique the flowing liquid must contain scatters in the form of particles or bubbles which will reflect the sound waves. These scatters should be traveling at the velocity of the liquid. A Doppler method is applied by transmitting sound waves along the flow path and measuring the frequency shift in the returned signal from the scatters in the process fluid. This frequency shift is proportional to liquid velocity.

Both measurement techniques respond to the average liquid velocity along the sound-wave path. Because the velocity profile changes with the Reynolds number, the meter coefficient which relates the measurement to the flow rate is also affected. In particular, there are two distinct meter coefficients for the laminar and turbulent regions. Errors in the flow-transition region from laminar to turbulent flow are much larger than in the regions on either side because of the unpredictability of flow patterns.

LEVEL MEASUREMENT

The measurement of level might be defined as the determination of the location of the interface between two fluids, separable by gravity, with respect to a fixed datum plane. The most common level measurement is that of the interface between a liquid and a gas. Other level measurements frequently encountered are the interface between two liquids, between a granular or fluidized solid and a gas, and between a liquid and its vapor.

A commonly used basis for classification of level devices is as follows: visual, float-actuated, displacer, and head devices, and a miscellaneous group which depend mainly on fluid characteristics.

Visual Devices This category includes such devices as dipstick, tape and plumb bob, open manometer, and gauge glass.

Glasses The most commonly used visual process-level device is the gauge glass. The gauge glass may be thought of as a manometer in which the process fluid level in the gauge seeks the same elevation as that in the vessel. The gauge glass is usually installed with valves which will permit the gauge to be isolated from the vessel and removed without depressurizing the vessel. The shutoff valves are frequently of an offset pattern, so that the tube or chamber may be cleaned without dismantling the gauge.

The tubular gauge glass is normally limited to 3103 kPa (450 lbf/in^2) and 240°C (400°F). Specialized construction is required for elevated pressures or temperatures. At higher temperatures and pressures the so-called flat-glass gauges, illustrated in Fig. 22-73, are employed.

When it is desirable to observe the color or other characteristics of the liquid as well as the interface between two liquids, the *transparent*-type gauge is used. This type has flat glass plates on both sides of the liquid chamber. Visibility of an interface in the transparent gauge is often enhanced by illuminating the gauge from the rear.

Reflex gauge glasses use only one glass plate which has reflecting prisms molded on the side of the glass in contact with the fluid. Incident light from the front of the gauge is internally reflected, giving the viewing slot a white, mirrorlike appearance if gas or vapor is in contact with the glass. Liquid in contact with the glass causes incident light to be transmitted through the glass and absorbed in the fluid chamber, giving the viewing slot a dark appearance. This construction results in a distinct level indication even with clear liquids.

Float-Actuated Devices Float-actuated devices are characterized by a buoyant member which floats at the interface between two fluids. Since a significant force is usually required to move the indicating mechanism, float-actuated devices are generally limited to liquid-gas interfaces. By properly weighting the float, they can be used to measure liquid-liquid interfaces.

Float-actuated devices may be classified on the basis of the method used to couple the float motion to the indicating system.

Chain or Tape Float Gauge In these types of gauges, the float is connected to the indicating mechanism by means of a flexible chain or tape. Figure 22-74 illustrates typical installations of tape float gauges. These gauges are commonly used in large atmospheric storage tanks.

The gauge-board type is provided with a counterweight to keep the tape or chain taut. A perforated tape is used to drive the level-indicating dials through a sprocket in the ground-level reading gauge. The tape is stored in the gauge head on a spring-loaded reel. The float is usually a pancake-shaped hollow metal float, with guide wires from top to bottom of the tank to constrain it.

Lever and Shaft Mechanisms In pressurized vessels, float-actuated lever and shaft mechanisms are frequently used for level measurement. As illustrated in Fig. 22-75, this type of mechanism consists of a hollow metal float and lever attached to a rotary shaft which transmits the float motion to the outside of the vessel through a rotary seal.

Magnetically Coupled Devices A variety of float-actuated level devices which transmit the float motion by means of magnetic coupling have been developed. Typical of this class of devices are

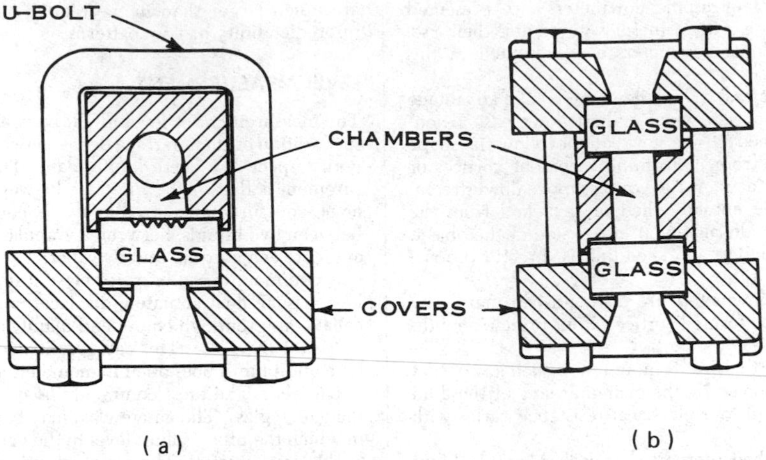

FIG. 22-73 Flat-glass gauge glasses. (*a*) Cross section of reflex gauge. (*b*) Cross section of transparent gauge.

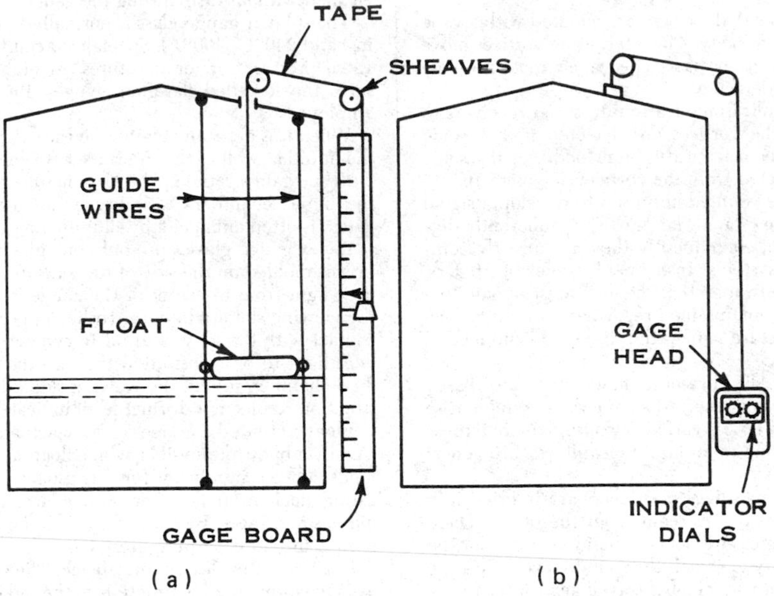

FIG. 22-74 Tape float gauges. (*a*) Gauge-board indicator. (*b*) Ground reading gauge.

magnetically operated level switches and magnetic-bond float gauges.

A typical magnetic-bond float gauge is illustrated in Fig. 22-76. This device consists of a hollow magnet-carrying float which rides along a vertical nonmagnetic guide tube. The follower magnet suspended by tape drives an indicating dial similar to that on a conventional tape float gauge. The float and guide tube, which come in contact with the measured fluid, are available in a variety of materials for resistance to corrosion and to withstand high pressures or vacuum. Weighted floats for liquid-liquid interfaces are available.

Magnetic coupling is frequently employed in level-sensing electrical switches. A typical device of this type is illustrated in Fig. 22-77.

Displacer Devices Displacer-actuated level-measuring devices use the buoyant force on a partially submerged displacer as a measure of the location of the interface along the axis of the displacer. Vertical motion of the displacer is usually restrained by some elastic member whose motion or distortion is directly proportional to the buoyant force and therefore to the interface level. The range is limited to the length of the displacer. Coupling of float motion to the indicating mechanism is almost always effected through some type of packless mechanism, which frequently also constitutes the elastic restraining member.

Since the buoyant force is dependent upon the difference in density between the measured fluids, accurate level measurement with displacement devices depends upon accurate knowledge of the fluid densities.

Torque-Tube Displacer One of the most frequently used level-measuring devices is the torque-tube displacer illustrated in Fig. 22-78. The displacer is suspended on a displacer rod attached to a torque tube. This is fixed at its outer end and supported on a knife-edge bearing at its inner end. The torque tube, in addition to being the elastic member, constitutes a packless pressuretight barrier. Inside the torque tube is a shaft which is fixed to the torque tube at its inner end. Rotation (5 to 10°) of the outer end of the shaft is proportional

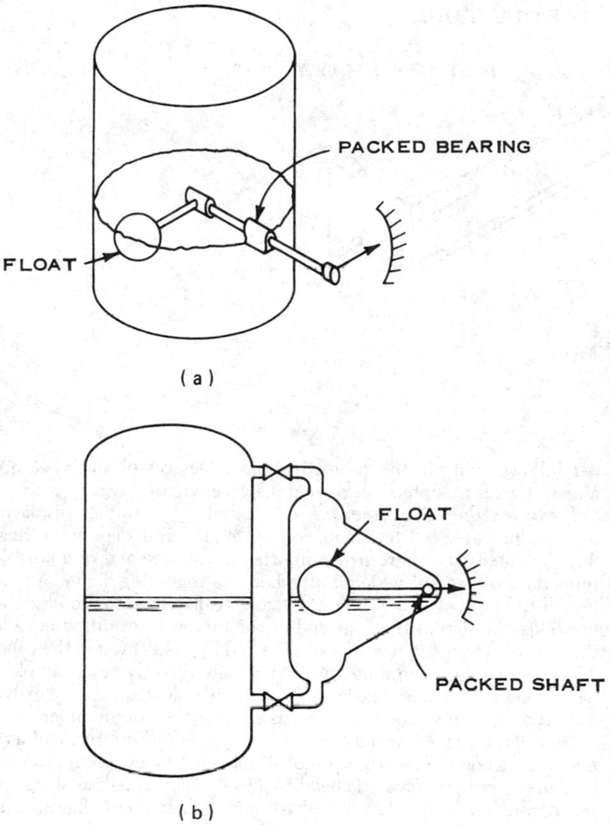

FIG. 22-75 Level and shaft float mechanisms. (*a*) Internal float gauge. (*b*) External float gauge.

Level and shaft float mechanisms. (*a*) Internal float gauge. (*b*) External float

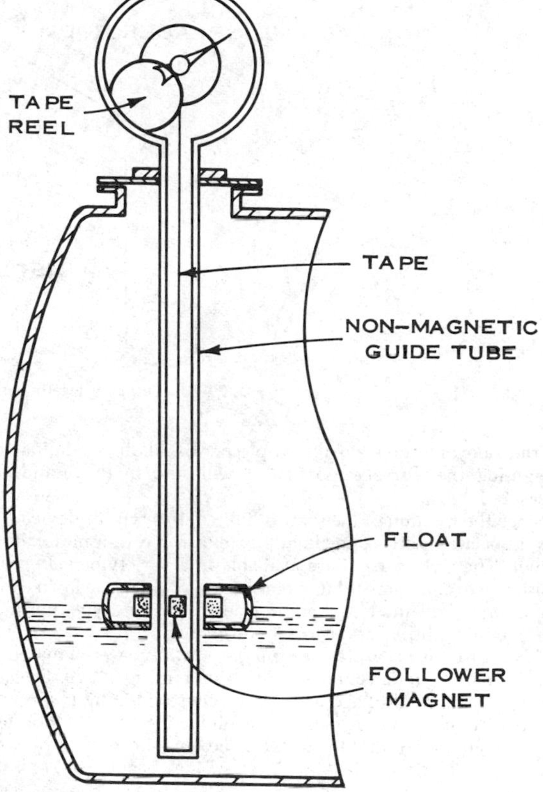

FIG. 22-76 Magnetic-bond float gauge.

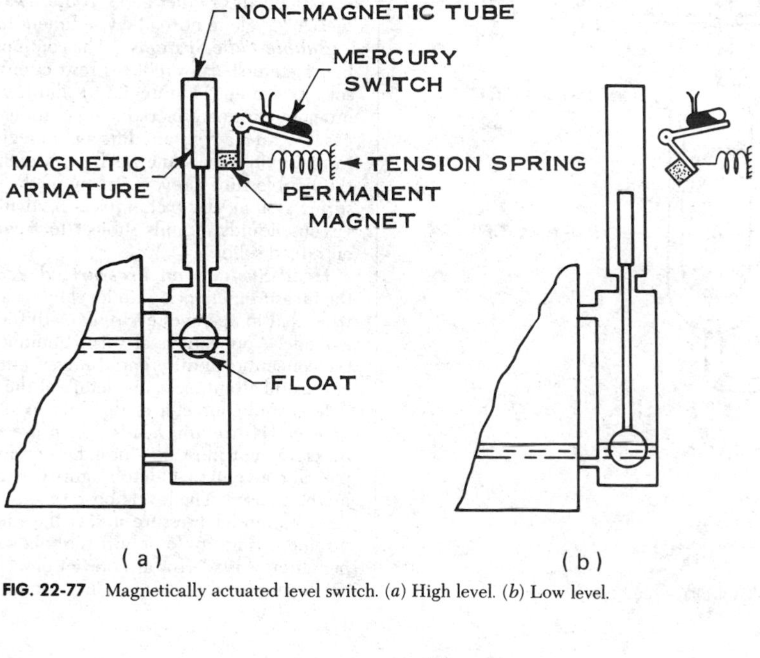

FIG. 22-77 Magnetically actuated level switch. (*a*) High level. (*b*) Low level.

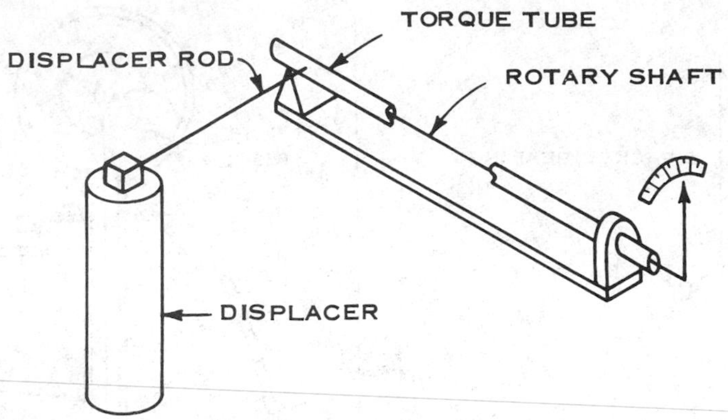

FIG. 22-78 Displacer level-measuring device.

to the buoyant force on the displacer. Backlash is eliminated by weighting the displacer so that it will sink in the liquid being measured.

Since the rotation of the shaft is limited, this type of device is usually associated with a pneumatic or electronic transmitter or controller. The wetted parts are available in a variety of materials for resistance to corrosion or high temperatures. Mounting in an external displacer cage is most common for devices of this type, but top- or side-mounted displacer devices are available.

Magnetically Coupled Displacer A displacer-actuated unit employing magnetic coupling is illustrated in Fig. 22-79. Devices of this type commonly have a displacer constrained by a spring and move a drive magnet enclosed in a protecting tube. Motion of the drive magnet is transmitted to the indicating mechanism by a magnet follower outside the protecting tube. Devices of this type are almost always mounted in external displacer cages.

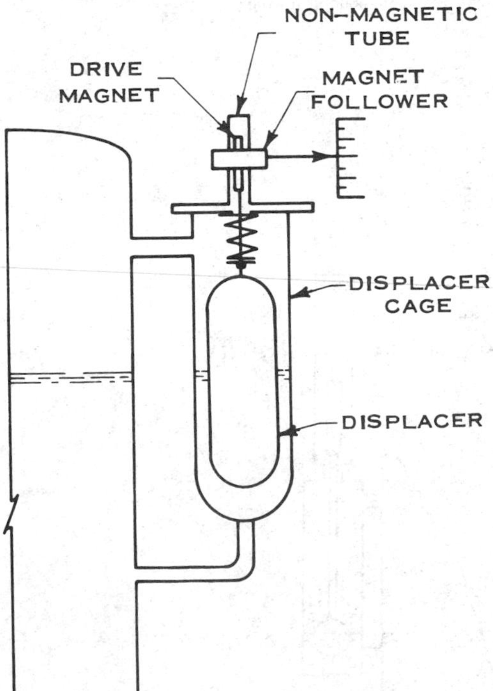

FIG. 22-79 Magnetically coupled displacer unit.

Flexure-Tube Displacer A comparatively simple displacer device, illustrated in Fig. 22-80, consists of an elliptical or cylindrical float mounted on a short arm connected to the free end of a flexible tube, the fixed end of which is attached to a mounting flange. A portion of the tube near the mounting flange is flattened to increase the flexibility. Motion of the float end of the tube is transmitted outside the float chamber by means of a rod extending out through the tube. These devices are commonly used to actuate directly either an electrical switch or a pneumatic pilot. A similar device uses a flexible disk, through which the float rod passes, as the elastic member.

Head Devices A variety of devices utilize hydrostatic head as a measure of level. As in the case of displacer devices, accurate level measurement by hydrostatic head requires an accurate knowledge of the densities of both heavier-phase and lighter-phase fluids. The majority of this class of systems utilize standard-pressure and differential-pressure measuring devices.

Pressure-Gauge Systems on Open Vessels The simplest application of head level measurement is the measurement of liquid level in an open vessel. The pressure-measuring element is located at or below the minimum operating level or process outlet connection. Pressure piping between the vessel and the measuring element must be sloped upward toward the vessel in order to prevent errors due to entrapped air or other gases. A drain valve should be provided at the measuring element to allow sediment to be flushed from the piping.

Bubble-Tube Systems The commonly used bubble-tube or bubbler system sharply reduces restrictions on the location of the measuring element. Figure 22-81 illustrates a typical installation. In order to eliminate or reduce variations in pressure drop due to the gas flow rate, a constant differential regulator is commonly employed to maintain a constant gas flow rate. Since the flow of gas through the bubble tube prevents entry of the process liquid into the measuring system, this technique is particularly useful with corrosive or viscous liquids, liquids subject to freezing, and liquids containing entrained solids.

Head Systems on Pressurized Vessels In pressurized vessels the measurement of liquid level by means of hydrostatic head differs from that in open vessels in that a differential-pressure measurement is made. Applications of this technique may employ almost any of the conventional differential-pressure-measuring devices.

Careful attention to the details of the installation is important. The density and character of the fluid in all parts of the system must be known. Hydrostatic heads which are not pertinent to the desired measurement must be eliminated or compensated for. As an example, a conventional diaphragm-type differential-pressure element might be used. The level above the lower connection is measured by the differential pressure across the measuring element. This measurement is accurate only if the following conditions are met: compensation is made for any deviation of the density of the liquid, the connection to the low-pressure side of the measuring element con-

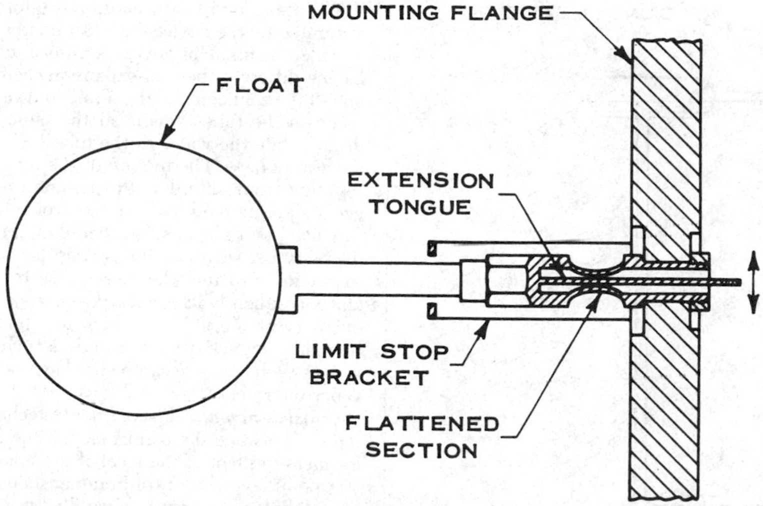

FIG. 22-80 Flexure-tube displacer unit.

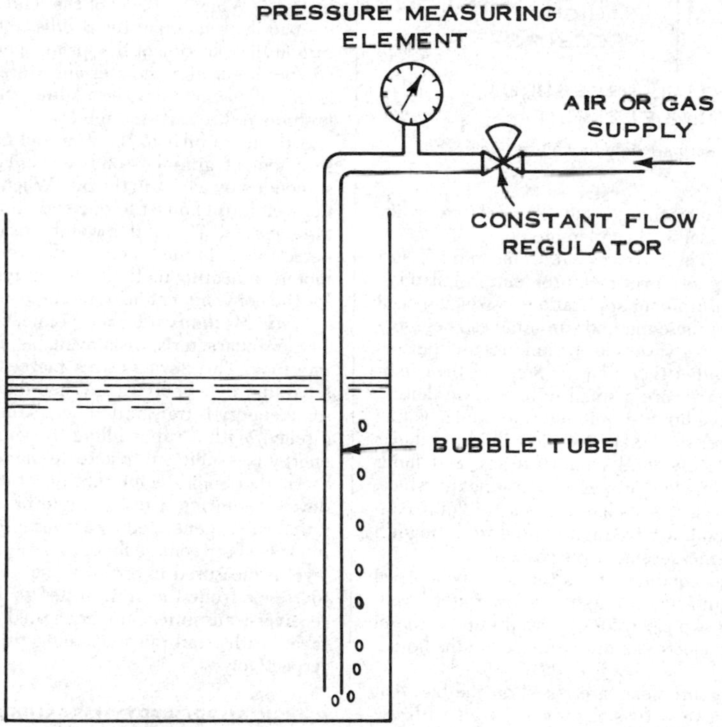

FIG. 22-81 Bubble-tube level-measuring system.

tains no liquid accumulated because of overflow or condensation, the density of the gas or vapor above the liquid is either negligible or compensated for, and the measuring element is located at the same elevation as the minimum level to be measured or suitable compensation is made.

Boiling-Liquid Service The conditions described for the simple head system will not be met if the liquid is at or near its boiling point and its temperature is higher than ambient temperature. For such applications, the system illustrated in Fig. 22-82 may be used.

Liquid condenses in the low-pressure side of the measuring element and in the condensate pot until it overflows back into the vessel. This constitutes a "reference leg" with a known constant head. This constant head can be suppressed in most differential-pressure-measuring elements by means of either the zero adjustment or a zero-suppression mechanism.

Systems with this configuration may have the measuring element placed below the bottom connection without further compensation, since the same hydrostatic head is added to both sides of the measuring element.

The condensate pot also provides a liquid seal which can be used to prevent corrosive or high-temperature fluids from coming in contact with the measuring element. If condensed process liquid is sub-

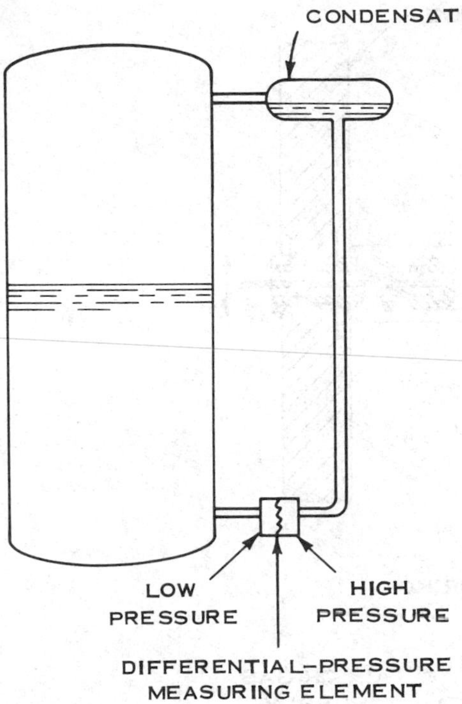

FIG. 22-82 Hydrostatic level-measuring system for boiling-liquid service.

ject to freezing at ambient temperatures, a suitable nonfreezing liquid is used to fill the reference leg and condensate pot.

Miscellaneous Methods The devices and techniques just described cover the majority of level-measurement installations. There are, however, level-measurement applications in which special problems and conditions make these methods unsuitable. These special problems have given rise to a variety of techniques and devices based on fluid characteristics other than density. Some of the conditions requiring special techniques are a small difference in density between the two fluids, such as a liquid-liquid interface and a boiling liquid near its critical pressure; extremely corrosive fluids; granular solids which do not exhibit truly fluid characteristics; and fluids which are miscible to a limited extent, giving an indistinct interface.

Electrical Methods Two electrical characteristics of fluids, conductivity and dielectric constant, are frequently used to distinguish between two phases for level-measurement purposes.

An application of electrical conductivity is the fixed-point level detection of a conductive liquid such as high and low water levels. A voltage is applied between two electrodes inserted into the vessel at different levels. When both electrodes are immersed in the liquid, a current flows.

Capacitance-type level measurements are based on the fact that the electrical capacitance between two electrodes varies with the dielectric constant of the material between them. A typical continuous level-measurement system consists of a rod electrode positioned vertically in a vessel, the other electrode usually being the metallic vessel wall. The electrical capacitance between the electrodes is a measure of the height of the interface along the rod electrode. The rod is usually conductively insulated from process fluids by a coating of plastic.

The dielectric constant of most liquids and solids is markedly higher than that of gases and vapors. The dielectric constant of water and other polar liquids is also higher than that of hydrocarbons and other nonpolar liquids.

Thermal Methods Level-measuring systems may be based on the difference in thermal characteristics between the fluids, such as temperature or thermal conductivity.

The use of temperature difference for level measurement may be illustrated by the device once frequently used for water-level control in boiler drums. This device consists of two concentric metallic tubes having different thermal-expansion coefficients with the ends of the outer tube connected to the liquid and vapor spaces of the drum. The steam in the tube remains at the same temperature as that in the drum, while the water in the tube cools because of heat transfer to the atmosphere. The motion of the inner tube relative to the outer, resulting from differential expansion, is transmitted through a linkage to operate the feedwater control valve.

A fixed-point level sensor based on the difference in thermal conductivity between two fluids consists of an electrically heated thermistor inserted into the vessel. The temperature of the thermistor and consequently its electrical resistance increase as the thermal conductivity of the fluid in which it is immersed decreases. Since the thermal conductivity of liquids is markedly higher than that of vapors, such a device can be used as a point level detector for liquid-vapor interface.

Consistency and Viscosity Detection of interface based on difference between the consistency of the fluids is frequently the basis for measurement of the level of granular solids or semisolids.

A point level sensor for granular solids based on the consistency of the solid phase consists of a paddle turned at a slow speed (5 to 10 r/min) by an electric motor through a tension spring or similar mechanism. The paddle is free to turn if it is not in contact with the solid material. When the level of the solid material rises to the location of the paddle, rotation of the paddle is stopped. The motor continues to turn until deflection of the spring operates a limit switch, which shuts off the motor and actuates an external alarm or control. When the level falls away from the paddle, the spring returns to its original position and restarts the motor.

A device similar to the float and tape gauge is available for measurement of granular-solids level. This device consists of a weight suspended by a tape in the bin. When a level measurement is desired, the weight is allowed to descend into the bin by a slip clutch on the tape sheave. When the weight reaches the solids level, a switch detects slack in the tape, and the weight is withdrawn by an electric motor. Indicating methods for this type of device are similar to those for the conventional float and tape device.

Sonic Methods A fixed-point level detector based on sonic-propagation characteristics is available for detection of a liquid-vapor interface. This device uses a piezoelectric transmitter and receiver, separated by a short gap. When the gap is filled with liquid, ultrasonic energy is transmitted across the gap, and the receiver actuates a relay. With a vapor filling the gap, the transmission of ultrasonic energy is insufficient to actuate the receiver.

Another sonic technique used for level measurement is a sonar device requiring a distinct interface between two fluids. A pulsed sound wave, generated by a transmitting crystal, is reflected from the interface between the fluids and returned to the receiver crystal. The level is measured in terms of the time required for the sound pulse to travel from the transmitter to the interface and return. The receiver-transmitter may be located either below or above the interface, with transmission being in the liquid or vapor phase respectively.

PHYSICAL-PROPERTY MEASUREMENT

Physical-property measurements are sometimes thought of as composition analyzers, because for binary or pseudobinary mixtures the composition can frequently be inferred from the measurement.

Density and Specific Gravity For binary or pseudobinary mixtures of liquids or gases or a solution of a solid or gas in a solvent, the density is a function of the composition at a given temperature and pressure. For nonideal solutions, empirical calibration will give the relationship between density and composition.

Liquid Column Density may be determined by measuring the gauge pressure at the base of a fixed-height liquid column open to the atmosphere. If the process system is closed, then a differential-pressure measurement is made between the bottom of the fixed-height liquid column and the vapor over the column. If vapor space is not always present, the differential-pressure measurement is made between the bottom and top of a fixed-height column with the top measurement being made at a point below the liquid surface.

Displacement There are a variety of density-measurement devices based on displacement techniques.

A hydrometer is a constant-weight, variable-immersion device. The degree of immersion, when the weight of the hydrometer equals the weight of the displaced liquid, is a measure of the density. The hydrometer is adaptable to manual or automatic usage.

A displacement float consists of a completely submerged float with which the liquid buoyance may be measured by the float position as in Fig. 22-83. The position of the float varies with the weight of the chain supported by the buoyant force acting on the float. A similar device measures, by means of a torque tube, the force exerted on a completely submerged float.

Another modification includes a magnetic float suspended below a solenoid, the varying magnetic field maintaining the float at a constant distance from the solenoid. Change in position of the float, resulting from a density change, excites an electrical system which increases or decreases the current through the solenoid.

Direct Mass Measurement One device of this type electrically forces a U tube as shown in Fig. 22-84 to vibrate in a plane perpendicular to the plane of the paper. The amplitude of vibration is measured to indicate changes in density. The amplitude decreases with density.

Weight of Fixed Volume A number of devices use force-balance mechanisms (see Fig. 22-85) to measure the weight of a fixed-volume chamber or tube. The chamber or tube is attached to the sample source and discharge by flexible couplings such as bellows.

Radiation-Density Gauges Gamma radiation may be used to measure the density of material inside a pipe or process vessel. The equipment is basically the same as for level measurement, except that here the pipe or vessel must be filled over the effective, irradiated sample volume. The source is mounted on one side of the pipe or vessel and the detector on the other side with appropriate safety radiation shielding surrounding the installation. Cesium 137 is usually used as the radiation source for path lengths under 610 mm (24 in) and cobalt 60 above 610 mm. The detector is usually an ionization gauge. The absorption of the gamma radiation is a function of density. Since the absorption path includes the pipe or vessel walls, an empirical calibration is used. Appropriate corrections must be made for the source intensity decay with time.

Viscous-Drag Gas-Density Sensor This device consists of a motor-driven vane-type impeller and a similar stationary impulse wheel located close together as illustrated in Fig. 22-86. These wheels are located within coaxial cylindrical baffles. The viscous-drag

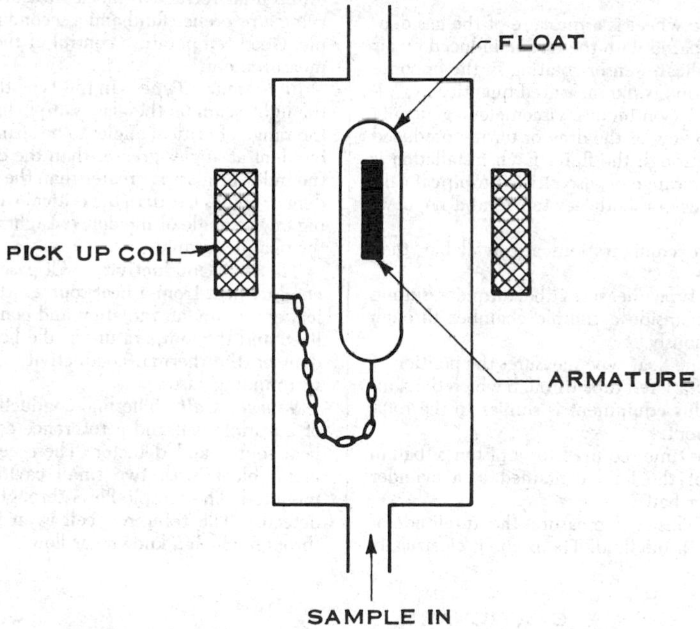

FIG. 22-83 Chain-balanced density meter in which the position of the float is determined by the weight of chain required to balance the buoyant force.

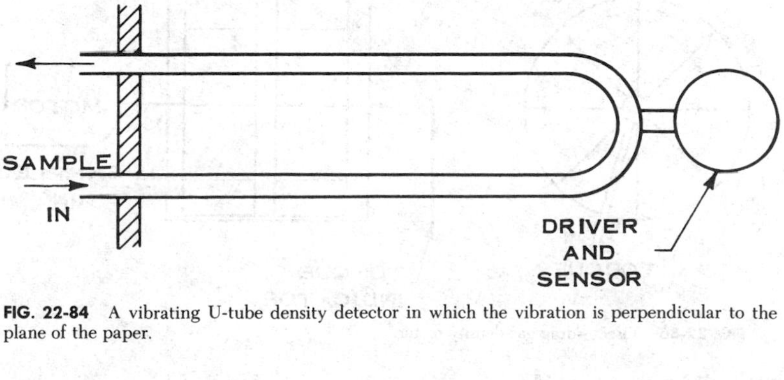

FIG. 22-84 A vibrating U-tube density detector in which the vibration is perpendicular to the plane of the paper.

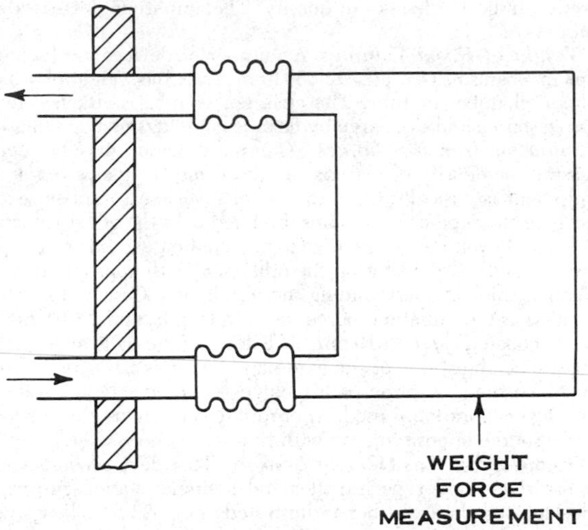

FIG. 22-85 Continuous-weighing determination of fluid density.

imparted torque on the impulse wheel is a measure of the gas density. This torque is usually compared with the torque induced by air or other standard gas on a duplicate sensor rotating in the opposite direction. The difference in torque is the measured quantity.

Viscosity and Consistency Continuous viscometers generally measure either the resistance to flow or the drag or torque produced by movement of an element through the fluid. Each installation is normally applied over a narrow range of viscosities. Empirical calibration over this range allows use on both newtonian and nonnewtonian fluids.

Hardware Many hardware configurations are available; they fall into four general categories:

1. *Rotational type.* This type measures the torque resulting from the rotation of a spindle inside a sample chamber through which the sample flows continuously.

2. *Float or piston type.* The float type measures the position of a specially shaped float inside a tapered tube through which the sample flows at a constant rate. This equipment is similar to the rotameters used for flow measurement.

The piston type measures the time required for a piston or ball to fall through a static sample of the fluid contained in a cylinder slightly larger than the piston or ball.

3. *Vibrating-probe type.* This type measures the amplitude of vibration of a probe immersed in the fluid. The probe is electrically excited at a constant frequency. Another method of excitation is to pulse the probe at a rate determined by the rate of decrease in the amplitude of vibration. In this case, the frequency of pulsation is the measured variable.

4. *Capillary type.* This type measures the pressure drop resulting from constant flow of the fluid through a capillary tube of specified diameter and length.

Refractive-Index Analyzers When light travels from one medium (e.g., air or glass) into another (e.g., a liquid), it udergoes a change of velocity and, if the angle of incidence is not 90°, a change of direction. For a given interface, angle, temperature, and wavelength of light the amount of deviation or refraction will depend on the composition of the liquid.

If the sample is transparent, the normal method is to measure the refraction of light transmitted through the glass-sample interface. If the sample is opaque, the reflectance near the critical angle at a glass-sample interface is measured.

Transmission Type Most process instruments of this type operate on a nulling principle. The refractive displacement of the light traversing the sample-filled prism is measured by rotating a plate of glass in the optical path. The glass plate or "refractor block" is part of a servo system which automatically keeps the measuring light beam centered on a phototube detector. The angle of rotation of the plate is the quantity measured. In some cases it is feasible to make a differential refractive-index measurement between one prism carrying a reference fluid and a second prism carrying the flowing sample. Good temperature control of the sample is required for precise measurement.

Reflectance Type In this type the average angle of incidence of the light beam on the glass-sample interface is arranged to be within the range of critical angles to be spanned by the sample. All the light incident at angles greater than the critical angle will be reflected if the index of glass is greater than the index of the sample. Light incident at angles less than the critical will be partially reflected, depending on the angle of incidence. Light not reflected will be absorbed in the opaque sample.

Thermal Conductivity All gases and vapor have the ability to conduct heat from a heat source. At a given temperature and physical environment, radiation and convection heat losses will be stabilized and the temperature of the heat source will be mainly dependent on the thermal conductivity and thus the composition of the surrounding gases.

Sample Cell Thermal-conductivity analyzers normally consist of a sample cell and a reference cell, each containing a combined heat source and detector. These cells are normally contained in a metal block with two small cavities in which the detectors are mounted. The sample flows through the sample-cell cavity past the detector. The reference cell is an identical cavity with a detector through which a known gas flows.

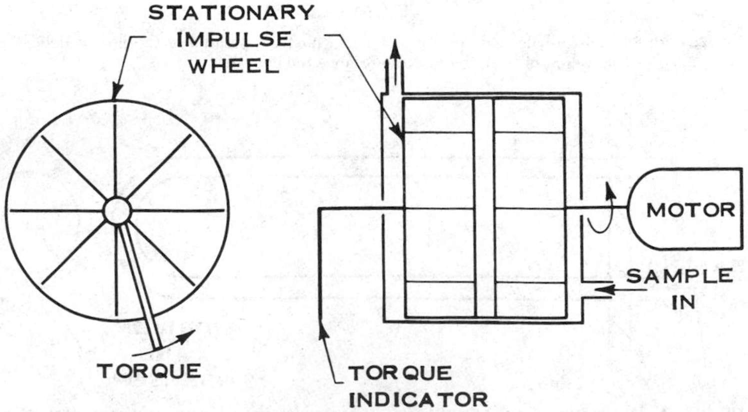

FIG. 22-86 Viscous-drag gas-density meter.

Heat-Source Detectors The combined heat source and detectors are normally either wire filaments or thermistors heated by a constant current. Since their resistance is a function of temperature, the sample-detector resistance will vary with sample composition while the reference-detector resistance will remain constant. The output from the detector bridge will be a function of sample composition.

Boiling-Point Analyzers Process analyzers for obtaining various boiling points (initial point, midpoint, or end point) of hydrocarbon streams are fairly common. These analyzers are miniature distillation processes in which the sample temperature is measured as distillation occurs. The various designs result from different methods used in determining the amount of sample distilled off and whether the method is a batch or continuous measurement.

Flash-Point Analyzers In these analyzers a liquid sample is heated, its vapor is mixed with a controlled flow of air, and the mixture is fed into a spark chamber. As the liquid-sample temperature and therefore the vapor concentration are increased, the mixture will finally be ignited by the spark. The temperature of the sample at this point is then recorded as the flash point.

PROCESS CHEMICAL-COMPOSITION ANALYZERS

A number of composition analyzers used for process monitoring and control require chemical conversion of one or more sample components preceding quantitative measurement. These reactions include formation of suspended solids for turbidimetric measurement, formation of colored materials for colorimetric detection, selective oxidation or reduction for electrochemical measurement, and formation of electrolytes for measurement by electrical conductance. Some nonvolatile materials may be separated and measured by gas chromatography after conversion into volatile derivatives.

Chromatographic Analyzers Chromatographic analyzers are widely used for the separation and measurement of volatile compounds and of compounds that can be quantitatively converted into volatile derivatives. These materials are separated by placing a portion of the sample in a chromatographic column and carrying the compounds through the column with a gas stream. As a result of the different affinities of the sample components for the column packing, the compounds emerge successively as binary mixtures with the carrier gas. A detector at the column outlet measures some physical property which can be related to the concentrations of the compounds in the carrier gas. Both the concentration peak height and the peak height–time integral, i.e., peak area, can be related to the concentration of the compound in the original sample.

Process chromatographs have three distinctive features:

1. Capability for multicomponent analyses ranging from one or two selected sample components to complete compositional analyses
2. Cyclic operation with information on composition presented on a discontinuous (sampled-data) basis requiring a programmer to control analyzer functions and data presentation
3. Instrumental-analysis lag (in addition to sample lag) determined by the degree to which column parameters such as length, packing material, efficiency, flow rate, and temperature have been optimized for minimum analysis time

The completeness of the analysis available by chromatography techniques enables the computation of physical properties of the sample when it is inconvenient or time-consuming to measure the property directly, for example, the octane number of a gasoline.

Columns Chromatographic columns are of two general types: packed columns and wall-coated open tubular columns. Packed columns usually contain either a granular adsorbent or a granular support material coated with a thin layer of high-boiling solvent (partitioning liquid). Open tubular columns contain a thin film of partitioning liquid on the column walls and have an open gas passage through the center of the column.

There is a wide choice of adsorbents and partitioning liquids available for chromatographic columns. For process chromatographs, selection of the column packing most suitable for specific separations is one of the most important factors in obtaining the short analysis time desired for automatic process control.

Detectors The two detectors most commonly used for process chromatographs are the thermal-conductivity detector (katharometer) and the hydrogen-flame ionization detector. Thermal-conductivity detectors, discussed earlier, require calibration for the thermal response of each compound.

Hydrogen-flame ionization detectors are more complicated than thermal-conductivity detectors but are capable of 100 to 10,000 times greater sensitivity for hydrocarbons and organic compounds. For ultrasensitive detection of trace impurities, carrier gases must be specially purified.

Programming The cyclic nature of process chromatographs requires a device to control addition of sample, switch columns, gate peak signals, and measure signal levels. These functions are provided by a mechanical or electronic programmer or by a digital computer. Electronic circuits for additional functions such as measurement of peak heights or peak areas and computation of sample-concentration ratios are frequently provided.

A digital computer may be used to program the functions of one or more chromatographs and calculate concentrations. More complex interfacing of chromatographs and computers allows online calculations such as combining partial analysis by several chromatographic columns into a total analysis, performing a mathematical resolution of unseparated peaks, and computing physical properties of the sample stream.

Data Presentation and Process Control For process monitoring, presentation of component concentrations in bar-graph form on a strip-chart recorder or logged digitally is usually adequate.

When the analyzer lag (sample lag plus analysis lag) is a small fraction of the process time constant, the output of a process chromatograph can be used as the measured variable in automatic closed-loop control of a process. Changes in the control signal are available only once per analyzer cycle, and this control signal must be maintained constant until a new analyzer output signal is available.

When the analyzer lag is long compared with the process time constant, the output from the chromatograph may be used to adjust the set point on a primary controller automatically in a cascaded loop or to update the value of a computed variable.

Trace Analysis Gas chromatographs are very useful for the measurement of trace concentrations of volatile materials. The hydrogen-flame ionization detector is extremely sensitive to most organic materials; special forms of these detectors are sensitized to halogen-containing compounds and may be used to measure trace concentrations of these compounds in the presence of large quantities of hydrocarbons and other non-halogen-containing organics. Other types of ionization detectors are very sensitive to nonorganic materials.

Infrared Analyzers Many gaseous and liquid compounds absorb infrared radiation to some degree. The degree of absorption at specific wavelengths depends on molecular structure and concentration.

Process instruments generally are of the nondispersive type. That is, the detector is exposed to a wide-wavelength band of radiation. This contrasts with the normal laboratory instrument which uses a scanning monochromator to isolate successive narrow-wavelength bands. The main components of a process instrument are the radiation source, sample cell, and detector.

Radiation Source This is commonly a heated wire which yields radiation over the analytically useful portion of the infrared spectrum, 1 to 10 μm. The main requirement is stability of emission. Dual sources are sometimes used.

Sample Cell This can vary from a few millimeters to a meter or more in length, depending on whether the sample is a liquid or a low-absorbing gas.

Detectors There are two common detector types for nondispersive analyzers. These analyzers normally have two beams of radiation, an analyzing and a reference beam. One type of detector consists of two gas-filled cells separated by a diaphragm. As the amount of infrared energy absorbed by the detector gas in one cell changes, the cell pressure changes. This causes movement in the diaphragm, which in turn causes a change in capacitance between the diaphragm and a reference electrode. This change in electrical capacitance is measured as the output. The second type of detector consists of two thermopiles or two bolometers, one in each of the two radiation beams. The infrared radiation absorbed by the detector is measured by a differential thermocouple output or a resistance-thermometer (bolometer) bridge circuit.

With gas-filled detectors, a chopped light system is normally used in which one side of the detector sees the source through the analyzing beam and the other side the reference beam, alternating at a frequency of a few hertz.

Ultraviolet and Visible-Radiation Analyzers Many gas and liquid compounds absorb radiation in the near-ultraviolet or visible region. For example, organic compounds containing aromatic and carbonyl structural groups are good absorbers in the ultraviolet region. Also many inorganic salts and gases absorb in the ultraviolet or visible region. In contrast, straight-chain and saturated hydrocarbons, inert gases, air, and water vapor are essentially transparent.

Process analyzers are designed to measure the absorbance in a particular wavelength band. The desired band is normally isolated by means of optical filters. When the absorbance is in the visible region, the term "colorimetry" is used.

The main components of a process instrument are the light source, sample cell, and detector. There are two common arrangements used in process instruments. In one, the light from a single source is split into two beams by means of a mirror system. One beam traverses the sample cell and the second a reference cell. The beams are then compared by use of a single detector or two detectors. The second arrangement is similar except that the beam is split after traversing the sample cell. In this case, one beam is made insensitive to the sample compound of interest by some type of filtering. In either case, the light beam may or may not be chopped mechanically to produce an ac input to the detector.

Light Source A tungsten lamp is a common source. The mercury lamp is also frequently used. The main requirement is strong, stable emission in the wavelength region of interest.

Detectors A phototube is the normal detector. Appropriate optical filters are used to limit the energy reaching the detector to the desired level and the desired wavelength region. Since absorption by the sample is logarithmic if a sufficiently narrow wavelength region is used, an exponential amplifier is sometimes used to compensate and produce a linear output.

Turbidimetry Turbidimetry is in a sense an extension of colorimetry in that it refers to absorption or scattering of visible light. Therefore, the same equipment is used. Wide-wavelength bands are normally suitable, since the particulate material absorbs or scatters over the entire visible region. Scattering is frequently measured at near 180° to the incident light.

Paramagnetism A few gases including O_2, NO, and NO_2 exhibit paramagnetic properties as a result of unpaired electrons. In a nonuniform magnetic field, paramagnetic gases, because of their magnetic susceptibility, tend to move toward the strongest part of the field, thus displacing diamagnetic gases. Paramagnetic susceptibility of these gases decreases with temperature. These effects permit measurement of the concentration of the strongest paramagnetic gas, oxygen.

Nephelometry Nephelometry differs from colorimetry in that it refers to measurement of light reflected or scattered from particles suspended in the sample, while colorimetry (and turbidimetry) refer to measurement of light transmitted through the sample. Nephelometers are similar to colorimeters and turbidimeters in the requirement of a light source, sample cell, and detector. Some process colorimeters are designed for simple conversion to nephelometers. Normally the reflected or scattered light is measured by a photocell placed at 90° to the transmitted light beam.

ELECTROANALYTICAL INSTRUMENTS

Conductometric Analysis Solutions of electrolytes in ionizing solvents, e.g., water, conduct current when an electrical potential is applied across electrodes immersed in the solution. Conductance is a function of ion concentrations, ionic charge, and ion mobility.

Conductance measurements are ideally suited for measurement of the concentration of a single strong electrolyte in dilute solutions. At higher concentrations conductance becomes a complex, nonlinear function of concentration requiring suitable calibration for quantitative measurements.

Gas analysis is commonly accomplished through conductance measurements. Acidic or reactive gases dissolved in liquids or quantitatively adsorbed from gas streams and gases which may be oxidized or reduced produce electrolytes that can be measured.

Conductometric analyzers utilize a conductivity cell and electronic measurement circuits which consist of the audio-frequency oscillator, an alternating-current Wheatstone bridge, and an electronic compensator for temperature variation.

Conductivity cells are simple in basic structure, consisting of two electrodes firmly spaced within an insulating chamber such that the measured impedance over the anticipated span of ion concentration will be in the range of 500 to 10,000 Ω.

Electrodeless Conductometric Measurements Special electrodeless systems for the measurement of conductance have been devised. The resistance of a closed solution loop is measured by the extent to which the loop couples two transformer coils.

Electrodeless conductance measurements are especially useful for solutions containing abrasive or fibrous solids and highly conductive and highly corrosive materials. Typical examples include hydrofluoric acid, 98 percent sulfuric acid, molten ammonium nitrate, cement slurry, and drilling mud.

Oscillometric Analysis In oscillometry, the properties measured are dielectric constant and conductance. For most measurements, a cell containing the sample is placed between the plates of a capacitor or inside a coil. The capacitor or coil is part of a radio-frequency resonant coil. Some oscillometric instruments have sensing heads that are pressed against bulk materials or are held to close tolerance within a specified distance of a moving web or film.

Oscillometry is frequently used for the measurement of water content, since water has a dielectric constant 15 to 20 times greater than most materials. Typical applications are moisture in granular materials, water in hydrocarbon and organic liquids, and water in fibers and paper.

Potentiometric Analysis Measurement of chemical electromotive force (emf) is the basis for a number of analysis methods. The emf measured is the potential difference between a sensing electrode and a reference electrode both immersed in an electrolytic solution. The potential is related to solution chemical activities through the Nernst equation. Activities are related to concentrations through activity coefficients.

Direct determination of concentration by emf measurement is somewhat inexact at concentrations above about 10^{-4} molar. The low accuracy is a result of inaccurately known activity coefficients for higher concentrations and the logarithmic relation between emf and activity. However, the extreme sensitivity of electrode response to trace concentrations makes the technique useful for dilute solutions.

Accurate potentials can be measured only under conditions approaching zero current, which minimize errors resulting from resistive potential drop and polarization effects. Voltage-measuring circuits are generally of two types: null-balance potentiometers with one or two stages of dc amplification or high-input resistance devices such as vacuum-tube voltmeters or vibrating-reed electrometers. Some form of temperature compensation is generally included.

A stable reference potential is a prime requirement for accurate measurement. The most commonly used reference electrodes are the calomel electrode and the silver–silver chloride electrode. Indicator electrodes are usually noble-metal electrodes or membrane electrodes. Membrane electrodes are widely used in pH measurement and specific ion measurement.

Measurement of pH The primary detecting element in pH measurement is the glass electrode. A potential is developed at the pH-sensitive glass membrane as a result of differences in hydrogen-ion activity in the sample and a standard solution contained within the electrode. This potential measured relative to the potential of the reference electrode gives a voltage which is expressed as pH.

Instrumentation for pH measurement is among the most widely used process-measurement devices. Rugged electrode systems and highly reliable electronic circuits have been developed for this use.

Specific-Ion Electrodes In addition to the pH glass electrode, specific for hydrogen ions, a number of electrodes which are selective for the measurement of other ions have been developed. This selectivity is obtained through the composition of the electrode membrane (glass, polymer, or liquid-liquid) and the composition of the electrode. These electrodes are subject to interference from other ions, and the response is a function of the total ionic strength of the solution. However, electrodes have been designed to be highly selective for specific ions, and when properly used these provide valuable process measurements.

Coulometric Analysis Coulometric analysis is based on exhaustive quantitative electrolysis of the species being measured or electrolytic generation of a reagent which reacts quantitatively with the measured species. Concentration of the measured component is related to total current by Faraday's law.

Coulometric process analyzers commonly are of continuous flow-through design in which current is determined by the number of moles per second entering the electrolysis cell. For a known flow rate, concentration can then be determined.

Required instrument components include the electrolysis cell, current-generating and -measuring circuits, and amplifiers with suitable output for display and control.

Electrolysis cells are of rugged construction and may frequently be placed directly in the process line or a bypass loop in the process line. In some cells, additional electrodes to control electrolysis current are incorporated. In some systems coulometric generation of reagent to react with the sample is obtained by including a base-metal electrode which is gradually consumed.

Polarographic Analysis In polarographic analysis, electrolysis current is proportional to the concentration of electrolyzed material. Current in the sample phase is limited by diffusion. In contrast to coulometric analysis, only a small fraction of the material is electrolyzed.

Polarographic cells consist of two electrodes, one of noble metal and one of base metal which is slowly consumed. The electrolyte is usually a cellulose-base potassium chloride gel held in place by and separated from the sample stream by a membrane of Teflon or silicone rubber. The sample diffuses into the cell and is electrolyzed. These cells are temperature-sensitive, and a temperature-compensating circuit is generally built into the sensor.

MOISTURE MEASUREMENT

Moisture measurements may be divided into two broad categories, absolute-moisture methods and relative-humidity methods. The absolute methods are those which provide a primary output which can be directly calibrated in terms of dew-point temperature, molar concentration, or weight concentration. Loss of weight on heating is the most familiar of these methods. The more specialized methods discussed are listed in approximate decreasing order of directness of measurement of moisture. The relative-humidity methods are those which provide a primary output which can be more directly calibrated in terms of percentage of saturation of moisture.

Dew-Point Method For many applications the dew point is the desired moisture measurement. When concentration is desired, the relation between water content and dew point is well known and available. The dew-point method requires an inert surface whose temperature can be adjusted and measured, a sample gas stream flowing past the surface, a control for adjusting the surface temperature to the dew point, and a means of detecting the onset of condensation.

Although the presence of condensate can be detected electrically, the original and most often used method is the optical detection of change in light reflection from an inert metallic-surface mirror. Some instruments measure the attenuation of reflected light at the onset of condensation. Others measure the increase of light dispersed and scattered by the condensate instead of, or in addition to, the reflected-light measurement. Surface cooling is obtained with an expendable refrigerant liquid, conventional mechanical refrigeration, or thermoelectric cooling. Surface-temperature measurement is usually made with a thermocouple or a thermistor.

Sharpness of onset of the dew point may be reduced or a shift in the apparent dew point may occur because of the presence in the sample of a component which condenses at a higher temperature than the anticipated dew-point range, gradual deposition of a low-vapor-pressure hygroscopic film, or graduation deposition of a low-vapor-pressure hydrophobic film, dust particles, or the accumulation of scratches on the mirror surface. The error of temperature measurement is about ± 0.56°C (1°F) in the higher humidity range, but the error increases to 2.2 to 2.8°C (4 to 5 °F) in the -62 to -73°C (-80 to -100°F) range. In the range of 0 to approximately -18°C (32 to approximately 0°F) the deposition may be of either supercooled liquid drops or frost. At low dew points the cooling-system capacity may be limiting.

Electrolysis Method In an electrolytic cell the water vapor from a flowing gas stream is quantitatively absorbed into a film of concentrated phosphoric acid. By means of electrodes embedded in the acid film, the absorbed water is simultaneously and quantitatively electrolyzed. The mass rate of water flowing through the cell is given directly by the electrolytic current. Since use of the instrument requires the calibration of only the ammeter and flowmeter and is based on the easily verifiable assumption that absorption and electrolysis are quantitative, the method is one of the most fundamental available.

The method requires a vapor sample through the detector cell. However, models are available which include a countercurrent gas-liquid stripper column for removing water from a liquid stream continuously, with the stripper gas transporting the water to the detector. The stripper's efficiency should be evaluated for samples with a high water solubility or of high viscosity.

The method is also applied to moisture in solids. A sample of the solid is manually placed in a drying chamber where it is heated and the chamber simultaneously purged with dry gas which transports the water vapor to the detector. Integration of the electrolysis current gives the amount of water driven off by heating.

The following are problems which should be considered in using this method:

1. The method should not be used under conditions in which the gas stream may at times reach saturation.
2. The presence of oil in the sample may eventually coat the acid surface in the cell.
3. The phosphoric acid will act as a polymerization catalyst for olefins, thus creating an oil film.
4. The platinum electrode and phosphoric acid act to re-form water from either hydrogen or oxygen in an excess of the other, giving high outputs in hydrogen- or oxygen-containing gas streams. This effect may be reduced with rhodium electrodes.
5. The method should not be used for sample streams that react with the acid film such as bases and alcohols.

The response time varies inversely with the moisture concentration and is less than 1 s in the 100-plus-ppm range, a few minutes in the 1- to 10-ppm range, and several hours in the fractional-ppm range. The fractional-ppm range is generally not usable without special operating techniques.

Karl Fischer Method This general laboratory method has been automated for routine use. It includes a moisture extraction, a reaction, and titration. The supplies of both extractant and reagent which must be maintained generally have short effective shelf lives.

The method may be used on any sample that does not react with the extractant, reagent, or handling equipment and in which water is considerably less soluble than in the extractant. This includes many vapors, liquids, and even solids which can be crushed or otherwise dispersed to permit extraction of the moisture. The water content may be computed directly from the sample size and titrant used. The volumes required and the dryness of the extractant limit the lower detection limit to from 5 to 10 ppm.

Dew-Point (Salt-Phase-Transition) Method In an environment of variable relative humidity, a hygroscopic salt will undergo one or more phase transitions at specific humidities. The transition causes a large change in electrical resistivity. Dew-point analyzers based on this phenomenon usually consist of a double helix of wire on a cylinder which is coated with a thin film of the salt solution. The analyzer contains a low-voltage power supply to the wire helix and a method of measuring the film temperature. In operation the wet film of low resistance permits the flow of high current through the film, thus heating and drying it until the transition is achieved. At this point the resistance increases, thus reducing heating. The device then is self-regulating, maintaining the film at the temperature required to maintain the gas near its surface at the relative humidity of the transition. Lithium chloride is the most frequently used salt. Noble-metal electrodes are normally used.

At any given sample temperature (dry-bulb temperature) the detector range is from saturation (100 percent relative humidity) to the relative humidity of the phase transition (approximately 11 percent relative humidity for lithium chloride). The detector is further limited to the dew-point range of −46 to +60 or 66°C (−50 to +140 or 150°F). The method is essentially independent of the actual sample-gas temperature as long as it is within the operating range.

The measurement is continuous, always near equilibrium, and the heat capacity of the detector is fairly low; so the time response should be satisfactory for most applications. The error of the method may approach 0.56°C (1°F) as long as the salt film is free of contamination.

General Analytical Methods Several more general analytical methods such as gas and liquid chromatography, infrared absorption, and microwave absorption may also be used for moisture analysis. See also "Process Chemical-Composition Analyzers" and "Oscillometric Analysis."

Piezoelectric Method A piezoelectric crystal in a suitable oscillator circuit will oscillate at a frequency dependent on its mass. If the crystal has a stable hygroscopic film on its surface, the equivalent mass of the crystal varies with the mass of water sorbed in the film. Thus the frequency of oscillation depends on the water in the film. The analyzer contains two such crystals in matched oscillator circuits. Typically, valves alternately direct the sample to one crystal and a dry gas to the other on a 30-s cycle. The oscillator frequencies of the two circuits are compared electronically, and the output is the difference between the two frequencies. This output is then representative of the moisture content of the sample. The output frequency is usually converted to a variable dc voltage for meter readout and recording. Multiple ranges are provided for measurement from about 1 ppm to near saturation.

The dry reference gas is preferably the same as the sample except for the moisture content of the sample. Other reference gases which are adsorbed in a manner similar to the dried sample gas may be used. The dry gas is usually supplied by an automatic dryer.

The method requires a vapor sample to the detector. Mist striking the detector destroys the accuracy of measurement until it vaporizes or is washed off the crystals. Water droplets or mist may destroy the hygroscopic film, thus requiring crystal replacement. Vaporization or gas-liquid strippers may sometimes be used for the analysis of moisture in liquids.

Capacitance Method Several analyzers utilize the high dielectric constant of water for its detection in solutions. The alternating electric current through a capacitor containing all or part of the sample between the capacitor plates is measured. Selectivity and sensitivity are enhanced by increasing the concentration of moisture in the cell by filling the capacitor sample cell with a moisture-specific sorbent as part of the dielectric. This both increases the moisture content and reduces the amount of other interfering sample components. Granulated alumina is the most frequently used sorbent. These detectors my be cleaned and recharged easily and with satisfactory reproducibility if the sorbent itself is uniform.

Heat-of-Sorption Method This analyzer detects moisture in vapors by measuring the heat of sorption of water onto a desiccant. Two cells containing desiccant and thermistors in a constant-temperature zone are used. The analyzer dries a part of the sample stream. The dried sample and the process sample are directed alternately through the two cells, where one cell dries the wet stream and the wet cell is dried by the dry stream on a 1- to 2-min cycle. The cell being dried is cooled by desorption of water, and the cell being wetted is warmed by sorption of water. The thermistor bridge measures the difference in temperature between the two cells.

The choice of desiccant for the cells is determined by compatibility with the sample including impurities and by the required sensitivity. Normal ranges are from 0 to 10 ppm up to 0 to 1000 ppm.

Although the full cycle time is 2 to 4 min, thermal and humidity equilibration is not reached for about 15 min at low-ppm concentrations.

Psychrometric Method The basic sling-psychrometer method compares the temperature depression of a wet-bulb thermometer with the dry-bulb or ambient temperature. The method has been automated with various modifications to furnish continuous wicking, cooling, and temperature sensors which provide an electrical output.

Resistance Method (Conductance) A number of resistive-type sensors which may be used to measure moisture are available. These sensors normally depend on the changes in resistance of surfaces exposed to the atmosphere, the resistance being measured by means of a low potential gradient so that the electric current flow through the surface produces negligible heating. Detector surfaces may be of a salt such as lithium chloride, polymeric films, bare nonconducting materials, or semiconductors. Any high-valued resistor is, in fact, a sensor of this type. Different combinations of surface-material component configurations, voltage, and current measurement may be required to obtain a sensor for the contemplated humidity range and sensitivity required. Such sensors are generally economical and structurally rugged, but they may not withstand a high-velocity or abrasive sample flow. The calibration will usually be dependent on temperature. The time response rate is generally slow at low temperatures, and the calibration will not be stable in the presence of materials which coat, dilute, react with, or dissolve into the surface. These detectors are normally used to measure relative humidity in gases.

SAMPLING SYSTEMS FOR PROCESS ANALYZERS

The sampling system consists of all the equipment required to present a process analyzer with a clean representative sample of a process stream and to dispose of that sample. When the analyzer is part of an automatic control loop, the reliability of the sampling system is as important as the reliability of the analyzer or the control equipment.

Sampling systems have several functions. The sample must be withdrawn from the process, transported, conditioned, introduced into the analyzer, and disposed of. Probably the most common problem in sample-system design is the lack of realistic information concerning the properties of the process material at the sampling point. Another common problem is the lack of information regarding the conditioning required so that the analyzer may utilize the sample without malfunction for long periods of time. Some samples require enough conditioning and treating that the sampling systems become equivalent to miniature online processing plants. These systems possess many of the same fabrication, reliability, and operating problems as small-scale pilot plants except that the sampling system must generally operate reliably for much longer periods of time.

Figure 22-87 illustrates schematically a typical dual-stream chromatographic-analyzer sampling system of the type usually employed when no sample conditioning is required beyond pressure regulation at the sampling points (pressure regulators are not shown).

Selecting the Sampling Point The selection of the sampling point is based primarily on supplying the analyzer with a sample whose composition or physical properties are pertinent to the control function to be performed. Other considerations include selecting locations that provide representative homogeneous samples with minimum transport delay, locations which collect a minimum of contaminating material, and locations which are accessible for test and maintenance procedures.

Sample Withdrawal from Process A number of considerations are involved in the design of sample-withdrawal devices which will provide representative samples. For example, in a horizontal pipe

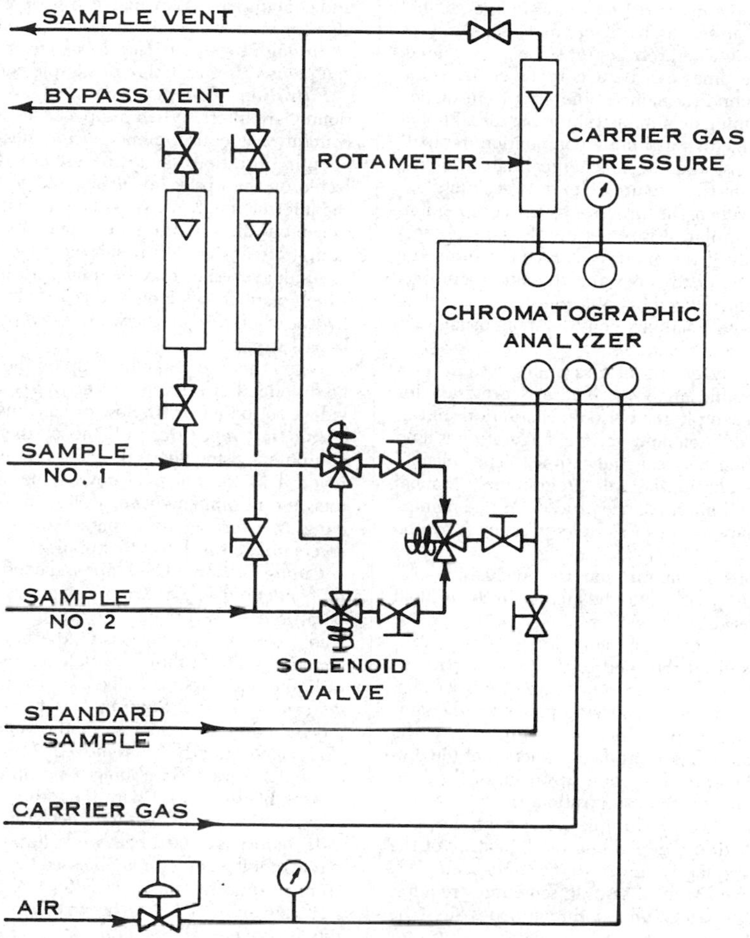

FIG. 22-87 Simple two-stream chromatographic-analyzer sampling system. The solenoid valves are operated by the chromatograph programmer to coincide with the chromatographic-analysis cycle.

carrying process fluid, a sample point on the bottom of the pipe will collect a maximum amount of rust, scale, or other solid materials being carried along by the process fluid. In a gas stream, such a location will also collect a maximum amount of liquid contaminants. A sample point on the top side of a pipe will, for liquid streams, collect a maximum amount of vapor contaminants being carried along. Bends in the piping which produce swirls or cause centrifugal concentration of the denser phase may cause maximum contamination to be at unexpected locations. Two-phase process materials are difficult to sample for a total-composition representative sample.

A typical method for obtaining a sample of process fluid well away from vessel or pipe walls is an eduction tube inserted through a packing gland. This sampling method withdraws liquid sample and vaporizes it for transporting to the analyzer location. The transport lag time from the end of the probe to the vaporizer is minimized by using tubing having a small internal volume compared with pipe and valve volumes.

This sample probe may be removed for maintenance and reinstalled without shutting down the process. The eduction tube is made of material which will not corrode so that it will slide through the packing gland even after long periods of service. There may be a small amount of process-fluid leakage until the tubing is withdrawn sufficiently to close the gate valve. A swaged ferrule on the end of the tube prevents accidental ejection of the eduction tube prior to removal of the packing gland.

The section of pipe surrounding the eduction tube and extending into the process vessel provides mechanical protection for the eduction tube.

Sample Transport Transport time, the time elapsed between sample withdrawal from the process and its introduction into the analyzer, should be minimized, particularly if the analyzer is an automatic analyzer-controller. Any sample-transport time in the analyzer-controller loop must be treated as equivalent to process dead time in determining conventional feedback controller settings or in evaluating controller performance. Reduction in transport time usually means transporting the sample in the vapor state.

Some of the things to be considered in selecting sample-line material are that the structural strength or protection must be compatible with the area through which the sample line runs; line size and length must be small enough to meet transport-time requirements without excessive pressure drop or excessive bypass of sample at the analyzer input; line size and internal-surface quality must be adequate to prevent clogging by the contaminants in the sample; the prevention of a change of state of the sample may require insulation, refrigeration, or heating of the sample line; and sample-line material must be such as to minimize corrosion due to sample or environment.

Some common solutions to vapor-handling sample-line problems are to use ⅛- to ¼-in-diameter stainless-steel tubing where feasible, limit sample-line length to 61 m (200 ft) or less, use sample flow rates in the 0.5- to 3.0-L/min range, limit sample temperature when the line is steam-traced by using low-pressure [less than 376 kPa (40 lbf/in^2)] wet steam and a steam trap for discharge of condensate, and use essentially vaportight insulation when refrigerant tracing is required.

Thermal tracing of sample lines can be accomplished by spiral winding of a copper tube around the sample line prior to insulation or by using duplex parallel tubing or suitable tubing bundles. Plastic-tubing sample lines are satisfactory if sample composition, temperature, and pressure limitations, and the mechanical nature of the installation and environment of the line are taken into account.

Multistream Sampling When the process-stream compositions at several sample points are similar, it is sometimes feasible to use a single analyzer for all of them. If the analyzer is used for automatic control, the added control-loop dead times and the multiloop overlapping reliability problems introduced by multiple-stream sampling must be considered. Timing and reliability considerations usually are not as important in monitoring applications.

Figure 22-87 shows the analyzer end of a sampling system for a monitoring gas chromatographic analyzer. There is provision for introduction of a calibration sample for periodic calibration. In this case a "double block and bleed" solenoid-valve configuration is used to prevent cross contamination between the samples. The solenoid valves are operated by the chromatograph programmer. Manual block valves for calibration and maintenance procedures and manual flow-adjusting valves are shown. Fresh sample is assured by the bypass vent.

In the double-block-and-bleed configuration, the process-side solenoid diverts any leakage to the vent, while the analyzer-side solenoid valve blocks any unvented leakage.

Sample-Return Point For reasons of economy or safety, all or part of the sample stream is often returned to the process, particularly if a large bypass stream is required to minimize transport lag. In order to circumvent pumping or pressurizing problems, the sample is usually returned to some low-pressure point downstream of the sample point. If the sample stream is a significant fraction of the process stream, its return to a low-pressure point upstream of the sampling point may introduce undesirable composition-time effects.

Other techniques for disposing of the sample are through the plant disposal system, flare line, or drain, depending on the nature of the sample. Vapor samples are frequently vented to the atmosphere. Disposal of contaminants removed from a sample sometimes requires special handling. In all cases safety and pollution codes must be observed.

Calibration Sample Most analytical instruments require the periodic introduction of a standardization sample for the purposes of recalibration or maintenance of the operator's confidence in the analytical results, particularly if the analyzer is part of an analyzer-controller. This periodic introduction may be handled by manual operation of the appropriate valves as indicated in Fig. 22-87.

If the standardization sample is to be introduced automatically for automatic recalibration or standardization of the analyzer, the automatic sample introduction and handling must be as reliable as the process-sampling system and an integral part of it.

Sample Conditioning Sample conditioning usually involves the removal of contaminants or some deleterious component from the sample mixture and/or the adjustment of temperature, pressure, and flow rate of the sample to values acceptable to the analyzer.

Some of the more common contaminants which must be removed are rust, scale, corrosion products, deposits due to chemical reactions, and tar. In sampling some process streams, the material to be removed may include the primary process product such as polymer or the main constituent of the stream such as oil. In other cases the material to be removed is present in trace quantities, for example, water in an online chromatograph sample when even minute quantities damage the chromatographic column packing.

It should be borne in mind that when contaminants or other materials which will hinder analysis represent a large percentage of the stream composition, their removal may significantly alter the integrity of the sample. In some cases, removal must be done as part of the analysis function so that removed material can be accounted for. In other cases, proper calibration of the analyzer output will suffice.

If reliable information on sample composition, physical properties, and contaminants is available, a wide variety of information is available for sample-conditioning equipment design. For example, Secs. 18 through 21 of this *Handbook* cover separation techniques which can be used in the design of sample-conditioning equipment.

Filtration The most commonly used elements in sample conditioning are filters. When sample conditions of viscosity, pressure, and contaminant entrainment permit, surface filtration, as opposed to "depth" filtration, should be used. That is, the contaminant is collected on the upstream surface of the filter rather than penetrating the filter element. A bypass flow is swirled over the filter element to sweep out the contaminant stopped by the filter. A smaller analyzer sample stream flows through the filter element.

Similar schemes may be used with hydrophobic filter elements in which water is selectively rejected. In removing free water from a hydrocarbon stream, the water is swept from the filter surface by the bypass stream.

When the filter cake built up on the filter element surface cannot be dislodged effectively by the bypass stream, it is necessary to provide a periodic reverse flow or backflush through the filter.

Inertial Separators Some of the more common conditioning equipment using inertial forces for separation are the simple flow reversal, "knockout pot" device, the more effective cyclone separators, and impingement separators in which material to be separated clings to a surface after contact and is drained off by gravity. Coalescers utilize this latter technique.

Condensation Cold taps and condensers are used to remove easily condensed vapors from a vapor sample. The required low temperature may be obtained by the use of such agents as cooling water, cold process fluids, refrigerant, thermoelectric cooling, or vortex tube (Ranque-Hilsch tube). In small condensers the countercurrent flow of liquid and vapor may be a problem.

Scrubbing and Stripping Scrubbing or stripping has the disadvantage of introducing another material into the sample which must subsequently be removed. Spray, bubbler, and wetted-wall devices are used. Scrubbing is useful for removing materials which are soluble in or react with the scrubbing liquid. Scrubbing may also provide a vehicle which will mechanically remove such hard-to-provide materials as tars, heavy oils, or polymers. Stripping with a countercurrent flow of gas may be used to remove the volatile components from a liquid sample. This latter technique has been used to sample parts-per-million levels of water in liquid hydrocarbons using a modified wet-wall technique.

Adsorption Techniques A group of techniques related to gas chromatography and gas adsorption are becoming increasingly useful. Vapor sample is introduced into an adsorption column which removes the unwanted components. Before the column becomes loaded with these unwanted materials, they are backflushed. These techniques may require considerable auxiliary equipment.

A simple example of this technique which has been used on a larger scale is the "heatless" dryer. Two parallel adsorbing beds alternately adsorb water from the high-pressure air stream and desorb water by backflush with low-pressure air which has been through the high-pressure adsorption part of the cycle.

Adaptation of Analytical Methods A number of analytical techniques may be used for sample conditioning provided proper precautions are taken to ensure quantitative removal, reaction, etc., or a consistent degree of separation. For example, distillation can be used when the degree of separation is controllably consistent and reproducible rather than quantitative.

Temperature, Pressure, and Flow Control of temperature, pressure, and flow rates in sampling systems is usually a simple matter if fouling can be prevented. Many forms of controllers are available in sizes appropriate for sampling systems. Steam heating is preferred to electrical heating in process areas requiring explosionproof equipment.

Flow adjustment is usually manual in sampling systems when pressures are stable or pressure regulators are employed. Glass-tube flow indicators or rotameters are usually sufficiently repeatable and give a visual indication of sample flow rate and contamination. The use of differential-pressure flow controllers is widespread. Successful flow control of sample streams usually requires a high degree of removal of solid- or tar-type contaminants.

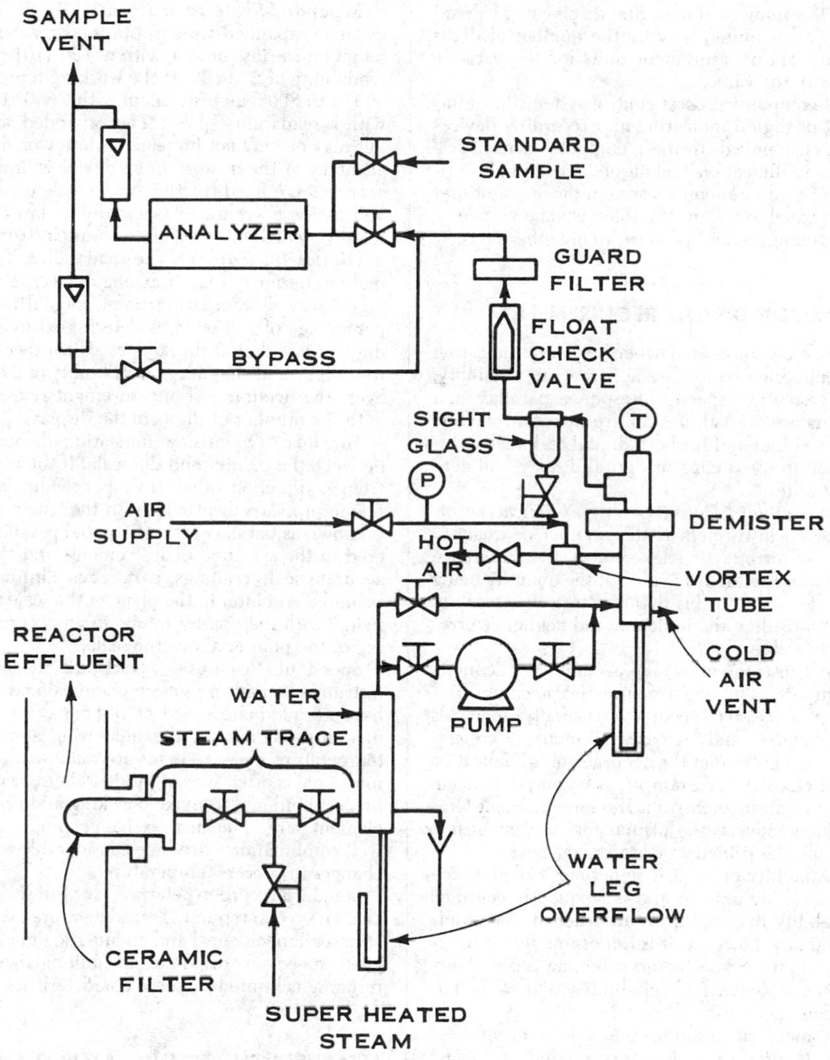

FIG. 22-88 A typical sampling system requiring a moderate amount of conditioning.

Pressure control is not a problem if there is sufficient differential pressure available to move the sample through the pressure regulator, the sample conditioner, and the analyzer and then return it to the process or disposal system. Problems arise when the pressure available is not adequate. Mechanical sample pump boosters are notoriously troublesome and expensive in the small sizes, and special materials are needed.

When applicable, one solution to low-pressure differential is to use small air, steam, or water aspirators to reduce the pressure at the analyzer sample outlet.

Example 1 Figure 22-88 illustrates a relatively simple sampling system. It is designed to provide a carbon-black reactor-effluent gas sample to a gas chromatograph. Carbon black, water, and water-soluble sulfur compounds are the materials to be removed. The fine-pore ceramic filter is cleared of embedded carbon black by periodic manual backflush with superheated steam.

Water, soluble sulfur compounds, and oil are removed by two stages of simple shell-and-tube heat exchangers using water and cold air as the cooling media. A float check valve protects the analyzer from accidental flooding in case of failure of the water-leg overflows which control sample pressure. The sight glass provides a manual inspection point for visual check of the sample condition.

INDICATING AND RECORDING INSTRUMENTS

Most display instruments can be classified as analog or digital. Analog display is characterized by a continuously varying response, for example, a pointer on a scale, an ink line on a chart, or a light trace on a cathode-ray tube. Digital display is characterized by numerical information. Typical examples include digital voltmeters, digital counters, and typed or printed log sheets. Digital data can also be recorded in coded form on magnetic or paper tape.

With analog display devices it is possible to tell at a glance the approximate value as well as the relative value with respect to full scale. However, analog devices are not as quickly and accurately read as digital devices. The numerical or digital display can be read as precisely as desired, being limited only by the number of digits contained in the display. The measurement units for the variable may also be displayed with the value.

The advent of digital-computer process-control systems has stimulated the development of digital indicating and recording devices that are functional and economical. In these computer control systems person-machine communication techniques are digitally oriented. In a typical digital-computer control system the operator uses a push-button operator's panel to call up variables, change set points, adjust process-control parameters, and perform many other tasks.

OPERATING-INFORMATION DISPLAY REQUIREMENTS

A number of factors must be considered when selecting indicating and recording instruments. Such factors are accuracy, repeatability, reliability, readability, resolution, speed of response, parallax, and cost. Not all these factors are critical, but all are important. While the same factors must be considered for both digital and analog display devices, the decision to use analog or digital display will generally be based on other criteria.

Accuracy *American Standard Definitions of Electrical Terms* states, "The accuracy of an instrument is the number or quantity which defines its limit of error. . . . The error is the difference between the indicated value and the true value of the quantity being measured." In defining accuracy in this manner, it is customary to assume negligible error in reading the indicator and negligible error in determining the "true" value.

It is important to select instruments whose accuracy is adequate. However, it is equally important to recognize practical accuracy limitations. Specification of excessive accuracy generally increases investment, decreases reliability, and increases maintenance costs.

It is also important to recognize that the accuracy of an indication on the control-room panelboard, for example, is a composite accuracy of the indicator, the transmission from the measurement location, the transducer which converts the primary measurement into the transmitted signal, and the primary measurement device.

Repeatability Repeatability or, as it is sometimes called, *precision* of measurement, is closely akin to and is frequently confused with accuracy. Repeatability may be up to an order of magnitude better than absolute accuracy; however, it is impossible for accuracy to be better than repeatability. Some factors affecting repeatability are hysteresis, dead band, stability, and reliability. As these factors improve, the cost generally goes up.

In many applications, major process objectives are to repeat operating conditions known to produce satisfactory product, to obtain information on trends, or to compare results under prescribed variations of operating conditions. For these objectives, repeatability rather than absolute accuracy is the decisive criterion.

As with accuracy, the repeatability of an indication on the panelboard is the composite of the repeatability of the entire chain of measurements and translations from the process to the control-room panelboard.

Readability Since the objective of indicating and recording instruments is to present information to process operators, readability is a major consideration in selection. The final reading of the display includes instrument errors as well as human errors. Readability requirements depend primarily on the use to be made of the information.

A field-mounted indicator for a process variable such as pressure, temperature, or flow needs to have a long scale with bold scale marking and a broad pointer so that it can be quickly read from several feet away. Control-room indicators usually have smaller scales and finer markings because of space limitations and a greater need for indicator precision. Test instruments such as the pressure gauges used for calibrating pneumatic instruments have a large-diameter circular dial with a long knife-edge pointer.

Readability may be defined as the smallest-scale increment to which the reading may be determined expressed as a percentage of full scale.

Expanded-Scale Instruments Readability can be increased by using an expanded-scale (suppressed-zero) instrument. A 0° to 1000° temperature instrument with a 4-in (101.6-mm) scale may have a readability of 2° to 3°. If the working-temperature range is between 600° and 800°, an instrument with a 600° to 800° scale may be used with a readability of ½°. The expanded scale increases readability but may or may not increase accuracy or repeatability. If the usable accuracy of the measurement device is limited by readability or the accuracy is a fixed fraction of the scale used, then an expanded scale will increase accuracy. Generally a zero check on accuracy is not readily available for expanded-scale instruments.

Digital Indicators The readability of digital indicators is normally superior to that of analog devices, although there is no meaningful way of comparing them. Digital readability, expressed as a percentage of full scale, is determined by the number of displayed digits. Most digital displays used in process operator's panels have four digits and therefore a readability of 0.01 percent of scale. However, the accuracy of measurement does not necessarily improve with the number of digits in the display.

Parallax On analog indicating devices there is a separation between the pointer and the scale. If the instrument reading is taken from a direction other than perpendicular to the scale face, the pointer appears displaced from the true reading. This displacement is known as parallax error. Potential parallax error should be considered in the selection of instruments and their locations. For highly accurate scale readings, parallax is eliminated by placing a mirror behind the pointer in the plane of the scale and matching the pointer image with the pointer to obtain an accurately perpendicular viewing of the pointer above the scale.

Speed of Response Nearly all the indicating and recording instruments used in modern chemical and related-process industries have an adequate speed of response. The limiting factor in many measuring systems is the measuring element. For instance, when temperature measurements are made with thermocouples, the thermocouple is often placed inside a thermal well to protect it from the process fluid and to make checking and replacing the thermocouple element safe. The heat-exchange properties of the thermocouple-well combination introduce considerable time lag in the response to changes in process temperature.

Standard recorder-pen travel for full scale is 2 to 3 s with ¼ s available on special request. Transducers are available which will convert almost all mechanical and pneumatic signals to electrical signals with good speed of response. In null-balance instruments, speed of response is limited by servo characteristics.

MEASUREMENT-TO-INDICATOR TRANSDUCERS

The measurement-display device is generally located at least somewhat remotely from the process-variable measurement point. Therefore, the actuating signal for the indicator, recorder, or controller is normally different from the process variable being measured. The process variable is sensed by some type of electrical, mechanical, chemical, or combination device that converts the process variable to a motion or signal that can be transmitted to the display device only a few inches or many feet away.

Some examples of the more common types of transducing devices are discussed here. In the case of electronic indicating equipment, the transition from vacuum tubes through solid-state discrete components to integrated circuits has produced many differences between older equipment currently in use and new equipment.

Rotational Motion or Angular Position

Gear Train Rotary motion and angular position are easily transduced by various types of gear arrangements. A gear train in conjunction with a mechanical counter, shown in Fig. 22-89, is a direct and effective way to obtain a digital readout of shaft rotations. The numbers on the counter can mean anything desired, depending on the gear ratio and the actuating device used to turn the shaft. A pointer attached to a gear train can be used to indicate a number of

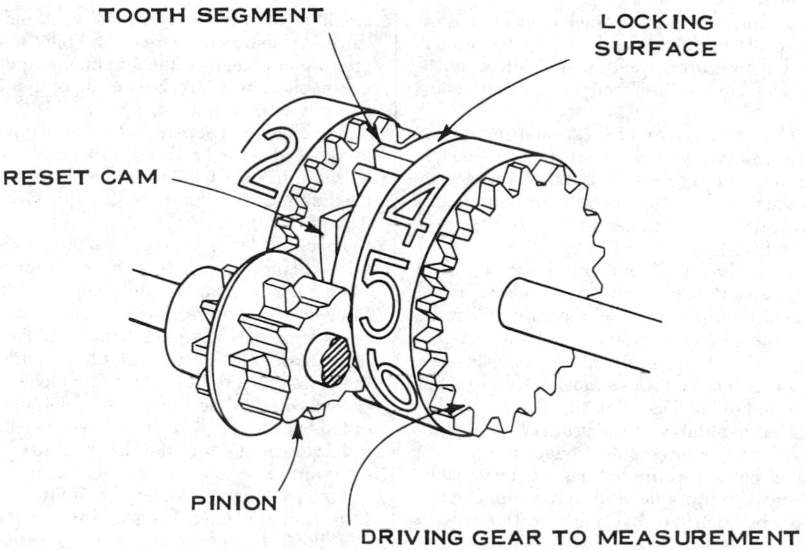

FIG. 22-89 Mechanical counter using interrupted-gear transfer between decade-counter wheels. The pinion and left wheel are held stationary by two of the pinion's long teeth and the right wheel's locking ring. To move the left wheel, the right wheel's tooth segment engages the pinion teeth, turning the pinion to the next locking position and the left wheel to the next figure.

revolutions or a small fraction of a revolution for any specified pointer rotation.

Lever or Pointer Lever arrangements are commonly used as direct couplings to convert a physical position to an indicating pointer. Direct-acting field-mounted indicating units are good examples.

Synchromotor Rotary position is quite often transduced, indicated, and transmitted by the synchromotor (Fig. 22-90). Motors with three-phase stators and two-phase rotors are connected to the same ac power line. Rotors R_1 and R_2 remain stationary unless moved by an external force. If primary rotor R_1 is turned, R_2 will follow with 1 to 3° lag.

Linear Position Linear-position transducers are similar to rotary transducers in that the motion or position is easily and economically converted to a pointer or display through the use of gears and levers. The forms these take depend on the ingenuity of the designer. Other transducers are more specifically used with linear position.

Differential Transformer These devices produce an ac electrical output from linear movement of an armature. They are very versatile in that they can be designed for a full range of output with any range of armature travel up to several inches. Figure 22-91 shows one used for long armature travel. The transformers have one or two primaries and two secondaries connected to oppose each other. With an ac voltage applied to the primary, the output voltage depends on

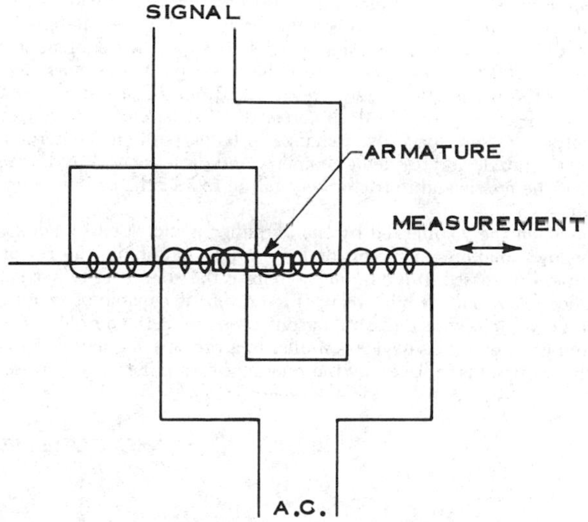

FIG. 22-91 Differential transformer used for large motions.

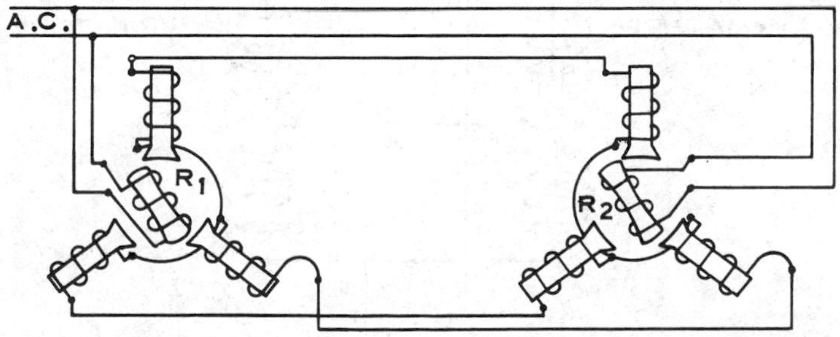

FIG. 22-90 Schematic synchromotor transmission and indicating system.

the position of the armature and the coupling. Such devices produce accuracies of 0.5 to 1.0 percent of full scale and are used to transmit forces, pressures, differential pressures, weights, and others up to 1524 m (5000 ft). They can also be designed to transmit rotary motion.

Inductance Bridge This device is used for transmitting indications from inside sealed instruments such as flowmeter manometers and rotameters. Transmitter and receiver coils are connected as shown in Fig. 22-92, and the motion of the soft-iron-armature transmitter creates magnetic fields in the coils such that the receiver armature positions itself similarly.

Capacitance Level positions of liquids or solids may be obtained with continuous capacitance probes. An insulated metal electrode probe is one side of the capacitor, and the other side is the vessel wall. Variations of the dielectric character of the material between the electrode and the wall are measured on a capacitance bridge as the level of the air-material interface moves. The output of the bridge is usually converted to an electric dc signal.

Force or Distortion A large number of the primary sensing elements used for process-variable measurements produce a force or a distortion of a diaphragm or bellows as the response to the variable. This force or distortion is usually transduced into usable pneumatic or electric signals. See also subsection "Strain Gauges" with reference to Fig. 22-93.

Bellows A common method of transducing a pneumatic signal for indicating or recording is by the use of a bellows. Figure 22-93 shows a typical bellows-driven indicating device.

Pneumatic Amplifier and Pilot Relay On pneumatic instruments using an orifice-flapper-nozzle combination to detect motion or force changes, it is necessary that the orifice and nozzle bleed air continuously. This combination constitutes a pneumatic amplifier as illustrated in Fig. 22-94. When the baffle is very near the nozzle ($\delta \approx 0$), the air pressure inside the nozzle is approximately that of the supply and the output pressure is high. As the baffle moves away from the nozzle, the nozzle pressure and output pressure decrease because of the pressure drop across the restriction. The change in output pressure for a given change in baffle position is determined by the nozzle and restrictor geometry and dimensions. It is desirable that the nozzle and restrictor be small to reduce the consumption of air.

When the volume fed by the amplifier is increased by adding a bellows, diaphragm motor, or long length of tubing, the response time is increased. It is customary to limit the volume fed by use of a pilot relay. Such a pilot relay (Fig. 22-95) is capable of feeding a large volume with the same output pressure as the amplifier or an amplified one. This device is similar to a pressure regulator in which the set point is replaced by the pneumatic amplifier. An increase in amplifier output first closes the vent and opens the air-supply-inlet valve. Air enters to increase the pilot-relay output until the output pressure just balances the amplifier output-signal pressure. If the signal diaphragm is larger than the output diaphragm, the pilot relay also acts as an amplifier.

The change in separation of the flapper and nozzle to give 20.7- to 103.4-kPa (3- to 15-lbf/in^2) relay output is typically less than 0.0254 mm (0.001 in) when feedback is used as in Fig. 22-96. The maximum air consumption for this type of relay is approximately 0.85 m^3/s (30 ft^3/s).

Force Balance Figure 22-96 shows a beam-type pneumatic force-balance transmitter. Force from any type of measurement (shown here as that of a differential-pressure bellows) is applied to a force-balanced beam. The beam obstructs the flow of air from nozzle N, which is part of a pneumatic amplifier and relay. If P_1 is assumed to increase, the beam will tilt to raise nozzle pressure, but this increases pressure in bellows B, which restores the beam to substantially its original position. Hence, output pressure is balanced against or fed back to the differential pressure, either exactly or in some multiple depending on the lever arms. The balancing pressure is transmitted.

Electronic force-balance transmitters perform the same functions as pneumatic transmitters and in the same sequence. As shown in Fig. 22-97, force from any type of measurement, such as differential-

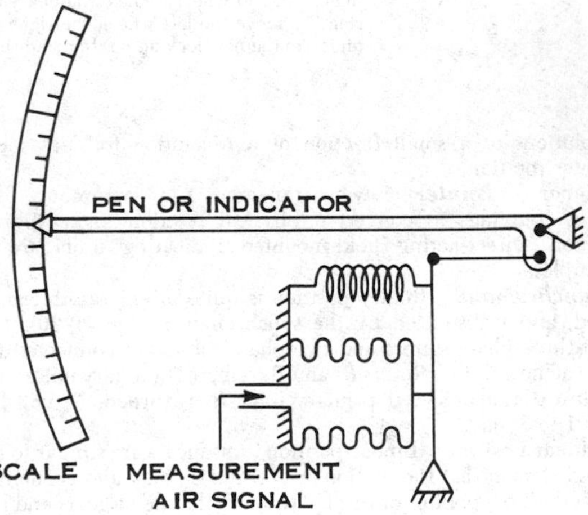

FIG. 22-93 Bellows-driven indicator.

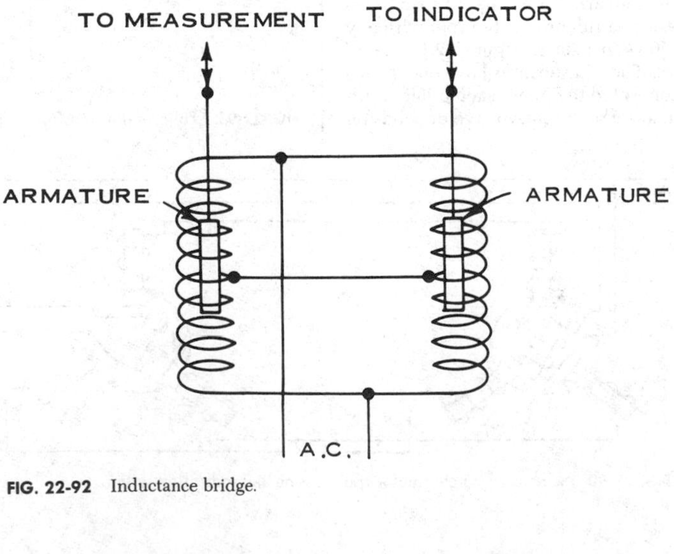

FIG. 22-92 Inductance bridge.

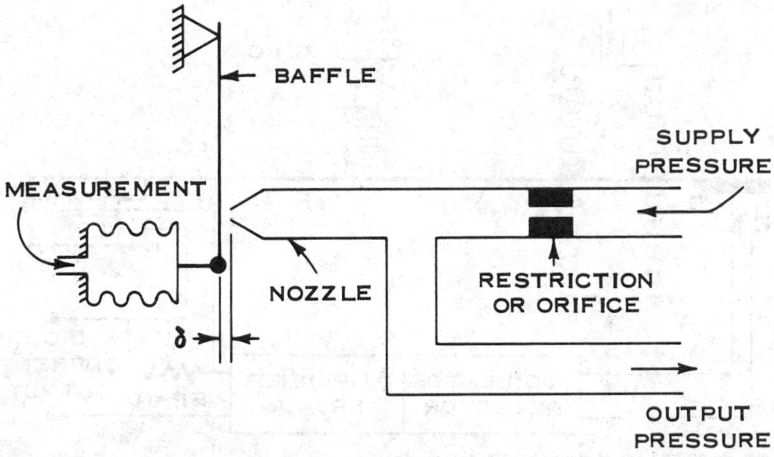

FIG. 22-94 Baffle-nozzle pneumatic amplifier in which the output pressure is dependent on the baffle-nozzle spacing δ.

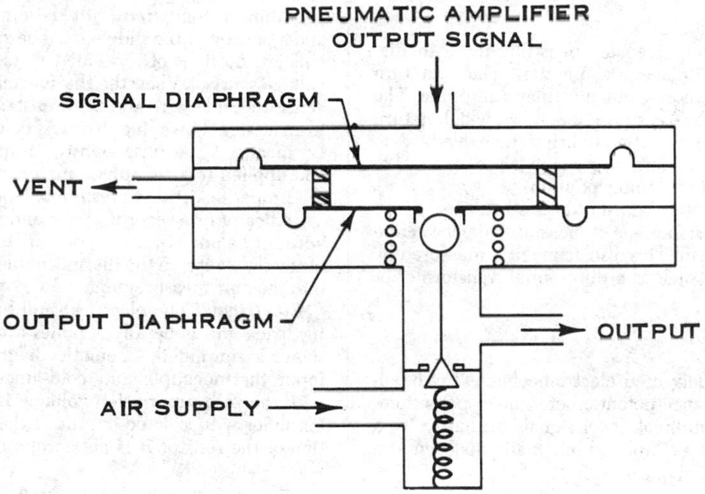

FIG. 22-95 Pneumatic pilot relay.

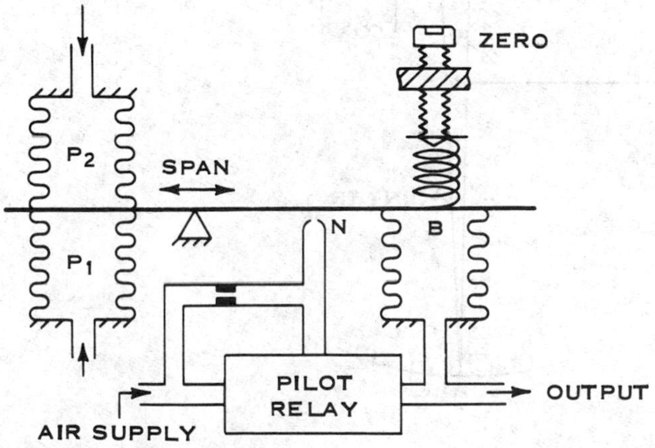

FIG. 22-96 Pneumatic force-balance transmitter.

pressure bellows, is applied to the force-balanced beam. The beam deflection produced by the force from the differential bellows is sensed by detector *D*, which changes the direct current in the output circuit of an oscillator-amplifier unit. The output current is applied to the magnet unit *M*, which supplies a rebalancing force to the beam to complete the feedback loop. The output-signal current is proportional to the measurement force.

Portions of these two types of transmitters may be combined to convert electric current signals into pneumatic signals.

Hall-Effect Sensors Some semiconductor materials exhibit a phenomenon in the presence of a magnetic field which is adaptable to sensing devices. When a current is passed through one pair of wires attached to a semiconductor, such as germanium, another pair of wires properly attached and oriented with respect to the semiconductor will develop a voltage proportional to the magnetic field present and the current in the other pair of wires. Holding the exciting current constant and moving a permanent magnet near the semiconductor produce a voltage output proportional to the movement of the magnet. The magnet may be attached to a process-variable measurement device which moves the magnet as the variable changes.

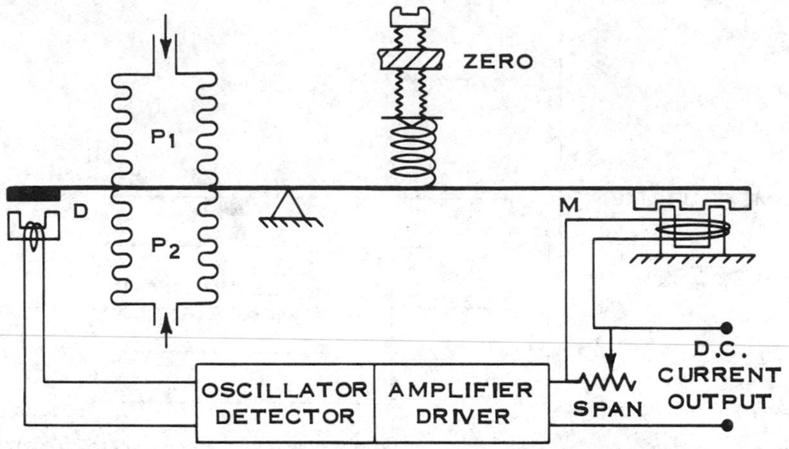

FIG. 22-97 Electronic force-balance transmitter.

Hall-effect devices provide high speed of response, excellent temperature stability, and no physical contact.

Motion Balance

Pneumatic A motion-balance mechanism is illustrated in Fig. 22-98. A measuring device displaces the pointer, which in turn moves the baffle obstructing the nozzle of a pneumatic amplifier. The amplifier output through a pilot relay moves a bellows, which in turn moves the movable fulcrum in a direction to bring the baffle back to near its original position. Thus the output pressure is proportional to the measurement displacement or pointer position.

Electrical In electrical motion-transmitter instruments, a displacement change results from some type of measuring device acting through a beam or linkage system. This displacement is measured by an electronic position detector such as a differential transformer or magnetic detector.

Null Balance

Electromechanical A widely used electromechanical null-balance transducer is the null-balance potentiometer. Such potentiometers are used for measuring millivolt-level signals primarily from thermocouples. Figure 22-99 is a simplified schematic diagram of a typical potentiometer using simplified zero suppression and span adjustments.

The thermocouple output with the necessary cold-junction compensation is compared with the emf across the zero-adjust resistor and a portion of the slide wire. The zero-adjust resistor and slide-wire emf are portions of a divided voltage supplied by the zener-diode voltage source. When the thermocouple emf is not equal to the emf across the slide wire and zero resistance, the error signal is converted to an ac signal by a line-driven synchronous switch (dc-ac converter). By means of the phase-sensive amplifier the ac signal is amplified and applied to a two-phase motor to drive the slide-wire contact in the proper direction to restore balance. The motor may also position an indicator or a recording pen and may drive a retransmitting slide wire. At balance (null), no net current flows in the left-hand portion of the slide wire; so the thermocouple circuit resistance does not enter into the emf measurement.

Electrical The electrical null-balance system uses an electrical feedback and generally supplies a dc output signal. Figure 22-100 shows a simplified schematic diagram of such a transducer. The input thermocouple and cold-junction millivolt signal, combined with the zero-suppression voltage, is applied to a chopper input dc amplifier with isolated output. If the net current through the E portion of the resistor R is not zero, a current flows in the input trans-

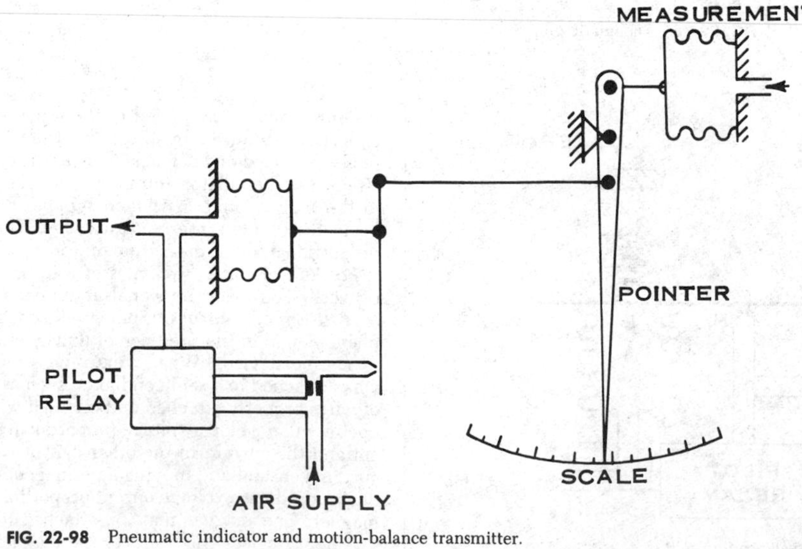

FIG. 22-98 Pneumatic indicator and motion-balance transmitter.

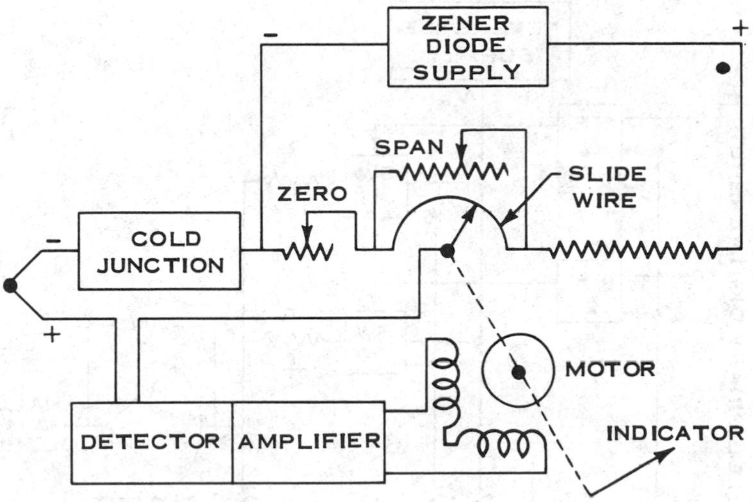

FIG. 22-99 An electromechanical null-balance potentiometer.

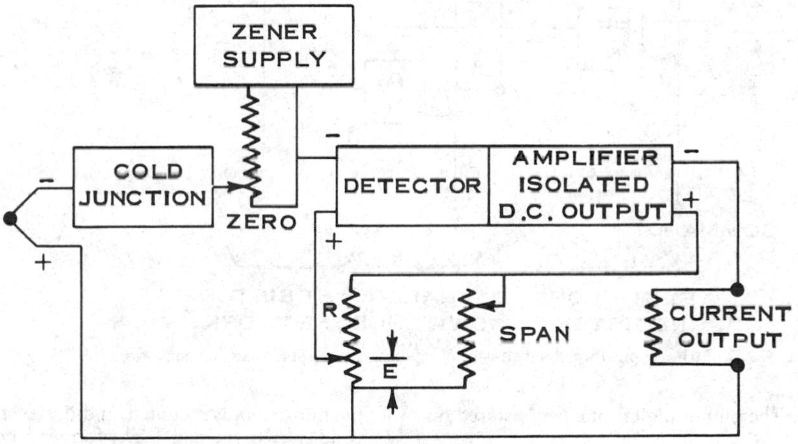

FIG. 22-100 An electronic null-balance potentiometer.

former of the low-level high-gain amplifier, with the result that the amplifier output is adjusted to move the net current in E to zero. The amplifier output current, say 4 to 20 mA, is indicated on an appropriate meter, indicator, or recorder.

Electrical Measurements The most common type of electrical indicating meters used in the process industries are the d'Arsonval dc voltmeter and ammeter. The segmented-circle face on which the pointer moves in the same plane as the scale and the edgewise type in which the pointer moves around an axis located behind a cylindrical scale are typical. While most recorder-pen drivers are servo-operated, one utilizes a magnetic-deflection motor operating directly from a dc signal to position the pen.

Analog and Digital Conversions Many analog and digital signals from measuring devices must be converted to the other form before they can be used. For example, pulse outputs from turbine flowmeters and some tachometers must sometimes be converted to continuous signals for analog-type indicating or recording. Pulse information may be used directly by a digital indicator for totaling purposes.

When digital-computer controllers are used, the analog process-variable input signals must be converted to electric digital signals to be handled by the computer. Likewise the digital signals generated by the computer must be converted to electric analog signals before being applied to most control-valve actuators or other final control devices. Digitally operated final control elements are currently finding limited use, even in digital-computer control systems, because of cost and performance.

Digital-to-Analog Conversion A typical digital-to-analog converter (DAC) is the weighted-resistor parallel converter illustrated in Fig. 22-101. It consists of a flip-flop register, analog switches, a resistor network, a reference power supply, and an output operational amplifier. This type of operational amplifier is described in greater detail in connection with Fig. 22-103.

On the load-command signal, the parallel digital input (a conductor for each bit in the binary digital signal) sets the logic flip-flops to true or not true (1 or 0). These in turn set the analog switches to ON or OFF. The switches connect the appropriate resistors between the reference supply and the summing junction. The operational amplifier then produces an output voltage or current corresponding to the digital input signal.

Analog-to-Digital Conversion The purpose of the analog-to-digital converter (ADC) is to provide the digital computer with a digital signal which is representative of the value of the analog signal. One widely used method is the successive-approximation ADC. As indicated in Fig. 22-102, a parallel output is used; that is, each bit signal of the digital output number is carried on a separate conduct-

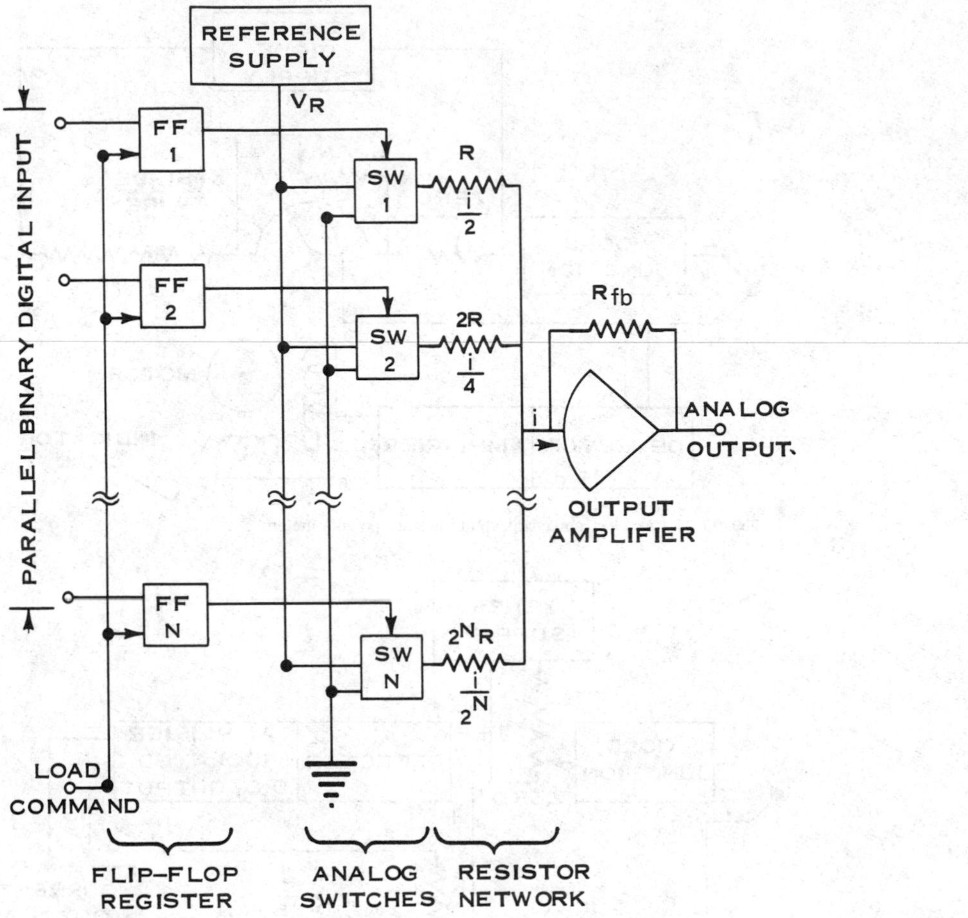

FIG. 22-101 Typical weighted-resistor N-input parallel digital-to-analog converter.

ing line to the computer. The output digital number is stored on a flip-flop register.

Upon receipt of a "start-conversion" pulse the control unit sets the highest-level flip-flop to true and all others to untrue. This through circuitry, similar to the DAC described in Fig. 22-101, produces a voltage V_f. Then V_f is compared with the input voltage V_x. If $V_x - V_f$ is positive, the control unit leaves FF-1 set. If $V_x - V_f$ is negative, the flip-flop is reset to untrue. In the next clock interval, the control unit sets FF-2 and again compares V_f and V_x. This process continues until all flip-flops in the register have been set or reset and an "end-of-conversion" pulse is generated to instruct the computer to read the register.

INDICATING INSTRUMENTS

Analog Analog-type indicators fall into two main categories, moving-pointer and moving-scale. Moving-pointer indicators are the most widely used; moving-scale indicators are generally associated with servo-operated devices. Most moving-scale and moving-pointer indicators are deflection-type devices.

The indicator mechanisms may be powered directly, pneumatically, or electrically, depending somewhat on their proximity to the process and the nature of the measurement. Multiple pointers utilizing one scale are fairly common. Indicators, such as pressure gauges, which use process fluid to provide power are employed quite frequently when the indicator is on or near a process vessel.

Deflection Devices The moving-pointer indicators usually encountered in process instrumentation fall into two categories. In one type the pointer moves in a plane parallel to the panel in which the instrument is mounted, and in the other it moves in a plane perpendicular to the panel. The former type requires a large amount of panel space compared with scale length. The minimum required rectangular panel area is wider than the scale and higher than the length of the pointer. In modern process systems, this type of meter is rarely used on control panels, although it is frequently used as a field-mounted indicator.

The more usable pointer movement for panel mounting is the edgewise type. Here the pointer rotates in a plane perpendicular to the panel around an axis located behind a cylindrical scale. The pointer can move in vertical or horizontal planes. By using a pointer that is long compared with the scale length, the indicator motion approximates a straight line. Compact displays can be provided, since the length of the pointer does not affect the panel space required. This type of indicator lends itself particularly to displaying related variables in close proximity to each other.

Straight-line pointer motion is available on several models of display devices. Straight-line display is nearly always associated with recording instruments and is usually obtained by attaching the pointer to a servo-driven cable.

One advantage of the servo-driven moving scale over moving-pointer indicators is that longer scales may be used in the same instrument-panel space. For example, miniature panel-mounted high-density instruments are available which make use of a movable scale 190.5 (7.5 in) long using only 76.2 mm (3 in) of panel space. In console-type temperature indicators, extensive use has been made of moving scales having an effective length of over 711.2 mm (28 in).

Direct-Powered Indicating devices that require no external power to make them function are considered to be direct- or self-

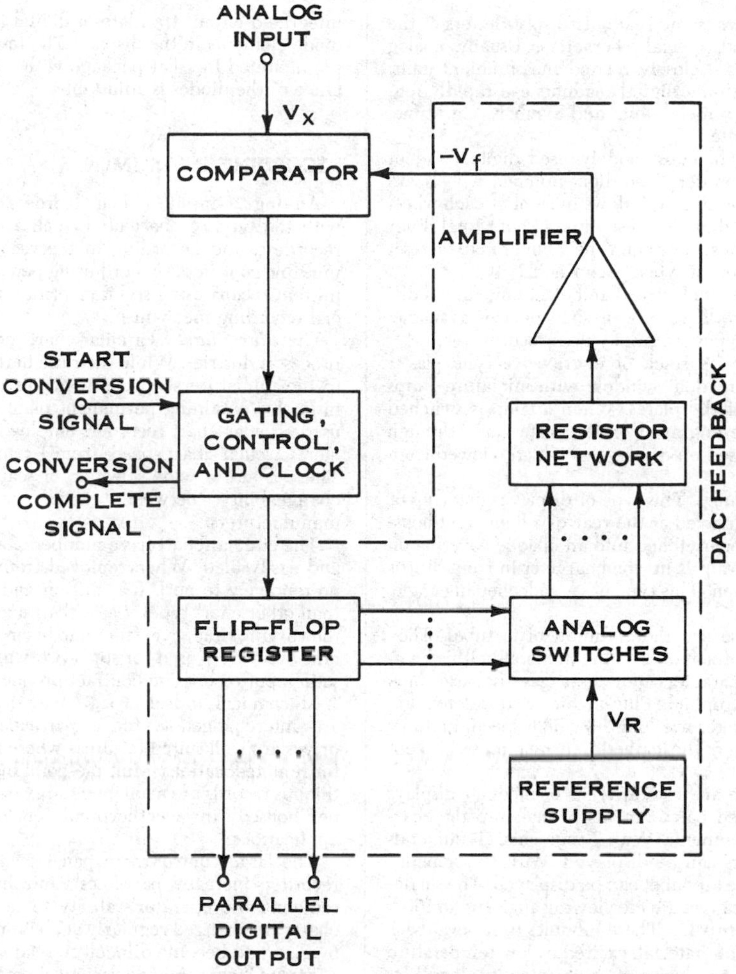

FIG. 22-102 Successive-approximation parallel feedback analog-to-digital converter.

powered. Such devices include Bourdon-tube pressure indicators, fluid-filled temperature indicators, direct-driven tachometers, and rotameters.

Pneumatic-Powered In the process industries the most prevalent deflection-type indicators are pneumatic-powered. The indicator motion comes from air pressure acting directly through a bellows or diaphragm plus a linkage or lever system to an indicator or indirectly through a pneumatic servo to the indicating device.

Electric-Powered The common d'Arsonval dc meter movement accounts for a major part of all process electrical indicating devices. It is generally the basic measuring element whether the process signal is a direct current or a dc voltage. Few ac voltmeter-type process-variable indicators are used in the process industries. They generally are associated with electric power generation or consumption. Electrically driven servo indicators are usually associated with recording indicators.

The process-variable indicators in some modern control rooms are of the deviation type, a zero-center type in which a deflection from the midscale point is the deviation from the set point. The scale can be either a dial or a movable scale.

Multiple Pointer Watthour meters and other totalizing instruments often use several separate pointers for units, tens, hundreds, etc. This display is simple and inexpensive and puts little load on the driving mechanism. However, it has the disadvantage of being prone to being misread by persons with little experience.

Moving-Material Indicators Many process measurements use the process fluid itself to provide the indication. Flow–no-flow gauges and liquid-level indicators are characterized by local visual indication in which level or movement of the process fluid or float is viewed through transparent tubes or sight glasses.

Digital Formerly it was usually cheaper and easier to obtain process data for computation and records from analog indicators by way of manual notation than from digital indicators. The subsequent expanding use of digital computers made great masses of already processed information in digital form available for essentially instantaneous display. This, coupled with solid-state electronic developments in analog-to-digital conversion, led to increased digital readout or display.

Digital Readout The direct display of numerical characters or coded equivalents offers the following advantages compared with analog readout:

1. An in-line digital display requires no translation. Analog meter readings frequently involve verniers, multiple scales, calibration curves, and other complications.
2. Electronic digital displays are as accurate as the information fed to them. In addition, electronic digital displays reduce chances of mechanical malfunction and human reading error.
3. Most digital readouts register data as fast as the human eye can follow. Also the characters snap into position to prevent blurring.
4. Many digital readouts can be used as either counters or as indi-

cators. In addition to numbers, some readouts display letters of the alphabet, full messages, and special characters. Usually analog meters handle only one or two closely related measurement units. With proper switching, electronic digital readouts can rapidly register widely different measurement units and symbols, e.g., time, temperature, pressure, flow, etc.

Mechanical Counters The most widely used digital readout device is the mechanical counter. Ten-digit number wheels are mechanically mounted side by side. Windows in front of each wheel present an in-line display of digits representing a count total. Four to six digit counters are the most common. The counter may be reset to zero by a push button, lever, or wheel. See Fig. 22-89.

Electrical Developments in electrical and electronic digital display devices have made a wide variety of such devices available. Some of the more common types are briefly described:

1. *Edge-lighted display.* A stack of engraved acrylic plastic plates is arranged behind a readout window with miniature lamps arrayed to edge-light each of the plates. When a lamp is switched on, the character or message engraved on the plate glows through the other plates. These are used on displays which are viewed from close up and head on.
2. *Projected-image display.* This type of display is made up of modules which have lamps arrayed at the rear of a light-tight housing behind a transparent integer etched into an opaque covering on a condensing lens. When a lamp is lit, the character in front of it is projected onto a viewing screen. This type has a wide viewing angle, up to 150°.
3. *Register tubes.* These are the common nixie tubes. They have stacked elements in the form of metallic numerals with a common anode. The selected characters glow as a gas-discharge tube when a negative voltage is applied. Plug-in binary translators are available. The major advantages are long life, high speed, and low power consumption. However, the cathode current must be kept within tight limits.
4. *Bar segments.* There are several types of modular displays which incorporate illuminated bar segments to make up the characters, using either 7 or 16 segments. With 7 segments, all numerals and some alphabet characters can be displayed. With 16 segments, all numerals and the complete alphabet can be displayed. The single-plane display of bar segments gives a wide viewing angle, up to 150°.
5. *Low-voltage vacuum tubes.* These modules have seven segmented bars of phosphorescent material, excited by low-temperature discharge. Their advantages are low cost, high intensity, long life, high speed, and low power consumption. An integrated-circuit package is available to supply decoding storage and counting functions.
6. *Electroluminescent panels.* The lamps for these panel segments are flat-plate luminous capacitors individually excited by a high-frequency or high-voltage signal applied across its electrodes. The advantages of this display are cool operation, low current drain, and few sudden failures. The disadvantages are low light level, the need for extensive external translation circuitry, and high-voltage, high-frequency power packs. Each segment has its own lamp source.
7. *Solid-state bar segments.* These are similar to incandescent bar segments except that the light source is a gallium arsenide phosphide. Their life is almost infinite, but brightness is limited.
8. *Matrices.* Matrices consist of neon lamps in optical reflectors banked in a solid mosaic pattern behind a viewing screen. A solid-state miniature integral counter and decoder network translates sequential pulses directly into digital presentation by turning on the proper lamps to form an image.
9. *High-voltage vacuum tubes.* These units combine the rear-projection-type display with a cathode-ray tube. A 10-gun cathode-ray tube projects any one of the 10 numerals through a grid onto a fluorescent screen. The major disadvantage is the need for a high-voltage (2.5-kV) dc power supply.
10. *Cathode-ray tubes.* Cathode-ray-tube displays are usually custom-engineered for highly specialized applications. Complete systems provide alphanumeric information positioned on predetermined formats. Use of cathode-ray-tube displays in process control is currently limited.
11. *Light-emitting diodes.* Each digit or character consists of a matrix of light-emitting gallium arsenide phosphide diodes. An integrated circuit translates a digital input to drive the appropriate diode elements in the display. The integrated circuit and the diodes are mounted in a flat package with a glass top. The luminous emittance of the diodes is adjustable.

RECORDING INSTRUMENTS

Analog Graphic analog recorders may be grouped in accordance with the type of chart used, such as circular-chart and strip-chart recorders, and according to the recording mechanism: single and multiple continuous record using pen and fluid ink, multipoint intermittent record using stylus or print wheel and semifluid ink, or special recording mechanisms.

Circular Chart Circular-chart recorders are widely used in the process industries. While their use in the control room has given way to the miniature strip chart, they are effective and useful mechanisms for obtaining permanent plant records. The most commonly used circular-chart recorders employ a 305-mm (12-in) chart diameter. Circular-chart speeds from 1 r/min to 1 r/month are available. There is also a wide selection of scales. There is little or no interchangeability between charts for instruments of different manufacturers.

The circular chart has a number of advantages. It is easily handled and easily filed. When removed from the recorder, it lies flat with no tendency to curl. It is put on and taken off much more simply than other chart types. Basic chart accuracy is very good. In spite of the varying time scale from hub to circumference, circular-chart records are widely used for supervisory and accounting purposes and as settlement records on contract product sales or transfers.

Mechanical and electrical chart drives are available and usually are interchangeable for a particular manufacturer. Mechanical drives are well suited for areas where there is an explosion hazard or for remote locations. Multiple-point instruments with up to four continuous records are available. Large-case circular-chart recorders can be housed in weatherproof enclosures for remote or field applications.

The large control-room panel space necessary for circular-chart recorders increases panel costs and disperses the display, materially reducing the operator's ability to monitor the process. Unless the charts are changed regularly at about the same time, the records may overlap and become difficult to read and interpret.

Strip Chart A wide variety of strip-chart recorders are available. The most commonly used chart widths use 11-, 6-, and 4-in (279-, 152-, and 102-mm) scales. Miniature strip-chart recorders can generally be placed in two categories: the square case [usually a 6- by 6-in (152- by 152-mm) enclosure] and the narrow case [approximately 3 in (76 mm) wide and generally used in conjunction with a control station].

1. *Wide chart.* The 11-in scale strip-chart recorder is invariably electrical and servo-operated, has rectilinear coordinates, and is available with one or two continuous records or with a variety of multipoint (discontinuous) recording arrangements. This type of recorder was developed for thermocouple temperature measurement and is used for recording electrical process variables and electrical output from transducers of nonelectrical variables.

The instrument is accurate and reliable; it is very flexible, with provision for auxiliary functions including control, alarm, and signal contacts and retransmission systems.

A variety of chart speeds is available from a few inches per day up to feet per minute. Pen speeds of 1 or 2 s for full scale are standard, with speeds up to ¼ s also available. Several manufacturers have servo drives available for the chart, providing a short-length record which is a function of two variables, one driving the pen and one driving the chart.

Wide strip-chart recorders are used extensively in industrial-process operations as temperature recorders, in pilot-plant operations, in analysis and test recording, and in specialty applications for which the accuracy of the instrument and the readability of the rectangular-coordinate wide-strip record are important. For central control rooms, the applications are generally limited to multipoint thermocouple measurements.

2. *Medium chart.* The 6-in-scale strip-chart recorder is used in much the same type of application as the wide, 11-in recorder. The basic input to medium-sized recorders is a millivolt-level signal. Overall dimensions of approximately 9 in (229 mm) wide and 10 in (254 mm) high provide considerable space saving on a panelboard. Also, the recorder is made with a greater number of options and with pneumatic or electronic control functions.

3. *Miniature strip chart.* Of the variety of sizes available, the most common is approximately 6 by 6 in (152 by 152 mm) with a chart-scale length of 4 in (102 mm). These miniature recorders normally appear as recorder-receiver elements only, but some models can operate as a recorder-control station. Actuating signals may be 20.7- to 103.4-kPa (3- to 15-lbf/in^2) pneumatic signals or 1- to 5-, 4- to 20-, and 10- to 50-mA dc electric signals. Up to four continuous records are available. Chart speed is usually in the vicinity of ¾ in/h (19 mm/h). Most miniature strip-chart recorders have linear pen motion and rectilinear charts. The pen-drive mechanisms are usually servo-operated. Miniature recorders are generally designed for indoor panel-mounted operation.

The narrow [approximately 3-in- (76 mm-) wide] miniature strip-chart recorders perform the same functions as the 6- by 6-in recorders. The recorder may be on the same chassis with an automatic or manual control station. The narrow recorders were developed to conserve panel space and reduce control-room costs as well as operating labor.

The manufacturers of both the 6- by 6-in recorders and the narrow 3-in recorders [some indicating control stations are 1 in (25 mm) wide] provide multi-instrument mounting shelves so that several instruments may be placed side by side for high instrument density.

Digital Digital recording is not used extensively in the process industries unless a digital-computer control system is used. In processing plants using conventional instrumentation, the operator is interested in maintaining process variables at fixed or specified operating points on a continuous basis. Thus, the history or trend of the process variables is an important consideration in the action that the operator takes. Analog recording is ideally suited for this. A digital record such as a long column of numbers is normally more difficult to interpret.

Computer-Associated Teletypewriter In processing plants where digital-computer controllers are used for direct digital control or digital set-point control, the recording of digital information is a basic function of the computer control system. A teletypewriter unit capable of communicating with a computer, i.e., sending and receiving digital information, is a standard and essential item in the computer control system.

1. *Periodic log.* Teletypewriters can be connected to the computer under program control and can be directed to turn on and print a list of process variables and their values on a predetermined periodic basis. Process logs can also be made on demand by the operator.

2. *Data logger.* Data logging has been applied in a large number of processing plants. The term "data logging" generally refers to the scanning and printout of process-variable values utilizing equipment designed specifically for logging data. Only scaling and minor calculations are performed on the input signals. The values (in engineering units) of a large group of process variables are typed or printed sequentially and at regular intervals with provisions for a printout on demand or when values are outside tolerance limits. This type of digital record is used primarily for process studies when the log sheets are collected and analyzed by process engineers.

3. *Strip printer.* Small digital printers (some with integrated circuits) are available which will convert, scan, and print out a limited number (under 50) of electric input signals. Printout is columnar on narrow, 2-in- (51-mm-) wide paper with each variable identified by number. Some in-line blending operations utilize this type of printer for totaled flow readout.

CONTROL CENTER

Information Requirements A major purpose of a control center in a processing plant is to supply all the information required for operating the process. There are three principal users of this information: operators, management, and the automatic equipment. Operators are responsible for the performance of the process units under their supervision. To fulfill this responsibility, operators must have current information on all process variables. They need to have alarms which call their attention to dangerous situations, potential equipment damage or failure, off-specification product, and situations which jeopardize scheduled throughputs.

Operators also must have current controller information such as set points, tuning constants, and controller outputs or valve positions. They sometimes need process-variable trend information and calculated operating guides based on material balances and heat balances. In addition, they need some information from operators in other portions of the plant, particularly if those portions of the plant receive process material from or supply process material to the processing units for which they are responsible. Operators also need information about process goals, equipment availability, and process results from their management.

Management requires information concerning processing results such as feed rates, product rates and quality, yields, and operating costs for accounting purposes. Process performance evaluation and engineering studies require this same information and frequently some additional information.

The automatic equipment, like a person, requires both information and instructions. If a high degree of computerized automation of process control as illustrated in Fig. 22-151 is assumed, the equipment requires process-variable information inputs, set points and economic information, timing and logic, alarm limits, controller-tuning parameters, and emergency procedures. In order that one process unit may be coordinated with others, the automatic equipment needs information from other control centers or other parts of a large control center.

Equipment Functions A second major purpose of the control center is to provide the means for controlling the process. Control during normal operation is usually automatic. The supervision required for the automatic control equipment depends on the level of automation discussed in connection with Fig. 22-151. Manual control is always an available option for operators. Another option, when enough is known about the process, is automatic optimal control.

Emergency operation can be accomplished either manually or automatically. Operators tend to put the process on manual control during upsets. This often defeats one important purpose of automatic equipment, that is, to compensate for the effects of the disturbances. As controller reliability and capabilities improve, as better control algorithms are developed, and as operator confidence increases, the control loops are in manual mode less and less frequently.

Automatic startup can be accomplished either by programmed control or manually. Digital control computers make it possible to automate all the known decisions that must be made.

Shutdowns can be accomplished in much the same way as startups. Although shutdowns are frequently emergency situations, the procedures are generally better defined. Many control rooms have an arrow diagram for emergency shutdowns.

System Display Well-engineered display, important in individual indicating and recording instruments, becomes crucial in control-room design when data on large and often critical process operations must be used by a human operator. Indicators and recorders must be coordinated with controls, switches, alarms, and auxiliary equipment so as to present a clear, easily grasped display of the process condition.

Developments in process-control centers have evolved through several stages: full graphic panel, semigraphic panel with separately mounted control instruments, and semigraphic panel with control instruments mounted in multiunit shelves. Processing plants employing digital computers for direct digital control or set-point control may have a combination of special person-computer communication equipment, conventional instruments, and special-purpose control stations linking the computer to the process.

Full Graphic The full graphic panel is essentially a pictorial flow diagram of the process painted on the control panel with miniature instruments mounted in appropriate locations on the diagram itself. This presents a clear and comprehensible picture of each process and control element in its relation to the whole. It also simplifies operator training and could reduce mistakes from a mix-up in instruments. However, the disadvantages of full graphic displays generally outweigh the advantages.

The large panel space required is costly and adds to the cost of the control room. It must be fully custom-engineered. Any changes after the initial installation are expensive. It is difficult for the operator to observe quickly all the process variables, particularly when several unit processes are grouped in one control room. Full graphic panels are now seldom used in new processing plants.

Semigraphic Semigraphic panels provide a better use of panel space than full graphic panels. Indicators, recorders, controllers, and auxiliary equipment are grouped together as dictated by the process. The graphic portion of the display is usually a small-scale pictorial flow diagram mounted above the instruments. This may be simply a line drawing, or colors and lights may be included to locate the operating instruments, switches, and valves on the diagram.

Early semigraphic displays utilized miniature instruments mounted in separate cases. The demand for greater instrument density on control panels led to the development of narrow-front instruments with provisions for mounting them in multiunit housings.

These steps have greatly increased the number of instruments that one operator can survey from one location. They have also changed control-room layout. The increased instrument concentration has made it necessary to utilize auxiliary equipment racks to mount transducers, switches, annunciators, and auxiliary devices.

When a digital-computer controller is used online to perform individual loop control functions directly, the control panel normally contains indicating control stations linking the computer to each individual process-control loop. Such direct digital control (DDC) systems normally have an operator's panel containing digital displays and means for changing loop set points and controller tuning.

Control Stations The primary function of a process controller is, of course, to maintain one or more process variables at or near desirable set points. However, since human operators are responsible for the performance of the process, some means must be provided to enable them to verify that the controller is doing its job and to enable them to take over the control of the process if necessary. The control station provides the link between operators and the process.

There are five essential parts to every control station: (1) a process-variable indicator, (2) a set-point adjusting mechanism, (3) an adjustment device (usually called manual) that directly manipulates the signal to the control valve, (4) an output-signal indicator, and (5) a device for switching between automatic mode and manual mode of control. The implementation of the five control-station essentials will vary among manufacturers, but they must be present in any control loop whether it be pneumatic, electronic, or computerized.

In conventional pneumatic or electronic control panels, the control station is normally located adjacent to the controller. If the process variable is being recorded, the recorder is adjacent to the control station and controller or at least near them.

In current DDC computer control systems, all five of the control-station functions are provided at the operator's console where communication with a particular control loop is established through a push-button keyboard. All the control-station functions except the controller set point are also duplicated at individual loop control stations. Depending on the redundancy required, in some designs a backup electronic analog controller and its set-point adjustment are included in the individual loop control station.

Process-Variable Indicator Process-variable indicators appear in many forms. In large-case instruments, the indicator may be a pointer on a circular or horizontal scale. It may be part of a controller, a recorder, or both. Current configurations separate recorders and controllers. The control stations are joined with the controllers to become indicating controllers.

Set-Point Adjusting Mechanism For each configuration of recorder or indicator, there is a different way to manipulate the set point. In pneumatic instruments the set-point mechanism generally adjusts a small transmitter which supplies a 20.7- to 103.4-kPa (3- to 15-lbf/in^2) signal to the controller. In electronic instruments the adjustment generally consists of a potentiometer which generates a voltage of opposite polarity to the process-variable signal.

Manual Adjustment This manually operated device directly manipulates the output signal to the valve or valve actuator. It is used only when the automatic controller is not in service. Most manual outputs are directly proportional to the position of the adjusting device. However, in some of the newer controllers manual adjustments change the output by actuating an "increase" or "decrease" mechanism.

Signal Indicator This device indicates what signal level is being applied to the control valve. The types of indicators used for this are similar to those used for process-variable indicators. This meter normally will have the ends of the scale labeled as "open" or "close," referring to the position of the control valve.

Automanual Transfer Switch The purpose of the automanual transfer switch is to change the source of the signal to the valve. To do this without changing the signal size, that is, with bumpless transfer, older controllers require a balance procedure. When switching from the normal or automatic mode to manual, the transfer device is first moved to an intermediate or balance position and the manual output is matched with the automatic controller output. The transfer lever is then moved to the manual position to complete the bumpless transfer. Then the output is adjusted to the desired value. When going from manual to automatic, the balance position is used to match the set point with the process variable before switching to automatic. Then the set point is adjusted to the desired value.

The balance position is being removed from currently available process-control stations, each manufacturer generally using a different scheme. The most common method is to force the manual-adjust output to track the controller output while in automatic mode and to make the set-point-adjustment output track the measurement while in manual. Thus when switching, no balancing or alignment is necessary. In computerized control systems, the computer controller automatically makes the adjustments required.

Remote-Local Set Point In addition to the five essential parts of a control station, most manufacturers furnish models which have a remote-local switch. "Remote-local" refers to the source of the set point for the controller. In the remote position, the set-point signal comes from another controller, a remotely located set-point station, or a computer. In the local position, it is generated by the set-point adjustment on the control station. The remote-local switch becomes an essential part of the control loop when the controller is the secondary of a cascade system as illustrated in Fig. 22-49. Special procedures are provided by each manufacturer to switch between remote and local.

New Control Center: Some Design Considerations The potential variations possible for control centers are limited only by human imagination. Economic considerations reduce the number of such variations, and some standardization of centers is appearing. A few basic assumptions appear reasonable in considering new control-center designs:

1. Hybrid computer–conventional-controller control centers are practical with small digital computers.
2. Several process units can be operated from one control center by utilizing multiple small computers or a large computer system.
3. Each process unit may have a separate console or area on a master console.
4. All process equipment inside the plant fence can be automatically computer-controlled online.
5. Control centers controlling 1000 to 2000 individual loops are feasible.

Possible computerized tasks include DDC, set-point control, unit optimization, control-loop trend data collection and recall, data collection and recall for accounting and process engineering, interunit optimization, and plant optimization.

With the increased amount of processing equipment supervised by a single operator strong efforts are being exerted to simplify the person-plant interface—the control panel. Some of the objectives of this effort are reduction in investment per control loop, improved interface between operator and control computers, implementation of operator involvement only in "control by exception" situations, and elimination of separate alarm systems and temperature consoles.

In semigraphic flow diagrams, a space-saving device is optical projection of small sections of the diagram on a screen in front of the operator. The particular section to be projected may be selected by the operator through a random-access slide-storage system.

Apparatus is available to carry out this projection on the back of the front surface of a cathode-ray tube. Values of set points, process variables, valve positions, control-loop automanual status, and alarm conditions can be superimposed on this graphic display in digital form and in engineering units. Inputs for changing the loop status, changing set points, changing valve position, or requesting a cathode-ray-tube analog display can be done by a keyboard and/or joystick movable electronic locator.

AUTOMATIC CONTROLLERS

Industrial automatic process controllers vary from simple on-off devices to special-purpose computing instruments. They are normally used to couple the controlled process variable to the manipulated variable, commonly by means of a process-control valve, forming a feedback control system.

Commercial models differ greatly in many respects. In some instances the controller input is mechanically linked to a controlled variable indicator, the latter being powered by the output of the measuring element. Some controllers are manufactured and housed as an integral part of a recorder-controller or an indicator-controller. Other controllers are housed separately to facilitate field-mounting the controller at the process and locating all recorders and indicators in a centralized control room.

For some applications, the sensing element may not be able to supply sufficient power to operate both the error detector and the final control element. In these instances it is necessary to provide a means for amplifying the power supplied by the sensing element. In essence, the power amplifier (often called power booster or power relay) uses the small amount of power provided by the sensing element to control larger amounts of power supplied from an external source. Types of boosters include electronic, pneumatic, mechanical, and hydraulic. Examples of these will be described in the discussion of the corresponding controller type.

SELF-OPERATED CONTROLLERS

Some control systems obtain the power required to operate the error detector and final control element from the controlled medium of the process by way of the sensing element. Such controllers are termed "self-operated controllers." Typically, they are simple devices and are frequently used in the chemical-process industries for pressure, temperature, and liquid-level control and safety devices. Most self-operated controllers fall into one of two categories: on-off (two-position) or proportional-action.

On-Off Controllers The significant characteristic of on-off control action is that the amount of control action applied at the process input is either zero or the maximum available. On-off action is commonly used in temperature-control systems in which, for example, the fuel rate to the furnace is zero or maximum or, similarly, the current through an electric heating element is zero or maximum. On-off controllers are also frequently used in safety devices such as pressure-relief valves and thermocouple-operated safety valves.

Proportional Controllers As opposed to the on-off action described above, the action of a proportional controller is smooth and continuous over its operating range. Under steady-state conditions the controller assumes an intermediate position where the measured variable equals the set point except for offset. The controller output varies in an amount proportional, either directly or inversely, to the deviation of the measured process variable from the set point. For example, in a steam-heated process vessel if the temperature to be controlled increases above the set point, the controller output will increase by a proportional amount. This output may reduce the opening of the steam valve to reduce the amount of heat being added. Self-operating proportional controllers are rather widely used. The linearity and range of the proportional action are dependent on the construction details of the manipulated devices.

ELECTRONIC CONTROLLERS

There is widespread use of electronic controllers for process control. Among numerous reasons for increased usage is the development of solid-state circuitry, which has led to greater reliability, easier servicing, and smaller physical size. These factors make electronic controllers' overall cost competitive with the cost of pneumatic controllers. In addition, electronic controllers can be more easily linked with process-control computers.

Another advantage of electronic controllers is that lengthy pneumatic transmission lines may be replaced by electrical wiring to reduce substantially or eliminate the pneumatic lags that often degrade the performance of pneumatic control loops.

When it is necessary to convert from electric voltage or current signals to pneumatic signals (for example, to operate a pneumatic valve actuator) or the reverse, the required transducers may be field-mounted, i.e., located near the valve or pneumatic signal source with short pneumatic lines. Examples of such transducers are described under "Indicating and Recording Instruments" and "Telemetering and Transmission."

Electronic controllers may be more accurately tuned, setting will generally show less drift, and there are fewer moving parts than in the pneumatic counterparts.

The principal disadvantage of electronic controllers is that special equipment designs are required to meet safety requirements for hazardous environments at additional cost.

Frequently, electronic controllers are housed with an instrument for recording or indicating the value of the controlled process variable to form a recorder-controller.

Operational Amplifier The heart of the conventional electronic controller is a high-gain operational amplifier. Present-day controllers accept dc signals at the inputs and supply dc signals at the outputs. If ac amplifiers are used as part of the operational amplifier, a means must be included to change the dc input signal to an ac signal. Then, at the amplifier output, the ac signal must be changed back to a dc signal. An operational amplifier for amplifying dc signals by using an ac amplifier is illustrated in Fig. 22-103.

The oscillator-amplifier section generates an ac signal that is of fixed frequency and has an amplitude proportional to the magnitude of the dc input signal. The amplifier-detector stage rectifies the ac signal, thus generating a dc signal that is proportional to the input signal. This signal then passes through an output power-driver stage to provide the dc output signal. One type of power driver is shown in Fig. 22-143. The essential features of operational amplifiers are a very high voltage gain, a very high input impedance, and very low input-current requirements.

For illustrative purposes, the operational amplifier (regardless of whether it is built around an ac or a dc amplifier) may be considered as shown in Fig. 22-104, where R_s, the input impedance of the amplifier, shown inside the amplifier symbol, is usually omitted. Normally the ground symbols are also omitted. Z_1 and i_1 represent the input impedance and current respectively in the network external to the amplifier. Z_f and i_f represent the feedback impedance and current respectively.

The relation between e_s and e_{out} (the input and output voltages of the operational amplifier) is given by

$$e_{out} = -Ge_s \tag{22-23}$$

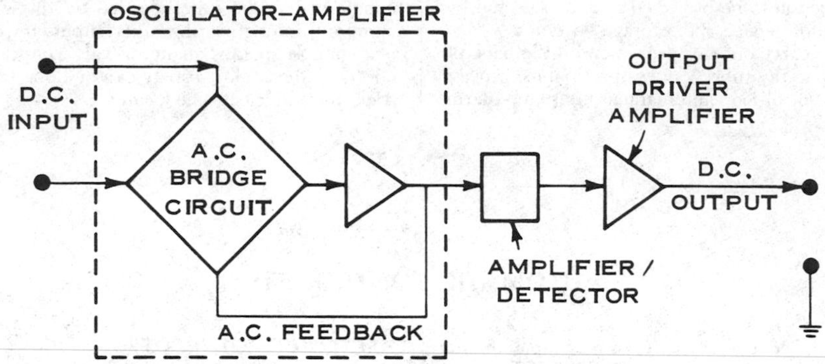

FIG. 22-103 Functional diagram of an operational amplifier utilizing alternating-current amplification.

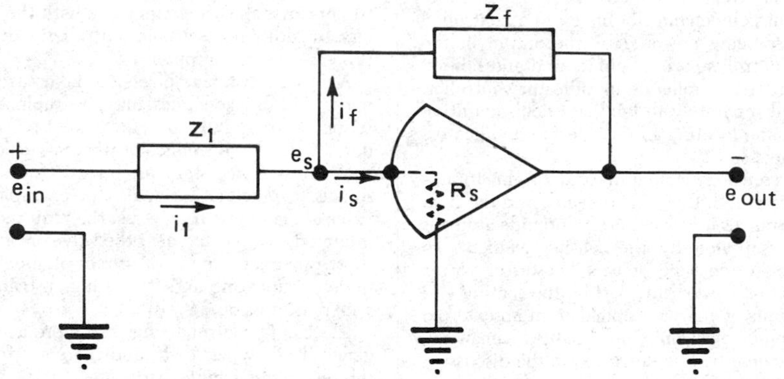

FIG. 22-104 Operational amplifier with input and feedback impedances.

where G is the gain of the operational amplifier. The subscript s denotes the summing junction or amplifier-input connection, and the minus sign indicates a polarity reversal from amplifier input to output. From the circuit of Fig. 22-104 and Eq. (22-23), the relationship between e_{in} and e_{out} may be determined.

$i_1 = i_f + i_s$ Kirchhoff's current law (22-24)

$i_1 = (e_{in} - e_s)/Z_1$
$i_f = (e_s - e_{out})/Z_f$ Ohm's law (22-25)

$i_s = e_s/R_s \approx 0$ an essential factor of operational amplifiers

Substituting in Eq. (22-24) for i_1, and i_s for e_s from Eq. (22-23) yields

$$e_{out} = \frac{-(Z_f/Z_1)e_{in}}{1 + (1/G)(1 + Z_f/Z_1)} \tag{22-26}$$

When G is very large,

$$e_{out} = -Ke_{in} \qquad K = Z_f/Z_1 \tag{22-26a}$$

The impedances Z_1 and Z_f may be resistors, capacitors, or resistor-capacitor networks, depending upon the desired relationship between e_{in} and e_{out}. Frequently, Z_1 and Z_f are both resistors, in which case $Z_1 = R_1$, $Z_f = R_f$, and $K = R_f/R_1$. The value of K may be in the range of 0.01 to 100 for typical use in electronic controllers.

Electronic-Controller Inputs and Outputs The signals from electronic controllers have not been standardized. For 0 to 100 percent of the input-signal range, typical outputs are 1 to 5 mA, 4 to 20 MA, 10 to 50 mA, and 1 to 5 V. Most controllers will accept a variety of input signals by properly sizing input resistors, but the output is usually limited to a single range. The instruction manual should be consulted for each manufacturer.

For a specific controller application (for example, a pressure controller) a transducer converts the actual physical measurement (pressure in this instance) to an electric signal. Similarly, if the final control element is a pneumatic control valve, a transducer is used to convert the controller electric output signal to a pneumatic signal. Compatibility of tthe electronic controller with measurement devices, final control elements, and transducers should be carefully checked in a new installation.

Two-Position (On-Off) Control Mode The two-position controller (sometimes called on-off or bang-bang) is, as its name implies, a controller with an output level that can assume only two values. There are numerous ways to implement a two-position controller. A typical one is depicted in Fig. 22-105.

The operation of this controller is described as follows. When the difference between the value of the set point r and the value of the process variable c is negative, the amplifier output e_0 will tend to increase until e_0 is slightly greater than E_2. At this point diode D_2 will begin to conduct and thereby prevent e_0 from increasing further. On the other hand, if $r - c$ is positive, the amplifier output will tend to decrease until e_0 is slightly less than $-E_1$, at which time diode D_1 begins to conduct, thus limiting e_0. There is a very narrow range of values of the difference $r - c$ for which the amplifier output voltage takes on some intermediate value between $-E_1$ and $+E_2$. The width of this range is determined by the ratio of the diode resistance (in reverse-biased state) to the resistance of the input resistor R.

The output of the controller in Fig. 22-105 may be used to drive directly the final control element, the output transducer, or an electrical relay. The relay is usually capable of handling greater power, and its action is analogous to that of pneumatic and hydraulic pilots and relays.

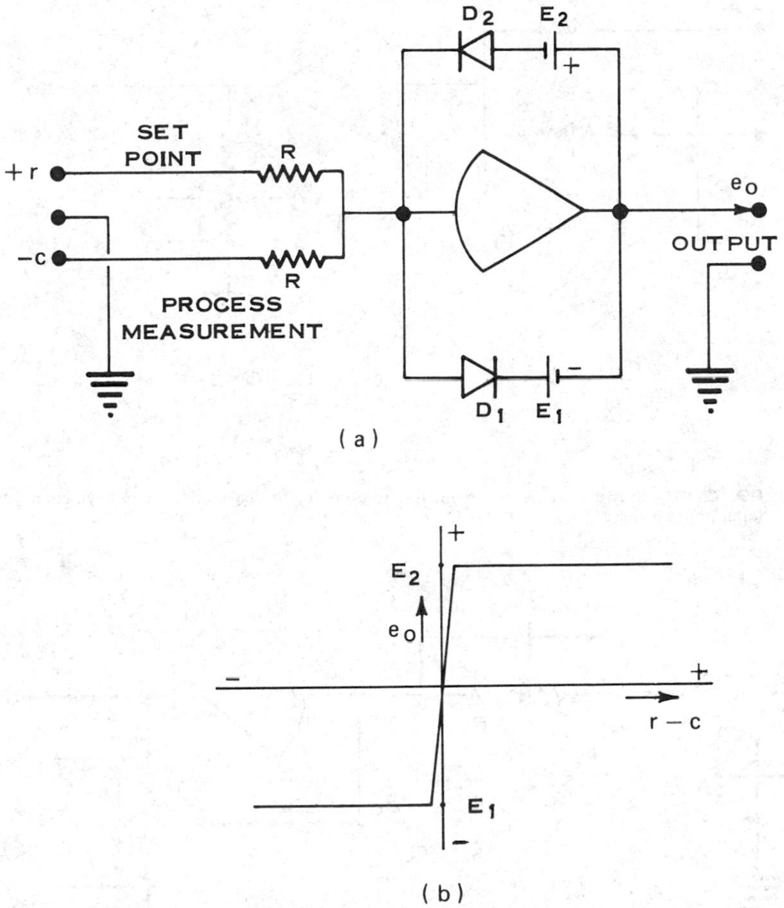

FIG. 22-105 Typical two-position controller using an operational amplifier where (*a*) is the basic circuit and (*b*) the response of the output to input deviations from set point.

Proportional Control Mode (P) A proportional controller supplies an output signal that is proportional to the difference between set point (reference signal) and the controlled-variable measurement. Some typical basic feedback circuits for operational amplifiers used for proportional control are shown in Figs. 22-106 and 22-107. In each instance the controller output is $e_0 = -K(r - c)$, where K is defined by Eq. (22-26). In Fig. 22-106 the input-impedance component is a capacitor, and the feedback impedance is a resistor-capacitor network in which the resistance is adjustable.

In Fig. 22-107 all impedance components are resistors. The impedance components could be capacitors. Although all circuits mentioned will provide proportional control, the network of Fig. 22-107 is most widely used because capacitance circuits tend to possess drift problems and are proportional only for transient signals.

Proportional-Integral Controllers (PI) The reasons for adding other control modes to the basic proportional controller are discussed under "Fundamentals of Automatic Control." The descriptions provided here are intended to show how these additional modes may be implemented.

Integral, or reset, action can be added to the proportional network in numerous ways. Figures 22-108 and 22-109 show some typical basic circuits for implementing a PI controller.

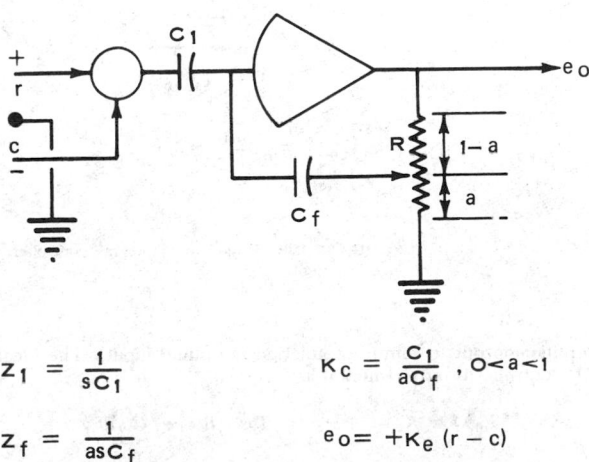

Proportional-Integral-Derivative Controllers (PID) Derivative, or rate, action may be added to a PI controller by adding a resistance-capacitance network in the feedback path from operational-

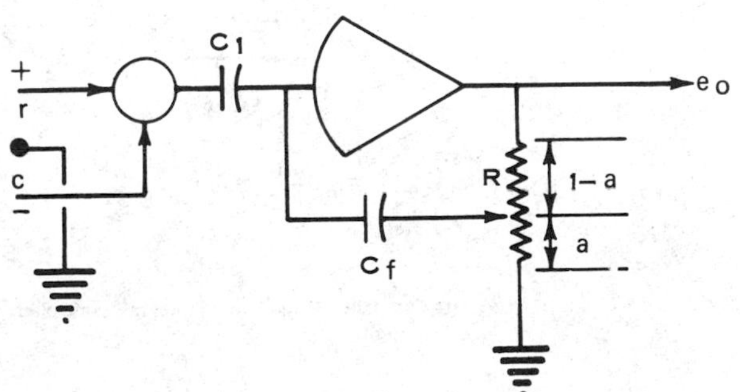

$$Z_1 = \frac{1}{sC_1} \qquad K_c = \frac{C_1}{aC_f}, \quad 0<a<1$$

$$Z_f = \frac{1}{asC_f} \qquad e_o = +K_e\,(r - c)$$

FIG. 22-106 Proportional controller utilizing capacitances for the feedback and summing-junction input impedances. K_c is the proportional gain and R the proportional adjustment resistor.

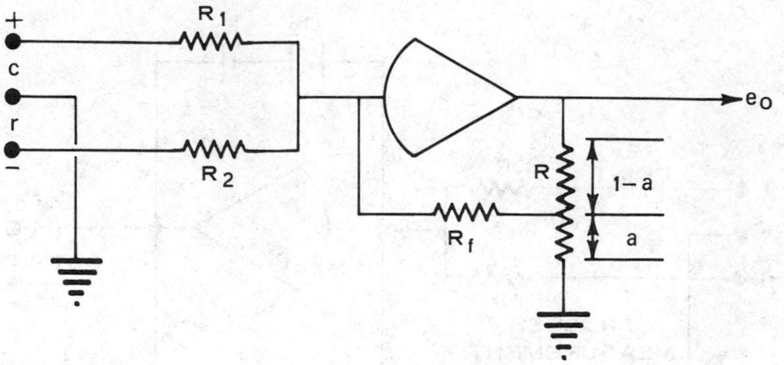

$$Z_1 = R_1 \qquad K_c = \frac{R_f}{aR},\ 0<a<1$$

$$Z_f = \frac{R_f}{a} \qquad e_o = +K\,(r-c)$$

FIG. 22-107 Proportional controller utilizing resistors for feedback and summing-junction input impedances.

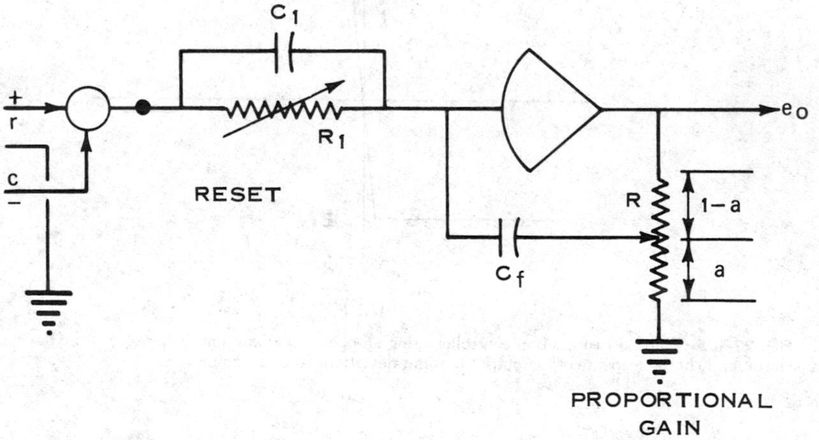

$$Z_1 = \frac{\frac{1}{s_1}}{1+\frac{1}{sC_1R_1}} \qquad e_o = \frac{Z_f}{Z_1}\,(r-c) = K_c\left[(r-c)+\frac{1}{T_i}\int(r-c)\,dt\right]$$

$$Z = \frac{1}{asC_f} \qquad K_c = \frac{C_1}{aC_f},\ 0<a<1$$

$$T_i = R_1C_1$$

FIG. 22-108 Proportional-integral controller utilizing input capacitor-resistor networks for reset adjustment.

amplifier output to summing junction (amplifier input). The ideal PID controller transfer function is

$$E_0(s) = K_c\left(1+\frac{1}{T_i s}+T_d s\right)[R(s)-C(s)]$$

In practice, true derivative action is not easily achieved; in addition, true derivative action may not be desired since it amplifies noise. To minimize the noise problem, circuit modifications that result in approximate derivative action are made. Generally, the modified action may be described by

$$T_d s \rightarrow \frac{T_d s + 1}{(T_d s/\gamma)+1} \qquad \text{and} \qquad \gamma \gg 1$$

Figure 22-110 shows a basic circuit for adding rate action to the PI controller.

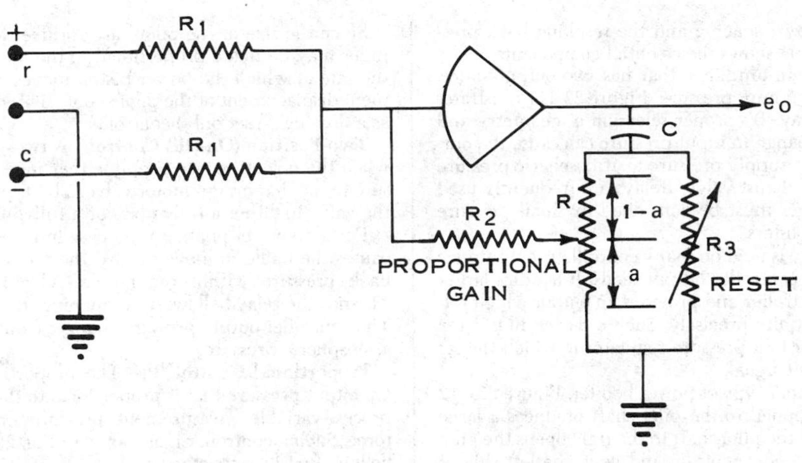

$$K_C = \frac{R_2}{aR_1},\ 0<a<1 \qquad T_i = R_3C$$

FIG. 22-109 Proportional-integral controller utilizing a feedback capacitive-resistive network for adjustments of proportional gain and reset.

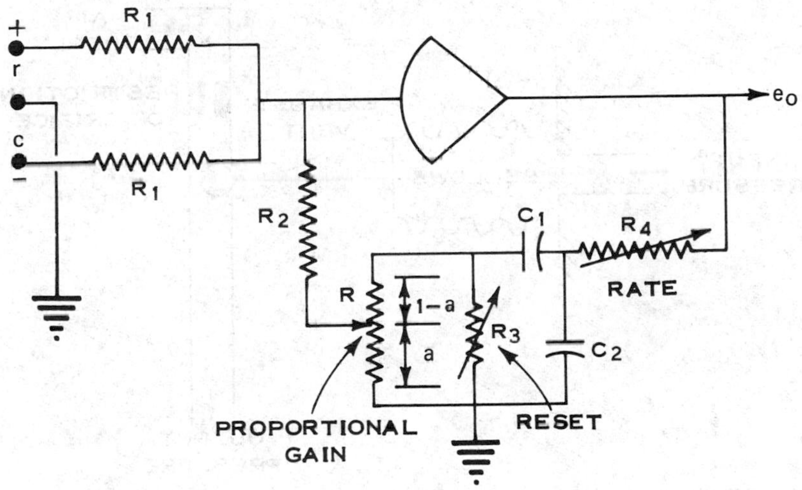

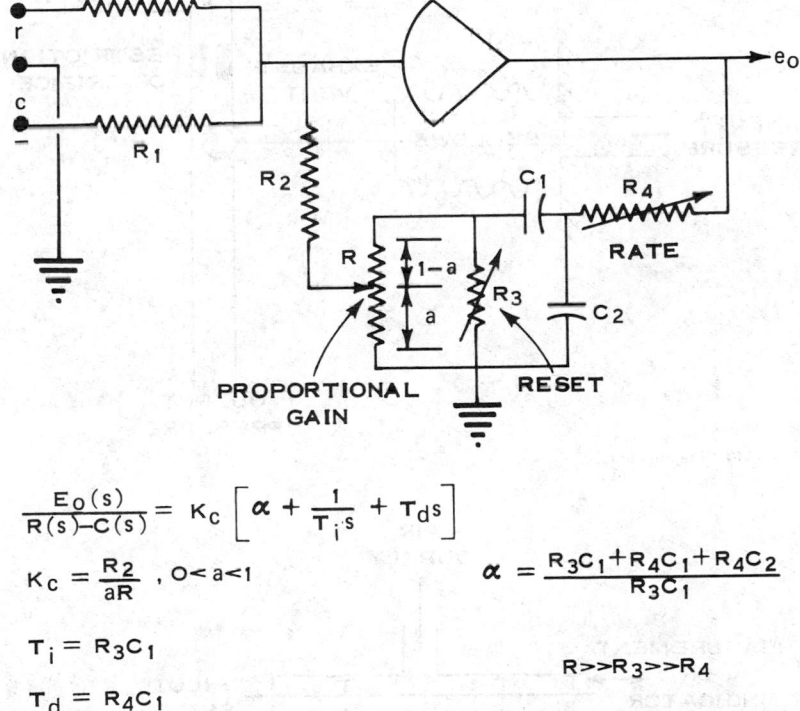

$$\frac{E_O(s)}{R(s)-C(s)} = K_C\left[\alpha + \frac{1}{T_i s} + T_d s\right]$$

$$K_C = \frac{R_2}{aR},\ 0<a<1 \qquad \alpha = \frac{R_3C_1+R_4C_1+R_4C_2}{R_3C_1}$$

$$T_i = R_3C_1 \qquad R \gg R_3 \gg R_4$$

$$T_d = R_4C_1$$

FIG. 22-110 Proportional-integral-derivative controller based on the PI controller of Fig. 22-109.

PNEUMATIC CONTROLLERS

The basic purpose of a pneumatic controller is to supply controlled pressurized air to a pneumatic valve actuator in response to an error signal based on the deviation of the measured variable from the set point. This air pressure is usually in the 20.7- to 103.4-kPa (3- to 15-lbf/in^2) range, standardized by agreement between instrument manufacturers and users. The valve actuator is usually a pneumatic diaphragm actuator as described under "Final Control Elements."

A variety of pneumatic-equipment configurations are available which will produce controller outputs corresponding to on-off, proportional, reset (integral), and rate (derivative) control modes. Since each manufacturer has its own designs, instruction books should be consulted for details of a particular controller.

Pneumatic Amplifiers and Relays The baffle-nozzle amplifier, relay, and pilot are typical basic building blocks for pneumatic controllers.

The operation of the baffle-nozzle amplifier is based on the relationship between baffle-nozzle spacing and the resulting back pressure developed. Figure 22-94 shows the essential components.

The pneumatic relay is an amplifier that has two output states: maximum pressure and minimum pressure. Figure 22-111 illustrates the operation of such a relay. By proper selection of geometry and dimensions, a very small change in input pressure can cause the output pressure to change from supply pressure to atmospheric pressure by closing or opening the exhaust valve. Relays are frequently used when large pressure changes must be controlled by small pressure signals; relays are power boosters.

The pneumatic pilot relay is used on many controllers to minimize the volume into which the controller output feeds. It handles larger volumes of air than the controller and provides an output which follows the controller output quite precisely. Such a device illustrated in Fig. 22-95 is very similar to a pressure regulator in which the set point is the controller output signal.

Pneumatic pilots are another type of power booster. Figure 22-112 shows how a small force applied to the pilot shaft produces a large force. A small movement of the pilot shaft to the right opens the pilot port to the right side of the power piston and vents the left side of the piston, resulting in a leftward piston movement. Similarly, moving the pilot shaft to the left produces a rightward piston movement. The force supplied by the piston is determined by the air-supply pressure and piston area.

Since the size of the pilot-valve orifices leading to the power cylinder may be made proportional to the measurement displacement, the rate at which the power piston moves is related to the measurement displacement of the pilot spools. This combination may be used as a floating (reset only) controller.

Two-Position (On-Off) Control A two-position controller is used when the objective is to supply either maximum or minimum pressure to the diaphragm motor valve. This type of control action moves the valve to either a fully open or a fully closed position.

For a given set point, an increase in measurement signal pressure causes the baffle to move toward the nozzle, thereby increasing the back pressure within the nozzle. This increased back pressure expands the relay bellows, thus opening the exhaust port of the relay. The controller output pressure changes from supply pressure to near-atmospheric pressure.

Proportional Control (P) The proportional controller produces an output pressure that is proportional to the difference between the process-variable measurement pressure and the set-point spring force. Such a controller is as shown in Fig. 22-113 but without a reset bellows and its restrictor.

A decrease in the measurement air pressure causes the balance beam to move away from the nozzle opening, thereby reducing the nozzle back pressure. The reduced nozzle pressure decreases the controller output pressure and the force applied to the beam by the pro-

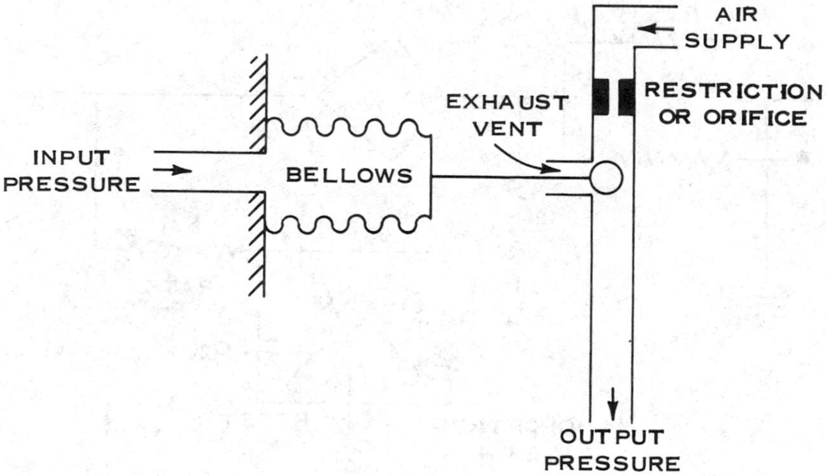

FIG. 22-111 Pneumatic relay.

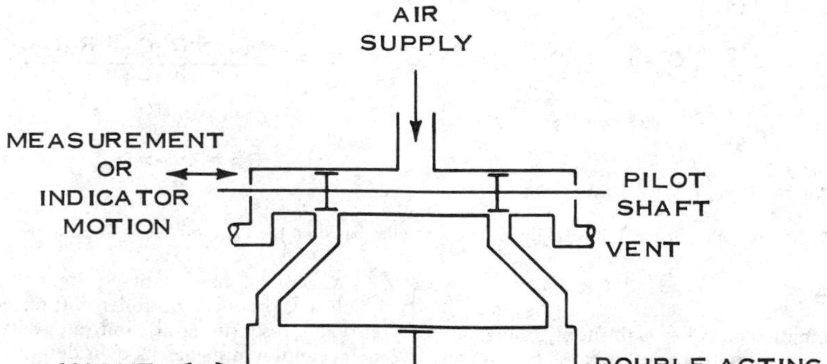

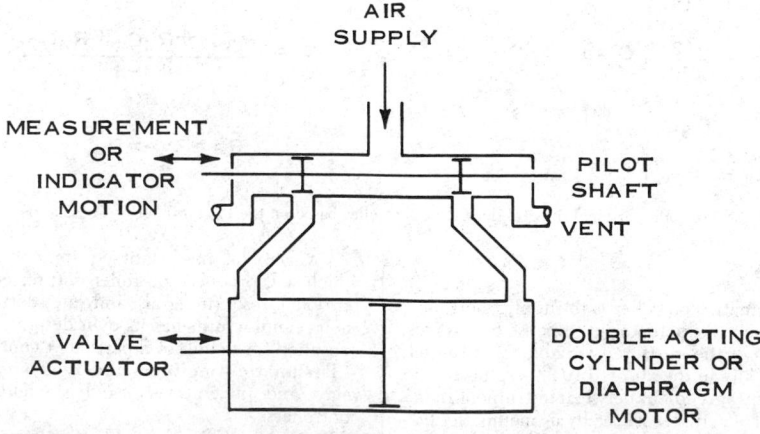

FIG. 22-112 Pneumatic pilot valve and double-acting cylinder.

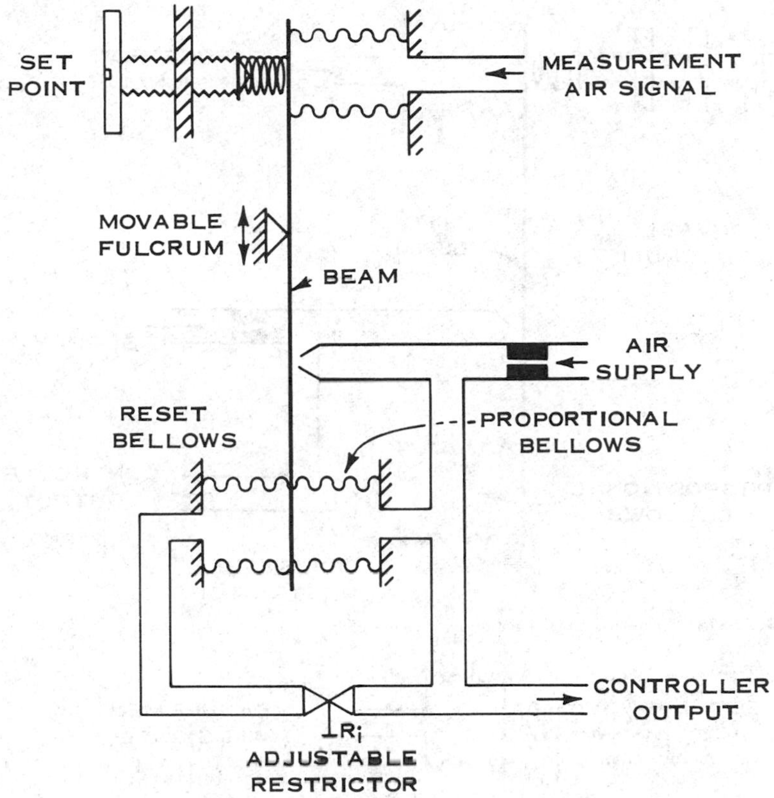

FIG. 22-113 Pneumatic proportional-integral controller.

portional bellows, thus offsetting some of the beam movement caused by the measurement bellows. This is negative feedback; therefore, the beam assumes a position such that the controller output pressure varies in proportion to the difference in forces exerted by the measurement bellows and the set-point spring. The size of the change in output pressure is determined by the position of the movable fulcrum on the beam. The greater the distance between the fulcrum and the nozzle, the greater will be the output-pressure change; that is, the controller has a higher proportional gain.

Proportional and Reset Control (PI) The addition of reset (or integral) action to a proportional controller involves the addition of a reset bellows and an adjustable restrictor. A typical proportional plus reset controller is shown in Fig. 22-113. The action of a proportional plus reset controller is as follows. If the measurement pressure increases, the balance beam moves toward the nozzle, increasing the controller output pressure. The increased output pressure causes the proportional bellows to reposition the beam and stabilize the output at an increased pressure. There is now a pressure drop across the restriction R_i. As a small flow of air enters the reset bellows, it forces the beam toward the nozzle, further increasing the output pressure.

If this controller is in a closed loop, the output pressure will remain above the steady-state value until the measurement pressure is driven to equal the set point. The output pressure stabilizes at whatever pressure is required to hold the measurement at the set point, eliminating offset following load or set-point changes.

If the process does not return to the set point or the measurement feedback loop is broken, the increase in measurement bellows pressure causes the output to increase proportionally, then slowly increase to the limit of the air supply. This situation is called "reset windup." Before the automatic control mode can be restored, this pressure must bleed back through R_i. Bleed-down time may be reduced by opening the restrictor until the pressures in the reset and proportional bellows have been equalized and the output pressure has decreased to be within the normal control range, then closing it to the original restriction setting.

Proportional plus Rate Control (PD) Proportional plus rate (derivative) action may be obtained with the arrangement shown in Fig. 22-114. By placing an adjustable restrictor R_d between the proportional bellows and the controller output, a delayed proportional action results. This operation is described in the following way. Assume that the measurement signal increases linearly with time (a ramp). The beam moves slowly toward the nozzle, increasing the output pressure. Air flows slowly through the restrictor R_d to restore the beam-nozzle position. Because of the length of time required for the air to flow through the restrictor, proportional action is temporarily reduced and delayed. The output is therefore higher and more advanced in time than it would be with proportional control only. To this extent the controller output leads the measurement input. The magnitude of the output is proportional to the rate of change of the measurement signal.

Proportional plus Reset plus Rate Control (PID) As the name suggests, the PID controller is a combination of the three control modes discussed previously. Figure 22-115 shows a typical three-mode arrangement.

The arrangements shown in Figs. 22-113 through 22-115 are not standard. Each manufacturer has its own design. Some common variations are as follows: (1) In some designs a pneumatic pilot relay is used between the controller and the output to the valve actuator to reduce the volume of air which must be supplied through the restrictions within the controller. (2) Some applications require a reverse-acting controller. That is, the controller output decreases when the measurement signal increases. (3) In some designs the set point is provided by using a set-point bellows and adjustable restrictor. (4) The proportional-gain (or proportional-band) adjustment may be made with an adjustable restrictor instead of a movable fulcrum.

Stack-Type Controller The pneumatic controllers described in

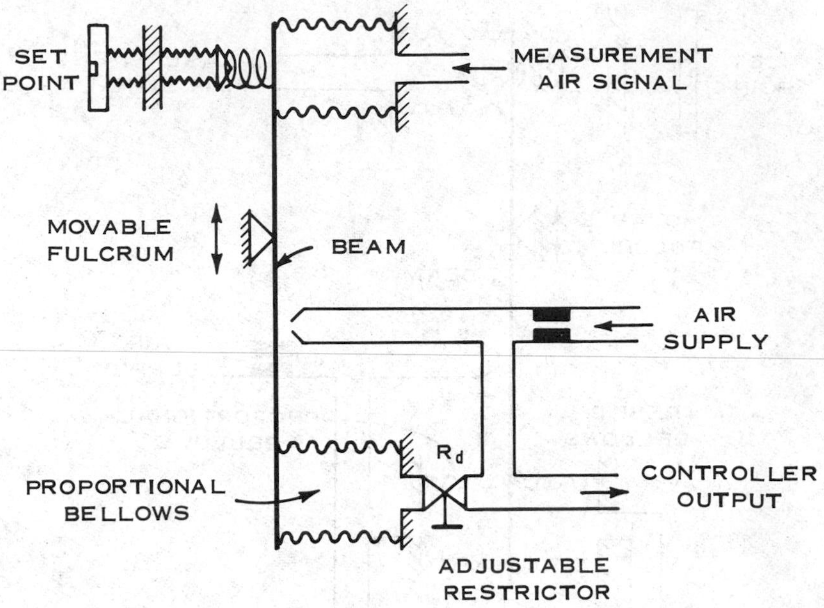

FIG. 22-114 Pneumatic proportional-derivative controller.

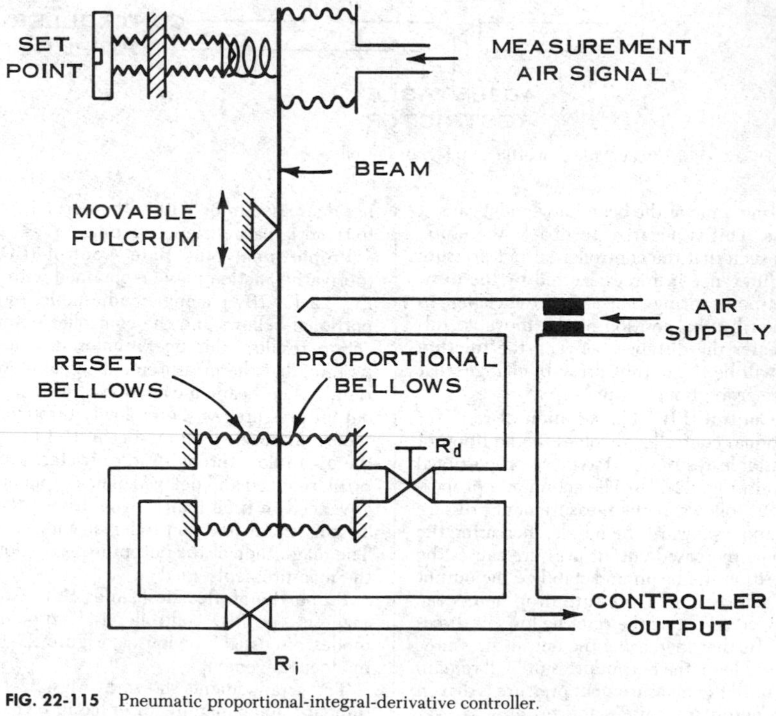

FIG. 22-115 Pneumatic proportional-integral-derivative controller.

the preceding paragraphs may be implemented by any of numerous hardware configurations. A typical configuration is the stack-type controller illustrated in Fig. 22-116. Its name is derived from the fact that the controller is assembled by literally stacking one module on top of another, each module performing a segment of the overall control action. The order of the stacking may vary from that shown. This type of equipment is used particularly when the controller is located far from the control room and near the processing equipment.

In this illustration the proportional-gain adjustment is accomplished as suggested in variation 4 in the preceding paragraph, and the set-point signal is supplied as suggested in variation 3. Reset action is produced by the reset tank, reset adjustment, and restrictions R_1 and R_2.

Consider an increase in set-point pressure that causes a reduction in nozzle pressure and an increase in controller output pressure. The increased output pressure, retarded by the reset-adjustment restrictor, forces the diaphragm in the reset tank to close the vent. This supplies air through R_2 and R_1 to increase the pressure in the proportional positive-feedback bellows until the pressure between R_2

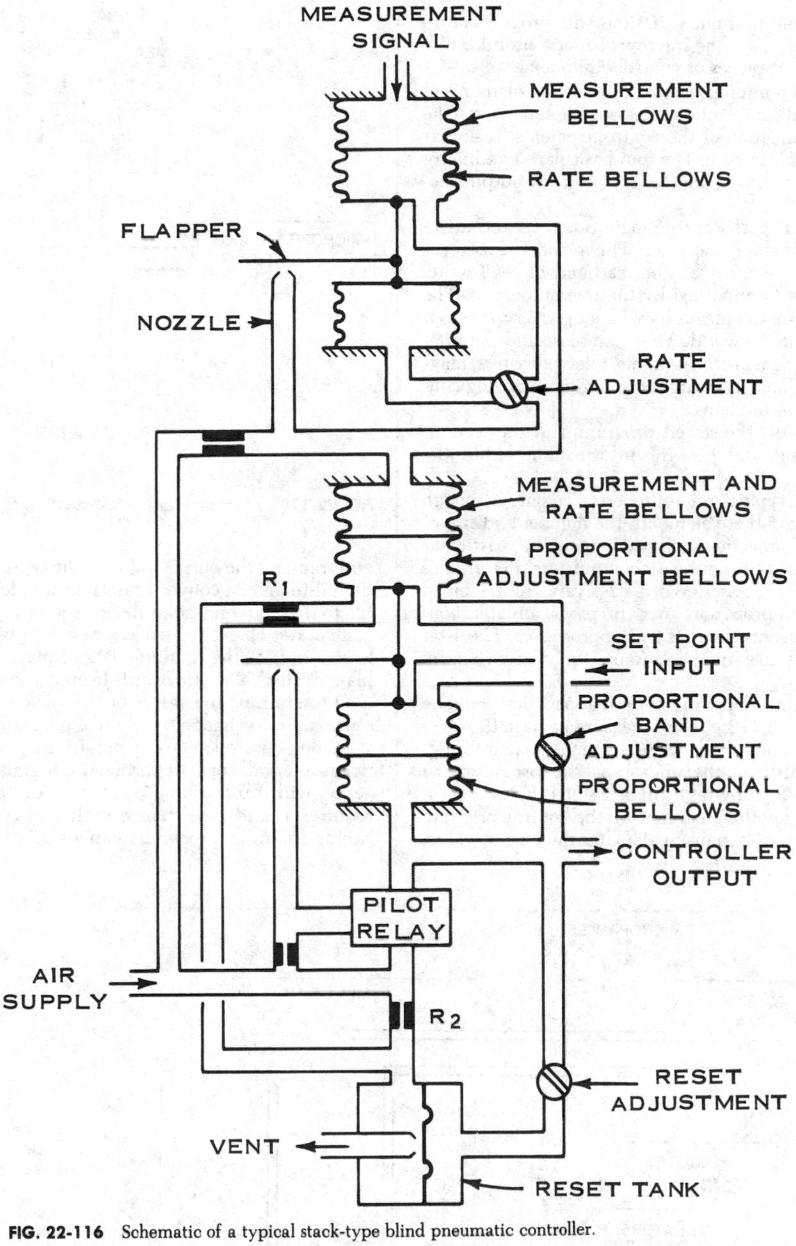

FIG. 22-116 Schematic of a typical stack-type blind pneumatic controller.

and R_1 in the reset tank balances the reset-tank pressure supplied by the controller output through the reset-adjustment restriction. In Fig. 22-116 bellows are shown to correspond to earlier illustrations. However, this type of controller is usually built around a series of diaphragms instead of bellows.

MICROPROCESSOR CONTROLLERS

Since the introduction of microprocessors in 1971, there has been increasingly widespread use of them in numerous products including many applications in process control. In the majority of such applications, the microprocessor is incorporated into a product (for example, mass-flow metering, data loggers, batching controllers, and continuous-process controllers) in such a way that the end user does not require a knowledge of microprocessors and may not even be aware that a microprocessor is used. In other applications, the user may purchase a microprocessor system and develop the software (i.e., the control programs) for a specific use.

General Concepts Before discussing process-control applications of microprocessors, we must first define what a microprocessor is. A microprocessor is a semiconductor device that can perform arithmetic, logic, and decision-making operations under the control of a set of instructions stored in memory. Essentially it is a general-purpose digital data manipulator (computer) with a stored program. In a microprocessor all the elements of a minicomputer have been reproduced in large-scale integrated circuits. Unlike other types of digital logic circuits that are inflexible because they are programmed by the way in which the circuits are wired, a microprocessor's logic functions are programmed by an instruction set stored in memory.

Microprocessors can respond to input variations with program alternatives. This flexibility has made the microprocessor a useful building block for many different pieces of control equipment.

It was stated earlier that a microprocessor contains the elements of a minicomputer. In addition, microprocessor architecture (the method used for the organization of the control elements) is similar to classic minicomputer architecture. The four basic parts of a microprocessor are the arithmetic logic unit, memory, input-output sections, and control unit.

The arithmetic logic unit performs the logic functions and arithmetic operations within the microprocessor. The memory is used for program and data storage. Memory can be read-only or read-write. Read-only memory cannot be modified by the microprocessor. The input and output sections are the means by which the microprocessor communicates with the outside world. They can sense and manipulate a variety of devices such as switches, lights, teletypewriters, magnetic- or paper-tape units, analog-to-digital or digital-to-analog converters, and communication modems.

The control unit interprets the stored program, and the sections also coordinate the timing and flow of information within the microprocessor.

Figure 22-117 shows a typical microprocessor layout. Although most microprocessors contain these elements, the number and size of registers and their interconnection will vary with the particular microprocessor. The size of the registers (accumulators) determines the word size of the microprocessor. Word sizes vary from 1 bit in some special-purpose microprocessors used in programmable controllers to 16 bits in some general-purpose microprocessors. The 4-bit and 8-bit microprocessors are most common for process-control applications.

A microprocessor-based controller in a typical application could be organized as shown in Fig. 22-118. Process measurements in the form of electrical analog signals are applied as inputs to the analog-to-digital converter. Under control of the microprocessor program, the analog data are converted to a digital form and stored in the microprocessor memory. Later another segment of the control program selects the needed data from memory, calculates the output value, and transfers the output value to the digital-to-analog converter. The digital-to-analog converter outputs an electrical analog signal suitable to drive an end-point device such as a control valve.

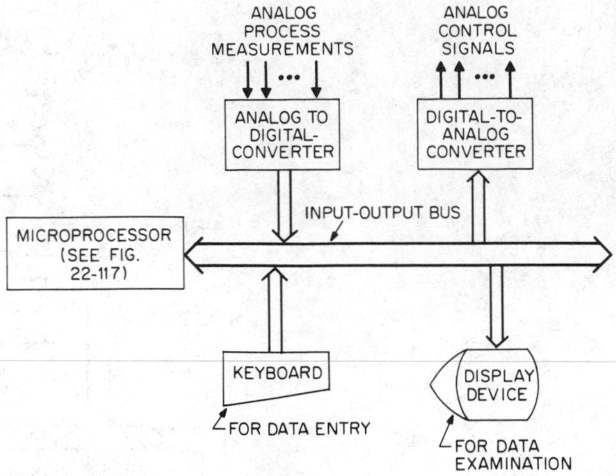

FIG. 22-118 Microprocessor-based controller.

Operator changes, such as a new set-point value, are made via the keyboard with the resulting data displayed on the alphanumeric display device. The keyboard-display combination can typically be used to examine the values of the analog inputs and outputs and, in some cases, to change the control program.

Analog controllers are generally limited to proportional, proportional-integral, or proportional-integral-derivative (PID) control action with fixed tuning constants. Any change in control strategy requires a hardware change with analog equipment. Control algorithms for microprocessors can be written for any control action

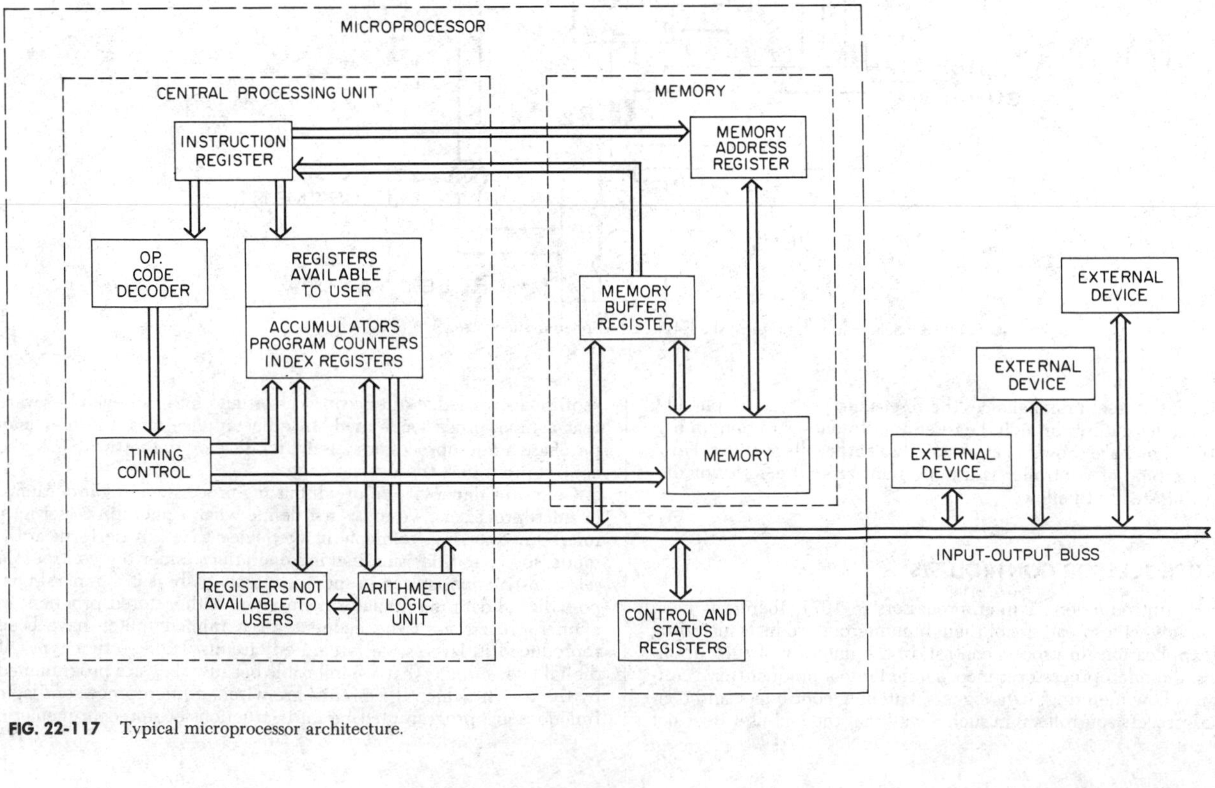

FIG. 22-117 Typical microprocessor architecture.

desired. However, the standard PID and a few variations of the PID algorithm are usually all that are available in standard software packages. A derivation of the PID algorithm as it could be implemented in a microprocessor-based controller is discussed in the following paragraphs.

Algorithms Since the microprocessor is a digital device operating on discrete data sampled at a known rate, equations can be written to mimic the action of analog control circuitry.

The algorithm describing the action of a PID controller can be derived from the definition of PID. By definition, the output is effected by three terms: one is proportional to the error signal, one is the integral of the error signal, and one is the derivative of the error signal. To be consistent with the analog circuitry which applies the gain to all three error terms, our algorithm will also be written so that the integral and reset actions are affected by the gain setting. A similar algorithm could easily be written when the reset and rate were independent of the gain term. On the basis of this definition, the output of an analog controller equals the sum of three terms, or

$$\text{Output} = \text{proportional term} + \text{integral term} + \text{derivative term}$$

$$= G_1 e + G_2 \int e\,dt + G_3 \frac{de}{dt} \qquad (22\text{-}27)$$

where output = signal sent to valve or final control element
G_1 = controller gain term
G_2 = controller reset or integral term
G_3 = controller rate or derivative term
e = error or present difference between process signal and controller set point

Since the microprocessor operates on discrete numbers, the integral is replaced by a summation and the differential is replaced by a difference. Because the sample rate is small compared with the process time constants, this does not produce any significant errors. Equation (22-27) then becomes

$$\text{Output} = G_1\, e_n + G_2 \sum_{k=0}^{n} e_k + G_3\,(e_n - e_{n-1}) \qquad (22\text{-}28)$$

where $e_n = s_n - p_n$ error at time n
p_n = process signal at time n
s_n = set-point value at time n

Therefore $e_n - e_{n-1} = p_{n-1}$ since $s_n = s_{n-1}$.

A constant Q_k may be included in the integral term to prevent the reset windup problem that can occur in analog controllers:

$$\text{Output} = G_1\, e_n + G_2 \sum_{K=0}^{n} (Q_k\, e_k) + G_3\,(p_n - p_{n-1}) \qquad (22\text{-}29)$$

where Qk = 1 if output is within defined limits
= 0 if output exceeds limits

The gain terms are now rewritten to correspond with the gain, reset, and rate terms as used on analog controllers:

$$\text{Output} = K_g \left[e_n + \frac{T_s}{T_i} \sum_{K=0}^{n} (Q_k\, e_k) + \frac{T_d}{T_s}\,(p_n - p_{n-1}) \right] \qquad (22\text{-}30)$$

where p_n = process signal at mth sample period
$e_n = S - P_n$
S = set point
Q_k = 1 if output is within limits
= 0 if output exceeds limits
T_s = sampling time
T_i = reset term
T_d = rate term
K_g = proportional-gain term

This is only one of many possible control algorithms. Among other popular algorithms are (1) PID with error squared on integral, (2) PID with error squared on gain, (3) gap controllers in which no output change is made until predefined error limits have been exceeded, and (4) gap controllers in which only the integral action is controlling until a predefined error limit has been exceeded.

Most algorithms in existing software packages are based on the three-mode PID controller. However, a wide variety of input conditioning (ratio, difference, square-root, multiplier, etc.) is available.

Adaptive Tuning The capability of a microprocessor-based controller for adaptive tuning is a major advantage in some control loops. Adaptive tuning is the ability to change automatically the tuning constants of a feedback controller. The most common form is adaptive gain control, which previously was limited to control loops on a process-control computer. The microprocessor-based controller has extended the possibility of adaptive gain control to all control loops.

Four major application areas for adaptive gain control are:

1. Control loops requiring tighter, more responsive control at a specific set point
2. Control loops requiring an overdampened control at a specific set point
3. Control loops requiring gain changes based on an external input to provide startup, event, or time-based changes in the tuning constants
4. Control loops when the process gain is nonlinear

Tighter, More Responsive Control The reasons for desiring tighter control for some loops are obvious. It was on loops such as these that adaptive gain control was first used. Process deviations from set point can usually be decreased by increasing the controller gain, but stability also decreases as the gain increases. If the controller gain is increased as a function of process deviation from the set point, a compromise in which the deviation is reduced without forcing the loop into oscillation can be reached.

Overdampened Control While at first it may not be obvious why overdampened control is desirable, there are several applications for which it is beneficial. Two possible applications are the reduction of controller action caused by process or measurement noise and the reduction of control-loop interactions.

Previous steps used to reduce process and measurement noise include dampening the transmitter and tuning the controller for low gains and high reset rates. These steps reduce controller action due to noise, but they also reduce the response to actual process deviations. Adaptive tuning allows controller gain to be low in the region around the set point with a higher gain as the deviation increases.

Adaptive gain control can be used to reduce control-loop interaction caused by process coupling in both parallel and series processes. A common problem in processing plants in which equipment is operated in parallel (heaters, boilers, generators, refrigeration compressors, or an entire processing train) is that disturbances in one unit are coupled into all the units. One method of decoupling the control loops to prevent disturbances from cycling between parallel units is to slow down controller action. However, as in the preceding example, overdampened control results in larger deviations and slow recoveries. With adaptive tuning reducing the dampening as the deviation increases, the size of the maximum deviation can be reduced.

Another example of how control-loop interaction can be reduced is feed control into a processing unit such as a fractionation tower. For smooth unit operation the feed rate should be as nearly constant as possible. A common method of controlling the feed rate to a fractionation tower employs the level controller on the bottom of the preceding tower. The bottom of this tower can be used as a surge tank to smooth any variation in feed rate if the level is allowed to fluctuate over a specified range. In this example, adaptive tuning is used as a gap controller with controller gain being a function of the level signal.

Gain Changes Based on External Inputs The ability to change automatically controller gain on the basis of external startup, event, or time-based inputs has few applications in the continuous-process industries, but it has major potential in batch processes. It might be applied to a batch process that requires nonoverpeaking startup and consistent tight control throughout the batch.

Nonlinear Process Gains Adaptive tuning has applications on control loops when the process transfer function is nonlinear. This nonlinearity can be caused by nonlinearities in process characteristics such as pH control (acid or base neutralization), or it may be the result of a wide range of loads or set points. The object of adaptive tuning in this application is to maintain the product of process gain and controller gain at a fairly constant value.

An example of nonlinear process gain is the application of split-range valves. The flow characteristics and sizes of the two valves are often quite different. To compensate for the changes in valve characteristics controller gain can be adjusted as a function of controller output.

Further Applications Microprocessors can also be applied to control the operation of a complete system within a process unit. For example, a specially designed microprocessor system could control the operation of a boiler. It could be responsible for monitoring the status of the safety switches, modulate the control valves to control steam pressure, drum level, etc., optimize the efficiency of the boiler by adjusting the fuel-to-air ratio, and control sequencing and timing for startup. Such specially designed unit controllers would be purchased as a microprocessor system, not as a piece of hardware containing a microprocessor. In this type of system the end user has an input into the programming of the unit and may be responsible for generating the control program.

CONTROL-SYSTEM LOGIC SIMPLIFICATION

A fault tree is a list of criteria having logical connections which must be met before a specified goal is reached. If your goal is to start your car and your car has an automatic transmission, then your fault tree would consist of placing the transmission in neutral or park, placing the key in the ignition, and turning it to start. There are other advisable things to be done, such as to depress the accelerator partially and keep your foot on the brake, but you can start a car without these items; they are not, therefore, on your car's start fault tree. A fault tree will be made up of binary functions such as a light switch being on or off, speed, pressure, level, or flow switches being above or below the set point, or a vibration switch being closed or tripped. Fault trees being made up of binary functions, lead to notations and solutions by boolean algebra.

Boolean Algebra Boolean equations are written by using + for OR and · for AND. They have only two possible answers, 0 and 1. In an OR equation, when any input is 1, the output is 1. In an AND equation, all inputs must be 1 before the output is 1. Your car's start fault tree then becomes (neutral + park) · turn to start = start car. If we assign N to transmission neutral, P to transmission park, S to ignition turned to start, and SC to start car, our equation is $(N + P) \cdot S = SC$. Our functions being binary, we assign 0 to each function that is not in the state required to give our desired output, i.e., ignition switch is not turned to start $S = 0$. A 1 is assigned to each function in the desired state. If the transmission is in park and the ignition turned to start, $N = 0$, $P = 1$, and $S = 1$. By logic addition $0 + 1 = 1$ and $1 \cdot 1 = 1$; therefore, the car should start. A bar (line) over the function denotes the function is NOT; i.e., if $N = 1$, then $\overline{N} = 0$, $\overline{0} = 1$, and $\overline{1} = 0$. The following boolean laws are used to simplify a fault tree:

$$A + A = A$$
$$A \cdot A = A$$
$$A + B = B + A$$
$$A \cdot B = B \cdot A$$
$$(A + B) + C = A + (B + C) = (A + B + C)$$
$$(A \cdot B) + C = A \cdot (B \cdot C) = (A \cdot B \cdot C)$$
$$A \cdot (B + C) = A \cdot B + A \cdot C$$
$$A + B \cdot C = (A + B) \cdot (A + C)$$
$$A + 1 = 1$$
$$A \cdot 1 = A$$
$$A + 0 = A$$
$$A \cdot 0 = 0$$
$$\overline{A} + A = 1$$
$$\overline{A} \cdot A = 0$$
$$\overline{A \cdot B} = \overline{A} + \overline{B}$$
$$\overline{A + B} = \overline{A} \cdot \overline{B}$$
$$A \cdot B = \overline{A} + \overline{B}$$
$$A + B = \overline{A} \cdot \overline{B}$$
$$\overline{\overline{A}} = A$$

In the earlier example we could have said that to start the car the transmission must be in neutral and not in park or in park and not in neutral and that the ignition must be turned to start:

$$SC = N \cdot \overline{P} \cdot S + \overline{N} \cdot P \cdot S$$
$$SC = S \cdot (\overline{P} \cdot N + \overline{N} \cdot P)$$

We know that the car transmission cannot be in park and neutral at the same time. Another way of stating this is that when $P = 1$, $N = 0$, and when $N = 1$, $P = 0$, or $\overline{P} = N$, $\overline{N} = P$. When this is applied to the equation, it changes to $SC = S \cdot (N \cdot N + P \cdot P)$. From the boolean law $A \cdot A = A$, the equation reduces to $SC = S \cdot (N + P)$. This equation is reduced to the same equation as our earlier example even though we have more completely defined the original conditions. We can add the other possible positions of the shift lever to the initial equation and still reduce the final equation to $SC = S \cdot (N + P)$.

Assume that a pump is to operate when there is a level above a fixed reference point in tank A and below a fixed reference point in tank B: P (pump) $= A \cdot \overline{B}$. This pump also needs to run when the level in tank A is above a fixed reference point ($P = A \cdot C$). When the levels in tanks B and C are both above their respective fixed reference points, we require the pump to run ($P = B \cdot C$). Accumulating the running requirements for the pump gives the equation $P = A \cdot \overline{B} + A \cdot C + B \cdot C$, which is reduced by use of the boolean laws to $P = A \cdot \overline{B} + B \cdot C$. Without the use of the laws, it may not be apparent that the function $A \cdot C$ in this equation may be eliminated. The level in tank B must be either above or below its fixed reference point ($B + \overline{B} = 1$). If $\overline{B} = 1$, then $A \cdot \overline{B} = 1$ as soon as $A = 1$. If $B = 1$, then $B \cdot C = 1$ as soon as $C = 1$. $A \cdot \overline{B} + B \cdot C$ will equal 1 before or at the same time that the $A \cdot C$ function does not change the equation. Direct application of the laws without reference to the problem yields

$$\begin{aligned} A \cdot \overline{B} + A \cdot C + B \cdot C &= A \cdot \overline{B} + A \cdot C \cdot (B + \overline{B}) + B \cdot C \\ &= A \cdot \overline{B} + A \cdot C \cdot B + A \cdot C \cdot \overline{B} + B \cdot C \\ &= A \cdot \overline{B} + A \cdot \overline{B} \cdot C + B \cdot C \cdot A + B \cdot C \\ &= A \cdot \overline{B} \cdot (1 + C) + B \cdot C\,(A + 1) \\ &= A \cdot \overline{B} + B \cdot C \end{aligned}$$

Why should we be concerned with reducing an equation such as $A \cdot \overline{B} + A \cdot C + B \cdot C$ to $A \cdot \overline{B} + B \cdot C$? Each + (OR) and (AND) in the final equation increases the cost of the control system in equipment and programming time. By reducing the equation, we have reduced from five to three the number of AND and OR functions required. If AND and OR are assumed to cost the same, this is a 40 percent cost reduction with no change in how the system functions.

As another example, a request has been received for the installation of a single alarm which will sound when the flow in a pipe line is below normal flow and the pipe-line booster pump is running. This alarm is also to sound at any time when the flow rate is above normal flow. The normal flow rate will be detected by a flow switch which operates on a single point (no dead space). If A is assigned to flow above normal and B to the booster pump running, then the equation for the alarm is $AL = \overline{A} \cdot B + A$. The boolean laws can be used to show that $\overline{A} \cdot B + A = A + B$. This simplification shows that the initial request was not realistic and will result in an alarm whenever the pump is running or the flow is above normal. For the alarm to function on low flow as intended, the flow switch must have two set points with a dead space between them. If A and B are assigned as before and C is assigned to the new below-normal flow, the equation becomes $AL = C \cdot B + A$, which will function as intended.

Programmable Control The control system, after the boolean equations have been simplified, can be designed by using any of a number of binary control systems. The easiest system for the user to install initially and modify after installation is the programmable controller.

A programmable controller is a logic system which issues output commands when a preset combination of input conditions are met.

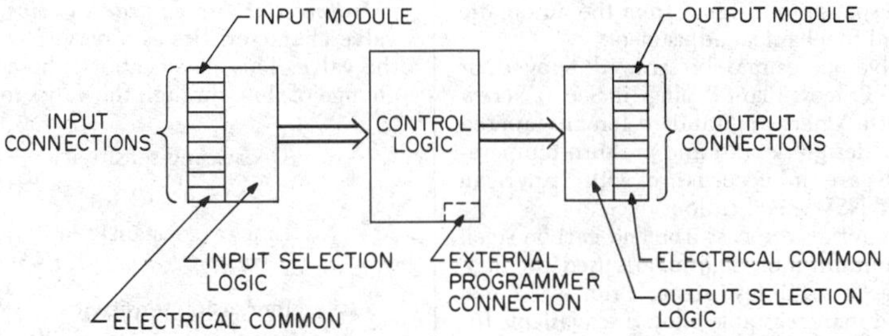

FIG. 22-119 Programmable controller.

Hardware consists of a specialized form of computer that has limited arithmetic ability and is generally designed for the harsher industrial environment of an industrial plant. Basically, the programmable controller as shown in Fig. 22-119 consists of three sections: input modules, control logic (programmable) section, and output modules. The input modules convert the incoming signals to a voltage level acceptable to the logic section. The logic section polls the information from the input modules, compares this information with the programmed requirements, and issues commands to the output modules on the basis of this comparison. The output modules convert the logic-level signal from the logic section to a power level compatible with the utilization equipment. The voltage to the input module may range from 24 to 230 V ac or dc, depending on the input module used. The logic-level voltage will usually be in the range of 5 to 15 V dc. The output voltage of the output modules will have generally the same range as the input voltage to the input modules.

The cost of equipment, maintenance, and programming, along with the frequency of control-concept changes, will dictate when a programmable controller should be used in place of a less flexible nonprogrammed control system. Normally, programmable controllers become more cost-effective than relays when more than 20 relays are required. However, the cost of the external programmer used to generate the control program can increase this number if only a few systems are involved.

FINAL CONTROL ELEMENTS

GENERAL REFERENCE: Beard, *Final Control Elements*, Chilton, Philadelphia, 1969.

The final control element is the mechanism which alters the value of the manipulated variable in response to the output signal from an automatic controller, the output signal from a manually manipulated control device, or direct manual manipulation. In automatic-control installations, it normally consists of two parts: (1) an actuator which translates the output signal of the controlling device into an action involving a large force or the manipulation of large power and (2) a device responsive to the actuator force which adjusts the value of the manipulated variable. For example, the actuator may be used to change the position of a valve plug in an orifice, the velocity of a rotating device, or the amount of power being delivered to an electrical load.

In automatic process control, the most frequently used final control element is a diaphragm motor valve (DMV). It consists of a pneumatic diaphragm motor actuator and a process-fluid control valve. Figure 22-120 illustrates a typical diaphragm motor valve.

Each device that is used to make up the final control element possesses its own dynamic-lag characteristics or time constants. This is to say, the devices will not respond instantaneously to changes in control signals or load disturbances. The significance of the effect of the lags depends on the process in which the device is used. In some instances, these lags may seriously degrade the control-system performance and hence lead to decreased process profit. They may also lead to increased operator attention and intervention.

CONTROL VALVES

When discussing control valves and valve characteristics, two parts of the valve need to be considered: first, the valve body, its geometry and materials of construction, and second, the valve plug, its geometry and materials of construction. The combined geometry of the body and plug determines the flow properties of the valve.

Most control valves are operated by a linear-position actuator or some modification of such an actuator. These actuators position the

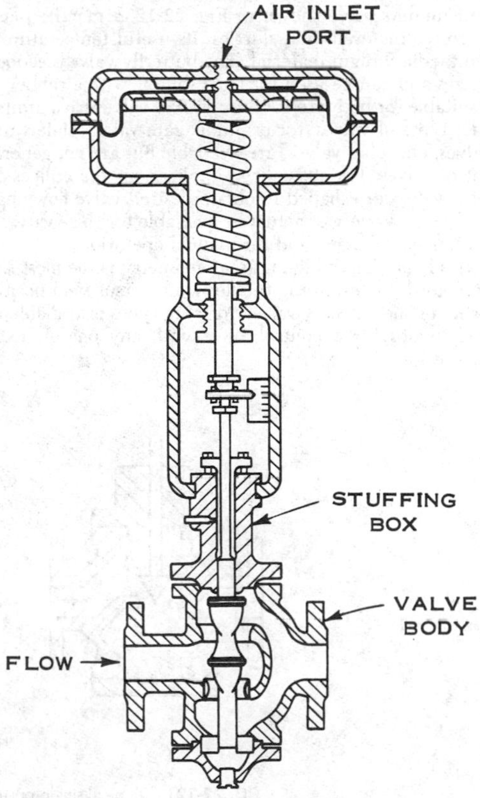

FIG. 22-120 Schematic of a typical control valve and pneumatic actuator.

valve plug in the orifice in response to a signal from the automatic controller or through a manual mechanical adjustment.

Valve Bodies Control-valve bodies may be screwed, flanged, or welded into the flow line; see "Process-Plant Piping" in Sec. 6. Screw ends usually are threaded with American Standard female tapered pipe threads. The dimensions, design details, and pressure-temperature ratings of flanged ends are in accordance with American National Standards Institute (ANSI) specifications.

The most common body materials are cast iron and carbon steel. Low-alloy steels, such as chromium-molydenum, are used for high temperatures. Various grades of stainless steels, bronze, Monel, nickel, Inconel, Hastelloys, and many castable alloys are available for corrosive chemical service or for very high- or low-temperature applications. When it is not practical to cast valve bodies, they may be machined from solid bar or forged material, particularly in small sizes.

The construction material for control-valve trim, i.e., those parts which must retain close machined tolerances for sealing, metering, or moving, must be selected with care. It must generally be more resistant to corrosion, erosion, galling, and distortion than the body material.

Several of the common types of control-valve design and construction are illustrated schematically in Figs. 22-121 through 22-123. When valves illustrated in Fig. 22-121 are operated with flows in the direction of the arrows, the actuator must overcome flow forces when closing. When flow is reversed, these valves have a tendency to slam on closing. They are all adaptable to tight-shutoff (i.e., no-leakage) service when plug and orifice are properly designed for this use. Soft valve seats may also be provided for tight shutoff.

Valves using the configurations indicated in Fig. 22-122 are not tight-closing unless used with soft seats or O rings for sealing. The Fig. 22-122*a* configuration has balanced forces on closing, therefore reducing the force requirements of the actuator. The other two configurations have essentially the same unbalanced forces as single-ported valves. Flow directions may be reversed from that shown. Valve-plug orientation may also be reversed from that shown.

The Saunders valve shown in Fig. 22-123*a* may be plastic-lined for corrosive materials and slurries. Its useful temperature range is limited by diaphragm material. The butterfly valve is economical in large sizes and can be used for tight shutoff with a rubber lining. It is not suitable for high pressure drops. Ball valves have limited usage.

Other types of valves, for example, gate valves, slide valves, poppet valves, and plug valves, are available but are not generally used as control valves. For specialized applications the orifices and seats for these valves are shaped to obtain desired valve flow characteristics. There is a variety of actuators available for these valves both for automatic-control usage and for manual operation.

Valve Plugs The plug-and-seat geometry is the most significant single factor in determining the flow rate versus stem position characteristics of the valve. A variety of plug types is available and may be used to provide a control valve with any one of several flow characteristics.

Inherent Flow Characteristics There are three basic types of valve characteristics as observed with constant pressure drop across the valve. This classification is based on the sensitivity of the rate of change of flow through the valve to the valve-stem position.

$$\text{Decreasing sensitivity } \left.\frac{dQ}{dL}\right|_{Q=0} > n > \left.\frac{dQ}{dL}\right|_{Q=\max}$$

$$\text{Linear (constant) sensitivity } \frac{dQ}{dL} = n$$

$$\text{Increasing sensitivity } \left.\frac{dQ}{dL}\right|_{Q=0} < n < \left.\frac{dQ}{dL}\right|_{Q=\max}$$

Here L is a percentage of maximum valve-stem travel (valve lift), Q is a percentage of maximum flow, and n is the valve-flow coefficient. These three basic types of characteristics are illustrated schematically in Fig. 22-124.

Examples of the decreasing-sensitivity type are quick-opening globe valves, weir valves, and poppet valves. Equal-percentage, logarithmic, and parabolic valves are of the increasing-sensitivity type. The inherent flow characteristics are determined by the valve orifice and plug geometry. They describe the flow rate through the valve as a function of the stem position with the pressure drop across the valve kept constant. Valve manufacturers provide this information, referring to it as "sizing coefficient" (C_V, C_G, or C_S) data, generally referenced to the standard condition of a constant pressure drop of 6.9 kPa (1 lbf/in^2). Figure 22-125 shows a variety of typical control-valve-plug configurations. Figure 22-126 illustrates typical inherent flow characteristics for these plug configurations in a conventional valve seat.

The inherent flow characteristic of a valve may be determined by the shape of the seat as well as the plug. Figure 22-127 illustates the conventional seat with a parabolic plug and a quick-opening plug in a contoured seat. In addition to a large change in the flow characteristic, the contoured seat causes the unbalanced forces on the plug to change less rapidly as the valve is opened or closed than with the conventional seat. This leads to a smoother actuator action and generally better actuator positioning with small changes in controller output, reducing or eliminating the need for the inclusion of force-compensation devices in the valve design.

Besides valve sensitivity, two other valve properties are important. The first is rangeability R, defined as

$$R = \frac{\text{maximum controllable flow}}{\text{minimum controllable flow}}$$

and the second is turndown T, defined as

$$T = \frac{\text{normal maximum flow}}{\text{minimum controllable flow}}$$

Generally, valves are sized so that $T \approx 0.7R$, where R lies, depending on the type of valve, between 20 and 50.

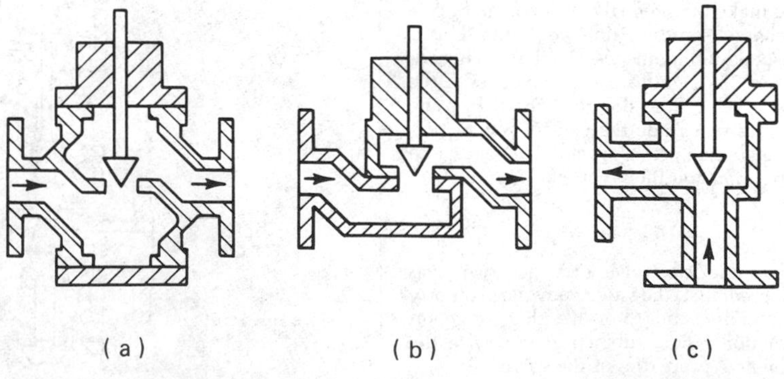

(a) (b) (c)

FIG. 22-121 Typical single-ported valves of different body designs. (*a*) Globe. (*b*) Split-body. (*c*) Angle.

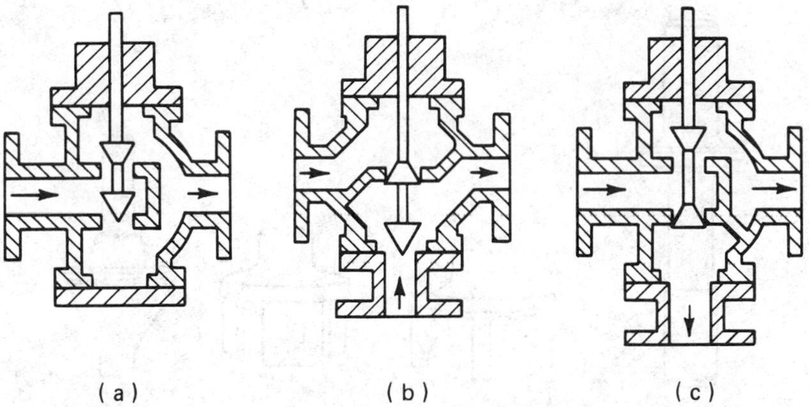

FIG. 22-122 Typical double-ported-valve configurations for (*a*) through flow, (*b*) blending, and (*c*) stream splitting.

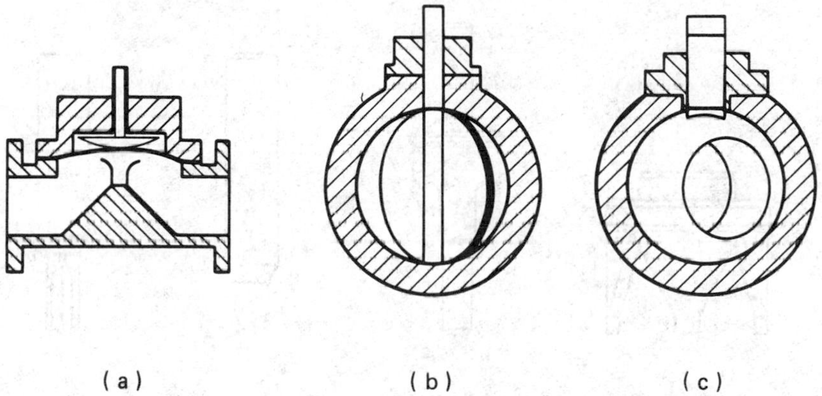

FIG. 22-123 Special-purpose control-valve bodies. (*a*) Saunders diaphragm valve. (*b*) Butterfly valve. (*c*) Ball valve.

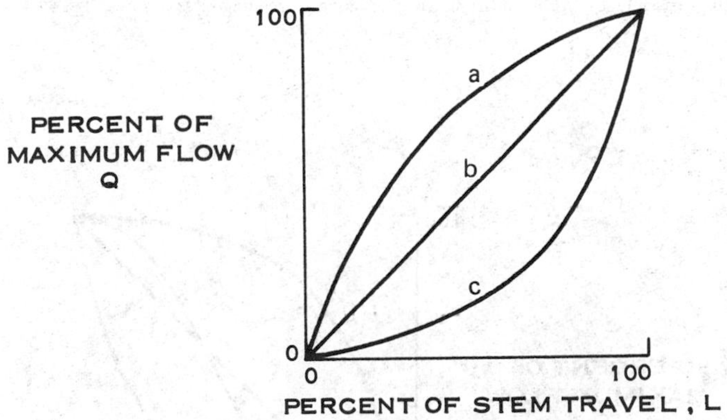

FIG. 22-124 Inherent flow characteristics. (*a*) Decreasing sensitivity. (*b*) Linear or constant sensitivity. (*c*) Increasing sensitivity.

The usefulness of rangeability and turndown may be indicated by a specific control-valve application. For example, if the design of a tank-level control system calls for a 25:1 change in inlet flow rate to maintain the level because of a changing outlet flow rate, then the indicated control-valve turndown ratio is at least 25 and the rangeability ratio at least 35. Typically, increasing-sensitivity-type valves have higher rangeability ratios.

Seldom is the pressure drop across an installed valve constant as the stem position is adjusted. As a result, inherent flow characteristics are not sufficient to determine the installed flow rate for any given stem position.

Installed Flow Characteristics When control valves are combined with other fluid-handling equipment in processing systems, the composite flow-rate characteristics differ from the characteristics of

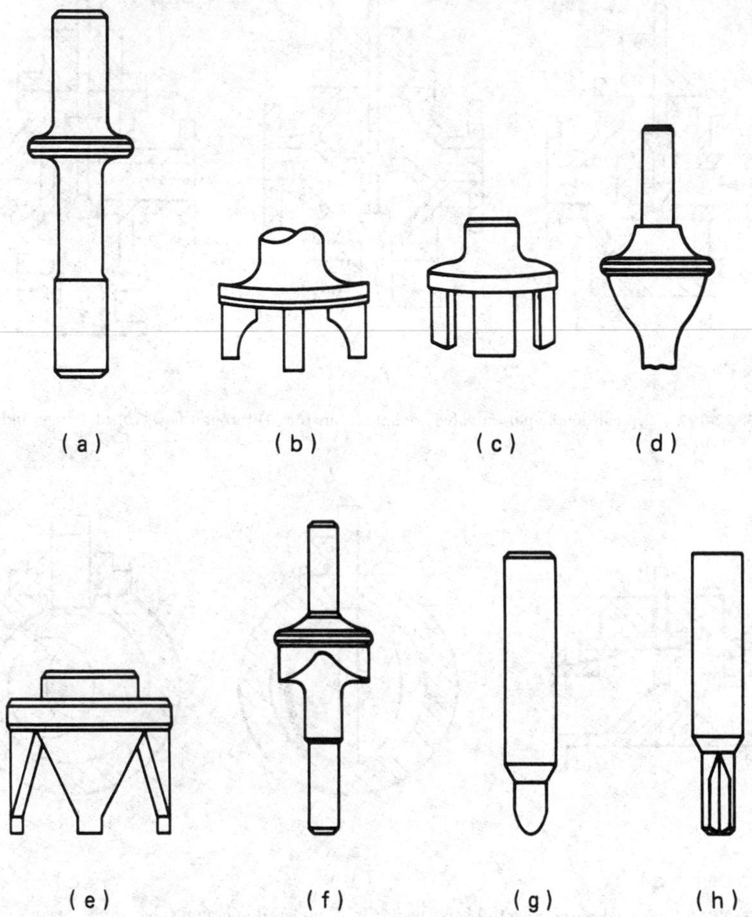

FIG. 22-125 Typical valve-plug shapes. (*a*) Top-and-bottom-guided single-port quick-opening. (*b*) Port-guided quick-opening. (*c*) Rectangular (linear) port. (*d*) Throttle plug (modified linear). (*e*) V port (modified linear). (*f*) Equal-percentage V port. (*g*) Miniature throttle plug (equal-percentage). (*h*) Miniature fluted plug (equal-percentage).

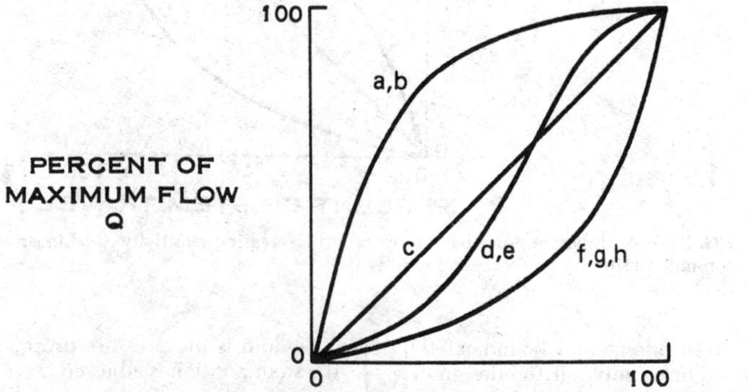

FIG. 22-126 Generalized inherent flow characteristics of the valve plugs shown in Fig. 22-125.

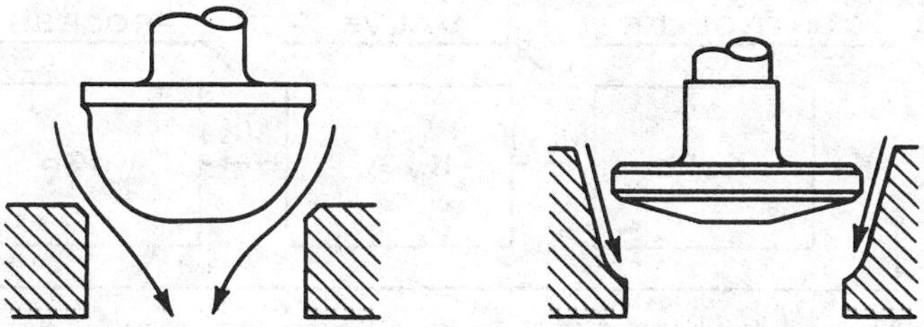

FIG. 22-127 Conventional and contoured valve seats with parabolic and quick-opening plugs.

any single component in the system. Flow rates through the valve are no longer determined solely by the geometry of the valve body and plug.

The effects of resistances resulting from pipe lines, orifices, or other equipment in series with the control valve and the variation of available head with flow rate affect the flow versus stem position relationship. This effect may be described for a linear valve by the equation

$$Q = \frac{L}{[\alpha + (1 - \alpha)L^2]^{1/2}} \tag{22-31}$$

where Q and L are the fractions of maximum flow and stem travel respectively; the term α is defined as

$$\alpha = \frac{\text{valve head differential at maximum flow}}{\text{valve head differential at zero flow}}$$

Installed flow characteristics for a linear control valve installed in flow systems having different values of α are shown in Fig. 22-128. Decreasing values for α indicate increasing flow restrictions external to the valve. For α equal to 1 the installed flow characteristic is identical to the inherent flow characteristic.

The installed flow characteristics of the increasing-sensitivity type of valve (parabolic or equal-percentage) with series resistance in the flow system are described by the equation

$$Q = \frac{L^2}{[\alpha + (1 - \alpha)L^4]^{1/2}} \tag{22-32}$$

and are also illustrated in Fig. 22-128.

Valve Selection

Valve-Type Selection One of the objectives of control-system design is to obtain a specified gain relationship between the controller and some process output variable. This relationship is usually a linear one. That is, using the generalized block diagram (Fig. 22-129) and transfer-function nomenclature, the product $K_vK_pK_c$ is constant independent of $M(s)$, $A(s)$, and $C(s)$, and K_c is also constant. In this block diagram $M(s)$ is the valve-stem-positioning signal and $A(s)$ the flow of process fluid resulting from the instantaneous position of the

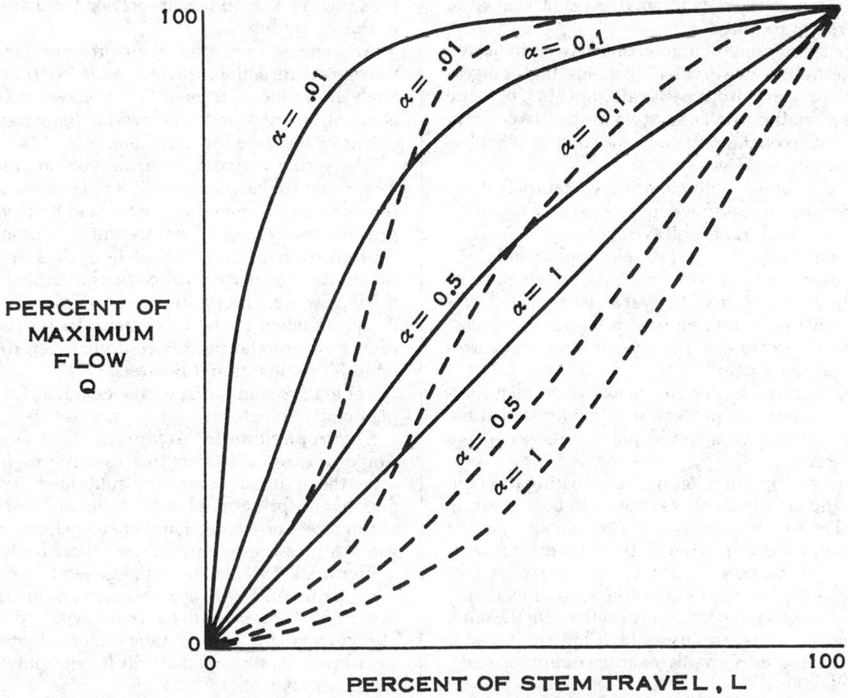

FIG. 22-128 Generalized installed flow characteristics of linear (solid curves) and parabolic (dashed curves) control valves as α is varied with constant available head across the portion of the process which influences the installed characteristic.

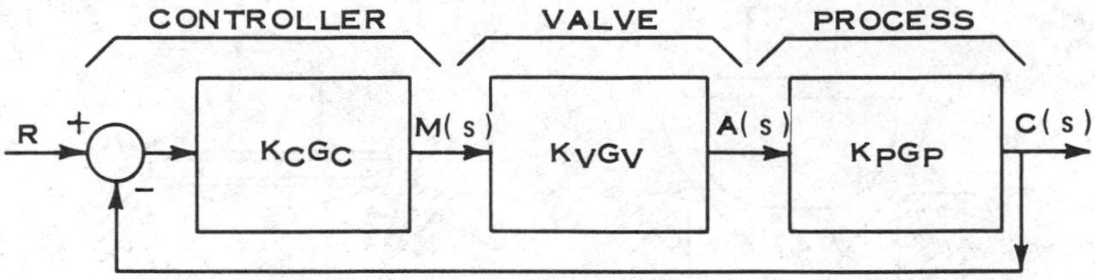

FIG. 22-129 Generalized block diagram where *M(s)* is the valve-stem position and *A(s)* the flow of process fluid resulting from the instantaneous position of the valve stem as in Fig. 22-28.

valve stem. (See also Fig. 22-28.) This calls for a valve having a linear installed characteristic or at least linear over the normal operating range if the process gain K_p is constant. If K_p is not constant over the normal operating range, then K_v is proportional to $1/K_p$. That is, the valve gain characteristics must be the inverse of the process gain characteristics if the combination is to be characterized by constant gain.

In any case, the installed characteristic for a control valve should be governed by the nature of the process and the desired overall characteristics. For example, it may be desired that K_vK_p be a function of $A(s)$ or $C(s)$; for example, $K_vK_p = f(C)$. In this case, $K_v = f(C)/K_p$ may, for example, call for an increasing-sensitivity valve.

Linear-inherent-characteristic valves are indicated when more than 40 percent of the pressure drop in the system occurs across the control valve, high rangeability is not required, valve supply and discharge pressures are stable, combined valve and actuator time constants dominate the control loop, the control loop is stable with controller proportional gains greater than 10, major disturbances in the process are load changes, or there is a linear relationship between the controlled variable and the measured variable.

Some conditions which indicate the choice of decreasing-sensitivity-inherent-characteristic (quick-opening) valves are that control can be accomplished with on-off action, maximum valve capacity must be rapidly available, the process time constant is small but not smaller than the valve time constant, and small rates of change of flow with stem travel are not required.

An increasing-sensitivity-inherent-characteristic valve is indicated when good control is desired at small valve openings, high rangeability is required, supply pressure drops with valve opening, pressure drop across the valve is a small portion of system drop at large valve openings, and the response to changes in controller output is nonlinear in a decreasing-sensitivity process.

Modified linear valve inherent characteristics exemplified by curve *d*, *e* of Fig. 22-126 are indicated when good control is desired at small valve openings and high rangeability is not required.

Other factors entering into valve-type selection include the amount and nature of suspended material in the process stream, the tightness of shutoff desired on closure of the valve, balancing of process fluid forces on the valve plug desired, total pressure drop occurring across the valve, vapor pressure of the process fluid, and corrosive properties of the processing fluid.

Valve Sizing In addition to selecting the proper valve-plug characteristics to match the valve to the process, it is important that the control valve be properly sized. One would not usually use a 2-in valve in a 2-in line, for example.

The control valve performs its function of manipulating throughput rate by virtue of being an adjustable resistance to flow, creating a pressure drop from valve inlet to valve outlet. Controllability of the process flow is determined by the fraction of the total process pressure drop that occurs across the valve. If the valve is too large, sensitivity is low and rangeability too high so that the valve is used only in the almost-closed position, giving poor control action. On the other hand, if the valve is too small, sensitivity is high but the range is limited so that the valve operates normally near maximum opening, giving restricted controllability.

Formulas and procedures are available from valve manufacturers to aid in selecting valve sizes. However, they must be used with caution. These selection procedures are based on modifications of the flow equation $Q = K\sqrt{\Delta P}$, where Q = flow rate, ΔP = pressure drop across the valve or orifice, and K is a constant associated with the valve coefficient C_r. C_r includes the valve-aperture size and shape. The various modifications take into account the properties of the fluid being controlled and the geometry of the particular manufacturer's valve.

LINEAR-POSITION ACTUATORS

According to the definition stated for a final control element, the actuator is a transducer. It translates the control signal from one form or level of energy or power to another, e.g., from a pneumatic signal into a mechanical action which is used to manipulate a process variable. For the most part, a valve actuator may be termed a linear-position actuator. It provides an output position that is proportional to its input signal. The motion of the actuator output element is generally translational (as opposed to rotational), although the translational motion may be transformed into rotary motion to operate some valves.

Pneumatic Actuators The pneumatic actuator is the most common of all process-control actuators. There are several different hardware configurations available for a variety of applications, each possessing its own particular advantages. The pneumatic actuator shown in Fig. 22-120 is a spring actuator in which the controlled input-air pressure provides sufficient force on the diaphragm to offset a portion of the spring force.

An increase in controlled air pressure moves the valve plug in the downward direction, thereby reducing the flow through the valve. Such an action is termed "air-to-close" (see Fig. 22-130*a*). If the actuator air-inlet port is below the diaphragm, the actuator action is "air-to-open" (see Fig. 22-130*b*).

The spring actuator is capable of providing sufficient thrust for many control-valve applications. However, valve-stem friction or forces resulting from the process fluid flowing through the valve may prevent the actuator from assuming a position proportional to the controlled air-pressure input. In critical applications, a valve positioner may be used to overcome this difficulty (see Fig. 22-130*c* and *d*). In essence, the positioner is a high-gain position-feedback controller attached to the valve stem. Its set point is the controlled air pressure from the process controller. See discussion of Fig. 22-135 under "Positioners and Boosters."

Table 22-9 summarizes some characteristics of the more common pneumatic actuators.

Electropneumatic Actuators Increased installation of electronic process controllers that usually provide dc electric signals to drive the actuator system and continued use of the widely accepted pneumatic spring-diaphragm actuator have given rise to a number of interfaces or connections between electronic and pneumatic hardware. A pressure-to-current transducer is illustrated in Fig. 22-97.

Figure 22-131 illustrates the general principle of an electropneumatic actuator. In essence, this is a combination of a current-to-pressure transducer, a feedback positioner, and a pneumatic spring-diaphragm actuator. If a stepping motor actuates the nozzle flapper, the electropneumatic actuator will handle pulse signals from a digital-computer controller.

Hydraulic Actuators Hydraulic actuators usually take the form of a hydraulic piston. The piston may be moved by the application of pressurized hydraulic fluid to either side of the piston (double-

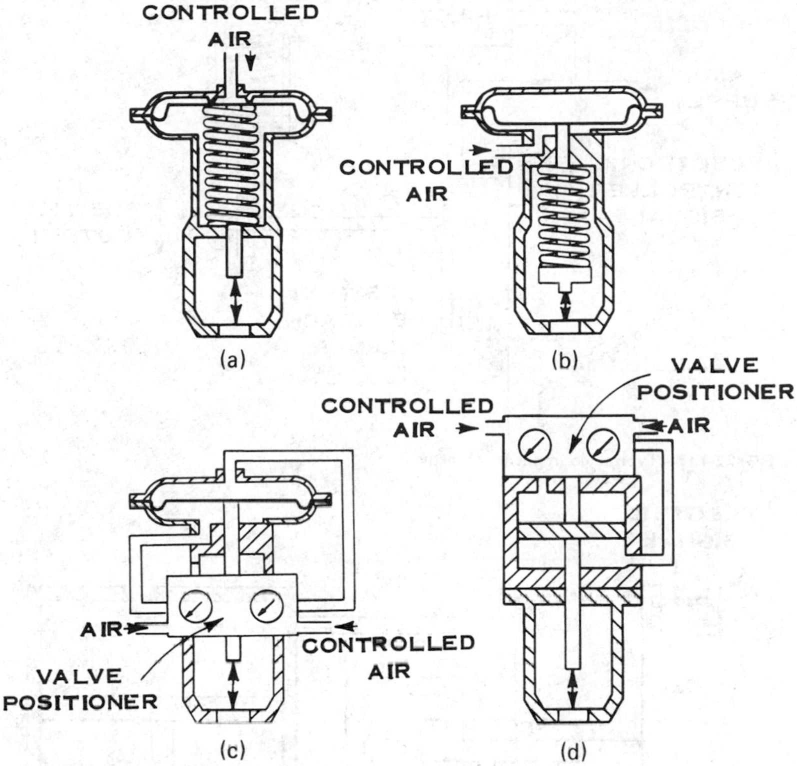

FIG. 22-130 Pneumatic linear actuators.

TABLE 22-9 Properties of Typical Pneumatic Actuators of the Types Illustrated in Fig. 22-130

Schematic diagram	Type of construction	Positioner used	Power hp.	Stroke, in.	Thrust, lb.	Design limitations	Advantages and most common applications
a	Spring and diaphragm; air moves stem downward	No	0.05	To 3	To 800	Thrust is limited. Spring absorbs thrust	Simple and dependable. Fails upward on loss of air. Widely used in process control
b	Spring and diaphragm; air moves stem upward	No	0.05	To 3	To 800	As *a*	Fails downward on loss of air, otherwise as *a*
a or *b*	Spring and diaphragm	Yes	0.15	To 3	To 800	As *a* or *b*	As *a* or *b* with improved control dynamics
c	Pressure-balanced or springless diaphragm	Required	0.15	To 3	To 2000	May not fail safe	Has more thrust than spring types
d	Pressure-balanced piston	Required	1.0	To 36	To 5000	May not fail safe	Uses high-pressure air. Widely used in chemical processes

acting) or to only one side with an opposing spring providing for motion in the other direction. These different arrangements are analogous to the pneumatic actuators illustrated in Fig. 22-130.

Electrohydraulic Actuator The typical electrohydraulic actuator consists of an electric-to-hydraulic transducer, a feedback positioner, and a hydraulic piston actuator. The hydraulic-piston-actuator operation is basically the same as that of the pneumatic piston actuator. Both require some means of piston-position feedback. A typical electric-to-hydraulic transducer consists of a coil-magnet mechanism and a hydraulic pilot spool. Figure 22-132 illustrates an electrohydraulic piston actuator utilizing such a transducer.

Electric Actuators There are two basic types of electric actuators for linear-positioning applications. These are the electric motor and the electrical solenoid. Both types are easily used for two-position actuators. The motor may be provided with a forward and a reverse winding in order to achieve the two-position operation. The reverse winding may be eliminated if some type of spring-return feature is provided. Both motor configurations are shown schematically in Fig. 22-133.

There are a variety of ways to translate motor rotation into linear actuation. For example, one may use a gear train or a reversible hydraulic pump and a hydraulic motor or cylinder. The rate at which the actuator moves from one position to the other is basically determined by the power of the electric-motor driver.

The addition of position feedback and appropriate signal sensor-switching equipment permits the actuator to be started or stopped at intermediate positions. The actuator then becomes equivalent to an electropneumatic or electrohydraulic actuator.

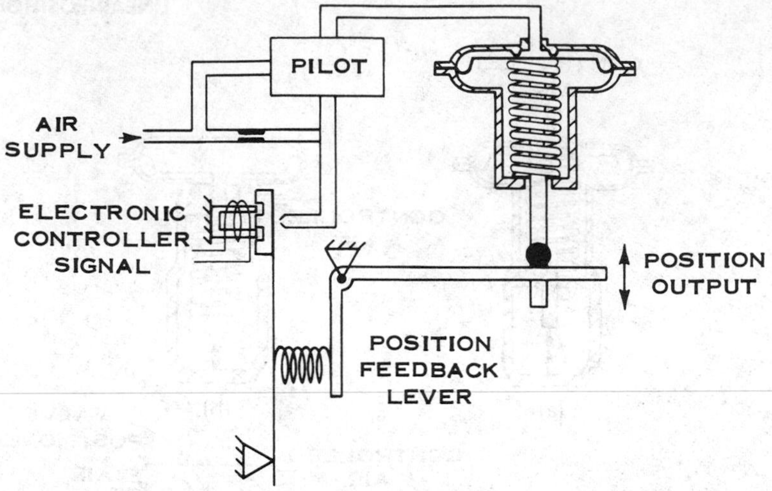

FIG. 22-131 Typical electropneumatic actuator.

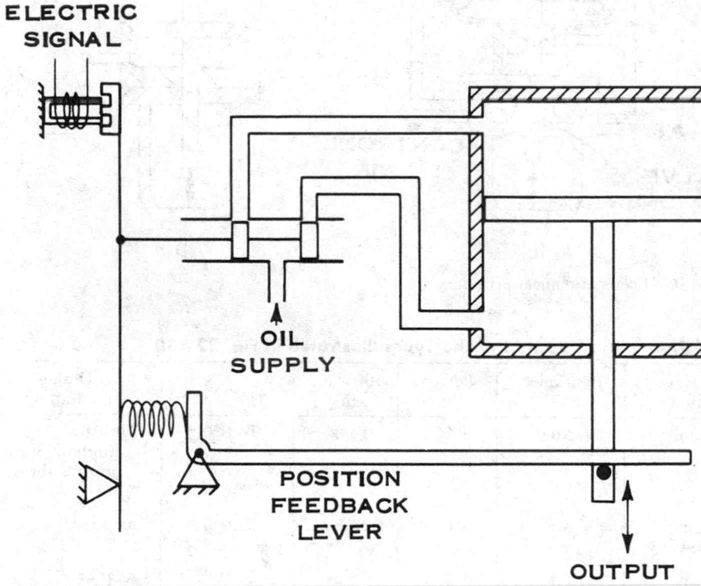

FIG. 22-132 Electrohydraulic actuator.

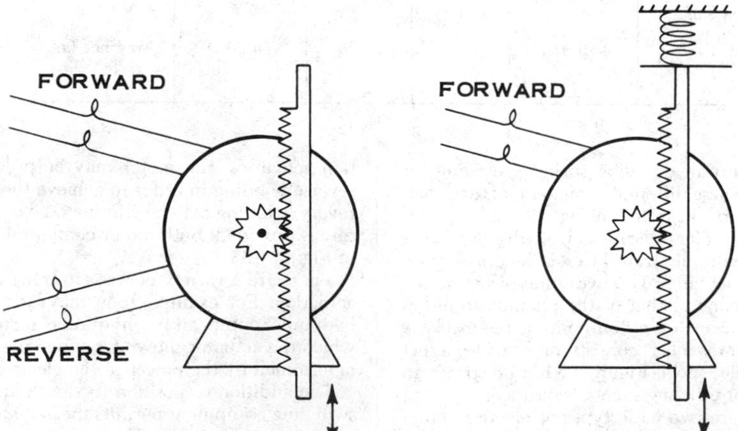

FIG. 22-133 Electric-motor actuators.

The electrical solenoid is a two-position device. When electric current is supplied to the coil, the solenoid armature moves to the energized extreme position; when the current is removed, the armature is returned to the deenergized position by a spring. Increased thrust capability may be obtained by attaching a hydraulic or pneumatic pilot valve or an electric-motor switch to the solenoid armature.

Mechanical Actuators Handwheels and manual actuators are common examples of mechanical actuators. The actual hardware configuration may be quite varied, ranging from a handwheel mounted on the top or side of a spring-diaphragm actuator to a manually operated mechanical lever-linkage arrangement.

POSITIONERS AND BOOSTERS

Positioners were mentioned in connection with some linear-position actuators and hydraulic automatic controllers. Another auxiliary device that is used to improve the performance of certain types of final control elements is the booster or pilot relay. Only pneumatic positioners and boosters are discussed here.

Operating Principles The positioner is a high-gain (10 to 100), mechanical-pneumatic feedback amplifier. The booster is a pneumatic-power feedforward amplifier. Figure 22-134 shows schematically the normal configurations for using the positioner (*b*) and booster (*a*). The particular configuration selected should be based on the requirements of the application. A pilot is normally an integral part of the positioner as illustrated in Fig. 22-135. It usually operates over the normal 20.7- to 103.4-kPa (3- to 15-lbf/in^2) pressure range. If additional pressure is required to operate the valve actuator, then a separate booster, not usually integral with the positioner, is used with higher-pressure air. The booster or pilot relay may be similar to that illustrated in Fig. 22-94.

The typical positioner operation is to compare a mechanical position signal with a pneumatic control signal and to adjust the positioner's output to reduce any difference between these two signals. This operation is illustrated schematically in Fig. 22-135. An increase in control-signal air pressure expands the input bellows, thereby increasing the pressure in the nozzle. This allows more air to flow through the pilot to the spring-diaphragm actuator. See discussion of Figs. 22-94 and 22-95 under "Measurement-to-Indicator Transducers." The stem is caused to move down, thereby moving the feedback cam to increase the force on the feedback spring. The actuator stem is properly positioned when the cam-spring force is equal to the bellows force.

The use of the feedback cam provides the positioner with a useful feature. By properly selecting the cam contour, the control signal versus output position relationship may be "characterized." The result is indicated schematically in Fig. 22-124 if flow is replaced by stem travel and the abscissa is the control signal. To a limited extent the positioner-cam contour can be selected either to offset or to reinforce valve and/or process nonlinearities as desired. Generally, the cams are easily changed to provide various characteristics.

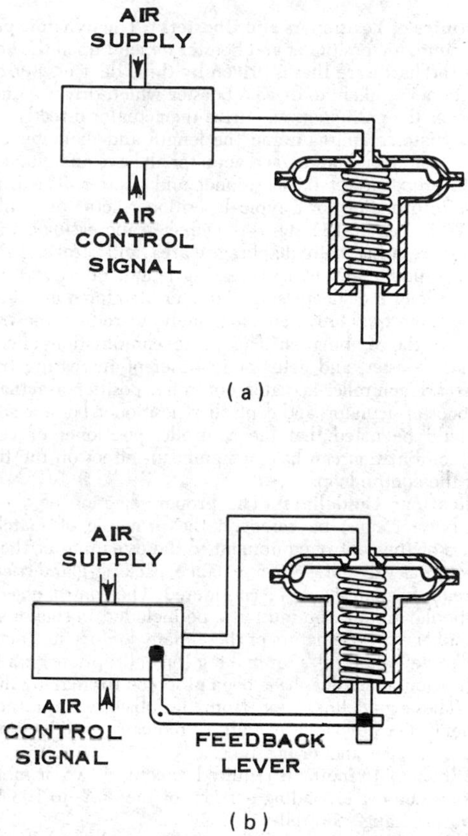

FIG. 22-134 Pneumatic booster and positioner configurations.

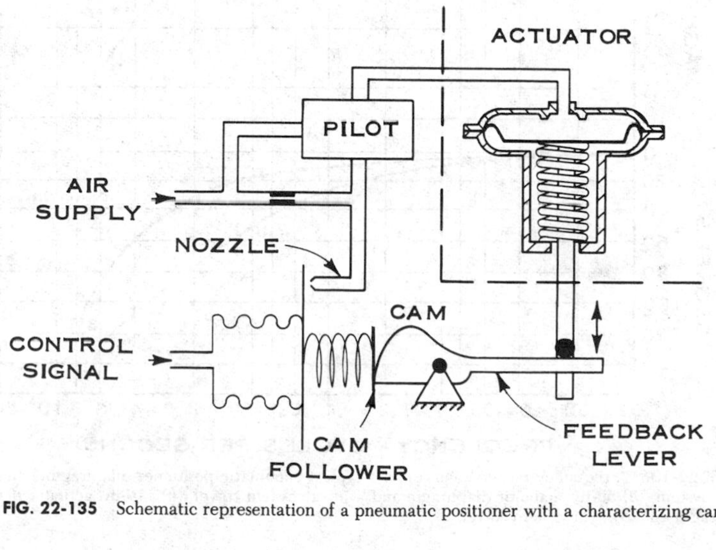

FIG. 22-135 Schematic representation of a pneumatic positioner with a characterizing cam.

Dynamics of Positioners and Boosters The dynamic properties of the pneumatic positioner and booster depend upon the characteristics of the hardware that is driven by these devices. For example, the positioner is likely to drive a booster which drives a diaphragm actuator, or the positioner may drive the actuator directly. The area of the actuator diaphragm and the length and diameter of tubing between positioner, booster, and actuator all have an influence on the dynamic properties of the positioner and booster. The frequency-response information for a typical positioner-actuator combination, illustrated in Fig. 22-136, shows the effect of increasing actuator volume by increasing both diaphragm area and stroke. In general, increasing diaphragm area, increasing length of valve travel, and increasing length or decreasing diameter of interconnecting tubing have the same qualitative effect, namely, to reduce the frequency bandpass of the combination. The several combinations of controller, positioner, booster, and actuator in order of increasing frequency bandpass are controller-actuator, controller-positioner-actuator, controller-booster-actuator, and controller-positioner-booster-actuator.

It should be noted that the controller-positioner or controller-booster combination can have a significant effect on the time constant of the control loop.

Applications Guidelines The proper selection of a positioner and/or booster must be based on the principle of matching the dynamics of the control equipment to the dynamics of the process. Such factors as valve-pressure unbalance, packing-gland friction, and valve-steam friction should be considered. The control objectives for the particular installation must also be included in such a selection. Because of the diverse nature of these many factors, it is not possible to state any absolute rules for making the appropriate choice.

Applications guidelines have been proposed for making the proper choice. These guidelines stem from the objectives for the control equipment. The following conditions indicate consideration of the use of a positioner and/or booster:

1. When split ranging is required to control two or more valves sequentially on corresponding portions of the 20.7- to 103.4-kPa (3- to 15-lbf/in^2) gauge controller output
2. When the controller output signal must be amplified to provide additional actuator thrust to overcome friction, obtain accurate valve positioning, or overcome unbalanced valve forces
3. When improved transient response is required to recover from upsets such as response to slow and small controller output changes or when pneumatic transmission lines from the controller are long
4. When a change in the valve flow characteristic is required without a change in valve construction
5. When positioning of a springless cylinder or diaphragm actuator is required

Once the decision has been made to use one of these auxiliary devices, the process dynamic characteristics are the key to selecting the proper device. Generally, positioners are recommended for relatively "slow" processes such as level, thermal, and mixing processes. Boosters are suggested for "fast" processes such as flow, liquid-pressure, and gas-pressure processes. A booster may also be desirable in the positioner loop if a large actuator is involved.

SOLIDS-METERING VALVES

Linear-position actuators and the control valves described earlier are primarily used for the control of fluid (liquid or gas) flow. Other types of equipment are required to handle the transfer of solid materials. The particular solids-metering-valve hardware depends on the volume, density, particle shape, and coarseness of the solids to be handled.

The throughput rate of such valves is adjusted by driving them with variable-speed motors. This makes them equivalent to the valve-and-actuator configurations in which the speed controller is the equivalent of a booster or a positioner.

OTHER FINAL CONTROL ELEMENTS

Variable-Speed Drives A common variable-speed drive is the variable-speed motor. Such a motor may be a universal electric motor with an adjustable supply voltage or an ac electric motor which is supplied by an adjustable frequency source. For special applications the variable-speed motor may be a gas- or air-turbine-type motor in which the source pressure is varied to change the speed.

Other variable-speed drives include variable-ratio belt drives, variable-ratio cone drives, and a variable-displacement hydraulic pump

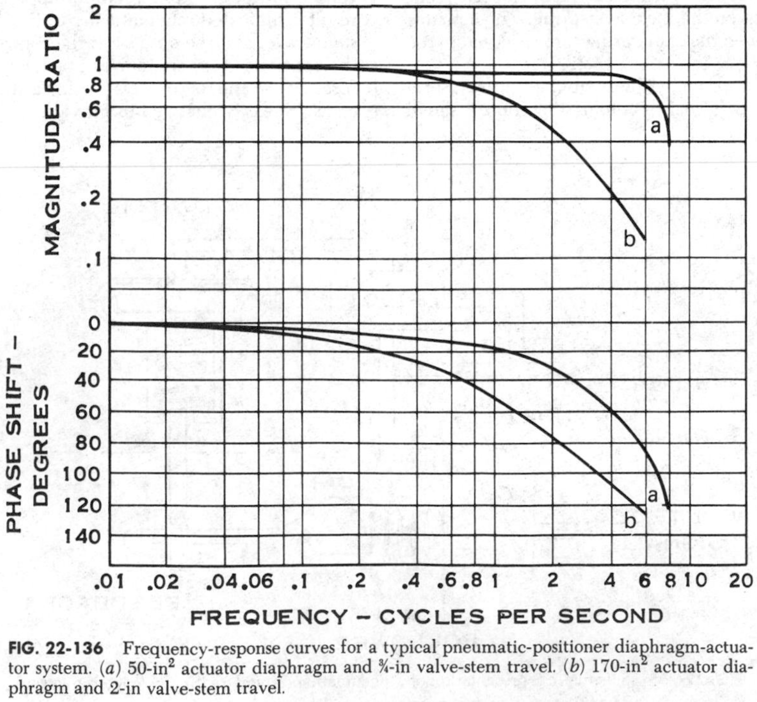

FIG. 22-136 Frequency-response curves for a typical pneumatic-positioner diaphragm-actuator system. (*a*) 50-in^2 actuator diaphragm and ¾-in valve-stem travel. (*b*) 170-in^2 actuator diaphragm and 2-in valve-stem travel.

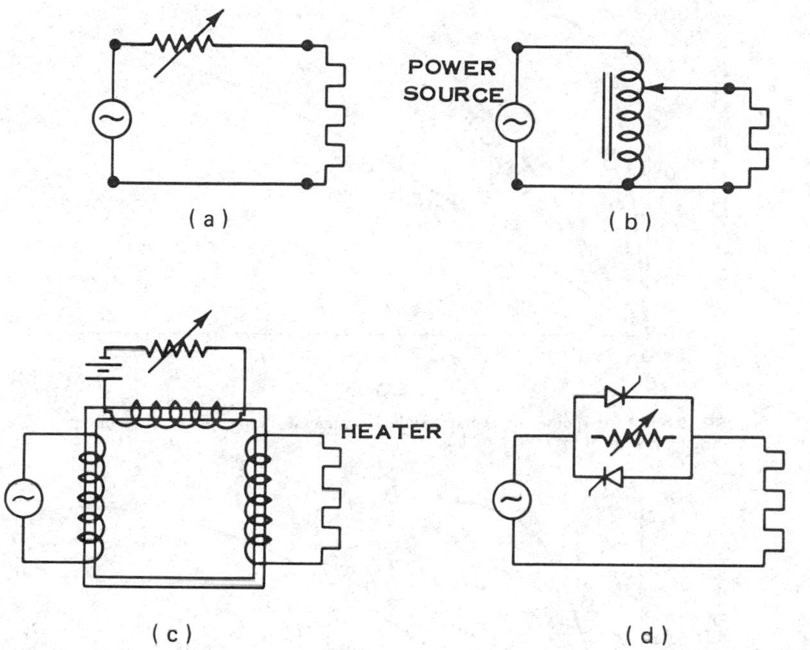

FIG. 22-137 Electric power control devices. (*a*) Series resistance. (*b*) Variable autotransformer. (*c*) Saturable reactor. (*d*) Silicon-controlled rectifiers.

driving a hydraulic motor. These devices are all driven by a constant-speed motor, usually electric.

Variable-Electric-Power Actuators A variety of devices is available for adjusting electric power. Figure 22-137 illustrates several of the more common methods as applied to heating elements.

Variable-Output Pumps There are several types of variable-output pumps. The more common ones are constant-volume pumps which are operated by variable-speed drivers or variable-volume pumps driven by constant-speed motors. The variable-volume pumps are usually adjusted by changing the piston stroke length. The piston may pump the process fluid directly by using multiple check valves to assure reproducible pumping rates. The piston may also intrude into a sealed hydraulic chamber which activates the diaphragm of a diaphragm pump.

TELEMETERING AND TRANSMISSION

Telemetering is usually thought of as the transmission of a measurement signal over some distance. A telemetering system usually consists of five components: a measuring device, a transmitter which converts the measurement signal into a signal suitable for transmission and which presents this signal to the transmission medium, the transmission path, a receiving device which translates the transmitted signal into a signal which is acceptable to the end device, and the end device or display.

For the most part ratio, microwave, and video telemetering links are used only in the upper-hierarchy levels of process control, i.e., off line, and they will therefore not be discussed here. They are, however, used online in such areas as pipe-line control, scientific work, and business systems.

Almost all telemetering systems associated with process-control systems are of a fixed location-to-location type. Most of them involve fixed electrical or pneumatic conductors as the transmission medium. The distance of transmission may vary from a few feet to a few thousand feet.

Thermocouples constitute the largest class of process-measurement devices whose output is transmitted over some distance from point of measurement to point of use. For the most part their millivolt dc signals are transmitted untransformed via extension wires to their point of usage. However, with the increasing use of solid-state and integrated-circuit electronics these measurements are being transformed and transmitted in other forms.

This discussion will be concerned with such things as the transmission of control signals to a remote valve actuator, the transmission of a measurement signal to a remote receiver, and the transfer of information between computers. Signal-handling hardware such as transmitters, transducers, controllers, and computer interface equipment are discussed elsewhere in this section.

ANALOG SIGNAL TRANSMISSION

An analog signal is continuously variable over a given range, and the size of the signal represents the numerical value of the transmitted variable. The most common type of process analog transmission is pneumatic.

Pneumatic Signal Transmission Pneumatic transmission normally uses 20.7 to 103.4 kPa (3 to 15 lbf/in^2) of air pressure for the representation of process variables. The signal delays in long lines and the effects of the volume of the terminating device such as the pneumatic valve actuator are of particular importance in pneumatic transmission. The quality of control in a pneumatic control loop can be greatly affected when the signal time constant is significant compared with the process time constant.

The curves in Fig. 22-138 show approximate pressure responses to a step change in a pneumatic control signal at the inlet of typical pneumatic transmission lines with different terminal volumes. Terminal volumes in some control loops may be greater than 0.016 m^3

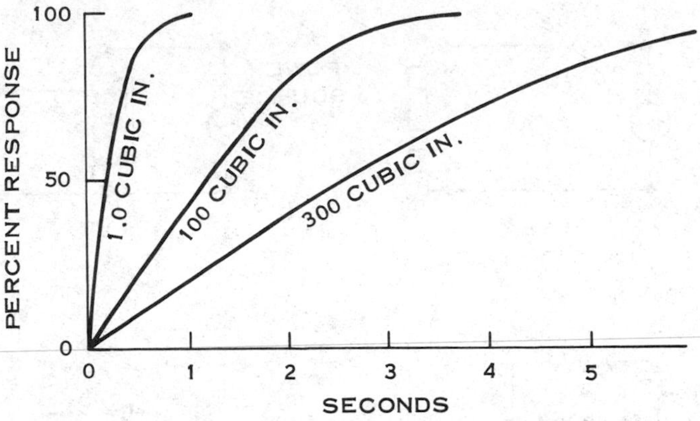

FIG. 22-138 Response of several terminal volumes to a pneumatic step input through 200 ft (61 m) of ¼-in tubing.

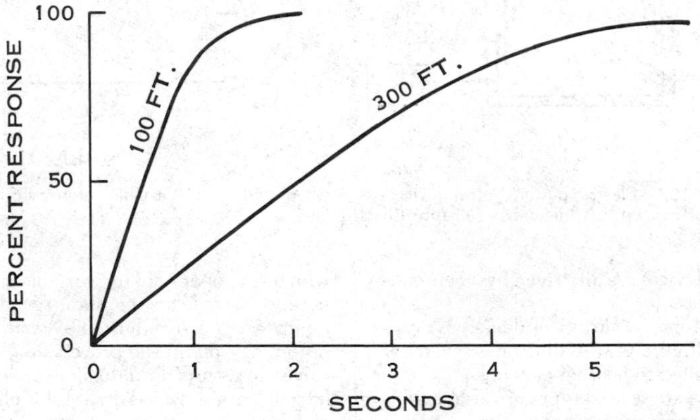

FIG. 22-139 Response of a 100-in^3 terminal volume to a pneumatic step input through the indicated length of ¼-in tubing.

(1000 in^2). Figure 22-139 shows the effect of pneumatic transmission-line length.

A method of avoiding long transmission time constants is to use field-mounted controllers and valve positioners. The field-mounted controller will eliminate the long transmission distance within the closed loop as illustrated in Fig. 22-140. A valve positioner can be used to further reduce the time constant of a large terminal volume of a valve diaphragm actuator.

Closed-loop transmission delay can often be eliminated in computer control systems by the use of predictive-type calculations. Consider the control of distillation-column internal reflux flow with a pneumatic computer. One control method computes internal reflux R_I for use as a controller measurement as shown by the feedback configuration in Fig. 22-141. This places the remotely located computer within the feedback loop, putting the long measurement and control-signal transmission lines and their time constants in the flow control loop. Thus corrections for disturbances which affect the flow must all be handled through the long transmission time constants.

The predictive method is illustrated in the cascade configuration in Fig. 22-141. The computer predicts the external reflux flow rate and is outside the field-mounted secondary flow control loop with its much shorter transmission lines. Thus disturbances other than those affecting ΔT are handled by a control loop with short time constants, the secondary loop of a cascaded control system.

Electronic Analog Transmission Methods Electronic analog transmission systems use some form of electric signal to represent the transmitted value. The most common types of representation use current, voltage, phase shift, or frequency.

Current Transmission An electric current transmission system normally uses a direct current proportional to the transmitted variable. Common process-control current ranges are 1 to 5, 4 to 20, and 10 to 50 mA. The transmission path is a two-wire circuit which can be over a relatively long distance. The current-transmitting device can be a flow, pressure, level, or temperature transmitter, an analog computer, or a digital-to-analog converter. The power supply can be part of the transmitter or receiver or be an independent unit.

A typical control loop may have an electronic automatic controller with a current output and a control valve operated by a pneumatic diaphragm-motor valve actuator. In this type of installation the current-to-pressure transducer-transmitter should be located close to the actuator, as in Fig. 22-142, so that a long pneumatic transmission is avoided.

An illustration of an analog-computer-output current-transmitting circuit is shown in Fig. 22-143. An operational amplifier is used as a voltage-to-current converter since the current output is proportional to the voltage applied to the amplifier. The current output is not affected by variations in line impedance and receiver input impedance. For the circuit show, if e_1 = 0 to 10 V, i_b = 1 mA, and R_1 = 2500 Ω, the output current i is 1 to 5 mA. See discussion of Fig. 22-104 under "Electronic Controllers" for a more detailed description.

In general, current-transmission methods are very effective since they are not affected by small line and receiver impedance changes. The high-level current signal is less affected by noise pickup than some of the other electrical systems.

Voltage Signal Transmission In most analog voltage-transmission systems, the signal is represented by a dc voltage proportional to

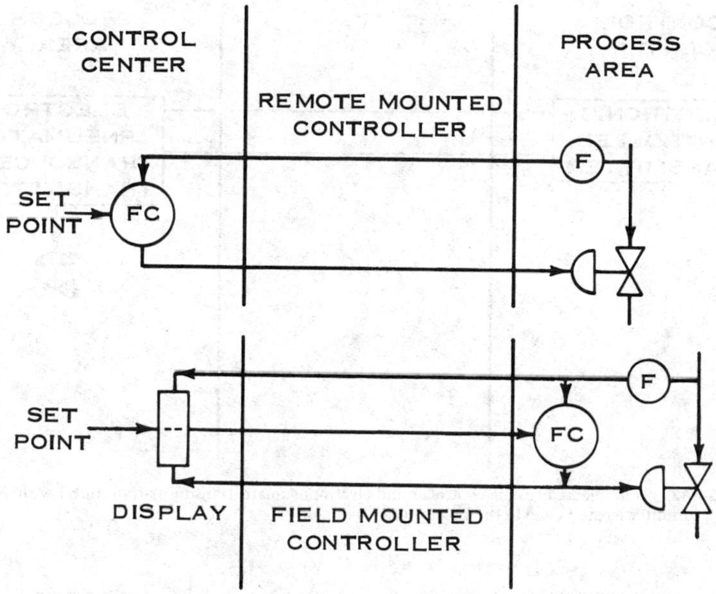

FIG. 22-140 Reduction of closed-loop pneumatic transmission delays by using a field-mounted controller to replace a remote controller.

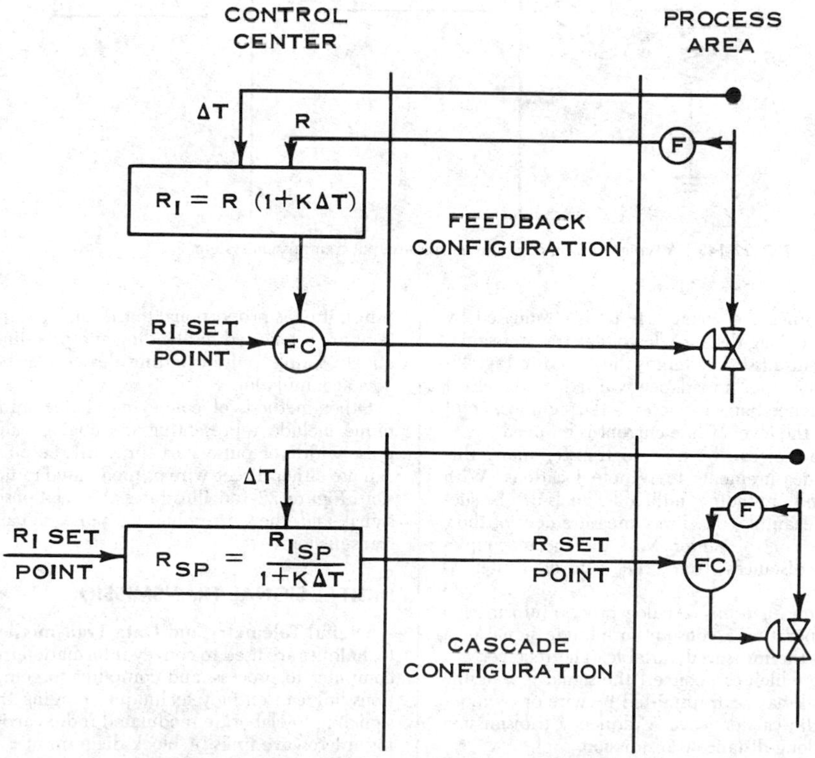

FIG. 22-141 Reduction of closed-loop pneumatic transmission delays in a pneumatic computer controller used for internal reflux flow-rate control on a fractionator by placing the computer in the cascade configuration rather than the computed-variable feedback configuration.

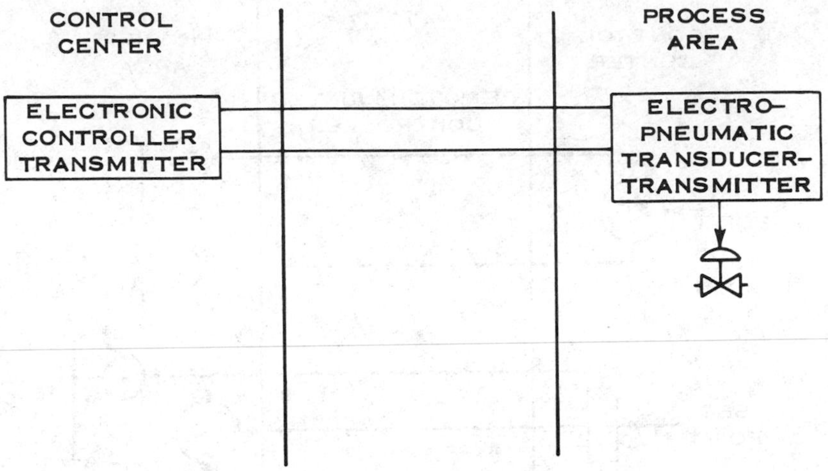

FIG. 22-142 Combined electronic controller and electropneumatic transducer-transmitter with a pneumatic transmitter located close to the process valve actuator.

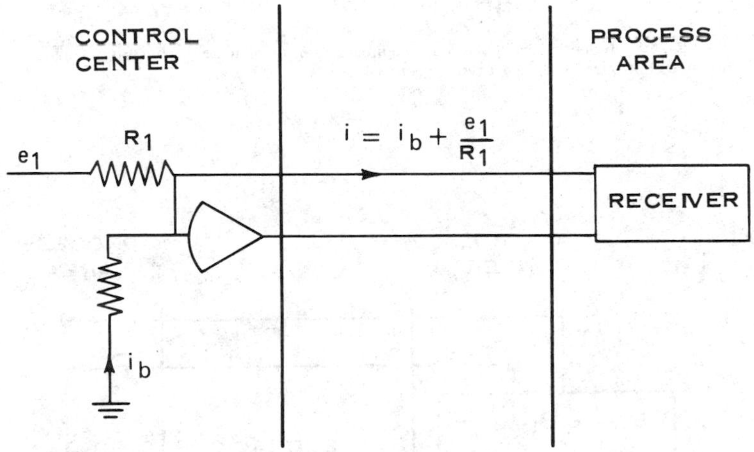

FIG. 22-143 A typical analog-computer current-output transmission system.

the transmitted variable. Since the voltage signal is attenuated by transmission-line impedance, long-distance dc voltage transmission is not advisable. A voltage-transmission system is illustrated in Fig. 22-144. Note that high receiver input impedance will reduce the effect of transmission-line impedance but will increase the significance of stray-current pickup since the level of line current is reduced.

Voltage-transmitting devices can be used to transfer many different types of process measurements to remote locations. With proper care in shielding and grounding, millivolt signals can be successfully transmitted. For example, the direct measurement of thermocouple voltages in the control center with the thermocouples located at considerable distances from the control center is commonplace.

Frequency-Transmission Systems Analog process information can be transmitted by generating a constant-amplitude signal of a frequency proportional to the measured variable. The receiver is a frequency-sensitive device which can convert the signal to a usable form. The frequency signal may be transmitted by wire or by modulated high-frequency radio carrier wave. Frequency transmission systems are well suited to long-distance transmission.

Pulse-Transmission Systems Analog process information can also be transmitted as voltage or current pulses. The most common method is pulse-frequency transmission, which is closely related to frequency-transmission methods mentioned earlier. The pulse rate transmitted is proportional to the analog signal. Pulse-frequency systems are often used in blending and pipe-line operations, since a relatively simple pulse-counting device can be used to indicate total flow accumulation.

Other methods of conveying analog information by pulse technique include representing the analog value by pulse amplitude, pulse width, or pulse start time. All the pulse-transmission methods can use either direct-wire or modulated radio-carrier-wave transmission. Figure 22-145 illustrates the relationships between an analog signal and the corresponding pulse signals that are suitable for transmission.

DIGITAL SIGNAL TRANSMISSION

Digital Telemetry and Data Transmission Digital transmission techniques are used to convey information from process to computer, computer to process, and computer to computer. The communications link can employ techniques ranging from telegraphic-type dc signaling to elaborate modulated radio-carrier-wave systems including microwave links. A block diagram of a digital transmission system is shown in Fig. 22-146.

The simplest, most common form of digital transmission code is binary. Three methods of transmitting binary information are amplitude-shift keying, frequency-shift keying, and phase-shift keying.

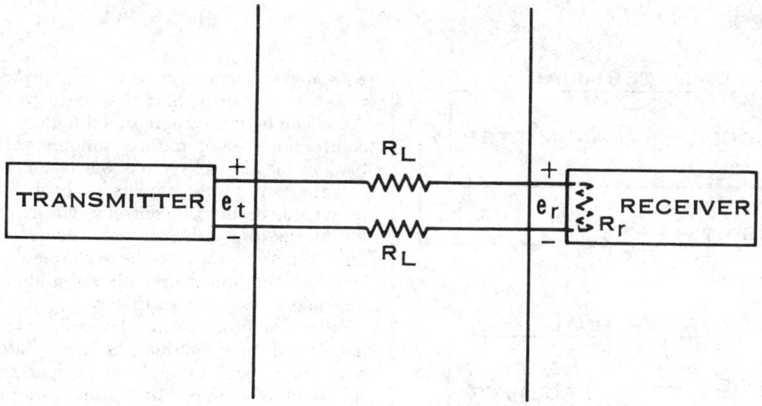

$$e_r = e_t \left[\frac{R_r}{R_r + 2R_L} \right]$$

FIG. 22-144 Direct-current voltage transmission where e_t is the transmitted voltage, e_r the received voltage, R_L the transmission-line resistance, and R_r the receiver input resistance.

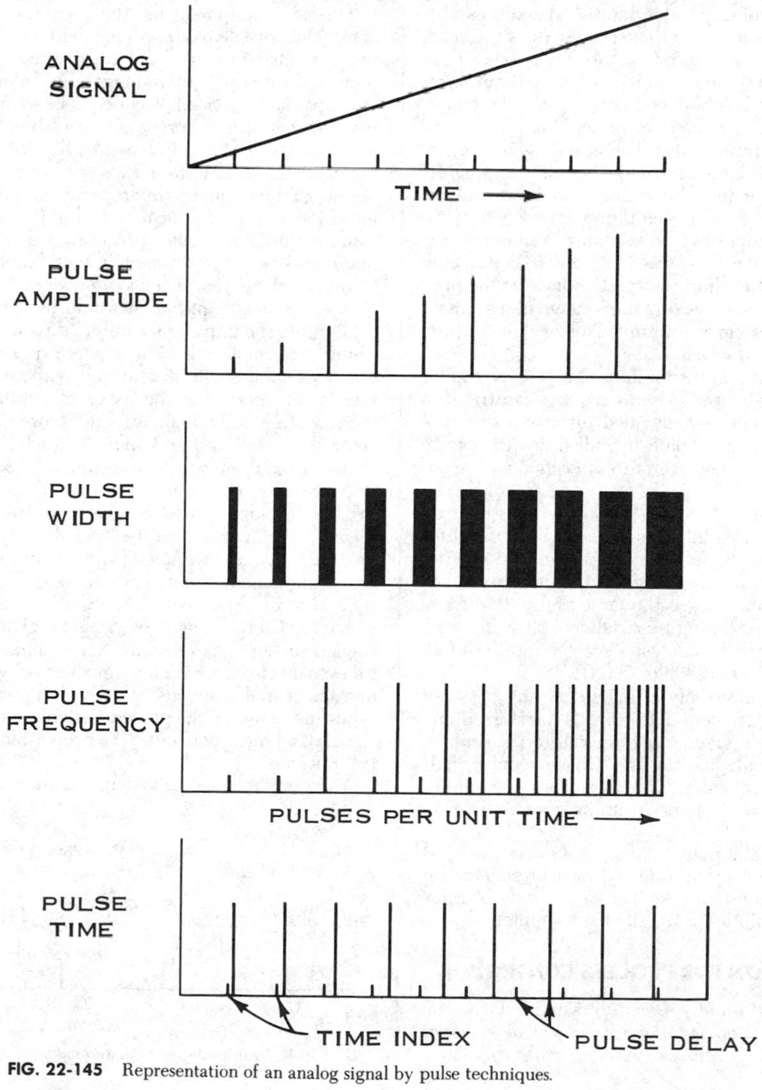

FIG. 22-145 Representation of an analog signal by pulse techniques.

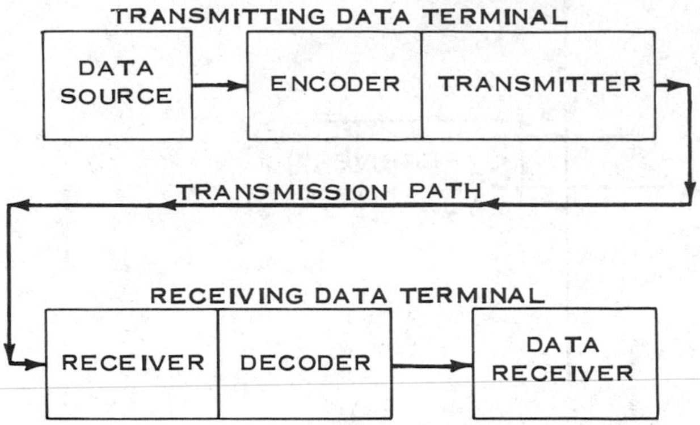

FIG. 22-146 Digital-data-transmission system.

Amplitude-shift keying uses the presence or absence of the signal to designate a 1 or 0. Frequency-shift keying uses two different frequencies to represent a 1 or 0, and phase-shift keying uses a change in signal phase to denote a 1 or 0.

In addition to the two-state binary codes, multilevel codes can be used for digital data transmission. In multilevel codes, the signal can assume three or more discrete levels as opposed to only two levels in binary representation. Although some transmission systems have been used with as many as 32 levels, decoding and noise problems become severe, and fewer levels are more common.

A binary decoder must determine only whether the signal is present or not. A signal below 0.5 is designated as 0; above 0.5, as 1. Only severe noise can cause an error in interpretation; however, only a single binary bit is transmitted during each time interval t_1 to t_2. A four-level code requires more sophisticated decoding. A signal below 0.25 is 0, between 0.25 and 0.50 is 1, between 0.50 and 0.75 is 2, and above 0.75 is 3. Since the decoder must interpret with twice the precision of a binary decoder, noise pickup may cause many more errors. However, the four-level signal transmits 2 binary bits during each time interval and hence is twice as fast.

Computer-to-Computer Data Transmission The use of multicomputer hierarchical process-control systems requires the transfer of information between computers. A detailed discussion of computer-to-computer communications should include a description of computer input-output hardware; however, this subsection will cover only the basic concepts.

Data exchange between computers can be broadly classified as parallel or serial transmission. A parallel link will consist of one line for each bit of the digital word plus additional lines for send-receive commands. All bits of a data word are transferred simultaneously so that the transfer time for a data word is the same as the time for a single bit. Parallel data transfer through coaxial cable is a fast and reliable method; however, parallel cable costs become significant at distances greater than 120 or 150 m (400 or 500 ft).

Serial data transmission requires only a single data link between computers but is much slower, since a data word is transferred one bit at a time. Serial systems are divided into three categories: simplex system (data and control signals flow in one direction only), half duplex system (signals flow in either direction but not simultaneously), and full duplex system (simultaneous two-way transmission).

Data transmission between computers is classified as synchronous when data are transferred at a constant rate and asynchronous when data are not transferred at a constant rate. The rate of data transfer is always limited by the capabilities of the slowest computer.

EQUIPMENT ORGANIZATION FOR PROCESS CONTROL

Process-control equipment is rapidly being transformed by the advances in microprocessor technology into a variety of new shapes and functional organizations. A process variable with the latest equipment is more apt to be digitally displayed and entered than recorded on a strip-chart recorder or varied by turning a knob. One "box" can be user-programmed to do several unique and/or related control functions or, more commonly, can be manufacturer-designed to allow the easy selection of one or more preprogrammed functions.

Significant improvements in control-system performance have been made in difficult control situations by using digital control computers. Because of their initial expense control computers are kept as busy as possible through the assignment of as many tasks as possible. In effect, the proper operation of a large part of the plant becomes dependent upon the computer.

When the process-control computer fails or malfunctions during a manufacturing operation, the immediate result is usually a process upset and a large expenditure of operations-personnel time to undo the results of the upset and place the affected part of the plant back into normal operation. Often manual control of equipment becomes necessary for protracted periods of time with a resultant loss of operation attention given to normal operator duties.

The relatively high reliability of a digital control computer tends to compound the failure problem. When equipment which has been doing an excellent job of controlling the process with little or no operator intervention suddenly fails, the immediate response is more likely to be panic than action because the operator has forgotten what to do owing to a lack of recent experience.

This new equipment has the capability of overcoming the major operational problem of applying digital computers to process control; i.e., a single-component failure in the computer can render large parts of the control system inoperable unless a backup system of similar capability has been provided. The solution results from the much reduced cost of microprocessors plus their capability to control communication interfaces independently. In effect, it has become economical and feasible to use a very large number of quasi-independent digital computers (microprocessors) to control the process with interprocessor communication techniques employed to assure overall coordination. Each major processing element would contain its own microprocessor, and consequently its failure need not affect the performance of the rest of the system.

In effect, such equipment would have the desirable properties of (1) limiting the impact of a failure due to the small extent of responsibility, (2) rendering failures more quickly repairable since functional assemblies can be removed without disturbing the rest of the system, (3) increasing the usage of common parts while reducing the cost of an individual part, and, more important, (4) retaining the computational and time-handling capabilities of a digital computer.

Background material consisting of the techniques evolved for improving the availability and repairability of analog equipment will first be developed. Next will be presented digital-control-computer techniques and their related problems while achieving the same goals. Finally, an example of the techniques available for microprocessor implementation to achieve better digital equipment availability and repairability constitutes the last subsection.

Analog Control Systems In general, analog controls are built of single-function blocks which are connected either by pneumatic tubes that convey air analog signals or by wires which convey current or voltage analog signals. System design consists of determining the types and ranges of the process variables, scaling the function module signals, and interconnecting the proper blocks to accomplish the control purpose.

An example of such a system is shown in Fig. 22-147. The control

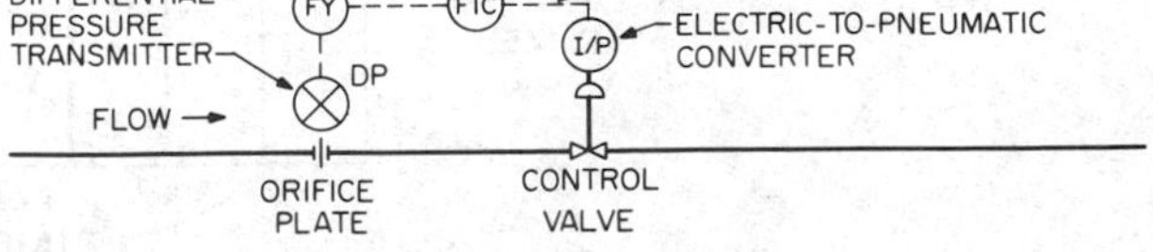

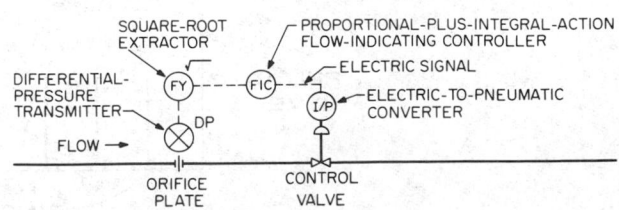

FIG. 22-147 Flow control with an analog controller.

loop is intended to maintain a constant flow rate within the ability of the valve to control the flow.

An orifice plate or restriction is placed in the flow path, and a differential-pressure transmitter is used to measure the pressure drop across the orifice plate as a function of the flow rate. From other process data it is known that the flow will be varying from 0 to 200 volumes per minute. For illustration purposes, let the orifice-plate pressure drop range from 0 to 125 units of water for this amount of flow.

The signal actually developed by the orifice plate will appear as shown in Fig. 22-148. Consequently, the differential-pressure transmitter when sized for 0 to 150 units of water pressure drop (some overrange is needed) will yield an electrical analog signal which is not linear with flow.

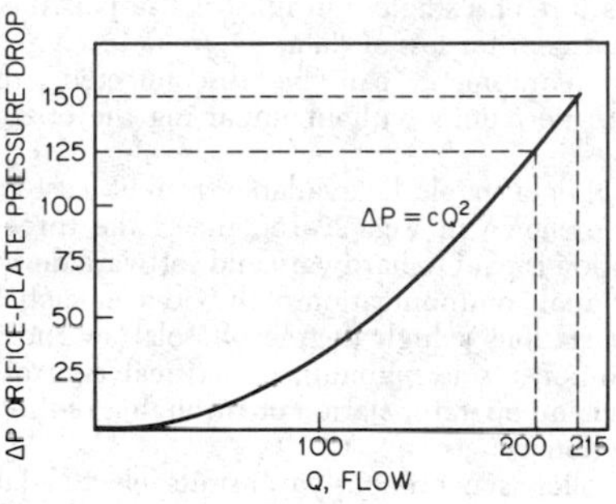

FIG. 22-148 Flow versus orifice-plate pressure drop.

A square-root extractor is used to linearize the differential-pressure signal so that the square-root-extractor output is linear with flow. The full-scale output of the square-root extractor will correspond to 219 volumes per minute of flow for this example.

Finally, the square-root signal is applied as input to the controller, which in turn manipulates an electric-to-pneumatic converter. The converter supplies the air signal to the valve actuator.

The scaling of the signals was determined when the range of the differential-pressure transmitter was selected and the value of the resulting flow computed. The analog scale on the face of the controller will have linear graduations, and a full-scale range will be 0 to 219 volumes per minute (the set-point range will be the same). Normally, an operator-convenient range such as 0 to 250 volumes per minute would be used and the differential-pressure transmitter calibrated accordingly.

The controller output will assume whatever output level is necessary to drive the flow measurement into correspondence with the set point. Externally caused flow increases result in the valve's being driven toward close with flow decreases causing the valve to open. It may occur that the valve is driven full-open or full-closed when the controller attempts to maintain flow-measurement and set-point equivalence.

The following questions can now be asked of the control system design:

1. How are the differential-pressure transmitter, square-root extractor, controller, and electric-to-pneumatic converter to be calibrated or worked upon?

2. What is the best way for the control system to behave if a module fails?

Analog control systems have uniformly answered question 1 by providing each controller with an independent control and signal-display source for manipulating the controller output signal. This independent control source is known as an automanual station, and whenever a controller is placed in manual, the valve signal becomes an operator adjustment and equipment upstream of the manual station, including the controller, can be worked upon. The electric-to-pneumatic converter would require that the control valve be taken out of service for the converter to be worked upon.

Question 2 has been answered in analog control systems by minimizing the detrimental sense of adverse failures. In practice, an attempt is made to cause the control valve to go to the safe position in the event of a module failure. For instance, if instrument air is lost and the safe direction is for the valve to go closed, a valve-actuator set which requires an increasing air signal to open the valve should be purchased. A like criterion can be applied to the other components since the function modules can be purchased with direct- or reverse-acting outputs.

In summary, analog control systems have limited the instrumental effects of failures by using self-contained replaceable modules to compose control loops, by providing a final independent control station, and by using a fail-safe strategy when practical.

Digital-Computer Control Systems The digital-computer control system directly comparable to the analog control system of Fig. 22-147 is shown in Fig. 22-149. The differential-pressure-transmitter signal goes directly to the computer-control-system analog-to-digital converter. The converted signal is linearized internally through a square-root-extraction routine and supplied as input to a proportional plus integral control algorithm (see subsection "Microprocessor Controllers"). The output of the algorithm is sent to the digital-to-analog converter, where it is converted to a representative electrical signal for driving the pneumatic-to-electric converter.

Process measurement, set-point value, and controller output values are obtained via the operator-interface equipment. This control system is termed "direct digital control" since the computer system directly manipulates the valve signal.

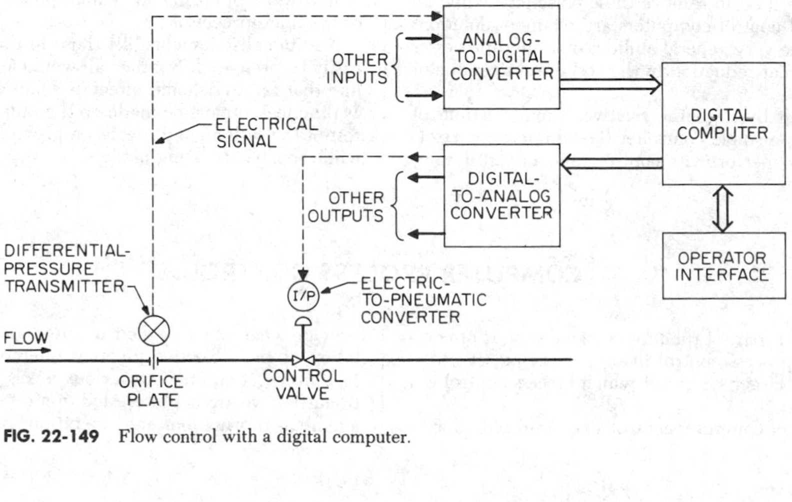

FIG. 22-149 Flow control with a digital computer.

Comparison of the direct digital control system with the capabilities of the earlier analog control system demonstrates that the same capabilities are present except for the manual station for control of the valve in the event that the computer or any equipment preceding the computer needs to be calibrated or worked upon. This capability is supplied in three ways:

1. A backup manual station for valve control is provided.
2. A backup analog controller for valve control is provided.
3. Control loops which directly manipulate valves are constrained to be analog controllers with the computer supplying set-point values only.

Option 3 is normally referred to as a "supervisory control system." In each case means is provided to override the computer-generated signal.

The basic control loop discussed thus far does not really indicate the significance of equipment failure in computer control systems. Computer control systems are installed to solve control problems which are not manageable with analog equipment. Consequently, when a computer is lost, even though backup controllers continue to hold the process variables at the last value, effective handling of process changes is lost since the more complex part of the control system is lost.

In general, repair of a digital computer requires that the computer be made completely available to the service worker while repairs are made. This requirement follows because computer-related equipment is normally passive except under the command of a stored program. Consequently, to exercise equipment to check for failure requires the use of the central processing unit. Most diagnostic programs are built to be run as the only programs resident in the computer.

To counter the loss of capability, backup computers have been installed. These backup units may or may not be dedicated to control. In either case, current process data must be constantly transmitted to the backup unit and a safe cold-start strategy adopted for it, or dual data acquisition, with the hardware interfaces accessible to both computers, must be practiced. In particular, automatic strategies for transfer to the backup equipment are reliable for only the most obvious types of failures.

From the standpoint of development and maintenance costs these approaches are unsatisfactory because of the presence of redundant or unused equipment and the programming required to handle the additional equipment.

Microprocessor Distributed Control Systems Operational objections to the use of digital control computers arose for these reasons:

1. A large loss of control capability can result from the failure of a single component.
2. A backup system to retain complete control capability is almost as expensive as the primary system.
3. Effective repairs require that the computer be removed from service.

Microprocessors can be used to remove these objections while all the capabilities of digital control computers are retained. Solutions are possible because of the very general abilities of a microprocessor as a controller and the great reduction in the cost of semiconductor devices.

Objection 1 can be met by assigning relatively small sections of each controlled system to a single controller (i.e., a microprocessor) which would continue to perform its control function until commanded to change. Cascade and interacting control loops are assigned to the same controller. Consequently, failure of one microprocessor would affect only the loops controlled by that microprocessor.

The cost of backup systems (objection 2) would be much reduced because the hardware would be identical for all controllers, specifics being determined by the program. Consequently, only one type of device need be stocked, and this device would be relatively inexpensive as compared with the whole system.

The problem of repair raised in objection 3 is also solved. The ability to substitute controllers readily allows off-line repair without disabling the control system.

In this discussion the questions of controller intercommunications and operator interfaces have been ignored. Some guidelines to interconnecting microprocessor-driven units need to be developed. One set of guidelines is:

1. The failure of a single unit must not impair the operations of another unit except for loss of data.
2. Failed equipment can be disconnected, removed, and replaced with new units without impairing the operation of other units.
3. Critical data should be available from at least two sources.

The system shown in Fig. 22-150 meets the three guidelines if proper attention is paid to hardware and software design. By providing an individual communication path between each controller and the operator stations a high degree of isolation and replaceability between components is maintained. Critical control data can be transferred via an operator station or through a redundant measurement connection.

Each controller asynchronously transmits selected data to the operator stations. Time is left between each controller transmission to allow operator-station transmissions. Since only small amounts of data are involved for each communication line, the data-transmission rates per line are low with a corresponding increase in noise immunity.

All commands from either operator station are echoed back to each operator station to allow simultaneous updates of status. Destination and source codes are unnecessary since addresses are wire-connection-determined. However, error check bits should be appended to each message. Bad data would be flagged on the operator-station display with one type of special character and nonupdated data flagged with another special character.

Operator-station software allows bad data from the controllers to be overridden by an operator-keyboard entry. Also, reception of data from any controller can be halted with an operator keyboard command.

Each controller should be equipped with a keyboard and display which allows its memory contents to be displayed. This equipment is primarily for use by the service worker when calibrating and repairing the equipment.

The manual station of each controller need not be a separate piece of hardware, but it should be manipulatable by a means independent of the microprocessor.

Additional statements like those in the last few paragraphs could easily be presented. But they all would arise from meeting the guideline that failures should affect as small a part of the control system as practical, repairs be made on the failed part without affecting the rest of the system, and means be provided for operation to continue while repairs are being made.

COMPUTER PROCESS CONTROL

Process computer control is part of the more general subject of computer control. To place process control in proper perspective, it is necessary to consider the larger system of which process control is a part.

The ultimate objective of computer control, overly simplified, is to provide what may be called an automated company. The implementation of this objective involves intercommunicating computers or sections of computers at various levels of the organization. In this discussion we are not interested in most of the data-processing, management-information, and operations-programming aspects of this

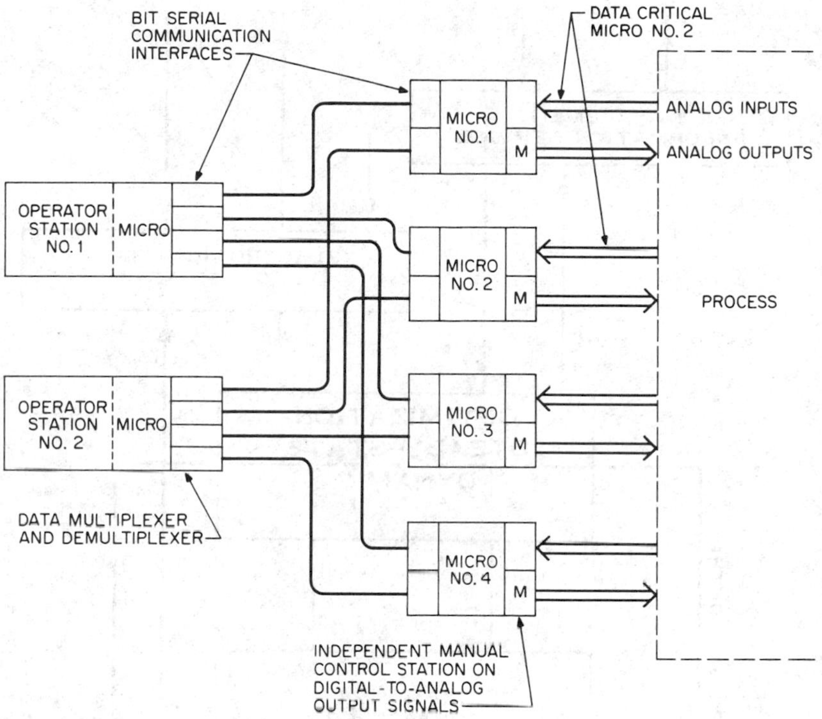

FIG. 22-150 Distributed control system.

ultimate objective except as they may be related to process control in the engineering rather than the business sense.

To gain the perspective required, one can look at the concept of intercommunicating computers or sections of computers at various levels as a "hierarchy of computer control."

HIERARCHY OF COMPUTER CONTROL

In the hierarchy of control, five levels can be identified.

1. *Unit-operation control.* This pertains to control of an individual item of process equipment such as a heat exchanger, a reactor, or a dryer.

2. *Unit-process control.* This is the control of a collection of unit operations whose individual operations must be coordinated to meet a specific objective, for example, a gasoline reformer which includes catalytic reactors, a product fractionator, and heat exchangers.

3. *Plant control.* This level includes control of the collection of all unit processes at one physical location.

4 and 5. *Department and corporate control levels.* Above the plant level are the department and corporate levels of a company. Process control is not concerned with the operations of either of these levels, except as some of the outputs of these levels are plant inputs which are of concern in process control and some data generated within the plant, such as nonstandard operations or capacity limits, are inputs to the upper two levels.

Information originating at a lower level is communicated to the next higher level, which in turn produces decisions or actions that are transmitted down to the lower level. An important consequence of using a control system is the reduction of the degrees of freedom. Thus, as control at a particular level becomes better or more complete, less information is needed by the next higher level of control and it is required less frequently. To accomplish its control function, an upper level normally requires only the current values of the performance variables from the immediately lower level.

PROCESS CONTROL

The Realm of Process Control Process control is concerned with the control of unit operations, unit processes, and those aspects of plant control which pertain to processing objectives and effectiveness. Process control is not directly concerned with such plant-level decisions as plant maintenance, equipment replacement, product shipment, and the like except as they affect the performance variables of the lower levels because of the availability status of online processing equipment, availability of feedstocks, and availability of product-storage facilities. Likewise process control is not usually concerned with the decisions at the department or corporate levels.

Computer-Operator-Process Configurations Computers may be used in a variety of configurations with the process and operator, depending on the level of automation desired or justifiable. These configurations cover a wide range: the use of a computing device as an off-line aid to the operator, the use of a computing device as an auxiliary to a conventional controller, online data logging and computation serving as a guide to the operator, an online automatic computer control system setting the set points of conventional controllers under the supervision of the operator, and the inclusion of all controllers within the computer with the system under the supervision of the operator.

Of these arrangements the first three represent variations of computer-aided manual process control. The last two represent automatic computer control. The first three represent computer control in the sense that all the data are available for feedback control but one link in the chain is the responsive action of the operator. This represents computer control in the business sense but not in the engineering sense. The last two represent computer control in the engineering sense.

Functional Hierarchy of Computerized Company Control System If we again look at the hierarchy of control but include the computer or computers in our analysis, we find that a diagram such as Fig. 22-151 represents a functional description of the computer-

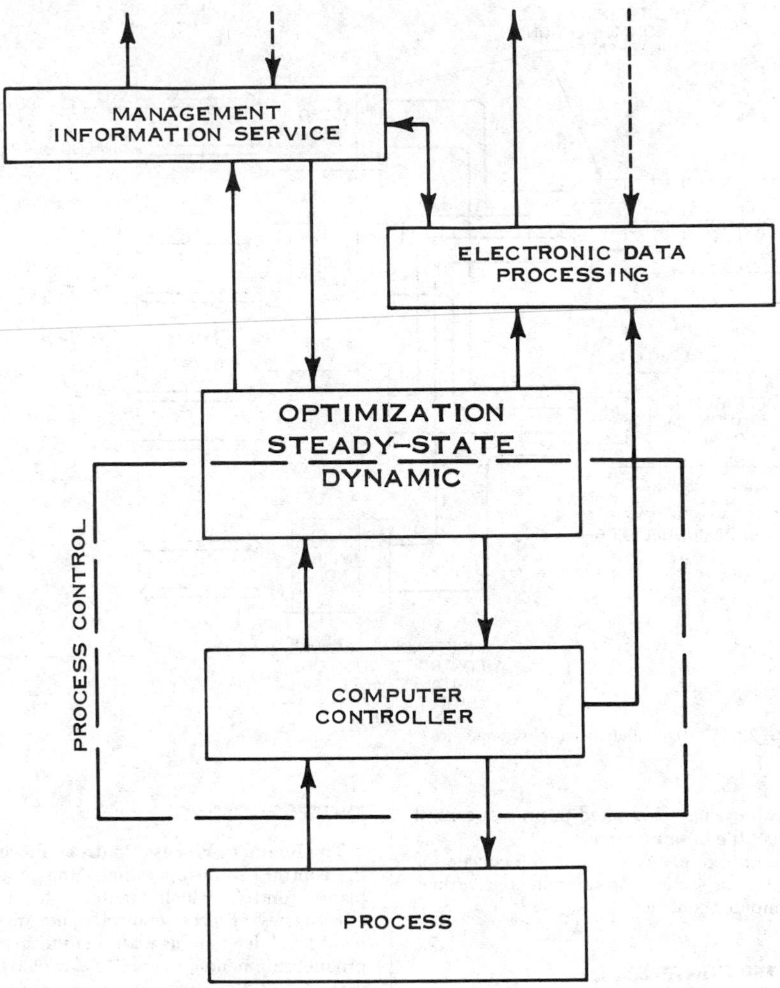

FIG. 22-151 Simplified diagram of the various functions of a fully computerized control system.

ized system. The portion of the system indicated by the dashed box incorporates what is usually designated, on an engineering basis, as computer process control. The division indicated between steady-state and dynamic optimization is a separation which depends on the process being controlled and the frequency of making adjustments to the process. The division may vary from the unit-operation level to the plant level, depending on the economic and technical conditions and the viewpoint of the system designer. The blocks may represent separate communicating computers or sections of a single computer.

In general, the control equations and models used in the various levels of the control hierarchy are characterized by a decreasing number of differential equations or dynamic characteristics as one moves up the hierarchy. By the time that the unit-process level is reached, the need for dynamic control equations has normally disappeared. In future processes in which the design of the process is such that computer control is required for operability, the dynamic characteristics will extend farther up the hierarchy.

Sometimes the functional description of an operating computer control system as illustrated in Fig. 22-151 is much more in evidence than a hierarchical arrangement. One of the reasons for this may be that the implementation team designing the computer control system is computer-equipment-oriented without adequate process, control, and measurement experience. Such a situation may lead to the replacement of difficult measurement and more direct control at a lower hierarchical level by a process model which spans more than one level.

In building such models, it is sometimes difficult to relate control action at a particular valve to an upper-level performance variable. This difficulty arises from such things as dead time, high-order response at the performance variable, and multiple disturbances affecting the performance variable in similar manners. Such difficulties have led to computer-control-system failure and large computer-program development costs exceeding budgetary allowances.

The Performance Variable Each level of the hierarchy exists for a reason. Its reason for existing usually establishes its performance variables. Performance variable, performance index, operational objective, and operational criteria are frequently used interchangeably. At the production levels in the hierarchy, the basic purpose in operating any process is to produce, at minimum cost, a specific product at a required rate consistent with the operational constraints of the unit. Usually, the manufacturing costs, product qualities, and production rate are the performance variables of the process. Control at any level of the hierarchy should be such as to reduce the degrees of freedom of the system, with respect to its performance variables, toward zero.

COMPUTER CONTROL OF UNIT OPERATIONS

Computer control improves performance of the unit operation when conventional control systems are characterized by any or all of the following conditions: (1) not all controllable disturbances are kept out of the system, (2) there are uncontrollable disturbances for which there is no compensation, and (3) manually adjusted controller set points are on variables of only indirect interest. With computer control systems, it is possible to calculate, specify, and control performance variables which previously could be neither measured nor controlled.

Reduction in Degrees of Freedom As an example, consider the chemical reactor illustrated later in Fig. 22-191. With conventional controllers the operator attempts to maintain the production rate (product-formation rate) by observing the temperature rise of the coolant in flowing through the reactor jacket and adjusts the flow of catalyst accordingly. This rise in coolant temperature is an indication of production rate but is not the whole story. As regards product quality, the only information available to the operator is from a laboratory analysis usually run once a shift.

In this example, using conventional controls, the operation of the reactor is a function of 18 independent variables. An analysis of most conventionally controlled unit operations will reveal the existence of a similar number of independent variables or degrees of freedom. With the typical existence of so many uncontrolled variables, considerable fluctuation is observed in the performance variables for unit operations.

The technique of reducing the degrees of freedom for this reactor example, described in detail in connection with Fig. 22-191, can be accomplished only through computer control.

Simplification of Control Equations Another feature of reduction of degrees of freedom by control is that process variables are made constants through control. For example, in the calculation of the concentration of product in the chemical reactor illustrated later in Fig. 22-191 and Table 22-13, the material-balance equation may be written as

$$PR - F_0 W_P = (1/V)(DW_P/dt) \tag{22-33}$$

where the product-formation rate or production rate $PR = f(F_{CSP}, E_C, C_P, W_P, \ldots)$ and the flow leaving the reactor $F_0 = f(PR, \ldots)$, giving a nonlinear differential equation. The volume of the reactor is V, and the other variables are indicated in Table 22-13.

When the production rate is controlled automatically, PR is equal to PR_{SP}, the production-rate set point. Under these conditions, the material-balance equation becomes

$$PR_{SP} - F_0 W_P = (1/V)(dW_P/dt) \tag{22-34}$$

where $F_0 = f(PR_{SP}, \ldots)$ and PR_{SP} is a constant. This is now a linear differential equation which is normally explicitly solvable. This simplification of the dynamic equation describing the concentration of product in the reactor also simplifies any control equation or model in the upper levels of the control hierarchy which involves online information about product concentration.

Selection and Design of Control Loops In the selection and design of control loops, there are essentially four steps:

1. Establish the performance variables of the particular unit. This is usually straightforward. The reason that a unit was built and is operated serves to establish the performance variables.

2. List all the variables that affect the performance variables. These will include those variables which can be manipulated and those which cannot. Those which cannot be manipulated are the process disturbances. Some of these disturbance variables can be measured directly, while others cannot. The list of all variables establishes the degrees of freedom of the system.

3. Reduce the number of variables (degrees of freedom) by defining and controlling variables that are more fundamental to the performance variables of the unit. The more the effects of disturbances on the performance variables can be eliminated by so doing, the more precise will be the final control.

4. Establish possible ways to control the performance variables. The number of possible ways of controlling the performance variables is usually large. This number can normally be reduced to a few plausible ways by consideration of time constants, steady-state gains, and range over which the variables can be changed. From the plausible ways, a final selection can be made by simulation if knowledge of the process is not sufficient to evaluate final choices by other means.

In the development of process-control systems, particular attention should be given to defining the end result desired and determining when this result has been achieved. The design of the control system for any unit should be directed toward employing the performance variables as set points as in the chemical-reactor example. Although this example is of a unit operation, the same procedure applies at any level of the control hierarchy.

The basic operating objectives of this reactor are to produce product of a specified property, product index, and production rate and at a catalyst productivity of not less than a specified value. These objectives constitute the reason for building and operating the reactor. Therefore, the computer control set points should be production rate, product property, and product index, since these are the performance variables. The desired control will be achieved when and only when the degrees of freedom of the reactor, with respect to these set points, are zero, subject to the catalyst productivity constraint and any other operational constraints that may exist. If several control systems are possible, the one selected should normally be based upon performance-variable control quality as the first consideration and any local process optimization that may be possible as a secondary consideration.

Computer–Operator–Unit-Operation Configuration In computer control of a unit operation, the computer controller must contain the equations relating the measurement inputs and the output desired whether that output is a valve position or a conventional controller set point. These equations are sometimes called the control equations or the process model. Such equations may be simple cause-and-effect, simultaneous algebraic equations, simultaneous differential equations, or rather complex simulations of the process being controlled.

If the process is well understood and quantitative data are available, the control equations may be written explicitly or a mathematical model which represents the process accurately may be constructed. These equations or models incorporated into the computer controller then form the basis upon which the control signal is generated from the measurement signals.

When the process is too complex or is not understood sufficiently to write explicit equations or construct explicit models, empirical ones are used. When empirical equations or models are used, it may be difficult to evaluate the equation or model parameters accurately, or the equations or models may be difficult to implement by the computation equipment available for wide ranges of operating conditions.

Some feedback from performance variables to the control mechanism is required because of inaccurate empirical relationships, inaccuracies in process online measurements and in their conversion to inputs compatible with the computer, inaccuracies in process responses to control signals, presence of uncontrolled variables, and accumulative discrepancies in calculation procedures.

One example occurs when the operator must observe the qualities of the feedstock and products through laboratory tests or other means not easily automated to input directly to the computer controller. From these observations, the operator makes minor adjustments to the process model or control equations on an hourly or shift basis. These minor adjustments usually consist of modifying the value of a constant or resetting the value of a set point of a computerized controller.

One consequence of accepting such a design premise is that the system designer will use the apparent continuous availability of operator action as a substitute for becoming sufficiently knowledgeable about the process to formulate adequate models or control equations. This results in a less effective process.

One approach to the design of computer control systems is to put in a grossly oversimplified control system requiring considerable operator attention and action with hardware for future projected use.

The expectation is gradually to accumulate sufficient process knowledge to build up the automatic control system. This may lead to future labor commitments of uncontrollable size, failure to reach the objective of adequate economically justifiable control, and ineffective use of the computer. This situation becomes particularly important when it is intended to implement computer control all the way from unit operations through the plant level.

Another condition occurs when all the normal operating control information is automatically collected by the computer controller at a frequency considered adequate for automatic control except for, say, one performance variable. This performance variable is available automatically but with some delay. In this case, the performance variable is calculated on the basis of automatically available process data, and control action is taken on the basis of this calculated value. When the automatically determined variable value is available, it is compared with the calculated value used, and any required model or equation adjustment is made to equate the calculated and measured values. For example, in Fig. 22-191 the process dynamics are such that the process can move out of specification operating range before the change in concentration of the reactant can be measured by a process chromatograph. The essentially instantaneous reactant concentration, calculated by a dynamic material balance around the reactor, is used for control of reactant flow to the reactor. Appropriate computational adjustments are automatically made to bring the calculated value into coincidence with the measured value when it becomes available a few minutes later.

In the usual fully automated unit-operation computer control system all the information normally required is automatically available to the computer on a time basis which is adequate for control. The operator enters the control circuit only infrequently for control-parameter modification on the basis of observation of process performance, information from outside the process such as a change in processing objectives or a maintenance procedure, or correction of an operational situation which the computer controller is unable to handle. For example, in the fractionator control system illustrated later in Fig. 22-175, the operator observes the quality of the distillate produced by means of an online analyzer or laboratory reports. From this information the operator makes infrequent minor adjustments of some computational parameter in the predictive computer, say, a constant in the material-balance equation.

Unit-Operations Optimization The optimization of a continuous-process unit operation is a local optimization, usually to produce the required product or specified quality at minimum cost. For example, the operation of a single distillation column is optimized when specification products are produced at minimum boil-up rate, i.e., minimum heat input. See also Fig. 22-188 and associated discussion in the subsection "Maximizing Continuous-Unit Operation Performance."

When a unit operation is the process bottleneck, keeping the unit loaded is the responsibility of the computer at the unit-process level. In the case of fractionation, the unit-process computer would continue to increase the feed or accept all the feed available until some other operational constraint was reached. This would in no way alter optimization for the individual column, since specification product at minimum boil-up is optimum no matter what the feed rate to the column.

In addition to the explicit mathematical formulation of optimum operating conditions, there are two basic online procedures for obtaining optimum or near-optimum operating conditions for continuous unit operations. The first procedure is automatically to test the process periodically by moving the process slightly from its operating point by means of the manipulated variables and to determine the new value of the performance variable to be optimized. By keeping track of the value of the performance variable obtained and comparing this value with a subsequent value, the process can be moved toward an optimum operating condition or operating constraints which constitute optimum attainable conditions. The second procedure is to test a model of the process and translate the results of these tests into corresponding operating conditions for the unit operation.

Either of these two procedures may be operated in the automatic mode or in the manual mode. In the usual process implementation, the manual mode, the testing is done automatically but the operator makes the decision as to what moves are actually made on the process.

These techniques are rather specialized, and as such their details are beyond the scope of this *Handbook*. A number of simple automatic, single-variable, peak-seeking control devices have been reported. However, the digital-computer controller is the most useful tool for this type of control, and practical examples of such usage appear in the literature. These procedures can be extended to other levels of the control hierarchy, and a number of manipulatable variables may be simultaneously optimized to obtain a near-optimum value of the performance variable being considered.

Adaptive Control Computer control with the conventional controllers built into the computer system provides a mechanism for partially optimizing controller tuning. In conventional controller usage, the tuning quality is gradually reduced to a condition which represents stable responses under the most adverse operational conditions that the operator will tolerate. This generally represents very poor responses under normal operating conditions.

In general, optimum controller-tuning changes with process throughput, since process gains and time constants change. If the controllers are included in the computer controller, the controller tuning constants can be changed automatically to correspond to the current process-operating conditions. This is handled by setting up the controller-tuning constants as an explicit function of a process variable such as flow rate. Another way is to tabulate, within the computer memory, controller settings versus process variable and match the settings to process-variable value.

Examples of Computer Control of Unit Operations Further examples will be found in the subsection "Unit-Operations Control." Specifically, reference is made to the discussions associated with Figs. 22-167 through 22-169.

COMPUTER CONTROL OF UNIT PROCESSES

The unit process is the next higher level of the control hierarchy. For control purposes, it is usually a collection of unit operations whose individual operations must be coordinated to meet a specified set of objectives. Conventional controllers provide very little automatic control at the level of the unit processes. The operation of unit processes without computer control consists of the analysis of daily operating data or of data averaged over whatever time is considered an operational period by supervisory personnel. They decide how to operate the various unit operations of the unit process over the next operational period. These decisions then are transmitted to the unit-operations operator. The operator, in turn, determines what set points to use on all the individual conventional controllers or the unit-operations computer controllers as discussed under "Computer Control of Unit Operations." In many plants, both of these levels of control hierarchy may be operated by the same person; so this procedure may not be apparent.

With conventional control and with computer control limited to unit operations, the unit-operation performance objectives tend to become fixed over the standard plant operational period, usually at least 1 day. With computer control of the unit-process level, some of the unit-operation performance variables become essentially minute-by-minute functions of the performance variables of the unit process. The principles of unit-process control are normally a direct extension of those for unit operations. However, the unit-process control system is usually one step further removed from the operation of valves, and the feedback or feedforward input control signals are frequently computed performance variables rather than direct process measurements. In addition, the calculation of the unit-process control outputs which are set points is usually in steady-state rather than dynamic format.

There is a wide range of complexity of processing units which may be designated as unit processes. As mentioned earlier, the simpler unit processes can be handled by extension of the unit-operation computer control techniques. The more complex unit processes can be handled by the techniques discussed under "Computer Process Control of Plants."

Fractionation System An example of a combined unit-process and unit-operation computer controller on a fractionator complex is

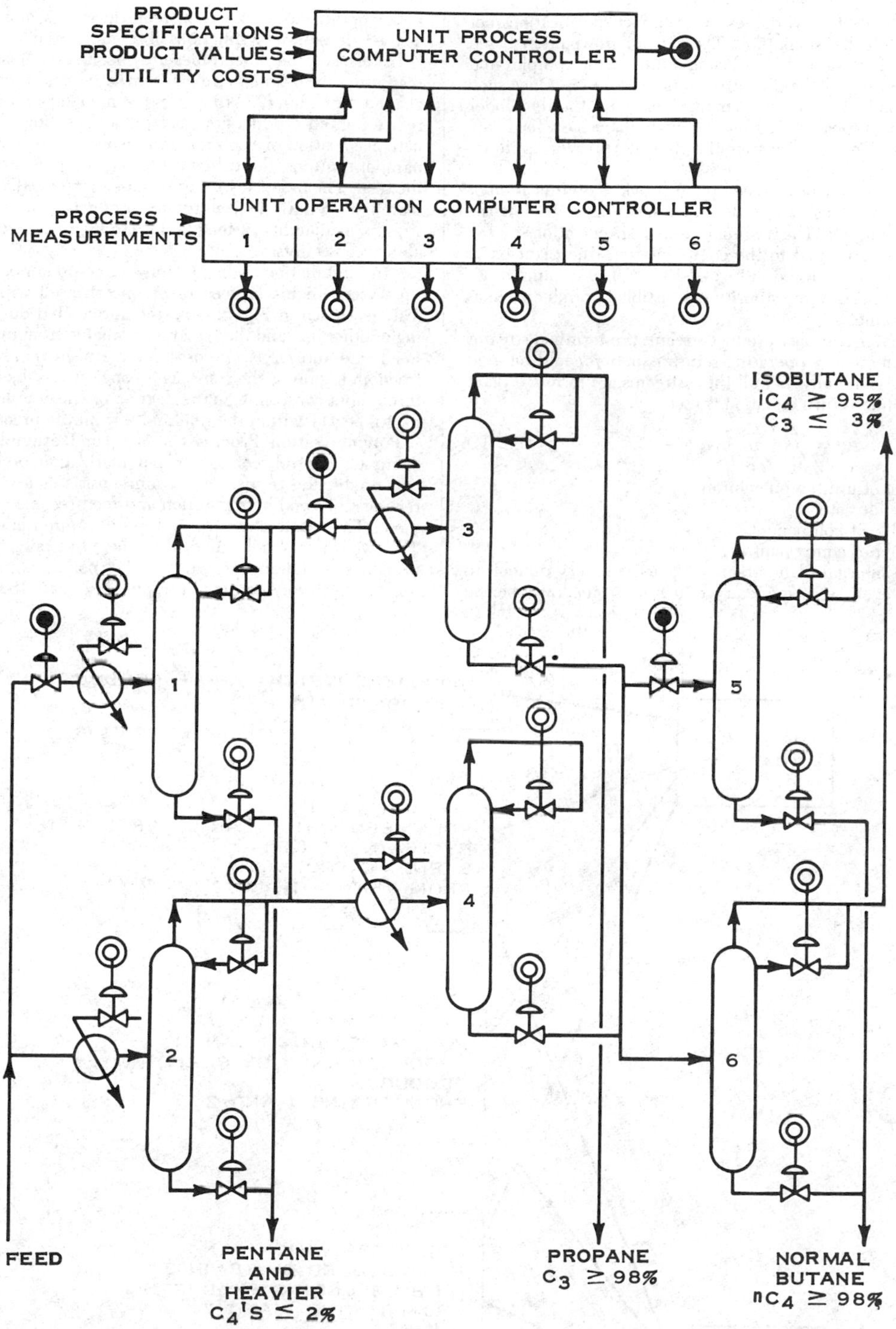

FIG. 22-152 Unit-process computer control of a fractionation system showing the division of the normal control actions between the unit-process and unit-operations computer controllers.

illustrated in Fig. 22-152. Only the computer-controller outputs to the process are indicated. Further details concerning process-measurement requirements and control of the individual fractionators may be obtained in subsections "Distillation-Column Control" and "Maximizing Continuous-Unit-Operation Performance" concerning Figs. 22-175 and 22-188. A typical enthalpy control as used on the column feeds is illustrated later in Fig. 22-168.

Without the unit-process computer controller, the plant supervisor would be faced daily with determining the product specifications of each column as well as the feed division between the parallel columns. This is a total of 15 variables, two product specifications per column and three feed distributions. These decisions would be based upon anticipated feed rate and feed composition to the complex and anticipated operations of the downstream unit processes. Once such decisions were made, they would remain in force for the remainder of the day or operational period even though changes in feed rate and feed composition would change the characteristics of optimum operation.

Installation of the unit-process computer provides optimization of the product specifications of the individual columns as well as the feed rates to the columns. The terminal-product specifications for the complex would be specified to the computer from the plant level. Typical specifications are listed in Fig. 22-152 as the minimum product purity and/or maximum impurity permissible in product, i.e., as inequality constraints.

If the fractionator-complex product streams are terminal streams for the plant, a monetary operating-return equation could be used for optimization. Generally, not all these streams are terminal plant streams; so a pseudo-return equation is used.

$$P_S = \sum_j \left(\sum_i V_i Y_i \right)_j - C_j \qquad (22\text{-}35)$$

where C_j = cost of utilities on column i
P_s = pseudo return
V_i = value of component i
Y_i = yield of component i

The value of an internal plant stream is sometimes very difficult to determine. However, since monetary return must generally be the major optimization criterion, realistic values must be determined. These values may be functions of several variables rather than fixed.

In this example, the values are specified by plant personnel and readjusted from time to time as marketing and processing conditions change. Equation (22-35) relates the monetary return to the internal reflux, bottom-product flow, heating and cooling costs, and the product specification of each column, as well as the feed split between parallel columns, the feed rate to the complex, and composition of the feed. The monetary return can be maximized by a computerized search procedure subject to the product-purity specification constraints, availability of feed, ability to use or sell products, and operability constraints.

After installation of the unit-process computer controller, the plant supervision regularly need only concern itself with the four specifications shown in Fig. 22-152, the anticipated downstream processing or other use, and the feed rate available from upstream processes. Feed rate and feed composition have been removed as degrees of freedom insofar as the complex is concerned, subject to a maximum-throughput constraint. So the degrees of freedom have been reduced by 13, i.e., 11 within the unit and 2 from the preceding process.

Polymerization Process As a second example of unit-process computer control, consider a polymerization process consisting of two production trains. Each train consists of a number of parallel reactors followed by separation and feed-recycle purification equipment. The reactors are of the type illustrated later in Fig. 22-176. The two trains may be used to produce two polymers with different properties simultaneously, or the two polymers may be blended to form a third polymer as illustrated in Fig. 22-153. For purposes of

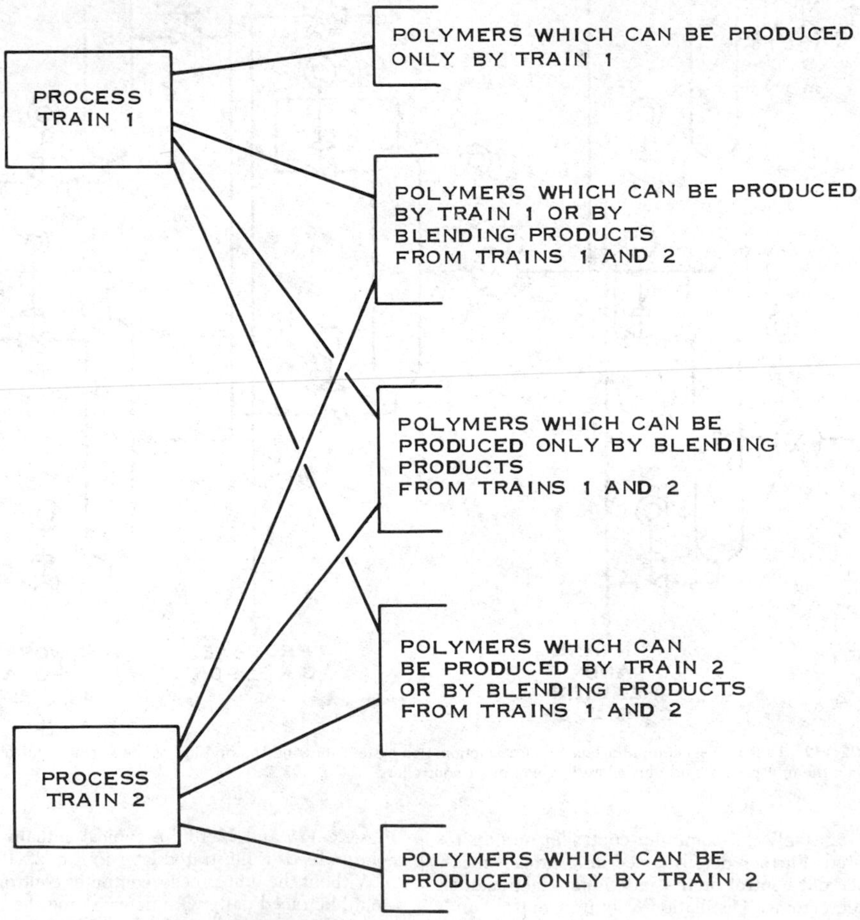

FIG. 22-153 Products available from a two-train polymerization unit process.

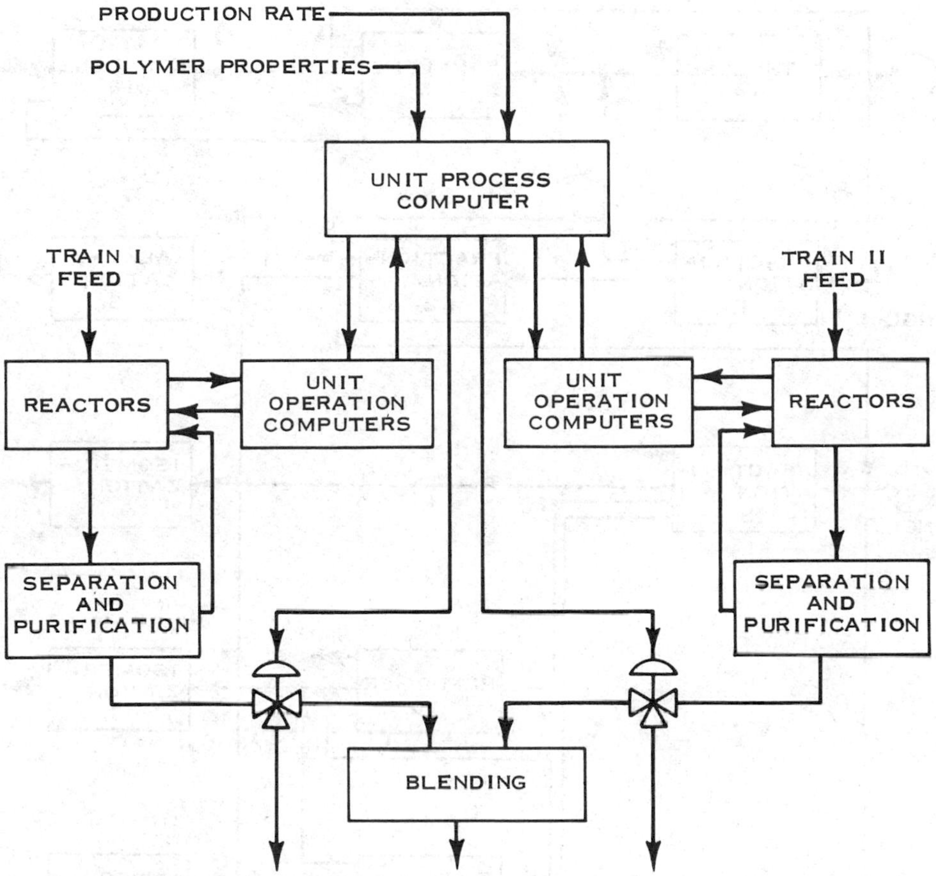

FIG. 22-154 Simplified schematic showing the relationship between the unit-process control computer and unit-operation control computer (reactors only) and the multiple-product polymerization process.

this illustration, details of the feed preparation, separation process, and recycle feed purification will not be considered.

This process can produce polymers with a wide range of properties. The spectrum of polymers which can be produced consists of those polymers which can be produced only by one train, those polymers which can be produced only by blending polymers from the two trains, and those polymers which can be produced either by one of the trains or by blending the products from both trains.

An optimization problem arises since the cost of producing a given quantity of a particular polymer is not the same for both trains, the production capacity of the trains varies with the type of polymer being made, and the properties of the polymers do not add linearly on blending. The purpose of the unit-process computer is to meet the polymer demand at the lowest production cost. The unit-process computer determines the production rate as well as the polymer type for each reactor as shown in Fig. 22-154.

COMPUTER PROCESS CONTROL OF PLANTS

The plant is the next level in the hierarchy of control. It usually consists of the collection at one location of all unit processes whose operation must be coordinated to meet a specified set of objectives. Conventional controllers provide essentially no automatic control at this level.

Without computer control at the plant or unit process levels, the "steady-state" plant-operating period tends to become several days. That is, no daily adjustments in throughput or product-specification goals for the individual unit processes are normally given to the unit-process operator or supervisor. If a disturbance affects a unit process which in turn affects a second unit process, particularly in product quality, the second unit process is locally readjusted to minimize off-specification product until such time as different instructions are obtained.

With computer control through the plant level, such a disturbance would quickly cause automatic adjustments to any unit process affected by quality or throughput to maintain performance variables. If the disturbance persisted, the plant-product throughput distribution would be altered at the next plant-level control-computation cycle to, say, minimize plant-operating costs.

Development of Control Project One of the characteristics of a large continuous-processing plant is that it represents a network of interacting units as illustrated in Fig. 22-155. Normally when it is desired to limit the computer control project to a part of the plant, the plant may be decomposed into several unit processes which can be treated essentially independently for the purposes of online control and optimization. The decomposition also aids in organizing the control and optimization concepts to be used if the entire plant is to be put under computer control with either a single computer or a group of unit-process computers. It also aids in identifying the interface between noninteracting unit processes and the interactions which cause interface problems between interacting unit processes.

Special attention must be given the interacting units to devise control systems which take into account the interactions between them. All other units may be treated independently of each other except that one unit may feed to or receive from other units.

An alternative strategy leading to computer control of the entire plant is to start with unit processes or even unit operations and add

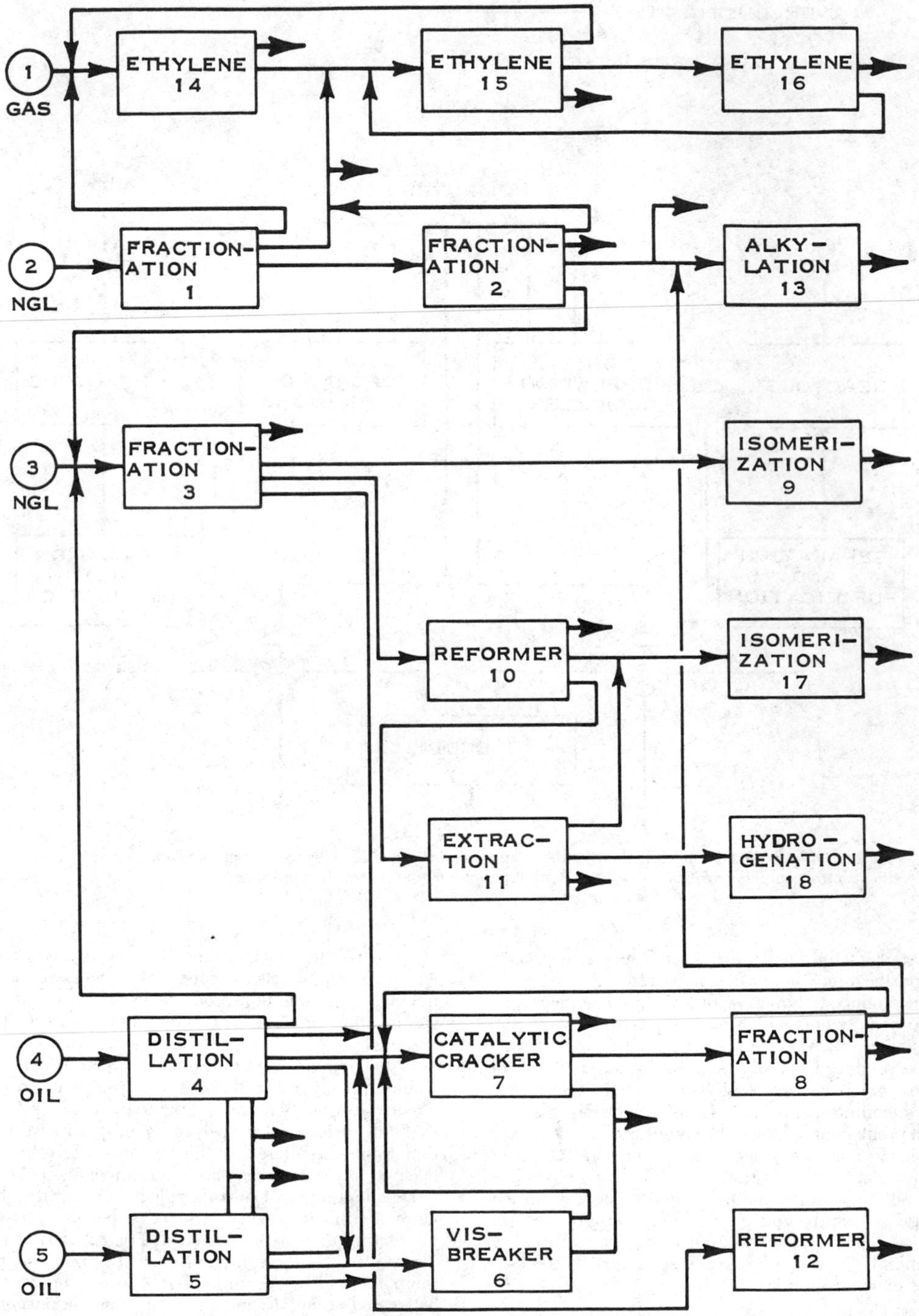

FIG. 22-155 Abbreviated flow diagram of a typical refinery-petrochemical complex or plant. ①, ②, ③, ④, ⑤ and → indicate feed and product streams respectively.

them to the plant control system one at a time. If the control at the unit process is good, then the amount of information required across the process-unit interface to or from interacting units is minimum and the job of adding the next hierarchy level is minimum. That is, if the degrees of freedom in the process unit are reduced to the lowest practical level by the control system, the number of inputs to the plant controller have been reduced and changed in character to inputs normally representing unit-process performance variables.

Other considerations, in the decision of what units to control, are posed by units in which the volumetric and, therefore, monetary throughput is large, the price differential between feed and product is large, the units are process bottlenecks, the unit is unstable or difficult to control by conventional means, and the unit produces final process products.

Decomposition of Large Process Complex As an example of the decomposition of a large system, consider a typical refinery-petrochemical complex (Fig. 22-155). This is an abbreviated flow diagram showing the unit processes as they might be identified by the plant management. The plant is fed by a gas stream, natural-gas liquids (NGL) from two sources, and crude oil from two sources. The variety of products go to pipe-line, storage, or blending facilities.

Putting Fig. 22-155 into signal-flow graph form by combining units tied together by major recycle streams and omitting product outputs gives Fig. 22-156. Here some terminal products are as much as six process stages from the plant feeds. There is also a wide variation in the size and complexity of the units shown in this diagram.

Further simplification by combining similar functions and multiple sources to other units results in Fig. 22-157. This reduces the entire complex to four process stages or less from feeds to terminal products. There is an even wider variation in the size and complexity of the units in this diagram.

The plant illustrated in Fig. 22-155 is large, and computer control is complex. Likewise, some of the unit processes are large and may be too large to be tackled at one time in view of the resources available to the organization contemplating computerized process control.

The large unit processes can be considered as plants, and the unit-process concept can be reduced in scale. For example, units 14, 15, and 16 can be considered a plant composed of three smaller unit processes with three major feed streams. The three streams are the outside gas feed and two feed streams from unit 1. Even here, the first two units are large from a control viewpoint when compared with the third, and for limited-resource situations they could be considered as separate plants. In the latter situation, the designation of unit process in the ethylene plant could be as follows: the parallel pyrolysis furnaces and quench-heat boiler as one unit process, the cracked-gas compression system as another, and fractionation as yet another.

As larger unit processes or plants are decomposed into smaller units, the recycle and minor interunit streams become more important in setting the individual unit-processing objectives. The mone-

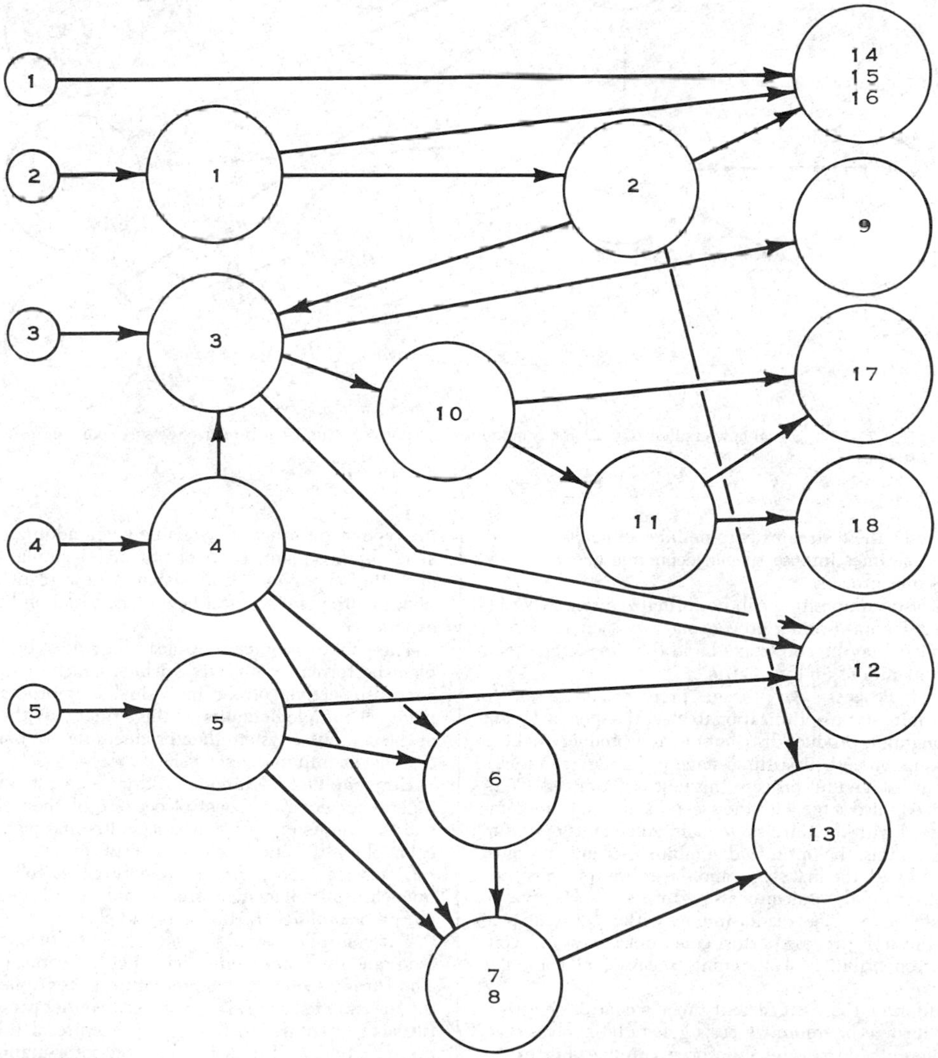

FIG. 22-156 Signal-flow graph of a refinery–petrochemical plant combining those units interacting through recycle streams.

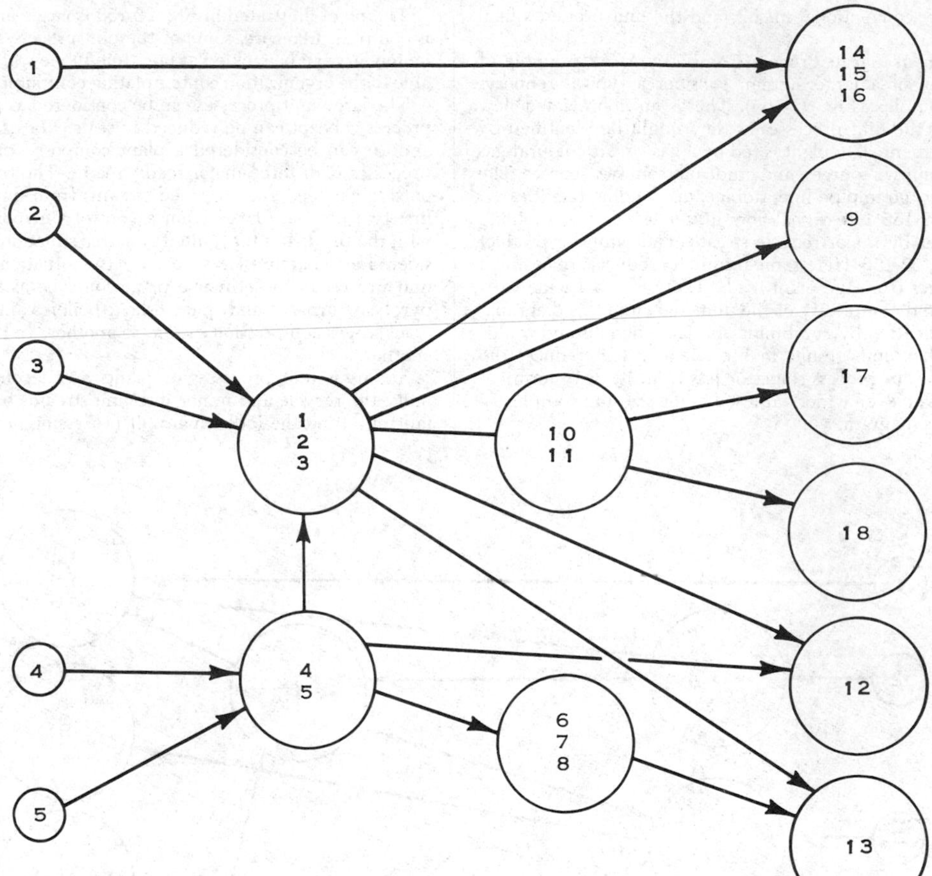

FIG. 22-157 Signal-flow graph of Fig. 22-156 combining those units of a similar nature or possessing common feed streams.

tary value of some of these streams is sometimes difficult to determine, and they sometimes impose special economic and technical constraints on the unit process.

The technique of decomposition can be formalized somewhat by the use of boolean algebra. From a process diagram such as Fig. 22-155 a boolean matrix may be constructed and the process analyzed by mathematical operations on this matrix.

Optimization of Process An n-stage process lends itself to dynamic-programming-type optimization to meet the specified plant objectives of throughput, product distribution, and product qualities at minimum production cost. In a three-stage process, starting with the terminal or third-stage unit process, this unit is optimized. Combining the optimized third stage with the second-stage unit, the combination is optimized with the third stage maintained at its optimum condition. Following this, the optimized combined second and third stages are combined with the first stage, and the entire system is optimized in terms of setting the unit-process performance objectives to meet the plant objectives. The readjustments within the unit processes to satisfy the unit-process performance objectives and constraints are the responsibility of the unit-process-level computer controller.

The determination of the n-stage unit-process control objectives which maximize income or minimize costs generally involves optimization techniques other than dynamic programming. For example, if the unit process is a batch-type operation, variation-calculus procedures may be used. There may also be significant differences in the type of information available when adding computer control to an established plant, a new plant which is similar to a plant for which operating experience is available, a new plant based on extensive pilot-plant operation, or a new plant based on little or no operating experience.

The plant computer control strategy must be based upon realistic plant-performance objectives which can be used to determine the control objectives of the individual unit processes making up the plant. The implementation of this strategy by the plant-level portion of the computer system then produces the set points of the unit-process-level computer controller.

Ethylene-Plant Example Ethylene plants are worthwhile candidates for computer control because of their size and complexity and the frequency of disturbances affecting plant operation. A typical naphtha-fed ethylene plant consists of several parallel-feed pyrolysis furnaces and a few recycle furnaces followed in order by a quench-heat boiler, main fractionator, cracked-gas compressor, and a train of product fractionators and cleanup reactors.

The controlled variables on each of the furnaces are hydrocarbon feed rate, steam feed rate, coil outlet temperature, and coil pressure. The furnace-coil outlet temperature is controlled by manipulating the fuel rate to each furnace. The coil outlet pressure, common to all furnaces, is controlled by suction pressure of the cracked-gas compressor. For N furnaces, the computer-manipulatable variables account for $3N + 1$ degrees of freedom.

The main process response to these controlled variables is the concentration distribution of the furnace-effluent components, ranging from hydrogen and methane to gasoline and fuel oil. The furnace unit-process model will express the individual yields as a function of the four furnace control variables. The furnace and quench disturbances include feed composition, quality of steam to furnace, coke laydown in furnace and quench-heat boiler tubes affecting coil pressure and heat-transfer ambient temperature, and cooling-water availability.

There are a number of physical and safety constraints upon the plant beyond which it cannot operate. These constraints can be divided into two groups, furnace and downstream. Constraints on the furnaces are maximum coil outlet temperature, maximum and minimum steam rates, maximum refractory temperature, maximum quench-boiler outlet temperature, maximum and minimum hydrocarbon feed rates, minimum steam-to-feed ratio, and maximum tube-skin temperature. Downstream constraints include maximum power available for the cracked-gas compressor, maximum gas flow through the compressor, and maximum refrigeration capacity.

Plant-performance criteria are formulated in terms of the yields of the several products, product values, and product markets for specified feed characteristics.

CONTROL EQUATIONS AND MODELS

Most computer control systems utilize some sort of model or control equation. These models may be divided into categories corresponding to four levels of computerized control: replacement of conventional controllers or regulation, dynamic computer control, steady-state computer control, and off-line computer control.

Computerized Conventional Controllers The objective of the computational portion of a conventional one- to three-mode controller is to correct the error in the measured variable in an optimal manner by adjusting some process variable which affects the measured variable. The optimal adjustment is to get to the desired value in minimum time with minimum total error. This led to the tuning procedures and controller settings discussed under "Fundamentals of Automatic Control."

The conventional controller contains an empirical process model in that it uses the error measurement to predict how much the manipulated variable should be adjusted to cause the error to go toward zero. Since the model in the controller is empirical and is nearly always too simple, it depends on the feedback signal to keep the process adjustment updated. The tuning procedures developed for such controllers take this simplification into account. They limit the control response to the maximum that will keep the response from excessively overshooting the desired value of the controlled variable prior to readjustment by the feedback signal.

The model in the conventional controller is some modification of the ideal three-mode controller represented by

$$m = k_c e + \frac{k_c}{T_i}\int e\,dt + k_c T_d \frac{de}{dt}$$

or

$$M = k_c\left(1 + \frac{1}{T_i s} + T_d s\right)E \qquad (22\text{-}36)$$

The modifications are usually the result of physical limitations of the mechanical or electrical implementation of this idealized form. See also text with Eq. (22-20).

In a computerized implementation, the computerized controller algorithm can fit Eq. (22-36) quite precisely. Unlike the mechanical implementations k_c, T_i, and T_d, adjustments are noninteracting, making tuning procedures somewhat simpler. In addition, other modifications can be made to the model such as the addition of dead-time compensation as in Fig. 22-55, or very nonlinear responses may be added. For example, in a proportional-only controller, k_c can be a nonlinear function of the error.

With computerized controllers, it is also relatively easy to change the model automatically with process-operating conditions. For example, the tuning constants can be made functions of some process variable making the controller adaptive, maintaining near optimum tuning over wide ranges of a process variable. Then Eq. (22-36) might become

$$M = f_c[C(s)]\left\{1 + \frac{1}{f_i[C(s)]s} + f_d[C(s)]s\right\}E$$

In digital-computer implementations of conventional controllers, one frequently encounters the terms "velocity algorithm" and "position algorithm." These terms are designations of how the controller output signal m is presented to the valve-actuating system. The velocity algorithm indicates that the change in the signal from its most recently presented value is presented to the valve-actuating system. If the position algorithm is used, then the new position is presented rather than the incremental position change.

The tuning settings for the digitally computerized controller may vary slightly from conventional controllers because the digital systems use periodic sampling of inputs (i.e., sampled-data inputs) rather than continuous inputs.

Multivariable-Input Computerized Conventional Controllers In many feedback controller configurations utilizing several measurements to compute a process variable, the algebraic determination is an instantaneous operation transmitted directly to a conventional controller or a computerized controller as the feedback-measurement signal or as a set point. Figure 22-176 is a rather complex example of this type of control system in which the calculated production rate is fed to the production-rate controller as the process-variable measurement.

In this case, for a completely computerized system, the model is divided into three parts, one part being the calculation of the input to the computerized controller, the second being the model required to implement the dead-time compensation of the computerized controller, and the third being the computerized conventional controller.

In feedforward control systems, as illustrated by Fig. 22-3, the model required for the controller is usually a quite realistic model of a portion of the process. In this case, the flow and temperature of the process fluid give instantaneously the heat required to maintain constant process output temperature. That is, $M = k(\text{flow})\,(T - T_0)$, where T_0 is the desired temperature of the process output stream and T the temperature of the incoming fluid. This is a proportional-only controller.

In more complex feedforward situations as illustrated in the subsection "Distillation-Column Control" with reference to Fig. 22-175, the model will need some of the dynamic characteristics of the process in order to maintain smooth operation of the process.

Steady-State Models As one moves up the hierarchy of control, the frequency of modification of control signals decreases. The models represent the computation of variables which are fixed until recalculated, at which time a variable is moved immediately to the new value. These are known as steady-state models. Constraints may be put on these models so that specific limits or specified rates of change cannot be exceeded. Such constraints may be explicit in a model or may result from some limitation in the process such that further movement of the variable is not possible even though called for in the computation.

Steady-state models are used at the conventional-controller level. They are frequently used for determining set points of controllers whose measured variables are dynamic multiple-input computed variables such as reactor production rate, fractionator reflux rate, and fractionator product rate as described in "Control of Unit Operations." Steady-state models are normally used at the unit-process level to determine and adjust the set points of unit-operations performance variables, for example, the purity specifications and feed splits for each pair of fractionators in Fig. 22-152 and the production-rate and polymer-specification set points for the individual reactors in Fig. 22-154.

The steady-state equations or models may be derived from process-design information, theoretical relationships, or experimental information obtained from process operation or simulation. One of the hazards of using design or theoretical information in developing the models is that too much detail may be included. On the other hand, if experimental data alone are used, important relationships may be obscured because of the difficulty of performing definitive experiments in the plant with plant-scale equipment. Simulation techniques offer a good compromise between theoretical and experimental development of steady-state equations. Further details on developing control equations or models are presented under "Control-System Analysis."

Dynamic Models Dynamic models or control equations are those which give a time-varying control signal or a signal which is dependent on the time variation of a measured process variable. Dynamic models are usually differential equations or their equivalent in such terms as integral equations and transfer functions.

An illustration of the use of dynamic models is the dead-time compensation illustrated in Fig. 22-55. Another is the predictive computer controller for fractionation control illustrated in Fig. 22-175. A third illustration is the calculation and control of the reactant concentration in the reactor shown in Fig. 22-191. In each of these illustrations, the dynamic characteristics enter into the model in a somewhat different fashion.

In the dead-time compensation system, the dynamic model provides a simulated response to control action with the dead time removed from the feedback signal. This simulated response is then compared with the process response on a process real-time basis for further control action.

In the fractionator predictive control system, the dynamic model provides control action in process real time to compensate for the change in feed rate and composition as these changes move through the fractionator.

In the reactant-concentration control system, the dynamic character of the model is in providing real-time reactant concentrations for control action by continuous integration of flows in, flows out, and reaction rate. Concentration computation is compared with on-stream analysis automatically obtained sometime after control action has taken place.

There is little need for dynamic process-control models above the unit-process level of control hierarchy. If the unit-operation controls are designed to control the unit-operation performance variables by reducing their degrees of freedom to near zero, the need for dynamic process-control models above the unit-operation level is low.

The techniques for obtaining dynamic process-control models are discussed under "Controller Tuning and "Control-System Analysis."

PROCESS-CONTROL COMPUTERS

Both analog and digital computers are used for process control. Analog computers represent the numbers being manipulated by the magnitude of a physical quantity such as volts or pressure (pascals or pounds-force per square inch). Mathematical operations are carried out in a continuous manner. If the computer network has several branches, several operations are carried out in parallel or simultaneously. In analog computers the computational precision available is limited by the precision to which the computer output can be measured. The accuracy is limited by the accuracy of the process measurements and their conversion into the physical medium with which the computer operates.

Digital computers are counting devices that operate directly on numbers to perform the four fundamental arithmetical operations: addition, subtraction, multiplication, and division. These mathematical operations are done by discrete serial machine operations. The instructions telling the computer to perform these operations are stored in the memory of the computer as programs. The computational precision available is essentially unlimited. The accuracy is limited by the accuracy of the process measurements and their conversion into digitized signals.

Both types of machines can be used to solve arithmetic equations and differential equations. Therefore, the choice of computer type is governed by the control job to be done and the cost of doing it. Figure 22-158 illustrates the basic cost versus job size characteristics of digital- and analog-computer control systems. That is, analog-computer-control-system costs are dependent on job size starting at a low level, while digital-computer-control-system costs start at a higher level with a smaller dependence on job size. The large number of varying-sized digital computers becoming available and the rapidly changing prices make it difficult to make general price versus job size statements. When job size does not give a clear-cut choice, other factors become significant. Some of these factors are listed in Table 22-10.

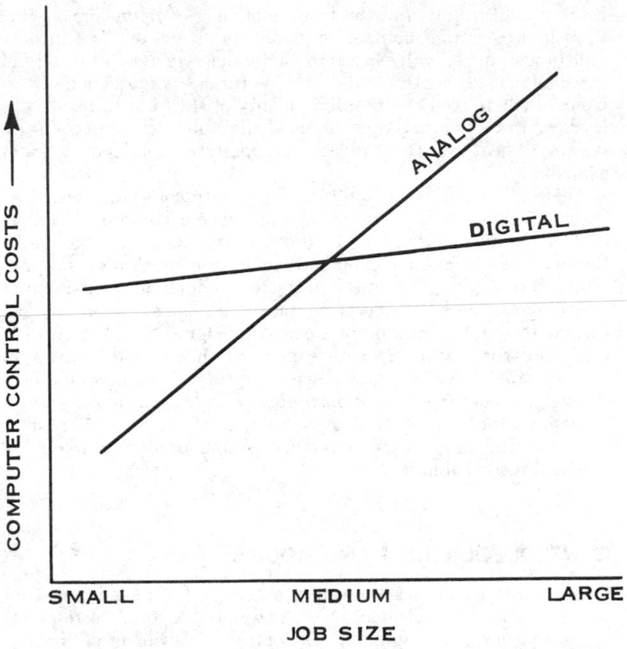

FIG. 22-158 Relative costs of analog- and digital-computer control systems.

TABLE 22-10 Special Features Which Help in Selection of Control-Computer Type

Analog Advantages	Digital Advantages
Lower cost for small systems	Lower cost for large systems
More flexible in small systems	More flexibility in large systems
Techniques employed more familiar to plant instrumentation personnel	Techniques employed more familiar to process engineers
More familiar troubleshooting procedures	More computational precision available
Parallel computation; therefore, component failure may not shut down entire system	Able to handle complex procedures such as optimization
More immune to noise	Drift-free data storage
Less complex interfacing with analog measurements and other instrumentation	Lower expansion cost

General Hardware (Equipment) Requirements There are a few major hardware requirements that must be met by a process-control computer whether that computer be analog or digital. Since a control system must operate essentially 24 h a day, year after year, an online availability of greater than 99 percent is a normal requirement for the computer and its input-output equipment.

An equally important requirement is that all components of the computer control system meet the safety requirements of the process area in which they are located. This requirement can be met for the small pieces of electronic equipment by conventional methods. Larger items can be handled by air purging or intrinsically safe design or by locating them in a general-purpose area.

A third requirement is that the system be designed so that it can be maintained quickly by normally available personnel. Modular design and good diagnostic procedures contribute to meeting this requirement.

Sometimes a computer-control-system measurement device is shared with an indicator or a backup controller. In these cases the computer's process-instrument interface must be designed so that the computer's operation does not affect the indicator or backup controller.

Computer control systems that are to be kept in operation during electric power interruptions require a backup power source. If alternating current is required, an inverter must also be installed.

Analog Equipment Analog-computer control equipment is usually electronic, electromechanical, or pneumatic. The pneumatic and electromechanical systems are usually limited to a few computational operations, with electronic devices being used for the more complex computations. Electronic analog computation is discussed under "Control-Systems Analysis" and "Automatic Controllers," as well as in Sec. 2.

Programming of analog computers for control is essentially the same as programming for analog simulation. See "Analog Simulation" under "Control-Systems Analysis."

Examples of applications of analog-computer control can be found under "Unit-Operations Control." All the examples of computerized control illustrated there have been implemented with analog equipment. Many have been implemented with pneumatic equipment.

Parallel processing units may be controlled with a single analog computer as illustrated in Fig. 22-159. Here block *A* represents amplification of low-level process-measurement signals, and block *H* represents a holding device to retain the output signal until a new one is available.

Digital Equipment The hardware or equipment required for a digital-computer control system can be divided into three parts: main-frame or central processor unit, CPU; interface; and process input and output, I/O.

The hardware requirements differ for the various levels of control. Direct digital control (DDC) systems require many inputs, rapid scanning, relatively fast computation, many outputs, backup equipment, and the highest reliability economically feasible.

Set-point control systems, sometimes referred to as supervisory control systems, generally require fewer inputs, little backup equipment, slower scanning, and relatively slower computation speeds.

An online optimizing system may require extensive computation based on relatively few process variables (see Fig. 22-151). Since new optimum control-parameter values are usually required at infrequent intervals, very-high-speed computation is not normally required. However, an extensive program and, therefore, a correspondingly large memory may be required.

Combinations of DDC, set-point, and optimizing control can therefore lead to quite large computer systems in large plants.

The major difference between a process-control computer and a typical business-scientific computer is the real-time environment in which the control computer works as compared with the batch environment in which the business-scientific computer operates. Actions taken by the computer are usually time-dependent or event-dependent. Time-dependent actions are scheduled by the real-time clock. Event-dependent actions are scheduled by priority rank by the interrupt system. The interrupt system and its associated software have the capability of stopping the computer in the midst of a program, having it carry out some other action or computation, and then returning to its original program, restarting from the point at which it was stopped.

Another difference is that the control computer can usually perform its computations with a shorter word length than the business-scientific computer. This is a result of the limited accuracy and resolution of the process-measurement and process-manipulative equipment.

Software There are two classes of programs in process-control-computer software: system programs and application programs. The system programs, commonly called the *operating system* or *executive program*, deal with the real-time operation of the computer con-

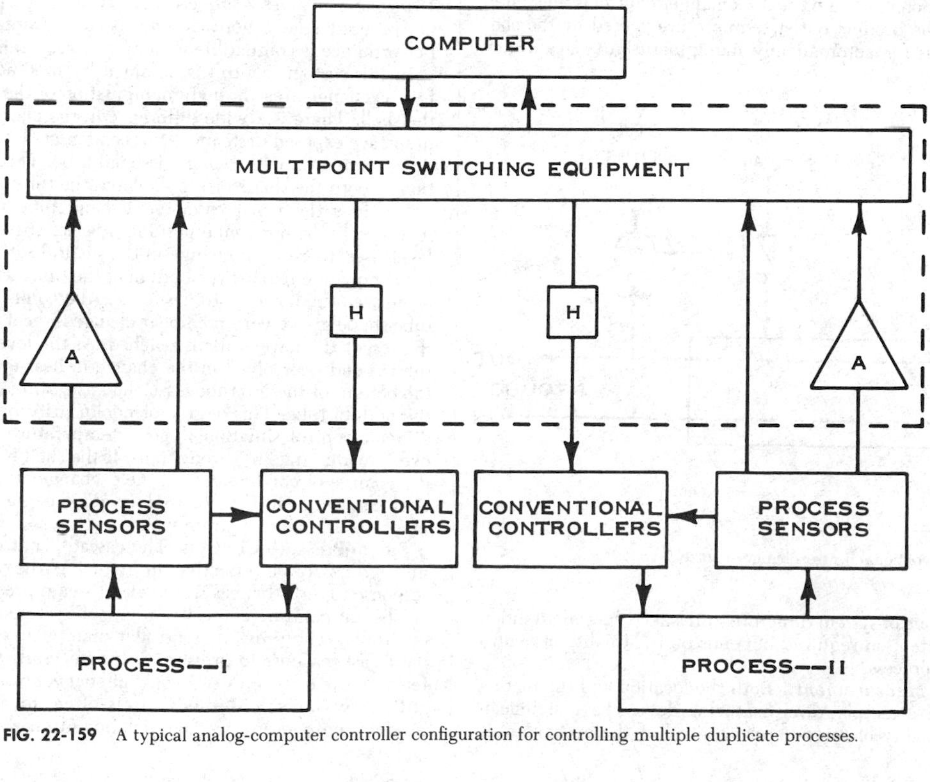

FIG. 22-159 A typical analog-computer controller configuration for controlling multiple duplicate processes.

trol system. They organize the hardware operations into orderly sequences which provide efficient use of the arithmetic unit, memory unit, peripheral equipment, and process-control equipment.

The application programs perform the specific functions which justified the design and installation of the system, for example, computing controller set points, monitoring process variables, and calling for an alarm if a variable exceeds some preset value. The complexity of this software and the amount of process and control knowledge required to implement it should not be underestimated. The application software may be largely custom-made for each process and installation. It is also dependent on the system software available for the computer used.

A major element in the implementation of application software is the language used to write the software. Languages used to communicate with the computer may be divided into three levels: machine languages, symbolic assembly languages, and higher-level languages.

Higher-level languages allow one to state the solution of the problem in a form more understandable to the programmer. These statements are then translated by another program in the computer into a sequence of machine-language instructions. One of the first of these languages was FORTRAN. It is widely used to allow mathematicians, scientists, engineers, and others to utilize computers in their everyday work.

Higher-level languages enable one to write programs in less time. However, there is a tradeoff in small systems between the cost of additional working memory required for utilizing higher-level languages and the cost of programming in machine or assembly language. A small control system may not be able to justify economically the extra memory required for using higher-level languages.

UNIT-OPERATIONS CONTROL

GENERAL REFERENCES: D. E. Lupfer, "Distillation Column Control for Utility Economy," presented at 53d Annual GPA Convention, Denver, March 25–27, 1974. F. G. Shinskey, *Distillation Control*, McGraw-Hill, New York, 1977.

Examples of the control of unit operations have been selected from which *Handbook* users can draw ideas for the solution of their unit-operation control problems, rather than providing a comprehensive survey of unit operations and applicable control systems.

CONTROL OF HEAT EXCHANGERS

Direct Control of Steam A number of equipment configurations are used in heat exchangers. We have used, for illustrative purposes, a simple tube-and-shell design. It contains a single U-bend tube in the shell for systems using steam as the heating medium, with an added shell-side baffle for liquid-liquid heat-exchange applications.

The most common method of heat-exchanger control is illustrated in Fig. 22-160. The product outlet temperature is used by the controller as the basis for automatically manipulating the steam flow rate.

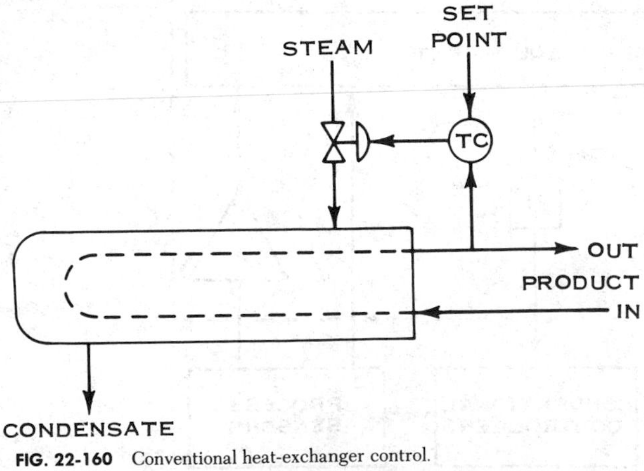

FIG. 22-160 Conventional heat-exchanger control.

Some of the control-system design features need special attention. The degree of attention required depends on the quality of control required by the process.

Temperature Measurement Both the location and the method of installation of a temperature-measuring device have significant effects on the quality of temperature control of the exchanger effluent. A few feet of piping between the exchanger and temperature measurement adds a dead time which may be as large as or larger than the exchanger time constant. Placing the temperature-measuring device in a thermal well may add another time constant in the feedback control loop which limits the quality of control obtainable.

If high performance is required of the control system, the temperature-measuring device should be placed directly in the process fluid unless maintenance, safety, or process problems indicate that a thermal well is required.

Control Valves Heat exchangers normally operate over a large range of throughputs. This large range requires accurate positioning of the control valve and nearly constant gain over the entire range of the valve. Thus, valve positioners are used, and the choices of size and type of valve trim must be carefully considered. Equal-percentage trim is normally used.

Condensate Control The control scheme illustrated in Fig. 22-161 generally gives relatively slow dynamic performance. It should not be used unless there is some special advantage in its use or its performance is compatible with process requirements.

In this system, control is obtained by modification of the steam heat-exchange area through manipulation of the condensate level in the shell. There is a wide difference in the character of the change in surface exposed to steam versus condensate level when a tube bundle is compared with a single U-bend tube. This variation is a function of both the shape of the shell and the tube geometry.

Consider the worst case: the U-bend tube with the exchanger mounted in the horizontal position (Fig. 22-161*a*). As the condensate level rises, there is no change in the steam heat-exchange area until the condensate reaches the bottom of the tube, where a sharp change occurs. As the level continues to rise, the combined shell shape and tube shape give a very nonlinear change in heat-exchange area until the top of the lower tube is reached. As the level rises further, only the U bend is involved in the change in heat-exchange area. When the bottom of the top tube is reached, the same situation exists as for the bottom tube. This degree of nonlinearity presents an extremely difficult control situation if good temperature control is expected over a wide range of throughputs. If the shell is shaped to minimize the volume of condensate and the exchanger is mounted in the vertical position as in Fig. 22-161*b* or at a steep angle, control performance is dramatically improved.

Steam-Pressure Control The cascade control system illustrated in Fig. 22-162 is effective in reducing the process-temperature response to disturbances which affect steam pressure.

The temperature controller adjusts the set point of a shell pressure controller. The pressure controller manipulates the steam rate and has a fast response to pressure disturbances. As with all cascade systems, this one requires two conventional controller loops.

Bypass Control Methods It is often advantageous to control product temperature by manipulation of an exchanger bypass flow

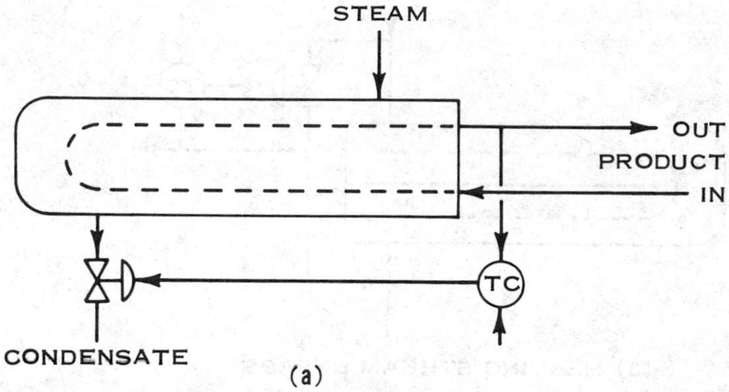

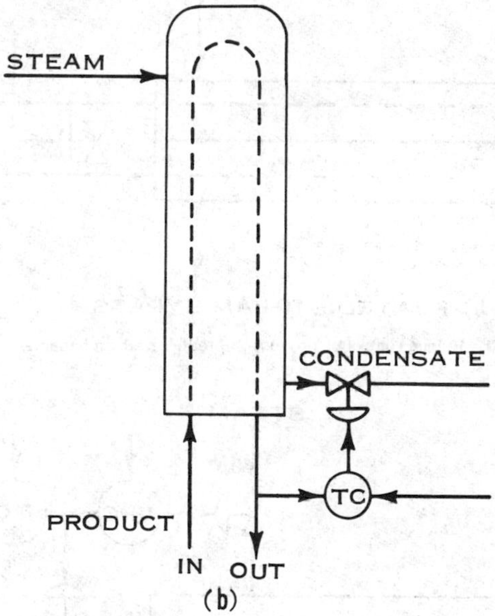

FIG. 22-161 Condensate throttling control. (*a*) Conventional construction and maintenance-oriented installation. (*b*) Control-oriented installation.

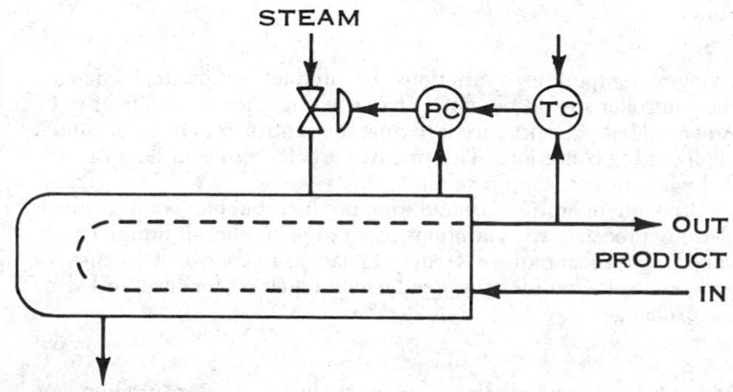

FIG. 22-162 Temperature-pressure cascade control.

rate. This approach is used when the heating medium is a process stream whose flow cannot be manipulated by the temperature controller. The bypass line can be on either side of the heat exchanger as shown in Fig. 22-163. The configuration in *b*, a heated-stream bypass, will usually give better temperature control because of the shorter time constant in the control loop.

An even better, but more expensive, bypass control arrangement for steam heat exchangers is shown in Fig. 22-164. The temperature controller manipulates the three-way valve to regulate the product bypass rate and maintain outlet temperature. The fast response of the bypass arrangement permits good temperature control. To keep control quality high over a wide range of throughput, the temperature-controller output, which also represents the three-way valve position, is used as the measured-variable input by the valve-position controller (VPC). The VPC, tuned for a slow response relative to the temperature controller, manipulates steam flow to hold the three-way valve near a midpoint position.

Feedforward Control All heat-exchanger temperature-control methods are subjected to disturbances in product flow rate, product inlet temperature, and steam quality. Well-designed systems can

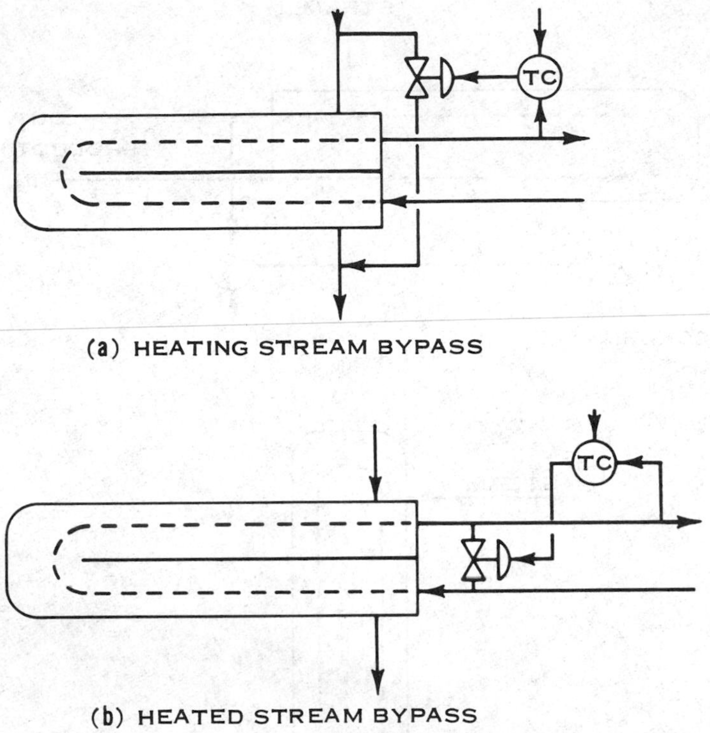

FIG. 22-163 Bypass control of product-to-product heat exchangers.

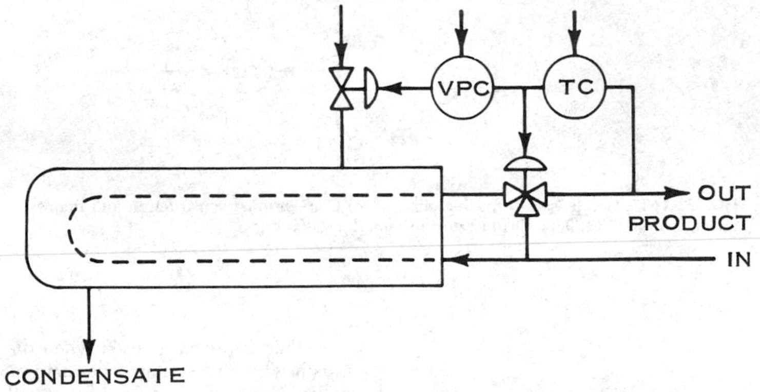

FIG. 22-164 Bypass control of a steam heat exchanger.

control outlet temperature well if the feedback controller is properly tuned for the existing conditions. However, the transfer function of a heat exchanger can change significantly with operating conditions. An extreme change in transfer function occurs when the product outlet temperature is near the bubble point or dew point as illustrated in Fig. 22-165.

The process steady-state gain G_p is the change in temperature per unit change in steam rate, or $G_p = dT/dS$. Hence, the process gain at any operating point is given by the slope of the curve in Fig. 22-165.

If the operating point is in the all-liquid region, at A, the process gain is large and the temperature controller will require low gain. However, if the operating point is changed to the partially vaporized region, at B, higher proportional gain is required for good control. Since the product physical state can change because of set-point changes, temperature variations, or product-composition changes, the controller should be tuned after each significant change of state. A normal but poor practice is to tune the controller to be stable under all operating conditions. This practice usually results in sluggish control when process gain is small.

Although operation around the product bubble point presents obvious process gain variations, operation in the all-liquid or all-vapor region also requires frequent changes in controller settings for good control. Consider the steady-state equation for heat added by the exchanger,

$$h_s S = FC_p(T_2 - T_1) \tag{22-37}$$

where h_s = steam enthalpy, S = steam flow rate, F = product flow rate, C_p = product specific heat, T_2 = product outlet temperature, and T_1 = product inlet temperature. Taking the derivative of Eq.

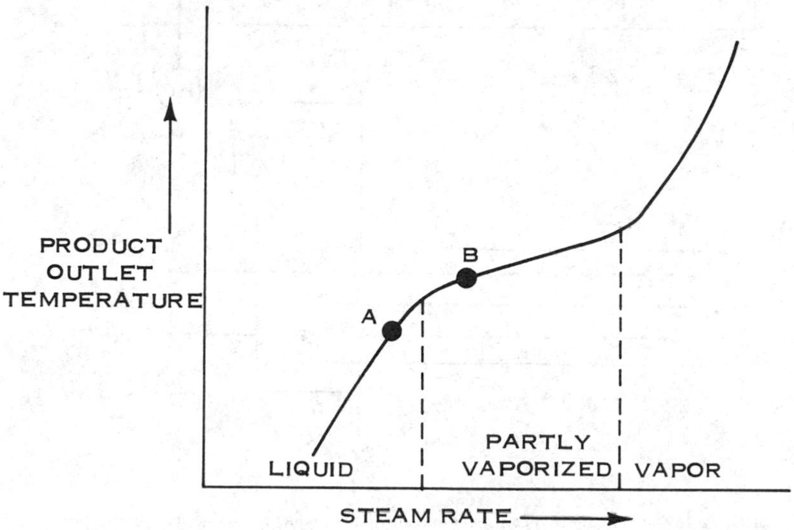

FIG. 22-165 Exchanger outlet temperature versus steam rate for constant product rate near the vaporization temperature.

(22-37) with respect to steam flow, the process steady-state gain is $G_p = dT_2/dS = h_s/fC_p$. Hence, the steady-state gain is inversely proportional to the product flow rate F. The process gain with respect to the steam-to-product ratio is a constant given by $G_p = dT_2/d(S/F) = h_s/C_p$.

Process gain changes can be compensated for by automatic adjustment of controller gain as a function of product flow rate or by using the temperature controller to manipulate the steam-to-product ratio as shown in Fig. 22-166. This is an elementary form of combined feedforward and feedback control.

Feedforward Control of Enthalpy The object of heat-exchanger control is to maintain a specified product outlet enthalpy. Simple computerized feedforward control systems can be designed to achieve this goal. The three examples presented here require only a small amount of computational equipment, and a conventional controller operates the final control element. These systems can also be included in larger computer control systems.

The heat-balance equation for a product-to-product heat exchanger is

$$H_p = C_{p1}(T_1 - T_0) + C_{p2}(F_2/F_1)(T_3 - T_2)$$

where H_p = heated-stream outlet enthalpy, C_{p1} = specific heat of heated stream at inlet, T_1 = inlet temperature of heated stream, T_0 = reference temperature, C_{p2} = specific heat of heating stream, F_2 = flow rate of heating stream, F_1 = flow rate of heated stream, T_2 = outlet temperature of heating stream, and T_3 = inlet temperature of heating stream. This equation can be mechanized and the calculated outlet enthalpy used as a measurement signal to a standard controller as illustrated in Fig. 22-167.

The heat-balance equation for a steam-heated heat exchanger is

$$H_p = C_p(T_1 - T_0) + h_s(S/F) \qquad (22\text{-}38)$$

where H_p = product outlet enthalpy, C_p = specific heat of product, T_1 = product inlet temperature, T_0 = reference temperature, h_s = steam heat content, S = steam flow rate, and F = product flow rate. If the steam flow rate is an input to the feedforward controller and Eq. (22-38) is mechanized, the resultant control system is illustrated in Fig. 22-168. Here the desired product enthalpy H is the set point, and H_p is used as the calculated variable for the conventional enthalpy controller HC.

Solving Eq. (22-38) for the steam flow rate

$$S = (F/h_s)[H - C_p(T_1 - T_0)]$$

where H_p becomes H, the desired enthalpy. In the implementation illustrated in Fig. 22-169, S, the steam-flow rate, is the calculated feedforward variable and is used as the cascade set point of the

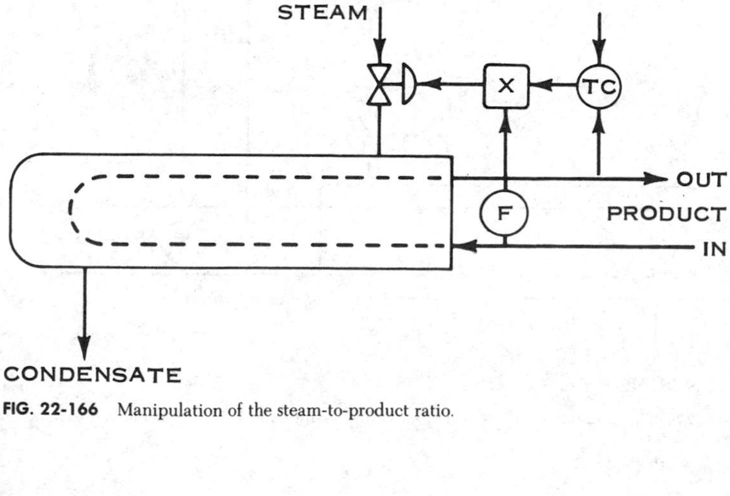

FIG. 22-166 Manipulation of the steam-to-product ratio.

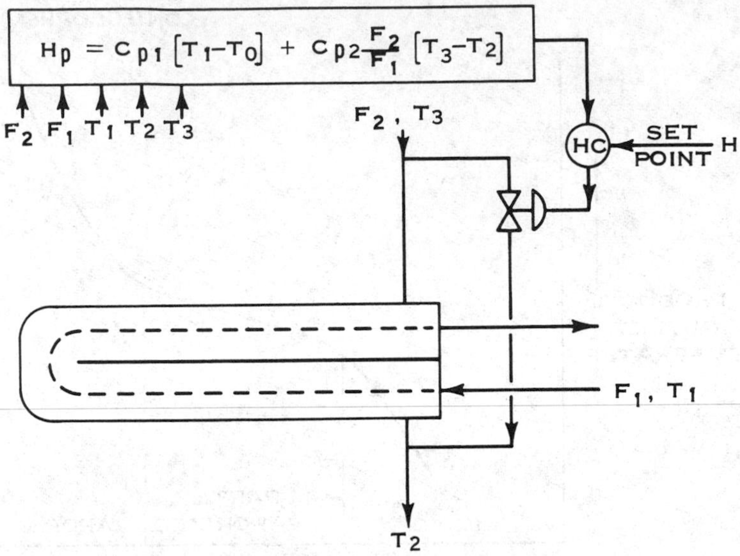

FIG. 22-167 Computerized control of enthalpy on a product-to-product heat exchanger.

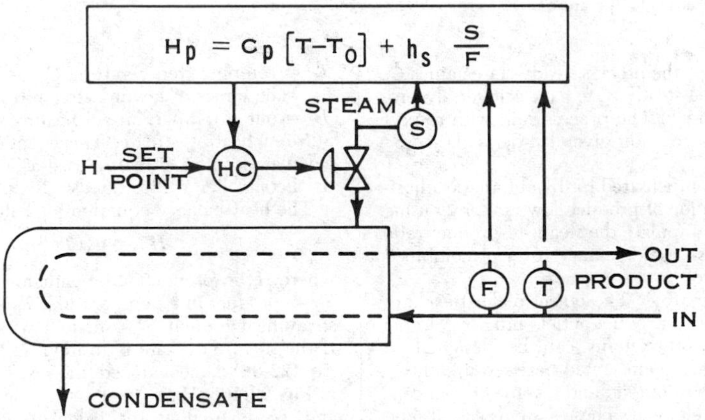

FIG. 22-168 Computerized control of enthalpy. Steam flow rate is an input to the controller.

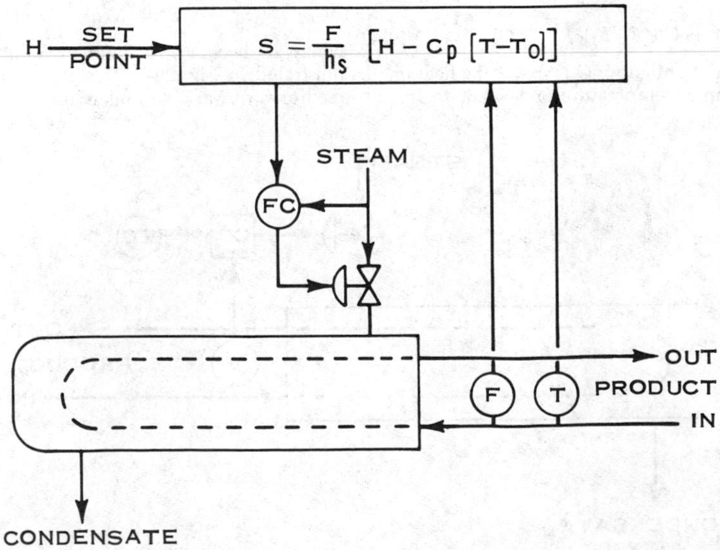

FIG. 22-169 Computerized control of enthalpy. Steam flow rate is the calculated variable as set point.

steam-flow controller. H, the desired product enthalpy, is the feed-forward process set point.

These computerized enthalpy-control systems may require the addition of dynamic compensation in order to perform the control action at the correct time to prevent feeding momentary enthalpy disturbances downstream. For example, it may be desirable to delay steam response to an inlet-temperature change by inserting a lag in the computer or in the temperature measurement. Such lag time constants can be estimated from dynamic tests on the heat exchanger or from the differential equations or simulations which describe the heat-exchanger operation.

DISTILLATION-COLUMN CONTROL

The ability of an effective distillation control system to reduce utility cost, increase capacity, or key component recovery is well known. While distillation columns are a major part of many processes, each of them has a unique set of operating objectives and constraints. It is for this reason that no single distillation control system will suffice for all applications. Presented here, therefore, are concepts and techniques which should assist those involved in the design and operation of control systems for the more common continuous two-product type of distillation column. Note that the more advanced control systems detailed here will only affect an operation which produces the specified separation. Space does not permit a detailed review of control systems which would affect the optimum (least-cost) operation.

Basic Controls The major objective of a control system for any process is stability. Stability, when related to distillation-column operation, means the ability to prevent disturbances which can result from changes in the independent variables to which the operation is subject. Table 22-11 contains a list of the more common variables which affect distillation-column operation.

TABLE 22-11 Classification of Distillation Variables

Input (independent) variables		Output (dependent) variables
Uncontrolled	Manipulated	
Feed composition	Reflux rate	
Feed rate	Boil-up rate	
Feed enthalpy	Distillate rate	
Reflux temperature	Bottom rate	Distillate composition
Heat-medium enthalpy	Column pressure	Bottom composition

The most stable distillation operation results when product withdrawal and energy input (or removal) are manipulated independently to produce the specified separation. This requires one product flow rate to be set in a manner which will maintain the material balance. Reflux or heat, then, is manipulated to regulate the amount of energy which resides within the process.

One of the more common distillation control systems is shown in

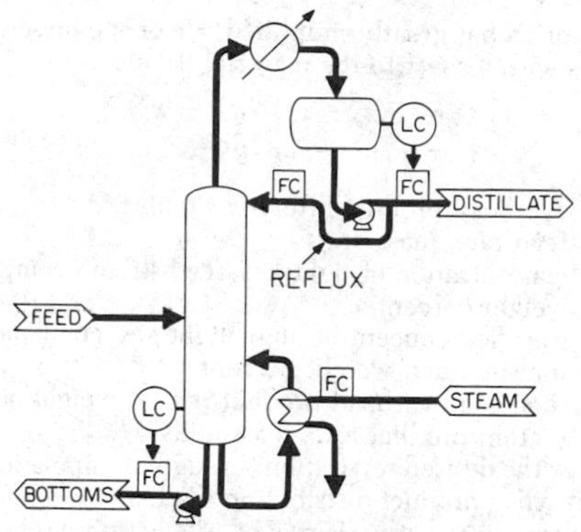

FIG. 22-170 Arrangement of controls which promotes instability. Pressure controls are not illustrated.

Fig. 22-170. Note that overhead and bottom products are set at rates which will maintain levels in the accumulator and the column base respectively. This arrangement of controls promotes instability because product rates become functions of the amount of energy within the system. Energy disturbances resulting from changes in feed, reflux and heat-medium enthalpies, and feed rate will each have an effect on product distribution. Large variations in product quality result in such an operation because product quality is most sensitive to product distribution.

Recovery from disturbances is slow because the operator is forced to manipulate both reflux and heat input independently for product quality control. This further complicates the operation because the amount of energy required to produce a given separation is a function of many variables, each of which must be taken into consideration by the operator.

Acceptable basic distillation control systems are illustrated in Fig. 22-171. Note that in each system one of the product flow rates is directly manipulated for product distribution (i.e., material-balance control). Reflux rate is the variable manipulated for energy control when bottom product is set from the material balance. Heat, then, is manipulated for energy control when the flow of overhead product is set with respect to the material balance.

The preferred control systems tend to be self-regulating. Disturbances in heat-medium enthalpy, for example, are detected by the base-level controller, which provides the required compensation when bottom product is manipulated for material-balance control. Variations in overhead vapor rate which are the result of changes in reflux subcooling are compensated for in like manner by the accumulator-level controller when overhead product is set for material-balance control.

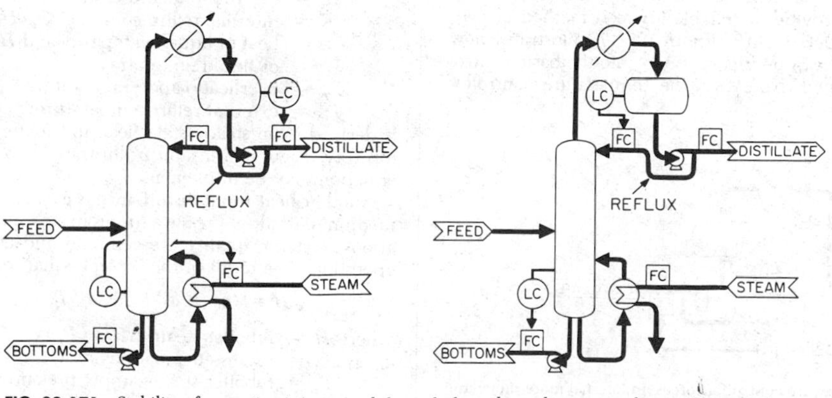

FIG. 22-171 Stability of operation is improved through the independent manipulation of material energy.

The operator's job is greatly simplified. He or she merely has to set product flows with respect to the material balance:

$$D = F(z - x^\circ)/(y^\circ - x^\circ) \quad (22\text{-}39)$$

or

$$B = F(y^\circ - z)/(y - x^\circ)$$

where D = overhead product rate, mass units
F = feed rate, mass units
z = concentration of light-key-feed-stream component, weight percent
x° = specified concentration of light key component in bottom product, weight percent
y° = specified overhead product purity, weight percent
B = bottom product rate, mass units

and then force the desired separation through manipulation of either reflux or heat when product distribution is set. The operator's adjustments are fewer because the degree of interaction has been considerably reduced by the properly arranged control system.

The correct product for manipulation is normally the one with the smaller flow rate, if for no other reason than accuracy: the same relative measurement error will produce less disturbance to the material balance. Other variables to be considered include reflux ratio and reboiler dynamics, especially when bottom product is contemplated for manipulation by material-balance methods.

Product Quality Controls Product quality control means control of the concentration of product-stream impurities. Consider, for example, the typical deisobutanizer operation which separates a feed mixture containing propane, isobutane, and butane into relatively pure isobutane (distillate) and butane (bottom) products. If the objective of the distillation is to provide 96 percent pure isobutane, the sum of the propane and normal butane impurities may not exceed 4 percent in the distillate stream. The concentration of propane in the overhead cannot be controlled by direct means and will appear as

$$C_3D = C_3F/(D/F)$$

where C_3D = propane in overhead product, weight percent
C_3F = propane in feed, weight percent
D/F = weight ratio, overhead product-to-feed

in the distillate. This leaves butane as the component which can be controlled for overhead product purity. Butane, in this case, is referred to as the "heavy key component" in the separation. Isobutane, on the other hand, is referred to as the "light key component," the bottoms-stream concentration of which can also be maintained at specified levels by adequate control instrumentation.

Product quality measurements are both specific and nonspecific. A specific composition measurement is one which directly presents stream-quality data (i.e., percentage of component concentrations) to the observer. Nonspecific composition measurements, then, present data which infer the composition of the monitored stream from measurements of other stream properties.

Tray temperature has been used for many years as the principal nonspecific indication of distillation-product quality. Temperature control is still used in many operations in which the separation is binary and column pressure is held constant. Tray-temperature control, in such instances, provides a reliable low-cost method of controlling distillation-product quality. Figure 22-172 illustrates how tray-temperature control can be incorporated into the basic control systems previously outlined. Note that the temperature controller manipulates product distribution to correct the material balance. Attendant response by the accumulator-level controller then forces energy to assume the proper value.

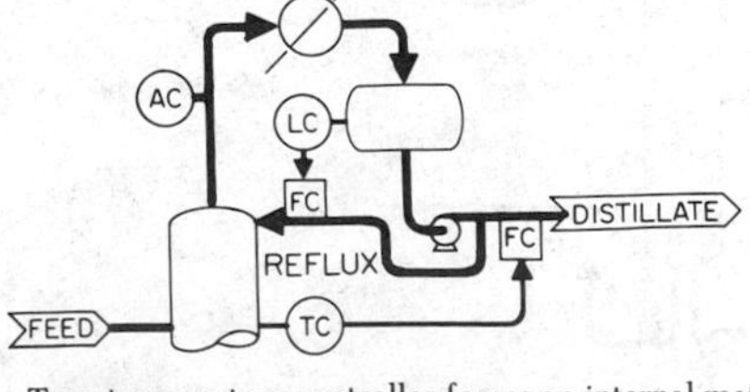

FIG. 22-172 Tray-temperature controller forces an internal material balance by manipulating the external material flow rate.

The temperature sensor should be located at a point on the column which exhibits a significant temperature variation when tray composition changes. Ideally it would be located in the vapor stream which ascends the selected tray. This location ensures a quick response to temperature variation.

Other nonspecific indications of product quality include density, viscosity, refractive index, and boiling point. Each of these properties has been used, with varying degrees of success, in composition control.

Specific product quality measurements include the use of mass spectrometers and gas chromatographs, the latter being the most widely accepted in recent years. Excellent results have been realized when the chromatographic-analysis system is properly designed, installed, and maintained.

Sample-system design and installation are probably the most important aspects of gas chromatography. The sample system must be of a quality which will provide the analyzer with a sample representative of the monitored stream in the shortest possible time. This means that the sample must be drawn from a point upstream of such lag-producing elements as the overhead accumulator and reboiler.

Analyzer-controller feedback is introduced in a manner similar to that described for temperature control. Figure 22-173 illustrates how composition control may be added to the basic control systems.

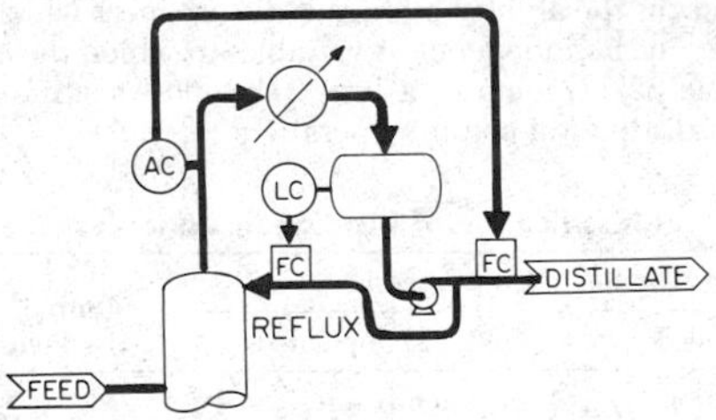

FIG. 22-173 Analyzer-controller provides corrective response to changes in overhead heavy-key-component concentration.

Heat-Input Controls Variations in distillation heat input create instability by disturbing internal liquid and vapor rates. Variations in reflux temperature, for example, affect internal liquid downflow. One proven method of eliminating this disturbance is to apply a special-purpose internal reflux controller, the objective of which is to maintain the liquid which descends from the top tray at a specified rate. The control equation

$$R_i = R_e\, 1. + K^\circ(T_{ov} - T_{re})$$

where R_i = internal reflux rate, mass units
R_e = external reflux rate, mass units
K = internal reflux constant, C_p/H, where C_p = specific heat of liquid on top tray and H = heat of vaporization of liquid on top tray
T_{ov} = overhead vapor temperature
T_{re} = external reflux temperature

is derived from steady-state heat and material balances around the top tray. Figure 174*a* and *b* illustrates two methods by which this equation may be implemented.

Another heat-input disturbance is caused by variations in heating-medium enthalpy. Pressure surges in steam-supply systems and variations in steam quality are common and can upset the distillation operation. The result of the control equation

$$H = (a_0 + a_1{}^\circ P_s + a_2{}^\circ T_s) - (a_3 + a_4{}^\circ T_r)$$

where H = reboiler steam ΔH
$a_0, a_1 \ldots$ = regressed equation constants
P_s = reboiler-steam-supply pressure
T_s = reboiler-steam-supply temperature
T_r = reference temperature

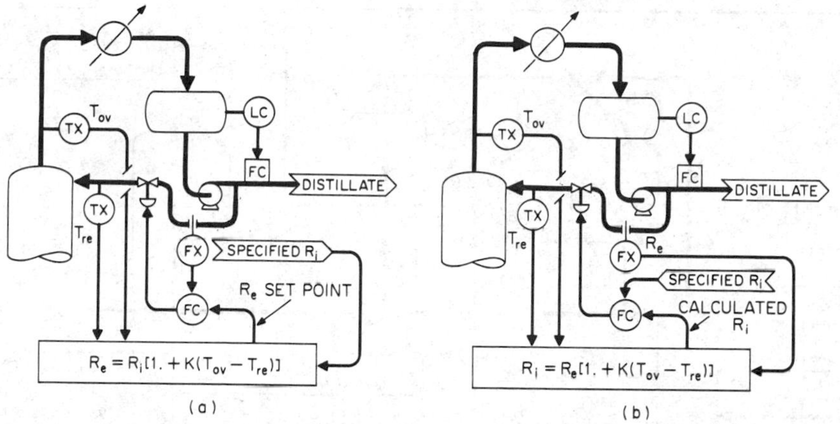

FIG. 22-174 (*a*) Internal reflux computer calculates set point of external reflux controller. (*b*) Internal reflux computer calculates current internal reflux rate.

has been used in a control system which set the heat-medium rate in a manner which compensated for these disturbances.

Feed-enthalpy disturbances are serious for the same reasons that heat-input disturbances are. The effects of feed-enthalpy disturbances can, however, be minimized through implementation of the control techniques discussed in the subsection "Control of Heat Exchangers."

Predictive Control A predictive-control system will provide response to changes in key independent variables before the effects of those changes can significantly disturb the process. A predictive-control system for distillation, then, will shift material and energy as required to maintain the specified separation in the presence of variations in feed rate, composition, temperature, tray efficiency, and other variables to which the separation is sensitive. The predictive distillation control system is an extension of the basic systems previously described because it independently manipulates product withdrawal and energy input to produce the specified separations. Product is withdrawn with respect to the material balance:

$$D = F_{\text{lagged}}\,(z - x^\circ)/(y^\circ - x) \qquad \text{or}$$

$$B = F_{\text{lagged}}\,(y^\circ - z)/(y^\circ - x)$$

where F_{lagged} = lagged feed rate, mass units. The feed-rate signal is lagged so that changes in control-equation input will have an effect on the manipulated variable (*D* or *B*) at the proper time; i.e., there will be a smooth transition from one operating state to the next. This "dynamic compensation" is normally in the form of dead time plus a first- or second-order lag.

The predicted values for reflux and heat are not as easily attained. Operating equations must be developed for each case. The steps involved in equation development are (1) base-case-operation simulation, (2) sensitivity study, (3) independent-variable perturbations, and (4) curve-fitting data to the assumed equation form.

The following is an example of equations which are generated by this technique:

$$\begin{aligned} R_i/F = {} & (a_1 + b_1/(c_1 + E)))^\circ(a_2 + (b_2/(c_2 + (100. \\ & - y^\circ))))^\circ \ldots (a_3 + b_3/(c_3 + x^\circ)))^\circ(a_4 + (a_5{}^\circ T_f) \\ & + (a_6{}^\circ LKF) + \ldots (a_7{}^\circ HKF) + (a_8{}^\circ P_c)) \end{aligned} \qquad (22\text{-}40)$$

where R_i/F = weight ratio, internal reflux-to-feed
$b_1, b_2, \ldots b_n$ = regressed equation constants
$c_1, c_2, \ldots c_n$ = regressed equation constants
E = separation efficiency, percent
LKF = light-key-feed-stream-component concentration, weight percent
HKF = heavy-key-feed-stream component

The reflux rate required to satisfy R_i/F is then calculated:

$$R_e = (F_{\text{lagged}}{}^\circ R_i/F)/[1. + K^\circ(T_{ov} - T_{re})]$$

The external reflux set point is maintained at the value R_e.

Note that the form of the operating-heat equation for the prediction of Q/F (reboiler heat to feed) is essentially the same as that of Eq. (22-40).

Absolute and rate-of-change limits should be placed on the outputs of all predictive-control equations. These limits will protect the operation from upsets created by failures in equipment which supply equation inputs.

The accuracy of the prediction depends on the accuracies of the inputs to the equation. Since absolute accuracy is assumed to be unobtainable, feedback control is necessary. The techniques previously mentioned still apply.

Figure 22-175 illustrates a computerized distillation control system in which the basic controls are assisted by predictive, product quality, and heat-input controls. Table 22-12 explains the symbols used on Figure 22-175 not previously defined.

CONTROL OF CHEMICAL-REACTOR PRODUCTION RATE

An example of the use of computer control based on a computed variable is the production-rate control of an exothermic reaction in a stirred reactor. By the online computing of a dynamic heat balance, the production rate can be determined for use as the controlled process variable. [Tolin and Fluegel, *J. Instrum. Soc. Am.*, **6**, 32–38 (October 1959).]

Typical Reactor System A typical reactor, with some simplifications, is shown schematically in Fig. 22-176. The solvent, reactant, and catalyst may be considered as continuous inputs to a stirred vessel. The reactant and product are dissolved in the solvent. Coolant circulated through the cooling coils and the reactor jacket maintains the reactor temperature.

The production rate of product is proportional to the heat generated by the reaction and can be controlled by catalyst-rate adjustment. The heat generated by the reaction can be computed by performing an online heat balance around the reactor. The information required for such an online heat-balance computation includes temperature and mass flows of all input streams, coolant flow and temperature, reactor temperature, the energy added by the stirrer, the heat of solution of the reactant in the solvent, and the heat of reaction.

Reactor Heat Balance The equations for the heat balance can be written with reactor temperature as the reference so that dynamic computation of reactor-effluent mass flow and the resulting heat removal is not required. In order to simplify the example calcula-

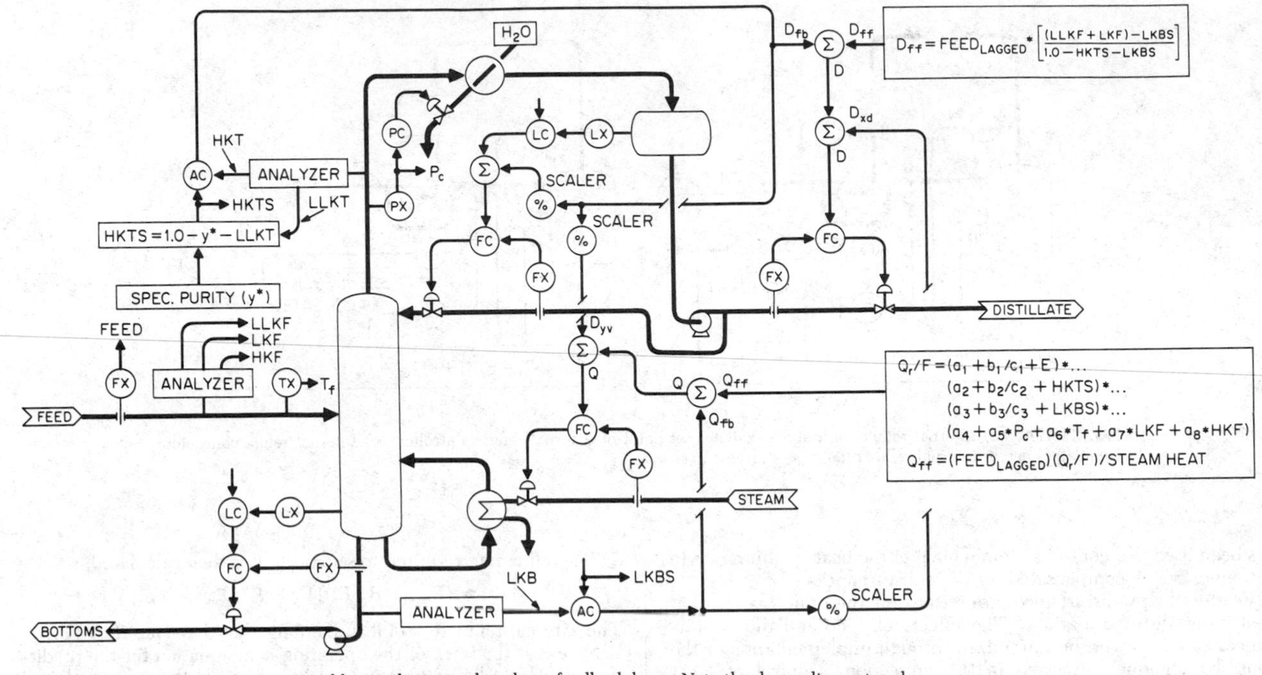

FIG. 22-175 Basic controls are assisted by predictive and analyzer feedback loops. Note the decoupling network.

TABLE 22-12 Nomenclature Summary for Fig. 22-175

Symbol	Explanation
D_{fb}	Contribution of feedback control to distillate-flow set point
D_{ff}	Contribution of predictive control
D_{xd}	Contribution of noninteractive decoupler to distillate-flow set point
D_{yv}	Contribution of noninteractive decoupler to reboiler-heat set point
HKF	Heavy key feed-stream component
HKT	Heavy key overhead-product-stream-component concentration, weight percent
HKTS	Specified concentration of heavy key component in overhead product, weight percent
LKBS, $x°$	Specified concentration of light key component in bottoms product, weight percent
LKF, z	Light-key-feed-stream-component concentration, weight percent
LLKF	Lighter-than-light-key-feed-stream-component concentration, weight percent
LLKT	Lighter-than-light-key-overhead-product-stream-component concentration, weight percent
Q	Reboiler heat, BTU/unit time
Q_{fb}	Contribution of feedback control to steam-flow set point
Q_{ff}	Contribution of predictive control to steam-flow set point
Q_r/F	Weight ratio, reboiler-heat-to-feed
$x°$, *LKBS*	Specified concentration of light key component in bottoms product, weight percent
z, *LKF*	Light-key-feed-stream-component concentration, weight percent

tions, mass-flow measurements will be assumed. Specific heat of each stream is also assumed to be constant over the range of temperature encountered. The overall heat balance, based on the heat inputs and removals listed in Table 22-13, is

$$Q_s + Q_m + Q_c + Q_w + Q_r + Q_a + Q_L + Q_p = 0$$

$$\text{Thus } PR = -(1/h_p)(Q_s + Q_m + Q_c + Q_w + Q_r + Q_a + Q_L)$$

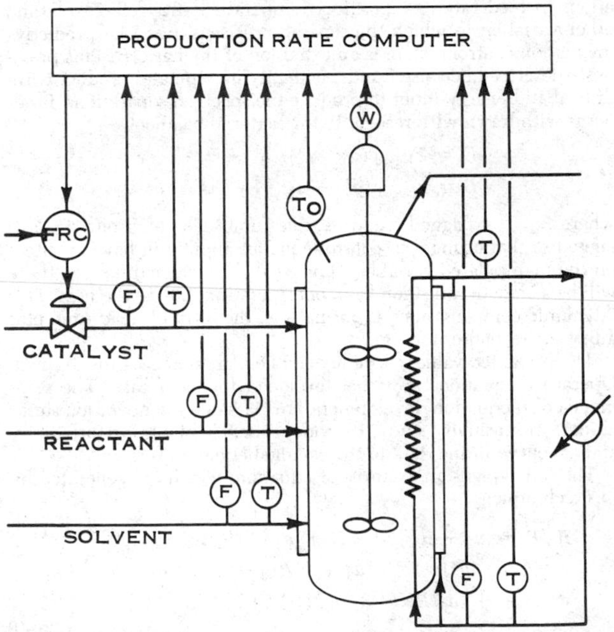

FIG. 22-176 Control of production rate in a chemical reactor.

h_p and all terms in the equations for the Q's making up the functional equation for *PR* either are known or can be automatically computed from online process measurements.

The computed *PR* can then be used as the measurement of the controlled variable for automatic control of catalyst addition as illustrated in Fig. 22-176. In the actual reaction system from which this example is drawn, the catalyst has an induction time which is a significant fraction of its residence time in the reactor. Therefore, dead-time compensation as illustrated in Fig. 22-55 is required for any kind of reasonable controller performance.

TABLE 22-13 Heat Input and Removal from Chemical Reactor

Source	Equation	Variables
Solvent	$Q_s = F_s C_{ps}(T_s - T_o)$	T_o = reactor temperature Q_s = solvent heat input F_s = solvent mass flow C_{ps} = solvent specific heat T_s = solvent temperature
Reactant	$Q_m = F_m C_{pm}(T_m - T_o) + F_m h$	Q_m = reactant heat input F_m = reactant mass flow C_{pm} = reactant specific heat T_m = reactant temperature h = reactant heat of solution
Catalyst	$Q_c = F_c C_{pc}(T_c - T_o)$	Q_c = catalyst heat input F_c = catalyst mass flow C_{pc} = catalyst specific heat T_c = catalyst temperature
Mixer	$Q_w = K(W - W_o)$	Q_w = mixer heat input K = heat-power conversion W = mixer power W_o = mixer no-load power
Coolant	$Q_r = F_r C_{pr}(T_r - T_i)$	$-Q_r$ = coolant heat removal F_r = coolant mass flow C_{pr} = coolant specific heat T_r = coolant temperature out T_i = coolant temperature in
Reactor	$Q_a = M_a C_a \frac{dT_o}{dt}$	Q_a = reactor heat accumulation M_a = mass of reactor and contents C_a = heat capacity, reactor and contents
Ambient	$Q_L = f(T_L, V_L)$	$-Q_L$ = ambient-heat loss $f(T_L, V_L)$ = function of ambient temperature and wind velocity
Reaction	$Q_p = PRh_p$	PR = product-production rate Q_p = reaction heat input h_p = heat of reaction

CONTROL OF DRYING OPERATIONS

The control of drying operations is often difficult because online direct measurement of product quality or drying effectiveness is difficult. Therefore, the drying process must be controlled from measurements of environmental conditions and by changing the environmental conditions when an off-line or laboratory analysis indicates the need. If direct measurement of input-material quality is possible, a predictive-control system is desirable.

There are many types of solids dryers which have a variety of different measurement and control problems. Although the mechanical configurations differ, the basic control problems are essentially the same for rotary, traveling-screen, fluidized-bed, moving-sheet, tray, and screen-conveyor dryers.

Rotary Dryers The countercurrent rotary dryer illustrated in Fig. 22-177 is the basic form of this type of dryer. It consists of a tilted cylindrical shell, with attached mixing blades, which rotates around its longitudinal axis. The wet material enters the elevated end and is mixed with the hot air as it travels through the cylinder. The hot air, flowing countercurrently to the material being dried, is heated by a steam-heated heat exchanger in this example. The air flow rate is regulated by a damper, and a second damper is located in the exhaust-air outlet to regulate the pressure in the dryer. For this discussion, we will assume that the drive speed is fixed and that the feed condition and rate are determined by the unit operation just preceding the dryer.

Dryer Variables The independent input variables for the rotary dryer may be divided into two categories, uncontrolled and manipulated. The uncontrolled inputs include feed moisture content and feed rate. The independent manipulated inputs are variables which are adjusted to control the drying process. These variables include heat applied, air input, and exhaust-damper position.

The dependent output variables for the rotary dryer and air heater include dryer pressure, dryer inlet-air temperature, and exhaust-air wet- and dry-bulb temperature; however, the primary output variable, which is usually unmeasurable by automatic apparatus, is product moisture content. The product moisture content is sometimes indirectly determined online by measurement of product temperature, but frequently this is an unreliable measurement.

Control System The normal control system for a rotary dryer uses heat applied to regulate inlet-air temperature; the pressure at one point is controlled by the exhaust-damper setting, and the inlet-air rate is used to control the outlet-air wet-bulb temperature. A slightly different arrangement is to use the air rate to control the difference between dry- and wet-bulb exhaust temperatures, employing this temperature difference as a measure of the rate of evaporation. This latter modification in control configuration is

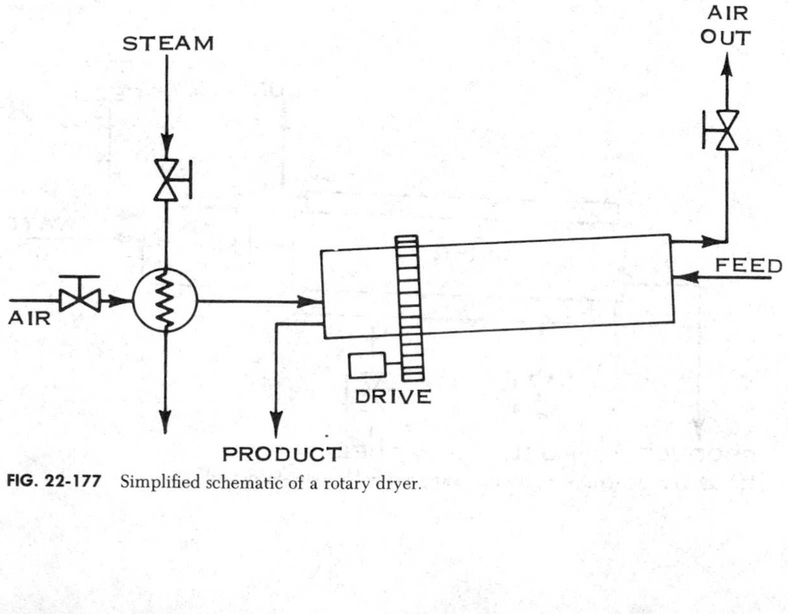

FIG. 22-177 Simplified schematic of a rotary dryer.

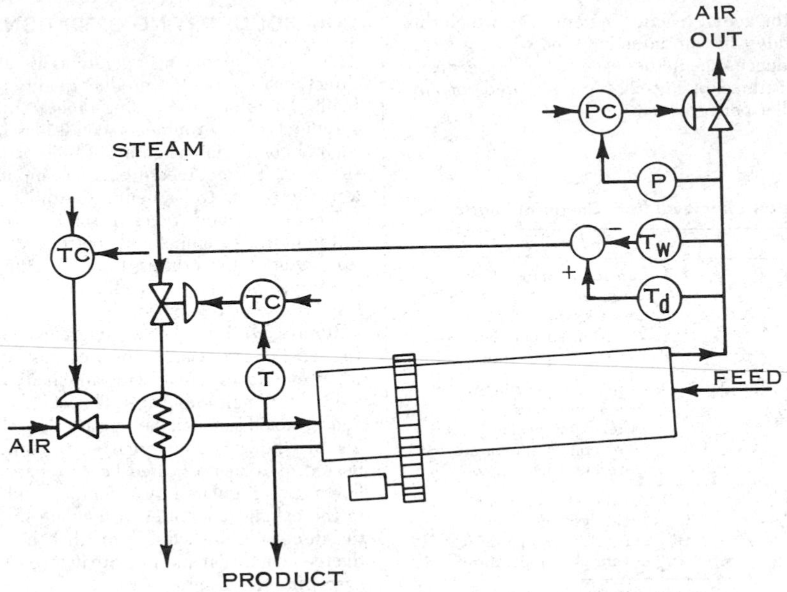

FIG. 22-178 A rotary-dryer control configuration.

shown in Fig. 22-178. Another modification is to control the metering of the feed in response to the exhaust temperature or metering to hold the feed constant.

Predictive Control Improved control of a drying operation is possible if more pertinent measurements are available. As an example, a predictive-control system may be applied to the dryer illustrated in Fig. 22-179 when the load is measurable. The feed is a dry powder which is mixed with water and a binder to form pellets. The water is added in proportion to the powder feed rate so that the dryer-feed moisture content is reasonably constant. Two sets of burners are used to apply heat to the rotating shell.

The control objective is to maintain the final product temperature within specified limits above the boiling point of water. Feedback control of this temperature by simultaneous manipulation of fuel-gas rate to both sets of burners to compensate for variations in feed rate is difficult. The process-temperature response to feed changes includes a large dead time, and the steady-state relationship between water injection and fuel-gas rate is very nonlinear.

One solution is to measure the incoming-feed disturbances and begin to adjust the fuel-gas rate before the disturbances affect the final product temperature as illustrated in Fig. 22-180. Such a predictive-control system employs the measured water-injection rate to make adjustments in the fuel-gas rates to the two furnaces at the proper times to minimize product-temperature variation. The dynamic portion of the feedforward control is essentially dead time, and the two fuel-gas rates are changed in sequence. A product-temperature measurement is used by a conventional controller to trim the predictive-control signal to keep the product temperature within the desired operating limits.

Drum Dryers The control of a drum dryer also is dependent upon the measurements available. In the drum dryer illustrated in Fig. 22-181, heat is applied to the inside of the rotating drum, and the wet feed clings to the outside of the drum. The material is dried during part of a revolution and is removed by a blade or a scraper.

The operation of the drum dryer may be controlled by the adjustment of drum speed and heat supplied. The variables to be controlled depend upon the product quality measurements available such as product moisture content or temperature. Figure 22-181 illustrates control of the steam rate on the basis of a continuous measurement of moisture content. Drum speed is usually fixed.

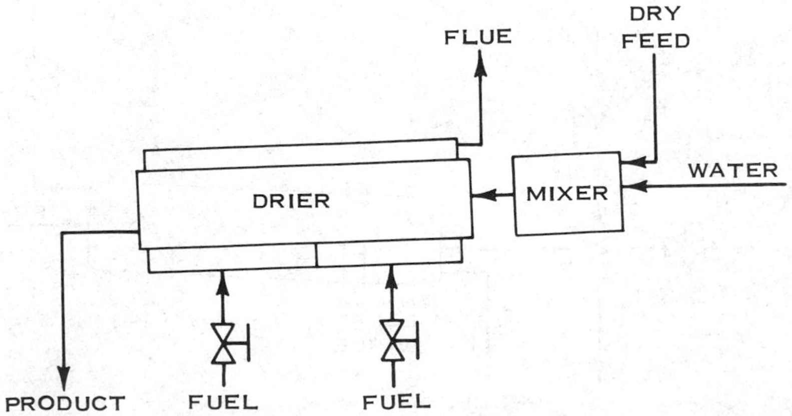

FIG. 22-179 Mixer-dryer pelletizer with measurable water-binder input.

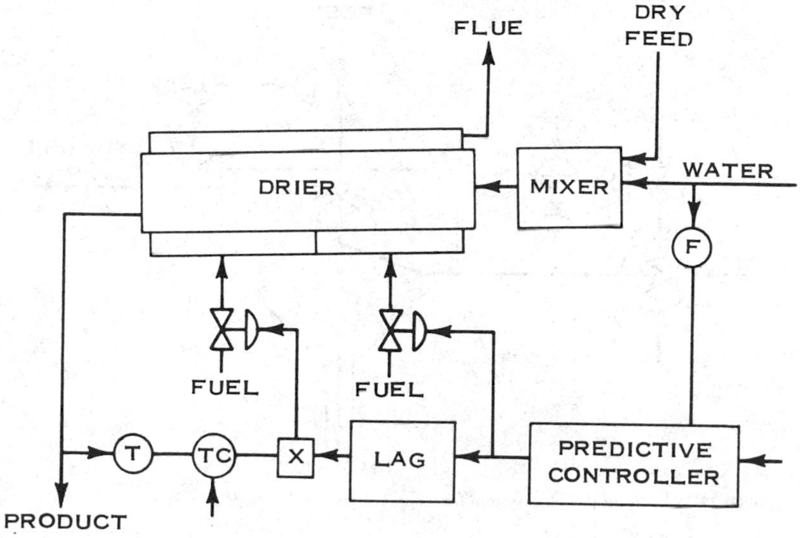

FIG. 22-180 Predictor control of a mixer-dryer pelletizer.

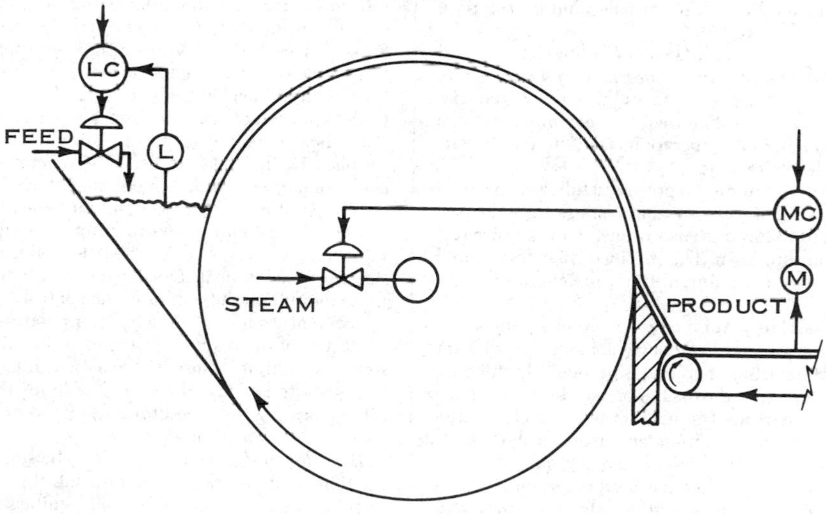

FIG. 22-181 Typical control system for a drum dryer utilizing moisture measurement and controller for product quality control.

Spray Dryers Liquid mixtures are often dried by spraying the material into a stream of hot gases. The small droplets offer a large area of exposed surface, so drying is rapid. The short process dead time due to rapid drying will often allow good automatic control.

A common control configuration for a spray dryer is shown in Fig. 22-182. The entering-air temperature is maintained by automatic manipulation of the heat rate. The exhaust-air-temperature measurement is used to adjust the dryer feed rate.

CONTROL OF BATCH OR START-STOP OPERATIONS

Many process operations can be classified as start-stop operations. In addition to batch processes, startup and shutdown of a normally continuous unit operation require batch-process control techniques. Also, the changing of a continuous process from one steady-state operating level to another may be similar to batch processing. The automatic control of batch or start-stop processes can be difficult because of rapid changes in operating conditions. Computer control systems can often be employed to improve the control of start-stop-process operation.

Special control procedures may be required to handle the wide changes in operating conditions in, for example, a batch reactor. The operation may include charging the reactor with reactants and catalyst, applying the heat required to reach the reaction temperature, maintaining a desired level of operation until the reaction reaches the desired stage of completion, stopping the reaction, removing the product, and preparing the reactor for another batch. Similar problems exist in the control of batch drying and distillation.

Start-Stop Performance Criteria Performance criteria for start-stop operations may be stated in several forms. Such operations are often evaluated by the time required to complete a process change without exceeding processing constraints during the change. For example, a batch dryer may be operated to obtain a specified product moisture content in minimum time without exceeding a specified maximum dryer temperature.

Another common control performance criterion is to complete the

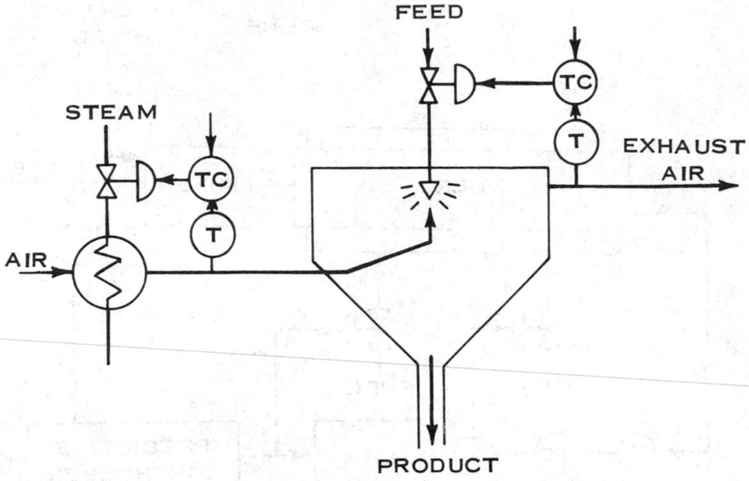

FIG. 22-182 Typical control system for a spray dryer.

required operation at minimum cost. An example is a batch reactor in which a given quantity of product is produced for minimum cost of heating, material, and catalyst with no specific limit on the time required.

A batch process may also be operated to obtain maximum yield within a given time period when the time constraint is set by the rate at which product must be produced or the rate at which raw material is made available. A shorter batch time than the maximum may be of no benefit if the frequency of the operation is determined by circumstances arising outside the process.

A control criterion can also require the process to follow a specified trajectory or path during a change in operation. For example, a batch reactor may be operated to follow a predetermined temperature trajectory during startup and shutdown. The specified trajectory would be normally designed as a minimum-time or minimum-cost operation.

Continuous-Process Startups Automatic startup of a continuous operation in many industrial processes does not have a significant economic appeal unless the startup operation is particularly difficult, some process condition frequently throws the process into the equivalent of a startup situation, there are frequent start-stop cycles, or the control system has been designed to minimize startup problems. To date, few unit operations have been automated for an optimized startup. However, if the startup problem has been taken into consideration in the design of the control system, startup time may approach optimum with little automatic control.

Exothermic Polymerization Reactor The control system illustrated in Fig. 22-183 uses a minimum of instrumentation and requires some operator attention to obtain adequate temperature control during startup. [Mayer and Spencer, *J. Instrum. Soc. Am.*, 8(7), 58–64 (1961).] The reaction is exothermic, and feed temperatures are lower than the reactor temperature. Startup procedure calls for having diluent in the reactor and at reaction temperature prior to the addition of reactant feed and catalyst. Prior to reaction initiation, heat is removed by the coil and the same amount of heat is supplied by the jacket to keep the reactor temperature constant. As cool reactant and diluent feed and catalyst are introduced, the reactor temperature starts to fall and more heat is automatically supplied to raise the incoming feed to reaction temperature. As the reaction rate increases, less and less heat is required, and finally the jacket starts to remove heat. During the time when the reactor is coming up to the full production rate, the control system continuously makes adjustments to hold the reactor temperature at the set point.

Startup of the reactor illustrated in Fig. 22-176 requires somewhat similar conditions. Here the reactor is filled liquid-full and pressurized prior to heating. If the reactor fluid-removal rate is limited, the rate of heating must be limited to prevent reactor pressure buildup due to fluid expansion.

Distillation-Column Startup The startup and shutdown of distillation columns are a common control problem. Automatic control of column startup can often yield significant economic benefits from operating the column only at full capacity when maximum tray ef-

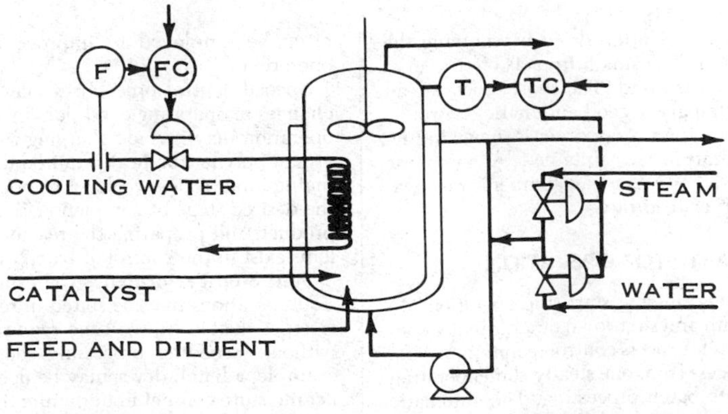

FIG. 22-183 Typical temperature-control system for an exothermic polymerization reactor requiring little operator attention during startup.

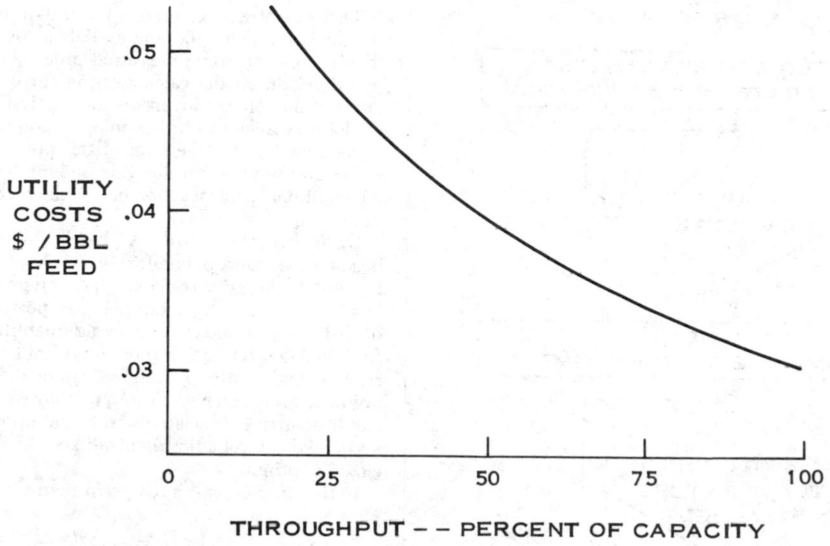

FIG. 22-184 Distillation-column utility costs at constant product purity.

ficiency is achieved. If feed-storage capacity is available for a feed-limited situation, an automatic startup and shutdown system will permit an on-off type of operation that is more efficient than continuous operation at low throughput.

Figure 22-184 illustrates the variation in utility costs with throughput for a typical distillation column operating at fixed product compositions. If the column is operated continuously at partial load, the operating costs may be higher than operating the column part-time at full load. At full load, the processing costs include the utilities plus startup, shutdown, and storage costs. Therefore, an efficient automatic startup-control system could be profitable.

A distillation-column computer-controlled startup system is of about the same complexity as putting a chemical reactor on line. The startup routine is an automatic sequencing of the tasks normally performed by an operator. It can be a subroutine in a multicolumn computer control system. The objective of the program is to reach the desired operating level within a reasonable time at minimum costs.

Control of Batch Reactors The control of a batch reactor in a simple case consists of charging the reactor, controlling the reactor temperature to meet some processing criterion, and shutting down and emptying the reactor. For an exothermic reaction, heat may be required to obtain the desired reaction temperature, and then cooling is used to maintain the proper reaction temperature.

The use of conventional instrumentation for batch-reactor control usually requires close operator attention. For example, the operator must manually adjust flows to charge the reactor and then adjust the temperature controls to obtain the desired reaction temperature. The latter is accomplished by increasing the reactor temperature in steps well below the maximum rate.

Batch Programming One technique for reducing the attention required from the operator is to use a batch programmer—a mechanism that automatically advances the batch through its process stages to a preselected time schedule. To ensure completion of each operating stage, the time sequences are normally set longer than necessary.

End-Point Programmer The use of an end-point programmer which employs some measurement of process performance for triggering the next operating sequence cuts down on batch time. For example, in an esterification process, limiting values of temperature and pressures are used to sequence the various stages of the batch process. To the extent that the end points used describe the products obtained, the end-point programmer reduces processing time compared with the fixed-time programmer. When a desired end point is reached, the processing switches to the next step with no loss of time. Production is thus increased.

Computer Control Computer control can sometimes be used to advantage in batch reactors. The amount of reactants charged can be closely monitored to ensure the proper mixture. The temperature controls can be preprogrammed to obtain a minimum time for the temperature transients. With computerized regulators, the controller can be automatically adjusted to maintain the proper controller tuning at all levels of operation. An example of an operator-action flow diagram is shown in Fig. 22-185. This flow diagram can also serve as the outline for a computer program to operate the batch reactor.

Addition of Reactant Sometimes a small amount of automation of the batch reactor can result in substantial improvement of throughput if more complete information concerning reaction conditions and equipment constraints is available. Consider Fig. 22-186, a batch reactor in which an exothermic, irreversible reaction of feed Y to form product Z takes place in the presence of a catalyst. [Schrock, *J. Instrum. Soc. Am.*, **12,** 75–82 (October 1965).] A batch is considered complete when a fixed amount of Y has been added and the reaction approaches equilibrium. A totalizer integrates the rate of addition of Y, and at the preset total the addition is stopped. For startup, a small initiator batch including catalyst is added to the empty reactor, the initiator is brought up to operating temperature, and the reactor is pressurized with an inert gas. As the reaction proceeds, the inerts are vented. Reactor temperature is automatically controlled by the cascade control of jacket-water temperature. The jacket-water temperature and reactor pressure are both bounded by upper and lower limits.

The control problem is to operate the process in minimum time, subject to temperature and pressure constraints. Manual operation consists of the operator's adjusting the vent flow rate and mass flow of Y so that jacket-water temperature and reactor pressure do not exceed the limits.

In order to obtain more information about the process and to increase automatic control, a simulation study was made. This simulation revealed that for minimum batch time the process must operate at maximum pressure and reactor venting must be started at the start of the run. The simulation also revealed that the minimum run time was less than one-half of the manual operating time, this minimum time being limited by the vent condenser capacity. The simulation led to the installation of the pressure controller, selector relay, and multiplying relay, shown in Fig. 22-186, to control addition of Y. In this control scheme, the selector relay selects the lower of the

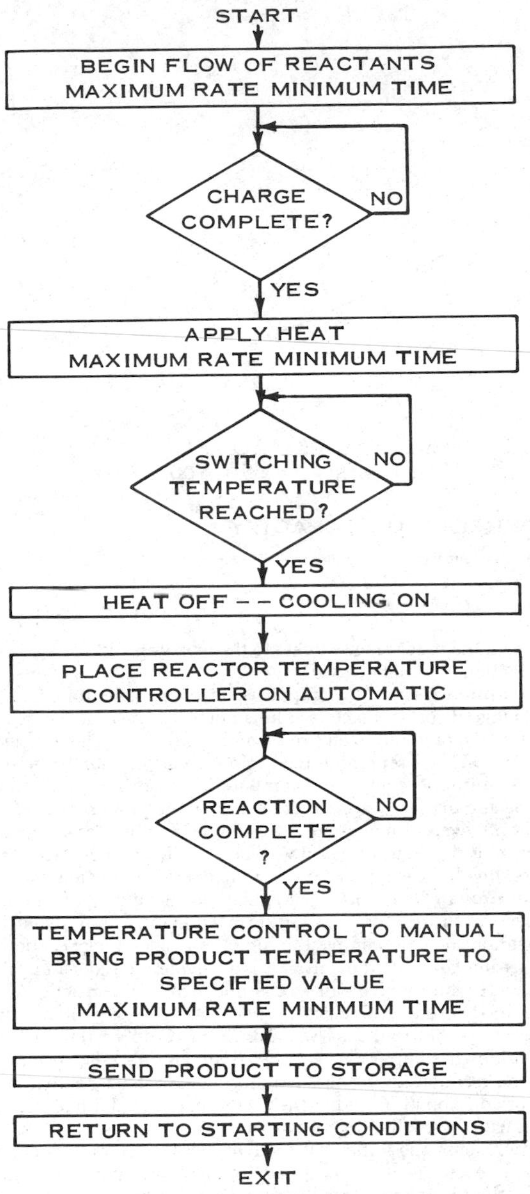

FIG. 22-185 Outline of a batch-reactor control program.

reactor pressure set point or the product of B times the jacket-temperature set point.

Predictive Control An analog computer, functioning as an online predictive controller, serves as another example of control techniques available for use on batch reactors. This example is a pilot-plant reactor producing a synthetic rubber by an exothermic reaction. [Adams and Schooley, *Instrum. Technol.*, **16,** 57–62 (1969).] Heat-transfer characteristics of the reaction mixture deteriorate as reaction proceeds and temperature variations produce large fluctuations in the reaction rate. Product properties are highly dependent on reaction-temperature history. Low or high monomer concentrations yield poor-quality product; additionally high monomer concentration can cause a rapid secondary reaction. The control problem is to make a batch of rubber within a small range of reaction temperature and stop the reaction at a predetermined conversion.

The control scheme used is illustrated in Fig. 22-187. The model of reaction kinetics and energy balance is operated in parallel with the real reactor, using measured process temperatures as inputs. In real time, the model calculates the current state of the reaction in terms of monomer conversion and catalyst activity. Periodically, the model is switched to fast-time operation to predict the future state of the reaction. On the basis of this prediction, the reactor-temperature error in the controller is biased by the future temperature, giving additional control action before the reaction moves out of desired limits.

Optimizing Control A batch hydrogenation reaction has become the classic laboratory example of dynamic optimization of batch-unit operation [Eckman and Lefkowitz, *Control Eng.*, **4,** 197–204 (1957)]. In this example the performance variable is best described by an integral involving quantities which vary with time. By utilizing information about a typical batch, an optimum path for reaction and control parameters can be determined. By using this as a guide, each batch is closely monitored with a computer. As the batch progresses, deviations from the optimum are used to predict and make control adjustments to keep the batch on a reaction path close to optimum.

In the simplest case such performance variables are of the form

$$I = \int_0^T f(\dot{x}, x, t)\, dt$$

where x is the process variable, $\dot{x}$ is dx/dt, t is time, and T is the time required for the completion of the batch. The problem is to find the variation of x with time which will maximize the performance variable I. This form of integral objective function can usually be employed to obtain a solution for the optimum path by the use of variational calculus. Extension of the problem to several variables and using lagrangian multipliers to handle the introduction of constraints permit use of this technique on real control problems. These techniques are useful when dynamics are essential in the optimization of a batch through computer control.

MAXIMIZING CONTINUOUS-UNIT-OPERATION PERFORMANCE

The control of the more complex continuous unit operations, such as distillation columns, may be divided into three levels: dynamic regulation, set-point control, and performance optimization. Dynamic regulation, or first-level control, includes the function of conventional controllers which manipulate valves to regulate such variables as temperatures, flows, levels, and pressures. Set-point or second-level control adjusts the set points of the first-level controllers to achieve the specified values for the performance variables or process outputs. The third level, unit-performance optimization, computes the specific process outputs which will result in optimum operation for the unit, that is, meet the performance specification with maximum profit. This hierarchical control structure is shown in Fig. 22-188.

In line with the previous discussion of computer control hierarchy, the information entering the performance-optimization block of Fig. 22-188 from the top comes from outside the unit-operations control system. It either is a higher-level control-system input to the unit-operation level or is a manual input. It should also be pointed out that both the dynamic regulation and set-point control levels may involve cascaded control loops.

As an example, consider a distillation column with predictive computer control which must accept whatever feed it receives. The set-point control computer maintains the specified product compositions: fraction light key component in the bottom product and fraction heavy key component in the distillate product at minimum boil-up rate. The third level of control must supply the product specifications which will result in maximum profit.

The equation for profit of a distillation column may be written as

$$P = DV_D + BV_B - FV_F - C$$

where P = profit rate, D = distillate rate, V_D = value of the distillate product, B = bottom-product rate, V_B = value of the bottom

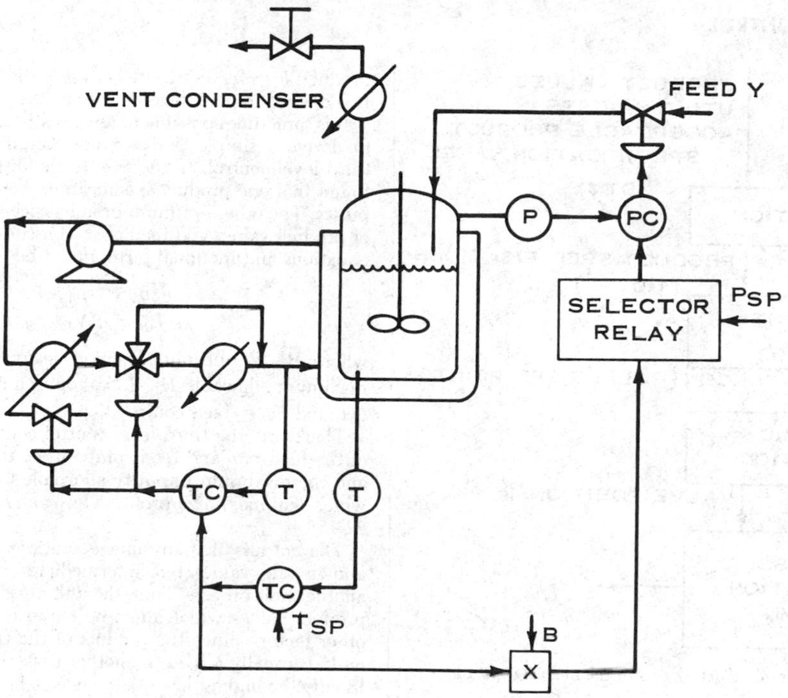

FIG. 22-186 Controlled addition of feed to a batch reactor to maintain reactor pressure and temperature within preset limits.

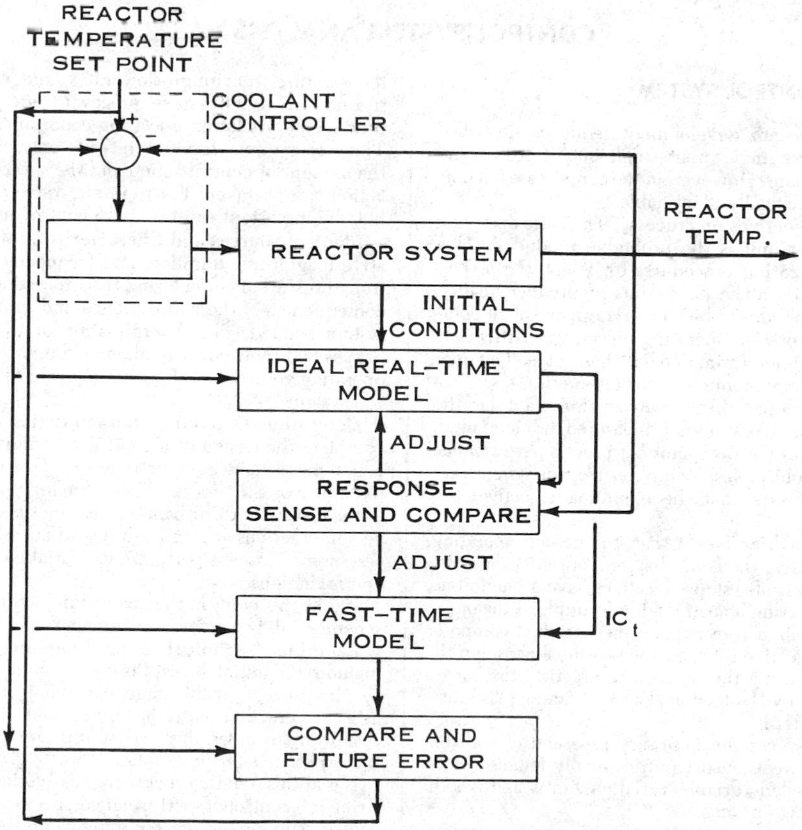

FIG. 22-187 Predictive-type analog-computer control of an exothermic-polymerization batch reactor.

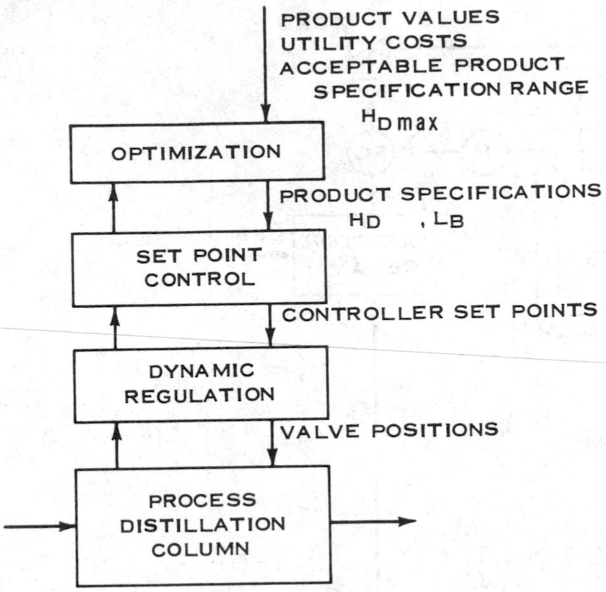

FIG. 22-188 Hierarchical control structure of a distillation unit operation.

product, F = feed rate, V_F = cost of the feed, and C = utility costs. It is desired to maximize the profit P.

It is sometimes possible to use a mathematical model of the process to derive a simple performance-optimization equation for use in third-level control. In the case of the distillation column, it is often found that one product specification should be held at its minimum purity. The other optimum product specification is a simple function of product values and feed characteristics. The third-level controller equations, in functional form, might be

$$H_D^\circ = H_{D\max}$$

$$L_B^\circ = f(V_D, V_B, F, F_C)$$

where H_D° = optimum fraction heavy key in distillate, $H_{D\max}$ = maximum allowable H_D, L_B° = optimum light key in bottom product, and F_C = feed composition.

Therefore, the third-level control computer needs measurements of the feed rate and feed composition, the specified product values, and the maximum impurity allowable in one product specifications which will maximize profits. This configuration is illustrated in Fig. 22-188.

The outputs of many unit operations are not final products with known sale values but intermediate process streams which feed another unit process. Thus, the unit must be optimized with assumed product values which must be realistic and which may vary with other factors. Since the product of the fractionator discussed previously is usually a feed to another unit, the second level of control is usually the highest level that needs to be operated on line. The third level requires infrequent adjustment and may be done off line.

CONTROL-SYSTEM ANALYSIS

DEVELOPMENT OF A CONTROL SYSTEM

The generalized control-system development sequence described here fits both the very simple single-input–single-output process and the computer control of a plant. However, in the simple case, all nine of the steps may not be specifically identifiable.

Establish Operating Objectives of Process The basic operating objective of any processing unit is the profitable production of a specified product. The operating procedures for which the process yields maximum profit, within the constraints of product quality, constitute the best set of operator's objectives. However, plant operators often cannot specify precise operating objectives, particularly economic ones. Thus, the determination of these objectives is a significant step toward development of a control system.

The establishment of the operating objectives also establishes the process performance variables. Whether measured or computed, these variables give a measure of the quality of process performance in meeting the operating objectives. Computer control systems are sometimes necessary to ensure that the operating objectives are followed.

Define Control-System Objectives After the process-operating objectives have been specified, the control-system objectives may be established. Ideally, the control-system objectives have a one-to-one correspondence to the operating objectives. For example, if the operating objective is to maintain a constant specified product composition, the control-system objective is to control, say, the flow-rate ratio between two materials entering the process. In practice, the correspondence may be limited by the technical and economic feasibility of the required control system.

Another objective is to meet the necessary precision of control. Specifying the necessary precision of control usually entails determining permissible controlled-variable variation limits as to both their value and their time variation.

Determine Constraints on the Process These constraints include such items as minimum product purity, maximum process temperature, maximum flow rates, and maximum pressure. Constraints are usually some processing-equipment operability limit, product quality limit, processing-condition limit, or safety constraint. These constraints impose limits not only on the process but also on the amount of control action and the speed with which such control action can be taken. For example, valve sizing and surge capacity may be important constraints on control action.

Identify Sources and Characteristics of Disturbances The relative magnitude, duration, and frequency of occurrence of uncontrolled disturbances entering the processing system and the process consequences of these disturbances indicate the nature of the control system needed. The determination of the nature of these disturbances and their effect on the performance variables along with the operating and control objectives is the basis for the synthesis of a control system.

Determine Dynamic Characteristics of Process In order to complete the design of a control system and to specify the control-action performance required, it is necessary to have some idea of the dynamics of the process. These dynamics include the time response of the process to disturbances, the time-dependent description of process interactions arising from disturbances, and the time-dependent response of process performance variables to both disturbances and control actions.

It is at this point in the analysis procedure that alternative control schemes will begin to become evident.

Examine Technical and Economic Feasibility This step includes the initial description of a complete proposed control system for the process. At this stage, each control loop should be evaluated as to its contribution to the control and operating objectives and its cost. In many cases, this analysis must be qualitative and be based on the experience of the evaluator.

It is at this step that such things as overinstrumentation, excessively stringent controller performance specifications, trade-offs between control and processing equipment, and overcompensation for the consequences of control-system reliability show up.

Consider Alternative Solutions For a given process, there will probably be a best solution from a technical point of view. However, there may be several designs which give adequate technical performance but improved economic performance when compared with the best technical solution.

Leads as to some of the alternatives to be considered will normally show up in some of the preceding steps. However, new concepts may need to be introduced at this stage.

One of the more difficult comparisons that needs to be made is reliability versus investment, operating, and maintenance costs between various manufacturers of both processing and control equipment. The provisions which must be made for maintenance procedures are currently more widely understood for processing equipment than for control equipment. However, both types of equipment are integral parts of the processing system.

Through this point the development of a control system is an iterative process to arrive at the desired solution. However, in the interests of costs, this iteration cannot be unrestrained. In simple cases, the iteration will not be apparent.

Finalize Control-System Design The final stages of development of a control system appear obvious but are important. The finalization of the design puts the results of the preceding steps into a complete system which can be implemented. From the viewpoint of a process operator, the design is not complete until the tuning constants of controllers are specified. This means that in some cases the design is not entirely complete until the control system is running.

With the exception of the adjustments required to make the control system handle the process, the final design permits initiation of equipment ordering, construction planning, operator-training planning, computer programming, and other activities required to get the system installed.

Complete Control-System Development The last stages of the development of the control system include construction, checkout, and adjustment; operator training and education; and complete documentation. These final stages must be carried out at an appropriate time, since the economic benefits of the control system are not available until the system is operating and operating personnel have sufficient confidence to make maximum use of its capabilities.

TOOLS FOR CONTROL-SYSTEMS ANALYSIS

Some of the basic tools and techniques for control-systems analysis have been covered under "Fundamentals of Automatic Control: Controller Tuning; Additional Topics on Control." Other tools and techniques which may be used for systems more complex than the single-input–single-output process systems are discussed in the following subsections. The particular tools used will vary with their availability; availability of experienced personnel; and the stage of development of the process, that is, new process emerging from pilot plant, new plant using a proven process, or modernization of an existing process.

Frequency-Response Analysis Frequency-response analysis is a method for investigating the characteristics of a closed-loop process on the basis of its open-loop response to sinusoidal inputs. Given a transfer function or experimental frequency-response data, the method allows the estimation of controller settings and predicts the limits of stability of the closed-loop system. See "Controller Tuning" with reference to Figs. 22-41 through 22-44.

The frequency response of a process can be defined as the relationship in amplitude and phase between a process output and a sinusoidal input after any transients in the response have disappeared. Mathematically, the frequency response is the particular solution of the process-systems differential equations for a sinusoidal forcing function. The stipulation that the transients must die out requires that the system be stable.

Since frequency-response analysis generally makes extensive use of Laplace-transform methods, the system to be analyzed must also be linear or such that it can be approximated by a linear system.

Bode Diagram Frequency-response data are most conveniently displayed on a Bode diagram.

Amplitude data are often given in decibels (dB), where dB = 20 $\log_{10} AR$. ω is usually given in radians per second and ϕ in degrees. See Fig. 22-41.

It can be shown that when a transfer function $G(s)$ is available, amplitude and phase data for a Bode diagram can be obtained by substituting $s = j\omega$, where $j = \sqrt{-1}$, and writing $G(j\omega)$ in polar form.

$$G(j\omega) = |G(j\omega)|\, e^{j\angle G(j\omega)}$$
$$AR = |G(j\omega)|$$
$$\phi = \angle G(j\omega)$$

Values of AR and ϕ can be computed for various values of ω to obtain the necessary information, but if $G(s)$ is not a simple function, these calculations can be quite tedious. An approximate graphical solution is quite easy and is often sufficiently accurate to be used in place of the true frequency-response curves.

For a first-order system $G(s) = 1/(Ts + 1)$. Substituting $s = j\omega$ and separating into real and imaginary parts,

$$G(j\omega) = \frac{1}{1 + (T\omega)^2} - j\frac{T\omega}{1 + (T\omega)^2}$$

Using the right-triangle rule to determine magnitude and phase (Fig. 22-189),

$$AR = |G| = \frac{1}{\sqrt{1 + (T\omega)^2}} \qquad \text{and} \qquad \phi = \tan^{-1}(-T\omega)$$

See Table 22-4. $\log AR = -\frac{1}{2}\log[1 + (T\omega)^2]$. Therefore, as $T\omega \to 0$, $AR \to 1$; so the low-frequency asymptote of the amplitude plot is a horizontal line at $AR = 1$. See Fig. 22-42.

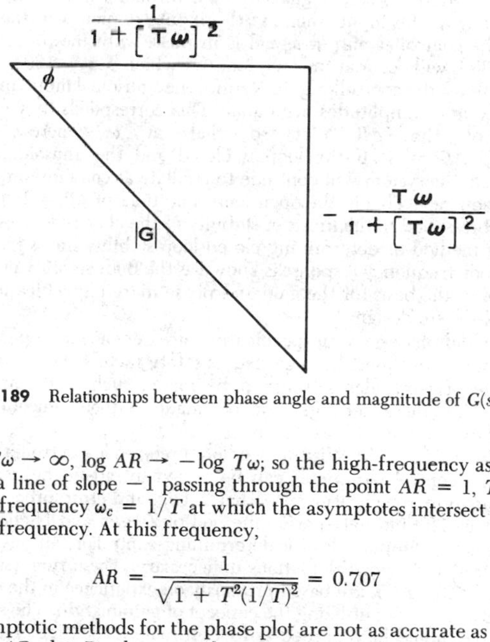

FIG. 22-189 Relationships between phase angle and magnitude of $G(s)$.

As $T\omega \to \infty$, $\log AR \to -\log T\omega$; so the high-frequency asymptote is a line of slope -1 passing through the point $AR = 1$, $T\omega = 1$. The frequency $\omega_c = 1/T$ at which the asymptotes intersect is the corner frequency. At this frequency,

$$AR = \frac{1}{\sqrt{1 + T^2(1/T)^2}} = 0.707$$

Asymptotic methods for the phase plot are not as accurate as those for the AR plot. By observing that $\phi \to 0$ at low frequencies and $\phi \to -90°$ at high frequencies and that the curve is symmetric about ω_c, the following values can be used to approximate the curve: $\omega = 1/T$, $1/2T$, and $1/10T$, and $\phi = -45$, -27, and $-6°$, respectively.

For a second-order system, as shown in Table 22-4, there is an additional parameter ξ. Thus, there is a different Bode plot for each value of ξ as noted in Fig. 22-41.

At high frequencies $\log AR \to -2 \log T\omega$, and the high-frequency AR asymptote for the second-order system has a slope of -2. The peak in the AR curve near $T\omega = 1$ is due to the phenomenon of resonance that can occur in underdamped systems. Note also that at high frequencies $\phi \to -180°$ and that at $T\omega = 1$, $\phi_c = -90°$.

When a second-order system is overdamped, $\xi > 1.0$, as it is normally in chemical processes, it can be represented as two real first-order systems in series. Since multiplication of complex numbers is effected by multiplying the magnitude and adding the phase angles, the Bode diagram for the overall system can be obtained by graphically adding the curves for the two first-order systems.

A transportation delay contributes nothing to the amplitude ratio but introduces a phase lag which increases without bound (see Table 22-4).

True integration $G(s) = 1/T_i s$ and differentiation $G(s) = T_d s$ exhibit frequency responses as shown by extensions of the straight-line portions of Fig. 22-44. Thus, an ideal mixed-mode controller, e.g., proportional-integral-derivative with transfer function

$$G_c(s) = K_c\left(1 + \frac{1}{T_i s}\right)(1 + T_d s)$$

will yield the Bode plot which would be a composite of the solid-line curves of Fig. 22-44.

An actual controller has limiting effects which cause the true frequency response to deviate from the ideal at both low and high frequencies. A typical real controller might have a transfer function of the form

$$G_c(s) = K_c \frac{T_i s + 1}{T_i s + 1/K_i} \frac{T_d s + 1}{T_d s/K_d + 1}$$

These data are usually furnished by the controller manufacturer.

Bode Stability Criterion Consider a control loop opened between the controller and the control valve, with a low-frequency sinusoidal signal applied to the valve. After the transient response has disappeared, the process signals to the controller will be of the same frequency as the input. Increase the input frequency to the valve until the controller output signal is in phase with the input to the valve. This will occur at the frequency at which $\phi = -180°$. At this point, adjust the controller gain K_c until the sinusoidal input and controller output amplitudes are equal. This corresponds to a vertical shift of the AR plot so that at ω where $\phi = -180°$, $AR = 1$. If the loop is closed and the sinusoidal input removed, the system will continue to oscillate at constant amplitude and frequency. That is, the open-loop conditions of $AR = 1$, $\phi = -180°$ correspond to the limits of stability of the closed-loop response.

This method of determining closed-loop stability limits from the open-loop frequency response is known as the Bode stability criterion and forms the basis for the gain and phase-margin specifications for control-system design.

Gain and phase-margin specifications are dependent on particular applications and must be regarded as safety factors. They allow an estimate of controller settings to be made, such that reasonable changes in these settings can be made without encountering instability.

Ultimate Period With a two- or three-mode controller, more than one control parameter can be varied to satisfy the gain and phase-margin specifications, so that a trial-and-error procedure is indicated. The ultimate-period method of Ziegler and Nichols prescribes a few simple rules for determining controller settings which make trial-and-error calculations unnecessary. These rules, summarized in Table 22-5, are based primarily on experience in the control of typical processes utilizing the concept of gain margin. The settings they give should be regarded as first estimates. The Ziegler-Nichols controller settings are determined from the open-loop system response. The ultimate gain K_u is defined as the gain margin, i.e., the reciprocal of AR at $\phi = -180°$. The ultimate period P_u is defined as $2\pi/\omega_u$, where ω_u is the frequency in radians per minute at $\phi = -180°$.

Controller settings for the proportional-derivative mode cannot be estimated by this method. A procedure commonly used in this case is to select T_d such that K_c is maximized when the phase margin is on specification, thus minimizing offset.

Pulse Testing Frequency-response and reaction-curve testing times are usually long compared with process response times. Pulse tests are executed by changing the open-loop process input from a steady-state value to some "appropriate" value for a "proper" length of time, and then returning the input to the steady-state value while observing the output response. Thus, the duration of tests is normally of the same order as process response times, and the process output is returned to its initial steady-state value at the end of the test.

Theoretically, the pulse test is executed once. The Fourier transform of a pulse, which contains energy at many frequencies, shows that this should be sufficient. However, selection of the appropriate magnitude and the proper pulse duration is important because a poor choice of either may necessitate additional tests. This method requires substitution of input pulse $x(t)$ and output response $y(t)$ into

$$G(j\omega) = \left[\int_0^\infty e^{-j\omega t} y(t)\,dt\right] \Big/ \left[\int_0^\infty e^{-j\omega t} x(t)\,dt\right]$$

and integrating for a range of frequency ω. Each integration (usually numerical) provides one point for the "gain versus frequency" and "phase shift versus frequency" plots. The main disadvantages of the dynamic pulse-testing procedure are the voluminous data storage, manipulation, and computation that are needed; its principal asset is the small number of test runs needed for a study.

An improvement of the pulse test is to time the removal of the pulse input and return to the initial steady-state value when the first derivative of the process response is a maximum. [Pemberton, "An Improved Pulse Testing Method," *Instrum. Technol.*, 17, 61–67 (December 1970).] In the Laplace notation, a second-order plus dead-time process response to a pulsed input is represented by

$$C(s) = \frac{KX(s)\,e^{-Ls}}{T_1 T_2 s^2 + (T_1 + T_2)s + 1}$$

where $X(s)$ is the input pulse. In the time domain, this is the differential equation

$$T_1 T_2 \frac{d^2c(t)}{dt^2} + T_1 T_2 \frac{dc(t)}{dt} + c(t) = Kx(t - L)$$

The input pulse, amplitude A, illustrated in Fig. 22-190 is described by

$$x(t) = A[U(t) - U(t - t_1)]$$

where $U(t) = \begin{cases} 0 & t \le 0 \\ 1 & t > 0 \end{cases}$ and $U(t - t_1) = \begin{cases} 0 & t \le t_1 \\ 1 & t > t_1 \end{cases}$

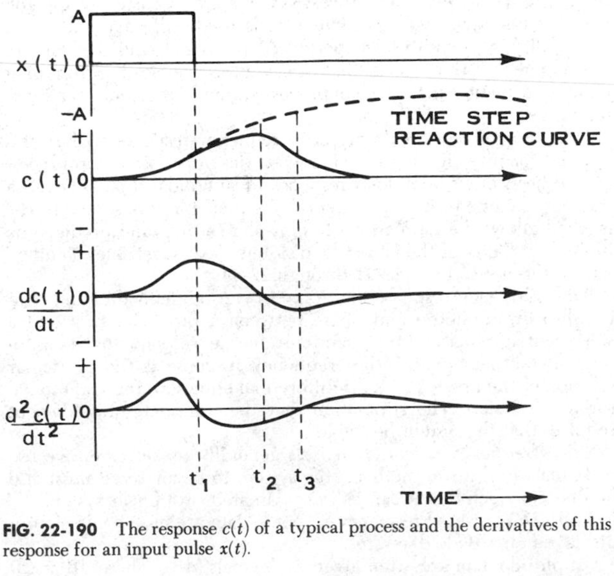

FIG. 22-190 The response $c(t)$ of a typical process and the derivatives of this response for an input pulse $x(t)$.

From the differential equation and the expression for $x(t)$, one can calculate the process characteristics. That is,

$$K = \frac{1}{A}\,\frac{c(t_1)}{[1 - c(t_3)/c(t_2)]}$$

$$L = t_1 + t_2 - t_3$$

$$T_2 \cong \left[0.925\,\frac{KA}{c(t_1)} - 2.3\right](t_3 - t_2)$$

$$T_1 \cong \left[\frac{3c(t_1)}{KA}\right]^3 T_2$$

where t_1 = time at which $dc(t)/dt$ is a maximum
t_2 = time at which $c(t)$ is a maximum
t_3 = time at which $dc(t)/dt$ is a minimum

Similar expressions are available for a two-sided pulse when a pulse of amplitude $-A$ immediately follows the initial pulse of amplitude A, the second pulse being terminated at t_3 when $dc(t)/dt$ is a minimum.

Analysis by Reduction in Degrees of Freedom An illustration of a typical analysis of a unit operation by degrees of freedom (independent variables) indicates how degrees of freedom are reduced by controls based on performance variables.

Chemical Reactor Consider a reactor used to carry out an exothermic reaction when the feed consists of reactant, diluent, and catalyst. Such a reactor, illustrated schematically in Fig. 22-191, is a simplification of an actual process put under computer control.

The major simplifications are involved in the following statements of process characteristics. Two major properties of the product are determined by the temperature of the reactor and the concentration of reactant in the reactor. The catalyst leaves the process with the product, and one specification of product acceptability is a maximun concentration of catalyst in product. Productivity of the catalyst (product-catalyst ratio) is affected by concentration of product in the reactor. The feed streams are added continuously to the reactor. Two automatic process analyzers are used to obtain the reactant concentration in the reactor effluent and the density of the reactor fluid.

The controls shown in Fig. 22-191 represent conventional automatic control of such reactors. The flow rate of cooling water is controlled on a volumetric basis. The inlet cooling-water temperature is controlled by regulating the volumetric split of cooling water to two parallel heat exchangers, one heating and one cooling, as in Fig. 22-

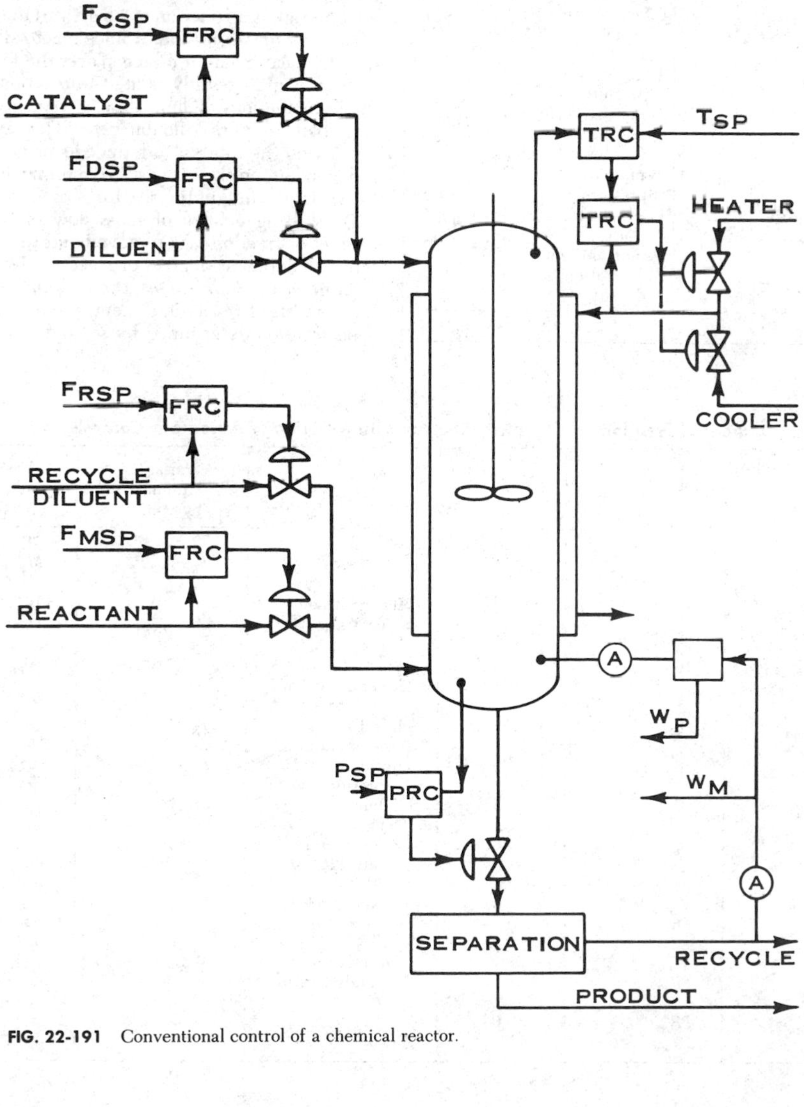

FIG. 22-191 Conventional control of a chemical reactor.

48. Reactor temperature is controlled by cascading to manipulate the set point of the coolant inlet-temperature controller. Reactant, catalyst, and diluent are on volumetric rate-of-flow control. Reactor pressure is controlled by manipulating the rate of discharge of reaction mixture from the reactor.

Variables The performance variables for this reactor are the production rate of product and its properties. These performance variables are functions of the independent variables (degrees of freedom) shown in Table 22-14. Of the 18 degrees of freedom, 6 are controller set points specified by the operator and 12 are free to wander uncontrolled. A change in any one of the uncontrolled variables will change at least one performance variable.

TABLE 22-14 List of Independent Variables or Degrees of Freedom for Chemical Reactor

	Manipulated variables		Uncontrolled variables	
Source	Controller set points	Symbol	Variable	Symbol
Catalyst	Flow rate	F_{CSP}	Catalyst feed concentration	E_C
Catalyst poisons	. . . .	. . . . . .	Effective rate of poisons addition	C_P
Fresh diluent . .	Flow rate	F_{DSP}	Temperature, pressure	T_D, P_D
Recycle diluent.	Flow rate	F_{RSP}	Temperature, pressure, composition	T_R, P_R, C_R
Reactant	Flow rate	F_{MSP}	Temperature, pressure	T_M, P_M
Reactor.	Temperature, pressure	T_{SP}, P_{SP}	Product concentration, reactant concentration, density	W_P, W_M, ρ
Total	6		12	

The primary sources of disturbances are those which affect the effective concentration of catalyst in the reactor and those which affect the reaction-mixture composition and heat balance. Those disturbances which affect the effective concentration of catalyst include variations in catalyst poisons and variations in catalyst feed. Catalyst poisons are present in all feed streams.

Those disturbances which affect reaction-mixture composition and heat balance are disturbances which affect mass flow of feed (these include variations in volumetric flow, feed temperatures, reactor pressure, and feed pressures), the variations in the amount of reactant in the recycle diluent, and variations in catalyst concentration.

The conventional controls indicated in Fig. 22-191 require nothing special in the way of a control algorithm; that is, direct measurements can be made of the variables being regulated, and conventional three-mode controllers suffice. In the computer control system computed variables and some special control algorithms are used.

The performance variables may be written as functions of the degrees of freedom listed in Table 22-14. The performance variables are production rate PR, a temperature-dependent product property, product index I, and productivity of catalyst PY. Table 22-15 illustrates the stepwise reduction in degrees of freedom as controls are added. In Table 22-15 the individual independent variables are not listed for performance variables if the degrees of freedom number more than three; only the total number is listed. For example, production rate is a function of all 18 variables.

Control of Mass Flow Since reactor-fluid composition regulates the chemical reaction taking place, the mass flow of the feed components is most readily related to reaction conditions. Uncontrolled small fluctuations of fluid flows carrying catalyst poisons introduce disturbances in production rate. This arises because poisons act quickly on the catalyst to reduce its productivity. High productivity is dependent on the catalyst's remaining active for a large portion of its residence time in the reactor.

By shifting control of mass flow of all the feed streams, nine degrees of freedom are removed and the three respective mass flow rates are added as degrees of freedom, for a reduction of six.

Control of Reaction-Mixture Composition The reaction mixture consists of reactant, diluent, product, and catalyst. The reactant concentration determined by solving a dynamic mass balance (dif-

TABLE 22-15 Reduction in Degrees of Freedom of Chemical Reactor with Addition of Automatic Controls

Performance variables as functions of the independent variables			Degrees of freedom	Control added	Variables eliminated by control	Variables added by control	Degrees of freedom removed
PR	I	PY					
18	6	18	18	Mass flow of feeds M	$T_D, F_{DSP}, P_D, T_R, F_{RSP}, P_R, T_M, F_{MSP}, P_M$	$M_{DSP}, M_{RSP}, M_{MSP}$	6
12	4	12	12	Reactant concentration in reactor	M_{MSP}, W_M, C_R	W_{MSP}	2
10	W_{MSP}	10	10	Product concentration in reactor	M_{RSP}, W_P	W_{PSP}	1
9	W_{MSP}, T_{SP}, P_{SP}	9	9	Reactor temperature	T_{SP}	. . . .	1
8	W_{MSP}, P_{SP}	8	8	Reactor pressure	P_{SP}	. . . .	1
7	W_{MSP}	7	7	Diluent flow	M_{DSP}	. . . .	1
6	W_{MSP}	6	6	Production rate PR	E_C, C_P, F_{CSP}, W_C	PR_{SP}, not degree of freedom	4
$W_{PSP}, W_{MSP}, PR_{SP}$	W_{MSP}	$W_{PSP}, W_{MSP}, PR_{SP}$	2	Product index I	W_{MSP}	I_{SP}, not degree of freedom	1
W_{PSP}, PR_{SP}	I_{SP}	W_{PSP}, PR_{SP}	1	Product-to-catalyst ratio PY	W_{PSP}	PY_{SP}, not degree of freedom	1
PR_{SP}, PY_{SP}	I_{SP}	PR_{SP}, PY_{SP}	0				

ferential equation) is the measured variable used for automatic control. The computed reactant concentration compared against a set point W_{MSP} generates a control signal that resets the reactant mass flow rate for a reduction in degrees of freedom by one. The concentration is also determined by the automatic analyzer sampling the reactor effluent. Since the reactant concentration can move outside the operating range before analyzer results are available, the analyzer results are used to update the reactant-concentration calculation automatically.

The product concentration in the reactor can be controlled by manipulating the recycle-diluent flow rate. The computed product concentration is based in part on an automatic density measurement. It is compared with the concentration set point to generate a control signal that manipulates the recycle flow, thus reducing the degrees of freedom by two. At this point, the concentrations of product and reactant in the reactor are on set-point control.

Reactor temperature, pressure, and fresh diluent-flow set points are determined by product properties determined in the laboratory, reactor operability, and recycle surge capacity. These criteria for set-point determination are dependent upon information supplied at least in part by sources outside the immediate reactor environment. Therefore, they are put under manual set-point control.

Control of Performance Variables In the preceding paragraphs the degrees of freedom have been reduced (thus making the system simpler to operate), and variables that are more fundamental have become the set-point variables. For further reduction of degrees of freedom, control considerations are shifted to the performance variables.

The production rate, determined by solving a dynamic heat balance, is a computed variable which when controlled eliminates four degrees of freedom associated with the catalyst performance. PR_{SP}, the set point for the performance variable, is normally determined by conditions external to the unit operation, i.e., sales and inventory. Therefore, it is set manually. An example of this type of control is discussed in connection with Fig. 22-176.

In this process the production rate is a function of a number of the variables eliminated in the tabulated listings in the rows of Table 22-15 above the row where PR_{SP} is introduced. However, catalyst-concentration manipulation will produce the desired production rate. Therefore, catalyst flow rate is chosen as the manipulated variable.

The remaining two degrees of freedom can be eliminated by making them manual set points. However, if sufficiently good process models are available, these two variables can be calculated. They would then become controllable variables whose set points could be determined from steady-state performance as indicated by periodic laboratory analyses.

The concentration of the reactant set point W_{MSP} is removed as a degree of freedom by designating it as an operator-adjusted set point based on operating results reported by the control laboratory. In this reaction system manipulation of the reactant concentrations in the reactor may be used as the primary online control of the product index. Adding a product-index controller would remove W_{MSP} as a degree of freedom and replace it by the product-index set point I_{SP}, a performance variable.

The last remaining degree of freedom W_{PSP} is eliminated by being set by the operator to a fixed value determined by operation results when productivity is above minimum specification. When productivity drops below specification, W_{PSP} can be eliminated by introducing a productivity controller which manipulates the product-concentration and production-rate set points.

Mathematical Models In the analysis and development of control systems a variety of mathematical process models and control equations may be used. Usually the term "model" is used to cover all kinds of process models, descriptions of controllers, and control equations. The extent of the variety of models used is indicated in this *Handbook* section in "Control Equations and Models" under "Computer Process Control," in "Additional Topics on Control" under "Fundamentals of Automatic Control," and in the examples cited under "Unit-Operations Control." Techniques for obtaining some of the simpler empirical process models are indicated in "Controller Tuning" under "Fundamentals of Automatic Control."

The accuracy and complexity of the models needed in developing a control system depend upon the processing objectives, control objectives, and degree of automation required to enforce these objectives. The models can be developed from theoretical considerations or empirical expressions based on experimental results. Usually they are a combination of theoretical and empirical results. Mathematical models range from a single algebraic equation to sets of simultaneous partial differential equations.

The more complex process models may be reduced in complexity by making use of the processing and control objectives to eliminate some of the variables. This is a judgment matter and must be done realistically, particularly if the reduction in variables depends upon control in some other part of the plant not directly under the supervision of personnel operating the portion of the plant being analyzed. Sometimes a change in process-equipment specifications such as minimum insulation thickness or maximum length of piping between vessel and measurement point will reduce the significance of some variables. Other techniques of reducing the complexity of a process model involve the reduction of degrees of freedom discussed previously and the decomposition of the processing system into smaller units as illustrated under "Computer Process Control of Plants" under "Computer Process Control."

As an example of developing a mathematical model, consider the stirred-tank reactor shown in Fig. 22-192, in which a mixture of materials A and B enters the reactor. Under the conditions existing in the reactor, A is transformed into B with a reaction-rate constant k, $A \xrightarrow{k} B$. The processing objective is to produce a constant ratio of B to A in the effluent, taking the feed at whatever rate it is produced by some previous processing step.

The mathematical-model development uses the basic concepts of mass and energy balances described by

$$\begin{pmatrix}\text{Accumulation}\\ \text{in reactor}\end{pmatrix} = \begin{pmatrix}\text{transfer}\\ \text{into reactor}\end{pmatrix} - \begin{pmatrix}\text{transfer out}\\ \text{of reactor}\end{pmatrix} + \begin{pmatrix}\text{generation}\\ \text{in reactor}\end{pmatrix}$$

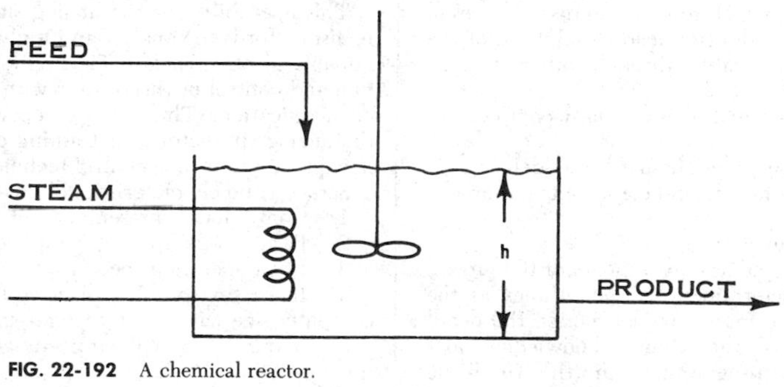

FIG. 22-192 A chemical reactor.

For Fig. 22-192, the overall material balance is

$$(dh/dt(a\rho) = F_1 - F_2$$

The material balance on material A is

$$(dx_A/dt)(a\rho h) = F_1x_{A1} - F_2x_A - kx_Aa\rho h$$

The heat balance is

$$(dT/dt)(a\rho hC) = F_1C_1(T_1 - T_{\text{ref}}) - F_2C(T - T_{\text{ref}}) + kx_A\Delta H + a_sH_s(T_s - T)$$

Other algebraic relationships which apply are

$$x_A + x_B = x_{A1} + x_{B1} = 1$$

$$k = k_0e^{-EA/RT}$$

$$H_s = f(\rho, C, T)$$

In these equations a is the cross-sectional area of the reactor; ρ the density, F the mass flow into or out of the reactor, h the level of fluid in the reactor, x_A and x_B the weight fraction of components A and B, T the temperature, and C the specific heat. The subscript 1 refers to the feed material and subscript 2 to the effluent. The steam temperature is T_s, the heat-exchange area a_s, and the transfer coefficient H_s. ΔH is the heat of reaction for $A \rightarrow B$. k_0 is the Arrhenius frequency factor and E_A the energy of activation (see Sec. 4).

These equations form the mathematical model of the process. Three obvious assumptions made to reduce the number of equations were that the reaction vessel was perfectly mixed and the heat loss to surroundings and heat generated by the mixer were negligible. It is important that the assumptions made are known so that their effect on the anticipated control-system performance may be taken into account.

The mathematical model normally consists of ordinary differential equations and algebraic expressions. The differential equations may be partial differential equations, and the algebraic expressions may be simultaneous nonlinear equations.

Inspection of the differential equations in the model will suggest control strategies to be studied. For example, if temperature of the reactor is to be controlled at a fixed value, then the residence time and in turn the level must be controlled in such a way that x_A/x_B in the effluent product meets the quality standard. The knowledge about the process gained in the formulation of mathematical models will often lead to better synthesis of solutions to the overlapping process modification and control-system problems.

Simulation Electronic computers have become valuable tools for control-system analysis through simulation. A process with its control system is simulated by programming an electronic computer so that it contains a mathematical model of the process and control system. This mathematical model responds to disturbances and adjustments or modifications in the same way as the real process. However, the responses may be observed as automatically plotted process or control variables, digital-computer-data printouts, or cathode-ray-tube displays in a laboratory environment rather than in a plant environment.

Simulation of a process and its control system is divided into two parts: the preparation of the model and the use of the model to study control of the process. Knowledge about the process developed in the course of preparation of a model, which responds to disturbances in the same manner as the process, will often lead to solutions of the control problems not readily available through other analysis techniques.

The development of a process simulation is normally composed of four basic steps:

1. Define the process and derive a mathematical model.
2. Determine the model scale factors and constraints required.
3. Program the computer.
4. Check out and run the simulation.

The simulation must be considered as only a model of the process it represents with limitations determined by such things as the assumptions made to simplify the mathematical model, the detail with which the process is described, the extent of knowledge about the process, the experience of the programmer, and the size of the computer.

The use of simulation in control-system analysis is normally a trial-and-error procedure. Results from other techniques of control-system analysis may assist in the analysis by providing a starting point or guidelines for the simulation analysis. The simulation study proceeds by implementing different control methods, making simulation runs testing these methods, and evaluating the relative performance of the methods. In some cases simulation is used for confirmation or demonstration of control techniques developed by other methods.

The usefulness of simulation is increased by time scaling. Depending on the problem requirements, simulation can be operated at speeds many times faster or slower than the actual process. Thus no restriction is imposed on the application of simulation to processes that operate at extreme speeds, either fast or slow.

Control-system analysis by simulation is characterized by the ease with which runs are made and parameters can be varied. Simulation also has the capability of computing variables that may not be accessible as direct measurements in the real process, for example, the reaction rate in a chemical process. Simulation of a computer control system would permit experimental testing of such computed variables for control along with various control algorithms, controller configurations, and controller tunings. The course of the simulation study depends on what control criteria are to be met.

The use of simulation can reduce the background in control theory required of the analyst, albeit at the expense of an increase in the cases studied by simulation.

Computerized process simulation allows one to study the control of the process in an online manner, that is, to conduct experiments with the process in much the same manner as they would be conducted on the real process. Some advantages of simulation over plant experimentation are that the economic and physical risks involved in manipulating the real process are circumvented, the time scale may be reduced or increased from the actual process depending on the needs of the experimenter, processing-condition excursions outside the normal operating range may be made, and modification of the process by programming changes rather than modifying process equipment permits studying the interaction of control and process design.

The simulation laboratory may contain analog computers, a digital computer, or an integrated hybrid analog-digital computer system.

In evaluating the relative merits of the different types of computers for simulation, the following factors need to be considered:

1. The run time of different computers to achieve a given simulation and the resulting cost
2. The time of programming and reprogramming for initial implementation and later (online or off-line) modifications of the program
3. The convenience and importance of person-machine interaction between the operator and the computer

Analog Simulation The electronic analog computer has become a valuable tool for control-system analysis through simulation (see also Sec. 2 of this *Handbook*). A process with its control system is simulated on an analog computer by wiring the computer components into an electrical network which represents the mathematical models of the process and control system. An analog process simulation is an operable electronic process which can be operated in a manner similar to the real process.

This operability of the analog simulation is an asset in that the realism affords the analyst an insight into the process which is unattainable by other means. The analyst has a sense of hands-on operation and control of the process with instant response to adjustments or modifications. This hands-on operation can be extended to utilize the analog simulator as a training device for process operators for new processes or new control techniques.

Some of the characteristics of analog-computer simulation are:

1. Continuous representation of variables
2. Parallel operation of components
3. High operation speed
4. Increasing problem size requiring a proportional increase in computer size without an increase in solution time
5. Accuracy and dynamic range limited by physical-measurement capability
6. Continuous integration

7. Difficulty in handling and storing discrete data
8. Solution of individual simulation case progressing at a rate proportional to the time scale of the system being studied
9. Ease in including components of actual control system in simulation
10. Programming by hand patching of components

An analog simulation is often easier and faster to implement than a digital simulation. Analog simulation is particularly useful in process-control studies in which interactions or nonlinearities in the process make an analytical study difficult. Since the maximum simulation size is limited by the number of computer components available, the detail with which the simulation may be made is limited. This may lead to approximations which limit the accuracy of the analysis.

The analog computer is equipped with a number of linear devices such as potentiometers, operational amplifiers, summers, and integrators. It normally has a lesser number of nonlinear devices such as multipliers, function generators, and comparators. The function generators are of both fixed and variable types. The variable types permit generation of almost any desired function.

Analog computers are available in a wide range of sizes. The smallest-sized commercially available units are useful for simulation of a simple, single control loop. The largest installations can handle models of rather complex processing units in their entirety.

The development of the wiring diagram for a simulation involves a number of techniques for translating the mathematical model into electric circuitry. If the operator is experienced in analog simulation, the mathematical model may not be explicitly formulated in conventional mathematical terms.

To illustrate the programming, consider the process model

$$(d^2y/dt^2) + a_1(dy/dt) + a_2y = kx$$

This model represents a second-order process whose input and output are x and y, respectively, and whose transfer function is

$$G(s) = k/(s^2 + a_1s + a_2)$$

Transformation of the model into *state variables* $y_1 = y$ and $y_2 = dy/dt$ gives two simultaneous first-order differential equations,

$$\begin{aligned} dy_1/dt &= y_2 \\ dy_2/dt &= kx - a_1y_2 - a_2y_1 \end{aligned} \qquad (22\text{-}41)$$

Process measurements usually have units of flow rates, composition, degrees, etc., whereas the output of the analog simulation is in volts. A typical range is -10 to $+10$ V. By defining reduced variables whose range is -1 to $+1$, corresponding to -10 and $+10$ V, the units conversion is simplified. In Eqs. (22-41) to determine the maximum and minimum absolute values of x, y_1, and y_2, define reduced variables $\bar{x}$, $\bar{y}_1$, and $\bar{y}_2$ in the form

$$\bar{x} = \frac{x - |x_{\min}|}{|x_{\max}| - |x_{\min}|}$$

Solving these reduced variable expressions for x, y_1, and y_2 and substituting in Eqs. (22-35)

$$\frac{d\bar{y}_1}{dt} = \frac{|y_{2\min}| + (|y_{2\max}| - |y_{2\min}|)}{|y_{1\max}| - |y_{1\min}|}$$

$$\frac{d\bar{y}_2}{dt} = \frac{k|x_{\min}| - a_1|y_{2\min}| - a_2|y_{1m}}{|y_{2\max}| -}$$

$$- \frac{a_2(|y_{1\max}| - |y_{1\min}|}{|y_{2\max}| - |y_{2\min}|} \bar{y}_2 - a$$

Since there is normally a difference in the time scale at which the process operates and at which the simulator operates, *time scaling* is accomplished by defining simulator time in seconds $\bar{t}$ in terms of process time t. $\bar{t} = \beta t$. Thus

$$\frac{d\bar{y}_1}{d\bar{t}} = \frac{1}{\beta}\frac{d\bar{y}_1}{dt} \quad \text{and} \quad \frac{d\bar{y}_2}{d\bar{t}} = \frac{1}{\beta}\frac{d\bar{y}_1}{dt}$$

The magnitudes of the constant coefficients in the expressions for $d\bar{y}_1/dt$ and $d\bar{y}_2/dt$ often indicate an appropriate value for β. β is usually chosen so that potentiometer settings are larger than 10 percent of scale and amplifier gains are less than 10.

The integral equations that need to be solved are

$$\bar{y}_1(\bar{t}) = \bar{y}_1(0) + \int_0^{\bar{t}} \frac{1}{\beta}\frac{d\bar{y}}{dt}\, d\bar{t}$$

$$\bar{y}_2(\bar{t}) = \bar{y}_2(0) + \int_0^{\bar{t}} \frac{1}{\beta}\frac{d\bar{y}_2}{dt}\, d\bar{t}$$

The solution may be obtained by the analog-computer circuit illustrated in Fig. 22-193. In this figure the potentiometer settings are

$$P1 = \bar{y}_2(0) \text{ the initial condition for } \bar{y}_2$$

$$P2 = \frac{|x_{\min}| - a_1|y_{2\min}| - a_2|y_{1\min}|}{\beta(|y_{2\max}| - |y_{2\min}|)}$$

$$P3 = \frac{|x_{\max}| - |x_{\min}|}{\beta(|y_{2\max}| - |y_{2\min}|)}$$

$$P4 = \frac{a_2}{\beta}\frac{|y_{1\max}| - |y_{1\min}|}{|y_{2\max}| - |y_{2\min}|}$$

$$P5 = a_1/\beta$$

$$P6 = \bar{y}_1(0) \text{ the initial condition for } \bar{y}$$

$$P7 = \frac{|y_{2\min}|}{\beta(|y_{1\max}| - |y_{1\min}|)}$$

$$P8 = \frac{|y_{2\max}| - |y_{2\min}|}{\beta(|y_{1\max}| - |y_{1\min}|)}$$

Digital Simulation Digital computers are capable of performing arithmetic and logic operations at very high speeds and of storing many programs and large amounts of data (see also Sec. 2 of this *Handbook*). They are superior to analog computers when the nature of the task to be performed requires these specific characteristics. But digital computers can be inefficient for the solution of certain classes of problems involving ordinary or partial differential equations. The expense of these inefficiencies is decreasing with the increased speed of newer generations of computers.

Some of the characteristics of digital-computer simulation are:

1. Variables represented by discrete numbers
2. Operations performed in sequence
3. Slower operation speed than analog
4. Increasing problem size requiring an increase in memory size and time of solution
5. Accuracy and dynamic range extendable to any desired degree
6. Integration by finite-difference calculus
7. Ease in handling and storing discrete data
8. Ordinarily no correspondence between rate of solution and rate of system studied
9. Difficulty in including part of actual system
10. Programming by computer language

Programming a digital computer for simulation is generally achieved by using either a general-purpose scientific language, such as FORTRAN or ALGOL, or by using a specialized programming language specifically developed for simulation.

General-purpose digital-simulation languages are designed for the programming of digital computers for the simulation of wide classes of systems. There are basically two categories of general-purpose simulation languages: continuous-systems simulation languages and discrete-events simulation languages.

Special-purpose digital-simulation languages are designed so that engineers in a specialized field can write a program using a vocabulary of terms and definitions directly related to the system to be simulated. For example, ECAP (*e*lectronic *c*ircuit *a*nalysis *p*rogram) for electric circuit analysis permits steady-state ac, steady-state dc, as well as transient response of circuits to be obtained by digital simulation.

Continuous-systems simulation languages, developed for the simulation of dynamic systems, have been derived from programming techniques long used in analog computation. A number of them make use of block-diagram-type notation and provide the analyst

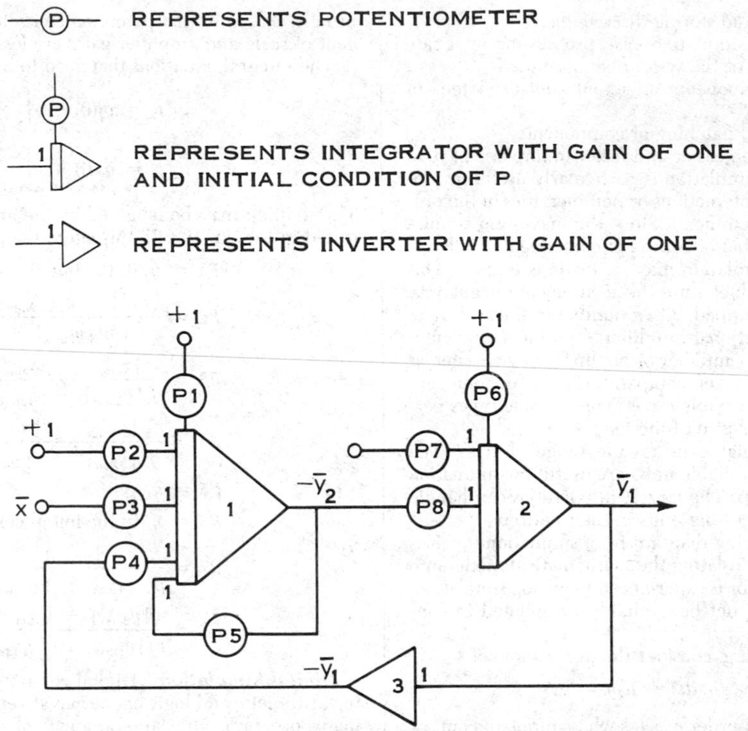

FIG. 22-193 Analog-computer program simulating the time- and magnitude-scaled process represented by Eqs. (22-35).

with conveniences previously available only to users of analog computers. A program developed by the use of such a simulation language might run more slowly than if it had been developed through use of a scientific programming language. However, less effort is involved in its preparation. Because of the number of programming languages for continuous dynamic simulation developed in just a few years, the Simulation Councils, Inc. (a professional society devoted to simulation) established a committee for the establishment of standards in simulation programming. One result of this committee's work was a new digital-simulation program, the CSSL (*c*ontinuous *s*ystems *s*imulation *l*anguage), which contains features found in its more than 20 ancestors.

There has been a proliferation of discrete-events simulation languages. Among the best known are SIMSCRIPT and TPSS.

Hybrid Simulation Hybrid computers used for process-control simulation consist of one or more analog computers and a general-purpose digital computer linked by a special interface into a single computing system as illustrated in Fig. 22-194. The interconnection linkage includes the following functions:

1. Multichannel analog-to-digital conversion equipment transforming analog signals to digital data
2. Multichannel digital-to-analog conversion equipment, transforming digital data to analog signals
3. Logic linkage, performing bilateral transfer of logic information
4. Monitoring and control-system linkage, permitting the analog and digital computer to be controlled by signals originating from either machine

The highly specialized hybrid computer is being applied to the solution of engineering problems in the process industries that formerly were deemed too expensive or impractical. In general, the engineering approach to the solution of a process or control problem using hybrid simulation does not change from that used in analog or digital simulation. The advantage of using the hybrid computer lies in the ability of the analyst to perform hundreds or even hundreds of thousands of complete solutions in a relatively short time, to evaluate several completely different control schemes with little additional effort, and to solve problems involving complex mathematical models which would not be practical or possible by using other techniques.

The hybrid is capable of retaining the hands-on feature of analog simulation, in which the operator observes the behavior of the system rather than simply performing calculations. Also, it is possible to have complete problem automation, in which the digital portion of the hybrid system handles operating control of the analog computer, makes potentiometer settings, controls output devices, changes initial conditions from run to run, and terminates automatically when all the desired data have been processed.

The completely integrated hybrid computer makes use of features of both the analog and digital computers for its capability to:

1. Solve partial differential equations at high speed by methods using continuous integration in the analog machine and the storage and playback functions of the digital machine
2. Solve dynamic optimization problems requiring the iterative solution of state and adjoint differential equations and the storage of intermediate iterative results
3. Preprogram sequences of solutions to be obtained at high speed by the analog computer and recorded or operated upon by the digital
4. Store in the digital computer numerical data to be used by the analog computer in data fitting, parameter-identification problems, solution of modeling, or in a data-processing function
5. Program the analog computer, set up the attenuators, and perform checking operations automatically from the digital

INSTRUMENTATION COSTS

Instrumentation used in chemical-process control is a wide subject with many variables; however, most control is based on measurement of temperature, pressure, differential pressure, and stream composi-

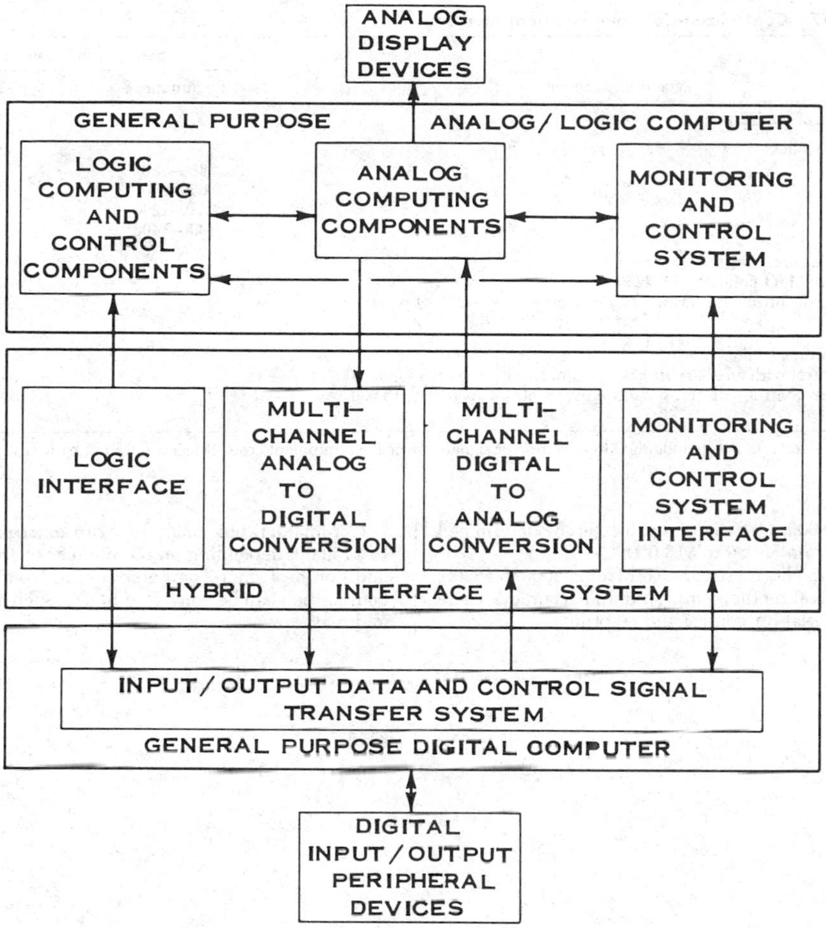

FIG. 22-194 Typical hybrid computer interfacing connecting analog and digital computers into a single hybrid machine.

tion. This subsection will concentrate on the cost of basic instruments to record and control these variables. The costs presented are list prices for 1977. Actual purchase can vary outside the cost ranges presented because of differences in quality, complexity of control, and other instrument features.

Local-Mounted Instrumentation Process instruments mounted at the point of measurement have many applications in simple processes and at remote locations such as compressor and pipe-line pump stations and product terminals. Presented in Table 22-16 are the local-mounted instruments commonly used in all processes.

The costs for the pneumatic instruments include the actual measurement device with the instrument. The electronic-instrument costs include a converter to produce a suitable electronic signal. In each case the cost presented covers the hardware cost at both sensing and measurement-control points.

Control-Room Instrumentation As the process complexity increases, a point is reached at which total control and measurement of the process is concentrated in a centralized control room. Presented in Table 22-17 are the cost spreads for the basic control-room instruments.

Plant Computer Control The third level of instrumentation currently being applied in plant control is achieved by computer monitor and control of the individual control-room instruments. A 50-loop monitor will cost from $85,000 to $175,000. The difference in price is due to the magnitude of software requirements, the hardware costs remaining constant. The computer is expanded by adding another unit at this cost.

TABLE 22-16 Local-Mounted Single-Loop Instrumentation

Type of instrument	Pneumatic, $	Electronic, $
Temperature		
Indicator	300–700	600–1500
Recorder	600–1100	1150–1650
Controller	600–1800	1700–2350
Recorder-controller	650–1850	1700–1450
Pressure		
Indicator	40–200	720–1830
Recorder	450–1490	1300–1860
Indicator-controller	470–2300	1910–2700
Recorder-controller	1000–2400	1930–2700
Flow measurement by differential pressure		
Indicator	560–1520	750–2000
Recorder	790–1550	1390–2100
Indicator-controller	500–1000	1700–3000
Recorder-controller	800–2000	1900–3000
Chromatographic stream analysis	5000–8500	7000–11,000

Process Composition Analysis Composition analysis is a variable used to control chemical processes and to establish product quality. Gas quality can be determined by specific-gravity instruments

TABLE 22-17 Centralized-Control Instrumentation

Type of instrument	Range of unit costs, 1977$	
	Pneumatic, $	Electronic, $
Multipoint recorder (12 points)	. . . .	1100–4000
Recorder-controllers		
One-pen	460–580	730–1400
Two-pen	650–900	1140–2000
Three-pen	820–1200	1500–2300
Four-pen	1000–1400	. . . .
Programmable controllers		
Relay control; I/O points = 250/250	. . . .	1000–10,000
Basic process control with no auxiliary support equipment; I/O points = 500/500	. . . .	8000–15,000
Microprocessor controllers		
Process control with auxiliary support equipment; I/O points = 500/500	. . . .	10,000–25,000
Multiprocess control with full support equipment; I/O points = 15,000/15,000	. . . .	25,000–150,000

NOTE: These costs do not include signal-generation or signal-conversion equipment costs. Prices are for the basic instrument up to full options.

costing from $1750 to $8000. Calorific gas value can be determined by calorimeters costing from $6000 to $13,000.

Densitometers measuring liquid streams cost from $3000 to $8000, depending on the analytical method employed and accuracy. These instruments transmit a signal for control and recording.

Chromatographic analysis of process streams will cost from $7500 to $15,000, depending on the number of streams monitored. By the addition of a microcomputer costing from $10,000 to $13,000, the composition can be converted to data such as calorific value, density, and mass flow.

Section 23

Materials of Construction*

Albert S. Krisher, *B.S. Ch.E., Oklahoma State University; Senior Fellow, Monsanto Company; member, American Institute of Chemical Engineers, National Association of Corrosion Engineers, American Society of Metals, American Society for Testing and Materials, Electrochemical Society*

Oliver W. Siebert, *B.S. M.E., Washington University; Senior Fellow, Monsanto Company; Adjunct Professor of Mechanical Engineering, Washington University; Member, American Society of Mechanical Engineers, American Institute of Chemical Engineers, National Association of Corrosion Engineers, American Welding Society, American Society for Testing and Materials, American Society of Metals, Society of The Plastics Industries, Electrochemical Society*

CORROSION AND ITS CONTROL

PROPERTIES OF MATERIALS

*The contribution of Mr. R. B. Norden to material used from the fifth edition is acknowledged.

HIGH- AND LOW-TEMPERATURE MATERIALS

CORROSION AND ITS CONTROL

GENERAL REFERENCES: Ailor (ed.), *Handbook on Corrosion Testing and Evaluation*, McGraw-Hill, New York, 1971. Bordes (ed.), *Metals Handbook*, 9th ed., vols. 1, 2, and 3, American Society for Metals, Metals Park, Ohio, 1978–1980; other volumes in preparation. Dillon (ed.), *Process Industries Corrosion*, National Association of Corrosion Engineers, Houston, 1975. Dillon and associates, *Guidelines for Control of Stress Corrosion Cracking of Nickel-Bearing Stainless Steels and Nickel-Base Alloys*, MTI Manual No. 1, Materials Technology Institute of the Chemical Process Industries, Columbus, 1979. Evans, *Metal Corrosion Passivity and Protection*, E. Arnold, London, 1940. Evans, *Corrosion and Oxidation of Metals*, St. Martin's, New York, 1960. Fontana and Greene, *Corrosion Engineering*, 2d ed., McGraw-Hill, New York, 1978. Gackenbach, *Materials Selection for Process Plants*, Reinhold, New York, 1960. Hamner (comp.), *Corrosion Data Survey: Metals Section*, National Association of Corrosion Engineers, Houston, 1974. Hamner (comp.), *Corrosion Data Survey: Non-Metals Section*, National Association of Corrosion Engineers, Houston, 1975. Hanson and Parr, *The Engineer's Guide to Steel*, Addison-Wesley, Reading, Mass., 1965. LaQue and Copson, *Corrosion Resistance of Metals and Alloys*, Reinhold, New York, 1963. Lyman (ed.), *Metals Handbook*, 8th ed., vols. 1–11, American Society for Metals, Metals Park, Ohio, 1961–1976. Mantell (ed.), *Engineering Materials Handbook*, McGraw-Hill, New York, 1958. Shreir, *Corrosion*, George Newnes, London, 1963. Speller, *Corrosion—Causes and Prevention*, McGraw-Hill, New York, 1951. Uhlig (ed.), *The Corrosion Handbook*, Wiley, New York, 1948. Uhlig, *Corrosion and Corrosion Control*, 2d ed., Wiley, New York, 1971. Wilson and Oates, *Corrosion and the Maintenance Engineer*, Hart Publishing, New York, 1968. Zapffe, *Stainless Steels*, American Society for Metals, Cleveland, 1949.

FLUID CORROSION

In the selection of materials of construction for a particular fluid system, it is important first to take into consideration the **characteristics of the system,** giving special attention to all factors that may influence corrosion. Since these factors would be peculiar to a particular system, it is impractical to attempt to offer a set of hard-and-fast rules that would cover all situations.

The **materials** from which the system is to be fabricated are the second important consideration; therefore, knowledge of the characteristics and general behavior of materials when exposed to certain environments is essential.

In the absence of factual corrosion information for a particular set of fluid conditions, a reasonably good selection would be possible from data based on the resistance of materials to a very similar environment. These data, however, should be used with some reservations. Good practice calls for applying such data for preliminary screening. Materials selected thereby would require further study in the fluid system under consideration.

FLUID CORROSION: GENERAL

Metallic Materials Pure metals and their alloys tend to enter into **chemical** union with the elements of a corrosive medium to form stable compounds similar to those found in nature. When metal loss occurs in this way, the compound formed is referred to as the corrosion product and the metal surface is spoken of as being corroded.

Corrosion is a complex phenomenon that may take any one or more of several forms. It is usually confined to the metal surface, and this is called general corrosion. But it sometimes occurs along grain boundaries or other lines of weakness because of a difference in resistance to attack or local electrolytic action.

In most aqueous systems, the corrosion reaction is divided into an anodic portion and a cathodic portion, occurring simultaneously at discrete points on metallic surfaces. Flow of electricity from the anodic to the cathodic areas may be generated by local cells set up either on a single metallic surface (because of local point-to-point differences on the surface) or between dissimilar metals.

Nonmetallics As stated, corrosion of metals applies specifically to chemical or electrochemical attack. The deterioration of plastics and other nonmetallic materials, which are susceptible to swelling, crazing, cracking, softening, etc., is essentially **physiochemical** rather than electrochemical in nature. Nonmetallic materials can either be rapidly deteriorated when exposed to a particular environment or, at the other extreme, be practically unaffected. Under some conditions, a nonmetallic may show evidence of gradual deterioration. However, it is seldom possible to evaluate its chemical resistance by measurements of weight loss alone, as is most generally done for metals.

FLUID CORROSION: LOCALIZED

Pitting Corrosion Pitting is a form of corrosion that develops in highly localized areas on the metal surface. This results in the development of cavities or pits. They may range from deep cavities of small diameter to relatively shallow depressions. Pitting examples: aluminum and stainless alloys in aqueous solutions containing chloride. **Inhibitors** are sometimes helpful in preventing pitting.

Crevice Corrosion Crevice corrosion occurs within or adjacent to a crevice formed by contact with another piece of the same or another metal or with a nonmetallic material. When this occurs, the intensity of attack is usually more severe than on surrounding areas of the same surface.

This form of corrosion can result because of a deficiency of oxygen in the crevice, acidity changes in the crevice, buildup of ions in the crevice, or depletion of an inhibitor.

Oxygen-Concentration Cell The oxygen-concentration cell is an electrolytic cell in which the driving force to cause corrosion results from a difference in the amount of oxygen in solution at one point as compared with another. Corrosion is accelerated where the oxygen concentration is least, for example, in a stuffing box or under gaskets. This form of corrosion will also occur under solid substances that may be deposited on a metal surface and thus shield it from ready access to oxygen. Redesign or change in mechanical conditions must be used to overcome this situation.

Galvanic Corrosion Galvanic corrosion is the corrosion rate above normal that is associated with the flow of current to a less active metal (cathode) in contact with a more active metal (anode) in the same environment. Table 23-1 shows the **galvanic** series of various metals. It should be used with caution, since exceptions to this series in actual use are possible. However, as a general rule when dissimilar metals are used in contact with each other and are exposed to an electrically conducting solution, combinations of metals that are as close as possible in the galvanic series should be chosen. Coupling two metals widely separated in this series generally will produce accelerated attack on the more active metal. Often, however, protective oxide films and other effects will tend to reduce galvanic corrosion. Galvanic corrosion can, of course, be prevented by **insulating** the metals from each other. For example, when plates are bolted together, specially designed plastic washers can be used.

Potential differences leading to galvanic-type cells can also be set up on a single metal by differences in temperature, velocity, or concentration (see subsection "Crevice Corrosion").

Area effects in galvanic corrosion are very important. An unfavorable area ratio is a large cathode and a small anode. Corrosion of the anode may be 100 to 1000 times greater than if the two areas were the same. This is the reason why stainless steels are susceptible to rapid pitting in some environments. Steel rivets in a copper plate will corrode much more severely than a steel plate with copper rivets.

TABLE 23-1 Galvanic Series of Metals and Alloys

Corroded end (anodic, or least noble)
Magnesium
Magnesium alloys
Zinc
Aluminum alloys
Aluminum
Alclad
Cadmium
Mild steel
Cast iron
Ni-Resist
13% chromium stainless (active)
50-50 lead-tin solder
18-8 stainless type 304 (active)
18-8-3 stainless type 316 (active)
Lead
Tin
Muntz Metal
Naval brass
Nickel (active)
Inconel 600 (active)
Yellow brass
Admiralty brass
Aluminum bronze
Red brass
Copper
Silicon bronze
70-30 cupronickel
Nickel (passive)
Inconel 600 (passive)
Monel 400
18-8 stainless type 304 (passive)
18-8-3 stainless type 316 (passive)
Silver
Graphite
Gold
Platinum
Protected end (cathodic, or most noble)

Intergranular Corrosion Selective corrosion in the grain boundaries of a metal or alloy without appreciable attack on the grains or crystals themselves is called intergranular corrosion. When severe, this attack causes a loss of strength and ductility out of proportion to the amount of metal actually destroyed by corrosion.

The **austenitic stainless steels** that are not stabilized or that are not of the extra-low-carbon types, when heated in the temperature range of 450 to 843°C (850 to 1550°F), have chromium-rich compounds (chromium carbides) precipitated in the grain boundaries. This causes grain-boundary impoverishment of chromium and makes the affected metal susceptible to intergranular corrosion in many environments. Hot nitric acid is one environment which causes severe intergranular corrosion of austenitic stainless steels with grain-boundary precipitation. Austenitic stainless steels stabilized with niobium (columbium) or titanium to decrease carbide formation or containing less than 0.03 percent carbon are normally not susceptible to grain-boundary deterioration when heated in the given temperature range. Unstabilized austenitic stainless steels or types with normal carbon content, to be immune to intergranular corrosion, should be given a solution anneal. This consists of heating to 1090°C (2000°F), holding at this temperature for a minimum of 1 h/in of thickness, followed by rapidly quenching in water (or, if impractical because of large size, rapidly cooling with an air-water spray).

Stress-Corrosion Cracking Corrosion can be accelerated by stress, either residual internal stress in the metal or externally applied stress. Residual stresses are produced by deformation during fabrication, by unequal cooling from high temperature, and by internal structural rearrangements involving volume change. Stresses induced by rivets and bolts and by press and shrink fits can also be classified as residual stresses. Tensile stresses at the surface, usually of a magnitude equal to the yield stress, are necessary to produce stress-corrosion cracking. However, failures of this kind have been known to occur at lower stresses.

Virtually every alloy system has its specific environment conditions which will produce stress-corrosion cracking, and the time of exposure required to produce failure will vary from minutes to years. Typical examples include cracking of cold-formed brass in ammonia environments, cracking of austenitic stainless steels in the presence of chlorides, cracking of Monel in hydrofluosilicic acid, and caustic-embrittlement cracking of steel in caustic solutions.

This form of corrosion can be prevented in some instances by eliminating high stresses. Stresses developed during fabrication, particularly during welding, are frequently the main sources of trouble. Of course, temperature and concentration are also important factors in this type of attack.

Presence of **chlorides** does not generally cause cracking of austenitic stainless steels when temperatures are below about 50°C (120°F). However, when temperatures are high enough to concentrate chlorides on the stainless surface, cracking may occur when the chloride concentration in the surrounding media is a few parts per million. Typical examples are cracking of heat-exchanger tubes at the crevices in rolled joints and under scale formed in the vapor space below the top tube sheet in vertical heat exchangers. The cracking of stainless steel under insulation is caused when chloride-containing water is concentrated on the hot surfaces. The chlorides may be leached from the insulation or may be present in the water when it enters the insulation. Improved design and maintenance of insulation weatherproofing, coating of the metal prior to the installation of insulation, and use of chloride-free insulation are all steps which will help to reduce (but not eliminate) this problem.

Serious stress-corrosion-cracking failures have occurred when chloride-containing hydrotest water was not promptly removed from stainless-steel systems. Use of potable-quality water and complete draining after test comprise the most reliable solution to this problem. Use of chloride-free water is also helpful, especially when prompt drainage is not feasible.

In handling caustic, as-welded steel can be used without developing caustic-embrittlement cracking if the temperature is below 50°C (120°F). If the temperature is higher and particularly if the concentration is above about 30 percent, cracking at and adjacent to non-stress-relieved welds frequently occurs.

Liquid-Metal Corrosion Liquid metals can also cause corrosion failures. The most damaging are liquid metals which penetrate the metal along grain boundaries to cause catastrophic failure. Examples include mercury attack on aluminum alloys and attack of stainless steels by molten zinc or aluminum. A fairly common problem occurs when galvanized-structural-steel attachments are welded to stainless piping or equipment. In such cases it is mandatory to remove the galvanizing completely from the area which will be heated above 260°C (500°F).

Erosion Erosion is the destruction of a metal by abrasion or attrition caused by the flow of liquid or gas (with or without suspended solids). The use of harder materials and changes in velocity or environment are methods employed to prevent erosion attack.

Impingement Corrosion This phenomenon is sometimes referred to as erosion-corrosion or velocity-accelerated corrosion. It occurs when damage is accelerated by the mechanical removal of corrosion products (such as oxides) which would otherwise tend to stifle the corrosion reaction.

Corrosion Fatigue Corrosion fatigue is a reduction by corrosion of the ability of a metal to withstand **cyclic or repeated stresses.** The surface of the metal plays an important role in this form of damage, as it will be the most highly stressed and at the same time subject to attack by the corrosive media. Corrosion of the metal surface will lower fatigue resistance, and stressing of the surface will tend to accelerate corrosion.

Under cyclic or repeated stress conditions, rupture of protective oxide films that prevent corrosion takes place at a greater rate than that at which new protective films can be formed. Such a situation frequently results in formation of anodic areas at the points of rupture; these produce pits that serve as stress-concentration points for the origin of cracks that cause ultimate failure.

Cavitation Formation of transient voids or vacuum bubbles in a liquid stream passing over a surface is called cavitation. This is often encountered around propellers, rudders, and struts and in pumps. When these bubbles collapse on a metal surface, there is a severe impact or explosive effect that can cause considerable mechanical damage, and corrosion can be greatly accelerated because of the destruction of protective films. Redesign or a more resistant metal is generally required to avoid this problem.

Fretting Corrosion This attack occurs when metals slide over each other and cause mechanical damage to one or both. In such a case, frictional heat oxidizes the metal and this oxide then wears away; or the mechanical removal of protective oxides results in exposure of fresh surface for corrosive attack. Fretting corrosion is minimized by using harder materials, minimizing friction (via lubrication), or designing equipment so that no relative movement of parts takes place.

Hydrogen Attack At elevated temperatures and significant hydrogen partial pressures, hydrogen will penetrate carbon steel, reacting with the carbon in the steel to form methane. The pressure generated causes a loss of ductility (hydrogen embrittlement) and failure by cracking or blistering of the steel. The removal of the carbon from the steel (decarburization) results in decreased strength. Resistance to this type of attack is improved by alloying with molybdenum or chromium. Accepted limits for the use of carbon and low-alloy steels are shown in Fig. 23-1, which is adapted from American Petroleum Institute (API) Publication 941, *Steels for Hydrogen Service at Elevated Temperatures and Pressures in Petroleum Refineries and Petrochemical Plants*.

Hydrogen damage can also result from hydrogen generated in electrochemical corrosion reactions. This phenomenon is most commonly observed in solutions of specific weak acids. H_2S and HCN are the most common, although other acids can cause the problem. The atomic hydrogen formed on the metal surface by the corrosion reaction diffuses into the metal and forms molecular hydrogen at microvoids in the metal. The result is failure by embrittlement, cracking, and blistering.

FLUID CORROSION: STRUCTURAL

Graphitic Corrosion Graphitic corrosion usually involves **gray cast iron** in which metallic iron is converted into corrosion products, leaving a residue of intact graphite mixed with iron-corrosion products and other insoluble constituents of cast iron.

When the layer of graphite and corrosion products is impervious to the solution, corrosion will cease or slow down. If the layer is porous, corrosion will progress by galvanic behavior between graphite and iron. The rate of this attack will be approximately that for the maximum penetration of steel by pitting. The layer of graphite formed may also be effective in reducing the galvanic action between cast iron and more noble alloys such as bronze used for valve trim and impellers in pumps.

Low-alloy cast irons frequently demonstrate a superior resistance to graphitic corrosion, apparently because of their denser structure and the development of more compact and more protective graphitic coatings. Highly alloyed **austenitic cast irons** show considerable superiority over gray cast irons to graphitic corrosion because of the more noble potential of the austenitic matrix plus more protective graphitic coatings.

Carbon steels heated for prolonged periods at temperatures above 455°C (850°F) may be subject to the segregation of carbon, which is transformed into graphite. When this occurs, the structural strength of the steel will be affected. Killed steels or low-alloy steels of chromium and molybdenum or chromium and nickel should be considered for elevated-temperature services.

Parting, or Dealloying, Corrosion This type of corrosion occurs when only one component of an alloy is removed by corrosion. The most common type is dezincification of brass.

Dezincification Dezincification is corrosion of a brass alloy containing zinc in which the principal product of corrosion is metallic copper. This may occur as plugs filling pits (plug type) or as continuous layers surrounding an unattacked core of brass (general type). The mechanism may involve overall corrosion of the alloy followed by redeposition of the copper from the corrosion products or selective corrosion of zinc or a high-zinc phase to leave copper residue. This form of corrosion is commonly encountered in brasses that contain more than 15 percent zinc and can be either eliminated or reduced by the addition of small amounts of **arsenic, antimony,** or **phosphorus** to the alloy.

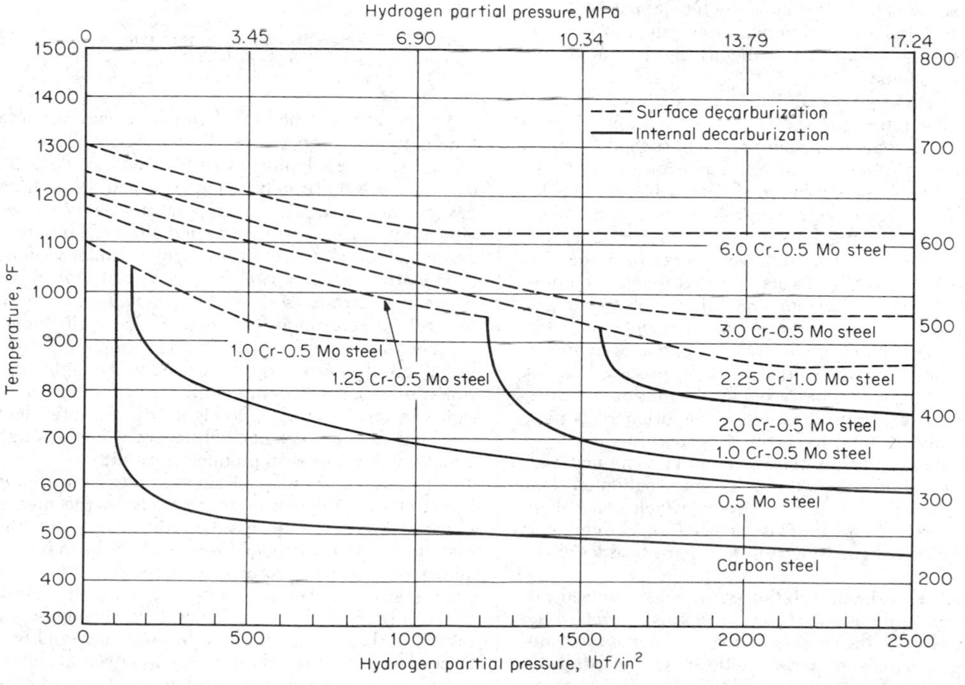

FIG. 23-1 Operating limits for steels in hydrogen service. Each steel is suitable for use under hydrogen-partial-pressure–temperature conditions below and to the left of its respective curve. *(Courtesy of National Association of Corrosion Engineers.)*

Biological Corrosion The metabolic activity of microorganisms can either directly or indirectly cause deterioration of a metal by corrosion processes. Such activity can (1) produce a corrosive environment, (2) create electrolytic-concentration cells on the metal surface, (3) alter the resistance of surface films, (4) have an influence on the rate of anodic or cathodic reaction, and (5) alter the environment composition.

Microorganisms associated with corrosion are of two types, aerobic and anaerobic. **Aerobic** microorganisms readily grow in an environment containing oxygen, while the **anaerobic** species thrive in an environment virtually devoid of atmospheric oxygen.

The manner in which many of these bacteria carry on their chemical processes is quite complicated and in some cases not fully understood. The role of **sulfate-reducing bacteria** (anaerobic) in promoting corrosion has been extensively investigated. The sulfates in slightly acid to alkaline (pH 6 to 9) soils are reduced by these bacteria to form calcium sulfide and hydrogen sulfide. When these compounds come in contact with underground iron pipes, conversion of the iron to iron sulfide occurs. As these bacteria thrive under these conditions, they will continue to promote this reaction until failure of the pipe occurs.

Several instances of serious biological corrosion occurred when hydrotest water was not promptly removed from stainless-steel systems. These cases involved both potable and nonpotable waters. Biological activity caused perforation of the stainless steel in a few months. Use of potable-quality water and prompt and complete draining after a test constitute the most reliable solution to this problem.

FACTORS INFLUENCING CORROSION

Solution pH The corrosion rate of most metals is affected by pH. The relationship tends to follow one of three general patterns:

1. Acid-soluble metals such as iron have a relationship as shown in Fig. 23-2*a*. In the middle pH range ($\approx$4 to 10), the corrosion rate is controlled by the rate of transport of oxidizer (usually dissolved O_2) to the metal surface. Iron is weakly amphoteric. At very high temperatures such as those encountered in boilers, the corrosion rate increases with increasing basicity, as shown by the dashed line.
2. Amphoteric metals such as aluminum and zinc have a relationship as shown in Fig. 23-2*b*. These metals dissolve rapidly in either acidic or basic solutions.
3. Noble metals such as gold and platinum are not appreciably affected by pH, as shown in Fig. 23-2*c*.

Oxidizing Agents In some corrosion processes, such as the solution of zinc in hydrochloric acid, hydrogen may evolve as a gas. In others, such as the relatively slow solution of copper in sodium chloride, the removal of hydrogen, which must occur so that corrosion may proceed, is effected by a reaction between hydrogen and some oxidizing chemical such as oxygen to form water. Because of the high rates of corrosion which usually accompany hydrogen evolution, metals are rarely used in solutions from which they evolve hydrogen at an appreciable rate. As a result, most of the corrosion observed in practice occurs under conditions in which the oxidation of hydrogen to form water is a necessary part of the corrosion process. For this reason, oxidizing agents are often powerful accelerators of corrosion, and in many cases the **oxidizing power of a solution** is its most important single property insofar as corrosion is concerned.

Oxidizing agents that accelerate the corrosion of some materials may also retard corrosion of others through the formation on their surface of oxides or layers of adsorbed oxygen which make them more resistant to chemical attack. This property of chromium is responsible for the principal corrision-resisting characteristics of the stainless steels.

It follows, then, that oxidizing substances, such as dissolved air, may accelerate the corrosion of one class of materials and retard the corrosion of another class. In the latter case, the behavior of the material usually represents a balance between the power of oxidizing compounds to preserve a protective film and their tendency to accelerate corrosion when the agencies responsible for protective-film breakdown are able to destroy the films.

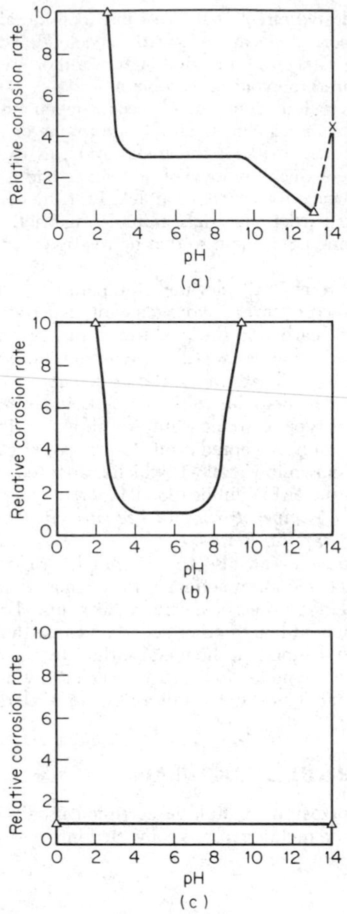

FIG. 23-2 Effect of pH on the corrosion rate. (*a*) Iron. (*b*) Amphoteric metals (aluminum, zinc). (*c*) Noble metals.

Temperature The rate of corrosion tends to increase with rising temperature. Temperature also has a secondary effect through its influence on the solubility of air (oxygen), which is the most common oxidizing substance influencing corrosion. In addition, temperature has specific effects when a temperature change causes phase changes which introduce a corrosive second phase. Examples include condensation systems and systems involving organics saturated with water.

Velocity An increase in the velocity of relative movement between a corrosive solution and a metallic surface frequently tends to accelerate corrosion. This effect is due to the higher rate at which the corrosive chemicals, including oxidizing substances (air), are brought to the corroding surface and to the higher rate at which corrosion products, which might otherwise accumulate and stifle corrosion, are carried away. The higher the velocity, the thinner will be the films which corroding substances must penetrate and through which soluble corrosion products must diffuse.

Whenever corrosion resistance results from the accumulation of layers of insoluble corrosion products on the metallic surface, the effect of high velocity may be either to prevent their normal formation or to remove them after they have been formed. Either effect allows corrosion to proceed unhindered. This occurs frequently in small-diameter tubes or pipes through which corrosive liquids may be circulated at high velocities (e.g., condenser and evaporator tubes), in the vicinity of bends in pipe lines, and on propellers, agitators, and centrifugal pumps. Similar effects are associated with cavitation and impingement corrosion.

Films Once corrosion has started, its further progress very often is controlled by the nature of films, such as passive films, that may form or accumulate on the metallic surface. The classical example is the thin **oxide film** that forms on stainless steels.

Insoluble corrosion products may be completely impervious to the corroding liquid and, therefore, completely protective; or they may be quite permeable and allow local or general corrosion to proceed unhindered. Films that are nonuniform or discontinuous may tend to localize corrosion in particular areas or to induce accelerated corrosion at certain points by initiating electrolytic effects of the concentration-cell type. Films may tend to retain or absorb moisture and thus, by delaying the time of drying, increase the extent of corrosion resulting from exposure to the atmosphere or to corrosive vapors.

It is agreed generally that the characteristics of the **rust films** that form on steels determine their resistance to atmospheric corrosion. The rust films that form on low-alloy steels are more protective than those that form on unalloyed steel.

In addition to films that originate at least in part in the corroding metal, there are others that originate in the corrosive solution. These include various salts, such as carbonates and sulfates, which may be precipitated from heated solutions, and insoluble compounds, such as "beer stone," which form on metal surfaces in contact with certain specific products. In addition, there are films of oil and grease that may protect a material from direct contact with corrosive substances. Such oil films may be applied intentionally or may occur naturally, as in the case of metals submerged in sewage or equipment used for the processing of oily substances.

Other Effects Stream **concentration** can have important effects on corrosion rates. Unfortunately, corrosion rates are seldom linear with concentration over wide ranges. In equipment such as distillation columns, reactors, and evaporators, concentration can change continuously, making prediction of corrosion rates rather difficult. Concentration is important during plant shutdown; presence of moisture that collects during cooling can turn innocuous chemicals into dangerous corrosives.

As to the effect of time, there is no universal law that governs the reaction for all metals. Some corrosion rates remain constant with time over wide ranges, others slow down with time, and some alloys have increased corrosion rates with respect to time. Situations in which the corrosion rate follows a combination of these paths can develop. Therefore, extrapolation of corrosion data and corrosion rates should be done with utmost caution.

Impurities in a corrodent can be good or bad from a corrosion standpoint. An impurity in a stream may act as an inhibitor and actually retard corrosion. However, if this impurity is removed by some process change or improvement, a marked rise in corrosion rates can result. Other impurities, of course, can have very deleterious effects on materials. The chloride ion is a good example; small amounts of chlorides in a process stream can break down the passive oxide film on stainless steels. The effects of impurities are varied and complex. One must be aware of what they are, how much is present, and where they come from before attempting to recommend a particular material of construction.

HIGH-TEMPERATURE ATTACK

Physical Properties The suitability of an alloy for high-temperature service [425 to 1100°C (800 to 2000°F)] is dependent upon properties inherent in the alloy composition and upon the conditions of application. Crystal structure, density, thermal conductivity, electrical resistivity, thermal expansivity, structural stability, melting range, and vapor pressure are all physical properties basic to and inherent in individual alloy compositions.

Of usually high relative importance in this group of properties is **expansivity.** A surprisingly large number of metal failures at elevated temperatures are the result of excessive thermal stresses originating from constraint of the metal during heating or cooling. Such constraint in the case of hindered contraction can cause rupturing.

Another important property is alloy **structural stability.** This means freedom from formation of new phases or drastic rearrangement of those originally present within the metal structure as a result of thermal experience. Such changes may have a detrimental effect upon strength or corrosion resistance or both.

Mechanical Properties Mechanical properties of wide interest include creep, rupture, short-time strengths, and various forms of ductility, as well as resistance to impact and fatigue stresses. Creep strength and stress rupture are usually of greatest interest to designers of stationary equipment such as vessels and furnaces.

Corrosion Resistance Possibly of greater importance than physical and mechanical properties is the ability of an alloy's chemical composition to resist the corrosive action of various hot environments. The forms of high-temperature corrosion which have received the greatest attention are **oxidation** and **scaling.**

Chromium is an essential constituent in alloys to be used above 550°C (1000°F). It provides a tightly adherent oxide film that materially retards the oxidation process. Silicon is a useful element in imparting oxidation resistance to steel. It will enhance the beneficial effects of chromium. Also, for a given level of chromium, experience has shown oxidation resistance to improve as the nickel content increases.

Aluminum is not commonly used as an alloying element in steel to improve oxidation resistance, as the amount required interferes with both workability and high-temperature-strength properties. However, the development of high-aluminum surface layers by various methods, including spraying, cementation, and dipping, is a feasible means of improving heat resistance of low-alloy steels.

Contaminants in fuels, especially alkali-metal ions, vanadium, and sulfur compounds, tend to react in the combustion zone to form molten fluxes which dissolve the protective oxide film on stainless steels, allowing oxidation to proceed at a rapid rate. This problem is becoming more common as the high cost and short supply of natural gas and distillate fuel oils force increased usage of residual fuel oils and coal.

COMBATING CORROSION

Material Selection The objective is to select the material which will most economically fulfill the process requirements. The best source of data is well-documented experience in an identical process unit. In the absence of such data, other data sources such as experience in pilot units, corrosion-coupon tests in pilot or bench-scale units, laboratory corrosion-coupon tests in actual process fluids, or corrosion-coupon tests in synthetic solutions must be used. The data from such alternative sources (which are listed in decreasing order of reliability) must be properly evaluated, taking into account the degree to which a given test may fail to reproduce actual conditions in an operating unit. Particular emphasis must be placed on possible composition differences between a static laboratory test and a dynamic plant as well as on trace impurities (chlorides in stainless-steel systems, for example) which may greatly change the corrosiveness of the system. The possibility of severe localized attack (pitting, crevice corrosion, or stress-corrosion cracking) must also be considered.

Permissible **corrosion rates** are an important factor and differ with equipment. Appreciable corrosion can be permitted for tanks and lines if anticipated and allowed for in design thickness, but essentially no corrosion can be permitted in fine-mesh wire screens, orfices, and other items in which small changes in dimensions are critical.

In many instances use of **nonmetallic materials** will prove to be attractive from an economic and performance standpoint. These should be considered when their strength, temperature, and design limitations are satisfactory.

Proper Design Design considerations with respect to minimizing corrosion difficulties should include the desirability for free and complete drainage, minimizing crevices, and ease of cleaning and inspection. The installation of baffles, stiffeners, and drain nozzles and the location of valves and pumps should be made so that free drainage will occur and washing can be accomplished without holdup. Means of access for inspection and maintenance should be provided whenever practical. Butt joints should be used whenever possible. If lap joints employing fillet welds are used, the welds should be continuous.

The use of dissimilar metals in contact with each other should generally be minimized, particularly if they are widely separated in their nominal positions in the galvanic series (see Table 23-1). If they are to be used together, consideration should be given to insulating them from each other or making the anodic material area as large as possible.

Equipment should be supported in such a way that it will not rest in pools of liquid or on damp insulating material. Porous insulation should be weatherproofed or otherwise protected from moisture and spills to avoid contact of the wet material with the equipment. Specifications should be sufficiently comprehensive to ensure that the desired composition or type of material will be used and the right condition of heat treatment and surface finish will be provided. Inspection during fabrication and prior to acceptance is desirable.

Altering the Environment Simple changes in environment may make an appreciable difference in the corrosion of metals and should be considered as a means of combating corrosion. **Oxygen** is an important factor, and its removal or addition may cause marked changes in corrosion. The treatment of boiler feedwater to remove oxygen, for instance, greatly reduces the corrosiveness of the water on steel. Inert-gas purging and blanketing of many solutions, particularly acidic media, generally minimize corrosion of copper and nickel-base alloys by minimizing air or oxygen content. Corrosiveness of acid media to stainless alloys, on the other hand, may be reduced by aeration because of the formation of passive oxide films. Reduction in temperature will almost always be beneficial with respect to reducing corrosion if no corrosive phase changes (condensation, for example) result. Velocity effects vary with the material and the corrosive system. When pH values can be modified, it will generally be beneficial to hold the acid level to a minimum. When acid additions are made in batch processes, it may be beneficial to add them last so as to obtain maximum dilution and minimum acid concentration and exposure time. Alkaline pH values are less critical than acid values with respect to controlling corrosion. Elimination of moisture can and frequently does minimize, if not prevent, corrosion of metals, and this possibility of environmental alteration should always be considered.

Inhibitors The use of various substances or inhibitors as additives to corrosive environments to decrease corrosion of metals in the environment is an important means of combating corrosion. This is generally most attractive in closed or recirculating systems in which the annual cost of inhibitor is low. However, it has also proved to be economically attractive for many once-through systems, such as those encountered in petroleum-processing operations. Inhibitors are effective as the result of their controlling influence on the cathode- or anode-area reactions.

Typical examples of inhibitors used for minimizing corrosion of iron and steel in aqueous solutions are the chromates, phosphates, and silicates. Organic sulfide and amine materials are frequently effective in minimizing corrosion of iron and steel in acid solution.

The use of inhibitors is not limited to controlling corrosion of iron and steel. They frequently are effective with stainless steel and other alloy materials. The addition of copper sulfate to dilute sulfuric acid will sometimes control corrosion of stainless steels in hot dilute solutions of this acid, whereas the uninhibited acid causes rapid corrosion.

The effectiveness of a given inhibitor generally increases with an increase in **concentration,** but inhibitors considered practical and economically attractive are used in quantities of less than 0.1 percent by weight.

In some instances the amount of inhibitor present is critical in that a deficiency may result in localized or pitting attack, with the overall results being more destructive than when none of the inhibitor is present. Considerations for the use of inhibitors should therefore include review of experience in similar systems or investigation of requirements and limitations in new systems.

Cathodic Protection This electrochemical method of corrosion control has found wide application in the protection of carbon steel underground structures such as pipe lines and tanks from external soil corrosion. It is also widely used in water systems to protect ship hulls, offshore structures, and water-storage tanks.

Two methods of providing cathodic protection for minimizing corrosion of metals are in use today. These are the sacrificial-anode method and the impressed-emf method. Both depend upon making the metal to be protected the cathode in the electrolyte involved.

Examples of the **sacrificial-anode** method include the use of zinc, magnesium, or aluminum as anodes in electrical contact with the metal to be protected. These may be anodes buried in the ground for protection of underground pipe lines or attachments to the surfaces of equipment such as condenser water boxes or on ship hulls. The current required is generated in this method by corrosion of the sacrificial-anode material. In the case of the **impressed emf,** the direct current is provided by external sources and is passed through the system by use of essentially nonsacrificial anodes such as carbon, noncorrodible alloys, or platinum buried in the ground or suspended in the electrolyte in the case of aqueous systems.

The requirements with respect to current distribution and anode placement vary with the resistivity of soils or the electrolyte involved.

Anodic Protection This electrochemical method relies on an external potential control system (potentiostat) to maintain the metal or alloy in a noncorroding (passive) condition. Practical applications include acid coolers in sulfuric acid plants and storage tanks for sulfuric acid.

Coatings and Linings The use of nonmetallic coatings and lining materials in combination with steel or other materials has and will continue to be an important type of construction for combating corrosion.

Organic coatings of many kinds are used as linings in equipment such as tanks, piping, pumping lines, and shipping containers, and they are often an economical means of controlling corrosion, particularly when freedom from metal contamination is the principal objective. One principle that is now generally accepted is that thin nonreinforced paintlike coatings of less than 0.75-mm (0.03-in) thickness should not be used in services for which full protection is required in order to prevent rapid attack of the substrate metal. This is true because most thin coatings contain defects or holidays and can be easily damaged in service, thus leading to early failures due to corrosion of the substrate metal even though the coating material is resistant. Electrical testing for continuity of coating-type linings is always desirable for immersion-service applications in order to detect holiday-type defects in the coating.

The most dependable barrier linings for corrosive services are those which are bonded directly to the substrate and are built up in multiple-layer or laminated effects to thicknesses greater than 2.5 mm (0.10 in). These include flake-glass-reinforced resin systems and elastomeric and plasticized plastic systems. Good surface preparation and thorough inspections of the completed lining, including electrical testing, should be considered as minimum requirements for any lining applications.

Linings of this type are slightly permeable to many liquids. Such permeation, while not damaging to the lining, may cause failure by causing disbonding of the lining owing to pressure buildup between the lining and the steel.

Ceramic or carbon-brick linings are frequently used as facing linings over plastic or membrane linings when surface temperatures exceed those which can be handled by the unprotected materials or when the membrane must be protected from mechanical damage. This type of construction permits processing of materials that are too corrosive to be handled in low-cost metal constructions.

Glass-Lined Steel By proprietary methods, special glasses can be bonded to steel, providing an impervious liner 1.5 to 2.5 mm (0.060 to 0.100 in) thick. Equipment and piping lined in this manner are routinely used in severely corrosive acid services. The glass lining can be mechanically damaged, and careful attention to details of design, inspection, installation, and maintenance is required to achieve good results with this system.

The **cladding** of steel with an alloy is another approach to this problem. There are a number of cladding methods in general use. In one, a sandwich is made of the corrosion-resistant metal and carbon steel by hot rolling to produce a **pressure weld** between the plates.

Another process involves **explosive bonding.** The corrosion-resistant metal is bonded to a steel backing metal by the force generated by properly positioned explosive charges. Relatively thick sections of metal can be bonded by this technique into plates.

In a third process, a **loose liner** is fastened to a carbon steel shell by welds spaced so as to prevent collapse of the liner. A fourth method is **weld overlay**, which involves depositing multiple layers of alloy weld metal to cover the steel surface.

All these methods require careful design and control of fabrication methods to assure success.

Metallic Linings for Mild Environments **Zinc coatings** applied by various means have good corrosion resistance to many atmospheres. Such coatings have been extensively used on steel. Zinc has the advantage of being anodic to steel and therefore will protect exposed areas of steel by electrochemical action.

Steel coated with tin **(tinplate)** is used to make food containers. Tin is more noble than steel; therefore, well-aerated solutions will galvanically accelerate attack of the steel at exposed areas. The comparative absence of air within food containers aids in preserving the tin as well as the food. Also the reversible potential which the tin-iron couple undergoes in organic acids serves to protect exposed steel in food containers.

Cadmium, being anodic to steel, behaves quite similarly to zinc in providing corrosion protection when applied as a coating on steel. Tests of zinc and cadmium coatings should be conducted when it becomes necessary to determine the most economical selection for a particular environment.

Lead has a good general resistance to various atmospheres. As a coating, it has had its greatest application in the production of terneplate, which is used as a roofing, cornicing, and spouting material.

Aluminum coatings on steel will perform in a manner similar to zinc coatings. Aluminum has good resistance to many atmospheres; in addition, being anodic to steel, it will galvanically protect exposed areas. Aluminum-coated steel products are quite serviceable under high-temperature conditions, for which good oxidation resistance is required.

CORROSION-TESTING METHODS

There is no standard or preferred way to carry out a corrosion test; the method must be chosen to suit the purpose of the test. The principal **types of tests** are, in decreasing order of reliability:

1. Actual operating experience with full-scale plant equipment exposed to the corroding medium.
2. Small-scale plant-equipment experience, under either commercial or pilot-plant conditions.
3. Sample tests in the field. These include coupons, stressed samples, electrical-resistance probes exposed to the plant corroding medium, or samples exposed to the atmosphere, to soils, or to fresh, brackish, or saline waters.
4. Laboratory tests on samples exposed to "actual" plant liquids or simulated environments.

Plant or field corrosion tests are useful for:

1. Selection of the most suitable material to withstand a particular environment and to estimate its probable durability in that environment
2. Study of the effectiveness of means of preventing corrosion

CORROSION TESTING: LABORATORY TESTS

Metals and alloys do not respond alike to all the influences of the many factors that are involved in corrosion. Consequently, it is impractical to establish any universal standard laboratory procedures for corrosion testing except for inspection tests. However, some details of laboratory testing need careful attention in order to achieve useful results.

In the selection of materials for the construction of a chemical plant, resistance to the corroding medium is often the determining factor; otherwise, the choice will fall automatically on the cheapest material mechanically suitable. Laboratory corrosion tests are frequently the quickest and most satisfactory means of arriving at a **preliminary selection** of the most suitable materials to use. Unfortunately, however, it is not yet within the state of the art of laboratory tests to predict with accuracy the behavior of the selected material under plant-operating conditions. The outstanding difficulty lies not so much in carrying out the test as in interpreting the results and translating them into terms of plant performance. A laboratory test of the conventional type gives mainly one factor—the chemical resistance of the proposed material to the corrosive agent. There are numerous other factors entering into the behavior of the material in the plant, such as dissolved gases, velocity, turbulence, abrasion, crevice conditions, hot-wall effects, cold-wall effects, stress levels of metals, trace impurities in corrodent that act as corrosion inhibitors or accelerators, and variations in composition of corrodent.

One method of determining the chemical-resistance factor, the so-called **total-immersion test,** represents an unaccelerated method that has been found to give reasonably concordant results in approximate agreement with results obtained on the large scale when the other variables are taken into account. Various other tests have been proposed and are in use, such as salt-spray, accelerated electrolytic, alternate-immersion, and aerated-total-immersion; but in view of the numerous complications entering into the translation of laboratory results into plant results the simplest test is considered the most desirable for routine preliminary work, reserving special test methods for special cases. The total-immersion test serves quite well to eliminate materials that obviously cannot be used; further selection among those materials which apparently can be used can be made on the basis of a knowledge of the properties of the materials concerned and the working conditions or by constructing larger-scale equipment of the proposed materials in which the operating conditions can be simulated.

The National Association of Corrosion Engineers (NACE) has established Standard TM-01-69, Laboratory Corrosion Testing of Metals for the Process Industries [*Mater. Prot.*, 13–24 (May 1969)], covering the total-immersion test. The description that follows is based on this standard.

Test Piece The size and the shape of specimens will vary with the purpose of the test, nature of the material, and apparatus used. A large surface-to-mass ratio and a small ratio of edge area to total area are desirable. These ratios can be achieved through the use of rectangular or circular specimens of minimum thickness. Circular specimens should be cut preferably from sheet and not bar stock to minimize the exposed end grain.

A circular specimen of about 38-mm (1.5-in) diameter is a convenient shape for laboratory corrosion tests. With a thickness of approximately 3 mm (⅛ in) and an 8- or 11-mm- (5/16- or 7/16-in-) diameter hole for mounting, these specimens will readily pass through a 45/50 ground-glass joint of a distillation kettle. The total surface area of a circular specimen is given by the equation:

$$A = \frac{\pi}{2}(D^2 - d^2) + t\pi D + t\pi d$$

where t = thickness, D = diameter of the specimen, and d = diameter of the mounting hole. If the hole is completely covered by the mounting support, the final term ($t\pi d$) in the equation is omitted.

Strip coupons [50 by 25 by 1.6 or 3.2 mm (2 by 1 by 1/16 or ⅛ in)] may be preferred as corrosion specimens, particularly if interface or liquid-line effects are to be studied by the laboratory test.

All specimens should be measured carefully to permit accurate calculation of the exposed areas. An area calculation accurate to plus or minus 1 percent is usually adequate.

More uniform results may be expected if a substantial layer of metal is removed from the specimens to eliminate variations in condition of the original metallic surface. This can be done by chemical treatment (pickling), electrolytic removal, or grinding with a coarse abrasive paper or cloth, such as No. 50, using care not to work-harden the surface. At least 2.5×10^{-3} mm (0.0001 in) or 1.5 to 2.3 mg/cm^2 (10 to 15 mg/in^2) should be removed. If clad alloy specimens are to be used, special attention must be given to ensure that excessive metal is not removed. After final preparation of the specimen surface, the specimens should be stored in a desiccator until exposure if they are not used immediately.

Specimens should be finally degreased by scrubbing with bleach-free scouring powder, followed by thorough rinsing in water and in a suitable solvent (such as acetone, methanol, or a mixture of 50 percent methanol and 50 percent ether), and air-dried. For relatively soft metals such as aluminum, magnesium, and copper, scrubbing with abrasive powder is not always needed and can mar the surface of the specimen. The use of towels for drying may introduce an error through contamination of the specimens with grease or lint. The dried specimen should be weighed on an analytic balance.

Apparatus A versatile and convenient apparatus should be used, consisting of a kettle or flask of suitable size (usually 500 to 5000 mL), a reflux condenser with atmospheric seal, a sparger for controlling atmosphere or aeration, a thermowell and temperature-regulating device, a heating device (mantle, hot plate, or bath), and a specimen-support system. If agitation is required the apparatus can be modified to accept a suitable stirring mechanism such as a magnetic stirrer. A typical resin-flask setup for this type of test is shown in Fig. 23-3. Open-beaker tests should not be used because of evaporation and contamination.

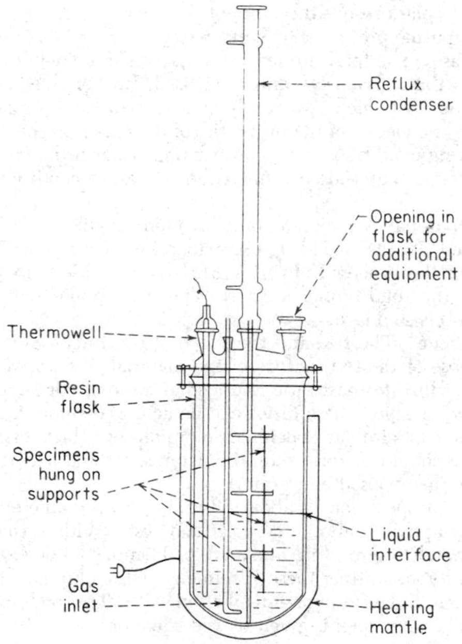

FIG. 23-3 Laboratory-equipment arrangement for corrosion testing. *(Based on NACE Standard TM-01-69.)*

In more complex tests, provisions might be needed for continuous flow or replenishment of the corrosive liquid while simultaneously maintaining a controlled atmosphere.

Apparatus for testing materials for heat-transfer applications is shown in Fig. 23-4. Here the sample is at a higher temperature than the bulk solution.

If the test is to be a guide for the selection of a material for a particular purpose, the limits of controlling factors in service must be determined. These factors include oxygen concentration, temperature, rate of flow, pH value, and other important characteristics.

The **composition of the test solution** should be controlled to the fullest extent possible and be described as thoroughly and as accurately as possible when the results are reported. Minor constituents should not be overlooked because they often affect corrosion rates. Chemical content should be reported as percentage by weight of the solution. Molarity and normality are also helpful in defining the concentration of chemicals in the test solution. The composition of the test solution should be checked by analysis at the end of the test to determine the extent of change in composition, such as might result from evaporation.

Temperature of Solution Temperature of the corroding solution should be controlled within ±1°C (±1.8°F) and must be stated in the report of test results.

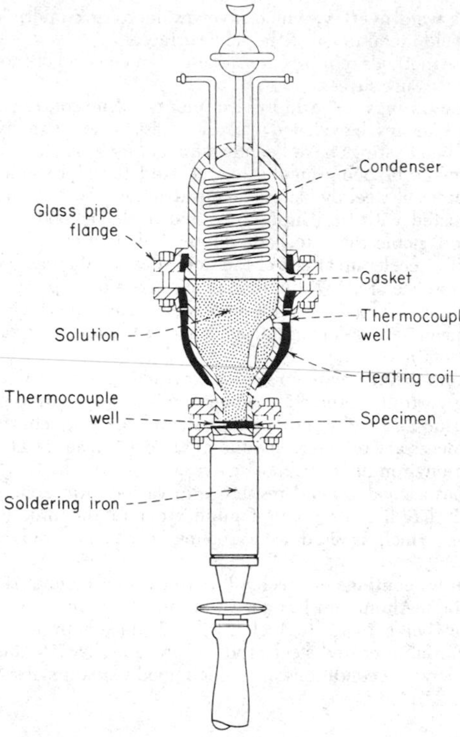

FIG. 23-4 Laboratory setup for the corrosion testing of heat-transfer materials.

For tests at ambient temperatures, the tests should be conducted at the highest temperature anticipated for stagnant storage in summer months. This temperature may be as high as 40 to 45°C (104 to 113°F) in some localities. The variation in temperature should be reported also (e.g., 40°C ± 2°C).

Aeration of Solution Unless specified, the solution should not be aerated. Most tests related to process equipment should be run with the natural atmosphere inherent in the process, such as the vapors of the boiling liquid. If aeration is used, the specimens should not be located in the direct air stream from the sparger. Extraneous effects can be encountered if the air stream impinges on the specimens.

Solution Velocity The effect of velocity is not usually determined in laboratory tests, although specific tests have been designed for this purpose. However, for the sake of reproducibility some velocity control is desirable.

Tests at the boiling point should be conducted with minimum possible heat input, and boiling chips should be used to avoid excessive turbulence and bubble impingement. In tests conducted below the boiling point, thermal convection generally is the only source of liquid velocity. In test solutions of high viscosities, supplemental controlled stirring with a magnetic stirrer is recommended.

Volume of Solution Volume of the test solution should be large enough to avoid any appreciable change in its corrosiveness through either exhaustion of corrosive constituents or accumulation of corrosion products that might affect further corrosion.

A suitable volume-to-area ratio is 20 mL (125 mL) of solution/cm^2 (in^2) of specimen surface. This corresponds to the recommendation of ASTM Standard A262 for the Huey test. The preferred volume-to-area ratio is 40 mL/cm^2 (250 mL/in^2) of specimen surface, as stipulated in ASTM Standard G31, Laboratory Immersion Testing of Materials.

Method of Supporting Specimens The supporting device and container should not be affected by or cause contamination of the test solution. The method of supporting specimens will vary with the apparatus used for conducting the test but should be designed to insulate the specimens from each other physically and electrically and to insulate the specimens from any metallic container or supporting device used with the apparatus.

Shape and form of the specimen support should assure free contact of the specimen with the corroding solution, the liquid line, or the vapor phase, as shown in Fig. 23-3. If clad alloys are exposed, special procedures are required to ensure that only the cladding is exposed (unless the purpose is to test the ability of the cladding to protect cut edges in the test solution). Some common supports are glass or ceramic rods, glass saddles, glass hooks, fluorocarbon plastic strings, and various insulated or coated metallic supports.

Duration of Test Although the duration of any test will be determined by the nature and purpose of the test, an excellent procedure for evaluating the effect of time on corrosion of the metal and also on the corrosiveness of the environment in laboratory tests has been presented by Wachter and Treseder [*Chem. Eng. Prog.*, 315–326 (June 1947)]. This technique is called the **planned-interval test.** Other procedures that require the removal of solid corrosion products between exposure periods will not measure accurately the normal changes of corrosion with time.

Materials that experience severe corrosion generally do not need lengthy tests to obtain accurate corrosion rates. Although this assumption is valid in many cases, there are exceptions. For example, lead exposed to sulfuric acid corrodes at an extremely high rate at first while building a protective film; then the rate decreases considerably, so that further corrosion is negligible. The phenomenon of forming a protective film is observed with many corrosion-resistant materials, and therefore short tests on such materials would indicate high corrosion rates and would be completely misleading.

Short-time tests also can give misleading results on alloys that form passive films, such as stainless steels. With borderline conditions, a prolonged test may be needed to permit breakdown of the passive film and subsequently more rapid attack. Consequently, tests run for long periods are considerably more realistic than those conducted for short durations. This statement must be qualified by stating that corrosion should not proceed to the point at which the original specimen size or the exposed area is drastically reduced or the metal is perforated.

If anticipated corrosion rates are moderate or low, the following equation gives a suggested test duration.

$$\text{Duration of test (h)} = \frac{78{,}740}{\text{corrosion rate (mm/year)}} = \frac{2000}{\text{corrosion rate (mils/year)}}$$

Cleaning Specimens after Test Before specimens are cleaned, their appearance should be observed and recorded. Locations of deposits, variations in types of deposits, and variations in corrosion products are extremely important in evaluating localized corrosion such as pitting and concentration-cell attack.

Cleaning specimens after the test is a vital step in the corrosion-test procedure and, if not done properly, can give rise to misleading test results. Generally, the cleaning procedure should remove all corrosion products from specimens with a minimum removal of sound metal. Set rules cannot be applied to cleaning because procedures will vary with the type of metal being cleaned and the degree of adherence of corrosion products.

Mechanical cleaning includes scrubbing, scraping, brushing, mechanical shocking, and ultrasonic procedures. Scrubbing with a bristle brush and a mild abrasive is the most widely used of these methods; the others are used principally as supplements to remove heavily encrusted corrosion products before scrubbing. Care should be used to avoid the removal of sound metal.

Chemical cleaning implies the removal of material from the surface of the specimen by dissolution in an appropriate chemical agent. Solvents such as acetone, carbon tetrachloride, and alcohol are used to remove oil, grease, or resin and are usually applied prior to other methods of cleaning. Various chemicals are chosen for application to specific materials; some of these treatments in general use are outlined in the NACE standard.

Electrolytic cleaning should be preceded by scrubbing to remove loosely adhering corrosion products. One method of electrolytic cleaning that has been found to be useful for many metals and alloys is as follows:

Solution: 5 percent (by weight) H_2SO_4
Anode: carbon or lead
Cathode: test specimen
Cathode current density: 20 A/dm^2 (129 A/in^2)
Inhibitor: 2 cm^3 organic inhibitor per liter
Temperature: 74°C (165°F)
Exposure period: 3 min

Precautions must be taken to ensure good electrical contact with the specimen, to avoid contamination of the solution with easily reducible metal ions, and to ensure that inhibitor decomposition has not occurred. Instead of using 2 mL of any proprietary inhibitor, 0.5 g/L of inhibitors such as diorthotolyl thiourea or quinoline ethiodide can be used.

Whatever treatment is used to clean specimens after a corrosion test, its effect in removing metal should be determined, and the weight loss should be corrected accordingly. A "blank" specimen should be weighed before and after exposure to the cleaning procedure to establish this weight loss.

Evaluation of Results After the specimens have been reweighed, they should be examined carefully. Localized attack such as pits, crevice corrosion, stress-accelerated corrosion, cracking, or intergranular corrosion should be measured for depth and area affected.

Depth of localized corrosion should be reported for the actual test period and not interpolated or extrapolated to an annual rate. The rate of initiation or propagation of pits is seldom uniform. The size, shape, and distribution of pits should be noted. A distinction should be made between those occurring underneath the supporting devices (concentration cells) and those on the surfaces that were freely exposed to the test solution. An excellent discussion of pitting corrosion has been published [*Corrosion*, 25t (January 1950)].

The specimen may be subjected to simple bending tests to determine whether any **embrittlement** has occurred.

If it is assumed that localized or internal corrosion is not present or is recorded separately in the report, the **corrosion rate** or penetration can be calculated alternatively as

$$\frac{\text{Weight loss} \times 534}{\text{(Area)(time)(metal density)}} = \text{mils/year (mpy)}$$

$$\frac{\text{Weight loss} \times 13.56}{\text{(Area)(time)(metal density)}} = \text{mm/year (mmpy)}$$

where weight loss is in milligrams, area is in square inches of metal surface exposed, time is in hours exposed, and density is in grams per cubic centimeter. Densities for alloys can be obtained from the producers or from various metal handbooks.

The following **checklist** is a recommended guide for reporting all **important information and data:**

Corrosive media and concentration (changes during test)
Volume of test solution
Temperature (maximum, minimum, average)
Aeration (describe conditions or technique)
Agitation (describe conditions or technique)
Type of apparatus used for test
Duration of each test (start, finish)
Chemical composition or trade name of metals tested
Form and metallurgical conditions of specimens
Exact size, shape, and area of specimens
Treatment used to prepare specimens for test
Number of specimens of each material tested and whether specimens were tested separately or which specimens were tested in the same container
Method used to clean specimens after exposure and the extent of any error expected by this treatment
Actual weight losses for each specimen
Evaluation of attack if other than general, such as crevice corro-

sion under support rod, pit depth and distribution, and results of microscopic examination or bend tests

Corrosion rates for each specimen expressed as millimeters (mils) per year

Effect of Variables on Corrosion Tests It is advisable to apply a factor of safety to the results obtained, the factor varying with the degree of confidence in the applicability of the results. Ordinarily, a factor of from 3 to 10 might be considered normal.

Among the more important points that should be considered in attempting to base plant design on laboratory corrosion-rate data are the following.

Galvanic corrosion is a frequent source of trouble on a large scale. Not only is the use of different metals in the same piece of equipment dangerous, but the effect of cold working may be sufficient to establish potential differences of objectionable magnitude between different parts of the same piece of metal. The mass of metal in chemical apparatus is ordinarily so great and the electrical resistance consequently so low that a very small voltage can cause a very high current. Welding also may leave a weld of a different physical or chemical composition from that of the body of the sheet and cause localized corrosion.

Local variations in temperature and crevices that permit the accumulation of corrosion products are capable of allowing the formation of **concentration cells**, with the result of accelerated local corrosion.

In the laboratory, the **temperature** of the test specimen is that of the liquid in which it is immersed, and the measured temperature is actually that at which the reaction is taking place. In the plant (heat being supplied through the metal to the liquid in many cases), the temperature of the film of (corrosive) liquid on the inside of the vessel may be a number of degrees higher than that registered by the thermometer. As the relation between temperature and corrosion is a logarithmic one, the rate of increase is very rapid. Like other chemical reactions, the speed ordinarily increases twofold to threefold for each 10°C temperature rise, the actual relation being that of the equation $\log K = A + (B/T)$, where K represents the rate of corrosion and T the absolute temperature. This relationship, although expressed mathematically, must be understood to be a qualitative rather than strictly a quantitative one.

Cold walls, as in coolers or condensers, usually have somewhat decreased corrosion rates for the reason just described. However, in some cases, the decrease in temperature may allow the formation of a more corrosive second phase, thereby increasing corrosion.

The effect of **impurities** in either structural material or corrosive material is so marked (while at the same time it may be either accelerating or decelerating) that for reliable results the actual materials which it is proposed to use should be tested and not types of these materials. In other words, it is much more desirable to test the actual plant solution and the actual metal or nonmetal than to rely upon a duplication of either. Since as little as 0.01 percent of certain organic compounds will reduce the rate of solution of steel in sulfuric acid 99.5 percent and 0.05 percent bismuth in lead will increase the rate of corrosion over 1000 percent under certain conditions, it can be seen how difficult it would be to attempt to duplicate here all the significant constituents.

CORROSION TESTING: PLANT TESTS

It is not always practical or convenient to investigate corrosion problems in the laboratory. In many instances, it is difficult to discover just what the conditions of service are and to reproduce them exactly. This is especially true with processes involving changes in the composition and other characteristics of the solutions as the process is carried out, as, for example, in evaporation, distillation, polymerization, sulfonation, or synthesis.

With many natural substances also, the exact nature of the corrosive is uncertain and is subject to changes not readily controlled in the laboratory. In other cases, the corrosiveness of the solution may be influenced greatly by or even may be due principally to a constituent present in such minute proportions that the mass available in the limited volume of corrosive solution that could be used in a laboratory setup would be exhausted by the corrosion reaction early in the test, and consequently the results over a longer period of time would be misleading.

Another difficulty sometimes encountered in laboratory tests is that contamination of the testing solution by corrosion products may change its corrosive nature to an appreciable extent.

In such cases, it is usually preferable to carry out the corrosion-testing program by exposing specimens in operating equipment under **actual conditions of service.** This procedure has the additional advantages that it is possible to test a large number of specimens at the same time and that little technical supervision is required.

In certain cases, it is necessary to choose materials for equipment to be used in a process developed in the laboratory and not yet in operation on a plant scale. Under such circumstances, it is obviously impossible to make plant tests. A good procedure in such cases is to construct a pilot plant, using either the cheapest materials available or some other materials selected on the basis of past experience or of laboratory tests. While the pilot plant is being operated to check on the process itself, specimens can be exposed in the operating equipment as a guide to the choice of materials for the large-scale plant or as a means of confirming the suitability of the materials chosen for the pilot plant.

Test Specimens In carrying out plant tests it is necessary to install the test specimens so that they will not come into contact with other metals and alloys; this avoids having their normal behavior disturbed by galvanic effects. It is also desirable to protect the specimens from possible mechanical damage.

There is no single standard size or shape for corrosion-test coupons. They usually weigh from 10 to 50 g and preferably have a large surface-to-mass ratio. Disks 40 mm (1½ in) in diameter by 3.2 mm (⅛ in) thick and similarly dimensioned square and rectangular coupons are the most common. Surface preparation varies with the aim of the test, but machine grinding of surfaces or polishing with a No. 120 grit is common. Samples should not have sheared edges, should be clean (no heat-treatment scale remaining unless this is specifically part of the test), and should be identified by stamping. See Fig. 23-5 for a typical plant test assembly.

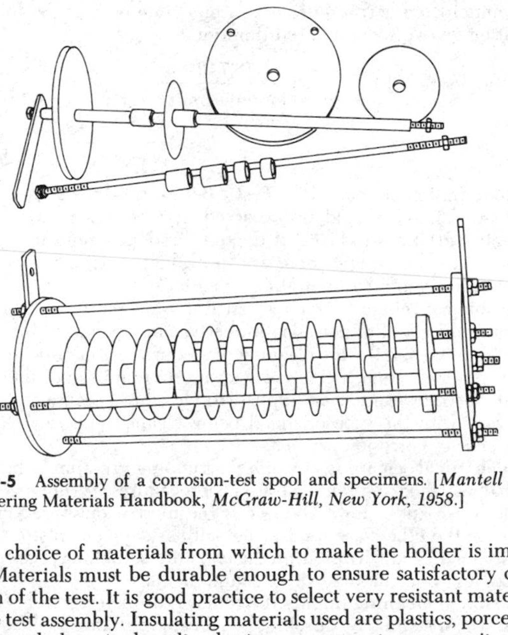

FIG. 23-5 Assembly of a corrosion-test spool and specimens. [*Mantell (ed.),* Engineering Materials Handbook, *McGraw-Hill, New York, 1958.*]

The choice of materials from which to make the holder is important. Materials must be durable enough to ensure satisfactory completion of the test. It is good practice to select very resistant materials for the test assembly. Insulating materials used are plastics, porcelain, Teflon, and glass. A phenolic plastic answers most purposes; its principal limitations are unsuitability for use at temperatures over 150°C (300°F) and lack of adequate resistance to concentrated alkalies.

The method of supporting the specimen holder during the test period is important. The preferred position is with the long axis of the holder horizontal, thus avoiding dripping of corrosion products from one specimen to another. The holder must be located so as to cover the conditions of exposure to be studied. It may have to be submerged, or exposed only to the vapors, or located at liquid level, or holders may be called for at all three locations. Various means have been utilized for supporting the holders in liquids or in vapors. The simplest is to suspend the holder by means of a heavy wire or light metal chain. Holders have been strung between heating coils, clamped to agitator shafts, welded to evaporator tube sheets, etc. The best method is to use test racks.

In a few special cases, the standard "spool-type" specimen holder is not applicable and a suitable special test method must be devised to apply to the corrosion conditions being studied.

For conducting tests in **pipe lines** of 75-mm (3-in) diameter or larger, a spool holder as shown in Fig. 23-6, which employs the same

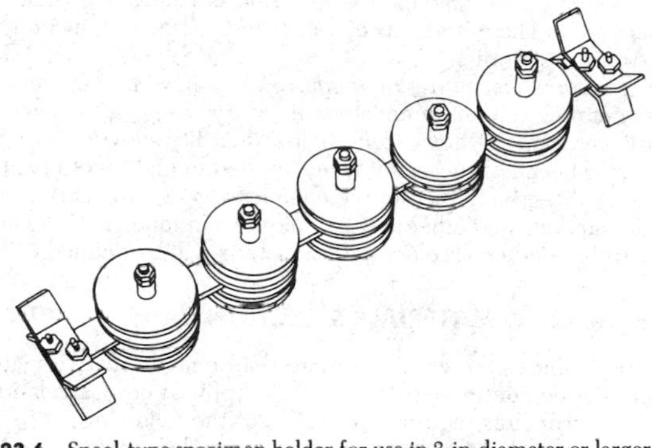

FIG. 23-6 Spool-type specimen holder for use in 3-in-diameter or larger pipe. [*Mantell (ed.),* Engineering Materials Handbook, *McGraw-Hill, New York, 1958.*]

disk-type specimens used on the standard spool holder, has been used. This frame is so designed that it may be placed in a pipe line in any position without permitting the disk specimens to touch the wall of the pipe. As with the strip-type holder, this assembly does not materially interfere with the fluid through the pipe and permits the study of corrosion effects prevailing in the pipe line.

Another way to study corrosion in pipe lines is to install in the line short sections of pipe of the materials to be tested. These test sections should be insulated from each other and from the rest of the piping system by means of nonmetallic couplings. It is also good practice to provide insulating gaskets between the ends of the pipe specimens where they meet inside the couplings. Such joints may be sealed with various types of "dope" or cement. It is desirable in such cases to paint the outside of the specimens so as to confine corrosion to the inner surface.

It is occasionally desirable to expose corrosion-test specimens in operating equipment without the use of specimen holders of the type described. This can be accomplished by attaching specimens directly to some part of the operating equipment and by providing the necessary insulation against galvanic effects as shown in Fig. 23-7. The suggested method of attaching specimens to racks has been found to be very suitable in connection with the exposure of specimens to corrosion in seawater.

Test Results The methods of cleaning specimens and evaluating results after plant corrosion tests are identical to those described earlier for laboratory tests.

CORROSION TESTING: NEW METHODS

Electrical-Resistance Method If a corrosion-test sample is in the form of a thin wire or strip, its electrical resistance increases as corrosion decreases its cross section. Therefore, a periodic or continuous

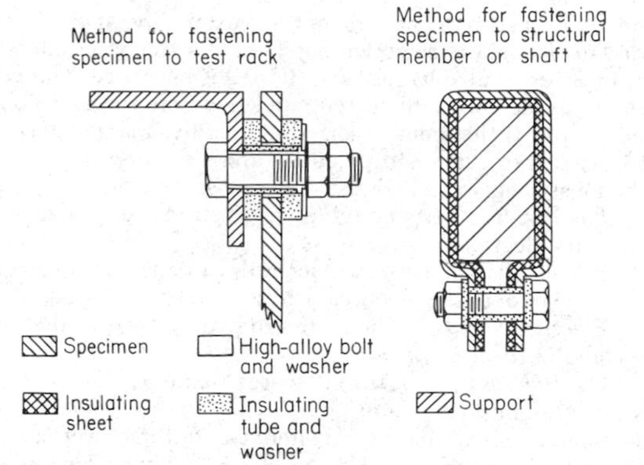

FIG. 23-7 Methods for attaching specimens to test racks and to parts of moving equipment. [*Mantell (ed.),* Engineering Materials Handbook, *McGraw-Hill, New York, 1958.*]

measurement of the longitudinal resistance of the specimen can be used to monitor the corrosion. The electrical-resistance measurement has nothing to do with the electrochemistry of the corrosion reaction. It merely measures a bulk property that is dependent upon the specimen's cross-section area. Commercial instruments are available (Fig. 23-8).

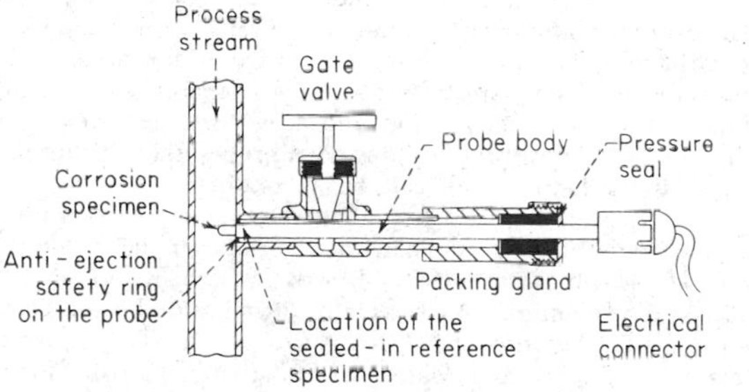

FIG. 23-8 Typical retractable corrosion probe.

Advantages of the electrical-resistance technique are:

1. A corrosion measurement can be made without having to see or remove the test sample.

2. Corrosion measurements can be made quickly—in a few hours or days, or continuously. This enables sudden increases in corrosion rate to be detected. In some cases, it will be possible then to modify the process to decrease the corrosion.

3. The method can be used to monitor a process to indicate whether the corrosion rate is dependent on some critical process variable.

4. Corrodent need not be an electrolyte (in fact, need not be a liquid).

5. The method can detect low corrosion rates that would take a long time to detect with weight-loss methods.

Limitations of the technique are:

1. It is usually limited to the measurement of uniform corrosion only and is not generally satisfactory for localized corrosion.

2. The probe design includes provisions to compensate for temperature variations. This feature is not totally successful. The most reliable results are obtained in constant-temperature systems.

Linear-Polarization Method The linear-polarization (or polarization-resistance) technique gives an indication of the corrosion rate in a minute or so without making any prior measurements. This method can be used to monitor corrosion in the plant. Commercial instruments are available, and some of these give a direct readout (semicontinuous recording, if desired).

The technique is dependent upon the corrosion's being electrochemical in nature. The amount of externally applied current needed to change the corrosion potential of a freely corroding specimen by a few millivolts (usually 10 mV) is measured. This current is related to the corrosion current, and therefore the corrosion rate, of the sample. If the metal is corroding rapidly, a large external current is needed to change its potential, and vice versa.

The measuring system consists of four basic elements:

1. *Electrodes.* Test and reference electrodes, and in some cases, an auxiliary electrode.
2. *Probe.* It connects the electrodes in the corrodent on the inside of a vessel to the electrical leads.
3. *Electrical leads.* They run from the probe to the current source and instrument panel.
4. *Control system.* Current source (batteries), ammeter, voltmeter, instrument panel, etc.

Commercial instruments have either two or three electrodes. Also, there are different types of three-electrode systems. The application and limitations of the instruments are largely dependent upon these electrode systems.

Electrochemical Corrosion Testing The application of electrochemical methods to practical corrosion problems (other than cathodic protection systems, in which its use has been well established for years) is a relatively recent phenomenon. It is now well established that the activity of pitting, crevice corrosion, and stress-corrosion cracking is strongly dependent upon the corrosion potential (i.e., the potential difference between the corroding metal and a suitable reference electrode). By using readily available electronic equipment, the quantity and direction of direct current required to control the corrosion potential in a given solution at a given selected value can be measured. A plot of such values over a range of potentials is called a polarization diagram. By using proper experimental techniques, it is possible to define approximate ranges of corrosion potential in which pitting, crevice corrosion, and stress-corrosion cracking will or will not occur. With properly designed probes, these techniques can be used in the field as well as in the laboratory.

For further details, the following references are recommended: Martin, "Potentiodynamic Polarization Studies in the Field," *Mater. Prot.*, **18**, 41–50 (March 1979). Morris, "Use of Rapid-Scan Potentiodynamic Techniques to Evaluate Pitting and Crevice Corrosion Resistance," *Galvanic and Pitting Corrosion*, STP 576, American Society for Testing and Materials, Philadelphia, 1976. Morris and Scarberry, "Predicting Corrosion Rates with the Potentiostat," *Corrosion*, **28**, 444–452 (December 1972).

Stress-Corrosion-Cracking Tests Prestressed samples, such as are shown in Fig. 23-9, have been used for laboratory and field SCC testing. The variable observed is "time to failure or visible cracking." Unfortunately, such tests do not provide acceleration of failure.

Since SCC frequently shows a fairly long induction period (months to years), such tests must be conducted for very long periods before reliable conclusions can be drawn.

Much work is under way to evaluate the newer concepts listed here, which show promise of providing reliable data in shorter (more practical) time periods.

1. *Slow-strain-rate-testing.* This involves a very slow tensile test conducted with the tensile specimen exposed to the environment of interest. Typical strain rates used range from $1 \times 10^{-4}\ s^{-1}$ to $1 \times 10^{-9}\ s^{-1}$. SCC is detected by deviation of the stress-strain curve from that observed in a noncorrosive environment. This technique has been effectively applied to a number of practical problems. (See *Stress Corrosion Cracking, The Slow Strain Rate Technique*, STP 665, American Society for Testing and Materials, Philadelphia, 1979, for further details.)
2. *Electrochemical methods.* These involve the use of corrosion potential monitoring or control of corrosion potential to a selected value (see subsection "Electrochemical Corrosion Testing") in conjunction with various methods of stress application. They have proved useful in acceleration tests and in defining the range of corrosion potentials in which stress-corrosion cracking can occur in a given system. These methods can also be used in conjunction with slow-strain-rate testing.
3. *Fracture mechanics methods.* These have proved very useful for defining the minimum stress intensity (K_{ISCC}) at which stress-corrosion cracking of high-strength, low-ductility alloys occurs. They have so far been less successful when applied to high-ductility alloys, which are extensively used in the chemical-process industries.

Work on these and other new techniques continues, and it is hoped that a truly reliable, accelerated test or tests will be defined.

ECONOMICS IN MATERIALS SELECTION

In most instances, there will be more than one alternative material which may be considered for a specific application. Calculation of true long-term costs requires estimation of the following:

1. Total cost of fabricated equipment and piping
2. Total installation cost
3. Service life
4. Maintenance costs: amount and timing
5. Time and cost requirements to replace or repair at the end of service life
6. Cost of downtime to replace or repair
7. Cost of inhibitors, extra control facilities, etc., required to assure achievement of predicted service life
8. Time value of money
9. Factors which impact taxation, such as depreciation and tax rates
10. Inflation rate

Proper economic analysis will allow comparison of alternatives on a sound basis. Detailed calculations are beyond the scope of this section. The reader should review the material in Sec. 25. NACE Standard RP-02-72 (Direct Calculation of Economic Appraisals of Corrosion Control Measures, National Association of Corrosion Engineers, Houston) presents one sound method.

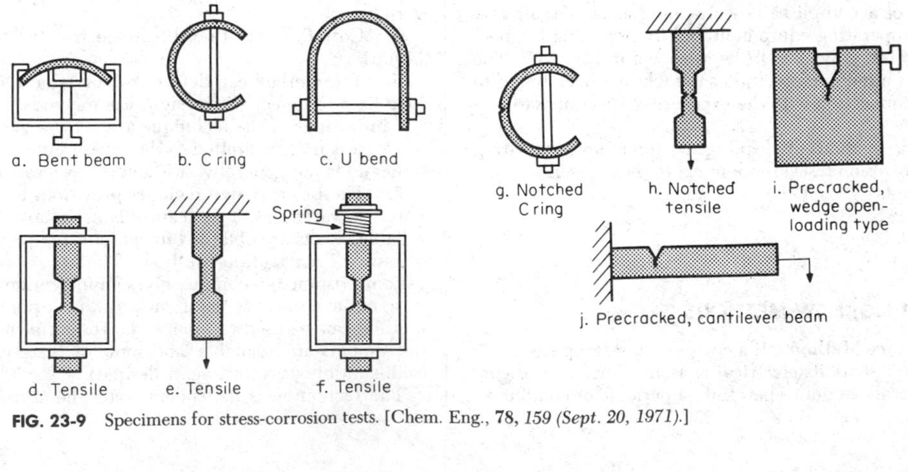

FIG. 23-9 Specimens for stress-corrosion tests. [Chem. Eng., **78**, 159 *(Sept. 20, 1971).*]

PROPERTIES OF MATERIALS

GUIDE TO TABULATED DATA

In keeping with the theme set by the preceding subsection on corrosion and its control, the emphasis in this subsection, dealing with properties of materials, is on their corrosion resistance. Since many other properties enter into consideration when selecting materials of construction, physical and mechanical properties are covered also.

Following is a general discussion of properties of various materials, arranged by type. The bulk of the information presented here on properties of materials, however, is contained in an extensive series of tables, Tables 23-2 through 23-33. Additional information specifically related to materials for low and high temperatures is found later in Sec. 23.

MATERIALS STANDARDS AND SPECIFICATIONS

There are obvious benefits to be derived from consensus standards which define the chemistry and properties of specific materials. Such standards allow designers and users of materials to work with confidence that the materials supplied will have the expected minimum properties. Designers and users can also be confident that comparable materials can be purchased from several suppliers. Producers are confident that materials produced to an accepted standard will find a ready market and therefore can be produced efficiently in large factories.

While a detailed treatment is beyond the scope of this section, a few of the organizations which generate standards of major importance to the chemical-process industries in the United States are listed here. An excellent overview is presented in the *Encyclopedia of Chemical Technology* (3d ed., Wiley, New York, 1978–1980).

1. **American National Standards Institute (ANSI),** formerly American Standards Association (ASA). ANSI promulgates the piping codes used in the chemical-process industries.
2. **American Society of Mechanical Engineers (ASME).** This society generates the Boiler and Pressure Vessel Codes.
3. **American Society for Testing and Materials (ASTM).** This society generates specifications for most of the materials used in the ANSI Piping Codes and the ASME Boiler and Pressure Vessel Codes.
4. **International Organization for Standardization (ISO).** This organization is engaged in generating standards for worldwide use. It has 80 member nations.

FERROUS METALS AND ALLOYS

Steel Carbon steel is the most common, cheapest, and most versatile metal used in industry. It has excellent ductility, permitting many cold-forming operations. Steel is also very weldable.

The grades of steel most commonly used in the chemical-process industries have tensile strength in the 345- to 485-MPa (50,000 to 70,000- lbf/in^2) range, with good ductility. Higher strength levels are achieved by cold work, alloying, and heat treatment.

Carbon steel is easily the most commonly used material in process plants despite its somewhat limited corrosion resistance. It is routinely used for most organic chemicals and neutral or basic aqueous solutions at moderate temperatures. It is also used routinely for the storage of concentrated sulfuric acid and caustic soda [up to 50 percent and 55°C (130°F)]. Because of its availability, low cost, and ease of fabrication steel is frequently used in services with corrosion rates of 0.13 to 0.5 mm/year (5 to 20 mils/year), with added thickness **(corrosion allowance)** to assure the achievement of desired service life. Product quality requirements must be considered in such cases.

Low-Alloy Steels Alloy steels contain one or more alloying agents to improve mechanical and corrosion-resistant properties over those of carbon steel.

A typical low-alloy grade [American Iron and Steel Institute (AISI) 4340] contains 0.40 percent C, 0.70 percent Mn, 1.85 percent Ni, 0.80 percent Cr, and 0.25 percent Mo. Many other alloying agents are used to produce a large number of standard AISI and proprietary grades.

Nickel increases toughness and improves low-temperature properties and corrosion resistance. Chromium and silicon improve hardness, abrasion resistance, corrosion resistance, and resistance to oxidation. Molybdenum provides strength at elevated temperatures.

The addition of small amounts of alloying materials greatly improves corrosion resistance to atmospheric environments but does not have much effect against liquid corrosives. The alloying elements produce a tight, dense adherent rust film, but in acid or alkaline solutions corrosion is about equivalent to that of carbon steel. However, the greater strength permits thinner walls in process equipment made from low-alloy steel.

Cast Irons Generally, cast iron is not a particularly strong or tough structural material, although it is one of the most economical and is widely used industrially.

Gray cast iron, low in cost and easy to cast into intricate shapes, contains carbon, silicon, managanese, and iron. Carbon (1.7 to 4.5 percent) is present as combined carbon and graphite; combined carbon is dispersed in the matrix as iron carbide (cementite), while free graphite occurs as thin flakes dispersed throughout the body of the metal. Various strengths of gray iron are produced by varying size, amount, and distribution of graphite.

Gray iron has outstanding "damping" properties—i.e., ability to absorb vibration—as well as wear resistance. However, gray iron is brittle, with poor resistance to impact and shock. Machinability is excellent.

With some important exceptions, gray-iron castings generally have corrosion resistance similar to that of carbon steel. They do resist atmospheric corrosion as well as attack by natural or neutral waters and neutral soils. However, dilute acids and acid-salt solutions will attack this material.

Gray iron is resistant to concentrated acids (nitric, sulfuric, phosphoric) as well as to some alkaline and caustic solutions. Caustic fusion pots are usually made from gray cast iron with low silicon content; cast-iron valves, pumps, and piping are common in sulfuric acid plants.

White cast iron is brittle and difficult to machine. It is made by controlling the composition and rate of solidification of the molten iron so that all the carbon is present in the combined form. Very abrasive- and wear-resistant, white cast iron is used as liners and for grinding balls, dies, and pump impellers.

Malleable iron is made from white cast iron. It is cast iron with free carbon as dispersed nodules. This arrangement produces a tough, relatively ductile material. Total carbon is about 2.5 percent. Two types are produced: standard and pearlitic (combined carbon plus nodules). Standard malleable iron is easily machined; pearlitic, less so. Both types will withstand bending and cold working without cracking. Large welded areas are not recommended with fusion welding because welds are brittle. Corrosion resistance is about the same as for gray cast iron.

Ductile cast iron includes a group of materials with good strength, toughness, wear resistance, and machinability. This type of cast iron contains combined carbon and dispersed nodules of carbon. Composition is about the same as gray iron, with more carbon (3.7 percent) than malleable iron. The spheroidal graphite reduces the notch effect produced by graphite flakes, making the material more ductile.

There are a number of grades of ductile iron; some have maximum toughness and machinability; others have maximum resistance to oxidation.

Generally, corrosion resistance is similar to gray iron. But ductile iron can be used at higher temperatures—up to 590°C (1100°F) and sometimes even higher.

Alloy Cast Irons Cast iron is not usually considered corrosion-resistant, but this condition can be improved by the use of various cast-iron alloys. A number of such materials are commercially available.

TABLE 23-2 Detailed Corrosion Data on Construction Materials*

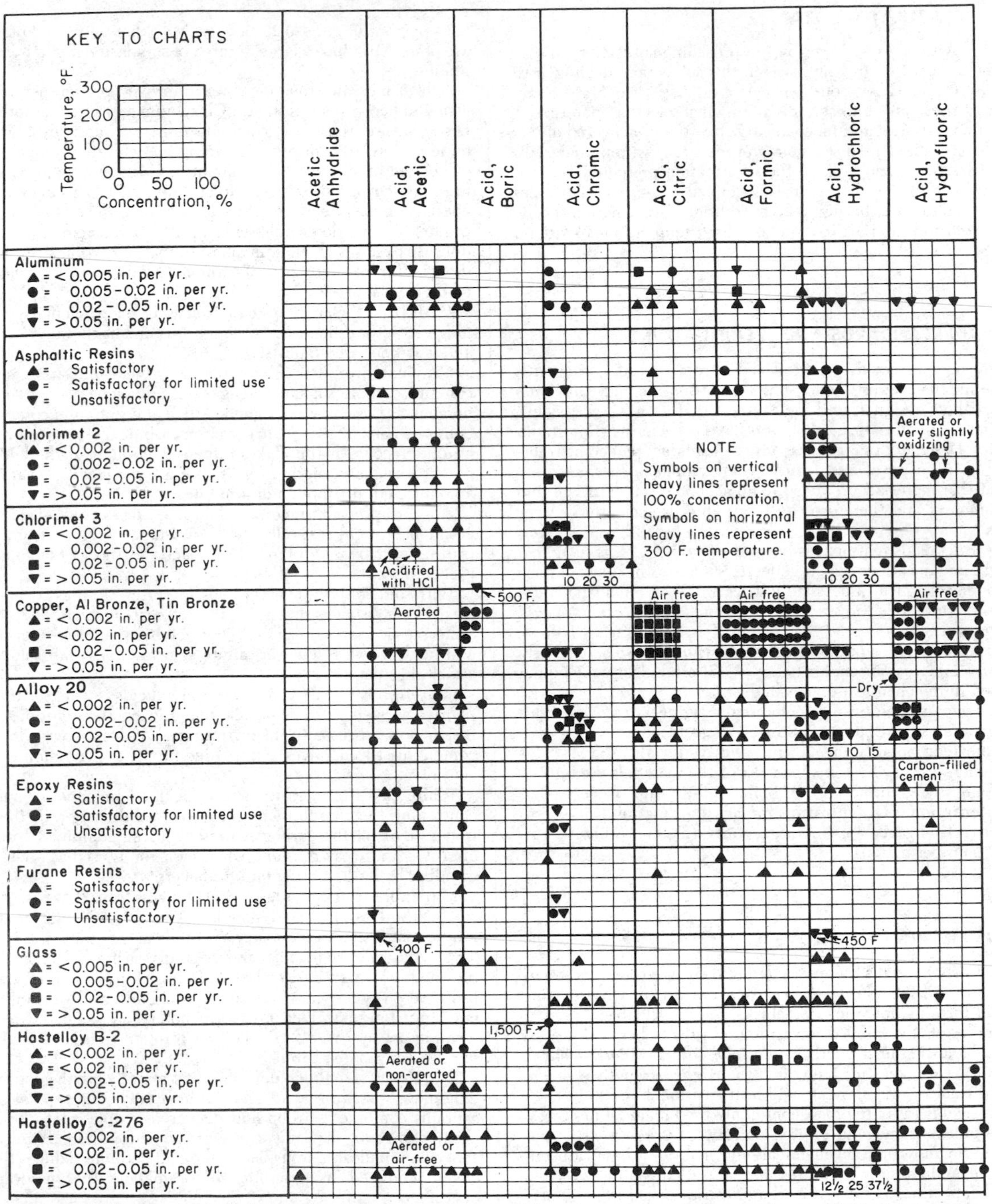

*Based on information published in *Chemical Engineering* magazine. These charts are to assist in narrowing your field of choice. They should not be used for anything more. Effects of contaminants, aeration, galvanic coupling, erosion, etc., must be taken into account. Field tests are best for determining suitability. To convert from inches to millimeters, multiply by 25.4.

TABLE 23-2 Detailed Corrosion Data on Construction Materials* *(Continued)*

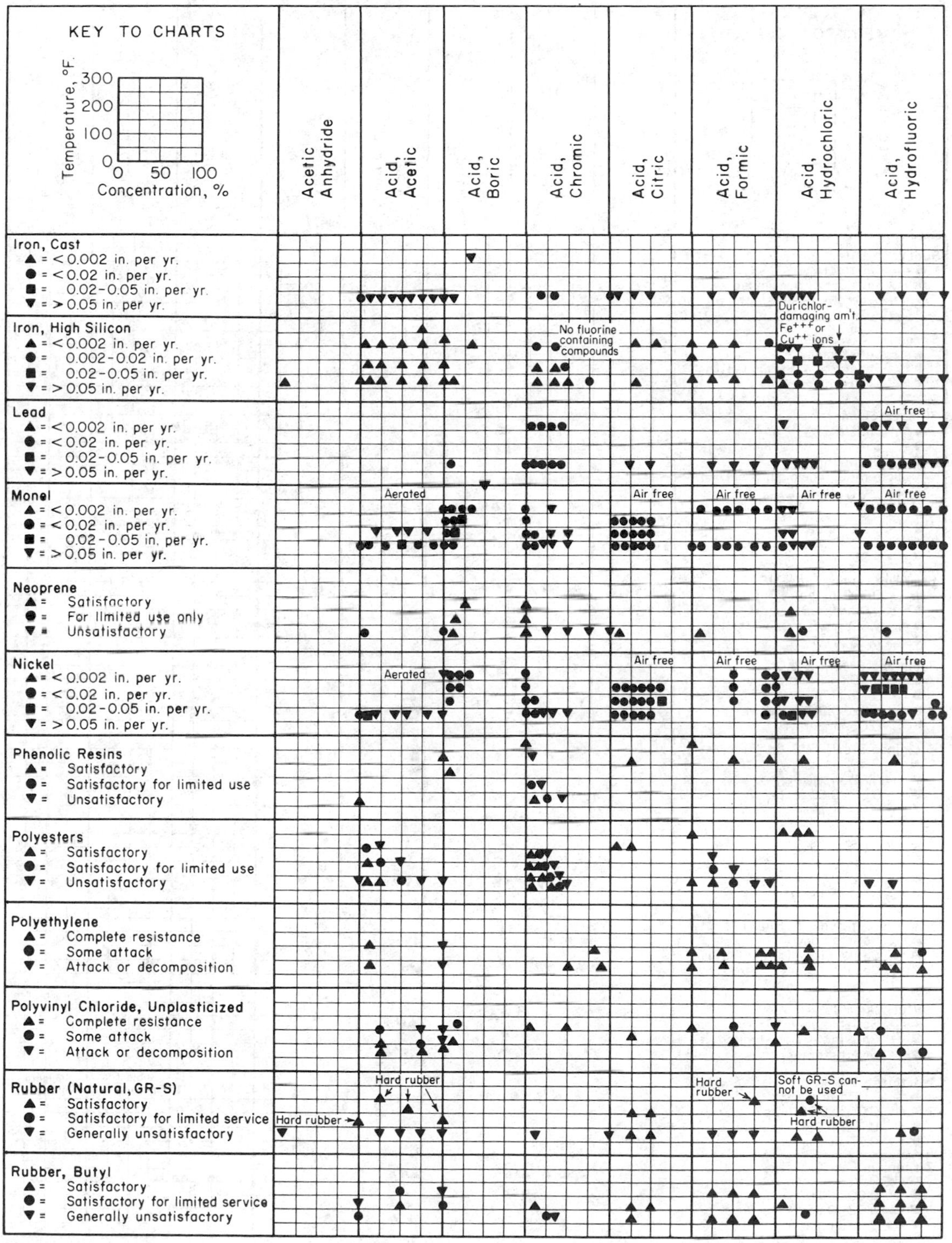

*To convert from inches to millimeters, multiply by 25.4.

TABLE 23-2 Detailed Corrosion Data on Construction Materials* *(Continued)*

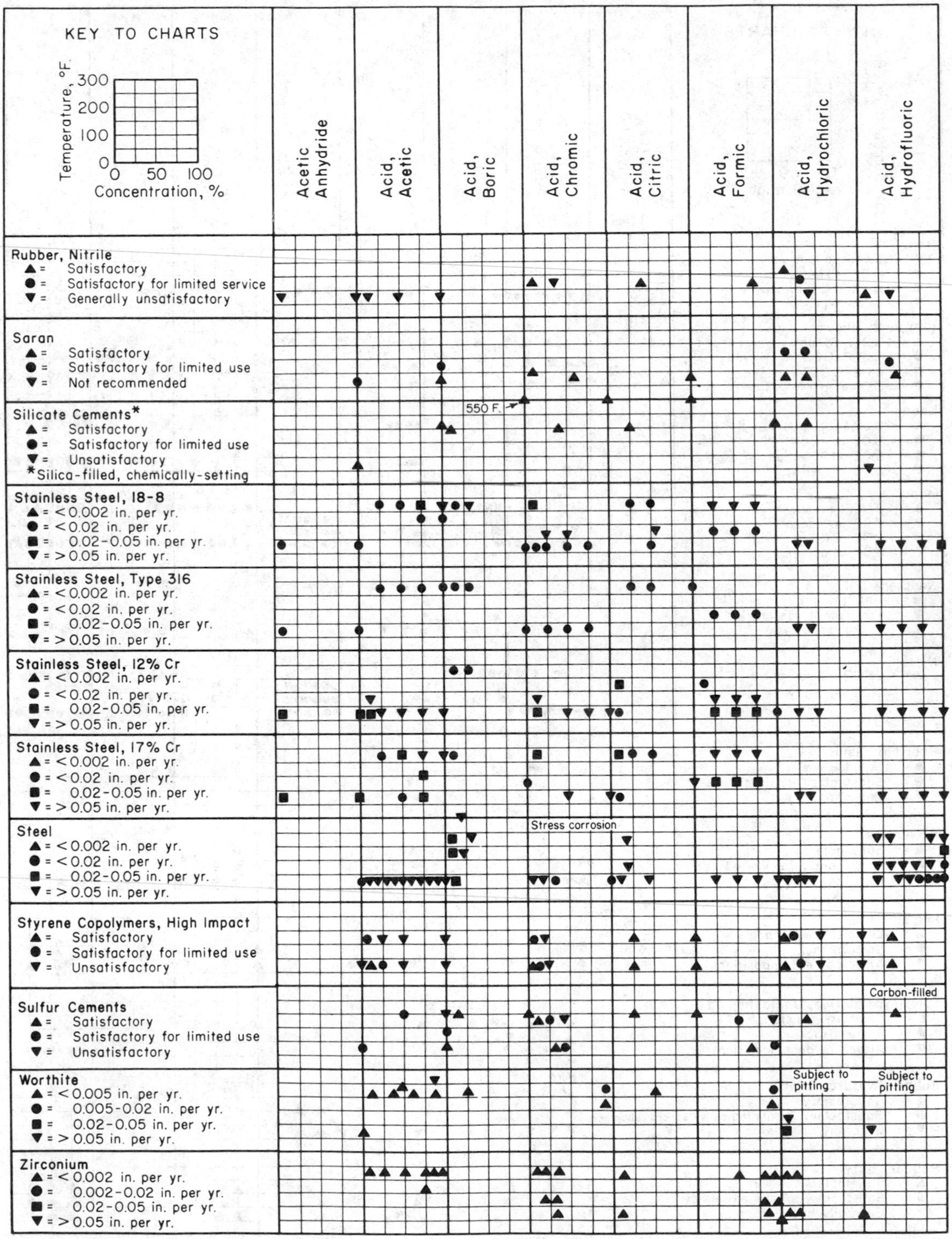

*To convert from inches to millimeters, multiply by 25.4.

TABLE 23-2 Detailed Corrosion Data on Construction Materials* *(Continued)*

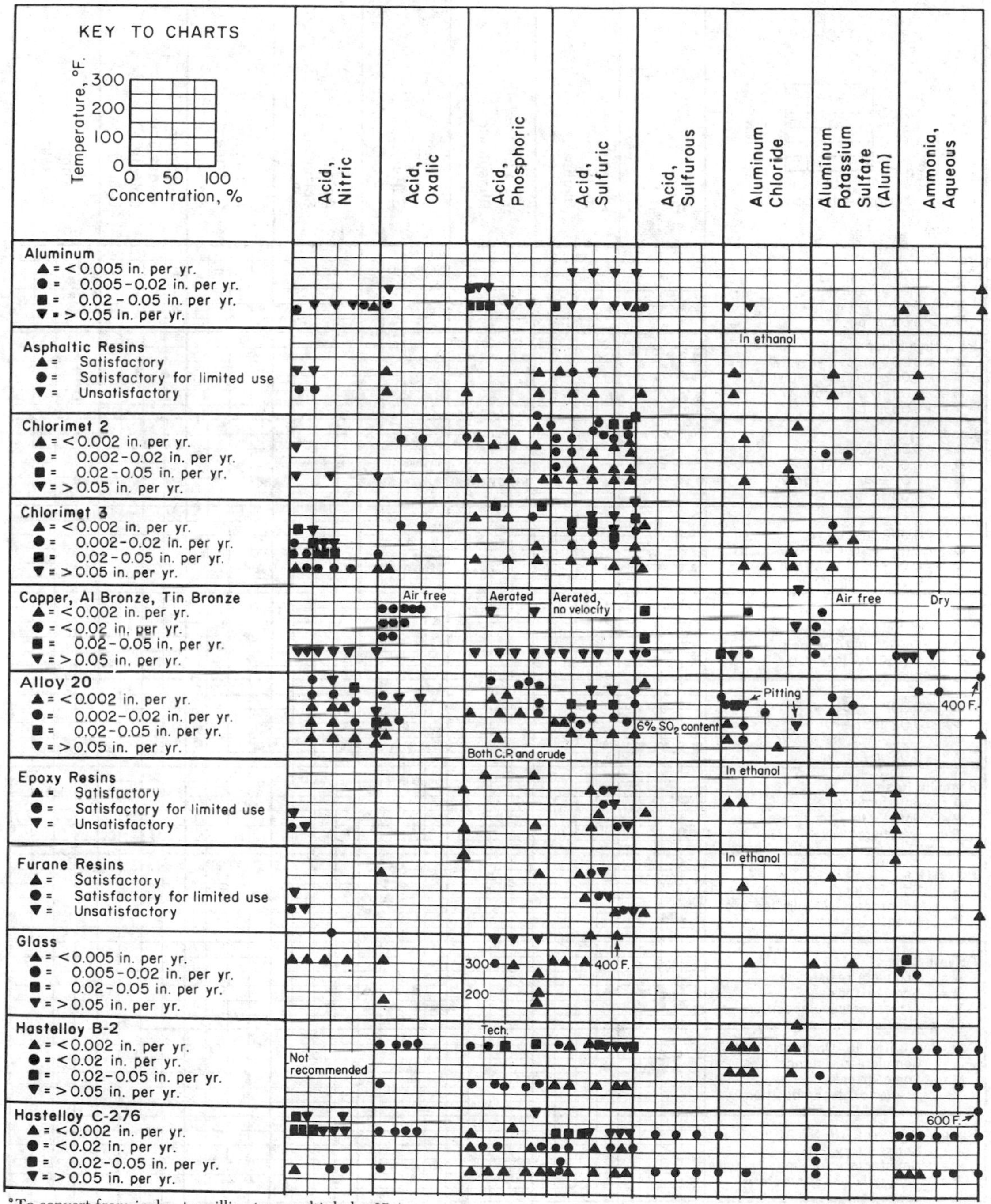

*To convert from inches to millimeters, multiply by 25.4.

TABLE 23-2 Detailed Corrosion Data on Construction Materials* *(Continued)*

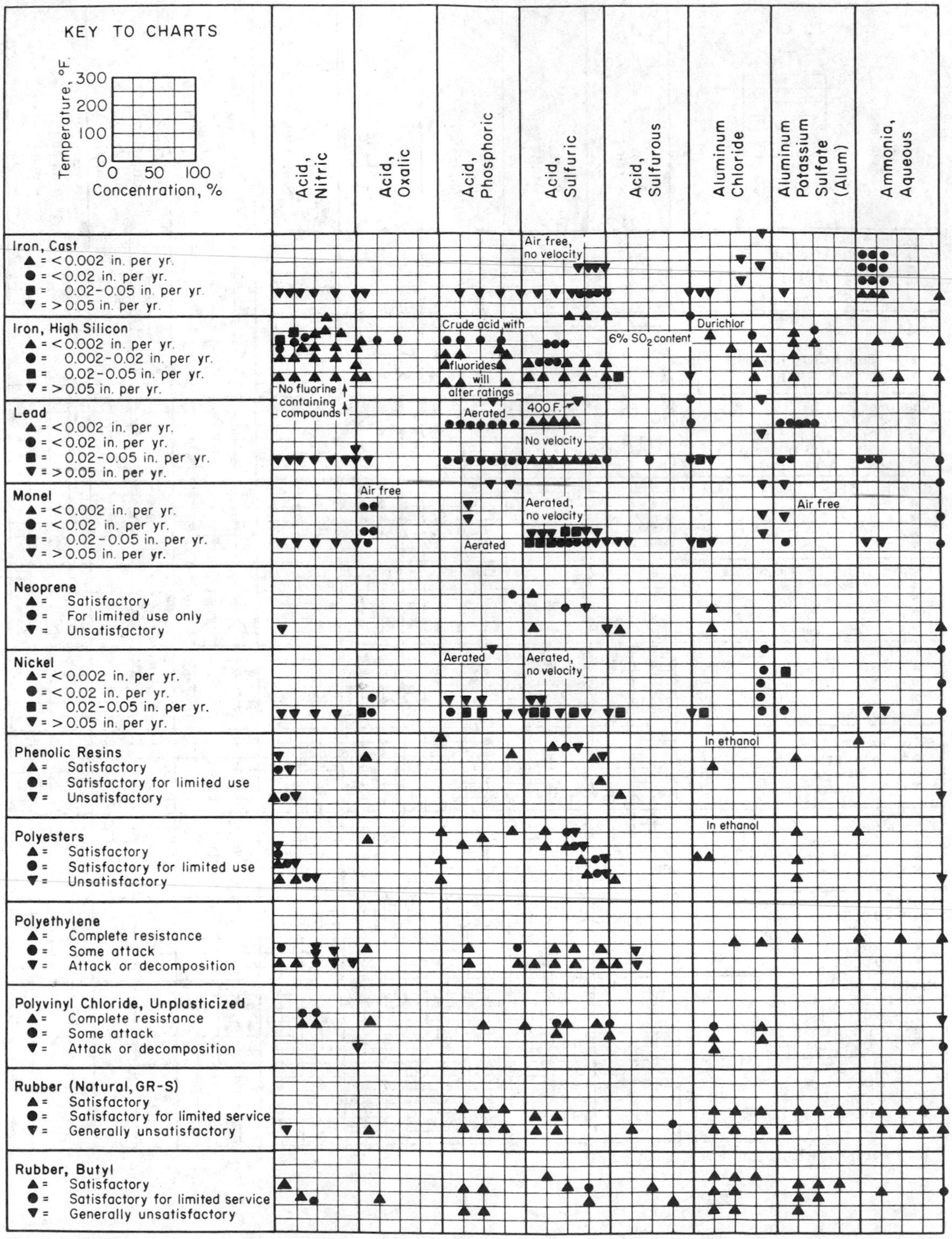

*To convert from inches to millimeters, multiply by 25.4.

TABLE 23-2 Detailed Corrosion Data on Construction Materials* *(Continued)*

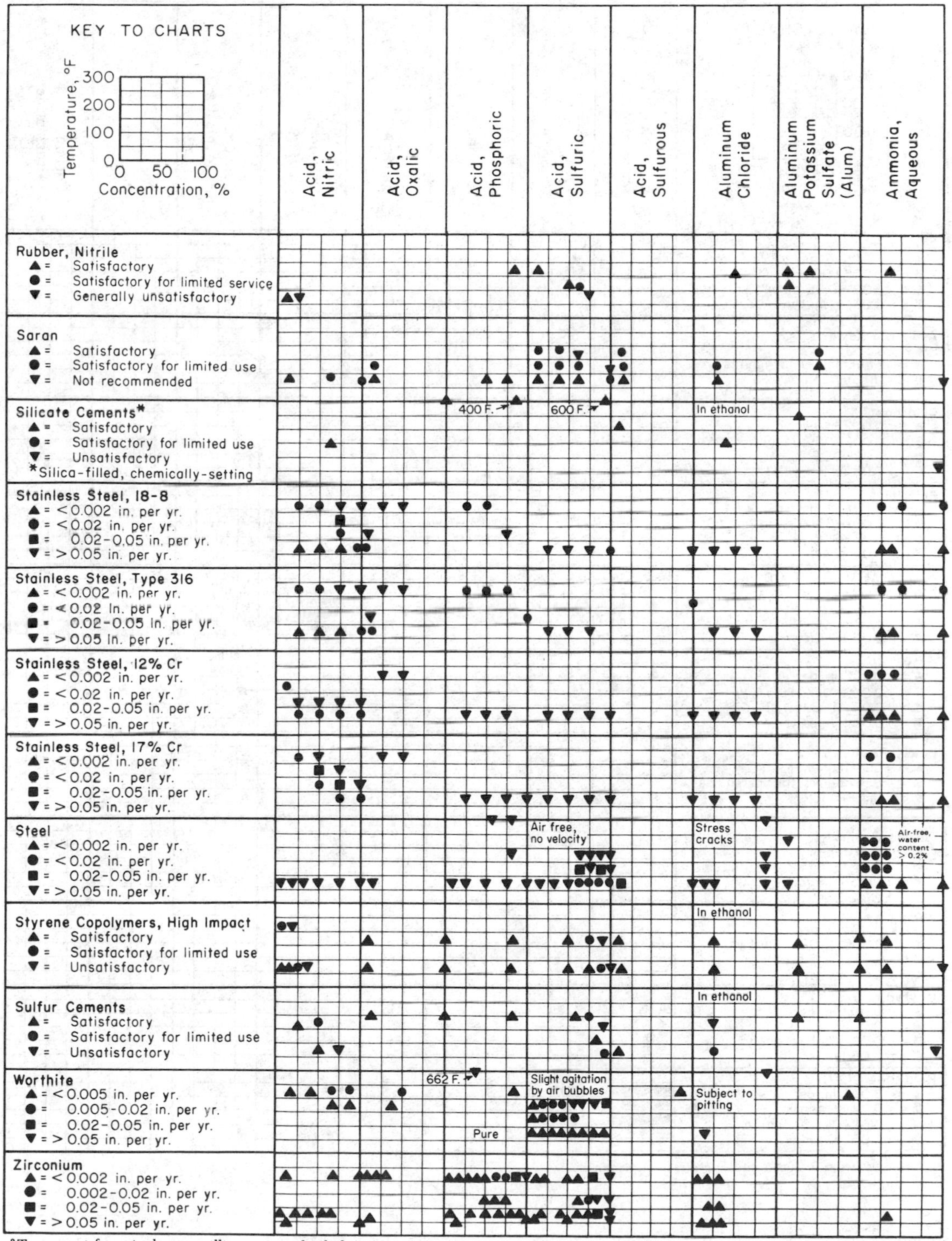

*To convert from inches to millimeters, multiply by 25.4.

TABLE 23-2 Detailed Corrosion Data on Construction Materials* *(Continued)*

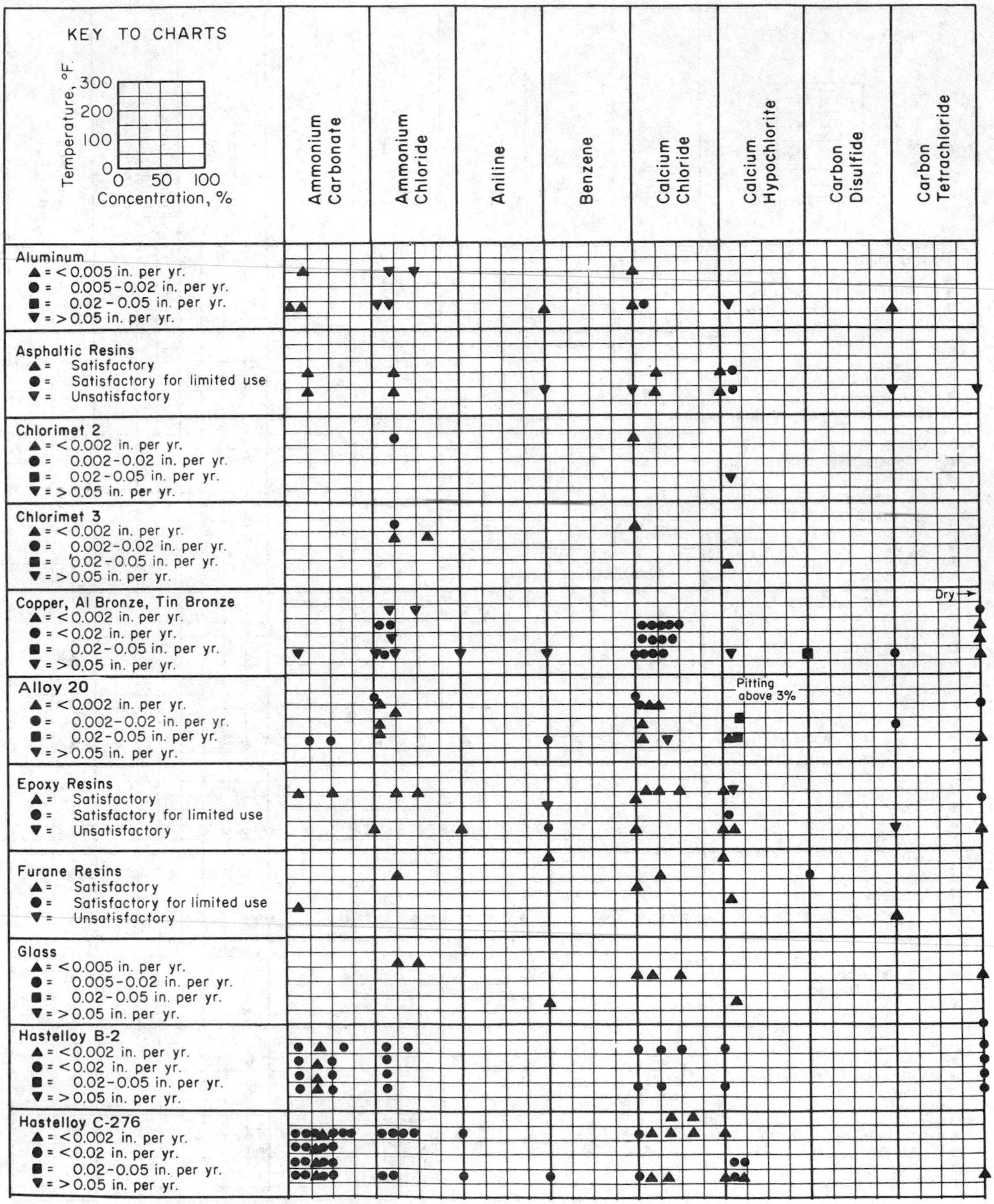

*To convert from inches to millimeters, multiply for 25.4.

TABLE 23-2 Detailed Corrosion Data on Construction Materials* *(Continued)*

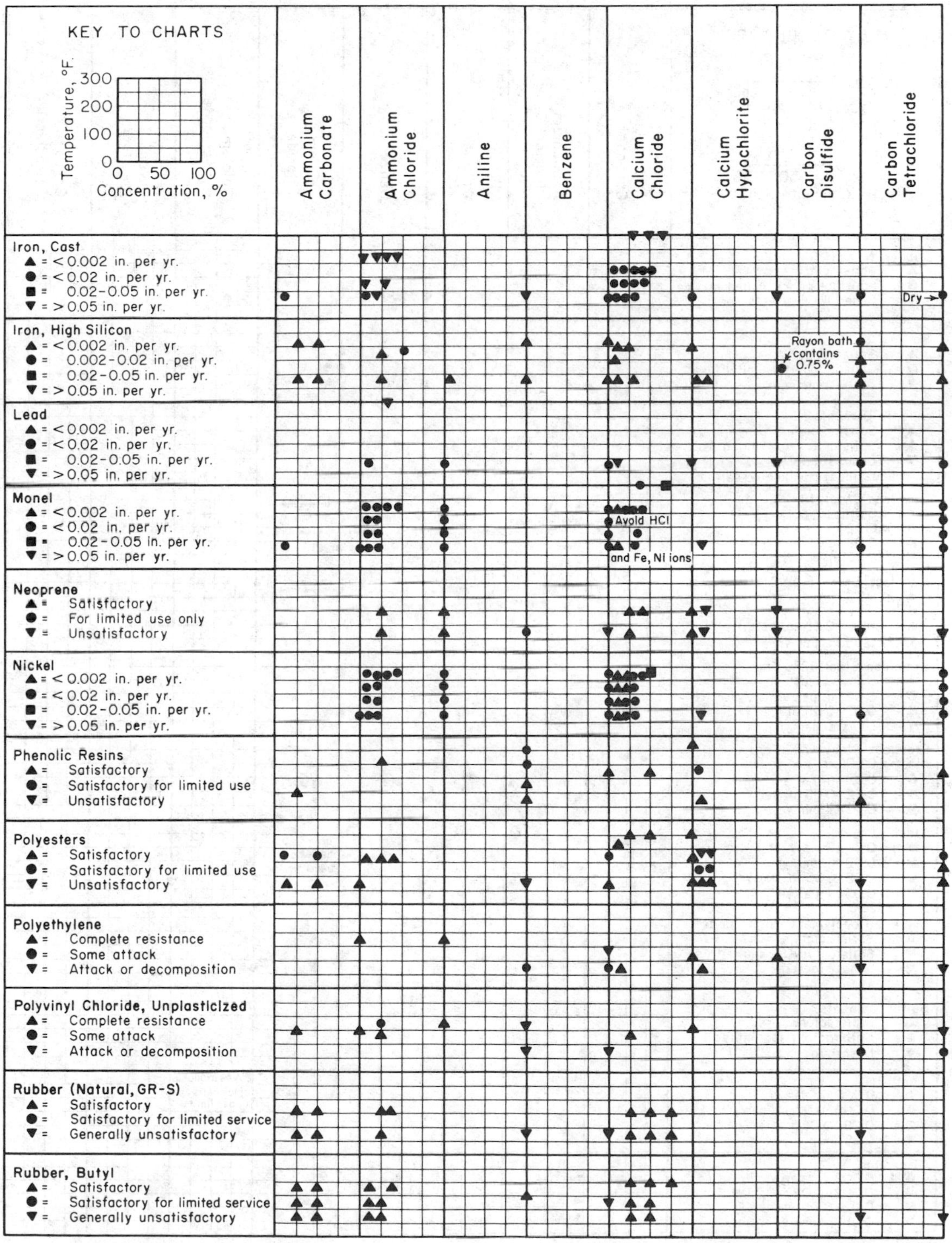

*To convert from inches to millimeters, multiply by 25.4.

TABLE 23-2 Detailed Corrosion Data on Construction Materials* *(Continued)*

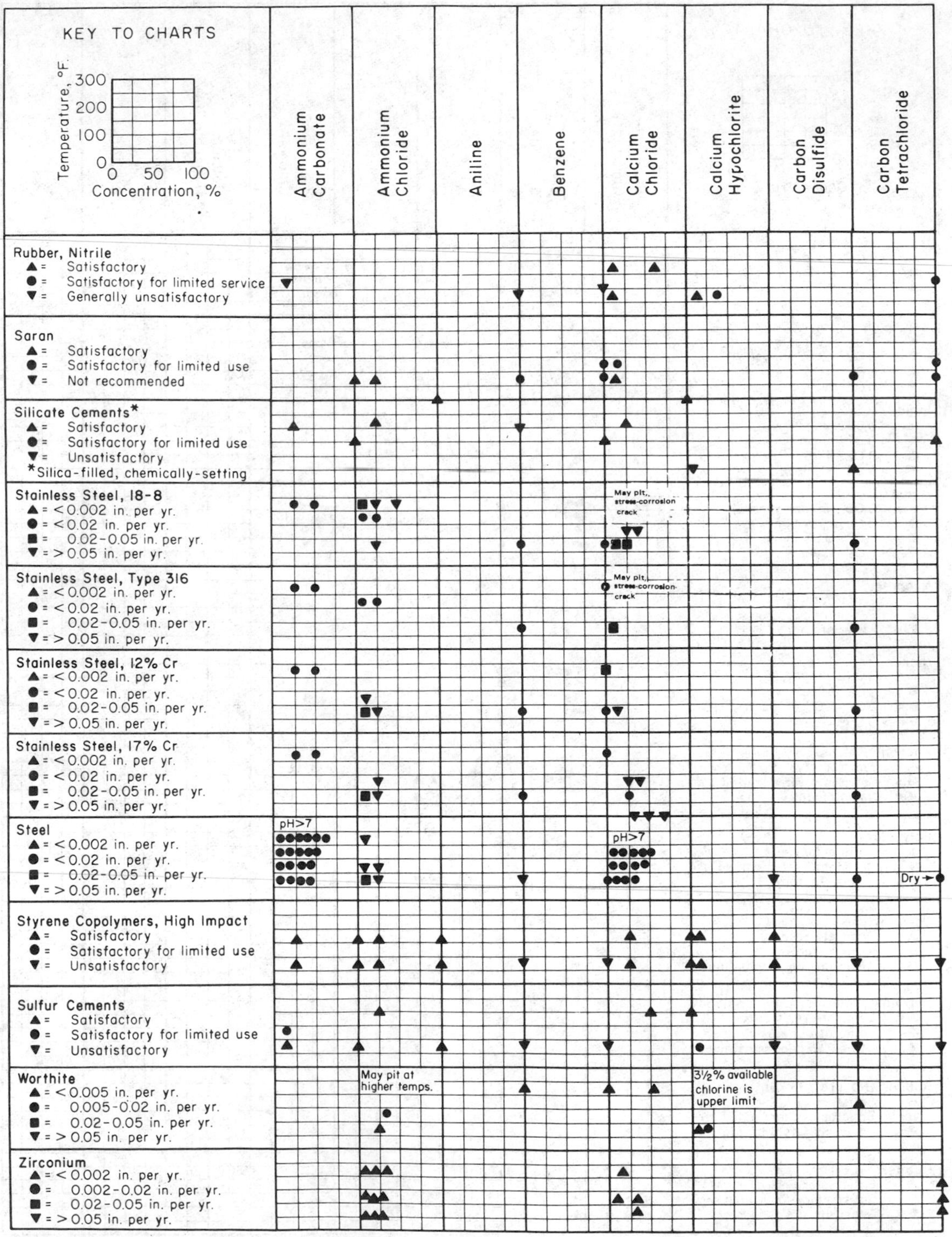

*To convert from inches to millimeters, multiply by 25.4.

TABLE 23-2 Detailed Corrosion Data on Construction Materials* *(Continued)*

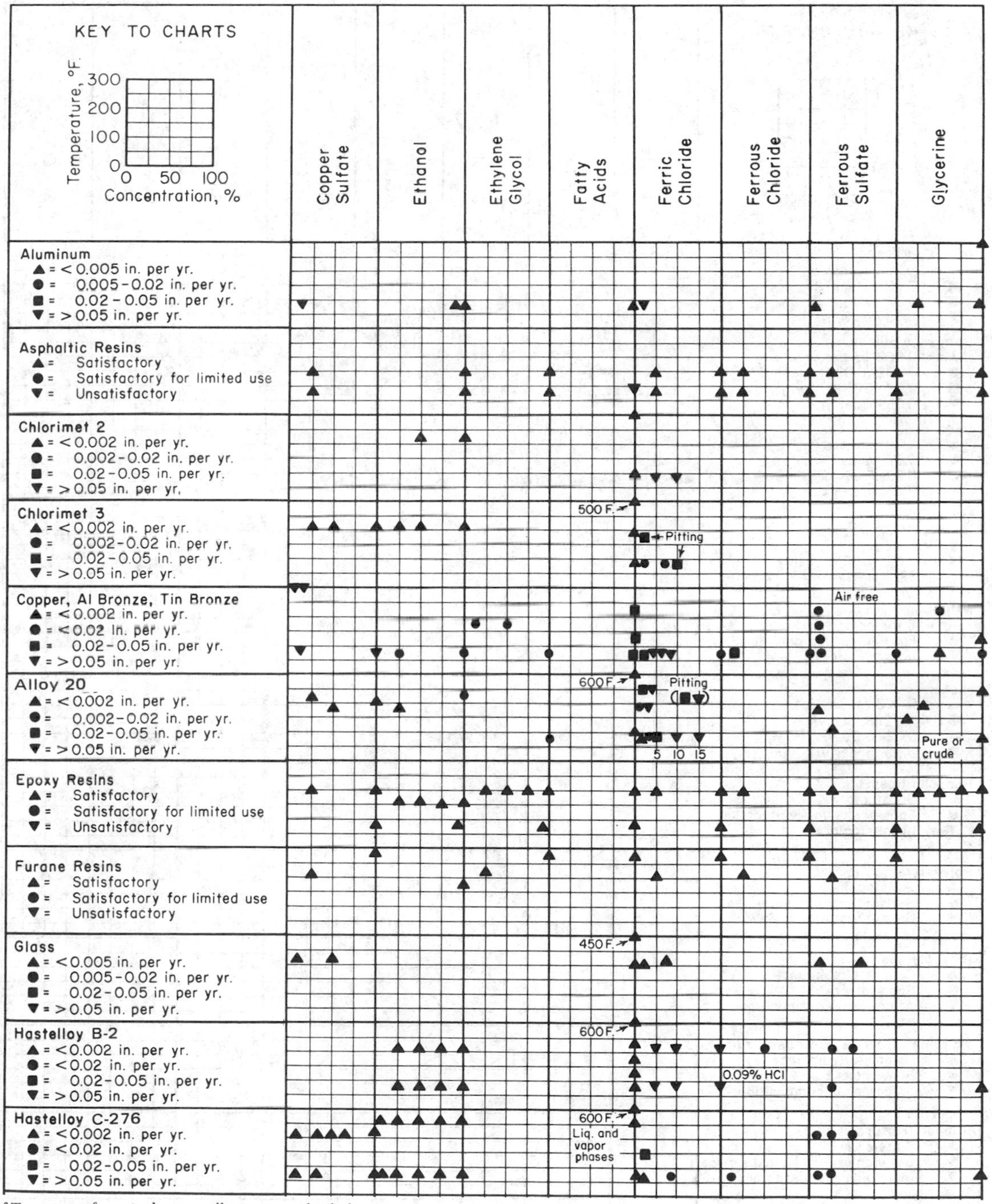

*To convert from inches to millimeters, multiply by 25.4.

TABLE 23-2 Detailed Corrosion Data on Construction Materials* *(Continued)*

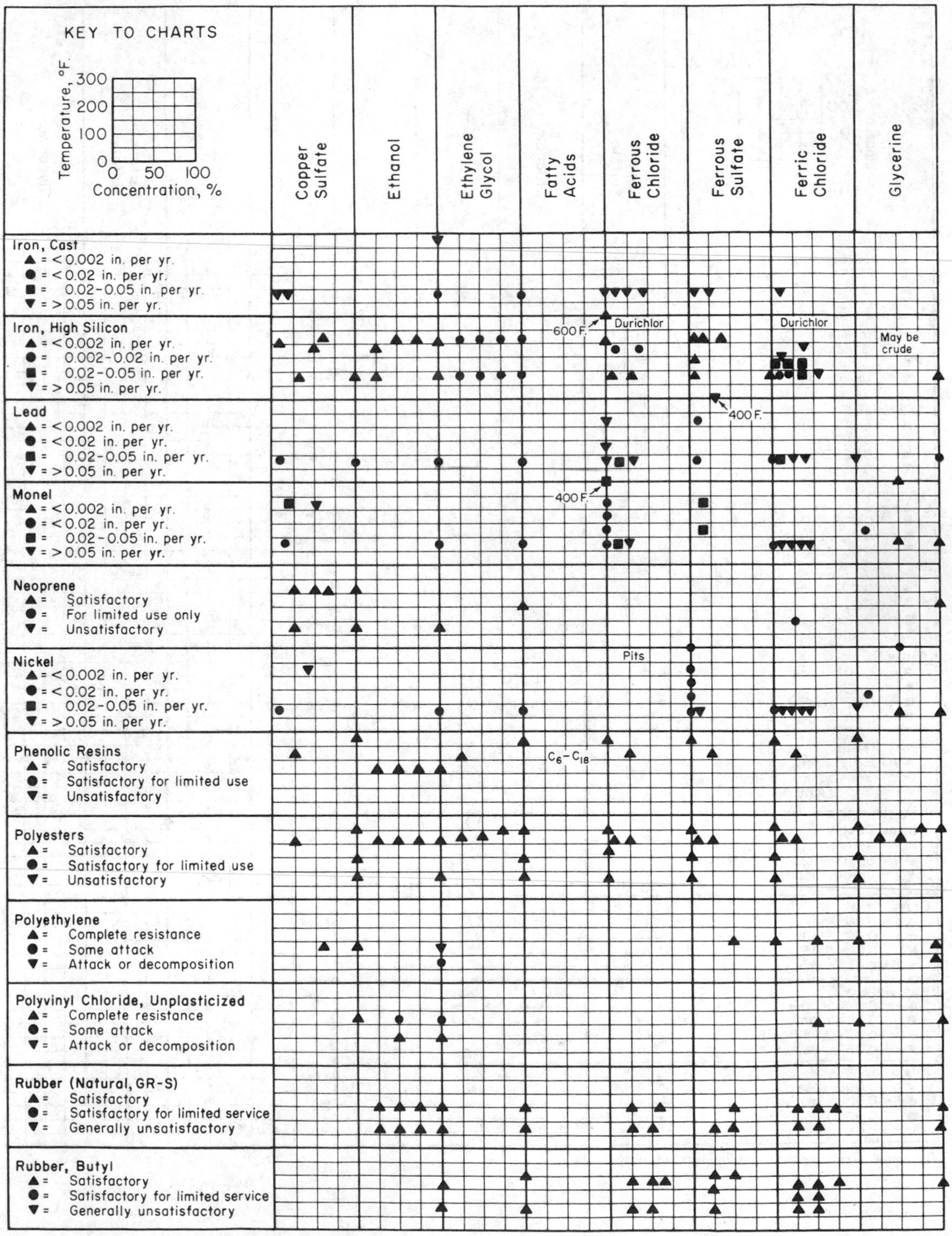

*To convert from inches to millimeters, multiply by 25.4.

TABLE 23-2 Detailed Corrosion Data on Construction Materials* *(Continued)*

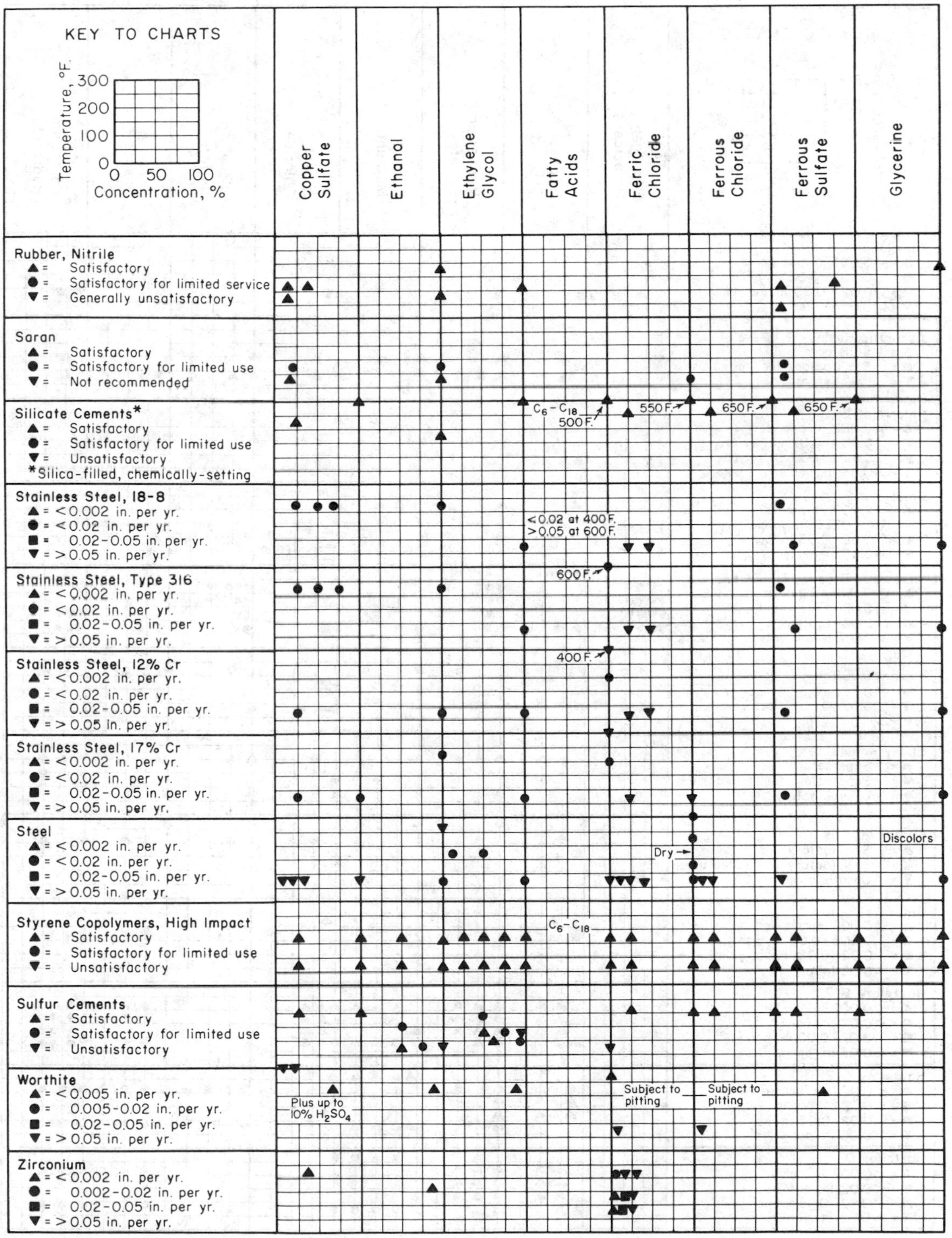

°To convert from inches to millimeters, multiply by 25.4.

TABLE 23-2 Detailed Corrosion Data on Construction Materials* *(Continued)*

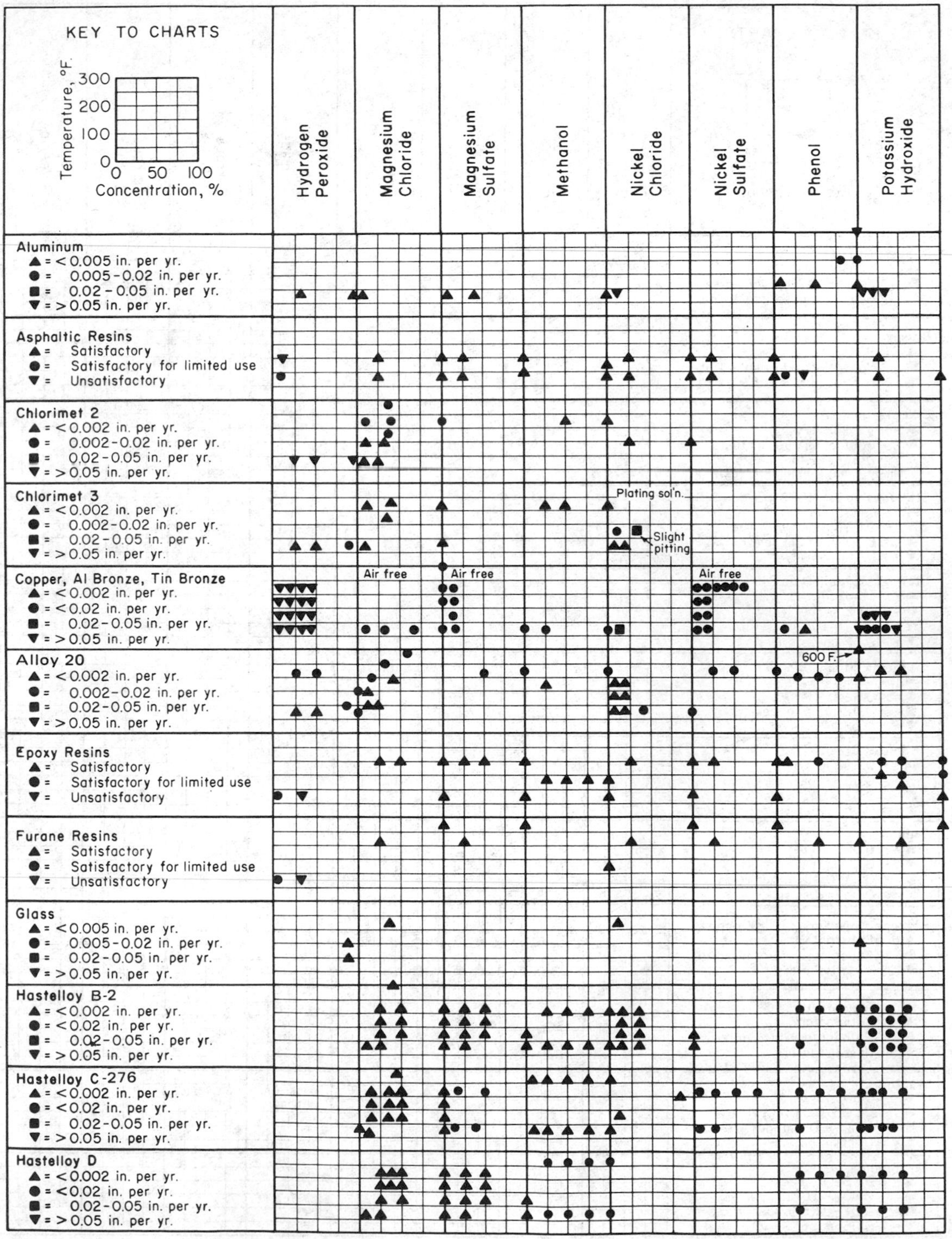

*To convert from inches to millimeters, multiply by 25.4.

TABLE 23-2 Detailed Corrosion Data on Construction Materials* *(Continued)*

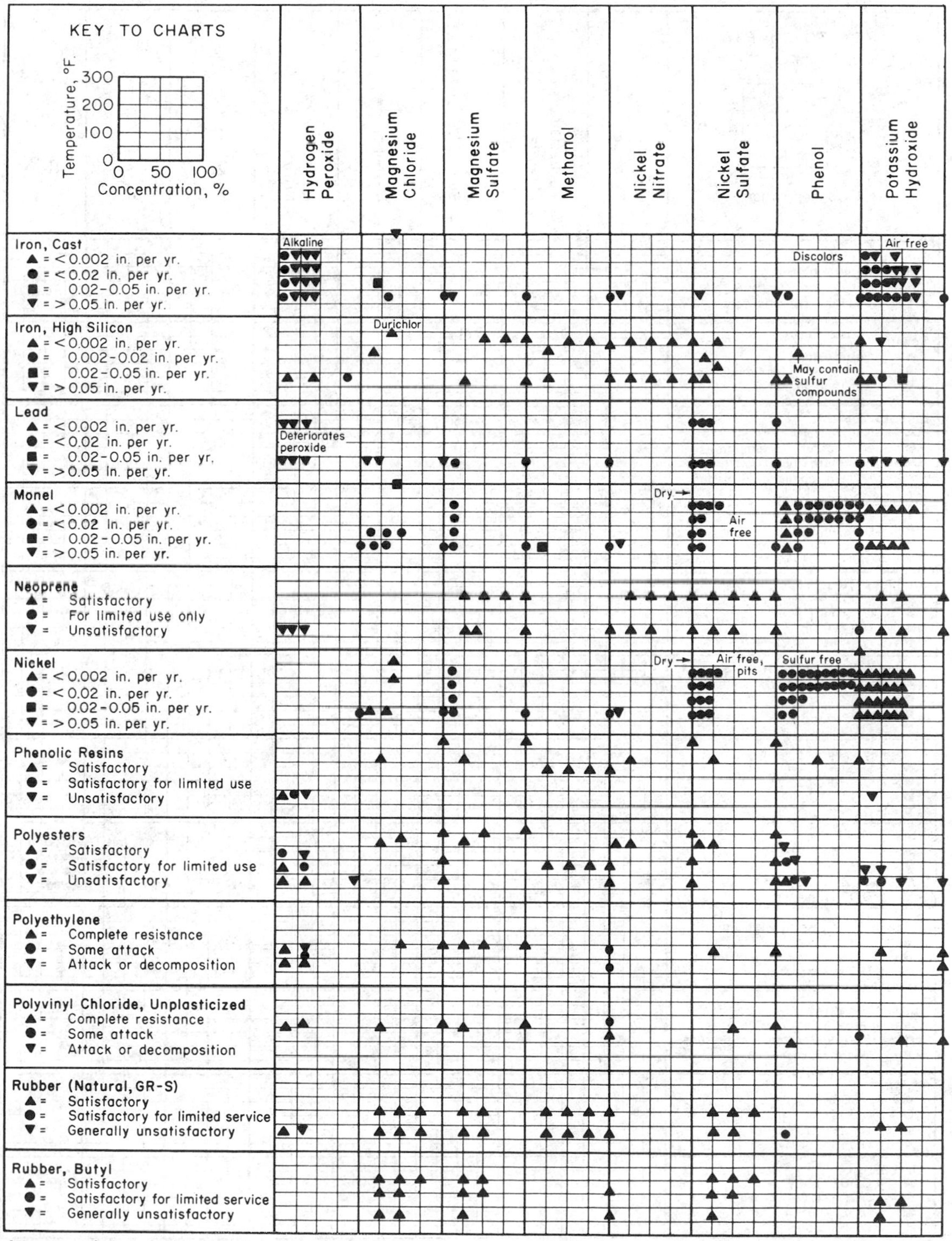

*To convert from inches to millimeters, multiply by 25.4.

TABLE 23-2 Detailed Corrosion Data on Construction Materials* *(Continued)*

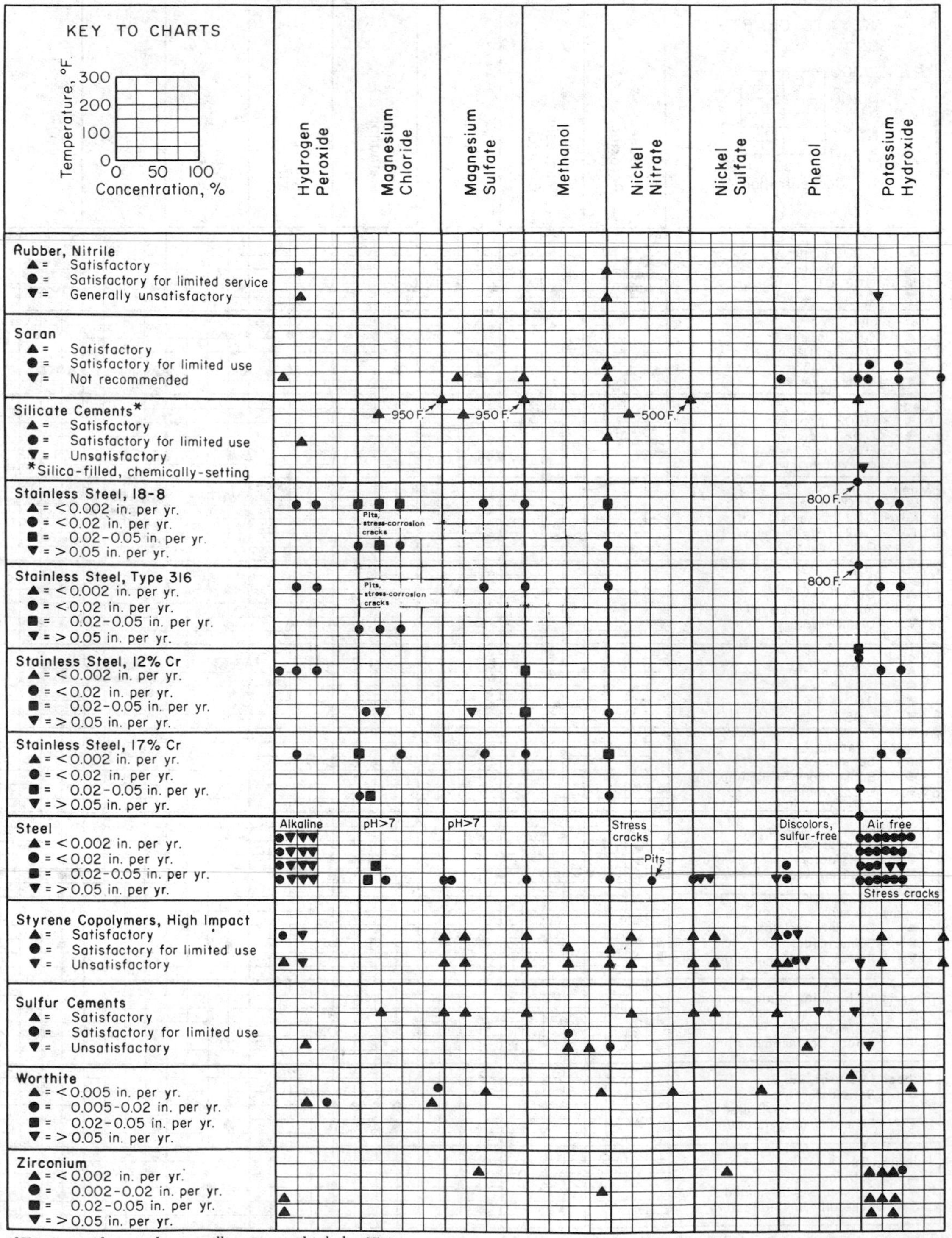

*To convert from inches to millimeters, multiply by 25.4.

TABLE 23-2 Detailed Corrosion Data on Construction Materials* *(Continued)*

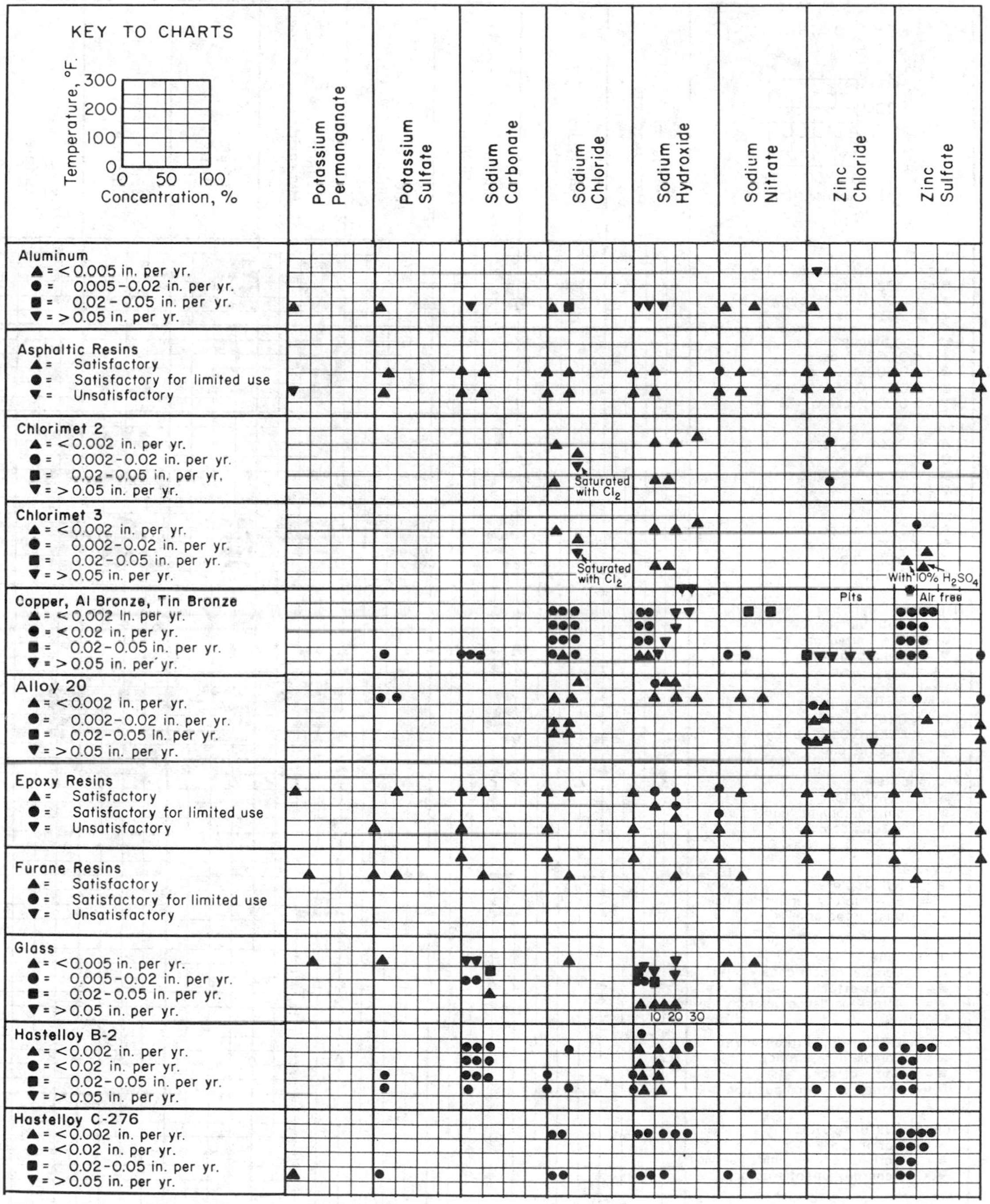

*To convert from inches to millimeters, multiply by 25.4.

TABLE 23-2 Detailed Corrosion Data on Construction Materials* *(Continued)*

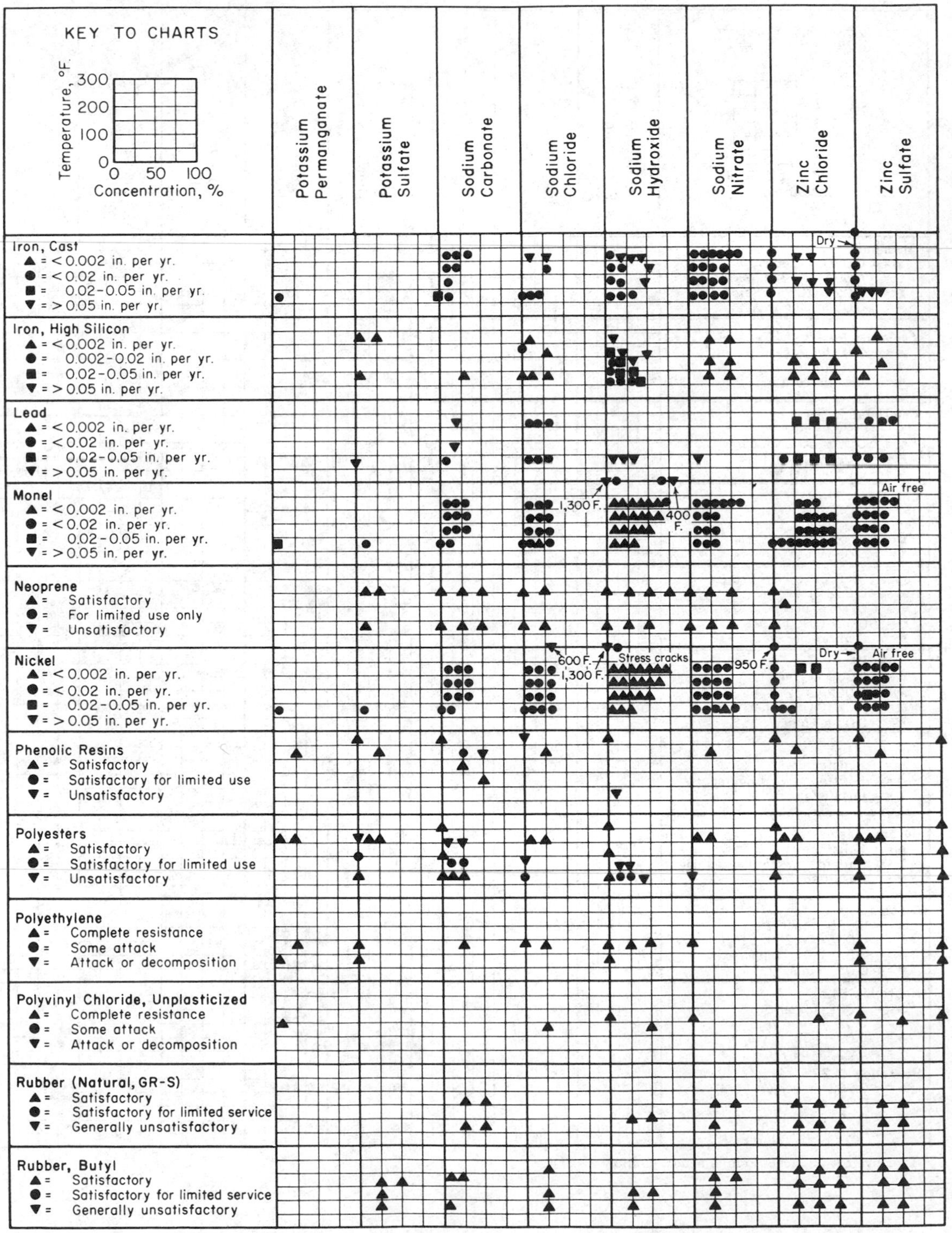

*To convert from inches to millimeters, multiply by 25.4.

TABLE 23-2 Detailed Corrosion Data on Construction Materials* *(Concluded)*

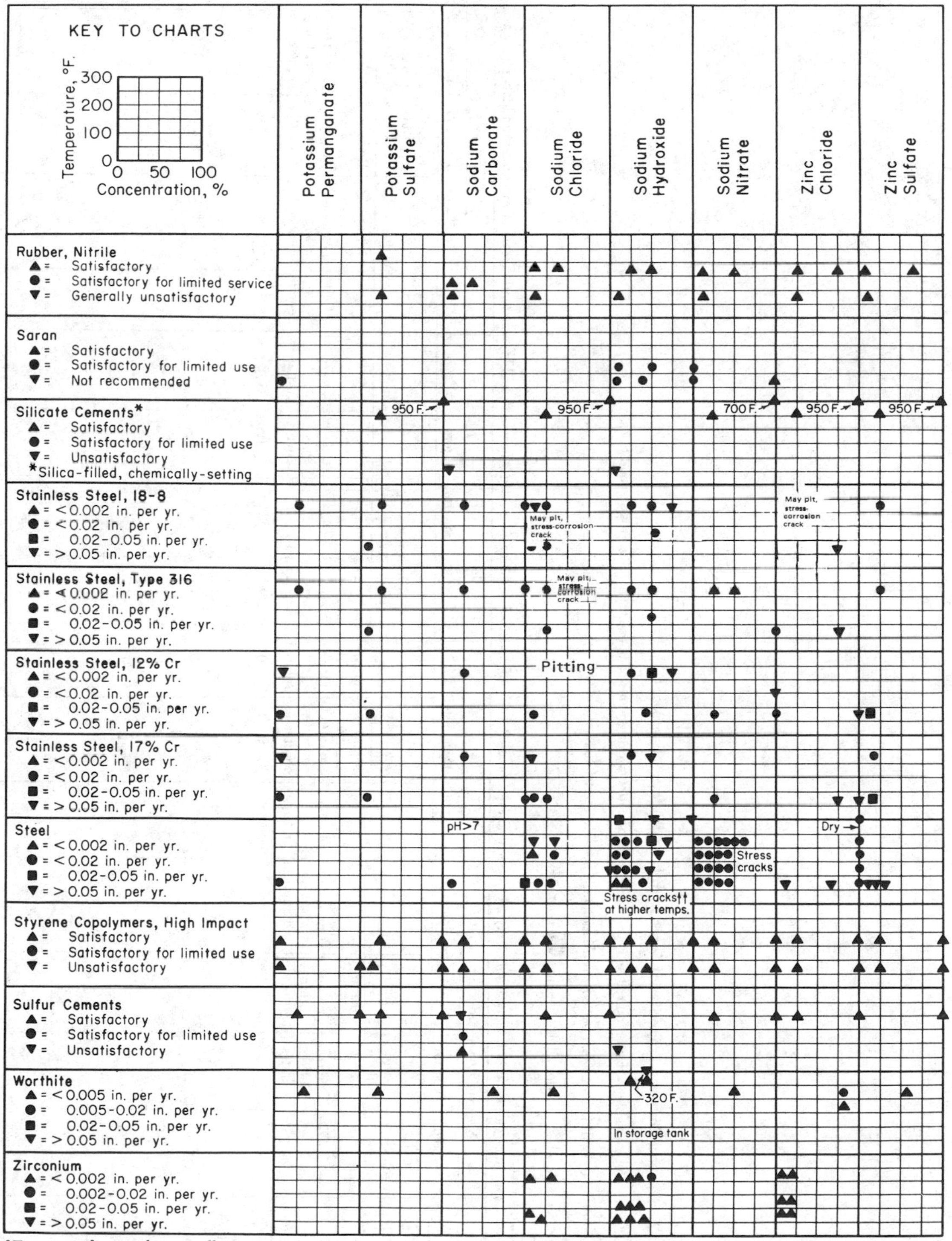

°To convert from inches to millimeters, multiply by 25.4.

TABLE 23-3 General Corrosion Properties of Some Metals and Alloys* *(Continued)*

Ratings:
0: Unsuitable. Not available in form required or not suitable for fabrication requirements or not suitable for corrosion conditions.
1: Poor to fair.
2: Fair. For mild conditions or when periodic replacement is possible. Restricted use.
3: Fair to good.
4: Good. Suitable when superior alternatives are uneconomic.
5: Good to excellent.
6: Normally excellent.
Small variations in service conditions may appreciably affect corrosion resistance. Choice of materials is therefore guided wherever possible by a combination of experience and laboratory and site tests.

	Liquids											Gases				
	Nonoxidizing or reducing media				Oxidizing media			Natural waters				Common industrial media				
			Alkaline solutions, e.g.					Fresh-water supplies		Seawater		Steam		Furnace gases with incidental sulfur content		
Material	Acid solutions, excluding hydrochloric, e.g., phosphoric, sulfuric, most conditions, many organics	Neutral solutions, e.g., many nonoxidizing salt solutions, chlorides, sulfates	Caustic and mild alkalies, excluding ammonium hydroxide	Ammonium hydroxide and amines	Acid solutions, e.g., nitric	Neutral or alkaline solutions, e.g., persulfates, peroxides, chromates	Pitting media,† acid ferric chloride solutions	Static or slow-moving	Turbulent	Static or slow-moving	Turbulent	Moist, condensate	Dry at high temperature, promoting slight dissociation	Reducing e.g., heat-treatment furnace gases	Oxidizing, e.g., flue gases	Ambient air, city or industrial
Cast iron, flake graphite, plain or low-alloy	1	3	4	5	0	4	0	4	3	4	2	4	4	1	1	3
Ductile iron (higher strength and hardness may be attained by composition and heat treatment or both)	1	3	4	5	0	4	0	4	4	4	3	4	4	1	1	3
Ni-Resist corrosion-resistant cast irons	4	5	5	5	0	5	0	5	5	5	5	5	5	3	2	4
14% silicon iron	6	6	2	5	6	6	3	5	5	5	5	6	4	4	3	6

Mild steel, also low-alloy irons and steels	1	3	4	5	0	4	0	4	3	4	2	4	4	1	1	3
Stainless steel, ferritic 17% Cr type	2	4	4	6	5	6	0	4	6	1	4	5	6	3	2	4
Stainless steel, austenitic 18 Cr; 8 Ni type	3	4	5	6	6	6	0	6	6	2	5	6	6	2	3	5
Stainless steel, austenitic 18 Cr; 12 Ni; 2.5 Mo type	4	5	5	6	5	6	1	6	6	3	5	6	6	2	4	6
Stainless steel, austenitic 20 Cr; 29 Ni; 2.5 Mo; 3.5 Cu type	5	6	5	6	5	6	2	6	6	4	6	6	6	2	4	6
Incoloy 825 nickel-iron-chromium alloy (40 Ni; 21 Cr; 3 Mo; 1.5 Cu; balance Fe)	6	6	5	6	5	6	2	6	6	4	6	6	6	2	5	6
Hastelloy alloy C-276 (55 Ni; 17 Mo; 16 Cr; 6 Fe; 4 W)	5	6	5	6	4	6	5	6	6	6	6	6	6	3	4	6
Hastelloy alloy B-2 (61 Ni; 28 Mo; 6 Fe)	6	5	4	4	0	3	0	6	6	4	4	6	5	3	2	5
Inconel 600 (78 Ni; 15 Cr; 7 Fe)	3	6	6	6	3	6	1	6	6	4	6	6	6	2	4	6

TABLE 23-3 General Corrosion Properties of Some Metals and Alloys* *(Continued)*

Ratings:
0: Unsuitable. Not available in form required or not suitable for fabrication requirements or not suitable for corrosion conditions.
1: Poor to fair.
2: Fair. For mild conditions or when periodic replacement is possible. Restricted use.
3: Fair to good.
4: Good. Suitable when superior alternatives are uneconomic.
5: Good to excellent.
6: Normally excellent.
Small variations in service conditions may appreciably affect corrosion resistance. Choice of materials is therefore guided wherever possible by a combination of experience and laboratory and site tests.

	Liquids											Gases			
	Nonoxidizing or reducing media											Common industrial media			
			Alkaline solutions, e.g.		Oxidizing media			Natural waters							
								Fresh-water supplies		Seawater		Steam		Furnace gases with incidental sulfur content		
Material	Acid solutions, excluding hydrochloric, e.g., phosphoric, sulfuric, most conditions, many organics	Neutral solutions, e.g., many nonoxidizing salt solutions, chlorides, sulfates	Caustic and mild alkalies, excluding ammonium hydroxide	Ammonium hydroxide and amines	Acid solutions, e.g., nitric	Neutral or alkaline solutions, e.g., persulfates, peroxides, chromates	Pitting media,† acid ferric chloride solutions	Static or slow-moving	Turbulent	Static or slow-moving	Turbulent	Moist, condensate	Dry at high temperature, promoting slight dissociation	Reducing, e.g., heat-treatment furnace gases	Oxidizing, e.g., flue gases	Ambient air, city or industrial
Copper-nickel alloys up to 30% nickel	4	5	5	0	0	4	1	6	6	6	6	6	5	2	2	5
Monel 400 nickel-copper alloy (66 Ni; 30 Cu; 2 Fe)	5	6	6	1	0	5	1	6	6	4	6	6	6	2	3	5
Nickel 200—commercial (99.4 Ni)	4	5	6	1	0	5	0	6	6	3	5	6	6	2	2	4
Copper and silicon bronze	4	4	4	0	0	4	0	6	5	4	1	6	5	2	2	5
Aluminum brass (76 Cu; 22 Zn; 2 Al)	3	4	2	0	0	3	0	6	6	4	5	6	5	2	2	5
Nickel-aluminum bronze (80 Cu; 10 Al; 5 Ni; 5 Fe)	4	4	2	0	0	3	0	6	6	4	5	6	5	2	3	5
Bronze, type A (88 Cu; 5 Sn; 5 Ni; 2 Zn)	4	5	4	0	0	4	0	6	6	5	5	6	5	2	2	5
Aluminum and its alloys	1	3	0	6	0–5	0–4	0	4	5	0–5	4	5	2	5	4	5
Lead, chemical or antimonial	5	5	2	2	0	2	0	6	5	5	3	2	0	4	3	5
Silver	4	6	6	0	0	2	0	6	6	5	5	6	5	4	4	4
Titanium	3	6	2	6	6	6	6	6	6	6	6	6	5	3	5	6
Zirconium	3	6	2	6	6	6	2	6	6	6	6	6	6	3	5	6

Material	Gases (continued): Halogens and derivatives: Halogens: Moist, e.g., chlorine below dew point	Halogens: Dry, e.g., fluorine above dew point	Halide acids, moist, e.g., hydrochloric hydrolysis products of organic halides	Hydrogen halides, dry,‡ e.g., dry hydrogen chloride, °F	Available forms	Cold formability in wrought and clad form	Weldability	Maximum strength annealed condition × 1000 lb/in²	Coefficient of thermal expansion, millionths per °F, 70–212 °F	Remarks¶
Cast iron, flake graphite, plain or low alloy	0	2	0	2 < 400 1 < 750	Cast	No	Fair§	45	6.7	
Ductile iron (higher strength and hardness may be attained by composition and heat treatment or both)	0	2	0	2 < 400 1 < 750	Cast	No	Good§	67	7.5	
Ni-Resist corrosion-resistant cast irons	0	2	3	3 < 400 2 < 750	Cast	No	Good§	22–31	10.3	
Durion—14% silicon iron	0	0	4	1 < 400	Cast	No	No	22	7.4	Very brittle, susceptible to cracking by mechanical and thermal shock
Mild steel, also low-alloy irons and steels	0	3	0	3 < 400 1 < 750	Wrought, cast	Good	Good	67	6.7	High strengths obtainable by alloying, also improved atmospheric corrosion resistance. See ASTM specifications for particular grade
Stainless steel, ferritic 17% Cr type	0	2	0	2 < 400	Wrought, cast, clad	Good	Good§	78	6.0	AISI type 430 ASTM corrosion- and heat-resisting steels
Stainless steel, austenitic 18 Cr; 8 Ni types	0	2	0	3 < 400	Wrought, cast, clad	Good	Good	90	9.6	AISI type 304 ASTM corrosion- and heat-resisting steels; stabilized or LC types used for welding
Stainless steel, austenitic 18 Cr; 12 Ni; 2.5 Mo type	0	3	2	4 < 400 3 < 750	Wrought, cast, clad	Good	Good	90	8.9	AISI type 316 ASTM corrosion- and heat-resisting steel; LC type used for welding
Stainless steel, austenitic 20 Cr; 29 Ni; 2.5 Mo; 3.5 Cu type	1	3	3	4 < 400 3 < 750	Wrought, cast	Good	Good	90	9.4	ACI CH-7M; good resistance to sulfuric, phosphoric, and fatty acids at elevated temperatures
Incoloy 825 nickel-iron-chromium alloy (40 Ni; 21 Cr; 3 Mo; 1.5 Cu; bal. Fe)	2	3	3	4 < 400 3 < 750	Wrought, cast, clad	Good	Good	100	7.3	Special alloy with good resistance to sulfuric, phosphoric, and fatty acids; resistant to chlorides in some environments
Hastelloy alloy C-276 (55 Ni; 17 Mo; 16 Cr; 6 Fe; 4 W)	5	4	4	4 < 750 3 < 900	Wrought, cast, clad	Fair	Good	145	6.3	Excellent resistance to wet chlorine gas and sodium hypochlorite solutions
Hastelloy alloy B-2 (61 Ni; 28 Mo; 6 Fe)	1	3	5	4 < 750 3 < 900	Wrought, cast, clad	Fair	Good	135	5.6	Resistant to solutions of hydrochloric and sulfuric acids
Inconel 600 (78 Ni; 15 Cr; 7 Fe)	2	5	3	5 < 400 4 < 900	Wrought, cast, clad	Good	Good	90	8.9	Wide application in food and pharmaceutical industries

TABLE 23-3 General Corrosion Properties of Some Metals and Alloys* *(Concluded)*

Ratings:
0: Unsuitable. Not available in form required or not suitable for fabrication requirements or not suitable for corrosion conditions.
1: Poor to fair.
2: Fair. For mild conditions or when periodic replacement is possible. Restricted use.
3: Fair to good.
4: Good. Suitable when superior alternatives are uneconomic.
5: Good to excellent.
6: Normally excellent.
Small variations in service conditions may appreciably affect corrosion resistance. Choice of materials is therefore guided wherever possible by a combination of experience and laboratory and site tests.

Material	Gases *(continued)*				Available forms	Cold formability in wrought and clad form	Weldability	Maximum strength annealed condition × 1000 lb/in²	Coefficient of thermal expansion millionths per °F, 70–212°F	Remarks¶
	Halogens and derivatives									
	Halogens		Halide acids, moist, e.g., hydrochloric, hydrolysis products of organic halides	Hydrogen halides, dry,‡ e.g., dry hydrogen chloride °F						
	Moist, e.g., chlorine below dew point	Dry, e.g., fluorine above dew point								
Copper-nickel alloys up to 30% nickel	1	5	2	4 < 400 3 < 750	Wrought, cast, clad	Good	Good	38–62	9.3–8.5	High-iron types excellent for resisting high-velocity effects in condenser tubes
Monel 400 nickel-copper alloy (66 Ni; 30 Cu; 2 Fe)	2	6	3	6 < 400 3 < 750 2 < 900	Wrought, cast, clad	Good	Good	77	7.5	Widely used for sulfuric acid pickling equipment; also for propeller shafts in motor boats; precautions needed to avoid sulfur attack during fabrication
Nickel 200—commercial (99.4 Ni)	2	6	2	6 < 400 5 < 750 4 < 900	Wrought, cast, clad	Good	Good	54	6.6	Widely used for hot concentrated caustic solutions; precautions needed to avoid sulfur attack during fabrication
Copper and silicon bronze	0	5	2	3 < 400 2 < 750	Wrought, cast, clad	Excellent	Fair	29	9.3–9.5	Unsuitable for hot concentrated mineral acids or for high-velocity HF
Aluminum brass (76 Cu; 22 Zn; 2 Al)	0	4	2	2 < 400	Wrought, cast	Good	Fair	60	10.3	Possibility of developing localized corrosion in seawater
Nickel-aluminum bronze (80 Cu; 10 Al; 5 Ni; 5 Fe)	0	4	3	3 < 400 2 < 750	Wrought, cast	Good	Fair	60–80	9.4	Ship propellers an excellent application
Bronze, type A (88 Cu; 5 Sn; 5 Ni; 2 Zn)	0	4	3	3 < 400 2 < 750	Cast	No	§	45	11.0	High strengths obtained by heat treatment; not susceptible to dezincification
Aluminum and its alloys	0	6	0	3 < 400 1 < 750	Wrought, cast, clad	Good	Good	9–90	11.5–13.7	Extent of corrosion dependent upon type and concentration of acidic ions; wide range of mechanical properties obtainable by alloying and heat treatment
Lead, chemical or antimonial	0	1	3	0	Wrought, cast, clad	Excellent	Good	2	16.4–15.1	High purity "chemical lead" preferred for most applications
Silver	5	5	3	4 < 400 2 < 750	Wrought, cast, clad	Excellent	Good	21	10.6	Used as a lining
Titanium	6	0	1	0	Wrought, cast	Fair	Good§	6–90	5.0	Possibility of red fuming HNO_3 initiating explosions; good resistance to solutions containing chlorides
Zirconium	6	1	6	0	Wrought, cast	Fair	Good§			

*Data courtesy of International Nickel Co..
‡On unsuitable materials these media may promote potentially dangerous pitting.
‡Temperatures are approximate.
§Special precautions required.
¶Many of these materials are suitable for resisting dry corrosion at elevated temperatures.

TABLE 23-4 Unified Alloy Numbering System (UNS)

UNS was established in 1974 by ASTM and SAE to reduce the confusion involved in the labeling of commercial alloys. Metals have been placed into 15 groups, each of which is given a code letter. The specific alloy is identified by a five-digit number following this code letter.

Nonferrous metals and alloys	
A00001–A99999	Aluminum and aluminum alloys
C00001–C99999	Copper and copper alloys
E00001–E99999	Rare-earth and rare-earth-like metals and alloys
L00001–L99999	Low-melting metals and alloys
M00001–M99999	Miscellaneous nonferrous metals and alloys
N00001–N99999	Nickel and nickel alloys
P00001–P99999	Precious metals and alloys
R00001–R99999	Reactive and refractory metals and alloys
Ferrous metals and alloys	
D00001–D99999	Specified-mechanical-properties steels
F00001–F99999	Cast irons and cast steels
G00001–G99999	AISI and SAE carbon and alloy steels
H00001–H99999	AISI H steels
K00001–K99999	Miscellaneous steels and ferrous alloys
S00001–S99999	Heat- and corrosion-resistant (stainless) steels
T00001–T99999	Tool steels

When possible, earlier widely used three- or four-digit alloy numbering systems such as those developed by the Aluminum Association (AA), Copper Development Association (CDA), American Iron and Steel Institute (AISI), etc., have been incorporated by the addition of the appropriate alloy-group code letter plus additional digits. For example:

Alloy description	Former designation System	Former designation No.	UNS designation
Aluminum + 1.2% Mn	AA	3003	A93003
Copper, electrolytic tough pitch	CDA	110	C11000
Carbon steel, 0.2% C	AISI	1020	G10200
Stainless steel 18 Cr, 8 Ni	AISI	304	S30400

Proprietary alloys are assigned numbers by the AA, AISI, CDA, ASTM, and SAE, which maintain master listings at their headquarters. Handbooks describing the system are available. (Cf. ASTM publication DS-56AC.)

SOURCE: ASTM DS-56A. Courtesy of National Association of Corrosion Engineers.

High-silicon cast irons have excellent corrosion resistance. Silicon content is 13 to 16 percent. This material is known as Durion. Adding 4 percent Cr yields a product called Durichlor, which has improved resistance in the presence of oxidizing agents. These alloys are not readily machined or welded.

Silicon irons are very resistant to oxidizing and reducing environments, and resistance depends on the formation of a passive film. These irons are widely used in sulfuric acid service, since they are unaffected by sulfuric at all strengths, even up to the boiling point.

Because they are very hard, silicon irons are good for combined corrosion-erosion service.

Another group of cast-iron alloys are called **Ni-Resist.** These materials are related to gray cast iron in that they have high carbon contents (3 percent), with fine graphite flakes distributed throughout the structure. Nickel contents range from 13.5 to 36 percent, and some have 6.5 percent Cu.

Generally, nickel-alloy castings have superior toughness and impact resistance compared with gray irons. The nickel-alloy castings can be welded and machined.

Corrosion resistance of nickel alloys is superior to that of cast irons but less than that of pure nickel. There is little attack from neutral or alkaline solutions. Oxidizing acids such as nitric are highly detrimental. Cold, concentrated sulfuric acid can be handled.

Ni-Resist has excellent heat resistance, with some grades serviceable up to 800°C (1500°F). Also, a ductile variety is available, as well as a hard variety (Ni-Hard).

TABLE 23-5 Coefficient of Thermal Expansion of Common Alloys*

	UNS	10^{-6} in/(in·°F)	10^{-6} mm/(mm·°C)	Temperature range, °C
Aluminum alloy AA1100	A91100	13.1	24.	20–100
Aluminum alloy AA5052	A95052	13.2	24.	20–100
Aluminum cast alloy 43	A24430	12.3	22.	20–100
Copper	C11000	9.4	16.9	20–100
Red brass	C23000	10.4	18.7	20–300
Admiralty brass	C44300	11.2	20.	20–300
Muntz Metal	C28000	11.6	21.	20–300
Aluminum bronze D	C61400	9.0	16.2	20–300
Ounce metal	C83600	10.2	18.4	0–100
90-10 copper nickel	C70600	9.5	17.1	20–300
70-30 copper nickel	C71500	9.0	16.2	20–300
Carbon steel, AISI 1020	G10200	6.7	12.1	0–100
Gray cast iron	F10006	6.7	12.1	0–100
4-6 Cr, ½ Mo steel	S50100	7.3	13.1	20–540
Stainless steel, AISI 410	S41000	6.1	11.0	0–100
Stainless steel, AISI 446	S44600	5.8	10.4	0–100
Stainless steel, AISI 304	S30400	9.6	17.3	0–100
Stainless steel, AISI 310	S31000	8.0	14.4	0–100
Stainless steel, ACI HK	J94224	9.4	16.9	20–540
Nickel alloy 200	N02200	7.4	13.3	20–90
Nickel alloy 400	N04400	7.7	13.9	20–90
Nickel alloy 600	N06600	7.4	13.3	20–90
Nickel-molybdenum alloy B-2	N10665	5.6	10.1	20–90
Nickel-molybdenum alloy C-276	N10276	6.3	11.3	20–90
Titanium, commercially pure	R50250	4.8	8.6	0–100
Titanium alloy T1-6A1-4V	R56400	4.9	8.8	0–100
Magnesium alloy AZ31B	M11311	14.5	26.	20–100
Magnesium alloy AZ91C	M11914	14.5	26.	20–100
Chemical lead	......	16.4	30.	0–100
50-50 solder	L05500	13.1	24.	0–100
Zinc	Z13001	18.	32.	0–100
Tin	L13002	12.8	23.	0–100
Zirconium	R60702	2.9	5.2	0–100
Molybdenum	R03600	2.7	4.9	20–100
Tantalum	R05200	3.6	6.5	20–100

*Courtesy of National Association of Corrosion Engineers.

Stainless Steel There are more than 70 standard types of stainless steel and many special alloys. These steels are produced in the wrought form (AISI types) and as cast alloys [Alloy Casting Institute (ACI) types]. Generally, all are iron-based, with 12 to 30 percent chromium, 0 to 22 percent nickel, and minor amounts of carbon, niobium (columbium), copper, molybdenum, selenium, tantalum, and titanium. These alloys are very popular in the process industries. They are heat- and corrosion-resistant, noncontaminating, and easily fabricated into complex shapes.

There are three groups of stainless alloys: (1) martensitic, (2) ferritic, (3) austenitic.

The **martensitic alloys** contain 12 to 20 percent chromium with controlled amounts of carbon and other additives. Type 410 is a typical member of this group. These alloys can be hardened by heat treatment, which can increase tensile strength from 550 to 1380 MPa (80,000 to 200,000 lbf/in^2).

TABLE 23-6 Melting Temperatures of Common Alloys*

	UNS	Melting range °F	Melting range °C
Aluminum alloy AA1100	A91100	1190–1215	640–660
Aluminum alloy AA5052	A95052	1125–1200	610–650
Aluminum cast alloy 43	A24430	1065–1170	570–630
Copper	C11000	1980	1083
Red brass	C23000	1810–1880	990–1025
Admiralty brass	C44300	1650–1720	900– 935
Muntz Metal	C28000	1650–1660	900– 905
Aluminum bronze D	C61400	1910–1940	1045–1060
Ounce metal	C83600	1510–1840	854–1010
Manganese bronze	C86500	1583–1616	862– 880
90-10 copper nickel	C70600	2010–2100	1100–1150
70-30 copper nickel	C71500	2140–2260	1170–1240
Carbon steel, AISI 1020	G10200	2760	1520
Gray cast iron	F10006	2100–2200	1150–1200
4-6 Cr, ½ Mo Street	S50100	2700–2800	1480–1540
Stainless steel, AISI 410	S41000	2700–2790	1480–1530
Stainless steel, AISI 446	S44600	2600–2750	1430–1510
Stainless steel, AISI 304	S30400	2550–2650	1400–1450
Stainless steel, AISI 310	S31000	2500–2650	1400–1450
Stainless steel, ACI HK	J94224	2550	1400
Nickel alloy 200	N02200	2615–2635	1440–1450
Nickel alloy 400	N04400	2370–2460	1300–1350
Nickel alloy 600	N06600	2470–2575	1350–1410
Nickel-molybdenum alloy B-2	N10665	2375–2495	1300–1370
Nickel-molybdenum alloy C-276	N10276	2420–2500	1320–1370
Titanium, commercially pure	R50250	3100	1705
Titanium alloy T1-6A1-4V	R56400	2920–3020	1600–1660
Magnesium alloy AZ 31B	M11311	1120–1170	605– 632
Magnesium alloy HK 31A	M13310	1092–1204	589– 651
Chemical lead	......	618	326
50-50 solder	L05500	361– 421	183– 216
Zinc	Z13001	787	420
Tin	Z13002	450	232
Zirconium	R60702	3380	1860
Molybdenum	R03600	4730	2610
Tantalum	R05200	5425	2996

*Courtesy of National Association of Corrosion Engineers.

Corrosion resistance is inferior to that of austenitic stainless steels, and martensitic steels are generally used in mildly corrosive environments (atmospheric, fresh water, and organic exposures).

Ferritic stainless contains 15 to 30 percent Cr, with low carbon content (0.1 percent). The higher chromium content improves its corrosive resistance. Type 430 is a typical example. The strength of ferritic stainless can be increased by cold working but not by heat treatment. Fairly ductile ferritic grades can be fabricated by all standard methods. They are fairly easy to machine. Welding is not a problem, although it requires skilled operators.

Corrosion resistance is rated good, although ferritic alloys are not good against reducing acids such as HCl. But mildly corrosive solutions and oxidizing media are handled without harm. Type 430 is widely used in nitric acid plants. In addition, it is very resistant to scaling and high-temperature oxidation up to 800°C (1500°F).

Austenitic stainless steels are the most corrosion-resistant of the three groups. These steels contain 16 to 26 percent chromium and 6 to 22 percent nickel. Carbon is kept low (0.08 percent maximum) to minimize carbide precipitation. These alloys can be work-hardened, but heat treatment will not cause hardening. Tensile strength in the annealed condition is about 585 MPa (85,000 lbf/in^2), but work-hardening can increase this to 2000 MPa (300,000 lbf/in^2). Austenitic stainless steels are tough and ductile.

They can be fabricated by all standard methods. But austenitic grades are not easy to machine; they work-harden and gall. Rigid machines, heavy cuts, and high speeds are essential. Welding, however, is readily performed, although welding heat may cause chromium carbide precipitation, which depletes the alloy of some chromium and lowers its corrosion resistance in some specific environments, notably nitric acid. The carbide precipitation can be eliminated by heat treatment (solution annealing). To avoid precipitation, special stainless steels stabilized with titanium, niobium, or tantalum have been developed (types 321, 347, and 348). Another approach to the problem is the use of low-carbon steels such as types 304L and 316L, with 0.03 percent maximum carbon.

The addition of molybdenum to the austenitic alloy (types 316, 316L, 317, and 317L) provides generally better corrosion resistance and improved resistance to pitting.

In the stainless group, nickel greatly improves corrosion resistance over straight chromium stainless. Even so, the chromium-nickel steels, particularly the 18-8 alloys, perform best under **oxidizing conditions,** since resistance depends on an oxide film on the surface of the alloy. Reducing conditions and chloride ions destroy this film and bring on rapid attack. Chloride ions tend to cause pitting and crevice corrosion; when combined with high tensile stresses, they can cause stress-corrosion cracking.

Cast stainless alloys are widely used in pumps, valves, and fittings. These casting alloys are designated under the ACI system. All corrosion-resistant alloys have the letter C plus a second letter (A to N) denoting increasing nickel content. Numerals indicate maximum carbon. While a rough comparison can be made between ACI and AISI types, compositions are not identical and analyses cannot be used interchangeably. Foundry techniques require a rebalancing of the wrought chemical compositions. However, corrosion resistance is not greatly affected by these composition changes. Typical members of this group are CF-8, similar to type 304 stainless; CF-8M, similar to type 316; and CD-4M Cu, which has improved resistance to nitric, sulfuric, and phosphoric acids.

In addition to the C grades, there is a series of heat-resistant grades of ACI cast alloys, identified similarly to the corrosion-resistant grades, except that the first letter is H rather than C. Mention should also be made of precipitation-hardening (PH) stainless steels, which can be hardened by heat treatments at moderate temperatures. Very strong and hard at high temperatures, these steels have but moderate corrosion resistance. A typical PH steel, containing 17 percent Cr, 7 percent Ni, and 1.1 percent Al, has high strength, good fatigue properties, and good resistance to wear and cavitation corrosion. A large number of these steels with varying compositions are commercially available. Essentially, they contain chromium and nickel with added alloying agents such as copper, aluminum, beryllium, molybdenum, nitrogen, and phosphorus.

Medium Alloys A group of (mostly) proprietary alloys with somewhat better corrosion resistance than stainless steels are called medium alloys. A popular member of this group is the **20 alloy,** made by a number of companies under various trade names. Durimet 20 is a well-known cast version, containing 0.07 percent C, 29 percent Ni, 20 percent Cr, 2 percent Mo, and 3 percent Cu. The ACI designation of this alloy is CN-7M. A wrought form is known as Carpenter 20 (Cb3). Worthite is another proprietary 20 alloy with about 24 percent Ni and 20 percent Cr. The 20 alloy was originally developed to fill the need for a material with sulfuric acid resistance superior to the stainless steels.

Other members of the medium-alloy group are **Incoloy 825** and **Hastelloy G-3.** Wrought Incoloy 825 has 40 percent Ni, 21 percent Cr, 3 percent Mo, and 2.25 percent Cu. Hastelloy G-3 contains 44 percent Ni, 22 percent Cr, 6.5 percent Mo, and 0.05 percent C maximum.

These alloys have extensive applications in sulfuric acid systems. Because of their increased nickel and molybdenum contents they are more tolerant of chloride-ion contamination than standard stainless steels. The nickel content decreases the risk of stress-corrosion cracking; molybdenum improves resistance to crevice corrosion and pitting.

High Alloys The group of materials called high alloys all contain relatively large percentages of nickel. **Hastelloy B-2** contains 61 percent Ni and 28 percent Mo. It is available in wrought and cast forms. Work hardening presents some fabrication difficulties, and machining is somewhat more difficult than for type 316 stainless. Conventional welding methods can be used. The alloy has unusually high resistance to all concentrations of hydrochloric acid at all temperatures in the absence of oxidizing agents. Sulfuric acid attack is low

TABLE 23-7 Carbon and Low-Alloy Steels*

Steel type	ASTM	UNS	Composition, %[a]	Mechanical properties[b]: Yield strength, kip/in^2 (MPa)	Tensile strength, kip/in^2 (MPa)	Elongation, %
C-Mn	A53B	K03005	0.30 C, 1.20 Mn	35 (241)	60 (415)	
C-Mn	A106B	K03006	0.30 C, 0.29–1.06 Mn, 0.10 min. Si	35 (241)	60 (415)	30
C	A285A	K01700	0.17 C, 0.90 Mn	24 (165)	45–55 (310–380)	30
HSLA	A517F	K11576	0.08–0.22 C, 0.55–1.05 Mn, 0.13–0.37 Si, 0.36–0.79 Cr, 0.67–1.03 Ni, 0.36–0.64 Mo, 0.002–0.006 B, 0.12–0.53 Cu, 0.02–0.09 V	100 (689)	115–135 (795–930)	16
HSLA	A242(1)	K11510	0.15 C, 1.00 Mn, 0.20 min Cu, 0.15 P	42–50 (290–345)	63–70 (435–480)	21
2¼Cr, 1Mo	A387(22)	K21590	0.15 C, 0.30–0.60 Mn, 0.5 Si, 2.00–2.50 Cr, 0.90–1.10 Mo	30 (205)[c] 45 (310)[d]	60–85 (415–585)[c] 75–100 (515–690)[d]	18[c] 18[d]
4–6Cr, ½Mo	A335 (P5)	K41545	0.15 C, 0.30–0.60 Mn, 0.5 Si, 4.00–6.00 Cr, 0.45–0.65 Mo	30 (205)	60 (415)	
9Cr, 1Mo	A335 (P9)	K81590	0.15 C, 0.30–0.6 Mn, 0.25–1.00 Si, 8.00–10.00 Cr, 0.90–1.10 Mo	30 (205)	60 (415)	
9Ni	A333(8), A353(1)	K81340	0.13 C, 0.90 Mn, 0.13–0.32 Si, 8.40–9.60 Ni	75 (515)	100–120 (690–825)	20
	AISI 4130	G41300	0.28–0.33 C, 0.80–1.10 Mn, 0.15–0.3 Si, 0.8–1.10 Cr, 0.15–0.25 Mo	120 (830)[e]	140 (965)[e]	22[e]
	AISI 4340	G43400	0.38–0.43 C, 0.60–0.80 Mn, 0.15–0.3 Si, 0.70–0.90 Cr, 1.65–2.00 Ni, 0.20–0.30 Mo	125 (860)[f]	148 (1020)[f]	20[f]

*Courtesy of National Association of Corrosion Engineers. To convert megapascals to pounds-force per square inch, multiply by 145.04.
[a]Single values are maximum values unless otherwise noted.
[b]Room-temperature properties. Single values are minimum values.
[c]Class 1.
[d]Class 2.
[e]1-in-diameter bars water-quenched from 1575°F (860°C) and tempered at 1200°F (650°C).
[f]1-in-diameter bars oil-quenched from 1550°F (845°C) and tempered at 1200°F (650°C).

for all concentrations at 65°C (150°F), but the rate goes up with temperature. Oxidizing acids and salts rapidly corrode Hastelloy B. But alkalies and alkaline solutions cause little damage.

Chlorimet 2 has 63 percent Ni and 32 percent Mo and is somewhat similar to Hastelloy B-2. It is available only in cast form, mainly as valves and pumps. This is a tough alloy, very resistant to mechanical and thermal shock. It can be machined with carbide-tipped tools and welded with metal-arc techniques.

Hastelloy C-276 is a nickel-based alloy containing chromium (15.5 percent), molybdenum (15.5 percent), and tungsten (3 percent) as major alloying elements. It is available only in wrought form. This alloy is a low-impurity modification of Hastelloy C, which is still available in cast form. The low impurity level substantially reduces the risk of intergranular corrosion of grain-boundary precipitation in weld-heat-affected zones. This alloy is resistant to strong oxidizing chloride solutions, such as wet chlorine and hypochlorite solutions. It is one of the very few alloys which are totally resistant to seawater.

A still newer variation, **Hastelloy C-4,** is almost totally immune to selective intergranular corrosion in weld-heat-affected zones.

Chlorimet 3 is an alloy, available only in cast form, which is similar in alloy content and corrosion resistance to Hastelloy C.

Inconel 600 (80 percent Ni, 16 percent Cr, and 7 percent Fe) should also be mentioned as a high alloy. It contains no molybdenum. The corrosion-resistant grade is recommended for reducing-oxidizing environments, particularly at high temperatures. When heated in air, this alloy resists oxidation up to 1100°C (2000°F). The alloy is outstanding in resisting corrosion by gases when these gases are essentially sulfur-free.

The alloys discussed are typical examples of the large number of proprietary high alloys used in the chemical industry. For more comprehensive lists and data, refer to the listed references.

NONFERROUS METALS AND ALLOYS

Nickel and Nickel Alloys Nickel is available in practically any mill form as well as in castings. It can be machined easily and joined by welding. Generally, oxidizing conditions favor corrosion, while reducing conditions retard attack. Neutral alkaline solutions, seawater, and mild atmospheric conditions do not affect nickel. The metal is widely used for handling alkalies, particularly in concentrating, storing, and shipping high-purity caustic soda. Chlorinated solvents and phenol are often refined and stored in nickel to prevent product discoloration and contamination.

A large number of nickel-based alloys are commercially available. Many have been mentioned in the preceding discussion of alloy castings and high alloys. One of the best known of these is **Monel 400,** 67 percent Ni and 30 percent Cu. It is available in all standard forms. This nickel-copper alloy is ductile and tough and can be readily fabricated and joined. Its **corrosion resistance** is generally superior to that of its components, being more resistant than nickel in reducing environments and more resistant than copper in oxidizing environments. The alloy can be used for relatively dilute sulfuric acid (below 80 percent), although aeration will result in increased corrosion. Monel will handle hydrofluoric acid up to 92 percent and 115°C (235°F). Alkalies have little effect on this alloy, but it will not stand up against very highly oxidizing or reducing environments.

Aluminum and Alloys Aluminum and its alloys are made in practically all the forms in which metals are produced, including castings. Thermal conductivity of aluminum is 60 percent of that of pure copper, and unalloyed aluminum is used in many heat-transfer applications. Its high electrical conductivity makes aluminum popular in electrical applications. Aluminum is one of the most workable of metals, and it is usually joined by inert-gas-shielded arc-welding techniques.

Commercially pure aluminum has a tensile strength of 69 MPa (10,000 lbf/in^2), but it can be strengthened by cold working. One limitation of aluminum is that strength declines greatly above 150°C (300°F). When strength is important, 200°C (400°F) is usually considered the highest permissible safe temperature for aluminum. However, aluminum has excellent low-temperature properties; it can be used at −250°C (−420°F).

Aluminum has high resistance to atmospheric conditions as well as to industrial fumes and vapors and fresh, brackish, or salt waters. Many mineral acids attack aluminum, although the metal can be used with concentrated nitric acid (above 82 percent) and glacial acetic acid. Aluminum cannot be used with strong caustic solutions.

It should be noted that a number of **aluminum alloys** are available (see Table 23-17). Many have improved mechanical properties over pure aluminum. The wrought heat-treatable aluminum alloys have

TABLE 23-8 Properties of Low-Alloy AISI Steels

Typical physical properties[a]				
AISI type	Melting temperature, °F	Thermal conductivity, Btu/[(h·ft²)(°F/ft)] (212°F)	Coefficient of thermal expansion (0–1200°F) per °F	Specific heat (68–212°F), Btu/(lb·°F)
13XX	....	27	7.9×10^{-6}[b]	0.10–0.11
23XX	2600–2620	38.3[c]	8.0×10^{-6}	0.11–0.12
25XX	2610–2620	34.5–38.5[c]	7.8×10^{-6}	0.11–0.12
40XX	....	27	8.3×10^{-6}[b]	0.10–0.11
41XX	....	24.7[d]	....	0.11
43XX	2740–2750	21.7[c]	8.1×10^{-6}	0.107
46XX	....	27[d]	6.3×10^{-6}[e]	0.10–0.11
48XX	2750	26[f]	8.6×10^{-6}	
51XX	2720–2760	27–34[g]	7.4×10^{-6}[h]	0.10–0.11
61XX	....	27	8.1×10^{-6}[b]	0.10–0.11
86, 87XX	2745–2755	21.7[c]	8.2×10^{-6}	0.107
92, 94XX	....	27	8.1×10^{-6}[b]	0.10–0.12

XX = nominal percent carbon.
[a]Density for all low-alloy steels is about 0.28 lb/in³.
[b]68 to 1200°F.
[c]120°F.
[d]68°F.
[e]0 to 200°F.
[f]75°F.
[g]32 to 212°F.
[h]100 to 518°F.

tensile strengths of 90 to 228 MPa (13,000 to 33,000 lbf/in²) as annealed; when they are fully hardened, strengths can go as high as 572 MPa (83,000 lbf/in²). However, aluminum alloys usually have lower corrosion resistance than the pure metal. The **alclad** alloys have been developed to overcome this shortcoming. Alclad consists of an aluminum layer metallurgically bonded to a core alloy.

The corrosion resistance of aluminum and its alloys tends to be very sensitive to trace contamination. Very small amounts of metallic mercury, heavy-metal ions, or chloride ions can frequently cause rapid failure under conditions which otherwise would be fully acceptable.

When alloy steels do not give adequate corrosion protection—particularly from sulfidic attack—steel with an **aluminized surface coating** can be used. A spray coating of aluminum on a steel is not likely to spall or flake, but the coating is usually not continuous and may leave some areas of the steel unprotected. Hot-dipped "aluminized" steel gives a continuous coating and has proved satisfactory in a number of applications, particularly when sulfur or hydrogen sulfide is present. It is also used to protect thermal insulation and as weather shields for equipment. The coated steel resists fires better than solid aluminum.

Copper and Alloys Copper and its alloys are widely used in chemical processing, particularly when heat and electrical conductivity are important factors. The thermal conductivity of copper is twice that of aluminum and 90 percent that of silver. A large number of copper alloys are available, including brasses (Cu-Zn), bronzes (Cu-Sn), and cupronickels.

Copper has excellent low-temperature properties and is used at −200°C (−320°F). Brazing and soldering are common joining methods for copper, although welding, while difficult, is possible. Generally, copper has high resistance to industrial and marine atmospheres, seawater, alkalies, and solvents. Oxidizing acids rapidly corrode copper. However, the alloys have somewhat different properties than commercial copper.

Brasses with up to 15 percent Zn are ductile but difficult to machine. Machinability improves with increasing zinc up to 36 percent Zn. Brasses with less than 20 percent Zn have corrosion resistance equivalent to that of copper but with better tensile strengths. Brasses with 20 to 40 percent Zn have lower corrosion resistance and are subject to dezincification and stress-corrosion cracking, especially when ammonia is present.

TABLE 23-8 Properties of Low-Alloy AISI Steels *(Continued)*

Typical mechanical properties[a]						
AISI type	Tensile strength, 1000 lbf/in²	Yield strength (0.2% offset), 1000 lbf/in²	Elongation (in 2 in), %	Reduction of area, %	Hardness, Brinell	Impact strength (Izod), ft·lbf
1330[b]	122	100	19	52	248	
1335[c]	126	105	20	59	262	
1340[c]	137	118	19	55	285	
2317[c]	107	72	27	71	222	84
2515[c]	113	94	25	69	233	85
E2517[c]	120	100	22	66	244	80
4023[d]	120	85	20	53	255	
4032[c]	210	182	11	49	415	
4042[f]	235	210	10	42	461	
4053[g]	250	223	12	40	495	
4063[h]	269	231	8	15	534	
4130[i]	200	170	16	49	375	25
4140[j]	200	170	15	48	385	16
4150[k]	230	215	10	40	444	12
4320[d]	180	154	15	50	360	32
4337[k]	210	140	14	50	435	18
4340[k]	220	200	12	48	445	16
4615[d]	100	75	18	52	...	42
4620[d]	130	95	21	65	...	68
4640[l]	185	160	14	52	390	25
4815[d]	150	125	18	58	325	44
4817[d]	...	...	15	52	355	36
4820[i]	...	...	13	47	380	28
5120[d]	143	114	13	45	302	6
5130[m]	189	175	13	51	380	
5140[m]	190	170	13	43	375	16
5150[m]	224	208	10	40	444	
6120[n]	125	94	21	56	...	28
6145[o]	176	169	16	52	429	20
6150[o]	187	179	13	42	444	13
8620[p]	122	98	21	63	245	76
8630[p]	162	142	14	54	325	42
8640[p]	208	183	13	43	420	18
8650[p]	214	194	12	41	423	
8720[p]	122	98	21	63	245	76
8740[p]	208	183	13	43	420	18
8750[p]	214	194	12	41	423	
9255[p]	232[q]	215	9	21	477	6
9261[p]	258[r]	226	10	30	514	12

[a]Properties are for materials hardened and tempered as follows: [b]water-quenched from 1525°F, tempered at 1000°F; [c]oil-quenched from 1525°F, tempered at 1000°F, [d]pseudocarburized 8 h at 1700°F, oil-quenched, tempered 1 h at 300°F, [e]water-quenched from 1525°F, tempered at 600°F, [f]oil-quenched from 1500°F, tempered at 600°F, [g]oil-quenched from 1475°F, tempered at 600°F, [h]oil-quenched from 1450°F, tempered at 600°F, [i]water-quenched from 1500 to 1600°F, tempered at 800°F, [j]oil-quenched from 1550°F, tempered at 800°F, [k]oil-quenched from 1525°F, tempered at 800°F, [l]normalized at 1650°F, reheated to 1475°F, oil-quenched, tempered at 800°F, [m]normalized at 1625°F, reheated to 1550°F, water-quenched, tempered at 800°F; [n]carburized 10 h at 1680°F., pot-cooled, oil-quenched from 1525°F, tempered at 300°F; [o]normalized at 1600°F, oil-quenched from 1575°F, tempered at 1000°F, [p]oil-quenched tempered at 800°F; [q]normalized at 1650°F, reheated to 1625°F, quenched in agitated oil, tempered at 800°F, [r]normalized at 1600°F, reheated to 1575°F, quenched in agitated oil, tempered at 800°F.

NOTE: °C = (°F − 32) × 5/9. To convert British thermal units per hour-foot-degrees Fahrenheit to watts per meter-degrees Celsius, multiply by 0.8606; to convert British thermal units per pounds-force-degrees Fahrenheit to kilojoules per kilogram-degrees Celsius, multiply by 0.2388; to convert pounds-force per square inch to megapascals, multiply by 0.006895; and to convert foot-pounds-force to joules, multiply by 0.7375.

Bronzes are somewhat similar to brasses in mechanical properties and to high-zinc brasses in corrosion resistance (except that bronzes are not affected by stress cracking). **Aluminum and silicon bronzes** are very popular in the process industries because they combine good strength with corrosion resistance.

TABLE 23-9 Cast-Iron Alloys*

Alloy	ASTM	UNS	Composition, %†	Condition	Mechanical properties‡ Yield strength, kip/in² (MPa)	Tensile strength, kip/in² (MPa)	Elongation, %	Hardness, HB
Gray cast iron	A159 (G3000)	F10006	3.1–3.4 C, 0.6–0.9 Mn, 1.9–2.3 Si	As cast	......	30 (207)	..	187–241
Malleable cast iron	A602 (M3210)	F20000	2.2–2.9 C, 0.15–1.25 Mn, 0.9–1.90 Si	Annealed	32 (229)	50 (345)	12	130
Ductile cast iron	A395 (60-40-18)	F32800	None specified	Annealed	40 (276)	60 (414)	18	170
Cast iron	A436(1)	F41000	3.0 C, 1.5–2.5 Cr, 5.5–7.5 Cu, 0.5–1.5 Mn, 13.5–17.5 Ni, 1.0–2.8 Si	As cast	......	25 (172)	..	150
Cast iron	A436(2)	F41002	3.0 C, 1.5–2.5 Cr, 0.50 Cu, 0.5–1.5 Mn, 18–22 Ni, 1.0–2.8 Si	As cast	......	25(172)	..	145
Cast iron	A436(5)	F41006	2.4 C, 0.1 Cr, 0.5 Cu, 0.5–1.5 Mn, 34–36 Ni, 1.0–2.0 Si	As cast	......	20 (138)	..	110
Ductile austenitic cast iron	A439(D-2)	F43000	3.0 C, 1.75–2.75 Cr, 0.7–1.25 Mn, 18–22 Ni, 1.5–3.0 Si	As cast	30 (207)	58 (400)	..	170
Ductile austenitic cast iron	A439 (D-5)	F43006	2.4 C, 0.1 Cr, 1.0 Mn, 34–36 Ni, 1.0–2.8 Si	As cast	30 (207)	55 (379)	..	155
Silicon cast iron	A518	F47003	0.7–1.1 C, 0.5 Cr, 0.5 Cu, 1.50 Mn, 0.5 Mo, 14.2–14.75 Si	As cast	......	16 (110)	..	520

*Courtesy of National Association of Corrosion Engineers. To convert megapascals to pounds-force per square inch, multiply by 145.04.
†Single values are maximum values.
‡Typical room-temperature properties.

Cupronickels (10 to 30 percent Ni) have become very important as copper alloys. They have the highest corrosion resistance of all copper alloys and find application as heat-exchanger tubing. Resistance to seawater is particularly outstanding.

Lead and Alloys Chemical leads of 99.9+ percent purity are used primarily in the chemical industry in environments that form thin, insoluble, and self-repairable protective films, e.g., salts such as sulfates, carbonates, or phosphates. More soluble films such as nitrates, acetates, or chlorides offer little protection.

Alloys of antimony, tin, and arsenic offer limited improvement in mechanical properties, but the usefulness of lead is limited primarily because of its poor structural qualities. It has a low melting point and a high coefficient of expansion, and it is a very ductile material that will creep under a tensile stress as low as 1 MPa (145 lbf/in²).

Titanium Titanium has become increasingly important as a construction material. It is strong and of medium weight. Corrosion resistance is very superior in oxidizing and mild reducing media (a Ti-Pd alloy has a superior resistance in reducing environments). Titanium is usually not bothered by impingement attack, crevice corrosion, and pitting attack in seawater. Its general resistance to seawater is excellent. Titanium is resistant to nitric acid at all concentrations except with red fuming nitric. The metal also resists ferric chloride, cupric chloride, and other hot chloride solutions. However, there are a number of disadvantages to titanium which have limited its use.

TABLE 23-10 Standard Wrought Martensitic Stainless Steels*

AISI type	UNS	Composition, %† Cr	Ni	Mo	C	Other	Mechanical properties‡ Yield strength, kip/in² (MPa)	Tensile strength, kip/in² (MPa)	Elongation, %	Hardness, HB
403	S40300	11.5–13.0	....	....	0.15	.....	40 (276)	75 (517)	35	155
410	S41000	11.5–13.5	....	....	0.15	.....	35 (241)	70 (483)	30	150
414	S41400	11.5–13.5	1.25–2.5	....	0.15	.....	90 (621)	115 (793)	20	235
416	S41600	12–14	....	0.6	0.15	0.15S§	40 (276)	75 (517)	30	155
416Se	S41623	12–14	....	....	0.15	0.15Se§	40 (276)	75 (517)	30	155
420	S42000	12–14	....	....	0.15	.....	50 (345)	95 (655)	20	195
420F	S42020	12–14	....	0.6	0.15§	0.155§	55 (379)	95 (655)	22	220
422	S42200	11–13	0.5–1.0	0.75–1.25	0.20–0.25	0.15–0.30 V, 0.75–1.25 W	125 (862)	145 (1000)	18	320
431	S43100	15–17	1.25–2.5	....	0.20	.....	95 (665)	125 (862)	20	260
440A	S44002	16–18	....	0.75	0.6–0.75	.....	60 (414)	105 (724)	20	210
440B	S44003	16–18.0	....	0.75	0.75–0.95	.....	62 (427)	107 (738)	18	215
440C	S44004	16–18	....	0.75	0.95–1.20	.....	65 (448)	110 (758)	14	220
501	S50100	4–6	....	0.40–0.65	0.10§	.....	30 (207)	70 (483)	28	160
502	S50200	4–6	....	0.40–0.65	0.10	.....	25 (172)	65 (448)	30	150

*Courtesy of National Association of Corrosion Engineers. To convert megapascals to pounds-force per square inch, multiply by 145.04.
†Single values are maximum values unless otherwise noted.
‡Typical room-temperature properties of annealed plates.
§Minimum.

TABLE 23-11 Standard Wrought Ferritic Stainless Steels*

AISI type	UNS	Composition, %† Cr	C	Mn	Si	P	S	Other	Mechanical properties‡ Yield strength, kip/in² (MPa)	Tensile strength, kip/in² (MPa)	Elongation, %	Hardness, HB
405	S40500	11.5–14.5	0.08	1.0	1.0	0.04	0.03	0.1–0.3 Al	40 (276)	65 (448)	30	150
409	S40900	10.5–11.75	0.08	1.0	1.0	0.045	0.045	(6 × C) Ti§	35 (241)	65 (448)	25	137
429	S42900	14–16	0.12	1.0	1.0	0.04	0.03	.....	40 (276)	70 (483)	30	163
430	S43000	16–18	0.12	1.0	1.0	0.04	0.03	.....	40 (276)	75 (517)	30	160
430F	S43020	16–18	0.12	1.25	1.0	0.06	0.15¶	0.6 Mo	55 (379)	80 (552)	25	170
430FSe	S43023	16–18	0.12	1.25	1.0	0.06	0.06	0.15 Se¶	55 (379)	80 (552)	25	170
434	S43400	16–18	0.12	1.0	1.0	0.04	0.03	0.75–1.25 Mo	53 (365)	77 (531)	23	160
436	S43600	16–18	0.12	1.0	1.0	0.04	0.03	0.75–1.25 Mo (5 × C)(Cb + Ta)§	53 (365)	77 (531)	23	160
442	S44200	18–23	0.20	1.0	1.0	0.04	0.03	.....	45 (310)	80 (552)	20	185
446	S44600	23–27	0.20	1.5	1.0	0.04	0.03	0.25N	55 (379)	85 (586)	25	160

*Courtesy of National Association of Corrosion Engineers. To convert megapascals to pounds-force per square inch, multiply by 145.04.
†Single values are maximum values unless otherwise noted.
‡Typical temperature properties of annealed plates.
§0.70 maximum.
¶Minimum.

Titanium is not easy to form, it has a high springback and tends to gall, and welding must be carried out in an inert atmosphere.

Zirconium Zirconium was originally developed as a construction material for atomic reactors. Reactor-grade zirconium contains very little hafnium, which would alter zirconium's neutron-absorbing properties. Commercial-grade zirconium, for process applications, however, contains 2.5 percent hafnium. Zirconium resembles titanium from a fabrication standpoint. All welding must be done under an inert atmosphere. Zirconium has excellent resistance to reducing environments. Oxidizing agents frequently cause accelerated attack. It resists all chlorides except ferric and cupric. There are a number of alloys of titanium and zirconium, with mechanical properties superior to those of the pure metals. The zirconium alloys are referred to as Zircaloys.

TABLE 23-12 Standard Wrought Austenitic Stainless Steels*

AISI type	UNS	Composition, %† Cr	Ni	Mo	C	Si	Mn	Other	Mechanical properties† Yield strength, kip/in² (MPa)	Tensile strength, kip/in² (MPa)	Elongation, %	Hardness, HB
201	S20100	16–18	3.5–5.5	...	0.15	1.0	5.5–7.5	0.25 N	55 (379)	115 (793)	55	185
202	S20200	17–19	4–6	...	0.15	1.0	7.5–10.	0.25 N	55 (379)	105 (724)	55	185
301	S30100	16–18	6–8	...	0.15	1.0	2.0	...	40 (276)	105 (724)	55	165
302	S30200	17–19	8–10	...	0.15	1.0	2.0	...	35 (241)	90 (621)	60	150
302B	S30215	17–19	8–10	...	0.15	2.0–3.0	2.0	...	40 (276)	90 (621)	50	165
303	S30300	17–19	8–10	0.6	0.15	1.0	2.0	0.15 S,§ 0.2 P	35 (241)	90 (621)	50	160
303Se	S30323	17–19	8–10	...	0.15	1.0	2.0	0.15 Se,§ 0.2 P	35 (241)	90 (621)	50	160
304	S30400	18–20	8–10.5	...	0.08	1.0	2.0	...	35 (241)	82 (565)	60	149
304L	S30403	18–20	8–12	...	0.03	1.0	2.0	...	33 (228)	79 (545)	60	143
304N	S30451	18–20	8–10.5	...	0.08	1.0	2.0	0.10–0.16 N	48 (331)	90 (621)	50	180
308	S30800	19–21	10–12	...	0.08	1.0	2.0	...	30 (207)	85 (586)	55	150
309	S30900	22–24	12–15	...	0.20	1.0	2.0	...	40 (276)	95 (655)	45	170
309S	S30908	22–24	12–15	...	0.08	1.0	2.0	...	40 (276)	95 (655)	45	170
310	S31000	24–26	19–22	...	0.25	1.5	2.0	...	45 (310)	95 (655)	50	170
310S	S31008	24–26	19–22	...	0.08	1.5	2.0	...	45 (310)	95 (655)	50	170
314	S31400	23–26	19–22	...	0.25	1.5–3.0	2.0	...	50 (345)	100 (609)	45	180
316	S31600	16–18	10–14	2.0–3.0	0.08	1.0	2.0	...	36 (248)	82 (565)	55	149
316L	S31603	16–18	10–14	2.0–3.0	0.03	1.0	2.0	...	34 (234)	81 (558)	55	146
316N	S31651	16–18	10–14	2.0–3.0	0.08	1.0	2.0	0.10–0.16 N	42 (290)	90 (621)	55	180
317	S31700	18–20	11–15	3.0–4.0	0.08	1.0	2.0	...	40 (276)	85 (586)	50	160
317L	S31703	18–20	11–15	3.0–4.0	0.03	1.0	2.0	...	35 (241)	85 (586)	55	150
321	S32100	17–19	9–12	...	0.08	1.0	2.0	(5 × C) Ti§	30 (207)	85 (586)	55	160
329	S32900	25–30	3–6	1.0–2.0	0.10	1.0	2.0	...	80 (552)	105 (724)	25	230
347	S34700	17–19	9–13	...	0.08	1.0	2.0	(10 × C)(Cb + Ta)§	35 (241)	90 (621)	50	160
348	S34800	17–19	9–13	...	0.08	1.0	2.0	(10 × C)(Cb + Ta)¶ 0.20 Co	35 (241)	90 (621)	50	160

*Courtesy of National Association of Corrosion Engineers. To convert megapascals to pounds-force per square inch, multiply by 145.04.
†Single values are maximum values unless otherwise noted.
‡Typical room-temperature properties of solution-annealed plates.
§Minimum.
¶Minimum except Ta = 0.1 maximum.

TABLE 23-13 Special Stainless Steels*

Alloy	UNS	Composition, %† Cr	Ni	Mo	C	Mn	Si	Other	Mechanical properties‡ Yield strength, kip/in^2 (MPa)	Tensile strength, kip/in^2 (MPa)	Elongation, %	Hardness, HB
A-286	K66286	13.5–16	24–27	1.0–1.5	0.08	2.0	1.0	1.90–2.35 Ti, 0.1–0.5 V, 0.001–0.01 B,	100 (690)	140 (970)	20	
20Cb-3	N08020	19–21	32–38	2.0–3.0	0.07	2.0	1.0	8 × C–1.0 Cb 3.0–4.0 Cu	53 (365)	98 (676)	33	185
20Mod	N08320	21–23	25–27	4.0–6.0	0.05	2.5	1.0	(4.0 × C) Ti min.	43 (296)	84 (579)	42	160
PH13-8Mo	S13800	12.25–13.25	7.5–8.5	2.0–2.5	0.05	0.2	0.1	0.90–1.35 Al	120 (827)	160 (1100)	17	300
PH14-8Mo	S14800	13.75–15.0	7.75–8.75	2.0–3.0	0.05	1.0	1.0	0.75–1.5 Al, 0.15–45 Cb	55–210 (380–1450)	125–230 (860–1540)	2–25	200–450
15-5PH	S15500	14.0–15.5	3.5–5.5	....	0.07	1.0	1.0	2.5–4.5 Cu	145 (1000)	160 (1100)	15	320
PH15-7Mo	S15700	14.0–16.0	6.5–7.75	2.0–3.0	0.09	1.0	1.0	0.75–1.5 Al, 0.15–0.45 Cb	55–210 (380–1450)	130–220 (900–1520)	2–35	200–450
17-4PH	S17400	15.5–17.5	3.0–5.0	....	0.07	1.0	1.0	3.0–5.0 Cu, 0.4 Al	145 (1000)	160 (1100)	15	320
W	S17600	16.0–17.5	6.0–7.5	....	0.08	1.0	1.0	0.4–1.20 Ti	90–200 (620–1380)	135–210 (930–1450)	3–12	260–420
17-7PH	S17700	16.0–18.0	6.5–7.75	.....	0.09	1.0	1.0	0.75–1.5 Al	40 (276)	130 (710)	10	185
216	S21600	17.5–22.0	5.0–7.0	2.0–3.0	0.08	7.5–9.0	1.0	0.25–0.5 N	70 (480)	115 (790)	45	200
Nitronic 60	S21800	16.0–18.0	8.0–9.0	....	0.10	7.0–9.0	3.5–4.5	0.08–0.18 N	60 (410)	103 (710)	62	210
21-6-9	S21900	18.0–21.0	5.0–7.0	....	0.08	8.0–10.0	1.0	0.15–0.40 N	68 (470)	112 (770)	44	220
AM350	S35000	16.0–17.0	4.0–5.0	2.5–3.25	0.07–0.11	0.5–1.25	0.5	0.07–0.13 N	60–173 (410–1200)	145–206 (1000–1420)	13.5–40	200–400
AM355	S35500	15.0–16.0	4.0–5.0	2.5–3.25	0.10–0.15	0.5–1.25	0.5	......	182 (1250)	216 (1490)	19	402–477
Almar 362	S36200	14.0–15.0	6.0–7.0	....	0.05	0.5	0.3	0.55–9.0 Ti	105–185 (124–286)	120–188 (827–1300)	15–13	250–400
18-18-2	S38100	17.0–19.0	17.5–18.5	....	0.08	2.0	1.5–2.5	0.20 × (C + N)	40 (280)	80 (550)	55	165
Stab. 18-2	S44400	17.5–19.5	1.0	1.75–2.5	0.025	1.0	1.0	min.–0.8 Ti + Cb 0.015 N	45 (310)	60 (414)	20	210
26-1	S44625	25.0–27.5	0.5	0.75–1.50	0.01	0.4	0.4	0.5 Ni + Cu	50 (345)	70 (480)	30	165
Stab. 26-1	S44626	25.0–27.0	0.5	0.75–1.50	0.06	0.75	0.75	7 × (C + Ni)–1.0 Ti, 0.15 Cu	50 (345)	70 (480)	30	165
28-4	S44700	28.0–30.0	0.15	3.5–4.2	0.010	0.3	0.2	0.02 N, 0.15 Cu	70 (480)	90 (620)	25	210
28-4-2	S44800	28.0–30.0	2.0–2.5	3.5–4.2	0.010	0.3	0.2	0.02 N, 8 × C Cb min.	85 (590)	95 (650)	25	230
Custom 450	S45000	14.0–16.0	5.0–7.0	0.5–1.0	0.05	1.0	1.0	1.25–1.75 Cu	117–184 (800–1270)	144–196 (990–1350)	14	270–400
Custom 455	S45500	11.0–12.5	7.5–9.5	0.5	0.05	0.5	0.5	1.5–2.5 Cu, 0.8–1.4 Ti	115–220 (790–1500)	140–230 (970–1600)	10–14	290–460

*Courtesy of National Association of Corrosion Engineers. To convert megapascals to pounds-force per square inch, multiply by 145.04.
†Single values are maximum values unless otherwise noted.
‡Typical room-temperature properties.

TABLE 23-14 Standard Cast Heat-Resistant Stainless Steels*

ACI	Equivalent AISI	UNS	Composition, %[a] Cr	Ni	C	Mn	Si	Other	Mechanical properties at 1600°F, Short term: Tensile strength, kip/in² (MPa)	Elongation, %	Stress to rupture in 1000 h: kip/in²	MPa
HA	. . .	. . . . .	8–10	. . . .	0.2	0.35–0.65	1.0	0.9–1.2 Mo.	44 (303)[b]	36[b]	27	186[c]
HC	446	J92605	26–30	4	0.5	1.0	2.0	. . . .	. . . .		1.3	9.0
HD	327	J93005	26–30	4–7	0.5	1.5	2.0	. . . .	23 (159)	18	7.0	48[d]
HE	. . .	J93403	26–30	8–11	0.2–0.5	2.0	2.0					
HF	302B	J92603	18–23	9–12	0.2–0.4	2.0	2.0	. . .	21 (145)	16	4.4	30
HH[e]	. . .	J93503	24–28	11–14	0.2–0.5	2.0	2.0	0.2 N	18.5 (128)	30	3.8	26
HH[f]	309	J93503	24–28	11–14	0.2–0.5	2.0	2.0	0.2 N	21.5 (148)	18	3.8	26
HI	. . .	J94003	26–30	14–18	0.2–0.5	2.0	2.0	. . . .	26 (179)	12	4.8	33
HK	310	J94224	24–28	18–22	0.2–0.6	2.0	2.0	. . . .	23 (159)	16	6.0	41
HL	. . .	J94604	28–32	18–22	0.2–0.6	2.0	2.0	. . . .	30 (207)			
HN	. . .	J94213	19–23	23–27	0.2–0.5	2.0	2.0	. . . .	20 138)	37	7.4	51
HP	. . .	J95705	24–28	33–37	0.35–0.75	2.0	2.5	. . . .	26 (179)	27	7.5	52
HT	330	J94605	13–17	33–37	0.35–0.75	2.0	2.5	. . . .	19 (131)	26	5.8	40
HU	. . .	J95405	17–21	37–41	0.35–0.75	2.0	2.5	. . . .	20 (138)	20	5.2	36
HW	. . .	. . . . . .	10–14	58–62	0.35–0.75	2.0	2.5	. . . .	19 (131)		4.5	31
HX	. . .	. . . . . .	15–19	64–68	0.35–0.75	2.0	2.5	. . . .	20.5 (141)	48	4.0	28

*Courtesy of National Association of Corrosion Engineers. To convert megapascals to pounds-force per square inch, multiply by 145.04.
[a]Single values are maximum values; S and P are 0.04 maximum; Mo is 0.5 maximum.
[b]At 1100°F (593°C).
[c]At 1000° (538°C).
[d]At 1400°F (760°C).
[e]Type I; partially ferritic.
[f]Type II; wholly austenitic.

TABLE 23-15 Standard Cast Corrosion-Resistant Stainless Steels*

ACI	Equivalent AISI	UNS	Composition, %[a] Cr	Ni	Mo	C	Mn	Si	Other	Mechanical properties[b]: Yield strength, kip/in² (MPa)	Tensile strength, kip/in² (MPa)	Elongation, %	Hardness, HB
CA-15	410	J91150	11.5–14	1.0	0.5	0.15	1.00	1.50	. . . . .	150 (1034)[c]	200 (1379)[c]	7[c]	390[c]
CA-15M	. . .	J91151	11.5–14	1.0	0.15–1.0	0.15	1.00	1.50	. . . . .	150 (1034)[c]	200 (1379)[c]	7[c]	390[c]
CA-6NM	. . .	J91540	11.5–14	3.5–4.5	0.4–1.0	0.06	1.00	1.00	. . . . .	100 (690)[d]	120 (827)[d]	4[d]	269[d]
CA-40	420	J91153	11.5–14	1.0	0.5	0.20–0.40	1.00	1.50	. . . . .	165 (1138)[c]	220 (1517)[c]	1[c]	470[c]
CB-30	431	J91803	18.21	2.0	. . . .	0.30	1.00	1.50	. . . . .	60 (414)[e]	95 (655)[e]	15[e]	195[e]
CC-50	446	J92615	26–30	4.0	. . . .	0.50	1.00	1.50	. . . . .	65 (448)[f]	97 (669)[f]	18[f]	210[f]
CE-30	312	J93423	26–30	8–11	. . . .	0.30	1.50	2.00	. . . . .	63 (434)	97 (669)	18	190
CB-7Cu	(17–4PH)	. . . . . .	(16)	(4)	. . . .	0.07	. . . .	. . . .	(3) Cu	165 (1138)	. . . .	3	418
CD-4MCu	. . .	. . . . . .	25–26.5	4.75–6.0	1.75–2.25	0.04	1.00	1.00	2.75–3.25 Cu	82 (565)	108 (745)	25	253
CF-3	304L	J92500	17–21	8–12	. . . .	0.03	1.50	2.00	. . . . .	36 (248)	77 (531)	60	140
CF-8	304	J92600	18–21	8–11	. . . .	0.08	1.50	2.00	. . . . .	37 (255)	77 (531)	55	140
CF-20	302	J92602	18–21	8–11	. . . .	0.20	1.50	2.00	. . . . .	36 (248)	77 (531)	50	163
CF-3M	316L	J92800	17–21	9–13	2.0–3.0	0.03	1.50	1.50	. . . . .	38 (262)	80 (552)	55	150
CF-8M	316	J92900	18–21	9–12	2.0–3.0	0.08	1.50	2.00	. . . . .	42 (290)	80 (552)	50	160
CF-12M	. . .		18–21	9–12	2.0–3.0	0.12	1.50	2.00	. . . . .	42 (290)	80 (552)	50	160
CG-12	317	J93000	18–21	9–13	3.0–4.0	0.08	1.50	1.50	. . . . .	44 (303)	83 (572)	45	170
CF-8C	347	J92710	18–21	9–12	. . . .	0.08	1.50	2.00	(8 × C) Cb[g]	38 (262)	77 (531)	39	149
CF-16F	303	J92701	18–21	9–12	1.50	0.16	1.50	2.00	. . . . .	40 (276)	77 (531)	52	150
CG-12	. . .	J93001	20–23	10–13	. . . .	0.12	1.50	2.00	. . . . .	28 (193)	. . . .	35	
CH-20	309	J93402	22–26	12–15	. . . .	0.20	1.50	2.00	. . . . .	50 (345)	88 (607)	38	190
CK-20	310	J94202	23–27	19–22	. . . .	0.20	1.50	2.00	. . . . .	38 (262)	76 (524)	37	144
CN-7M	. . .	J95150	19–22	27.5–30.5	2.0–3.0	0.07	1.50	1.50	3–4 Cu	32 (221)	69 (476)	48	130

*Courtesy of National Association of Corrosion Engineers. To convert megapascals to pounds-force per square inch, multiply by 145.04.
[a]Single values are maximum values except those in parentheses, which are minimum values. P and S values are 0.04 maximum.
[b]Typical room-temperature properties for solution-annealed material unless otherwise noted.
[c]For material air-cooled from 1800°F and tempered at 600°F.
[d]For material air-cooled from 1750°F and tempered at 1100 to 1150°F.
[e]For material annealed at 1450°F, furnace-cooled to 1000°F, then air-cooled.
[f]Air-cooled from 1900°F.
[g]1.0 maximum.

TABLE 23-16 Nickel Alloys*

Alloy	UNS	Composition, %† NI(+Co)§	Cr	Fe	Mo	C	Other	Condition	Mechanical properties‡ Yield strength, kip/in² (MPa)	Tensile strength, kip/in² (MPa)	Elongation, %	Hardness, HB
200	N02200	99.	.....	0.4	.....	0.15	....	Annealed	15–30 (103–207)	55–80 (379–552)	55–40	90–120
201	N02201	99.	.....	0.4	.....	0.02	....	Annealed	10–25 (69–172)	50–60 (345–414)	60–40	75–102
400	N04400	63–70	.....	1.0–2.5	.....	0.3	28–34 Cu	Annealed	25–50 (172–345)	70–90 (483–621)	60–35	110–149
K-500	N05500	63–70	.....	2.0	.....	0.25	2.3–3.15 Al 0.35–0.85 Ti, 30 Cu	Age-hardened Annealed	85–120 (586–827)	130–165 (896–1138)	35–20	250–315
600	N06600	72.	14–17	6–10	.....	0.15	....	Annealed	30–50 (207–345)	80–100 (552–690)	55–35	120–170
601	N06601	58–63	21–25	Bal.	.....	0.10	1.0–1.7 Al	Annealed	30–60 (207–414)	80–115 (552–793)	70–40	110–150
625	N06625	Bal.	20–23	5	8–10	0.10	3.15–4.15 (Cb + Ta)	Annealed	60–95 (414–655)	120–150 (827–1034)	60–30	145–220
706	N09706	39–44	14.5–17.5	Bal.	.....	0.06	....	Solution-treated and aged	161 (1110)	193 (1331)	20.	371.
718	N07718	50–55	17–21	Bal.	2.8–3.3	0.08	4.75–5.5 (Cb + Ta)	Special heat treatment	171 (1180)	196 (1351)	17.	382.
X-750	N07750	70	14–17	5–9	.....	0.08	0.65–1.15 Ti, 0.2–0.8 Al 0.7–1.2 (Cb + Ta) 2.25–2.75 Ti, 0.4–1.0 Al	Special heat treatment	115–142 (793–979)	162–193 (1117–1331)	30–15	300–390
800	N08800	30–35	19–23	Bal.	.....	0.10	0.15–0.6 Al, 0.15–0.6 Ti	Annealed	30–60 (207–414)	75–100 (517–690)	60–30	120–184
800H	N08800	30–35	19–23	Bal.	.....	0.05–0.10	0.15–0.6 Al, 0.15–0.6 Ti	Solution-treated	20–50 (138–345)	65–95 (448–655)	50–30	100–184
801	N08801	30–34	19–22	Bal.	.....	0.10	0.75–1.5 Ti	Special heat treatment	79.5 (548)	129 (889)	29.5	
825	N08825	38–46	19.5–23.5	Bal.	2.5–3.5	0.05	1.5–3.0 Cu, 0.6–1.2 Ti	Annealed	35–65 (241–448)	85–105 (586–724)	50–30	120–180
B-2	N10665	Bal.	1.0	2	26–30	0.02	....	Annealed	76 (524)	139 (958)	53.	210
C-276	N10276	Bal.	14.5–16.5	4–7	15–17	0.02	3.0–4.5 W	Annealed	52 (358)	115 (793)	61.	194
C-4	N06455	Bal.	14–18	3	14–17	0.015	0.7 Ti	Annealed	61 (421)	116 (800)	54.	194
G	N06007	Bal.	21–23	18–21	5.5–7.5	0.05	1.0–2.0 Mn, 1.5–2.5 Cu 1.75–2.5 (Cb + Ta)	Annealed	46 (317)	102 (703)	61.	161
X	N06002	Bal.	20.5–23	17–20	8–10	0.05–0.15	0.2–1.0 W	Annealed	56 (386)	110 (758)	45.	178

*Courtesy of National Association of Corrosion Engineers. To convert megapascals to pounds-force per square inch, multiply by 145.04.
†Single values are maximum unless otherwise noted.
‡Typical room-temperature properties.
§Single values are minima.

TABLE 23-17 Aluminum Alloys

AA designation	UNS	Composition, %* Cr	Cu	Mg	Mn	Si	Other	Condition‡	Mechanical properties† Yield strength, kip/in² (MPa)	Tensile strength, kip/in² (MPa)	Elongation in 2 in, %	Hardness, HB
Wrought												
1060	A91060	....	....	....	....	....	99.6 Al min.	0	4 (28)	10 (69)	43	19
1100	A91100	....	0.05–0.2	....	....	....	99.0 Al min.	0	5 (34)	13 (90)	45	23
2024	A92024	0.1	3.8–4.9	1.2–1.8	0.3–0.9	0.5	....	T4	47 (324)	68 (469)	19	120
3003	A93003	....	0.05–0.2	....	1.0–1.5	0.6	....	H14	21 (145)	22 (152)	16	40
5052	A95052	0.15–0.35	0.1	2.2–2.8	0.1	....	....	0	13 (90)	2.8 (193)	30	47
5083	A95083	0.05–0.25	0.1	4.0–4.9	0.4–1.0	0.4	....	0	21 (145)			
5086	A95086	0.05–0.25	0.1	3.5–4.5	0.2–0.7	0.4	....	0	17 (117)	38 (262)	30	
5154	A95154	0.05–0.35	0.1	3.1–3.9	0.1	0.25	....	0	17 (117)	35 (241)	27	58
6061	A96061	0.04–0.35	0.15–0.4	0.8–1.2	0.15	0.4–0.8	....	T6	40 (276)	45 (310)	17	95
6063	A96063	0.1	0.1	0.45–0.9	0.1	0.2–0.6	....	T6	31 (214)	35 (241)	18	73
7075	A97075	0.18–0.28	1.2–2.0	2.1–2.9	0.3	0.40	5.1–6.1 Zn	T6	73 (503)	63 (572)	11	150
Cast												
242.0	A02420	0.25	3.5–4.5	1.2–1.8	0.35	0.7	1.7–2.3 Ni	S-T571	....	29 (200)		
295.0	A02950	....	4.0–5.0	0.03	0.35	0.7–1.5	....	S-T4	....	29 (200)	6	
A332.0	A13320	....	0.5–1.5	0.7–1.3	0.35	11–13	2.0–3.0 Ni	P-T551	....	31 (214)		
B443.0	A24430	....	0.15	0.05	0.35	4.5–6.0	....	S-F	....	17 (117)	3	
514.0	A05140	....	0.15	3.5–4.5	0.35	0.35	....	S-F	....	22 (152)	6	
520.0	A05200	....	0.25	9.5–10.6	0.15	0.25	....	S-T4	22 (152)	42 (290)	12	

SOURCE: Aluminum Association. Courtesy of National Association of Corrosion Engineers. To convert megapascals to pounds-force per square inch, multiply by 145.04.

*Single values are maximum values.

†Typical room-temperature properties.

‡S = sand-cast; P = permanent-mold-cast; other = temper designations.

Tantalum The physical properties of tantalum are similar to those of mild steel except that tantalum has a higher melting point. Tantalum is ductile and malleable and can be worked into intricate forms. It can be welded by using inert-gas-shielded techniques. The metal is practically inert to many oxidizing and reducing acids (except fuming sulfuric). It is attacked by hot alkalies and hydrofluoric acid. Its cost generally limits use to heating coils, bayonet heaters, coolers, and condensers operating under severe conditions. When economically justified, larger items of equipment (reactors, tanks, etc.) may be fabricated with tantalum liners, either loose (with proper anchoring) or explosion-bonded-clad . Since tantalum linings are usually very thin, very careful attention to design and fabrication details is required.

INORGANIC NONMETALLICS

Glass and Glassed Steel Glass is an inorganic product of fusion which is cooled to a rigid condition without crystallizing. With unique properties compared with metals, they require special considerations in their design and use.

Glass has excellent resistance to all acids except hydrofluoric and hot, concentrated H_3PO_4. It is also subject to attack by hot alkaline solutions. Glass is particularly suitable for piping when transparency is desirable.

The chief drawback of glass is brittleness, and it is also subject to damage by thermal shock. However, glass armored with epoxy-polyester fiberglass can readily be protected against breakage. On the other hand, glassed steel combines the corrosion resistance of glass with the working strength of steel. Accordingly, **glass linings** are resistant to all concentrations of hydrochloric acid to 120°C (250°F), to dilute concentrations of sulfuric to the boiling point, to concentrated sulfuric to 230°C (450°F), and to all concentrations of nitric acid to the boiling point. Acid-resistant glass with improved alkali resistance (up to 12 pH) is available.

A nucleated crystalline ceramic-metal composite form of glass has superior mechanical properties compared with conventional glassed steel. Controlled high-temperature firings chemically and physically bond the ceramic to steel, nickel-based alloys, and refractory metals. These materials resist corrosive hydrogen chloride gas, chlorine, or sulfur dioxide at 650°C (1200°F). They resist all acids except HF up to 180°C (350°F). Their impact strength is 18 times that of safety glass; abrasion resistance is superior to that of porcelain enamel. They have 3 to 4 times the thermal-shock resistance of glassed steel.

Porcelain and Stoneware Porcelain and stoneware materials are about as resistant to acids and chemicals as glass, but with the advantage of greater strength. This is offset somewhat by poor thermal conductivity, and the materials can be damaged by thermal shock fairly easily. **Porcelain enamels** are used to coat steel, but the enamel has slightly inferior chemical resistance. Some refractory coatings, capable of taking very high temperatures, are also available.

Brick Construction Brick-lined construction can be used for many severely corrosive conditions under which high alloys would fail. Common bricks are made from carbon, red shale, or acidproof refractory materials. Red-shale brick is not used above 175°C (350°F) because of spalling. Acidproof refractories can be used up to 870°C (1600°F).

A number of **cement** materials are used with brick. Standard are phenolic and furan resins, polyesters, sulfur, silicate, and epoxy-based materials. Carbon-filled polyesters and furanes are good against nonoxidizing acids, salts, and solvents. Silica-filled resins should not be used against hydrofluoric or fluosilicic acids. Sulfur-based cements are limited to 93°C (200°F), while resins can be used to about 180°C (350°F). The sodium silicate–based cements are good against acids to 400°C (750°F).

Differential thermal expansion of the brick, its joints, and the vessel substrate necessitates an intermediate lining of lead, asphalt, rubber, or plastic. This membrane functions as a barrier to protect the substrate from corrosion damage. A special prestressed-brick design that maintains the brick in compression by using a controlled-expansion resinous mortar and brick bedding material precludes the use of an elastomeric membrane.

Cement and Concrete Concrete is an aggregate of inert reinforcing particles in an amorphous matrix of hardened cement paste. Concrete made of portland cement has limited resistance to acids and bases and will fail mechanically following absorption of crystal-forming solutions such as brines and various organics. Concretes made of corrosion-resistant cements (such as calcium aluminate) can be selected for specific chemical exposures.

TABLE 23-18 Copper Alloys*

Alloy	CDA	UNS	Composition, %†						Mechanical properties‡		
			Cu	Zn	Sn	Al	Ni	Other	Yield strength, kip/in² (MPa)	Tensile strength, kip/in² (MPa)	Elongation in 2 in, %
Wrought											
Copper	110	C11000	99.90	....	....	....	....	....	10 (69)	32 (221)	55
Commercial bronze	220	C22000	89–91	Rem.	....	....	....	......	10 (69)	37 (255)	50
Red brass	230	C23000	84–86	Rem.	....	....	....	....	10 (69)	40 (276)	55
Cartridge brass	260	C26000	68.5–71.5	Rem.	....	....	....	....	11 (76)	44 (303)	66
Yellow brass	270	C27000	63–68.5	Rem.	....	....	....	....	14 (97)	46 (317)	65
Muntz Metal	280	C28000	59–63	Rem.	....	....	....	....	21 (145)	54 (372)	52
Admiralty brass	443	C44300	70–73	Rem.	0.9–1.2	....	....	0.02–0.1 As	18 (124)	48 (331)	65
Admiralty brass	444	C44400	70–73	Rem.	0.9–1.2	....	....	0.02–0.1 Sb	18 (124)	48 (331)	65
Admiralty brass	445	C44500	70–73	Rem.	0.9–1.2	....	....	0.02–0.1 P	18 (124)	48 (331)	65
Naval brass	464	C46400	59–62	Rem.	0.5–1.0	....	....	....	25 (172)	58 (400)	50
Phosphor bronze	510	C51000	Rem.	0.3	4.2–5.8	....	....	0.03–0.35 P	19 (131)	47 (324)	64
Phosphor bronze	524	C52400	Rem.	0.2	9.0–11.0	....	....	0.03–0.35 P	28 (193)	66 (455)	70
Aluminum bronze	613	C61300	86.5–93.8	....	0.2–0.5	6–8	0.5	3.5 Fe	30 (207)	70 (483)	42
Aluminum bronze D	614	C61400	88.0–92.5	0.2	....	6–8	....	1.5–3.5 Fe, 1.0 Mn	33 (228)	76 (524)	45
Nickel-aluminum bronze	630	C63000	78–85	0.3	0.2	9–11	4.0–5.5	2.0–4.0 Fe, 1.5 Mn, 0.25 Si	36 (248)	90 (620)	10
High-silicon bronze	655	C65500	94.8	1.5	....	....	0.6	0.8 Fe, 0.5–1.3 Mn, 2.8–3.8 Si	21 (145)	56 (386)	63
Manganese bronze	675	C67500	57–60	Rem.	0.5–1.5	0.25	....	0.05–0.5 Mn, 0.8–2.0 Fe	30 (207)	65 (448)	33
Aluminum brass	687	C68700	76–79	Rem.	....	1.8–2.5	....	0.02–0.1 As	27 (186)	60 (414)	55
90-10 copper nickel	706	C70600	86.5	1.0	....	....	9.0–11.0	1.0–1.8 Fe, 1.0 Mn	16 (110)	44 (303)	42
70–30 copper nickel	715	C71500	Rem.	1.0	....	....	29–33	0.4–1.0 Fe, 1.0 Mn	20 (138)	54 (372)	45
65-18 nickel silver	752	C75200	63–66.5	Rem.	....	....	16.5–19.5	0.25 Fe, 0.5 Mn	25 (172)	56 (386)	45
Cast											
Ounce metal	836	C83600	84–86	4–6	4–6	0.005	1.0	4–6 Pb	17 (117)	37 (255)	30
Manganese bronze	865	C86500	55–65	36–42	1.0	0.5–1.5	1.0	0.4–2.0 Fe; 0.1–1.5 Mn	28 (193)	71 (490)	30
G bronze	905	C90500	86–89	1.0–3.0	9–11	0.005	1.0	....	22 (152)	45 (310)	25
M bronze	922	C92200	86–90	3.0–5.0	5.5–6.5	0.005	1.0	1.0–2.0 Pb	20 (138)	40 (276)	30
Ni-Al-Mn bronze	957	C95700	71	....	....	7.0–8.5	1.5–3.0	2.0–4.0 Fe; 11–14 Mn	45 (310)	95 (665)	26
Ni-Al bronze	958	C95800	79	....	....	8.5–9.5	4.0–5.0	3.5–4.5 Fe; 0.8–1.5 Mn	38 (262)	95 (655)	25
Copper nickel	964	C96400	65–69	....	....	....	28–32	0.5–1.5 Cb; 0.25–1.5 Fe, 1.5 Mn	37 (255)	68 (469)	28

*Courtesy of National Association of Corrosion Engineers. To convert megapascals to pounds-force per square inch, multiply by 145.04.
†Single values are maximum values except for Cu, which is minimum.
‡Typical room-temperature properties of annealed or as-cast material.

Soil Clay is the primary construction material for settling basins and waste-treatment evaporation ponds. Since there is no single type of clay even within a given geographical area, shrinkage, porosity, absorption characteristics, and chemical resistance must be checked for each application.

ORGANIC NONMETALLICS

Plastic Materials In comparison with metallic materials, the use of plastics is limited to relatively moderate temperatures and pressures [230°C (450°F) is considered high for plastics]. Plastics are also less resistant to mechanical abuse and have high expansion rates, low strengths (thermoplastics), and only fair resistance to solvents. However, they are lightweight, are good thermal and electrical insulators, are easy to fabricate and install, and have low friction factors.

Generally, plastics have excellent resistance to weak mineral acids and are unaffected by inorganic salt solutions—areas where metals are not entirely suitable. Since plastics do not corrode in the electrochemical sense, they offer another advantage over metals: most metals are affected by slight changes in pH, or minor impurities, or oxygen content, while plastics will remain resistant to these same changes.

The important thermoplastics used commercially are polyethylene, acrylonitrile butadiene styrene (ABS), polyvinyl chloride (PVC), cellulose acetate butyrate (CAB), vinylidene chloride (Saran), fluorocarbons (Teflon, Kel-F, Kynar), polycarbonates, polypropylene, nylons, and acetals (Delrin). Important thermosetting plastics are general-purpose polyester glass reinforced, bisphenol-based polyester glass, epoxy glass, vinyl ester glass, furan and phenolic glass, and asbestos reinforced.

THERMOPLASTICS

The most chemical-resistant plastic commercially available today is **tetrafluoroethylene** or **TFE** (Teflon). This thermoplastic is practically unaffected by all alkalies and acids except fluorine and chlorine gas at elevated temperatures and molten metals. It retains its properties up to 260°C (500°F). **Chlorotrifluoroethylene** or **CTFE** (Kel-

TABLE 23-19 Miscellaneous Alloys*

Alloy	Designation	UNS	Composition, %‡	Condition	Mechanical properties† Yield strength, kip/in² (MPa)	Tensile strength, kip/in² (MPa)	Elongation, %	Hardness, HB
			Refractory alloys					
Niobium R04210 (columbium)	.....	204–210	99.6 Cb	Annealed	37 (255)	53 (365)	26	80
Molybdenum	.....	R03600	0.01–0.04 C					
Molybdenum, low C	.....	R03650	0.01 C					
Molybdenum alloy	.....	R03630	0.01–0.04 C, 0.40–0.55 Ti, 0.06–0.12 Zn					
Tantalum	.....	R05200	99.8 min. Ta.	Annealed	....	50 (345)	40	45
Tungsten	.....	R07030	99.9 min. W	Annealed	....	270 (1862)		
Zirconium	.....	R60702	4.5 Hf, 0.2 Fe + Cr, 99.2 Zi + Hf	Annealed	16 (110)	36 (248)	31	77
			Precious metals and alloys					
Gold	.....	P00020	99.95 min. Au	Annealed	....	19 (131)	45	25
Silver	.....	P07015	99.95 min. Ag	Annealed	8 (55)	18 (124)	54	27
Sterling silver	.....	......	7.5 Cu, 92.5 Ag	Annealed	20 (138)	41 (283)	26	65
Platinum	.....	P04955	99.95 min. Pt	Annealed	....	18 (124)	38	39
Palladum	.....	P03980	99.80 min. Pd	Annealed	....	25 (172)	27	38
			Lead alloys					
Chemical lead	.....	.....	99.9 min. Pb	Rolled	1.9 (13)	2.5 (17)	50	5
Antimonial lead	.....	.....	90 Pb, 10 Sb	Rolled	....	4.1 (28)	47	13
Tellurium lead	.....	.....	99.85 Pb, 0.04 Te, 0.06 Cu	Rolled	2.2 (15)	3 (21)	45	6
50-50 solder	.....	L05500	50 Pb, 50 Sn, 0.12 max. Sb	Cast	......	6.8 (47)	50	14
			Magnesium alloys					
Wrought alloy	AZ31B	M11311	2.5–3.5 Al, 0.20 min. Mn, 0.6–1.4 Zn	Annealed	15–18 (103–124)	32 (220)	9–12	56
Cast alloy	AZ91C	M11914	8.1–9.3 Al, 0.13 min. Mn, 0.4–1.0 Zn	As cast	11 (76)	23 (159)	....	60
Cast alloy	EZ33A	M12330	2.0–3.1 Zn, 0.5–1.0 Zr	Aged	14 (97)	20 (138)	2	50
Wrought alloy	HK31A	M13310	0.3 Zn, 2.5–4.0 Th, 0.4–1.0 Zr	Stress hard-annealed	24–26 (165–179)	33–34 (228–234)	4	57
			Titanium alloys					
Commercial pure	Gr. 1	R50250	0.20 Fe, 0.18 O	Annealed	35 (241)	48 (331)	30	120
Commercial pure	Gr. 2	R50400	0.30 Fe, 0.25 O	Annealed	50 (345)	63 (434)	28	200
Ti-Pd	Gr. 7	R52400	0.30 Fe, 0.25 O, 0.12–0.25 Pd	Annealed	50 (345)	63 (434)	28	200
Ti-6Al-4V	Gr. 5	R56400	5.5–5.6 Al, 0.40 Fe, 0.20 O, 3.5–4.5 V	Annealed	134 (924)	144 (993)	14	330
Low alloy	Gr. 12	....	0.2–0.4 Mo, 0.6–0.9 Ni	Annealed	65 (448)	75 (517)	25	
			Cobalt alloys					
	N-155	R30155	0.08–0.16 C, 0.75–1.25 Cb, 18.50–21.0 Co, 20.0–22.5 Cr, 1.0–2.0 Mn, 2.5–3.5 Mo, 19–21 Ni, 1.0 Si, 2.0–3.0 W					
	MP35N	R30036	0.025 C, 19–21 Cr, 1.0 Fe, 0.15 Mn, 9.0–10.5 Mo, 33.37 Ni, 0.15 Si, 1.0 Ti	Annealed	60 (414)	135 (931)	70	
	Stelite 6	R30006	0.9–1.4 C, 27–31 Cr, 3 Fe, 1.0 Mn, 1.5 Mo, 3.0 Ni, 1.5 Si, 3.5–5.5 W	As cast	....	105 (724)	1	

*Courtesy of National Association of Corrosion Engineers. To convert megapascals to pounds-force per square inch, multiply by 145.04.
†Typical room-temperature properties.
‡Single values are maximum values unless otherwise noted.

F, Plaskon) also possesses excellent corrosion resistance to almost all acids and alkalies up to 180°C (350°F). A Teflon derivative has been developed from the copolymerization of tetrafluoroethylene and hexafluoropropylene. This resin, **FEP,** has similar properties to TFE except that it is not recommended for continuous exposures at temperatures above 200°C (400°F). Also, FEP can be extruded on conventional extrusion equipment, while TFE parts must be made by complicated powder-metallurgy techniques. Another version is **polyvinylidene fluoride,** or **PVF_2** (Kynar), which has excellent resistance to alkalies and acids to 150°C (300°F). It can be extruded. A more recent development is a copolymer of CTFE and ethylene (Halar). This material has excellent resistance to strong inorganic acids, bases, and salts up to 150°C. It also can be extruded.

Perfluoroalkoxy, or **PFA** (Teflon), has the general properties and chemical resistance of FEP at a temperature approaching 300°C (600°F).

Polyethylene is the lowest-cost plastic commercially available. Mechanical properties are generally poor, particularly above 50°C (120°F), and pipe must be fully supported. Carbon-filled grades are resistant to sunlight and weathering.

Unplasticized polyvinyl chlorides (type I) have excellent resistance to oxidizing acids other than concentrated and to most nonoxidizing acids. Resistance is good to weak and strong alkaline materials. Resistance to chlorinated hydrocarbons is not good. Polyvinylidene chloride, known as **Saran,** has good resistance to chlorinated hydrocarbons.

Acrylonitrile butadiene styrene (ABS) polymers have good resistance to nonoxidizing and weak acids but are not satisfactory with oxidizing acids. The upper temperature limit is about 65°C (150°F).

Acetals have excellent resistance to most organic solvents but are not satisfactory for use with strong acids and alkalies.

Cellulose acetate butyrate is not affected by dilute acids and alkalies or gasoline, but chlorinated solvents cause some swelling. **Nylons** resist many organic solvents but are attacked by phenols, strong oxidizing agents, and mineral acids.

Polypropylene has a chemical resistance about the same as that of polyethylene, but it can be used at 120°C (250°F). **Polycarbonate** is a relatively high-temperature plastic. It can be used up to 150°C (300°F). Resistance to mineral acids is good. Strong alkalies slowly decompose it, but mild alkalies do not. It is partially soluble in aromatic solvents and soluble in chlorinated hydrocarbons. **Polyphenylene oxide** has good resistance to aliphatic solvents, acids, and bases but poor resistance to esters, ketones, and aromatic or chlorinated solvents.

Polyphenylene sulfide (PPS) has no known solvents below 190 to 205°C (375 to 400°F); mechanical properties of PPS are unaffected by exposures in air at 230°C (450°F). It is resistant to aqueous inorganic salts and bases.

Polysulfone can be used to 170°C (340°F); it is highly resistant to mineral acid, alkali, and salt solutions as well as to detergents, oils, and alcohols. It is attacked by such organic solvents as ketones, chlorinated hydrocarbons, and aromatic hydrocarbons.

Polyamide or polyimide polymers are resistant to aliphatic, aromatic, and chlorinated or fluorinated hydrocarbons as well as to many acidic and basic systems but are degraded by high-temperature caustic exposures.

Thermosetting Plastics Among the thermosetting materials are phenolic plastics filled with asbestos, carbon or graphite, glass, and silica. Relatively low cost, good mechanical properties, and chemical resistance (except against strong alkalies) make phenolics popular for chemical equipment. Furan plastics filled with asbestos and glass have much better alkali resistance than phenolic resins. They are more expensive than the phenolics but also offer somewhat higher strengths.

Polyester resins, reinforced with fiberglass, have good strength and good chemical resistance except to alkalies. Some special materials in this class, based on bisphenol and vinyl esters are more alkali-resistant. The temperature limit for polyesters is about 90 to 150°C (200 to 300°F), depending upon exposure conditions.

Epoxies reinforced with fiberglass have very high strengths and resistance to heat. The chemical resistance of the epoxy resin is excellent in nonoxidizing and weak acids but not good against strong acids. Alkaline resistance is excellent in weak solutions. Resistance is poor to such organic solvents as ketones, chlorinated hydrocarbons, and aromatic hydrocarbons.

The thermoset polyimides are a family of heat-resistant polymers with acceptable properties up to 260°C (500°F). They are unaffected by dilute acids, aromatic and aliphatic hydrocarbons, esters, ethers, and alcohols but are attacked by dilute alkalies and concentrated inorganic acids.

Chemical resistance of thermosetting-resin-glass-reinforced laminates may be affected by any exposed glass in the laminate.

Phenolic asbestos, general-purpose polyester glass, Saran, and CAB are adversely affected by alkalies. And thermoplastics generally show poor resistance to organics.

The lack of homogeneity and the friable nature of FRP composite structures dictate that caution be followed in mechanical design, vendor selection, inspection, shipment, installation, and use.

FRP code vessels for pressure service over 0.1 MPA (15 lbf/in^2) may be designed and built under ASME Sec. X. Equipment for service from full vacuum through 0.1-MPA (15-lbf/in^2) pressure, while not presently covered by an ASME code designation, can be designed or fabricated in accordance with the SPI-MTI Quality Assurance Practices and Procedures Report for RTP equipment.

Rubber and Elastomers Rubber and elastomers are widely used as lining materials. To meet the demands of the chemical industry, rubber processors are continually improving their products. A number of synthetic rubbers have been developed, and while none has all the properties of natural rubber, they are superior in one or more ways. The isoprene and polybutadiene synthetic rubbers are duplicates of natural.

The ability to bond natural rubber to itself and to steel makes it ideal for lining tanks. Many of the synthetic elastomers, while more chemically resistant than natural rubber, have very poor bonding characteristics and hence are not well suited for lining tanks.

Natural rubber is resistant to dilute mineral acids, alkalies, and salts, but oxidizing media, oils, and most organic solvents will attack it. **Hard rubber** is made by adding 25 percent or more of sulfur to natural or synthetic rubber and, as such, is both hard and strong. **Chloroprene** or **neoprene rubber** is resistant to attack by ozone, sunlight oils, gasoline, and aromatic or halogenated solvents but is easily permeated by water, thus limiting its use as a tank lining. **Styrene rubber** has chemical resistance similar to that of natural. **Nitrile rubber** is known for resistance to oils and solvents. **Butyl rubber**'s resistance to dilute mineral acids and alkalies is exceptional; resistance to concentrated acids, except nitric and sulfuric, is good. **Silicone rubbers,** also known as polysiloxanes, have outstanding resistance to high and low temperatures as well as against aliphatic solvents, oils, and greases. **Chlorosulfonated polyethylene,** known as **Hypalon,** has outstanding resistance to ozone and oxidizing agents except fuming nitric and sulfuric acids. Oil resistance is good. **Fluoroelastomers (Viton A, Kel-F, Kalrez)** combine excellent chemical and temperature resistance. **Polyvinyl chloride elastomer (Koroseal)** was developed to overcome some of the limitations of natural and synthetic rubbers. It has excellent resistance to mineral acids and petroleum oils.

The **cis-polybutadiene, cis-polyisoprene,** and **ethylene-propylene** rubbers are close duplicates of natural rubber. The newer ethylene-propylene rubbers (EPR) have excellent resistance to heat and oxidation.

Asphalt Asphalt is used as a flexible protective coating, as a brick-lining membrane, and as a chemical-resisting floor covering and road surface. Resistant to acids and bases, alphalt is soluble in organic solvents such as ketones, most chlorinated hydrocarbons, and aromatic hydrocarbons.

Carbon and Graphite The chemical resistance of impervious carbon and graphite depends somewhat on the type of resin impregnant used to make the material impervious. Generally, impervious graphite is completely inert to all but the most severe oxidizing conditions. This property, combined with excellent heat transfer, has made impervious carbon and graphite very popular in heat exchangers, as brick lining, and in pipe and pumps. One limitation of these

TABLE 23-20 Properties of Glass and Silica*

	Pyroceram	96% silica	Borosilicate	Glass lining
Specific gravity, 77°F	2.60	2.18	2.23	2.56
Water absorption, %	0.00	0.00	0.00	
Gas permeability	Gastight	Gastight	Gastight	
Softening temperature, °F (°C)	2282 (1250)	2732 (1500)	1508 (1820)	
Specific heat, 77°F Btu/(lb·°F)[J/(kg·K)]	0.185 (775)	0.178 (746)	0.186 (779)	
Mean specific heat (77–752°F)	0.230	0.224	0.233	
Thermal conductivity, mean temperature, 77°F, Btu/(ft²·h·°F)/in [W/(m·K)]	25.2 (3.6)	....	7.5 (1.1)	
Linear thermal expansion, per °F (77–572°F; (per °C), $\times 10^{-6}$	3.2 (5.8)	0.44 (0.79)	1.8 (3.2)	
Modulus of elasticity, kip/in² (MPa) $\times 10^3$	17.3 (119)	9.6 (66)	9.5 (66)	6–9 (40–60)
Poisson's ratio	0.245	0.17	0.20	
Modulus of rupture, kip/in²	20 (140)	5–9 (35–63)	6–10 (42–70)	
Knoop hardness, 100 g	698	532	481	480
Knoop hardness, 500 g	619	477	442	
Adhesion strength kip/in² (MPa)	....	....	....	5–10 (35–70)
Maximum operating temperature, °F (°C)	....	....	....	500 (260)
Thermal shock resistance, temperature difference, °F (°C)	....	....	....	305 (152)

*Courtesy of National Association of Corrosion Engineers.

materials is low tensile strength. Threshold oxidation temperatures are 350°C (660°F) for carbon and 400°C (750°F) for graphite.

Several types of resin impregnates are employed in manufacturing impervious graphite. The standard impregnant is a phenolic resin suitable for service in most acids, salt solutions, and organic compounds. A modified phenolic impregnant is recommended for service in alkalies and oxidizing chemicals. Furan and epoxy thermosetting resins are also used to fill structural voids. The chemical resistance of the impervious graphite is controlled by the resin used. However, no type of impervious graphite is recommended for use in over 60 percent hydrofluoric, over 20 percent nitric, and over 96 percent sulfuric acids and in 100 percent bromine, fluorine, or iodine.

Wood While fairly inert chemically, wood is readily dehydrated by concentrated solutions and hence shrinks badly when subjected to the action of such solutions. It is also slowly hydrolyzed by acids and alkalies, especially when hot. In tank construction, if sufficient shrinkage once takes place to allow crystals to form between the staves, it becomes very difficult to make the tank tight again.

A number of manufacturers offer wood impregnated to resist acids or alkalies or the effects of high temperatures.

HIGH- AND LOW-TEMPERATURE MATERIALS

LOW-TEMPERATURE METALS

The low-temperature properties of metals have created some unusual problems in fabricating cryogenic equipment.

Most metals lose their ductility and impact strength at low temperatures, although in many cases yield and tensile strengths increase as the temperature goes down.

TABLE 23-21 Chemical Resistance of Important Plastics

	Polypropylene polyethylene	CAB*	ABS†	PVC‡	Saran§	Polyester glass¶	Epoxy glass	Phenolic asbestos	Fluorocarbons	Chlorinated polyether (Penton)	Polycarbonate
10% H_2SO_4	Excel.	Good	Excel.	Excel.	Excel.	Excel.	Excel.	Excel.	Excel.	Excel.	Excel.
50% H_2SO_4	Excel.	Poor	Excel.	Excel.	Excel.	Good	Excel.	Excel.	Excel.	Excel.	Excel.
10% HCl	Excel.	Excel.	Excel.	Excel.	Excel.	Excel.	Excel.	Excel.	Excel.	Excel.	Excel.
10% HNO_3	Excel.	Poor	Good	Excel.	Excel.	Good	Good	Fair	Excel.	Excel.	Excel.
10% Acetic	Excel.	Good	Excel.	Excel.	Excel.	Excel.	Excel.	Excel.	Excel.	Excel.	Excel.
10% NaOH	Excel.	Fair	Excel.	Good	Fair	Fair	Excel.	Poor	Excel.	Excel.	Excel.
50% NaOH	Excel.	Poor	Excel.	Excel.	Fair	Poor	Good	Poor	Excel.	Excel.	Excel.
NH_4OH	Excel.	Poor	Excel.	Excel.	Poor	Fair	Excel.	Poor	Excel.	Excel.	Excel.
NaCl	Excel.	Excel.	Excel.	Excel.	Excel.	Excel.	Excel.	Excel.	Excel.	Excel.	Excel.
$FeCl_3$	Excel.	Excel.	Excel.	Excel.	Excel.	Excel.	Excel.	Excel.	Excel.	Excel.	Excel.
$CuSO_4$	Excel.	Excel.	Excel.	Excel.	Excel.	Excel.	Excel.	Excel.	Excel.	Excel.	Excel.
NH_4NO_3	Excel.	Excel.	Excel.	Excel.	Excel.	Excel.	Excel.	Good	Excel.	Excel.	Excel.
Wet H_2S	Excel.	Excel.	Excel.	Excel.	Excel.	Excel.	Excel.	Excel.	Excel.	Excel.	
Wet Cl_2	Poor	Poor	Excel.	Good	Poor	Poor	Poor	Excel.	Excel.	Excel.	
Wet SO_2	Excel.	Poor	Excel.	Excel.	Good	Excel.	Excel.	Excel.	Excel.	Excel.	
Gasoline	Poor	Excel.	Excel.	Excel.	Excel.	Excel.	Excel.	Excel.	Excel.	Excel.	Excel.
Benzene	Poor	Poor	Poor	Poor	Fair	Good	Excel.	Excel.	Excel.	Fair	Fair
CCl_4	Poor	Poor	Poor	Fair	Fair	Excel.	Good	Excel.	Excel.	Fair	Poor
Acetone	Poor	Poor	Poor	Poor	Fair	Poor	Good	Poor	Excel.	Good	Good
Alcohol	Poor	Poor	Excel.	Excel.	Excel.	Excel.	Excel.	Excel.	Excel.	Excel.	Excel.

NOTE: Ratings are for long-term exposures at ambient temperatures [less than 38°C (100°F)].
*Cellulose acetate butyrate.
†Acrylonitrile butadiene styrene polymer.
‡Polyvinyl chloride, type I.
§Chemical resistance of Saran-lined pipe is superior to extruded Saran in some environments.
¶Refers to general-purpose polyesters. Special polyesters have superior resistance, particularly in alkalies.

TABLE 23-22 Typical Property Ranges for Plastics

Thermosets[a]	Specific gravity	Tensile strength, kip	Tensile strength, MPa	Modulus of elasticity, tension, 10^2kip/in^2	Modulus of elasticity, tension, 10^2MPa	Impact strength, Izod[b], ft·lb	Impact strength, Izod[b], J	Maximum use temperature (no load), °F	Maximum use temperature (no load), °C	HDT at 254 lbf/in^2[c], °F	HDT at 254 lbf/in^2[c], °C	Chemical resistance[d]: Weather resistance	Weak acid	Strong acid	Weak alkali	Strong alkali	Solvents
Alkyds																	
Glass-filled	2.12–2.15	4–9.5	28–66	20–28	138–193	0.6–10	0.8–14	450	230	400–500	200–260	R	A	A	A	A	A
Mineral-filled	1.60–2.30	3–9	21–62	5–30	34–207	0.3–0.5	0.4–0.7	300–450	150–230	350–500	180–260	R	R	A	A	D	A
Asbestos-filled	1.65	4.5–7	31–48	.....	.....	0.4–0.5	0.6–0.7	450	230	315	160	R	R	S	R	S	R
Synthetic fiber–filled	1.24–2.10	4.5–7	31–48	20	138	0.5–4.5	0.7–6.1	300–430	150–220	245–430	120–220	R	R	S	R	S	A
Alkyl diglycol carbonate	1.30–1.40	5–6	34–41	3.0	21	0.2–0.4	0.3–0.5	212	100	140–190	60–90	R	R	A[e]	R	R-S	R
Diallyl phthalates																	
Glass-filled	1.61–1.78	6–11	41–76	14–22	97–152	0.4–15	0.5–20	300–400	150–200	330–540	165–280	R	R	S	R-S	S	R
Mineral-filled	1.65–1.68	5–9	34–62	12–22	83–152	0.3–0.5	0.4–1	300–400	150–200	320–540	160–280	R	R	S	R-S	S	R
Asbestos-filled	1.55–1.65	7–8	48–55	12–22	83–152	0.4–0.5	0.5–0.7	300–400	150–200	320–540	160–280	R	R	S	R-S	S	R
Epoxies (bis-A)																	
No filler	1.06–1.40	4–13	28–90	2.15–5.2	15–36	0.2–1.0	0.3–1.4	250–500	120–260	115–500	45–260	R	R	A	R	S	R-S
Graphite-fiber reinforced	1.37–1.38	185–200	1280–1380	118–120	814–827	.....	.....	.....	.....	.....	.....	S	R	R	R	R	R-S
Mineral-filled	1.6 –2.0	5–15	34–103	.....	.....	0.3–0.4	0.4–0.5	300–500	150–260	250–500	120–260	S	R	R	R	R	R-S
Glass-filled	1.7 –2.0	10–30	69–207	30	207	10–30	14–41	300–500	150–260	250–500	120–260	S	R	R-S	R	R	R-S
Epoxies (novolac): no filler	1.12–1.24	5–11	34–76	2.15–5.2	15–36	0.3–0.7	0.4–0.9	400–500	200–260	450–500	230–260	R	R	R	R	R	R
Epoxies (cycloaliphatic): no filler	1.12–1.18	10–17.5	69–121	5–7	34–48	.....	.....	480–550	250–290	500–550	260–290	R	R	R-A	R	R-A	R
Melamines																	
Cellulose-filled	1.45–1.52	5–9	34–62	11	76	0.2–0.4	0.3–0.5	250	120	270	130	S	R-S	D	R	D	R
Flock-filled	1.50–1.55	7–9	48–62	.....	.....	0.4–0.5	0.5–0.7	250	120	270	130	S	R-S	D	R	D	R-S
Asbestos-filled	1.70–2.0	5–7	34–48	20	138	0.3–0.4	0.4–0.5	250–400	120–200	265	130	S	R-S	D	S	S	R
Fabric-filled	1.5	8–11	55–76	14–16	97–110	0.6–1.0	0.8–1.4	250	120	310	150	S	R	D	R	A	R-S
Glass-filled	1.8 –2.0	5–10	34–69	24	165	0.6–18	0.8–24	300–400	150–200	400	200	S	R	D	R	R-S	R
Phenolics																	
Wood-flour-filled	1.34–1.45	5–9	34–62	8–17	55–117	0.2–0.6	0.3–0.8	300–350	150–180	300–370	150–190	S	R-S	S-D	S-D	A	R-S
Asbestos-filled	1.45–2.00	4.5–7.5	31–52	10–30	69–207	0.2–0.4	0.3–0.5	350–500	180–260	300–500	150–260	S	R-S	S-D	S-D	A	R-S
Mica-filled	1.65–1.92	5.5–7	38–48	25–50	172–345	0.3–0.4	0.4–0.5	250–300	120–150	300–350	150–180	S	R-S	S-D	S-D	A	R-S
Glass-filled	1.69–1.95	5–18	34–124	19–33	131–228	0.3–18	0.4–24	350–550	180–290	300–600	150–320	S	R-S	S-D	S-D	A	R-S
Fabric-filled	1.36–1.43	3–9	21–62	9–14	62–97	0.8–8	1.1–11	220–250	100–120	250–330	120–170	S	R-S	S-D	S-D	A	R-S
Polybutadienes																	
Very high vinyl (no filler)	1.00	8	55	2	14	1.1	1.5	500	260	.....	.....	S	R	R	R	R	R
Polyesters																	
Glass-filled BMC	1.7 –2.3	4–10	28–69	16–25	110–172	1.5–16	2.0–22	300–350	150–180	400–450	200–230	R-E	R-A	S-A	S-A	S-D	A-D
Glass-filled SMC	1.7 –2.1	8–20	55–138	16–25	110–172	8–22	11–30	300–350	150–180	400–450	200–230	R-E	R-A	S-A	S-A	S-D	A-D
Glass-cloth reinforced	1.3 –2.1	25–50	172–345	19–45	131–310	5–30	7–41	300–350	150–180	400–450	200–230	R-E	R-A	S-A	S-A	S-D	A-D
Silicones																	
Glass-filled	1.7 –2.0	4–6.5	28–45	10–15	69–103	3–15	4–20	600	320	600	320	R-S	R-S	R-S	S	S-A	R-A
Mineral-filled	1.8 –2.8	4–6	28–41	13–18	90–124	0.3–0.4	0.4–0.5	600	320	600	320	R-S	R-S	R-S	S	S-A	R-A
Ureas																	
Cellulose-filled	1.47–1.52	5.5–13	38–90	10–15	69–103	0.2–0.4	0.3–0.5	170	80	260–290	130–140	S	R-S	A-D	S-A	D	R-S
Urethanes																	
No filler	1.1 –1.5	0.2–10	1–69	1 –10	7–69	5–NB	7	129–250	90–120	.....	.....	R-S	S	A	S	S-A	R-S

TABLE 23-22 Typical Property Ranges for Plastics *(Continued)*

Thermoplastics	Specific gravity	Tensile strength kip/in²	Tensile strength MPa	Modulus of elasticity tension 10²kip/in²	Modulus of elasticity tension 10²MPa	Impact strength, Izod[b] ft·lb	Impact strength, Izod[b] J	Maximum-use temperature (no load) °F	Maximum-use temperature (no load) °C	HDT at 66 lbf/in² °F	HDT at 66 lbf/in² °C	HDT at 264 lbf/in² °F	HDT at 264 lbf/in² °C	Chemical resistance[d] Weather resistance	Weak acid	Strong acid	Weak alkali	Strong alkali	Solvents
ABS																			
GP	1.05–1.07	5.9	41	3.1	21	6	8	160–200	70–90	210–225	100–110	190–206	90–95	R-E	R	A^e	R	R	A^fR
High-impact	1.01–1.06	4.8	33	2.4	17	7.5	10	140–210	60–100	210–225	100–110	188–211	85–100	R-E	R	A^f	R	R	A^fR
Heat-resistant	1.06–1.08	7.4	51	3.9	27	2.2	3.0	190–230	90–110	225–252	110–120	226–240	110–115	R-E	R	A^f	R	R	A^fR
Trans.	1.07	5.6	39	2.9	20	5.3	7.1	130	55	180	80	165	75	R-E	R	A^f	R	R	A^fR
	1.20	6.0	41	3.2	22	2.5	3.4	130–180	55–80	210–220	100–105	195	90	R-E	R	A^e	R	R	A^fR
Acetals																			
Homopolymers	1.42	10	69	5.2	36	1.4	1.9	195	90	338	170	255	125	R	R	A	R	A-D	R
Copolymers	1.41	8.8	61	4.1	28	1.2–1.6	1.6–2.2	212	100	316	160	230	110	R	R	A	R	R	R
Acrylics																			
GP	1.11–1.19	5.6–11.0	39–76	2.25 4.65	16–32	0.3–2.3	0.4–3.1	130–230	55–110	175–225	80–110	165–210	75–100	R	R	A^e	R	A	A^fR
High-impact	1.12–1.16	5.8–8.0	40–55	2.3–3.3	16–23	0.8–2.3	1.1–3.1	140–195	60–90	180–205	80–95	165–190	75–90	R	R	A^e	R	R	A^fR
	1.21–1.28	8.0–12.5	55–86	3.5–4.8	24–33	0.3–0.4	0.4–0.5	125–200	50–90	170–200	75–95	155–205	70–95	R	R	A^e	R	A	A^fR
Cast	1.18–1.28	9.0–12.5	62–86	3.7–5.0	26–34	0.4–1.5	0.5–2.0	140–200	60–90	165–235	75–115	160–215	70–100	R	R	A^e	R	A	A^fR
Multipolymer	1.09–1.14	6–8	41–55	3.1–4.3	21–30	1–3	1–4	165–175	75–80	....	....	185–195	85–90	E	R	A^e	H	S	A^f
Cellulosics																			
Acetate	1.23–1.34	3.0–8.0	21–55	1.05–2.55	7–18	1.1–6.8	1.5–9	140–220	60–105	120–209	50–100	111–195	45–90	S	S	D	S	D	D-S
Butyrate	1.15–1.22	3.0–6.9	21–48	0.7–1.8	5–12	3.0–10.0	4–14	140–220	60–105	130–227	55–110	113–202	45–95	S	S	D	S	D	D-S
E cellulose	1.10–1.17	3–8	21–55	0.5–3.5	3–24	1.7–7.0	2.3–9.5	115–185	45–85	....	....	115–190	45–90	S	S	D	R	S	D
Nitrate	1.35–1.40	7–8	48–55	1.9–2.2	13–15	5–7	7–9	140	60	....	....	140–160	60–70	E	S	D	S	D	D
Propionate	1.19–1.22	4.0–6.5	28–45	1.1–1.8	8–12	1.7–9.4	2.3–13	155–220	70–105	147–250	65–120	111–228	45–110	S	S	D	S	D	D-S
Chloro polyether	1.4	5.4	37	1.5	10	0.4	0.5	290	140	285	140	....	....	R-S	R	A^e	R	R	R
Ethylene copolymers																			
EEA	0.93	2.0	14	0.05	0.3	NB	....	190	90	....	....	....	....	S	R	A^e	R	R	A-D
EVA	0.94	3.6	25	0.02–0.12	0.14–0.8	NB	....	....	....	140–147	60–65	93	35	S	R	A	R	R	A-D
Fluoropolymers																			
FEP	2.14–2.17	2.5–3.9	17–27	0.5–0.7	3–5	NB	....	400	208	158	70	....	....	R	R	R	R	R	R
PTFE	2.1–2.3	1.4	7–28	0.38–0.65	2.6–4.5	2.5–4.0	3.4–5.4	550	290	250	120	....	....	R	R	R	R	R	R
CTFE	2.10–2.15	4.6–5.7	32–39	1.8–2.0	12–14	3.5–3.6	4.7–4.9	350–390	180–200	258	125	....	....	R	R	R	R	R	S^g
PVF_2	1.77	7.2	50	1.7	12	3.8	5.2	300	150	300	150	195	90	S	R	A^h	R	R	R
ETFE and ECTFE	1.68–1.70	6.5–7.0	45–48	2–2.5	14–17	NB	....	300	150	220	105	160	70	R	R	R	R	R	
Methylpentene	0.83	3.3–3.6	23–25	1.3–1.9	10–13	0.95–3.8	1.3–5.2	275	135	....	....	....	....	E	R	A^h	R	R	A
Nylons																			
6/6	1.13–1.15	9–12	62–83	3.85	27	2.0	27	180–300	80–150	360–470	180–240	150–220	165–105	R	R	A	R	R	R-D^i
6	1.14	12.5	86	....	....	1.2	1.6	180–250	80–170	300–365	150–185	140–155	60–70	R	R	A	R	R	R-A^i
6/10	1.07	7.1	49	2.8	19	1.6	2.2	180	80	300	150	....	....	R	R	A	R	R	R-A^i
8	1.09	3.9	27			>16	>22							R	R	A	R	R	R-A^i
12	1.01	6.5–8.5	45–59	1.7–2.1	12–14	1.2–4.2	1.6–5.7	175–260	80–125	....	....	120–130	50–55	R	R	A	R	R	R-A^i

Copolymers	1.08–1.14	7.5–11.0	52–76	....	....	1.5–19	2–26	180–250	80–120	....	....	130–350	55–180	R	R	A	R	R	R-A[i]
Polyesters																			
PET	1.37	10.4	72	....	....	0.8	1.1	175	80	240	115	185	85	R	R	A[e]	R	A	R-A[i]
PBT	1.31	8.0–8.2	55–57	3.6	25	1.2–1.3	1.6–1.8	280	140	310	155	130	55	R	R	R	R	A	R
PTMT	1.31	8.2	57	....	....	1.0	1.4	270	130	302	150	122	50	R	R	R	R	A	R
Copolymers	1.2	7.3	50	....	....	1.0	1.4					154	70						
Polyaryl ether	1.14	7.5	52	3.2	22	10	14	250	120	320	100	300	150	E	R	R	R	R	A
Polyaryl sulfone	1.36	13	90	3.7	26	2	2.7	500	260	....	....	525	275	Darkens	R	R	R	R	R
Polybutylene	0.910	3.8	26	0.26	1.8	NB		225	105	215	100	130	55	E	R	A[e]	R	R	
Polycarbonate	1.2	9	62	3.45	24	12–16	16–22	250	120	270–290	130–145	265–285	130–140	R	R	A[e]	A	A	A
PC-ABS	1.14	8.2	57	3.7	26	10	14	220	105	235	115	220	105	R-E	R	A[e]	R	S	A
Polyethylenes																			
LD	0.91–0.93	0.9–2.5	6–17	0.20–0.27	1.4–1.9	NB	....	180–212	80–100	100–120	40–50	90–105	30–40	E	R	A[e]	R	R	R
HD	0.95–0.96	2.9–5.4	20–37	....	....	0.4–14	0.5–19	175–250	80–120	140–190	60–90	110–130	45–55	E	R	R-A[e]	R	R	R
HMW	0.945	2.5	17	1	7	NB	....	....	....	155–180	70–80	105–180	40–80	E	R	A[e]	R	R	R
Ionomer	0.94–0.95	3.4–4.5	23–31	0.3–0.7	2–5	6-NB	8	160–180	70–80	110	45	100–120	40–50	E	A	A[e]	R	R	R
Phenylene oxide-based materials	1.06–1.10	7.8–9.6	54–66	3.5–3.8	24–26	5.0	68	175–220	80–105	230–280	110–140	212–265	100–130	R	R	R	R	R	R-A
Polyphenylene sulfide	1.34	10	69	4.8	33	0.3	0.4	500	260	....	....	278	135	R	R	A[e]	R	R	R
Polyimide	1.43	5–7.5	34–52	5.4	37	5–7	7–9	500	260	....	....	680	360		R	R	A	A	R
Polypropylenes																			
GP	0.90–0.91	4.8–5.5	33–38	1.6–2.2	11–15	0.4–2.2	0.5–3.0	225–300	105–150	200–230	95–110	125–140	50–60	E	R	A[e]	R	R	R
High-impact	0.90–0.91	3–5	21–34	1.3	9	1.5–12	2–16	200–250	95–120	160–200	70–95	120–135	50–60	E	R	A[e]	R	R	A
Propylene copolymer	0.91	4	28	1.0–1.7	7–12	1.1	1.5	190–240	90–115	185–230	85–110	115–140	45–60	E	R	A[e]	R	R	R
Polystyrenes																			
GP	1.04–1.07	6.0–7.3	41–50	4.5	31	0.3	0.4	150–170	65–80	....	....	180–220	80–105	S	R	A[e]	R	R	D
High-impact	1.04–1.07	2.8–4.6	20–32	2.9 4.0	20–28	0.7–1.0	0.9–1.4	140–175	60–80	....	....	175–210	80–100	S	R	A[e]	R	R	D
Polysulfone	1.24	10.2	70	3.6	25	1.2	1.6	300	150	360	180	345	175	S	R	R	R	R	R-A
Polyurethanes	1.11–1.25	4.5–8.4	31–58	0.1–3.5	0.7–24	NB	....	190	90	....	....	....	....	R-S	S-D	S-D	S-D	S-D	R
Vinyl, rigid	1.3–1.5	5–8	34–55	3– 5	21–34	0.5–20	0.7–27	150–175	65–80	135–180	60–80	130–175	55–80	R	R	R-S	R	R	R-A
Vinyl, flexible	1.2–1.7	1–4	7–28	....	....	0.5–20	0.7–27	140–175	60–80	....	....	....	....	S	R	R-S	R	R	R-A
Rigid CPVC	1.49–1.58	7.5–9.0	52–62	3.6–4.7	25–32	1.0–5.6	1.4–7.6	230	110	215–245	100–120	200–235	95–115	R	R	R	R	R	R
PVC-acrylic	1.30–1.35	5.5–6.5	38–45	2.75–3.35	19–23	15	20	....	....	180	80	170	80	R	R	S	R	R	A
PVC-ABS	1.10–1.21	2.6–6.0	18–41	0.8–3.4	6–23	10–15	14–20	....	....	....	....	....	....	S	R	R-S	R	R	R-D
SAN	1.08	10–12	69–83	5.0–5.6	34–39	0.4–0.5	0.5–0.7	140–200	60–95	....	....	190–220	90–105	S-E	R	A	R	R	A

SOURCE: *Plastics Engineering Handbook*, 4th ed., Van Nostrand Reinhold, New York, 1976. Courtesy of National Association of Corrosion Engineers. To convert megapascals to pounds-force per square inch, multiply by 145.04.

[a]All values at room temperature unless otherwise listed.
[b]Notched samples.
[c]Heat-deflection temperature.
[d]Ac = acid, and Al = alkali; R = resistant; A = attached; S = slight effects; E = embrittles; D = decomposes.
[e]By oxidizing acids.
[f]By ketones, esters, and chlorinated and aromatic hydrocarbons.
[g]Halogenated solvents cause swelling.
[h]By fuming sulfuric.
[i]Dissolved by phenols and formic acid.

TABLE 23-23 Chemical Resistance of Coatings for Immersion Service (Room Temperatures)

	Asphalt, unmodified	Coal tar: Hot-applied	Coal tar: Cold-applied	Coal tar–epoxy	Coal tar–urethanes	Epoxy: phenolic-baked	Epoxy: amine-cured	Epoxy ester	Furfuryl alcohol	Phenolics, baked	Polyesters (unsaturated)	Polyvinyl chloracetates	Vinyl ester	Urethanes: Air-dried	Urethanes: Baked	Vinylidene chloride	Chlorinated rubber
Acids																	
Sulfuric, 10%	R	LR	NR	R	R	R	R	LR	R	R	R	R	R	LR	LR	R	R
Sulfuric, 80%	. .	NR	NR	. .	NR	NR	NR	NR	LR	NR	R	NR	R	NR	LR	LR	R
Hydrochloria, 10%	R	LR	NR	LR	LR	R	R	LR	R	R	R	R	R	LR	LR	R	R
Hydrochloric, 35%	R	NR	NR	. .	. .	NR	NR	NR	R	R	R	LR	R	LR	LR	R	R
Nitric, 10%	NR	LR	NR	. .	. .	LR	NR	NR	NR	NR	R	R	R	LR	LR	R	R
Nitric, 50%	NR	NR	NR	. .	. .	NR	NR	NR	NR	NR	NR	NR	. .	NR	NR	LR	NR
Acetic, 100%		NR	NR	NR	NR	NR	NR	NR	LR	LR	NR	NR	LR	NR	NR	NR	NR
Water																	
Distilled	R	R	R	R	LR	R	R	R	R	R	R	R	R	LR	LR	. .	R
Salt water	R	R	R	R	LR	R	R	R	R	R	R	R	R	LR	LR	R	R
Alkalies																	
Sodium hydroxide, 10%	R	R	LR	R	R	R	R	NR	R	NR	R	R	R	LR	LR	LR	R
Sodium hydroxide, 70%	. .	NR	NR	. .	LR	R	R	NR	LR	NR	NR	LR	R	LR	LR	NR	R
Ammonium hydroxide, 10%	R	R	LR	R	R	R	LR	LR	R	NR	R	R	R	LR	R	NR	R
Sodium carbonate, 5%	R	R	R	. .	. .	R	. .	. .	. .	. .	. .	R	R	R	R	R	R
Gases																	
Chlorine	R	NR	NR	LR	NR	LR	LR	LR	NR	NR	R	LR	R	LR	R	LR	R
Ammonia	. .	LR	LR	NR	NR	LR	LR	R	R	NR	NR	LR	R	LR	R	NR	NR
Hydrogen sulfide	. .	R	R	R	. .	R	R	. .	R	R	. .	LR	R	R	R	R	R
Organics																	
Alcohols	R	LR	LR	NR	NR	R	R	LR	R	R	R	R	R	NR	R	R	LR
Aliphatic hydrocarbons	NR	LR	LR	LR	LR	R	R	R	R	R	R	R	R	R	R	R	LR
Aromatic hydrocarbons	NR	NR	NR	NR	LR	R	R	R	R	R	R	NR	LR	R	R	LR	NR
Ketones	NR	NR	NR	NR	NR	LR	LR	NR	LR	R	NR	NR	NR	NR	R	NR	NR
Ethers	NR	NR	NR	. .	. .	LR	LR	NR	LR	R	. .	NR	NR	R	R	NR	NR
Esters	NR	NR	NR	NR	NR	LR	LR	NR	R	R	LR	NR	LR	NR	R	NR	NR
Chlorinated hydrocarbons	NR	NR	NR	NR	LR	LR	LR	NR	LR	R	NR	LR	LR	LR	R	LR	NR
Maximum temperature (dry conditions), °F	150	. . .	. . .	200	200	250	250	250	300	250–300	. . .	160	350	. . .	. . .	. . .	160
Maximum temperature (wet conditions), °F	. . .	120	120	150	150	150	150	150	190	160–250	250	150	210	. . .	. . .	150	140

SOURCE: NACE TPC-2, *Coatings and Linings for Immersion Service*. Courtesy of National Association of Corrosion Engineers.

NOTE: Chemical resistance data are for coatings only. Thin coatings generally are not suitable for substrates such as carbon steel which are corroded significantly (e.g., >20 mils/year) in the test environment. R = recommended; LR = limited recommendation; NR = no recommendation.

TABLE 23-24 Properties of Coatings for Atmospheric Service

	Physical properties	Water resistance	Acid resistance	Alkali resistance	Solvent resistance	Temperature resistance	Weathering	Recoating
Alkyd								
Short-oil alkyd	Hard	Fair	Fair	Poor	Fair	Good	Fair	Easy
Long-oil alkyd	Flexible	Fair	Poor	Poor	Poor	Good	Good	Easy
Silicone alkyd	Tough	Good	Fair	Poor	Fair	Best of group	Very good	Fair
Vinyl alkyd	Tough	Good	Best of group	Poor	Fair	Fair	Very good	Difficult
Vinyl								
Polyvinyl chloride acetate copolymers	Tough	Very good	Excellent	Excellent	(Aliphatic hydrocarbon, good; aromatic hydrocarbon, poor)	Fair, 150°F	Very good	Easy
Vinyl acrylic copolymers	Tough	Good	Very good	Very good	(Alphatic, good; aromatic, poor)	Fair, 150°F	Excellent	Easy
Chlorinated rubber								
Resin-modified	Hard	Very good	Very good	Very good	(Alphatic, good; aromatic, poor)	Fair	Good	Easy
Alkyd-modified	Tough	Good	Fair	Fair	(Aliphatic, good; aromatic, poor)	Fair	Very good	Easy
Water base								
Polyvinyl acetate	Scrub-resistant	Poor	Poor	Poor	Poor	Fair	Very good	Easy
Acrylic polymers	Scrub-resistant	Poor	Poor	Poor	Poor	Fair	Excellent	Easy
Epoxy	Tough	Good	Good	Good	Good	Good	Fair	Difficult
Epoxy								
Epoxyamine	Hard	Good	Good	Good	Very good	Very good	Fair; chalks	Difficult
Epoxy polyamide	Tough	Very good	Fair	Excellent	Fair	Good	Good; chalks	Difficult
Epoxy coal tar	Hard	Excellent	Good	Good	Poor	Good	Poor	Difficult
Epoxyester	Flexible	Good	Fair	Poor	Fair	Good	Good; chalks	Reasonable
Polyurethane								
Air-drying polyurethane varnish	Very tough	Fair	Fair	Fair	Fair	Good	Yellowing	Requires care
Two-package-reactive polyurethane	Tough; hard	Good	Fair	Fair	Good	Good	Some yellowing and chalking	Difficult
Moisture-reactive polyurethane	Very tough; abrasion-resistant	Fair	Fair	Fair	Good	Good	Fades in light; yellows in shade	Difficult
Nonyellowing polyurethane	Fairly hard to rubbery	Good	Fair	Fair	Good	Good	Very good	Difficult
Inorganic zinc								
Water base (sodium or potassium silicate)	Tough; abrasion-resistant; excellent chemical bond	Good	Poor	Poor	Excellent	Excellent	Excellent; unaffected by weather	Easy
Organic base (ethyl silicate)	Tough; hard; excellent bond	Good	Poor	Poor	Good	Excellent	Excellent	Easy

SOURCE: F. L. LaQue, *Marine Corrosion: Causes and Prevention*, Wiley, New York, 1975, pp. 302–305. Courtesy of National Association of Corrosion Engineers.

TABLE 23-25 Typical Physical Properties of Surface Coatings for Concrete

		Polyester		Epoxy		
	Concrete	Isophthalic	Bisphenol	Polyamide	Amine	Urethane*
Tensile strength (ASTM C307), lbf/in^2	200–400	1200–2500	1200–2500	600–4000	1200–2500	200–1200
MPa	1.4–2.8	8.3–17	8.3–17	4.0–28	8.3–17	1.4–8.3
Thermal coefficient of expansion (ASTM C531)						
Maximum in/(in·°F)	6.5×10^6	20×10^6	20×10^6	40×10^6	40×10^6	†
Maximum mm/(mm·°C)	11.7×10^6	36×10^6	36×10^6	72×10^6	72×10^6	
Compressive strength, (ASTM) C579), lbf/in^2	3500	10,000	10,000	4000	6000	†
MPa	24	70	70	28	42	
Abrasion resistance, Taber abraser—weight loss, mg, 1000-g load/1000 cycles	.....	15–27	15–27	15–27	15–27	†
Shrinkage, ASTM C531, %	.....	2–4	2–4	0.25–0.75	0.25–0.75	0–2
Work life, min	.....	15–45	15–45	30–90	30–90	15–60
Traffic limitations, h after application	Light	16	16	24	24	24
	Heavy	36	36	48	48	48
	Ready for service	48	48	72	72	72
Adhesion characteristics‡	.....	Poor	Fair	Excellent	Good	Fair
Flexural strength (ASTM C580), lbf/in^2	.....	1500	1500	1000	1500	†
MPa	.....	10	10	7	10	

SOURCE: NACE RP-03-76, *Monolithic Organic Corrosion Resistant Floor Surfacing,* 1976. Courtesy of National Association of Corrosion Engineers.
NOTE: All physical values depend greatly on reinforcing. Values are for ambient temperatures.
*Type of urethane used is one of three: (1) Type II, moisture-cured; (2) Type IV, two-package catalyst; or (3) Type V, two-package polyol. (Ref. ASTM C16.)
†Urethanes not shown because of great differences in physical properties, depending on formulations. Adhesion characteristics should be related by actual test data. Any system which shows concrete failure when tested for surfacing adhesion should be rated excellent with decreasing rating for systems showing failure in cohesion or adhesion below concrete failure.
‡Adhesion to concrete: primers generally are used under polyesters and urethanes to improve adhesion.

TABLE 23-26 Chemical Resistance of Rubbers

Type of rubber	Features
Butadiene styrene	General-purpose; poor resistance to hydrocarbons, oils, and oxidizing agents
Butyl	General-purpose; relatively impermeable to air; poor resistance to hydrocarbons and oils
Chloroprene	Good resistance to aliphatic solvents; poor resistance to aromatic hydrocarbons and many fuels
Chlorosulfonated polyethylene	Excellent resistance to oxidation, chemicals, and heat; poor resistance to aromatic oils and most fuels
cis-Polybutadiene	General-purpose; poor resistance to hydrocarbons, oils, and oxidizing agents
cis-Polyisoprene	General-purpose; poor resistance to hydrocarbons, oils, and oxidizing agents
Ethylene propylene	Excellent resistance to heat and oxidation
Fluorinated	Excellent resistance to high temperature, oxidizing acids, and oxidation; good resistance to fuels containing up to 30% aromatics
Natural	General-purpose; poor resistance to hydrocarbons, oils, and oxidizing agents
Nitrile (butadiene acrylonitrile)	Excellent resistance to oils, but not resistant to strong oxidizing agents; resistance to oils proportional to acrylonitrile content
Polysulfide	Good resistance to aromatic solvents; unusually high impermeability to gases; poor compression set and poor resistance to oxidizing acids
Silicone	Excellent resistance over unusually wide temperature range [−100 to 260°C (−150 to 500°F)]; fair oil resistance; poor resistance to aromatic oils, fuels, high-pressure steam, and abrasion
Styrene	Synonymous with butadiene-styrene

Materials selection for low-temperature service is a specialized area. In general, it is necessary to select materials and fabrication methods which will provide adequate toughness at all operating conditions. It is frequently necessary to specify Charpy V-notch (or other appropriate) qualification tests to demonstrate adequate toughness of carbon and low-alloy steels at minimum operating temperatures.

Stainless Steels Chromium-nickel steels are suitable for service at temperatures as low as −250°C (−425°F). Type 304 is the most popular. The original cost of stainless steel may be higher than that of another metal, but ease of fabrication (no heat treatment) and welding, combined with high strength, offsets the higher initial cost. Sensitization or formation of chromium carbides can occur in several stainless steels during welding, and this will affect impact strength. However, tests have shown that impact properties of types 304 and 304L are not greatly affected by sensitization but that the properties of 302 are impaired at −185°C (−300°F).

Nickel Steel Low-carbon 9 percent nickel steel is a ferritic alloy developed for use in cryogenic equipment operating as low as −195°C (−320°F). ASTM specifications A 300 and A 353 cover low-carbon 9 percent nickel steel (A 300 is the basic specification for low-temperature ferritic steels). Refinements in welding and (ASME code–approved) elimination of postweld thermal treatments make 9 percent steel competitive with many low-cost materials used at low temperatures.

Aluminum Aluminum alloys have unusual ability to maintain strength and shock resistance at temperatures as low as −250°C (−425°F). Good corrosion resistance and relatively low cost make these alloys very popular for low-temperature equipment. For most welded construction the 5000-series aluminum alloys are widely used. These are the aluminum-magnesium and aluminum-magnesium-manganese materials.

Copper and Alloys With few exceptions the tensile strength of copper and its alloys increases quite markedly as the temperature goes down. However, copper's low structural strength becomes a problem when constructing large-scale equipment. Therefore, alloys must be used. One of the most successful for low temperatures is silicon bronze, which can be used to −195°C (−320°F) with safety.

HIGH-TEMPERATURE MATERIALS

Metals Successful applications of metals in high-temperature-process service depend on an appreciation of certain engineering factors. The important alloys for service to up 1100°C (2000°F) are shown in Table 23-35. Among the most important properties are creep, rupture, and short-time strengths (see Figs. 23-10 and 23-11). **Creep** relates initially applied stress to rate of plastic flow. **Stress rup-**

TABLE 23-27 Properties of Elastomers

Property	NR Natural rubber (cis-polyisoprene)	SBR Butadiene styrene (GR-S)	IR Synthetic (polyisoprene)	COX Butadiene acrylonitrile (nitrile)	CR Chloroprene (neoprene)	ITR Butyl (isobutylene isoprene)	BR Polybutadiene	T Polysulfide	Silicone (polysiloxane)
Physical properties									
Specific gravity (ASTM D 792)	0.93	0.91	0.93	0.98	1.25	0.90	0.91	1.35	1.1–1.6
Thermal conductivity, Btu/[(h·ft²)(°F/ft)] (ASTM C 177)	0.082	0.143	0.082	0.143	0.112	0.053	.....	.....	0.13
Coefficient of thermal expansion (cubical), 10^{-5}°F (ASTM D 696)	37	37	.....	39	34	32	37.5	.....	45
Electrical insulation	Good	Good	Good	Fair	Fair	Good	Good	Fair	Excellent
Flame resistance	Poor	Poor	Poor	Poor	Good	Poor	Poor	Poor	Good
Minimum, recommended service temperature, °F	−60	−60	−60	−60	−40	−50	−150	−60	−178
Maximum, recommended service temperature, °F	180	180	180	300	240	300	200	250	600
Mechanical properties									
Tensile strength, lbf/in²									
Pure gum (ASTM D 412)	2500–3500	200–300	2500–3500	500–900	3000–4000	2500–3000	200–1000	250–400	600–1300
Black (ASTM D 412)	3500–4500	2500–3500	3500–4500	3000–4500	3000–4000	2500–3000	2000–3000	>1000	
Elongation, %									
Pure gum (ASTM D 412)	750–850	400–600		300–700	800–900	750–950	400–1000	450–650	100–500
Black (ASTM D 412)	550–650	500–600	300–700	300–650	500–600	650–850	450–600	150–450	
Hardness (durometer)	A30–90	A40–90	A40–80	A40–95	A20–95	A40–90	A40–90	A40–85	A30–90
Rebound									
Cold	Excellent	Good	Excellent	Good	Very good	Bad	Excellent	Good	Very good
Hot	Excellent	Good	Excellent	Good	Very good	Very good	Excellent	Good	Very good
Tear resistance	Excellent	Fair	Excellent	Good	Fair to good	Good	Fair	Poor	Fair
Abrasion resistance	Excellent	Good to excellent	Excellent	Good to excellent	Good	Good to excellent	Excellent	Poor	Poor
Chemical resistance									
Sunlight aging	Poor	Poor	Fair	Poor	Very good	Very good	Poor	Very good	Excellent
Oxidation	Good	Good	Excellent	Good	Excellent	Excellent	Good	Very good	Excellent
Heat aging	Good	Very good	Good	Excellent	Excellent	Excellent	Good	Fair	Excellent
Solvents									
Aliphatic hydrocarbons	Poor	Poor	Poor	Excellent	Good	Poor	Poor	Excellent	Fair
Aromatic hydrocarbons	Poor	Poor	Poor	Good	Fair	Poor	Poor	Excellent	Poor
Oxygenated, alcohols	Good	Good	Good	Good	Very good	Very good	.....	Very good	Excellent
Oil, gasoline	Poor	Poor	Poor	Excellent	Good	Poor	Poor	Excellent	Poor
Animal, vegetable oils	Poor to good	Poor to good	.....	Excellent	Excellent	Excellent	Poor to good	Excellent	Excellent
Acids									
Dilute	Fair to good	Fair to good	Fair to good	Good	Excellent	Excellent	.....	Good	Very good
Concentrated	Fair to good	Fair to good	Fair to good	Good	Good	Excellent	.....	Good	Good
Permeability to gases	Low	Low	Low	Very low	Low	Very low	Low	Very low	High
Water-swell resistance	Fair	Excellent	Excellent	Excellent	Fair to excellent	Excellent	Excellent	Excellent	Excellent

ture is another important consideration at high temperatures since it relates stress and time to produce rupture. As the figures show, ferritic alloys are weaker than austenitic compositions, and in both groups molybdenum increases strength. Austenitic castings are much stronger than their wrought counterparts. And higher strengths are available in the superalloys. Other properties which become important at high temperatures include thermal conductivity, thermal expansion, ductility at temperature, alloy composition, and stability.

Actually, in many cases strength and mechanical properties become of secondary importance in process applications, compared with resistance to the corrosive surroundings. All common heat-resistant alloys form oxides when exposed to **hot oxidizing environments.** Whether the alloy is resistant depends upon whether the oxide is stable and forms a protective film. Thus mild steel is seldom used above 480°C (900°F) because of excessive scaling rates. Higher temperatures require **chromium** (see Fig. 23-12). Thus type 502 steel, with 4 to 6 percent Cr, is acceptable to 620°C (1150°F). A 9 to 12 percent Cr steel will handle 730°C (1350°F); 14 to 18 percent Cr extends the limit to 800°C (1500°F); and 27 percent Cr to 1100°C (2000°F).

The well-known austenitic stainless steels have excellent oxidation resistance: up to 900°C (1650°F) for 18-8; and up to 1100°C (2000°F) for 25-12 (and Inconel 600 and Incoloy 800). The cobalt-based alloys, of which Stellite 25 is an example, show excellent strengths up to 1100°C (2000°F).

Another useful element in imparting oxidation resistance to steel is **silicon** (complementing the effects of chromium). In the lower-chromium ranges, silicon in the amounts of 0.75 to 2 percent is more effective than chromium on a weight-percentage basis. The influence of 1 percent silicon in improving the oxidation rate of steels with varying chromium contents is shown in Fig. 23-13.

Aluminum also improves the resistance of iron to oxidation as well as sulfidation. But use as an alloying agent is limited because the amount required interferes with the workability and high-temperature strength properties of the steel. However, development of high-aluminum surface layers by spraying, dipping, and cementation is a feasible means of improving the heat resistance of low-alloy steels.

Hydrogen Atmospheres Austenitic stainless steels, by virtue of their high chromium contents, are usually resistant to hydrogen atmospheres.

TABLE 23-27 Properties of Elastomers *(Continued)*

Property	ECO, CO Epichlrohydrin homopolymer and copolymer	Fluorosilicone	EPDM Ethylene propylene	CSM Chlorosulfonated polyethylene	FPM Fluorocarbon elastomers
Physical properties					
Specific gravity	1.32–1.49	1.4	0.86	1.1–1.26	1.4–1.95
Thermal conductivity, Btu/[(h·ft^2)(°F/ft)]	....	0.13	....	0.065	0.13
Coefficient of thermal expansion, 10^{-5}/°F	....	45	....	27	8.8
Flame resistance	Fair	Poor	Poor	Good	Excellent
Colorability	Good	Good	Excellent	Excellent	Good
Mechanical properties					
Hardness (Shore A)	30–95	40–70	30–90	45–95	65–90
Tensile strength, kip/in^2					
Pure gum	....	1	<1	4	<2
Reinforced	2–3	<2	0.8–3.2	1.5–2.5	1.5–3
Elongation, % reinforced	320–350	200–400	200–600	250–500	100–450
Resilience	Poor to excellent	Good to fair	Good	Good	Fair
Compression-set resistance	Very good		Good	Fair to good	Good to excellent
Hysteresis resistance	Good	Good	Good	Good	Good
Flexcracking resistance	Very good	Good	Good	Good	Good
Slow rate	Very good	Good	Good	Good	Good
Fast rate	Good	Good	Good	Good	Good
Tear strength	Good	Fair	Poor to fair	Fair to good	Poor to fair
Abrasion resistance	Fair to good	Poor	Good	Excellent	Good
Electrical properties					
Dielectric strength	Fair	Good	Excellent	Excellent	Good
Electrical insulation	Fair	Good	Very good	Good	Fair to good
Thermal properties					
Service temperature, °F					
Minimum, for continuous use	−15 to −80	−90	−60	−40	−10
Maximum, for continuous use	300	400	<350	<325	<500
Corrosion resistance					
Weather	Excellent	Excellent	Excellent	Excellent	Excellent
Oxidation	Very good	Excellent	Excellent	Excellent	Outstanding
Ozone	Good to excellent	Excellent	Excellent	Excellent	Excellent
Radiation	....	Good	Excellent	Fair to good	Fair to good
Water	Good	Excellent	Good to excellent	Good	Good
Acids	Good	Very good to excellent	Good to excellent	Excellent	Good to excellent
Alkalies	Good	Very good	Good to excellent	Excellent	Poor to good
Aliphatic hydrocarbons	Excellent	Excellent	Poor	Fair	Excellent
Aromatic hydrocarbons	Very good	Excellent	Fair	Poor to fair	Excellent
Halogenated hydrocarbons	Good	....	Poor	Poor to fair	Good
Alcohol	Good	....	Good	Very good	Excellent
Synthetic lubricants (diester)	Fair to good	Excellent	Poor to fair	Poor	Fair to good
Hydraulic fluids					
Silicates	Very good	Excellent	Fair to good	Good	Good
Phosphates	Poor to fair	Excellent	Good to excellent	Poor to fair	Poor

SOURCE: C. H. Harper, *Handbook of Plastics and Elastomers,* McGraw-Hill, New York, 1975, Table 35. Courtesy of National Association of Corrosion Engineers. °C = (°F − 32) × 5/9; to convert British thermal units per hour–foot–degrees Fahrenheit to watts per meter–degrees Celsius, multiply by 0.861; to convert pounds-force per square inch to megapascals, multiply by 6.895 × 10^{-3}.

Sulfur Corrosion Chromium is the most important material in imparting resistance to sulfidation (formation of sulfidic scales similar to oxide scales). The austenitic alloys are generally used because of their superior mechanical properties and fabrication qualities, despite the fact that nickel in the alloy tends to lessen resistance to sulfidation somewhat.

Halogens (Hot, Dry Cl_2, HCl) Pure nickel and nickel alloys are useful with dry halogen gases. But even with the best materials, corrosion rates are relatively high at high temperature. There are cases in which equipment for high-temperature halogenation has used platinum-clad nickel-base alloys. These materials have high initial cost but long life. Platinum and gold have excellent resistance to dry HCl even at 1100°C (2000°F).

Refractories Refractories are selected to accomplish four objectives:

1. Resist heat
2. Resist high-temperature chemical attack
3. Resist erosion by gas with fine particles
4. Resist abrasion by gas with large particles

Refractories are available in three general physical forms: solids in the form of brick and monolithic castable ceramics and as ceramic fibers.

The primary method of selection of the type of refractory to be used is by gas velocity:

< 7.5 m/s (25 ft/s): fibers
7.5–60 m/s (25–200 ft/s): monolithic castables
> 60 m/s (200 ft/s): brick

Within solids the choice is a trade-off because, with brick, fine particles in the gas remove the mortar joints and, in the monolithic castables, while there are no joints, the refractory is less dense and less wear-resistant.

Internal Insulation The practice of insulating within the vessel (as opposed to applying insulating materials on the equipment exterior) is accomplished by the use of fiber blankets and lightweight

TABLE 23-28 Important Properties of Gasket Materials*

Material	Max. service temp., °F.	Important properties
Rubber (straight):		
Natural	225	Good mechanical properties. Impervious to water. Fair to good resistance to acids, alkalies. Poor resistance to oils, gasoline. Poor weathering, aging properties
Styrene-butadiene (SBR)	250	Better water resistance than natural rubber. Fair to good resistance to acids, alkalies. Unsuitable with gasoline, oils and solvents
Butyl	300	Very good resistance to water, alkalies, many acids. Poor resistance to oils, gasoline, most solvents (except oxygenated)
Nitrile	300	Very good water resistance. Excellent resistance to oils, gasoline. Fair to good resistance to acids, alkalies
Polysulfide	150	Excellent resistance to oils, gasoline, aliphatic and aromatic hydrocarbon solvents. Very good water resistance, good alkali resistance, fair acid resistance. Poor mechanical properties
Neoprene	250	Excellent mechanical properties. Good resistance to non-aromatic petroleum, fatty oils, solvents (except aromatic, chlorinated, or ketone types). Good water and alkali resistance. Fair acid resistance
Silicone	600	Excellent heat resistance. Fair water resistance; poor resistance to steam at high pressures. Fair to good acid, alkali resistance. Poor (except fluorosilicone rubber) resistance to oils, solvents
Acrylic	450	Good heat resistance but poor cold resistance. Good resistance to oils, aliphatic and aromatic hydrocarbons. Poor resistance to water, alkalies, some acids
Chlorosulfonated polyethylene (Hypalon)	250	Excellent resistance to oxidizing chemicals, ozone, weathering. Relatively good resistance to oils, grease. Poor resistance to aromatic or chlorinated hydrocarbons. Good mechanical properties
Fluoroelastomer (Viton, Fluorel 2141, Kel-F)	450	Can be used at high temperatures with many fuels, lubricants, hydraulic fluids, solvents. Highly resistant to ozone, weathering. Good mechanical properties
Asbestos:		
Compressed asbestos-rubber sheet	To 700	Large number of combinations available; properties vary widely depending on materials used
Asbestos-rubber woven sheet	To 250	Same as above
Asbestos-rubber (beater addition process)	400	Same as above
Asbestos composites	To 1000	Same as above
Asbestos-TFE	500	Combines heat resistance and sealing properties of asbestos with chemical resistance of TFE
Cork compositions	250	Low cost. Truly compressible materials which permit substantial deflections with negligible side flow. Conform well to irregular surfaces. High resistance to oils; good resistance to water, many chemicals. Should not be used with inorganic acids, alkalies, oxidizing solutions, live steam
Cork rubber	300	Controlled compressibility properties. Good conformability, fatigue resistance. Chemical resistance depends on kind of rubber used
Plastics:		
TFE (solid) (Tetrafluoroethylene, Teflon)	500	Excellent resistance to almost all chemicals and solvents. Good heat resistance; exceptionally good low-temperature properties. Relatively low compressibility and resilience

*From *Materials in Design Engineering*, Reinhold, New York, December 1959, pp. 111–126. °C = (°F − 32) × 5/9.

TABLE 23-28 Important Properties of Gasket Materials* *(Continued)*

Material	Max. Service temp., °F.	Important properties
TFE (filled)	To 500	Selectively improved mechanical and physical properties. However, fillers may lower resistance to specific chemicals
TFE composites	To 500	Chemical and heat resistance comparable with solid TFE. Inner gasket material provides better resiliency and deformability
CFE (Chlorotrifluoroethylene, Kel-F)	350	Higher cost than TFE. Better chemical resistance than most other gasket materials, although not quite so good as TFE
Vinyl	212	Good compressibility, resiliency. Resistant to water, oils, gasoline, and many acids and alkalies. Relatively narrow temperature range
Polyethylene	150	Resists most solvents. Poor heat resistance
Plant fiber:		
Neoprene-impregnated wood fiber	175	Nonporous; recommended for glycol, oil, and gasoline to 175°F.
SBR-bonded cotton	230	Good water resistance
Nitrile rubber-cellulose fiber	......	Resists oil at high temperatures
Vegetable fiber, glue binder	212	Resists oil and water to 212°F.
Vulcanized fiber	......	Low cost, good mechanical properties. Resists gasoline, oils, greases, waxes, many solvents
Inorganic fibers	To 2200°F.	Excellent heat resistance, poor mechanical properties
Felt:		
Pure felt	......	Resilient, compressible and strong, but not impermeable. Resists medium-strength mineral acids and dilute mineral solutions if not intermittently dried. Resists oils, greases, waxes, most solvents. Damaged by alkalies
TFE-impregnated	300	Good chemical and heat resistance
Petrolatum or paraffin-impregnated	......	High water repellency
Rubber-impregnated	......	Many combinations available; properties vary widely depending on materials used
Metal:		
Lead	500	Good chemical resistance. Best conformability of metal gaskets
Tin	.....	Good resistance to neutral solutions. Attacked by acids, alkalies
Aluminum	800	High corrosion resistance. Slightly attacked by strong acids, alkalies
Copper, brass	......	Good corrosion resistance at moderate temperatures
Nickel	1400	High corrosion resistance
Monel	1500	High corrosion resistance. Good against most acids and alkalies, but attacked by strong hydrochloric and strong oxidizing acids
Inconel	2000	Excellent heat, oxidation resistance
Stainless steel	......	High corrosion resistance. Properties depend on type used
Metal composites	......	Many combinations available; properties vary widely depending on materials used
Leather	220	Low cost. Limited chemical and heat resistance. Not recommended against pressurized steam, acid or alkali solutions
Glass fabric	......	High strength and heat resistance. Can be impregnated with TFE for high chemical resistance
Packing and Sealing Materials		
Rubber (straight)	To 600	See Gasket Materials for properties. Mainly used for ring-type seals, although some types are available as spiral packings
Rubber composites:		
Cotton-reinforced	350	High strength. Chemical resistance depends on type of rubber used; however, most types are noted for high resistance to water, aqueous solutions
Asbestos-reinforced	450	High strength combined with good heat resistance
Asbestos:		
Plain, braided asbestos	500	Heat resistance combined with resistance to water, brine, oil, many chemicals. Can be reinforced with wire

TABLE 23-28 Important Properties of Gasket Materials* *(Continued)*

Material	Max. service temp., °F.	Important properties
Packing and Sealing Materials *(Cont.)*		
Asbestos, *(Cont.)*		
Impregnated asbestos	To 750	Environmental properties vary widely depending on type of asbestos and impregnant used. Neoprene-cemented type resists hot oils, gasoline, and solvents. Oil and wax-impregnated type resists caustics. Wax-impregnated blue asbestos type has high acid resistance. TFE-impregnated type has good all-around chemical resistance
Asbestos composities	To 1200	End properties vary widely depending on secondary material used
Metals:		
Copper	To 1500	Properties depend on other construction materials and form of copper used. Packing made of copper foil over asbestos core resists steam and alkalies to 1000°F. Packing of braided copper tinsel resists water, steam, and gases to 1500°F.
Aluminum	To 1000	Resists hot petroleum derivatives, gases, footstuffs, many organic acids

TABLE 23-28 Important Properties of Gasket Materials* *(Concluded)*

Material	Max. service temp., °F.	Important properties
Packing and Sealing Materials *(Cont.)*		
Lead	550	Many types are available
Organic fiber:		
Flax	300	Good water resistance
Jute	300	Good water resistance
Ramie	300	Good resistance to water, brine, cold oil
Cotton	300	Good resistance to water, alcohol, dilute aqueous solutions
Rayon	300	Good resistance to water, dilute aqueous solutions
Felt	300	See Gasket Materials
Leather	To 210	Good mechanical properties for sealing. Resistant to alcohol, gasoline, many oils and solvents, synthetic hydraulic fluids, water
TFE	To 500	Available in many forms, all of which have high chemical resistance
Carbon-graphite	700	Good bearing and self-lubricating properties. Good resistance to chemicals, heat

TABLE 23-29 Properties of Graphite and Silicon Carbide

	Graphite	Impervious graphite	Impervious silicon carbide
Specific gravity	1.4–1.8	1.75	3.10
Tensile strength, lbf/in^2 (MPa)	400–1400 (3–10)	2600 (18)	20,650 (143)
Compressive strength, lbf/in^2 (MPa)	2000–6000 (14–42)	10,500 (72)	150,000 (1000)
Flexural strength, lbf/in^2 (MPa)	750–3000 (5–21)	4700 (32)	
Modulus of elasticity ($\times 10^6$), lbf/in^2 (MPa)	0.5–1.8 (0.3–12 $\times 10^4$)	2.3 (1.6 $\times 10^4$)	56 (39 $\times 10^4$)
Thermal expansion, in/(in·°F $\times 10^{-6}$) [mm/(mm·°C)]	0.7–2.1 (1.3–3.8)	2.5 (4.5)	1.80 (3.4)
Thermal conductivity, Btu/[(h·ft^2)(°F/ft)] [(W/(m·K)]	15–97 (85–350)	85 (480)	60 (340)
Maximum working temperature (inert atmosphere), °F (°C)	5000 (2800)	350 (180)	4200 (2300)
Maximum working temperature (oxidizing atmosphere), °F (°C)	660 (350)	350 (180)	3000 (1650)

SOURCE: Carborundum Co. Courtesy of National Association of Corrosion Engineers.

TABLE 23-30 Properties of Stoneware and Porcelain

	Stoneware	Porcelain
Specific gravity	2.2–2.7	2.4–2.9
Hardness, Mohs scale	6.5	7.5
Modulus of rupture, lb./sq. in.	3–7,000	8–15,000
Modulus of elasticity, lb./sq. in.	5–10 $\times 10^6$	10–15 $\times 10^6$
Compressive strength, lb./sq. in.	40–60,000	60–90,000
Pore volume, %	1.5	0.2–0.5
Water absorption, %	0.5–4.0	0–0.5
Linear thermal expansion, per °F.	2.4 $\times 10^{-6}$	2.5 $\times 10^{-6}$
Thermal conductivity, Btu./(sq. ft.)(hr.)(°F./in.)	8–22	8–10

aggregates in ceramic cements. Such construction frequently incorporates a thin, high-alloy shroud (with slip joints to allow for thermal expansion) to protect the ceramic from erosion. In many cases this design is more economical than externally insulated equipment because it allows use of less expensive lower-alloy structural materials.

Refractory Brick Nonmetallic refractory materials are widely used in high-temperature applications in which the service permits the appropriate type of construction. The more important classes are described in the following paragraphs.

Fireclays can be divided into plastic clays and hard flint clays; they may also be classified as to alumina content. **Firebricks** are usually made of a blended mixture of flint clays and plastic clays which is formed, after mixing with water, to the required shape. Some or all of the flint clay may be replaced by highly burned or calcined clay, called grog. A large proportion of modern brick production is molded by the dry-press or power-press process, in which the forming is carried out under high pressure and with a low water content. Extruded and hand-molded bricks are still made in large quantities.

The dried bricks are burned in either periodic or tunnel kilns at temperatures ranging between 1200 and 1500°C (2200 and 2700°F). Tunnel kilns give continuous production and a uniform burning temperature.

Fireclay bricks are used in kilns, malleable-iron furnaces, incinerators, and many portions of metallurgical furnaces. They are resistant to spalling and stand up well under many slag conditions but are not generally suitable for use with high-lime slags or fluid-coal-ash slags or under severe load conditions.

High-alumina bricks are manufactured from raw materials rich in alumina, such as diaspore. They are graded into groups with 50, 60, 70, 80, and 90 percent alumina content. When well fired, these bricks contain a large amount of mullite and less of the glassy phase than is present in firebricks. Corundum is also present in many of these bricks. High-alumina bricks are generally used for unusually severe temperature or load conditions. They are employed extensively in lime kilns and rotary cement kilns, in the ports and regenerators of glass tanks, and for slag resistance in some metallurgical furnaces; their price is higher than that of firebrick.

Silica bricks are manufactured from crushed ganister rock containing about 97 to 98 percent silica. A bond consisting of 2 percent lime is used, and the bricks are fired in periodic kilns at temperatures of 1500 to 1540°C (2700 to 2800°F) for several days until a stable volume is obtained. They are especially valuable when good strength is required at high temperatures. Superduty silica bricks are finding

TABLE 23-31 Wood for Chemical Equipment

Condition of Woods after 31 Days Immersion in Cold Solutions
Examined after 7 days drying

	Fir	Oak	Oregon pine	Yellow pine	Spruce	Redwood	Maple	Cypress
Hydrochloric acid, 5%	NAC	NAC	NAC	SS	SS	SS	NAC	NAC
Hydrochloric acid, 10%	NAC	NAC	NAC	SS	SS	SS	NAC	NAC
Hydrochloric acid, 50%	SS,SB,SWF	SS,WF	S,WF	S,WF	S,WF	S,WF	S,WF	S,WF
Sulfuric acid, 1%	NAC	NAC	NAC	SS	SS	NAC	NAC	SS,SB
Sulfuric acid, 5%	SS	SS	SS	SS	SS,SB	SS,SB	NAC	SS,SB
Sulfuric acid, 10%	S,FSD	S,FSD	S,FSD	S,FSD	S,FSD	S,FSD	S,FSD	S,FSD
Sulfuric acid, 25%	SSp,FSD	SSp,FSD	SSp,FSD	SSp,FSD	SSp,FSD	SSp,FSD	SSp,FSD	SSp,FSD
Caustic soda, 5%	S,NAC	MSh,SWp	SS	SS,FSD	SSp,FSD	SSp,FSD	MSh	SSp,FSD
Caustic soda, 10%	S,FSD	MSh,WF,Horny	SS	SS,SB,FSD	SS,SB,FSD	SS,SB,FSD	MSh	S,SB,FSD
Alum, 13%	NAC	NAC	NAC	NAC	NAC	NAC	NAC	NAC
Sodium carbonate, 10%	SB,GC	NAC	GC	SB,GC	SB,GC	SB,GC	GC	SB,GC
Calcium chloride, 25%	NAC	NAC	NAC	NAC	NAC	NAC	NAC	NAC
Common salt, 25%	NAC	NAC	NAC	SS,GC	SS,GC	SS,GC	NAC	NAC
Water	NAC	NAC	NAC	NAC	NAC	NAC	NAC	NAC
Sodium sulfide	SS,SB	MSh,WF	SB	SB	SB	SB	MSh,FSD	FSD

Condition of Woods after 8 Hr. Boiling in Solutions
Examined after 7 days drying

	Fir	Oak	Oregon pine	Yellow pine	Spruce	Redwood	Maple	Cypress
Hydrochloric acid, 10%	SB,S	FSD	FSD	FSD	FSD	FSD	FSD	FSD
Hydrochloric acid, 50%	FD,Ch,B,S,NG	FD,Ch,B,S,NG	FD,Ch,B,S,NG	FD,Ch,B,S,NG	FD,Ch,B,S,NG	FD,Ch,B,S,NG	FD,Ch,B,S,NG	FD,Ch,B,S,NG
Sulfuric acid, 4%	SB,GC	SB,GC	SB,GC	SB,GC	SB,GC	SB,GC	SB,GC	SB,GC
Sulfuric acid, 5%	SS,GC	SB,GC	SB,GC	SB,GC	SB,FSD	SB,GC	SB,GC	SB,FSD
Sulfuric acid, 10%	SS,GC	BFD,Wpd,NG	Sp,FD,NG	B,Sp,FD,NG	B,Sp,FD,NG	SB,FSD	SB,FSD	B,FD
Caustic soda, 5%	SS	MSh	S	GC	S,GC	S,GC	Sh	SSp
Alum, 13%	SB,GC	NAC	NAC	SB,GC	SB,GC	SB,GC	NAC	SB,GC
Sodium carbonate, 10%	SB,GC	GC	GC	GC	GC	GC	GC	SB,GC
Calcium chloride, 25%	SB,GC	SB,SS,GC	NAC	SB,GC	SB,GC	NAC	NAC	SB,GC
Common salt, 25%	NAC	NAC	NAC	SB,GC	NAC	SB,GC	NAC	NAC
Water	NAC	NAC	NAC	SB,GC	NAC	NAC	NAC	NAC

TABLE 23-32 Comparison of Properties of Refractory Metals

Melting point, °F	Element	Advantages	Disadvantages
6180	Tungsten	Highest melting point; nonvolatile oxide to at least 2500°F	Highest density; oxidizing rapidly; brittle at low temperatures
5425	Tantalum	Very high melting point; nonvolatile oxide; ductile	High density; oxidizing rapidly; least abundant
4730	Molybdenum	High melting point; less dense than tungsten or tantalum; moderately ductile at room temperature	Extremely high oxidation rate (volatile oxide)
4380	Niobium (columbium)	High melting point; nonvolatile oxide; ductile; moderate density	Oxidizing rapidly
3435	Chromium	Extremely oxidation-resistant; lighest of refractory metals	Lowest melting point of refractory metals, brittle at low temperatures

some use in the steel industry. They have a lowered alumina content and often a lowered porosity.

Silica bricks are used extensively in coke ovens, the roofs and walls of open-hearth furnaces, and the roofs and sidewalls of glass tanks and as linings of acid electric steel furnaces. Although silica brick is readily spalled (cracked by a temperature change) below red heat, it is very stable if the temperature is kept above this range and for this reason stands up well in regenerative furnaces. Any structure of silica brick should be heated up slowly to the working temperature; a large structure often requires 2 weeks or more.

Magnesite bricks are made from crushed magnesium oxide, which is produced by calcining raw magnesite rock to high temperatures. A rock containing several percent of iron oxide is preferable, as this permits the rock to be fired at a lower temperature than if pure materials were used. Magnesite bricks are generally fired at a comparatively high temperature in periodic or tunnel kilns. A large proportion of magnesite brick made in the United States uses raw material extracted from seawater.

Magnesite bricks are basic and are used whenever it is necessary to resist high-lime slags, as in the basic open-hearth steel furnace. They also find use in furnaces for the lead- and copper-refining industries. The highly pressed unburned bricks find extensive use as linings for cement kilns. Magnesite bricks are not so resistant to spalling as fireclay bricks.

Chrome bricks are manufactured in much the same way as magnesite bricks but are made from natural chromite ore. Commercial ores always contain magnesia and alumina. Unburned hydraulically pressed chrome bricks are also available.

Chrome bricks are very resistant to all types of slag. They are used as separators between acid and basic refractories, also in soaking pits and floors of forging furnaces. The unburned hydraulically pressed bricks now find extensive use in the walls of the open-hearth furnace. Chrome bricks are used in sulfite-recovery furnaces and to some extent in the refining of nonferrous metals. Basic bricks combining various properties of magnesite and chromite are now made in large quantities and have advantages over either material alone for some purposes.

The **insulating firebrick** is a class of brick that consists of a highly porous fire clay or kaolin. Such bricks are light in weight (about one-half to one-sixth of the weight of fireclay), low in thermal conductivity, and yet sufficiently resistant to temperature to be used successfully on the hot side of the furnace wall, thus permitting thin walls of low thermal conductivity and low heat content. The **low heat content** is particularly valuable in saving fuel and time on heating up, allows rapid changes in temperature to be made, and permits rapid cooling. These bricks are made in a variety of ways, such as mixing organic matter with the clay and later burning it out to form pores; or a bubble structure can be incorporated in the clay-water mixture which is later preserved in the fired brick. The insulating firebricks are classified into several groups according to the maximum use limit; the ranges are up to 870, 1100, 1260, 1430, and above 1540°C (1600, 2000, 2300, 2600, and above 2800°F).

Insulating refractories are used mainly in the heat-treating industry for furnaces of the periodic type. They are also used extensively in stress-relieving furnaces, chemical-process furnaces, oil stills or

TABLE 23-33 General Physical and Chemical Characteristics of Refractory Brick*

Type of brick	Typical composition	Approx. bulk density, lb./cu.ft.	Fusion point, °F.	Chemical nature	Deformation under hot loading	Apparent porosity, %	Permeability	Hot strength	Thermal shock resistance	Chemical resistance: To acid	Chemical resistance: To alkali
Silica	SiO_2, 95%	115	3100	Acid	Excellent	21	High	Excellent	Poor‡	Good	Good at low temperatures
High-duty fire clay	SiO_2, 54% Al_2O_3, 40%	134	3125	Acid	Fair	18	Moderate	Fair	Fair	Good	Good at low temperatures
Superduty fire clay	SiO_2, 52% Al_2O_3, 42%	140	3170	Acid	Good	15	High	Fair	Good	Good	Good at low temperatures
Acid-resistant (type H)	SiO_2, 59% Al_2O_3, 34%	142	3040	Acid	Poor	7	Low	Poor	Good	Insoluble in acids except HF and boiling phosphoric	Very resistant in moderate concentrations
Insulating brick	Varies	30–75	Varies	...	Poor	65–85	High	Poor	Excellent	Poor	Poor
High-alumina	Al_2O_3, 50–85%	170	3200–3400	Slightly acid	Good	20	Low	Good	Good	Good except for HF and aqua regia	Very slight attack with hot solutions
Extra-high-alumina	Al_2O_3, 90–99%	185	3000–3650	Neutral	Excellent	23	Low	Excellent	Good		
Mullite	Al_2O_3, 71%	153	3290	Slightly acid	Excellent	20	Low	Good	Good	Insoluble in most acids	Slight reaction
Chrome-fired	Chrome ore, 100%	195	Varies	Neutral	Fair	20	Low	Good	Poor	Fair to good	Poor
Magnesite-chrome bonded§	MgO, 50–80% Cr_2O_3, 5–18% Fe_2O_3, 3–13% Al_2O_3, 6–11% SiO_2, 1.2–5%	190	Varies	Basic	Good	12	Very low	Good	Excellent	Fair except to strong acids	Fair resistance at low temperatures
Magnesite-chrome fired		180			Excellent	20	High	Good	Excellent		
Magnesite-chrome high-fired		180			Excellent	18	High	Excellent	Excellent		
Magnesite-bonded§	MgO, 95%	181	3900	Basic	Good	11	Low	Good	Good	Soluble in most acids	Good resistance at low temperatures
Magnesite-fired		178			Good	19	Moderate	Good	Good		
Zircon	ZrO_2, 67% SiO_2, 33%	200	3100†	Acid	Excellent	25	Very low	Excellent	Good	Very slight	Very slight
Zirconia (stabilized)	ZrO_2, 94% CaO, 4%	245	4800	Slightly acid	Excellent	23	Low	Excellent	Excellent	Very slight	Very slight
Silicon-carbide	SiC, 80–90%	160	4175	Slightly acid	Excellent	15	Very low	Excellent	Excellent	Slight reaction with HF	Attacked at high temperatures
Graphite	C, 97%	105	6400	Neutral	Excellent	16	Low	Excellent	Excellent	Insoluble	Insoluble

*From *Chem. Eng.*, 100 (July 31, 1967). To convert pounds per cubic foot to kilograms per cubic meter, multiply by 0.0624; °C = (°F − 32) × 5/9.
†Dissociates above 1700°C (3100°F).
‡Good above 650°C (1200°F).
§Chemically bonded.

heaters, and the combustion chambers of domestic-oil-burner furnaces. They usually have a life equal to that of the heavy brick that they replace. They are particularly suitable for constructing experimental or laboratory furnaces because they can be cut or machined readily to any shape. They are not resistant to fluid slag.

There are a number of types of special brick obtainable from individual producers. **High-burned kaolin refractories** are particularly valuable under conditions of severe temperature and heavy load or severe spalling conditions, as in the case of high-temperature oil-fired boiler settings or piers under enameling furnaces. Another brick for the same uses is a high-fired brick of Missouri aluminous clay.

There are on the market a number of bricks made from **electrically fused materials,** such as fused mullite, fused alumina, and fused magnesite. These bricks, although high in cost, are particularly suitable for certain severe conditions.

Bricks of **silicon carbide,** either recrystallized or clay-bonded, have a high thermal conductivity and find use in muffle walls and as a slag-resisting material.

Other types of refractory that find use are forsterite, zirconia, and zircon. Acid-resisting bricks consisting of a dense body like stoneware are used for lining tanks and conduits in the chemical industry. Carbon blocks are used as linings for the crucibles of blast furnaces, very extensively in a number of countries and to a limited extent in the United States. Fusion-cast bricks of mullite or alumina are largely used to line glass tanks.

Ceramic-Fiber Insulating Linings Ceramic fibers are produced by melting the same alumina-silica china (kaolin) clay used in conventional insulating firebrick and blowing air to form glass fibers. The fibers, 50.8 to 101.6 mm (2 to 4 in) long by 3 μm in diameter, are interlaced into a mat blanket with no binders or chopped into shorter fibers and vacuum-formed into blocks, boards, and other shapes. Ceramic-fiber linings, available for the temperature range of

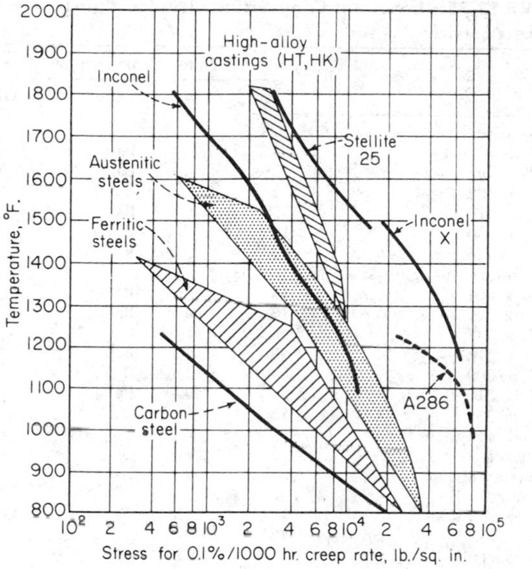

FIG. 23-10 Effect of creep on metals for high-temperature use. °C = (°F − 32) × 5⁄9; to convert pounds-force per square inch to megapascals, multiply by 6.895×10^{-3}. [Chem. Eng., *139 (Dec. 15, 1958).*]

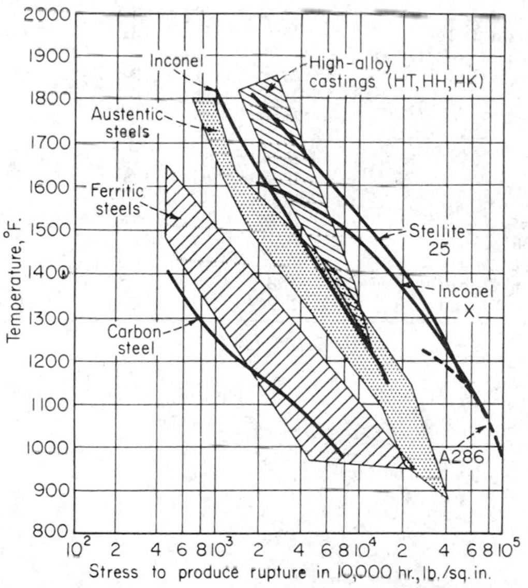

FIG. 23-11 Rupture properties of metals as a function of temperature. °C = (°F − 32) × 5⁄9; to convert pounds-force per square inch to megapascals, multiply by 6.895×10^{-3}. [Chem. Eng., *139 (Dec. 15, 1958).*]

650 to 1430°C (1200 to 2600°F), are more economical than brick in the 650- to 1230°C- (1200- to 2250°F-) range. Savings come from reduced first costs, lower installation labor, 90 to 95 percent less weight, and a 25 percent reduction in fuel consumption.

Because of the larger surface area (compared with solid-ceramic refractories) the chemical resistance of fibers is relatively poor. Their acid resistance is good, but they have less alkali resistance than solid materials because of the absence of resistant aggregates. Also, because they have less bulk, fibers have lower gas-velocity resistance. Besides the advantage of lower weight, since they will not hold heat,

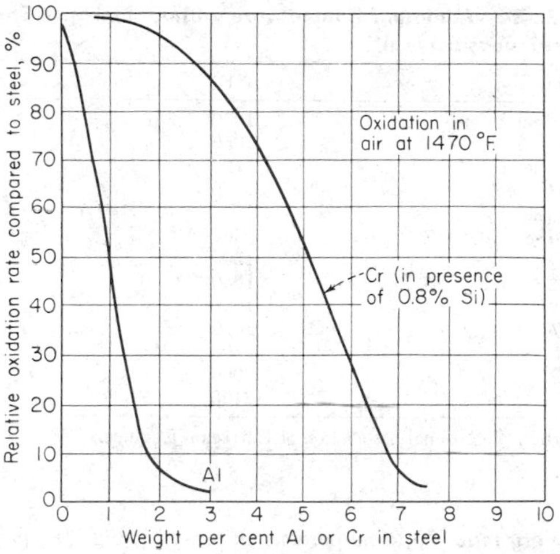

FIG. 23-12 How chromium and aluminum reduce steel oxidation. °C = (°F − 32) × 5⁄9.

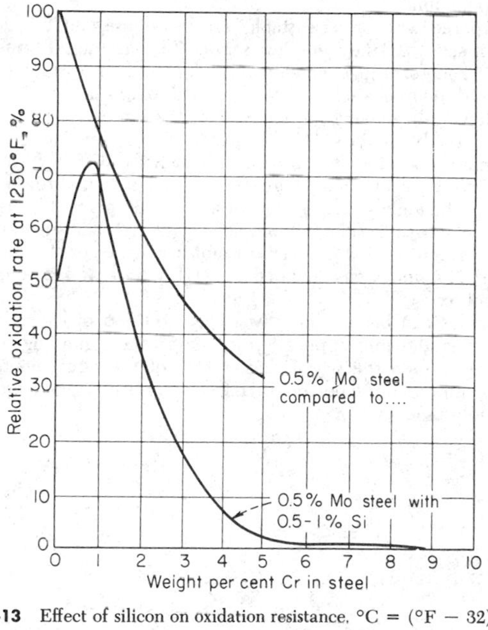

FIG. 23-13 Effect of silicon on oxidation resistance. °C = (°F − 32) × 5⁄9.

fibers are more quickly cooled and present no thermal-shock structural problem.

Castable Monolithic Refractories Standard portland cement is made of calcium hydroxide. In exposures above 427°C (800°F) the hydroxyl ion is removed from portland (water removed); below 427°C (800°F), water is added. This cyclic exposure results in spalling. Castables are made of calcium aluminate (rather than portland); without the hydroxide they are not subject to that cyclic spalling failure.

Castable refractories are of three types:

1. *Standard.* 40 percent alumina for most applications at moderate temperatures.
2. *Intermediate purity.* 50 to 55 percent alumina. The anorthite (needle-structure) form is more resistant to the action of steam exposure.

TABLE 23-34 Minimum Temperature without Excessive Scaling in Air (Continuous Service)*

Alloy	°F	°C
Carbon steel	1050	565
½Mo steel	1050	565
1Cr ½Mo steel	1100	595
2¼Cr 1Mo steel	1150	620
5Cr ½Mo steel	1200	650
9Cr 1Mo steel	1300	705
AISI 410	1300	705
AISI 304	1600	870
AISI 321	1600	870
AISI 347	1600	870
AISI 316	1600	870
AISI 309	2000	1090
AISI 310	2100	1150

*Courtesy of National Association of Corrosion Engineers.

3. *Very pure.* 70 to 80 percent alumina for high temperatures. Under reducing conditions the iron in the ceramic is controlling, as it acts as a catalyst and converts the CO to CO_2 plus carbon, which results in spalling. The choice among the three types of castables is generally made by economic considerations and the temperature of the application.

Compared with brick, castables are less dense, but this does not really mean that they are less serviceable, as their cements can hydrate and form gels which can fill the voids in castables. Extra-large voids do indicate less strength regardless of filled voids and dictate a lower allowable gas velocity. If of the same density as a given brick, a castable will result in less permeation.

Normally, castables are 25 percent cements and 75 percent aggregates. The aggregate is the more chemically resistant of the two components. The highest-strength materials have 30 percent cement, but too much cement results in too much shrinkage. The standard insulating refractory, 1:2:4 LHV castable, consists of 1 volume of cement, 2 volumes of expanded clay (Haydite), and 4 volumes of vermiculite.

Castables can be modified by a clay addition to keep the mass intact, thus allowing application by air-pressure gunning (gunite). Depending upon the size and geometry of the equipment, many castable linings must be reinforced; wire and expanded metal are commonly used.

TABLE 23-35 Important Commercial Alloys for High-Temperature Process Service

	Nominal composition, %			
	Cr	Ni	Fe	Other
Ferritic steels				
Carbon steel	....	....	Bal.	
2¼ chrome	2¼	....	Bal.	Mo
Type 502	5	....	Bal.	Mo
Type 410	12	....	Bal.	
Type 430	16	....	Bal.	
Type 446	27	....	Bal.	
Austenitic steels				
Type 304	18	8	Bal.	
Type 321	18	10	Bal.	Ti
Type 347	18	11	Bal.	Cb
Type 316	18	12	Bal.	Mo
Type 309	24	12	Bal.	
Type 310	25	20	Bal.	
Type 330	15	35	Bal.	
Nickel-base alloys				
Nickel	....	Bal.		
Incoloy 800	21	32	Bal.	
Hastelloy B	....	Bal.	6	Mo
Hastelloy C	16	Bal.	6	W, Mo
60-15	15	Bal.	25	
Inconel 600	15	Bal.	7	
80-20	20	Bal.		
Hastelloy X	22	Bal.	19	Co, Mo
Multimet	21	20	Bal.	Co
Rene 41	19	Bal.	5	Co, Mo, Ti
Cast irons				
Ductile iron	....	....	Bal.	C, Si, Mg
Ni-Resist, D-2	2	20	Bal.	Si, C
Ni-Resist, D-4	5	30	Bal.	Si, C
Cast stainless (ACI types)				
HC	28	4	Bal.	
HF	21	11	Bal.	
HH	26	12	Bal.	
HK	15	20	Bal.	
HT	15	35	Bal.	
HW	12	Bal.	28	
Superalloys				
Inconel X	15	Bal.	7	Ti, Al, Cb
A 286	15	25	Bal.	Mo, Ti
Stellite 25	20	10	Co-base	W
Stellite 21 (cast)	27.3	2.8	Co-base	Mo
Stellite 31 (cast)	25.2	10.5	Co-base	W

Section 24

Process Machinery Drives

Frank L. Evans, Jr., B.S.M.E., L.L.B., *Editor, Hydrocarbon Processing, Gulf Publishing Co.; Member, American Society of Mechanical Engineers, National Association of Corrosion Engineers; Registered Professional Engineer (Texas). (Section Editor)*

Carl R. Olson, M.S.E.E., *Staff Planning Manager, Industrial Projects Marketing, Westinghouse Electric Corp.; Member, Institute of Electrical and Electronics Engineers; Registered Professional Engineer (Pennsylvania). (Electric Motors and Motor Controls)*

H. Steen-Johnsen, M.S.M.E., *Chief Staff Engineer (Retired), Elliott Co.; Member, Life Fellow, American Society of Mechanical Engineers. (Steam Turbines)*

J. S. Swearingen, Ph.D., *President, Rotoflow Corp.; Member, American Institute of Chemical Engineers, American Society of Mechanical Engineers, American Chemical Society. (Expansion Turbines)*

Eric Jenett, M.S.Ch.E., *Manager Process Engineering, Brown & Root, Inc.; Associate Member, American Institute of Chemical Engineers, Project Management Institute; Registered Professional Engineer (Texas). (Power Recovery from Liquid Streams)*

ELECTRIC MOTORS AND AUXILIARIES

RECIPROCATING ENGINES

STEAM TURBINES

COMBUSTION GAS TURBINES

EXPANSION TURBINES

POWER RECOVERY FROM LIQUID STREAMS

MECHANICAL POWER-TRANSMISSION EQUIPMENT

General References: Bartlett, *Steam Turbine Performance and Economics*, McGraw-Hill, New York, 1958. Baumeister, *Standard Handbook for Mechanical Engineers*, 7th ed., McGraw-Hill, New York, 1967. Collins and Canaday, *Expansion Machines for Low Temperature Processes*, Oxford, Fair Lawn, N.J., 1958. Csanady, *Theory of Turbomachines*, McGraw-Hill, New York, 1964. Fink and Carroll, *Standard Handbook for Electrical Engineers*, 10th ed., McGraw-Hill, New York, 1968. Jennings and Rogers, *Gas Turbines Analysis and Practice*, McGraw-Hill, New York, 1953. Katz et al., *Handbook of Natural Gas Engineering*, McGraw-Hill, New York, 1959. Rase and Barrow, *Project Engineering of Process Plants*, Wiley, New York, 1957. Salisbury, *Steam Turbines and Their Cycles*, Wiley, New York, 1950. Scott, *Cryogenic Engineering*, Van Nostrand, Princeton, N.J., 1959. Shepherd, *Principles of Turbomachinery*, Macmillan, New York, 1956. Stepanoff, *Centrifugal and Axial Flow Pumps*, 2d ed., Wiley, New York, 1957. Stodola, *Steam and Gas Turbines*, Peter Smith, New York, 1945.

Nomenclature and Units

Symbol	Definition	SI units	U.S. customary units
b	Belt width	m	in
C	Constant		
C_1	Velocity of steam flow	m/s	ft/s
c	Clearance	m	ft
d	Bearing diameter	m	ft
E	Applied voltage	V	V
E	Modulus of elasticity	N/m^2	lbf/ft^2
e	Gross pump efficiency	Dimensionless	Dimensionless
e	Base of natural logarithm		
e	Efficiency	Dimensionless	Dimensionless
e_h	Hydraulic efficiency	Dimensionless	Dimensionless
F	Load		
F_f	Friction force	N	lbf
f	Friction coefficient		
f	Frequency	Hz	Hz
g	Gravitational constant	m/s^2	ft/s^2
H	Power	W	hp
H_e	Enthalpy change, ideal	J	Btu
H_i	Reversible enthalpy change	J/kg	Btu/lb
H_p	Total head (pump)	m	ft
H_t	Total head (turbine)	m	ft
h	Oil thickness	m	ft
Δh	Enthalpy change per mass	J/kg	Btu/lb
I	Line current	A	A
K_A	Constant		
K_L	Constant		
K_S	Constant		
K_T	Constant		
k	Constant		
Mol. wt.	Molecular weight	kg	lb
N	Force	N	lbf
N_c	Compressor efficiency	Dimensionless	Dimensionless
N_e	Expander efficiency	Dimensionless	Dimensionless
n	Speed	r/s	r/min
n_s	Specific speed (turbine or pump)		
P	Pressure	kPa	lbf/in^2
P	Mean bearing pressure	N/m^2	lbf/ft^2
P	Power	kW	kW
P_{max}	Contact pressure, maximum	N/m^2	lbf/ft^2
p	Number of poles		
Q_e	Quantity of heat removed	J	Btu
Q_p	Capacity, pump	m^3/h	ft^3/h
Q_t	Capacity, turbine	m^3/h	ft^3/h
R	Operating load	kg	lb
R	Gas constant	J/(mol·K)	Btu/(mol·°R)
R	Armature resistance	Ω	Ω
r	Radius	m	ft
S	Apparent power	kVA	kVA
S_{max}	Stress, maximum	N/m^2	lbf/ft^2
T	Tension	N	lbf
T	Torque	N·m	lbf·ft
T	Temperature	K	°R
T_a	Average temperature	K	°R
T_c	Centrifugal tension	N	lbf
t	Time	s	s
t	Belt thickness	m	in
V	Applied voltage	V	V
V	Counterelectromotive force	V	V
V_2	Belt speed	m/s	ft/min
W	Net work to compressor	J	Btu
W_c	Compressor work	J	Btu
W_{theor}	Theoretical work	J	Btu
WK^2	Inertia	$kg \cdot m^2$	$lb \cdot ft^2$
y	Power factor	Dimensionless	Dimensionless
Z	Viscosity	$N \cdot s/m^2$	cP
Z_a	Average compressibility factor	Dimensionless	Dimensionless

Greek symbols

Symbol	Definition	SI units	U.S. customary units
θ	Arc of contact		
φ	Belt density	kg/m^3	lb/in^3
ϕ	Magnetic-field flux	Wb	Wb

ELECTRIC MOTORS AND AUXILIARIES

All electric motors operate on the same basic principle regardless of type or size. When a wire carries electric current in the presence of a magnetic field (at least partially perpendicular to the current), a force on the wire is produced perpendicular to both the current and the magnetic field. In a motor the magnetic field radiates either in toward or outward from the motor axis (shaft) across the air gap, which is the annular space between the rotor and stator. Current-carrying conductors parallel to the axis (shaft) then have a force on them tangent to the rotor circumference. The force on the wire opposes an equal force (or reaction) on the magnetic field. It makes no difference whether the magnetic field is created in the rotor or the stator; the net result is the same: the shaft rotates.

Within these basic principles there are many types of electric motors. Each has its own individual operating characteristics peculiarly suited to specific drive applications. Equations (24-1) through (24-9), presented in Table 24-1, describe the general operating characteristics of alternating-current motors. When several types are suitable, selection is based on initial installed cost and operating costs (including maintenance and consideration of reliability).

TABLE 24-1 Useful Formulas for Alternating-Current Motors

Power output:

$$H = Tn/5250 \tag{24-1}$$

$$P = 0.00173Vlye \text{ (three-phase)} \tag{24-2}$$

$$P = 0.001Vlye \text{ (single-phase)} \tag{24-3}$$

Power input:

$$P = 0.00173Vly \text{ (three-phase)} \tag{24-4}$$

$$P = 0.001Vly \text{ (single-phase)} \tag{24-5}$$

$$P = 0.746H/e \tag{24-6a}$$

$$S = P/y = 0.746H/ye \tag{24-6b}$$

Line current and power factor:

$$I = \frac{0.746H \text{ (output)}}{1.73Vye} \text{ (three-phase)} \tag{24-7}$$

$$I = \frac{0.746H \text{ (output)}}{Vye} \text{ (single-phase)} \tag{24-8}$$

$$y = \frac{P \text{ (input)}}{S} \tag{24-9}$$

where e = efficiency, decimal
H = power, hp
I = line current, A
n = speed, r/min
P = power, kW
S = apparent power, kVA
T = torque, lbf·ft
V = applied voltage, V
y = power factor, decimal

NOTE: To convert horsepower to watts, multiply by 746; to convert pound-force-feet to newton-meters, multiply by 1.356; and to convert revolutions per minute to radians per second, multiply by 0.1047.

ALTERNATING-CURRENT MOTORS, CONSTANT-SPEED

The majority of industrial drives are constant-speed. Typical applications include:

Pumps
Compressors
Fans
Conveyors
Crushers and mills

Alternating-Current Squirrel-Cage Induction Motors These motors are by far the most common constant-speed drives. They are relatively simple in design and therefore both low in cost and highly reliable. Representative prices are shown in Fig. 24-1 for various speeds and horsepowers.

The typical three-phase squirrel-cage motor has stator windings which are connected to the power source. The rotor is a cylindrical magnetic structure mounted on the shaft with slots in the surface, parallel (or slightly skewed) to the shaft; either bars are inserted into these slots or molten metal is cast in place and connected by a short-circuiting end ring at both ends of the rotor. The name "squirrel-cage" derives from this rotor-bar construction. In operation, current passing through the stator winding creates a rotating magnetic field which cuts the rotor winding unless the rotor is turning in exact synchronism with the stator field. This cutting action induces a voltage, and hence a current, in the rotor which in turn reacts with the magnetic field to produce torque.

The typical medium-sized squirrel-cage motor is designed to operate at 2 to 3 percent slip (97 to 98 percent of synchronous speed). The **synchronous speed** is determined by the power-system frequency and the stator-winding configuration. If the stator is wound to produce one north and one south magnetic pole, it is a two-pole motor; there is always an even number of poles (2, 4, 6, 8, etc.). The synchronous speed is

$$n = 120\, f/p \tag{24-10}$$

where n = speed, r/min
f = frequency, Hz (cycles/s)
p = number of poles

The actual operating speed will be slightly less by the amount of slip. **Slip** depends upon motor size and application. Typically, the larger the motor, the less slip; an ordinary 7460-W (10-hp) motor may have 2½ percent slip, whereas motors over 746 kW (1000 hp) may have less than ½ percent. High-slip motors (as much as 13 percent slip) are used for applications with high inertia and requiring high starting torque; typical applications are punch presses and some crushers. Typical speed versus torque curves for various National Electrical Manufacturers Association (NEMA) design motors up to 149.2 kW (200 hp) are shown in Fig. 24-2. Typical characteristics and applications for these motors are given in Table 24-2.

Control or **starting** of squirrel-cage induction motors normally consists of applying full voltage to the motor terminals. The speed-torque curves in Fig. 24-2 are based on full voltage throughout the speed range from start to run. The specific motor design determines the amount of starting current. However, if the motor is a typical standard (NEMA A or B) design, the starting current may be estimated at 6 to 6.5 times normal full-load current with full voltage applied. Particularly for large motors, this starting inrush current may cause an undesirable voltage dip which can shut down other equipment, temporarily dim lights, or even initiate malfunctions in sophisticated control on the power system. For these conditions various alternatives exist.

1. *Reduced-voltage starting.* A reactor, resistor, or transformer is temporarily connected ahead of the motor during start to reduce the current inrush and limit voltage dip. This is accompanied by reduced starting torque. For reactor or resistor start, the torque decreases as the square of current; for transformer start, the torque decreases directly with line current. The reactor, resistor, or transformer can be adjusted to give a proper balance between torque and current.

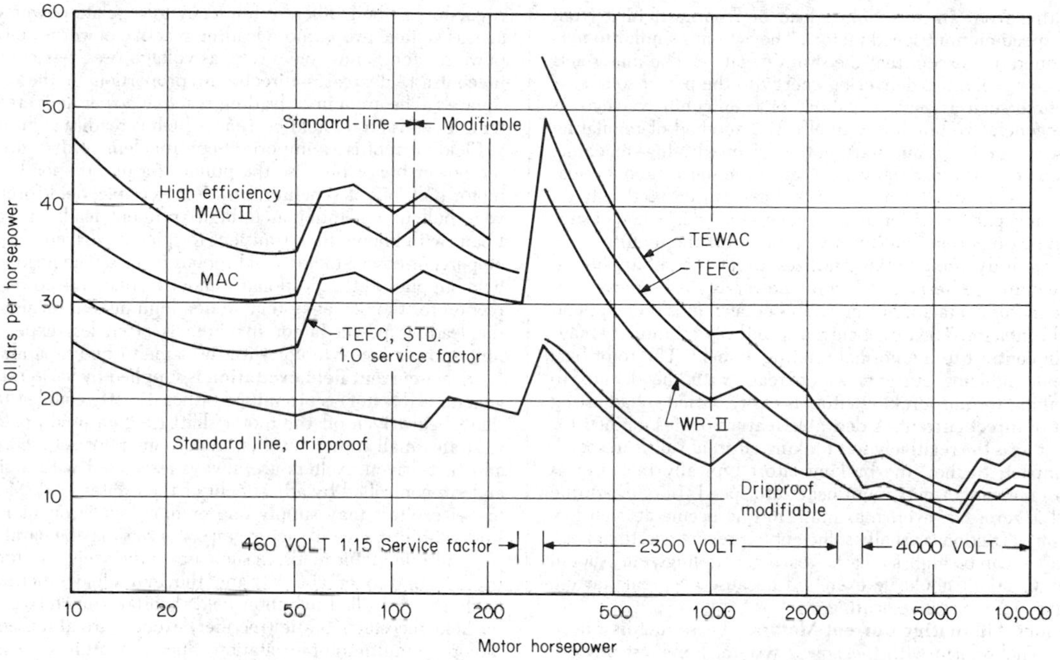

FIG. 24-1 Motor prices in dollars per horsepower for 1800-r/min squirrel-cage induction motors from 10 to 10,000 hp. Except for TEFC, Standard, motors shown from 10 to 250 hp have service factor; above 250 hp the standard service factor is 1.0. The basis of these data is June 1979. To convert dollars per horsepower to dollars per kilowatt, multiply by 1.340; to convert horsepower to kilowatts, multiply by 0.746.

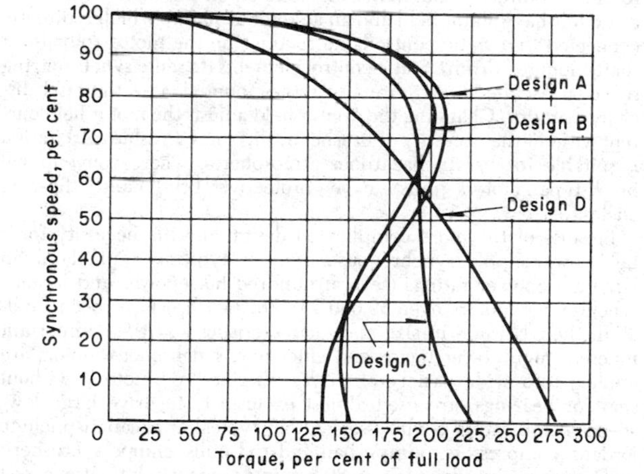

FIG. 24-2 Typical speed versus torque curves for various NEMA-design squirrel-cage induction motors. (See Table 24-2 for an explanation of design types.)

TABLE 24-2 Characteristics and Typical Applications for Squirrel-Cage Induction Motors

	NEMA A and B	NEMA C	NEMA D
Starting torque	Normal	High	High
Running slip	Low	Low to medium	High
Applications	Pumps Compressors Fans Machine tools General use	Electrical stairways Pulverizers Conveyors	Punch presses Crushers

2. *Star-delta starting.* A delta-connected motor is reconnected in Y form for starting, thus applying 57.7 percent voltage to each phase winding. This results in a developed torque of $(0.577)^2$, or only 33 percent. There is no means of adjustment; therefore, this method is useful only for loads requiring less than one-third of the motor's normal starting and accelerating torques. It is rarely used in the United States.

3. *Part-winding starting.* This method employs a motor with two sets of windings, only one of which is energized during start. Torque and current are both roughly 50 percent. Two small contactors (starting switches) are used instead of one large one, and no reactors or transformers are required. The disadvantages are the fixed value of available torque and the harmonic disturbances from possible winding unbalance, causing deviations in the speed-torque curve and therefore possible failure to accelerate.

Braking and regeneration are possible with squirrel-cage motors. The direction of rotation is determined by the sequence or phase rotation of the power supply. If two leads on a three-phase motor are interchanged, the rotation reverses. If this occurs during operation, the motor will come to a rapid stop and reverse. Power is removed at standstill for effective braking. For estimating only, this plug-stop torque is approximately equal to starting torque. The braking time can be estimated by

$$t = WK^2\, n/308T \tag{24-11}$$

where t = time, s
WK^2 = inertia, lb·ft²
n = running speed, r/min
T = torque, lbf·ft

To convert pound–square feet to kilogram–square meters, multiply by 0.0421; to convert revolutions per minute to radians per second, multiply by 0.1047; and to convert pound-force-feet to newton-meters, multiply by 1.356.

These estimates are frequently inaccurate because of second-order effects such as rotor saturation and harmonics. If the application is at all critical, the motor manufacturer should be consulted.

Regenerative braking occurs at speeds above synchronous-motor speed resulting from an overhauling load or from switching from high to low speed on multispeed motors. The action is similar to normal motor operation except that the slip is negative. The motor acts as an induction generator, delivering energy to the power source. If a power source such as a gas expander or a downhill conveyor is available, regenerative braking is an effective method of regulating speed, conserving energy, and starting the driven (driving) machine. An induction generator can deliver power to the source about equal to its rating as a motor. Regenerative braking can be used only on power systems capable of absorbing the generated energy and of supplying magnetizing excitation (reactive power) for the motor.

Direct-current dynamic braking utilizes direct current applied to the stator winding. Alternating-current power is first removed by opening the motor contactor or starter; direct current is then applied by a second contactor. The direct current produces a stationary magnetic flux, in contrast to the normal rotating ac field. The rotor bars cut this field, inducing currents which react with the dc flux to develop braking torque. Braking effort is easily varied by adjusting the amount of direct current. A desirable feature of this method for standard motors is the relatively soft braking effort at full-load speed, reducing impact; further, the braking effort typically increases as speed drops, reaching a maximum near zero speed. Braking torque at standstill is zero; however, maximum torque occurs at such low speed that static friction is usually sufficient to prevent coasting. Peak braking torques can be high; so shafts, gearing, couplings, etc., should be checked. Caution should be exercised because frequent starting and stopping cause excessive heating.

Synchronous Alternating-Current Motors These motors run in exact clock synchronism with the power system. For most modern power systems, these are truly constant-speed motors.

In the conventional synchronous motor a rotating magnetic field is developed by the stator currents as in induction motors. The rotor, however, is different, consisting typically of pairs of electromagnets (poles) spaced around the rotor periphery. The rotor field corresponds to the field produced by the ac stator having the same number of poles. The rotor or field coils are supplied with direct current; the magnetic field is therefore stationary with respect to the rotor structure. Torque is developed by the interaction of the rotor magnetic field and the stator current (in-phase component). Under no-load conditions and with appropriate dc field current, rotor and stator magnetic-field centers coincide. The voltage applied to the stator winding is balanced by an opposing voltage generated in the stator by the rotor field (induced), and no ac power current flows. As load is applied, the rotor tends to decelerate, causing a shift of rotor position with respect to the ac field. This shift produces a difference between the applied and induced voltages; the voltage difference causes current to flow; the current reacts with the rotor magnetic flux, producing torque.

Synchronous motors cannot be started with the dc field applied. Instead they are started as induction motors; bars, acting like a squirrel-cage rotor, are embedded in the field-pole surface and connected by end rings at both ends of the rotor. These damper bars also serve to damp out oscillations under normal running conditions. When the motor is at approximately 95 percent speed (depending upon application and motor design), direct current is applied to the field and the motor pulls into step (synchronism). Because the damper bars do not affect the synchronous-speed characteristics, they are designed for starting performance. This provides flexibility in the accelerating characteristics to meet specific application requirements without affecting running efficiency and other synchronous-speed characteristics. The rotor design of a squirrel-cage motor, on the other hand, must be a compromise between starting and running performance. The dc field is usually shorted by a resistor during starting and contributes accelerating torque, particularly near synchronous speed.

Power-factor correction is an important feature of synchronous motors. Conventional synchronous-motor power factors are either 100 or 80 percent leading. Leading-power-factor machines are used frequently to correct for the lagging power factor of the remaining plant load (such as induction motors), preventing penalty charges on power bills. Even 100 percent power-factor motors can be operated leading at reduced loads. An advantage of synchronous motors over capacitors is their inherent tendency to regulate power-system voltage; as voltage drops, more leading reactive power is delivered to the power system, and, conversely, as voltage rises, less reactive power, in contrast, decreases directly in proportion to the voltage drop squared. The amount of leading reactive power delivered to the system depends on dc field current, which is readily adjustable.

Field current is an important control element. It controls not only the power factor but also the pullout torque (the load at which the motor pulls out of synchronism). For example, field forcing can prevent pullout on anticipated high transient loads or voltage dips. Loads with known high transient torques are driven frequently with 80 percent power-factor synchronous motors. The needed additional field supplies both additional pullout torque and power-factor correction for the power system. When high pullout torque is required, the leading power-factor machine is often less expensive than a unity-power-factor motor with the same torque capability.

Direct-current field excitation is supplied by various means. A dc generator (exciter) is often used either directly coupled to the motor shaft, belt-driven off the motor shaft (seldom used), or driven by a separate small motor (exciter motor-generator set). Direct-coupled and belt-driven exciters are always associated with a single motor and are controlled by adjustment of the exciter field. Motor-generator-set exciters may supply one or more synchronous motor fields. For reliability, several motor-generator sets may be paralleled to supply multiple motor fields; in such a case the exciter voltage is usually fixed (e.g., 125 or 250 V), and the individual synchronous-motor fields are controlled by motor field rheostats (much larger than exciter field rheostats). Static (rectifier) exciters are also used for single-motor or multimotor excitation. Special rectifiers are required to avoid damage from surge voltages on pullout. These exciters, rotating or static, require brushes and slip rings to conduct direct current to the rotating field structure.

Another concept is **brushless excitation,** in which an ac generator (exciter) is directly coupled to or mounted on the motor shaft. The ac exciter has a stator field and an ac rotor armature which is directly connected to a static controllable rectifier on the motor rotor (or a shaft-mounted drum). Static control elements (to sense synchronizing speed, phase angle, etc.) are also rotor-mounted, as is the field discharge resistor. Changing the exciter field adjusts the motor field current without the necessity of brushes or slip rings. Brushless excitation is suitable for use in hazardous atmospheres, where conventional brush-type motors must have protective brush and slip-ring enclosures.

Because of the more complicated design and the necessity for a field power supply, synchronous motors are typically applied only in large-horsepower ratings (several hundred horsepower and larger); synchronous motors over 59,680 kW (80,000 hp) have been built. With their latitude in size and characteristics and their important inherent high power factor and efficiency, synchronous motors are applied to a wide variety of drives. Engine-type motors (without shaft or bearings) are used almost exclusively to drive large low-speed reciprocating compressors. Other typical applications include jordans, compressors, pumps, ball and rod mills, chippers, crushers, and grinders. Speeds as low as 80 r/min are practical; the top speed is limited by the rotor structure and is dependent on horsepower. The approximate limit for 1800 r/min is 2238 kW (3000 hp); for 1200 r/min it is 29,840 kW (40,000 hp).

Synchronous speeds are calculated by Eq. (24-10). Speeds above the limits given are obtained through step-up gears; large high-speed centrifugal compressors are examples. Two-pole (3600 r/min at 60 Hz) synchronous motors can be built but are uneconomical in comparison with geared drives.

ALTERNATING-CURRENT MOTORS, MULTISPEED

Squirrel-cage induction motors are inherently single-speed machines, but multispeed operation can be obtained by reconnecting the stator windings of motors designed for this purpose.

Two-Winding Motors These motors illustrate the simplest concept. The two separate stator windings (three-phase or two-phase only) are designed and wound for a different number of poles. For example, one winding may be four poles (1800 r/min at 60 Hz) and the other six poles (1200 r/min at 60 Hz). Only one winding is connected at a time. This method is used for speed ratios other than 2:1. Since the two windings are independent, a large number of speed combinations is possible. The two windings are not necessarily of equal capacity.

Two-winding motors may be built for constant torque, variable torque, or constant horsepower. Constant-horsepower motors are capable of handling the same horsepower at both speeds (i.e., higher torque at the low speed). Constant-torque motors can handle the same load torque at either speed (e.g., conveyor drives). Variable-torque motors are designed for loads in which load torque varies as the square of speed and horsepower varies as the cube of speed. Typical applications are as follows:

Variable torque	Constant torque	Constant horsepower
Fans	Conveyors	Machine tools
Centrifugal pumps	Feeders	
	Reciprocating compressors	

Single-Winding Consequent-Pole Motors These motors can be used when the low speed is one-half of the high speed. They are available as three-phase only. The specially designed winding is regrouped by external reconnection (motor control) to obtain the desired speed. A 2:1 speed ratio only is obtainable by this method; speeds such as 3600/1800, 1800/900, and 1200/600 are obtainable. Variable-torque, constant-torque, and constant-horsepower designs are available with torque characteristics as discussed under "Two-Winding Motors." The control for two-speed single-winding motors is more complicated than for the two-winding control.

Four-Speed, Two-Winding Squirrel-Cage Motors These motors are built by combining the preceding two methods. The stator winding is composed of two consequent-pole windings. Each winding gives two speeds with a relation of 2:1 to each other. The standard 60-Hz speed combinations are 1800/1200/900/600 r/min and 1200/900/600/450 r/min. The three torque-capability designs of variable torque, constant torque, and constant horsepower are also available in these four-speed motors.

Pole-Amplitude-Modulated Induction Motors These are single-winding squirrel-cage motors with any combination of poles or speeds (e.g., 8/10 poles or 900/720 r/min or as wide as 4/20 poles or 1800/360/min). They are smaller and lighter than equivalent two-winding machines. The entire winding works at both high and low speed, resulting in greater thermal capacity and higher efficiency. The basic principle is that one frequency acted on (modulated) by another produces new frequencies equal to the sum and difference of the two. Thus, a six-pole field modulated by a two-pole field produces a four- and an eight-pole field. The four-pole field can be eliminated by proper winding geometry. Such a motor runs at six-pole speed (1200 r/min, 60 Hz) when connected normally and at eight-pole speed (900 r/min) when half of the coils are reversed to produce the two-pole modulation.

In the preceding discussion of multispeed ac motors note that only induction motors are considered. These have no discrete physical rotor poles, so that only the stator-pole configuration need be modified to change speed. To operate multispeed, a synchronous motor would require a distinct rotor structure for each speed. Thus multispeed is practical only for squirrel-cage induction motors.

ALTERNATING-CURRENT MOTORS, WOUND-ROTOR INDUCTION

Wound-rotor induction motors operate on the same principle as squirrel-cage motors. However, as the name implies, the rotor has windings rather than bars, and these windings are connected to shaft-mounted slip rings. Brushes riding on the slip rings are connected to external resistance or short-circuited. Wound-rotor motors have an additional dimension of flexibility in the **variability of external rotor resistance.** Rotor resistance affects the shape of the speed-torque curve; increasing resistance decreases the speed at which maximum torque occurs. Figure 24-3 illustrates this effect as resistance is increased from zero external resistance (at top) to a very high value (extreme left).

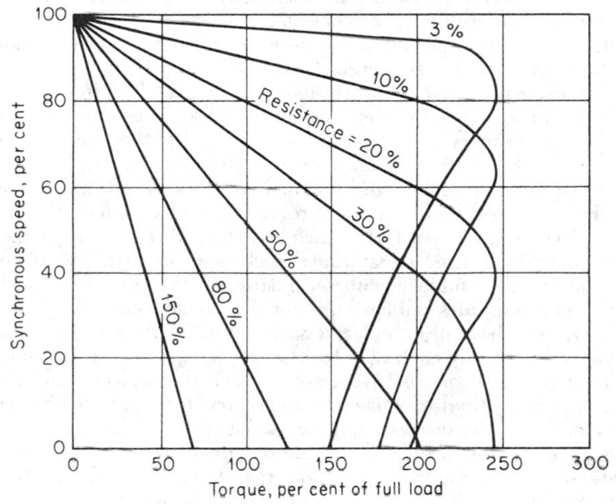

FIG. 24-3 Typical speed versus torque curves for a wound-rotor induction motor with varying amounts of external secondary (rotor) resistance. Resistance values are based on resistance at 100 percent torque and zero speed = 100 percent.

Reduced-speed operation reduces efficiency. **Efficiency** is approximately equal to speed expressed as a percentage of synchronous speed. Thus at 75 percent speed, about three-fourths of the motor input goes to the load; the other quarter is dissipated in the rotor resistance. The external resistance is sized to get rid of this heat. Further, in accelerating a load to operating speed, the heat generated in the rotor resistance is equal to the energy required to accelerate the load (inertia plus friction). The rotor resistance (internal or external) must have capacity to store this heat, since accelerating time is typically too short to allow any significant heat dissipation.

These characteristics of wound-rotor motors determine their **scope of application.** They can accelerate very-high-inertia loads, such as crushers, by using a large external resistance to absorb the heat. Loads sensitive to shock-accelerating torques may also be accelerated softly by inserting a high starting rotor resistance; this effect is used, for example, to take up slack in gears. Wound rotors are also used to handle loads like punch presses and car crushers, for which extreme transient peak loads are supplied by the mechanical-system inertia, allowing the system to slow down during these peaks; permanent external rotor resistance provides this soft characteristic. Wound-rotor motors are also used to provide adjustable-speed drives for pumps, cranes, and other loads when precise speed regulation is not required. Reduced-speed losses are not very significant for pump loads; the percent efficiency is low at reduced speed, but the torque and load horsepower are dropping rapidly. If torque is proportional to speed squared, the maximum rotor-resistance losses never exceed 10.5 percent of the full-speed load (occurs at 70 percent speed and efficiency, 50 percent torque, 35 percent load horsepower, and 30 percent losses, thus 10.5 percent losses based on full load at full speed).

Control of wound-rotor motors, as discussed, can be effected by adjusting the external secondary (rotor) resistance either in steps or continuously by liquid rheostat (this method is seldom used). Commonly when secondary resistance is varied to adjust speed or torque or to control acceleration, multiple resistance steps are used. These steps may be switched manually (typically a drum switch) or electrically by contactor.

In addition to secondary resistance control, other devices such as reactors and thyristors (solid-state controllable rectifiers) are used to control wound-rotor motors. Fixed secondary reactors combined with resistors can provide very constant accelerating torque with a minimum number of accelerating steps. The change in slip frequency with speed continually changes the effective reactance and hence the value of resistance associated with the reactor. The secondary reactors, resistors, and contacts can be varied in design to provide the proper accelerating speed-torque curve for the protection of belt conveyors and similar loads.

Saturable reactors, which are adjustable by a small dc signal, have also been used for both primary (stator) and secondary (rotor) control. In the primary they control motor voltage and therefore torque. In combination with fixed secondary resistors and feedback from a tachometer, this system can be used for precise speed and torque control of cranes, hoists, etc. Even reversing can be accomplished by using two saturable reactors in each of two (of three) phases. Other combinations of fixed or saturable reactors in the primary and/or secondary, all combined with secondary resistors, provide a wide range of capabilities and flexibility for the wound-rotor motor.

Thyristors have been replacing saturable reactors; they are small, efficient, and easily controlled by a wide variety of control systems. A modern crane control drive uses fixed secondary resistors and two sets of primary thyristors (one set for hoist, one for lower). With tachometer feedback for speed sensing, the control for the motor provides speed regulation and torque limiting in both directions, all with static devices. A wide variety of control systems is possible; the control should be designed for the specific application.

DIRECT-CURRENT MOTORS

Direct-current motors are adjustable in speed over a wide range. Further, efficiency is high over the entire speed range, unlike wound-rotor motors, in which efficiency is roughly proportional to speed. This flexibility is attained at the expense of additional complexity and cost.

Direct-current motor fields are on the stator. The rotor is the armature. The magnetic field does not rotate like the field in ac machines. Current in the armature reacts with the stator field to produce torque.

The armature windings generate a voltage opposite to the applied voltage as they cut the magnetic field. The difference between the terminal voltage and the generated voltage (counterelectromotive force) applied to the armature resistance produces armature current. Torque is proportional to armature current and magnetic flux. Counterelectromotive force is almost equal to the applied voltage, so the speed can be changed by changing the applied voltage; the speed can also be changed by varying the field current.

$$E = V + IR \qquad \text{or} \qquad I = (E - V)/R \tag{24-12}$$

$$V = kn\phi \tag{24-13}$$

where E = applied voltage, V
V = counterelectromotive force (generated voltage), V
R = armature resistance, Ω
I = armature current, A
k = constant dependent on motor design
n = speed, r/min
ϕ = magnetic-field flux

Generated voltage is proportional to the magnetic-flux cut; so a motor must change speed to generate the same counterelectromotive force if the field current is changed.

Direct-current motors are connected in several ways. **Shunt motors** have armature and field connected in parallel. This connection provides almost constant speed regardless of the applied voltage. If the voltage drops, the counterelectromotive force also drops because of the reduction in field strength. Speed can be changed by varying the field current and/or armature voltage independently. Increasing armature voltage increases speed; increasing field current decreases speed.

Series motors have their armature and field windings connected in series; both carry the same current. The speed depends on both voltage and load. Torque is the product of armature current and magnetic field; both are dependent on current. For any specific load torque, the current is constant and speed is proportional to applied voltage. If, however, the load changes, the speed also changes. A 50 percent drop in load torque reduces the motor current to 70 percent, making the product of armature and field current 50 percent. The reduced field current decreases counterelectromotive force, and the motor speed increases by 40 percent. Because of this characteristic, series motors overspeed severely if unloaded. Series motors are suitable for constant-horsepower applications with a wide speed range, such as machine tools. They are also used for traction drives (e.g., shuttle cars and locomotives), some cranes, hoists, and elevators.

Compound-wound dc motors have both series and shunt fields. The addition of a small series field helps provide the proper amount of no-load to full-load speed regulation or droop. Shunt or compound-wound motors are applied widely to many adjustable-speed drives. They are important for drives requiring accurate speed regulation and adjustment.

A dc motor's inherent speed-torque curve can be varied widely by adjusting the relative amounts of shunt and series fields. The series field may also be connected to aid or buck the shunt field. The usual practice is to connect the series field so that it adds to the shunt field (cumulative compound), which gives a stable, drooping speed with increasing load.

The great flexibility of dc motors both through inherent design characteristics and through the way in which they are operated makes them ideally suited to adjustable-speed drives, particularly regulated drive systems.

ADJUSTABLE-SPEED DRIVES

One of the oldest adjustable-speed drives is the **Ward-Leonard system.** This consists of an ac to dc motor-generator set and a shunt or compound-wound dc motor. Speed is adjusted by changing the generator voltage. A functional equivalent of this drive uses an adjustable-voltage rectifier feeding a dc motor. This system has only one rotating machine in contrast to the three of a conventional Ward-Leonard system.

Modern static controllable rectifiers such as thyristors respond almost instantaneously to control signals and are adaptable as the power supply to the most critical regulated-drive systems. The sensing and regulating system can be designed to hold speed, speed differential, tension, torque, current, acceleration and deceleration, etc., or any required combination. For example, a drive may be speed-regulated with torque limit or speed-regulated with acceleration and/or deceleration regulated or limited.

Typical applications for adjustable-voltage, adjustable-speed dc drives include winders, paper machines and auxiliaries, blending systems, feeders, extruders, calenders, machine tools, range and slasher drives, cranes, hoists, shovels and draglines, and an almost unlimited variety of drives requiring the flexibility and efficiency possible with direct current.

Mechanical adjustable-speed drives are used when a high degree of regulation is not required. One drive consists of a constant-speed ac motor driving the load through V belts and variable-pitch pulleys. The speed range can be as high as 8:1. It is available up to 18.6 kW (25 hp). Speed adjustment is either manual or remote with a motor drive for the adjustment. Speed regulation from no load to full load is normally 3 to 6 percent. Efficiency is high over the entire speed range since there are no slip losses.

Electromagnetic drives are simple adjustable-speed ac drives with efficiency comparable with that of wound-rotor motors. This drive uses a magnetic slip coupling driven by a squirrel-cage motor. The slip is determined by the excitation current and the load. Efficiency is proportional to speed. Therefore, these drives are uneconomical for continuous low-speed high-torque operation but are ideal for fans and centrifugal pumps requiring little speed adjustment when torque decreases rapidly with speed and for controlling acceleration. Electromagnetic drives are functionally equivalent to hydraulic couplings.

RectiFlow adjustable-speed drives use both a wound-rotor motor and a dc motor connected to the same shaft. The rotor winding of the wound-rotor motor is connected to a rectifier. The dc rectifier output supplies the dc motor. Semiconductor-rectifier developments make this a practical, high-efficiency adjustable-speed drive for applications up to several hundred horsepower. Since the rotor losses of the wound-rotor motor are used to produce shaft power and are not dissipated in a resistor, the efficiency of these drives is high over the entire speed range. A typical 223.8-kW (300-hp) RectiFlow drive operating over a 3:1 speed range has an efficiency of more than 83 percent over its operating range. RectiFlow adjustable-speed drives are suitable for high-torque low-speed drives such as extruders, mixers, pumps, fans, and kilns.

A modification of this basic drive system uses solid-state rectifiers and thyristors to convert the wound-rotor, variable-frequency slip power first to direct current and then to line-frequency power (60 Hz in the United States). This in turn is fed back to the power system as useful energy.

Adjustable-frequency alternating current can be used with squirrel-cage motors for adjustable-speed drives. A typical application is small synthetic-fiber spinning drives which require multiple motors operating at constant speed. When precise synchronism between drives is required, synchronous-reluctance motors are used. These are squirrel-cage motors with flats or grooves on the rotor which form magnetic poles because of the change in magnetic path as the rotor position moves with respect to the stator field. This causes the rotor to rotate in exact synchronism with the stator field for light loads. The power source for these drives is either an ac generator driven by any adjustable-speed drive or a static inverter.

MOTOR ENCLOSURES

Except for areas with fire or explosion hazards (hazardous areas), motor enclosures are designed to provide protection to the internal working parts. The development of improved insulating materials and finishes has affected the required degree of protection and consequently the design and classification of enclosures.

Open, dripproof is the standard enclosure for induction, high-speed synchronous, and industrial dc motors. This design is useful for most indoor and many outdoor applications. Dripproof construction provides good mechanical protection to the internal working parts of the motor and prevents the entrance of dropping liquids and heavy dirt particles. However, it does not protect against airborne moisture, dust, or corrosive fumes. **Guarded** machines have all openings protected to prevent objects more than 12.7 mm (½ in) in diameter from entering the motor. **Splashproof** motors are not affected by water or by solid particles striking or entering the enclosure at an angle less than 100° from the vertical.

Weather-protected, type I is the next degree of protection for larger motors. Such a motor is defined as "an open machine with its ventilating passages so constructed as to minimize the entrance of rain, snow, and airborne particles to the electric parts" (NEMA Standard MG-1, "Motors and Generators"). All openings are restricted against passage of a 19-mm- (¾-in-) diameter rod. Some modern insulation systems are completely satisfactory for most outdoor applications.

Weather-protected, type II motors are recommended for large sizes when a higher degree of protection and longer life are desired. They have extensive baffling of the ventilating system so that the air must turn at least three 90° corners before entering the active motor parts [maximum air velocity, 3.05 m/s (600 ft/min)]; thus, rain, snow, and dirt carried by driving winds are blown through the motor housing without entering the active parts.

Totally enclosed motors offer the greatest protection against moisture, corrosive vapors, dust, and dirt. Totally enclosed fan-cooled (TEFC) motors are the obvious choice rather than *weather-protected* below 186.5 kW (250 hp). Their internal and external ventilating air are kept separate; external air never gets inside except for the small amount that enters by breathing.

TEFC motors have both an internal fan for circulating air within the motor and an external fan for forcing the air through or over the motor frame or heat exchanger. Small motors [approximately 2.238 kW (3 hp) and below] do not require ventilating fans; these totally enclosed nonventilated motors are similar to TEFC with the fans omitted.

Separate forced ventilation is required for some applications (for example, adjustable-speed drives which operate at low speed); these must depend on an external ventilation. This classification includes *open externally ventilated* machines, *open pipe-ventilated* machines, and *totally enclosed pipe-ventilated* machines. In corrosive or hazardous areas, safe or clean air ventilates the motor.

Enclosed motors with air-to-water coolers cost much less than TEFC motors above 373 kW (500 hp); in large synchronous-motor ratings, they cost even less than weather-protected, type II. Large enclosed synchronous machines with coolers for mounting in the motor foundation are frequently supplied at lower cost than motors with integral-mounted coolers.

Fire or explosion hazards require special motor enclosures. Hazards include combustible gases and vapors such as gasoline; dust such as coal, flour, or metals that can explode when suspended in air; and fibers such as textile lint. The kind of motor enclosure used depends on the type of hazard, the type and size of motor, and the probability of a hazardous condition occurring. Some available enclosures are explosionproof motors, which can withstand an internal explosion; force-ventilated motors cooled with air from a safe location; and totally enclosed motors cooled by air-to-water heat exchangers and pressurized with safe air, instrument air, or inert gas.

MOTOR CONTROL

The basic functions of motor starters are:

1. Normal "start-stop" control of the motor.
2. Protection of the motor.
3. Protection of the electrical supply system in the event of a motor or motor-feeder short circuit. The fault must be cleared from the rest of the system to prevent further trouble.
4. Electrical isolation to provide accessibility for maintenance.
5. Provision for other control such as master sequence control, protective shutdown devices (e.g., bearing overtemperature, overtravel, pump high pressure, remote control, etc.).

Types of Starters

High Voltage and Low Voltage The electrical industry has standardized the distinction between high voltage and low voltage at 600 V. Below 600 V the common system voltages in use in the United States are 120, 208, 240, 480, and 600 V. Above 600 V, the standard nominal system voltages commonly in use are 2400, 4160, and 6900 or 7200 V. Higher voltages are available, but the motor cost is usually prohibitive.

A group of low-voltage starters is shown in Fig. 24-4 in a motor-control-center configuration. The same starters are used for any voltage below 600. The insulation is the same for all voltages, since for this equipment there is only one insulation class.

For high-voltage motor-starting applications, there are several classes of insulation: 2500, 5000, 7500, and 15,000 V. The conventional control-type high-voltage motor starter is available for 2500- or 5000-V service. For voltages higher than this, switchgear must be used.

Control-type high-voltage starters are shown in Fig. 24-5. The construction used here employs much greater clearances and provides additional safety features such as grounded barriers between the high- and low-voltage sections of the starter. Extensive mechanical and electrical interlocking is also used for additional safety.

One of the major differences between high- and low-voltage starters is the amount of power handled. An approximate dividing line is 149.2 kW (200 hp). This, however, is not a fixed and rigid rule.

Line Starters and Combination Starters A line starter consists of a contactor (motor-starting switch) and motor-overload relays. Contactors are capable of carrying and interrupting normal motor-starting and -running currents; they are not, however, normally

FIG. 24-4 Motor control center for a group of low-voltage ac motors. Note the modular construction. Starters are draw-out.

FIG. 24-5 Typical high-voltage induction motor starters with air-break contactors for normal start-stop control and fuses for short-circuit protection. Note the protective relays on the upper panels.

capable of interrupting short-circuit currents. They must be backed up by fuses or a circuit breaker for this function.

When a disconnect switch, circuit breaker, or set of fuses is included in the same enclosure as the contactor, the starter is then called a **combination starter.** In addition to the fault-current-interrupting function, the breaker or fuses serve as the disconnecting device. Figure 24-6 illustrates schematically combination starters of various types. The starters pictured in Figs. 24-4 and 24-5 are combination starters. The latch is arranged to open the disconnect before the door can be swung open. There are also provisions for padlocking the disconnect open with the door closed so that maintenance work on the motor may proceed in safety.

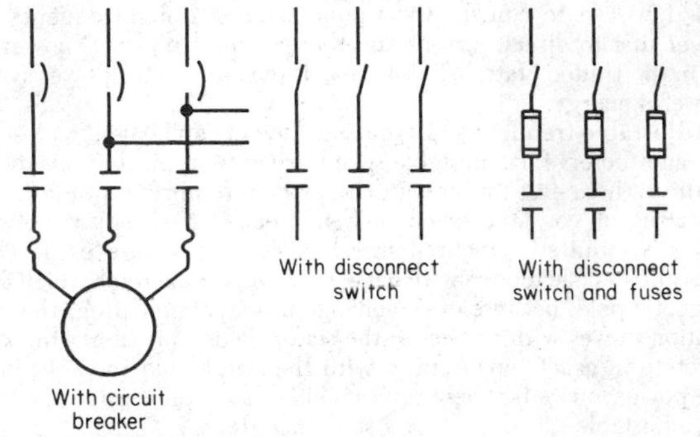

FIG. 24-6 Simplified schematic diagram of a combination line starter with a circuit breaker as the fault interrupter and disconnect. Alternative fuses and disconnect switch are shown as substitutes for the circuit breaker.

Manual and Magnetic Starters Manual motor starters are operated by hand. The simplest type of manual starter is a snap switch with no overload protection, used only for motors of 1.492 kW (2 hp) and smaller, usually single-phase motors with integral overload protection.

Magnetic motor starters are similar in function to manual starters except that they are solenoid-operated. They are available up to 3730 kW (5000 hp). One of the main advantages is the convenience of electrical operation. Start-stop push buttons can be located anywhere. When automatic or remote operation is needed, magnetic starters are essential.

Comparison of Switchgear and Contactor-Type Control Frequently switchgear is used for motor control, particularly for large high-voltage motors. Switchgear (Fig. 24-7) must be used for motors larger than 3357 kW (4500 hp)* at 4160 or 4600 V or 1865 kW (2500 hp)* at 2300 V and for all motors above 5000 V. Switchgear consists of circuit breakers and protective relaying. Circuit breakers are electrical switches designed primarily for their ability to interrupt short-circuit currents. This is one of the major differences from contactors, which are designed principally to handle starting and running currents. Contactors normally depend on a set of fuses or a circuit breaker to handle major faults (short circuits).

Contactors are designed for frequent operation. Circuit breakers are designed for far fewer operations and therefore are never used as motor starters when repetitive operation is required. A typical example of frequent operation is mine-hoist service, in which the motor must be reversed at the end of every hoisting or lowering operation; contactors would be used.

High-voltage ac control-type (contactor) motor starters use fuses to provide short-circuit interrupting capacity. One disadvantage of fuses is that only one fuse may blow. This leaves single phase applied to the motor. Motors will continue to operate with single-phase power but can overheat even with less than rated current flowing. In

*3730 kW (5000 hp) and 2051 kW (2750 hp) for unity power factor synchronous motors.

FIG. 24-7 Typical lineup metal-clad switchgear including motor starters and protective relaying.

contrast to contactors, circuit breakers are three-pole devices: a fault on one phase will trip all three, minimizing the single-phasing problem.

Centralized Control As mentioned previously, motor starters may be located either at the motor or at some remote point. Frequently they are grouped at a location convenient to the source of power. The feeders radiate from this point to the individual motor loads. A convenient method is the control-center (Fig. 24-4) modular structure for low-voltage control, into which are assembled motor starters and other control devices. The individual starters can be drawn out of the structure for rapid, easy maintenance and adjustment. With this construction it is easy to change starter size or add additional starters. All the starters are in one location, so that interwiring is simple and easy to check. Auxiliary relays, control transformers, and other special control devices can also be included. See Fig. 24-8.

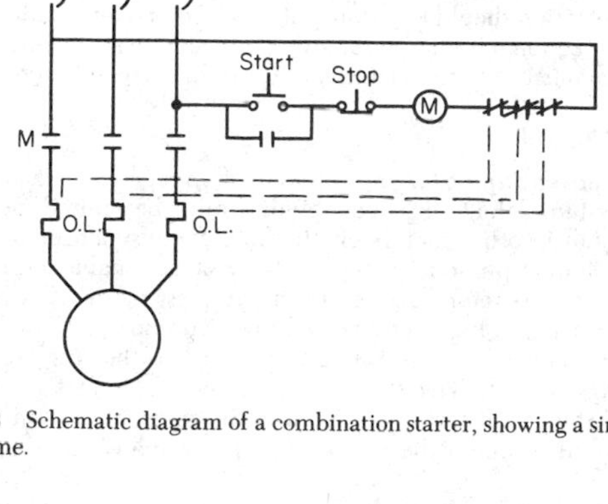

FIG. 24-8 Schematic diagram of a combination starter, showing a simple control scheme.

Motor Protection Money spent for motor-protective devices can be compared to insurance, in which premiums depend on the protected value when the protected value is the cost of the motor, the cost of anticipated repairs, or the cost of downtime, lost production, and, in some cases, contingent damage to other equipment.

Overload Protection Overload relays for protecting motor insulation against excessive temperature are located either in the motor control or in the motor itself. The most common method is to use thermal overcurrent relays in the starter. These relays have heating characteristics similar to those of the motor which they are intended to protect. Either motor current or a current proportional to motor-line current passes through the relays so that relay heating is comparable to motor heating.

Standard thermal overcurrent relays located in the starter have some disadvantages. They cannot detect abnormal temperatures in the motor caused by blocked ventilation passages or high ambient temperature at the motor. They are also likely to trip out unnecessarily in locations where the control enclosure is at a higher temperature than the motor. Motors are normally ventilated with external air so that their ambient temperature is the ambient temperature of the surrounding air. However, control enclosures are not freely ventilated, so their internal temperature can become quite high if they are located in a sunny location High-current relays are sometimes used to avoid this difficulty. This prevents the motor from being tripped out unnecessarily because of high ambients inside the control enclosure, but the motor will be improperly protected during cool weather and overcast days and at night. Ambient-temperature-compensated relays should be used in these situations.

Some overload-protection schemes measure motor-winding temperature directly; various methods are used. Small single-phase motors are available with built-in overload protection. A thermostat built into the motor senses motor-winding temperature directly. When the motor overheats, the thermostat opens, interrupting motor-line current. Pilot thermostats mounted on the windings of larger motors trip the motor starter rather than interrupt line current. This method gives good protection for sustained overloads, but because of the thermal time lag between the copper winding and the thermostat it does not provide adequate protection for stalled conditions or severe overloads.

Temperature detectors embedded in the motor winding give close, accurate indication of motor temperature. Both conventional resistance temperature detectors (RTD) and special thermistors (highly temperature-sensitive nonlinear resistors) are used. With appropriate auxiliaries these devices can indicate or record motor temperature, alarm, and/or shut down the motor.

Short-Circuit Protection Short circuits must be removed promptly to avoid severe damage at the fault and to avoid disturbances to the rest of the electrical system. Short-circuit protection should be set as low as possible so that tripping action is initiated quickly. Motor-starting inrush current sets a limit on how low short-circuit devices may be set. For squirrel-cage ac motors, instantaneous short-circuit tripping should be initiated at about 7 to 10 times full-load running current. This gives an adequate margin above the normal inrush of approximately 6 times full-load current. Modern low-voltage combination starters are available with adjustable instantaneous circuit breakers which can be set just above motor-starting current.

High-voltage contactor-type motor controls depend on power fuses for short-circuit protection. The fuses are coordinated with the overload relays to protect the motor circuit over the full range of fault conditions from overload conditions to solid maximum-current short circuits.

Locked-Rotor Protection Under locked-rotor (stalled) conditions the rotors of large synchronous and squirrel-cage motors are the most likely motor elements to be damaged by overheating. The rotor's heating is not related to stator heating during startup. Therefore for large motors it is common practice to use separate devices or characteristics to protect against running overloads and locked-rotor conditions if the overload and short-circuit protective devices cannot be coordinated to handle this condition for the specific motor characteristics.

Synchronous-motor rotor frequency can be detected because the rotor field circuit is available. Special control schemes have been devised which take into account both speed and induced rotor current in providing locked-rotor and accelerating protection.

Undervoltage Protection If a power outage occurs, it is necessary to remove motors from the line to prevent excessive starting current surges on the electrical system when voltage is reestablished. It is also unsafe to have drives starting indiscriminately when electrical service is reestablished. Conversely, it may be desirable to leave the motors connected during short voltage dips; this is time-delay undervoltage protection. Instantaneous undervoltage protection disconnects the motor as soon as the voltage drops appreciably. This is satisfactory if continuity of operation is relatively unimportant. It is inherent in low-voltage magnetic starters when a power loss drops out all contactors as soon as a voltage dip occurs. If time-delay undervoltage protection is desired for these controls, time-delay relays must be added to the standard control circuit. Because circuit breakers do not drop out on a voltage dip, undervoltage relays are necessary.

Reverse-Phase Protection Reverse-phase relays are used on some large motors to prevent their starting when the electrical-system phase rotation is reversed because of improper wiring or maintenance. They are also used as undervoltage and voltage-balance relays. Individual relays may be applied to each motor in place of the undervoltage relay, or one relay may be operated off a bus for several motors. Individual relays are more expensive but more reliable, particularly when motor circuits are changed frequently. This type of protection is normally supplied only on high-voltage switchgear-type starters.

Phase-Current Balance Protection Three-phase ac motors will usually continue to operate on single phase. Single phase is serious on large ac motors because of the severe rotor heating it causes. Single-phase conditions cannot be detected by measuring voltage; a running motor acts as a generator so that, even under single-phase conditions, motor terminal voltage is nearly normal. Current-balance relays give a positive indication of system current unbalance and single-phase operation. Normally one three-phase relay is used for each motor. The use of these relays is restricted to large motors [approximately 1119 kW (1500 hp) and larger] when the value of the equipment protected justifies the cost of this protection.

Adequate single-phase protection is provided on low-voltage ac motor starters by three overload relays, which are now standard. Rotor heating is not particularly a problem on smaller motors which have more thermal capacity, but it is important to protect the stator windings of these machines against burnout.

Differential Protection Differential protection is applied to detect internal motor faults quickly and limit damage. The cost of this protection is justified on large motors [1119 kW (1500 hp) and above], for which limiting the motor damage may save the cost of this additional protection many times over.

Motor differential protection is one of the most sensitive forms of large-motor protection available. Figure 24-9 illustrates the basic principles involved. All six leads (both ends of all three windings) are brought out to terminals. The electric current entering each winding and the current leaving that winding pass through the same current transformer in opposite directions. If everything is normal, these currents are equal and no current is induced to the current transformer winding. If a phase-to-phase (winding-to-winding) or winding-to-ground short circuit occurs, the currents do not balance, current is induced in the current-transformer winding, and the differential relay operates instantaneously, shutting down the motor. Because of its sensitivity and speed, this system limits motor damage, minimizing repair costs and downtime.

Ground-Fault Protection High-voltage motors (2300 V and above) should be protected with ground-fault relays if the power source is grounded (see Fig. 24-9). This scheme includes a large-diameter current transformer (CT) encircling all three motor leads. Short-circuit current to ground flows through the CT to ground and returns to the power source external to the CT; this unbalance induces current in the CT and ground relay to shut down the motor. With this protection only two overload relays and two line CTs

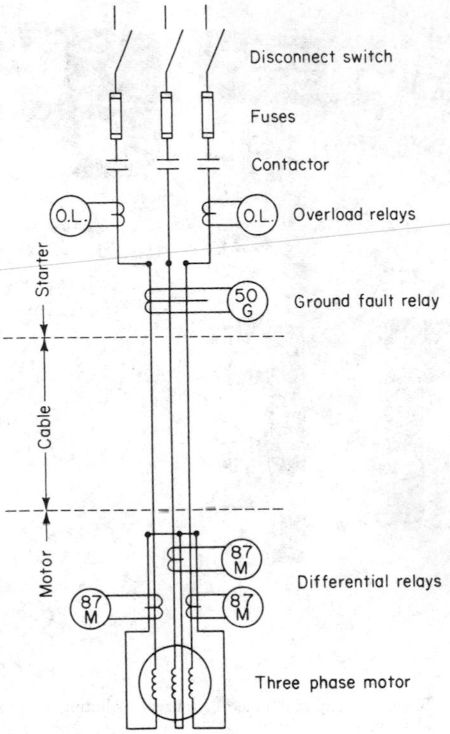

FIG. 24-9 Typical high-voltage ac motor starter illustrating several protective schemes: fuses, overload relays, ground-fault relays, and differential relays with the associated current transformer which act as fault-current sensors. In practice, the differential protection current transformers are located at the motor, but the relays are part of the starter.

(rather than the standard three) are required, so the additional protection is very economical. It cannot be used, however, unless the power source is grounded.

Both differential and ground relaying detect ground faults. Ground-fault protection is located at the starter and protects the cable and the motor; differential CTs are located at the motor and protect the motor only. Economic priorities indicate ground-fault protection first, adding differential protection when justified by potential savings in downtime and repair costs.

Surge Protection High-voltage motors should be equipped with surge-protection apparatus consisting of a set of three lightning arresters and three surge capacitors. Potentially damaging voltage surges or spikes can be generated on the power system by switching operations, certain faults, or lightning. The surge capacitors slope off these steep front voltage spikes, and the lightning arresters limit the peak voltage; both functions are essential for adequate protection. Surge protection should be located at each motor's terminals for maximum protection, although in many instances one set of surge equipment is connected to the electrical bus serving several motors.

Special Control

Reduced-Voltage Starting Reduced-voltage starting is used to reduce system voltage dip. Voltage dips must be limited; otherwise, they may drop other motors off the line, cause synchronous motors on the system to pull out of step, or cause objectionable lamp flicker.

Resistor and reactor starting are the simplest methods of reduced-voltage starting. These systems require two contactors or breakers and a set of reactors or resistors, in contrast to the single-contactor full-voltage starter. The starting contactor closes first, connecting power to the motor terminals through the reactors. The impedance in the circuit reduces the motor terminal voltage and the starting current. As the motor approaches full speed, the running contactor closes, shorting out or bypassing the reactors, applying line voltage to the motor terminals. Starting current is reduced in proportion to the reduction in motor voltage. However, torque is proportional to the square of motor voltage, so starting torque is reduced far more than starting current.

If a greater reduction in line current is required for starting ac motors than is possible with reactor starting, autotransformers may be used. Because of transformer action, the reduction in motor-starting torque is directly proportional to the reduction in line current. Table 24-3 compares reactor and autotransformer starting with respect to line current and torque. Other, less commonly used methods of reduced-voltage starting include part-winding starting and star-delta starters.

TABLE 24-3 Effects of Reduced Voltage Starting*

Starter type	Motor voltage	Motor current	Line or source current	Motor torque	Source voltage dip
Design	100	100	100	100	0
Actual full voltage	80	80	80	64	20
Reactor:					
0.8 tap	67	67	67	45	17
0.65 tap	56	56	56	31	14
0.5 tap	44	44	44	20	11
Autotransformer:					
0.8 tap	69	69	55	48	14
0.65 tap	59	59	38	35	10
0.5 tap	47	47	24	22	6

*Values shown are in percent of design or normal starting values and are calculated for an arbitrary hypothetical power source whose voltage would dip by 20 percent if full-voltage starting were used.

Synchronous-Motor Starters Except for the addition of the synchronous-motor field-application panel, control schemes are identical for both synchronous and induction motors. Excitation is not applied to synchronous motors until they reach approximately 95 percent speed. Field current should be applied when the field poles are in proper space relationship to the stator's rotating magnetic field. Both speed and position are indicated by the ac voltage generated in the field winding. The frequency is directly proportional to slip and therefore indicates speed; the magnitude and polarity of the generated wave indicate position relative to the armature field. When the proper speed and position are detected, field current is applied. When reduced-voltage starting is employed, the ac starting sequence is completed before the application of field current.

Multispeed Alternating-Current Starters Multispeed induction motors are either two-winding motors, single-winding motors with consequent-pole connection or pole-amplitude-modulated motors (see subsection "Alternating-Current Motors Multispeed"). The starters for two-winding, two-speed motors are quite simple; they consist of two standard single-pole, single-speed starters in the same enclosure with appropriate mechanical and electrical interlocks so that the two contactors cannot be closed simultaneously.

Two-speed, single-winding motors, either consequent-pole or pole-amplitude-modulated, require a three-pole and a five-pole contactor mechanically and electrically interlocked. Three- and four-speed, two-winding motors require a combination of two-speed, single-winding and two-speed, two-winding starters. Further modifications are possible by making these multispeed-control-reversing.

Secondary Control of Wound-Rotor Motors Wound-rotor motors may be effectively reduced-voltage-started or have their speed controlled by using external secondary resistance. The addition of resistance into the secondary circuit of a wound-rotor motor reduces the starting current and affects the speed under load conditions.

When external secondary resistance is used for improved starting characteristics, short-time-rated resistors are employed. As the motor accelerates, steps of resistance are cut out on a time or current basis to give the desired accelerating torque and current characteristics.

When external secondary resistance is used for speed adjustment, the resistors may be either infinitely adjustable (e.g., liquid rheostats) or adjustable in steps (if fine speed adjustment is not required).

Direct-Current Motor Control Control for dc motors runs the gamut from simple manual line starters to elaborate regulating systems. Only the starting problems are considered here since variable-speed drives and regulating systems are discussed elsewhere.

The major differences between ac and dc starters are necessitated by the commutation limitation of dc motors, which is the ability of the individual commutator segments to interrupt their share of armature current as each segment moves away from the brushes. Normally 250 to 275 percent of rated current can be commutated safely. Since motor-starting current is limited only by armature resistance, line starting can be used only for very small [approximately 1492-W (2-hp)] dc motors. Otherwise, the commutator would flash over and destroy the motor. External resistance to limit the current must be used in starting to prevent this.

Manual rheostats can be used in series with the motor armature for the current-limiting function. If the rheostat has ample thermal capacity, it can also be used to vary speed. If this system is used, interlocks should be included to prevent closing of the contactor unless maximum resistance is in the circuit.

Magnetic starters short out the starting resistance in one or several steps based on time, current, or speed. The number of steps depends on the size of the motor and the application. Current-limit acceleration is used frequently for high-inertia drives which require a long accelerating time. Motor current is sensed by a current relay which actuates the shorting contactors in sequence as the current drops. Time-limit acceleration is more common. The motor accelerates in a definite time by shorting out the starting resistor steps in timed sequence.

RECIPROCATING ENGINES

STEAM ENGINES

The advent of electric motors, steam turbines, and other drivers has relegated the steam engine to a minor position as an industrial driver. It does have the advantages of reliability and operating characteristics that are not obtainable with other drivers but also the disadvantage of bulkiness and oily exhaust steam.

In the simple **nonexpanding engine** as used with direct-acting reciprocating pumps, steam is admitted over the entire stroke and does not expand in the cylinder, resulting in relatively low efficiency. Control is simple, and pump speed is regulated by steam throttling. By proper selection of the steam and pump piston sizes, these pumps can deliver high shutoff pressures, which can be used to overcome temporary blockages of pipe lines or for other situations requiring high pressure of short duration.

The higher-efficiency **expanding steam engines** use cutoff valves to limit steam admission to the cylinder during the initial part of the stroke, and the expansion occurs during the remainder of the stroke. Larger engines use several cylinders in series to achieve full expansion. Although this type of engine can be controlled with throttling valves, the preferred method is to change the cutoff point, thus eliminating throttle-valve losses and permitting change in output from zero to maximum design power. Thus almost complete expansion is achieved at part loads and overloads, resulting in efficient operation for the full range of loadings.

Although this type of engine is efficient, it is limited by its inability to utilize low-vacuum exhaust and/or high steam pressures and temperatures commonly used by steam turbines. With low steam pressures [2068 kN/m² (300 psig)] and low vacuums [88 kN/m² (26 inHg)] a steam engine will have a better efficiency than a steam turbine of the same rated power. For each cutoff setting an engine will develop the same torque at all speeds, with steam consumption and power output directly proportional to speed. All other drivers, except certain dc electric motors, require the same input for constant torque at varying speeds.

Changing the cutoff-point setting will allow an engine designed for one gas to operate efficiently on any other gas (limited only by corrosiveness, fouling, etc.). This characteristic favors the use of reciprocating engines when it is necessary to expand gases efficiently in process applications in which the composition is variable. A further advantage is that an engine is the only driver with essentially zero gas consumption at zero speed while developing full torque and maintaining full process pressures. A combination of an expansion engine driving an oil brake provides a high-efficiency refrigeration effect over a wide range of process conditions, with the process controllers throttling the oil flow in the oil brake.

The **uniflow design** reduces cylinder condensation and also allows greater expansion ratios per cylinder (see Figs. 24-10 and 24-11).

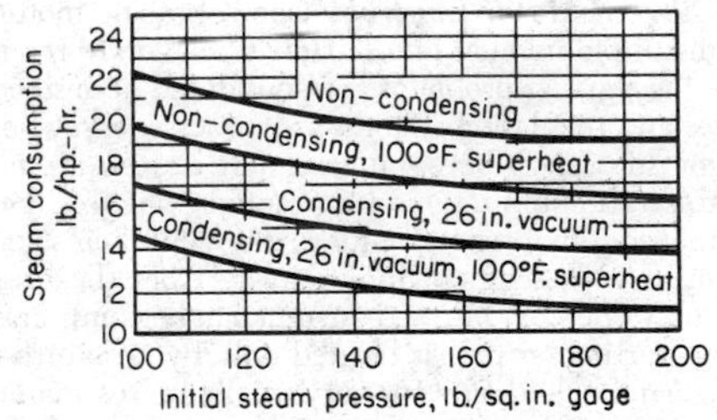

FIG. 24-10 Steam consumption of a 400-hp uniflow engine. To convert pounds per horsepower-hour to kilograms per kilowatthour, multiply by 0.6084; to convert pounds-force per square inch to megapascals, multiply by 6.89 × 10^{-3}.

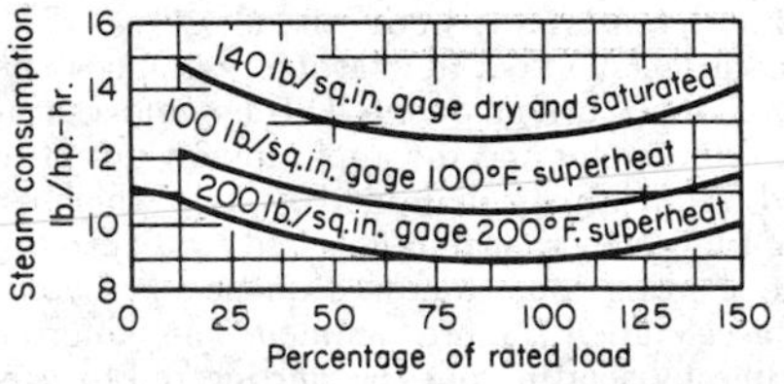

FIG. 24-11 Steam consumption of a uniflow engine with 27.5-in vacuum. To convert pounds per horsepower-hour to kilograms per kilowatthour, multiply by 0.6084; to convert pounds-force per square inch to megapascals, multiply by 6.89 × 10^{-3}.

Steam is admitted during the start of the power stroke and after cutoff is expanded down to a pressure slightly higher than the exhaust pressure. At the end of the stroke the piston uncovers the exhaust ports of the cylinder with partial steam and discharges into the exhaust system. The steam remaining in the cylinder is compressed during the return piston stroke, maintaining higher average cylinder temperatures in order to reduce steam condensation during admission.

To reduce friction and cylinder wear oil is injected into the cylinders of engines, and to maintain lubricity cylinders of engines on low-temperature services are warmed. Oil causes foaming in boilers and can contaminate low-temperature process streams. Therefore, in steam plants oil is usually removed from the condensate. By using carbon or plastic rings similar to those used in oil-free reciprocating compressors, oil for lubrication can be omitted. But these rings are not as reliable nor do they have as good a life as lubricated cast-iron piston rings.

INTERNAL-COMBUSTION ENGINES

Internal-combustion engines range in size from small portable gasoline engines to large [14,914-kW (20,000-hp)] diesels for ship propulsion. They are usually designed for particular industrial applications and to meet specific objectives as to weight per horsepower, reliability, and operating conditions.

All internal-combustion engines fall into two main types, namely, four-cycle and two-cycle engines. These engines may be further classified as (1) gasoline or gas engines (Otto cycle), in which a spark plug is used to ignite a premixed fuel-air mixture; (2) diesel engines (diesel cycle), in which high-pressure compression raises the air temperature to the ignition temperature of the injected fuel oil; (3) dual-fuel or gas-diesel engines, in which the fuel is a combination of gas and oil in any desired ratio, provided that at least 5 percent oil is used at all times; and (4) trifuel engines, which can operate as dual-fuel or as straight gas engines by replacing the oil-injection system with a spark plug for ignition.

Design Characteristics Internal-combustion engines involve consideration of the following design features: (1) **Compression ratio,** an increase of which usually increases engine efficiency but also results in higher average cycle temperatures and therefore hotter cylinders, piston heads, and rings which increases the difficulty of piston lubrication. (2) **Piston speed,** which is a major overall criterion of engine design since power rating is proportional to piston speed. Reciprocating forces and lubrication problems increase with piston speed. (3) **Brake mean effective pressure** (bmep), which is an overall measure of the output of an engine frame; bmep increases with compression ratio and degree of supercharging, and, in general, high values are associated with modern high-efficiency industrial engines. It is also the overall criterion for bearing loadings and average piston-head and cylinder temperature. (4) **Engine rating,** which is proportional to the product of piston speed and bmep. Acceptable values of piston speed and bmep are more dependent on industrial usage than on the type of engine. (5) **Supercharging,** which connotes means for increasing the inlet manifold air pressure above ambient pressure. (6) **Turbocharging,** which applies to the expansion of the hot exhaust gases through a turbine to drive the supercharging compressor. Since power output is proportional to inlet manifold air pressure, supercharging provides increased power output and usually higher efficiency. Highly supercharged engines may require an air cooler between the compressor and inlet manifold.

Engine size is usually in terms of power rating (horsepower) rather than the physical size of the engine frame. Frame size is determined by the diameter of the cylinder bore, length of stroke, and number of cylinders. The power rating of a given frame varies with industrial practice and usage; accordingly, automobile engines are designed to develop between 0.54 and 0.95 hp/in³ of piston displacement, while industrial diesels are limited to 0.045 to 0.175 hp/in³.

Engine rating reflects industrial practice. Automobile engine rating is the peak horsepower developed on a test stand, whereas industrial engine rating is usually in terms of continuous load.

Industrial engines are made in a wide variety of frame sizes, each offered at several power and speed ratings. The same engine frame might drive either a 900-kW generator at 900 r/min or a 600-kW generator at 600 r/min by using lighter pistons and special parts for the higher-speed application. An engine originally designed for 746 kW (1000 hp) may be uprated to 820 kW (1100 hp) by higher compression heads and further to 984 kW (1320 hp) by supercharging. Thus one frame can cover a power range from 447 kW (600 hp) (60 percent speed) to 984 kW (1320 hp). An engine frame can also be designed for different output by varying the number of cylinders; for example, a 10-cylinder 2610-kW (3500-hp) engine will develop 1566 kW (2100 hp) and 2088 kW (2800 hp) at the same efficiency with six and eight cylinders respectively. This enables a manufacturer to use the same basic design, tools, and fixtures for a variety of ratings.

The published power rating of American industrial engines is usually at 4.57 m (1500 ft) above sea level and at 32°C (90°F) air temperature. It is important that atmospheric conditions at the installation site be specified, since engines operate with very little excess combustion air and consequently maximum power output is proportional to air density.

Operating Characteristics The operating characteristics of internal-combustion engines are basically the same regardless of fuel used or whether the engine is two- or four-cycle. To vary the speed and/or load-carrying capacity, which can range from zero to full torque for all speeds within the operating range, it is necessary only to vary the fuel input. Large engines use a governor to control the fuel-input rate and maintain constant speed under variations of load. With auxiliary instrumentation the governor can be used as a controller of power or process. Starting, stopping, and operation from a remote location are made possible by instrumentation that will also shut down an engine in the event of loss of cooling-water or lubricating-oil flow.

Starting is accomplished by rotating the engine at a speed sufficient to achieve ignition and self-sustained operation. Small engines are started with electric motors or smaller hand-starting engines, and large engines are provided with special valving whereby some of the engine cylinders can be operated as air motors, utilizing high-pressure air to rotate the engine. Starting motors are usually sized at 5 to 10 percent of the engine rating. Starting-air requirements are 0.014 m^3 (0.5 ft^3) (free air)/hp stored at a pressure of 1723 kN/m^2 (250 psig) to 2068 kN/m^2 (300 psig). Usually three starts per successful firing are assumed for sizing the starting air compressor and air-storage vessel.

Operating characteristics of engines are also influenced by **service requirements.** Automotive engines must develop maximum torque at lower speeds for hill climbing; marine engines are required to develop full torque only when the propeller is at full speed; generator drivers run at a single speed and therefore require maximum possible efficiency at all loads for this speed; and reciprocating compressors at constant pressure require full torque over the entire operating-speed range, thus presenting an engine-instability problem at reduced speeds. The engine must be selected to meet these service requirements.

Supercharging is employed primarily to improve engine efficiency or power output, but part-load performance and rate of response to load changes depend on the type of supercharging system used. Some systems result in supercharged engines having the same torque-speed characteristics and rate of response to load changes as nonsupercharged engines, while other systems result in poorer or better response and in different speed-torque characteristics.

Maintenance and Reliability Preventive maintenance requires that all engines be shut down at periodic intervals for inspection and repair. For properly maintained heavy-duty engines availability is over 97 percent, with maintenance costs of $1.50 to $3 per horsepower-year and lubricating-oil consumption of 1 to 2 gal/hp·year.* While this represents a high degree of reliability, outages of heavy-duty engines are more frequent than those of electric motors or steam turbines.

Satisfactory engine performance is assured by maintaining a log that allows comparison of present characteristics with past records of (1) cylinder exhaust, cooling water, and supercharger exhaust temperature for similar loadings, (2) running sounds and smokiness of exhaust, and (3) engine efficiency and losses.

The minimum safety devices for an industrial engine are a governor and separate overspeed and low-oil-pressure trips. Engines operating in nonsupervised areas should be arranged to shut down in the event of cooling-water or lubricating-oil failures and for excessive exhaust or jacket-water temperatures. Engines operating in supervised areas should be provided with instruments for the operator to check performance.

*For further details of industrial operating costs see *Annual Oil and Gas Engine Power Costs*, published by the American Society of Mechanical Engineers. Engine manufacturers will supply recommended schedules for preventive maintenance.

Fuel Characteristics Fuels used in industrial engines of the internal-combustion type are usually derivatives of petroleum or else natural or manufactured gases. Alcohols and mixtures of gasoline and alcohol or benzol can also be used. A gas engine will operate satisfactorily on any gas which is free of dust, noncorrosive (i.e., less than 0.6 grains/ft^3), does not detonate, does not preignite during compression stroke, and produces enough heat on burning to develop power.

In general, the fuel must have a heat capacity of over 600 Btu/ft^3. Gasoline engines require, in addition, that the fuel will vaporize in the carburetor. Diesels will burn any fuel that can be injected, provided that it will burn under controlled conditions, possesses sufficient lubricity to lubricate the injection plungers, will supply enough heat, and is grit-free, containing less than 3 percent sulfur, 70 ppm vanadium, and 125 ppm vanadium pentoxide. Most diesel engines use either No. 2 or No. 5 fuel oil. The latter must be heated to a viscosity of 50 to 70 SSU [121°C (250°F) approximately] for proper injector lubrication and injection characteristics.

Gaseous fuels containing fractions whose ignition temperature is lower than that of methane may require the use of low-compression heads and a resulting derating of the gas engine.

The method of reporting fuel consumption varies among different industries and also among countries. Trade associations usually have recommended procedures. Thus the Diesel Engine Manufacturers Association (United States) calculates efficiencies based on the lower heating value (LHV) for gas fuels and the higher heating value for oil fuels. It is general practice to report gas-engine performance in terms of British thermal units per horsepower-hour (LHV) and oil-engine performance in terms of pounds of fuel consumed per horsepower-hour. For electric power plants, fuel consumption is reported in terms of kilowatts. Auxiliaries included with engine-efficiency calculations vary with industry practice.

Fuel Economy and Heat Recovery The high efficiencies obtained by modern high-compression supercharged diesels or gas engines can be approached only by very large high-pressure reheat steam plants or by very complicated gas-turbine cycles. The following efficiencies (LHV) based on methane fuel for gas engines and oil fuel for diesel engines can be used for estimating fuel consumption: 63 kW (85 hp) to 298 kW (400 hp), 28 percent; 328 kW (440 hp) to 597 kW (800 hp), 32 percent; 597 kW (800 hp) to 2237 kW (3000 hp), 36 percent; and over 2461 kW (3300 hp), 38 percent.

Plant fuel costs chargeable to power production can be reduced if heat losses can be utilized to provide process or other heating. Engine heat losses as percentages of heat input (LHV) are (1) lubricating-oil cooling, 5 to 7 percent available at 82°C (165°F); (2) jacket cooling, 17 to 30 percent available as 82°C water or with vapor-phase cooling as 103 kN/m^2 (15 psig) steam; (3) engine exhaust gases, 26 to 30 percent available with approximately half of this at sufficient temperature to generate 689.5 kN/m^2 (100 psig) steam in waste-heat boilers. Table 24-4 gives some heat-balance data for gas engines.

Vapor-phase cooling reduces the cost of the cooling system, increases heat recovery, and may result in improved engine efficiency.

At full speed fuel consumption decreases linearly with reduction in load, becoming at zero load almost one-third to one-fourth of full-load consumption and with no potential for heat recovery in the exhaust. Jacket-water heat losses decrease by 20 to 40 percent at zero load. Fuel consumption at rated torque is almost proportional to speed. At half speed exhaust recovery decreases to less than half, and jacket-water cooling becomes more than half of full-torque values.

Installation and Costs An engine installation includes auxiliary equipment necessary for operation, such as lubrication pumps and attendant storage, filtering and cooling equipment; jacket-water pumps with expansion tank and cooler; starting-air tanks and compressors; inlet-air piping and screens; exhaust-air piping and silencers or waste-heat boilers; a fuel system, which in the case of diesels would include a day tank, filters, pumps, heaters, and a main storage tank; ignition system or fuel-injection plungers; and cooling towers or radiators. Diesel engines operating on heavy residual fuel oils would have two-fuel systems, a heavy-fuel-oil system for normal operation and a light-fuel-oil system for starting and stopping.

Pipe-line and marine installations are frequently arranged so that

TABLE 24-4 Approximate Heat Balance for Gas Engines*

Manifold type	Two cycle: Wet† B.t.u./hp.	%	Two cycle: Dry† B.t.u./hp.	%	Four cycle: Atmospheric wet B.t.u./hp.	%	Four cycle: Supercharged dry B.t.u./hp.	%
Heat input (LHV)	7200	....	6900	....	7700	....	6200	
Work	2545	35.4	2545	36.9	2545	33.1	2545	41.0
Jacket	1700	23.6	1200	17.4	2200	28.6	1250	20.2
Lube oil	460	6.4	500	7.2	450	5.8	300	4.8
Exhaust	2095	29.1	1795	26.0	2300	29.9	1685	27.2
Radiation	400	5.5	860	12.5	205	2.6	200	3.2
Combustion	....	....	....	....	....	....	220	3.5

*Courtesy of Cooper-Bessemer Corp. To convert British thermal units per horsepower to kilojoules per kilowatt, multiply by 1.4148.
†Wet type refers to water-cooled manifolds, and dry type refers to air-cooled manifolds.

the engine drives all its auxiliaries from the crankshaft by means of chains and V belts. But process-plant practice is to have all the auxiliaries independently driven, using standby pumps to minimize engine downtime.

Foundations should be designed to control **vibrating motion** resulting from reciprocating masses. Engine manufacturers will recommend the size of the foundation, but usually their recommendations do not take soil properties into account and are based on making the combined engine and foundation weight sufficiently large to limit vibration. When possible, foundations should be separate from the building structure. In many cases vibration of the engine will cause no damage; nevertheless, it is good practice to reduce it whenever possible. Even though vibration does not increase any forces in the engine, it can loosen pipe joints, nuts, etc.

Torsional vibration can also be a problem and results from pressure variations in the cylinders which can produce cyclic torques with harmonics ranging from half speed to 10 or 12 times running speed. To avoid operating difficulties, the torsional critical frequencies of the combined engine and driven equipment should be calculated or measured to assure that operating speeds are removed from these criticals or that vibration dampers are provided or that the equipment is designed for the resulting cyclic stresses.

The costs of both engines and auxiliaries are reasonably consistent on the basis of dollars per horsepower as long as essential details are the same. Published figures on **installed engine costs** are often misleading, since with supercharging more power output can be obtained from the same size of engine, which also reduces cooling-water and foundation requirements. Pipe-line compressor costs frequently include piping, buildings, etc.; and in some process plants, cooling water which has been used and charged against process operation can be reused for engine cooling at no cost. It is obvious that general cost predictions must be used with caution unless their detailed basis is known. However, as preliminary figures, the following are suggested (1981 basis):

Integral engine compressors:	
Uninstalled and without auxiliaries	$200/hp
Installed cost with auxiliaries and cooling water	$338/hp
Installed costs of large units with cooling water supplied from process units	$313/hp
Diesel or gas-engine generators:	
Uninstalled	$169/kW
Installed	$300–$338/kW

Table 24-5 gives costs for engine installation, which can be prorated to make preliminary estimates of installation costs.

FREE-PISTON GAS GENERATOR

The free-piston gas generator is a relatively simple device that uses "bounce cylinders" to return two power pistons to their starting positions. The power pistons form a highly supercharged variable-stroke opposed-piston type of diesel engine in which each power piston drives an air-compressor piston to produce pressurized air from which a turbine develops power. Industrial installation began in 1951 with an electric power plant using free-piston gas generators to drive a turbine. Succeeding applications include pumping stations, petrochemical-plant drivers, locomotive drivers, and marine-propulsion units.

TABLE 24-5 Comparative Installation Costs of Integral-Engine Compressors for Pipe-Line Stations and Process Plants*

	Pipe-line station	Process plant
A. Land and improvements	$ 328,000	$ 48,500
B. Structures	648,000	346,000
C. Testing	25,000	25,000
D. Equipment	4,900,000	3,625,000
Subtotal	$5,900,000	$4,045,000
Add 10% for overhead and undistributed field costs	590,000	404,500
Subtotal	$6,490,000	$4,449,500
Add 5% for contingencies	324,500	Not used
Total	$6,814,500	$4,449,500
Cost per horsepower	487	318

*10 units are assumed for a total of 10.4 MW (14,000 hp). Basis year is 1981.

Design Characteristics The characteristics of the power turbine are similar to those of steam turbines, as discussed later in this section. The air compressor of the gasifier is similar to a reciprocating air compressor and differs only in its higher piston speeds. The power cylinders form a two-cycle opposed-piston diesel engine, differing only in elimination of the crankshaft.

Absence of a crankshaft provides the free-piston gas generator with certain unique characteristics, namely, variable piston stroke and piston speed that is independent of the turbine or other power output. While crankshaft elimination avoids high bearing loads, it does not avoid the problem of high average cycle temperatures present in highly supercharged engines. In present designs, the power cylinders are water-jacketed and large amounts of oil are circulated to cool the pistons and minimize ring wear.

Potentially, a free-piston turbine plant can operate at an efficiency higher than that of a diesel; however, this is accomplished only at extremely high supercharging pressures, which result in poor ring life. Early failures were probably due to seeking ultimate efficiencies. Present models have a fuel consumption of 0.38 to 0.40 lb/(bhp·h), which is considered a good efficiency, although higher than corresponding values for diesel engines.

Operating Characteristics The operating cycle is described in Fig. 24-12. In essence, a free-piston gasifier (or compressor) is a spring-mass system, with the air in the bounce cylinder providing the spring effect and the pistons the mass. The spring-mass ratio determines the speed in strokes per minute. Idling speeds of 400 cycles/min are attained at low bounce-cylinder pressures and speeds of 600 cycles/min at full pressure. Varying the fuel input changes the available forces and results in a change of the piston stroke and speed, which determines the amount of air that is compressed.

In a simple system consisting of one gasifier and one turbine, the amount of fuel can be decreased from full power requirements down to a point at which the free-piston compressor becomes unstable. Any further reductions in power can be obtained only by using a split-range controller that vents gas to the atmosphere. The split-range controller may operate on turbine speed or on some process factor such as flow or pressure.

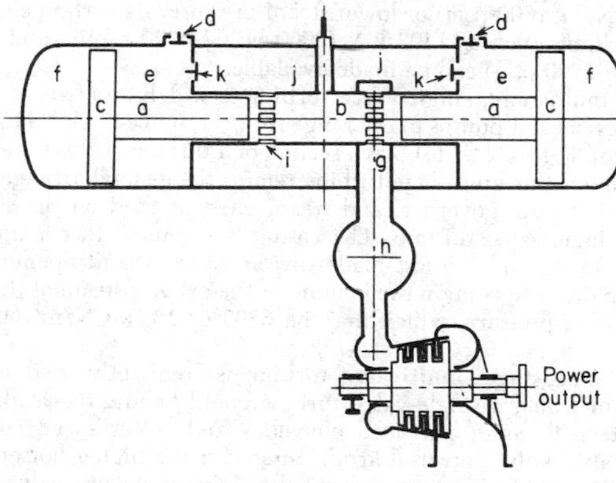

FIG. 24-12 Free-piston compressor. When pistons (*a*) are at their innermost position, fuel oil is injected into the power cylinder (*b*). The resulting rise in pressure in the power cylinder (*b*) forces the power pistons (*a*) and the air-compressor systems (*c*) in an outward direction, drawing air through valves (*d*) into the air-compressor cylinder (*e*) and at the same time compressing air in the bounce cylinder (*f*). Near the end of the piston stroke, the exhaust ports (*g*) are uncovered, allowing products of combustion to flow outward through pipe (*h*) to the power turbine. Further movement of the power piston uncovers the air-inlet ports (*i*), allowing air to flow from the chamber (*j*) into the power cylinder (*b*) through the exhaust ports and to the power turbine via the exhaust pipe (*h*). The power cylinder (*b*) now has a charge of fresh air. The pressure in the bounce cylinders (*f*) returns the two power pistons (*a*) and their connected air-compressor cylinders (*c*) toward the center, the inward movement of the air-compressor piston (*c*) discharging the air through valves (*k*) into the chamber (*j*). The inward movement of the power pistons (*a*) compresses the air in the power cylinder (*b*), making it ready for a new power stroke.

In large single-turbine plants, each gasifier is usually connected to a separate group of first-stage turbine nozzles. To decrease power output, fuel input can be lowered to all the gasifiers, or successive gasifiers can be shut down to maintain high-efficiency performance from the gasifiers left in operation. Blowoff valves are used for transient part-load operation.

In large multiturbine installations, the gasifiers discharge into a common header whose pressure is maintained by a split-range pressure controller which positions the fuel racks and the blowoff valves. Each turbine is controlled by its governor, which throttles the gas as required.

In each case, operating characteristics depend on the control arrangement. When the gasifier is controlled by a back-pressure controller and the turbine by its own governor, performance of the power turbine is similar to that of a steam turbine operating on constant-pressure steam. Gasifiers controlled by a turbine governor respond quickly to power swings as changes in fuel setting vary the length of the next stroke. The only lag in response is the interval required to pressurize the gas header.

Reduction of gasifier output pressure results in reduced turbine output power and at the same time in lower mean effective pressures, which increase ring life. Conversely, an increase in pressure results in more turbine power and shorter ring life but will not stall the gasifiers until high back pressures are attained. Almost all the work output of the power pistons is absorbed by the air-compressing pistons. At normal output pressures the air compressor discharges excess air through the power cylinders. At high back pressure the volumetric efficiency of the compressor decreases, resulting in less air to the power cylinders. The engine continues to operate with increasing back pressure until a stalling pressure is reached.

Installation and Operation Inertia forces are balanced by the two pistons moving simultaneously in opposite directions, thus minimizing foundation requirements, and without a crankshaft mechanical torsional vibrations do not exist. The pulsating gas outflow involves the same problems as are encountered with reciprocating compressors in exciting vibrations with possible resonance in connected piping.

With a single gasifier serving a turbine, pressure fluctuation can excite torsional vibrations in the turbine and driven equipment. To avoid resonance the torsional critical frequencies of the turbine and driven equipment must differ from the frequencies of the exciting forces which are multiples of the strokes per minute. The exciting frequencies of multigasifier installations range from a condition in which all units fire simultaneously through various patterns involving spaced firing between units.

Gasifiers must be shut down for periodic maintenance, and the percentage availability of a single unit appears to be lower than that of a diesel. However, the drivers have the high availabilities usual with turbines, and the use of standby gasifiers will therefore result in a high degree of availability of a complete plant.

Fuels and Heat Recovery The products of combustion from the gasifier contain as much as 15 percent oxygen (by volume). This oxygen can be used in firing a pressurized boiler ahead of the turbine or in an atmospheric-pressure boiler in the turbine exhaust. It is also possible simply to heat the gases and thus achieve higher turbine-power output. These arrangements result in higher-efficiency plants but at an increased investment cost.

Heat recovery from cooling-water and lubricating-oil circuits is limited to low temperatures [49°C (120°F.)], which are much lower than for diesels.

Most installations use No. 2 diesel fuels, and some report successful use of heavier fuels. Higher air pressure favors more complete combustion, and some authorities therefore contend that a gasifier can burn heavier oils than a diesel. This favorable pressure condition is partly offset, however, by the short combustion time resulting from high piston speeds.

STEAM TURBINES

Steam turbines are divided into two broad categories: those used for generating **electric power** and general-purpose units used for driving pumps, compressors, etc., and frequently called **mechanical-drive turbines.**

Figure 24-13 illustrates in general the relationship of capability versus speed. At 1800 and 3600 r/min are the turboelectric generator drives with capability limits above the top of the chart. The majority of mechanical-drive applications are within the shaded area; capabilities above the solid line are special and unusual.

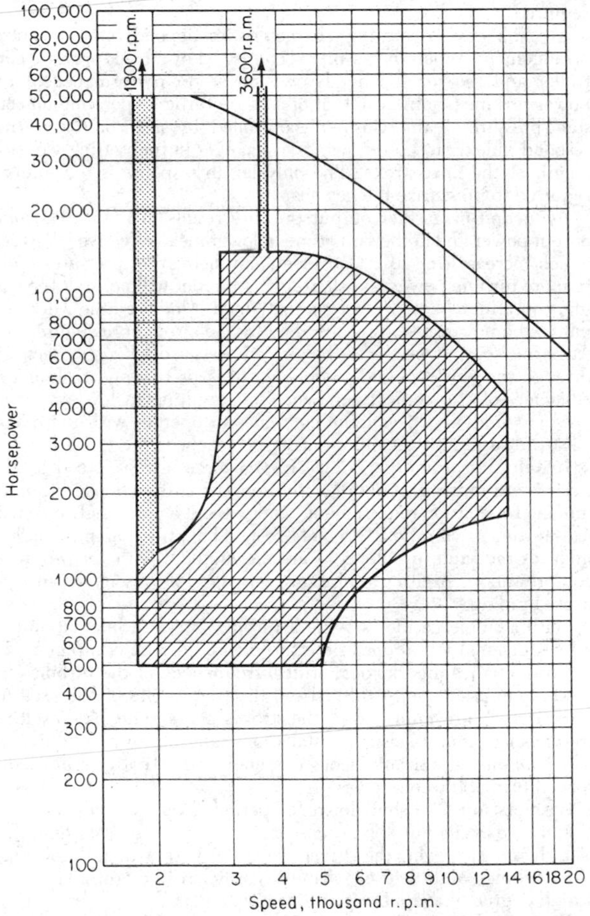

FIG. 24-13 Steam-turbine capability versus speed. To convert horsepower to watts, multiply by 745.7.

Inlet-steam pressure is usually in the range of 1723 kN/m^2 (250 psig) at zero superheat to 5860 kN/m^2 (850 psig) at 482°C (900°F). Some turbines have been built to operate at 35 kN/m^2 (5 psig) with zero superheat from a process exhaust. Pressures of 10,342, 12,410, and 16,547 kN/m^2 (1500, 1800, and 2400 psig) are common for large turbine generators, and some operate at supercritical pressures of 24,131 kN/m^2 (3500 psig) and 34,474 kN/m^2) (5000 psig). Power plants that generate steam with nuclear reactors generate saturated steam in the range of 1379 to 6895 kN/m^2 (200 to 1000 psig). Early units were 125 MW, but currently 250 to 1000 MW are the most common size. These units have multiple casings and 1.32-m-(52-in-) long blades in 1800-r/min exhaust stages.

TYPES OF STEAM TURBINES

Straight Condensing Turbine All the steam enters the turbine at one pressure, and all the steam leaves the turbine exhaust at a pressure below atmosphere.

Straight Noncondensing Turbine All the steam enters the turbine at one pressure, and all the steam leaves the turbine exhaust at a pressure equal to or greater than atmosphere.

Nonautomatic-Extraction Turbine, Condensing or Noncondensing Steam is extracted from one or more stages, but without means for controlling the pressures of the extracted steam.

Automatic-Extraction Turbine, Condensing or Noncondensing Steam is extracted from one or more stages with means for controlling the pressures of the extracted steam.

Automatic-Extraction-Induction Turbine, Condensing or Noncondensing Steam is extracted from or inducted into one or more stages with means for controlling the pressures of the extraction and/or induction steam.

Mixed-Pressure Turbine, Condensing or Noncondensing Steam enters the turbine at two or more pressures through separate inlet openings with means for controlling the inlet-steam pressures.

Reheat Turbine After the steam has expanded through several stages, it leaves the turbine and passes through a section of the boiler, where superheat is added. The superheated steam is then returned to the turbine for further expansion.

STAGE AND VALVE OPTIONS

The single-stage, single-valve turbine is the simplest turbine, and it sees the most varied application. There is a single governor valve in a steam chest, operated directly from a mechanical flyball governor. After passing through the valve, steam is expanded through the nozzles, where it gains velocity and momentum for driving the wheels by impulse action against the blades. The shaft is sealed by carbon rings. The bearings are ring-oiled. The majority of applications are below 1119 kW (1500 hp) and at speeds below 5500 r/min with 4137-kN/m^2 (600-psig) or lower steam pressure. By certain changes higher limits such as 1492 kW (2000 hp), 10,000 r/min, and 8274 kN/m^2 (1200 psig) can be made available.

The multistage, single-valve turbine is widely used for driving compressors and pumps in the range from 1119 to 4474 kW (1500 to 6000 hp). Figure 24-14 shows a section of a turbine of this type. The inlet end of this multistage turbine retains the general arrangement of bearing case, governor, and steam chest as used on the single-stage, single-valve turbine. The casing is extended to contain the added stages, and the last blade row and the exhaust opening are large in order to contain the volume of the exhaust steam at the low condensing pressure, which may be 6895 or 13,790 N/m^2 (1 or 2 psia).

The multistage, multivalve turbine is frequently used in the range in which a single-valve turbine could handle the load. The advantage is better part-load efficiency and better speed control. Multivalve is the universal arrangement for the higher-horsepower outputs to avoid a large governing valve that might cause valve problems. There is no exact dividing line; special situations may allow single-valve units to be used at double the limit given here. Different valve arrangements are used on multivalve turbines, and complexity increases with steam conditions and volume. The following are the principal arrangements that are used as the steam flow increases:

1. Steam chest in upper half with bar-lift valves
2. Steam chest in upper half with cam-lift valves
3. Steam chest in both upper and lower halves with cam-lift
4. Separate side-mounted steam chest

The changeover from one arrangement to the next varies with the several manufacturers. The arrangement of separate side-mounted steam chest and valves is characteristic of very large generator drives. Also, as the flow increases, it becomes too much for one row of exhaust blades, and double-flow or multiple-exhaust casings are used.

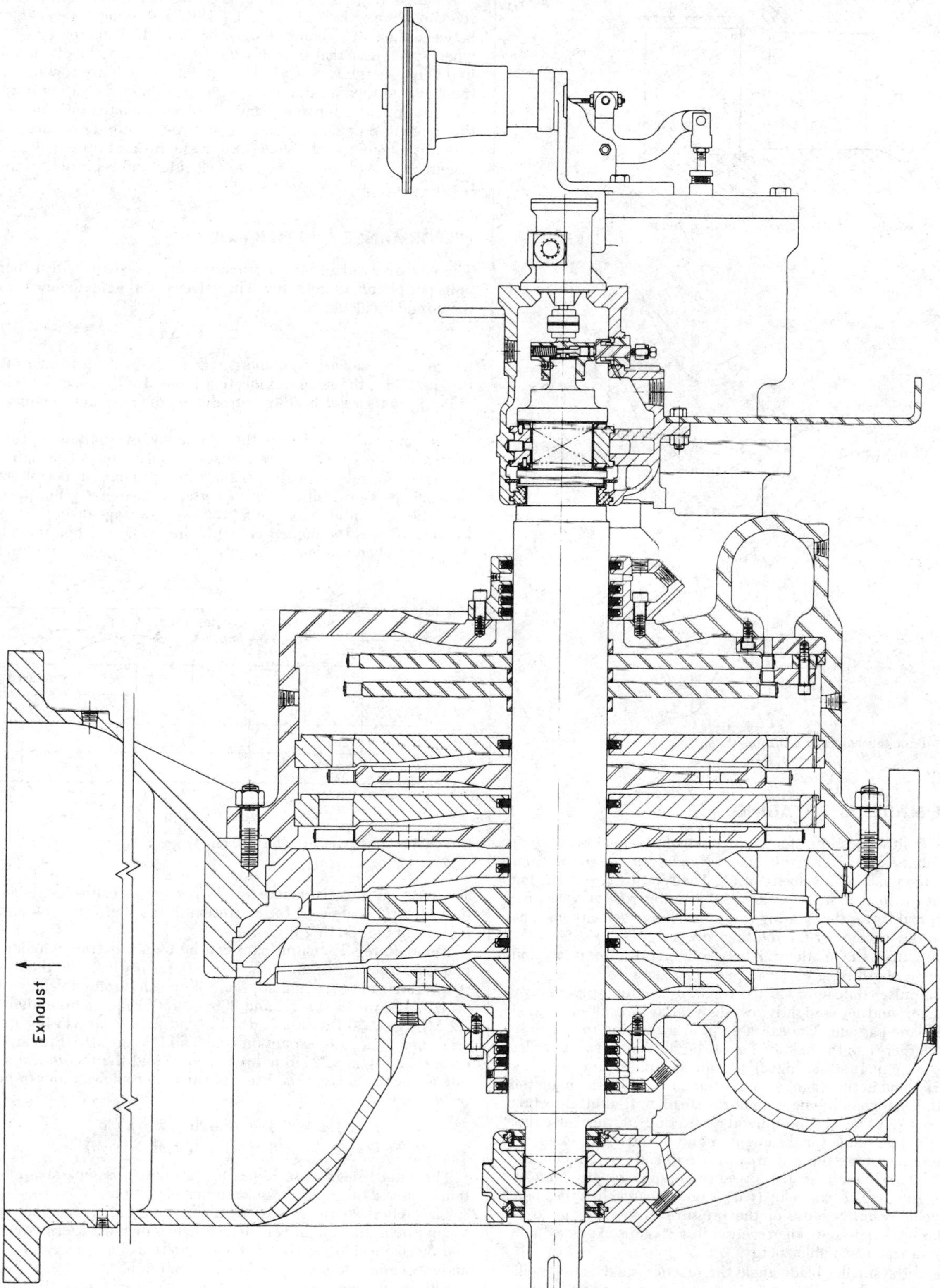

FIG. 24-14 Single-valve, multistage steam turbine. (*Elliott.*)

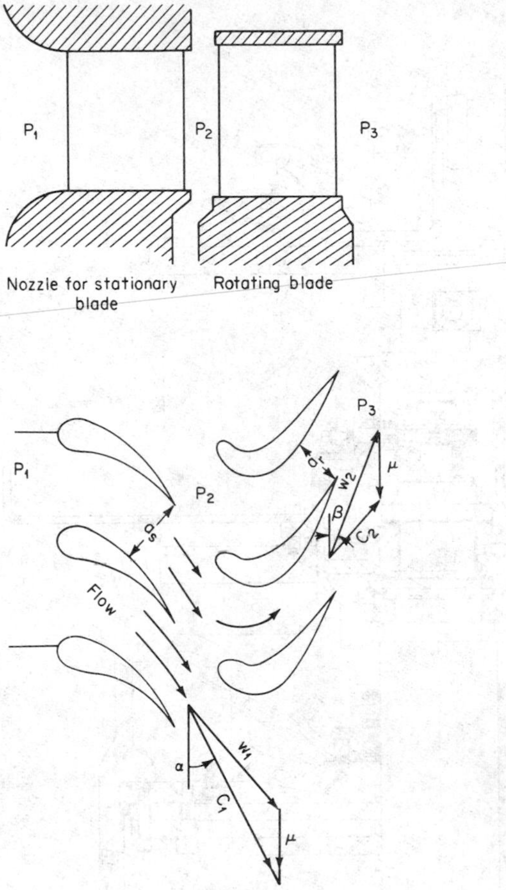

FIG. 24-15 Basic mechanics of a turbine stage.

TYPES OF BLADES AND STAGING

Figure 24-15 illustrates a turbine stage in which steam at pressure p_1 enters the nozzle or stationary blade and expands to a lower pressure p_2, leaving the nozzle at a velocity of C_1. The rotating row is moving at a velocity μ so that steam enters the rotating row at a relative velocity w_1 and leaves the rotating row at a relative velocity w_2. The pressure on the exit side of the rotating row is p_3.

Depending upon the relationship between the pressures p_1, p_2, and p_3, the stage is classified as either impulse or reaction.

For an **impulse** (Rateau or Curtis) **stage,** p_2 is equal to p_3 or only slightly higher, and w_2 is slightly less than w_1 as a result of friction loss in the blade passage. The exit area a_r of a rotating row is 50 to 74 percent larger than the exit area a_s of the stationary row in order to pass the same quantity of higher-specific-volume steam.

The work done in the stage, which is the push of the steam on the blades, is the change in momentum of the steam as it alters direction from C_1 to C_2, using the peripheral projection of the velocities. Therefore, it is desirable for the angles α and β to be very small.

For a **reaction stage** the exit area a_r is reduced by reducing the angle β. This will increase the pressure p_2, possibly to midway between p_1 and p_3. The exit velocity w_2 is now greater than the blade entrance velocity w_1 because of the pressure drop from p_2 to p_3 through the blade passage. The reaction in a stage is expressed as a percentage of stage available energy.

Because of the smaller blade angle the reaction stage is more efficient than the impulse stage, but it requires more stages for the same pressure drop. This will increase the losses and the leakage, and the choice is generally close to a standoff.

A **Curtis stage** is an impulse stage that makes use of two rotating rows to absorb the energy in C_1 with a stationary reversing row between them. This comes about when C_1 is high compared with the wheel speed μ, so that the exit velocity C_2 still has a lot of energy left in it. This energy is dissipated by adding a stationary row to reverse the flow and then putting it through an additional rotating row. It is universal practice for single-stage turbines when the ratio p_1/p_3 is in the range of 2 to 2.5 or greater. The two-row impulse (Curtis) stage is not as efficient as the single-row stage, owing to the higher losses encountered with the high steam velocities and repeated turning of the stream.

PERFORMANCE AND EFFICIENCY

The energy available in the steam is expressed in British thermal units per pound, or enthalpy. The velocity of the steam flow through the nozzle is calculated from

$$C_1 = 223.7\sqrt{\Delta h} \qquad (24\text{-}14)$$

where C = velocity, ft/s, and Δh = enthalpy drop from p_1 to p_2, Btu/lb. This is the same formula that is used for the spouting velocity of liquid in terms of head by introducing the mechanical equivalent of heat.

For calculation purposes the **efficiency of individual turbine stages** is plotted as efficiency μ versus velocity ratio μ/C_1. Figure 24-16 shows the relation between the three principal types of stages; each one has a peak efficiency at a certain ratio of μ/C_1. For the same revolution per minute, of course, the two-row stage takes the highest value of C_1 and the highest enthalpy drop. The reaction stage takes the lowest. The single-row impulse (Rateau) stage is in the middle.

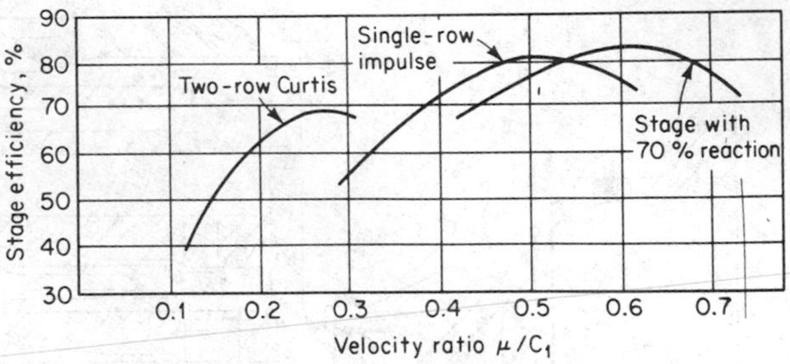

FIG. 24-16 Stage efficiency for different types of stages.

The reaction stage is denoted as 70 percent, because 30 percent of the enthalpy is allowed for expansion in the stationary row and 70 percent in the rotating row.

Steam Rate Enthalpy data can be obtained from Mollier diagrams or from steam tables (see Sec. 3), from which the **theoretical steam rate** can be calculated. For example, a throttle inlet condition of 4137 kN/m² (600 psig) and 399°C (750°F) gives an enthalpy of 3.2 MJ/kg (1380 Btu/lb), and if the end point is at 348 kN/m² (50 psig), then adiabatic expansion is to 2.69 MJ/kg (1157 Btu/lb). This gives 0.52 MJ/kg (223 Btu/lb) available, and the theoretical steam rate is calculated from the Btu equivalent per kilowatthour or horsepower-hour:

$$3412.4/223 = 15.3 \text{ lb steam/kWh}$$
$$2544/223 = 11.4 \text{ lb steam/(hp·h)}$$

Theoretical-steam-rate tables are available as separate publications. Table 24-6 covers some common conditions.

The **actual steam rate** is obtained by dividing the theoretical steam rate by the engine efficiency, which includes thermodynamic and mechanical losses. Alternatively, internal efficiency can be used, and mechanical losses applied in a second step.

Efficiency varies over a wide range, dependent upon the number of stages in the turbine. If steam conditions are assumed to be 4.14 MN/m² (600 psig) at 399°C (750°F) inlet and 13.8 kN/m² (2 psia)

TABLE 24-6 Theoretical Steam Rates for Steam Turbines at Some Common Conditions, lb/kWh

Exhaust pressure	Inlet conditions							
	150 lb./sq. in. gage, 366°F., saturated	200 lb./sq. in. gage, 388°F., saturated	250 lb./sq. in. gage, 500°F., 94°F. superheat	400 lb./sq. in. gage, 750°F., 302°F. superheat	600 lb./sq. in. gage, 750°F., 261°F. superheat	600 lb./sq. in. gage, 825°F., 336°F. superheat	850 lb./sq. in. gage, 825°F., 298°F. superheat	850 lb./sq. in. gage, 900°F., 373°F. superheat
2 in. Hg	10.52	10.01	9.07	7.37	7.09	6.77	6.58	6.28
4 in. Hg	11.76	11.12	10.00	7.99	7.65	7.28	7.06	6.73
0 lb./sq. in. gage	19.37	17.51	15.16	11.20	10.40	9.82	9.31	8.81
10 lb./sq. in. gage	23.96	21.09	17.90	12.72	11.64	10.96	10.29	9.71
30 lb./sq. in. gage	33.6	28.05	22.94	15.23	13.62	12.75	11.80	11.07
50 lb./sq. in. gage	46.0	36.0	28.20	17.57	15.36	14.31	13.07	12.21
60 lb./sq. in. gage	53.9	40.4	31.10	18.75	16.19	15.05	13.66	12.74
70 lb./sq. in. gage	63.5	45.6	34.1	19.96	17.00	15.79	14.22	13.25
75 lb./sq. in. gage	69.3	48.5	35.8	20.59	17.40	16.17	14.50	13.51

NOTE: To convert pounds-force per square inch to megapascals, multiply by 6.8948×10^{-3}; to convert pounds per kilowatthour to kilograms per kilowatthour, multiply by 0.4536; °C = 5/9(°F − 32).

exhaust, Table 24-6 shows a theoretical steam rate of 3.47 kg/kWh (7.65 lb/kWh). With efficiencies that might be experienced for the given steam conditions with single-stage, five-stage, seven-stage, and nine-stage turbines, the actual steam rates would be as in Table 24-7.

From this table note that efficiency increases with the number of stages and that the increased number of stages corresponds to larger horsepower values. For each stage, as characterized by diameter and speed, there is a Btu drop that gives the best efficiency provided there is enough steam to fill the stage so that it operates with minimum friction and windage loss. Thus the nine-stage turbine is fine for 7.46 MW (10,000 hp), but it would not show up as well as a five-stage turbine at 0.746 MW (1000 hp) because the losses would increase with the light flow.

From the velocity diagram in Fig. 24-15 it is apparent that an increase in wheel speed μ permits an increase in nozzle exit velocity C_1 without increasing C_2. Accordingly, a high-speed turbine can use more Btu per stage and will have fewer stages than a slow-speed turbine.

The leaving velocity C_2 is a measure of the unused energy. For best efficiency C_2 should have no radial component; C_2 should be straight axial. For all stages except the last one, C_2 represents a carry-over to the next stage. For the last stage, C_2 is the velocity into the exhaust hood and is referred to as the leaving loss or exhaust loss.

The curves in Figs. 24-17 and 24-18 can be used for estimating **steam rates of single-stage turbines** by proceeding according to the following example. For steam conditions of 2.76 MN/m^2 (400 psig) and 399°C (750°F) inlet and 5.17 N/m^2 (75 psig) exhaust, Table 24-6 gives theoretical steam rate as 20.59 lb/kWh. If the turbine is 300 hp and 4000 r/min, enter the top of Fig. 24-17 at 4000 and for a trial stop at 18-in-base diameter of turbine blading, then drop to TSR 20.59, and the base steam rate is 37 lb/hp·h. Next enter the top of Fig. 24-18 at 4000 r/min, and intersect 75 psig, then drop to 18-in diameter and find 8-hp loss. Total horsepower then is 308.5, and steam required is (308.5) (37.0) = 11,400 lb/h. By reading values for other diameters a selection table can be prepared, shown as Table 24-8.

From this table the most efficient unit can be selected and balanced against price. It is apparent that for 300 hp the 28-in diameter achieves no gain over the 22-in diameter because of the increase in horsepower loss. For 22 in versus 18 in the gain is small and may be offset by the higher price.

Steam rates for multistage turbines depend upon many more variables than do single-stage turbines and require extensive computation. Depending upon the type of turbine, single-valve or multivalve, general-purpose or generator-drive, condensing or noncondensing, with or without extraction, the manufacturers have shortcut procedures for estimating performance in their bulletins for the different types. As a general approximation the curve in Fig. 24-19 may be used, if one keeps in mind that an actual turbine may be several points above or below the curves, depending upon the use of optimum staging for efficiency or a compromise to meet price. Speed is also important; high speed at 373 kW (500 hp) may have high losses, while 746 MW (10,000 hp) at 12,000 r/min may be above the curve. And steam pressure affects performance. The most efficient turbine is one in which speed, pressure, and steam flow combine to fill the blade path so that there are no partial-admission stages. For partial admission the nozzles do not fill the entire 360° arc because there is not enough steam for that many nozzles. Those portions of the blades which are spinning outside of the nozzle arc create friction and windage.

TURBINE CONTROL

A turbine may be speed-, pressure-, or process-controlled. Some of the terms used are defined as follows:

Speed-governing system includes the speed governor, the speed changer, the servomotor that moves the valves, and the governor-controlled valves.

Speed governor includes only those elements which are directly responsive to speed and position the other elements.

The **speed changer** is a device by means of which the set point may be varied.

Steady-state regulation is the change in sustained speed or pressure (expressed as a percentage of rated) when power or flow output is gradually reduced from rated value to zero.

Speed variation is the total variation in speed from the set point and includes both dead band and oscillation.

Proportional-action governor is a governor with inherent regulation and a continuous linear relation between the input (speed

TABLE 24-7 Typical Stage Efficiencies for Steam Turbines

Turbine design	Turbine hp.	Internal efficiency, %	Exhaust enthalpy, B.t.u./lb.	Δh,* B.t.u./lb.	Steam rate
Single-stage	500	30	1245	135	7.65/0.30 = 25.5 lb./kw.-hr.
5-stage	1,000	55	1135	245	7.65/0.55 = 13.9 lb./kw.-hr.
7-stage	4,000	65	1090	290	7.65/0.65 = 11.75 lb./kw.-hr.
9-stage	10,000	75	1020	360	7.65/0.75 = 10.02 lb./kw.-hr.

NOTE: To convert horsepower to kilowatts, multiply by 0.7457; to convert British thermal units per pound to kilojoules per kilogram, multiply by 2.33.
*Based on inlet enthalpy = 1380 Btu/lb.

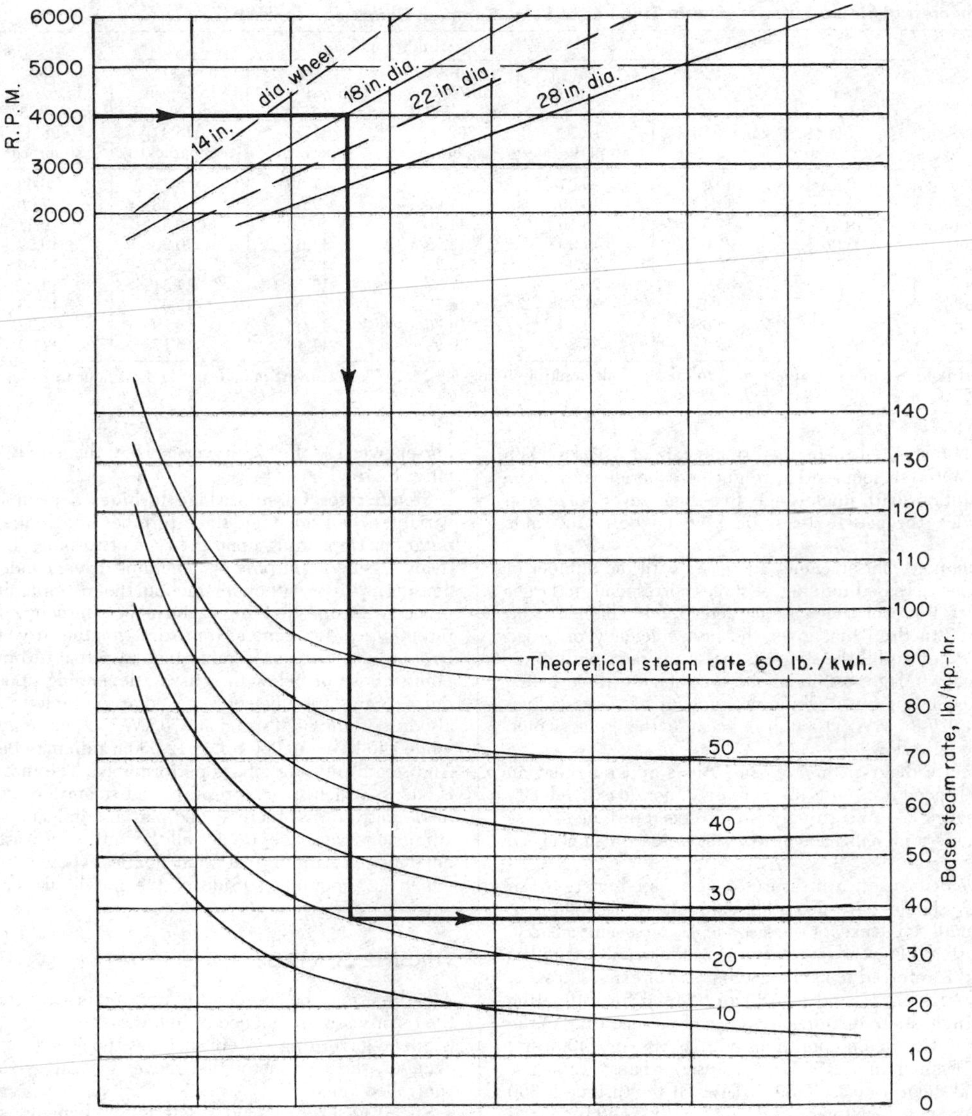

FIG. 24-17 Approximate steam rate for single-stage turbines. To convert pounds per kilowatthour to kilograms per kilowatthour, multiply by 0.4537; to convert inches to meters, multiply by 0.0254; and to convert pounds per horsepower-hour to kilograms per kilowatthour, multiply by 0.6084.

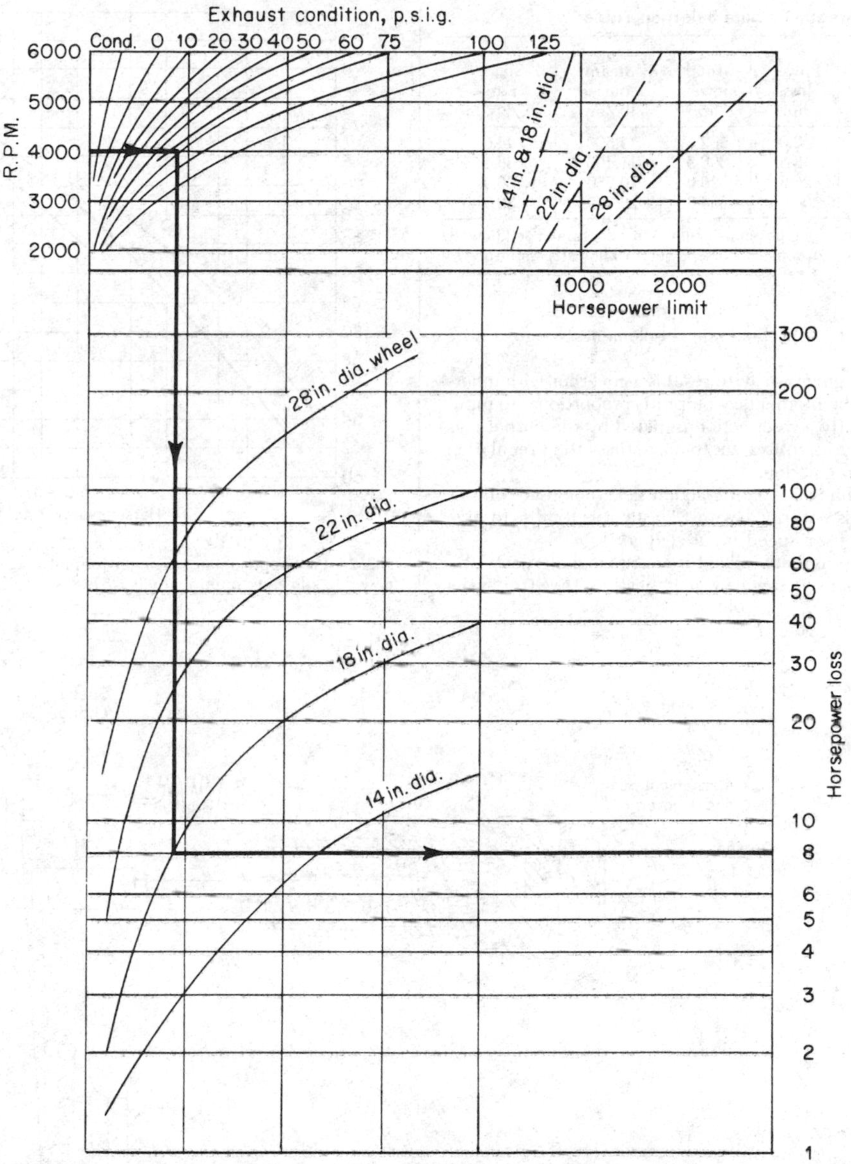

FIG. 24-18 Approximate horsepower loss for single-stage turbines. To convert horsepower to kilowatts, multiply by 0.7457; to convert inches to meters, multiply by 0.0254; and to convert pounds per square inch gauge to kilopascals, multiply by 6.895.

TABLE 24-8 Typical Steam-Turbine Selection Table

Wheel diameter, in.	Base steam rate, lb./hp.-hr.	Power loss, hp.	Total power. hp.	Steam required, lb./hr.	Steam rate, lb./hp.-hr.
14	44.5	3.0	303	13,500	44.6
18	37.0	8.5	308.5	11,400	38.1
22	33.0	26.0	326.0	10,750	36.9
28	29.5	64.0	364.0	10,750	36.9

NOTE: To convert pounds per horsepower-hour to kilograms per kilowatt-hour, multiply by 0.6084; to convert horsepower to kilowatts, multiply by 0.7457.

change) and the output of the final control element, the governing valve.

Proportional-action governor with reset is a governor with inherent regulation so that the momentary output is proportional to input change, and subsequently a reset action initiated by the output acts on the speed changer or its equivalent to make the settled regulation less than the inherent regulation.

Isochronous governor is a floating-action governor that controls for constant speed. It is equipped with a dashpot or buffer to give momentary regulation for a speed-input change.

Control-System Components The three principal elements of a control system are the sensing device which measures the error as the

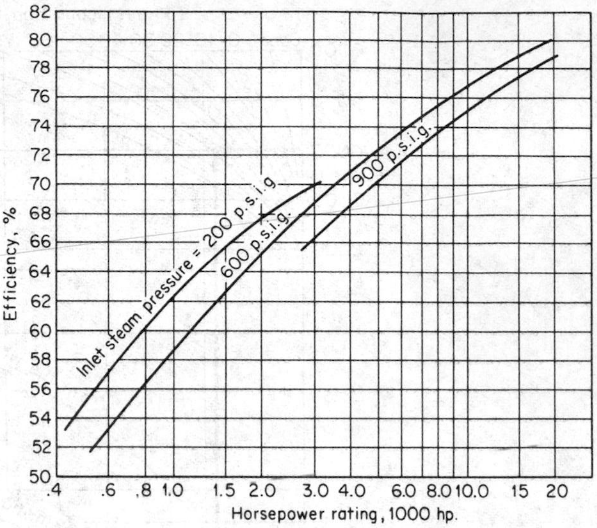

FIG. 24-19 Approximate efficiency for multistage turbines. To convert horsepower to kilowatts, multiply by 0.7457.

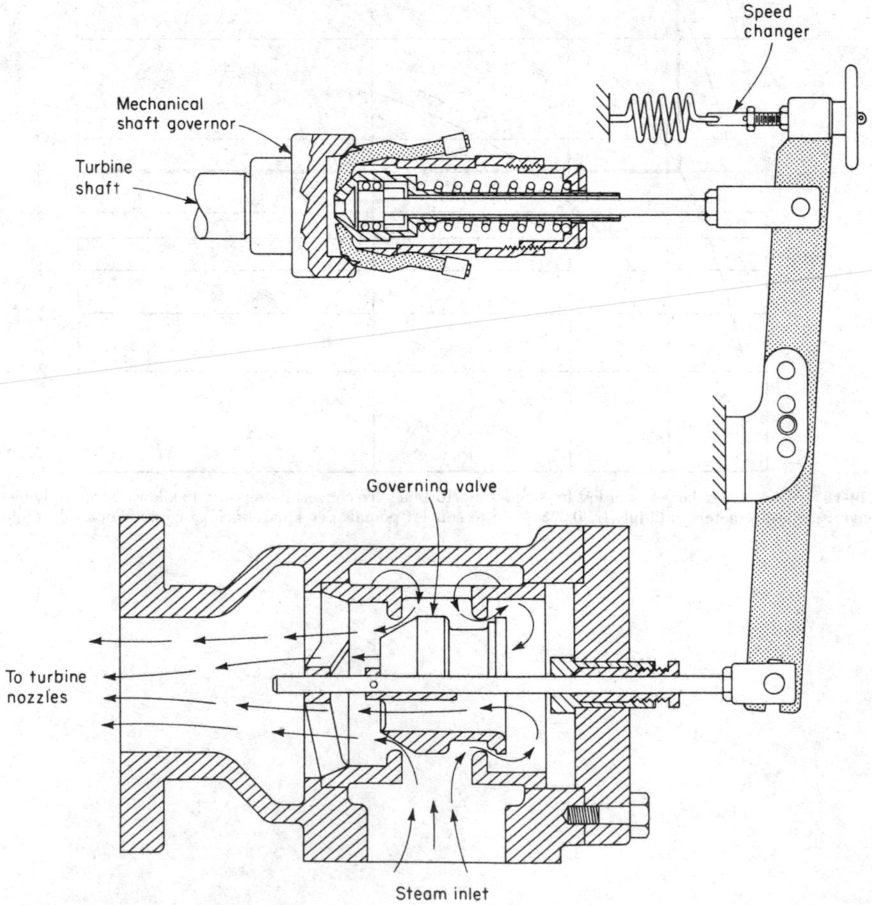

FIG. 24-20 Direct-acting flyball governor.

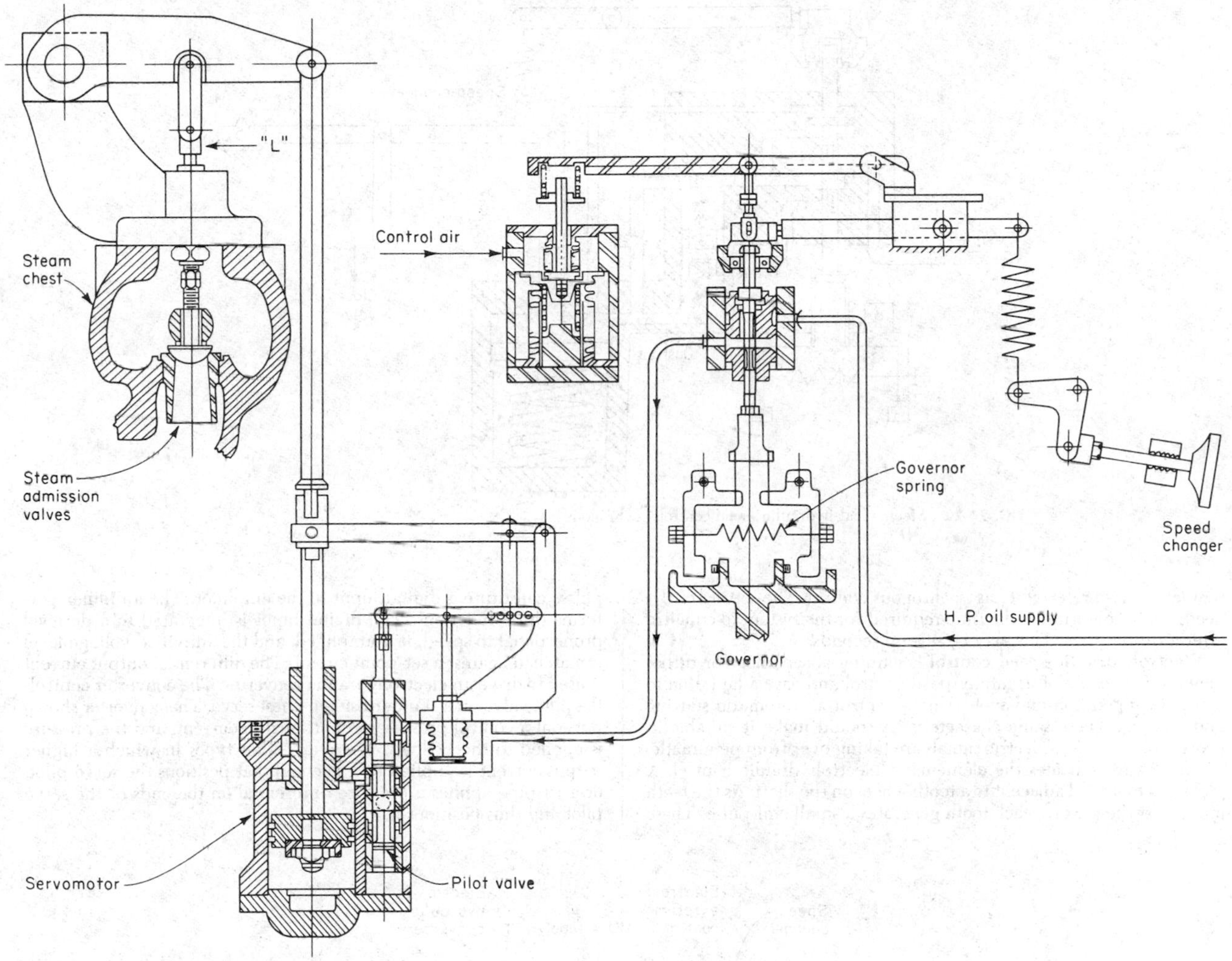

FIG. 24-21 Mechanical-hydraulic speed control: proportional.

deviation from the set point, means for transmission and amplification of the error signal, and the control output device in the form of a servo-operated valve. In the case of the **direct-acting flyball governor** (Fig. 24-20) these three elements are combined in the flyball element and the linkage that connects to the valve.

The centrifugal force of the weights is continually compared against the set point as established by the governor spring which opposes the force from the governor weights. For an increase in load the speed drops, and the weight force is reduced, which allows the spring to push the governor spindle to the left. The lever then pivots, and the valve moves to the right and opens to increase steam flow and torque. The feedback in the control is the resultant increase in speed to match the set point of the governor spring again and eliminate the error. This governor corresponds to NEMA Class A, 10 percent speed regulation, that is, 10 percent speed rise from full load to no load, and the governor weights must move out to close the valve.

The force required to position the valve and the specified regulation puts a practical limit on the direct-acting governor. Beyond this limit a **servo** is required. The servo requirement has three levels. The lowest level is for a single valve of the balanced type in Fig. 24-20 which may be controlled with less than 889.6-N (200-lb) force. The next level is a single valve with single seat of the venturi type; partially balanced it requires 3558.4-N (800-lbf) to 4448-N (1000-lbf) force. The top level is the multivalve bar-lift or the cam-lift valve gear, in which 8- and 10-in and larger oil servos are used in developing several thousand pounds-force.

Speed-Control Systems The most common sensing element is mechanical, although some systems use hydraulic and a more recent arrival is electronic. For valve positioner they all have a hydraulic servo as first choice, with an occasional choice of pneumatic for the lighter loads.

Figures 24-21 and 24-22 illustrate two different **mechanical-hydraulic systems.** Figure 24-21 is a bar-lift steam chest with a heavy-duty hydraulic servo. The speed-sensing element is a flyball assembly attached to a rotating pilot. This rotating pilot sends a control-pressure signal that is proportional to speed to a bellows on the servo. A change in control pressure initiated through the rotating pilot by either speed or speed changer deflects the bellows and servo pilot valve. The servopiston position is proportional to the control pressure.

Figure 24-22 has a spring-return servo that finds application on single-valve turbines with a moderate valve force. For a speed change the rotating pilot sends an error signal through the dashpot to the servopiston. This is a dashpot-type isochronous governor; the error signal sees an instantaneous regulation due to the springs in the dashpot, but after the pressures have equalized through the needle valve, there is no resultant force change on the governor pilot. Control pressure is not proportional to speed; so the governor has zero

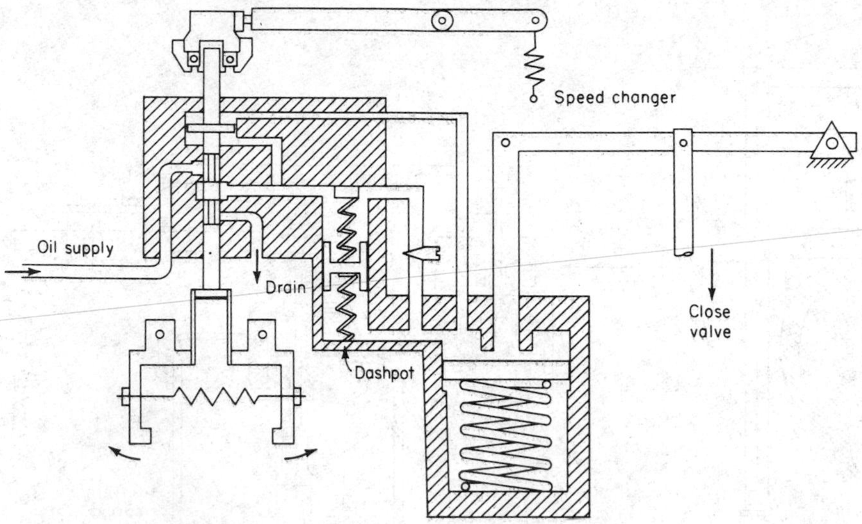

FIG. 24-22 Mechanical-hydraulic speed control: isochronous.

regulation, also referred to as isochronous control. The setting of the needle valve determines the time required for the system to equalize after a disturbance. This may be several seconds.

Electrohydraulic speed control is gaining acceptance for turbogenerators because of accurate speed control and easy adaptation to computer operation and such remote control as automatic starting and loading. These same characteristics should make it suitable in process plants where electric signals are taking over from penumatic. Figure 24-23 indicates the elements of electrohydraulic control. A pickup is mounted adjacent to a tooth wheel on the shaft. As the teeth move past the pickup, each tooth generates a small emf pulse. These pulses constitute a digital input to the amplifier. The amplifier performs three functions. The digital input is integrated to a dc level proportional to speed, it is amplified, and the amplified voltage level is matched against a set-point circuit. The differential output current is used to drive an electrohydraulic converter. The converter controls the pilot valve on a standard steam-chest servo. The converter shown puts out a control pressure proportional to current, and the pressure is applied to the bellows. There are other types in which a higher output current is applied to a solenoid that positions the servo pilot, or a jet pipe applies a pressure differential on the ends of the servo pilot and thus positions it.

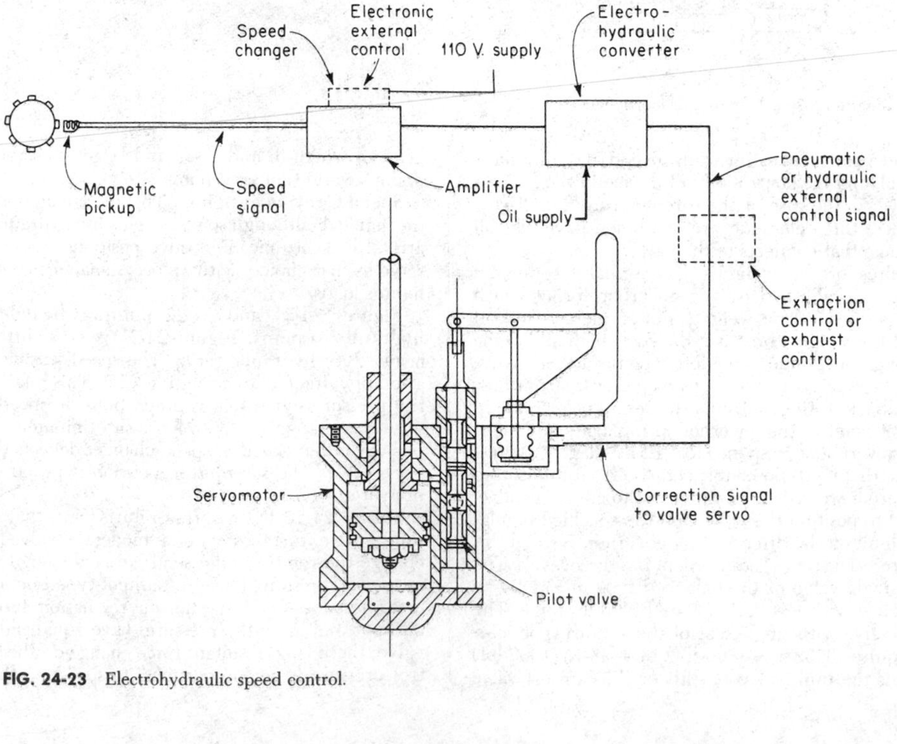

FIG. 24-23 Electrohydraulic speed control.

Remote speed control or **process speed control** takes one of two forms. The remote signal may position the valve directly by acting on the valve stem or on the servo pilot. This action is independent of the governor. For this form of operation the governor is set for maximum operating speed and will take over in an emergency. The operation is referred to as preemergency. In the second type the remote signal acts on the speed changer or its equivalent to adjust the set point. The governor is now always in the circuit, and the unit is always speed-responsive.

The speed control operates the governing valve to maintain steam flow commensurate with load demand while holding speed essentially constant. For sudden load changes there will be a short-time overshoot, and a special case is the instantaneous loss of load, load dump at full load. The usual specification states that the overshoot on load dump must not exceed 9 to 10 percent of rated speed. The settled speed rise will of course be equal to the regulation, 4 or 6 percent for a NEMA Class C or B governor and less than 1 percent for Class D.

The **trip valve** is provided as a second line of defense in case of overspeed, and the trip is adjusted in the range of 10 to 15 percent above rated speed. The trip valve is frequently equipped with a trip-actuating solenoid which can be operated by push button, by low oil pressure, or by some other process upset. When the speed control functions as described above, the trip will not be actuated by load dump.

Extraction-Pressure Control An extraction turbine equipped with a regulator so that the extraction pressure will be automatically controlled is provided with two sets of steam-chest valves as shown in the schematic in Fig. 24-24. Each of the two sets of steam-chest valves is operated by a servomotor. The throttle flow is illustrated as the total of two flow streams: *A*, which travels the length of the turbine and leaves through the exhaust opening; and *B*, which leaves through the extraction opening. The shaft output is the sum of the power generated by the two streams. If the process demand increases, flow *B* increases and develops more power. For constant output, flow *A* must be reduced, and this is the function of the three-arm linkage: to open the governing valves and close the extraction valves, which will increase throttle flow ($A + B$) and decrease condenser flow *A* for more extraction flow at constant load. For a reduction the opposite happens.

For an increase or decrease in load the governor moves the three-arm link parallel to itself, and both sets of valves move in the same direction.

This principle of the three-arm lever is the same for all so-called compensated extraction control. It is applied to mechanical, hydraulic, or electric turbines to produce opposite valve motion for changes in steam demand and parallel valve motion for changes in load.

The ratio of the three-arm lever depends upon the ratio of zero to total extraction flow for a certain load.

SELECTING A TURBINE

The **major variables** that affect turbine selection are as follows:

1. Horsepower and speed of the driven machine
2. Steam pressure and temperature available or to be decided
3. Steam needed for process, so that a back-pressure turbine should be considered
4. Steam cost and value of turbine efficiency, so that consideration can be given to stage and valve options
5. Use of speed-reducing or speed-increasing gears
6. Extraction for feedwater heating
7. Condensing turbine with extraction for process
8. Control system, speed control, pressure control, and process control, so that consideration can be given to pneumatic or electric remote control, speed or pressure variation that can be tolerated, and system response speed
9. Safety features such as overspeed trip, low-oil trip, remote-solenoid trip, vibration monitor, or other special monitoring of temperature, temperature changes, and casing and rotor expansion
10. Price range from the minimum single-stage turbine to the most efficient multistage turbine

The **initial step in selection** could logically be to make an estimate of the steam flow at various steam pressures by using Fig. 24-19 for a rough estimate of efficiency. Unfortunately, there are no rigid standards for steam pressure and temperature as electrical-voltage steps are fixed, and many engineers pick a pressure and a temperature that look good to them. In general, however, manufacturers prefer working with the standards proposed some years ago by a joint ASME-IEEE committee. The values are 2.76 MN/m^2 (400 psig) and 399°C (750°F), 4.14 MN/m^2 (600 psig) and 441°C (825°F), 5.86 MN/m^2 (850 psig) and 482°C (900°F), and 8.62 MN/m^2 (1250 psig) and 510°C (950°F) or 538°C (1000°F). The values fall on line *A* in Fig. 24-25. For a 1.5-inHg absolute exhaust pressure, line *A* corresponds to 9 percent moisture. Operating to the left of this state line (78 to 80 percent efficiency), last-stage moisture increases rapidly, which means more erosion; also less heat is available. Moving to the right of this line, the temperature lines become quite flat, the pressure drops at constant temperature, and heat available is reduced. It should be noted that 538°C (1000°F) is a good upper limit for steam turbines without a sharp increase in cost because of special materials. Experience has indicated that maintenance and initial cost exceed

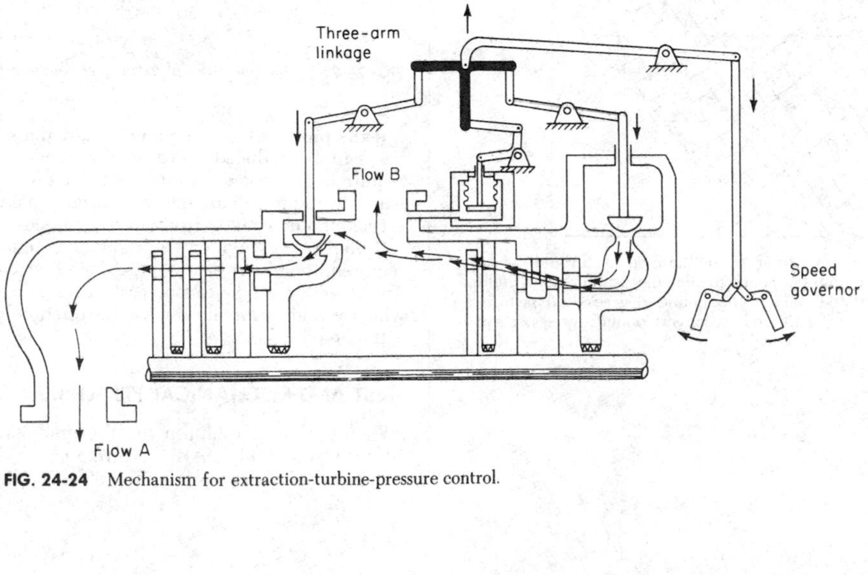

FIG. 24-24 Mechanism for extraction-turbine-pressure control.

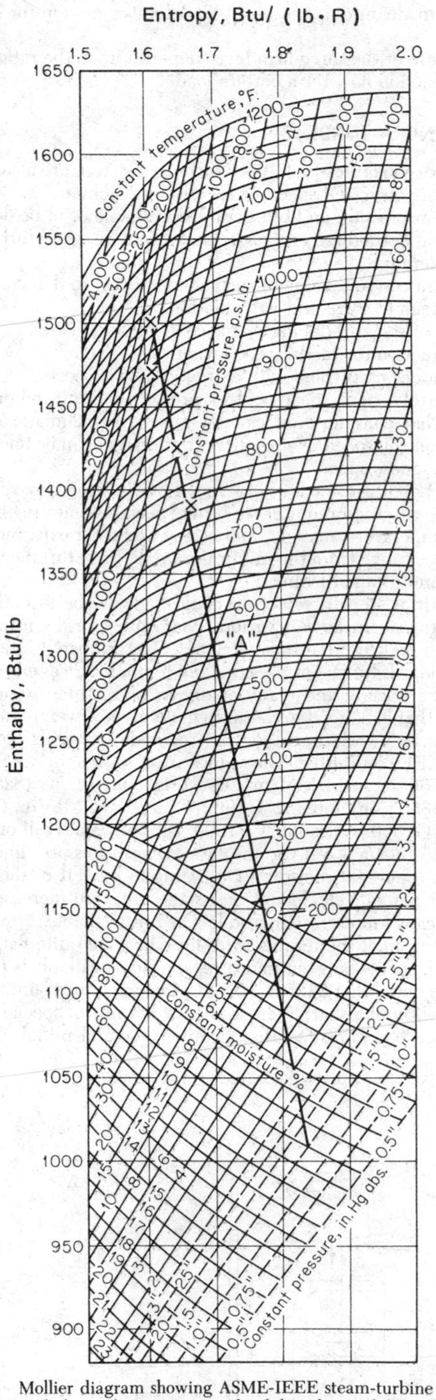

FIG. 24-25 Mollier diagram showing ASME-IEEE steam-turbine standards. To convert British thermal units per pound to kilojoules per kilogram, multiply by 2.328; to convert British thermal units per pound–degrees Rankine to joules per gram-kelvin, multiply by 4.19; and to convert pounds-force per square inch to megapascals, multiply by 6.895^{-3}

the gain in performance with temperatures in excess of the range of 538 to 566°C (1000 to 1050°F).

Example 1 If a turbine is to be operated with exhaust to a condenser vacuum that will give 3 inHg absolute in the summer and 1 inHg absolute in the winter, what vacuum should be specified?

The turbine for 3 inHg will have shorter exhaust blades and a smaller exhaust opening and will lose 16 Btu/lb when operated at 1 inHg. It will also have a high pressure drop over the last-stage blading. Sonic velocity and shock waves will emanate from the last-stage blading, inducing blade loading and oscillation that may lead to fatigue failure. For operation at 1 inHg the last stage should operate with a diffuser which will limit the annulus discharge area of the stage. A turbine designed for 1 inHg will have a lot of windage in the last stage because the blade annulus cannot be filled by the higher-density steam.

The best answer is an exhaust designed for 2 inHg and equipped with exhaust-blade diffuser. This will protect at 1 inHg and give some pressure recovery at 3 inHg by virtue of its venturi action.

When an **extraction-condensing turbine** is decided upon, it may be specified in three different ways, depending upon process steam and power demand. Referring to Fig. 24-26, the usual purchase is a unit in which rated capability can be carried either straight-condensing or total-extraction. The zero-extraction line terminates at *A*, and the total-extraction line terminates at *B*.

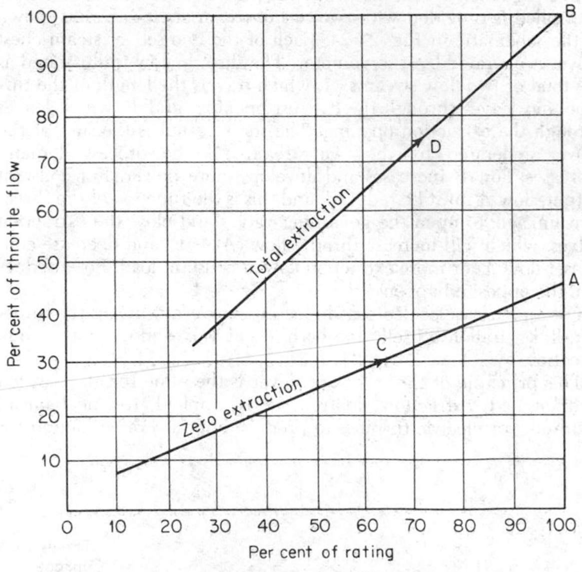

FIG. 24-26 Characteristics of extraction-condensing steam turbines.

If the process-steam demand is high and steady, then the exhaust size can be reduced, because not much condensing capacity is required. The choice would be to save cost by a smaller exhaust which would terminate the zero-extraction line at *C*, while the total-extraction line would extend to *B* for rated capability.

If the process demand is light and high-extraction flow is not required, then most of the power will be from the condensing flow. The choice would be to save cost by a smaller inlet and steam chest, which would terminate the total-extraction line at *D* and the zero-extraction line at *A*.

TEST AND MECHANICAL PERFORMANCE

When testing to establish the **thermodynamic performance** of a steam turbine, the ASME Performance Test Code 6 should be followed as closely as possible. The effect of deviations from code procedure should be carefully evaluated. The flow measurement is particularly critical, and Performance Test Code 19 gives details of flow nozzles and orifices. The test requirements should be carefully studied when the piping is designed to ensure that a meaningful test can be conducted.

Mechanical performance is generally checked by a running test at the factory before shipment and again when the turbine is installed on the site. The following is an enumeration of items that may provide smooth and vibration-free operation. The rotor must be in dynamic balance, and at-speed balancing is the most effective. The bearings must be in line, and the seal clearance must be correct. This alignment must be maintained when both cold and hot and during the transient from cold to hot. Disturbance of alignment may originate by unequal heat expansion of supports, from pipe expansion, or from binding and lack of freedom to expand. Excessive pipe expansion and high forces on the turbine have caused many vibration problems. Flexible couplings that do not flex also cause problems. This may result in torque lock or in unloading of bearings, and driver and driven machine bearings may be shifted out of line. Then there is the question of resonance and critical speed. The foundation enters into this relationship, and slender columns may contribute problems of resonance. If there is a gear in the lineup, it should be checked carefully for bearing load, alignment, and resonance frequencies that may be multiples of the gear ratio. Also, bearing-oil supply must be adequate.

From this it can be seen that **vibration** is the universal manifestation that something is wrong. Therefore, many units are equipped with instruments that continuously monitor vibration. Numerous new instruments for vibration analysis have become available. Frequency can be accurately determined and compared with computations, and by means of oscilloscopes the waveform and its harmonic components can be analyzed. Such equipment is a great help in diagnosing a source of trouble.

OPERATING PROBLEMS

While most turbines have a 10-year-availability record in the range of 95 to 99 percent, troubles may develop in any number of places. The most common are vibration, cycling governor, sticking valve stems, leaky packing, temperature bow, erosion of blading, loss of power, and bearing problems.

The causes of **vibration** have already been discussed. An increase in vibration over a period of time is generally caused by loss of alignment, settling of the foundation, or sticking of some expansion feature such as a pipe or a pedestal. Other causes are wear in the teeth of a flexible coupling, an internal rub in the unit, loss of bearing oil, and bearing wear. (Startup vibration is discussed later under temperature bow.)

If a **governor** starts to **cycle** after operating for some time, this is generally the result of wear which causes dead band or sticking. Also, all pilot valves should be inspected for the effects of dirt in the oil.

Sticking of valve stems is common if solids are present in the steam. The steam must be without solids. (Note comments later under loss of power.) It is important that units operating on a steady load for long periods be checked for sticking stems at regular intervals. The records show that in several cases deposits have caused the stem of both the governor valve and the trip valve to stick when there was a loss of load. The effect of the loss of load was destructive overspeed.

Wear and increased **leakage from the glands** are common. Carbon rings may need replacement after 1 or 2 years of operation. A unit with labyrinth packing may never need packing replacement. This depends upon operation. If a unit is started quickly with a temperature bow in the shaft, the result is a rub in the labyrinth, and then all packing may need replacement.

It is important to understand the reasons for a **temperature bow.** When a turbine is shut down and starts to cool, the lower half, particularly on a condensing unit, will cool faster than the top half. After the rotor has stopped turning, the temperature difference increases, and in 20 min there may be a 28 to 83°C (50 to 150°F) difference between top and bottom. Both the casing and the shaft bow up because of this temperature difference in the vertical plane. If the throttle is opened now and the bowed shaft starts to turn in the bowed casing, the packing may wipe out in a few revolutions, and at 200 r/min and up there will be a heavy thumping. The packing rub serves to increase the temperature bow by heating the high side of the shaft.

Other causes of a temperature bow are leaky valves and damaged sleeves. If steam is leaking into a stopped turbine from either an exhaust valve or a stop valve that is leaking, the upper half will be warmer than the lower half, and it bows. If a shaft sleeve does not contact uniformly, there will be a transient difference in heat transfer to the shaft, and a bow will result. Also, turning sealing steam on before the shaft rotates may cause a bow.

Leaky valves are also a cause of **erosion.** Most turbine erosion-corrosion problems come from damage that takes place when the unit is not running. A slight steam leak into the turbine will let the steam condense inside the turbine, and salt from the boiler water will settle on the inside surfaces and cause pitting, even of the stainless blading. There must be two valves with a drain between them, i.e., a block valve on the header and an open drain in the line before it reaches the closed trip-throttle valve.

In a turbine that is running, erosion-corrosion is pretty much confined to units that are operating on saturated steam with inadequate boiler-water treatment. This type of erosion takes place behind the nozzle ring and around the diaphragms where they fit in the casing.

Loss of power is another item generally tied to water treatment. With dissolved salts in the steam these salts stay in solution while the steam is superheated. After the steam has expanded through several stages and become saturated, the salts condense out with the moisture. Silica and other salt deposits build up on the blading and the nozzles. The stage pressures increase, and the load drops. The thrust load increases, and the thrust bearing may fail. Depending upon the nature of the salts, it is possible to have corrosion associated with the deposits or corrosion only in the region where the steam changes from superheated to steam with moisture.

Turbine bearings show very little sign of wear as long as there is an adequate oil film. Wiping of the bearing is generally traced to dirt in the oil or a block in the oil supply. Thus filtration should be adequate and retain particles that may exceed the oil-film thickness. A checkerboard cracking of the babbitt is sometimes observed. This may have either of two origins. The shaft transports a lot of heat from the steam parts into the journal, and the oil flow in the bearing must be sufficient to lubricate and to remove this heat. Otherwise, the heat is conducted into the babbitt and the babbitt will soften and crack. This cracking may take place if oil flow is stopped too soon when a unit is shut down. The second origin for cracking of the bearing surface is pounding of the journal caused by shaft bow or oil whirl.

COMBUSTION GAS TURBINES

Gas turbines are usually rated according to power output at sea level and 26.7°C (80°F) ambient air (some European designs are rated at 15.7°C (60°F) and when burning a specified fuel. Power output and efficiency will be larger for fuels which produce larger volumes of products of combustion, since the compressor does not do any work on additional volume. Gas turbines are further classified by the physical arrangements of the component parts, such as (1) single-shaft, (2) two-shaft, (3) regenerative (a heat exchanger is used to recover exhaust losses and heat the air to the combustors), (4) intercooled (heat is removed between compressors), and (5) reheat (heat is added

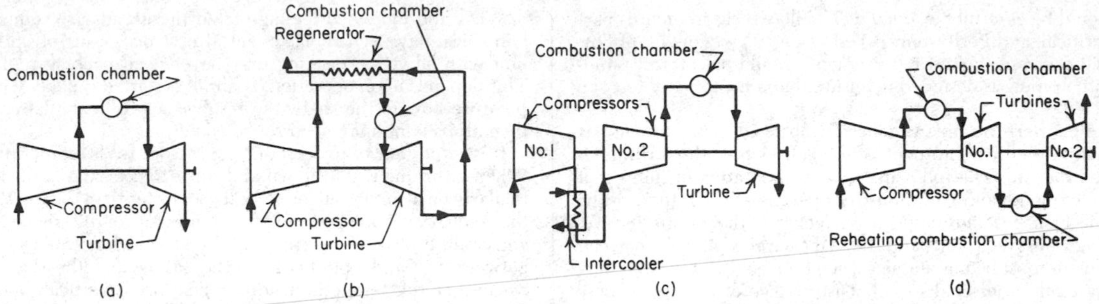

FIG. 24-27 Gas-turbine cycles. (*a*) Basic Brayton or Joule cycle. (*b*) Gas turbine with regeneration. (*c*) Gas turbine with intercooling. (*d*) Gas turbine with reheating. (*From Baumeister,* Standard Handbook for Mechanical Engineers, *7th ed., McGraw-Hill, New York, 1967.*)

between turbines). Thus an installation may be a two-shaft 7.46-MW (10,000-hp) turbine with one intercooler, one reheat, and a regenerator (see Figs. 24-27 and 24-28).

Efficiency The **overall efficiency** of a gas turbine is a function of compressor and turbine efficiencies, ambient-air temperature, nozzle inlet temperature, and the type of cycle used (i.e., reheat, etc.). The compressor and turbine are designed for high efficiency, and the first-stage gas temperature establishes material and stress conditions for the first set of rotating blades. To the gas temperature at these blades is added the temperature drop across the first-stage nozzles to determine the inlet temperature of the turbine, which may vary from 704 to 816°C (1300 to 1500°F) for industrial gas turbines and higher for aviation gas turbines. The higher values are generally used in impulse turbines.

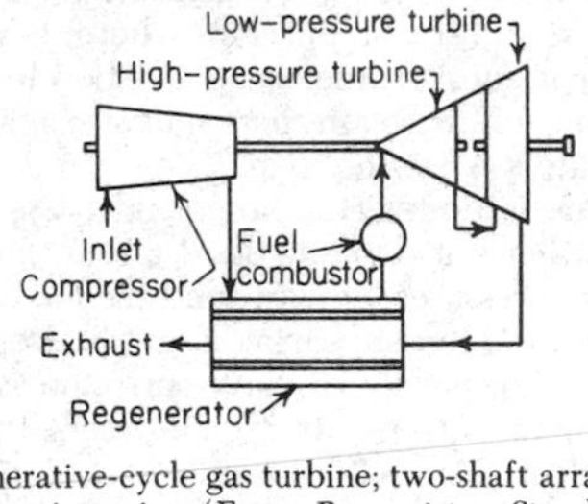

FIG. 24-28 Regenerative-cycle gas turbine; two-shaft arrangement with separate power turbines in series. (*From Baumeister,* Standard Handbook for Mechanical Engineers, *7th ed., McGraw-Hill, New York, 1967.*)

In a simple-cycle turbine there is, for each turbine inlet temperature, an optimum **pressure ratio** producing the highest possible efficiency. The efficiency and optimum pressure ratio increase with increasing turbine inlet temperatures. These pressure ratios vary from 4 at 704°C (1300°F) up to 6 for 816°C (1500°F) turbine inlet temperature.

Regenerative cycles favor lower pressure ratios, which result in low compressor discharge temperatures, thus allowing greater recovery of heat from the turbine exhaust gases. High-ratio regenerative plants use intercoolers in the compressor circuit to lower the compressor discharge air temperature.

Any type of efficient compressor can be used, e.g., positive-displacement (Lysholm), centrifugal, and axial-flow; however, most industrial gas turbines use axial-flow compressors. The turbine may have impulse or reaction blading, and to minimize losses air from the compressor discharge flows through the combustor directly into the turbine nozzle. Throttle valves are not used because the resulting pressure drop decreases overall efficiency.

A gas turbine has a large amount of excess air, and the combustor is designed with an inner portion burning only part of the air to achieve high combustion temperatures and efficiency. The products of combustion are effectively mixed with the remainder of the air to minimize temperature stratification. Each turbine may have one large combustor or several smaller combustors operating in parallel.

Design Options Most gas-turbine installations are of the **open-cycle** type using atmospheric air as the working medium and burning relatively clean fuels. For "dirty" fuels it has been suggested that the burner be located in the gas-turbine discharge, using a heat exchanger to heat the air discharged by the compressor.

In **closed-cycle** plants (see Figs. 24-29 and 24-30) it may be desirable to use other gases, since plant efficiency increases as the specific heat ratio c_p/c_v decreases. Optimum plant efficiency occurs at increasingly higher pressure ratios with decreasing values of c_p/c_v. However, most closed systems use air for convenience. Closed systems can provide a high plant efficiency over a power range from 25 to 100 percent by varying the turbine exhaust and compressor inlet pressure from atmospheric to 414 kN/m^2 (60 psig). These plants require expensive heaters, located between the compressor discharge

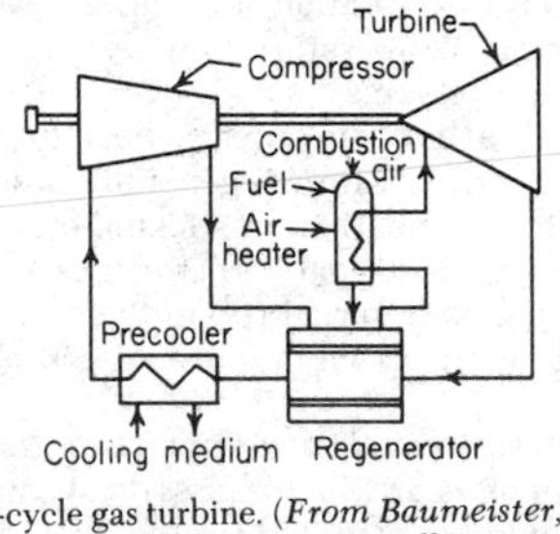

FIG. 24-29 Closed-cycle gas turbine. (*From Baumeister,* Standard Handbook for Mechanical Engineers, *7th ed., McGraw-Hill, New York, 1967.*)

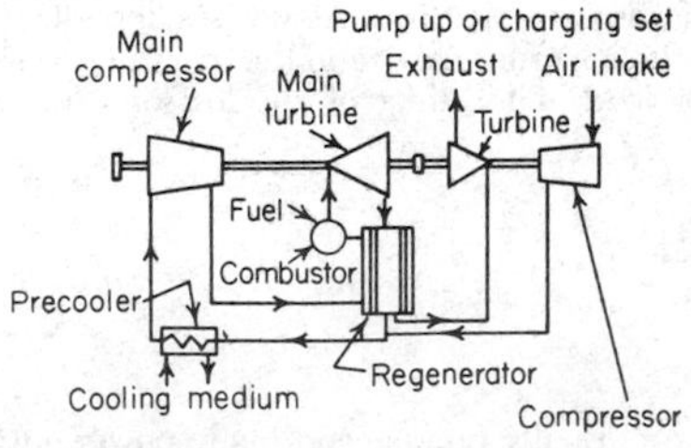

FIG. 24-30 Semiclosed, internally fired gas-turbine cycle. (*From Baumeister,* Standard Handbook for Mechanical Engineers, *7th ed., McGraw-Hill, New York, 1967.*)

and the turbine inlet, and large coolers, located between the turbine exhaust and the compressor suction. Usually, combustion of a fuel provides the heat source and cooling water the coolant medium.

A large overload, even if only temporary in nature, can cause a single-shaft gas turbine to shut down, since its fuel input is limited by the inlet overtemperature protective system. If the torque requirements of the driven machine do not decrease sufficiently with speed reduction, the gas turbine will continue to slow down, resulting in higher exhaust temperatures. The high-exhaust-temperature protective system will either shut off the fuel valve or further reduce fuel input, causing the turbine to decrease its speed and finally to shut down. To prevent such an occurrence the load characteristics of the driven equipment must be suited to those of the driver.

The popularity of the **single-shaft gas turbine** is due to its low cost and compactness in terms of power output per volume of machinery space (see Fig. 24-31). Its disadvantage is a relatively low operating-speed range and its sensitivity to atmospheric temperatures. The low operating-speed range is caused by the following factors: (1) the quantity of air flow induced by the compressor is proportional to its speed, and (2) the back pressure produced by the turbine nozzles is proportional to air flow. At low speeds the turbine power is decreased by low air flows and secondarily by the effect of low pressures on allowable inlet-air temperatures. At low flows the decreased pressure at the turbine inlet may require a reduction of turbine inlet temperature to keep the exhaust temperature within design limitations, resulting in a further reduction in power. In most applications it is necessary to unload the turbine during startup.

A wider operating-speed range is provided by the more expensive **two-shaft machine,** which consists of a high-pressure turbine driving the air compressor and a low-pressure turbine on a separate shaft to provide output power (see Fig. 24-32). A variable-area nozzle can be used in the low-pressure turbine to increase the operating-speed range. Change in the fuel input to the high-pressure turbine causes the speed and quantity of air flow to change. The low-pressure turbine power output is changed by varying the quantity of air flow and the nozzle area of the power turbine (see also Fig. 24-33).

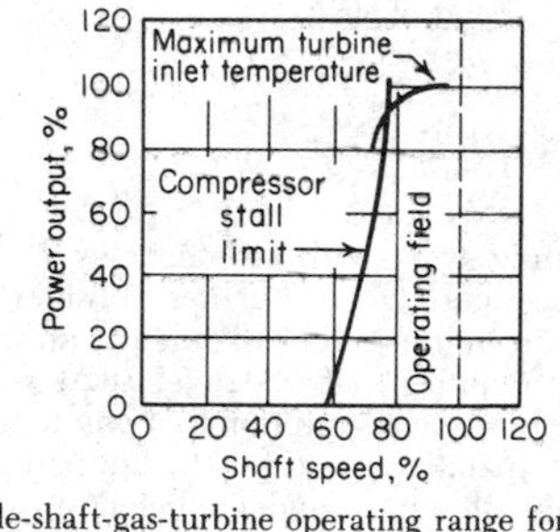

FIG. 24-31 Single-shaft-gas-turbine operating range for various speeds and loads. (*From Baumeister,* Standard Handbook for Mechanical Engineers, *7th ed., McGraw-Hill, New York, 1967.*)

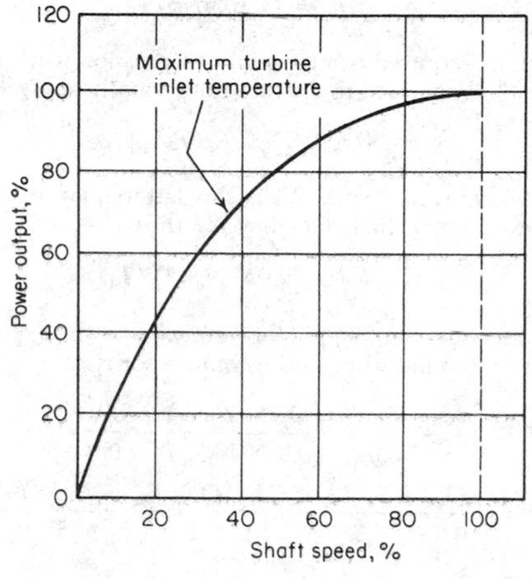

FIG. 24-32 Two-shaft-gas-turbine operating range for various speeds and loads. (*From Baumeister,* Standard Handbook for Mechanical Engineers, *7th ed., McGraw-Hill, New York, 1967.*)

Operating Characteristics The air flow in weight per hour to a gas turbine is inversely proportional to the absolute air temperature at the compressor inlet, and the compressor discharge pressure is set by the turbine nozzles (proportional to flow), resulting in decreased turbine power output during hot weather and increased power during cold days. In dry areas, it is possible to cool the hot incoming air by evaporation by using water injection. In most locations maximum summer temperature is of short duration, and it may be possible to obtain rated power by increasing turbine temperature for this short period without appreciably shortening turbine life. In extreme winter temperatures, high air pressures will exist at the turbine inlet and must be considered in the design of a gas turbine.

Since the power required by the compressor is approximately twice as great as the shaft output, a 1 percent change in compressor efficiency will result in a 2 percent change in shaft power, and a 1 percent change in turbine efficiency produces a 3 percent change in shaft power. It is therefore important that all losses be minimized and that ample-sized inlet and exhaust piping or ducts be used.

A gas turbine is started by bringing it up to **starting** speed and maintaining this speed for several minutes in order to purge the casing. Some machines require that the casing or rotor be heated slowly by burning a nominal amount of fuel in the combustors for several minutes. The turbine inlet temperature is then increased rapidly to a value above the design temperature, thus producing enough power in the turbine to bring the set up to full speed. Some installations will require a blowoff valve to prevent surging during startup. The starting power requirements of an unloaded gas turbine will be between 5 and 10 percent of the rating, and the starting speed will be between 20 and 30 percent of full-load speed. Two-shaft turbines will require slightly more starting power than single-shaft machines, but by opening the nozzles of the low-pressure turbine, the load is not driven during startup.

Gas turbines can use a wide variety of fuels. Major **fuel limitations** are that the fuel does not (1) form ashes which deposit on the blades and interfere with operation, (2) contain dust which will erode the blades, and (3) contain uninhibited vanadium. Gas turbines are now operating on fuel gas (natural and refinery), blast-furnace gases, and fuel oils (including heavy residuals); at least one experimental coal-burning gas turbine has been built.

The simple-cycle gas turbine is relatively inefficient, and almost all its losses are in the hot exhaust gases. When exhaust gases can be used in a boiler or for process heating, the combination of turbine and heat-recovery apparatus results in a high-efficiency plant. Another method that results in high efficiency is to integrate the gas turbine with process requirements.

Following are some examples of gas turbines that have been used in industry: (1) In the Houdry process a gas turbine is used to drive the regeneration-air compressor; (2) the air compressor of a gas turbine can be made oversize to supply pressurized air for blast-furnace operation; (3) removal of the last stages of the turbine supplies hot regeneration air at nominal pressures; (4) in a pressurized-boiler system the turbine drives the compressor and acts as an economizer by reducing the boiler exhaust-gas temperature; (5) in large central stations gas turbines have been used to drive boiler-feed pumps and supply hot combustion air to a boiler; and (6) the use of waste-heat boilers installed in the exhaust is the most common and easiest way of improving the efficiency of a simple-cycle gas turbine.

If the heat in the exhaust gases cannot be used for process applications, the gas-turbine plant can be made efficient only by using intercoolers in the compressors, reheat in the turbine system, and regeneration to recover exhaust heat. Such plants are expensive and usually achieve efficiencies of 30 to 34 percent.

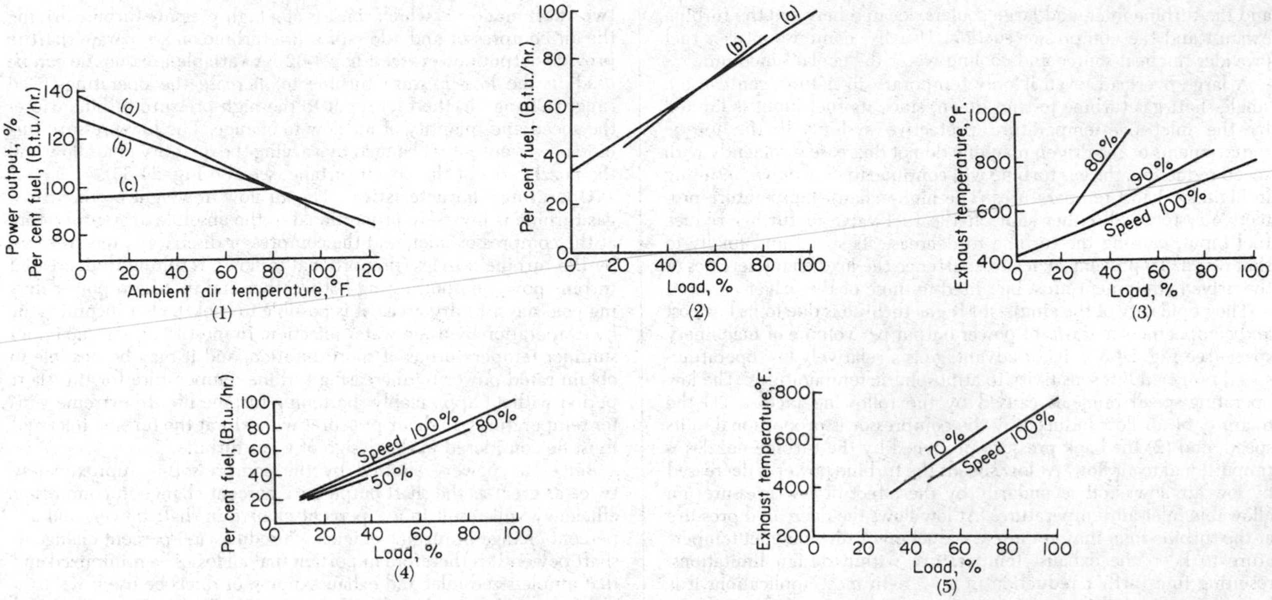

FIG. 24-33 Gas-turbine performance at various speeds, temperatures, and loads. (1) At 100 percent speed, output *a* and fuel consumption *b* are shown as functions of ambient temperature; fuel at 100 percent power is shown as *c*. (2) Fuel consumption with a single-shaft gas turbine at 80°F and (*a*) 100 percent and (*b*) 90 percent speed. (3) Exhaust temperatures of a single-shaft gas turbine at ambient temperatures of 80°F and speeds as shown. (4) Fuel consumption of a two-shaft gas turbine at various loads and speeds. (5) Exhaust temperature of a two-shaft gas turbine at an ambient temperature of 80°F. °C = 5/9(°F − 32).

EXPANSION TURBINES

Fundamentally, an expansion turbine is a device for converting the pressure energy of a gas or vapor stream into mechanical work as the gas or vapor expands through the turbine. The mechanical work so produced, however, is generally a by-product, the primary objective of the turboexpander being to chill the process gas. Turboexpanders are in wide use in the cryogenic field to produce the refrigeration required for the separation and liquefaction of gases.

By common usage, the terms "turboexpanders" and "expansion turbines" specifically exclude steam turbines and combustion gas turbines, which are covered elsewhere in Sec. 24.

Any work developed by the turboexpander is at the expense of the enthalpy of the process stream, and the latter is correspondingly cooled. A low inlet temperature means a correspondingly lower outlet temperature, and the lower the temperature range, the more effective the expansion process becomes.

FUNCTIONAL DESCRIPTION

The turboexpander in combination with a compressor and a heat exchanger functions as a heat pump and is analyzed as follows: In Fig. 24-34 consider the compressor and aftercooler as an isothermal

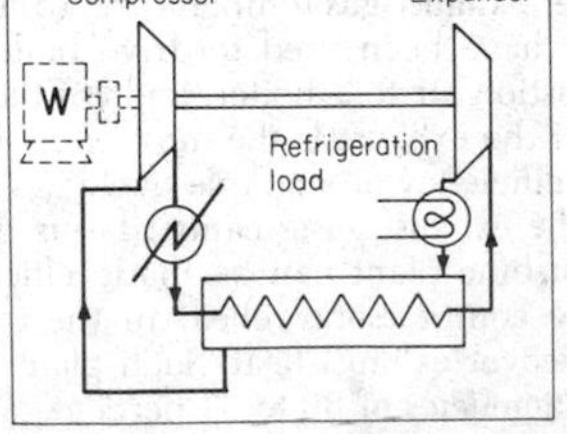

FIG. 24-34 Turboexpander system functioning as a refrigeration machine.

compressor operating at T_2 with an efficiency E_c, and assume the working fluid to be a perfect gas. Further, consider the removal of a quantity of heat Q_e by the turboexpander at an average low temperature T_1. This requires that it deliver shaft work equal to Q_e. Now, make the reasonable assumption that one-tenth of the temperature drop in the expander is used for the temperature difference in the heat exchanger. If the expander efficiency is N_e and this efficiency is multiplied by 0.9 to include the effect of the temperature difference in the heat exchanger, the needed ideal enthalpy drop across the expander is

$$H_e = Q_e/0.9N_e \tag{24-15}$$

The theoretical required (isothermal) compression work in the compressor, which is assumed to operate isothermally at T_2, is

$$(Q_e/0.9N_e)(T_2/T_1) \tag{24-16}$$

The actual compressor work W_c is this latter quantity, divided by the compressor isothermal efficiency N_c; thus,

$$W_c = (Q_e/0.9N_eN_c)(T_2/T_1) \tag{24-17}$$

Mechanical work equal to $Q_e/0.9$ is returned by the expander to the compressor, so the net work to the compressor is

$$W = W_c - \frac{Q_e}{0.9} = \left(\frac{Q_e}{0.9}\frac{T_2}{N_eN_cT_1}\right) - \frac{Q_e}{0.9}$$
$$W = \left(\frac{Q_e}{0.9}\frac{T_2}{N_eN_cT_1}\right) - 1 = \frac{Q_e}{0.9}\left(\frac{T_2 - N_eN_cT_1}{N_eN_cT_1}\right) \tag{24-18}$$

The second-law theoretical work is

$$W_{\text{theor}} = Q_e\frac{T_2 - T_1}{T_1} \tag{24-19}$$

Hence, the second-law efficiency of the expander–heat-exchanger–compressor system is

$$\frac{W_{\text{theor}}}{W} = \frac{Q_e \dfrac{T_2 - T_1}{T_1}}{\dfrac{Q_e}{0.9}\left(\dfrac{T_2 - N_e N_c T_1}{N_e N_c T_1}\right)} = \frac{0.9(T_2 - T_1)N_e N_c}{T_2 - N_e N_c T_1} \tag{24-20}$$

A plot of this efficiency in which commonly available equipment is assumed is shown by the expander curve in Fig. 24-35.

The family of short curves in Fig. 24-35 shows the power efficiency of conventional refrigeration systems. The curves for the latter are taken from the *Engineering Data Book*, Gas Processors Suppliers Association, Tulsa, Oklahoma. The data refer to the evaporator temperature as the point at which refrigeration is removed. If the refrigeration is used to cool a stream over a temperature interval, the efficiency is obviously somewhat less. The short curves in Fig. 24-35 are for several refrigeration-temperature intervals. A comparison of these curves with the expander curve shows that the refrigeration power requirement by expansion compares favorably with mechanical refrigeration below 360°R (−100°F). The expander efficiency is favored by lower temperature at which heat is to be removed.

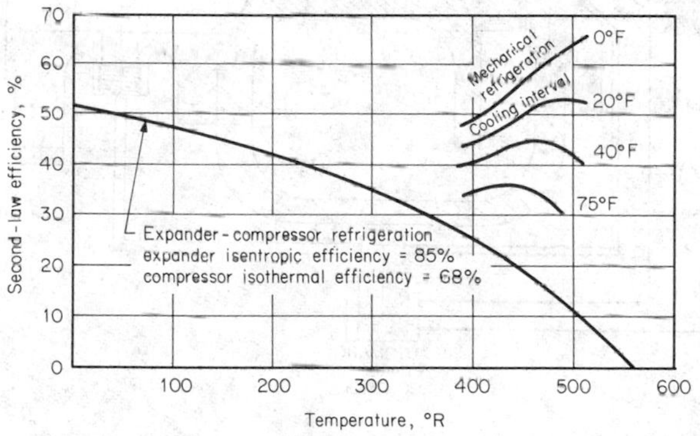

FIG. 24-35 Mechanical versus turboexpander refrigeration. K = 5/9°R; °C = 5/9(°F − 32).

Another conclusion that can be drawn from Fig. 24-35 is that if the process can justify the complexity, it is more efficient, powerwise, to use conventional means rather than expanders to absorb heat at moderate temperatures in the range of ambient to 360°R, although for expediency expanders are frequently used in any case.

SPECIAL CHARACTERISTICS

An example of a typical turboexpander is shown in Fig. 24-36. Radial-flow turbines are normally single-stage and have combination impulse-reaction blades, and the rotor resembles a centrifugal-pump impeller. The gas is jetted tangentially into the outer periphery of the rotor and flows radially inward to the "eye," from which the gas is jetted backward by the angle of the rotor blades so that it leaves the rotor without spin and flows axially away.

Radial-flow turbines have been developed primarily for the production of low temperatures, but they also may be used as power-recovery devices.

The characteristics of these machines include the following:

1. High efficiency: 75 to 88 percent
2. Operation usually at a very low temperature
3. Operation often on small or moderate streams, dictating a rather high rotating speed
4. Supports having low heat conductivity
5. Effective shaft seals to conserve the process stream

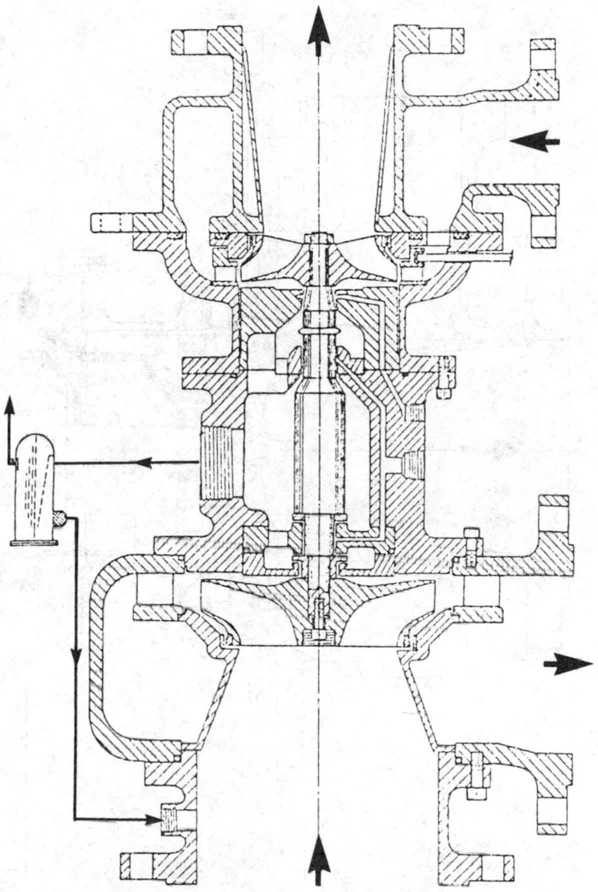

FIG. 24-36 Typical radial-flow turboexpander. Rotate above figure 90° counterclockwise; continue figure on page 24-34. *(Rotoflow Corp.)*

6. Heavy-duty construction resistant to abuse
7. High reliability

Commonly established operating limitations for turboexpanders without special design features are an enthalpy drop of 93 to 116.3 kJ/kg (40 to 50 Btu/lb) per stage of expansion and a rotor-tip speed of 304.8 m/s (1000 ft/s). Commercial turboexpanders are available for inlet pressure up to 20.68 MN/m^2 (3000 lbf/in^2) and inlet temperatures from near absolute zero to 538°C (1000°F). The permissible liquid condensation in the expanding stream varies with discharge pressure; it may be 50 weight percent or higher in the discharge, provided the turboexpander has been specially designed to handle condensation.

RADIAL INFLOW DESIGN

The radial reaction design has been selected for turboexpanders primarily because it attains the highest efficiency of all turbine designs. However, it has several additional features which favor this application:

1. In the single-stage configuration, it is usual to have the rotor on the end of the shaft (overhung); this provides a convenient opportunity for thermally insulating the cold turbine portion and is an ideal arrangement for axial discharge.
2. This design permits variable primary nozzles, which enable the attainment of high efficiency over a wide flow range.
3. It applies lower axial thrust to the shaft than a single-stage axial reaction turbine.

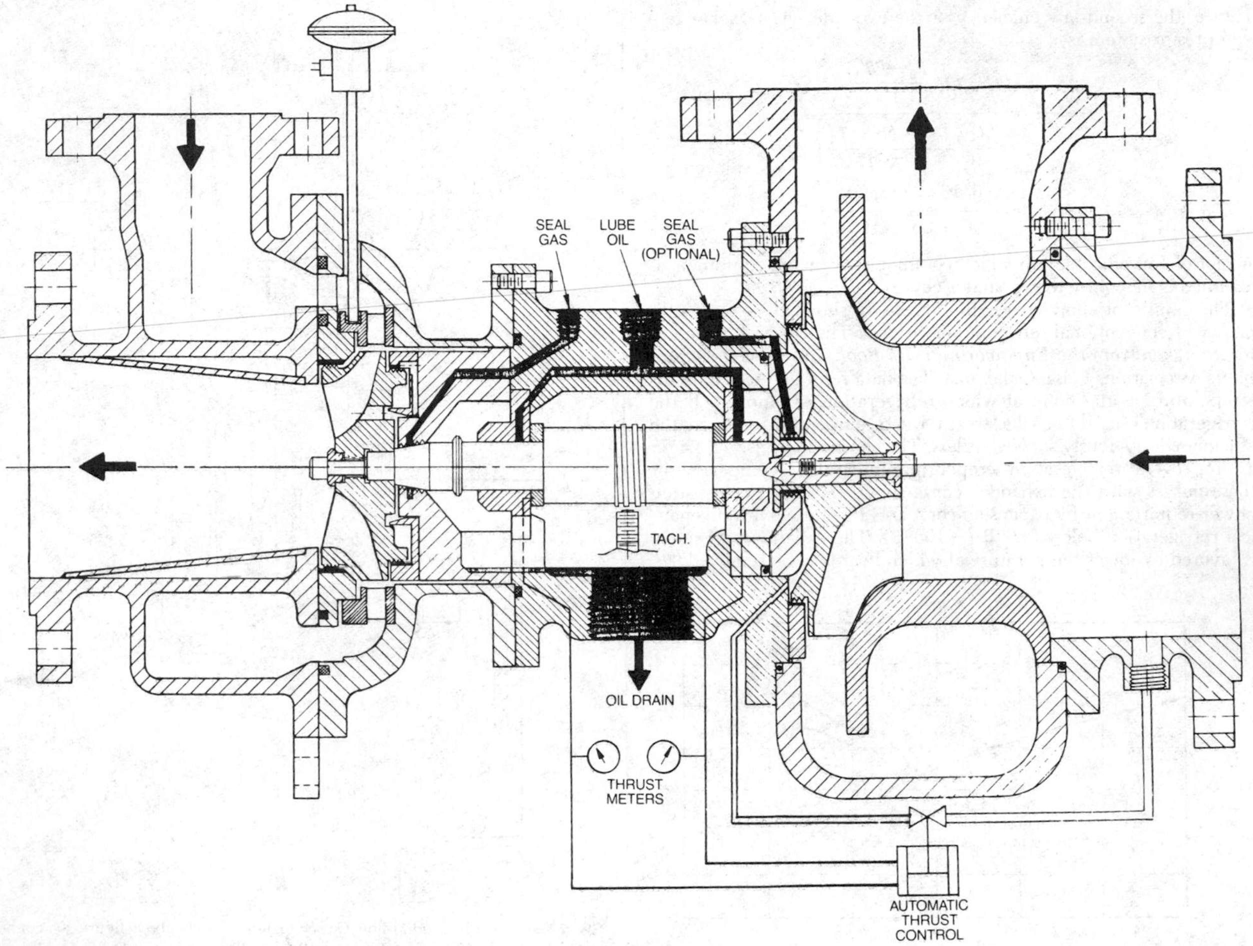

FIG. 24-36 *(continued)*

4. It has shaft bearings on only one side of the expander rotor, and heat-barrier insulation between the warm, lubricated bearings and the cold turbine is convenient to arrange.
5. Its speed is reasonably acceptable for suitable loading devices.

EFFICIENCY

Efficiency for a turboexpander is calculated on the basis of isentropic rather than polytropic expansion even though its efficiency is not 100 percent. This is done because the losses are largely introduced at the discharge of the machine in the form of seal leakages and disk friction which heats the gas leaking past the seals and in exducer losses. (The exducer acts to convert the axial-velocity energy from the rotor to pressure energy.)

BEARINGS

Radial Bearings Antifriction bearings are largely unsuitable for these applications, chiefly because of the special attention and maintenance which they demand.

Virtually all expanders have lubricated sleeve bearings or tilting-shoe bearings. The advantage of sleeve bearings is that they can be designed to support the shaft sufficiently rigidly that the oil-film critical (first critical) can be designed to be safely above the design running speed. This is desirable because it eliminates all shaft vibrations in the full operating range; more important, if the critical were below the design speed, then at design speed the shaft assembly would be rotating about its center of gravity. In turboexpanders, there frequently are ice deposits or other reasons for rotor imbalance, and such imbalance would cause gyration and damage to the shaft seals.

In arrangements in which the shaft is running with its first critical (oil-film critical) below the design speed, then not far above design speed is the oil swirl (half-speed gyration). Substitution of tilting-shoe bearings for sleeve bearings moves the oil swirl to a higher speed, but it does not prevent the rotor from rotating about its center of gravity. Tilting-shoe bearings do not lubricate well at this high rubbing speed and small shaft diameter.

The sleeve bearing has the further advantage that it functions as a lubricated, pressurized shaft seal to contain the pressure on the process gas (see subsection "Shaft Seals").

Thrust Bearings Turboexpanders often have process upsets or ice plugging or the like, which can cause serious thrust-bearing load variations. In applications above 506.6 to 1013.2 kPa (5 or 10 atm), the best available thrust bearing usually is insufficient to protect against such high thrust loads. Various indications, such as the differential pressure across the rotor and thrust-bearing temperatures, are available to protect the unit at least to some extent.

Thrust-bearing-load meters for protection against excessive loading are available on thrust bearings, and automatic thrust control, which functions by controlling the pressure behind a balancing drum, also is available.

SEALS

Shaft Seals Mechanical shaft seals generally are not acceptable in turboexpanders for the same general reason that antifriction bearings are not: they require periodic replacement and careful attention. In their stead, close-clearance labyrinth-type seals are generally used. Such a labyrinth seal generally has an injection point intermediate from its two ends into which a suitable buffer gas is injected to prevent the escape of the process gas; instead, buffer gas escapes. If the escaping seal gas is inexpensive and nontoxic, it may be allowed to leak to the atmosphere. However, it is possible to enclose the expander housing to the bearing housing and allow the journal bearing, acting as an oil seal, to contain the outleaking seal gas and collect it in a float-operated drainer for suitable disposal or reuse.

In refrigeration applications in which the refrigerant must be completely conserved, the expander housing and bearing housing can be hermetically sealed through a speed-reducing gearbox. The low-speed shaft of the gearbox is sealed with a low-speed mechanical seal. Then any refrigerant which leaks out of the labyrinth seal is totally contained in the gearbox and in the closed lubrication system for complete collection and reuse.

Rotor Seals To balance the thrust on the rotor, usually there are one or two labyrinth-type seals on the rotor. These seals often are damaged if there is dust in the incoming fluid or gas, and wear on the backside seal causes serious upsets in thrust-bearing loads. Provisions are available for collecting and disposing of the dust which tends to accumulate in the seal so as to protect the seal from serious erosion.

VARIABLE NOZZLES

The pressurized process stream is guided radially into the rotor by the primary nozzles, which are a series of vanes forming nozzles jetting the gas tangentially and inwardly into the rotor (see Fig. 24-37). These nozzle vanes are clamped between two flat rings and usually are pivoted so that they can be rotated in unison to open or close the spaces between them in order to vary the nozzle-throat areas. This is quite an important function because it can be used to vary the flow widely through the expander without wasteful throttling: all the expansion energy in the nozzles is recovered in the rotor. The variable nozzles, from a control standpoint, act just as a throttle valve would act in controlling the flow (but without throttling loss), and conventional flow-control instrumentation can be used to operate them.

FIG. 24-37 The primary nozzles and the rotor.

ROTOR RESONANCE

Variable nozzles produce a series of jets of gas entering the rotor, and these impulses add up to form a frequency equal to the blade-passing frequency: the number of revolutions per second multiplied by the number of nozzle vanes, which is of the order of thousands of cycles per second. Frequently the rotor will resonate at this frequency, and if it does, it will be fatigued and crack and break up; thus these frequencies must be avoided, and the manufacturer should be asked to supply information to the customer on this subject.

CONDENSING STREAMS

It is advantageous to have the expander generate its refrigeration at the lowest possible temperature in the process (see Fig. 24-35), and this frequently encounters the condensation temperature of the process stream. Steam-turbine practice advises against operating on condensing streams because efficiency is deteriorated and there also usually is erosion of the rotor blades.

Another advantage of the radial reaction turbine is that it can be designed to accept condensation in any amount without efficiency deterioration or erosion.* This is possible because there are two forces acting on suspended fog particles, the deceleration force and the centrifugal force, and these two forces can be balanced against each other to prevent the droplets from impinging on specially shaped blades. The process is explained as follows:

This expansion of a condensing vapor is highly desirable thermodynamically, but the liquid must not bombard and erode the rotor blades, and, in particular, it must not accumulate in the rotor, since that would cause efficiency loss.

If liquid droplets form as the gas is expanded in the turboexpander, one's first thought may be that a radial inflow design is the last thing to use, but the following explanation will show that this is the only design that can accomplish expansion efficiently.

Figure 24-37 shows the primary nozzles and the rotor. About half of the pressure drop takes place in the primary nozzles, which jet the gas tangentially into the periphery of the rotor. Cooling takes place during this expansion, and the jetted stream entering the rotor may be foggy. This foggy gas flows radially inward within the rotor as the latter rotates, and at the outlet, which is near the center of the rotor, the fluid is discharged by being jetted backward out of the rotor so as to leave without rotary motion. The second half of the expansion energy is spent by the gas passing radially through the rotor against centrifugal force, and further precipitation of liquid takes place.

The stream from the nozzles enters the rotor with a tangential velocity of about 152.4 or 304.8 m/s (500 or 1000 ft/s; see Fig. 24-38) and follows a path through the rotor of such curvature that the centrifugal force acting on an element of the stream and, therefore, on suspended droplets is of the order of 75,000 G (75,000 times the force of gravity). Also, the stream, because it is moving radially inward in the rotor, is decelerated in the rotor from this tangential velocity of 304.8 m/s (1000 ft/s) down to zero tangential velocity in about a half a revolution of the rotor. This deceleration force amounts to something like 10,000 G. The vector sum of these two forces, therefore, amounts to 75,000 or 100,000 G in a direction 5 to 15° from the radial direction (see Fig. 24-38). This is an acceptable

*This same principle applies also to the expansion of flashing liquids in which the bubbles are guided away from the blades.

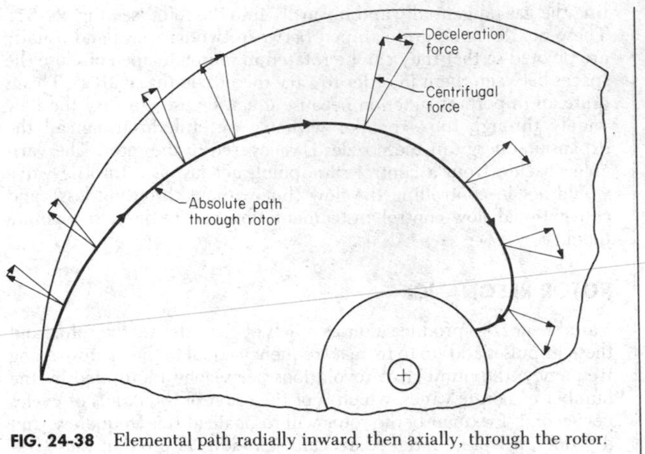

FIG. 24-38 Elemental path radially inward, then axially, through the rotor.

direction for the blades to lie, so the problem of avoiding bombardment of the blades by the droplets is solved by shaping the expander rotor blades parallel to this resultant vector. Then there is no force causing droplets to drift in the direction of any surface. They do drift back upstream slightly but nevertheless are carried on through by the mainstream and discharged. By this method any amount of condensing liquid can pass through, or, in the case of flashing liquids being expanded, the bubbles can pass through without efficiency loss. It would be impossible to construct a turbine blade meeting this requirement without two vector forces.

APPLICATIONS

The important uses of turboexpanders are in:

1. Air separation
2. Recovery of condensables from natural gas
3. Liquefaction of gases, including helium
4. Augmenting refrigeration in various cryogenic processes such as the recovery of ethylene
5. Power generation, sometimes referred to as power recovery

A potential application for the turboexpander for power recovery exists whenever a large flow of gas is reduced from a high pressure to some lower pressure or when high-temperature process streams (waste heat) are available to boil a secondary liquid. When such conditions exist, they should be examined to see if use of a turboexpander is justified. In such cases a turboexpander can be used to drive a pump, compressor, or electric generator, thus recovering a large portion of otherwise wasted energy. In applications of this type, careful consideration should be given to the temperature drop which will occur in the expander. It may sometimes be necessary to heat or to dry the inlet gas to avoid low exhaust temperatures that cause the formation of ice or liquids.

Expanders are used because they are in an advanced state of development and reliability, attain high efficiency, and are relatively inexpensive.

LUBRICATION

Expander bearings are usually high-speed, and they should have full film lubrication. This is best assured by using force-feed lubrication at a pressure of the order of 689.5 kN/m^2 (100 lbf/in^2) or more. There is no special objection to using pressures as high as 6.895 MN/m^2 (1000 lbf/in^2) or higher, if for some reason it is desirable to do so.

Usually, a journal bearing and a thrust bearing are combined in one assembly, and oil is injected so as to feed both of them. The rate of flow usually is adjusted so as to carry the heat away with a temperature rise of the order of 11 to 17°C (20 to 30°F).

The smallest expanders usually use oil with a viscosity at 38°C (100°F) of 60 to 100 SSU, and large machines up to 500 SSU. If the oil is kept in a totally enclosed system in contact with hydrocarbon or another partly soluble gas, which would dissolve and reduce the viscosity of the oil, then a compensating higher viscosity should be used so that the working viscosity after ultimate equilibrium with such gas is suitable for the bearings.

The lubrication system, for reliability reasons, usually has an operating and a standby pump and dual switchable filters. If there is a cooling-water scaling problem, coolers may also be switchable.

BUFFER-GAS SYSTEM

The shaft seal (see subsection "Shaft Seals") generally is a close-clearance labyrinth-type seal. It is desirable that there be available a suitable pressurized buffer gas for injection into the intermediate point in the seal, such gas to be available at an absolute pressure well above the highest shaft pressure to be sealed. Then the seal-gas system may consist of only a filter, a flow-indicating device, and a throttle valve or other flow or pressure control, usually a pressure regulator and a graduated needle valve.

If the available pressure is not far above that of the pressure to be sealed, then with simple throttling the flow may be insufficient when the two pressures come too near together. Then more precise control, such as by differential pressure between the process side and the seal-gas pressure, may be required.

SIZE SELECTION

Size, rotating speed, and efficiency correlate well with the available isentropic head, the volumetric flow at discharge, and the expansion ratio across the turboexpander. The head and the volumetric flow and rotating speed are correlated by the specific speed. Figure 24-39 shows the efficiency at various specific speeds for various sizes of rotor. This figure presumes the expansion ratio to be less than 4:1. Above 4:1, certain supersonic losses come into the picture and there is an additional correction on efficiency, as shown in Fig. 24-40.

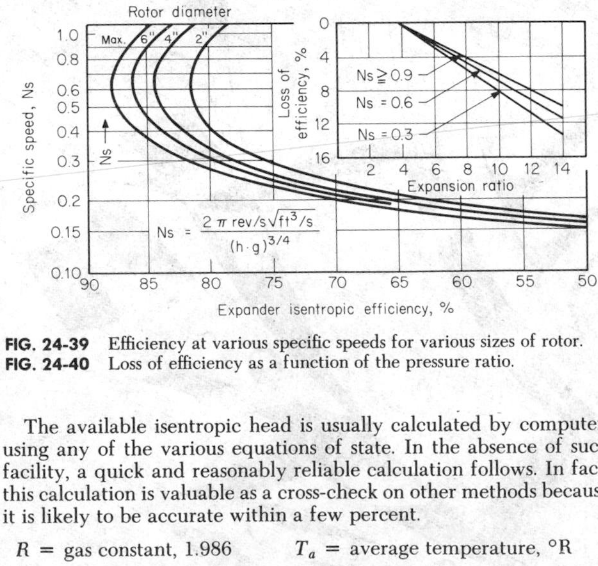

FIG. 24-39 Efficiency at various specific speeds for various sizes of rotor.
FIG. 24-40 Loss of efficiency as a function of the pressure ratio.

The available isentropic head is usually calculated by computer, using any of the various equations of state. In the absence of such facility, a quick and reasonably reliable calculation follows. In fact, this calculation is valuable as a cross-check on other methods because it is likely to be accurate within a few percent.

R = gas constant, 1.986 $\quad$ T_a = average temperature, °R

Z_a = compressibility at T_a $\quad$ H_i = isenthalpic work,

$$H_i = \frac{RT_aZ_a}{\text{mol. wt.}} \ln \frac{P_1}{P_2} \qquad \text{Btu/lb} \qquad (24\text{-}21)$$

where H_i = reversible incremental enthalpy drop, Btu/lb
R = gas constant, 1.986, Btu/(lb·mol·°R)
T_a = average temperature for the increment, °R
Z_a = average compressibility for the increment
P_1/P_2 = pressure ratio

The use of this equation requires that an average P and T, based on an assumed increment, be used to find Z.

INSTRUMENTATION

Process-flow control and buffer-gas control have been discussed under "Variable Nozzles" and "Buffer-Gas System" respectively. Speed is usually self-controlled by a matching speed-sensitive load such as a compressor or a pump. If the load is an induction or synchronous generator feeding into a stable ac system, the system frequency fixes the speed. Otherwise, the speed can be controlled by a conventional governor.

Various protective instruments are used to provide a shutdown signal (to a fast-acting trip valve at the expander inlet) that senses various things, such as overspeed, lubricant pressure, bearing temperature, lubricant temperature, shaft runout, icing, lubricant level, thrust-bearing load, and process variables such as sensitive temperatures, levels, pressures, etc. However, too many safety shutdown devices may lead to excessive nuisance shutdowns.

POWER RECOVERY FROM LIQUID STREAMS

BASIC PRINCIPLES

The potential for power recovery from liquid streams exists whenever a liquid flows from a high-pressure source to one of lower pressure in such a manner that throttling to dissipate pressure occurs. Such throttling represents a system potential for power that is the reverse of a pump—in other words, a potential for power extraction. Just as in a pump, there exists a hydraulic horsepower and a brake horsepower, except that in the recovery they are generated or available horsepowers.

Basically, power recovery from liquids is achieved in industrial installations as shaft horsepower. While this potentially could appear either as reciprocating or as rotating power, most larger applications are rotating. Consideration of power recovery from liquids involves a choice among several possible uses, and usually this choice involves as alternatives (1) driving of a few large-horsepower services versus driving of more smaller-horsepower services; (2) driving of essential versus nonessential services or of spared versus nonspared services; (3) driving as sole driver versus partial driver or full horsepower versus partial horsepower for the selected service; and (4) converting power-recovery energy to some other intermediate energy form, as by driving an electric generator.

In applying power recovery, three basic problems are (1) limitations in designing equipment to recover the power, (2) operating reluctance to consider rotating equipment that is not absolutely necessary, and (3) the way in which the economics of the installed system is evaluated. It is important to recognize that there has always been an operable, acceptable alternative to power recovery from liquid streams in the form of the throttling or letdown valve, whereas no such simple, cheap, foolproof substitute exists for the pump.

Basic to establishing whether power recovery is even feasible, let alone economical, are considerations of the flowing-fluid capacity available, the differential pressure available for the power recovery, and corrosive or erosive properties of the fluid stream. A further important consideration in feasibility and economics is the probable physical location, with respect to each other, of fluid source, power-production point, and final fluid destination. In general, the tendency has been to locate the power-recovery driver and its driven unit where dictated by the driven-unit requirement and pipe the power-recovery fluid to and away from the driver. While early installations were in noncorrosive, nonerosive services such as rich-hydrocarbon absorption oil, the trend has been to put units into mildly severe services such as amine plants, hot-carbonate units, and hydrocracker letdown.

Economics Power-recovery units have no operating costs; in essence, the energy is available free. Furthermore, there is no incremental capital cost for energy supply. Incremental installed energy-system costs for a steam-turbine driver and supply system amount to about \$250 per kilowatt, and the incremental cost of an electric-motor driver plus supply system is about \$25 per kilowatt. By contrast, even the highest-inlet-pressure, largest-flow power-recovery machines will seldom have an equipment cost of more than \$45 per kilowatt, and costs frequently are as low as \$20 per kilowatt. However, at bare driver costs (not including power supply) of \$20 to \$45 per kilowatt for the power-recovery driver versus about \$8 to \$25 per kilowatt for steam turbines or \$15 to \$20 per kilowatt for electric motors, operating costs must be considered to make power-recovery units attractive. Using commonly accepted values for power costs, turbine steam rates, and steam selling prices, operating costs for either motors or steam turbines approximate \$80 per year for 746 W (1 hp).

Thus, barring technical difficulties or operational considerations in application, power-recovery units ought to show payouts of less than 6 months. Actual project payouts run from 1 to 3 years. This difference is due principally to (1) the fact that while the incremental costs just presented are valid in comparing large systems, specific designs encounter frame-size breaks, standardized capacities and horsepowers, code requirements, etc.; and (2) operating requirements, sparing considerations, and a certain lack of confidence in power-recovery units stemming from lack of extensive experience produce equipment-selection schemes that deviate from the straightforward comparison.

Development The following discussion relates specifically to the use of what could be called radial-inflow, centrifugal-pump power-recovery turbines. It does not apply to the type of unit nurtured by the hydroelectric industry for the large-horsepower, large-flow, low- to medium-pressure differential area of hydraulic water turbines of the Pelton or Francis runner type. There seems to have been little direct transfer of design concepts between these two fields; the major manufacturers in the hydroelectric field have thus far made no effort to sell to the process industries, and the physical arrangement of their units, developed from the requirements of the hydroelectric field, is not suitable to most process-plant applications.

Despite a rather slow start, centrifugal power-recovery pump-turbines have built a respectable record of process-plant installations extending back to the middle 1950s. Applications have included drives for the following services: cooling-tower fans; reciprocating recycle compressors; gas-treating-solution circulation pumps, as sole drive and as tandem with a steam turbine or motor helper; refinery-unit charge-stock pumps with a helper driver; and floating-online electric generators.

In general, early experiences were in the small-horsepower, nonessential or spared services, using sole drivers at full horsepower. The present trend is more and more toward the few large drivers in essential services, usually supplying only partial horsepower but not spared. If a plant is based on electric drivers and the economics of rate structure, demand charges, etc., permits, the use of a power-recovery unit driving a generator electrically floating on the line has found increasing favor. In general, operating experience in regard to reliability, serviceability, and maintainability has shown the units to be comparable with centrifugal pumps and has resulted in increasing acceptance even as drivers in large-horsepower units on essential service equipment. Presently accepted industrial limits are shown in Fig. 24-41.

Hydraulic Behavior The basic hydraulic behavior of centrifugal pumps operating as power-recovery units (turbines) is not much different from that of centrifugal pumps and follows the same sort of affinity laws over narrow ranges. Typical generalized curves are shown in Figs. 24-42 and 24-43. Note particularly that both torque and horsepower go negative (turn to values indicating power con-

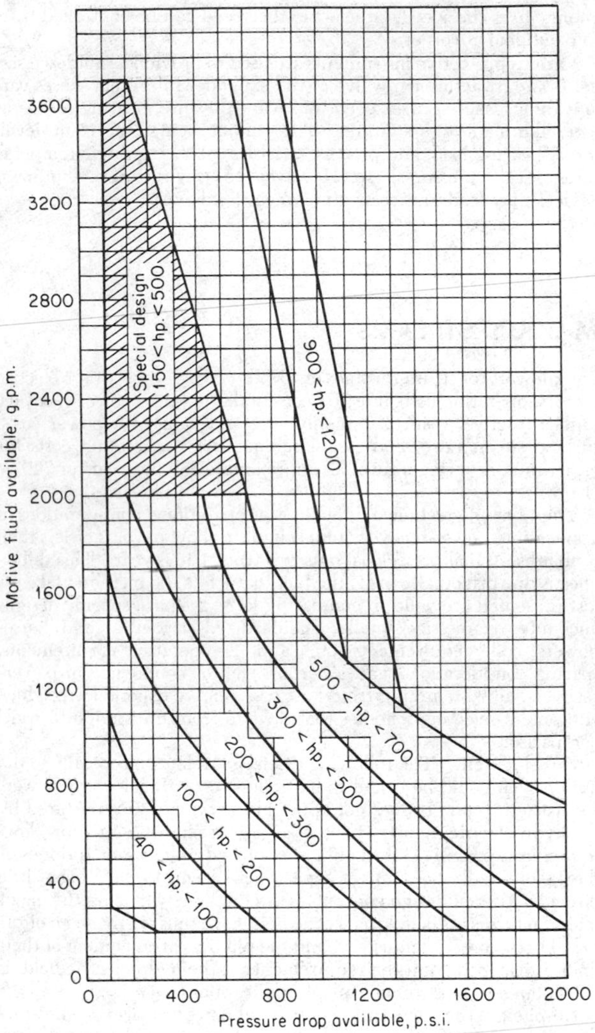

FIG. 24-41 Application areas for centrifugal-pump turbines. Curves apply between the following minimum and maximum limits: inlet pressure, 100 to 3000 psig; pressure differential, 100 to 2800 lbf/in^2; flow of motive fluid, 200 to 4000 gal/min; horsepower, 50 to 3000 hp. Curve horsepower is based on a speed of 3600 r/min and a fluid of 1.0 specific gravity; for other fluids, multiply the curve horsepower by specific gravity to get the actual horsepower. To convert pounds-force per square inch to megapascals, multiply by 6.89×10^{-3}; to convert gallons per minute to cubic meters per minute, multiply by 3.79×10^{-3}; and to convert horsepower to kilowatts, multiply by 0.7457.

sumption) when head and capacity are within a fairly wide range representing at least startup and shutdown conditions if not also part load. Note also that even if head goes to 125 percent of design, speed (r/min) at zero torque for this unit does not exceed about 130 to 150 percent of design.

Tests conducted to operate centrifugal pumps as hydraulic turbines throughout the head-capacity-speed range show that a good centrifugal pump generally makes an efficient hydraulic turbine. From theoretical considerations it is possible to state that at the same speed

$$H_t = H_p/e_h^2 \tag{24-22}$$

$$Q_t = Q_p/e_h \tag{24-23}$$

$$n_{st} = n_{sp}e_h \tag{24-24}$$

where H = total head at best efficiency point
Q = capacity
n_s = specific speed
e_h = hydraulic efficiency, taken as the same for the turbine and the pump
t, p = subscripts denoting turbine and pump respectively

Since the exact value of the **hudraulic efficiency** e_h is never known, $\sqrt{e}$ can be taken as an approximation where e is the gross (hydraulic horsepower/brake horsepower) pump efficiency. Efficiencies of pump designs running as turbines are usually 5 to 10 efficiency points lower than those as pumps at the best efficiency point.

OPERATING BEHAVIOR

By considering the flow as stopped but the turbine casing full of liquid, it is intuitively obvious that to rotate the wheel or impeller in either direction power will have to be put in. As the flow increases from the no-flow conditions, the fluid velocity through the wheel gradually approaches such a rate that it imparts enough energy to the wheel not only to overcome internal friction but also to permit some net power output for consumption; this point usually occurs at about 30 to 40 percent of design flow or capacity. As in any turbine driver, the machine will speed up until the load imposed on the shaft coupling by the driven unit equals the power entering from the power-recovery wheel. Like a pump, a power-recovery unit will ride its characteristic curve and seek a point at which its particular head-capacity-speed–power-output relationship is satisfied. In most applications, the head available to the unit, being largely composed of a static-pressure difference, is nearly constant and varies only to the extent that inlet and exit piping-friction losses vary with flow through the unit. Thus the unit finally acts as an orifice in a relatively fixed differential system, meaning that it has a definite flow limit which also produces a torque and horsepower limit. This can be seen by following the 100 percent head curve of Fig. 24-42.

Performance Characteristics Performance of the power-recovery unit **operating as the sole driver** (Fig. 24-44) is shown in Fig. 24-45. If it is assumed that more liquid at the available head is presented to the power-recovery unit than is needed to generate the horsepower required by the pump, the turbine unit will speed up to handle the liquid, and at the same time the pump speed must go up. In speeding up, the turbine will generate more horsepower, which the pump must absorb while it is at the new speed. Finally, a balance point on horsepower is reached with the driven unit, but the number of revolutions per minute may be off design. If speed control of the driven unit is necessary, throttling some of the available capacity across a valve bypassing the unit permits the unit to satisfy its horsepower-capacity-speed relationship at the desired number of revolutions. A similar problem occurs when the capacity available to the unit is less than that needed at available head and design revolutions. The unit will slow down, try shedding load, and attempt to come to peace with its head-capacity-speed curve sets. Here speed control can be achieved by throttling the available pressure so that the unit sees only that portion of the available head needed to satisfy its head-capacity-speed relationship at the desired number of revolutions.

Performance of the power-recovery unit **operating with a makeup driver** (Fig. 24-46) is shown in Fig. 24-47; specific percentage values are shown, but the general characteristics and curve shapes are typical. It should be noted that the flow scheme, the selection of equipment, and the design of that equipment have produced the relatively inflexible system pattern shown in the curves, in which (1) except at a single point the recovery unit always requires either flow bypassing or inlet-pressure throttling (see bottommost curve); and (2) the horsepower output of the recovery unit is reduced at any point away from design (note the horsepower-difference curve), which, combined with the characteristics of steam turbines, produces the unusual turbine throttle steam-flow curve (second from the top in Fig. 24-47).

DESIGN CONSIDERATIONS

Involved in producing the curves for Figs. 24-45 and 24-47 is a calculation of the so-called **balance point** at which the flow and revo-

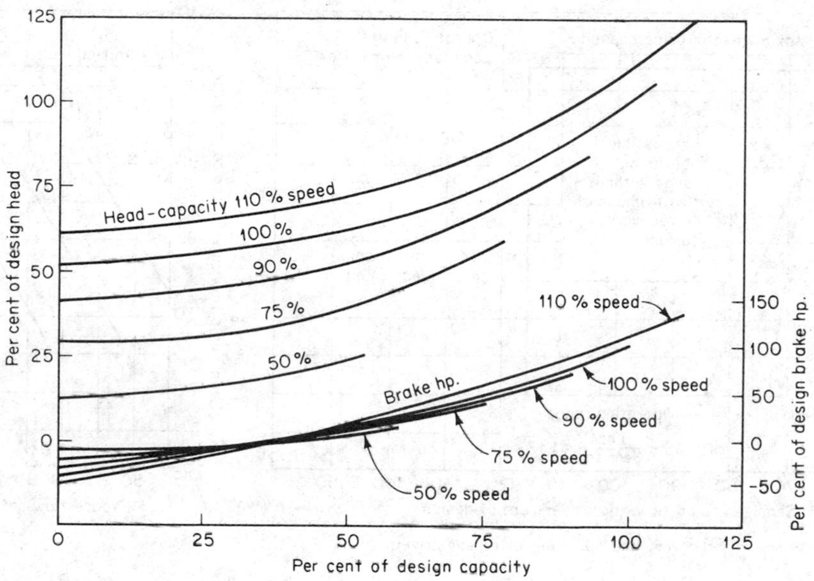

FIG. 24-42 Generalized curves showing hydraulic behavior of centrifugal pumps operating as power-recovery turbines.

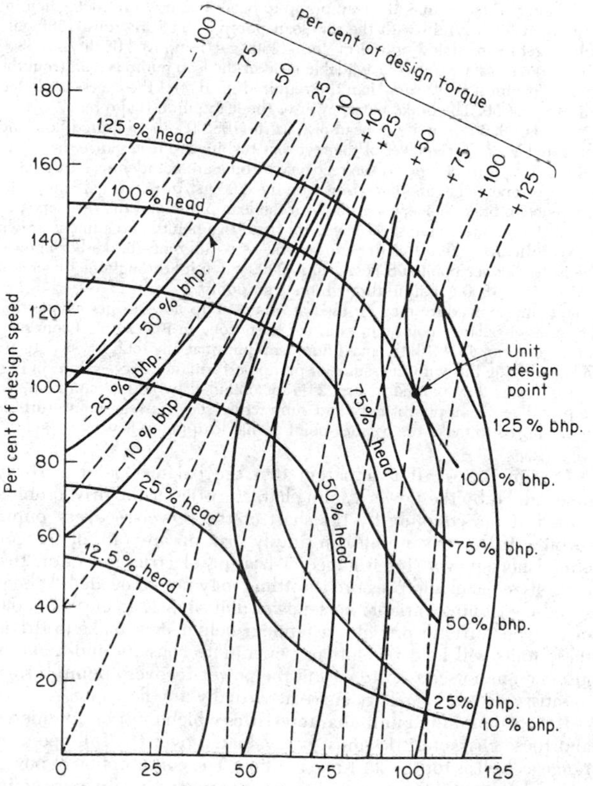

FIG. 24-43 Generalized curves for centrifugal pumps operating as power-recovery turbines.

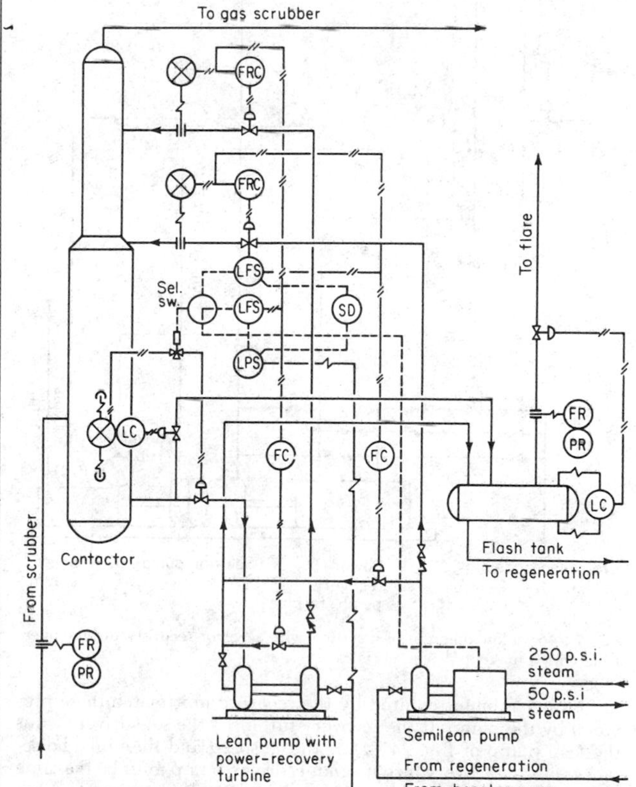

FIG. 24-44 Flow diagram of a power-recovery unit operating as the sole driver.

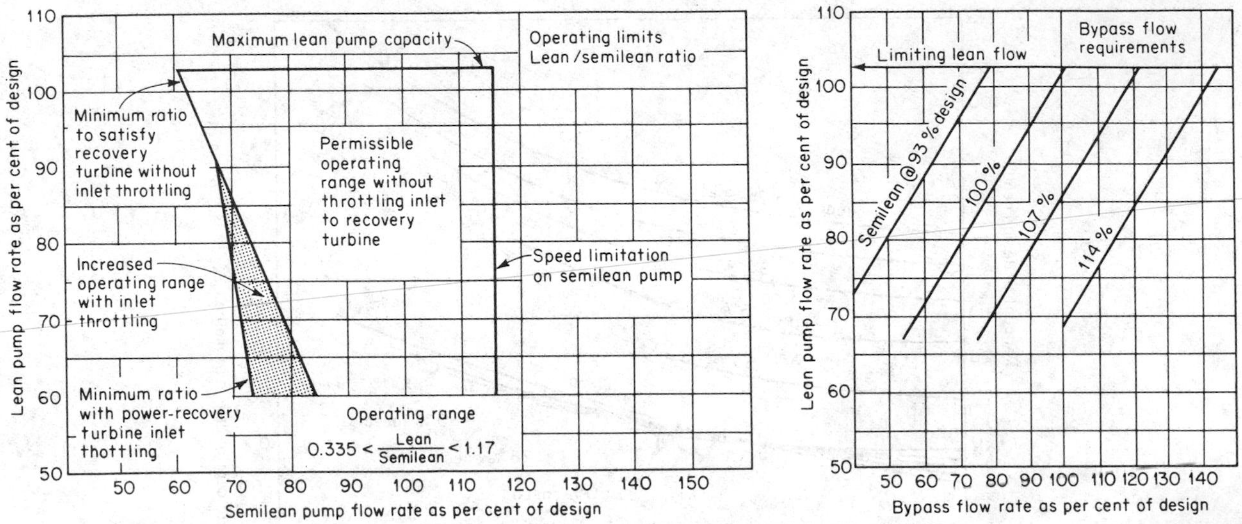

FIG. 24-45 Performance of a power-recovery unit operating as the sole driver.

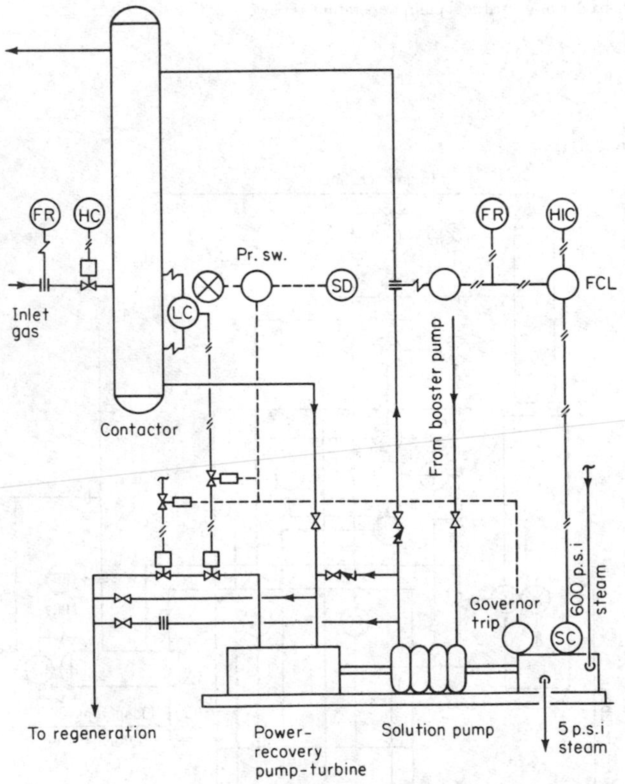

FIG. 24-46 Flow diagram of a system with a power-recovery turbine operating with a makeup driver.

lutions per minute required by the recovery unit match those provided by the pump. If the recovery turbine is the sole driver (as for the lean pump of Fig. 24-44), both the speed and the brake horsepower of the recovery turbine and its driven pump must be the same at the so-called balance point. If there is a makeup driver and the recovery unit has available to it just the flow from the pump that it is driving, as for the pump of Fig. 24-46, then the speed and capacity must match at the balance point.

Example 2 The scheme of Fig. 24-44 and the actual units supplied (Figs. 24-48 and 24-49) will be used for purposes of illustration. Since this case has the recovery unit as the sole driver, the balance point is set by speed and horsepower.

For purposes of example, assume a flow of 8.71 m^3/min (2300 gal/min) through the tower. The maximum head available to the recovery turbine was calculated to be 604 m (1982 ft); this value will be slightly in error when part of the flow is bypassed since frictional losses into and out of the recovery unit will change. First, assume the lean pump to be at 3.03 m^3/min (800 gal/min) running at 3900 r/min with the semilean pump at 5.68 m^3/min (1500 gal/min) to get the total flow of 8.71 m^3/min (2300 gal/min). At 3.03 m^3/min (800 gal/min) and 3900 r/min the available head of the lean pump is read from the curve. This must be greater than the required head, and the excess is plotted as in Fig. 24-50. The brake horsepower of the lean pump is also read.

Now, at 3900 r/min and a head of 6.04 m (1982 ft), the required flow and generated brake horsepower of the recovery turbine are read. Since the horsepower of the lean pump and the recovery turbine are not identical, this entire process is repeated at another speed with the 3.03 m^3/min (800 gal/min). The difference in brake horsepower between the lean pump and the recovery turbine is then plotted against the speed for these two points, and a line is drawn between them. Where this line crosses the zero-difference brake-horsepower line is the balance point at 8.71 m^3/min (2300 gal/min) through the tower and 3.03 m^3/min (800 gal/min) through the lean pump.

The same procedure may be used at other pump flows to permit plotting the series of balance-point curves as has been done in Fig. 24-51. From such curves, one can establish the maximum lean pump at any total tower outflow, and combining this with the semilean-pump performance curve results in Fig. 24-45. Bypass flow plotted in Fig. 24-45 is obtained by adding simultaneous lean- and semilean-pump flows and subtracting the recovery pump-turbine flow required to make the balance point at that lean-pump flow.

Design Bases It is apparent that the balance point is always determined by the power r/min characteristics of the driven unit as sensed at the coupling by the shaft of the power-recovery pump-turbine. If the driven unit can simply soak up any (all) of the generated horsepower, for instance, a floating electric generator, then capacity control and pressure throttling may not be needed. When a speed-controlling variable-horsepower unit such as an electric motor or a steam turbine provides a tandem helper or a makeup driver, these units will hold revolutions per minute constant and make up just enough horsepower to permit the power-recovery pump-turbine to satisfy its head-capacity curve at virtually any flow rate.

It is the **combined unit characteristics** which must be considered, and these characteristics must be evaluated over the full operating range as well as for the startup condition. The consumption of power, up to almost 40 percent of design output, on starting up and coming up to speed and the fact that under a relatively fixed head condition the maximum speed at zero torque is about 140 percent of design have both already been noted. These, of course, bear particular sig-

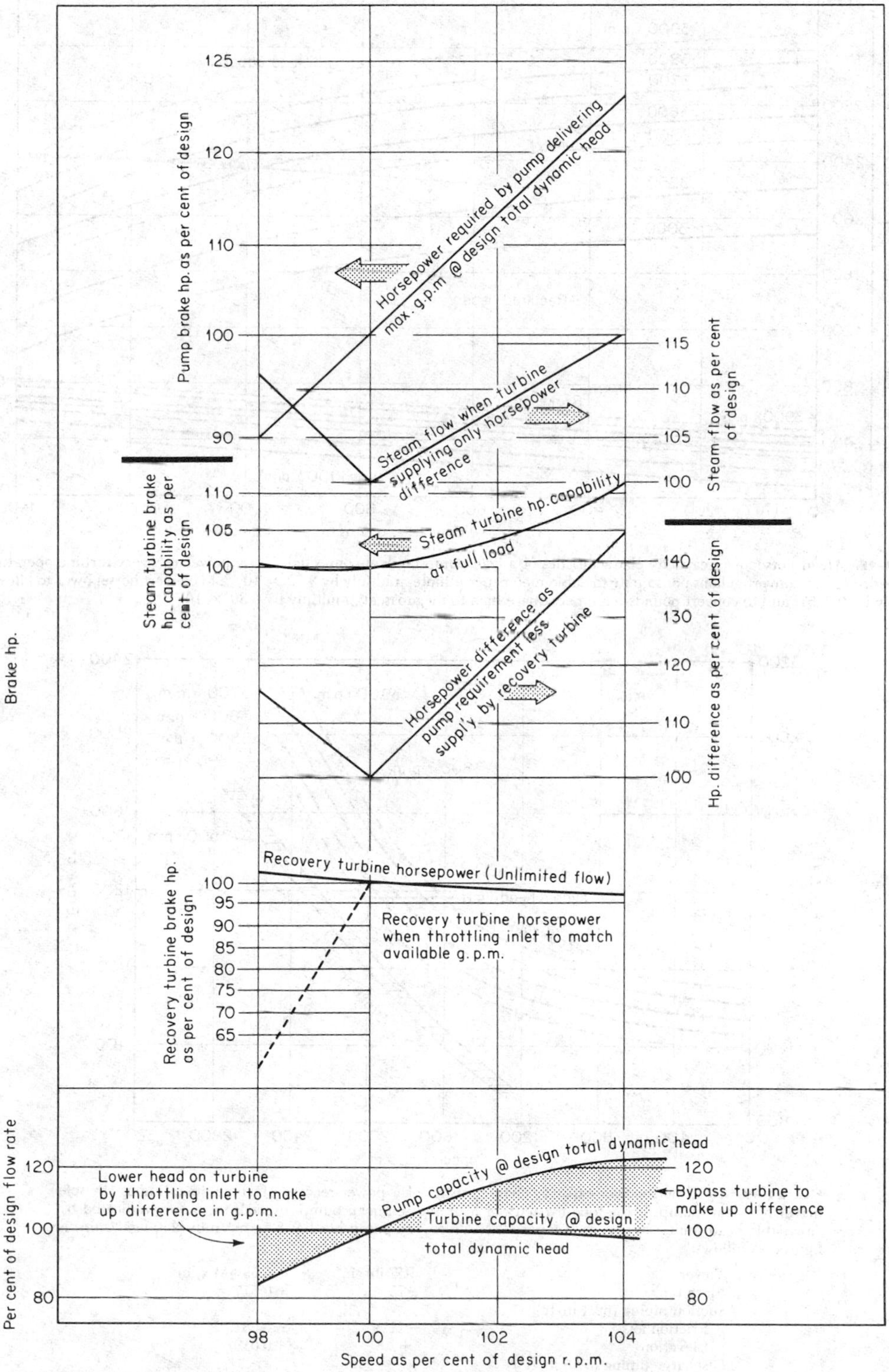

FIG. 24-47 Performance of a power-recovery turbine operating with a makeup driver.

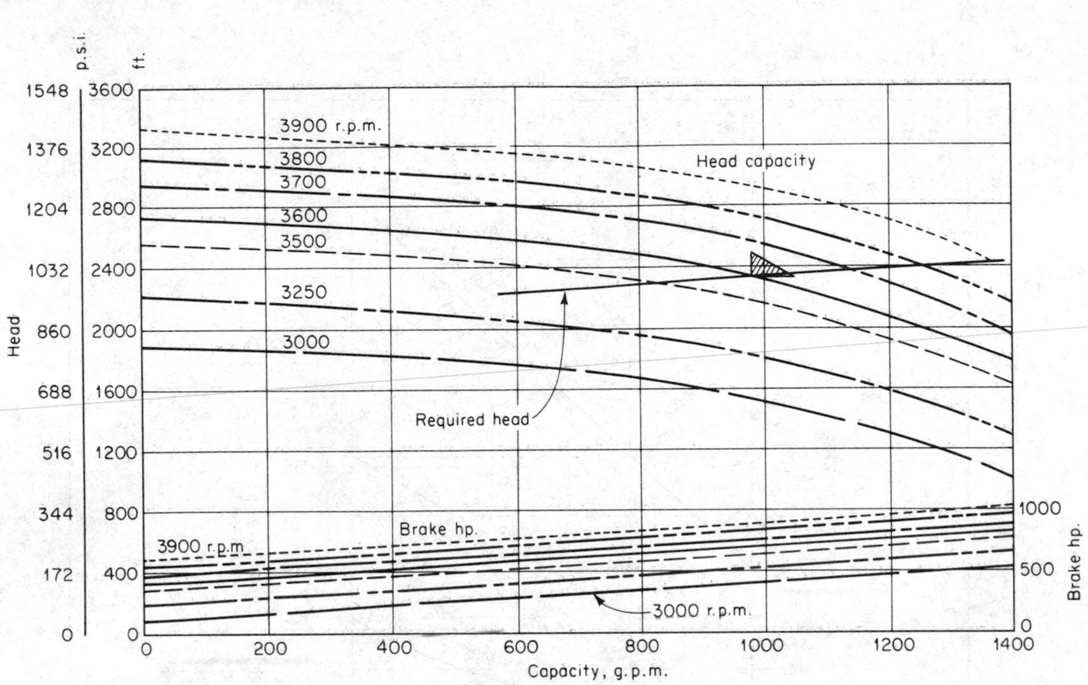

FIG. 24-48 Head-horsepower-capacity characteristics of a lean pump tandem-connected with a power-recovery turbine operating as the sole driver. To convert gallons per minute to cubic meters per minute, multiply by 3.79×10^{-3}; to convert horsepower to kilowatts, multiply by 0.7457; and to convert pounds-force per square inch to megapascals, multiply by 6.89×10^{-3}.

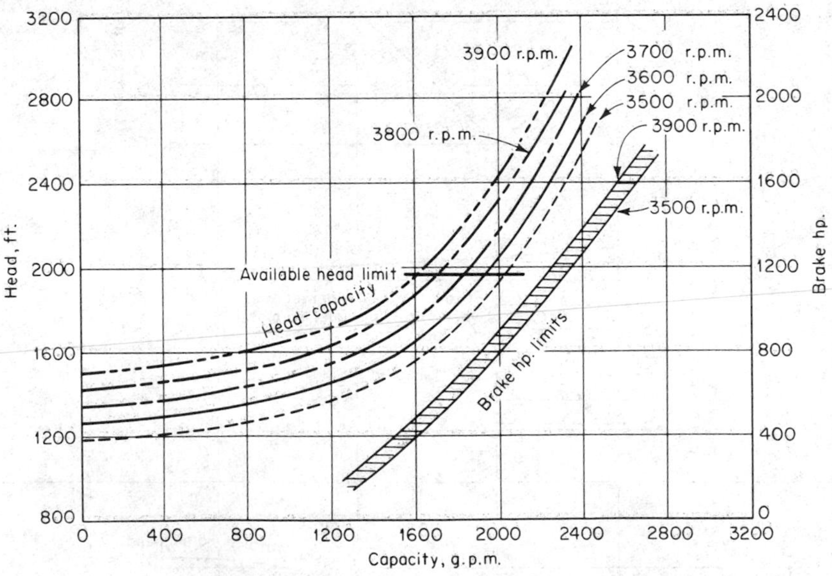

FIG. 24-49 Head-horsepower-capacity characteristics of a power-recovery turbine operating as the sole driver of a lean pump. If the total capacity of lean and semilean pumps exceeds the values indicated by "available head limit," bypass must be used. Net recovery-pump head at 8.71 m^3/min (2300 gal/min) is figured as follows:

Tower	957 lbf/in^2	6.598 MN/m^2
Flash tank	−75	−0.517
Suction piping (62.4 lb/ft^3)		
Friction loss	−11.8	−0.081
Elevation	+2.8	+0.019
Discharge piping (57.8 lb/ft^3)		
Friction loss	−8.8	−0.061
Elevation	−5.6	−0.038
Net	859 lbf/in^2	5.923 MN/m^2

NOTE: To convert gallons per minute to cubic meters per minute, multiply by 3.79×10^{-3}; to convert horsepower to kilowatts, multiply by 0.7457; and to convert pounds-force per square inch to megapascals, multiply by 6.89×10^{-3}.

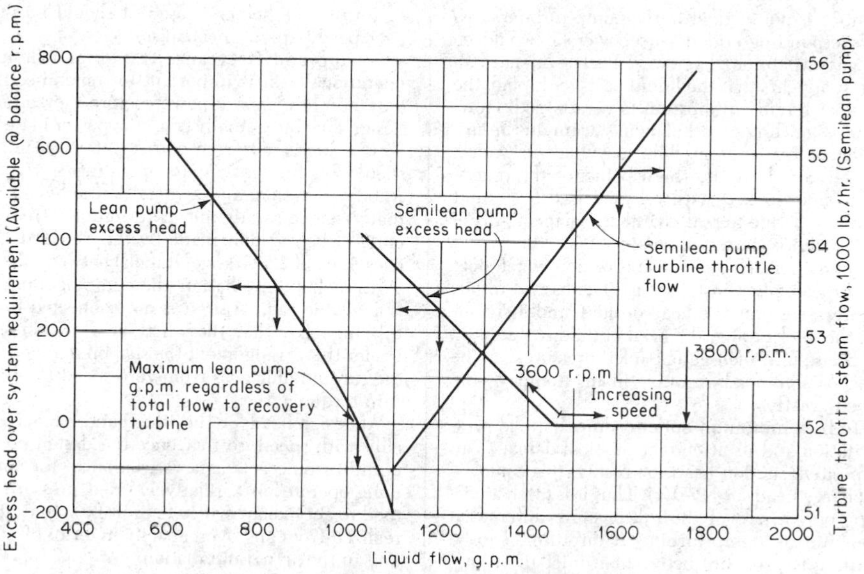

FIG. 24-50 Excess head developed by lean and semilean pumps and the steam-throttle flow for a semilean-pump turbine. To convert gallons per minute to cubic meters per minute, multiply by 0.00379; to convert pounds per hour to kilograms per second, multiply by 1.260×10^{-4}.

nificance for starting the unit up or shutting it down and must be considered. Failure to perform a complete system analysis can frequently lead to a process design that proves, upon installation, to be an operating trap.

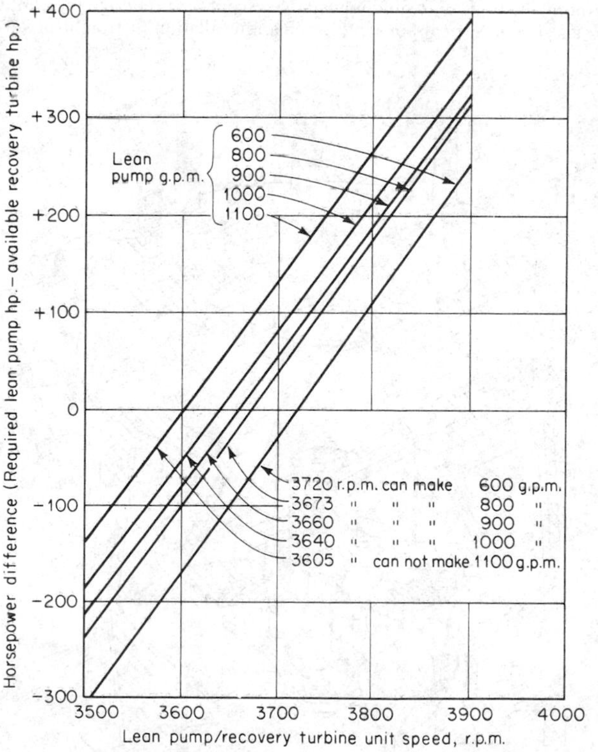

FIG. 24-51 Horsepower-r/min balance for a lean pump tandem-connected with a power-recovery turbine operating as the sole driver. Horsepower differences are calculated from excess head requirements as typically shown in Fig. 24-50. To convert gallons per minute to cubic meters per hour, multiply by 0.2271; to convert horsepower to kilowatts, multiply by 0.746.

In realizing the advantages of competitive designs for hydraulic power-recovery systems, there is usually an investment premium which must be paid for the **operating flexibility** illustrated in Fig. 24-45. This takes the form of additional reduced-capacity pumps and steam-turbine drivers, as well as some sacrifice in power recovery that results from bypassing or throttling even at the design point. If investment is to be minimized, a tightly designed system for full power recovery with a single pump-turbine helper, as in Fig. 24-46, may be worth considering. In such systems, the only fluid available to the pump-turbine is that provided by the pump it drives, and no separate pump with auxiliary driver is available for startup. Operating personnel will immediately see that such systems are relatively inflexible and difficult to operate (Fig. 24-47). Thus, while the tightly designed system has a minimum investment and more power recovery, it may not be the most desirable if operation away from the design point is anticipated, since only at that one point is neither throttling nor bypassing needed. Furthermore, it becomes apparent that the maximum steam requirements may be set by a partial-load condition rather than by design conditions or overload.

While the foregoing examples have dealt with applications on centrifugal pumps, the same sort of analysis can be made for reciprocating pumps, reciprocating compressors, or other rotary users like cooling-tower fans.

INSTALLATION FEATURES

In addition to performing the system analysis, a number of details or peculiarities of the units must be considered with respect to:

1. Vaporization, flashing, or cavitation
2. Fluid volumes
3. Process-stream controls
4. Speed control
5. Startup and overcapacity
6. Electrical-system characteristics if the recovered power is used to generate electricity

Vaporizing Fluids Many pump-turbines are installed on gas-saturated liquid streams, and loss in pressure can cause problems whenever this occurs across balancing drums or pressure-reducing labyrinth seals. Piping that carries bleed streams from the drums and seals to the low-pressure (outlet) side of the pump-turbine must be sized generously to allow for some gas evolution.

In general, gas evolution is not evident in the pump-turbine's casing proper, for the corresponding added horsepower (as would be anticipated from the gas expansion) has never been evidenced in reports on field testing. It appears that the liquid passage through the casing is too fast for vapor-liquid equilibrium to be attained. However, slower shearing passage through a balancing drum and return line does permit gas evolution, and this line may become vapor-locked if it is undersized. Similarly, the high points of the pump-turbine casing may be vapor-locked with released gas if the unit stands idle; this can cause damage to seal chamber, balancing-drum chambers, etc., when the unit is started up again.

There is another potential hazard due to vaporization which does not generally occur in process-plant installations. The hazard results from the fact that a pump sees only the head of fluid, and if significant flashing occurs in the inlet piping, the head of fluid represented by the pounds-force-per-square-inch-gauge inlet pressure can be many times greater than design head, resulting in an attempt by the unit to increase its speed greatly.

There also appears to be a minimum outlet or impeller eye pressure below which cavitation and its attendant physical damage can occur (similar to net positive suction head, or NPSH, this could be called net positive discharge head, or NPDH). Thus it is often advisable to design by using only a part of the full pressure differential available in the process for the pump-turbine. If throttling is to be provided, outlet throttling is probably better than inlet throttling, and if used, any mechanical seals, as well as the unit casing outlet flange, must be adequate to withstand the full inlet pressure when the throttle valve is closed.

Fluid Volumes Many process-plant installations of these units are made by handling *rich liquid* out of an absorber. In most cases both liquid volume and liquid density will change from input to output. While such changes are not normally significant, the sensitivity of the balance point to the volume of flow makes it mandatory to consider volumetric swell by absorption in checking the suitability and adequacy of the pump-turbine unit and the controls and bypassing arrangemens as part of a system analysis.

Process Controls If the inlet or outlet liquid to a pump-turbine is regulated by a level controller on the liquid-supply vessel, a falling liquid level inside the vessel will cause this controller to throttle a valve, reducing the differential pressure available to the pump-turbine; or if the liquid level is rising, the control valve tends to open wide, so that the pump-turbine sees its full available head. Since under this latter condition the head is at its maximum, no more liquid will flow through the pump, and if the level continues to rise, the system goes off level control. One performance like this inevitably leads to a request from operating personnel for a bypass around the pump-turbine. On a startup it is frequently necessary to have a bypass from a point upstream of the driven pump's discharge check and block valves to the associated pump-turbine.

Speed Speed control can be critical with pump-turbines. Considering the characteristics shown in Fig. 24-47 (and the curves on which it is based), a 4 percent change in speed, from 98 to 102 percent, produces a 33 percent change in pump flow, from 82 to 115 percent, and about a 22 percent horsepower change, from 118 to 100 to 122 percent. A NEMA Class A steam-turbine mechanical governor has about 10 percent steady-state regulation, so that for the 22 percent horsepower speed change one should not expect better than a 2 percent speed regulation. Since 4 percent is the total speed change being considered, it becomes apparent that a problem exists in governing the speed of the steam turbine. Since this unit is supposed to serve to hold pump speed by supplying horsepower as needed to get a match of pump and system curves at the desired flow rate, this is a serious concern. In this case the problem can be solved by eliminating the steam-turbine speed governor, leaving only the overspeed trip, and placing a control valve actuated directly by pump-flow measurement in the steam-supply line to the turbine.

In contrast to steam turbines, in which runaway overspeeding is always a problem, pump-turbines operating at design head go to zero torque at about 130 to 140 percent of design speed. Thus, overspeed protection may not be necessary if the pump-turbine can withstand 140 to 150 percent of design speed and it is the sole driver. When a steam-turbine helper is used, it should be provided with the usual overspeed trip-out mechanism.

Startup and Overcapacity From a design standpoint and also operationally, it is important to remember that pump-turbines not only do not generate power before they attain about 40 percent of design flow but actually consume power in decreasing amounts as the flow is increased from zero to 40 percent. This means that they should be brought up to operating speed as rapidly as possible. Another solution, and a more desirable one from an operating and maintenance standpoint, is to install a free-wheeling or overriding clutch (such as that made by the Marland Division of Zurn Industries, Inc., of La Grange, Illinois) between the turbine and the pump. With such an installation, the pump does not have to turn until fluid is available to it. Also, it is not connected to the pump until it tries running faster than the pump, at which time it is putting out power. Under this arrangement the startup sequence can be selected so that the turbine unit goes from zero speed to operating speed along the zero-torque curve.

At design head, on the other hand, capacity does not change markedly with speed, so that once the design point has been passed the pump-turbine acts as a restriction in the line. Since most of these units operate on a relatively fixed pressure differential, they then tend to act like an orifice to limit flow, and little or no benefit can be realized from any overcapacity in terms of fluid flow available to the unit in the actual installation.

Electrical Generation When pump-turbines are used to generate electricity, the units should be tied into electrically strong networks of such a size that the pump-turbines cannot swing the system frequency but are governed by that frequency and become constant-speed machines. Inlet or outlet throttling, as well as bypassing, must be used for such intallations. Consequently, with speed fixed and available head essentially constant because of the process, the maximum amount of recoverable power is established by the design. Once installed, a pump-turbine cannot be pushed into generating more power if more fluid is available except by a redesign of its internals. The electrical portion of such an installation is straightforward,

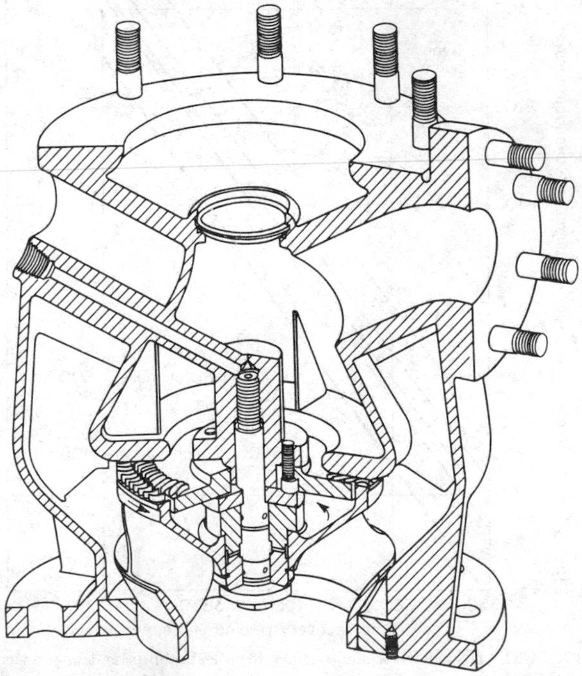

FIG. 24-52 Integral or packaged pump-turbine unit. *(Worthington Division, McGraw-Edison Co.)*

controls are simply inlet throttling and a bypass (if needed) for the pump-turbine, and these controls can be operated on a split range from the same level controller on the liquid-source vessel.

Integral Units A relatively recent development is the integral or packaged pump-turbine unit assembled in a single casing, having a common discharge port and designed for mounting directly in the piping. A cutaway view of one such design is shown in Fig. 24-52. It is obvious that the use of such a unit presupposes the compatibility and desirability of having the pump discharge material contaminated by all the powering fluid which passed through the driving turbine impeller. So far, this unit has seen little service in process plants, but its simplicity, its independence of normal power sources, and its ease of installation make it an attractive candidate for consideration when the process-flow-scheme can be adapted to its characteristics. It has been proposed for use in place of liquid eductors or ejectors, for it has a much higher efficiency than such units, and furthermore throttling motive fluid flow will cause the pumped fluid flow to follow according to balance-point requirements.

MECHANICAL POWER-TRANSMISSION EQUIPMENT

BELT DRIVES

Belts are used to transmit power from one shaft to another. At the same time, a reduction or increase in speed can be obtained.

A belt drive is fundamentally a friction drive. The most important factor is the relationship between the belt tension and the pulley. This produces a pressure and develops pulley torque.

For static conditions, the friction coefficient f is used to define the limits of the friction force F_f available to prevent a sliding action between the two flat surfaces and the force N which holds the two surfaces together.

$$F_f = fN \tag{24-25}$$

For belt speeds below 609 m/min (200 ft/min) a general equation which relates belt tension to the friction coefficient can be written:

$$T_1/T_2 = e^{f\theta} \tag{24-26}$$

in which T_1 is the tension on the tight side of the belt, lb; T_2 is the tension on the slack side of the belt, lb, e is the base of the natural logarithms (2.71828); f is the friction coefficient between the belt and the pulley; and θ is the arc of contact between the belt and pulley, rad.

For belt speeds in the range of 914 to 1524 m/min (3000 to 5000 ft/min), the centrifugal stress within the belt becomes an appreciable portion of the total stress. The significance of the centrifugal tension T_c is that a part of tension T_1 and T_2 is used up in forcing the belt to follow a curved path around the pulley. Therefore, this latter tension is not available to produce a pressure between the belt and the pulley.

An equation for centrifugal tension is

$$T_c = \rho btV_2/300g \tag{24-27}$$

where T_c is in pounds; ρ is the belt density, lb/in^3 (0.035 to 0.045); b is the belt width, in; t is the belt thickness, in; V is the belt speed, ft/min; and g is the gravitational constant, 32.2 ft/s^2.

The tension on the tight side of the belt should not be so great as to exceed a design stress of 1.38 to 2.76 MN/m^2 (200 to 400 lbf/in^2) for leather belts.

Belts can reach a **critical speed** which is noted when the pulleys move from side to side or by violent flapping on the slack side of the belt. This critical speed is a function of the installation geometry, belt characteristics, and tension. When the critical condition occurs, it can be corrected by changing these parameters.

The **power capacity** of a belt can be estimated from the following:

At 183 m/min (600 ft/min), use 1.2 hp/in of belt width.

At 1828 m/min (6000 ft/min), use 8.9 hp/in of belt width.

V Belts The wedging action of a V belt against the pulley grooves increases the contact load. Therefore, the friction of V belts is 2 or 3 times that of flat belts, which results in proportionately greater power transmission. In addition, V belts can use smaller pulley diameters than flat belts for the same power rating.

The load-carrying capacity T_1 of a V belt varies between 15.9 and 181.4 kg (35 and 400 lb).

The theoretical speed ratio between the driver and the driven equipment is essentially proportional to the pulley diameter for both flat and V belts. However, because of slippage and creep, the actual speed ratio is not constant. Slippage between the pulley and the belt can be as much as 2 percent for flat belts but is almost zero for V belts. Creep is caused by elongation of the belt under tension, which results in a greater belt velocity on the driving side than on the slack side.

Adjustable cones are used to obtain variable pulley diameters, and therefore adjustable speed can be obtained by using V-belt drives.

CHAIN DRIVES

Steel roller chains and inverted-tooth chains are classified as power-transmission chains. The roller chain is a series of alternately assembled roller links and pin links in which the rollers are free to turn. The chain may be a single strand having one row of roller links or a multiple strand having more than one row. Chain and sprockets eliminate slip that is often noted in using belt drives. They can also carry a much greater load. The speed reduction equals exactly the ratio of the number of teeth on the driver and driven sprockets. Most chain drives run at speeds to 1828.8 m/min (6000 ft/min). Tensile strengths are as high as 444.8 kN per chain (100,000 lb per chain).

GEAR DRIVES

Meshing teeth formed with special cutters provide a much more compact drive than either belts or chain drives. They can operate at higher speeds and power. The most common gears are spur gears. As gear speeds increase, friction increases the heat released from the sliding contact between the gear teeth, and special tooth profiles must be used. These special profiles, such as the cycloidal, provide a rolling action to minimize friction. However, the gears and shafts must be very accurately aligned and spaced. Another profile, the involute, does not provide as good a rolling action as the cycloidal profile, and some sliding between the teeth occurs. But the involute profile is not as sensitive to shaft alignment or gear spacing.

Helical gears are formed by cutters that produce an angle which allows several teeth to mesh simultaneously.

Herringbone gears have two helices of opposite angles on the same gear. This eliminates the axial thrust of a single helix.

Planetary Gears This type of gearing consists of three sets of gears. A pinion or sun gear meshes with two or more planet gears. These, in turn, rotate within an internal gear. The planetary gears are connected to one shaft and the sun gear to another. This system is compact and achieves a large speed reduction in a limited space. It also makes possible the design of the input and output shaft on the same centerline.

Bevel Gears Bevel gears are used most frequently for 90° drives. However, other angles can be used. The most typical application is the driving of a vertical pump with a horizontal driver.

Worm Gears Speed reductions in the range of 500 to 1 can be obtained by using worm gears. The contact surface of the screw on the worm slides along the gear teeth. Slightly less efficiency is obtained with this type of gear than with precision spur gears. Also, heat removal is a problem, and this limits worm gears to low-speed applications. Large helix angles on the gear teeth produce higher efficiencies. The gear can be driven from either the worm or the gear teeth. This type of gear can be self-locking. Thus, a torque on the gear will not cause the worm to rotate.

The worm may have one or more screw threads. One revolution of the worm advances the gear teeth in direct proportion to the number of worm threads. The speed-reduction ratio is equal to the total number of worm-gear teeth divided by the number of worm threads.

VARIABLE-SPEED DRIVES

Every machine has an optimum speed at which it is most efficient. Within the same machine, this optimum speed may differ for various circumstances or products handled by the machine, and this optimum speed can seldom be accurately predetermined. Speed changes are also often necessary during the operation of the machine.

One mechanical type of variable-speed drive is the PIV gear. This device uses a chain drive enclosed in a housing. This chain drive transmits power from the input to the output shaft. It is shown in Fig. 24-53.

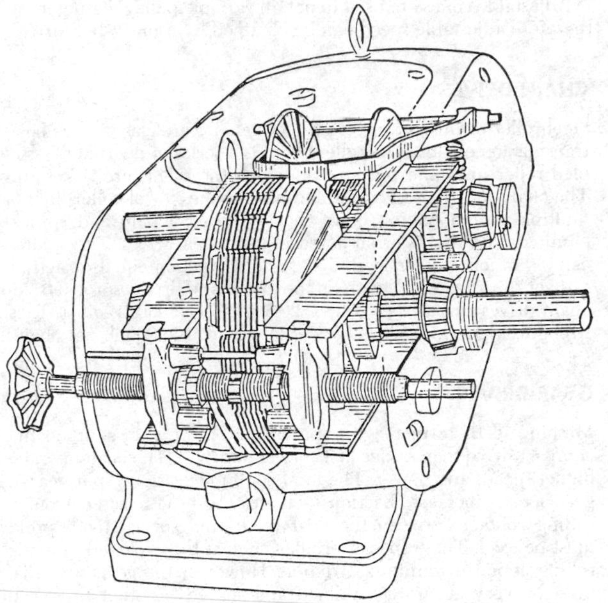

FIG. 24-53 PIV speed changer. (*From Kent,* Mechanical Engineers' Handbook, *12th ed., Wiley, New York, 1961.*)

By rotating the handwheel, control levers move in and out of pivots. This changes the effective diameter of the opposing wheels and provides a no-slip transmission with a continuously variable speed. The entire gear assembly is encased and operates in an oil bath. Speed ratios of 2:1 to 6:1 can be obtained with this device.

Variable-Pitch Pulleys The Reeves variable-speed transmission (Fig. 24-54) has a capacity range up to 74.6 kW (100 hp). Two pairs of cone-shaped disks are mounted on spline shafts. The handwheel is rotated, producing a movement in the levers which rotate about *P*. This causes the disks of one pair to approach and those of the other pair to recede from each other. By varying the distance between the disks, any desired speed ratio can be obtained.

The Reeves Vari-Speed pulley, shown in Fig. 24-55, uses a V belt to transmit power. Two cone-shaped disks are used in the device. One of the disks is stationary, and the other can slide laterally. The entire assembly can be mounted on a standard motor shaft. When the handwheel is rotated, the motor-and-disk assembly is moved forward or backward. When the motor is nearest the driven pulley, maximum speed is developed. Capacities range from 193 W to 11.2

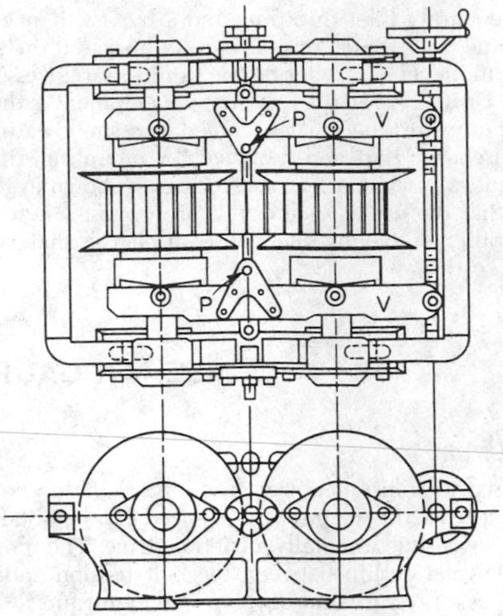

FIG. 24-54 Reeves variable-speed transmission. (*From Kent,* Mechanical Engineers' Handbook, *12th ed., Wiley, New York, 1961.*)

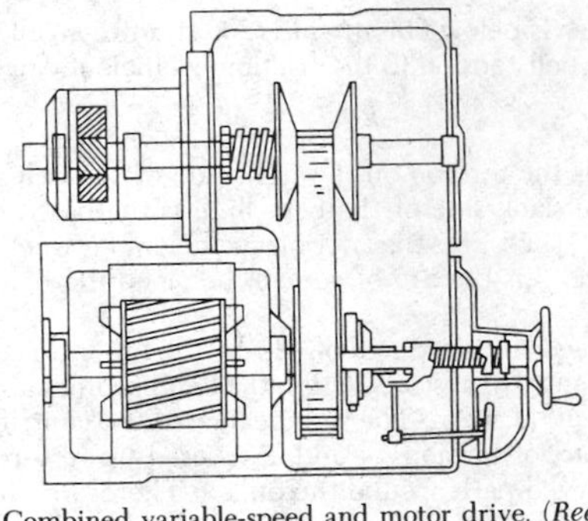

FIG. 24-55 Combined variable-speed and motor drive. (*Reeves Pulley Co.; from Kent,* Mechanical Engineers' Handbook, *12th ed., Wiley, New York, 1961.*)

kW (⅛ to 15 hp) at 1800-r/min motor speed with speed ratios up to 4:1.

Vari-Pitch Drive is a V-belt drive with one or two sheaves designed so that the grooves can be adjusted for a larger or smaller pitch diameter. This increases or decreases the speed of the driven shaft. The capacity is between 746 W and 93.2 kW (1 and 125 hp) and larger.

Several other mechanical variable-speed drives are available. They operate primarily on the principle of varying the pitch of a V-belt disk.

Hydraulic Variable-Speed Drives The hydrostatic type uses a variable-displacement pump. This type also uses a constant or variable-displacement hydraulic motor.

In this type of drive, the input shaft is driven at a constant speed by an electric motor or other prime mover. The output shaft can be varied in speed, and the rated torque can be transmitted throughout the speed range.

In the hydrostatic type, oil is pumped through a piston which regulates a piston on the output side. The output piston, in turn, positions a universal joint or other device that regulates the speed.

The hydrodynamic type of hydraulic variable-speed drive uses the turbine action of oil to transmit power. In one of these types, shown in Fig. 24-56, two rotors with straight radial vanes form a torus ring.

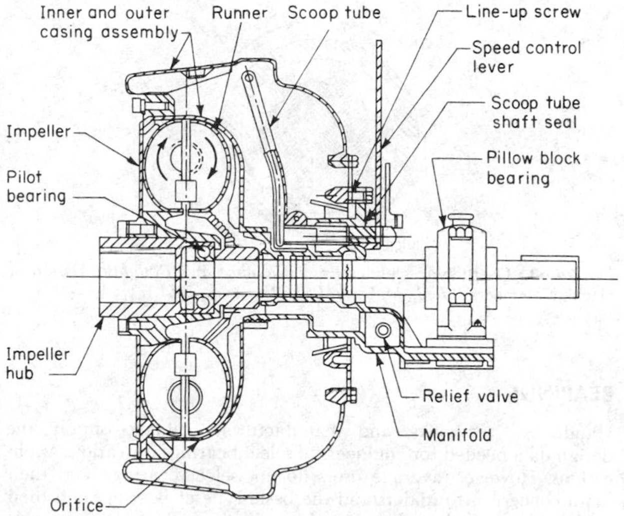

FIG. 24-56 Variable-speed fluid drive. (*American Blower Corp.; from Kent,* Mechanical Engineers' Handbook, *12th ed., Wiley, New York, 1961.*)

An impeller attached to the driving motor acts as a pump and sets up a vortex of oil which drives the runner like a turbine wheel. The runner is connected to the driven machine. By changing the amount of oil in the vortex, output speed can be adjusted.

In the lever-controlled fluid drive shown in Fig. 24-56, the rotating casing is attached to the impeller and stores centrifugally the oil not in the vortex. This type of drive is used for applications between 3.7 and 261 kW (5 and 350 hp).

In the pump-controlled fluid drive, an external reservoir is used to store the oil. To increase the speed, a small oil pump transfers the oil from the reservoir to the vortex. To decrease the speed, oil is diverted to the external reservoir by a valve or by reversing the pump. The pump-controlled fluid drive has applications for fan and centrifugal pump drives between 74.6 kW and 1.49 MW (100 and 2000 hp).

The enclosed-fluid-drive design uses a stationary housing. There are neither nozzles to regulate the oil flow from the vortex nor an oil pump to supply oil to the rotors. The oil is distributed centrifugally throughout the vortex, and a speed-control tube which skims off the rotating ring of oil according to its position is inserted between the casings. The speed controller sets the level of the oil in the vortex as well as the casing and controls the output speed and torque. This type of fluid drive is designed for applications at 74.6 kW (100 hp) and above.

Electrical and Electronic Speed Control A great variety of electrical methods of speed control are available. The most recent is the use of the silicon rectifier to control motor speed. Other methods for dc motors include series resistance, field shunt, two-motor series, and field control. For ac motors, frequency and voltage change are used to control speed along with two windings and consequent poles, secondary resistance, variable voltage, field control, and brush shifting (see subsection "Electric Motors").

Hydraulic Torque Converters The hydraulic torque converter produces an output speed that varies with torque requirements. The speed ratio is controlled automatically by the torque. To obtain accurate control of output speed, the torque converter must be used with a variable-speed driver, such as an internal-combustion engine.

Simply, the hydraulic torque converter consists of a pump, a turbine, and a reaction stator (see Fig. 24-57). The pump acts like a centrifugal pump and delivers hydraulic fluid to a turbine in which the fluid flows radially inward. Fluid leaves the turbine through a series of guide vanes in the stationary reaction member or stator and again enters the pump.

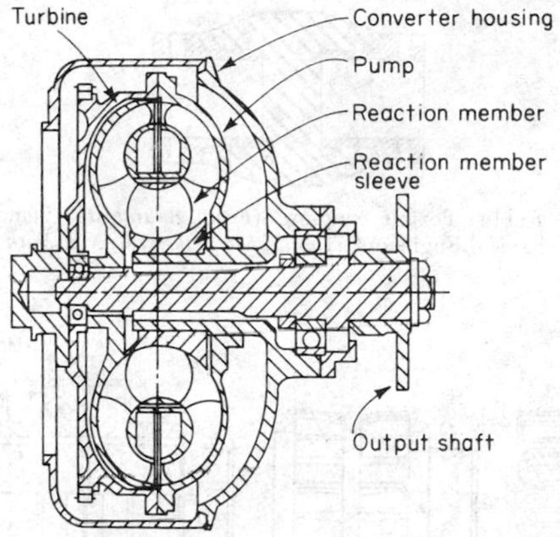

FIG. 24-57 Three-element, single-stage hydraulic torque converter. (*From Kent,* Mechanical Engineers' Handbook, *12th ed., Wiley, New York, 1961.*)

MECHANICAL COUPLINGS

Connections between the driver shaft and driven machine must be capable of transmitting torque.

Solid couplings consist of rigid flanges on each shaft bolted together solidly to form a continuous shaft. When the driver and the driven equipment are operated at the same speed, these couplings can be used effectively. Bearings can be eliminated. However, shaft misalignment can result in high shaft stresses, bearing overloading, or vibration. If misalignment can occur, flexible couplings must be used.

Flexible Couplings It is difficult to make an exact alignment of the driver and the driven shaft. To compensate for shaft misalignment, flexible couplings are often used (see Figs. 24-58 to 24-60).

Flexible couplings are also used to reduce the effect of torque variations. Rubber bushings or rubber inserts are used to obtain torsional flexibility.

To compensate for misalignment, a design is used that allows a sliding action between the meshing parts of the coupling. This sliding action creates friction and limits the allowable misalignment of high-speed shafts.

For high-speed shafts, the gear-type flexible coupling is used (see Fig. 24-58). In this type of coupling, the two coupling halves are bolted together. Each half contains an internal gear meshing with gear teeth on the shaft flanges. At operating speeds, any misalignment is corrected by a movement between the mating gear teeth. These couplings are filled with lubricating oil or grease to prevent damage by movement of the teeth at operating speeds.

Disconnect Couplings At times, it is desirable to connect or disconnect the driver from the driven equipment while in operation. Various types of friction clutches allow the shafts to be engaged or disengaged at operating speeds (see Figs. 24-61 and and 24-62).

To prevent reverse rotation, an overrunning clutch is used (Fig.

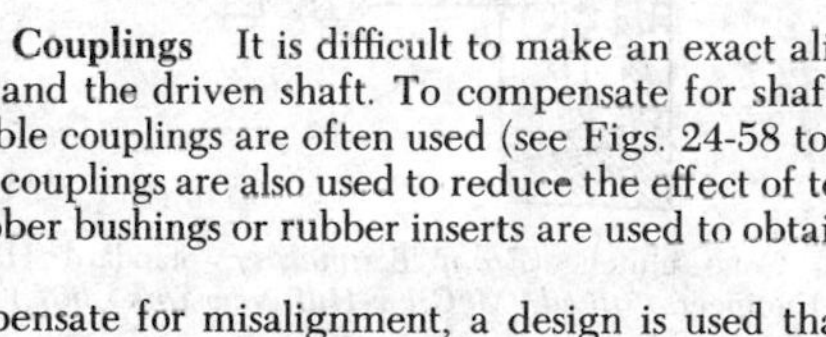

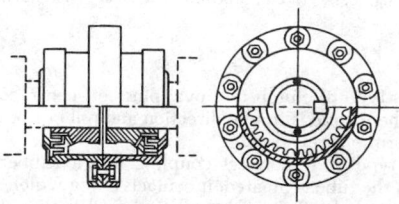

FIG. 24-58 Gear and dental flexible coupling. (*From Baumeister,* Standard Handbook for Mechanical Engineers, *7th ed., McGraw-Hill, New York, 1967.*)

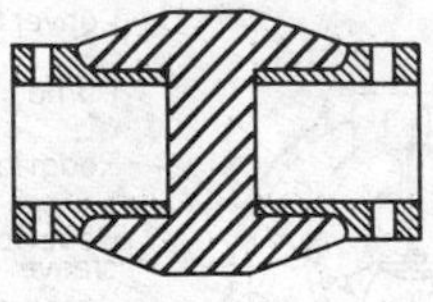

FIG. 24-59 Rubber flexible coupling. (*From Baumeister,* Standard Handbook for Mechanical Engineers, *7th ed., McGraw-Hill, New York, 1967.*)

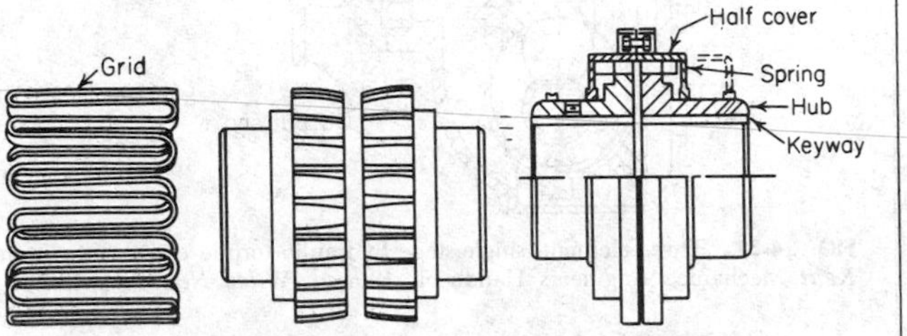

FIG. 24-60 Falk Steelflex coupling. (*From Baumeister,* Standard Handbook for Mechanical Engineers, *7th ed., McGraw-Hill, New York, 1967.*)

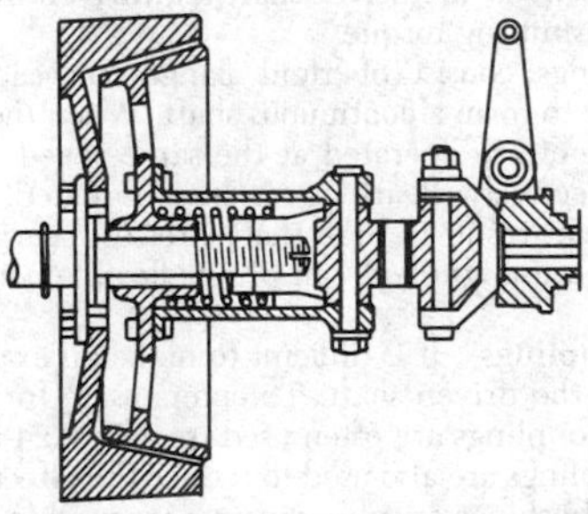

FIG. 24-61 Cone clutch. (*From Baumeister,* Standard Handbook for Mechanical Engineers, *7th ed., McGraw-Hill, New York, 1967.*)

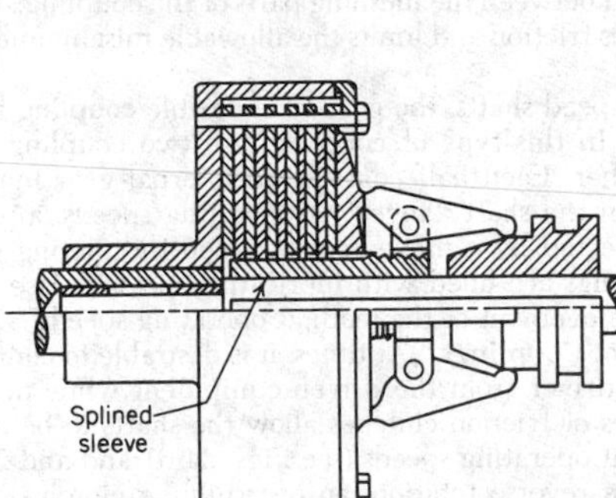

FIG. 24-62 Multidisk clutch. (*From Baumeister,* Standard Handbook for Mechanical Engineers, *7th ed., McGraw-Hill, New York, 1967.*)

24-63). This type of clutch uses oval pins between raceways which jam when the torque is in one direction and roll out of engagement when the torque is reversed.

Another type of disconnect coupling uses a rubberized canvas tube. When the tube is inflated, it contacts a ring which is connected to the output shaft. By simply inflating or deflating the tube, the driver can be disconnected or connected to the driven machine while in operation.

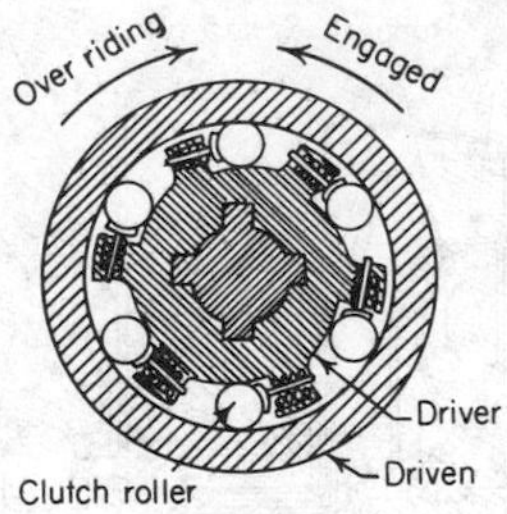

FIG. 24-63 Overunning clutch. (*From Vallance and Doughtie,* Design of Machine Members, *3d ed., McGraw-Hill, New York, 1951.*)

BEARINGS

Handbooks, textbooks, and manufacturers' catalogs supply the design data needed for engineers to select bearings for various applications. However, average users do not select bearings, and their main concern is to understand the basic type of bearings and their relative merits and weaknesses.

Types of Bearings Bearings are classified as low-speed sliding, high-speed oil-lubricated, and antifriction. The last-named includes ball and roller bearings.

The velocity between the elements of a low-speed sliding bearing is so slow that an oil wedge does not form to separate the sliding surfaces. Lubrication lowers static and sliding friction, which can be further reduced by carefully polished surfaces and/or by lubricant additions (such as graphite, talc, or soapstone) to fill the surface cavities. The essential characteristics of such bearings are the slow wearing of the sliding surfaces and increased machine clearances, with ample warning of excessive wear and little hazard of sudden disastrous damage. Heat generation is low enough to be readily conducted away from the rubbing surfaces by the machine parts.

The load-carrying capacity of a high-speed oil-lubricated bearing results from the pressure buildup caused by the moving surface drawing oil into a region of diminishing oil-film thickness to form an oil wedge. Oil wedges in journal bearings cause the shaft not to be concentric with the bearings (see Fig. 24-64). In stationary thrust bearings this wedge may be obtained by machining a tapered depression into the stationary thrust collars; in tilting-pad thrust bearings, such as those of the **Kingsbury type,** the angle assumed under rotation forms the oil wedge.

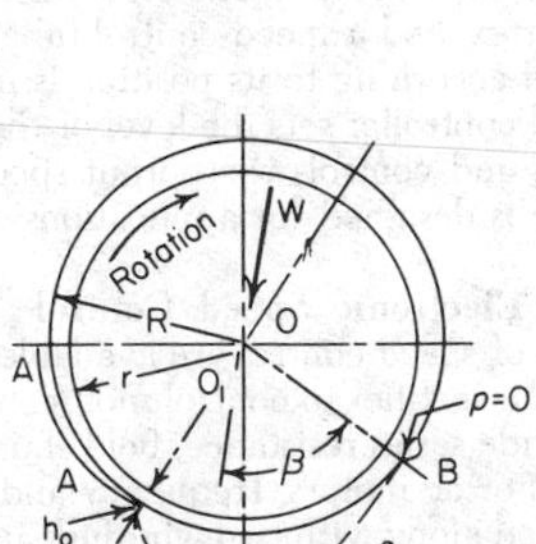

FIG. 24-64 Journal bearing with perfect lubrication. (*From Baumeister,* Standard Handbook for Mechanical Engineers, *7th ed., McGraw-Hill, New York, 1967.*)

Design parameters of various industries differ because bearings must be designed not only to carry load but also to satisfy other conditions under which specific machines must operate. Thus, turbine design favors small clearances to minimize steam leakage; whereas machines subject to fouling, such as those operating in dusty areas, must have increased clearances to allow ample oil flow and to pass dust particles. Locomotives must transmit high traction loads with attendant high unit bearing loads. To minimize the hazard of damage, such bearings are hand-fitted to shafts and are under continual inspection.

At nominal speeds, lightly loaded bearings which can dissipate friction heat through the bearing housing need be supplied only with sufficient oil to make up "end leakage." Ring oilers, oil wicks, etc., can be used to supply this makeup from a reservoir, which in turn receives the end leakage. Grease-packed bearings can be used for slow speeds. The grease blocks oil loss, prevents the ingress of dust, and supplies lubricating oil under the melting action of the frictional heat. High-speed bearings are pressure-flow-lubricated to assure sufficient oil flow to make up end leakages, to maintain lubricant and metal temperatures in bearing areas, and to remove heat conducted along the shaft and the housing. Bearing materials of high conductivity are frequently selected to minimize the formation of localized hot spots. Properly designed and lubricated bearings will continue to run for many years without requiring replacement.

Load capacity of a bearing increases with surface velocity, viscosity, and wedge angle. A generalized equation defines the load-carrying ability of journal bearings in terms of the ratio of minimum oil-film thickness h to the bearing clearance c and is usually plotted, for given ratios of bearing length L to bearing diameter d, as follows:

$$h = ZN\,d^2/Pc \tag{24-28}$$

where Z is viscosity, cP; P is the mean bearing pressure (load/Ld); and N is r/min.

The frictional power dissipated in an oil film is proportional to the film thickness, velocity, density, and bearing length. The temperature rise t is proportional to the heat dissipated divided by the product of specific heat c_p of the oil and the oil-flow quantity. High oil temperatures can decrease oil viscosity sufficiently to break down the oil film, and as a result each bearing design has a maximum load-carrying capability depending on Z, N, and d/c.

Although different industries use different values of c/d, ZN/P, and L/d, bearings are always designed to ensure that the minimum film thickness is above an acceptable value and that the temperature rise is within allowable limits. Values of c/d = 0.001 and L/d less than unity are commonly used for turbine and centrifugal compressors. Higher values of c/d = 0.002 and L/d = 2 are used in slower-speed machines and less frequently for high-speed machinery. Values of P usually vary from 0.34 to 2.0 MN/m^2 (50 to 300 lbf/in^2) and for special conditions up to 3.79 MN/m^2 (550 lbf/in^2). Higher values can be obtained by the use of tilting pads that promote effective oil cooling.

Reciprocating machines (crankshafts and crankpins) have high bearing unit loads [in the order of 12.41 MN/m^2 (1500 lbf/in^2)], since during the unloaded part of the cycle fresh oil flows into the bearing and pressure buildup occurs as the oil is forced out. Thus the load is always carried by cool oil and the bearings are thermally stable.

Oil whip is a bearing phenomenon encountered on high-speed machinery and vertical bearings and is a result of lightly loaded bearings (i.e., high values of ZN/P). The action, for a horizontal bearing carrying a vertical load, is as follows:

If a shaft starts to whirl (move in a circular path) in the direction of rotation, thus decreasing the oil-film thickness, the resulting bearing pressures will raise the shaft until the vertical component of the bearing oil-film reaction is less than the bearing load. At this stage the horizontal component of the bearing oil-film reaction will force the shaft toward the bearing center, increasing the oil-film thickness and allowing the shaft to fall back to the starting position. Under certain conditions, the pumping action of the shaft will develop enough energy to overcome the losses and the whirling motion of the shaft in the bearing will persist. The whirl frequency is about half of the revolutions per minute of the shaft.

The amplitude of the whip of a stiff shaft (i.e., critical speed above operating speed) is limited to the amplitude of whirl in the bearing, and its occurrence may not be noticed. The whipping amplitude of a shaft increases as the running speed approaches twice the critical frequency of the rotor, at which time resonance occurs with resultant magnified amplitudes of vibration, which can lead to a fatigue failure.

Mild cases of oil whip have been stopped by changing oil viscosity through either increased oil temperature or reduced oil flow or in a few cases by the reverse (i.e., decrease in oil temperatures). More severe cases require that the bearing length be decreased to produce higher mean bearing pressures.

Grit-particle size in the oil and/or surface irregularities determine the minimum practical oil-film thickness and may result in low unit pressures. To prevent oil whip under these conditions, specially designed bearings are used, such as **segmental or tilting pads,** which change the bearing contour and decrease the lifting capacity of the bearing as it whirls. Or else passages in the bearing housing can be designed to change pressure distribution for snubbing the whirling action.

Thrust Bearings Low axial loads can be carried by a simple thrust collar on a shaft that runs against a flat and parallel stationary bearing surface in which lubrication to the center flows radially outward. The flat and parallel surfaces prevent the formation of an oil wedge and thus limit this type of bearing to low axial loads. The increased capacity obtained by several collars on a shaft is not proportional to the number of collars used, since one collar usually carries most of the load.

Radial grooves in the surface can be chamfered to allow wedge formation and increase the load-carrying capacity of a simple thrust collar. Manufacturing tolerances and errors in chamfer angle can result in some segments of the bearing carrying more load than other segments, thus affecting the load-carrying capacity of this type of thrust bearing. Tilting-pad thrust bearings were developed to overcome the limitations of fixed-wedge bearings by permitting the individual pads to tilt and distribute the load among them.

Thrust bearings have the same thermal and minimum film-thickness limitations as sleeve bearings. In addition, tilting pads may deform under loads, causing hot spots; or the pumping action of the thrust collar may cause the inner portion of the thrust bearing to run dry. Tilting-pad bearings are preferred for heavy-duty thrust bearings. They are made by several specialist companies and by most manufacturers of heavy-duty equipment according to the basic designs developed by Kingsbury in the United States and Michelle in France.

Ball and Roller Bearings These are often referred to as antifriction bearings, since a shaft rotating on rolling balls or rollers appears to be frictionless. Their actual performance does not attain this ideal state, as will be explained later for ball bearings. Roller bearings, including the needle and tapered varieties, use different equations.

The maximum unit contact pressure P_{max} resulting from a sphere of radius r carrying a load F against a flat plate of the same material as the sphere (modulus of elasticity E) is

$$P_{max} = 0.388\sqrt[3]{FE^2/r^2} \tag{24-29}$$

and the stress is

$$S_{max} = 0.31P_{max} \tag{24-30}$$

By associating this stress with the fatigue properties of the material, it is possible to calculate the life of an ideal bearing for an assumed load. The relative size of adjacent balls and their spacing will influence the life of a ball bearing. The effect of these parameters is determined by testing a large number of bearings to a statistically defined life such that not more than 10 percent of the bearings will fail at the selected design life and less than 50 percent will fail at 5 times this design life. Some of the bearings may reach a life of 20 or 30 times design. These tests are performed at rated speed and load.

The catalog **bearing capacity** C is adjusted to service conditions by multiplying the actual operating load R by several factors, which are K_L, a life factor to correct from catalog life to the required life; K_S, a speed factor to correct from catalog speed to required speed; K_A, an application factor to correct for shock possibilities; and K_T, a thrust factor to account for the application of both thrust and radial loads.

The equation relating these factors is written as

$$C = K_A K_L K_S K_T R \tag{24-31}$$

where $K_L = \sqrt[3]{(\text{required life})/(\text{catalog speed})}$
$K_S = \sqrt[3]{(\text{required speed})/(\text{catalog speed})}$
$K_A =$ 1 for steady loads and increases to 3 for ball bearings or to 2 for roller bearings under extreme shock loads
$K_T =$ 1, usually, at design load conditions and is different for each bearing design for the various thrust radial-load conditions

Bearings of different manufacturers are interchangeable as a result of dimensional standardization as to width and outer and inner diameters, with standard series rated as light-, medium-, and heavy-duty. To assure selection of comparable bearings it is necessary that ratings be corrected to common operating conditions.

Bearing friction is partly due to rubbing between elements and the retaining cages (whose function is to keep the elements apart) but is mostly due to the uphill motion that results from the impression of the loaded elements on the raceways. For slow-speed bearings effective lubrication is obtained by oil spray, flow, flooding, or grease packing. As the speed increases, the heat generated by the elements churning in the lubricant becomes critical, and consequently high-speed ball-bearing lubrication systems are designed to provide only the exact quantity of lubricant required; the lubricant is usually applied in the form of a spray or mist.

Because of their ease of installation and availability, roller and ball bearings are popular among manufacturers and users. Additional advantages are that they can provide accurate centering and location of the shaft with respect to the housing, runout can be kept to a minimum, and low starting torques can be obtained. Their major disadvantage as compared with journal bearings is that life is less assured and a certain percentage fails with time.

LUBRICATING SYSTEMS

Lubrication may involve the simple manual filling of oil cups or grease cups, or pressure application of lubricant to fittings, or automatic lubrication from sumps with ring oilers, chains, or wicks. However, high-speed bearings require more complicated systems to assure a continuous supply of clean oil.

Large machines require an oil cooler to remove the heat generated in their bearings; usually the oil is contained in a reservoir and pumped to the bearing through oil coolers and filters. Oil temperature is controlled to maintain constant viscosity, and sufficient oil pressure is maintained to ensure flow of oil through the orifices into the individual bearings. Oil-return lines slope from the bearing to the reservoir and are generously sized to allow the passage of foam with the return oil. Lubricating oils are selected to minimize foaming, and when excessive, antifoaming additives are used. Lubricating systems must be cleaned by flushing or oil circulation prior to operation and kept clean thereafter to prevent damage.

Oil filters are placed as near the bearings as possible and should be capable of removing all particles which are large compared with the minimum oil-film thickness. **Underfiltering** will result in scratched bearings and may lead to more serious damage. **Overfiltering** risks the plugging of filters and either collapse with sudden release of their contents or else stoppage of oil flow.

Some machines will produce small particles (i.e., carbon particles in diesels) whose size is not large enough to damage the lubricating system. It is good practice, however, to prevent their accumulation; for this purpose a microfilter is effective, and when installed in a side stream it will not endanger the main bearings.

To minimize condensation and consequent accumulation of moisture from the atmosphere or as a product of combustion, lubricating systems are maintained at approximately 71°C (60°F). Centrifuges will remove sludge, moisture, and heavier-than-oil contaminants and should be used at periodic intervals. Magnetic filters will remove magnetic impurities.

Hazards from lubricating systems fall into several categories, viz., failure of the lubrication system itself, oil leakages, vapors, and carryover and accumulation of oil.

Unless the machine is stopped immediately, failure of its lubrication system will result in damaged bearings and shafts. To minimize this danger most lubricating circuits are provided with automatic controls to stop drivers immediately when lubricating-oil pressure decreases.

Oil leakages commonly represent a housekeeping and maintenance problem (to protect employees from slipping) and, when leaking on hot parts, a serious fire hazard. Oil leakages result from (1) inadequate or defective lubricating-oil piping connections, bearings, and bearing housings; (2) undersize oil-return lines that allow foam to build up and leak out of the bearings; and (3) certain rotating parts, such as oil and grit flingers, which induce air flow that blows foam and/or droplets out of bearing housings.

The air space above the lubricating-oil reservoirs always contains oil vapor and oil droplets. Some reciprocating machines, such as diesel and gas engines, cool the underside of pistons with lubricating oil, and the action of the reciprocating parts increases the quantity of oil droplets in the crankcase. Recommendations of manufacturers should be followed to minimize the hazards of crankcase explosions which result from ignition of oil droplets.

The oil reservoirs of turbines and of axial or centrifugal compressors are enclosed and vented to the atmosphere and are usually considered as nonhazardous. However, when the lubricating oil or seal-oil systems of a compressor or turbine are in contact with hydrocarbons, the oil reservoir must be vented to a safe location.

The oil droplets carried by the discharge air of lubricated reciprocating air compressors will deposit in the discharge pipes or accumulate in low spots and in the air-receiver tanks. High discharge temperatures will carbonize the oil, resulting in considerable carbon deposits. Explosions have resulted from oil deposits, and fires have taken place on carbon deposits. These hazards are minimized by thorough draining of the discharge system, the use of oil separators, and the selection of compressor stages that limit discharge temperature. Nonflammable lubricants and the use of nonlubricated compressors will eliminate this hazard, but at higher initial and operating costs.

Section 25

Process Economics

F. A. Holland, D.Sc., Ph.D., *Chairman, Department of Chemical Engineering, University of Salford, Salford, England; Partner in Salchem Associates, Consulting Engineers; Fellow, Institution of Chemical Engineers, London. (Section Editor)*

F. A. Watson, M.Sc., *Senior Lecturer, Department of Chemical Fngineering, University of Salford; Partner in Salchem Associates, Consulting Engineers; Fellow, Institution of Chemical Engineers, London*

J. K. Wilkinson, M.Sc., *Senior Lecturer, Department of Chemical Engineering, University of Salford; Partner in Salchem Associates, Consulting Engineers; Fellow, Institution of Chemical Engineers, London*

Nomenclature and Units

Symbol	Definition	Units
a	Empirical constant in general equations	Various
A	Annual income or expenditure particularized by the subscript	$/year
A_A	Annual allowances against tax other than for depreciation of fixed assets	$/year
A_D	Annual writing down (depreciation) of fixed assets, allowable against tax	$/year
(ATR)	Asset-turnover ratio defined by Eq. (25-131)	Dimensionless
b	Empirical constant in general equations	Various
b_c	Deviation from budgeted capacity	Dimensionless
B	Parametric constant in Eq. (25-204)	Dimensionless
c	Empirical constant in general equations	Various
c	Cost (or income) per unit of sales or production particularized by the subscript	$/unit
c_B	Cost of base heat supply	$/unit
c_D	Cost of heat energy delivered by a heat pump defined by Eq. (25-240)	$/GJ
c_I	Cost of high-grade energy supplied to the compressor of a vapor compression heat pump	$/GJ
c_L	Cost of labor per unit of production	$/hour
c°	Standard cost particularized by the subscript	$/hour
C	Cost particularized by the subscript	$
C_{CT}	Installed cost of a cooling tower	$
C_{DS}	Installed cost of a demineralized-water system	$
$(C_{EQ})_{DEL}$	Delivered-equipment cost	$
C_K	Capitalized cost of a fixed asset defined by Eq. (25-47)	$
C_L	Cost of land and other nondepreciable assets	$
C_{RS}	Installed cost of a refrigeration system	$
C_{RW}	Installed cost of a river-water supply system	$
C_{WS}	Installed cost of a water-softening system	$
(CI)	Cost index as used in Eq. (25-246)	Dimensionless
$(COP)_A$	Actual coefficient of performance of a heat pump	Dimensionless
(CR)	Capital ratio defined by Eq. (25-134)	Year
(CRR)	Capital-rate-of-return ratio defined by Eq. (25-56)	Year
(CSR)	Contribution-sales ratio defined by Eq. (25-236)	Dimensionless
d	Empirical constant in general equations	Various
d	Symbol indicating differentiation	Dimensionless
(DR)	Debt ratio defined by Eq. (23-139)	Dimensionless
(DCFRR)	Discounted-cash-flow rate of return	$Year^{-1}$
e	Empirical constant in general equations	Various
e	Base of natural logarithms, 2.71828	Dimensionless
exp (a)	Exponential function of a, e^a	Dimensionless
(EMIP)	Equivalent maximum investment period defined by Eq. (25-55)	Year
f_{AF}	Annuity future-worth factor, $i[(1+i)^n - 1]^{-1}$	Dimensionless
f_{AP}	Annuity present-worth factor, $f_{AF}(1+i)^n$	Dimensionless
f_d	Discount factor, $(1+i)^{-n}$	Dimensionless
f_i	Compound-interest factor, $(1+i)^n$	Dimensionless
f_k	Capitalized-cost factor, f_{AP}/i	Dimensionless
f_p	Piping-cost factor defined by Eq. (25-249)	Dimensionless
$f(x)$	Distribution function of x variously defined	Dimensionless
F	Future value of a sum of money	$
F_n	Sum of f_d for Years 1 to n	Dimensionless
i	Interest rate per period, usually annual, often the cost of capital	Dimensionless
i_e	Effective interest rate defined by Eq. (25-111)	Dimensionless
i_m	Minimum acceptable interest rate defined by Eq. (25-107)	Dimensionless
i_r	Entrepreneurial-risk interest rate	Dimensionless
i'	Nominal annual interest rate	Dimensionless
I	Value of inventory particularized by the subscript	$
k_n	Constants in Eq. (25-81)	Various
K	Effective value of the first unit of production	$/unit, time/unit, etc.
ln (a)	Logarithm to the base e of a	Dimensionless
log (a)	Logarithm to the base 10 of a	Dimensionless
m	Number of interest periods due per year	Dimensionless
m	Number of units removed from inventory	Dimensionless
(MSF)	Measured-survival function defined by Eq. (25-106)	Dimensionless
n	Number of years, units, etc.	Dimensionless
N	Slope of the learning curve defined by Eq. (25-64)	Dimensionless
N	Number of inventory orders per year	Dimensionless
(NPV)	Net present value	$
$p(x)$	Probability of the variable having the value x	Dimensionless
P	Present value of a sum of money	$
P_a	Production time worked	Hour
P_b	Budgeted production	Standard hour
P_e	Production efficiency defined by Eq. (25-216)	Dimensionless
P_1	Level of productive activity defined by Eq. (25-217)	Dimensionless
P_s	Actual production rate	Standard hour
P_s'	Book value of asset at the end of year s'	$
P_w	Budgeted working time	Hour
(PBP)	Payback period defined by Eq. (25-30)	Year
(PM)	Profit margin defined by Eq. (25-127)	Dimensionless
(PSR)	Profit-sales ratio defined by Eq. (25-235)	Dimensionless
q	Quantity defining the scale of operation	Various
Q_D	Process-heat-rate requirement	GJ/hour
r	Fraction of range of the independent variable	Dimensionless
R	Production rate	Units/year
R°	Standard production rate	Units/year
R_B	Breakeven production rate	Units/year

Nomenclature and Units *(continued)*

Symbol	Definition	Units
R_0	Scheduled production rate	Units/year
R_S	Sales rate	Units/year
(ROA)	Return on assets defined by Eq. (25-129)	Dimensionless
(ROE)	Return on equity defined by Eq. (25-130)	Dimensionless
(ROI)	Return on investment defined by Eq. (25-128)	Dimensionless
s	Scheduled number of productive years	
s'	Number of productive years to date	
s°	Sample standard deviation	Various
S	Scrap value of a depreciable asset	$
t	Fractional tax rate payable on adjusted income	Dimensionless
t_C	Time taken to construct plant	Years
t_{SU}	Time taken to start up plant	Years
T	Auxiliary variable defined by Eq. (25-92)	Various
U	Size of inventory order	Units
V	Variable cost of inventory order	$/unit
W	Power supplied at shaft of a heat pump	GJ/hour
x	General variable	
$\bar{x}$	Mean value of x,	Various
X	Cumulative production from startup	Units
y	Cumulative probability	Dimensionless
y	Operating time of a heat pump	Hours/year
Y	Cumulative average cost, production time, etc.	$/unit, hour/unit, etc.
Y	Operating-labor rate in Eq. (25-204)	labor-hour/ton
$\bar{Y}$	Cumulative-average batch cost, etc.	$/unit, etc.
z	Standard score defined by Eq. (25-73)	Dimensionless
	Greek symbols	
α	Proportionality factor in Eq. (25-168)	Dimensionless
β	Proportionality factor in Eq. (25-171)	Dimensionless
β	Exponent in Eqs. (25-106) and (25-117)	Dimensionless
δ	Symbol indicating partial differentiation	Dimensionless
Δ	Symbol indicating a difference of like quantities	Dimensionless
η	Contribution efficiency defined by Eq. (25-119)	Dimensionless
η	Margin of safety defined by Eq. (25-229)	Dimensionless
θ	Time taken to produce a given amount of product	Hour
σ	Population standard deviation	Various
Σ	Symbol indicating a sum of like quantities	Dimensionless
ϕ	Fractional increase in production rate	Dimensionless
ϕ_P	Parameter defined with Eq. (25-249)	Dimensionless
ψ	Parameter defined with Eq. (25-241)	Dimensionless
χ	Plant capacity in Eq. (25-204)	Tons/day
χ	Weight of product per unit of raw material	Dimensionless
	Subscripts	
A	Allowance against tax other than for capital depreciation	
BD	Depreciation allowance shown in company balance sheet	

Symbol	Definition	Units
BL	Within project boundary limits	
BOH	Budgeted overhead	
CF	Cash flow after payment of tax and expenses	
CI	Cash income after payment of expenses	
DCF	Discounted cash flow	
DME	Direct manufacturing expense	
FC	Fixed capital	
FE	Fixed expense	
FGE	Fixed general expense	
FIFO	On a first-in–first-out basis	
FIN	Financial-resources inventory	
FME	Fixed manufacturing expense	
FOH	Fixed overhead	
GE	General expense	
GP	Gross profit	
IME	Indirect manufacturing expense	
INV	Inventory	
IO	Inventory-orders cost	
IT	Income tax payable	
IW	Inventory working cost	
L	Labor-earnings index	
L	Lower-quartile value of the variable	
LIFO	Last-in–first-out basis	
max	Maximum value	
M	Median value of the variable	
ME	Manufacturing expense	
N	At agreed normal production rate	
NCI	Net cash income after payment of tax	
NOH	Overhead cost at agreed normal production rate	
NNP	Net profit after payment of tax	
NP	Net profit before payment of tax	
OH	Overhead cost	
P	Profit	
RM	Raw material	
s'	In the s'th productive year	
S	From sales and other income	
SAV	On a simple-average basis	
ST	Steel-price index	
$SVOH$	Semivariable overhead	
TC	Total capital	
TE	Total expense	
TFE	Total fixed expense	
TVE	Total variable expense	
U	Utilities	
U	Upper-quartile value of the variable	
VE	Variable expense	
VGE	Variable general expense	
VME	Variable manufacturing expense	
VOH	Variable overhead expense	
W	Weighted value	
WC	Working capital	
WAV	On a weighted-average basis	
1, 2, j, n	lst, 2d, jth, nth item, year, etc.	

General References: Allen, D. H., *A Guide to the Evaluation of Projects,* Institution of Chemical Engineers, Rugby, England, 1972. Aries, R. S., and R. D. Newton, *Chemical Engineering Cost Estimation,* McGraw-Hill, New York, 1955. Baasel, W. D., *Preliminary Chemical Engineering Plant Design,* Elsevier, Amsterdam, 1976. Barish, N. N., and S. Kaplan, *Economic Analysis for Engineering and Managerial Decision Making,* 2d ed., McGraw-Hill, New York, 1978. Bauman, H. A., *Fundamentals of Cost Engineering in the Chemical Industry,* Van Nostrand Reinhold, New York, 1964. Bierman, H., Jr., and S. Smidt, *The Capital Budgeting Decision, Economic Analysis and Financing of Investment Projects,* Collier-Macmillan, London, 1975. Canada, J. R., *Intermediate Economic Analysis for Management and Engineering,* Prentice-Hall, Englewood Cliffs, N.J., 1971. Carsberg, B., and A. Hope, *Business Investment Decisions under Inflation,* Macdonald & Evans, London, 1976. Chemical Engineering (ed.), *Modern Cost Engineering,* McGraw-Hill, New York, 1979. Garvin, W. W., *Introduction to Linear Programming,* McGraw-Hill, New York, 1960. Gass, S. I., *Linear Programming,* McGraw-Hill, New York, 1958. Granger, C. W. J., *Forecasting in Business and Economics,* Academic Press, New York, 1980. Hackney, J. W., *Control and Management of Capital Projects,* Wiley, New York, 1965. Happel, J., W. H. Kapfer, B. J. Blewitt, P. T. Shannon, and D. G. Jordan, *Process Economics,* American Institute of Chemical Engineers, New York, 1974. Holland, F. A., F. A. Watson, and J. K. Wilkinson, *Introduction to Process Economics,* Wiley, London, 1983. Institution of Chemical Engineers, London (ed.), *A New Guide to Capital Cost Estimating,* Institution of Chemical Engineers, Rugby, England, 1977. Jelen, F. C., *Cost and Optimization Engineering,* McGraw-Hill, New York, 1970. Jordan, R. B., *How to Use the Learning Curve,* Materials Management Institute, Boston, 1965. Kharbanda, O. P., *Process Plant and Equipment Cost Estimation,* Sevak Publications, Bombay, 1977. Kirkman, P. R. A., *Inflation Accounting: A Guide for Non-Accountants,* Associated Business Programmes, London, 1975. Koopmans, T. C. (ed.), *Activity Analysis of Production and Allocation,* Wiley, New York, 1951. Liddel, C. J., and A. M. Gerrard, *The Application of Computers to Capital Cost Estimation,* Institution of Chemical Engineers, Rugby, England, 1975. Loomba, N. P., *Linear Programming,* McGraw-Hill, New York, 1964. Merrett, A. J., and A. Sykes, *The Finance and Analysis of Capital Projects,* Longman Group, London, 1963. Merrett, A. J., and A. Sykes, *Capital Budgeting and Company Finance,* Longmans, London, 1966. Ostwald, P. F., *Cost Estimating for Engineering and Management,* Prentice-Hall, Englewood Cliffs, N.J., 1974. Park, W. R., *Cost Engineering Analysis,* Wiley, New York, 1973. Peters, M. S., and K. D. Timmerhaus, *Plant Design and Economics for Chemical Engineers,* 2d ed., McGraw-Hill, New York, 1968. Pilcher, R., *Appraisal and Control of Project Costs,* McGraw-Hill, Maidenhead, England, 1973. Popper, H. (ed.), *Modern Cost Estimating Techniques,* McGraw-Hill, New York, 1970. Ridge, W. J., *Value Analysis for Better Management,* American Management Association, New York, 1969. Rockley, L. E., *Capital Investment Decisions,* Business Books, London, 1968. Rose, L. M., *Engineering Investment Decisions: Planning under Uncertainty,* Elsevier, Amsterdam, 1976. Rudd, D. F., and C. C. Watson, *The Strategy of Process Engineering,* Wiley, New York, 1968. Schlaifer, R., *Probability and Statistics for Business Decisions,* McGraw-Hill, New York, 1959. Tocher, K. D., *The Art of Simulation,* rev. ed., English Universities Press, London, 1967. Vilbrandt, F. C., and C. E. Dryden, *Chemical Engineering Plant Design,* McGraw-Hill, New York, 1959. Weaver, J. B., "Project Selection in the 1980's," *Chem. Eng. News,* 37–46 (Nov. 2, 1981). Wells, G. L., *Process Engineering with Economic Objectives,* Wiley, New York, 1973. Wilkes, F. M., *Capital Budgeting Techniques,* Wiley, London, 1977. Wood, E. G., *Costing Methods for Managers,* Business Books, London, 1974. Woods, D. R., *Financial Decision Making in the Process Industry,* Prentice-Hall, Englewood Cliffs, N.J., 1975. Wright, M. G., *Financial Management,* McGraw-Hill, London, 1970.

NOMENCLATURE

An attempt has been made to bring together most of the methods currently available for project evaluation and to present them in such a way as to make the methods amenable to modern computational techniques. To this end the practices of accountants and others have been reduced, where possible, to mathematical equations which are usually solvable with an electronic hand calculator equipped with scientific function keys. To make the equations suitable for use on high-speed computers an attempt has been made to devise a nomenclature which is suitable for machines using ALGOL, COBOL, or FORTRAN compilers. The number of letters and numbers used to define a variable has usually been limited to five. The letters are mnemonic in English wherever possible and are derived in two ways. First, when a standard accountancy phrase exists for a term, this has been abbreviated in capital letters and enclosed in parentheses, e.g., (ATR), for assets-to-turnover ratio; (DCFRR), for discounted-cash-flow rate of return. Clearly, the parentheses are omitted when the letter group is used to define the variable name for the computer. Second, a general symbol is defined for a type of variable and is modified by a mnemonic subscript, e.g., an annual cash quantity A_{TC}, annual total capital outlay, $/year. Clearly, the symbols are written on one line when the letter group is used to define a variable name for the computer. In other cases, when well-known standard symbols exist, they have been adopted, e.g., z for the standard score as used in the normal distribution. Also, $a,b,c,d,$ and e have been used to denote empirical constants and x and y to denote general variables where their use does not clash with other meanings of the same symbols.

The coverage in this section is so wide that nomenclature has sometimes proved a problem which has required the use of primes, asterisks, and other symbols not universally acceptable in the naming of computer variables. However, it is realized that each individual will program only his or her preferred methods, which will release some symbols for other uses. Also, it is not difficult to replace a forbidden symbol by an acceptable one; e.g., c_{RM} might be rendered CARM and P_S' as PSP by using A for asterisk and P for prime. For compilers which recognize only one alphabetical case, an extra prefix can be used to distinguish between uppercase and lowercase letters, for which purpose the letters U and L have been used only in a restricted way in the nomenclature.

It is, of course, impossible to allow for all possible variations of equation requirements and machine capability, but it is hoped that the nomenclature in the table presented at the beginning of the section will prove adequate for most purposes and will be capable of logical extension to other more specialized requirements.

INVESTMENT AND PROFITABILITY

In order to assess the profitability of projects and processes it is necessary to define precisely the various parameters.

Annual Costs, Profits, and Cash Flows To a large extent, accountancy is concerned with annual costs. To avoid confusion with other costs, annual costs will be referred to by the letter A.

The revenue from the annual sales of product A_S, minus the total annual cost or expense required to produce and sell the product A_{TE}, excluding any annual provision for plant depreciation, is the annual cash income A_{CI}:

$$A_{CI} = A_S - A_{TE} \tag{25-1}$$

Net annual cash income A_{NCI} is the annual cash income A_{CI}, minus the annual amount of tax A_{IT}:

$$A_{NCI} = A_{CI} - A_{IT} \tag{25-2}$$

Taxable income is $(A_{CI} - A_D - A_A)$, where A_D is the annual writing-down allowance and A_A is the annual amount of any other allowances. A distinction is made between the writing-down allowance permissible for the computation of tax due, the actual depreciation in value of an asset, and the book depreciation in value of that asset as shown in the company position statement. There is no necessary connection between these values unless specified by law, although the first two or all three are often assigned the same value in practice. Some governments give cash incentives to encourage companies to build plants in otherwise unattractive areas. Neither A_D nor A_A involves any expenditure of cash, since they are merely book transactions. The annual amount of tax A_{IT} is given by

$$A_{IT} = (A_{CI} - A_D - A_A)t \tag{25-3}$$

where t is the fractional tax rate. The value of t is determined by the appropriate tax authority and is subject to change. For most developed countries the value of t is about 0.5, or 50 percent.

The annual amount of tax A_{IT} included in Eq. (25-2) does not necessarily correspond to the annual cash income A_{CI} in the same year. The tax payments in Eq. (25-2) should be those actually paid in that year. In the United States, companies pay about 80 percent of the tax on estimated current-year earnings in the same year. In the United Kingdom, companies do not pay tax until at least 9 months after the end of the accounting period, which, for the most part, amounts to paying tax on the previous year's earnings. When assessing projects for different countries, engineers should acquaint themselves with the tax situation in those countries.

In modern methods of profitability assessment, cash flows are more meaningful than profits, which tend to be rather loosely defined. The net annual cash flow after tax is given by

$$A_{CF} = A_{NCI} - A_{TC} \tag{25-4}$$

where A_{TC} is the annual expenditure of capital, which is not necessarily zero after the plant has been built. For example, working capital, plant additions, or modifications may be required in future years.

The total annual expense A_{TE} required to produce and sell a product can be written as the sum of the annual general expense A_{GE} and the annual manufacturing cost or expense A_{ME}:

$$A_{TE} = A_{GE} + A_{ME} \tag{25-5}$$

Annual general expense A_{GE} arises from the following items: administration, sales, shipping of product, advertising and marketing, technical service, research and development, and finance.

The terms gross annual profit A_{GP} and net annual profit A_{NP} are commonly used by accountants and misused by others. Normally, both A_{GP} and A_{NP} are calculated before tax is deducted. Gross annual profit A_{GP} is given by

$$A_{GP} = A_S - A_{ME} - A_{BD} \tag{25-6}$$

where A_{BD} is the balance-sheet annual depreciation charge, which is not necessarily the same as A_D used in Eq. (25-3) for tax purposes. Net annual profit A_{NP} is simply

$$A_{NP} = A_{GP} - A_{GE} \tag{25-7}$$

Equation (25-7) can also be written as

$$A_{NP} = A_{CI} - A_{BD} \tag{25-8}$$

Net annual profit after tax A_{NNP} can be written as

$$A_{NNP} = A_{NCI} - A_{BD} \tag{25-9}$$

The relationships among the various annual costs given by Eqs. (25-1) through (25-9) are illustrated diagrammatically in Fig. 25-1. The top half of the diagram shows the tools of the accountant; the bottom half, those of the engineer. The net annual cash flow A_{CF}, which excludes any provision for balance-sheet depreciation A_{BD}, is used in two of the more modern methods of profitability assessment: the net-present-value (NPV) method and the discounted-cash-flow-rate-of-return (DCFRR) method. In both methods, depreciation is inherently taken care of by calculations which include capital recovery.

Annual general expense A_{GE} can be written as the sum of the fixed and variable general expenses:

$$A_{GE} = A_{FGE} + A_{VGE} \tag{25-10}$$

Similarly, annual manufacturing expense A_{ME} can be written as the sum of the fixed and variable manufacturing expenses:

$$A_{ME} = A_{FME} + A_{VME} \tag{25-11}$$

A variable expense is considered to be one which is directly proportional to the rate of production R_P or of sales R_S as is most appro-

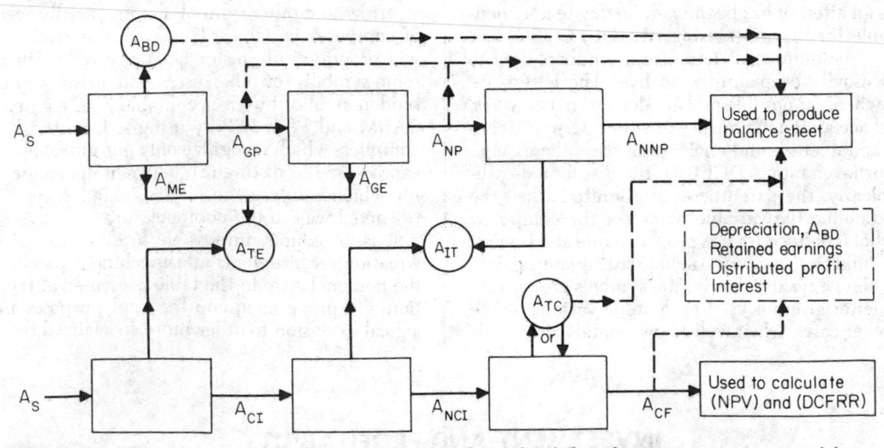

FIG. 25-1 Relationship between annual costs, annual profits, and cash flows for a project. A_{BD} = annual depreciation allowance; A_{CF} = annual net cash flow after tax; A_{CI} = annual cash income; A_{GE} = annual general expense; A_{GP} = annual gross profit; A_{IT} = annual tax; A_{ME} = annual manufacturing cost; A_{NCI} = annual net cash income; A_{NNP} = annual net profit after taxes; A_{NP} = annual net profit; A_S = annual sales; A_{TC} = annual total cost; (DCFRR) = discounted-cash-flow rate of return; (NPV) = net present value.

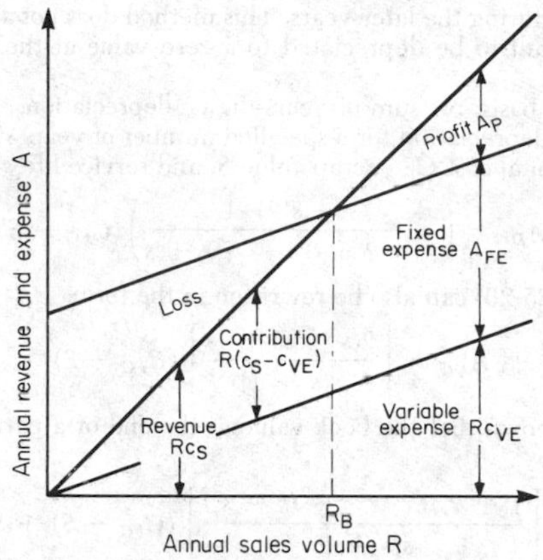

FIG. 25-2 Conventional breakeven chart.

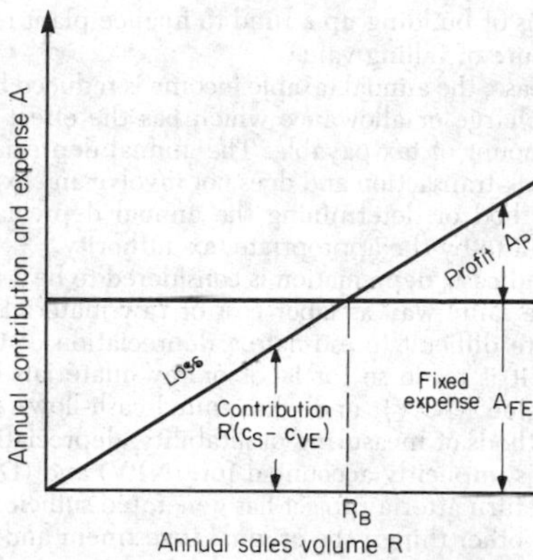

FIG. 25-4 Breakeven chart showing relationship between contribution and fixed expense.

priate to the case under consideration. Unless the variation in finished-product inventory is large when compared with the total production over the period in question, it is usually sufficiently accurate to consider R_P and R_S to be represented by the same-numerical-value R units of sale or production per year. A fixed expense is then considered to be one which is not directly proportional to R, such as overhead charges. Fixed expenses are not necessarily constant but may be subject to stepwise variation at different levels of production. Some authors consider such steps as included in a semivariable expense, which is less amenable to mathematical analysis than the above division of expenses.

Contribution and Breakeven Charts These can be used to give valuable preliminary information prior to the use of the more sophisticated and time-consuming methods based on discounted cash flow. If the sales price per unit of sales is c_S and the variable expense is c_{VE} per unit of production, Eq. (25-7) can be rewritten as

$$A_{NP} = R(c_S - c_{VE}) - A_{FE} \tag{25-12}$$

where $R(c_S - c_{VE})$ is known as the annual contribution. The net annual profit is zero at an annual production rate

$$R_B = A_{FE}/(c_S - c_{VE}) \tag{25-13}$$

where R_B is the breakeven production rate.

Breakeven charts can be plotted in any of the three forms shown in Figs. 25-2, 25-3, and 25-4. The abscissa shown as annual sales volume R is also frequently plotted as a percentage of the designed production or sales capacity R_0. In the case of ships, aircraft, etc., it is then called the percentage utilization. The percentage margin of safety is defined as $100(R_0 - R_B)/R_0$.

A decrease in selling price c_S will decrease the slope of the lines in Figs. 25-2, 25-3, and 25-4 and increase the required breakeven value R_B for a given level of fixed expense A_{FE}.

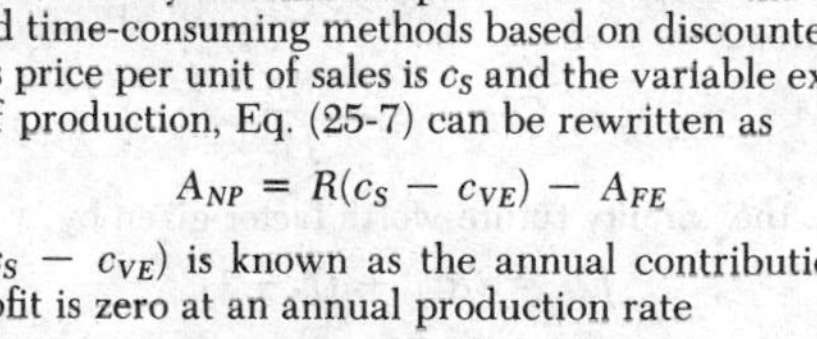

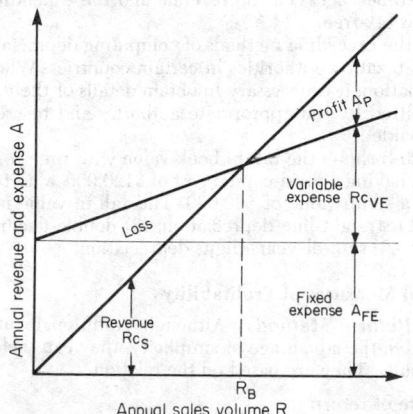

FIG. 25-3 Breakeven chart showing fixed expense as a burden cost.

Capital Costs The total capital cost C_{TC} of a project consists of the fixed-capital cost C_{FC} plus the working-capital cost C_{WC}, plus the cost of land and other nondepreciable costs C_L:

$$C_{TC} = C_{FC} + C_{WC} + C_L \tag{25-14}$$

The project may be a complete plant, an addition to an existing plant, or a plant modification.

The working-capital cost of a process or a business normally includes the items shown in Table 25-1. Since working capital is completely recoverable at any time, in theory if not in practice, no tax allowance is made for its depreciation. Changes in working capital arising from varying trade credits or payroll or inventory levels are usually treated as a necessary business expense except when they exceed the tax debt due. If the annual income is negative, additional working capital must be provided and included in the A_{TC} for that year. The value of land and other nondepreciables often increases over the working life of the project. These are therefore not treated in the same way as other capital investments but are shown to have made a (taxable) profit or loss only when the capital is finally recovered.

Working capital may vary from a very small fraction of the total capital cost to almost the whole of the invested capital, depending on the process and the industry. For example, in jewelry-store operations, the fixed capital is very small in comparison with the working capital. On the other hand, in the chemical-process industries, the working capital is likely to be in the region of 10 to 20 percent of the value of the fixed-capital investment.

Depreciation The term "depreciation" is used in a number of different contexts. The most common are:

1. A tax allowance
2. A cost of operation

TABLE 25-1 Working-Capital Costs

Raw materials for plant startup
Raw-materials, intermediate, and finished-product inventories
Cost of handling and transportation of materials to and from stores
Cost of inventory control, warehouse, associated insurance, security arrangements. etc.
Money to carry accounts receivable (i.e., credit extended to customers) less accounts payable (i.e., credit extended by suppliers)
Money to meet payrolls when starting up
Readily available cash for emergencies
Any additional cash required to operate the process or business

3. A means of building up a fund to finance plant replacement
4. A measure of falling value

In the first case, the annual taxable income is reduced by an annual depreciation charge or allowance which has the effect of reducing the annual amount of tax payable. The annual depreciation charge is merely a book transaction and does not involve any expenditure of cash. The method of determining the annual depreciation charge must be agreed to by the appropriate tax authority.

In the second case, depreciation is considered to be a manufacturing cost in the same way as labor cost or raw-materials cost. However, it is more difficult to estimate a depreciation cost per unit of product than it is to do so for labor or raw-materials costs. In the net-present-value (NPV) and discounted-cash-flow-rate-of-return (DCFRR) methods of measuring profitability, depreciation, as a cost of operation, is implicitly accounted for. (NPV) and (DCFRR) give measures of return after a project has generated sufficient income to repay, among other things, the original investment and any interest charges that the invested money would otherwise have brought into the company.

In the third case, depreciation is considered as a means of providing for plant replacement. In the rapidly changing modern chemical-process industries, many plants will never be replaced because the processes or products have become obsolete during their working life. Management should be free to invest in the most profitable projects available, and the creation of special-purpose funds may hinder this. However, it is desirable to designate a proportion of the retained income as a fund from which to finance new capital projects. These are likely to differ substantially from the projects that originally generated the income.

In the fourth case, a plant or a piece of equipment has a limited useful life. The primary reason for the decrease in value is the decrease in future life and the consequent decrease in the number of years for which income will be earned. At the end of its life, the equipment may be worth nothing, or it may have a salvage or scrap value S. Thus a fixed-capital cost C_{FC} depreciates in value during its useful life of s years by an amount that is equal to $(C_{FC} - S)$. The useful life is taken from the startup of the plant.

On the basis of straight-line depreciation, the average annual amount of depreciation A_D over a service life of s years is given by

$$A_D = (C_{FC} - S)/s \tag{25-15}$$

The book value after the first year P_1 is given by

$$P_1 = C_{FC} - A_D \tag{25-16}$$

The book value at the end of a specified number of years s' is given by

$$P_{s'} = C_{FC} - s'A_D \tag{25-17}$$

The principal use of a particular depreciation rate is for tax purposes. The permitted annual depreciation is subtracted from the annual income before the latter is taxed. The basis for depreciation in a particular case is a matter of agreement between the taxation authority and the company, in conformity with tax laws.

Other commonly used methods of computing depreciation are the declining-balance method (also known as the fixed-percentage method) and the sum-of-years-digits method.

On the basis of declining-balance (fixed-percentage) depreciation, the book value at the end of the first year is given by

$$P_1 = C_{FC}(1 - r) \tag{25-18}$$

where r is a fraction to be agreed with the taxation authority.

The book value at the end of specified number of years s' is given by

$$P_{s'} = C_{FC}(1 - r)^{s'} \tag{25-19}$$

When the fraction r is chosen to be $2/s$, i.e., twice the reciprocal of the service life s, the method is called the double-declining-balance method.

The declining-balance method of depreciation allows equipment or plant to be depreciated by a greater amount during the earlier years than during the later years. This method does not allow equipment or plant to be depreciated to a zero value at the end of the service life.

On the basis of sum-of-years-digits depreciation, the annual amount of depreciation for a specified number of years s' for a plant of fixed-capital cost C_{FC}, scrap value S, and service life s is given by

$$A_{Ds'} = \left(\frac{s - s' + 1}{1 + 2 + 3 + \cdots + s}\right)(C_{FC} - S) \tag{25-20}$$

Equation (25-20) can also be rewritten in the form

$$A_{Ds'} = \left[\frac{2(s - s' + 1)}{s(s + 1)}\right](C_{FC} - S) \tag{25-21}$$

It can be shown that the book value at the end of a particular year s' is

$$P_{s'} = 2\left[\frac{1 + 2 + \cdots + (s - s')}{s(s + 1)}\right](C_{FC} - S) + S \tag{25-22}$$

The sum-of-years-digits depreciation allows equipment or plant to be depreciated by a greater amount during the early years than during the later years.

A fourth method of computing depreciation (now seldom used) is the sinking-fund method. In this method, the annual depreciation A_D is the same for each year of the life of the equipment or plant. The series of equal amounts of depreciation A_D, invested at a fractional interest rate i and made at the end of each year over the life of the equipment or plant of s years, is used to build up a future sum of money equal to $(C_{FC} - S)$. This last is the fixed-capital cost of the equipment or plant minus its salvage or scrap value and is the total amount of depreciation during its useful life. The equation relating $(C_{FC} - S)$ and A_D is simply the annual cost or payment equation, written either as

$$C_{FC} - S = A_D\left[\frac{(1 + i)^s - 1}{i}\right] \tag{25-23}$$

or

$$C_{FC} - S = \frac{A_D}{f_{AF}} \tag{25-24}$$

where f_{AF} is the annuity future-worth factor given by

$$f_{AF} = i/[(1 + i)^s - 1]$$

Values of the future-worth factor f_{AF} are given in tables (see Sec. 1).

In the sinking-fund method of depreciation, the effect of interest is to make the annual decrease of the book value of the equipment or plant less in the early than in the later years with consequent higher tax due in the earlier years when recovery of the capital is most important.

It is preferable not to think of annual depreciation as a contribution to a fund to replace equipment at the end of its life but as part of the difference between the revenue and the expenditure, which difference is tax-free.

Some of the preceding methods of computing depreciation are not allowed by taxation authorities in certain countries. When calculating depreciation, it is necessary to obtain details of the methods and rates permitted by the appropriate authority and to use the information provided.

Figure 25-5 shows the fall in book value with time for a piece of equipment having a fixed-capital cost of $120,000, a useful life of 10 years, and a scrap value of $20,000. This fall in value is calculated by using (1) straight-line depreciation, (2) double-declining depreciation, and (3) sum-of-years-digits depreciation.

Traditional Measures of Profitability

Rate-of-Return Methods Although traditional rate-of-return methods have the advantage of simplicity, they can yield very misleading results. They are based on the relation

Percent rate of return

$$= [(\text{annual profit})/(\text{invested capital})]100 \tag{25-25}$$

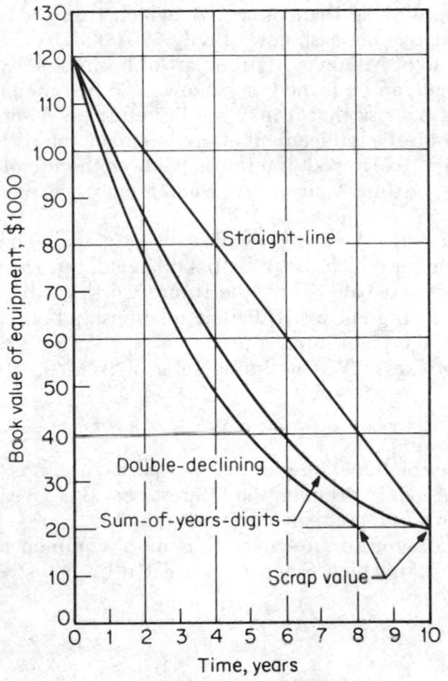

FIG. 25-5 Book value against time for various depreciation methods.

Since different meanings are ascribed to both annual profit and invested capital in Eq. (25-25), it is important to define the terms precisely. The invested capital may refer to the original total capital investment, the depreciated investment, the average investment, the current value of the investment, or something else. The annual profit may refer to the net annual profit before tax A_{NP}, the net annual profit after tax A_{NNP}, the annual cash income before tax A_{CI}, or the annual cash income after tax A_{NCI}.

The fractional interest rate of return based on the net annual profit after tax and the original investment is

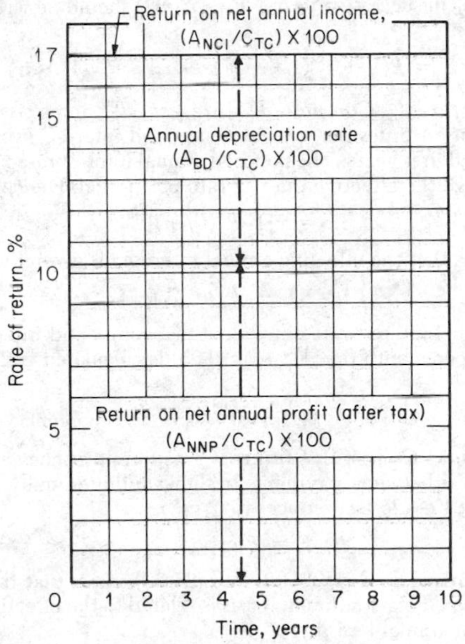

FIG. 25-6 Effect of straight-line depreciation on rate of return for a project. A_{BD} = annual depreciation allowance; A_{NCI} = annual net cash income after tax; A_{NNP} = annual net profit after payment of tax; C_{TC} = total capital cost.

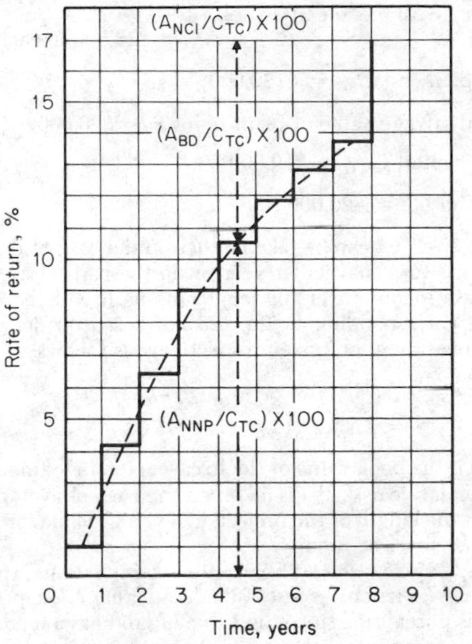

FIG. 25-7 Effect of double-declining depreciation on rate of return for a project.

$$i = A_{NNP}/C_{TC} \tag{25-26}$$

which can be written in terms of Eq. (25-9) as

$$i = (A_{NCI}/C_{TC}) - (A_{BD}\ /\ C_{TC}) \tag{25-27}$$

where A_{BD} is the balance-sheet annual depreciation. The main disadvantage of using Eq. (25-27) is that the fractional depreciation rate A_{BD}/C_{TC} is arbitrarily assessed. Its value will affect the fractional rate of return considerably and may lead to erroneous conclusions when making comparisons between different companies. This is particularly true when making international comparisons.

Figures 25-6, 25-7, and 25-8 show the effect of the depreciation method on profit for a project described by the following data:

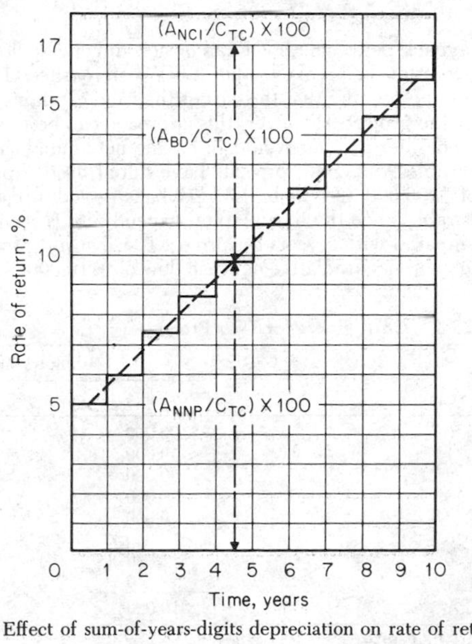

FIG. 25-8 Effect of sum-of-years-digits depreciation on rate of return for a project.

Net annual cash income after tax A_{NCI} = \$25,500 in each of 10 years

Fixed-capital cost C_{FC} = \$120,000

Estimated salvage value of plant items S = \$20,000

Working capital C_{WC} = \$10,000

Cost of land C_L = \$20,000

In Eq. (25-27), i can be taken either on the basis of the net annual cash income for a particular year or on the basis of an average net annual cash income over the length of the life of the project. The equations corresponding to Eq. (25-26) based on depreciated and average investment are given respectively as follows:

$$i = A_{NNP}/(P_{s'} + C_{WC} + C_L) \qquad (25\text{-}28)$$

and

$$i = 2A_{NNP}/(C_{FC} + S + 2C_{WC} + 2C_L) \qquad (25\text{-}29)$$

where $P_{s'}$ is the book value of the fixed-capital investment at the end of a particular year s'. If i is taken on the basis of average values for A_{NNP} over the length of the project, an average value for the working capital C_{WC} must be used.

In Eqs. (25-28) and (25-29), the computations are based on unchanging values of the cost of land and other nondepreciable costs C_L. This is unrealistic, since the value of land has a tendency to rise. In such circumstances, the accountancy principle of conservatism requires that the lowest valuation be adopted.

Payback Period Another traditional method of measuring profitability is the payback period or fixed-capital-return period. Actually, this is really a measure not of profitability but of the time it takes for cash flows to recoup the original fixed-capital expenditure.

The net annual cash flow after tax is given by

$$A_{CF} = A_{NCI} - A_{TC} \qquad (25\text{-}4)$$

where A_{TC} is the annual expenditure of capital, which is not necessarily zero after the plant has been built. The payback period (PBP) is the time required for the cumulative net cash flow taken from the startup of the plant to equal the depreciable fixed-capital investment $(C_{FC} - S)$. It is the value of s' that satisfies

$$\sum_{s'=0}^{s'=(\text{PBP})} A_{CF} = C_{FC} - S \qquad (25\text{-}30)$$

The payback-period method takes no account of cash flows or profits received after the breakeven point has been reached. The method is based on the premise that the earlier the fixed capital is recovered, the better the project. However, this approach can be misleading.

Let us consider projects A and B, having net annual cash flows as listed in Table 25-2. Both projects have initial fixed-capital expenditures of \$100,000. On the basis of payback period, project A is the more desirable since the fixed-capital expenditure is recovered in 3 years, compared with 5 years for project B. However, project B runs for 7 years with a cumulative net cash flow of \$110,000. This is obviously more profitable than project A, which runs for only 4 years with a cumulative net cash flow of only \$10,000.

TABLE 25-2 Cash Flows for Two Projects

	Cash flows A_{CF}	
Year	Project A	Project B
0	−\$100,000	−\$100,000
1	50,000	0
2	30,000	10,000
3	20,000	20,000
4	10,000	30,000
5	0	40,000
6	0	50,000
7	0	60,000
ΣA_{CF}	\$10,000	\$110,000
Payback period (PBP)	3 years	5 years

Time Value of Money A large part of business activity is based on money that can be loaned or borrowed. When money is loaned, there is always a risk that it may not be returned. A sum of money called interest is the inducement offered to make the risk acceptable. When money is borrowed, interest is paid for the use of the money over a period of time. Conversely, when money is loaned, interest is received.

The amount of a loan is known as the principal. The longer the period of time for which the principal is loaned, the greater the total amount of interest paid. Thus, the future worth of the money F is greater than its present worth P. The relationship between F and P depends on the type of interest used.

Simple Interest When simple interest is used, F and P are related by

$$F = P(1 + ni) \qquad (25\text{-}31)$$

where i is the fractional interest rate per period and n is the number of interest periods. Normally, the interest period is 1 year, in which case i is known as the effective interest rate.

Annual Compound Interest It is more common to use compound interest, in which F and P are related by

$$F = P(1 + i)^n \qquad (25\text{-}32)$$

or

$$F = Pf_i \qquad (25\text{-}33)$$

where the compound-interest factor $f_i = (1 + i)^n$. Values for compound-interest factors are readily available in tables.

The present value P of a future sum of money F is

$$P = F/(1 + i)^n \qquad (25\text{-}34)$$

or

$$F = P/f_d \qquad (25\text{-}35)$$

where the discount factor f_d is

$$f_d = 1/f_i = 1/[(1 + i)^n]$$

Values for the discount factors are readily available in tables which show that it will take 7.3 years for the principal to double in amount if compounded annually at 10 percent per year and 14.2 years if compounded annually at 5 percent per year.

For the case of different annual fractional interest rates $(i_1, i_2, \ldots, i_n$ in successive years), Eq. (25-32) should be written in the form

$$F = P(1 + i_1)(1 + i_2)(1 + i_3) \cdots (1 + i_n) \qquad (25\text{-}36)$$

Short-Interval Compound Interest If interest payments become due m times per year at compound interest, mn payments are required in n years. The nominal annual interest rate i' is divided by m to give the effective interest rate per period. Hence,

$$F = P[1 + (i'/m)]^{mn} \qquad (25\text{-}37)$$

It follows that the effective annual interest i is given by

$$i = [1 + (i'/m)]^m - 1 \qquad (25\text{-}38)$$

The annual interest rate equivalent to a compound-interest rate of 5 percent per month (i.e., $i'/m = 0.05$) is calculated from Eq. (25-38) to be

$$i = (1 + 0.05)^{12} - 1 = 0.796, \text{ or } 79.6 \text{ percent/year}$$

Continuous Compound Interest As m approaches infinity, the time interval between payments becomes infinitesimally small, and in the limit Eq. (25-37) reduces to

$$F = P \exp(i'n) \qquad (25\text{-}39)$$

A comparison of Eqs. (25-32) and (25-39) shows that the nominal interest rate i' on a continuous basis is related to the effective interest rate i on an annual basis by

$$\exp(i'n) = (1 + i)^n \qquad (25\text{-}40)$$

Numerically, the difference between continuous and annual compounding is small. In practice, it is probably far smaller than the errors in the estimated cash-flow data. Annual compound interest conforms more closely to current acceptable accounting practice. However, the small difference between continuous and annual compounding may be significant when applied to very large sums of money.

Let us suppose that $100 is invested at a nominal interest rate of 5 percent. We then compute the future worth of the investment after 2 years and also compute the effective annual interest rate for the following kinds of interest: (1) simple, (2) annual compound, (3) monthly compound, (4) daily compound, and (5) continuous compound. The following tabulation shows the results of the calculations, along with the appropriate equation to be used:

Interest type	Equation	Future worth F	Effective rate i, %	Equation
1	(25-31)	$110.000	5	(25-31)
2	(25-32)	$110.250	5	(25-38)
3	(25-37)	$110.495	5.117	(25-38)
4	(25-37)	$110.516	5.1267	(25-38)
5	(25-39)	$110.517	5.1271	(25-38)

When computing the effective annual rate for continuous compounding, the first term of Eq. (25-38), $[1 + (i'/m)]^m$, approaches $e^{i'}$ as m approaches infinity.

Annual Cost or Payment A series of equal annual payments A invested at a fractional interest rate i at the end of each year over a period of n years may be used to build up a future sum of money F. These relations are given by

$$F = A\left[\frac{(1+i)^n - 1}{i}\right] \tag{25-41}$$

or

$$F = A/f_{AF} \tag{25-42}$$

where the annuity future-worth factor is

$$f_{AF} = i/[(1 + i)^n - 1]$$

Values for f_{AF} are readily available in tables.

Equation (25-41) can be combined with Eq. (25-34) to yield

$$P = A\left[\frac{(1+i)^n - 1}{i(1+i)^n}\right] \tag{25-43}$$

$$P = A/f_{AP} \tag{25-44}$$

where P is the present worth of the series of future equal annual payments A and the annuity present-worth factor is

$$f_{AP} = [i(1 + i)^n]/[(1 + i)^n - 1]$$

Values for f_{AP} are also available in tables.

Alternatively, the annual payment A required to build up a future sum of money F with a present value of P is given by

$$A = Ff_{AF} \tag{25-45}$$

$$A = Pf_{AP} \tag{25-46}$$

Equation (25-41) represents the future sum of a series of uniform annual payments that are invested at a stated interest rate over a period of years. This procedure defines an ordinary annuity. Other forms of annuities include the annuity due, in which payments are made at the beginning of the year instead of at the end; and the deferred annuity, in which the first payment is deferred for a definite number of years.

Capitalized Cost A piece of equipment of fixed-capital cost C_{FC} will have a finite life of n years. The capitalized cost of the equipment C_K is defined by

$$(C_K - C_{FC})(1 + i)^n = C_K - S \tag{25-47}$$

C_K is in excess of C_{FC} by an amount which, when compounded at an annual interest rate i for n years, will have a future worth of C_K less the salvage or scrap value S. If the renewal cost of the equipment remains constant at $(C_{FC} - S)$ and the interest rate remains constant at i, then C_K is the amount of capital required to replace the equipment in perpetuity.

Equation (25-41) may be rewritten as

$$C_K = \left[C_{FC} - \frac{S}{(1+i)^n}\right]\left[\frac{(1+i)^n}{(1+i)^n - 1}\right] \tag{25-48}$$

or

$$C_K = (C_{FC} - Sf_d)f_k \tag{25-49}$$

where f_d is the discount factor and f_k, the capitalized-cost factor, is

$$f_k = [(1 + i)^n]/[(1 + i)^n - 1]$$

Values for each factor are available in tables.

Example 1 A piece of equipment has been installed at a cost of $100,000 and is expected to have a working life of 10 years with a scrap value of $20,000. Let us calculate the capitalized cost of the equipment based on an annual compound-interest rate of 5 percent.

Therefore, we substitute values into Eq. (25-48) to give

$$C_K = \left[\$100{,}000 - \frac{\$20{,}000}{(1+0.05)^{10}}\right]\left[\frac{(1+0.05)^{10}}{(1+0.05)^{10} - 1}\right]$$

$$C_K = [\$100{,}000 - (\$20{,}000/1.62889)](2.59009)$$

$$C_K = \$227{,}207$$

Modern Measures of Profitability An investment in a manufacturing process must earn more than the cost of capital for it to be worthwhile. The larger the additional earnings, the more profitable the venture and the greater the justification for putting the capital at risk. A profitability estimate is an attempt to quantify the desirability of taking this risk.

The ways of assessing profitability to be considered in this section are (1) discounted-cash-flow rate of return (DCFRR), (2) net present value (NPV) based on a particular discount rate, (3) equivalent maximum investment period (EMIP), (4) interest-recovery period (IRP), and (5) discounted breakeven point (DBEP).

Cash Flow Let us consider a project in which C_{FC} = $1,000,000, C_{WC} = $90,000, and C_L = $10,000. Hence, C_{TC} = $1,100,000 from Eq. (25-14). If all this capital expenditure occurs in Year 0 of the project, then A_{TC} = $1,100,000 in Year 0 and $-A_{TC}$ = −$1,100,000. From Eq. (25-4), it is seen that any capital expenditure makes a negative contribution to the net annual cash flow A_{CF}.

Let us consider another project in which the fixed-capital expenditure is spread over 2 years, according to the following pattern:

$$C_{FC} = C_{FC0} + C_{FC1}$$

Year 0	Year 1
C_{FC0} = $400,000	C_{FC1} = $600,000
C_L = 10,000	C_{WC} = 90,000
A_{TC} = 410,000	A_{TC} = 690,000

In the final year of the project, the working capital and the land are recovered, which in this case cost a total of $100,000. Thus, in the final year of the project, A_{TC} = −$100,000 and $-A_{TC}$ = +$100,000. From Eq. (25-4), it is seen that any capital recovery makes a positive contribution to the net annual cash flow.

During the development and construction stages of a project, A_{CI} and A_{IT} are both zero in Eqs. (25-2) and (25-4). For this period, the cash flow for the project is negative and is given by

$$A_{CF} = -A_{TC} \tag{25-50}$$

Figure 25-9 shows the cash-flow stages in a project. The expenditure during the research and development stage is normally relatively small. It will usually include some preliminary process design and a market survey. Once the decision to go ahead with the project has been taken, detailed process-engineering design will commence, and the rate of expenditure starts to increase. The rate is increased

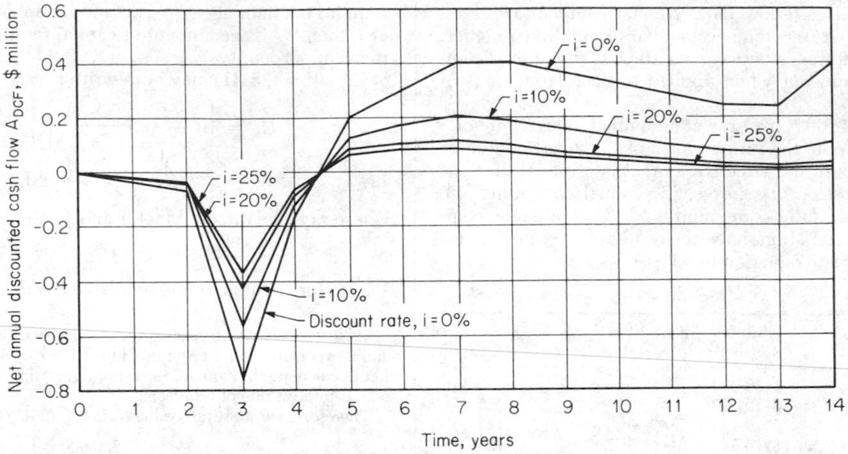

FIG. 25-9 Effect of discount rate on cash flows.

still further when equipment is purchased and construction gets under way. There is no return on the investment until the plant is started up. Even during startup, there is some additional expenditure. Once the plant is operating smoothly, an inflow of cash is established. During the early stages of a project, there may be a tax credit because of the existence of expenses without corresponding income.

Discounted Cash Flow The present value P of a future sum of money F is given by

$$P = Ff_d \tag{25-51}$$

where $f_d = 1/(1 + i)^n$, the discount factor. Values for this factor are readily available in tables. For example, $90,909 invested at an annual interest rate of 10 percent becomes $100,000 after 1 year. Similarly, $38,554 invested at 10 percent becomes $100,000 after 10 years.

Thus, cash flow in the early years of a project has a greater value than the same amount in the later years of a project. Therefore, it pays to receive money as soon as possible and to delay paying out money for as long as possible.

Time is taken into account by using the annual discounted cash flow A_{DCF}, which is related to the annual cash flow A_{CF} and the discount factor f_d by

$$A_{DCF} = A_{CF}f_d \tag{25-52}$$

Thus, at the end of any year n,

$$(A_{DCF})_n = (A_{CF})_n/(1 + i)^n$$

The sum of the annual discounted cash flows over n years, ΣA_{DCF}, is known as the net present value (NPV) of the project:

$$(\text{NPV}) = \sum_0^n (A_{DCF})_n \tag{25-53}$$

The value of (NPV) is directly dependent on the choice of the fractional interest rate i. An interest rate can be selected to make (NPV) = 0 after a chosen number of years. This value of i is found from

$$\sum_0^n (A_{DCF})_n = \frac{(A_{CF})_0}{(1 + i)^0} + \frac{(A_{CF})_1}{(1 + i)^1} + \cdots + \frac{(A_{CF})_n}{(1 + i)^n} = 0 \tag{25-54}$$

Equation (25-54) may be solved for i either graphically or by an iterative trial-and-error procedure. The value of i given by Eq. (25-54) is known as the discounted-cash-flow rate of return (DCFRR). It is also known as the profitability index, true rate of return, investor's rate of return, and interest rate of return.

Cash-Flow Curves Figure 25-9 shows the cash-flow stages in a project together with their discounted-cash-flow values for the data given in Table 25-3. In addition to cash-flow and discounted-cash-flow curves, it is also instructive to plot cumulative-cash-flow and cumulative-discounted-cash-flow curves. These are shown in Fig. 25-10 for the data in Table 25-3.

The cost of capital may also be considered as the interest rate at which money can be invested instead of putting it at risk in a manufacturing process. Let us consider the process data listed in Table 25-3 and plotted in Fig. 25-10. If the cost of capital is 10 percent, then the appropriate discounted-cash-flow curve in Fig. 25-10 is *abcdef*. Up to point *e*, or 8.49 years, the capital is at risk. Point *e* is the discounted breakeven point (DBEP). At this point, the manufacturing process has paid back its capital and produced the same return as an equivalent amount of capital invested at a compound-interest rate of 10 percent. Beyond the breakeven point, the capital is no longer at risk and any cash flow above the horizontal baseline, $\Sigma A_{DCF} = 0$, is in excess of the return on an equivalent amount of capital invested at a compound-interest rate of 10 percent. Thus, the greater the area above the baseline, the more profitable the process.

When (NPV) and (DCFRR) are computed, depreciation is not considered as a separate expense. It is simply used as a permitted writing-down allowance to reduce the annual amount of tax in accordance with the rules applying in the country of earning. The tax payable is deducted in accordance with Eq. (25-2) in the year in which it is paid, which may differ from the year in which the corresponding income was earned.

A (DCFRR) of, say, 15 percent implies that 15 percent per year will be earned on the investment, in addition to which the project generates sufficient money to repay the original investment plus any interest payable on borrowed capital plus all taxes and expenses.

It is not normally possible to make a comprehensive assessment of profitability with a single number. The shape of the cumulative-cash-flow and cumulative-discounted-cash-flow curves both before and after the breakeven point is an important factor.

D. H. Allen [*Chem. Eng.*, **74,** 75–78 (July 3, 1967)] accounted for the shape of the cumulative-undiscounted-cash-flow curve up to the breakeven point e_0 in Fig. 25-10 by using a parameter known as the equivalent maximum investment period (EMIP), which is defined as

$$(\text{EMIP}) = \frac{\text{area } (a_0 \text{ to } e_0)}{(\Sigma A_{CF})_{\max}} \quad \text{for} \quad A_{CF} \leq 0 \tag{25-55}$$

where the area (a_0 to e_0) refers to the area below the horizontal baseline ($\Sigma A_{CF} = 0$) on the cumulative-cash-flow curve in Fig. 25-10. The sum $(\Sigma A_{CF})_{\max}$ is the maximum cumulative expenditure on the project, which is given by point d_0 in Fig. 25-10. (EMIP) is a time in years. It is the equivalent period during which the total project debt would be outstanding if it were all incurred at one instant and all

TABLE 25-3 Annual Cash Flows and Discounted Cash Flows for a Project

			Discounted at 10%			Discounted at 20%			Discounted at 25%		
Year	A_{CF}, $	ΣA_{CF}, $	f_d	A_{DCF}, $	ΣA_{DCF}, $	f_d	A_{DCF}, $	ΣA_{DCF}, $	f_d	A_{DCF}, $	ΣA_{DCF},$
0	−10,000	−10,000	1.00000	−10,000	−10,000	1.00000	−10,000	−10,000	1.00000	−10,000	−10,000
1	−30,000	−40,000	0.90909	−27,273	−37,273	0.83333	−25,000	−35,000	0.80000	−24,000	−34,000
2	−60,000	−100,000	0.82645	−49,587	−86,860	0.69444	−41,666	−76,666	0.64000	−38,400	−72,400
3	−750,000	−850,000	0.75131	−563,483	−650,343	0.57870	−434,025	−510,691	0.51200	−384,000	−456,400
4	−150,000	−1,000,000	0.68301	−102,452	−752,795	0.48225	−72,338	−583,029	0.40960	−61,440	−517,840
5	+200,000	−800,000	0.62092	+124,184	−628,611	0.40188	+80,376	−502,653	0.32768	+65,536	−452,304
6	+300,000	−500,000	0.56447	+169,341	−459,270	0.33490	+100,470	−402,183	0.26214	+78,642	−373,662
7	+400,000	−100,000	0.51316	+205,264	−254,006	0.27908	+111,632	−290,551	0.20972	+83,888	−289,774
8	+400,000	+300,000	0.46651	+186,604	−67,402	0.23257	+93,028	−197,523	0.16777	+67,108	−222,666
9	+360,000	+660,000	0.42410	+152,676	+85,274	0.19381	+69,772	−127,751	0.13422	+48,319	−174,347
10	+320,000	+980,000	0.38554	+123,373	+208,647	0.16151	+51,683	−76,068	0.10737	+34,358	−139,989
11	+280,000	+1,260,000	0.35049	+98,137	+306,784	0.13459	+37,685	−38,383	0.08590	+24,052	−115,937
12	+240,000	+1,500,000	0.31863	+76,471	+383,255	0.11216	+26,918	−11,465	0.06872	+16,493	−99,444
13	+240,000	+1,740,000	0.28966	+69,518	+452,773	0.09346	+22,430	+10,965	0.05498	+13,195	−86,249
14	+400,000	+2,140,000	0.26333	+105,332	+558,105	0.07789	+31,156	+42,121	0.04398	+17,592	−68,657

NOTE: A_{CF} is net annual cash flow, A_{DCF} is net annual discounted cash flow, f_d is discount factor at stated interest, ΣA_{CF} is cumulative cash flow, and ΣA_{DCF} is cumulative discounted cash flow.

repaid at one instant. Clearly, the shorter the (EMIP), the more attractive the project.

Allen accounted for the shape of the cumulative-cash-flow curve beyond the breakeven point by using a parameter known as the interest-recovery period (IRP). This is the time period (illustrated in Fig. 25-11) that makes the area (e_0 to f_0) above the horizontal baseline equal to the area (a_0 to e_0) below the horizontal baseline on the cumulative-cash-flow curve.

C. G. Sinclair [*Chem. Process. Eng.*, **47**, 147 (1966)] has considered similar parameters to the (EMIP) and (IRP) based on a cumulative-discounted-cash-flow curve.

Consideration of the cash-flow stages in Fig. 25-10 shows the factors that can affect the (EMIP) and (IRP). If the required capital investment is increased, it is necessary to increase the rate of income after startup for the (EMIP) to remain the same. In order to have the (EMIP) small, it is necessary to keep the research and development, design, and construction stages short.

Example 2 The following data describe a project. Revenue from annual sales and the total annual expense over a 10-year period are given in the first three columns of Table 25-4. The fixed-capital investment C_{FC} is $1,000,000. Plant items have a zero salvage value. Working capital C_{WC} is $90,000, and cost of land C_L is $10,000. There are no tax allowances other than depreciation; i.e., A_A is zero. The fractional tax rate t is 0.50.

We shall calculate for these data the net present value (NPV) for the following depreciation methods and discount factors:

a. Straight-line, 10 percent
b. Straight-line, 20 percent
c. Double-declining, 10 percent
d. Sum-of-years-digits, 10 percent
e. Straight-line, 10 percent; income tax delayed for 1 year

In addition, we shall calculate the discounted-cash-flow rate of return (DCFRR) with straight-line depreciation.

a. We begin the calculations for this example by finding the total capital cost C_{TC} for the project from Eq. (25-14). Here, C_{TC} = $1,100,000. In Year 0, this amount is the same as the net annual capital expenditure A_{TC} and is listed in Table 25-4.

The annual rate of straight-line depreciation of the fixed-capital investment C_{FC}, from $1,000,000 at startup to a salvage value S, of zero at the end of a productive life s of 10 years, is given by

$$A_D = (C_{FC} - S)/s$$

$$A_D = (\$1{,}000{,}000 - \$0)/10 \text{ years} = \$100{,}000/\text{year}$$

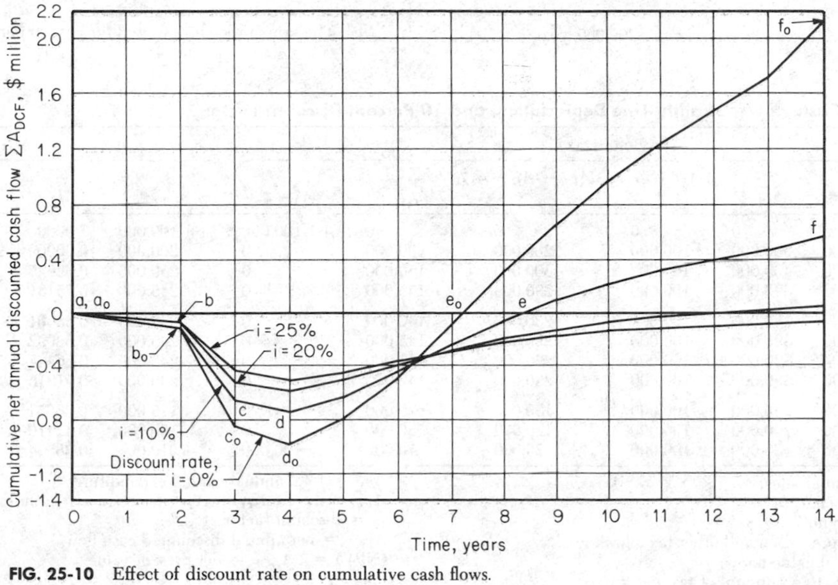

FIG. 25-10 Effect of discount rate on cumulative cash flows.

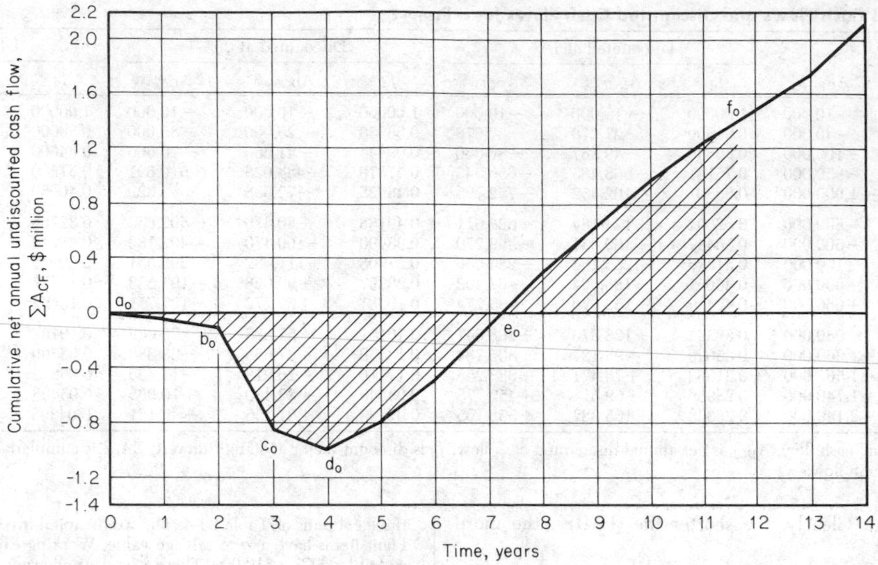

FIG. 25-11 Cumulative cash flow against time, showing interest recovery period.

The annual cash income A_{CI} for Year 1, when A_S = $400,000 per year and A_{TE} = $100,000 per year, is, from Eq. (25-1), $300,000 per year. Values for subsequent years are calculated in the same way and listed in Table 25-4.

Annual amount of tax A_{IT} for Year 1, when A_{CI} = $300,000 per year, A_D = $100,000 per year, A_A = $0 per year, and $t = 0.5$, is found from Eq. (25-3) to be

$$A_{IT} = [(\$300{,}000 - \$100{,}000 - \$0)/\text{year}](0.5)$$

$$= \$100{,}000/\text{year}$$

Values for subsequent years are calculated in the same way and listed in Table 25-4.

Net annual cash flow (after tax) A_{CF} for Year 0, when A_{CI} = $0 per year, A_{IT} = $0 per year, and A_{TC} = $1,100,000 per year, is found from Eq. (25-4) to be

$$A_{CF} = \$0/\text{year} - \$1{,}100{,}000/\text{year} = -\$1{,}100{,}000/\text{year}$$

Net annual cash flow (after tax) A_{CF} for Year 1, when A_{CI} = $300,000 per year, A_{IT} = $100,000 per year, and A_{TC} = $0 per year, is found from Eqs. (25-2) and (25-4) to be

$$A_{CF} = \$200{,}000/\text{year} - \$0/\text{year} = \$200{,}000/\text{year}$$

Values for the years up to and including Year 9 are calculated in the same way and listed in Table 25-4.

At the end of Year 10, the working capital (C_{WC} = $90,000) and the cost of land (C_L = $10,000) are recovered, so that the annual expenditure of capital A_{TC} in Year 10 is −$100,000 per year. Hence, the net annual cash flow (after tax) for Year 10 must reflect this recovery. By using Eq. (25-4),

$$A_{CF} = \$110{,}000/\text{year} - (-\$100{,}000/\text{year})$$

$$= \$210{,}000/\text{year}$$

The net annual discounted cash flow A_{DCF} for Year 1, when A_{CF} = $200,000 per year and f_d = 0.90909 (for i = 10 percent), is found from Eq. (25-52) to be

$$A_{DCF} = (\$200{,}000/\text{year})(0.90909) = \$181{,}820/\text{year}$$

Values for subsequent years are calculated in the same way and listed in Table 25-4.

The net present value (NPV) is found by summing the values of A_{DCF} for each year, as in Eq. (25-53). The net present value is found to be $276,210, as given by the final entry in Table 25-4.

b. The same procedure is used for i = 20 percent. The discount factors to

TABLE 25-4 Annual Cash Flows, Straight-Line Depreciation, and 10 Percent Discount Factor

	Before tax							After tax			
Year	A_S, $	A_{TE}, $	A_{CI}, $	$A_D + A_A$, $	$A_{CI} - A_D - A_A$, $	A_{IT}, $	A_{TC}, $	A_{CF}, $	f_d	A_{DCF}, $	(NPV), $
0	0	0	0	0	0	0	+1,100,000	−1,100,000	1.0000	−1,100,000	−1,100,000
1	400,000	100,000	300,000	100,000	200,000	100,000	0	200,000	0.90909	181,820	−918,180
2	500,000	100,000	400,000	100,000	300,000	150,000	0	250,000	0.82645	206,610	−711,570
3	500,000	110,000	390,000	100,000	290,000	145,000	0	245,000	0.75131	184,070	−527,500
4	500,000	120,000	380,000	100,000	280,000	140,000	0	240,000	0.68301	163,920	−363,580
5	520,000	130,000	390,000	100,000	290,000	145,000	0	245,000	0.62092	152,120	−211,460
6	520,000	130,000	390,000	100,000	290,000	145,000	0	245,000	0.56447	138,300	−73,160
7	520,000	140,000	380,000	100,000	280,000	140,000	0	240,000	0.51316	123,160	+50,000
8	390,000	140,000	250,000	100,000	150,000	75,000	0	175,000	0.46651	81,640	+131,640
9	350,000	150,000	200,000	100,000	100,000	50,000	0	150,000	0.42410	63,610	+195,250
10	280,000	160,000	120,000	100,000	20,000	10,000	−100,000	210,000	0.38554	80,960	+276,210

A_S = revenue from annual sales.
A_{TE} = total annual expense.
A_{CI} = annual cash income.
$A_D + A_A$ = annual depreciation and other tax allowances.
$A_{CI} - A_D - A_A$ = taxable income.
$A_{IT} = (A_{CI} - A_D - A_A)t$ = amount of tax at t = 0.5.
A_{TC} = total annual capital expenditure.
$A_{CF} = A_{CI} - A_{IT} - A_{TC}$ = net annual cash flow.
f_d = discount factor at 10%.
A_{DCF} = net annual discounted cash flow.
(NPV) = ΣA_{DCF} = net present value.

be used in a table similar to Table 25-4 must be those for 20 percent. The (NPV) is found to be −$151,020.

c. The calculations are similar to those for subexample *a* except that depreciation is computed by using the double-declining method of Eq. (25-19). The net present value is found to be $288,530.

d. Again, the calculations are similar to those for subexample *a* except that depreciation is computed by using the sum-of-years-digits method of Eq. (25-20). The net present value is found to be $316,610.

e. The calculations follow the same procedure as for subexample *a*, but the annual amount of tax A_{IT} is calculated for a particular year and then deducted from the annual cash income A_{CI} for the following year. The net present value for Year 11 is found to be $341,980.

The discounted-cash-flow rate of return (DCFRR) can readily be obtained approximately by interpolation of the (NPV) for i = 10 percent and i = 20 percent:

$$(\text{DCFRR}) = 0.100 + [(\$276{,}210)(0.20 - 0.10)]/[\$276{,}210 - (-\$151{,}020)]$$

$$(\text{DCFRR}) = 0.164, \text{ or } 16.4 \text{ percent}$$

The calculation of (DCFRR) usually requires a trial-and-error solution of Eq. (25-57), but rapidly convergent methods are available [N. H. Wild, *Chem. Eng.*, **83**, 153–154 (Apr. 12, 1976)]. For simplicity linear interpolation is often used.

A comparison of the (NPV) values for a 10 percent discount factor shows clearly that double-declining depreciation is more advantageous than straight-line depreciation and that sum-of-years-digits depreciation is more advantageous than the double-declining method. However, a significant advantage is obtained by delaying the payment of tax for 1 year even with straight-line depreciation.

This example is a simplified one. The cost of the working capital is assumed to be paid for in Year 0 and returned in Year 10. In practice, working capital increases with the production rate. Thus there may be an annual expenditure on working capital in a number of years subsequent to Year 0. Except in loss-making years, this is usually treated as an expense of the process. In loss-making years the cash injection for working capital is included in the A_{TC} for that year.

Analysis of Techniques Both the (NPV) and the (DCFRR) methods are based on discounted cash flows and in that sense are variations of the same basic method. However, when ranking different projects on the basis of profitability, they can produce different results.

Discounted-cash-flow rate of return (DCFRR) has the advantage of being unique and readily understood. However, when used alone, it gives no indication of the scale of the operation. The (NPV) indicates the monetary return, but unlike that of the (DCFRR) its value depends on the base year chosen for the calculation. Additional information is needed before its significance can be appreciated. However, when a company is considering investment in a portfolio of projects, individual (NPV)s have the advantage of being additive. This is not true of (DCFRR)s.

Increasing use is being made of the capital-rate-of-return ratio (CRR), which is the net present value (NPV) divided by the maximum cumulative expenditure or maximum net outlay, $-(\Sigma A_{CF})_{max}$

$$(\text{CRR}) = (\text{NPV})/(\Sigma A_{CF})_{max} \quad \text{for} \quad A_{CF} \le 0 \qquad (25\text{-}56)$$

The maximum net outlay is very important, since no matter how profitable a project is, the matter is academic if the company is unable to raise the money to undertake the project.

An (NPV) or (DCFRR) estimation will be no better than the accuracy of the projected cash flows over the life of the project. Clearly, one is likely to predict cash flows more accurately for 2 or 3 years ahead than, say, for 9 or 10 years ahead. However, since the cash flows for the later years are discounted to a greater extent than the cash flows for the earlier years, the latter have less effect on the overall estimation. Nevertheless, the difficulty of predicting cash flows in later years and the inherent lack of confidence in these predictions are serious disadvantages of the (DCFRR) method. In this respect (NPV)s are more useful since they are calculated for each year of a project. Thus, a project with a favorable (NPV) in the early years is a promising one.

One way of overcoming these disadvantages of the (DCFRR) method is to make estimates of the times required to reach certain values of (DCFRR). For example, how many years will it take to reach (DCFRR)s of 10 percent, 15 percent, 20 percent per year, etc.?

Although (DCFRR) trial-and-error calculations and (NPV) calculations are tedious if done manually, computer programs which are suitable for programmable pocket calculators can readily be written to make calculations easier.

It is possible for some projects to reach a stage at which repairs, replacements, etc., can exceed net earnings in a particular year. In this case the cumulative-discounted-cash-flow or net-present-value curve plotted against time has a genuine maximum.

It is important when appraising by (NPV) and (DCFRR) not to consider the past in profitability estimations. Good money should never follow bad. It is unwise to continue to put money into a project if a more profitable project exists, even though this course may involve scrapping an expensive plant. Other considerations may, however, outweigh purely financial criteria in a particular case.

No single value for a profitability estimate should be accepted without further consideration. An intelligent consideration of the cumulative-cash-flow and cumulative-discounted-cash-flow curves such as those shown in Fig. 25-10, together with experience and good judgment, is the best way of assessing the financial merit of a project.

When considering future projects, top management will most likely require the discounted-cash-flow rate of return and the payback period. However, the estimators should also supply management with the following:

Cumulative discounted-cash-flow or (NPV) curve for a discount rate of 10 percent per year or other agreed aftertax cost of capital

Maximum net outlay, $(\Sigma A_{CF})_{max}$, for $A_{CF} \le 0$

Discounted breakeven point (DBEP)

Plot of capital-return ratio (CRR) against time over the life of the project for a discount rate at the cost of capital

Number of years to reach discounted-cash-flow rates of return of, say, 15 and 25 percent per year respectively

Comparisons on the basis of time can be summarized by the following:

Duration of the project

Breakeven point (BEP)

Discounted breakeven point (DBEP)

Equivalent maximum investment period (EMIP)

Interest-recovery period (IRP)

Payback period (PBP)

Comparisons on the basis of cash can be summarized by the following:

Maximum cumulative expenditure on the project, $(\Sigma A_{CF})_{max}$, for $A_{CF} \le 0$

Maximum discounted cumulative expenditure on the project

Cumulative net annual cash flow ΣA_{CF}

Cumulative net annual discounted cash flow ΣA_{DCF} or net present value (NPV)

Capitalized cost C_K

Comparisons on the basis of interest can be summarized as (1) the net present value (NPV) and (2) the discounted-cash-flow rate of return (DCFRR), which from Eqs. (25-53) and (25-54) is given formally as the fractional interest rate i which satisfies the relationship

$$(\text{NPV}) = \sum_{0}^{n} (A_{DCF})_n = 0 \qquad (25\text{-}57)$$

When comparing project profitability, the ranking on the basis of net present value (NPV) may differ from that on the basis of discounted-cash-flow rate of return (DCFRR). Let us consider the data for two projects:

Cost of capital	Project C	Project D
i, %	(NPV), $	(NPV), $
4	+100,000	+62,000
8	+41,000	+28,000
12	−2,000	+10,000
16	−32,000	−4,000

These (NPV) data are plotted against the cost of capital, as shown in Fig. 25-12. The discounted-cash-flow rate of return is the value of

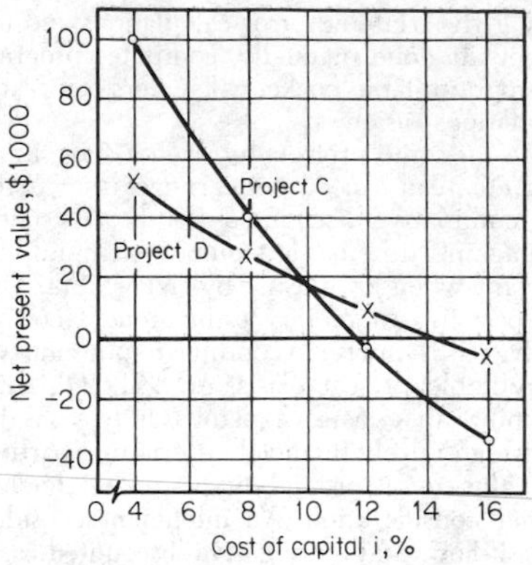

FIG. 25-12 Effect of cost of capital on net present value.

i that satisfies Eq. (25-5). From Fig. 25-12, (NPV) = 0 at a (DCFRR) of 11.8 percent for project C and 14.7 percent for project D. Thus, on the basis of (DCFRR), project D is more profitable than project C.

The (NPV) of project C is equal to that of project D at a cost of capital i = 9.8 percent. If the cost of capital is greater than 9.8 percent, project D has the higher (NPV) and is, therefore, the more profitable. If the cost of capital is less than 9.8 percent, project C has the higher (NPV) and is the more profitable.

Benefit of Early Cash Flows It pays to receive cash inflows as early as possible and to delay cash outflows as long as possible.

Let us consider the net annual cash flows (after tax) A_{CF} for projects E, F, and G, listed in Table 25-5. The cumulative annual cash flows ΣA_{CF} and cumulative discounted annual cash flows ΣA_{DCF}, using a discount of 10 percent for these projects, are also listed in Table 25-5. We notice that the cumulative annual cash flow for each project is +\$1000.

The (DCFRR) is the discount rate that satisfies Eq. (25-57) in the final year of the project. We can approximate the (DCFRR) for each project as follows:

For project E,

$$\Sigma A_{CF} = +\$1000 \text{ in Year 3 for } i = 0 \text{ percent}$$
$$\Sigma A_{DCF} = +\$131 \text{ in Year 3 for } i = 10 \text{ percent}$$
$$\Sigma A_{DCF} = \$0 \text{ in Year 3 for } i = \text{(DCFRR)}$$

TABLE 25-5 Cash-Flow Data for Projects E, F, and G

			Discounted at 10%		
Year	A_{CF}, \$	ΣA_{CF}, \$	f_d	A_{DCF}, \$	ΣA_{DCF} = (NPV), \$
			Project E		
0	−5000	−5000	1.0000	−5000	−5000
1	+3000	−2000	0.90909	+2727	−2273
2	+2000	0	0.82645	+1653	−620
3	+1000	+1000	0.75131	+751	+131
			Project F		
0	−5000	−5000	1.0000	−5000	−5000
1	+1000	−4000	0.90909	+909	−4091
2	+2000	−2000	0.82645	+1653	−2438
3	+3000	+1000	0.75131	+2254	−184
			Project G		
0	−5000	−5000	1.0000	−5000	−5000
1	+2000	−3000	0.90909	+1818	−3182
2	+2000	−1000	0.82645	+1653	−1529
3	+2000	+1000	0.75131	+1503	−26

TABLE 25-6 Cash-Flow Data for Projects H and I

		Discounted at 10%		
Year	A_{CF}, \$	f_d	A_{DCF}, \$	ΣA_{DCF} = (NPV), \$
		Project H		
0	−50,000	1.0000	−50,000	−50,000
1	+10,000	0.90909	9,091	−40,909
2	+10,000	0.82645	8,265	−32,644
3	+10,000	0.75131	7,513	−25,131
4	+10,000	0.68301	6,830	−18,301
5	+10,000	0.62092	6,209	−12,092
6	+10,000	0.56447	5,645	−6,447
7	+10,000	0.51316	5,132	−1,315
8	+10,000	0.46651	4,665	+3,350
9	+10,000	0.42410	4,241	+7,591
		Project I		
3	−40,000	0.75131	−30,052	−30,052
4	+20,000	0.68301	+13,660	−16,392
5	+20,000	0.62092	+12,418	−3,974
6	+20,000	0.56447	+11,289	+7,315
7	+20,000	0.51316	+10,263	+17,578
8	+20,000	0.46651	+9,330	+26,908
9	+20,000	0.42410	+8,482	+35,390

Therefore,

$$1000/(1000 - 131) \cong \text{(DCFRR)}/10$$
$$\text{(DCFRR)} \cong 11.5 \text{ percent}$$

Similarly for project F,

$$1000/(1000 + 184) \cong \text{(DCFRR)}/10$$
$$\text{(DCFRR)} \cong 8.4 \text{ percent}$$

Similarly for project G,

$$1000/(1000 + 26) \cong \text{(DCFRR)}/10$$
$$\text{(DCFRR)} \cong 9.7 \text{ percent}$$

In terms of net present value (NPV), the projects in order of merit are E, G, and F, with (NPV)s of +\$131, −\$26, and −\$184 respectively. In terms of (DCFRR), the projects in order of merit are also E, G, and F, with (DCFRR) values of 11.5 percent, 9.7 percent, and 8.4 percent respectively.

When to Scrap an Existing Process Let us suppose that a company invests \$50,000 in a manufacturing process that has positive net annual flows (after tax) A_{CF} of \$10,000 in each year. During the third year of operation, an alternative process becomes available. The new process would require an investment of \$40,000 but would have positive net annual cash flows (after tax) of \$20,000 in each year. The cost of capital is 10 percent, and it is estimated that a market will exist for the product for at least 6 more years. Should the company continue with the existing process (project H), or should it scrap project H and adopt the new process (project I)?

The net annual cash flows A_{CF} and cumulative discounted annual cash flow ΣA_{DCF} for a discount factor of 10 percent are listed in Table 25-6 for the two projects. At the end of Year 9, the net present values are

$$\text{(NPV)} = +\$35{,}390 \text{ for project I}$$
$$\text{(NPV)} = +\$7591 \text{ for project H}$$

The difference is +\$27,779, which is numerically greater than the money lost by the end of Year 3 for project H. Thus project H should be scrapped, and the new project I adopted if only economic reasons need to be considered. Recovery of working capital and the cost of land have been neglected since the latter is the same for each project and the former would also favor project I.

Incremental Comparisons A company may have the choice of, say, investing \$10,000 in project J, which will give a (DCFRR) of 16 percent, or $7000 in project K, which will give a (DCFRR) of 18 percent. Should it spend $10,000 on project J or spend only $7000 on project K and invest the difference of $3000 elsewhere?

Both projects have lives of 10 years and constant positive net annual cash flows A_{CF} of $2069 and $1558 for projects J and K respectively. The corresponding (NPV)s at a discount factor of 10 percent are +$2710 and +$2560 respectively. These data are summarized as follows:

	Project J	Project K	Project (J − K)
A_{CF}, $, in Year 0	−10,000	−7,000	−3,000
A_{CF}, $, in each of Years 1–10	+2,069	+1,558	+511
(NPV), i = 10 percent, $	+2,710	+2,560	+150
(DCFRR), percent	16	18	12.4

From the difference in cash flows between the projects, the discounted-cash-flow rate of return (DCFRR) for project (J-K) can be shown as 12.4 percent. This is significantly lower than for either project J or project K. Thus, if the $3000 can be invested to give a return greater than 12.4 percent, project K should be chosen in preference to project J.

Comparisons on the Basis of Capitalized Cost A machine in a process generates a positive net cash flow of $1000. Two alternatives are available: machine L, costing $2000, requires replacement every 4 years, and machine M, costing $3000, requires replacement every 6 years. Neither machine has any scrap value. The cost of capital is 10 percent. Which machine is the more profitable to operate?

In this case, the lives of the machines are unequal, and the comparison is conveniently made on the basis of capitalized cost. This puts lives on the same basis, which is an infinite number of years. The net annual cash flows generated by each machine are equal.

The capitalized cost C_K of a piece of fixed-capital cost C_{FC} is the amount of capital required to ensure that the equipment may be renewed in perpetuity. For a piece of equipment with no scrap value, C_K is given by

$$C_K = C_{FC}\left[\frac{(1+i)^n}{(1+i)^n - 1}\right] \qquad (25\text{-}58)$$

For machine L,

$$C_K = (\$2000)(3.15471) = \$6309.42$$

For machine M,

$$C_K = (\$3000)(2.29607) = \$6888.21$$

Thus, machine L with the lower capitalized cost is the more profitable to operate.

Relationship between (PBP) and (DCFRR) For the case of a single lump-sum capital expenditure C_{FC} which generates a constant annual cash flow A_{CF} in each subsequent year, the payback period is given by the equation

$$(\text{PBP}) = C_{FC}/A_{CF} \qquad (25\text{-}59)$$

if the scrap value of the capital outlay may be taken as zero.

For this simplified case the net present value (NPV) after n years with money invested at a required aftertax compound annual fractional interest rate i is given by the equation

$$(\text{NPV}) = C_{FC} - A_{CF}F_n \qquad (25\text{-}60)$$

where

$$F_n = \sum_{1}^{n} \frac{1}{(1+i)^n}$$

When (NPV) = 0, the value of i given by Eq. (25-60) is the discounted-cash-flow rate of return (DCFRR), and in this case Eqs. (25-59) and (25-60) can be combined to give:

$$(\text{PBP}) = F_n \qquad (25\text{-}61)$$

Figure 25-13 is a plot of Eq. (25-61) in the form of the number of years n required to reach a certain discounted-cash-flow rate of return (DCFRR) for a given payback period (PBP). The figure is a modification of plots previously published by A. G. Bates [*Hydrocarbon Process.*, **45,** 181–186 (March 1966)], C. Estrup [*Br. Chem. Eng.*, **16,** 171 (February–March 1971)], and F. A. Holland and F. A. Watson [*Process Eng. Econ.*, **1,** 293–299 (December 1976)].

In the limiting case when n approaches infinity, Eq. (25-61) can be written as

$$(\text{DCFRR})_{\max} = 1/(\text{PBP}) \qquad (25\text{-}62)$$

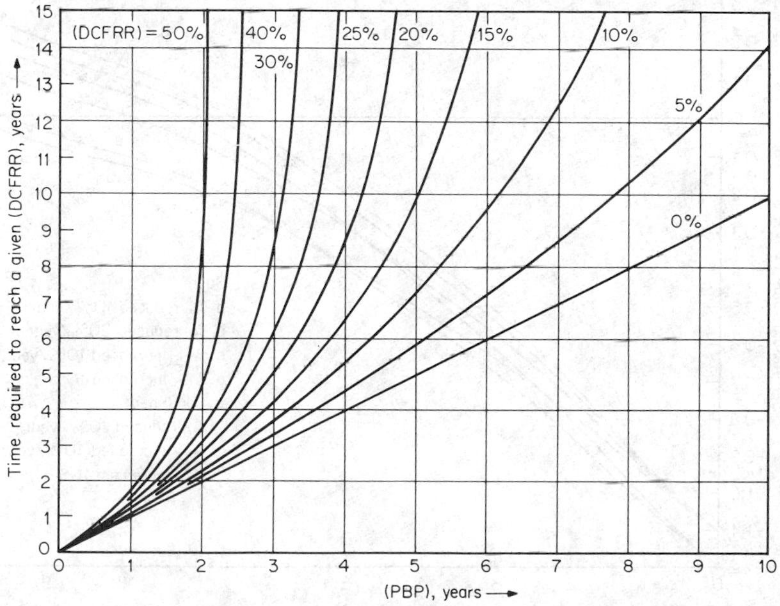

FIG. 25-13 Relationship between payback period and discounted-cash-flow rate of return.

which means, for example, that if the payback period is 4 years, the maximum possible discounted-cash-flow rate of return which can be reached is 25 percent. The corresponding (DCFRR) for (PBP) = 10 years is 10 percent.

Equations (25-59), (25-60), (25-61), and (25-62) may be used as they stand to assess expenditure on energy-conservation measures since a constant amount of energy is saved in each year subsequent to the capital outlay. However, the annual cash flows A_{CF} corresponding to the energy savings remain constant only if there is no inflation or if the money values are corrected to their purchasing power at the time of the capital expenditure.

Sensitivity Analysis An economic study should pinpoint the areas most susceptible to change. It is easier to predict expenses than either sales or profits. Fairly accurate estimates of capital costs and processing costs can be made. However, for the most part, errors in these estimates have a correspondingly smaller effect than changes in sales price, sales volume, and the costs of raw materials and distribution.

Sales and raw-materials prices may be affected by any of the following: discounts and allowances, availability of substitutes, contract pricing, government regulations, quality and form of the materials, and competition. Sales volume may be affected by any of the following: new uses for the product, new markets, advertising, quality, overcapacity, replacement by another product, competition, and timing of entry into the market.

Distribution costs depend on plant location, physical state of the material (whether liquid, gas, or solid), nature of the material (whether corrosive, explosive, flammable, perishable, or toxic), freight rates, and labor costs. Distribution costs may be affected by any of the following: new methods of materials handling, safety regulations, productivity agreements, wage rates, transportation systems, storage systems, quality, losses, and seasonal effects.

It is worthwhile to make tables or plot curves that show the effect of variations in costs and prices on profitability. This procedure is called **sensitivity analysis.** Its purpose is to determine to which factors the profitability of a project is most sensitive. Sensitivity analysis should always be carried out to observe the effect of departures from expected values.

For many years, companies and countries have lived with the problem of inflation, or the falling value of money. Costs—in particular, labor costs—tend to rise each year. Failure to account for this trend in predicting future cash flows can lead to serious errors and misleading profitability estimates.

Another important factor is the tendency of product prices to fall as the total national or international volume of production increases. Sales prices may fall by 20 percent for a doubling in volume or production.

No profitability estimate is better than the inherent accuracy of the data.

Example 3 The following data describe a project. Revenue from annual sales and total annual expense over a 10-year period are given in the first three columns of Table 25-4. The fixed-capital investment C_{FC} is $1 million. Plant items have a zero salvage value. Working capital C_{WC} is $90,000, and the cost of land C_L is $10,000. There are no tax allowances other than depreciation; i.e., A_A is zero. The fractional tax rate t is 0.50. For this project, the net present value for a 10 percent discount factor and straight-line depreciation was shown to be $276,210 and the discounted-cash-flow rate of return to be 16.4 percent per year.

We shall use these data and the accompanying information of Table 25-4 as the base case and calculate for straight-line depreciation the net present value (NPV) with a 10 percent discount factor and the discounted-cash-flow rate of return (DCFRR) for the project with the following situations.

Case	Modification
a	Revenue A_S reduced by 10 percent per year
b	Revenue A_S reduced by 20 percent per year
c	Total expense A_{TE} increased by 10 percent per year
d	Fixed-capital investment increased by 10 percent
e	A_S reduced by 10 percent per year, A_{TE} increased by 10 percent per year, and C_{FC} increased by 10 percent

The results are shown in Figs. 25-14 and 25-15 and Tables 25-7 and 25-8.

Learning Curves It is usual to learn from experience. Consequently, the time taken to produce an article, the number of spoiled

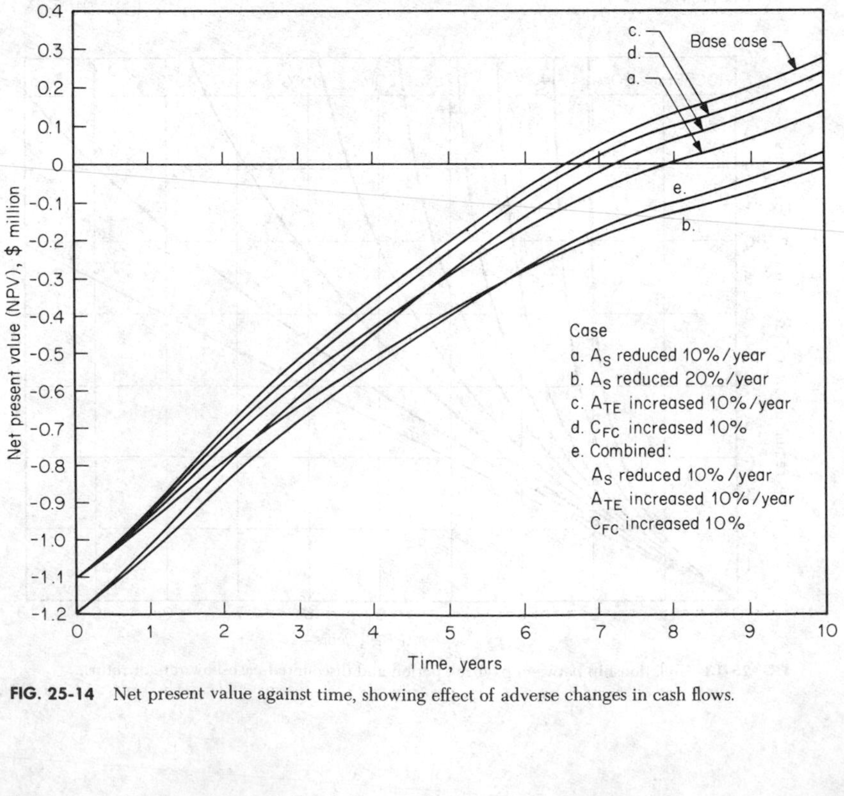

FIG. 25-14 Net present value against time, showing effect of adverse changes in cash flows.

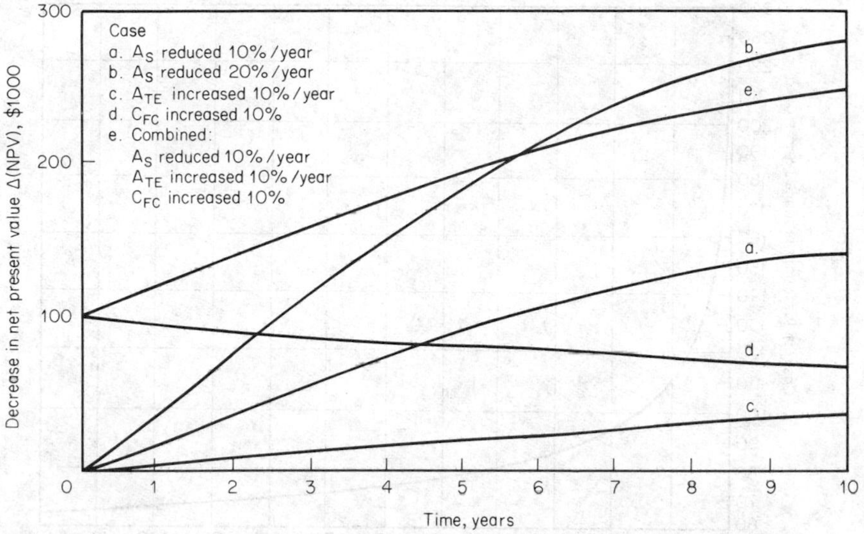

FIG. 25-15 Decrease in net present value against time resulting from adverse changes in cash flows.

TABLE 25-7 Annual Cash Flows, Straight-Line Depreciation, and 10 Percent Discount Factor When Revenue Is Reduced by 10 Percent per Year

	Base case							
Year	A_S, $	A_{TE}, $	(NPV), $	ΔA_S, $	f_d	ΔA_{DCF}, $	Δ(NPV), $	Reduced (NPV), $
0	0	0	−1,100,000	0	1.0000	0	0	−1,100,000
1	400,000	100,000	−918,180	40,000	0.90909	18,180	18,180	−936,360
2	500,000	100,000	−711,570	50,000	0.82645	20,660	38,840	−750,410
3	500,000	110,000	−527,500	50,000	0.75131	18,780	57,620	−585,120
4	500,000	120,000	−363,580	50,000	0.68301	17,070	74,690	−438,270
5	520,000	130,000	−211,460	52,000	0.62092	16,140	90,830	−302,290
6	520,000	130,000	−73,160	52,000	0.56447	14,680	105,510	−178,670
7	520,000	140,000	+50,000	52,000	0.51316	13,340	118,850	−68,850
8	390,000	140,000	+131,640	39,000	0.46651	9,100	127,950	+3,690
9	350,000	150,000	+195,250	35,000	0.42410	7,420	135,370	+59,880
10	280,000	160,000	+276,210	28,000	0.38554	5,390	140,760	+135,450

A_S = base revenue from annual sales before tax.
A_{TE} = base total annual expense before tax.
(NPV) = base net present value after tax.
ΔA_S = decrease in annual revenue.
f_d = discount factor at 10%.
ΔA_{DCF} = decrease in net discounted cash flow at income tax rate = 0.5.
Δ(NPV) = $\Sigma \Delta A_{DCF}$ = decrease in net present value.
Reduced (NPV) = ΣA_{DCF} = reduced net present value after tax.

batches, the cost per unit of production, etc., tend to decrease with the number of units produced. The relationships are expressed for the ideal case by

$$Y = KX^N \tag{25-63}$$

where Y = cumulative-average cost, production time, etc., per unit
X = cumulative production, units
K = effective value of first unit produced
N = slope of straight-line plot of Y versus X on log-log paper

The particular learning curve is usually characterized by the percentage reduction in the cumulative average value Y when the number of units X is doubled. From this definition it follows that

$$N = \log(\text{characteristic}/100)/\log 2 \tag{25-64}$$

The cost c_{ME} of the last unit of a block bringing the cumulative production to X units is, from Eq. (25-63),

$$c_{ME} = K[X^{N+1} - (X-1)^{N+1}] \tag{25-65}$$

These unit costs, or the time taken to produce the last unit, etc., may be plotted on cartesian coordinates against the number of units produced to provide a standard against which the performance of a new employee, a new machine, etc., can be judged. Figure 25-16 shows such a plot for the subsequent example.

In general, cost data will be available for multiple units. Typically, the cost of production for 1 week or of a specific order is computed and an average cost per unit obtained. This average value $\overline{Y}$ for the

TABLE 25-8 Summary of Results of Sensitivity Analysis

Case	(NPV), $ i = 10%	(DCFRR), %
Base case	276,210	16.4
A_S reduced 10% per year	135,450	12.9
A_S reduced 20% per year	−5,330	9.8
A_{TE} increased 10% per year	238,430	15.0
C_{FC} increased 10%	206,890	14.0
Combined: A_S reduced 10% per year; A_{TE} increased 10% per year; C_{FC} increased 10%	28,420	10.6

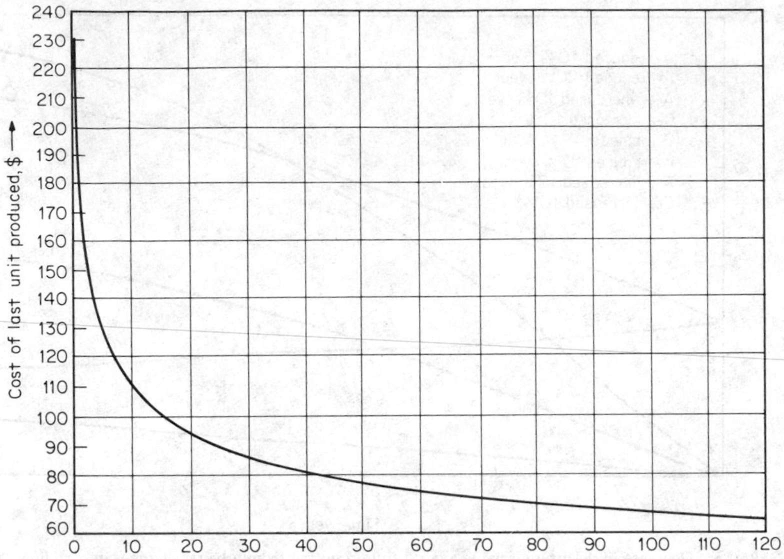

FIG. 25-16 Cartesian plot of learning curve.

batch should be plotted against the corresponding learning-curve value $\overline{X}$ calculated by Eq. (25-66):

$$\overline{X}^N = (X_2^{N+1} - X_1^{N+1})/(X_2 - X_1) \tag{25-66}$$

where X_1 and X_2 are the cumulative production before and after the batch. This form of the equation is useful when only the previous production history of the process is known, from the serial numbers or otherwise.

A straight line may be fitted to the (X,Y) or $(\overline{X},\overline{Y})$ pairs of data when plotted on log-log graph paper from which the slope N and the intercept log K with $X = 1$ may be read. Alternatively, the method of least squares may be used to estimate the values of K and N, giving the best fit to the available data.

It will be noted that a value of $N = 0$, corresponding to a characteristic of 100 percent for the learning curve, implies that the value of Y is independent of X. This would imply that learning by experience was not possible and thus corresponds to an optimally designed process or one for which the costs are determined by external factors. Similarly, a value of $N = -1$, corresponding to the 50 percent learning curve, implies that the cost of production is inversely proportional to the number produced, which is absurd. Projects having characteristics less than 70 percent are impractical. Low characteristics are typical of hasty entry into a market in an attempt to preempt it. Characteristics tend to increase with experience, so that established and mature projects are likely to have characteristics around 95 percent. Characteristics close to 100 percent are unlikely to be achieved because of random factors such as changes in personnel, accidents, supply delays, etc. Figure 25-17 represents a typical practical case, from which it can be seen that the curve has a point of inflexion but eventually settles down to an approximately straight line of lower slope than that of the conventionally defined learning curve. At some point it is useful to change to the equation of this mature project line.

Significant changes in working, such as the introduction of new equipment, the influx of a large number of inexperienced workers, or a temporary reduction in skills after a long shutdown, may produce a sudden increase in all the cumulative-average curves. The simplest way to handle this, when the next accurate costing is avail-

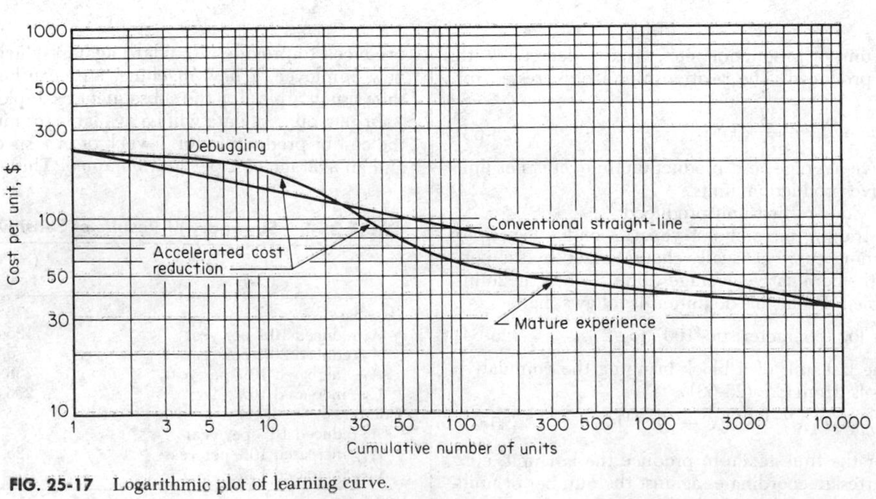

FIG. 25-17 Logarithmic plot of learning curve.

able, is to deduct the value of X obtained from the curve from that actually achieved and to use this value as a constant correction to X until the next break in the curve is reached. If the causes of such steps recur, the size of the step can often be related to a particular cause. In such cases the estimated step change can be used for predictions until the next accurately determined values are obtained.

Applications for the learning curve are already extensive, and new uses can often be found. Care is needed in applying the techniques to ensure that it is possible for learning to take place. In projecting prices, etc., unusual items, such as the cost of the special setting up of tools or factory rearrangements, should be excluded from the production costs used to establish the learning curve. In times of inflation, costs should be corrected for the effects of inflation in the manner to be shown subsequently. Production times or spoilage rates are not affected by cost allocations or inflation and may prove to be better standards of performance where appropriate. However, the learning curve is often required when preparing quotations for batch production runs, particularly when competition is likely to be keen. In such cases the average cost of the production run $\overline{Y}$ between cumulative production totals of X_1 and X_2 may be estimated by Eq. (25-67) when the previous cumulative-average cost Y_1 is known:

$$\overline{Y} = Y_1[(X_2/X_1)^{N+1} - 1]/[(X_2/X_1) - 1] \qquad (25\text{-}67)$$

In process engineering, fractional units can often be produced so that the learning curve can be treated as being continuous. When only discrete numbers of units can be produced, the learning curve is strictly a histogram. In order to allow for this it is sufficient to increase the value of X by half a unit before applying the above equations. The difference is significant at small values of X, such as may be used for the initial estimates of K and N. As the project matures, it is better to use the equations as presented, as the cost of the first unit K is an entirely notional one. Major technological changes should, of course, be treated as the start of a new project.

R. B. Jordan (*How to Use the Learning Curve*, Materials Management Institute, Boston, 1965) discusses the uses of the learning curve extensively and provides many tables of factors. The uses considered include estimating starting costs, determining labor requirements, establishing factory cost targets, checking employee-training progress, the make-or-buy decision, aid in purchasing negotiations, and aid in establishing a selling price.

Example 4 The cost of an initial batch of 21 units, exclusive of special tools and setting-up costs, averaged \$120 per unit. The average cost of the next batch of 80 units was \$75.81. Let us establish the learning curve implied by these data and hence estimate the probable average cost of the next 50 units. We shall establish also the unit-cost curve to be used as a control during follow-up orders.

If the batch units are capable of continuous subdivision, we proceed as follows. We substitute the given values of the cumulative-average cost Y and cumulative production X for the first batch into Eq. (25-63) to give, by taking logarithms of each side,

$$\log 120 = \log K + N \log 21$$

The cost of the first batch is $120 \times 21 = \$2520$, and that of the second batch is $75.81 \times 80 = \$6065$. The total cost of the first 101 units is therefore \$8585, with a cumulative-average unit cost of \$85. We substitute as before to give

$$\log 85 = \log K + N \log 101$$

From these equations it follows that $K = \$234.15$ and $N = -0.2196$. This line is plotted in Fig. 25-18.

From Eq. (25-64) it follows that the value of the characteristic of this learning curve $= 100$ antilog $(-0.2196 \log 2) = 85.0$ percent. From Eq. (25-65) the production cost of the third unit is

$$c_{ME} = (\$234.15)(3^{0.7804} - 2^{0.7804}) = \$149.70$$

Values calculated in this way are plotted in Fig. 25-16 and also in Fig. 25-18. It will be noted that after about 10 units this latter curve becomes parallel to the cumulative-average-cost curve and that the Y values are $(N + 1)$ times those obtained from the latter curve.

Since the cumulative-average cost Y_2 of the first 101 units was \$85, it follows from Eq. (25-67) that the average cost of the third batch of 50 units, bringing the cumulative total to 151, is given by

$$\overline{Y}_3 = (\$85)[(151/101)^{0.7804} - 1]/[(151/101) - 1]$$
$$= \$63.30 \text{ per unit}$$

This may be used as a cost guide when quoting the order.

If the units of production may not be subdivided, the procedure is similar except that all X values are increased by 0.5 unit in establishing the curves. The results are not sufficiently different to be significant for estimation purposes.

To the above costs must be added back any unit costs omitted from those to which learning might bring improvement. These will normally include overheads and specific charges on the project such as the unit cost of special tools, jigs, etc.

Risk and Uncertainty Discounted-cash-flow rates of return (DCFRR) and net present values (NPV) for future projects can never be predicted absolutely because the cash-flow data for such projects are subject to uncertainty. Therefore, when stating predicted values of (DCFRR) and (NPV) for projects, it is also desirable to give a measure of confidence in the predictions.

For example, for a particular project it may be estimated that there is a 90 percent chance of the (DCFRR) being greater than 10 percent, a 50 percent chance of its being greater than 16 percent, and only a 10 percent chance of its being greater than 20 percent. Management retains the power of decision to proceed with the project or not, but the probability data provide desirable information for the decision.

The estimation of probabilities requires the use of statistics. Thus statistical methods play an increasing role in decision making.

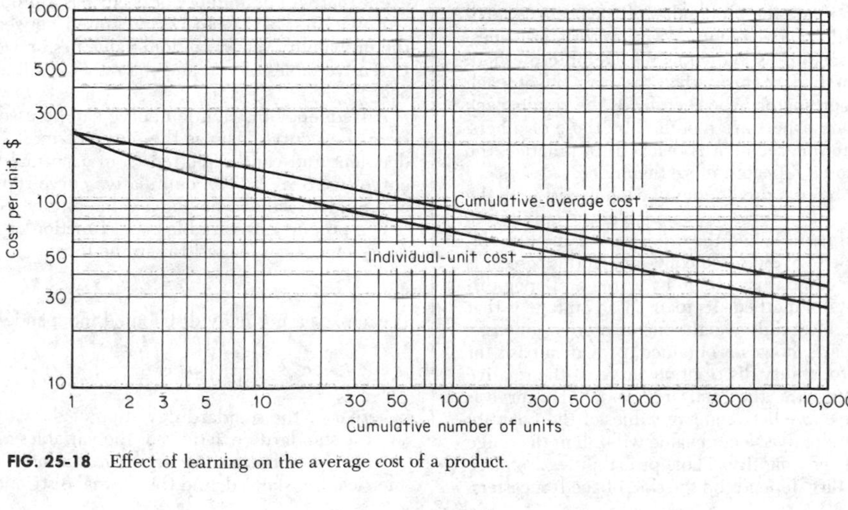

FIG. 25-18 Effect of learning on the average cost of a product.

Predictions from Limited Data Predictions of future sales price, sales volume, etc., are normally based on a very limited amount of data about past events. Furthermore, it would not be convenient to use the entire population of past events even if it were available. A statistic is a measure, based on limited information from a sample, that allows the corresponding parameter of the population to be estimated.

The mean value $\bar{x}$ of a property x is a statistic based on a sample of n items defined by

$$\bar{x} = (x_1 + x_2 + x_3 + \cdots x_n)/n \tag{25-68}$$

The mean $\bar{x}$ is the statistic corresponding to the population parameters μ, which is the arithmetic average of all the items in the population. In many cases, not all the x values will be different. In such circumstances, Eq. (25-68) can be written as

$$\bar{x} = [\Sigma x_i f(x_i)]/[\Sigma f(x_i)] \tag{25-69}$$

where $f(x_i)$ is the frequency with which a particular value x_i occurs. It is often convenient to divide the frequency of occurrence by the total number of items. In this case, $f(x_i)$ becomes the relative frequency of occurrence of the value x_i, and $\Sigma f(x_i) = 1$.

The values of x may be either discrete or continuous. The number of sales of, say, automobiles in any one day must be an integer. If a business sells 4 automobiles, this represents all possible values of x in the range of 3.5 to 4.5.

When x represents a continuous variable quantity, it is sometimes convenient to take the total or relative frequency of occurrences within a given range of x values. These frequencies can then be plotted against the midvalues of x to form a histogram. In this case, the ordinate should be the frequency per unit of width x. This makes the area under any bar proportional to the probability that the value of x will lie in the given range. If the relative frequency is plotted as ordinate, the sum of the areas under the bars is unity.

If x is a continuous variable and the interval ranges are made smaller and smaller, a smooth curve will eventually result. The area under such a curve between x_1 and x_2 represents the probability that a randomly selected item will have a value of x lying in the range x_1 to x_2. This is the information that is desired.

Data available from past experience can be used to generate frequency distribution curves. It is essential for a company to have an efficient commercial-intelligence system to assess market conditions.

Accuracy of sales forecasting can also be increased by a careful study of past sales records, price trends, etc. However, the uncertainty of an estimate increases the farther into the future that the estimate is projected.

Estimates of sales income and other types of forecasts are usually based on the opinions of experts. Experts should be able to estimate maximum, minimum, and most likely, or modal, values for a quantity. The modal value is not necessarily midway between the minimum and maximum values, since many distributions are skewed. An expert may be asked to estimate the probability of the occurrence of certain values on each side of the mode. When experts are questioned separately, the procedure is known as the **Delphic method.** Strictly speaking, this method requires that the opinion of each expert be assessed by a coordinator, who then feeds the results back to see if the opinions of one expert are modified by those of others. The process is repeated until agreement is reached. In practice, the procedure is too tedious to be repeated more than once.

It is useful to compare the past predictions of each expert with the results obtained in practice. This information enables the opinions to be weighted by the coordinator. When the experts work in close collaboration, it is not possible to avoid some collusion. In this case, it is better to arrive at a single consensus opinion by a free and open discussion. This is the think-tank method. Its main disadvantage is that rank or aggressiveness might unduly weight one or more opinions.

The opinions of the experts, however obtained, provide a basis for plotting a frequency or probability distribution curve. If the relative frequency is plotted as ordinate, the total area under the curve is unity. The area under the curve between two values of the quantity is the probability that a randomly selected value will fall in the range between the two values of the quantity. These probabilities are mere estimates, and their reliability depends on the skill of the forecasters.

The estimated (DCFRR) and the estimated (NPV) are both functions of the estimated cumulative revenue from annual sales ΣA_S, the estimated cumulative total annual cost or expense ΣA_{TE}, and the estimated fixed capital cost C_{FC} of the plant. The revenue from annual sales for each year is in turn the product of the sales price and sales volume. Initially it is desirable to select those values from the distribution curves of ΣA_S, ΣA_{TE}, and C_{FC} which enable the maximum and minimum (DCFRR) and (NPV) to be calculated.

If the maximum values of (DCFRR) and (NPV) are not acceptable to the company, the project should promptly be rejected. If the minimum values of (DCFRR) and (NPV) are acceptable, a detailed assessment should be made. If the maximum values of (DCFRR) and (NPV) are acceptable but the minimum values are not, the feasibility study should be continued.

Mathematical Models for Distribution Curves Mathematical models have been developed to fit the various distribution curves. It is most unlikely that any frequency distribution curve obtained in practice will exactly fit a curve plotted from any of these mathematical models. Nevertheless, the approximations are extremely useful, particularly in view of the inherent inaccuracies of practical data. The most common are the binomial, Poisson, and normal, or gaussian, distributions.

A normal distribution curve is bell-shaped (see Sec. 2). The curve obeys the relationship

$$f(x) = \frac{\exp - [(x - \mu)^2/2\sigma^2]}{\sigma(2\pi)^{0.5}} \tag{25-70}$$

where σ is known as the true standard deviation. The standard deviation $s°$ from a sample is given by

$$s° = \left[\frac{\Sigma(x_i - \bar{x})^2 f(x_i)}{\Sigma f(x_i) - 1}\right]^{0.5} \tag{25-71}$$

The standard deviation $s°$ for the sample corresponds to the true standard deviation σ for the whole population in the same way that the mean $\bar{x}$ of the sample corresponds to the arithmetic average μ for the whole population. Equation (25-70) can be written more compactly as

$$f(z) = [\exp(-z^2/2)/[(2\pi)^{0.5}] \tag{25-72}$$

where the standard score z is

$$z = (x - \mu)/\sigma \tag{25-73}$$

The area under the curve of $f(z)$ is unity if the abscissa extends from minus infinity to plus infinity. The area under the curve between z_1 and z_2 is the probability that a randomly selected value of x will lie in the range z_1 and z_2, since this is the relative frequency with which that range of values would be represented in an infinite number of trials.

An event that will definitely occur has a probability of unity. An event that will definitely not occur has a probability of zero.

Equation (25-72) can be integrated between limits to determine the probability that a random value lies between the selected limits. Extensive tables of $f(z)$ and the associated integral are available (see Sec. 1).

A frequency distribution curve can be used to plot a cumulative-frequency curve. This is the curve of most importance in business decisions and can be plotted from a normal frequency distribution curve (see Sec. 2). The cumulative curve represents the probability of a random value z having a value of, say, z_1 or less.

If a property or variable c is a function of several other variables x_1, x_2, etc., it can be written in the form

$$c = \phi(x_1, x_2, \ldots x_n) \tag{25-74}$$

If each x is a normally distributed independent variable, then

$$\sigma_c^2 = \left(\frac{\partial c}{\partial x_1}\right)^2 \sigma_1^2 + \left(\frac{\partial c}{\partial x_2}\right)^2 \sigma_2^2 + \cdots \left(\frac{\partial c}{\partial x_n}\right)^2 \sigma_n^2 \tag{25-75}$$

where σ_c is the standard deviation of the variable c and σ_1, σ_2, etc., are the standard deviations of the variables x_1, x_2, etc.

Many distributions occurring in business situations are not symmetrical but skewed, and the normal distribution curve is not a good fit. However, when data are based on estimates of future trends, the accuracy of the normal approximation is usually acceptable. This is particularly the case as the number of component variables x_1, x_2, etc., in Eq. (25-74) increases. Although distributions of the individual variables (x_1, x_2, etc.) may be skewed, the distribution of the property or variable c tends to approach the normal distribution.

Let us consider an event that must have one of two outcomes. It must either occur with probability p_1 or fail to occur with probability p_2. Since these are exclusive events and the probability that something will happen is unity, it follows that

$$p_1 + p_2 = 1 \tag{25-76}$$

Provided that no learning process is involved (so that the value of p_1 is not influenced by previous results), the probability of x successes in n trials is given by the term containing p_1^x in the expansion of the binomial:

$$(p_1 + p_2)^n = p_1^n + \cdots \frac{n!}{x!(n-x)!} p_1^x p_2^{(n-x)} \cdots + p_2^n \tag{25-77}$$

where x and n are integers and $x!$ (read as x factorial) is the product of all integers from unity to x.

Example 5 If a six-sided die marked with the numbers 1, 2, 3, 4, 5, and 6 is thrown, the probability that any given number will be uppermost is 1/6. If the die is thrown twice in succession, then the probability of a given sequence of numbers occurring, say, 5 followed by 6, is (1/6)(1/6) = 1/36. The chance of any particular number occurring 0, 1, 2, 3, or 4 times in four throws of the die (or in a simultaneous throw of four dice) is given by the successive terms of Eq. (25-77), expanded as

$$(5/6 + 1/6)^4 = (1)(5/6)^4(1/6)^0 + (4)(5/6)^3(1/6)^1 + (6)(5/6)^2(1/6)^2 + (4)(5/6)^1(1/6)^3$$
$$+ (1)(5/6)^0(1/6)^4 = 0.4823 + 0.3858 + 0.1157 + 0.0154 + 0.0008 = 1$$

The distribution of the number of successes is skewed toward the low numbers. In particular, there is only a slightly better than even-money chance that any given number will occur even once in four throws. Such highly unsymmetrical distributions cannot be approximated by the normal distribution curve.

However, an increasing number of throws will result in totals that are close to the normal distribution. This fact can be used to approximate such a distribution without the enormous labor of the calculations required by the use of Eq. (25-77).

Possible values of the total of four throws of a die are integers from 4 to 24 and hence represent values in the range from 3.5 to 24.5. The mean value $\bar{x}$ of this range is given by Eq. (25-68) as $\bar{x}$ = (3.5 + 24.5)/2 = 14.0.

The cumulative probability of a normally distributed variable lying within 4 standard deviations of the mean is 0.49997. Therefore, it is more than 99.99 percent (0.49997/0.50000) certain that a random value will be within $\pm 4\sigma$ from the mean. For practical purposes, σ may be taken as one-eighth of the range of certainty, and the standard deviation can be obtained:

$$s^\circ \cong (24.5 - 3.5)/8 = 2.625$$

From Eq. (25-73) the standard score becomes

$$z = (x - \mu)/\sigma \cong (x - \bar{x})/s^\circ$$

For a total score of 4 (i.e., x = 4), the standard score is approximately z = (4 − 14)/2.625 = −3.81. Since the normal curve is symmetrical about z = 0, the height of the ordinate at z = −3.81 is the same as that at z = +3.81. From tables of values of cumulative probabilities of the normal distribution (see Sec. 2), the height of the ordinate is 0.0003 in units of $1/\sigma$. The relative frequency of 4 occurring is thus approximately 0.0003/2.625 = 0.0001.

This concept can be used to translate Delphic or other opinions into probability distributions and hence into useful decision-making tools.

Example 6 A store that is open 5 days a week is to promote a new product. The manager believes that not more than 5 units will be sold in any one day, but he cannot be more precise about the probable sales pattern. Stocks are delivered once per week. What size should the first order be to give a 95 percent certainty of meeting demand?

Since the product is sold in units, the possible range of weekly sales is from −0.5 to +25.5 units. Therefore, the mean of the sales distribution will be

$$\bar{x} \cong [25.5 - (-0.5)]/2 = 13$$

The standard deviation for this example will be

$$s^\circ \cong [25.5 - (-0.5)]/8 = 3.25$$

From this, the approximate frequency distribution of daily sales can be derived by using Eqs. (25-70), (25-72), or (25-73). The desired area to the right of z = 0 for the normal probability distribution curve is 0.95 − 0.50 = 0.45. For this value the standard score z = 1.645.

$$x \cong 13 + (1.645)(3.25) = 18.35$$

Hence, to be 95 percent certain of meeting demand, 19 units should be purchased.

If the value of n in Eq. (25-77) is large and neither p_1 nor p_2 is too close to zero, the binomial distribution can be approximated by

$$z = \frac{x - np_1}{\sqrt{np_1p_2}} \tag{25-78}$$

The approximation of Eq. (25-78) is good enough for most purposes if np_1 and np_2 are each greater than 5.

Example 7 The records of a business show that never more than 1 item is sold in a day and that 2 sales per week can be expected. What is the probability of selling between 90 and 120 items in a 300-day year?

In a year consisting of 50 weeks of 6 days, the mean or expected value of the distribution is 100 items. The probability of a sale of an item on a given day is p_1 = 100/300 = 1/3, and of no sale is p_2 = 2/3. From Eq. (25-78),

$$z = \frac{x - (300)(1/3)}{\sqrt{(300)(1/3)(2/3)}} = \frac{x - 100}{8.165}$$

The integral range of 90 to 120 items contains all possible values of x from 89.5 to 120.5. For x = 89.5, z = −1.286; and for x = 120.5, z = 2.511.

The cumulative probability of a standard score of 1.286 is 0.11, while that of a standard score of 2.511 is 0.99. Therefore, the probability of annual sales in the range of 90 to 120 items is (0.99 − 0.11) = 0.88, or 88 percent.

There are times when the frequency measurement is an integral number of events in a given segment of a continuum, for example, the number of automobiles passing a given point in 1 h or the number of leaks in a given length of hosepipe. In such cases, the correct frequency distribution is the Poisson distribution, in which the probability of x events per unit of a continuum occurring is given by

$$f(x) = \lambda^x e^{-\lambda}/x! \tag{25-79}$$

where x is an integer, e is the base of natural logarithms, and λ is a parameter of the system $\lambda = \mu = \sigma^2$.

As λ increases, the Poisson distribution approaches the normal distribution, with the relationship

$$z = (x - \lambda)/\sqrt{\lambda} \tag{25-80}$$

When the value of p_1 is very close to zero in Eq. (25-77), so that the occurrence of the event is rare, the binomial distribution can be approximated by the Poisson distribution with $\lambda = np_1$ when $n > 50$ while $np_1 < 5$.

Example 8 The daily chance of a breakdown in a production line operated continuously for 300 days per year is estimated at 1 percent from past performance. Let us estimate the probability of 4 or more breakdowns in the coming year.

For n = 300 and p_1 = 0.01, $\lambda \cong np_1 = 3 < 5$, the probability of no breakdown is found from Eq. (25-79) to be

$$f(0) = (3)^0 e^{-3}/0! = 1/e^3 = 0.0498$$

Similarly,

Breakdowns	Probability
1	$f(1) = (3)^1 e^{-3}/1! = 0.1496$
2	$f(2) = (3)^2 e^{-3}/2! = 0.2240$
3	$f(3) = (3)^3 e^{-3}/3! = 0.2240$

Since something must happen, the probability of 4 or more breakdowns is

$$1 - 0.0498 - 0.1496 - 0.2240 - 0.2240 = 0.3526$$

A simple trial will show how much more easily the preceding calculation is carried out than direct use of Eq. (25-77).

The necessary value of λ may often be established as in the following example.

Example 9 In a production period of 100 days, 0, 1, 2, 3, and 4 machine failures occurred in a single day on 41, 37, 15, 6, and 1 occasions respectively. Let us fit a Poisson distribution to the data and estimate the maximum number of machine failures likely to occur in 1 day of a 300-day year.

The mean number of failures is found from Eq. (25-69) to be

$$\bar{x} = \frac{0(41) + 1(37) + 2(15) + 3(6) + 4(1)}{41 + 37 + 15 + 6 + 1} = 0.89$$

The standard deviation is found from Eq. (25-71):

$$s^\circ = \left[\frac{\Sigma(x_i - \bar{x})^2 f(x_i)}{\Sigma f(x_i) - 1}\right]^{0.5}$$

The steps for calculating the numerator and denominator for this equation are tabulated as follows:

$(x_i - \bar{x})^2$	$f(x_i)$	$(x_i - \bar{x})^2 f(x_i)$
$(0 - 0.89)^2$	41	32.48
$(1 - 0.89)^2$	37	0.45
$(2 - 0.89)^2$	15	18.48
$(3 - 0.89)^2$	6	26.71
$(4 - 0.89)^2$	1	9.67
	$\Sigma(x_i - \bar{x})^{-2} f(x_i)$	= 87.79
	$\Sigma f(x_i) - 1$	= 99

Therefore, $s^\circ = (87.79/99)^{0.5} = 0.9417$.

The Poisson distribution is a good fit since

$$\lambda = \mu \cong \bar{x} = 0.8900$$

and

$$\lambda = \sigma^2 \cong (s^\circ)^2 \cong 0.8868$$

The Poisson distribution is found from Eq. (25-79) to be

$$f(x) = [(0.89)^x e^{-0.89}]/x!$$

By substituting the appropriate values of x for this example into the preceding equation, we find $f(0) = 0.4107$, $f(1) = 0.3655$, $f(2) = 0.1627$, $f(3) = 0.0483$, $f(4) = 0.0107$, $f(5) = 0.0019$, and $f(6) = 0.0003$. Hence, in 300 days the expected maximum number of breakdowns in 1 day is 5 since $(300)f(6) = 0.09$ occurrence.

In many business applications, Eq. (25-74) can be reduced to the linear relationship

$$c = k_1x_1 + k_2x_2 + \cdots + k_nx_n \tag{25-81}$$

where the k's are constants. Equation (25-75) then becomes

$$\sigma_c^2 = k_1^2\sigma_1^2 + k_2^2\sigma_2^2 + \cdots + k_n^2\sigma_n^2 \tag{25-82}$$

On the other hand, for a product function such as

$$c = x_1x_2 \tag{25-83}$$

Eq. (25-75) can be written in the form

$$\sigma_c^2/c^2 = \sigma_1^2/x_1^2 + \sigma_2^2/x_1^2 \tag{25-84}$$

The discounted-cash-flow rate of return (DCFRR) and net present value (NPV) are functions of the cumulative revenue from annual sales ΣA_{TE} and the fixed-capital cost of the plant C_{FC}, among other factors.

Equation (25-75) can be written for (DCFRR) and for (NPV) as

$$\sigma^2_{(\text{DCFRR})} = \left[\frac{\partial(\text{DCFRR})}{\partial \Sigma A_S}\right]^2 \sigma^2_{\Sigma AS} + \left[\frac{\partial(\text{DCFRR})}{\partial \Sigma A_{TE}}\right]^2 \sigma^2_{\Sigma ATE} + \left[\frac{\partial(\text{DCFRR})}{\partial C_{FC}}\right]^2 \sigma^2_{CFC} \tag{25-85}$$

$$\sigma^2_{(\text{NPV})} = \frac{\partial(\text{NPV})^2}{\partial \Sigma A_S}\sigma^2_{\Sigma AS} + \frac{\partial(\text{NPV})^2}{\partial \Sigma A_{TE}}\sigma^2_{\Sigma ATE} + \frac{\partial(\text{NPV})^2}{\partial C_{FC}}\sigma^2_{CFC} \tag{25-86}$$

The revenue from annual sales A_S of a product at an annual production rate R and sales price of c_s per unit of production is

$$A_S = Rc_S \tag{25-87}$$

Equation (25-84) can be written as:

$$\sigma^2_{AS} = (A_S/R)^2\sigma_R^2 + (A_S/c_s)^2\sigma^2_{c_s} \tag{25-88}$$

An extensive example illustrating the use of Eqs. (25-81) through (25-86) in establishing the probability of attaining a given value of the net present value or less in a particular year of a project was presented by Holland et al. [F. A. Holland, F. A. Watson, and J. K. Wilkinson, *Chem. Eng.*, **81**, 105–110 (Jan. 7, 1974)]. The result is shown in Fig. 25-19.

Decision makers often prefer to have graphs showing the probability of attaining a value greater than a given value. Such curves are easily obtained by subtracting the probability of achieving a given value or less from 100 percent. Figure 25-20 was obtained in this way and shows the probability of attaining a (DCFRR) greater than a given value.

Monte Carlo Method The Monte Carlo method makes use of random numbers. A digital computer can be used to generate pseudo-random numbers in the range from 0 to 1. To describe the use of random numbers, let us consider the frequency distribution curve of a particular factor, e.g., sales volume. Each value of the sales volume has a certain probability of occurrence. The cumulative probability of that value (or less) being realized is a number in the range from 0 to 1. Thus, a random number in the same range can be used to select a random value of the sales volume.

In the same way, random values of the other factors can be obtained. These can then be combined to give random values of (DCFRR) and (NPV) and, in turn, used to plot cumulative-probability curves for (DCFRR) and (NPV). The computer may be required to perform some 10,000 to 50,000 calculations.

The use of the Monte Carlo method in project appraisal was illustrated by Holland et al. [F. A. Holland, F. A. Watson, and J. K. Wilkinson, *Chem. Eng.*, **81**, 76–79 (Feb. 4, 1974)]. The cumulative-probability curves of (DCFRR) and (NPV) can never be more accurate than the opinions on which they are based, and comparable accuracy can be obtained by the use of S-shaped curves with relatively small computational effort.

S-Shaped Curves K. D. Tocher (*The Art of Simulation*, rev. ed., English Universities Press, London, 1967) presented a comprehensive treatment of the generation of random and pseudo-random numbers and their use in a wide range of simulated processes. He also considered sampling techniques from the various statistical distributions and the design of simulated processes. It will be noted that the cumulative distribution curves are S-shaped, and Tocher (op. cit., p. 16) recommended as a general equation for such curves

$$x = a + by + cy^2 + d(1 - y)^2 \ln y + ey^2 \ln(1 - y) \tag{25-89}$$

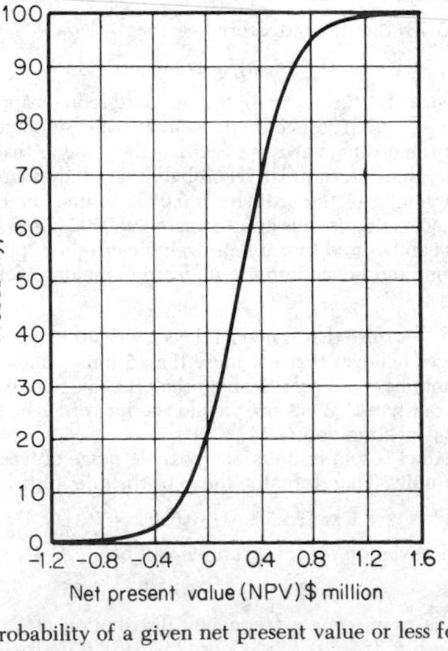

FIG. 25-19 Probability of a given net present value or less for a project.

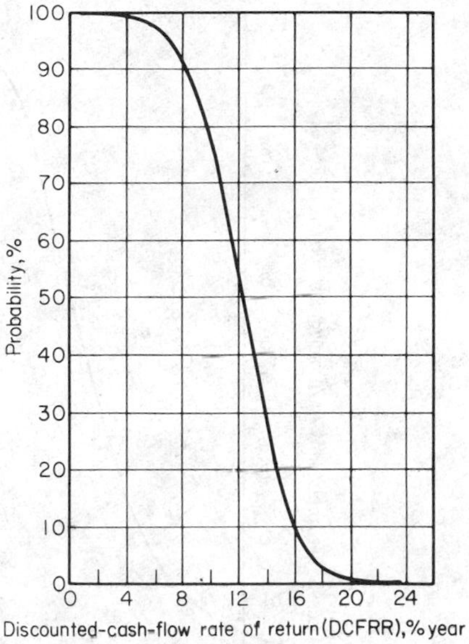

FIG. 25-20 Probability of a given discounted-cash-flow rate of return or more for a project.

in which x varies from $-\infty$ to $+\infty$ as y varies from 0 to 1. The underlying frequency curve corresponding to Eq. (25-89) is

$$\frac{1}{p(x)} = \frac{dx}{dy} = b + 2cy + d(1-y)\left(\frac{1-y}{y} - 2\ln y\right) + ey\left[2\ln(1-y) - \frac{y}{1-y}\right] \qquad (25\text{-}90)$$

If necessary, the fit can be improved by increasing the order of the polynomial part of Eq. (25-89), so that this approach provides a very flexible method of simulation of a cumulative-frequency distribution. The method can even be extended to J-shaped curves, which are characterized by a maximum frequency at $x = 0$ and decreasing frequency for increasing values of x, by considering the reflexion of the curve in the y axis to exist. The resulting single maximum curve can then be sampled correctly by Monte Carlo methods if the vertical scale is halved and only absolute values of x are considered.

When the data do not warrant the accuracy of Eq. (25-89) or Eq. (25-90), simpler curves will usually suffice if the frequency distribution may be assumed to have a single maximum value.

Let us consider a product which is sold entirely on the basis of personal recommendation. The rate of sale will depend on the number of people who have already bought the product. Thus initially sales will increase exponentially. Eventually the market will be saturated, and only replacement purchases will be made. If the frequency curve may be assumed to be symmetrical about a single maximum value, the cumulative distribution curve is known as the logistics curve and is defined by Eq. (25-91):

$$y = c/[1 + a\exp(-bx)] \qquad (25\text{-}91)$$

where y varies between zero and c as x ranges from $-\infty$ to $+\infty$. Although only three constants appear explicitly in Eq. (25-91), two further constants are implied by the choice of zero as the lower bound of y and the point of inflexion at $y = c/2$. The usual use of Eq. (25-91) is in sales forecasting, in which case y is sales demand and x is time. If such a curve already exists, the value of c can be read as the upper asymptote and a and b obtained by the use of an auxiliary variable T where

$$T = x_2\,(\text{at } y = r_2c) - x_1\,(\text{at } y = r_1c) \qquad (25\text{-}92)$$

$$b = [\ln(1/r_1 - 1) - \ln(1/r_2 - 1)]/T \qquad (25\text{-}93)$$

$$a = (1/r_1 - 1)\exp(bx_1) \qquad (25\text{-}94)$$

or

$$a = (1/r_2 - 1)\exp(bx_2)$$

If the values of a obtained from Eq. (25-94) differ significantly, the logistics curve is not a suitable representation of the data.

Example 10 We shall derive the logistics curve representing the cumulative-frequency distributions of the normal distribution curve defined by Eqs. (25-72) and (25-73). In this case, y varies between a cumulative probability of zero and unity as z varies from $-\infty$ to $+\infty$. Since the upper bound is unity, $c = 1$. From Table 25-9 the area under the right-hand side of the curve between $z = 0$ and $z = z$ may be read. Since the frequency curve is symmetrical about the mean, this is also the area between $z = 0$ and $z = z$. Hence, the area under the frequency curve, which represents the cumulative probability, is 0.50000 at $z = 0$ and the 80 percentile, for which the area is 0.80000, corresponds to the value $z = 0.842$. We substitute these values into Eqs. (25-92) through (25-94) to give

$$T = 0.842 - 0.000 = 0.842$$

$$b = [\ln(1/0.50 - 1) - \ln(1/0.80 - 1)]/0.842 = 1.6464$$

$$a = 1.0000 \textit{ or } 1.00000$$

From Eq. (25-91) the corresponding logistic curve is

$$y = [1 + \exp(-1.6464z)]^{-1}$$

The cumulative-frequency function calculated from this simple expression is compared with the precise value in Table 25-9.

When a cumulative-frequency curve can be satisfactorily represented by a logistics curve, the underlying frequency curve can be obtained by differentiation of Eq. (25-91) as

$$p(x) = \frac{dy}{dx} = \frac{abc\exp(-bx)}{[1 + a\exp(-bx)]^2} \qquad (25\text{-}95)$$

The probability-density function for the normal distribution curve calculated from Eq. (25-95) by using the values of a, b, and c obtained in Example 10 is also compared with precise values in Table 25-9. In such symmetrical cases the best fit is to be expected when

TABLE 25-9 Data for Normal Distribution Curve

Standard score, z	Ordinate of normal distribution curve, $p(z)$: Precise	Estimated	Area under normal curve, cumulative probability, $y = \int_0^z p(z)\,dz$: Precise	Estimated
0.000	0.3989	0.4116	0.0000	0.00000
0.100	0.3970	0.4088	0.03983	0.04017
0.200	0.3910	0.4006	0.07926	0.08158
0.253	0.3864	0.3943	0.10000	0.10265
0.300	0.3814	0.3875	0.11791	0.12103
0.400	0.3683	0.3700	0.15542	0.15894
0.500	0.3521	0.3491	0.19146	0.19492
0.524	0.3478	0.3436	0.20000	0.20323
0.600	0.3332	0.3255	0.22575	0.22866
0.700	0.3123	0.3003	0.25804	0.25996
0.800	0.2897	0.2744	0.28814	0.28870
0.842	0.2798	0.2634	0.30000	0.30000
0.900	0.2661	0.2484	0.31594	0.31484
1.000	0.2420	0.2231	0.34134	0.33840
1.200	0.1942	0.1761	0.38493	0.37822
1.282	0.1953	0.1587	0.40000	0.39194
1.400	0.1497	0.1358	0.41234	0.40929
1.600	0.1109	0.1029	0.44520	0.43303
1.645	0.1032	0.0964	0.45000	0.43752
1.800	0.0790	0.0769	0.46407	0.45092
2.000	0.0540	0.0569	0.47725	0.46418
2.500	0.0175	0.0260	0.49379	0.48395
3.000	0.0044	0.0116	0.49865	0.49289
4.000	0.0001	0.0023	0.49997	0.49862
00	0.0000	0.0000	0.50000	0.50000

the median or 50 percentile x_M is used in conjunction with the lower quartile or 25 percentile x_L or with the upper quartile or 75 percentile x_U. These statistics are frequently quoted, and determination of values of a, b, and c by using x_M with x_L and with x_U is an indication of the symmetry of the curve. When the agreement is reasonable, the mean values of b so determined should be used to calculate the corresponding value of a.

In practice most distribution curves are not symmetrical about the median but are inherently skewed. The effect of an advertising campaign is usually to increase the rate of sales in the early years. It may also increase the level of mature demand for the product, but this mature demand must be asymptotic to a finite upper limit of sales c. Such a curve is positively skewed since $(x_M - x_L) < (x_U - x_M)$. This situation can often be approximated by the Gompertz curve defined by Eq. (25-96):

$$\ln y = \ln c - a \exp(-bx) \qquad (25\text{-}96)$$

which has its point of inflexion at 0.3679 c. In terms of the upper and lower quartiles and the median,

$$b = 0.8794/(x_U - x_M) \qquad (25\text{-}97)$$

$$b = 0.6931/(x_M - x_L) \qquad (25\text{-}98)$$

$$a = 0.6931 \exp(bx_M) \qquad (25\text{-}99)$$

The suitability of the Gompertz fit to the curve can be assessed by comparing the values of b calculated from Eqs. (25-97) and (25-98), and, if suitable, the average value of b may be used in Eq. (25-99) to calculate the corresponding value of a to ensure a fit at the median and reasonable accuracy over the more important practical range within a couple of standard deviations on either side of the median.

The underlying frequency distribution curve of the Gompertz curve may be obtained by differentiation of Eq. (25-96) to give

$$p(x) = dy/dx = yab \exp(-bx) \qquad (25\text{-}100)$$

The logistic and Gompertz curves are of the general shape illustrated by Fig. 25-19. They may be adapted to fit curves of the general shape illustrated by Fig. 25-20 by a little mathematical manipulation. As an example, let us consider the current ratio, the ratio of current assets to current debts, as is quoted in Dun & Bradstreet statistics. A typical value for United States industrial chemical companies might be listed as $x_L = 1.82$, $x_M = 2.59$, and $x_U = 3.25$. First, we notice that $(x_M - x_L) > (x_U - x_M)$. This curve is, therefore, negatively skewed, or reversed S-shaped, and the logistics curve is not suitable. Nor can the Gompertz equation be used directly. However, it is clear that if the curve is drawn upside down and backward, the transformed curve will be positively skewed. Mathematically, this is equivalent to interchanging the upper and lower bounds and considering the dependent variable to be $(c - y)$. In the present case the quoted values represent the cumulative probabilities that the current ratio will be less than the quoted value and hence the value of y ranges between zero and unity. Hence, $c = 1$. In the transformed curve $x_L = 3.25$, $x_M = 2.59$, and $x_U = 1.82$. Hence, from Eq. (25-97)

$$b = 0.8794/(1.82 - 2.59) = -1.1421$$

and from Eq. (25-98)

$$b = 0.6931/(2.59 - 3.25) = -1.0502$$

The variation is within 5 percent of the mean value of $b = -1.0961$, and the transformed curve should be sufficiently accurate for many purposes. From Eq. (25-99)

$$a = 0.6931 \exp[(-1.0961)(2.59)] = 0.04054$$

Hence, from Eq. (25-96) the equation of the transformed curve is

$$\ln(1 - y) = -0.04054 \exp(1.0961x)$$

Since $d(1 - y) = -dy$, the corresponding underlying frequency distribution curve is from Eq. (25-100):

$$p(x) = +0.0444(1 - y) \exp(1.0961x)$$

Values of y and $p(x)$ calculated from last two equations are plotted in Figs. 25-21 and 25-22 respectively.

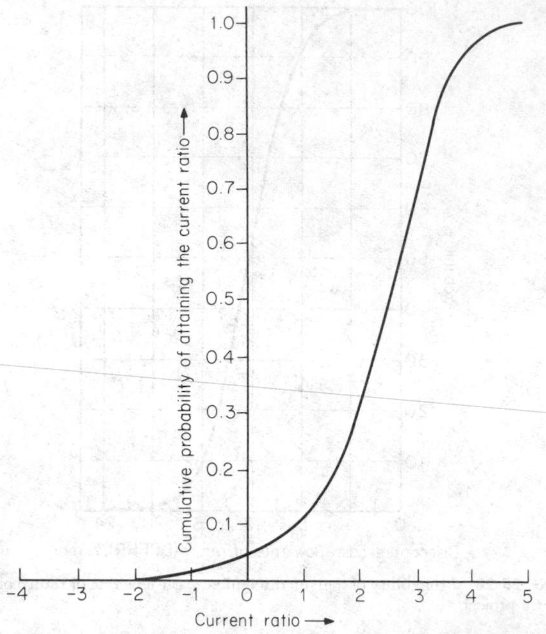

FIG. 25-21 Cumulative probability of a given current ratio.

In all such S-shaped curves the range of x is from $-\infty$ to $+\infty$, so that there is always a finite possibility of negative values of x occurring. In the present case the definition of the current ratio makes values of x below zero meaningless. The error of some 4 percent in the cumulative-probability curve implied by this factor may be tolerable in a given case.

It can be shown [J. J. Molder and E. G. Rogers, *Manage. Sci.*, **15**, B-76 (1968)] that for continuous events it is possible to estimate the mean and standard deviation of a skewed distribution from estimates

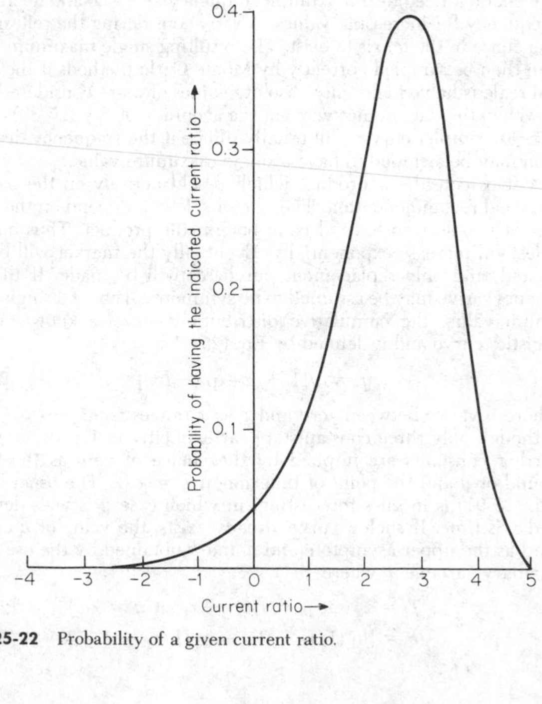

FIG. 25-22 Probability of a given current ratio.

TABLE 25-10 Data for Risk Analysis

	(NPV), \$/year				
Parameter	Low	Modal	High	Mean	Standard deviation
A_S	1,126,240	1,407,800	1,478,190	1,355,008	132,811
A_{TE}	−471,698	−377,358	−358,490	−396,226	−42,720
C_{FC}	−900,603	−692,772	−658,133	−736,070	−91,498
C_{WC}	−61,446	−61,446	−61,446	−61,446	0

of a low value, a most likely or modal value, and a high value. It is suggested, since it is difficult to make very fine subjective judgments as to probabilities, that the range most likely to be accurate is that for which there is a 10 percent chance of a value less than the low value and a 10 percent chance of a value greater than the high value. These values will usually imply a skewed distribution. For the suggested 80 percent confidence level the best available estimates are

$$\bar{x} = \text{(low value)} + [(2)\text{(modal value)}] + \text{(high value)} \qquad (25\text{-}101)$$

$$s^\circ = [\text{(high value)} - \text{(low value)}]/2.65 \qquad (25\text{-}102)$$

On this basis an alternative approach to risk analysis is the parameter method [D. O. Cooper and L. B. Davidson, *Chem. Eng. Prog.*, **72**, 73–78 (November 1976)].

Example 11 Let us consider the project outlined in Table 25-4. It is estimated that the basic data represent the most likely values and that there is a 10 percent chance that A_S will be reduced by more than 20 percent or will be increased by more than 5 percent. In the same way the low and high levels at 10 percent probability for A_{TE} are considered to be 5 percent below and 25 percent above the base figures respectively. The low and high values for C_{FC} are considered to be 5 percent below and 30 percent above the base figure, while changes in other parameters are considered to be immaterial.

With a cost of capital i of 10 percent the various cash flows can be discounted and summed. Thus for the base cases $\Sigma A_s f_d$ = \$2,815,600, $\Sigma A_{TE} f_d$ = \$754,716, $\Sigma A_D f_d$ = \$614,457, and $\Sigma C_{WC} f_d$ = \$61,446. With corporate taxes payable at 50 percent the aftertax cash flows of the first three items are (1 − 0.50) of the sums calculated above. The discounted working capital and the fixed-capital outlay are not subject to tax. These most probable values are listed and summed in Table 25-10 and, after adjustment for tax, give the modal value of the (NPV) as \$276,224.

A reduction of 20 percent in A_s for each year will result in a 20 percent reduction in $\Sigma A_s f_d$ below the modal value, i.e., a reduction of (0.20)(\$2,815,600) = \$563,120. The aftertax effect of this reduction on the contribution to (NPV) is (1 − 0.50)(\$563,120) = \$281,560, making the low value \$1,407,800 − \$281,560 = \$1,126,240 or, more directly, (0.8)(\$1,407,800). Other values in Table 25-10 are calculated in a similar manner.

The mean value of each of the distributions is obtained from these high, modal, and low values by the use of Eq. (25-101). If the distribution is skewed, the mean and the mode will not coincide. However, the mean values may be summed to give the mean value of the (NPV) as \$161,266. The standard deviation of each of the distributions is calculated by the use of Eq. (25-75). The fact that the (NPV) of the mean or the mode is the sum of the individual mean or modal values implies that Eq. (25-81) is appropriate with all the k's equal to unity. Hence, by Eq. (25-81) the standard deviation of the (NPV) is the root mean square of the individual standard deviations. In the present case s° = \$166,840 for the (NPV).

If the resulting distribution is assumed to be normal, then the cumulative distribution curve can immediately be generated. From Table 25-9, a standard score of 4 corresponds to a probability of 0.5 + 0.49997 = 0.99997 and one of −4 to a probability of 0.5 − 0.49997 = 0.00003, virtually unity and zero respectively. From Eq. (25-73) a standard score of 4 corresponds to an (NPV) of \$161,266 + (4)(\$166,840) = \$828,626 and one of −4 to an (NPV) of \$506,094. Values of (NPV) corresponding to other confidence limits may be calculated in the same way and plotted to give the curve of Fig. 25-23.

As has been stated, with the uncertainties attached to many business assessments of the range of various factors, the central-limit theorem implies that the assumption of a normal distribution of the main variable is sufficiently accurate provided that there are several factors contributing to that main variable. The results are as informative as most Monte Carlo estimates and have the advantage that they can be rapidly obtained without recourse to a digital computer, although a good desk calculator speeds the work. Strictly, the variables should be independent and additive. Thus it is better, for example, to treat ($A_S - A_{TVE}$) as a single variable since both sales income and total variable expense are related to the annual rate of sales R. In such cases the standard deviation of A_{TVE} would be added to or subtracted from that of A_S before squaring to obtain the variance according as the uncertainty of the group was greater or less than that of the individual factors. Also, when a product such as $A_S = Rc_S$ is involved, Eq. (25-84) should be used to estimate the variance rather than Eq. (25-82). When the predominant uncertainties are multiplied together, a log-normal distribution may provide a better final distribution. A similar technique may be applied to the (DCFRR) provided that Eq. (25-85) is used in place of Eq. (25-86) to estimate the overall variance of the main variable.

When the estimates are well founded, the skewness may be preserved by using a distribution such as the Gompertz. The median of that curve occurs as y = 0.5 c, while the point of inflexion corresponds to the mode at $y = C/\exp(1) = 0.3679\ c$. The statistician Karl Pearson suggested as a simple measure of skewness

$$\text{Skewness} = 3\ (\text{mean} - \text{median})/\sigma \qquad (25\text{-}103)$$

with an empirical approximation in terms of the mode given by

$$(\text{Mean} - \text{mode}) = 3\ (\text{mean} - \text{median}) \qquad (25\text{-}104)$$

Applying these equations to the present problem,

$$(\text{Mean} - \text{mode}) = \$161{,}266 - \$276{,}224 = -\$114{,}958$$

$$\text{Skewness} = -\$114{,}958/\$166{,}840 = -0.6890$$

For symmetrical distributions, such as the logistic or normal, the skewness should be zero.

The Gompertz distribution requires the distribution to be positively skewed, which can be achieved by treating −(NPV) as the independent variable and ($c - y$) as the dependent variable. From Eq. (25-104) the median of the distribution is given approximately as

$$\text{Median} = [\$161{,}266 - (-\$114{,}958)]/3 = \$199{,}585$$

Substituting values into Eq. (25-96) with −(NPV) as the independent variable to give, since the range of y is zero to unity,

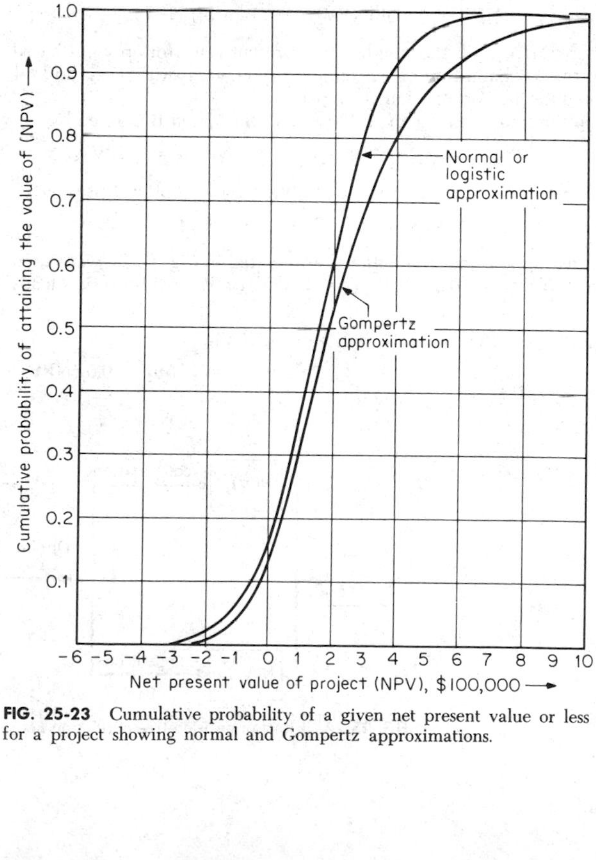

FIG. 25-23 Cumulative probability of a given net present value or less for a project showing normal and Gompertz approximations.

$$\ln(1 - 0.5) = \ln(1) - a \exp[(-b)(-\$199{,}585)]$$

$$\ln(1 - 1/e) = \ln(1) - a \exp[(-b)(-\$276{,}224)]$$

$$\text{whence } b = \frac{-\ln[\ln(0.5)/\ln(1 - 1/e)]}{[(-\$199{,}585) - (-\$276{,}224)]} = -5.388 \times 10^{-6}/\$$$

$$a = \frac{-\ln 0.5}{\exp[(5.388 \times 10^{-6}/\$)(-\$199{,}585)]} = 2.0315$$

The Gompertz curve of the distribution is then, in terms of (NPV),

$$\ln y = -2.0315 \exp[-5.388 \times 10^{-6}(\text{NPV})]$$

For the same degree of certainty as before, the minimum value of the (NPV) is likely to be

$$\ln\left[\frac{\ln(0.00003)}{-2.0315}\right] - 5.388 \times 10^{-6} = -\$303{,}365$$

and the maximum of \$2,064,569 calculated in the same way for $y = 0.99997$. Other values are calculated in the same way and are plotted as in Fig. 25-23.

Decision Trees In a typical decision tree, illustrated in a very simplified form by Fig. 25-24, each node represents a decision point (DP) at which one or more alternatives are available. Some quantifiable result of each alternative is chosen as a basis for comparison: for example, the net present value (NPV). A value is assigned to the probability of attaining each result, either cumulative or not as required. These may be obtained by the methods just described or otherwise. The estimates are subject to the restriction that the sum of the probabilities for all branches leaving each node shall be unity since some decision must be taken there.

In considering two investments, we shall let option B be a safe investment having a base net present value $(\text{NPV})_B$ that is independent of any competition. We shall let option A yield a net present value $(\text{NPV})_{A1}$ if no competition exists and $(\text{NPV})_{A2}$ if competition exists. We shall then let the probabilities of no competition and competition be p_1 and p_2 respectively. Then p_2 must equal $(1 - p_1)$.

The expected (NPV) for option A can be written, from Eq. (25-105), which follows, as

$$(\text{NPV})_{WA} = p_1(\text{NPV})_{A1} + (1 - p_1)(\text{NPV})_{A2}$$

where $(\text{NPV})_{WA}$ is the weighted net present value for option A based on the probabilities of encountering no competition p_1 and of encountering competition $(1 - p_1)$.

In the same way the expected (NPV) for option B is given by

$$(\text{NPV})_{WB} = p_1(\text{NPV})_B + (1 - p_1)(\text{NPV})_B = (\text{NPV})_B$$

The gain in the expected value of option A over option B is thus

$$\Delta(\text{NPV})_W = (\text{NPV})_{WA} - (\text{NPV})_{WB}$$

Let us suppose that the options represented in Fig. 25-24 were such that $(\text{NPV})_B = (0.5)(\text{NPV})_{A1} = 2(\text{NPV})_{A2}$. Then substitution leads to

$$\Delta(\text{NPV})_W = [2p_1 + 0.5(1 - p_1) - 1](\text{NPV})_B$$
$$= (1.5p_1 - 0.5)(\text{NPV})_B$$

The choice is immaterial when $\Delta(\text{NPV})_W = 0$, i.e., when $p_1 = 1/3$. If the probability of no competition is greater than 1/3, option A should be chosen; otherwise option B should be chosen.

The technique is based on the methods of linear algebra and the theory of games. When the problem contains many multibranched decision points, a computer may be needed to follow all possible paths and list them in order of desirability in terms of the quantitative criterion chosen. The decision maker may then concentrate on the routes at the top of the list and choose from among them by using other, possibly subjective criteria. The technique has many uses which are well covered in an extensive literature and will not be further considered here.

Numerical Measures of Risk Without risk and the reward for successfully accepting risk, there would be no business activity. In estimating the probabilities of attaining various levels of net present value (NPV) and discounted-cash-flow rate of return (DCFRR), there was a spread in the possible values of (NPV) and (DCFRR). A number of methods have been suggested for assessing risks and rewards to be expected from projects.

Let us consider a proposed project in which there is a probability p_1 that a net present value $(\text{NPV})_1$ will result, a probability p_2 that $(\text{NPV})_2$ will result, etc. A weighted average $(\text{NPV})_w$, known as the expected value, can then be calculated from

$$(\text{NPV})_w = p_1(\text{NPV})_1 + p_2(\text{NPV})_2 + \cdots \qquad (25\text{-}105)$$

where $p_1 + p_2 + \cdots = 1.0$.

Analogous equations may be written for other additive measures of profitability such as net profit.

Example 12 Let us consider a contractor who stands to make a net profit of \$100,000 on a contract. The cost of preparing the bid on the contract is \$10,000. There are four competing contractors, each with a probability $p_1 = 0.25$ of obtaining the contract. Thus, each contractor has a probability $p_2 = 0.75$ of not obtaining the contract. Therefore, the expected value of the project is

$$0.25(\$100{,}000) + 0.75(-\$10{,}000) = \$17{,}500$$

In this case, the potential gain is 10 times greater than the potential loss.

If the potential loss can bankrupt the company, then decisions are not necessarily made on the basis of expected value even though the potential gain may be very high. Also, decisions are not necessarily made on the basis of expected value if the potential loss represents a relatively small amount of money to the company. Between these two extremes, expected value can be a very useful criterion, particularly for a company with a large number of projects.

A company may be considering a project with a very high potential rate of return and a low risk, but it may prove impossible to raise the money to start the project. Conversely, the company may be prepared to undertake an extremely risky project if the investment is trivial. Thus, the attitude of a company to risk depends on the circumstances.

Money does not hold the same value for each company or each individual. A dollar may keep a pauper from starvation while being a trivial amount to the person who gave it. Attempts have been made to quantify a company's attitude to money, risk, and uncertainty by

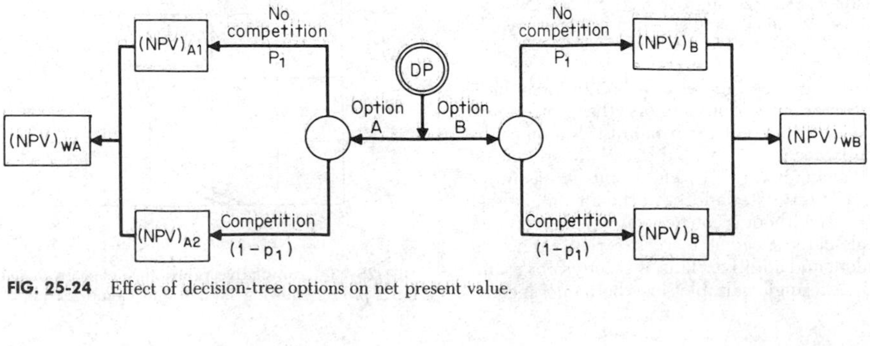

FIG. 25-24 Effect of decision-tree options on net present value.

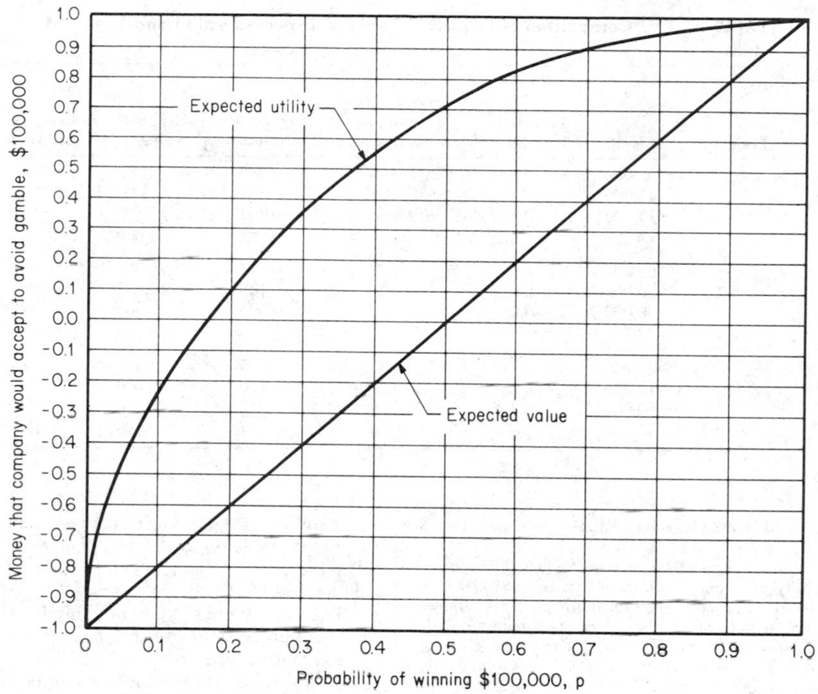

FIG. 25-25 Utility-function plot for $100,000.

asking business executives a number of questions such as the following:

"Your company has signed a business contract with potential after-tax proceeds of $\$P$. The probability of achieving the net gain of $\$P$ is, say, $p_1 = 0.75$, and the probability of a net loss of $\$P$ is $p_2 = 0.25$. If you would rather keep the contract, how much cash would you accept for your interest in it? If you would rather be released from the contract, how much cash would you pay to be released from it?"

The same questions may then be asked for different values of the probabilities p_1 and p_2. The answers to these questions can give an indication of the importance to the company of $\$P$ at various levels of risk and are used to plot the utility curve in Fig. 25-25. Positive values are the amounts of money that the company would accept in order to forgo participation. Negative values are the amounts the company would pay in order to avoid participation. Only when the utility value and the expected value (i.e., the straight line in Fig. 25-25) are the same can net present value (NPV) and discounted-cash-flow rate of return (DCFRR) be justified as investment criteria.

Since the utility curve has such a subjective basis, most companies prefer the objectivity of (NPV) and (DCFRR) over the range of the normal income and expenditure budget. Subjective methods tend to be reserved for exceptionally high risk projects.

A utility curve such as that in Fig. 25-25 is specific to a certain sum of money. The curve is likely to be different for, say, $P =$ $10,000. Figure 25-25 can only be used to consider projects that fall within the range of −$100,000 to +$100,000. Other utility curves must be used to cover projects that lie outside this range.

R. O. Swalm ["Utility Theory—Insight into Risk Taking," *Harv. Bus. Rev.*, **44**, 123–136 (November–December 1966)] found that many business executives had difficulty in appreciating fine shades of odds and confined his considerations to even-money bets. He asked various executives to state what guaranteed sum of money they considered equivalent to a gamble related to the toss of a coin. If the coin fell on one side, they would win a given sum of money; if the coin fell on the other side, they would get nothing.

Swalm started by considering a sum of money equivalent to twice the maximum expenditure that the executive could authorize in 1 year. This was used to obtain a further utility. In this way, a utility curve could be sketched. Swalm chose an arbitrary utility scale based on a range of −120 utiles to +120 utiles. (NOTE: It is as incorrect to compare utiles by ratio as it is to imply that an object at 30°C is twice as hot as an object at 15°C.)

Swalm found that most executives are conservative in their expenditure and that the patterns of utility curves are very similar if plotted with an ordinate range of ±1 unit. The unit, in this case, is the maximum authorized annual expenditure of the executive. Such curves may appear to differ quite widely when plotted in terms of absolute money values. The curves also show that executives tend to be more conservative when considering a loss than they do when considering a reduced gain.

Example 13 A company is considering investment in one or more of three projects, A, B, and C. We wish to evaluate the investment priorities if the probabilities of attaining various net present values (NPV) are as listed in the third column of Table 25-11. Equation (25-105) gives the expected value for $(NPV)_w$. Hence for project A, $(NPV)_w$ is computed from the data in Table 25-11 and found to be

$$(NPV)_w = 0.1(\$95{,}000) + 0.8(\$45{,}000) + 0.1(-\$75{,}000)$$

$$(NPV)_w = \$9{,}500 + \$36{,}000 - \$7{,}500$$

$$(NPV)_w = \$38{,}000$$

Corresponding values for projects B and C are calculated in the same way and are listed in Table 25-11.

In project A, the probability $p = 0.1$ for (NPV) = $95,000. Figure 25-25 shows that $95,000 is the amount of money that this company would pay for a 0.8 probability of gaining $100,000. There is, therefore, a 0.2 probability of losing $100,000. In this case, a probability of $p = 0.1$ of attaining $95,000 is equivalent to a probable utility of $(0.1)(0.8) = 0.08$ of gaining $100,000.

Equation (25-105) can be used to calculate the expected utility if the probabilities p_1, p_2, etc., are replaced by the probable utilities and if the net present values $(NPV)_1$, $(NPV)_2$, etc., are each replaced by $100,000. For project A, the expected utility U_w is

$$U_w = [0.1(0.8) + 0.8(0.34) + 0.1(0.03)]\$100{,}000$$

$$U_w = \$8{,}000 + \$27{,}200 + \$300$$

$$U_w = \$35{,}500$$

TABLE 25-11 Comparison of Projects in Terms of Expected Value and Expected Utility

Project	(NPV), $	Probability, p	Equivalent probability of winning $100,000	Probable utility	Expected value, $	Expected utility, $
	95,000	0.1	0.80	0.080	9,500	8,000
A	45,000	0.8	0.34	0.272	36,000	27,200
	−75,000	0.1	0.03	0.003	−7,500	300
		1.0		0.355	38,000	35,500
	50,000	0.2	0.37	0.074	10,000	7,400
B	20,000	0.6	0.23	0.138	12,000	13,800
	−60,000	0.2	0.04	0.008	−12,000	800
		1.0		0.220	10,000	22,000
	45,000	0.1	0.35	0.035	4,500	3,500
C	10,000	0.6	0.20	0.120	6,000	12,000
	−60,000	0.3	0.04	0.012	−18,000	1,200
		1.0		0.167	−7,500	16,700

Corresponding values for projects B and C are calculated in the same way and are listed in Table 25-11.

The straight line in Fig. 25-25 represents the situation in which the expected value and the expected utility are equal over the range of −$100,000 to +$100,000. In this case, decisions can be taken on the basis of the highest expected value as a routine matter. In other cases, decisions should be made on the basis of the highest expected utility. The utility curve in Fig. 25-25 represents the present attitude of management to $100,000. This curve should be updated as the company's business position changes. In this example, the utility curve is above the straight line. This represents a tendency on the part of the company's decision makers to gamble. When it is below the straight line, the utility curve implies conservatism. The investment priorities should be to implement project A and then, if finance is available, project B.

It might appear that project C should also be considered in view of the expected utility of $16,700. However, it is better to do nothing than to implement project C. The utility of doing nothing, which is equivalent to paying $0, is read from Fig. 25-25 to be 0.17. This gives a corresponding probable utility of (1.0)(0.17)($100,000), or $17,000. This is a better result than investing in project C.

In this example, the order of priorities based on expected utilities is the same as that based on expected values. However, the order of priorities is clear-cut on the basis of expected value but much less so on the basis of expected utility.

Capital is at risk until the breakeven point has been reached. It is common practice to give consideration to the discounted breakeven point (DBEP), the time at which the (NPV) is zero when discounting at the cost of capital. At any time after the (DBEP), the project will have recovered its cost and provided a greater return on the capital than the cost of capital. It is customary for management to spread risk by diversifying the activities of a company among a portfolio of projects.

R. L. Reul [*Chem. Eng. (London)*, **238,** CE 120–125 (May 1970)] has defined a parameter, which he calls the measured-survival function (MSF), given by

$$(\text{MSF}) = 1 - (1 - p)^{\beta} \qquad (25\text{-}106)$$

where (MSF) is the probability that a portfolio of bets with a similar strategy will at least break even, and β is the amount of one win divided by the amount of each bet. Reul has applied Eq. (25-106) to the research and development activities of a company. Equation (25-106) is based on the simplified assumption that a project either succeeds with probability p and achieves the expected reward or fails completely with probability $(1 - p)$. Therefore, (MSF) is the probability of at least one success when β similar projects are undertaken and represents a conservative measure of risk. It follows that $\beta > 1$ and hence that (MSF) $> p$. Many projects may result in greater returns or have an increased probability of attaining a given return if more money is spent. Each alternative derivable result from a given project is treated as a separate risk in the portfolio.

Research and development activities do not, in themselves, produce a salable product. Thus, they cannot directly generate a return on capital outlay. A successful research and development project is one that results in an activity that earns revenue for the company. The life cycle of the revenue from an individual product may be as shown in Fig. 25-26.

This revenue has to pay not only for the successful project but for all the unsuccessful research and development activities. It is common practice to consider all R&D as a portfolio. Disbursements for R&D are relatively flexible and can be switched from less favorable to more favorable projects at short notice.

When considering individual projects, β should be taken as the lesser of

$$\beta = \frac{\text{expected proceeds if project is successful}}{\text{disbursement on project}}$$

or

$$\beta = \frac{\text{total expenditure on all projects over budget period}}{\text{expenditure on project over budget period}}$$

Because the projects in a portfolio will usually have different probabilities of success and different rewards for success, β and p in Eq. (25-106) are conservatively estimated as follows:

$$\beta = \frac{\text{total annual proceeds if all projects are successful}}{\text{total annual disbursements on all projects}}$$

$$p = \frac{\text{total expected value of all projects}}{\text{total proceeds if all projects are successful}}$$

The expected value can be calculated from Eq. (25-105).

The relationship between (MSF), p, and $1/\beta$ in Eq. (25-104) is shown graphically in Fig. 25-27. It is the responsibility of management to decide on an acceptable value of the (MSF) for its company.

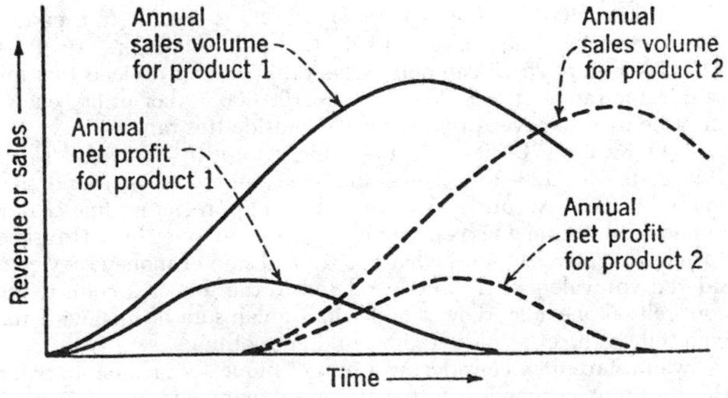

FIG. 25-26 Life cycle of products.

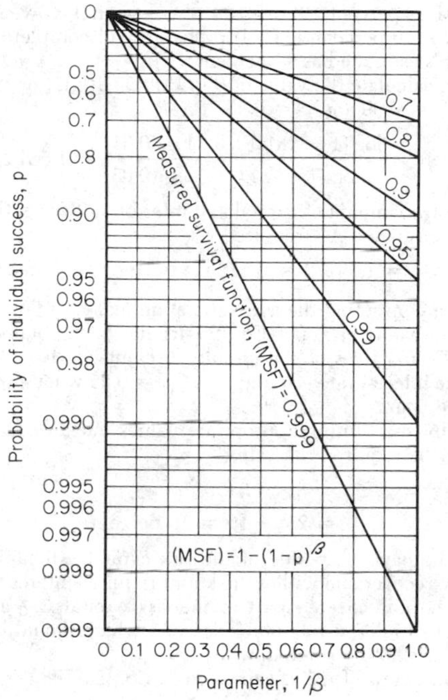

FIG. 25-27 Measured-survival-function plot.

The value chosen will depend on the company's attitude to risk that can be quantified in the form of a utility curve such as the one shown in Fig. 25-25, from which a value of equivalent (MSF) can be obtained.

It is also the responsibility of management to estimate the probabilities for the success of individual projects after due consideration of all the data provided by the various departments. The rate of return on investment that is acceptable to management is a function of these responsibilities. Each industry has a reasonably well defined return on investment that reflects the degree of risk inherent in that industry. If management decisions are faulty, the company either will overspend or will miss opportunities.

With a disbursement of \$1000 in Year 0, the discounted breakeven point (DBEP) will be reached in 3 years at a compound-interest rate of 30 percent if the annual net profit A_{NP} = \$550.63 per year. Thus, a discounted-cash-flow rate of return (DCFRR) of 30 percent corresponds to

$$\frac{1}{\beta} = \frac{\$1000}{(3\text{ years})(\$550.63/\text{year})} = 0.61$$

For $1/\beta = 0.61$ and (MSF) = 0.999, the probability of individual success is read from Fig. 25-27 to be $p = 0.985$. Similarly, it can be deduced that if (MSF) = 0.999 and $p = 0.95$, a (DCFRR) of 45 percent is required; if breakeven in 20 years is acceptable, then a (DCFRR) of only 10 percent is needed.

Example 14 Details of the estimates for the current research and development program of a company are given in Table 25-12. We shall estimate the probability that this portfolio will at least break even.

The total annual proposed disbursement for R&D is \$500,000. The effective total annual income if all projects reached their anticipated income would be \$1,300,000. Therefore,

$$\beta = \$1{,}300{,}000/\$500{,}000 = 2.600$$

Project A has an expected value of (0.95)(\$500,000/year) = \$475,000/year; project B has an expected value of (0.90)(\$400,000/year) = \$360,000/year; and so on. We sum these values to obtain the total expected value of the portfolio as \$1,109,500 per year. Hence,

$$p = (\$1{,}109{,}500/\text{year})/(\$1{,}300{,}000/\text{year}) = 0.8535$$

TABLE 25-12 Example of a Portfolio of Projects for a Research and Development Program

Project	Proposed disbursement for coming year	Annual aftertax income if successful	Probability of success
A	\$125,000	\$ 600,000	0.95
B	100,000	400,000	0.90
C	100,000	125,000	0.80
D	80,000	100,000	0.75
E	50,000	60,000	0.70
F	20,000	30,000	0.65
G	20,000	70,000	0.50
H	5,000	15,000	0.20
Totals	\$500,000	\$1,400,000	

From Fig. 25-27, for a probability of success $p = 0.8535$ and a value $1/\beta = 1/2.600 = 0.3846$, the (MSF) is 99.3 percent. This is the probability that this portfolio will at least break even.

Alternatively, we can substitute the values for p and β into Eq. (25-106) to get

$$(\text{MSF}) = 1 - (1 - 0.8535)^{2.6} = 0.9932, \text{ or } 99.32 \text{ percent}$$

The (MSF) and utility curves can be related.

Example 15 Let us sketch a utility-function curve that is equivalent to the following pattern of measured-survival functions (MSF), which expresses the observed strategy of a particular manager when spending an authorized annual budget of \$1,000,000:

Case	Potential proceeds annually, \$	(MSF), %
a	Above 600,000	99.9
b	300,000–600,000	95.0
c	0–300,000	65.0
d	Losses	75.0

We shall plot the resultant curve on a utility scale of ±120 utiles against a potential gain of ±\$1,000,000.

The required axes range from −\$1,000,000 per year to +\$1,000,000 per year, and from −120 utiles to +120 utiles. Utiles can be compared by ratio on an absolute scale only. Hence, for purposes of calculation the axes are moved to provide a working range of \$0 per year to \$2,000,000 per year and 0 to 240 utiles as in Fig. 25-28. On these axes, a potential gain of \$600,000 per year corresponds to an absolute amount of (600,000 + 1,000,000) = \$1,600,000 per year, and a potential loss of \$200,000 per year to an absolute amount of (−200,000 + 1,000,000) = \$800,000 per year.

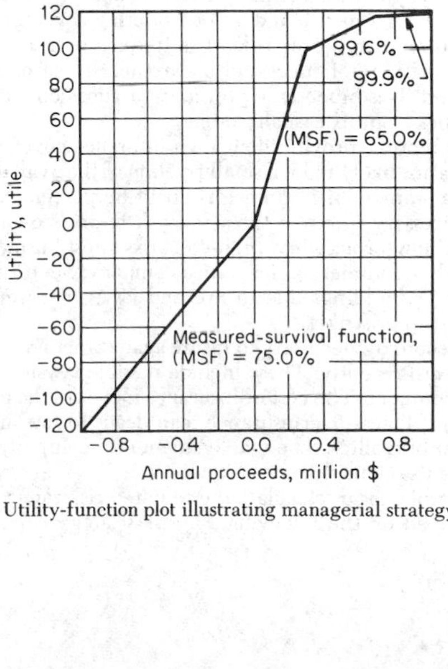

FIG. 25-28 Utility-function plot illustrating managerial strategy.

a. For annual proceeds above $600,000 per year, (MSF) is 99.9 percent. If the certainty of an annual gain of $600,000 has to be abandoned in an effort to obtain an annual gain of $1,000,000, then on an absolute scale

$$\beta = (\$2{,}000{,}000/\text{year})/(\$1{,}600{,}000) = 1.2500$$

With $1/\beta = 0.8000$ and (MSF) = 99.9 percent, we find the required probability of success by solving Eq. (25-106) for p:

$$p = 1 - (1 - 0.999)^{0.800} = 0.996$$

The utility of an amount of money is its utility when it is certain to be obtained, multiplied by its probability of being attained. On a scale in which an absolute annual income of $2,000,000 per year has a utility of 240 utiles, the utility of $1,600,000 is (0.996)(240), or 239 utiles.

b. For annual proceeds between $300,000 and $600,000, (MSF) = 95 percent. If the certainty of an annual gain of $300,000 has to be abandoned to obtain an annual gain of $600,000, then, as before,

$$1/\beta = \$1{,}300{,}000/\text{year}/\$1{,}600{,}000/\text{year} = 0.8125$$

$$p = 1 - (1 - 0.95)^{0.8125} = 0.912$$

Since, to this manager, the utility of an absolute income of $1,600,000 is 239 utiles, the value of $1,300,000 is (0.912)(239) = 218 utiles. On the original scales, potential annual proceeds of $300,000 have a utility of (218 − 120), or 98 utiles.

c and *d.* Values of utility at other potential annual gains are calculated in the same way and shown graphically in Fig. 25-28.

This strategy is extremely conservative when high gains are possible but becomes less so for smaller potential gains. If potential losses are involved, the strategy is a fair one for which (NPV) would be an accurate guide for choosing alternatives.

Insurance and Risk In the venture-premium method of assessment, risky investments are required to yield a rate of return that adds a premium to the cost of finance. D. F. Rudd and C. C. Watson (*The Strategy of Process Engineering*, Wiley, New York, 1968, p. 91) consider this relationship:

$$i_m = i + i_r \tag{25-107}$$

where i is the cost of capital, i_m is the minimum acceptable interest rate of return on the investment, and i_r is known as the risk rate.

They suggested that each project should pay an insurance premium i_r to guarantee the expected profits. The magnitude of i_r is proportional to the amount of capital to be risked. It is also a function of the degree of risk involved. Working capital and capital for auxiliary facilities are assumed to be risk-free. Thus, the risk rate is applied only to the fraction of the capital investment likely to be lost if the project is unexpectedly terminated.

The main objection to the venture-premium method is that the assessment of the riskiness of a project may be too subjective. This could lead to the rejection of potentially attractive proposals and the acceptance of projects that merely appear to be risk-free.

Insurance is protection against risk. Commercial insurance companies minimize their own risks by covering a large number of individuals against a given risk and also by offering coverage on a wide variety of different types of risk. It is frequently quite difficult to assess the probability of success of a particular research and development project. It is much easier for an insurance company to assess its probabilities from its casualty tables.

Businesses tend to provide their own insurance cover when individual claims are likely to be a small fraction of the available capital. The cost of commercial insurance is about 30 percent higher than would be necessary to cover the same risk in one's own company. However, for low-probability, high-cost risks, most businesses prefer to insure with a commercial insurance company. Such risks include loss of plant or buildings due to fire and losses of revenue due to delays in startup or strikes.

It is also becoming necessary to insure against factors not normally considered until recently. These include possible lawsuits for polluting the environment. The cost of insurance increases the annual total expense A_{TE}. Thus, overinsurance can lead to an unnecessary decrease in profitability. The management of a company must ultimately judge its own risks.

As an example, let us calculate the required risk rate for a project that is described by the following: (1) risk strategy is equivalent to an (MSF) of 99 percent, (2) payback of risk capital is 3 years, (3) cost of capital i is 10 percent, and (4) probability of complete success of the project is estimated as 95 percent.

First, we calculate the value of β in Eq. (25-106). For this project, with (MSF) = 0.99 and $p = 0.95$,

$$\beta = \frac{\log\,[1 - (\text{MSF})]}{\log\,(1 - p)} = \frac{\log\,(0.01)}{\log\,(0.05)} = 1.537$$

To recover this amount of capital and interest in 3 years, the average net annual cash flow A_{CF} required is

$$A_{CF} = 1.537/3 = 0.5124\ (\$/\text{year})/(\$\ \text{invested})$$

In effect, in computing the average net annual cash flow per dollar invested, the value of f_{AP} of Eq. (25-46) has been obtained for this example. From tables of the annuity present-worth factor f_{AP} the value of the interest rate is found to be $i_m = 0.25$ when $f_{AP} = 0.5124$ with $n = 3$ years.

Hence, by substituting appropriate values into Eq. (25-107) and solving for the required risk rate,

$$\begin{aligned} i_r &= i_m - i \\ &= 25 - 10 = 15 \text{ percent} \end{aligned}$$

based on the payback period of the risk capital. All capital C_{WC} is completely recoverable without risk and requires interest only at 10 percent. The unrecovered part of the risk capital C_{FC} attracts the additional risk interest rate of 15 percent, which should be reduced as the risk capital is written down.

A different view of risk is expressed in Eq. (25-108):

$$[1 + (\text{DCFRR})] = (1 + i)(1 + i_r') \tag{25-108}$$

The (DCFRR) represents the return on all capital invested after such capital has been paid back, together with any interest incurred by borrowing it, and after payment of all expenses, including taxes, associated with the project. It thus represents the entrepreneurial return to the company for managing the total capital employed. If the cost of capital i is set at the best risk-free use of that capital, such as the interest rate on a bank deposit or on government bonds, etc., i_r' represents the increased entrepreneurial return on the capital for taking the risks involved. This is a useful concept since the probability of achieving a given (DCFRR), and hence of a particular value of i_r', may be estimated by the methods detailed previously. We notice that i, as so defined, implies that all taxes and interest have been paid. Thus, $100 deposited in a bank at a rate of 10 percent with half of the money borrowed at 15 percent and corporation tax at 40 percent would result in a risk-free income after tax of

$$[(\$100)(0.10/\text{year}) - (0.5)(\$100)(0.15/\text{year})](1 - 0.40) = \$1.5/\text{year}$$

The same money invested in a project with a (DCFRR) of 10 percent would, by Eq. (25-108), obtain an entrepreneurial return $i_r' =$ 8.37 percent on the whole investment, i.e., $8.37/$100. Investment of the entrepreneur's own money would only achieve an aftertax return of (0.1)(1 − 0.40) = 6 percent on $50, or $3/$100 of total investment. The incentive to the entrepreneur to manage the project thus corresponds to a tax-free income of $5.37/$100 of total investment. In practice, money is borrowed from more than one source at different interest rates and at different tax liabilities. The effective cost of capital in such cases can be obtained by an extension of the above reasoning and is treated in detail by A. J. Merrett and A. Sykes (*Capital Budgeting and Company Finance*, Longmans, London, 1966, pp. 30–48).

Inflation It is currently necessary to evaluate the profitability of proposed investments whose future earnings are virtually certain to be eroded by inflation. It has been common practice to ignore the effects of inflation. This is done on the reasonable grounds that predicting the market rate of interest, and thus the appropriate discount rate for future cash flows, is difficult enough without having to worry about inflation as well. But failure at least to try to predict inflation rates and take them into account can greatly distort a project's economics, especially at the double-digit rates that have been found

TABLE 25-13 (NPV) Calculations with No Inflation

Year, n	Net capital expenditure, A_{TC}	Revenue from sales, A_S	Total expenses, A_{TE}	Cash income, A_{CI} (= A_S − A_{TE})	Depreciation charge, A_D	Taxable income, $(A_{CI} - A_D)$	Amount of tax at t = 0.5, A_{IT} [= $(A_{CI} - A_D)t$]	Net cash flow after tax, A_{CF} (= $A_{CI} - A_{IT} - A_{TC}$)	Discount factor at i = 10%, f_d $\left[= \frac{1}{(1+0.1)^n}\right]$	Discounted net cash flow, A_{DCF} [= $A_{CF}(f_d)$]	Net present value (NPV), $\left(= \sum_0^n A_{DCF}\right)$
0	$1,100,000	0	0	0	0	0	0	−$1,100,000	1.00000	−$1,100,000	−$1,100,000
1	0	$500,000	$100,000	$400,000	$200,000	$200,000	$100,000	300,000	0.90909	272,727	−827,273
2	0	500,000	100,000	400,000	200,000	200,000	100,000	300,000	0.82645	247,935	−579,338
3	0	500,000	100,000	400,000	200,000	200,000	100,000	300,000	0.75131	225,393	−353,945
4	0	500,000	100,000	400,000	200,000	200,000	100,000	300,000	0.68301	204,903	−149,042
5	−100,000	500,000	100,000	400,000	200,000	200,000	100,000	400,000	0.62092	248,368	+99,326

TABLE 25-14 (NPV) Calculations with Inflation Present But Not Allowed For

Year, n	Net capital expenditure, A_{TC}	Revenue from sales, A_S	Total expenses, A_{TE}	Cash income, A_{CI}	Depreciation charge, A_D	Taxable income, $(A_{CI} - A_D)$	Amount of tax at t = 0.5, A_{IT}	Net cash flow, A_{CF}	Discount factor at i = 10%, f_d	Discounted net cash flow, A_{DCF}	Net present value (NPV)
0	$1,100,000	0	0	0	0	0	0	−$1,100,000	1.00000	−$1,100,000	−$1,100,000
1	0	$500,000	$100,000	$400,000	$200,000	$200,000	$100,000	300,000	0.90909	272,727	−827,273
2	0	600,000	120,000	480,000	200,000	280,000	140,000	340,000	0.82645	280,993	−546,280
3	0	720,000	144,000	576,000	200,000	376,000	188,000	388,000	0.75131	291,508	−254,772
4	0	864,000	172,800	691,200	200,000	491,200	245,600	445,600	0.68301	304,349	+49,577
5	−100,000	1,036,800	207,360	829,440	200,000	629,440	314,720	614,720	0.62092	381,692	+431,269

throughout the world. It is the common experience that a given amount of money buys less and less of goods and services as time goes by. The problem is to express this experience quantitatively.

Published figures for inflation rates are based on some particular mixture of goods and services that is chosen to represent the material wants of the average citizen. If a given quantity of this specific mixture cost $100 last year and now costs $120, then the mix has suffered a 20 percent rate of inflation. The purchasing power of the currency (i.e., of the $120) in respect of these goods and services has consequently fallen by a factor of ($120 − 100)/$120, or 16.7 percent.

Two kinds of inflation can be considered: general, or open, inflation and repressed, or differential, inflation. In the first case, all costs and prices increase at a uniform rate. Thus, the same rate of inflation will be calculated regardless of the particular mixture of goods and services chosen. In the second case, the rate of inflation will depend on the spending spectrum of the individual or company. For instance, a given company's labor costs and material costs may inflate at different rates. To quite a large extent, inflation becomes repressed, or differential, in such fields as taxation, import control, and price restriction.

The effect of inflation on the real value of future earnings from a project should not be confused with the effect of the market rates of interest on those earnings. Strictly speaking, the market interest rate and the inflation rate are not fully independent, at least according to some economic theorists. However, they are here treated as being separate. Because of each effect, a dollar of project income next year has a smaller true value than does a dollar in hand today. The interest-rate effect could be offset because a dollar could be financially invested at the prevailing interest rate and the dollar plus interest earnings recouped in a year. By contrast, the inflation effect comes about simply because a dollar can buy more now than a year hence because of an irreversible rise in prices. The distinction is clarified in the following subsections.

Effect of Inflation on (NPV) When computing the (NPV) for a proposed project, error arises if the actual cash flows are simply added together instead of adjusting all the values to their purchasing power in a particular year. The reason lies in the basis of (NPV) calculations. We shall rewrite Eq. (25-57) to give

$$(\text{NPV}) = A_{CF0} + \sum_1^n \frac{A_{CFn}}{(1+i)^n} \tag{25-109}$$

Equation (25-109) is valid for the case of no inflation. In the case of general inflation at a fractional rate i_i, this equation can be written in the modified form

$$(\text{NPV}) = A_{CF0} + \sum_1^n \frac{A_{CF0}}{(1+i)^n(1+i_i)^n} \tag{25-110}$$

Equation (25-110) enables all the net annual cash flows to be corrected to their purchasing power in Year 0. If the inflation rate is zero, Eq. (25-110) becomes identical with Eq. (25-109).

The following example illustrates the effect of inflation on (NPV) as well as on the taxes the company pays.

Example 16 Let us consider a simplified project in which $1,100,000 of capital is spent in Year 0, $1,000,000 for fixed-capital items and $100,000 for working capital. The fixed capital is depreciated on a straight-line basis to a book value of zero at the end of Year 5. The annual sales revenue in Years 1 through 5 is $500,000. There is no inflation. The $100,000 of working capital is recovered at the end of Year 5. The taxation rate is 50 percent, and the market interest rate is 10 percent. Table 25-13 lists the cash-flow data for this project, showing that the (NPV) at the end of Year 5 is $99,326 by using Eq. (25-109).

Let us modify this example by assuming that there is a general inflation rate of 20 percent per year and that the project analyst ignores the inflation and (inappropriately) applies Eq. (25-109). The revenue and expense data for this case are shown in Table 25-14, yielding an (NPV) of $431,269. When Eq. (25-109) is (inappropriately) used for the same example with various other rates of inflation, the resulting (NPV)s can be plotted as the upper line in Fig. 25-29.

If the inflation is correctly taken into account by applying Eq. (25-110), the results are strikingly different. By further discounting the discounted cash flows A_{DCF} of Table 25-14 by the f_d factors corresponding to an inflation rate of 20 percent before summing, it can be seen that the project actually incurs a negative (NPV) of $208,733 in uninflated-money terms. The lower line in Fig. 25-29 extends the example by assuming other rates of inflation. Figure 25-29 shows that the effect of inflation, if not taken into account, is to make a project seem more profitable than it actually is.

Table 25-14 shows that the total amount of tax actually paid over the 5-year period was $988,320. This becomes $534,272 in uninflated-money terms when the tax for each year is corrected to its purchasing power in Year 0, using f_d factors for the 20 percent inflation rate employed for the example. Calculations for other rates of inflation can also be made, and the results plotted as in Fig. 25-30.

This confirms that although the tax paid will increase with inflation, the gain to the government is more apparent than real. It is interesting to note that although the tax paid corrected to its purchasing power in Year 0 is almost constant irrespective of the inflation rate, it does go through a maximum at an inflation rate of about 17 percent in this example.

Effect of Inflation on (DCFRR) A net annual cash flow A_{CF} will have a cash value of $A_{CF}(1 + i)$ 1 year later if invested at a fractional

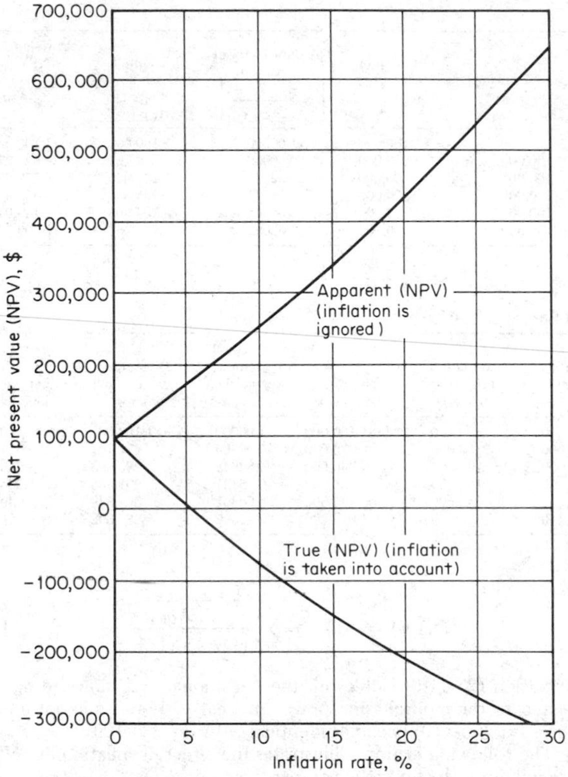

FIG. 25-29 Effect of inflation rate on net present value for a project.

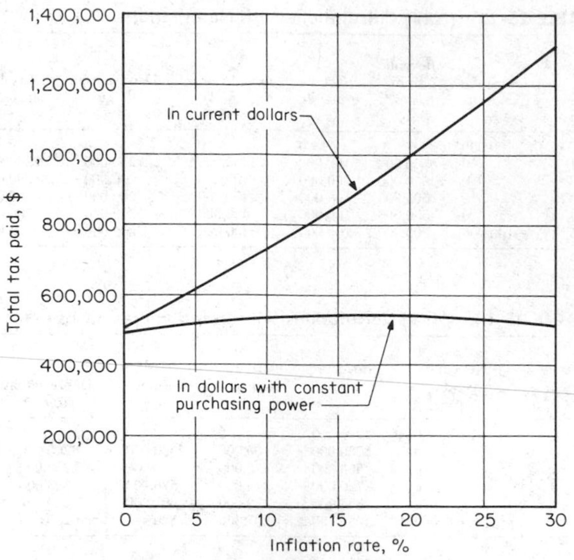

FIG. 25-30 Effect of inflation rate on taxes paid for a project.

interest rate i. If there is inflation at an annual rate i_i, then an effective rate of return or interest rate i_e can be defined by the equation

$$A_{CF}(1 + i_e) = [A_{CF}(1 + i)]/(1 + i_i) \quad (25\text{-}111)$$

which can be simplified and rewritten to give

$$i_e = i - i_i - i_e i_i \quad (25\text{-}112)$$

In the context of the discounted-cash-flow rate of return, Eq. (25-112) becomes

$$i_e = (\text{DCFRR}) - i_i - i_e i_i \quad (25\text{-}113)$$

In this equation, (DCFRR) can be viewed as the nominal discounted-cash-flow rate of return uncorrected for inflation and i_e can be thought of as the true or real discounted-cash-flow rate of return.

Instead of using Eq. (25-113), it is unfortunately common practice to try to obtain the true or effective rate of return by calculating the nominal (DCFRR), based on actual net annual cash flows uncorrected for inflation, and then subtracting the inflation rate from it as if

$$i_e = (\text{DCFRR}) - i_i \quad (25\text{-}114)$$

Equation (25-113) shows that Eq. (25-114) is only approximately true and should be used, if at all, solely for low interest rates. Let us consider the case of a nominal (DCFRR) of 5 percent and an inflation rate of 3 percent. Equation (25-14) yields an approximate effective return rate of 2 percent, compared with the real effective rate of 1.94 percent given by Eq. (25-113); i.e., there is an error of 3.1 percent. Now let us consider the case of a nominal (DCFRR) of 25 percent and an inflation rate of 23 percent. Equation (25-114) yields an approximate effective return rate of 2 percent, compared with 1.63 percent from Eq. (25-113); in this case, the error that results is 22.7 percent.

Inflation, (DCFRR), and Payback Period More insight into the effect of inflation on (DCFRR) calculations can be gained by considering the payback period (PBP), which is defined as the elapsed time necessary for the positive aftertax cash flows from the project to recoup the original fixed-capital expenditure. In this definition, the cash flows are not discounted to allow for the market rate of interest or for the inflation rate, so that a project with a given (PBP) could show various values for its (DCFRR) and a given (DCFRR) could pertain to projects with various payback periods.

We shall consider the simple case of (1) a single capital expenditure made immediately before the start of production and (2) equal positive net annual cash flows A_{CF} in all the productive years of the project. For this case, Eq. (25-109) can be rewritten in terms of the payback period and the (DCFRR) as follows:

$$(\text{PBP}) = \sum_{1}^{n} \frac{1}{[(1 + (\text{DCFRR})]^n} \quad (25\text{-}115)$$

The relationship set out in Eq. (25-115) can also be viewed via a different chain of causality with (DCFRR) as a given parameter, (PBP) as the independent variable, and n as the variable whose value is being sought. Such an approach is the basis for the lines in Fig. 25-31, each of which shows the number of years of project life required to achieve an effective interest rate or a (DCFRR) of 20 percent by projects having various payback periods. The three lines differ from each other with respect to the matter of inflation.

If there is no inflation, then the middle line pertains. Because there is no inflation, the nominal (DCFRR) is equal to or identical with i_e, the real discounted-cash-flow rate of return, as can be seen from the relationship expressed in Eq. (25-113).

When inflation does exist, the relevant parameter is i_e, which is different from the nominal (DCFRR). Equation (25-113), manipulated into equivalent form,

$$(\text{DCFRR}) = (1 + i_e)(1 + i_i) - 1$$

shows that in order to achieve an i_e of 20 percent when the general inflation rate is likewise 20 percent, a project must generate a nominal (DCFRR) of 44 percent. This is the basis for the uppermost line in Fig. 25-31. Other lines pertaining to other rates of inflation could be plotted in the same way.

Let us assume that 20 percent inflation prevails but that the analyst ignores it and mistakenly takes a (DCFRR) of the project at its nom-

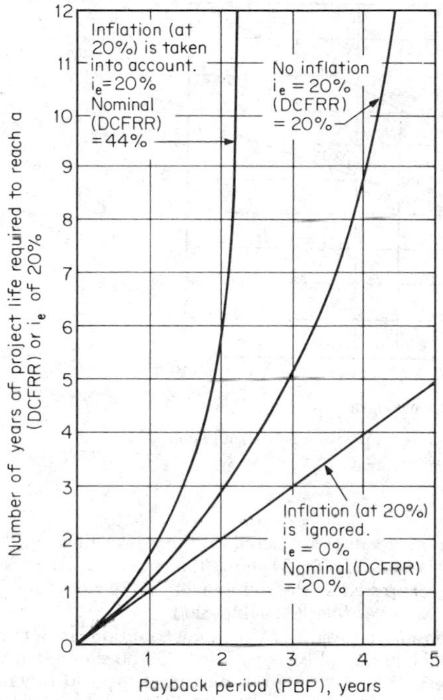

FIG. 25-31 Effect of inflation rate on the relationship between the payback period and the discounted-cash-flow rate of return.

inal value instead of converting it to an i_e. Equation (25-115) rearranged into the form

$$i_e = [1 + (\text{DCFRR})]/(1 + i_i) - 1 \quad (25\text{-}116)$$

shows that with a nominal (DCFRR) of 20 percent and a general inflation rate of 20 percent, the true or effective rate of interest is zero. This is the basis for the lowest line in Fig. 25-31. Points for lines corresponding to other rates of inflation could be plotted onto that figure. Plots similar to Fig. 25-31 can be drawn for other (DCFRR) values.

Figure 25-31 shows that the elapsed time necessary to reach a nominal (DCFRR) for a given project decreases sharply with inflation. This figure, like Fig. 25-29, shows that the effect of inflation is to make a project seem more profitable than it actually is.

The magnitude of the effect comes through even more clearly in Fig. 25-32, a plot of the time needed to reach a nominal (DCFRR) of 20 percent against the inflation rate for various values of (PBP). This plot also shows that the longer the payback period, the greater the increase in apparent profitability of the project.

The true rates of return i_e can be calculated from Eq. (25-116) to be 20, 9.09, 0, and −7.69 percent respectively for general inflation rates of 0, 10, 20, and 30 percent. Thus, although the time required for a project with a payback period of 4 years to reach a nominal (DCFRR) of 20 percent is reduced from almost 9 years under conditions of no inflation to less than 3½ years for 30 percent inflation, the true rate of return that prevails for the latter condition is −7.69 percent, implying that the project loses money in real terms.

It is interesting to note that, in order to reach a real (DCFRR) or i_e of 20 percent within a reasonable project lifetime when the general inflation rate is 20 percent, it follows from Fig. 25-31 that the payback period for the project must not be much in excess of 2 years.

Although it is difficult to carry out economic-feasibility studies on projects in a time of high inflation, it is important to try to predict inflation rates and allow for them in such studies.

When different people talk about inflation, they often adopt different concepts without realizing it. The area of conceptual uncertainty can be said to lie somewhere between the upper and lower lines shown on Fig. 25-31 in most cases.

Inflation and the (MSF) By applying the measured-survival-function concept to manufacturing projects rather than to research and development, we can define a modified (MSF) for a given project as

$$(\text{MSF}) = 1 - (1 - \eta)^{\beta} \quad (25\text{-}117)$$

Here, β is the number of payback periods that have elapsed since the project started to generate positive net annual cash flows A_{CF} up to any given year n since project startup. It is given by

$$\beta = \frac{\sum_{n=0}^{n=n} (A_{CF})_n}{C_{FC} - S} \quad (25\text{-}118)$$

If all the net annual cash flows in Eq. (25-118) are based on their purchasing power in Year 0, then β is independent of inflation.

As for the contribution efficiency η, it is the ratio of (1) the annual profit that can actually be achieved in a given year for a given sales volume to (2) the profit that could be obtained if no repayment of capital or interest were required and all fixed-expense items were credited free to the project. It is defined by

$$\eta = [R(c_S - c_{VE}) - A_{FE}]/[R(c_S - c_{VE})] \quad (25\text{-}119)$$

where R is the annual production rate or sales volume in physical units, c_S is the sales price per unit, c_{VE} is the variable production and selling cost per unit, and A_{FE} is the annual fixed cost.

If the project gets a "free ride," i.e., if A_{FE} is zero, then η takes on its maximum possible value of unity. Conversely, if the project and its production rate are only at the breakeven point, then η becomes zero. Therefore, contribution efficiency can be regarded as a measure of the probability of success for the project.

The relationship between the number of payback periods, the contribution efficiency, and the measured-survival function as set out in Eq. (25-117) is plotted in Fig. 25-33.

The contribution efficiency defined by Eq. (25-119) may vary

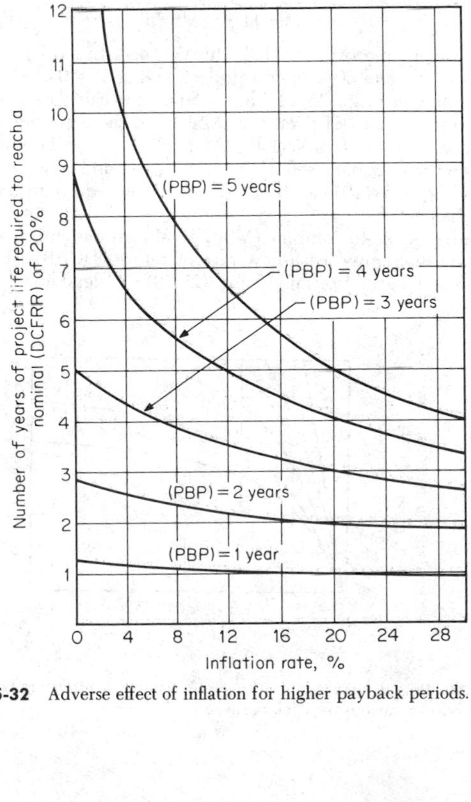

FIG. 25-32 Adverse effect of inflation for higher payback periods.

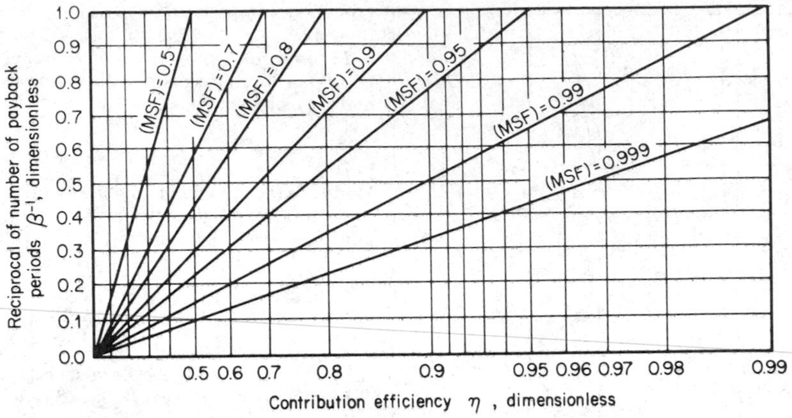

FIG. 25-33 Relationship between measured-survival function, number of payback periods, and contribution efficiency.

from year to year. In that case, Eq. (25-117) can be written in the modified form

$$(\text{MSF}) = 1 - [(1 - \eta_1)(1 - \eta_2) \cdot \cdot \cdot (1 - \eta_n)]^{\beta/n} \qquad (25\text{-}120)$$

where $\eta_1, \eta_2, \ldots \eta_n$ are the contribution efficiencies in Years 1, 2, . . . n respectively.

As in the case of the (MSF) defined by Reul for research and development projects, it is the responsibility of management in a particular manufacturing company to decide on an acceptable level of (MSF) for manufacturing projects. That decision reflects and helps quantify the company's attitude toward risk.

Thus, (MSF) should in practice be regarded as a given or predetermined variable, and Eq. (25-117) accordingly becomes more useful if it is rearranged. For instance, the values of contribution efficiency for a given value of (MSF) are related to the number of elapsed payback periods by

$$\eta = 1 - [1 - (\text{MSF})]^{1/\beta} \qquad (25\text{-}121)$$

If the acceptable (MSF) is 0.9, this can be satisfied by a project having $\eta = 0.9$ and $\beta = 1$, or a project having $\eta = 0.684$ and $\beta = 2$, and so on. Once Eq. (25-121) has been used to calculate a required contribution efficiency [given the (MSF) and the expected number of payback periods of project life], Eq. (25-119) can be applied to determine the necessary selling price if R, c_{VE}, and A_{FE} are known. Similarly, Eq. (25-119) can be used to find the required production rate if c_S is known.

It is also possible to combine (MSF) considerations with evaluation of the true discounted-cash-flow rate of return (DCFRR) by using Eq. (25-62). The relationship of Eq. (25-59) is independent of inflation if all money values are based on those prevailing in the startup year. For this case, Fig. 25-34 shows the true (DCFRR) reached in a given time, expressed as the number of elapsed payback periods β for various values of the payback period.

Let us consider a project having a contribution efficiency of 0.684 and a payback period of 3 years. Figure 25-33 shows that when two payback periods have elapsed, a measured-survival function of 0.9 has been attained. In addition, Fig. 25-34 shows that the discounted-cash-flow rate of return reached at that time is 24 percent.

Effects of Differential Inflation Inflation can be general or differential. In the first case, all costs and prices increase at a uniform rate. In the second, government controls and other factors cause the various costs and prices to inflate at different rates.

The onset of general inflation does not change the value of the contribution efficiency η, as can be seen from Eq. (25-119), and it does not affect the value of β if the cash flows in Eq. (25-118) are converted to their purchasing power in Year 0. Thus, general inflation does not cause the measured-survival function to change.

Differential inflation, on the other hand, can affect the measured-survival function. We shall assume, for instance, that the sales price per unit product c_S in Eq. (25-119) is frozen at a constant level while some or all of the production costs are allowed to rise. This causes the value of η to decrease; therefore, (MSF) likewise decreases, as can be seen from Eq. (25-117).

Let us consider the effect of differential inflation on the overall profitability of the project of the last example. The effect of general

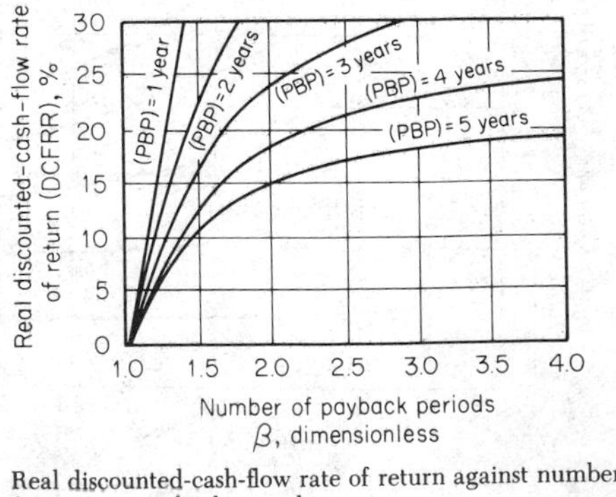

FIG. 25-34 Real discounted-cash-flow rate of return against number of payback periods for various payback periods.

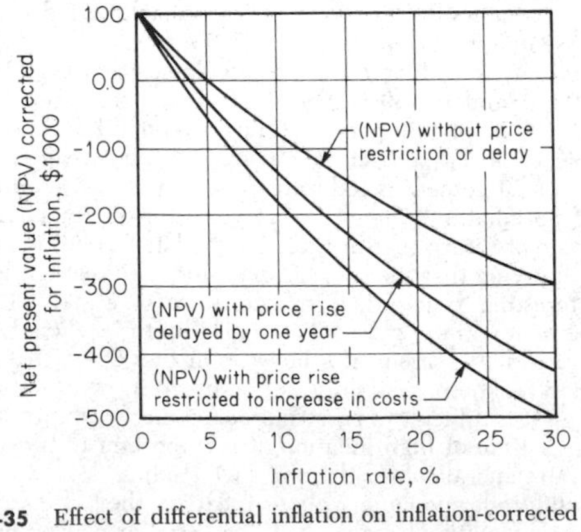

FIG. 25-35 Effect of differential inflation on inflation-corrected net present value.

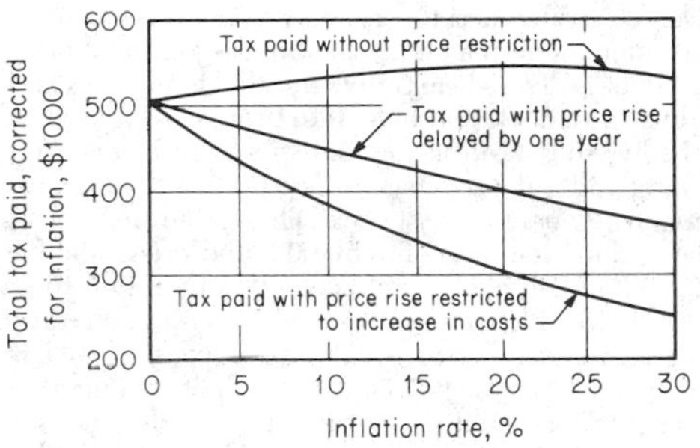

FIG. 25-36 Effect of differential inflation on inflation-corrected tax revenue.

inflation on this project showed that the apparent profitability rises sharply, to an (NPV) of $431,269 at a general inflation rate of 20 percent. However, when the cash flows of the (NPV) are properly corrected to their purchasing power in Year 0, the (NPV) instead becomes $208,733.

The effect of differential inflation on this project emerges in Fig. 25-35, with all (NPV)s corrected to their purchasing power in Year 0. The top line shows (NPV) for various rates of general inflation. The bottom line shows (NPV) for the differential-inflation case in which only the costs are allowed to increase while product selling price and thus cash income remain constant from year to year. The middle line shows the effect of general inflation when the price rises are delayed by 1 year. The figure confirms that both of these situations take away from the attractiveness of the project.

The effect upon total taxes paid, when they are corrected to their purchasing power in Year 0, is shown in Fig. 25-36. Differential inflation not only decreases the profitability of the project to its owner but also decreases the revenue received by the taxing authority. The method of calculation is identical to that of the earlier example.

Another instance of differential inflation occurs when the prices of goods and services rise uniformly but the cost of borrowing money, the interest rate charged on a loan, does not rise.

If the fractional inflation rate is i_i, a fractional interest rate i_L on a loan can be corrected to an effective rate of interest by Eq. (25-116) with i_L substituted for (DCFRR). The effect of various amounts of loan, borrowed at various interest rates i_L, on the net present value of a particular, fairly simple project is shown in Fig. 25-37. Thus, if $25,000 were borrowed at an interest rate of 15 percent for the project, the (NPV) would be about $43,000 at a zero inflation rate. But if the inflation for goods and services i_i is 10 percent, the effective interest rate for that loan can be calculated from Eq. (25-116) to be only 4.55 percent. It is seen from Fig. 25-37 that this increases the (NPV) of the project to $48,000. This confirms the economic advantage of borrowing at a fixed interest rate in a time of general inflation.

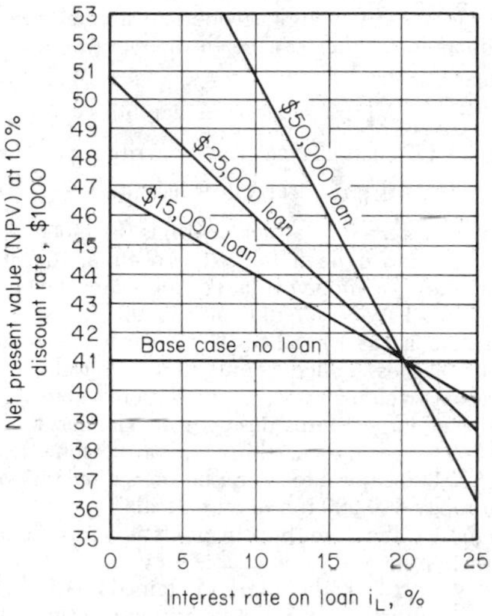

FIG. 25-37 Effect of loan interest rate on the net present value of a project.

A topical aspect of differential inflation is the question of energy costs. Will the cost of a particular fuel rise or fall in relation to prices in general, and if so, what effect will this have on the economics of a project?

Example 17 A process unit is heated by gas. We assume that $100 spent on energy-conservation measures for this particular unit at the end of 1980 would save 200 therms (21.1 GJ) of gas energy in each subsequent year. If the cost of gas in 1980 is $x per therm, the annual dollar savings at 1980 prices is $200x. The (NPV) at the end of year n for this project is

$$(\text{NPV}) = -100 + \sum_{1}^{n} \frac{(200)x}{(1+i)^n}$$

if the appropriate discount factor is i.

This is independent of inflation provided that the cost of gas rises in line with any general rate of inflation. However, if the real cost of gas rises at a fractional annual rate r over and above the general inflation rate, it should be modified into the form

$$(\text{NPV}) = -100 + \sum_{1}^{n} \frac{(200x)(1+r)^n}{(1+i)^n}$$

This equation confirms that as the gas price rises because of inflation, the attractiveness of the conservation project also rises.

ACCOUNTING AND COST CONTROL

Principles of Accounting Accounting is the art of recording business transactions in a systematic manner. Financial statements are both the basis for and the result of management decisions. Such statements can tell managers or engineers a great deal about their company, provided that they can interpret the information correctly.

Since a fair allocation of costs requires considerable technical knowledge of operations in the chemical-process industries, a close liaison between the senior process engineers and the accountants in a company is desirable. Indeed, the success of a company depends on a combination of financial, technical, and managerial skills.

Accounting is also the language of business, and the different departments of management use it to communicate within a broad context of financial and cost terms. Engineers involved in feasibility studies and detailed process evaluations are dependent for financial information on the company accountants, especially for information on the way in which the company intends to allocate its overhead costs. It is vital that engineers correctly interpret such information and that they can, if necessary, make the accountants understand the effect of the chosen method of allocation.

The method of allocating overheads can seriously affect the assigned costs of a project and hence the apparent cash flows for that project. Since these cash flows are used to assess profitability by the net-present-value (NPV) and discounted-cash-flow-rate-of-return (DCFRR) methods, unfair allocation of overhead costs can result in a wrong choice between alternative projects.

In addition to understanding the principles of accountancy and obtaining a working knowledge of its practical techniques, engineers should be aware of possible inaccuracies of accounting information in the same way that they allow for errors in any technical data.

At first acquaintance, the language of accountancy appears illogical to most engineers. Although accountants normally express themselves in tabular form, the basis of all their practice can be simply expressed by

$$\text{Capital} = \text{assets} - \text{liabilities} \tag{25-122}$$

Equation (25-122) can alternatively be written as

$$\text{Assets} = \text{capital} + \text{liabilities} \tag{25-123}$$

Capital, often referred to as net worth, is the money value of the business, since assets are the money values of things the business owns while liabilities are the money values of the things the business owes.

Most engineers have great difficulty in thinking of capital (also known as ownership) as a liability. This is easily overcome once it is realized that a business is a legal entity in its own right, owing money to the individuals who own it. This realization is absolutely essential when considering large companies with stockholders and is used for consistency even for sole ownerships and partnerships. If an individual puts up $10,000 capital to start a business, then that business has a liability to repay $10,000 to the individual.

It is even more difficult to think of profit as being a liability. Profit is the increase in money value available for distribution to the owners and effectively represents the interest obtained on the capital. If the profit is not distributed, it represents an increase in capital by the normal concept of compound interest. Thus, if the individual's business makes a profit of $5000, the liability to the individual is increased to $15,000. With this concept in mind, Eq. (25-123) can be expanded to

$$\text{Assets} = \text{capital} + \text{liabilities} + \text{profit} \tag{25-124}$$

where the capital is considered as the cash investment in the business and is distinguished from the resultant profit in the same way that principal and interest are separated.

Profit (as referred to above) is the difference between the total cash revenue from sales and the total of all costs and other expenses incurred in making those sales. With this definition, Eq. (25-124) can be further expanded to

$$\text{Assets} + \text{expenses} = \text{capital} + \text{liabilities} + \text{revenue from sales} \tag{25-125}$$

Engineers usually have the greatest difficulty in regarding an expense as being equivalent to an asset, as is implied by Eq. (25-125). Let us consider a one-person business. We assume for a given period a profit of $5000 and total expenses excluding the individual's earnings of $8000. Also we assume that the individual's labor to the business in this period is worth $12,000. The revenue required from sales would be $25,000. Effectively, the individual has made a personal income of $17,000 in the period but has apportioned it to the business as $12,000 expense for the individual's labor and $5000 return on capital. In larger businesses, there will also be those who receive salaries but do not hold stock and, therefore, receive no profits and stockholders who receive profits but no salaries. Thus, the difference between expenses and profits is very practical.

The period covered by the published accounts of a company is usually 1 year, but the details from which these accounts are compiled are entered daily in a journal. The journal is a chronological listing of every transaction of the business, with details of the corresponding income or expenditure. For the smallest businesses, this may provide sufficient documentation, but in most cases the unsystematic nature of the journal can lead to computational errors. Therefore, the usual practice is to keep accounts that are listings of transactions related to a specific topic such as "purchase-of-oil account." This account would list the cost of each purchase of oil, together with the date of purchase, as extracted from the journal.

Principles of Double-Entry Accounting Many of the accounts involve both income and expenditure. The general practice is to keep accounts by the double-entry system, which may be summarized by

$$\text{Debits} = \text{credits} \tag{25-126}$$

The principle of double entry dates from the fifteenth century and is based on the premise that every transaction involves a giver and a receiver of value. Double entry requires that each transaction be entered into two accounts, the convention being that the account of the giver is credited and the account of the receiver is debited with the same amount of money, as noted in the journal. For convenience, each account is divided centrally, and the debit items are entered on the right-hand side. It is also usual to provide a cross-reference to the journal entry so that errors and omissions can be checked.

Let us consider the purchase of $50,000 worth of plant equipment by company A, paid for by check. The accounting entries are: debit the plant-equipment account $50,000, and credit the bank account $50,000. The plant-equipment account is then said to have a debit balance of $50,000, and the bank account a credit balance of $50,000, if these happen to be the only entries.

If company A then sells $100,000 worth of product that is paid for by check, the accounting entries are: credit the sales account $100,000, and debit the bank account $100,000. The bank account will now have a debit balance of ($100,000 − $50,000) = $50,000, and the sales account a credit balance of $100,000, if this happens to be the only sale to date in the accounting period.

In principle, the debiting and crediting of accounts are relatively straightforward. However, a great deal of practice is essential in order to achieve proficiency. Although it is not at all necessary for engineers to compete with professional accountants in this field, engineers should appreciate what accountants do and why they do it.

Of the accounts considered in the preceding illustrations, the plant-equipment and bank accounts are asset accounts, and the sales account is a liability account. To increase an asset, debit the asset account; to increase a liability, credit the liability account. Conversely, to decrease an asset, credit the asset account; to decrease a liability, debit the liability account.

Closing the Books At the end of the accounting period, the individual accounts are closed by balancing each in accordance with Eq. (25-126). The balances are transferred either to the balance sheet in the case of capital expenditure or to the income statement in the case of revenue expenditure. An alternative name for the balance sheet is the position statement; the income statement is also called the trading and profit-and-loss account.

The purpose of capital expenditure, such as the purchase of a piece of plant equipment for $50,000, is to earn future revenue. In contrast, the purpose of revenue expenditure is to maintain existing business.

Revenue expenditure includes the direct material costs and direct labor costs incurred in the manufacture of a product, together with the associated overheads that include maintenance of the plant. Since these expenses are debits, the debit balance for a given accounting period is obtained by adding up the debit balances from each individual expenditure account. Similarly, since revenues from sales and other income are credits, the credit balance for a given accounting period is obtained by adding up the credit balances from each individual income or revenue account.

To ascertain profit or loss (calculated as income minus expenditure for a given accounting period), income and expenditure must be matched. For example, any rent paid in advance beyond the current accounting period should not be included in the profit or loss calculation. Similarly, goods sold but not yet paid for in a given accounting period should not be included in the revenue total for that period.

An income statement such as the one shown in Table 25-15 is used to obtain the profit or loss for a given period. The debit and credit balances of all the accounts that do not represent expenditure or income for a given accounting period are entered as assets and liabilities in a balance sheet such as that shown in Table 25-16.

There is no rigid format for either the income statement or the balance sheet. Tables 25-15 and 25-16 show common layouts for the income statement and balance sheet respectively, but these are not the only forms. For example, vertical balance sheets, with the assets listed above the liabilities and equity, are also popular.

Some expenditures are partly capital and partly revenue. For example, repair and improvement work may be done on a plant simultaneously. In this case, the repair work should be classified as revenue expenditure and the plant-improvement work as capital expenditure.

Accounting Concepts and Conventions Accounting is based on the following concepts: (1) money measurement, (2) business entity, (3) going concern, (4) cost, and (5) matching.

TABLE 25-15 Income Statement for ABC Company

Revenue		
Sales revenue	$1,900,000	
Other revenue	100,000	
		$2,000,000
Expenses		
Raw materials	953,000	
Wages	185,000	
Utilities	44,000	
Depreciation	68,000	
Other expenses	376,000	
Income taxes	194,000	
		1,820,000
Net profit (after tax)		$ 180,000

Concept 1. "Money measurement" means that only those facts that can be represented in monetary terms are recorded. The balance sheet and income statement for a company give no indication as to what might happen in the future. The company may be about to be successfully sued for a large sum of money, or a competitor may be launching a new product that will seriously reduce future sales of the company's products.

Concept 2. "Business entity" means that accounts are kept for the company quite independently of the people who may own the company. For example, if an individual puts an additional $10,000 into a one-person business, the accounts show that the business is $10,000 richer. They do not show that the individual's personal wealth has been depleted by $10,000.

Concept 3. "Going concern" means that the accounting is based on the premise that the business will continue indefinitely. It is most unlikely that the values of the assets shown in the balance sheet are what the assets would realize if sold. No attempt is made in normal accounting to measure the value of the business to a potential buyer.

Concept 4. "Cost" means that the assets are normally shown in the balance sheet at cost price together with their subsequent depreciation. Some assets such as land may be considerably more valuable than when originally purchased, but no indication of this is given in the balance sheet. However, some governments now require a note giving the current estimated value of the land.

Concept 5. "Matching" means that the revenue in a given accounting period should correspond to the expenses for that accounting period.

Accounting is also based on the following conventions: (1) materiality, (2) conservatism, or prudence, and (3) consistency.

Materiality deals with determining whether certain expenditures will have a significant effect on a company's accounting procedures. This is a matter of judgment that is to be made by each company. Obviously, the purchase of a vehicle is a material item, but writing paper or tools for maintenance are less obvious. Although such items may last well beyond the current accounting period, it may not be worth the accounting effort to treat them as material items. Some companies will treat a particular item as capital; other companies, as expenditure. Clearly, the purchase of a piece of equipment costing, say, $1000, will be regarded as less material by a giant company than by a small one.

Conservatism, or prudence, means monetary values that tend to understate rather than overstate the profit are taken.

Consistency means that accounting items are normally treated in the same way over an indefinite number of years. For example, an individual item would not be treated as an expenditure during one year and as a capital item during the next year without good reason being given.

Balance Sheet The balance sheet, also called the position statement, presents an accounting view of the financial status of a company at a particular point in time. A typical balance sheet is shown in Table 25-16. Although a balance sheet has two sides that balance, it is not part of the double-entry system. In fact, it is not an account but rather a statement listing all the assets of a company and the various claims against these assets on the last day of the accounting period. The assets must be equal to the claims against them at all times. Those who have claims against the assets are the owners (stockholders in a business corporation) and the people to whom the company owes money. In the case of the latter, the company is said to have liabilities to its creditors. The total claim against the assets is often labeled "liabilities and owners' equity."

Assets are classified as current or fixed, and liabilities as current or long-term. Fixed assets are material items that have a relatively long life and normally include land, buildings, plant, vehicles, etc. They are held for the specific purpose of earning revenue and are not for sale in the normal course of business. Current assets include cash and those items that can be fairly easily converted into cash, such as raw-materials inventories, etc. In contrast to fixed assets, current assets are acquired for the specific purpose of conversion into cash in the normal course of business. However, what is regarded as a fixed asset by one type of company might be regarded as a current asset by another. For example, a chemical company would normally classify its vehicles as a fixed asset. However, a company whose primary business was to sell vehicles would classify them as a current asset.

Similarly, the distinction between current and long-term liabilities is also not clear-cut. Current liabilities include accounts payable

TABLE 25-16 Balance Sheet for XYZ Company

Assets (thousands of dollars)		Liabilities and stockholders' equity (thousands of dollars)	
Current assets		**Current liabilities**	
Cash	$ 38,893	Notes payable	$ 34,507
Notes and accounts receivable	110,740	Accounts payable and accrued liabilities	106,433
Inventories:		Accrued taxes	7,264
Finished products	17,396	Total current liabilities	148,204
Work in process	56,690		
Raw materials and supplies (at cost)	35,790	**Long-term liabilities**	67,677
Total inventories	109,876	Deferred income taxes	13,225
		Other deferred credits	2,307
Total current assets	259,509		
		Stockholders' equity	
Investments and long-term receivables (at cost)	94,009	Common stock, $20 par value	
		Shares authorized, 7,750,000	
Property, plant, and equipment (at cost)		Shares issued, 4,794,450	95,889
Land	6,110	Capital in excess of par value of common stock	31,798
Buildings	63,848	Retained earnings	101,492
Machinery and equipment	106,185	Total stockholders' equity	229,179
	176,143		
Less accumulated depreciation	75,163		
Net property, plant and equipment	100,980		
Prepaid and deferred charges	6,094		
Total assets	$460,592	Total liabilities and stockholders' equity	$460,592

TABLE 25-17 Provision for Depreciation of Plant-Equipment Account

1978			
		Jan. 1 Balance brought down	0
Dec. 31 Balance carried down	$100,000	Dec. 31 Debited to income statement	$100,000
1979			
		Jan. 1 Balance brought down	$100,000
Dec. 31 Balance carried down	$200,000	Dec. 31 Debited to income statement	100,000
	$200,000		$200,000
1980			
		Jan. 1 Balance brought down	$200,000
Dec. 31 Balance carried down	$300,000	Dec. 31 Debited to income statement	100,000
	$300,000		$300,000

(money owed to creditors), taxes payable, dividends payable, etc., if due within a year. Long-term liabilities include deferred income taxes, bonds, notes, etc., that do not have to be paid within a year. The owners' equity includes the par, or face, value of the capital received from stockholders and any retained earnings. The balance sheet shows only the nominal value and not the current or real value of this capital.

A balance sheet includes items that are not regarded as assets or liabilities in normal language, such as expenditures carried forward and accumulated profits.

Accountants regard assets as resources that have not yet been used up. Assets are normally shown on the balance sheet at cost minus accumulated depreciation. In this sense, the depreciation charge for an accounting period is the means of converting a part of an asset into a current expenditure that is then listed as an expense in the income statement.

Let us consider plant equipment costing $1 million and purchased on Jan. 1, 1978. Table 25-17 shows the provision for the depreciation account for 1978, 1979, and 1980 for straight-line depreciation, assuming a service life of 10 years and zero scrap value. The credit entries of $100,000 for the depreciation in each year are balanced by the depreciation charge of $100,000 debited to the income statement (or trading and profit-and-loss account) in each year. Table 25-18 shows the corresponding entries in the balance sheets for the years 1978, 1979, and 1980. Entries for subsequent years are made in the same way.

A balance sheet is true only for one particular point in time; it tells nothing about the trends in a company. However, by comparing balance sheets for successive years, management can follow changes in the various items. If the observed trend is undesirable, management can take corrective action. Since the accounting period of 1 year is long for most businesses, it is usual to draw up balance sheets at more frequent intervals for control purposes. These may be less formal than those issued annually to the stockholders. In general, balance sheets are less useful to management than are income statements.

Income Statement Income statements range from the very simple presentation shown in Table 25-15 to the more informative and more complex presentation shown in Table 25-19. The income statement shows the revenue and the corresponding expenses that were incurred to earn that revenue over a period of time. It is the most obvious measure of the efficiency of a business. Although published income statements are normally for 1-year periods, many companies use monthly income statements for internal purposes.

TABLE 25-18 Balance-Sheet Entries

As of Dec. 31, 1978		
Plant equipment at cost	$1,000,000	
Less depreciation to date	100,000	
		$900,000
As of Dec. 31, 1979		
Plant equipment at cost	$1,000,000	
Less depreciation to date	200,000	
		$800,000
As of Dec. 31, 1980		
Plant equipment at cost	$1,000,000	
Less depreciation to date	300,000	
		$700,000

Income statements are very useful tools to assist management in controlling a business and planning for the future. Since management needs to follow the trends of the normal expenses, extraordinary expenses such as those incurred as a result of a major fire or flood should be shown separately.

If revenue and expenses are not properly matched, an understatement or an overstatement of profit may occur. If raw materials were previously purchased at a lower cost than their current cost, profit will be overstated. Any overstatement of profit will mean that more tax will be paid.

One of the most important items in an income statement is depreciation expense. Although depreciation should not be thought of as a means to build up a fund to replace plant, it nevertheless does enable money to be retained in the business by reducing the profit available for distribution to stockholders. It is of course a duty of both accountants and management to see that sufficient money is retained in the business to replace assets and to invest such money in other processes or outside investment.

A further duty of accountants and management is to ensure that the company always has sufficient working capital to enable it to carry on its business.

Types of Accountancy The traditional work of accountants has been to prepare balance sheets and income statements. Nowadays, accountants are becoming increasingly concerned with forward planning. Modern accountancy can roughly be divided into two branches, financial accountancy and management or cost accountancy.

Financial accountancy is concerned with stewardship. This involves the preparation of balance sheets and income statements that represent the interest of stockholders and are consistent with existing legal requirements. Taxation is an important element of financial accounting.

Management accounting is concerned with decision making and control. This is the branch of accountancy closest to the interest of most process engineers. Management accounting is concerned with standard costing, budgetary control, and investment decisions.

Accounting statements present only facts that can be expressed in financial terms. They do not indicate whether a company is developing new products that will ensure a sound business future. A company may have impressive current financial statements and yet be heading for bankruptcy in a few years' time if provision is not being made for the introduction of sufficient new products or services.

Financing Assets by Equity and Debt

Financial Ratios Probably the most commonly mentioned ratio is the profit margin (PM), defined as

$$(\text{PM}) = \frac{\text{net annual profit}}{\text{revenue from annual sales}}\,100 \qquad (25\text{-}127)$$

TABLE 25-19 Income Statement for a Mature Year for a New Chemical Product, Produced at 10 Million lb/Year

			Unit values, cents/lb	% sales revenue, %
Revenue from annual sales A_S		$2,000,000	20.00	100.0
Direct manufacturing expense A_{DME}				
Raw materials	$ 884,000		8.84	44.2
Catalysts and solvents	69,000		0.69	3.4
Operating labor	102,000		1.02	5.1
Operating supervision	20,000		0.20	1.0
Utilities	22,000		0.22	1.1
Operating maintenance	21,000		0.21	1.1
Operating supplies	4,000		0.04	0.2
Royalties and patents	10,000		0.10	0.5
Total A_{DME}	$1,132,000	$1,132,000	11.32	56.6
Indirect manufacturing expense A_{IME}				
Payroll overhead	28,000		0.28	1.4
Central laboratory	10,000		0.10	0.5
General plant overhead	52,000		0.52	2.6
Packaging and storage	22,000		0.22	1.1
Property taxes	14,000		0.14	0.7
Insurance	6,000		0.06	0.3
Total A_{IME}	$ 132,000	$ 132,000	1.32	6.6
Total manufacturing expense (excluding depreciation) A_{ME}		$1,264,000	12.64	63.2
Depreciation A_{BD}		68,000	0.68	3.4
Other expenses				
Administration	74,000		0.74	3.7
Sales and shipping	124,000		1.24	6.2
Advertising and marketing	40,000		0.40	2.0
Technical service	10,000		0.10	0.5
Research and development	60,000		0.60	3.0
Total other expenses	308,000	308,000	3.08	15.4
Total expense A_{TE}		$1,640,000	16.40	82.0
Net annual profit A_{NP}		$360,000	3.60	18.0

Another common ratio is the return on investment (ROI), defined as

$$(\mathrm{ROI}) = \frac{\text{net annual profit}}{\text{investment}} 100 \qquad (25\text{-}128)$$

In both Eq. (25-127) and Eq. (25-128), the net annual profit can be either before or after tax. It can also include interest and dividends receivable, etc.

Obviously, the net annual profit must be clearly defined before comparisons are made with other companies. Similarly the term "investment" in Eq. (25-128) can have a variety of meanings. The two most common ones (used when assessing the profitability of companies as opposed to projects) are total assets and owners' equity or capital employed. In the first case, Eq. (25-128) can be written as

$$(\mathrm{ROA}) = \frac{\text{net annual profit}}{\text{total assets}} 100 \qquad (25\text{-}129)$$

where (ROA) is called the return on assets. In the second case, Eq. (25-128) can be written as

$$(\mathrm{ROE}) = \frac{\text{net annual profit}}{\text{stockholders' equity}} 100 \qquad (25\text{-}130)$$

where (ROE) is the return on equity.

Asset-turnover ratio (ATR) is a commonly used measure of company performance, defined as

$$(\mathrm{ATR}) = \frac{\text{revenue from annual sales}}{\text{total assets}} 100 \qquad (25\text{-}131)$$

A comparison between Eqs. (25-127), (25-129), and (25-131) shows that

$$(\mathrm{ROA}) = (\mathrm{ATR})(\mathrm{PM}) \qquad (25\text{-}132)$$

Thus (ROA) can be improved by increasing either (ATR) or (PM). A variation of Eq. (25-131) is the fixed-asset turnover ratio (FATR), defined as

$$(\mathrm{FATR}) = \frac{\text{revenue from annual sales}}{\text{fixed assets}} 100 \qquad (25\text{-}133)$$

Clearly, (FATR) is of less value than (ATR) when applied to companies that use relatively large amounts of working capital. The (FATR) is the inverse of the capital ratio (CR) for single projects. (CR) is defined as

$$(\mathrm{CR}) = C_{FC}/A_S \qquad (25\text{-}134)$$

where C_{FC} is the fixed-capital cost for a green-fields (grass-roots) site and A_S is the revenue from annual sales.

The fixed assets in Eq. (25-133) and those included in the total assets in Eqs. (25-129) and (25-131) are usually taken at their written-down, or book, value, which may differ significantly from their market value. This is one disadvantage in using Eqs. (25-129), (25-131), and (25-133).

The revenue from annual sales referred to in Eqs. (25-127), (25-131), and (25-132) is normally taken to be the gross turnover, which includes intergroup sales. However, intergroup sales are eliminated in consolidated or group accounts. Again, revenue from annual sales must be clearly defined before comparisons are made with other companies.

Let us consider the simplified balance-sheet or position statement shown in Table 25-20. Essentially, total assets are related to liabilities and stockholders' equity by

$$\text{Total assets} = \text{stockholders' equity} + \text{total debt} \qquad (25\text{-}135)$$

Equation (25-135) can also be written as

$$\text{Stockholders' equity} = \text{total assets} - \text{total debt} \qquad (25\text{-}136)$$

Equations (25-130) and (25-136) can be combined to give

$$(\mathrm{ROE}) = \frac{\text{net annual profit}}{\text{total assets} - \text{total debt}} 100 \qquad (25\text{-}137)$$

Equation (25-137) can also be written to include a quantity called the debt ratio (DR), which gives

$$(\mathrm{ROE}) = \frac{\text{net annual profit}}{\text{total assets}} \left[\frac{100}{1 - (\mathrm{DR})}\right] \qquad (25\text{-}138)$$

where (DR) is the debt ratio as given by

$$(\mathrm{DR}) = \frac{\text{total debt}}{\text{total assets}} \qquad (25\text{-}139)$$

Return on assets (ROA) can be related to the return on equity (ROE) by combining Eqs. (25-129) and (25-138):

$$(\mathrm{ROA}) = (\mathrm{ROE})/[1 - (\mathrm{DR})] \qquad (25\text{-}140)$$

TABLE 25-20 Simplified Balance Sheets for Companies X and Y

X Company balance sheet		
	Total debt	0
	Stockholders' equity	$100,000
Total assets $100,000	Total liabilities and stockholders' equity	$100,000
Y Company balance sheet		
	Total debt	$ 50,000
	Stockholders' equity	$ 50,000
Total assets $100,000	Total liabilities and stockholders' equity	$100,000

TABLE 25-21 Return on Equity after Tax for Companies X and Y

(ROA) before tax	5%	10%	15%	20%
		X company		
A_{NP}	$5,000	$10,000	$15,000	$20,000
Less tax at 50%	($2,500)	($ 5,000)	($ 7,500)	($10,000)
A_{NNP}	$2,500	$ 5,000	$ 7,500	$10,000
(ROE) after tax	2½%	5%	7½%	10%
		Y company		
A_{NP} before interest	$5,000	$10,000	$15,000	$20,000
Less interest	($5,000)	($ 5,000)	($ 5,000)	($ 5,000)
A_{NP} after interest	0	$ 5,000	$10,000	$15,000
Less tax at 50%	0	($ 2,500)	($ 5,000)	($ 7,500)
A_{NNP}	0	$ 2,500	$ 5,000	$ 7,500
(ROE) after tax	0	5%	10%	15%

(ROE) can also be related to the asset-turnover ratio (ATR) and the profit margin (PM) by combining Eqs. (25-132) and (25-140):

$$(\text{ROE}) = [(\text{ATR})(\text{PM})]/[1 - (\text{DR})] \qquad (25\text{-}141)$$

Financing by Debt, or Leverage The debt ratio (DR) is also known as the leverage, or gearing, ratio. Highly levered companies have a high proportion of debt to total assets. At first glance, it may appear that the use of leverage is a simple way of increasing the return on equity (ROE). However, interest charges have to be paid on the debt. Whether leverage is a good thing or not will depend on exactly what the interest charges are in relation to the return on assets and the return on equity.

Let us consider the simplified balance sheets of two companies, X and Y, shown in Table 25-20. Companies X and Y have a debt, or leverage, ratio of zero and 0.5 respectively. Let us assume that the debt is of the debenture type for tax purposes and that the interest rate is 10 percent per annum. The return on equity (ROE) after tax is given in Table 25-21 for companies X and Y for various values of net annual profit A_{NP} before tax. A_{NNP} is the net annual profit after tax. The data for Table 25-21 are plotted in Fig. 25-38. This figure shows that leverage has no effect on the (ROE) when the interest rate charged for the borrowed money is equal to the return on assets (ROA) before tax. Leverage provides increased (ROE) values when the (ROA) is greater than the interest rate charged for the borrowed money and decreased (ROE) values when it is less.

The greater the debt, or leverage, ratio (DR), the more sensitive the (ROE) is to a change in (ROA) and the steeper the slope of the line in Fig. 25-38. Dividends to stockholders are paid out of the net annual profit after tax A_{NNP}, from which the (ROE) after tax in Fig. 25-38 is calculated. Thus, the higher the leverage, the greater the financial risk to the stockholder. This risk is not the same as the business risk of the company, which is a function of its overall prospects in its particular industry. Leverage increases the return to the stockholders when the (ROA) is higher than the interest rate on debt and decreases the return when the (ROA) is lower than the interest rate.

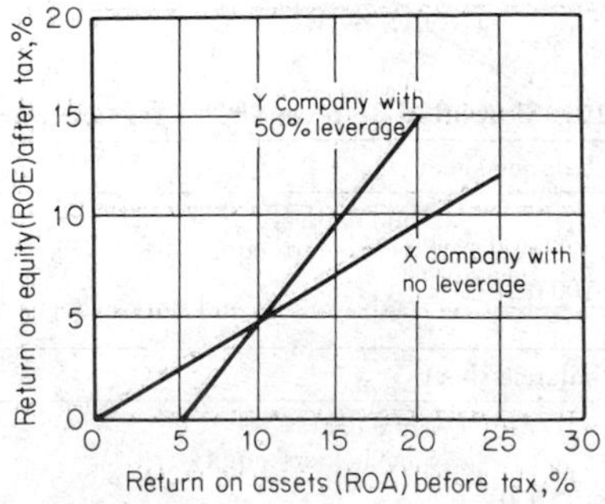

FIG. 25-38 Effect of leverage on the return on equity.

Whether the assets of a company are financed largely by stockholders' equity (also called net worth), or largely by debt, or by some combination of the two depends on a number of factors. If sales do not fluctuate, a company is in a good position to pay the fixed interest charges on debt. This is also the case if the revenue from sales is steadily increasing. In this case, any new common stock issued by the company is likely to command a good price, and it also increases the attractiveness of equity financing.

The attitude of management is also an important factor in determining how much debt financing is used. In a small firm in which management owns most of the equity, management may be very reluctant to issue further amounts of common stock that would lead to a dilution of its control. Furthermore, if management has great confidence in future prospects, it will wish to ensure the maximum return for itself. In contrast, the equity in a large company is widely distributed, and the issue of further amounts of common stock has little effect on the control of the company.

The difference between equity financing and debt financing is not always clear-cut. For example, preferred stock can be classified as stockholders' equity or debt, depending on who is doing the financial analysis.

Equity Financing Typically, the company balance sheet will show the stockholders' equity and list the preferred stock, common stock, and retained earnings as in Table 25-22.

The issue of common stock is the basic method of financing a company. Common stockholders take the ultimate risk in a business because they have no right to a return on their investment. However, they have the right to elect the directors of the company, who in turn are responsible for the management of the business. Stockholders are likely to vote the board of directors out if adequate dividends are not paid. Usually the liability of stockholders is limited to the nominal, or par, value of their stock, and hence they can lose only what they have already paid for the stock. If the liability is not limited by law, the personal assets of the stockholders are at risk in the event of company bankruptcy, in proportion to the amount of stock held.

Preferred stock is often used as an alternative to debt when companies do not wish to issue additional common stock or to incur the fixed interest charges required to finance debt. Preferred stockholders are not normally allowed to vote for the board of directors. They have the right to receive fixed amounts of dividends before common stockholders are paid any dividends. However, a company does not have to pay dividends. The board of directors may decide to pay small or no dividends in a particular year. Holders of cumulative preferred stock are entitled to receive compensation for the previous underpayment of dividends when the company again pays dividends.

Common stockholders have a right to the residual assets of a company in the event of dissolution or liquidation but only after all the

TABLE 25-22 Stockholders' Equity as Shown in Section of a Company's Balance Sheet

Preferred stock, par value $100 per share			
Authorized 2000 issued and outstanding	1,500	$ 150,000	
Less discount on preferred stock		(10,000)	
Preferred-stock equity			$ 140,000
Common stock, par value $10 per share			
Authorized and issued 100,000 shares	$1,000,000		
Amount paid in excess of par	100,000		
		$1,100,000	
Retained earnings		100,000	
Common-stock equity			$1,200,000
Total stockholders' equity			$1,340,000

creditors and then any liabilities to the preferred stockholders have been paid. The larger the proportion of debt financing in a company, the smaller the amount the common stockholders are likely to receive if the company is liquidated.

Common stockholders normally have a preemptive right to the first option to purchase any additional issues of common stock. This prevents management from using an additional issue of common stock to override the control exercised by existing stockholders. Preemptive rights also protect existing stockholders from having the value of their shares decreased by such dilution, since the same net earnings would be spread over more units of stock.

Let us consider the very simplified case of a company with 100,000 shares of common stock, each with a market value of \$10, giving a total market value of \$1,000,000. If a further 50,000 shares are sold at \$4 each, the total market value of the 150,000 shares is \$1,200,000, or \$8 each. This means that the new stockholders have gained at the expense of the original ones. The preemptive right is designed to prevent this. In practice, the situation is rather more complex than is indicated here.

Both common and preferred stocks normally have a par, or nominal, value. In the case of common stock, the market value at the time of issue usually differs from the par value. Stock can be issued either at a premium or at a discount, depending on prevailing economic conditions and the strength of the company. The difference between the actual amount paid and the par value is listed in the stockholders'-equity section of the balance sheet, as shown in Table 25-22. The issuance of stock at a premium or a discount is done to protect existing stockholders.

In the case of preferred stock, the par value has more meaning than with common stock, since it is the amount due preferred stockholders if the company goes into liquidation, provided that this is a condition of issue.

The advantage of using common stock to finance assets is that it does not incur fixed interest charges. Furthermore, there is no maturity date, as there is with all loans and most preference issues. Common stock can often be issued more easily than debt can be financed. However, the flotation costs of common stock can be quite high, especially when stock values are depressed, so that large discounts for the stock are needed to induce purchase.

Stockholders' equity in a company is made up of the capital contributed by the stockholders and the capital generated from retained earnings. The presence of retained earnings on a balance sheet, as shown in Table 25-22, does not necessarily mean that they are matched by an equal amount of cash. In fact, there may be little or no cash available. The retained earnings shown on a balance sheet may be largely fictitious. For example, the assets on a balance sheet may be worth less than shown by at least the value of the retained earnings.

Purchase and Sale of Equities Stockholders usually require an adequate return on their investment, and the quoted price of the stock reflects the consensus opinion of investors as to the current health of the company. Purchases or sales are normally made through stockbrokers.

Most stock transactions are completed through organized security exchanges on which the stock is listed. Such exchanges have physical existence in the form of buildings located in different regions of the country. Each exchange has members who are often the nominated representatives of large brokerage firms having offices in various cities. These offices are in constant telephone and telegraph communication with the members at the exchange, passing on requests to buy or sell specified stocks. Since brokers live by commissions and charges on transactions, they attempt to match such requests either directly or by dealings with other brokers. In the United Kingdom, brokers must deal through an independent "jobber," similar in function to a specialist broker, who quotes a low price for sales and a higher one for purchases before the jobber knows whether the broker is buying or selling. The difference represents the jobber's margin, or "turn." If requests to buy exceed offers for sale, the price of the stock rises until someone is tempted to sell. Conversely, if an excess of stock is offered for sale, the price is likely to fall.

It is an advantage to a company to be listed on a stock exchange since its investors can more easily sell their stock if they decide to do so. This increased liquidity makes investors more willing to accept a lower rate of return, which effectively lowers the cost of capital to the company.

Because dealings in the stock of a listed company are published, a healthy company engenders confidence that makes it easier to obtain other forms of finance. In the absence of a regular market, stock transactions are necessarily infrequent, and prices are liable to wide fluctuation, which may make creditors wary and possibly lead to bankruptcy proceedings. Such dealings are usually referred to as "over-the-counter" and are confined to the relatively few specialist brokers who hold inventories of such stock and are prepared to "make a market" in them or are limited to private transactions.

Retained Earnings Much confusion is caused by the practice of dividing retained earnings under various headings such as reserve for replacement of plant, reserve for contingencies, etc. This procedure also restricts the flexibility of management in expenditure decisions.

The amount of retained earnings shown on a balance sheet should not be taken as a measure of the amount of future dividends that the company is likely to pay. A contract may exist that specifies a minimum balance of retained earnings, which is then not available for dividends until bonds issued by the company have been retired.

Dividends can be paid either as cash or in the form of an additional issue of stock. A stock split is really a stock dividend, and both are used to reduce the price of stock when management considers that it is too high. A stock dividend is essentially a transfer of retained earnings to the common-stock account and makes the amount transferred unavailable for future dividends. A stock dividend may be used in place of a cash dividend when a company is short of cash.

Debt Financing In practice, debt financing covers a variety of fixed-income securities, both long-term and short-term. The most common forms of long-term debt are bonds, mortgages, and debentures.

A bond is simply a long-term promissory note. It is a contract established between borrower and lender in a document called an indenture. A bond indenture includes a detailed description of assets that are pledged, together with any protective clauses and provisions for redemption. A trustee is appointed to look after the interest of the bondholders. The trustee is normally a commercial bank. Bonds may be issued with a call provision that enables a company to redeem its bonds at any date earlier than scheduled. Obviously, this would be an advantage to a company in times of falling interest rates. However, a company has to pay more than the par value of the bond for this privilege. The additional amount is called the bond premium.

Sometimes a company uses a sinking fund to retire a bond. A series of equal annual payments A, invested at a fractional interest rate i and made at the end of each year over a period of n years, is equivalent to a sum of money of present value P, given by

$$A = Pf_{AP} \tag{25-46}$$

where f_{AP} is the annuity present-worth factor, which is

$$f_{AP} = [i(1 + i)^n]/[(1 + i)^n - 1]$$

A company may use a sinking fund in a variety of ways, but the simplest is to pay a fixed amount A at the end of each year to buy and retire bonds until after n years all the bonds have been retired. This annual payment may prove a significant strain on the resources of a company. Failure to make the payment could result in bankruptcy. In the case of income bonds, a company is required only to pay interest when it earns it.

A mortgage is a bond in which specific real assets are pledged as security. A senior mortgage has a prior claim on assets. A junior mortgage is normally a second mortgage on the residual value of the assets. A blanket mortgage is a pledge on all real property owned by a company.

A debenture is an unsecured bond. Strong companies are in a better position to issue debentures than weak companies since they have less need to pledge specific assets. Debenture holders are really general creditors. Subordinated debenture holders have claims on assets only after the claims of certain other claimants have been met. The issue of subordinated debentures provides a tax advantage for a company compared with the issue of preferred stock because the interest payable is a tax-deductible expense.

A financial analyst looking at a company from a potential common stockholder's point of view is likely to classify preferred stock as debt. In contrast, bondholders and general creditors are likely to regard preferred stock as additional equity. Since preferred stock is a hybrid type of security, it may be issued by a company whose management is divided over the question of whether to use equity or debt to finance additional assets. However, preferred stock does have the disadvantage that the dividends are not allowed as a tax-deductible expense.

Comparative Company Data Table 25-23 gives comparative company data that have been compiled by Dun & Bradstreet for various types of processing industries. The median value for each ratio is given together with the lower- and upper-quartile values respectively in parentheses.

Row 1 in Table 25-23 is the

$$\text{Current ratio} = \frac{\text{current assets}}{\text{current liabilities}} \qquad (25\text{-}142)$$

Compare

$$\text{Quick ratio} = \frac{\text{liquid assets}}{\text{current liabilities}} \qquad (25\text{-}143)$$

Row 2 in Table 25-23 is the profit margin (PM) of Eq. (25-127). In this case, the net profit referred to is the net annual profit after tax and depreciation A_{NNP}. The net sales is the revenue from annual sales A_S after deductions for returns, allowances, and discounts for gross sales.

Row 3 in Table 25-23 is the return on equity (ROE) of Eq. (25-130). In this case, the net worth is the tangible net worth representing the sum of the preferred and common stocks and the surplus and undistributed profits or retained earnings, less any intangible items such as goodwill, etc.

Row 7 in Table (25-23) is the

Average collection period

$$= \frac{\text{average value of accounts receivable}}{\text{revenue from sales per day}} \qquad (25\text{-}144)$$

The funded debt (referred to in row 14) consists of mortgages, bonds, debentures, serial notes, or other obligations with maturity of more than 1 year from the statement date.

Robert Morris Associates also compiles extensive comparative company data for various industries. In addition to ratios similar to the Dun & Bradstreet ratios shown in Table 25-23, Robert Morris Associates gives very useful breakdowns of assets and liabilities for various industries. Table 25-24 shows a breakdown of assets and lia-

TABLE 25-23 Comparative Ratios for Selected United States Industry Groups for 1979*

Row	Industry / Ratio	Agricultural chemicals	Chemicals: alkalies and chlorines	Paints and allied products	Petroleum refining	Plastic materials and resins	Soap and other detergents
	Number of companies	43	5	196	45	91	56
1	Current assets/Current debt, ratio	1.47 (1.01 2.43)	2.62 (1.05 9.91)	2.40 (1.64 3.76)	1.43 (1.10 2.18)	2.07 (1.47 2.90)	2.71 (1.97 5.42)
2	Net profit × 100/Net sales, %	1.45 (.54 3.76)	2.55 (1.90 3.21)	3.90 (2.10 6.64)	2.12 (1.54 3.43)	4.44 (1.91 7.63)	3.37 (2.61 6.47)
3	Net profit × 100/Net worth, %	10.91 (2.05 27.05)	14.76 (10.43 19.09)	20.43 (9.73 35.29)	15.88 (10.54 41.00)	16.78 (11.33 26.17)	17.91 (12.41 34.16)
4	Net profit × 100/Net working capital, %	6.92 (3.64 58.31)	54.09 (11.68 96.50)	22.76 (12.05 41.46)	43.06 (14.89 69.31)	27.48 (14.06 45.18)	23.27 (13.10 42.48)
5	Net sales/Net worth, ratio	4.61 (2.10 8.51)	3.64 (2.10 7.03)	4.63 (3.44 7.17)	6.31 (3.54 10.87)	4.43 (2.78 7.86)	4.48 (3.32 6.97)
6	Net sales/Net working capital, ratio	5.10 (3.17 12.91)	3.66 (2.66 27.20)	5.59 (4.21 8.98)	16.91 (4.77 21.60)	7.00 (3.58 10.93)	6.29 (4.28 8.28)
7	Collection period, days	49 (27 91)	58 (45 63)	40 (26 53)	33 (15 51)	48 (36 60)	33 (22 51)
8	Net sales/Inventory, ratio	7.3 (3.8 16.5)	16.1 (8.2 42.7)	8.3 (6.2 13.00)	18.4 (8.7 34.8)	10.6 (8.1 18.4)	8.9 (7.7 13.1)
9	Fixed assets × 100/Net worth, %	74.1 (26.7 98.5)	58.0 (25.7 79.2)	36.1 (19.6 55.00)	91.1 (48.2 147.5)	50.6 (26.8 116.0)	30.5 (19.2 56.4)
10	Current debt × 100/Net worth, %	90.2 (39.2 161.0)	40.8 (15.4 108.1)	55.1 (28.1 98.7)	101.6 (45.6 187.2)	54.3 (27.8 119.2)	46.8 (22.2 79.4)
11	Total debt × 100/Net worth, %	104.8 (46.7 224.1)	84.7 (24.2 148.0)	79.6 (38.3 150.2)	137.8 (75.1 335.6)	89.5 (37.8 194.5)	55.6 (26.5 111.2)
12	Inventory × 100/Net working capital, %	81.1 (30.5 139.4)	17.8 (4992.6 75.0)	69.6 (45.4 91.3)	57.5 (14.2 104.0)	53.0 (23.4 81.7)	58.9 (42.9 88.9)
13	Current debt × 100/Inventory, %	140.4 (78.3 311.8)	137.8 (67.3 806.5)	94.9 (56.9 154.5)	240.8 (156.4 503.0)	154.3 (97.5 277.9)	96.5 (41.7 144.0)
14	Funded debt × 100/Net working capital, %	44.7 (48.6 95.6)	32.3 (4991.9 63.5)	40.6 (16.8 66.7)	75.5 (24.1 168.7)	37.4 (9.9 102.0)	25.4 (14.4 64.8)

*Reprinted with the special permission of *Dun's Review*, October 1979, copyright 1979, Dun & Bradstreet Publications Corporation.

NOTE: Numbers in parentheses are lower- and upper-quartile values respectively. Numbers above are median value.

TABLE 25-24 Balance Sheet for United States Manufacturers of Industrial Inorganic Chemicals, 1972*

Assets	%	Liabilities	%
Cash	4.3	Short-term due to banks	5.6
Marketable securities	2.2	Due to trade	14.5
Net receivables	26.2	Income taxes	3.1
Net inventory	24.6	Current maturities long-term debt	2.1
All other current assets	0.9	All other current liabilities	5.0
Total current assets	58.3	Total current debt	30.3
Fixed assets	36.3		
All other noncurrent assets	5.4	Noncurrent debt unsubordinated	18.3
		Total unsubordinated debt	48.6
		Subordinated debt	3.1
		Tangible net worth	48.3
Total assets	100.0	Total liabilities and stockholders' equity	100.0

*Abridged from *Annual Statement Studies*, 1973 ed., copyright 1973 by Robert Morris Associates, Philadelphia. The composite figures for each industry shown in the RMA studies may not be representative of that entire industry, and they should not be automatically considered as representative norms.

bilities for United States manufacturers of industrial inorganic chemicals for 1972.

Application of Overall Company Ratios The various ratios for a hypothetical company are listed in Table 25-25. The balance sheet shown in Table 25-26 has been built up from the ratios in Table 25-25 in terms of the revenue from net annual sales A_S.

Let us calculate the following values for the right-hand side of the balance sheet as follows:

TABLE 25-25 Ratios for a Typical Industrial Chemical Company

No.	Ratio
1	$\dfrac{\text{Current assets}}{\text{Current debt}} = 2.60$
2	$\left(\dfrac{\text{Net profit}}{\text{Net sales}}\right) 100 = 4.00$
3	$\left(\dfrac{\text{Net profit}}{\text{Net worth}}\right) 100 = 10.0$
4	$\left(\dfrac{\text{Net profit}}{\text{Net working capital}}\right) 100 = 18.18$
5	$\dfrac{\text{Net sales}}{\text{Net worth}} = 2.50$
6	$\dfrac{\text{Net sales}}{\text{Net working capital}} = 4.50$
7	$\text{Collection period} = \dfrac{\text{accounts receivable}}{\text{sales per day}} = 61 \text{ days}$
8	$\dfrac{\text{Net sales}}{\text{Inventory}} = 7.14$
9	$\left(\dfrac{\text{Fixed assets}}{\text{Net worth}}\right) 100 = 74.00$
10	$\left(\dfrac{\text{Current debt}}{\text{Net worth}}\right) 100 = 35.00$
11	$\left(\dfrac{\text{Total debt}}{\text{Net worth}}\right) 100 = 65.00$
12	$\left(\dfrac{\text{Inventory}}{\text{Net working capital}}\right) 100 = 63.00$
13	$\left(\dfrac{\text{Current debt}}{\text{Inventory}}\right) 100 = 100.00$
14	$\left(\dfrac{\text{Funded debt}}{\text{Net working capital}}\right) 100 = 76.50$

From ratio 5

$$\text{Net worth} = A_S/2.50 = 0.4\ A_S$$

From ratio 11

$$\text{Total debt} = (0.4\ A_S)(0.65) = 0.26\ A_S$$

From ratio 10

$$\text{Current debt} = (0.4\ A_S)(0.35) = 0.14\ A_S$$

$$\text{Long-term debt} = \text{total debt} - \text{current debt}$$

$$\text{Long-term debt} = 0.26\ A_S - 0.14\ A_S = 0.12\ A_S$$

We calculate the following values for the left-hand side of the balance sheet:

From ratio 9

$$\text{Fixed assets} = (0.4\ A_S)(0.74) = 0.29\ A_S$$

From ratio 1

$$\text{Current assets} = (0.14\ A_S)(2.60) = 0.36\ A_S$$

From ratio 8

$$\text{Inventory} = A_S/7.14 = 0.14\ A_S$$

From ratio 7

$$\text{Accounts receivable} = (A_S/365)(61) = 0.167\ A_S$$

$$\text{Cash and short-term investments} = \text{total current assets} - (\text{inventory} + \text{accounts receivable})$$

$$\text{Cash and short-term investments} = 0.364\ A_S - 0.307\ A_S = 0.057\ A_S$$

In addition to the data for the balance sheet, we calculate the net annual profit (after tax), i.e., ratio 2, to be $A_{NNP} = 0.04\ A_S$.

In practice, the ratios are obtained from the information published in the balance sheet. The advantage of the above presentation is that it relates everything to the revenue from net annual sales and hence underlines the importance of sales.

Careful study of the ratios can produce many inferences as to the health of the company. For example, the leverage, or debt, ratio (DR) for this example is

$$(\text{DR}) = \frac{\text{total debt}}{\text{total assets}} = \frac{0.260\ A_S}{0.660\ A_S} = 0.40$$

This value is quite low and does not present any problems of control by debtors, such as can arise when (DR) is greater than 1.

From Table 25-26 we calculate the ratio for

TABLE 25-26 Balance Sheet for a Typical Industrial Chemical Company, Dec. 31, 1981

Assets		Liabilities and stockholders' equity	
Current assets		Liabilities	
Cash	$0.057\ A_S$	Current debt	$0.140\ A_S$
Accounts receivable	$0.167\ A_S$	Long-term debt	$0.120\ A_S$
Inventory	$0.140\ A_S$		
Total current assets	$0.364\ A_S$	Total debt	$0.260\ A_S$
Fixed assets	$0.296\ A_S$	Net worth	$0.400\ A_S$
Total assets	$0.660\ A_S$	Total liabilities and stockholders' equity	$0.660\ A_S$

$$\frac{\text{Current debt}}{\text{Cash} + \text{short-term investments}} = \frac{0.140\ A_S}{0.057\ A_S} = 2.45$$

Therefore, requests for early repayment by more than 40 percent of the debtors could be met. Hence, no liquidity problems are likely to arise, and advantage can be taken of discounts for early payment. Also, the current debt could be met by sale of the inventory, which takes $(0.140\ A_S/A_S)(365)$, or 51 days. The quick ratio is $1/2.45 = 0.407$.

If it is assumed that current debtors are due for payment within 61 days, the same time as that allowed to creditors, no bankruptcy petitions are likely.

The profit of 10 percent, indicated by ratio 3 in Table 25-25, will be reduced by any dividend due to preferred stockholders, because such payments are not part of fixed-debt expenses; the residue is shared among the ordinary stockholders. If all the long-term debts were in redeemable 6 percent preferred shares, then (from ratio 3) the net annual profit (after tax) is $A_{NNP} = 0.10(0.40\ A_S)$, or $0.04\ A_S$. Interest due on preferred shares is $0.06(0.12\ A_S)$, or $0.0072\ A_S$. Therefore, the earnings for the ordinary shares are

$$(0.04\ A_S - 0.0072\ A_S)/(0.4\ A_S - 0.12\ A_S) = 0.1171$$

This value corresponds to 11.71 cents per dollar of common stockholders' equity.

If it is assumed that available interest rates offered by banks, government, etc., for no-risk investment of capital are 10 percent, then the maximum economic market price of $100 stock units in this hypothetical company is about $117. If all the debt is in bonds, etc., earnings on ordinary stock would be 10 cents per dollar of net worth, and the maximum economic price of the stock would be about $100 unless stock prices were expected to rise.

Other ratios can easily be deduced from those listed. For example, the return on assets (ROA) and the asset-turnover ratio (ATR) are

$$(\text{ROA}) = \frac{\text{net annual profit}}{\text{total assets}}\,100 = \frac{0.04\ A_S}{0.66\ A_S} = 6.06$$

$$(\text{ATR}) = \frac{\text{revenue from annual sales}}{\text{total assets}}\,100$$

$$(\text{ATR}) = (A_S/0.66\ A_S)100 = 151.5$$

The lower quartile, median, and upper quartile listed in Table 25-23 are those at which the cumulative probabilities of achieving those values are 25 percent, 50 percent, and 75 percent respectively. Since such curves become asymptotic to plus or minus infinity at 0 percent and 100 percent probability, the probability curves for each ratio can be sketched in by the methods described earlier.

Cost of Capital The value of the interest rate of return used in calculating the net present value (NPV) of a project is usually referred to as the cost of capital. It is not a constant value since it depends on the financial structure of the company, the policy of the company toward a particular project, the local method of assessing taxation, and, in some cases, the measure of risk associated with the particular project. The last-named factor is best dealt with by calculating the entrepreneur's risk allowance inherent in the project i_r' from Eq. (25-108), written in the form

$$i_r' = [1 + (\text{DCFRR})]/(1 + i) - 1$$

where i is the cost of capital exclusive of the risk allowance. The value of i_r' should be compared with the probability of exceeding or of failing to achieve an (NPV) of zero when using that value of i. The decision to proceed can then be made with a full knowledge of the odds against success. The decision can be related to the company attitude to budgets of the relevant size by the use of probable utilities, as has already been discussed. Cash flows used in calculating (NPV) and (DCFRR) should, of course, be corrected for the anticipated rates of inflation, preferably to the time when the utility curve was obtained. This is important since inflation is likely to have a distorting effect on utility curves obtained at different times. This may be due to an unconscious wish to protect against inflation by achieving higher rewards while assigning less importance to any losses incurred, thus tending toward a gambling outlook.

In the absence of a risk allowance the cost of capital becomes a technical financial computation based on sources of funds and company policy. As such it will usually be presented as a figure specified for use in a particular appraisal and is therefore of little concern to the project assessor. However, the following résumé indicates the kinds of factors to be considered.

In most companies the objective of company policy is to maximize the financial return to the equity stockholders. This is not invariably the case, since a young company will often plow back an unusually large proportion of its profits to encourage growth. Also, it is increasingly the case that projects are undertaken to restore or preserve an environmental amenity or to bring work into a particular locality. In such circumstances a low value of the cost of capital might be assigned to the project. In many government projects a limited loss is acceptable, in which case the value of i would be negative.

When the objective is to maximize the aftertax return to the stockholders, a balance must be struck between the proportion of aftertax company profits which are retained to permit growth of the company assets and the proportion which are distributed to provide an income for the stockholders in the form of dividends. The latter will usually be subject to personal income taxes, sometimes at higher-than-normal rates. The growth potential should be reflected in an increased value of the stock as quoted on the stock exchange. Such growth may result in the imposition of capital gains or inheritance taxes. The selection of the right proportion of earnings to be retained is crucial since this affects the appeal of the company to investors and hence its credit worthiness in the eyes of creditors. The optimum split is influenced by the type of investor since institutional tax rates and exemptions often differ from those applied to private investors. It is for this reason that the optimum split is sensitive to local taxation policy.

Most companies can maintain a given level of business only by continuous reinvestment in plant and equipment. If company growth is required, additional investment is essential. In general, a company has only three sources of new money, namely, cash received from the sale of newly issued shares, retained earnings, and debt capital of all kinds including deferred taxes. In certain circumstances cash grants may be forthcoming from government sources. Each of these sources has its own effective rate of interest, and it is the weighted average of these rates which constitutes the cost of capital exclusive of risk allowance.

There is no interest payable directly on equity stocks, but there is a concealed rate expected by investors. Without the expectation of a certain return on their investment they would not invest in a new issue, nor would they retain existing holdings of stock. The sale of stocks on the stock exchange does not affect the cash holding of the company, but new issues must be at prices lower than existing values quoted on the exchange unless great confidence exists that the new money will produce an increased income greatly in excess of the reduction in earnings per share caused by the new issue. Stock carrying a fixed interest rate normally has the interest treated as an allowable expense before tax in the same way as a bank overdraft, which is a relatively short-term source of debt. Deferred taxes carry an interest rate which, like an overdraft, is normally compounded daily at a nominal annual rate but naturally is not an allowable expense for tax purposes. Cash owing on outstanding bills carries a notional rate of interest since in many cases prompt settlement of bills would attract a cash discount.

Example 18 A company requires an investment of \$100,000 in new plant to maintain its present sales. Let us determine the current cost of capital to the company and the risk-free cost of capital that it should assign to the plant-replacement project, given the following data.

Company assets: from stock sales	\$ 300,000
from retained earnings	200,000
as bills due	100,000
as deferred taxes	200,000
as bank overdraft	200,000
Total assets	\$1,000,000
Current annual income	\$ 200,000

Bills are due on monthly account with a 2 percent discount for cash. Overdraft and deferred-tax interest are compounded daily at nominal annual interest rates of 15 and 9 percent respectively. Corporation tax, capital gains tax, and personal income tax rates are 50, 40, and 30 percent respectively. The current rate of inflation is at 8 percent per year. The traditional return expected by investors is 7 percent per year net of all taxes in real terms.

The interest-rate equivalent of the cash discounts is 2 percent per month, since this discount could be obtained every month if payment were to be made at the beginning of the month rather than, as at present, at its end. Since the bills are settled monthly, the notional interest is paid monthly and should not be compounded. The discount is equivalent to 12 monthly simple-interest payments per year. Hence, from Eq. (25-31) the effective annual interest rate on discounts = (12)(0.02) = 0.24 = 24 percent. It would, therefore, be a good use of surplus cash to reduce this debt as quickly as possible. This would require cash equivalent to one-sixth of the annual bills due, or \$16,700, to be available. It can, therefore, be assumed that this level of liquidity is not available for capital projects, either as working capital to reduce the debt or for fixed-capital projects. Further, since the new project will not increase sales, it cannot generate further debt of this kind. Hence, this source is not available to capitalize the new project.

Since the overdraft is payable daily at a nominal annual interest rate of 15 percent, it follows from Eq. (25-38) that the effective annual interest rate on overdraft = $(1 + 0.15/365)^{365} - 1$ = 16.18 percent. Similarly, the effective annual interest rate on deferred tax = $(1 + 0.09/365)^{365} - 1$ = 9.42 percent.

The new plant will not increase sales and will therefore not increase the tax debt, so that this source is not available to capitalize the project. An increase in overdraft may be available, subject to a maximum imposed by the acceptable gearing of the company.

Since the liquidity of the company is so low, it is possible that it is already extended to its maximum debt, in which case the gearing

$$\frac{\text{Total equity}}{\text{Total debt}} = \frac{\$300{,}000 + \$200{,}000}{\$100{,}000 + \$200{,}000 + \$200{,}000} = 1.00$$

Since neither increased bills due nor increased tax debt is available to finance the new project, this implies that the required \$100,000 of new capital will be available as \$50,000 from increased overdraft and \$50,000 from increased equity. The effective interest rate on the equity involved must therefore be calculated.

Equity is available from two sources. First, the company can sell new stock which, if in the form of ordinary shares, carries no interest payment. Although this course appears cheap, its use for projects which do not increase earnings, at least to a compensatory level, is usually inadvisable. This leaves retained earnings as the most likely source of equity for the present project.

Equity holders require a real return on their outlay, which they assume to be at the stock-market price if this differs from the face value of the stock, of 7 percent net of all taxes. Retained earnings attract a 40 percent capital gains tax; hence the actual interest rate required on distribution forgone is 7/(1 − 0.40) = 11.67 percent. This is in real terms and at a time of 8 percent inflation rate must be increased in cash terms to (1 + 0.1167)(1.08) − 1 = 20.60 percent.

In the same way the effective interest rate on distributed earnings, on which an income tax of 30 percent is payable, may be calculated to be [0.07/(1 − 0.30) + 1](1.08) − 1 = 18.80 percent. If the shares are currently valued at par by the stock exchange, this would require distribution, on the \$300,000 of issued equity, of (\$300,000)(0.188) = \$56,400. This amount is required after corporation tax has been paid on the earnings of \$200,000. Thus the earnings which can be retained are (\$200,000)(1 − 0.50) − \$56,400 = \$43,600. This is close to the \$50,000 required. If the company were well regarded on the stock exchange, a slight reduction in distributed dividends might not reduce share values since the purpose is to maintain future earnings. However, it would be possible for share values to be reduced by up to (\$56,400 − \$50,000)/\$56,400 = 0.1135 = 11.35 percent. If this happened, the effective interest rate on the retained earnings should be increased to 20.60/(1 − 0.1135) = 23.24 percent. These rates are net of corporation tax, which is the correct form for use in (NPV) calculations.

The interest rate due on deferred tax is also net of corporation tax at 9.42 percent. The interest payable on overdrafts is an expense fully allowable against tax, so that the effective aftertax rate is reduced to (16.18)(1 − 0.50) = 8.09 percent. Similarly, as the advantage forgone on the discounts would tend to increase company profits and hence tax due, the effective aftertax gain is reduced to (24)(1 − 0.50) = 12.00 percent.

The present cost of capital to the company is the weighted-average interest payable on the various sources of funds on an after-corporation-tax basis. This is readily calculable to be at least [(\$300,000)(0.00) + (\$200,000)(20.60) + (\$100,000)(12.00) + (\$200,000)(9.42) + (\$200,000)(8.09)]/(\$1,000,000) = 8.82 percent. The cost of capital to the new project, with only two sources, should be [(\$50,000)(20.60) + (\$50,000)(8.09)]/(\$100,000) = 14.35 percent.

Since this project is essential if current production is to be maintained, many companies would assess the cost of capital at somewhere near the lower value. Values of cost of capital in the region of 10 percent are to be expected in developed countries at the present time.

We notice in particular that inflation does not affect quoted interest rates when assessing present values of cost of capital. It must, however, be taken into account in assessing the interest rate on the dividend which will be expected by investors.

As has been stated, it is alternatively possible to assign to the cost of capital the best risk-free return available on the money. The assessment then proceeds as discussed in connection with Eq. (25-108).

Management and Cost Accounting In any given time period, cost may be divided into expired and unexpired cost. An expired cost is an expense; an unexpired cost is an asset. This division is the basis for income statements and balance sheets.

Cost accounting is the name traditionally given to accounting for manufacturing costs. The manufacturing cost of a product is traditionally taken as the sum of the costs for (1) direct materials, (2) direct labor, (3) manufacturing overheads, and (4) administration, selling, and finance.

Two methods are in general use in accounting for manufacturing costs, absorption costing and marginal costing. In absorption costing, which is the traditional method, all manufacturing overhead costs are included in the cost of sales. In marginal costing only variable manufacturing overhead costs are included in the cost of sales. Marginal costing is more valuable than absorption costing in decision making. However, it is sometimes quite difficult to separate costs and particularly manufacturing overhead costs into fixed and variable components. In the long term virtually all costs are variable. The difference between the two methods assumes great importance in inventory evaluation. In cost accounting, costs are identified with cost centers. These are accounting devices which may or may not have a physical existence. In the simplest case of a plant manufacturing a single product, the entire plant may be the cost center.

In practice there are two major classifications of cost accounting systems, job costing and process costing. In the former, costs are collected for each job or batch irrespective of the accounting period. This system is normally used in construction work. Process costing is normally used in continuous and semicontinuous processes. Costs are collected for a specific accounting period.

Allocation of Overheads How overheads are allocated can affect the total cost of a product and, hence, the estimated future cash flows for a project. Since these cash flows are used in the net-present-value (NPV) and discounted-cash-flow-rate-of-return (DCFRR) methods for estimating profitability, erroneous allocations could result in the wrong choice of project.

The modern trend is for overhead costs to become an increasing

TABLE 25-27 Profits of Products with Different Methods of Overhead Allocation

	Product A—basis of overhead allocation			Product B—basis of overhead allocation		
	31.25% of direct materials cost, \$/unit	125% of direct labor cost, \$/unit	25% of prime cost, \$/unit	31.25% of direct materials cost, \$/unit	125% of direct labor cost, \$/unit	25% of prime cost, \$/unit
Direct materials cost	8.00	8.00	8.00	6.000	6.000	6.000
Direct labor cost	2.00	2.00	2.00	4.000	4.000	4.000
Prime or direct cost	10.00	10.00	10.00	10.000	10.000	10.000
Overhead cost	2.50	2.50	2.50	1.875	5.000	2.500
Total cost	12.50	12.50	12.50	11.875	15.000	12.500
Selling price	14.00	14.00	14.00	14.000	14.000	14.000
Profit/(loss)	1.50	1.50	1.50	2.125	(1.000)	1.500

proportion of total product costs. This results from the ever-greater sophistication of process plants. Therefore, it is highly desirable that chemical engineers should have some say in the allocation of overheads and that this should not be left entirely to accountants.

Direct costs are those that can be directly charged to a single product. The most obvious direct cost is for raw materials, of which the quantity consumed is directly proportional to the amount of product manufactured. Direct process labor is also considered to be a direct cost.

However, many costs cannot be directly charged to an individual product. These so-called indirect, burden, or overhead costs range from the lighting and heating required for the plant and offices to the cafeteria and medical facilities provided. When several products are made in a plant, it becomes increasingly difficult to allocate overheads correctly among the various products.

A number of different methods are commonly used to estimate the amount of overhead to be allocated to an individual product. These methods are necessary because accountancy costs become prohibitive for charging all costs directly to an individual product. Unfortunately, there is always an arbitrary element inherent in the process of allocation.

Overheads in the chemical-process industries are commonly calculated as a percentage of (1) direct materials cost, (2) direct labor cost, or (3) prime or direct costs. Other methods of allocating overheads are on the basis of (1) plant area, (2) number of employees, (3) capital value, and (4) electric power.

These listings do not include all the methods in use. The validity of a particular method depends on the process and the industry. An inappropriate method can lead to misleading and even absurd results.

Let us consider the manufacture of metal ornaments. The processing cost, exclusive of material, may vary very little for a wide range of materials. However, the direct materials cost will be much greater for precious than for base metals. In this case, an overhead allocation on the basis of direct material costs could be very misleading, while one based on direct labor cost could be quite accurate.

Problems can also arise when allocating overheads on the basis of direct labor cost. Let us consider a company that evaluates overheads at 125 percent of direct labor cost. A process plant employs seven operators, each with a direct cost of \$10,000 per budget period. As a result of a works-study exercise, it is found that the plant can operate satisfactorily with six operators. The actual cost saving is likely to be far nearer to the direct labor savings of \$10,000 per period than to the calculated saving of \$10,000 + \$10,000(125/100) = \$22,500 per period. The \$22,500 calculated saving is the direct labor cost plus overheads taken as 125 percent of the direct labor cost.

A thorough analysis should be made before production is stopped on a product that is losing money. Although direct costs of the discontinued product will be saved, overheads are not eliminated, as might be inferred from taking overheads as a percentage of direct material, direct labor, or prime costs. The plant is still there, together with its associated costs for interest charges, insurance, painting, some maintenance, etc. Continued production can still make a useful contribution to such overheads. This contribution is lost if production is stopped and must then be borne by other products.

Problems can arise with each of the methods used for allocating overheads. Two process plants may occupy similar areas yet have vastly different material or labor costs. Problems can also arise with an individual plant that can be used to make different products, as Example 19 will show.

Example 19 Let us consider a plant that can make either product A or product B. At normal capacity, the overhead cost is known to be \$2.50 per unit. Product A has a direct materials cost of \$8 per unit and a direct labor cost of \$2 per unit. For simplicity, the prime cost is here taken as the sum of these two costs, i.e., \$10 per unit.

In Table 25-27, a correct overhead cost of \$2.50 per unit at normal capacity is calculated by taking either 31.25 percent of the direct materials cost, 125 percent of the direct labor cost, or 25 percent of the prime cost. All these methods give a total cost of \$12.50 per unit and a profit of \$1.50 per unit for a selling price of \$14 per unit.

The alternative, product B, has a direct materials cost of \$6 per unit and a direct labor cost of \$4 per unit. In Table 25-27 overhead costs of \$1.875 per unit, \$5 per unit, and \$2.50 per unit are calculated by taking 31.25 percent of direct materials cost, 125 percent of direct labor cost, and 25 percent of prime cost respectively. Total costs are \$11.875 per unit, \$15 per unit, and \$12.50 per unit respectively and profits of \$2.125 per unit, −\$1 per unit, and \$1.50 per unit respectively, for a selling price of \$14 per unit.

An alternative to allocating overheads by using a single method is to classify the various overheads into groups and to use the most appropriate allocation for each group. For example, depreciation would be allocated on the basis of capital cost, while indirect labor might be allocated either on the basis of direct labor cost or on the number of employees. Clearly, this alternative method is more complex, increases the associated accountancy costs, and is prone to misinterpretation and possibly abuse.

Inventory Evaluation and Control

Inventory Effect on Cash Income and Profit When the annual production rate is equal to the annual sales volume, the revenue from annual sales A_S is

$$A_S = Rc_S \tag{25-87}$$

where R is the production rate, units per year, and c_S is the sales price per unit. In this case, the annual cash income A_{CI} and the net annual profit A_{NP} before tax are given respectively from Eqs. (25-1), (25-8), and (25-12) by

$$A_{CI} = R(c_S - c_{TVE}) - A_{TFE} \tag{25-145}$$

$$A_{NP} = R(c_S - c_{TVE}) - A_{TFE} - A_{BD} \tag{25-146}$$

where c_{TVE} is the total variable expense per unit of production, A_{TFE} is the total annual fixed expense required to produce and sell a product but excluding any annual provision for plant depreciation, and A_{BD} is the balance-sheet annual depreciation charge.

However, in a given accounting period the sales volume may differ from the volume of production. In this case, the inventory of finished product I_1 at the beginning of the accounting period will differ from that at the end, I_2, and Eqs. (25-145) and (25-146) need to be written in modified form as

$$A_{CI} = R(c_S - c_{TVE}) - A_{TFE} + (I_1 - I_2) \quad (25\text{-}147)$$

$$A_{NP} = R(c_S - c_{TVE}) - (A_{TFE} + A_{BD}) + (I_1 - I_2) \quad (25\text{-}148)$$

If the annual sales volume exceeds the annual production rate R by an amount ΔR, then Eqs. (25-147) and (25-148) can be written as

$$A_{CI} = R(c_S - c_{TVE}) - A_{TFE} + \Delta R(c_S - c_{INV}) \quad (25\text{-}149)$$

$$A_{NP} = R(c_S - c_{TVE}) - (A_{TFE} + A_{BD}) + \Delta R(c_S - c_{INV}) \quad (25\text{-}150)$$

where c_{INV} is the value per unit of inventory of the finished product.

Clearly, the value of c_{INV} affects both the annual cash income and the net annual profit. Since annual cash incomes are the basic data for (NPV) and (DCFRR) methods of estimating profitability, the actual value per unit of inventory is of direct importance for chemical engineers engaged in economic assessments.

Let us divide Eq. (25-150) by Eq. (25-87) to give

$$\frac{A_{NP}}{A_S} = \left(1 - \frac{c_{TVE}}{c_S}\right) - \frac{(A_{TFE} + A_{BD})}{A_S} + \left(\frac{\Delta R}{R}\right)\left(1 - \frac{c_{INV}}{c_S}\right) \quad (25\text{-}151)$$

$$\frac{A_{NP}}{A_S} = (\text{CSR}) - \frac{(A_{TFE} + A_{BD})}{A_S} + \left(\frac{\Delta R}{R}\right)\left(1 - \frac{c_{INV}}{c_S}\right) \quad (25\text{-}152)$$

where $(\text{CSR}) = 1 - (c_{TVE}/c_S)$ is the contribution to the sales-price ratio. When the ratio of net annual profit to revenue from annual sales A_{NP}/A_S is expressed as a percentage it is known as the profit margin.

The terms (CSR) and $(A_{TFE} + A_{BD})/A_S$ have a similar order of magnitude in chemical processing. For example, (CSR) values of 0.1 to 0.4 are typical, and these are quite close to a value of 0.2 that is common in general chemical processing for $(A_{TFE} + A_{BD})/A_S$. Thus, the profit margin is very sensitive to the value c_{INV} per unit of inventory.

For example, let us consider that both (CSR) and $(A_{TFE} + A_{BD})/A_S$ are equal to 0.3. In this case, Eq. (25-152) can be written as

$$A_{NP}/A_S = (\Delta R/R)(1 - c_{INV}/c_S) \quad (25\text{-}153)$$

If the sales volume exceeds the annual production rate by 10 percent and the inventory is valued at the sales price, then Eq. (25-153) shows that the profit margin is $(A_{NP}/A_S)100 = 0$ percent. If the inventory is valued at the total variable cost, then the profit margin $(A_{NP}/A_S)100 = (0.1)(1 - 0.7)(100) = 3$ percent. Hence, the value of the inventory is of vital importance.

Unfortunately, there is no universally accepted method for valuing inventory. The value of c_{INV} per unit can be taken on any of the following bases:

1. Direct material plus direct labor cost.
2. Direct material plus direct labor plus other direct production expenses. (This is the total variable production cost.)
3. Total variable production cost plus fixed production overhead cost.
4. Total variable cost. (This includes both production and general expenses.)
5. Total cost. (This includes variable and fixed production and general expenses.)

Methods 1, 2, and 4 are termed direct costing, variable costing, and marginal costing respectively. Although direct costing is being increasingly used for internal accounting and control purposes, it is not acceptable to tax authorities as a basis for calculating profit. Tax authorities and most accountants favor method 3, which is known as absorption costing. We have already seen that this method has the disadvantage that the fixed-overhead cost per unit is determined for a particular normal production rate. If the production rate exceeds the normal, there is overabsorption of fixed overheads. Conversely, if the production rate falls below the normal, there is underabsorption of fixed overheads. Method 5 is known as full-absorption costing.

Most people agree that general expenses incurred in administration, selling, distribution, etc., should not be included in the cost of inventory. In fact, many feel that no costs should be absorbed before they have been incurred. In general, method 2 is favored by engineers and method 3 by most accountants. However, the accountancy convention is to value at either cost or market value, whichever is the lower. In the methods considered, either actual or standard costs can be used. Note that method 3 shows a higher profit than method 2 when sales volume exceeds the production rate and a lower profit when the production rate exceeds sales volume.

The total profits (before tax) over the life of a project are independent of the method used to value inventory. Over a project life of n years, Eq. (25-148) can be written as

$$\sum_0^n A_{NP} = \sum_0^n R(c_S - c_{TVE}) - \sum_0^n (A_{TFE} + A_{BD}) + \sum_0^n (I_m - I_{m+1}) \quad (25\text{-}154)$$

where the last term in Eq. (25-154) becomes $\sum_0^n (I_m - I_{m+1}) = I_0 - I_{n+1}$. Since there will be no material in inventory in the year before the project starts or in the year after it terminates, $I_0 = I_{n+1} = 0$. Hence, total profits do not depend on individual values for I.

However, the annual profit A_{NP} (before tax) does depend on the value of the inventory. Since the tax payable in any individual year is based on A_{NP}, the net annual profit A_{NNP} (after tax) is also dependent on the method chosen for valuing inventory. Frequently, a particular method for valuing inventory is chosen to delay payment of tax as long as it is legally possible to do so.

So far, only the inventory of finished product has been considered. There are also inventories of raw materials and work in process, i.e., partially processed materials or intermediate products, to be considered. It is necessary to modify Eqs. (25-147) through (25-154) accordingly to take these inventories into account.

Effect of Raw-Materials Prices Raw materials for the chemical-process industries are subject to relatively wide variations in price. These effects on profits will now be considered.

When the price of raw materials varies from week to week, not all the units in storage will have been purchased at the same price. Let us consider χ units in storage at the start of the inventory period, purchased at a price c_1 per unit. Additional quantities χ_2, χ_3, etc., are purchased at prices c_2, c_3, etc., per unit respectively until finally χ_n units are purchased at the latest price of c_n per unit at the end of the inventory period. The total value of the inventory C_{INV} at the end of the inventory period (in the absence of any withdrawal) is given by

$$C_{INV} = \sum_1^n \chi_j c_j \quad (25\text{-}155)$$

The value of the inventory I at any given time depends on the values ascribed to the units withdrawn from inventory. There are five methods for valuing inventory: (1) FIFO (first-in–first-out), (2) LIFO (last-in–first-out), (3) average cost, (4) standard cost, and (5) market value.

In the FIFO method, the units taken out of storage are valued at their purchase price beginning with the earliest item purchased. If a number of units m are removed from inventory during the period, the total cost of these items on a FIFO basis is given by:

$$C_{FIFO} = \sum_1^p \chi_j c_j + \left(m - \sum_1^p \chi_j\right) c_{p+1} \quad (25\text{-}156)$$

providing that the value of m satisfies

$$m < \sum_0^n \chi_j$$

and where p is the largest integer, such that

$$\sum_1^p \chi < m$$

Hence, the value of the inventory, I_{FIFO} at any given time is

$$I_{FIFO} = C_I - C_{FIFO} \tag{25-157}$$

In terms of Eqs. (25-155) and (25-156), Eq. (25-157) can be written as

$$I_{FIFO} = \sum_1^n \chi_j c_j - \sum_1^p \chi_j c_j - \left(m - \sum_1^p \chi_j\right) c_{p+1} \tag{25-158}$$

In the LIFO method, the m units taken out of storage are valued at their purchase price, beginning with the latest item purchased. In a similar manner, the value of the material C_{LIFO} taken out of inventory is given by

$$C_{LIFO} = \sum_p^n \chi_j c_j + \left(m - \sum_p^n \chi_j\right) c_{p-1} \tag{25-159}$$

where p is the smallest integer, such that

$$\sum_p^n \chi_j \leq m$$

Hence, the value of the inventory I_{LIFO} at any given time is

$$I_{LIFO} = C_I - C_{LIFO} \tag{25-160}$$

In terms of Eqs. (25-155) and (25-159), Eq. (25-160) can be written as

$$I_{LIFO} = \sum_1^n \chi_j c_j - \sum_p^n \chi_j c_j - \left(m - \sum_p^n \chi_j\right) c_{p-1} \tag{25-161}$$

Example 20 Let us consider 10 successive batches of raw materials, each of 1000 units, purchased in a time of rising prices in which c_1 = \$0.10 per unit, c_2 = \$0.11 per unit, etc., as listed in Table 25-28. The total cost of the purchases in the inventory is found from Eq. (25-155) to be \$1450.

Let us calculate the value of the raw-materials inventory after, say, 5500 units have been withdrawn from inventory, first by using FIFO and then by using LIFO.

We substitute the appropriate quantities into Eq. (25-158) for FIFO, keeping in mind that $p = 5$, $n = 10$, and $m = 5500$, to get

$$I_{FIFO} = \$1450 - \$600 - (5500 - 5000)(\$0.15)$$
$$I_{FIFO} = \$775$$

In a similar manner, we substitute values into Eq. (25-161) for LIFO (except that p is now 6) to get

$$I_{LIFO} = \$1450 - \$850 - (5500 - 5000)(\$0.14)$$
$$I_{LIFO} = \$530$$

Thus, in a time of rising raw-materials prices, the FIFO method gives a higher value for the remaining inventory than will LIFO. In a time of falling prices, the FIFO method will give a lower value for the remaining inventory than will LIFO.

Average-Cost Basis for Inventory Either a simple average or a weighted average can be used to value inventory cost.

TABLE 25-28 Costs of Inventory with Rising Prices

p	x_j, units	c_j, \$/unit	x_jc_j, \$	$\sum_1^p x_j$, units	$\sum_1^p x_jc_j$, \$	$\sum_p^n x_j$, units	$\sum_p^n x_jc_j$, \$
1	1,000	0.10	100	1,000	100	10,000	1,450
2	1,000	0.11	110	2,000	210	9,000	1,350
3	1,000	0.12	120	3,000	330	8,000	1,240
4	1,000	0.13	130	4,000	460	7,000	1,120
5	1,000	0.14	140	5,000	600	6,000	990
6	1,000	0.15	150	6,000	750	5,000	850
7	1,000	0.16	160	7,000	910	4,000	700
8	1,000	0.17	170	8,000	1,080	3,000	540
9	1,000	0.18	180	9,000	1,260	2,000	370
10	1,000	0.19	190	10,000	1,450	1,000	190

Using the simple-average method, the value C_{SAV} of the material taken out of inventory is given by

$$C_{SAV} = m(c_1 + c_n)/2 \tag{25-162}$$

The value of the inventory I_{SAV} at any given time is

$$I_{SAV} = C_I - C_{SAV} \tag{25-163}$$

In terms of Eqs. (25-155) and (25-162), Eq. (25-163) can be written as

$$I_{SAV} = \sum_1^n \chi_j c_j - [m(c_1 + c_n)/2] \tag{25-164}$$

We shall consider the value of the inventory after 5500 units have been withdrawn, using the data listed in Table 25-28. On the basis of a simple average, the materials withdrawn are priced at (\$0.10 + \$0.19)/2 = \$0.145 per unit. Since the total cost of the purchases in the raw-materials inventory is found from Eq. (25-155) to be \$1450, the value of the inventory after 5500 units have been withdrawn is calculated from Eq. (25-164) to be

$$I_{SAV} = \$1450 - (5500)(\$0.145) = \$652.50$$

For the weighted-average method, the value of the material C_{WAV} taken out of inventory is given by

$$C_{WAV} = m \sum_1^n \chi_j c_j \Big/ \sum_1^n \chi_j \tag{25-165}$$

$$I_{WAV} = C_I - C_{WAV} \tag{25-166}$$

In terms of Eqs. (25-153) and (25-164), Eq. (25-166) can be written

$$I_{WAV} = \sum_1^n \chi_j c_j \left[1 - \left(m \Big/ \sum_1^n \chi_j\right)\right] \tag{25-167}$$

Let us use Eq. (25-167) to value the inventory, after 5500 units have been withdrawn, by employing the data of Table 25-28:

$$I_{WAV} = \$1450\,[1 - (5500/10{,}000)] = \$652.50$$

For this example, the values of I_{SAV} and I_{WAV} are the same because the batches purchased are of equal size and the prices are linearly progressive. This is a combination rarely found in practice.

A more realistic example, in which the buyer seeks to purchase at the lowest price, is provided by the data of Table 25-29. The quantities bought vary according to price, but some may have been made at high prices to maintain production.

On the basis of Table 25-29, when 5500 items have been removed from inventory, the value of the inventory by using the FIFO, LIFO, simple-average, and weighted average methods respectively is

$$I_{FIFO} = \$1175 - \$560 - (5500 - 5200)(\$0.10)$$
$$I_{FIFO} = \$573.00$$
$$I_{LIFO} = \$1175 - \$645 - (5500 - 5000)(\$0.10)$$
$$I_{LIFO} = \$480.00$$
$$I_{SAV} = \$1175 - 5500\,[(\$0.10 + \$0.15)/2]$$
$$I_{SAV} = \$487.50$$
$$I_{WAV} = \$1175[1 - (5000/10{,}000)]$$
$$I_{WAV} = \$528.75$$

Pros and Cons of Inventory Valuation In the standard-price method of inventory valuation, all materials are taken out at the same price. In addition to simplicity, the method has the advantage that the efficiency of raw-materials purchase is constantly checked.

In both the average-cost and the standard-cost methods of valuing inventory, materials are not charged out at actual cost. Thus, the amount of profit or loss for the period may be varied by the method chosen to value the inventory. For this reason, accountants usually insist that the method of inventory valuation be consistent from period to period. This causes inertia but does not prevent a change of method when it can be justified. In such cases, it is usual to inform

TABLE 25-29 Costs of Inventory with Fluctuating Prices

p	x_j, units	c_j, \$/unit	x_jc_j, \$	$\sum_1^p x_j$, units	$\sum_1^p x_jc_j$, \$	$\sum_p^n x_j$, units	$\sum_p^n x_jc_j$, \$
1	1,000	0.10	100	1,000	100	10,000	1,175
2	1,500	0.11	165	2,500	265	9,000	1,075
3	500	0.13	65	3,000	330	7,500	910
4	2,000	0.10	200	5,000	530	7,000	845
5	200	0.15	30	5,200	560	5,000	645
6	1,000	0.14	140	6,200	700	4,800	615
7	2,500	0.11	275	8,700	975	3,800	475
8	300	0.15	45	9,000	1,020	1,300	200
9	100	0.20	20	9,100	1,040	1,000	155
10	900	0.15	135	10,000	1,175	900	135

stockholders of the change because the influence on declared profits can be large.

Unfortunately, there is no right or wrong way to value inventory, although certain methods are not allowed in certain countries for tax-assessment purposes. For example, LIFO is not allowed in the United Kingdom. As a general rule, the method used should be the one that gives the lowest tax liability. However, it is generally accepted that consistency is also a virtue in inventory valuation.

It is important to realize that the method used to value inventory for cost accounting purposes is not necessarily the one used to draw up the balance sheet and financial accounts. In this case, inventory is valued either at the cost given by another method or at the market value, whichever is lower.

Inventory Control The optimum size of inventory depends on the type of industry and on the skills available to the individual company. Inventories are high in the tobacco industry and low in perishable-foods businesses. The larger the inventory, the larger the warehousing and associated costs. These costs include insurance, taxes, depreciation, handling and security charges, etc., and can be taken as roughly proportional to the value of the inventory I. The annual cost A_{IW} of maintaining an inventory of value I is given by

$$A_{IW} = \alpha I \qquad (25\text{-}168)$$

where α is the proportionality factor which is of the order of 0.25 for many industries.

In contrast, some costs can be reduced as a result of larger inventories. For example, larger discounts can be obtained on bulk purchases and deliveries. In addition, larger inventories reduce the risk of losing sales and goodwill through interruptions to production and consequently running out of stocks of finished goods.

The cost of placing an order for materials is partly fixed and partly variable. The annual cost of ordering A_{IO} is given by

$$A_{IO} = FN + VR \qquad (25\text{-}169)$$

where F is the fixed cost per order, N is the number of orders per year, V is the variable cost of ordering per unit of production, and R is the annual production rate. Although the administrative cost of placing an order will be more or less fixed, the shipping costs are proportional to the size of the order and, hence, for the year are proportional to the annual production rate. The total annual cost of inventory A_I is the sum of Eqs. (25-168) and (25-169):

$$A_I = \alpha I + FN + VR \qquad (25\text{-}170)$$

For a given number of orders per year N, the value of the inventory is proportional to the magnitude of the average individual order U. Hence, Eq. (25-170) can be written as

$$A_I = \beta U + (FR/U) + VR \qquad (25\text{-}171)$$

where β is a proportionality factor and the number of orders per year N has been written as R/U.

By differentiating Eq. (25-171) with respect to U, the optimum size of order can be estimated. The differential is:

$$dA_I/dU = \beta - (FR/U^2) \qquad (25\text{-}172)$$

Setting the right-hand side of Eq. (25-172) equal to zero yields

$$U = \sqrt{FR/\beta} \qquad (25\text{-}173)$$

where U is now the optimum size of order.

A number of models have been developed to enable managers to handle inventories in the most profitable manner. These models can be applied to other elements of working capital, such as cash.

Working Capital The amount and disposition of working capital and the efficiency of its use determine the immediate prospects for future growth in a company. The bulk of managerial effort in a company is directly or indirectly concerned with the manipulation of working capital. Insufficient or misused working capital is the commonest cause of business failure.

Engineers concerned with cost estimations tend to make estimates of the fixed-capital cost of a project, leaving considerations of working capital to the accountants. Although the estimation of fixed-capital cost is more straightforward from an engineering point of view, the estimation of working capital is of vital importance both for an individual project and for the company as a whole.

Working capital can range from about 10 percent to almost 100 percent of the invested capital, depending on the industry, and is an important factor in the profitability index of a business. For this reason, it is best to compare the performance of an individual company with that of others that are as similar as possible.

Gross working capital is normally defined as total current assets, while net working capital is current assets minus current liabilities. Current assets normally amount to more than half of the total assets of a company. When accountants wish to emphasize working capital in a company, they present the balance sheet in the vertical form, as shown in Table 25-30. In this table, the stockholders' equity of \$91,650 funds the sum of \$38,650 (net working capital) and the sum of \$53,000 (fixed assets less the long-term loan). The flow of working capital is diagrammatically illustrated in Fig. 25-39.

The necessary working capital varies with sales volume or production rate. For example, sales and hence accounts receivable will double for a doubling in sales volume. In addition, an increase in sales volume normally requires increased inventories of raw materials, work in progress, and finished goods, all of which tie up capital. An increase in sales volume may lead to a relative shortage of working capital. In turn, this may mean that accounts payable cannot be paid in time and that valuable cash discounts may be lost or interest and penalty charges incurred. Creditors may take legal action to obtain payments and thereby put an additional strain on the current assets of the company to provide legal fees. Often, such action leads other creditors to take similar steps, which may lead a fundamentally sound company into bankruptcy.

A shortage of cash may prevent a company from taking advantage of large discounts available for bulk purchase of raw materials. The

TABLE 25-30 Balance Sheet for BCD Company, Dec. 31, 1981

Current assets		
Cash	\$14,575	
Accounts receivable	35,575	
Inventories		
Raw materials	6,000	
Work in process	3,750	
Finished product	10,000	
Gross working capital		\$69,900
Current liabilities		
Accounts payable	25,000	
Notes payable	3,000	
Bank loans	3,000	
Accruals payable	250	
		31,250
Net working capital		\$38,650
Fixed assets	75,000	
Long-term loan	22,000	
		\$53,000
Stockholders' equity		\$91,650

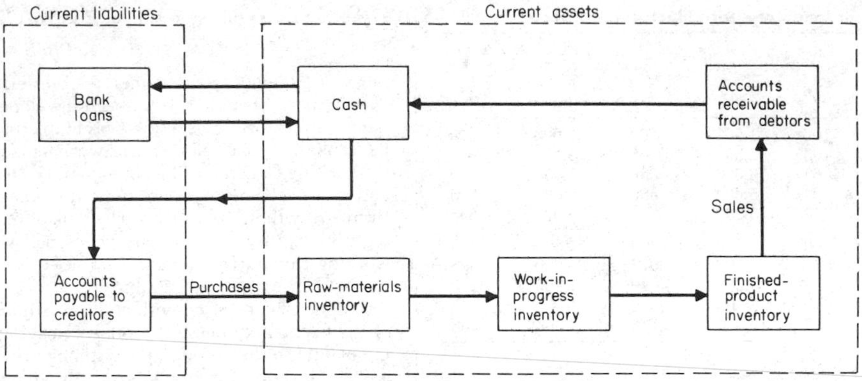

FIG. 25-39 Flow of working capital showing relationship between current liabilities and current assets.

importance of the availability of adequate cash or near cash can be seen by considering an account payable within 28 days, with a 2 percent discount allowed if paid within 7 days. If cash is not available to pay the account within 7 days, this is then equivalent to paying 2 percent interest on the money for the remaining 21-day period, or an annual compound-interest rate of more than 41 percent.

Adequate cash and a history of prompt payment of accounts strengthen the credit standing of a company and make it easier to obtain bank loans, etc. A company needs additional cash as a contingency against fires, floods, strikes, etc., as well as for additional advertising required to counteract the activities of competitors. This additional money is normally held as interest-bearing investments that can be turned into cash on short notice.

Working-Capital Ratios Financial analysts make extensive use of ratios in assessing the economic health of a company. For evaluating the ability of a company to successfully maintain and develop its immediate business activities, analysts apply a current ratio and a quick (or acid-test) ratio, as given by

$$\text{Current ratio} = \frac{\text{current assets}}{\text{current liabilities}} \qquad (25\text{-}142)$$

$$\text{Quick ratio} = \frac{\text{liquid assets}}{\text{current liabilities}} \qquad (25\text{-}143)$$

Liquid assets are those that can be realized almost immediately, such as cash, accounts receivable, and marketable securities. Although inventories are current assets, they must not be regarded as liquid assets because they cannot usually be converted into cash without winding up the business.

Although a high current ratio is desirable, this may be achieved by having unnecessarily high inventories that bring no profit except when commodity prices are rising rapidly. The quick ratio is less misleading in this respect.

Good management practice will hold inventories at the lowest possible levels consistent with customer satisfaction and efficient plant operation. Excessive inventories are unproductive and are an investment having little or no rate of return. Excessive inventories should be maintained only when supplies are erratic or rising in price. Management should normally aim for a high inventory-turnover ratio, as given by:

$$\text{Inventory-turnover ratio} = \frac{\text{revenue from annual sales}}{\text{average value of inventory}} \qquad (25\text{-}174)$$

Similarly, good management practice is to hold accounts receivable at a low level and to have a high accounts-receivable-turnover ratio, as given by

Accounts-receivable-turnover ratio

$$= \frac{\text{revenue from annual sales}}{\text{average value of accounts receivable}} \qquad (25\text{-}175)$$

Alternatively, it is good practice to have a low average collection period, as given by

Average collection period

$$= \frac{\text{average value of accounts receivable}}{\text{revenue from sales per day}} \qquad (25\text{-}144)$$

Equation (25-144) gives the number of days during which sales are tied up in receivables.

An accounts-receivable-turnover ratio of 12 is considered fairly good for a manufacturing company. This implies an average collection period of about 1 month. The price obtained for the goods should include an allowance for interest (otherwise obtainable) on the money tied up for such a period.

Since ratios, like balance sheets, refer to a particular point in time, they have a limited use unless they are compared with previous values. A study of ratio trends indicates whether or not a company is approaching a working-capital or a liquidity crisis and may enable management to compare the performance of the company with that of competitors.

Funds Statement A typical funds statement is shown in Table 25-31. It displays the change in net working capital and can be obtained from a statement of changes in working capital, such as the one shown in Table 25-32. A funds statement shows where the cash came from and how it was used.

Changes in working capital ΔC_{WC} in an annual accounting period can be represented by

$$\Delta C_{WC} = A_S - A_{TE} - \Sigma\Delta C_{FC} - \Sigma\Delta C_L + \Sigma\Delta C_{FIN} \qquad (25\text{-}176)$$

where A_S is the revenue from annual sales of a product, A_{TE} is the total cost or expense required to produce and sell the product but excluding any annual provision for plant depreciation, $\Sigma\Delta C_{FC}$ is the sum of the changes in depreciable fixed assets, $\Sigma\Delta C_L$ is the sum of the changes in nondepreciable fixed assets, and $\Sigma\Delta C_{FIN}$ is the sum of the changes in financial resources such as loans, bonds, preferred stock, common stock, etc.

TABLE 25-31 Funds Statement for Year Ending Dec. 31, 1981

Sources of funds		
Cash income from operations		$15,000
New finance		10,000
Sale of plant equipment		2,000
Total sources of funds		$27,000
Applications of funds		
Purchase of land	$3,000	
Cash dividend on stock	9,000	
Total applications of funds		$12,000
Increase in net working capital		$15,000

TABLE 25-32 Statement of Changes in Working Capital

	Jan. 1, 1981	Jan. 1, 1982	Change
Cash	$ 80,000	$ 94,000	+$14,000
Accounts receivable	20,000	35,000	+ 15,000
Inventories	30,000	25,000	− 5,000
Total current assets	$130,000	$154,000	+$24,000
Current liabilities	($ 60,000)	($ 69,000)	(+$ 9,000)
Net working capital	$ 70,000	$ 85,000	+$15,000
Current ratio	2.17	2.23	
Quick ratio	1.67	1.87	

Equation (25-176) can also be written as

$$\Delta C_{WC} = A_{CI} - \Sigma\Delta C_{FC} - \Sigma\Delta C_L + \Sigma\Delta C_{FIN} \qquad (25\text{-}177)$$

where A_{CI} is the annual cash income, which is the main source of funds for most companies. In this case, the annual cash income excludes all noncash expenses such as the balance-sheet annual depreciation charge A_{BD}, which is purely a book transaction.

A positive value of any term in Eq. (25-177) implies an increase in working capital, and a negative value a decrease. For example, the sale of fixed assets such as plant, buildings, land, etc., is a source of cash, and the purchase of fixed assets uses up cash. Similarly, an increase in financial resources in the form of loans and stock and bond issues is a source of cash, and a decrease in financial resources in the form of repayment of loans, retirement of stocks and bonds, and the payment of cash dividends uses up cash. (Note that a stock dividend as opposed to a cash dividend does not use up cash.)

The relation between (1) net annual cash flow A_{CF} after tax for individual projects, (2) the annual amount of tax A_{IT}, and (3) the annual expenditure of capital A_{TC} is

$$A_{CF} = A_{CI} - A_{IT} - A_{TC} \qquad (25\text{-}178)$$

Equation (25-178) for a single project is really analogous to Eq. (25-177) for a company.

An income statement or profit-and-loss account gives the net annual profit A_{NP} before tax. In order to assess the annual cash income A_{CI} as a source of funds from the value of the net annual profit A_{NP} given in the income statement, it is necessary to add back all noncash expenses such as the balance-sheet annual depreciation charge A_{BD}. This practice sometimes erroneously suggests that depreciation is a source of funds, whereas cash income is the only source of funds.

Although A_{BD} does not affect working capital in any way, the annual depreciation charge A_D does affect the annual amount of tax A_{IT} given by

$$A_{IT} = (A_{CI} - A_D - A_A)t \qquad (25\text{-}3)$$

where t is the fractional tax rate and A_A is the annual amount of any other allowances. Thus, the net annual cash income is affected by depreciation allowances, as follows:

$$A_{NCI} = A_{CI} - A_{IT} \qquad (25\text{-}2)$$

In this sense, depreciation makes working capital available by reducing the cash outflow for taxes.

When a fixed asset is sold at a price that differs from its book value, an accounting gain or loss is recorded. This gain or loss does not affect working capital, which has simply been increased by the amount of cash received from the sale. For example, if an item of plant is sold for $40,000 (whatever its book value), the increase in working capital resulting from the sale is $40,000. If the book value of the plant had been $50,000, the accounting loss of $10,000 on the sale would be included as an expense when calculating the cash income. This $10,000 must be added back to the cash income in order to get the cash income excluding noncash expenses as required for A_{CI}. Conversely, any accounting gain on the sale of a fixed asset must be subtracted if it has been included in the cash income.

Book values of fixed assets are determined by the balance-sheet annual depreciation charges A_{BD}, which do not affect working capital. Although the accounting gain or loss on the sale of a fixed asset is based on its book value, working capital is not affected by depreciation assessments.

Transactions that change the character of the net working capital but do not affect its value occur in a company. For example, a cash payment of $10,000 for accounts payable reduces both the current asset of cash by $10,000 and the current liability of accounts payable by $10,000, leaving the net working capital unchanged. However, this transaction affects both the current and the quick ratios.

After the balance sheet and the income statement or profit-and-loss account, the funds statement is generally regarded as the most important financial document. However, many financial managers regard a statement showing changes in cash as being of equal importance.

As with balance sheets and income statements, there is no rigid format for funds statements. These vary in the amount of detail given and also in their layout. Funds statements help management in planning for the future, for example, in timing and making financial provision for future expenditures.

A typical management accountant's statement for changes in working capital and sources and applications of funds is shown in Table 25-33. This is based on the following relation: an increase in application of funds equals an increase in sources of funds. The relation can also be expresssed as follows: an increase in assets plus a

TABLE 25-33 Statement of Changes in Working Capital and Sources and Applications of Funds for BCD Company

	Balance sheets		Change in funds	
	Dec. 31, 1981	Dec. 31, 1982	Applications	Sources
Cash	$ 14,575	$ 13,000		$ 1,575
Accounts receivable	35,575	36,000	$ 425	
Inventories	19,750	20,750	1,000	
Total current assets	$ 69,900	$ 69,750		
Accounts payable	$ 25,000	$ 27,000		2,000
Notes payable	3,000	2,500	500	
Bank loans	3,000	2,000	1,000	
Accruals payable	250	300		50
Total current liabilities	$ 31,250	$ 31,800		
Net working capital	$ 38,650	$ 37,950		
Fixed assets	75,000	85,000	10,000	
Net assets	$113,650	$122,950		
Long-term loan	$ 22,000	$ 22,000		
Stockholders' equity	$ 91,650	$100,950		9,300
			$12,925	$12,925

decrease in liabilities equals an increase in liabilties plus a decrease in assets.

In Table 25-33 the cash income for the year has been absorbed in the increase in stockholders' equity, which consists of issued stock plus retained profits or income.

General Considerations Many people regard the management of working capital as essentially a cash-flow problem. Certainly, expansion involves increased investment in fixed assets and in the various items that comprise the current assets, all of which require cash. If this leads to a shortage of cash so that a company cannot pay its liabilities, then a situation called overtrading results, and the company may ultimately be forced into liquidation. Therefore, it is essential that a company have access to readily available cash or sources of short-term financing, preferably without having to specify a particular asset as collateral, since such collateral can only be pledged to an agreed realizable value that is much less than its true value.

Accounts payable, also called trade credit, are the major source of short-term financing. Accounts payable normally amount to about 40 percent of the current liabilities of a manufacturing company. Such short-term financing is relatively expensive when available discounts are lost.

The second most important source of short-term financing is notes payable from commercial banks. Banks normally require a borrower to maintain a compensating balance. For example, if a company requires a loan of $100,000, it must borrow more than this, say, $120,000 (on which it pays interest), in order to maintain a minimum checking-account balance of $20,000. Commercial banks also provide a wide variety of other services that can be of great help to companies in temporary financial difficulties.

The third most important source of short-term financing is the commercial paper, or promissory notes, of large companies. This is the cheapest form of finance. However, the amount of money available is a function of the excess liquidity of the large companies at a given time, and the money may have to be repaid at relatively short notice.

Sometimes, fixed assets are purchased via short-term loans, which can lead to liquidity problems. For the most part, fixed assets should be financed from long-term or permanent capital such as stocks or bonds. The proven ability of management to handle working capital efficiently will put a company in a better position to obtain such long-term capital when required, because the confidence of bankers and stockholders will have been obtained.

Budgets and Cost Control R. J. Bull (*Accounting in Business*, 2d ed., Butterworth, London, 1972, p. 163) defined a budget as a comprehensive and coordinated plan, expressed in monetary terms, directing and controlling the resources and trading activities of an enterprise for some specified period in the future. A budget is not a forecast. A forecast is an estimate of the future which may or may not be attained. A budget is an overall objective based on a forecast. In addition, a budget defines the detailed objectives to be achieved by various levels of management in an organization. Since the achievement of these objectives requires the complete cooperation of management, the targets set must be realistic in terms of resources and past performance.

A comparison of actual with budgeted results can be used as the basis for control at the company, departmental, plant, or project level. In addition, a continuing record of performance should be maintained to provide the data for the preparation of further budgets.

Since company accounts are normally published annually, 1 year is commonly taken as a budget period. However, budget periods can vary widely depending on the nature of the operation. For example, a sales budget may be for a period of, say, 3 or 6 months, while the budget period for the installation of, say, a nuclear power station would extend over many years.

The basic objectives of budgets are planning and control. The first step is to determine the limiting factor. For example, budgeted sales cannot exceed the maximum productive capacity of the available plant. Since all the activities in the plan are interrelated, the extent of the plan is determined by the limiting factor.

After the plan is put into operation, actual progress is monitored against established standards. These data may subsequently lead to the plan's being modified in order to achieve the objectives more effectively.

R. Pilcher (*Appraisal and Control of Project Costs*, McGraw-Hill, Maidenhead, England, 1973, p. 233) stated the main purposes of a cost control system to be:

1. To provide immediate warning of uneconomic operations in both the long and the short terms
2. To provide the relevant feedback, carefully qualified in detail by all the conditions under which the work has been carried out, to the estimator who is responsible for establishing the standards in the past and in the future
3. To provide data to assist in the valuation of those variations that will arise during the course of the work
4. To promote cost consciousness
5. To summarize progress

Budgeted income statements are identical in form to actual income statements. However, the budgeted numbers are objectives rather than achievements. Budgetary models based on mathematical equations are increasingly being used. These may be used to determine rapidly the effect of changes in variables. Variance analysis is discussed in the treatment of manufacturing-cost estimation.

MANUFACTURING-COST ESTIMATION

The annual manufacturing cost or expense A_{ME} can be written as the sum of the direct manufacturing or prime cost A_{DME} and the indirect manufacturing or overhead cost A_{IME}:

$$A_{ME} = A_{DME} + A_{IME} \tag{25-179}$$

The determination of direct or prime costs is more straightforward than the determination of indirect or overhead costs. When more than one product is involved, the question arises as to the correct distribution of overhead costs between the various products.

In addition to fixed and variable costs there are mixed or semivariable costs. These have a fixed and a variable element. The variable element can vary with production linearly, stepwise, or in a curvilinear manner. However, it is a convenient simplification to divide all manufacturing costs into fixed and linearly variable costs.

General Considerations Manufacturing costs are best considered in the context of the manufacturing, trading, and profit-and-loss accounts. Typical examples of these are shown in Tables 25-34, 25-35, and 25-36 respectively. These are based on the conventional accountancy period of 1 year.

The gross annual profit A_{GP} in Table 25-35 is dependent on the balance-sheet annual depreciation charge A_{BD}, which is not necessarily the same as the depreciation allowance used for tax purposes. Since A_{BD} is arbitrarily chosen, it can be used to make the gross annual profit A_{GP} high or low according to the company policy.

The gross annual profit A_{GP} is also dependent on the method of valuing the inventory. For example, raw materials may have been purchased at the beginning of the accounting year at, say, 9 cents per kilogram. The purchase price may have risen to, say, 12 cents per kilogram at the end of the accounting year. If valuation of the inventory is made at the higher purchase price, the production cost is lower and the gross profit higher than if the valuation is made at the lower purchase price. Although in this case the profit looks better, a higher tax is payable.

TABLE 25-34 Manufacturing Account

Inventory of raw materials, Jan. 1, 1981	\$ 35,000	
Add purchases	870,000	
Add carriage inward	25,000	
	\$ 930,000	
Less inventory of raw materials Dec. 31, 1981	51,000	
Cost of raw materials consumed	\$ 879,000	
Direct wages	122,000	
Direct utilities	22,000	
Other direct expenses	104,000	
Prime cost or direct manufacturing expense, A_{DME}	\$1,127,000	\$1,127,000
Payroll overhead	28,000	
General plant overhead	52,000	
Other indirect expenses	52,000	
Indirect manufacturing expense, A_{IME}	\$ 132,000	
Depreciation, A_{BD}	68,000	
	\$ 200,000	200,000
		\$1,327,000
Add work in progress, Jan. 1, 1981		30,000
		\$1,357,000
Less work in progress, Dec. 31, 1981		35,000
Production cost of products		\$1,322,000

The profit is also dependent on the method of valuing the work in progress. We shall consider the manufacture of 100,000 kg of product with a prime or direct manufacturing cost of 10 cents per kilogram and an additional indirect manufacturing expense of 5 cents per kilogram. We assume that 90,000 kg of the product is sold and 10,000 kg is stored in inventory. The value of the inventory is \$1000 on the basis of prime or direct cost and \$1500 when the indirect manufacturing expense is included. It is still a controversial question as to whether manufacturing overheads should be absorbed in the cost. The latter is known as absorption costing and is the traditional accounting method. Direct costing is being increasingly used and is particularly favored by engineers.

The annual cash income A_{CI} before tax, in terms of the revenue from annual sales A_S of a product and the components of the total annual cost or expense required to produce and sell the product (excluding any allowance for plant depreciation), is expressed by

$$A_{CI} = [A_S - (A_{VGE} + A_{VME})] - (A_{FGE} + A_{FME}) \qquad (25\text{-}180)$$

where A_{VGE} and A_{VME} are the annual variable general and variable manufacturing expenses respectively and A_{FGE} and A_{FME} are the annual fixed general and fixed manufacturing expenses respectively.

Revenue from annual sales is

$$A_S = Rc_S \qquad (25\text{-}181)$$

where R is the production rate and c_S the sales price per unit of production.

Similarly, A_{VGE} can be taken as proportional to the annual production rate R and the variable general expense per unit of production c_{VGE}:

$$A_{VGE} = Rc_{VGE} \qquad (25\text{-}182)$$

Likewise, for A_{VME},

$$A_{VME} = Rc_{VME} \qquad (25\text{-}183)$$

TABLE 25-35 Trading Account

Inventory of finished products, Jan. 1, 1981	\$ 200,000
Add production cost of products	1,322,000
	\$1,522,000
Less inventory of finished products, Dec. 31, 1981	190,000
	\$1,332,000
Gross profit	668,000
Sales	\$2,000,000

TABLE 25-36 Profit-and-Loss Account

Administration	\$ 74,000	
Sales and shipping	124,000	
Advertising and marketing	40,000	
Technical service	10,000	
Research and development	60,000	
	\$308,000	\$308,000
Net profit before taxes		360,000
Gross profit		\$668,000

In practice, annual direct variable costs such as raw materials, utilities, etc., are not always proportional to the production rate.

Substituting Eqs. (25-182) and (25-183) into the first term on the right of Eq. (25-180) yields

$$A_S - (A_{VGE} + A_{VME}) = R[c_s - (c_{VGE} + c_{VME})] \qquad (25\text{-}184)$$

In Eq. (25-184), the term $[c_S - (c_{VGE} + c_{VME})]$ is the contribution to cash income.

For simultaneous production of more than one product, Eq. (25-184) can be written as

$$A_S - (A_{VGE} + A_{VME}) = R_1 \{(c_S)_1 - [(c_{VGE})_1 + (c_{VME})_1]\} + R_2 \{(c_S)_2 - [(c_{VGE})_2 + (c_{VME})_2]\} + \dots \qquad (25\text{-}185)$$

where the subscripts 1, 2, etc., refer to the various coproducts.

The contribution to cash income made by a particular product depends on the method of accounting. Widely different values for the contribution can be calculated by using different methods of assigning manufacturing expenses.

The cost of each item in a cost estimate should be presented in such a way that the estimate can be modified and updated at any time in the future when revised data become available.

Published data and shortcut estimating methods can be used to calculate the approximate manufacturing cost of a new product. However, most companies have extensive data on various items of cost such as overheads, property taxes, etc. These data should be used whenever possible to give the estimate that is most valid for a particular company.

For a new product, the ratio of manufacturing expense to sales price c_{ME}/c_S should be compared with the ratio of total manufacturing cost or expense to sales revenue for the company as a whole. If the ratio c_{ME}/c_S is less than or equal to the ratio for the company, then the proposed sales price appears to be reasonable and the product is probably commercially viable. This comparison is, of course, used only as an approximate guide in preliminary assessments.

One Main Product Plus By-Products We shall let one unit of raw material yield χ_1, χ_2, etc., weights of products 1, 2, etc., respectively. The variable general expense per unit of raw material will be c_{VGE} and the variable manufacturing expense per unit of raw material will be c_{VME}. The unit of raw material may be either a single material or a mixture of several components. I. Leibson and G. A. Trischman [*Chem. Eng.*, **78**, 69–74 (May 31, 1971)] showed the effects on manufacturing expenses of alternate feedstocks having different compositions and costs for producing the same products.

If product 1 is the main product, then this carries all the variable expenses, so that

$$(c_{VGE})_1 = c_{VGE}/\chi_1 \qquad (25\text{-}186)$$

$$(c_{VME})_1 = c_{VME}/\chi_1 \qquad (25\text{-}187)$$

where $(c_{VGE})_1$ and $(c_{VME})_1$ are the variable general and variable manufacturing expenses respectively per unit of product 1.

For by-product 2, etc., $(c_{VGE})_2 = 0$ and $(c_{VME})_2 = 0$, etc. For this case, Eq. (25-184) becomes

$$A_S - (A_{VGE} + A_{VME}) = R_1 [(c_S)_1 - (c_{VGE} + c_{VME})/\chi_1] + R_2(c_S)_2 + \dots \qquad (25\text{-}188)$$

If R is the annual rate at which the raw material is consumed, then

$$R_1 = \chi_1 R \qquad (25\text{-}189)$$

$$R_2 = \chi_2 R \qquad (25\text{-}190)$$

In terms of one unit of raw material, Eq. (25-185) can be combined with Eq. (25-189) and (25-190) and written as

$$[A_S - (A_{VGE} + A_{VME})]/R = \chi_1(c_S)_1 - (c_{VGE} + c_{VME}) + \chi_2(c_S)_2 + \cdots \quad (25\text{-}191)$$

Equation (25-191) gives the contribution of products 1, 2, etc., per unit of raw material. In this particular case, the contribution of the main product per unit weight of raw material is $\chi_1(c_S)_1 - (c_{VGE} + c_{VME})$. The contribution of product 2 is $\chi_2(c_S)_2$, i.e., its selling price per unit weight of raw material.

Two Main Products

By Weight We shall let one unit of raw material yield χ_1 and χ_2 weights of products 1 and 2 respectively. The variable general expense per unit of raw material will be c_{VGE} and the variable manufacturing expense per unit of raw material c_{VME}. In practice, it is rare for $\chi_1 + \chi_2$ to be exactly unity.

If the variable expenses are shared by weight, then

$$(c_{VGE})_1 = \chi_1 c_{VGE}/[(\chi_1 + \chi_2)\chi_1] \quad (25\text{-}192)$$

$$(c_{VME})_1 = \chi_1 c_{VME}/[(\chi_1 + \chi_2)\chi_1] \quad (25\text{-}193)$$

$$(c_{VGE})_2 = \chi_2 c_{VGE}/[(\chi_1 + \chi_2)\chi_2] \quad (25\text{-}194)$$

$$(c_{VME})_2 = \chi_2 c_{VME}/[(\chi_1 + \chi_2)\chi_2] \quad (25\text{-}195)$$

where $(c_{VGE})_1$ and $(c_{VME})_1$ are the variable general and variable manufacturing expenses per unit of product 1 respectively and $(c_{VGE})_2$ and $(c_{VME})_2$ are the comparable expenses for product 2.

For this example, Eq. (25-185) becomes

$$A_S - (A_{VGE} + A_{VME}) = R_1\left[(c_S)_1 - \frac{(c_{VGE} + c_{VME})}{(\chi_1 + \chi_2)}\right] + R_2\left[(c_S)_2 - \frac{(c_{VGE} + c_{VME})}{(\chi_1 + \chi_2)}\right] \quad (25\text{-}196)$$

In Eqs. (25-192) through (25-196) the sum of χ_1 and χ_2 may be equal to or less than 1. The above analysis can be extended for any number of coproducts.

In terms of one unit of raw material, Eq. (25-196) can be combined with Eqs. (25-189) and (25-190) and written as

$$\frac{A_S - (A_{VGE} + A_{VME})}{R} = \chi_1\left[(c_S)_1 - \frac{(c_{VGE} + c_{VME})}{(\chi_1 + \chi_2)}\right] + \chi_2\left[(c_S)_2 - \frac{(c_{VGE} + c_{VME})}{(\chi_1 + \chi_2)}\right] \quad (25\text{-}197)$$

Equation (25-197) gives the contribution of products 1 and 2 per unit weight of raw material.

By Value We shall let one unit of raw material yield χ_1 and χ_2 weights of products 1 and 2 respectively, with values of $\chi_1(c_S)_1$ and $\chi_2(c_S)_2$ respectively; $(c_S)_1$ and $(c_S)_2$ are the sales prices of products 1 and 2 per unit of production.

We shall let the variable general expense per unit of raw material be c_{VGE} and the variable manufacturing expense per unit of raw material be c_{VME}. If the variable expenses are shared by value then

$$(c_{VGE})_1 = \frac{\chi_1(c_S)_1 c_{VGE}}{[\chi_1(c_S)_1 + \chi_2(c_S)_2]\chi_1} \quad (25\text{-}198)$$

$$(c_{VME})_1 = \frac{\chi_1(c_S)_1 c_{VME}}{[\chi_1(c_S)_1 + \chi_2(c_S)_2]\chi_1} \quad (25\text{-}199)$$

$$(c_{VGE})_2 = \frac{\chi_2(c_S)_2 c_{VGE}}{[\chi_1(c_S)_1 + \chi_2(c_S)_2]\chi_2} \quad (25\text{-}200)$$

$$(c_{VME})_2 = \frac{\chi_2(c_S)_2 c_{VME}}{[\chi_1(c_S)_1 + \chi_2(c_S)_2]\chi_2} \quad (25\text{-}201)$$

where $(c_{VGE})_1$ and $(c_{VME})_1$ are the variable general and variable manufacturing expenses per unit of product 1 respectively and $(c_{VGE})_2$ and $(c_{VME})_2$ are the comparable expenses for product 2.

For this case, Eq. (25-185) becomes

$$A_S - (A_{VGE} + A_{VME}) = R_1\left\{(c_S)_1 - \frac{(c_S)_1(c_{VGE} + c_{VME})}{[\chi_1(c_S)_1 + \chi_2(c_S)_2]}\right\} + R_2\left\{(c_S)_2 - \frac{(c_S)_2(c_{VGE} + c_{VME})}{[\chi_1(c_S)_1 + \chi_2(c_S)_2]}\right\} \quad (25\text{-}202)$$

This analysis can be extended for any number of coproducts.

In terms of one unit of raw material, Eq. (25-202) can be combined with Eqs. (25-189) and (25-190) and written as

$$\frac{A_S - (A_{VGE} + A_{VME})}{R} = \chi_1\left\{(c_S)_1 - \frac{(c_S)_1(c_{VGE} + c_{VME})}{[\chi_1(c_S)_1 + \chi_2(c_S)_2]}\right\} + \chi_2\left\{(c_S)_2 - \frac{(c_S)_2(c_{VGE} + c_{VME})}{[\chi_1(c_S)_1 + \chi_2(c_S)_2]}\right\} \quad (25\text{-}203)$$

Equation (25-203) gives the contribution of products 1 and 2 per unit weight of raw material.

Example 21 One kilogram of raw material is used to manufacture $\chi_1 = 0.32$ kg of product 1 and $\chi_2 = 0.64$ kg of product 2. The balance of the raw material goes to waste. Product 1 sells at $(c_S)_1 = 40$ cents per kilogram, and product 2 sells at $(c_S)_2 = 12$ cents per kilogram. The variable general and variable manufacturing expenses, including raw materials, $c_{VGE} + c_{VME}$, total 10 cents per kilogram.

Let us calculate (*a*) the total contribution to cash income per kilogram of raw material and (*b*) the individual contribution of each product to cash income per kilogram of raw material for the following:

Case	Condition
1	Product 1 as the main product charged for all the variable expenses
2	Product 2 as the main product charged for all the variable expenses
3	Products 1 and 2 sharing the variable expenses on the basis of weight
4	Products 1 and 2 sharing the variable expenses on the basis of value

Case 1. The total contribution per kilogram of raw material is obtained by substituting the appropriate values into Eq. (25-191). For these conditions,

$$\begin{aligned}\chi_1(c_S)_1 + \chi_2(c_S)_2 - (c_{VGE} + c_{VME}) &= 0.32(40 \text{ cents/kg}) + 0.64\,(12 \text{ cents/kg}) - 10 \text{ cents/kg} \\ &= 12.8 \text{ cents/kg} + 7.68 \text{ cents/kg} - 10 \text{ cents/kg} \\ &= 10.48 \text{ cents/kg}\end{aligned}$$

The individual contribution of product 1 as the main product is 12.8 cents − 10 cents = 2.8 cents per kilogram of raw material. The individual contribution of product 2 as the by-product is 7.68 cents per kilogram of raw material.

Case 2. The total contribution when product 2 is the main product is also obtained from Eq. (25-191) and as in Case 1 is found to be 10.48 cents per kilogram.

The individual contribution of product 2 as the main product is 7.68 cents − 10 cents = −2.32 cents per kilogram of raw material. The individual contribution of product 1 as the by-product is 12.8 cents per kilogram of raw material.

Case 3. When products 1 and 2 share the variable expenses on the basis of weight, the total contribution per kilogram of raw material is found by substituting the unit costs into Eq. (25-197). Values for each term are

$$\begin{aligned}\chi_1(c_S)_1 &= 12.8 \text{ cents/kg} \quad \text{(same as Case 1)} \\ \chi_2(c_S)_2 &= 7.68 \text{ cents/kg} \quad \text{(same as Case 1)} \\ [\chi_1/(\chi_1 + \chi_2)](c_{VGE} + c_{VME}) &= (0.32/0.96)(10 \text{ cents/kg}) = 3.34 \text{ cents/kg} \\ [\chi_2/(\chi_1 + \chi_2)](c_{VGE} + c_{VME}) &= (0.64/0.96)(10 \text{ cents/kg}) = 6.66 \text{ cents/kg}\end{aligned}$$

Therefore, the total contribution becomes 12.8 + 7.68 − 3.34 − 6.66 = 10.48 cents per kilogram.

The individual contribution of product 1 is 12.8 cents − 3.34 cents = 9.46 cents per kilogram of raw material. The individual contribution of product 2 is 7.68 cents − 6.66 cents = 1.02 cents per kilogram of raw material.

Case 4. When products 1 and 2 share the variable expenses on the basis of value, the total contribution per kilogram of raw material is found by substituting the unit costs into Eq. (25-203). Values of each term are

$$\chi_1(c_S)_1 = 12.8 \text{ cents/kg} \qquad \text{(same as Case 1)}$$
$$\chi_2(c_S)_2 = 7.68 \text{ cents/kg} \qquad \text{(same as Case 1)}$$
$$\frac{\chi_1(c_S)_1(c_{VGE} + c_{VME})}{\chi_1(c_S)_1 + \chi_2(c_S)_2} = \frac{(12.8)(10)}{12.8 + 7.68} = 6.24 \text{ cents/kg}$$
$$\frac{\chi_2(c_S)_2(c_{VGE} + c_{VME})}{\chi_1(c_S)_1 + \chi_2(c_S)_2} = \frac{(7.68)(10)}{12.8 + 7.68} = 3.76 \text{ cents/kg}$$

Therefore, the total contribution becomes 12.8 + 7.68 − 6.24 − 3.76 = 10.48 cents per kilogram.

The individual contribution of product 1 is found as 12.8 cents − 6.24 cents = 6.56 cents per kilogram of raw material. The individual contribution of product 2 then becomes 7.68 cents − 3.76 cents = 3.92 cents per kilogram of raw material.

Direct Manufacturing Costs Direct manufacturing costs include raw materials, operating labor, utilities, and some miscellaneous items. A summary of the characteristics of each follows.

Raw Materials The cost of raw materials is normally the largest item of expense in the manufacturing cost of a product. The quantities of raw materials consumed can be calculated from material balances.

Material costs are conveniently presented in tables that give the following: name of material, form and grade, method of delivery, unit of measure, cost per unit, source of cost, annual consumption, annual cost, fractional consumption per unit of production, and cost per unit of production.

Net consumption of materials should be used for catalysts, solvents, filter aids, etc., that may have a recovery value. Current prices of chemicals are published in various trade journals. However, quotations from suppliers should be used whenever possible.

It may be possible for a company to negotiate the purchase of a material at a cost per unit that is significantly lower than the current published price. This is particularly true if large quantities are involved. Thus, estimates should be presented for both minimum and maximum costs. Price trends, availability, and quality are other factors that should be considered. A knowledge of price trends is particularly important for a product that a company may not manufacture for several years.

The yield in a chemical reaction determines the quantities of materials in the material balance. Assumed yields are used to obtain approximate exploratory estimates. In this case, possible ranges should be given. Firmer estimates require yields based on laboratory or, preferably, pilot-plant work.

Operating Labor The cost of operating labor is the second largest item of expense in the manufacturing cost. Labor requirements for a process can be estimated from an intelligent study of the equipment flow sheet, paying careful attention to the various primary process steps such as fractionation, filtration, etc. The hourly wage rate should be that currently paid in the company. Once the number of persons required per shift has been estimated for a particular production rate, the annual labor cost and the labor cost per unit of production can be estimated.

H. E. Wessel [*Chem. Eng.*, **59**, 209–210 (July 1952)] made a study of the operating-labor requirements in the United States chemical industry and presented the data as a plot of labor-hours per ton per processing step versus plant capacity in tons per day. These data can be represented by:

$$\log_{10} Y = 0.783 \log_{10} \chi + 1.252 + B \qquad (25\text{-}204)$$

where Y is the operating-labor–hours per ton per processing step; χ is plant capacity, tons per day; and B is a constant having values of 0.132 when multiple units are used to increase capacity or when the process is completely batch, of 0 for the average chemical-processing plant, and of −0.167 for large, highly automated plants or plants concerned with fluid processing.

Wessel's data for the United States chemical industry refer to the short ton equal to 2000 lb, or 907.2 kg. Labor requirements are higher in countries with lower productivities.

The approximate cost of supervision for operating labor is equivalent to 10 percent of the labor cost for simple operations and 25 percent for complex operations.

Utilities These include steam, cooling water, process water, electricity, fuel, compressed air, and refrigeration. The consumption of utilities can be estimated from the material and energy balances for the process, together with the equipment flow sheet.

Let us consider a cooler in the equipment flow sheet. The required rate of heat removal is known from the balances, and the rate of cooling water can be calculated once the inlet and outlet temperatures of the water have been specified. The calculation of the consumption of other utilities is also straightforward. Allowances should be made for wastage.

The current cost per unit for each utility is usually well known in a company. Thus, the annual cost for utilities and the utilities cost per unit of production can be estimated. The latter is normally much smaller than the raw-materials and labor costs. However, a great deal more work is involved in calculating the utilities cost than for any other item in the manufacturing cost.

Unfortunately, there are no satisfactory shortcut methods for doing this. When the utilities cost is relatively small, it may be possible to make an intelligent guess on the basis of known costs for similar processes in the company. Alternatively, published data for the consumption of utilities per unit of production for various processes may be used.

Miscellaneous Direct Costs Estimates for the cost of maintenance and repairs, operating supplies, royalties, and patents are best based on company records for similar processes. A rough average value for the annual cost of maintenance is 6 percent of the capital cost of the plant. This percentage can vary from 2 to 10 percent, depending on the severity of plant operation. Approximately half of the maintenance costs are for materials and half for labor. Royalty and patents costs are in the order of 1 to 5 percent of the sales price of the product.

Indirect Manufacturing Costs Estimates for the cost of payroll overhead, control laboratory, general plant overhead, packaging, and storage facilities are best based on company records for similar processes.

Payroll overhead includes the cost of pensions, holidays, sick pay, etc., and is normally between 15 and 20 percent of the operating-labor cost. Laboratory work is required for product quality control, and its cost is approximately 10 to 20 percent of the operating-labor cost.

Plant overhead includes the cost of medical, safety, recreational, effluent-disposal, and warehousing facilities, etc. In general, the larger the plant, the lower the overhead per unit of production. Plant-overhead costs can vary between 15 and 150 percent of the operating-labor cost. Packaging costs depend on the physical and chemical nature of the product as well as on its use and value. The cost of packaging is as high as one-third of the selling price for soaps and pharmaceuticals.

Rapid Manufacturing-Cost Estimates Fixed manufacturing costs are a function of the fixed-capital investment and are independent of the production rate of the plant. Property taxes or rates depend on location. They may be taken as 2 percent of the fixed-capital cost of the plant in the absence of specific data. The cost of insurance depends on both location and the hazardous nature of the materials handled. This cost is normally of the order of 1 percent of the fixed-capital cost of the plant.

The manufacturing cost of a product is the sum of the processing or conversion cost and the cost of raw materials. The processing cost can be roughly broken down into three parts: investment-related cost, labor-related cost, and utility cost.

Companies usually include in the charge for overhead the following items: operating supplies, supervision, indirect payroll expenses, plant protection, plant office, general plant overhead, and control laboratory. This overhead charge is frequently taken as an equivalent percentage of the direct labor cost.

The percentage is best obtained from company records. Although it can vary over a wide range, a reasonable value is 125 percent. In this case, the labor-related cost would be 2.25 times the direct labor cost. For a 6000-h year and N persons per shift earning c_L\$ per hour, the annual labor-related cost would be 13,500 ($c_L N$).

Let us consider a plant of fixed-capital cost C_{FC}. If the annual property taxes are taken as 0.02 C_{FC}, insurance as 0.01 C_{FC}, and maintenance as 0.06 C_{FC}, the annual investment-related cost would be 0.09 C_{FC}. Annual utilities cost is A_U. The annual processing cost A_p can be represented by

$$A_p = \eta_1 C_{FC} + \eta_2 c_L N + A_U \tag{25-205}$$

where the factors η_1 and η_2 can be obtained from the data available in a particular company and have the dimensions of year^{-1} and hours per year respectively.

Substituting the information previously given into Eq. (25-205) yields the relationship

$$A_p = 0.09 C_{FC} + 13{,}500 c_L N + A_U \tag{25-206}$$

Equation (25-206) represents very closely the manufacturing costs of a particular company and is typical of the coefficients to be expected.

The annual processing cost A_{P2} for a similar plant of a different size designed for an annual production rate R_2 can be approximately calculated from an equation of the form

$$A_{P2} = \eta_1 C_{FC1}(R_2/R_1)^{0.7} + \eta_2 c_L N_1 (R_2/R_1)^{0.25} + A_{U1}(R_2/R_1) \tag{25-207}$$

(F. A. Holland, F. A. Watson, and J. K. Wilkinson, *Introduction to Process Economics*, Wiley, London and New York, 1974, p. 152).

The processing cost per unit of production for a plant with an annual production R_2 can be approximately calculated from

$$\frac{A_{P2}}{R_2} = \frac{100}{R_2}\left[\eta_1 C_{FC}\left(\frac{R_2}{R_1}\right)^{0.7} + \eta_2 c_L N_1\left(\frac{R_2}{R_1}\right)^{0.25} + A_{U1}\left(\frac{R_2}{R_1}\right)\right] \tag{25-208}$$

where A_{P2}/R_2 is in cents per kilogram.

Equation (25-208) can be used to compute data for plots such as Fig. 25-40, which shows the decrease in processing cost per unit of production A_{P2}/R_2 with increasing plant size.

Manufacturing Cost as a Basis for Product Pricing Pricing on the basis of cost plus a fair profit has the disadvantage of ignoring demand. The modern approach is to price on the basis of market research. However, the classic cost-plus-fair-profit approach can still give useful complementary information. This can be done by any of the following three methods:

1. Absorption pricing
2. Rate-of-return pricing
3. Marginal pricing

The gross annual profit A_{GP} for a product is given by

$$A_{GP} = A_S - A_{ME} - A_{BD} \tag{25-6}$$

where A_S is the revenue from annual sales, A_{ME} the annual manufacturing cost, and A_{BD} the balance-sheet annual depreciation charge.

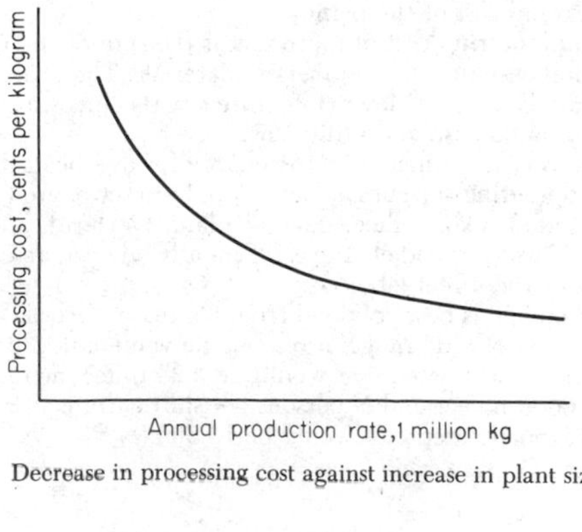

FIG. 25-40 Decrease in processing cost against increase in plant size.

Equation (25-6) can also be rewritten in the form

$$A_S = A_{GP} + A_{VME} + A_{FME} + A_{BD} \tag{25-209}$$

where A_{VME} and A_{FME} are the annual variable and fixed manufacturing costs or expenses respectively.

Equation (25-209) can also be rewritten as

$$c_S = c_{VME} + (A_{GP}/R) + (A_{FME} + A_{BD})/R \tag{25-210}$$

where R is the annual sales volume taken as equal to the annual production rate, c_S is the sales price per unit of production, and c_{VME} is the variable manufacturing expense per unit of production.

Absorption pricing is based on a normal annual production rate R. The gross profit per unit A_{GP}/R is taken as a fixed percentage χ of the fixed plus variable manufacturing costs given by the equation

$$\frac{A_{GP}}{R} = \left(\frac{\chi}{100}\right)\left[c_{VME} + \left(\frac{A_{FME} + A_{BD}}{R}\right)\right] \tag{25-211}$$

We combine Eqs. (25-210) and (25-211) to give

$$c_S = \left(\frac{100 + \chi}{100}\right)\left[c_{VME} + \left(\frac{A_{FME} + A_{BD}}{R}\right)\right] \tag{25-212}$$

Equations (25-210), (25-211), and (25-212) are based on a fixed normal annual production rate R.

Let us consider a change in annual production rate to $R + \Delta R$. In order to maintain the gross profit per unit as A_{GP}/R, the sales price per unit of production would need to be $c_S - \Delta c_S$. For this case Eq. (25-210) can be written in the modified form

$$c_S - \Delta c_S = c_{VME} + (A_{GP}/R) + [(A_{FME} + A_{BD})/(R + \Delta R)] \tag{25-213}$$

We subtract Eq. (25-213) from Eq. (25-212) to give

$$\Delta c_S = \frac{\Delta R}{R}\left(\frac{A_{FME} + A_{BD}}{R + \Delta R}\right) \tag{25-214}$$

Equation (25-214) gives the overpricing Δc_S per unit of production for an increase in annual production rate ΔR. Equation (25-214) also gives the underpricing Δc_S per unit of production for a decrease in annual production rate ΔR. In the first case the fixed costs or overheads are said to be overabsorbed and in the second case underabsorbed.

Absorption pricing is rigid and arbitrary and may result in business being turned away if the fixed sales price c_S cannot be obtained even though the business may give a useful contribution to fixed costs.

Rate-of-return pricing is a modified form of absorption pricing. It is based on the equation

$$\frac{A_{GP}}{A_{ME}} = \left(\frac{C_{TC}}{A_{ME}}\right)\left(\frac{A_{GP}}{C_{TC}}\right) \tag{25-215}$$

where C_{TC} is the total capital employed. Equation (23-215) can also be written as

Percentage markup on cost

= (capital-turnover ratio)(projected rate of return on capital)

The percentage markup on cost is calculated for a known capital-turnover ratio and a desired rate of return on capital. As with absorption pricing, the percentage markup on manufacturing cost per unit of production is calculated for a normal annual production rate. If this production rate is exceeded, the rate of return on capital will be higher than projected because of the decrease in unit cost. Conversely, if the production rate is lower than normal, the rate of return on capital will be lower than projected because of the increase in unit cost. For production rates both higher and lower than the normal production rate, the percentage markup is based on the normal unit cost. Thus the method is strictly valid only for the normal production rate.

With marginal pricing a company chooses its selling prices so as to maximize the total contribution $\Sigma R(c_S - c_{VME})$ from its various products. The method is particularly useful for large multiproduct companies with extensive existing facilities since it is the marginal

TABLE 25-37 Variances

Revenue from sales $Rc_S - R^\circ c_S^\circ$	=	Sales price $R(c_S - c_S^\circ)$	+	Sales volume $c_S^\circ(R - R^\circ)$		
Profit $Rc_{NP} - R^\circ c_{NP}^\circ$	=	Unit profit $R(c_{NP} - c_{NP}^\circ)$	+	Quantity $c_{NP}^\circ(R - R^\circ)$		
Total cost $Rc_{TE} - R^\circ c_{TE}^\circ$	=	Unit cost $R(c_{TE} - c_{TE}^\circ)$	+	Production rate $c_{TE}^\circ(R - R^\circ)$		
Direct material cost $Rc_{RM} - R^\circ c_{RM}^\circ$	=	Material price $R(c_{RM} - c_{RM}^\circ)$	+	Material usage $c_{RM}^\circ(R - R^\circ)$		
Direct labor cost $\theta c_L = \theta^\circ c_L^\circ$	=	Wage rate $\theta(c_L - c_L^\circ)$	+	Labor efficiency $c_L^\circ(\theta - \theta^\circ)$		
Overhead cost $C_{OH} - \theta^\circ c_{NOH}^\circ$	=	Budgeted cost $(C_{OH} - \theta c_{BOH})$	+	Volume $\theta(c_{BOH} - c_{NOH}^\circ)$	+	Efficiency $c_{NOH}^\circ(\theta - \theta^\circ)$

c_{BOH} = Flexible budgeted overhead cost, \$/h.
c_L = Actual labor cost, \$/h.
c_{NOH}° = Standard overhead cost based on normal production rate, \$/h.
c_{NP} = Actual net profit before tax, \$/unit.
C_{OH} = Actual overhead cost, \$/period.
c_{RM} = Actual raw-materials cost, \$/unit.
c_S = Actual selling price, \$/unit.
c_{TE} = Actual total cost, \$/unit.
R = Actual quantity, units/period.
R° = Standard quantity, units/period.
θ = Actual time to produce a given quantity, h.
θ° = Standard time to produce a given quantity, h.
$c_{NP} = c_S - c_{TE}$.
NOTE: The asterisk on all items not otherwise defined indicates the standard cost for that item.

cost that must be considered as the base case when entering into competition with another company. The lowest acceptable price for a product is that which gives the lowest worthwhile contribution to fixed costs. Marginal pricing enables a company to develop a more aggressive pricing policy than when using absorption or rate-of-return pricing.

Absorption, rate-of-return, and marginal pricing have been considered here on the basis of manufacturing cost. Total cost, which is the sum of manufacturing and general costs, can also be considered as the basis. In this case the appropriate profit to consider is the net annual profit rather than the gross annual profit.

Standard Costs for Budgetary Control For convenience and simplicity, we shall consider the total cost of a manufactured product to be the sum of the material, labor, and overhead costs. Standard costs are those that have been predetermined and budgeted for the manufacture of a given amount of product in a given time. The deviation of the actual cost from the standard cost is called the variance. It is far easier to make comparisons between periods by using variances than by using actual production data. The different variances for material, labor, and overhead costs are listed in Table 25-37.

Standard costing is extensively used in budgetary-control systems. Criteria for the establishment of standards range from the maximum possible under ideal conditions to those expected under normal conditions. Past or historical costs are not always the best basis for setting up standards because past performance may have been unnecessarily inefficient.

Static and Flexible Budgets Overhead cost can significantly affect the profitability of a project and is the only cost outside the control of the project manager. The project is expected to contribute a definite amount toward the expenses of the company and will be charged this amount even if the production rate is zero. This is the fixed component of the overhead cost and will include directly allocable costs such as depreciation and a proportion of general costs such as office salaries and heating.

Other nonproduction costs such as indirect labor may vary linearly with the production rate and represent the variable component of the overhead. Costs that are neither fixed nor variable but occur in discrete steps at various production levels (such as supervisory labor) are the semivariable component of the overhead cost. It is an easy matter to determine these various components for various production rates and list them as shown in Table 25-38.

Two types of overhead budget are currently in use. The static (often referred to as the fixed) budgeted overhead cost is related to the standard budgeted production rate. The flexible budgeted overhead cost is that shown as the total cost in Table 25-38. Values for intermediate production rates are often obtained by interpolation. This is justifiable only when semivariable costs are a negligible part of overhead costs.

Flexible budgeting is more widely used than static budgeting despite certain logical difficulties. This is so because production in many cases is seasonal and the use of a static production norm might distort evaluation of performance. Variances are the difference between the actual costs expended and the budgeted costs expected. Variances are unfavorable if positive and favorable if negative. Any variance should be explained and, if necessary, controlled; the largest variance should be considered first.

Let us consider the overhead-cost data for Table 25-38 with 10 million kg per month as the standard production rate. The static budgeted overhead is then \$150,000 per month, or 1.5 cents per kilogram. We assume that the actual overhead is \$186,000 for a month in which 12 million kg was produced. Then, the static budgeted overhead cost would be 12 million(1.5), or \$180,000 per month. Therefore, the variance is \$186,000 − \$180,000 = +\$6000, which is unfavorable because \$6000 more was spent than was anticipated.

From Table 25-38 we find that the flexible budgeted overhead cost for a production rate of 12 million kg per month is \$190,000. The corresponding variance is \$186,000 minus \$190,000, or −\$4,000, which is favorable because \$4,000 less was spent than was antici-

TABLE 25-38 Flexible Budget for Overhead Costs

Overhead cost, \$/month	Production, 1 million lb/month 8	9	10	11	12
Fixed	40,000	40,000	40,000	40,000	40,000
Variable	40,000	45,000	50,000	55,000	60,000
Semivariable	40,000	40,000	60,000	60,000	90,000
Total	120,000	125,000	150,000	155,000	190,000

pated. Thus, the use of flexible budgeting makes this particular performance look better without changing either the production rate or a single cost of the planned budget.

The Standard Hour The standard hour can be defined as the number of units of output expected to be produced in 1 h. It is often used as a measure of output rate by cost accountants.

Let us consider a batch processing unit that can produce either 1000 kg of product A in a cycle time of 5 h or 900 kg of product B in a cycle time of 3 h. Thus, for this processing unit a standard hour is 200 kg of product A or 300 kg of product B. In a budget period of, say, 1000 h, it is possible to produce 200,000 kg of product A, or 300,000 kg of product B, or any appropriate combination of the two products.

For example, let us assume that production requirements are twice as great for product A as for product B, i.e., a ratio of 600 kg, or 3 standard hours, of product A to 300 kg, or 1 standard hour, of product B. On this basis, for a budget period of 1000 h, 750 standard hours [(3/4)1000] would be used to produce 750(200) = 150,000 kg of product A, and 250 standard hours [(1/4)1,000] would be used to produce 250(300) = 75,000 kg of product B.

Production efficiency P_e can be calculated from

$$P_e = (P_s/P_a)100 \qquad (25\text{-}216)$$

where P_s is the actual production rate in standard hours and P_a is the actual hours worked.

The level of production activity P_ℓ can be calculated from

$$P_\ell = (P_s/P_b)100 \qquad (25\text{-}217)$$

where P_b is the budgeted production in standard hours.

The deviation from budgeted capacity b_c can be calculated from

$$b_c = (P_a/P_w)100 \qquad (25\text{-}218)$$

where P_w is the budgeted number of working hours.

Actual Costs versus Standard Costs Let us consider the sales, profits, and manufacturing-cost data in Table 25-39. The gross profit is $33,129 per period better than expected. Clearly, there is less incentive to investigate overall costs when the profit variance is favorable than if the profit were less than expected. However, standard costing enables an objective analysis of the data, whether good or bad, to be made.

The individual variances in Table 25-39 show that the increased profit is due to reduced material costs, which affect manufacturing costs to a greater extent than increased labor and overhead costs. The individual variances also show that to an even greater extent the increased profit is due to the increase in sales revenue.

Clearly, management will wish to investigate both labor and overhead costs for any inefficiencies and to ascertain the reasons for the improved sales revenue. If necessary, the standard values can be revised.

We notice that profit is obtained as the difference between two large cash sums and that variances of some 3 percent in manufacturing costs and sales revenue have resulted in a variance of some 33 percent in gross profit.

Table 25-39 is a very simplified presentation. In a full standard costing system, the direct material, direct labor, and overhead variances are broken down into component parts to enable an even closer look at the operation. Standard costing is an invaluable aid to management for controlling a business.

TABLE 25-39 Sales, Profits, and Manufacturing Costs

Component	Actual, \$/period	Standard, \$/period	Variance, \$/period
Direct materials cost	325,080	350,000	−24,920
Direct labor cost	55,001	51,600	+ 3,401
Overhead cost	75,000	68,370	+ 6,630
Manufacturing cost	455,081	469,970	−14,889
Revenue from sales	588,240	570,000	+18,240
Gross profit	133,159	100,030	+33,129

TABLE 25-40 Cost Data for Problems

Factor	Actual value	Standard value
Raw-materials cost, \$/unit	$c_{RM} = 0.18$	$c^\circ_{RM} = 0.20$
Direct labor cost, \$/unit	$c_L = 8.45$	$c^\circ_L = 8.00$
Production rate, units/period	$R = 1{,}806{,}000$	$R^\circ = 1{,}750{,}000$
Production time h/period	$\theta = 6509$	$\theta^\circ_N = 6250$
Fixed and semivariable overhead cost, \$/period	. . . .	$C^\circ_{FOH} = 60{,}000$
Variable overhead cost, \$/h	. . . .	$C^\circ_{VOH} = 1.00$
Overhead cost, \$/period	$C_{OH} = 75{,}000$	

Variances: Direct Material Cost Since the variance in direct material cost $\Delta\,(Rc_{RM})$ is the difference between actual cost and standard cost,

$$\Delta\,(Rc_{RM}) = Rc_{RM} - R^\circ c^\circ_{RM} \qquad (25\text{-}219)$$

where R is the actual quantity and R° is the standard quantity, units per period, c_{RM} is the actual price, and c°_{RM} is the standard price, \$ per unit.

Equation (25-219) can be written in an expanded form:

$$\Delta\,(Rc_{RM}) = R(c_{RM} - c^\circ_{RM}) + c^\circ_{RM}(R - R^\circ) \qquad (25\text{-}220)$$

where $R(c_{RM} - c^\circ_{RM})$, known as the direct-material-price variance, is the actual quantity multiplied by the deviation in unit price, and $c^\circ_{RM}(R - R^\circ)$, known as the direct-material-usage variance, is the standard unit price multiplied by the deviation in quantity.

By using the data of Table 25-40, let us calculate the direct materials cost and the standard direct materials cost as

$$Rc_{RM} = 1{,}806{,}000(0.18) = \$325{,}080/\text{period}$$

$$R^\circ c^\circ_{RM} = 1{,}750{,}000(0.20) = \$350{,}000/\text{period}$$

From Eq. (25-219) we calculate the direct-material-cost variance as −$24,920 per period. This variance is favorable. However, by using the relations of Eq. (25-220) we calculate a direct-material-price variance of (1,806,000)(0.18 − 0.20) = −$36,120 per period. This variance is favorable. Likewise, we calculate a direct-material-usage variance of (0.20)(1,806,000 − 1,750,000) = $11,200 per period. This variance is unfavorable.

In this case, the favorable direct-material-cost variance was achieved because of a lower unit price despite an inefficient material usage, which needs to be investigated. (There is no room for complacency since the lower unit price may well be temporary.)

In the case of mixtures of raw materials, the direct-material-usage variance can be further subdivided into (1) a direct-material-mixture variance and (2) a direct-material-yield variance. The former is due to the difference between the actual and standard mixture compositions, and the latter to the difference between the actual and standard yields. Here, the standard yield is the output expected from the standard input of material. The yield variance denotes the extent of loss of material. The direct-material-mixture variance can be illustrated by Example 22.

Example 22 A standard mixture of 100 units of material contains 70 percent of material A at $0.08 per unit and 30 percent of material B at $0.12 per unit. The standard mixture cost is 70(0.08) + 30(0.12) = $9.20.

Now let us consider a mixture of 100 units containing 75 percent of material A and 25 percent of material B. The cost of this mixture at standard prices is 75(0.08) + 25(0.12) = $9.00. The direct-material-mixture variance is $9.00 − $9.20 = −$0.20 and is favorable. The favorable variance has been brought about by using more of the lower-priced material A and less of the higher-priced material B.

The direct-material-yield variance is illustrated as follows. Let us assume that the standard mixture (cost $9.20 for 100 units) has a standard loss of 20 percent, making the cost $9.20 for 80 units, or $0.115 per unit of output. Now let us consider the actual loss to be 30 percent, leaving 70 units of output for each 100 units of input. The direct-material-yield variance is 0.115(80 − 70) = $1.15 and is unfavorable.

Variances: Direct Labor Cost Since the variance in direct labor cost $\Delta\,(\theta c_L)$ is the difference between actual cost and standard cost,

$$\Delta\,(\theta c_L) = \theta c_L - \theta^\circ c_L^\circ \tag{25-221}$$

where c_L is the actual pay or wage rate, \$ per hour; c_L° is the standard pay or wage rate, \$ per hour; θ is the actual time taken to produce a given quantity of product in a given period, hours; and θ° is the standard time taken to produce a given quantity of product in a given period, hours.

Equation (25-221) can also be written in expanded form:

$$\Delta\,(\theta c_L) = \theta(c_L - c_L^\circ) + c_L^\circ(\theta - \theta^\circ) \tag{25-222}$$

where $\theta(c_L - c_L^\circ)$, known as the direct pay, or wage-rate, variance, is the actual time taken to produce a given output multiplied by the deviation in wage rate, and $c_L^\circ(\theta - \theta^\circ)$, known as the direct-labor-efficiency variance, is the standard wage rate multiplied by the deviation in time taken to produce a given output.

By using the data of Table 25-40, we calculate that 1 standard hour corresponds to

$$(1{,}750{,}000/6250) = 280 \text{ units/standard hour}$$

Standard time to actual production is

$$\theta^\circ = (1{,}806{,}000/280) = 6450 \text{ standard hours/period}$$

Direct labor cost is

$$\theta c_L = 6509(8.45) = \$55{,}001/\text{period}$$

Standard direct labor cost is

$$\theta^\circ c_L^\circ = 6450(8.00) = \$51{,}600/\text{period}$$

From Eq. (25-219) we calculate the direct-labor-cost variance as \$3401 per period. This variance is unfavorable. However, by using the relations of Eq. (25-220), we calculate a direct pay, or wage-rate, variance of $\theta(c_L - c_L^\circ) = 6509(8.45 - 8.00) = \$2929$ per period. This direct pay variance is unfavorable. Likewise, we calculate the direct-labor-efficiency variance as $c_L^\circ(\theta - \theta^\circ) = 8.00(6509 - 6450) = \$472$ per period. This variance is also unfavorable.

For this example, the adverse direct-labor cost variance of \$3401 is due to both a higher wage rate per hour and a higher number of labor-hours.

The direct-labor-cost variance can, if necessary, be broken down into a direct-labor-idle-time variance in addition to the direct-wage-rate and direct-labor-efficiency variances. The direct-labor-idle-time variance is simply the number of idle labor-hours in the period multiplied by the standard wage rate. This is rarely relevant to the conditions existing in process plants except when maintenance is involved.

Variances: Overhead Cost The variance in overhead cost ΔC_{OH} is the difference between actual overhead cost and static standard overhead cost,

$$\Delta C_{OH} = C_{OH} - \theta^\circ c_{NOH}^\circ \tag{25-223}$$

where C_{OH} is the actual overhead cost incurred in a given period, \$ per period; c_{NOH}° is the static standard overhead cost based on the normal production rate, \$ per hour; θ is the actual time taken to produce a given quantity of product in a given period; and θ° is the standard time taken to produce a given quantity of product in a given period, hours.

Equation (25-223) can also be written in expanded form

$$\Delta C_{OH} = (C_{OH} - \theta c_{BOH}) + \theta(c_{BOH} - c_{NOH}^\circ) + c_{NOH}^\circ(\theta - \theta^\circ) \tag{25-224}$$

where c_{BOH} is the flexible budgeted overhead cost at the actual production rate or operating capacity.

In Eq. (25-223), $(C_{OH} - \theta c_{BOH})$ is known as the budgeted overhead-cost variance, $\theta(c_{BOH} - c_{NOH}^\circ)$ as the overhead-volume variance, and $c_{NOH}^\circ(\theta - \theta^\circ)$ as the overhead-efficiency variance. The last is analogous to the labor-efficiency variance and is the standard overhead rate multiplied by the deviation in time taken to produce a given output.

Also in Eq. (25-224), c_{BOH} is simply the flexible budgeted overhead cost in dollars per hour for the actual production rate, and the overhead-volume variance $\theta(c_{BOH} - c_{NOH}^\circ)$ is the actual time taken to produce a given output multiplied by the difference between the flexible budgeted overhead cost and the standard overhead cost in dollars per hour. The budgeted overhead-cost variance $(C_{OH} - \theta c_{BOH})$ is the difference between the actual overhead cost and the actual time (in hours) required to produce the given output multiplied by the flexible budgeted overhead cost (in dollars per hour).

We shall write the fixed overhead cost for the budget period as C°_{FOH}, the semivariable overhead cost as C_{SVOH}°, and the standard hours to produce the agreed normal production as θ_N°. The standard overhead cost at the agreed normal production rate can then be calculated from

$$c_{NOH}^\circ = [(C_{FOH}^\circ + C_{SVOH}^\circ)/\theta_N^\circ] + c_{VOH}^\circ \tag{25-225}$$

where c_{VOH}° is the standard variable overhead cost, \$ per hour.

For production rates that differ from the agreed normal rate, the flexible budgeted overhead cost is given by

$$c_{BOH} = [(C_{FOH}^\circ + C_{SVOH}^\circ)/\theta] + c_{VOH}^\circ \tag{25-226}$$

where θ is the actual hours taken to produce a given amount of product.

For production rates lower than normal, the fixed overheads are underused, and the flexible budgeted overhead cost c_{BOH} is greater than the standard overhead cost c_{NOH}°. For production rates higher than normal, c_{BOH} is less than c_{NOH}°.

It is common practice in cost accountancy to treat the standard semivariable cost C_{SVOH}° at the normal production rate as part of the standard fixed cost. In this case, Eqs. (25-225) and (25-226) can be written respectively as

$$c_{NOH}^\circ = (C_{FOH}^\circ/\theta_N^\circ) + c_{VOH}^\circ \tag{25-227}$$

$$c_{BOH} = (C_{FOH}^\circ/\theta) + c_{VOH}^\circ \tag{25-228}$$

By using the data of Table 25-40 in Eq. (25-227), we calculate the standard overhead cost to be

$$c_{NOH}^\circ = (60{,}000/6250) + 1.00 = \$10.60/\text{h}$$

From Eq. (25-228) and the data in Table 25-40 we calculate the flexible budgeted overhead cost to be

$$c_{BOH} = (60{,}000/6509) + 1.00 = \$10.22/\text{h}$$

By substituting into Eq. (25-223), we calculate the overhead cost variance:

$$\Delta c_{OH} = 75{,}000 - 6450(10.60) = \$6630$$

This variance is unfavorable. (Note that standard time for actual production was previously calculated to be 6450 h per period.)

The overhead cost variance comprises (1) budgeted overhead-cost variance, (2) overhead-volume variance, and (3) overhead-efficiency variance. The calculations for each follow:

$$c_{OH} - \theta c_{BOH} = 75{,}000 - 6509(10.22) = \$8478$$

Budgeted overhead-cost variance is positive and, therefore, unfavorable.

$$\theta(c_{BOH} - c_{NOH}^\circ) = 6509(10.22 - 10.60) = -\$2473$$

Overhead volume variance is negative and favorable.

$$c_{NOH}^\circ(\theta - \theta^\circ) = 10.60(6509 - 6450) = \$625$$

Overhead efficiency variance is positive and unfavorable.

The total variances in each category are listed in Table 25-39.

Chemical engineers usually make detailed evaluations of costs rather than evaluations for profits or sales. However, the latter can be analyzed in a similar manner to costs by using the equations shown in Table 25-37. For this purpose, the sign convention will be reversed because an increase in sales or profits would be considered favorable, whereas an increase in cost would be considered unfavorable. The equations can be applied to both batch and continuous processes.

Budgets can be used for both forward planning and control. Variances show managers what their costs should have been and how near they came to meeting budgeted values. Managers will be able to assess, over a number of budget periods, the rate of improvement in performance in their areas of responsibility. A good budgetary system not only should provide detailed information and an appraisal of performance but also should motivate people to improve performance.

Contribution Analysis Contribution analysis can be used to make rapid assessments of the effect of changes in manufacturing costs on profitability. A dimensionless contribution efficiency η can be defined by rewriting Eq. (25-12) in the form

$$\eta = \frac{R(c_S - c_{VE}) - A_{FE}}{R(c_S - c_{VE})} \tag{25-229}$$

[F. A. Holland and F. A. Watson, *Eng. Process Econ.*, **1**, 135–143 (1976)].

This represents the ratio of the net annual profit A_{NP} actually achieved divided by the profit which could be obtained if no repayment of capital or interest were required and all fixed-expense items were credited free to the project. The contribution efficiency η is also the profit per unit of contribution. A value for η of unity would be obtained for a very high production rate R whether c_S is greater or less than c_{VE}. For the unusual case of c_S being equal to c_{VE}, the value of η would become negatively infinite for a finite annual fixed expense A_{FE} or positively infinite if A_{FE} became negative because of excessive subsidy of expenses. However, for most projects which are intended to pay their own expenses and taxes, A_{NP} must be positive, and hence c_S is usually greater than c_{VE}, so that η will normally have values in the range of zero to unity. For projects which are not intended to make a profit but are provided for their social or amenity value, the aim should be to bring the value of η as near to zero as possible.

The breakeven production rate R_B is defined by Eq. (25-13) as the production rate at which the project makes neither a profit nor a loss. Equation (25-13) and (25-229) can be combined to give

$$\eta = (R - R_B)/R \tag{25-230}$$

which shows that the contribution efficiency η is a function of the production rate R and that η has the value zero when the production rate is the breakeven production rate R_B. For all real projects R_B will be positively finite while R cannot be less than zero, and hence the practical range of η is from negative infinity to unity.

At first glance it might appear that it is desirable to have a value of η as near to unity as possible. However, this is not necessarily so. Reference to Eq. (25-227) will show that if the unit contribution $(c_S - c_{VE})$ is positive, as it must be if the project is ever to make a profit, a value of unity for the contribution efficiency η implies a negligible value of the annual fixed expense A_{FE} when compared with the annual contribution $R(c_S - c_{VE})$. In such cases either $R(c_S - c_{VE})$ is very large, thus attracting competition in spite of high capital charges, or the fixed expenses are very low, so that it is easy for many small competitors to enter the market. The result of such competition often leads to a rapid reduction in sales price c_S. Ball-point pens and electronic calculators were both drastically reduced in price as a result of competitors entering the market.

Since the variable expense per unit of production c_{VE} is by definition independent of production rate, the unit contribution $(c_S - c_{VE})$ and hence the value of η will be reduced. On the other hand, a very large company may be in a stronger and more stable position with a modest contribution efficiency and relatively high fixed costs which will deter competitors from entering the market and thereby depressing the sales price. In this argument it is implicit that c_S is also independent of output from the project under consideration. In many cases this will be the case since if many small buyers can choose from many alternative producers, the individual producer cannot adjust the price to suit its output, while at the opposite extreme a group of producers that are in a position to make such adjustments are also likely to attract the attention of antitrust legislation.

It is of interest to be able to examine the effect of changes in productivity on the profitability of projects. Historically, labor costs have been regarded as variable costs, implying that if workers doubled their output their net wages also doubled. This may have been the case for some piecework rates, but it is generally not true today. It is not normally possible to reduce the work force in step with falling demand or to recruit and train labor in step with increasing demand. In general it is better to consider labor as a fixed cost, with any part of a production bonus which is truly proportional to output included in the variable expense c_{VE}. If the annual fixed expense A_{FE} varies significantly with production rate R owing to this factor, then the breakeven chart will consist of curves, the simplicity of the method is lost, and it will be assumed that a particular change in productivity agreements implies a step change in A_{FE} and/or in c_{VE}. Let us consider an increase in productivity for the same fixed labor cost, with other fixed costs remaining the same. We shall let the original production rate R be increased by an increase in productivity by a fraction ϕ. Therefore,

$$\Delta R = R\phi \tag{25-231}$$

where ΔR is the increase in sales volume or production rate. By substitution into Eq. (25-229), the resulting increase in profit ΔA_{NP} is given by

$$A_{NP} + \Delta A_{NP} = A_S + \Delta A_S - (A_{VE} + \Delta A_{VE}) - (A_{FE} + 0) \tag{25-232}$$

We shall subtract A_{NP} from Eq. (25-232) to give

$$\Delta A_{NP} = \Delta A_S - \Delta A_{VE} = \Delta R(c_S - c_{VE}) \tag{25-233}$$

It follows from Eqs. (25-12), (25-229), and (25-233) that

$$\Delta A_{NP}/A_{NP} = \phi/\eta \tag{25-234}$$

Thus a change of productivity ϕ of 10 percent will result in a 10 percent increase in profit when $\eta = 1$ and a very large increase in profit when η is close to zero. If η is negative, increased productivity reduces the profit (or increases the loss).

Equation (25-234) illustrates the enormous influence that go-slow tactics can have on the profitability of companies and processes which have low contribution efficiencies, since a slowdown has little effect on A_{FE}. It is sometimes the case that in different countries productivity per worker varies considerably in similar industries. When poor productivity is not the result of technical or capital inadequacy, it should be possible to increase profitability without a proportionate increase in A_{FE}.

Breakeven charts present a snapshot of the present situation by means of graphs which are generally drawn in the manner shown in Figs. 25-2, 25-3, and 25-4. Since the lines are straight, this implies that c_S, c_{VE}, and A_{FE} will remain constant over the range of variation of R, which is of interest. The values would be based on the production rate currently achieved (or scheduled), since all the data are available from the financial analysis of current production, so that c_S would be $(A_S)_0/R_0$, and so on.

Two well-known ratios used by financial analysts are the profit-to-sales ratio (PSR) and the contribution-to-sales ratio (CSR). The profit-to-sales ratio is simply Eq. (25-127) for profit margin (PM) rewritten as the ratio

$$(\text{PSR}) = A_{NP}/Rc_S \tag{25-235}$$

The contribution-to-sales ratio

$$(\text{CSR}) = (A_S - A_{VE})/A_S = (c_S - c_{VE})/c_S \tag{25-236}$$

We substitute these values into Eq. (25-229) to give

$$\eta = (\text{PSR})/(\text{CSR}) \qquad (25\text{-}237)$$

The contribution efficiency at the scheduled output η_0 is given by substituting the value of the scheduled output into Eq. (25-229) to give

$$\eta_0 = \frac{R_0(c_S - c_{VE}) - A_{FE}}{R_0(c_S - c_{VE})}$$

The characteristic shape of a given breakeven chart of the type represented by Figs. (25-2), (25-3), and (25-4) can be defined by the two ratios (CSR) and η_0, while the scale of the project can be defined by a single annual cost such as A_{FE}. This information may be used for the rapid investigation of the likely effect on current profits obtainable by changes in various factors such as prices, expenses, and throughput. It should be noted that this technique is not intended to replace discounted methods of investment appraisal but to provide a rapid assessment of the probable effect of changes in current conditions. If the current profitability is always maximized, then the discounted-cash-flow present value will always be made as great as conditions in a changing world will permit.

Valuation of Recycled Heat Energy The rising cost of energy is having an inflationary effect on manufacturing costs. One obvious way to reduce energy costs is to recycle heat energy whenever possible [S. A. K. El-Meniawy, F. A. Watson, and F. A. Holland, *Indian Chem. Eng.*, **22** (July–September 1980)].

Heat pumps are particularly suitable for recycling heat energy in the chemical-process industries. For the outlay of an additional fixed-capital expenditure C_{FC} on a heat-pump system, a considerable reduction in the annual heating cost can be effected.

Let us consider a process unit requiring heat at the rate of Q_D GJ/h operating for y h in a year. We shall let the unit cost of this base heating requirement be c_B \$ per gigajoule. Therefore the annual heating cost for this unit is $Q_D y c_B$ \$ per year.

We then consider the use of a heat pump to supply this heat so that

$$Q_D = W(\text{COP})_A \qquad (25\text{-}238)$$

where W is the rate of energy input to the compressor in gigajoules per hour and $(\text{COP})_A$ is the actual coefficient of performance of the heat pump.

When interest charges are involved, the fixed-capital expenditure on a heat-pump system C_{FC} can be related to an annual cost A_{FC} for the estimated life of the heat pump in years by the equation

$$A_{FC} = C_{FC} f_{AP} \qquad (25\text{-}239)$$

where $f_{AP} = [i(1 + i)^n]/[(1 + i)^n - 1]$, the annuity present-worth factor, and i is the fractional interest rate per year payable on the borrowed money.

A given value for A_{FC} enables a cost in, say, dollars per gigajoule, to be assigned to the heat energy made available by a heat pump.

We shall consider a heat-pump system which operates for y h/year and consumes W GJ/h of high-grade energy to drive the compressor. We shall let the unit cost of the input energy to the compressor be c_I \$ per gigajoule. The annual amount of heat delivered by the heat pump is $Q_D y$, which in terms of Eq. (25-238) can be written as $W(\text{COP})_A y$ in gigajoules per year. The annual cost of this delivered heat, neglecting any maintenance cost, is $(W c_I y + A_{FC})$ in dollars per year. Therefore the unit cost c_D of the heat energy delivered by a compressor-driven pump is

$$c_D = (W c_I y + A_{FC})/[W(\text{COP})_A y] \qquad \$/\text{GJ} \qquad (25\text{-}240)$$

The ratio of the unit costs of the delivered heat energy and the input energy can be obtained by combining Eqs. (25-239) and (25-240) to give

$$(c_D/c_I) = [(1 + \psi)/(\text{COP})_A] \qquad (25\text{-}241)$$

where $\psi = (C_{FC}/W)(f_{AP}/y c_I)$. ψ is a dimensionless parameter which contains cost and usage data for a particular heat pump; it should have as low a value as possible in order to minimize the unit cost of the delivered heat c_D. The cost C_{FC}/W is the fixed-capital cost per unit of input energy in \$ $(\text{GJh}^{-1})^{-1}$, and this should also be as low as possible consistent with a long life and good reliability for the heat pump. Clearly y, the number of operating hours per year, should approach as closely as possible to the maximum value of 8760 h for a 365-day year.

Since all costs refer to a given year, Eqs. (25-239), (25-240), and (25-241) are independent of inflation.

The annual cost of heat delivered by the heat pump is $Q_D y c_D$, where the unit cost c_D is given by Eq. (25-240). Therefore, the annual saving on heating costs in dollars per year is $Q_D y(c_B - c_D)$, which can also be written in terms of Eq. (25-238) as $W(\text{COP})_A y(c_B - c_D)$.

The payback period in years for a heat-pump system is the additional fixed-capital cost C_{FC} divided by the annual saving on heating costs. This can be written as

$$(\text{PBP}) = \left(\frac{C_{FC}}{W}\right)\left[\frac{1}{(\text{COP})_A y(c_B - c_D)}\right] \qquad (25\text{-}242)$$

which can also be written in terms of Eq. (25-241) as

$$(\text{PBP}) = \left(\frac{C_{FC}}{W}\right)\left\{\frac{1}{y[(\text{COP})_A c_B - c_I(1 + \psi)]}\right\} \qquad (25\text{-}243)$$

For the special case of the unit cost of the input energy to the compressor being the same as the unit cost of the base heat supply, i.e., $c_B = c_I$, Eq. (25-243) simplifies to

$$(\text{PBP}) = \left(\frac{C_{FC}}{W}\right)\frac{1}{y c_I[(\text{COP})_A - (1 + \psi)]} \qquad (25\text{-}244)$$

Equation (25-244) can be used to calculate the payback period when electricity, oil, or gas, etc., is used to drive the compressor and also to provide the base heating.

Equation (25-244) shows that, to have a low payback period (PBP), C_{FC}/W and ψ should be small and y, $(\text{COP})_A$, and c_I large. Clearly as the unit cost of input energy c_I increases, the economics of heat pumps becomes more favorable.

The value of ψ will for most cases be less than 0.2 and with the right application may well be less than 0.1. Values for the annuity present-worth factor will in most cases be less than 0.15.

Since 1 bbl (0.159 m^3) of oil is normally quoted as having a thermal-energy value of 6.12 GJ, a world oil price of, say, US\$40 per barrel is equivalent to US\$6.54 per gigajoule.

For simplicity, we substitute c_I = US\$6⅔ per gigajoule and a conservative value of $[(\text{COP})_A - (1 + \psi)] = 3$ into Eq. (25-244) to give

$$(\text{PBP}) = (C_{FC}/W)(1/20y) \qquad (25\text{-}245)$$

where (PBP) is the payback period in years, y is the operating hours per year, and (C_{FC}/W) is the fixed-capital cost in US\$ $(\text{GJh}^{-1})^{-1}$ of primary-energy input.

Equation (25-245) shows that in this particular case the fixed-capital cost per unit of input energy (C_{FC}/W) must not exceed \$160,000 $(\text{GJh}^{-1})^{-1}$, or \$576 per kilowatt, to have a 1-year payback period if the heat pump is operational for 8000 h/year. For this case the corresponding value of ψ is about 0.12 for a heat pump with an operating life of 10 years purchased with money borrowed at a 10 percent rate of interest.

Equation (25-245) also shows that the fixed-capital cost per unit of energy input (C_{FC}/W) must not exceed \$40,000 $(\text{GJh}^{-1})^{-1}$, or \$144 per kilowatt, to have a 1-year payback period if the heat pump is operational for only 2000 h/year. For this case the corresponding value of ψ is also about 0.12 for a heat pump with an operating life of 10 years purchased with money borrowed at a 10 percent rate of interest.

FIXED-CAPITAL-COST ESTIMATION

Total Capital Cost The installed cost of the fixed-capital investment C_{FC} is obviously an essential item which must be forecast before an investment decision can be made. It forms part of the total capital investment C_{TC}, defined by Eq. (25-14). The fixed-capital investment is usually regarded as the capital needed to provide all the depreciable facilities. It is sometimes divided into two classes by defining battery limits and auxiliary facilities for the project. The boundary for battery limits includes all manufacturing equipment but excludes administrative offices, storage areas, utilities, and other essential and nonessential auxiliary facilities.

Cost Indices The value of money will change because of inflation and deflation. Hence cost data can be accurate only at the time when they are obtained and soon go out of date. Data from cost records of equipment and projects purchased in the past may be converted to present-day values by means of a cost index. The present cost of the item is found by multiplying the historical cost by the ratio of the present cost index divided by the index applicable at the previous date. Ideally each cost item affected by inflation should be forecast separately. Labor costs, construction costs, raw-materials and energy prices, and product prices all change at different rates. Composite indices are derived by adding weighted fractions of the component indices. Most cost indices represent national averages, and local values may differ considerably.

Table 25-41 presents information on some cost indices for the United States. *Engineering News-Record* updates its construction-cost index in March, June, September, and December. The *Oil and Gas Journal* gives the Nelson refinery indices in the first issue of each quarter. The *Chemical Engineering* plant-cost index and Marshall and Swift equipment-cost index are given in each issue of the publication *Chemical Engineering*. Derivation of the base values is referred to in the respective publications.

Table 25-42 is based on the method suggested by J. Cran [*Eng. Process Econ*., **2**, 89–90 (1977)]. He showed that reasonably accurate plant-cost indices for various countries could be derived by using two component indices in the equation

$$(\mathrm{CI})_P = 0.327(\mathrm{CI})_{ST} + 0.673(\mathrm{CI})_L \qquad (25\text{-}246)$$

where $(\mathrm{CI})_{ST}$ is the steel-price index and $(\mathrm{CI})_L$ the earnings index for labor in the particular country. Most of the data required can be obtained from the *United Nations Monthly Bulletin of Statistics* or the Organization for Economic Cooperation and Development (OECD) annual review of the iron and steel industry. In Table 25-42, the plant-cost indices have been brought to a common base of 1970 = 100. The values given do not relate costs in one country to those in another country, as this involves many complex and difficult problems. However, the table indicates the inflationary trends in plant costs since 1970 for each of the countries listed.

Types and Accuracy of Estimates Capital-cost estimates may be required for a variety of reasons, among others to enable feasibility studies to be carried out, to enable a manufacturing company to select from alternative investments, to assist in selection from alternative designs, to provide information for planning the appropriation of capital, and to enable a contractor to bid on a new project. It is therefore essential to achieve the greatest accuracy of estimation with a minimum expenditure of time and money.

Two simple rules are invaluable in aiding the production of consistently accurate estimates:

1. Check the completeness of the project scope.
2. Reduce the effect of bias by using statistically proven methods of estimation based on experience.

Estimates which are lower than actual project costs are often the result of sizable omissions of equipment, services, or auxiliary facilities rather than of errors in pricing or estimation methods. To avoid this, the use of a checklist of items involved in a new project as given in Table 25-43 can be invaluable.

The first stage toward producing an accurate estimate is to use a standard cost code for all construction projects. Table 25-44 shows a suitable numerical cost code, and Table 25-45 shows a typical alphabetical-numerical code. The cost-code system can be used throughout the estimating and construction stages for the collection of cost data by manual or computer methods. There are numerous types of fixed-capital-cost estimates, but in 1958 the American Association of Cost Engineers defined five types as follows:

1. *Order-of-magnitude estimate (ratio estimate)*. Rule-of-thumb method based on cost data for previous similar types of plant; probable error within 10 to 50 percent.
2. *Study estimate (factored estimate)*. Better than order-of-magnitude; requires knowledge of major items of equipment; used for feasibility surveys; probable error up to 30 percent.
3. *Preliminary estimate (budget-authorization estimate)*. Requires more detailed information than study estimate; probable error up to 20 percent.
4. *Definitive estimate (project-control estimate)*. Based on considerable data prior to preparation of completed drawings and specifications; probable error within 10 percent.
5. *Detailed estimate (firm or contractor's estimate)*. Requires

TABLE 25-41 Cost Indices for the United States

	Price indices*				Construction-cost indices				
Year	Consumer	Wholesale commodity (all)	Total, chemical and allied industries	Industrial chemicals	Chemical equipment†	Process plants‡	Petroleum refinery§	General construction¶	General building¶
1969	110	107	100	100	283	119	115	119	118
1970	116	110	102	101	301	126	127	129	124
1971	121	114	104	102	319	132	141	148	141
1972	125	119	104	101	330	137	153	164	156
1973	133	135	110	103	344	144	163	177	169
1974	147	161	147	152	398	165	182	189	179
1975	161	175	181	207	444	182	200	207	194
1976	170	182	187	219	472	192	215	224	212
1977	181	194	192	224	505	204	223	241	230
1978	195	209	199	226	545	219	244	259	249
1979	217	234	222	264	599	239	260	281	271
1980	247	269	260	324	660	261	286	301	289
1981	272	293	288	363	721	297	315	329	310
1982	287(May)	299(June)	294(June)	353(June)	746	314	340	356	331
1983	. . .	. . .	. . .	. . .	761	. . .	. . .	379	353

*Bureau of Labor Statistics, 1967 = 100. †Marshall and Swift, *Chem. Eng.*, 1926 = 100. ‡*Chem. Eng.*, plant index, 1957–1959 = 100. §Nelson refinery index, *Oil Gas J.*, 1967 = 100. ¶*Eng. News Rec.*, 1967 = 100.

TABLE 25-42 International Plant-Cost Indices*

Year	Australia	Austria	Belgium	Canada	Denmark	France	West Germany	Ireland	Nether-lands	South Africa	Spain	Sweden	United Kingdom	United States
1971	106	107	105	107	109	109	110	113	108	112	112	99	113	104
1972	116	114	117	115	116	120	118	126	129	115	124	103	126	107
1973	125	126	133	123	137	141	132	145	138	128	142	105	132	113
1974	143	156	162	135	170	168	144	176	154	147	175	108	148	119
1975	186	199	188	165	204	197	176	240	173	188	217	154	205	149
1976	214	195	201	189	211	219	174	287	191	219	267	171	245	149
1977	245	217	211	207	240	256	186	344	195	263	337	182	285	163
1978	270	233	227	224	260	277	194	356	210	284	397	197	302	176
1979	286	257	243	244	280	311	203	447	219	328	457	212	348	193

*From J. Cran, *Eng. Process Econ.*, **2**, 89–90 (1977).
NOTE: Figures apply for January of each year (1970 = 100).

completed drawings, specifications, and site surveys; probable error within 5 percent.

Greater accuracy of estimation may be achieved, within limits, by the expenditure of more time and money. The greater the accuracy required, the greater the time and effort needed to obtain the design and cost data prior to making the estimate.

W. R. Park investigated the cost and accuracy of estimates for a project with a total cost of $1 million as shown in Fig. 25-41 (*Cost Engineering Analysis,* Wiley, New York, 1973, p. 133). Table 25-46 shows typical average costs for producing estimates [A. Pikulik and H. E. Diaz, *Chem. Eng.*, **84,** 106–122 (Oct. 10, 1977)].

Rapid Estimations

Ratio Methods J. E. Haselbarth [*Chem. Eng.*, **74,** 214–215 (Dec. 4, 1967)] published data giving the total capital investment per unit of annual production capacity C_{TC}/R. Table 25-47 lists data for many processes involving production units constructed on a previously developed site. Plants built on a green-field site would cost about 30 to 40 percent more, but enlargements of an existing plant would cost about 20 to 30 percent less than the values given in Table 25-47. Total fixed-capital investments for installations within the battery limits are given in Table 25-47. These refer to North American values corresponding to a Marshall and Swift index of 1000 and are taken from the data of D. R. Woods (*Financial Decision Making in the Process Industry,* Prentice-Hall, Englewood Cliffs, N.J., 1975, pp. 288–290). L. Lynn and R. F. Howland [*Chem. Eng.*, **67,** 131–136 (Feb. 8, 1960)] studied the capital ratios for 17 process industries and summarized data for more than 1000 processes. The capital ratio (CR) for a plant erected on a green-field site is defined as the ratio of the fixed-capital investment C_{FC} to the annual sales revenue A_S:

$$(CR) = C_{FC}/A_S \qquad (25\text{-}134)$$

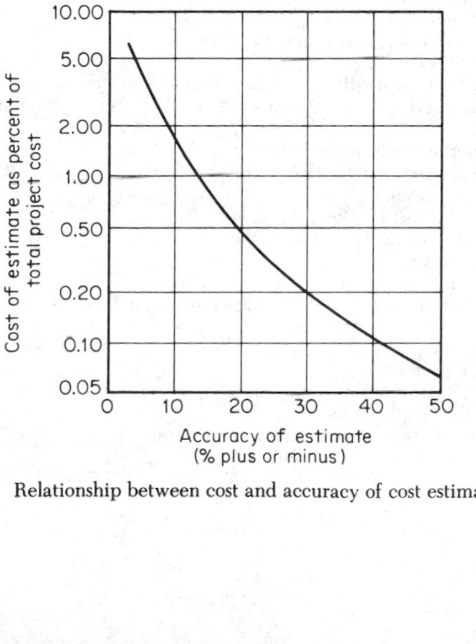

FIG. 25-41 Relationship between cost and accuracy of cost estimations.

However, Lynn and Howland included in the fixed-capital cost not only money invested in production and storage facilities but also that invested in land, research and development costs, and any auxiliary facilities necessary to support the process. Typical values of capital ratios for the year 1958 are listed in Table 25-48.

Both of the preceding methods are relatively inaccurate and can be used only for rough screening. They have the advantage that an estimate can be made in a few minutes, and they do not require design work or process flow sheets.

Exponential Methods Rapid capital-cost estimates can be made by using capacity-ratio exponents based on existing cost data of a company or drawn from published correlations.

If the cost of a piece of equipment or plant of size or capacity q_1 is C_1, then the cost of a similar piece of equipment or plant of size or capacity q_2 can be calculated from

$$C_2 = C_1(q_2/q_1)^n \qquad (25\text{-}247)$$

where the value of the exponent n depends on the type of equipment or plant. Cost indices should be used to bring the cost data to a common year. Table 25-47 gives typical values of n for various processes along with the cost of a plant of given capacity at a particular time and the capacity range of applicability. For process plants, capacity is expressed in terms of annual production capacity in metric tons per year.

Exponential cost correlations have been developed for individual items of equipment. Care must be taken in determining whether the cost of the equipment has been expressed as free on board (FOB), delivered (DEL), or installed (INST), as this is not always clearly stated. In many cases the cost must be correlated in terms of parameters related to capacity such as surface area for heat exchangers or power for grinding equipment. There are four main sources of error in such cost correlations:

1. Oversimplification by correlating the cost of equipment in terms of a single variable
2. Representation of data by using a simple exponential relationship
3. Failure to include the effects of technological improvements
4. Errors incurred because of special circumstances

Table 25-49 gives typical values of the exponent n for many types of equipment. Prices are North American with a Marshall and Swift index of 1000, mainly for carbon steel equipment.

Factor Estimations Most factor methods for estimating the total installed cost of a process plant are based on a combination of materials, labor, and overhead cost components. These can be conveniently grouped as

1. Cost of major items of equipment
2. Cost of complete installation of equipment
3. Auxiliary equipment to make the process work
4. Engineering and field expenses
5. Contractor's fees and contingencies

A great variety of factors are in use, depending on the time available and the accuracy expected. Normally the input information required is the base cost. Determination of this cost usually requires a knowledge of equipment sizes, probably using mass and energy balances for the proposed process.

Table 25-43 Checklist of Items for Fixed-Capital-Cost Estimates

Land:
- Surveys
- Fees
- Property cost

Site development:
- Site clearing
- Grading
- Roads, access and on-site
- Walkways
- Railroads
- Fence
- Parking areas
- Other paved areas
- Wharves and piers
- Recreational facilities
- Landscaping

Process buildings:
- (List as required) Include in each as required substructure, superstructures, platforms, supports, stairways, ladders, access ways, cranes, monorails, hoists, elevators

Auxiliary buildings:
- Administration and office
- Medical or dispensary
- Cafeteria
- Garage
- Product warehouse(s)
- Parts or stores warehouse
- Maintenance shops—electric, piping, sheet metal, machine, welding, carpenters, instrument
- Guard and safety
- Hose houses
- Change houses
- Smoking stations (in hazardous plants)
- Personnel building
- Shipping office and platforms
- Research laboratory
- Control laboratories

Building services:
- Plumbing
- Heating
- Ventilation
- Dust collection
- Air conditioning
- Sprinkler systems
- Elevators, escalators
- Building lighting
- Telephones
- Fire alarm
- Paging
- Intercommunication systems
- Painting

Process equipment:
- (List carefully from checked flow sheets)

Non-process equipment:
- Office furniture and equipment
- Cafeteria equipment
- Safety and medical equipment
- Shop equipment
- Automotive heavy maintenance and yard material-handling equipment
- Laboratory equipment
- Lockers and locker-room benches
- Garage equipment
- Shelves, bins, pallets, hand trucks
- Housekeeping equipment
- Fire extinguishers, hoses, fire engines

Process appurtenances:
- Piping—carbon steel, alloy, cast iron, lead-lined, aluminum, copper, asbestos-cement, ceramic, plastic, rubber, reinforced concrete
- Pipe hangers, fittings, valves
- Insulation—piping, equipment
- Instruments
- Instrument panels
- Electrical—panels, switches, motors, conduit, wire, fittings, feeders, grounding, instrument and control wiring

Utilities:
- Boiler plant
- Incinerator
- Ash disposal
- Boiler feed-water treatment
- Electric generation
- Electrical substations
- Refrigeration plant
- Air plant
- Wells
- River intake
- Primary water treatment—filtration, coagulation, aeration
- Secondary water treatment—deionization, demineralization, pH and hardness control
- Cooling towers
- Water storage
- Effluent outfall
- Process-waste sewers
- Process-waste pumping stations
- Sanitary-waste sewers
- Sanitary-waste pumping stations
- Impounders, collection basins
- Waste treatment, including gases
- Storm sewers

Yard distribution and facilities (outside battery limits):
- Process pipe lines—steam, condensate, water, gas, fuel oil, air, fire, instrument, and electric lines
- Raw-material and finished-product handling equipment—elevators, hoists, conveyors, airveyors, cranes
- Raw-material and finished-product storage—tanks, spheres, drums, bins, silos
- Fuel receiving, blending, and storage
- Product loading stations
- Track and truck scales

Miscellaneous:
- Demolition and alteration work
- Catalysts
- Chemicals (initial charge only)
- Spare parts and non-installed equipment spares
- Surplus equipment, supplies and equipment allowance
- Equipment rentals (for construction)
- Premium time (for construction)
- Inflation cost allowance
- Freight charges
- Taxes and insurance
- Duties
- Allowance for modifications and extra construction work during startup

Engineering costs:
- Administrative
- Process, project, and general engineering
- Drafting
- Cost engineering
- Procurement, expediting, and inspection
- Travel and living expense
- Reproductions
- Communications
- Scale model
- Outside architect and engineering fees

Construction expense:
- Construction, operation, and maintenance of temporary sheds, offices, roads, parking lots, railroads, electrical, piping, communication, and fencing
- Construction tools and equipment
- Warehouse personnel and expense
- Construction supervision
- Accounting and timekeeping
- Purchasing, expediting, and traffic
- Safety and medical
- Guards and watchmen
- Travel and transportation allowance for craft labor
- Fringe benefits
- Housekeeping
- Weather protection
- Permits, special licenses, field tests
- Rental of off-site space
- Contractor's home office expense and fees
- Taxes and insurance, interest

Table 25-44 Standard Cost Code (Numerical) Summary

Code Group	Subdivision
Direct costs:	
0–349	Process section (process equipment listed in flow-sheet sequence 1, 2, 3, 4, 5, . . . , *N*, preceded by process identification number)
350–399	Site development
400–449	Process buildings
450–489	Auxiliary buildings
490–499	Non-process equipment
500–599	Process appurtenances (piping, insulation, instrumentation, electrical, etc.)
600–699	Utilities and yard services (boiler plant, refrigeration, compressed air, water supply and treatment, effluents, fire protection, yard piping, yard electrical, yard materials handling, raw and finished-product storage)
700–729	Substructures
730–749	Superstructures
750	Painting
760–769	Building services
770–799	Demolition and alteration to existing structure
800–819	Surplus equipment, supplies and materials, royalty payments
820–830	Design modifications, construction modifications, and extra work during startup
Indirect costs:	
850–874	Home office engineering
875–889	Architect's and engineer's charges and fees
900–969	Construction expenses
970–994	Taxes and insurance
995	Contractor's home office expenses
996	Contractor's fees

Equipment or Base Cost The total cost of the main-plant items is generally used as the base cost. Again, care must be taken with equipment costs which may be quoted as installed (INST), delivered to site (DEL), or free on board the delivery vehicle at the place of manufacture or other specified location (FOB).

Base equipment includes all equipment within the battery limits whose cost is as significant as the cost of a pump. For example, storage tanks, knockout drums, accumulators, heat exchangers, and pumps are classed as main-plant items (MPI). Early in the development of the process-flow diagram, it is advisable to increase the estimated (MPI) cost by 10 to 20 percent to allow for later additions. When the scope of the process has been well defined, (MPI) costs should be increased by 1 to 10 percent.

For order-of-magnitude estimates the cost of equipment delivered $(C_{EQ})_{DEL}$ varies approximately from 1.1 to 1.25 times the FOB cost $(C_{EQ})_{FOB}$. The factor would be at the lower end of the range for domestic purchases and at the higher end for imports. Installation costs include unpacking, mounting, and connecting up to existing auxiliaries or utilities. The cost of equipment installed $(C_{EQ})_{INST}$ varies with type and size but generally ranges from 1.4 to 2.2 times the delivered-equipment cost $(C_{EQ})_{DEL}$.

Single-factor methods collect the various items of expenditure into one factor, which is usually used to multiply the total cost of delivered equipment $\Sigma(C_{EQ})_{DEL}$ to give the fixed-capital cost for plant within the battery limits:

$$(C_{FC})_{BL} = f\Sigma(C_{EQ})_{DEL} \tag{25-248}$$

Typical values for single factors f for battery-limit-plant costs (for a carbon steel plant including auxiliaries but not land) are as follows:

Solids processing (S)	3.8
Solids-fluid processing (S-F)	4.1
Fluid processing (F)	4.8

Thus the factors vary with the type of processing, although the boundaries between the classifications are not clear-cut and considerable judgment is required in selection of the correct factor.

Multiple-factor methods include the cost contributions for each given activity, which can be added together to give an overall factor. This factor can be used to multiply the total cost of delivered equipment $\Sigma(C_{EQ})_{DEL}$ to produce an estimate of the total fixed-capital investment either for grass-roots or for battery-limit plants. The costs may be divided into four groups:

1. Cost of plant within battery limits
2. Cost of auxiliaries
3. Cost of engineering and field expenses
4. Cost of contractor's fees plus contingency allowance

Table 25-50 gives typical values of such factors for carbon steel installations taken from the data of D. R. Woods (*Financial Decision Making in the Process Industry*, Prentice-Hall, Englewood Cliffs, N.J., 1975, p. 184). Auxiliaries and site preparation are given as factors of the delivered-equipment cost in Table 25-50, whereas C. A. Miller [*Chem. Eng.*, **72,** 226–236 (Sept. 13. 1965)] expresses auxiliary costs as factors of the battery-limit (BL) cost. Table 25-51 gives the factors from the breakdown of Miller, which is more detailed than that of Woods.

Example 23 Let us estimate the total installed cost for a grass-roots plant producing an organic chemical (S-F process) on a continuous basis. We assume

Table 25-45 Standard Cost Code (Alphabetical-Numerical)

A	Site work and foundations
B	Buildings (less foundations)
C	Steel structures and platforms (other than buildings)
D	Heat exchangers
E	Fractionating towers
F	Tanks and drums
G	Pumps and pump drivers
H	Compressors and blowers
J	Reactors and converters
K	Grinding, crushing, and classifying equipment
L	Materials-handling equipment
M	Fired heaters
N	Catalysts and chemicals
O	Laboratory equipment
P	Piping
Q	Instruments and controls
R	Electrical
S	Insulation and painting
T	Utility equipment (boilers, generators, refrigeration)
U	Plant and building accessories (railroads, fence, etc.)
V	Laboratory equipment
W	Safety equipment
X	Warehouse spares
100	Equipment and materials delivered and installed (A through X)
200	Sales and use taxes
300	Temporary facilities
320	Construction equipment, tools, and supplies
340	Construction supervision
360	Field office expense
380	Warehousing expense
400	Pay-roll taxes and insurance
500	Home office engineering costs
600	Procurement costs
700	Resident engineering
800	Royalty payments
900	Engineering administrative overhead and profit
950	Constructor's administrative overhead and profit

TABLE 25-46 Typical Cost Ranges for Preparing Estimates (1976 Prices)*

Type of estimate	Cost of project: Less than \$1 million \$ thousand	\$1 to \$5 million \$ thousand	\$5 to \$50 million \$ thousand
Study estimate	5–15	12–30	20–40
Preliminary estimate (scope or authorization)	15–35	30–60	50–90
Definitive estimate (project control)	25–60	60–120	100–230

*From A. Pikulik and H. E. Diaz, "Cost Estimating for Major Process Equipment," *Chem. Eng.*, **84,** 106–122 (Oct. 10, 1977).

TABLE 25-47 Capital-Cost Data for Processing Plants

Product	Process route	Size, 1000 metric tons/year	Approximate cost, $ × 10^6	Cost $/annual metric tons	Size range, 1000 metric tons/year	Exponent n
Acetaldehyde	Ethanol	23	4.0	174		
Acetaldehyde	Ethylene	50	13.7	274	25–100	0.70
Acetic acid	Methanol	10	77	7700	2–50	0.68
Acetone	Propylene	10	25	2500	2–50	0.45
Acetone	Isopropanol	25	23.4	936		
Acetylene	Natural gas or petrochemical	15	23.4	1560	2–150	0.70
Acrylic acid	Propylene gives methyl acrylate by-product	1.1	10.7	9730		
Acrylic fiber		10	27.4	2740	4–20	0.69
Acrylonitrile	Acetylene/HCN	100	107	1070	10–500	0.60
Adipic acid		10	21.7	2170	5–50	0.53
Alkylbenzene (linear)		40	9.4	235	10–120	0.70
Alum (liquid		10	3.0	300	7–350	0.71
Alumina	Bauxite	200	124	620	100–400	0.66
Aluminum	Alumina	50	177	3540	25–200	0.80
Aluminum sulfate		50	6.0	120	10–500	0.71
Ammonia (anhydrous liquefied)	Steam reforming of naphtha	150	37	247	30–330	0.70
	Steam reforming of natural gas	150	30	200	30–330	0.70
	Partial oxidation of natural gas	150	37	247	30–330	0.70
	Naphtha	150	44	293	30–330	0.70
	Fuel oil	150	47	313	30–330	0.70
	Coal	150	63	420	30–330	0.70
Ammonium nitrate	Ammonia/nitric acid	100	9.4	94	15–300	0.65
Propylene oxide	Chlorhydrin	40	14.4	358	25–60	0.90
Propylene oxide	Propylene plus isobutane	60	42.7	711		
	Electrochemical	60	39	650		
Protein	Paraffinic concentrates	60	33.4	556		
Pseudocumene	Distillation separation	32	6.0	188		
Pulp: acid sulfite	Bleached with recovery	50	66.7	1334	8–200	0.78
	Without recovery	50	60	1200	8–200	0.86
Resorcinol	Sulfonation	1.8	3.7	2040		
Soap		3	0.87	289	1.5–7	0.23
Sodium		18	26.4	1463		
Sodium carbonate (soda ash)	Solvay process	120	1.27	10.6	60–200	0.55
	Natural brine	360	130	361		
Sodium chlorate		30	16.7	556	15–60	0.66
Sodium hydroxide	Electrolysis of brine	30	33.4	1110	3–300	0.38
Sorbitol	Corn sugar	4.5	2.7	593		
Steel	Integrated	500	767	1534	150–4000	0.65
Styrene	Benzene/ethylene	100	35	350	20–400	0.67
Styrene	Ethyl benzene	40	11.7	292		
Styrene	Reformate aromatics	32	21.7	677		
Sugar	Sugarcane: operates 120 days/year	10	10.7	1070	6–30	0.41

NOTE: All costs are North American values with M & S = 1000, assuming 330 operating days per year.

TABLE 25-48 Capital Ratios for Process Industries

Industry	Capital ratio,* 1958
Chemicals, general	2.02
Carbon black	3.98
Explosives	1.64
Glass	1.46
Fibers, synthetic	3.44
Foodstuffs, processed	0.66
Inorganics, heavy	2.24
Nonferrous metals	3.31
Petroleum	3.08
Pharmaceuticals	0.92
Pigments, paints, and inks	1.04
Pulp and paper	2.01
Resins and plastics	1.90
Rubber	1.04
Soap and detergents	0.69
Steel	2.78
Sulfur	1.97
Average	2.01

*Capital ratio = (fixed-capital investment)/(annual sales revenue).

that the total cost of delivered equipment $\Sigma(C_{EQ})_{DEL}$ is $1 million and use suitable factors from Table 25-50.

The estimated values for the various contributions are given in Table 25-52, resulting in an estimate of $4,280,000 for the total fixed-capital investment, including a contingency factor.

A newer multiple-factor method for predesign cost estimating has been put forward by D. H. Allen and R. C. Page [*Chem. Eng.*, **82**, 142–150 (Mar. 3, 1975)] for fluid-type plants (F) that include some vapor processing. The method requires the following input information:

1. Plant flow sheet giving main-plant items and process streams
2. Total process-stream input per year
3. Extreme temperature and pressure conditions, if any
4. Materials of construction for main-plant items
5. Operating phases for each main-plant item
6. Expectation of any unusually high or low, direct or indirect initial costs

By means of 12 procedural steps involving the input information, several equations, graphs, and tables, the total cost of delivered equipment $\Sigma(C_{EQ})_{DEL}$ is estimated. This is then converted into a grass-roots investment estimate by dividing $\Sigma(C_{EQ})_{DEL}$ by a single factor ranging from 15 to 30 percent (average value, 21 percent).

TABLE 25-49 Typical Exponents for Equipment Cost versus Capacity

Equipment	Size	Unit	Approximate cost, $000	Size range	Exponent
Agitator, turbine, top entry, open, FOB	10 (7.5)	hp (kW)	7.0	2–30 (1.5–22.4)	0.45
Agitator, turbine, top entry, closed, FOB	10 (7.5)	hp (kW)	10.7	2–200 (1.5–150)	0.56
Blower, centrifugal, 4 lbf/in^2 (27.6 kN/m^2), DEL, excluding motor	10 (4.72)	10^3 sft^3/min (sm^3/s)	67	0.5–150 (0.24–71)	0.60
Cone crusher, FOB, crusher only	100 (74.6)	hp (kW)	130	30–300 (22.4–224)	0.92
Jaw crusher, FOB, excluding motor	10 (7.5)	hp (kW)	34	1–60 (0.75–44.7)	0.65
Jaw crusher, FOB, exluding motor	100 (74.6)	hp (kW)	284	60–400 (44.7–300)	0.81
Centrifugal pump, C/S, FOB, excluding motor	10 (7.5)	hp (kW)	1.6	0.5–40 (0.37–30)	0.30
Centrifugal pump, C/S, FOB, exluding motor	100 (74.6)	hp (kW)	4.4	40–400 (30–300)	0.67
Conveyor, belt, C/S, FOB, excluding motor	100 (9.3)	ft^2 (m^2)	6.7	60–200 (5.6–18.6)	0.50
Conveyor, screw, C/S, DEL, excluding motor	70 (540)	ft × m diameter (m × mm diameter)	10	50–100 (390–780)	0.46
Centrifuge, automatic batch, horizontal, C/S, FOB	20 (1.86)	Filter area, ft^2 (m^2)	100	7–80 (0.65–7.43)	0.65
Compressor, reciprocating, < 1000 lbf/in^2, FOB, including motor	300 (224)	hp (kW)	133	1–20000 (0.75–1490)	0.84
Crystallizer, forced circulation, C/S, FOB	100 (91)	ton/day (Mg/day)	283	10–1000 (9.1–970)	0.59
Dryer, drum, C/S, FOB, excluding motor	100 (9.3)	ft^2 (m^2)	73	10–400 (0.9–37)	0.52
Dryer, vacuum, shelf, C/S, FOB, excluding trays, vacuum equipment	100 (9.3)	ft^2 (m^2)	17	15–1000 (1.4–93)	0.56
Dust collector, cloth, shaker type, FOB, including motors	10^4 (4.7)	sft^3/min (m^3/s)	17	10^3–5 × 10^4 (0.47–23.6)	0.79
Dust collector, multicyclones, FOB	10^4 (4.7)	sft^3/min (m^3/s)	7	10^3–1.5 × 10^5 (0.47–70.8)	0.66
Electrostatic precipitator, FOB	10^4 (4.7)	ft^3/min at 40°C	77	10^3–8 × 10^4 (0.47–73.8)	0.39
	2 × 10^5 (94)	(m^3/s)	383	8 × 10^4–10^6 (37.8–472)	0.81
Ejector, single-stage, 100 psig, steam, FOB	3 (10^{-2})	lb/h (air/mmHg absolute)	2.7	0.2–30 (6.8 × 10^{-4}–0.1)	0.50
Ejector, two-stage, FOB, including condenser, piping	1 (3.4 × 10^{-3})	[kg/h/(N/m^2)]	6.3	0.2–10 (6.8 × 10^{-4}–3.4 × 10^{-2})	0.43
Ejector, multistage, FOB, including condenser, piping	10 (3.4 × 10^{-2})	[kg/h/(N/m^2)]	16.7	0.2–100 (6.8 × 10^{-4}–0.34)	0.26
Filter, vertical-pressure leaf, C/S, DEL	100 (9.3)	ft^2 (m^2)	17	30–1500 (2.8–140)	0.57
Filter, plate and frame, C/S, DEL	100 (9.3)	ft^2 (m^2)	5.7	10–1000 (0.9–93)	0.55
Filter, vacuum rotary drum, C/S, FOB, including motor	100 (9.3)	ft^2 (m^2)	63.3	10–1500 (0.9–140)	0.48
Heat exchanger, shell-tube, floating head, C/S, DEL; fixed tube × 0.85; U tube × 0.87; kettle × 1.35	1000 (9.3)	ft^2 (m^2)	21.7	20–20,000 (1.9–1860)	0.59
Heat exchanger, thermal screw, C/S, FOB, excluding motor	100 (9.3)	ft (m^2)	33	10–400 (0.9–37)	0.78
Kettle, jacketed, glass-lined, FOB	100 (0.38)	U.S. gal (m^3)	53	50–1000 (0.2–3.8)	0.48
Motors, ac induction, wound rotor, TEFC, FOB	10 (7.5)	hp (kW)	12.3	10–25 (7.5–18.6)	0.56
Motors, ac induction, wound rotor, TEFC, FOB	70 (52)	hp (kW)	19.3	25–200 (18.6–149)	0.77
Piping, typical straight run, C/S, FOB, $/ft					
Installed: $/ft × 6 to 7	6 (152)	Nominal diameter in (mm)	0.0093	1–24 (25–610)	1.33
Complex network: FOB $/ft × 2					
Installed: $/ft × 13					
Pressure vessel horizontal drum (150 psig), C/S	1000 (3.8)	U.S. gal (m^3)	6.3	100–80000 (0.4–302)	0.62
Jacketed reactors, including mixer, FOB	100 (0.38)	U.S. gal (m^3)	9.3	10–4000 (0.04–15.1)	0.53
Refrigeration, packaged mechanical, INST	100 (351.7)	U.S. tons (kW)	133	10–1000 (35.2–3520)	0.73
Screen, vibrating, single-deck, DEL, including motor	500 (46)	ft^2 (m^2)	10	150–700 (14–65)	0.62
Stack, carbon steel	. . . . . .	ft (m)	—	20–150 (6.1–45.7)	1.00
Tanks: atm, horizontal cylinder, C/S, FOB	1000 (3.8)	U.S. gal (m^3)	4.7	100–40000 (0.4–151)	0.57
Vertical cylinder, C/S, FOB	1000 (3.8)	U.S. gal (m^3)	3.3	100–20000 (0.4–76)	0.30
Vertical jacketed, C/S, FOB	1000 (3.8)	U.S. gal (m^3)	15	70–1500 (0.26–5.7)	0.57
Vertical agitated, C/S, FOB, including motor	1000 (3.8)	U.S. gal (m^3)	12.3	100–20000 (0.4–76)	0.50
Towers, distillation including internals, INST	4000 (trays)	$\left(\frac{\text{feed, lb/year}}{10^6}\right)^{0.65}$	3300	300–30000	1.00

TABLE 25-50 Factors to Convert Delivered-Equipment Costs into Fixed-Capital Investment

Details	Grass-roots plants: Solids processing	Grass-roots plants: Solids-fluid processing	Grass-roots plants: Fluid processing	Battery-limit installations: Solids processing	Battery-limit installations: Solids-fluid processing	Battery-limit installations: Fluid processing
Equipment, delivered	1.00	1.00	1.00	1.00	1.00	1.00
Installed	0.19–1.23	0.39–0.43	0.76	0.45	0.39	0.27–0.47
Piping	0.07–0.23	0.30–9.39	0.33	0.16	0.31	0.66–1.20
Structural steel foundations, reinforced concrete			0.28			0 –0.13
Electrical	0.13–0.25	0.08–0.17	0.09	0.10	0.10	0.09–0.11
Instruments	0.03–0.12	0.13	0.13	0.09	0.13	
Battery-limits building and service	0.33–0.50	0.26–0.35	0.45	0.25	0.39	0.18–0.34
Excavation and site preparation	0.03–0.18	0.08–0.22	....	0.13	0.10	0.10
Auxiliaries	0.14–0.30	0.48–0.55	Included above	0.40	0.55	0.70
Total physical plant	2.37	2.97	3.04	2.58	2.97	3.50
Field expense	0.10–0.12	0.35–0.43	....	0.39	0.34	0.41
Engineering	....	0.35–0.43	0.41	0.33	0.32	0.33
Direct plant costs	2.48	3.73	3.45	3.30	3.63	4.24
Contractor's fees, overhead, profit	0.30–0.33	0.09–0.17	0.17	0.17	0.18	0.21
Contingency	0.26	0.39	0.36	0.34	0.36	0.42
C_{FC}: total fixed-capital investment	3.06	4.27	3.98	3.81	4.17	4.87

TABLE 25-51 Factor Method of Miller (Based on Delivered-Equipment Costs = 100)*

		Battery-limit costs (range of factors in percent of basic equipment); average unit cost of main-plant item (MPI)						
		Under $3000	$3000 to $5000	$5000 to $7000	$7000 to $10,000	$10,000 to $13,000	$13,000 to $17,000	Over $17,000
Field erection of basic equipment	High percentage of equipment involving high field labor	23/18	21/17	19.5/16	18.5/15	17.5/14.2	16.5/13.5	15.5/13
	Average (mild steel) equipment	18/12.5	17/11.5	16/10.8	15/10	14.2/9.2	13.5/8.5	13/8
	High percentage of corrosion materials and other high-unit-cost equipment involving little field erection	12.5/7.5	11.5/6.7	10.8/6	10/5.5	9.2/5.2	8.5/5	8/4.8
Equipment foundations and structural supports	High: predominance of compressors or mild steel equipment requiring heavy foundations			17/12	15/10	14/9	12/8	10.5/6
	Average: for mild steel fabricated-equipment solids			12.5/7	11/6	9.5/5	8/4	7/3
	Average: for predominance of alloy and other high-unit-price fabricated equipment	7/3	8/3	8.5/3	7.5/3	6.5/2.5	5.5/2	4.5/1.5
	Low: equipment more or less sitting on floor	5/0	4/0	3/0	2.5/0	2/0	1.5/0	1/0
	Piling or rock excavation	Increase above values by 25 to 100%						
Piping, including ductwork but excluding insulation	High: gases and liquids, petrochemicals, plants with substantial ductwork	105/65	90/58	80/48	70/40	58/34	50/30	42/25
	Average for chemical plants: liquids, electrolytic plants	65/33	58/27	48/22	40/16	34/12	30/10	25/9
	Liquids and solids	33/13	27/10	22/8	16/6	12/5	10/4	9/3
	Low: solids	13/5	10/4	8/3	6/2	5/1	4/0	3/0
Insulation of equipment only	Very high: substantial mild steel equipment requiring lagging and very low temperatures	13/10	11.5/8.5	10/7.4	9/6.2	7.8/5.3	6.8/4.5	5.8/3.5
	High: substantial equipment requiring lagging and high temperatures (petrochemicals)	10.3/7.5	9/6.3	7.8/5.2	6.7/4.2	5.7/3.4	4.7/3.8	4.8/2.5
	Average for chemical plants	7.8/3.4	6.5/2.6	5.5/2.1	4.5/1.7	3.6/ 1.4	2.9/1.1	2.2/.8
	Low	3.5/0	2.7/0	2.2/0	1.8/0	1.5/0	1.2/0	1/0
Insulation of piping only	Very high: substantial mild steel piping requiring lagging and very low temperatures	22/16	19/13	16/11	14/9	12/7	9/5	6/3.5
	High: substantial piping requiring lagging and high temperatures (petrochemicals)	18/14	15/12	13/10	11/8	9/6	7/4	4.5/2.5
	Average for chemical plants	16/12	14/10	12/8	10/6	8/4	6/2	4/2
	Low	14/8	12/6	10/5	8/4	6/3	4/2	2/1

TABLE 25-51 Factor Method of Miller (Based on Delivered-Equipment Costs = 100)* *(Continued)*

		Battery-limit costs (range of factors in percent of basic equipment); average unit cost of main-plant item (MPI)						
		Under $3000	$3000 to $5000	$5000 to $7000	$7000 to $10,000	$10,000 to $13,000	$13,000 to $17,000	Over $17,000
All electrical except building, lighting, and instrumentation	Electrolytic plants, including rectification equipment		55/42	50/38	45/33	40/30	35/26	
	Plants with mild steel equipment, heavy drives, solids	26/17	22.5/15	19.5/12.5	17/10	14/8.5	12/7	10/6
	Plants with alloy or high-unit-cost equipment. chemical and petrochemical plants	18/9.5	15.5/8.5	13/6.5	11/5.5	9/4.5	7.3/3.5	6/2.5
Instrumentation°	Substantial instrumentation, central control panels, petrochemicals		58/31	46/24	37/18	29/13	23/10	18/7
	Miscellaneous chemical plants		32/13	26/10	20/7	15/5	11/3	8/2
	Little instrumentation, solids		21/9	17/7	13/5	10/3	7/2	5/1
Miscellaneous, including site preparation, painting, and other items not accounted for above	Top of range—large complicated processes; bottom of range—smaller, simple processes	Range for all values of basic equipment is 6 to 1%						

Building evaluation when most process units are located inside buildings

Buildings—architectural and structural, excludes building services†

	High, brick and steel	Medium		Economical	Evaluation
Quality of construction	+4	+2		0	
	Very high unit cost equipment	Mostly alloy steel	Mixed materials	Costly carbon steel	
Type of equipment	−3	−2	−1	0	
	Very high	Intermediate		Atmosphere	
Operations pressures	−2	−1		0	

cost equipment, chemical Building class = algebraic sum =

	Average unit cost of MPI							
	Building class	Under $3000	$3000 to $5000	$5000 to $7000	$7000 to $10,000	$10,000 to $13,000	$13,000 to $17,000	Over $17,000
Cost of process	+2	92/68	82/61	74/56	67/49	59/44	52/39	46/33
Units inside buildings	+1 to −1	72/49	62/43	56/38	51/33	45/29	41/26	36/21
	−2	50/37	44/33	40/29	35/25	30/21	27/18	23/15
Open-air plants with minor buildings		37/16	32/13	28/11	24/8	20/6	17/4	14/2

Building services‡	High	Normal	Low
Compressed air for general service only	4	1½	0.5
Electric lighting	18	9	5
Sprinklers	10	6	3
Plumbing	20	12	3
Heating	25	16	8
Ventilation:			
Without air conditioning	18	8	0
With air conditioning	45	35	25
Total overall average§	85	55	20

The above factors apply to those items normally classified as building services. They do not include (1) services located outside the building such as substations, outside sewers, and outside water lines, all of which are considered to be outside the battery limit as well as outside the building; and (2) process services.

°Courtesy C. A. Miller of Canadian Industries Ltd. and the American Association of Cost Engineers.

NOTE: The average unit cost of the main-plant items is the total cost of the MPI divided by the total number of items. Figures include up to 3 percent for BL outside lighting, which is not covered in building services.

°Total instrumentation cost does not vary a great deal with size and hence is not readily calculated as a percentage of basic equipment. This is particularly true for distillation systems. If in doubt, detailed estimates should be made.

†When building specifications and dimensions are known, a high-speed building-cost estimate is recommended, especially if buildings are a significant item of cost. If a separate estimate is not possible, evaluate the buildings as shown before selecting the factors.

‡The following factors are for battery-limit (process) buildings only and are expressed in percentage of the building architectural and structural cost. They are not related to the basic equipment cost.

§The totals provide the ranges for the type of building involved and are useful when individual service requirements are not known. Note that the overall averages are not the sum of the individual columns.

TABLE 25-52 Estimate Using Factors from Table 25-50

Details (solids-fluid, grass-roots plant)	Factor assumed	Cost, $	Percentage of total
Equipment, delivered	1.00	1,000,000	23.4
Installed	0.41	410,000	9.6
Piping	0.34	340,000	8.0
Electrical	0.13	130,000	3.0
Instruments	0.13	130,000	3.0
Battery-limit building and service	0.30	300,000	7.0
Excavation and site preparation	0.15	150,000	3.5
Auxiliaries	0.52	520,000	12.2
Total physical plant	2.98	2,980,000	69.7
Field expense	0.39	390,000	9.1
Engineering	0.39	390,000	9.1
Direct plant costs	3.76	3,760,000	87.9
Contractor's fees, overhead, profit	0.13	130,000	3.0
Contingency	0.39	390,000	9.1
Total fixed-capital investment	4.28	4,280,000	100.0

The method is rapid and is claimed to be accurate within −20 to +25 percent, but it has only been tested by using data published in the literature for eight plants.

Multiple-Factor Methods That Separate Materials and Labor These methods have become increasingly popular. While they are similar to the preceding methods, labor and materials costs are considered separately. Hence it is possible to allow for variations in efficiency and labor costs in different localities or countries. H. C. Bauman (*Fundamentals of Cost Engineering in the Chemical Industry*, Van Nostrand Reinhold, New York, 1964, p. 295) divides most of the components of Table 25-50 into material and labor components, quoting the data as ranges and medians of the percentage of the total fixed-capital investment. In Table 25-53, Bauman's data have been converted to factors of the delivered-equipment cost for a grass-roots installation.

A study has been made by A. V. Bridgwater [*Chem. Eng.*, **86**, 119–121 (Nov. 5, 1979)] of the geographical variations in capital costs. He concluded that because of trade and competition basic equipment costs do not vary significantly in the industrialized countries of the western world. The main differences in construction costs at various international locations are due to variations in labor costs and productivity, the use of specialized equipment, and sundry local factors. Table 25-54 gives location factors for the construction of chemical plants of similar function in various countries (1977 values). The factors have been corrected by Bridgwater for location variations in labor costs and efficiency and converted at the average value of the exchange rate.

TABLE 25-53 Typical Factors with Separation of Materials and Labor*

	Total factor	Materials factor	Labor factor
Equipment delivered	1.00		
Installation	0.09		0.09
Instruments installed	0.13	0.09	0.04
Piping	0.29	0.155	0.135
Foundations and steel	0.18	0.08	0.10
Insulation painting	0.11	0.025	0.085
Electrical	0.18	0.06	0.12
Battery-limit building	0.21	0.13	0.08
Site preparation	0.08		
Auxiliaries	0.55		
Physical-plant cost	2.82		
Engineering and home office	0.31	0.01	0.30
Field expense	0.43	0.30	0.13
Direct plant cost	3.56		
Contractor's fees	0.17		
Contingency	0.39		
Fixed-capital cost	4.12		

*Based on the data of H. C. Bauman, *Fundamentals of Cost Engineering in the Chemical Industry*, Van Nostrand Reinhold, New York, 1964, p. 295, for essentially carbon steel equipment.

TABLE 25-54 Location Factors for Chemical Plants of Similar Function (1977 Values)

Location	Factor (United Kingdom = 1.0)	Factor (United States = 1.0)
Australia	1.4	1.3
Austria	1.1	1.0
Belgium	1.1	1.0
Canada	1.25	1.15
Central Africa	2.0	2.0
Central America	1.1	1.0
China: imported element	1.2	1.1
Indigenous element	0.6	0.55
Denmark	1.1	1.0
Finland	1.3	1.2
France	1.05	0.95
Germany, West	1.1	1.0
Greece	1.0	0.9
India: imported element	2.0	1.8
Indigenous element	0.7	0.65
Ireland	0.9	0.8
Italy	1.0	0.9
Japan	1.0	0.9
Malaysia	0.9	0.8
Middle East	1.2	1.1
Netherlands	1.1	1.0
Newfoundland	1.3	1.2
New Zealand	1.4	1.3
North Africa: imported element	1.2	1.1
Indigenous element	0.8	0.75
Norway	1.2	1.1
Portugal	0.8	0.75
South Africa	1.25	1.15
South America, northern	1.5	1.35
South America, southern	2.5	2.25
Spain: imported element	1.3	1.2
Indigenous element	0.8	0.75
Sweden	1.2	1.1
Switzerland	1.2	1.1
Turkey	1.1	1.0
United Kingdom	1.0	0.9
United States	1.1	1.0
Yugoslavia	1.0	0.9

NOTE:
1. Increase a factor by 10 percent for each 1000 mi or part of 1000 mi that the new plant is distant from a major manufacturing or import center, or both.
2. When materials or labor, or both, are obtained from more than a single source, prorate the appropriate factors.
3. Investment incentives have been ignored.

Factor Methods Using the Modular Approach These are methods used for estimating the cost of major-equipment units and have been proposed by several authors. Perhaps the most comprehensive is the method suggested by K. M. Guthrie [*Chem. Eng.*, **76**, 114–142 (Mar. 24, 1969)]. Table 25-55 gives average factors for major-equipment items based on a $(C_{EQ})_{FOB}$ cost for carbon steel units. To the FOB cost of the item is added, by means of factors, the total materials cost to complete the module M. Erection and setting costs L are added as a factor or calculated from the L/M cost ratio to give $M + L = X$, the direct module cost. Indirect costs, such as freight, taxes, insurance, engineering, and field expense, are added to $(M + L)$ to give the total module cost. This excludes contingency allowances, contractor's fees, auxiliaries, site development, land, and industrial buildings, which may have to be added when applicable. The factors in Table 25-55 were based on mid-1968 prices for a United States Gulf Coast location.

More recently A. Pikulik and H. E. Diaz [*Chem. Eng.*, **84**, 106–122 (Oct. 10, 1977)] presented a graphical method for estimating the fabricated cost of distillation columns and pressure vessels, storage

TABLE 25-55 Factors for Individual Items*

Details	Furnaces	Exchangers: Shell and tube	Exchangers: Air-cooled	Vessels: Vertical	Vessels: Horizontal	Pump and driver	Compressor and driver	Tanks
FOB equipment	1.00	1.00	1.00	1.00	1.00	1.00	1.00	1.00
Piping	0.18	0.46	0.18	0.61	0.42	0.30	0.21	
Concrete	0.10	0.05	0.02	0.10	0.06	0.04	0.12	
Steel	....	0.03	....	0.08				
Instruments	0.04	0.10	0.05	0.12	0.06	0.03	0.08	
Electrical	0.02	0.02	0.12	0.05	0.05	0.31	0.16	
Insulation	....	0.05	....	0.08	0.05	0.03	0.03	
Paint	....	....	0.01	0.01	0.01	0.01	0.01	
Total materials = M	1.34	1.71	1.38	2.05	1.65	1.72	1.61	1.20
Erection and setting (L)	0.30	0.63	0.38	0.95	0.59	0.70	0.58	0.13
X, excluding site preparation and auxiliaries ($M + L$)	1.64	2.34	1.76	3.00	2.24	2.42	2.19	1.33
Freight, insurance, taxes, engineering, home office, construction	....	0.08	....	0.08	0.08	0.08	0.08	0.08
Overhead or field expense	0.60	0.95	0.70	1.12	0.92	0.97	0.97	
Total module factor	2.24	3.37	2.46	4.20	3.24	3.47	3.24	1.41

*From K. M. Guthrie, *Chem. Eng.*, **76**, 114–142 (Mar. 24, 1969). Based on FOB equipment cost = 100 (carbon steel).

tanks, fired heaters, pumps and drivers, compressors and drivers, and vacuum equipment.

Piping Estimation The cost of fabrication and installation of process-plant piping appears to range from 18 to 61 percent of the FOB equipment cost as indicated in Table 25-55. This would normally represent about 7 to 15 percent of the installed plant cost and is obviously a significant item. The various available piping-estimation methods are as follows:

1. Detailed pricing for piping drawings
2. Guthrie method
3. Dickson N method
4. Pricing by weight of specific types of pipe
5. Price estimation by cost per joint
6. Pricing as a factor of equipment cost
7. Pricing as a factor of total plant installed cost

The first five methods are applicable only after rigorous circuit analysis and when piping layouts and isometric drawings or scale models are available for quantity takeoff (e.g., pipe size, length, and specification, flanges and valve count, etc.).

Guthrie's method [K. M. Guthrie, *Chem. Eng.*, **76**, 201–216 (Apr. 14, 1969)] is mainly graphical, using average mid-1968 costs for a United States Gulf Coast location.

The Dickson N method [R. A. Dickson, *Chem. Eng.*, **54**, 121–123 (November 1947)] is a variation of the detailed price takeoff. Various circuits for each type of pipe are completely priced for a base size. Another chart gives an N factor for all other pipe sizes. Multiplying the cost of the circuit for the base size by the appropriate N factor yields the estimated cost of the new circuit of the desired pipe size. The method depends for its accuracy on periodic repricing of the base-size circuits in order to keep the base charts up to date.

Estimating by weight requires virtually complete takeoff, including weight calculations and a full record of past costs on this basis. Its only advantage lies in the time saved in the detailed estimates of the cost of piping components.

Estimating by cost per joint depends on the accumulation of past data, analyzed and conveniently correlated for use. The main advantage of the method lies in the fact that good engineering flow sheets can be used for the estimation.

Figure 25-42 is a plot of the number of labor-hours of field erection time per joint against the nominal pipe size of shop-fabricated carbon steel and low-alloy pipe. The unit of work measurement used in this method is the pipe joint, requiring two joints for couplings and valves, three for tees, etc., as most of the labor-hours involved in pipe erection are expended in making connections. The additional costs of handling, suspending, and placing lengths of pipe in position are included in the chart.

It should be noted that Fig. 25-42 gives labor-hours only. Material costs must be obtained by price takeoff from drawings on which all valves and instrument connections are shown. Pipe lengths and fittings are taken off by referring to the equipment-layout plan and elevation drawing. The graph of Fig. 25-42 can be updated by using actual costs for a specific job, in which case the labor cost per joint represents a total labor cost including all the factors applicable to labor shown in Table 25-56. It should be possible to analyze statistically uniform data from a number of complete jobs to determine the value of each factor for various project locations.

Methods 6 and 7 are simpler procedures, using factors for estimating piping costs when neither flow sheets nor detailed piping drawings are available. Tables 25-50, 25-51, 25-53, and 25-55 include typical values of piping factors based on total equipment cost, detailed or FOB, as indicated in the particular table. These methods require some degree of judgment in selecting the appropriate factor,

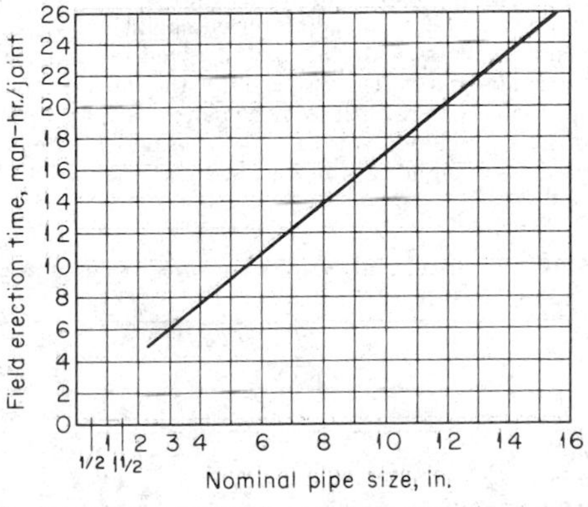

FIG. 25-42 Labor-hours required to erect large quantities of shop-fabricated steel and low-alloy piping.

TABLE 25-56 Components of Total Installed Piping Cost

Material	Pipe, valves, fittings, nuts, bolts, gaskets, and hangers
Labor	Cut, erect, align, fit, bolt, thread or weld, and test
Indirect costs	Handle and haul, store, scaffold, lost time, tools and rentals, contractor's overhead and profit
Factors applicable to labor-hours	Craft rate, productivity, height, and complexity
Crafts involved in piping erection	Pipefitters, laborers, carpenters, warehouse workers, teamsters, and operating engineers

based on experience gained by comparing piping costs for similar previously installed process plants.

A rough method of estimating the piping factor as a percentage of the total delivered cost of major process equipment (excluding instruments and electrical items) was presented by E. S. Sokullu in the form

$$f_P = 11\phi_P^{1.6} \tag{25-249}$$

where ϕ_P = (number of actual pipes on flow diagram)/(number of major process equipment units) [*Chem. Eng.*, **76,** 148–150 (Feb. 10, 1969)].

The equipment-unit method would appear to give more accuracy than the preceding methods, particularly for unfamiliar process arrangements. It requires the accumulation of piping costs for various sizes of main-plant items such as pumps, heat exchangers, evaporators, tanks, and columns. Basically it is assumed that piping designs for specific items are similar for most projects. Statistical analysis of such data shows good agreement with the more detailed takeoff pricing methods. Since for most processes the length of pipe used is a small proportion of the total piping cost, the assumption of an average length of piping per main-plant item, based on actual costs for several previous jobs, should give sufficient accuracy. Correction for escalation of costs can be carried out by using a single cost index, unlike methods 1 to 5.

Most of the factorial methods of estimation given previously, with the exception of the method of Allen and Page (loc. cit.), tend to estimate costs which are based on carbon steel equipment or installations. Table 25-57 gives typical multiplying factors for converting carbon steel costs to equivalent-alloy costs for a few items of equipment. (Adapted from *A New Guide to Capital Cost Estimating,* Institution of Chemical Engineers, Rugby, England, 1977, p. 18.)

Electrical and Instrumentation Estimation These costs usually range from 4 to 10 percent of the total installed plant cost, with a median value of about 7.5 percent. As with piping estimation, the process design must be almost completed before detailed drawings and specifications can be prepared for estimating purposes. However, actual electrical costs can be up to 100 percent higher than estimated costs, and so it is important to attempt to maintain the accuracy range within reasonable limits.

During the design stages, frequent changes in the type and sizes of equipment lead to delays in establishing electrical requirements. Hence it is very difficult to obtain a detailed estimate of the cost of the electrical part of the project. For order-of-magnitude and study estimates, an appropriate factor in the range 4 to 10 percent of the total installed plant cost can be used. However, for budget-authorization or preliminary estimates requiring an accuracy within 5 percent more accurate methods are necessary.

The methods available for electrical estimates are as follows:

1. Detailed takeoff
2. Factored electrical cost as a percentage of total installed plant cost for specific types of plant
3. Unit pricing

The detailed-takeoff method can rarely if ever be used. When detailed drawings are available, costs may be estimated by pricing materials and components from suppliers' catalogs or, for special items, from quotations. Handbooks are available which give typical values of the labor-hours required to perform units of installation work, such as installation of switches, starters, motors, conduit wiring, and push buttons of various sizes, for both hazardous and nonhazardous areas. Labor rates can be obtained from various government statistical sources or elsewhere. For the United States the National Electrical Contractors Association publishes an excellent manual of electrical costs. From the complete plans and specifications, the estimator can take off materials, estimate the labor cost, apply appropriate factors for labor efficiency, productivity, and local conditions, and achieve good results.

The factor estimate, if based on tested actual data, gives good results in the study estimate and often proves adequate at the preliminary estimate stage. It is essential to accumulate from past experience data showing actual electrical costs (1) as a percentage of total installed plant cost and (2) as a percentage of installed equipment costs. Studies of electrical installations for more than 100 plants (H. C. Bauman, *Fundamentals of Cost Engineering in the Chemical Industry,* Van Nostrand Reinhold, New York, 1964, p. 134) showed electrical costs ranging from about 4 to 11 percent of total plant cost, with a median for battery-limit process plants of 7.5 percent. The corresponding range based on installed equipment costs was 15 to 40 percent, with a median of 26 percent. Thus, it appears that there is a better correlation between electrical costs and total installed plant cost than with installed equipment costs. Table 25-58, taken from Bauman's data, gives typical values of electrical costs as a percentage of total installed plant cost. Cost ranges for installed instrumentation costs are also included in Table 25-58, as these would form part of electrical costs. The ranges of values are rather wide, depending on the degree of automatic control required.

TABLE 25-57 Typical Factors for Converting Carbon Steel Cost to Equivalent-Alloy Costs

Material	Pumps, etc.	Other equipment
All carbon steel	1.00	1.00
Stainless steel, Type 410	1.43	2.00
Stainless steel, Type 304	1.82	2.50
Stainless steel, Type 316	2.00	2.86
Stainless steel, Type 310	2.00	3.33
Rubber-lined steel	1.43	1.25
Bronze	1.54	
Monel	3.33	

Material	Heat exchangers
Carbon steel shell and tubes	1.00
Carbon steel shell, aluminum tubes	1.25
Carbon steel shell, monel tubes	2.08
Carbon steel shell, 304 stainless tubes	1.67
304 stainless steel shell and tubes	2.86

Electrical costs involve four main components: (1) power wiring, (2) lighting, (3) transformation and service, and (4) instrument and control wiring. A breakdown of these component costs as a percentage of total electrical cost is given in Table 25-59.

The unit-cost method can give a quick and accurate estimation, provided it is based on accumulated data from many jobs on various types of plant. The actual data are analyzed to provide unit-cost information for electrical components as follows:

1. Total installed cost per motor
2. Total installed cost per lighting outlet by type
3. Total installed cost of receptacles by type (incandescent, fluorescent, etc.)
4. Total installed cost for each wired instrumentation point
5. Total installed cost for each unit of transformation
6. Total installed cost per lineal foot of distribution by type (overhead bare and insulated, underground)
7. Total installed cost of each interlock point

Each unit cost contains all the costs involved in the installation of that unit. For motors installed costs include the starter, conduit, wire, and a proportionate share of the service panelboard and busbars. The motor cost is not included since this will be part of the equipment cost. In the case of lighting, the installed cost includes the lighting

TABLE 25-58 Electrical and Instrumentation as Percentage of Total Installed Plant Cost

	Electrical (process and service)		Instrumentation	
Type of plant	Range, %	Median,	Range, %	Median, %
Solids plants	3.7–10.7	5.4	0.3–6.0	0.8
Grass-roots process	4.0–7.9	5.9	1.9–4.3	3.2
Battery-limit process	4.3–10.1	7.5	0.1–7.9	3.7

TABLE 25-59 Component Electrical Costs as Percentage of Total Electrical Cost

Component	Range, %	Median, %
Power wiring	25–50	40
Lighting	7–25	12
Transformation and service	9–65	40
Instrument-control wiring	3–8	5

fixtures, the conduit and wire, and a proportional share of the lighting panelboard and service switching costs.

Auxiliaries Estimation Chemical-plant auxiliaries normally include all structures, equipment, and services which are not directly involved in the process. Within this broad range there are two major classifications, utilities and service facilities.

The typical cost range for auxiliaries is from 20 to 40 percent of the total installed plant cost. For a small continuous-process plant making a single product, the cost of auxiliaries would lie in the lower part of the range, while for large multiprocess grass-roots plants the factor would tend to be near the upper limit of the range.

Auxiliary Buildings Typical variations in the cost of auxiliaries for a variety of process plants are given in Table 25-60. The widest variation is shown for auxiliary buildings, which is not surprising in view of the many types and quality of materials and the wide variation in methods of construction. For example, amenities buildings such as offices, cafeteria, first-aid rooms, gatehouses, and control rooms would necessitate fairly expensive brick and plaster-wall construction. On the other hand, services buildings such as substations, switch rooms, and pump or compressor houses would cost about 5 to 10 percent less. Provision of air conditioning, furniture, and equipment for cafeteria, laboratory, and office buildings would add about 50 percent to the basic cost of the building.

Steam-Generating Facilities These form the second largest investment item for chemical-plant auxiliary equipment. Variations in capacity, location indoors or outdoors, the type of fuel used, pressure and temperature levels, and the type of process served have an important effect on actual cost as well as on cost relative to other auxiliary items. Package boiler installations can be purchased as shop-built units which are assembled, piped, and wired ready to be erected on the owner's foundations. They are available in units up to about 136,000 kg/h (300,000 lb/h), although units larger than about 45,360 kg/h (100,000 lb/h) may be available only on a semierected basis. It is usually necessary to obtain firm price quotations that take into account all the factors involved. Housing the boiler installations in buildings will generally increase the cost by about 7 to 9 percent

TABLE 25-60 Typical Ranges of Auxiliary Facilities as Percentage of Total Installed Plant Cost

Grass roots and large additions

	Range, %	Median, %
Auxiliary buildings	3–9	5.0
Steam generation	2.6–6	3.0
Refrigeration, including distribution	1–3	2.0
Water supply, cooling, and pumping	0.4–3.7	1.8
Finished-product storage	0.7–2.4	1.5
Process-waste systems	0.4–1.8	1.1
Raw-materials storage	0.3–3.2	1.1
Steam distribution	0.2–2	1.0
Electrical distribution	0.4–2.1	1.0
Air compressor and distribution	0.2–3.0	1.0
Water distribution	0.1–2	0.9
Fire protection system	0.3–1.0	0.7
Water treatment	0.2–1.1	0.6
Railroads	0.3–0.9	0.6
Roads and walks	0.2–1.2	0.6
Gas supply and distribution	0.2–0.4	0.3
Sanitary-waste disposal	0.1–0.4	0.3
Communications	0.1–0.3	0.2
Yard and fence lighting	0.1–0.3	0.2

per kilogram-hour (15 to 20 percent per pound-hour) of steam-generating capacity over the cost for outdoor installations or for installations in existing buildings.

For most chemical plants, process steam is used at pressures of 1.825 MN/m^2 (250 psig), saturated or lower. When combined heat and power generation is economically justified, the steam may be generated at about 5.96 MN/m^2 (850 psig) appropriately superheated and used to drive back-pressure steam turbines passing out process steam at the required pressure level.

Electricity A reliable and adequate electricity supply is usually available through government or private enterprises. Owing to the increasing cost of purchased electricity, many companies have investigated the economics of installing combined steam and electricity generation systems. A cogeneration plant may (1) be owned and operated by the industrial user or the utility, (2) serve or be isolated from one or more industrial users, or (3) form an integral part of the local utility grid. A study [J. Javetski, *Power*, **122**, 35–40 (April 1978)] predicted that typical 1982 average cogeneration steam costs would be $11.42 per 1000 kg ($5.18 per 1000 lb) after deducting $2.82 per 1000 kg ($1.28 per 1000 lb) of steam as credit for electric power produced. The figures assumed the use of western United States coal of a heat content from 18.6 to 19.8 MJ/kg (8000 to 8500 Btu/lb) and a sulfur content of less than 0.5 percent.

For most plants, electric distribution systems start at the power company's service point on the plant's property. The choice of an electric distribution system will depend on many factors such as lightning hazards and other environmental conditions. This will result in the use of various types of underground and overhead distribution systems as required. An overhead system incurs only 30 to 50 percent of the cost of an underground system. The system itself includes main substations, distribution substations, feeders, switches, and ancillary equipment.

With 30.5-m (100-ft) spans, overhead systems consisting of insulated aerial cable supported on messengers strung between wooden poles would cost from $97 to $125 per meter ($30 to $38 per foot) of pole line installed at 1979 prices, depending on the size of conductor, ranging from No. 6 AWG (4.1-mm diameter) to 500,000 cmil (18-mm diameter). For this size range, an equivalent system of bare conductors on wooden poles, including lightning arresters on alternate spans, would cost from $27 to $84 per meter ($8 to $26 per foot) of overhead line installed.

Underground systems would require fiber-duct banks encased in concrete, manholes, substations, and insulated cables. This would cost from $350 to $840 per meter ($110 to $255 per foot) installed, depending on the number of ducts in a bank, the length of the system, the voltage and capacity, and soil and temperature conditions. The cost of underground distribution only, in fiber concrete-encased ducts laid in average soil and including manholes, less substations, varies from about $210 to $350 per meter ($65 to $105 per foot) of duct bank. Direct-burial cable for underground systems can sometimes be used. Costs for this type of system, less substations, range from $85 to $175 per meter ($25 to $55 per foot). Substations, according to type, capacity, and voltage, cost about $55 to $130 per kilovolt-ampere installed, all the above being at 1979 prices (M & S = 600).

Water Systems These systems usually form the third highest cost item in chemical-plant auxiliaries, with cooling towers representing the largest part of the investment. Although the installed cost increases with the terminal temperature range, an approximate cost correlation is given by

$$C_{CT} = 100q^{0.87} \tag{25-250}$$

where C_{CT} is the installed cost of the cooling tower in United States dollars for a Marshall and Swift (M & S) index of 1000 and q is the capacity in United States gallons per minute over the range from $(1)(10^3)$ to $(1)(10^5)$ U.S. gal/min.

River-water pumping and filtering installations can be approximately correlated by

$$C_{RW} = 0.65q^{0.81} \tag{25-251}$$

where C_{RW} is the installed cost of the river-water system in United States dollars for an M & S index of 1000 and q is the capacity over the range from $(4)(10^5)$ to $(1)(10^7)$ U.S. gal/day.

Similarly, installed costs of water-softening systems can be correlated in United States dollars (M & S = 1000) as follows:

$$C_{WS} = 1380q^{0.44} \tag{25-252}$$

over the range of capacity from $(3)(10^7)$ to $(1)(10^9)$ U.S. gal/day and of demineralizing systems by

$$C_{DS} = 0.17q^{1.9} \tag{25-253}$$

over the range of q values from $(1)(10^4)$ to $(4)(10^5)$ U.S. gal/day. Actual water-treatment costs may vary widely from the above, depending on the quality of the water, the percentage of dissolved solids, and the total hardness.

Refrigeration Systems These systems are being used increasingly in chemical processing. Installed costs of packaged mechanical units in United States dollars (M & S = 1000) can be approximately correlated by

$$C_{RS} = 4630q^{0.73} \tag{25-254}$$

where q is the capacity in tons of refrigeration over the range from 10 to 1000 tons. One ton of refrigeration is equivalent to a rate of heat removal of 3.517 kW (12,000 Btu/h).

Roads and Walks The cost of roads and walks in chemical plants is difficult to estimate since these vary with type of construction and thickness of applied cover. Some typical unit costs for roads are as follows: gravel and asphalt, \$10.20 to \$12.70 per square meter (\$8.50 to \$10.60 per square yard); concrete with a 152-mm (6-in) base, \$12.70 to \$15.20 per square meter (\$10.60 to \$12.70 per square yard); and concrete with a 203-mm (8-in) base, \$15.20 to \$19.00 per square meter (\$12.70 to \$15.90 per square yard). Installed costs for railroads, including switches and frogs, can be estimated as follows (July 1979 prices):

Linear meters	Linear feet	\$ per foot	\$ per meter
152–305	500–1000	125.00	39.00
305–915	1000–3000	118.00	36.00
915–3050	3000–10,000	112.00	34.00
Above 3050	Above 10,000	105.00	32.00

Use of Computers in Cost Estimation A large part of estimation consists of the collection and storage of data obtained from records of actual plant costs. The data then must be correlated and updated and the required information rapidly retrieved for use in further cost estimations. A comprehensive survey (C. J. Liddle and A. M. Gerrard, *The Application of Computers to Capital Cost Estimation*, Institution of Chemical Engineers, London, 1975, pp. 6–17) suggests that large chemical manufacturers, equipment vendors, and some contractors are using the computer increasingly for data retrieval, followed by simple correlation and the application of factorial methods to cost estimation.

In the case of equipment vendors, the computer's contribution appears to be particularly worthwhile owing to the elimination of estimating errors in producing price quotations. Several companies have developed an automated quotation system to overcome delay and inaccuracy in estimating and bidding. Such systems appear to have been developed by firms already possessing significant computing facilities, since the cost of computer time is small compared with the cost of the computer. Qualitatively, operating costs for an automated quotation system appear to be about half of those of a manual system of price quotation. The methods of estimation used are based on the manual methods described previously.

Several of the larger chemical manufacturers, particularly those in the petrochemicals field, have developed computer packages based on manual methods of factorial estimating. Usually the input data consists of the cost of each main-plant item (MPI) obtained from quotations or historical records. The program then estimates the costs of erection, piping, instrumentation, electricals, civil engineering, and lagging for each (MPI) in turn by adding a series of factors. These account for the complexity of the process and the constructional difficulties for each (MPI) to produce an estimate of the overall plant cost. It is obviously necessary to introduce appropriate inflation indices to bring the estimated costs up to date.

For process plants, it is often possible to use these cost-estimation programs with a design or flow-sheet program to optimize on a particular component or even over the whole plant (*A New Guide to Capital Cost Estimating*, Institution of Chemical Engineers, Rugby, England, 1977, pp. 20–21). However, it must be remembered that optimization is expensive on computer time, although there appear to be no data available on the cost effectiveness of the computer in this area. It is also possible to incorporate the capital-cost estimate in an investment evaluation involving forecasts of expenses and revenue from sales. Thus, by means of the computer design and costing can be brought together. There is an immediate feedback of information, resulting in improved design and lower costs. In some types of plants, costing data can be fed in as a subroutine to the design programs. All these possibilities assume that the total cost of using the computer is not unreasonable.

Startup Costs Startup problems can reduce aftertax earnings during the early years, the most serious effect being to delay the startup of production, causing a loss of earnings. An accurate estimate of startup time and cost can help in (1) predicting the availability of new products, (2) planning market entry, and (3) estimating the overall profitability because of more accurate cash-flow forecasts and (NPV) calculations [R. P. Feldman, *Chem. Eng.*, **76**, 87–90 (Nov. 3, 1969)].

Startup costs are defined as the total of those costs directly related to bringing a new production facility into operation. They should not include the costs of entering or expanding a business. Hence startup costs include the following:

1. All expenses due to changes in process and equipment after completion of construction but excluding those due to changes in project scope
2. All labor costs after completing construction, especially those incurred in checking the functioning of equipment
3. All costs incurred during the startup period but excluding normal operating expenses
4. Expenses for training plant personnel even if incurred before startup has officially begun
5. All research and development costs incurred during startup

The following expenses should *not* be included:

1. Marketing costs
2. Expense of training sales representatives
3. Penalties for shipping outside optimum freight areas
4. Costs associated with starting a new company
5. Lost sales unless there is a contract with a penalty
6. Profit lost due to timing

Startup time may be defined as the time span between the end of construction and the beginning of normal operation. Hence it should start when the contractor finishes the whole plant or a specified section of it to enable comparisons to be made with other startup times. It is usual to define "normal" operation as (1) operations at a certain percentage of design capacity, (2) a specified number of days of continuous operation, or (3) the capability of making products of a specified purity.

It is essential for project and production management to agree beforehand on the definition to be applied. Obtaining agreement on the definition of "normal" operation is important since (1) it sets a target for field personnel, (2) it ensures that everyone is striving for the same target, (3) it permits comparisons with other plants, and (4) it determines a cutoff point for completion of startup. It may be necessary to wait until the plant is running well to obtain the actual total cost of startup.

For control purposes it is advisable to estimate startup cost and time beforehand and then try to stay within the estimates. The general parameters which can be used to estimate startup cost C_{SU}, which are usually between 2 and 20 percent of the battery-limit fixed-capital cost, are as follows:

1. Direct fixed-capital cost for plant (battery-limit capital), $(C_{FC})_{BL}$
2. Newness of process and technology, b
3. Newness of type and size of equipment, c
4. Labor quality and quantity, d
5. Interplant dependency, e

Hence startup cost may be expressed as

$$C_{SU} = (C_{FC})_{BL}[0.10 + b + c + d + ne] \qquad (25\text{-}255)$$

When applied to large air-separation and ammonia plants (1000 to 1400 metric tons/day), the following values for the parameters can be used:

$b =$ 0.05 for a radically new process
$=$ 0.02 for a relatively new process
$= -0.02$ for an old process
$c =$ 0.07 if radically new
$=$ 0.04 if very new
$=$ 0.02 if relatively new
$= -0.03$ if old
$d =$ 0.04 if labor is in very short supply
$=$ 0.02 if labor is in short supply
$= -0.1$ if labor is in surplus supply
$e =$ 0.04 if plant is very dependent on another
$=$ 0.02 if moderately dependent on another
$= -0.02$ if independent

and

n = number of plants or sections making up the process chain

Startup time t_{SU} for these plants may be estimated from construction time t_c by developing an equation similar to Eq. (25-255):

$$t_{SU} = t_C(0.15 + b + c + d + ne) \qquad (25\text{-}256)$$

For the same type of plant the values of the parameters are

$b =$ 0.15 for a radically new process
$=$ 0.05 for a relatively new process
$= -0.01$ for an old process
$c =$ 0.15 if radically new
$=$ 0.08 if very new
$=$ 0.05 if relatively new
$= -0.01$ if old
$d =$ 0.15 if labor is in very short supply
$=$ 0.05 if labor is in short supply
$= -0.01$ if labor is in surplus supply
$e =$ 0.25 if plant is very dependent on another
$=$ 0.10 if moderately dependent on another
$= -0.02$ if independent

and

n = number of plants or sections making up the process chain

It should be noted that these values are based on previous experience with certain types of plants, but appropriate values which apply to other processes and locations could be selected.

Construction Time The duration of construction is difficult to estimate owing to the large number of variables involved. In general, estimates of construction time tend to be overoptimistic, especially for larger projects. Usually projects costing less than \$2 million at 1979 prices can be completed in 10 to 18 months, while those costing more than \$10 million may take from 18 to 42 months to complete. Delays of up to 12 months behind schedule are quite possible, particularly when there are labor problems. As mentioned previously, such delays will usually result in increased construction costs. Often, a more serious effect is loss of earnings resulting from a delayed startup. Both of these factors increase the payback period and reduce the attainable net present value and discounted-cash-flow rate of return of the project.

Project Control Having made a good estimate of the capital cost and the expected construction time, it is essential to introduce an effective system for controlling expenditure of time and money during construction. Good capital-cost control can cut down expenditures even when the definitive estimate is not very accurate. It is most important for management to receive early warnings if overruns in expenditure or time are likely to occur.

Effective cost control should start from the beginning of the project at the research and development stage and continue through the design and estimating stages to initial operation of the plant (J. W. Hackney, *Control and Management of Capital Projects*, Wiley, New York, 1965, p. 4). The stages discussed here are the later steps after authorization of funds and during project construction. After the purchasing department has placed the orders for equipment and materials, speed and efficiency during the construction stage is most important in ensuring the financial success of the project. Field expenditure during construction can amount to 30 to 60 percent of the fixed-capital cost and includes the costs of all labor, installed equipment, and materials together with associated process piping, electrical instrumentation, and insulation. Construction therefore requires efficient execution and prompt feedback of progress information, necessitating a good cost-control system.

Figure 25-43 shows the flow of information needed for cost control. The chart assumes a definitive estimate which has been linked to a standard code of accounts. As construction proceeds, up-to-date cost-control reports are supplied to the field cost engineer. From the home office the engineer receives monthly reports of engineering and drafting labor-hours used and money expended, together with a list of drawings and specifications completed up to that time. Monthly expenditures and current commitments come in coded detail from the job ledgers of the accountants. Timekeepers' records give details of craft and nonmanual labor-hour expenditures. Quantities of equipment and material held on site are reported daily by quantity surveyors to the construction superintendent. All purchase orders are posted in the ledgers as current commitments, whether they are placed at the home office or in the field, and an up-to-date warehouse inventory is maintained.

At the end of each month, the field cost engineer collects all current information on a detailed cost report form. As these are actual costs, they can be used to estimate future job costs to completion. Daily reports of unit-cost progress for concrete, excavation, masonry, steel, piping, and electrical work, etc., are then used to predict possible overruns or underruns for the various items. Analysis and comparison with the original estimate point out trouble spots for early attention. If an item is running into difficulty, it is red-flagged to the resident and project engineers for remedial action.

In practice, the existence of a tight cost-control system tends to spread a cost consciousness among the personnel involved in the project. Such an awareness, even in construction-equipment maintenance and job housekeeping, can lead to efficient cost control throughout.

Cost reports should be brief but informative, preferably in summary form. They should report expenditures and commitments, estimated costs to complete, and expected overruns or underruns of the authorized budget for each important item of cost. Brief notes should emphasize significant deviations from predicted cost. Any large, persistent overrun should have already been investigated and reported to the project and construction managers for immediate attention. If an expected overrun cannot be avoided, the current summary cost report should serve as justification for a request for additional funds.

When organized efficiently, the cost-control system should require no more paperwork than for the normal construction procedure. The cost of cost control appears to vary between 0.2 and 0.5 percent of the total project value. Proper use of the normal records available for craft-labor time, warehouse-inventory control, and the usual accounting purposes should be adequate. The savings achieved by good cost control should far exceed the additional costs of operating

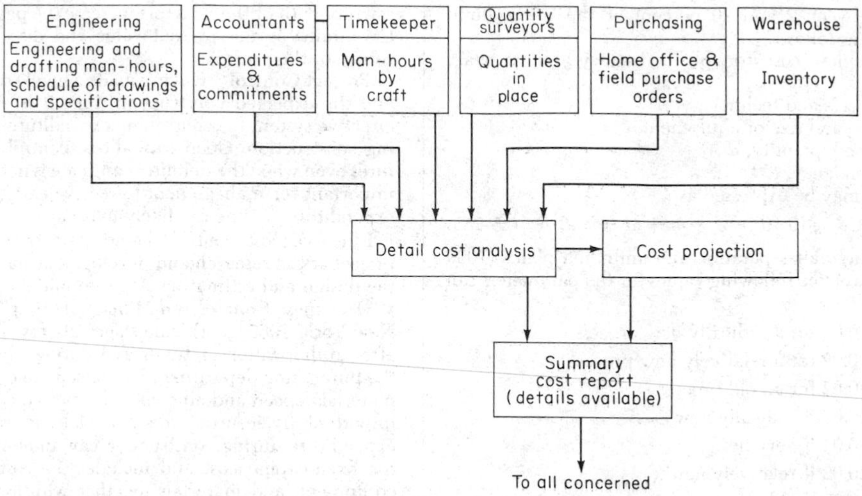

FIG. 25-43 Information flow for cost control.

the system. Additional details on the technique are given by H. C. Bauman (*Fundamentals of Cost Engineering in the Chemical Industry,* Van Nostrand Reinhold, New York, 1964, pp. 190–196).

Scheduling construction to ensure that the project is completed in the shortest possible time is an essential part of project control. The project-control estimate defines to a large extent the construction-time schedule. It is then possible to prepare a master schedule from the control estimate by carefully sequencing and synchronizing the installation work according to past experience. Drawings are usually completed in predictable order owing to the dependence of certain designs on preceding work. The normal order of completion of drawings and specifications is (1) site work, (2) substructures, (3) equipment and building superstructures, (4) equipment layouts, (5) piping, (6) insulation, (7) instrumentation, and (8) electrical work.

Detailed planning and scheduling then involve establishing the items of work required and determining the correct sequence of work and the number of persons required to perform each item of work. From this information it is possible to prepare bar charts by using a 4-week month and noting on the chart the interdependence of the various functions. The starting time for each class of work is fixed on the chart, and the duration is calculated from the labor-hours allocated to that work from the control estimate. Work should progress smoothly as time elapses, but the operations must be linked by the order of necessary precedence. Starting times for the various items of work will be staggered as drawings are released and also to smooth out labor requirements.

Sketching the bar chart is commenced by inserting the arrival dates of key items, observing any necessary precedence. Estimates of the duration of erection time can be made to obtain the starting date for process piping. Since the supporting structure must be in place when the key item arrives, it is possible to work back along the bar chart to the preparation of the foundations. From the complete bar chart, built up in a similar manner, a tentative startup date can be set after allowing a few weeks for tidying up bits and pieces. Some activities can be speeded up, but it is necessary to estimate the increased cost of so doing.

Actual progress made with construction work can be indicated on the bar chart by filling in the open bars according to the percentage toward completion. Comparison of the actual progress bar for the whole project with the cumulative labor-hour curve indicates whether the job is ahead of schedule or not. If corrective action is required, effort should be concentrated on the key or critical items.

Large projects will usually require network analysis using the critical-path method (CPM) or program evaluation and review technique (PERT) in the planning, scheduling, and progress-control stages. Examples of bar charts and a fuller description of network analysis are given by J. W. Hackney (*Control and Management of Capital Projects,* Wiley, New York, 1965, pp. 118–131). A detailed treatment of the use of PERT and CPM techniques as applied to contract bidding strategy and to project control is given by L. A. Swanson and H. L. Pazer (*Pertsim: Text and Simulation,* International Textbook, Scranton, Pa., 1969), who present a hand simulation technique based on probabilistic methods.

Overseas Construction Costs Although Table 25-54 gives location factors for the construction of chemical plants of similar function in various countries at 1977 values, these may vary differentially over a period of time owing to local changes in labor costs and productivity. Hence, it is often necessary to estimate the various components of overseas construction costs separately. Equipment and material prices will depend on local labor costs and the availability of raw materials. If the basic materials have to be imported, costs in the source area become important and import duties and freight charges must be added.

Equipment and material normally amount to about 40 to 45 percent of the costs of a typical chemical plant. Table 25-61 shows the relative costs of some commonly used equipment and materials for several countries. In general, equipment and material costs are slightly cheaper in European countries and Japan, whereas in Mexico and Canada they are nearer the United States average.

Construction labor makes up about 20 to 35 percent of total costs for a chemical plant. Table 25-62 compares average 1979 hourly rates for various types of construction labor in several countries with those for the United States. Most of the rates, except those for Canada and West Germany, are considerably below United States average values. The data are published quarterly in *Engineering News-Record* and include all applicable fringe benefits.

Fringe benefits are known in countries other than the United States as "social charges"; they vary considerably in degree of coverage from country to country. Typical allowances in these countries include family benefits based on number of children, health service, maternity benefits, disability allowances, grants for funeral espenses, old-age and war pensions, unemployment benefits, and pension schemes. Additional fringe benefits may include paid holidays, starting allowances for new workers, relocation grants, severance pay, profit sharing, production bonuses, special gratuities, and sometimes housing allowances. It is essential to investigate the local situation thoroughly to determine the benefits payable and the additional cost on the basic hourly wage rate.

TABLE 25-61 Relative Equipment and Materials Costs, United States and Other Countries, 1979

	United States	United Kingdom	Italy	Mexico	West Germany	France	Canada	Japan
Equipment								
Tanks, vessels	1.00	0.95	0.90	0.90	0.95	0.95	1.05	0.85
Pumps	1.00	0.90	0.85	0.95	0.90	0.90	1.00	0.90
Filters, centrifuges	1.00	0.90	0.90	1.00	0.90	0.90	1.05	0.90
Compressors	1.00	0.85	0.80	0.95	0.90	0.90	1.05	1.00
Heat exchangers	1.00	0.95	0.90	1.00	0.90	1.00	1.05	1.10
Dryers, boilers	1.00	0.95	1.00	1.05	0.90	1.00	1.05	0.95
Dust collectors	1.00	1.05	1.00	1.05	0.95	0.95	1.05	0.90
Refrigeration units	1.00	1.00	0.95	1.10	0.95	0.95	1.05	0.90
Material								
Piping	1.00	0.90	0.90	1.05	0.90	0.95	1.05	0.90
Insulation	1.00	0.90	0.90	0.95	0.90	0.95	1.05	0.90
Electrical	1.00	1.10	0.85	1.05	0.95	0.95	1.10	0.85
Instruments	1.00	1.05	1.00	1.20	1.00	1.05	1.05	0.95

TABLE 25-62 Hourly Construction Labor Rates in Various Countries Compared with United States, December 1979*

Craft	United States	Canada	England	West Germany	Brazil	France	Italy	Australia	Mexico	Japan	India
Laborers	$11.11	$ 9.64	$3.96	$11.02	$0.48	$3.17	$8.43	$3.23	$0.75	$ 6.07	$0.16
Carpenters	14.20	12.42	4.60	16.70	0.76	4.82	9.15	4.17	1.52	9.11	0.31
Bricklayers	14.41	13.14	4.60	12.15	0.76	4.82	8.97	4.17	1.10	7.84	0.31
Ironworkers	15.24	13.50	4.60	11.84	0.74	4.82	8.73	4.30	1.74	10.12	0.32
Average for above workers	13.74	12.17	4.44	12.93	0.69	4.41	8.82	3.97	1.28	8.29	0.28
Ratio to the United States	1.00	0.89	0.32	0.94	0.05	0.32	0.64	0.29	0.09	0.60	0.02

*From *Engineering News-Record,* fourth quarterly cost roundup, Dec. 20, 1979.
NOTE: Fringe benefits are included.

Labor productivity is very much dependent on the health and well-being of the workers and also on the availability of laborsaving tools and construction equipment. The frequency of strikes, holidays, slowdowns, and political unrest will also depress productivity. Closed-shop practices or demarcation disputes will also affect the productivity of labor. The use of standard equipment, parts, and methods tends to improve productivity.

In a particular country, productivity will depend largely on the number of hours worked per week. Production will increase with the number of hours worked during the week, but as more overtime is worked, fatigue will produce a falloff in productivity.

In the United States, construction craft labor usually work a normal 35 h/week. The United Kingdom operates a 40-h schedule, although there is strong pressure to reduce this to 35 h/week. European countries tend to work a normal 42 h/week, and other countries, such as Japan, may work up to 45 h/week.

Productivity of local craft labor also depends on the use and availability of modern mechanical tools and construction equipment. Normally, the low cost of labor in certain countries tends to cut out the purchase or hire of sophisticated laborsaving equipment and to encourage the employment of large pools of labor, particularly in developing countries such as India, Pakistan, southeast Asian countries, and many African countries. In turn, this usually leads to higher construction costs. The use of laborsaving equipment is prevalent in Canada, western Europe, Japan, and, to an increasing extent, the Middle East.

Complete Plant Costs It is difficult to compare costs of domestic and overseas plants owing to the wide variation in types of plants and sizes and the rapid changes in technology. Useful data are scarce, and the following comparisons must be used with caution and then only for order-of-magnitude estimates of fixed-capital costs.

The method uses a breakdown of costs for a typical chemical plant installed in the United States, as shown in Fig. 25-44. Costs of equipment, appurtenances, construction, and engineering with material and labor separate are given as a percentage of total installed United States costs. The four components of cost are defined as follows:

Equipment includes all prefabricated machines, appliances, or systems such as tanks, heat exchangers, pumps, motors, switchgear, and boilers.

Appurtenances are auxiliary items which cover materials, such as pipes, valves, fittings, conduit, wire, tubing, and insulation.

Construction expense includes the cost of construction equipment, tools, sheds, railroad trackage, road materials, welding machines, scaffolding, and timber, which are all used in construction but do not form a permanent part of the plant.

Engineering is mainly labor but has a small component cost which can be classified with equipment and materials, such as tools, paper, pencils, and reproduction costs.

In total, labor amounts to 34 percent and material to 66 percent of total installed costs.

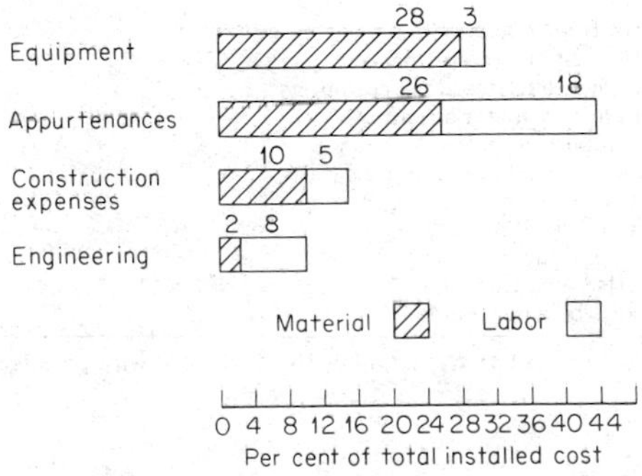

FIG. 25-44 Typical breakdown of chemical-plant costs by major component.

TABLE 25-63 Relative Plant Construction Costs in Various Countries Compared with the United States (1979 Values)

Country	Equipment	Material	Labor	Engineering	Total
United States	0.28	0.38	0.26	0.08	1.00
England	0.26	0.41	0.18	0.05	0.90
Italy	0.22	0.32	0.31	0.05	0.90
Mexico	0.26	0.35	0.35	0.04	1.00
Australia	0.38	0.54	0.29	0.09	1.30
Canada	0.32	0.44	0.31	0.08	1.15
France	0.27	0.40	0.22	0.06	0.95
West Germany	0.26	0.36	0.32	0.06	1.00
Japan	0.29	0.35	0.22	0.04	0.90

Table 25-63 uses the data of Fig. 25-44 to compare the relative fixed-capital costs for plant construction in other countries with those for the United States. The relative cost ratios were developed from data similar to those in Tables 25-61 and 25-62. Labor ratios were corrected for the different local rates and hours per working week, job duration, and degree of mechanization available in other countries. Some of these factors are difficult to estimate, and the final "total" ratios give a reasonable order-of-magnitude value for relative construction costs for equivalent plants in the countries indicated.

The choice of an overseas manufacturing site involves the consideration of many political and economic factors in addition to costs. Table 25-64 gives a list of 92 items which should be taken into account when choosing a plant location for manufacturing abroad.

TABLE 25-64 Factors in Choosing a Foreign Manufacturing Site*

Economic factors
- Size of GNP and rate of growth
- Is there a working development plan?
- Resistance to recession
- Relative dependence on imports and exports
- Foreign-exchange position
- Balance-of-payments outlook
- Stability of currency; convertibility
- Remittance and repatriation regulations
- Balance of economy (industry-agriculture-trade)
- Size of market for your products; rate of growth
- Size of population; rate of growth
- Per capita income; rate of growth
- Income distribution
- Current or prospective membership in a customs union
- Price levels; rate of inflation

Political factors
- Stability of government; its form
- Presence or absence of class antagonism
- Special political, ethnic, and social problems
- Attitude toward private and foreign investment
- Acceptability of United States investment by government
- Acceptability of United States investment by customers and competitors
- Presence or absence of nationalization threat
- Presence or absence of state industries
- Do state industries receive favored treatment?
- Concentration of influence in small groups
- Treaty of friendship or establishment with United States?

Government factors
- Are fiscal and monetary policies sound?
- Freedom from bureaucratic red tape
- Fairness and honesty of administrative procedures
- Degree of antiforeign or anti–United States discrimination
- Fairness of courts
- Clear and modern corporate investment laws
- Patentability of your products
- Presence or absence of price controls
- Restrictions on 100 percent United States or foreign ownership

Geographic factors
- Efficiency of transport (railways, waterways, highways)
- Port facilities
- Free ports, free zones, bonded warehouses
- Proximity of site to export markets
- Proximity of site to suppliers, customers
- Proximity to raw-material sources
- Existing supporting industry
- Availability of local raw materials
- Availability of power, water, gas
- Reliability of utilities
- Waste-disposal facilities
- Can exports be easily made?
- Can imports be easily made?

Geographic factors (continued)
- Are plant sites readily available?
- Cost of suitable land

Labor factors
- Availability of English-speaking managerial, technical, office personnel
- Availability of skilled labor
- Availability of semiskilled and unskilled labor
- Level of worker productivity
- Training facilities
- Outlook for increase in labor supply
- Degree of skill and discipline at all levels
- Tranquillity of labor relations
- Presence or absence of militant or Communist-dominated unions
- Degree of labor voice in management
- Freedom to hire and fire
- Compulsory and voluntary fringe benefits
- Social security taxes
- Total cost, including fringes, compared wtih alternative sites
- Compulsory or customary profit sharing

Tax factors
- Tax rates (corporate and personal income, capital, withholding, turnover, excise, payroll, capital gains, customs, other indirect and local taxes)
- General tax morality
- Fairness and incorruptibility of tax authorities
- Long-term trend for taxes
- Taxation of export income and income earned abroad
- Tax incentives for new businesses
- Depreciation rates
- Tax-loss carry-forward and carry-back
- Joint tax treaties
- Duty and tax drawbacks when imported goods are exported
- Availability of tariff protection

Capital-sources factors
- Availability of local capital
- Costs of local borrowing
- Normal terms for local borrowing
- Availability of convertible currencies locally
- Modern banking system
- Government credit aids to new businesses
- Availability and cost of export financing, insurance
- Do United States or European capital sources favor loans here?

Business factors
- Availability of United States government investment insurance
- General business morality
- State of marketing and distribution system
- Are administrative procedures simple and effective?
- Normal profit margins in general, in your industry
- Competitive situation in your industry; is it cartelized?
- What are antitrust and restrictive practices laws, and do they conflict with United States laws?
- Availability of amenities for United Stated expatriate executives and families

*From *Business International*, May 18, 1962. Used with permission.

Section 26

Waste Management

Anthony J. Buonicore, M.CH.E., P.E., *Diplomate, AAEE, President, Buonicore-Cashman Associates, Inc.; Member, American Institute of Chemical Engineers, Air Pollution Control Association. (Section Editor, Air Pollution Management)*

Louis Theodore, Sc.D., *Professor of Chemical Engineering, Manhattan College; Member, Air Pollution Control Association. (Section Coeditor)*

Ross E. McKinney, Sc.D., *N. T. Veatch Professor of Environmental Engineering, University of Kansas; Member, American Academy of Engineering. (Wastewater Management)*

George Tchobanoglous, *Professor of Environmental Engineering, University of California at Davis; Member, American Society of Civil Engineers, Water Pollution Control Federation. (Management of Industrial Solid Wastes)*

INTRODUCTION TO WASTE MANAGEMENT

AIR POLLUTION MANAGEMENT OF STATIONARY SOURCES

INDUSTRIAL-WASTEWATER MANAGEMENT

MANAGEMENT OF INDUSTRIAL SOLID WASTES

INTRODUCTION TO WASTE MANAGEMENT

In this section a number of references are made to laws and procedures that have been formulated in the United States with respect to waste management. An engineer handling waste-management problems in another country would be well advised to know the specific laws and regulations of that country. Nevertheless, the treatment given here is believed to be useful as a general guide.

Multimedia Approach to Environmental Regulations in the United States Among the most complex problems to be faced by industry during the 1980s are the proper control and use of the natural environment. In the 1970s the engineering profession became acutely aware of its responsibility to society, particularly for the protection of public health and welfare. The decade saw the formation and rapid growth of the U.S. Environmental Protection Agency (EPA) and the passage of federal and state laws governing virtually every aspect of the environment. The end of the decade, however, brought a realization that only the more simplistic problems had been addressed. A limited number of large sources had removed substantial percentages of a few readily definable air pollutants from their emissions. The incremental costs to increase the removal percentages would be significant and would involve increasing numbers of smaller sources, and the health hazards of a host of additional toxic pollutants remained to be quantified and control techniques developed.

Early in the 1970s, air, water, and land were treated as separate problem areas, to be governed by their own statutes and regulations. Toward the latter part of the decade, however, it became obvious that environmental problems were closely interwoven and should be treated in concert. The traditional process-standard type of regulation—command and control—had severely restricted compliance options.

The 1980s began with EPA efforts redirected to take advantage of the case-specific knowledge, technical expertise, and imagination of those being regulated. Providing plant engineers with an incentive to find more efficient ways of abating pollution should greatly stimulate innovation in control technology. This is a principal objective, for example, of EPA's "controlled trading" air pollution program, established in the Offsets Policy Interpretative Ruling issued by the EPA in 1976, with statutory foundation given by the Clean Air Act Amendments of 1977.

The rapidly expanding body of federal regulation presents an awesome challenge to traditional practices of corporate decision making, management, and long-range planning. Those responsible for new plants must take stock of the emerging requirements and construct a fresh approach.

The full impact of the Clean Air Act Amendments of 1977, the Clean Water Act, the Safe Drinking Water Act, the Resource Conservation and Recovery Act, and the Toxic Substances Control Act is still not generally appreciated. The combination of all these requirements, sometimes imposing conflicting demands or establishing differing time schedules, makes the task of obtaining all regulatory approvals extremely complex.

One of the dominant impacts of environmental regulations is that the lead time required for the planning and construction of new plants will be substantially increased. When new plants may generate major environmental complexities, the implications can be profound. Of course, the exact extent of additions to lead time will vary widely from one case to another, depending on which permit requirements apply and on what difficulties are encountered. For major expansions in any field of heavy industry, however, the delay resulting from federal requirements could conceivably add 2 to 3 years to total lead time. Moreover, there is always the possibility that regulatory approval will be denied. So, contingency plans for fulfilling production needs must be developed.

Any company planning a major expansion must concentrate on environmental factors from the outset. Since many environmental approvals require a public hearing, the views of local elected officials are extremely important. To an unprecedented degree, the political acceptability of a project can now be critical.

Plant Strategies At the plant level, a number of things can be done to minimize the impact of environmental-quality requirements. These include:

1. Maintaining an accurate source-emission inventory
2. Continually evaluating process operations to identify potential modifications that might reduce or eliminate environmental impacts
3. Ensuring that good housekeeping and strong preventive-maintenance programs exist and are followed
4. Investigating available and emerging pollution-control technologies
5. Keeping well informed of the regulations and the directions in which they are moving
6. Working closely with the appropriate regulatory agencies and maintaining open communications to discuss the effects that new regulations may have

It is unrealistic to expect that at any point in the foreseeable future Congress will reverse direction, reduce the effect of regulatory controls, or reestablish the preexisting legal situation in which private companies were free to construct major industrial facilities with little or no restraint by federal regulation.

Corporate Strategic Planning Contingency plans represent an essential component of sound environmental planning for a new plant. The environmental uncertainties surrounding a large capital project should be specified and related to other contingencies (such as marketing, competitive reactions, politics, foreign trade, etc.) and mapped out in the overall corporate strategy.

Environmental factors should also be incorporated into a company's technical or research and development program. Since the planning horizons for new projects may now extend to 5 to 10 years, R&D programs can be designed for specific projects. These may include new process controls or end-of-pipe technologies.

Another clear need is to integrate environmental factors into financial planning for major projects. It must be recognized that strategic environmental planning is as important to the long-range goal of the corporation as is financial planning. Trade-off decisions regarding financing may have to change as the project goes through successive stages of environmental planning and permit negotiations. For example, requirements for the use of more expansive pollution-control technology may significantly increase total project costs; or a change from end-of-pipe to internal-process technology may preclude use of industrial-revenue-bond financing under Internal Revenue Service (IRS) rules. Regulatory delays can affect assumptions as to both the rate of expenditure and inflation factors. Investment, production, environmental, and legal factors are all interrelated and can have a major impact on corporate cash flow.

Most companies must learn to deal more creatively with local officials and public opinion. The social responsibility of companies can become an extremely important issue. Companies should apply thoughtfulness and skill to the timing and conduct of public hearings. Management must recognize that local officials have views and constituencies that go beyond attracting new jobs.

From all these factors, it is clear that the approval and construction of major new industrial plants or expansions is a far more complicated operation than it has been in the past, even the recent past.

Stringent environmental restrictions are likely to preclude construction of certain facilities at locations where they otherwise might have been built. In other cases, acquisition of required approvals may generate a heated technical and political debate that can drag out the regulatory process for several years.

In many instances, new requirements may be imposed while a company is seeking approval for a proposed new plant. Thus, companies intending to expand their basic production facilities should anticipate their needs far in advance, begin preparation to meet the regulatory challenge they will eventually confront, and select sites with careful consideration of environmental attributes. It is the objective of this section to assist the engineer in meeting this environmental regulatory challenge.

AIR POLLUTION MANAGEMENT OF STATIONARY SOURCES

General References: Billings and Wilder, *Fabric Filter Handbook*, U.S. EPA, NTIS Publ. PB 200-648 (vol. I), PB 200-649 (vol. II), PB 200-651 (vol. III), and PB 200-650 (vol. IV), 1970. Buonicore, "Air Pollution Control," *Chem. Eng.*, **87**(13), 81 (June 30, 1980). Buonicore and Theodore, *Industrial Control Equipment for Gaseous Pollutants*, vols. I and II, CRC Press, Boca Raton, Fla., 1975. Calvert, *Scrubber Handbook*, U.S. EPA, NTIS Publ. PB 213-016 (vol. I) and PB 213-017 (vol. II), 1972. Danielson, *Air Pollution Engineering Manual*, EPA Publ. AP-40, 1973. Davis, *Air Filtration*, Academic, New York, 1973. Kleet and Galeski, *Flare Systems Study*, EPA-600/2-76-079 (NTIS), 1976. Lund, *Industrial Pollution Control Handbook*, McGraw-Hill, New York, 1971. Oglesby and Nichols, *Manual of Electrostatic Precipitator Technology*, U.S. EPA, NTIS Publ. PB 196-380 (vol. I), PB 196-381 (vol. II), PB 196-370 (vol. III), and PB 198-150 (vol. IV), 1970. *Package Sorption System Study*, EPA-R2-73-202 (NTIS), 1973. Rolke et al., *Afterburner Systems Study*, U.S. EPA, NTIS Publ. PB 212-500, 1972. Slade, *Meteorology and Atomic Energy*, AEC (TID-24190), Oak Ridge, Tenn., 1969. Stern, *Air Pollution*, Academic, New York, 1974. Strauss, *Industrial Gas Cleaning*, Pergamon, New York, 1966. Theodore and Buonicore, *Industrial Air Pollution Control Equipment for Particulants*, CRC Press, Boca Raton, Fla., 1976. Theodore and Buonicore, *Air Pollution Control Equipment—Selection, Design, Operation and Maintenance*, Prentice-Hall, Englewood Cliffs, N.J., 1982. Treybal, *Mass Transfer Operations*, 3d ed., McGraw-Hill, New York, 1980. Turner, *Workbook of Atmospheric Dispersion Estimates*, U.S. EPA Publ. AP-26, 1970. White, *Industrial Electrostatic Precipitation*, Addison-Wesley, Reading, Mass., 1963.

INTRODUCTION

Air pollutants may be classified into two broad categories: (1) natural and (2) human-made. Natural sources of air pollutants include:

- Windblown dust
- Volcanic ash and gases
- Ozone from lightning and the ozone layer
- Esters and terpenes from vegetation
- Smoke, gases, and fly ash from forest fires
- Pollens and other aeroallergens
- Gases and odors from natural decomposition
- Natural radioactivity

Such sources constitute background pollution and that portion of the pollution problem over which control activities can have little, if any, effect.

Human-made sources cover a wide spectrum of chemical and physical activities and are the major contributors to urban air pollution. Air pollutants in the United States pour out from over 100 million vehicles, from the refuse of over 200 million people, the generation of billions of kilowatts of electricity, and the production of innumerable products demanded by everyday living. Almost 300 million tons of air pollutants are generated annually in the United States alone. The five main classes of pollutants are particulates, carbon monoxide, hydrocarbons, nitrogen oxides, and sulfur oxides. Emissions from stationary combustion systems in the United States are summarized for the year 1976 in Table 26-1.

Air pollutants may also be classified as to origin and state of matter:

1. Origin

a. Primary—emitted to the atmosphere from a process

b. Secondary—formed in the atmosphere as a result of a chemical reaction

2. State of matter

a. Gaseous—true gases such as sulfur dioxide, nitrogen oxides, ozone, carbon monoxide, etc.; vapors such as gasoline, paint solvents, dry cleaning agents, etc.

b. Particulate—finely divided solids or liquids; solids such as dust, fumes, and smoke; and liquids such as droplets, mists, fogs, and aerosols

Gaseous Pollutants Gaseous pollutants may be classified as inorganic or organic. Inorganic pollutants consist of:

1. Sulfur gases: sulfur dioxide, sulfur trioxide, hydrogen sulfide
2. Oxides of carbon: carbon monoxide, carbon dioxide
3. Nitrogen gases: nitrous oxide, nitric oxide, nitrogen dioxide, other nitrous oxides
4. Halogens, halides: hydrogen fluoride, hydrogen chloride, chlorine, fluorine, silicon tetrafluoride
5. Photochemical products: ozone, oxidants
6. Cyanides: hydrogen cyanide
7. Ammonium compounds: ammonia

Organic pollutants consist of:

1. Hydrocarbons

a. Paraffins: methane, ethane, octane

b. Acetylene

c. Olefins: ethylene, butadiene

d. Aromatics: benzene, toluene, benzpyrene

2. Aliphatic oxygenated compounds

a. Aldehydes: formaldehyde

b. Ketones: acetone

TABLE 26-1 Emissions from Stationary Combustion Systems in the United States*

	Air							
						Organics†		
	Particles, %	Sulfur oxides, SO^x	Nitrogen oxides, NO^x	Hydrocarbons, HC, %	Carbon monoxide, CO, %	BSO, %	PPOM, %	BaP, %
Electric generation	63.8	72.5	64.8	34.0	33.6	8.8	0.3	0.2
Industrial	28.3	14.5	24.7	22.3	14.9	20.0	0.5	1.3
Commercial and institutional	4.9	6.7	7.3	12.2	7.7	16.0	0.2	0.4
Residential	3.0	6.3	3.2	31.5	44.7	55.2	99.0	98.1
Total, 10,000 tons/year	7060	22,100	10,950	353	1070	125	4.14	0.40

*Source: GCA Corp., 1976.

†BSO = benzene-soluble organics; PPOM = particulate polycylic organic material; BaP = benzo-α-pyrene.

c. Organic acids
d. Alcohols
e. Organic halides: cyanogen chloride, bromobenzyl cyanide
f. Organic sulfides: dimethyl sulfide
g. Organic hydroperoxides: peroxyacyl nitrite or nitrate (PAN)

The most common gaseous pollutants and their major sources and significance are presented in Table 26-2.

Particulate Pollutants Particulates may be defined as solid or liquid matter whose effective diameter is larger than a molecule but smaller than approximately 1000 μm. Particulates dispersed in a gaseous medium are collectively termed an "aerosol." The terms "smoke," "fog," "haze," and "dust" are commonly used to describe particular types of aerosols, depending on the size, shape, and characteristic behavior of the dispersed particles. Aerosols are rather difficult to classify on a scientific basis in terms of their fundamental properties such as settling rate under the influence of external forces, optical activity, ability to absorb an electric charge, particle size and structure, surface-to-volume ratio, reaction activity, physiological action, etc. In general, particle size and settling rate have been the most characteristic properties for many purposes. For example, particles larger than 100 μm may be excluded from the category of dispersions because they settle too rapidly. On the other hand, particles on the order of 1 μm or less settle so slowly that, for all practical purposes, they are regarded as permanent suspensions. Despite possible advantages of scientific classification schemes, the use of popular descriptive terms such as smoke, dust, and mist, which are essentially based on the mode of formation, appears to be a satisfactory and convenient method of classification. In addition, this approach is so well established and understood that it undoubtedly would be difficult to change.

Dust is typically formed by the pulverization or mechanical disintegration of solid matter into particles of smaller size by processes such as grinding, crushing, and drilling. Particle sizes of dusts range from a lower limit of about 1 μm up to about 100 or 200 μm and

Table 26-2 Typical Gaseous Pollutants and Their Principal Sources and Significance

Air pollutants	From manufacturing sources such as these	In typical industries	Cause these damaging effects
Sulfur dioxide	Fuel combustion (coal, oil), smelting and casting, manufacture of paper by sulfite process	Primary metals (ferrous and nonferrous); pulp and paper	Sensory and respiratory irritation, vegetation damage, corrosion, possible adverse effect on health
Carbon monoxide	Fuming of metallic oxides, gas-operated fork trucks	Primary metals; steel and aluminum	Reduction in oxygen-carrying capacity of blood
Aldehydes	Results from thermal decomposition of fats, oil, or glycerol	Food processing, light process	An irritating odor, suffocating, pungent, choking; not immediately dangerous to life; can become intolerable in a very short time
Ammonia	Used in refrigeration, chemical processes such as dye making, explosives, lacquer, fertilizer	Textiles, chemicals	Corrosive to copper, brass, aluminum, and zinc; high concentration producing chemical burns on wet skin
Arsine	Any soldering, pickling, etching, or plating process involving metals or acids containing arsenic	Chemical processing, smelting	Breakdown of red cells in blood
Chlorine	Manufacture by electrolysis, bleaching cotton and flour, organic-chemicals by-product	Textiles, chemicals	Attacking entire respiratory tract and mucous membrane of eye
Hydrogen cyanide	From metal plating, blast furnaces, dyestuff works	Metal fabricating, primary metals, textiles	Capable of affecting nerve cells
Hydrogen fluoride	Catalyst in some petroleum refining, etching glass, silicate extraction; by-product in electrolytic production of aluminum	Petroleum, primary metals, aluminum	Strong irritant and corrosive action on all body tissue; damage to citrus plants; effect on teeth and bones of cattle from plants
Hydrogen sulfide	Refinery gases, crude oil, sulfur recovery, various chemical industries using sulfur compounds	Petroleum and chemicals	Foul odor of rotten eggs; irritating to eyes and respiratory tract; darkening exterior paint
Nitrogen oxides	High-temperature combustion: metal cleaning, fertilizer, explosives, nitric acid; carbon-arc combustion; manufacture of H_2SO_4	Metal fabrication, heavy chemicals	Irritating gas affecting lungs; vegetation damage
Phosgenes	Thermal decomposition of chlorinated hydrocarbons, degreasing, manufacture of dyestuffs, pharmaceuticals, organic chemicals	Metal fabrication, light chemicals, textiles	Damage capable of leading to pulmonary edema, often delayed
Suspended particles (iron oxide, fly ash, soot, smoke)	Combustion and industrial processes (basic oxygen processes)	Primary metal, plant powerhouse	Causing soiling and reducing visibility
Odors	Slaughtering and rendering animals, tanning animal hides, canning, smoking meats, roasting coffee, brewing beer, processing toiletries	Food processing, allied industries	Objectionable odors

larger. Dust particles are usually irregular in shape, and particle size refers to some average dimension for any given particle. Common examples include fly ash, rock dusts, and ordinary flour. Smoke implies a certain degree of optical density and is typically derived from the burning of organic materials such as wood, coal, and tobacco. Smoke particles are very fine, ranging in size from less than 0.01 μm up to 1 μm. They are usually spherical in shape if of liquid or tarry composition and irregular in shape if of solid composition. Owing to their very fine particle size, smokes can remain in suspension for long periods of time and exhibit lively brownian motion. Fumes are typically formed by processes such as sublimination, condensation, or combustion, generally at relatively high temperatures. They range in particle size from less than 0.1 μm to 1 μm. Similar to smokes, they settle very slowly and exhibit strong brownian motion. Mists or fogs are typically formed either by the condensation of water or other vapors on suitable nuclei, giving a suspension of small liquid droplets, or by the atomization of liquids. Particle sizes of natural fogs and mists usually lie between 2 and 200 μm. Droplets larger than 200 μm are more properly classified as drizzle or rain. Many of the important properties of aerosols which depend on particle size are presented in Sec. 20 (Fig. 20-102.).

When a liquid or solid substance is emitted to the air as particulate matter, its properties and effects may be changed. As a substance is broken up into smaller and smaller particles, more of its surface area is exposed to the air. Under these circumstances, the substance, whatever its chemical composition, tends to combine physically or chemically with other particulates or gases in the atmosphere. The resulting combinations are frequently unpredictable. Very small aerosol particles (from 0.001 to 0.1 μm) can act as condensation nuclei to facilitate the condensation of water vapor, thus promoting the formation of fog and ground mist. Particles less than 2 or 3 μm in size (about half by weight of the particles suspended in urban air) can penetrate the mucous membrane and attract and convey harmful chemicals such as sulfur dioxide. By virtue of the increased surface area of the small aerosol particles and as a result of the adsorption of gas molecules or other such properties that are able to facilitate chemical reactions, aerosols tend to exhibit greatly enhanced surface activity. Many substances that oxidize slowly in their massive state will oxidize extremely fast or possibly even explode when dispersed as fine particles in air. Dust explosions, for example, are often caused by the unstable burning or oxidation of combustible particles, brought about by their relatively large specific surfaces. Adsorption and catalytic phenomena can also be extremely important in analyzing and understanding the problems of particulate pollution. The conversion of sulfur dioxide to corrosive sulfuric acid assisted by the catalytic action of iron oxide particles, for example, demonstrates the catalytic nature of certain types of particles in the atmosphere. Finally, aerosols can absorb radiant energy and rapidly conduct heat to the surrounding gases of the atmosphere. These are gases that ordinarily would be incapable of absorbing radiant energy by themselves. As a result, the air in contact with the aerosols can become much warmer.

Estimating Emissions from Sources Knowledge of the types and rates of emissions is fundamental to evaluation of any air pollution problem. A comprehensive material balance on the process can often assist in this assessment. Estimates of the rates at which pollutants are discharged from various processes can also be obtained by utilizing published emission factors (see *Compilation of Air Pollution Emission Factors*, 2d ed., AP-42, U.S. EPA, Research Triangle Park, N.C., April 1973, with all succeeding supplements). The emission factor is a statistical average of the rate at which pollutants are emitted from the burning or processing of a given quantity of material or on the basis of some other meaningful parameter. Emission factors are affected by the techniques employed in the processing, handling, or burning operations, by the quality of material used, and by the efficiency of air pollution control. Since the combination of these factors tends to be unique to a source, emission factors for one source may not be satisfactory for another source. Hence, care and good judgment must be exercised in identifying appropriate emission factors. If appropriate emission factors cannot be found or if air-pollution-control equipment is to be designed, specific source sampling should be conducted. The major industrial sources of pollutants, the air contaminants emitted, and typical control techniques are summarized in Table 26-3.

Effects of Air Pollutants There seems to be little question that during many of the more serious pollution episodes air pollution can have a significant effect on health. Hundreds of excess deaths have been attributed to incidents in London in 1952, 1956, 1957, and 1962, in Donora, Pennsylvania, in 1948, and in New York City in 1953, 1963, and 1966. Many of the people were in failing health and generally were suffering from lung conditions. In addition, hundreds of thousands of persons have suffered from serious discomfort and inconvenience, including eye irritation and chest pains, during these and other such incidents. These acute problems are actually the lesser of the health problems. There is considerable evidence of a chronic threat to human health from air pollution. This evidence ranges from the rapid rise of emphysema as a major health problem through the identification of carcinogenic compounds in smog to statistical evidence that people exposed to polluted atmospheres over extended periods of time suffer from a number of ailments and a reduction in their life span.

Humans Sufficient evidence is available to indicate that atmospheric pollution in varying degrees does affect health adversely. [Amdur, Melvin, and Drinker, "Effect of Inhalation of Sulfur Dioxide by Man," *Lancet*, London, **2,** 758 (1953); Barton, Corn, Gee, Vasallo, and Thomas, "Response of Healthy Men to Inhaled Low Concentrations of Gas-Aerosol Mixtures," *Arch. Environ. Health*, Chicago, **18,** 681 (1969); Bates, Bell, Burnham, Hazucha, and Mantha, "Problems in Studies of Human Exposure to Air Pollutants," *Can. Med. Assoc. J.*, Toronto, **103,** 833 (1970); Ciocco and Thompson, "A Follow-Up of Donora Ten Years After: Methodology and Findings," *Am. J. Public Health*, New York, **51,** 155 (1961); Daly, "Air Pollution and Causes of Death," *Br. J. Prev. Soc. Med.*, London, **13,** 14 (1959); Jaffe, "The Biological Effect of Photochemical Air Pollutants on Man and Animals," *Am. J. Public Health*, New York, **57,** 1269 (1967); New York Academy of Medicine, Committee on Public Health, "Air Pollution and Health," *Bull. N.Y. Acad. Med.*, **42,** 588 (1966); Pemberton and Goldberg, "Air Pollution and Bronchitis," *Br. Med. J.*, London, **2,** 567 (1954); Snell and Luchsinger, "Effect of Sulfur Dioxide on Expiratory Flowrates and Total Respiratory Resistance in Normal Human Subjects," *Arch. Environ. Health*, Chicago, **18,** 693 (1969); Speizer and Frank, "A Comparison of Changes in Pulmonary Flow Resistance in Healthy Volunteers Acutely Exposed to SO_2 by Mouth and by Nose," *J. Ind. Med.*, **23,** 75 (1966); Stocks, "Cancer and Bronchitis Mortality in Relation to Atmospheric Deposit and Smoke," *Br. Med. J.*, London, **1,** 74 (1959); Toyama, "Air Pollution and Its Health Effects in Japan," *Arch. Environ. Health*, Chicago, **8,** 153 (1963); U.K. Ministry of Health, "Mortality and Morbidity during London Fog of December 1952," Report on Public Health and Medical Subjects No. 95, London, 1954; U.S. Public Health Service, "Air Pollution in Donora, Pa.: Preliminary Report," Public Health Bull 306.] It contributes to excesses of deaths, increased morbidity, and earlier onset of chronic respiratory diseases. There is evidence of a relationship between the intensity of the pollution and the severity of attributable health effects and a consistency of the relationship between these environmental stresses and the diseases of the target organs. Air pollutants can both initiate and aggravate a variety of respiratory diseases including asthma. In fact, the clinical presentation of asthma may be considered an air pollution host-defense disorder brought on by specific airborne irritants: pollens, infectious agents, and gaseous and particulate chemicals. The bronchopulmonary response to these foreign irritants is bronchospasm and hypersecretion; the airways are intermittently and reversibly obstructed.

Air-pollutant effects on neural and sensory functions in humans vary widely. Odorous pollutants cause only minor annoyance; yet, if persistent, they can lead to irritation, emotional upset, anorexia, and mental depression. Carbon monoxide can cause death secondary to the depression of the respiratory centers of the central nervous system. Short of death, repeated and prolonged exposure to carbon monoxide can alter sensory protection, temporal perception, and higher mental functions. Lipid-soluble aerosols can enter the body

TABLE 26-3 Control Techniques Applicable to Unit Processes at Important Emission Sources

Industry	Process of operation	Air contaminants emitted	Control techniques
Aluminum reduction plants	Materials handling Buckets and belt Conveyor or pneumatic Conveyor	Particulates (dust)	Exhaust systems and baghouse
	Anode and cathode electrode preparation		Exhaust systems and mechanical collectors
	Cathode: baking	Hydrocarbon emissions from binder	
	Anode: grinding and blending	Particulates (dust)	
	Baking	Particulates (dust), CO, SO_2, hydrocarbons, and fluorides	High-efficiency cyclone, electrostatic precipitators, scrubbers, catalytic combustion or incinerators, flares, baghouse
	Pot charging	Particulates (dust), CO, HF, SO_2, CF_4, and hydrocarbons	High-efficiency cyclone, baghouse, spray towers, floating-bed scrubber, electrostatic precipitators, chemisorption, wet electrostatic precipitators
	Metal casting	Cl_2, HCl, CO, and particulates (dust)	Exhaust systems and scrubbers
Asphalt batch plants	Materials handling, storage and classifiers: elevators, chutes, vibrating screens	Particulates (dust)	Local exhaust systems with a cyclone precleaner and a scrubber or baghouse
	Drying: rotary oil- or gas-fired	Particulates and smoke	Proper combustion controls, fuel-oil preheating where required; local exhaust system, cyclone and a scrubber or baghouse
	Truck traffic	Dust	Wetting down truck routes
Cement plants	Quarrying: primary crusher, secondary crusher, conveying, storage	Particulates (dust)	Wetting; exhaust systems with mechanical collectors
	Dry processes; materials handling, air separator (hot-air furnace)	Particulates (dust)	Local exhaust system and mechanical collectors and baghouse
	Grinding	Particulates (dust)	Local exhaust system with cyclones and baghouse
	Pneumatic, conveying and storage	Particulates (dust)	
	Wet process: materials handling, grinding, storage	Wet materials, no dust	
	Kiln operations: rotary kiln	Particulates (dust), CO, SO_x, NO_x, hydrocarbons, aldehydes, ketones	Electrostatic precipitators and baghouses, scrubber, flare
	Clinker cooling: materials handling	Particulates (dust)	Local exhaust system and mechanical collectors
	Grinding and packaging: air separator, grinding, pneumatic conveying, materials handling, packaging	Particulates (dust)	Local exhaust systems and mechanical collectors
Coal-preparation plants	Materials handling: conveyors, elevators, chutes	Particulates (dust)	Local exhaust systems and cyclones
	Sizing: crushing, screening, classifying	Particulates (dust)	Local exhaust systems and cyclones
	Dedusting	Particulates (dust)	Local exhaust system, cyclone precleaners, and baghouse
	Storing coal in piles	Blowing particulates (dust)	Wetting, plastic-spray covering
	Refuse piles	H_2S, particulates, and smoke from burning storage piles	Digging out fire, pumping water onto fire area, blanketing with incombustible material
	Coal drying: rotary, screen, suspension, fluid-bed, cascade	Dust, smoke, particulates, sulfur oxides, H_2S	Exhaust systems with cyclones and venturi scrubbers
Coke plants	By-product-ovens charging	Smoke, particulates (dust)	Pipe-line charging, careful charging techniques, portable hooding and scrubber or baghouses
	Pushing	Smoke, particulates (dust), SO_2	Minimizing green-coke pushing, scrubbers and baghouses
	Quenching	Smoke, particulates (dust and mists), phenols, and ammonia	Baffles and spray tower
	By-product processing	CO, H_2S, methane, ammonia, H_2, phenols, hydrogen cyanide, N_2, benzene, xylene, etc.	Electrostatic precipitator, scrubber, flaring
	Material storage (coal and coke)	Particulates (dust)	Wetting, plastic spray, fire-prevention techniques
Fertilizer industry (chemical)	Phosphate fertilizers: crushing, grinding, and calcining	Particulates (dust)	Exhaust system, scrubber, cyclone, baghouse
	Hydrolysis of P_2O_5	PH_3, $P_2O_5PO_4$ mist	Scrubbers, flare
	Acidulation and curing	HF, SiF_4	Scrubbers
	Granulation	Particulates (dust) (product recovery)	Exhaust system, scrubber, or baghouse
	Ammoniation	NH_3, NH_4Cl, SiF_4, HF	Cyclone, electrostatic precipitator, baghouse, high-energy scrubber
	Nitric acid acidulation	NO_x, gaseous fluoride compounds	Scrubber, addition of urea
	Superphosphate storage and shipping	Particulates (dust)	Exhaust system, cyclone, or baghouse
	Ammonium nitrate reactor	NH_3, NO_x	Scrubber
	Prilling tower	NH_4, NO_3	Proper operation control, scrubbers

TABLE 26-3 Control Techniques Applicable to Unit Processes at Important Emission Sources *(Continued)*

Industry	Process of operation	Air contaminants emitted	Control techniques
Foundries	Melting (cupola)		
Iron	Charging	Smoke and particulates	Closed top with exhaust system, CO afterburner, gas-cooling device and scrubbers, baghouse or electrostatic precipitator, wetting to extinguish fire
	Melting	Smoke and particulates, fume	
	Pouring	Oil, mist, CO	
	Bottom drop	Smoke and particulates	
Brass and bronze	Melting		
	Charging	Smoke particulates, oil mist	Low-zinc-content red brass: use of good combustion controls and slag cover; high-zinc-content brass: use of good combustion controls, local exhaust system, and baghouse or scrubber
	Melting	Zinc oxide fume, particulates, smoke	
	Pouring	Zinc oxide fume, lead oxide fume	
Aluminum	Melting: charging, melting, pouring	Smoke and particulates	Charging clean material (no paint or grease); proper operation required; no air-pollution-control equipment if no fluxes are used and degassing is not required; dirty charge requiring exhaust system with scrubbers and baghouses
Zinc	Melting		
	Charging	Smoke and particulates	Exhaust system with cyclone and baghouse; charging clean material (no paint or grease)
	Melting	Zinc oxide fume	Careful skimming of dross
	Pouring	Oil mist and hydrocarbons from die-casting machines	Use of low-smoking die-casting lubricants
	Sand-handling shakeout	Particulates (dust), smoke, organic vapors	
	Magnetic pulley, conveyors and elevators, rotary cooler, screening, crusher-mixer	Particulates (dust)	Exhaust system, cyclone, and baghouse
	Coke-making ovens	Organic acids, aldehydes, smoke, hydrocarbons	Use of binders that will allow ovens to operate at less than 204°C (400°F) or exhaust systems and afterburners
Galvanizing operations	Hot-dip-galvanizing-tank kettle: dipping material into the molten zinc; dusting flux onto the surface of the molten zinc	Fumes, particulates (liquid), vapors: NH_4Cl, ZnO, $ZnCl_2$, Zn, NH_3, oil, and carbon	Close-fitting hoods with high in-draft velocities (in some cases the hood may not be able to be close to the kettle, so the in-draft velocity must be very high), baghouses, electrostatic precipitators
Kraft pulp mills	Digesters: batch and continuous	Mercaptans, methanol (odors)	Condensers and use of lime kiln, hog fuel boiler, or furnaces as afterburners
	Multiple-effect evaporators	H_2S, other odors	Caustic scrubbing and thermal oxidation of noncondensables
	Recovery furnace	H_2S, mercaptans, organic sulfides, and disulfides	Paper combustion controls for fluctuating load and unrestricted primary and secondary air flow to furnace and scrubber or electrostatic precipitator
	Weak and strong black-liquor oxidation	H_2S	Packed tower and cyclone
	Smelt tanks	Particulates (mist or dust)	Demisters, venturi, packed tower, or impingement-type scrubbers
	Lime kiln	Particulates (dust), H_2S	Venturi scrubbers
Municipal and industrial incinerators	Single-chamber incinerators	Particulates, smoke, volatiles, CO, SO_x, ammonia, organic acids, aldehydes, NO_x, hydrocarbons, odors, HCl	
	Flue-fed		Settling chambers, scrubbers, afterburner
	Multiple-chamber incinerators: retort, inline	Particulates, smoke, and combustion contaminants	Operating at rated capacity, using auxiliary fuel as specified, and good maintenance, including timely cleanout of ash
	Flue-fed	Particulates, smoke, and combustion contaminants	Use of charging gates and automatic controls for draft
	Wood waste	Particulates, smoke, and combustion contaminants	Continuous-feed systems; operation at design load and excess air; limit of charging of oily material
	Municipal incinerators: 50 tons and up per day	Particulates, smoke, volatiles, CO, ammonia, organic acids, aldehydes, NO_x, hydrocarbons, SO_x, hydrogen chloride, odors	Preparation of materials, including weighing, grinding, shredding; control of tipping area, furnace design with proper automatic controls; proper startup techniques; maintenance of design operating temperatures; use of electrostatic precipitators, scrubbers and baghouses; proper ash cleanout

TABLE 26-3 Control Techniques Applicable to Unit Processes at Important Emission Sources *(Continued)*

Industry	Process of operation	Air contaminants emitted	Control techniques
	Pathological incinerators	Odors, hydrocarbons	Proper charging
	Wood waste and industrial waste	Particulates, smoke, and combustion contaminants	Modified fuel feed, auxiliary fuel and dryer systems, cyclones, scrubbers
	Box type	Particulates, smoke, and combustion contaminants	Allowing proper startup, charging material slowly, no overloading
Nonferrous smelters, primary			
Copper	Roasting	SO_2, particulates, fume	Exhaust system, settling chambers, cyclones or scrubbers and electrostatic precipitators for dust and fumes and sulfuric acid plant for SO_2
	Reverberatory furnace	Smoke, particulates, fume, SO_2	Exhaust system, settling chambers, cyclones or scrubbers and electrostatic precipitators for dust and fumes and sulfuric acid plant for SO_2
	Converters: charging, slag skim, pouring, air or oxygen blow	Smoke, fume, SO_2	Exhaust system, settling chambers, cyclones or scrubbers, electrostatic precipitators for dust and fumes and sulfuric acid plant for SO_2
Lead	Sintering	SO_2, particulates, smoke	Exhaust system, cyclones and baghouse or precipitators for dust and fumes, sulfuric acid plant for SO_2
	Blast furnace	SO_2, CO, particulates, lead oxide, zinc oxide	Exhaust system, settling chambers, afterburner and cooling device, cyclone, and baghouse
	Dross reverberatory furnace	SO_2, particulates, fume	Exhaust system, settling chambers, cyclone and cooling device, baghouse
	Refining kettles	SO_2, particulates	Local exhaust system, cooling device, baghouse or precipitator
Cadmium	Roasters, slag, fuming furnaces, deleading kilns	Particulates	Local exhaust system, baghouse or precipitator
Zinc	Roasting	Particulates (dust) and SO_2	Exhaust system, humidifier, cyclone scrubber, electrostatic precipitator, and acid plant
	Sintering	Particulates (dust) and SO_2	Exhaust system, humidifier, electrostatic precipitator, and acid plant
	Calcining Retorts: electric arc	Zinc oxide fume, particulates, SO_2, CO	Exhaust system, baghouse
Nonferrous smelters, secondary	Blast furnaces and cupolas—recovery of metal from scrap and slag	Dust, fumes, particulates, oil vapor, smoke, CO	Exhaust systems, cooling devices, CO burners and baghouses or precipitators
	Reverberatory furnaces	Dust, fumes, particulates, smoke, gaseous fluxing materials	Exhaust systems and baghouses or precipators, or venturi scrubbers
	Sweat furnaces	Smoke, particulates, fumes	Precleaning metal and exhaust systems with afterburner and baghouse
	Wire reclamation and autobody burning	Smoke, particulates	Scrubbers and afterburners
Paint and varnish manufacturing	Resin manufacturing: closed reaction vessel	Acrolein, other aldehydes and fatty acids (odors), phthalic anhydride (sublimed)	Exhaust systems with scrubbers and fume burners
	Varnish: cooking—open or closed vessels	Ketones, fatty acids, formic acids, acetic acid, glycerine, acrolein, other aldehydes, phenols and terpenes; from tall oils, hydrogen sulfide, alkyl sulfide, butyl mercaptan, and thiofene (odors)	Exhaust system with scrubbers and fume burners; close-fitting hoods required for open kettles
	Solvent thinning	Olefins, branched-chain aromatics and ketones (odors), solvents	Exhaust system with fume burners
Rendering plants	Feedstock storage and housekeeping	Odors	Quick processing, washdown of all concrete surfaces, paving of dirt roads, proper sewer maintenance, packed towers
	Cookers and percolators	SO_2, mercaptans, ammonia, odors	Exhaust system, condenser, scrubber, or incinerator
	Grinding	Particulates (dust)	Exhaust system and scrubber
Roofing plants (asphalt saturators)	Felt or paper saturators: spray section, asphalt tank, wet looper	Asphalt vapors and particulates (liquid)	Exhaust system with high inlet velocity at hoods [3658 m/s (>200 ft/min)] with either scrubbers, baghouses, or two-stage low-voltage electrostatic precipitators
	Crushed rock or other minerals handling	Particulates (dust)	Local exhaust system, cyclone or multiple cyclones

TABLE 26-3 Control Techniques Applicable to Unit Processes at Important Emission Sources *(Continued)*

Industry	Process of operation	Air contaminants emitted	Control techniques
Steel mills	Blast furnaces: charging, pouring	CO, fumes, smoke, particulates (dust)	Good maintenance, seal leaks; use of higher ratio of pelletized or sintered ore; CO burned in waste-heat boilers, stoves, or coke ovens; cyclone, scrubber, electrostatic precipitator, or venturi scrubber
	Electric steel furnaces: charging, pouring, oxygen blow	Fumes, smoke, particulates (dust), CO	Segregating dirty scrap; proper hooding, baghouses, venturi scrubbers, or electrostatic precipitator
	Open-hearth furnaces: oxygen blow, pouring	Fumes, smoke, SO_x, particulates (dust), CO, NO_x	Proper hooding, settling chambers, waste-heat boiler, baghouse, electrostatic precipitator, or venturi scrubber
	Basic oxygen furnaces: oxygen blowing	Fumes, smoke, CO, particulates (dust)	Proper hooding (capturing of emissions and dilute CO), scrubbers, or electrostatic precipitator
	Raw-material storage	Particulates (dust)	Wetting or application of plastic spray
	Pelletizing	Particulates (dust)	Proper hooding, cyclone, baghouse
	Sintering	Smoke, particulates (dust), SO_2, NO_x	Proper hooding, cyclones, venturi scrubbers, baghouse, or precipitator

and be absorbed in the lipids of the central nervous system. Once there, their effects may persist long after the initial contact has been removed. Examples of agents of long-term chronic effects are organic phosphate pesticides and aerosols carrying the metals lead, mercury, and cadmium.

The acute toxicological effects of most air contaminants are reasonably well understood, but the effects of exposure to heterogeneous mixtures of gases and particulates at very low concentrations are only beginning to be comprehended. Two general approaches can be used to study the effects of air contaminants on humans: epidemiology, which attempts to associate the effect in large populations with the cause; and laboratory research, which begins with the cause and attempts to determine the effects. Ideally, the two methods should complement each other.

Epidemiology is the more costly of the two, requires great care in planning, and often suffers from incomplete data and lack of controls. One great advantage, however, is that moral barriers do not limit its application to humans as they do with some kinds of laboratory research. The method is therefore highly useful and has produced considerable information. Laboratory research is less costly than epidemiology, and its results can be checked against controls and verified by experimental repetition.

Animals Considerable work continues to be performed on the effects of pollutants on animals, including, for a few species, experiments involving mixed pollutants and mixed gas-aerosol systems. In general, such work has shown that mixed pollutants may act in several different ways. They may produce an effect that is additive, amounting to the sum of the effects of each contaminant acting alone; they may produce an effect that is greater than the simply additive (synergistic) or less than the simply additive (antagonistic); or they may produce an effect that differs in some other way from the simply additive.

The mechanism by which an animal can become poisoned in many instances is completely different from that by which humans are affected. As in humans, inhalation is an important route of entry in acute air pollution exposures such as occurred in the Meuse Valley and Donora incidents. However, probably the most common exposure for herbivorous animals grazing within a zone of pollution will be the ingestion of feed contaminated by air pollutants. In this case, inhalation is of secondary importance.

Air pollutants that present a hazard to livestock, therefore, are those which are taken up by vegetation or deposited on the plants. Only a few pollutants have been observed to cause harm to animals. These include arsenic, fluorides, lead, and molybdenum.

Vegetation Vegetation is more sensitive than animals to many air contaminants, and methods have been developed that use plant response to measure and identify contaminants. The effects of air pollution on vegetation can appear as death, stunted growth, reduced crop yield, and degradation of color. It is interesting to note that in some cases of color damage such as the silvering of leafy vegetables by oxidants, the plant may still be used as food without any danger to the consumer; however, the consumer usually will not buy such vegetables on aesthetic grounds, so the grower still sustains a loss. Among the pollutants that can harm plants are sulfur dioxide, hydrogen fluoride, and ethylene. Plant damage caused by constituents of photochemical smog has been studied extensively. Damage has been attributed to ozone and peroxyacyl nitrates, to higher aldehydes, and to products of the reaction of ozone with olefins. However, none of these cases precisely duplicates all features of the damage observed in the field, and the question remains open to some debate and further study.

Materials The damage that air pollutants can do to some materials is well known: ozone in photochemical smog cracks rubber, weakens fabrics, and fades dyes; hydrogen sulfide tarnishes silver; smoke dirties laundry; acid aerosols ruin nylon hose. Among the most important effects are discoloration, corrosion, the soiling of goods, and impairment of visibility.

1. *Discoloration.* Many air pollutants accumulate on and discolor buildings. Not only does sooty material blacken buildings, but it can accumulate and become encrusted. This can hide lines and decorations and thereby disfigure structures and reduce their aesthetic appeal. Another common effect is the discoloration of paint by certain acid gases. A good example is the blackening of white paint with a lead base by hydrogen sulfide.

2. *Corrosion.* A more serious effect and one of great economic importance is the corrosive action of acid gases on building materials. Such gases can cause stone surfaces to blister and peel; mortar can be reduced to powder. Metals are also damaged by the corrosive action of some pollutants. Another common effect is the deterioration of tires and other rubber goods; cracking and apparent "drying" occur when these goods are exposed to ozone and other oxidants.

3. *Soiling of goods.* Wearing apparel and household goods can easily be soiled by air contaminants, and the more frequent cleaning thus required can become expensive. Also, more frequent cleaning often leads to a shorter life span for materials and to the need to purchase new goods more often.

4. *Impairment of visibility.* The impairment of atmospheric visibility (i.e., decreased visual range through a polluted atmosphere) is caused by the scattering of sunlight by particles suspended in the air. It is not a result of sunlight being obscured by materials in the air. Since light scattering, and not obscuration, is the main cause of the reduction in visibility, reduced visibility due to the presence of air pollutants occurs primarily on bright days. On cloudy days or at night there may be no noticeable effect, although the same particulate concentration may exist at these times as on sunny days. Reduction in visibility creates several problems. The most significant are

the adverse effects on aircraft, highway, and harbor operations. Reduced visibility can also cause adverse aesthetic impressions which can seriously restrict the growth and development of any area. Extreme conditions such as dust storms or sandstorms can actually cause physical damage by themselves.

UNITED STATES LEGISLATION, REGULATIONS, AND GOVERNMENTAL AGENCIES

Although considerable federal legislation dealing with air pollution has been enacted since the 1950s, the basic statutory framework now in effect was established by the Clean Air Act of 1970, amended in 1974 to deal with energy-related issues and again in 1977, when a number of amendments containing particularly important provisions associated with the approval of new industrial plants were adopted.

Clean Air Act of 1970 The Clean Air Act of 1970 was founded on the concept of attaining National Ambient Air Quality Standards (NAAQS). Data were accumulated and analyzed to establish what the air quality actually was, identify sources of pollution, determine how pollutants disperse and interact in the ambient air, and define reductions and controls necessary to achieve air-quality objectives.

EPA promulgated the basic set of current ambient-air-quality standards in April 1971. The specific regulated pollutants were particulates, sulfur dioxide, photochemical oxidants, hydrocarbons, carbon monoxide, and nitrogen oxides. In 1978 lead was added. Table 26-4 enumerates the present (1983) standards.

To provide basic geographic units for the air-pollution-control program, the United States was divided into 247 air-quality-control regions (AQCRs). By a standard rollback approach, the total quantity of pollution in a region was estimated, the quantity of pollution that could be tolerated without exceeding standards was then calculated, and the degree of reduction called for was determined. States were required by EPA to develop state implementation plans (SIPs) to achieve compliance.

The act also directed EPA to set new-source performance standards (NSPS) for specific industrial categories. New plants were required to use the best system of emission reduction available. EPA gradually issued these standards, which now cover a number of basic industrial categories (as listed in Table 26-5). The 1977 amendments to the Clean Air Act directed EPA to accelerate the NSPS program and included a regulatory program to prevent significant deterioration in those areas of the country where the NAAQS were being attained.

Finally, Sec. 112 of the Clean Air Act required that EPA promulgate National Emission Standards for Hazardous Air Pollutants (NESHAPs). Standards have been promulgated for asbestos, beryllium, mercury, and vinyl chloride (see Table 26-6). A NESHAP proposal for benzene is in preparation. Cadmium, arsenic, copper, and polycyclic organic matter also are being considered.

Prevention of Significant Deterioration (PSD) Of all the federal laws placing environmental controls on industry (and, in particular, on new plants), perhaps the most confusing and restrictive are the limits imposed for the prevention of significant deterioration (PSD) of air quality. These limits apply to areas of the country that already are cleaner than required by ambient-air-quality standards. This regulatory framework evolved from judicial and administrative action under the 1970 Clean Air Act and subsequently was given full statutory foundation by the 1977 Clean Air Act Amendments.

EPA established an area-classification scheme to be applied in all such regions. The basic idea was to allow a moderate amount of industrial development but not enough to degrade air quality to a point at which it barely complied with standards. In addition, states were allowed to designate certain areas where pristine air quality was especially desirable. All air-quality areas were categorized as Class I, Class II, or Class III. Class I areas were pristine areas subject to the tightest control. Permanently designated Class I areas included international parks, national wilderness areas, and memorial parks exceeding 5000 acres, and national parks exceeding 6000 acres. Although the nature of these areas is such that industrial projects would not be located within them, their Class I status could affect projects in neighboring areas where meteorological conditions might result in the transport of emissions into them. Class II areas were areas of moderate industrial growth. Class III areas were areas of major industrialization. Under EPA regulations promulgated in December 1974, all areas were initially categorized as Class II. States were authorized to reclassify specified areas as Class I or Class III.

The EPA regulations also established another critical concept, known as the "increment." This was the numerical definition of the amount of additional pollution that may be allowed through the combined effects of all new growth in a particular locality (see Table 26-7). To assure that the increments would not be used up hastily, EPA specified that each major new plant must install best-available control technology (BACT) to limit emissions. This reinforced the same policy underlying the NSPS; and where an NSPS had been promulgated, it would control determinations of BACT. Where such standards had not been promulgated, an ad hoc determination was called for in each case.

To implement these controls, EPA required that every new source undergo a preconstruction review. The regulations prohibited a company from commencing construction on a new source until the review had been completed and provided that, as part of the review procedure, public notice should be given and an opportunity provided for a public hearing on any disputed questions.

Sources Subject to Prevention of Significant Deterioration (PSD) Sources subject to PSD regulations (40 CFR, Sec. 52.21, Aug. 7, 1980) are major stationary sources and major modifications located in attainment areas and unclassified areas. A major stationary source is defined as any source listed in Table 26-8 with the potential to emit 100 tons per year or more of any pollutant regulated under the Clean Air Act (CAA) or any other source with the potential to emit 250 tons per year or more of any CAA pollutant. The "potential to emit" is defined as the maximum capacity to emit the pollutant under applicable emission standards and permit conditions (after application of any air-pollution-control equipment) excluding secondary emissions. A "major modification" is defined as any physical or operational

TABLE 26-4 National Primary and Secondary Ambient-Air-Quality Standards, 1983

Pollutant	Averaging time	Primary standards	Secondary standards
Sulfur oxides	Annual arithmetic mean	80 $\mu g/m^3$ (0.03 ppm)	
	24 h	365 $\mu g/m^3$ (0.14 ppm)	
	3 h		1,300 $\mu g/m^3$ (0.5 ppm)
Particulate matter	Annual geometric mean	75 $\mu g/m^3$	60 $\mu g/m^3$
	24 h	260 $\mu g/m^3$	150 $\mu g/m^3$
Carbon monoxide	8 h	10 mg/m^3	Same as primary standards
	1 h	40 mg/m^3 (35 ppm)	
Ozone (corrected for NO_2 and SO_2)	1 h	240 $\mu g/m^3$ (0.12 ppm)	Same as primary standard
Hydrocarbons (corrected for methane)	3 h	160 $\mu g/m^3$ (0.24 ppm)	Same as primary standard
Nitrogen oxides	Annual arithmetic mean	100 $\mu g/m^3$ (0.05 ppm)	Same as primary standard
Lead	3 months	1.5 $\mu g/m^3$	Same as primary standard
Photochemical oxidants (expressed as ozone)	1 h	160 $\mu g/m^3$ (0.08 ppm)	Same as primary standard

NOTE: National standards, other than those based on annual arithmetic means or annual geometric means, are not to be exceeded more than once a year.

TABLE 26-5 Standards of Performance, 40 CFR Part 60

Source category	Affected facility	Pollutant	Emission level	Monitoring requirement[a]
Subpart D: **Fossil-fuel-fired steam generators** > 73 MW heat input (>250 million Btu/h)	Coal-fired boilers[b]	Particulate	0.10 lb/million Btu	No requirement
		Opacity	20% (27% for 6 min/h)	Continuous
		SO_2	1.20 lb/million Btu	Continuous
Proposed 8/17/71 (36FR 15703) Promulgated 12/23/71 (36 FR24876)		NO_x		Continuous
		Anthracite, bituminous, or subbituminous coal	0.70 lb/million Btu	
		Lignite[c]	0.60 lb/million Btu	
Revised 7/26/72 (37 FR 14877) 6/14/74 (39 FR 20790) 1/16/75 (40 FR 2803) 10/6/75 (40 FR 46250) 8/15/77 (42 FR 41122) 12/15/77 (42 FR 61537) 3/7/78 (43 FR 9276)		More than 25% coal refuse	Exempt	
	Oil- or gas-fired boilers	Particulate	0.10 lb/million Btu	No requirement
		Opacity	20% (27% for 6 min/h)	Continuous
		SO_2—oil	0.80 lb/million Btu	Continuous
		NO_x—oil	0.30 lb/million Btu	Continuous
		NO_x—gas	0.20 lb/million Btu	Continuous
Subpart D*a*: **Electric utility steam> generating units** > 73 MW heat input (>250 million Btu/h)	Coal-fired boilers (and coal-derived fuels)	Particulate	0.03 lb/million Btu	No requirement
		Opacity	20% (27% for 6 min/h)	Continuous
Proposed 9/19/78 (43 FR 42154) Promulgated 6/11/79 (44 FR 33580)		SO_2	1.20 lb/million Btu[d] and 90% reduction,[e] except 70% reduction when emissions are less than 0.60 lb/million Btu[e]	Continuous compliance[f]
		NO_x[g]	0.60 lb/million Btu	Continuous compliance[f]
		Anthracite, bituminous, and lignite[c]		
		Subbituminous coal	0.50 lb/million Btu	
		Coal-derived fuels and shale oil	0.50 lb/million Btu	
		More than 25% coal refuse	Exempt	
	Oil- or gas-fired boilers	Particulate	0.03 lb/million Btu	No requirement
		Opacity	20% (27% for 6 min/h)	Continuous
		SO_2	0.80 lb/million Btu and 90% reduction[e] or 0.20 lb/million Btu (no reduction requirement)	Continuous compliance[f]
		NO_x—oil	0.30 lb/million Btu	Continuous compliance[f]
		NO_x—gas	0.20 lb/million Btu	Continuous compliance[f]
Subpart E: **Incinerators** (>50 tons/day) Proposed 8/17/71 (36 FR 15703) Promulgated 12/23/71 (36 FR 24876) Revised 6/14/74 (39 FR 20790)	Incinerators	Particulate	0.08 gr/dscf corrected to 12% CO_2	No requirement
Subpart F: **Portland cement plants** Proposed 8/17/71 (36 FR 15703)	Kiln	Particulate	0.30 lb/ton	No requirement
		Opacity	20%	No requirement
	Clinker cooler	Particulate	0.10 lb/ton	No requirement
Promulgated 12/23/71 (36 FR 24876)		Opacity	10%	No requirement
Revised 6/14/74 (39 FR 20790) 11/12/74 (39 FR 39874) 10/6/75 (40 FR 46250)	Fugitive Emission points	Opacity	10%	No requirement

TABLE 26-5 Standards of Performance, 40 CFR Part 60 *(Continued)*

Source category	Affected facility	Pollutant	Emission level	Monitoring requirement[a]
Subpart G: **Nitric acid plants** Proposed 8/17/71 (36 FR 15703) Promulgated 12/23/71 (36 FR 24876) Revised 5/23/73 (38 FR 13562) 6/14/74 (39 FR 20790) 10/6/75 (40 FR 46250)	Process equipment	Opacity NO_x	10% 3.0 lb/ton	No requirement Continuous
Subpart H: **Sulfuric acid plants** Proposed 8/17/71 (36 FR 15703) Promulgated 12/23/71 (36 FR 24876) Revised 5/23/73 (38 FR 13562) 6/14/74 (39 FR 20790) 10/6/75 (40 FR 46250)	Process equipment	SO_2 Acid mist Opacity	4.0 lb/ton 0.15 lb/ton 10%	Continuous No requirement No requirement
Subpart I: **Asphalt concrete plants** Proposed 6/11/73 (38 FR 15406) Promulgated 3/8/74 (39 FR 9308) Revised 10/6/75 (40 FR 46250)	Dryers; screening and weighing systems; storage, transfer, and loading systems; and dust-handling equipment	Particulate Opacity	0.04 gr/dscf (90 mg/dscm) 20%	No requirement No requirement
Subpart J: **Petroleum refineries** Proposed 6/11/73 (38 FR 15406) Promulgated 3/8/74 (39 FR 9308) Revised 10/6/75 (40 FR 46250) 6/24/77 (42 FR 32426) 3/15/78 (43 FR 10866) 3/12/79 (44 FR 13480)	Fluid-catalytic-cracking-unit catalyst regenerator	Particulate Opacity CO	1.0 lb/1000 lb of coke burn-off 30% (6-min exemption) 0.05%	No requirement Continuous Continuous[h]
	Fuel-gas combustion devices	SO_2 H_2S	0.10 gr/dscf 230 mg/dscm	Continuous Continuous[h]
	Claus sulfur-recovery plants	SO_2 Reduced sulfur compounds plus H_2S	0.025% (at 0% oxygen) 0.030% (at 0% oxygen) 0.0010% (at 0% oxygen)	Continuous Continuous[h] Continuous[h]
Subpart K: **Storage vessels for petroleum liquids** constructed after 6/11/73 and prior to 5/19/78 Proposed 6/11/73 (38 FR 15406) Promulgated 3/8/74 (39 FR 9308) Revised 4/17/74 (39 FR 13776) 6/14/74 (39 FR 20790) 4/4/80 (45 FR 23373)	Storage tanks >40,000-gal but not >65,000-gal capacity, constructed after 3/8/74 and prior to 5/19/78 >65,000 gal, constructed after 6/11/73 and prior to 5/19/78	VOC	Equipment standard[m]	No requirement
Subpart Ka: **Storage vessels for petroleum liquids** constructed after May 18, 1978 Proposed 5/18/78 (43 FR 21615) Promulgated 4/4/80 (45 FR 23373)	Storage vessel >40,000-gal capacity constructed after 5/18/78 Storage vessel > 420,000-gal capacity for petroleum or condensate stored, processed, or treated prior to custody transfer *not* an affected facility and exempted	VOC	Equipment standard[n]	None

TABLE 26-5 Standards of Performance, 40 CFR Part 60 *(Continued)*

Source category	Affected facility	Pollutant	Emission level	Monitoring requirement[a]
Subpart L:				
Secondary lead smelters	Reverberatory and blast furnaces	Particulate	0.022 gr/dscf (50 mg/dscm)	No requirement
Proposed 6/11/73 (38 FR 15406)		Opacity	20%	No requirement
	Pot furnaces	Opacity	10%	No requirement
Promulgated 3/8/74 (39 FR 9308)				
Revised 4/17/74 (39 FR 13776) 10/6/75 (40 FR 46250)				
Subpart M:				
Secondary brass and bronze plants	Reverberatory furnaces	Particulate	0.022 gr/dscf (50 mg/dscm)	No requirement
Proposed 6/11/73 (38 FR 15406)		Opacity	20%	No requirement
	Blast and electric furnaces	Opacity	10%	No requirement
Promulgated 3/8/74 (39 FR 9308)				
Revised 10/6/75 (40 FR 46250)				
Subpart N:				
Iron and steel plants	Basic-oxygen-process furnaces	Particulate	0.022 gr/dscf (50 mg/dscm)	No requirement
Proposed 6/11/73 (38 FR 15406)		Opacity	10%	Scrubber pressure loss
Promulgated 3/8/74 (39 FR 9308)			20% once each production cycle	Water pressure
Revised 4/13/78 (43 FR 15600)				
Subpart O:				
Sewage-treatment plants	Sludge incinerators	Particulate	1.30 lb/ton	Mass or volume of sludge; mass of municipal solid waste, if any
Proposed 6/11/73 (38 FR 15406)		Opacity	20%	No requirement
Promulgated 3/8/74 (39 FR 9308)				
Revised 4/17/74 (39 FR 13775) 5/3/74 (39 FR 15396) 10/6/75 (40 FR 46250) 11/10/77 (42 FR 58520)				
Subpart P:				
Primary copper smelters	Dryer	Particulate	0.022 gr/dscf (50 mg/dscm)	No requirement
Proposed 10/16/74 (39 FR 37039)		Opacity	20%	Continuous
	Roaster, smelting furnace,° copper converter	SO_2	0.065%	Continuous
Promulgated 1/15/75 (41 FR 2331)		Opacity	20%	No requirement
Revised 2/26/76 (41 FR 8346) 11/1/77 (42 FR 125)	°Reverberatory furnaces that process high-impurity feed materials are exempt from SO_2 standard			
Subpart Q:				
Primary zinc smelters	Sintering machine	Particulate	0.022 gr/dscf (50 mg/dscm)	No requirement
Proposed 10/16/74 (39 FR 37039)		Opacity	20%	Continuous
Promulgated 1/15/76 (41 FR 2331)	Roaster	SO_2	0.065%	Continuous
		Opacity	20%	No requirement
Subpart R:				
Primary lead smelters	Blast or reverberatory furnace, sintering-machine discharge end	Particulate	0.022 gr/dscf (50 mg/dscm)	No requirement
Proposed 10/16/74 (39 FR 37039)		Opacity	20%	Continuous
Promulated 1/15/76 (41 FR 2331)	Sintering machine, electric smelting furnace, converter	SO_2	0.065%	Continuous
		Opacity	20%	No requirement

TABLE 26-5 Standards of Performance, 40 CFR Part 60 *(Continued)*

Source category	Affected facility	Pollutant	Emission level	Monitoring requirement[a]
Subpart S: **Primary aluminum-reduction plants**	Potroom group (*a*) Soderberg plant	(*a*) Total fluorides	2.0 lb/ton	No requirement
		Opacity	10%	No requirement
Proposed 10/23/74 (39 FR 37729)	(*b*) Prebake plant	(*b*) Total fluorides	1.9 lb/ton	No requirement
Promulgated 1/26/76 (41 FR 3825)		Opacity	10%	No requirement
	Anode bake plants	Total fluorides	0.1 lb/ton	No requirement
6/30/80 (45 FR 44296)		Opacity	20%	No requirement
Subpart T: **Phosphate-fertilizer plants** Proposed 10/22/74 (39 FR 37601) Promulgated 8/6/75 (40 FR 33152)	Wet-process phosphoric acid	Total fluorides	0.02 lb/ton	Total pressure drop across process scrubbing system
Subpart U:	Superphosphoric acid	Total fluorides	0.01 lb/ton	Total pressure drop across process scrubbing system
Subpart V:	Diammonium phosphate	Total fluorides	0.06 lb/ton	Total pressure drop across process scrubbing system
Subpart W:	Triple superphosphate	Total fluorides	0.2 lb/ton	Total pressure drop across process scrubbing system
Subpart X:	Granular triple superphosphate	Total fluorides	$(5.0)(10^{-4})$ lb/(h·ton)	Total pressure drop across process scrubbing system
Subpart Y **Coal-preparation plants** Proposed 10/24/74 (39 FR 37921 Promulgated 1/15/76 (41 FR 2232)	Thermal dryer	Particulate	0.031 gr/dscf (0.070 g/dscm)	Temperature scrubber pressure loss Water pressure
		Opacity	20%	No requirement
	Pneumatic coal-cleaning equipment	Particulate	0.018 gr/dscf (0.040 g/dscm)	No requirement
		Opacity	10%	No requirement
	Processing and conveying equipment, storage systems, transfer and loading systems	Opacity	20%	No requirement
Subpart Z: **Ferroalloy-production facilities** Proposed 10/21/74 (39 FR 37469)	Electric-submerged-arc furnaces	Particulate	0.99 lb/MWh (0.45 kg/MWh) ("high-silicon alloys"); 0.51 lb/MWh (0.23 lb/MWh) (chrome and manganese alloys)	No requirement
Promulgated 5/4/75 (41 FR 18497)			No visible emissions may escape furnace capture system.	Flow-rate monitoring in hood
Revised 5/20/76 (41 FR 20659)			No visible emission may escape tapping system for >40% of each tapping period.	Flow-rate monitoring in hood
		Opacity	15%	Continuous
		CO	20% volume basis	No requirement
	Dust-handling equipment	Opacity	10%	No requirement
Subpart AA: **Iron and steel plants** Proposed 10/21/74 (39 FR 37465) Promulgated 9/23/75 (40 FR 43850)	Electric-arc furnaces	Particulate	0.0052 gr/dscf (12 mg/dscm)	No requirement
		Opacity		
		(*a*) control device	3%	Continuous
		(*b*) shop roof	0, except 20%—charging 40%—tapping	Flow-rate monitoring in capture hood

TABLE 26-5 Standards of Performance, 40 CFR Part 60 *(Continued)*

Source category	Affected facility	Pollutant	Emission level	Monitoring requirement[a]
				Pressure monitoring in DSE system
	Dust-handling equipment	Opacity	10%	No requirement
Subpart BB: **Kraft pulp mills** (kraft pulping operations within neutral sulfite semichemical pulping mills)	Digester, washer, evaporator, condensate stripper, or black-liquor oxidation systems	Total reduced sulfur (TRS)	5 ppm by volume on a dry basis, corrected to a specific oxygen content	Continuous[h] Continuous[h]
Proposed 9/24/76 (41 FR 42012)	Straight kraft recovery furnaces	TRS	5 ppm by volume on a dry basis corrected to 8% oxygen	
Promulgated 2/23/78 (43 FR 7568) Revised	Cross-recovery furnaces	TRS	25 ppm by volume on a dry basis, corrected to 8% oxygen	Continuous[h]
8/7/78 (43 FR 34784) 1/12/79 (44 FR 2578)	Smelt tanks	TRS	0.0084 g/kg black-liquor solids (dry weight) [0.0168 lb/ton liquor solids (dry weight)]	Continuous[h]
	Lime kilns	TRS	8 ppm by volume on a dry basis, corrected to 10% oxygen	Continuous[h]
	Any recovery furnace	Particulate	0.10 g/dscm (0.044 gr/dscf), corrected to 8% oxygen	No requirement
		Opacity	35%	Continuous
	Smelt tanks	Particulate	0.1 g/kg black-liquor solids (dry weight) [0.2 lb/ton black-liquor solids (dry weight)]	Continuous measurement of pressure loss when using a scrubber emission-control device
	Lime kilns	Particulate	0.15 g/dscm (0.067 gr/dscf), corrected to 10% oxygen when gaseous fossil fuel is burned	Continuous measurement of pressure loss when using a scrubber emission-control device
			0.30 g/dscm (0.13 gr/dscf), corrected to 10% oxygen when liquid fossil fuel is burned	
Subpart CC: **Glass manufacturing plants** Proposed 6/15/79 (44 FR 34840) Promulgated 10/7/80 (45 FR 66751) **Petitions for reconsideration have been filed.**	Glass-melting furnace	Particulate	Rates vary.	None
Subpart DD: **Grain elevators**[i,j] Proposed	All facilities listed below	Particulate	0.01 gr/dscf	No requirement
1/13/77 (42 FR 2842) 6/24/77 (42 FR 32264)	Truck-loading stations	Opacity	10%	No requirement
Proposal suspended	Truck-unloading stations	Opacity	5%	No requirement
8/3/78 (43 FR 34349) Proposal reinstated	Barge or ship-loading stations	Opacity	20%	No requirement
Promulgated 8/3/78 (43 FR 34340)	Barge, or ship-unloading stations		Equipment standard[k]	No requirement
	Railcar-loading stations	Opacity	5%	No requirement
	Railcar-unloadng stations	Opacity	5%	No requirement
	Grain dryers	Opacity	0%	No requirement
	Column dryers which have perforated plates with hole sizes larger than 0.094-in diameter			
	Rack dryers with screen filters coarser than 50-mesh			
	Grain-handling operations	Opacity	0%	No requirement

TABLE 26-5 Standards of Performance, 40 CFR Part 60 *(Continued)*

Source category	Affected facility	Pollutant	Emission level	Monitoring requirement[a]
Subpart GG: **Gas turbines**—heat input at peak load equal to or greater than 10.7 GJ/h (~1000 hp) Proposed Oct. 3, 1977 (42 FR 53788) Promulgated Sept. 10, 1979 (44 FR 52798) Standard under petition to review°	Simple and regenerative-cycle gas turbines and the gas-turbine portion of a combined cycle steam-electric generating system°	NO_x SO_2	See 44FR 52798	Continuous monitoring of water-fuel ratio Monitoring of sulfur and nitrogen content of fuel fired
°Proposed revision 4/15/81 (46 FR 22005). Revises the applicability of the standard with respect to industrial gas turbines.				
Subpart HH: **Lime-manufacturing plants** Proposed 5/3/77 (42 FR 22506) Promulgated 3/7/78 (43 FR 9452)	Rotary lime kilns	Particulate	0.30 lb/ton	Scrubber pressure loss; scrubber-liquid supply pressure
		Opacity	10%	Continuous
	Lime hydrators	Particulate	0.15 lb/ton	Scrubber-liquid flow rate; electric energy used by scrubber
Subpart MM: **Automobile and light-duty truck surface-coating operations** Proposed: 10/5/79 (44 FR 57792) Promulgated: 12/24/80 (45 FR 85414)	Prime coat operations Guide coat operations Topcoat operations Exempted from these provisions: operations to coat plastic body components or all-plastic automobiles or light-duty truck bodies on separate coating lines	VOC	0.16 kg of VOC per liter of applied coating solids from each prime-coat operation 1.40 kg of VOC per liter of applied coating solids from each guide-coating operation 1.47 kg of VOC per liter of applied coating solids from each topcoat operation	Incinerator temperature (if applicable)
Subpart PP: **Ammonium sulfate manufacturing** Proposed 2/4/80 (45 FR 7758) Promulgated 11/12/80 (45 FR 74846)	Ammonium sulfate dryer within ammonium-sulfate-manufacturing plant in caprolactam by-product, synthetic, and coke-oven by-product sectors	Particulate Opacity	0.15 kg/Mg (0.30 lb/ton) 15%	Provide flow monitoring devices or weigh scales for feed-material streams. Monitor pressure loss across the control device.

[a]Continuous monitors are used to determine excess emissions only, unless noted as "continuous compliance."

[b]Includes boilers firing coal-wood mixtures.

[c]If more than 25% lignite which was mined in North Dakota, South Dakota, or Montana is fired in a slap-tag furnace, the standard is 0.80 lb/million Btu.

[d]For SRC-I an 85% reduction requirement applies (24-h average).

[e]Percent reduction requirement does not apply to facilities firing 100% anthracite, resource-recovery facilities firing less than 25% fossil fuel (90-day average), or facilities located in noncontinental areas.

[f]30-day rolling average, except where noted.

[g]Commercial demonstration permits are available for SRC-I (SO_2) 1.20 lb/million Btu and 80% reduction (24-h); FBC (SO_2) 1.20 lb/million Btu and 85% reduction; coal liquefaction (NO_x) 0.70 lb/million Btu.

[h]Not effective until monitor performance specifications have been proposed and promulgated.

[i]Grain-elevator terminals (i.e., grain elevators which have permanent grain-storage capacity of over 2,500,000 bu), which handle or process wheat, corn, sorghum, rice, rye, oats, barley, or soybeans.

[j]Grain-storage elevators at wheat-flour mills, wet-corn mills, dry-corn mills (human consumption), rice mills, and soybean-oil-extraction plants, which handle or process wheat, corn sorghum, rice, rye, oats, barley, or soybeans and which have a permanent grain-storage capacity of over 1 million bu.

[k]Marine leg enclosed from top to bottom of leg, with ventilation flow rate of both leg and receiving hopper of 40 ft^3 of air per bushel of grain unloaded.

[m]For vapor pressure 78 to 570 mmHg, equip with floating roof, vapor-recovery system, or equivalent; for vapor pressure >570 mmHg, equip with vapor-recovery system or equivalent.

[n]For vapor pressure 10.3 kPa (1.5 psia) but not 75.6kPa (11.1 psia) equipped with external floating roof with primary and secondary seals (see *Federal Register* for seal-gap requirements).

or For fixed roof with internal floating-type cover equipped with continuous-closure device.

or For vapor-pressure 76.6 kPa, equipped with a vapor-recovery system which collects all VOC vapors and gases and a vapor-return or -disposal system designed for 95% removal by weight.

or Equivalent to one of the above.

TABLE 26-6 Hazardous-Air-Pollutant Standards

Asbestos	No visible emissions from milling, manufacturing, spraying, stripping, fabricating, or waste disposal, or must use fabric filter at specific setting of air-flow, pressure drop, and weight, or equivalent method of control.
Beryllium	Emissions must not exceed 10 g/24 h, or with administrator's approval must not exceed an ambient concentration of 0.01 $\mu g/m^3$ averaged over 30 days.
Mercury	1. Mercury-ore-processing facilities and mercury-cell chloralkali plant emissions must not exceed 2,300 g/24 h. 2. Mercury emissions from sludge incineration and drying must not exceed 3,200 g/24 h.
Vinyl chloride	Specific allowable emissions of vinyl chloride (and polyvinyl chloride) in all phases of production in ethylene dichloride purification plants, oxychlorination reactors, and vinyl chloride and polyvinyl chloride formation and purification plants may be found in 40 CFR Sec. 61.

change of a major stationary source producing a "significant net emissions increase" of any CAA pollutant (see Table 26-9).

Continuous monitoring is required of all CAA pollutants with emissions greater than or equal to Table 26-9 values for which there are NAAQS (except hydrocarbons). Continuous monitoring is also required for other CAA pollutants for which the EPA or the state determines that monitoring is necessary. The EPA or the state may exempt any CAA pollutant from these monitoring requirements if the maximum air-quality impact of the emissions increase is less than the values in Table 26-10 or if present concentrations of the pollutant in the area that the new source would affect are less than the Table 26-10 values. The EPA or the state may accept representative existing monitoring data collected within 3 years of the permit application to satisfy monitoring requirements.

EPA regulations provide exemption from BACT and ambient-air-impact analysis if the modification that would increase emissions is accompanied by other changes within the plant that would net a zero increase in total emissions. This exemption is referred to as the "bubble" or "no net increase" exemption.

A full PSD review would include a case-by-case determination of the controls required by BACT, an ambient-air-impact analysis to determine whether the source might violate applicable increments or air-quality standards, an assessment of effect on visibility, soils, and vegetation, submission of monitoring data, and full public review.

EPA regulations exempted smaller sources from the major elements of PSD review and, in particular, relieved those sources from compliance with BACT (though they still had to comply with applicable NSPS as well as with requirements under the SIP program). Smaller sources were also exempted from conducting ambient-air-impact analysis and submitting data supporting an ambient-air-quality analysis. Smaller sources, however, were not exempted from the program altogether. They remained subject to the statutory requirements to obtain preconstruction approval, including procedures for public review, and they still might be required, at EPA request, to submit data supporting their applications. Also, if emissions from a smaller source would affect a Class I area or if an applicable incre-

TABLE 26-8 Sources Subject to PSD Regulation If Their Potential to Emit Equals or Exceeds 100 Tons per Year

Fossil-fuel-fired steam electric plants of more than 250 million Btu/h heat input
Coal-cleaning plants (with thermal dryers)
Kraft-pulp mills
Portland-cement plants
Primary zinc smelters
Iron and steel mill plants
Primary aluminum-ore-reduction plants
Primary copper smelters
Municipal incinerators capable of charging more than 250 tons of refuse per day
Hydrofluoric, sulfuric, and nitric acid plants
Petroleum refineries
Lime plants
Phosphate-rock-processing plants
Coke-oven batteries
Suflur-recovery plants
Carbon-black plants (furnace process)
Primary lead smelters
Fuel-conversion plants
Sintering plants
Secondary metal-production plants
Chemical-process plants
Fossil-fuel boilers (or combinations thereof) totaling more than 250 million Btu/h heat input
Petroleum-storage and -transfer units with total storage capacity exceeding 300,000 bbl
Taconite-ore-processing plants
Glass-fiber-processing plants
Charcoal-production plants

TABLE 26-9 Significant-Net-Emissions Increase

Pollutant	Tons/year
CO	100
NO_x (as NO_2)	40
SO_2	40
Particulate matter	25
Ozone	40 of volatile organic compounds
Lead	0.6
Asbestos	0.007
Beryllium	0.0004
Mercury	0.1
Vinyl chloride	1
Fluorides	3
Sulfuric acid mist	7
Hydrogen sulfide	10
Total reduced sulfur	10
Reduced-sulfur compounds	10
Other CAA pollutants	> 0

TABLE 26-7 Prevention-of-Significant-Deterioration (PSD) Air-Quality Increments

	Maximum allowable increase over baseline air quality, $\mu g/m^3$			Primary ambient-air-quality standard, $\mu g/m^3$
	Class I	Class II	Class III	
Particulate matter				
Annual geometric mean	5	19	37	75
24-h maximum	10	37	75	260
SO_2				
Annual arithmetic mean	2	20	40	80
24-h maximum	5	91	182	365
3-h maximum	25	512	700	1300*

*Secondary standard rather than primary standard.

TABLE 26-10 Concentration Impacts below Which Ambient Monitoring May Not Be Required

	$\mu g/m^3$	Average time
CO	575	8-h maximum
NO_2	14	24-h maximum
TSP	10	24-h maximum
SO_2	13	24-h maximum
Lead	0.1	24-h maximum
Mercury	0.25	24-h maximum
Beryllium	0.0005	24-h maximum
Fluorides	0.25	24-h maximum
Vinyl chloride	15	24-h maximum
Total reduced sulfur	10	1-h maximum
H_2S	0.04	1-h maximum
Reduced-sulfur compounds	10	1-h maximum

ment were already being violated, the full PSD requirements for ambient-air-impact analysis would apply.

Nonattainment (NA) Those areas of the United States failing to attain compliance with ambient-air-quality standards were considered nonattainment areas. New plants could be constructed in nonattainment areas only if stringent conditions were met. Emissions had to be controlled to the greatest degree possible, and more than equivalent offsetting emission reductions had to be obtained from other sources to assure progress toward achievement of the standards. Specifically, (1) the new source must be equipped with pollution controls to assure lowest-achievable-emission rate (LAER), which in no case can be less stringent than any applicable NSPS; (2) all existing sources owned by an applicant in the same region must be in compliance with applicable state-implementation-plan requirements or be under an approved schedule or an enforcement order to achieve such compliance; (3) the applicant must have sufficient offsets to more than make up for the emissions to be generated by the new source (after application of LAER); and (4) the emission offsets must provide "a positive net air quality benefit in the affected area."

LAER was deliberately a technology-forcing standard of control. The statute stated that LAER must reflect (1) the most stringent emission limitation contained in the implementation plan of any state for such category of sources unless the applicant can demonstrate that such a limitation is not achievable, or (2) the most stringent limitation achievable in practice within the industrial category, whichever is more stringent. In no event could LAER be less stringent than any applicable NSPS. While the statutory language defining BACT directed that "energy, environmental and economic impacts and other costs" be taken into account, the comparable provision on LAER provided no instruction that costs be considered.

For existing sources emitting pollutants for which the area is nonattainment, reasonable available control technology (RACT) would be required. EPA is in the process of defining RACT by industrial category.

Controlled-Trading Program The legislation enacted under the Clean Air Act Amendments of 1977 provided the foundation for EPA's controlled-trading program, the essential elements of which include:

- Bubble policy (or bubble exemption under PSD)
- Offsets policy (under nonattainment)
- Banking and brokerage (under nonattainment)

While these different policies vary broadly in form, their objective is essentially the same: to substitute flexible economic-incentive systems for the current rigid, technology-based regulations that specify exactly how companies must comply. Although still in the early stages, these market mechanisms could make regulating easier for EPA and less burdensome and costly for industry.

Bubble Policy The bubble concept introduced under PSD provisions of the Clean Air Act Amendments of 1977 was formally proposed as EPA policy on Jan. 18, 1979, the final policy statement being issued on Dec. 11, 1979. The bubble policy allows a company to find the most efficient way to control a plant's emissions as a whole rather than by meeting individual point-source requirements. If it is found less expensive to tighten control of a pollutant at one point and relax controls at another, this would be possible as long as the total pollution from the plant would not exceed the sum of the current limits on individual point sources of pollution in the plant. Properly applied, this approach should promote greater economic efficiency and increased technological innovation.

There are some restrictions, however, in applying the bubble concept:

1. The bubble may only be used for pollutants in an area where the state implementation plan has an approved schedule to meet air-quality standards for that pollutant.
2. The alternatives used must ensure that air-quality standards will be met.
3. Emissions must be quantifiable, and trades among them must be even. Each emission point must have a specific emission limit, and that limit must be tied to enforceable testing techniques.
4. Only pollutants of the same type may be traded, that is, particulates for particulates, and hydrocarbons for hydrocarbons, etc.
5. Control of hazardous pollutants cannot be relaxed through trades with less toxic pollutants.
6. Development of the bubble plan cannot delay enforcement of federal and state requirements.

Some additional considerations must be noted:

1. The bubble may cover more than one plant within the same area.
2. In some circumstances, states may consider trading open dust emissions for particulates (although EPA warns that this type of trading will be difficult).
3. EPA may approve compliance-date extensions in special cases. For example, a source may obtain a delay in a compliance schedule to install a scrubber if such a delay would have been permissible without the bubble.

EPA will closely examine particulate-size distribution in particulate-emission trades because finer particulates disperse more widely and remain in the air longer.

It will be the responsibility of industry to suggest alternative control approaches and demonstrate satisfactorily that the proposal is equivalent in pollution reduction, enforceability, and environmental impact to existing individual process standards.

Offsets Policy Offsets were EPA's first application of the concept that one source could meet its environmental-protection obligations by getting another source to assume additional control actions. In nonattainment areas, pollution from a proposed new source, even one that controls its emissions to the lowest possible level, would aggravate existing violations of ambient-air-quality standards and trigger the statutory prohibition. The offsets policy provided these new sources with an alternative. The source could proceed with construction plans, provided that:

1. The source would control emissions to the lowest achievable level.
2. Other sources owned by the applicant were in compliance or on an approved compliance schedule.
3. Existing sources were persuaded to reduce emissions by an amount at least equal to the pollution that the new source would add.

Banking and Brokerage Policy EPA's new banking policy is aimed at providing companies with incentives to find more offsets. Under the original offset policy, a firm shutting down or modifying a facility could apply the reduction in emissions to new construction elsewhere in the region only if the changes were made simultaneously. However, with banking a company can "deposit" the reduction for later use or sale. Such a policy will clearly establish that clean air (or the right to use it) has direct economic value.

APPROACH TO A SOURCE-CONTROL-PROBLEM STRATEGY

Strategy Control technology is self-defeating if it creates undesirable side effects in meeting objectives. Air pollution control must be considered in terms of both total technological systems (equipment and processes) and ecological consequences, such as the problems of treatment and disposal of collected pollutants.

The control strategy for environmental-impact assessment often focuses on five alternatives whose purpose would be the reduction and/or elimination of pollutant emissions:

1. Elimination of the operation entirely or in part
2. Modification of the operation
3. Relocation of the operation
4. Application of appropriate control technology
5. Combinations thereof

In view of the relatively high costs often associated with pollution-control systems, engineers are directing considerable effort toward process modification to eliminate as much of the pollution problem as possible at the source. This includes evaluating alternative manufacturing and production techniques, substituting raw materials, and improving process-control methods. Unfortunately, if there is no alternative, the application of pollution-control equipment must be considered. In view of the relatively high costs, proper selection of this equipment is essential. The equipment must be designed to comply with regulatory emission limitations on a continual basis, interruptions being subject to severe penalty depending upon the circumstances. The requirement for design performance on a continual basis places very heavy emphasis on operation and maintenance practices. The rapidly escalating costs of energy, labor, and materials can make operation and maintenance considerations even more important than the original capital cost.

Factors in Control-Equipment Selection A number of factors must be considered prior to selecting a particular piece of air-pollution-control equipment. In general, they can be grouped into three categories: environmental, engineering, and economic.

Environmental Factors These include (1) equipment location, (2) available space, (3) ambient conditions, and (4) availability of adequate utilities (i.e., power, water, etc.) and ancillary-system facilities (i.e., waste treatment and disposal, etc.), (5) maximum allowable emission (air pollution codes), (6) aesthetic considerations (i.e., visible steam or water-vapor plume, etc.), (7) contributions of the air-pollution-control system to wastewater and land pollution, and (8) contribution of the air-pollution-control system to plant noise levels.

Engineering Factors These include:

1. Contaminant characteristics (i.e., physical and chemical properties, concentration, particulate shape and size distribution (in the case of particulates), chemical reactivity, corrosivity, abrasiveness, toxicity, etc.)

2. Gas-stream characteristics (i.e., volume flow rate, temperature, pressure, humidity, composition, viscosity, density, reactivity, combustibility, corrosivity, toxicity, etc.)

3. Design and performance characteristics of the particular control system (i.e., size and weight, fractional efficiency curves (in the case of particulates), mass-transfer and/or contaminant-destruction capability (in the case of gases or vapors), pressure drop, reliability, turndown capability, power requirements, utility requirements, temperature limitations, maintenance requirements, and flexibility toward complying with more stringent air pollution codes.

Economic Factors These include capital cost (equipment, installation, engineering, etc.), operating cost (utilities, maintenance, etc.), and life-cycle cost over the expected equipment lifetime.

TABLE 26-11 Advantages and Disadvantages of Cyclone Collectors

Advantages
1. Low cost of construction
2. Relatively simple equipment with few maintenance problems
3. Relatively low operating pressure drops (for degree of particulate removal obtained) in the range of approximately 2- to 6-in water column
4. Temperature and pressure limitations imposed only by the materials of construction used
5. Dry collection and disposal
6. Relatively small space requirements

Disadvantages
1. Relatively low overall particulate collection efficiencies, especially on particulates below 10 μm in size
2. Inability to handle tacky materials

TABLE 26-12 Advantages and Disadvantages of Wet Scrubbers

Advantages
1. No secondary dust sources
2. Relatively small space requirements
3. Ability to collect gases as well as particulates (especially "sticky" ones)
4. Ability to handle high-temperature, high-humidity gas streams
5. Capital cost low (if wastewater treatment system, not required)
6. For some processes, gas stream already at high pressures (so pressure-drop considerations may not be significant)
7. Ability to achieve high collection efficiencies on fine particulates (however, at the expense of pressure drop)

Disadvantages
1. Possible creation of water-disposal problem
2. Product collected wet
3. Corrosion problems more severe than with dry systems
4. Steam plume opacity and/or droplet entrainment possibly objectionable
5. Pressure-drop and horsepower requirements possibly high
6. Solids buildup at the wet-dry interface possibly a problem
7. Relatively high maintenance costs

TABLE 26-13 Advantages and Disadvantages of Electrostatic Precipitators

Advantages
1. Extremely high particulate (coarse and fine) collection efficiencies attainable (at a relatively low expenditure of energy)
2. Dry collection and disposal
3. Low pressure drop (typically less than 0.5-in water column)
4. Designed for continuous operation with minimum maintenance requirements
5. Relatively low operating costs
6. Capable of operation under high pressure (to 150 lbf/in^2) or vacuum conditions
7. Capable of operation at high temperatures [to 704°C (1300°F)]
8. Relatively large gas flow rates capable of effective handling

Disadvantages
1. High capital cost
2. Very sensitive to fluctuations in gas-stream conditions (in particular, flows, temperatures, particulate and gas composition, and particulate loadings)
3. Certain particulates difficult to collect owing to extremely high- or low-resistivity characteristics
4. Relatively large space requirements required for installation
5. Explosion hazard when treating combustible gases and/or collecting combustible particulates
6. Special precautions required to safeguard personnel from the high voltage
7. Ozone produced by the negatively charged discharge electrode during gas ionization
8. Relatively sophisticated maintenance personnel required

Comparing Control-Equipment Alternatives The final choice in equipment selection is usually dictated by the equipment capable of achieving compliance with regulatory codes at the lowest uniform annual cost (amortized capital investment plus operation and maintenance costs). To compare specific control-equipment alternatives, knowledge of the particular application and site is essential. A preliminary screening, however, may be performed by reviewing the advantages and disadvantages of each type of air-pollution-control equipment. General advantages and disadvantages of the most popular types of air-pollution-control equipment for gases and particulates are presented in Tables 26-11 through 26-19. Other activities that must be accomplished before final compliance is achieved are presented in Table 26-20.

DISPERSION FROM STACKS

Stacks discharging to the atmosphere have long been the most common industrial method of disposing of waste gases. The concentrations to which humans, plants, animals, and structures are exposed at ground level can be reduced significantly by emitting the waste gases from a process at great heights. Although tall stacks may be effective

TABLE 26-14 Advantages and Disadvantages of Fabric-Filter Systems

Advantages
1. Extremely high collection efficiency on both coarse and fine (submicrometer) particulates
2. Relatively insensitive to gas-stream fluctuation; efficiency and pressure drop relatively unaffected by large changes in inlet dust loadings for continuously cleaned filters
3. Filter outlet air capable of being recirculated within the plant in many cases (for energy conservation)
4. Collected material recovered dry for subsequent processing or disposal
5. No problems with liquid-waste disposal, water pollution, or liquid freezing
6. Corrosion and rusting of components usually not problems
7. No hazard of high voltage, simplifying maintenance and repair and permitting collection of flammable dusts
8. Use of selected fibrous or granular filter aids (precoating), permitting the high-efficiency collection of submicrometer smokes and gaseous contaminants
9. Filter collectors available in a large number of configurations, resulting in a range of dimensions and inlet and outlet flange locations to suit installation requirements
10. Relatively simple operation

Disadvantages
1. Temperatures much in excess of 288°C (550°F) requiring special refractory mineral or metallic fabrics that are still in the developmental stage and can be very expensive
2. Certain dusts possibly requiring fabric treatments to reduce dust seeping or, in other cases, assist in the removal of the collected dust
3. Concentrations of some dusts in the collector (~50 g/m^3) forming a possible fire or explosion hazard if a spark or flame is admitted by accident; possibility of fabrics burning if readily oxidizable dust is being collected
4. Relatively high maintenance requirements (bag replacement, etc.)
5. Fabric life possibly shortened at elevated temperatures and in the presence of acid or alkaline particulate or gas constituents
6. Hygroscopic materials, condensation of moisture, or tarry adhesive components possibly causing crusty caking or plugging of the fabric or requiring special additives
7. Replacement of fabric possibly requiring respiratory protection for maintenance personnel
8. Medium pressure-drop requirements, typically in the range 4- to 10-in water column

TABLE 26-15 Advantages and Disadvantages of Absorption Systems (Packed and Plate Columns)

Advantages
1. Relatively low pressure drop
2. Standardization in fiberglass-reinforced plastic (FRP) construction permiting operation in highly corrosive atmospheres
3. Capable of achieving relatively high mass-transfer efficiencies
4. Increasing the height and/or type of packing or number of plates capable of improving mass transfer without purchasing a new piece of equipment
5. Relatively low capital cost
6. Relatively small space requirements
7. Ability to collect particulates as well as gases

Disadvantages
1. Possibility of creating water (or liquid) disposal problem
2. Product collected wet
3. Particulates deposition possibly causing plugging of the bed or plates
4. When FRP construction is used, sensitive to temperature
5. Relatively high maintenance costs

in lowering the ground-level concentration of pollutants, they do not in themselves reduce the amount of pollutants released into the atmosphere. However, in certain situations their use can be the most practical and economical way of dealing with an air pollution problem.

Preliminary Design Considerations To determine the acceptability of a stack as a means of disposing of waste gases, the acceptable ground-level concentration (GLC) of the pollutant or pollutants must first be determined. The topography of the area must also be considered so that the stack can be properly located with respect to buildings and hills which might introduce a factor of air turbulence into the operation of the stack (see Fig. 26-1). Awareness of the meteorological conditions prevalent in the area, such as the prevailing winds, humidity, and rainfall, is also essential. Finally, an accurate knowledge of the constituents of the waste gas and their physical and chemical properties is paramount.

Wind Direction and Speed Wind direction is measured at the height at which the pollutant is released, and the mean direction will

TABLE 26-16 Comparison of Plate and Packed Columns

Packed column
1. Lower pressure drop
2. Simpler and cheaper to construct
3. Preferable for liquids with high-foaming tendencies

Plate column
1. Less susceptible to plugging
2. Less weight
3. Less of a problem with channeling
4. Temperature surge resulting in less damage

TABLE 26-17 Advantages and Disadvantages of Adsorption Systems

Advantages
1. Possibility of product recovery
2. Excellent control and response to process changes
3. No chemical-disposal problem when pollutant (product) recovered and returned to process
4. Capability of systems for fully automatic, unattended operation
5. Capability to remove gaseous or vapor contaminants from process streams to extremely low levels

Disadvantages
1. Product recovery possibly requiring an exotic, expensive distillation (or extraction) scheme
2. Adsorbent progressively deteriorating in capacity as the number of cycles increases
3. Adsorbent regeneration requiring a steam or vacuum source
4. Relatively high capital cost
5. Prefiltering of gas stream possibly required to remove any particulate capable of plugging the adsorbent bed
6. Cooling of gas stream possibly required to get to the usual range of operation [less than 49°C (120°F)]
7. Relatively high steam requirements to desorb high-molecular-weight hydrocarbons

TABLE 26-18 Advantages and Disadvantages of Combustion Systems

Advantages
1. Simplicity of operation
2. Capability of steam generation or heat recovery in other forms
3. Capability for virtually complete destruction of organic contaminants

Disadvantages
1. Relatively high operating costs (particularly associated with fuel requirements)
2. Potential for flashback and subsequent explosion hazard
3. Catalyst poisoning (in the case of catalytic incineration)
4. Incomplete combustion possibly creating potentially worse pollution problems

TABLE 26-19 Advantages and Disadvantages of Condensers

Advantages
1. Pure product recovery (in the case of indirect-contact condensers)
2. Water used as the coolant in an indirect-contact condenser (i.e., shell-and-tube heat exchanger) not in contact with contaminated gas stream and so reusable after cooling

Disadvantages
1. Relatively low removal efficiency for gaseous contaminants (at concentrations typical of pollution control applications)
2. Coolant requirements possibly extremely expensive

TABLE 26-20 Compliance Activity and Schedule Chart

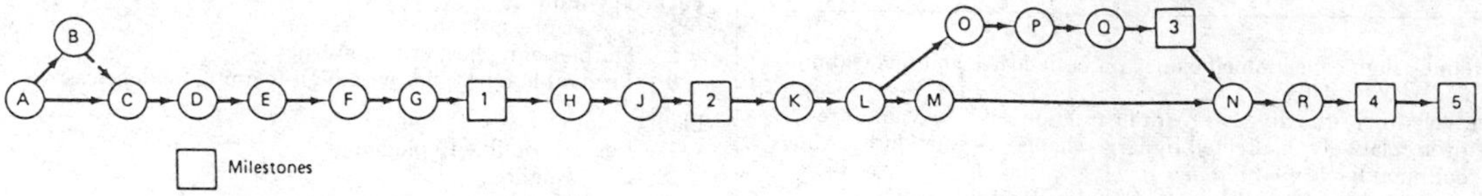

Milestones
1. Date of submittal of final control plan to appropriate agency
2. Date of award of control-device contract
3. Date of initiation of on-site construction or installation of emission-control equipment
4. Date by which on-site construction or installation of emission-control equipment is completed
5. Date by which final compliance is achieved

Activities

Designation	Activity	Designation	Activity
A-C	Preliminary investigation.	K-L	Review and approval of assembly drawings.
A-B	Source tests, if necessary.	L-M	Vendor prepares fabrication drawings.
C-D	Evaluate control alternatives.	M-N	Fabricate control device.
D-E	Commit funds for total program.	L-O	Prepare engineering drawings.
E-F	Prepare preliminary control plan and compliance schedule for agency.	O-P	Procure construction bids.
F-G	Agency review and approval.	P-Q	Evaluate construction bids.
G-1	Finalize plans and specifications.	Q-3	Award construction contract.
I-H	Procure control-device bids.	3-N	On-site construction.
H-J	Evaluate control-device bids.	N-R	Install control device.
J-2	Award control-device contract.	R-4	Complete construction (system tie-in).
2-K	Vendor prepares assembly drawings.	4-5	Startup, shakedown, source test.

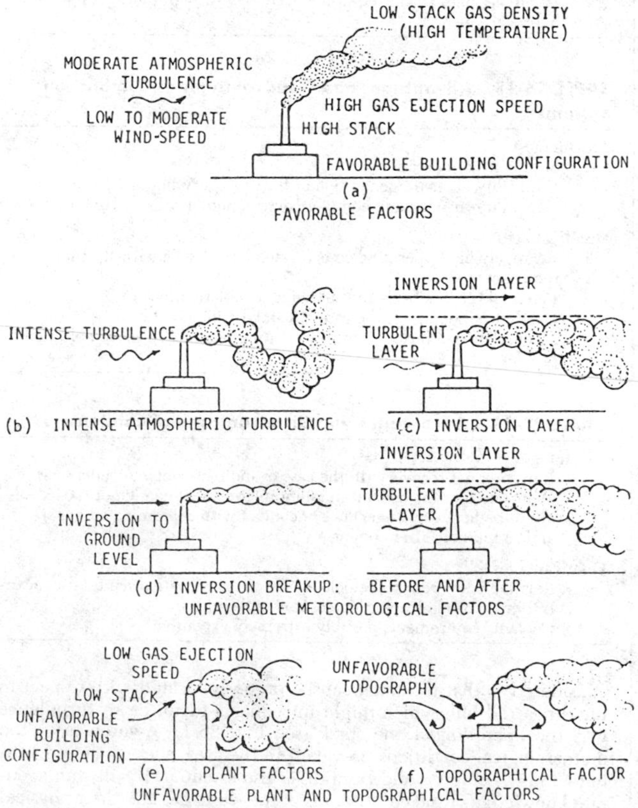

FIG. 26-1 Favorable and unfavorable aspects of factors important in air pollution from stacks.

indicate the direction of travel of the pollutants. In meteorology it is conventional to consider the wind direction as the direction from which the wind blows; therefore a northwest wind will move pollutants to the southeast of the source.

The effect of wind speed is twofold: (1) wind speed will determine the travel time from a source to a given receptor; and (2) wind speed will affect dilution in the downwind direction. Generally, the concentration of air pollutants downwind from a source is inversely proportional to wind speed.

Wind speed has velocity components in all directions, so that there are vertical motions as well as horizontal ones. These random motions of widely different scales and periods are essentially responsible for the movement and diffusion of pollutants about the mean downwind path. These motions can be considered atmospheric turbulence. If the scale of a turbulent motion, i.e., the size of an eddy, is larger than the size of the pollutant plume in its vicinity, the eddy will move that portion of the plume. If an eddy is smaller than the plume, its effect will be to diffuse or spread out the plume. This diffusion caused by eddy motion is widely variable in the atmosphere, but even when the effect of this diffusion is least, it is on the order of three orders of magnitude greater than diffusion by molecular action alone.

Mechanical turbulence is the induced-eddy structure of the atmosphere due to the roughness of the surface over which the air is passing. Therefore, the existence of trees, shrubs, buildings, and terrain features will cause mechanical turbulence. The height and spacing of the elements causing the roughness will affect the turbulence. In general, the higher the roughness elements, the greater the mechanical turbulence. In addition, mechanical turbulence increases as wind speed increases.

Thermal turbulence is turbulence induced by the stability of the atmosphere. When the earth's surface is heated by the sun's radiation, the lower layer of the atmosphere tends to rise and thermal turbulence becomes greater, especially under conditions of light wind. On clear nights with wind, heat is radiated from the earth's surface, resulting in the cooling of the ground and the air adjacent to it. This results in extreme stability of the atmosphere near the earth's surface. Under these conditions turbulence is at a minimum.

Attempts to relate different measures of turbulence of the wind (or stability of the atmosphere) to atmospheric diffusion have been made for some time. The measurement of atmospheric stability by temperature-difference measurements on a tower is frequently utilized as an indirect measure of turbulence, particularly when climatological estimates of turbulence are desired.

Lapse Rate and Atmospheric Stability Apart from mechanical interference with the steady flow of air caused by buildings and other obstacles, the most important factor which influences the degree of turbulence and hence the speed of diffusion in the lower air is the variation of temperature with height above the ground, referred to as the "lapse rate." The dry-adiabatic lapse rate (DALR) is the temperature change for a rising parcel of dry air. The dry-adiabatic lapse rate can be approximated as −one°C per 100 m, or $dT/dz = -10^{-2}$°C/m or −5.4°F/1000 ft. If the rising air contains water vapor, the cooling due to adiabatic expansion will result in the relative humidity being increased, and saturation may be reached. Further ascent would then lead to condensation of water vapor, and the latent heat thus released would reduce the rate of cooling of the rising air. The buoyancy force on a warm-air parcel is caused by the difference between its density and that of the surrounding air. The perfect-gas law shows that at a fixed pressure (altitude) the temperature and density of an air parcel are inversely related: temperature is normally used to determine buoyancy because it is easier to measure than density. If the temperature gradient (lapse rate) of the atmosphere is the same as the adiabatic lapse rate, a parcel of air displaced from its original position will expand or contract in such a manner that its density and temperature remain the same as its surroundings. In this case there will be no buoyancy forces on the displaced parcel, and the atmosphere is termed "neutrally stable."

If the atmospheric temperature decreases faster with increasing altitude than the adiabatic lapse rate (superadiabatic), a parcel of air displaced upward will have a higher temperature than the surrounding air. Its density will be lower, giving it a net upward buoyancy force. The opposite situation exists if the parcel of air is displaced downward, and the parcel experiences a downward buoyancy force. Once a parcel of air has started moving up or down, it will continue to do so, causing unstable atmospheric conditions. If the temperature decreases more slowly with increasing altitude than the adiabatic lapse rate, a displaced parcel of air experiences a net restoring force. The buoyancy forces then cause stable atmospheric conditions (see Fig. 26-2).

Strongly stable lapse rates are commonly referred to as "inversions." The strong stability inhibits mixing across the inversion layer. Normally these conditions of strong stability extend for only several hundred meters vertically. The vertical extent of the inversion is

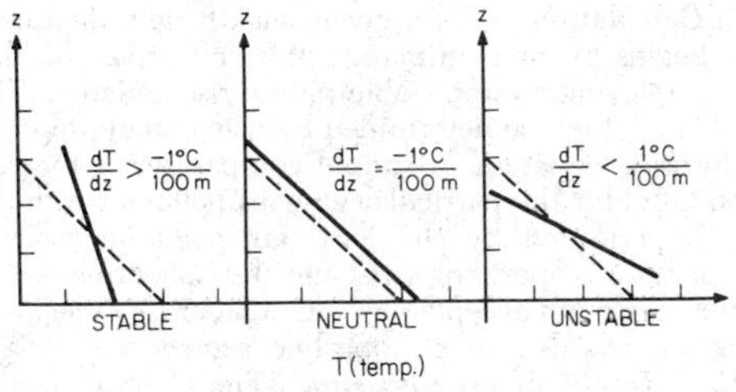

FIG. 26-2 Stability criteria with measured lapse rate.

referred to as the "inversion depth." Two distinct types are observed: the **ground-level inversion,** caused by radiative cooling of the ground at night, and **inversions aloft,** occurring between 500 and several thousand meters above the ground (see Fig. 26-3).

Some of the more common lapse-rate profiles with their corresponding effect on stack plumes are presented in Fig. 26-4.

From the viewpoint of air pollution, both stable surface layers and low-level inversions are undesirable because they minimize the rate of dilution of contaminants in the atmosphere. Even though the surface layer may be unstable, a low-level inversion will act as a barrier to vertical mixing, and contaminants will accumulate in the surface layer below the inversion. Stable atmospheric conditions tend to be more frequent and persist longest in the autumn, but inversions and stable lapse rates are prevalent at all seasons of the year.

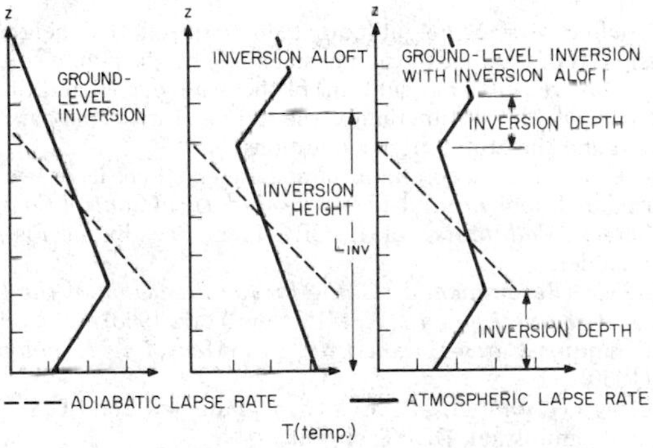

FIG. 26-3 Characteristic lapse rates under inversion conditions.

Temperature gradient	Observation	Description

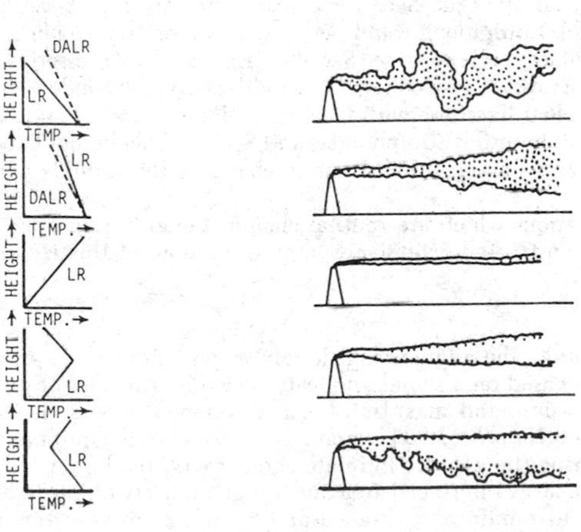

Strong lapse—looping (unstable)
This is usually a fair-weather daytime condition, since strong solar heating of the ground is required. Looping is not favored by cloudiness, snow cover, or strong winds.

Weak lapse—coning (slightly unstable or neutral)
This is usually favored by cloudy and windy conditions and may occur day or night. In dry climates it may occur infrequently, and in cloudy climates it may be the most frequent type observed.

Inversion—fanning (stable)
This is principally a nighttime condition. It is favored by light winds, clear skies, and snow cover. The condition may persist in some climates for several days at a time during winter, especially in the higher latitudes.

Inversion below, lapse aloft—lofting (transition from unstable to stable)
This condition occurs during transition from lapse to inversion and should be observed most frequently near sunset; it may be very transitory or persist for several hours. The shaded zone of strong effluent concentration is caused by trapping by the inversion of effluent carried into the stable layer by turbulent eddies that penetrate the layer for a short distance.

Lapse below, inversion aloft—fumigation (transition from stable to unstable)
This occurs when the nocturnal inversion is dissipated by heat from the morning sun. The lapse layer usually starts at the ground and works its way upward (less rapidly in winter than in summer). Fumigation may also occur in sea-breeze circulations during late morning or early afternoon. The shaded zone of strong concentration is that portion of the plume which has not yet been mixed downward.

FIG. 26-4 Lapse-rate characteristics of atmospheric-diffusion transport of stack emissions.

Design Calculations For a given stack height, the calculational sequence begins by first estimating the "effective" height of the emission, employing an applicable plume-rise equation. The maximum GLC may then be determined by using an appropriate atmospheric-diffusion equation. A simple comparison of the calculated maximum GLC for the particular gaseous pollutant with the maximum GLC permitted by the local air pollution codes dictates whether the stack is operating satisfactorily. Conversely, with knowledge of the maximum acceptable GLC standards, a stack which will satisfy these standards can be properly designed.

Effective Height of an Emission The effective height of an emission rarely corresponds to the physical height of the stack. If the plume is caught in the turbulent wake of the stack or of buildings in the vicinity of the stack, the effluent will be mixed rapidly downward toward the ground. If the plume is emitted free of these turbulent zones, a number of emission factors and meteorological factors influence the rise of the plume. The emission factors include the gas flow rate and temperature of the effluent at the top of the stack and the diameter of the stack opening. The meteorological factors influencing plume rise include wind speed, air temperature, shear of the wind speed with height, and atmospheric stability. No theory on plume rise presently takes into account all these variables. Most of the equations that have been formulated for computing the effective height of an emission are semiempirical. When considering any of these plume-rise equations, it is important to evaluate each in terms of the assumptions made and the circumstances existing at the time that the particular correlation was formulated. The formulas generally are not applicable to tall stacks [above 305-m (1000 ft) effective height].

The effective stack height (equivalent to the effective height of the emission) is the sum of the actual stack height, the plume rise due to the exhaust velocity (momentum) of the issuing gases, and the buoyancy rise, which is a function of the temperature of the gases being emitted and the atmospheric conditions.

Some of the more common plume-rise equations have been summarized by Buonicore and Theodore (*Industrial Control Equipment for Gaseous Pollutants*, vol. II, CRC Press, Boca Raton, Fla., 1975) and include:

- ASME (*Recommended Guide for the Prediction of the Dispersion of Airborne Effluents*, ASME, New York, 1968)
- Bosanquet-Carey-Halton [*Proc. Inst. Mech. Eng. (London)*, **162,** 355 (1950)]
- Briggs (*Plume Rise*, AEC Critical Review ser., U.S. Atomic Energy Commission, Div. Tech. Inf.)
- Carson and Moses [*J. Air Pollut. Control Assoc.*, **18,** 454 (1968), and **19,** 862 (1969)]
- CONCAWE (Brummage et al., *The Calculation of Atmospheric Dispersion from a Stack*, CONCAWE, The Hague, 1966)
- Csanady [*Int. J. Air Water Pollut.*, **4,** 47 (1961)]
- Davison-Bryant (*Trans. Conf. Ind. Wastes*, 14th Ann. Meet. Ind. Hyg. Found. Am., **39,** 1949)
- Holland (*Workbook of Atmospheric Dispersion Estimates*, U.S. EPA Publ. AP-26, 1970)
- Lucas, Moore, and Spurr [*Int. J. Air Water Pollut.*, **7,** 473 (1963)]
- Stone and Clarke (*British Experience with Tall Stacks for Air Pollution Control on Large Fossil-Fueled Power Plants*, American Power Conference, Chicago, 1967)
- Stumke [*Staub*, **23,** 549 (1963)]
- TVA [Montgomery et al., *J. Air Pollut. Control Assoc.*, **22** (10), 779 (1972)]

Maximum Ground-Level Concentrations The effective height of an emission having been determined, the next step is to study its path downward by using the appropriate atmospheric-dispersion formula. Some of the more popular atmospheric-dispersion calculational procedures have been summarized by Buonicore and Theodore (op. cit.) and include:

- Bosanquet-Pearson model [*Trans. Faraday Soc.*, **32,** 1249 (1936)]
- Pasquill-Gifford model [Pasquill, *Meteorol. Mag.*, **90,** 33, 1063 (1961), and Gifford, *Nucl. Saf.*, **2,** 4, 47 (1961)]
- Sutton model [*Q.J.R. Meteorol. Soc.*, **73,** 257 (1947)]
- TVA model [Carpenter et al., *J. Air Pollut. Control Assoc.*, **21** (8) (1971), and Montgomery et al., *J. Air Pollut. Control Assoc.*, **23** (5), 388 (1973)]

Miscellaneous Effects

Evaporative Cooling When effluent gases are washed to absorb certain constituents prior to emission, the gases are cooled and become saturated with water vapor. Upon release of the gases, further cooling due to contact with cold surfaces of ductwork or stack is likely. This cooling causes water droplets to condense in the gas stream. Upon release of the gases from the stack, the water droplets evaporate, withdrawing the latent heat of vaporization from the air and cooling the plume. The resulting negative buoyancy reduces the effective stack height. The result may be a plume (with a greater density than that of the ambient atmosphere) which will fall to the ground. If any pollutant remains after scrubbing, its full effect will be felt on the ground in the vicinity of the stack.

Aerodynamic Downwash Should the stack exit velocity be too low as compared with the speed of the crosswind, some of the effluent can be pulled downward by the low pressure on the lee side of the stack. This phenomenon, known as "downwash," can be minimized by keeping the exit velocity greater than the mean wind speed, i.e., typically twice the mean wind speed. Another way to minimize downwash is to fit the top of the stack with a flat disk that extends for at least one stack diameter outward from the stack.

If it becomes necessary to increase the stack-gas exit velocity to avoid downwash, it may be necessary to remodel the stack exit. A venturi-nozzle design has been found to be most effictive. This design also keeps pressure losses to a minimum.

Multiple Stacks The effect of multiple-stack sources has been handled in the past by simply treating each as a distinct source and the resulting concentrations added to obtain the total concentration. However, an improved method is available for estimating the effect of multiple sources [Montgomery, *J. Air Pollut. Control Assoc.* **23** (5), 388 (1973)].

SOURCE CONTROL OF GASEOUS EMISSIONS

There are four chemical engineering unit operations commonly used for the control of gaseous emissions:

1. *Absorption.* Sec. 4: "Thermodynamics"; and Sec. 18: "Liquid-Gas Systems." For plate columns, see Sec. 18: "Gas-Liquid Contacting: Plate Columns." For packed columns, see Sec. 18: "Gas-Liquid Contacting: Packed Columns."
2. *Adsorption.* See Sec. 16: "Adsorption and Ion Exchange."
3. *Combustion.* See Sec. 9: "Heat Generation" and "Fired Process Equipment." For incinerators, see material under these subsections.
4. *Condensation.* See Sec. 10: "Heat Transmission"; Sec. 11: "Heat-Transfer Equipment"; and Sec. 12: "Evaporative Cooling." For direct-contact condensers, see Sec. 11: "Evaporator Accessories"; and Sec. 12: "Evaporative Cooling." For indirect-contact condensers, see Sec. 10: "Heat Transfer with Change of Phase" and "Thermal Design of Heat-Transfer Equipment"; and Sec. 11: "Shell-and-Tube Heat Exchangers" and "Other Heat Exchangers for Liquids and Gases."

These operations which are routine chemical engineering operations, have been treated extensively in other sections of this *Handbook*.

Absorption

Introduction The engineering design of gas-absorption equipment must be based on a sound application of the principles of diffusion, equilibrium, and mass transfer as developed in Secs. 4, 14, and 18 of the *Handbook*. The main requirement in equipment design is to bring the gas into intimate contact with the liquid, that is, to provide a large interfacial area and a high intensity of interface renewal, and to minimize resistance and maximize driving force. This contacting of the phases can be achieved in many different types of equipment, the most important being packed and plate columns. The final choice between them rests with the various criteria which must be met. For example, if the pressure drop through the column is large enough that compression costs become significant, a packed column may be preferable to a plate-type column because of the lower pressure drop.

In most processes involving the absorption of a gaseous pollutant from an effluent gas stream, the gas stream is the processed fluid; hence, its inlet condition (flow rate, composition, and temperature) are usually known. The temperature and composition of the inlet liquid and the composition of the outlet gas are usually specified. The main objectives in the design of an absorption column, then, are the determination of the solvent flow rate and the calculation of the principal dimensions of the equipment (column diameter and height to accomplish the operation). These objectives can be attained by evaluating, for a selected solvent at a given flow rate, the number of theoretical separation units (stages or plates) and converting them into practical units of column heights or number of actual plates by means of existing correlations.

The general design procedure consists of a number of steps to be taken into consideration. These include:

1. Solvent selection
2. Equilibrium-data evaluation
3. Estimation of operating data (usually consisting of a mass and energy balance in which the energy balance decides whether the absorption process can be considered isothermal or adiabatic)
4. Column selection (should the column selection not be obvious or specified, calculations must be carried out for the different types of columns and the final selection based on economic considerations)
5. Calculation of column diameter (for packed columns, this is usually based on flooding conditions, and for plate columns on the optimum gas velocity or the liquid-handling capacity of the plate)
6. Estimation of column height or number of plates (for packed columns, column height is obtained by multiplying the number of transfer units, obtained from a knowledge of equilibrium and operating data, by the height of a transfer unit; for plate columns, the number of theoretical plates determined from the plot of equilibrium and operating lines is divided by the estimated overall plate efficiency to give the number of actual plates, which in turn allows the column height to be estimated from the plate spacing)
7. Determination of pressure drop through the column (for packed columns, correlations dependent on packing type, column-operating data, and physical properties of the constituents involved are available to estimate the pressure drop through the packing; for plate columns, the pressure drop per plate is obtained and multiplied by the number of plates)

Solvent Selection The choice of a particular solvent is most important. Frequently, water is used, as it is very inexpensive and plentiful, but the following properties must also be considered:

1. *Gas solubility.* A high gas solubility is desired since this increases the absorption rate and minimizes the quantity of solvent necessary. Generally, solvents of a chemical nature similar to that of the solute to be absorbed will provide good solubility.
2. *Volatility.* A low solvent vapor pressure is desired since the gas leaving the absorption unit is ordinarily saturated with the solvent and much may thereby be lost.
3. *Corrosiveness.*
4. *Cost.*
5. *Viscosity.* Low viscosity is preferred for reasons of rapid absorption rates, improved flooding characteristics, lower pressure drops, and good heat-transfer characteristics.
6. *Chemical stability.* The solvent should be chemically stable and, if possible, nonflammable.
7. *Toxicity.*
8. *Low freezing point.* If possible, a low freezing point is favored since any solidification of the solvent in the column makes the column inoperable.

Equipment The principal types of gas-absorption equipment may be classified as follows:

1. Packed columns (continuous operation)
2. Plate columns (staged operation)
3. Miscellaneous

Of the three categories, the packed column is by far the most commonly used for the absorption of gaseous pollutants. Additional information may be found by referring to the appropriate sections in this *Handbook* and the many excellent texts available, e.g., McCabe and Smith, *Unit Operations of Chemical Engineering*, 3d ed., McGraw-Hill, New York, 1976; Sherwood and Pigford, *Absorption and Extraction*, 2d ed., McGraw-Hill, New York, 1952; Smith, *Design of Equilibrium Stage Processes*, McGraw-Hill, New York, 1963; and Treybal, *Mass Transfer Operations*, 3d ed., McGraw-Hill, New York, 1980.

Adsorption

Introduction The design of gas-adsorption equipment is in many ways analogous to the design of gas-absorption equipment, with a solid adsorbent replacing the liquid solvent (see Secs. 16 and 19). Similarity is evident in the material- and energy-balance equations as well as in the methods employed to determine the column height. The final choice, as one would expect, rests with the overall process economics.

Selection of Adsorbent Industrial adsorbents are usually capable of adsorbing both organic and inorganic gases or vapors. However, their preferential adsorption characteristics and other physical properties make each of them more or less specific for a particular application. General experience has shown that, for the adsorption of vapors of an organic nature, activated carbon has superior properties, having hydrocarbon-selective properties and high adsorption capacity for such materials. Inorganic adsorbents, such as activated alumina or silica gel, can also be used to adsorb organic materials, but difficulties can arise during regeneration. Activated alumina, silica gel, and molecular sieves will also preferentially adsorb any water vapor with the organic contaminant. At times this may be a considerable drawback in the application of these adsorbents for organic-contaminant removal.

The normal method of regeneration of adsorbents is by use of steam, inert gas (i.e., nitrogen), or other gas streams, and in the majority of cases this can cause at least slight decomposition of the organic compound on the adsorbent. Two difficulties arise: (1) incomplete recovery of the adsorbate, although this may be unimportant; and (2) progressive deterioration in capacity of the adsorbent as the number of cycles increases owing to blocking of the pores from carbon formed by hydrocarbon decomposition. With activated carbon a steaming process is used, and the difficulties of regeneration are thereby overcome. This is not feasible with silica gel or activated alumina because of the risk of breakdown of these materials when in contact with liquid water.

In some cases, none of the adsorbents has sufficient retaining capacity for a particular contaminant. In these applications, a large-surface-area adsorbent can be impregnated with an inorganic compound or, in rare cases, with a high-molecular-weight organic compound which can react chemically with the particular contaminant. For example, iodine-impregnated carbons are used for removal of mercury vapor and bromine-impregnated carbons for ethylene or propylene removal. The action of these impregnants is either catalytic conversion or reaction to a nonobjectionable compound or to a more easily adsorbed compound. For this case, general adsorption theory no longer applies to the overall effects of the process. For example, mercury removal by an iodine-impregnated carbon proceeds faster at a higher temperature, and a better overall efficiency can be obtained than in a low-temperature system.

Since adsorption takes place at the interphase boundary, the adsorbent surface area becomes an important consideration. Generally, the higher the adsorption surface area, the greater its adsorption capacity. However, the surface area has to be "available" in a particular pore size within the adsorbent. At low partial pressure (or concentration) a surface area in the smallest pores in which the adsorbate can enter is the most efficient. At higher pressures the larger pores become more important, while at very high concentrations capillary condensation will take place within the pores, and the total micropore volume becomes the limiting factor.

The action of molecular sieves is slightly different from that of other adsorbents in that selectivity is determined more by the pore-size limitations of the particular molecular sieve. In selecting molecular sieves, it is important that the contaminant to be removed be smaller than the available pore size. Hence, it is important that the particular adsorbent not only have an affinity for the contaminant in question but also have sufficient surface available for adsorption.

Design Data The adsorbent having been selected, the next step is to calculate the quantity of adsorbent required and eventually to consider other factors such as the temperature rise of the gas stream due to adsorption and the useful life of the adsorbent under operating conditions. The sizing and overall design of the adsorption system depend on the properties and characteristics of both the feed gas to be treated and the adsorbent. The following information should be known or available for design purposes:

1. Gas stream.
a. Adsorbate concentration.
b. Temperature.
c. Temperature rise during adsorption.
d. Pressure.
e. Flow rate.
f. Presence of adsorbent contaminant material.
2. Adsorbent.
a. Adsorption capacity as used on stream.
b. Temperature rise during adsorption.
c. Isothermal or adiabatic operation.
d. Life, if presence of contaminant material is unavoidable.
e. Possibility of catalytic effects causing an adverse chemical reaction in the gas stream or the formation of solid polymerizates on the adsorbent bed, with consequent deterioration.
f. Bulk density.
g. Particle size, usually reported as a mean equivalent particle diameter. The dimensions and shape of particles affect both the pressure drops through the adsorbent bed and the diffusion rate into the particles. All things being equal, adsorbent beds consisting of smaller particles, although causing a higher pressure drop, will be more efficient.
h. Pore data, which are important because they may permit elimination from consideration of adsorbents whose pore diameter will not admit the desired adsorbate molecule.
i. Hardness, which indicates the care that must be taken in handling adsorbents to prevent the formation of undesirable fines.
j. Regeneration information.

The design techniques used include both stagewise and continuous-contacting methods and can be applied to batch, continuous, and semicontinuous operations.

Adsorption Phenomena The adsorption process involves three necessary steps. The fluid must first come in contact with the adsorbent, at which time the adsorbate is preferentially or selectively adsorbed on the adsorbent. Next, the unadsorbed fluid must be separated from the adsorbent-adsorbate, and, finally, the adsorbent must be regenerated by removing the adsorbate or by discarding used adsorbent and replacing it with fresh material. Regeneration is performed in a variety of ways, depending on the nature of the adsorbate. Gases or vapors are usually desorbed by either raising the temperature (thermal cycle) or reducing the pressure (pressure cycle). The more popular thermal cycle is accomplished by passing hot gas through the adsorption bed in the direction opposite to the flow during the adsorption cycle. This ensures that the gas passing through the unit during the adsorption cycle always meets the most active adsorbent last and that the adsorbate concentration in the adsorbent at the outlet end of the unit is always maintained at a minimum.

In the first step, in which the molecules of the fluid come in contact with the adsorbent, an equilibrium is established between the adsorbed fluid and the fluid remaining in the fluid phase. Figures 26-5 through 26-7 show several experimental equilibrium adsorption isotherms for a number of components adsorbed on various adsorbents. Consider Fig. 26-5, in which the concentration of adsorbed gas on the solid is plotted against the equilibrium partial pressure $p°$ of the vapor or gas at constant temperature. At 40°C, for example, pure propane vapor at a pressure of 550 mmHg is in equilibrium with an adsorbate concentration at point P of 0.04-lb adsorbed propane per pound of silica gel. Increasing the pressure of the propane will cause

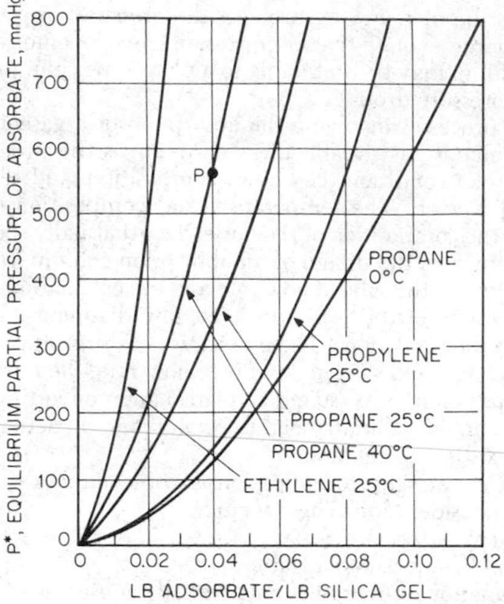

FIG. 26-5 Equilibrium partial pressures for certain organics on silica gel.

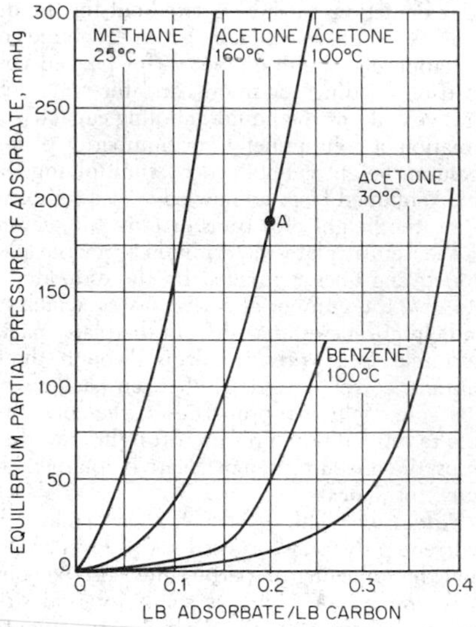

FIG. 26-6 Equilibrium partial pressures for certain organics on carbon.

more propane to be adsorbed, while decreasing the pressure of the system at P will cause propane to be desorbed from the carbon.

The adsorptive capacity of activated carbon for some common solvent vapors is shown in Table 26-21.

Adsorption-Control Equipment If a gas stream must be treated for a short period, usually only one adsorption unit is necessary, provided, of course, that a sufficient time interval is available between adsorption cycles to permit regeneration. However, this is usually not the case. Since an uninterrupted flow of treated gas is often required, it is necessary to employ one or more units capable of operating in this fashion. The units are designed to handle gas flows without interruption and are characterized by their mode of contact, staged or continuous. By far the most common type of adsorption system used to remove an objectionable pollutant from a gas stream consists of a number of fixed-bed units operating in such a sequence that the gas flow remains uninterrupted. A two- or three-bed system is usually employed, with one or two beds bypassed for regeneration while one is adsorbing. A typical two-bed system is shown in Fig. 26-8, while a

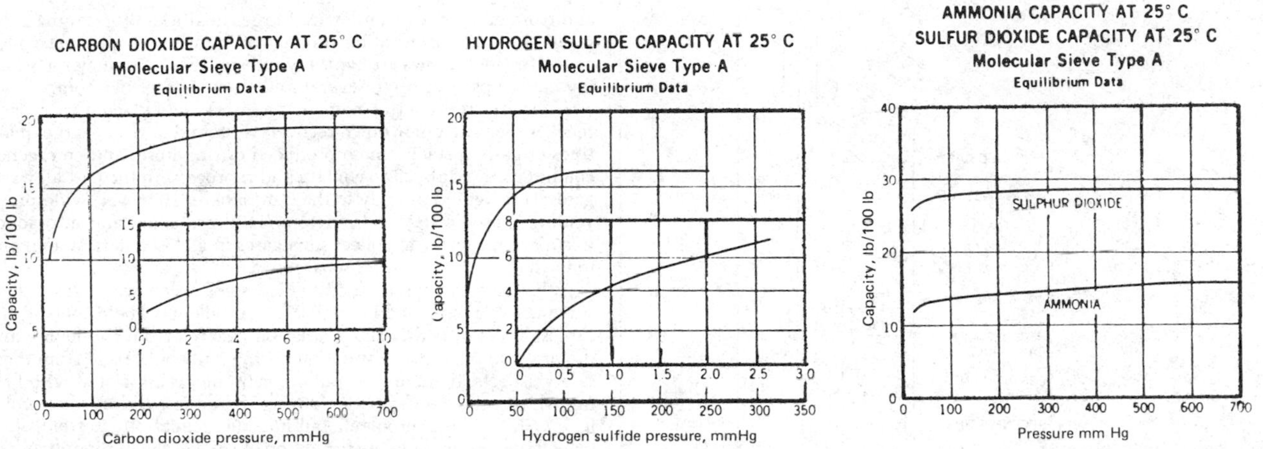

FIG. 26-7 Equilibrium partial pressures for certain gases on molecular sieves. (*A. J. Buonicore and L. Theodore,* Industrial Control Equipment for Gaseous Pollutants, *vol. I, CRC Press, Boca Raton, Fla., 1975.*)

TABLE 26-21 Adsorptive Capacity of Common Solvents on Activated Carbons*

Solvent	Carbon bed weight, %†
Acetone	8
Heptane	6
Isopropyl alcohol	8
Methylene chloride	10
Perchloroethylene	20
Stoddard solvent	2–7
1,1,1-Trichloroethane	12
Trichloroethylene	15
Trichlorotrifluoroethane	8
VM&P naphtha	7

*Assuming steam desorption at 5 to 10 psig.
†For example, 8 lb of acetone adsorbed on 100 lb of activated carbon.

typical three-bed system is shown in Fig. 26-9. The type of system best suited for a particular job is determined from several factors, including the amount and rate of material being adsorbed, the time between cycles, the time required for regeneration, and the cooling time, if required.

Typical of continuous-contact operation for gaseous-pollutant adsorption is the use of a fluidized bed. During steady-state staged-contact operation, the gas flows up through a series of successive fluidized-bed stages, permitting maximum gas-solid contact on each stage. A typical arrangement of this type is shown in Fig. 26-10 for multistage countercurrent adsorption with regeneration. In the upper part of the tower, the solids are contacted countercurrently on perforated trays in relatively shallow beds with the gas stream containing the pollutant, the adsorbent solids moving from tray to tray through downspouts. In the lower part of the tower, the adsorbent is regenerated by similar contact with hot gas, which desorbs and carries off the pollutant. The regenerated adsorbent is then recirculated by an air lift to the top of the tower.

Although the continuous-countercurrent type of operation has found limited application in the removal of gaseous pollutants from process streams (for example, the removal of carbon dioxide and sulfur compounds such as hydrogen sulfide and carbonyl sulfide), by far the most common type of operation presently in use is the fixed-bed adsorber. The relatively high cost of continuously transporting solid particles as required in steady-state operations makes fixed-bed adsorption an attractive, economical alternative. If intermittent or batch operation is practical, a simple one-bed system, cycling alternately between the adsorption and regeneration phases, will suffice.

Additional information may be found by referring to the appro-

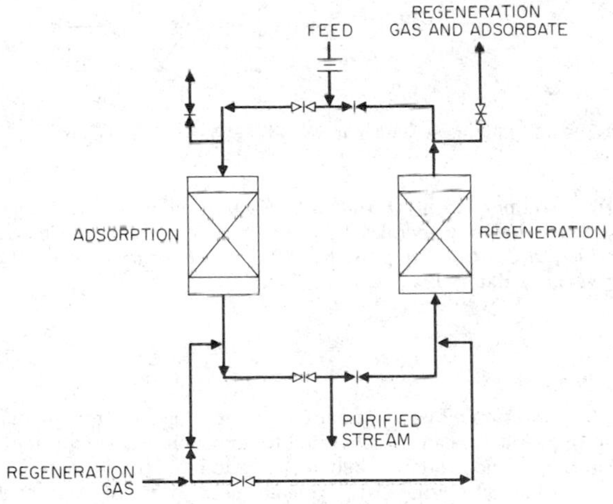

FIG. 26-8 Typical two-bed adsorption system.

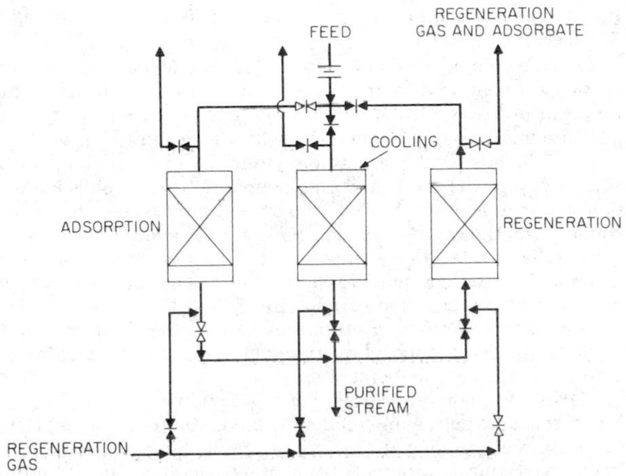

FIG. 26-9 Typical three-bed adsorption system.

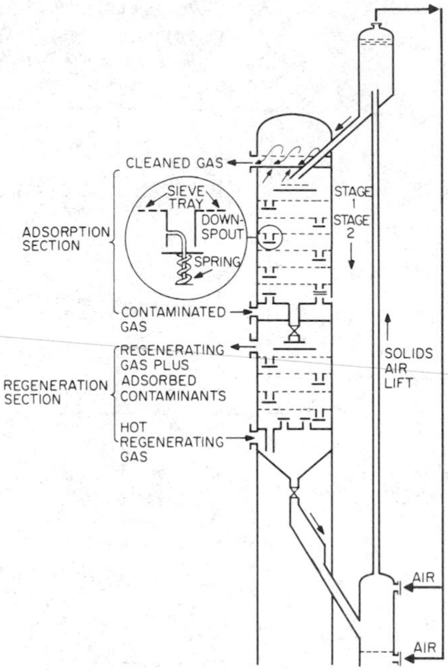

FIG. 29-10 Multistage countercurrent adsorption with regeneration.

priate sections of this *Handbook*. A comprehensive treatment of adsorber design principles is given in Buonicore and Theodore, *Industrial Control Equipment for Gaseous Pollutants*, vol. 1, CRC Press, Boca Raton, Fla., 1975.

Combustion

Introduction Many organic compounds released from manufacturing operations can be converted to innocuous carbon dioxide and water by rapid oxidation (chemical reaction): combustion. Three rapid oxidation methods are typically used to destroy combustible contaminants: (1) flares (direct-flame combustion), (2) thermal combustors, and (3) catalytic combustors. The thermal and flare methods are characterized by the presence of a flame during combustion. Catalytic combustors utilize a metallic catalyst to promote rapid oxidation and are characterized by flameless-type combustion. The combustion process is also commonly referred to as "afterburning" or "incineration."

To achieve complete combustion, i.e., the combination of the combustible elements and compounds of a fuel with all the oxygen which they can utilize, sufficient space, time, and turbulence and a temperature high enough to ignite the constituents must be provided.

The "three T's" of combustion—time, temperature, and turbulence—govern the speed and completeness of the combustion reaction. For complete combustion, the oxygen must come into intimate contact with the combustible molecule at sufficient temperature and for a sufficient length of time for the reaction to be completed. Incomplete reactions may result in the generation of aldehydes, organic acids, carbon, and carbon monoxide.

Combustion-Control Equipment Combustion-control equipment can be divided into three types: (1) flares, (2) thermal incinerators, and (3) catalytic incinerators.

Flares In many industrial operations and particularly in chemical plants and petroleum refineries, large volumes of combustible waste gases are produced. These gases result from undetected leaks in the operating equipment, from upset conditions in the normal operation of a plant in which gases must be vented to avoid dangerously high pressures in operating equipment, from plant startups, and from emergency shutdowns. Large quantities of gases may also result from off-specification product or from excess product which cannot be sold. Flows are typically intermittent, with flow rates during major upsets of up to several million cubic feet per hour.

The preferred control method for excess gases and vapors is to recover them in a blowdown recovery system. However, large quantities of gas, especially those produced during upset and emergency conditions, are difficult to contain and reprocess. In the past all waste gases were vented directly to the atmosphere. However, widespread venting caused safety and environmental problems, and in practice it is now customary to collect such gases in a closed flare system and to burn them as they are discharged.

Although flares can be used to dispose of excess waste gases, such systems can present additional safety problems. These include the explosion potential, thermal-radiation hazards from the flame, and the problem of toxic asphyxiation during flameout. Aside from these safety aspects, there are several other problems associated with flaring which must be dealt with during the design and operation of a flare system. These problems fall into the general area of emissions from flares and include the formation of smoke, the luminosity of the flame, noise during flaring, and the possible emission of by-product air pollutants during flaring.

The heat content of the waste stream to be disposed is another important consideration. The heat content of the waste gas falls into two classes. The gases can either support their own combustion or not. In general, a waste gas with a heating value greater than 7443 kJ/m^3 (200 Btu/ft^3) can be flared successfully. The heating value is based on the lower heating value of the waste gas at the flare. Below 7443 kJ/m^3, enriching the waste gas by injecting another gas with a higher heating value may be necessary. The addition of such a rich gas is called "endothermic flaring." Gases with a heating value as low as 2233 kJ/m^3 (60 Btu/ft^3) have been flared but at a significant fuel demand. It is usually not feasible to flare a gas with a heating value below 3721 kJ/m^3 (100 Btu/ft^3). If the flow of low-Btu gas is continuous, thermal or catalytic incineration can be used to dispose of the gas. For intermittent flows, however, endothermic flaring may be the only possibility.

Although most flares are used to dispose of intermittent waste gases, some continuous flares are in use, but generally only for relatively small volumes of gases. The heating value of large-volume continuous-flow waste gases is usually too valuable to lose in a flare. Vapor recovery or the use of the vapor as fuel in a process heater is preferred over flaring. Since auxiliary fuel must be added to the gas in order to flare, large continuous flows of a low-heating-value gas are usually more efficient to burn in a thermal incinerator than in the flame of a flare.

Flares are mostly used for the disposal of hydrocarbons. Waste gases composed of natural gas, propane, ethylene, propylene, butadiene, and butane probably constitute over 95 percent of the material flared. Flares have been used successfully to control malodorous gases such as mercaptans and amines, but care must be taken when flaring these gases. Unless the flare is very efficient and gives good combustion, obnoxious fumes can escape unburned and cause a nuisance.

Flaring of hydrogen sulfide should be avoided because of its toxicity and low odor threshold. In addition, burning relatively small amounts of hydrogen sulfide can create enough sulfur dioxide to cause crop damage or a local nuisance. For gases whose combustion products may cause problems, such as those containing hydrogen sulfide or chlorinated hydrocarbons, flaring is not recommended.

Thermal Incinerators Thermal incinerators or afterburners can be used over a fairly wide but low range of organic vapor concentration. The concentration of the organics in air must be substantially below the lower flammable level (lower explosive limit). As a rule, a factor of 4 is employed for safety precautions. Reactions are conducted at elevated temperatures to ensure high chemical-reaction rates for the organics. To achieve this temperature, it is necessary to preheat the feed stream with auxiliary energy. Along with the contaminant-laden gas stream, air and fuel are continuously delivered to the incinerator (see Fig. 26-11), where the fuel is combusted with air in a firing unit (burner). The burner may utilize the air in the

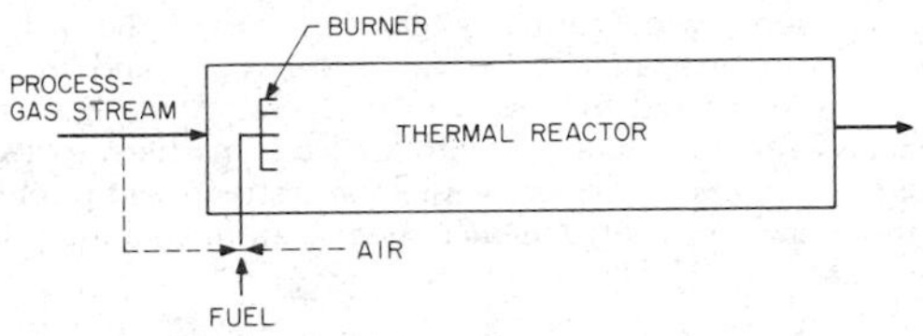

FIG. 26-11 Thermal-combustion device.

process-waste stream as the combustion air for the auxiliary fuel, or it may use a separate source of outside air. The products of combustion and the unreacted feed stream are intensively mixed and enter the reaction zone of the unit. The pollutants in the process-gas stream are then reacted at the elevated temperature. Thermal incinerators generally require operating temperatures in the range of 650 to 820°C (1200 to 1500°F) for combustion of most organic pollutants (see Table 26-22). A residence time of 0.2 to 1.0 is often recommended, but this factor is dictated primarily by kinetic considerations. The end products are continuously discharged at the outlet of the reactor. The averaged gas velocity can range from as low as 3 m/s (10 ft/s) to as high as 15 m/s (50 ft/s). These high velocities are required to prevent settling of particulates (if present) and to minimize the dangers of flashback and fire hazards.

The fuel is usually natural gas. The energy liberated by reaction may be directly recovered in the process or indirectly recovered by suitable external heat exchange (see Fig. 26-12).

Because of the high operating temperatures, the unit must be constructed of metals capable of withstanding this condition. Combustion devices are usually constructed with an outer steel shell that is lined with refractory material. Refractory-wall thickness is usually in the 0.05- to 0.23-m (2- to 9-in) range, depending upon temperature considerations.

Some of the advantages of the thermal incinerators are:

1. Removal of organic gases
2. Removal of submicrometer organic particles
3. Simplicity of construction
4. Small space requirements
5. Low maintenance costs

Some of the disadvantages are:

1. High operating costs
2. Fire hazards
3. Flashback possibilities

Catalytic Incinerators Catalytic incinerators are an alternative to thermal incinerators. For simple reactions, the effect of the presence of a catalyst is to (1) increase the rate of reaction, (2) permit the reaction to occur at a lower temperature, and (3) reduce the reactor volume.

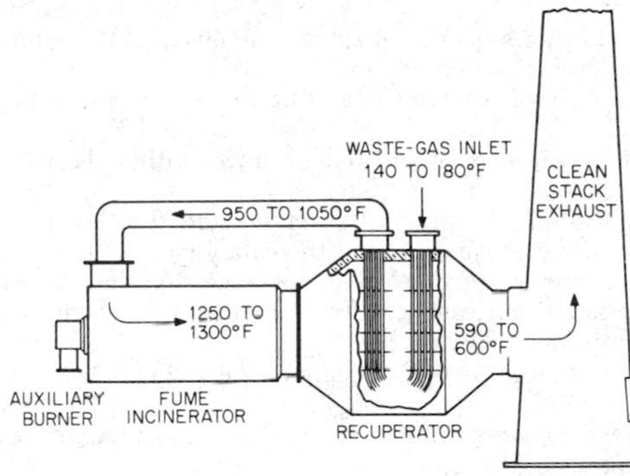

FIG. 26-12 Thermal combustion with energy (heat) recovery.

In a typical catalytic incinerator for the combustion of organic vapors, the gas stream is delivered to the reactor continuously by a fan at a velocity in the range of 3 to 15 m/s (10 to 30 ft/s), but at a lower temperature, usually in the range of 350 to 425°C (650 to 800°F), than the thermal unit. The gases, which may or may not be preheated, pass through the catalyst bed, where the combustion reaction occurs. The combustion products, which again are made up of water vapor, carbon dioxide, inerts, and unreacted vapors, are continuously discharged from the outlet at a higher temperature. Energy savings can again be effected by heat recovery from the exit stream.

Metals in the platinum family are recognized for their ability to promote combustion at low temperatures. Other catalysts include various oxides of copper, chromium, vanadium, nickel, and cobalt. These catalysts are subject to poisoning, particularly from halogens, halogen and sulfur compounds, zinc, arsenic, lead, mercury, and particulates. It is therefore important that catalyst surfaces be clean and active to ensure optimum performance.

Catalysts may be porous pellets, usually cylindrical or spherical in shape, ranging from 0.16 to 1.27 cm (1/16 to 1/2 in) in diameter. Small sizes are recommended, but the pressure drop through the reactor increases. Among other shapes are honeycombs, ribbons, and wire mesh. Since catalysis is a surface phenomenon, a physical property of these particles is that the internal pore surface is nearly infinitely greater than the outside surface.

TABLE 26-22 Thermal Afterburners: Conditions Required for Satisfactory Performance in Various Abatement Applications

Abatement category	Afterburner residence time, s	Temperature, °F
Hydrocarbon emissions: 90 + % destruction of HC	0.3–0.5	1100–1250*
Hydrocarbons + CO: 90 + % destruction of HC + CO	0.3–0.5	1250–1500
Odor		
50–90% destruction	0.3–0.5	1000–1200
90–99% destruction	0.3–0.5	1100–1300
99 + % destruction	0.3–0.5	1200–1500
Smokes and plumes		
White smoke (liquid mist)		
Plume abatement	0.3–0.5	800–1000†
90 + % destruction of HC + CO	0.3–0.5	1250–1500
Black smoke (soot and combustible particulates)	0.7–1.0	1400–2000

*Temperatures of 1400 to 1500°F (760 to 816°C) may be required if the hydrocarbon has a significant content of any of the following: methane, cellosolve, and substituted aromatics (e.g., toluene and xylenes).

†Operation for plume abatement only is not recommended, since this merely converts a visible hydrocarbon emission into an invisible one and frequently creates a new odor problem because of partial oxidation in the afterburner.

The following sequence of steps is involved in the catalytic conversion of reactants to products:

1. Transfer of reactants to and products from the outer catalyst surface
2. Diffusion of reactants and products within the pores of the catalyst
3. Activated adsorption of reactants and the desorption of the products on the active centers of the catalyst
4. Reaction or reactions on active centers on the catalyst surface

At the same time, energy effects arising from chemical reaction can result in the following:

1. Heat transfer to or from active centers to the catalyst-particle surface
2. Heat transfer to and from reactants and products within the catalyst particle
3. Heat transfer to and from moving streams in the reactor
4. Heat transfer from one catalyst particle to another within the reactor
5. Heat transfer to or from the walls of the reactor

Some of the advantages of catalytic incinerators are:

1. Lower fuel requirements as compared with thermal incinerators
2. Lower operating temperatures
3. Minimum insulation requirements
4. Reduced fire hazards
5. Reduced flashback problems

The disadvantages include:

1. Higher initial cost than thermal incinerators
2. Catalyst poisoning
3. Necessity of first removing large particulates
4. Catalyst-regeneration problems

Condensation

Introduction Frequently in air-pollution-control practice it becomes necessary to treat an effluent stream consisting of a condensable pollutant vapor and a noncondensable gas. One control method to remove such pollutants from process-gas streams that is often overlooked is condensation. Condensers can be used to collect condensable emissions discharged to the atmosphere, particularly when the vapor concentration is high. This is usually accomplished by lowering the temperature of the gaseous stream, although an increase in pressure will produce the same result. The former approach is usually employed by industry, since pressure changes (even small ones) on large volumetric gas flow rates are often economically prohibitive.

Condensation Equipment There are two basic types of condensers used for control: contact and surface. In contact condensers, the gaseous stream is brought into direct contact with a cooling medium so that the vapors condense and mix with the coolant (Fig. 26-13). The more widely used system, however, is the surface condenser (or heat exchanger), in which the vapor and the cooling medium are separated by a wall (Fig. 26-14). Since high removal efficiencies cannot be obtained with low condensable-vapor concentrations, condensers are typically used for pretreatment prior to some other more efficient control device such as an incinerator, absorber, or adsorber.

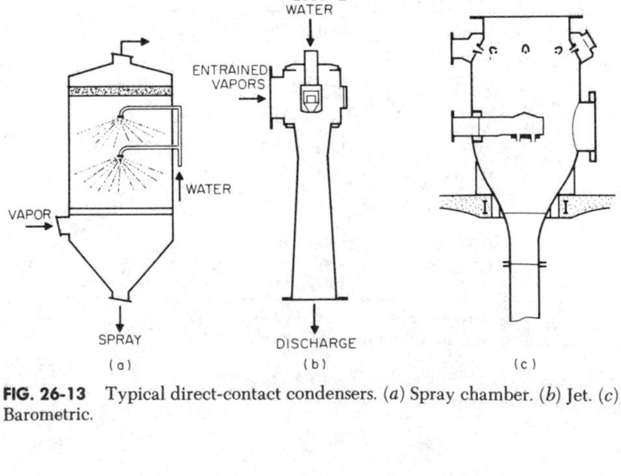

FIG. 26-13 Typical direct-contact condensers. (*a*) Spray chamber. (*b*) Jet. (*c*) Barometric.

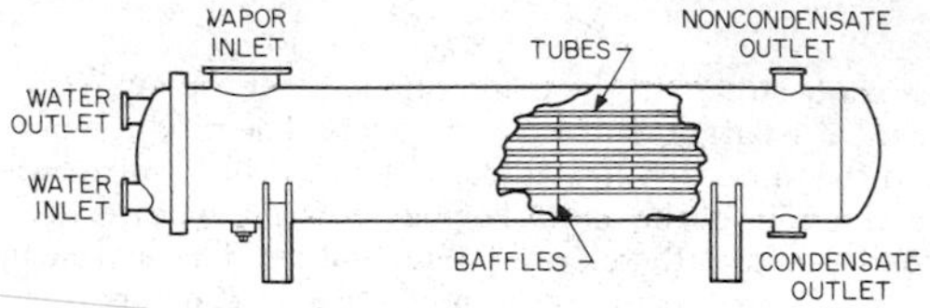

FIG. 26-14 Typical surface condenser (shell-and-tube).

Contact Condensers Spray condensers, jet condensers, and barometric condensers all utilize water or some other liquid in direct contact with the vapor to be condensed. The temperature approach between the liquid and the vapor is very small, so the efficiency of the condenser is high, but large volumes of the liquid are necessary. If the vapor is soluble in the liquid, the system is essentially an absorptive one. If the vapor is not soluble, the system is a true condenser, in which the temperature of the vapor must be below the dew point. Direct-contact condensers are seldom used for the removal of organic solvent vapors because the condensate will contain an organic-water mixture which must be separated or treated before disposal. They are, however, the most effective method of removing heat from hot gas streams when the recovery of organics is not a consideration.

In a direct-contact condenser a stream of water or other cooling liquid is brought into direct contact with the vapor to be condensed. The liquid stream leaving the chamber contains the original cooling liquid plus the condensed substances. The gaseous stream leaving the chamber contains the noncondensable gases and such condensable vapors as did not condense; it is reasonable to assume that the vapors in the leaving gas stream are saturated. It is then the temperature of the leaving gas stream which determines the collection efficiency of the condenser.

The advantages of contact condensers are that (1) they can be used to produce a vacuum, thereby creating a draft to remove odorous vapors and also reduce boiling points in cookers and vats; (2) they usually are simpler and less expensive than the surface type; and (3) they usually have considerable odor-removing capacity because of the greater condensate dilution (13 lb of 60°F water is required to condense 1 lb of steam at 212°F and cool the condensate to 140°F). The principal disadvantage is the large water requirement. Depending on the nature of the condensate, odor in the wastewater can sometimes be offset by using treatment chemicals.

Direct-contact condensers involve the simultaneous transfer of heat and mass. Design procedures available for absorption, humidification, cooling towers, etc., may be applied with some modifications.

Surface Condensers Surface condensers (indirect-contact condensers) are used extensively in the chemical-process industry. They are employed in the air-pollution-equipment industry for recovery, control, and/or removal of trace impurities or contaminants. In the surface type, coolant does not contact the vapor condensate. There are various types of surface condensers including the shell-and-tube, fin-fan, finned-hairpin, finned-tube-section, and tubular. The use of surface condensers has several advantages. Salable condensate can be recovered. If water is used for coolant, it can be reused, or the condenser may be air-cooled when water is not available. Also, surface condensers require less water and produce 10 to 20 times less condensate. Their disadvantage is that they are usually more expensive and require more maintenance than the contact type.

SOURCE CONTROL OF PARTICULATE EMISSIONS

There are four conventional types of equipment used for the control of particulate emissions:

1. Mechanical collectors
2. Wet scrubbers
3. Electrostatic precipitators
4. Fabric filters

Each is discussed in Sec. 20 of this *Handbook* under "Gas-Solids Separations." The effectiveness of conventional air-pollution-control equipment for particulate removal is compared in Fig. 26-15. These

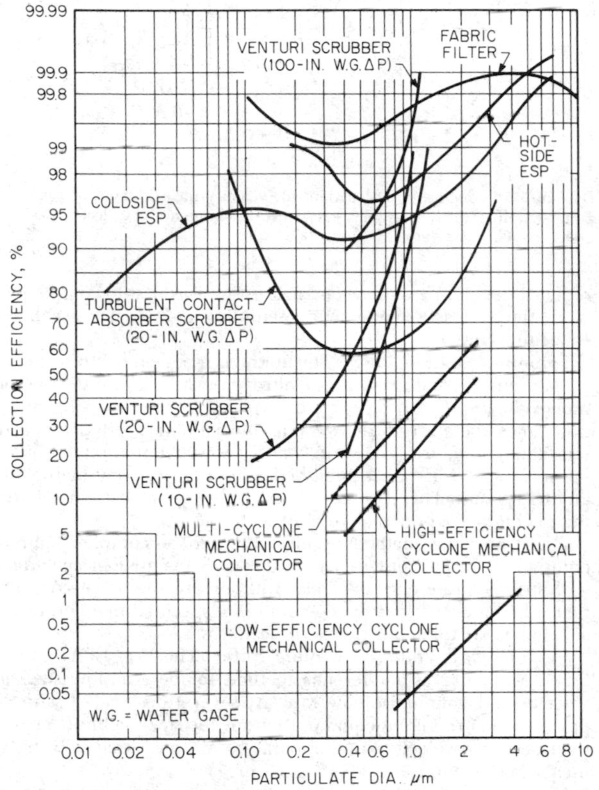

FIG. 26-15 Fractional efficiency curves for conventional air-pollution-control devices. [Chem. Eng., *87(13), 83 (June 30, 1980).*]

fractional efficiency curves indicate that the equipment is least efficient in removing particulates in the 0.1- to 1.0-μm range. For wet scrubbers and fabric filters, the very small particulates (0.1 μm) can be efficiently removed by brownian diffusion. The smaller the particulates, the more intense their brownian motion and the easier their collection by diffusion forces. Larger particulates (> 1 μm) are collected principally by impaction, and removal efficiency increases with particulate size. The minimum in the fractional efficiency curve for scrubbers and filters occurs in the transition range between removal by brownian motion and removal by impaction.

A somewhat similar situation exists for electrostatic precipitators. Particulates larger than about 1 μm have high mobilities because they are highly charged. Those smaller than a few tenths of a micrometer can achieve moderate mobilities with even a small charge because of aerodynamic slip. A minimum in collection efficiency usually occurs in the transition range between 0.1 and 1.0 μm. The situation is further complicated because not all particulates smaller than about 0.1 μm acquire charges in an ion field. Hence the efficiency of removal of very small particulates decreases after reaching a maximum in the submicrometer range.

The general trend in industry is toward the use, where possible, of baghouses for particulate-emission control. Baghouses provide extremely high collection efficiency, are dry collection systems, and are relatively easy to operate and maintain—a key to success in control-equipment design. An additional attraction is that the cleaned gas stream exhausted from the baghouse can sometimes be returned to the plant, reducing makeup-air and heating requirements.

Improvements in existing control technology for fine particulates and the development of advanced techniques are top-priority research goals. Conventional control devices have certain limitations. Precipitators, for example, are limited by the magnitude of charge on the particulate, by the electric field, and by dust reentrainment. Also, the resistivity of the particulate material may adversely affect both charge and electric field. Advances are needed to overcome resistivity and to extend the performance of precipitators not limited by resistivity (see Buonicore, Reynolds, and Theodore, "Control Technology for Fine Particulate Emissions," DOE Rep. ANL/ECT-5, Argonne National Laboratory, Argonne, Ill., October 1978). Recent design developments with the potential to improve precipitator performance include pulse energization, electron-beam ionization, wide plate spacing, and precharged units [Balakrishnan et al., "Emerging Technologies for Air Pollution Control," *Pollut. Eng.*, **11**, 28–32 (November 1979); "Pulse Energization," *Environ. Sci. Technol.*, **13**(9), 1044 (1974); and Midkaff, "Change in Precipitator Design Expected to Help Plants Meet Clean Air Laws," *Power*, **126** (10), 79 (1979)].

Fabric filters are limited by physical size and bag-life considerations. Some sacrifices in efficiency might be tolerated if higher air-cloth ratios could be achieved without reducing bag life (improved pulse-jet systems). Improvements in fabric filtration may also be possible by enhancing electrostatic effects which may contribute to rapid formation of a filter cake after cleaning.

Scrubber technology is limited by scaling and fouling, overall reliability, and energy consumption. The use of supplementary forces acting on particulates to cause them to grow or otherwise be more easily collected at lower pressure drops is being closely investigated. The development of electrostatic and flux-force–condensation scrubbers is a step in this direction.

The electrostatic effect can be incorporated into wet scrubbing by charging the particulates and/or the scrubbing-liquor droplets. Electrostatic scrubbers may be capable of achieving the same efficiency for fine-particulate removal as is achieved by high-energy scrubbers, but at substantially lower power input. The major drawbacks are increased maintenance of electrical equipment and higher capital cost.

Flux-force–condensation scrubbers combine the effects of flux force (diffusiophoresis and thermophoresis) and water-vapor condensation to contact hot, humid gas with subcooled liquid and/or inject steam into saturated gas.

Research has demonstrated that a number of these novel devices can remove fine particulates (see Fig. 26-16). Although limited in terms of commercialization, these systems may find application in many industries.

EMISSIONS MEASUREMENT

Introduction An accurate quantitative analysis of the discharge of pollutants from a process must be determined prior to the design and/or selection of control equipment. If the unit is properly engineered by utilizing the emission data as input to the control device and the code requirements as maximum-effluent limitations, most pollutants can be successfully controlled.

Sampling is the keystone of source analysis. Sampling methods and tools vary in their complexity according to the specific task; therefore, a degree of both technical knowledge and common sense is needed to design a sampling function. Sampling is done to measure quantities or concentrations of pollutants in effluent-gas streams, to measure the efficiency of a pollution-abatement device, to guide the designer of pollution-control equipment and facilities, and/or to appraise contamination from a process or a source. A complete measurement requires determination of the concentration and contami-

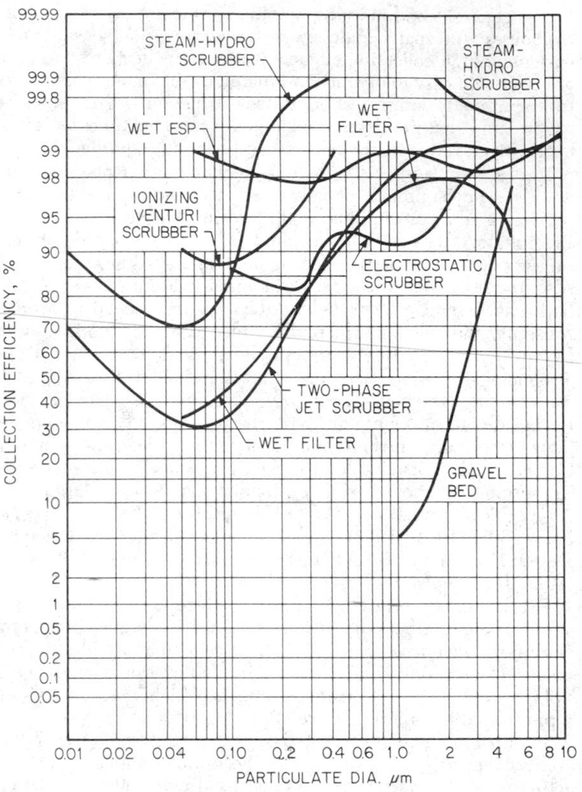

FIG. 26-16 Fractional efficiency curves for novel air-pollution-control devices. [Chem. Eng., *87(13), 85 (June 30, 1980).*]

nant characteristics as well as the associated gas flow. Most statutory limitations require mass rates of emission; both concentration and volumetric-flow-rate data are therefore required.

The selection of a sampling site and the number of sampling points required are based on attempts to get representative samples. To accomplish this, the sampling site should be at least eight stack or duct diameters downstream and two diameters upstream from any bend, expansion, contraction, valve, fitting, or visible flame.

Once the sampling location has been decided on, the flue cross section is laid out in a number of equal areas, the center of each being the point where the measurement is to be taken. For rectangular stacks, the cross section is divided into equal areas of the same shape, and the traverse points are located at the center of each equal area, as shown in Fig. 26-17*a*. The ratio of length to width of each elemental area should be selected. For circular stacks, the cross section is divided into equal annular areas, and the traverse points are located at the centroid of each area. The location of the traverse points as a percentage of stack diameter from the inside wall to the traverse point for circular-stack sampling is given in Table 26-23. The number of traverse points necessary on each of two perpendiculars for a particular stack may be estimated from Fig. 26-18.

Once these traverse points have been determined, velocity measurements are made to determine gas flow. The stack-gas velocity is usually determined by means of a pitot tube and differential-pressure gauge. When velocities are very low [less than 3 m/s (10 ft/s)] and when great accuracy is not required, an anemometer may be used. For gases moving in small pipes at relatively high velocities or pressures, orifice-disk meters or venturi meters may be used. These are valuable as continuous or permanent measuring devices.

Once a flow profile has been established, sampling strategy can be considered. Since sampling collection can be simplified and greatly

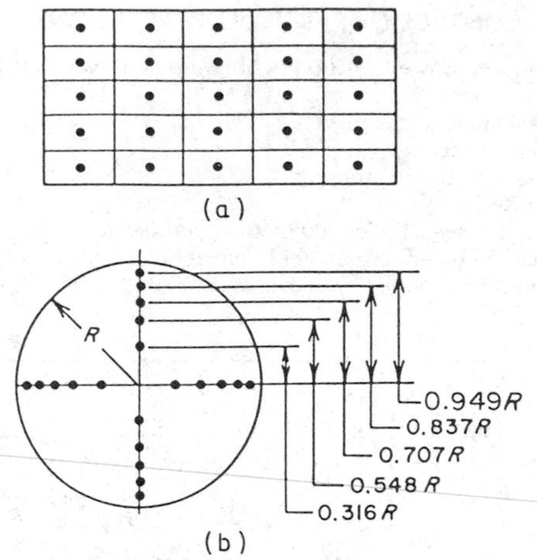

FIG. 26-17 Traverse-point locations for velocity measurement or for multipoint sampling. (*a*) Rectangular stack (measure at center of at least nine equal areas). (*b*) Circular stack (20-point traverse).

reduced depending on flow characteristics, it is best to complete the flow-profile measurement before sampling or measuring pollutant concentrations.

Sampling Methodology The methods specified for the sampling of particulates, sulfur dioxide, nitrogen oxides, fluorides, carbon monoxide, and hydrocarbons are reviewed in the following subsections since these are the most common pollutants tested for. In all sampling procedures, the main concern is to obtain a representative sample; the U.S. EPA has published reference sampling methods for all these pollutants so that uniform procedures can be applied in testing to obtain a representative sample.

Each sampling method requires the use of complex sampling equipment which must be calibrated and operated in accordance with specified reference methods. Additionally, the process or source that is being tested must be operated in a specified manner, usually at rated capacity, under normal procedures.

Velocity and Volumetric Flow Rate The U.S. EPA has published Method 2 as a reference method for determining stack-gas velocity and volumetric flow rate. At several designated sampling points, which represent equal portions of the stack volume (areas in the stack), the velocity and temperature are measured with instrumentation shown in Fig. 26-19.

Measurements to determine volumetric flow rate usually require approximately 30 min. Since sampling rates depend on stack-gas velocity, a preliminary velocity check is usually made prior to testing for pollutants to aid in selecting the proper equipment and in determining the approximate sampling rate for the test.

The volumetric flow rate determined by this method is usually within ±10 percent of the true volumetric flow rate.

Molecular Weight and CO_2 or O_2 EPA Method 3 is used to determine carbon dioxide or oxygen content and molecular weight of the stack-gas stream. Depending on the intended use of the data, these values also can be obtained with an integrated sample (see Fig. 26-20) or a grab sample (see Fig. 26-21).

Grab sampling is used primarily to determine the molecular weight of the gas stream. A sampling probe is placed at the center of the stack, and a sample is drawn directly into an Orsat analyzer or a Fyrite-type combustion-gas analyzer. The sample is then analyzed for carbon dioxide and oxygen content. With these data, the dry molecular weight of the gas stream can then be calculated.

Moisture Content EPA Method 4 is the reference method for determining the moisture content of the stack gas. A value for moisture content is needed in some of the calculations for determining pollutant-emission rates.

A sample is taken at several designated points in the stack, which

TABLE 26-23 Traverse-Point Locations in Circular Stacks

Traverse point number on a diameter	Number of traverse points on a diameter									
	6	8	10	12	14	16	18	20	22	24
1	4.4	3.3	2.5	2.1	1.8	1.6	1.4	1.3	1.1	1.1
2	14.7	10.5	8.2	6.7	5.7	4.9	4.4	3.9	3.5	3.2
3	29.5	19.4	14.6	11.8	9.9	8.5	7.5	6.7	6.0	5.5
4	70.5	32.3	22.6	17.7	14.6	12.5	10.9	9.7	8.7	7.9
5	85.3	67.7	34.2	25.0	20.1	16.9	14.6	12.9	11.6	10.5
6	95.6	80.5	65.8	35.5	26.9	22.0	18.8	16.5	14.6	13.2
7		89.5	77.4	64.5	36.6	29.8	23.6	20.4	18.0	16.1
8		96.7	85.4	75.0	63.4	37.5	29.6	25.0	21.8	19.4
9			91.8	82.3	73.1	62.5	38.2	30.6	26.1	23.0
10			97.5	88.2	79.9	71.7	61.8	38.8	31.5	27.2
11				93.3	85.4	78.0	70.4	61.2	39.3	32.3
12				97.9	90.1	83.1	76.4	69.4	60.7	39.8
13					94.3	87.5	81.2	75.0	68.5	60.2
14					98.2	91.5	85.4	79.6	73.9	67.7
15						95.1	89.1	83.5	78.2	72.8
16						98.4	92.5	87.1	82.0	77.0
17							95.6	90.3	85.4	80.6
18							98.6	93.3	88.4	83.9
19								96.1	91.3	86.8
20								98.7	94.0	89.5
21									96.5	92.1
22									98.9	94.5
23										96.8
24										98.9

NOTE: Figures in body of table are percentages of stack diameter from inside wall to traverse point.

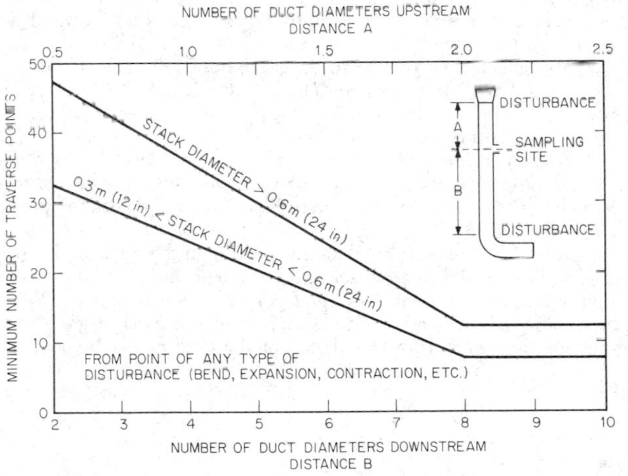

FIG. 26-18 Minimum number of sample points.

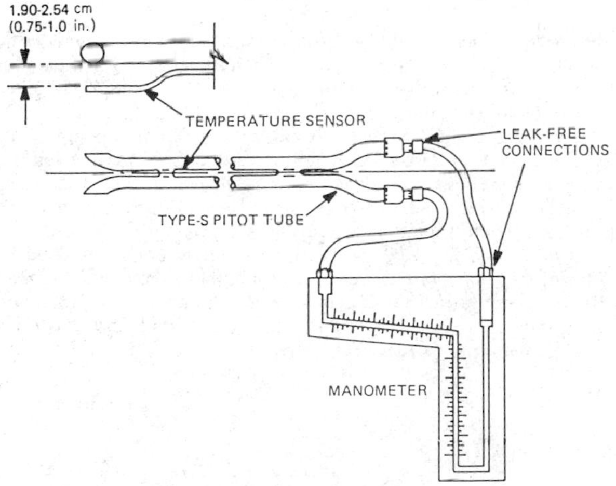

FIG. 26-19 Velocity-measurement system.

represent equal areas. The sampling probe is placed at each sampling point, and the apparatus is adjusted to take a sample at a constant rate. As the gas passes through the apparatus, a filter collects the particulate matter, the moisture is removed, and the sample volume is measured. The collected moisture is then measured, and the moisture content of the gas stream is calculated. A schematic of the sampling apparatus used in this reference method is shown in Fig. 26-22.

Particulates Procedures for testing a particulate source are more detailed than those used in sampling gases. Because particulates exhibit inertial effects and are not uniformly distributed within a stack, sampling to obtain a representative sample is more complex than for gaseous pollutants. EPA Method 5 (as shown in Fig. 26-23) is the most widely used procedure for determination of particulate emissions from a stationary source. In-stack sampling guidelines are presented in EPA Method 17.

According to Method 5 (except as applied to fossil-fuel-fired steam generators), a particulate is defined as any material collected at 121°C (250°F) on a filtering medium. The sampling apparatus used in Method 5 is designed to catch particulate matter at this specified temperature. Most states accept Method 5 even though they define particulate differently. The sampling apparatus, however, may have to be modified to conform with the state's definition of particulate. For example, a state may define particulate as any material collectible at stack conditions, a definition that would allow the filtering medium to be located in the stack.

In performing a particulate-source test, samples are taken at several designated sampling points in the stack, which represent equal areas. At each sampling point, the velocity, temperature, molecular weight, and static pressure of the particulate-laden gas stream are measured. The sampling probe is placed at the first sampling point, and the sampling apparatus adjusted to take a sample at the conditions measured at this point, and the process is repeated continuously

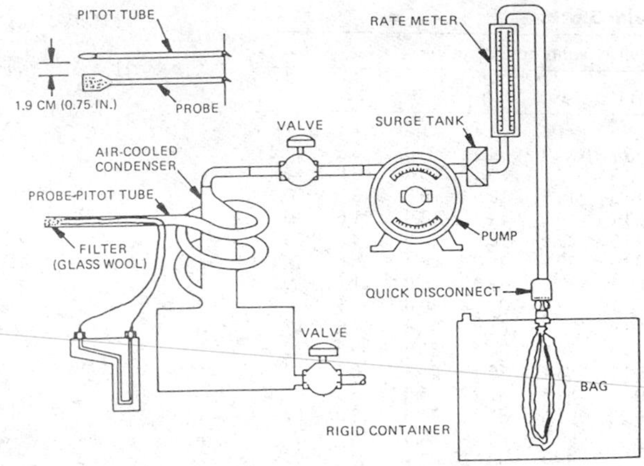

FIG. 26-20 Integrated-sample setup for molecular-weight determination.

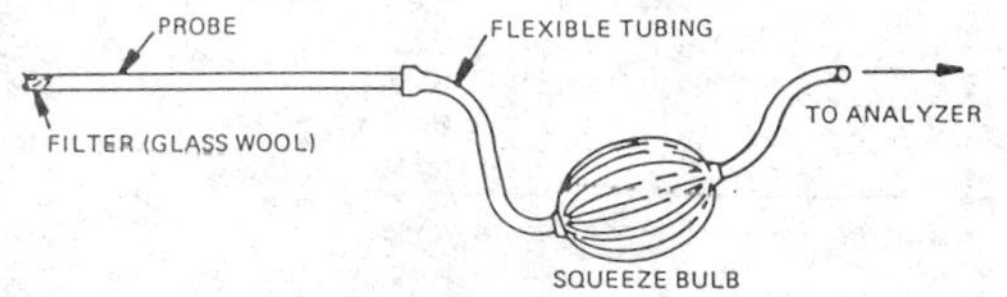

FIG. 26-21 Grab-sample setup for molecular-weight determination.

until a sample has been taken from each designated sampling point. To achieve valid results in a particulate-source test, the sample must be taken isokinetically. Measurement of stack conditions allows adjustment of the sampling rate to meet this requirement.

As the gas stream proceeds through the sampling apparatus, the particulate matter is trapped on a filter, the moisture is removed, and the volume of the sample is measured. Upon completion of sampling, the collected material is recovered and sent to a laboratory for a gravimetric determination or analysis.

Sulfur Dioxide EPA Method 6 is the reference method for determining emissions of sulfur dioxide (SO_2) from all stationary sources except sulfuric acid plants. As the gas goes through the sampling apparatus (see Fig. 26-24), the sulfuric acid mist and sulfur trioxide are removed, the SO_2 is removed by a chemical reaction

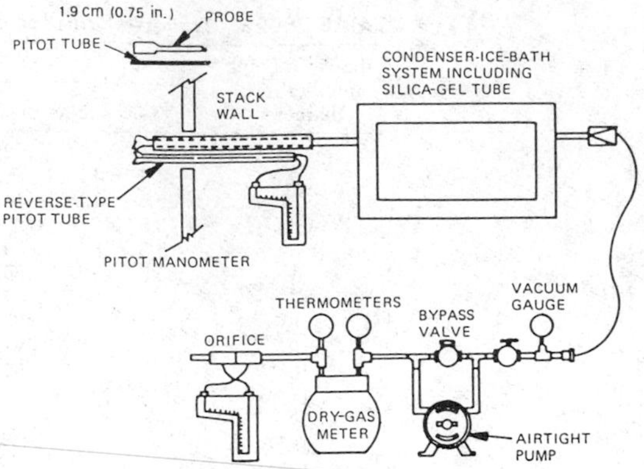

FIG. 26-22 Moisture-sample train.

with a hydrogen peroxide solution, and, finally, the sample-gas volume is measured. Upon completion of the run, the sulfuric acid mist and sulfur trioxide are discarded, and the collected material containing the SO_2 is recovered for analysis at the laboratory. The concentration of SO_2 in the sample is determined by a titration method.

For determination of the total mass-emission rate of SO_2, the moisture content and the volumetric flow rate of the exhaust-gas stream must also be measured.

For Method 6, the minimum sampling time is 20 min per sample, and two separate samples constitute a run. Three runs are required, resulting in six separate samples. An interval of 30 min is required between each sample. Longer sampling times may be required if a larger sample is needed.

Stack concentrations of 50 to 10,000 ppm of sulfur dioxide can be determined with this method. The minimum detectable limit has been determined to be 3.4 mg of SO_2 per cubic meter of gas (2.1×10^{-7} lb of SO_2 per cubic foot of gas).

Nitrogen Oxides (NO_x) EPA Method 7 is the reference method for determining emissions of nitrogen oxides from stationary sources. Sampling for NO_x by this method is relatively simple with the proper equipment.

A sampling probe is placed at any location in the stack, and a grab sample is collected in an evacuated flask. This flask contains a solution of sulfuric acid and hydrogen peroxide, which reacts with the NO_x. The volume and moisture content of the exhaust-gas stream must be determined for calculation of the total mass-emission rate.

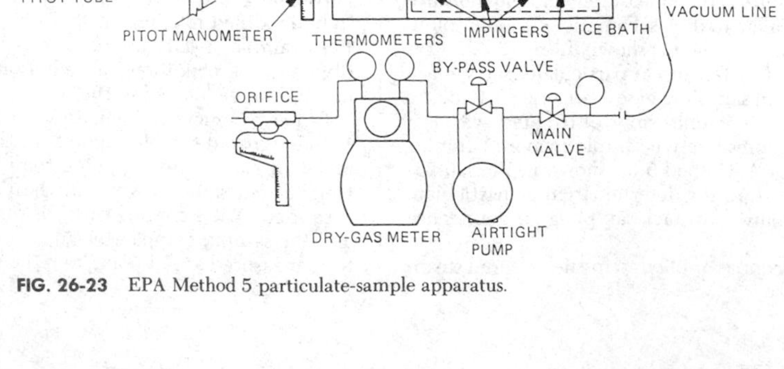

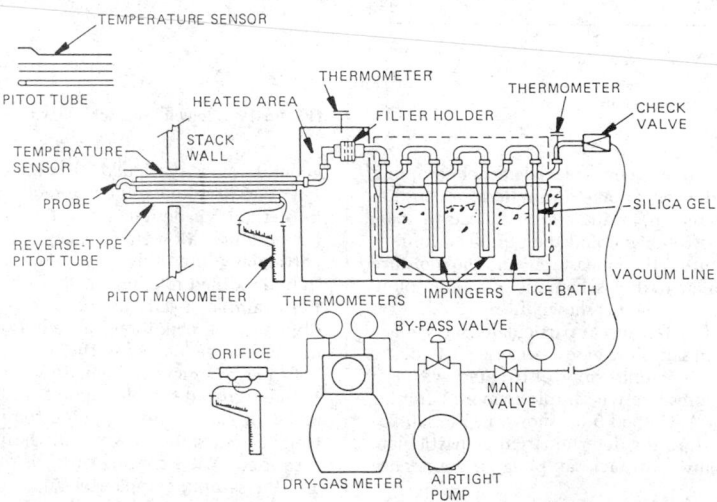

FIG. 26-23 EPA Method 5 particulate-sample apparatus.

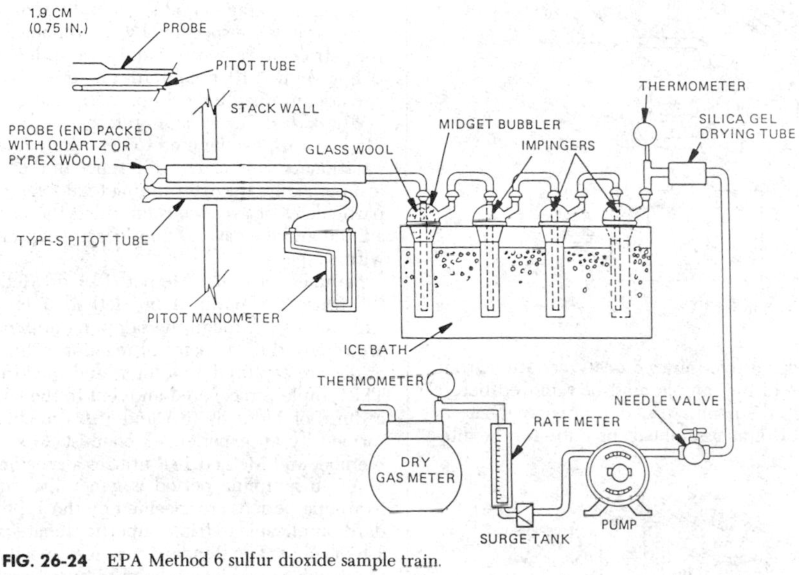

FIG. 26-24 EPA Method 6 sulfur dioxide sample train.

The sample is sent to a laboratory, where the concentration of nitrogen oxides, except nitrous oxide, is determined colorimetrically.

Each grab sample is obtained fairly rapidly (15 to 30 s), and 4 grab samples constitute one run; a total of 12 grab samples is required for a complete series of three runs. An interval of 15 min between grab samples is required. The range of this method has been determined to be 2 to 400 mg of NO_x (as NO_2) per dry standard cubic meter (without dilution). Figure 26-25 shows a schematic of the sampling apparatus for an NO_x source test.

Carbon Monoxide (CO) EPA Method 10 is the reference method for determining emissions of carbon monoxide from stationary sources. An integrated or a continuous gas sample may be required, depending on operating conditions.

When the operating conditions are uniform and steady (there are no fluctuations in flow rate or in concentration of CO in the gas stream), the continuous sampling method can be used. A sampling probe is placed in the stack at any location, preferably near the center. The sample can be extracted at any convenient and constant sampling rate. As the gas stream passes through the sampling apparatus, any moisture or carbon dioxide in the sample-gas stream is removed. The CO concentration is then measured by a nondispersive infrared analyzer, which gives direct readouts of CO concentrations.

Figure 26-26 is a schematic of an assembled sampling apparatus used to determine CO concentrations by the continuous sampling method.

An integrated sampling method is required when operation of the source is uniform but unsteady (fluctuations in flow rate can occur). For an integrated sample, the sampling probe is located at any point near the center of the stack, and the sampling rate is adjusted proportionately to the stack-gas velocity. As the stack gas passes through the sampling apparatus, moisture is removed and the sample gas is collected in a flexible bag. Analysis of the sample is then performed

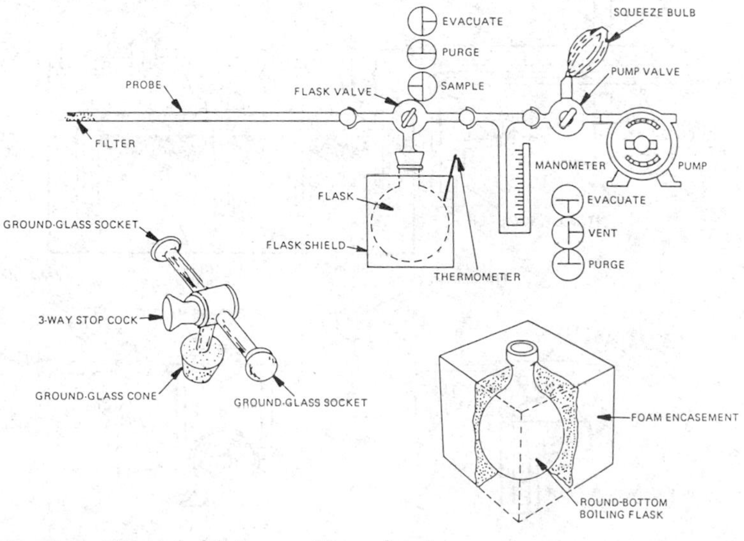

FIG. 26-25 EPA Method 7 nitrogen oxide sample train.

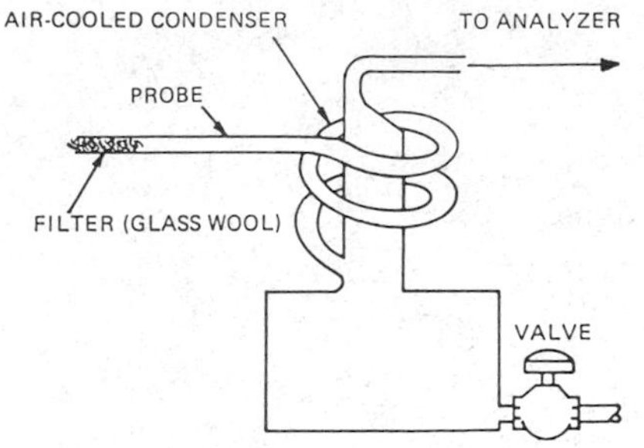

FIG. 26-26 Continuous sample train for CO.

in a laboratory with a nondispersive infrared analyzer. Any carbon dioxide or residual moisture in the sample must be removed before the sample is passed through the nondispersive infrared analyzer.

Figure 26-27 is a schematic of an assembled apparatus for the integrated sampling of CO.

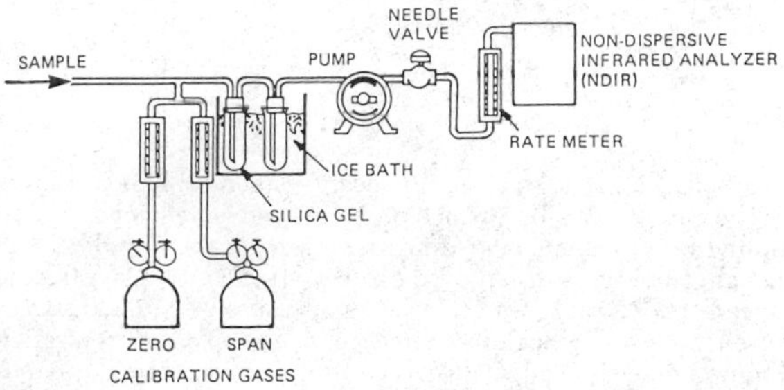

FIG. 26-27 Sampling apparatus for CO.

A 1-h sampling period is generally required for this method. Sampling periods are specified by the applicable standard; e.g., standards for petroleum refineries require sampling for 1 h or more.

For Method 10, the minimum detectable concentration of CO has been determined to be 20 ppm in a range of 1 to 1000 ppm.

Fluorides Two EPA reference methods, Method 13*A* and Method 13*B*, can be used to determine total fluoride emissions from a stationary source. The difference in the two methods is the analytical procedure for determining total fluorides. Fluorides can occur as particulates or as gaseous fluorides; the particulates are captured on a filter, and the gaseous fluorides are captured in a chemical reaction with water.

Samples for either Method 13*A* or Method 13*B* are obtained by the procedure outlined in Method 5 for particulates. As the gas stream passes through the sampling apparatus, the gaseous fluorides are removed by a chemical reaction with water, the particulate fluorides are captured on a filter, and the sample volume is measured. The sample is recovered and sent to the laboratory for analysis. Procedures of Methods 13*A* and 13*B* are complex and should be performed by an experienced chemist. Method 13*A* is a colorimetric method, and Method 13*B* utilizes a specific ion electrode.

A 1-h sampling period is generally required for both methods. Sampling periods are specified by the applicable standard; e.g., standards applicable to triple-superphosphate plants require sampling of 1 h or more. The standard may also specify a minimum sample volume, which will dictate the minimum length of the sampling period.

The determination range of Method 13*A* is 0 to 1.4 mg of fluoride per milliliter; the range of Method 13*B* is 0.2 to 2000 mg of fluoride per milliliter. Figure 26-28 is a schematic of an assembled fluoride-sampling apparatus used in Methods 13*A* and 13*B*.

Organics and Hydrocarbons The U.S. EPA has published two reference source-testing methods for organics and hydrocarbons, Method 23 for the determination of halogenated organics from stationary sources and Method 25 for the determination of total gaseous nonmethane organics (TGNMO).

The principle behind Method 23 is an integrated bag sample of

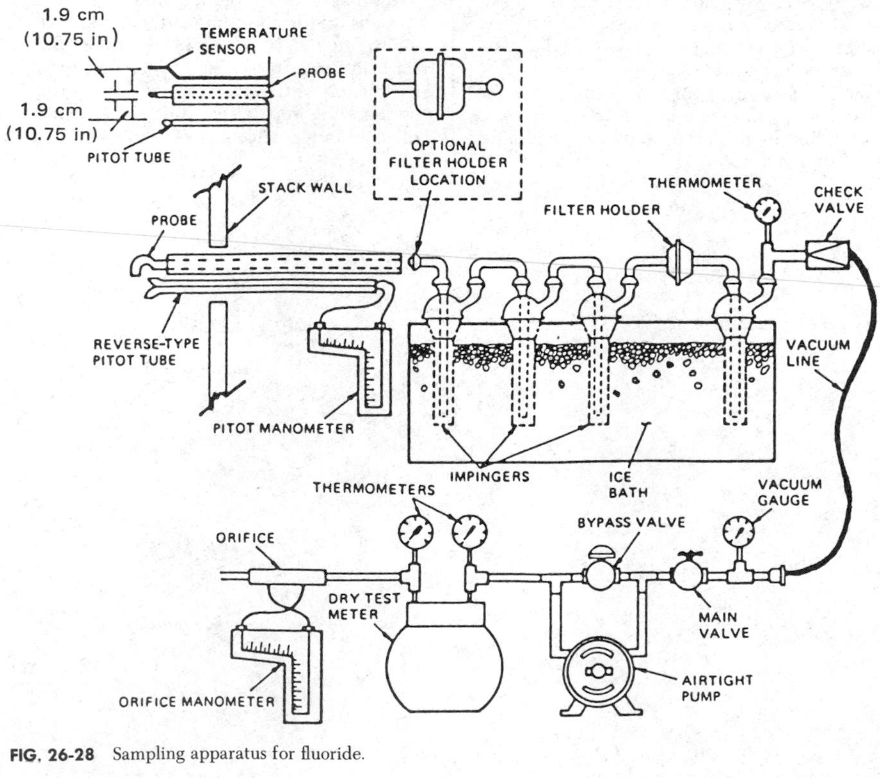

FIG. 26-28 Sampling apparatus for fluoride.

stack gas that is subjected to gas chromatographic analysis, using flame ionization detection. The range of this method is 0.1 to 200 ppm. The upper limit may be increased by extending the calibration range or by diluting the sample. The sampling train is shown schematically in Fig. 26-29.

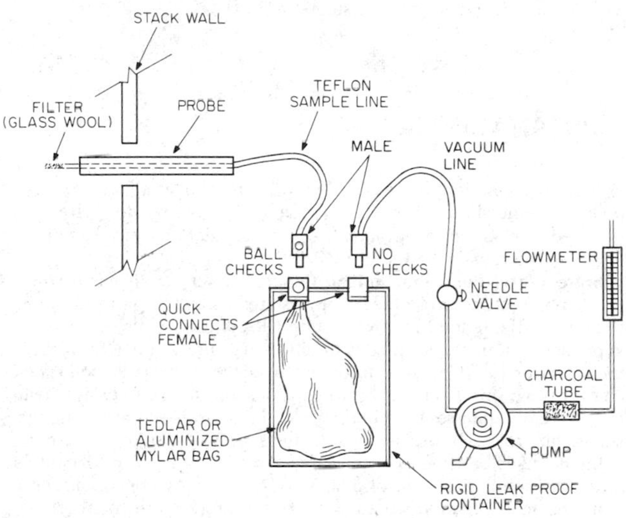

FIG. 26-29 Sampling system for organics and hydrocarbons.

Method 25 applies to the measurement of volatile organic compounds (VOC) as gaseous nonmethane organics (TGNMO), reported as carbon. Organic particulate matter will interfere with the analysis, and therefore in some cases an in-stack particulate filter will be required. The method requires an emission sample to be withdrawn at a constant rate through a chilled-condensate trap by means of an evacuated-sample tank.

TGNMO are determined by combining the analytical results obtained from independent analyses of the condensate-trap and sample-tank fractions. After sampling has been completed, the organic contents of the condensate trap are oxidized to carbon dioxide, which is quantitatively collected in an evacuated vessel; then a portion of the CO_2 is reduced to methane and measured by a flame ionization detector (FID). The organic content of the sample fraction collected in the sampling tank is measured by a chromatographic column to achieve separation of the nonmethane organics from carbon monoxide (CO), CO_2, and CH_4; the nonmethane organics (NMO) are oxidized to CO_2, reduced to CH_4, and measured by an FID. In this manner, the variable response of the FID associated with different types of organics is eliminated.

The sampling system consists of a condensate trap, flow-control system, and sample tank (Fig. 26-30). The analytical system consists of two major subsystems: an oxidation system for the recovery and conditioning of the condensate-trap contents and an NMO analyzer. The NMO analyzer is a gas chromatograph with backflush capability for NMO analysis and is equipped with an oxidation catalyst, a reduction catalyst, and an FID. The system for the recovery and conditioning of the organics captured in the condensate trap consists of a heat source, an oxidation catalyst, a nondispersive infrared (NDIR) analyzer, and an intermediate collection vessel.

Opacity Actual particulate-emission-rate determinations from industrial sources fall into two general categories: opacity measurements and mass measurements. Except in very special cases, there is no universally acceptable correlation between these two parameters.

Stack-opacity measurements are typically made according to EPA Method 9, which entails visual observation of the plume according to a specified procedure. In addition to the mass emissions of particulates, a number of other variables affect opacity. These variables can be divided into two categories: those that are a function of the control equipment and can be "controlled" by the operator or the designer and those beyond the control of the operator (see Table 26-24).

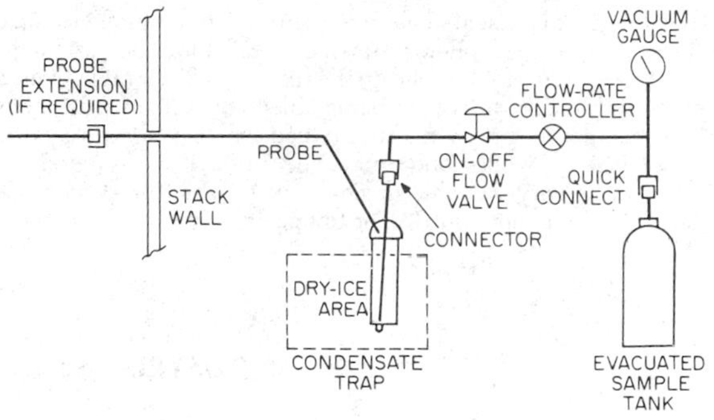

FIG. 26-30 EPA Method 25 sampling system.

TABLE 26-24 Variables Influencing Plume Opacity

Controllable factors
Mass emission of particulate matter
Mean particle size
Deviation from the mean size
Stack diameter
Stack-gas temperature
Stack velocity and other factors influencing plume dispersion
Uncontrollable factors related to type of fuel burned or process involved
Particle density
Particle index of refraction
Water vapor
"Color" of plume
Uncontrollable factors related to human observer, ambient weather conditions, and movement of earth about the sun
Wind speed
Wind direction
Wind turbulence
Ambient-air temperature and humidity
"Color" of sky
Distance of observer from stack
Nonlevel terrain
Observer offset angle
Time of day
Day of year
Longitude of stack
Latitude of stack
"EPA allowable" human error
Sun angle

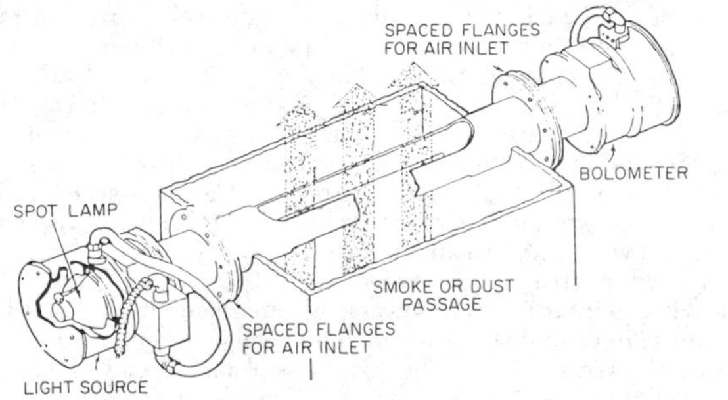

FIG. 26-31 Stack-density transmissometer.

To help quantify visual opacity measurements, a number of opacity-measuring devices (photometers) are available. The basic objective of these systems is to duplicate the readings of the trained human observer with the advantage of being able to operate on a 24-h basis without being affected by nighttime, bad weather, and overcast or other adverse viewing conditions. The in-stack measurement of opacity can be very valuable in situations in which suspended particulates are the major contributor to opacity. To provide such continuous opacity measurement, a large number of stacks have been equipped with in-stack transmissometers (transmission photometers). A typical installation is shown in Fig. 26-31. Usually a light or radiant-heat beam is projected across the stack. The intensity of radiation reaching the other side is proportional to the number and size of particles obstructing the beam at any time. Once the unit has been installed, it can be calibrated in place in terms of stack opacity with visual observations of the stack.

INDUSTRIAL-WASTEWATER MANAGEMENT

GENERAL REFERENCES: *Manual of Practice: Wastewater Treatment Plant Design*, MOP/8, Water Pollution Control Federation and American Society of Civil Engineers, 1977. Metcalf and Eddy, *Wastewater Engineering*, McGraw-Hill, New York, 1972.

INTRODUCTION

All industrial operations produce some wastewaters which must be returned to the environment. Wastewaters can be classified as (1) domestic wastewaters, (2) process wastewaters, and (3) cooling wastewaters. Domestic wastewaters are produced by plant workers, shower facilities, and cafeterias. Process wastewaters result from spills, leaks, and product washing. Cooling wastewaters are the result of various cooling processes and can be once-pass systems or multiple-recycle cooling systems. Once-pass cooling systems employ large volumes of cooling waters that are used once and returned to the environment. Multiple-recycle cooling systems have various types of cooling towers to return excess heat to the environment and require periodic blowdown to prevent excess buildup of salts.

Domestic wastewaters are generally handled by the normal sanitary-sewerage system to prevent the spread of pathogenic microorganisms which might cause disease. Normally, process wastewaters do not pose the potential for pathogenic microorganisms, but they do pose potential damage to the environment through either direct or indirect chemical reactions. Some process wastes are readily biodegraded and create an immediate oxygen demand. Other process wastes are toxic and represent a direct health hazard to biological life in the environment. Cooling wastewaters are the least dangerous, but they can contain process wastewaters as a result of leaks in the cooling systems. Recycle cooling systems tend to concentrate both inorganic and organic contaminants to a point at which damage can be created.

UNITED STATES LEGISLATION, REGULATIONS, AND GOVERNMENTAL AGENCIES

Federal Legislation Public Law 92-500, passed in 1972, is the primary federal legislation affecting wastewater treatment. It was responsible for the National Pollutant Discharge Elimination System (NPDES), with the objective of eliminating all discharges by 1985. NPDES allowed the federal government to gather the data needed to determine the magnitude of the wastewater-pollution problem by requiring every discharger to have a permit and to require regular reporting of data to the EPA. Public Law 95-217, the Clean Water Act of 1977, modified Public Law 92-500 somewhat by recognizing the need to modify the time schedule for wastewater-treatment construction and for greater emphasis on control of toxic materials discharged from industrial plants. Pretreatment regulations have been developed for various industries to reduce the potential disruption in municipal wastewater-treatment plants receiving industrial wastes. It is standard practice to develop new legislation every 5 years.

Environmental Protection Agency President Nixon created the EPA in 1970 to coordinate all environmental pollution-control activities at the federal level. The EPA was placed directly under the Office of the President so that it could be more responsive to the political process. In the succeeding decade the EPA produced a series of federal regulations increasing federal control over all wastewater-pollution-control activities. In January 1981, President Reagan reversed the trend of greater federal regulation and began to decrease the role of the federal EPA.

State Water-Pollution-Control Offices Every state has its own water-pollution-control office. Some states have reorganized along the lines of the federal EPA with state EPA offices, while others have kept their water-pollution-control offices within state health departments. Prior to 1965 each state controlled its own water-pollution-control programs. Conflicts between states and uneven enforcement of state regulations resulted in the federal government's assuming the leadership role. Unfortunately, conflicts between states shifted to being conflicts between the states and the federal EPA. By 1980 the state water-pollution-control offices were primarily concerned with handling most of the detailed permit and paper work for the EPA and in furnishing technical assistance to industries at the local level.

WASTEWATER CHARACTERISTICS

Organics Organic compounds in wastewaters have created most of the pollution problems as a result of their effect on oxygen resources in the environment. The low-molecular-weight water-soluble organics tend to be biodegraded by bacteria and fungi with the utilization of oxygen. As the complexity of organic molecules increases, their solubility and biodegradability decrease. The total chemical oxygen demand (COD) of organic compounds in wastewaters is measured by the dichromate COD test. A 2-h reflux with concentrated sulfuric acid and potassium dichromate with silver sulfate and mercuric sulfate catalysts is adequate for complete oxidation of all but a few aromatic organic compounds. In recent years the total organic carbon (TOC) apparatus has been developed to give a TOC measurement similar to the COD measurement. Unfortunately, the TOC data have not been correlated with COD data except for a few industrial wastes. The TOC measurements have found use in industries needing immediate information on organic concentrations, primarily soluble organics. The 5-day, 20°C, biochemical oxygen demand (BOD5) test has been the primary test for evaluating BOD in wastewaters. Unfortunately, the BOD5 test is subject to considerable error and requires careful evaluation for good results. Normally, BOD5 data are approximately 67 percent of the ultimate BOD and 58 percent of the biodegradable COD. Efforts have been made to replace the BOD5 test with the COD test, but the results have not been satisfactory. Both tests are normally required for proper evaluation of the organic fraction of wastewaters. By using a microbial seed acclimated to the specific industrial wastes, the BOD5 test will give a good measure of the biodegradability of the organics, while the COD test will give both the biodegradable COD and the nonbiodegradable COD. Soluble organics will be metabolized more easily than insoluble organics. Complex proteins and complex carbohydrates must be hydrolyzed to simple amino acids and simple sugars prior to metabolism. Complex hydrocarbons must be metabolized by direct contact with the microbes at the water-hydrocarbon interface. Thus it is that organics are generally analyzed as to their oxygen equivalence and as to their impact on the environmental oxygen resources. In a few instances direct analyses of organics by gas chromatography (GC) or gas-chromatography–mass-spectrometry (GC-MS) techniques have been used for unusual organics.

Inorganics The inorganics in most industrial wastes are the direct result of inorganic compounds in the carriage water. Soft-water sources will have lower inorganics than hard-water or saltwater sources. In a few instances the industrial processes add inorganic compounds in the wastewaters. While domestic wastewaters have a balance in organics and inorganics, many process wastewaters from industry are deficient in specific inorganic compounds. Biodegradation of organic compounds requires adequate nitrogen, phosphorus, iron, and trace salts. Ammonium salts or nitrate salts can provide the nitrogen, while phosphates supply the phosphorus. Either ferrous or ferric salts or even normal steel corrosion can supply the needed iron. Other trace elements needed for biodegradation are potassium, calcium, magnesium, cobalt, molybdenum, chloride, and sulfur. Carriage water or demineralizer wastewaters or corrosion products can supply the needed trace elements for good metabolism. Occasionally, it is necessary to add specific trace elements or nutrient elements.

pH and Alkalinity Wastewaters should have pH values between 6 and 9 for minimum impact on the environment. Wastewaters with pH values less than 6 will tend to be corrosive as a result of the excess hydrogen ions. On the other hand, raising the pH above 9 will cause some of the metal ions to precipitate as carbonates or as hydroxides at higher pH levels. Alkalinity is important in keeping pH values at the right levels. Bicarbonate alkalinity is the primary buffer in wastewaters. It is important to have adequate alkalinity to neutralize the acid waste components as well as those formed by partial metabolism of organics. Many neutral organics such as carbohydrates, aldehydes, ketones, and alcohols are biodegraded through organic acids which must be neutralized by the available alkalinity. If alkalinity is inadequate, sodium carbonate is a better form to add than lime. Lime tends to be hard to control accurately and results in high pH levels and precipitation of the calcium which forms part of the alkalinity. In a few instances, sodium bicarbonate may be the best source of alkalinity.

Temperature Most industrial wastes tend to be on the warm side. For the most part, temperature is not a critical issue below 37°C if wastewaters are to receive biological treatment. It is possible to operate thermophilic biological wastewater-treatment systems up to 65°C with acclimated microbes. Low-temperature operations in northern climates can result in very low winter temperatures and slow reaction rates for both biological treatment systems and chemical treatment systems. Increased viscosity of wastewaters at low temperatures makes solid separation more difficult. Efforts are generally made to keep operating temperatures between 10 and 30°C if possible.

Dissolved Oxygen Oxygen is a critical environmental resource in receiving streams and lakes. Aquatic life requires reasonable dissolved-oxygen (DO) levels. EPA has set minimum stream DO levels at 5 mg/L during summer operations, when the rate of biological metabolism is a maximum. It is important that wastewaters have maximum DO levels when they are discharged and have a minimum of oxygen-demanding components so that DO remains above 5 mg/L. DO is a poorly soluble gas in water, having a solubility around 9.1 mg/L at 20°C and 101.3-kPa (1-atm) air pressure. As the temperature increases and the pressure decreases with higher elevations above sea level, the solubility of oxygen decreases. Thus, DO is a minimum when BOD rates are a maximum. Lowering the temperature yields higher levels of DO saturation, but the biological metabolism rate decreases. Warm-wastewater discharges tend to aggravate the DO situation in receiving waters.

PRETREATMENT

Many industrial-wastewater streams should be pretreated prior to discharge to municipal sewerage systems or even to a central industrial sewerage system. Pretreatment of individual streams should be considered whenever these streams might have an adverse effect on the total treatment system.

Equalization Equalization is one of the most important pretreatment devices. The batch discharge of concentrated wastes is best suited for equalization. It may be important to equalize wastewater flows, wastewater concentrations, or both. Periodic wastewater discharges tend to overload treatment units. Flow equalization tends to level out the hydraulic loads on treatment units. It may or may not level out concentration variations, depending upon the extent of mixing within the equalization basin. Mechanical mixing may be adequate if the wastes are purely chemical in their reactivity. Biodegradable wastes normally require aeration mixing so that the microbes are kept aerobic and nuisance odors are prevented. Diffused aeration systems offer better mixing under variable load conditions than mechanical surface aeration equipment. Mixing and oxygen transfer are both important with biodegradable wastewaters. Operation on regular cycles determines the size of the equalization basin. There is no advantage in making the equalization basin any larger than necessary to level out wastewater variations. Industrial operation on a 5-day, 40-h week will normally make a 2-day equalization basin as large as needed for continuous operation of the wastewater-treatment system under uniform conditions.

Neutralization Acidic or basic wastewaters must be neutralized prior to discharge. If an industry produces both acidic and basic wastes, these wastes may be mixed together at the proper rates to obtain neutral pH levels. Equalization basins can be used as neutralization basins. When separate chemical neutralization is required, sodium hydroxide is the easiest base material to handle in a liquid form and can be used at various concentrations for in-line neutralization with a minimum of equipment. Yet, lime remains the most widely used base for acid neutralization. Limestone is used when reaction rates are slow and considerable time is available for reaction. Sulfuric acid is the primary acid used to neutralize high-pH wastewaters unless calcium sulfate might be precipitated as a result of the neutralization reaction. Hydrochloric acid can be used for neutralization of basic wastes if sulfuric acid is not acceptable. For very weak basic wastewaters carbon dioxide can be adequate for neutralization.

Grease and Oil Removal Grease and oils tend to form insoluble layers with water as a result of their hydrophobic characteristics. These hydrophobic materials can be easily separated from the water phase by gravity and simple skimming, provided they are not too well mixed with the water prior to separation. If the oils and greases form emulsions with water as a result of turbulent mixing, the emulsions are difficult to break. Separation of oil and grease should be carried out near the point of their mixing with water. In a few instances, air bubbles can be added to the oil and grease mixtures to separate the hydrophobic materials from the water phase by flotation. Chemicals have also been added to help break the emulsions. American Petroleum Institute (API) separators have been used extensively by the petroleum industry to remove oils from wastewaters. The food industries use grease traps to collect the grease prior to its discharge. Unfortunately, grease traps are designed for regular cleaning of the trapped grease. Too often they are allowed to fill up and discharge the excess grease into the sewer or are flushed with hot water and steam to fluidize the grease for easy discharge to the sewer. A grease trap should be designed for a specific volume of grease to be collected over specific time periods. Care should be taken to design the trap so that the grease can easily be removed and properly handled. Neglected or poorly designed grease traps are worse than no grease traps at all.

Toxic Substances Recent federal legislation has made it illegal for industries to discharge toxic materials in wastewaters. Each industry is responsible for determining if any of its wastewater components are toxic to the environment and to remove them prior to the wastewater discharge. The EPA has identified a number of priority pollutants which must be removed and kept under proper control from their origin to their point of ultimate disposal. Major emphasis has recently been placed on heavy metals and on complex organics that have been implicated in possible cancer production. Pretreatment is essential to reduce heavy metals below toxic levels and to prevent discharge of any toxic organics. Fortunately, toxic organics can ultimately be destroyed by various chemical oxidation systems. Incineration appears to be the most economical method for destroying toxic organics. To make incineration economical, the organics must be kept separated from the dilute wastewaters and treated in their concentrated form. If the heavy metals cannot be reused, they must be concentrated and placed into insoluble materials which will not leach the heavy metals. Toxic substances currently pose the greatest challenge to industries since very little attention has been paid to these materials in the past.

PRIMARY TREATMENT

Wastewater treatment is directed toward removal of pollutants with the least effort. Suspended solids are removed by either physical or chemical separation techniques and handled as concentrated solids.

Screens Fine screens such as hydroscreens are used to remove moderate-size particles that are not easily compressed under fluid flow. Fine screens are normally used when the quantities of screened particles are large enough to justify the additional units. Mechanically cleaned fine screens have been used for separating large particles. A few industries have used large bar screens to catch large solids that could clog or damage pumps or equipment following the screens.

Grit Chambers Industries with sand or hard, inert particles in their wastewaters have found aerated grit chambers useful for the rapid separation of these inert particles. Aerated grit chambers are relatively small, with total volume based on 3-min retention at maximum flow. Diffused air is normally used to create the mixing pattern shown in Fig. 26-32, with the heavy, inert particles removed by centrifugal action and friction against the tank walls. The air flow rate is adjusted for the specific particles to be removed. Floatable solids are removed in the aerated grit chamber. It is important to provide for regular removal of floatable solids from the surface of the grit chamber; otherwise, nuisance conditions will be created. The settled grit is normally removed with a continuous screw and buried in a landfill.

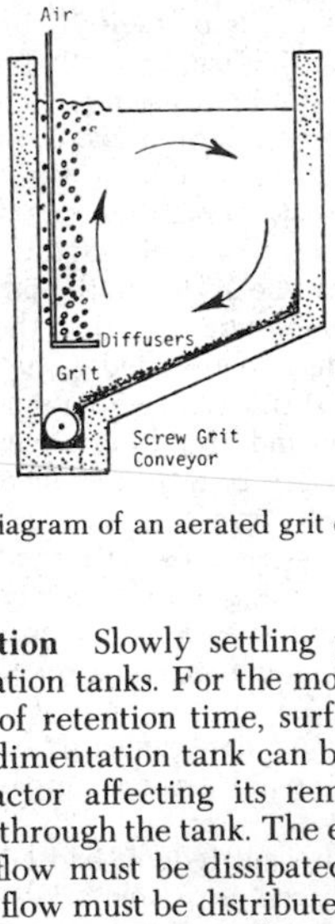

FIG. 26-32 Schematic diagram of an aerated grit chamber.

Gravity Sedimentation Slowly settling particles are removed with gravity sedimentation tanks. For the most part, these tanks are designed on the basis of retention time, surface overflow rate, and minimum depth. A sedimentation tank can be rectangular or circular. The important factor affecting its removal efficiency is the hydraulic flow pattern through the tank. The energy contained in the incoming-wastewater flow must be dissipated before the solids can settle. The wastewater flow must be distributed properly through the sedimentation volume for maximum settling efficiency. After the solids have settled, the settled effluent should be collected without creating serious hydraulic currents that could adversely affect the sedimentation process. Effluent weirs are placed at the end of rectangular sedimentation tanks and around the periphery of circular sedimentation tanks to ensure uniform flow out of the tanks. Once the solids have settled, they must be removed from the sedimentation-tank floor by scraping and hydraulic flow. Conventional sedimentation tanks have sludge hoppers to collect the concentrated sludge and to prevent removal of excess volumes of water with the settled solids. Cross-sectional diagrams of conventional sedimentation tanks are shown in Figs. 26-33 and 26-34.

Design criteria for gravity sedimentation tanks normally provide

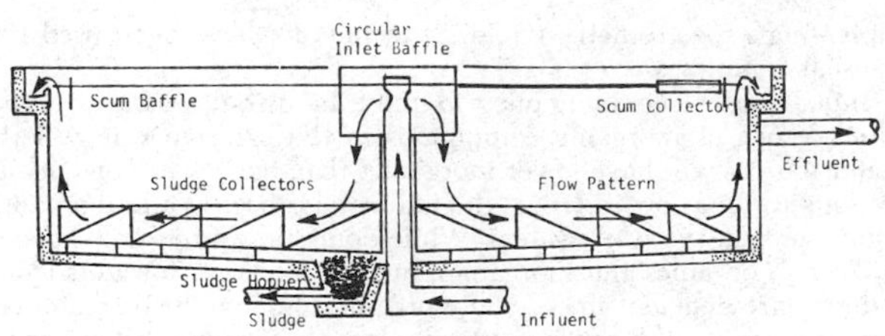

FIG. 26-33 Schematic diagram of a circular sedimentation tank.

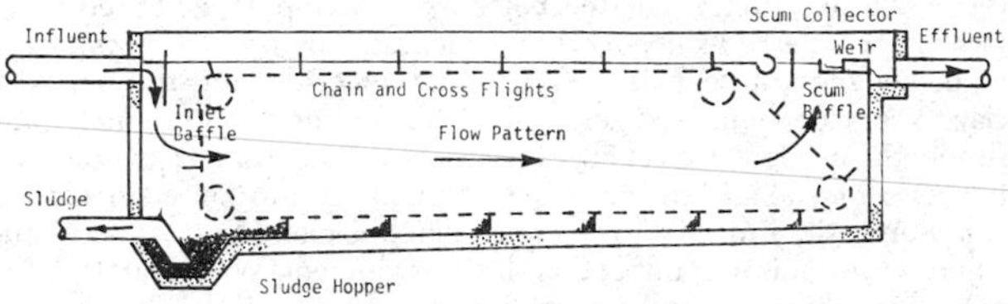

FIG. 26-34 Schematic diagram of a rectangular sedimentation tank.

for 2-h retention based on average flow, with longer retention periods used for light solids or inert solids that do not change during their retention in the tank. Care should be taken that sedimentation time is not too long; otherwise, the solids will compact too densely and affect solids collection and removal. Organic solids generally will not compact to more than 5 to 10 percent. Inorganic solids will compact up to 20 or 30 percent. Centrifugal sludge pumps can handle solids up to 5 or 6 percent, while positive-displacement sludge pumps can handle solids up to 10 percent. With solids above 10 percent the sludge tends to lose fluid properties and must be handled as a semisolid rather than a fluid. Circular sedimentation tanks have steel truss boxes with angled sludge scrapers on the lower side. As the sludge scrapers rotate, the solids are pushed toward the sludge hopper for removal on a continuous or semicontinuous basis. The rectangular sedimentation tanks employ chain-and-flight sludge collectors or rail-mounted sludge collectors. When floating solids can occur in primary sedimentation tanks, surface skimmers are mounted on the sludge scrapers so that the surface solids are removed at regular intervals.

The surface overflow rate (SOR) for primary sedimentation is normally held close to 40.74 $m^3/(m^2 \cdot day)$ [1000 $gal/(ft^2 \cdot day)$] for average flow rates, depending upon the solids characteristics. Lowering the SOR below 40.74 $m^3/(m^2 \cdot day)$ does not produce improved effluent quality in proportion to the reduction in SOR. Generally, the minimum depth of sedimentation tanks is 3.0 m (10 ft), with circular sedimentation tanks having a minimum diameter of 6.0 m (20 ft) and rectangular sedimentation tanks having length-to-width ratios of 5:1. Chain-and-flight limitations generally keep the width of rectangular sedimentation tanks to increments of 6.0 m (20 ft) or less. While hydraulic overflow rates have been limited on the effluent weirs, operating experience has indicated that the recommended limit of 186 $m^3/(m \cdot day)$ [15,000 $gal/(ft \cdot day)$] is lower than necessary for good operation. A circular sedimentation tank with a single-edge weir provides adequate weir length and is easier to adjust than one with a double-sided weir. More problems appear to be created from improper adjustment of the effluent weirs than from improper length.

Chemical Precipitation Lightweight suspended solids and colloidal solids can be removed by chemical precipitation and gravity sedimentation. In effect, the chemical precipitate is used to agglomerate the tiny particles into large particles that settle rapidly in normal sedimentation tanks. Aluminum sulfate, ferric chloride, ferrous sulfate, lime, and polyelectrolytes have been used as coagulants. The choice of coagulant depends upon the chemical characteristics of the particles being removed, the pH of the wastewaters, and the cost and availability of the precipitants. While the precipitation reaction results in removal of the suspended solids, it increases the amount of sludge to be handled. The chemical sludge must be considered along with the characteristics of the original suspended solids in evaluating sludge-processing systems.

Normally, chemical precipitation requires a rapid mixing system and a flocculation system ahead of the sedimentation tank. With a rectangular sedimentation tank, the rapid-mixer and flocculation units are added ahead of the tank. With a circular sedimentation tank the rapid-mixer and flocculation units are built into the tank. Schematic diagrams of chemical treatment systems are shown in Figs. 26-35 and 26-36. Rapid mixers are designed to provide 30-s retention at average flow with sufficient turbulence to mix the chemicals with the incoming wastewaters. The flocculation units are designed for slow mixing at 20-min retention. These units are designed to cause the particles to collide and increase in size without excessive shearing. Care must be taken to move the flocculated mixture from the flocculation unit to the sedimentation unit without disrupting the large floc particles.

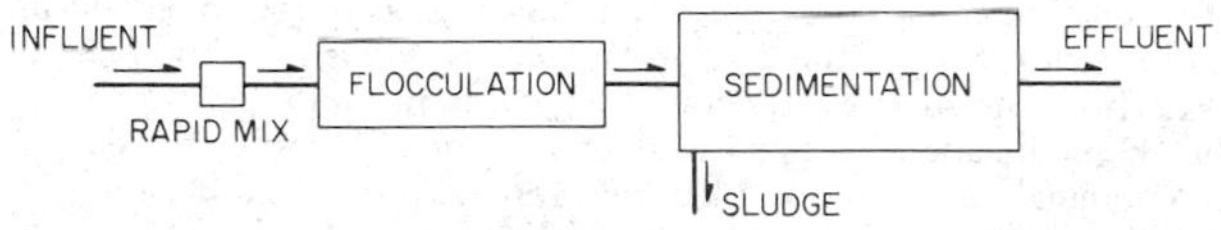

FIG. 26-35 Schematic diagram of a chemical precipitation system for rectangular sedimentation tanks.

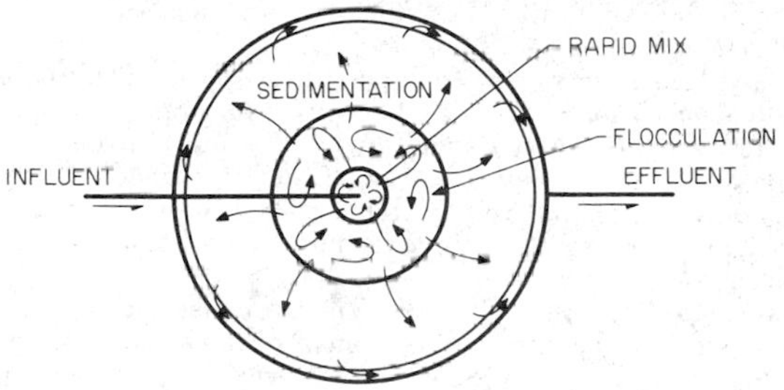

FIG. 26-36 Schematic diagram of a chemical precipitation system for circular sedimentation tanks.

SECONDARY TREATMENT

Concepts of Biological Treatment Secondary treatment is designed to remove colloidal and soluble organics through microbial metabolism. Biological treatment systems convert the biodegradable organics in solution into suspended organics which flocculate and are removed by gravity sedimentation. Colloidal solids tend to be adsorbed onto the microbial flocs.

Biological wastewater-treatment systems employ bacteria as the primary microorganisms responsible for removing excess bacteria and producing a clarified effluent. The bacteria average 0.5 to 1.0 μm in diameter and 1.0 to 2.0 μm in length. Protozoa range from 10 μm to several hundred micrometers in size. The bacteria cannot metabolize solid organics but rather must convert solid particles to soluble organics prior to metabolism. Fortunately, the bacteria have enzymes on their cell surface which can hydrolyze the complex organics to simple organic molecules. Bacteria are widely distributed throughout the environment with different biochemical characteristics. Soil is the best source of concentrated microbes, since the bacteria in the soil are responsible for the stabilization of waste materials in the soil. Normally, the desired bacteria are extracted from the soil by growing them in the desired wastewaters until sufficient numbers have been obtained for the desired reactions. While specialized bacteria are available for purchase, it is not normally necessary to do so as soil will supply the desired microorganisms more economically.

In order to grow, bacteria must have a suitable environment with all the proper nutrients. The environment must provide good mixing for adequate contact between the microorganisms and the pollutants being metabolized. It can be either aerobic with excess dissolved oxygen or anaerobic without any dissolved oxygen. The pH should be maintained between 6 and 9. There should be sufficient nitrogen, phosphorus, iron, and trace metals for good growth, but there should not be too high a concentration of heavy metals to be toxic. The temperature should be maintained between 5 and 35°C, although it is possible to operate thermophilic systems up to 65°C. If the environment is not balanced, improper microbial growth and inadequate treatment can occur. One of the major problems lies with filamentous microbes which adversely affect flocculation and sedimentation. In the absence of adequate oxygen and iron, filamentous bacteria can predominate over normal bacteria. Under low-pH conditions and low-nitrogen or high-carbohydrate wastes, filamentous fungi predominate over bacteria.

The best-quality effluent is produced under aerobic conditions. The bacteria metabolize the biodegradable organics, using dissolved oxygen as their terminal electron acceptor. Approximately one-third of the organics metabolized are oxidized to carbon dioxide and water to obtain the energy to convert the remaining two-thirds of the organics to microbial protoplasm. Microbial protoplasm is similar regardless of the bacteria species or the organics being metabolized. Chemical analysis indicates that microbial protoplasm contains 90 percent organics and 10 percent inorganics. The organic fraction contains approximately 50 percent C, 7 percent H, 31 percent O, and 12 percent N by weight. The major problem in aerobic treatment is the large volume of microbial solids to be processed. The microbial protoplasm is not stable but rather continues to be degraded with time in a process known as "endogenous respiration." Approximately 80 percent of microbial protoplasm can be oxidized, but 20 percent of the organic fraction is stable in aerobic treatment systems and must be processed further.

In the absence of dissolved oxygen the bacteria must find another electron acceptor. Chemically bound oxygen is the primary electron acceptor under anaerobic conditions. NO_3, NO_2, SO_4, CO_2, and organic oxygen compounds are used as electron acceptors. Nitrates are reduced to nitrites and to various intermediates before being reduced to nitrogen gas. It is important to recognize that denitrification results in the ultimate production of nitrogen gas that is insoluble in the wastewaters rather than in the production of ammonia or ammonium ions. If the nitrates are needed as a source of nitrogen for the bacteria, they will be reduced to ammonium ions in proportion to the nitrogen needs for protoplasm. A special group of bacteria use sulfates for their electron acceptor and can produce thiosulfates, free sulfur, or hydrogen sulfide. The methane bacteria can reduce carbon dioxide with the production of methane, which is relatively insoluble in water and can be used as a source of energy. Organic oxygen compounds such as carbohydrates are broken down to simple organic acids, aldehydes, ketones, and alcohols which can be metabolized by the methane bacteria to methane and carbon dioxide. Since it takes the same amount of energy to produce a unit of microbial protoplasm aerobically and anaerobically, less microbial cell mass is produced per unit of organic pollutants metabolized anaerobically than aerobically. The anaerobic reactions do not yield as much energy to the microbes as the aerobic reactions. The anaerobic end products contain considerable energy that is not available to the microbes. Current research into anaerobic wastewater-treatment systems is being stimulated by the production of methane, which can be used as an energy source, and by the lower microbial solids production. Approximately 80 to 95 percent of the organics can be converted to methane and carbon dioxide, with 5 to 20 percent being converted to microbial solids.

Time is a critical variable in sizing wastewater-treatment systems. A definite time period is required to metabolize a given amount of organic matter by a unit of cell mass. By retaining the microbes in the treatment system, the treatment time per unit of organic matter is reduced. Unfortunately, the time for aerobic treatment is controlled by oxygen transfer. Under anaerobic conditions contact between the microbes and the organic pollutants controls the total reaction time. With proper design more organic matter can be treated anaerobically than aerobically in the same time period. Unfortunately, the anaerobic system cannot produce as high-quality effluent as the aerobic system. The net result is that anaerobic systems are often used as the first stage of biological treatment, with aerobic systems being used as the second stage.

Lagoons Lagoons are low-cost, easy-to-operate wastewater-treatment systems capable of producing satisfactory effluents.

Facultative Lagoons These lagoons have been designed to use both aerobic and anaerobic reactions. Normally, facultative lagoons consist of two or more cells in series. The settleable solids tend to settle out in the first cell and undergo anaerobic metabolism with the production of organic acids and methane gas, which bubbles out to the atmosphere. Algae at the surface of the lagoon utilize sunlight for their energy in converting carbon dioxide, water, and ammonium ions into algal protoplasm with the release of oxygen as a waste product. Aerobic bacteria utilize the oxygen released by the algae to stabilize the soluble and colloidal organics. Thus, the bacteria and algae form a symbiotic relationship as shown in Fig. 26-37. The interesting aspect of facultative lagoons is that the organic matter in the incoming wastewaters is not stabilized but rather is converted to microbial protoplasm, which has a slower rate of oxygen demand. In fact, in some facultative lagoons inorganic compounds in the wastewaters are converted to organic compounds with a total increase in organics within the lagoon system.

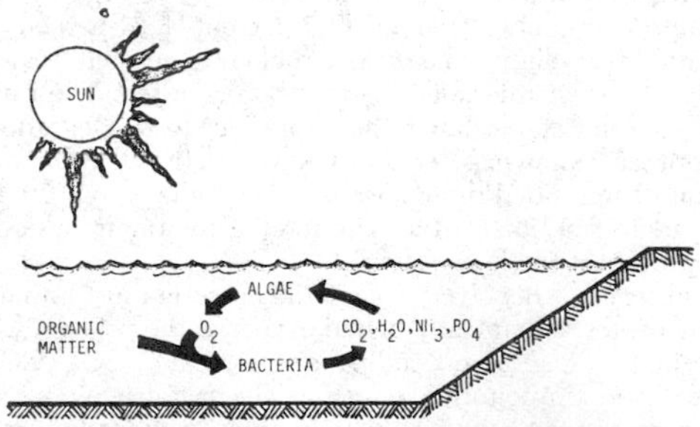

FIG. 26-37 Schematic diagram of oxidation-pond operations.

Facultative lagoons are designed on the basis of organic load in relationship to the potential sunlight availability. In the northern part of the United States facultative lagoons are designed on the basis of 2.2 g/(m^2·day) [20 lb BOD5/(acre·day)]. In the middle part of the United States the organic load can be increased to 3.4 to 4.5 g/(m^2·day) [30 to 40 lb BOD5/(acre·day)], while in the southern part the organic load can be increased to 6.7 g/(m^2·day) [60 lb BOD5/(acre·day)]. The depth of lagoons is normally maintained between 1.0 and 1.7 m (3 and 5 ft). A depth less than 1.0 m (3 ft) encourages the growth of aquatic weeds and permits mosquito breeding. In dry areas the maximum depth may be increased above 1.7 m (5 ft) depending upon evaporation. Most facultative lagoons depend upon natural wind action for mixing and should not be placed in screened areas where wind action is blocked.

Effluent quality from facultative lagoons is related primarily to the suspended solids created by living and dead microbes. The long retention period in the lagoons allows the microbes to die off, leaving a small particle that settles slowly. The release of nutrients from the dead microbes permits the algae to survive by recycling the nutrients. Thus, the algae determine the ultimate effluent quality. Use of series ponds with well-designed transfer structures between ponds permits maximum retention of algae within the ponds and the best-quality effluent. Normally the soluble BOD5 is under 5 or 10 mg/L with a total effluent BOD5 under 30 mg/L. The effluent suspended solids will vary widely during the different seasons of the year, being a maximum of 70 to 100 mg/L in the summer months and a minimum of 10 to 20 mg/L in the winter months. If suspended-solids removal is essential, chemical precipitation is the best method available at the present time. Slow sand filters and rock filters have been studied for suspended-solids removal; they work well as long as the effluent suspended solids are relatively low, 40 to 70 mg/L.

Aerated Lagoons These lagoons originated from efforts to control overloaded facultative lagoons. Since the lagoons were deficient in oxygen, additional oxygen was supplied by either mechanical surface aerators or diffused aerators. Mechanical surface aerators were quickly accepted as the primary aerators because they could be quickly added to existing ponds and moved to strategic locations. Unfortunately, the high-speed, floating surface aeration units were not efficient, and large numbers were required for existing lagoons. The problem was simply one of poor mixing in a very shallow lagoon.

Eventually, diffused aeration equipment was added to relatively deep lagoons [3.0 to 6.0 m (10 to 20 ft)]. Mixing became the most significant parameter for good oxygen transfer in aerated lagoons. From an economical point of view, it was found that a completely mixed aerated lagoon with 24-h retention provided the best balance between mixing and oxygen transfer. As the organic load increased, the fluid-retention time also increased. Short-term aeration permitted metabolism of the soluble organics by the bacteria, but time did not permit metabolism of the suspended solids. The suspended solids were combined with the microbial solids produced from metabolism and discharged from the aerated lagoon to a solids-separation pond. Data from the short-term aerated lagoon indicated that 50 percent BOD5 stabilization occurred, with conversion of the soluble organics to microbial cells. The problem was separation and stabilization of the microbial cells. Short-term sedimentation ponds permitted separation of the solids without significant algae growths but required cleaning at frequent intervals to keep them from filling with solids and flowing into the effluent. Long-term lagoons permitted solids separation and stabilization but also permitted algae to grow and affect effluent quality.

Aerated lagoons were simply dispersed microbial reactors which permitted conversion of the organic components in the wastewaters to microbial solids without stabilization. The residual organics in solution were very low, less than 5 mg/L BOD5. By adding oxygen and improving mixing, the microbial metabolism reaction was speeded up, but the stabilization of the microbial solids has remained a problem to be solved.

Anaerobic Lagoons These lagoons were developed when a major fraction of the organic contaminants consisted of suspended solids that could be removed easily by gravity sedimentation. The anaerobic lagoons are relatively deep [8.0 to 6.0 m (10 to 20 ft)], with a short fluid-retention time (3 to 5 days) and a high BOD5 loading rate, up to 3.2 kg/(m^3·day) [200 lb/(1000 ft^3·day)]. Microbial metabolism in the settled-solids layer produces methane and carbon dioxide, which quickly rise to the surface, carrying some of the suspended solids. A scum layer that retards oxygen transfer and release of obnoxious gases is quickly produced in anaerobic lagoons. Mixing with a grinder pump can provide a better environment for metabolism of the suspended solids. The key for anaerobic lagoons is adequate buffer to keep the pH between 6.5 and 8.0. Protein wastes have proved to be the best pollutants to be treated by anaerobic lagoons, with the ammonium ions reacting with carbon dioxide and water to form ammonium bicarbonate as the primary buffer. High-carbohydrate wastes are poor in anaerobic lagoons since they produce organic acids without adequate buffer, making it difficult to maintain a suitable pH for good microbial growth.

Anaerobic lagoons do not produce a high-quality effluent but are able to reduce the BOD load by 80 to 90 percent with a minimum of effort. Since anaerobic lagoons work best on strong organic wastes, their effluent must be treated by either aerated lagoons or facultative lagoons. An anaerobic lagoon is simply the first stage in the treatment of strong organic wastewaters.

Trickling Filters For years trickling filters were the mainstay of biological wastewater treatment systems because of their simplicity of design and operation. Trickling filters were displaced as the primary biological treatment system by activated sludge because of better effluent quality. Trickling filters are simply fixed-medium biological reactors with the wastewaters being spread over the surface of a solid medium where the microbes are growing. The microbes remove the organics from the wastewaters flowing over the fixed medium. Oxygen from the air permits aerobic reactions to occur at the surface of the microbial layer, but anaerobic metabolism occurs at the bottom of the microbial layer where oxygen does not penetrate.

Originally, the medium in trickling filters was rock, but rock has largely been replaced by plastic, which provides greater void space per unit of surface area and occupies less volume within the filter. A plastic medium permitted trickling filters to be increased from a medium depth of 1.8 m (6 ft) to one of 4.2 m (14 ft) and even 6.0 m (20 ft). The wastewaters are normally applied by a rotary distributor or a fixed-spray nozzle. The spraying or discharging of wastewaters above the trickling-filter medium permits better distribution over the medium and oxygen transfer before reaching the medium. The effluent from the trickling-filter medium is captured in a clay-tile underdrain system or in a tank below the plastic medium. It is important that the bottom of the trickling filter be open for air to move quickly through the filter and bring adequate oxygen for the microbial reactions.

If a high-quality effluent is required, trickling filters must be operated at a low hydraulic-loading rate and a low organic-loading rate. Low-rate trickling filters are operated at hydraulic loadings of 2.2×10^{-5} to 4.3×10^{-5} $m^3/(m^2 \cdot s)$ [2 million to 4 million gal/(acre·day)]. High-rate trickling filters are designed for 10.8×10^{-5} to 40.3×10^{-5} $m^3/(m^2 \cdot s)$ [10 million to 40 million gal/(acre·day)] hydraulic loadings and organic loadings up to 1.4 $kg/(m^3 \cdot day)$ [90 lb BOD5/(1000 $ft^3 \cdot day$)]. Plastic-medium trickling filters have been designed to operate up to 108×10^{-5} $m^3/(m^2 \cdot s)$ [100 million gal/(acre·day)] or even higher, with organic loadings up to 4.8 $kg/(m^3 \cdot day)$ [300 lb BOD5/(1000 $ft^3 \cdot day$)]. Low-rate trickling filters will produce better than 90 percent BOD5 and suspended-solids reductions, while high-rate trickling filters will produce from 65 to 75 percent BOD5 reduction. Plastic-medium trickling filters will produce from 59 to 85 percent BOD5 reduction depending upon the organic-loading rate. It is important to recognize that concentrated industrial wastes will require considerable hydraulic recirculation around the trickling filter to obtain the proper hydraulic-loading rate without excessive organic loads. With high recirculation rates the organic load is distributed over the entire volume of the trickling filter for maximum organic removal. The short fluid-retention time within the trickling filter is the primary reason for the low treatment efficiency.

Rotating Biological Contactors (RBC) The newest form of trickling filter is the rotating biological contactor with a series of circular plastic disks, 3.0 to 3.6 m (10 to 12 ft) in diameter, immersed to approximately 40 percent diameter in a shaped contact tank. The RBC disks rotate at 2 to 5 r/min. As the disks travel through the wastewaters, a small layer adheres to them. As the disks travel into the air, the microbes on the disk surface oxidize the organics. Thus, only a small amount of energy is required to supply the required oxygen for wastewater treatment. As the microbes build up on the plastic disks, the shearing velocity that is created by the movement of the disks through the water causes the excess microbes to be removed from the disks and discharged to the final sedimentation tank.

Rotating biological contactors have been very popular in treating industrial wastes because of their relatively small size and their low energy requirements. Unfortunately, there have occurred a number of problems which should be recognized prior to using RBCs. Strong industrial wastes tend to create excessive microbial growths which are not easily sheared off and which create high oxygen-demand rates with the production of hydrogen sulfide and other obnoxious odors. The heavy microbial growths have damaged some of the disks and caused some shaft failures. The disks are currently being covered with plastic shells to prevent nuisance odors from occurring. Air must be forced through the covered RBC systems and be chemically treated before being discharged back into the environment. Recirculation of wastewater flow around the RBC units can distribute the load over all the units and reduce the heavy initial microbial growths. RBC units also work best under uniform organic loads, requiring surge tanks for many industrial wastes. The net result has been for the cost of RBC units to approach that of other treatment units in terms of organic matter stabilized.

RBCs should be designed on both a hydraulic-loading rate and an organic-loading rate. Normally, hydraulic-loading rates of up to 0.16 $m^3/(m^2 \cdot day)$ [4 gal/($ft^2 \cdot day$)] of surface area are used with organic loading rates up to 44 $kg/(m^2 \cdot day)$ [9 lb BOD5/($ft^2 \cdot day$)]. Treatment efficiency is primarily a function of the fluid-retention time and the organic-loading rate. At low organic-loading rates the RBC units will produce nitrification in the same way as low-rate trickling filters.

Activated Sludge Activated sludge has been the most widespread biological wastewater-treatment system because of its high-quality effluent. The high energy requirements for activated sludge have caused some questions to be raised about the continued use of activated-sludge systems. There is no doubt that these systems are energy-intensive, but there does not appear to be any way to provide treatment without the expenditure of a definite amount of energy. Care should be taken with regard to activated-sludge systems providing a high degree of treatment with a minimum expenditure of energy.

Modifications The modifications of activated-sludge systems offer considerable choice in processes. Complete-mixing activated sludge is the most popular system for industrial wastes because of its ability to absorb shock loads better than other modifications. Contact stabilization is a modification of activated sludge that is best suited to wastewaters having high suspended solids and low soluble organics. Contact stabilization employs a short-term mixing tank to adsorb the suspended solids and metabolize the soluble organics, a sedimentation tank for solids separation, and a reaeration tank for stabilization of the suspended organics. Extended-aeration systems are actually long-term-aeration, completely mixed activated-sludge systems. They employ 24- to 48-h aeration periods and high mixed-liquor suspended solids to provide complete stabilization of the organics and aerobic digestion of the activated sludge in the same aeration tank. The oxidation ditch is a popular form of the extended-aeration system employing mechanical aeration. Pure-oxygen systems are designed to treat strong industrial wastes in series, completely mixed units having relatively short contact periods. One of the latest modifications of activated sludge employs powdered activated carbon to adsorb complex organics and assist in solids separation. Another modification employs a redwood-medium trickling filter ahead of a short-term aeration tank with mixed liquor recycled over the redwood-medium tower to provide heavy microbial growth on the redwood as well as in the aeration tank. Figure 26-38 shows schematic diagrams of the various modifications of activated sludge.

Each activated-sludge process employs an aeration tank for waste stabilization with conversion of the organics to new microbial solids and a sedimentation tank for solids separation. Solids are normally recycled from the sedimentation tank back to the aeration tank as a seed for microorganisms. Since more solids are produced than can be oxidized, there are always some solids that must be continuously wasted from the activated-sludge system. Short-term-aeration systems require more solids wasting than long-term-aeration systems.

Aeration Systems These systems control the design of aeration tanks. Aeration equipment has two major functions: mixing and oxygen transfer. Diffused-aeration equipment employs either a fixed-speed positive-displacement blower or a high-speed turbine blower for readily adjustable air volumes. Air diffusers can be located along one side of the aeration tank or spread over the entire bottom of the tank. They can be either fine-bubble or coarse-bubble diffusers. Fine-bubble diffusers are more efficient in oxygen transfer but require more extensive air-cleaning equipment to prevent them from clogging as a result of dirty air. Mechanical-surface-aeration equipment is more efficient than diffused-aeration equipment but is not as flexible. Economics has dictated the use of large-power aerators, but tank configuration has tended to favor the use of greater numbers of lower-power aerators. Oxidation ditches use horizontal rotor-type aerators. Mixing is a critical problem with mechanical-surface aerators since they are a point-source pump of limited capacity. Experience has indicated that bearings are a serious problem with mechanical-aeration equipment. Wave action generated within the aeration tank tends to produce lateral stresses on the bearings and has resulted in failures and increased maintenance costs. Slow-speed mechanical-surface-aeration units present fewer problems than the high-speed mechanical-surface-aeration units. Deep tanks, greater than 3.0 m (10 ft), require draft tubes to ensure proper hydraulic flow through the aeration tank. Short-circuiting is one of the major problems associated with mechanical aeration equipment. Combined mechanical-

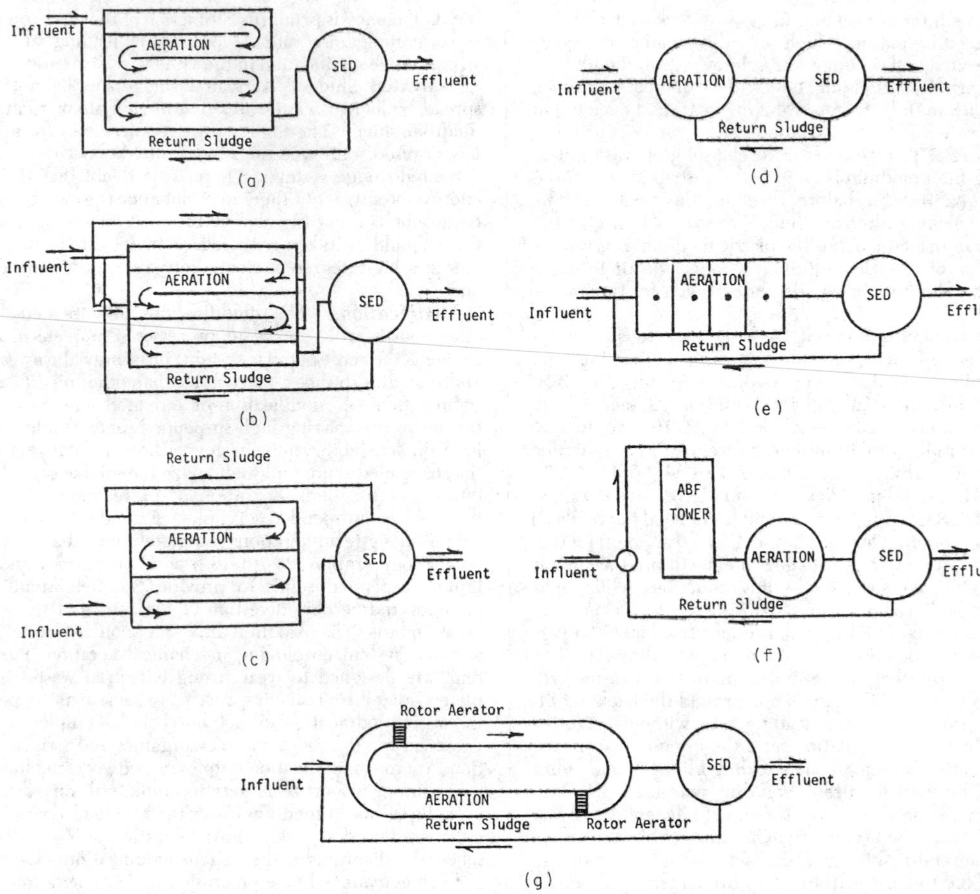

FIG. 26-38 Schematic diagrams of various modifications of the activated-sludge process. (*a*) Conventional activated sludge. (*b*) Step aeration. (*c*) Contact stabilization. (*d*) Complete mixing. (*e*) Pure oxygen. (*f*) Activated biofiltration (ABF). (*g*) Oxidation ditch.

and diffused-aeration systems have enjoyed some popularity for industrial-waste systems that treat variable organic loads. The mechanical mixers provide the fluid mixing with the diffused aeration varied for different oxygen-transfer rates.

Diffused-aeration systems transfer from 20 to 40 mg/(L $O_2 \cdot h$). Combined mechanical- and diffused-aeration systems can transfer up to 65 mg/(L $O_2 \cdot h$), while mechanical-surface aerators can provide up to 90 mg/(L $O_2 \cdot h$). Pure-oxygen systems can provide the highest oxygen-transfer rate, up to 150 mg/(L $O_2 \cdot h$). Aeration equipment must provide sufficient oxygen to meet the peak oxygen demand; otherwise, the system will fail to provide proper treatment. For this reason, the peak oxygen demand and the rate of transfer for the desired equipment determine the size of the aeration tank in terms of retention time. Economics dictates a balance between the size of the aeration tank and the size of the aeration equipment. As the cost of power increases, economics will favor constructing a larger aeration tank and smaller aerators. It is equally important to examine the hydraulic flow pattern around each aerator to ensure maximum efficiency of oxygen transfer. Improper spacing of aeration equipment can waste energy.

There is no standard aeration-tank shape or size. Aeration tanks can be round, square, or rectangular. Shallow aeration tanks are more difficult to mix than deeper tanks. Yet aeration-tank depths have ranged from 0.6 m (2 ft) to 18 m (60 ft). The oxidation-ditch systems tend to be shallow, while some high-rate diffused-aeration systems have used very deep tanks to provide more efficient oxygen transfer.

Regardless of the aeration equipment employed, oxygen-transfer rates must provide from 0.6 to 1.4 kg of oxygen/kg BOD5 (0.6 to 1.4 lb oxygen/lb BOD5) stabilized in the aeration tank for carbonaceous-oxygen demand. Nitrogen oxidation can increase oxygen demand at the rate of 4.3 kg (4.3 lb) of oxygen/kg (lb) of ammonia nitrogen oxidized. At low oxygen-transfer rates more excess activated sludge must be removed from the system than at high oxygen-transfer rates. Here again the economics of sludge handling must be balanced against the cost of oxygen transfer. The quantity of waste activated sludge will depend upon wastewater characteristics. The inert suspended solids entering the treatment system must be removed with the excess activated sludge. The soluble organics are stabilized by converting a portion of the organics into suspended solids, producing from 0.3 to 0.8 kg (0.3 to 0.8 lb) of volatile suspended solids/kg (lb) of BOD5 stabilized. Biodegradable suspended solids in the wastewaters will result in destruction of the original suspended solids and their conversion to a new form. Depending upon the chemical characteristics of the biodegradable suspended solids, the conversion factor will range from 0.7 to 1.2 kg (0.7 to 1.2 lb) of microbial solids produced/kg (lb) of suspended solids destroyed. If the suspended solids produced by metabolism are not wasted from the system, they will eventually be discharged in the effluent. While considerable efforts have been directed toward developing activated-sludge systems which totally consume the excess solids, no such system has proved to be practical. The concept of total oxidation of excess sludge is fundamentally unsound and should be recognized as such.

Sedimentation Tanks These tanks are an integral part of any activated-sludge system. It is essential to separate the suspended solids from the treated liquid if a high-quality effluent is to be produced. Circular sedimentation tanks with various types of hydraulic sludge collectors have become the standard secondary sedimentation system. Square tanks have been used with common-wall construction for compact design with multiple tanks. Most secondary sedimentation tanks use center-feed inlets and peripheral-weir outlets. Recently, efforts have been made to employ peripheral inlets with submerged-orifice flow controllers and either center-weir outlets or peripheral-weir outlets adjacent to the peripheral-inlet channel.

Aside from flow control, basic design considerations have centered on surface overflow rates, retention time, and weir overflow rate. Surface overflow rates have been slowly reduced from 33 $m^3/(m^2 \cdot day)$ [800 $gal/(ft^2 \cdot day)$] to 24 $m^3/(m^2 \cdot day)$ to 16 $m^3/(m^2 \cdot day)$ [600 $gal/(ft^2 \cdot day)$ to 400 $gal/(ft^2 \cdot day)$] and even to 12 m^3 $(m^2 \cdot day)$ [300 $gal/(ft^2 \cdot day)$] in some instances, based on average raw-waste flows. Operational results have not demonstrated that lower surface overflow rates improve effluent quality, making 33 $m^3/(m^2 \cdot day)$ [800 $gal/(ft^2 \cdot day)$] the design choice in most systems. Retention time has been found to be an important design factor, averaging 2 h on the basis of raw-waste flows. Longer retention periods tend to produce rising sludge problems, while shorter retention periods do not provide for good solids separation with high-return sludge flow rates. Effluent-weir overflow rates have been limited to 186 $m^3/(m \cdot day)$ [15,000 $gal/(ft \cdot day)$] with a tendency to reduce the rate to 124 $m^3/(m \cdot day)$ [10,000 $gal/(ft \cdot day)$]. Lower effluent-weir overflow rates are obtained by using dual-sided effluent weirs cantilevered from the periphery of the tank. Unfortunately, proper adjustment of dual-side effluent weirs has created more hydraulic problems than the weir overflow rate. Field data have shown that effluent quality is not really affected by weir overflow rates up to 990 $m^3/(m \cdot day)$ [80,000 $gal/(ft \cdot day)$] or even 1240 $m^3/(m \cdot day)$ [100,000 $gal/(ft \cdot day)$] in a properly designed sedimentation tank. A single peripheral weir, being easy to adjust and keep clean, appears to be optimal for secondary sedimentation tanks from an operational point of view.

Depth tends to be determined from the retention time and the surface overflow rate. As surface overflow rates were reduced, the depth of sedimentation tanks was reduced to keep retention time from being excessive. It was recognized that depth was a valid design parameter and was more critical in some systems than retention time. As mixed-liquor suspended-solids (MLSS) concentrations increase, the depth should also be increased. Minimum sedimentation-tank depths for variable operations should be 3.0 m (10 ft) with depths to 4.5 m (15 ft) if 3000 mg/L MLSS concentrations are to be maintained under variable hydraulic conditions. With MLSS concentrations above 4000 mg/L, the depth of the sedimentation tank should be increased to 6.0 m (20 ft). The key is to keep a definite freeboard over the settled-sludge blanket so that variable hydraulic flows do not lift the solids over the effluent weir.

Scum baffles around the periphery of the sedimentation tank and radial scum collectors are standard equipment to ensure that rising solids or other scum materials are removed as quickly as they form. Hydraulic sludge-collection tubes have replaced the center sludge well, but they have caused a new set of operational problems. These tubes were designed to remove the settled sludge at a faster rate than conventional sludge scrapers. To obtain good hydraulic distribution in the sludge-collection tubes, it was necessary to increase the rate of return sludge flow and decrease the concentration of return sludge. The higher total inflow to the sedimentation tank created increased forces that lifted the settled-solids blanket at the wall, causing loss of excessive suspended solids and lower effluent quality. Operating data tend to favor conventional secondary sedimentation tanks over hydraulic sludge-collection systems. Return-sludge rates normally range from 25 to 50 percent for MLSS concentrations up to 3300 mg/L. Most return-sludge pumps are centrifugal pumps with capacities up to 100 percent raw-waste flow.

Gravity settling can concentrate activated sludge to 10,000 mg/L, but hydraulic sludge-collecting tubes tend to operate best below 8,000 mg/L. The excess activated sludge can be wasted either from the return sludge or from a separate waste-sludge hopper near the center of the tank. The low solids concentrations result in large volumes of waste activated sludge in comparison with primary sludge. Unfortunately, the physical characteristics of waste activated sludge prevent significant concentration without the expenditure of considerable energy. Gravity thickening can produce 2 percent solids, while air flotation can produce 4 percent solids concentration. Centrifuges are able to concentrate activated sludge from 10 to 15 percent solids, but the capture is limited. Vacuum filters can equal the performance of centrifuges if the sludge is chemically conditioned. Filter presses and belt-press filters can produce cakes with 15 to 25 percent solids. It is very important that the excess activated sludge formed in the aeration tanks be wasted on a regular basis; otherwise, effluent quality will deteriorate. Care should be taken to ensure that sludge-thickening systems do not control activated-sludge operations. Alternative sludge-handling provisions should be available during maintenance on sludge-thickening equipment. At no time should final sedimentation tanks be used for the storage of sludge beyond that required by daily operational variations.

PHYSICAL-CHEMICAL TREATMENT (PCT)

Concepts of PCT Physical-chemical treatment processes have been developed to treat wastewaters that are either toxic to or difficult to treat with biological treatment. In a few instances PCT has been used to replace biological treatment. The first step in PCT is the use of chemical precipitants to remove the suspended solids by flocculation and gravity sedimentation. The clarified effluent is then passed through an activated-carbon bed to remove soluble organic compounds. If it is desired to remove soluble salts, the effluent can be passed through ion-exchange resins. PCT treatment can produce an effluent of sufficient quality for reuse. Sand media or multimedia filters have also been used after chemical precipitation and gravity sedimentation to produce a higher-quality effluent to be applied to the activated carbon. Reverse osmosis has also been used as a final treatment when a very high quality of water is desired for reuse.

Limitations of PCT Continuous chemical additions and disposal of chemical sludge have proved to be a major limitation of PCT. Chemical coagulants not only add suspended solids to the waste suspended solids but also add soluble salts to the effluent. Inadequate suspended-solids separation results in carry-over of solids into the activated-carbon bed, which acts as a filter as well as an adsorber of organics. Unfortunately, most activated-carbon beds are not designed for easy removal of suspended solids. For this reason multimedia filters have often been placed after chemical precipitation and sedimentation to ensure removal of the suspended solids prior to carbon adsorption. Reverse osmosis has proved to be a necessity when a high-quality effluent is desired for reuse, but it simply concentrates the contaminants into a smaller volume that must be handled as concentrated wastewaters.

SLUDGE PROCESSING

Objectives Sludges from primary and secondary treatment systems pose a major processing problem in waste-treatment systems. These sludges consist of concentrated unstabilized organics together with inert organics and inert inorganics. The inert sludges must be collected, concentrated, dewatered, and returned to the environment. The biodegradable organics in the sludges create problems and necessitate further processing. Normal processing of sludges consists of thickening to minimize the total volume to be handled and biological treatment or heat treatment for stabilization.

Thickening Gravity thickening has been used to concentrate solids. It is possible to thicken primary sludges to 6 or 8 percent and secondary sludges to 2 percent. Since these concentrations can be reached in a well-designed and -operated sedimentation tank, there is little value in gravity thickening and some negative benefits. Microbial activity creates odor nuisances along with hydrolyzing some of the organics. The return flow returns some solids and soluble organics to the treatment system for removal a second time.

Flotation Air flotation has proved to be successful in concentrating secondary sludges to about 4 percent concentration. The incoming solids are normally saturated with air at 275 to 350 kPa (40 to 50 psig) prior to being released in the flotation tank. As the air comes out of solution, the fine bubbles collect under the suspended solids and carry them to the surface of the tank. The air bubbles compact the floating solids to their maximum extent. Normally, the air-to-solids ratio is about 0.05 by weight. The thickened solids are scraped off the surface, while the effluent is drawn off the middle of the tank and returned to the treatment system. In large flotation tanks with high flows the effluent rather than the incoming solids is pressurized and recycled to the influent. The size of the flotation tanks is determined primarily by the solids-loading rate, directly or indirectly. A solids loading of 25 to 97 kg/(m^2·day) [5 to 20 lb/(ft^2·day)] has been found to be adequate. On a flow basis this translates into 0.14 to 2.7 L/(m^2·day) [0.2 to 4 gal/(ft^2·min)] surface area. For the most part, air-flotation equipment has been developed on a trial-and-error basis with various equipment-manufacturing companies supplying their own units.

Centrifugation Both basket and solid-bowl centrifuges have been used to concentrate waste sludges. Field data have shown that it is possible to obtain 10 to 15 percent solids with waste activated sludge, 15 to 25 percent solids with a mixture of primary and waste activated sludge, and up to 30 to 35 percent solids with primary sludge alone. Centrifuges result in 85 to 90 percent solids capture with good operation. The problem is that the centrate contains the fine solids not easily removed. The centrate is normally returned to the treatment process, where it may or may not be removed. Economics do not favor centrifuges unless the sludge cake produced is at least 20 to 25 percent solids. For the most part, centrifuges are designed by equipment manufacturers from field experience. With varying sludge characteristics centrifuge characteristics will also vary widely.

Anaerobic Digestion Since the organics in both primary and secondary sludges contain biodegradable compounds, concentrated sludges can be treated by anaerobic digestion. Anaerobic digesters are large covered tanks with detention times of 20 to 30 days, based on the volume of sludge added daily. Currently, two anaerobic digesters are constructed of equal size and operated in series so that each unit has a 10- to 15-day retention time. The first digester is heated with an external heat exchanger to 35 to 37°C to speed the rate of reaction. Mixing provides good contact between the microbes and the incoming organic solids. Gas mixing and mechanical mixers have been used to provide mixing in the anaerobic digester. The second digester is basically a solids-separation unit and is not normally equipped for either heating or mixing. The supernatant is recycled back to the treatment plant, while the settled sludge is allowed to concentrate to from 6 to 8 percent solids before being returned to the environment.

Anaerobic digestion results in the conversion of the biodegradable organics to methane, carbon dioxide, and microbial cells. Because of the energy in the methane, the production of microbial mass is quite low, less than 0.1 kg/kg [0.1 lb volatile suspended solids (VSS)/lb] BCOD metabolized except for carbohydrate wastes. The production of methane is 0.35 m^3/kg (5.6 ft^3/lb) BCOD destroyed. Digester gases range from 50 to 80 percent methane and 20 to 50 percent carbon dioxide, depending on the chemical characteristics of the waste organics being digested.

There are two major groups of bacteria in anaerobic digesters: acid-forming and methane-forming. Acid-forming bacteria break down the complex organics to organic acids, which are metabolized by the methane bacteria. It is important that the acid-forming bacteria do not produce excess acid; otherwise, the pH will fall to the toxic level for methane bacteria. Acid production is best controlled by the addition of organics to the anaerobic digester. A continuous, uniform addition of fresh solids will keep the system in good equilibrium. If continuous addition of solids is not possible, additions should be made at as short intervals as possible. Alkalinity levels are normally maintained at about 3000 to 5000 mg/L to keep the pH above 6.5 as a buffer against variable organic acid production with varying organic loads. Proteins will produce an adequate buffer, but carbohydrates will require the addition of alkalinity to provide a sufficient buffer.

Anaerobic Filter Anaerobic digestion is limited by the ability of the system to retain the methane bacteria at high levels. The anaerobic filter was developed to retain the methane bacteria on a fixed medium so that high volumes of relatively dilute organic wastes could be treated by anaerobic digestion with the production of methane gas. Considerable research has been done on anaerobic treatment of various types of soluble organic wastewaters. With a fixed medium fluid-retention time is reduced to a few hours or a few days, depending upon the strength of the wastewaters. Wastewaters containing up to 2000 mg/L BCOD can be treated in a few hours, while wastewaters containing up to 10,000 mg/L BCOD require several days' retention. Approximately 90 percent of the BCOD is metabolized with proper contact between the microbes and the wastewaters. Recirculation of wastewaters around the filter help provide increased organic reduction by furnishing optimum contact between the microbes and the wastes. Normal recycle rates are 6 to 10 times raw-wastewater flow rates.

Aerobic Digestion Waste activated sludge can be treated more easily in aerobic treatment systems than in anaerobic systems. The sludge has already been partially aerobically digested in the aeration tank. For the most part, only about 25 to 35 percent of the waste activated sludge can be digested. An additional aeration period of 15 to 20 days should be adequate to reduce the residual biodegradable mass to a satisfactory level for dewatering and return to the environment. One of the problems in aerobic digestion is the inability to concentrate the solids to levels greater than 2 percent. A second problem is nitrification. The high protein concentration in the biodegradable solids results in the release of ammonia, which can be oxidized during the long retention period in the aerobic digester. Limiting oxygen transfer to the aerobic digester appears to be the best method to handle nitrification and the resulting low pH. High power costs for aerobic digestion help keep the oxygen supply close to the amount required.

Chemical Conditioning Lime, alum, and various ferric salts have been used to condition sludge prior to dewatering. Lime reacts to form calcium carbonate crystals, which act as a solid matrix to hold the sludge particles apart and allow the water to escape during dewatering. Alum and iron salts help displace some of the bound water from hydrophilic organics and form part of the inorganic matrix. Chemical conditioning increases the mass of sludge to be ultimately handled from 10 to 25 percent, depending upon the characteristics of the individual sludge. Chemical conditioning can also help remove some of the fine particles by incorporating them into insoluble chemical precipitates.

Thermal Conditioning Heat conditioning of waste sludges was first applied toward the total oxidation of organics, but operating problems resulted in shifting heat treatment from a total-oxidation process to a heat-conditioning process. By raising the temperature to 180 to 230°C for 15 to 60 min, it is possible to dewater the remaining solids without the addition of chemicals. To keep the system fluid, the heat-treatment system must be operated at a pressure of 1380 to 2070 kPa (200 to 300 lbf/in^2). After the heat reaction, the solids are normally separated in a covered sedimentation tank. The gases over the tank are odorous and generally are passed through an incinerator for complete combustion. The liquid supernatant is returned to the treatment process and retreated. Approximately 40 to 50 percent of the organics are returned to the treatment process, creating a very heavy load on the biological units. Economics does not favor thermal conditioning of sludges at the present time.

Vacuum Filtration Vacuum filtration has been the most common method employed in dewatering sludges. Vacuum filters consist of a rotary drum covered with a cloth-filter medium. Various plastic fibers as well as wool have been used for the filter cloth. The filter operates by drawing a vacuum as the drum rotates into chemically conditioned sludge. The vacuum holds a thin layer of sludge, which is dewatered as the drum leaves the sludge vat, carrying the attached sludge into the air. As the drum rotates the cloth to the opposite side, air-pressure jets replace the vacuum, causing the sludge cake to separate from the cloth medium as the cloth moves away from the drum. The cloth travels over a series of rollers, with the sludge being separated by a knife edge and dropping onto a conveyor belt by gravity. The dewatered sludge is moved on the conveyor belt to the next concentration point, while the filter cloth is spray-washed and returned to the drum prior to entering the sludge vat. Vacuum filters yield the poorest results on waste activated sludge and the best results on primary sludge. Waste activated sludge will concentrate to between 12 and 18 percent solids at a rate of 4.9 to 9.8 kg dry cake/($m^2 \cdot h$) [1 to 2 lb/($ft^2 \cdot h$)]. Primary sludge can be dewatered to 25 to 30 percent solids at a rate of 49 kg dry cake/($m^2 \cdot h$) [10 lb/$ft^2 \cdot h$)].

Pressure Filtration Pressure filtration has been used increasingly since the early 1970s because of its ability to produce a drier sludge cake. The pressure filters consist of a series of plates and frames separated by a cloth medium. Sludge is forced into the filter under pressure, while the filtrate is drawn off. When maximum pressure is reached, the influent-sludge flow is stopped and the pressure filter is allowed to discharge the residual filtrate prior to opening the filter and allowing the filter cake to drop by gravity to a conveyor belt below the filter press. The pressure filter operates at a pressure between 689 and 1380 kPa (100 and 200 psig) and takes 1.5 to 4 h for the pressure cycle. Normally, 20 to 30 min is required to remove the filter cake. The sludge cakes will vary from 20 to 25 percent for waste activated sludge to 50 percent for primary sludge. Chemical conditioning is necessary to obtain good dewatering of the sludges.

Belt-Press Filters The newest filter for handling waste activated sludge is the belt-press filter. The belt press utilizes a continuous cloth-filter belt. Waste activated sludge is spread over the filter medium, and water is removed initially by gravity. The open belt with the sludge moves into contact with a second moving belt, which squeezes the sludge layer between rollers with ever-increasing pressure. The sludge cake is removed at the end of the filter press by a knife blade, with the sludge dropping by gravity to a conveyor belt. Belt-press filters can produce up to 20 percent solids.

Sand Beds Sand filter beds can be used to dewater either anaerobically or aerobically digested sludges. They work best on relatively small treatment systems located in relatively dry areas. The sand bed consists of coarse gravel graded to fine sand in a series of layers to a depth of 0.45 to 0.6 m (1.5 to 2 ft). The digested sludge is placed over the entire filter surface to a depth of 0.3 m (12 in) and allowed to sit until dry. Free water will drain through the sand bed to an open pipe underdrain system and be removed from the filter. Air drying will slowly remove the remaining water. The sludge must be cleaned from the bed by hand prior to adding a second layer of sludge. The sludge layer will drop from an initial thickness of 3 m (12 in) to about 0.006 m (¼ in). An open sand bed can generally handle 49 to 122 kg dry solids/($m^2 \cdot$ year) [10 to 25 lb/($ft^2 \cdot$ year)]. Covered sand beds have been used in wet climates as well as in cold climates, but economics does not favor their use.

SLUDGE DISPOSAL

Incineration Incineration has been used to reduce the volume of sludge after dewatering. The organic fractions in sludges lend themselves to incineration if they do not have too much water. Multiple-hearth and fluid-bed incinerators have been extensively used for sludge combustion.

A multiple-hearth incinerator consists of several hearths in a vertical cylindrical furnace. The dewatered sludge is added to the top hearth and is slowly pushed through the incinerator, dropping by gravity to the next lower layer until it finally reaches the bottom layer. The top layer is used for drying the sludge with the hot gases from the lower layers. As the temperature of the furnace increases, the organics begin to degrade and undergo combustion. Air is used to add the necessary oxygen and to control the temperature during combustion. It is very important to keep temperatures above 600°C to ensure complete oxidation of the volatile organics. One of the problems with the multiple-hearth incinerator is volatilization of odorous organics during the drying phase before the temperature reaches combustion levels. Even afterburners on the exhaust-gas line may not be adequate for complete oxidation. Air-pollution-control devices are required on all incinerators to remove fly ash and corrosive gases. The ash from the incinerator must be cooled, collected, and conveyed back to the environment, normally to a sanitary landfill for burial. The residual ash will weigh from 10 to 30 percent of the original dry weight of the sludge. Supplemental fuels are needed to start the incinerator and to ensure adequate temperatures with sludges containing excessive moisture, such as activated sludge. Heat recovery from wastes is being given more consideration. It is possible to combine the sludges with other wastes to provide a better fuel for the incinerator.

A fluid-bed incinerator uses hot sand as a heat reservoir for dewatering the sludge and combusting the organics. The turbulence created by the incoming air and the sand suspension requires the effluent gases to be treated in a wet scrubber prior to final discharge. The ash is removed from the scrubber water by a cyclone separator. The scrubber water is normally returned to the treatment process and diluted with the total plant effluent. The ash is normally buried.

Sanitary Landfills Dewatered sludge, either raw or digested, is often buried in a sanitary landfill to minimize the environmental impact. Increased concern over sanitary landfills has made it more difficult simply to bury dewatered sludge. Sanitary landfills must be made secure from leachate and be monitored regularly to ensure that no environmental damage occurs. The moisture content of most sludges makes them a problem at sanitary landfills designed for solid wastes, requiring separate burial even at the same landfill.

Land Spreading The nutrient content of most sludges makes them useful as fertilizers or as soil conditioners if properly mixed with the surface soil. Land spreading has gained in popularity in agricultural areas. Normally, the rate of application of sludge to land is controlled by the nitrogen content of the sludge. Since nitrogen uptake varies with different crops, nitrogen application is limited to approximately twice the annual uptake of nitrogen by the proposed crop. Approximately one-half of the nitrogen is readily available in sludge. Nutrient release with sludge is slower than with chemical fertilizers, allowing the nutrients to become available as the crop needs it. Activated sludge appears to be an excellent soil conditioner because the humus material in the sludge provides a good matrix for root growth, while the nutrient elements are released in approximately the right combination for optimal plant growth. There is a growing concern over heavy metals in some sludge, and care should be taken to minimize heavy-metal concentrations in sludges placed on the land. Since heavy metals cannot be easily removed from sludges, it is important to prevent them from entering the wastewater-treatment system. Greater concern will be placed on other potentially toxic or hazardous materials, including some organic compounds such as pesticides and PCBs. Land spreading of sludge requires careful application of the sludge at the surface and its mixing with the soil. Soil microbes will assist in further stabilization of any biodegradable organics remaining. Land spreading of sludge will become more popular as energy and nutrients become scarcer.

RECYCLING AND REUSE

Water Conservation Water shortages are increasing in some areas of the United States and will increase further as the population grows. Industries will be faced with increased water-conservation measures. Less water will be wasted, and waste concentrations will generally increase. In some areas dual water systems will be employed to permit use of treated wastewater effluents for lawn sprinkling and other nonpotable uses. Potable water will be carefully regulated to ensure that it is safe for use. Periodic droughts have provided data on methods to conserve limited water resources while minimizing the adverse impact on the industry involved. More and more industries are setting up and maintaining regular water-conservation programs.

Waste Segregation Greater efforts will be made to segregate toxic and hazardous wastes. Resource Conservation and Recovery Act (RCRA) requirements of the EPA are designed to encourage minimum loss of toxic and hazardous chemicals. Many industries will find that it is economical to segregate certain wastes and to reuse them in the process or to keep them segregated as future raw materials. Waste segregation may make it easier to handle certain wastes prior to mixing with carriage waters. There is no single approach to waste segregation. Each industry must examine its own processes and determine exactly where wastes are being generated and how each waste stream can best be controlled.

Zero Discharge Public Law 92-500 set zero discharge of pollutants in the rivers and streams of the United States. While zero discharge is an excellent concept, it is fundamentally unsound because there will always be some liquid wastes that must be treated and returned to the environment. Although zero discharge is not a practical goal, zero pollution is a practical goal that can be attained by every industry. Concern over water pollution has forced many industries to reexamine their basic processes and to develop modifications that could produce the same product with less waste. Less waste generally means greater profit. Water pollution control is a positive part of every industry and will play an even greater role as scarce resources must be conserved and used to their maximum extent. Enforcement of water-pollution-control regulations may vary with different governmental administrations, but they will continue to be a part of normal industrial operations.

MANAGEMENT OF INDUSTRIAL SOLID WASTES

General References: A. V. Bridgwater and C. J. Mumford, *Waste Recycling and Pollution Control Handbook*, Van Nostrand Reinhold, New York, 1979. V. Cavaseno (ed.), *Industrial Wastewater and Solid Waste Engineering*, McGraw-Hill, New York, 1980. EMCON Associates, *Methane Generation and Recovery from Landfills*, Ann Arbor Science Publishers/Butterworth Group, Woburn, Mass., 1981. A. Mallow, *Hazardous Waste Regulations: An Interpretive Guide*, Van Nostrand Reinhold, New York, 1981. N. L. Nemerow, *Industrial Solid Wastes*, Ballinger Publishing Co., Cambridge, Mass., 1983. R. B. Pojasek, (ed.), *Toxic and Hazardous Waste Disposal*, vols. 1–6, Ann Arbor Science Publishers/Butterworth Group, Woburn, Mass., 1979–1981. G. Tchobanoglous, H. Theisen, and R. Eliassen, *Solid Wastes: Engineering Principles and Management Issues*, McGraw-Hill, New York, 1977. P. A. Vesilind and A. E. Rimer, *Unit Operations in Resource Recovery Engineering*, Prentice-Hall, Englewood Cliffs, N.J., 1981. D. G. Wilson, (ed.), *Handbook of Solid Waste Management*, Van Nostrand Reinhold, New York, 1977.

INTRODUCTION

"Solid wastes" are all the wastes arising from human and animal activities that are normally solid and that are discarded as useless or unwanted. The term as used in this subsection is all-inclusive, and it encompasses the heterogeneous mass of throwaways from several industrial activities as well as the more homogeneous accumulations of a single industrial activity. To avoid confusion the term "refuse," often used interchangeably with the term solid wastes, is not used in this subsection.

Functional Elements The activities associated with the management of solid wastes from the point of generation to final disposal have been grouped into the six functional elements identified in Fig. 26-39. By considering each fundamental element separately it is possible to (1) identify the fundamental aspects and relationships involved in each element and (2) develop, when possible, quantifiable relationships for the purpose of making engineering comparisons, analyses, and evaluations.

Waste Generation Waste generation encompasses those activities in which materials are identified as no longer being of value and are either thrown away or gathered together for disposal. From the standpoint of economics, the best place to sort waste materials for recovery is at the source of generation.

On-Site Handling, Storage, and Processing This functional element encompasses those activities associated with the handling, storage, and processing of solid wastes at or near the point of generation. On-site storage is of primary importance because of the aesthetic considerations, public health, public safety, and economics involved.

Collection The functional element of collection includes the gathering of solid wastes and the hauling of wastes after collection to the location where the collection vehicle is emptied. As shown in Fig. 26-39, this location may be a transfer station, a processing station, or a landfill disposal site.

Transfer and Transport The functional element of transfer and transport involves two steps: (1) the transfer of wastes from the smaller collection vehicle to the larger transport equipment and (2) the subsequent transport of the wastes, usually over long distances, to the disposal site.

Processing and Recovery The functional element of processing and recovery includes all the techniques, equipment, and facilities used both to improve the efficiency of the other functional elements and to recover usable materials, conversion products, or energy from solids wastes.

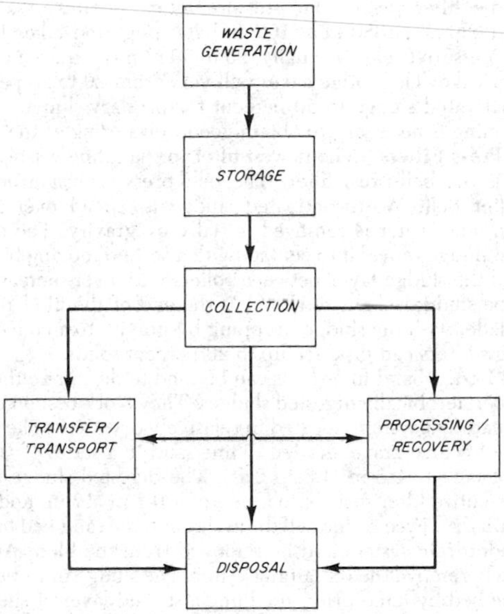

FIG. 26-39 Functional elements in a solid-waste-management system. *(From G. Tchobanoglous, H. Theisen, and R. Eliassen,* Solid Wastes: Engineering Principles and Management Issues, *McGraw-Hill, New York, 1977.)*

Disposal The final functional element in the solid-waste-management system depicted in Fig. 26-39 is disposal. Disposal is the ultimate fate of all solid wastes, whether they are wastes collected and transported directly to a landfill site, semisolid wastes (sludge) from industrial treatment plants, incinerator residue, compost, or other substances from various solid-waste-processing plants that are of no further use.

Solid-Waste-Management Systems Practical aspects associated with solid-waste-management systems not covered in this presentation include financing; operations; equipment management; personnel; reporting, cost accounting, and budgeting; contract administration; ordinances and guidelines; and public communications.

UNITED STATES LEGISLATION, REGULATIONS, AND GOVERNMENTAL AGENCIES

Much of the current activity in the field of solid-waste management, especially with respect to hazardous wastes and resources recovery, is a direct consequence of recent legislation. It, therefore, is important to (1) review the principal legislation that has affected the entire field of solid-waste management and (2) examine the role of several governmental agencies responsible for implementing the legislation.

Legislation What follows is a brief review of existing legislation that affects the management of solid wastes. The actual legislation must be consulted for specific details. Implementation of the legislation is accomplished through regulations adopted by federal, state, and local agencies. Because these regulations are revised continuously, they must be monitored continuously, especially when design and construction work is to be undertaken.

Rivers and Harbors Act, 1899 Passed in 1899, the Rivers and Harbors Act directed the U.S. Army Corps of Engineers to regulate the dumping of debris in navigable waters and adjacent lands.

Solid Waste Disposal Act, 1965 Modern solid-waste legislation dates from 1965, when the Solid Waste Disposal Act, Title II of Public Law 88-272, was enacted by Congress. The principal intent of this act was to promote the demonstration, construction, and application of solid-waste-management and resource-recovery systems which preserve and enhance the quality of air, water, and land resources.

National Environmental Policy Act, 1969 The National Environmental Policy Act (NEPA) of 1969 was the first federal act that required coordination of federal projects and their impacts with the nation's resources. The act specified the creation of the Council on Environmental Quality in the Executive Office of the President. This body has the authority to force every federal agency to submit to the council an environmental impact statement on every activity or project which it may sponsor or over which it has jurisdiction.

Resources Recovery Act, 1970 The Solid Waste Disposal Act of 1965 was amended by Public Law 95-512, the Resources Recovery Act of 1970. This act directed that the emphasis of the national solid-waste-management program should be shifted from disposal as its primary objective to that of recycling and reuse of recoverable materials in solid wastes or to the conversion of wastes to energy.

Occupational Safety and Health Act, 1970 Passed by Congress in 1970, the Occupational Safety and Health Act (OSHA) was designed to protect workers by improving the safety of their working environment. The chemical industries are most directly affected by Subpart 2, which deals with the exposure of workers to toxic and hazardous substances.

Resource Conservation and Recovery Act, 1976 The Resource Conservation and Recovery Act (RCRA), passed in 1976, was enacted to accomplish two major goals: (1) to ensure that all solid and hazardous wastes are managed in a manner that will protect public health and the environment, and (2) to conserve national resources directly and through the management, reuse, or recovery of solid and hazardous wastes. To meet these goals, three major programs are identified in the act. They are (1) the establishment of a hazardous-waste-control program to be administered by the states or, when the states choose not to do so, by EPA; (2) the establishment of a land-disposal regulatory program in each state; and (3) the initiation and support of resource-conservation programs by state and local governments to conserve resources and reduce the amount of solid waste requiring land disposal.

Toxic Substances Control Act, 1976 The two major goals of the Toxic Substances Control Act (TSCA), passed by Congress in 1976, are (1) the acquisition of sufficient information to identify and evaluate potential hazards from chemical substances and (2) the regulation of the manufacture, processing, distribution, use, and disposal of any substance that presents an unreasonable risk of injury to health or the environment.

Miscellaneous Laws Other laws that apply to the environmental control of solid-waste-management problems include the Noise Pollution and Abatement Act of 1970 and the Clean Air Act of 1970 (Public Law 91-604).

Governmental Agencies The various laws, regulations, and executive orders have created a divided responsibility among many federal departments and agencies for the regulation and financing of solid-waste management. Some of the more significant agencies are as follows.

Environmental Protection Agency The EPA was created by presidential order in 1970 to become the central or lead agency for the control of pollution of the nation's air, water, and land resources. It took over the responsibilities of the U.S. Public Health Service for air pollution control, quality of water supply, and solid-waste management. The Federal Water Pollution Control Administration was abolished and incorporated into the EPA.

U.S. Army Corps of Engineers The responsibility of the U.S. Army Corps of Engineers in the field of solid-waste management has been mentioned previously.

Department of Labor The Occupational Safety and Health Act of 1970 directed the secretary of labor to set mandatory standards to protect the occupational health and safety of all employers and employees of businesses engaged in interstate commerce. Occupational safety is a significant problem in the design of solid-waste-management facilities, which in the past have been subject to high accident rates.

Department of Transportation Stringent regulations have been placed on the transportation of hazardous wastes. Containers must be labeled, and the wastes must be placed in specially designed and approved containers. The Coast Guard controls marine shipment of solid wastes of a hazardous nature.

Interstate Commerce Commission Freight rates are controlled by the Interstate Commerce Commission. Differences now exist in the rates for the shipment of virgin (unprocessed) materials as opposed to higher rates for manufactured and reclaimed materials.

Department of Health and Human Services Public-health effects of solid-waste disposal facilities, particularly land disposal sites where vector control is necessary for the prevention of disease transmission, come within the province of the Department of Health and Human Services. Cooperation with state departments of public health is maintained through regional offices.

GENERATION OF SOLID WASTES

Solid wastes, as noted previously, include all solid or semisolid materials that are no longer considered of sufficient value to be retained in a given setting. (The wastes that are discharged may be of significant value in another setting.) The types and sources of solid wastes, the physical and chemical composition of solid wastes, and typical solid-waste-generation rates are considered in this subsection.

Types of Solid Wastes The term solid wastes is all-inclusive and encompasses all sources, types of classifications, compositions, and properties. As a basis for subsequent discussions, it will be helpful to define the various types of solid wastes that are generated.

Conventional Solid Wastes It is important to note that the definitions of solid-waste terms and the classifications vary greatly in practice and in the literature. Consequently, the use of published data requires considerable care, judgment, and common sense. The following definitions are intended to serve as a guide.

1. *Food wastes.* Food wastes are the animal, fruit, or vegetable residues (also called garbage) resulting from the handling, preparation, cooking, and eating of foods. The most important characteristic of these wastes is that they are putrescible and will decompose rapidly, especially in warm weather.
2. *Rubbish.* Rubbish consists of combustible and noncombustible solid wastes, excluding food wastes or other putrescible materials. Typically, combustible rubbish consists of materials such as paper, cardboard, plastics, textiles, rubber, leather, wood, furniture, and garden trimmings. Noncombustible rubbish consists of items such as glass, crockery, tin cans, aluminum cans, ferrous and other nonferrous metals, dirt, and construction wastes.
3. *Ashes and residues.* These are the materials remaining from the burning of wood, coal, coke, and other combustible wastes. Residues from power plants normally are not included in this category. Ashes and residues are normally composed of fine, powdery materials, cinders, clinkers, and small amounts of burned and partially burned materials.
4. *Demolition and construction wastes.* Wastes from razed buildings and other structures are classified as demolition wastes. Wastes from the construction, remodeling, and repair of commercial and industrial buildings and other similar structures are classified as construction wastes. These wastes may include dirt, stones, concrete, bricks, plaster, lumber, shingles, and plumbing, heating, and electrical parts.
5. *Special wastes.* Wastes such as street sweepings, roadside lit-

TABLE 26-25 Sources and Types of Industrial Wastes*

Code	SIC group classification	Waste-generating processes	Expected specific wastes
19	Ordnance and accessories	Manufacturing, assembling	Metals, plastic, rubber, paper, wood, cloth, chemical residues
20	Food and kindred products	Processing, packaging, shipping	Meats, fats, oils, bones, offal, vegetables, fruits, nuts and shells, cereals
22	Textile mill products	Weaving, processing, dyeing, shipping	Cloth and filter residues
23	Apparel and other finished products	Cutting, sewing, sizing, pressing	Cloth, fibers, metals, plastics, rubber
24	Lumber and wood products	Sawmills, millwork plants, wooden containers, miscellaneous wood products, manufacturing	Scrap wood, shavings, sawdust; in some instances, metals, plastics, fibers, glues, sealers, paints, solvents
25*a*	Furniture, wood	Manufacture of household and office furniture, partitions, office and store fixtures, mattresses	Those listed under Code 24; in addition, cloth and padding residues
25*b*	Furniture, metal	Manufacture of household and office furniture, lockers, springs, frames	Metals, plastics, resins, glass, wood, rubber, adhesives, cloth, paper
26	Paper and allied products	Paper manufacture, conversion of paper and paperboard, manufacture of paperboard boxes and containers	Paper and fiber residues, chemicals, paper coatings and fillers, inks, glues, fasteners
27	Printing and publishing	Newspaper publishing, printing, lithography, engraving, bookbinding	Paper, newsprint, cardboard, metals, chemicals, cloth, inks, glues
28	Chemicals and related products	Manufacture and preparation of inorganic chemicals (ranging from drugs and soaps to paints and varnishes and explosives)	Organic and inorganic chemicals, metals, plastics, rubber, glass, oils, paints, solvents, pigments
29	Petroleum refining and related industries	Manufacture of paving and roofing materials	Asphalt and tars, felts, asbestos, paper, cloth, fiber
30	Rubber and miscellaneous plastic products	Manufacture of fabricated rubber and plastic products	Scrap rubber and plastics, lampback, curing compounds, dyes
31	Leather and leather products	Leather tanning and finishing, manufacture of leather belting and packing	Scrap leather, thread, dyes, oils, processing and curing compounds
32	Stone, clay, and glass products	Manufacture of flat glass, fabrication or forming of glass; manufacture of concrete, gypsum, and plaster products; forming and processing of stone and stone products, abrasives, asbestos, and miscellaneous nonmineral products	Glass, cement, clay, ceramics, gypsum, asbestos, stone, paper, abrasives
33	Primary metal industries	Melting, casting, forging, drawing, rolling, forming, extruding operations	Ferrous and nonferrous metals scrap, slag, sand, cores, patterns, bonding agents
34	Fabricated metal products	Manufacture of metal cans, hand tools, general hardware, nonelectrical heating apparatus, plumbing fixtures, fabricated structural products, wire, farm machinery and equipment, coating and engraving of metal	Metals, ceramics, sand, slag, scale, coatings, solvents, lubricants, pickling liquors
35	Machinery (except electrical)	Manufacture of equipment for construction, mining, elevators, moving stairways, conveyors, industrial trucks, trailers, stackers, machine tools, etc.	Slag, sand, cores, metal scrap, wood, plastics, resins, rubber, cloth, paints, solvents, petroleum products
36	Electrical	Manufacture of electric equipment, appliances, and communication apparatus, machining, drawing, forming, welding, stamping, winding, painting, plating, baking, firing operations	Metal scrap, carbon, glass, exotic metals, rubber, plastics, resins, fibers, cloth residues
37	Transportation equipment	Manufacture of motor vehicles, truck and bus bodies, motor-vehicle parts and accessories, aircraft and parts, ship and boat building, repairing motorcycles and bicycles and parts, etc.	Metal scrap, glass, fiber, wood, rubber, plastics, cloth, paints, solvents, petroleum products
38	Professional scientific controlling instruments	Manufacture of engineering, laboratory, and research instruments and associated equipment	Metals, plastics, resins, glass, wood, rubber, fibers, abrasives
39	Miscellaneous manufacturing	Manufacture of jewelry, silverware, plated ware, toys, amusement, sporting, and athletic goods, costume novelties, buttons, brooms, brushes, signs, advertising displays	Metals, glass, plastics, resins, leather, rubber, composition, bone, cloth, straw, adhesives, paints, solvents

*From C. L. Mantell (ed.), *Solid Wastes: Origin, Collection, Processing, and Disposal*, Wiley-Interscience, New York, 1975.

ter, catch-basin debris, dead animals, and abandoned vehicles are classified as special wastes.

6. *Treatment-plant wastes*. The solid and semisolid wastes from water, wastewater, and industrial waste-treatment facilities are included in this classification.

7. *Agricultural wastes*. Wastes and residues resulting from diverse agricultural activities, such as the planting and harvesting of row, field, and tree and vine crops, the production of milk, the production of animals for slaughter, and the operation of feedlots, collectively are called agricultural wastes.

Hazardous Wastes Wastes that pose a substantial danger immediately or over a period of time to human, plant, or animal life are classified as hazardous wastes. A waste is classified as hazardous if it exhibits the following characteristics: (1) ignitability, (2) corrosivity, (3) reactivity, and (4) toxicity. A detailed definition of these terms was first published in the *Federal Register* on May 19, 1980, pages 33, 121–122.

In the past, hazardous wastes were often grouped into the following categories: (1) radioactive substances, (2) chemicals, (3) biological wastes, (4) flammable wastes, and (5) explosives. The chemical category included wastes that were corrosive, reactive, and toxic. The principal sources of hazardous biological wastes are hospitals and biological-research facilities.

Sources of Industrial Wastes Knowledge of the sources and types of solid wastes, along with data on the composition and rates of generation, is basic to the design and operation of the functional elements associated with the management of solid wastes.

Conventional Wastes Sources and types of industrial solid wastes generated by Standard Industrial Classification (SIC) group classification are reported in Table 26-25. The expected specific wastes in the table are those that are most readily identifiable.

Hazardous Wastes Hazardous wastes are generated in limited amounts throughout most industrial activities. In terms of generation, concern is with the identification of the amounts and types of hazardous wastes developed at each source, with emphasis on those sources where significant waste quantities are generated. Unfortunately, very little information is available on the quantities of hazardous wastes generated in various industries.

The generation of hazardous wastes by spillage must also be considered. The quantities of hazardous wastes that are involved in spillages usually are not known. After a spill, the wastes requiring collection and disposal are often significantly greater than the amount of spilled wastes, especially when an absorbing material, such as straw, is used to soak up liquid hazardous wastes or when the soil into which a hazardous liquid waste has percolated must be excavated. Both the straw and the liquid and the soil and the liquid are classified as hazardous wastes.

Properties of Solid Wastes Information on the properties of solid wastes is important in evaluating alternative equipment needs, systems, and management programs and plans.

Physical Composition Information and data on the physical composition of solid wastes including (1) identification of the individual components that make up industrial and municipal solid wastes, (2) density of solid wastes, and (3) moisture content are presented below.

1. *Individual components*. Components that typically make up most industrial and municipal solid wastes and their relative distribution are reported in Table 26-26. Although any number of components could be selected, those listed in the table have been chosen because they are readily identifiable, are consistent with component categories reported in the literature, and are adequate for the characterization of solid wastes for most applications.

TABLE 26-26 Typical Data on Distribution of Industrial Wastes Generated by Major Industries and Municipalities*

	Percent by mass									
SIC code	Food wastes†	Paper	Wood	Leather	Rubber	Plastics	Metals	Glass	Textiles	Miscellaneous
20 Food and kindred products	15–20	50–60	5–10	0–2	0–2	0–5	5–10	4–10	0–2	5–15
22 Textile mill products	0–2	40–50	0–2	0–2	0–2	3–10	0–2	0–2	20–40	0–5
23 Apparel and other finished products	0–2	40–60	0–2	0–2	0–2	0–2	0–2	0–2	30–50	0–5
24 Lumber and wood products	0–2	10–20	60–80	0–2	0–2	0–2	0–2	0–2	0–2	5–10
25a Furniture, wood	0–2	20–30	30–50	0–2	0–2	0–2	0–2	0–2	0–5	0–5
25b Furniture, metal	0–2	20–40	10–20	0–2	0–2	0–2	20–40	0–2	0–5	0–10
26 Paper and allied products	0–2	40–60	10–15	0–2	0–2	0–2	5–15	0–2	0–2	10–20
27 Printing and publishing	0–2	60–90	5–10	0–2	0–2	0–2	0–2	0–2	0–2	0–5
28 Chemicals and related products	0–2	40–60	2–10	0–2	0–2	5–15	5–10	0–5	0–2	15–25
29 Petroleum refining and related industries	0–2	60–80	5–15	0–2	0–2	10–20	2–10	0–12	0–2	2–10
30 Rubber and miscellaneous plastics products	0–2	40–60	2–10	0–2	5–20	10–20	0–2	0–2	0–2	0–5
31 Leather and leather products	0–2	5–10	5–10	40–60	0–2	0–2	10–20	0–2	0–2	0–5
32 Stone, clay, and glass products	0–2	20–40	2–10	0–2	0–2	0–2	5–10	10–20	0–2	30–50
33 Primary metal industries	0–2	30–50	5–15	0–2	0–2	2–10	2–10	0–5	0–2	20–40
34 Fabricated metal products	0–2	30–50	5–15	0–2	0–2	0–2	15–30	0–2	0–2	5–15
35 Machinery (except electrical)	0–2	30–50	5–15	0–2	0–2	1–5	15–30	0–2	0–2	0–5
36 Electrical	0–2	60–80	5–15	0–2	0–2	2–5	2–5	0–2	0–2	0–5
37 Transportation equipment	0–2	40–60	5–15	0–2	0–2	2–5	0–2	0–2	0–2	15–30
38 Professional scientific controlling instruments	0–2	30–50	2–10	0–2	0–2	5–10	5–15	0–2	0–2	0–5
39 Miscellaneous manufacturing	0–2	40–60	10–20	0–2	0–2	5–15	2–10	0–2	0–2	5–15
Municipal	10–20	40–60	1–4	0–2	0–2	2–10	3–15	4–16	0–4	5–30

*Adapted in part from D. G. Wilson (ed.), *Handbook of Solid Waste Management*, Van Nostrand Reinhold, New York, 1977.

†With the exception of food and kindred products, food wastes are from company cafeterias, canteens, etc.

2. *Density.* Typical densities for various wastes as found in containers are reported by source in Table 26-27. Because the densities of solid wastes vary markedly with geographic location, season of the year, and length of time in storage, great care should be used in selecting typical values.

3. *Moisture content.* The moisture content of solid wastes usually is expressed as the mass of moisture per unit mass of wet or dry material. In the wet-mass method of measurement, the moisture in a sample is expressed as a percentage of the wet mass of the material; in the dry-mass method, it is expressed as a percentage of the dry mass of the material. In equation form, the wet-mass moisture content is expressed as follows:

$$\text{Moisture content } (\%) = \left(\frac{a-b}{a}\right) 100 \qquad (26\text{-}1)$$

where a = initial mass of sample as delivered
b = mass of sample after drying

Typical data on the moisture content for the solid-waste components are given in Table 26-27. For most industrial solid wastes, the moisture content will vary from 10 to 25 percent.

Chemical Composition Information on the chemical composition of solid wastes is important in evaluating alternative processing and recovery options. If solid wastes are to be used as fuel, the four most important properties to be known are:

1. Proximate analysis
   a. Moisture (loss at 105°C for 1 h)
   b. Volatile matter (additional loss on ignition at 950°C)
   c. Ash (residue after burning)
   d. Fixed carbon (remainder)
2. Fusing point of ash
3. Ultimate analysis, percent of C (carbon), H (hydrogen), O (oxygen), N (nitrogen), S (sulfur), and ash
4. Heating value

TABLE 26-27 Typical Density and Moisture-Content Data for Domestic, Commercial, and Industrial Solid Waste

Item	Density, kg/m³		Moisture content, % by mass	
	Range	Typical	Range	Typical
Residential (uncompacted)				
Food wastes (mixed)	130–480	290	50–80	70
Paper	40–130	85	4–10	6
Cardboard	40–80	50	4–8	6
Plastics	40–130	65	1–4	2
Textiles	40–100	65	6–15	10
Rubber	100–200	130	1–4	2
Leather	100–260	160	8–12	10
Garden trimmings	60–225	100	30–80	60
Wood	130–320	240	15–40	20
Glass	160–480	195	1–4	2
Tin cans	50–160	90	2–4	3
Nonferrous metals	65–240	160	2–4	2
Ferrous metals	130–1150	320	2–4	2
Dirt, ashes, etc.	320–1000	480	6–12	8
Ashes	650–830	745	6–12	6
Rubbish (mixed)	90–180	130	5–20	15
Residential (compacted)				
In compactor truck	180–450	300	15–40	20
In landfill (normally compacted)	360–500	450	15–40	30*
In landfill (well-compacted)	590–740	600	15–40	30*
Commercial				
Food wastes (wet)	475–950	535	50–85	75
Appliances	150–200	180	0–5	...
Wooden crates	110–160	110	10–30	20
Tree trimmings	100–180	150	20–80	50
Rubbish (combustible)	50–180	120	5–25	15
Rubbish (noncombustible)	180–360	300	5–15	10
Rubbish (mixed)	140–180	160	5–20	12
Construction; demolition				
Mixed demolition (noncombustible)	1000–1600	1420	2–10	4
Mixed demolition (combustible)	300–400	360	4–15	8
Mixed construction (combustible)	180–360	260	4–15	8
Broken concrete	1200–1800	1540	0–5	...
Industrial wastes				
Chemical sludges (wet)	800–1100	1000	75–99	80
Fly ash	700–900	800	2–10	4
Leather scraps	100–250	160	6–15	10
Metal scrap (heavy)	1500–2000	1780	0–5	...
Metal scrap (light)	500–900	740	0–5	...
Metal scrap (mixed)	700–1500	900	0–5	...
Oils, tars, asphalts	800–1000	950	0–5	2
Sawdust	100–350	290	10–40	15
Textile wastes	100–220	180	6–15	10
Wood (mixed)	400–675	500	10–40	20
Agricultural wastes				
Agricultural (mixed)	400–750	560	40–80	50
Fruit wastes (mixed)	250–750	360	60–90	75
Manure (wet)	900–1050	1000	75–96	94
Vegetable wastes (mixed)	200–700	360	50–80	65

*Depends on degree of surface-water infiltration.

TABLE 26-28 Typical Proximate-Analysis and Energy-Content Data for Components in Domestic, Commercial, and Industrial Solid Waste*

	Proximate analysis, % by mass				Energy content, kJ/kg		
Component	Moisture	Volatile matter	Fixed carbon	Noncombustible	As collected	Dry	Moisture- and ash-free
Food and food products							
Fats	2.0	95.3	2.5	0.2	37,530	38,296	38,374
Food wastes (mixed)	70.0	21.4	3.6	5.0	4,175	13,917	16,700
Fruit wastes	78.7	16.6	4.0	0.7	3,970	18,638	19,271
Meat wastes	38.8	56.4	1.8	3.1	17,730	28,970	30,516
Paper products							
Cardboard	5.2	77.5	12.3	5.0	16,380	17,278	18,240
Magazines	4.1	66.4	7.0	22.5	12,220	12,742	16,648
Newsprint	6.0	81.1	11.5	1.4	18,550	19,734	20,032
Paper (mixed)	10.2	75.9	8.4	5.4	15,815	17,611	18,738
Waxed cartons	3.4	90.9	4.5	1.2	26,345	27,272	27,615
Plastics							
Plastics (mixed)	0.2	95.8	2.0	2.0	32,000	32,064	32,720
Polyethylene	0.2	98.5	<0.1	1.2	43,465	43,552	44,082
Polystyrene	0.2	98.7	0.7	0.5	38,190	38,266	38,216
Polyurethane	0.2	87.1	8.3	4.4	26,060	26,112	27,316
Polyvinyl chloride	0.2	86.9	10.8	2.1	22,690	22,735	23,224
Wood, trees, etc.							
Garden trimmings	60.0	30	9.5	0.5	6,050	15,125	15,316
Green wood	50.0	42.3	7.3	0.4	4,885	9,770	9,848
Hardwood	12.0	75.1	12.4	0.5	17,100	19,432	19,542
Wood (mixed)	20.0	67.9	11.3	0.8	15,444	19,344	19,500
Leather, rubber, textiles, etc.							
Leather (mixed)	10	68.5	12.5	9.0	18,515	20,572	22,858
Rubber (mixed)	1.2	83.9	4.9	9.9	25,330	25,638	28,493
Textiles (mixed)	10	66.0	17.5	6.5	17,445	19,383	20,892
Glass, metals, etc.							
Glass and mineral	2	....	....	96–99+	196†	200	200
Metal, tin cans	5	....	....	94–99+	1,425†	1,500	1,500
Metal, ferrous	2	....	....	96–99+	....	....	....
Metal, nonferrous	2	....	....	94–99+	....	....	....
Miscellaneous							
Office sweepings	3.2	20.5	6.3	70	8,535	8,817	31,847
Municipal wastes	20 (15–40)	53 (30–60)	7 (5–15)	20 (9–30)	10,470	13,090	17,450
Industrial wastes	15 (10–30)	58 (30–60)	7 (5–15)	20 (10–30)	11,630	13,682	17,892

*Adapted in part from D. G. Wilson (ed.), *Handbook of Solid Waste Management*, Van Nostrand Reinhold, New York, 1977, and G. Tchobanoglous, H. Theisen, and R. Eliassen, *Solid Wastes: Engineering Principles and Management Issues*, McGraw-Hill, New York, 1977.

†Energy content is from coatings, labels, and attached materials.

Typical proximate-analysis data for the combustible components of industrial and municipal solid wastes are presented in Table 26-28.

Typical data on the inert residue and energy values for solid wastes are also reported in Table 26-28. As-discarded energy values may be converted to a dry basis by using Eq. (26-2):

$$\text{kJ/kg (dry basis)} = \text{kJ/kg (as discarded)} \left(\frac{100}{100 - \%\ \text{moisture}}\right) \quad (26\text{-}2)$$

The corresponding equation on an ash-free dry basis is

$$\text{kJ/kg (ash-free dry basis)} = \text{kJ/kg (as discarded)} \left(\frac{100}{100 - \%\ \text{ash} - \%\ \text{moisture}}\right) \quad (26\text{-}3)$$

Representative data on the ultimate analysis of typical industrial- and municipal-waste components are presented in Table 26-29. If energy values are not available, approximate values can be determined by using Eq. (26-4), known as the modified Dulong formula, and the data in Table 26-29.

$$\text{kJ/kg} = 337\ C + 1428\ (H - 1/8\ O) + 95\ S \quad (26\text{-}4)$$

where C = carbon, percent
H = hydrogen, percent
O = oxygen, percent
S = sulfur, percent

Quantities of Solid Wastes Representative data on the quantities of solids wastes and factors affecting the generation rates are considered briefly in the following paragraphs.

Typical Generation Rates Typical unit waste-generation rates for selected industrial sources are reported in Table 26-30. Because waste-generation practices are changing so rapidly, the presentation of specific waste-generation data is meaningless.

Factors That Affect Generation Rates Factors that influence the quantity of industrial wastes generated include (1) the extent of salvage and recycle operations, (2) company attitudes, and (3) legislation. The existence of salvage and recycling operations within an industry definitely affects the quantities of wastes collected. Whether such operations affect the quantities generated is another matter. Until more information is available, no definite statement can be made on this issue. Significant reductions in the quantities of solid wastes that are generated will occur when and if companies are willing to change—on their own volition—to conserve national resources and to reduce the economic burdens associated with the manage-

TABLE 26-29 Typical Ultimate-Analysis Data for Components in Domestic, Commercial, and Industrial Solid Waste*

Components	Percent by mass (dry basis)					
	Carbon	Hydrogen	Oxygen	Nitrogen	Sulfur	Ash
Foods and food products						
Fats	73.0	11.5	14.8	0.4	0.1	0.2
Food wastes (mixed)	48.0	6.4	37.6	2.6	0.4	5.0
Fruit wastes	48.5	6.2	39.5	1.4	0.2	4.2
Meat wastes	59.6	9.4	24.7	1.2	0.2	4.9
Paper products	45.4	6.1	42.1	0.3	0.1	6.0
Cardboard	43.0	5.9	44.8	0.3	0.2	5.0
Magazines	32.9	5.0	38.6	0.1	0.1	23.3
Newsprint	49.1	6.1	43.0	<0.1	0.2	23.3
Paper (mixed)	43.4	5.8	44.3	0.3	0.2	6.0
Waxed cartons	59.2	9.3	30.1	0.1	0.1	1.2
Plastics						
Plastics (mixed)	60.0	7.2	22.8	...	...	10.0
Polyethylene	85.2	14.2	...	<0.1	<0.1	0.4
Polystyrene	87.1	8.4	4.0	0.2	...	0.3
Polyurethane†	63.3	6.3	17.6	6.0	<0.1	4.3
Polyvinyl chloride†	45.2	5.6	1.6	0.1	0.1	2.0
Wood, trees, etc.						
Garden trimmings	46.0	6.0	38.0	3.4	0.3	6.3
Green timber	50.1	6.4	42.3	0.1	0.1	1.0
Hardwood	49.6	6.1	43.2	0.1	<0.1	0.9
Wood (mixed)	49.5	6.0	42.7	0.2	<0.1	1.5
Wood chips (mixed)	48.1	5.8	45.5	0.1	<0.1	0.4
Glass, metals, etc.						
Glass and mineral‡	0.5	0.1	0.4	<0.1	...	98.9
Metals (mixed)	4.5	0.6	4.3	<0.1	...	90.5
Leather, rubber, textiles						
Leather (mixed)	60.0	8.0	11.6	10.0	0.4	10.0
Rubber (mixed)	69.7	8.7	...	...	1.6	20.0
Textiles (mixed)	48.0	6.4	40.0	2.2	0.2	3.2
Miscellaneous						
Office sweepings	24.3	3.0	4.0	0.5	0.2	68.0
Oils, paints	66.9	9.6	5.2	2.0	...	16.3
Refuse-derived fuel (RDF)	44.7	6.2	38.4	0.7	<0.1	9.9

*Adapted in part from D. G. Wilson (ed.), *Handbook of Solid Waste Management*, Van Nostrand Reinhold, New York, 1977, and G. Tchobanoglous, H. Theisen, and R. Eliassen, *Solid Wastes: Engineering Principles and Management Issues*, McGraw-Hill, New York, 1977.
†Remainder is chlorine.
‡Organic content is from coatings, labels, and other attached materials.

TABLE 26-30 Unit Solid-Waste-Generation Rates for Selected Industrial Sources

Source	Unit	Range
Canned and frozen foods	Metric tons/metric tons of raw product	0.04–0.06
Printing and publishing	Metric tons/metric tons of raw paper	0.08–0.10
Automotive	Metric tons/vehicle produced	0.6–0.8
Petroleum refining	Metric tons/(employee·day)	0.04–0.05
Rubber	Metric tons/metric tons of raw rubber	0.01–0.3

ment of solid wastes. Perhaps the most important factor affecting the generation of certain types of wastes is the existence of local, state, and federal regulations concerning the use and disposal of specific materials. Legislation dealing with hazardous wastes is an example.

ON-SITE HANDLING, STORAGE, AND PROCESSING

The handling, storage, and processing of solid wastes at the source before they are collected is the second of the six functional elements in the solid-waste-management system.

On-Site Handling On-site handling refers to the activities associated with the handling of solid wastes until they are placed in the containers used for their storage before collection. Depending on the type of collection service, handling may also be required to move loaded containers to the collection point and to return the empty containers to the point where they are stored between collections.

Conventional Solid Wastes In most office, commercial, and industrial buildings, solid wastes that accumulate in individual offices or work locations usually are collected in relatively large containers mounted on rollers. Once filled, these containers are removed by means of the service elevator, if there is one, and emptied into (1) large storage containers, (2) compactors used in conjunction with the storage containers (see Fig. 26-40), (3) stationary compactors that can compress the material into bales or into specially designed containers, or (4) other processing equipment.

Hazardous Wastes When hazardous wastes are generated, special containers are usually provided and trained personnel are responsible (or should be) for the handling of these wastes.

On-Site Storage Factors that must be considered in the on-site storage of solid wastes include (1) the type of container to be used, (2) the container location, (3) public health and aesthetics, and (4) the collection methods to be used.

Containers To a large extent, the types and capacities of the containers used depend on the characteristics of the solid wastes to

FIG. 26-40 Small compactor used in conjunction with a large portable container. Note the wheeled container used to collect wastes within the warehouse.

be collected, the collection frequency, and the space available for the placement of containers.

1. *Containers for conventional wastes.* The types and capacities of containers now commonly used for on-site storage of solid wastes are summarized in Table 26-31. The small containers are used in individual offices and work stations. The medium-size and large containers are used at locations where large volumes are generated.

2. *Containers for hazardous wastes.* On-site storage practices are a function of the types and amounts of hazardous wastes generated and the time period over which waste generation occurs. Usually, when large quantities are generated, special facilities that have sufficient capacity to hold wastes accumulated over a period of several days are used. When only small amounts of hazardous wastes are generated on an intermittent basis, they may be containerized, and limited quantities may be stored for periods covering months or years. General information on the storage containers used for hazardous wastes and the conditions of their use is presented in Table 26-32.

Container Location The location of containers at existing commercial and industrial facilities depends on both the location of available space and service-access conditions. In newer facilities, specific service areas have been included for this purpose. Often, because the containers are not owned by the commercial or industrial activity, the locations and types of containers to be used for on-site storage must be worked out jointly between the industry and the public or private collection agency.

On-Site Processing of Solid Wastes On-site-processing methods are used to (1) recover usable materials from solid wastes, (2) reduce the volume, or (3) alter the physical form. The most common on-site-processing operations as applied to large commercial and industrial sources include manual sorting, compaction, and incineration. These and other processing operations are considered in the portion of this section dealing with processing and resource recovery. Factors that should be considered in the selection of on-site-processing equipment are summarized in Table 26-33.

COLLECTION OF SOLID WASTES

Information on collection, one of the most costly functional elements, is presented in four parts dealing with (1) the types of collection ser-

TABLE 26-31 Data on the Types and Sizes of Containers Used for On-Site Storage of Solid Wastes

	Capacity			Dimensions*	
Container type	Unit	Range	Typical	Unit	Typical
Small capacity					
Plastic or metal (office type)	L	16–40	28	mm	(180 × 300) B × (260 × 380) T × 380 H
Plastic or galvanized metal	L	75–150	120	mm	510 D × 660 H
Barrel, plastic, aluminum, or fiber barrel	L	75–250	120	mm	510 D × 660 H
Disposable paper bags (standard, leak-resistant, and leakproof)	L	75–210	120	mm	380 W × 300 d × 1100 H
Disposable plastic bag					460 W × 380 d × 1000 H
Medium capacity					
Side or top loading	m^3	0.75–9	3	mm	1830 W × 1070 d × 1650 H
Large capacity					
Open-top, roll-off (also called debris boxes)	m^3	9–38	27	mm	2440 W × 1830 H × 6100 L
Used with stationary compactor	m^3	15–30	23	mm	2440 W × 1830 H × 5490 L
Equipped with self-contained compaction mechanism	m^3	15–30	23	mm	2440 W × 2440 H × 6710 L
Trailer-mounted					
Open-top	m^3	15–38	27	mm	2440 W × 3660 H × 6100 L
Enclosed, equipped with self-contained compaction mechanism	m^3	15–30	27	mm	2440 W × 3660 H × 7320 L

*B = bottom, T = top, D = diameter, H = height, L = length, W = width, and d = depth.

TABLE 26-32 Typical Data on Containers Used for Storage and Transport of Hazardous Wastes

	Container		
Waste category	Type	Capacity	Auxiliary equipment and conditions of use
Radioactive substances	Lead encased in concrete Lined metal drums	Varies with waste 210 L	Isolated storage buildings; high-capacity hoists and lighting equipment; special container markings
Corrosive, reactive, and toxic chemicals	Metal drums Lined metal drums Lined and unlined storage tanks	210 L 210 L Up to 20 m^3	Washing facilities for empty containers; special blending precautions to prevent hazardous reactions; incompatible wastes stored separately
Biological wastes	Sealed plastic bags Lined metal drums	120 L	Heat sterilization prior to bagging; special heavy-duty bags with hazard warning printed on sides
Flammable wastes	Metal drums Storage tanks	210 L Up to 20 m^3	Fume ventilation; temperature control
Explosives	Shock-absorbing containers	Varies	Temperature control; special container markings

TABLE 26-33 Factors That Should Be Considered in Evaluating On-Site Processing Equipment

Factor	Evaluation
Capabilities	What will the device or mechanism do? Will its use be an improvement over conventional practices?
Reliability	Will the equipment perform its designated functions with little attention beyond preventive maintenance? Has the effectiveness of the equipment been demonstrated in use over a reasonable period of time or merely predicted?
Service	Will servicing capabilities beyond those of the local building maintenance staff be required occasionally? Are properly trained service personnel available through the equipment manufacturer or the local distributor?
Safety of operation	Is the proposed equipment reasonably foolproof so that it may be operated by tenants or building personnel with limited mechanical knowledge or abilities? Does it have adequate safeguards to discourage careless use?
Ease of operation	Is the equipment easy to operate by a tenant or by building personnel? Unless functions and actual operations of equipment can be carried out easily, they may be ignored or short-circuited by paid personnel and most often by "paying" tenants.
Efficiency	Does the equipment perform efficiently and with a minimum of attention? Under most conditions, equipment that completes an operational cycle each time that it is used should be selected.
Environmental effects	Does the equipment pollute or contaminate the environment? When possible, equipment should reduce environmental pollution presently associated with conventional functions.
Health hazards	Does the device, mechanism, or equipment create or amplify health hazards?
Aesthetics	Do the equipment and its arrangement offend the senses? Every effort should be made to reduce or eliminate offending sights, odors, and noises.
Economics	What are the economics involved? Both first and annual costs must be considered. Future operation and maintenance costs must be assessed carefully. All factors being equal, equipment produced by well-established companies, having a proven history of satisfactory operation, should be given appropriate consideration.

*From G. Tchobanoglous, H. Theisen, and R. Eliassen, *Solid Wastes: Engineering Principles and Management Issues*, McGraw-Hill, New York, 1977.

vices, (2) the types of collection systems, (3) an analysis of collection systems, and (4) the general methodology involved in setting up collection routes.

Collection Services The various types of collection services now used for commercial-industrial sources are described in this subsection.

Commercial-Industrial The collection service provided to large commercial and industrial activities typically is centered in the use of large movable and stationary containers and large stationary compactors. Compactors are of the type that can be used to compress material directly into large containers (see Fig. 26-40) or to form bales that are then placed in large containers.

Hazardous Wastes Hazardous wastes for delivery to a treatment or disposal facility normally are collected by the waste producer or a specialized hauler. Typically, the loading of collection vehicles is completed in one of two ways: (1) wastes stored in large-capacity tanks are either drained or pumped into collection vehicles, and (2) wastes stored in sealed drums or other sealed containers are loaded by hand or by mechanical equipment onto flatbed trucks. To avoid accidents and possible loss of life, two collectors should always be assigned when hazardous wastes are to be collected.

Types of Collection Systems On the basis of their mode of operation, collection systems are classified into two categories: (1) hauled-container systems and (2) stationary-container systems.

Hauled-Container Systems (HCS) Collection systems in which the containers used for the storage of wastes are hauled to the processing, transfer, or disposal site, emptied, and returned to either their original location or some other location are defined as "hauled-container systems." In most hauled-container systems, a single collector is used. The collector is responsible for driving the vehicle, loading full containers and unloading empty containers, and emptying the contents of the container at the disposal site. In some cases, for safety reasons, both a driver and helper are used.

There are three main types of hauled-container systems: (1) hoist-truck, (2) tilt-frame-container, and (3) trash-trailer. Typical data on the containers and collection vehicles used with these systems are reported in Tables 26-34 and 26-35 respectively.

1. *Hoist-truck systems.* With the advent of self-loading compaction-type collection vehicles, hoist-truck systems (see Fig. 26-41) are being used in a limited number of cases, the most important of which are (1) the collection of wastes from only a few pickup points at which a considerable amount of waste is generated and (2) the collection of bulky items and industrial rubbish not suitable for collection with compaction vehicles.

2. *Tilt-frame-container systems.* Systems that use tilt-frame-loaded vehicles and large containers, often called "drop boxes" (see Fig. 26-42), are ideally suited for the collection of all types of solid waste and rubbish from locations where the generation rate warrants

TABLE 26-34 Typical Data on Container Capacities for Use with Various Collection Systems

Collection: Vehicle	Collection: Container type	Typical range of container capacities,* m^3
Hauled-container systems		
Hoist-truck	Open and covered, also used with stationary compactor	2–9
Tilt-frame	Open-top, also called debris boxes	8–40
	Used in conjunction with stationary compactor	10–30
	Equipped with self-contained compaction mechanism	15–30
Truck-tractor	Open-top trash trailers	10–30
	Enclosed trailer-mounted containers equipped with self-contained compaction mechanism	15–30
Stationary-container systems		
Compactor, mechanically loaded	Open-top and enclosed top- and side-loading	0.6–6
Compactor, manually loaded	Small plastic or galvanized metal containers, disposable paper and plastic bags	75–200 L†

*See Table 26-31 for typical dimensions.
†Loaded mass of container should not exceed 30 kg.

TABLE 26-35 Typical Data on Vehicles Used for Collection of Solid Wastes

Collection vehicle			Typical overall collection-vehicle dimensions				
Type	Available container or truck body capacities,* m^3	Number of axles	With indicated container or truck body capacity,* m^3	Width, mm	Height, mm	Length,* mm	Unloading method
Hauled-container systems							
Hoist-truck	2–9	2	7.6	2440	2030–2540	2800–3800	Gravity, bottom-opening
Tilt-frame	8–40	3	22.9	2440	2030–2290	5590–7620	Gravity, inclined-tipping
Truck-tractor trash trailer	10–30	3	30.6	2440	2290–3800	5590–11,430	Gravity, inclined-tipping
Stationary-container systems							
Compactor (mechanically loaded)							
Front-loading	15–35	3	22.9	2440	3560–3800	6100–7370	Hydraulic ejector panel
Side-loading	6–28	3	22.9	2440	3350–3000	5590–6600	Hydraulic ejector panel
Rear-loading	6–24	2	15.3	2440	3175–3430	5330–5824	Hydraulic ejector panel
Compactor (manually loaded)							
Side-loading	6–28	3	28.3	2440	3350–3800	6100–7620	Hydraulic ejector panel
Rear-loading	6–24	2	15.3	2440	3175–3430	5330–5840	Hydraulic ejector panel

*From front of truck to rear of container or truck body.

FIG. 26-41 Typical hoist-truck collection vehicle used in a hauled-container system.

FIG. 26-42 Typical tilt-frame collection vehicle used in a hauled-container system. The container being unloaded is also known as a drop box.

the use of large containers. Open-top containers are used routinely at warehouses and construction sites. Large containers used in conjunction with stationary compactors are common at commercial and industrial services and transfer stations. Because of the large volume that can be hauled, the use of tilt-frame hauled-container systems has become widespread, especially among private collectors servicing industrial accounts.

3. *Trash-trailer systems.* The application of trash trailers is similar to that for tilt-frame-container systems. Trash trailers are better for the collection of especially heavy rubbish, such as sand, timber, and metal scrap, and often are used for the collection of demolition wastes at construction sites.

Stationary-Container Systems (SCS) Collection systems in which the containers used for the storage of wastes remain at the point of waste generation, except for occasional short trips to the collection vehicle, are defined as "stationary-container systems." Labor requirements for mechanically loaded stationary-container systems are essentially the same as for hauled-container systems. There are two main types of stationary-container systems: (1) those in which self-loading compactors are used and (2) those in which manually loaded vehicles are used.

1. *Systems with self-loading compactors.* Container size and utilization are not as critical in stationary-container systems using self-loading collection vehicles equipped with a compaction mechanism (see Fig. 26-43 and Table 26-35) as they are in hauled-container systems. Trips to the disposal site, transfer station, or processing station are made after the contents of a number of containers have been collected and compacted and the collection vehicle is full. Because a variety of container sizes and types are available, these systems may be used for the collection of all types of wastes. Container sizes vary from relatively small sizes (0.6 m^3) to sizes comparable to those handled with a hoist truck (see Table 26-34).

FIG. 26-43 Typical self-loading compactor used in a stationary-container collection system.

2. *Systems with manually loaded vehicles.* The major application of manual transfer and loading methods is in the collection of residential wastes and litter. Manual methods are used for the collection of industrial wastes when pickup points are inaccessible to the collection vehicle.

Equipment for Hazardous-Wastes Collection The equipment used for collection varies with the characteristics of the wastes. For short-haul distances, drum storage and collection with a flatbed truck are often the preferred methods. As hauling distances increase, larger tank trucks, trailers, and railroad tank cars are used.

Determination of Vehicle and Labor Requirements By separating collection activities into unit operations, it is possible to develop design data and relationships that can be used to establish vehicle and labor requirements for the various collection systems.

Definition of Terms The operational tasks for the hauled-container and stationary-container systems are shown schematically in Fig. 26-44. The activities involved in the collection of solid wastes can be resolved into four unit operations: (1) pickup, (2) haul, (3) at site, and (4) off route.

1. *Pickup*. The definition of the term "pickup" (P) depends on the type of collection system used.

a. For hauled-container systems (see Fig. 26-44*a*), pickup (P_{hcs}) refers to the time spent in picking up the loaded container, the time required to redeposit the container after its contents have been emptied, and the time spent in driving to the next container.

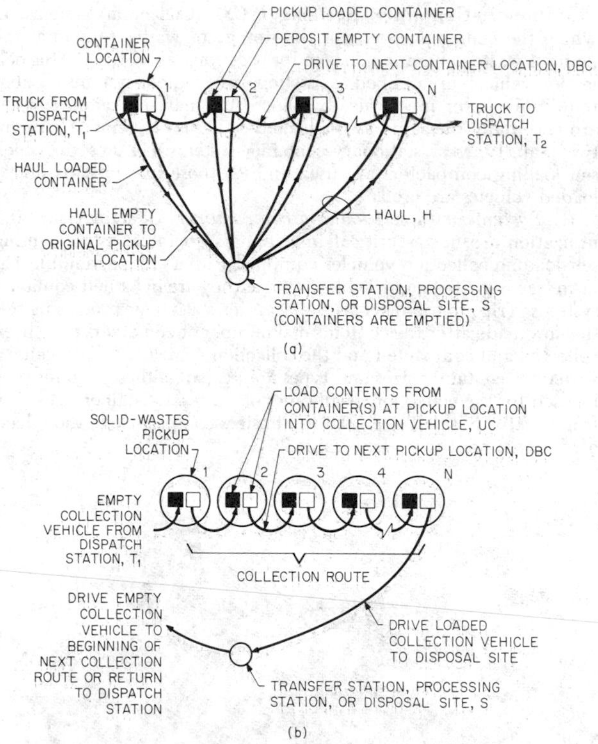

FIG. 26-44 Operation sequences for hauled- and stationary-container collection systems. (*a*) Hauled-container system. (*b*) Stationary-container system. (*Adapted from G. Tchobanoglous, H. Theisen, and R. Eliassen,* Solid Wastes: Engineering Principles and Management Issues, *McGraw-Hill, New York, 1977.*)

b. For stationary-container systems (see Fig. 26-44*b*), pickup (P_{scs}) refers to the time spent in loading the collection vehicle, beginning with the stopping of the vehicle prior to loading the contents of the first container and ending when the contents of the last container to be emptied have been loaded.

2. *Haul*. The definition of the term "haul" (h) also depends on the type of collection system used.

a. For hauled-container systems, haul represents the time required to reach the disposal site, starting after a container whose contents are to be emptied has been loaded on the truck, plus the time after leaving the disposal site until the truck arrives at the location where the empty container is to be redeposited. It does not include any time spent at the disposal site.

b. For stationary-container systems, haul refers to the time required to reach the disposal site, starting after the last container on the route has been emptied or the collection vehicle filled, plus the time after leaving the disposal site until the truck arrives at the location of the first container to be emptied on the next collection route. It does not include the time spent at the disposal site.

3. *At site*. The unit operation at site (s) refers to the time spent at the disposal site and includes the time spent in waiting to unload as well as the time spent in unloading.

4. *Off route*. The unit operation off route (W) includes all time spent on activities that are nonproductive from the point of view of the overall collection operation. Necessary off-route time includes (1) time spent in checking in and out in the morning and at the end of the day, (2) time lost due to unavoidable congestion, and (3) time spent on equipment repairs and maintenance, etc. Unnecessary off-route time includes time spent for lunch in excess of the stated lunch period and time spent on taking unauthorized coffee breaks, talking to friends, etc.

Hauled-Container Systems The time required per trip, which also corresponds to the time required per container, is equal to the sum of pickup, at-site, and haul times and a factor accounting for off-route activities and is given by the following equation:

$$T_{hcs} = (P_{hcs} + s + a + bx)/(1 - W) \qquad (26\text{-}5)$$

where T_{hcs} = time per trip for hauled-container system, h/trip
P_{hcs} = pickup time per trip for hauled container system, h/trip
s = at-site time per trip, h/trip
a = empirical haul constant, h/trip
b = empirical haul constant, h/km
x = round-trip haul distance, km/trip
W = off-route factor, expressed as a fraction

The pickup time per trip P_{hcs} is equal to

$$P_{hcs} = pc + uc + dbc \qquad (26\text{-}6)$$

where P_{hcs} = pickup time per trip, h/trip
pc = time required to pick up loaded container, h/trip
uc = time required to unload empty container, h/trip
dbc = average time spent driving between container locations, h/trip (determined locally)

The number of trips that can be made per vehicle per day with a hauled-container system is determined using Eq. (26-7):

$$N_d = [(1 - W)H - (t_1 + t_2)]/(P_{hcs} + s + a + bx) \qquad (26\text{-}7)$$

where N_d = number of trips per day, trip/day
H = length of work day, h/day
t_1 = time from garage to first container location, h
t_2 = time from last container location garage, h

Other terms are as defined previously.

Data that can be used in the solution of Eqs. (26-6) and (26-7) for various types of hauled container systems are given in Tables 26-36 and 26-37. The off-route factor in Eq. (26-7) varies from 0.10 to 0.25; a factor of 0.15 is representative for most operations.

TABLE 26-36 Typical Values for Haul Constant Coefficients *a* and *b* in Eqs. (26-5), (26-7), (26-8), and (26-12)

Speed limit, km/h (mi/h)	*a*, h/trip	*b*, h/km (h/mi)
88 (55)	0.016	0.011 (0.018)
72 (45)	0.022	0.014 (0.022)
56 (35)	0.034	0.018 (0.029)
40 (25)	0.050	0.025 (0.040)

Stationary-Container Systems For systems using mechanically self-loading compactors, the time per trip is

$$T_{scs} = (P_{scs} + s + a + bx)/(1 - W) \qquad (26\text{-}8)$$

TABLE 26-37 Typical Data for Computing Equipment and Labor Requirements for Hauled- and Stationary-Container Collection Systems

Collection vehicle	Loading method	Compaction ratio, r	Pick up loaded container and deposit empty container, h/trip*	Empty contents of loaded container, h/container†	At-site time, s, h/trip
Hauled-container systems					
Hoist-truck	Mechanical	....	0.067	....	0.053
Tilt-frame	Mechanical	....	0.40	....	0.127
Tilt-frame	Mechanical	2.0–4.0‡	0.40	....	0.133
Stationary-container systems					
Compactor	Mechanical	2.0–4.0		0.050	0.10
Compactor	Manual	2.0–4.0		....	0.10

*$pc + uc$ in Eq. (26-6).
†uc in Eq. (26-9).
‡Containers used in conjunction with stationary compactor (see Fig. 26-40).

where T_{scs} = time per trip for stationary-container systems, h/trip
P_{scs} = pickup time per trip for stationary-container system, h/trip
Other terms are as defined previously.

The pickup time for the stationary-container system is given by:

$$P_{scs} = C_t(uc) + (n_P - 1)(dbc) \qquad (26\text{-}9)$$

where P_{scs} = pickup time per trip for stationary-container systems, h/trip
C_t = number of containers emptied per trip, containers/trip
uc = average unloading time per container for stationary-container systems, h/container
n_p = number of container pickup locations per trip, locations/trip
dbc = average time spent in driving between container locations, h/location (determined locally)

The term $(n_p - 1)$ accounts for the fact that the number of times when the collection vehicle will have to be driven between container locations is equal to the number of container locations less 1.

The number of containers that can be emptied per collection trip is related directly to the volume of the collection vehicle and the compaction ratio that can be achieved. This number is given by

$$C_t = vr/(cf) \qquad (26\text{-}10)$$

where C_t = number of containers emptied per trip, container/trip
v = volume of collection vehicle, m^3/trip
r = compaction ratio
c = container volume, m^3/container
f = weighted container-utilization factor

The number of trips required per day is given by

$$N_d = V_d/(vr) \qquad (26\text{-}11)$$

where N_d = number of collection trips required per day, trips/day
V_d = daily waste-generation rate, m^3/day
Other terms are as defined previously.

When an integer number of trips is to be made each day, the proper combination of trips per day and the size of the vehicle can be determined by using Eq. (26-12) in conjunction with an economic analysis:

$$H = [(t_1 + t_2) + N_d(P_{scs} + s + a + bx)]/(1 - W) \qquad (26\text{-}12)$$

where N_d = number of collection trips per day, trips/day, and other terms are as defined previously.

To determine the required truck volume, two or three different values for N_d are substituted in Eq. (26-12), and the available pickup times per trip are determined. Then, by trial and error, the volume required for each value of N_d is determined by using Eqs. (26-9) and (26-10). From the available truck sizes, select the ones that most nearly correspond to the computed values. If the available truck sizes are smaller than the required values, compute the actual times per day that will be required in using these sizes. The most cost-effective combination then can be selected.

Collection Routes Once the equipment and labor requirements have been determined, collection routes must be laid out so that both work force and equipment are used effectively. In general, the layout of collection routes is a trial-and-error process: there are no fixed rules that can be applied to all situations.

Some of the factors that should be taken into consideration when laying out routes are as follows: (1) Existing company policies and regulations related to such items as the point of collection and frequency of collection must be identified. (2) Existing system conditions such as crew size and vehicle types must be coordinated. (3) Wastes generated at traffic-congested locations should be collected as early in the day as possible. (4) Sources at which extremely large quantities of wastes are generated should be serviced during the first part of the day. (5) Scattered pickup points where small quantities of solid wastes are generated that receive the same collection frequency should, if possible, be serviced during one trip or on the same day.

Layout of Routes The layout of collection routes is a four-step process. First, prepare location maps. On a relatively large-scale map of the industrial area, the following data should be plotted for each solid-waste pickup point: location, number of containers, collection frequency, and, if a stationary-container system with self-loading compactors is used, the estimated quantity of wastes to be collected at each pickup location. Second, prepare data summaries. Estimate the quantity of wastes to be collected from pickup locations serviced each day on which the collection operation is to be conducted. When a stationary-container system is used, the number of locations that will be serviced during each pickup cycle must also be determined. Third, lay out preliminary collection routes, starting from the dispatch station or the place where the collection vehicles are parked. A route that connects all the pickup locations to be serviced during each collection day should be laid out. This route should be laid out so that the last location is nearest the disposal site. Fourth, develop balanced routes. After the preliminary collection routes have been laid out, the haul distance for each route should be determined. Next, determine the labor requirements per day, and check against the available work times per day. In some cases it may be necessary to readjust collection routes to balance the workload and the distance traveled. After the balanced routes have been established, they should be drawn on the master map.

Schedules A master schedule for each collection route should be prepared for use by the engineering department and the transportation dispatcher. A schedule for each route, on which can be found the location and order of each pickup point to be serviced, should be prepared for the driver. In addition, a route book should be maintained by each truck driver.

TRANSFER AND TRANSPORT

The functional element of transfer and transport refers to the means, facilities, and appurtenances used to effect the transfer of wastes from relatively small collection vehicles to larger vehicles and to transport them over extended distances to either processing centers or disposal sites. Transfer and transport operations become a necessity when haul distances to available disposal sites or processing centers increase to a point at which direct hauling is no longer economically feasible.

Transfer Stations Important factors that must be considered in the design of transfer stations include (1) the type of transfer operation to be used, (2) capacity requirements, (3) equipment and accessory requirements, and (4) environmental requirements.

Types of Transfer Stations Depending on the method used to load the transport vehicles, transfer stations may be classified into three types: (1) direct-discharge, (2) storage-discharge, and (3) combined direct- and storage-discharge.

1. *Direct-discharge.* In a direct-discharge transfer station, wastes from the collection vehicles usually are emptied directly into the vehicle to be used to transport them to a place of final disposition. To accomplish this, these transfer stations usually are constructed in a two-level arrangement. Either the unloading dock or platform from which wastes from collection vehicles are discharged into the transport trailers is elevated, or the transport trailers are located in a depressed ramp (see Fig. 26-45). Direct-discharge transfer stations employing stationary compactors also are popular.

FIG. 26-45 Direct-discharge transfer station with open-top trailers located in a depressed ramp under the loading hoppers.

2. *Storage-discharge.* In the storage-discharge transfer station, wastes are emptied either into a storage pit or onto a platform from which they are loaded into transport vehicles by various types of auxiliary equipment. In a storage-discharge transfer station, storage volume varies from about one-half to 2 days' volume of wastes.

3. *Combined direct- and storage-discharge.* In some transfer stations, both direct-discharge and storage-discharge methods are used. Usually, these are multipurpose facilities designed to service a broader range of users than a single-purpose facility. In addition to serving a broader range of users, a multipurpose transfer station may house a materials-salvage operation.

Capacity Requirements The operational capacity of a transfer station must be such that collection vehicles do not have to wait long to unload. In most cases, it will not be cost-effective to design the station to handle the ultimate peak number of hourly loads. An economic trade-off analysis should be made between the annual cost for the time spent by the collection vehicles waiting to unload against the incremental annual cost of a larger transfer station and/or the use of more transport equipment. Because of the increased cost of transport equipment, a trade-off analysis must also be made between the capacity of the transfer station and the cost of the transport operation, including both equipment and labor components.

Equipment and Accessory Requirements The types and amounts of equipment required vary with the capacity of the station and its function in the waste-management system. Specifically, scales should be provided at all medium-size and large transfer stations both to monitor the operation and to develop meaningful management and engineering data.

Environmental Requirements Most large modern transfer stations are enclosed and are constructed of materials that can be maintained and cleaned easily. For direct-discharge transfer stations with open loading areas, special attention must be given to the problem of blowing papers. Windscreens or other barriers are commonly used. Regardless of type, the station should be designed and constructed so that all accessible areas where rubbish or paper can accumulate are eliminated.

Transfer Means and Methods Motor vehicles, railroads, and oceangoing vessels are the principal means used to transport solid wastes. Pneumatic and hydraulic systems have also been used. Still other systems have been suggested, but most have not been tested.

Motor-Vehicle Transport Motor vehicles used to transport solid wastes on highways should satisfy the following requirements: (1) Wastes must be transported at minimum cost. (2) Wastes must be covered during the haul operation. (3) Vehicles must be designed for highway traffic. (4) Vehicle capacity must be such that allowable weight limits are not exceeded. (5) Methods used for unloading must be simple and dependable. The maximum volume that can be hauled in highway transport vehicles depends on the regulations in force in the state in which they are operated.

1. *Trailers and semitrailers.* In recent years, because of their simplicity and dependability, open-top trailers and semitrailers have found wide acceptance (see Table 26-38 and Fig. 26-45). To maximize the payload, transport trailers are often designed so that they are higher than the legal limit when empty and lower when full. Some trailers are equipped with sumps to collect any liquids that accumulate from the solid wastes. The sumps are equipped with drains so that they can be emptied at the disposal site.

Methods used to unload the transport trailers may be classified as (1) self-emptying and (2) requiring the aid of auxiliary equipment. Self-emptying transport trailers are equipped with mechanisms such as hydraulic dump beds, powered diaphragms, and moving floors that are part of the vehicle. Moving floors are an adaptation of equipment used in the construction industry. An advantage of the moving-floor trailer is the rapid turnaround time (typically 6 to 10 min) achieved at the disposal site without the need for auxiliary equip-

TABLE 26-38 Typical Data on Haul Vehicles Used at Transfer Stations

	Capacity per trailer		Dimensions of single trailer			
Type	m^3	Metric tons	Width, m	Length, m	Approximate height, empty, m	Length of tractor and trailer units, m*
Tractor-trailer-trailer†	54	11.4	2.44	8.25	4.12	19.8
Tractor-trailer	54	10.0	2.44	8.50	4.12	18.3
	74	17.3	2.44	12.2	4.12	
Tractor-compactor-trailer	58	18	2.44	10.2	4.12	14.0

*Overall length will vary with the type of tractor (e.g., conventional or cab-over) and the turning radius of the trailers.
†Dimensions are for the tractor-trailer-trailer combination shown in Fig. 26-47.

ment. Unloading systems that require auxiliary equipment are usually of the "pull-off" type, in which the wastes are pulled out of the truck by either a movable bulkhead or wire-cable slings placed forward of the load (see Fig. 26-46). The disadvantage of requiring auxiliary equipment and work force at the disposal site to unload is relatively minor in view of the simplicity and reliability of the method.

FIG. 26-46 Unloading wastes from a transport trailer by using a cable pull-off system.

Another auxiliary unloading system that has proved to be very effective and efficient involves the use of movable, hydraulically operated tipping ramps located at the disposal site (see Fig. 26-47). Operationally, the semitrailer of a tractor-trailer-trailer combination is backed up onto one of the tipping ramps, and the tractor-trailer combination is backed up onto a second tipping ramp. The backs of the trailers are opened, and the units are then tilted upward until the wastes fall out by gravity. The time required for the entire unloading operation typically is about 5 min per trip.

FIG. 26-47 Unloading wastes from a trailer and tractor-trailer by using hydraulically operated tilting ramps.

2. *Compactors.* Large-capacity containers and container-trailers are used in conjunction with stationary compactors at transfer stations. In some cases, the compaction mechanism is an integral part of the container. When containers are equipped with a self-contained compaction mechanism, the movable bulkhead used to compress the wastes is also used to discharge the compacted wastes.

Railroad Transport Although railroads were commonly used for the transport of solid wastes in the past, they are now used by only a few communities. However, renewed interest is developing in the use of railroads for hauling solid wastes, especially to remote areas where highway travel is difficult and railroad lines exist.

Water Transport Barges, scows, and special boats have been used in the past to transport solid wastes to processing locations and to seaside and ocean disposal sites, but ocean disposal is no longer practiced by the United States. Although some self-propelled vessels (such as U.S. Navy garbage scows and other special boats) have been used, the most common practice is to use vessels towed by tugs or other special boats.

Pneumatic Transport Both low-pressure air and vacuum conduit transport systems have been used to transport solid wastes. The most common application is the transport of wastes from high-density apartments or commercial activities to a central location for processing or for loading into transport vehicles. The largest pneumatic system now in use in the United States is at the Walt Disney World amusement park in Orlando, Florida.

Location of Transfer Stations Whenever possible, transfer stations should be located (1) as near as possible to the weighted center of the individual solid-waste-production areas to be served, (2) within easy access of major arterial highways as well as near secondary or supplemental means of transportation, (3) where there will be a minimum of public and environmental objection to the transfer operations, and (4) where construction and operation will be most economical. Additionally, if the transfer-station site is to be used for processing operations involving material recovery and/or energy production, the requirements for those operations must be considered.

Transfer and Transport of Hazardous Wastes The facilities of a hazardous-waste transfer station are quite different from those of an industrial or municipal solid-waste transfer station. Typically, hazardous wastes are not compacted (mechanical volume reduction), discharged at differential levels, or delivered by numerous collection companies. Instead, liquid hazardous wastes are generally pumped from collection vehicles, and sludges or solids are reloaded without removal from the collection containers for transport to processing and disposal facilities.

It is unusual to find a hazardous-waste transfer facility at which wastes are simply transferred to larger transport vehicles. Some processing and storage facilities are often part of the materials-handling sequence at a transfer station. For example, neutralization of corrosive wastes will result in the use of lower-cost holding tanks on transport vehicles.

PROCESSING AND RESOURCE RECOVERY

The purpose of this subsection is to introduce the reader to the techniques and methods used to recover materials, conversion products, and energy from solid wastes. Topics to be considered include (1) processing techniques, (2) processing techniques for hazardous wastes, (3) materials-recovery systems, (4) recovery of biological conversion products, (5) recovery of chemical conversion products, and (6) recovery of energy from conversion products.

Because many of the techniques, especially those associated with the recovery of materials and energy and the processing of solid hazardous wastes, are in a state of flux with respect to application and design criteria, the objective here is only to introduce them to the reader. If these techniques are to be considered in the development of waste-management systems, current engineering design and performance data must be obtained from the records of operating installations, from field tests, from equipment manufacturers, and from the literature.

Processing Techniques Processing techniques are used in solid-waste-management systems to (1) improve the efficiency of the systems, (2) to recover resources (usable materials), and (3) to prepare materials for recovery of conversion products and energy. The more important techniques used for processing solid wastes are summarized in Table 26-39.

Manual Component Separation The manual separation of solid-waste components can be accomplished at the source where solid wastes are generated, at a transfer station, at a centralized processing station, or at the disposal site. Manual sorting at the source of generation is the most positive way to achieve the recovery and reuse of materials. The number and types of components salvaged or sorted (e.g., cardboard and high-quality paper, metals, and wood)

TABLE 26-39 Summary of Techniques Used for Processing Solid Wastes

Processing technique	Function	Representative equipment and/or facilities and applications
Manual component separation	Separation of recoverable materials, usually at point of generation	
Storage and transfer	Storage and transfer of wastes to be processed	Open storage pits for unprocessed wastes, storage bins and silos for processed wastes; transfer equipment including metal and rubber belt conveyors, vibratory conveyors with unprocessed wastes, pneumatic conveyors, and screw conveyors with processed wastes
Mechanical volume reduction	Reduction of solid-waste volume; alteration of shape of solid-waste components; all modern collection vehicles essentially equipped with compaction equipment	Hydraulic piston-type compactors for collection vehicles, on-site compactors, and transfer-station compactors; roll crushers used to fracture brittle materials and to crush tin and aluminum cans and other ductile materials
Chemical volume reduction	Reduction of volume of solid wastes through burning (incineration)	Mass-fired incinerators, with and without heat recovery, for unprocessed wastes
Mechanical size and shape alteration	Alteration of size and shape of solid-waste components	Equipment used to reduce the size of solid waste including hammer mills, shredders, roll crushers, grinders, chippers, jaw crushers, rasp mills, and hydropulpers
Mechanical component separation	Separation of recoverable materials, usually at a processing facility	(See Table 26-40)
Magnetic and electromechanical separation	Separation of ferrous and nonferrous materials from processed solid wastes	Magnetic separation for ferrous materials; eddy-current separation for aluminum; electrostatic separation for glass from wastes free of ferrous and aluminum scrap; magnetic fluid separation for nonferrous materials from processed wastes
Drying and dewatering	Removal of moisture from solid wastes	Convection, conduction, and radiation dryers used for solid wastes and sludge; centrifugation and filtration used to dewater treatment-plant sludge

depend on the location, the opportunities for recycling, and the resale market.

Storage and Transfer When solid wastes are to be processed for materials recovery, storage and transfer facilities should be considered an essential part of the processing operation. Important factors in the design of such facilities include (1) the size of the material before and after processing, (2) the density of the material, (3) the angle of repose before and after processing, (4) the abrasive characteristics of the material, and (5) the moisture content.

Mechanical Volume Reduction Mechanical volume reduction is perhaps the most important factor in the development and operation of solid-waste-management systems. Vehicles equipped with compaction mechanisms are used for the collection of most industrial solid wastes. To increase the useful life of landfills, wastes are compacted. Paper for recycling is baled for shipping to processing centers. When compacting industrial solid wastes, it has been found that the final density (typically about 1100 kg/m^3) is essentially the same regardless of the starting density and applied pressure. This fact is important in evaluating manufacturers' claims.

Chemical Volume Reduction Incineration has been the method most commonly used to reduce the volume of wastes chemically. One of the most attractive features of the incineration process is that it can be used to reduce the original volume of combustible solid wastes by 80 to 90 percent. Although the technology of incineration has advanced since 1960, air pollution control and cost remain major problems in implementation.

Mechanical Size Alteration The objective of size reduction is to obtain a final product that is reasonably uniform and considerably reduced in size in comparison to its original form. It is important to note that size reduction does not necessarily imply volume reduction. In some situations, the total volume of the material after size reduction may be greater than the original volume.

Mechanical Component Separation Component separation is a necessary operation in the recovery of resources from solid wastes and in instances when energy and conversion products are to be recovered from processed wastes. Mechanical separation techniques that have been used are reported in Table 26-40.

Magnetic and Electromechanical Separation Magnetic separation of ferrous materials is a well-established technique. More recently, a variety of electromechanical techniques have been developed for the removal of several nonferrous materials (see Table 26-39).

Drying and Dewatering In many solid-waste energy-recovery and incineration systems, the shredded light fraction is predried to decrease weight. Although energy requirements for drying wastes vary with local conditions, the required energy input can be estimated by using a value of about 4300 kJ/kg of water evaporated.

Processing of Hazardous Wastes As with conventional solid wastes, the processing of hazardous wastes is undertaken for three purposes: (1) to recover useful materials, (2) to reduce the amount of waste that must be disposed in landfills, and (3) to prepare the wastes for ultimate disposal.

Processing Techniques The processing of hazardous wastes on a batch basis can be accomplished by physical, chemical, thermal, and biological means. The various individual processes in each category are reported in Table 26-41. Clearly, the number of possible treatment-process combinations is staggering. In practice, the physical, chemical, and thermal treatment operations and processes are the ones most commonly used.

Identification of Waste Constituents In any processing (and disposal) scheme, the key item is knowledge of the characteristics of the wastes to be handled. Without this information, effective processing or treatment is impossible. For this reason, the characteristics of the wastes must be known before they are accepted and hauled to a treatment or disposal site. In most states, proper identification of the constituents of the waste is the responsibility of the waste generator.

Materials-Recovery Systems Paper, rubber, plastics, textiles, glass, ferrous metals, and organic and inorganic materials are the principal recoverable materials contained in industrial solid wastes.

TABLE 26-40 Mechanical Methods for Separating Solid-Waste Components

Method	Function	Equipment and/or facilities and applications
Screening	Used to separate solid waste components by size	Trommels and horizontal and vibrating screens for unprocessed and processed wastes; disk screens with processed wastes
Air separation	Used to separate light (organic) materials from heavy (inorganic) materials in solid waste	Zig zag-air, vibrating-air, rotary-air, and air-knife classifiers used with processed wastes
Jig separation	Used to separate light and heavy materials in solid waste by means of density separation	
Pneumatic separation (stoners)	Used to separate light and heavy materials in solid waste	
Sink-float separation	Used to separate light and heavy materials in solid wastes	
Inertial separation	Used to separate light and heavy materials in solid wastes	
Inclined-table separation	Used to separate light and heavy materials in solid wastes	
Shaking-table separation	Used to separate light and heavy materials in solid wastes	
Flotation	Used to separate light and heavy materials in solid wastes	
Optical sorting	Used to separate light and heavy materials in solid wastes	

TABLE 26-41 Treatment Operations and Processes for Hazardous Wastes[a]

Operation or process	Functions performed[b]	Types of wastes[c]	Forms of waste[d]
Physical treatment			
Adsorption[e]	Se	1, 2, 3, 4,	L
Aeration	Se	1, 2, 3, 4, 5	L
Ammonia stripping	VR, Se	1, 2, 3, 4	L
Carbon sorption	VR, Se	1, 2, 3, 4, 5	L, G
Centrifugation	VR, Se	1, 2, 3, 4, 5	L
Dialysis	VR, Se	1, 2, 3, 4	L
Distillation[e]	VR, Se	1, 2, 3, 4, 5	L
Electrodialysis	VR, Se	1, 2, 3, 4, 6	L
Encapsulation	St	1, 2, 3, 4, 6	L, S
Evaporation	VR, Se	1, 2, 5	L
Filtration[e]	VR, Se	1, 2, 3, 4, 5	L, G
Flocculation or setting	VR, Se	1, 2, 3, 4, 5	L
Flotation[e]	Se	1, 2, 3, 4	L
Reverse osmosis	VR, Se	1, 2, 4, 6	L
Screening	Se	1, 2, 3, 4, 5	L
Sedimentation[e]	VR, Se	1, 2, 3, 4, 5	L
Solar evaporation[e]	VR, Se	1, 2, 5	L
Solvent extraction	Se	1, 2, 3, 4, 5	L
Thickening	Se	1, 2, 3, 4	L
Ultrafiltration	Se	1, 2, 3, 4, 5	L
Vapor scrubbing	VR, Se	1, 2, 3, 4	L
Chemical treatment			
Calcination	VR	1, 2, 5	L
Chemical dechlorination	De	1, 3	L
Ion exchange	VR, Se, De	1, 2, 3, 4, 5	L
Neutralization[e]	De	1, 2, 3, 4	L
Oxidation	De	1, 2, 3, 4	L
Precipitation[e]	VR, Se	1, 2, 3, 4, 5	L
Reduction	De	1, 2	L
Sorption	De	1, 2, 3, 4	L
Stabilization or solidification[e]	De	1, 2, 3, 4	L
Thermal treatment			
Incineration[e]	VR, De	3, 5, 6, 7, 8	S, L, G
Pyrolysis	VR, De	3, 4, 6	S, L, G
Biological treatment			
Activated sludge[e]	De	3	L
Aerated lagoons	De	3	L
Anaerobic digestion	De	3	L
Anaerobic filters	De	3	L
Trickling filters	De	3	L
Waste-stablization ponds[e]	De	3	L

[a]Adapted from *Report to Congress: Disposal of Hazardous Wastes*, U.S. EPA Publ. SW-115, 1974.

[b]Functions: VR, volume reduction; Se, separation; De, detoxification; St, storage.

[c]Waste types: 1, inorganic chemical without heavy metals; 2, inorganic chemical with heavy metals; 3, organic chemical without heavy metals; 4, organic chemical with heavy metals; 5, radiological; 6, biological; 7, flammable; 8, explosive.

[d]Waste forms: S, solid; L, liquid; G, gas.

[e]Most widely used technologies for hazardous-waste management.

Once a decision has been made to recover materials and/or energy, process flow sheets must be developed for the removal of the desired components, subject to predetermined materials specifications. A typical flow sheet for the recovery of specific components and the preparation of combustible materials for use as a fuel source is presented in Fig. 26-48. The light conbustible materials are often identified as refuse-derived fuel (RDF).

The design and layout of the physical facilities that make up the processing-plant flow sheet are an important aspect in the implementation and successful operation of such systems. Important factors that must be considered in the design and layout of such systems include (1) process performance efficiency, (2) reliability and flexibility, (3) ease and economy of operation, (4) aesthetics, and (5) environmental controls.

Recovery of Biological Conversion Products Biological conversion products that can be derived from solid wastes include compost, methane, various proteins and alcohols, and a variety of other intermediate organic compounds. The principal processes that have been used are reported in Table 26-42. Composting and anaerobic digestion, the two most highly developed processes, are considered fur-

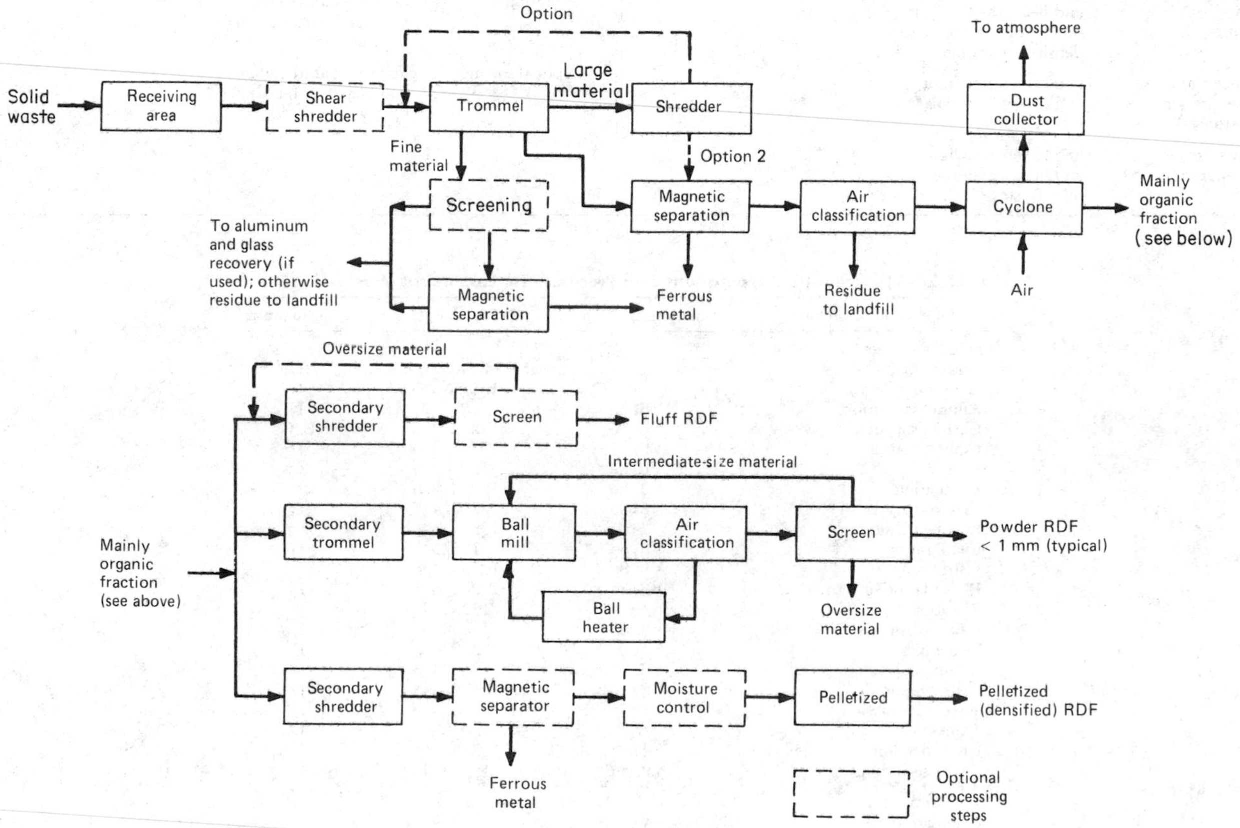

FIG. 26-48 Typical flow sheet for the recovery of materials and the production of refuse-derived fuels (RDF). Adapted in part from D. C. Wilson (ed.), *Waste Management: Planning, Evaluation, Technologies*, Oxford University Press, Oxford, 1981.

TABLE 26-42 Biological and Chemical Processes Used for Recovery of Conversion Products from Solid Wastes

Process	Conversion product	Preprocessing required	Comments
Biological			
Composting	Humuslike material	Shredding, air separation	Lack of markets primary shortcoming; technically proven in full-scale application
Anaerobic digestion	Methane gas	Shredding, air separation	Technology on laboratory scale only
Biological conversion to protein	Protein, alcohol	Shredding, air separation	Technology on pilot scale only
Biological fermentation	Glucose, furfural	Shredding, air separation	Used in conjunction with the hydrolytic process
Chemical			
Incineration with heat recovery	Energy in the form of steam	None	Markets for steam required; proven in numerous full-scale applications; air-quality regulations possibly prohibiting use
Supplementary fuel firing in boilers	Energy in the form of steam	Shredding, air separation, magnetic separation	If least capital investment desired, existing boiler required to be capable of modification; air-quality regulations possibly prohibiting use
Gasification	Energy in the form of low energy gas	Shredding, air separation, magnetic separation	Gasification also capable of being used for codisposal for industrial sludges
Pyrolysis	Energy in the form of gas or oil	Shredding, magnetic separation	Technology proven only in pilot applications; even though pollution is minimized, air-quality regulations possibly prohibiting use
Hydrolysis	Glucose, furfural	Shredding, air separation	Technology on pilot scale only
Chemical conversion	Oil, gas, cellulose acetate	Shredding, air separation	Technology on pilot scale only

ther. The recovery of gas from landfills is discussed in the portion of this section dealing with ultimate disposal.

Composting If the organic materials, excluding plastics, rubber, and leather, are separated from municipal solid wastes and subjected to bacterial decomposition, the end product remaining after dissimilatory and assimilatory bacterial activity is called "compost" or "humus." The entire process involving both separation and bacterial conversion of the organic solid wastes is known as "composting." Decomposition of the organic solid wastes may be accomplished either aerobically or anaerobically, depending on the availability of oxygen.

Most composting operations involve three basic steps: (1) preparation of the solid wastes, (2) decomposition of the solid wastes, and (3) product preparation and marketing. Receiving, sorting, separation, size reduction, and moisture and nutrient addition are part of the preparation step. Several techniques have been developed to accomplish the decomposition step. Once the solid wastes have been converted to a humus, they are ready for the third step of product preparation and marketing. This step may include fine grinding, blending with various additives, granulation, bagging, storage, shipping, and, in some cases, direct marketing. The principal design considerations associated with the biological decomposition of prepared solid wastes are presented in Table 26-43.

TABLE 26-43 Important Design Considerations for Aerobic-Composting Processes*

Item	Comment
Particle size	For optimum results the size of solid wastes should be between 25 and 75 mm (1 and 3 in).
Seeding and mixing	Composting time can be reduced by seeding with partially decomposed solid wastes to the extent of about 1 to 5 percent by weight. Sewage sludge can also be added to prepared solid wastes. When sludge is added, the final moisture content is the controlling variable.
Mixing or turning	To prevent drying, caking, and air channeling, material in the process of being composted should be mixed or turned on a regular schedule or as required. Frequency of mixing or turning will depend on the type of composting operation.
Air requirements	Air with at least 50 percent of the initial oxygen concentration remaining should reach all parts of the composting material for optimum results, especially in mechanical systems.
Total oxygen requirements	The theoretical quantity of oxygen required can be estimated.
Moisture content	Moisture content should be in the range between 50 and 60 percent during the composting process. The optimum value appears to be about 55 percent.
Temperature	For best results temperature should be maintained between 322 and 327 K (130 and 140°F) for the first few days and between 327 and 333 K (130 and 140°F) for the remainder of the active composting period. If temperature goes beyond 339 K (150°F), biological activity is reduced significantly.
Carbon-nitrogen ratio	Initial carbon-nitrogen ratios (by mass) between 35 and 50 are optimum for aerobic composting. At lower ratios ammonia is given off. Biological activity is also impeded at lower ratios. At higher ratios nitrogen may be a limiting nutrient.
pH	To minimize the loss of nitrogen in the form of ammonia gas, pH should not rise above about 8.5.
Control of pathogens	If the process is properly conducted, it is possible to kill all the pathogens, weeds, and seeds during the composting process. To do this, the temperature must be maintained between 333 and 344 K (140 and 160°F) for 24 h.

*Adapted from G. Tchobanoglous, H. Theisen, R. Eliassen, *Solid Wastes: Engineering Principles and Management Issues,* McGraw-Hill, New York, 1977.

Anaerobic Digestion Anaerobic digestion or anaerobic fermentation, as it is often called, is the process used for the production of methane from solid wastes. In most processes in which methane is to be produced from solid wastes by anaerobic digestion, three basic steps are involved. The first step involves preparation of the organic fraction of the solid wastes for anaerobic digestion and usually includes receiving, sorting, separation, and size reduction. The second step involves the addition of moisture and nutrients, blending, pH adjustment to about 6.7, heating of the slurry to between 327 and 333 K (130 and 140°F), and anaerobic digestion in a reactor with continuous flow, in which the contents are well mixed for a period of time varying from 8 to 15 days. The third step involves capture, storage, and, if necessary, separation of the gas components evolved during the digestion process. The disposal of the digested sludge is an additional task that must be accomplished. Some important design considerations are reported in Table 26-44. Because of the variability of the results reported in the literature, it is recommended that pilot-plant studies be conducted if the digestion process is to be used for the conversion of solid wastes.

TABLE 26-44 Important Design Considerations for Anaerobic Digestion*

Item	Comment
Size of material shredded	Wastes to be digested should be shredded to a size that will not interfere with the efficient functioning of pumping and mixing operations.
Mixing equipment	To achieve optimum results and to avoid scum buildup, mechanical mixing is recommended.
Percentage of solid wastes mixed with sludge	Although amounts of waste varying from 50 to 90+ percent have been used, 60 percent appears to be a reasonable compromise.
Hydraulic and mean cell residence time, $\Theta_h = \Theta_c$	Washout time is in the range of 3 to 4 days. Use 8 to 15 days for design or base design on results of pilot-plant studies.
Loading rate	0.6 to 1.6 kg/(m^3 · day) [0.04 to 0.10 lb/(ft^3 · day)]. Not well defined at present time. Significantly higher rates have been reported.
Temperature	Between 327 and 333 K (130 and 140°F).
Destruction of volatile solid wastes	Varies from about 60 to 80 percent; 70 percent can be used for estimating purposes.
Total solids destroyed	Varies from 40 to 60 percent, depending on amount of inert material present originally.
Gas production	0.5 to 0.75 m^3/kg (8 to 12 ft^3/lb) of volatile solids destroyed (CH_4 = 60 percent; CO_2 = 40 percent).

*From G. Tchobanoglous, H. Theisen, and R. Eliassen, *Solid Wastes: Engineering Principles and Management Issues,* McGraw-Hill, New York, 1977.

†Actual removal rates for volatile solids may be less, depending on the amount of material diverted to the scum layer.

Recovery of Chemical Conversion Products Chemical conversion products that can be derived from solid wastes include heat, gases, a variety of oils, and various related organic compounds. The principal chemical conversion processes that have been used for the recovery of usable conversion products from solid wastes are reported in Table 26-42.

Incineration with Heat Recovery Heat contained in the gases produced from the incineration of solid wastes can be recovered by conversion to steam. The low-level heat remaining in the gases after heat recovery can also be used to preheat the combustion air, boiler makeup water, or solid-waste fuel.

1. *In existing incinerators.* With existing incinerators, waste-heat boilers can be installed to extract heat from the combustion gases without introducing excess amounts of air or moisture. Typically, incinerator gases will be cooled from a range of 1250 to 1375 K (1800 to 2000°F) to a range from 500 to 800 K (600 to 1000°F) before being disharged to the atmosphere. Apart from the production of steam, the use of a boiler system is beneficial in reducing the volume of gas to be processed in the air-pollution-control equipment.

2. *In water-wall incinerators.* In these incinerators, the internal walls of the combustion chamber are lined with boiler tubes that are arranged vertically and welded together in continuous sections. When water walls are employed in place of refractory materials, they are not only useful for the recovery of steam but also extremely effective in controlling furnace temperature without introducing excess air; however, they are subject to corrosion by the hydrochloric acid produced from the burning of some plastic compounds.

Use of Refuse-Derived Fuels Prepared solid-waste fuels (refuse-derived fuels, RDF) can also be fired directly in large industrial boilers that are now used for the production of power with pulverized coal or oil. In addition, they can be fired in conjunction with coal or oil. Although the process is not well established with coal, it appears that about 15 to 20 percent of the heat input can be from prepared solid wastes. With oil as the fuel, about 10 percent of the heat input can be from solid wastes. Depending on the degree of processing, suspension, spreader-stoker, and double-vortex firing systems have been used.

Gasification The gasification process involves the partial combustion of a carbonaceous fuel to generate a combustible fuel gas rich in carbon monoxide and hydrogen. A gasifier is basically an incinerator operating under reducing conditions. Heat to sustain the process is derived from exothermic reactions, while the combustible components of the low-energy gas are primarily generated by endothermic reactions. The reaction kinetics of the gasification process are quite complex and still the subject of considerable debate.

When a gasifier is operated at atmospheric pressure with air as the oxidant, the end products of the gasification process are a low-energy gas typically containing (by volume) 10 percent CO_2, 20 percent CO, 15 percent H_2, and 2 percent CH_4, with the balance being N_2, and a carbon-rich char. Because of the diluting effect of the nitrogen in the input air, the low-energy gas has an energy content in the range of the 5.2 to 6.0 MJ/m^3 (140 to 160 Btu/ft^3). When pure oxygen is used as the oxidant, a medium-energy gas with an energy content in the range of 12.9 to 13.8 MJ/m^3 (345 to 370 Btu/ft^3) is produced.

Pyrolysis Of the many alternative chemical conversion processes that have been investigated, excluding incineration, pyrolysis has received the most attention. Depending on the type of reactor used, the physical form of the solid wastes to be pyrolyzed can vary from unshredded raw wastes to the finely ground portion of the wastes remaining after two stages of shredding and air classification. Upon heating in an oxygen-free atmosphere, most organic substances can be split through a combination of thermal cracking and condensation reactions into gaseous, liquid, and solid fractions. Pyrolysis is the term used to describe the process. In contrast to the combustion process, which is highly exothermic, the pyrolytic process is highly endothermic. For this reason, the term "destructive distillation" is often used as an alternative for pyrolysis.

The characteristics of the three major component fractions resulting from the pyrolysis are (1) a gas stream containing primarily hydrogen, methane, carbon monoxide, carbon dioxide, and various other gases, depending on the organic characteristics of the material being pyrolyzed; (2) a fraction that consists of a tar and/or oil stream that is liquid at room temperatures and has been found to contain chemicals such as acetic acid, acetone, and methanol; and (3) a char consisting of almost pure carbon plus any inert material that may have entered the process. It has been found that distribution of the product fractions varies with the temperature at which the pyrolysis is carried out. Under conditions of maximum gasification, the energy content of the resulting gas is about 26.1 MJ/kg (700 Btu/ft^3). The energy content of pyrolytic oils has been estimated to be about 23.2 MJ/kg (10,000 Btu/lb).

Recovery of Energy from Conversion Products Once conversion products have been derived from solid wastes by one or more of the biological and chemical methods listed in Table 26-42, the next step involves their storage and/or use. If energy is to be produced, an additional conversion step is required.

Energy-Recovery Systems The principal components involved in the recovery of energy from heat, steam, various gases and oils, and other conversion products are boilers for the production of steam, steam and gas turbines for motive power, and electric generators for the conversion of motive power into electricity.

Typical flow sheets for alternative energy-recovery systems are shown in Fig. 26-49. Perhaps the most common flow sheet for the production of electric energy involves the use of a steam-turbine–generator combination (see Fig. 26-49*a*). As shown, when solid wastes are used as the basic fuel source, four operational modes are possible. A flow sheet using a gas-turbine–generator combination is shown in Fig. 26-49*b*. The low-energy gas is compressed under high pressure so that it can be used more effectively in the gas turbine.

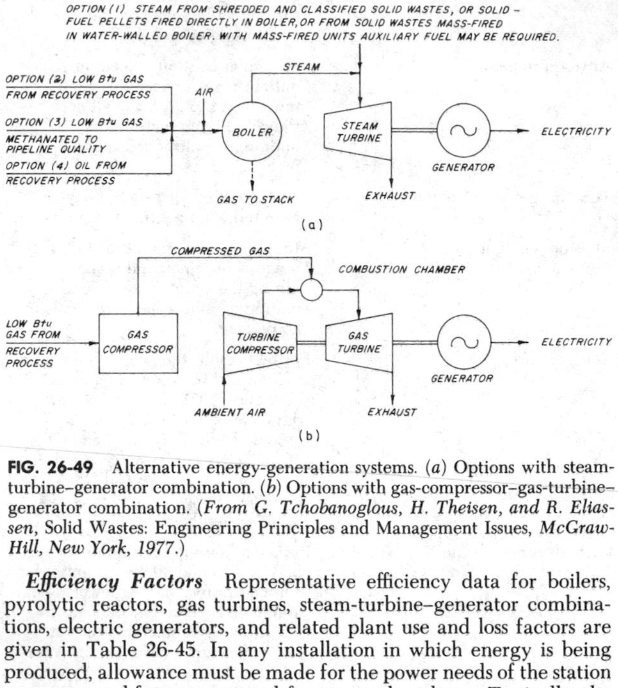

FIG. 26-49 Alternative energy-generation systems. (*a*) Options with steam-turbine–generator combination. (*b*) Options with gas-compressor–gas-turbine–generator combination. (*From G. Tchobanoglous, H. Theisen, and R. Eliassen,* Solid Wastes: Engineering Principles and Management Issues, *McGraw-Hill, New York, 1977.*)

Efficiency Factors Representative efficiency data for boilers, pyrolytic reactors, gas turbines, steam-turbine–generator combinations, electric generators, and related plant use and loss factors are given in Table 26-45. In any installation in which energy is being produced, allowance must be made for the power needs of the station or process and for unaccounted-for process-heat losses. Typically, the auxiliary power allowance varies from 4 to 8 percent of the power produced. Process-heat losses usually will vary from 2 to 8 percent.

Determination of Energy Output and Efficiency for Energy-Recovery Systems An analysis of the amount of energy produced from a solid-waste energy conversion system using an incinerator–boiler–steam-turbine–electric-generator combination with a capacity of 1000 metric tons per day is presented in Table 26-46. If it is assumed that 10 percent of the power generated is used for the front-end processing system (typical values vary from 8 to 14 percent), then the net power for export is 24,604 kW and the overall efficiency is 17.5 percent.

TABLE 26-45 Typical Thermal Efficiency and Plant Use and Loss Factors for Individual Components and Processes Used for Recovery of Energy from Solid Wastes

Component	Efficiency* Range	Efficiency* Typical	Comments
Incinerator-boiler	40–68	63	Mass-fired
Boiler			
Solid fuel	60–75	72	Processed solid wastes (RDF)
Low-Btu gas	60–80	75	Necessity to modify burners
Oil-fired	65–85	80	Oils produced from solid wastes possibly required to be blended to reduce corrosiveness
Gasifier	60–70	70	
Pyrolysis reactor			
Conventional	65–75	70	
Purox	70–80	75	
Turbines			
Combustion gas			
Simple cycle	8–12	10	
Regenerative	20–26	24	Including necessary appurtenances
Expansion gas	30–50	40	
Steam-turbine–generator system			
Less than 12.5 MW	24–40	29†‡	Including condenser, heaters, and all other necessary appurtenances but not boiler
Over 10 MW	28–32	31.6†‡	
Electric generator			
Less than 10 MW	88–92	90	
Over 10 MW	94–98	96	
			Plant use and loss factors
Station-service allowance			
Steam-turbine–generator plant	4–8	6	
Purox process	18–24	21	
Unaccounted heat losses	2–8	5	

*Theoretical value for mechanical equivalent of heat = 360 KJ/kWh.
†Efficiency varies with exhaust pressure. Typical value given is based on an exhaust pressure in the range of 50 to 100 mmHg.
‡Heat rate = 11,395 kJ/kWh = (3600 kJ/kWh)/0.316.

TABLE 26-46 Energy Output and Efficiency for 1000 Metric-Ton/Day Steam-Boiler–Turbine–Generator Energy-Recovery Plant Using Unprocessed Industrial Solid Wastes with Energy Content of 12,000 kJ/kg

Item	Value
Energy available in solid wastes, million kJ/h [(1000 metric tons/day × 1000 kg/metric ton × 12,000 kJ/kg)/(24 h/day × 10^6 kJ/million kJ)]	500
Steam energy available, million kJ/h (500 million kJ/h × 0.7)	350
Electric power generation, kW (350 million kJ/h)/(11,395 kJ/kWh)*	30,715
Station-service allowance, kW [30,715 (0.06)]	−1,843
Unaccounted heat losses, kW [30,715 (0.05)]	−1,536
Net electric power for export, kW	27,336
Overall efficiency, percent {(27,336 kW)/[(500,000,000 Btu/h)/(3,600 kJ/kWh)]}(100)	19.7

*11,395 kJ/kWh = (3600 kJ/kWh)/0.316.

ULTIMATE DISPOSAL

Disposal on or in the earth's mantle is, at present, the only viable method for the long-term handling of (1) solid wastes that are collected and are of no further use, (2) the residual matter remaining after solid wastes have been processed, and (3) the residual matter remaining after the recovery of conversion products and/or energy has been accomplished. The three land disposal methods used most commonly are (1) landfilling, (2) landfarming, and (3) deepwell injection. Although incineration is often considered a disposal method, it is, in reality, a processing method. Recently, the concept of using muds in the ocean floor as a waste-storage location also has received some attention.

Landfilling of Solid Wastes Landfilling involves the controlled disposal of solid wastes on or in the upper layer of the earth's mantle. Important aspects in the implementation of sanitary landfills include (1) site selection, (2) landfilling methods and operations, (3) occurrence of gases and leachate in landfills, (4) movement and control of landfill gases and leachate, and (5) landfill design. The landfilling of hazardous wastes is considered separately.

Site Selection Factors that must be considered in evaluating potential solid-waste-disposal sites are summarized in Table 26-47. Final selection of a disposal site usually is based on the results of a preliminary site survey, results of engineering design and cost studies, and an environmental-impact assessment.

Landfilling Methods and Operations To use the available area at a landfill site effectively, a plan of operation for the placement of solid wastes must be prepared. Various operational methods have been developed primarily on the basis of field experience. The principal methods used for landfilling dry areas may be classified as (1) area, (2) trench, and (3) depression.

1. *Area method.* The area method is used when the terrain is unsuitable for the excavation of trenches in which to place the solid wastes. The filling operation usually is started by building an earthen levee against which wastes are placed in thin layers and compacted (see Fig. 26-50). Each layer is compacted as the filling progresses until the thickness of the compacted wastes reaches a height varying from 2 to 3 m (6 to 10 ft). At that time and at the end of each day's operation, a 150- to 300-mm (6- to 12-in) layer of cover material is placed over the completed fill. The cover material must be hauled in by truck or earth-moving equipment from adjacent land or from borrow-pit areas. In some newer landfill operations, the daily cover material is omitted. A completed lift, including the cover material, is called a "cell" (see Fig. 26-51). Successive lifts are placed on top of one another until the final grade called for in the ultimate development plan is reached. A final layer of cover material is used when the fill reaches the final design height.

TABLE 26-47 Important Factors in Preliminary Selection of Landfill Sites

Factor	Remarks
Available land area	In selecting potential land disposal sites, it is important to ensure that sufficient land area is available. Sufficient area to operate for at least 1 year at a given site is needed to minimize costs.
Impact of processing and resource recovery	It is important to project the extent of resource-recovery-processing activities that are likely to occur in the future and determine their impact on the quantity and condition of the residual materials to be disposed of.
Haul distance	Although minimum haul distances are desirable, other factors must also be considered. These include collection-route location, types of wastes to be hauled, local traffic patterns, and characteristics of the routes to and from the disposal site (condition of the routes, traffic patterns, and access conditions).
Soil conditions and topography	Because it is necessary to provide material for each day's landfill and a final layer of cover after the filling has been completed, data on the amounts and characteristics of the soils in the area must be obtained. Local topography will affect the type of landfill operation to be used, equipment requirements, and the extent of work necessary to make the site usable.
Climatological conditions	Local weather conditions must also be considered in the evaluation of potential sites. Under winter conditions where freezing is severe, landfill cover material must be available in stockpiles when excavation is impractical. Wind and wind patterns must also be considered carefully. To avoid blowing or flying papers, windbreaks must be established.
Surface-water hydrology	The local surface-water hydrology of the area is important in establishing the existing natural drainage and runoff characteristics that must be considered. Other conditions of flooding must also be identified.
Geologic and hydrogeologic conditions	Geologic and hydrogeologic conditions are perhaps the most important factors in establishing the environmental suitability of the area for a landfill site. Data on these factors are required to assess the pollution potential of the proposed site and to establish what must be done to the site to control the movement of leachate or gases from the landfill.
Local environmental conditions	The proximity of both residential and industrial developments is extremely important. Great care must be taken in their operation if they are to be environmentally sound with respect to noise, odor, dust, flying paper, and vector control.
Ultimate uses	Because the ultimate use affects the design and operation of the landfill, this issue must be resolved before the layout and design of the landfill are started.

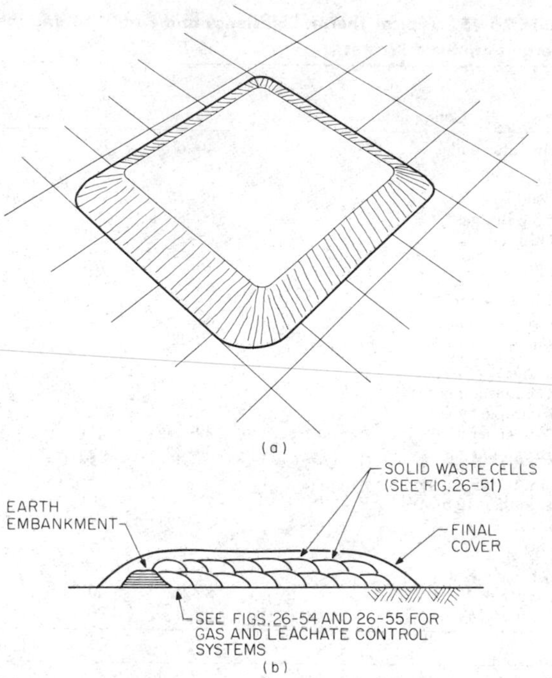

FIG. 26-50 Area method for landfilling solid wastes. (*a*) Pictorial view of completed landfill. (*b*) Section through landfill.

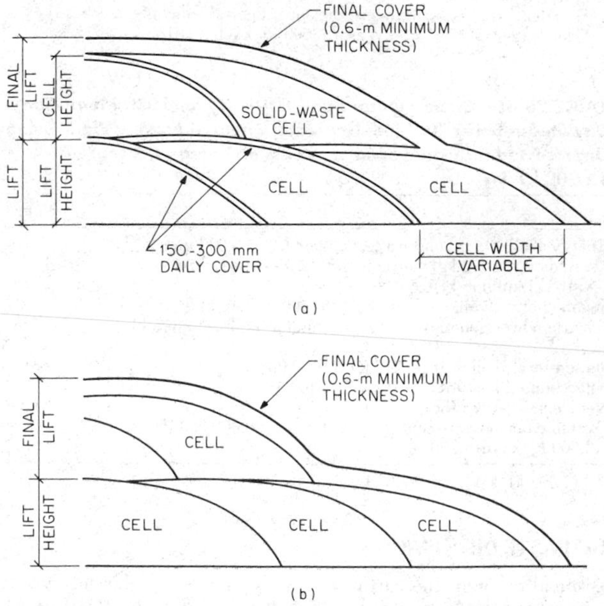

FIG. 26-51 Typical section through a landfill. (*a*) With daily or intermediate cover. (*b*) Without daily or intermediate cover.

2. *Trench method.* The trench method (see Fig. 26-52) of landfilling is ideally suited to areas where an adequate depth of cover material is available at the site and where the water table is well below the surface. To start the process, a portion of the trench is dug and the dirt is stockpiled to form an embankment behind the first trench. Wastes are then placed in the trench, spread into thin layers and compacted. The operation continues until the desired height is reached. Cover material is obtained by excavating an adjacent trench or continuing the trench that is being filled.

3. *Depression method.* At locations where natural or artificial depressions exist, it is often possible to use them effectively for land-

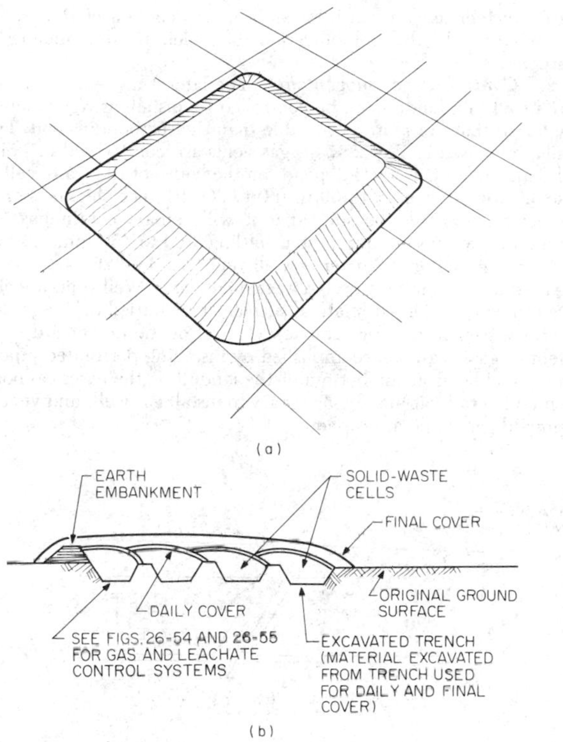

FIG. 26-52 Trench method for landfilling solid wastes. (*a*) Pictorial view of completed landfill. (*b*) Section through landfill.

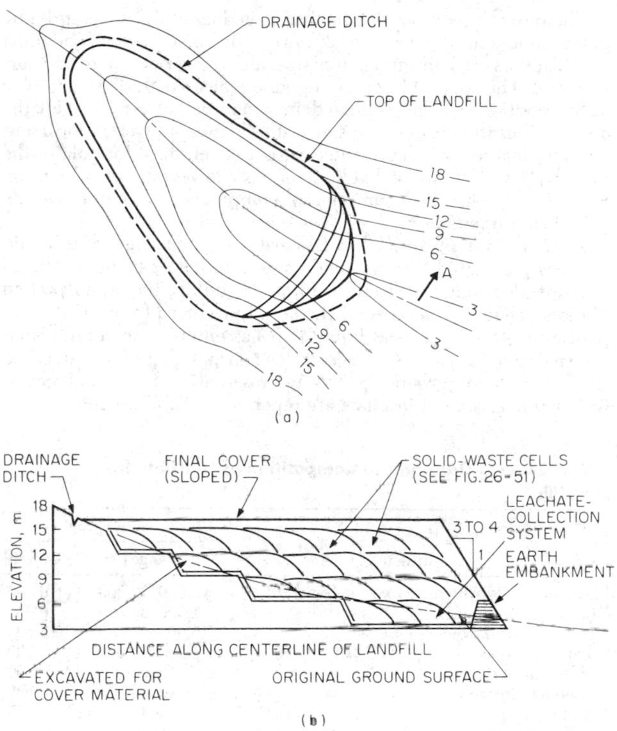

FIG. 26-53 Depression method for landfilling solid wastes. (*a*) Plan view: canyon-site landfill. (*b*) Section through landfill.

filling operations. Canyons, ravines, dry borrow pits, and quarries have all been used for this purpose. The techniques to place and compact solid wastes in depression landfills vary with the geometry of the site, the characteristics of the cover material, the hydrology and geology of the site, and access to the site.

In a canyon site filling starts at the head end of the canyon (see Fig. 26-53) and ends at the mouth. This practice prevents the accumulation of water behind the landfill. Wastes usually are deposited on the canyon floor and from there are pushed up against the canyon face at a slope of about 2 to 1. In this way, a high degree of compaction can be achieved.

4. *Landfills in wet areas.* Because of the problems associated with contamination of local groundwaters, the development of odors, and structural stability, landfills are seldom used in wet areas. If wet areas such as swamps and marshes, tidal areas, and ponds, pits, or quarries must be used as landfill sites, special provisions must be made to contain or eliminate the movement of leachate and gases from completed cells. Usually this is accomplished by first draining the site and then lining the bottom with a clay liner or other appropriate sealants. If a clay liner is used, it is important to continue operation of the drainage facility until the site is filled to avoid the creation of uplift pressures that could cause the liner to rupture from heaving.

Occurrence of Gases and Leachate in Landfills The following biological, physical, and chemical events occur when solid wastes are placed in a sanitary landfill: (1) biological decay of organic materials, either aerobically or anaerobically, with the evolution of gases and liquids; (2) chemical oxidation of waste materials; (3) escape of gases from the fill; (4) movement of liquids caused by differential heads; (5) dissolving and leaching of organic and inorganic materials by water and leachate moving through the fill; (6) movement of dissolved material by concentration gradients and osmosis; and (7) uneven settlement caused by consolidation of material into voids.

With respect to item 1, bacterial decomposition initially occurs under aerobic conditions because a certain amount of air is trapped within the landfill. However, the oxygen in the trapped air is exhausted within days, and long-term decomposition occurs under anaerobic conditions.

1. *Gases in landfills.* Gases found in landfills include air, ammonia, carbon dioxide, carbon monoxide, hydrogen, hydrogen sulfide, methane, nitrogen, and oxygen. Data on the molecular weight and density of these gases are presented in Table 3-30 in Sec. 3. Carbon dioxide and methane are the principal gases produced from the anaerobic decomposition of the organic solid-waste components.

The anaerobic conversion of organic compounds is thought to occur in three steps: the first involves the enzyme-mediated transformation (hydrolysis) of higher-weight molecular compounds into compounds suitable for use as a source of energy and cell carbon; the second is associated with the bacterial conversion of the compounds resulting from the first step into identifiable lower-molecular-weight intermediate compounds; and the third step involves the bacterial conversion of the intermediate compounds into simpler end products, such as carbon dioxide (CO_2) and methane (CH_4). The overall anaerobic conversion of organic industrial wastes can be represented with the following equation:

$$C_aH_bO_cN_d \rightarrow nC_wH_xO_yN_z + mCH_4 + sCO_2 + rH_2O + (d - nz)NH_3 \qquad (26\text{-}13)$$

where $s = a - nw - m$

$r = c - ny - 2s$

The terms "$C_aH_bO_cN_d$" and "$C_wH_xO_yN_z$" are used to represent on a molar basis the composition of the material present at the start end of the process. If it is assumed that the organic wastes are stabilized completely, the corresponding expression is

$$C_aH_bO_cN_d + \left(\frac{4a - b - 2c + 3d}{4}\right) H_2O \rightarrow \left(\frac{4a + b - 2c - 3d}{8}\right) CH_4 + \left(\frac{4a - b + 2c + 3d}{8}\right) CO_2 + dNH_3 \qquad (26\text{-}14)$$

The rate of decomposition in unmanaged landfills, as measured by gas production, reaches a peak within the first 2 years and then slowly tapers off, continuing in many cases for periods up to 25 years or more. The total volume of the gases released during anaerobic decomposition can be estimated in a number of ways. If all the organic constituents in the wastes (with the exception of plastics, rubber, and leather) are represented with a generalized formula of the form $C_aH_bO_cN_d$, the total volume of gas can be estimated by using Eq. (26-14), with the assumption of complete conversion to carbon dioxide and methane.

2. *Leachate in landfills.* Leachate may be defined as liquid that has percolated through solid waste and has extracted dissolved or suspended materials from it. In most landfills, the liquid portion of the leachate is composed of the liquid produced from the decomposition of the wastes and liquid that has entered the landfill from external sources, such as surface drainage, rainfall, groundwater, and water from underground springs. Representative data on the chemical characteristics of leachate are reported in Table 26-48.

TABLE 26-48 Data on the Composition of Leachate from Landfills*

Constituent	Value†, mg/L	
	Range‡	Typical
BOD_5 (5-day biochemical oxygen demand)	2,000–30,000	10,000
TOC (total organic carbon)	1,500–20,000	6,000
COD (chemical oxygen demand)	3,000–45,000	18,000
Total suspended solids	200–1,000	500
Organic nitrogen	10–600	200
Ammonia nitrogen	10–800	200
Nitrate	5–40	25
Total phosphorus	1–70	30
Ortho phosphorus	1–50	20
Alkalinity as $CaCO_3$	1,000–10,000	3,000
pH	5.3–8.5	6
Total hardness as $CaCO_3$	300–10,000	3,500
Calcium	200–3,000	1,000
Magnesium	50–1,500	250
Potassium	200–2,000	300
Sodium	200–2,000	500
Chloride	100–3,000	500
Sulfate	100–1,500	300
Total iron	50–600	60

*From G. Tchobanoglous, H. Theisen, and R. Eliassen, *Solid Wastes: Engineering Principles and Management Issues*, McGraw-Hill, New York, 1977.
†Except pH.
‡Representative range of values. Higher maximum values have been reported in the literature for some of the constituents.

Gas and Leachate Movement and Control Under ideal conditions, the gases generated from a landfill should be either vented to the atmosphere or, in larger landfills, collected for the production of energy. The leachate should be either contained within the landfill or removed for treatment.

1. *Gas movement.* In most cases, over 90 percent of the gas volume produced from the decomposition of solid wastes consists of methane and carbon dioxide. Although most of the methane escapes to the atmosphere, both methane and carbon dioxide have been found in concentrations of up to 40 percent at lateral distances of up to 120 m (400 ft) from the edges of landfills. If vented into the atmosphere in an uncontrolled manner, methane can accumulate (because its specific gravity is less than that of air) below buildings or in other enclosed spaces on or close to a sanitary landfill. With proper venting, methane should not pose a problem.

Because carbon dioxide is about 1.5 times as dense as air and 2.8 times as dense as methane, it tends to move toward the bottom of the landfill. As a result, the concentration of carbon dioxide in the lower portions of landfill may be high for years. Ultimately, because of its density, carbon dioxide will also move downward through the underlying formation until it reaches the groundwater. Because carbon dioxide is readily soluble in water, it usually lowers the pH, which in turn can increase the hardness and mineral content of the groundwater through the solubilization of calcium and magnesium carbonates.

2. *Control of gas movement.* The lateral movement of gases produced in a landfill can be controlled by installing vents made of materials that are more permeable than the surrounding soil. Typically, as shown in Fig. 26-54*a*, gas vents are constructed of gravel. The spacing of cell vents depends on the width of the waste cell but usually varies from 18 to 60 m (60 to 200 ft). The thickness of the gravel layer should be such that it will remain continuous even though there may be differential settling; 300 to 450 mm (12 to 18 in) is recommended. Barrier or well vents (see Fig. 26-54*b*) also can be used to control the lateral movement of gases. Well vents are often used in conjunction with lateral surface vents buried below grade in a gravel trench (see Fig. 26-54*c*). Control of the downward movement of gases can be accomplished by installing perforated pipes in the gravel layer at the bottom of the landfill. If the gases cannot be vented laterally, it may be necessary to install gas wells and vent the pumped gas to the atmosphere.

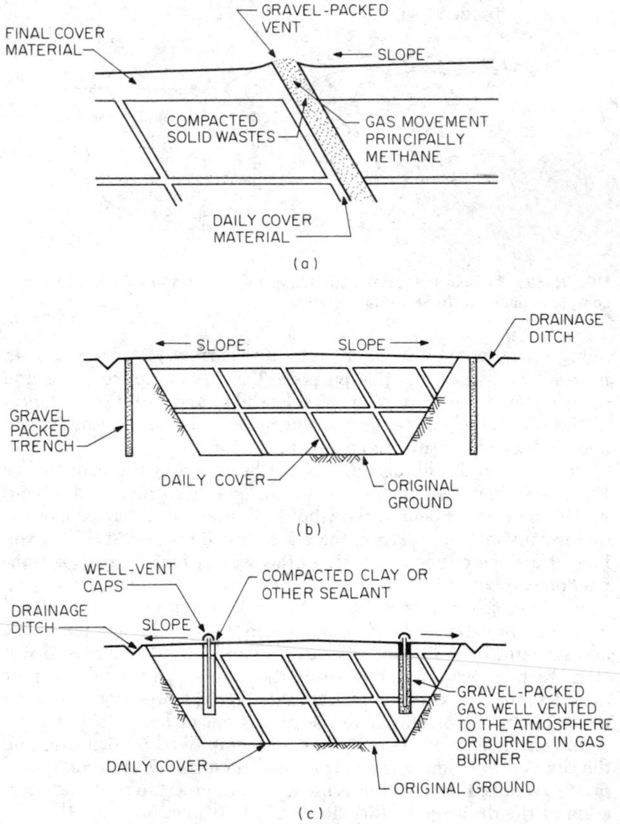

FIG. 26-54 Vents used to control the lateral movement of gases in landfills. (*a*) Cell. (*b*) Barrier. (*c*) Well. (*From G. Tchobanoglous, H. Theisen, and R. Eliassen,* Solid Wastes: Engineering Principles and Management Issues, *McGraw-Hill, New York, 1977.*)

The movement of landfill gases through adjacent soil formations can be controlled by constructing barriers of materials that are more impermeable than the soil (see Fig. 26-55*a*). Some of the landfill sealants that are available for this use are identified in Table 26-49. Of these, the use of compacted clays is the most common. The thickness will vary depending on the type of clay and the degree of control required; thicknesses ranging from 0.15 to 1.25 m (6 to 48 in) have been used.

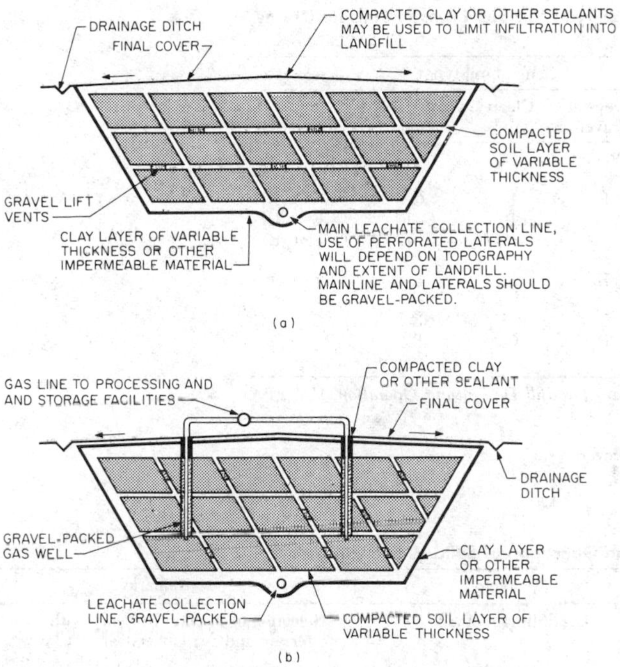

FIG. 26-55 Use of an impermeable liner to control the movement of gases and leachate in landfills. (*a*) Without gas recovery. (*b*) With gas recovery. (*From G. Tchobanoglous, H. Theisen, and R. Eliassen,* Solid Wastes: Engineering Principles and Management Issues, *McGraw-Hill, New York, 1977.*)

TABLE 26-49 Landfill Sealants for the Control of Gas and Leachate Movement

Sealant		
Classification	Representative types	Remarks
Compacted soil		Should contain some clay or fine silt.
Compacted clay	Bentonites, illites, kaolinites	Most commonly used sealant for landfills; layer thickness varies from 6 to 48 in; layer must be continuous and not be allowed to dry out and crack.
Inorganic chemicals	Sodium carbonate, silicate, or pyrophosphate	Use depends on local soil characteristics.
Synthetic chemicals	Polymers, rubber latex	Experimental; use not well established.
Synthetic membrane liners	Polyvinyl chloride, butyl rubber, Hypalon, polyethylene, nylon-reinforced liners	Expensive; may be justified where gas is to be recovered.
Asphalt	Modified asphalt, asphalt-covered polypropylene fabric, asphalt concrete	Layer must be thick enough to maintain continuity under differential settling conditions.
Others	Gunite concrete, soil cement, plastic soil cement	

°From G. Tchobanoglour, H. Theisen, and R. Eliassen, *Solid Wastes: Engineering Principles and Management Issues*, McGraw-Hill, New York, 1977.

3. *Control of gas movement by recovery.* The movement of gases in landfills can also be controlled by installing gas-recovery wells in completed landfills (see Fig. 26-55*b*). Clay and other liners are used when landfill gas is to be recovered. In some gas-recovery systems, leachate is collected and recycled to the top of the landfill and reinjected through perforated lines located in drainage trenches. Typically, the rate of gas production is greater in leachate-recirculation systems.

Although gas-recovery systems have been installed in some large municipal landfills, the economics of such operations are, at present, not well defined. The cost of the gas-cleanup and -processing equipment may limit the recovery of landfill gases, especially from small landfills.

4. *Leachate movement.* Under normal conditions, leachate is found in the bottom of landfills. From there it moves through the underlying strata, although some lateral movement may also occur, depending on the characteristics of the surrounding material. The rate of seepage of leachate from the bottom of a landfill can be estimated by Darcy's law by assuming that the material below the landfill to the top of the water table is saturated and that a small layer of leachate exists at the bottom of the fill. Under these conditions the leachate discharge rate per unit area is equal to the value of the coefficient of permeability K, expressed in meters per day. The computed value represents the maximum amount of seepage that would be expected, and this value should be used for design purposes. Under normal conditions, the actual rate would be less than this value because the soil column below the landfill would not be saturated.

5. *Control of leachate movement.* As leachate percolates through the underlying strata, many of the chemical and biological constituents originally contained in it will be removed by the filtering and adsorptive action of the material composing the strata. In general, the extent of this action depends on the characteristics of the soil, especially the clay content. Because of the potential risk involved in allowing leachate to percolate to the groundwater, best practice calls for its elimination or containment. Ultimately, it may be necessary to collect and treat the leachate.

The use of clay has been the favored method of reducing or eliminating the percolation of leachate (see Fig. 26-55 and Table 26-49). Membrane liners have also been used, but they are expensive and require care so that they will not be damaged during the filling operations. Equally important in controlling the movement of leachate is the elimination of surface-water infiltration, which is the major contributor to the total volume of leachate. With the use of an impermeable clay layer, an appropriate surface slope (1 to 2 percent), and adequate drainage, surface infiltration can be controlled effectively. Generalized ratings for the suitability of various types of soil for use as a landfill cover are reported in Table 26-50.

6. *Settlement and structural characteristics of landfills.* The settlement of landfills depends on the initial compaction, characteristics of wastes, degree of decomposition, and effects of consolidation when the leachate and gases are formed in the landfill. The height of the completed fill will also influence the initial compaction and degree of consolidation. The degree of consolidation can be modeled with a first-order equation.

Design and Operation of Landfills Important design considerations in the design and operation of landfills include (1) land requirements, (2) types of wastes that must be handled, (3) evaluation of seepage potential, (4) design of drainage and seepage-control facilities, (5) development of a general operation plan, (6) design of solid-waste-filling plan, and (7) determination of equipment requirements. The more important individual factors that must be considered in the design of a landfill are reported in Table 26-51. The last three items are considered further in the following discussion.

1. *Landfill-operation plan.* The layout of the site and the development of a workable operating schedule are the main features of a landfill-operation plan. In planning the layout of a landfill site, the location of the following must be determined: (1) access roads, (2) equipment shelters, (3) scales, if used, (4) storage sites for special wastes, (5) topsoil-stockpile sites, (6) landfill areas, and (7) plantings.

2. *Solid-waste-filling plan.* The specific method of filling will

TABLE 26-50 Generalized Ratings of Suitability of Various Types of Soils for Use as Landfill-Cover Material*

	General soil type†					
Function	Clean gravel	Clayey-silty gravel	Clean sand	Clayey-silty sand	Silt	Clay
Prevents rodents from burrowing or tunneling	G	F–G	G	P	P	P
Keeps flies from emerging	P	F	P	G	G	E‡
Minimizes moisture entering fill	P	F–G	P	G–E	G–E	E‡
Minimizes landfill-gas venting through cover	P	F–G	P	G–E	G–E	E‡
Provides pleasing appearance and controls blowing paper	E	E	E	E	E	E
Supports vegetation	P	G	P–F	E	G–E	F–G
Vents decomposition gas (is permeable)§	E	P	G	P	P	P

*From D. R. Brummer and D. J. Keller, *Sanitary Landfill Design and Operation*, U.S. EPA Publ. SW-65ts, 1972.

†E, excellent; G, good; F, fair; P, poor.

‡Except when cracks extend through the entire cover.

§Only if well drained.

TABLE 26-51 Important Factors That Must Be Considered in Design and Operation of Solid-Waste Landfills

Factor	Remarks
Design	
Access	Paved all-weather access roads to landfill site; temporary roads to unloading areas.
Cell design and construction	Will vary depending on terrain, landfilling method, and whether gas is to be recovered.
Cover material	Maximize use of on-site earth materials; approximately 1 m^3 of cover material will be required for every 4 to 6 m^3 of solid wastes; mix with sealants to control surface infiltration. In some designs, intermediate cover is not used.
Drainage	Install drainage ditches to divert surface-water runoff; maintain 1 to 2 percent grade on finished fill to prevent ponding.
Equipment requirements	Vary with size of landfills.
Fire prevention	Water on site; if nonpotable, outlets must be marked clearly; proper cell separation prevents continuous burn-through if combustion occurs.
Groundwater protection	Divert any underground springs; if required, install sealants for leachate control; install wells for gas and groundwater monitoring.
Land area	Area should be large enough to hold all wastes for a minimum of 1 year but preferably for 5 to 10 years.
Landfilling method	Selection of method will vary with terrain and available cover.
Litter control	Use movable fences at unloading areas; crews should pick up litter at least once per month or as required.
Operation plan	With or without the codisposal of treatment-plant sludges and the recovery of gas.
Spread and compaction	Spread and compact waste in 0.6-m (2-ft) layers.
Unloading area	Keep small, generally under 30 m (100 ft).
Operation	
Communications	Telephone for emergencies.
Days and hours of operation	Usual practice is 5 to 6 days/week and 8 to 10 h/day.
Employee facilities	Rest rooms and drinking water should be provided.
Equipment maintenance	A covered shed should be provided for field maintenance of equipment.
Operational records	Tonnage, transactions, and billing if a disposal fee is charged.
Salvage	No scavenging; salvage should occur away from the unloading area; no salvage storage on site.
Scales	Essential for record keeping.

depend on the characteristics of the site, such as the amount of available cover material, the topography, and local hydrology and geology. To assess future development plans, it will be necessary to prepare a detailed plan for the layout of the individual solid-waste cells. On the basis of the characteristics of the site or the method of operation (e.g., gas recovery), it may be necessary to incorporate special features for the control of the movement of gases and leachate from the landfill.

3. *Equipment requirements.* The types of equipment that have been used at sanitary landfills include both crawler and rubber-tired tractors, scrapers, compactors, draglines, and motor graders (see Fig. 26-56). The size and amount of equipment required will depend primarily on local site conditions, the size of the landfill operation, and the method of operation.

Landfilling of Hazardous Wastes In most states, the only disposal option available for most hazardous wastes is landfilling. The basis for the management of hazardous-wastes landfills is set forth in the Resource Conservation and Recovery Act of 1976. In general, disposal sites for hazardous wastes should be separate from sites for municipal solid wastes. If separate sites are not possible, great care must be taken to ensure that separate disposal operations are maintained.

Requirements The requirements for a hazardous-waste landfill are detailed in RCRA and the regulations developed to implement

FIG. 26-56 Equipment used to compact and place wastes in landfills. (*a*) Conventional crawler tractor. (*b*) Articulated metal wheel compactor.

the act. From a design standpoint, two of the most important requirements are (1) complete leachate containment, and (2) control of the surface water on and around the site.

Site Selection Factors that must be considered in evaluating potential sites for the disposal of hazardous waste are currently in a state of flux. In California, landfills where hazardous wastes can be received are referred to as Class I disposal sites. To qualify as a Class I site, it must be shown that:

1. Geological conditions are naturally capable of preventing vertical hydraulic continuity between liquids and gases emanating from the waste in the site and usable surface or groundwaters.
2. Geological conditions are naturally capable of preventing lateral hydraulic continuity between liquids and gases emanating from wastes in the site and usable surface or groundwaters, or the disposal area has been modified to achieve such capability.
3. Underlying geological formations which contain rock fractures or fissures of questionable permeability must be permanently sealed to provide a competent barrier to the movement of liquids or gases from the disposal site to usable water.
4. Inundation of disposal areas shall not occur until the site is closed in accordance with requirements of the regional board.
5. Disposal areas shall not be subject to washout.
6. Leachate and subsurface flow into the disposal areas shall be contained within the site unless other disposition is made in accordance with requirements of the regional board.
7. Sites shall not be located over zones of active faulting or where other forms of geological change would impair the competence of natural features or artificial barriers which prevent continuity with usable waters.
8. Sites made suitable for use by human-made physical barriers shall not be located where improper operations or maintenance of such structures could permit the waste, leachate, or gases to contact usable groundwater or surface water.
9. Sites which comply with the above-noted clauses but would be subject to inundation by a tide or a flood of greater than 100-year frequency may be considered by the regional board as limited Class I disposal sites.

Landfilling Methods and Operations Operation of a landfill for hazardous wastes is quite different from that of a conventional landfill. The specific details will vary depending on whether the wastes are containerized, liquid, or solid. When containerized hazardous wastes are to be disposed of, they are unloaded and placed in position individually to avoid rupturing the containers. To avoid the codisposal of incompatible wastes, separate storage areas within the total landfill site should be designated for various classes of compatible wastes. Liquid wastes are usually placed in containment areas and allowed to dry by solar evaporation. At many sites, liquid wastes are processed before disposal. In some cases, liquid wastes are injected into the soil. For dry wastes, conventional landfilling methods are used.

Design of Hazardous-Waste Landfills At the present time, many of the regulations governing the design of hazardous-waste landfills are unresolved. Although specific requirements will vary, the factors identified in Table 26-51 can be used as a design guide. Some special precautions that can be taken to prevent contamination of underlying strata are shown in Fig. 26-57.

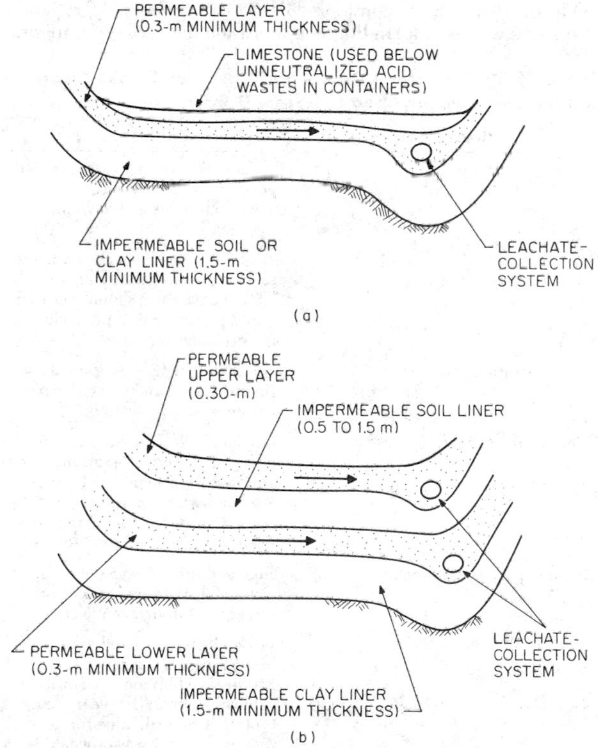

FIG. 26-57 Special design features for hazardous-waste landfills to prevent the contamination of the underlying strata. (*a*) Single-liner system. (*b*) Double-liner system.

Landfarming Landfarming is a waste-disposal method in which the biological, chemical, and physical processes that occur in the surface of the soil are used to treat biodegradable industrial wastes. Wastes to be treated are either applied on top of the land which has been prepared to receive the wastes or injected below the surface of the soil (see Fig. 26-58).

Process Description When organic wastes are added to the soil, they are subjected simultaneously to the following processes: (1) bacterial and chemical decomposition, (2) leaching of water-soluble

FIG. 26-58 Typical rig used for the subsurface injection of liquid wastes.

components in the original wastes and from the decomposition products, and (3) volatilization of selected components in the original wastes and from the products of decomposition.

Factors that must be considered in evaluating the biodegradability of organic wastes in a landfilling application include (1) composition of the waste; (2) compatibility of wastes and soil microflora; (3) environmental requirements including oxygen, temperature, pH, and inorganic nutrients, and (4) moisture content of soil-waste mixture.

Although most of the volatile components are released to the atmosphere, a small fraction is dissolved and/or carried away with the water in the soil matrix. Leached wastes are carried with the water as it percolates through the underlying soil strata. Most of the organic constituents contained in the leachate receive additional treatment as they pass through the soil column. Leached wastes can also be lost in surface runoff.

Ultimately, a portion of the wastes that are added is incorporated into the soil matrix. For this reason, it is important to biodegrade the added organic wastes to the maximum extent possible. It is for this reason that inorganic constituents such as cadmium, chromium, copper, and lead must also be controlled in wastes to be disposed of by landfarming.

Applications Landfarming is suitable for wastes that contain organic constituents that are biodegradable and are not subject to significant leaching while the bioconversion process is occurring. For example, petroleum oily wastes and oily sludges are ideally suited for disposal by landfarming. A variety of other organic wastes with similar characteristics are also suitable. Properly managed landfarming sites can be reused at frequent intervals with no adverse effects.

Design and Operation Important considerations in the design and operation of landfarming systems include (1) site selection, (2) site preparation, (3) waste characteristics, (4) method of waste application, (5) waste-application rate, (6) site management, and (7) monitoring. Important factors related to these design and operation considerations are reported in Table 26-52.

TABLE 26-52 Important Design and Operation Considerations for Landfarming Systems Used for Waste Disposal

Item	Remarks
Site selection location	Proximity to critical areas specified in government regulations, accessibility, site geology and hydrology.
Site selection: soil characteristics	Adequate area soil cover and depth to groundwater usually greater than 1.5 m (4 ft). Slope should not exceed 5 to 8 percent. Soil type, including ion-exchange capacity.
Site preparation	Area should be fenced, graded for runoff control, and disked or plowed before waste application.
Waste characterization	Suspended solids, organic content, nitrogen (all forms), phosphorus pH, and inorganic metals including arsenic, barium, cadmium, chromium, copper, lead, mercury, selenium, silver, sodium, and zinc.
Method of waste application	Ridge and furrow, sprinkling (fixed or portable systems), tank-truck spreading, subsurface injection.
Waste-application rates	For petroleum crude oil and lubricating oils the range is from 250 to 1250 bbl/(ha · year) with a value of 400 bbl/(ha · year) being typical. A typical value for general refinery oils and wastes would be about 150 bbl/(ha · year).
Site management	Wastes spread on the surface should be disked or plowed into the soil soon after application (1 to 7 days). To promote aerobic conditions and rapid bioconversion of the wastes the soil-waste mixture should be cultivated periodically.
Monitoring	Periodic samples should be taken to assess the extent of completion of the bioconversion process. Core samples should be taken annually to monitor the movement of leached wastes in the underlying strata.

Deep-Well Injection Deep-well injection for the disposal of liquid solid wastes involves injecting the wastes deep in the ground into

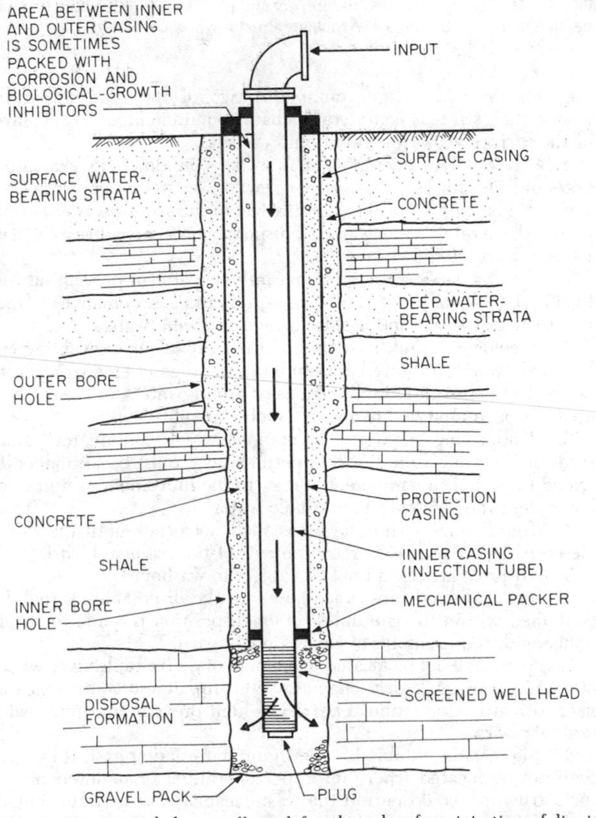

FIG. 26-59 Typical deep well used for the subsurface injection of liquid wastes.

TABLE 26-53 Important Design and Operation Considerations for Deep Wells Used for Waste Injection

Item	Remarks
Well-site selection	Criteria for assessing the feasibility of a deep-well-injection site include (1) uniformity, (2) large extent, (3) substantial thickness, (4) high porosity and permeability, (5) low pressure, (6) saline aquifer, (7) separation from potable-water horizons, (8) adequate overlying and underlying aquicludes, (9) no poorly plugged wells nearby, and (10) compatibility between the mineralogy and fluids of the reservoir and the injected wastes.
Waste pretreatment	Suspended solid less than 10 to 15 mg/L; particle sizes equal to or less than 1 to 5 μm (depends on injection formation). Adjustment of pH and buffering of the waste may be necessary.
Deep-well installation	Well depths vary from 550 to 3660 m (1800 to 1200 ft); well-injection rates vary from 4 to 60 L/s; rates in the range from 15 to 20 L/s are typical. Operation pressures up to 27,600 kPa (4000 psig) are used.
Monitoring	Continuous monitoring facilities should be installed when wells are put into operation. Irregularities in the pressure may require changes in operating procedures.

permeable rock formations (typically limestone or dolomite) or underground caverns.

Process Description The installation of deep wells for the injection of wastes closely follows the practices used for the drilling and completion of oil and gas wells. To isolate and protect potential water-supply aquifers, the surface casing must be set well below such aquifers and cemented to the surface of the well (see Fig. 26-59). The drilling fluid should not be allowed to penetrate the formation that is to be used for waste disposal. To prevent clogging of the formation, the drilling fluid is replaced with a compatible solution. Also,

TABLE 26-54 Short- and Long-Term Actions for Effective Industrial Solid-Waste Management*

Actions	Remarks
Short-term	
1. Inventory wastes.	Document all types, quantities, and sources of wastes (both nonhazardous and hazardous wastes).
2. Inventory inactive sites.	All inactive sites where wastes have been disposed of in the past should be inventoried. Data should be gathered on buried wastes, including types, quantity, and sources. A groundwater-monitoring program should be developed.
3. Characterize wastes.	In addition to general information on the characteristics of the wastes, all hazardous wastes should be individually characterized.
4. Assign responsibilities.	Assign responsibilities and authority at plant and headquarters for the storage, collection, treatment, and disposal of all types of hazardous wastes.
5. Track the movement of wastes.	Develop a logging system for hazardous wastes containing the date, waste description, source, volume shipped or hauled, name of hauler, and destination. Follow through to be sure that wastes reach destination.
6. Develop emergency procedures.	Develop procedures for dealing with emergency situations involving the storage, collection, treatment, and disposal of hazardous wastes.
7. Obtain permits.	Start obtaining the necessary waste-disposal permits as soon as possible.
Long-term	
1. Remove all accumulated wastes.	Develop a systematic program for removing all accumulated wastes stored on the plant site.
2. Separate wastes.	Institute a long-term program to separate wastes at the source of production.
3. Reduce wastes.	A systematic program should be undertaken to examine all sources of waste production and to develop alternative operations and processes to reduce waste generation.
4. Improve facilities.	Upgrade facilities to meet RCRA requirements. A data-collection program should be instituted to obtain any needed data.
5. Review all waste-management agreements.	Develop detailed contracts with outside waste-management firms. Define clearly the duties and responsibilities of plant personnel and waste-collection personnel.
6. Review and develop disposal-site options.	Develop long-term projections for landfill requirements and initiate a program to secure the needed sites.
7. Secure appropriate engineering and consulting services.	Make sure that your engineering departments are involved early in the process. Retain outside consultants for specific tasks.
8. Monitor legislative programs.	Develop a program for monitoring new regulations and for inputting to appropriate federal, state, and local agencies on the modification and development of new regulations.

*Adapted in part from R. Sobel, "How Industry Can Prepare for RCRA," *Chem. Eng.*, 86(1), 82 (Jan. 29, 1979).

in some cases it may be necessary to acid-treat the formation before the injection of wastes is initiated.

Applications Deep-well injection has been used principally for liquid wastes that are difficult to treat and dispose of by more conventional methods and for hazardous wastes. Chemical, petrochemical, and pharmaceutical wastes are those most commonly disposed of with this method. The waste may be liquid, gases, or solids. The gases and solids are either dissolved in the liquid or are carried along with the liquid.

Design and Operation Important design and operation considerations for deep-well injection are related to (1) well-site selection, (2) pretreatment, (3) installation of an injection well, and (4) monitoring. Important factors related to these design and operation considerations are reported in Table 26-53. As noted in the table, wastes are usually treated prior to injection to prevent clogging of the formation and damage to equipment. Particles greater than about 1 to 5 μm must be removed. Typically, treated wastes must be filtered prior to injection. Wastes must also be compatible with the characteristics of the aquifer. This may require pH adjustment and the use of compatible buffers.

Ocean Disposal of Solid Wastes Although ocean dumping of municipal solid wastes was abandoned in the United States in 1933, the concept has persisted throughout the years and is still frequently discussed today. Some industrial wastes are still discharged at sea. Within the past few years, the idea that the ocean is a gigantic sink, into which an infinite amount of pollution of all types can be dumped, has been discarded. On the other hand, it is argued that many of the wastes now placed in landfills or on land could be used as fertilizers to increase the productivity of the ocean. It is also argued that the placement of wastes in ocean-bottom trenches where tectonic folding is occurring is an effective method of waste disposal.

PLANNING

Because of the ever-growing number of federal regulations governing the disposal of nonhazardous and hazardous solid wastes, it is prudent to develop both short-term and long-term action programs to deal with all aspects of solid-waste management. Important short- and long-term actions are identified in Table 26-54.

Section 27

Biochemical Engineering

Henry R. Bungay, P.E., Ph.D., *Professor of Chemical and Environmental Engineering, Rensselaer Polytechnic Institute; Member, American Institute of Chemical Engineers, American Chemical Society, American Society for Microbiology, American Society for Engineering Education. (Section Editor)*

George T. Tsao, Ph.D., *Director, Laboratory of Renewable Resource Engineering, Purdue University; Member, American Institute of Chemical Engineers, American Chemical Society, American Society for Microbiology*

Arthur E. Humphrey, Ph.D., *Provost, Lehigh University; Member, U.S. National Academy of Engineering; Member, American Institute of Chemical Engineers, American Chemical Society, American Society for Microbiology*

GENERAL REFERENCES: Aiba, S., A. E. Humphrey, and N. F. Millis, *Biochemical Engineering,* 2d ed., University of Tokyo Press, Tokyo, 1973. Atkinson, B., *Biochemical Reactors,* Pion, London, 1974. Bailey, J. E., and D. F. Ollis, *Biochemical Engineering Fundamentals,* McGraw-Hill, New York, 1977. Bungay, H. R., *Energy: The Biomass Options,* Wiley, New York, 1981. Bushell, M. E., and J. H. Slater (eds.), *Mixed Culture Fermentations,* Academic, New York, 1981. Calcott, P. H., *Continuous Cultures of Cells,* vols. I and II, CRC Press, Boca Raton, Fla., 1981. Chibata, I., S. Fukiu, and L. B. Wingard, *Enzyme Engineering,* vol. 6 of series, Plenum, New York, 1982. Goldstein, I. S. (ed.), *Organic Chemicals from Biomass,* CRC Press, Boca Raton, Fla., 1981. Grady, C. P. L., Jr., and H. C. Lim, *Biological Wastewater Treatment: Theory and Applications,* Marcel Dekker, New York, 1980. Hollaender, A., R. Rabson, P. Rogers, A. San Pietro, R. Valentine, and R. Wolfe, *Trends in the Biology of Fermentation for Fuels and Chemicals,* Plenum, New York, 1982. Kirk, T. K., *Lignin Biodegradation: Microbiology, Chemistry, and Potential Applications,* vols. I and II, CRC Press, Boca Raton, Fla., 1980. Klass, D. L. (ed.), *Biomass as a Nonfossil Fuel Source,* CRC Press, Boca Raton, Fla., 1981. Moo-Young and various coauthors, *Advances in Biotechnology: Proceedings of the 6th International Fermentation Symposium, London, Canada,* 4 vols., Pergamon, New York, 1981. Peppler, H. J., and D. Perlman, *Microbial Technology,* Academic, New York, 1979. Rose, A. H. (ed.), *Economic Microbiology Series,* Academic, New York, annually. San Pietro, A. (ed.), *Biochemical and Photosynthetic Aspects of Energy Production,* Academic, New York, 1980. Smith, J. E., D. R. Berry, and B. Kristiansen (eds.), *Fungal Biotechnology,* Academic, New York, 1980. Sofer, S. S., and O. R. Zaborsky (eds.), *Biomass Conversion Processes for Energy and Fuels,* Plenum, New York, 1981. Solomons, G. L., *Materials and Methods in Fermentation,* Academic, New York, 1969. Stafford, D. A., D. L. Hawkes, and H. R. Horton, *Methane Production from Waste Organic Matter,* CRC Press, Boca Raton, Fla., 1980. Vogt, F. (ed.), *Energy Conservation and the Use of Renewable Energies in the Bio-Industries,* Pergamon, New York, 1981. Wang, D. I. C., C. L. Cooney, A. L. Demain, P. Dunnill, A. E. Humphrey, and M. D. Lilly, *Fermentation and Enzyme Technology,* Wiley, New York, 1979. Wise, D. L. (ed.), *Fuel Gas Production from Biomass,* vols. I and II, CRC Press, Boca Raton, Fla., 1981. Wiseman, A., *Principles of Biotechnology,* Surrey University Press, Glasgow, 1982.

Nomenclature and Units

Symbol	Definition	SI units	U.S. customary units
A	Empirical constant	Dimensionless	Dimensionless
C	Concentration (mass)	kg/m^3	lb/ft^3
C	Concentration	mol/m^3	$(lb \cdot mol)/ft^3$
D	Diameter	m	ft
D	Effective diffusivity	m^2/s	ft^2/h
D	F/V	s^{-1}	h^{-1}
DRT	Decimal reduction time for sterilization	s	h
E	Activation energy	cal/mol	$Btu/(lb \cdot mol)$
F	Flow or feed rate	m^3/s	ft^3/h
H	Concentration of host organisms	kg/m^3	lb/ft^3
K	Rate coefficient	Units dependent on order of reaction	Units dependent on order of reaction
K_M	Michaelis constant	kg/m^3	lb/ft^3
K_1a	Lumped mass-transfer coefficient	s^{-1}	h^{-1}
K_d	Death-rate coefficient	s^{-1}	h^{-1}
K_s	Monod coefficient	kg/m^3	lb/ft^3
k	Kinetic constants	Dependent on reaction order	Dependent on reaction order
M	Coefficient for maintenance energy	Dimensionless	Dimensionless
N	Numbers of organisms or spores	Dimensionless	Dimensionless
P	Product concentration	kg/m^3	lb/ft^3
Q_{O_2}	Specific-respiration-rate coefficient	$kg\ O_2/(kg\ organism \cdot s)$	$lb\ O_2/(lb\ organism \cdot h)$
R	Universal-gas-law constant	$8314\ J/(mol \cdot K)$	$0.7299\ (ft^3)\ (atm)/(lb \cdot mol \cdot R)$
r	Radial position	m	ft
S	Substrate concentration	kg/m^3	lb/ft^3
S	Shear	N/m^2	lbf/ft^2
S_o	Substrate concentration in feed	kg/m^3	lb/ft^3
T	Temperature	K	°F
t	Time	s	h
V	Velocity of reaction	mol/s	$(lb \cdot mol)/h$
V_m	Maximum velocity of reaction	mol/s	$(lb \cdot mol)/h$
V	Air velocity	m/s	ft/h
V	Fermenter volume	m^3	ft^3
V	Derivative of product concentration [Eq. (27-30)]	$kg/(m^3 \cdot s)$	$lb/(ft^3 \cdot h)$
VVM	Volume of air/volume of fermentation broth per minute	Dimensionless	Dimensionless
X	Organism concentration	kg/m_3	lb/ft^3
Y	Yield coefficient	$kg/kg\ O_2$	$lb/lb\ O_2$
		Greek symbols	
β	Dimensionless Michaelis constant	Dimensionless	Dimensionless
ΔE	Activation energy	cal/mol	$Btu/(lb \cdot mol)$
μ	Specific-growth-rate coefficient	s^{-1}	h^{-1}
$\hat{\mu}$	Maximum-specific-growth-rate coefficient	s^{-1}	h^{-1}
ω	Recycle ratio	Dimensionless	Dimensionless
ϕ	Thiele modulus	Dimensionless	Dimensionless

INTRODUCTION TO BIOCHEMICAL ENGINEERING

The differences between biochemical engineering and chemical engineering lie not in the principles of unit operations and unit processes but in the nature of living systems. The commercial exploitation of cells or of enzymes taken from cells is restricted to conditions at which biological systems can function; thus the mild parameters that are the great advantage for biosystems are also a major problem. At exactly the correct pH, temperature, redox potential, and medium, a given strain should achieve overwhelming predominance. However, operation at other conditions may impair rates and allow contaminating organisms to thrive. In mixed-culture systems, especially those for biological waste treatment, there is an ever-shifting interplay between populations and their environments that influences performance and control. The optimization of the complicated biochemical activities of isolated strains, of aggregated cells, of mixed populations, and of cell-free enzymes or components presents engineering challenges that are sophisticated and difficult. Stability of biological systems may be compromised by any of the many biochemical steps functioning in concert, and genetic controls are subject to mutation. Offspring of specialized mutants that yield high concentrations of product tend to revert during propagation to less productive strains—a phenomenon called rundown.

This section emphasizes microbial processes and enzymatic catalysis; medical, animal, and agricultural engineering systems have been omitted because of space limitations. Engineering aspects of biological waste treatment are covered in Sec. 26.

Although fermented beverages have been produced for several thousand years, biochemical engineering is not yet fully mature. Recent developments such as immobilized enzymes and cells have been partially exploited, but many exciting advances should be forthcoming. Genetic engineering is leading to practical processes for molecules that previously could be found only in trace quantities in plant or animal tissues. Fermentation processes that are now important for relatively valuable products may be overshadowed by those for cheaper bulk chemicals. As prices for petrochemicals soar, there are large profits to be made with fermentations to produce equivalent compounds.

BIOLOGICAL CONCEPTS

Cells The cell is the unit of life. Cells in multicellular organisms function in association with other specialized cells, but many organisms are free-living single cells. Although differing in size, shape, and functions, they have basic common features. Every cell contains cytoplasm, a colloidal system of large biochemicals in a complex solution of smaller organic molecules and inorganic salts. The cytoplasm is bounded by a semielastic, selectively permeable cell membrane which controls the transport of molecules into and out of the cell. Genes containing information that controls cellular activity are in sequence along a threadlike chromosome. As the units of heredity, genes determine the cellular characteristics passed from one generation to the next.

In most cells, the chromosomes are surrounded by a membrane to form a conspicuous nucleus. Cells with organized nuclei are described as eukaryotic. Other intracellular structures serve as specialized sites for cellular activities. For example, photosynthesis is carried out by organelles called chloroplasts.

In bacteria and blue-green algae the chromosomes are not surrounded by a membrane, and there is little apparent subcellular oganization. The chlorophyll of blue-green algae is associated with loosely arranged membranes within the cytoplasm. When bacteria have chlorophyll, it is a unique type located in vesicular chromatophores. Lacking a discrete nucleus, these organisms are said to be prokaryotic.

Plant cells, bacteria, and blue-green algae are protected by rigid cell walls external to the cell membranes. Certain algae and protozoa have gelatinous sheaths of inorganic materials such as silica.

Three groups of microorganisms are of special concern to biochemical engineering: bacteria, algae, and fungi. A fourth group, the protozoa, can feed on smaller organisms in natural waters and in waste-treatment processes. Certain viruses called phages are also important in that they can infect microorganisms and may destroy a culture.

Bacteria Bacteria are tiny single-cell organisms ranging from 0.5 to 20 μm in size, although some may be smaller and a few exceed 100 μm in length. The cell wall imparts a characteristic round or ovoid, rod-shaped, or spiral shape to the cell. Some bacteria can vary in shape, depending on culture conditions; this is termed "pleomorphism." Certain species are further characterized by the arrangement of cells in clusters, chains, or discrete packets. Pigments which impart a characteristic color may be formed. The cytoplasm of bacteria may also contain numerous granules of storage materials such as carbohydrates and lipids. Many bacteria exhibit motility by means of one or more hairlike appendages called flagella. Bacteria reproduce by dividing into equal parts—a process termed "binary fission."

Under adverse conditions, certain microorganisms produce spores which germinate upon return to a favorable environment. Spores may survive dryness and temperature extremes. Other microorganisms form spores at a stage in their normal life cycle.

Many species may, under appropriate circumstances, become surrounded by gelatinous material that provides a means of attachment and some protection from other organisms. If many cells share the same gelatinous covering, it is called a slime; otherwise, each is said to have a capsule.

Each species of microorganism grows best within certain temperature and pH ranges, commonly between 20 and 40°C and not too far from neutral pH. There are some species, however, which thrive at extremes.

Algae Algae are a very diverse group of photosynthetic organisms that range from microscopic size to giant kelp that may reach lengths of 20 m (66 ft). Some commercial biochemicals come from algal seaweeds, and algae supply oxygen and consume nutrients in several different processes for biological waste treatment. Although their rapid growth rates relative to other green plants offer great potential for producing biomass as an energy or chemical feedstock, there is little use of algae for industrial fermentations. A new process is being developed with *Dunaliella,* a species that grows in high salinity and produces glycerol. Outdoor ponds are good for growing algae because vast surfaces and high illumination are needed.

Fungi As a group, fungi are characterized by simple vegetative bodies from which reproductive structures are elaborated. All fungal cells possess distinct nuclei and, at some stage in their life cycles, produce spores in specialized fruiting bodies. The fungi contain no chlorophyll and therefore require sources of complex organic molecules. Many species grow on dead organic material, while others live as parasites. Many can live on carbohydrate, inorganic nitrogen, and salts.

Yeasts are unicellular organisms surrounded by a cell wall and possessing a distinct nucleus. With very few exceptions, yeasts reproduce by a process known as budding; a small new cell is pinched off the parent cell, but under certain conditions an individual yeast cell may become a fruiting body, producing four spores.

Viruses Viruses, particles of a size below the resolution of the light microscope, are not cellular in structure and are composed mainly of nucleic acid surrounded by a protein sheath. Lacking metabolic machinery, viruses exist only as intracellular, highly host-specific parasites. Many bacteria and certain molds are subject to invasion by virus particles. Those which attack bacteria are called bacteriophages and may be either virulent or temperate. Virulent bacteriophages divert cellular resources to the manufacture of phage particles. As new phage particles are released to the medium, the host cell dies and lyses. Temperate bacteriophages have no immediate effect upon the host cell. They become attached to the bacterial chromosome and may be carried through many generations before being triggered to virulence by some physical or chemical event.

Biochemistry All organisms require sources of carbon, nitrogen, sulfur, phosphorus, water, and trace elements. Some have specific vitamin requirements as well. Green plants need only carbon dioxide, nitrate or ammonium ions, dissolved minerals, and water to manufacture all their cellular components. Photosynthetic bacteria require specific sources of hydrogen ions, and chemosynthetic bacteria must have an oxidizable substrate. Some microorganisms have the ability to "fix" atmospheric nitrogen by reduction. Organisms which use only simple inorganic compounds as nutrients are said to be autotrophic (self-nourishing).

Organisms which require compounds that have been manufactured by other organisms are called heterotrophs (other-nourishing). Many heterotrophs secrete enzymes (exoenzymes) which hydrolyze large molecules to smaller units which can readily enter the cell.

Proteins are macromolecules that play roles such as serving as components of cell membranes and of muscle. The antibodies which protect organisms against invasion by foreign substances are themselves proteins. There are about 20 amino acids which are found regularly in naturally occurring proteins as polypeptides. Because of the great length of protein chains and the various sequences of amino acids, the number of possible proteins is astronomical. The amino-acid sequence is referred to as the primary structure of a protein. The polypeptide chain is usually folded to provide secondary structure to the molecule, and linkages through other functional groups form the tertiary structure. For some protein molecules, there may be a spatial arrangement forming defined aggregates known as the quaternary structure of proteins. For a polypeptide polymer to have biological activities including the catalytic activities of enzymes, usually a certain molecular structure, including not only primary and secondary but also tertiary and sometimes quaternary structure, is necessary. Such a strict structural requirement provides the basis of explanation of the high specificity of proteins. In the presence of certain chemical reagents, excessive heat, radiation, unfavorable pH, etc., the protein structure may become disorganized. This is called denaturation and may be reversible if it is not too severe.

A special class of proteins, the enzymes, are biological catalysts which expedite reactions by lowering the amount of activation required. Enzymes function in conjunction with another special class of compounds known as coenzymes. Coenzymes are not proteins. Many of the known coenzymes include vitamins, such as niacin and riboflavin, as part of their molecular structure. Coenzymes carry reactant groups or electrons between substrate molecules in the course of a reaction. Because coenzymes serve merely as carriers and are constantly recycled, only small amounts are needed to produce large amounts of biochemical product.

Hundreds of reactions take place simultaneously in cells. There are branched and parallel pathways, and a biochemical may participate in several distinct reactions. Through mass action, concentration changes caused by one reaction may affect the kinetics and equilibrium concentrations of another. In order to prevent the accumulation of too much of a biochemical, the product or an intermediate in the pathway may slow the production of an enzyme or may inhibit the activation of existing enzymes. This is termed "feedback control" and is shown in Fig. 27-1. More complicated examples in which two biochemicals act in concert to inhibit an enzyme are known. As accumulation of excessive amounts of a certain biochemical may be the key to economic success, creating mutant cultures with defective metabolic controls has great value.

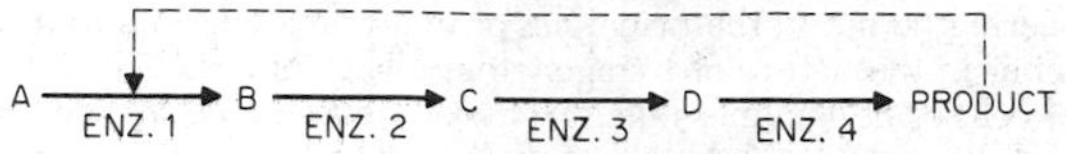

FIG. 27-1 Feedback control. Product inhibits the first enzyme.

Cell efficiency is improved by omitting the synthesis of unneeded enzymes, so there are two classes of enzymes, those that are constitutive and are always produced and those that are synthesized in response to an inducer, usually the initial substrate in a pathway. Enzymes that are induced in one organism may be constitutive in another.

Microorganisms exhibit nutritional preferences. The enzymes for common substrates such as glucose are constitutive, as are the enzymes for common metabolic pathways. Furthermore, the synthesis of enzymes for attack on less common substrates such as galactose is blocked by the presence of appreciable amounts of common substrates or metabolites. This is logical for cells to conserve their resources for enzyme synthesis as long as their usual substrates are readily available. If presented with mixed substrates, those that are in the main metabolic pathways are consumed simultaneously while the other substrates are consumed later after the common substrates have been depleted. A diauxic growth curve exhibits a plateau, while the enzymes needed for the uncommon substrates are synthesized (see Fig. 27-2). There may also be preferences for the less common substrates such that a mixture shows a sequence of one being exhausted before the start of metabolism of the next.

Energy Many metabolic reactions, once activated, proceed spontaneously with a net release of energy. Hydrolysis and molecular rearrangements are examples of spontaneous reactions. The hydro-

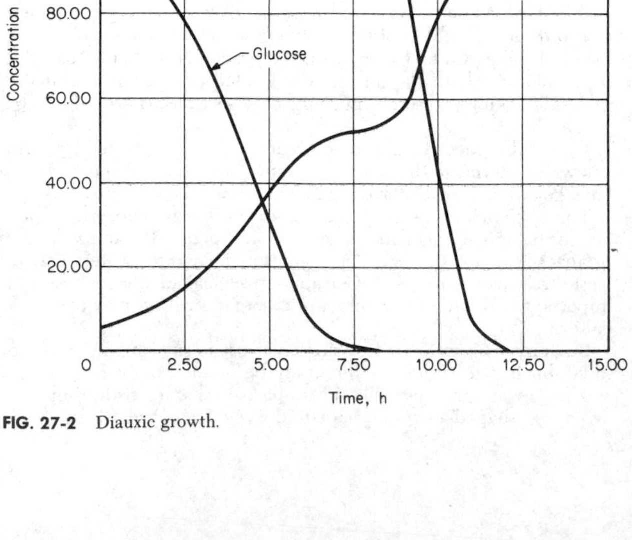

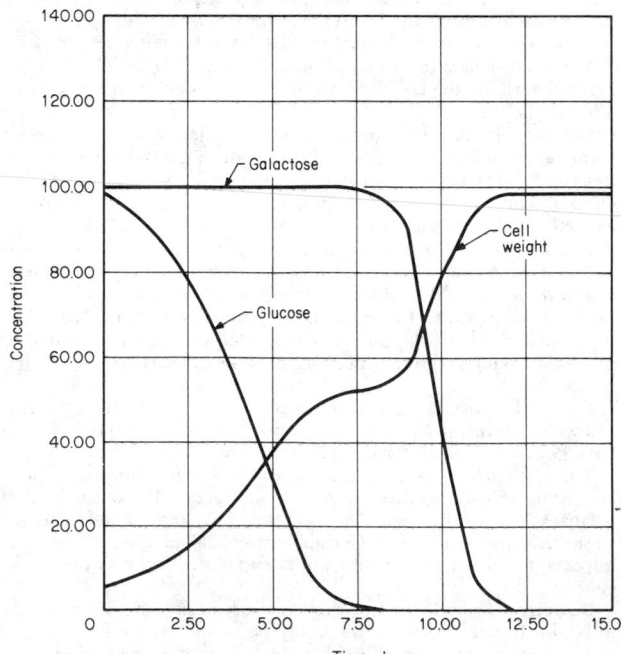

FIG. 27-2 Diauxic growth.

lytic splitting of starch to glucose, for instance, results in a net release of energy. A great many biochemical reactions are not spontaneous and therefore require an energy input. In living systems this requirement is met by coupling an energy-requiring reaction with an energy-releasing reaction. If a sufficient amount of energy is produced by a metabolic reaction, it may be used to synthesize a high-energy compound such as adenosinetriphosphate (ATP). When the terminal phosphate linkage is broken, adenosinediphosphate and inorganic phosphate are formed and energy is provided. When sufficient energy becomes available, ATP is re-formed.

In biological systems, the most frequent mechanism of oxidation is the removal of hydrogen, and, conversely, the addition of hydrogen is the common method of reduction. Nicotinamide–adenine dinucleotide (NAD) and nicotinamide–adenine dinucleotide phosphate (NADP) are two coenzymes which assist in oxidation and reduction. Most oxidation reactions are coupled to reduction reactions so that one can drive the other. However, stepwise release or consumption of energy requires driving forces and losses at each step such that overall efficiency suffers.

The overall redox potential of a system determines the amount of energy that cells can derive from their nutrients. When oxygen is present to be the ultimate acceptor of electrons, complete oxidation of organic molecules yields maximum energy. However, inside animals, in polluted waters, in the benthos (bottom region) of natural waters, and elsewhere there is little or no free oxygen. In these environments, organisms develop that can partially oxidize substrates or can derive a small amount of energy from reactions in which some products are oxidized while others are reduced. The pathways for complete oxidation may be absent, and the presence of oxygen can disrupt the mechanisms for anaerobic metabolism so that the cell is quickly killed. The differences in efficiency are striking. Aerobic metabolism of 1 molecule of glucose can generate bond energy as 33 molecules of ATP, while anaerobic metabolism yields only 2 molecules of ATP. Natural anaerobic processes accumulate compounds such as ethanol, acetoin, acetone, butanol, lactate, and malate. Products of natural aerobic metabolism are water and carbon dioxide. Aerobic processes of commercial importance produce valuable cell mass or use special mutants to accumulate products such as antibiotics.

Mutation and Genetic Engineering Exposing organisms to agents such as mustard chemicals, ultraviolet light, and x-rays increases mutation rate by damaging chromosomes. There are few survivors of an intensive treatment, and some of these cells are mutants. Most of the mutations are harmful to the cell, but a very small number may have economic importance in that some controls are impaired, resulting in better yields of product. Screening of many strains to find the very few worthy of further study is tedious and expensive.

Whereas mutagenic agents delete genes, recombinant-DNA techniques add desirable features of genetic material from very different cells. The genes may come from plant, animal, or microbial cells, or in a few instances they may be synthesized in the laboratory by knowing nucleic-acid sequences in natural genes. Opening a chromosome and splicing in foreign DNA are simple in concept, but there are complications. Genes in fragments of DNA must have control signals from other nucleic-acid sequences in order to function. Both the gene and its controls must be spliced into the chromosomes of the receiving culture. Bacterial chromosomes (circular DNA molecules called plasmids) are cut open with enzymes, mixed with the new fragments to be incorporated, and closed enzymatically. Each plasmid is reproduced many times during cell division; thus the organism has acquired new traits.

There are tricks and much art for genetic engineering, such as using bacteriophage infection to introduce a gene or an enzyme to a cell. Certain strains of *E. coli*, *Bacillus subtilis*, yeast, and streptomyces are the usual working organisms (cloning vectors) to which genes are added because their own genetics is well understood and the methodology has become fairly routine.

FURTHER READING: S. Broome and W. A. Gilbert, *Proc. Nat. Acad. Sci. (U.S.A.)*, **72**, 3961 (1978); S. N. Cohen, A. C. Y. Chang, H. W. Boyer, and R. B. Hellings, *Proc. Nat. Acad. Sci. (U.S.A.)*, **70**, 3240 (1973); H. A. Erlich, S. N. Cohen, and H. O. McDevitt, *Cell*, **13**, 681 (1978); D. O. Nathans, *Science*, **206**, 903 (1979); A. Skalka and L. Shapiro, *Gene*, **1**, 65 (1976).

Photosynthesis All living cells synthesize ATP, but only green plants and a few microorganisms can drive biochemical reactions to form ATP with radiant energy through the process of photosynthesis. The number per cell of membrane-surrounded organelles called chloroplasts varies with species and environmental conditions. In higher plants numerous chloroplasts are found in each cell of the mesophyll tissue of leaves, while an algal cell may contain a single chloroplast. A chloroplast has a sandwich of many layers alternating between pigments and enzymatic proteins such that electromagnetic excitation from light becomes chemical-bond energy. Prokaryotic organisms—bacteria and blue-green algae (Chlorophyta)—do not possess chloroplasts. Instead, their photosynthetic systems are associated with the cell membrane or with lamellar structures located in organelles known as chromatophores. Chromatophores, unlike chloroplasts, are not surrounded by a membrane.

All photosynthetic organisms contain one or more of the group of green pigments called chlorophylls. In addition, many contain accessory pigments which impart characteristic colors to the cells.

The net result of photosynthesis is to reduce carbon dioxide to form carbohydrates. A key intermediate is phosphoglyceric acid, and various simple sugars are reacted and disproportionated as the carbon from carbon dioxide is incorporated.

BIOLOGICAL REACTORS

Fermenters Commercial fermentations are usually performed aseptically in vessels held under positive pressure of sterile air to resist entry of contaminating microorganisms. However, the production of pathogenic organisms for medical purposes or for biological warfare operates below atmospheric pressure because the safety of the plant operators is more important than the integrity of the product. Biological waste treatment employs elective cultures of microorganisms in relatively crude, open equipment. Fermentation formerly distinguished processes from which oxygen was absent, but the term has now been extended to aerobic processes.

The functions desired in fermentation are gas-liquid contacting, online sensing of concentrations, mixing, heat transfer, foam control, and feed of nutrients or reagents such as those for pH control. The workhorse of the fermentation industry is the conventional batch fermenter as shown in Fig. 27-3. There is extensive process piping, and copper or brass fittings are avoided because of the highly deleterious effects of copper on many biological systems. For example, more than 50 percent reduction in yield was noted in penicillin fermentations when a bronze valve was in a feed line. Cooling coils or circulation through external heat exchangers must be used for larger tanks because the heat-transfer area of a jacket is inadequate for cooling from sterilization temperature to operating temperature in a reasonable time. Jackets may be inadequate for fermentations with intensive mixing to satisfy aeration demand plus metabolic heat from rapid respiration and growth. Some features of interest for a conventional fermenter are that (1) a bypass valve in the air system allows diversion of air so that foaming is not excessive and the redox potential is not too high during the early stage of fermentation when the inoculum is becoming established; (2) antifoam is added when excessive foam reaches a conductive or capactitive electronic probe; (3) all piping is protected from contamination by the use of steam; (4) the level of liquid when filling the vessel is determined by reference

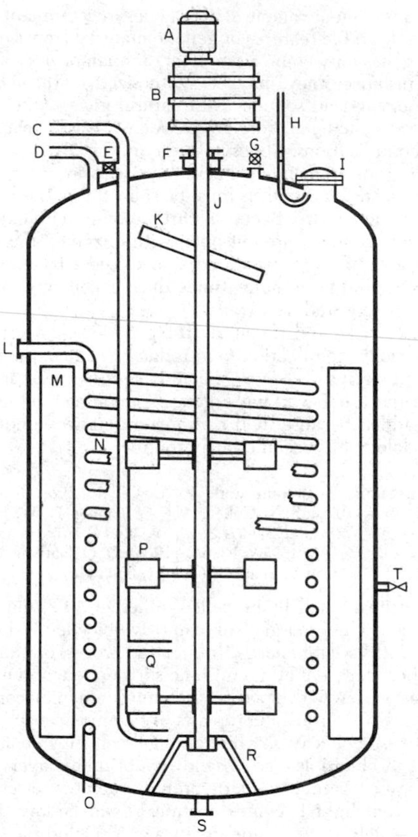

FIG. 27.3 Conventional batch fermenter. A = agitator motor; B = speed-reduction gears; C = air inlet; D = air outlet; E = air bypass valve; F = shaft seal; G = sight glass with light; H = sight-glass clean-off line; I = manhole with sight glass; J = agitator shaft; K = paddle to break foam; L = cooling-water outlet; M = baffle; N = cooling coils; O = cooling-water inlet; P = mixer; Q = sparger; R = shaft bearing and bracket; S = outlet (steam seal not shown); T = sample valve (steam seal not shown). Not shown are ladder rungs for maintenance, antifoam probe, antifoam system, and sensors (pH, dissolved oxygen, temperature, etc.). Note that (1) coils may be between baffles and tank wall, (2) coils may connect to top to minimize openings below water level, and (3) bottom-entering mixers are feasible.

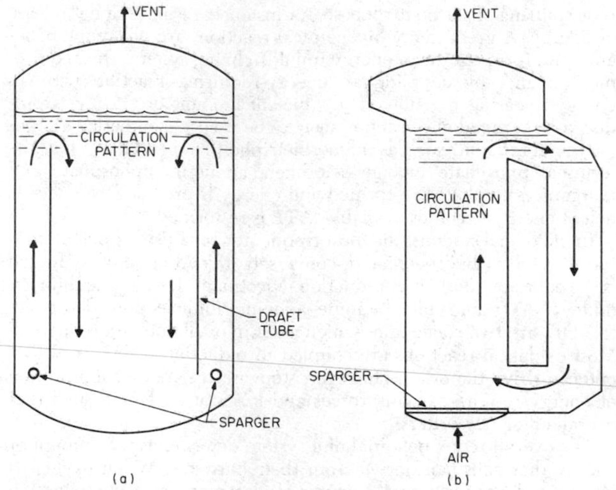

FIG. 27-4 Air-lift fermenters. (*a*) Concentric cylinder. (*b*) External recycle.

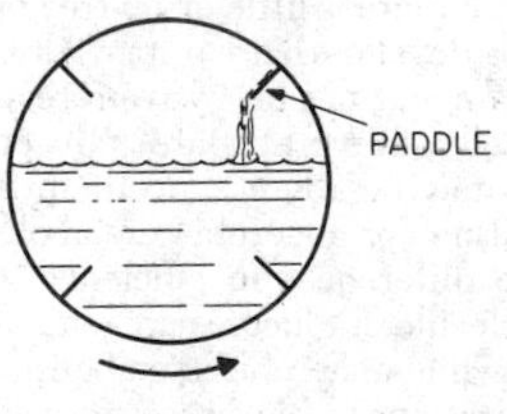

FIG. 27-5 Rotating fermenter.

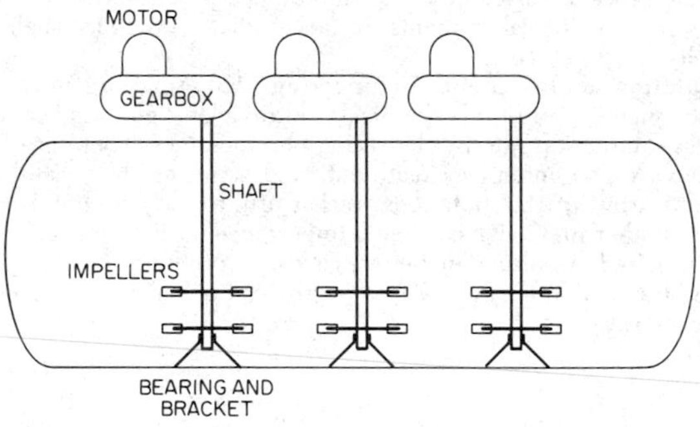

FIG. 27-6 Horizontal fermenter.

to a calibration chart based on points in the tank such as a rung on the ladder; (5) the weight of the tank contents can be determined by the hydrostatic balance against air bubbled slowly through the sparger; and (6) pumps are very uncommon because it is easy to force fluid from a pressurized vessel.

The need for highly efficient oxygen transfer in fermentations such as those with hydrocarbon feedstocks has led to air-lift fermenters as shown in Fig. 27-4. The world's largest fermenter was designed for producing single-cell protein from hydrocarbons at Billingham, England. It is 100 m (328 ft) in height and 10 m (32.8 ft) in diameter.

A few variations on the standard fermenter have been attempted, but none has become popular. An obsolete design in which the fermenter was rotated to aerate the medium is shown in Fig. 27-5. Performance was unsatisfactory, and the units were turned on end with spargers and agitation added. One of the largest fermenters used for antibiotics is a horizontal cylinder with several agitators as in Fig. 27-6. Multiple agitators have also been used with vertical cylindrical vessels.

Ethanol fermentation is a particularly good example of product accumulation inhibiting microbial culture. Most strains of yeast have a much slower alcohol production rate when ethanol reaches about 10 percent, and the wine or saki strains that achieve over 20 percent by volume of ethanol are very, very slow. A system known as the Vacuferm for the removal of alcohol by distillation as it is formed is shown in Fig. 27-7. The vacuum is adjusted to the vapor pressure of the alcohol-water solution at the fermentation temperature [30 to 40°C) (86 to 104°F)]. Volumetric productivity is far better than that of a conventional fermenter, but there is the killing disadvantage of having to recompress large volumes of vapor so that alcohol can be condensed with normal cooling water instead of expensive cold brine. Furthermore, the large amounts of carbon dioxide generated by fermentation are evacuated and recompressed along with the alcohol and water vapors. A far better design is shown in Fig. 27-8, in which the fermenter operates at normal pressure so that carbon dioxide escapes. Broth is circulated through the flash pot for vaporization of the ethanol. Although this system may seem to have attractive energy economy because metabolic heat is removed as the vapor flashes, the initial investment and the cost of pumping the vapors are high. Operating the ethanol fermentation at higher temperatures

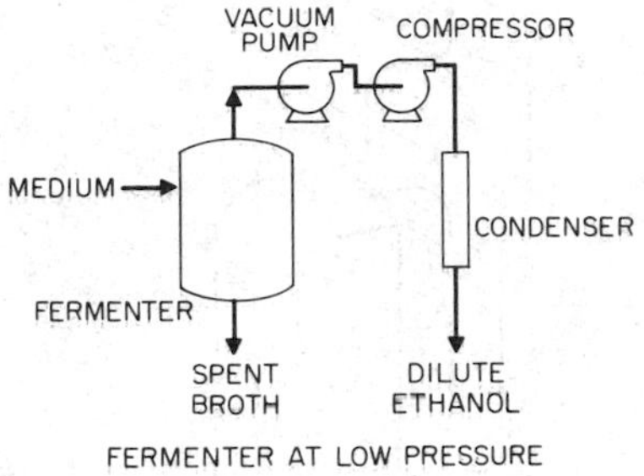

FIG. 27-7 Vacuferm.

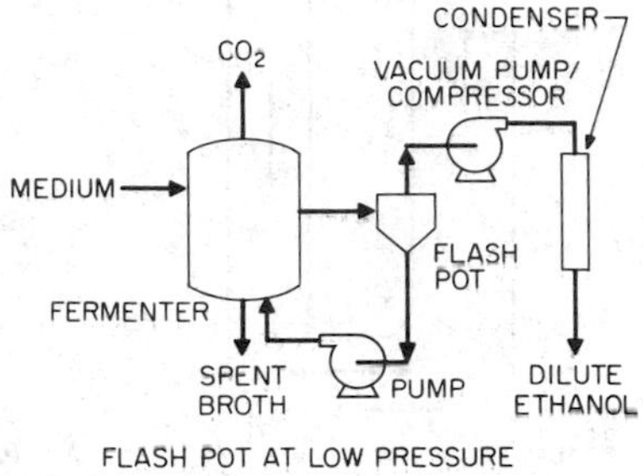

FIG. 27-8 Flash-pot fermenter.

with thermophilic organisms has better economics in terms of milder vacuum and less recompression because of the higher vapor pressure. Other alternatives for overcoming inhibition by the product are extraction from the fermentation broth with an immiscible solvent and/or operating with very dense cultures so that low productivity per cell is compensated for by having many more cells.

Denser cell concentrations can be achieved by separating cells from the effluent and recycling them to the fermenter. This has been standard practice for many years in biological waste treatment when dilute feed streams result in slow growth of the culture. Producing cell flocs that are collected easily by sedimentation is aided by recycling the cells that do settle. This is a selective advantage that may allow those cells to dominate. Industrial fermentations can afford greater expense than can waste treatment, and centrifuges for collecting cells are not uncommon. Recycle of yeast cells to an ethanol fermentation can lower the fermentation time greatly. Heavily coagulated cells can be retained in the fermenter. The tower fermenter shown in Fig. 27-9 uses this principle with yeast strains that flocculate naturally. Cells can also be retained in the reactor by attachment to a support. Vinegar is sometimes produced in a "generator" filled

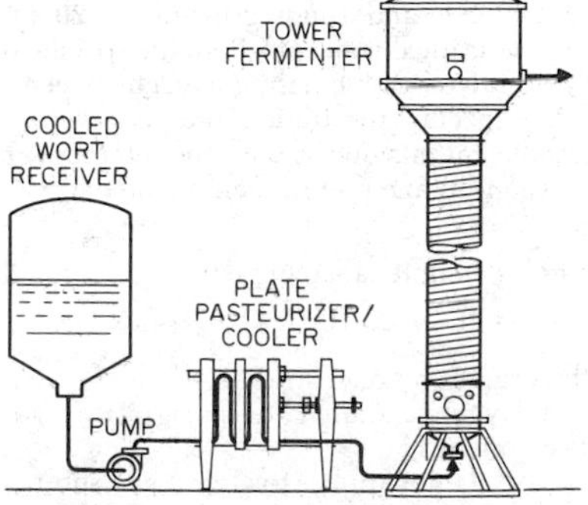

FIG. 27-9 Tower fermenter. *(Compliments of APV Corp.)*

with wood shavings to which bacteria attach. Rocks or plastic support materials are used in trickling filters for waste treatment (see Sec. 26). Chemical agents can link cells to the support materials when simple adsorption does not hold them tightly enough. Gel entrapment can provide extremely high cell concentrations because the cells continue to multiply within the gel. Comparisons of results with fermentation of ethanol are given in Table 27-1.

TABLE 27-1 Comparison of Ethanol Fermenters

System	Typical time, h	Typical ethanol concentration, %
Conventional	72	10
Cell recycle	12	8
Tower fermenter	3	8
Gel immobilization	1	10

Fermentation can be combined with other operations. For example, feedback inhibition of enzymatic hydrolysis of cellulose can be relieved by removal of the product glucose by fermentation as it forms.

Valves and pumps that have a potential path for contaminating organisms are taboo for aseptic operations. Rising-stem valves could bring organisms to the sterile side by in-and-out motion as the valve operates. Diaphragm valves are still commonly used, but heating, cooling, and abrasion by solids in the nutrient media are somewhat severe conditions leading to occasional rupture of a diaphragm and contamination of a run. Ball valves or plug valves do not have an absolute seal to the outside, but the direction of motion does not tend to bring organisms into the system. Contamination is seldom attributed to these valves, they are designed for easy maintenance in place, and there is the very nice human advantage that a glance at the handle tells easily whether the valve is open or closed. Many runs have been spoiled or impaired because a manual valve had been left in the wrong position. For plant operations, pumps with diaphragms are satisfactory. In the laboratory or in a pilot plant, peristaltic pumps (also known as tubing squeezers) predominate.

Transfer of fluid in a fermentation plant usually makes use of air-pressure differences. One or more manifold headers may interconnect many vessels. As transfers may have to be aseptic, headers are pressurized with steam until they are needed. A typical arrangement of steam seals is shown in Fig. 27-10.

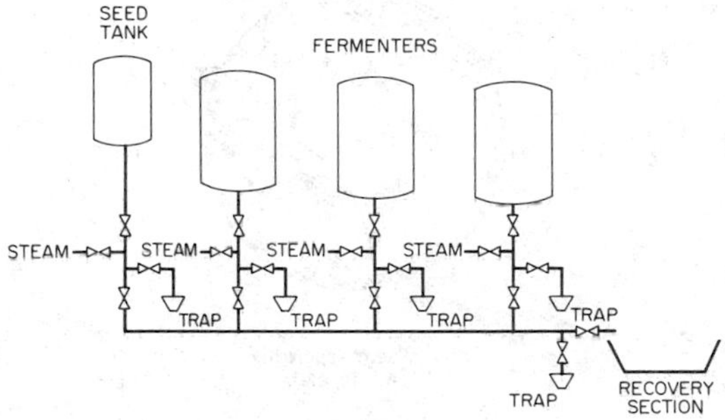

FIG. 27-10 Inoculation and harvest header.

Sample lines also commonly have steam seals. A typical layout is shown in Fig. 27-11. In the closed position, steam provides an absolute barrier to contamination. To take a sample, the steam line and the trap line are closed, and fermentation medium is flowed to waste until the pipes are cool to the touch so that sensitive products do not give false assays because of thermal destruction. Cooling takes up to 5 L of medium if not done carefully, and bad practices can waste considerably more. Pilot-sized tanks have less massive fittings that are easier to cool. Less medium is wasted, but oversampling to a point at

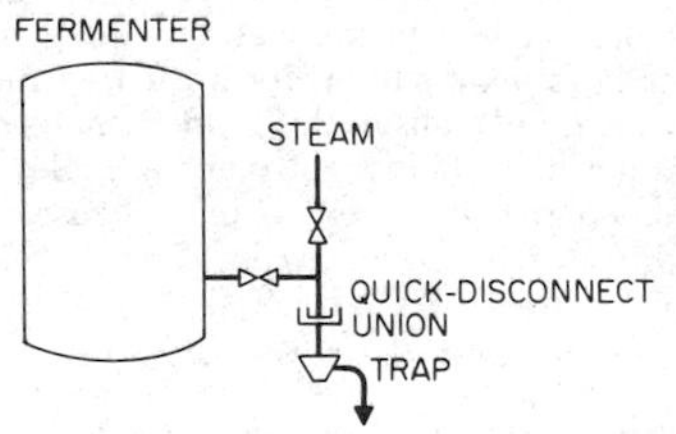

FIG. 27-11 Sample-line piping. (It may have a valve to the sewer, allowing bypass of the trap while cooling the line.)

which fermenter volume is low can be a problem. For this reason, alternative sampling methods have been devised. For example, a sterile syringe and needle may be used to sample through a rubber diaphragm in the wall of the tank. Although such methods appear reliable, there is a tendency to scrap all innovations and to return to tried-and-true steam seals when a factory encounters any period of contamination.

All piping to a fermenter is flushed with steam during the sterilization period. A clever means for sight-glass clean-off uses steam condensate that is naturally sterile in a dead leg. Steam pressure behind the condensate forces this water to the sight glass. Without cleaning, splashing and spray can quickly cover the sight glass with a thick coating of microorganisms and medium.

While it is easy to add materials to a fermentation, removal is difficult. Membrane devices have been placed in the fermenter or in external recycle loops to dialyze away a soluble component. Cells release wastes or metabolites that can be inhibitory; these are sometimes referred to as staling factors. Removal by dialysis has allowed cell concentrations to reach from 10 to 100 times the concentration of control cultures.

Another innovative design is the toroidal fermenter shown in Fig. 27-12. Motion in an axial direction allows intimate mixing with air. A special case of the air-lift fermenter is shown in Fig. 27-13. The feed enters at moderate pressure and is drawn downward to regions of very high hydrostatic pressure that provide a great driving force for gas transfer. In the other leg, lowering the pressure allows gases to expand, providing the impulse for circulation. Experimental units have been built in elevator shafts.

Solid substrates such as wood cannot be stirred when slurry con-

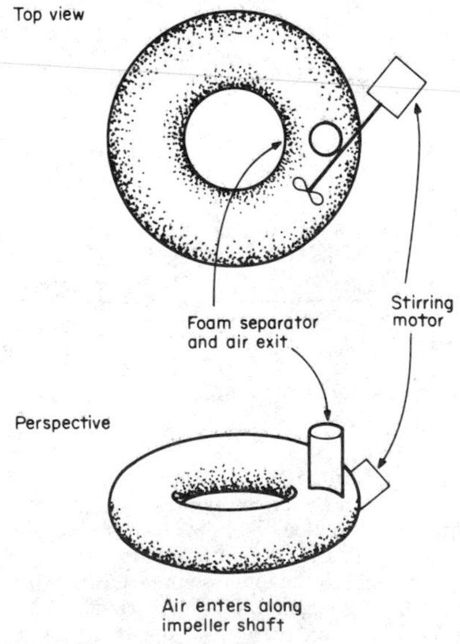

FIG. 27-12 Toroidal fermenter.

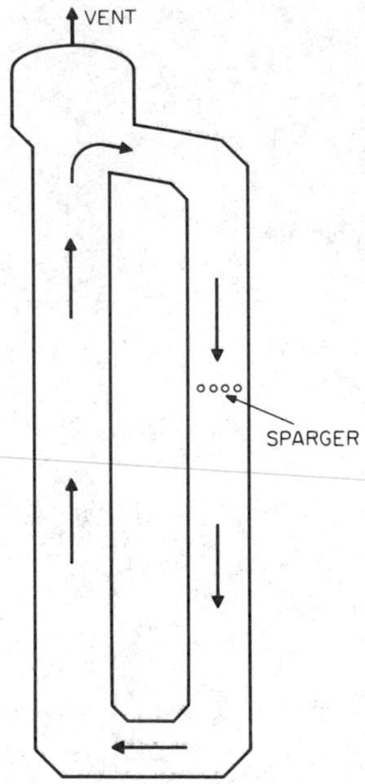

FIG. 27-13 Deep-shaft fermenter.

centration exceeds about 10 percent. For saccharification prior to ethanol fermentation, keeping the sugar concentration high can avoid an evaporation step. In a batch reactor, mixing limitations with the wood result in a dilute sugar solution. This has been circumvented by placing the wood in a column and percolating the solution through it. As wood dissolves, more is added. Simultaneous fermentation of the sugars formed in the column is possible.

Oxygen Transfer Supplying sufficient oxygen can be a very challenging engineering problem for some aerobic fermentations. Oxygen is sparingly soluble in water. Saturation with air at room temperature provides only 6 or 7 mg/L of oxygen. A vigorous process can deplete the dissolved oxygen in several seconds when aeration is stopped. The mass transfer of gases to liquids is covered in Sec. 18. Emphasis is somewhat different for biological systems that commonly have bubble aeration. Because the number and size of the bubbles are very difficult to estimate, transfer area is usually lumped with the mass-transfer coefficient as a $K_l a$ term. The l subscript in K_l signifies that liquid-film resistance should be controlling for a sparingly soluble gas such as oxygen. The relationship between oxygen concentration and growth is of a Michaelis-Menten type (see Fig. 27-26 later). When a process is rate-limited by oxygen, specific respiration rate (Q_{O2}) also increases steeply with dissolved-oxygen concentration until a plateau is reached. The concentration below which respiration is severely limited is termed the "critical oxygen concentration," which typically ranges from 0.5 to 2.0 ppm for well-dispersed bacteria, yeast, and fungi growing at 20 to 30°C (86 to 104°F). Above this critical concentration, the specific oxygen uptake increases only slightly with increasing oxygen concentrations.

A plot of the specific respiration rate Q_{O2} versus the specific-growth-rate coefficient μ is linear, with the intercept on the ordinate equal to the oxygen-uptake rate for cell maintenance. A formulation of this is

$$\begin{aligned}\text{Uptake rate} &= \text{maintenance uptake} + \text{growth uptake}\\ Q_{O2}X &= (Q_{O2})_M X + \mu X\,(1/Y_G)\end{aligned} \qquad (27\text{-}1)$$

where X is the organism concentration, Y_G is the yield of cell mass per unit mass of oxygen, and the Q terms signify oxygen-uptake rates in mass O_2/mass organism.

This type of correlation applies to almost any substrate involved in cellular-energy metabolism and is supported by experimental data and energetic considerations. However, it is based on assumptions that are true at or near steady-state equilibrium conditions and may not be valid during transient states. The oxygen-uptake equation should be modified when other cellular activities requiring oxygen can be identified. For example, utilization of oxygen for product formation would be represented by

$$\begin{matrix} \text{Uptake rate} & = & \text{maintenance uptake} & + & \text{growth uptake} & + & \text{product uptake} \\ Q_{O_2}X & = & (Q_{O_2})_M X & + & \dfrac{dX}{dt}\dfrac{1}{Y_G} & + & \dfrac{dP}{dt}\dfrac{1}{Y_P} \end{matrix} \quad (27\text{-}2)$$

where P is product concentration and Y_P is yield of product mass per unit mass of oxygen uptake. Oxygen uptake is distributed between the uptake for growth and that for cellular activities dependent on cell concentration.

As the oxygen-transfer rate under steady-state conditions must equal the oxygen uptake, $K_l a$ may be calculated from

$$K_l a = \text{overall oxygen-uptake rate}/(C° - C)_{\text{mean}} \quad (27\text{-}3)$$

where C is the concentration of oxygen in gas bubbles, $C°$ is the concentration of oxygen in the liquid that would be in equilibrium with the gas-bubble concentration, and $K_l a$ is the volumetric oxygen-transfer rate.

A convenient method for measuring oxygen-transfer rates in microbial systems depends on dissolved-oxygen electrodes with relatively fast response times. Quite inexpensive oxygen electrodes are available for use with open systems, and steam-sterilizable electrodes are available for aseptic systems. There are two basic types: one develops a voltage from an electrochemical cell based on oxygen, and the other is a polarographic cell whose current depends on the rate at which oxygen arrives (see Fig. 27-14). Measurement of oxygen-transfer properties requires only a brief interruption of oxygen supply. A mass balance for oxygen is

$$dO/dt = \text{rate of supply} - \text{uptake rate} \quad (27\text{-}4)$$

A tracing of the electrode signal during a cycle of turning aeration off and on is shown in Fig. 27-15. The rate of supply is zero (after bubbles have escaped) in the first portion of the response curve; thus the slope equals the uptake rate by the organisms. When aeration is resumed, both the supply-rate and the uptake-rate terms apply. The

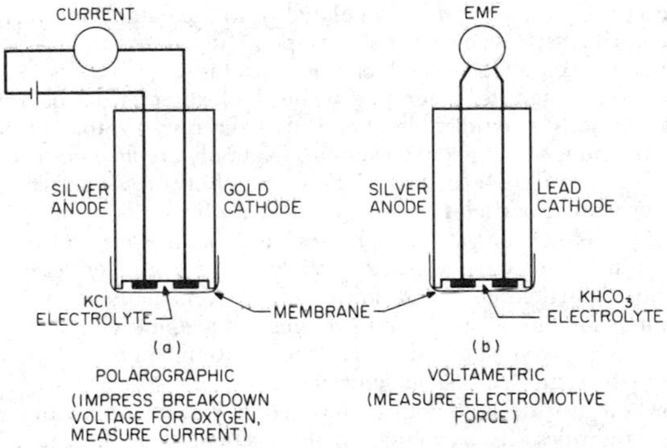

FIG. 27-14 Dissolved-oxygen electrodes. (*a*) Polarographic (impress breakdown voltage for oxygen; measure content). (*b*) Voltametric (measure electromotive force).

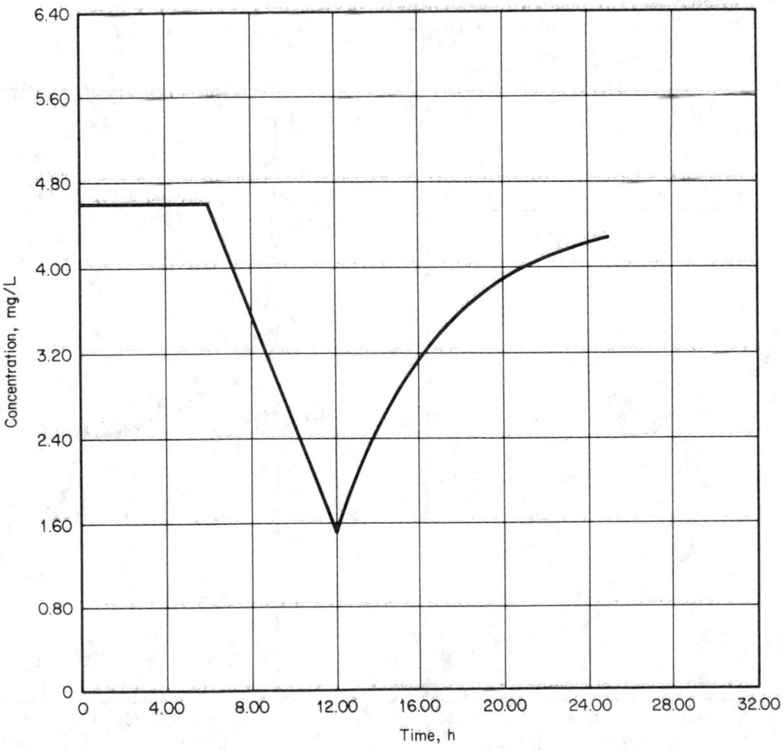

FIG. 27-15 Dynamic measure of $K_l a$.

values for $(C^\circ - C)$ can be calculated from the data. Thus, from the slope of the response curve at a given point all values for the equation are known except $K_l a$, which can be solved for.

Studies of the rate of aeration when the fermenter has been filled with solutions of oxidizable chemicals have questionable value for assessing rates with biological systems. Not only are flow patterns and bubble sizes different for biological systems, but surface-active agents and suspended particles can impair gas transfer seriously. Fig. 27-16 shows the effects of various particles. In general, spherical particles have a small effect, elongated particles have a greater effect, and entangled particles markedly impair transfer. As many mold cultures are intertwined and lipids and proteins are present with strong surface activity, oxygen transfer to a fermentation can be much slower than that to simple aqueous solutions.

Carbon dioxide is seldom considered in fermentations but plays important roles. Its participation in carbonate equilibrium affects pH. Removal of carbon dioxide by photosynthesis can force the pH above 10 in dense, well-illuminated algal cultures. Several biochemical reactions involve carbon dioxide; so their kinetics and equilibrium concentrations depend on gas concentrations, and metabolic rates of associated reactions may also change. Attempts to increase oxygen-transfer rates by elevating pressure to get more driving force sometimes encounter poor process performance which might be attributed to excessive dissolved carbon dioxide.

Sparger Systems Gas spargers are discussed in detail in Sec. 18. Large openings are desirable for spargers in industrial fermentations to avoid clogging by microbial growth. Relatively small holes or diffusers are used in activated-sludge units for biological waste treatment, but there is commonly a means for swinging a section of the aerator out of the vessel for cleaning. Newer designs for fermenters were conceived as answers to the problems of oxygen transfer. The air-lift fermenter (Fig. 27-4) creates intimate mixing of air and medium while using the buoyancy of the gas to mix the fluid.

Surface-active substances also lead to foaming that can be so bad that most of the contents of the fermenter are lost. Mechanical antifoam devices are helpful but cannot function alone except when there is little propensity for foaming. The mechanical foam breakers rupture the large, weak bubbles while allowing tiny, rugged bubbles to accumulate. Surface-active antifoam agents tend to reduce the elasticity of the bubbles so that mechanical shocks are easily transmitted to encourage rupture. Several antifoam delivery systems are shown in Fig. 27-17. Some lipids used as antifoams are metabolized by the culture and must be replaced. The nutrition supplied by these oils may be beneficial, but they are much more expensive than their equivalents in carbohydrate nutrients. Furthermore, it is troublesome to have nutrition coupled to foam control. Several antifoam agents are not nutrients and tend to persist. Their tendency to be lost by coating solid surfaces in the fermenter means that more agents must be added occasionally. Various synthetic antifoam agents are toxic to some organisms, but one of the many types available is usually satisfactory.

Scale-Up Fermenters ranging from about 2 to more than 100 L (0.07 to 3.5 ft^3) have been used for research and development. The smaller sizes provide too little volume for sampling and are difficult to replicate, while large vessels are too expensive and use too much

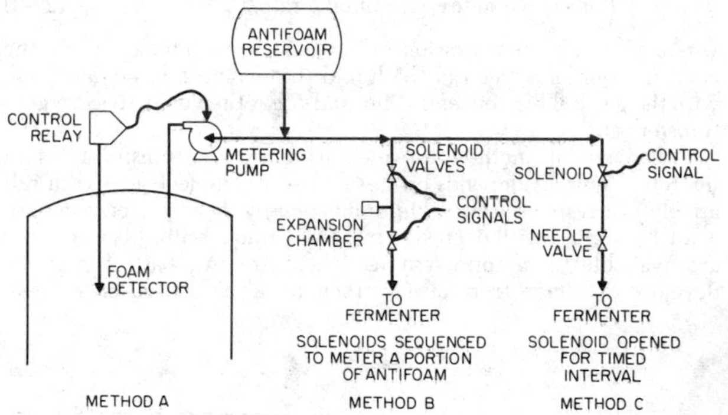

FIG. 27-17 Antifoam systems.

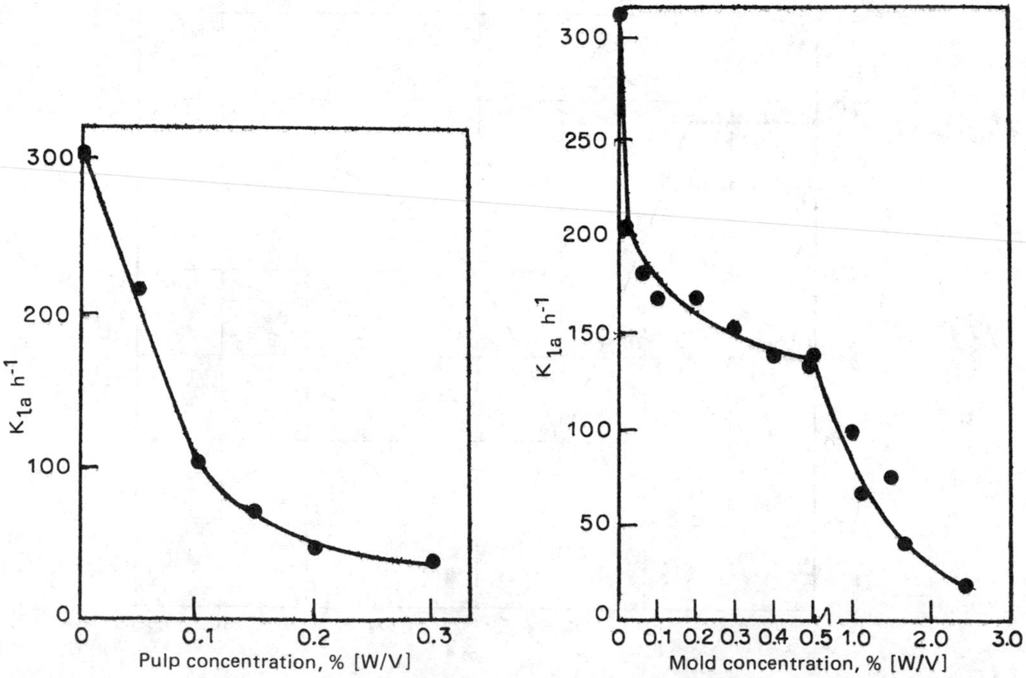

FIG. 27-16 Effect of solids on $K_l a$. Operating conditions: agitator speed, 800 r/min; air flow, 2.5 L/min. [*M. R. Brierly and R. Steel,* Appl. Microbiol., *7, 57 (1959). (Courtesy of American Society for Microbiology.)*]

medium. The goal in using small equipment is to provide insights for the design of full-sized units. Minor refinements are best performed in production equipment, whereas the pilot plant is used to screen broad ranges of variables.

An effective means of scale-up for aerobic processes is to measure the dissolved-oxygen level that is adequate in small equipment and to adjust conditions in the plant until this level of dissolved oxygen is reached. However, some antibiotic fermentations as well as the production of fodder yeast from hydrocarbon substrates have very severe requirements, and designers are hard-pressed to supply enough oxygen.

Older methods of fermentation scale-up were based on geometric similarity by using proportional physical dimensions. It was thought that applying the same power per unit volume as in the pilot equipment would give an equivalent process performance in large fermenters. Antibiotic fermentations aim for mixer power in the range of 0.2 to 2 kW/m^3 (0.1 to 1 hp/100 gal). As mixing devices have areas (dimensions squared) to supply a volume (dimensions cubed), methods based on dimensional similarity are fundamentally unsound. Scale-up based on equivalent-oxygen-transfer coefficient $k_l a$ has been reasonably successful.

Impeller Reynolds number and equations for mixing power for particle suspensions are presented in Sec. 19. Dispersion of gases into liquids is discussed in Sec. 18. Usually, an increase in mechanical agitation is more effective than is an increase in aeration rate for improving mass transfer.

Other scale-up factors are shear, mixing time, Reynolds number, momentum, and the mixing provided by rising bubbles. Shear is maximum at the tip of the impeller and may be estimated from Eq. (27-5) [R. Steel and W. D. Maxon, **Biotechnol. Bioeng., 4**, 231 (1962)].

$$S_s = S_l\,(D_{is}/D_{il})^{1/3} \tag{27-5}$$

where S_s and S_l are shears in the smaller and larger vessels respectively, and D_{is} and D_{il} are impeller diameters of the smaller and larger vessels.

Some mycelial fermentations exhibit early sporulation, breakup of mycelium, and low yields if the shear is excessive. A tip speed of 250 to 500 cm/s (8 to 16 ft/s) is considered permissible. Mixing time has been proposed as a scale-up consideration, but little can be done to improve it in a large fermenter because excessively large motors would be required to achieve rapid mixing.

Constant Reynolds number is not used for fermentation scale-up; it is only one factor in the aeration task. This is also true for considering the impeller as a pump and attempting scale-up by constant momentum. As mechanical mixing tends to predominate over bubble effects in improving aeration, equations including bubble effects have had little use.

Fermentation biomass productivities usually range from 2 to 5 g/(L·h) [0.12 to 0.31 lb/(ft^3·h)]. This represents an oxygen demand in the range of 1.5 to 4 g O_2/(L·h) [0.093 to 0.25 lb/(ft^3·h)]. In a 500-m^3 (18,000-ft^3) fermenter this means the achievement of a volumetric oxygen-transfer coefficient in the range of 250 to 400 h^{-1}. Such oxygen-transfer capabilities can be achieved with aeration rates of the order of 0.5 VVM (volume of air at STP/volume of broth) and rather excessive mechanical-agitation power inputs of 2.4 to 3.2 kW/m^3 (1.2 to 1.6 hp/100 gal).

Often heat removal causes design problems for scale-up. Mechanical agitation coupled with metabolic heat from the growing biomass overwhelms the cooling capacity of a large fermenter equipped with only a jacket. External circulation through a heat exchanger or extensive coils inside the fermenter must be used. In highly viscous fermentations, internal cooling coils are usually not desirable because of interference with mixing patterns. Numerous schemes exist for heat removal in large fermenters such as half-coil baffles (plate coils) and draft tubes.

Evaporation of the medium provides a little cooling. The inlet air to particulate filters must not be near saturation because condensation of moisture on the filter medium mobilizes contaminating microorganisms so that their chances of penetration are greatly increased. Sometimes humidified air is used, and the filter unit is heated to prevent condensation. However, this is common only for small equipment with which the extra operations are relatively easy to install and maintain.

Once a plant has been built, the conditions of agitation, aeration, oxygen transfer, and heat transfer are more or less set, and sterilization cycles are defined. Those environmental conditions achievable in plant-scale equipment should be scaled down to the pilot plant and laboratory equipment (shaken flasks) to ensure that results can be translated.

Sterilization Some old, traditional fermentations such as those for alcohol and pickles are conducted by organisms that are hardy and help their own cause by creating conditions that are unfavorable for competitors. Yeast, for example, lower pH by producing acids from sugars, and tolerance to alcohol is another powerful advantage that allows their domination. Nevertheless, modern factories use aseptic techniques or extreme care to minimize contamination that can jeopardize product quality by affecting taste, texture, aroma, or appearance. Bioprocesses such as tissue culture to produce vaccines are very easily contaminated because there is an abundance of nutrients and no inherent protection against foreign organisms. Thus, there is a range from relatively good self-protection to practically none. The value of the product and the need for quality control determine the extent to which precautions must be taken. Sterility in the practical sense is statistical in that foreign organisms are almost always present, but generally they do not propagate rapidly enough to damage a run. Good defense against contamination is relatively inexpensive, while absolute protection is impossible and attempts to achieve it can be inordinately costly. Production of agents for biological warfare takes extreme pains to keep the organisms away from the workers, yet people operating the fermentations are occasionally killed. The most expensive and best-protected industrial fermentations are sometimes contaminated. Sterilization and aseptic techniques to keep bioprocesses uncontaminated can be crucial.

Common ways of sterilization are (1) removal of microorganisms by filtration and (2) killing them with heat or chemicals. Sterilization by filtration follows a standard unit operation that is covered in Sec. 19. The differences for biochemical engineering are that (1) the filter medium and the downstream lines are steamed at a pressure at which the temperature kills all organisms and spores, (2) the size of the particles being removed is in the micrometer range, and (3) the filter medium should be reusable and not be degraded by repeated heating.

Sterilization by Filtration Air is almost always sterilized by filtration. Small-scale employment of heat has been successful, but large equipment for heating an air stream to sterilizing temperatures has not been sufficiently reliable. While it seems simple enough to maintain a section of the air-supply pipe line at high temperature, automatic control is needed to adjust for varying heat transfer as the flow rate changes, and an air-cooling section is needed to prevent excessive heat load on the fermenter. Furthermore, energy costs are now much higher than when heating seemed a promising alternative to filtration. For exit gases from a fermentation that has hazardous organisms, heating is a reasonable precaution and cost is not the key factor.

Sterilization of liquids by filtration has performed very well since the advent of membrane filters of small pore size. When heat would damage the ingredients, filtration is an ideal choice. However, the extra handling and equipment required mitigate against filtration because heat sterilization is easy and relatively inexpensive. A tank must be steam-sterilized in any case, so it is convenient to fill it first and to sterilize the contents as well. When some constituents must not be subjected to heat, it is customary to sterilize the rest of the medium with heat and to filter concentrated solutions of the delicate ingredients.

The magnitude of the air-sterilization problem is seen from the usual needs of a highly aerobic fermentation in which roughly one volume of air per volume of medium per minute may be used. For a factory with 20 fermenters of 100,000 L each, 2 million L/min (70,000 ft^3/min) of air is handled. Very large compressors are used, and at least two are required so that one can be down for maintenance.

Typical performance data for the removal of bacteria from air streams by fibrous filters are presented in Fig. 27-18. For a particular filter there is an intermediate air velocity at which filtration efficiency is at a minimum because different collection mechanisms predominate at different ranges of velocity. At low velocities, gravitational, diffusional, and electrostatic forces on the particle are important, and their effect is inversely proportional to air velocity.

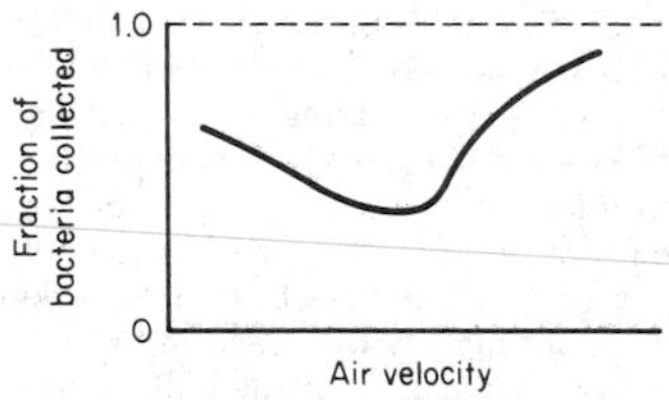

FIG. 27-18 Typical collection of bacteria by a fibrous filter.

At high velocities, inertial forces directly proportional to air velocity come into play. The nature of inertial effects is such that below a certain air velocity, their effect on collection is zero. If the filter design is based upon a performance observed at an operating velocity other than that at which minimum efficiency occurs, surges or brief power failures could create periods of operation at lower efficiencies. For the collection of 1-μm bacterial particles with specific gravity of 1.0 from air streams at room temperature and air pressure, this velocity is equal (in SI units) to

$$V_{\text{min,eff}} = 0.02\, d_f \tag{27-6}$$

where velocity $V_{\text{min,eff}}$ is in meters per second and the fiber diameter d_f is in micrometers.

Currently, there is a pronounced trend to the use of membrane filters for air sterilization in order to obtain excellent performance with units of relatively small size. Although membranes are quite efficient for filtering air and tend to capture particles larger than pore size, moisture may help particles to pass. To ensure safety, a pore size of 0.2 to 0.3 μm is recommended. Hydrophilic membranes should not be used because moisture is held tightly in the pores and not dislodged unless quite high pressure drops are created across the membrane. Moisture tends to drain from hydrophobic membranes and collect in a sump. Sizing of a membrane unit for air filtration is based on the number of cartridges needed. Only 60 percent of the available pressure drop should be used in the calculation to allow for increased resistance as particles collect on the membrane. Figure 27-19 shows a housing for 12 membrane cartridges in parallel and a graph of typical flow rates versus head loss.

Sterilization of Media First-order kinetics may be assumed for heat destruction of living matter, and this leads to a linear relationship when the logarithm of the fraction surviving is plotted against time. However, nonlogarithmic kinetics of death are quite often found for bacterial spores. One model for such behavior assumes inactivation of spores via a sensitive intermediate state by the mechanism:

$$C_R \rightarrow C_S \rightarrow C_D \tag{27-7}$$

where C_R, C_S, and C_D are concentrations of resistant, sensitive, and dead spores respectively. Typical plots are shown in Fig. 27-20.

Relative thermal resistance for the different types of microorganisms encountered in typical environments associated with fermentation broths is shown in Table 27-2. Bacterial spores are far more resistant to moist heat than are any other type of microbial contaminants; thus a sterilization cycle based on the destruction of bacterial spores should destroy all life.

As predicted by the Arrhenius equation (Sec. 4), a plot of microbial death rate versus the reciprocal of temperature is usually linear with slope that is a measure of the susceptibility of microorganisms to heat. Correlations other than the Arrhenius equation are used, particularly in the food-processing industry. A common temperature relationship of the thermal resistance is "decimal reduction time" (DRT), defined as the time required to reduce the microbial population by one-tenth. Over short temperature intervals [e.g., 6°C (10°F)], DRT is useful, but extrapolation over a wide temperature interval produces serious errors.

(a)

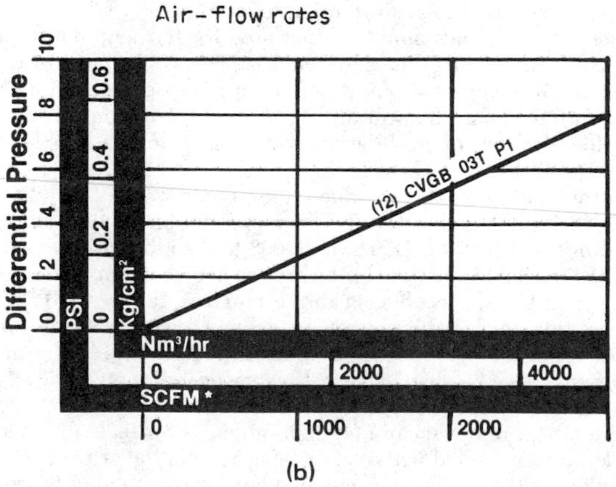

(b)

FIG. 27-19 Unit for 12 cartridges and pressure data for one cartridge. (*a*) Housing for 12 cartridges. (*b*) Air flow through 12 cartridge filters at 20°C discharging to the atmosphere. *Cubic feet of air at 1 atm and 20°C. Cartridges are 76 cm long and 7.36 cm in diameter of polyvinylidene difluoride and have 0.22-μm pores. (*Courtesy of Millipore Corp.*)

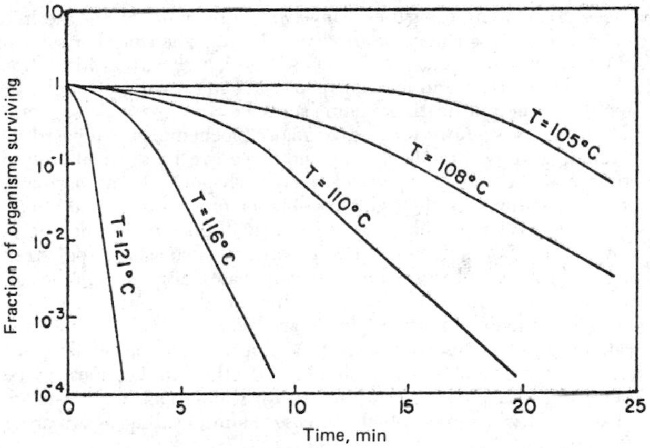

FIG. 27-20 Typical death-rate data for bacterial spores (*B. stearothermophilus*). To convert °C to °F, multiply by 1.8 and add 32. (*Wang et al.*, Fermentation and Enzyme Technology, *Wiley-Interscience, New York, 1979, p. 140.*)

TABLE 27-2 Relative Resistance to Killing by Moist Heat

Microorganism	Relative resistance
Vegetative bacteria or yeast	1.0
Bacterial spores	3,000,000
Mold spores	2–10
Virus and bacteriophage	1–5

The activation energy (ΔE) associated with microbial death is larger than the thermal inactivation of chemical compounds in fermentation broths (see Table 27-3). Thus, by sterilizing at high temperatures for short times (HTST), overcooking of nutrients is minimized.

TABLE 27-3 Activation Energies for Thermal Destruction

	Activation energy, cal/mol*
Folic acid	16,800
d-Panthothenyl alcohol	21,000
Cyanocobalamin	23,100
Thiamine hydrochloride	22,000
Bacillus stearothermophilus	67,700
Bacillus subtilus	76,000
Clostridium botulinum	82,000
Putrefactive anaerobe NCE 3679	72,400

*To convert calories per mole to British thermal units per pound-mole, multiply by 1.8.

Batch Sterilization If it is assumed that the presence of one single contaminating organism could cause ultimate failure of the desired fermentation, it is necessary to assign some probability of success. For example, if 1 contaminated fermentation per 1000 can be tolerated, the design calculation for the sterilization cycle should use 0.001 organism per fermentation. However, in batch sterilization, the heating, holding, and cooling portions of the cycle all contribute toward the reduction of the microbial contaminants. Furthermore, the specific-death-rate constant varies because the temperature of the medium changes. Therefore, the design criterion ∇ total is composed of

$$\nabla \text{ total} = \nabla \text{ heating} + \nabla \text{ holding} + \nabla \text{ cooling}$$

$$\nabla \text{ heating} = \ln \frac{N_0}{N_1} = A \int_0^{t1} \exp(-\Delta E/RT)\, dt \qquad (27\text{-}8)$$

$$\nabla \text{ holding} = \ln \frac{N_1}{N_2} = A \int_0^{t2} \exp(-\Delta E/RT)\, dt \qquad (27\text{-}9)$$

$$\nabla \text{ cooling} = \ln \frac{N_2}{N} = A \int_0^{t3} \exp(-\Delta E/RT)\, dt \qquad (27\text{-}10)$$

where ∇ is the incremental killing contribution for each phase of sterilization, N_0, N_1, etc., are numbers of organisms, T is absolute temperature, R is the gas constant, A is an empirical constant, and E is the activation energy.

Other parameters affecting the temperature profile include the viscosity of the medium and the amount of suspended solids or insoluble materials such as vegetable oils. Rate of heat transfer depends on heat-transfer coefficients and film coefficients. Fouling can be important in heat transfer. The temperature-time profile during the cooling portion of the sterilization cycle includes the effects of cooling coils or the fermenter jacket and evaporative cooling if sterile air is injected to remove heat. Releasing part of the pressure gives flash cooling. Most of the sterilization is derived from the holding portion of the cycle. The cooling cycle contributes little to the overall process. An extremely long heating cycle should be avoided because its contribution toward microbial destruction is far outweighed by its detrimental effects.

Continuous Sterilization Continuous sterilization permits short detention times at high temperature to avoid overcooking and has potential for improvement in yield. Tubular or plate-and-frame heat exchangers are used for continuous sterilization instead of steam injection. Although heating is almost instantaneous, steam injection produces violent noise and offers few opportunities for economical heat exchange between process streams.

In the holding section of a continuous sterilizer, correct exposure time and temperature must be maintained. Because of the distribution of residence times, the actual reduction of microbial contaminants in the holding section is significantly lower than that predicted from plug-flow assumption. The difference between actual and predicted reduction in microorganisms can be of several orders of magnitude; therefore, a design based on ideal flow conditions may fail.

PRODUCT RECOVERY

Recovery of products can be more difficult and expensive than the fermentation step. As an example, for one antibiotic production plant the investment for recovery facilities was about 4 times that for fermenter vessels and their auxiliary equipment. As much as 60 percent of the fixed costs of fermentation plants for organic acids or amino acids is attributable to the recovery section. Most of the purification equipment in a large factory is the same as that used throughout the chemical-process industries.

Certain products are contained inside the cells and are not released or only partially released to the medium. It may be possible to wash the cells to remove impurities before breaking them to get the product. Cell disruption is a unit operation peculiar to biochemical engineering. The equipment, however, may be borrowed from other industries. Colloid mills and shear devices such as the Manton-Gaulin homogenizer used for manufacturing paint and other products effectively rupture walls of many types of cells. Cells with high resistance to shear can be passed through several times, but heat generation can be a problem. Ultrasonic energy is commonly used on a small scale for cell disintegration but is impractical for large batches. Grinding with sand or beads, high-pressure pumping through a tiny orifice, freezing and thawing, dessication, adding lytic enzymes, inducing autolysis with a chemical such as chloroform, and various means of creating shear are alternative or synergistic means of rupturing cells. There are encouraging results with special mutants of cells with impaired ability to form cell walls at temperatures slightly above their normal growth temperatures. When cells are shifted to the elevated temperature, cell division gives damaged walls and lets the cell contents leak out.

Although it is sometimes possible to proceed directly to ion exchange, solvent extraction, or some other step with whole broth, there is usually a removal of biological cells and other solids. Centrifugation can be considered, but rotary-drum filters with string discharge are commonly used. Whereas very large amounts of filter aid were used in the past because many fermentation broths were slimy and hard to filter, present practice employs polymeric bridging agents to agglomerate the solids. This allows good filtration with only small amounts of filter aid.

Ion exchange or solvent extraction is desirable for recovery from bioprocesses because selectivity is good, costs are reasonable, and a large scale is feasible. These and other purification steps can be affected by modifications in fermentation. For example, adding excess lipids or antifoam oils in fermentation can aggravate emulsion problems for solvent extraction or impair ion exchange by coating the resin. Stability of biochemical products can be troublesome. Penicillin fermentation broth is acidified just prior to contact with the extracting solvent (esters, ketones, or halogenated hydrocarbons are commonly used) because low pH causes very rapid destruction of penicillin in water. Penicillin is extracted back into an aqueous phase at pH 7.5 to 8. Bicarbonate is used to buffer the pH because harsher agents are difficult to control. The Podbelniak design of centrifuge (see Sec. 21) has been widely used in the United States for the extraction of fermentation broths, while the Westphalia design is common in Europe. There may be lead-and-trail operation of the centifuges to improve yields because emulsions result in less than the number of theoretical stages predicted by the manufacturer. Extraction back into water often has no emulsion problem, and Delaval separators work well.

Of the various types of purification steps based on sorption to a solid phase, ion exchange is the most straightforward (see Sec. 19 for a discussion of techniques). Carbon adsorption has considerable importance to biochemical engineering, primarily for the removal of traces of colored impurities. However, adsorption of the desired material on carbon may be the least of evils when neither solvent extraction nor ion exchange can be used. Although carbon adsorption may be unselective and low in capacity, it can provide a "roughing" step to get the purification scheme started.

Isolation procedures for a great many biochemicals are based on chromatography. Practically any substance can be selected from a crude mixture and eluted at relatively high purity by a chromatographic column with the right combination of adsorbent, conditions, and reagents. For bench scale or for a small pilot plant, such chromatography has made alternative procedures such as electrophoresis nearly obsolete. Unfortunately, as size increases, dispersion in the column ruins resolution. To produce a few hundred grams or even a few kilograms, chromatography is an excellent choice. When the scale-up problem is solved, these procedures will revolutionize purification of biochemicals and should find other applications in the chemical-process industries. Prospects for scale-up are good because it has recently been possible to increase column loadings by an order of magnitude.

Affinity chromatography uses ligands with high specificity for certain compounds. There are several types of affinity that can be employed: antigen-antibody, enzyme-substrate, enzyme-cofactor, or special biochemical attractions such as the protein avidin for the vitamin biotin. Numerous purifications in which affinity chromatography is able to isolate quite pure product from a very crude mixture have been devised. Expensive affinity agents are regenerated and reused many times. In some cases, the attraction is so strong that the adsorbent can be added batchwise. This scales up well but adds laborious steps for collection, elution, and regeneration of the affinity agent.

Crystallization is the preferred method of forming the final product because very high purification is possible. High-purity antibiotic crystals can be produced from colored, rather impure solutions if the cake is uniform and amenable to good washing to remove the mother liquor. When a sterile pharmaceutical product is desired, crystals are produced from liquid streams that have been sterilized by filtration.

PROCESS MODELING

It is generally assumed that the properties of very large numbers of cells can be treated as continuous functions having average properties because so many cell divisions occur that the overall rates follow smooth curves. There are exceptions in which the cells can all be induced to divide at the same time because events such as illumination or temperature changes slow or halt a step in division. The cells can be triggered to proceed together from that point with overall numbers that are stepwise with time. This is termed a "synchronous culture." The steps are seldom distinct for more than a few generations unless the triggering event continues to be applied periodically.

Mass balances for common, unsynchronized batch cultures give

$$dX/dt = \mu X - K_d X \tag{27-11}$$

$$dS/dt = -\mu X/Y \tag{27-12}$$

$$\mu = f(S) \tag{27-13}$$

where K_d is the death-rate coefficient, μ is the specific-growth-rate coefficient, S is the substrate concentration, and other terms are as defined earlier.

Various functional relationships between μ and S have been proposed, but the Monod equation is used almost exclusively:

$$\mu = \hat{\mu} S/(K_s + S) \tag{27-14}$$

where $\hat{\mu}$ is the maximum-specific-growth-rate coefficient and K_s is the Monod coefficient.

The death-rate coefficient is usually relatively small unless inhibitory substances accumulate, so Eq. (27-14) shows an exponential rise until S becomes depleted to reduce μ. This explains the usual growth curve (Fig. 27-21) with its lag phase, logarithmic phase, resting phase, and declining phase as the effect of K_d takes over.

Structured Models Meaningful details can be added to culture models in several ways. Cells can be compartmentalized according

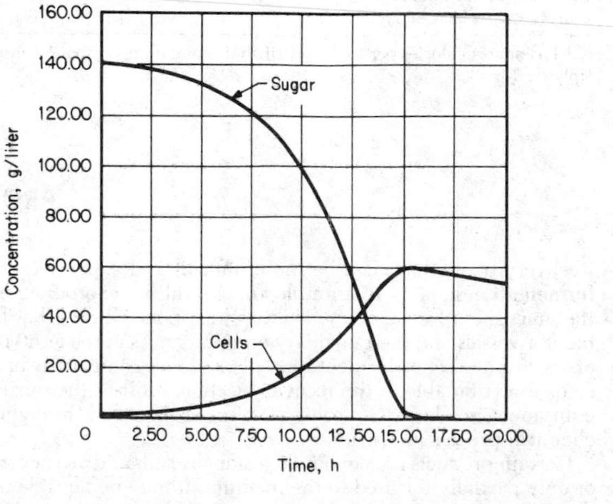

FIG. 27-21 Microbial growth curve. $dx/dt = \mu X - K_d X$; $ds/dt = -\mu X/Y$; $\mu = \hat{\mu} S/(K_s + S)$. $\hat{\mu} = 0.35$; $K_d = 0.025$; $K_s = 12.0$; $Y = 0.48$.

to biochemical functions, and the components can interact. For example, there can be a group of equations for carbohydrate metabolism, a group for protein synthesis, another for nucleic-acid synthesis, etc. This permits a much more intricate description of cell activities, but at the expense of having so many rate constants that assigning values to them may end up essentially as guesswork. For cells with distinct life cycles, a structured model may have compartments corresponding to each stage in the cycle. In addition, each compartment may be subdivided into the biochemical functions mentioned earlier. Such complicated models have had limited practical use but have great value for directing research toward areas where information is lacking.

Continuous Culture Continuous culture has been the goal of bioengineers for several decades because batch culture has inherent downtime for cleaning and sterilization and long lags before the organisms enter a brief period of high productivity. Continuous runs can last many weeks, but there must be stoppages for cleaning and maintenance. The nutrition and the product mix can be advantageously manipulated as functions of dilution rate. A serious problem, however, is the instability of the culture itself. The main successes with continuous fermentation have been achieved with rugged strains that produce either cell mass for cattle feed or a simple enzyme or metabolite. When a single stage is used, it is difficult to optimize cell growth, product production, and good utilization of substrates. Molds used for antibiotic fermentations are particularly messy and form voluminous coating on the shaft, coils, and any protuberances in the fermenter. Although this may limit the length of a run, a more important problem is genetic because there is a tendency to revert to less productive strains that quickly replace the finely tuned mutants which achieve high titers of product.

In view of the limited industrial application, continuous culture receives what may be viewed as a disproportionate amount of attention from academicians. As a research tool, batch culture suffers from changing concentrations of products and reactants, varying pH and redox potential, and a complicated mix of growing, dying, and dead cells. Data from continuous cultures are much easier to interpret because steady states are achieved or there are repeatable excursions from the steady state. The usual explanations for the limited use of continuous culture in industry are culture instability, difficulty of maintaining asepsis, insufficient knowledge of microbial behavior, and reluctance to convert existing factories. Another, more subtle reason is the cost of each research station. Rapid progress in research and development requires multiple vessels for screening many variables, but there are usually only one or two continuous fermenters because the cost of pumps, reservoirs, sterilizers, and controls is relatively high.

Conventional means for continuous culturing are the chemostat, in which nutrient is fed to a reactor at constant rate, and the turbidostat, which employs feedback control of pumping rate to maintain a fixed turbidity of the culture. Another alternative with feedback control of a nutrient or of product concentration has been termed "controlled, concentration-coupled, continuous culture," or "C5." Proportional control of the pumping rate is desirable because continuous cultures can have oscillatory responses induced by turning the feed pump on or off.

Mathematical Analysis The concept of a limiting nutrient is essential to the theory of continuous culture. There will be exact stoichiometric balance of all the ingredients going into the cells only when a very deliberate and time-consuming effort has been made to determine the details of cell nutrition. Even then there may be a different balance if the growth rate is changed, or kinetic rather than stoichiometric limitations may apply. The ingredient in short supply relative to the other ingredients will be exhausted first and will thus limit cellular growth or product synthesis. The other ingredients may exhibit toxicity or may influence cellular activities, but there will not be acute shortage as in the case of the limiting nutrient.

Mass balances for vessel n in a series of continuous fermenters give

Rate of change = rate in − rate out + rate of production

$$V\,dX_n/dt = FX_{n-1} - FX + V\,\mu_n X_n \qquad (27\text{-}15)$$

where V is the fermenter volume, F is the feed rate, and other terms are as previously defined. Subscript n refers to the fermenter-vessel number.

Dividing through by V and substituting $D = F/V$,

$$dX_n/dt = D\,(X_{n-1} - X_n) + \mu_n X_n \qquad (27\text{-}16)$$

and

$$V\,dS_n/dt = F\,S_{n-1} - F\,S_n - V\,\mu_n X_n/Y - VMX_n \qquad (27\text{-}17)$$

$$dS_n/dt = D\,(S_{n-1} - S_n) - \mu_n X_n/Y - MX_n \qquad (27\text{-}18)$$

where M = coefficient for maintenance energy (metabolism with no growth). For a single vessel with sterile feed, this reduces to

$$dX/dt = \mu X - D\,X \qquad (27\text{-}19)$$

$$dS/dt = D\,(S_0 - S) - \mu\,X/Y - M\,X \qquad (27\text{-}20)$$

This is an analysis that applies to any continuous culture that meets the assumptions of perfect mixing and constant volume. The equations are fundamental except for the Monod equation [Eq. (27-14)], which has no time dependency and should be applied with caution to transient states in which there may be a time lag as μ responds to changing S. At steady state, the rates of change become zero, and $\mu = D$. Substituting,

$$D = \hat{\mu} S/(K_s + S) \qquad (27\text{-}21)$$

$$S = D\,K_s/(\hat{\mu} - D) \qquad (27\text{-}22)$$

where $\hat{\mu}$ is the maximum-specific-growth-rate constant. Solving for X gives

$$X = D\,Y\,(S - S_0)/(D + M\,Y) = Y\,(S_0 - S) \qquad (27\text{-}23)$$

From these equations, the behavior of X and S as functions of dilution rate can be plotted as in Fig. 27-22. The interesting features are that (1) X goes to zero and S reaches S_0 as D approaches $\hat{\mu}$; (2) S is not a function of S_0 when D is less than $\hat{\mu}$; and (3) the maintenance coefficient is very important, but only at low dilution rates.

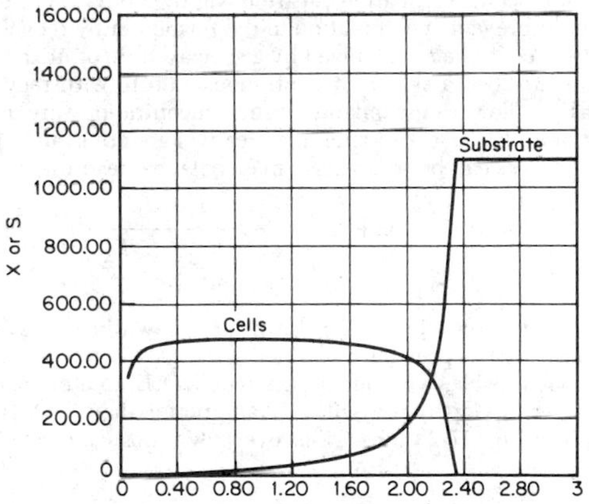

FIG. 27-22 Effects of dilution rate in continuous culture; $\hat{\mu}$ = 2.38; S_0 = 1100; Y = 0.45; K_s = 35; E_M = 0.05.

Mixing has been shown to be critical at low dilution rates because the uptake of substrate is extremely rapid for cells in a starved condition. Quite vigorous agitation is required in small vessels to ensure homogeneous distribution of the feed. Such intense agitation is probably impractical in large vessels, but it might be possible to distribute the feed from many fine openings throughout the tank.

It is easy to postulate advantages for multistage continuous culture but very difficult to conduct all the research and development of the many parameters that should be optimized. Each stage could have its feed streams, control of pH and other conditions, and recycle of cells or fluids from other steps in the process. Not only are there many parameters to study for each stage, but changes in one stage can markedly affect other stages. It can be quite troublesome to get representative conditions and cultures in a given stage to begin research because of the complicated interactions with other stages. Time delays in lines and in separators for recycle plus complexities from nonideal-flow regimes cause theoretical analysis to be faulty. An optimized multistage continuous fermentation system with recycle and control is one of the most difficult engineering feats. Furthermore, the dynamics of microbial responses to upsets are poorly understood.

Recycle Separation and recycle of cells result in much longer residence times for the cells than for the fluid and permit relatively high cell concentrations. In waste treatment, the dilute feed leads to slow growth rates. Therefore, more rapid processing is attained by achieving higher populations through cell recycle, although a higher percentage of the cells may be dead because all are in a starved state. High rates of production are also important in industrial fermentations with cell recycle, and there is the added advantage of reusing cells instead of diverting expensive substrate to producing more cells.

In activated sludge, organisms not associated with flocs are not collected and tend to leave the system as recycle increases the proportions of flocculating types. Recycle of collectible algae to outdoor ponds has profound influence on the population, but seasonal changes can develop small algae despite the retention of large algae. Thus, in such a case recycle fails as the few large algae die or escape collection.

Recycle of fermentation fluids has quite different objectives from those for cell recycle that aims at population control or greater productivity. Since spent broths have leftover nutrients, recycle can save on costs of nutrients and makeup water while greatly reducing the volumes sent to waste treatment. Of course, total recycle is not desirable because unwanted materials build up in concentration and can poison the fermentation. This buildup determines the amount of recycle, but purification steps may be added to remove toxic substances. For example, acetic acid is a by-product of alcohol fermentation, but its low volatility means that much is in the stillage (bottoms from alcohol distillation). Its removal by a physical or chemical step would prevent accumulation in the fermenter by recycle. Alternatively, it could be metabolized by a special strain or mixed culture.

Figure 27-23 is a sketch of continuous culture with recycle. The symbols for flow rates and organism concentrations are F and X respectively. If perfect mixing and steady state are assumed so that the derivatives can be set to zero, mass balances lead to

$$\mu = D\left[1 + \omega\left(1 - \frac{1+\omega}{1 + \omega - \dfrac{F_e}{F}}\right)\right] \qquad (27\text{-}24)$$

where ω is the recycle ratio, F terms are shown in Fig. 27-23, and other terms are as previously defined.

Without recycle, washout occurs when D is greater than $\hat{\mu}$, but recycle permits operation with D far greater than $\hat{\mu}$. A family of curves is in Fig. 27-24 for various overflow ratios F_e/F relating μ/D

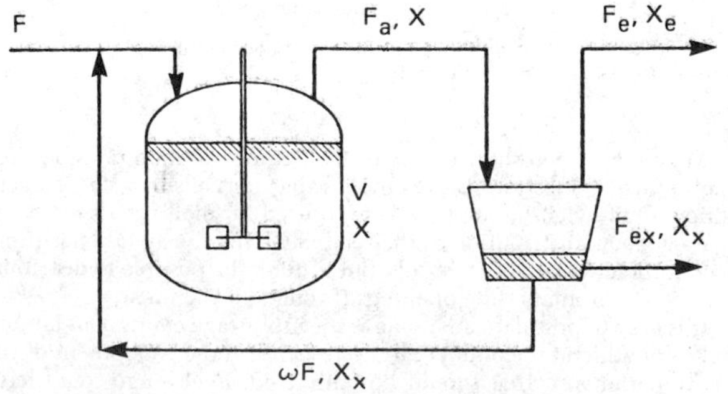

FIG. 27-23 Continuous culture with recycle. (*A. E. Humphrey, "Biochemical Engineering,"* in Encyclopedia of Chemical Processing and Design, *vol. 4, July 1977, pp. 359–394.*)

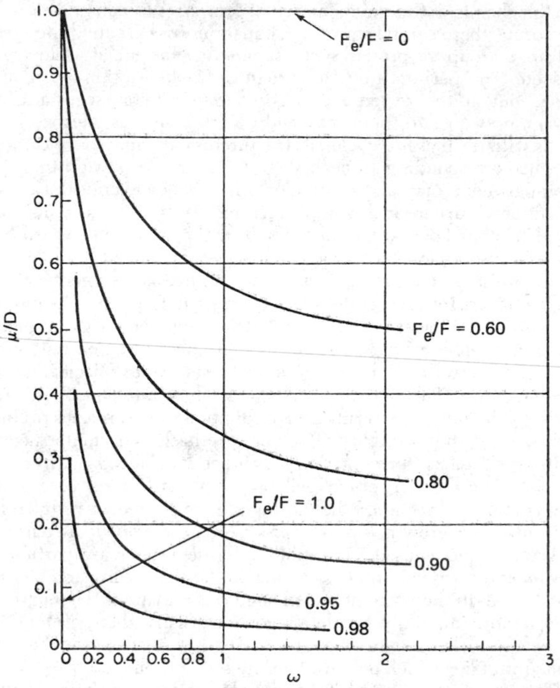

FIG. 27-24 Graphical solutions to recycle equation. (*A. E. Humphrey, "Biochemical Engineering,"* in Encyclopedia of Chemical Processing and Design, *vol. 4, July 1977, pp. 359–394.*)

to the recycle ratio. The curves show that cell concentration in the recycle stream has a large effect when ω is small, but the effect levels off. When $F_e/F = 0$, there is no recycle and $\mu = D$.

Mixed Cultures Mixtures of microorganisms characterize the processes shown in Table 27-4. Processes for biological waste treatment have elective cultures, and the proportions of different species can shift dramatically in response to changing nutrition or physiological conditions. There is an interesting area of research on defined mixtures of microorganisms, but there has been little practical application of the results. The definitions of various interactions are given in Table 27-5. These definitions are difficult to apply to real systems in which there is highly complicated interaction among organisms that play various roles with respect to each other.

Two types of interaction, competition and predation, are so important that it is worthwhile to gain insight from considering mathe-

TABLE 27-4 Mixed-Culture Processes

Process	Types of organisms
Commercial	
Alcoholic beverages	Various yeasts, molds, and bacteria
Sauerkraut	*L. plantarum* plus other bacteria
Pickles	*L. plantarum* plus other bacteria
Cheeses	Propionibacteria, molds, and possibly many other microorganisms
Lactic acid	Two *Lactobacillus* species
Beta carotene	Opposites sexes of *Blakelea trispora*
Waste treatment	
Trickling filters	Zoogloea, protozoa, algae, and fungi
Activated sludge	Zoogloea, *Sphaerotilus*, yeasts, molds, and protozoa
Sludge digestion	Cellulolytic and acid-forming bacteria; methanogenic bacteria
Sewage lagoons	Many types from most microbial families

TABLE 27-5 Definitions of Microbial Interactions

Competition	Race for nutrients and space
Predation	One feeding on another
Commensalism	One living off another with negligible help or harm
Mututalism	Each benefiting the other
Synergism	Combination with cooperative metabolism
Antibiosis	One excreting a factor harmful to the other

matical formulations. If it is assumed that specific-growth-rate coefficients are different, no steady state can be reached in a well-mixed continuous culture with both organisms of type A and type B present because if one were at steady state with $\mu_A = D$, the other would have μ_B unequal to D and a rate of change unequal to zero. The net effect is that the faster-growing type takes over while the other declines to zero. In real systems, even those that approximate well-mixed continuous cultures, there may be profound changes in relative numbers of the various organisms present, but a complete takeover by one type is extremely uncommon. Survival of a broad range of species is highly advantageous in natural systems because a needed type will be present should an uncommon nutrient (pollutant?) be added or if the conditions change.

Prey-predator or host-parasite systems can be analyzed by mass-balance equations:

$$dH/dt = \mu_H H - D\,H \tag{27-25}$$

$$dP/dt = \mu_P P - D\,P \tag{27-26}$$

$$dS/dt = D\,(S_0 - S) - \mu_H H/Y \tag{27-27}$$

where H = concentration of hosts (prey), P = concentration of predators, S = substrate concentration (food for prey), S_0 = concentration of feed stream, μ_H = Monod function of S, and μ_P = Monod function of H.

Computer simulation of these equations is shown in Fig. 27-25. Real systems do have this type of oscillating behavior, but frequencies and amplitudes are erratic.

Another interaction with grave consequences is attack on a species by a phage (microbial virus) that is usually highly specific. Infection of a cell by a virulent phage results in the production of from 10 to several hundred new phage particles as phage nucleic acid takes over control of cellular activities. The cell disintegrates and releases phage that infects other cells to reach high-phage titers quickly. A few cells of the host species may be resistant to phages. Such resistance can be acquired through mutation. These cells have fewer competitors and may thrive. However, mutations also occur in phage, so highly complicated behavior occurs as the hosts mutate and mutate further as the phage mutates to counter host resistance.

Commercial fermentation groups usually maintain different strains of cultures suitable for production so that phage attacks can be thwarted by substituting a nonsusceptible culture. After a period of time for the phage to dissipate, it may be possible to return the most desirable production strain.

Computers and Instrumentation A fermenter is a fairly sophisticated device with control of temperature, aeration rate, and per-

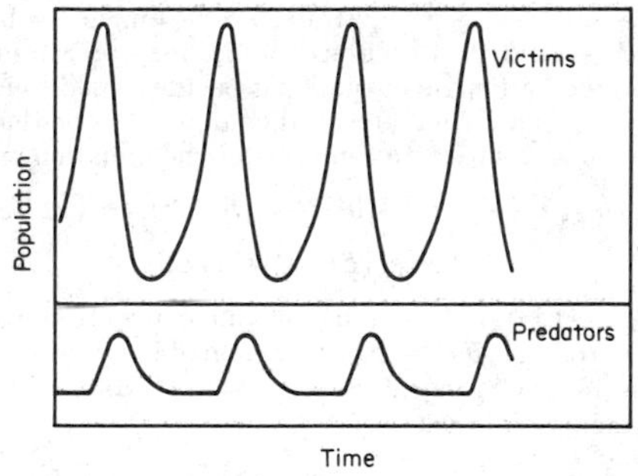

FIG. 27-25 Prey-predator kinetics.

haps pH, concentration of dissolved oxygen, or some nutrient concentration. There has been a strong trend toward automated data collection and analysis. Analog control is still very common, but when a computer is available for online data collection, it often is good practice to use it for control as well.

Many of the sensor problems, such as those with steam-sterilizable pH electrodes and dissolved-oxygen probes, have been solved, but there is still an urgent need for a method of quickly and accurately measuring the amount of cellular material. Indirect methods, such as measuring protein produced by cells or monitoring nucleic acids, are sometimes valuable, but their proportionality to cell mass may vary during fermentation. The development of computer models that can be used to interpret measured variables to calculate cell mass or product concentration that may be difficult or impractical to measure online has been an important advance.

Although dynamic responses of microbial systems are poorly understood, models with some basic features and some empirical features have been found to correlate with actual data fairly well. Real fermentations take days to run, but many variables can be tried in a few minutes by using computer simulation. Optimization of fermentation with models and real-time dynamic control is in its early infancy. However, bases for such work are advancing steadily. The foundations for all such studies are accurate material balances.

The common indices of the physical environment are temperature, pressure, shaft-power input, impeller speed, foam level, gas flow rate, liquid feed rates, broth viscosity, turbidity, pH, oxidation-reduction potential, dissolved oxygen, and exit-gas concentrations. A wide variety of chemical assays can be performed. Product concentration, nutrient concentration, and product-precursor concentration are important. Indices of respiration were mentioned earlier with regard to oxygen transfer; they are particularly useful in tracking fermentation behavior. Computer control schemes for fermentation can focus on high productivity, high-product titer, or minimum cost. Computer systems may perform online optimization of fermentation. Progress has been slow by using empirical methods because there are many variables and because statistical techniques suffer from the relatively poor reproducibility of fermentations.

ENZYMATIC-REACTION KINETICS

Enzymes are considered to be excellent catalysts for two reasons: high specificity and high turnover rates. With few exceptions, all reactions in biological systems are catalyzed by enzymes, and each enzyme usually catalyzes only one reaction. For a few enzymes and other proteins, the amino-acid sequences and three-dimensional structures have been determined. As the molecular structure of an enzyme is seldom known, a precise molecular weight is not available for measuring the quantity of an enzyme in gram-moles. Instead, the amount is usually expressed in terms of catalytic activity. An international unit (IU) of an enzyme is defined as the amount capable of producing one micromole of its reaction product in one minute under its optimal (or some defined) reaction conditions. Specific activity is an index of enzyme purity.

Although the mechanisms may be complicated and varied, some simple equations can often describe the reaction kinetics of common enzymatic reactions quite well. Each enzyme molecule is considered to have an active site that must first encounter the substrate (the reagent) to form a complex so that the enzyme can function. Accordingly, the following reaction scheme is written:

$$E + S \underset{2}{\overset{1}{\rightleftharpoons}} ES \overset{3}{\rightarrow} P + E \tag{27-28}$$

where E = enzyme, S = substrate, ES = enzyme-substrate complex, and P = product. Reactions 1 and 2 may be assumed to be in equilibrium soon after the enzyme has been exposed to its substrate. The rate of product formation will then depend upon the concentration of the enzyme-substrate complex for the equation

$$k_2 [ES] = k_1 [E][S] = k_1 [S] ([E]_T - [ES]) \quad (27\text{-}29)$$

and

$$V = d[P]/dt = k[ES] \quad (27\text{-}30)$$

where $[E]_T = [E] + [ES]$ = total enzyme concentration, $[S]$ = substrate concentration, $[ES]$ = concentration of enzyme-substrate complex, $[P]$ = product concentration, t = time, and k_1, k_2, k_3 are = kinetic constants. This leads to

$$V = \frac{V_{max}[S]}{K_M + [S]} \quad (27\text{-}31)$$

where $K_M = k_2/k_1$ = Michaelis constant. This equation successfully describes the kinetic behavior of a surprisingly large number of reactions of different enzymes (see Fig. 27-26). By taking reciprocals of both sides, the following results:

$$1/V = (K_M/V_{max})/S + 1/V_{max} \quad (27\text{-}32)$$

A linear plot of $1/V$ versus $1/S$ (the Lineweaver-Burk plot) will allow the determination of K_M and V_{max} from experimental data.

This derivation illustrates how the kinetics of enzymatic reactions are usually treated. Kinetic behavior becomes complicated when there are two chemical species which can both complex with the enzyme molecules. One of the species might behave as an inhibitor of the enzyme reaction with the other as the substrate. Depending upon the nature of the complex, different inhibition patterns will yield different kinetic equations. For example,

$$E + S \rightleftharpoons ES \rightarrow P + S \quad (27\text{-}33)$$

$$E + I \rightleftharpoons EI \quad (27\text{-}34)$$

where I = inhibitor.

Since the EI complex does not yield product P and since I competes with S for E, this is known as competitive inhibition. By analogy to the Michaelis-Menten equation,

$$\frac{1}{V} = \frac{K_M}{V_{max}}\frac{1}{S}\left(1 + \frac{[I]}{K_i}\right) + \frac{1}{V_{max}} \quad (27\text{-}35)$$

where I = concentration of the competitive inhibitor and K_i = inhibition constant.

Enzyme reactions are also sensitive to pH and temperature changes. In characterizing an enzyme, its optimal pH and optimal temperature are conditions at which the enzyme has its highest catalytic activity.

For a somewhat more extensive exposure to enzyme reaction kinetics, consult standard biochemistry texts and also M. Dixon and E. C. Webb, *Enzymes,* 2d ed., Academic, New York, 1964; and I. H. Segal, *Enzyme Kinetics,* Wiley, New York, 1975.

Enzyme Immobilization and Immobilized-Enzyme Engineering One factor that usually impedes the development of wide industrial application of enzymes is high cost. Immobilization is a technique to retain enzyme molecules for repeated use. The method of immobilization can be adsorption, covalent bonding, or entrapment. Semipermeable membranes in the form of flat sheets or hollow fibers are one way to restrain the enzyme while allowing smaller molecules to pass. Polyacrylamide gel, silica gel, and similar materials have been used to entrap biologically active materials including enzymes. Encapsulation is another means of capture by coating liquid droplets containing enzymes with some semipermeable materials formed in situ. Generally speaking, entrapment does not involve a chemical or physical chemical reaction directly with the enzyme molecules, and the enzyme molecules are not altered. Physical adsorption on active carbon particles and ionic adsorption on ion-exchange resins are important for enzyme immobilization. A method with a myriad of possible variations is covalent bonding of the enzyme to a selected carrier. Materials such as glass particles, cellulose, silica, etc., have been used as carriers for immobilization. Enzymes immobilized by entrapment and adsorption may be subject to loss due to leakage and desorption respectively. On the other hand, the chemical treatment in forming the covalent bond between an enzyme and its carrier may permanently damage some enzyme molecules. In enzyme immobilization, two efficiency terms are often used. "Immobilization yield" can be used to describe the percentage of enzyme activity that is immobilized:

Percent yield = 100 × activity immobilized/starting activity

"Immobilization efficiency" describes the percentage of enzyme activity that is observed:

Percent efficiency

= 100 × observed activity/activity immobilized

When an enzyme molecule is attached to a carrier, its active site might be sterically blocked and thus its activity becomes unobservable (inactivated).

One of the most important parameters of an immobilized-carrier complex is stability of its activity. Catalytic activity of the complex diminishes with time because of leakage, desorption, deactivation, and the like. The half-life of the complex is often used to describe the activity stability. Even though there may be frequent exceptions, linear decay is often assumed in treating the kinetics of activity decay of an immobilized complex.

Immobilization by adsorption or by covalent bonding often helps

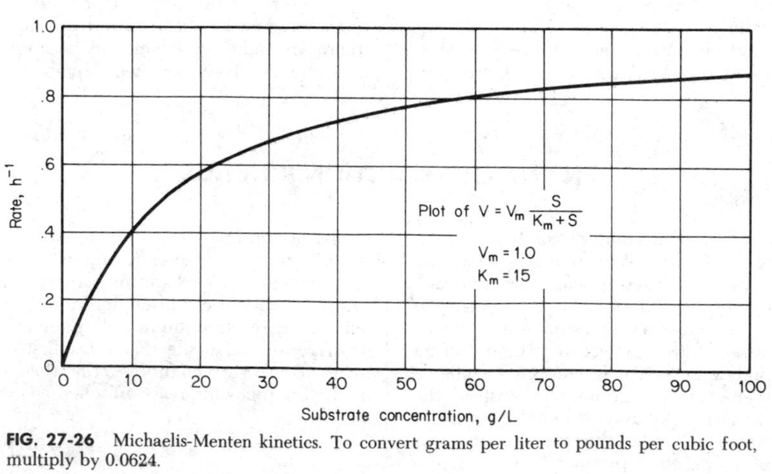

FIG. 27-26 Michaelis-Menten kinetics. To convert grams per liter to pounds per cubic foot, multiply by 0.0624.

to stabilize the molecular configurations of an enzyme against alterations including those that may cause thermal deactivation. Immobilized enzymes tend to be less sensitive to pH changes than are free enzymes. Most carriers are designed to have high porosity and large internal-surface areas so that a relatively large amount of enzymes can be immobilized onto a given volume or given weight of the carrier. Therefore, in an immobilized enzyme-carrier complex, the enzyme molecules are subject to the effect of the microenvironment in the pores of the complex. Surface charges and other microenvironmental effects can create a shift up or down of optimal pH of the enzyme activity.

An immoblized enzyme-carrier complex is a special case for which the methodology developed for evaluation of a heterogeneous catalytic system can be employed. The enzyme complex also has external diffusional effects, pore diffusional effects, and an effectiveness factor. When carried out in aqueous solutions, heat transfer is usually good, and it is safe to assume that isothermal conditions prevail for an immobilized enzyme complex.

The Michaelis-Menten equation and similar nonlinear expressions characterize immobilized-enzyme kinetics. Therefore, for a spherical porous carrier particle with enzyme molecules immobilized on its external as well as internal surfaces, a material balance on the substrate results in the following:

$$2\frac{D_e}{r}\frac{dS}{dr} + D_e\frac{d^2S}{dr^2} = \frac{V_{max}S}{K_M + S} \qquad (27\text{-}36)$$

with also the usual boundary conditions

$$\text{at } r = R, S = S \qquad \text{and} \qquad \text{at } r = 0, dS/dr = 0$$

where R = radius of sphere, r = distance from sphere center, S = substrate concentration, and D_e = effective diffusivity.

$$\frac{d^2y}{dx^2} + \frac{2}{x}\frac{dy}{dx} - \phi^2\beta\left(\frac{y}{\beta} + y\right) = 0 \qquad (27\text{-}37)$$

where y is dimensionless concentration, x is dimensionless distance, and ϕ β and are dimensionless constants. The boundary conditions are at $x = 1$, $y = 1$ and at $x = 0$, $dy/dx = 0$. The parameter ϕ is the Thiele modulus of the immobilized-enzyme complex. Graphical solutions are available in standard tests. Two meaningful asymptotic conditions have analytical solutions. In one extreme, $\beta \rightarrow 0$, meaning $S \gg K_M$, and accordingly the Michaelis-Menten equation reduces to a zero-order reaction with $V = V_{max}$. This is the condition of saturation; i.e., the substrate supply is high and saturates all the active sites of the enzyme molecules. In the other extreme, $\beta \rightarrow \infty$, $K_M \gg S$, and accordingly the Micaelis-Menten equation approaches that of a first-order reaction with $V = (V_{max}S/K_M)$. This is the condition of complete substrate control.

ADDITIONAL READING: *Enzyme immobilization.* R. A. Messing, *Immobilized Enzymes for Industrial Reactors*, Academic, New York, 1975. O. R. Zaborsky, *Immobilized Enzymes*, CRC Press, Boca Raton, Fla., 1973. *Review of heterogeneous catalysis.* C. N. Satterfield, *Mass Transfer in Heterogeneous Catalysis*, M.I.T., Cambridge, Mass., 1970. T. K. Sherwood, R. L. Pigford, and C. R. Wilke, *Mass Transfer*, McGraw-Hill, New York, 1975.

Enzymatic Reactors An enzyme immobilized on beads of a support material or captured in a gel droplet is essentially a catalytic particle. When it is mounted in a packed column, there may be upflow or downflow of the feed solution, and a fluidized bed may be feasible except that particle collision often endangers stability of the enzyme. A serious problem is growth of microorganisms on the particles because enzymes are proteins that are nutritious. As immobilized enzymes often have more thermal stability than do free enzymes, the columns can be run at elevated temperatures of 50 to 65°C (122 to 149°F) to improve reaction rate and inhibit most but not all contaminating organisms. Sterile feed solutions and aseptic technique can minimize contamination. More commonly, antiseptics are added to the feed, and there is occasional treatment with a toxic chemical to wash organisms from the column. Particles with immobilized enzymes are sometimes added to a reactor and recovered later by filtration or by a unique method such as using magnets to collect enzymes attached to iron.

Cellulose is hydrolyzed by a complex of several enzymes. The mix of enzyme activities produced by mold cultures can have insufficient amounts of the enzyme beta glucosidase to maintain a commercially acceptable hydrolysis rate. This enzyme can be produced with a different microbial culture and used to supplement the original enzyme mix, but the cost is high. It is logical to immobilize the beta glucosidase for multiple use. Handling is minimized by circulating fluid from the main reactor through an external packed column of immobilized enzyme.

Reactors that employ membranes to retain enzymes can take various forms. It is possible to add free enzyme and to recover it by ultrafiltration, but the requirement for sufficient membrane surface area to obtain good rates and the necessary auxiliary equipment cause this approach to be expensive. A hollow fiber device packs a vast amount of membrane area into a small volume. The enzyme may be immobilized inside or outside of the fiber, and it is easy to flush and replace the enzyme. Drawbacks to this design are that (1) stability of the enzyme is not improved as in chemical immobilization; (2) there are two mass-transfer steps, as the substrate must diffuse through the fiber to reach the enzyme and the product must diffuse back; and (3) diffusion is poor on the outside of packed fibers. Diffusional resistances are minimized when the enzyme is immobilized on or in the membrane to provide excellent contact as the feed is forced through.

ADDITIONAL READING: *Enzyme Engineering*, vol. 2, E. K. Pye and L. B. Wingard (eds.), 1974; vol. 3, E. K. Pye and H. H. Weetall (eds.), 1978; vol. 4, G. B. Broun and G. Manecke (eds.), 1978; vol. 5, H. H. Weetall and G. P. Royer (eds.), 1980, Plenum, New York. *Enzyme Engineering: Future Directions*, L. B. Wingard, I. V. Berezin, and A. A. Klyosov (eds.), Plenum, New York, 1980. Y. Y. Lee and G. T. Tsao, "Engineering Problems of Immobilized Enzymes," *J. Food Technol.*, **39**, 667 (1974).

Index

ABOUT THE EDITORS

The late **Robert H. Perry** served as chairman of the Department of Chemical Engineering at the University of Oklahoma and program director for graduate research facilities at the National Science Research Foundation. He was a consultant to various United Nations and other international organizations. From 1973 until his death in 1978 Dr. Perry devoted his time to a study of the cross impact of technologies within the next half century. The subjects under his investigation on a global basis were energy, minerals and metals, transportation and communications, medicine, food production, and the environment.

Don W. Green is the Conger-Gabel distinguished professor of chemical and petroleum engineering and codirector of the Tertiary Oil Recovery Project at the University of Kansas in Lawrence, Kansas, where he has taught since 1964. He received his doctorate in chemical engineering in 1963 from the University of Oklahoma, where he was Dr. Perry's first doctoral student. Dr. Green has won several teaching awards from the University of Kansas and the National Distinguished Achievement Award of the Society of Petroleum Engineers. He is the author of numerous articles in technical journals.

[illegible] and the environment.

[illegible] the [illegible] of the [illegible] of the [illegible]

TABLE 1-5 Metric Conversion Factors as Exact Numerical Multiples of SI Units

The first two digits of each numerical entry represent a power of 10. For example, the entry "—02 2.54" expresses the fact that 1 in $= 2.54 \times 10^{-2}$ m.

To convert from	To	Multiply by
abampere	ampere	+01 1.00
abcoulomb	coulomb	+01 1.00
abfarad	farad	+09 1.00
abhenry	henry	−09 1.00
abmho	mho	+09 1.00
abohm	ohm	−09 1.00
abvolt	volt	−08 1.00
acre	meter^2	+03 4.046 856
ampere (international of 1948)	ampere	−01 9.998 35
angstrom	meter	−10 1.00
are	meter^2	+02 1.00
astronomical unit	meter	+11 1.495 978
atmosphere	newton/meter^2	+05 1.013 25
bar	newton/meter^2	+05 1.00
barn	meter^2	−28 1.00
barrel (petroleum 42 gal)	meter^3	−01 1.589 873
barye	newton/meter^2	−01 1.00
British thermal unit (ISO/TC 12)	joule	+03 1.055 06
British thermal unit (International Steam Table)	joule	+03 1.055 04
British thermal unit (mean)	joule	+03 1.055 87
British thermal unit (thermochemical)	joule	+03 1.054 350
British thermal unit (39°F)	joule	+03 1.059 67
British thermal unit (60°F)	joule	+03 1.054 68
bushel (U.S.)	meter^3	−02 3.523 907
cable	meter	+02 2.194 56
caliber	meter	−04 2.54
calorie (International Steam Table)	joule	+00 4.1868
calorie (mean)	joule	+ 00 4.190 02
calorie (thermochemical)	joule	+00 4.184
calorie (15°C)	joule	+00 4.185 80
calorie (20°C)	joule	+00 4.181 90
calorie (kilogram, International Steam Table)	joule	+03 4.186 8
calorie (kilogram, mean)	joule	+03 4.190 02
calorie (kilogram, thermochemical)	joule	+03 4.184
carat (metric)	kilogram	−04 2.00
Celsius (temperature)	kelvin	$t_K = t_c + 273.15$
centimeter of mercury (0°C)	newton/meter^2	+03 1.333 22
centimeter of water (4°C)	newton/meter^2	+01 9.806 38
chain (engineer's)	meter	+01 3.048
chain (surveyor's or Gunter's)	meter	+01 2.011 68
circular mil	meter^2	−10 5.067 074
cord	meter^3	+00 3.624 556
coulomb (international of 1948)	coulomb	−01 9.998 35
cubit	meter	−01 4.572
cup	meter^3	−04 2.365 882
curie	disintegration/second	+10 3.70
day (mean solar)	second (mean solar)	+ 04 8.64
day (sidereal)	second (mean solar)	+04 8.616 409
degree (angle)	radian	−02 1.745 329
denier (international)	kilogram/meter	−07 1.00
dram (avoirdupois)	kilogram	− 03 1.771 845
dram (troy or apothecary)	kilogram	−03 3.887 934
dram (U.S. fluid)	meter^3	−06 3.696 691
dyne	newton	−05 1.00
electron volt	joule	−19 1.602 10
erg	joule	−07 1.00
Fahrenheit (temperature)	kelvin	$t_K = (5/9)(t_F + 459.67)$
Fahrenheit (temperature)	Celsius	$t_c = (5/9)(t_F - 32)$
farad (international of 1948)	farad	−01 9.995 05
faraday (based on carbon 12)	coulomb	+04 9.648 70
faraday (chemical)	coulomb	+04 9.649 57
faraday (physical)	coulomb	+04 9.652 19
fathom	meter	+00 1.828 8
fermi (femtometer)	meter	−15 1.00

To convert from	To	Multiply by
fluid ounce (U.S.)	meter^3	−05 2.957 352
foot	meter	−01 3.048
foot (U.S. survey)	meter	−01 3.048 006
foot of water (39.2°F)	newton/meter^2	+03 2.988 98
footcandle	lumen/meter^2	+01 1.076 391
footlambert	candela/meter^2	+00 3.426 259
furlong	meter	+02 2.011 68
gal (galileo)	meter/second^2	−02 1.00
gallon (U.K. liquid)	meter^3	−03 4.546 087
gallon (U.S. dry)	meter^3	−03 4.404 883
gallon (U.S. liquid)	meter^3	−03 3.785 411
gamma	tesla	−09 1.00
gauss	tesla	−04 1.00
gilbert	ampere turn	−01 7.957 747
gill (U.K.)	meter^3	−04 1.420 652
gill (U.S.)	meter^3	−04 1.182 941
grad	degree (angular)	−01 9.00
grad	radian	−02 1.570 796
grain	kilogram	−05 6.479 891
gram	kilogram	−03 1.00
hand	meter	−01 1.016
hectare	meter^2	+04 1.00
henry (international of 1948)	henry	+00 1.000 495
hogshead (U.S.)	meter^3	−01 2.384 809
horsepower (550 ft lbf/s)	watt	+02 7.456 998
horsepower (boiler)	watt	+03 9.809 50
horsepower (electric)	watt	+02 7.46
horsepower (metric)	watt	+02 7.354 99
horsepower (U.K.)	watt	+02 7.457
horsepower (water)	watt	+02 7.460 43
hour (mean solar)	second (mean solar)	+03 3.60
hour (sidereal)	second (mean solar)	+03 3.590 170
hundredweight (long)	kilogram	+01 5.080 234
hundredweight (short)	kilogram	+01 4.535 923
inch	meter	−02 2.54
inch of mercury (32°F)	newton/meter^2	+03 3.386 389
inch of mercury (60°F)	newton/meter^2	+03 3.376 85
inch of water (39.2°F)	newton/meter^2	+02 2.490 82
inch of water (60°F)	newton/meter^2	+02 2.4884
joule (international of 1948)	joule	+00 1.000 165
kayser	1/meter	+02 1.00
kilocalorie (International Steam Table)	joule	+03 4.186 74
kilocalorie (mean)	joule	+03 4.190 02
kilocalorie (thermochemical)	joule	+03 4.184
kilogram mass	kilogram	+00 1.00
kilogram-force (kgf)	newton	+00 9.806 65
kilopond-force	newton	+00 9.806 65
kip	newton	+03 4.448 221
knot (international)	meter/second	−01 5.144 444
lambert	candela/meter^2	+04 $1/\pi$
lambert	candela/meter^2	+03 3.183 098
langley	joule/meter^2	+04 4.184
lbf (pound-force, avoidupois)	newton	+00 4.448 221
lbm (pound-mass, avoirdupois)	kilogram	−01 4.535 923
league (British nautical)	meter	+03 5.559 552
league (international nautical)	meter	+03 5.556
league (statute)	meter	+03 4.828 032
light-year	meter	+15 9.460 55
link (engineer's)	meter	−01 3.048
link (surveyor's or Gunter's)	meter	−01 2.011 68
liter	meter^3	−03 1.00
lux	lumen/meter^2	+00 1.00
maxwell	weber	−08 1.00
meter	wavelengths Kr 86	+06 1.650 763
micrometer	meter	−06 1.00
mil	meter	−05 2.54
mile (U.S. statute)	meter	+03 1.609 344
mile (U.K. nautical)	meter	+03 1.853 184
mile (international nautical)	meter	+03 1.852
mile (U.S. nautical)	meter	+03 1.852
millibar	newton/meter^2	+02 1.00
millimeter of mercury (0°C)	newton/meter^2	+02 1.333 224